The Columbia Gazetteer of the World

The
Columbia
Gazetteer
of the World

VOLUME 2 H TO O

EDITED BY SAUL B. COHEN

SECOND EDITION

COLUMBIA UNIVERSITY PRESS

New York

Columbia University Press

Publishers since 1893

New York   Chichester, West Sussex

Copyright © by Columbia University Press

2008

Copyright under the Berne Convention

Library of Congress Cataloging-in-Publication
Data

The Columbia gazetteer of the world / edited
by Saul B. Cohen.—2nd ed.

    p. cm.

  ISBN 978-0-231-14554-1 (alk. paper)

1. Gazetteers. I. Cohen, Saul Bernard.
II. Title.

G103.5.C645   2008

910.3—dc22

2008009181

Casebound editions of Columbia University
Press books are printed on permanent and
durable acid-free paper.

Printed in the United States of America.

The crown logo is a registered trademark of
Columbia University Press.

The Columbia Gazetteer of the World

H

Ha (HAH), village, W BHUTAN, on Ha Chu tributary of the WANG Chu River, on main YADONG-PUNAKHA trading route through the CHUMBI VALLEY of TIBET; 27°20′N 89°11′E. Location of a current encampment of the Indian army. Near the site of a battle between CHINA and INDIA in the 1960s.

Haabai, TONGA: see HA'APAI.

Haacht (HAHGT), commune (2006 population 13,665), Leuven district, BRABANT province, central BELGIUM, 7 mi/11.3 km NNW of LEUVEN; 50°59′N 04°38′E. Brewing; agriculture. Also spelled Haecht.

Haad Yai, THAILAND: see HAT YAI.

Haag (HAHG), town, W LOWER AUSTRIA, 8 mi/12.9 km NE of STEYR; 48°07′N 14°34′E. Pig breeding. Renaissance castle Salaberg.

Haag am Hausruck (HAHG ahm HOUS-ruk), township, W central UPPER AUSTRIA, 7 mi/11.3 km ESE of RIED IM INNKREIS, in HAUSRUCK MOUNTAINS; 48°11′N 13°38′E. Railroad terminus; cattle breeding; Starhemberg castle.

Haag, Den, NETHERLANDS: see HAGUE, THE.

Haag in Oberbayern (HAHG in O-buhr-bei-ern), village, BAVARIA, S GERMANY, in Upper BAVARIA, 16 mi/26 km WSW of Mühldorf; 48°10′N 12°11′E. Has 13th-century watchtower. Chartered 1324.

Haakon (HAW-kin), county (□ 1,827 sq mi/4,750.2 sq km; 2006 population 1,864), central SOUTH DAKOTA, ⊙ PHILIP; 44°17′N 101°31′W. Agricultural and cattle-raising region drained in S by BAD RIVER and in N by Plum Creek; bounded N by CHEYENNE RIVER. Wheat; ranching; cattle, poultry. Cheyenne River Indian Reservation to N in ZIEBACH county. Formed 1914.

Haaksbergen (HAHKS-ber-khuhn), town, OVERIJSSEL province, E NETHERLANDS, 8 mi/12.9 km SSW of HENGELO, and 8 mi/12.9 km SW of ENSCHEDE; 52°09′N 06°46′E. German border 5 mi/8 km to SE. Dairying; cattle, hogs, poultry; grain, vegetables; manufacturing (construction materials, electronics, clothing).

Haaltert (HAHL-tuhrt), commune (2006 population 17,356), Aalst district, EAST FLANDERS province, N central BELGIUM, 3 mi/4.8 km SSW of AALST; 50°54′N 04°00′E.

Haamstede (HAHM-ste-duh), village, ZEELAND province, SW NETHERLANDS, at center of SCHOUWEN region, and 8 mi/12.9 km WNW of Zierikee; 51°42′N 03°45′E. NORTH SEA 3 mi/4.8 km to W, N end of OOSTERSCHELDE DAM 3 mi/4.8 km to SSW, S end of Brouwers Dam, on GREVELINGEN channel, 4 mi/6.4 km to NE. Dairying; livestock; vegetables, grain, sugar beets. Airport. Castle, lighthouse to W.

Haan (HAHN), town, NORTH RHINE–WESTPHALIA, GERMANY, just NW of SOLINGEN; 51°12′N 07°00′E. Manufacturing includes electronics, plastics, and various small industries.

Haan, de (HAHN, duh), commune, Ostend district, WEST FLANDERS province, W BELGIUM, 6 mi/10 km NE of OSTENDE, on NORTH SEA. Summer resort.

Haan Tayshiriin Hüryee, MONGOLIA: see ALTAY.

Ha'apai (HAH-pei), island group (2006 population 7,572), central TONGA, S PACIFIC OCEAN, mainly aligned with N-S reef; 19°42′S 174°29′W. Comprised of 36 small low-lying coral islands of which LIFUKA, site of Pangai village (capital of Ha'apai group), is most important; a W group of andesitic volcanoes, including KAO and TOFUA. Also Haabai, Hapai.

Haapajärvi (HAH-pah-YAHR-vee), village, OULUN province, W central FINLAND, on the KALAJOKI (river), and 65 mi/105 km E of KOKKOLA; 63°45′N 25°20′E. Elevation 330 ft/100 m. Railroad junction.

Haapakoski (HAH-pah-KOS-kee), village, MIKKELIN province, SE central FINLAND, 35 mi/56 km SSW of KUOPIO; 62°27′N 27°10′E. Elevation 495 ft/125 m. In SAIMAA lake region. Metalworks, machine shops.

Haapamäki (HAH-pah-MAH-kee), village, KESKI-SUOMEN province, W FINLAND, 40 mi/64 km W of JYVÄSKYLÄ; 62°15′N 24°28′E. Elevation 495 ft/150 m. In lumbering region.

Haapaniemi (HAH-pah-NEE-e-mee), town, KUOPION province, S central FINLAND, on KALLAVESI (lake); 62°50′N 27°40′E. Elevation 330 ft/100 m. Primarily residential suburb of KUOPIO.

Haapsalu (HAHP-sah-loo), German *Hapsal*, city, W ESTONIA, shallow port on BALTIC SEA inlet, 55 mi/89 km SW of TALLINN; 58°56′N 23°32′E. Railroad terminus. Meatpacking; fisheries; ship repair, textile manufacturing. Summer resort (swimming beach, mud baths). Ruins of 13th century episcopal castle; was summer home of Russian imperial family. Founded by Danes (1228); passed to Livonian Knights (1346), to SWEDEN (1561); occupied 1710 by RUSSIA.

Haar (HAHR), suburb of MUNICH, BAVARIA, S GERMANY, in UPPER BAVARIA, 8 mi/12.9 km ESE of city center. Metalworking. Has 13th-century church.

Haardt Mountains, GERMANY: see HARDT MOUNTAINS.

Haari, Har (hah-ah-REE, HAHR), Upper GALILEE (3,435 ft/1,047 m), N ISRAEL, 8 mi/12.9 km SW of ZEFAT; 32°57′N 35°22′E. Formerly Jebel Heidar.

Haarlem (HAHR-luhm), city (2000 population 189,698), ⊙ NORTH HOLLAND province, W NETHERLANDS, on the Spaarne River, near the NORTH SEA, 12 mi/19 km W of AMSTERDAM; 52°23′N 04°38′E. Zaandvoort (North Sea resort) 4 mi/6.4 km to W; NORTH SEA CANAL passes to N, RINGVAART Canal to SE. Dairying; cattle, sheep, poultry; flowers, fruit, vegetables; manufacturing (pharmaceuticals, machinery, signs, paper products; food processing). Chartered in 1245. In 1573 it was sacked by the Spanish during the revolt of the Netherlands. During the sixteenth and seventeenth century, Haarlem was a center of Dutch painting; Frans Hals, Jacob van Ruisdael, and Adriaen van Ostade worked here. Among Haarlem's numerous historic buildings are the Church of St. Bavo, or Grote Kerk (fifteenth century), which has a world-famous organ; the Stadhuis [Dutch=city hall], formerly a palace of the counts of Holland, begun in 1250; and many medieval gabled houses. Has a number of museums. Nearby, at Spaarndam, is a monument commemorating the legendary boy of Haarlem who stopped a leak in the dike with his finger. Headquarters of Netherlands Geological Service. Also spelled Harlem.

Haarlemmermeer (HAHR-luhm-uhr-MIR), region (□ 71.5 sq mi/185.2 sq km; 1994 population 103,684), NORTH HOLLAND province, W NETHERLANDS, SW of AMSTERDAM and SE of HAARLEM. Reclaimed land (polder) in former shallow lake, surrounded by RINGVAART Canal, crossed by network of smaller canals, supplemented by pumping stations. Dairying; cattle, poultry; flowers, vegetables. Formed by flood (sixteenth century), completely drained by 1852. Towns of NIEUW-VENNEP in SW, HOOFDDORP in center, Badhoevedorp (suburb of Amsterdam) in NE. Airport to NE.

Haarsteeg (HAHR-staikh), village, NORTH BRABANT province, S central NETHERLANDS, 5 mi/8 km WNW of 's-HERTOGENBOSCH; 51°43′N 05°12′E. MAAS RIVER (pumping station) 2 mi/3.2 km to N. Dairying; livestock; grain, vegetables.

Haase River (HAH-se), 80 mi/129 km long, NW GERMANY; rises in TEUTOBURG FOREST NW of BORGHOLZHAUSEN; flows N and W, past OSNABRÜCK and Quackenbrück (head of navigation), to DORTMUND-EMS CANAL at MEPPEN; source at 52°07′N 08°16′E. Canalized in lower course. Also spelled HASE RIVER.

Haast Pass, between S WESTLAND and the Queenstown-Lakes districts, S SOUTH ISLAND, NEW ZEALAND.

Lowest-level pass (1,850 ft/564 m) across SOUTHERN ALPS, over 50 mi/80 km S of MOUNT COOK, connecting dense rain forests of Westland with open grasslands of glacially sculptured lakes in E lee of ranges. Low level probably due to W Haast glacier flowing E to Makarora valley.

Haastrecht (HAHS-trekht), village, SOUTH HOLLAND province, W NETHERLANDS, on HOLLANDSCHE IJSSEL River, at mouth of Vlist River, 3 mi/4.8 km E of GOUDA; 52°00′N 04°47′E. Dairying; cattle; grain, vegetables.

Haast River, 60 mi/97 km long, WESTLAND district, SOUTH ISLAND, NEW ZEALAND; rises in SOUTHERN ALPS SW of MOUNT SEFTON; flows SW to TASMAN SEA 22 mi/35 km NE of JACKSON BAY. Lower Valley followed by HAAST PASS route across ranges between E and W of South Island.

Ha Bac, province (□ 1,781 sq mi/4,630.6 sq km), N VIETNAM, ⊙ BAC GIANG; 21°20′N 106°30′E. Bordered N by LANG SON province, SE by QUANG NINH province, S by HAI HUNG province, W by HANOI, NW by BAC THAI province. In complex region of mountains, midlands, and riverine lowlands. Province exhibits range of regional habitats and localized economies. Its once-forested mountains and hills have suffered serious deforestation in recent decades. Diverse soils and mineral resources (limestone, copper, coal), shifting cultivation, wet rice farming, livestock raising, agroforestry, and commercial agriculture (tea, mulberry, cassava, vegetables, fruits). Sawmilling, handicrafts, various industries (sericulture and silk weaving; textiles, furniture, beer and soft drinks, plastics, cement, fertilizers, brick and tile; food and meat processing, boatbuilding). Kinh population with Nung, Tay, San Chay, and other minorities. In 1997 province was scheduled to be divided into two new provinces: BAC GIANG (⊙ Bac Giang) and BAC NINH (⊙ BAC NINH).

Habahe (HAH-BAH-HUH), town, ⊙ Habahe county, northernmost XINJIANG UYGUR AUTONOMOUS REGION, NW CHINA, on Kaba River, and 90 mi/145 km W of ALTAY, near KAZAKHSTAN border; 47°53′N 86°12′E. Livestock; logging; food processing. Also called Kaba.

Haba, La (AH-vah, lah), town, BADAJOZ province, W SPAIN, 30 mi/48 km E of MÉRIDA. Cereals, chickpeas, tubers, melons, olives; livestock.

Habana, CUBA: see HAVANA.

Habana, La, CUBA: see HAVANA.

Habartov, German *Habersbirg*, town, ZAPADOCESKY province, W BOHEMIA, CZECH REPUBLIC, 4 mi/6.4 km WNW of SOKOLOV; 50°12′N 12°33′E. Manufacturing of textiles; in lignite-mining area.

Habay (ah-BAI), agricultural commune (2006 population 7,830), Virton district, LUXEMBOURG province, SE BELGIUM, in the ARDENNES, 8 mi/12.9 km WNW of ARLON. Has eighteenth-century castle.

Habban (huh-BAN), township, S YEMEN, on the Wadi Habban, and 75 mi/121 km WNW of BALHAF, 15 mi/24 km NE of Yashbum. Agricultural oasis. Formerly a separate Wahidi sultanate.

Habbaniya (huhb-buh-NEE-yuh), town, ANBAR province, central IRAQ, on the W bank of the EUPHRATES RIVER, and 45 mi/72 km W of BAGHDAD, near NE shore of LAKE HABBANIYA. Seaplane base on Lake Habaniya. Was British military and air base until transferred to the Iraqi government in 1955. In the spring of 1941, the scene of fighting between the British garrison and the army of a pro-German government that seized power in Baghdad.

Habbaniya, Lake (huhb-buh-NEE-yuh), salt lake (□ 54 sq mi/140.4 sq km), ANBAR province, central IRAQ, just S of the EUPHRATES RIVER, in a natural depression between AL FALLUJA and AR RAMADI, 50 mi/80 km W of BAGHDAD; 15 mi/24 km long, 8 mi/13 km wide. Since 1911 attempts have been made to use it for flood control and as a storage reservoir. The Ramadi Dam regulating apparatus and a number of canals were completed in 1956, flooding the basin and creating an artificial lake within it.

Habelschwerdt, POLAND: see BYSTRZYCA KLODZKA.

Habelschwerdt Mountains, German *Habelschwerdter Gebirge,* Polish *Góry Bystrzyckie,* range of the SUDETES Mountains, along LOWER SILESIA, SW POLAND and N CZECH REPUBLIC; extend c.20 mi/32 km between PO-LANICA ZDRÓJ (Poland; NW) and MEZHYRICH (Uk-raine; SE); rise to 3,205 ft/977 m, 5 mi/8 km SW of BYSTRZYCA KŁODZKA (Poland). The parallel Ad-lergerbirge are largely in Czech Republic, and sepa-rated from Habelschwerdt Mountains by upper DIVOCHA ORLICE RIVER.

Habern, CZECH REPUBLIC: see HABRY.

Habersbirg, CZECH REPUBLIC: see HABARTOV.

Habersham (HA-buhr-sham), county (□ 283 sq mi/735.8 sq km; 2006 population 41,112), NE GEORGIA on SOUTH CAROLINA state line; CLARKESVILLE; 34°38'N 83°32'W. BLUE RIDGE (N) and piedmont (S) area. Farming (cotton, hay, sweet potatoes, apples, peaches; poultry, cattle, hogs; timber). Textile manufacturing. Part of CHATTAHOOCHEE National Forest in N. Formed 1818.

Habichtswald (HAH-bichts-vahlt), village, HESSE, central GERMANY, 6 mi/9.7 km W of KASSEL; 51°22'N 09°21'E.

Habiganj (ho-bee-gawnj), town (2001 population 55,476), SYLHET district, E EAST BENGAL, BANGLA-DESH, in SURMA VALLEY, on branch of KUSIYARA RIVER, and 45 mi/72 km SW of SYLHET; 24°25'N 91°19'E. Trades in rice, tea, oilseeds, hides. Rice research station; tea processing nearby. Railroad station 7 mi/11.3 km SSE, at Shaistaganj.

Habikino (hah-BEE-kee-no), city, OSAKA prefecture, S HONSHU, W central JAPAN, 9 mi/15 km S of OSAKA; 34°33'N 135°36'E.

Habla, Arab village, TULKARM district, 2 mi/3 km SE of QALQILYA, in the Samarian Highlands, WEST BANK; 32°09'N 34°58'E. Agriculture (olives, fruit).

Habo (HAH-BOO), town, SKARABORG county, S SWE-DEN, near S end of LAKE VÄTTERN, 8 mi/12.9 km NNW of JÖNKÖPING; 57°55'N 14°05'E. Has seventeenth-century church.

Habor, SYRIA and TURKEY: see KHABUR.

Haboro (hah-BO-ro), town, Rumoi district, W HOK-KAIDO prefecture, N JAPAN, on SEA OF JAPAN, 90 mi/145 km N of SAPPORO; 44°21'N 141°42'E. Protected seafowl breeding ground and yew forest.

Habra (ah-BRAH), irrigated lowland in MOSTAGANEM and MASCARA wilaya, NW ALGERIA, between MO-HAMMADIA (Mascara wilaya; S) and the Gulf of ARZEW (N); water supplied by the Oued HAMMAM. Grows citrus fruit, cotton, and vegetables.

Hab River, c.250 mi/402 km long, in SE BALUCHISTAN province, SW PAKISTAN; rises in N PAB RANGE S of KHUZDAR; flows SSE, just W of KIRTHAR RANGE, and SSW, along SIND province border, to ARABIAN SEA N of Cape MONZE. Source for part of KARACHI water supply.

Habry, German *Habern,* town, VYCHODOCESKY prov-ince, E BOHEMIA, CZECH REPUBLIC, 11 mi/18 km SSE of CASLAV; 49°45'N 15°29'E. Agriculture (sugar beets, potatoes); poultry. Historical town. In 1100, it was a customs station on the road from Bohemia to MOR-AVIA.

Habsburg (HAHBS-boorg), castle, AARGAU canton, N SWITZERLAND, near the AARE RIVER. Built c.1030, it served during the 12th and 13th century as the seat of the counts of Habsburg, or Hapsburg, whose name derives from the castle [German *Habichtsburg* =hawk's castle].

Habsheim (ahb-ZEM), German (HAPS-heim), com-mune (□ 6 sq mi/15.6 sq km), HAUT-RHIN de-partment, in ALSACE, E FRANCE, 4 mi/6.4 km ESE of MULHOUSE; 47°44'N 07°25'E. Hosiery manufacturing. Has 16th-century town hall.

Habsor, ISRAEL: see ESHKOL.

Hacarí (ah-kahr-EE), town, ⊙ Hacarí municipio, NORTE DE SANTANDER department, N COLOMBIA, 50 mi/80 km NW of CÚCUTA; 08°19'N 73°08'W. Eleva-tion 2,946 ft/897 m. Coffee, corn; livestock.

Hachado Pass, ARGENTINA and CHILE: see PINO HA-CHADO PASS.

Hachem, El (ah-SHEM, el), village, MASCARA wilaya, NW ALGERIA, in the TELL ATLAS Mountains, 20 mi/32 km E of MASCARA. Cereals. Formerly Dombasle.

Hachenburg (HAH-khuhn-boorg), town, RHINELAND-PALATINATE, W GERMANY, 17 mi/27 km SW of SIEGEN; 50°40'N 07°49'E. Industry includes printing and brewing. Has 17th-century castle. Chartered in 1247.

Hachijo (hah-CHEE-jyo), town, on HACHIJO-SHIMA island, TOKYO prefecture, SE JAPAN, 16 mi/25 km S of SHINJUKU; 33°06'N 139°47'E. Fish (*kusaya*) drying, silk (*kihachijo*) dyeing. Dropworts (*ashitaba*).

Hachijo-shima (hah-chee-JO–shee-mah), island (□ 40 sq mi/104 sq km), TOKYO prefecture, SE JAPAN, in the PACIFIC OCEAN. HACHIJO is the chief town of this mountainous island, which has two active volcanoes. Dairying, fishing, and weaving are the main occupa-tions.

Hachikai (hah-chee-KAH-ee), village, Ama district, AICHI prefecture, S central HONSHU, central JAPAN, 12 mi/20 km W of NAGOYA; 35°12'N 136°42'E.

Hachiman (hah-CHEE-mahn), town, Gujo district, GIFU prefecture, central HONSHU, central JAPAN, 28 mi/45 km N of GIFU; 35°44'N 136°57'E. Ayu, natural mineral (*Sogi*) water; dyeing (*aizome*); pongee.

Hachimori (hah-CHEE-mo-ree), town, Yamamoto district, Akita prefecture, N HONSHU, NE JAPAN, 47 mi/75 km N of AKITA city; 40°21'N 140°01'E.

Hachinohe (hah-chee-no-HE), city (2005 population 269,144), Aomori prefecture, N HONSHU, N JAPAN, on the Oirase River, and the PACIFIC OCEAN, 43 mi/70 km S of AOMORI; 40°30'N 141°29'E. Fishing and com-mercial port and industrial and commercial center. Paper; marine-product processing. Breeding ground of black-tailed gull. Stone Age ruins at Korekawa.

Hachioji (hah-chee-O-jee), city (2005 population 563,077), Tokyo prefecture, E central HONSHU, E central JAPAN, 16 mi/25 km W of SHINJUKU; 35°39'N 139°19'E. Railroad center. Manufacturing includes cameras, computer components, ties. Seat of many colleges. Bombed (1945) in World War II. Absorbed former town of Komiya in early 1940s.

Hachirogata (hah-chee-RO-gah-tah), town, South Akita district, Akita prefecture, N HONSHU, NE JAPAN, 16 mi/25 km N of AKITA city; 39°56'N 140°04'E. On reclaimed land of HACHIRO-GATA lagoon.

Hachiro-gata (hah-chee-RO–gah-tah), lagoon (□ 87 sq mi/226.2 sq km), Akita prefecture, N HONSHU, NE JAPAN, at base of OGA PENINSULA, 11 mi/18 km NNW of AKITA; connected (S) with SEA OF JAPAN; 16 mi/26 km long, 8 mi/12.9 km wide.

Hachiryu (hah-CHEE-ryoo), town, Yamamoto district, Akita prefecture, N HONSHU, NE JAPAN, 28 mi/45 km N of AKITA city; 40°05'N 140°00'E. Melons.

Hachita (ah-CHEE-tuh), unincorporated village, GRANT county, SW NEW MEXICO, 38 mi/61 km WSW of DEMING, at crossroads just N of New Mexico bootheel, 12 mi/19 km NW of MEXICO (CHIHUA-HUA) border. In desert area. Little Hatchet Mountains to W; Cedar Mountain Range to E; CONTINENTAL DIVIDE passes to N. Port of entry at ANTELOPE WELLS c.46 mi/74 km to SSW.

Hacho, Monte, MOROCCO: see CEUTA.

Hacienda Cartavio, PERU: see CARTAVIO.

Hacienda Heights (hah-see-EN-dah), unincorporated city (2000 population 53,122), LOS ANGELES county, S CALIFORNIA; residential suburb 16 mi/26 km ESE of downtown LOS ANGELES, NE of WHITTIER, in Puente Hills; 34°00'N 117°58'W. Manufacturing (electrical and electronic equipment).

Hacin, TURKEY: see SAIMBEYLI.

Hacine (hah-SEEN), village, MASCARA wilaya, NW AL-GERIA, on railroad, and 11 mi/18 km NW of MASCARA. Citrus groves; brick manufacturing. Formerly Du-blineau.

Hackberry, unincorporated town (2000 population 1,699), CAMERON parish, LOUISIANA, 15 mi/24 km N of CAMERON; 29°59'N 93°20'W. Storage facility of the National Strategic Petroleum Reserve. A recreational fishing area, it claims to be the Crab Capital of the South.

Hackensack (HA-kin-sak), city (2006 population 43,671), ⊙ BERGEN county, NE NEW JERSEY, on the HACKENSACK River, a residential and industrial sub-urb of NEW YORK city; 40°53'N 74°02'W. Manu-facturing. Trading post here established 1647 by Dutch settlers from Manhattan. During the American Re-volution the city served as camping ground for armies of both sides. Grew as a commercial and shipping center in the early 1800s. Of interest are the Church on the Green (First Dutch Reformed; built 1696, rebuilt 1728) and the von Steuben House (1739), a state his-toric site and the headquarters of the county historical society. A campus of Fairleigh Dickinson University is here. Settled 1647, incorporated as a city 1921.

Hackensack (HA-kin-sak), village (2000 population 285), CASS county, central MINNESOTA, 12 mi/19 km SSE of WALKER on E shore of Birch Lake; 46°55'N 94°31'W. In wooded lakes region. Poultry; oats; wild rice; dairying; timber; manufacturing (signs, septic tanks). Part of Foothills State Forest to W; Chippewa National Forest to N.

Hackensack (HA-kin-sak), river, c.45 mi/72 km long, NEW YORK and NEW JERSEY; rises in SE New York; flows S through the Jersey Meadowlands (wetlands), NE New Jersey, to NEWARK BAY. The lower Hack-ensack is heavily industrialized and economically tied to the ports on Newark Bay and to the industrial development on the nearby PASSAIC River. It is nav-igable by oceangoing vessels in New Jersey to KEARNY, and by tugs and barges to HACKENSACK. Upper course is dammed to form three reservoirs that supply water to ROCKLAND (New York) and BERGEN (New Jersey) counties.

Hacketstown (HA-kuhts-TOUN), Gaelic *Baile an Droichid,* town (2006 population 606), NE CARLOW county, SE IRELAND, 16 mi/26 km E of CARLOW; 52°52'N 06°33'W. Agricultural market. Two battles were fought here in the rising of 1798.

Hackett, village (2000 population 694), SEBASTIAN county, W ARKANSAS, 13 mi/21 km S of FORT SMITH, near OKLAHOMA state line; 35°11'N 94°24'W.

Hackettstown, town (2000 population 10,403), WAR-REN county, NW NEW JERSEY, in fertile Musconetcong Valley, 14 mi/23 km W of DOVER; 40°51'N 74°49'W. Headquarters for several large corporations. State fish hatcheries located here. Seat of Centenary College. Incorporated 1853.

Hacking, Port (HACK-eeng), inlet of PACIFIC OCEAN, E NEW SOUTH WALES, AUSTRALIA, just S of BOTANY BAY, between Cape Baily and Big Jibbon Point; 4 mi/6 km wide at mouth, 4 mi/6 km long; semi-circular; 34°04'S 151°08'E. Drowned valley, estuary 22 mi/35 km S of SYDNEY, easily reached by public and private trans-port, especially from Sydney's S and SW suburbs. Good water quality; wading birds, amateur fishing (commercial fishing ban). CRONULLA, resort town, on W shore; bathing beach.

Hackleburg, town (2000 population 1,527), Marion co., NW Alabama, 13 mi/21 km NE of Hamilton. Apparel; mobile homes. William B. Bankhead National Forest is E. Named for the thick plants growing in the area. Inc. in 1909.

Hackletons Cliff, 1,000-ft/305-m limestone escarpment, c.1 mi/1.6 km inland from E coast of BARBADOS; 1.5 mi/2.4 km long.

Hackney (HAK-nee), inner borough (□ 7 sq mi/18.2 sq km; 2001 population 202,824) of GREATER LONDON, SE

ENGLAND, on the LEA RIVER; 51°25'N 00°15'W. Clothing manufacturing (in Hackney); printing and furniture making (in Shoreditch). London's first theater was built in SHOREDITCH (c.1575). The parish church of St. Mary, in Stoke Newington, is one of the few remaining Elizabethan churches. The writer Daniel Defoe lived and Edgar Allen Poe went to school here. The London College of Furniture, the Shoreditch College for the Clothing Industry, and Cordwainer's Technical College are in the borough. Hackney Marshes, a large sports and recreation area intersected by the Lea, lies just outside the borough. The borough includes Clapton, Dalston, Hoxton, and Stamford Hill.

Hacoi, VIETNAM: see QUANG HA.

Haçreş Dag, peak (8,822 ft/2,689 m), E central TURKEY, in BITLIS MOUNTAINS, 6 mi/9.7 km SSW of Museum. Sometimes called Kurtik Dag.

Hadagalli (HUH-dah-gah-lee), town, BELLARY district, KARNATAKA state, S INDIA, 35 mi/56 km SW of HOSPET; 16°49'N 75°57'E. Peanut milling; date palms. Also spelled Huvvinahadagalli, Huvinahadgalli.

Hadali (huh-DAH-lee), town, SARGODHA district, W central PUNJAB province, central PAKISTAN, 31 mi/50 km WNW of SARGODHA; 32°18'N 72°12'E.

Hadama, ETHIOPIA: see NAZRĒT.

Hadamar (HAH-dah-mahr), town, HESSE, W GERMANY, 4 mi/6.4 km N of Limburg; 50°27'N 08°02'E. Glass industry. Has 17th-century town hall and former castle. Site of concentration camp during Hitler regime.

Hadano (hah-DAH-no), city (2005 population 168,587), KANAGAWA prefecture, E central HONSHU, E central JAPAN, 25 mi/40 km N of YOKOHAMA; 35°22'N 139°13'E. Computers. Peanuts. Tanzawa area in nearby Tanzawa Oyama quasi-national park known for its scenic beauty.

Hadar (HAI-duhr), village (2006 population 326), PIERCE county, NE NEBRASKA, 7 mi/11.3 km SSE of PIERCE, and on branch of ELKHORN RIVER; 42°06'N 97°27'W.

Hadar Hacarmel (hah-DAHR hah-kahr-MEL), central section of HAIFA, NW ISRAEL, on N slope of MOUNT CARMEL, overlooking Bay of Acre of the MEDITERRANEAN SEA; 32°48'N 35°00'E. Elevation 150 ft/45 m. Business and entertainment center. Formerly a residential neighborhood to the SW before expansion. First modern Jewish settlement. Site of Haifa municipal buildings and law courts. Founded 1912.

Hadashville, hamlet, SE MANITOBA, W central CANADA, 57 mi/92 km from WINNIPEG, and in REYNOLDS rural municipality; 49°40'N 95°54'W.

Hadd, township (□ 232 sq mi/603.2 sq km; 2001 population 11,637), BAHRAIN, E of MUHARRAQ island; 26°14'N 50°39'E. Also called Al-Hadd or Hedd.

Hadda (HAHD-dah), village, NANGARHAR province, AFGHANISTAN, 6 mi/9.7 km S of JALALABAD; 34°22'N 70°29'E. Inhabited largely by Mohmands. Archaeological site; excavations (1920s) have uncovered Buddhist monasteries and sculpture of the Greco-Buddhist school of Gandhara. The stupas discovered at Hadda were partially destroyed in military operations in the 1980s or looted during the subsequent civil war.

Hadda (HAH-duh), village, S HEJAZ, SAUDI ARABIA, in the WADI FATIMA, 17 mi/27 km W of MAKKA, and on highway to JIDDA. Boundary between TRANSJORDAN (now JORDAN) and Saudi Arabia was partly defined here in 1925 by the British and Ibn Saud.

Haddaj, Al (hah-DAHJ, el), village, QABIS province, SE TUNISIA, 3 mi/4.8 km N of MATMATAH. Berber troglodyte (underground) dwellings here; the largest troglodyte village in the JABAL DAHAR, it was abandoned after 1969 floods.

Haddam, town, MIDDLESEX county, S CONNECTICUT, bisected by the CONNECTICUT River (here bridged), just S of MIDDLETOWN; 41°28'N 72°32'W. Includes the HIGGANUM section of Haddam. Agriculture. Industries include lumberyards; manufacturing of ma-

chinery, electronic equipment, wire products, plastics, molding; extruding and offset printing. Atomic energy plant. State forest. Settled 1662, incorporated 1668.

Haddam (HAD-uhm), village (2000 population 169), WASHINGTON county, N KANSAS, on small affluent of LITTLE BLUE RIVER, and 13 mi/21 km WNW of WASHINGTON; 39°51'N 97°17'W. Grain; livestock.

Haddenham (HAD-uhn-uhm), agricultural village (2001 population 8,368), central BUCKINGHAMSHIRE, S central ENGLAND, 6 mi/9.7 km SW of AYLESBURY; 51°46'N 00°55'W. Has 13th-century church.

Haddenham (HAD-uhn-uhm), village (2001 population 5,731), in Isle of ELY, N CAMBRIDGESHIRE, E ENGLAND, 6 mi/9.7 km SW of ELY; 52°21'N 00°09'E. Agricultural market. Has 13th-century church, built on site of a church founded 673.

Haddington (HAH-ding-tuhn), town (2001 population 8,851), ⊙ EAST LOTHIAN, E Scotland, on TYNE RIVER, and 16 mi/26 km E of EDINBURGH; 55°56'N 02°46'W. This market town once had a large corn exchange. Hides are traded. Previously manufacturing of farm machinery and textiles. Giffordgate, a suburb, was the birthplace of Protestant reformer John Knox. French troops sacked the town for eighteen months (1547–1549). Formerly in STRATHCLYDE, abolished 1996.

Haddock (HAD-uhk), town, JONES county, GEORGIA, 17 mi/27 km NE of MACON; 33°02'N 83°25'W. Manufacturing of canned food products.

Haddon (HA-duhn), township, CAMDEN county, S NEW JERSEY, 4 mi/6.4 km SE of CAMDEN; 39°54'N 75°03'W. Suburban residential community. Incorporated 1865.

Haddonfield (HA-duhn-feeld), borough (2006 population 11,515), CAMDEN county, SE NEW JERSEY, a residential suburb just SE of CAMDEN; 39°53'N 75°01'W. Of interest are Indian King Tavern (1750), where the first state legislature met in 1777, and the Haddonfield Historical Society; downtown area is historic preservation district with many 19th- and some 18th-century structures. Site of first complete dinosaur fossil ever found, "Haddonfield Hadrosaur," made national historical landmark 1993. Settled c.1713, incorporated 1875.

Haddon Heights (HA-duhn), residential borough (2006 population 7,365), CAMDEN county, SW NEW JERSEY, 5 mi/8 km SE of CAMDEN; 39°52'N 75°04'W. Laid out c.1891, incorporated 1904.

Hadd, Ras Al (HAHD, RAHS ahl), village, E OMAN, 110 mi/177 km SE of MUSCAT, on RAS AL HADD (E cape of Arabian Peninsula); 22°33'N 59°48'E. Formerly a military airfield; buildings converted into housing units.

Hadeija (hah-dai-JAH), river, NE NIGERIA; rises in KANO state, S of KANO; flows NE across JIGAWA state, past Hadeija, across N YOBE state, to the NIGER border, where it merges with the Komadugu Gana River to form the KOMADOUGOU YOBE River W of DAMASAK (BORNO state). Receives the Jama'are River N of GORGORAM. The Hadeija-Jama'are system produces 6,000 tons/5,442 metric tons of fish annually. Part of the flood plain area receives 24 in/60 cm–28 in/70 cm of rain per year during the 3- to 4-month rainy season. The area supports dry savanna vegetation; grazing by Fulani pastoralists. Millet farming in the area depends on rainwater.

Hadejia (hah-dai-JAH), town, JIGAWA state, N NIGERIA, on Hadejia River (branch of the KOMADUGU YOBE River), and 110 mi/177 km NE of KANO; 12°27'N 10°03'E. Agricultural trade center; cotton, peanuts, millet, durra. Sometimes spelled Hadeija.

Hademarschen, GERMANY: see HANERAUHADE-MARSCHEN.

Hadera (khah-DE-rah), city (2006 population 76,300), W ISRAEL, on the plain of SHARON, near the MEDITERRANEAN SEA; 32°26'N 34°55'E. Elevation 180 ft/54 m. Initially completely agricultural, there are still carp ponds, beehives, cattle, poultry, and flower and banana fields, but industry has taken over much of the

city's economy. Manufacturing includes tires, paper, and processed foods. NW of the town, one of Israel's largest electric power plants, a combination coal and oil plant with its own port. Founded in 1890 by Zionist immigrants from Russia and Eastern Europe, who drained vast malarial swamps and planted eucalyptus woods, groves of citrus fruit, and fields of grain. The ruins of Caesarea Palestinae are nearby. Hometown of Israel's first Olympic gold medalist, windsurfer Gal Friedman.

Hadera Stream, ISRAEL; rises in the Samarian Highlands, W of TUBAS; flows W, reaching the MEDITERRANEAN SEA N of the town HADERA. Only its lower part is perennial.

Haderslev (HA-dthuhr-slev), German *Hadersleben*, city (2000 population 21,114), SØNDERJYLLAND county, S DENMARK, a seaport on the HADERSLEV FJORD, an inlet of the LILLE BÆLT; 55°14'N 09°30'E. It is a commercial and industrial center and railroad junction. Held by Prussia 1864–1920. Of note is the Church of Saint Mary (thirteenth century).

Haderslev Fjord (HA-dthuhr-slev), 13 mi/21 km long, narrow inlet of the LILLE BÆLT, S JUTLAND, DENMARK. HADERSLEV city and port are on it.

Hadgaon (HUHD-goun), town, tahsil headquarters, NANDED district, MAHARASHTRA state, W INDIA, 32 mi/51 km NE of NANDED; 19°30'N 77°40'E. Millet, cotton, wheat.

Hadhal, MONGOLIA: see HATGAL.

Hadháztéglás (HAHD-hahz-taig-lahsh), city, HAJDU-BIHAR county, Hungary, 7 mi/11.3 km N of DEBRECEN. Wheat, corn, sugar beets; cattle, hogs; manufacturing (consumer durables, handicrafts).

Hadhdhunmathi Atoll, S central group of MALDIVES, in INDIAN OCEAN; 01°46'N 73°14'E–02°08'N 73°45'E. Has commercial airport, fish-refrigeration plant. Also spelled Haddummati.

Hadhr, Al (HUH-der, ahl), ancient *Hatra*, village, NINEVEH province, N IRAQ, 55 mi/89 km SW of MOSUL; 35°34'N 42°43'E. Conquered C.E. 115 by Trajan, its ruins constitute some of the few remaining stone monuments of IRAQ. Sometimes spelled Al-Hadr.

Hadhramaut (HAH-der-MOT), region, S ARABIA, on the Gulf of ADEN and the ARABIAN SEA, occupying the E part of S YEMEN. Historically, the name refers to the former Hadhramaut states, a collective term for the QUAITI and KATHIRI sultanates. The chief port and city of the region is MUKALLA. The Hadhramaut extends c.400 mi/644 km from E to W. It consists of a narrow, arid coastal plain, a broad plateau averaging 4,500 ft/1,370 m high, a region of deeply sunk wadis (watercourses), and an escarpment fronting the desert. The sedentary population, the Hadranis, live in towns and villages built along the wadis and harvest crops of wheat, maize, alfalfa, millet, dates, coconuts, and coffee. On the plateau the Bedouins raise sheep and goats. The Hadhramaut is believed by some scholars to be identical with the biblical name Hazarmaveth. Also spelled Hadramaut.

Hadhrami (huh-DRAH-mee), S YEMEN; 15°15'N 44°17'E. Ash SHIBR was its main center. Came under British protection in 1903. Since 1967 in S. YEMEN. Also spelled Hadrami.

Hadibu (HAH-de-boo), town, ⊙ SOCOTRA island and adjoining islands, on island's N coast, YEMEN; 12°39'N 54°01'E. Former capital of entire MAHRI sultanate of QISHN and Socotra. Fisheries. Airfield. In the 1970s became important Soviet base in the ARABIAN SEA. Sometimes spelled Tamarida, Tamrida, and Tamridah.

Hadim, township, S central TURKEY, 60 mi/97 km S of KONYA; 36°58'N 32°27'E. Cereals.

Haditha (hah-DEETH-uh), town, ANBAR province, NW central IRAQ, on the W bank of the EUPHRATES RIVER, and 70 mi/113 km NW of AR RAMADI; 34°08'N 42°22'E. Oil refinery; hydroelectric power station. Dates; livestock. Just W, the oil pipelines from the N Iraqi oilfields leading to LEBANON and SYRIA on the

---

Cross-references are shown in SMALL CAPITALS. The pronunciation guide is shown on page xix. The sources of population figures are shown on page xvii.

MEDITERRANEAN coast and closed down following the PERSIAN GULF War.

Hadja, YEMEN: see HAJJA.

Hadjab al Ayyūn (hah-JEB el ah-YOON), town (2004 population 9,648), AL QAYRAWAN province, central TUNISIA, on Susah-Tawzar railroad, and 36 mi/58 km SW of AL QAYRAWAN; 35°23'N 09°32'E. Sheep market; esparto, olives. Lead mines nearby.

Hadjadj (hah-JAHJ), village, MOSTAGANEM wilaya, NW ALGERIA, 17 mi/27 km NE of MOSTAGANEM. Wine. Formerly called Bosquet.

Hadjar, El (hah-JAHR, el), town, ANNABA wilaya, NE ALGERIA, 7 mi/12 km S of ANNABA. Major center of Algerian cast-iron and steel production; connected by railroad to iron mines at OUENZA and Annaba port. Formerly called Duzerville.

Hadjer-Lamis, administrative region, W CHAD, on shore of Lake CHAD (to NW); ⊙ MASSAKORY. Borders LAC (NW), KANEM (N), BATHA (E), GUERA (SE), and CHARI-BAGUIRMI (S) administrative regions and CAMEROON (W). Livestock; agriculture. Major centers include Massakory and BOKORO. Formed following a decree in October 2002 that reorganized Chad's administrative divisions from twenty-eight departments to eighteen regions. This area is roughly coterminous with the N portion of former CHARI-BAGUIRMI prefecture.

Hadjout (hah-JOOT), town, TIPAZA wilaya, N central ALGERIA, at W edge of the MITIDJA plain, 23 mi/37 km W of BLIDA; 36°33'N 02°20'E. Agricultural trade center in vegetable-gardening region. Formerly called Marengo.

Had Kourt (hahd KOORT), village, Sidi Kacem province, Gharb-Chrarda-Beni Hssen administrative region, N MOROCCO, in GHARB lowland, 16 mi/26 km ESE of SOUK EL ARBA; 34°37'N 05°44'W. Livestock.

Hadleigh (HAD-lee), town (2001 population 18,300), ESSEX, SE ENGLAND, 4 mi/6.4 km W of SOUTHEND-ON-SEA; 51°33'N 00°36'E. Remains of Norman castle.

Hadley (HAD-lee), town (2001 population 5,100), central SHROPSHIRE, W ENGLAND, 2 mi/3.2 km E of WELLINGTON; 52°42'N 02°29'W. Light industries.

Hadley, town, HAMPSHIRE county, W central MASSACHUSETTS, on CONNECTICUT RIVER opposite NORTHAMPTON; 42°22'N 72°34'W. Paper mills. Settled 1659, incorporated 1661.

Hadley, village (2000 population 81), MURRAY county, SW MINNESOTA, 5 mi/8 km W of SLAYTON, near Beaver Creek; 44°00'N 95°51'W. Dairying.

Hadley, ENGLAND: see BARNET.

Hadleys Purchase, land purchase, COOS county, N central NEW HAMPSHIRE, 25 mi/40 km SSW of BERLIN. Wilderness area in WHITE MOUNTAIN National Forest.

Hadlock, unincorporated town, JEFFERSON county, NW WASHINGTON, 6 mi/9.7 km S of PORT TOWNSEND, on Port Townsend Bay, arm of PUGET SOUND; 48°03'N 122°46'W. Fish, crabs, oysters. Resort area. Old Fort Townsend State Park to N; Anderson Lake State Park to SW; Jefferson County International Airport to W. Bridge to Indian Island Naval Reservation 1 mi/1.6 km ESE. Also called PORT HADLOCK.

Hadlow (HAD-lo), village (2001 population 3,908), W KENT, SE ENGLAND, 4 mi/6.4 km NE of TONBRIDGE; 51°13'N 00°20'E. Agricultural market in hop-growing region. Site of 19th-century castle.

Hadlyme, village and postal section of EAST HADDAM and LYME, CONNECTICUT, in N Lyme and S East Haddam. Gillette Castle State Park, with rock mansion of actor William Hooker Gillette (noted for creating role of Sherlock Holmes in 1899) overlooking CONNECTICUT River. Seasonal ferry to CHESTER.

Ha Dong (HAH DOUNG), city, ⊙ HA TAY province, N VIETNAM, 5 mi/8 km S of HANOI; 20°58'N 105°46'E. Silk-spinning, food-processing, and trading center; light manufacturing (wood products, lace, embroi-

dery). Transportation and commercial complex. Fast-growing city with much emphasis from government development. Formerly Ha Dong.

Hadong (HAH-DONG), town, SOUTH KYONGSANG province, SOUTH KOREA, on SOMJIN RIVER, and 20 mi/32 km WSW of CHINJU. Agricultural center (rice, barley, soybeans, hemp, fruit). Serves as a S entrance to CHIRI MOUNTAIN NATIONAL PARK.

Hadr, Al, IRAQ: see HADHR, AL.

Hadramawt, YEMEN: see HADHRAMAUT.

Hadrami, YEMEN: see HADHRAMI.

Hadres (HAH-dres), township, N LOWER AUSTRIA, in the WEINVIERTEL, on Pulkau River, and 10 mi/16 km SSE of ZNOJMO (Czech Republic). Near CZECH border; 48°43'N 16°08'E. Vineyards; corn.

Hadrianopolis, TURKEY: see EDIRNE.

Hadrian's Wall, wood pallisade on earth berm barrier, with approximately 20 forts, constructed by Roman forces in Britain in C.E. 122, along the border with the Scots tribes; approximately 60 mi/97 km long, from modern CARLISLE to modern NEWCASTLE UPON TYNE, running N of the TYNE RIVER. Named for Emperor Hadrian (ruled C.E. 117-A.D. 138), who supervised part of the construction. The wall was not intended to keep out raiders, but restrict their movement, trap them on the near side if they should cross, and shield movement of limited Roman garrisons in the area. A modern tourist attraction.

Hadrumetum, TUNISIA: see SUSAH.

Hadseløya (HAHT-suhl-uh-yah), island (□ 39 sq mi/101.4 sq km), in NORTH SEA, NORDLAND county, N NORWAY, in the VESTERÅLEN group, between LANGØYA (N), HINNØYA (E), and AUSTVÅGØY (S), 20 mi/32 km NNE of SVOLVÆR; 10 mi/16 km long (E-W), 5 mi/8 km wide; rises to 1,683 ft/513 m. On it are MELBU (S) and STOKKMARKNES (N) villages. Highway circles island (W) is tourist attraction. Many Viking relics found here.

Hadspen (HAD-spen), town, TASMANIA, SE AUSTRALIA, 11 mi/18 km SW of LAUNCESTON, on Bass Highway; 41°30'S 147°05'E. Nineteenth-century buildings.

Hadsten (HAHS-duhn), city (2000 population 6,794), ÅRHUS county, E JUTLAND, DENMARK, 13 mi/21 km NNW of ÅRHUS; 56°20'N 10°05'E. Meatpacking; hogs; dairying; barley; manufacturing (electronic parts).

Hadsund (HAH-soon), town (2000 population 4,779), NORDJYLLAND county, N JUTLAND, DENMARK, on MARIAGER FJORD, and 23 mi/37 km SSE of ÅLBORG; 56°46'N 10°10'E. Fishing (cod, eels); cement, limestone; meatpacking; cattle; dairying.

Hadyach (HAH-dyahch), (Russian *Gadyach*), city, N POLTAVA oblast, UKRAINE, on PSEL RIVER, and 60 mi/97 km NNW of POLTAVA; 50°22'N 34°00'E. Elevation 360 ft/109 m. Railroad terminus. Raion center; metalworking, manufacturing (asphalt-cement, bricks, cheese, food flavoring; beeswax application), flour and feed milling, broiler packing. Beekeeping research station, educational and professional technical schools, sanatorium, heritage museum. Known since the beginning of the 17th century, city since 1634.

Had Yai, THAILAND: see HAT YAI.

Hadzhidimitrovo (hahd-JEE-dee-MEE-tro-vo), village, LOVECH oblast, Svishtov obshtina, N BULGARIA, 9 mi/15 km SE of SVISHTOV; 43°31'N 25°28'E. Grain, vegetables; livestock. Sometimes spelled Khadzhi Dimitrovo. Formerly called Saryar.

Hadzhidimovo (hahd-JEE-DEE-mo-vo), village, SOFIA oblast, ⊙ Hadzhidimovo obshtina (1993 population 12,117), BULGARIA; 41°31'N 23°52'E.

Hadzhi Eles, Bulgaria: see PURVOMAI.

Haebaru (HAH-e-bah-roo), town, Shimajiri district, OKINAWA prefecture, SW JAPAN, 3.1 mi/5 km S of NAHA; 26°11'N 127°43'E.

Haecht, BELGIUM: see HAACHT.

Haedo, Cuchilla de (HAH-e-do, koo-CHEE-yah dah), hill range, N URUGUAY, extends c.200 mi/322 km S

and SW from the BRAZIL border 30 mi/48 km SW of RIVERA; rises to 1,000 ft/305 m; 31°40'S 56°18'W. Crossed by railroad from Brazilian border through TACUAREMBO SE toward PIEDRA.

Hægebostad Tunnel, NORWAY: see KVINESHEI TUNNEL.

Haeju (HAI-JOO), city, ⊙ SOUTH HWANGHAE province, central NORTH KOREA, fishing port on inlet of YELLOW SEA, 65 mi/105 km S of PYONGYANG; 38°02'N 125°42'E. Commercial center for gold mining and agricultural area; gold refining.

Haenam (HAI-NAHM), town, SOUTH CHOLLA province, SW SOUTH KOREA, 90 mi/145 km SSW of KWANGJU, southernmost tip of Korean peninsula; 34°34'N 126°35'E. Rice, barley, soybeans.

Ha'erbin, CHINA: see HARBIN.

Ha-erpin, CHINA: see HARBIN.

Hafat (HUH-fet), village, SE YEMEN, on coast of the ARABIAN SEA, 6 mi/9.7 km ENE of QISHN. Sometimes used as a port.

Hafelekarspitze (HAH-fe-le-KAHR-shpi-tse), peak (7,114 ft/2,168 m) of the NORDKETTE (TYROLEAN-BAVARIAN LIMESTONE ALPS), in TYROL, W AUSTRIA, overlooking INNSBRUCK (just S); 47°19'N 11°23'E. Reached by cable car; superior view.

Hafetz Hayim or **Hafetz Haim** (khah-FETS khah-YEEM), kibbutz, 8.5 mi/13.7 km ESE of ASHDOD, ISRAEL; 31°47'N 34°48'E. Elevation 209 ft/63 m. Mixed farming, guest house. Founded in 1933 by orthodox Jews.

Haffe (HAH-fai), village, LATTAKIA district, W SYRIA, 15 mi/24 km ENE of LATTAKIA; 35°35'N 36°01'E. Tobacco, cereals. Just S is the citadel of SAHYUN.

Hafford (HA-fuhrd), town (2006 population 360), central SASKATCHEWAN, W CANADA, 50 mi/80 km NW of SASKATOON, near Redberry Lake (9 mi/14 km long, 6 mi/10 km wide); 52°43'N 107°21'W. Wheat.

Hafik, village, central TURKEY, on the KIZIL IRMAK, and 21 mi/34 km ENE of SIVAS; 39°53'N 37°17'E. Wheat, barley. Formerly Kochisar.

Hafizabad (HUH-fee-zah-bahd), town, GUJRANWALA district, E PUNJAB province, central PAKISTAN, 30 mi/48 km WSW of GUJRANWALA; 32°04'N 73°41'E. Trade center.

Haflong (HAH-fluhng), town, ⊙ NORTH CACHAR HILL district, central ASSAM state, NE INDIA, in BARAIL RANGE, on railroad, 27 mi/43 km NNE of SILCHAR; 25°11'N 93°02'E. Trades in rice, cotton, sugarcane, tobacco. Silk growing nearby. Its landscape and climate are similar to the once-popular resort of SHILLONG, and it is considered a smaller version. Also spelled Haflang.

Hafnarfjörur (HAHP-nahr-FYUH-[th]uhr), town (2000 population 19,688), SW ICELAND, S of REYKJAVIK. Distribution, industrial, and fishing center with an excellent harbor. During the 15th and 16th century German and English traders fought over the port. The town was chartered in 1908.

Hafnerzell, GERMANY: see OBERNZELL.

Hafrsfjord (HAHFSH-fyawr), inlet of the NORTH SEA, ROGALAND county, SW NORWAY, near STAVANGER. Harald I won (872) a decisive victory here that made him king of all Norway. Also spelled Hafsfjord.

Hafslo (HAHFS-law), village, SOGN OG FJORDANE county, W NORWAY, on N shore of Hafslovatnet, near W shore of LUSTRAFJORDEN (an arm of SOGNEFJORDEN), 6 mi/9.7 km NE of SOGNDAL. Cattle raising, lumbering nearby.

Haft Gel (HAHFT gel), oil town, Khuzestān province, SW IRAN, 50 mi/80 km ENE of AHVAZ; 31°25'N 49°32'E. Pipeline to ABADAN refinery; oil field opened 1928. Also spelled Haftquel or Haft Kel.

Hafun, town, NE SOMALIA, on N shore of Hafun peninsula; easternmost point of AFRICA. Extensive evaporation basin in bay for salt production. Linked by aerial cable to neighboring town of HURDIO for salt hauling. Renamed DANTE by Italians.

Hafun (hah-FOON), peninsula of NE SOMALIA, on INDIAN Ocean, 90 mi/145 km S of Cape GUARDAFUI; 10°27′N 51°23′E. Hafun, at its E tip, is AFRICA's easternmost headland. Peninsula consists of a rocky island (17 mi/27 km long, 13 mi/21 km wide, 390 ft/119 m high), linked to mainland by sandy isthmus (19 mi/31 km long, up to 2.5 mi/4 km wide). Encloses bay (N), largely transformed by the Italian colonial administration into an extensive evaporation basin for salt. HURDIO, on mainland, and Hafun, on peninsula's N shore, are linked by aerial cable (15 mi/24 km long) for salt hauling.

Haga (HAH-gah), town, Shiso district, HYOGO prefecture, S HONSHU, W central JAPAN, 47 mi/76 km N of KOBE; 35°09′N 134°32′E.

Haga, town, Haga district, TOCHIGI prefecture, central HONSHU, N central JAPAN, 9 mi/15 km E of UTSUNOMIYA; 36°32′N 140°03′E. Pears.

Hag Abdalla (HAG ahb-DUHL-lah), township, Sinnar state, E central SUDAN, on left bank of the BLUE NILE River, S of influx of DINDER RIVER, on railroad, and 30 mi/48 km S of WAD MADANI; 13°58′N 33°35′E. Cotton, wheat, barley, corn, fruits, and durra; livestock. Also called Hagg Abd-Allah.

Hagaman, village (□ 1 sq mi/2.6 sq km; 2006 population 1,320), MONTGOMERY county, E central NEW YORK, 2 mi/3.2 km N of AMSTERDAM; 42°58′N 74°09′W.

Hagan (HAI-guhn), city (2000 population 898), EVANS county, E central GEORGIA, 3 mi/4.8 km W of CLAXTON; 32°09′N 81°56′W.

Hagar (HAI-gahr), unincorporated village (□ 41 sq mi/106.6 sq km; 2001 population 874), S central ONTARIO, E central CANADA, 22 mi/35 km from FRENCH RIVER, and included in town of MARKSTAY-WARREN; 46°27′N 80°24′W.

Hagari, INDIA: see BELLARY.

Hagari River (HUH-gah-ree), c.100 mi/161 km long, in KARNATAKA and ANDHRA PRADESH states, S INDIA; rises in BABA BUDAN RANGE in two headstreams joining NE of KADUR; flows (as the Vedavati River) NE, past HIRIYUR, and (here becoming the Hagari River) c.170 mi/274 km generally N to TUNGABHADRA RIVER 25 mi/40 km WNW of ADONI, on Karnataka–Andhra Pradesh state line. Dammed 10 mi/16 km SW of Hiriyur to form irrigation reservoir. Sometimes spelled Hagary or Haggari.

Hagarty and Richards (HA-gahr-tee, RI-chuhrdz), former township (2001 population 1,831), SE ONTARIO, E central CANADA. Amalgamated in 2000 into KILLALOE, HAGARTY AND RICHARDS township. Also written Hagarty & Richards.

Hagaström (HAH-ga-STRUHM), village, GÄVLEBORG county, E SWEDEN, on GAVLEÅN River, 3 mi/4.8 km W of GÄVLE; 60°41′N 17°04′E. Railroad junction.

Hagbourne, East (HAG-bawn), agricultural village (2001 population 2,708), OXFORDSHIRE, S central ENGLAND, 5 mi/8 km W of WALLINGFORD; 51°35′N 01°14′W. Has 13th-century church.

Hage (HAH-ge), village, LOWER SAXONY, NW GERMANY, in EAST FRIESLAND, 18 mi/29 km N of EMDEN; 53°37′N 07°17′E.

Hagemeister Island, FRENCH POLYNESIA: see APATAKI.

Hagen (HAH-gen), city, NORTH RHINE–WESTPHALIA, W GERMANY, on the ENNEPE river; 51°22′N 07°38′E. Industrial center in the RUHR district. Manufacturing includes iron and steel, chemicals, machinery, paper, and textiles. Chartered in 1746; became famous for its textiles in the late 18th century. Main industrial growth dates from 1870. City contains parks, theaters, and museum. Also called HAGEN IN WESTFALEN.

Hagen am Teutoburger Wald (HAH-gen ahm TOI-tuh-boor-ger VAHLT), suburb of OSNABRÜCK, LOWER SAXONY, NW GERMANY, 7 mi/11.3 km SW of city center; 52°12′N 07°52′E. Climatic health resort.

Hagenau, FRANCE: see HAGUENAU.

Hagenbach (HAH-guhn-bahkh), village, RHINELAND-PALATINATE, W GERMANY, near the RHINE RIVER, 6 mi/9.7 km W of KARLSRUHE; 49°01′N 08°15′E.

Hagenbach, GERMANY: see BAD FRIEDRICHSHALL.

Hagenburg (HAH-gen-boorg), village, LOWER SAXONY, NW GERMANY, on S shore of STEINHUDER MEER Lake, 19 mi/31 km E of HANOVER; 52°24′N 09°20′E. Tourism.

Hagendingen, FRANCE: see HAGONDANGE.

Hagengebirge (HAH-gen-ge-BIR-ge), plateau of SALZBURG–UPPER AUSTRIAN LIMESTONE ALPS, extending c.15 mi/24 km S of HALLEIN, along Austro-Germany border, W of the SALZACH; rises (N) to 7,163 ft/2,183 m in the Kahlersberg. KÖNIGSSEE (in Germany) is at W foot.

Hagen in Westfalen, GERMANY: see HAGEN.

Hagenow (HAH-ge-nou), town, MECKLENBURG–WESTERN POMERANIA, N GERMANY, 17 mi/27 km SW of SCHWERIN; 53°27′N 11°11′E. Manufacturing includes food processing; bricks.

Hagen Range (HAH-guhn), E central PAPUA NEW GUINEA, SW PACIFIC OCEAN; highest point circa 13,000 ft/3,962 m.

Hagensborg (HAH-guhnz-borg), community, unincorporated village, W central BRITISH COLUMBIA, W CANADA, at E end of Bella Coola Valley, 13 mi/21 km E of BELLA COOLA, in CENTRAL COAST regional district; 52°23′N 126°34′W. Settled by Norwegians from MINNESOTA (1874), attracted to area for its similarity to NORWAY. Mixed farming; fishing; timber.

Hager City (HAI-guhr), village, PIERCE county, W WISCONSIN, 3 mi/4.8 km N (by bridge) of RED WING (MINNESOTA), on MISSISSIPPI RIVER. Barley, soybeans; livestock. Manufacturing (power transmission poles).

Hagere Hiywot, ETHIOPIA: see AMBO.

Hageremariam (HAH-guhr-MAH-ree-um), township (2007 population 23,871), OROMIYA state, S ETHIOPIA, on main road to KENYA, and 40 mi/64 km NNE of YABELO; 05°38′N 38°14′E. In livestock raising region. Also called Alga.

Hägere Selam (HAH-ger-ai sai-LAHM), town (2007 population 8,896), SOUTHERN NATIONS state, S ETHIOPIA, on road to SOMALIA, and 22 mi/35 km SSE of YIRGA-ALEM; elevation c.9,200 ft/2,800 m; 06°29′N 38°31′E. Commercial center (coffee; hides and livestock). Coptic churches. Also called Hula; also spelled Aghere-Salam.

Hagere Selam, town (2007 population 7,113), TIGRAY state, N ETHIOPIA, 25 mi/40 km NW of MEK'ELE; 13°43′N 39°11′E.

Hagerhill (HAI-guhr-hil), unincorporated town, JOHNSON county, E KENTUCKY, 3 mi/4.8 km SSE of PAINTSVILLE, near Levisa Fork River. Bituminous coal. Manufacturing (machinery).

Hagerman (HAI-guhr-muhn), former township, S ONTARIO, E central CANADA, 19 mi/30 km from PARRY SOUND; 45°36′N 79°55′W. Amalgamated into WHITESTONE township in 2000.

Hagerman, town (2006 population 1,174), CHAVES county, SE NEW MEXICO, near PECOS River, near mouth of Rio Felix, 23 mi/37 km SSE of ROSWELL; 33°06′N 104°19′W. In cotton and alfalfa region.

Hagerman, village (2000 population 656), GOODING county, S IDAHO, 15 mi/24 km SW of GOODING, and on SNAKE RIVER, near mouth of BIG WOOD RIVER; elevation 2,964 ft/903 m; 42°49′N 114°54′W. Watermelons; corn, wheat, alfalfa; winery. U.S. fish hatchery. Hagerman Valley Historical Society Museum here. Bliss Dam (Snake River) to N; HAGERMAN FOSSIL BEDS NATIONAL MONUMENT to S.

Hagerman, village, SUFFOLK county, SE NEW YORK, just E of PATCHOGUE; 40°46′N 72°59′W.

Hagerman Fossil Beds National Monument (□ 7 sq mi/18.1 sq km), TWIN FALLS county, S IDAHO, 23 mi/37 km WNW of TWIN FALLS. Fossils dating from the Pliocene era, with over 125 prehistoric zebra-like horse skeletons excavated. Authorized 1988.

Hagerman Pass, Colorado: see SAWATCH MOUNTAINS.

Hagerstown (HAI-guhrz-toun), city (2006 population 39,008), ⊙ WASHINGTON county, NW MARYLAND, on Antietam Creek, near its junction with the POTOMAC River, in the fertile CUMBERLAND VALLEY; 39°38′N 77°43′W. Shipping and processing center for agriculture products. Its diverse manufacturing includes pipe organs, aircraft, and furniture. The first settler was Jonathan Hager, a Westphalian German, who built a home here in 1737, which is now a museum. Most of the other settlers were also German. Occupied both by Northern and Southern troops during the Civil War; the bloody battle of Antietam (Sharpsburg) was fought nearby at the ANTIETAM NATIONAL BATTLEFIELD SITE. The BALTIMORE AND OHIO RAILROAD reached here in 1867, and for many decades the town was the junction of four railroads. Fort Ritchie, 13 mi/21 km NE on PENNSYLVANIA state line now closed. Was major center for Intelligence Training in World War II. Incorporated 1791.

Hagerstown, town (2006 population 1,661), WAYNE county, E INDIANA, on W Fork of WHITEWATER RIVER, and 16 mi/26 km WNW of RICHMOND; 39°55′N 85°10′W. Trading center in livestock and grain area; manufacturing (candy, consumer goods, wood products).

Hagerstown Valley, N MARYLAND. The valley and ridge extend approximately 65 mi/105 km W from SOUTH MOUNTAIN and ELK RIDGE (E) to DANS (local name for the Allegheny front), POWELL, and Fairview mountains (W). The two distinct subdivisions of the region are the Valley-Ridge section and the GREAT APPALACHIAN VALLEY, also known as HAGERSTOWN VALLEY in Maryland, SHENANDOAH VALLEY in VIRGINIA, and CUMBERLAND VALLEY in PENNSYLVANIA. The two rivers draining into Hagerstown Valley are the Antietam to the E and Conococheague to the W.

Hagersville (HAI-guhrz-vil), unincorporated village, S ONTARIO, E central CANADA, 22 mi/35 km SW of HAMILTON, and included in HALDIMAND; 42°57′N 80°03′W. Dairying; grain and seed milling. Stone quarrying; natural-gas wells.

Hagetmau (ah-zhet-MO), town (□ 11 sq mi/28.6 sq km), LANDES department, AQUITAINE region, SW FRANCE, 17 mi/27 km SSW of MONT-DE-MARSAN; 43°40′N 00°35′W. Woodworking (chairs); market center for grain and hams of the CHALOSSE agricultural district. The crypt of St. Girons is only evidence of an old abbey; visited by pilgrims en route to SANTIAGO DE COMPOSTELA (SPAIN).

Hagfors (HAHG-FORSH), town, VÄRMLAND county, W SWEDEN, in BERGSLAGEN region, on small tributary of Klarälven River, 25 mi/40 km NW of FILIPSTAD; 60°02′N 13°41′W. In former iron-mining region.

Haggari River, INDIA: see HAGARI RIVER.

Hagi (HAH-gee), city, YAMAGUCHI prefecture, SW HONSHU, W JAPAN, on the delta of the Abu River, 16 mi/25 km N of YAMAGUCHI; 34°24′N 131°24′E. Oranges. Was a castle (ruins remain) town of the Mori clan during the Tokugawa era.

Hagia, Hagion, and **Hagios** [Greek=saint], for Greek names beginning thus and not found here: see AYIA AYION and AYIOS.

Ha Giang, province (□ 3,023 sq mi/7,859.8 sq km), N VIETNAM, in northernmost mountain region, on CHINA (N and NW) border; ⊙ HA GIANG, 22°50′N 105°05′E. Bordered E by CAO BANG province, SE by TUYEN QUANG province, S by YEN BAI province, SW by LAO CAI province. Highland province in an isolated area where ecologically complex forests still survive despite recent deforestation. Diverse soils and mineral resources, shifting cultivation, opium growing; forest products (medicinals, honey, beeswax, resins, foraged foods), and some commercial agriculture (tea, cardamom, vegetables, fruits). Lumbering, sawmilling; handicrafts. Tay, H'mong, Dao, and other minorities.

Ha Giang (HAH YAHNG), city, ⊙ HA GIANG province, N VIETNAM, near CHINA border, on Song Lo (CLEAR RIVER), and 135 mi/217 km NNW of HANOI (linked by road); 22°50′N 104°59′E. Transportation hub; market

---

Cross-references are shown in SMALL CAPITALS. The pronunciation guide is shown on page xix. The sources of population figures are shown on page xvii.

complex; administrative center. Cardamoms; cinnabar deposits; shifting cultivation, coffee and tea plantations. Tay, H'mong, Dao, and other minority peoples. Formerly Hagiang.

Ha Giao (HAH YOU), stream, BINH DINH province, central VIETNAM, major coastal stream; 13°58′N 108°53′E.

Hagihara, JAPAN: see HAGIWARA.

Hagiwara (hah-gee-WAH-rah), or **Hagihara**, town, Mashita district, GIFU prefecture, central HONSHU, central JAPAN, 43 mi/70 km N of GIFU; 35°52′N 137°12′E.

Hagley (HAG-lee), village, N TASMANIA, AUSTRALIA, 15 mi/24 km WSW of LAUNCESTON; 41°32′S 146°55′E.

Hagley (HAG-lee), suburb (2001 population 5,600), Worcestershire, central ENGLAND, 2 mi/3.2 km S of STOURBRIDGE; 52°26′N 02°08′W. Site of Hagley Hall, 18th-century Gothic structure, built by Sanderson Miller.

Hagondange (ah-gon-DAHNZH), German *Hagendingen* (HAH-gen-DING-guhn), town (□ 2 sq mi/5.2 sq km), MOSELLE department, LORRAINE region, NE FRANCE, near left bank of MOSELLE RIVER, 10 mi/16 km N of METZ; 49°15′N 06°10′E. In iron-mining district; steel mill.

Hagonoy (hah-GO-noi), town, BULACAN province, S central LUZON, PHILIPPINES, 23 mi/37 km NW of MANILA, on PAMPANGA delta; 14°48′N 120°43′E. Agricultural center (rice, sugar, corn).

Hagoshrim (hah-gosh-REEM), kibbutz, 2 mi/3.2 km E of KIRYAT SHMONA, Upper Galilee, N ISRAEL; 33°13′N 35°37′E. Elevation 288 ft/87 m. Mixed farming and some industry. Guest house. Dan stream (see DAN village) runs through the kibbutz, and nearby is Hurshat Tal national park, where prehistoric artifacts have been found. Also, Bronze Age and Hellenistic cemeteries and ruins from Roman, Byzantine, and later times. Founded in 1948.

Hag's Head, cape on the ATLANTIC OCEAN, W CLARE county, W IRELAND, 8 mi/12.9 km W of ENNISTYMON; 52°57′N 09°29′W. Site of ruined O'Brien's Tower, built 1835 as a tea house for visitors to the Cliffs of Moher.

Hague (HAIG), village (2006 population 707), central SASKATCHEWAN, W CANADA, 30 mi/48 km NNE of SASKATOON; 52°30′N 106°25′W. Mixed farming; dairying.

Hague (HAIG), resort village, WARREN county, E NEW YORK, in the ADIRONDACK MOUNTAINS, on W shore of Lake GEORGE, 8 mi/12.9 km SSW of TICONDEROGA; 43°42′N 73°32′W.

Hague, village (2006 population 77), EMMONS co., S NORTH DAKOTA, 20 mi/32 km SSE of LINTON; 46°01′N 100°00′W. Rice Lake to W. Founded 1902 and named for THE HAGUE, NETHERLANDS, to honor the Dutch settlers that eventually moved further west.

Hague (HAIG), unincorporated village, WESTMORELAND county, E VIRGINIA, 50 mi/80 km ESE of FREDERICKSBURG; 38°04′N 76°39′W. Agriculture (grain, soybeans; cattle). Nearby is Yeocomico Church, originally built 1655; rebuilt 1706.

Hague, La (AHG, lah), cape of MANCHE department, BASSE-NORMANDIE region, NW FRANCE, forming NW extremity of COTENTIN PENINSULA, on ENGLISH CHANNEL, separated from ALDERNEY island (10 mi/16 km W) by Race of ALDERNEY, and 15 mi/24 km WNW of CHERBOURG; 49°43′N 01°57′W. Lighthouse. A nuclear power plant (capacity, 2,600 MW) operates (since 1985) 10 mi/16 km S of La Hague, at FLAMANVILLE, on W Coast of the Cotentin Peninsula (at foot of cliffs overlooking the English Channel).

Haguenau (AH-guh-no), German *Hagenau*, town, subprefecture of BAS-RHIN department, E FRANCE, in ALSACE, on the MODER RIVER, and 16 mi/26 km N of STRASBOURG; 48°49′N 07°47′E. Commercial center (trade in grains, especially Alsatian hops); manufacturing (electrical and mechanical equipment, footwear, carpets, furniture, pharmaceuticals, and

candies). Oil drilled nearby. Has large military camp. Heavily damaged in World War II (including 12th-century church of St. George and 12th–13th-century church of St. Nicholas, only partly restored). Just N, Forest of Haguenau (50 sq mi/130 sq km), one of the largest in France, was scene of heavy fighting (1945). The Regional Park of the N Vosges Mountains (VOSGES DU NORD NATURAL REGIONAL PARK) extends N to German border.

Hague's Peak, Colorado: see MUMMY RANGE.

Hague, The (HAIG), Dutch *'s-Gravenhage* or *Den Haag*, French *La Haye*, city (2007 population 474,680), W NETHERLANDS, on the NORTH SEA, administrative and governmental seat of the Kingdom of the Netherlands, also ⊙ SOUTH HOLLAND province, 15 mi/24 km NW of ROTTERDAM, and 35 mi/56 km SW of AMSTERDAM; 52°05′N 04°18′E. Entrance to NEW WATERWAY, shipping channel for Rotterdam, 10 mi/16 km to SW. Railroad terminus; airport to E. Although it has some industries (the manufacturing of computer software, petroleum products, electrical equipment; food processing), as well as dairying, agriculture (flowers, fruit, vegetables), and livestock raising. The Hague's economy revolves around government administration. It is the seat of the Dutch legislature, the Dutch supreme court, the International Court of Justice, and foreign embassies. Headquarters of numerous companies, including the Royal Dutch Shell petroleum company. Also of economic importance are banking, insurance, and trade. Site of a hunting lodge of the counts of Holland (*'s Gravenhage* means "the count's hedge") in the thirteenth century William, count of Holland, began (c.1250) the construction of a palace, around which a town grew in the fourteenth and fifteenth century. In 1586 the States-General of the United Provinces of the Netherlands convened here; later (seventeenth century) it became the residence of the stadtholders and the capital of the Dutch republic. In the seventeenth century, The Hague rose to be one of the chief diplomatic and intellectual centers of Europe. In the early nineteenth century, after Amsterdam had become the constitutional Dutch capital, The Hague received its own charter. Was (1815–1830) the alternative meeting place, with Brussels, of the legislature of the United Netherlands. As the Dutch royal residence 1815–1948, the city was greatly expanded and beautified in the mid-nineteenth century by King William II. In 1899 the First Hague Conference met here; ever since, The Hague has been a center for the promotion of international justice and arbitration. Among the numerous landmarks are the Binnenhof, which grew out of the thirteenth-century palace and houses both chambers of the legislature, and the thirteenth-century Hall of Knights (Du. *Ridderzaal*) within the Binnenhof. Nearby is the Gevangenenpoort, the fourteenth-century prison where Jan and Cornelius de Witt were murdered in 1672. The Mauritshuis, a seventeenth-century structure built as a private residence for John Maurice of Nassau, is an art museum and contains several of the greatest works of Rembrandt and Vermeer. The Peace Palace (Du. *Vredespaleis*), which was financed by Andrew Carnegie and opened in 1913, houses the Permanent Court of Arbitration and, since 1945, the International Court of Justice. Among the other notable buildings are the former royal palace; the Groote Kerk, a Gothic church (fifteenth and sixteenth century); the Nieuwe Kerk, containing Spinoza's tomb; the sixteenth-century town hall; and the Netherlands Conference Center (1969). Educational institutions in The Hague include schools of music and international law. Museums include the Royal Picture Gallery, Netherlands Post and Telecommunications Museum, Museum of Education, and State Archives. NW of the city is SCHEVENINGEN, a popular North Sea resort and a fishing port with a vehicle ferry to Yarmouth (U.K.). William III (Wil-

liam of Orange), stadtholder of Holland and other Dutch provinces as well as king of England (1689–1702), was born here. Another local attraction is the Madurodam, a miniature city built to reflect the different architectural forms of the country's urban landscape.

Haguro (HAH-goo-ro), town, East Tagawa district, YAMAGATA prefecture, N HONSHU, NE JAPAN, 43 mi/70 km N of YAMAGATA city; 38°42′N 139°54′E. Mt. Haguro is nearby.

Ha Ha Bay, inlet of the SAGUENAY River, S central QUEBEC, E CANADA, 12 mi/19 km ESE of CHICOUTIMI; 7 mi/11 km long, 3 mi/5 km wide; 48°21′N 70°49′W. At its head are BAGOTVILLE, PORT ALFRED, and GRANDE-BAIE.

Haha-jima (hah-HAH–jee-mah) [=mother island], volcanic island (□ 8 sq mi/20.8 sq km), northernmost of BAILEY ISLANDS, BONIN ISLANDS, TOKYO prefecture, SE JAPAN, in W PACIFIC OCEAN, 32 mi/51 km ESE of CHICHI-JIMA; 9 mi/14.5 km long, 1.5 mi/2.4 km wide; rises to 1,000 ft/300 m. Sugar refinery. In World War II, site of Japanese air base. Formerly Coffin Island.

Haha-jima-retto, JAPAN: see BAILEY ISLANDS.

Hahaya (hah-HAH-yah), village, Njazidja island and district, NW Comoros Republic, 9 mi/14.5 km N of Moroni, on W coast of island; 11°35′S 43°15′E. Fish; livestock; ylang-ylang, vanilla, coconuts, bananas. Hahaya Intl. Airport is 1 mi/1.6 km to N. Also spelled Hahaia.

Hahira (hai-HEI-ruh), town (2000 population 1,626), LOWNDES county, S GEORGIA, 12 mi/19 km NNW of VALDOSTA; 30°59′N 83°22′W. Tobacco market; logging, wood production; center of production for honey and beekeeping supplies.

Hahnbach (HAHN-bahkh), village, BAVARIA, central GERMANY, in UPPER PALATINATE, 33 mi/53 km E of NUREMBERG; 49°32′N 11°47′E.

Hahndorf (HAHN-dorf), town, SE SOUTH AUSTRALIA, 20 mi/32 km SE of ADELAIDE on South East Highway, in ADELAIDE HILLS; 35°02′S 138°48′E. Settled by Germans; promoted as AUSTRALIA's oldest German town. Dairying; beef cattle, poultry.

Hahnenkamm, AUSTRIA: see KITZBÜHEL.

Hahnville (HAHN-vil), unincorporated town (2000 population 2,792), ⊙ SAINT CHARLES parish, SE LOUISIANA, on the MISSISSIPPI RIVER, and 19 mi/31 km W of NEW ORLEANS; 29°58′N 90°25′W. Vegetables, sugarcane; cattle; alligators, catfish, crabs; manufacturing (jambalaya and dirty rice). Petroleum refining nearby.

Haho, prefecture (2005 population 194,917), PLATEAUX region, S TOGO ⊙ NOTSÉ; 07°05′N 01°15′E.

Hai'an (hei-AHN), town, SW GUANGDONG province, SE CHINA; port on HAINAN STRAIT, at S tip of LEIZHOU PENINSULA, 45 mi/72 km S of HAIKANG; 20°15′N 110°10′E. Boat connection with Haikou in Hainan province across strait.

Hai'an (HEI-AHN), town, ⊙ Hai'an county, central JIANGSU province, E CHINA, 12 mi/19 km NNW of RUGAO; 32°33′N 120°27′E. Rice, cotton, oilseeds. Food processing; textiles, machinery, building materials, chemicals.

Haibach (HEI-bahkh), suburb of ASCHAFFENBURG, BAVARIA, central GERMANY, in LOWER FRANCONIA, 3 mi/4.8 km SE of city center; 49°58′N 09°14′E.

Haibak, AFGHANISTAN: see AIBAK.

Haibara (hah-ee-BAH-rah), town, Uda district, NARA prefecture, S HONSHU, W central JAPAN, 13 mi/21 km S of NARA; 34°31′N 135°57′E.

Haibara (HAH-ee-bah-rah), town, Haibara district, SHIZUOKA prefecture, central HONSHU, E central JAPAN, 19 mi/30 km S of SHIZUOKA; 34°44′N 138°13′E. Tea.

Haicheng (HEI-CHENG), city (□ 1,056 sq mi/2,735 sq km; 1994 estimated urban population 235,900; estimated total population 1,058,800), S LIAONING province, NE CHINA, 80 mi/129 km SW of SHENYANG;

40°53′N 122°45′E. Heavy industry and agriculture are the largest economic sectors. Manufacturing (textiles, iron and steel, machinery).

Haicheng (HEI-CHENG), town, S FUJIAN province, SE CHINA, 13 mi/21 km W of XIAMEN, on LONG RIVER estuary; 24°24′N 117°53′E. Rice, sugarcane. Building materials, furniture, chemicals, machinery; food processing; papermaking.

Haid, CZECH REPUBLIC: see BOR.

Haida, CZECH REPUBLIC: see NOVY BOR.

Haida Gwaii, CANADA: see QUEEN CHARLOTTE ISLANDS.

Haidarabad, PAKISTAN: see HYDERABAD.

Haidargarh (HEI-duhr-guhr), town, tahsil headquarters, BARA BANKI district, central UTTAR PRADESH state, N CENTRAL INDIA, on branch of SARDA CANAL, and 25 mi/40 km SSE of BARA BANKI; 26°37′N 81°22′E. Rice, gram, wheat, oilseeds.

Haidarkan, KYRGYZSTAN: see KHAYDARKAN.

Haidarnagar, INDIA: see NAGAR.

Haidar Pasha, TURKEY: see HAYDARPASA.

Haidian (HEI-DYEN), NW suburban district of BEIJING, NE CHINA. Its Xishan (West Mountains) on the W and N are covered with rich vegetation and provide a variety of natural scenery. Over seventy significant historic sites are located in the mountains, among which are numerous royal gardens and palaces of the Qing (Ch'ing) dynasty (1644–1911) such as the Fragrant Hills, Jade Spring Hills, and Summer Palace. The ruined residence of the late Qing writer Cao Xueqin is also here, along with many Buddhist and Taoist temples. Also a cultural and scientific center. There are 138 research institutes that are part of the Chinese Academy of Sciences, and fifty-eight universities, including the well-known Beijing and Qinghua universities and People's University of China. It is also the site of numerous national art and performing arts institutions, and the new Beijing library. The core is the Zhongguancun (JUNG-GWAHN-ZWUN) area, where twenty-eight Chinese Academy of Science research institutes and nearly 200 other national research institutions, nearly thirty universities, and 5,000 high-tech companies are located. The Zhongguancun area is known as "The Silicon Valley of Beijing" and "Science Town."

Hai Duong (HEI DUH-uhng), city, HAI HUNG province, N VIETNAM, in RED RIVER delta, on THAI BINH RIVER, on HANOI–HAI PHONG railroad, and 30 mi/48 km ESE of Hanoi; 20°56′N 106°19′E. Transportation and trading center in intensive agricultural (rice, tobacco, corn, tomatoes) area; administrative complex. Currently shifting investment toward agro-processing industries, including large-scale export tomato-processing zone, meat-, lychee-, and longan-export processing. Limestone, kaolin, clay. Tourism. Formerly Haiduong.

Hai Duong, VIETNAM: see HAI HUNG.

Haifa (HEI-fah), city (2006 population 267,000), NW ISRAEL, the country's major port on the MEDITERRANEAN SEA, at the foot of MOUNT CARMEL; 32°48′N 34°59′E. Elevation 331 ft/100 m. One of the chief cities of N Israel. The city extends up the N slope of Mount Carmel and into the N part of the Carmel Range, where the majority of the population now lives. The other two sections lie along the coastal plain and the Hadar (the slope of Mount Carmel). The country's principal oil-refining center. Along with ASHDOD, Haifa handles oceangoing vessels, including oil tankers. Industries include steel, textiles, chemicals, high-tech electronics; shipbuilding, military research, and food processing and storing. It also services large naval vessels, and is a port of leave for the U.S. Sixth Fleet. Along the coast on S outskirts is the Center for Scientific Industries, a complex of high-tech industries known as "Matan." The old commercial center is on the Hadar. A large, new shopping center is located in the "Kraygoth," the residential, industrial, and commercial areas in the ZEBULUN VALLEY NNE of the

city. Haifa is known to have existed by the 3rd century C.E. (Sycaminum) but was of little importance during early Muslim times. The Crusaders, who called it Caiffa or Caiphas, developed it commercially. Destroyed by Saladin in 1191, it began to revive in the late 18th century. The city's main growth occurred in the 20th century with the development of its port. Haifa was contested by Jews and Arabs in the 1948–1949 war because of its industrial importance. By the late 20th century the city's population was 90% Jewish, although Muslims, Christians, and Druse continued to live in the area. Haifa was a target of Iraqi missiles during the Persian Gulf War (1991). The Technion (Israel Institute of Technology; established 1924) and Haifa University (established 1963) are here. Known also as "Red Haifa" due to its longtime loyalty to the Labor Party. World center of Bahaism and the site of the shrine of Bab and a Bahai temple. On the coast S of the city (at ancient Shikmona) is the Israel Oceanographic Research Center.

Haifa, Bay of (hai-FAH) inlet (7 mi/11.3 km–10 mi/16 km wide) of MEDITERRANEAN SEA, NW ISRAEL; 32°52′N 35°02′E. Extends SW-NE between HAIFA and AKKO. Formerly Bay of Acre.

Haifeng (HEI-FUHNG), town, ⊙ Haifeng county, SE GUANGDONG province, S CHINA, 60 mi/97 km SE of HUIZHOU; 22°58′N 115°20′E. Rice, sugarcane, oilseeds. Fisheries; saltworks; food processing; textiles.

Haifif, YEMEN: see GHAIDHA.

Haiger (HEI-ger), town, HESSE, W GERMANY, in the WESTERWALD, on the DILL RIVER, 3 mi/4.8 km W of DILLENBURG; 50°45′N 08°12′E. Manufacturing of steel products, rubber goods, electronics, and furniture. Iron ore formerly mined here. Has late-Gothic church. Chartered 914; incorporated thirteen surrounding villages in 1977.

Haigerloch (HEI-ger-lokh), town, BADEN-WÜRTTEMBERG, SW GERMANY, 14 mi/23 km SW of TÜBINGEN; 48°22′N 08°46′E. Brewing, bottling of mineral water. Baroque castle and church. Was nuclear research center during the Nazi regime.

Haigler, village (2006 population 197), DUNDY county, S NEBRASKA, 20 mi/32 km W of BENKELMAN, and on REPUBLICAN RIVER, at KANSAS state line, near COLORADO state line; 40°00′N 101°56′W. Livestock; grain. Rock Creek State Recreation Area and fish hatchery to NE.

Hai Hung, province (□ 985 sq mi/2,561 sq km), N VIETNAM, in RED RIVER Delta; ⊙ HAI DUONG; 20°50′N 106°15′E. Bordered N by HA BAC province, NE by QUANG NINH province, E by HAI PHONG, S by THAI BINH and NAM HA provinces, W by HA TAY province, and NW by HANOI. Framed by the RED RIVER and its distributaries and interlaced by irrigation canals, Hai Hung is an area of fertile alluvial soils and agriculture (wet rice cultivation, vegetable farming, tobacco and fruit growing, poultry raising). Fishing and aquaculture; mining (limestone, kaolin, clay); various industries (cement, brick and tile manufacturing, meat and food processing, light manufacturing). Domestic tourism. Predominantly Kinh population. In 1997 the province was divided into the provinces: Hai Duong (⊙ Hai Duong) and HUNG YEN (⊙ HUNG YEN).

Haikang (HEI-KAHNG), town, ⊙ Haikang county, SW GUANGDONG province, SE CHINA, 25 mi/40 km SW of ZHANJIANG; 20°55′N 110°04′E. Economic center of LEIZHOU PENINSULA; sugarcane, rice, oilseeds; food processing; pharmaceuticals; fisheries; saltworks. Also Leizhou.

Haikou (HEI-KO), city (□ 84 sq mi/218 sq km; 1994 estimated non-agrarian population 364,700; estimated total population 462,900), ⊙ HAINAN province, SE CHINA; seaport on HAINAN STRAIT, and c.80 mi/128 km S of ZHANJIANG; 20°05′N 110°25′E. Largest city on Hainan Island. The opening of the city to foreign trade and tourism has stimulated the economy and rapid population growth. Light industry and com-

merce are the largest sectors of the city's economy. Crop growing; fishing; animal husbandry. Grain; manufacturing (food and beverages, textiles, leather, utilities, pharmaceuticals, synthetic fibers, rubber, electronics). Airport. Jinhiu and Wanlh parks are points of interest. Sometimes appears as Hoihow.

Haiku (HAH-ee-KOO), town, on N coast of MAUI island, MAUI county, HAWAII, 9 mi/14.5 km E of KAHULUI, 3 mi/4.8 km inland from Pauwela Point. Sugarcane, pineapples; cattle. Hookipa Beach Park to NW, Koolau Forest Reserve to SE.

Hail (HAH-il), city (2004 population 267,005), ⊙ HAIL province, N central SAUDI ARABIA; 27°33′N 41°42′E. The city grew because of its location on a pilgrimage route from IRAQ to MAKKA. It was capital of the independent emirate of JABAL SHAMMAR, which Ibn Saud conquered in 1921. Important communications, commerce, and services center. Airport. Also spelled Hayel.

Hailakandi (hei-LAH-kahn-dee), district (□ 512 sq mi/1,331.2 sq km), ASSAM state, NE INDIA; ⊙ HAILAKANDI.

Hailakandi (hei-LAH-kahn-dee), town, ⊙ HAILAKANDI district, S ASSAM state, NE INDIA, in SURMA VALLEY, on tributary of BARAK (Surma) River, and 17 mi/27 km WSW of SILCHAR; 24°41′N 92°34′E. Trades in rice, tea, sugarcane, rape, and mustard; tea processing. Extensive tea gardens nearby. Railroad spur terminus 10 mi/16 km SSE, at Lalaghat.

Hailar (HEI-LAH-UHR), city (□ 556 sq mi/1,445.6 sq km), E INNER MONGOLIA, N CHINA, on railroad, on Hailar River (a tributary of the ERGUN); 49°15′N 119°41′E. Industry, commerce, and agriculture (especially animal husbandry and crop growing) are the main economic activities. Main agricultural products include milk, eggs, beef, and lamb. The main industry is food processing. Formerly called Hulun.

Hailey, city (2000 population 6,200), ⊙ BLAINE county, S central IDAHO, on BIG WOOD RIVER, and c.100 mi/161 km E of BOISE; 43°31′N 114°18′W. Elevation 5,342 ft/1,628 m. Trade center for mining (silver, lead) and livestock area; manufacturing (printing and publishing); lumber milling. Summer resort, headquarters for Sawtooth National Forest (NW and NE). Friedman Memorial Airport. Clarendon Hot Springs to NW. SAWTOOTH MOUNTAINS are N. Incorporated as village 1903, as city 1909.

Hailey (HAI-lee), agricultural village (2001 population 1,635), W OXFORDSHIRE, S central ENGLAND, 2 mi/3.2 km N of WITNEY; 51°46′N 00°01′W.

Haileybury (HAI-lee-buh-ree), town (□ 35 sq mi/91 sq km; 2001 population 4,543), ⊙ TIMISKAMING district, E ONTARIO, E central CANADA, on Lake TIMISKAMING, 80 mi/129 km N of NORTH BAY; 47°27′N 79°39′W. Distributing and residential center for rich mining region, with pulp, lumber milling. Ski resort. Airfield. Silver, gold, cobalt are mined in immediate vicinity. Site of mining school. Developed after discovery (1903) of silver at nearby COBALT; incorporated 1905. Town was almost wholly destroyed (1922) by fire. Forms part of city of Temiskaming Shores.

Haileyville, town (2006 population 904), PITTSBURG county, SE OKLAHOMA, 12 mi/19 km ESE of MCALESTER and adjoining HARTSHORNE (E); 34°51′N 95°34′W. In cattle-raising area. Jack Fork Mountains to S; headwaters of Lake Arrowhead reservoir to N. Settled c.1890.

Hailin (HEI-LIN), town, ⊙ Hailin county, SE HEILONGJIANG province, NE CHINA, 10 mi/16 km W of MUDANJIANG, and on railroad; 44°35′N 129°25′E. Tobacco, sugar beets; sugar refining; lumbering.

Haillan, Le (ei-YAHN, luh), town (□ 4 sq mi/10.4 sq km), GIRONDE department, AQUITAINE region, SW FRANCE; 6 mi/10 km NW of BORDEAUX. Electronics.

Haillicourt (ai-yee-KOOR), town (□ 1 sq mi/2.6 sq km), PAS-DE-CALAIS department, NORD PAS-DE-CALAIS region, N FRANCE, 5 mi/8 km SSW of BÉTHUNE; 50°28′N 02°35′E. In depleted coal-mining area.

Hailong, CHINA: see MEIHEKOU.

Hailsham (hahl-SHUHM), town (2001 population 19,658), East SUSSEX, SE ENGLAND, 7 mi/11.3 km N of EASTBOURNE; 50°51′N 00°15′E. Formerly an agricultural market, now primarily a commuter town. Has 15th-century church.

Hailun (HEI-LWUN), city (□ 1,699 sq mi/4,400 sq km; 1994 estimated urban population 149,300; estimated total population 787,400), E central HEILONGJIANG province, NE CHINA, on railroad, and 120 mi/193 km N of HARBIN; 47°29′N 126°58′E. Agriculture and light industry are the main sources of income. Agriculture (grain, tobacco, jute, sugar beets; milk; hogs); manufacturing (food processing; plastics).

Hailuoto (HEI-loo-o-to), Swedish *Karlö*, island (□ 68 sq mi/176.8 sq km) in GULF OF BOTHNIA, OULUN province, W FINLAND, 12 mi/19 km W of OULU; 14 mi/23 km long (E-W), 1 mi/1.6 km–9 mi/14.5 km wide; 65°02′N 24°42′E. Elevation 66 ft/20 m. Hailuoto fishing village (S). The island's isolated position makes it of ethnographical interest. Popular summer resort.

Haimen (HEI-MEN), town, E ZHEJIANG province, SE CHINA; outer port 20 mi/32 km SE of LINHAI, on EAST CHINA SEA, at mouth of LING RIVER; 28°40′N 121°47′E.

Haimen (HEI-MUHN), town, in Haimen county, S JIANGSU province, E CHINA, 50 mi/80 km NNW of SHANGHAI, across estuary of CHANG JIANG (Yangzi River; N); 31°54′N 121°10′E. Cotton, rice, oilseeds; textiles and clothing, machinery; food processing.

Haina (HEI-nah), village, HESSE, central GERMANY, 17 mi/27 km NE of MARBURG; 52°02′N 08°59′E. Founded around former Cistercian monastery from 13th century with early-Gothic church from 1250.

Hainan (hei-NAHN) or **Qiong** (CHYONG), island and province (□ 13,100 sq mi/34,060 sq km; 2000 population 7,559,035), SE CHINA, in the SOUTH CHINA SEA; ⊙ HAIKOU; 19°12′N 109°36′E. Created in 1988, the province is coextensive with Hainan Island, which was designated (1988) a special economic zone to spur the development of its considerable natural resources. The second-largest island off the Chinese coast (after TAIWAN), Hainan is separated from the mainland (LEIZHOU PENINSULA) by HAINAN STRAIT (c.30 mi/48 km wide). Haikou is its largest city and major port. Other major cities include SANYA, QIONGHAI, QIONGSHAN, and TONGSHI.

The year-round growing season and monsoon climate favor the cultivation of rice, coconuts, palm oil, sisal, tropical fruit, coffee, tea, and sugarcane; the island also produces most of China's rubber. Major rubber farms are concentrated around Haikou and Sanya. Tropical agriculture has become a leading industry in Hainan's economy. The mountainous interior is thickly forested (nearly 50% of the island is covered with forests) yielding tropical hardwoods, including teak and sandalwood. Hainan is rich in minerals, notably high-grade iron and tungsten, but also rich in titanium, manganese, salt, copper, bauxite, molybdenum, gold, silver, coal, cobalt, graphite, and crystal. Hainan's rich offshore fishing grounds provide shrimp, scallops, tuna, and Spanish mackerel, and pearls are harvested in the shallow bays surrounding the island. The growth of Hainan's industries, which include the production of textiles and farm equipment, has been hindered by a lack of energy resources. Tropical climate and many beaches are making Hainan a popular resort site. Recreational villages have been established in Haikou, Sanya and Xinglong. Xinglong is well known for its forest park and its farm operated by overseas Chinese. The southernmost tip of the island, Jimu Jao, is a famous tourist point.

The many aboriginal Li, who inhabit the forested interior, have been constituted with the Miao into a large Li-Miao autonomous district. Hainan has a large population of Chinese born outside the country. Under Chinese control since the 1st century C.E., Hainan was not fully incorporated into China until the 13th century. It became part of GUANGDONG province in the late 14th century. In World War II it was occupied (1939) by the Japanese, who developed the industries and exploited the great iron-ore deposits. Taken over (1945) by the Nationalists. The Chinese Communists landed in April 1949, and gained control in 1950. The Yulin naval base, a natural harbor developed by the Japanese, has been expanded under the Communist government.

Hainan Strait (hei-NAHN), in SOUTH CHINA SEA, CHINA, joins GULF OF TONKIN (W) and South China Sea proper (E), separating LEIZHOU PENINSULA (N) and HAINAN island (S); 10 mi/16 km–15 mi/24 km wide, 50 mi/80 km long. HAIKOU is on S shore. Main shipping route between HONG KONG and HAI PHONG (Vietnam). Also called Qiongzhou Strait.

Hainasch, LATVIA: see AINAŽI.

Hainaut (ai-NO), Flemish *Henegouwen*, province (□ 1,437 sq mi/3,736.2 sq km; 2006 population 1,291,850), S BELGIUM, on border of FRANCE (S); ⊙ MONS; 50°27′N 03°56′E. The chief cities of the predominately French-speaking province are Mons, CHARLEROI, and TOURNAI. It lies mainly on the low-lying sedimentary plains that slope gently toward the NORTH SEA, but its panhandle, the *Botte de Hainaut*, extends S to include part of the ARDENNES Mountains. Its largely agricultural landscape became one of the earliest industrialized regions of Europe when the coalfields of the BORINAGE began to be exploited. The small and broken coal seams became economically unexploitable in the 1960s. Conversion of the economy, though subsidized by the government, has only slowly reduced the widespread unemployment and general economic stagnation of the region. Manufacturing now includes chemicals and electrical equipment. Drained by the SCHELDT, DENDER, and SAMBRE rivers; served by a dense railroad network and the CHARLEROI-BRUSSELS CANAL. The province of Hainaut was created in the late ninth century, and in the divisions of the Carolingian empire became a fief of Lotharingia. In 1433, Philip the Good of Burgundy added Hainaut and Holland to his dominions. Hainaut remained under the house of Burgundy until the death (1482) of Mary of Burgundy, when its history became that of the Austrian Netherlands. By the treaties of the Pyrenees (1659) and of Nijmegen (1678) parts of Hainaut, including the city of VALENCIENNES, were permanently annexed by France; they form part of the present NORD department.

Hainburg (HEIN-boorg), suburb of Frankfurt region, HESSE, central GERMANY, on Main River, 4 mi/6.4 km S of HANAU; 50°04′N 08°55′E. Printing, metalworking, manufacturing of bricks and cigars. Formed in 1977 from unification of Hainstadt and Klein-Krotzenburg.

Hainburg an der Donau (HEIN-boorg ahn der DOnou), town, E LOWER AUSTRIA, and 8 mi/12.9 km W of BRATISLAVA (SLOVAKIA); 48°09′N 16°57′E. Medieval town walls and gates are well-preserved; ruins of a large castle mentioned in the Nibelungen epic, place of the marriage between King Ottokar of BOHEMIA and the Babenberg Duchess Margarethe of AUSTRIA. Bridge across the DANUBE River; Danube Flood Plains National Park is nearby. Site has been a town since 1244.

Haindorf, CZECH REPUBLIC: see HEJNICE.

Haine River (EN), Flemish *Henne*, 40 mi/64 km long, SW BELGIUM and N FRANCE; rises near ANDERLUES (Belgium); flows N and W, through HAINAUT coal-mining area, past SAINT-GHISLAIN, to SCHELDT RIVER at CONDÉ-SUR-L'ESCAUT (France).

Haines, village (2000 population 1,811), SE ALASKA, on CHILKOOT INLET, arm of LYNN CANAL, and 75 mi/121 km NNW of JUNEAU, 15 mi/24 km SSW of SKAGWAY; 59°14′N 135°26′W. Tourism; fishing and fish processing. Site of state fair; Chilkoot dancers. During World War II a highway, the HAINES CUT-OFF, was built connecting Haines with the ALASKA HIGHWAY at HAINES JUNCTION (BRITISH COLUMBIA). N terminus of Alaska ferry system. Docks near former Chilkoot Barracks, previously site of Fort William H. Seward, U.S. army post.

Haines, village (2006 population 400), BAKER county, NE OREGON, 10 mi/16 km NNW of BAKER CITY, on POWDER RIVER, at mouth of ROCK CREEK BUTTE; 44°54′N 117°56′W. Elevation 3,333 ft/1,016 m. Dairy products; wheat, potatoes; sheep, cattle. Timber. Site of Eastern Oregon Museum. Anthony Lakes Ski Area to W. Wallowa-Whitman National Forest to W.

Haines, borough (□ 2,357 sq mi/6,128.2 sq km; 2006 population 2,257), SE ALASKA, on CANADA (BRITISH COLUMBIA; E and NW) border, at head of Long Canal inlet (which forms part of E boundary). Main towns are HAINES and SKAGWAY. Chilkat Range to S; part of the COAST MOUNTAINS to W. Drained by CHILKAT and Klehini rivers. Part of TONGASS NATIONAL FOREST in S and E. Haines State Forest and Chilkat State Park in center. KLONDIKE GOLD RUSH HISTORICAL PARK at Skagway. Tourism. Fishing; timber.

Haines City (HAINZ), town (□ 8 sq mi/20.8 sq km; 2000 population 13,174), POLK county, central FLORIDA, c.50 mi/80 km ENE of TAMPA; 28°06′N 81°37′W. Citrus-fruit shipping center, with packing houses and canneries; manufacturing of fertilizer, citrus oil, and pulp feed. Seat of military institute.

Haines Cut-off, SE ALASKA and NW CANADA (NW BRITISH COLUMBIA and SW YUKON), branch (150 mi/241 km long) of the ALASKA HIGHWAY; extends N from HAINES (Alaska), via CHILKAT PASS (British Columbia), to junction with Alaska Highway at HAINES JUNCTION (Yukon Territory). Alaska Highway outlet to Gulf of ALASKA and SE Alaska towns on the Alaska Panhandle.

Haines Falls, resort village, GREENE county, SE NEW YORK, in the CATSKILL MOUNTAINS, at upper end of KAATERSKILL CLOVE, 12 mi/19 km W of CATSKILL; 42°12′N 74°06′W. Scenic waterfall here. NORTH LAKE is E.

Haines Junction, village (□ 13 sq mi/33.8 sq km; 2006 population 589), SW YUKON, NW CANADA, 120 mi/193 km NW of SKAGWAY, and on ALASKA HIGHWAY, at junction with Haines Cut-off (road to HAINES, ALASKA); 60°45′N 137°32′W. Administrative center for Kluane National Park Reserve; eco-tourism.

Haines Landing, MAINE: see RANGELEY.

Hainesport, township, BURLINGTON county, W NEW JERSEY, 2 mi/3.2 km W of MOUNT HOLLY; 39°58′N 74°50′W. Rancocas State Park to E.

Hainesville, village, Lake county, NE ILLINOIS, 10 mi/16 km W of WAUKEGAN; 42°20′N 88°04′W.

Hainesville, village, SUSSEX county, NW NEW JERSEY, 2 mi/3.2 km E of the DELAWARE RIVER, and 14 mi/23 km N of NEWTON; 41°15′N 74°48′W. In hilly recreation region; Delaware Water Gap National Recreation Area just W.

Hainfeld (HEIN-feld), town, central LOWER AUSTRIA, 13 mi/21 km SE of Sankt Pölten, on Gölsen River; 48°02′N 15°46′E. Manufacturing of metals; brewery. Here, the Austrian Social Democratic Party was founded December 30, 1888–January 1, 1889.

Hainichen (HEI-ni-khen), town, SAXONY, E central GERMANY, 13 mi/21 km NE of CHEMNITZ; 50°58′N 13°07′E. Manufacturing includes furniture, construction materials. Poet Christian Gellert born here 1715.

Haining (HEI-NEENG), city (□ 263 sq mi/683.8 sq km; 2000 population 617,623), N ZHEJIANG province, SE CHINA, 25 mi/40 km ENE of HANGZHOU, on Hangzhou Bay; 30°31′N 120°35′E. Agriculture and light industry are the largest sectors of the city's economy. Rice, tobacco, jute, sugarcane, oilseeds; manufacturing (food processing; textiles, leather, chemicals, plastics, machinery, transportation equipment, electrical equipment, and electronics).

Hai Ninh (HEI NIN), town, QUANG NINH province, N VIETNAM, on CHINA border (opposite Tunghing), near Gulf of BAC BO coast, 145 mi/233 km ENE of HANOI; 21°32′N 107°58′E. Fisheries; antimony deposits. Formerly MON CAI.

Hainspach, CZECH REPUBLIC: see LIPOVA.

Hai Phong (HEI PONG), city (2005 population 820,700), NE VIETNAM, on a large branch of the RED RIVER delta, c. 10 mi/16 km from the Gulf of BAC BO; 20°48′N 106°43′E. Connected with the sea by a narrow access channel that requires continual dredging. A major port of Vietnam and one of the largest ports in SOUTHEAST ASIA, Hai Phong was developed (1874) by the French and became the chief naval base of FRENCH INDOCHINA. A shipbuilding industry and cement, glass, porcelain, and textile works were established by the French and remain important. At the beginning of the French-Indochina War (Nov. 1946), French naval vessels shelled the city, killing c. 6,000 Vietnamese. After the French departed and the new state of Democratic Republic of Vietnam (also NORTH VIETNAM) was created (1954), the silted-up harbor was reconstructed with Chinese and Soviet aid and the docks and shipbuilding yards were repaired and modernized. The old French cement plant was enlarged, and fish-canning, chemical-fertilizer, machine-tool, and additional textile industries were established. During the Vietnam War, Hai Phong was heavily bombed by the U.S.; the shipyards and the industrial section of the city were devastated, railroad connections with HANOI were disrupted, and thousands of homes were destroyed. The harbor was mined by U.S. naval planes in May 1972, and effectively sealed until the mines were swept by U.S. forces after the cease-fire agreement in early 1973. Reconstruction, while slow, was aided by the fact that many of the factories had been dismantled during the bombings and relocated in rural areas; when returned to Hai Phong, much of the machinery was able to function in ruined structures. The city experienced a major exodus of its Chinese population in the period leading up to the 1979 Sino-Vietnamese conflict, leaving little of the traditional fishing fleet in place. In recent years, Hai Phong has undergone considerable development, including a number of export processing zones (e.g., Dinh Vu Zone, Nomura–Hai Phong Industrial Park, Taiwanese–Hai Phong Industrial Park). Industries undergoing modernization include the following sectors: chemicals, fertilizers, machine tools, plastics, textiles, agriculture, and mariculture processing; shipbuilding. In 1994, a loan treaty between JAPAN and Vietnam provided funds to improve the Hai Phong port and transport network (especially Highway 5). Thuong Li railroad station is located in the W part of city. Tourism. Formerly Haiphong.

Hairedin (HEI-re-deen), village (1993 population 2,487), MONTANA oblast, ⊙ Hairedin obshtina, NW BULGARIA, on the OGOSTA RIVER, 17 mi/27 km SW of ORYAHOVO; 43°37′N 23°40′E. Vineyards; fruit, grain; livestock. Sometimes spelled Khayredin or Hayredin. Formerly Yeredin.

Hairy Hill (HE-ree HIL), former village (2001 population 30), central ALBERTA, W CANADA, 19 mi/31 km NNE of VEGREVILLE; 53°46′N 111°58′W. Tanning, dairying; mixed farming; grain. Dissolved in 1996; amalgamated into Two Hills County No. 21.

Hais (HAH-is), township, W YEMEN, on Tihama coastal plain, 70 mi/113 km SE of HODEIDA, and on road to TAIZ. Noted pottery-manufacturing center; jewelry manufacturing; sheepskin tanning. Sometimes also spelled Hays, Heis, Heys, or Hes.

Hais, SOMALIA: see HEIS.

Hai, Shatt al, IRAQ: see GHARRAF, SHATT AL.

Haisnes (EN), town (□ 2 sq mi/5.2 sq km), PAS-DE-CALAIS department, NORD PAS-DE-CALAIS region, N FRANCE, 6 mi/9 km N of LENS; 50°31′N 02°48′E.

Haiterbach (HEI-tuhr-bahkh), town, BADEN-WÜRTTEMBERG, GERMANY, in BLACK FOREST, 4 mi/6.4 km SW of NAGOLD; 48°32′N 08°39′E. Manufacturing includes plastics; lumbering.

Haiti (HAI-tee), French *Haiti* (ah-ee-TEE), independent republic (□ 10,700 sq mi/27,713 sq km; 2004 estimated population 7,656,166; 2007 estimated population 8,706,497), WEST INDIES, on the W one-third of the island of HISPANIOLA, on the ATLANTIC OCEAN (N) and the CARIBBEAN SEA (S); ⊙ PORT-AU-PRINCE.

Geography
Bordered E by the DOMINICAN REPUBLIC. JAMAICA lies to the W and CUBA to the NW. Important cities include Port-au-Prince, CAP-HAÏTIEN, and GONAÏVES. The offshore islands of TORTUGA and GONÂVE also belong to Haiti. Mostly mountainous; c.one-third of the land is arable.

Population
About 95% of the inhabitants are descendants of African slaves who still follow West African cultural patterns. Since the mid-19th century, however, Haiti has been dominated by the mulatto minority, which clings to French cultural traditions; the official languages of Haiti are Haitian Creole and French, although the vast majority of the people speak the former. Roman Catholicism is the predominant religion, but African nature gods are still worshiped and *vodun* (voodoo) rites are practiced. Economic hardship and political upheavals have caused numerous Haitians to emigrate, especially to the UNITED STATES.

Economy
Agriculture is the principal economic activity. Subsistence crops include cassava, rice, sugarcane, sorghum, yams, corn, and plantains. Most Haitians own and farm tiny plots of land, and great population density has caused great poverty. Coffee is the major export; other exports include cotton, sugar, sisal, bauxite, and essences. Spiny lobsters constitute an important share of Haitian exports to the U.S., the country's leading trading partner. Industry in Haiti consists largely of light manufacturing; products include foodstuffs, liquors, essential oils, leather goods, soap, and footwear. Some bauxite and copper are mined but other mineral deposits have barely been tapped. Haiti is the most densely populated country in Latin America and has the lowest per capita income with 80% of the population living in abject poverty. Need for more farmland, fuel, and construction materials has led to serious deforestation. Prolonged economic inequality, political instability, and a near total lack of medical care continue to plague Haiti.

History to 1795
The island of Hispaniola was inhabited by the Arawaks prior to the arrival of Columbus in 1492. Disease, ill treatment, and execution by the Spaniards decimated the Arawaks, who gave Haiti ("land of mountains") its name. While establishing plantations in E Hispaniola (now the Dominican Republic), however, the Spanish largely ignored the W part of the island, which by the 17th century became a base for French and English buccaneers. Gradually French colonists, importing African slaves, developed sugar plantations on the N coast. SPAIN ceded Haiti (then called Saint-Domingue) to FRANCE in 1697. It became France's most prosperous colony in the Americas and one of the world's chief coffee and sugar producers. Haitian society became stratified into Frenchmen, Creoles, freed blacks, and black slaves, with the mulattoes, whose social status was in-between. When Creole planters sought to prevent mulatto representation in the French National Assembly and in local assemblies, the mulattoes revolted, destroying the rigid structure of Haitian society. The blacks formed guerrilla bands led by Toussaint L'Ouverture, a former slave who had been made an officer of the French forces on Hispaniola.

History: 1795 to 1915
In 1795, Spain ceded its part of the island to France, and in 1801 Toussaint conquered it, abolished slavery, and proclaimed himself governor general of an autonomous government over all Hispaniola. Napoleon sent his brother-in-law, General Charles Leclerc, in an unsuccessful effort to reconquer the island. Toussaint, taken by trickery, died in a French prison; but the revolt continued and forced the French troops, already ravaged by yellow fever, to withdraw. The rebels received unexpected aid from U.S. President Thomas Jefferson, who feared that Napoleon would use Saint-Domingue as a base to invade Louisiana. In 1804, Haiti became the second nation in the Western Hemisphere, after the U.S., to win complete independence. The remaining French and Creoles were expelled, and Jean-Jacques Dessalines, an ex-slave, proclaimed himself emperor. Haiti's last emperor (1847–1859) was Faustin Soulouque. Since the end of his reign, the country has been a republic. Political and social conflict persisted, intensified by the mulatto-black hostility, and Haiti's economy, which had never recovered from the violent struggle for independence, declined further.

History: 1915 to 1990
After the dictator Guillaume Sam was killed in a popular uprising in 1915, the U.S. took the opportunity to invade Port-au-Prince. Although financial and general material progress advanced under American military occupation, Haiti protested, and a U.S. Senate investigation in 1921 found that the avowed purpose of preparing Haiti for responsible self-government had been ignored. In 1930 a U.S. presidential commission recommended that Haiti be allowed to elect a legislature that would, in turn, name a president. The marines were finally withdrawn in 1934, although U.S. fiscal control was maintained until 1947. Political instability persisted in Haiti after World War II. François ("Papa Doc") Duvalier, elected president in 1957, suppressed opposition and in 1964 he proclaimed himself president for life. When he died in 1971 he was succeeded by his 19-year-old son, Jean-Claude ("Baby Doc"), who also became president for life. In 1986, popular discontent became great enough to induce him to flee the country.

History: 1990 to Present
In December 1990, Jean-Bertrand Aristide was elected president but was exiled in a military coup led by Lieutenant General Raoul Cedras. The U.S. and the OAS responded with economic sanctions. In September 1994, Cedras agreed to a U.S.-backed plan for Aristide to return to power, after a show of U.S. military force. UN troops (mostly from the U.S.) landed unopposed in Haiti in September 1994, U.S. sanctions were lifted, and Cedras resigned on October 10, paving the way for Aristide's return on October 15. Command of U.S. forces was turned over to the UN on March 31, 1995, and on December 17, 1995, Haitians voted for Aristide's successor, René Préval, who took office February 7, 1996. Irregularities marred the elections in 2000 causing nearly all international aid to Haiti to be suspended. Aristide was sworn in as president again in February 2001, but his second term was marked by continued civil unrest; escalating violence between political parties erupted into armed rebellion in January 2004, forcing Artistide's resignation from office after which he left the country. Boniface Alexander became interim president on February 29, 2004, and on March 12, 2004 Gerald Latortue was named interim prime minister. In September 2004, Haiti was devastated by Tropical Storm Jeanne which caused widespread death and destruction. Elections held in early 2006 were marred by violence and irregularities but

---

resulted in the election of René Préval as president. UN peacekeepers remain in the country.

Government

Haiti is governed under the constitution of 1987, which was suspended and reinstated several times between 1988 and 2006, when the country returned to constitutional rule. The president is the head of state; the prime minister, who is appointed by the president and confirmed by the legislature, is the head of government. Most power resides with the president. Haiti has a bicameral legislature, the National Assembly, with a thirty-seat Senate, whose members are elected to six-year terms, and a ninety-nine-seat Chamber of Deputies, whose members are elected to four-year terms. Administratively, the country is divided into 10 departments. The head of state is President René Préval; the current head of government is Prime Minister Jacques-Édouard Alexis.

Haiti, sugar-mill village, CAMAGÜEY province, E CUBA, 45 mi/72 km SSE of CAMAGÜEY, 2 mi/3 km from CARIBBEAN SEA; 20°46′N 77°52′W. Formerly Francisco.

Haiti, WEST INDIES: see HISPANIOLA.

Hai Van, Deo, VIETNAM: see CLOUDS, PASS OF THE.

Haiya Junction, SUDAN: see TAQATU HAYYA.

Haiyan (HEI-YAHN), town, ⊙ Haiyan county, NE QINGHAI province, W CHINA, on upper XINING RIVER, and 60 mi/97 km NW of XINING, near QINGHAI LAKE; 36°58′N 100°50′E. Oil crops; livestock. Also called Sanjiaocheng.

Haiyan, town, ⊙ Haiyan county, NE ZHEJIANG province, SE CHINA, 50 mi/80 km NE of HANGZHOU, on Hangzhou Bay; 30°31′N 120°57′E. Cotton, rice, oilseeds, jute; textiles and clothing; food and beverages; building materials. The Qinshan Power Station (nuclear), the first in China, began operating in 1994.

Haiyang (HEI-YAHNG), town, ⊙ Haiyang county, NE SHANDONG province, NE CHINA; port on S coast of SHANDONG PENINSULA, on YELLOW SEA, 65 mi/105 km NE of QINGDAO; 36°46′N 121°09′E. Grain, oilseeds. Also called Dongcun.

Haiyou, CHINA: see SANMEN.

Haiyuan, town, ⊙ Haiyuan county, SW NINGXIA HUI AUTONOMOUS REGION, N central CHINA, 105 mi/169 km ENE of LANZHOU; 36°35′N 105°40′E. In mountain region. Grain, oilseeds; livestock. Population is largely Muslim.

Haiyuan (HEI-yuh-WAHN), town, SW GUANGXI ZHUANG, S CHINA, on road, and 60 mi/97 km SW of NANNING; 22°10′N 107°35′E. Food processing; grain, sugarcane.

Hajab, Al (hah-JEB, el), oil and gas field, Safaqis prov., SE TUNISIA, 12 mi/19 km W of SAFAQIS. Crude oil pipeline to As-Sakhirra terminal.

Hajagos, SLOVAKIA: see KLOKOCOV.

Hajar (HAH-jer), mountainous region of OMAN, extending 300 mi/483 km parallel to the coast of the GULF OF OMAN between OMAN PROMONTORY (N) and RAS AL HADD (S). Consisting of a series of limestone ridges and tablelands (rising to 9,950 ft/3,033 m in the JABAL AKHDAR), the HAJAR region is divided by the Wadi Sama'il into the WESTERN HAJAR and EASTERN HAJAR. The Western Hajar is separated from the Gulf of Oman by the fertile, populous BATINA coastal plain, while the Eastern Hajar's limestone formations hug the coast between MUSCAT and Ras al Hadd. Sometimes spelled Hajr.

Hajdu-Bihar (HAH-yuh-doo-BI-hahr), county (□ 998 sq mi/2,594.8 sq km), E HUNGARY; ⊙ DEBRECEN; 47°25′N 21°30′E. Includes the HORTOBÁGY puszta (NW; part of the Alföld); drained by BERETTYÓ, Kösely, and HORTOBÁGY rivers. Level agricultural region (grain, potatoes; hogs, cattle, sheep; spices, honey, fruit, fodder crops). First large natural gas discoveries in Hungary. Two of the country's three largest irrigation canals pass through the county, N-S. Manufacturing (textiles, furniture, footwear, pharmaceuticals; tobacco and food processing). Major center of Hun-

garian Protestantism through history; Debrecen is the seat of the largest bishopric of the Calvinist church. Major cities include Debrecen, Hajdúböszörmény, and HAJDUSZOBOSZLÓ (has largest medicinal spa on the Great Plain).

Hajdúböszörmény (HAH-yuh-doo-buh-suhr-mainyuh), city (2001 population 31,993), HAJDU-BIHAR county, E HUNGARY, 12 mi/19 km NNW of DEBRECEN; 47°40′N 21°31′E. Wheat, corn; hogs, sheep; manufacturing (textiles, furniture, dairy products, aluminum articles and implements). Almost perfectly round, the city's layout retains the marking of its origin, as one of the giant plain villages built for protection in the decades of Ottoman warfare.

Hajdudorog (HAH-yuh-doo-do-rog), city, HAJDU-BIHAR county, E HUNGARY, 10 mi/16 km N of Hajdúböszörmény; 47°49′N 21°30′E. Market center; wheat, corn, tobacco, sunflowers; hogs, cattle; manufacturing (textiles, clothing).

Hajduhadház (HAH-yuh-doo-hahd-hahz), city, HAJDU-BIHAR county, E HUNGARY, 10 mi/16 km N of DEBRECEN; 47°41′N 21°40′E. Wheat, paprika, sugar beets; cattle, hogs. Small textile industry; food processing.

Hajdunánás (HAH-yuh-doo-nah-nahsh), city, HAJDU-BIHAR county, E HUNGARY, 15 mi/24 km SW of NYIREGYHÁZA; 47°51′N 21°26′E. Wheat, tobacco, corn, sugar beets; cattle, hogs; manufacturing (home industries, bricks; milling); export of straw hats.

Hajdusámson (HAH-yuh-doo-shahm-shon), village, HAJDU-BIHAR county, E HUNGARY, 7 mi/11 km NE of DEBRECEN; 47°36′N 21°46′E. Wheat, corn; cattle, hogs.

Hajduszoboszló (HAH-yuh-doo-so-bos-lo), city (2001 population 23,425), HAJDU-BIHAR county, E HUNGARY, on Kösely River, and 12 mi/19 km SW of DEBRECEN; 47°27′N 21°24′E. Grain, tobacco; cattle, hogs; brickmaking. Mineral springs, natural gas nearby; largest mineral spa on Great Alföld.

Hajiganj, town (2001 population 42,806), COMILLA district, SE EAST BENGAL, BANGLADESH, 28 mi/45 km SW of COMILLA; 23°19′N 90°52′E. Trade and administrative center.

Hajipur (HAH-jee-puhr), town (2001 population 119,276), ⊙ VAISHALI district, central BIHAR state, E INDIA, on GANGA RIVER, at GANDAK RIVER mouth, and 7 mi/11.3 km N of PATNA 25°41′N 85°13′E. Railroad and road junction; trade center (rice, wheat, barley, corn, sugarcane, tobacco); rice and sugar milling. Important in late-sixteenth-century struggle between Akbar and Afghan governors of BENGAL.

Hajipur (HAH-jee-poor), INDIA: see VAISHALI.

Hajja (HAH-juh), town, N central YEMEN, near the W edge of the Yemeni highlands, 50 mi/80 km NW of SANA; 15°41′N 43°55′E. Trade center. Sometimes spelled Hadja.

Hajo (HAH-jo), town, KAMRUP district, W ASSAM state, NE INDIA, near the BRAHMAPUTRA RIVER, 15 mi/24 km WNW of GUWAHATI; 26°15′N 91°32′E. Rice, mustard, jute. Pilgrimage center. Has sixteenth-century Vishnuite temple (erection said to be consecrated by human sacrifice).

Hajós (HAH-yosh), Hungarian *Hajós*, village, BÁCS-KISKUN county, S central HUNGARY, 16 mi/26 km NNE of BAJA; 46°24′N 19°07′E. Corn, wheat, apples, tobacco; hogs; vineyards. Shrine of Saint George.

Hajr, OMAN: see HAJAR.

Hajr, Wadi (HUH-jer, WAH-dee), intermittent, partly seasonal, or sporadic coastal river, 60 mi/97 km long, S YEMEN; formerly formed border between Wahidi sultanate of BIR ALI and the Quaiti state; flows SE to Gulf of ADEN at MEIFA. Irrigates Hajr valley, one of chief agricultural areas of S YEMEN.

Haj ve Slezsku, Czech *Háj ve Slezsku*, German *Freiheitsau*, village, SEVEROMORAVSKY province, central SILESIA, CZECH REPUBLIC, on OPAVA RIVER, on railroad, and 9 mi/14.5 km ESE of OPAVA; 49°54′N 18°06′E. Agriculture (wheat, barley); poultry. Has rest home. Part of GERMANY until 1920.

Haka, MYANMAR: see HAKHA.

Hakalau (HAH-ka-LOU), village, E HAWAII island, HAWAII county, HAWAII, 12 mi/19 km N of HILO, near Hakalau Bay, on Hamakua Coast (NE). Hilo Forest Reserve to SW.

Hakari, TURKEY: see HAKKARI.

Hakata (hah-KAH-tah), town, Ochi district, EHIME prefecture, NW SHIKOKU, W JAPAN, 31 mi/50 km N of MATSUYAMA; 34°12′N 133°06′E. Prawns; salt. Boats.

Hakata, JAPAN: see FUKUOKA.

Hakata Bay (hah-KAH-tah), SE inlet of GENKAI SEA, off FUKUOKA prefecture, N KYUSHU, SW JAPAN, between Cape Nishiura (W) and SHIKANO-SHIMA island (E); 12 mi/19 km long E-W, 4 mi/6.4 km wide N-S; contains small island Noko-shima. FUKUOKA is on SE shore.

Hakata-jima (hah-KAH-tah–JEE-mah), island (□ 8 sq mi/20.8 sq km), EHIME prefecture, W JAPAN, in HIUCHI SEA (central section of INLAND SEA), 8 mi/12.9 km off N coast of SHIKOKU, just NE of O-SHIMA island; 5 mi/8 km long, 2.5 mi/4 km wide. Mountainous, fertile; sweet potatoes, tobacco, oranges, melons. Some fishing.

Hakgala (HUHK-gah-luh), peak (7,127 ft/2,172 m) in PIDURU RIDGES, CENTRAL PROVINCE, near UVA PROVINCE border, central SRI LANKA, 5 mi/8 km SE of NUWARA ELIYA; last 1,600 ft/488 m rises in bare rock; 06°55′N 80°49′E. Noted Hakgala botanical gardens on NE slope; meteorological observatory.

Hakha (HAHK-ah), township, CHIN STATE, MYANMAR, 20 mi/32 km SSW of FALAM, near INDIA border.

Haki (HAH-kee), town, Asakura district, FUKUOKA prefecture, N central KYUSHU, SW JAPAN, on CHIKUGO RIVER, and 28 mi/45 km S of FUKUOKA; 33°21′N 130°48′E. Persimmons.

Hakirya (hah-kir-YAH), a quarter of government offices and public services in the heart of TEL AVIV, W ISRAEL. Formerly German colony of Sarona (founded 1871), it was renamed 1948 when it became provisional seat of most government offices of Israel. While JERUSALEM became capital early in 1950, some government offices remain here. Also spelled Kirya.

Hakkari (Turkish=*Hakkâri*) village (2000 population 58,145), SE TURKEY, near Great ZAB River, 200 mi/322 km E of DIYARBAKIR, 24 mi/39 km from IRAQ border, and 36 mi/58 km from IRAN border; 37°36′N 43°45′E. Grain. Formerly spelled Hakari. Formerly called Colemerik (Turkish=*Çölemerik*) or Julamerk.

Hakkari Mountains (Turkish=*Hakkâri*) SE TURKEY, cover area 100 mi/161 km by 50 mi/80 km, with IRAN border on E, IRAQ frontier on S, and BOTAN RIVER on N; rise to 13,675 ft/4,168 m in CILO DAG.

Hakluyt Island (HAHK-loot), off NW GREENLAND, in N BAFFIN BAY, at mouth of INGLEFIELD GULF; 3 mi/4.8 km long, 1 mi/1.6 km wide; 77°N 72°35′W. Northernmost point reached (1616) by Baffin.

Hakmana, town, SOUTHERN PROVINCE, SRI LANKA; 06°04′N 80°38′E. Trades in rice, coconuts, rubber, cinnamon, coffee, and tea.

Hakodate (hah-ko-DAH-te), city (2005 population 294,264), extreme SW HOKKAIDO prefecture, N JAPAN, on the TSUGARU STRAIT, 96 mi/155 km S of SAPPORO; 41°45′N 140°43′E. Opened (1854) to U.S. ships and a little later (1857) to general foreign trade, it was the chief port of the island until recently replaced by Sapporo. Linked with AOMORI on HONSHU island by the Seikan Tunnel. A commercial and industrial center, the city's main industries are food processing (fish), fish-oil manufacturing, and the production of robots. Potatoes are grown in the area. Of interest is the Goryaku, the fort where the last Tokugawa shogun made his final stand, and nearby Mt. Hakodate.

Hakone (hah-KO-ne), town and resort area, Ashigarashimo district, KANAGAWA prefecture, E central HONSHU, E central JAPAN, 34 mi/55 km SW of Yokohama; 35°13′N 139°06′E. Famous for its mountains, lakes, and hot springs, it is included in FUJI-HAKONE-

IZU NATIONAL PARK, of which the central feature is FUJI-SAN. Wooden mosaics. The Hakone Shrine (built 757), a major shrine of central Japan, is one of the many religious monuments that dot the area. There is also a Tokugawa-era tollgate here.

Hakone, Mount (hah-KO-ne), Japanese *Hakone-yama* (hah-KO-ne–YAH-mah), extinct volcano in KANAGAWA prefecture, E central HONSHU, S central JAPAN, near HAKONE; has six central cones (highest is c.4,700 ft/1,433 m) and a crater (c.8 mi/12.9 km long, c.4 mi/6.4 km wide). Part of high volcanic range cutting across central Honshu. During period of Tokugawa shogunate, there was a gate (at S end of crater) forming the barrier to HAKONE Pass connecting TOKYO and KYOTO.

Håksberg (HOKS-ber-yuh), village, KOPPARBERG county, central SWEDEN, in BERGSLAGEN region, 2 mi/3.2 km NNE of LUDVIKA; 60°12′N 15°12′E.

Hakuba (HAH-koo-bah), village, North Azumi district, NAGANO prefecture, central HONSHU, central JAPAN, 19 mi/30 km W of NAGANO; 36°41′N 137°51′E. CHUBU-SANGAKU NATIONAL PARK nearby; skiing area. Alpine plant (*hakuba renzan*) zone. Traditional *minshuku*-style guesthouse originated here.

Hakui (HAH-koo-ee), city, ISHIKAWA prefecture, central HONSHU, central JAPAN, on W NOTO PENINSULA, on SEA OF JAPAN, 25 mi/40 km N of KANAZAWA; 36°53′N 136°46′E. Watermelons.

Hakusan (hahk-SAHN), town, Ichishi district, MIE prefecture, S HONSHU, central JAPAN, 12 mi/20 km S of TSU; 34°38′N 136°20′E. Mushrooms (shiitake, *shimeji*), strawberries, bamboo shoots; rice; bracken. Vegetable canning.

Hakushiu (hah-koo-SHYOO), town, North Koma district, YAMANASHI prefecture, central HONSHU, central JAPAN, 16 mi/25 km N of KOFU; 35°48′N 138°20′E.

Hakusui (HAHK-swee), village, Aso district, KUMAMOTO prefecture, W KYUSHU, SW JAPAN, near source of Shirakawa River, 19 mi/30 km E of KUMAMOTO; 32°49′N 131°05′E. Habitat of *oruri shijimi* butterfly.

Hakuta (HAHK-tah), town, Nogi district, SHIMANE prefecture, SW HONSHU, W JAPAN, 15 mi/24 km S of MATSUE; 35°20′N 133°16′E.

Hal, BELGIUM: see HALLE.

Hala (HAH-lah), town, HYDERABAD district, central SIND province, SE PAKISTAN, near INDUS River, 28 mi/45 km N of HYDERABAD; 25°49′N 68°25′E.

Halab, SYRIA: see ALEPPO.

Halabja (hah-LUHB-zhuh), town, SULAIMANIYA province, NE IRAQ, in the mountains of KURDISTAN, near IRAN border, 35 mi/56 km SE of SULAIMANIYA. Tobacco, fruit; livestock. Chief town of the Jaf Kurds. Center of Kurdish rebellious activities in the late 1980s, surprised by the Iraqi army allegedly by the use of poison gas. Also spelled Alabja.

Halachó (ah-lah-CHO), town, Halachó municipio, YUCATÁN, SE MEXICO, on railroad, and 45 mi/72 km SW of MÉRIDA; 20°28′N 90°04′W. Agricultural center (henequen, sugarcane, corn, fruit); tropical wood.

Ha-la-ha Ho, river, CHINA and MONGOLIA: see HALHIN GOL.

Halaib (hah-lah-EEB), township, RED SEA state, NE SUDAN, port on RED SEA, on road, and 185 mi/298 km NNW of PORT SUDAN; 22°13′N 36°38′E. Fishing. Lies N of 22°00′N lat. (official border with EGYPT), but within SUDAN under the de facto border between both countries.

Halamish, settlement, in SAMARIAN HIGHLANDS, 8 mi/13 km NW of RAMALLAH, WEST BANK. High-tech industry; candy factory. Poultry farming; hothouse flowers. Founded in 1977 by Jewish settlers (previously named Neve Tzuf).

Halas, HUNGARY: see KISKUNHALAS.

Halawa (HAH-LAH-vah), city, near E end of MOLOKAI island, MAUI county, HAWAII, 17 mi/27 km ENE of KAUNAKAKAI, on Halawa Bay, at mouth of Halawa Stream; 21°22′N 157°55′W. Cattle. Landing strip. Halawa Point 2 mi/3.2 km E. Halawa Beach Park to N;

Molokai Forest Reserve to W. Mokuhooniki Island Bird Refuge to SE.

Halawa Heights (HAH-LAH-vah), town, S OAHU island, HONOLULU county, HAWAII, residential suburb 7 mi/11.3 km NW of downtown HONOLULU, between South Halawa and Aiea streams. PEARL HARBOR to SW; Ewa Forest Reserve to NE. New Interstate Highway H3 crosses KOOLAU Mountains, to NE, from here.

Halba (HEL-buh), township, N LEBANON, near SYRIAN border, 15 mi/24 km ENE of TRIPOLI; 34°33′N 36°05′E. Cereals, citrus.

Halberry Head, Scotland: see CLYTH NESS.

Halberstadt (HAHL-ber-shtaht), city, SAXONY-ANHALT, central GERMANY; 51°55′N 11°03′E. Industrial center and railroad junction; has sugar-refining, metal-processing, and electrical equipment plants; other manufacturing includes machinery, rubber, and processed food. Made an episcopal see in 814. Burned (1179) by Henry the Lion; after Henry's fall (1180) the bishopric of Halberstadt was given in temporal fief to the bishops by Emperor Frederick I. Became (13th–14th century) a flourishing trade center. In 1648 it was annexed by BRANDENBURG. Severely damaged in World War II. Noteworthy buildings include the Cathedral of Saint Stephen (13th–17th century) and the Liebfrauenkirche, a 12th-century church.

Halberton (hal-BUH-tuhn), village (2001 population 1,851), E DEVON, SW ENGLAND, 3 mi/4.8 km E of TIVERTON; 50°55′N 03°25′W. Has 15th-century church.

Halbstadt, CZECH REPUBLIC: see MEZIMESTI.

Halbstadt, UKRAINE: see MOLOCHANS'K.

Halbturn (HAHLB-toorn), township, N BURGENLAND, E AUSTRIA, in the SEEWINKEL, near Hungarian border, 21 mi/34 km E of EISENSTADT; 47°52′N 16°59′E. Wine; Baroque castle built (1101–1111) by Lukas von Hildebrandt; fresco by Franz Maulbertsch; exhibitions.

Halbur, town (2000 population 202), CARROLL county, W central IOWA, 7 mi/11.3 km SW of CARROLL; 42°00′N 94°58′W. In agricultural area.

Halcott, Mount, 3,537 ft/1,078 m, GREENE county, SE NEW YORK, in the CATSKILL MOUNTAINS, 28 mi/45 km NW of KINGSTON; 42°11′N 74°26′W.

Halda, CZECH REPUBLIC: see SEZEMICE.

Haldaur (HUHL-dour), town, BIJNOR district, N UTTAR PRADESH state, N CENTRAL INDIA, 11 mi/18 km SE of BIJNOR; 29°18′N 78°17′E. Rice, wheat, gram, sugarcane.

Haldefjäll, FINLAND: see HALTIATUNTURI.

Haldeman (HAWL-duh-muhn), unincorporated village, ROWAN county, NE KENTUCKY, 7 mi/11.3 km NE of MOREHEAD, near source of TYGARTS CREEK, in Daniel Boone National Forest. Agriculture (tobacco; cattle; timber).

Halden (HAHL-luhn), city (2007 population 27,835), ØSTFOLD county, SE NORWAY, a port on the IDDEFJORD (an arm of the SKAGERRAK), near the Swedish border; 59°09′N 11°23′E. Manufacturing includes forest products and textiles. New transport terminal for import/export of goods. The first atomic reactor plant in SCANDINAVIA was built here for research purposes. The town was chartered in 1665 and developed around Fredrikssten Fortress, which three times (1716, 1718, 1814) withstood Swedish attacks. Karl XII fell before its walls in 1718. From 1665 to 1928 Halden was known as Fredrikshald.

Haldensleben (hahl-duhns-LAI-ben), town, SAXONY-ANHALT, central GERMANY, port city on WESER-ELBE CANAL, 14 mi/23 km NW of MAGDEBURG; 52°18′N 11°24′E. Sugar refining; manufacturing of machinery, pottery. Has late-Gothic church, remains of medieval town walls. Formed 1938 by union of towns of NEU-HALDENSLEBEN and ALTHALDENSLEBEN.

Haldia (huhl-dee-YAH), city, MEDINIPUR district, WEST BENGAL state, E INDIA; 21°57′N 86°21′E. Being developed as a major trade port for CALCUTTA, intended mainly for bulk cargoes. Fertilizer factory and a refinery; various light manufacturing (batteries, soap,

cement, salt, miscellaneous instruments, industrial phosphates, and pesticides). A joint public-private petrochemical complex is being set up for rapid industrialization of WEST BENGAL.

Haldibari (huhl-DEE-bah-ree), town, KOCH BIHAR district, NE WEST BENGAL state, E INDIA, 42 mi/68 km W of KOCH BIHAR; 26°20′N 88°46′E. Jute trade center; rice, tobacco, oilseeds, sugarcane.

Haldimand County (HOL-dim-uhnd), city (☐ 483 sq mi/1,255.8 sq km; 2001 population 43,728), S ONTARIO, E central CANADA, on Lake ERIE, and on GRAND RIVER; 42°55′N 79°50′W. Retains pre-1973 borders; since 2001, forms E half of former Haldimand-Norfolk regional municipality.

Haldi River, INDIA: see KASAI RIVER.

Haldummulla (HUHL-thum-mul-luh), town, UVA PROVINCE, S central SRI LANKA, in SRI LANKA HILL COUNTRY, 20 mi/32 km SSW of BADULLA; 06°46′N 80°53′E. Tea, rice, vegetables.

Haldwani (huhld-WAH-nee), town, NAINI TAL district, N UTTAR PRADESH state, N CENTRAL INDIA, 12 mi/19 km SSE of NAINI TAL, at foot of SHIWALIK RANGE; 29°13′N 79°31′E. Match manufacturing, sugar milling; trade center for hill goods, rice, wheat, oilseeds. Founded 1834. Railroad spur terminus 4 mi/6.4 km N, at KATHGODAM.

Hale, county (☐ 656 sq mi/1,705.6 sq km; 2006 population 18,236), W central Alabama; ⊙ Greensboro. In Black Belt; bounded on W by Black Warrior River. Cattle; corn, soybeans, grain; catfish; timber. Formed 1867. Named for Stephen F. Hale, a prominent lawyer who had lived at Eutaw in Greene County and was killed at Gaines Mill, VA while leading the Eleventh Alabama Infantry Regiment of the Confederate army.

Hale (HAIL), county (☐ 1,004 sq mi/2,610.4 sq km; 2006 population 36,317), NW TEXAS; ⊙ PLAINVIEW; 34°04′N 101°49′W. On the LLANO ESTACADO; elevation 3,250 ft/990 m–3,500 ft/1,070 m. Drained by WHITE RIVER, Running Water Draw (continuation of White River) and Blackwater Draw. A leading Texas agricultural county, with large irrigated acreage; grain sorghum, cotton, alfalfa, soybeans, corn, sunflowers, sugar beets, vegetables; beef cattle. Oil, clay, and natural gas deposits. Has small lakes (waterfowl hunting). Formed 1876.

Hale, city (2000 population 473), CARROLL county, NW central MISSOURI, near GRAND RIVER, 17 mi/27 km SE of CHILLICOTHE; 39°36′N 93°20′W. Wheat, corn, soybeans; hogs, cattle; feed and fertilizer.

Hale (HAIL), village (2001 population 1,898), N CHESHIRE, W ENGLAND, just S of ALTRINCHAM; 53°23′N 02°21′W.

Haleakala National Park (HAH-lai-AH-ka-LAH) (☐ 45 sq mi/117 sq km), on MAUI island, MAUI county, HAWAII; park area extends SE through Kipahulu Valley to Puhilele Point on Haleakala Coast, including Waimoku Falls and the Seven Pools. Originally established 1916 as part of Hawaii National Park, which included unit on Hawaii island, renamed Hawaii Volcanoes National Park, 1960; separated in 1961 as Haleakala National Park. Haleakala volcano, PUU ULAULA, 10,023 ft/3,055 m high, has been dormant since the mid-1700s. Its crater, 2,720 ft/829 m deep with an area of 19 sq mi/49 sq km, is one of the largest in the world. In 1976, 19,270 acres/7,799 ha were designated as a wilderness area. Main entrance and visitors center via Haleakala Crater Road from PUKALANI. Rare silversword plants and many native and migratory birds are found here. Biosphere Reserve (1980).

Haleb, SYRIA: see ALEPPO.

Halebid, INDIA: see BELUR.

Haleburg, town (2000 population 108), Henry co., SE Alabama, near Chattahoochee River, 14 mi/23 km SE of Abbeville. Named for Jonathan Hale, an early settler. Inc. in 1911.

Hale Center (HAIL), town (2006 population 2,186), HALE county, NW TEXAS, on the LLANO ESTACADO, 10 mi/16 km SW of PLAINVIEW; 34°03′N 101°50′W. Elevation 3,423 ft/1,043 m. In cattle and irrigated agricultural area (cotton; sunflowers, sugar beets, vegetables). Established 1893, incorporated 1921.

Haledon (HAIL-duhn), borough (2006 population 8,358), PASSAIC county, NE NEW JERSEY, just NW of PATERSON; 40°56′N 74°11′W. Light manufacturing. Incorporated 1908.

Haleiwa (HAH-lai-EE-vah), town (2000 population 2,225), on NW coast of OAHU, HONOLULU county, HAWAII, 23 mi/37 km NW of HONOLULU, at mouth of Anahulu Stream; 21°36′N 158°06′W. Coral gardens. Two beach parks (including Haleiwa Beach Park) and a harbor border Waialua Bay. Manufacturing (clothing).

Halemaumau (HAH-lai-MAH-oo-MAH-oo), crater in W part of Kilauea Caldera, S central HAWAII, HAWAII county, HAWAII, 27 mi/43 km SSW of HILO; 250 ft/76 m wide, 400 ft/122 m long, with floor covering 95 acres/38 ha; elevation of crater rim, 3,640 ft/1,109 m; elevation of crater floor 3,412 ft/1,040 m; rim of Kilauea Caldera, 4,078 ft/1,243 m. Periodic eruptions send lava over the rim.

Halen (HAH-luhn), agricultural commune (2006 population 8,648), Hasselt district, LIMBURG province, NE BELGIUM, 4 mi/6.4 km SE of DIEST; 50°57′N 05°06′E.

Hales Bar Dam, TENNESSEE: see NICKAJACK LAKE.

Hales Corners, city (2006 population 7,566), MILWAUKEE county, SE WISCONSIN, a suburb 10 mi/16 km SW of downtown MILWAUKEE; 42°56′N 88°02′W. In dairy and farm area. Fox farm; manufacturing (wood and fabricated metal products). Speedway. Jeremiah Curtin House State Historic Site and Experimental Aircraft Association Aviation Museum.

Halesite, village (2000 population 2,582), SUFFOLK county, SE NEW YORK, on HUNTINGTON HARBOR, on N shore of W LONG ISLAND, just N of HUNTINGTON; 40°53′N 73°24′W. In summer recreational and residential area.

Hales Location, locality, CARROLL county, E central NEW HAMPSHIRE, 7 mi/11.3 km NNW of CONWAY, in WHITE MOUNTAIN National Forest. Drained by Lucy Brook. Timber.

Halesowen (hailz-O-win), town (2001 population 57,918), WEST MIDLANDS, W central ENGLAND, just SW of BIRMINGHAM; 52°27′N 02°03′W. Listed in the *Domesday Book* as Hala. Former site of nail making and coal mining. Manufacturing of machine tools, electronic equipment, and iron goods.

Halesworth (hailz-WURHTH), town (2001 population 4,637), NE SUFFOLK, E ENGLAND, on BLYTH RIVER, and 15 mi/24 km SW of LOWESTOFT; 52°21′N 01°30′E. Agricultural market. Originally settled by Romans. Tourist attraction. Has 15th-century church.

Halethorpe, village, BALTIMORE county, N MARYLAND, 2 mi/3.2 km SW of BALTIMORE. The name meaning, "Healthy Town," was suggested by BALTIMORE AND OHIO RAILROAD officials, many of whom lived here.

Halewood (hail-WUD), town (2001 population 18,283), MERSEYSIDE, NW ENGLAND, 7 mi/11.3 km ESE of LIVERPOOL; 53°22′N 02°49′W. Motor vehicle assembly plant to SW.

Haley's Island, MAINE: see ISLES OF SHOALS.

Haley Station (HAI-lee), hamlet, unincorporated area, SE ONTARIO, E central CANADA, 8 mi/12 km from RENFREW, and included in Whitewater Region township; 45°33′N 76°46′W. Established 1845–1850.

Haleyville, city (2000 population 4,182), Winston co., NW ALABAMA, 15 mi/24 km WNW of Double Springs, in William B. Bankhead National Forest. Furniture and mobile home manufacturing. During the Civil War, Haleyville seceded from Alabama and c.2,000 Haleyville and Winston co. residents joined the Union army. Originally called 'Ark,' it was renamed in 1891 for C. L. Haley, the settlement's first merchant.

Halfa, SUDAN: see WADI HALFA.

Half Assini (ah-SEE-nee), town, district headquarters, WESTERN REGION, GHANA, on Gulf of GUINEA, close to border of Côte d'IVOIRE, 45 mi/72 km WNW of AXIM; 05°03′N 02°53′W. Fishing center; rice, cassava.

Halfaya Pass (hahl-FAH-yuh), in coastal hills of NW EGYPT, 5 mi/8 km SSE of SALUM, near Libyan border. Scene of heavy fighting (1941–1942) in World War II.

Half Dome, California: see YOSEMITE NATIONAL PARK.

Half Moon (HAF MOON), unincorporated town (□ 4 sq mi/10.4 sq km; 2000 population 6,645), ONSLOW county, E NORTH CAROLINA, residential suburb 5 mi/8 km N of downtown JACKSONVILLE, near NEW RIVER; 34°49′N 77°28′W.

Half Moon Bay, city (2000 population 11,842), SAN MATEO county, W CALIFORNIA; suburb 23 mi/37 km S of downtown SAN FRANCISCO, on picturesque Half Moon Bay of PACIFIC OCEAN, at mouth of Pilarcitos Creek; 37°28′N 122°27′W. Sheltered on N by Pillar Point. Artichokes, brussel sprouts; grain; Christmas trees, ornamentals, flowers, nursery products; fishing; light manufacturing. Annual Pumpkin Festival. Half Moon Bay Airport to NW; SANTA CRUZ MOUNTAINS and San Francisco State Fish and Game Refuge to NE; part of Half Moon Bay State Beach is to N.

Half Moon Bay (HAF MOON), summer village (2001 population 37), S central ALBERTA, W CANADA, in Lacombe County; 52°20′N 114°10′W. Incorporated 1978.

Halfmoon Bay, NEW ZEALAND: see OBAN.

Half Moon Cay National Monument, wildlife refuge, on LIGHTHOUSE REEF, BELIZE, E of TURNEFFE ISLANDS; 17°12′N 87°32′W. Extends over 45 acres/18.2 ha. Features one of two Caribbean colonies of red-footed boobies.

Halfmoon Lake (HAF-MOON), hamlet (2006 population 228), central ALBERTA, W CANADA, included in municipality of Strathcona County. Rural, residential. HALFMOON LAKE (c.1 mi/2 km long, 820 ft/250 m wide, and 30 ft/9 m deep; 54°04′N 113°21′W); summer resort.

Half Tree Hollow, settlement and local administrative district (2006 population 785), on the island of ST. HELENA, in the British dependency of St. Helena, in the S ATLANTIC OCEAN; 15°56′S 05°43′W. Elevation 961 ft/292 m. Former military fort on the outskirts of JAMESTOWN, protecting the N approaches to the island. Much of the fortification and cannon still stand.

Halfway, village (2000 population 10,065), WASHINGTON county, W MARYLAND, just SW of HAGERSTOWN; 39°37′N 77°46′W.

Halfway, village (2006 population 317), BAKER county, NE OREGON, on W Fork Pine Creek, near IDAHO state line, 36 mi/58 km ENE of BAKER CITY; 44°52′N 117°06′W. Dairy products; wheat, potatoes; sheep, cattle. Oxbow Dam, on Snake River, 15 mi/24 km to NE. Brownlee Dam, on Snake River, 11 mi/18 km to E. HELLS CANYON National Recreation Area to NE. Umatilla National Forest, including Eagle Cap Wilderness Area, to N.

Half Way House, unincorporated town, MONTGOMERY county, SE PENNSYLVANIA, residential suburb 3 mi/4.8 km N of POTTSTOWN; 40°17′N 75°38′W. Also spelled Halfway House.

Halfway Mountain (1,400 ft/427 m), SW NEWFOUNDLAND AND LABRADOR, E CANADA, 8 mi/13 km SW of BUCHANS, near N shore of RED INDIAN LAKE.

Halfway Rock Light, SW MAINE, small lighthouse island midway between PORTLAND HEAD and Seguin Island Light, CASCO BAY, and 10 mi/16 km E of PORTLAND. Completed 1871.

Half Way Tree, N suburb of KINGSTON, ⊙ SAINT ANDREW parish, SE JAMAICA, in LIGUANEA plain, c.3.5 mi/5.6 km N of Kingston; 18°00′N 76°48′W. Small agricultural region (mangoes, vegetables, coffee; cattle); manufacturing of cigars and cigarettes; dairying. Township of shopping plazas; historic clock tower as landmark. In N outskirts is King's House, since 1872

the official residence of governors and, beginning 1996, governor general of Jamaica. Historical Saint Andrew parish church, burial place of governors.

Halfweg (HAH-luh-vai), village, NORTH HOLLAND province, W NETHERLANDS, on the RINGVAART Canal, and 7 mi/11.3 km W of AMSTERDAM; 52°23′N 04°45′E. Amerikahaven port facility to N. Dairying; cattle, poultry; flowers, fruit, vegetables; manufacturing (cans).

Halgan, NEW CALEDONIA: see OUVÉA.

Halhal (hahl-hahl), village, ANSEBA region, N ERITREA, 20 mi/32 km NW of KEREN; 15°54′N 38°17′E. Fruit, vegetables, grain. Axumite ruins at nearby Aratu.

Halhin Gol (HAHL-hin GOL), Chinese *Ha-la-ha Ho*, river, 145 mi/233 km long, on CHINA (Manchuria)–MONGOLIA border; rises in the Great Khingan Mountains of E Wenchüan, N China; flows NW near or along international border to the BUYR NUR (lake), forming a delta at 47°46′N 118°45′E. Drains an area of c.6,600 sq mi/17,000 sq km. NOMONHAAN (Chinese *No-men-han*), on right bank, was scene (May–September 1939) of Soviet-Japanese border incident. Also spelled Khalkhin Gol.

Halhul, Arab township, HEBRON district, 3.1 mi/5 km N of Hebron, in the JUDEAN HIGHLANDS, WEST BANK; 3,281 ft/1,000 m; 31°35′N 35°07′E. The township is well known for its vineyards and fruit. According to the Islamic tradition the prophet Jonah was buried here.

Haliacmon River, Greece: see ALIÁKMON RIVER.

Haliartos (ah-LEE-ahr-tos), ancient town of BOEOTIA, now in Boeotia prefecture, central GREECE, on S edge of drained Lake COPAIS, 15 mi/24 km W of THEBES; 38°22′N 23°06′E. Here the Spartan general Lysander was killed in battle (395 B.C.E.) against Thebes. Destroyed (171 B.C.E.) by Romans. On site is modern village of Haliartos or Aliartos, formerly called Krimpas.

Haliburton (HA-li-buhr-tuhn), county (□ 1,554 sq mi/4,040.4 sq km; 2001 population 15,085), S ONTARIO, E central CANADA, on Burnt River; ⊙ MINDEN; 44°10′N 78°30′W.

Haliburton (HA-li-buhr-tuhn), unincorporated village, S ONTARIO, E central CANADA, near N end of KASHAGAWIGAMOG LAKE, 50 mi/80 km NNE of LINDSAY, and included in township of Dysart and Others; 45°03′N 78°30′W. Lumbering. Center of Haliburton Highlands lakes region (sport fishing, hunting; seasonal cabins).

Halibut Cove, fishing and tourist settlement, S ALASKA, SW KENAI PENINSULA, on KACHEMAK BAY, 19 mi/31 km NE of SELDOVIA. Artist colony reached by ferry from HOMER.

Halicarnassus, ancient city of CARIA, SW ASIA MINOR, on the Ceramic Gulf, present-day Gulf of GOKOVA (Turkish) or KOS (Greek), on the site of the modern city of BODRUM, TURKEY; 37°03′N 27°28′E. Halicarnassus was Greek in origin, but there were Carian inhabitants. Except for a brief period in the 5th century B.C.E., the city was not intimately concerned with Greek affairs. As a Persian vassal it was ruled by tyrants and participated in Xerxes' invasion of Greece (480 B.C.E.), but after the expulsion of the tyrants (460 B.C.E.–455 B.C.E.) it joined the DELIAN LEAGUE. A dynasty of Carian kings in the 4th century B.C.E. was made famous by Mausolus, whose wife, Artemisia, built him a magnificent tomb, considered one of the Seven Wonders of the World. Alexander the Great conquered the city (c.334 B.C.E.). Herodotus and Dionysius of Halicarnassus born here.

Halicz (HAH-leech), Russian *Galich*, peak (4,380 ft/1,335 m), in the CARPATHIAN Mountains, SE POLAND, 13 mi/21 km SW of TURKA (UKRAINE).

Halicz, UKRAINE: see HALYCH.

Halidon Hill (HAIL-eh-DAHN), hill, in N NORTHUMBERLAND, NE ENGLAND, 2 mi/3.2 km NW of BERWICK-UPON-TWEED, overlooking the TWEED. Elevation 537 ft/164 m. The English defeated the Scots here in 1333.

Halifax (HA-li-fax), county (□ 2,063 sq mi/5,363.8 sq km; 2001 population 359,183), S NOVA SCOTIA, E CANADA, on the ATLANTIC OCEAN; ⊙ was HALIFAX; part of HALIFAX REGIONAL MUNICIPALITY (HRM). Was largest county of province; disolved 1996 and amalgamated as part of HRM.

Halifax (HA-luh-faks), county (□ 731 sq mi/1,900.6 sq km; 2006 population 55,521), NE NORTH CAROLINA; ⊙ HALIFAX; 36°15′N 77°39′W. Bounded N and E by ROANOKE RIVER (forms ROANOKE RAPIDS and GASTON reservoirs in NW), S by FISHING CREEK. In PIEDMONT region. Manufacturing at Weldon, some manufacturing at Roanoke Rapids and Scotland Neck; service industries; agriculture (tobacco, peanuts, cotton, corn, wheat, soybeans; hay, sweet potatoes; poultry, hogs; some dairying); timber. County border in NW, within 3 mi/4.8 km of VIRGINIA state line. Medoc State Park in SW. Formed 1758 from Edgecombe County. Named for George Montagu, second Earl of Halifax (1716–1771), and president of the Board of Trade and Plantations. Original residents were Tuscarora Indians. Early settlers were from the British Islands and migrated to Halifax after settling first in New Jersey and other northern states.

Halifax (HA-li-faks), county (□ 824 sq mi/2,142.4 sq km; 2006 population 36,149), S VIRGINIA; ⊙ HALIFAX; 36°46′N 78°56′W. Independent city of SOUTH BOSTON reverted to town status in 1995. Bounded S by NORTH CAROLINA state line, N and E by ROANOKE (Staunton) River; drained by DAN, BANISTER, and Hyco rivers (hydroelectric plants). Manufacturing at South Boston and Halifax; agriculture (a leading Virginia tobacco-growing county; also wheat, corn, hay, soybeans; cattle). Staunton River State Park in E. Formed 1752.

Halifax, (HA-li-fax) city (1991 population 114,455; 1996 population 113,910; 2001 pop. 119,292—note that census area for Halifax changed during 1996 reorganization), ⊙ NOVA SCOTIA, E CANADA, on the Atlantic Ocean, part of the HALIFAX REGIONAL MUNICIPALITY; 44°39′N 63°36′W. Elevation 150 ft/45 m. Largest city in the Maritime provinces and one of Canada's principal ice-free Atlantic ports. E terminus of Canada's two great railroad systems and of its transcontinental highway. Its many industries include commercial fishing, oil refining, manufacturing. Home port of the Canadian Atlantic fleet and the headquarters of its E army. Founded in 1749 as Chebucto meaning Chief Harbour; later renamed for the second earl of Halifax, Geroge Montagu Dunk, then president of the Board of Trade and Plantations (1748–1761). It was intended originally to be a British naval stronghold comparable to that of France at LOUISBURG. It served as a naval base for the expedition against Louisburg in 1758, against the American colonies in the American Revolution, and against the U.S. in the War of 1812. The first transatlantic steamship service, from Halifax to Great Britain, began in 1840. During both world wars the port was an important naval and air base, convoy terminal, and embarkation center. In 1917 a French munitions vessel carrying explosives was rammed in the harbor by a Belgian relief vessel, causing an explosion that killed about 1,800 people, wounded about 9,000 more (20% of the population), and destroyed the N part of the city. Places of interest include the Citadel fortress (1856); Province House (1818); St. Paul's Church, the oldest (1750) Anglican church in Canada; and Point Pleasant Park. Seat of Dalhousie University (1818), the University of Kings College, Mount St. Vincent University, St. Mary's University, and technical and art schools. Incorporated in 1841; corporate entity disolved on April 1, 1996 when city was amalgamated into Halifax Regional Municipality.

Halifax (HAL-ih-FAKS), town (2001 population 82,056), WEST YORKSHIRE, central ENGLAND, on the Hebble River, a small tributary of the CALDER RIVER; 53°40′N 01°29′W. Adjoins COPLEY 2 mi/3.2 km S, on Calder River. An industrial town centered around the production of woolen goods, carpets, and machine tools. Other industries include the manufacturing of confections, cotton, silk and synthetics, and iron and steel. Noteworthy are the Bankfield Museum, the 18th-century Piece Hall, the 15th-century parish church of St. John the Baptist, the Renaissance town hall designed by Sir Charles Barry, and Heath Grammar School (1585).

Halifax, rural town, PLYMOUTH county, SE MASSACHUSETTS, 10 mi/16 km WNW of PLYMOUTH; 42°00′N 70°52′W. Settled c.1670, incorporated 1734.

Halifax, town, WINDHAM CO., SE VERMONT, on MASSACHUSETTS state line, 11 mi/18 km SW of BRATTLEBORO; 42°46′N 72°45′W. Named for George Montagu Dunk (1716–1771), second Earl of Halifax.

Halifax (HA-li-faks), town (2006 population 1,286), ⊙ HALIFAX county, S VIRGINIA, 5 mi/8 km N of SOUTH BOSTON, on BANISTER RIVER (Banister Dam and hydroelectric plant); 36°46′N 78°55′W. Manufacturing (textiles, shoes, pet food); in agricultural area (tobacco, corn, soybeans, wheat, hay; cattle). Incorporated 1875.

Halifax (HA-luh-faks), village (2006 population 322), ⊙ HALIFAX county, NE North Carolina, 28 mi/45 km NNE of ROCKY MOUNT, and on ROANOKE RIVER; 36°19′N 77°35′W. Service industries; agriculture (tobacco, cotton, peanuts, grain; poultry, livestock). Site of first North Carolina constitutional convention (1776). Settled c.1750. Named for George Montagu, second Earl of Halifax. Inc. in 1760.

Halifax (HA-li-faks), borough (2006 population 835), DAUPHIN county, S central PENNSYLVANIA, 14 mi/23 km NNW of HARRISBURG, on SUSQUEHANNA RIVER. Agriculture (grain, soybeans, apples; livestock, poultry; dairying); manufacturing (plastic products, stoves and furnaces, leather products). APPALACHIAN TRAIL passes to S on PETERS MOUNT ridge; Clemson Island to W.

Halifax Bay (HA-luh-faks), inlet of CORAL SEA, E QUEENSLAND, AUSTRALIA, between LUCINDA POINT (NW) and MAGNETIC ISLAND (SE); 50 mi/80 km long, 15 mi/24 km wide; 18°35′S 146°17′E. PALM ISLANDS at N entrance.

Halifax Harbour, inlet of the ATLANTIC OCEAN, S NOVA SCOTIA, E CANADA; 15 mi/24 km long, 6 mi/10 km wide at entrance; 44°35′N 63°31′W. On W shore is HALIFAX; on E is DARTMOUTH; two bridges and ferry link the two cities at the Narrows. N part of inlet is called Bedford Basin. One of largest natural harbors in North America. Major container port facilities. Autoport receives 120,000 imported motor vehicles annually. Formerly called Chebucto Harbour.

Halifax Regional Municipality (HAL-i-faks), municipality (2001 population 359,111) ⊙ NOVA SCOTIA province, CANADA. Largest population center in the province. Created in 1996, combining former cities of HALIFAX and DARTMOUTH, and the town of BEDFORD, essentially comprising most of HALIFAX county.

Halifax River (HAL-i-FAKS), narrow lagoon, c.25 mi/40 km long, in VOLUSIA county, E central FLORIDA, sheltered from the ATLANTIC OCEAN by barrier beach (site of DAYTONA BEACH and other resorts); extends N from PONCE DE LEON INLET (at N end of HILLSBOROUGH RIVER lagoon). Followed by INTRACOASTAL WATERWAY.

Haliimaile (HAH-LEE-ee-MEI-lai), town (2000 population 895), MAUI island, MAUI county, HAWAII, 8 mi/12.9 km ESE of KAHULUI, 5 mi/8 km inland from N coast; 20°52′N 156°20′W. Sugarcane, pineapples; cattle. Maunaloa College 1.5 mi/2.4 km N.

Halil River (hah-LEEL), 200 mi/322 km long, SE IRAN; rises in one of ZAGROS ranges S of KERMAN; flows SE, past Sabzevaran, to JAZ MURIAN Salt-lake depression. Also called Halilrud.

Halin (HAH-leen), village, NE SOMALIA, 100 mi/161 km NE of LAS ANOD. Livestock raising.

Halisahar (HAH-lee-suh-huhr), city, NORTH 24-PARAGNAS district, SE WEST BENGAL state, E INDIA, on HUGLI RIVER, and 25 mi/40 km N of CALCUTTA city center; 22°56′N 88°25′E. Paper milling. Noted religious center, where a 19th-century saint-singer (Ramprosad) worshipped the goddess Kali. Formerly called KUMARHATA.

Halita, UKRAINE: see YALTA, city.

Haliyal (HUH-lee-yahl), town, UTTAR KANNAD district, KARNATAKA state, S INDIA, 19 mi/31 km WSW of DHARWAD; 15°20′N 74°46′E. Trades in teak, blackwood, bamboo, rice. British frontier post in eighteenth century. Also spelled Halyal.

Halkett, Cape, N ALASKA, on ARCTIC OCEAN, 110 mi/177 km ESE of BARROW; 70°47′N 152°05′W. Inuit trading post of Cape Halkett is 10 mi/16 km WNW.

Halkirk (HAL-kuhrk), village (2001 population 117), S central ALBERTA, W CANADA, 23 mi/37 km E of STETTLER, in Paintearth County No. 18; 52°17′N 112°09′W. Coal mining. Mixed farming; wheat; cattle; dairying. Established in 1912.

Halkirk (HAL-kuhrk), agricultural village (2001 population 923), HIGHLAND, N Scotland, on THURSO RIVER, and 6 mi/9.7 km S of THURSO; 58°30′N 03°30′W. There are ruins of ancient Brawl Castle, adjoining modern structure. Previously whiskey distilling.

Halkis, Greece: see KHALKÍS.

Halkyn (HAL-kin), village (2001 population 2,876), FLINTSHIRE, NE Wales, 3 mi/4.8 km WSW of FLINT; 53°13′N 03°11′W. Formerly in CLWYD, abolished 1996.

Hall, county (□ 426 sq mi/1,107.6 sq km; 2006 population 173,256), NE GEORGIA; ⊙ GAINESVILLE; 34°19′N 83°49′W. Piedmont area drained by CHATTAHOOCHEE and OCONEE rivers. Agricultural products include soybeans, hay, sweet potatoes; cattle, hogs, poultry; eggs. Also producer of granite and marble. Formed 1818.

Hall, county (□ 552 sq mi/1,435.2 sq km; 2006 population 55,555), S central NEBRASKA; ⊙ GRAND ISLAND; 40°52′N 98°30′W. Irrigated agricultural and industrial region drained by PLATTE, Wood, and SOUTH LOUP rivers. Manufacturing at Grand Island; corn, soybeans, wheat, sorghum, vegetables; cattle, hogs; dairy and poultry products. In area are Cheyenne and War Axe (SW) and Mormon Island (SE) state waysides. Formed 1859.

Hall (HAHL), county (□ 904 sq mi/2,350.4 sq km; 2006 population 3,668), NW TEXAS; ⊙ MEMPHIS; 34°31′N 100°41′W. In plains region below CAPROCK ESCARPMENT of LLANO ESTACADO; elevation 2,200 ft/670 m–3,300 ft/1,000 m. Crossed by PRAIRIE DOG TOWN FORK OF RED RIVER. Agriculture, chiefly cotton; also grain and sweet sorghum, wheat, peanuts; beef and dairy cattle, hogs. Formed 1876.

Hall, village, GRANITE county, W MONTANA, on FLINT CREEK, 55 mi/89 km W of HELENA. Hay; cattle, horses. In area are Lolo (W) and Deerlodge (S and SE) national forests. Garnet Range to N.

Hall, Austria: see HALL IN TIROL or BAD HALL.

Hällabrottet (HEL-lah-brot-tet), village, ÖREBRO county, S central SWEDEN, 13 mi/21 km S of ÖREBRO; 59°07′N 15°14′E.

Halladale River (HAWL-uh-DAIL), 22 mi/35 km long, HIGHLAND, N Scotland; rises 18 mi/29 km NW of HELMSDALE; flows N to the ATLANTIC OCEAN 3 mi/4.8 km E of Strathy in Melvich Bay.

Hallam (HAWL-uhm), village (2006 population 575), LANCASTER county, SE NEBRASKA, area drainage of North Fork of BIG NEMAHA RIVER, 18 mi/29 km S of LINCOLN; 40°32′N 96°47′W. Olive Creek Lake State Recreation Area to NW.

Hallam (HA-lahm), borough (2006 population 2,790), YORK county, S PENNSYLVANIA, 7 mi/11.3 km ENE of YORK. Post office name is Hellam; borough surrounded by Hellam township. Manufacturing (apparel, paper products). Agricultural area (apples, soybeans, grain; poultry, livestock; dairying). Samuel S. Lewis State Park to E.

Halla, Mount, Korean *Halla-san,* volcanic peak (6,398 ft/1,950 m), central CHEJU Island, SOUTH KOREA, 10 mi/16 km S of CHEJU. Highest peak of South Korea; part of HALLA MOUNTAIN NATIONAL PARK.

Halla Mountain National Park, Korean *Halla-san Kungnip Kongwon,* CHEJU province, SOUTH KOREA, in the center of Cheju Island (largest in Korea), S of KOREAN Peninsula. Extinct volcanic mountain (made of basalt) is the highest (6,398 ft/1,950 m) in South Korea. Sitting on its summit is Paengnoktam (White Deer), a crater lake 1.2 mi/2 km in circumference. Many deep valleys, rare rocks, and a variety of forests at different altitudes with 1,800 species of flora. International tourism. Ferry from PUSAN, MOKPO, and WAN ISLAND. Established 1970.

Hallam Peak (HA-luhm) (10,560 ft/3,219 m), SE BRITISH COLUMBIA, W CANADA, near HAMBER PROVINCIAL PARK, 55 mi/89 km SW of JASPER (ALBERTA); 52°11′N 118°46′W.

Halland (HAHL-lahnd), county (□ 1,901 sq mi/4,942.6 sq km), SW SWEDEN, on KATTEGATT STRAIT; ⊙ HALMSTAD; 56°45′N 13°00′E. Metalworking; agriculture (grain); fish (especially salmon). Almost co-extensive with historical province of Halland, conquered from DENMARK by Charles X of Sweden in 1658. Low and undulating, drained by LAGAN, NISSAN, Viskan, and ÄTRAN rivers. Cities are HALMSTAD, LAHOLM, FALKENBERG, VARBERG, and KUNGSBACKA.

Hallandale (HAH-luhn-dal), city (□ 4 sq mi/10.4 sq km; 2000 population 34,282), extreme SE BROWARD county, SE FLORIDA, on the ATLANTIC coast and the INTRACOASTAL WATERWAY; 25°59′N 80°08′W. Retirement center, especially for French Canadians. Horse and greyhound racetracks are major sources of employment. A principal tourist attraction is Gulf Stream Park, site of the annual Florida Derby. Settled 1897, incorporated 1927.

Hallaniya (HEL-uh-NEE-yuh), chief island (□ 22 sq mi/57.2 sq km) of the KURIA MURIA group, off SE OMAN; 7.5 mi/12.1 km long, 3.5 mi/5.6 km wide; rises to 1,647 ft/502 m; 17°31′N 56°05′E. Fishing; goat raising. Telegraph station operated here temporarily (1859–1860).

Hall Basin, channel, in the ARCTIC OCEAN, between NE ELLESMERE ISLAND (Canada) and NW GREENLAND, joining ROBESON (N) with KENNEDY (S) channels; part of passage between the ATLANTIC OCEAN and LINCOLN SEA of the Arctic Ocean; 40 mi/64 km long, 30 mi/48 km wide; 81°30′N 62°30′W.

Hall Beach, Inuktitut *Sanirajak* [=along the coast], hamlet (□ 7 sq mi/18.2 sq km; 2001 population 609), BAFFIN region, NUNAVUT territory, N CANADA, on E shore of MELVILLE PENINSULA, on W side of FOXE BASIN; 68°46′N 81°13′W. Rises to c.27 ft/8.2 m. The land is covered with sand and gravel beaches, lakes, and tundra ponds. Fishing, hunting; soapstone. Scheduled air service from IQALUIT; access by boat.

Hallbergmoos (hahl-berg-MOS), suburb of MUNICH, BAVARIA, S GERMANY, in UPPER BAVARIA, 16 mi/26 km NE of city center; 48°20′N 11°45′E.

Hällbybrunn (HEL-BEE-BRUN), residential village, SÖDERMANLAND county, E SWEDEN, 3 mi/4.8 km WNW of ESKILSTUNA.

Halle (HAH-luh), French *Hal,* commune, ⊙ Halle-Vilvoorde district (2006 population 34,947), together with VILVOORDE, in BRABANT province, central BELGIUM, on the CHARLEROI-BRUSSELS CANAL; 50°44′N 04°14′E. Commercial and industrial center. Manufacturing. Halle's Gothic Church of Our Lady (fourteenth–fifteenth century), a popular pilgrimage site, contains a celebrated miraculous image of the Virgin.

Halle (HAH-le), city (2005 population 237,198), SAXONY-ANHALT, central GERMANY, on the SAALE River; 51°28′N 12°00′E. Industrial center and major transportation hub. Manufacturing includes chemicals, refined sugar and other food products, machinery,

rubber, cement, and electrical and chemical products. Lignite and potash are mined in the region; during the Middle Ages, the city was raised to prominence for mining and trading salt. Industrialization has caused Halle and the region surrounding it to become one of the most polluted areas in EUROPE. Located on the site of Bronze- and Iron-age settlements, Halle was first mentioned in the 9th century; possessed by (from 968) the archbishops of MAGDEBURG, who frequently resided here. Member (1281–1478) of the HANSEATIC LEAGUE; accepted the Reformation 1544. Annexed by BRANDENBURG in 1648. The famous University of Halle was founded in 1694, and in 1817 it absorbed the University of WITTENBERG; in 1933, the name of the university changed to Martin Luther University Halle-Wittenberg. Here, in 1695 the philanthropist A. H. Francke founded a school for paupers, the first of the Francke Institutes. The first Bible Society was founded here in 1710. Noteworthy buildings include the Gothic Red Tower (1418–1506) and the Marienkirche, a 16th-century church. Composer George Friedric Handel born here (1685).

Hällefors (HEL-le-FORSH), town, ÖREBRO county, S central SWEDEN, in BERGSLAGEN region, 13 mi/21 km ENE of FILIPSTAD; 59°47′N 14°31′E.

Hälleforsnäs (HEL-le-FORSH-NES), town, SÖDERMANLAND county, E SWEDEN, 15 mi/24 km S of ESKILSTUNA; 59°09′N 16°30′E.

Hallein (HAHL-lein), town, in SALZBURG province, W AUSTRIA, in the Tennergau, on the SALZACH River, near the German border; 47°41′N 13°06′E. Manufacturing includes pulp and paper, machines, chemicals, shoes, cosmetics, foodstuffs; brewery. Market center of the Tennergau. Hydropower station on the Salzach River. An ancient Celtic settlement, Hallein was first mentioned in the 12th century. Salt mines in nearby Dürmberg and saltworks (abandoned 1989) made it a prosperous town in the Middle Ages. A museum of Celtic history displays findings from Dürmberg.

Halle in Westfalen (HAHL-le in vest-FAH-len), town (2006 population 21,300), NORTH RHINE–WESTPHALIA, NW GERMANY, on S slope of TEUTOBURG FOREST, 8 mi/12.9 km NW of BIELEFELD; 52°04′N 08°22′E. Manufacturing of candy, textiles, pharmaceuticals, packaging materials; metalworking. Annual Gerry Weber Open tennis tournament prepares players for Wimbledon. Has 16th-century castle.

Hällekis (HEL-le-SHEES), village, SKARABORG county, S SWEDEN, on E shore of LAKE VÄNERN, 14 mi/23 km WSW of MARIESTAD; 58°38′N 13°26′E.

Hallenberg (HAHL-len-berg), town, North NORTH RHINE–WESTPHALIA, W GERMANY, 22 mi/35 km N of MARBURG; 51°07′N 08°37′E. Forestry; metalworking; manufacturing of furniture.

Hallenberg, GERMANY: see STEINBACH-HALLENBERG.

Hallennes-lez-Haubourdin (ah-LEN–laiz–o-boor-DAN), commune (□ 2 sq mi/5.2 sq km), NORD department, NORD PAS-DE-CALAIS region, N FRANCE; W suburb of LILLE; 50°37′N 02°58′E.

Hallett (HA-let), village (2006 population 171), PAWNEE county, N OKLAHOMA, 14 mi/23 km ESE of PAWNEE; 36°13′N 96°34′W. Agricultural center.

Hallett, Cape, on the ROSS SEA, at the foot of MOUNT SABINE, on the BORCHGREVINCK COAST of VICTORIA LAND, EAST ANTARCTICA; 72°19′S 170°16′E.

Hallettsville, city (2006 population 2,538), ⊙ LAVACA county, on LAVACA RIVER, and 17 mi/27 km NE of YOAKUM; elevation 232 ft/71 m; 29°26′N 96°56′W. In agricultural area (cattle, hay; corn, rice, milo); manufacturing (natural-gas processing, printing; canned and bottled beverages, livestock trailers). Annual Kolache Festival in September.

Hällevik (HEL-le-VEEK), fishing village, BLEKINGE county, S SWEDEN, on S shore of LISTERLANDET peninsula, on BALTIC SEA, 4 mi/6.4 km SE of SÖLVESBORG; 56°01′N 14°41′E. Seaside resort.

Hälleviksstrand (HEL-le-VEEKS-STRAHND), fishing village, GÖTEBORG OCH BOHUS county, SW SWEDEN, on W coast of ORUST island, on SKAGERRAK, 10 mi/16 km S of LYSEKIL; 58°08′N 11°26′E. Seaside resort.

Halley (HA-lee), village, DESHA county, SE ARKANSAS, 6 mi/9.7 km E of DERMOTT, between Big Bayou and BOEUF RIVER.

Halley Station, ANTARCTICA, British station on BRUNT ICE SHELF; 75°35′S 26°15′W. Replaced Halley Bay Station, February 1984.

Hall, Fort, Idaho: see FORT HALL.

Halliday (HA-luh-dai), village (2006 population 215), DUNN CO., W central NORTH DAKOTA, 27 mi/43 km WNW of BEULAH, and on Spring Creek; 47°21′N 102°20′W. Fort Berthold Indian Reservation to N.

Hallig Islands (HAHL-lig), German *Die Halligen,* island group of the N FRISIAN group, NW GERMANY, in the NORTH SEA, off SCHLESWIG-HOLSTEIN coast; 54°25′N 08°27′E–54°40′N 08°57′E. There are eleven islands, including NORDSTRAND, PELLWORM, NORDMARSCH-LANGENESS, and HOOGE. Rising barely above sea level, and without protection from dikes and jetties, the islands are often submerged during storms. Noted for excellent pastures. Houses are built on piles.

Hallikhed, town, BIDAR district, KARNATAKA state, S INDIA, 18 mi/29 km WSW of BIDAR. Millet, cotton, oilseeds, rice.

Halling (HA-ling), town (2001 population 2,698), N KENT, SE ENGLAND, on MEDWAY RIVER, and 4 mi/6.4 km SW of ROCHESTER; 51°20′N 00°26′E. Agricultural market. Has 13th-century church. An Ice Age human skeleton was found here.

Hallingdal (HAHL-ling-dahl), valley of Hallingdal River, upper course of DRAMMENSELVA River, BUSKERUD county, S NORWAY; 70 mi/113 km long. Extends ENE from S slope of HALLINGSKARVET mountains to GOL, then SSE to the lake KRØDEREN. The HARDANGERVIDDA rises on W side of valley. Important barley- and oat-growing region; livestock raising; lumbering. The rivers are major producers of hydroelectric power. Main villages are NESBYEN, GEILO, HOL, Ål, Gol, and FLÅ. There are several ancient stave churches. Valley is noted for its peasants' customs and dances. Tourism is important here year-round.

Hallingskarvet (HAHL-lings-kahr-vaht), mountain range in BUSKERUD, HORDALAND, and SOGN OG FJORDANE counties, S NORWAY; forms NE section of the Hardangerfjell. Extends E-W c.60 mi/97 km; rises to 6,342 ft/1,933 m in the FOLARSKARDNUTEN, 10 mi/16 km E of FINSE. Tourist area.

Hall in Tirol (HAHL in ti-ROL), town, TYROL, W AUSTRIA, on INN River, 5 mi/8 km E of INNSBRUCK; 47°17′N 11°30′E. Market center; manufacturing of metals, plastics, textiles, foodstuffs; hydroelectric power station; tourism. Many medieval buildings and fortifications; 15th century Gothic church rebuilt in rococo style in the 18th century; salt mining from the 13th century to 1967; saltworks and salt shipping on Inn River; mint from 1477 to 1809. Monasteries and castles.

Hall Island, NUNAVUT territory, N CANADA, in the Gulf of BOOTHIA, near W entrance of FURY AND HECLA STRAIT, off NW BAFFIN ISLAND; 40 mi/64 km long, 6 mi/10 km wide; 70°00′N 87°00′W.

Hall Island, NUNAVUT territory, N CANADA, in the ATLANTIC OCEAN, off HALL PENINSULA, SE BAFFIN ISLAND; 6 mi/10 km long; 62°32′N 64°10′W.

Hall Island, in S FRANZ JOSEF LAND, ARCHANGEL oblast, extreme N European Russia, in the ARCTIC OCEAN, E of MACCLINTOCK ISLAND; separated from WILCZEK LAND (NE) by the AUSTRIAN SOUND; 25 mi/40 km long, 25 mi/40 km wide.

Hall Island, W ALASKA, in BERING SEA, 3 mi/4.8 km NW of SAINT MATTHEW ISLAND, and 180 mi/290 km WNW of NUNIVAK Island; 6 mi/9.7 km long, 2 mi/3.2 km wide; 60°33′N 172°42′W. Rugged.

Hall Island, GILBERT ISLANDS, KIRIBATI: see MAIANA.

Hall Islands, group, State of CHUUK, E CAROLINE ISLANDS, Federated States of MICRONESIA, W PACIFIC, c.70 mi/113 km N of CHUUK ISLANDS; include NOMWIN and MURILO atolls.

Hall Land, peninsula, NW GREENLAND, on HALL BASIN, opposite ELLESMERE ISLAND; 81°30'N 61°00'W. A great outlet from the inland ice, PETERMANN GLETSCHER, forms the SW border of the peninsula. On coast is POLARIS BAY. The area is named after the American explorer Charles Francis Hall, who was buried on Hall Land (1871). Investigations of his body in 1968 suggested that he was poisoned by arsenic.

Hallmundarhraun, ICELAND: see EIRÍKSJÖKULL.

Hällnäs (HEL-NES), village, VÄSTERBOTTEN county, N SWEDEN, on VINDELÄLVEN RIVER, 40 mi/64 km NW of UMEÅ; 64°19'N 19°37'W. Railroad junction.

Hallock (HA-luhk), town (2000 population 1,196), ⊙ KITTSON county, extreme NW MINNESOTA, c.60 mi/97 km N of GRAND FORKS (NORTH DAKOTA), on South Branch TWO RIVERS, opposite mouth of Middle Branch Two Rivers, near CANADA (MANITOBA) border; elevation 827 ft/252 m; 48°46'N 96°56'W. Grain, potatoes, sunflowers, flax, sheep, alfalfa, sugar beets; livestock; manufacturing (concrete products). Hunting and fishing in vicinity. Plotted 1879, incorporated 1887.

Hallowell (HAH-luh-wel), city (2000 population 2,467), KENNEBEC county, S MAINE, on the KENNEBEC RIVER, just S of AUGUSTA; 44°17'N 69°48'W. Boatbuilding. Hallowell granite used for state capitol. Settled c.1754; town incorporated 1771, including present Augusta; city incorporated 1850.

Hall Park, town (2000 population 1,088), CLEVELAND county, central OKLAHOMA, residential suburb 17 mi/27 km SSE of downtown OKLAHOMA CITY, and 3 mi/4.8 km NE of downtown NORMAN; 35°14'N 97°24'W. Surrounded by city of Norman.

Hall Peninsula, SE BAFFIN ISLAND, BAFFIN region, NUNAVUT territory, N CANADA, extends 150 mi/241 km SE into DAVIS STRAIT, between CUMBERLAND SOUND and FROBISHER BAY; 100 mi/161 km wide at base; 62°40'N 65°10'W. SE extremity is Blunt Peninsula. IQALUIT on S shore.

Halls, town (2006 population 2,214), LAUDERDALE county, W TENNESSEE, 10 mi/16 km S of DYERSBURG; 35°53'N 89°24'W. Veterans museum here. Was main base for B-17 Flying Fortress bombers in mid-1940s.

Hallsberg (HAHLS-ber-yuh), town, ÖREBRO county, S central SWEDEN, 14 mi/23 km SSW of ÖREBRO; 59°04'N 15°08'E. Railroad junction; manufacturing (metalworking, tanning; footwear, foods).

Hallsboro (HAHLZ-buhr-o), unincorporated village, COLUMBUS county, SE North Carolina, 6 mi/9.7 km E of WHITEVILLE; 34°19'N 78°35'W. Lake Waccamaw State Park to SE. Named for an early settler, it was settled in 1888 and inc. in 1889.

Hall, Schwäbisch, GERMANY: see SCHWÄBISCH HALL.

Halls Creek (HAHLZ KREEK), village, NE WESTERN AUSTRALIA, 1,850 mi/2,977 km from PERTH, 190 mi/306 km S of WYNDHAM; 18°16'S 127°46'E. Mining center of KIMBERLEY GOLDFIELD. Copper and zinc deposits.

Halls Gap (HAHLZ GAP), town, VICTORIA, SE AUSTRALIA, 156 mi/251 km NW of MELBOURNE, in Fyan's Valley, at edge of GRAMPIANS NATIONAL PARK; 37°07'S 142°31'E. Tourism. Wildlife park and zoo, Aboriginal heritage center, lakes nearby. Several annual arts and culture festivals.

Hall's Stream, c.20 mi/32 km long, NEW HAMPSHIRE, VERMONT, and CANADA (QUEBEC); rises in NW New Hampshire; flows SSW to the CONNECTICUT RIVER at CANAAN (Vermont); forms international border above Beecher Falls (Vermont).

Hallstadt (HAHL-shtaht), town, BAVARIA, central GERMANY, in UPPER FRANCONIA, on Main River, 2 mi/3.2 km NW of BAMBERG; 49°55'N 10°53'E. Meat processing; brickworks; brewing. Has 16th-century town hall and 18th-century palace. Chartered in 1954.

Hallstahammar (HAL-stah-HAHM-mer), town, VÄSTMANLAND county, central SWEDEN, on KOLBÄCKSÅN RIVER, 10 mi/16 km W of VÄSTERÅS; 59°37'N 16°13'E. Manufacturing (steel, machine tools, electrical equipment).

Hallstatt (HAHL-shtaht), township, S UPPER AUSTRIA province, W central AUSTRIA, in the SALZKAMMERGUT, on the Lake of HALLSTATT; 47°34'N 13°39'E. A tourist center and the oldest salt mines in the world are here. One of the oldest and most picturesque settlements in Austria. The term Hallstatt now refers to late Bronze- and early Iron-Age culture in central and W Europe. During excavations in the latter half of the 19th century, over 2,000 graves were discovered in an ancient cemetery near Hallstatt. Most of the graves dated to two time periods, an earlier (c.1100 B.C.E.–c.700 B.C.E.) and a later (c.700 B.C.E.–450 B.C.E.). Near the cemetery, preserved in a prehistoric salt mine, the bodies of miners have been discovered, as well as their implements and clothing. Location of a 15th century church with a famous altar. Corpus Christi procession on the lake takes place here.

Hallstatt, Lake of (HAHL-shtaht), German *Hallstätter See* (HUL-shte-tuhr-se), S UPPER AUSTRIA, in the SALZKAMMERGUT, at the N foot of the DACHSTEIN massif (mountain), 7 mi/11.3 km S of BAD ISCHL; 3.6 mi/5.8 km long, 1.4 mi/2.3 km wide; maximum depth 381 ft/116 m; elevation 1,548 ft/472 m. Both fed and drained by the TRAUN RIVER; beautiful alpine scenery and tourist area; 47°35'N 13°40'E. HALLSTATT on SW; STEEG on NW; OBERTRAUN on SE shore.

Hallstavik (HAHL-stah-VEEK), town, STOCKHOLM county, E SWEDEN, on small bay of GULF OF BOTHNIA, 35 mi/56 km ENE of UPPSALA; 60°03'N 18°36'E. Seaport. Has thirteenth-century church.

Hallstead (HAL-sted), borough (2006 population 1,163), SUSQUEHANNA county, NE PENNSYLVANIA, 13 mi/21 km SE of BINGHAMTON (New York), on SUSQUEHANNA RIVER (bridged to GREAT BEND, 1 mi/1.6 km N), near New York state line. Agriculture (grain; livestock; dairying); manufacturing (electronics). Salt Springs State Park to SW. Founded 1787, incorporated 1874.

Hall Summit, village (2000 population 264), RED RIVER parish, NW LOUISIANA, 30 mi/48 km SE of SHREVEPORT; 32°11'N 93°15'W. In agricultural area (cotton, peaches, soybeans; cattle). Logging.

Hallsville, town (2000 population 978), BOONE county, central MISSOURI, 13 mi/21 km NNE of COLUMBIA; 39°07'N 92°13'W. Corn, hay; cattle.

Hallsville (HAHLS-vil), town (2006 population 2,957), HARRISON county, E TEXAS, 12 mi/19 km W of MARSHALL, and 10 mi/16 km E of LONGVIEW, near SABINE RIVER; 32°30'N 94°34'W. In agricultural area (cattle, hogs; nurseries; timber.

Halltown, town (2000 population 189), LAWRENCE county, SW MISSOURI, 18 mi/29 km W of MOUNT VERNON; 37°11'N 93°37'W. Plotted 1887.

Hallûf, Jabal, TUNISIA: see SUQ AL KHAMIS.

Halluin (ahl-WAN), town (□ 4 sq mi/10.4 sq km), NORD department, NORD-PAS-DE-CALAIS region, N FRANCE, on Belgian border opposite MENEN (French *Menin*) (Belgium), 5 mi/8 km NNW of TOURCOING, near the LYS River; 50°47'N 03°08'E. Textile center (linen and cotton cloth, rugs). Diverse manufacturing (beer, furniture, biscuits).

Hallwilersee (HAHL-veel-uhr-zai), lake (□ 4 sq mi/10.4 sq km), N SWITZERLAND, bordering on LUCERNE and AARGAU cantons, traversed by AA RIVER; 5 mi/8 km long, maximum depth 154 ft/47 m. Elevation 1,473 ft/449 m. Hallwil castle (9th century) is N.

Hallwood (HAHL-wud), village (2006 population 287), ACCOMACK county, E VIRGINIA, 12 mi/19 km N of ACCOMAC; 37°52'N 75°35'W. CHESAPEAKE BAY to W. Agriculture (grain, vegetables; livestock; poultry).

Hallyo Marine National Park, Korean *Hallyo Haesang Kungnip Kongwon* (□ 197,035 sq mi/512,291 sq km), SOUTH KOREA. National sea park, E KOREA STRAIT. Better known as Hallyo-sudo, a 93-mi/150-km-long waterway stretching from KOJE ISLAND in SOUTH KYONGSANG province, to Odong Island in SOUTH CHOLLA province. Has 400 islands and islets (many uninhabited); many beaches and unusual rock formations. Seascape around Haegum River, Pijin Island, Sejon Island, Pipe Organ Valley, Ssangyonggul Cave and Nakhwaam Rock. Historic sites include Chesungdang and Chungnyolsa Shrine, associated with Admiral Yi Soon Shin, one of Korea's greatest war heroes during the 1592 Japanese invasion. Established 1968.

Halma (HAL-muh), village (2000 population 78), KITTSON county, NW MINNESOTA, 18 mi/29 km SE of HALLOCK, in RED RIVER valley; 48°39'N 96°35'W. Grain, sunflowers. Twin Lakes Wildlife Area to SE.

Hǎlmagiu (huhl-mah-JYOO), Hungarian *Nagyhalmágy*, village, ARAD county, W ROMANIA, on Crişul Alb River, on railroad, and 63 mi/101 km E of ARAD; 46°16'N 22°35'E. Lumbering and agriculture center. Nearby Gaina is noted for picturesque peasant costumes, as well as for traditional yearly fairs at which marriageable girls of the district offer themselves for bids to prospective husbands.

Halmahera (HAHL-mah-hai-rah), island (□ 6,928 sq mi/17,944 sq km; 1990 population 138,000), MALUKU province, E INDONESIA, between NEW GUINEA and SULAWESI islands, on the equator; 01°00'N 128°00'E. The largest of the MALUKU and irregular in shape, it consists of two intersecting mountain ranges (rising to c.5,000 ft/1,520 m), which form four rocky peninsulas separated by three deep bays. Has several active volcanos, lush forests, streams, and a few lakes. The indigenous population, mostly Malays, engage in subsistence farming, hunting, and fishing. Chief products are spices, resin, sago, rice, tobacco, and coconuts. Anchorages at JAILOLO and Weda. Known to the Portuguese and the Spaniards as early as 1525, Halmahera came under Dutch influence in 1660. Taken by the Japanese (1942) in World War II, it was frequently bombed by the Allies. The island was called Jailolo (Djailolo) by the Dutch.

Halmstad (HAHLM-stahd), town, ⊙ HALLAND county, SW SWEDEN, seaport on KATTEGATT STRAIT at mouth of NISSAN RIVER; 56°40'N 12°52'E. Manufacturing (textiles, glass, paper; brewing); engineering. Summer resort. Important Danish fortified city conquered by Sweden (1645). Gothic church (fourteenth century), seventeenth-century castle. Chartered 1307. Airport.

Halmyros, Greece: see ALMIROS.

Halol (HAH-lol), town, PANCH MAHALS district, GUJARAT state, W INDIA, 20 mi/32 km SSW of GODHRA; 22°30'N 73°28'E. Trades in grain, timber, hardware, tobacco, cotton cloth; markets corn, wheat, millet; sawmills; cotton ginning.

Ha Long Bay, Gulf of BAC BO, NE VIETNAM, c. 34 mi/55 km from HAI PHONG; 20°55'N 107°05'E. Entire bay, including 3,000 islands, extends 932 mi/1,500 km along the border between Hai Phong urban region and QUANG NINH province. It is a major tourist site whose name, Ha Long, refers to the Legend of the Descending Mountain Dragon. Beaches, impressive grottos, and limestone define the area. Its delicate ecosystem and landscape is threatened by industrial pollution, as well as the tampering of coral reefs and limestone outcrops by tourists and others. Formerly Halong (or ALONG) Bay.

Halonnesos, Greece: see ALONNISSOS.

Haloze (HAH-lo-ze), hilly region, NE SLOVENIA, S of DRAVINJAT, on Croatian border. Vineyards.

Halq al Wadi (HAHLK el WE-dee), town (2004 population 28,407), TUNIS province, N TUNISIA, 6 mi/9.7 km E of TUNIS; 36°49'N 10°18'E. Principal outport for TUNISIA, at LAKE OF TUNIS (W) and GULF OF TUNIS (E), at lake's outlet. Linked with TUNIS by a deep-sea channel (for ships drawing 20 ft/6 m) and by a causeway (electric trolley) across lake. Sheltered anchorage; fishing port and iron-ore shipping center. A favorite summer resort of Tunis residents. Captured by Spaniards, town was fortified by Charles V. Taken by Turks in 1574 after a memorable siege, it was used as a base for corsair raids. Formerly called La Goulette.

Hals (hals), town and port, NORDJYLLAND county, N JUTLAND, DENMARK, at E entrance of Lim Fjord (the Langrrak), 15 mi/24 km E of ÅLBORG; 57°05′N 10°10′E. Fisheries (cod, shrimp).

Halsa (HAHL-sah), village, MØRE OG ROMSDAL county, W NORWAY, near mouth of HALSAFJORDEN, 16 mi/26 km E of KRISTIANSUND. Tannery, sawmills nearby.

Halsa, ISRAEL: see KIRYAT SHMONA.

Halsafjorden (HAHL-sah-fyawr-uhn), inlet of NORTH SEA, in MØRE OG ROMSDAL CO., W NORWAY, E of KRISTIANSUND, and extending SE c.30 mi/48 km to Todal. On it are HALSA and Stangvik.

Halsey (HAHL-zee), village (2006 population 782), LINN county, W OREGON, 17 mi/27 km S of ALBANY, 6 mi/9.7 km E of WILLAMETTE RIVER; 44°22′N 123°06′W. Dairy products; grain, fruit; cattle; timber; pulp mills.

Hälsingland (HELS-eeng-lahnd), historic province (□ 5,933 sq mi/15,425.8 sq km), E central SWEDEN, included largely in GÄVLEBORG county.

Halsnøy (HAHL-suhn-uh-oo), island (□ 15 sq mi/39 sq km) in HARDANGERFJORDEN, HORDALAND county, SW NORWAY, 30 mi/48 km NNE of HAUGESUND; 9 mi/14.5 km long, 5 mi/8 km wide. Main villages are Saebøvik in W and Toftevåg in N. On a NW peninsula are remains of medieval Augustinian abbey.

Halsö (HAHLS-UH), fishing village, GÖTEBORG OCH BOHUS county, SW SWEDEN, on island (309 acres/125 ha) of Halsö in SKAGERRAK, 12 mi/19 km WNW of GÖTEBORG; 57°44′N 11°39′E.

Halstad (HAL-stuhd), village (2000 population 622), NORMAN county, NW MINNESOTA, on RED RIVER, and 32 mi/51 km N of FARGO (NORTH DAKOTA); 47°21′N 96°49′W. Agricultural area (grain, potatoes, sunflowers, sugar beets); manufacturing (printing).

Halstead (HAWL-sted), town (2001 population 14,312), N ESSEX, SE ENGLAND, on COLNE RIVER, and 12 mi/19 km WNW of COLCHESTER; 51°57′N 00°38′E. Agricultural market, with tanneries, artificial-silk mills, flour mills. Has 15th-century church.

Halstead (HAWL-sted), town (2000 population 1,873), HARVEY county, S central KANSAS, on LITTLE ARKANSAS RIVER, and 10 mi/16 km WSW of NEWTON; 38°00′N 97°30′W. Market and shipping point for wheat and livestock region; flour milling. Manufacturing (animal feeds, lumber, mobile homes). Health Museum. Incorporated 1877.

Halstenbek (HAHL-stuhn-bek), town, SCHLESWIG-HOLSTEIN, NW GERMANY, 2 mi/3.2 km NE of PINNEBERG; 53°38′N 09°50′E. Seed nurseries; woodworking.

Halsteren (HAHL-stuh-ruhn), town, NORTH BRABANT province, SW NETHERLANDS, 3 mi/4.8 km NNW of BERGEN OP ZOOM; 51°32′N 04°17′E. MARKIEZA-ATSMEER, impounded SE end of EASTERN SCHELDT estuary, 2 mi/3.2 km to SW. Scheldt-Rhine Canal passes 3 mi/4.8 km to W. Dairying; cattle, hogs; agriculture (vegetables, sugar beets); manufacturing (fabricated metal products, building materials).

Haltdalen (HAHLT-dah-luh), village, SØR-TRØNDELAG county, central NORWAY, on GAULA River, on railroad, and 42 mi/68 km SE of TRONDHEIM. Cattle raising; lumbering.

Haltemprice, ENGLAND: see HULL.

Halten, NORWAY: see FROAN.

Haltern (HAHL-tern), town, NORTH RHINE–WESTPHALIA, W GERMANY, in the RUHR industrial district, on LIPPE RIVER, 9 mi/14.5 km N of RECKLINGHAUSEN; 51°44′N 07°11′E. Woodworking; limestone quarrying. Since 1985, coal mining. Nearby is Haltern reservoir (18.3 million cu yd/14 million cu m), which is the

principal water supply for the Ruhr district. Roman excavations in vicinity.

Haltiatunturi (HAHL-tee-ah-TOON-too-ree), Swedish *Haldefäll*, NORWEGIAN *Reisduoddarhaldde*, mountain peak, LAPIN province, NW FINLAND, on Norwegian border; 69°17′N 21°15′E. At 4,343 ft/1,324 m, it is the highest peak in Finland.

Haltom City (HAHL-tuhm), city (2006 population 39,987), TARRANT county, N TEXAS, suburb 6 mi/9.7 km NE of downtown FORT WORTH, near West Fork TRINITY RIVER; 32°47′N 97°16′W. Drained by Big Fossil Creek. Manufacturing (transportation equipment, building materials, light poles, foods). Incorporated after 1940.

Halton (HAHL-tuhn), region, former county (□ 373 sq mi/969.8 sq km; 2001 population 375,229), S ONTARIO, E central CANADA, on Lake ONTARIO; 43°30′N 79°53′W. One of six regional governments of Ontario. Composed of the city of BURLINGTON and the towns of Halton Hills, MILTON, and OAKVILLE. Oakville and Burlington constitute the region's S urban area, while Milton and Halton Hills make up its largely rural N area. Agriculture. Incorporated 1974.

Halton (HAWL-tun), village (2006 population 2,200), central BUCKINGHAMSHIRE, central ENGLAND, 4 mi/6.4 km SE of AYLESBURY; 51°47′N 00°43′W. Former agricultural market.

Halton Hills (HAHL-tuhn HILZ), town (□ 107 sq mi/278.2 sq km; 2001 population 48,184), S ONTARIO, E central CANADA, 12 mi/19 km W of metropolitan TORONTO; 43°37′N 79°57′W. Created by 1974 amalgamation of ACTON, GEORGETOWN, and Esquising townships. Manufacturing (paper products, electric and electronic products; weather stripping); construction; agriculture; mining.

Haltwhistle (HAWLT-wihsuhl), town (2001 population 3,811), SW NORTHUMBERLAND, NE ENGLAND, on SOUTH TYNE RIVER, and 14 mi/23 km W of HEXHAM; 54°59′N 02°26′W. Formerly a major site of paint works. Tourist attraction. Nearby are Hadrian's Wall, Featherstonehaugh (Fanshaw) Castle, and Unthank Hall (Bishop Ridley born here).

Halvad (HUHL-vuhd), town, SURENDRANAGAR district, GUJARAT state, W INDIA, 38 mi/61 km NW of WADHWAN; 23°01′N 71°11′E. Railroad terminus; local market for cotton, millet, salt; hand-loom weaving. Has fine palace built on small lake.

Halve Maan (HAHL-vuh MAHN), provincial recreation center near DIEST, BRABANT province, central BELGIUM, 17 mi/27 km NE of LEUVEN.

Halver (HAHL-ver), town, NORTH RHINE–WESTPHALIA, W GERMANY, 6 mi/9.7 km WSW of LÜDENSCHEID; 51°12′N 07°30′E. Metalworking (iron).

Halyal, INDIA: see HALIYAL.

Halych (HAH-lich) (Russian *Galich*), (Polish *Halicz*), city, N IVANO-FRANKIVS′K oblast, UKRAINE, on the DNIESTER River, and 13 mi/21 km N of IVANO-FRANKIVS′K; 49°07′N 24°44′E. Elevation 793 ft/241 m. Food processing (cheese, fruit, vegetables, cereals, vegetable oils); manufacturing (building materials). Has several old churches with medieval relics. Ruins of an 18th-century fortress nearby. On site of the outport of the former princely Halych. The walled princely Halych was located 3 mi/4.8 km S, on a hill at Krylos. The present Halych arose after princely Halych, former capital of duchy of Halych (since 1144), was razed by the Mongols in 1241. The new Halych passed to Poland in about 1366, acquired (1375) Latin-rite Roman Catholic diocese. In the 17th century, the castle was destroyed by the Tartars. Passed to Austria in 1772. Scene of Ukrainian Sich Rifelmen (of the Austrian army) battles with invading Russian army during World War I. Part of Western Ukrainian Republic (1918). Reverted to Poland (1919); ceded to USSR in 1945. Part of independent Ukraine since 1991.

Halychyna, UKRAINE and POLAND: see GALICIA.

Halys, TURKEY: see KIZIL IRMAK.

Ham (HAHM), commune (2006 population 9,754), Hasselt district, LIMBURG province, BELGIUM.

Ham (AHM), town, SOMME department, PICARDIE region, N FRANCE, on the SOMME RIVER (canalized), and 12 mi/19 km SW of SAINT-QUENTIN; 49°45′N 03°04′E. Road center. Metalworking (wire drawing, manufacturing of bakery equipment and metal plumbing fixtures; sugar milling (at Eppeville), canning. Its 15th-century castle (state prison famed for detention 1840–1846 and escape of Louis Napoleon) is now in ruins. Changed hands several times in World War I, was rebuilt and again damaged in 1939–1945.

Ham, ENGLAND: see RICHMOND UPON THAMES.

Hama (HAH-mah), district, W SYRIA, ⊙ HAMA. Mainly desert except for its W part, which is semi-arid and where a large area is irrigated by the ORONTES River; cotton, corn, millet, oranges, apples, pears, apricots. Main urban centers are HAMA, Suqayilibiya, MASYAF, SALAMIYA.

Hama (HAH-mah), city, ⊙ HAMA district, W central SYRIA, on the ORONTES River; 35°08′N 36°44′E. Market center for an irrigated farm region where cotton, wheat, barley, millet, and maize are grown. Manufacturing includes cotton and woolen textiles, silk, carpets, and dairy products. Famous old waterwheels, some as much as 90 ft/27 m in diameter, used to bring water up from the Orontes River for irrigation. Road and railroad center; airport nearby. The city has a long history, having been settled as far back as the Bronze and Iron ages. In the 2nd millennium B.C.E., it was a center of the Hittites. Often mentioned in the Bible as HAMATH. The Assyrians under Shalmaneser III captured the city in the mid-9th century B.C.E. Later included in the Persian Empire, it was conquered by Alexander the Great and, after his death (323 B.C.E.), was claimed by the Seleucid kings, who renamed it EPIPHANIA, after Antiochus IV (Antiochus Epiphanes). Later came under the control of ROME and of the BYZANTINE Empire. In C.E. 638 it was captured by the Arabs. Christian Crusaders held Hama briefly (1108), but in 1188 it was taken by Saladin, in whose family it remained until it passed to Egyptian Mamluk control in 1299. An early Mamluk governor of Hama was Abd al-Fida (reigned 1310–1330), the historian and geographer. In the early 16th century the city came under the OTTOMAN Empire. After World War I it was made part of the French Levant States League of Nations mandate, and in 1946 it became part of independent Syria. Political insurgency by the Muslim brotherhood beginning in the early 1980s culminated in an uprising in February 1982. Govt. forces quelled the revolt after months of fighting, but destroyed much of the city in the process, including some of its finest mosques; estimated deaths numbered over 20,000. Points of interest include the remains of the Roman aqueduct (still in use) and the Great Mosque of Djami al-Nuri (until 638 a Christian basilica).

Hama'apil (hah-mah-uh-PEEL), kibbutz, in SHARON plain, 5 mi/8 km SE of HADERA, ISRAEL; 32°23′N 34°59′E. Elevation 121 ft/36 m. Agriculture and light industry (plastic fabrication; clothing). Founded in 1945 by Polish and German settlers.

Hamachtesh Hagadol (hah-mahkh-TESH hah-gah-DOL) [Hebrew=The Big Crater] or **Wadi Hatira**, cirque or valley, in the NEGEV mountains, S ISRAEL, between DIMONA and Yeruham; 9.3 mi/15 km long, 4.3 mi/7 km wide; maximum depth 1,312 ft/400 m. Glass sands, phosphate nearby.

Hamachtesh Hakatan (hah-mahkh-TESH hah-kah-TAHN) [Hebrew=The Little Crater] or **Wadi Hadhira**, cirque or valley, in the NEGEV mountains, S ISRAEL, 12.5 mi/20 km SE of DIMONA; 5 mi/8 km long, 3.7 mi/6 km wide; max. depth 1,476 ft/450 m.

Hamad, town (2001 population 52,7189), BAHRAIN, on PERSIAN GULF, SW of MANAMAH; 26°06′N 50°30′E.

Area is shown by the symbol □, and capital city or county seat by ⊙.

new town built by the Ministry of Housing to attract population out of Manamah. Target population is 60,000. Also called Madinat Hamad.

Hamada (HAH-mah-dah), city, SHIMANE prefecture, SW HONSHU, W JAPAN, on the SEA OF JAPAN, 68 mi/109 km S of MATSUE; 34°53′N 132°04′E. Fishing and commercial port. Dried globefish. Iwami Kagura temple; ruins of Iwami Kokubunji temple.

Hamada el Homra, LIBYA: see HAMMADA AL HAMRA.

Hamadān, province (□ 7,639 sq mi/19,861.4 sq km), W central IRAN; ⊙ Hamadān; 34°50′N 48°35′E. Bordered N by Zanjān province, E by Markazī province, S by Lorestān province, and W by Kermānshāhān and Kordestān provinces. ZAGROS Mountains in SW; section of interior plateau in NE. Produces a diversity of agriculture crops (wheat, grapes, barley, cotton); livestock herding. Hamadān and MALAYER are large urban centers.

Hamadān (hah-mahd-DAN), city (2006 population 479,640), ⊙ Hamadān province, W IRAN, at the foot of Mount Alvand; 34°47′N 48°30′E. Elev. 6,000 ft/1,830 m. Trade center for a fertile farm region where fruit and grain are grown. Noted for its rugs, leatherwork, textiles, chemicals, and wood and metal products. Important road junction. Airport. In ancient times, as Hangmatana or Agbatana, it was capital of MEDIA. Known to the Greeks as ECBATANA. In the 7th century Hamadan passed to the Arabs, and it was later held by the Seljuk Turks (12th–13th century) and the Mongols (13th–14th century). The city has had a Jewish colony for many years; the reputed tombs of Mordecai and Esther are here. Avicenna (980–1037), the physician and philosopher, is buried here.

Hamajima (hah-MAH-jee-mah), town, Shima district, MIE prefecture, S HONSHU, central JAPAN, port on AGO BAY, 34 mi/55 km S of TSU; 34°17′N 136°45′E.

Hamakita (hah-MAH-kee-tah), city, SHIZUOKA prefecture, central HONSHU, E central JAPAN, 37 mi/60 km S of SHIZUOKA; 34°47′N 137°47′E. Engines. Bonsai technology.

Hamale (hah-MAH-lai), town (2000 population 5,245), UPPER WEST region, GHANA, 40 mi/64 km N of LAWRA, at NW corner close to border with BURKINA FASO, on MOUHOUN RIVER; 10°59′N 02°44′W. Livestock; groundnuts; shea-nut butter.

Hamam (hah-MAHM), Beduin village, 4.3 mi/7 km NW of TIBERIAS, N ISRAEL. Founded in 1948.

Hamam, TURKEY: see HAYMANA.

Hamamasu (hah-MAH-mahs), village, Ishikari district, HOKKAIDO prefecture, N JAPAN, 37 mi/60 km N of SAPPORO; 43°35′N 141°23′E.

Hamamatsu (hah-MAH-mahts), city (2005 population 804,032), SHIZUOKA prefecture, S central HONSHU, E central JAPAN, 40 mi/65 km S of SHIZUOKA; 34°42′N 137°43′E. Important transportation hub. Chief products are motorcycles, engines, musical instruments. Local agriculture includes potatoes, onions, melons; eels. Castle town in the 16th century.

Haman (HAHM-AHN), county (□ 161 sq mi/418.6 sq km), S SOUTH KYONGSANG province, SOUTH KOREA. Bordered by UIRYONG (across Nam River) and Changnyong (across NAKDONG RIVER). Mountainous in E and S, fields in basin of Nakdong and Nam rivers in N and W. Agriculture (rice, barley, cotton, watermelon, pepper, garlic, fruit). Kyongchon railroad; NAMHAE and Kuma expressways.

Hamanaka (hah-MAH-nah-kah), town, Kushiro district, SE HOKKAIDO prefecture, N JAPAN, on the PACIFIC OCEAN, 192 mi/310 km E of SAPPORO; 43°04′N 145°08′E. Kombu; dairying. Kiritappu Peninsula (Kiritappu marshland) nearby.

Hamana, Lake (HAH-mah-nah), Japanese *Hamana-ko* (hah-mah-NAH-ko), lagoon (□ 28 sq mi/72.8 sq km), SHIZUOKA prefecture, central HONSHU, central JAPAN, 5 mi/8 km W of HAMAMATSU; connected with PHILIPPINE SEA by narrow, shallow outlet; 8 mi/12.9 km

long, 5 mi/8 km wide; many small coves. Fish hatcheries and resorts on shores. Excursion boats.

Hamaoka (hah-MAH-o-kah), town, Ogasa district, SHIZUOKA prefecture, central HONSHU, E central JAPAN, 28 mi/45 km S of SHIZUOKA; 34°38′N 138°07′E. Strawberries. Nuclear power plant.

Hamar (HAH-mahr), city (2007 population 27,909), ⊙ HEDMARK county, SE NORWAY, on Lake MJØSA. It is a commercial, industrial, and winter-sports center. Was an Olympic city during the 1994 LILLEHAMMER Games, hosting the speedskating events. Founded in 1152 as an episcopal see by Nicholas Breakspear (later Pope Adrian IV) and is now a Lutheran episcopal see.

Hamasa, INDONESIA: see TANAHMASA.

Hamasaka (hah-MAH-sah-kah), town, Mikata district, HYOGO prefecture, S HONSHU, W central JAPAN, on SEA OF JAPAN, 75 mi/121 km N of KOBE; 35°37′N 134°27′E.

Hamatama (hah-MAH-tah-mah), town, East Matsuura district, SAGA prefecture, N KYUSHU, SW JAPAN, 22 mi/35 km N of SAGA; 33°26′N 130°02′E. Mandarin oranges.

Hamath, SYRIA: see HAMA.

Hamatonbetsu (hah-mah-TON-bets), town, Soyashi district, HOKKAIDO prefecture, N JAPAN, 152 mi/245 km N of SAPPORO; 45°07′N 142°22′E. KUTCHARO LAKE is nearby.

Hambach (ahm-bahk), commune (□ 6 sq mi/15.6 sq km), MOSELLE department, LORRAINE region, NE FRANCE, 4 mi/6.4 km S of SARREGUEMINES and German border, on E-W expressway between PARIS and the RHINE valley; 49°04′N 07°02′E. Site of Franco-German motor vehicle assembly plant for low-cost European car.

Hambach (HAHM-bahkh), suburb of NEUSTADT AN DER WEINSTRASSE, RHINELAND-PALATINATE, W GERMANY, on E slope of HARDT MOUNTAINS, 1.5 mi/2.4 km S of city center; 49°20′N 08°09′E. Ruins of 11th-century castle were scene (1832) of German liberal meeting known as Hambacher Fest. Independent village until 1969.

Hambantota, district (□ 964 sq mi/2,506.4 sq km; 2001 population 526,414), SOUTHERN PROVINCE, on S coast of SRI LANKA; ⊙ HAMBANTOTA; 06°15′N 81°10′E.

Hambantota (HUHM-buhn-tho-tuh), town (2001 population 11,134), ⊙ HAMBANTOTA district, SOUTHERN PROVINCE, SRI LANKA, on S coast, 65 mi/105 km E of GALLE; 06°07′N 81°07′E. Coconut palms. Meteorological observatory. Major government salterns of Sri Lanka, consisting of natural depressions called *lewayas*, here. Largest concentration of Malays in Sri Lanka.

Hamberg, village (2006 population 25), WELLS CO., central NORTH DAKOTA, 10 mi/16 km NNE of FESSENDEN; 47°45′N 99°30′W. Near New Rockford Canal.

Hamberg, SOUTH AFRICA: see ROODEPOORT.

Hambergen (HAHM-ber-gen), village, LOWER SAXONY, NW GERMANY, 15 mi/24 km N of BREMEN; 53°18′N 08°49′E.

Hamber Provincial Park (HAM-buhr) (□ 3,800 sq mi/9,880 sq km), SE BRITISH COLUMBIA, W CANADA, on ALBERTA border, in the ROCKY and SELKIRK MOUNTAINS, NE of REVELSTOKE; extends along upper reaches of the COLUMBIA River and borders on JASPER and BANFF national parks; 52°22′N 117°52′W. Peaks more than 10,000 ft/3,048 m high include Sir Sanford, LAUSSEDAT, MUMMERY, BRYCE, SHACKLETON, ICONOCLAST, ADAMANT, and GHOST mountains. Numerous small headwaters of Columbia River rise here.

Hamble (HAM-buhl), town (2001 population 4,147), S HAMPSHIRE, S central ENGLAND, on SOUTHAMPTON WATER, at mouth of HAMBLE RIVER estuary, 5 mi/8 km SE of SOUTHAMPTON; 50°51′N 01°19′W. Also referred to as Hamble-le-Rice. Aircraft works. Yachting and boatbuilding. Site of air-training school. Has 14th–15th-century church.

Hambleden (HAM-buhl-duhn), agricultural village (2001 population 2,617), SW BUCKINGHAMSHIRE, central ENGLAND, near the THAMES RIVER, 3 mi/4.8 km NNE of HENLEY-ON-THAMES; 51°34′N 00°52′W. Flour milling. Has 14th-century church.

Hambledon (HAM-buhl-duhn), village (2001 population 765), SE HAMPSHIRE, S central ENGLAND, 10 mi/16 km N of PORTSMOUTH; 50°55′N 01°04′W. Agricultural market. Church dates from 13th century. The home of cricket; local club founded in 18th century.

Hambledon Junction, AUSTRALIA: see EDMONTON.

Hamblen (HAM-blin), county (□ 174 sq mi/452.4 sq km; 2006 population 61,026), NE TENNESSEE; ⊙ MORRISTOWN; 36°13′N 83°16′W. In GREAT APPALACHIAN VALLEY region; BAYS MOUNTAIN along SE border. Bounded N by CHEROKEE Reservoir (HOLSTON River), S by NOLICHUCKY River. Agriculture; diversified manufacturing. Formed 1870.

Hamble River (HAM-buhl), 9 mi/14.5 km long, HAMPSHIRE, S central ENGLAND; rises 5 mi/8 km N of Fareham; flows SSW, past Botley and Hamble, to SOUTHAMPTON WATER just S of HAMBLE.

Hambleton, village (2006 population 245), TUCKER county, NE WEST VIRGINIA, on Laurel Fork River, in Monongahela National Forest, 2 mi/3.2 km SE of PARSONS; 39°04′N 79°39′W. Includes Fernow Experimental Forest to SW.

Hamborn (HAHM-born), industrial outer suburb of DUISBURG, W GERMANY, on right bank of the RHINE RIVER, and 5 mi/8 km N of city center, adjoining (N) RUHRORT and MEIDERICH; 51°26′N 06°47′E. Steel mills; zinc refining. Commune until 1911, when it incorporated numerous suburbs and was chartered; incorporated 1929 into DUISBURG, which was subsequently called Duisburg-Hamborn until 1935.

Hambrücken (hahm-BRYOOK-ken), village, BADEN-WÜRTTEMBERG, SW GERMANY, 13 mi/21 km NNE of KARLSRUHE; 49°11′N 08°33′E.

Hamburg (HAHM-boorg), officially, Freie und Hansestadt Hamburg (German=Free and Hanseatic City of Hamburg), city (2005 population 1,743,627), coextensive with, and ⊙ Hamburg state (□ 290 sq mi/755 sq km), N GERMANY, on the ELBE River near its mouth in the NORTH SEA, and on the ALSTER River; 53°33′N 10°00′E. The economic center of Northern Germany and the country's second-largest city, Hamburg is the nation's busiest port and its major industrial city. Manufacturing includes copper, vegetable and mineral oils, machinery, electrotechnical goods, and cigarettes. Its harbor handles approximately half of Germany's imports (foodstuffs, tea, coffee, and petroleum) and exports (machinery, processed copper, copper, and pharmaceuticals). It is a very wealthy city, and the city's inhabitants have a very high standard of living: the city is distinct for having the largest average amount of living space per person of any major city in the world. Hamburg originated (early 9th century) in the Carolingian castle of Hammaburg, probably built by Charlemagne as a defense against the Slavs. It became (834) an archepiscopal see (united in 847 with the archdiocese of Bremen) and a missionary center for N EUROPE. Grew to commercial importance and in 1241 formed an alliance with LÜBECK, which later became the basis of the HANSEATIC LEAGUE. Accepted the Reformation in 1529. In 1558 the first German stock exchange was founded here; with the arrival of Dutch Protestants, Portuguese Jews, and English cloth merchants (expelled from ANTWERP), and with the expansion of commercial ties with the UNITED STATES after 1783, Hamburg continued to prosper. Occupied by FRANCE in 1806 and in 1815 joined the GERMAN CONFEDERATION. In 1842 a fire destroyed much of the city. After World War I Hamburg was briefly (1918–1919) a socialist republic. In 1937 the city ceded CUXHAVEN, its outlying port, to PRUSSIA, but incorporated the neighboring towns of ALTONA, HARBURG, and

WANDSBEK. During World War II (especially in 1943) Hamburg was severely damaged by aerial bombardment, and some 55,000 persons were killed. Rebuilt after 1945, Hamburg today is an elegant, modern city and a cultural center, widely known for its opera, theaters, magazine- and book-publishing houses, radio and television broadcasting centers, and film studios. At its center are two lakes, the *Binnenalster* (Inner Alster) and the *Aussenalster* (Outer Alster). The ST. PAULI district, with its well-known street, the Reeperbahn, includes numerous places of entertainment. Seat of a university (founded 1919), several museums, and medical and technical institutes. There are extensive zoological and botanical gardens. Noteworthy buildings include the baroque St. Michael's Church (1750–1762), rebuilt (1907–1912) after a fire; the Church of St. Jacobi (begun in the 14th century); and the Renaissance-style city hall (1886–1897). Composers Felix Mendelssohn and Johannes Brahms born here.

Hamburg, city (2000 population 1,240), FREMONT county, extreme SW IOWA, on NISHNABOTNA RIVER, near MISSOURI state line, and 18 mi/29 km SW of SHENANDOAH; 40°36′N 95°39′W. Manufacturing (foods and beverages, feed, alcohol products, bricks); has large nursery. Waubonsie State Park to NW. Incorporated 1867.

Hamburg, town (2000 population 3,039), ⊙ ASHLEY county, SE ARKANSAS, c.48 mi/77 km E of EL DORADO; 33°13′N 91°47′W. In lumbering (pine, oak) and agricultural area (cotton, rice, hay). Manufacturing (lumber, apparel, radio loudspeaker cones). Overflow Wildlife Refuge to SE.

Hamburg, village (2000 population 126), CALHOUN county, W ILLINOIS, on the MISSISSIPPI RIVER, and 7 mi/11.3 km NW of HARDIN; 39°13′N 90°43′W. Apple growing.

Hamburg, village, LIVINGSTON county, SE MICHIGAN, 10 mi/16 km N of ANN ARBOR; 42°26′N 83°48′W. Located in small lakes district. Manufacturing (electrical equipment, fabricated metal products). Pinckney State Recreation Area to W.

Hamburg, village (2000 population 538), CARVER county, S MINNESOTA, c.40 mi/64 km WSW of MINNEAPOLIS; 44°43′N 93°57′W. Poultry, livestock; soybeans, alfalfa; dairying; manufacturing (feeds, fertilizers).

Hamburg, village, FRANKLIN county, SW MISSISSIPPI, 20 mi/32 km E of NATCHEZ.

Hamburg, village (□ 2 sq mi/5.2 sq km; 2006 population 9,495), ERIE county, W NEW YORK, S of BUFFALO; 42°43′N 78°49′W. Part of a township of 48,000 people, Hamburg is a residential and industrial suburb of Buffalo. Its manufacturing includes rubber goods and optical products. It is thought to be the site of the invention or development, by the Menches brothers in 1885, of the American hamburger sandwich, although other locales make similar claims. Seat of Hilbert College. Settled c.1808, incorporated 1874.

Hamburg, village, AIKEN county, W SOUTH CAROLINA, on SAVANNAH River, opposite AUGUSTA (GEORGIA) and now part of NORTH AUGUSTA (S.C.). Founded as W terminus of state's first railroad, completed 1833 to CHARLESTON.

Hamburg, village, MARATHON county, central WISCONSIN, 14 mi/23 km NW of WAUSAU; 45°06′N 89°53′W.

Hamburg, borough (2006 population 3,554), SUSSEX county, NW NEW JERSEY, on WALLKILL RIVER, 12 mi/19 km NE of NEWTON; 41°08′N 74°34′W. "Gingerbread Castle" here has scenes from various fairy tales. In suburbanizing area. Incorporated 1920.

Hamburg, borough (2006 population 4,197), BERKS county, E central PENNSYLVANIA, 15 mi/24 km NNW of READING, on SCHUYLKILL RIVER. Agricultural area (grain, soybeans; livestock; poultry; dairying); manufacturing (food products, fabricated metal products, apparel, machinery; printing and publishing; iron

foundry). Pennsylvania Dutch Folk Culture Center to E at LENHARTSVILLE. Weiser State Forest and APPALACHIAN TRAIL, in BLUE MOUNTAIN ridge to N. Founded 1779, incorporated 1837.

Hamburg, SOUTH AFRICA: see KEISKAMMA RIVER.

Hamburg, CONNECTICUT: see LYME.

Hambye (ahn-BEE), commune (□ 11 sq mi/28.6 sq km), MANCHE department, BASSE-NORMANDIE region, NW FRANCE, 10 mi/16 km SW of SAINT-LÔ; 48°57′N 01°16′E. Livestock raising. Has nearby ruins of Benedictine abbey founded c.1145.

Hamdallaye (HAHM-dah-lai), village, FIFTH REGION/MOPTI, MALI, S of MOPTI. Tourist center. Seat of the early 19th-century Caliphate of Hamdallahi. Also spelled Hamdallay.

Hamdaniya (HAHM-dah-NEE-yuh), township, IDLIB district, NW SYRIA, on railroad, 55 mi/89 km S of ALEPPO; 35°58′N 32°07′E. Cotton, cereals, fruit.

Hamden, town, NEW HAVEN county, S CONNECTICUT; 41°23′N 72°55′W. The town was named for John Hampden, the English Puritan patriot. A residential and manufacturing suburb of NEW HAVEN, of which it was once a part, Hamden makes machinery, electrical products, metal goods, computer products, wire and cable, building materials, fabricated metals and rolled steel; other industries are construction, business services, and retail trade. The town's industrial development dates back to 1798, when Eli Whitney set up an arms factory using techniques of mass production. A plaque marks the site. Has many early mill sites, pre-Revolutionary and Civil War houses, and a restored opera house. Seat of Quinnipiac College. Settled c.1638, incorporated 1786.

Hamden (HAM-duhn), village (2006 population 946), VINTON county, S OHIO, 27 mi/43 km ESE of CHILLICOTHE; 39°10′N 82°31′W. In agricultural area.

Hamdh, Wadi (HAH-mid, WAH-dee), largest wadi of HEJAZ, SAUDI, extending c.249 mi/400 km from area of MEDINA to the RED SEA S of WEJH along 26°00′N latitude.

Hämeen (HA-main), Swedish *Tavastehus*, province (□ 8,590 sq mi/22,334 sq km), S central FINLAND; ⊙ HÄMEENLINNA. Low and undulating, interspersed with numerous lakes; Lake PÄIJÄNNE forms E border of province. Lumber and timber-processing industries; agriculture (rye, oats, barley, potatoes); livestock raising; dairy farming. Manufacturing of furniture, machinery, glass (RIIHIMÄKI), leather and rubber products. Major cities are TAMPERE, LAHTI, and Hämeenlinna.

Hämeenkyrö (HA-main-kuh-RUH), Swedish *Tavastkyro*, village, HÄMEEN province, SW FINLAND, 20 mi/32 km NW of TAMPERE; 61°38′N 23°12′E. Elevation 198 ft/60 m. In lake region; lumbering.

Hämeenlinna (HA-main-LIN-nah), Swedish *Tavastehus*, city, ⊙ HÄMEEN province, S FINLAND; 61°00′N 24°27′E. Elevation 248 ft/75 m. A lake port and manufacturing town with plywood mills, tanneries, spool mills, and rubber works. Built around a thirteenth-century castle on Lake Vanajavesi, the city was chartered in 1638. Composer Jean Sibelius born (1865) and raised here.

Hamel, village (2000 population 570), MADISON county, SW ILLINOIS, 8 mi/12.9 km NE of EDWARDSVILLE, just beyond SAINT LOUIS urban fringe; 38°53′N 89°50′W. Corn, soybeans; manufacturing (concrete, metal tools).

Hamel, MINNESOTA: see MEDINA.

Hamelin, GERMANY: see HAMELN.

Hamelin Pool, AUSTRALIA: see SHARK BAY.

Hameln (HAH-meln), city, LOWER SAXONY, N central GERMANY, a port on the WESER RIVER; 52°07′N 09°22′E. Industrial center and railroad junction. Manufacturing includes carpets, chemicals, electrical goods, machinery, and food products. Tourist center, known as the scene of the legend of the Pied Piper of Hamelin. Frescoes illustrating the tale adorn the

so-called Ratcatcher's House (built 1602–1603). An ancient Saxon settlement, Hameln became a missionary outpost c.750, received city rights c.1200, and, while frequently changing hands, acquired considerable independence. Was a member of the HANSEATIC LEAGUE 1426–1572. Passed to HANOVER in 1814 and to PRUSSIA in 1866. Retained many historic buildings, including an early Gothic church (14th century), the Rattenkrug (built 1568), and the Wedding House (1610–1617); now the city hall). Sometimes spelled HAMELIN.

Hamer (HAI-mur), unincorporated village, DILLON county, NE SOUTH CAROLINA, 6 mi/9.7 km NE of DILLON, at NORTH CAROLINA state line. "South of the Border," a large tourist attraction with a Mexican theme nearby; well-known landmark of Interstate 95.

Hamersley Range (HA-muhrz-lee), N WESTERN AUSTRALIA, S of FORTESCUE RIVER; extends 160 mi/257 km ESE from Robe River; 21°53′S 116°46′E. Highest peak (2,798 ft/853 m), Mount Margaret. Pilbara iron-ore boom since 1960s has led to creation of the new towns of NEWMAN, TOM PRICE, PARABURDOO. Hamersley Range National Park (KARIJINI NATIONAL PARK) at center of range.

Hamersley Range National Park, AUSTRALIA: see KARIJINI NATIONAL PARK.

Hamersville (HA-muhrz-vil), village (2006 population 519), BROWN county, SW OHIO, 32 mi/51 km ESE of CINCINNATI; 38°55′N 83°59′W. In agricultural area.

Hamgyong-namdo, NORTH KOREA: see SOUTH HAMGYONG.

Hamgyong-pukdo, NORTH KOREA: see NORTH HAMGYONG.

Hamhung (HAHM-HUNG), city; ⊙ SOUTH HAMGYONG province, E central NORTH KOREA; 40°00′N 127°40′E. The 2nd-largest city in North Korea, it is a leading port for Korean foreign trade as well as a major industrial center. Textiles, metalware, machinery, and chemicals are produced; oil is refined, and food is processed. Coal mines are nearby. The founder of the Yi dynasty, the last imperial line of Korea, was born here. Airport; connects primary railroads.

Hami (HAH-MEE), city (□ 32,832 sq mi/85,035 sq km; 1994 estimated urban population 179,000; estimated total population 314,300), E XINJIANG UYGUR, NW CHINA, in oasis, 300 mi/483 km ESE of URUMQI, on SILK ROAD, and on the road to the HEXI CORRIDOR; 42°48′N 93°27′E. Home to twenty-four ethnic groups, the largest being the Uygur, Kazakh, Hui (Muslim), and Han Chinese. Has 270 mosques, three Protestant churches, and a Roman Catholic church. Well known for its melon production. Agriculture, heavy industry, and transportation are the largest sources of income; light industry is also important. Crop growing; animal husbandry. Main industries include coal mining, salt extraction, and chemical manufacturing.

Hami (HEM-mee), village, SE YEMEN, 14 mi/23 km ENE of SHIHR, on Gulf of ADEN; 14°48′N 49°50′E.

Hamidiya (HAHM-e-DEE-yah), township, LATTAKIA district, W SYRIA, on the MEDITERRANEAN Sea, 5 mi/8 km N of the Lebanese border, 55 mi/89 km S of Lattakia; 34°42′N 35°56′E. Sericulture; cereals.

Hamidiye, TURKEY: see CEYHAN.

Hamidiye, TURKEY: see DEVREK.

Hamidiye, TURKEY: see MESUDIYE.

Hamill Peak (HA-mil) (10,640 ft/3,243 m), SE BRITISH COLUMBIA, W CANADA, in SELKIRK MOUNTAINS, 60 mi/97 km NE of NELSON; 50°13′N 116°38′W.

Hamilton (HAM-uhl-tuhn), county (□ 519 sq mi/1,349.4 sq km; 2006 population 14,215), N central FLORIDA, on GEORGIA state line (N), on SUWANNEE (S,E) and WITHLACOOCHEE (W) RIVERS; ⊙ JASPER; 30°29′N 82°57′W. Flatwoods area with swamps in E; drained by ALAPAHA RIVER. Farming (corn, peanuts, cotton, tobacco, vegetables), livestock raising (hogs, cattle), and lumbering. Formed 1827.

Area is shown by the symbol □, and capital city or county seat by ⊙.

Hamilton, county (□ 435 sq mi/1,131 sq km; 2006 population 8,335), SE ILLINOIS; ⊙ MCLEANSBORO; 38°04′N 88°33′W. Agriculture (livestock; fruit, wheat, corn, redtop seed). Drained by North Fork of SALINE RIVER. Formed 1821.

Hamilton, county (□ 402 sq mi/1,045.2 sq km; 2006 population 250,979), central INDIANA; ⊙ NOBLESVILLE; 40°04′N 86°03′W. Diversified manufacturing, including the processing of farm products, at Noblesville and SHERIDAN; soybeans, corn, wheat; dairy products; cattle, hogs. Drained by West Fork of WHITE RIVER, and by CICERO (MORSE RESERVOIR at center of county), and small Prairie and Duck creeks. Part of INDIANAPOLIS metropolitan area (residential). GEIST RESERVOIR, on Fall Creek, in SE. Formed 1823.

Hamilton, county (□ 577 sq mi/1,500.2 sq km; 2006 population 16,087), central IOWA; ⊙ WEBSTER CITY, 42°22′N 93°43′W. Prairie agricultural area (hogs, poultry, cattle; corn, soybeans, oats) drained by BOONE and SKUNK rivers; bituminous-coal deposits. Little Wall Lake in S. Widespread flooding in 1993. Formed 1856.

Hamilton, county (□ 997 sq mi/2,592.2 sq km; 2006 population 2,594), SW KANSAS, on COLORADO (W) state line; ⊙ SYRACUSE; 38°01′N 101°40′W. Prairie region; drained by ARKANSAS RIVER. Wheat, sorghum; cattle. One of four KANSAS counties in Mountain time zone; Central-Mountain time zone boundary follows S and E county border. Formed 1886.

Hamilton, county (□ 546 sq mi/1,419.6 sq km; 2006 population 9,490), SE central NEBRASKA; ⊙ AURORA; 41°52′N 97°59′W. Irrigated agricultural region bounded N and NW by PLATTE River; drained by BIG BLUE and W Fork BIG BLUE rivers. Cattle, hogs; dairying; corn, soybeans, sorghum. Formed 1870.

Hamilton, county (□ 1,807 sq mi/4,698.2 sq km; 2006 population 5,162), NE central NEW YORK; ⊙ LAKE PLEASANT; 43°39′N 74°30′W. Situated entirely in the ADIRONDACK MOUNTAINS; drained by tributaries of the HUDSON River and by RAQUETTE, BLACK, and SACANDAGA rivers. Well-known resorts on INDIAN, LONG, RAQUETTE, PISECO, and Pleasant lakes, and FULTON CHAIN OF LAKES; many skiing areas; hunting, fishing. Some logging and light industry. Formed 1816.

Hamilton (HAM-uhl-tuhn), county (□ 414 sq mi/1,076.4 sq km; 2006 population 822,596), extreme SW OHIO; ⊙ CINCINNATI; 39°11′N 84°32′W. Bounded W by INDIANA state line, S by OHIO RIVER (here forming KENTUCKY state line); drained by GREAT MIAMI, Little Miami, and WHITEWATER rivers and by small Mill Creek. In the Till Plains physiographic region. Entirely within metropolitan district of Cincinnati. Oil and gas extraction. Some coal mining. Agricultural products include vegetables, nursery and greenhouse crops; dairying; manufacturing of food products and beverages, clothing, plastics, consumer goods, metal products, electrical products, transportation equipment. Formed 1790. Population has gradually declined since 1990.

Hamilton, county (□ 576 sq mi/1,497.6 sq km; 2006 population 312,905), SE TENNESSEE, on GEORGIA (S) state line; ⊙ CHATTANOOGA; 35°11′N 85°10′W. Crossed N-S by TENNESSEE RIVER. Includes parts of Hales Bar and CHICKAMAUGA reservoirs. WALDEN RIDGE is in W and NW, parts of LOOKOUT MOUNTAIN and of CHICKAMAUGA AND CHATTANOOGA NATIONAL MILITARY PARK in S. Fertile farm lands; manufacturing; tourism. Nuclear power plants SEQUOYAH 1 (initial criticality July 5, 1980) and Sequoyah 2 (initial criticality November 5, 1981) are 10 mi/16 km NE of Chattanooga. They use cooling water from Chickamauga Lake, and each has a maximum dependable capacity of 1,122 MW. Formed 1819; absorbed James county in 1919.

Hamilton, county (□ 836 sq mi/2,173.6 sq km; 2006 population 8,186), central TEXAS; ⊙ HAMILTON; 31°41′N 98°06′W. Mainly prairies; drained by LEON, BOSQUE, and LAMPASAS rivers. Livestock, especially cattle, also sheep, goats (wool, mohair); dairying; agriculture (wheat, oats, barley, corn; hay). Natural gas and oil fields; sand and gravel. Formed 1842.

Hamilton, city, ⊙ BERMUDA, in PEMBROKE parish, on Bermuda Island; 32°17′N 64°47′W. Port at the head of GREAT SOUND, a huge lagoon and deep-water harbor protected by coral reefs. Focus of Bermuda's commercial and social life and a major tourist resort and port-of-call of cruise ships. Duty-free shopping center.

Hamilton (HA-mil-tuhn), city (□ 431 sq mi/1,120.6 sq km; 2001 population 490,268), S ONTARIO, E central CANADA, at the W end of Lake ONTARIO; 43°15′N 79°50′W. On a narrow plain between its harbor (connected by canal with the lake) and the Niagara escarpment. Important port, transportation center, and manufacturing city. Canada's leading producer of iron and steel; other manufacturing includes motor vehicles, heavy machinery, chemicals, and electrical, paper, and textile products. Settled in 1778, it became an important port city with the opening (1830) of the Burlington Canal, which linked Hamilton Harbor with Lake Ontario. Places of interest include the Royal Botanical Gardens, the open-air market, the historical museum in Dundern Park, and the Canadian Football Hall of Fame. Seat of McMaster University (1887). Restructured in 2001; the communities of ANCASTER, DUNDAS, Flamborough, Glanbrook, Hamilton, and STONEY CREEK merged to form the new city of Hamilton.

Hamilton, city (□ 36 sq mi/94 sq km; 2006 population 184,838), ⊙ WAIKATO region (□ 13,472 sq mi/34,892 sq km; 2006 population 184,838), N central NORTH ISLAND, NEW ZEALAND, in WAIKATO RIVER basin, between AUCKLAND and WELLINGTON; 37°47′S 175°17′E. Transportation, industrial, and commercial center of a densely populated dairying, prime lamb-raising, and fruit-producing area. The University of Waikato is a major regional university; agricultural research centers are nearby.

Hamilton, city (2000 population 3,029), HANCOCK county, W ILLINOIS, on the MISSISSIPPI RIVER, opposite KEOKUK (IOWA; connected by bridge), 12 mi/19 km W of CARTHAGE; 40°23′N 91°21′W. Trade and shipping center in agricultural area (corn, wheat, soybeans; livestock). Stone quarry. Nearby is Lake Keokuk, or Lake Cooper, formed by Keokuk Dam in Mississippi River, with recreational and resort facilities. Incorporated 1859.

Hamilton, city (2000 population 1,813), CALDWELL county, NW MISSOURI, 24 mi/39 km W of CHILLICOTHE; 39°44′N 94°00′W. Corn, soybeans, oats, wheat; cattle, hogs; manufacturing (shoes); lumber. Founded 1855, incorporated 1868. Boyhood home of James Cash (J.C.) Penney, retail merchant.

Hamilton (HAM-uhl-tuhn), city (□ 22 sq mi/57.2 sq km; 2006 population 62,130), ⊙ BUTLER county, SW OHIO, on GREAT MIAMI RIVER; 39°23′N 84°34′W. An agricultural trading and manufacturing center, Hamilton has paper and pulp mills and many factories that make a variety of products, including safes, machinery, chemicals, textiles, and motor-vehicle parts. Employment sectors include steel and insurance. Settled on the site of Fort Hamilton, built 1791. Author William Dean Howells was raised here. Points of interest include the Soldiers, Sailors, and Pioneers Monument and the county historical society. Miami University of Ohio has a branch here. Incorporated 1857.

Hamilton (HA-mil-tuhn), town, SW VICTORIA, AUSTRALIA, 160 mi/257 km W of MELBOURNE; 37°45′S 142°04′E. Railroad junction; dairying and agriculture center; wool.

Hamilton (HA-mil-tuhn), township (□ 99 sq mi/257.4 sq km; 2001 population 10,785), SE ONTARIO, E central CANADA, 18 mi/29 km from PETERBOROUGH; 44°03′N 78°12′W. Forms part of the Oak Ridges Moraine.

Hamilton (HAM-il-tuhn), town (2001 population 48,546), ⊙ SOUTH LANARKSHIRE, S central Scotland, at the confluence of the AVON and the CLYDE rivers, 11 mi/18 km SE of GLASGOW; 55°46′N 04°02′W. Once known for its coal mining, Hamilton's industries now include light engineering, textile manufacturing, and food processing. Market town for fruits, vegetables, and dairy goods. Racecourse. Formerly in STRATHCLYDE, abolished 1996.

Hamilton, town (2000 population 6,786), ⊙ Marion co., NW Alabama, 45 mi/72 km NW of Jasper, near Mississippi state line. Lumber; cotton. Mobile homes; textiles. Settled c.1818. Originally called 'Toll Gate' because a station for collecting tolls from vehicles traveling on the road between Washington, D.C. and New Orleans was located here. Renamed for Albert J. Hamilton, an early settler. Inc. in 1896.

Hamilton, town (2000 population 1,233), STEUBEN county, NE INDIANA, on small Hamilton Lake, 8 mi/12.9 km SSE of ANGOLA; 41°32′N 84°55′W. In agricultural and resort area; manufacturing (fabricated metal products, electrical products, transportation equipment, plastics).

Hamilton, town (2000 population 144), MARION county, S central IOWA, near CEDAR CREEK, 15 mi/24 km SE of KNOXVILLE; 41°10′N 92°54′W. In bituminous-coal-mining and agricultural area.

Hamilton, rural upper-income residential town, ESSEX county, NE MASSACHUSETTS, 9 mi/14.5 km W of GLOUCESTER; 42°38′N 70°52′W. Site of Myopia Hunt and Polo Clubs. Settled 1638, incorporated 1793.

Hamilton, town (2000 population 3,705), ⊙ RAVALLI county, W MONTANA, 45 mi/72 km S of MISSOULA, and on BITTERROOT RIVER, near BITTERROOT RANGE and IDAHO state line (to W); 46°15′N 114°10′W. Trading point for irrigated agricultural and mining region; silver, lead, zinc mines; lumber mills; cattle, sheep, horses, llamas. Manufacturing (apparel, animal feeds, log homes, wood products). Rocky Mountain Laboratory (focussing upon Rocky Mountain Spotted Fever); Private Biotechnology Laboratory, fish hatchery to SE. Incorporated 1894. Daly Mansion (1890), forty-two-room mansion of "Copper King" Marcus Daly, on 50 acres/20 ha N of town.

Hamilton, mining ghost town, WHITE PINE county, E NEVADA, in WHITE PINE MOUNTAINS, 32 mi/51 km W of ELY, in Humboldt National Forest. During its silver boom (c.1865–1873), its population reached c.10,000. Mt. HAMILTON to W.

Hamilton, township, MERCER county, central NEW JERSEY, 5 mi/8 km E of TRENTON; 40°12′N 74°40′W. Incorporated 1842.

Hamilton, town (2006 population 2,946), ⊙ HAMILTON county, central TEXAS, c.60 mi/97 km W of WACO; 31°42′N 98°07′W. Elevation 1,154 ft/352 m. In agricultural area (cattle; wheat, sorghum); dairying. A natural gas field is W; sand and gravel; manufacturing (wood products).

Hamilton (HAM-il-tuhn), village, S central TASMANIA, AUSTRALIA, 32 mi/51 km NW of HOBART; 42°33′S 146°50′E. Sheep. Declared 1863.

Hamilton, village, SIERRA LEONE, minor port on ATLANTIC OCEAN, on SIERRA LEONE PENINSULA, and 7 mi/11.3 km SSW of FREETOWN; 08°22′N 13°15′W. Fishing.

Hamilton, village, W ALASKA, on arm of YUKON River delta, 70 mi/113 km SW of SAINT MICHAEL; 62°53′N 163°51′W. Fur-trading post, supply point, steamer landing. School. Sometimes called Old Hamilton. NEW FORT HAMILTON village is 10 mi/16 km S.

Hamilton, village (2000 population 307), ⊙ HARRIS county, W GEORGIA, 21 mi/34 km NNE of COLUMBUS; 32°46′N 84°53′W.

Hamilton, village (2000 population 334), GREENWOOD county, SE KANSAS, 12 mi/19 km NNE of EUREKA; 37°58′N 96°09′W. In livestock-raising, dairying, and grain-growing region.

Hamilton, village (□ 1 sq mi/2.6 sq km; 2006 population 3,749), MADISON county, central NEW YORK, 25 mi/40 km SW of UTICA; 42°49′N 75°32′W. In dairying and farming region. Seat of Colgate University. Settled 1795, incorporated 1816.

Hamilton (HAM-il-tuhn), village (2006 population 487), MARTIN county, E North Carolina, on ROANOKE RIVER (head of navigation), and 18 mi/29 km ENE of TARBORO. Service industries; manufacturing (fabrics); agriculture (tobacco, cotton, peanuts, grain; poultry, livestock). Historic Hamilton Visitor Center; Fort Branch Battlefield State Historic Site. Incorporated 1804. Earlier known as Milton, it was renamed in 1804.

Hamilton, village (2006 population 66), PEMBINA CO., NE NORTH DAKOTA, 9 mi/14.5 km E of CAVALIER; 48°48′N 97°27′W. Founded 1882.

Hamilton (HA-mil-tuhn), village (2006 population 742), LOUDOUN county, N VIRGINIA, 6 mi/10 km W of LEESBURG, in E foothills of the BLUE RIDGE MOUNTAINS; 39°08′N 77°40′W. Light manufacturing; agriculture (dairying); livestock; grain, soybeans, apples.

Hamilton, village (2006 population 361), SKAGIT county, NW WASHINGTON, on SKAGIT River, and 18 mi/29 km ENE of Mount Vernon; 48°31′N 122°60′W. Agricultural and logging region. Mount Baker–Snoqualmie National Forest to NE and SE; Lower Baker Dam (forms Lake Shannon). MOUNT BAKER (10,775 ft/3,284 m) to NE; Mount Baker National Recreation Area on S slopes.

Hamilton, parish, N BERMUDA, S of SAINT George's parish; 32°20′N 64°43′W.

Hamilton (HAM-il-tuhn), W suburb of NEWCASTLE, E NEW SOUTH WALES, AUSTRALIA; 32°55′S 151°45′E. Horse and dog racing. Extensively damaged in 1989 earthquake. Mixed industry.

Hamilton, CANADA: see CHURCHILL.

Hamilton, Rhode Island: see NORTH KINGSTOWN.

Hamilton Air Force Base, California: see SAN RAFAEL.

Hamilton Beach (HA-mil-tuhn), unincorporated village, S ONTARIO, E central CANADA, included in city of HAMILTON, on narrow spit between Hamilton Bay and Lake ONTARIO; 43°17′N 79°47′W. Resort.

Hamilton, Cape, SW VICTORIA ISLAND, NORTHWEST TERRITORIES, N CANADA, on W coast of WOLLASTON PENINSULA; 69°31′N 116°25′W.

Hamilton City, unincorporated town (2000 population 1,903), GLENN county, N central CALIFORNIA, 10 mi/16 km W of CHICO, on SACRAMENTO RIVER, near mouth of STONY CREEK. Sugar beets, citrus, prunes, plums, olives, walnuts, pistachios, almonds, grain, rice, wheat; dairying. Tehama-Colusa Canal passes to W.

Hamilton Cove, CANADA: see PORTNEUF-SUR-MER.

Hamilton, Fort, New York: see FORT HAMILTON.

Hamilton Grange, memorial, in Hamilton Heights section of W HARLEM, Upper MANHATTAN, SE NEW YORK, on W 195th Street between Amsterdam Avenue and BROADWAY. The only home ever owned by Alexander Hamilton (and one of the few memorials to him), it is a rare example of John McComb residential architecture. Hamilton lived here until his 1804 death in duel with Aaron Burr. The home was moved here from its original site four blocks away in 1889, and it is expected to be moved to nearby St. Nicholas Park by c.2010. Memorial site also includes 7-ft/2.1-m bronze statue of Hamilton, resting atop 8-ft/2.4-m granite pier. Originally built for Hamilton Club in Brooklyn, the statue was moved here after the club's closure in 1936. Authorized 1962 as historic site.

Hamilton Inlet, CANADA: see MELVILLE, LAKE.

Hamilton Island (HA-mil-tuhn EI-luhnd), island, in Whitsunday island group, QUEENSLAND, NE AUSTRALIA, between MACKAY and TOWNSVILLE, on GREAT BARRIER REEF; 20°22′S 148°57′E. Most commercially developed of the Whitsundays. Resort; walking tracks, Hamilton Island Fauna Park. Airport.

Hamilton, Lake, GARLAND county, in W central ARKANSAS, on OUACHITA RIVER, 5 mi/8 km S of HOT SPRINGS; c.25 mi/40 km long; 34°25′N 93°01′W. Extends W and N to Blakely Mountain Dam (Lake

OUACHITA). Formed by Carpenter Dam. Ouachita National Forest on NW shore; fish hatchery on S shore.

Hamilton, Mount, peak (4,213 ft/1,284 m), SANTA CLARA county, W CALIFORNIA, in DIABLO RANGE, 15 mi/24 km E of SAN JOSE. Site of LICK OBSERVATORY (built 1876–1888), directed by the University of California. The California Institute of Technology installed a 120-in/305-cm telescope in 1959.

Hamilton, Mount, Nevada: see WHITE PINE MOUNTAINS.

Hamilton-on-Forth, AUSTRALIA: see FORTH.

Hamilton Square, village, MERCER county, W NEW JERSEY, 5 mi/8 km E of TRENTON; 40°13′N 74°39′W. Pottery, rubber products. Largely residential.

Hamina (HAH-mi-nah), Swedish *Fredrikshamn*, town, KYMEN province, SE FINLAND, port on GULF OF FINLAND; 60°34′N 27°12′E. Military center; timber and wood products exported. Originally named Veckelaks, it was a noted trade center in the Middle Ages. The Treaty of Fredrikshamn (1809), by which Sweden ceded all of Finland to RUSSIA, was signed here.

Hamiota (ha-mi-O-tuh), town (2001 population 858), SW MANITOBA, W central CANADA, 36 mi/58 km NW of BRANDON, and in HAMIOTA rural municipality; 50°11′N 100°36′W. Lumbering; wheat. Incorporated 1907.

Hamiota (ha-mi-O-tuh), rural municipality (□ 1 sq mi/2.6 sq km; 2001 population 479), SW MANITOBA, W central CANADA, 36 mi/58 km NW of BRANDON; 50°09′N 100°35′W. Agriculture and related businesses. Includes village of HAMIOTA and the communities of OAKNER, DECKER, and MCCONNELL.

Hamira (huh-MEE-rah), village, KAPURTHALA district, PUNJAB state, N INDIA, 6 mi/9.7 km NNE of KAPURTHALA. Sugar milling, liquor distilling, manufacturing of agricultural implements and chemicals. Was in N part of former in PATIALA AND EAST PUNJAB STATES UNION.

Hamirpur (hah-MIR-puhr), district (□ 432 sq mi/1,123.2 sq km), HIMACHAL PRADESH state, N INDIA; ⊙ HAMIRPUR.

Hamirpur, district (□ 1,582 sq mi/4,113.2 sq km), S UTTAR PRADESH state, N central INDIA; ⊙ HAMIRPUR. Bounded NE by the YAMUNA RIVER; drained by BETWA RIVER. Agriculture (gram, jowar, wheat, sesame, pearl millet, barley, cotton). Main towns are CHARKHARI, MAHOBA, RATH, MAUDAHA, Hamirpur. Formerly part of BRITISH BUNDELKHAND. Original district was enlarged by incorporated 1950 of several former petty states.

Hamirpur (hah-MIR-puhr), town, ⊙ HAMIRPUR district, HIMACHAL PRADESH state, N INDIA, 38 mi/61 km SSE of DHARMSHALA; 31°41′N 76°31′E. Wheat, corn; leather goods.

Hamirpur, town, ⊙ HAMIRPUR district, S UTTAR PRADESH state, N central INDIA, on YAMUNA RIVER, just above BETWA RIVER mouth, and 37 mi/60 km WSW of KANPUR; 25°57′N 80°09′E. Trades in gram, jowar, wheat, sesame, pearl millet. Has eleventh-century Rajput fort ruins. Founded eleventh century by a Karchuli Rajput.

Hamitabat, TURKEY: see ISPARTA.

Hamiz Dam, BOUMERDES wilaya, N ALGERIA, 4.3 mi/7 km S of KHEMIS EL KHECHNA. Dam raised and strengthened 1934–1935; has a capacity of 36,620,000 cu yd/28,000,000 cu m for hydroelectric power and irrigation of over 19 sq mi/49 sq km.

Ham Lake, town (2000 population 12,710), ANOKA county, E MINNESOTA, suburb 17 mi/27 km N of downtown MINNEAPOLIS; 45°15′N 93°12′W. Drained by Coon Creek in S; Ham Lake in center, Lake Netta and part of Coon Lake in N. Manufacturing (commercial fixtures, building materials, wood products; machining). Carlos Avery Wildlife Area to E.

Hamler (HAM-luhr), village (2006 population 671), HENRY county, NW OHIO, 16 mi/19 km SSE of NAPOLEON; 41°14′N 84°02′W.

Hamlet, town (2000 population 820), STARKE county, NW INDIANA, 6 mi/9.7 km N of KNOX; 41°23′N 86°35′W. Ships farm produce.

Hamlet (HAM-luht), town (□ 5 sq mi/13 sq km; 2006 population 5,749), RICHMOND county, S North Carolina, 5 mi/8 km SE of ROCKINGHAM, near SOUTH CAROLINA state line; 34°53′N 79°42′W. Manufacturing (transportation equipment, paper products, apparel, plastic products, fabricated metal products, consumer goods; machining, poultry processing); agriculture (cotton, tobacco, grain, sweet potatoes); poultry, livestock. Railroad junction; once had 42 passenger trains per day. National Railroad Museum. Seat of Richmond Community College. Settled in 1875 and incorporated in 1897, it was named by John Shortridge, a local settler.

Hamlet, village (2006 population 52), HAYES county, SW NEBRASKA, 35 mi/56 km WNW of MCCOOK, and on FRENCHMAN CREEK; 40°22′N 101°13′W.

Hamlet, hamlet, CHAUTAUQUA county, extreme W NEW YORK, 13 mi/21 km SE of DUNKIRK; 43°19′N 77°55′W.

Hamletsburg, village, POPE county, extreme SE ILLINOIS, on OHIO River, and 17 mi/27 km S of GOLCONDA; 37°08′N 88°27′W.

Hamley Bridge (HAM-lee BRIJ), town, S SOUTH AUSTRALIA, 40 mi/64 km N of ADELAIDE, and on LIGHT RIVER; 34°22′S 138°41′E. Agriculture center.

Hamlin, county (□ 537 sq mi/1,396.2 sq km; 2006 population 5,616), E SOUTH DAKOTA, ⊙ HAYTI; 44°40′N 97°12′W. Dairying and livestock area drained by BIG SIOUX RIVER. Several lakes in S; LAKE POINSETT (largest) on border with BROOKINGS county (has Lake Poinsett State Recreation Area). Corn, wheat, soybeans, flax, potatoes; cattle, hogs, poultry. Formed 1873.

Hamlin, town (2006 population 1,988), on border between FISHER and JONES counties, W central TEXAS, 37 mi/60 km NNW of ABILENE; 32°53′N 100°07′W. In cotton area; wheat; cattle. Oil, gypsum, sand and gravel. Manufacturing (grain and feed, gasoline). Incorporated 1907.

Hamlin, town (2006 population 1,096), ⊙ LINCOLN county, W WEST VIRGINIA, on MUD RIVER, 20 mi/32 km ESE of HUNTINGTON; 38°16′N 82°05′W. Coal, natural gas, and oil region. Agriculture (corn, tobacco, potatoes); cattle. Light manufacturing.

Hamlin, village (2000 population 53), BROWN county, NE KANSAS, 6 mi/9.7 km NW of HIAWATHA, near NEBRASKA state line; 39°55′N 95°37′W. Corn; livestock, poultry; dairying.

Hamlin, plantation, AROOSTOOK county, NE MAINE, on SAINT JOHN RIVER, and 29 mi/47 km NE of PRESQUE ISLE; 47°05′N 67°52′W.

Hamlin Lake, MICHIGAN: see BIG SABLE RIVER.

Hamlyn Heights (HAM-lin HEITS), suburb of GEELONG, VICTORIA, SE AUSTRALIA. Grew residentially during the 1950s and 1960s.

Hamm (HAHM), city, NORTH RHINE–WESTPHALIA, W GERMANY, on the LIPPE RIVER, in the RUHR industrial district; 51°41′N 07°49′E. Important railroad junction. Has iron and steel foundries; manufacturing includes textiles, machinery, wire and cable. Coal is mined here. Was capital of the county of MARK until 1809. An active member of the HANSEATIC LEAGUE, it passed to CLEVES in the 14th century and later (1614) to BRANDENBURG. Founded 1226. Also HAMM IN WESFALEN.

Hamm (HAHM), village, Luxembourg commune, S LUXEMBOURG, near LUXEMBOURG city; 49°36′N 06°10′E. Five thousand American soldiers, including General George S. Patton, are buried in the large U.S. military cemetery here.

Hamma, Al (hah-MAH, el), ancient *Aquae Tacapitanae*, market town (2004 population 34,835), QABIS province, SE central TUNISIA, 17 mi/27 km W of QABIS; 33°54′N 09°48′E. Oasis with date palms and olive trees. Mineral springs and remains of Roman thermae.

Hamma Bouziane (hah-MAH boo-ZYAHN), village, CONSTANTINE wilaya, NE ALGERIA, near the Oued

RHUMEL, 4 mi/6.4 km N of CONSTANTINE. Thermal springs; paper and flour milling. Formerly Hamma Plaisance.

Hammada al Hamra (huh-MAH-duh el HUHM-ruh) [Arabic=red rocky desert], desolate rocky region (□ 19,300 sq mi/50,180 sq km) of the SAHARA DESERT, FAZZAN and TRIPOLITANIA regions, NW LIBYA; c.170 mi/274 km long, 160 mi/257 km wide; 28°00′N–30°30′N. Rises in JABAL AS-SAWDA (E) to c.2,700 ft/823 m. Crossed by roads connecting Fazzan region with Tripolitanian coast. Also spelled Hamada el Homra.

Hamma-Jarīd, Al (hah-MAH-je-REED, el), village (2004 population 6,259), TAWZAR province, SW TUNISIA, in the BILAD AL JARID oasis, at SE edge of the SHATT AL GHARSAH, 6 mi/9.7 km NNE of TAWZAR. Date palms. Roman thermae.

Hammam (huhm-MAHM), village, central YEMEN, 45 mi/72 km S of SANA, and on road to HODEIDA. Hot-springs resort.

Hammamat, Al (hah-mah-MET, el), town (2004 population 63,116), NABUL province, NE TUNISIA, port on the GULF OF HAMMAMAT, 40 mi/64 km SE of TUNIS; 36°24′N 10°37′E. Citrus-growing center (chiefly lemons). Fisheries; tourism. Played important role in piracy in 15th century Recent site of resort development.

Hammamat, Gulf of, inlet of the central MEDITERRANEAN SEA, off NE coast of TUNISIA, between CAPE BON peninsula (N) and MUNASTIR (S); 25 mi/40 km long, 50 mi/80 km wide. On it are towns of SUSAH, AL HAMMAMAT, and NABUL. Tuna and anchovy fisheries.

Hammam Bou Hadjar (hah-mAHM boo ah-JAHR), town, AÏN TÉMOUCHENT wilaya, NW ALGERIA, 29 mi/47 km SW of ORAN; 35°22′N 00°58′W. Railroad spur terminus, health resort; vineyards.

Hammam, El, township, ALEPPO district, NW SYRIA, on Turkish border, 34 mi/55 km WNW of ALEPPO; 36°22′N 36°36′E. Cotton, cereals. Border post of the road between Aleppo and ISKENDERUN (ALEXANDRETTA).

Hammam, El (HAHM-mahm, el), village, WESTERN DESERT province, N EGYPT, on coastal railroad, and 35 mi/56 km SW of ALEXANDRIA.

Hammam-Lif (hah-MAHM–LEEF), ancient *Naro*, town (2004 population 38,401), BIN ARUS province, N TUNISIA, on S shore of GULF OF TUNIS, 10 mi/16 km ESE of TUNIS; 36°43′N 10°20′E. Bathing and health resort (medicinal hot springs). Cement- and brickworks; industry. Vineyards. Was winter residence of the Ottoman Bey of Tunis.

Hammam Mélouane, ALGERIA: see BOUGUERRA.

Hammam Meskoutine (hahm-MAHM mes-koo-TEEN), village and noted spa with hot springs (160°F/71°C–205°F/96°C), GUELMA wilaya, NE ALGERIA, in CONSTANTINE MOUNTAINS, on railroad, and 9 mi/14.5 km W of GUELMA. Olive oil milling.

Hammam, Oued (hahm-MAHM, WED), stream, c.160 mi/257 km long, in NW ALGERIA; rises in several headstreams in High Plateaus; flows N to the HABRA lowland (adjoining the Gulf of ARZEW), which its waters irrigate. Storage and flood-control dams at BOU HANIFIA (177 ft/54 m high; 20 mi/32 km SSW of MOHAMMADIA) and at influx of Oued FERGOUG.

Hammam Rirha (hahm-MAHM ree-GAH), village, AÏN DEFLA wilaya, N central ALGERIA, 11 mi/18 km NE of MILIANA, in coastal range of TELL ATLAS (elev. 1,700 ft/518 m). Spa and winter resort, amidst Aleppo pine forests; noted for its curative waters. Ruins of Roman *Aquae Calidae*. Also spelled Hammam Righa.

Hammarby (HAHM-mahr-BEE), village, GÄVLEBORG county, E SWEDEN, on small lake, 8 mi/12.9 km SW of SANDVIKEN; 60°33′N 16°34′E. Old manor house.

Hammar, Hor al (he-MAHR, HOR ahl), lake (□ 750 sq mi/1,950 sq km), SE IRAQ, NW of BASRA, just S of junction of the EUPHRATES with the TIGRIS rivers, both of which feed it; c.70 mi/113 km long, c.15 mi/24 km wide. Largely surrounded by marshes. Also spelled Hur al-Hammar.

Hammarland (HAH-mahr-LAHND), fishing village, Åland province, SW FINLAND, on W shore of ÅLAND island, opposite ECKERÖ island, and 12 mi/19 km NW of MAARIANHAMINA; 60°13′N 19°45′E.

Hammarön (HAHM-mahr-UHN), island (□ 21 sq mi/54.6 sq km), VÄRMLAND county, W SWEDEN, in N part of LAKE VÄNERN, 4 mi/6.4 km S of KARLSTAD; 7 mi/11.3 km long, 1 mi/1.6 km–5 mi/8 km wide; 59°19′N 13°33′E. On NW coast is SKOGHALL village. E shore has fourteenth-century church.

Hammarstrand (HAHM-mahr-STRAHND), village, JÄMTLAND county, N central SWEDEN, on INDALSÄLVEN RIVER (falls), 30 mi/48 km W of SOLLEFTEÅ; 63°07′N 16°21′E. Tourist resort. Hydroelectric station. Has thirteenth-century church.

Hamme (HAH-muh), commune (2006 population 23,351), in Dendermonde district, EAST FLANDERS province, NW BELGIUM, on branch of SCHELDT RIVER, and 5 mi/8 km N of DENDERMONDE; 51°06′N 04°08′E.

Hammel (HAH-muhl), city (2000 population 5,729), ÅRHUS county, E central JUTLAND, DENMARK, 15 mi/24 km N of SKANDERBORG; 56°15′N 09°50′E. Cement works, machine shops; meat processing; agriculture (dairying; hogs; barley, rye).

Hammelburg (HAHM-mel-boorg), town, LOWER FRANCONIA, NW BAVARIA, GERMANY, on the SAALE RIVER, 10 mi/16 km SW of BAD KISSINGEN; 50°07′N 09°53′E. Various small industries; vineyards. Has 14th-century church; nearby are ruins of 11th-century castle. Johann Froben, printer, born here.

Hammels, a section of S QUEENS borough of NEW YORK city, SE NEW YORK, on ROCKAWAY PENINSULA; 40°35′N 73°49′W. Also spelled Hammel.

Hammenhög (HAHM-men-HUHG), village, SKÅNE county, S SWEDEN, 9 mi/14.5 km WSW of SIMRISHAMN; 55°30′N 14°10′E.

Hammer an See, CZECH REPUBLIC: see DOKSY.

Hammerdal (HAHM-mer-DAHL), village, JÄMTLAND county, N central SWEDEN, on Hammerdalssjön Lake (5 mi/8 km long), 35 mi/56 km NE of ÖSTERSUND; 63°35′N 15°21′E.

Hammerfest (HAHM-muhr-fest), town, FINNMARK county, N NORWAY, on KVALØYA island. Europe's northernmost town, but its harbor is always ice-free. Tourists are attracted by its uninterrupted daylight May 17–July 29. Has fish-processing plants. Chartered c.1795. Heavily damaged by British naval bombardment in 1809, by fire in 1890, and by retreating German forces in 1944.

Hamme River (HAHM-me), c.30 mi/48 km long, NW GERMANY; rises 10 mi/16 km NW of OSTERHOLZ-SCHARMBECK; flows S and SW to the WÜMME RIVER just N of BREMEN; source at 53°18′N 08°45′E.

Hammersmith and Fulham (HAM-uhr-SMITH and FUHL-ham), inner borough (□ 6 sq mi/15.6 sq km; 2001 population 165,242) of GREATER LONDON, SE ENGLAND, on the THAMES RIVER; 51°30′N 00°15′W. It has various industries (such as wharves and pottery kilns) and is the principal television center of the British Broadcasting Corporation (BBC). Fulham Palace, with 0.06 sq mi/0.16 sq km of grounds, is the residence of the bishop of London. William Morris's Kelmscott Press was in Hammersmith, and St. Paul's School for boys was moved here in 1884.

Hammerstein, POLAND: see CZARNE.

Hammerum (HAH-muh-room), town, RINGKJØBING county, W JUTLAND, DENMARK, 32 mi/51 km E of Ringkjøbing; 56°08′N 09°04′E. Agriculture (cattle, hogs; potatoes); textiles.

Hamminkeln (hahm-MING-keln), town, North NORTH RHINE–WESTPHALIA, NW GERMANY, on ISSEL RIVER, 6 mi/9.7 km N of WESEL; 51°46′N 06°36′E. Agricultural center; manufacturing of building materials. Has remains of Augustinian monastery from 13th century.

Hamm in Westfalen, GERMANY: see HAMM.

Hammon (HA-muhn), village (2006 population 453), ROGER MILLS county, W OKLAHOMA, 15 mi/24 km N of ELK CITY; 35°37′N 99°22′W. In cattle-raising, dairying, agricultural area (wheat).

Hammonasset Point, S CONNECTICUT, peninsula, on LONG ISLAND SOUND near MADISON; c.2 mi/3.2 km long. Here are Hammonasset Beach State Park and mouth of Hammonasset River (c.20 mi/32 km long), which rises S of MIDDLETOWN, flows SSE to Clinton Harbor just E of Hammonasset Point.

Hammond, city (2000 population 83,048), Lake county, extreme NW INDIANA, bounded by Lake MICHIGAN, the ILLINOIS state line, and the Little CALUMET RIVER, and traversed by the Grand Calumet River; 41°37′N 87°29′W. Originally important as a slaughterhouse site, Hammond was a meatpacking town until its great packing house was destroyed by fire in 1901. Manufacturing here includes food products, fabricated metal products, chemicals, machinery, building materials, transportation equipment, foods, petroleum products; secondary steel processing; fire brick refractories. A campus of Purdue University is here. Settled 1851. Laid out 1875. Incorporated 1884.

Hammond, city, TANGIPAHOA parish, SE LOUISIANA, 43 mi/69 km NNW of NEW ORLEANS; 30°31′N 90°28′W. Railroad junction. In agricultural area (vegetables; cattle; dairying); timber; manufacturing (building materials, consumer goods, paper products, fabricated metal products, machinery; meat processing). Seat of Southeastern Louisiana University. Has major highway crossroads, I-55 (N-S) and I-12 (E-W). Zemurry Gardens and Global Wildlife Park. Home of the Black Heritage Festival. Incorporated 1888.

Hammond, town (2006 population 1,855), ST. CROIX county, W WISCONSIN, 16 mi/26 km E of HUDSON; 44°58′N 92°26′W. Dairying; manufacturing (polyester film).

Hammond (HA-muhnd), unincorporated village, SE ONTARIO, E central CANADA, 18 mi/29 km from OTTAWA, and included in Clarence-Rockland; 45°26′N 75°14′W.

Hammond, village (2000 population 518), PIATT county, central ILLINOIS, 18 mi/29 km E of DECATUR; 39°47′N 88°35′W. In grain area.

Hammond, village (2000 population 198), WABASHA county, SE MINNESOTA, 15 mi/24 km NNE of ROCHESTER, on ZUMBRO RIVER, in Richard J. Dorer Memorial Hardwood State Forest; 44°13′N 92°22′W. Grain; livestock.

Hammond, village (2006 population 292), ST. LAWRENCE county, N NEW YORK, between BLACK LAKE and SAINT LAWRENCE River, 19 mi/31 km SW of OGDENSBURG; 44°27′N 75°41′W. Timber; agricultural commerce.

Hammond, village, CLATSOP county, NW OREGON, 5 mi/8 km W of ASTORIA, on Columbia River estuary, 1 mi/1.6 km E of PACIFIC OCEAN; 46°12′N 123°57′W. Terminus of railroad spur from WARRENTON. Smoked, fresh, frozen seafood. Fort Stevens State Park to W.

Hammond, plantation, AROOSTOOK county, E MAINE, just NW of HOULTON; 46°14′N 67°57′W. In agricultural area.

Hammond, CANADA: see PORT HAMMOND.

Hammond-Harwood House National Historic Site (c.1774), ANNAPOLIS, ANNE ARUNDEL county, central MARYLAND. The house exemplifies American Georgian architecture in refinement of detail and excellence of design. The symmetrical brick building has two wings with polygonal bays. The architect, William Buckland, one of few of the period who can be identified, designed it for Matthais Hammond, a young member of the Maryland Assembly who played a role in events leading up to the American Revolution. Ownership of the house eventually passed to William Harwood, a grandson of William Buckland. Preserved as a historical landmark since 1924, first by St. Johns College, later by the Hammond-Harwood House Association.

Hammondsport, village (2006 population 701), STEU-BEN county, S NEW YORK, at S end of KEUKA LAKE, 20 mi/32 km NNW of CORNING; 42°24′N 77°13′W. In grape-growing area; wine-making center; light manufacturing. Summer residential and recreational area. Birthplace of Glenn Curtiss, who made aviation experiments in the village (museum). Incorporated 1856.

Hammondville, town (2000 population 486), De Kalb co., NE Alabama, near Valley Head, c.10 mi/16 km NE of FortPayne.

Hammonia, Brazil: see IBIRAMA.

Hammonton, town (2006 population 13,572), ATLANTIC county, S NEW JERSEY, a residential and manufacturing suburb of PHILADELPHIA; 39°38′N 74°46′W. Shipping and processing center for fruit; light manufacturing. Site of a winery. Incorporated 1866.

Ham-Nord (AHM–NOR), canton (□ 39 sq mi/101.4 sq km; 2006 population 928), Centre-du-Québec region, S QUEBEC, E CANADA, 5 mi/8 km from Saint-Fortunat; 45°54′N 71°39′W.

Hamochi (HAH-mo-chee), town, Sado district, NII-GATA prefecture, N central JAPAN, 40 mi/65 km W of NIIGATA; 37°50′N 138°19′E.

Hamoir (ahm-WAHR), commune (2006 population 3,613), Huy district, LIÈGE province, E BELGIUM, 14 mi/23 km S of LIÈGE; 50°26′N 05°32′E. Dairying center; tourist resort.

Hamois (ahm-WAH), commune (2006 population 6,645), Dinant district, NAMUR province, S central BELGIUM, 12 mi/19 km NW of MARCHE-EN-FAMENNE, in FAMENNE depression of the ARDENNES Mountains; 50°20′N 05°08′E.

Hamônia, Brazil: see IBIRAMA.

Hamont-Achel (HAH-mon–AH-guhl), commune (2006 population 13,802), Maaseik district, LIMBURG province, NE BELGIUM, 15 mi/24 km S of EINDHOVEN, near NETHERLANDS border.

Hámor, Lake (HAH-mor), Hungarian *Hámor*, NE HUNGARY, in Bükk Mountains, 7 mi/11 km W of MISKOLC; 1 mi/1.6 km long, 360 ft/110 m wide; 48°06′N 20°37′E. Village of Hámor and resort of Lillafüred on its shore.

Hamoukar, Tell, ancient city and archaeological site, NE SYRIA, near IRAQ border, 5 mi/8 km W of Tell KOJAK. Excavations begun in 1999 revealed the city to have been a flourishing center of trade in MESOPOTAMIA c.3,500 B.C.E., making it one of the oldest cities outside ancient SUMER; in 2005 evidence was found that the city may have been destroyed in a large-scale battle, one of the earliest such battle sites discovered. Also Tall Hamoukar.

Hampden (HAM-duhn), canton (□ 43 sq mi/111.8 sq km; 2006 population 163), Estrie region, S QUEBEC, E CANADA, 19 mi/30 km from LAC-MÉGANTIC; 45°30′N 71°15′W.

Hampden, county (□ 634 sq mi/1,648.4 sq km; 2006 population 460,520), SW MASSACHUSETTS, on CONNECTICUT state line; ⊙ SPRINGFIELD; 42°07′N 72°40′W. Bisected by CONNECTICUT RIVER, which supplies power to its industrial cities. The Connecticut Valley Lowland formed in middle of county. Varied manufacturing (paper and paper products, foods, rubber and plastic products, metal products, industrial machinery and equipment, electronic equipment, consumer goods; printing and publishing). Vegetables, fruit; livestock; dairying, nursery products. Formed 1812.

Hampden, township, WAITAKI district, OTAGO region, E SOUTH ISLAND, NEW ZEALAND, 22 mi/35 km S of OA-MARU. Service center for agriculture and fishing; resort with beaches and Moeraki boulders tourist attraction.

Hampden, town, PENOBSCOT county, S MAINE, on the PENOBSCOT RIVER, just S of BANGOR; 44°43′N 68°53′W. Park commemorates Dorothea Dix (born here). Settled 1767, incorporated 1794.

Hampden, town, HAMPDEN county, S MASSACHUSETTS, near SCANTIC RIVER, 8 mi/12.9 km ESE of SPRING-

FIELD; 42°04′N 72°25′W. Dairying; vegetables. Settled c.1740, incorporated 1878.

Hampden, village (2006 population 56), RAMSEY co., NE NORTH DAKOTA, 21 mi/34 km NNE of DEVILS LAKE; 48°32′N 98°39′W.

Hampden Sydney (HAM-duhn SID-nee), unincorporated town (2000 population 1,264), PRINCE EDWARD county, central VIRGINIA, 7 mi/11 km SSW of FARM-VILLE; 37°14′N 78°27′W. Manufacturing (wine); agriculture (tobacco, grain; hogs; dairying); timber. Seat of Hampden-Sydney College.

Hampi (huhm-PEE), village, BELLARY district, KARNA-TAKA state, S INDIA, on right bank of TUNGABHADRA RIVER, and 33 mi/53 km NW of BELLARY. Site of famous ruined city of Vijayanagar [Sanskrit/Hindi=city of victory]; flourished from 1336 until sacked 1565 by Deccan sultans after battle of Talikota. Ruins of city's main ramparts enclose remains of several palaces, Shivaite and Vishnuite temples, and other granite monuments, providing notable panorama of Dravidian architecture. Tourists' bungalow just S, at village of Kamalapuram.

Hampshire or **Hants** (HAMP-SHI-uhr), county (□ 1,503 sq mi/3,907.8 sq km; 2001 population 1,644,249), S central ENGLAND; ⊙ WINCHESTER; 51°05′N 01°15′W. The terrain is undulating and is crossed by two chalk downs, rising in places to over 800 ft/245 m. Fronts on The SOLENT, facing across to the Isle of WIGHT. In the SW is the NEW FOREST. The principal rivers are the TEST, the ITCHEN, and the AVON. Agriculture (grain production; dairy farming; market gardening); oil refining at Fawley; aircraft engineering at Farnborough. Gosport, Southampton, and Portsmouth are three of Britain's leading ports. Other towns are Aldershot (with its military associations), Basingstoke, and Eastleigh. Evidence of prehistoric and Roman settlement is found here. Hampshire was once part of the Anglo-Saxon kingdom of WESSEX and has numerous historical and literary associations.

Hampshire, county (□ 545 sq mi/1,417 sq km; 2006 population 153,471), W central MASSACHUSETTS; ⊙ NORTHAMPTON; 42°20′N 72°40′W. Bisected by CONNECTICUT RIVER; drained by WESTFIELD RIVER and other small streams. Connecticut Valley Lowland bisects county. The BERKSHIRES touch W section; Mounts Tom and Holyoke are near the CONNECTICUT border. Generally forested and agricultural (tobacco, hay, vegetables, fruit; poultry, livestock; dairying); manufacturing (textile goods; paper, plastic, and metal products; machinery; photographic equipment; consumer goods). Formed 1662.

Hampshire, county (□ 645 sq mi/1,677 sq km; 2006 population 22,480), NE WEST VIRGINIA, in EASTERN PANHANDLE; ⊙ ROMNEY; 39°18′N 78°36′W. Bounded N, in part, by POTOMAC River (MARYLAND state line), E by VIRGINIA; drained by South Branch of Potomac, CACAPON, and North rivers. Traversed by valleys and ridges of the APPALACHIAN MOUNTAINS. Agriculture (corn, hay, alfalfa, grains, potatoes, sorghum, honey, fruit); livestock, poultry. Some manufacturing at Romney. Timber. Small part of George Washington National Forest in SE; Edwards Run Wildlife Management Area in E; Short Mountain and Nathanial Mountain wildlife areas in S; Fort Mill and Springfield wildlife management areas in W. Formed 1753.

Hampshire, village (2000 population 2,900), KANE county, NE ILLINOIS, 18 mi/29 km NNW of GENEVA; 42°06′N 88°31′W. In agricultural area (dairy products; livestock; grain).

Hampstead (HAMP-sted), city, Montréal administrative region, S QUEBEC, E CANADA, 4 mi/6 km SW of downtown MONTREAL, on MONTREAL ISLAND; 45°28′N 73°39′W. Part of the Metropolitan Community of Montreal (*Communauté Metropolitaine de Montréal*).

Hampstead (HAMP-sted), town (2000 population 5,060), CARROLL county, N MARYLAND, 25 mi/40 km

NW of BALTIMORE; 39°37′N 76°52′W. Power tool manufacturing. Laid out in 1786.

Hampstead (HAMP-stuhd), town, ROCKINGHAM county, SE NEW HAMPSHIRE, 6 mi/9.7 km N of SALEM; 42°52′N 71°10′W. Manufacturing (plastic products, computer equipment, electronic equipment; printing and publishing); agriculture (nursery crops, vegetables; poultry, cattle; dairying). Island Pond (2 mi/3.2 km long) in W; Angle and Wash ponds in N. Includes village of EAST HAMPSTEAD.

Hampstead (HAMP-stid), village (2001 population 312), S NEW BRUNSWICK, E CANADA, on ST. JOHN RIVER (ferry), and 24 mi/39 km N of SAINT JOHN; 45°36′N 66°10′W. Granite quarrying.

Hampstead, ENGLAND: see CAMDEN.

Hampton, county (□ 562 sq mi/1,461.2 sq km; 2006 population 21,268), S SOUTH CAROLINA; ⊙ HAMPTON; 32°46′N 81°08′W. Bounded SW by SAVANNAH River; drained by the COOSAWHATCHIE RIVER. Hunting. Timber; agriculture includes hogs, cattle; corn, wheat, soybeans, peanuts, cotton, watermelons, peaches. Formed 1878.

Hampton (HAMP-tuhn), city (2000 population 4,218), ⊙ FRANKLIN county, N central IOWA, 29 mi/47 km S of MASON CITY; 42°45′N 93°12′W. Railroad junction. Agricultural trade and processing center (corn, packed poultry, feed). Wood, metal, stone, and concrete products; fertilizers; motor home components; nursery. Limestone quarries, sand and gravel pits nearby. Beeds Lake State Park is NW. Founded 1856, incorporated 1870.

Hampton (HAMP-tuhn), independent city (□ 57 sq mi/148.2 sq km; 2006 population 145,017), SE VIRGINIA, 8 mi/11 km NNW of NORFOLK, port on HAMPTON ROADS, at mouth of JAMES RIVER, connected to Norfolk by Hampton Roads Bridge-Tunnel; 37°02′N 76°17′W. Separate from adjoining county; bounded W by independent city of NEWPORT NEWS, N by independent city of POQUOSON. Railroad terminus; manufacturing (machinery, chemicals, lumber, building materials, consumer goods, wood products, foods, transportation equipment; machining, steel fabrication, electronic instrumentation); large seafood packing and shipping industry (fish, crabs, and oysters). LANGLEY AIR FORCE BASE (established 1917), which includes NASA's Langley Research Center, are in N; historic FORT MONROE Military Reservation (built 1819–1834) is in SE on CHESAPEAKE BAY. One of the oldest continuous English settlements in the country, Hampton was founded on the site of the Native American village Kecoughtan. Attacked by pirates in late 17th century (Blackbeard captured off the coast), shelled in the Revolutionary War, sacked by British in 1813, and nearly burned to the ground by evacuating Confederates in 1861 to prevent Union possession. Seat of Hampton University, Thomas Nelson Community College, St. John's Episcopal Church (1728; original church established 1610), nearby reproduction of a Native American village. Plum Tree Island National Wildlife Reservation to N. Big Bethel Reservoir in NW corner. Formerly Elizabeth City county. Settled 1610 by colonists from Jamestown, incorporated 1849.

Hampton (HAMP-tuhn), town (2000 population 1,579), ⊙ CALHOUN county, S ARKANSAS, on Champagnolle Creek, 21 mi/34 km E of CAMDEN; 33°32′N 92°28′W. In agricultural area. Manufacturing (apparel).

Hampton, town, WINDHAM county, E CONNECTICUT, on LITTLE RIVER, and 9 mi/14.5 km NE of WILLIMANTIC; 41°47′N 72°03′W. In hilly region. Agriculture. Has 18th- and 19th-century houses; part of state forest.

Hampton, town (2000 population 3,857), HENRY county, N central GEORGIA, 9 mi/14.5 km N of GRIF-FIN; 33°23′N 84°17′W. Manufacturing includes building materials, machinery parts, consumer goods, fabricated metal goods, electrical equipment; egg production. NASCAR racing at Atlanta Motor Speedway; Atlanta regional air-traffic control center.

Hampton, town, ROCKINGHAM county, SE NEW HAMP-SHIRE, on ATLANTIC OCEAN (E) and HAMPTON HARBOR (S), 9 mi/14.5 km S of PORTSMOUTH; 42°56′N 70°49′W. Manufacturing (chemicals, electronic equipment, tools, aerospace products, leather; architectural millwork; printing and publishing). Agriculture (nursery crops, vegetables, apples; cattle, poultry; dairying). Henry Dearborn born here. Hampton Beach and North Beach, on coast, are resort areas. Hampton Beach State Park and State Pier in SE. NORTH HAMPTON and HAMPTON FALLS were formerly in Hampton. Settled 1638, incorporated 1639.

Hampton, township, ALLEGHENY county, W PENNSYL-VANIA, residential suburb 9 mi/14.5 km NNW of PITTSBURGH; 40°35′N 79°57′W. Manufacturing includes tool and die and ornamental iron. Includes the communities of ALLISON PARK, DeHaven, Wildwood, and Hardy.

Hampton, town (2006 population 2,786), ⊙ HAMPTON county, SW SOUTH CAROLINA, 55 mi/89 km N of SA-VANNAH; 32°52′N 81°06′W. Manufacturing of wood products, chemicals, plastics, and laminates. In agricultural area producing cotton, corn, sweet potatoes, watermelons, peanuts, peaches; timber; livestock.

Hampton (HAMP-tuhn), village (2001 population 3,997), ⊙ Kings county, S NEW BRUNSWICK, E CA-NADA, on KENNEBECASIS RIVER, and 20 mi/32 km NE of SAINT JOHN; 45°30′N 65°50′W. In dairying and farming region. Popular summer residence for Saint John residents.

Hampton (HAMP-tuhn), village (2000 population 1,626), ROCK ISLAND county, NW ILLINOIS, suburb of ROCK ISLAND and MOLINE, on the MISSISSIPPI RIVER; 41°32′N 90°24′W. Metal electroplating.

Hampton, village (2000 population 434), DAKOTA county, SE MINNESOTA, 24 mi/39 km S of ST. PAUL, and 11 mi/18 km SW of HASTINGS; 44°36′N 93°00′W. Grain, soybeans; livestock, poultry; dairying; manufacturing (meat processing; fertilizers).

Hampton, village (2006 population 432), HAMILTON county, SE central NEBRASKA, 5 mi/8 km E of AURORA; 40°52′N 97°53′W.

Hampton, village, CARTER county, NE TENNESSEE, 5 mi/8 km SE of ELIZABETHTON; 36°17′N 82°10′W.

Hampton, borough (2006 population 1,658), HUNTER-DON county, W NEW JERSEY, near MUSCONETCONG River, 13 mi/21 km E of PHILLIPSBURG; 40°42′N 74°58′W.

Hampton (HAMP-tuhn), neighborhood (2001 population 9,429) in the GREATER LONDON outer borough of RICHMOND UPON THAMES, SE ENGLAND, on the THAMES RIVER; 51°25′N 00°22′W. Site of Hampton Court Palace, begun by Cardinal Wolsey in 1514 as his private residence. After his downfall it was taken by Henry VIII and remained a royal residence until the time of George II. William III had part of it torn down and redesigned by Christopher Wren. Although much of the palace is open to the public, many of its rooms are occupied by royal pensioners. Part of Richmond upon Thames since 1965.

Hampton (HAMP-tuhn), suburb 9 mi/14 km SSE of MELBOURNE, VICTORIA, SE AUSTRALIA, between BRIGHTON and SANDRINGHAM suburbs; 37°57′S 145°00′E. Railway station.

Hampton, NORTH CAROLINA: see RUTH.

Hampton Bays, summer-resort village (□ 21 sq mi/54.6 sq km; 2000 population 12,236), SUFFOLK county, SE NEW YORK, on shore of SE LONG ISLAND, 7 mi./11.3 km W of SOUTHAMPTON; 40°51′N 72°31′W. In diversified-farming area; boatyards; commercial fishing. Until 1922 called Good Ground. Pop. figure includes Rampasture, Squiretown, and Tiana.

Hampton Beach, New Hampshire: see HAMPTON.

Hampton Falls, town, ROCKINGHAM county, SE NEW HAMPSHIRE, 5 mi/8 km SSW of HAMPTON; 42°55′N 70°53′W. Bounded on E by HAMPTON HARBOR. Agriculture (dairying; cattle, poultry; vegetables;

nursery crops); manufacturing (commercial printing). Set off from Hampton 1726.

Hampton Harbor, SE NEW HAMPSHIRE, small bay (1 mi/1.6 km long, 0.5 mi/0.8 km wide) S of HAMPTON, near MASSACHUSETTS state line; receives two tidal streams.

Hampton National Historic Site, BALTIMORE county, NE MARYLAND. Late-18th-century Georgian mansion (1783–1790) with a portrait gallery and terraced gardens. Built by George Ridgely. Authorized 1948.

Hampton Roads (HAMP-tuhn), harbor, SE VIRGINIA, in JAMES RIVER estuary at its entrance to CHESAPEAKE BAY, East Elizabeth River estuary enters from S, Willoughby Bay on S, just within entrance; 4 mi/6 km long and 40 ft/12 m deep; 36°58′N 76°21′W. One of the world's finest natural harbors, it has been a major anchorage point since colonial times and has extensive harbor facilities and shipyards; NEWPORT NEWS, HAMPTON on the N shore; NORFOLK, PORTSMOUTH on the S. The Port of Hampton Roads, established 1926 under the State Port Authority of Virginia, is one of the busiest U.S. seaports. Hampton Roads has long been important to the U.S. Navy; Norfolk is head-quarters for the Atlantic Fleet. Norfolk Navy Base on SE side; CRANEY ISLAND Army Corps of Engineers Base on S. A vehicular tunnel (7,479 ft/2,280 m) under the roads opened in 1957. Hampton Roads was the site of the Civil War battle (March 1862) between the ironclads *Monitor* and *Merrimack*.

Hampyong (HAHM-PYOUNG), county (□ 111 sq mi/288.6 sq km), NW SOUTH CHOLLA province, SOUTH KOREA, on YELLOW SEA (W). Adjacent to YONGGWANG and CHANGHUNG on N and MUAN on S. Noryong Mountains in N. Agriculture (rice, barley, sweet potatoes, onions, peppers, garlic, strawberries, peaches, tobacco) in wide, fertile flatland created by small mountain rivers running to the Yongsan River. Sedge mats; oxen market. Honam railroad and freeway between KWANGJU and MOKPO pass in S. Beach.

Hamra, El (HAHM-rah, el), village, HAMA district, W SYRIA, 22 mi/35 km NE of HAMA; 35°17′N 37°04′E. Cotton, cereals. Sometimes called Al-HUMAYRAH.

Hamrah, Al (HUHM-ruh, el), terminal, TRIPOLITANIA region, W central LIBYA, at S edge of GHUDAMIS BASIN, 250 mi/402 km S of AZ-ZAWIYAH; 26°43′N 13°54′E. S terminus of oil pipeline to Az-Zawiyah.

Hamra, Jazirat al (HUHM-ruh, zhuh-ZEE-ret el), island, off the coast of the UNITED ARAB EMIRATES, in PERSIAN GULF, belonging to RAS AL-KHAIMAH emirate; 25°43′N 55°47′E.

Hamrångefjärden (HAHM-RONG-e-FYER-den), village, GÄVLEBORG county, E SWEDEN, near GULF OF BOTHNIA, 18 mi/29 km N of GÄVLE; 60°54′N 17°04′E.

Hamrat Al-Sheikh (HAHM-ruh ahl-SHAIK), village, N KURDOFAN state, SUDAN, at S edge of LIBYAN DESERT, on the road from KHARTOUM to AL-FASHER, and 180 mi/290 km NW of AL UBAYYID; 14°35′N 27°58′E. Gum arabic; livestock. Also spelled Hamrat Ash Sheikh.

Hamrat Ash Sheikh, SUDAN: see HAMRAT AL-SHEIKH.

Hamrun (hahm-ROON), industrial town (2005 population 9,541), E MALTA, 1.5 mi/2.4 km SW of VAL-LETTA; 35°53′N 14°29′E. Adjoined (NE) by FLORIANA at base of Mount Sceberras Peninsula. Manufacturing of buttons, bricks, tiles, candles, mirrors, beer. Though many houses were wrecked during World War II air raids, its monuments have suffered little.

Hamry, CZECH REPUBLIC: see HLINSKO.

Hams Fork, river, c.100 mi/161 km long, SW WYOMING; rises in S tip of SALT RIVER RANGE, central LINCOLN county; flows S through Lake Viva Naughton Reservoir, past KEMMERER, then E, past OPAL, and SE to BLACKS FORK of GREEN RIVER at GRANGER.

Ham-Sud (ham–SYOOD), unincorporated village, S QUEBEC, E CANADA, at foot of Ham Mountain (2,325 ft/709 m), 30 mi/48 km NE of SHERBROOKE, and included in Saint-Joseph-de-Ham-Sud; 45°45′N 71°36′W. Dairying; cattle raising; potato growing. Was seat of historic WOLFE county.

Ham-sur-Heure-Nalinnes (AHM–syoor–uhr–nah-LEEN), commune (2006 population 13,387), Thuin district, HAINAUT province, SW BELGIUM, 7 mi/11 km SSE of CHARLEROI, on EAU D'HEURE RIVER; 50°19′N 04°23′E.

Hamta Pass (huhm-TAH) (c.14,050 ft/4,282 m), in SE PIR PANJAL RANGE of PUNJAB HIMALAYAS, in KANGRA district, HIMACHAL PRADESH state, NE INDIA, 60 mi/97 km E of DHARMSHALA. Also spelled Hamtah.

Hamtramck (ham-TRA-mik), city (2000 population 22,976), WAYNE county, SE MICHIGAN, suburb 4 mi/6.4 km N of downtown DETROIT; 42°23′N 83°03′W. Manufacturing (meat processing; machinery, foods, motor vehicles and transportation equipment, chemicals, plastic products). The site was settled by Frenchmen in the late 18th-century. The city grew quickly after the coming of the automobile industry c.1910. Points of interest include St. Florian's Church (a prime example of Gothic architecture); and the memorial and grave of John F. Hamtramck, 1st U.S. commander of the Detroit garrison. The city has a large Polish-American community and various Polish cultural events. NW corner touches city of HIGHLAND PARK; both cities are completely surrounded by city of Detroit. Incorporated as a city 1922.

Hamun-i-Lora (HAH-moon–e–LO-rah), closed salt-marsh depression in CHAGAI district, N BALUCHISTAN province, SW PAKISTAN; 36 mi/58 km long (N-S), 3 mi/4.8 km–10 mi/16 km wide. Receives the PISHIN LORA (SE) and drainage of CHAGAI HILLS (W); dry for most of year.

Hamun-i-Mashkel (HAH-moon–e–MUSH-kahl), closed salt-marsh depression in CHAGAI district, W BALUCHISTAN province, SW PAKISTAN, just W of Sandy Desert; 55 mi/89 km long-(E-W), 8 mi/12.9 km–20 mi/32 km wide. Receives Mashkel River (E) and drainage of hills (N); dry for most of year. Thick deposits of brine salt worked.

Hamun-i-Puzak, depression forming a small lake, E NIMRUZ province, AFGHANISTAN; extends slightly into S FARAH province at the Iranian border; 15 mi/24 km long, c.11 mi/18 km wide (measured in 1884), although size changes seasonally with the flow of water from the Kashrud and Khuspas rivers; at times completely dry; 31°30′N 61°45′E. Also called Hamnun-i-Helmand.

Hamurre, El (hah-MOOR-re, el), village, E central SOMALIA, 80 mi/129 km SW of EIL. Road junction. Also called Girriban.

Hamusit (hah-MOO-seet) [=Thursday market], town (2007 population 12,684), AMHARA state, NW ETHIOPIA, near SE shore of LAKE TANA, and 3 mi/4.8 km S of WERETA; 11°47′N 37°33′E. Market town.

Hamyang (HAHM-YAHNG), county (□ 280 sq mi/728 sq km), NW SOUTH KYONGSANG province, SOUTH KOREA. Adjacent to KOCHANG and SANCHONG (E), HADONG (S), and NORTH CHOLLA (W) provinces. Sobaek mountains surround county; fields along Imchon River basin in NE. Agriculture (cotton, sesame, vegetables); sericulture; husbandry. Served by 88 Expressway; CHIRI MOUNTAINand Tokyu Mountain national parks.

Hamza (HUHM-zuh), town, QADISSIYA province, SE central IRAQ, on the HILLA (canal of the EUPHRATES RIVER), and 17 mi/27 km SE of DIWANIYA; 37°05′N 43°25′E. Rice, dates, corn, millet, sesame. Also called Imam el Hamza.

Hamzoren, Bulgaria: see BEZMER.

Han (HAHN), river, S CHINA, 210 mi/338 km long; rises in Wuyi mountains of W FUJIAN province; flows S through E GUANGDONG province past Dabu, Chaozhou, and Chenghai to the SOUTH CHINA SEA near Shantou; navigable for about 100 mi/161 km upstream. The densely populated delta is a rich agricultural area; two crops of rice are grown annually. Other crops are oilseeds and sugarcane. Manganese and tungsten are mined in the upper valley.

Cross-references are shown in SMALL CAPITALS. The pronunciation guide is shown on page xix. The sources of population figures are shown on page xvii.

Hana (HAH-nah), village (2000 population 709), E MAUI, MAUI county, HAWAII, 32 mi/51 km ESE of KAHULUI, at E end of Maui, on Hana Bay; 20°46′N 155°59′W. Harbor; airport. Secluded and beautiful area. Charles A. Lindbergh buried nearby. Hana Airport to NW. Waianapanapa State Park to N, on coast. Extension to coast of Haleakala National Park 7 mi/11.3 km to SW. Hana Forest Reserve to SW. Hana Cultural Center and Museum Hana Bay Beach Park is here.

Hana (HAH-nah), Czech *Haná*, fertile agricultural region of undulating lowlands (elevation 632 ft/193 m–832 ft/254 m), JIHOMORAVSKY and SEVEROMORAVSKY provinces, central MORAVIA, CZECH REPUBLIC. Noted for its strong ethnographic and agricultural characteristics; lies approximately between LITOVEL (N), Prostějov (central), VYSKOV (SW), HOLESOV (SE), and LIPNIK NAD BECVOU (NE). Chief industries include sugar milling, brewing, dairying; grows well-known malt and barley. Drained by MORAVA RIVER.

Hanábana River (ah-NAH-bah-nah), c.50 mi/111 km long, W CUBA; flows along border between MATANZAS and CIENFUEGOS provinces; 22°28′N 80°34′W. Drains 406 sq mi/1,051 sq km of plains and foothills W of Alturas de Santa Clara, emptying into the ZAPATA marshes. Formerly called Amarillas River.

Hanabanilla River (ah-nah-bah-NEE-yah), c.20 mi/32 km long, VILLA CLARA province, central CUBA; small affluent of the ARIMAO RIVER, which it joins at Arimao town, 15 mi/24 km E of CIENFUEGOS; 22°07′N 80°06′W. Known for its waterfall. Headwaters held in a dam of same name.

Hanaford, village (2000 population 55), FRANKLIN county, S ILLINOIS, 5 mi/8 km SE of BENTON; 37°57′N 88°50′W. In bituminous-coal and agricultural area. Also called Logan.

Hanagal, INDIA: see HANGAL.

Hanagita Peak (HA-nuh-GEE-duh) (8,504 ft/2,592 m), S ALASKA, in CHUGACH MOUNTAINS, 80 mi/129 km ENE of CORDOVA; 61°05′N 143°43′W.

Hanahan (HA-nuh-han), city (2006 population 13,846), BERKELEY county, SE SOUTH CAROLINA suburb 9 mi/14.5 km NNW of downtown CHARLESTON; 32°55′N 80°00′W. Manufacturing includes machining, metal fabricating, manufacturing of plastic products. Charleston Naval Base to NE; Charleston Air Force Base to W. Charleston International Airport to SW. Goose Creek Reservoir to N.

Hana-izumi (hah-NAH–ee-ZOO-mee), town, West Iwai district, IWATE prefecture, N HONSHU, NE JAPAN, 59 mi/95 km S of MORIOKA; 38°49′N 141°11′E.

Hanakpinar, TURKEY: see CINAR.

Hanalei (HAH-nah-LAI), village (2000 population 478), N KAUAI, KAUAI county, HAWAII, near head of Hanalei Bay and Hanalei River, 18 mi/29 km NW of LIHUE, on Kuhio Highway, between Hanalei River (N) and Waioli Stream (W); 22°12′N 159°30′W. Hanalei Museum Hanalei National Wildlife Refuge to E. Halelea Forest Reserve to S, including Mount Kaliko (4,201 ft/1,280 m). Waioli Mission. Waioli Beach Park is here. Hanalei Beach Park at mouth of Hanalei River, to N.

Hanam (HAH-NAHM), city (2005 population 122,337), central KYONGGI province, SOUTH KOREA, just E of SEOUL. Hilly in E and S; wide valley of HAN RIVER forms in NE (rice farming). Bedroom community to Seoul. Manufacturing of chemicals and textiles for export. Chungbu Expressway and Olympic Freeway intersect here. Paldang dam and a boat race lake are located here.

Ha Nam (HAH NAHM), city, ⊙ NAM HA province, N VIETNAM, on the Song Dai (arm of RED RIVER delta), 35 mi/56 km S of HANOI; 20°33′N 105°56′E. Transportation and market center; light industry; administrative complex. Silk spinning; rice, vegetable, fruit, and coffee growing; food processing.

Ha Nam, VIETNAM: see NAM HA.

Hanamaki (hah-NAH-mah-kee), city, IWATE prefecture, N HONSHU, NE JAPAN, on KITAKAMI RIVER, and 22 mi/35 km S of MORIOKA; 39°23′N 141°07′E. Electric machinery. Hot springs nearby.

Hanamaulu (HAH-nah-MAH-OO-loo), village (2000 population 3,272), SE KAUAI, KAUAI county, HAWAII, 2 mi/3.2 km N of LIHUE, on Kuhiō Highway, and on Hanamaulu Bay, at mouth of Hanamaulu Stream; 22°00′N 159°20′W. County Correctional Facility to N. Kalepa Forest Reserve and Wailua River State Park to N. Ahukini State Recreational Park to E on coast; museum.

Hanamsagar (huh-nuhm-SAH-guhr), town, RAICHUR district, KARNATAKA state, S INDIA, 45 mi/72 km NW of GANGAWATI; 15°53′N 76°04′E. Millet, oilseeds.

Hanang Mountain (HAHN-ahng), extinct volcano, Manyara region, N central TANZANIA, 50 mi/80 km NE of SINGIDA; 04°28′S 35°22′E. Elevation 11,215 ft/3,418 m. Lake Balangida to NW; near GREAT RIFT VALLEY.

Hanapepe (HAH-nah-PAI-PAI), town (2000 population 2,153), S KAUAI, KAUAI county, HAWAII, at head of Hanapepe Bay, at mouth of Hanapepe River, NW of Port Allen, on Kaumualii Highway; 21°55′N 159°35′W. Manufacturing (ice cream); Port Allen Airport (Burns Field) at Puolo Point to S; Salt Pond Beach Park to SW. Hanapepe Stadium; Hoary Head Stadium to W.

Hanau (HAH-nou), city, HESSE, central GERMANY, on the Main and KINZIG rivers; 50°08′N 08°55′E. Important railroad and road junction and center of the German jewelry industry. Other manufacturing includes rubber goods, nuclear technology, and precious-metal works. Chartered in 1303 and in the 16th century accepted refugees from the Low Countries who contributed significantly to the city's economic growth. Hanau passed to HESSE-KASSEL in 1736, and with it to PRUSSIA in 1866. The philologists Jakob and Wilhelm Grimm and the composer Paul Hindemith born here. Since almost complete destruction in World War II, the city has been rebuilt where possible according to its historical character.

Hanawa (hah-NAH-wah), village, East Shirakawa district, FUKUSHIMA prefecture, N central HONSHU, NE JAPAN, 53 mi/85 km S of FUKUSHIMA city; 36°57′N 140°72′E.

Hanayama (hah-NAH-yah-mah), village, Kurihara district, MIYAGI prefecture, N HONSHU, NE JAPAN, 37 mi/60 km N of SENDAI; 38°47′N 140°50′E.

Hanazono (hah-NAH-zo-no), town, Osato district, SAITAMA prefecture, E central HONSHU, E central JAPAN, 31 mi/50 km N of URAWA; 36°07′N 139°13′E. Pigs.

Hanazono (hah-NAH-zo-no), village, Ito district, WAKAYAMA prefecture, S HONSHU, W central JAPAN, 22 mi/36 km S of WAKAYAMA; 34°07′N 135°31′E.

Hancacato (ahn-kah-KAH-to), town and canton, CHALLAPATA O AVAROA province, ORURO department, W central BOLIVIA, 12 mi/20 km NE of Challapata, on the Tacagua River; 18°50′S 66°40′W. Elevation 12,188 ft/3,715 m. Tin-bearing lodes in area. Tin, lead, zinc, silver, and iron mining; clay, limestone. Agriculture (potatoes, yucca, bananas, wheat); cattle.

Hânceşti, town (2004 population 15,281), S central MOLDOVA, 18 mi/29 km SW of Chişinău (KISHINEV). Agricultural center; flour and oilseed milling, distilling. Until 1944, called Gancheshty; renamed Kotovskoye for Bolshevik military leader, Kotovski (b. here). Population 50% Jewish until World War II.

Hanceville (HANS-vil), town (2000 population 2,951), Cullman co., N central Alabama, 10 mi/16 km SSE of Cullman. Farm shipping center. Animal feeds, fertilizer, building materials, wood products. Seat of Wallace State Community College. First named 'Gilmer,' it was renamed for Horace Kinney, the first postmaster, who had come from Ireland. Inc. in 1878.

Hancheng (HAHN-CHENG), city (□ 626 sq mi/1,627.6 sq km; 2000 population 345,502), E SHAANXI province, NW central CHINA, 120 mi/193 km NE of XI'AN, and on the HUANG HE (Yellow River), near SHANXI border; 35°28′N 110°26′E. Heavy industry and agriculture are the largest sources of income. Crop growing, forestry, and animal husbandry. Main industries include coal mining; utilities; iron and steel. Tangjiacun (a village within Hancheng) is famous for its well-maintained and highly stylized residential complexes built between 1331 and 1851; each of them is square and built with gray bricks and shingles.

Han Chiang, CHINA: see HAN RIVER.

Hanchuan (HAHN-CHWAHN), town, ⊙ Hanchuan co., E central Hubei province, 30 mi/48 km Wof Wuhan, and on Han River; 30°39′N 113°46′E. Rice, cotton, sugarcane, oilseeds. Textiles; papermaking; food processing.

Hanchung, CHINA: see HANZHONG.

Hancock, county (□ 485 sq mi/1,261 sq km; 2006 population 9,677), E central GEORGIA, ⊙ SPARTA; 33°16′N 83°00′W. Bounded E by OGEECHEE RIVER, W by OCONEE River, intersected by fall line. Agriculture includes cotton, corn, forage, pecans; cattle; lumber and wood products. Formed 1793.

Hancock, county (□ 814 sq mi/2,116.4 sq km; 2006 population 19,091), W ILLINOIS; ⊙ CARTHAGE; 40°31′N 91°09′W. Bounded W by the MISSISSIPPI RIVER, here dammed into Lake Keokuk; drained by LA MOINE RIVER and BEAR CREEK. Agricultural area (corn, wheat, soybeans, fruit; cattle, hogs; dairying). Limestone quarries, sand pits. Some manufacturing. Formed 1825.

Hancock, county (□ 306 sq mi/795.6 sq km; 2006 population 65,050), central INDIANA; ⊙ GREENFIELD; 39°49′N 85°46′W. Drained by SUGAR CREEK, small Brandywine Creek, and BIG BLUE RIVER. Agricultural area (vegetables, corn, wheat; hogs, cattle; dairying). Part of INDIANAPOLIS metro area. Formed 1827.

Hancock, county (□ 573 sq mi/1,489.8 sq km; 2006 population 11,680), N IOWA; ⊙ GARNER, 43°05′N 93°43′W. Prairie agricultural area (cattle, hogs, poultry; corn, oats) drained by branches of IOWA RIVER and LIME CREEK. Contains small lakes (glacial origin); sand and gravel pits. Crystal Lake State Park in NW; Pilot Knob Lake and State Park in NE; Eagle Lake in center; Twin Lakes in S. General flooding here in 1993. Formed 1851.

Hancock, county (□ 198 sq mi/514.8 sq km; 2006 population 8,636), NW KENTUCKY; ⊙ HAWESVILLE; 37°51′N 86°46′W. Bounded N and NE by OHIO River (INDIANA state line); drained by Blackford Creek, forms part of W border. Agricultural area (burley tobacco, corn, hay, alfalfa, soybeans, wheat; cattle, hogs). Cannelton Locks and Dam on Ohio River in NE. Formed 1829.

Hancock, county (□ 2,351 sq mi/6,112.6 sq km; 2006 population 53,797), S and SE MAINE; ⊙ ELLSWORTH; 44°32′N 68°22′W. Gateway to MOUNT DESERT ISLAND and ACADIA NATIONAL PARK. Manufacturing (wood and paper products, machinery). Agriculture; dairying; granite quarrying; hunting and fishing. Bays, islands, and inland lakes are resort sites. Drained by PENOBSCOT RIVER.

Hancock, county (□ 552 sq mi/1,435.2 sq km; 2006 population 40,421), SE MISSISSIPPI, on MISSISSIPPI SOUND of Gulf of MEXICO (S); ⊙ BAY ST. LOUIS; 30°23′N 89°28′W. PEARL RIVER (Louisiana state line) on W; drained by Jordan and Wolf rivers. Agriculture (corn, cotton, pecans; timber); seafood. Beach resorts and casinos. Buccaneer State Park on coast in SE. Stennis Space Center (NASA) in W. Suffered widespread and catastrophic damage in Hurricane Katrina in August 2005. Formed 1812.

Hancock (HAN-kahk), county (□ 532 sq mi/1,383.2 sq km; 2006 population 73,824), NW OHIO; ⊙ FINDLAY; 41°01′N 83°40′W. Intersected by BLANCHARD RIVER. In the Till Plains physiographic region, except for the NW portion (in the Lake Plains region). Hogs, sheep, corn, soybeans, vegetables. Diversified manufacturing, especially at Findlay (food products, plastic products, household appliances, transportation equipment); limestone quarries. Includes Van Buren State Park. Formed 1820.

Hancock, county (□ 231 sq mi/600.6 sq km; 2006 population 6,713), NE TENNESSEE, on VIRGINIA (N) state line; ⊙ SNEEDVILLE; 36°32′N 83°13′W. Traversed by POWELL MOUNTAIN and other ridges of the APPALACHIAN MOUNTAINS; drained by CLINCH and POWELL rivers. Agriculture; diversified industry; natural resources. Formed 1844.

Hancock, county (□ 88 sq mi/228.8 sq km; 2006 population 30,911), N WEST VIRGINIA, at tip of NORTHERN PANHANDLE; ⊙ NEW CUMBERLAND; 40°31′N 80°34′W. Bounded N and W by OHIO River (OHIO state line), E by PENNSYLVANIA Industrial region, especially steel milling. Manufacturing at WEIRTON, NEWELL, and Chester. Clay and glass-sand pits. Agriculture (corn, wheat, oats, alfalfa, nursery crops). Includes Tomlinson Run State Park in NW; Hillcrest Wildlife Management Area in NE. Formed 1848.

Hancock, city (2000 population 4,323), HOUGHTON county, NW UPPER PENINSULA, N MICHIGAN, opposite HOUGHTON, on KEWEENAW WATERWAY (port facilities); 47°07′N 88°35′W. Light manufacturing; meat processing. Tourism; resort. Lift bridge connects it to Houghton. Houghton County Airport to NE. Seat of Suomi College. Historic Arcadian Copper Mines (tours). F. J. McLain State Park to N. Plotted 1859; incorporated as village 1875, as city 1903.

Hancock, town (2000 population 207), POTTAWATTAMIE county, SW IOWA, 26 mi/42 km ENE of COUNCIL BLUFFS; 41°23′N 95°21′W.

Hancock, resort town, HANCOCK county, S MAINE, on FRENCHMAN BAY, N of MOUNT DESERT ISLAND, and 8 mi/12.9 km E of ELLSWORTH; 44°31′N 68°16′W.

Hancock, town (2000 population 1,725), WASHINGTON county, W MARYLAND, on the POTOMAC River (bridged), and 25 mi/40 km W of HAGERSTOWN, at narrowest point of Maryland (c.2 mi/3.2 km wide); 39°42′N 78°10′W. Center for agriculture and timber. Hunting preserves nearby. Incorporated 1853.

Hancock, town, BERKSHIRE county, NW MASSACHUSETTS, 7 mi/11.3 km NNW of PITTSFIELD, near NEW YORK state line; 42°31′N 73°19′W. Resort; dairying.

Hancock, town, HILLSBOROUGH county, S NEW HAMPSHIRE, 28 mi/45 km W of MANCHESTER; 42°58′N 71°59′W. Drained by CONTOOCOOK RIVER. Manufacturing (wood products); timber; agriculture (vegetables, fruit; livestock, poultry; dairying). Powder Mill Pond in E; Nubanusit Lake on W border.

Hancock, town, ADDISON CO., central VERMONT, on WHITE RIVER, and 18 mi/29 km ESE of MIDDLEBURY, in GREEN MOUNTAINS; 43°55′N 72°55′W. Winter sports. Forest products. Named for John Hancock, signer of the American Declaration of Independence.

Hancock, village (2000 population 717), STEVENS county, W MINNESOTA, 9 mi/14.5 km SE of MORRIS; 45°30′N 95°47′W. Grain, sunflowers, soybeans; cattle, hogs; manufacturing (concrete products). Page Lake to N.

Hancock, summer-resort village (□ 1 sq mi/2.6 sq km; 2006 population 1,120), DELAWARE county, S NEW YORK, in the CATSKILL MOUNTAINS, at junction of East and West branches here forming the DELAWARE RIVER, 35 mi/56 km ESE of BINGHAMTON; 41°57′N 75°16′W. Manufacturing (medical supplies, electrical equipment, fabricated metal products, wood products, consumer goods, furniture). Incorporated 1888.

Hancock, village (2006 population 440), WAUSHARA county, central WISCONSIN, 24 mi/39 km SE of WISCONSIN RAPIDS; 44°07′N 89°31′W. In dairying and farming area, on Pine Lake. Agricultural research station and wildlife area nearby.

Hancock, Fort, NEW JERSEY: see SANDY HOOK.

Hancock, Lake (HAN-kahk), POLK county, central FLORIDA, 3 mi/4.8 km N of BARTOW; c.4 mi/6.4 km long, 3 mi/4.8 km wide; source of PEACE RIVER.

Hancock, Mount, peak (4,430 ft/1,350 m), GRAFTON county, central NEW HAMPSHIRE, 9 mi/14.5 km NE

of NORTH WOODSTOCK, in WHITE MOUNTAINS, in White Mountain National Forest.

Hancock, Mount, peak (10,214 ft/3,113 m), ROCKY MOUNTAINS, in S YELLOWSTONE NATIONAL PARK, TETON county, NW WYOMING, 10 mi/16 km S of YELLOWSTONE LAKE.

Hancocks Bridge, village, SALEM county, SW NEW JERSEY, on ALLOWAY'S CREEK, and 5 mi/8 km S of SALEM; 39°30′N 75°27′W. Hancock House (1734), site of massacre of Revolutionary troops by Tories, here.

Hand, county (□ 1,440 sq mi/3,744 sq km; 2006 population 3,323), central SOUTH DAKOTA; ⊙ MILLER; 44°32′N 99°00′W. Level agricultural region watered by Wolf, Turtle, and Sand creeks. Wheat, oats, rye, barley, corn; dairy products; cattle, hogs. Lake Louise State Recreation Area at center. Formed 1873.

Handa (HAHN-DAH), city, AICHI prefecture, S central HONSHU, central JAPAN, on the CHITA PENINSULA and CHITA BAY, 19 mi/30 km S of NAGOYA; 34°53′N 136°56′E. Dairying; vinegar. Was a commercial port during the Tokugawa era. Known for Handa-no-Anda Spring Festival.

Handa (HAHN-DAH), town, Mima district, E central SHIKOKU, W JAPAN, 31 mi/50 km W of TOKUSHIMA; 34°01′N 134°02′E.

Handan (HAHN-DAHN), city (□ 176 sq mi/457.6 sq km; 2000 population 1,769,315), SW HEBEI province, NE CHINA, on Beijing-Wuhan railroad, 35 mi/56 km S of XINTAI; 36°35′N 114°29′E. Its position as a communication and transportation center has led to significant industrial growth. Cotton, wheat, oilseeds; cotton textiles, electric and textile machinery, medicinal chemicals and chemical pesticides, instrument and meters; coal and iron-ore mining and processing. Sometimes spelled Hantan.

Handawor (hahn-DAH-wor), village, BARAMULA (or Kashmir North) district, JAMMU AND KASHMIR state, N INDIA, 14 mi/23 km NNW of BARAMULA. Rice, corn, wheat, oilseeds. Also spelled Handwara.

Handegg (HAHND-egg), waterfall (150 ft/46 m high), BERN canton, S central SWITZERLAND, on AARE RIVER, just N of GRIMSEL pass. Rapids.

Handen (HAHND-en), suburb, STOCKHOLM county, SWEDEN, 10 mi/16 km SSE of STOCKHOLM; 59°10′N 18°09′E.

Handeni (hahn-DAI-nee), town, TANGA region, NE TANZANIA, 80 mi/129 km WSW of TANGA; 05°27′S 38°01′E. Road junction. Rice, grain, sisal; livestock.

Handewitt (HAHN-duh-vit), village, SCHLESWIG-HOLSTEIN, N GERMANY, near Danish border, 6 mi/9.7 km WSW of FLENSBURG; 54°45′N 09°18′E.

Handia (HUHND-yah), town, ALLAHABAD district, SE UTTAR PRADESH state, N CENTRAL INDIA, 22 mi/35 km ESE of ALLAHABAD. Rice, gram, barley, wheat, oilseeds, sugarcane.

Handies Peak (14,048 ft/4,282 m), HINSDALE county, SW COLORADO, in SAN JUAN MOUNTAINS, 11 mi/18 km NE of SILVERTON.

Handley, village (2006 population 342), KANAWHA county, W central WEST VIRGINIA, on KANAWHA River, 17 mi/27 km SE of CHARLESTON; 38°11′N 81°22′W.

Handlova (hahn-DLO-vah), Slovak *Handlová,* Hungarian *Nyitrabánya,* town, STREDOSLOVENSKY province, W central SLOVAKIA, on railroad, 35 mi/56 km ESE of TRENČÍN; 48°44′N 18°46′E. Plastic manufacturing; power plants. Important lignite mines. Gothic church.

Handöl (HAHND-UHL), village, JÄMTLAND county, NW SWEDEN, on small lake, 12 mi/19 km ESE of STORLIEN; 63°16′N 12°25′E.

Handsboro, village, in E part of GULFPORT, HARRISON county, SE MISSISSIPPI, 8 mi/12.9 km W of BILOXI, on MISSISSIPPI SOUND. In shore-resort area. Now part of city of Gulfport.

Handsworth, ENGLAND: see BIRMINGHAM.

Handsworth, ENGLAND: see SHEFFIELD.

Handwara, INDIA: see HANDAWOR.

Haneda Airport (hah-NAI-dah), officially Tokyo International Airport, serving TOKYO, OTA ward, in the southest corner of Tokyo city, Tokyo prefecture, east central HONSHU, east central JAPAN, on TOKYO BAY, 10 mi/16 km south of Tokyo city center (road, rail links); elevation 36 ft/11m; 35°33′N 139°47′E. Haneda has three runways and three passenger terminals (over sixty-two million passengers annually), and is the fourth busiest airport in the world. The airport was opened in 1931, but with the opening of NARITA AIRPORT, Haneda has handled mostly domestic air traffic. Airport Code HND.

Hanerau-Hademarschen (HAH-ne-rou–HAH-de-mahr-shuhn), village, SCHLESWIG-HOLSTEIN, NW GERMANY, 14 mi/23 km ESE of HEIDE; 54°08′N 09°25′E. Residence (1879–1888) of Theodor Storm.

Haneti (hah-NAI-tee), village, DODOMA region, central TANZANIA, 50 mi/80 km N of DODOMA; 05°31′S 36°00′E. Cattle, sheep, goats; corn, wheat. Also called Waneta.

Haney, CANADA: see MAPLE RIDGE.

Hanford, city (2000 population 41,686), ⊙ KINGS county, central CALIFORNIA, 21 mi/34 km W of VISALIA; 36°20′N 119°39′W. Railroad junction. Agricultural, trade, and processing center of the San Joaquin Valley. Along with FRESNO, Hanford has prospered from the commercial enterprises and agricultural-related business. Rubber and oil companies here. Cotton; wheat, barley, pistachios, almonds, olives, cantaloupes, tomatoes, plums, peaches, nectarines; dairying; turkeys; flour mill. Lemoure Naval Air Station to W. TULARE LAKE irrigation reservoir to SW. Incorporated 1891.

Hanford Works, WASHINGTON: see RICHLAND.

Hang, MONGOLIA: see HANGAYN NURUU.

Hangae (HAHNG-GA), village, NARATHIWAT province, S THAILAND, in MALAY PENINSULA, on railroad, and 12 mi/19 km SW of NARATHIWAT (linked by highway). Locally known as Tanyong Mas or Tanyong Mat.

Hangal (HAHNG-guhl), town, DHARWAD district, KARNATAKA state, S INDIA, 50 mi/80 km S of DHARWAD; 14°46′N 75°08′E. Rice, millet, betel leaf, cotton. Irrigation canal nearby. Also spelled Hanagal.

Hangam, IRAN: see HENGAM.

Hanga Roa (AHNG-gah RO-ah), town, ⊙ Isla de Pascua comuna and Isla de Pascua province, VALPARAISO region, N central CHILE; 27°09′S 109°26′W. Principal settlement of EASTER ISLAND, which is all included in the Parque Nacional RAPA NUI.

Hangay Mountains, MONGOLIA: see HANGAYN NURUU.

Hangayn Nuruu (KHAHN-gein NOO-roo), massive series of mountain ranges in W central MONGOLIA, extending E-W for c.500 mi/800 km, with a width of about 125 mi/200 km, between 96°00′E–103°00′E and 46°00′N–49°00′N. Rises to 13,192 ft/4,021 m in Otgon-Tenger-Uul (peak). The N slopes are forested (larch, pine, birch, cedar). The S slopes are generally dry steppes. Many rivers, notably the ORHON and the SELENGA and their tributaries, rise on its N slopes. It forms part of the main continental divide of ASIA between rivers that drain to the ARCTIC OCEAN to the N and those that end in the interior drainage basins of central Asia to the S. Also spelled Khangai, Hangay Mountains.

Hangchow, CHINA: see HANGZHOU.

Hanggin Houqi (HAHNG-JEEN HO-CHEE), town, ⊙ Hanggin Houqi county, SW INNER MONGOLIA AUTONOMOUS REGION, N CHINA, 25 mi/40 km NW of LINHE, in HETAO oasis; 40°52′N 107°04′E. Agricultural settlement; cattle raising; grain, oilseeds, sugar beets.

Hangha (HAHNGH-hah), village, EASTERN province, E SIERRA LEONE, 7 mi/11.3 km NNE of KENEMA; 07°56′N 11°08′W.

Han Giang (HAHN YAHNG), river, MEKONG Delta, S VIETNAM, S branch of Mekong River; 10°00′N 105°46′E.

Hanging Hills, S central CONNECTICUT, small trap ridge just NW of MERIDEN, rising to 1,024 ft/312 m in West Peak. Area is state park, with recreational facilities.

Hanging Rock, village (2006 population 293), LAWRENCE county, S OHIO, on OHIO RIVER, 4 mi/6 km NW of IRONTON; 38°33′N 82°43′W.

Hangklip, Cape, WESTERN CAPE province, SOUTH AFRICA, on the ATLANTIC OCEAN, between FALSE (W) and SANDOWN (E) bays, 35 mi/56 km SE of CAPE TOWN, 20 mi/32 km E of CAPE OF GOOD HOPE; 34°24′S 18°49′E. Lighthouse situated on point. Notorious hideaway for runaway slaves and deserters in the late 18th century.

Hangö, FINLAND: see HANKO.

Hanguranketa, village, CENTRAL PROVINCE, SRI LANKA, 18 mi/29 km SE of KANDY; 07°10′N 80°46′E. Second residence of 17th-century Kandyan king. Ancient Buddhist temple with old manuscripts.

Hangzhou (HAHNG-JO), city (□ 166 sq mi/430 sq km; 1994 estimated urban population 1,184,300; estimated total population 1,412,700), ⊙ ZHEJIANG province, E CHINA, on the Fuchun (Qiantang) River, at the head of Hangzhou Bay, and c.100 mi/160 km SW of SHANGHAI; 30°15′N 120°10′E. Railroad hub with important links to Shanghai; handles river traffic through its port. A historically famous silk-producing center in a rich agricultural area, Hangzhou has recently been developed into a major industrial complex. Manufacturing includes silk and cotton textiles; iron and steel products; motor vehicles; pharmaceuticals; cement, rubber, paper, and bamboo products; chemicals; machine tools; electronic equipment; processed tea. One of the seven ancient capitals of China. Its charming natural setting on the shore of scenic Xihu (West Lake) amid high, wooded hills attracts many tourists. Other major points of interest include the Xi'an River, Thousand Island Lake, pagodas built during the Wu Kingdom (C.E. 220–280), Cliff Temple built during the Yuan dynasty (1206–1308), Lingyin Temple, and Running Tiger springs. Also famous for its Longjing [=dragon well] tea. Together with SUZHOU in JIANGSU province, Hangzhou has been called "paradise on earth" since ancient times, due to its ample agricultural supply, long history of civilization, and natural beauty.

Established as a township during the Qin, (or Chin) dynasty (221–206 B.C.E.); power base for seven regimes during the Southern Song dynasty (1132–1276). Was capital of both the Wu kingdom and the Southern Song dynasty. Many of the city's picturesque monasteries and shrines date from this period. Sacked by Kublai Khan during the Yuan dynasty. In the Southern Song period Hangzhou, rich with a thriving silk trade, was a center of art, literature, and scholarship and a cosmopolitan city with a large colony of foreign merchants—Arabs, Persians, and Nestorian Christians. Marco Polo, who visited it then, described it as the finest and noblest city in the world. Famous for its splendid buildings before its near destruction (1861) by Taiping rebels; it was subsequently rebuilt along mainly modern lines. Its modern prosperity dates from the opening of the Shanghai-Hangzhou-Ningbo railroad in 1909. Occupied by the Japanese 1937–1945; fell to the Communists in 1949. Seat of Zhejiang and Hangzhou universities, an agricultural institute, a medical college, and an institute of fine arts. Botanical gardens and an astronomical observatory also here. Hangzhou Bay, an arm of the EAST CHINA SEA, begins at the mouth of the Fuchun River. When the tide is coming in, the funnel shape of the bay creates a spectacular bore, 5 ft/1.5 m–15 ft/4.6 m high, which sweeps past Hangzhou, menacing shipping. Zhoushan Archipelago lies across the S entrance of the bay. Its ancient name is Liu'an. Sometimes spelled Hangchow.

Hanham Mount, ENGLAND: see KINGSWOOD.

Hä̈nheim (HUHN-heim), outer N suburb (□ 1 sq mi/ 2.6 sq km), of STRASBOURG, BAS-RHIN department, in ALSACE, E FRANCE, on MARNE-RHINE CANAL; 48°37′N 07°45′E. Brick- and tileworks.

Hania, Greece: see KHANIÁ.

Hanifa, Wadi (HAH-ni-fuh, WAH-dee), one of chief wadies of ARABIAN PENINSULA, SAUDI ARABIA, in former Nejd region, extending c.200 mi/322 km from NW of RIYADH, SE into the DAHNA desert.

Hanigovsky, SLOVAKIA: see LIPANY.

Hanish Islands (huhn-EESH), archipelago in RED SEA, 70 mi/113 km SSW of HODEIDA, belonging to YEMEN, also claimed by ERITREA; 13°45′N 42°45′E. Includes the larger Hanish Kabir (12 mi/19 km long) and the Hanish Saghir (4 mi/6.4 km long). Yemen deployed troops here for several years and announced plans to promote tourism. Eritrean troops attacked and seized the islands in 1995.

Hanita (khah-NEE-tah), kibbutz, 7 mi/11.3 km NE of NAHARIYA, in Upper Galilee, near Lebanese border, N ISRAEL; 35°05′N 35°10′E. Elevation 980 ft/298 m. Founded in 1938 and named for ancient Hanita, it was the first Jewish settlement in region. Stone tools and other relics were found from the ancient settlement dating back to Chalcolithic times. Also, Roman burial caves and remains of large late-Byzantine buildings, including a church. Agriculture; light manufacturing (metals factory; optical lenses). Just E of the kibbutz is the Hanita observation point (on Jabel el-Marad) and the Hanita nature reserve; to the W is the Nahal Hanita (Hanita stream) nature reserve.

Hanjam, IRAN: see HENGAM.

Hanjiaji, CHINA: see LINXIA.

Hanjiang (HAHN-JYAHNG), town, E FUJIAN province, SE CHINA, 3 mi/4.8 km SE of PUTIAN, on XINGHUA BAY; 25°33′N 119°08′E. Rice; processing of marine products.

Han Kiang, CHINA: see HAN RIVER.

Hankins, hamlet, SULLIVAN county, SE NEW YORK, on the DELAWARE RIVER (here forming PENNSYLVANIA state line), and 17 mi/27 km W of LIBERTY; 41°49′N 75°05′W. In summer recreational area.

Hankinson, town (2006 population 1,018), Richland co., SE North Dakota, 20 mi/32 km SW of Wahpeton. Dairy products; livestock; grain; 46°04′N 96°53′W. Railroad junction. Site of convent and Academy of Sisters of St. Francis. Willard and Grass lakes to SW. (Sisseton) Wahpeton Indian Reservation to S. Founded in 1886 and named for Col. Richard Henry Hankinson (1841–1911).

Hanko (HAHNG-ko), Swedish *Hangö* (HAHNG-uh), city, UUDENMAAN province, SW FINLAND, at tip of 15-mi/24-km-long Hanko peninsula, on BALTIC SEA, and 50 mi/80 km SE of TURKU; 59°50′N 22°57′E. Important seaport kept ice-free and used year round. Manufacturing town and railroad terminus. Popular seaside resort. Offshore are remains of fortress (1789). Leased to former USSR for thirty years as a naval base after the Finnish-Russian War (1939–1940). The USSR exchanged it for a fifty-year lease on the PORKKALA district in 1944 but evacuated in 1956. City incorporated 1874.

Hankou (HAHN-ko), former city, E HUBEI province, central CHINA. Part of the WUHAN conurbation since 1950. Built on an alluvial plain on the left banks of both the HAN RIVER and CHANG JIANG (Yangzi River), it was the largest city in the conurbation and contains its port, a major facility handling oceangoing vessels. Has many industries. Hankou owes much of its development to the Beijing-Guangzhou railroad, which crosses the Chang Jiang here. The city opened as a treaty port in 1862, was held (1938–1945) by the Japanese, and passed to the Chinese Communists in 1949. Linked by bridges and ferries with HANYANG and WUCHANG.

Han Krum (HAHN KROOM), village, VARNA oblast, VELIKI PRESLAV obshtina, E BULGARIA, on the GOLYAMA KAMCHIYA river, at the mouth of the Vrana River, 5 mi/8 km S of SHUMEN; 43°03′N 27°54′E. Railroad junction; railroad construction, wine making; vineyards. Tobacco experimental station. Formerly called Chatali. Sometimes spelled Khan Krum.

Hanks, village, WILLIAMS CO., NW NORTH DAKOTA, 33 mi/53 km NNW of WILLISTON; 48°36′N 103°47′W. Ranching in vicinity. Lake Zahl Wildlife Refuge to E. Founded in 1916 and named for banker, W.F. Hanks. Post office closed in1964.

Hanksville, unincorporated village (2006 population 203), WAYNE county, S central UTAH, 50 mi/80 km SSW of GREEN RIVER town, and 50 mi/80 km W of LOA, on FREMONT RIVER S of its junction with MUDDY RIVER, where the two form the DIRTY DEVIL RIVER; elevation c.4,300 ft/1,310 m. Cattle; tourism. Settled 1882. Crossroads for national park area, with Capitol Reef (W) and CANYONLANDS (E) national parks in area, Glen Canyon National Recreation Area to SE, and Goblin Valley State Park to N.

Hanley (HAN-lee), town (2006 population 464), S central SASKATCHEWAN, W CANADA, 35 mi/56 km SSE of SASKATOON; 51°37′N 106°26′W. Grain elevators; dairying.

Hanley, ENGLAND: see STOKE-ON-TRENT.

Hanley Falls, village (2000 population 323), YELLOW MEDICINE county, SW MINNESOTA, 9 mi/14.5 km SW of GRANITE FALLS, on YELLOW MEDICINE RIVER; 44°41′N 95°37′W. Railroad junction. Corn, oats, soybeans; livestock. Wood Lake to E.

Hanley Hills, town (2000 population 2,124), SAINT LOUIS county, E MISSOURI, residential suburb 8 mi/12.9 km NW of downtown ST. LOUIS; 38°40′N 90°19′W.

Hanlontown, town (2000 population 229), WORTH county, N IOWA, near LIME CREEK, 14 mi/23 km SW of NORTHWOOD; 43°16′N 93°22′W. Peat mining and processing.

Hanmer Springs, N CANTERBURY region, SOUTH ISLAND, NEW ZEALAND, 92 mi/148 km N of CHRISTCHURCH. The largest source of thermal water in South Island, in upland intermontane basin, used as vacation and health resort (for nervous disorder treatment); springs on fault line.

Hann (AHN), suburb, DAKAR administrative region, W SENEGAL, on Hann Bay, on Dakar-Niger railroad, and 3 mi/4.8 km NE of DAKAR; 14°43′N 17°26′E. Beach; botanic and zoological gardens.

Hanna (HA-nuh), town (□ 3 sq mi/7.8 sq km; 2001 population 2,986), SE central ALBERTA, W CANADA, 35 mi/56 km ENE of DRUMHELLER; 51°38′N 111°55′W. Coal mining; dairying; beef cattle, hogs, poultry, sheep, lamb; mixed farming; flour milling; wheat, rye, oats; oil and gas. Incorporated as a village in 1912, and as a town in 1914.

Hanna, town (2006 population 857), CARBON county, S WYOMING, 35 mi/56 km E of RAWLINS; 41°52′N 106°33′W. Elevation c.6,777 ft/2,066 m. Coal mines, one of most productive in area. Nearby is site of Fort Halleck, military post established 1862 on Overland stage route. SEMINOE RESERVOIR to NW. Settled 1887, incorporated 1890.

Hanna, village, LA PORTE county, NW INDIANA, 13 mi/21 km SSW of LA PORTE. Railroad junction. Dairying; fruit. Laid out 1858.

Hanna, village (2006 population 134), MCINTOSH county, E OKLAHOMA, 20 mi/32 km NNW of MCALESTER, on CANADIAN River (upper reach of EUFAULA LAKE); 35°12′N 95°53′W. In agricultural area.

Hanna City, village (2000 population 1,013), PEORIA county, central ILLINOIS, 11 mi/18 km W of PEORIA; 40°41′N 89°47′W. In agricultural (corn, soybeans; cattle) and bituminous-coal-mining area.

Hannaford, village (2006 population 157), GRIGGS CO., E central NORTH DAKOTA, 10 mi/16 km S of COOPERSTOWN, and on Bald Hill Creek; 47°18′N 98°11′W. Railroad junction. Lake Ashtabula Reservoir to SE. Founded in 1886 and incorporated in 1906. Named for Jules M. Hannaford, a president of the Northern Pacific Railroad.

Hannah, village (2006 population 17), CAVALIER CO., N NORTH DAKOTA, port of entry 25 mi/40km NW of LANGDON, near Canadian (border; 48°58'N 98°41'W. Railroad terminus. Rush Lake to S. Founded in 1884 and named for Frank Hannah, the postmaster's father-in-law.

Hannastown (HA-nuhz-toun), unincorporated village, WESTMORELAND county, SW PENNSYLVANIA, 5 mi/8 km NNE of GREENSBURG, on Crabtree Creek; 40°21'N 79°29'W. Colonial-era county seat and center of dispute between PENNSYLVANIA and VIRGINIA over possession of WESTMORELAND territory, resolved 1779 by extension of MASON-DIXON LINE W. Destroyed 1782 by Native American raid. Old Hanna's Town historic site.

Hannibal (HAN-nuh-buhl), city (2000 population 17,757), MARION and RALLS counties, NE MISSOURI, on the MISSISSIPPI RIVER; 39°42'N 91°22'W. River port and shipping center. Industries include meat canning, printing, and the manufacturing of boats, electronics, and lumber and metal products. Corn, soybeans; dairying; cattle, hogs. Major tourist destination. Railroad junction; railroad and highway bridges. Famous as the boyhood home of Mark Twain; his house has been preserved, and a museum, statue, lighthouse, and a bridge across the Mississippi commemorate him; the famous Mark Twain cave is S of city. Seat of Hannibal–La Grange College. City spared during 1993 floods. Plotted 1819, incorporated 1845.

Hannibal, village (□ 1 sq mi/2.6 sq km; 2006 population 525), OSWEGO county, N central NEW YORK, 10 mi/16 km SW of OSWEGO; 43°19'N 76°34'W. In dairy and fruit area.

Hannibal Lock and Dam, WETZEL county, N WEST VIRGINIA, and MONROE county, SE OHIO, on OHIO River, at Hannibal, Ohio, and 2 mi/3.2 km N (upstream) from NEW MARTINSVILLE, West Virginia; 42 ft/13 m high; the reservoir created has a maximum water storage capacity of 130,000 acre-ft. Built (1974) by the Army Corps of Engineers for navigational purposes; originally called Lock and Dam No. 14.

Hanno (hahn-NO), city, SAITAMA prefecture, E central HONSHU, E central JAPAN, on the Naguri River, 19 mi/30 km W of URAWA; 35°51'N 139°19'E. Lumber.

Hannover, GERMANY: see HANOVER.

Hannoversch-Münden, town, LOWER SAXONY, central GERMANY, 9 mi/14.5 km NE of KASSEL; 51°26'N 09°39'E. Wood- and metalworking, manufacturing (synthetic fiber); tourism. WERRA and FULDA rivers meet here, forming WESER RIVER. Town first mentioned 1183; has 16th-century castle, church (13th–15th century), town hall with Renaissance facade (1603–1619).

Hannsdorf, CZECH REPUBLIC: see HANUSOVICE.

Hannut (ah-NOO), Flemish *Hannuit*, commune (2006 population 14,336), Waremme district, LIÈGE province, E central BELGIUM, 13 mi/21 km NW of HUY; 50°40'N 05°05'E. Market center.

Hano, Indian pueblo, NAVAJO county, NE ARIZONA, atop a mesa in Hopi Indian Reservation, 3 mi/4.8 km N of Polacca, on Wepo Wash, c.65 mi/105 km NNE of WINSLOW; elevation c.6,200 ft/1,890 m. Founded in 17th century by immigrant Tewa-speaking Native Americans who have since adopted Hopi customs.

Hanö Bay (HAHN-UH), inlet of BALTIC SEA, S SWEDEN. Chief ports are SIMRISHAMN (at S end) and ÅHUS; 35 mi/56 km long, 15 mi/24 km wide. Receives Helgeån River.

Hanoi (ha-NOI), city, ⊙ VIETNAM, in N part of the country, on the right bank of the RED RIVER; 21°05'N 105°51'E. Transportation hub and the administrative center of the country and former capital of NORTH VIETNAM. Has two airports and railroad connections to KUNMING (CHINA), as well as to the main Chinese system centering on BEIJING; also linked by railroad with HAI PHONG and HO CHI MINH CITY. Manufacturing includes both light and heavy industry such as machine tools, plywood, textiles, chemicals, matches, tires, construction materials, food products, and handicrafts. Known for its European-style public squares and tree-lined boulevards. Became (7th century) the seat of the Chinese rulers of Vietnam. Its Chinese name, Dong Kinh or TONG KING, became TONKIN and was applied by Europeans to the entire region of N Vietnam. Occupied briefly by the French in 1873; passed to them ten years later. Became capital of FRENCH INDOCHINA after 1887. The French developed Hanoi industrially, centering railroad repair shops and small processing industries here. Occupied by the Japanese 1940–1945; then became the seat of the Vietnam government. From 1946 to 1954, it was the scene of heavy fighting between the French and Viet Minh forces. After the French evacuated Hanoi in accordance with the GENEVA Conference (July 1954), the city became the North Vietnamese capital. Under the N Vietnamese, it was greatly expanded industrially. During the Vietnam War its transportation facilities were continually disrupted by the bombing of bridges and railroads, which were, however, immediately repaired. The city remained remarkably intact despite heavy U.S. bombings, although widespread destruction occurred during the massive attacks of December 18–30, 1972, when many non-military targets, including the French embassy and large residential areas, were hit. Much of the civilian population had been evacuated and factories had been dismantled and reassembled in rural areas. After the cease-fire, much of the machinery was returned and functioned again in ruined structures. Hanoi became capital of the Socialist Republic of Vietnam on July 2, 1976. In the 1980s, Hanoi underwent significant change. Its administrative boundaries were extended to include a number of rural districts (e.g., GIA LAM) that together comprise the Thanh Po Ha Nai (Hanoi urban region), allowing for coordinated development plans of the entire metropolitan area. The population has considerably increased and is likely to continue to do so as rural migrants make their way to the capital complex. A new road network between Hanoi and the Gia Lam airport has been built, in addition to ongoing efforts to regrade existing transportation networks. The West Lake (Ho Tay) area has undergone enormous change in the 1990s, with the development of the West Lake Resort and wealthy housing enclave. The city's industrial base is quite diversified, with a number of joint venture industrial parks completed (e.g., Daewoo Industrial Park) and more being planned. Alongside these changes, the booming tourist industry has stimulated the construction of an increasing number of luxury hotels, along with the restoration of still-impressive French colonial buildings for growing numbers of resident foreign business people. Residential housing for locals remains a problem, but plans to expand housing along the outer perimeter of the city center are underway. Cultural center; seat of the National University (formerly Hanoi University), National History Museum, Revolution Museum, Temple of Literature (Van Mieu), Mot Cot Pagoda, and the Temple of the Trung Sisters. Many of the key cultural sites here are scheduled for renovation and historic preservation.

Hanoi Hilton, historic site, HANOI city, N VIETNAM. Nickname given to Hoa Lo Prison, where American prisoners of war were held during the Vietnam War (1965–1975) Conditions were poor with bad food and nickname refers to the Hilton Hotel chain in the U.S. It was built in 1904 by the French to house political prisoners when Vietnam was part of French Indochina (1887–1954). Most of the prison was demolished for a high rise but a section was left as a museum.

Hanoura (hah-NO-oo-rah), town, Naka district, TOKUSHIMA prefecture, E SHIKOKU, W JAPAN, 9 mi/15 km S of TOKUSHIMA; 33°57'N 134°37'E. Ayu; cucumbers, strawberries. Lumber products, house fittings.

Hanover (hahn-NO-fer), German *Hannover*, former independent kingdom and former province of Germany, LOWER SAXONY, NW GERMANY. Very irregular in outline, Hanover stretched from the DUTCH border and the NORTH SEA in the NW to the HARZ Mountains in the SE. Most of the territory was included in the duchy of BRUNSWICK, which the house of the Guelphs retained after 1180. In 1692 Duke Ernest Augustus of Brunswick-Lüneburg was raised to the rank of elector, choosing Hanover city as his capital; the totality of his holdings became known as the electorate of Hanover. The marriage of Ernest Augustus to Sophia, granddaughter of James I of ENGLAND, brought (1714) the English throne to his son, Elector George Louis (George I of England). In 1815 the Congress of Vienna raised Hanover to a kingdom, with membership in the GERMAN CONFEDERATION. At the accession (1837) of Queen Victoria in England, Hanover was separated from the British crown because of the Salic law of succession. Refusal to support PRUSSIA in the Austro-Prussian War (1866) resulted in Prussian annexation; became province. After World War II the province was incorporated into Lower Saxony.

Hanover (HA-no-vuhr), rural municipality (□ 286 sq mi/743.6 sq km; 2001 population 10,789), SE of WINNIPEG, SE MANITOBA, W central CANADA; 49°28'N 96°50'W. Land set aside in 1873 as East Reserve for Mennonite settlements. Diversified agriculture. Commuter community for Winnipeg. Includes communities of Mitchell, Blumenort, Grunthal; surrounds city of STEINBACH. Incorporated 1881.

Hanover (HA-no-vuhr), county (□ 474 sq mi/1,232.4 sq km; 2006 population 98,983), E central VIRGINIA; ⊙ HANOVER; 37°45'N 77°29'W. Partly in the PIEDMONT region (W), with coastal plain (E); bounded NE by NORTH ANNA and PAMUNKEY rivers, S by CHICKAHOMINY River; drained by SOUTH ANNA River. Agricultural area (tobacco, soybeans, barley, wheat, corn, peanuts, hay; cattle, poultry; dairying). Many Civil War battle sites (at COLD HARBOR, Gaines's Mill, MECHANICSVILLE). Cold Harbor National Cemetery, part of RICHMOND NATIONAL BATTLEFIELD PARK in SE. Formed 1720.

Hanover (hahn-NO-fer), German *Hannover*, city (2005 population 515,729), ⊙ LOWER SAXONY, N GERMANY, on the LEINE RIVER and the MIDLAND CANAL; 52°22'N 09°44'E. Major industrial, commercial, and trans-shipment center, also serving as a vital railroad and road junction in N Germany. Manufacturing includes iron and steel, tires, machinery, and motor vehicles. The world's largest annual industrial fair (*Hannover Messe*) and computer expo (*CeBIT*) held here every April. Chartered in 1241 and in 1369 passed to BRUNSWICK. In 1386 it joined the HANSEATIC LEAGUE. In 1692 it became capital of the electorate (from 1815 kingdom; from 1866 province) of Hanover. Badly damaged in World War II, but after 1945 numerous old buildings were reconstructed and many modern structures were erected. Points of interest include the Gothic former city hall (15th century); the Marktkirche (14th century), a red-brick church with a high (318 ft/97 m) tower; the Leineschloss (17th century), a château that now houses the parliament of Lower Saxony; and the remains of Herrenhausen castle (17th century). Seat of technical, medical, and veterinary university and several museums. Has numerous parks and gardens. Elector Ernest Augustus, his wife Sophia, and their son, George I of ENGLAND, are buried here.

Hanover (HA-no-vuhr), town (□ 4 sq mi/10.4 sq km; 2001 population 6,869), S ONTARIO, E central CANADA, on SAUGEEN RIVER, and 28 mi/45 km S of OWEN SOUND; 44°09'N 81°02'W. Woodworking, light manufacturing. Dairying.

Hanover, town, NORTHERN CAPE province, SOUTH AFRICA, 40 mi/64 km SE of DE AAR, on junction of N10 and main route (N1) between CAPE TOWN and JOHANNESBURG; elevation 4,922 ft/1,500 m; 31°04'S

24°12′E. Sheep; feed crops, fruit. Airfield on farm by that name.

Hanover, town (2000 population 2,834), JEFFERSON county, SE INDIANA, near OHIO River, 5 mi/8 km WSW of MADISON; 38°43′N 85°28′W. Agricultural area. Manufacturing (medical supplies), meat processing. Seat of Hanover College. Laid out 1832.

Hanover, town, OXFORD county, W MAINE, on the ANDROSCOGGIN RIVER, and 21 mi/34 km NNW of SOUTH PARIS; 44°29′N 70°44′W.

Hanover, town, PLYMOUTH county, E MASSACHUSETTS, on NORTH RIVER, and 20 mi/32 km SE of BOSTON; 42°07′N 70°52′W. Manufacturing rubber, tools, machinery; binding and publishing. Dairying; poultry. Settled 1649, incorporated 1727.

Hanover, town (2000 population 1,355), HENNEPIN and WRIGHT counties, E MINNESOTA, suburb 24 mi/39 km NW of downtown MINNEAPOLIS, on CROW RIVER; 45°09′N 93°39′W. Grain; livestock, poultry; dairying; manufacturing (fabricated metal and wood products). Crow Hassan Park Reserve to NE; several small natural lakes in area.

Hanover, town, GRAFTON county, SW NEW HAMPSHIRE, 4 mi/6.4 km NW of LEBANON, bounded W by CONNECTICUT RIVER (VERMONT state line). Light manufacturing. Seat of Dartmouth College and a cultural and recreational center. Dartmouth Hitchcock Medical Center. North Peak (2,300 ft/701 m) in NE; APPALACHIAN TRAIL crosses town NE to W center (crosses bridge to Vermont). Includes village of Etna, 3 mi/4.8 km E of town center, manufacturing (mobile classrooms). Settled 1765, incorporated 1769.

Hanover, township, MORRIS county, N NEW JERSEY, 5 mi/8 km NE of MORRISTOWN; 40°49′N 74°25′W. Incorporated 1798.

Hanover (HA-no-vuhr), township, LUZERNE county, NE central PENNSYLVANIA, suburb 2 mi/3.2 km SW of downtown WILKES-BARRE, on SUSQUEHANNA RIVER, in WYOMING VALLEY. In declining anthracite-coal-mining region. Surrounds boroughs of WARRIOR RUN and SUGAR NOTCH.

Hanover, village (2000 population 836), JO DAVIESS county, NW ILLINOIS, on APPLE RIVER (bridged here), and 13 mi/21 km SSE of GALENA; 41°58′N 88°08′W. In agricultural area (dairying; cattle, hogs, ducks; oats); manufacturing (transportation equipment). Savanna Army Depot to SW on MISSISSIPPI RIVER.

Hanover, village (2000 population 653), WASHINGTON county, NE KANSAS, on LITTLE BLUE RIVER, and 11 mi/18 km NW of MARYSVILLE; 39°53′N 96°52′W. Railroad junction and Pony Express Station to E.

Hanover, village (2000 population 424), JACKSON county, S MICHIGAN, 13 mi/21 km SW of JACKSON; 42°06′N 84°32′W. In diversified farming area. Machining.

Hanover, village, FERGUS county, central MONTANA, on Big Spring Creek, and 7 mi/11.3 km NW of LEWISTOWN. Gypsum rock deposits.

Hanover, unincorporated village, GRANT county, SW NEW MEXICO, in foothills of PINOS ALTOS MOUNTAINS, 11 mi/18 km ENE of SILVER CITY; elevation 6,340 ft/1,932 m. Gold, silver, copper mines in vicinity. BLACK PEAK is 8 mi/12.9 km NNW, Gila National Forest to N. Santa Rita Open Pit Copper Mine is here.

Hanover (HAN-o-vuhr), village (□ 1 sq mi/2.6 sq km; 2006 population 975), LICKING county, central OHIO, 8 mi/13 km E of NEWARK; 40°04′N 82°16′W.

Hanover (HA-no-vuhr), unincorporated village, ☉ HANOVER county, E central VIRGINIA, near PAMUNKEY River, 15 mi/24 km N of RICHMOND; 37°45′N 77°22′W. Agriculture (grain, soybeans; poultry, cattle). Patrick Henry lived here and pleaded (1763) his first important case in the county courthouse (c.1735; extant). Nearby hamlet of Hanovertown nearly became capital of Virginia in 1751.

Hanover (HA-no-vuhr), borough (2006 population 15,015), YORK county, SE PENNSYLVANIA, 17 mi/27 km SW of YORK; 39°48′N 76°59′W. Railroad junction. Large and varied industrial base includes food processing (especially pretzels) and manufacturing (apparel, machinery, tool and die, fabricated metal products, food products, plastics products, chemicals, paper products; printing and publishing). Agriculture (grain, apples, soybeans; livestock, poultry; dairying). Standardbred horses raised here (many famous trotters have "Hanover" in their names). A cavalry action preceding the battle of Gettysburg was fought here in June 1863. Nearby are Devener (N) and HANOVER (W) airports. Codorus State Park, on Lake Marburg reservoir, to E. Incorporated 1815.

Hanover, parish (□ 177 sq mi/460.2 sq km; 2001 population 67,037), CORNWALL county, NW JAMAICA, on W promontory of the island; ☉ LUCEA; 18°25′N 78°08′W. E boundary along Great River. Principal agricultural crops (rice, bananas, lime, yams, breadfruit). Lucea and GREEN ISLAND ship bananas. Near Lucea are phosphate deposits.

Hanover, CONNECTICUT: see SPRAGUE.

Hanover Island (HAN-o-vuhr), off coast of S CHILE, between JORGE MONTT (S) and CHATHAM (NE) islands, 95 mi/153 km NW of PUERTO NATALES; 40 mi/64 km long, 5 mi/8 km—22 mi/35 km wide; 51°00′S 74°40′W.

Hanover Park, village (2000 population 38,278), COOK and DU PAGE counties, NE ILLINOIS, suburb 27 mi/43 km WNW of downtown CHICAGO, 2 mi/3.2 km SW of SCHAUMBURG; 41°59′N 88°08′W. Manufacturing (fabricated metal products, electrical and electronic equipment, water softeners).

Hanoverton (ha-NO-vuhr-tuhn), village (2006 population 387), COLUMBIANA county, E OHIO, 23 mi/37 km E of CANTON; 40°45′N 80°56′W.

Hanovertown, Virginia: see HANOVER.

Han Pijesak (HAHN PYE-sahk), town, central BOSNIA, BOSNIA AND HERZEGOVINA, 7 mi/11.3 km S of VLASENICA, and S of JAVOR MOUNTAINS; 44°05′N 18°57′E. Also spelled Khan Piyesak.

Han River (HAHN), river, central CHINA, c.700 mi/1,127 km long; rises in SW SHAANXI province, flows E between the QINLING and the DABA mountains, then SE through HUBEI province, past Shiyan and Xiangfan, to join CHANG JIANG (Yangzi River) at WUHAN; navigable for about 300 mi/483 km upstream. The river floods its fertile lower valley in summer. There is a hydroelectric power station near Xiangfan (Hubei province). The name sometimes appears as Han Chiang, Han Kiang, or Han Shui.

Han River, Korean *Han-gang*, central SOUTH KOREA, 292 mi/470 km long; rises in mountains c. 100 mi/161 km E of SEOUL; flows SW, and turns generally NW past CHUNGJU, yoju, and Seoul to YELLOW SEA at KANGHWA Island (nearly connected to mainland). Navigable 185 mi/298 km by small craft. Has hydroelectric plant in upper course; drains large agricultural area. It was an important defense line in the seesaw fighting (1950–1951) during the Korean War. Principal tributary, called Pukhan (or North Han) River, is c. 110 mi/177 km long.

Hansa, Brazil: see CORUPÁ.

Hanság (HAHN-shahg), Hungarian *Hanság*, central swampy district (□ 218 sq mi/566.8 sq km), of the Little Alföld, NW HUNGARY; 34 mi/55 km long, 11 mi/18 km wide; fed by IKVA and RÉPCE rivers, drained by Fertő Canal. Peat cutting.

Hansa-Park, GERMANY: see SIERKSDORF.

Hansboro, village (2006 population 7), TOWNER CO., N NORTH DAKOTA, port of entry 33 mi/53 km N of CANDO, near Canadian (Man.) border; 48°57′N 99°22′W. Founded in 1905 and incorporated in 1917. Named for Henry Clay Hansbrough (1848–1933) who served as U.S. Senator (1891–1909).

Hanseatic League (han-see-A-tik), initially an alliance of BALTIC SEA seaports and trading states formed in the mid-13th century at LÜBECK. Empowered by control of Baltic fisheries, Europe's richest; ties to the Teutonic Order, expanding into central European hinterland; and redirection of Russian trade from CONSTANTINOPLE to Baltic Sea. Over ninety member states exercised a trade monopoly from the NORTH SEA to RUSSIA at its mid-14th century peak. Then in four divisions: Prussia/Livonia (chief town, DANZIG), Wendic (chief town, Lübeck, includes Mecklenburg and Pomerania), Westphalia (chief town, COLOGNE), and Saxony (chief town, BRUNSWICK). Major trade stations at LONDON, BRUGES (Belgium), BERGEN (Norway), STOCKHOLM (Sweden), and NOVGOROD (Russia). The League fell into decline by the mid-15th century as fish stocks moved (possibly for climatological reasons) from the Baltic to North Sea areas it did not control and as the power of the Teutonic Order waned.

Hansell, town (2000 population 96), FRANKLIN county, N central IOWA, 5 mi/8 km E of HAMPTON; 42°45′N 93°05′W. Livestock; grain.

Hansen, town (2000 population 970), TWIN FALLS county, S IDAHO, 8 mi/12.9 km E of TWIN FALLS, and on SNAKE RIVER; 42°32′N 114°18′W. Elevation 4,000 ft/1,220 m. Irrigated farming (sugar beets, potatoes, fruit, hay, grain; cattle).

Hansen Dam, California: see TUJUNGA CREEK.

Hansford, county (□ 920 sq mi/2,392 sq km; 2006 population 5,237), extreme N TEXAS, on OKLAHOMA (N) state line; ☉ SPEARMAN; 36°16′N 101°20′W. In high grassy plains of the PANHANDLE; elevation 3,000 ft/914 m–3,800 ft/1,158 m. Drained by Coldwater and PALO DURO creeks. Wheat-producing area; also grain, sorghums, corn, wheat; beef cattle, some sheep. Natural gas wells, some oil; stone; helium. Palo Duro reservoir in E. Formed 1876.

Hanshou (HAHN-SHO), town, ☉ Hanshou county, N HUNAN province, S central CHINA, on DONGTING LAKE, at delta of YUAN River, and 25 mi/40 km ESE of CHANGDE; 28°55′N 111°58′E. Rice, oilseeds, jute, cotton.

Han Shui, CHINA: see HAN RIVER.

Hansi (HAHN-see), town, HISAR district, HARYANA state, N INDIA, 14 mi/23 km ESE of HISAR. Trades in cotton, oilseeds, salt, grain; cotton ginning, handloom weaving; metalwork. Captured 1192 by Kutb-ud-din, general of Muhammad Ghori; at end of eighteenth century held briefly by George Thomas, noted adventurer.

Hansjö (HAHN-SHUH), village, KOPPARBERG county, central SWEDEN, near N end of Lake Orsasjön, 10 mi/16 km NNE of MORA; 61°09′N 14°36′E.

Hanska (HAN-skuh), village (2000 population 443), BROWN county, S MINNESOTA, 12 mi/19 km S of NEW ULM; 44°08′N 94°29′W. Grain, soybeans, peas; livestock, poultry; dairying. Manufacturing (fertilizers and feeds). Lake HANSKA to W, Linden Lake to E.

Hanska, Lake (HAN-skuh), BROWN county, S MINNESOTA, 13 mi/21 km SW of NEW ULM; 7.5 mi/12.1 km long, 0.5 mi/0.8 km wide; 44°08′N 94°36′W. Dam at SE end raises lake level; drains 6 mi/9.7 km through channel to Butterfield Creek, then 2 mi/3.2 km, to WATONWAN RIVER. Lake Hanska County Park at SE end.

Hans Lollik Islands (HANZ LAW-lik), two islets and several rocks just off N SAINT. THOMAS Island, U.S. VIRGIN ISLANDS, 4 mi/6.4 km NNE of CHARLOTTE AMALIE. Hans Lollik Island proper (489 acres/198 ha; elevation 713 ft/217 m) is at 18°24′N 64°54′W; Little Hans Lollik is just N.

Hanslope (HANZ-luhp), agricultural village (2001 population 2,215), Milton Keynes, central ENGLAND, 4 mi/6.4 km N of WOLVERTON; 52°06′N 00°49′W. Has 15th-century church.

Hanson, county (□ 435 sq mi/1,131 sq km; 2006 population 3,690), SE central SOUTH DAKOTA; ☉ ALEXANDRIA; 43°39′N 97°47′W. Agricultural and livestock-raising region drained by JAMES RIVER. Dairy products; corn, wheat; poultry, cattle, hogs. Formed 1871.

Hanson, town (2000 population 625), HOPKINS county, W KENTUCKY, 7 mi/11.3 km N of MADISONVILLE; 37°25′N 87°28′W. Coal mining; agriculture (tobacco, grain; livestock).

Hanson, residential town, PLYMOUTH county, E MASSACHUSETTS, 25 mi/40 km SSE of BOSTON; 42°03′N 70°53′W. Manufacturing (flourescent lighting); cranberries. Settled 1632, set off from PEMBROKE 1820.

Hansot (HAHN-sot), town, BHARUCH district, GUJARAT state, W INDIA, 13 mi/21 km SW of BHARUCH; 21°35′N 72°48′E. Local market for wheat, millet, fish.

Hanspach, CZECH REPUBLIC: see LIPOVA.

Hanstedt (HAHN-stet), village, LOWER SAXONY, N GERMANY, 20 mi/32 km S of HAMBURG; 53°31′N 10°00′E.

Hanston, village (2000 population 259), HODGEMAN county, SW central KANSAS, on Buckner Creek, and 31 mi/50 km W of LARNED; 38°07′N 99°42′W. Grain; livestock.

Han-sur-Lesse (HAHN–syoor–LES), village, in commune of ROCHEFORT, Dinant district, NAMUR province, SE BELGIUM, on LESSE RIVER, and 15 mi/24 km SE of DINANT; 50°08′N 05°11′E. Tourist resort. Here, river flows through series of limestone caves. The Domain of the Grottoes of Han theme park.

Hansweert (HAHN-svairt), town, ZEELAND province, SW NETHERLANDS, in SOUTH BEVELAND region, 6 mi/9.7 km SE of GOES, on the EASTERN SCHELDT estuary, at S end of SOUTH BEVELAND CANAL; 51°27′N 04°00′E. Vehicle ferry to Perkpolder, on S shore, 2 mi/3.2 km to SE. Dairying; cattle; grain, vegetables, sugar beets. Light manufacturing.

Hantan, CHINA: see HANDAN.

Hantengri (HAHN-TENG-GREE), peak (22,949 ft/6,995 m), in central TIANSHAN mountain system, on CHINA-KAZAKHSTAN-KYRGYZSTAN border, 100 mi/161 km E of ISSYK-KOL (lake); 42°13′N 80°11′E. Thought to be highest peak of Tianshan until discovery of POBEDA PEAK. Sometimes called Khan Tengri.

Hanthawaddy (hahn-thah-WAH-dee), formerly southernmost district (□ 1,927 sq mi/5,010.2 sq km) of BAGO division, MYANMAR, on GULF OF MARTABAN; ⊙ YANGON. Split by RANGOON RIVER; drained by numerous creeks; mud flats, scrub vegetation on coast. Intensive rice-growing area; fisheries; forests (government reserves) in W. Served by Thongwa-Pegu railroad and TWANTE CANAL. Separated in 1879 from Rangoon district.

Hants, county (□ 1,229 sq mi/3,195.4 sq km; 2001 population 40,513), central NOVA SCOTIA, E CANADA, on the BAY OF FUNDY; ⊙ WINDSOR. Split from Kings county in 1781.

Hants, ENGLAND: see HAMPSHIRE.

Hantsport, town (2001 population 1,202), W central NOVA SCOTIA, E CANADA, on AVON RIVER, near its mouth on the MINAS BASIN, and 5 mi/8 km NNW of WINDSOR; 45°04′N 64°11′W. Paper manufacturing, fruit packing; gypsum quarrying.

Hanumangarh (huh-noo-MAHN-guhr), town, GANGANAGAR district, N RAJASTHAN state, NW INDIA, 125 mi/201 km NNE of BIKANER; 29°35′N 74°19′E. In canal-irrigated area; market center for millet, wheat, gram, wool, livestock; hand-loom cotton and woolen weaving. Railroad junction just NW. Has noted fort. Formerly called BHATNAIR.

Hanumannagar (huh-noo-MAHN-nuh-guhr), village, SE NEPAL, in the TERAI on SAPTA KOSI RIVER, and 11 mi/18 km E of RAJBIRAJ; 26°30′N 86°51′E. Rice, jute, corn, oilseeds. Annual fair.

Hanusovce nad Toplou (hah-NUH-shou-TSE NAHD top-LYOO), Slovak *Hanušovce nad Topl'ou*, Hungarian *Tapolyhanusfalva*, town, VYCHODOSLOVENSKY province, E SLOVAKIA, on TOPLA RIVER, on railroad, and 12 mi/19 km ENE of PREŠOV; 49°02′N 21°30′E. Manufacturing of building materials; wheat, barley, rape. Has three churches; one museum.

Hanusovice (HAH-nuh-SHO-vi-TSE), Czech *Hanušovice*, German *Hannsdorf*, town, SEVEROMORAVSKY province, NW MORAVIA, CZECH REPUBLIC, on MORAVA RIVER, and 8 mi/12.9 km NNW of Šumperk; 50°05′N 16°57′E. Railroad junction. Manufacturing (machinery, tractors, building materials); spinning mill (flax) and brewery (established 1874). Has a 17th-century church.

Hanuy Gol, river, 260 mi/421 km long, W central MONGOLIA; rises in the HANGAYN NURUU (Khangai Mountains); flows NNE joining the SELENGA River at 49°20′N 102°24′E. Has a drainage basin of 5,640 sq mi/14,600 sq km.

Hanwell, ENGLAND: see EALING.

Hanworth, ENGLAND: see HOUNSLOW.

Hanyang (HAHN-YAHNG), former city, now part (since 1950) of WUHAN conurbation, E HUBEI province, central CHINA, on right bank of HAN RIVER, at its junction with CHANG JIANG (Yangzi River). Heavy industrial center. Founded during the Sui dynasty (C.E. 581–618). Linked by bridge and ferries with Hankou.

Hanyin (HAHN-YEEN), town, ⊙ Hanyin county, S SHAANXI province, NW central CHINA, 32 mi/51 km NW of ANKANG; 32°53′N 108°37′E. Rice, wheat, oilseeds. Textiles, chemicals, electronics; food processing.

Hanyu (HAHN-yoo), city, SAITAMA prefecture, E central HONSHU, E central JAPAN, 22 mi/35 km N of URAWA; 36°10′N 139°33′E.

Hanyuan (HAHN-yuh-WAHN), town, ⊙ Hanyuan county, SW SICHUAN province, SW CHINA, 50 mi/80 km SE of KANGDING, and on highway; 29°21′N 102°43′E. Rice, wheat. Food and beverages; iron-ore and non-ferrous ore mining.

Hanzan (HAHN-zahn), town, Ayauta district, KAGAWA prefecture, NE SHIKOKU, W JAPAN, 12 mi/20 km S of TAKAMATSU; 34°15′N 133°51′E.

Hanzhong (HAHN-JUNG), city (□ 215 sq mi/559 sq km; 2000 population 441,706), SW SHAANXI province, NW central CHINA, on the HAN RIVER, near SICHUAN border; 33°09′N 107°03′E. Heavy industry and agriculture. Crop growing (grain); animal husbandry (hogs). Manufacturing includes food processing, pharmaceuticals and machinery. N terminal of the ancient plank roads that connect Shaanxi and Sichuan. Formerly called Hancheng.

Hao (HOU), atoll, central TUAMOTU ARCHIPELAGO, FRENCH POLYNESIA, S PACIFIC; 18°13′S 140°54′W. Used and modified as logistic support base for French nuclear testing operations on nearby MURUROA atoll, starting in 1962; under the Rarotonga and Comprehensive Nuclear Test Ban treaties, facilities have since been dismantled, except for those that may be reused, such as the freshwater production system at Hao. Formerly Bow Island, Harp Island (so named for its shape).

Haofel (hah-fel), part of JERUSALEM, ISRAEL. Site of the original Yebusite settlement captured by David.

Haora (huh-O-rah), district (□ 566 sq mi/1,471.6 sq km), S WEST BENGAL state, E INDIA; ⊙ HAORA. Bounded E by HUGLI RIVER, W by RUPNARAYAN RIVER; drained by the DAMODAR RIVER. Alluvial soil (swamps toward W); rice, jute, pulses, potatoes, coconuts, betel nuts. Heavy industrial area centered around Haora. Many iron- and steel-rolling works, factories, repair shops, paper factories; cotton and jute milling, glass manufacturing; motor vehicle manufacturing and railroad workshops at LILUAH. Engineering college, Indian Botanical Gardens at SIBPUR.

Haora (huh-O-rah), city, ⊙ HAORA district, WEST BENGAL state, E INDIA, on the HUGLI RIVER, opposite CALCUTTA; 22°35′N 88°20′E. At the region's main railroad terminus. Nerve center of trade and manufacturing; second only to CALCUTTA in the Hugliside industrial complex. Produces textiles, glass, hosiery, cigarettes, paper, batteries, and products made from

jute, cotton, and wood. Numerous engineering works. Acute traffic congestion; recent planning has improved only a small section of the city, near the railroad station (one of INDIA's largest). Also spelled HOWRAH.

Haora Salkia, INDIA: see SALKHIA.

Haouz (HAWZ), lowland of SW MOROCCO, just N of the High ATLAS mountains, surrounding MARRAKECH. A former saline lake, it was filled by alluvium and is drained by the Oued TENSIFT. Olive and date palm groves, vineyards irrigated from Oued N'FISS; Lalla Takerkoust barrage is 19 mi/31 km SSW of Marrakech.

Hapai, TONGA: see HA'APAI.

Haparanda (HAH-PAH-rahn-dah), town, NORRBOTTEN county, NE SWEDEN, on GULF OF BOTHNIA, at mouth of TORNEÄLVEN RIVER, 55 mi/89 km ENE of LULEÅ; 65°50′N 24°06′E. Frontier station on Finnish border, opposite TORNIO (FINLAND). Railroad junction. Manufacturing (wood products, foods, metal goods; printing). Meteorological observatory. Reception station for refugees, wounded, and prisoners of war in both world wars. Incorporated as town 1842.

Hapchon (HAHP-CHUHN), county (□ 380 sq mi/988 sq km), NW SOUTH KYONGSANG province, SOUTH KOREA. Mountainous in W; Hwang River in E flows into NAKDONG RIVER. Agriculture (rice, barley, potatos, cotton, vegetables, medical herbs, sedge); pottery; tile. Expressway 88 passes through county in N KAYA MOUNTAIN NATIONAL PARK; Haein Buddhist Temple. Hapchon dam on upstream Hwang River.

Hapert (HAH-puhrt), village, NORTH BRABANT province, S NETHERLANDS, 10 mi/16 km SE of EINDHOVEN; 51°22′N 05°16′E. Dairying; cattle, hogs; grain, vegetables; manufacturing (metalware, concrete, trailers, furniture, dairy products).

Hapeville (HAIP-vil), railroad town and ATLANTA suburb (2000 population 6,180), FULTON county, NW central GEORGIA, 7 mi/11.3 km SSW of Atlanta; 33°40′N 84°25′W. Motor vehicle-assembly and airplane-rebuilding plants; printing; sawmilling. Adjacent to Atlanta Hartsfield International Airport, which has expanded into the community, acquiring many homes and businesses. Incorporated 1891.

Happisburgh (HAPS-buh-ruh), village (2001 population 2,290), NE NORFOLK, E ENGLAND, on NORTH SEA, 17 mi/27 km NE of NORWICH; 52°49′N 01°32′E. Seaside resort and small fishing port. Has 15th-century church.

Happurg (HAHP-poorg), village, BAVARIA, S central GERMANY, in MIDDLE FRANCONIA, 18 mi/29 km E of NUREMBERG; 49°29′N 11°30′E.

Happy, village (2006 population 619), SWISHER county, NW TEXAS, on LLANO ESTACADO, c.40 mi/64 km N of PLAINVIEW; 34°44′N 101°51′W. In wheat, corn, cotton, and cattle area.

Happy Valley, town, S central NEWFOUNDLAND AND LABRADOR, NE CANADA, 5 mi/8 km SE of GOOSE BAY, on N bank of CHURCHILL RIVER; 53°18′N 60°18′W. Connected by road to Goose Bay and air base. Established 1970 as residential support center for military base. Has also served to support hydro projects on Churchill River Amalgamated as Happy Valley–Goose Bay, 1975.

Happy Valley, town (2006 population 9,945), CLACKAMAS county, NW OREGON, residential suburb 7 mi/11.3 km SE of downtown PORTLAND; 45°26′N 122°32′W. Agriculture to E and SE. Dairying; poultry; fruit, grain; nurseries.

Happy Valley-Goose Bay, town, E NEWFOUNDLAND AND LABRADOR, CANADA, on Goose Bay, inlet of Lake Melville, at mouth of Churchill (Hamilton) River; 53°19′N 60°25′W. The large air base and radio station, built here in World War II (1941) as military and ferrying base, is now used by Canadian commercial aircraft on transatlantic routes. Lumbering. Staging center for Churchill River hydro plan. Goose Bay consolidated with Happy Valley in 1975.

Hapsal, ESTONIA: see HAAPSALU.

Hapsburg, SWITZERLAND: see HABSBURG.

Hapton (HAP-tuhn), village (2001 population 6,084), E LANCASHIRE, N ENGLAND, 3 mi/4.8 km W of BURNLEY; 53°47′N 02°19′W. Some manufacturing.

Hapur (hah-POOR), city (2001 population 211,983), GHAZIABAD district, NW UTTAR PRADESH state, N CENTRAL INDIA, 18 mi/29 km SSE of MEERUT; 28°43′N 77°47′E. Railroad and road junction. Trades in wheat, millet, sugarcane, oilseeds, cotton, bamboo; cotton ginning, manufacturing of chemical fertilizer. Has 17th-century mosque. Harpur-made "Papad" (food) nationally known.

Haputale (HUH-puh-THUH-lee), town, UVA PROVINCE, S central SRI LANKA, in UVA BASIN, 18 mi/29 km SE of NUWARA ELIYA; 06°45′N 80°57′E. Tea-transport center; tea processing. Just W is scenic Haputale Gap (elev. 4,583 ft/1,397 m).

Haquira (hah-KEE-rah), town, COTABAMBAS province, APURÍMAC region, S central PERU, in the ANDES Mountains, 65 mi/105 km SE of ABANCAY; 14°13′S 72°11′W. Elevation 11,975 ft/3,650 m.

Hara (hah-RAH), village, Suwa district, NAGANO prefecture, central HONSHU, central JAPAN, 47 mi/75 km S of NAGANO; 35°57′N 138°13′E. Vegetables.

Hara, JAPAN: see HARAMACHI.

Haraa Gol (KHAH-rah GOL), river, 180 mi/290 km long, N MONGOLIA; rises in the HENTIYN NURUU (Kentei Mountains), 25 mi/40 km WNW of ULAANBAATAR; flows generally N, past DARHAN, to ORHON RIVER, 50 mi/80 km SSW of SÜHBAATAR. Used for logging for 100 mi/160 km above mouth. Also spelled Hara Gol, Khara Gol; also called Hara River.

Har Adar, Jewish settlement, 11 mi/18 km WNW of JERUSALEM, 3.1 mi/5 km N of ABU GHOSH, in JUDEAN HIGHLANDS, WEST BANK. Name derives from adjacent Givat Haradar (Radar Hill), a Jordanian outpost in the 1948 war. Founded in 1986, it is a commuter suburb of Jerusalem.

Haradh (HUH-ruhd), village, NW YEMEN, 15 mi/24 km E of Midi, at ASIR border; 16°26′N 43°05′E.

Haradh (HAHR-uhd), oil field in AL AHSA region, EASTERN PROVINCE, SAUDI ARABIA, 160 mi/257 km SSW of DHAHRAN, and on railroad and highway to RIYADH; 24°18′N 49°04′E. Discovered in 1949.

Haragauli (hah-rah-gah-OO-lee), town, W central GEORGIA, on railroad, and 30 mi/48 km SE of KUTAISI. Food processing. Formerly ORDZHONIKIDZE.

Hara Gol, MONGOLIA: see HARAA GOL.

Harahan (HA-ruh-han), city (2000 population 9,885), JEFFERSON parish, SE LOUISIANA, a suburb 8 mi/13 km W of downtown NEW ORLEANS, on the MISSISSIPPI RIVER; 29°57′N 90°12′W. Agricultural shipping center; manufacturing (fabricated metal products, building materials, machinery, soft drinks, paper products, foods, chemicals). Huey P. Long Bridge over the MISSISSIPPI is nearby.

Hara Hira, MONGOLIA: see HARHIRA UUL.

Haraiya (huh-REI-yah), town, tahsil headquarters, BASTI district, NE UTTAR PRADESH state, N central INDIA, 17 mi/27 km W of BASTI; 26°48′N 82°28′E. Rice, wheat, barley, sugarcane. Sometimes spelled HARRAIYA.

Haralson (HER-uhl-suhn), county (□ 285 sq mi/741 sq km; 2006 population 28,616), NW GEORGIA; ⊙ BUCHANAN; 33°47′N 85°13′W. Bounded W by ALABAMA state line. Piedmont area drained by TALLAPOOSA River. Agriculture includes corn; cattle, hogs, poultry. Textile manufacturing and sawmilling. Formed 1856.

Haralson (HER-uhl-suhn), village (2000 population 144), COWETA county, W GEORGIA, 16 mi/26 km SE of NEWNAN; 33°14′N 84°34′W. Manufacturing of surveillance equipment.

Haramachi (hah-RAH-mah-chee), city, FUKUSHIMA prefecture, N central HONSHU, NE JAPAN, on the PACIFIC OCEAN, 31 mi/50 km E of FUKUSHIMA city; 37°38′N 140°57′E. Watches. Sometimes called Hara.

Haramosh Range, PAKISTAN and INDIA: see KAILAS-KARAKORAM RANGE.

Haramsøya (HAH-rahms-uh-yah), island (□ 5 sq mi/13 sq km) in NORTH SEA, MØRE OG ROMSDAL county, W NORWAY, one of the Nord Islands, 12 mi/19 km NNE of Ålesund; c.5 mi/8 km long, 2 mi/3.2 km wide. Fisheries; some manufacturing in village of Austnes.

Haramukh (HUH-ruh-mook), peak (c.16,000 ft/4,875 m) of GREATER HIMALAYAS, JAMMU AND KASHMIR state, N INDIA, 22 mi/35 km N of SRINAGAR. WULAR LAKE is to the W.

Haran, village, SE TURKEY, 24 mi/39 km SE of SANLIURFA; 36°51′N 39°00′E. Nearby hydroelectric developments of the Southeast Anatolia Project, the centerpiece of which is the ATATÜRK DAM AND RESERVOIR, have led to some growth in the area. Factories have been built recently to produce cotton textiles; also processing of irrigated fruits and vegetables. Known for villagers' unusual beehive-shaped dwellings made of mud. An important ancient Mesopotamian center on the trade route from NINEVEH to CARCHEMISH and the seat of the Assyrian moon god. The Babylonians defeated the Assyrian army here in 609 B.C.E. Frequently mentioned in the Bible, it was the home of Abraham's family after the migration from Ur. The Greek form of the name is Charan or Charran; in Roman times it was Carrhae. Also known as Helenopolis; also spelled Harran.

Haranhalli (HAH-ruhn-huh-lee), town, HASSAN district, KARNATAKA state, S INDIA, 5 mi/8 km SSW of ARSIKERE; 13°15′N 76°14′E. Cotton ginning. Kaolin workings nearby. Also spelled Harnahalli.

Hara Nuur, MONGOLIA: see HAR NUUR.

Haraoti (hah-ROU-tee), the country of the Haras, a tribe of Chauhan Rajputs, in SE RAJASTHAN state, NW INDIA; comprises area of former RAJPUTANA STATES of BUNDI, KOTAH, and JHALAWAR.

Harappa (hah-RAH-pah), village, SAHIWAL district, SE PUNJAB province, central PAKISTAN, on old bed of RAVI River (S of present course), and 13 mi/21 km W of SAHIWAL; 30°38′N 72°52′E. Noted site of several successive cities of Indus Civilization (4000 B.C.E.–1500 B.C.E.). Prehistoric nature estimated early 1920s; excavations have uncovered brick walls and dwellings, cemetery, copper, bronze implements, pictographic seals, many terra-cotta objects; antiquities in nearby museum.

Harar (HAH-rahr), city (2007 population 127,000), E central ETHIOPIA; ⊙ HARARI state, in highlands at edge of the GREAT RIFT VALLEY, 225 mi/362 km E of ADDIS ABABA; 09°19′N 42°07′E. Elevation c.6,000 ft/1,830 m. Commercial and trade center for coffee, cereals, and cotton. Fruit, tobacco, hides, and oilseed pressing are also important. An ancient walled city, it was probably founded in the 7th century. After 1520, the Somali conqueror Ahmad Gran made it the capital of a large Muslim state, but an invasion by the Oromo people brought an end to its political power (1577). The city maintained a precarious independence until its occupation by Egypt (1875–1885). Incorporated 1887 into ETHIOPIA by Menelik II. The Harari inhabitants of the city are a distinctive Ethiopian group who speak a Semitic language, but write in Arabic. Long a center of Muslim learning; considered Islam's 4th-holiest city. Bombed and occupied by Italians in 1936 in the Italo-Ethiopian War, the city was taken by British forces during World War II (1941). Newer sections have grown outside the walls along the road to DIRE DAWA. Major tourist destination. Also spelled Harer and Harrar.

Harar, ETHIOPIA: see HARERGE.

Harardera (hah-RAHR-de-rah), village, E central SOMALIA, 12 mi/19 km from INDIAN Ocean, 65 mi/105 km SW of OBBIA. Wood carving. Has native forts.

Harare (hah-RAH-rai), city (1992 urban population 1,184,169; 2002 urban population 1,444,534) and province (□ 337 sq mi/876.2 sq km; 2002 population

1,444,534), ⊙ ZIMBABWE, NE Zimbabwe; 17°49′S 31°02′E. Elevation 4,865 ft/1,483 m. Zimbabwe's largest city and its administrative, commercial, and communications center. With a major commercial airport, it is also a major railroad and highway junction, connected by railroad with BULAWAYO, BEIRA (MOZAMBIQUE), and SOUTH AFRICA. Has a mild climate; trade center for an agricultural region (tobacco, maize, cotton, citrus fruits, vegetables, coffee, tea, soybeans; cattle, sheep, goats, hogs, poultry; dairying). Manufacturing (clothing, processed food, pharmaceuticals, farm supplies, motor vehicles, tobacco products, photographic supplies, leather products, aircraft engines, steel, beverages, furniture, construction materials; printing and publishing). Gold is mined in the area. Founded in 1890 as a fort by the Pioneer Column, a mercenary force organized by Cecil J. Rhodes to seize MASHONALAND. Originally named Salisbury after River A. Salisbury, then British prime minister. Became a municipality in 1897 and a city in 1935. Was capital of the Federation of RHODESIA AND NYASALAND (1953–1963). After World War II the population grew as people migrated here. Continued as Rhodesian capital after the breakup of the federation in 1964. The name was changed to Harare in 1980 when Zimbabwe gained independence from BRITAIN. Seat of the University of Zimbabwe, Gwebi College of Agriculture, the School of Social Work, Harare School of Art, Harare Polytechnic College, and Zimbabwe College of Music. Site of the National Gallery, which has collections of African soapstone carvings, and of the National Museum and Zimbabwe Museum of Human Sciences, known for its archaeological holdings. There are several agriculture research institutes here. LAKE CHIVERO and LAKE MANYAME reservoirs (both recreational parks) are to W; Ewarigg Botanical Garden is to NE. Formerly part and capital of MASHONALAND EAST province, became its own province in 1997; formerly capital of MASHONALAND CENTRAL province.

Harari (HAH-rah-ree), state (2007 population 203,000), E ETHIOPIA; ⊙ HARAR; 09°19′N 42°07′E. It is a small state that only includes the city of HARAR and nearby surrounding areas. The ethnically distinct Harari people comprise most of the region's population, with almost 60% living in the city of HARAR.

Hara River, MONGOLIA: see HARAA GOL.

Harasta (hah-rahs-tah), village, DAMASCUS district, SW SYRIA, on Damascus-Homs road, and 5 mi/8 km NE of Damascus; 33°33′N 36°22′E. Located within the Damascus metropolitan area, in the fertile GHUTA belt; melons, olives, grapes.

Harat, AFGHANISTAN: see HERAT.

Harato, town (2007 population 6,254), OROMIYA state, W central ETHIOPIA, 45 mi/72 km NE of NEKEMTE; 09°22′N 37°07′E.

Hara Usa, MONGOLIA: see HAR US NUUR.

Haraz-Djombo (hah-RAHZ–jom-BO), town and former military outpost, BATHA administrative region, central CHAD, 90 mi/145 km NE of ATI.

Haraz Mangueigne (hah-RAHZ mahng-GAIN-YUH), village, SALAMAT administrative region, SE CHAD, 85 mi/137 km ESE of AM-TIMAN, near border with CENTRAL AFRICAN REPUBLIC; 09°55′N 20°48′E. Livestock; millet. Reserve de Faune de L'Aouk-Aoukale is here.

Harbel (HAHR-bel), town, MARGIBI county, W LIBERIA, on FARMINGTON RIVER, and 32 mi/51 km E of MONROVIA; 06°17′N 10°21′W. Headquarters of extensive rubber plantations. Agriculture research station. Operations at the rubber plantations have been affected by rebel warfare that started in 1989.

Harberton (HAHR-buh-tuhn), village (2001 population 1,250), S DEVON, SW ENGLAND, 2 mi/3.2 km SW of TOTNES; 50°23′N 03°43′W. Market center. Fine medieval church. Has 16th-century inn.

Harbertón, ARGENTINA: see PUERTO HARBERTÓN.

Area is shown by the symbol □, and capital city or county seat by ⊙.

Harbin (HAHR-bin), Russian *Kharbin*, city (□ 632 sq mi/1637 sq km; 1994 estimated urban population 2,505,200; estimated total population 2,888,100), ⊙ HEILONGJIANG province, NE CHINA, on the SONGHUA RIVER, and c.140 mi/225 km NNE of CHANGCHUN; 45°45′N 126°39′E. The major trade and communications center of the central NORTHEAST, the junction of the two most important railroads in the Northeast, and the main port on the Songhua River. Part of the great NE industrial complex of metallurgical, machinery, chemical, petroleum, and coal industries, Harbin also has railroad shops, food-processing establishments (soybeans are a major commodity), and plants making tractors, turbines, boilers, precision instruments, electrical and electronic equipment, cement, and fertilizer. Developed rapidly after RUSSIA was granted a concession (1896–1924) and built a modern section alongside the old Chinese town. Flooded by White Russian refugees after 1917, Harbin had one of the largest European population in East Asia. Most of the Europeans left the city following the rise to power of the Chinese Communists. Seat of Harbin Polytechnical University, a medical college, and several technical institutes. The Sun Island in the Songhua River is a famous recreational location. The name may appear as Ha'erbin or Ha-erpin.

Harbine (HAHR-bein), village (2006 population 55), JEFFERSON county, SE NEBRASKA, 13 mi/21 km WSW of BEATRICE; 40°11′N 96°58′W.

Harbison Canyon, unincorporated town (2000 population 3,645), SAN DIEGO county, S CALIFORNIA; residential suburb 19 mi/31 km NE of downtown SAN DIEGO; 32°49′N 116°50′W. Sycuan Indian Reservation to S.

Harbledown, ENGLAND: see CANTERBURY.

Harbor, unincorporated town (2000 population 2,622), CURRY county, SW OREGON, 53 mi/85 km SW of GRANTS PASS, and 1 mi/1.6 km E of BROOKINGS, on PACIFIC OCEAN, at mouth of Chetco River; 42°02′N 124°15′W. Fish. Manufacturing (seafood processing).

Harbor Beach, city (2000 population 1,837), HURON county, E MICHIGAN, 17 mi/27 km E of BAD AXE, on LAKE HURON; 43°51′N 82°39′W. Railroad terminus. Resort center; manufacturing (food processing, motor vehicle wiring; harnesses, chemicals). Settled 1837; incorporated as village 1882, as city 1909.

Harbor City, suburban section of LOS ANGELES, LOS ANGELES county, S CALIFORNIA, 18 mi/29 km S of downtown Los Angeles, in harbor district. Manufacturing (consumer goods, fabricated rubber products, dental equipment, plastic products, transportation equipment, chemicals). PALOS VERDES PENINSULA to SW.

Harbor Country, term for a group of 8 small towns, SW MICHIGAN, including Grand Union, MICHIANA, NEW BUFFALO (main center), THREE OAKS, Union Pier, LAKESIDE, Harbert, and Sawyul. Upscale summer communities, marinas, hotels. Area was formerly a farming and logging community; today, there are still many working farms and orchards.

Harborcreek, unincorporated town, ERIE county, NW PENNSYLVANIA, residential suburb 7 mi/11.3 km E of ERIE, and 1 mi/1.6 km SE of LAKE ERIE; 42°09′N 79°57′W. Manufacturing includes plastic molding. Agriculture includes dairying; livestock; hay, corn, apples, grapes. PRESQUE ISLE Wine Cellars and Moorhead Airport to NE.

Harbord (HAHR-buhrd), town, E NEW SOUTH WALES, AUSTRALIA, suburb of SYDNEY, on PACIFIC OCEAN; 33°47′S 151°18′E. Seaside resort.

Harbor Island, BAHAMAS: see HARBOUR ISLAND.

Harbor Island, Texas: see ARANSAS PASS.

Harborne, ENGLAND: see BIRMINGHAM.

Harbor Springs, town (2000 population 1,567), EMMET county, NW MICHIGAN, on LITTLE TRAVERSE BAY of LAKE MICHIGAN, and 4 mi/6.4 km NNW of PETOSKEY; 45°25′N 84°59′W. Lumbering; boatyards, fisheries; agriculture (potatoes, apples, grain; livestock); manufacturing (electrical enclosures, fabricated metal products). Boyne Highlands Ski Area to N; Nubs Nob Ski Area to NE. Year-round resort. Incorporated as village 1881, as city 1932.

Harbor View (HAHR-buhr VYOO), village (2006 population 2,000), LUCAS county, NW OHIO, on Maumee Bay, 9 mi/14.4 km E of TOLEDO, across MAUMEE RIVER; 41°42′N 83°26′W.

Harbour au Bouche, CANADA: see HAVRE BOUCHER.

Harbour Buffet (BUH-fit), former fishing village, SE NEWFOUNDLAND AND LABRADOR, E CANADA, on SE coast of LONG ISLAND, in PLACENTIA BAY, 17 mi/27 km NNW of Argentia; 47°32′N 54°04′W. Residents moved to Arnold's Cove in the 1960s as part of a government relocation program.

Harbour Grace, town (□ 21 sq mi/54.6 sq km; 2001 population 3,380), SE NEWFOUNDLAND AND LABRADOR, E CANADA, on CONCEPTION BAY; 47°41′N 53°15′W. Leading fishing port with fish-processing plants. Settled c.1550, one of the oldest towns in the province. Airport nearby.

Harbour Island, islet and district (□ 2 sq mi/5.2 sq km; 2000 population 1,639), central BAHAMAS, off NE tip of ELEUTHERA Island, 33 mi/53 km NE of NASSAU; 25°30′N 76°38′W. The small island is one of the most densely populated in the archipelago; produces tomatoes, pineapples, coconuts. The chief settlement, DUNMORE TOWN, has a fine harbor. Harbour Island was one of the first in the Bahamas to be settled by buccaneers. Its fine beaches attract many tourists. Also spelled Harbor Island.

Harbour Island, largest of PENGUIN ISLANDS, just off S Newfoundland, NEWFOUNDLAND AND LABRADOR, E CANADA; 1 mi/2 km long; 47°23′N 56°59′W.

Harbu, town (2007 population 13,548), AMHARA state, central ETHIOPIA, 25 mi/40 km SE of DESSIE; 10°47′N 39°50′E.

Harburg (HAHR-boorg), district of Hamburg, N GERMANY, port on the ELBE RIVER; 53°33′N 10°00′E. Refined petroleum and rubber goods are produced here. Formerly an independent town, incorporated 1937 into HAMBURG.

Harburg (HAHR-boorg), town, BAVARIA, S GERMANY, in SWABIA, on the Wörnitz River, 6 mi/9.7 km NW of DONAUWÖRTH; 48°47′N 10°41′E. Cement works. On nearby hill is a 17th-century castle with 13th-century watchtower. Limestone quarries in area. Chartered c.1250.

Harbury (HAH-buh-ree), village (2001 population 2,485), SE central WARWICKSHIRE, central ENGLAND, 5 mi/8 km SE of ROYAL LEAMINGTON SPA; 52°18′N 01°31′W. Cement works. Has 13th-century church.

Harchoka, INDIA: see BHARATPUR.

Harcourt, town (2000 population 340), WEBSTER county, central IOWA, 16 mi/26 km S of FORT DODGE; 42°15′N 94°10′W. In agricultural area.

Harcourt (HAHR-kort), village, central VICTORIA, AUSTRALIA, 105 mi/169 km NW of MELBOURNE, near CASTLEMAINE, at foot of Mount Alexander; 37°00′S 144°15′E. In fruit-growing area (especially apples); granite quarries; vineyards.

Harcuvar Mountains, in LA PAZ and YAVAPAI counties, W ARIZONA, c.40 mi/64 km W of WICKENBURG; rise to c.5,000 ft/1,525 m.

Hard (HAHRD), township, VORARLBERG, W AUSTRIA, on Lake Constance, between mouths of BREGENZERACH and RHINE rivers, and 3 mi/4.8 km W of BREGENZ; 47°30′N 09°41′E. Manufacturing of metals, foodstuffs (cheese); textiles (silk, velvet).

Harda (HUHR-dah), town (2001 population 61,712), ⊙ HARDA district, S central MADHYA PRADESH state, central INDIA, in fertile NARMADA RIVER valley, 50 mi/80 km SW of HOSHANGABAD; 22°20′N 77°06′E. Agriculture trade center (wheat, millet, cotton, oil-seeds); brass ware. Part of HOSHANGABAD district until 1998.

Hardangerfjorden (hahr-DAHNG-uhr-fyawr-uhn), second-largest fjord of NORWAY, penetrating 114 mi/183 km from the ATLANTIC OCEAN into HORDALAND county, SW NORWAY. A S branch, the SØRFJORD, cleaves the HARDANGERVIDDA, a barren and rocky plateau, for 25 mi/40 km; at its head are the village of ODDA and the famous SKJEGGEDALSFOSS, a waterfall 525 ft/160 m high. An E branch, the Eidfjord, extends 15 mi/24 km to the quaint village of VIK near the Vøringfoss, a waterfall 535 ft/163 m high. The valleys of the Hardangerfjord are fertile and dotted with picturesque villages. Embroidery and violin making are traditional home industries. The region is a favorite tourist area. Extending inland from the fjord is the Hardangerfjell, a mountain mass rising to 6,153 ft/1,875 m in the Hardangerjøkel.

Hardangervidda (hahr-DAHNG-uhr-VID-dah), mountain plateau (□ c.2,500 sq mi/6,475 sq km), HORDALAND and BUSKERUD counties, S NORWAY, between SØRFJORDEN (W), GEILO (NE), and Tinn (SE); average elevation c.3,500 ft/1,067 m. Europe's largest barren bedrock area, extending c.100 mi/161 km between head of HARDANGERFJORDEN and the HALLINGDAL. Grooved by deep, steep-sided valleys; contains numerous lakes; traversed by OSLO-BERGEN railroad and E-W highway. On the coast it is indented by Hardangerfjorden and its branches. Since 1981, central parts protected as the Hardangervidda National Park (1,324 sq mi/3,430 sq km; Norway's largest), half of which is privately owned. To the N is the Hardangerjøkulen glacier. The park is a major recreational area year round, and has a rich wildlife, including Europe's largest herd of wild reindeer (about 10,000 animals).

Hardap Region, administrative division (2001 population 68,249), S central NAMIBIA, extending from ATLANTIC OCEAN to BOTSWANA border. Includes most of Namib-Naukluft park, upper Fish River valley, and artesian basins of NOSSOB and AUOB rivers.

Hardburly (HAHR-buhr-lee), village, PERRY county, SE KENTUCKY, in CUMBERLAND foothills, 5 mi/8 km NE of HAZARD, and 2 mi/3.2 km E of Bulan. Railroad spur terminus. Bituminous coal.

Hardee (hahr-DEE), county (□ 638 sq mi/1,658.8 sq km; 2006 population 28,621), central FLORIDA; ⊙ WAUCHULA; 27°29′N 81°48′W. Rolling terrain, partly swampy, with many small lakes; drained by PEACE RIVER. Citrus and strawberry region, with cattle and poultry raising. Formed 1921.

Hardeeville, town (2006 population 1,850), JASPER county, S SOUTH CAROLINA, 18 mi/29 km W of HILTON HEAD, and 5 mi/8 km from GEORGIA state line; 32°16′N 81°04′W. Manufacturing includes fabricated metal products, lumber; sand processing. Agriculture includes timber; livestock; grain, soybeans.

Hardegg (hahr-DAIK), town, N LOWER AUSTRIA, N AUSTRIA, in the WALDVIERTEL, on Thaya River, and on CZECH border; 48°51′N 15°52′E. Elevation 939 ft/286 m. Border station. Town since 1290; medieval fortification well preserved; impressive castle high above Thaya valley.

Hardegsen (HAHR-deg-sen), town, LOWER SAXONY, W GERMANY, 9 mi/14.5 km NNW of GÖTTINGEN; 51°39′N 09°50′E. Health center; cement works. Has remains of ancient castle.

Hardelot-Plage (ahr-duh-lo–PLAHZH), swimming resort, in Neufchâtel-Hardelot commune, PAS-DE-CALAIS department, NORD-PAS-DE-CALAIS region, N FRANCE, on ENGLISH CHANNEL, 6 mi/9.7 km S of BOULOGNE-SUR-MER; 50°38′N 01°35′E. Forest-covered dunes extend behind sandy beach. Many sports activities.

Hardeman (HAHR-duh-muhn), county (□ 655 sq mi/1,703 sq km; 2006 population 28,176), SW TENNESSEE, on MISSISSIPPI (S) state line; ⊙ BOLIVAR; 35°12′N

89°00′W. Drained by HATCHIE River. Agriculture; diverse industries; natural resources; timbering; recreation. Formed 1823.

Hardeman, county (□ 697 sq mi/1,812.2 sq km; 2006 population 4,250), N TEXAS; ⊙ QUANAH; 34°17′N 99°45′W. Bounded in part and drained in S by PEASE RIVER; bounded by RED RIVER (here forming OKLAHOMA state line) on N. Wheat, cotton, peanuts; sheep, horses, angora goats, cattle; also grain. Gypsum mining; some oil and gas production. Drained by Wanderer's Creek, which forms LAKE PAULINE at center of county. Lake Copper Breaks and Copper Breaks State Park in S. Formed 1858.

Harden, ENGLAND: see BINGLEY.

Hardenberg (HAHR-duhn-berkh), town (2001 population 16,501), OVERIJSSEL province, E NETHERLANDS, on VECHT RIVER, near OVERIJSSEL CANAL, and 16 mi/26 km N of ALMELO; 52°34′N 06°38′E. Overijssel Canal passes to E; German border 4 mi/6.4 km to E. Dairying; cattle, hogs; grain, vegetables; manufacturing (radiators, fiberglass, packaging materials). Oldmeijer Recreational Park to SW; Shetland Ponypark to NW.

Hardenberg, GERMANY: see NÖRTEN-HARDENBERG.

Harden-Murrumburrah (HAHR-duhn–MUHR-uhm–BUHR-uh), municipality, S central NEW SOUTH WALES, AUSTRALIA, 65 mi/105 km NW of CANBERRA. Railroad terminus; dairying center; wheat, sheep; wine.

Harderwijk (HAHR-duhr-veik), city (2001 population 32,577), GELDERLAND province, central NETHERLANDS, 17 mi/27 km NW of APELDOORN; 52°21′N 05°38′E. At NE end of NULDERNAUW (channel between mainland and FLEVOLAND polders, at entrance to VELUWEMEER channel), lock and dam separate channels. Dairying; cattle, hogs, poultry, ducks; fruit, vegetables, grain; manufacturing (equestrian equipment, packaging machines; food processing). Marine aquarium and recreational center are here. Chartered in 1231; in Middle Ages a member of Hanseatic League. Seat of provincial university (1647–1811) where Linnaeus once taught.

Hardesty (HAHR-des-tee), village (2006 population 279), TEXAS county, central OKLAHOMA Panhandle, 16 mi/26 km ESE of GUYMON; 36°36′N 101°11′W. Optima Lake reservoir (on NORTH CANADIAN, or Beaver, River) and Optima National Wildlife Refuge to N.

Hardheim (HAHRD-heim), village, BADEN-WÜRTTEMBERG, SW GERMANY, on E slope of the ODENWALD, 11 mi/18 km S of Wertheim; 49°36′N 09°29′E. Metalworking. Has 16th-century castle, keep of 15th-century castle.

Hardin, county (□ 181 sq mi/470.6 sq km; 2006 population 4,585), extreme SE ILLINOIS; ⊙ ELIZABETHTOWN; 37°31′N 88°31′W. Bounded S and E by OHIO River; drained by short Big Creek. Agriculture (wheat, corn, sorghum; livestock; timber). Manufacturing (fluorspar products from imported fluorspars). About 90% of county in Shawnee National Forest, also part of Cave in Rock State Park. Toll ferry to KENTUCKY at CAVE IN ROCK. One of 17 Illinois counties to retain Southern-style commission form of county government. The last Illinois fluorspar mine closed in 1996. Formed 1839.

Hardin, county (□ 569 sq mi/1,479.4 sq km; 2006 population 17,791), central IOWA; ⊙ ELDORA; 42°23′N 93°15′W. Prairie agricultural area (hogs, cattle, poultry; corn, oats, soybeans) drained by IOWA RIVER. Manufacturing at IOWA FALLS. Limestone quarries; sand, clay, and gravel pits. Pine Lake and Steamboat Rock state parks in E. Widespread flooding of fields and rivers in 1993. Formed 1851.

Hardin, county (□ 629 sq mi/1,635.4 sq km; 2006 population 97,087), central and N KENTUCKY, on INDIANA (extreme N; across OHIO River) state line; ⊙ ELIZABETHTOWN; 37°42′N 85°58′W. Bounded E by SALT RIVER and ROLLING FORK; drained by ROUGH and NOLIN rivers. Gently rolling agricultural area (burley

tobacco, corn, wheat, hay, alfalfa, soybeans; hogs, cattle, poultry; dairying); limestone quarries, sand pits, asphalt deposits. Manufacturing at ELIZABETHTOWN. Part of FORT KNOX Military Reservation in E, includes U.S. Gold Bullion Depository. Vernon-Douglas State Nature Preserve in E. Formed 1792.

Hardin (HAHR-duhn), county (□ 467 sq mi/1,214.2 sq km; 2006 population 31,966), W central OHIO; ⊙ KENTON; 40°39′N 83°40′W. Intersected by SCIOTO, BLANCHARD, and OTTAWA rivers. In the Till Plains physiographic region. Agricultural area (hogs, poultry; wheat, corn, soybeans); manufacturing at Kenton, ADA, and FOREST (paper products, transportation equipment); limestone quarries, gravel pits. Formed 1833.

Hardin, county (□ 595 sq mi/1,547 sq km; 2006 population 26,089), SW TENNESSEE, on MISSISSIPPI and ALABAMA (both S) state lines; ⊙ SAVANNAH; 35°12′N 88°11′W. Bounded NE by TENNESSEE RIVER. Agriculture; manufacturing. Includes Pickwick Landing Dam and part of Pickwick Landing Reservoir in S, SHILOH NATIONAL MILITARY PARK in SW. Formed 1819.

Hardin, county (□ 897 sq mi/2,332.2 sq km; 2006 population 51,483), SE TEXAS; ⊙ KOUNTZE; 30°20′N 94°23′W. Bounded E by NECHES RIVER and drained by its tributaries. S part on Gulf coastal plains; rolling, wooded in N (lumbering). Agriculture (forage crops); livestock raising (cattle, hogs; egg production). Oil, natural gas fields; sand and gravel; timber. Formed 1858.

Hardin (HAHR-din), city (2000 population 614), RAY county, NW MISSOURI, near MISSOURI River, 8 mi/12.9 km E of RICHMOND; 39°16′N 93°49′W. Corn, wheat, soybeans; hogs, cattle; grain elevators. Town and surrounding farmland damaged in 1993 flood.

Hardin (HAHR-duhn), town (2000 population 3,384), ⊙ BIG HORN county, S MONTANA, on BIGHORN RIVER, at mouth of LITTLE BIGHORN River, and 45 mi/72 km E of BILLINGS; 45°44′N 107°37′W. Beet sugar; barley; livestock. Gas fields, coal mining; tourism. Trading point for Crow Indian Reservation to S and SE. County fairgrounds, Bighorn County Historical Museum, LITTLE BIGHORN BATTLEFIELD National Monument to SE; Big Horn Canyon National Recreation Area to SW, Grant Marsh Wildlife Management Area to N. Incorporated 1911.

Hardin, village (2000 population 959), ⊙ CALHOUN county, W ILLINOIS, on ILLINOIS River (bridged to East Hardin), and 30 mi/48 km NW of ALTON; 39°09′N 90°37′W. Center of apple- and grain-growing district; also sorghum; hogs; vinegar factory. Sustained heavy damage in floods of 1993.

Hardin (HAHR-duhn), village (2000 population 564), MARSHALL county, W KENTUCKY, 10 mi/16 km N of MURRAY, near East Fork Clarks River; 36°45′N 88°17′W. Railroad terminus. Agriculture (tobacco, grain, soybeans; hogs, cattle); manufacturing (fertilizer, boat docks, meats). Kenlake State Resort park on Kentucky Lake, to E.

Hardin, village (2006 population 799), LIBERTY county, SE TEXAS, 36 mi/58 km W of BEAUMONT, near TRINITY RIVER; 30°08′N 94°44′W. Timber; oil and natural gas. Agriculture (rice, soybeans). BIG THICKET NATIONAL PRESERVE to NE.

Harding (HAHR-deeng), community, SW MANITOBA, W central CANADA, 28 mi/44 km WNW of BRANDON, and in Woodworth rural municipality; 49°59′N 100°31′W.

Harding, county (□ 2,136 sq mi/5,553.6 sq km; 2006 population 718), NE NEW MEXICO; ⊙ MOSQUERO; 35°51′N 103°49′W. Hay, some alfalfa, wheat, oats, barley, millet; cattle, some sheep. Bounded W by CANADIAN River (Canadian River Canyon), drained by Ute, Tequesquite, and Carrizo creeks. Formed 1921.

Harding, county (□ 2,677 sq mi/6,960.2 sq km; 2006 population 1,205), NW SOUTH DAKOTA, on MONTANA (W) and NORTH DAKOTA (N) state lines; ⊙ BUFFALO;

45°35′N 103°30′W. Wheat; sheep-raising area watered by LITTLE MISSOURI RIVER and branches of GRAND and MOREAU rivers. Petroleum and large deposits of lignite coal; numerous small ranches. Cave Hills; Crow Butte (3,185 ft/971 m) has limestone ridges in N, SW, and E; part of Custer National Forest near each. Slim Buttes (3,672 ft/1,119 m) in SE; site of Battle of Slim Buttes (1876) in E, near Reva. Several units of Custer National Forest in SW, E, and N (remainder mostly in S Montana). Formed 1908.

Harding, village (2000 population 105), MORRISON county, central MINNESOTA, near PLATTE RIVER, 20 mi/32 km NE of LITTLE FALLS; 46°07′N 94°02′W. Agricultural area (grain; livestock; dairying). Incorporated 1938.

Hardinge Bridge, 5,900 ft/1,798 m long, spanning the GANGA (or PADMA) River, W central BANGLADESH; connects PAKSEY, PABNA district (E bank), 3 mi/4.8 km S of SARA with Veramara in the KUSHTIA district (W bank); strategic railroad link. Largest railroad bridge in the country; carries two broad-guage railroad tracts and one pedestrian walkway, and consists of fifteen spans. Water level readings are taken at PAKSEY for flood forecasting. BANGLADESH receives its share of the GANGA waters as outlined in the Farakka Agreement with INDIA. Provides a strategic railroad connection in W BANGLADESH. Opened on March 4, 1915.

Hardingstone, ENGLAND: see NORTHAMPTON.

Hardinsburg, town (2000 population 244), WASHINGTON county, S INDIANA, 13 mi/21 km SW of SALEM; 38°28′N 86°16′W. Agricultural area. Wood furniture. Laid out 1838.

Hardinsburg, town (2000 population 2,345), ⊙ BRECKINRIDGE county, NW KENTUCKY, 35 mi/56 km E of OWENSBORO; 37°46′N 86°27′W. Agriculture (burley tobacco, corn, wheat) and hardwood timber; manufacturing (linens, apparel, building materials, lumber, paper products, crushed stone). County Airport to E. Broadmoor Gardens and Conservatory to NE, at IRVINGTON. ROUGH RIVER LAKE reservoir and Rough River Dam State Resort Park to S, Mt. Laurel Lake to E.

Hardinxveld-Giessendam (HAHR-deenks-felt–GEE-suhn-DAHM), town, SOUTH HOLLAND province, SW NETHERLANDS, on LOWER MERWEDE RIVER, 8 mi/12.9 km W of DORDRECHT; 51°50′N 04°50′E. UPPER MERWEDE RIVER divides into NEW MERWEDE and Lower Merwede rivers 2 mi/3.2 km to E. Dairying; cattle, hogs, poultry; grain, vegetables, sugar beets; manufacturing (shipbuilding, food processing).

Hardisty (HAHR-dis-tee), town (□ 2 sq mi/5.2 sq km; 2001 population 743), E ALBERTA, W CANADA, on Battle River, and 22 mi/35 km SW of WAINWRIGHT, in FLAGSTAFF county; 52°41′N 111°18′W. Grain elevators, stockyards; mixed farming; oil and gas. Habitat for birds, deer, and moose. Established as a village in 1906; became a town in 1910.

Hardman, unincorporated village, MORROW county, N OREGON, 15 mi/24 km SSW of HEPPNER. Wheat. Umatilla National Forest to SE.

Hardo Daska, PAKISTAN: see DASKA.

Hardoi (HUHR-do-ee), district (□ 2,311 sq mi/6,008.6 sq km), central UTTAR PRADESH state, N central INDIA, on GANGA plain; ⊙ HARDOI. Bounded SW by GANGA RIVER, E by GOMATI RIVER; drained by the RAMGANGA RIVER; irrigated by SARDA CANAL system. Agriculture (wheat, gram, barley, pearl millet, rice, oilseeds, jowar, sugarcane, corn); a major Indian sugar-processing district; cotton weaving. Main towns include HARDOI, SHAHABAD, SANDILA, MALLANWAN.

Hardoi (HUHR-do-ee), town (⊙ HARDOI district, central UTTAR PRADESH state, N central INDIA, 35 mi/56 km SSE of SHAHJAHANPUR; 27°25′N 80°07′E. Road and trade (grains, oilseeds) center; sugar milling, woodworking, saltpeter processing.

Hä″rdt (HUHRT), commune (□ 6 sq mi/15.6 sq km), BAS-RHIN department, in ALSACE, NE FRANCE;

48°42′N 07°47′E. Agricultural village 8 mi/ 13 km N of STRASBOURG known for asparagus festival.

Hardt Forest, FRANCE: see HARTH FOREST.

Hardt Mountains (HAHRT), range in RHINELAND-PALATINATE, W GERMANY, extends c.25 mi/40 km N of QUEICH RIVER; rises to 2,240 ft/683 m in the KALMIT; 49°00′N 07°50′E–49°35′N 08°10′E. Densely forested (broadleaf); orchards and vineyards on E slopes. Geologically, range is a continuation of the VOSGES MOUNTAINS. Also spelled HAARDT MOUNTAINS.

Hardtner (HAHRT-nuhr), village (2000 population 199), BARBER county, S KANSAS, 9 mi/14.5 km W of KIOWA, near OKLAHOMA state line; 37°00′N 98°39′W. Cattle; wheat.

Harduaganj (huhrd-WAH-guhnj), town, ALIGARH district, W UTTAR PRADESH state, N central INDIA, near UPPER GANGA CANAL, 6 mi/9.7 km NE of ALIGARH; 27°57′N 78°10′E. Wheat, barley, pearl millet, gram, cotton. Hydroelectric plant.

Hardwick, town, BALDWIN county, central GEORGIA, 29 mi/47 km NE of MACON; 33°04′N 83°13′W. Georgia College nearby.

Hardwick, town, WORCESTER county, central MASSACHUSETTS, on branch of WARE RIVER, 21 mi/34 km WNW of Worcester, and near QUABBIN RESERVOIR; 42°21′N 72°13′W. Dairying. Settled 1737, incorporated 1739. Includes villages of Gilbertville and Wheelwright.

Hardwick, town, including Hardwick village, CALEDONIA CO., N central VERMONT, 19 mi/31 km WNW of St. Johnsbury, and on LAMOILLE RIVER; 44°31′N 72°20′W. Dairy products. Settled before 1800. Named for HARDWICK, MASSACHUSETTS.

Hardwick, village (2000 population 222), ROCK county, extreme SW MINNESOTA, 8 mi/12.9 km N of LUVERNE; 43°46′N 96°12′W. Corn, soybeans, oats; livestock; poultry; dairying. Blue Mounds State Park to S.

Hardwicke Bay (HAHRD-wik BAI), inlet of SPENCER GULF, SOUTH AUSTRALIA, between CORNY POINT of YORKE PENINSULA and S shore of WARDANG ISLAND; 33 mi/53 km long, 20 mi/32 km wide; 34°56′S 137°22′E.

Hardy, county (□ 585 sq mi/1,521 sq km; 2006 population 13,420), NE WEST VIRGINIA, in EASTERN PANHANDLE, on VIRGINIA (E and S) state line; ⊙ MOOREFIELD; 39°00′N 78°51′W. Drained by South Branch of POTOMAC River and its South Fork, and by CACAPON and Lost rivers. Traversed by ridges (including North and SHENANDOAH mountains) and valleys of the ALLEGHENY MOUNTAINS. Includes Lost River State Park in S center, and part of George Washington National Forest in E. Agriculture (corn, oats, grains, soybeans, potatoes, alfalfa, sorghum); livestock, poultry; dairying. Limestone and marble quarrying. Manufacturing at Moorefield (especially poultry processing). Formed 1786.

Hardy, town (2000 population 578), N ARKANSAS, 28 mi/45 km W of POCAHONTAS, and on SPRING RIVER; 36°19′N 91°28′W. Harold E. Alexander Wildlife Management Area to S. Growing retirement population. Former county seat (shared with EVENING SHADE); later replaced by ASH FLAT.

Hardy, town (2000 population 57), HUMBOLDT county, N central IOWA, 23 mi/37 km NNE of FORT DODGE; 42°48′N 94°02′W. In agricultural area.

Hardy, unincorporated town, PIKE county, E KENTUCKY, in the CUMBERLAND MOUNTAINS, near TUG FORK River (WEST VIRGINIA state line), 5 mi/8 km SSE of WILLIAMSON (W.Va.). In bituminous coal mining area.

Hardy, village (2006 population 167), NUCKOLLS county, S NEBRASKA, 7 mi/11.3 km E of SUPERIOR, at KANSAS state line, near REPUBLICAN RIVER; 40°00′N 97°55′W. Dairy and poultry products, livestock; grain.

Hardy, village, KAY county, N OKLAHOMA, near KANSAS state line, 25 mi/40 km NE of PONCA CITY. In agricultural area.

Hardy Dam, MICHIGAN: see MUSKEGON RIVER.

Hardy Dam Pond, reservoir, NEWAYGO county, central MICHIGAN, on MUSKEGON RIVER, between ROGERS

DAM and CROTON DAM ponds, in Manistee National Forest, 16 mi/26 km SSW of BIG RAPIDS; 43°29′N 85°38′W. Formed by HARDY DAM (100 ft/30 m high), built in 1931.

Hardy Lake, reservoir, SCOTT county, SE INDIANA, on Quick Creek, in Hardy Lake State Recreation Area, 8 mi/12.9 km NNE of SCOTTSBURG; 3 mi/4.8 km long; 38°46′N 85°42′W. Maximum capacity 14,700 acre-ft. Formed by Hardy Lake Dam (48 ft/15 m high), built (1970s) by the Indiana Department of Natural Resources for water supply. Also called Quick Creek Reservoir.

Hardy Peninsula, CHILE: see HOSTE ISLAND.

Hardyston, township, SUSSEX county, NW NEW JERSEY, 1 mi/1.6 km S of HAMBURG; 41°07′N 74°33′W. Incorporated 1798.

Hare Bay, inlet, E NEWFOUNDLAND AND LABRADOR, E CANADA; 20 mi/32 km long, 10 mi/16 km wide at entrance; 51°17′N 55°50′W.

Harefield, ENGLAND: see HILLINGDON.

Hare Hill (1,958 ft/597 m), SW NEWFOUNDLAND AND LABRADOR, E CANADA, at SW end of GRAND LAKE, 22 mi/35 km SSW of CORNER BROOK.

Hare Hill, (995 ft/303 m), S NEWFOUNDLAND AND LABRADOR, E CANADA, on W side of BURIN PENINSULA, near E side of FORTUNE BAY; highest point of Burin Peninsula.

Hareidlandet (HAHR-aid-LAHN-uht), island (□ 67 sq mi/174.2 sq km), in NORTH SEA, MØRE OG ROMSDAL county, W NORWAY, one of the SØRØYANE islands, SW of Ålesund; 13 mi/21 km long, 9 mi/14.5 km wide. Separated from mainland by narrow sound. Fishing and manufacturing of fish products. Hareid village, on NE coast, makes wood products.

Hare Indian River, 120 mi/193 km long, NORTHWEST TERRITORIES, N CANADA; rises W of GREAT BEAR LAKE; flows W to MACKENZIE RIVER at FORT GOOD HOPE; 66°17′N 128°37′W.

Hare Island (HER) or **Ile aux Lièvres** (EEL o LYE-vruh), island, SE central QUEBEC, E CANADA, in the SAINT LAWRENCE RIVER, opposite RIVIÈRE-DU-LOUP; 8 mi/13 km long, 1 mi/2 km wide; 47°51′N 69°43′W. Just E are the BRANDYPOT islets.

Harelbeke (HAH-ruhl-bai-kuh), commune (2006 population 26,231), Kortrijk district, WEST FLANDERS province, W BELGIUM, on Leie River, and 3 mi/4.8 km NNE of KORTRIJK; 50°51′N 03°18′E. Textile industry; market center for tobacco-growing, cattle-raising area. Has twelfth-century church. Town chartered 1153. Also spelled Harlebeke.

Haren (HAH-ren), town, LOWER SAXONY, N GERMANY, on the EMS RIVER (here joined by short HAREN-RÜTENBROCK CANAL), and 7 mi/11.3 km NNW of MEPPEN; 52°48′N 07°14′E. Health resort; metalworking, boatbuilding, plastics processing. Site of research for German supertrain "Transrapid."

Haren (HAH-ruhn), town, GRONINGEN province, N NETHERLANDS, 3 mi/4.8 km SSE of GRONINGEN; 53°10′N 06°37′E. NOORD-WILLEMS Canal passes to W, Paterswoldersmeer to W. Dairying; cattle, sheep; grain, vegetables, fruit; manufacturing (paper products).

Haren-Rütenbrock Canal (HAH-ren–RYOO-ten-brok), c.10 mi/16 km long, NW GERMANY, connects SÜD-NORD KANAL (at Rütenbrock) with the EMS River (at HAREN); 52°50′N 07°10′E.

Harer, ETHIOPIA: see HARAR.

Harerge, or **Harer**, former province, SE ETHIOPIA, on DJIBOUTI and SOMALIA borders; ⊙ was HARAR; 08°30′N 43°00′E. Was the largest province of ETHIOPIA, occupying nearly 24% of the country's area; it extended c.500 mi/805 km S from AWASH RIVER and c.600 mi/966 km E from the GENALE RIVER. Population was largely Somali and Oromo. Commercial centers included DIRE DAWA, HARAR, JIJIGA, and GINIR. Was crossed by Djibouti–Addis Ababa railroad. Its main road, paralleling the railroad, was

joined by roads from BERBERA (NW SOMALIA) and MOGADISHO (SE SOMALIA). Now part of SOMALI, OROMIYA, and AFAR states. Formerly spelled Harrar, or Harar.

Hareskov (HAH-ruh-skou), town, Copenhagen county, SJÆLLAND, DENMARK, 11 mi/18 km NW of COPENHAGEN; 55°46′N 12°25′E.

Harfleur (ahr-FLUR), town (□ 1 sq mi/2.6 sq km), SEINE-MARITIME department, HAUTE-NORMANDIE region, N FRANCE, at mouth of the SEINE RIVER, on the ENGLISH CHANNEL, 4 mi/6.4 km E of Le HAVRE; 49°30′N 00°12′E. Was a flourishing port during the late Middle Ages but declined because of silting in the 16th century. Has chemical and aeronautical plants; part of the Le Havre industrial port zone. The siege and capture (1415) of Harfleur by the English in the Hundred Years War is described in Shakespeare's *Henry V*.

Harford (HAHR-ford), county (□ 526 sq mi/1,367.6 sq km; 2006 population 241,402), NE MARYLAND, on PENNSYLVANIA (N) state line; ⊙ BEL AIR; 39°32′N 76°18′W. Bounded NE by the SUSQUEHANNA River, SE and S by CHESAPEAKE BAY. PIEDMONT agricultural area (N) produces dairy products, vegetables, fruit, grain; poultry. Stone quarries. Coastal plain (S fringe) occupied mostly by Federal reservations (Aberdeen Proving Ground, Army Chemical Center, and Edgewood Arsenal). Commercial fisheries, shore resorts, some industries (especially vegetable canneries, clothing factories). Includes many large estates. Formed 1773.

Hargeisa (hahr-GEI-zah), town, ⊙ WOQOOYI GALBEED region, NW SOMALIA. Commercial center and watering place for nomadic livestock herders. Transportation hub; has international airport. Former summer capital of BRITISH SOMALILAND; also, capital SOMALILAND, declared independent state from Somalia (1991) but not recognized. Heavily damaged by Somalian bombing. Also HARGHESSA or HARGEYSA.

Hargeysa, SOMALIA: see HARGEISA.

Harghessa, SOMALIA: see HARGEISA.

Harghita (hahr-GEE-tah), county, E central ROMANIA, in TRANSYLVANIA; ⊙ MIERCUREA-CIUC; 46°21′N 25°48′E. Hilly to mountainous topography; drained by MUREŞ and OLT rivers. Farming; grazing. Large Hungarian population.

Hargill, unincorporated town, HIDALGO county, S TEXAS, 22 mi/35 km NE of MCALLEN. Located in N margin of irrigated Rio Grande Valley; rich agricultural area (citrus, vegetables, cotton, sugarcane; cattle).

Hargraves (HAHR-graivz), town, NEW SOUTH WALES, SE AUSTRALIA, 186 mi/300 km NW of SYDNEY, and 22 mi/35 km SE of MUDGEE; 32°47′S 149°27′E. Old gold-mining town.

Hargshamn (HAHR-yuhs-hahm), village, UPPSALA county, E SWEDEN, port on BALTIC SEA, 6 mi/9.7 km SSW of ÖSTHAMMAR; 60°10′N 18°28′E.

Harhira Uul (KHAHR-khir-ah OOL), outlying massif of the MONGOLIAN ALTAY, SW MONGOLIA, 35 mi/55 km SW of ULAANGOM; 49°34′N 91°17′E. Highest elevation 13,245 ft/4,037 m. Also spelled Kharkhira, Kharkira, or Hara Hira.

Hari, AFGHANISTAN: see HARI RUD.

Haría (ah-REE-ah), village, LANZAROTE, CANARY ISLANDS, SPAIN, 13 mi/21 km N of ARRECIFE. Cereals, grapes, fruit, vegetables. Flour milling, wine making, embroidery manufacturing. Its ATLANTIC port ARRIETA is 2 mi/3.2 km E.

Hariana (huhr-YAH-nah), town, HOSHIARPUR district, N central PUNJAB state, N INDIA, 8 mi/12.9 km NW of HOSHIARPUR. Wheat, sugarcane, corn, gram; handloom woolen weaving.

Hariana (huhr-YAH-nah), level area in S PUNJAB region, HARYANA state, N INDIA; roughly between GHAGGAR RIVER (N), sandy tracts of THAR DESERT (W, S), and JIND and ROHTAK districts, Haryana state (E). Once scene of flourishing Hindu civilization. Was in S part of former old PATIALA AND EAST PUNJAB STATES UNION.

Harib, township, central YEMEN, c.100 mi/161 km ESE of SANA, at the edge of the desert; 14°56′N 45°30′E.

Haridwar (huhr-idh-WAHR), district (□ 911 sq mi/2,368.6 sq km), W Uttarakhand state, N central INDIA; ⊙ HARIDWAR. Name means "Gateway to God;" there are numerous famous temples scattered across the district. Approximately one third of Rajaji National Park-which contains deer, pea fowl, kingfishers, tigers, wild boar, elephants, king cobras, and sloth bears-is located here, with the rest of the park shared by DEHRADUN and PAURI GARHWAL districts. Popular tourist destination; known for mild climate and scenic beauty. In 2000, joined other districts to form Uttarakhand; was formerly a part of UTTAR PRADESH.

Haridwar (huhr-idh-WAHR), city (2001 population 175,010), ⊙ HARIDWAR district, W central Uttarakhand state, N central INDIA, on the GANGA RIVER, where the river descends onto the plains and at the headworks of the GANGA CANAL system; 29°58′N 78°10′E. Annual and duodecennial pilgrimages are associated with the town's Hindu temple and with the GANGA RIVER; it is one of seven major holy pilgrimage sites in Hinduism collectively referred to as moksapuri. Millions come each year to bathe and spread ashes of cremated relatives during the Kumbha Mela. The sacred bathing-ghat area, believed to have a footprint of the Hindu god Vishnu, is known as "Har ki Pauri" [=foot of Vishnu]. The Gurukul school, founded in 1902, is a center of Vedic studies. Serene and picturesque, it is a popular tourist site. Heavy-electrical equipment plant in nearby RANIPUR for steel castings and steel forging, in addition to manufacturing steam and water turbines, generators, and industrial electric motors.

Harigaon (HAH-ri-goun), village, AHMADNAGAR district, MAHARASHTRA state, W INDIA, 38 mi/61 km N of AHMADNAGAR. Railroad terminus; sugar milling.

Harihar, town, Chitaldurga district, KARNATAKA state, S INDIA, on TUNGABHADRA RIVER, on border of DHARWAD district, and 8 mi/12.9 km WNW of DAVANGERE; 14°31′N 75°48′E. Road center; manufacturing of cotton textiles, machine tools; known for fine handicrafts (leather footwear; woolen weaving, goldsmithing). Annual livestock fair. Large 13th-century Vishnuite-Shivaite temple contains substructure with inscriptions dating back to seventh century A.D.

Harij (hah-REEJ), town, tahsil headquarters, MAHESANA district, GUJARAT state, W INDIA, 31 mi/50 km WNW of MAHESANA; 23°42′N 71°54′E. Railroad spur terminus. Local market for millet, cotton, salt, oilseeds.

Harim (HAH-rim), town, IDLIB district, NW SYRIA, near the Turkish border, 38 mi/61 km W of ALEPPO; 36°12′N 36°30′E. Tobacco, cereals.

Harima, former province in S HONSHU, JAPAN; now part of HYOGO prefecture.

Harima (HAH-ree-mah), town, Kako district, HYOGO prefecture, S HONSHU, W central JAPAN, 17 mi/28 km W of KOBE; 34°42′N 134°52′E. Steel frames.

Harima Sea (HAH-ree-mah), Japanese *Harima-nada* (hah-REE-mah–NAH-dah), E section of INLAND SEA, W JAPAN, between HONSHU (N) and NE coast of SHIKOKU (S); merges W with HIUCHI SEA; bounded E by AWAJI-SHIMA; connected with PHILIPPINE SEA by NARUTO STRAIT and KII CHANNEL; c.50 mi/80 km long, 40 mi/64 km wide. Contains SHODO-SHIMA island. TAKAMATSU is on SW shore.

Haringey (HA-ring-gai), inner borough (□ 11 sq mi/28.6 sq km; 2005 population 224,500), of GREATER LONDON, SE ENGLAND; 51°35′N 00°05′W. Also called Tottenham. Primarily residential. Within the borough, Tottenham has furniture, light manufacturing, children's clothing, and printing industries. Seat of Middlesex University. Bruce Castle in Tottenham, built in the 16th century, houses a postal museum in

honor of Sir Rowland Hill, founder of the penny postage system. Alexandra Palace is a conference and exhibition center. Other districts include Crouch End, Finsbury Park, Highgate, Muswell Hill, Hornsey, Wood Green, and Tottenham.

Haringhata River, BANGLADESH: see MADHUMATI RIVER.

Haringvliet (HAH-ring-vleet), estuary, 21 mi/34 km long, SW NETHERLANDS, arm of NORTH SEA, N channel of WAAL and Maas (MEUSE) rivers; formed by dividing of HOLLANDS DIEP channel into the VOLKERAK (SW) and HARINGVLIET (W) channels 12 mi/19 km SW of DORDRECHT; flows WNW past MIDDELHARNIS and HELLEVOETSLUIS to North Sea; joined by SPUI RIVER channel from NE 14 mi/23 km SW of ROTTERDAM. Estuary is impounded 4 mi/6.4 km inside its North Sea entrance by Haringvliet Dam, part of the Delta Project. Bifurcated Volkerakdam-Haringvliet Bridge at E end; bridge joins dam at center of channel juncture. GOEREE-OVERFLAKKEE island to S.

Haringvliet Dam, NETHERLANDS: see HARINGVLIET.

Haripad (huh-ri-PAHD), city, ALAPPUZHA district, KERALA state, S INDIA, 30 mi/48 km NNW of KOLLAM; 09°18′N 76°28′E. Coir rope and mats; cashew nut processing.

Haripur (huh-REE-puhr), town, ABBOTTABAD district, NE NORTH-WEST FRONTIER PROVINCE, N PAKISTAN, 18 mi/29 km SW of ABBOTTABAD; 33°59′N 72°56′E. Markets maize, wheat, barley, rapeseeds; local trade in leather clothing; manufacturing and agriculture service center. Sometimes called Haripur Hazara.

Hariq (huh-REEK), town, S central NAJ'D, SAUDI ARABIA, 80 mi/129 km SSW of RIYADH; 23°37′N 46°31′E. Grain, vegetables, fruit; livestock raising.

Hari River (HAH-ree) or **Batang Hari** (BAH-tahng HAH-ree), c.250 mi/402 km long, S central SUMATRA, INDONESIA; rises in central BARISAN Highlands; flows SE, turning NE past JAMBI city, Jambi province, to BERHALA STRAIT 55 mi/89 km NE of Jambi city; 01°16′S 104°05′E. Navigable for seagoing vessels below Jambi city. Also called Jambi, or Djambi, River.

Hari Rud (HAH-ree ROOD), river, c.700 mi/1,127 km long, IRAN, AFGHANISTAN, and TURKMENISTAN; rises in the Kuh-e Baba range, central AFGHANISTAN; flows W, irrigating the fertile HERAT valley, then N, forming the Iran-Afghanistan border between ISLAM QALA and ZULFIQAR (both in Afghanistan); continues N as the TEJEN (formerly TEDZHEN) River forming the IRAN-TURKMENISTAN border, and turns NW into the steppes S of the KARA KUM desert in Turkmenistan (37°24′N 60°38′E), irrigating the TEJEN oasis, a wheat-, cotton-, and cattle-raising area. Also Hari River.

Harjavalta (HAHR-yah-VAHL-tah), town, TURUN JA PORIN province, SW FINLAND, on KOKEMÄENJOKI (rapids), and 17 mi/27 km SE of PORI; 61°19′N 22°08′E. Elevation 116 ft/35 m. Industrial town with large chemical and fertilizer industries; hydroelectric station.

Härjedalen, SWEDEN: see JÄMTLAND.

Harkány (HAHN-kah-nyuh), Hungarian *Harkány*, village, BARANYA county, S HUNGARY, at S foot of VILLANY MOUNTAINS, 16 mi/26 km S of PÉCS; 45°51′N 18°14′E. Railroad junction. Hot sulphur springs. Wine industry in region.

Harker Heights, city (2006 population 22,842), BELL county, central TEXAS, residential suburb 4 mi/6.4 km E of KILLEEN, and 17 mi/27 km WSW of TEMPLE; 31°03′N 97°39′W. Drained by Nolan Creek. Agriculture to S (grains, cotton; cattle). Large Fort Hood Military Reservation to N. STILLHOUSE HOLLOW LAKE reservoir to SE.

Harkers Island (HAHRK-uhrz EI-land), unincorporated town (□ 3 sq mi/7.8 sq km; 2000 population 1,525), CARTERET county, E NORTH CAROLINA, on HARKERS ISLAND in sheltered Core Sound, 4 mi/6.4 km E of BEAUFORT; 5 mi/8 km long, 1 mi/1.6 km wide; bridged to mainland; 34°42′N 76°33′W. Ferry to Cape

LOOKOUT Lighthouse, in CAPE LOOKOUT NATIONAL SEASHORE to SE. Mouth of North River inlet to W. Service industries; manufacturing (consumer goods).

Harkiko, Eritrea: see HIRGIGO.

Harkin, Mount (HAHR-kin) (9,788 ft/2,983 m), SE BRITISH COLUMBIA, W CANADA, near ALBERTA border, in ROCKY MOUNTAINS, on SE edge of KOOTENAY NATIONAL PARK, 30 mi/48 km SW of BANFF (Alberta); 50°47′N 115°52′W.

Harlan (HAHR-luhn), county (□ 467 sq mi/1,214.2 sq km; 2006 population 31,692), SE KENTUCKY, on Virginia (S and E) state line; ⊙ HARLAN; 36°51′N 83°13′W. In the CUMBERLAND MOUNTAINS; drained by CUMBERLAND RIVER and its POOR and CLOVER forks; includes Kentenia State Forest in NW, BLACK MOUNTAIN (4,139 ft/1,262 m; highest point in Kentucky) in E. One of leading coal-producing counties of Kentucky. Hardwood timber; some agriculture (burley tobacco; livestock). Labor conflicts between coal operators and miners here (nicknamed "Bloody Harlan") have been frequent and bitter; after 20 years of strife, mines were unionized in 1941. Part of Daniel Boone National Forest in N; Cranks Creek Wildlife Management Area in S; small part of CUMBERLAND GAP National Historical Park in SW corner; part of Kingdom Come State Park in NE.

Harlan, county (□ 574 sq mi/1,492.4 sq km; 2006 population 3,446), S NEBRASKA, on KANSAS (S) state line; ⊙ ALMA; 40°10′N 99°24′W. Agricultural region in S; drained by REPUBLICAN RIVER, here impounded by Harlan County Dam. Cattle, hogs; corn, wheat, sorghum; dairy and poultry products. Formed 1871.

Harlan, city (2000 population 5,282), ⊙ SHELBY county, W IOWA, on WEST NISHNABOTNA RIVER (hydroelectric plant), and 37 mi/60 km NE of COUNCIL BLUFFS; 41°38′N 95°19′W. Manufacturing (plastic products; pallet and hammer milling, seed corn and soybean processing; rendering plant, cement works, machine shop). Prairie Rose State Park to SE. Settled 1858, incorporated 1879.

Harlan (HAHR-luhn), town (2000 population 2,081), ⊙ HARLAN county, SE KENTUCKY, in the CUMBERLAND MOUNTAINS, 27 mi/43 km NE of MIDDLESBORO, at joining of CLOVER and Martin forks of CUMBERLAND RIVER; 36°50′N 83°19′W. Railroad junction. Coal-mining center in one of Kentucky's largest coalfields. Some timber; some agriculture (tobacco; livestock) in region; manufacturing (coal processing; machinery; lumber, foods; printing and publishing). Kentenia State Forest and part of Daniel Boone National Forest to N; Cranks Creek Wildlife Management Area to SE; Martins Fork Lake reservoir in S. City settled 1819 as Mt. Pleasant; grew as coal-shipping point after coming of railroad in 1911.

Harlan, village, ALLEN county, NE INDIANA, 13 mi/21 km NE of FORT WAYNE. Manufacturing (cabinets). Corn, soybeans.

Harlan County Lake, reservoir, HARLAN county, S NEBRASKA, on REPUBLICAN RIVER, 40 mi/64 km SSW of KEARNEY, near KANSAS state line; c.15 mi/24 km long; 40°03′N 99°12′W. Maximum capacity 850,000 acre-ft. PRAIRIE DOG CREEK enters from SW. Formed by Harlan County Dam (100 ft/30 m high), built (1952) by the Army Corps of Engineers for flood control and irrigation.

Harlaw (hahr-LAW), locality, ABERDEENSHIRE, NE Scotland, 2 mi/3.2 km NW of INVERURIE; 57°17′N 02°25′W. Scene of battle (1411) in which Alexander Stewart, earl of Mar, defeated Highlanders under Donald, lord of the Isles.

Harlebeke, BELGIUM: see HARELBEKE.

Harlech (HAR-luhk), town (2001 population 1,406), GWYNEDD, W Wales, 9 mi/14.4 km NNW of BARMOUTH; 52°52′N 04°06′W. Resort area with beautiful beaches on E side of TREMADOG BAY. The 13th-century castle rests on a cliff 200 ft/61 m above the modern seaside town. The heroic defense of the castle against the Yorkists (1468) is the theme of the Welsh battle song, "The March of the Men of Harlech." The Welsh fortress was the last to surrender (1647) to the Parliamentarians in the English Civil War. The marshland of Morfa Harlech lies to the N.

Harlem (HAHR-luhm), town (2000 population 2,730), HENDRY county, central FLORIDA, adjacent to CLEWISTON; 26°44′N 80°57′W.

Harlem, town (2000 population 1,814), COLUMBIA county, E GEORGIA, 19 mi/31 km WSW of AUGUSTA; 33°25′N 82°19′W. Manufacturing of building materials, fabricated metal products; printing and publishing.

Harlem, town (2000 population 848), BLAINE county, N MONTANA, on MILK River, and 43 mi/69 km E of HAVRE; 48°32′N 108°47′W. Cattle, sheep; sugar beets; gas, oil, and gold in area. Manufacturing (plastics products). Seat of Fort Belknap College. Headquarters of FORT BELKNAP AGENCY in Fort Belknap Indian Reservation, 3 mi/4.8 km SSE. Black Coulee National Wildlife Refuge to NE. Incorporated 1910.

Harlem, residential and business section of upper MANHATTAN, NEW YORK city, bounded roughly by 110th Street (S), the EAST RIVER (E) and HARLEM RIVER (NE), 168th Street (NW), Amsterdam Avenue (NW), and Morningside Park (SW). The Dutch settlement of Nieuw Haarlem was established 1658 by Peter Stuyvesant. To the W of Harlem, near the present site of Columbia University, British and Continental forces fought (Sept. 16, 1776) the Battle of Harlem Heights. Harlem remained rural until the 19th century when improved transportation facilities linked it with lower Manhattan. It then became a fashionable residential section of New York city. By the turn of the century Harlem had a large Jewish population; starting around 1910 Harlem became the scene of increasing African-American migration from the American South. It soon became the largest and most influential African-American community in the nation, one of the centers of innovation in jazz, and the home of such Harlem Renaissance authors as Langston Hughes, Countee Cullen, and Zora Neale Hurston. In East Harlem, a largely Italian neighborhood—the home of Mayor Fiorello H. LaGuardia—many Puerto Ricans and other Hispanic-Americans settled after World War II. The intersection of 7th Avenue and 125th Street is generally considered the heart of Harlem; Lenox Avenue, once internationally known for its entertainment spots, is now mainly lined with housing developments. Strivers' Row is a block of well-preserved turn-of-the-century townhouses. Site of the Abyssinian Baptist Church, headed for many years by Adam Clayton Powell, Jr., and the Apollo theater, noted for performances by African-American musicians and entertainers. An extensive scholarly collection is housed at the Schomburg Center for Research in Black Culture (part of the New York Public Library), which is adjacent to the Countee Cullen branch of the library. Harlem today has a mixture of poverty and some gentrification, in part due to a city policy favoring renovations of abandoned residential buildings. Increasingly popular as a tourist destination. Designated (1996) as an Enterprise Zone.

Harlem, NETHERLANDS: see HAARLEM.

Harlem River, navigable tidal channel, 8 mi./12.9 km long (including Spuyten Duyvil Creek), in NEW YORK city, SE NEW YORK, separating MANHATTAN from the BRONX. Connecting the HUDSON and EAST rivers, it is a shipping shortcut between LONG ISLAND SOUND and river ports N of New York city. Several railroad and many street bridges span the river. Clearing of the channel to create a shipping canal took place in 1895, the NW portion consisting of part of the Spuyten Duyvil (Dutch for "Devil's Spout") Creek, separating

MARBLE HILL from the Bronx. (The upper portion of the creek was filled in in 1914.) The waterway has suffered from severe pollution in the past but is today closely monitored, resulting in some improvements.

Harlesden, ENGLAND: see BRENT.

Harleston (HAHL-stuhn), town (2001 population 4,058), S NORFOLK, E ENGLAND, near WAVENEY RIVER, 17 mi/27 km S of NORWICH; 52°24′N 01°18′E. Agricultural market, with leather industry. Just ENE is agricultural village of Redenhall, with 14th-century church.

Hårlev (HOR-lev), town, STORSTRØM county, SJÆLLAND, DENMARK, 17 mi/27 km NNE of PRAESTO; 55°21′N 12°15′E. Railroad junction.

Harley (HAHR-lee), township (□ 36 sq mi/93.6 sq km; 2001 population 557), E central ONTARIO, E central CANADA, 8 mi/14 km N of NEW LISKEARD; 47°37′N 79°41′W. Agriculture; formerly logging.

Harleysville, unincorporated town (2000 population 8,795), MONTGOMERY county, SE PENNSYLVANIA, 27 mi/43 km NW of PHILADELPHIA; 40°16′N 75°23′W. Manufacturing includes machinery, fabricated metal products, food products, medical supplies, electronic equipment, chemicals. Agricultural includes dairying; livestock; grain, soybeans, apples.

Harleyville, village (2006 population 692), DORCHESTER county, SE central SOUTH CAROLINA, 20 mi/32 km NW of SUMMERVILLE; 33°12′N 80°27′W. Manufacturing of portland cement. Agriculture includes poultry; grain, soybeans, cotton.

Harlingen (HAHR-ling-uhn), city, FRIESLAND province, N NETHERLANDS, on the WADDENZEE, at W end of the VAN HARINXMA CANAL, 16 mi/26 km W of LEEUWARDEN; 53°10′N 05°25′E. Passenger ferries to VLIELAND, TERSCHELLING, and TEXEL islands; chief port for province NE end of AFSLUITDIJK, barrier dam forming IJSSELMEER, 7 mi/11.3 km to S. Dairying; fishing; cattle, sheep; vegetables, grain; manufacturing (food processing). Town destroyed by the sea in 1134. Town Hall (1730); museum railroad terminus.

Harlingen, city (2006 population 64,202), CAMERON county, extreme S TEXAS, 23 mi/37 km NW of BROWNSVILLE; 26°11′N 97°41′W. Railroad junction, in a shipping and processing center in the lower Rio Grande valley, an irrigated farming area yielding citrus and other fruits, grain, vegetables, sugarcane, and cotton. The city, which is linked to the Intracoastal Waterway by a barge channel (ARROYO COLORADO), has food processing, apparel manufacturing, and factories making various chemical, concrete, and metal products. Founded (c.1904) with the coming of the railroad and grew with the agriculture development of the surrounding area. Rio Grande campus of Texas Technical College and Marine Military Academy are both at Harlingen Industrial Air Park; Rio Grande Valley International Airport (NE); Iwo Jima War Memorial. Incorporated 1910.

Harlinger Trekvaart, NETHERLANDS: see VAN HARINXMA CANAL.

Harlington, ENGLAND: see HILLINGDON.

Harlow (HAH-lo), town (2001 population 78,768) and district, ESSEX, SE ENGLAND, about 20 mi/32.2 km NNE of LONDON; 51°47′N 00°08′E. Designated one of the New Towns in 1946 to alleviate overpopulation in London. Harlow grew rapidly to become a significant residential and industrial location (metallurgy, printing, manufacturing of furniture, and production of scientific and surgical goods).

Harlowton, town (2000 population 1,062), ⊙ WHEATLAND county, central MONTANA, on MUSSELSHELL River, at mouth of Antelope Creek, and 80 mi/129 km NW of BILLINGS; 46°26′N 109°50′W. Stone quarries; flour mill; dairying; wheat, barley, oats, alfalfa; cattle, sheep, hogs. Incorporated 1917. Upper Musselshell Historical Society Museum. Originally called Merino.

Harlu, RUSSIA: see KHARLU.

Harman, village (2006 population 125), RANDOLPH county, E WEST VIRGINIA, on Gandy Creek, in Monongahela National Forest, 17 mi/27 km E of ELKINS; 38°55′N 79°31′W. Canaan Valley State Park to NE.

Harmanec (HAHR-mah-nyets), Hungarian *Hermánd*, village, STREDOSLOVENSKY province, central SLOVAKIA, on railroad, and 7 mi/11.3 km NW of BANSKÁ BYSTRICA; 48°49′N 19°03′E. Large paper mills. Izbica (iz-BI-tsah) Cave (4,921 ft/1,500 m long) 2 mi/3.2 km NW.

Harmanli (HAHR-mahn-lee), city, HASKOVO oblast, ⊙ Harmanli obshtina, SE BULGARIA, on a right tributary of the MARITSA RIVER, 18 mi/29 km E of HASKOVO; 41°57′N 25°54′E. Railroad station. Sericulture center; manufacturing textiles (cotton, silk), ceramics, and electrotechnics, tobacco processing; mulberry trees. Has a school of sericulture. Sometimes spelled Kharmanli or Kharmanliy.

Harmanliiska (hahr-mahn-LEE-skah), river, HASKOVO oblast, SE BULGARIA. Tributary of the MARITSA RIVER.

Harmarville (HAHR-mahr-vil), unincorporated town, HARMAR township, ALLEGHENY county, SW PENNSYLVANIA, residential suburb 10 mi/16 km NE of downtown PITTSBURGH, on ALLEGHENY RIVER, opposite (1 mi/1.6 km NW of) OAKMONT; 40°31′N 79°50′W. Agricultural (corn, hay; livestock; dairying) to N.

Hármashatár, Mount (HAHR-mahsh-hah-tahr), Hungarian *Hármashatárhegy*, point (1,627 ft/496 m) in N range of BUDA MOUNTAINS, N central HUNGARY; 48°00′N 19°40′E. Residential districts on E slopes; stalactite cave of Pálvölgy nearby.

Hármas Körös, HUNGARY: see KÖRÖS.

Harmelen (HAHR-muh-luhn), town, UTRECHT province, W central NETHERLANDS, on the OLD RHINE RIVER, and 7 mi/11.3 km W of UTRECHT; 52°05′N 04°58′E. Dairying; cattle, hogs, poultry; vegetables, fruit, sugar beets. Castle to N.

Harmon, county (□ 538 sq mi/1,398.8 sq km; 2006 population 3,042), SW OKLAHOMA; ⊙ HOLLIS; 34°44′N 99°50′W. Bounded SW (RED RIVER) and W by TEXAS; drained by the SALT and PRAIRIE DOG TOWN forks of Red River. Hilly agricultural area (peanuts, sorghum, cotton, wheat, vegetables, black-eyed peas; cattle); mesquite. Originally claimed by both Texas and Oklahoma territories. Formed 1909.

Harmon, village (2000 population 149), LEE county, N ILLINOIS, 9 mi/14.5 km SSW of DIXON; 41°43′N 89°33′W. In rich agricultural area.

Harmon, section of CROTON-ON-HUDSON, WESTCHESTER county, SE NEW YORK, on E bank of the HUDSON RIVER, just N of OSSINING; an important stop on Amtrak and Metro-North railroad, at N end of its electrified section. Nearby is Croton Point Park (recreation).

Harmondsworth, ENGLAND: see HILLINGDON.

Harmony, town (2000 population 589), CLAY county, W INDIANA, suburb 3 mi/4.8 km E of BRAZIL; 39°32′N 87°04′W. Manufacturing (industrial machinery).

Harmony, town, SOMERSET county, central MAINE, 17 mi/27 km NE of SKOWHEGAN; 44°58′N 69°32′W. Wood products.

Harmony, town (2000 population 1,080), FILLMORE county, SE MINNESOTA, near IOWA state line, 9 mi/14.5 km SSE of PRESTON; 43°33′N 92°00′W. Grain, soybeans; livestock, poultry; dairying; manufacturing (machinery, feeds, transportation equipment). Hammervold Landing Field to N. Area known for its karst topography. Niagara Cave is 4 mi/6.4 km SSW, with 60 ft/18 m waterfall 200 ft/61 m below surface of earth.

Harmony (HAHR-muh-nee), town (□ 1 sq mi/2.6 sq km; 2006 population 609), IREDELL county, W central North Carolina, 14 mi/23 km NNE of STATESVILLE; 35°57′N 80°46′W. Manufacturing (stained glass windows, poultry feed); agriculture (tobacco, grain, soybeans; poultry, livestock). Originally known as Harmony Hill. Inc. in 1874. Home of the first high school in the country, which was built in 1907.

Harmony, township, BEAVER county, W PENNSYLVANIA, residential suburb 2 mi/3.2 km N of AMBRIDGE on OHIO RIVER; 40°36′N 80°13′W. Railroad yards.

Harmony, borough (2006 population 893), BUTLER county, W PENNSYLVANIA, 12 mi/19 km WSW of BUTLER, and 1 mi/1.6 km NE of ZELIENOPLE, on CONNOQUENESSING CREEK. Manufacturing (medical supplies, consumer goods). Agriculture (corn, hay; livestock; dairying). First settlement (1805) of Harmony Society.

Harnahalli, INDIA: see HARANHALLI.

Harnai (huhr-NEI), village, SIBI district, NE central BALUCHISTAN province, SW PAKISTAN, in N CENTRAL BRAHUI RANGE, 38 mi/61 km N of SIBI; 30°06′N 67°56′E. Limestone, coal, gypsum deposits worked in hills (S); woolen industry.

Harnamganj, INDIA: see KUNDA.

Harnes (AHRN), town (□ 4 sq mi/10.4 sq km), PAS-DE-CALAIS department, NORD-PAS-DE-CALAIS region, N FRANCE, 4 mi/6.4 km ENE of LENS; 50°27′N 02°54′E. Wholesale distribution center. In former coal-mining area.

Harnett (HAHR-nit) county (□ 601 sq mi/1,562.6 sq km; 2006 population 106,283), central NORTH CAROLINA; ⊙ LILLINGTON; 35°22′N 78°51′W. Forested sandhills area; drained by CAPE FEAR and SOUTH rivers; bounded S in part by Little River. Diversified manufacturing; service industries; agriculture (especially tobacco; cotton, corn, wheat, hay, sweet potatoes, peppers; poultry, cattle, hogs; catfish); timber. Sand and gravel. Raven Rock State Park in N center. Averasboro Battleground State Historical Site on S border, in SE. Formed 1855 from Cumberland County. Named for Revolutinary War patriot and delegate to the Contental Congress, Cornelius Harnett (1723–1781). First settlers came in the 1720s and then came the Highland Scots.

Harney (HAHR-nee), county (□ 10,227 sq mi/26,590.2 sq km; 2006 population 6,888), SE central OREGON; ⊙ BURNS; 43°04′N 118°58′W. Bounded S by NEVADA, drained by Silver River and DONNER UND BLITZEN RIVER. Land area forms ninth-largest county in U.S. Timber. Agriculture (alfalfa, wheat, oats, barley; sheep, cattle). Part of GREAT SANDY DESERT in W. STEENS MOUNTAIN, a ridge 40 mi/64 km long, is in SE. MALHEUR and HARNEY lakes in center. Burns Indian Reservation in N center, N of Burns. Units of Malheur National Wildlife Refuge in center, at Harney and Malheur lakes and Donner und Blitzen River. Alvord Desert is SE of Steens Mountain. Part of Malheur National Forest in N; part of Ochoco National Forest in NW. Squaw Butte Range Experimental Area in NW. Frenchglen Hotel State Wayside in S center. Founded 1889.

Harney (HAHR-nee), locality, HARNEY county, E central OREGON, 11 mi/18 km ENE of BURNS, near Rattlesnake Creek.

Harney, Lake (HAHR-nee), a shallow widening of ST. JOHNS RIVER, on border between VOLUSIA and SEMINOLE counties, E central FLORIDA, 12 mi/19 km ESE of SANFORD; c.4 mi/6.4 km long, 2 mi/3.2 km–3 mi/4.8 km wide; 28°45′N 81°03′W.

Harney Lake (HAHR-nee), SE central OREGON, alkali lake, surrounded by part of Malheur National Wildlife Refuge, linked by channel to MALHEUR LAKE to NE, 20 mi/32 km S of BURNS; c.10 mi/16 km wide. No outlet for two lakes. Receives Silver Creek from NW.

Harney Peak, mountain peak (7,242 ft/2,207 m) in the BLACK HILLS, SW SOUTH DAKOTA; highest point in state.

Härnösand (HERN-UH-sahnd), town, ⊙ VÄSTER-NORRLAND county, E SWEDEN, on GULF OF BOTHNIA, at mouth of ÅNGERMANÄLVEN RIVER; 62°38′N 17°57′E.

Manufacturing (machinery, processed tobacco, ships; pulp mill). Harbor icebound in winter. Long a cultural center of NORRLAND. Burned by Russians (1721). Chartered 1585.

Har Nuur (KHAHR NOOR), freshwater lake (□ 220 sq mi/572 sq km) in W MONGOLIA, 60 mi/100 km E of HOVD; 18 mi/29 km long, 15 mi/24 km wide; 48°06′N 93°12′E. Elevation 3,714 ft/1,132 m. Connected by 15-mi/24-km-long chain of lakes and straits with the salt lake DÖRGÖN NUUR (S) and with DZAVHAN GOL (river; E). The lake has low desert shores but is rich in fish. Also spelled Khara Nor, Khara Nur, or Hara Nur.

Haro (AH-ro), town, LA RIOJA province, N SPAIN, in CASTILE-LEÓN, near the EBRO RIVER, 22 mi/35 km WNW of LOGROÑO; 42°35′N 02°51′W. Chief wine-production center of La Rioja district; summer resort. Wine-related industries (barrel-making; chemicals); manufacturing of textiles, textile machinery; canning; food processing (sausage); pottery. Hydroelectric station. Has Gothic church (16th century), 18th-century town hall, and many fine mansions of 17th–18th century. Title of counts of Haro borne by lords of city since Middle Ages.

Harod, Well of (khah-ROD), Hebrew *Ein Harod*, (AIN khah-ROD), important well in E JEZREEL VALLEY, on NW slope of GILBOA mountains, ISRAEL. Due to its strategic location between DAMASCUS and EGYPT, has long been contested by armies, including Gideon's and Mameluke Sultan Baybars's (1260). Served as a base for Hagana fighters during 1936 Arab riots.

Haroekoe, INDONESIA: see HARUKU.

Harold, village, FLOYD county, E KENTUCKY, in CUMBERLAND foothills, on Levisa Fork River, and 8 mi/12.9 km WNW of PIKEVILLE. Bituminous coal, oil, and gas; manufacturing (commercial printing).

Haroldswick, Scotland: see UNST.

Haro Strait (HAI-ro), channel of the PACIFIC OCEAN, off W CANADA (SW BRITISH COLUMBIA) and U.S. (NW WASHINGTON), at SE end of VANCOUVER ISLAND, joins straits of GEORGIA (N) and JUAN DE FUCA (S); 48°35′N 123°19′W. Separates Vancouver and SATURNA islands (W) and SAN JUAN and Stuart islands (E). International border runs through center of strait.

Harøya (HAHR-uh-yah), island (5 sq mi/13 sq km) in NORTH SEA, MØRE OG ROMSDAL county, W NORWAY, one of the Nord Islands, 21 mi/34 km W of MOLDE; 5 mi/8 km long, 2 mi/3.2 km wide. Fisheries.

Harpalpur (huhr-PAHL-poor), village (2001 population 15,410), N CHHATARPUR district, MADHYA PRADESH state, central INDIA, 17 mi/27 km NNW of NOWGONG. Trades in wheat, cotton, sugar, millet; sugar and flour milling, hand-loom weaving.

Harpanahalli (huhr-PUH-nuh-huh-lee), town, BELLARY district, KARNATAKA state, S INDIA, 40 mi/64 km SW of HOSPET; 14°48′N 75°59′E. Road center; peanut milling, hand-loom cotton and woolen weaving; silk growing. Livestock raising.

Harpenden (HAH-puhn-duhn), town (2001 population 27,686), HERTFORDSHIRE, E central ENGLAND, 5 mi/8 km N of St. ALBANS; 51°48′N 00°21′W. Mainly residential, Harpenden is the site of Rothamsted Research, the largest agricultural research center in the United Kingdom.

Harper, county (□ 802 sq mi/2,085.2 sq km; 2006 population 5,952), S KANSAS, on OKLAHOMA (S) state line, in Red Hills region; ⊙ ANTHONY; 37°12′N 98°04′W. Drained (NE) by CHIKASKIA RIVER. Wheat, barley, oats; sheep, cattle. Food processing. Manufacturing at HARPER and ANTHONY. Formed 1873.

Harper, county (□ 1,041 sq mi/2,706.6 sq km; 2006 population 3,348), NW OKLAHOMA; ⊙ BUFFALO; 36°46′N 99°39′W. Bounded N by KANSAS; intersected by NORTH CANADIAN (Beaver) River and Buffalo Creek. CIMARRON RIVER forms NE border, also drains NW corner. Plains agricultural area (livestock; wheat, barley). Formed 1907.

Harper, town (2003 population 20,000), ⊙ MARYLAND county, SE LIBERIA, port on ATLANTIC OCEAN, on Cape PALMAS, 255 mi/410 km SE of MONROVIA; 04°25′N 07°43′W. Trade center; copra, cassava, rice; fishing. Road to rubber plantation (N). Airfield. Formerly called Cape Palmas.

Harper, town (2000 population 134), KEOKUK county, SE IOWA, 8 mi/12.9 km ENE of SIGOURNEY; 41°21′N 92°02′W. Feed milling.

Harper, town (2000 population 1,567), HARPER county, S KANSAS, 45 mi/72 km SW of WICHITA; 37°17′N 98°01′W. Railroad junction. In wheat area. Manufacturing (farm machinery; meatpacking). Settled 1877, incorporated 1880.

Harpers Ferry, town (2000 population 330), ALLAMAKEE county, extreme NE IOWA, on MISSISSIPPI RIVER, and 11 mi/18 km ESE of WAUKON; 43°12′N 91°09′W. Dairy products, concrete blocks.

Harpers Ferry, village (□ 4 sq mi/10.4 sq km; 2006 population 318), JEFFERSON county, NE WEST VIRGINIA, 7 mi/11.3 km NE of CHARLES TOWN; 39°19′N 77°44′W. On POTOMAC River (MARYLAND state line), at mouth of the SHENANDOAH River; VIRGINIA state line to E. The town is a tourist attraction, known for its history and its scenic beauty. John Brown's seizure of the U.S. arsenal here on October 16, 1859, and the town's subsequent strategic importance during the Civil War, when it was considered the key to the Shenandoah Valley, brought it into national prominence. In 1747, Robert Harper, a millwright, established a ferry at the junction of the two rivers—hence the town's name. The U.S. arsenal was located here in 1796, and by the mid-19th century, Harpers Ferry was an important arms-producing center, with mills, numerous gun factories, and huge stores of weapons and ammunition. The development of the Chesapeake and Ohio Canal and of the Baltimore & Ohio railroad increased its importance, making it a transportation link between the Ohio Valley and the E Coast. During the Civil War it was primarily held by Union soldiers, but changed hands a number of times. Its industrial plant was repeatedly destroyed by troops of both sides. Harpers Ferry never recovered economically, and a series of devastating floods in the late 19th century ended all hopes for revival. Despite continued flooding during the 20th century, many old buildings remain. Of interest are the fire engine house in which John Brown was captured; the John Brown Museum; and the old steps, hand-carved (early 1800s) into the natural stone, which lead to Robert Harper's house (1775–1782) and to Jefferson Rock. The HARPERS FERRY NATIONAL HISTORICAL PARK is here (□ 4 sq mi/10.4 sq km; est. 1955). Appalachian Trail passes through town. Incorporated 1763.

Harpers Ferry National Historical Park, WEST VIRGINIA, MARYLAND, VIRGINIA: see HARPERS FERRY.

Harpersville, town (2000 population 1,620), Shelby co., central Alabama, 15 mi/24 km NE of Columbiana.

Harper Woods, city (2000 population 14,254), WAYNE county, SE MICHIGAN, residential suburb 10 mi/16 km NE of downtown DETROIT, 3 mi/4.8 km W of LAKE SAINT CLAIR; 42°26′N 82°55′W. Manufacturing (fabricated metal products). Borders MACOMB county on N, city of DETROIT on W and S.

Harpeth River, 117 mi/188 km long, central TENNESSEE; rises 10 mi/16 km SW of MURFREESBORO, in RUTHERFORD county; meanders generally NW, past FRANKLIN, to CUMBERLAND RIVER 5 mi/8 km WNW of ASHLAND CITY; 36°18′N 87°09′W. The section in DAVIDSON county is designated a state scenic river.

Harp Island, TUAMOTU ARCHIPELAGO: see HAO.

Harplinge (HAHRP-leeng-e), village, HALLAND county, SW SWEDEN, near KATTEGAT strait, 6 mi/9.7 km NW of HALMSTAD; 56°44′N 12°43′E.

Harpstedt (HAHRP-stet), village, LOWER SAXONY, N GERMANY, 13 mi/21 km SW of BREMEN; 52°54′N 08°35′E. In peat region.

Area is shown by the symbol □, and capital city or county seat by ⊙.

Harpster (HAHRP-stuhr), village (□ 2 sq mi/5.2 sq km; 2006 population 195), WYANDOT county, N central OHIO, 11 mi/18 km NW of MARION; 40°44′N 83°15′W.

Harpswell, resort town, CUMBERLAND county, SW MAINE, on peninsula (Harpswell Neck; W) and islands (Orrs and Bailey islands, and Sebascodegan Island, also called Great, or East Harpswell, Island; all bridge-linked), in CASCO BAY, and 15 mi/24 km NE of PORTLAND; 43°46′N 69°58′W. Includes CUNDYS HARBOR, Orrs Island, and Harpswell Center villages, latter with church (1843) where Elijah Kellogg preached. Settled 1720, incorporated 1758.

Harpswell Sound, SW MAINE, arm of CASCO BAY extending c.12 mi/19 km between Harpswell Neck and ORRS and SEBASCODEGAN islands.

Harput, village, E central TURKEY, 3 mi/4.8 km NNE of ELAZIG, in the mountains near source of the TIGRIS RIVER; 38°44′N 39°15′E. Elevation c.4,200 ft/1,280 m. On an old trade route, it has been replaced by Elazig as trading center. Was known for its old Jacobite convent. Suffered heavily in Armenian massacre of 1895.

Harquahala Mountains, in LA PAZ and MARICOPA counties, W ARIZONA, S of HARCUVAR MOUNTAINS, c.35 mi/56 km WSW of WICKENBURG; rise to 5,672 ft/1,729 m.

Harra (HAHR-ruh), type of desert on ARABIAN PENINSULA, W SAUDI ARABIA, consisting of large areas of corrugated and fissured lava beds. It is common along the E part of Hejaz from the vicinity of the Jordanian boundary to SE of MAKKA.

Harrach, El (hah-RAHSH, el), suburb of ALGIERS, ALGIERS wilaya, N central ALGERIA, in the MITIDJA plain, 6 mi/9.7 km SE of city center. Important livestock market and site of Algeria's agricultural institute, with extensive experimental farms; chemical and metalworks, distilleries; manufacturing of building materials and flour products. Citrus groves, tobacco fields surround town. DAR EL BEIDA airport is just E. Formerly Maison Carrée.

Harrachov (HAH-rah-KHOF), town, VYCHODOCESKY province, N BOHEMIA, CZECH REPUBLIC, at WNW foot of the GIANT MOUNTAINS, 16 mi/26 km E of LIBEREC, near Polish border; 50°46′N 15°26′E. Noted glassworks; popular mountain resort. Village of NOVY SVET, Czech *Nový Svět* (NO-vee SVYET), and Novy Svet Pass are just NW.

Harrah (HA-ruh), town (2006 population 4,970), OKLAHOMA county, central OKLAHOMA, suburb 20 mi/32 km E of downtown OKLAHOMA CITY, and on NORTH CANADIAN River; 35°29′N 97°10′W. In agricultural area; manufacturing (meat processing, poultry processing and products). John Miskelly State Park to W.

Harrah (HAHR-ruh), village (2006 population 506), YAKIMA county, S WASHINGTON, 15 mi/24 km S of YAKIMA, in NE part of Yakima Indian Reservation; 46°24′N 120°32′W. Wheat, spearmint, peppermint, fruits (apples, pears, peaches), vegetables; manufacturing (fertilizers, peppermint extracts).

Harraiya, INDIA: see HARAIYA.

Harran, TURKEY: see HARAN.

Harrar, ETHIOPIA: see HARAR.

Harraseeket River (har-uh-SEE-ket), inlet of CASCO BAY, SW MAINE, near FREEPORT, just NE of PORTLAND; c.4 mi/6.4 km long.

Harrell (HAR-uhl), village (2000 population 293), CALHOUN county, S ARKANSAS, 26 mi/42 km E of CAMDEN; 33°30′N 92°24′W. Manufacturing (building materials). Lake Poinsett State Park to SE.

Harrells (HER-uhls), village (□ 3 sq mi/7.8 sq km; 2006 population 213), SAMPSON county, SE central North Carolina, 20 mi/32 km SSE of CLINTON; 34°43′N 78°12′W. Agriculture (tobacco, cotton, peanuts, sweet potatoes; poultry; livestock). Inc. in 1943 as Harrells Store, the name was changed in 1955.

Harrellsville (HER-uhls-vil), village (2006 population 96), HERTFORD county, NE North Carolina, 10 mi/16 km E of AHOSKIE, near CHOWAN RIVER; 36°17′N 76°47′W. Service industries; construction. Inc. in 1883. Originally named Bethel, it was renamed for Abner Harrell.

Harricanaw River (ha-ri-KAH-no), 250 mi/402 km long, W QUEBEC and NE ONTARIO, E CANADA; rises near VAL-D'OR; flows NW to JAMES BAY of HUDSON BAY, 50 mi/80 km SW of RUPERT HOUSE; crosses into Ontario near its mouth; 51°10′N 79°45′W. Navigable for 50 mi/80 km.

Harrietsham (HA-ree-uht-shuhm), village (2001 population 1,500), central KENT, SE ENGLAND, 7 mi/11.3 km ESE of MAIDSTONE; 51°15′N 00°41′E. Agricultural market. Has almshouses, built in 1642, and church dating partly from 14th century.

Harrietta (HA-ree-uht-uh), village (2000 population 169), WEXFORD county, NW MICHIGAN, 15 mi/24 km WNW of CADILLAC; 44°18′N 85°42′W. In farm area. Has state fish hatchery. In Manistee National Forest.

Harrietville (HA-ree-et-vil), town, VICTORIA, SE AUSTRALIA, 206 mi/332 km NE of MELBOURNE; 36°55′S 147°04′E. Old gold-mining center. Accommodation for and access to ski fields at Mount HOTHAM; walking tracks.

Harriman, city (2006 population 6,717), ROANE county, E TENNESSEE, near TENNESSEE RIVER, 35 mi/56 km W of KNOXVILLE; 35°56′N 84°33′W. In rolling hills; residential. WATTS BAR Reservoir is S. Plotted 1889; incorporated 1891.

Harriman, village (2006 population 2,273), ORANGE county, SE NEW YORK, 10 mi/16 km WSW of HIGHLAND FALLS; 41°18′N 74°08′W. Harriman section of PALISADES INTERSTATE PARK is just E.

Harriman Reservoir (HER-i-man), in towns of WHITINGHAM and WILMINGTON, WINDHAM CO., S VERMONT, on Deerfield River, 17 mi/27 km WSW of BRATTLEBORO; c.8 mi/12.9 km long; 42°52′N 72°52′W. Formed by Harriman Dam (earth-fill construction; 200 ft/61 m high), built (1924) for power generation. Green Mountain National Forest to W. Also called Davis Bridge Dam.

Harrington (HA-reeng-tuhn), canton (□ 94 sq km/244.4 sq km; 2006 population 805), LAURENTIDES region, S QUEBEC, E CANADA, 12 mi/20 km from BARKMERE; 45°50′N 74°40′W.

Harrington, town (2000 population 3,174), KENT county, central DELAWARE, 16 mi/26 km S of DOVER; 38°55′N 75°34′W. Elevation 52 ft/15 m. Railroad junction. Trading and shipping point in agricultural area; manufacturing. Delaware State Fairgrounds and Raceway are here to S; Killen's Pond State Park to NE. Incorporated 1869.

Harrington, town, WASHINGTON county, E MAINE, on PLEASANT BAY, and 19 mi/31 km SW of MACHIAS; 44°31′N 67°47′W. Fishing; lumbering; canneries. Resorts.

Harrington (HA-reeng-tuhn), resort village, NEW SOUTH WALES, SE AUSTRALIA, 217 mi/350 km N of SYDNEY, 21 mi/33 km NE of TAREE, and on MANNING RIVER; 31°52′S 152°42′E. Commercial, recreational fishing; whale-watching. Close to Crowdy Bay National Park.

Harrington, village (2006 population 416), LINCOLN county, E WASHINGTON, 13 mi/21 km SSW of DAVENPORT, on Coal Creek; 47°29′N 118°15′W. In COLUMBIA basin agricultural region; wheat, barley, oats, rye, alfalfa, potatoes; cattle. Coffeepot Lake and Twin Lakes to W.

Harrington Harbour (HA-reeng-tuhn HAHR-buhr), unincorporated village, E QUEBEC, E CANADA, on largest of the Harrington Islands, a group of 12 islets in Gulf of SAINT LAWRENCE; included in Côte-Nord-du-Golfe-du-Saint-Laurent; 50°30′N 59°29′W. Magnesite mining.

Harrington Lake, reservoir, PISCATAQUIS county, central MAINE, on sidestream of Soper Brook, 30 mi/48 km/64 km NW of MILLINOCKET; 4 mi/6.4 km long; 45°56′N 68°40′W. In wilderness recreational area.

Harrington Park, residential borough (2006 population 4,916), BERGEN county, NE NEW JERSEY, 11 mi/18 km NE of PATERSON; 40°59′N 73°58′W. Incorporated 1904.

Harrington Sound, landlocked lagoon, E BERMUDA Island; 2.25 mi/3.7 km long, 1.5 mi/2.4 km wide; has narrow, 200-ft/61-m entrance in NW.

Harris, county (□ 465 sq mi/1,209 sq km; 2006 population 28,785), W GEORGIA; ⊙ HAMILTON; 32°44′N 84°55′W. Bounded W by ALABAMA state line (formed here by CHATTAHOOCHEE RIVER). Piedmont livestock; agriculture (cotton, wheat, vegetables, fruit); cattle, hogs; and sawmilling area. Franklin D. Roosevelt State Park (NE). Langdale, Riverview, BARTLETTS FERRY, and GOAT ROCK dams create reservoirs on the Chattahoochee here. Formed 1827.

Harris, county (□ 1,777 sq mi/4,620.2 sq km; 2006 population 3,886,207), S TEXAS, on Gulf coast plains; ⊙ HOUSTON; 29°51′N 95°23′W. Seaport, industrial center. Important cities include Houston, PASADENA, BAYTOWN, and LA PORTE. Bounded on N by Spring Creek, on E in part by Cedar Creek, on S by Clear Creek and Clear Lake, on SE by GALVESTON BAY; drained by SAN JACINTO RIVER (forms LAKE HOUSTON reservoir in NE) and its tributaries. Forested in N; highly urbanized S and E; agriculture and timber in NW. Large oil, natural-gas production; also salt, sulphur, clay, lime, sand, and gravel; a leading cattle-raising county; irrigated agriculture (especially rice, corn, peanuts, vegetables, hay; some nurseries); some timber. Bay resorts. Houston Intercontinental Airport in N. Includes San Jacinto Battlefield State Historic Site and Battleship Texas State Historic Site both in E, on San Jacinto River; Lake Houston State Park in NE; Sheldon State Wildlife Management Area in E. Formed 1836.

Harris (HA-ris), township (□ 19 sq mi/49.4 sq km; 2001 population 518), E central ONTARIO, E central CANADA, 5 mi/8 km from NEW LISKEARD; 47°32′N 79°35′W.

Harris, town (2000 population 200), OSCEOLA county, NW IOWA, 15 mi/24 km E of SIBLEY; 43°27′N 95°25′W. In livestock and grain area.

Harris, town (2000 population 1,121), CHISAGO county, E MINNESOTA, 44 mi/71 km N of ST. PAUL, on Goose Creek; 45°36′N 92°59′W. Grain; cattle, poultry; dairying; light manufacturing. Small lakes in area.

Harris, town (2000 population 105), SULLIVAN county, N MISSOURI, 14 mi/23 km WNW of MILAN; 40°18′N 93°20′W.

Harris (HAR-uhs), village (2006 population 187), SE central SASKATCHEWAN, W CANADA, 45 mi/72 km SW of SASKATOON; 51°46′N 107°34′W. Flour milling; mixed farming.

Harris, village, E MONTSERRAT, WEST INDIES, 4 mi/6.4 km E of PLYMOUTH. Sea-island cotton, fruit. Abandoned in 1997 after eruption of SOUFRIÉRE Hills volcano; now lies in the exclusion zone.

Harris, village, ROUTT county, NW COLORADO, on YAMPA RIVER, near PARK RANGE, and 12 mi/19 km W of STEAMBOAT SPRINGS; elevation c.6,350 ft/1,935 m. Coal-mining point. Formerly Mount Harris.

Harris, village (2000 population 53), ANDERSON county, W KANSAS, 11 mi/18 km E of GARNETT; 38°19′N 95°26′W. Livestock; grain; dairying.

Harris, village, COVENTRY and WEST WARWICK towns, central RHODE ISLAND; 41°43′N 71°32′W.

Harris, Scotland: see LEWIS AND HARRIS.

Harrisburg, city (2000 population 9,860), ⊙ SALINE county, SE ILLINOIS, 25 mi/40 km ESE of WEST FRANKFORT; 37°44′N 88°32′W. Center of bituminous-coal-mining and agricultural area. Shawnee National Forest is S. Incorporated 1861. Annexed DORRISVILLE in 1923. Illinois Youth Center nearby.

Harrisburg, city (2006 population 47,164), ⊙ PENNSYLVANIA and DAUPHIN county, SE PENNSYLVANIA, 90 mi/145 km WNW of PHILADELPHIA, on the SUS-

QUEHANNA RIVER; 40°16'N 76°52'W. BLUE MOUNT RIDGE to N. Commercial, wholesale, administrative, and transportation center. Manufacturing (fabricated metal products, transportation equipment, food products, machinery, electrical and electronic equipment, building materials; steel fabricating, scrap metal processing, printing and publishing). Naval Ships Parts Center to W, Defense Distribution Center to SE. Harrisburg became the state capital in 1812 and grew as an inland transportation center with the opening of the Pennsylvania Canal in 1827 and the arrival of the railroad in 1836. Has numerous parks. Its sprawling Italian Renaissance state capitol (completed 1906) has a 272-ft/83-m dome modeled after St. Peter's in ROME. Other notable structures are the Education Building, which contains the state library; the Pennsylvania State Museum; the William Penn Memorial Museum; the John Harris Mansion (1766), founder of Harrisburg; and the Soldiers' and Sailors' Memorial Bridge. Seat of Harrisburg Area Community College, Penn State University Center; state hospital, county prison; Art Association of Harrisburg (exhibitions); State Farm Show Building; Pennsylvania National Race Course to NE; Harrisburg International Airport 8 mi/12.9 km SE at MIDDLETOWN; Capitol City Airport 3 mi/4.8 km to SSE at NEW CUMBERLAND; Olmstead Air Force Base to SE, at Middletown; Camp Hills State Correctional Institution to SW. THREE MILE ISLAND nuclear plant 10 mi/16 km to SE, site of a major nuclear accident in 1979. APPALACHIAN TRAIL passes to N, on PETERS MOUNTAIN RIDGE, and to W. Settled c.1710 by John Harris, who established a trading post and operated a ferry here; incorporated 1791.

Harrisburg, town (2000 population 2,192), ⊙ POINSETT county, NE ARKANSAS, 19 mi/31 km S of JONESBORO, on CROWLEY'S RIDGE; 35°33'N 90°43'W. In agricultural area; lumber milling; manufacturing (wood products, shoes, aggregate materials).

Harrisburg, town, BOONE county, central MISSOURI, 15 mi/24 km NNW of COLUMBIA; 39°08'N 92°27'W. Coal mining.

Harrisburg (HAR-uhs-buhrg), town (□ 6 sq mi/15.6 sq km; 2006 population 5,347), CABARRUS county, S central NORTH CAROLINA, 13 mi/21 km NE of CHARLOTTE, 7 mi/11.3 km SW of CONCORD, near ROCKY RIVER; 35°19'N 80°40'W. Service industries; manufacturing (transportation equipment, chemicals, building materials, fabricated metal products; galvanizing); agriculture (grain, soybeans; poultry, livestock; dairying).

Harrisburg, town (2006 population 3,405), LINN county, W OREGON, on WILLAMETTE RIVER. Manufacturing (hides, tallow, concrete); 44°16'N 123°09'W. Agriculture (fruit, vegetables, grain; cattle); dairy products. Timber. Washburn Wayside State Park to NW.

Harrisburg (HER-is-buhrg), village, ⊙ BANNER county, W NEBRASKA, 20 mi/32 km S of SCOTTSBLUFF. Grain; cattle. Wildcat Hills State Recreation Area to N.

Harrisburg (HER-uhs-buhrg), village (2006 population 312), on border between FRANKLIN and PICKAWAY counties, central OHIO, 13 mi/21 km SW of COLUMBUS; 39°49'N 83°10'W.

Harrisburg, village (2006 population 2,507), LINCOLN county, SE SOUTH DAKOTA, 8 mi/12.9 km S of SIOUX FALLS; 43°25'N 96°42'W. Manufacturing of cabinets.

Harrisfield (HA-ris-feeld), suburb 15 mi/24 km SE of MELBOURNE, VICTORIA, SE AUSTRALIA; 37°57'S 145°11'E.

Harris, Lake (HAR-uhs), Lake county, central FLORIDA; c.11 mi/18 km long, 6 mi/9.7 km wide; 28°44'N 81°45'W. Connected by canal with LAKE EUSTIS, it forms part of lake system drained by OKLAWAHA RIVER.

Harrislee (HAHR-ris-lai), town, SCHLESWIG-HOLSTEIN, N GERMANY, near Danish border, 2 mi/3.2 km NW of FLENSBURG; 54°48'N 09°22'E.

Harrismith, town, NE FREE STATE province, SOUTH AFRICA, 15 mi/24 km NW of KWAZULU-NATAL border, 10 mi/16km NE of STERKFONTEIN DAM, on WILGE RIVER, and 130 mi/209 km SE of VEREENIGING; 28°16'S 29°07'E. Elevation 5,321 ft/1,622 m. At foot of Platberg (7,462 ft/2,274 m). Maize production and wool-milling center; distributing point for E Free State province; stud sheep raising and horse breeding; agriculture market. Resort. Named for Sir Harry Smith, a governor of CAPE COLONY. Nearby are several caves with old Bushman paintings. Important road and railroad junction (junction of highways N3 and N5).

Harris Nuclear Power Plant, NORTH CAROLINA: see WAKE.

Harrison, county (□ 479 sq mi/1,245.4 sq km; 2006 population 36,992), S INDIANA; ⊙ CORYDON; 38°12'N 86°07'W. Bounded SE, S, and SW by OHIO River (here forming KENTUCKY state line), and W by BLUE River; drained by INDIAN CREEK and small Buck Creek. Agriculture (soybeans, corn, wheat; cattle, poultry); natural gas; limestone. Lumber milling; manufacturing of furniture, glass, dairy products. Stone quarries; timber. Corydon Capital State Memorial at Corydon. Part of Harrison Crawford State Forest in W. Formed 1808.

Harrison, county (□ 700 sq mi/1,820 sq km; 2006 population 15,745), W IOWA, on NEBRASKA state line (W; formed here by MISSOURI RIVER); ⊙ LOGAN; 41°41'N 95°48'W. Prairie agricultural area (cattle, hogs; corn, barley, oats) drained by BOYER, SOLDIER, and Little Sioux rivers. Wilson Island State Park in SW corner; Desoto Bend National Wildlife Refuge and Visitors' Center in SW. Several rivers flooded in 1993. Formed 1851.

Harrison, county (□ 309 sq mi/803.4 sq km; 2006 population 18,592), N KENTUCKY; ⊙ CYNTHIANA; 38°26'N 84°0'W. Bounded NE by LICKING River, S by Silas Creek; drained by South Fork of Licking River and several creeks. Gently rolling upland agricultural area, in BLUEGRASS REGION (burley tobacco, hay, alfalfa, soybeans, wheat, corn; hogs, cattle, poultry; dairying); timber; limestone quarries. Manufacturing at Cynthiana. Quiet Trails State Nature Preserve in N. Formed 1793.

Harrison, county (□ 976 sq mi/2,537.6 sq km; 2006 population 171,875), SE MISSISSIPPI; ⊙ GULFPORT and BILOXI, ports on MISSISSIPPI SOUND (S); 30°25'N 89°05'W. Drained by Biloxi, Wolf, and Tchoutacabouffa rivers and Bernard Bayou rivers. Agriculture (corn, pecans, citrus; cattle; timber); extensive seafood industries, especially at Biloxi. Includes part of De Soto National Forest in NW. Suffered widespread destruction, especially in coastal regions, during Hurricane Katrina in August 2005. Formed 1841.

Harrison, county (□ 720 sq mi/1,872 sq km; 2006 population 8,898), NW MISSOURI; ⊙ BETHANY; 40°21'N 93°58'W. Borders IOWA on N. Corn, wheat, hay; sheep, cattle; limestone. Light manufacturing at Bethany. Formed 1845.

Harrison (HER-i-suhn), county (□ 411 sq mi/1,068.6 sq km; 2006 population 15,799), E OHIO; ⊙ CADIZ; 40°17'N 81°04'W. Drained by STILLWATER and Conotton creeks. Includes Tappan Lake and CLENDENING LAKE reservoirs. In the Unglaciated Plain physiographic region. Agriculture (vegetables, nursery and greenhouse crops, hogs, grains, soybeans); manufacturing at Cadiz and SCIO (printing and publishing); limestone quarries. Some coal mining. Formed 1813.

Harrison, county (□ 984 sq mi/2,558.4 sq km; 2006 population 63,819), E TEXAS; ⊙ MARSHALL; 32°32'N 94°22'W. Commercial, industrial center. Bounded E by LOUISIANA state line, NE by CADDO LAKE (formed by BIG CYPRESS CREEK), SW by SABINE RIVER; drained by LITTLE CYPRESS BAYOU (forms part of N border). Hilly wooded region (extensive timber); agriculture (nurseries, hay); dairying; livestock (cattle,

hogs, horses). Large clay products industry; also oil, natural gas, coal; sand and gravel. Includes Caddo Lake State Park (recreation) on S shore of Caddo Lake, in NE. Formed 1839.

Harrison, county (□ 417 sq mi/1,084.2 sq km; 2006 population 68,745), N WEST VIRGINIA; ⊙ CLARKSBURG; 39°17'N 80°22'W. On ALLEGHENY PLATEAU; drained by the West Fork River, and Simpson and Tenmile creeks. Agriculture (corn, alfalfa, hay, nursery crops); livestock, poultry; dairying. Natural gas and oil wells; bituminous-coal mines. Manufacturing at Clarksburg and BRIDGEPORT. Watters Smith Memorial State Park in S; E part of North Bend State Trail in W. Formed 1784.

Harrison, city (2000 population 12,152), ⊙ BOONE county, N ARKANSAS, c.60 mi/97 km ENE of FAYETTEVILLE, in the OZARK MOUNTAINS; 36°14'N 93°07'W. Commercial center for farm area (fruit; cattle, hogs, poultry). Manufacturing (wood products, cheese, flour, clothing, fabricated metal products, consumer goods, building materials, transportation equipment, paper products, fiberglass products); die-casting plant, produce houses. Mystic Cave to S; Buffalo River National Park to S. Seat of Arkansas Community College. Growing retirement population. Plotted c.1860, incorporated 1876.

Harrison (HER-i-suhn), city (□ 4 sq mi/10.4 sq km; 2006 population 8,313), HAMILTON county, extreme SW OHIO, 19 mi/31 km WNW of CINCINNATI, on WHITEWATER RIVER, at INDIANA state line, contiguous to WEST HARRISON (Indiana); 39°14'N 84°48'W. Laid out 1813.

Harrison, town (2000 population 509), WASHINGTON county, E central GEORGIA, 12 mi/19 km SSE of SANDERSVILLE; 32°50'N 82°44'W.

Harrison, resort town, CUMBERLAND county, SW MAINE, 37 mi/60 km NW of PORTLAND, at N end of LONG LAKE; 44°06'N 70°38'W. Light manufacturing (machinery, building materials). Incorporated 1805.

Harrison, town (2000 population 2,108), ⊙ CLARE county, central MICHIGAN, 18 mi/29 km N of MOUNT PLEASANT, on W side of Budd Lake; 44°01'N 84°48'W. In agricultural area (livestock; potatoes, beans; dairy products); manufacturing (lumber, machinery). Resort area (lakes). Wilson State Park to E; Snowsnake Mountain Ski Area to S. Seat of Mid Michigan Community College Settled 1878, incorporated as city 1891.

Harrison, town (2006 population 13,942), HUDSON county, NE NEW JERSEY, an industrial suburb on the PASSAIC River, opposite NEWARK; 40°44'N 74°09'W. The town has several foundries. Manufacturing; important distributing hub. Incorporated 1869.

Harrison, township, ALLEGHENY county, W central PENNSYLVANIA, suburb 20 mi/32 km NE of PITTSBURGH; 40°37'N 79°43'W. Bounded E and S by ALLEGHENY River. Includes NATRONA and NATRONA HEIGHTS in S; rural in N. Agriculture (corn, hay; livestock; dairying).

Harrison, village (2000 population 267), KOOTENAI county, N IDAHO, 15 mi/24 km NW of SAINT MARIES, and on E shore of COEUR D'ALENE LAKE, at mouth of COEUR D'ALENE RIVER, 1 mi/1.6 km N of Coeur d'Alene Indian Reservation; 47°27'N 116°47'W. Agriculture; recreation. Coeur d'Alene National Forest to N.

Harrison, village, MADISON county, SW MONTANA, 36 mi/58 km W of BOZEMAN, and on North Willow and South Willow creeks, which closely parallel each other forming Willow Creek to NE. LEWIS AND CLARK CAVERNS State Park to N. Willow Creek Reservoir (HARRISON LAKE) to E, just NE of TOBACCO ROOT MOUNTAINS, in ranching and mining (gold, tungsten) region.

Harrison, village (2006 population 264), ⊙ SIOUX county, NW NEBRASKA, 45 mi/72 km WSW of CHADRON, near WYOMING state line; 42°41'N 103°52'W. In ranching region; livestock; poultry products; grain, potatoes. Relics of prehistoric man found nearby.

Area is shown by the symbol □, and capital city or county seat by ⊙.

Near source of WHITE RIVER. Oglala National Grasslands to N.

Harrison, residential village and town (□ 17 sq mi/44.2 sq km; 2006 population 26,337), WESTCHESTER county, SE NEW YORK, between MAMARONECK (SW) and RYE (NE), near LONG ISLAND SOUND; 41°01′N 73°43′W. Light manufacturing and commercial services. Village is coterminous with the town, and the two share a common government.

Harrison (HA-ri-suhn), rural municipality (□ 184 sq mi/478.4 sq km; 2001 population 837), SW MANITOBA, W central CANADA; 50°27′N 100°05′W. Includes SANDY LAKE, NEWDALE. Agriculture (rye, wheat, oats, barley, canola, alfalfa; livestock); tourism. Incorporated 1883.

Harrison Bay, N ALASKA, shallow inlet of BEAUFORT SEA, between CAPE HALKETT (W) and BEECHEY POINT (E), c.120 mi/193 km ESE of BARROW; 70°40′N 151°15′W. Receives COLVILLE River.

Harrisonburg (HA-ri-suhn-buhrg), independent city (□ 17 sq mi/44.2 sq km; 2006 population 40,885), ⊙ surrounding ROCKINGHAM county, NW VIRGINIA, in SHENANDOAH VALLEY; 38°26′N 78°52′W. Railroad junction. Manufacturing (computers, clothing, paper products, machinery, building materials, fabricated metal products, transportation equipment, chemicals, furniture, animal feeds, fertilizers; sheet metal fabrication, printing and publishing). Processing center in an agricultural area (poultry, livestock; dairying; also grain, apples, peaches, soybeans). General T. J. (Stonewall) Jackson ended his Valley Campaign just E in 1862. Seat of James Madison University, Eastern Mennonite University. Headquarters of George Washington National Forest. Limestone caverns in area: Endless Caverns to NE, Grand Caverns to S; SHENANDOAH NATIONAL PARK to SE. Settled 1739, incorporated 1916.

Harrisonburg, village (2000 population 746), ⊙ CATAHOULA parish, E LOUISIANA, on OUACHITA RIVER, and 30 mi/48 km NW of NATCHEZ (MISSISSIPPI); 31°46′N 91°49′W. In agricultural area (cotton, corn, fruit, sorghum, sweet potatoes, peas; cattle, horses); catfish. Sicily Island Hills State Wildlife Area to NE.

Harrison, Cape, promontory on the Atlantic Ocean, E NEWFOUNDLAND AND LABRADOR, E CANADA; 54°46′N 58°26′W. Site of navigation radio station, 130 mi/209 km NE of Happy Valley–Goose Bay. Offshore oil and gas exploration.

Harrison City, unincorporated town, WESTMORELAND county, SW PENNSYLVANIA, suburb 2 mi/3.2 km NWS of JEANNETTE; 40°21′N 79°38′W. Manufacturing of motor controls; agriculture includes dairying; livestock; corn, hay. Bushy Run Battlefield Historical Site to E.

Harrison Hot Springs (HA-ri-suhn HAHT SPREENGZ), village (□ 2 sq mi/5.2 sq km; 2001 population 1,343), SW BRITISH COLUMBIA, W CANADA, at S end of HARRISON LAKE, 12 mi/19 km NE of CHILLIWACK, and in FRASER VALLEY regional district; 49°18′N 121°47′W. Resort, with mineral springs.

Harrison Lake (HA-ri-suhn), (□ 87 sq mi/226.2 sq km), S BRITISH COLUMBIA, W CANADA, 12 mi/19 km NE of CHILLIWACK; 30 mi/48 km long, 1 mi/2 km–5 mi/8 km wide; 49°33′N 121°50′W. Contains Long Island (6 mi/10 km long). Receives LILLOOET RIVER (NW); drains S into FRASER River.

Harrison Mills (HA-ri-suhn), unincorporated village, S BRITISH COLUMBIA, W CANADA, on FRASER River, at mouth of Harrison River (outlet of HARRISON LAKE), 5 mi/8 km N of CHILLIWACK, in FRASER VALLEY regional district; 49°15′N 121°57′W. Lumbering.

Harrison's Cave, natural limestone and coral cavern, NE BRIDGETOWN, St. Thomas parish, central BARBADOS. Said to be largest caverns in WEST INDIES. Underground tours operated by Barbados National Trust.

Harrison Stickle, ENGLAND: see LANGDALE PIKES.

Harrisonville, city (2000 population 8,946), W MISSOURI, ⊙ CASS county, 32 mi/51 km SSE of KANSAS CITY; 38°39′N 94°20′W. Corn, wheat, sorghum; cattle. Manufacturing (consumer goods, machinery). Laid out 1837.

Harriston (HA-ri-stuhn), unincorporated town (□ 1 sq mi/2.6 sq km; 2001 population 2,034), S ONTARIO, E central CANADA, on Maitland River, 40 mi/64 km NW of GUELPH, and included in town of MINTO; 43°55′N 80°53′W. Meatpacking, dairying, woodworking.

Harristown, town (2000 population 1,338), MACON county, central ILLINOIS, 7 mi/11.3 km W of DECATUR; 39°50′N 89°03′W. Wheat, corn, soybeans. Lincoln Trail Homestead State Park to S.

Harrisville, town, CHESHIRE county, SW NEW HAMPSHIRE, 9 mi/14.5 km E of KEENE; 42°56′N 72°05′W. Agriculture (dairying; poultry, cattle, sheep; nursery crops); manufacturing (textiles, water coolers). Skatutakee Lake in S center; SILVER LAKE, Childs Bog in NW.

Harrisville, town (2006 population 5,247), WEBER county, N UTAH, residential suburb 4 mi/6.4 km NW of OGDEN; 41°16′N 111°58′W. Elevation c.4,400 ft/1,341 m. Agricultural area (vegetables, fruit, alfalfa, barley; dairying; cattle). WASATCH RANGE and National Forest to E. Ogden Defense Depot to S. Settled 1850.

Harrisville, town (2006 population 1,875), ⊙ RITCHIE county, NW WEST VIRGINIA, on North Fork of HUGHES RIVER, 27 mi/43 km E of PARKERSBURG; 39°12′N 81°02′W. Agriculture (corn); livestock, poultry. Manufacturing (clothing, building materials, textiles). Oil production. North Bend State Park to W; North Bend State Trail to NW. Plotted 1822.

Harrisville, village (2000 population 514), ⊙ ALCONA county, NE MICHIGAN, 30 mi/48 km SSE of ALPENA, on LAKE HURON; 44°39′N 83°17′W. Summer resort; nurseries; manufacturing (tool and die, concrete). Harrisville State Park on lake; Newegon State Park 21 mi/34 km N.

Harrisville, village (2006 population 612), LEWIS county, N central NEW YORK, on West Branch of OSWEGATCHIE RIVER, and 33 mi/53 km ENE of WATERTOWN; 44°08′N 75°19′W.

Harrisville (HER-uhs-vil), village (2006 population 263), HARRISON county, E OHIO, 8 mi/13 km SE of CADIZ; 40°11′N 80°53′W. In agricultural and former coal-mining area.

Harrisville, village (2000 population 1,561), BURRILLVILLE town, PROVIDENCE county, NW RHODE ISLAND, 15 mi/24 km NW of PROVIDENCE; 41°58′N 71°41′W.

Harrisville (HER-is-vil), borough (2006 population 900), BUTLER county, W PENNSYLVANIA, 4 mi/6.4 km ESE of GROVE CITY. Manufacturing (limestone processing; concrete); agricultural (potatoes, corn, hay; livestock; dairying).

Harrod (HER-uhd), village (2006 population 476), ALLEN county, W OHIO, 10 mi/16 km ESE of LIMA; 40°42′N 83°55′W. In agricultural area; wood products.

Harrodsburg (HA-ruhdz-buhrg), town (2000 population 8,014), ⊙ MERCER county, central KENTUCKY, S of FRANKFORT; 37°46′N 84°50′W. Trade center in BLUEGRASS REGION producing livestock; grain and tobacco; dairying; manufacturing (paper products, optical lenses, shipping goods, textiles). Tourist and resort city, with mineral springs. The oldest settlement W of the ALLEGHENIES, it was founded in 1774 by James Harrod. One of the settlement's early leaders was George Rogers Clark. OLD FORT HARROD STATE PARK to NE contains a replica of old fort (1775), site of state's first school, Historical Society Museum, Butaan War Memorial.

Harrodsburg Dam, KENTUCKY: see DIX RIVER.

Harrogate (HA-ruh-gait), town (2001 population 85,128), NORTH YORKSHIRE, N central ENGLAND; 53°59′N 01°32′W. Residential with a tourist economy and spa, having over eighty mineral springs. Popular trade exhibition and conference center with varied

light industries. Knaresborough Spa is 3 mi/4.8 km NE.

Harrogate (HA-ruh-gait), village, CLAIBORNE county, NE TENNESSEE, near KENTUCKY-VIRGINIA state line, 5 mi/8 km S of CUMBERLAND GAP; 36°34′N 83°39′W. Some industry. Seat of Lincoln Memorial University; museum on campus houses one of the largest Lincoln collections in the world.

Harrold (HA-ruhld), village (2001 population 2,780), NW BEDFORDSHIRE, central ENGLAND, on OUSE RIVER, and 8 mi/12.9 km NW of BEDFORD; 52°15′N 00°33′W. Leatherworks. Has 13th–14th-century church.

Harrold, village (2006 population 205), HUGHES county, central SOUTH DAKOTA, 22 mi/35 km ENE of PIERRE, and on Medicine Knoll Creek; 44°31′N 99°44′W.

Harrouch, El (hah-ROOSH, el), village, SKIKDA wilaya, NE ALGERIA, in irrigated SAF SAF valley, 16 mi/26 km S of SKIKDA. Olive-oil pressing, flour milling; tobacco. ZARDÉZAS DAM is 4 mi/6.4 km SE.

Harrow (HA-ro), former town (□ 1 sq mi/2.6 sq km; 2001 population 2,935), S ONTARIO, E central CANADA, near Lake ERIE, 17 mi/27 km S of WINDSOR; 42°02′N 82°55′W. In dairying, farming region. Amalgamated into the town of ESSEX in 1999.

Harrow (HA-ro), outer borough (□ 19 sq mi/49.4 sq km; 2001 population 206,814) of GREATER LONDON, SE ENGLAND; 51°35′N 00°20′W. Area previously grew foodstuffs for London. It is now mainly residential and contains parts of the Green Belt, areas set aside as parkland. Manufacturing of optical and photographic goods and glass. The famous Harrow public school, founded in 1572, is at Harrow on the Hill. Among its graduates were the writers George Byron and John Galsworthy and the statesmen Sir Robert Peel and Henry Palmerston. To the N lie Harrow Weald and Wealdstone. Other districts in the borough include Belmont, Pinner, and Stanmore.

Harrow (HA-ro), hamlet, VICTORIA, SE AUSTRALIA, 234 mi/376 km NW of MELBOURNE, in Wimmera district, and on GLENELG RIVER; 37°10′S 141°36′E. Historic buildings.

Harrowsmith (HA-ro-smith), unincorporated village, SE ONTARIO, E central CANADA, 15 mi/24 km NW of KINGSTON, and included in township of SOUTH FRONTENAC; 44°24′N 76°39′W. Dairying, mixed farming.

Harrow Weald, ENGLAND: see HARROW.

Harry S. Truman Historic Site, INDEPENDENCE, JACKSON county, W central MISSOURI. Home of U.S. President Harry S. Truman from 1919 until 1972. Authorized 1983. Truman Museum and Library nearby.

Harry S. Truman Reservoir (□ 87 sq mi/225 sq km), W central MISSOURI, on OSAGE RIVER, 31 mi/50 km SSW of SEDALIA; 38°16′N 93°24′W. Maximum capacity 5,202,000 acre-ft. Star shaped; has four main arms radiating from dam area. Fed by POMME DE TERRE, Horse, and South GRAND rivers. Formed by Harry S. Truman Dam (98 ft/30 m high), built (1978) by Army Corps of Engineers for power generation; also used for flood control and recreation. Harry S. Truman State Park near dam.

Harry Strunk Lake, reservoir (9 sq mi/23.3 sq km), FRONTIER county, S central NEBRASKA, on MEDICINE CREEK, 24 mi/39 km ENE of MCCOOK; 40°23′N 100°13′W. Maximum capacity 194,080 acre-ft. Formed by Medicine Creek Dam (165 ft/50 m high), built (1949) by the Bureau of Reclamation for irrigation; also used for flood control. Medicine Creek Reservoir State Recreational Area to W.

Harsefeld (HAHR-se-felt), village, LOWER SAXONY, NW GERMANY, 10 mi/16 km S of STADE; 53°27′N 09°30′E. Metalworking.

Harsens Island, SAINT CLAIR county, SE MICHIGAN, in delta of SAINT CLAIR RIVER, in Lower SAINT CLAIR, opposite WALPOLE ISLAND (ONTARIO); c.5 mi/8 km long, 3 mi/4.8 km wide. Summer resort, known for fishing. Sans Souci village (area noted for waterfowl) is on E shore. Settled c.1779.

Harsewinkel (hahr-se-VING-kel), town, NORTH RHINE–WESTPHALIA, W GERMANY, on EMS River, 14 mi/23 km W of BIELEFELD; 52°58′N 08°14′E. Industrial town (manufacturing of agricultural machines, metal- and woodworking, and meat processing). Has 12th-century church and 13th-century Cistercian abbey.

Harsin (hahr-SEEN), town (2006 population 51,636), Kermānshāhān province, W IRAN, 30 mi/48 km E of KERMANSHAH. Grain, fruit, cotton, tobacco; dairy products; sheep raising. Tribal Lur population. Also spelled HERSIN.

Harsit River, 83 mi/134 km long, NE TURKEY; rises in Gümüshane Mountains 20 mi/32 km W of GÜMÜSHANE; flows NW to BLACK SEA near TIREBOLU.

Hárs, Mount (HAHRSH), Hungarian *Hárshegy*, hill (1,502 ft/458 m) in BUDA MOUNTAINS, N central HUNGARY; residential district of Hüvösvölgy. Lipótmező Asylum at foot.

Harsovo, Bulgaria: see HURSOVO.

Harspränget, SWEDEN: see LULEÄLVEN.

Harstad (HAHR-stah), town (2007 population 23,261), TROMS county, NW NORWAY, on HINNØYA, the largest island of Norway. It is a fishing center and a bunkering place for coastal steamers and trawlers. Nearby is the fortified church of Trondenes (thirteenth century).

Harsud (huhr-SOOD), town (2001 population 15,869), EAST NIMAR district, SW MADHYA PRADESH state, central INDIA, 31 mi/50 km NE of KHANDWA; 22°02′N 76°42′E. Cotton ginning; millet, wheat, oilseeds.

Harsum (HAHR-sum), town, LOWER SAXONY, NW GERMANY, 4 mi/6.4 km N of HILDESHEIM; 52°13′N 09°58′E. Sugar refining.

Hart, county (□ 257 sq mi/668.2 sq km; 2006 population 24,276), NE GEORGIA; ⊙ HARTWELL; 34°21′N 82°58′W. Bounded E and N by SOUTH CAROLINA state line, formed here by SAVANNAH and TUGALOO rivers. Piedmont agricultural area (cotton, wheat, soybeans, corn, hay, sweet potatoes; cattle, hogs, poultry). Formed 1853.

Hart, county (□ 417 sq mi/1,084.2 sq km; 2006 population 18,547), central KENTUCKY; ⊙ MUNFORDVILLE; 37°17′N 85°53′W. Bounded NE by NOLIN RIVER (including part of Nolin River Lake reservoir); drained by GREEN RIVER. Rolling agricultural area (burley tobacco, corn, wheat, soybean). Includes part of MAMMOTH CAVE NATIONAL PARK in SW; many limestone caves in area; Kentucky, MAMMOTH ONYX, and HIDDEN RIVER caves; American Cave Museum in S, at Horse Cave town. Formed 1819.

Hart, town (2000 population 1,950), ⊙ OCEANA county, W MICHIGAN, 33 mi/53 km NNW of MUSKEGON, and on short Pentwater River; 43°42′N 86°21′W. Potatoes, fruit, vegetables (especially asparagus), beans; livestock, poultry; dairy; manufacturing (food processing, confections, and cherry products); resort. Charles Mears State Park to NW; Silver Lake State Park to SW; Manistee National Forest to E and N. Incorporated as village 1885, as city 1947.

Hart, town (2006 population 1,071), CASTRO county, NW TEXAS, 25 mi/40 km NW of PLAINVIEW; 34°23′N 102°06′W. Agricultural area (cattle, sheep, hogs; corn, wheat, cotton). Manufacturing (fertilizer).

Harta (HAHR-tah), village, Pest-Pilis-Solt-Kiskún county, central HUNGARY, on the DANUBE River, and 34 mi/55 km SW of KECSKEMÉT; 46°42′N 19°02′E. River port; barley, corn; cattle, poultry.

Hartberg (HAHRT-berg), town, E STYRIA, SE AUSTRIA, 28 mi/45 km NE of GRAZ, near the BURGENLAND border; 47°17′N 15°58′E. Market center; manufacturing includes metals, machines, implements, clothes, textiles; dairy products; drinks, vineyards; Medieval town walls; monastery; Romanesque bone house dating to before 1167.

Hartebees River, 70 mi/113 km long, NORTHERN CAPE province, SOUTH AFRICA; rises in Verneuk pan, 30 mi/ 48 km S of KENHARDT, joined by Zak River from SSW and short Mottels River from E; flows NW, past Kenhardt, to ORANGE RIVER 3 mi/4.8 km W of KAKAMAS. Drains Groot Vloer and Verneuk pans.

Hartel Canal (HAHR-tuhl kah-NAHL), NETHERLANDS, inland shipping connection between EUROPOORT and Old Meute, ROTTERDAM port area.

Hartfield (HAHT-feeld), village (2001 population 2,523), East SUSSEX, SE ENGLAND, on MEDWAY RIVER, and 6 mi/9.7 km ESE of East GRINSTEAD; 51°02′N 00°03′E. Agricultural market. Has 15th-century church.

Hartford, county (□ 750 sq mi/1,950 sq km; 2006 population 876,927), central and N CONNECTICUT, on MASSACHUSETTS state line, bisected by CONNECTICUT River; ⊙ HARTFORD; 41°48′N 72°43′W. Manufacturing (airplanes, machinery, hardware, tools, building materials, paper, clothing, food products, rubber products, leather products, furniture, fabricated metal products, textiles, wood products, consumer goods, electronic goods, transportation equipment, chemicals). Agriculture (tobacco, dairy products, poultry, vegetables, fruit, corn, potatoes, nursery products, seeds). Includes several state parks and forests. Drained by FARMINGTON, QUINNIPIAC, PEQUABUCK, HOCKANUM, and SCANTIC rivers. Constituted 1666.

Hartford, city (2000 population 121,578), ⊙ Connecticut and HARTFORD counties, central CONNECTICUT, on the W bank of the CONNECTICUT River; 41°46′N 72°40′W. Settled as Newtown 1635–1636 on the site of a Dutch trading post (1633; abandoned 1654). The second-largest city in the state, it is a port of entry and a world-famous insurance center. Its insurance business began in 1794, and the area remains home to the headquarters of several major companies (although mergers and downsizing during the 1990s diminished the insurance industry's importance somewhat). During the 1970s and 1980s, however, many insurance companies branched out of the city into the growing suburban locations. Manufacturing includes precision instruments, computers, transportation equipment, firearms, and electric equipment. One of the earliest and strongest colonial centers, Hartford and two other towns formed (1639) the Connecticut Colony, adopting the Fundamental Orders. From 1701 to 1875 it was joint capital with NEW HAVEN. It was an important military supply depot during the American Revolution, and in 1814–1815, it hosted the Hartford Convention. Landmarks include the Old State House (1796; designed by Charles Bulfinch), where the Hartford Convention met; the site of the Charter Oak; the capitol (completed 1878; designed by Richard M. Upjohn); and the famous Travelers Insurance tower. The Connecticut state library includes the Colt collection of firearms. Has a noted art museum (the Wadsworth Atheneum), a symphony orchestra, and an opera company. Other attractions are the Harriet Beecher Stowe House (1871), where Stowe lived 1873–1896, and the Mark Twain Memorial (1873–1874). Noah Webster, John Fiske, and the elder J. P. Morgan were born here; the theologian Horace Bushnell, the author Charles Dudley Warner, and the poet Wallace Stevens lived here. The *Hartford Courant*, founded in 1764, is one of the country's oldest newspapers. The city's many parks include Elizabeth Park, scene of an annual rose festival, and Colt Park. Among Hartford's institutions of higher education are Trinity College, Capital Community Technical College, the University of Hartford, Hartford College for Women, and a branch of the University of Connecticut. There is also the American School for the Deaf (in WEST HARTFORD) and the Connecticut Institute for the Blind. Constitution Plaza, a 15-acre/6-ha development project, was completed in 1964. Incorporated 1784.

Hartford, town (2000 population 2,369), Geneva co., SE Alabama, 12 mi/19 km ENE of Geneva, between Choctawhatchee River and Florida state line. Pecan shelling, cotton ginning, lumber milling. Founded 1894.

Hartford, town (2000 population 759), WARREN county, S central IOWA, 10 mi/16 km NE of INDIANOLA; 41°27′N 93°24′W. Sorghum mill.

Hartford, town, ⊙ OHIO county, W KENTUCKY, on ROUGH RIVER, and 25 mi/40 km SSE of OWENSBORO; "twin city" of BEAVER DAM 4 mi/6.4 km to SSE; 37°27′N 86°53′W. In coal, timber, limestone, and agricultural (corn, burley tobacco, hay; cattle, hogs) area; manufacturing (crushed limestone; apparel; lumber). Founded 1782.

Hartford, town, OXFORD county, W MAINE, 14 mi/23 km NE of SOUTH PARIS, and on branch of NEZINSCOT RIVER; 44°22′N 70°19′W. In farming, recreational area; wood products.

Hartford, town (2000 population 2,476), VAN BUREN county, SW MICHIGAN, 16 mi/26 km NE of BENTON HARBOR, and on PAW PAW RIVER; 42°12′N 86°10′W. In fruit-growing area; nurseries; vegetables; ships fruit; winery. Manufacturing (fruit and vegetable processing; asphalt). Incorporated 1877.

Hartford, town (2006 population 2,123), MINNEHAHA county, E SOUTH DAKOTA, 12 mi/19 km WNW of SIOUX FALLS; 43°37′N 96°56′W. Manufacturing (bulk conveyors); honey. Wild Game Farm.

Hartford, town, WINDSOR CO., E VERMONT, on the CONNECTICUT River, at mouth of WHITE RIVER, and 9 mi/14.5 km E of WOODSTOCK; 43°39′N 72°23′W. Includes residential villages of Hartford and Wilder, industrial and transportation center White River Junction, and Quechee village (E of Quechee Gorge whose top is 162 ft/49 m above the OTTAUQUECHEE RIVER; has a small woolen mill). Gateway to resort area (W). Large hydroelectric dam in the Connecticut Rivert at Wilder. Settled 1765.

Hartford, town (2006 population 13,265), WASHINGTON county, E WISCONSIN, on small Rubicon River (tributary of ROCK RIVER), near small Pike Lake (resort), and 30 mi/48 km NW of MILWAUKEE; 43°19′N 88°23′W. In dairy and farm area. Cheese, canned vegetables; manufacturing (consumer goods, wood products, plastic products, tool and die, electrostatic powder coating, transportation equipment, furniture, fabricated metal products, beverages; tanning, metal fabricating). Pike Lake State Park to E. Settled c.1844, incorporated 1883.

Hartford, village (2000 population 772), SEBASTIAN county, W ARKANSAS, 24 mi/39 km S of FORT SMITH; 35°01′N 94°22′W. In diversified agricultural area. Ouachita National Forest to S.

Hartford, village (2000 population 1,545), MADISON county, SW ILLINOIS, on the MISSISSIPPI RIVER, industrial suburb 15 mi/24 km NNE of downtown SAINT LOUIS, within Saint Louis metropolitan area; 38°49′N 90°05′W. Oil and copper refining. Lewis and Clark State Memorial on river, opposite confluence of Missouri River, starting point of Lewis and Clark Expedition to NW. Incorporated 1920.

Hartford, village (2000 population 500), LYON county, E central KANSAS, on NEOSHO River, and 14 mi/23 km SE of EMPORIA; 38°18′N 95°57′W. Grain; livestock. At W (upstream) end of John Redmond Reservoir.

Hartford (HAHRT-fuhrd), village (2006 population 398), LICKING county, central OHIO; 40°14′N 82°41′W.

Hartford, village, MASON county, W WEST VIRGINIA, on OHIO River, 14 mi/23 km NNE of POINT PLEASANT, and 3 mi/4.8 km SE of POMEROY (OHIO); 39°00′N 81°59′W. Agriculture (grain, tobacco); livestock; dairying. Also known as Hartford City.

Hartford, ENGLAND: see CRAMLINGTON.

Hartford City, city (2000 population 6,928), ⊙ BLACKFORD county, E INDIANA, 18 mi/29 km N of MUNCIE; 40°27′N 85°22′W. In rich agricultural area (livestock; dairy products; soybeans, grain). Natural gas and oil fields nearby. Manufacturing (glass, rubber products, transportation equipment, concrete products, canned goods, consumer goods, lumber products). Settled 1832, laid out 1839.

Area is shown by the symbol □, and capital city or county seat by ⊙.

Hartford City, WEST VIRGINIA: see HARTFORD, village.

Hartha (HAHR-tah), town, SAXONY, E central GERMANY, 7 mi/11.3 km WSW of DÖBELN; 51°06′N 12°58′E. Manufacturing (electric motors); textile milling (wool, linen, cotton).

Harth Forest (AHRT), German *Hardt*, HAUT-RHIN department, in ALSACE, E FRANCE, extending c.20 mi/32 km N-S near left bank of the RHINE RIVER, just E of MULHOUSE; 50°47′N 13°53′E. Traversed by sections of RHÔNE-RHINE CANAL.

Harthill, village, Lothian, Scotland, 6 mi/9.7 km SW of Bathgate. Service area on M8 motorway.

Harthill (HAHRT-hil), village (2001 population 3,575), NORTH LANARKSHIRE, central Scotland, on Almond River, and 2 mi/3.2 km W of WHITBURN; 55°51′N 03°45′W. Agriculture. Service area on M8 motorway. Formerly in STRATHCLYDE, abolished 1996.

Hartington, town (2006 population 1,541), ☉ CEDAR county, NE NEBRASKA, 30 mi/48 km SSE of YANKTON (SOUTH DAKOTA), near MISSOURI RIVER; 42°37′N 97°15′W. Grain; manufacturing (machinery, cheese products, printing). Incorporated 1883.

Hartington (HAHT-ing-tuhn), village (2006 population 1,600), W DERBYSHIRE, central ENGLAND, 9 mi/14.5 km SSE of BUXTON; 53°08′N 01°48′W. Market center in dairying region. Former site of limestone-quarrying. Has 14th-century church and 16th-century mansion.

Hart Island, New York: see HARTS ISLAND.

Hart Lake Reservoir (□ 12 sq mi/31 sq km), LAKE county, S central OREGON, in the Warner Valley, 25 mi/40 km NE of LAKEVIEW; 42°27′N 119°50′W. Maximum capacity 52,150 acre-ft. Formed by Hart Lake Reservoir Dam (50 ft/6 m high), built (1963) for irrigation. Hart Mountain National Antelope Refuge just E.

Hartland, town (2001 population 902), W NEW BRUNSWICK, E CANADA, on ST. JOHN RIVER (longest covered bridge in the world; 1,283 ft/391 m), and 10 mi/16 km W of WOODSTOCK, near U.S. (MAINE) border; 46°18′N 67°32′W. Agricultural market in Irish potato region; woodworking.

Hartland, town, HARTFORD county, N CONNECTICUT, in hilly region, on MASSACHUSETTS state line, and 21 mi/34 km NW of HARTFORD; 42°00′N 72°57′W. Includes East Hartland village. Agriculture and manufacturing of wood products. Part of BARKHAMSTED RESERVOIR (on East Branch FARMINGTON RIVER), state forests here.

Hartland, town, SOMERSET county, central MAINE, on the SEBASTICOOK river, and 15 mi/24 km NE of SKOWHEGAN; 44°53′N 69°30′W. In farming area; tannery. Settled c.1800, incorporated 1820.

Hartland, town, WINDSOR CO., E VERMONT, on the CONNECTICUT River, 10 mi/16 km SE of WOODSTOCK; 43°34′N 72°25′W. In dairying area. Includes villages of North Hartland, at mouth of OTTAUQUECHEE RIVER, and Hartland Four Corners. Settled 1763. Named Hertford until 1782.

Hartland, town (2006 population 8,689), WAUKESHA county, SE WISCONSIN, on BARK RIVER, and 22 mi/35 km W of MILWAUKEE; 43°06′N 88°20′W. In dairying and farming area with resort lakes nearby. Manufacturing of dairy products, wood products, steel products, medical equipment, microfiche, fiberglass, plastic products, teflon seals; machining.

Hartland (HAHT-luhnd), village (2001 population 1,600), NW DEVON, SW ENGLAND, 13 mi/21 km W of BIDEFORD; 50°59′N 04°29′W. Former agricultural market. On the Atlantic Ocean, 2 mi/3.2 km W, is fishing village of Hartland Quai. HARTLAND POINT is 3 mi/4.8 km N.

Hartland, village, LIVINGSTON county, SE MICHIGAN, 10 mi/16 km NE of HOWELL; 42°39′N 83°45′W. In farm area. Numerous lakes in area.

Hartland, village (2000 population 288), FREEBORN county, S MINNESOTA, 13 mi/21 km NNW of ALBERT LEA; 43°47′N 93°29′W. Dairying; poultry; grain.

Hartland Point (HAHT-luhnd), high promontory, NW DEVON, SW ENGLAND, on the Atlantic Ocean, at entrance to BRISTOL CHANNEL, 14 mi/23 km N of BUDE; 51°02′N 04°31′W. Lighthouse.

Hartlebury (HAH-tuhl-buh-ree), village (2001 population 2,549), Worcestershire, W ENGLAND, 4 mi/6.4 km S of KIDDERMINSTER; 52°23′N 02°14′W. Agricultural market. Has castle of Bishops of Worcester begun in early 12th century.

Hartlepool (HAH-tli-pool), town and county (□ 36 sq mi/93.6 sq km; 2001 population 88,611), NE ENGLAND, about 8 mi/12.9 km N of MIDDLESBROUGH; 54°41′N 01°12′W. A prominent seaport in England, Hartlepool imports timber, wood pulp, petroleum, and iron ore. Industries include shipbuilding, iron and steel manufacturing, marine engineering, and brewing. Servicing the NORTH SEA oil fields has become increasingly important. Also the home of a herring fleet. A convent founded on the site in 640 was famous under St. Hilda (649–657) and was destroyed by the Danes in 800. In the 12th and 13th century, Hartlepool was the chief port of the palatinate of Durham. The West Hartlepool dock developed in the 19th century as a port for coal export. At nearby Greythorp is the Hartlepool nuclear power plant (two units, 1,250 MW capacity), which went on-line in the mid-1980s.

Hartleton (HAHR-tel-tuhn), borough (2006 population 264), UNION county, central PENNSYLVANIA, 15 mi/24 km WSW of LEWISBURG. Corn, hay; dairying. Parts of Bald Eagle State Forest to N and S.

Hartley (HAHRT-lee), county (□ 1,463 sq mi/3,803.8 sq km; 2006 population 5,335), extreme N TEXAS; ☉ CHANNING; 35°50′N 102°36′W. Elevation 3,400 ft/1,036 m–4,400 ft/1,341 m. In high, grassy plains of the PANHANDLE, and bounded W by NEW MEXICO state line. Drained by RITA BLANCA, Carrizo, and Punta de Agua creeks. Large-scale cattle-ranching area; extensively irrigated region, wheat, corn, sorghum; natural-gas wells. Includes Rita Blanca Lake (recreational area) in N. Formed 1876.

Hartley (HAHRT-lee), township, NEW SOUTH WALES, SE AUSTRALIA, 83 mi/133 km from SYDNEY, below W escarpment of BLUE MOUNTAINS; 33°33′S 150°10′E. Established 1815. Administered by the New South Wales government; old courthouse open to the public.

Hartley, town (2000 population 1,733), O'BRIEN county, NW IOWA, 10 mi/16 km NE of PRIMGHAR; 43°10′N 95°28′W. Dairy, wood, and metal products. Incorporated 1888.

Hartley (HAHT-lee), village (2001 population 5,871), NW KENT, SE ENGLAND, 5 mi/8 km E of SWANLEY, on ROTHER RIVER; 51°22′N 00°19′E.

Hartley, ENGLAND: see SEATON DELAVAL.

Hartley State Historic Site (HAHRT-lee), E central NEW SOUTH WALES, AUSTRALIA, 85 mi/137 km W of SYDNEY, in W foothills of BLUE MOUNTAINS, at base of Victoria Pass; covers 25 acres/10 ha. Stopover for travelers during Australia's colonial period (early 1800s). Courthouse built by convicts, 1837, scene of many convict trials, is now a museum; visitors' center at old Roman Catholic presbytery. Established 1972.

Hartline, village (2006 population 143), GRANT county, E central WASHINGTON, 33 mi/53 km NE of EPHRATA; 47°42′N 119°07′W. Alfalfa; cattle. BANKS LAKE reservoir in GRAND COULEE, to NW.

Hartly, village, KENT county, W DELAWARE, 10 mi/16 km W of DOVER; 39°10′N 75°42′W. Elevation 45 ft/13 m. In agricultural area.

Hartman, village (2000 population 596), JOHNSON county, NW ARKANSAS, 8 mi/12.9 km WSW of CLARKSVILLE, near ARKANSAS RIVER (Lake DARDANELLE); 35°25′N 93°37′W.

Hartman, village (2000 population 111), PROWERS county, SE COLORADO, near ARKANSAS RIVER, and KANSAS state line, 22 mi/35 km E of LAMAR; 38°07′N 102°13′W. Elevation c.3,600 ft/1,100 m. Terminus of railroad spur from Lamar.

Hartmanice (HAHRT-mah-NYI-tse), German *Hartmanitz*, town, ZAPADOCESKY province, SW BOHEMIA, CZECH REPUBLIC, in SUMAVA mountains, 5 mi/8 km SSW of SUSICE; 49°10′N 13°28′E. Mountain resort. PRASILY, Czech *Prášily* (PRAH-shi-LI), summer resort, is 6 mi/9.7 km SSW.

Hartmannsdorf (HAHRT-mahns-dorf), town, SAXONY, E central GERMANY, 7 mi/11.3 km NW of CHEMNITZ; 50°44′N 12°48′E.

Hartmannswillerkopf (ahrt-mahnz-vee-LER-KOF) or **Vieil-Armand** (vee-ye–ahr-mahn), summit (3,136 ft/956 m) of the SE VOSGES MOUNTAINS, HAUT-RHIN department, E FRANCE, in ALSACE, 4 mi/6.4 km SSW of GUEBWILLER, commanding the valleys of the THUR (S) and the LAUCH (N); 47°52′N 07°13′E. Scene (1915) of fierce struggles in World War I. Numerous war monuments, including national monument dedicated to 10,000 French war dead.

Hart Mountain (HAHRT), (2,700 ft/823 m), W MANITOBA, W central CANADA, 27 mi/43 km NNW of SWAN RIVER; highest point of PORCUPINE MOUNTAIN; 52°28′N 101°25′W.

Hart Mountain, peak (7,710 ft/2,134 m), LAKE county, S OREGON, rising from high plateau of Harney Basin, in GREAT BASIN, c.80 mi/129 km SSW of BURNS. HART LAKE (6 mi/9.7 km long, 2 mi/3.2 km wide; semi-dry), one of numerous intermittent lakes in Warner Valley, which has geysers and sub-surface thermal activity, to W. Hart Mountain National Antelope Refuge in SW.

Hartnanitz, CZECH REPUBLIC: see HARTMANICE.

Hartney (HAHRT-nee), town (□ 1 sq mi/2.6 sq km; 2001 population 446), SW MANITOBA, W central CANADA, on SOURIS River, 35 mi/56 km SW of BRANDON, and surrounded by CAMERON rural municipality; 49°29′N 100°31′W. Grain elevators; livestock.

Harts, unincorporated town (2000 population 2,361), LINCOLN county, W WEST VIRGINIA, 16 mi/26 km W of MADISON, near GUYANDOTTE RIVER; 38°01′N 82°07′W. Agriculture (corn, tobacco); cattle. Timber. Manufacturing (lumber). Big Ugly Wildlife Management Area to NE.

Hartsburg, town, (2000 population 108), BOONE county, central MISSOURI, on MISSOURI River, 10 mi/16 km NW of JEFFERSON CITY. Pumpkins, corn, soybeans; cattle. Flooding in 1993 severely damaged town. Access to Katy Trail State Park.

Hartsburg, village (2000 population 358), LOGAN county, central ILLINOIS, 8 mi/12.9 km NNW of LINCOLN; 40°15′N 89°26′W. In agricultural area (cattle, hogs; corn, soybeans).

Hartsdale, residential village (□ 3 sq mi/7.8 sq km; 2000 population 9,830), WESTCHESTER county, SE NEW YORK, just SW of WHITE PLAINS; 41°01′N 73°48′W.

Hartsel, village, Park county, central COLORADO, in ROCKY MOUNTAINS, 55 mi/89 km WNW of COLORADO SPRINGS; elevation 8,864 ft/2,702 m. Hot springs and ANTERO RESERVOIR and Antero State Wildlife Area to W; SPINNEY MOUNTAIN Reservoir and State Park to E. Region nearly surrounded by Pike National Forest.

Hartselle, city (2000 population 12,019), Morgan co., N Alabama, 12 mi/19 km SSE of Decatur. In cotton, corn, and vegetable area. Copper wire and pipe, clothing manufacturing William B. Bankhead National Forest is SW. Founded 1870, incorporated 1875. Named for George S. Hartselle, who settled in the area before 1834.

Hartsfield-Jackson Atlanta International Airport, ATLANTA, GEORGIA (10 mi/16.2 km from downtown); elevation 1,026 ft/316 m; 33°28′N 84°26′W. Covers 4,700 acres/1,518 ha. Has five runways and its terminal complex covers 130 acres/52.6 ha with seven concourses connected by 3.5 mi/5.6 km long "people mover" rail system; handles over 80 million passengers and nearly 900,000 tonnes of cargo annually.

Opened in 1925 as Candler Field, named for the Coca-Cola magnate, Asa Candler, the airport became the Atlanta Municipal Airport in 1929. On February 22, 1971, the airport changed its name to William B. Hartsfield Atlanta Airport, to honor the former Mayor of Atlanta who did so much to enlarge it. On July 1, 1971, the airport added "International" to its name to become William B. Hartsfield Atlanta International Airport when the first international flights (to Mexico and Jamaica) were introduced. In 2003, to honor Maynard H. Jackson, another former Atlanta mayor, the airport became Hartsfield-Jackson Atlanta International Airport. Airport Code ATL.

Hartshill, ENGLAND: see NUNEATON AND BEDWORTH.

Hartshorne (HAHRTS-horn), town (2006 population 2,075), PITTSBURG county, SE OKLAHOMA, 13 mi/21 km SE of MCALESTER, and contiguous to HAILEYVILLE (to W); 34°50′N 95°33′W. In mining, cattle-raising, and agricultural area (grain); manufacturing (wood products, electronic equipment). Oil and natural-gas wells. Jack Fork Mountains to S; headwaters of Lake Arrowhead reservoir to N. Settled c.1890.

Harts Island, part of BRONX borough of NE NEW YORK city, SE NEW YORK, in LONG ISLAND SOUND, near CITY ISLAND and PELHAM BAY PARK; c.1 mi/1.6 km long; 40°51′N 73°46′W. Mainland ferry to City Island. The city's cemetery (potter's field) is here. Formerly site of city reformatory. Sometimes called Hart Island or Hart's Island.

Harts Location, town, CARROLL county, N central NEW HAMPSHIRE, 23 mi/37 km SSW of BERLIN, in WHITE MOUNTAINS. Drained by Saco River. Tourism. White Mountain National Forest in southern half; northern part is surrounded by White Mountain National Forest. CRAWFORD NOTCH State Park in N; APPALACHIAN TRAIL crosses N; Silver Cascade Falls in N.

Harts River, SOUTH AFRICA: see HARTZ RIVER.

Hartstene Island (HAHRT-steen), MASON county, W WASHINGTON, in PUGET SOUND, 9 mi/14.5 km N of OLYMPIA. Bounded E by Case Inlet, S by Dana Passage (bridged to mainland). In area are Jarrell Cove (N) and McMicken Island (E; off E shore) state parks; Squaxin Island Indian Reservation and State Park to SW. Village of Hartstene at N end. Timber. Fishing.

Hartsville, town (2000 population 376), BARTHOLOMEW county, S central INDIANA, on small Clifty Creek, and 12 mi/19 km ENE of COLUMBUS; 39°16′N 85°42′W. In agricultural area.

Hartsville, town (2006 population 7,473), DARLINGTON county, NE SOUTH CAROLINA, 21 mi/34 km NW of FLORENCE; 34°22′N 80°04′W. Manufacturing includes packaged foods, fertilizers, fabricated metal products, fiberglass boats, textiles, paper and paper products. Agriculture includes livestock, poultry; dairying; grain, soybeans. Seat of Coker College H. B. ROBINSON 2 NUCLEAR POWER PLANT on Lake Robinson immediately to NW. Experimental seed farm nearby.

Hartsville, town (2000 population 2,395), ⊙ TROUSDALE county, N TENNESSEE, 15 mi/24 km NE of LEBANON; 36°24′N 86°10′W. In agricultural area; manufacturing. In Civil War, Federal garrison here was defeated (1862) by General John H. Morgan. Settled in early 1800s; incorporated 1913.

Hartsville, Massachusetts: see NEW MARLBORO.

Hartville, city, S central MISSOURI, ⊙ WRIGHT county, in the OZARK MOUNTAINS, on GASCONADE RIVER, and 43 mi/69 km E of SPRINGFIELD; 38°25′N 93°55′W. Peaches, apples, hay; dairying; cattle; timber. Manufacturing of livestock feed and fertilizer. Settled 1832. Destroyed during Civil War.

Hartville (HAHRT-vil), industrial village (□ 2 sq mi/5.2 sq km; 2006 population 2,512), STARK county, E central OHIO, 14 mi/23 km SE of AKRON; 40°58′N 81°20′W. Rubber goods, plastic products.

Hartville, village (2006 population 74), PLATTE county, SE WYOMING, on North Platte River (GUERNSEY RE-

SERVOIR), and 22 mi/35 km NE of WHEATLAND; elevation c.4,750 ft/1,448 m; 42°19′N 104°43′W. Terminus of railroad spur from GUERNSEY. Region of prehistoric stone quarries to NE.

Hartwell (HAHRT-wel), suburb 6 mi/10 km ESE of MELBOURNE, VICTORIA, SE AUSTRALIA; 37°51′S 145°05′E.

Hartwell, town (2000 population 4,188), ⊙ HART county, NE GEORGIA, 16 mi/26 km NNW of ELBERTON, near SOUTH CAROLINA state line; 34°21′N 82°56′W. Manufacturing of athletic equipment, consumer goods, fabricated metal products, clothes, machinery parts, transportation equipment; egg production, mica mining and production, printing and publishing. Named for American Revolutionary heroine Nancy Hart. Adjacent to Lake Hartwell on SAVANNAH River. Incorporated 1856.

Hartwell Lake, reservoir, NE GEORGIA and NW SOUTH CAROLINA, on SAVANNAH River (Georgia-SOUTH CAROLINA state line), 13 mi/21 km SE of ANDERSON (South Carolina); 34°20′N 82°49′W. Lower reservoir 10 mi/16 km long; KEOWEE (SENECA) River enters from N, TUGALOO RIVER (Georgia-South Carolina state line) enters from NW, both forming arms each c.25 mi/40 km long. Formed by HARTWELL Dam (195 ft/59 m high), built (1960) by the Army Corps of Engineers for flood control, navigation, and power generation. Hartwell near SW shore. State parks on NE (South Carolina) and SW (Georgia) shores.

Hartwick, town (2000 population 83), POWESHIEK county, central IOWA, 12 mi/19 km NE of MONTEZUMA; 41°46′N 92°20′W. In agricultural area.

Hartz River, 270 mi/435 km long, NORTH-WEST and NORTHERN CAPE provinces, SOUTH AFRICA; rises in W WITWATERSRAND, NE of LICHTENBURG; flows SW, past SCHWEIZER RENEKE and TAUNG, to VAAL river at DELPORT'S HOPE. Dammed at Schweizer Reneke. Serves as border between the 2 provinces for 40 mi/64 km between Pampierstad and Spitkop Dam in Hartswater irrigation area. Also spelled Harts River.

Harue (hah-ROO-e), town, Sakai district, FUKUI prefecture, central HONSHU, W central JAPAN, 5 mi/8 km N of FUKUI; 36°07′N 136°13′E. Rice.

Haruhi (HAH-roo-hee), town, W Kasugai district, AICHI prefecture, S central HONSHU, central JAPAN, 5.6 mi/9 km N of NAGOYA; 35°14′N 136°50′E.

Haruku (HAH-roo-koo), largest island (10 mi/16 km long, 8 mi/12.9 km wide) of ULIASER Islands, INDONESIA, in BANDA SEA, just E of AMBON, and 4 mi/6.4 km S of SW coast of SERAM, across narrow Seram Strait; 03°35′S 128°31′E. Generally low, rising in SW to 1,926 ft/587 m. Coconuts, cloves, sago. Formerly spelled Haroekoe.

Harumukotan-kaikyo, RUSSIA: see KURILE STRAIT.

Harumukotan-to, RUSSIA: see KHARIMKOTAN ISLAND.

Haruna (HAH-roo-nah), town, Gumma district, GUMMA prefecture, central HONSHU, N central JAPAN, near Mt. Haruna, 12 mi/20 km W of MAEBASHI; 36°22′N 138°53′E.

Harunabad (HUH-roo-nah-BAHD), town, BAHAWALPUR district, PUNJAB province, central PAKISTAN, on railroad spur, 90 mi/145 km ENE of BAHAWALPUR (also called Bodruwala Mandi); 29°37′N 73°08′E.

Haruniya, township, NW YEMEN, E of SALIF, near RED SEA coast; 15°20′N 42°50′E. Small local port.

Harun, Jebel (huh-ROON, JE-bel) (huh-ROON, JE-bel), mountain (4,383 ft/1,336 m) of S central JORDAN, W of WADI MUSA, and overlooking ruins of PETRA (N). Some consider it the biblical Hor.

Haruno (HAH-roo-no), town, Agawa district, KOCHI prefecture, S SHIKOKU, W JAPAN, 4.3 mi/7 km S of KOCHI; 33°29′N 133°29′E. Cucumbers. Photocopier paper.

Haruno (HAH-roo-no), town, Syuchi district, SHIZUOKA prefecture, HONSHU, E central JAPAN, 28 mi/45 km W of SHIZUOKA; 34°58′N 137°54′E. Tea.

Harur (huh-ROOR), town, DHARMAPURI district, TAMIL NADU state, S INDIA, 22 mi/35 km ESE of DHARMAPURI; 12°04′N 78°30′E. In agricultural area. Magnetite deposits nearby.

Har Us Nuur (KHAHR OOS NOOR), freshwater lake (□ 715 sq mi/1,859 sq km) in W MONGOLIA, just E of HOVD; 50 mi/80 km long and 16 mi/26 km wide; 48°00′N 92°10′E. Elevation 3,796 ft/1,157 m. Receives HOVD GOL (Kobdo River) and drains into HAR NUUR (lake). Low marshy shore contains rocky Akbashi Island (□ 105 sq mi/274 sq km). Rich in fish and birds. A smaller lake (□ 24 sq mi/62 sq km) of the same name—also called NAMIRIIN USA for the Namir River (a small tributary)—is 80 mi/130 km N of Hovd (Kobdo). Formerly called Ikhe Aral Norwegian. Also spelled Khar-Us-Nur, Khara Usu, or Hara Usa.

Harut Rud (HAH-root ROOD), river, c.250 mi/402 km long, W AFGHANISTAN; rises in the SIAH KOH (outlier of the Hindu Kush mountains) 80 mi/129 km SE of HERAT; flows SW, past SHINDAND and ANARDARAH, to the Hamun-i-Sabari, one of the lagoons of the Seistan depression on IRAN border (31°35′N 61°18′E). Intermittent flow in lower course. Also called Adraskan or Adraskand, particularly in upper course.

Harvard, city (2000 population 7,996), MCHENRY county, N ILLINOIS, near WISCONSIN state line, 12 mi/19 km NW of WOODSTOCK; 42°25′N 88°37′W. Railroad town (with repair shops); trade center in dairying and resort area; manufacturing of dairy products, hardware, telecommunications. Incorporated 1867.

Harvard, residential town, WORCESTER county, E central MASSACHUSETTS, 19 mi/31 km NNE of WORCESTER; 42°31′N 71°35′W. Orchards. A Shaker house and cemetery, a Native American museum, and a Harvard University observatory are here. Nearby is a museum on the site of Fruitlands, a cooperative vegetarian community founded by Bronson Alcott. Includes village of Still River. Incorporated 1732.

Harvard, town (2006 population 916), CLAY county, S NEBRASKA, 15 mi/24 km E of HASTINGS. Dairy and poultry products, grain; livestock. Harvard Marsh, one of the declining number of Rainwater Basin ponds or lagoons remaining, protected as a wildlife preserve of U.S. Fish and Wildlife Service.

Harvard College Observatory, astronomical observatory located in CAMBRIDGE, MASSACHUSETTS, operated by Harvard University (Harvard College at the time of the observatory's founding in 1839). Its equipment includes a 61-in/155-cm reflecting telescope and 15-in/38-cm and 12-in/30-cm refracting telescopes. Programs of the Harvard Observatory include various aspects of solar physics, stellar and nebular spectroscopy and photometry, and theoretical cosmology. Among the noted directors of the observatory have been W. C. Bond, G. P. Bond, E. C. Pickering, and Harlow Shapley. In 1973 the research programs of the Harvard College Observatory were merged with those of the Smithsonian Astrophysical Observatory to form the Harvard-Smithsonian Center for Astrophysics; the observatory itself, however, maintains its separate status under the control of Harvard.

Harvard, Mount (14,420 ft/4,395 m), Chaffer county, central COLORADO, in Collegiate Range of SAWATCH MOUNTAINS, in San Isabel National Forest, E of CONTINENTAL DIVIDE, 23 mi/37 km S of LEADVILLE. Peak is third highest in ROCKY MOUNTAINS of U.S. Named in 1869 by a group of climbers from Harvard College.

Harve (AHRV), village in commune of MONS, Mons district, HAINAUT province, SW BELGIUM, 4 mi/6.4 km E of Mons.

Harvel, village (2000 population 235), in CHRISTIAN and MONTGOMERY counties, S central ILLINOIS, 20 mi/32 km SW of TAYLORVILLE; 39°21′N 89°31′W.

Harvest, village (2000 population 3,054), Madison co., N Alabama, 13 mi/21 km NW of Huntsville; 34°51′N

86°45′W. Originally named 'Kelly' for Thomas B. Kelly, it was renamed after 1905.

Harvey, county (□ 540 sq mi/1,404 sq km; 2006 population 33,643), S central KANSAS; ⊙ NEWTON; 38°02′N 97°25′W. Flat to gently rolling prairie, drained by LITTLE ARKANSAS RIVER and WALNUT CREEK (E). Wheat, corn, oats, barley, soybeans, apples; cattle, poultry. Millwork; industrial machinery. Formed 1872.

Harvey, city (2000 population 30,000), COOK county, NE ILLINOIS, suburb 20 mi/32 km SSW of downtown CHICAGO; 41°36′N 87°39′W. Manufacturing (steel forging; fabricated metal products, chemicals, machinery; electronic equipment). Harvey has an oil research center. Founded by Turlington W. Harvey, a wealthy lumberman, in 1890. South Suburban College of Cook County in neighboring SOUTH HOLLAND. Incorporated 1891.

Harvey (HAHR-vee), unincorporated city (2000 population 22,226), JEFFERSON parish, SE LOUISIANA, on W bank (levee) of the MISSISSIPPI RIVER, suburb opposite and 2 mi/3 km S of downtown NEW ORLEANS; 29°54′N 90°04′W. Manufacturing (building materials, paper products, fabricated metal products, mineral oil, chemicals, fiberglass fabrication, machinery; barge building, marine vessel repair); shrimp, crabmeat, oysters. Harvey Lock (425 ft/130 m long) links the MISSISSIPPI to INTRACOASTAL WATERWAY here. Also Cosmopolite City.

Harvey (HAHR-vee), town, SW WESTERN AUSTRALIA, 80 mi/129 km S of PERTH; 33°05′S 115°54′E. Dairying and agriculture center; butter; timber; orchards (citrus fruits). Harvey Weir and sawmills nearby. Museum housed in old railway station.

Harvey, town (2000 population 277), MARION county, S central IOWA, 10 mi/16 km E of KNOXVILLE, near DES MOINES RIVER; 41°19′N 92°55′W. Brick and tile plant. Limestone quarries, sand and gravel pits nearby.

Harvey, town (2006 population 1,705), WELLS co., central NORTH DAKOTA, on SHEYENNE River and 80 mi/129 km NNE of BISMARCK; 47°46′N 99°55′W. Sunflowers, dairy products, potatoes, wheat. Manufacturing (concrete). Founded 1893.

Harvey, village (2001 population 429), ALBERT county, SE NEW BRUNSWICK, E CANADA, near SHEPODY BAY, 2 mi/3 km SSE of RIVERSIDE-ALBERT, 10 mi/16 km SW of HOPEWELL CAPE; 45°42′N 64°43′W. Mining and fishing in region.

Harvey, village (2001 population 349), YORK county, SW NEW BRUNSWICK, E CANADA, on SE end of HARVEY LAKE, 24 mi/39 km SW of FREDERICTON. Mixed agriculture; dairying; potatoes, apples. Tourism. Formerly Harvey Station.

Harvey, village (2000 population 1,321), MARQUETTE county, on LAKE SUPERIOR, NW UPPER PENINSULA, N MICHIGAN, 4 mi/6.4 km SE of MARQUETTE; 46°29′N 87°20′W. Resort; fish hatchery. Cliffs Ridge Ski Area to W; Highland State Recreation Area to E.

Harvey Bay, AUSTRALIA: see TUMBY BAY.

Harvey Cedars, resort borough (2006 population 389), OCEAN county, E NEW JERSEY, on LONG BEACH island, 17 mi/27 km SSE of TOMS RIVER, and 30 mi/48 km NNE of ATLANTIC CITY; 39°42′N 74°08′W. Artists' summer colony here.

Harvey Lake (□ 3 sq mi/8 sq km), SW NEW BRUNSWICK, E CANADA, 23 mi/37 km SW of FREDERICTON; 3 mi/5 km long, 1 mi/2 km wide; 45°45′N 67°01′W. Village of HARVEY (YORK county) on SE end.

Harveysburg (HAHR-veez-buhrg), village (2006 population 624), WARREN county, SW OHIO, 20 mi/32 km SSE of DAYTON, on small Caesar Creek; 39°30′N 84°00′W.

Harveys Lake, borough (2006 population 2,900), LUZERNE county, NE central PENNSYLVANIA, residential community 12 mi/19 km NW of WILKES-BARRE; 41°21′N 76°01′W. Town surrounds Harveys Lake.

Harveyton, village, PERRY county, SE KENTUCKY, in CUMBERLAND foothills, 5 mi/8 km N of HAZARD. Bituminous coal.

Harveyville, village (2000 population 267), WABAUNSEE county, E central KANSAS, 23 mi/37 km SW of TOPEKA; 38°47′N 95°57′W. In cattle, poultry, and grain region.

Harvie Heights (HAHR-vee HEITS), hamlet, SW ALBERTA, W CANADA, near CANMORE, and in Bighorn No. 8 municipal district; 51°07′N 115°23′W.

Harviell (hahr-vee-EL), unincorporated town, BUTLER county, SE MISSOURI, 8 mi/12.9 km SW of POPLAR BLUFF.

Harwan (huhr-WAHN), village, ANANTNAG or KASHMIR SOUTH, district, JAMMU AND KASHMIR state, N INDIA, 7 mi/11.3 km NE of SRINAGAR. Headworks of water-supply system for Srinagar here. Buddhist ruins (some constructed in a peculiar pebble style) include stupa and extensive figured brick tiles; earliest ruins date from fourth century A.D.

Harwell (HAH-wuhl), village (2001 population 3,780), OXFORDSHIRE, S central ENGLAND, 2 mi/3.2 km W of DIDCOT; 51°37′N 01°18′W. Atomic energy research.

Harwich (HA-rich), town (2001 population 17,015), ESSEX, E central ENGLAND, on the estuary of the STOUR and the ORWELL rivers; 51°57′N 01°17′E. This increasingly important port imports foodstuffs, iron and steel, and machinery; it exports chemicals and motor vehicles. Serves as a port for passenger ships to Denmark, Germany, and Holland; Parkeston Quay is the ferry terminal. There is also container-ship and roll-on, roll-off service to Zeebrugge (Belgium). The town's other industries are boatbuilding, fishing, light engineering, and cement manufacturing. An ancient town, Harwich was known in the Middle Ages for its port and had an important shipbuilding industry in the 17th century.

Harwich (HAHR-wich), town, BARNSTABLE county, SE MASSACHUSETTS, on S coast of CAPE COD, 12 mi/19 km E of BARNSTABLE; 41°41′N 70°04′W. Summer resort; cranberries (birthplace of the industry), vegetables. Once whaling and shipbuilding center. Includes resort villages of Harwich Port (yachting), North Harwich, East Harwich (1990 population 3,828), South Harwich, and West Harwich. Pleasant Lake (c.1.5 mi/2.4 km long) is nearby. Historical museum. Settled c.1670, incorporated 1694.

Harwinton, town, LITCHFIELD county, NW CONNECTICUT, on NAUGATUCK River, and 4 mi/6.4 km SE of TORRINGTON; 41°45′N 73°03′W. In hilly region; agriculture. Retail services, construction, landscaping; tool and die. Settled 1730, incorporated 1737.

Harwood, town (2000 population 90), VERNON county, W MISSOURI, near OSAGE RIVER, 14 mi/23 km NE of NEVADA; 37°57′N 94°09′W. Hay, sorghum; cattle.

Harwood (HAHR-wud), unincorporated village, S ONTARIO, E central CANADA, on RICE Lake, 13 mi/21 km SE of PETERBOROUGH, and included in HAMILTON township; 44°08′N 78°11′W. Fruit; dairying; mixed farming.

Harwood, unincorporated village, GONZALES county, S central TEXAS, 9 mi/14.5 km E of LULING. Railroad spur junction. Cattle, poultry.

Harwood, Great (HA-wud), town (2001 population 11,220), E central LANCASHIRE, N ENGLAND, 4 mi/6.4 km NE of BLACKBURN; 53°47′N 02°25′W. Aircraft machinery and equipment; textiles. Has 16th-century church.

Harwood Heights, village (2000 population 8,297), COOK county, NE ILLINOIS, suburb 11 mi/18 km NW of downtown CHICAGO; 41°58′N 87°48′W. Manufacturing (fabricated metal products, machinery, tapes and adhesives, tools).

Harworth (HAH-wuhth), village (2005 population 7,400), N NOTTINGHAMSHIRE, central ENGLAND, 8 mi/12.9 km N of WORKSOP; 53°25′N 01°04′W. Has Norman church, rebuilt 19th century, and 18th-century mansion containing notable Flemish paintings.

Haryana (huhr-YAH-nah), state (□ 17,070 sq mi/44,382 sq km; 2001 population 21,144,564), N central INDIA; ⊙ CHANDIGARH. The terrain is mostly flat and dry but does support irrigated agriculture. Cotton is a major product. The state has a surplus in food-grain production, especially wheat, rice, bajra, maize, rape, mustard, barley, gram, and jowar. Other crops include sugarcane and oilseeds. Wheat farming depends on green revolution technology. Its proximity to DELHI state provides a huge market for agriculture and manufacturing products, including several durable goods (washing machines, motorcycles, tractors, cars, stoves). In the 2000s, the SE city GURGAON, a suburb of Delhi, has attracted investments from several multinational technology companies, and the area is experiencing rapid economic development. AMBALA, CHANDIGARH, KARNAL, PANIPAT, ROHTAK, BHIWANI, and PANCHKULA are important cities. Created in 1966 out of the Hindi-speaking portions of PUNJAB state. Governed by a chief minister and cabinet responsible to an elected, unicameral legislature and by a governor appointed by the president of India. Haryana has suffered from spillover of the violence associated with the Sikh separatist war in Punjab state. A prosperous state with a growing economy and industrial centers spread all over, producing maufactured goods. Although it is one of the wealthiest states in all of India, gauged by per capita income, it has the dubious distinction of having one of the lowest live birth sex ratios in the country. Haryana has been promised a new capital city if Chandigarh is transferred to Punjab.

Harz (HAHRTS), mountain range, N GERMANY, extending c.60 mi/97 km between the ELBE and LEINE rivers; 51°30′N 10°10′E– 51°55′N 11°30′E. The rugged mountains were once densely forested. They culminate in BROCKEN peak (3,747 ft/1,142 m high). The region has good water-power potential. The uranium and silver mines that opened after World War II are now closed. The Upper Harz has extensive wastelands and a severe, rainy climate. Noted for its mineral deposits (especially silver). GOSLAR is the chief town of the region, which also has some summer resorts. WERNIGERODE, the chief town of the Lower Harz, has a relatively mild climate; the surrounding region is predominantly agriculture (grain; cattle).

Harzburg, Bad, GERMANY: see BAD HARZBURG.

Harzgerode (hahrts-ge-RO-de), town, SAXONY-ANHALT, central GERMANY, in the Lower Harz, 10 mi/16 km S of QUEDLINBURG; 51°38′N 11°08′E. Metal- and woodworking; health resort. Has 16th-century palace. Silver and lead formerly mined here. Incorporates ALEXISBAD spa (NW), founded 1810.

Hasa, SAUDI ARABIA: see AHSA, AL.

Hasa, SAUDI ARABIA: see HOFUF.

Hasa, El (HUH-suh, el), village, central JORDAN, on HEJAZ RAILROAD, and 20 mi/32 km E of TAFILA, on Wadi el Hasa (affluent of DEAD Sea). Barley; camel raising. Phosphate mining. Formerly known as Qal'at el Hasa.

Hasakah, Al-, SYRIA: see HASEKE, El.

Hasaki (hah-SAH-kee) or **Hazaki**, town, Kashima district, IBARAKI prefecture, central HONSHU, E central JAPAN, on the PACIFIC OCEAN, at mouth of TONE RIVER, opposite CHOSHI, 47 mi/75 km S of MITO; 35°45′N 140°49′E. Sweet peppers, flowers; cane processing; game (go, shogi) board manufacturing.

Hasama (hah-SAH-mah), town, Tome district, MIYAGI prefecture, N HONSHU, NE JAPAN, 34 mi/55 km N of SENDAI; 38°41′N 141°11′E.

Hasama, town, Oita district, OITA prefecture, SE KYUSHU, SW JAPAN, 6 mi/10 km S of OITA; 33°11′N 131°30′E.

Cross-references are shown in SMALL CAPITALS. The pronunciation guide is shown on page xix. The sources of population figures are shown on page xvii.

Hasami (HAH-sah-mee), town, East Sonogi district, NAGASAKI prefecture, NW KYUSHU, SW JAPAN, 28 mi/ 45 km N of NAGASAKI; 33°08′N 129°53′E.

Hasanabad, INDIA: see TAKI.

Hasan Abdal (HUH-suhn UHB-dahl), town, ATTOCK district, NW PUNJAB province, SE PAKISTAN, 18 mi/29 km ENE of CAMPBELLPUR; 33°49′N 72°41′E. Wheat. Various Muslim and Sikh legends associated with nearby tomb-garden of Lala Rukh, with its shrine and spring. Also spelled HASSAN ABDAL.

Hasanah, Yemen: see DATHINA.

Hasan Dagi, Turkish *Hasan Daği*, peak (10,672 ft/3,253 m), central TURKEY, 29 mi/47 km WNW of NIGDE. Sometimes called Buyukhasan (Turkish=*Büyükhasan*).

Hasanganj (HUH-suhn-guhnj), town, tahsil headquarters, UNNAO district, central UTTAR PRADESH state, N central INDIA, on SAI RIVER and 18 mi/29 km NNE of UNNAO; 26°47′N 80°39′E. Wheat, barley, rice, gram. Fruit orchards.

Hasani, Greece: see ELLINIKON.

Hasani, Yemen: see DATHINA.

Hasankale, TURKEY: see PASINLER.

Hasan Kiadeh (hah-SAHN kee-yah-DE), village, GĪlān province, N IRAN, small CASPIAN SEA port, 33 mi/53 km NE of RASHT, at mouth of the SEFID; 37°24′N 49°58′E.

Hasan Kuli, TURKMENISTAN: see ESENGULY.

Hasanparti (HUH-suhn-PUHR-tee), town, WARANGAL district, ANDHRA state, S India, 6 mi/9.7 km NW of WARANGAL; 18°04′N 79°32′E. Rice, oilseeds; silk weaving. Also spelled Hassan Parthi.

Hasanpur (huh-SUHN-poor), town, MORADABAD district, N central UTTAR PRADESH state, N central INDIA, 31 mi/50 km WSW of MORADABAD. Wheat, rice, pearl millet, mustard, sugarcane. Founded 1634.

Hasayan (huh-SEIN), town, ALIGARH district, W UTTAR PRADESH state, N central INDIA, 13 mi/21 km E of HATHRAS. Wheat, barley, pearl millet, gram, corn. Also spelled HUSAIN.

Hasbaya, LEBANON: see HASBEYA.

Hasbergen (HAHS-ber-gen), suburb of OSNABRÜCK, LOWER SAXONY, NW GERMANY, on N slope of TEUTOBURG FOREST, 5 mi/8 km SW of city center; 52°13′N 07°58′E.

Hasbeya (has-BAI-yuh), township, S LEBANON, at NW foot of MOUNT HERMON, 22 mi/35 km SE of SAIDA; elevation 2,500 ft/760 m. Cotton, tobacco, cereals, almonds, olives, grapes. Iron deposits nearby (NE). An old town, it was center of the Druse from 13th to 19th century. One of the most important Druse shrines is the Khalwat el Bujad, or Khalwat el Biyad, on a hill nearby. Also spelled HASBAYA.

Hasbrouck Heights (HAZ-brook), borough (2006 population 11,621), BERGEN county, NE NEW JERSEY, a residential suburb adjoining HACKENSACK; 40°51′N 74°04′W. Settled c.1685, incorporated 1894.

Hascosay (HAS-ko-sai), uninhabited island (□ 1 sq mi/ 2.6 sq km) of the SHETLAND ISLANDS, extreme N Scotland, just off E coast of YELL; 60°37′N 00°59′W.

Hase (HAH-se), village, Kamina district, NAGANO prefecture, central HONSHU, central JAPAN, 59 mi/95 km S of NAGANO; 35°47′N 138°05′E.

Haseke, El, district, NE SYRIA, bordering TURKEY to the N and IRAQ to the W; ⊙ El HASEKE. Contains a large Kurdish population. Oil; herding.

Haseke, El (HAHS-e-ke, el), town, ⊙ HASEKE district, NE SYRIA, on KHABUR River, 85 mi/137 km NNE of DEIR; 36°30′N 40°44′E. The town and its surroundings mainly inhabited by Kurds; the Arabic name is a distortion of the Kurdish name, *Hassache*. It is now the urban center for the rich oil and gas fields to the E. Airport. Sheep raising. Also called Al-HASAKAH.

Haselberg, RUSSIA: see KRASNOZNAMENSK.

Haselünne (hah-se-LYOON-ne), town, LOWER SAXONY, NW GERMANY, on HAASE RIVER, 8 mi/12.9 km ESE of MEPPEN; 52°40′N 07°29′E. Industry includes

distilleries; sawmilling; metalworking. Has 14th-century church.

Hasenkamp (AH-sen-kahmp), town, W ENTRE RÍOS province, ARGENTINA, on railroad, and 45 mi/72 km NE of PARANÁ; 31°31′S 59°51′W. Grain, flax; livestock, poultry.

Hasenmatt (HAH-suhn-maht), mountain peak (4,741 ft /1,445 m) in the JURA, SOLOTHURN canton, NW SWITZERLAND, 5 mi/8 km NW of SOLOTHURN; highest point in Solothurn canton.

Hasenpoth, LATVIA: see AIZPUTE.

Hase River, GERMANY: see HAASE RIVER.

Hasharon, ISRAEL: see RAMAT DAVID.

Hashikami (hah-shee-KAH-mee), town, Sannohe district, Aomori prefecture, N HONSHU, N JAPAN, 53 mi/ 85 km S of AOMORI; 40°26′N 141°37′E.

Hashima (HAH-shee-mah), city, GIFU prefecture, central HONSHU, central JAPAN, on the Kijo River, 5 mi/8 km S of GIFU; 35°18′N 136°42′E. Textiles.

Hashimoto (hah-SHEE-mo-to), city, WAKAYAMA prefecture, S HONSHU, W central JAPAN, on N central KII PENINSULA, 27 mi/43 km E of WAKAYAMA; 34°18′N 135°36′E. Railroad junction.

Hashir, TURKEY: see PERVARI.

Hashmonaim (khahsh-mo-nah-EEM), urban settlement, ISRAEL, 7 mi/11.3 km E of RAMLA. Named for the House of Hasmonea (Beit Hashmonai in Hebrew), the priestly family that established an autonomous Jewish state after rebelling against the Seleucid kingdom; the Hasmonean kingdom lasted from 167 B.C.E. to 76 B.C.E.

Hashtgerd (hasht-GUHRD), town, Tehrān province, N IRAN, 45 mi/70 km NW of TEHRAN; 35°58′N 50°40′E. Railroad and highway between Tehran and QAZVIN.

Hashtpar (hasht-PAHR), town, GĪlān province, NW IRAN, 50 mi/80 km NW of RASHT, along SW coast of CASPIAN SEA; 37°48′N 48°55′E. Sometimes called Talesh.

Hashtrud (hasht-ROOD), town, in Āzerbāyjān-e Sharqi province, NW IRAN, on road, and 35 mi/56 km W of MIANEH. Center of Hashtrud agricultural area (grain growing; sheep raising), at E foot of SAHANDmassif. Also called SARASKAND.

Hasikiya (huhs-ik-EE-yuh), uninhabited westernmost island (□ 1 sq mi/2.6 sq km) of the KURIA MURIA group, off SW OMAN; rises to 501 ft/153 m; 17°29′N 55°38′E. Guano was worked here in 1850s.

Hasilpur (HUH-seel-puhr), town, BAHAWALPUR district, PUNJAB province, central PAKISTAN, 55 mi/89 km NE of BAHAWALPUR; 29°43′N 72°33′E.

Hasiye, SYRIA: see HISYA.

Haskeir (HAS-kir) or **Hyskier**, Gaelic *Ìgh-Sgeir*, rocky uninhabited islet (c.1 mi/1.6 km long), OUTER HEBRIDES, EILEAN SIAR, NW Scotland, 8 mi/12.9 km NW of NORTH UIST, and 12 mi/19 km N of HEISKER (or Monach Isles); 57°42′N 07°40′W. Seal sanctuary.

Haskell (HAS-kuhl), county (□ 577 sq mi/1,500.2 sq km; 2006 population 4,171), SW KANSAS; ⊙ SUBLETTE; 37°33′N 100°52′W. Flat to rolling prairie, with sand dunes in extreme N. CIMARRON RIVER in extreme SW. Wheat, corn, sorghum; cattle. Small natural-gas fields. Formed 1887.

Haskell, county (□ 625 sq mi/1,625 sq km; 2006 population 12,155), E OKLAHOMA; ⊙ STIGLER; 35°13′N 95°06′W. Bounded N by CANADIAN River (EUFAULA Dam and Lake in far NW corner) and ROBERT S. KERR LAKE SE corner in OUACHITA MOUNTAINS; Sansbois Mountains on S border. Drained by Sansbois Creek and NE by ARKANSAS RIVER (Robert S. Kerr Lake). Agriculture (barley, hay, soybeans, corn; cattle); timber; coal mines. Part of Sequoyah National Wildlife Refuge in N. Recreation. Formed 1907.

Haskell, county (□ 910 sq mi/2,366 sq km; 2006 population 5,438), NW central TEXAS; ⊙ HASKELL; 33°10′N 99°43′W. Drained by Double Mt. Fork of Brazos River and Paint Creek. Irrigated agriculture (cotton, sorghum, wheat, oats, barley, peanuts); livestock (beef

cattle). Oil, natural-gas wells. Lake Stamford reservoir in SE corner. Formed 1858.

Haskell, city (2000 population 2,645), SALINE county, central ARKANSAS, 5 mi/8 km SSW of BENTON; 34°30′N 92°38′W. Railroad junction; manufacturing (aluminum recycling).

Haskell (HAS-kuhl), town (2006 population 1,779), MUSKOGEE county, E OKLAHOMA, 18 mi/29 km WNW of MUSKOGEE, and 25 mi/40 km SE of TULSA, near ARKANSAS RIVER; 35°49′N 95°40′W. Trade center in agricultural area (corn, potatoes; livestock; dairying); manufacturing (hazardous waste-derived fuels, fabricated metal products, food products); oil and natural-gas wells. Founded 1903.

Haskell, town (2006 population 2,729), ⊙ HASKELL county, NW central TEXAS, c.50 mi/80 km N of ABILENE; 33°09′N 99°43′W. Elevation 1,553 ft/473 m. In cattle-ranching and agricultural area (cotton, wheat, peanuts); light manufacturing; oil and gas. Lake Stamford reservoir to SE. Settled 1882, incorporated 1907.

Haskell (HA-skuhl), village in WANAQUE borough, PASSAIC county, NE NEW JERSEY, on Wanaque River, and 10 mi/16 km NW of PATERSON. Largely residential.

Haskins (HA-skinz), village (□ 2 sq mi/5.2 sq km; 2006 population 635), Wood county, NW OHIO, 6 mi/10 km NNW of BOWLING GREEN, near MAUMEE RIVER; 41°28′N 83°42′W. In agricultural area.

Haskovo (HAHS-ko-vo), oblast (□ 5,097 sq mi/13,252.2 sq km), S central BULGARIA, on the border of GREECE and TURKEY (both to the S); 41°55′N 25°33′E. Bordered W by PLOVDIV oblast (includes the E part of the RODOPI Mountains), E by BURGAS oblast, N by LOVECH oblast and the E part of the STARA PLANINA; occupies part of the Thracian plain. Important crossroads between W Europe and ISTANBUL and across Stara Planina's Republika Pass to BUCHAREST. Large flat areas. Land along the ARDA RIVER and its tributaries suitable for tobacco. Comprised of former okrugs—Haskovo, Kurdzhali, and Stara Zagora; 25 cities. Provides 11.3% of nation's industrial and 11.9% of agricultural production. Manufacturing (electric and electronic equipment, foodstuffs, hydraulic elements, chemical industry equipment, wool and silk textiles; machine building). Agriculture (tobacco, cotton, grain, walnuts, hazelnuts, wine-grapes; sericulture). Mining (lignite, lead, zinc, marble, minerals, high quality clays) and non-ferrous metallurgy.

Haskovo (HAHS-ko-vo), city, ⊙ HASKOVO oblast and obshtina, S BULGARIA; 41°55′N 25°33′E. In an agricultural region noted for its tobacco. Has one of Bulgaria's largest cigarette factories. Other industries include textiles, furniture, foodstuffs, machines for metal processing and for meat and dairy factories; metalworking, tailoring. A mineral spa is located nearby. Sometimes spelled Khaskovo.

Haskovo Basin (HAHS-ko-vo), (□ 100 sq mi/260 sq km), S central BULGARIA, SE part of the THRACIAN PLAIN, between the MARITSA RIVER (N and E), RODOPI Mountains (S), and PLOVDIV BASIN (W); 41°55′N 25°33′E. Average elevation 550 ft/168 m. Black and alluvial soils. Extensive agriculture (cotton, silk, tobacco, sesame; vineyards). Main centers are HASKOVO and HARMANLI.

Haskovska (HAHS-kov-skah), river, S central BULGARIA; 41°55′N 25°33′E.

Hasköy, TURKEY: see ISTANBUL.

Haslach (HAHS-lahkh), town, BADEN-WÜRTTEMBERG, SW GERMANY, in BLACK FOREST, on the KINZIG RIVER, and 14 mi/23 km SSE of OFFENBURG; 48°17′N 08°05′E. Manufacturing of machinery, tools; lumber milling. Summer resort.

Haslach an der Mühl (HAHS-luhkh ahn der MYOOL), township, N UPPER AUSTRIA, in the Mühlviertel, on Grosse Mühl River, and 21 mi/34 km NW of LINZ. Near CZECH border; 48°35′N 14°02′E.

Area is shown by the symbol □, and capital city or county seat by ⊙.

Linen manufacturing; center of Mühlviertel's weaving industry.

Hasle (HAS-luh), town, BORNHOLM county, DENMARK, on W shore of BORNHOLM island; 55°11′N 14°43′E. Brickworks; pottery; shipbuilding. Granite quarry, kaolin quarries; fisheries.

Hasle bei Burgdorf (HAHS-luh bei-BOORG-dorf), commune, BERN canton, NW central SWITZERLAND, on EMME RIVER, and 3 mi/4.8 km SSE of BURGDORF.

Haslemere (HAI-zuhl-mir), town (2001 population 15,612), SW SURREY, SE ENGLAND, 12 mi/19 km SSW of GUILDFORD; 51°05′N 00°42′W. Noted arts and crafts center, and scene of music festivals. Church contains grave of John Tyndall and a Burne-Jones window in memory of Tennyson, who lived nearby. George Eliot lived at nearby Shottermill.

Haslet, town (2006 population 1,484), TARRANT county, N TEXAS, suburb 14 mi/23 km N of downtown FORT WORTH; 32°57′N 97°20′W. Agricultural area on fringe of Dallas–Fort Worth Metroplex. Manufacturing (plastics; machining).

Haslett (HAS-let), village (2000 population 11,283), INGHAM county, S central MICHIGAN, suburb 6 mi/9.7 km E of LANSING, near Lake Lansing (c.1.5 mi/2.4 km long, 1 mi/1.6 km wide); 42°45′N 84°24′W. Manufacturing (metal finishing). Part of incorporated city of MERIDIAN TOWNSHIP.

Haslev (HA-slev), city, VESTSJÆLLAND county, SJÆLLAND, DENMARK, 15 mi/24 km SE of SORØ; 55°20′N 11°58′E. Agriculture (dairy plant); manufacturing (chemical plant; pottery, furniture, liqueurs, machinery); meat cannery.

Hasliberg (HAHS-lee-berg), commune of many villages, BERN canton, central SWITZERLAND, N of MEIRINGEN, in lower HASLITAL valley. Elevation 3,287 ft/1,002 m–4,035 ft/1,230 m. Alpine tourist center.

Haslingden (HAZ-ling-duhn), town (2001 population 14,870), E central LANCASHIRE, N ENGLAND, 4 mi/6.4 km SSE of ACCRINGTON; 53°42′N 02°18′W. Footwear, chemicals, rubber goods. Slate quarrying nearby. In the borough (S) is town of Helmshore.

Haslington (HAZ-ling-tuhn), village (2001 population 6,430), CHESHIRE, W ENGLAND, 2 mi/3.2 km E of CREWE; 53°06′N 02°24′W. Trent-Mersey Canal to N.

Haslital (HAHS-lee-tahl), valley of upper AARE RIVER, BERN canton, S central SWITZERLAND, in BERNESE ALPS, between Grimsel Lake and Lake Brienz. OBERHASLI hydroelectric works here. Important resort area. Chief town, MEIRINGEN.

Hasparren (ahs-pah-RAWN), industrial town (□ 29 sq mi/75.4 sq km), PYRÉNÉES-ATLANTIQUES department, AQUITAINE region, SW FRANCE, 11 mi/18 km SE of BAYONNE; 43°23′N 01°18′W. Its predominantly Basque population specializes in heat-pump manufacturing and distribution. Town is also a historic tanning center that produces footwear.

Hasper, GERMANY: see ENNEPETAL.

Haspra (HAHS-prah) (Russian *Gaspra*), town, Republic of CRIMEA, UKRAINE, BLACK SEA resort, adjoining (E) KOREYIZ, 5 mi/8 km SW and under jurisdiction of YALTA; 44°26′N 34°06′E. Manufacturing of reinforced concrete building materials. Has eight sanatoria. The ruins of Charax, an ancient Roman fortress, are approximately 3 mi/4.8 km NW of Haspra.

Hasrun (has-ROON), village, N LEBANON, 15 mi/24 km SE of TRIPOLI; 34°14′N 35°59′E. Elevation 4,100 ft/1,250 m. Summer resort; cereals, fruit.

Hassa, township, S TURKEY, 45 mi/72 km NNE of ANTAKYA; 36°48′N 36°30′E. Grain, fruit.

Hassan (HUH-suhn), district (□ 2,631 sq mi/6,840.6 sq km), KARNATAKA state, S INDIA; ⊙ HASSAN. On DECCAN PLATEAU; bordered W by WESTERN GHATS (extensive coffee, tea, cardamom estates; timber, sandalwood); drained mainly by HEMAVATI RIVER. Agriculture includes rice (terrace farming), sugarcane, millet, cotton. Coffee curing, rice milling, cotton

ginning; chromite, asbestos, and kaolin working; handicrafts (biris, glass bangles, wickerwork); handloom weaving. Chief towns are Hassan, HOLE NARSIPUR, ARSIKERE. Noted archaeological sites at BELUR and Sravana Belgola.

Hassan (HUH-suhn), city, ⊙ HASSAN district, KARNATAKA state, S INDIA, 60 mi/97 km NW of MYSORE; 13°00′N 76°05′E. Road center. Manufacturing of agricultural implements, domestic cutlery, firebricks, electric stoves, road tar. Annual livestock fair.

Hassan Abdal, PAKISTAN: see HASAN ABDAL.

Hassan Parthi, INDIA: see HASANPARTI.

Hassayampa River, intermittent stream, c.60 mi/97 km long, W central ARIZONA; rises S of PRESCOTT, in YAVAPAI county, flows S, past WICKENBURG, to GILA River, 8 mi/12.9 km WSW of BUCKEYE. Crossed by HAYDEN-RHODES AQUEDUCT, c.25 mi/40 km N of its mouth.

Hasselborg, Lake (HA-suhl-borg), SE ALASKA, central ADMIRALTY ISLAND, 18 mi/29 km NE of ANGOON; 4 mi/6.4 km long; 57°43′N 134°16′W. Fishing, hunting.

Hässelby (HES-el-BEE), residential town, STOCKHOLM county, E SWEDEN, at E end of LAKE MÄLAREN, 10 mi/16 km WNW of STOCKHOLM. Has seventeenth-century castle.

Hasselfelde (HAHS-sel-fel-de), town, SAXONY-ANHALT, central GERMANY, in the lower HARZ, 10 mi/16 km SSE of WERNIGERODE; 51°42′N 10°51′E. Health resort; woodworking. In Middle Ages, silver and copper mined here.

Hassell, village (2006 population 71), MARTIN county, E North Carolina, 14 mi/23 km WNW of WILLIAMSTON; 35°54′N 77°16′W. Service industries; agriculture (tobacco, grain, cotton; livestock; poultry). Settled c. 1878 and incorporated in 1903. Earlier known as Dogville Crossroads until it was renamed for Baptist Elder Sylvester Hassell.

Hasselt (HAH-suhlt), city, ⊙ Hasselt district (□ 350 sq mi/910 sq km; 2006 population 70,236) and LIMBURG province, NE BELGIUM, in the KEMPENLAND (French *Campine*), a port on the ALBERT CANAL; 50°56′N 05°20′E. Commercial and industrial center and a railroad junction. The Dutch defeated the Belgians here in 1831. Chartered in 1232.

Hassfurt (HAHS-foort), town, BAVARIA, central GERMANY, in LOWER FRANCONIA, on the Main River, 20 mi/32 km NW of BAMBERG; 50°02′N 10°30′E. Manufacturing includes instruments for measurement, bricks. Has late-Gothic chapel and 15th-century town hall.

Hassi Mameche (hah-SEE mah-MESH), village, MOSTAGANEM wilaya, NW ALGERIA, 6 mi/9.7 km S of MOSTAGANEM. Vineyards. Formerly called Rivoli.

Hassi Messaoud (hah-SEE mes-sah-OOD), oil town, OUARGLA wilaya, E ALGERIA, E of OUARGLA; 31°52′N 05°57′E. Center for oil production in Ouargla wilaya, with pipelines to SKIKDA, ALGIERS, BEJAÏA, and ARZEW. Has a small refinery. Also, an important road junction, with airport. Since oil was discovered in 1956, the town has grown considerably and rapidly. Population consists mainly of engineers and their families.

Hassi R'Mel (hah-SEE ruh-MEL), town, LAGHOUAT wilaya, S ALGERIA; 33°10′N 03°12′E. Center of one of the world's largest gas fields, with 2/3 of Algeria's gas reserves. An important source for domestic gas consumption and natural gas is piped from here to ARZEW and SKIKDA for export. Also produces condensate for export. Source of Trans-MEDITERRANEAN pipeline (1,553 mi/2,500 km; built with Italian collaboration), which supplies ITALY with natural gas. A second pipeline, built in 1993, supplies SPAIN.

Hassi Zehana (hah-SEE ze-hah-NAH), village, SIDI BEL ABBÈS wilaya, NW ALGERIA, 18 mi/29 km SW of SIDI BEL ABBÈS. Vineyards. Formerly Tassin.

Hasslarp (HAHS-LAHRP), village, SKÅNE county, S SWEDEN, 7 mi/11.3 km NE of Hälsingborg; 56°08′N 12°50′E.

Hasslau, GERMANY: see WILKAU-HASSLAU.

Hässleholm (HES-le-HOLM), town, SKÅNE county, S SWEDEN, 17 mi/27 km NW of KRISTIANSTAD; 56°09′N 13°47′E. Railroad center; manufacturing (furniture, baby carriages). Chartered as town 1914.

Hassloch (HAHS-lohkh), town, RHINELAND-PALATINATE, W GERMANY, 6 mi/9.7 km E of NEUSTADT; 49°22′N 08°16′E. Metalworking; wine making. Amusement park nearby.

Hassmersheim (HAHS-muhrs-heim), village, BADEN-WÜRTTEMBERG, SW GERMANY, on NECKAR RIVER, 12 mi/19 km N of HEILBRONN; 49°19′N 09°08′E.

Hasson Heights (HA-SUHN), unincorporated town (2000 population 1,495), VENANGO county, NW PENNSYLVANIA, residential suburb 1 mi/1.6 km NE of OIL CITY; 41°27′N 79°40′W. Oil fields in vicinity.

Hastenbeck (HAHS-tuhn-bek), suburb of HAMELN, LOWER SAXONY, W GERMANY, 3 mi/4.8 km SE of city center; 52°07′N 09°22′E. Scene (1757) of French victory over duke of Cumberland.

Hasthigiripura, SRI LANKA: see KURUNEGALA.

Hastière (ahs-TYER), commune (2006 population 5,248), Dinant district, NAMUR province, S BELGIUM, on MEUSE RIVER, and 5 mi/8 km SW of DINANT. Tourist resort.

Hastings (HAI-steengz), county (□ 2,308 sq mi/6,000.8 sq km; 2001 population 125,915), SE ONTARIO, E central CANADA, on Lake ONTARIO; ⊙ BELLEVILLE; 44°45′N 77°35′W. It is the second-largest county in Ontario. Population (2001) excludes Census data for one or more incompletely counted Indian reserves or settlements. Incorporated 1850. Composed of the municipalities of CENTRE HASTINGS, HASTINGS HIGHLANDS, MARMORA and Lake, and TWEED; the towns of BANCROFT and DESERONTO; and the townships of Carlow/Mayo, FARADAY, LIMERICK, MADOC, Stirling-Rawdon, TUDOR & CASHEL, TYENDINAGA, and WOLLASTON.

Hastings, city (2000 population 18,204), ⊙ DAKOTA county, SE MINNESOTA, suburb 18 mi/29 km SE of downtown ST. PAUL, on the MISSISSIPPI RIVER, opposite and to W of its confluence with the ST. CROIX River; elevation 691 ft/211 m; 44°43′N 92°50′W. Vermillion River drains S part of city, and enters Mississippi River to S; WISCONSIN state line follows St. Croix River and Mississippi R, SE. Farm trade (poultry, livestock; grain, soybeans; dairying) and manufacturing center (flour, computer equipment, fertilizers and feeds; printing and publishing, other diversified light manufacturing). Minnesota Veterans Home is here. Afton State Park to N; Richard J. Dorer Memorial Hardwood State Forest to S; Gores Pool Number 3 Wildlife Area to SE, on Mississippi River; LOWER ST. CROIX NATIONAL SCENIC RIVERWAY to NE; Hastings Lock and Dam Number 2 is here, forms Spring Lake on Mississippi River. Incorporated 1857.

Hastings, city (2000 population 7,095), ⊙ BARRY county, SW MICHIGAN, 29 mi/47 km SE of GRAND RAPIDS, and on THORNAPPLE RIVER; 42°38′N 85°17′W. In agricultural area. Manufacturing (transportation equipment, machinery, rubber products, fire-fighting apparatus, crossbows; publishing). Resort; several lakes nearby. Native American mounds in vicinity. Airport to W. Settled c.1836; incorporated as village 1855, as city 1871.

Hastings, city (2006 population 25,144), ⊙ ADAMS county, S central NEBRASKA, 26 mi/42 km S of GRAND ISLAND, near headwaters of W Fork BIG BLUE RIVER; 40°35′N 98°23′W. Railroad junction. Manufacturing includes processed foods, construction materials; printing; railroad shops. Seat of Hastings College, Central Community College–Hastings Campus E of town. Has Hastings Museum, County Fairgrounds. Municipal Airport to W. D. Lake D. State Wayside to E. Incorporated 1874.

Hastings (HAI-stingz), town (□ 11 sq mi/28.6 sq km; 2001 population 85,029) and district, East SUSSEX, SE

ENGLAND; 50°51'N 00°36'E. Resort and residential town backed by cliffs with a 3-mi/4.8-km marine esplanade, parks, and bathing beaches. The site was occupied in Roman times. It was made famous by the Battle of Hastings, which took place at nearby Battle in 1066, between the Normans under William, duke of Normandy (later William I), and the Anglo-Saxons under Harold. The battle, one of the most celebrated in English history, was won by William's force after a single day's fighting. This was the first and most decisive victory of the Norman Conquest of England. Hastings became one of the Cinque Ports. Remains of Norman castle on cliff tops. Adjacent to resort of St. Leonards.

Hastings, major town ⊙ HASTINGS district (□ 2,014 sq mi/5,217 sq km), HAWKE BAY region (□ 2,014 sq mi/5,236.4 sq km), NORTH ISLAND, NEW ZEALAND, 13 mi/21 km from NAPIER (its twin city). It serves an extensive livestock and fruit and vegetable producing area, with food processing industries, including meatpacking, canning, and some dairy processing.

Hastings, town, SIERRA LEONE, on SIERRA LEONE PENINSULA, 12 mi/19 km SSE of FREETOWN; 08°22'N 13°08'W. Airfield for domestic flights. Founded 1819.

Hastings, town (2000 population 214), MILLS county, SW IOWA, on WEST NISHNABOTNA RIVER, and 13 mi/21 km E of GLENWOOD; 41°01'N 95°30'W. In agricultural area.

Hastings (HAI-steengz), unincorporated village (□ 1 sq mi/2.6 sq km; 2001 population 1,208), SE ONTARIO, E central CANADA, on TRENT RIVER, 18 mi/29 km E of PETERBOROUGH, and included in town of TRENT HILLS; 44°18'N 77°57'W. Light manufacturing.

Hastings, village (2006 population 146), JEFFERSON county, S OKLAHOMA, 8 mi/12.9 km NW of WAURIKA; 34°13'N 98°06'W. In agricultural area. WAURIKA LAKE reservoir to NE.

Hastings, SE residential suburb and seaside resort of BRIDGETOWN, SW BARBADOS.

Hastings, borough (2006 population 1,322), CAMBRIA county, W central PENNSYLVANIA, 25 mi/40 km NNE of JOHNSTOWN. Light manufacturing; bituminous coal; agricultural area (potatoes, grain; livestock, dairying). Prince Gallitzin State Park to E. Incorporated 1894.

Hastings (HAI-steengz), suburb of MELBOURNE, VICTORIA, SE AUSTRALIA; 38°18'S 145°11'E.

Hastings (HAI-steengz), settlement, TASMANIA, SE AUSTRALIA, 65 mi/104 km S of HOBART, on Huon Highway. Logging. Limestone caves, thermal pool nearby.

Hastings Highlands (HAI-steengz HEI-luhndz), township (□ 373 sq mi/969.8 sq km; 2001 population 3,992), SE ONTARIO, E central CANADA, 17 mi/27 km from BANCROFT; 45°17'N 77°51'W.

Hastings Lake (HAI-steengz), hamlet (2006 population 88), central ALBERTA, W CANADA, on shores of HASTINGS LAKE, and included in municipality of STRATHCONA COUNTY. Moose, white-tailed deer in area.

Hastings-on-Hudson, residential and industrial village (□ 2 sq mi/5.2 sq km; 2006 population 7,843), WESTCHESTER county, SE NEW YORK, on E bank of HUDSON RIVER, just N of YONKERS; 40°59'N 73°52'W. Light manufacturing and commercial services. Serves as a suburb of NEW YORK city. Incorporated 1879.

Hastings River (HAI-steengz), 108 mi/174 km long, E NEW SOUTH WALES, AUSTRALIA; rises in GREAT DIVIDING RANGE; flows S and E, past WAUCHOPE, to PACIFIC OCEAN at PORT MACQUARIE; 31°24'S 152°51'E.

Hästveda (HEST-VE-dah), village, SKÅNE county, S SWEDEN, 11 mi/18 km NE of HÄSSLEHOLM; 56°17'N 13°56'E. Railroad junction.

Hasuda (hah-SOO-dah) or **Hasuta**, city, SAITAMA prefecture, E central HONSHU, E central JAPAN, 9 mi/15 km N of URAWA; 35°59'N 139°39'E. Peas.

Hasumi (hah-SOO-mee), village, Ochi district, SHIMANE prefecture, SW HONSHU, W JAPAN, 47 mi/75 km S of MATSUE; 34°52'N 132°40'E.

Hasunuma (hah-SOO-noo-mah), village, Sanbu district, CHIBA prefecture, E central HONSHU, E central JAPAN, 16 mi/25 km E of CHIBA; 35°35'N 140°30'E.

Hasuta, JAPAN: see HASUDA.

Haswell, village (2000 population 84), KIOWA county, E COLORADO, 20 mi/32 km W of EADS; elevation c.4,538 ft/1,383 m; 38°27'N 103°09'W. Cattle; wheat, sunflowers, sorghum. ADOBE CREEK RESERVOIR (Blue Lake) to S.

Hata (HAH-tah), town, East Chikuma district, NAGANO prefecture, central HONSHU, central JAPAN, 34 mi/55 km S of NAGANO; 36°11'N 137°51'E. Watermelons.

Hata (HAH-tah), village, GORAKHPUR district, E UTTAR PRADESH state, N central INDIA, 24 mi/39 km E of GORAKHPUR. Rice, wheat, barley, oilseeds, sugarcane.

Hatano (HAH-tah-no), town, E central SADO island, Sado district, NIIGATA prefecture, N central JAPAN, 34 mi/55 km N of NIIGATA; 37°59'N 138°23'E. Bamboo works; pottery; stone processing (gems).

Hatashio (hah-TAH-shyo), town, Echi district, SHIGA prefecture, S HONSHU, central JAPAN, 25 mi/40 km N of OTSU; 35°10'N 136°14'E.

Ha Tay, province (□ 829 sq mi/2,155.4 sq km), N VIETNAM, in RED RIVER delta; ⊙ HA DONG; 20°55'N 105°45'E. Bordered N and NW by VINH PHU province, NE by HANOI, E by HAI HUNG province, S by NAM HA province, W by HOA BINH province. Partly framed by RED RIVER and its distributaries and interlaced by irrigation canals, this is an area of fertile alluvial soils and agriculture (wet rice farming, vegetables, fruits, tobacco; livestock raising). Ha Dong is presently undergoing rapid urbanization. Aquaculture; industries (silk spinning, textiles, brick and tile; meat and food processing), light manufacturing, handicrafts. Predominantly Kinh population.

Hatay Mountains, TURKEY: see AMANOS MOUNTAINS.

Hatboro (HAT-buhr-o), borough (2006 population 7,204), MONTGOMERY county, SE PENNSYLVANIA, suburb 15 mi/24 km NNE of downtown PHILADELPHIA, on Pennypack Creek. Manufacturing (machinery, medical equipment, machinery, fabricated metal products, consumer goods, tool and die, foods, building materials; commercial printing). Agricultural (grain, soybeans, apples; livestock; dairying). U.S. Naval Air Station (Willow Grove) to NW. Settled in early 18th century, incorporated 1871.

Hatch, town (2006 population 1,649), DONA ANA county, SW NEW MEXICO, on RIO GRANDE, and 39 mi/63 km NW of LAS CRUCES; 32°40'N 107°09'W. In irrigated agricultural region (cattle, sheep; vegetables, grain, chilies, jalpenos, alfalfa); light manufacturing. Railroad junction to E. Leasburg Dam State Park and Fort Selden State Monument to SE; Percha Dam and Caballo Lake state parks to NW. Sierra Caballo to N; Sierra de las Uvas to S.

Hatch, village (2006 population 116), GARFIELD county, SW UTAH, 15 mi/24 km S of PANGUITCH; 37°38'N 112°25'W. Alfalfa; cattle. Units of Dixie National Forest to E and W; Utah State Fish Hatchery is here.

Hatch 1 and 2 Nuclear Power Plants, GEORGIA: see APPLING.

Hatchet Bay, town, central BAHAMAS, on N ELEUTHERA Island, 20 mi/32 km NW of Governor's Harbour, 55 mi/89 km ENE of NASSAU; 25°20'N 76°28'W. Dairying; poultry farming.

Hatchie River (HACH-ee), c.175 mi/282 km long, in MISSISSIPPI and TENNESSEE; rises in NE UNION county, N Mississippi; flows NNW into W Tennessee, and WNW to MISSISSIPPI RIVER through Hatchis National Wildlife Refuge, passes N of COVINGTON, 30 mi/48 km N of MEMPHIS. Receives TUSCUMBIA RIVER from SE, 6 mi/9.7 km E of MIDDLETON.

Hatchineha Lake (ha-chuh-NEE-hah), POLK county, central FLORIDA, 11 mi/18 km ESE of HAINES city; c.7 mi/11.3 km long, 2 mi/3.2 km wide; 28°01'N 81°24'W. KISSIMMEE RIVER connects it with LAKE KISSIMMEE (S) and with chain of lakes (including TOHOPEKALIGA LAKE) to N.

Hatchville, Massachusetts: see FALMOUTH.

Hațeg (HAH-tseg), German *Hötzing*, Hungarian *Hátszeg*, town, HUNEDOARA county, W central ROMANIA, in TRANSYLVANIA, on railroad, and 10 mi/16 km S of HUNEDOARA. Agricultural center; livestock market; tanning, flour milling, and vinegar making. Base for excursions into RETEZAT MOUNTAINS. Still preserves colorful regional costumes.

Hatfield (HAIT-feeld), town (2001 population 27,883), HERTFORDSHIRE, SE ENGLAND; 7 mi/11.3 km N of BARNET; 51°46'N 00°13'W. Designated one of the New Towns in 1948 to alleviate overpopulation in London; plans for it were coordinated with those of nearby WELWYN GARDEN CITY. Aircraft, engineering, and electronics manufacturing. Seat of University of Hertfordshire. Hatfield House (17th century) to E.

Hatfield (HAIT-feeld), town (2001 population 16,184), SOUTH YORKSHIRE, N ENGLAND, DON RIVER to W, 7 mi/11.3 km NE of DONCASTER; 53°34'N 00°59'W. Has 12th–15th-century church.

Hatfield, town, HAMPSHIRE county, W MASSACHUSETTS, on CONNECTICUT RIVER, just above NORTHAMPTON; 42°23'N 72°37'W. Wood products; barium-sulphate mine; tobacco. Settled 1661, set off from HADLEY 1670.

Hatfield, unincorporated town, FAYETTE county, SW PENNSYLVANIA, residential suburb 2 mi/3.2 km S of UNIONTOWN; 39°52'N 79°44'W.

Hatfield, village (2000 population 402), POLK county, W ARKANSAS, 10 mi/16 km SW of MENA, near OKLAHOMA state line, near MOUNTAIN FORK River; 34°29'N 94°22'W. Sawmilling, manufacturing (posts and poles, lumber, wood products). Ouachita National Forest to E and NW.

Hatfield, village, SPENCER county, SW INDIANA, 18 mi/29 km ESE of EVANSVILLE. Livestock, poultry.

Hatfield, village (2000 population 47), PIPESTONE county, SW MINNESOTA, near ROCK RIVER, 7 mi/11.3 km ESE of PIPESTONE; 43°57'N 96°11'W. Grain; livestock; dairying.

Hatfield, borough (2006 population 2,847), MONTGOMERY county, SE PENNSYLVANIA, suburb 22 mi/35 km N of PHILADELPHIA, on West Branch of Neshaminy Creek. Manufacturing (machinery, building materials, fabricated metal products, plastic products, apparel, wood products, food products; library bindery; printing). Settled 1860, incorporated 1898.

Hatfield Peverel (HAT-feeld PE-vuh-ruhl), village (2001 population 4,384), central ESSEX, SE ENGLAND, 6 mi/9.7 km ENE of CHELMSFORD; 51°47'N 00°35'E. Agricultural market. Has ancient priory church, restored in 19th century.

Hatgal (KHAHD-khahl), village, HÖVSGÖL province, NW MONGOLIA, landing at S end of HÖVSGÖL NUUR (Lake Khubsugul), at outlet of the EGIYN GOL (river), 55 mi/90 km N of MÖRÖN; 50°26'N 100°09'E. Freight transshipment point on RUSSIA-Mongolia trade route; wool-washing plant. Also spelled Khatkhyl, Hadhal, Khadkhal.

Hat Head (HAT HED), coastal village, NEW SOUTH WALES, SE AUSTRALIA, 285 mi/459 km N of SYDNEY, 20 mi/32 km NE of KEMPSEY, and in Hat Head National Park; 31°03'S 153°03'E. Surfing beaches, fishing. Whales, dolphins, birdlife.

Hatherleigh (HA&hardTH;-uh-lee), village (1991 population 1,086; 2001 population 1,673), W central DEVON, SW ENGLAND, 6 mi/9.7 km NNW of OKEHAMPTON; 50°49'N 04°04'W. Agricultural market; makes electrical appliances, agricultural implements. Has 15th-century church.

Hathern (HA&hardTH;-uhn), village (1991 population 1,750; 2004 population 1,866), N LEICESTERSHIRE, central ENGLAND, 3 mi/4.8 km NW of LOUGHBOROUGH; 52°48'N 01°15'W. Some light industry. Has 14th-century church.

Hathersage (HA&hardTH;-uh-sij), village (1991 population 1,323; 2001 population 2,751), N DERBYSHIRE, central ENGLAND, 8 mi/12.9 km SW of SHEFFIELD; 53°19′N 01°38′W. Has 14th–15th-century church.

Hatherton Glacier, E ANTARCTICA, entering DARWIN GLACIER, at Junction Spur, flowing from the Polar Plateau, along the S side of the Darwin Mountains; 79°55′S 157°35′E.

Hathras (HAH-thruhs), city, ALIGARH district, W UTTAR PRADESH state, N central INDIA, 20 mi/32 km S of ALIGARH; 27°36′N 78°03′E. Road and trade center (wheat, barley, pearl millet, gram, cotton, corn, sugarcane); cotton milling, ginning, and baling, glass and cutlery manufacturing, oilseed milling. Strong Jat fort in early 19th century. Saltpeter processing 5 mi/8 km SE, at town of LAKHNAU.

Hathwa, INDIA: see SIWAN.

Hatia Island (hah-tee-ah), E island of GANGA DELTA, NOAKHALI district, SE EAST BENGAL, BANGLADESH, in BAY OF BENGAL, 20 mi/32 km S of NOAKHALI; 23 mi/37 km long, 4 mi/6.4 km–8 mi/12.9 km wide. Separated from SANDWIP ISLAND (E) by Hatia Channel, and from DAKHIN SHAHBAZPUR ISLAND (W) by SHAHBAZPUR RIVER. Hatia village is in N central section. Steamer service with mainland. Subject to severe cyclones (over ½ of population killed in 1876). Large island just to the NW is also called Hatia.

Ha Tien (HAH tee-EN), town, KIEN GIANG province, S VIETNAM, on Gulf of SIAM, and 155 mi/250 km WSW of SAIGON, on CAMBODIA border; 10°25′N 104°30′E. Small port and charming town with rich architectural heritage. Exports pepper, dried and salted fish, straw bags, betel nuts, tortoise shells. Long a Chinese port-of-trade and population still largely Chinese. Linked with CHAU DOC by VINHTE CANAL. Limestone quarries. Tourism; nearby white-sand beaches. Former Khmer territory; passed in 1798 to Vietnamese. Chinese, Vietnamese, and Cambodian (Khmer) population. Formerly Hatien.

Hatiguanico River (ah-tee-gwah-NEE-ko), sluggish stream in MATANZAS province, W CUBA; rises near Lake Tesoro; flows W through mangrove swamps of ZAPATA to the GULF OF BATABANÓ; 22°31′N 81°36′W.

Hatillo (ah-TEE-yo), urban district, COSTA RICA, just SW of SAN JOSÉ. Residential with a large component of public housing units.

Hatillo (ah-TEE-yo), town (2006 population 42,483), NW PUERTO RICO, on the coast, 7 mi/11.3 km W of ARECIBO. Dairying center (one-third of milk consumed in Puerto Rico produced here). Coffee-growing area. Light manufacturing (food, textiles, wood and cement products). Famous for its costume parade on Innocents' Day (Dec. 28).

Hatillo (ah-TEE-yo), village, DUARTE province, central DOMINICAN REPUBLIC, on YUNA RIVER, and 23 mi/37 km S of SAN FRANCISCO DE MACORÍS. Iron and gold mining.

Ha Tinh (HAH TIN), province (☐ 2,337 sq mi/6,076.2 sq km), N central VIETNAM; ☉ HA TINH; 18°20′N 105°50′E. Bordered N by NGHE AN province, E by VINH BAC BO, S by QUANG BINH province, W by LAOS province. An amalgam of coastal lowlands, hilly midlands, and a mountainous interior; exhibits a range of habitats and localized economies. Once densely forested, its mountains and hills have been badly overcut in recent decades. Diverse soils and mineral resources (zinc, iron ore, marble), wet rice farming, shifting cultivation; forest products (gums, resins, medicinals, rattan, foraged foods); agro-forestry, fishing, and aquaculture; and commercial agriculture (coffee, tea, pineapples, and other fruits, vegetables). Handicrafts, light manufacturing, fish and food processing. Predominantly Kinh population.

Ha Tinh (HAH TIN), city, ☉ HA TINH province, N central VIETNAM, 25 mi/40 km SE of VINH, near SOUTH CHINA SEA coast; 18°19′N 105°55′E. Transportation and trading center; administrative complex. Head of road via Cha Lo to Tha Khek (LAOS) over N TRUONG SON RANGE. Area remains underdeveloped, but govern-

ment plans underway to expand fisheries, commercial farming, agro-processing, aquaculture; forest products. Zinc and iron mining. Formerly Hatinh.

Hatirkul, INDIA: see KONNAGAR.

Hat Island, FIJI: see VATU VARA.

Hatkalangda (haht-kuh-LUHNG-dah), town, KOLHAPUR district, MAHARASHTRA state, W INDIA, 13 mi/21 km ENE of KOLHAPUR; 16°44′N 74°25′E. Sugarcane, tobacco. Sometimes spelled Hatkanagla.

Hatley (HAT-lee), canton (☐ 28 sq mi/72.8 sq km; 2006 population 1,676), ESTRIE region, S QUEBEC, E CANADA; 45°14′N 72°01′W.

Hatley (HAT-lee), village (☐ 24 sq mi/62.4 sq km; 2006 population 728), ESTRIE region, S QUEBEC, E CANADA, 19 mi/30 km from ORFORD; 45°11′N 71°56′W.

Hatley (HAT-lee), village (2000 population 476), MONROE county, E MISSISSIPPI, 4 mi/6.4 km E of AMORY; 33°58′N 88°25′W. Cotton, corn; cattle; dairying; timber.

Hatley, village (2006 population 481), MARATHON county, central WISCONSIN, 15 mi/24 km ESE of WAUSAU, on Placer River; 44°53′N 89°20′W. In dairying region. Manufacturing (hardwood veneers).

Hato (AH-to), town, ☉ Hato municipio, SANTANDER department, N central COLOMBIA, 35 mi/56 km SW of BUCARAMANGA, in the Cordillera ORIENTAL. Coffee, corn; livestock.

Hato (HAH-to), village and airport (Aeropuerto Internashonal Hato), E central CURAÇAO, NETHERLANDS ANTILLES, 5 mi/8 km N of WILLEMSTAD; 12°11′N 68°17′W.

Hato Corozal (AH-to ko-RO-sahl), town, ☉ Hato Corozal municipio, CASANARE department, S COLOMBIA, in the Oriente, 53 mi/85 km NE of YOPAL; 06°09′N 71°45′W. Plantains, sugarcane; livestock.

Hatogaya (hah-TO-gah-yah), city, SAITAMA prefecture, E central HONSHU, E central JAPAN, 5 mi/8 km S of URAWA; 35°49′N 139°44′E. Residential suburb of TOKYO.

Hatolia (HAH-to-lee-yah), town, in central TIMOR ISLAND, 21 mi/34 km SW of DILI; 08°48′S 125°22′E. Cinnamon, coffee. Formerly spelled Hato-Lia.

Hato-Lia, EAST TIMOR: see HATOLIA.

Hato Mayor, province (☐ 514 sq mi/1,336.4 sq km; 2002 population 87,631), E DOMINICAN REPUBLIC, along S shore of SAMANA BAY; ☉ HATO MAYOR; 18°50′N 69°20′W. Agricultural area (cocoa, rice, beans; cattle). Fishing along Samana Bay.

Hato Mayor (AH-to mei-YOR), town (2002 population 34,006), SEIBO province, E DOMINICAN REPUBLIC, 22 mi/35 km N of SAN PEDRO DE MACORÍS; ☉ HATO MAYOR province; 18°42′N 69°20′W. Agricultural center (sugarcane, coffee, cacao, rice, fruit). First settled 1520.

Hato Rey (AH-to RAI), N residential suburb of Central SAN JUAN, N PUERTO RICO, 5 mi/8 km SE of OLD SAN JUAN. Island's main business district, includes many financial institutions.

Hatoyama (hah-TO-yah-mah), town, Hiki district, SAITAMA prefecture, E central HONSHU, E central JAPAN, 19 mi/30 km N of URAWA; 35°58′N 139°20′E.

Hatra, IRAQ: see HADHR, AL.

Hatsa (HAHT-sah), village, PHONG SALI province, N LAOS, on the Nam Ou (head of navigation), 6 mi/9.7 km ENE of PHONG SALI; 21°44′N 102°12′E. Shifting cultivation; forest products, opium. Lolo, H'mong, and other minority peoples.

Hatsukaichi (hahts-KAH-ee-chee), city and port, HIROSHIMA prefecture, SW HONSHU, W JAPAN, on HIROSHIMA BAY, 9 mi/15 km W of HIROSHIMA; 34°20′N 132°20′E. Oysters. Woodwork (toys, sports items); furniture. Tallest (6.76 ft/2.06 m high) wooden Buddha statue in Japan; carved out of one tree.

Hátszeg, ROMANIA: see HAȚEG.

Hatta (HAHT-tah), village (2001 population 28,508), DAMOH district, N MADHYA PRADESH state, central INDIA, on SONAR RIVER, and 60 mi/97 km ENE of SAGAR; 24°07′N 79°36′E. Wheat, millet, oilseeds.

Hatta (HAHT-tah), village, Nakakoma district, YAMANASHI prefecture, central HONSHU, central JAPAN, 5 mi/8 km W of KOFU; 35°39′N 138°29′E.

Hattah-Kulkyne National Park (☐ 185 sq mi/479 sq km), NW VICTORIA, SE AUSTRALIA, 330 mi/531 km NW of MELBOURNE, 25 mi/40 km S of NEW SOUTH WALES border, and on SW banks of MURRAY RIVER; 34°40′S 142°30′E. Lies E of Calder Highway. Floodwaters of Murray back into Chalka Creek, filling numerous lakes, largest being Lake Hattah. Lakes contain large carp, attracting variety of birds (pelicans, cockatoos, mountain ducks), also emus, choughs, mallee fowl. Kangaroos, wallabies; giant goannas. River red gums near lakes; black box, cypress pine, bull oak among sand plains. Visitor center. Camping, picnicking, hiking. Established 1960.

Hattem (HAH-tuhm), town, GELDERLAND province, N central NETHERLANDS, on IJSSEL RIVER, at mouth of Grote Wetering River, and 3 mi/4.8 km S of ZWOLLE; 52°28′N 06°04′E. N terminus of Apeldoorns Canal. Dairying; cattle; grain, vegetables; light manufacturing Castle to S; nature area to SW.

Hatten (HAHT-ten), commune, LOWER SAXONY, NW GERMANY, 12 mi/19 km WSW of DELMENHORST; 53°01′N 08°21′E.

Hatteras (HA-tuh-ruhs), unincorporated village (2000 population 2,642), DARE county, E North Carolina, 48 mi/77 km S of MANTEO, on PAMLICO SOUND, in CAPE HATTERAS NATIONAL SEASHORE, near SW end of HATTERAS ISLAND, in North Carolina's OUTER BANKS sand barrier; ATLANTIC OCEAN to S; 35°13′N 75°41′W. Resort; boatbuilding, fishing. Manufacturing (commercial boats). Free ferry to NE end of Ocracoke Island, across HATTERAS INLET (passage). Cape Hatteras (lighthouse) 10 mi/16 km to E. Beach recreation area. Inc. in 1931 but no longer active in municipal affairs.

Hatteras, Cape (HA-tuh-ruhs), promontory, DARE county, E NORTH CAROLINA, on HATTERAS ISLAND, in OUTER BANKS, sand barrier between the ATLANTIC OCEAN (E and S) and PAMLICO SOUND (NW), in CAPE HATTERAS NATIONAL SEASHORE; 35°13′N 75°31′W. Called the Graveyard of the Atlantic, the cape experiences frequent storms that drive ships landward toward its dangerous shallow and changing depths. Cape Hatteras Lighthouse (built 1870) was removed in 1936 due to heavy beach erosion, but is again exposed to the ocean.

Hatteras Inlet (HA-tuh-ruhs), NORTH CAROLINA, passage c.1 mi/1.6 km wide, connecting PAMLICO SOUND (NW) with the ATLANTIC OCEAN (SE), between HATTERAS (NE) and OCRACOKE (SE) islands, in CAPE HATTERAS NATIONAL SEASHORE. Border between DARE and HYDE counties passes through inlet. Free ferry carries State Highway 12 traffic across inlet on PAMLICO SOUND side.

Hatteras Island (HA-tuh-ruhs), DARE CO., E NORTH CAROLINA, section of the OUTER BANKS, extending S from NEW INLET (now blocked; joined to PEA ISLAND to N) to CAPE HATTERAS and then SW to HATTERAS INLET, lying between PAMLICO SOUND (W) and the ATLANTIC OCEAN (E and S); 40 mi/64 km long, 1 mi/1.6 km–3 mi/4.8 km wide. Site of fishing, resort villages (including HATTERAS, AVON, BUXTON), several coast guard stations. In CAPE HATTERAS SEASHORE. Section of State Highway 12 extends length of island except SW end.

Hattersheim am Main (HAHT-ters-heim ahm MEIN), town, HESSE, central GERMANY, on right bank of Main River, 10 mi/16 km W of FRANKFURT; 50°05′N 08°29′E. Manufacturing of sweets and packaging material; rose breeding. Chartered 1970.

Hattes, Les, FRENCH GUIANA: see FRANÇAISE, POINTE.

Hattiesburg (HA-teez-buhrg), city (2000 population 44,779), ☉ FORREST county, Forrest and LAMAR counties, SE MISSISSIPPI, 83 mi/134 km SE of JACKSON, on the LEAF RIVER, at mouth of Bowie River; 31°18′N 89°18′W. Major railroad, trade, and industrial center in an agricultural area (cotton, corn; poultry; timber); manufacturing (sand and gravel processing, steel fabrication; lumber, industrial machinery, signs, resins, furniture, apparel, rubber, consumer goods, building materials, food and beverages, chemicals). Pine Belt Regional Airport to N; Municipal Airport to SE. Seat of the University of Southern Mississippi and of William Carey College. Camp Shelby National Guard base to SE (museum); All-American Rose Garden; Hattiesburg Area Historical Society Museum; Hattiesburg Arts Council Gallery; zoo at Kamper Park. De Soto National Forest to SE; Paul B. Johnson State Park to S. Incorporated 1884.

Hattieville, village, BELIZE district, BELIZE, 16 mi/26 km W of BELIZE CITY; 17°26′N 88°24′W. Formerly a temporary camp for refugees from 1961 hurricane in Belize City, it is now a permanent settlement.

Hattingen (HAHT-ting-uhn), city, NORTH RHINE–WESTPHALIA, W GERMANY, on the RUHR RIVER, 5 mi/8 km S of BOCHUM; 51°24′N 07°11′E. Manufacturing includes machinery, steel products. Was member of HANSEATIC LEAGUE. Nearby are historical locks on Ruhr from 1830.

Hatto (HAHT-to), town, Yazu district, TOTTORI prefecture, S HONSHU, W JAPAN, 11 mi/18 km S of TOTTORI; 35°21′N 134°21′E. Fruit (persimmons, pears, apples), shiitake mushrooms. Bamboo work.

Hatton (HAT-tuhn), town, CENTRAL PROVINCE, SRI LANKA, on HATTON PLATEAU, 28 mi/45 km S of KANDY; 06°53′N 80°35′E. Major tea-trade center of Sri Lanka.

Hatton (HA-tuhn), town (2006 population 668), TRAILL CO., E NORTH DAKOTA, 30 mi/48 km SW of GRAND FORKS; 47°38′N 97°27′W. Grain; livestock. Site of Carl Ben Eielson Memorial Arch. Founded in 1881 and named for Frank Hatton (1846–1894) Assistant Postmaster General.

Hatton (HAH-tun), village (2001 population 2,451), S DERBYSHIRE, central ENGLAND, on DOVE RIVER, and 9 mi/14.5 km WSW of DERBY, 5 mi/8 km N of BURTON UPON TRENT; 52°52′N 01°40′W. Some light manufacturing, but mainly residential.

Hatton, village (2006 population 97), ADAMS county, SE WASHINGTON, 32 mi/51 km SW of RITZVILLE, in COLUMBIA basin agricultural region; 46°47′N 118°50′W. Wheat; cattle.

Hatton Fields, village, MONTEREY county, W CALIFORNIA, near CARMEL.

Hatton Plateau (HAT-tuhn), in SRI LANKA HILL COUNTRY, CENTRAL PROVINCE, S central SRI LANKA, consists of parallel ridges (SE-NW) rising to crest line of 3,000 ft/914 m–4,000 ft/1,219 m; 23 mi/37 km long, 18 mi/29 km wide; average rainfall, 150 in/381 cm. Main settlements are HATTON, TALAWAKELE, MASKELIYA.

Hattorf am Harz (HAHT-torf ahm HAHRTS), village, LOWER SAXONY, W GERMANY, at S foot of the upper HARZ, 5 mi/8 km S of OSTERODE; 51°38′N 10°15′E. Woodworking.

Hattula (HAHT-too-lah), village, HÄMEEN province, S FINLAND, 6 mi/9.7 km NNW of HÄMEENLINNA; 61°04′N 24°23′E. Elevation 198 ft/60 m. In lake region. Grain, potatoes; livestock. Has church from 1250, once a Roman Catholic shrine.

Hatuey (ah-too-WAI), town, CAMAGÜEY province, E CUBA, on railroad, and 27 mi/43 km ESE of CAMAGÜEY; 21°12′N 77°32′W. Sugarcane; cattle; lumber. Alfredo Alvarez Mola sugar mill is 2 mi/3.2 km S. Named after valiant indigenous warrior who fought Spaniards.

Hatu Iti, FRENCH POLYNESIA: see MOTU ITI.

Hatutaa (HAH-TOO-tah), uninhabited island, most northerly of MARQUESAS ISLANDS, FRENCH POLYNESIA, S PACIFIC; 4 mi/6.4 km long, 1 mi/1.6 km wide; rises to 1,083 ft/330 m; 07°55′S 140°34′W. Numerous ground doves unknown elsewhere in Pacific. Also called Hatutu.

Hatutu, FRENCH POLYNESIA: see HATUTAA.

Hatvan (HAHT-vahn), city (2001 population 22,906), HEVES county, N HUNGARY, on ZAGYVA RIVER, and 30 mi/48 km ENE of BUDAPEST; 47°40′N 19°41′E. Railroad center; manufacturing (textiles, farinaceous food products, frozen foods, wine); sugar and alcohol refineries.

Hat Yai (HAHD YEI), village (2000 population 187,920), SONGKHLA province, S THAILAND, in MALAY PENINSULA, 15 mi/24 km SW of SONGKHLA; 07°01′N 100°28′E. Railroad center on SINGAPORE-BANGKOK railroad; junction for Malaysian E coast line and spur to SONGKHLA. S THAILAND's chief commercial and entertainment center. Airport. Sometimes spelled HAAD YAI or HAD YAI.

Hatzerim (khah-TZEH-rim) or **Hazerim**, kibbutz, 4 mi/6.4 km W of BEERSHEBA, in NW NEGEV, S ISRAEL; 31°14′N 34°43′E. Elevation 790 ft/240 m. Mixed farming; light manufacturing (irrigation systems); law firm. On the grounds are remains of two ancient settlements, the early-Iron Age Hurbat Sufa and Roman-Byzantine Be'er Sufa. Also, remains of a Byzantine bathhouse with parts of a mosaic floor. Nearby one of Israel's main military air bases. Founded in 1946 and named for biblical Hatzerim.

Hatzeva (khah-tse-VAH) or **Hazeva**, moshav, 44 mi/71 km SE of BEERSHEBA, in Arava Valley, S ISRAEL; 30°48′N 35°15′E. Elevation 472 ft/143 m below sea level. Includes a field school which serves as the starting point for tours in the desert. As part of the peace treaty with Jordan, Hatzeva was granted the right to farm lands in Jordanian territory. Just S of Hatzeva is the Nahal Shizef (Shizef stream) nature reserve. Nearby are the ruins of ancient Hatzeva, a station on the main road in Nabatean, Roman, Byzantine, and British periods. Includes the spring, Ein Hatzeva. Other ruins date from the Iron Age and Roman-Byzantine and Hasmonean times. Several fortresses from 10th century B.C.E. on, and a 7th-century B.C.E. Edomite temple. Early 20th century studies of the ruins and excavations performed between 1987 and 1995 found that King Solomon built the settlement of Tamar here. Founded in 1965.

Hatzfeld, ROMANIA: see JIMBOLIA.

Hatzic (HAT-sik), unincorporated village, SW BRITISH COLUMBIA, W CANADA, on FRASER River, 14 mi/23 km W of CHILLIWACK, and included in district of MISSION; 49°09′N 122°15′W. Lumbering; dairying; fruit, vegetables.

Hatzor Ashdod or **Hatsor Ashdod** (khah-TZOR ahsh-DOD), kibbutz, SW ISRAEL, in coastal plain, 10 mi/16 km SW of REHOVOT; 31°46′N 34°43′E. Elevation 101 ft/30 m. Metal and clothing industries; manufacturing of radio equipment. Mixed farming. Heavily shelled by Arabs, 1948.

Hauaria, Al (hah-WAHR-yah, el), village (2004 population 9,273), NABUL province, NE TUNISIA, 64 mi/103 km NE of TUNIS, on CAPE BON. Falcon nesting grounds. Roman quarries in limestone caves along the coast (the largest is Ghar al Qabir).

Haubourdin (o-boor-DAN), outer WSW suburb (□ 2 sq mi/5.2 sq km), of LILLE, NORD department, NORD-PAS-DE-CALAIS region, N FRANCE; 50°36′N 02°59′E. Textile milling, food processing; ceramics, chemicals.

Haubstadt, town (2000 population 1,529), GIBSON county, SW INDIANA, 10 mi/16 km S of PRINCETON; 38°12′N 87°34′W. In grain-growing area. Manufacturing (meatpacking; specialty machinery).

Haud (HOUD), arid plateau in SE ETHIOPIA and NW SOMALIA; 08°00′N 46°30′E. Inhabited by nomadic Somali tribesmen who graze camels, sheep, and goats.

Haud Plateau, SOMALIA and ETHIOPIA: see HAUD and SOMALIA.

Hauenstein (HOU-uhn-shtein), village, RHINELAND-PALATINATE, W GERMANY, in HARDT MOUNTAINS, 11 mi/18 km E of PIRMASENS; 49°11′N 07°51′E. Health resort. Manufacturing of shoes.

Hauenstein (HAH-oo-uhn-shtain), mountain range, in the JURA mountains, NW SWITZERLAND, 2 mi/3.2 km NNW of OLTEN; rising from the SW to the NE, its highest point is the Geissfluh (3,159 ft/963 m). Crossed by several highways leading from BASEL to the MITTELLAND. Oberer Hauenstein Pass (2,398 ft/731 m) leads to BALSTHAL; Unterer Hauenstein Pass (2,267 ft/691 m) leads to Olten; and Basel-Bern superhighway goes through a 2-mi/3.2-km tunnel at an elevation of 1,854 ft/565 m. Two railroad lines link Basel and Olten: one goes through a 1-mi/1.6-km tunnel, the other uses the Hauenstein Tunnel (opened 1916), which is 5 mi/8 km long at an elevation of 1,476 ft/450 m.

Haugen (HOU-gan), village (2006 population 278), BARRON county, NW WISCONSIN, SE of Bear Lake, 8 mi/12.9 km N of RICE LAKE; 45°36′N 91°46′W. Dairying. W terminus of Tuscobia State Trail to S.

Haugesund (HOU-guh-soon), city (2007 population 32,303), ROGALAND county, S NORWAY, a port on the NORTH SEA; 59°24′N 05°16′E. Has large fisheries and industries producing processed fish and aluminum. Also has a 928-ft/283-m drydock, which, at its completion in 1979, was the largest in SCANDINAVIA. Nearby are numerous Viking monuments, including the grave of Harald I (9th century).

Haughton (HO-tuhn), town (2000 population 2,792), BOSSIER parish, NW LOUISIANA, 15 mi/24 km E of SHREVEPORT; 32°30′N 93°31′W. Cotton; timber; manufacturing (pulpwood, propane). Lake Bistineau State Park to SE. Barksdale Air Force Base to W.

Hau Giang (HOU YAHNG), river, MEKONG Delta, S VIETNAM. S distributary of Mekong River; 10°00′N 105°52′E.

Haugsdorf (HOUGS-dorf), township, N LOWER AUSTRIA, on Pulkau River, and 10 mi/16 km S of ZNOJMO (CZECH REPUBLIC), near Czech border; 48°42′N 16°05′E. Wine.

Haukaban, YEMEN: see SHIBAM.

Haukelifjell (HOUK-li-fyel), mountains in SW NORWAY, extend from the valley BRATTLANDSDAL and the glacier BREIDFONN, in SE HORDALAND county, E c.40 mi/64 km to the Grungedal in TELEMARK county. Rises to 5,110 ft/1,558 m in the Store Nup, 11 mi/18 km ENE of RØLDAL. Traversed by highway; tourist area.

Haukipudas (HOU-ki-POO-dahs), village, OULUN province, W FINLAND, near GULF of BOTHNIA, 13 mi/21 km NNW of OULU; 65°11′N 25°21′E. Elevation 33 ft/10 m. Lumber and pulp milling. Historic church.

Haulbowline Island (hawl-BO-lin), in CORK HARBOUR, SE CORK county, SW IRELAND, just S of CÓBH; 1 mi/1.6 km long; 51°50′N 08°18′W. Headquarters of the Irish navy, site of naval dockyard and repair installations.

Hauppauge (HAW-pawg), unincorporated village (□ 10 sq mi/26 sq km; 2000 population 20,100), SUFFOLK county, SE NEW YORK, on central LONG ISLAND, 2 mi/3.2 km SW of VILLAGE OF THE BRANCH; 40°49′N 73°12′W. A central Long Island manufacturing-industrial region, stemming from long association with aircraft and aerospace industries. Today the economy is more diversified, and includes a strong service-industry segment.

Haura (HO-ruh), former small sheikdom, S YEMEN, on the Gulf of ADEN, 110 mi/177 km WSW of MUKALLA; consists of HAURA coastal village and environs. Protectorate treaty concluded in 1902; abolished in 1967 after SOUTH YEMEN gained its independence. Also spelled Haurah or Hawrah.

Hauraki Gulf, ramifying inlet of the PACIFIC OCEAN, NE NORTH ISLAND, E of AUCKLAND region and E of Coromandel Peninsula and GREAT BARRIER Island. It forms a heavily used commercial and favored recreational waterway.

Hauraki Gulf Maritime Park (HOW-RA-kee), includes much of the gulf area from Cape Home near the BAY OF ISLANDS to Whangamata on the E side of the COROMANDEL Peninsula, NEW ZEALAND; over forty islands and two mainland sites are included to conserve marine ecology, shoreline, scenery, and historic locations.

Hauraki Plains, NEW ZEALAND, a geographically distinct region of flat alluvial soil, partly derived from past wanderings of the WAIKATO River (which in the recent geological past discharged into the FIRTH OF THAMES), and partly from the contemporary PIAKO, WAIHOU, and other rivers. Originally swampy, it has been generally (but not completely) drained by canals and occupied by a close-knit settlement pattern with rural service townships often based around large dairy factories, or, to S and W, on prime-lamb production. To the E this region is sharply defined by the fault line marking the COROMANDEL-KAIMAI ranges; to the W by a less sharply cut sequence of hills separating it from present Waikato Basin.

Hauran [Hebrew=hollow or cavernous land], region, SW SYRIA, E of the JORDAN River The name derives from the numerous caverns in the mountainous NE, where rise the JEBEL ed DRUZ. A largely treeless area with rich lava soil. Most of the inhabitants are Druze, who migrated from LEBANON. Hauran belonged to the biblical kingdom of BASHAN, which the Israelites conquered. Later it was part of the Roman province of AURANITIS. There are many ancient towns (e.g., BUSRA, Der'a, Izra'), ruins, and Greek, Latin, and Arabic inscriptions. The region, formerly a Syrian province, is divided at present into several districts: SUWEIDA, Der'a, and QUNAYTIRAH. Main crops include wheat, barley, lentils, corn, sesame, tobacco, and fruit.

Hauran, Wadi (HO-ran, WAH-dee), dry stream with sporadic flows during the winter, W IRAQ, in SYRIAN DESERT, extending c.200 mi/322 km ENE from Jebel 'Unaiza, near the meeting area of the JORDANIAN, SAUDI ARABIAN, and Iraqi borders, to the EUPHRATES RIVER between HADITHA and HIT.

Hausa (HOU-sah), name referring to a group of former indigenous states in W Afr. and to its Muslim inhabitants of West Afr. stock. Region now included in N NIGERIA and in adjacent territories of MALI and CAMEROON. The term is now strictly an ethnological one. Also spelled Haussa.

Hausach (HOUS-ahkh), town, BADEN-WÜRTTEMBERG, SW GERMANY, in the BLACK FOREST, on the KINZIG RIVER, 16 mi/26 km SE of OFFENBURG; 48°17′N 08°10′E. Manufacturing of garments; metalworking; lumber milling; distilling. Summer resort.

Hauser, village (2000 population 668), KOOTENAI county, N IDAHO, 12 mi/19 km WNW of COEUR D'ALENE, near WASHINGTON state line; 47°46′N 117°01′W. Small Hauser Lake is N.

Hauser Dam (HOU-suhr), LEWIS AND CLARK county, W MONTANA, on the MISSOURI RIVER, 13 mi/21 km NE of HELENA; 125 ft/38 m high. Built (1911) by the Montana Power Company for hydroelectric power. Impounds LAKE HAUSER reservoir with maximum capacity of 109,470 acre-ft.

Hauser, Lake (HOU-suhr), reservoir, LEWIS AND CLARK county, W central MONTANA, on MISSOURI River, 10 mi/16 km NE of HELENA; c.15 mi/24 km long; 46°43′N 111°43′W. CANYON FERRY Dam at SE end. Channel (4 mi/6.4 km long), widening of Silver Creek, connects with Lake HELENA (SW). Formed by Hauser Dam (125 ft/38 m high). Helena National Forest on E side; Hauser Lake State Park on W side.

Haushabi (ho-SHAH-bee), tribal area, S YEMEN, between ABDALI and AMIRI areas. The township MU-SEIMIR is its main center. Until 1967, a powerful tribe under feudal sultan's government, controlling upper Wadi TIBAN trade route to Yemen. Farming; livestock raising. Became a British protectorate in 1895. Also spelled Hawshabi.

Hausham (HOUS-hahm), village, UPPER BAVARIA, BAVARIA, GERMANY, in Bavarian Alps, near the lake Schliersee, 14 mi/23 km WSW of ROSENHEIM; 47°45′N 11°50′E. Metalworking, plastics processing, printing. Lignite was mined here until 1966.

Hausruck Mountains (HOUS-ruk), low wooded range in SW central UPPER AUSTRIA, between basins of the INN and the AGER rivers; extends c. 20 mi/32 km W from HAAG AM HAUSRUCK, rising to 2,441 ft/744 m in the Göblberg. Crossed by a railroad (Attnang–Puchheim–Ried im Innkreis) and several roads.

Haussa, NIGERIA: see HAUSA.

Hauta (HO-tah), village and oasis, S YEMEN, 12 mi/19 km NNE of AZZAN; 13°27′N 46°46′E. Also spelled Hautah or Hawtah.

Hauta, SAUDI ARABIA: see HAWTAH.

Hauta, YEMEN: see QATN, AL.

Haut Atlas or **High Atlas**, mountains, MOROCCO: see ATLAS MOUNTAINS.

Haute Cime, SWITZERLAND: see DENTS DU MIDI.

Hautecombe, FRANCE: see BOURGET, LAC DU.

Haute-Corse (ot–kors) [French=upper Corsica], NE administrative department (□ 1,809 sq mi/4,703.4 sq km; 1999 population 141,584), of the French island of CORSICA in the MEDITERRANEAN SEA, c.100 mi/160 km SE of mainland FRANCE; ⊙ BASTIA; 42°30′N 09°00′E. The department is surrounded on three sides by the Mediterranean, and by CORSE-DU-SUD, Corsica's other administrative department, to the S. Monte CINTO (8,891 ft/2,710 m high) in the department, is the island's highest point. Formed in 1975, Haute-Corse corresponds to the former department of Golo. Its 236 communes are divided among three sub-prefectures: CALVI, CORTE, and Bastia, Corsica's largest city and commercial center, on the TYR-RHENIAN SEA.

Haute-Côte-Nord, La (ot–kot–NOR, lah), county (□ 4,830 sq mi/12,558 sq km; 2006 population 12,457), Côte-Nord region, E QUEBEC, E CANADA; ⊙ Les ESCOUMINS; 48°34′N 69°13′W. Composed of nine municipalities. Formed in 1982.

Haute-Garonne (OT–gah-RON) [French=upper Garonne], department (□ 2,436 sq mi/6,333.6 sq km), partly in LANGUEDOC and partly in GASCONY, SW FRANCE; ⊙ TOULOUSE; 43°25′N 01°30′E. It touches on SPAIN (S) in the central PYRENEES Mountains. The upper valleys of the GARONNE and ARIÈGE rivers make up most of the land area, physiographically situated within the E part of the AQUITAINE BASIN. Primarily agricultural, it has extensive wheat and corn fields, and grows wine grapes, fruits, and vegetables. Horses, poultry, and cattle are raised in the valleys, sheep in the uplands. Marble is quarried in the foothills. Manufacturing is chiefly of the food-processing type but the Toulouse metropolitan area is also the center of the French aircraft and aerospace industry. Electrometallurgical plants are scattered in the Pyrenean foothills in proximity to a network of hydroelectric plants. Tourism and winter sports at BAGNÈRES-DE-LUCHON and SUPERBAGNÈRES, near Spanish border. Part of the administrative region of MIDI-PYRÉNÉES (headquarters in Toulouse).

Haute-Gaspésie, La (OT–gahs-pai-ZEE, lah), county (□ 1,980 sq mi/5,148 sq km; 2006 population 12,708), Gaspésie—Îles-de-la-Madeleine region, E QUEBEC, E CANADA; ⊙ SAINTE-ANNE-DES-MONTS. Composed of ten municipalities. Formed in 1981.

Haute-Goulaine (OT–goo-LEN), commune (□ 7 sq mi/18.2 sq km), LOIRE-ATLANTIQUE department, PAYS DE LA LOIRE region, W FRANCE; 47°12′N 01°26′W.

Haute-Guinée, geographic region, E central GUINEA; ⊙ KANKAN. Bordered N and E by MALI, SE tip by CÔTE D'IVOIRE, S by Guinée-Forestière geographic region (W by Faranah and central and E by N'Zérékoré administrative regions), SW by SIERRA LEONE, and W by Moyenne-Guinée geographic region (N by Labé and S by Mamou administrative regions). Savanna. Includes the administrative regions of Kankan in the E and all but the SE section of Faranah (Dabola, Dinguiraye, and Faranah prefectures) in the W.

Haute, Île (OT, eel), islet in the BAY OF FUNDY, off N NOVA SCOTIA, E CANADA, in entrance of MINAS CHANNEL, 6 mi/10 km SW of CAPE CHIGNECTO; 45°15′N 65°00′W.

Haute-Kotto (OT–KO-to), prefecture (□ 33,447 sq mi/ 86,962.2 sq km; 2003 population 90,316), E CENTRAL AFRICAN REPUBLIC; ⊙ BRIA. Bordered N by VAKAGA prefecture, NE by SUDAN, SE by HAUTE-M'BOMOU prefecture (CHINKO RIVER, border), S by M'BOMOU prefecture, W by OUAKA prefecture, and NW by BAMINGUI-BANGORAN prefecture. Drained by Bolou, CHINKO, Dji, and KOTO rivers. Includes BONGOS Massif. Agriculture (coffee, cotton); cotton ginning; diamond mining; tin and copper deposits. Main centers are BRIA, Mouka, OUADDA, OUANGO, and YALINGA.

Haute-Loire (ot–LWAHR) [French=upper Loire], department (□ 1,922 sq mi/4,997.2 sq km), in former AUVERGNE and LANGUEDOC provinces, S central FRANCE; ⊙ Le PUY-EN-VELAY; 45°05′N 04°00′E. It lies wholly within the rugged MASSIF CENTRAL, bounded by the granitic Monts de la MARGERIDE (W) and the Monts du VIVARAIS (E). Its central portion is occupied by the Monts du Velay, the basin of Le Puy, and an arid tableland topped by extinct volcanic summits. Drained S-N by the upper courses of the LOIRE and ALLIER rivers whose valleys support agriculture (cereals, fruit, vegetables, especially lentils, and potatoes). Sheep and cattle are raised in the uplands, chiefly around Mont MÉZENC. Manufacturing is represented by metalworking shops in the mid-sized communities. Lacemaking (handmade in outlying villages and machine-made in Le Puy) is still a source of employment overshadowed, however, by modern factories making equipment for TGV (high-speed) trains and rubber tires for heavy machinery. Part of the administrative region of AUVERGNE.

Haute-Marne (ot–MAHRN) [French=upper Marne], department (□ 2,398 sq mi/6,234.8 sq km), NE FRANCE; ⊙ CHAUMONT; 48°05′N 05°10′E. Its S part is occupied by the Plateau of LANGRES, the N part lies within the region of CHAMPAGNE. Drained S-N by the upper MEUSE, MARNE, and AUBE rivers, all of which rise here; it is also traversed N-S by the MARNE-SAÔNE CANAL. Generally a poor agricultural area given to livestock raising, and with large forested tracts. Metalworking is chief industry, ranging from agricultural machinery to cutlery manufacturing. Woodworking and food processing are dispersed among the communities located chiefly in the river valleys, with Chaumont and SAINT-DIZIER being the main urban centers. The department has lost population to the larger cities of the PARIS BASIN. Part of the administrative region of CHAMPAGNE-ARDENNE (headquarters at CHÂLONS-EN-CHAMPAGNE).

Haute-M'bomou (OT–uhm-BO-moo), prefecture (□ 21,435 sq mi/55,731 sq km; 2003 population 57,602), SE corner of CENTRAL AFRICAN REPUBLIC; ⊙ OBO. Bordered N and E by SUDAN, S by CONGO, W by M'BOMOU prefecture, and NW by HAUTE-KOTTO prefecture (Ouaka Chinko River, border). Drained by OUARRA, Kerre, Goangoa, and Vovodo rivers. DJ NGOUA Mountain is located here. Fishing. Includes Zemongo animal reserve. Main centers are Obo, DJÉMA, GOUBÉRE, and ZEMIO.

Haute-Normandie (OT–nor-mahn-DEE) [French=upper Normandie], administrative region (□ 4,756 sq mi/12,365.6 sq km) of NW central FRANCE; ⊙ ROUEN; 49°30′N 01°00′E. It comprises the departments of SEINE-MARITIME and EURE, which contain the lower valley and estuary of the SEINE RIVER. The region essentially covers the E part of historic NORMANDY. The W part is administratively assigned to BASSE-NORMANDIE [French=lower Normandy].

Hauterive (o-TREEV), commune, NEUCHÂTEL canton, W SWITZERLAND; NE suburb of NEUCHÂTEL.

Hauterives (o-tuh-REEV), commune (□ 11 sq mi/28.6 sq km), DRÔME department, RHÔNE-ALPES region, in SE FRANCE; 45°15′N 05°02′E. Site of *le Palais idéal*, a fantastic structure erected (1879–1912) by postman Ferdinand Cheval, with stones collected on his route over the years.

Haute-Saint-Charles, La (OT–san–SHAHRL, lah), N central borough of QUEBEC city, Capitale-Nationale region, S QUEBEC, E CANADA.

Hautes-Alpes (ot–ZAHLP) [French=upper Alps], department (□ 2,142 sq mi/5,569.2 sq km), in old province of DAUPHINÉ, SE FRANCE; ⊙ GAP; 44°40′N 06°30′E. Bounded by ITALY (NE) and the DURANCE RIVER valley (S), it is mountainous throughout, containing the ÉCRINS massif (rising to 13,461 ft/4,103 m at the BARRE DES ÉCRINS). Alpine peaks dominate the QUEYRAS and Ubaye valleys (E), and several lesser ranges of the DAUPHINÉ ALPS (W). Traversed NE-SW by the Durance River valley, the only important artery of communication, with access to Italy via MONTGENÈVRE PASS. The department is also drained by the upper DRAC RIVER (which flows N) and by minor tributaries of the Durance (GUISANE, GUIL, BÜECH). Agriculture is limited to warmer valleys with dairying and cheese making in QUEYRAS district and around Gap. Other economic activities include lumber trade and woodworking, some fruit growing and distilling of perfume essences. Tourism is very active from the ROMANCHE valley (N), to the BRIANÇON area, the Queyras, and the mountain valleys surrounding the ÉCRINS NATIONAL PARK. Hydroelectric power drives an aluminum-reduction plant. The SERRE-PONÇON dam and its impounded lake (□ 12 sq mi/31 sq km) provide electricity and flood control for the lower Durance valley; it is one of the largest man-made impoundments in W Europe (completed 1960); suitable for sailing and waterskiing despite fluctuating water levels. Briançon, an old stronghold against invasions from Italian side of the ALPS, and Gap, a commercial and administrative center, are the main towns; population of department has been declining. Forms part of the administrative region of PROVENCE-ALPES-CÔTE D'AZUR, which encompasses the entire SE corner of France.

Haute-Sanaga, department (2001 population 115,305), CENTRAL province, CAMEROON; ⊙ NANGA-EBOKO.

Haute-Sangha (OT–SAHNG-gah), prefecture, CENTRAL AFRICAN REPUBLIC: see MAMBÉRE-KADÉÏ.

Haute-Saône (ot–son) [French=upper Saône], department (□ 2,070 sq mi/5,382 sq km), in FRANCHE-COMTÉ province, E FRANCE; ⊙ VESOUL; 47°40′N 06°10′E. Bounded by the VOSGES MOUNTAINS (NE) and the BELFORT GAP (E), the department is drained NE-SW by the SAÔNE and OGNON rivers. Dairying; cereal, potato, and fruit (especially cherry) growing and lumbering are resource-based activities. Industry includes cotton milling, manufacturing of hardware and parts of textile machinery, woodworking. Principal towns are Vesoul (agriculture market and small-scale manufacturing), LURE (textile center), HÉRICOURT (metalworks), and GRAY. LUXEUIL-LES-BAINS is known for its mineral springs and FOUGEROLLES for its kirsch liqueur. Tourism is significant in the S Vosges Mountains. Part of the regional park of VOSGES summits lies in NE section of Haute-Saône department. Administratively included in the Franche-Comté region (headquarters in BESANÇON).

Haute-Savoie (OT–sah-VWAH), department (□ 1,694 sq mi/4,404.4 sq km), in SAVOY, SE FRANCE; ⊙ ANNECY; 46°00′N 06°20′E. Bounded by SWITZERLAND (E), ITALY (SE), Lake GENEVA (N), and RHÔNE RIVER (W). Central part is occupied by the outer ranges of the Savoy Alps (ALPES FRANÇAISES); the MONT BLANC massif, in the SE, contains the highest peak of the ALPS (15,771 ft/4,807 m). Agriculture (orchards, vineyards; cereals) and population are concentrated in FAUCIGNY valley, in the broad depression occupied by Lake of ANNECY, and in the French part of the Lake Geneva (French *Lac Léman*) lowland. Cattle raising and cheese manufacturing are found principally in the

uplands. The rapid mountain streams that issue from the Alps have long been harnessed for hydroelectric power, activating aluminum and electrochemical plants. Watchmaking and woodworking are still traditional industries. The tourist industry is drawn to the Chamonix Valley and its satellites and ski areas, and to the picturesque shores of Lake of Annecy and Lake Geneva. On the latter, ÉVIAN-LES-BAINS and THONON-LES-BAINS are fashionable and popular spas. Department was formed in 1860 after the French annexation of Savoy. The economic vitality of the Geneva canton in neighboring Switzerland has spilled over into Haute-Savoie. For many years an international vacation area, the department has gained further recognition as a winter sports region from the 1992 Winter Olympics that took place in the N Alps, centered on ALBERTVILLE (just S of Haute-Savoie). Part of the administrative region of RHÔNE-ALPES (headquarters in LYON).

Hautes Fagnes, BELGIUM: see HOHES VENN.

Hautes-Pyrénées (OT–pee-rai-NAI) [French=high Pyrenees], department (□ 1,724 sq mi/4,482.4 sq km), SW FRANCE, bordering on SPAIN (S); ⊙ TARBES; 43°00′N 00°10′E. S half lies in the central range of the PYRENEES mountains, accessible only through the narrow valleys of the Gave de PAU, NESTE, and upper ADOUR rivers. Area around Tarbes is in Adour River lowland separated from the lowlands by the LANNEMEZAN PLATEAU (E center). Agriculture (corn, wheat, grapes, tobacco) is limited to irrigated lowlands and mountain valleys. Lumbering, woodworking; grazing; slate and marble quarrying are activities found in the Pyrenean foreland known as BIGORRE. Waterpower activates electrometallurgical and chemical industry. Department has popular spas (BAGNÈRES-DE-BIGORRE, BARÈGES, CAUTERETS) and attracts visitors for mountaineering and winter sports. Chief towns are Tarbes (electrical and aeronautical industry), LOURDES (famous pilgrimage center), and Bagnères-de-Bigorre (spa; manufacturing of electrical equipment). The PYRENEES NATIONAL PARK occupies the W part of the principal Pyrenean range in this department, including the well-known headwall and waterfalls of the Cirque de GAVARNIE. Administratively, this department is included in the region of MIDI-PYRÉNÉES.

Hautes-Rivières, Les (OT–ree-VYER, laiz), commune (□ 12 sq mi/31.2 sq km), ARDENNES department, CHAMPAGNE-ARDENNE region, N FRANCE, on the SEMOIS RIVER (a tributary of the MEUSE RIVER), on Belgian border, 10 mi/16 km NE of CHARLEVILLE-MÉZIÈRES, in the Ardenne hills; 49°53′N 04°51′E. Metalworks.

Haute-Vienne (ot–VYEN) [French=upper Vienne], department (□ 2,131 sq mi/5,540.6 sq km), W central FRANCE, occupying parts of MARCHE (N) and LIMOUSIN (center and S); ⊙ LIMOGES; 45°50′N 01°10′E. Bordered by W outliers of the MASSIF CENTRAL, drained by VIENNE RIVER and its tributary, the GARTEMPE RIVER. Agriculture is limited to potatoes, buckwheat, rye, and barley, but department is noted for livestock raised on natural pastures in a humid climate. Here are the kaolin quarries (at SAINT-YRIEIX-LA-PERCHE), which supply the famous porcelain industry of Limoges and SÈVRES, and the uranium deposits, which supply the nuclear industry. Harnessed waters of Vienne River power paper mills and metalworks. Leatherworking (especially glove manufacturing) and wool spinning are traditional industries. Half the department's population lives in the greater Limoges area. BELLAC is the main center of the Marche district. Administratively, Haute-Vienne forms part of the Limousin region.

Hauteville-Lompnès (ot–veel–lomp-NE), town (□ 19 sq mi/49.4 sq km) and spa, AIN department, RHÔNE-ALPES region, E FRANCE, in the S JURA MOUNTAINS, 15 mi/24 km NNW of BELLEY; elevation 2,674 ft/815 m; 45°58′N 05°36′E. Health resort with modern sanatoria,

in dairying area (BUGEY district). Quarries for stone cutting nearby.

Haute-Yamaska, La (OT–yuh-MA-skuh, lah), county (□ 290 sq mi/754 sq km; 2006 population 83,659), MONTÉRÉGIE region, S QUEBEC, E CANADA; ⊙ GRANBY; 45°23′N 72°43′W. Composed of ten municipalities. Formed in 1982.

Haut-Koenigsbourg (O–kuh-nigz-BOOR), German *Hoch-Königsburg* (HOKH–KUH-nigz-boorg), former large feudal castle, in BAS-RHIN department, ALSACE, E FRANCE, atop rocky height (c.2,500 ft/762 m) of the E VOSGES MOUNTAINS, overlooking SÉLESTAT (5 mi/8 km ENE); 48°14′N 07°22′E. Built in 15th century by German princes and partly destroyed in 17th century, it was restored at public expense in the style of late 19th-century improvements as a residence for Emperor William II (who never spent a night here). Major regional tourist attraction, with a panoramic view of the Vosges and the Alsatian lowland extending to the RHINE RIVER.

Haut Languedoc Natural Regional Park (O LAHN-guh-dok) (□ 579 sq mi/1,505.4 sq km), TARN and HÉRAULT departments, in LANGUEDOC region, S FRANCE; 46°30′N 03°05′E.

Haut Limousin, FRANCE: see LIMOUSIN.

Haut-Médoc, FRANCE: see MÉDOC.

Hautmont (o-MON), town (□ 4 sq mi/10.4 sq km), Nord department, NORD-PAS-DE-CALAIS region, N FRANCE, on the SAMBRE RIVER, suburb 3 mi/4.8 km SW of MAUBEUGE; 50°15′N 03°56′E. Metallurgical center (blast furnaces, rolling mills).

Haut-Nkam, department (2001 population 203,251), West province, CAMEROON; ⊙ BAFANG.

Haut-Nyong, department (2001 population 216,768), East province, CAMEROON; ⊙ Abong Mbang. Dja National Game Reserve in W central part of department.

Haut-Ogooué (O–oo-goo-ai), province (□ 14,111 sq mi/36,688.6 sq km; 2002 population 134,500), SE GABON; ⊙ FRANCEVILLE; 01°10′S 13°51′E. Bounded E and S by CONGO, N by OGOOUÉ-IVINDO province, W by OGOOUÉ-LOLO province FRANCEVILLE is largest city. Area contains major mineral resources, including uranium and manganese. Part forest, part savannah (coffee is an important crop).

Haut-Rhin (o-RAN) [French=upper Rhine], department (□ 1,361 sq mi/3,538.6 sq km), in ALSACE, NE FRANCE; ⊙ COLMAR; 48°00′N 07°20′E. Bounded by GERMANY (E), SWITZERLAND (S), BELFORT GAP (SW), and crest of the VOSGES MOUNTAINS (W). The department occupies S part of the Alsatian plain and also contains highest point of the Vosges, the Ballon de GUEBWILLER, or Grand Ballon (4,672 ft/1,424 m). Drained by the RHINE RIVER (which forms Franco-German border) and the ILL RIVER and its tributaries. Traversed by RHÔNE-RHINE CANAL. Chief agricultural crops include potatoes, vegetables (especially cabbage for sauerkraut), hops, cereals, tobacco, and fruits (especially cherries for kirsch distilling). Alsace wines are produced from grapes grown on E slopes of Vosges along the Route des Vins (the wine road). Noted cheese made in MUNSTER valley. Department has France's leading potash mines (N of MULHOUSE), which supply local chemical plants. Chief industry is textile milling, with main centers at Mulhouse, Colmar, Guebwiller, and ALTKIRCH. Textile and printing machinery, electrical equipment, and paper are also made here. Power is provided by a series of hydroelectric stations along the Alsace canal (parallel to the Rhine) and by the FESSENHEIM nuclear power plant. Mulhouse is the leading industrial center of S Alsace. Colmar and the picturesque smaller towns along the front of the Vosges attract the tourist trade. As part of ALSACE-LORRAINE, it was under German administration, 1871–1918, and again 1940–1944. Now forms part of the administrative region of Alsace.

Area is shown by the symbol □, and capital city or county seat by ⊙.

Haut-Richelieu, Le (O–reesh-LYU, luh), county (□ 360 sq mi/936 sq km; 2006 population 107,229), Montérégie region, S QUEBEC, E CANADA; ☉ SAINT-JEAN-SUR-RICHELIEU; 45°12′N 73°13′W. Composed of fourteen municipalities. Formed in 1982.

Haut-Saint-François, Le (O–san-frahn-SWAH, luh), county (□ 879 sq mi/2,285.4 sq km; 2006 population 21,853), ESTRIE region, S QUEBEC, E CANADA; ☉ COOKSHIRE-EATON; 45°28′N 71°37′W. Composed of fourteen municipalities. Formed in 1982.

Haut-Saint-Laurent, Le (O–san-lo-RAHN, luh) (French=upper Saint Lawrence), county, SW QUEBEC, E CANADA, bounded on N, NW by the SAINT LAWRENCE RIVER; ☉ HUNTINGDON; 45°04′N 74°10′W. Composed of thirteen municipalities and the former county of Huntingdon; formed 1982.

Haut-Sassandra, region (□ 5,870 sq mi/15,262 sq km; 2002 population 1,186,600), W central CÔTE D'IVOIRE; ☉ DALOA; 07°00′N 06°35′W. Bordered N by Worodougou region, E by Marahoué region, SE by Fromager region, SW by Bas-Sassandra region, W by Moyen-Cavally region, and WNW by Dix-Huit Montagnes region (border formed by SASSANDRA RIVER). Part of Lake Buyo in SW; Dé River in NE. Towns include DALOA, ISSIA, and VAVOUA. Regional airport S of Daloa.

Hauts-Bassins, region (□ 9,785 sq mi/25,441 sq km; 2005 population 1,348,441), W BURKINA FASO; ☉ BOBO-DIOULASSO. Borders BOUCLE DU MOUHOUN (N), SUD-OUEST (E), and CASCADES (S) regions and MALI (W). Composed of HOUET, KÉNÉDOUGOU, and TUY provinces.

Hauts-de-Seine (O–duh–SEN), department (□ 68 sq mi/176.8 sq km), N central FRANCE, just W and SW of PARIS; ☉ NANTERRE; 48°50′N 02°10′E. Extends along the city's periphery from SCEAUX (S) N, occupying the entire area contained within the first N bend of the SEINE below Paris. It also reaches E of that bend to include the inner suburbs of BOULOGNE-BILLANCOURT, NEUILLY-SUR-SEINE, LEVALLOIS-PERRET, and CLICHY. Almost entirely urbanized, the department includes residential, commercial, and industrial suburbs of Paris. The predominantly residential communities are RUEIL-MALMAISON, SÈVRES, CHAVILLE, MEUDON, and Sceaux, with parks or greenbelts buffering these densely settled towns. Industries are located chiefly along the Seine, while the port of Paris, at the N extremity of the river bend, is part of GENNEVILLIERS. The modern office complex of La Défense, built 1957–1989, houses the headquarters of many international business enterprises and has a center for the development of new industries and technologies. Its great arch, visible from Paris (along with its skyscrapers) was placed on the axis of the CHAMPS-ÉLYSÉES and the ARC DE TRIOMPHE [French=triumphal arch], both of which run NW from the Place de la CONCORDE. This department was carved (1964) out of the former SEINE department. Part of the administrative region of ÎLE-DE-FRANCE.

Hauts-Plateaux, department (2001 population 117,008), WEST province, CAMEROON.

Haut-Zaïre, DEMOCRATIC REPUBLIC OF THE CONGO: see ORIENTALE province.

Hauula (HOU-OO-lah), town (2000 population 3,651), N OAHU island, HONOLULU county, HAWAII, on NE coast, 20 mi/32 km N of HONOLULU; 21°35′N 157°55′W. Hauula Beach Park here; Hauula Forest Reserve to SW, Sacred Falls State Park to S.

Hauwara (hah-WAH-ruh), locality, FAIYUM province, Upper EGYPT, SE of Faiyum; 28°27′N 30°40′E. Site of pyramid of Amenemhet III of XII dynasty. Also spelled Hawara.

Hauzen, ETHIOPIA: see HAWZEN.

Hauzenberg (HOU-tsen-berg), town, BAVARIA, GERMANY, in LOWER BAVARIA, in BOHEMIAN FOREST, 10 mi/16 km NE of PASSAU; 48°39′N 13°37′E. Metalworking; granite quarrying, graphite mining. Summer resort. Chartered c.1359 and again in 1978.

Havana, Spanish *La Habana*, city (2002 population 2,201,610), ☉ both Cuba and Ciudad de La Habana province, W Cuba; c.100 mi/161 km SSW of Key West (Fla.); 23°08′N 82°21′W. The largest city and chief port of the West Indies, and the political, economic, and cultural center of Cuba. Havana's climate is humid and subtropical (mean annual temperature is 76°F/24.5°C and average rainfall is 43 in/109 cm) but moderated by seawinds. Subject to occasional hurricanes. With one of the best natural harbors in the Caribbean Sea, it has long been strategically and commercially important. An important hub of air and maritime transportation, it is also the focal point of Cuban commerce and tourism, exporting sugar, tobacco, and fruits. Imports passing through its port include consumer durables, foodstuffs, cotton, machinery, and technical equipment Local industries include shipbuilding, light industries (mostly food processing and canning), biotechnology; also, assembly plants, rum distilleries, and factories making the famous Havana cigars. One of few Latin Amer. capitals with light industry (e.g., tobacco factories) so close to government buildings (here, near the *Capitolio*, former capitol building). Tourism has been greatly revived in the 1990s as Cuba redirects its economic model from central planning toward a mixed economy. Havana and Varadero (56 mi/90 km E) are now major tourist destinations. The collapse of trade and aid from the Soviet bloc in the late 1980s led to a new, receptive attitude toward foreign (Western) investment. Founded 1st in 1516 on S coast of Cuba, then on N coast c.4.3 mi/7 km from current site near mouth of Almendares River, later re-established on W side of Havana Bay (1519). One of 7 original settlements (*villas*) of Diego Velázquez, Spanish conquistador and 1st colonial governor of Cuba. Havana became capital in the late 16th century. Spanish galleons assembled in Havana's harbor for their return voyage to Spain, combining cargo from Mexico, Panama, and Colombia. Privateers from other European countries preyed on these galleons and raided many Cuban towns during the 17th and 18th century Havana fell to Anglo-Amer. forces in 1762, but was returned to Spain the following year, partly in exchange for Florida. By the early 19th century, Havana was among the wealthiest commercial centers in the Western Hemisphere. It benefited greatly from the out-migration of Haitian French and Creole sugar barons fleeing the 1792 slave revolt. As the 19th century progressed, the city was caught up in the anti-Span. independence movement. The Spanish-Amer. War was precipitated by the destruction of the U.S. battleship *Maine* in Havana harbor in 1898, and Amer. troops occupied the city. The U.S. set up administrative headquarters here (1898–1902), modernizing ports, roads (including the W extension of the seaside promenade, the Malécon), public lighting, communications, and sanitary conditions (eliminating yellow fever). Until 1959, the close relations between the U.S. and Cuba were strongly reflected in the commercial and cultural life of the city and Havana's hotels and entertainment made it a popular winter resort for Amer. tourists until that year. Although almost entirely a city of hard-working merchants and civil servants, Havana was also part of an illicit triangle of gambling, prostitution, and political corruption that included Las Vegas (Nev.) and Miami (Fla.). After the government of Fidel Castro took control, the Amer. presence was replaced by that of the USSR, which provided favorable terms of trade and foreign aid through the late 1980s. Castro's policy of directing economic resources toward rural areas and smaller urban centers led to the deterioration of Havana, especially the old city (Habana Vieja). Restoration efforts have been spotty, despite the fact that in 1982, UNESCO declared Habana Vieja and an adjacent network of fortresses (El Morro, La Punta, La Cabaña, La Fuerza Real, Príncipe) a World Heritage Site. In 1994, a joint-venture firm (*empresa mixta*) called Habanaguex, began controlling about a dozen restaurants and small hotels in and around Habana Vieja in an attempt to generate hard currency that, in turn, would be used for improvements to the old quarters. Havana harbor is one of the most polluted in the Americas due to the activities of oil refineries in the back bay district of Regla. The city has a diverse collection of colonial, baroque, neoclassical, Art Deco, and Modern buildings Most of the city's housing stock, however, reflects modest 20th-cent. buildings, giving the city a remarkably uniform skyline. Although much in need of repair, Havana is expected to be refurbished through foreign investment and international tourism. Since 1993, over 100 types of private-sector employment jobs have been approved by the government, stimulating small-scale business growth both here and in the rest of the country.

Havana, city (2000 population 3,577), ☉ MASON county, central ILLINOIS, on ILLINOIS RIVER (shipping), opposite mouth of SPOON RIVER, and 32 mi/51 km SW of PEORIA; 40°17′N 90°03′W. Trade, shipping, and industrial center in agricultural area; manufacturing (food products, metal products). Important river port in early nineteenth century. Site of Lincoln-Douglas debate is marked. Nearby are resorts on lakes along Illinois River, Chautauqua National Wildlife Refuge, and DICKSON MOUNDS MUSEUM. Founded 1827, incorporated 1853.

Havana (huh-VAN-nah), town (□ 1 sq mi/2.6 sq km; 2000 population 1,713), GADSDEN county, NW FLORIDA, near GEORGIA state line, 15 mi/24 km NW of TALLAHASSEE; 30°37′N 84°25′W. Canned foods (fruit, vegetables); feed, fertilizer. Settled 1904, incorporated 1906.

Havana (huh-VA-nuh), village (2000 population 392), YELL county, W central ARKANSAS, 25 mi/40 km WSW of RUSSELLVILLE; 35°06′N 93°31′W. MAGAZINE MOUNTAIN, highest point in Arkansas (2,753 ft/839 m), 4 mi/6.4 km N, in LOGAN county. Nearby are Ozark (N) and Ouachita (S) national forests.

Havana, village (2000 population 86), MONTGOMERY county, SE KANSAS, 15 mi/24 km SW of INDEPENDENCE; 37°05′N 95°56′W. In livestock and grain area. Oil field here.

Havana, village (2006 population 88), SARGENT CO., SE NORTH DAKOTA, 11 mi/18 km S of FORMAN, at SOUTH DAKOTA state line; 45°57′N 97°37′W. (Sisseton) Wahpeton Indian Reservation to E. Founded in 1887 and named for HAVANA, ILLINOIS.

Havannah, VANUATU: see EFATE.

Havant (HAV-uhnt), town (2001 population 45,435), HAMPSHIRE, S ENGLAND, 20 mi/32 km SE of SOUTHAMPTON; 50°51′N 00°59′W. Manufacturing includes pharmaceuticals, kitchen equipment, electronic components, motor vehicles, and toys. Includes BEDHAMPTON.

Havasu City, ARIZONA: see LAKE HAVASU CITY.

Havasu Lake, on COLORADO RIVER, on the border between W ARIZONA and SE CALIFORNIA, formed by PARKER DAM, 60 mi/97 km S of KINGMAN (ARIZONA); c.45 mi/72 km long. Its headwaters extend c.50 mi/80 km from dam, to near NEEDLES (California). Maximum capacity 717,000 acre-ft. Formed by Parker Dam. Chemehueva Indian Reservation on W shore; parts of Havasu National Wildlife Refuge on NE shore and E of dam. LAKE HAVASU CITY (Arizona) on E shore. BILL WILLIAMS RIVER enters from E, near dam.

Havdhem (HAHVD-HEM), village, GOTLAND county, SE SWEDEN, in S part of Gotland island, 30 mi/48 km S of VISBY; 57°10′N 18°20′E.

Havel (HAH-fel), river, c.215 mi/350 km long, N GERMANY; rises in the lake region of MECKLENBURG–WESTERN POMERANIA; 53°29′N 12°58′E; flows generally S through BERLIN, to POTSDAM where it turns W, past BRANDENBURG, where it turns NW and enters the

ELBE RIVER near HAVELBERG. Navigable for most of its length. The SPREE RIVER, its chief tributary, joins it at SPANDAU. The HAVEL is linked with the ODER river by the ODER-SPREE CANAL. During the Soviet blockade of Berlin (1948) the Havel was used as a runway for amphibious aircraft.

Havelange (AH-vuh-lawn-zhuh), commune (2006 population 4,884), Dinant district, NAMUR province, S central BELGIUM, 9 mi/14 km S of HUY, in FAMENNE depression of the ARDENNES Mountains; 50°23′N 05°14′E.

Havelberg (HAH-fel-berg), town, SAXONY-ANHALT, E GERMANY, on the HAVEL River, near its mouth on the ELBE River (partly developed on island in Havel), 18 mi/29 km SE of WITTENBERGE; 52°51′N 12°05′E. Ship repairing; manufacturing of furniture. Has cathedral (1170; rebuilt in 15th century; late-Gothic church. Seat of bishopric (948–983; 1150–1548). In Thirty Years War, sacked (1627) by Danes.

Havelock (HA-ve-LOK), canton (□ 34 sq mi/88.4 sq km; 2006 population 823), MONTÉRÉGIE region, S QUEBEC, E CANADA, 21 mi/33 km to HUNTINGDON; 45°02′N 73°45′W.

Havelock (HAV-lahk), city (□ 17 sq mi/44.2 sq km; 2006 population 21,906), CRAVEN county, E North Carolina, 18 mi/29 km SE of NEW BERN, near NEUSE RIVER estuary (to N), in Croatan National Forest; 34°54′N 76°53′W. Railroad junction. Service industries; manufacturing (printing and publishing; apparel). Service center for Cherry Point Marine Corps Air Station, to N. Swampy area with several natural lakes to W, including Catfish, Long, Great, and Ellis Simon lakes. Incorporated 1971. Settled prior to 1857 and named for Sir Henry Havelock, British Major General and layman.

Havelock (HA-ve-LOK), township, N SOUTH ISLAND, NEW ZEALAND, in extreme SW of PELORUS SOUND, c.12 mi/20 km W of PICTON. Vacation resort; agricultural and tourist services. Gold discoveries in 1864.

Havelock, town (2000 population 177), POCAHONTAS county, N central IOWA, 6 mi/9.7 km N of POCAHONTAS; 42°49′N 94°42′W. Livestock; grain.

Havelock (HAV-lahk), unincorporated village (2001 population 1,318), S ONTARIO, E central CANADA, 23 mi/37 km ENE of PETERBOROUGH, and included in HAVELOCK-BELMONT-METHUEN township; 44°26′N 77°52′W. Dairying; manufacturing of fishing tackle.

Havelock (HAHV-luk), village, HHOHHO district, N SWAZILAND, 25 mi/40 km N of MBABANE, just W of BULEMBU, at SOUTH AFRICA border; 25°57′S 31°08′E. Connected by 12.5-mi/20-km aerial cableway to BARBERTON (South Africa) to NNW. Site of one of world's larges asbestos mines. First discovered in 1886, but commercially mined from 1939.

Havelock (HA-vuh-lawk), N suburb, 4 mi/6.4 km NE of downtown LINCOLN, LANCASTER county, SE Nebraska Industrial base. City annexed to Lincoln in 1930.

Havelock-Belmont-Methuen (HA-ve-lahk–BEL-mahnt–me-THOO-uhn), township (□ 203 sq mi/ 527.8 sq km; 2001 population 4,479), S ONTARIO, E central CANADA, 29 mi/46 km from PETERBOROUGH, between Belville and PETERBOROUGH; 44°34′N 77°54′W. Tourism.

Havelte (HAH-vuhl-tuh), village, DRENTHE province, N central NETHERLANDS, W of the SMILDERVAART, 6 mi/9.7 km NNE of MEPPEL; 52°46′N 06°14′E. Dairying; cattle raising; grain, vegetables. Castle.

Haven (HAI-ven), town (2000 population 1,175), RENO county, S central KANSAS, 13 mi/21 km SE of HUTCHINSON, near ARKANSAS RIVER; 37°53′N 97°46′W. In wheat region. Manufacturing (fabricated metal products). Cheney Reservoir and Cheney State Park to S.

Havensville (HAI-venz-vil), village (2000 population 146), POTTAWATOMIE county, NE KANSAS, 18 mi/29 km NE of WESTMORELAND; 39°30′N 96°04′W. Cattle; grain.

Haverford (HA-vuh-fuhrd), township, DELAWARE county, SE PENNSYLVANIA, W residential suburb 7 mi/11.3 km W of downtown PHILADELPHIA; 40°00′N 75°17′W. Bounded E by Cobbs Creek. Some manufacturing Includes villages of HAVERTOWN, Llanerch, Beechwood, Manoa, Preston, PENFIELD, Brookline, South Ardmore, Oakmont, and part of DREXEL HILL (partly in Upper Darby township). Seat of Haverford College Arboretum in N. Village of Haverford is to E in MONTGOMERY county.

Haverfordwest, Welsh *Hwlffordd*, town, Pembrokeshire, SW Wales, on the Western Cleddau River, 13 mi/20.8 km S of FISHGUARD; 51°49′N 04°58′W. Formerly in DYFED, abolished 1996.

Haverhill (HAI-vruhl), city (2000 population 58,969), ESSEX county, NE MASSACHUSETTS, on the MERRIMACK RIVER; 42°47′N 71°05′W. Formerly one of the nation's leading shoe producers, Haverhill processes leather and makes leather, textile, and paper products. High-technology computer industries in the area and the manufacturing of electronic components add to the city's economic base. Skiing at Ward Hill. Points of interest are John Greenleaf Whittier's birthplace (the house dates from c.1688) and the home of Hannah Dustin. Seat of the Haverhill branch of Northeastern University and a community college Bradford College closed June 2000. Includes section of Bradford. Incorporated as a town 1641, as a city 1870.

Haverhill (HAV-uh-ril), town (2001 population 22,010), SUFFOLK, E ENGLAND, STOUR RIVER to E, 16 mi/26 km SE of CAMBRIDGE; 52°05′N 00°26′E. Some industry and manufacturing.

Haverhill (HAI-vruhl), town, GRAFTON county, W NEW HAMPSHIRE, 22 mi/35 km W of LITTLETON; 44°04′N 72°00′W. Bounded by CONNECTICUT RIVER (VERMONT state line); drained by Oliverian and Clark brooks. Railroad terminus. Agriculture (dairying; cattle, poultry; nursery crops; timber); manufacturing (wood products). Includes WOODSVILLE village in NW; site of Bedell Bridge State Park is here (covered bridge); Black Mountain State Forest in NE. Incorporated 1763.

Haverhill, village (2000 population 170), MARSHALL county, central IOWA, 8 mi/12.9 km SSW of MARSHALLTOWN; 41°56′N 92°57′W. Agricultural area (corn, oats; cattle, hogs).

Haveri (HAH-vai-ree), town, DHARWAD district, KARNATAKA state, S INDIA, 50 mi/80 km SSE of DHARWAD; 14°48′N 75°24′E. Market center for cardamom, cotton, grain, chili; cotton ginning. Agricultural school 4 mi/6.4 km W, at Devihosur.

Havering (HAI-vuhr-ING), outer borough (□ 43 sq mi/111.8 sq km; 2001 population 224,248) of GREATER LONDON, SE ENGLAND; 51°35′N 00°15′E. Largely residential, the borough has expanded greatly with the creation of electrified suburban railroads and added housing. Machinery and equipment manufacturing and manufacturing of plastics, chemicals, clothing, and beer. A market has been held in Romford, within the borough, since 1247. Until 1892, Romford was the capital of Havering-atte-Bower, a group of parishes united since the time of Edward the Confessor. Other districts include Hornchurch, Rainham, Romford, and Upminister.

Haverskerque (ah-verz-kerk), commune (□ 3 sq mi/7.8 sq km), NORD department, NORD-PAS-DE-CALAIS region, N FRANCE, 5 mi/8 km S of HAZEBROUCK, and on LYS River; 50°38′N 02°32′E. British battlefield cemetery from both world wars.

Haverstraw (HA-vuhr-straw), village (□ 5 sq mi/13 sq km; 2006 population 10,672), ROCKLAND county, SE NEW YORK, on W bank of the HUDSON RIVER, and 6 mi/9.7 km NW of OSSINING; 41°11′N 73°57′W. Formerly a noted brickmaking center, today it is home to a variety of light manufacturing and service operations along with stone quarrying. Commuter suburb. Incorporated 1854.

Havertown (HA-vuhr-toun), unincorporated town, DELAWARE county, SE PENNSYLVANIA, residential suburb 7 mi/11.3 km W of PHILADELPHIA, near Cobbs Creek; 39°58′N 75°18′W. Manufacturing includes printing and publishing, and light diversified manufacturing.

Haviland (HA-vi-land), village (2000 population 612), KIOWA county, S KANSAS, 20 mi/32 km W of PRATT; 37°37′N 99°06′W. In grain and livestock region.

Haviland (HAV-i-luhnd), village (2006 population 170), PAULDING county, NW OHIO, 10 mi/16 km N of VAN WERT; 41°01′N 84°35′W.

Havířov (HAH-vee-RZHOF), city (2001 population 85,855), SEVEROMORAVSKY province, in SILESIA, CZECH REPUBLIC; 49°48′N 18°24′E. Manufacturing (machinery, rubber; tanning, food processing); mining town. Power plant. Part of the Ostrava-Karviná industrial complex. A planned city founded in 1947, it contains large blocks of workers' housing.

Havixbeck (HAH-viks-bek), town, NORTH RHINE–WESTPHALIA, NW GERMANY, 9 mi/14.5 km W of MÜNSTER; 51°58′N 07°25′E. Tourism; has 14th-century church and several moated castles in vicinity.

Havizeh, IRAN: see HOWAIZEH.

Havlíčkův Brod (HAHV-leech-KOOF BROT), city (2001 population 24,375), VYCHODOCESKY province, BOHEMIA, CZECH REPUBLIC, on SAZAVA RIVER; 49°37′N 15°35′E. Railroad center. Potato trade; food processing, distilling; brewery (established 1880); broadcloth manufacturing. Military base. Has a 14th-century cathedral and Renaissance town hall. Site of Zizka's victory (1422) over Emperor Sigismund. Until 1945, it was called Nemecky Brod, Czech *Nĕmecký Brod*, German *Deutschbrod*.

Havøysund (HAHV-uh-oo-soond), fishing village, FINNMARK county, N NORWAY, on fjord of NORWEGIAN SEA, 30 mi/48 km NE of HAMMERFEST.

Havre (HAI-vuhr), city (2000 population 9,621), ⊙ HILL county, N MONTANA, on the MILK River (forms FRESNO RESERVOIR to NW); 48°32′N 109°41′W. Wheat, hay; cattle, sheep, hogs. Manufacturing (agricultural equipment, building materials); gas and oil field to SE (BLAINE county). Fort Assinniboine (1879) and Agricultural Research Center to SW. Rocky Boy's Indian Reservation to S. Wahkpa Chu'qn Archaeological Site and Lake Thibadeau National Wildlife Refuge to N. Earl H. Clack Memorial Museum and units of Rookery Wildlife Management Area to W. Founded in 1891 with the coming of the railroad. The area is served by the Milk River project. Seat of Montana State University-Northern. Originally called Bull Hook Bottoms. Incorporated 1892.

Havre (AH-vruh), village in commune of MONS, Mons district, HAINAUT province, SW BELGIUM, 5 mi/8 km E of Mons; 50°28′N 04°02′E. Has sixteenth-century church, Gothic castle.

Havre-Aubert (leel–dyoo–AH-vruh–o-BER) or **Amherst** (AM-uhrst), former village, E QUEBEC, E CANADA, on E AMHERST ISLAND, one of the MAGDALEN ISLANDS; 47°15′N 61°50′W. Fishing port. Forms part of the ÎLES-DE-LA-MADELEINE agglomeration.

Havre-aux-Maisons, CANADA: see HOUSE HARBOUR.

Havre Boucher (AH-vruh boo-SHAI) or **Harbour au Bouche** (HAHR-buhr o BOOSH), village, E NOVA SCOTIA, E CANADA, on GEORGE BAY, 22 mi/35 km ENE of ANTIGONISH; 45°39′N 61°31′W. Fishing.

Havre de Grace, city (2000 population 11,331), HARFORD county, NE MARYLAND, on CHESAPEAKE BAY, at mouth of the SUSQUEHANNA River (bridged 1940), and 33 mi/53 km NE of BALTIMORE; 39°33′N 76°06′W. Trade center for agricultural area (vegetables, especially tomatoes; fruit, corn; dairy products; poultry), with granite quarries. Canneries; commercial fisheries. Resort center; sport fishing, yachting (annual July regatta), duck hunting (on nearby SUSQUEHANNA FLATS). Nearby are Federal reservations (Aberdeen Proving Ground, Army Chemical Center, Perry Point

Area is shown by the symbol □, and capital city or county seat by ⊙.

Veterans' Hospital). First settled in 1650 and known as Susquehanna Lower Ferry, the town was established in 1785.

Havre-de-Grâce, FRANCE: see HAVRE, LE.

Havre, Le (AH-vruh, luh), port city (□ 18 sq mi/46.8 sq km), SEINE-MARITIME department, HAUTE-NORMANDIE region, N FRANCE, at mouth of SEINE RIVER, on the ENGLISH CHANNEL, and 110 mi/177 km WNW of PARIS; 49°30′N 00°08′E. One of France's leading commercial seaports (first in export tonnage and container-ship traffic) and a major industrial center (based on such imported commodities as petroleum, coal, chemical raw materials, coffee, cotton, tobacco). It was completely rebuilt 1950s–1990 with extensive modern docks and ship-repair facilities. The 1944 Allied invasion of Normandy took place on the beaches some 25 mi/40 km WSW; was Europe's most heavily damaged port. The rebuilt port area, extending 10 mi/16 km along N shore of the Seine estuary (and sheltered by 20 long breakwaters), has created industrial sites for metal and petrochemical works, automotive and airplane parts factories, cement and fertilizer plants, new shipyards, as well as an oil refinery and a coal-fired electric power plant. Modern railroad facilities link the port to the hinterland. The Normandy road bridge spanning the Seine estuary (1.3 mi/2.1 km long; 160 ft/49 m along the river) between Le Havre and HONFLEUR was completed in 1995 to facilitate travel from the CHANNEL TUNNEL to W and SW France. As an example of postwar urban planning and architecture, the area of Le Havre rebuilt by Auguste Perret was in 2005 designated a UNESCO World Heritage site. Founded 1516 as Havre-de-Grâce by Francis I, the city has a modern museum of fine arts named for André Malraux, and a national college for merchant marine trainees.

Havre–Saint-Pierre (AH-vruh–san–PYER), village (□ 1,460 sq mi/3,796 sq km), ⊙ MINGANIE county, CÔTE-NORD region, E QUEBEC, E CANADA, on the SAINT LAWRENCE RIVER, and 19 mi/31 km ESE of MINGAN; 50°15′N 63°35′W. Titanium center; trading post.

Havrylivka (hahv-RI-lif-kah) (Russian *Gavrilovka*), town, N central KHARKIV oblast, UKRAINE, on railroad (Shpakivka station), and 9 mi/14.5 km W of KHARKIV; 50°02′N 36°01′E. Elevation 570 ft/173 m. Was surpassed and absorbed by the town of Solonytsivka after about 1955.

Havrylivs'kyy Zavod, UKRAINE: see DRUZHKIVKA.

Havstenssund (HAHV-STENS-SUND), fishing village, GÖTEBORG OCH BOHUS county, SW SWEDEN, on SKAGERRAK strait, 12 mi/19 km S of STRÖMSTAD; 58°45′N 11°11′E. Seaside resort.

Havza, township, N TURKEY, on Samsun-Omasya railroad, 40 mi/64 km SW of SAMSUN; 40°57′N 35°40′E. Cereals. Hot springs.

Hawa (HAH-wah), village, ORIENTALE province, NE CONGO, 90 mi/145 km NE of IRUMU, near UGANDA border. Center of sericulture and apiculture; manufacturing of fishing tackle.

Hawai (hah-WEI), town, Tohaku district, TOTTORI prefecture, S HONSHU, W JAPAN, 22 mi/35 km W of TOTTORI; 35°29′N 133°52′E. Fruit (pears, grapes, melons, kiwis). Hot springs.

Hawaii (hah-WEI-ee), state (□ 10,931 sq mi/28,311 sq km; 2000 population 1,211,537; 2006 estimated population 1,285,498), central PACIFIC, admitted to the Union in 1959 as the 50th state; ⊙ HONOLULU (on OAHU); 21°13′N 156°56′W. Hawaii is often referred to as the "Aloha State."

Geography
It consists of a group of eight major islands and numerous islets in the Pacific Ocean, c.2,100 mi/3,380 km SW of SAN FRANCISCO. HAWAII island, referred to locally as "the Big Island," is the largest and geologically the youngest of the group, and Oahu is the most populous and economically important. The other principal islands are KAHOOLAWE, KAUAI, LANAI,

MAUI, MOLOKAI, and NIIHAU. The PALMYRA atoll and KINGMAN REEF, which were within the boundaries of Hawaii when it was a U.S. territory, were excluded when statehood was achieved. The Hawaiian Islands are of volcanic origin and are edged with coral reefs. Generally fertile with a mild climate, they are sometimes called "the paradise of the Pacific" because of their spectacular beauty; abundant sunshine; acres of green plants and gaily colored flowers; coral beaches with rolling white surf and fringed with palms; and, rising with sober majesty to solitary heights, cloud-covered volcanic peaks. Some of the world's largest active and inactive volcanoes are found on Hawaii and Maui; eruptions of the active volcanoes have provided spectacular displays but their lava flows have occasionally caused great property damage. MAUNA KEA and MAUNA LOA are volcanic mountains on Hawaii island; Haleakala volcano is on Maui in HALEAKALA NATIONAL PARK. Vegetation is generally luxuriant below 6,500 ft/1,981 m elevation on the windward NE exposures, with giant fern forests in HAWAII VOLCANOES NATIONAL PARK. Seasonally arid growth prevails on the leeward SW slope and low-lying islands, NIIHAU, KAHOOLAWE, and W MOLOKAI. Mountain slopes above 6,500 ft/1,981 m are very arid. Although many species of birds and domestic animals have been introduced on the islands, there are few wild animals other than feral boars and goats, and there are no snakes. The coastal waters abound with fish.

Economy
Sugarcane and pineapples, grown chiefly on large company-owned plantations, have long been the major agricultural products and the basis of the islands' principal industry, food processing, but are declining from competition from other countries. Other products include macadamia nuts, bananas, avocados, and other fruits and vegetables, coffee, dairy products; cattle and calves. Commercial fishing is also prevalent; tuna is the principal species caught. U.S. military defense installations at PEARL HARBOR and elsewhere in the state are extremely important to Hawaii's economy. Tourism is the leading source of income.

History: to 1820
The first known settlers of the Hawaiian Islands were Polynesian voyagers (the date of first migration is believed to be C.E. c.750). The islands were first visited by Europeans in 1778 by the English explorer Captain James Cook, who named them the Sandwich Islands for the English earl of Sandwich. At that time the islands were under the rule of warring native kings. In 1810, Kamehameha I became the sole sovereign of all the islands, and, in the peace that followed, agriculture and commerce were promoted. As a result of Kamehameha's hospitality, American traders were able to exploit the islands' sandalwood, which was much valued in China at the time. Trade with China reached its peak during this period. However, the period of Kamehameha's rule was also one of decline. Europeans and Americans brought with them devastating infectious diseases, and over the years the native population was greatly reduced. The adoption of Western ways contributed to the decline of native cultural tradition. This period also marked the breakdown of the traditional Hawaiian religion; years of religious unrest followed.

History: 1820 to 1893
When missionaries arrived from Boston in 1820 they found a less idyllic Hawaii than the one Captain Cook had discovered. Kamehameha III, who ruled 1825–1854, relied on the missionaries for advice and allowed them to preach Christianity. The missionaries established schools, developed the Hawaiian alphabet, and used it for translating the Bible into Hawaiian. In 1839, Kamehameha III issued a guarantee of religious freedom, and the following year a constitutional monarchy was established. From 1842 to 1854 an American,

G. P. Judd, held the post of prime minister, and under his influence many reforms were carried out. In the following decades commercial ties between Hawaii and the U.S. increased. In 1848 the islands' feudal land system was abolished, making private ownership possible and thereby encouraging capital investment in the land. By this time the sugar industry, which had been introduced in the 1830s, was well established. Hawaiian sugar gained a favored position in U.S. markets under a reciprocity treaty made with the U.S. in 1875. The treaty was renewed in 1884 but not ratified. Ratification came in 1887 when an amendment was added giving the U.S. exclusive rights to establish a naval base at Pearl Harbor. The amount of sugar exported to the U.S. increased greatly, and American businessmen began to invest in the Hawaiian sugar industry. Along with the Hawaiians in the industry, they came to exert powerful influence over the islands' economy and government, a dominance that was to last until World War II.

History: 1893 to 1941
Toward the end of the 19th century, agitation for constitutional reform in Hawaii led to the overthrow (1893) of Queen Liliuokalani, who had ruled since 1891. A provisional government was established and John Lake Stevens, the U.S. minister to Hawaii, proclaimed the country a U.S. protectorate. President Grover Cleveland, however, refused to annex Hawaii since most Hawaiians did not support a revolution; the Hawaiians and Americans in the sugar industry had aggravated the overthrow of the monarchy to serve their business needs. The U.S. tried to bring about the restoration of Queen Liliuokalani, but the provisional government on the islands refused to give up power and instead established (1894) a republic with Sanford B. Dole as president. Cleveland's successor, President William McKinley, favored annexation, which was finally accomplished in 1898. In 1900 the islands were made a territory, with Dole as governor. In this period, Hawaii's pineapple industry expanded as pineapples were first grown for canning purposes. In 1937 statehood for Hawaii was proposed and refused by the U.S. Congress—the territory's mixed population and distance from the U.S. mainland were among the obstacles.

History: 1941 to Present
On December 7, 1941, Japanese aircraft made a surprise attack on Pearl Harbor, plunging the U.S. into World War II. During the war the Hawaiian Islands were the chief Pacific base for U.S. forces and were under martial law (December 7, 1941–March 1943). The post-war years ushered in important economic and social developments. There was a dramatic expansion of labor unionism, marked by major strikes in 1946, 1949, and 1958. The International Longshoremen's and Warehousemen's Union organized the waterfront, sugar, and pineapple workers. The tourist trade, which had grown to major proportions in the 1930s, expanded further with post-war advances in air travel and with further investment and development. The building boom brought about new construction of luxury hotels and housing developments. After having sought statehood for many decades, Hawaii was finally admitted to the Union on August 21, 1959. In 1969 the construction of a new state capitol was completed. The University of Hawaii is located at Honolulu. In 1961 the Center for Cultural and Technical Interchange between East and West was dedicated at the university and drew graduate students and technical trainees from Asia and the Pacific area.

Multiculturalism
More ethnic and cultural groups are represented in Hawaii than in any other state. Chinese laborers, who came to work in the sugar industry, were the first of the large groups of immigrants to arrive (starting in 1852), and Filipinos and Koreans were the last (after

1900). Other immigrant groups—including Portuguese, Germans, Japanese, and Puerto Ricans—came in the latter part of the 19th century. Intermarriage with other races has brought a further decrease in the number of pure-blooded Hawaiians, who comprise a very small percentage of the population.

Government

Hawaii's constitution was drafted in 1950 and became effective in 1959 upon attainment of statehood. A governor elected every four years heads the executive. The current governor is Linda Lingle. The legislature has a senate with twenty-five members elected for four-year terms and a house of representatives with fifty-one members elected for two-year terms. The state elects two representatives and two senators to the U.S. Congress and has four electoral votes. Hawaii has long been known as a Democratic state. John A. Burns, a Democrat, was elected governor in 1962 and reelected in 1966 and 1970. Daniel Inouye, a Democratic Hawaiian senator of Japanese descent elected in 1962, was the chairman of the Iran-Contra Committee 1987–1988.

Hawaii (hah-WEI-ee), county (□ 5,086 sq mi/13,173 sq km; 1990 population 120,317; 2000 population 148,677), coextensive with island of HAWAII, SE HAWAII; ⊙ HILO; 19°36′N 155°30′W. Largest county in Hawaii. Includes several small coastal islets, especially along NE coast. Divided into 12 administrative districts; there are no incorporated cities.

Hawaii (hah-WEI-ee), island (□ 4,034 sq mi/10,448 sq km; 1990 population 120,217), largest and southernmost island of the state of HAWAII and coextensive with HAWAII county (which also includes a few coastal islets). Hamakua Coast on NE, Kona Coast on W, Kohala Coast on NW, KOHALA PENINSULA in N. Geologically the youngest of the Hawaiian group, Hawaii is made up of 5 volcanic mountain masses rising from the floor of the PACIFIC OCEAN—MAUNA KEA (13,800 ft/4,205 m above sea level, the highest point in the state); (N) MAUNA LOA (13,677 ft/4,169 m), center of island; HUALALAI (8,275 ft/2,522 m) in W; Kohala (5,489 ft/1,673 m) in NW; and Kilauea (4,090 ft/1,247 m) in SE. KA LAE (South Point) is southernmost point in the UNITED STATES. Lava flows, some of which reach the sea, and volcanic ash cover parts of the island. The N and NE coasts are rugged with high cliffs; the W and S coasts are generally low, with some good beaches. An unusual black-sand beach of volcanic origin lies on the SE coast at PUNALUU. Has numerous (over 20) forest reserves, largest being Kohala (N); Mauna Kea, Mauna Loa, and Kapapala (center); Kau (S); Hilo, Upper Waiakea, and Puna (E). Short rivers radiate from the major summits; Wailuku River, the longest, flows into Hilo Bay. Many waterfalls are on the island. Much of Hawaii has a tropical-rainy climate, with the N and E slopes receiving the most rain. The W and S slopes are much drier; the Kau Desert is in S Hawaii. Temperatures decrease with elevation; Mauna Loa and Mauna Kea are usually snow-covered in winter. Vegetation varies from tropical rain forest to grasslands to barren volcanic areas. Sugarcane, the island's principal product, is no longer harvested. Macadamia nuts and other fruits, vegetables, flowers, coffee, and beef cattle remain important (Parker Ranch in N; 313 sq mi/811 sq km). The KONA district of W Hawaii is the coffee belt of the U.S. and is also known for its resorts and offshore deep-sea fishing. HILO, on the E coast, is the island's largest city, second largest city in Hawaii, chief port, and is the county seat. Hawaii Belt Road, a highway linking the coastal towns, encircles the island. At KEALAKEKUA BAY on W coast there is a monument to Captain James Cook, the first English explorer to visit (1778) the Hawaiian islands HAWAII VOLCANOES NATIONAL PARK in the SE and City of Refuge (Pu'uhonua o Honaunau) National Historical Park are on Hawaii. All over the island *heiaus* (ancient temples) are found.

Puukohola Heiau National Historic Site in NW. Wailuku River and Wailoa River state parks at Hilo, in E; Lapakahi State Historic Park in NW; Lava Tree State Monument in E; Kalopa State Recreational Area in N; Old Kona Airport State Park on W coast; Kealakekua State Underwater Park on W coast; Mauna Kea State Recreational Area in N center; Kilauea State Recreational Area in SE center; Kaloko-Honokokau National Historic Park on W (Kona) coast; Pohakuloa Military Training Area in N center; the Puna District 10 mi/16 km S of Hilo has several semirural residential developments, including Hawaiian Acres, Hawaiian Beaches, Orchid Land Estates, Hawaiian Paradise Parks.

Hawaiian Gardens, city (2000 population 14,779), LOS ANGELES county, S CALIFORNIA; residential suburb 16 mi/26 km SE of downtown LOS ANGELES, near SAN GABRIEL RIVER; 33°50′N 118°04′W. Los Alamitos Naval Air Station to S. Manufacturing (confections, aircraft parts). City finances depend largely on income from tax-free bingo parlor.

Hawaii Ocean View Estates, village, S HAWAII, 57 mi/92 km SW of HILO, 6 mi/9.7 km inland from Ka Lae (SW) Coast, on Hawaii Belt Road, opposite Hawaiian Ranchos. Kapua-Manuka Forest Reserve to NW. Located on barren lava at S end of Southwest Rift Zone of MAUNA LOA. Residential community.

Hawaii Volcanoes National Park (□ 355 sq mi/919 sq km), on HAWAII island, HAWAII county, HAWAII, generally 20 mi/32 km–25 mi/40 km SW of HILO, extending near center of island to SE coast; includes smaller unit 15 mi/24 km SW of Hilo. Established 1916 as Hawaii National Park. The NW extension of park includes MAUNA LOA at W end, reached by trail from Hawaii Belt Road; SE coastal section includes Kilauea in N, the Great Crack, and Kau Desert in W, Chain of Craters in E. The park contains two of the most active volcanoes in the world—KILAUEA with Halemaumau Crater, and MAUNA LOA with the active MOKUAWEOWEO crater on its summit. Active lava flows just beyond park's E boundary have destroyed coastal access to park on Chain of Craters Road. The vegetation around Kilauea is varied—a few miles W of the arid Kau Desert is a lush fern jungle. Park renamed 1960. The HALEAKALA section (Maui Island) was made a separate park in 1961.

Hawalli, governate (2005 population 487,514), KUWAIT, at head of PERSIAN GULF; 29°17′N 47°59′E. Smallest governate in land area. Site of first water well in Kuwait.

Hawal River, NIGERIA: see GONGOLA RIVER.

Hawamdiya, El (HAH-wahm-DAI-yuh, el), village, GIZA province, Upper EGYPT, on railroad, and 6 mi/9.7 km NW of HELWAN; 39°54′N 31°15′E. Sugar-refining center.

Hawar, island group, in PERSIAN GULF, between QATAR peninsula and the island of BAHRAIN, 15 mi/24 km–20 mi/32 km from the S tip of Ras' Ak Bar, 1 mi/1.6 km–3 mi/4.8 km W of Qatar. Consists of 16 small islands Strategic importance; sovereignty is disputed between Bahrain and Qatar. Oil rich; fresh water.

Hawara, Arab village, Nablus district, 5 mi/8 km S of NABLUS, in the Samarian Highlands, WEST BANK. Agriculture (olives, fruit, wheat, barley).

Hawara, EGYPT: see HAUWARA.

Hawarden (HAHR-duhn), town (2001 population 13,539), FLINTSHIRE, NE Wales, 5 mi/8 km W of CHESTER; 53°11′N 03°02′W. Home of William Gladstone, 19th-century statesman. Formerly in CLWYD, abolished 1996.

Hawarden (hai-WAHR-duhn), village (2006 population 75), S central SASKATCHEWAN, W CANADA, 50 mi/80 km S of SASKATOON; 51°25′N 106°36′W. Grain elevators.

Hawarib, YEMEN: see HUWAIRIB.

Hawash River, ETHIOPIA: see AWASH RIVER.

Hawatka, El (hah-WAHT-kah, el), township, ASYUT province, central Upper EGYPT, on W bank of the NILE River, on railroad, and 4 mi/6.4 km SE of MANFALUT; 27°16′N 31°01′E. Pottery making, wood and bone carving. Cereals, dates, sugarcane.

Hawdon, Lake (HAW-duhn) (□ 54 sq mi/140.4 sq km), SE SOUTH AUSTRALIA, 165 mi/266 km SSE of ADELAIDE, near Cape JAFFA; 16 mi/26 km long, 5 mi/8 km wide; 37°10′S 139°54′E. Shallow.

Hawea, Lake (HAH-wee-ah), glacial lake (□ 48 sq mi/124.8 sq km), Queenstown-Lakes district, SOUTH ISLAND, NEW ZEALAND, 100 mi/161 km NW of DUNEDIN, E slope of SOUTHERN ALPS, in headwaters of CLUTHA RIVER; 20 mi/32 km long, 5 mi/8 km wide.

Hawera (HAH-wuhr-ah), town, ⊙ S Taranaki district (□ 1,381 sq mi/3,590.6 sq km), TARANAKI region, NEW ZEALAND, 40 mi/64 km SSE of NEW PLYMOUTH. Cooperative dairy plants; sawmills. KAPUNI natural-gas field nearby.

Hawerib, YEMEN: see HUWAIRIB.

Hawes (HAWZ), town (2001 population 1,323), NORTH YORKSHIRE, N ENGLAND, on URE RIVER, in Wemsleydale, and 17 mi/27 km N of SETTLE; 54°18′N 02°12′W. Resort, with dairy market.

Hawesville (HAWZ-vil), town (2000 population 971), ⊙ HANCOCK county, NW KENTUCKY, 23 mi/37 km ENE of OWENSBORO, on the OHIO RIVER (bridge to CANNELTON, Indiana); 37°53′N 86°45′W. Shipping point for agricultural area (burley and dark tobacco, corn, wheat); manufacturing (primary aluminum production; fabricated metal products, paper and pulp). County Museum in historic depot (1903). Cannelton Locks and Dam to E, on Ohio River. Established 1836.

Haweswater (HAWZ-wat-uhr), lake, in the Lake District, CUMBRIA, NW ENGLAND, 9 mi/14.5 km S of PENRITH; 3 mi/4.8 km long, 0.5 mi/0.8 km wide, 200 ft deep (at maximum point); 54°31′N 02°49′W. Dammed to form reservoir. Golden eagles form their habitat nearby.

Hawi (HAH-VEE), town (2000 population 938), N HAWAII island, HAWAII county, HAWAII, near UPOLU POINT, NW tip of KOHALA PENINSULA, 59 mi/95 km NW of HILO, 1.5 mi/2.4 km inland from N coast; 20°14′N 155°49′W. Upolu Airport, at Upolu Point, to NW. Mookini Heiau (Temple) State Monument and King Kamehameha I Birthplace State Memorial to W at Limukoko Point. Puuokumau Reservoir to S.

Hawick (HAI-wik), town (2001 population 14,573), SCOTTISH BORDERS, S Scotland, on the TEVIOT RIVER, and 12 mi/19.2 km S of GALASHIELS; 55°26′N 02°47′W. The largest Scottish town near the English border, Hawick became famous for its woolens and tweeds. Current industries include dye works and light machinery plants. Also livestock markets. The house of the barons of Drumlanrig was the only building not burned by the English during a border raid in 1570. St. Mary's Church (1763) stands on the site of a 7th-century Celtic church.

Hawke Bay, broad indentation of S PACIFIC, E NORTH ISLAND, NEW ZEALAND, extending from MAHIA PENINSULA (NE) to CAPE KIDNAPPERS (S); 50 mi/80 km long, 35 mi/56 km wide. Receives WAIROA RIVER on N and wide HERETAUNGA delta plains in S. NAPIER city on SW shore, WAIROA town on N.

Hawke, Cape (HAWK), E NEW SOUTH WALES, AUSTRALIA, in PACIFIC OCEAN, near Lake WALLIS; 32°13′S 152°35′E. Rises to 777 ft/237 m. Historic lighthouse (1875), 20 mi/32 km S at Seal Rocks.

Hawker (HAW-kuhr), village, S SOUTH AUSTRALIA, 90 mi/145 km NNE of PORT PIRIE; 31°53′S 138°25′E. At junction of roads from PORT AUGUSTA, MARREE, ORROROO, and Wilpena Pound. Wheat, wool. Established 1880.

Hawke's Bay, region (□ 8,177 sq mi/21,260.2 sq km), NEW ZEALAND; as defined in 1989, fairly closely approximates former Hawke's Bay province and provincial district, with modifications to politically unify courses of rivers. Four parallel zones characterize the geomorphic pattern. The inland margin incorporates much of the NORTH ISLAND Main (Axial) Range trending NE from COOK STRAIT, with substantial areas, especially Waikaremoana National Park, in forest. These hard-rock ranges are paralleled by soft-rock hill country, narrow to the S but broadening to reach the sea in WAIROA district to the N. Though often eroded, this carries sheep and beef cattle stations. The economic heart of the region is composed of an interconnecting series of lowlands with easy SW-NE road and railroad transport, a succession of agricultural service centers, and both intensive farming and major urbanization in the alluvial HERE-TAUNGA PLAINS around HASTINGS and NAPIER. Here processing of fruit, vegetables, and livestock products is crucial. Another belt of hill country parallels the coast S of CAPE KIDNAPPERS. The rivers cross-cutting the region from ranges to sea in the N are prone to flood. But in the S their parallel lowlands, occupying a warm rain shadow in the lee of the Axial Ranges, are very productive.

Hawkesbury (HAWKS-buh-ree), town (□ 4 sq mi/10.4 sq km; 2001 population 10,314), SE ONTARIO, E central CANADA, on the OTTAWA River; 45°37′N 74°36′W. Lumber and paper mills; manufacturing of clothing, glass, and prefabricated homes.

Hawkesbury Island (HAWKS-buh-ree) (□ 159 sq mi/413.4 sq km), W BRITISH COLUMBIA, W CANADA, in Douglas Channel (N arm of HECATE STRAIT), just N of GRIBBELL ISLAND; 27 mi/43 km long, 2 mi/3 km–12 mi/19 km wide; 53°35′N 129°04′W.

Hawkesbury River (HAWKS-buh-ree), 293 mi/472 km long, E NEW SOUTH WALES, AUSTRALIA; rises in GREAT DIVIDING RANGE N of Lake George; flows NE, past GOULBURN, PENRITH, and WINDSOR, to BROKEN BAY; 33°30′S 151°10′E. Navigable 70 mi/113 km below Windsor by small craft. Called Wollondilly River between Goulburn and junction with Cox River, and Warragamba River to junction with Nepean River. Tourism.

Hawke's Harbour, settlement, on Hawke Island (5 mi/8 km long, 5 mi/8 km wide), just off SE Labrador, NEWFOUNDLAND AND LABRADOR, NE CANADA; 53°01′N 55°50′W. Lumbering. Explored by Captain Cook.

Hawkesville (HAWKS-vil), unincorporated village, WATERLOO region, S ONTARIO, E central CANADA, 13 mi/21 km from KITCHENER, and included in WELLESLEY township; 43°33′N 80°38′W.

Hawkeye, town (2000 population 489), FAYETTE county, NE IOWA, 7 mi/11.3 km WSW of WEST UNION; 42°56′N 91°57′W. In agricultural area.

Hawkeye State; see IOWA.

Hawkhurst (HAWK-huhst), village (2001 population 4,360), S KENT, SE ENGLAND, 17 mi/27 km WSW of ASHFORD; 51°02′N 00°30′E. Has 15th-century church. Formerly important center in Wealden iron industry.

Hawkins, county (□ 494 sq mi/1,284.4 sq km; 2006 population 56,850), NE TENNESSEE; ⊙ ROGERSVILLE; 36°27′N 82°57′W. Bordered N by VIRGINIA; traversed by CLINCH and BAYS mountains, ridges of the APPALACHIAN MOUNTAINS; drained by HOLSTON River. Includes part of CHEROKEE Reservoir. Agriculture; manufacturing. Dimension marble from here was used in the Washington Monument. Formed 1786.

Hawkins, town (2006 population 1,495), Wood county, NE TEXAS, on SABINE River, 17 mi/27 km N of TYLER; 32°35′N 95°12′W. In oil and agricultural area; manufacturing (gasoline). Lake Hawkins to NW.

Hawkins, village (2006 population 304), RUSK county, N WISCONSIN, 19 mi/31 km ENE of LADYSMITH; 45°30′N 90°42′W. Dairying; livestock raising; farming.

Woodworking; manufacturing (wood products). FLAMBEAU RIVER State Forest to N.

Hawkins Peak, mountain (10,024 ft/3,055 m), ALPINE county, E CALIFORNIA, in the SIERRA NEVADA, c.15 mi/24 km S of LAKE TAHOE; flanks EBBETTS PASS on S.

Hawkinsville, town (2000 population 3,280), ⊙ PULASKI county, S central GEORGIA, on OCMULGEE River, and 39 mi/63 km SSE of MACON; 32°17′N 83°28′W. Manufacturing includes meatpacking; confections; textiles; lumber mills. Incorporated 1830.

Haw Knob, TENNESSEE and NORTH CAROLINA: see UNICOI MOUNTAINS.

Hawk Peak (10,627 ft/3,239 m), GRAHAM county, SE ARIZONA, in PINALENO MOUNTAINS, near Mount GRAHAM, 13 mi/21 km SW of SAFFORD.

Hawk Point, town (2000 population 459), LINCOLN county, E MISSOURI, near West Fork of CUIVRE RIVER, 8 mi/12.9 km W of TROY; 38°58′N 91°07′W. Meat processing.

Hawk Run, unincorporated town, CLEARFIELD county, central PENNSYLVANIA, 2 mi/3.2 km NE of Phillipsburg, near Moshannon Creek; 40°55′N 78°12′W.

Hawksbill Mountain (HAWKS-bil), peak (4,049 ft/1,234 m) of the BLUE RIDGE MOUNTAINS, N VIRGINIA, 8 mi/13 km SSE of LURAY; highest point in SHENANDOAH NATIONAL PARK and of Blue Ridge (N of WAYNESBORO).

Hawksburn (HAHKS-buhrn), suburb 3 mi/5 km SE of MELBOURNE, VICTORIA, SE AUSTRALIA; 37°50′S 145°00′E.

Hawkshaw, ENGLAND: see TOTTINGTON.

Hawkshead (HAWK-shed), village (2001 population 1,703), CUMBRIA, NW ENGLAND, in the LAKE DISTRICT, 13 mi/21 km NNE of ULVERSTON; 54°22′N 03°00′W. Has 16th-century church with 13th-century tower, and grammar school that Wordsworth attended. Beatrix Potter Gallery here. Just S is lake of ESTHWAITE WATER.

Hawks Nest (HAWKS NEST), township, NEW SOUTH WALES, SE AUSTRALIA, 49 mi/79 km NE of NEWCASTLE; 32°40′S 152°10′E. Separated from its twin city, Tea Gardens, by Myall River; the two towns are connected by "Singing Bridge," named for the sound of the wind passing through its rails. Seafood; timber; sand mining; tourism. Dolphin viewing; koala colony on Tilligerry Peninsula. Deep-sea fishing, diving at nearby Broughton Island. The Worimi Aboriginal people inhabited the area before European settlement.

Hawks Nest State Park, WEST VIRGINIA: see GAULEY BRIDGE.

Hawkwell (HAWK-wel), village (2001 population 11,231), SE ESSEX, SE ENGLAND, 5 mi/8 km NNW of SOUTHEND-ON-SEA; 51°36′N 00°40′E.

Hawley (HAW-lee), former township, E central ONTARIO, E central CANADA; 46°24′N 80°37′W. Amalgamated into the town of MARKSTAY-WARREN in 1999.

Hawley, town, FRANKLIN county, NW MASSACHUSETTS, 15 mi/24 km W of GREENFIELD; 42°35′N 72°55′W. In hilly area. Fruit. Dubuque Memorial State Forest nearby.

Hawley, town (2000 population 1,882), CLAY county, W MINNESOTA, 23 mi/37 km E of FARGO, NORTH DAKOTA, near BUFFALO RIVER; 46°52′N 96°19′W. Manufacturing (printing and publishing; foods, machinery). Barnsville Wildlife Area to S; Buffalo River State Park to W; small lakes to E; state game refuge nearby. Settled c.1870.

Hawley, village (2006 population 598), JONES county, W central TEXAS, 14 mi/23 km NNW of Abilene, on CLEAR FORK OF BRAZOS RIVER; 32°36′N 99°48′W. Agricultural area (grain, cotton, watermelons; cattle). LAKE FORT PHANTOM HILL to E.

Hawley, borough (2006 population 1,299), WAYNE county, NE PENNSYLVANIA, 8 mi/12.9 km SSE of HONESDALE, on LACKAWAXEN RIVER, at mouth of Middle Creek. Manufacturing (printing and publish-

ing, sheet metal fabricating; consumer goods). Lake WALLENPAUPACK reservoir, resort area, to SW. Numerous residential developments in area, especially W and SW; Tanglewood Ski Area to S; part of Delaware State Forest to SE. Settled 1803, incorporated 1884.

Hawmat as-Suq (haw-MET es–SOOK), town (2004 population 44,555), MADANIYINA province, SE TUNISIA, on N shore of JARBAH ISLAND, 40 mi/64 km E of QABIS. Chief market center (olives, figs, pomegranates, pears, grapes, dates); sponge fishing; handicraft industries (colorful silk and woolen cloth, jewelry); tourism. International airport 7 mi/11.3 km W. Settled by Phoenicians in 6th century B.C.E.; conquered by Arabs in C.E. 667; razed to ground in 11th century. Became base for Muslim corsairs in S TUNIS in 15th century. Also called Jarbah or Jerba.

Haworth (HA-wuhth), town, WEST YORKSHIRE, N central ENGLAND, 8 mi/12.9 km NNW of HALIFAX; 53°50′N 01°57′W. Light industry. Patrick Brontë and his daughters, Charlotte and Emily, lived at the parsonage (now a museum and library) here.

Haworth (HAI-wuhrth), village (2006 population 350), MCCURTAIN county, extreme SE OKLAHOMA, 10 mi/16 km ESE of IDABEL; 33°50′N 94°39′W. Located in unit of Ouachita National Forest.

Haworth (HAW-wuhrth), borough (2006 population 3,433), BERGEN county, NE NEW JERSEY, 10 mi/16 km ENE of PATERSON; 40°57′N 74°00′W. Incorporated 1894.

Haw Par Village (HAH PAHR), amusement park, S Singapore island, SINGAPORE, at BUONA VISTA, 5 mi/8 km W of downtown SINGAPORE; covers 24 acres/10 ha. For many years it was known as Tiger Balm Gardens, a landscaped garden noted for its bizarre sculptures. The site was refurbished as an amusement park with a Chinese mythology theme in 1990.

Hawrah, YEMEN: see HAURA.

Haw River (HAH), town (□ 2 sq mi/5.2 sq km; 2006 population 1,992), ALAMANCE county, N central NORTH CAROLINA, on HAW RIVER, and suburb 4 mi/6.4 km E of BURLINGTON; 36°05′N 79°21′W. Manufacturing (textiles, rubber products); agriculture (tobacco, grain; chickens, hogs; dairying). Haw River Museum. Settled 1747.

Haw River (HAH), c.110 mi/177 km long, N central NORTH CAROLINA; rises NW of GREENSBORO in NW GUILFORD county; flows E and SSE, past town of HAW RIVER, E of BURLINGTON, through B. EVERETT JORDAN LAKE (Jordan Lake) reservoir, where New Hope River joins it from N, forming larger part of reservoir, before joining DEEP RIVER near HAYWOOD to form CAPE FEAR RIVER; 36°08′N 80°02′W.

Hawshabi, YEMEN: see HAUSHABI.

Hawsh 'Īsā, town, BEHEIRA province, N EGYPT, 30 mi/50 km SE of ALEXANDRIA, in W section of the NILE Delta; 30°55′N 30°17′E. Also spelled Hosh Isa.

Hawtah (HOU-tah), town, S central NAJ'D, Riyadh province, SAUDI ARABIA, 80 mi/129 km S of RIYADH. Trading center; grain, vegetables, fruit; livestock raising. Also spelled Hauta or Hautah; also called Hilla or Hillah.

Hawtah, YEMEN: see HAUTA.

Hawthorn (HAW-thorn), borough (2006 population 556), CLARION county, W central PENNSYLVANIA, 14 mi/23 km SSE of CLARION, on RED BANK CREEK. Agricultural (corn, hay, potatoes; dairying).

Hawthorn (HAW-thorn), inner E residential suburb of MELBOURNE, S VICTORIA, AUSTRALIA; 37°50′S 145°02′E. Glenferrie Road shopping strip.

Hawthorne, city (2000 population 84,112), LOS ANGELES county, S CALIFORNIA; suburb 10 mi/16 km SW of downtown LOS ANGELES, 4 mi/6.4 km E of PACIFIC OCEAN coast (Manhattan State Beach); 33°55′N 118°21′W. In an oil- and gas-producing area. Has large-scale manufacturing of navigation systems, solar panels, electronic components, silicon instruments, and transportation equipment. Hawthorne Municipal Airport, LOS ANGELES INTERNATIONAL AIRPORT to NW. Incorporated 1922.

Hawthorne (HAW-thorn), town (2000 population 1,415), ALACHUA county, N central FLORIDA, 12 mi/19 km SE of GAINESVILLE; 29°35′N 82°05′W.

Hawthorne, unincorporated town (2000 population 3,311), ⊙ MINERAL county, W NEVADA, near WALKER Lake, 90 mi/145 km SE of RENO; 38°31′N 118°37′W. Elev. 4,320 ft/1,317 m. U.S. naval arsenal and ammunition depot surrounds town, except on SW. Tourist center; gold, silver, sand and gravel; manufacturing (drilling services); cattle. Old mining town of AURORA is c.40 mi/64 km SW. WASSUK Range W. YOSEMITE National Park (CALIFORNIA) 50 mi/80 km to SW. Part of Toiyabe National Forest and Wassuk Range to SW. WALKER LAKE to NW; Walker Lake State Recreational Area on W shore. Incorporated c.1940.

Hawthorne, unincorporated village (□ 1 sq mi/2.6 sq km; 2000 population 5,083), WESTCHESTER county, SE NEW YORK, 6 mi/9.7 km N of WHITE PLAINS, near KENSICO RESERVOIR (E); 41°06′N 73°47′W. Light manufacturing and commercial services. Gate of Heaven cemetery is here; it is the burial place of Babe Ruth, James Cagney, and other notables.

Hawthorne (HAW-thorn), borough (2006 population 18,166), PASSAIC county, NE NEW JERSEY; 40°57′N 74°09′W. Residential suburb, with some light manufacturing. Settled 1850, incorporated 1898.

Hawthorn East (HAH-thorn EEST), suburb 6 mi/9 km E of MELBOURNE, VICTORIA, SE AUSTRALIA. Between two shopping strips. Several parks.

Hawthorn Woods, village (2000 population 6,002), Lake county, NE ILLINOIS, residential suburb 33 mi/53 km NW of downtown CHICAGO, 4 mi/6.4 km NE of LAKE ZURICH town; 42°13′N 88°03′W.

Hawzen, town (2007 population 5,901), TIGRAY state, N ETHIOPIA, 20 mi/32 km S of ADIGRAT; 13°58′N 39°26′E. Trade center (cereals, salt, honey, cotton goods). In 1988, bombings in the town killed 2,500. Also spelled Hauzen.

Haxby (HAKS-bee), village (2001 population 8,754), YORK, N ENGLAND, on OUSE RIVER, 4 mi/6.4 km N of York; 54°01′N 01°04′W.

Haxey, ENGLAND: see AXHOLME, ISLE OF.

Haxtun, town (2000 population 982), PHILLIPS county, NE COLORADO, near SOUTH PLATTE RIVER, 17 mi/27 km WNW of HOLYOKE; 40°38′N 102°37′W. Elevation 4,028 ft/1,228 m. Shipping point in irrigated grain, sugar beet, and cattle region. In Sand Hills area.

Hay (HAI), municipality, S NEW SOUTH WALES, AUSTRALIA, on N bank of MURRUMBIDGEE RIVER, 265 mi/426 km SE of BROKEN HILL, and in RIVERINA region; 34°30′S 144°51′E. Railroad terminus; dairying, sheep center (merino wool), cattle; grains, legumes (irrigation agriculture).

Hay (HAI), former township (□ 86 sq mi/223.6 sq km; 2001 population 2,187), S ONTARIO, E central CANADA; 43°24′N 81°36′W. Amalgamated into town of BLUEWATER in 2001.

Hay (HAI), river, c.530 mi/850 km long, W CANADA; rises in several headstreams in NE BRITISH COLUMBIA and NW ALBERTA; 60°00′N 116°56′W. Flows generally NE through NW Alberta, over Alexander Falls, and into GREAT SLAVE LAKE. Its valley, a principal N-S route, is followed by a highway and a railroad.

Hayakawa (hah-yah-KAH-wah), town, South Koma district, YAMANASHI prefecture, central HONSHU, central JAPAN, 22 mi/35 km S of KOFU; 35°24′N 138°21′E. Inkstone.

Hayakita (hah-YAH-kee-tah), town, Iburi district, HOKKAIDO prefecture, N JAPAN, 34 mi/55 km S of SAPPORO; 42°45′N 141°49′E. Horse breeding, dairy farming.

Hayama (hah-YAH-mah), town, Miura district, KANAGAWA prefecture, E central HONSHU, E central JAPAN, on NW MIURA PENINSULA, on NE shore of

SAGAMI BAY, 12 mi/20 km S of YOKOHAMA; 35°16′N 139°35′E. Emperor's villa at Hayama Goyote. Japanese yachting originated here.

Hayama (HAH-yah-mah), village, Takaoka district, KOCHI prefecture, S SHIKOKU, W JAPAN, 22 mi/35 km S of KOCHI; 33°26′N 133°12′E.

Hayange (ah-YAHNZH), German *Hayingen*, town (□ 4 sq mi/10.4 sq km), MOSELLE department, LORRAINE region, NE FRANCE, 6 mi/9.7 km WSW of THIONVILLE; 49°20′N 06°04′E. Metallurgical industry; oldest iron-working city in Lorraine.

Hayarkon (hah-yahr-KON) or **Yarkon River**, ISRAEL; rises in the Samarian Highlands at ROSH HAAYIN; flows W to the MEDITERRANEAN SEA just N of TEL AVIV. Along its course lie ancient settlements and mills. The waters of the river are diverted into the National Water Carrier to meet water requirements of the arid S.

Hayashima (hah-YAH-shee-mah), town, Tsukubo district, SW HONSHU, W JAPAN, 6 mi/10 km W of OKAYAMA; 34°35′N 133°49′E. Tatami mats; flower matting.

Hayatim, El (ha-YAH-tim, el), village, GHARBIYA province, Lower EGYPT, 6 mi/9.7 km SW of El MAHALLA EL KUBRA; 30°55′N 31°06′E. Cotton.

Hayato (HAH-yah-to), town, Aira district, KAGOSHIMA prefecture, S KYUSHU, JAPAN, 9 mi/15 km NE of KAGOSHIMA; 31°44′N 130°44′E. Peaches. Has large Shinto shrine believed to occupy site of palace of Emperor Jimmu, first emperor of Japan.

Haybes (AIB), commune (□ 10 sq mi/26 sq km), ARDENNES department, CHAMPAGNE-ARDENNE region, N FRANCE, near Belgian border, in the Ardennes hills, 17 mi/27 km N of CHARLEVILLE-MÉZIÈRES, on entrenched MEUSE RIVER; 50°00′N 04°43′E. Metalworks. Leveled in World War I. Also called Haybes-sur-Meuse.

Haybes-sur-Meuse, FRANCE: see HAYBES.

Hay, Cape, N extremity of BAFFIN ISLAND, BAFFIN region, NUNAVUT territory, N CANADA, on LANCASTER SOUND; 73°53′N 79°49′W.

Hay, Cape, NW extremity of BYLOT ISLAND, NUNAVUT territory, N CANADA, at E entrance of LANCASTER SOUND; 73°53′N 79°49′W.

Hay, Cape, S extremity of MELVILLE ISLAND, NORTHWEST TERRITORIES, N CANADA, on MCCLURE STRAIT; 74°24′N 113°08′W.

Haycock, village, W ALASKA, SE SEWARD PENINSULA, near KOYUK RIVER, 18 mi/29 km N of KOYUK.

Haycuri (ai-KOO-ree), town and canton, General Bernardin BILBAO province, in extreme N of POTOSÍ department, W central BOLIVIA; 17°55′S 66°04′W. Elevation 9,974 ft/3,040 m. No roads. Lead-bearing lodes and limestone deposits. Agriculture (potatoes, yucca, bananas); cattle.

Haydarpasa (Turkish=*Haydarpaşa*), district of ISTANBUL, NW TURKEY, part of KADIKOY, at entrance to the BOSPORUS, on Sea of MARMARA. Terminus of Anatolian railroad. The station, Turkey's largest, was built in the early 20th century by German architects Otto Ritter and Helmuth Cuno. A monument to the close Turkish-German relations of the time, the station is in neo-renaissance style and has a U-plan. Its façade is sandstone. Sometimes spelled Haidar Pasha.

Hayden, town (2000 population 470), Blount co., N central Alabama, 18 mi/29 km WSW of Oneonta. Formerly known as 'Rockland,' it was renamed in honor of an early settler. Inc. in 1949.

Hayden, town (2000 population 892), GILA and PINAL counties, SE central ARIZONA, on GILA River, 3 mi/4.8 km NW of confluence with SAN PEDRO RIVER, 55 mi/89 km NNE of TUCSON; 33°00′N 110°46′W. Copper smelting, lime; cattle, sheep. San Carlos Indian Reservation to E; Ray Mine (open pit copper mine) to N.

Hayden, town (2000 population 1,634), ROUTT county, NW COLORADO, on YAMPA river, W of PARK RANGE, and 22 mi/35 km W of STEAMBOAT SPRINGS; 40°29′N 107°15′W. Elevation 6,337 ft/1,932 m. Coal mining.

Shipping point in sheep and cattle region. Parts of Routt National Forest to N and S; ELKHEAD RESERVOIR to NW.

Hayden, town (2000 population 9,159), KOOTENAI county, N IDAHO, suburb 4 mi/6.4 km N of COEUR D'ALENE, near HAYDEN LAKE; 47°46′N 116°48′W. Manufacturing (embroidered logos, electrical and electronic equipment, fabricated metal products, lumber).

Hayden Lake, village (2000 population 494), KOOTENAI county, N IDAHO, 8 mi/12.9 km NNE of COEUR D'ALENE; 47°46′N 116°45′W. Coeur d'Alene National Forest to E.

Hayden Lake, reservoir, KOOTENAI county, N IDAHO, on SPOKANE RIVER tributary, 4 mi/6.4 km NNE of COEUR D'ALENE, and 6 mi/9.7 km N of COEUR D'ALENE LAKE; 7 mi/11.3 km long, 2 mi/3.2 km wide; 47°45′N 116°47′W. Used for irrigation. Extends E into Coeur d'Alene National Forest. Towns of HAYDEN and DALTON GARDENS at SW end, HAYDEN LAKE village on N shore.

Hayden Peak (12,479 ft/3,804 m), in UINTA MOUNTAINS, on border between SUMMIT and DUCHESNE counties, NE UTAH, in High Uintas Wilderness Area, 45 mi/72 km NNW of DUCHESNE.

Hayden-Rhodes Aqueduct, c.210 mi/338 km long, W and central ARIZONA, in LA PAZ and MARICOPA counties; begins at LAKE HAVASU Reservoir, near PARKER DAM; flows SE then E, passing N of PHOENIX to SALT RIVER at Granite Reef Dam, 22 mi/35 km E of Phoenix. Diverts water from COLORADO RIVER to Salt River and GILA RIVER valleys for irrigation and urban use. SALT-GILA AQUEDUCT continues S to SANTA CRUZ RIVER.

Haydenville, Massachusetts: see WILLIAMSBURG.

Haydock (HAI-duk), town (2001 population 11,962), MERSEYSIDE, NW ENGLAND, 3 mi/4.8 km ENE of St. HELENS; 53°28′N 02°40′W. Haydock Park Racecourse just NE.

Haydon Bridge (HAI-dun), town (2006 population 2,000), S NORTHUMBERLAND, N ENGLAND, on SOUTH TYNE RIVER, and 6 mi/9.7 km W of HEXHAM; 54°58′N 02°14′W. Nearby is 14th-century Langley Castle.

Haye-Descartes, FRANCE: see DESCARTES, LA.

Haye-du-Puits, La (AI—dyoo—PWEE, lah), commune (□ 2 sq mi/5.2 sq km), MANCHE department, NW FRANCE, in NORMANDY, BASSE-NORMANDIE administrative region, on COTENTIN PENINSULA, 18 mi/29 km NNW of COUTANCES; 49°18′N 01°33′W. In dairying country; woodworking. Scene of heavy fighting (July 1944) in Normandy campaign of World War II, prior to Allied breakthrough at SAINT-LÔ.

Hayel, SAUDI ARABIA: see HAIL.

Hayes, county (□ 713 sq mi/1,853.8 sq km; 2006 population 1,029), SW NEBRASKA; ⊙ HAYES CENTER; 40°31′N 101°03′W. Agricultural area drained by FRENCHMAN CREEK and other branches of REPUBLICAN River. Cattle, hogs; corn, wheat, beans. Central/Mountain time zone boundary follows W and NW county line. Formed 1877.

Hayes, town, CLARENDON parish, S JAMAICA, in irrigated VERE plain, 4 mi/6.4 km S of MAY PEN; 17°53′N 77°15′W. Sugarcane.

Hayes (HAIZ), river, c.300 mi/480 km long, MANITOBA, W central CANADA; rises in a lake NE of LAKE WINNIPEG, central Manitoba; flows NE to HUDSON BAY; 57°03′N 92°10′W. Chief route used by Hudson's Bay Company traders from Hudson Bay to Lake Winnipeg and the interior; YORK FACTORY, an important establishment of the company, is at its mouth.

Hayes, ENGLAND: see BROMLEY.

Hayes, ENGLAND: see HILLINGDON.

Hayes Center, village (2006 population 226), ⊙ HAYES county, SW NEBRASKA, 30 mi/48 km NW of MCCOOK, and on branch of REPUBLICAN RIVER; 40°30′N 101°01′W. Dairy and poultry products; livestock; grain, alfalfa pellets.

Hayes, Fort, Ohio: see COLUMBUS.

Hayes, Mount (13,832 ft/4,216 m), E ALASKA, in ALASKA RANGE, 90 mi/145 km SSE of FAIRBANKS; 63°38′N 146°43′W.

Hayes Peninsula, NW GREENLAND, extends c.100 mi/161 km into BAFFIN BAY and SMITH SOUND; 75°54′N 64°00′W–79°10′N 73°05′W; c.200 mi/322 km wide, between KANE BASIN (N) and MELVILLE BAY (S). Deeply indented by INGLEFIELD BAY and WOLSTENHOLME FJORD. THULE and ETAH settlements are on peninsula; PRUDHOE LAND and INGLEFIELD LAND forms its N part. Name suggested in 1867 by A. Petermann (German geographer). It appears still on some maps but is generally discarded.

Hayesville, unincorporated city (2000 population 18,222), MARION county, NW OREGON, residential suburb 3 mi/4.8 km NE of downtown SALEM; 44°58′N 122°58′W. State Fairgrounds to SW.

Hayesville, town (2000 population 64), KEOKUK county, SE IOWA, 10 mi/16 km SW of SIGOURNEY; 41°15′N 92°15′W. Limestone quarries nearby.

Hayesville (HAIZ-vil), village (2006 population 469), ⊙ Clay county, W North Carolina, 80 mi/129 km WSW of ASHEVILLE, on HIWASSEE RIVER (forms Chatuga Lake reservoir to SE); 35°02′N 83°49′W. Resort area. Service industries; manufacturing (apparel, wire and cable; printing and publishing); agriculture (tobacco, fruit; cattle). Nantahala National Forest to W, N, and E. Chattahoochee National Forest (GEORGIA) to S. Named for George W. Hayes, member of the General Assembly. Inc. in 1891.

Hayesville (HAIZ-vil), village (2006 population 347), ASHLAND county, N central OHIO, 7 mi/11 km SE of ASHLAND; 40°46′N 82°16′W.

Hayfield (HAI-feeld), town (2001 population 2,164), NW DERBYSHIRE, central ENGLAND, 4 mi/6.4 km S of GLOSSOP, in THE PEAK DISTRICT; 53°23′N 01°57′W. Former industrial town. Just W is town of BIRCH VALE.

Hayfield, town (2000 population 1,325), DODGE county, SE MINNESOTA, 20 mi/32 km SW of ROCHESTER; 43°53′N 92°50′W. Agricultural area (poultry, livestock; grain, soybeans, peas; dairying); manufacturing (windows and doors, feeds, fiber cans).

Hayfield (HAI-feeld), unincorporated town, FAIRFAX county, NE VIRGINIA, residential suburb 6 mi/10 km SW of ALEXANDRIA, 12 mi/19 km SSW of WASHINGTON, D.C.; 38°45′N 77°08′W. FORT BELVOIR Military Reservation to S, U.S. Coast Guard Radio Station to E.

Hayfork, unincorporated town (2000 population 2,315), TRINITY county, NW CALIFORNIA, 17 mi/27 km SW of WEAVERVILLE, on Hayfork Creek; 40°34′N 123°08′W. Area surrounded by Shasta-Trinity National Forest. Cattle; hay, timber.

Hayik, town (2007 population 14,998), AMHARA state, N central ETHIOPIA; 11°15′N 39°41′E. Roadside town just W of LAKE HAYK, 10 mi/16 km NNE of DESSIE. Sometimes spelled Hayk.

Hayingen, FRANCE: see HAYANGE.

Hayk, Lake, AMHARA state, NE ETHIOPIA, near LAKE ARDIBBO, 17 mi/27 km NE of DESSIE; 4 mi/6.4 km long, 3 mi/4.8 km wide; 11°21′N 39°43′E. Elevation of 6,660 ft/2,030 m.

Haykota (hei-KO-tuh), village, GASH-BARKA region, W ERITREA, on GASH RIVER, and 36 mi/58 km W of BARENTU; 15°10′N 37°05′E. Elevation c.2,000 ft/610 m. Road junction; cattle raising, dom nut gathering. Also spelled Aicota or Aikota.

Hay Lakes (HAI), village (2001 population 346), central ALBERTA, W CANADA, on small Little Hay Lake, 30 mi/48 km SE of EDMONTON, in CAMROSE COUNTY; 53°12′N 113°03′W. Mixed farming, dairying. Established 1928.

Hayle (HAI-lee), town (2001 population 7,474), W CORNWALL, SW ENGLAND, on St. IVES Bay of the Atlantic Ocean, and 7 mi/11.3 km NE of PENZANCE; 50°11′N 05°24′W. Fishing port. Seaside resort.

Haÿ-les-Roses, L' (LAI-lai-ROZ), town (□ 1 sq mi/2.6 sq km), sub-prefecture of VAL-DE-MARNE depart-ment, ÎLE-DE-FRANCE region, N central FRANCE, a residential S suburb of PARIS, 5 mi/8 km from Notre Dame Cathedral, on main highway to the S and to ORLY airport. Rose gardens. Brickworks.

Hayling Island (HAI-ling), just off SE coast of HAMPSHIRE, S ENGLAND, in the Channel at SE end of the SPITHEAD, in CHICHESTER HARBOUR; separated from mainland by narrow channel (bridged); 4 mi/6.4 km long, 1 mi/1.6 km–4 mi/6.4 km wide; 50°47′N 00°58′W. PORTSEA ISLAND is just W. Chief town, SOUTH HAYLING. Activity of windsurfing was invented here. Sailing. Tourist site.

Haymana, village, central TURKEY, 40 mi/64 km SSW of ANKARA; 39°26′N 32°30′E. Grain, fruit; mohair goats. Formerly Hamam or Sivrihamam.

Hayman Island (HAI-muhn), island in Whitsunday group, QUEENSLAND, NE AUSTRALIA, on GREAT BARRIER REEF; 20°03′S 148°53′E. Upscale resort. Walking tracks.

Haymarket (HAI-mahr-ket), town (2006 population 1,234), PRINCE WILLIAM county, NE VIRGINIA, 10 mi/16 km WNW of MANASSAS; 38°48′N 77°38′W. In agricultural area (grain, soybeans, nursery stock; livestock; dairying). Manassas National Battlefield Park to E.

Haymock Lake (HA-muhk), PISCATAQUIS county, N central MAINE, 50 mi/80 km NNW of MILLINOCKET, in wilderness recreational area; 2.5 mi/4 km long, 1 mi/1.6 km wide. Drains W into EAGLE LAKE.

Hay, Mount (8,870 ft/2,704 m), on UNITED STATES (ALASKA)–CANADA (BRITISH COLUMBIA) border, SAINT ELIAS MOUNTAINS, 75 mi/121 km ESE of YAKUTAT; 59°15′N 137°36′W.

Haynau, POLAND: see CHOJNOW.

Haynes (HAINZ), unincorporated village, S central ALBERTA, W CANADA, 18 mi/29 km E of RED DEER, in LACOMBE COUNTY; 52°19′N 113°24′W. Coal mining; oil and gas. Cattle; wheat, oats.

Haynes (HAINZ), village (2006 population 17), ADAMS co., SW NORTH DAKOTA, 8 mi/12.9 km E of HETTINGER; 45°58′N 102°28′W. Founded in 1908 and incorporated in 1910. Named for George B. Haynes, a president of the Milwaukee Road Railroad.

Haynesville (HAINZ-vil), town (2000 population 2,679), CLAIBORNE parish, N LOUISIANA, 50 mi/80 km NE of SHREVEPORT, near ARKANSAS state line; 32°58′N 93°08′W. Oil and natural-gas wells, oil refineries; varied agriculture (watermelons; cattle, hogs, poultry; dairying); timber; manufacturing (consumer goods, apparel). Incorporated 1861.

Haynesville, town, AROOSTOOK county, E MAINE, on the MATTAWAMKEAG RIVER, and 22 mi/35 km SSW of HOULTON; 45°49′N 67°56′W.

Hayneville, village (2000 population 1,177), ⊙ Lowndes co., S central Alabama, 21 mi/34 km SW of Montgomery. Originally named 'Big Swamp' for the nearby creek, it was renamed after Robert Y. Hayne, governor of SC and a U.S. senator. Inc. in 1841.

Hayogev (hah-yo-GEV), moshav, 3.7 mi/6 km W of AFULA, in PLAIN OF JEZREEL, N ISRAEL; 32°36′N 35°11′E. Elevation 249 ft/75 m. Livestock; poultry farming; hothouse flowers. Founded in 1949.

Hay-on-Wye (HAI-ON-WEI), town (2001 population 1,469), POWYS, E Wales, on WYE RIVER, and 15 mi/24 km NE of BRECON; 52°04′N 03°07′W. Has Norman castle. Formerly a walled town.

Hay Point (HAI), port, QUEENSLAND, NE AUSTRALIA, c.25 mi/40 km S of MACKAY; 21°17′S 149°18′E. Coal export.

Hayrabolu, township, European TURKEY, 26 mi/42 km NW of TEKIRDAG; 41°14′N 27°04′E. In rich grain-producing district. Wheat, spelt, barley, rye, oats, canary-grass, corn; also flax, sugar beets. Sometimes spelled Airobol.

Hayredin, Bulgaria: see HAIREDIN.

Hay River, town (2000 population 3,835), NORTHWEST TERRITORIES, N CANADA, on SW shore of GREAT SLAVE LAKE, at mouth of Hay River, 70 mi/113 km WSW of FORT RESOLUTION; 60°51′N 118°44′W. Terminus of Great Slave railroad (completed 1964) and of paved highway from ALBERTA. Transfer point and lake port for supply barges that go to points on lake and MACKENZIE RIVER. Hudson Bay Company fur trading post (established 1868); transshipment point for YELLOWKNIFE region, at N end of road from railhead at GRIMSHAW (Alberta). Airfield, government radio, TV, and meteorological stations; site of Anglican and Roman Catholic mission and hospital. Fishing, trapping; service industries; oats and vegetables are grown. Scheduled air service. Former lead-zinc mine at nearby PINE POINT (closed 1987). Town has seventeen-story apartment building, tallest building in Northwest Territories.

Hay River, c.50 mi/80 km long, NW WISCONSIN; rises in Beaverdam Lake (BARRON county); flows S to RED CEDAR RIVER (Tainter Lake), 8 mi/12.9 km N of MENOMONIE.

Hays, county (□ 679 sq mi/1,765.4 sq km; 2006 population 241), S central TEXAS; ⊙ SAN MARCOS; 30°03′N 98°01′W. Crossed SW-NE by BALCONES ESCARPMENT, dividing prairies in SE from hilly N and W, part of EDWARDS PLATEAU; drained by SAN MARCOS and BLANCO rivers. Agriculture: livestock (especially in N); cotton, corn, wheat, grain sorghum, oats, hay, fruit, and vegetables; cattle, sheep, goats (wool, mohair marketed). Limestone, sand and gravel. Tourism: springs, scenic hills, hunting, fishing. Formed 1848.

Hays, city (2000 population 20,013), ⊙ ELLIS county, W central KANSAS; 38°52′N 99°19′W. Elevation 1,997 ft/609 m. Railroad, trade, and medical center in a grain, cattle, and oil area. Manufacturing (electronic equipment, building materials, plastics products, feeds, medical supplies, consumer goods, aircraft, motorcycles; meatpacking). Fort Hays established 1865, 14 mi/23 km SE of the city, on a stagecoach road to DENVER. The fort was abandoned in 1889 and the land turned over to the state with the understanding that it be used for a school, an agricultural experiment station. The school has grown into Fort Hays State University; the agricultural experiment station (laid out 1901) is one of the world's largest; and Frontier Historical Park, a state historic site, contains Old Fort Hays surviving buildings. Incorporated 1885.

Hays (HAIZ), unincorporated town (□ 6 sq mi/15.6 sq km; 2000 population 1,731), WILKES county, NW NORTH CAROLINA, 8 mi/12.9 km NNE of WILKESBORO; 36°15′N 81°07′W. Retail trade; agriculture (tobacco, soybeans; poultry; dairying).

Hays (HAIZ), village (2000 population 702), BLAINE county, N MONTANA, 45 mi/72 km, SW of MALTA, in S part of Fort Belknap Indian Reservation, on Little Peoples Creek; 48°00′N 108°39′W. Wheat, barley, oats, beans, alfalfa; cattle, sheep; gas, oil, and gold in area. Little Rocky Mountains, an outline of Main ROCKY MOUNTAINS, to S. Old Mission; Natural Bridge State Monument. Upper MISSOURI WILD AND SCENIC RIVER to SW, Charles M. Russell National Wildlife Refuge to S.

Hays, YEMEN: see HAIS.

Hays Glacier, in NE ANTARCTICA, flowing into the head of Spooner Bay in ENDERBY LAND; 67°40′S 46°18′E.

Hay, Shatt al, IRAQ: see GHARRAF, SHATT AL.

Haysi (HAI-sei), town (2006 population 180), DICKENSON county, SW VIRGINIA, on RUSSELL FORK RIVER, 10 mi/16 km NE of CLINTWOOD; 37°12′N 82°17′W. Agriculture (cattle; tobacco). Breaks Interstate Park (KENTUCKY and Virginia) to N; FLANNAGAN RESERVOIR (Pound River) to W.

Hay Springs, village (2006 population 575), SHERIDAN county, NW NEBRASKA, 20 mi/32 km SE of CHADRON; 42°40′N 102°41′W. Livestock; dairy products; grain, potatoes in irrigated Mirage Flats. Collection of prehistoric bones is here. Pine Ridge unit of Nebraska National Forest to W (in Butte county); Walgren Lake State Recreation Area to SE.

Haystack, Mount (4,918 ft/1,499 m), ESSEX COUNTY, NE NEW YORK, a peak of the High Peaks section of the ADIRONDACK MOUNTAINS, just SE of Mount MARCY, and 14 mi/23 km SSE of LAKE PLACID village; 44°06′N 73°54′W.

Haystack Mountain, ski area, WINDHAM CO., S. VERMONT, in Wilmington, and 15 mi/24 km W of BRATTLEBORO.

Haystack Peak (12,020 ft/3,664 m), highest point in DEEP CREEK MOUNTAINS, NW JUAB county, W UTAH, near NEVADA state line.

Haysville, city (2000 population 8,502), SEDGWICK county, S central KANSAS, suburb 10 mi/16 km S of downtown WICHITA, near ARKANSAS RIVER; 37°34′N 97°20′W. Agriculture to S (wheat, sorghum; cattle). Manufacturing (plastics products).

Haysville, village, DUBOIS county, SW INDIANA, 7 mi/11.3 km N of JASPER, near East Fork WHITE RIVER. Cattle; soybeans.

Haysville (HAIZ-vil), borough (2006 population 74), ALLEGHENY county, SW PENNSYLVANIA, residential suburb 10 mi/16 km NW of downtown PITTSBURGH, and 2 mi/3.2 km E of SEWICKLEY, on OHIO RIVER, opposite CORAOPOLIS.

Haysyn (HEI-sin), (Russian *Gaysin*), city, E VINNYTSYA oblast, UKRAINE, 50 mi/80 km SE of VINNYTSYA; 48°48′N 29°24′E. Elevation 672 ft/204 m. Sugar refining, distilling, fruit canning, flour milling; butter, cheese, meat. Manufacturing (machines, bricks), metal- and woodworking, sewing. Medical school. Established in 1600; city since 1795. A sizable Jewish community since the 18th century, reaching its peak in 1939 (5,190), exterminated by the Nazis in 1941.

Hayter (HAI-tuhr), hamlet, SE ALBERTA, W CANADA, 6 mi/10 km E of PROVOST, in Provost No. 52 municipal district; 52°22′N 110°07′W.

Hayti (HAI-TEI), city (2000 population 3,207), PEMISCOT county, in bootheel of extreme SE MISSOURI, near MISSISSIPPI RIVER, 7 mi/11.3 km NW of CARUTHERSVILLE; 36°13′N 89°45′W. Railroad junction. Cotton, rice, soybeans; manufacturing (fabricated metal products; aircraft rebuilding).

Hayti (hai-TEI), town (2006 population 351), ⊙ HAMLIN county, E SOUTH DAKOTA, 18 mi/29 km SSW of WATERTOWN; 44°39′N 97°12′W. Dairy products; livestock, poultry; grain, potatoes.

Hayti Heights (HAI-TEI), town (2000 population 771), PEMISCOT county, in the bootheel of extreme SE MISSOURI, 1 mi/1.6 km E of HAYTI; 36°13′N 89°46′W. Residential.

Hayton (HAI-tun), village (2001 population 2,031), CUMBRIA, NW ENGLAND, 7 mi/11.3 km ENE of CARLISLE; 54°55′N 02°46′W. Agricultural village. Has remains of Hayton Castle.

Hayvoron (HEI-vo-ron) (Russian *Gayvoron*), city, W KIROVOHRAD oblast, UKRAINE, on the Southern BUH RIVER, and 30 mi/48 km SSW of UMAN′; 48°21′N 29°51′E. Elevation 498 ft/151 m. Raion center. Railroad terminus and narrow-gauge railroad junction; locomotive repair. Granite quarry, flour mill, dairy (butter, cheese); hydroelectric power station. Machine-building and professional technical schools. Established in the first half of the 18th century; city since 1949.

Hayward (HAI-wuhrd), city (□ 44 sq mi/114.4 sq km; 2000 population 140,030), ALAMEDA county, W CALIFORNIA; suburb 13 mi/21 km SE of downtown OAKLAND, and E of SAN FRANCISCO BAY; 37°38′N 122°06′W. Bounded by SAN LORENZO CREEK to N. Important commercial and distributing center for farm products; manufacturing (wire, plastics, and screw-machine products; fabricated metal products, paper products, machinery, motor vehicles). The city has profited from development in the San Francisco

Bay area. In the 1970s–1980s, Hayward was the site of active middle-income housing growth that spurred a population increase. Hayward Executive Airport just SW. Seat of Chabot College (two-year) and California State University, East Bay. Walpert ridge to E. City is E terminus of the San Mateo–Hayward Bridge (toll), to SW, across San Francisco Bay. Olympic gold-medalist figure skater Kristi Yamaguchi born here. Settled 1851, incorporated 1876.

Hayward, town, ⊙ SAWYER county, N WISCONSIN, on NAMEKAGON RIVER, St. Croix National Scenic Riverway, and 55 mi/89 km SE of SUPERIOR. In wooded lake region. Dairy plants, boatyards; manufacturing (coffee, furniture, wood products, consumer goods). Railroad terminus. Lac Courte Oreilles Indian Reservation to SE; National Freshwater Fishing Hall of Fame. Settled c.1881, incorporated 1915.

Hayward, village, FREEBORN county, S MINNESOTA, near IOWA state line, 6 mi/9.7 km E of ALBERT LEA; 46°00′N 91°28′W. Dairying; manufacturing (steel fabrication; machine tools). DODGE CENTER Airport to SE. Myre Big Island State Park to W, on ALBERT LEA LAKE.

Haywards Heath (HAI-wuhdz HEETH), town (2001 population 27,052), West SUSSEX, S ENGLAND, 13 mi/21 km N of BRIGHTON; 50°59′N 00°06′W. Largely residential but with agricultural market, flour mills.

Haywood (HAI-wud), county (□ 554 sq mi/1,440.4 sq km; 2006 population 56,447), W NORTH CAROLINA; ⊙ WAYNESVILLE; 35°32′N 82°58′W. Partly (SE) in the BLUE RIDGE MOUNTAINS; bounded NW by TENNESSEE state line, drained by PIGEON RIVER. Largely forested; agricultural area (tobacco, hay, corn; dairying; cattle); timber. Resort area. Service industries; some manufacturing at CANTON and Waynesville. BLUE RIDGE (National) Parkway follows SW and SE county borders. Parts of Pisgah National Forest in N and SE; WATERVILLE LAKE (Pigeon River) reservoir in N; part of GREAT SMOKY MOUNTAINS National Park in NW. Formed 1808 from Buncombe County. Named after Johy Haywood (1755–1827), state treasurer at the time county was formed. When it was first created Haywood county encompassed the present day counties of Macon, Jackson, Clay, Cherokee, Graham, and Swain.

Haywood, county (□ 519 sq mi/1,349.4 sq km; 2006 population 19,405), W TENNESSEE; ⊙ BROWNSVILLE; 35°35′N 89°17′W. Drained by HATCHIE River and South Fork of FORKED DEER River. Was leading cotton producer; manufacturing now mainstay of economy. Formed 1823.

Haywood (HAI-wud), unincorporated village, CHATHAM county, central North Carolina, 27 mi/43 km SW of RALEIGH, on HAW RIVER, near its confluence with DEEP RIVER to form CAPE FEAR RIVER; 35°37′N 79°03′W. Railroad junction. B. EVERETT JORDAN LAKE (Jordan Lake) (N) and Harris Lake (E) reservoirs nearby. Earlier named Haywoodsborough, it was changed to Haywood in 1800. Named for John Haywood, state treasurer.

Haywood (HAI-wud), hamlet (2006 population 150), S MANITOBA, W central CANADA, between St. Claude and Elm Creek, 49 mi/79 km WSW of WINNIPEG, and in Grey rural municipality; 49°40′N 98°11′W. Agriculture (livestock).

Hayydra (hah-YEE-drah), village, AL KAF province, W TUNISIA, in Tibassa Mountains, 50 mi/80 km NW of AL KAF; 35°34′N 08°28′E. Railroad junction and customhouse near ALGERIAN border. Lumber, esparto grass. Ruins of Roman city (Amaedara; 1st century C.E.) include a triumphal arch, mausoleums, and a Byzantine fort.

Hazak, TURKEY: see IDIL.

Hazara (huh-ZAHR-ah), division (□ 3,000 sq mi/7,800 sq km), NORTH-WEST FRONTIER PROVINCE, N PAKISTAN, at W end of Punjab Himalayas; ⊙ ABBOTTABAD. Includes ABBOTTABAD and MANSEHRA districts.

Bounded by INDUS (SW) and JHELUM (SE) rivers. Mostly mountainous (forests have pine, cedar, fir, spruce); agriculture in valleys (maize, wheat, fruit, vegetables, rice, rapeseeds, tobacco; poultry). Pakistan Military Academy at KAKUL; medical college in ABBOTTABAD. Asokan rock edict near MANSAHRA. Local mother tongue is Hindki.

Hazarajat (ha-zah-RAH-jaht), region, mountainous area of central AFGHANISTAN; 33°45′N 66°00′E. Inhabited by the Hazaras, a Shiite Persian-speaking people of Mongolian descent; main town is PANJAO. A little-developed region of high ranges and deep, narrow gorges, the Hazarajat is crossed by central Kabul-Bamian-Herat road; not passable in winter. Long independent, the Hazaras fell under Afghan rule in 1890s.

Hazara Toghai, AFGHANISTAN: see PATA KESAR.

Hazard, city (2000 population 4,806), ⊙ PERRY county, SE KENTUCKY, 90 mi/145 km SE of LEXINGTON, in CUMBERLAND foothills, on North Fork KENTUCKY RIVER; 37°15′N 83°12′W. Elevation 867 ft/264 m. Trade, shipping, and industrial center for bituminous-coal-mining area; agriculture (tobacco); livestock; timber. Manufacturing (machining, coal processing; machinery, lumber, building materials, apparel). Wendel H. Ford Airport to N, Davis Park and Museum Part of Daniel Boone National Forest to SW.

Hazard, village (2006 population 61), SHERMAN county, central NEBRASKA, 14 mi/23 km SSW of LOUP CITY, and on MUD CREEK; 41°05′N 99°04′W.

Hazardville, CONNECTICUT: see ENFIELD.

Hazaribag (huh-ZAH-ree-bahg) or **Hazaribagh**, district (□ 1,949 sq mi/5,067.4 sq km; 2001 population 2,277,108), central BIHAR state, E INDIA, in CHOTA NAGPUR division; ⊙ HAZARIBAG. Foothills of CHOTA NAGPUR PLATEAU in SW; drained by DAMODAR RIVER and tributaries of the GANGA RIVER. Major mica-mining area N (center at Kodarma); coal-mining area in SE (centers at GIRIDIH and BERMO). Agriculture (rice, rape and mustard, oilseeds, corn, sugarcane, barley, cotton); sal and mahua in forested areas. The national park in Hazaribag is famous for its wildlife reserve. Health resort; name means "a thousand gardens." Was part of BIHAR state until 2000, when it joined other districts to form Jharkhand.

Hazaribag (huh-ZAH-ree-bahg), town (2001 population 127,243), ⊙ HAZARIBAG district, Jharkhand state, E central INDIA; 23°59′N 85°21′E. Located on the CHOTA NAGPUR PLATEAU, elevation 2,000 ft/610 m, the town is market for rice, maize, and mustard.

Hazaribagh, INDIA: see HAZARIBAG district and HAZARIBAG town.

Hazar, Lake (Turkish=*Hazargölü*) (□ 27 sq mi/70 sq km), E central TURKEY, 15 mi/24 km SE of ELAZIG; 13 mi/21 km long, 3 mi/4.8 km wide; elevation 3,790 ft/1,155 m. TIGRIS RIVER rises here. Formerly Lake Golcuk.

Hazar Masjid Range (hah-ZAHR muhs-JEED), one of the Turkmen-Khurasan ranges, in NE Iran, a SE continuation of the Kopet Mountains; rises to 10,000 ft/3,048 m N of Mashhad; 36°52′N 59°26′E. Also Hezar Masjed Range.

Hazebrouck (ahz-BROOK), town (□ 10 sq mi/26 sq km), NORD department, PAS-DE-CALAIS region, in FLANDERS, N FRANCE, 24 mi/39 km WNW of LILLE and SSE of DUNKERQUE; 50°43′N 02°32′E. Railroad junction and agricultural trade center in dairying, flax- and potato-growing district. Textile industry; food processing, appliance manufacturing.

Hazega (hah-ZAI-guh), Italian *Azzega*, village, central ERITREA, on road, and 9 mi/14.5 km NW of ASMARA; 15°24′N 38°50′E. Grain, fruit, vegetables; livestock; tree nursery. Gold mining nearby.

Hazel, village (2000 population 440), CALLOWAY county, SW KENTUCKY, at TENNESSEE state line, 7 mi/11.3 km S of MURRAY, near East Fork CLARKS RIVER;

36°30′N 88°19′W. Agriculture (tobacco, grain; livestock; dairying).

Hazel, village (2006 population 101), HAMLIN county, E SOUTH DAKOTA, 17 mi/27 km SW of WATERTOWN; 44°45′N 97°22′W.

Hazelbrook (HAI-zuhl-bruk), town, NEW SOUTH WALES, SE AUSTRALIA, 55 mi/89 km from SYDNEY, in BLUE MOUNTAINS; 33°44′S 150°27′E. Walking tracks, waterfalls. Gloria Park to W.

Hazel Crest, village (2000 population 14,816), COOK county, NE ILLINOIS, S suburb of CHICAGO; 41°34′N 87°41′W. Incorporated 1911.

Hazel Green, unincorporated town, WOLFE county, E central KENTUCKY, on RED RIVER, and 45 mi/72 km ESE of WINCHESTER, in the CUMBERLAND MOUNTAINS. Manufacturing (textiles).

Hazel Green, town (2006 population 1,139), GRANT county, extreme SW WISCONSIN, near ILLINOIS state line, 12 mi/19 km E of DUBUQUE (IOWA); 42°31′N 90°26′W. In livestock area. Formerly an important lead-mining center.

Hazel Green, village (2000 population 3,805), Madison co., N Alabama, near Flint River, and Tennessee state line, 15 mi/24 km N of Huntsville; 34°55′N 86°34′W. Name is probably descriptive of the hazelnut trees in the area.

Hazel Grove and Bramhall (HAI-zel GROV and BRAM-uhl), town (2001 population 45,651), GREATER MANCHESTER, W central ENGLAND, 3 mi/4.8 km SE of STOCKPORT; 53°23′N 02°07′W. Has 15th-century mansion, Bramhall Hall.

Hazelhurst, village, ONEIDA county, N WISCONSIN, 19 mi/31 km NW of RHINELANDER, in wooded lake region. Manufacturing (electronic equipment, concrete products). American Legion State Forest to E; on Bearskin State Trail. Formerly a lumbering town.

Hazel Park, city (2000 population 18,963), OAKLAND county, SE MICHIGAN, a suburb 9 mi/14.5 km NNW of downtown DETROIT; 42°27′N 83°05′W. Has varied light manufacturing (machinery, metal products; metal plating) and Hazel Park Racetrack. Most of the early settlers were German. Ottawa chief Pontiac made his headquarters here. Borders Detroit city and WAYNE county in S, MACOMB county on E. Incorporated 1942.

Hazel Run, village (2000 population 64), YELLOW MEDICINE county, SW MINNESOTA, 9 mi/14.5 km WSW of GRANITE FALLS, near YELLOW MEDICINE RIVER; 44°45′N 95°43′W. Grain; livestock.

Hazelton (HAI-zuhl-tuhn), village (□ 1 sq mi/2.6 sq km; 2001 population 345), W central BRITISH COLUMBIA, W CANADA, on SKEENA RIVER, at mouth of BULKLEY RIVER, and 130 mi/209 km NE of PRINCE RUPERT, in Kitimat-Stikine regional district; 55°15′N 127°40′W. Silver, lead, zinc, uranium mining; logging; cattle. Established 1868 as Hudson's Bay Company trading post. Historic Indian Village at 'Ksan, 4.5 mi/7.2 km N. New Hazelton (2001 population 750) is 4 mi/6 km E. Riverboats once plied the Skeena River to the coast.

Hazelton, village (2000 population 687), JEROME county, S IDAHO, 15 mi/24 km WNW of BURLEY; 42°35′N 114°08′W. Elevation 4,068 ft/1,240 m. Beans, potatoes, sugar beets; barley, corn; fertilizers. WILSON LAKE reservoir to N; Miller Dam (SNAKE RIVER) to E.

Hazelton, village (2000 population 144), BARBER county, S KANSAS, 7 mi/11.3 km NE of KIOWA; 37°05′N 98°24′W. In cattle area.

Hazelton, village (2006 population 203), EMMONS co., S NORTH DAKOTA, 33 mi/53 km SE of BISMARCK; 46°28′N 100°16′W. Founded in 1903 and incorporated in 1916. Named for Hazel Roop, daughter of local landowner.

Hazelton, INDIANA: see HAZLETON.

Hazelwood, city (2000 population 26,206), SAINT LOUIS county, E MISSOURI, a residential, commercial, and industrial suburb 18 mi/29 km NW of downtown ST. LOUIS; 38°46′N 90°21′W. On N side of Lambert–St.

Louis Airport. Has a diverse manufacturing base (aircraft, motor vehicles, microbiology instrumentation, plastic products, utility receptacles, power tools, paper goods, foods and beverages, medical equipment, chemicals; publishing). Incorporated as a village 1949, city charter approved 1969.

Hazelwood (HAIZ-uhl-wud), town, HAYWOOD county, W North Carolina, 1 mi/1.6 km SW of WAYNESVILLE; 35°28′N 83°00′W. Railroad terminus. In mountain-resort area. Manufacturing (machining; footwear, fertilizers); agriculture (cattle; dairying). Part of Pisgah National Forest to SW and SE. Incorporated into Waynesville in 1995. Named for hazelnuts growing in the area.

Hazen (HAIZ-uhn), town (2000 population 1,637), PRAIRIE county, E central ARKANSAS, 19 mi/31 km N of STUTTGART; 34°47′N 91°34′W. Manufacturing (powder metal parts). Wattensaw Wildlife Management Area to NE.

Hazen, town (2006 population 2,319), MERCER CO., central NORTH DAKOTA, 52 mi/84 km NW of BISMARCK, and on KNIFE RIVER; 47°17′N 101°37′W. Lignite mines; wheat. Sakakawea State Park and Reservoir to N. Founded in 1885.

Hazen, Lake, N ELLESMERE ISLAND, BAFFIN region, NUNAVUT territory, N CANADA, at foot of UNITED STATES RANGE, WNW of LADY FRANKLIN BAY; 55 mi/89 km long, 3 mi/5 km–12 mi/19 km wide; 81°50′N 70°00′W.

Hazen Strait, NORTHWEST TERRITORIES, N CANADA, arm of the ARCTIC OCEAN, between South BORDEN and MELVILLE islands; 60 mi/97 km long, 50 mi/80 km wide; 77°15′N 110°00′W. Connects PRINCE GUSTAF ADOLF SEA and BYAM MARTIN CHANNEL.

Hazerswoude-Dorp (HAH-zuhrs-VOU-duh–DORP), town, SOUTH HOLLAND province, W NETHERLANDS, 6 mi/9.7 km SE of LEIDEN; 52°06′N 04°36′E. OLD RHINE RIVER passes 2 mi/3.2 km to N. Dairying; cattle, poultry; flowers, vegetables, fruit.

Hazlehurst (HAI-zuhl-huhrst), town (2000 population 3,787), ⊙ JEFF DAVIS county, SE central GEORGIA, c.45 mi/72 km NNW of WAYCROSS, near ALTAMAHA RIVER; 31°52′N 82°36′W. Tobacco market; manufacturing of clothes, paper goods, machinery, textiles, lumber, consumer goods. Settled late 1850s, incorporated 1891.

Hazlehurst (HAI-zuhl-huhrst), town (2000 population 4,400), ⊙ COPIAH county, SW MISSISSIPPI, 38 mi/61 km SSW of JACKSON; 31°51′N 90°23′W. In agricultural (cotton, corn, soybeans; poultry, cattle; dairying) and timber area; ships tomatoes, fruit; manufacturing (fabricated metal products, lumber, plastics; poultry processing). Homochitto National Forest to SW. Founded 1857, incorporated 1865.

Hazlemere (HAI-zuhl-mir), suburb (2001 population 18,700), BUCKINGHAMSHIRE, central ENGLAND, 2 mi/3.2 km NE of HIGH WYCOMBE; 51°39′N 00°42′W.

Hazlet, township, MONMOUTH county, NE NEW JERSEY, 10 mi/16 km E of PERTH AMBOY; 40°25′N 74°10′W. Major industry is flavors and fragrance company. Incorporated 1848.

Hazleton (HAI-zel-tuhn), city (2006 population 22,037), LUZERNE county, E PENNSYLVANIA, 20 mi/32 km SSW of WILKES-BARRE, near Black Creek; 40°57′N 75°58′W. Railroad junction. Once a major anthracite-coal-producing region, it now has a diverse economy. Manufacturing (apparel, food products, asphalt, fabricated metal products, fiberglass windows, plastic products; printing and publishing). Agricultural area (grain, potatoes; livestock; dairying). Settled c.1809. Its name derives from the hazel bushes that grew in the swamp called Haselschwamm by the early German settlers. The settlement increased in size and development after coal was discovered nearby in 1826. Coal production reached its peak during the first half of the 20th century, but declined afterward. Pennsylvania State University (Hazleton campus, two-year). State

hospital, with a school of nursing. Hazleton Municipal Airport to N; Eagle Rock Ski Area to E; Eckley Miners' Village, 19th-century mining town, to E. Incorporated as a borough 1856, as a city 1892.

Hazleton, town (2000 population 288), GIBSON county, SW INDIANA, on WHITE RIVER, and 9 mi/14.5 km N of PRINCETON; 38°29′N 87°32′W. Oil wells. Also spelled Hazelton.

Hazleton, town (2000 population 950), BUCHANAN county, E IOWA, 10 mi/16 km N of INDEPENDENCE; 42°37′N 91°54′W. Feed milling. Limestone quarries nearby.

Hazo, TURKEY: see KOZLUK.

Hazor, town, the modern Khirbat Hazzar, JORDAN, 4 mi/6.4 km NW of JERUSALEM.

Hazorea (hah-zo-REE-ah), kibbutz, 11 mi/18 km SW of NAZARETH in SW JEZREEL VALLEY, N ISRAEL; 32°38′N 35°07′E. Elevation 725 ft/220 m. Some farming; manufacturing includes furniture, plastics, quality control systems. Ruins from Chalcolithic to middle Bronze Age and from Hellenistic to Roman times. Founded in 1934, it suffered heavy damage during the 1936–1939 Arab revolt; later, prior to 1948, it was a base for the Palmach.

Hazor Haglilit (khah-TSOR hah-glee-LEET), township, N GALILEE, 3 mi/4.8 km NE of ZEFAT, and 1.9 mi/3.1 km N of ROSH PINA, ISRAEL. Founded 1958 near Tel Hazor River (from where it gets its name). Strategically located on the road leading from ancient EGYPT to SYRIA and ASIA MINOR, it was occupied from the early Bronze Age to Hellenistic times. Hazor-Rosh Pina Industrial Center nearby. Joshua destroyed it because it was the center of the league of Canaanite kingdoms. Solomon rebuilt it as one of his strongholds in the N; later destroyed by Tiglath Pileser III of Assyria. Excavations have revealed both the Canaanite and Israelite cities and have confirmed the biblical data.

Hazor'im (hah-zor-EEM), religious Zionist moshav, NE ISRAEL, Lower Galilee, 2 mi/3.2 km SW of TIBERIAS; 32°45′N 35°30′E. Elevation 229 ft/69 m below sea level. Mixed agriculture. Founded 1939.

Hazrat Imam (hahz-RAHT ee-MAHM), village, KUNDUZ province, NE AFGHANISTAN, on PANJ RIVER (TAJIKISTAN border), 30 mi/48 km N of KUNDUZ; 37°11′N 68°55′E.

Hazro (hah-ZAH-ro), town, ATTOCK district, NW PUNJAB province, central PAKISTAN, 11 mi/18 km NNE of CAMPBELLPUR; 33°54′N 72°29′E.

Hazu (HAHZ), town, Hazu district, AICHI prefecture, S central HONSHU, JAPAN, on ATSUMI BAY, 28 mi/45 km S of NAGOYA; 34°47′N 137°07′E. Mikawa Bay quasi-national park nearby.

Hazua (hah-ZWAH), town, TAWZAR province, SW TUNISIA, 68 mi/109 km SW of TAWZAR. Border post to ALGERIA.

Heacham (HEE-chuhm), village (2001 population 4,611), NW NORFOLK, E ENGLAND, near The WASH (a prominent estuary), 2 mi/3.2 km S of NEW HUNSTANTON; 52°55′N 00°29′E. Agricultural market. The 14th–15th-century church contains an alabaster statue of Pocahontas, who, along with her husband John Rolfe, are buried here.

Head, Clara and Maria (HED, KLA-ruh, muh-REE-uh), township (□ 281 sq mi/730.6 sq km; 2001 population 228), SE ONTARIO, E central CANADA, and 51 mi/81 km from PEMBROKE; bordered by the OTTAWA River on the N and ALGONQUIN PROVINCIAL PARK on the S; 46°10′N 78°02′W. Also written Head, Clara & Maria.

Headford, Gaelic *Lios na gCeann*, town (2006 population 760), W central GALWAY county, W IRELAND, near NE shore of LOUGH CORRIB, 14 mi/23 km NNW of GALWAY; 53°28′N 09°06′W. Agricultural market. Area rich in ancient remains, with ruins of Ross Abbey nearby.

Headingley (HE-deeng-lee), rural municipality (□ 41 sq mi/106.6 sq km; 2001 population 1,907), SE MANI-

TOBA, W central CANADA, just W of WINNIPEG; 49°52′N 97°23′W. Includes Headingley, South HEADINGLEY. Formerly part of Winnipeg; incorporated 1993.

Headingley (HE-deeng-lee), community, S MANITOBA, W central CANADA, 12 mi/19 km from WINNIPEG, in HEADINGLEY rural municipality; 49°52′N 97°24′W.

Headington, ENGLAND: see OXFORD.

Headland, town, Henry co., SE Alabama, 17 mi/27 km SSW of Abbeville. Peanut shelling; peanut and cottonseed products, fertilizer, lumber. State agr. experiment station here. Named for James J. Head, founder of the settlement. Inc. in 1893.

Headlands, town, MANICALAND province, E ZIMBABWE, 18 mi/29 km NNW of RUSAPE, on railroad; 18°17′S 32°03′E. Elevation 5,147 ft/1,569 m. Cattle, sheep, goats, hogs; tobacco, corn, peanuts. Copper and gold mining to NE.

Headley (HED-lee), village (2001 population 5,459), E HAMPSHIRE, S ENGLAND, 9 mi/14.5 km S of ALDERSHOT; 51°07′N 00°49′W. Former agricultural market.

Head of Jeddore (je-DOR), village, S NOVA SCOTIA, E CANADA, at head of JEDDORE HARBOUR, 30 mi/48 km ENE of HALIFAX. Fishing port; lumbering.

Head of Passes, section of lower MISSISSIPPI RIVER, 18 mi/29 km SE of TRIUMPH, LOUISIANA; 29°10′N 89°15′W. Area where the Mississippi River branches into small channels, including North, South, and Southwest passes. Because of sedimentation, navigation was difficult. The first naval battle on the Mississippi River during the Civil War occurred here.

Head of the Harbor, upper-income residential village (□ 3 sq mi/7.8 sq km; 2006 population 1,494), SUFFOLK county, SE NEW YORK, on Stony Brook Harbor on N shore of LONG ISLAND, 14 mi/23 km E of HUNTINGTON; 40°53′N 73°09′W. In summer-resort area.

Headquarters, unincorporated village, CLEARWATER county, N IDAHO, 22 mi/35 km ENE of OROFINO; elevation 3,136 ft/956 m; 46°37′N 115°48′W. Railroad terminus. Timber; cattle. Clearwater National Forest to E; Bald Mountain Ski Area to SW; Dworshak Reservoir to W.

Headrick, village (2006 population 122), JACKSON county, SW OKLAHOMA, 11 mi/18 km E of ALTUS, near NORTH FORK OF RED RIVER; 34°37′N 99°08′W. In cotton area.

Head-Smashed-In Buffalo Jump (HED–SMASHT–IN BUH-fuh-lo JUHMP), historical Native American hunting ground and UNESCO World Heritage Site, SW ALBERTA, CANADA, 11 mi/24 km W of Fort MCLEOD; 49°43′N 113°39′W. 36-ft/11-m cliff on N side of OLDMAN River valley; one of 150 sites in valley where buffalo (bison) were stampeded in large herds and plunged to their deaths by Plains Indians. Spear points date to 7000 B.C.; horses were in use by 1830s; final drive mid-1800s. Museum built into cliff face.

Heads Nook, ENGLAND: see WETHERAL.

Head Tide, MAINE: see ALNA.

Heage (HEEJ), village (2001 population 4,743), central DERBYSHIRE, central ENGLAND, 2 mi/3.2 km NE of BELPER; 53°03′N 01°15′W.

Healdsburg, city (2000 population 10,722), SONOMA county, W CALIFORNIA, 13 mi/21 km NNW of SANTA ROSA, on RUSSIAN RIVER; 38°37′N 122°52′W. Apples, grapes, grain, vegetables, nursery products; dairying; poultry. Manufacturing (dehydrated fruits and vegetables; wineries, electronic components, machinery, motor vehicles).

Healdton (HEELD-tuhn), city (2006 population 2,773), CARTER county, S OKLAHOMA, 21 mi/34 km WNW of ARDMORE; 34°13′N 97°29′W. In oil and natural-gas, agriculture, dairying, and livestock-raising area; manufacturing (lingerie; printing); oil wells. Oil museum.

Healesville (HEELZ-vil), town, S central VICTORIA, SE AUSTRALIA, on YARRA RIVER, and 32 mi/51 km ENE of MELBOURNE, in GREAT DIVIDING RANGE; 37°39′S 145°31′E. Railroad terminus; mountain resort; famous wildlife sanctuary where the first platypus bred in captivity was born, 1943.

Healing Spring (HEE-leeng SPREENG), unincorporated village, BATH county, NW VIRGINIA, in ALLEGHENY MOUNTAINS, 14 mi/23 km NNE of COVINGTON, in George Washington National Forest. Agriculture (cattle); timber. Mineral springs.

Healy (HEE-lee), village, S central ALASKA, near MOUNT MCKINLEY NATIONAL PARK, on Alaska railroad, 50 mi/80 km S of NENANA; 63°58′N 144°43′W. Dall sheep. Major coal mine; thermal electric power plant; coal is exported to KOREA and is main source of electricity in FAIRBANKS NORTH STAR borough.

Heanor (HEE-nuh), town (2001 population 16,040), SE DERBYSHIRE, central ENGLAND, 8 mi/12.9 km NE of DERBY; 53°00′N 01°20′W. Former site of coal mines. Light manufacturing. Has technical college and church with 15th-century tower.

Heany Junction, town, MATABELELAND NORTH province, SW ZIMBABWE, 15 mi/24 km ENE of BULAWAYO; 20°05′S 28°47′E. Elevation 4,462 ft/1,360 m. Junction of railroad spur to WEST NICHOLSON. Inxele Hills (4,652 ft/1,418 m) to N. Gold mining in area. Cattle, sheep, goats; dairying; tobacco, corn, cotton, soybeans.

Heard (HUHRD), county (□ 301 sq mi/782.6 sq km; 2006 population 11,472), W GEORGIA, on ALABAMA state line; ⊙ FRANKLIN; 33°18′N 85°08′W. Piedmont area intersected by CHATTAHOOCHEE RIVER Agriculture (hay, vegetables, fruit; cattle, poultry). Gravel; apparel and textiles. Formed 1830.

Heard and McDonald Islands (HIMI) (HUHRD AND muhk-DAHN-uhld), Australian territory, remote subantarctic islands, 2,486 mi/4,000 km SW of AUSTRALIA and 932 mi/1,500 km N of ANTARCTICA, in S INDIAN (Southern) Ocean; 53°00′S 73°00′E. HEARD and MCDONALD islands, dominated by Big Ben, an active volcano (Mawson Peak summit, 9,006 ft/2,745 m) on principal Heard Island, are set against glacial ice, snow, and black volcanic rocks surrounded by stormy waters. Penguins and seals occupy the beaches; a range of seabirds visit the islands. Sighted 1853 by American navigators. As the world's only volcanically active sub-Antarctic islands, and with an intact ecosystem undisturbed by humans, the islands were declared a UNESCO World Heritage Area in 1997.

Heard Island (HUHRD), subantarctic islet, part of the Australian Territory of Heard Island and McDonald Islands, in S INDIAN (Southern) Ocean, c.300 mi/483 km SE of KERGUELEN Islands; c.25 mi/40 km long, 10 mi/16 km wide; 53°06′S 72°31′E. Volcanic island, rising in Big Pen Peak to c.11,000 ft/3,353 m, is largely covered by snow and glaciers. Meteorological station. Discovered (1853) by U.S. Captain John J. Heard. Formally annexed by Australia in December 1947. Just N is SHAG ISLAND. Together with the MCDONALD ISLANDS (to the W), the only other volcanically active sub-Antarctic islands in the world, and with an intact ecosystem undisturbed by humans, it was declared a UNESCO World Heritage Area in 1997.

Hearne (HUHRN), town (2006 population 4,743), ROBERTSON county, E central TEXAS, in the BRAZOS valley, c.50 mi/80 km SSE of WACO; 30°52′N 96°35′W. Railroad junction in rich agricultural area (cotton, sorghum, grains, watermelons; cattle, hogs, poultry). Manufacturing (vitreous china, fabricated metal products, wood products, machinery). Settled 1868, incorporated 1871.

Hearst (HUHRST), town (□ 38 sq mi/98.8 sq km; 2001 population 5,825), N central ONTARIO, E central CANADA, 60 mi/97 km WNW of KAPUSKASING; 49°42′N 83°40′W. Farming; lumbering.

Hearst Island, off ANTARCTICA, in WEDDELL SEA off E coast of ANTARCTIC PENINSULA; 36 mi/58 km long, 7 mi/11.3 km wide; 69°25′S 62°10′W. Elevation 1,200 ft/366 m. Discovered 1928 by Sir Hubert Wilkins, who thought it was edge of Antarctic continent and named

it Hearst Land. Rediscovered 1940 by U.S. expedition, which proved it was an island. Also known as Wilkins Island.

Heart, river, 180 mi/290 km long, NORTH DAKOTA; rises in the low prairie country near the LITTLE MISSOURI RIVER, BILLINGS co., SW North Dakota, near the S unit of the Theodore Roosevelt National Park at 46°56′N 103°13′W; flows E through Patterson Reservoir, past DICKINSON, through Lake Tschida and Heart Butte Dam, to the MISSOURI River at MANDAN. The Heart Butte and Dickinson dams, irrigation and flood control units built by the U.S. Bureau of Reclamation as part of the Missouri River Basin Project, have created the region's largest lakes, which are major recreation areas. Main tributaries are Antelope and Big Muddy creeks and Green River

Heart Butte (BYOOT), village (2000 population 698), PONDERA county, N MONTANA, on the S fork of Whitetail Creek, 33 mi/53 km SW of CUT BANK, in S part of Blackfeet Indian Reservation; 48°17′N 112°50′W. Hogs; wheat, barley, oats, alfalfa. Lewis and Clark National Forest to SW.

Heart Butte Dam, North Dakota: TSCHIDA, LAKE.

Heart Island, small island, of the THOUSAND ISLANDS, JEFFERSON county, N NEW YORK, in the SAINT LAWRENCE River, just NW of ALEXANDRIA BAY; 2 mi/3.2 km long; 44°21′N 75°55′W. Boldt Castle here attracts tourists.

Heart's Content, coastal town (□ 24 sq mi/62.4 sq km; 2001 population 495), SE NEWFOUNDLAND AND LABRADOR, E CANADA, on NW coast of AVALON Peninsula, on inlet of TRINITY Bay, 37 mi/60 km NW of St. JOHN'S; 47°53′N 53°22′W. Fishing. Hydroelectric plant. Lumbering region. First trans-ATLANTIC cable landed here in 1866.

Heartwell, village (2006 population 80), KEARNEY county, S NEBRASKA, 10 mi/16 km NE of MINDEN, near PLATTE River; 40°34′N 98°47′W.

Heath (HEETH), city (□ 10 sq mi/26 sq km; 2006 population 8,892), LICKING county, central OHIO, near LICKING RIVER, 3 mi/5 km S of NEWARK; 40°01′N 82°26′W. Former Newark Air Force Base nearby that specialized in guidance of navigation systems; closed base scheduled for privatization.

Heath, town, FRANKLIN county, NW MASSACHUSETTS, 12 mi/19 km WNW of GREENFIELD; 42°42′N 72°50′W. Agricultural area.

Heath (HEETH), town (2006 population 6,853), ROCKWALL county, N TEXAS, residential suburb 18 mi/29 km ENE of downtown DALLAS, on E shore of LAKE RAY HUBBARD (E Fork TRINITY RIVER); 32°51′N 96°28′W. Agricultural area (cattle, horses; wheat). Recreation.

Heath (HEETH), village (2001 population 3,715), NE DERBYSHIRE, central ENGLAND, 5 mi/8 km ESE of CHESTERFIELD; 53°11′N 01°20′W.

Heath (HEETH), village, FERGUS county, central MONTANA, on the E fork of Big Spring Creek, 8 mi/12.9 km ESE of LEWISTOWN. Gypsum deposits and processing plants. Big Spring Trout Hatchery to W. Formerly called Gypsum.

Heathcote (HEETH-kot), town, central VICTORIA, AUSTRALIA, 65 mi/105 km NNW of MELBOURNE, at foot of Mount Ida; 36°54′S 144°42′E. In sheep-raising, old gold-mining area; also timber, honey, wine.

Heathcote, NEW ZEALAND: see CHRISTCHURCH.

Heatherdale (HE-thur-dail), residential locality, suburb of MELBOURNE, VICTORIA, SE AUSTRALIA.

Heatherton (HE-thur-tuhn), agricultural district, residential suburb 12 mi/20 km SE of MELBOURNE, VICTORIA, SE AUSTRALIA, between MOORABBIN and Dingley Village; 37°58′S 145°06′E. Originally called Kingston.

Heathfield (HEETH-feeld), town (2001 population 11,406), East SUSSEX, SE ENGLAND, 14 mi/23 km N of EASTBOURNE; 50°58′N 00°16′E. Agricultural market.

Has 13th-century church. Heathfield was important cannon-manufacturing center in Middle Ages.

Heathmont (HEETH-mahnt), residential suburb 15 mi/24 km E of MELBOURNE, VICTORIA, SE AUSTRALIA; 37°50′S 145°15′E. Dandenong Ranges, Puffing Billy scenic railway nearby.

Heath River, c. 100 mi/161 km long, ITURRALDE province, LA PAZ department; rises on BOLIVIA-PERU border at 13°47′S 69°00′W; flows N along international border to MADRE DE DIOS RIVER at Puerto Pardo and Puerto HEATH; 12°31′S 68°38′W.

Heathrow (HEETH-ro), town, SEMINOLE county, central FLORIDA, 15 mi/24 km N of Orlando; 28°46′N 81°22′W. Headquarters of American Automobile Association (AAA).

Heathrow Airport (HEETH-ro) or **London Airport**, is located in HILLINGDON, Greater LONDON, ENGLAND, 15 mi/24 km W of LONDON (subway (tube), rail, and motorway links); elevation 78 ft/24 m; 51°28′N 00°27′W. Two runways, four passenger terminals (over 67 million passengers annually) and air cargo facilities (over 1,300,000 tonnes annually) London's premier airport, Heathrow opened in 1946 on the site of World War I era Royal Air Corps training facility. The airport has expanded considerably since inception. A brand new fifth terminal is scheduled to open on March 30, 2008. The airport has other expansions planned through 2015. Airport code LHR.

Heath Springs, town (2006 population 862), LANCASTER county, N SOUTH CAROLINA, 9 mi/14.5 km SE of LANCASTER; 34°35′N 80°40′W. Fish hatcheries; manufacturing of chemicals, machinery. Agriculture includes turkeys, cattle; soybeans.

Heathsville (HEETHS-vil), unincorporated village, ⊙ NORTHUMBERLAND county, E VIRGINIA, 65 mi/105 km SE of FREDERICKSBURG; 37°55′N 76°28′W. Manufacturing (printing and publishing; oyster processing); agriculture (tomatoes, grain, soybeans; poultry, cattle). Entrance to POTOMAC RIVER estuary to NE.

Heaton, ENGLAND: see NEWCASTLE UPON TYNE.

Heaton Norris, ENGLAND: see MANCHESTER.

Heavener (HEEV-nuhr), town (2006 population 3,265), LE FLORE county, SE OKLAHOMA, 10 mi/16 km S of POTEAU, just N of the Ouachita Mountains railroad junction; 34°53′N 94°36′W. In farm and recreation area (corn, potatoes); manufacturing (apparel, machinery). Heavener-Runestone State Park here. Ouachita National Forest to S and E; Lake WISTER reservoir and State Park to W. A state fish hatchery is here.

Hebardville, village, WARE county, SE GEORGIA, just NW of WAYCROSS; 31°14′N 82°22′W. Also spelled Hebardsville.

Hebbronville (HE-brawn-vil), town (2000 population 4,498), ⊙ JIM HOGG county, extreme S TEXAS, c.55 mi/89 km E of LAREDO; elevation 550 ft/168 m; 27°19′N 98°41′W. In oil- and gas-producing region; cattle ranching; sorghum; manufacturing (lumber).

Hebburn (HE-buhn), town (2006 population 27,000), TYNE AND WEAR, NE ENGLAND, on the TYNE RIVER, 4 mi/6.4 km E of GATESHEAD; 54°58′N 01°31′W. Electrical industries; machinery works.

Hebden Bridge or **Hebden Royd** (HEB-den), town (2001 population 9,092), WEST YORKSHIRE, N central ENGLAND, on CALDER RIVER and Rochdale Canal, 7 mi/11.3 km W of HALIFAX; 53°45′N 02°01′W. Textiles, machine tools.

Hebei (HUH-bai) [Chinese=north of the river] or **Ji** (JEE), province (□ 78,900 sq mi/205,140 sq km; 2000 population 66,684,419), NE CHINA, on the BOHAI (an arm of the YELLOW SEA); ⊙ SHIJIAZHUANG; 39°00′N 116°00′E. Contains two autonomous municipalities—BEIJING and TIANJIN—administered directly by the central government. QINHUANGDAO is a large port on the relatively unindented coast line. Other important cities are TANGSHAN and BAODING. Hebei is mountainous in the N and W, where rich iron and coal deposits are extensively mined. The province also has extensive deposits of oil and natural gas. S Hebei is part of the NORTH CHINA PLAIN. The land is fertile and rainfall is adequate, but until water-conservation programs were instituted, the province was subject to severe drought and flooding due to the fluctuating monsoon climate. These recent improvements, along with the enlargement of farms and the expansion of mechanization, have greatly increased agricultural output. Hebei is a major cotton-producing province and an important producer of wheat. Other crops include rice, millet, kaoliang, potatoes, sweet potatoes, barley, corn, soybeans, and fruit. Livestock raising is important, and fishing and salt production are significant along the coast. Heavy industry (mainly metallurgical; iron and steel, machinery, and textiles) is concentrated in and around Beijing, Tianjin, and Tangshan. Light manufacturing includes ceramics, paper, and processed foods. With many good roads and railroad systems centering on Beijing, Tianjin, and Shijiazhuang, and with the GRAND CANAL and other excellent waterways, Hebei has one of the best communications systems in China. One of the earliest regions of Chinese settlement, Hebei has many prehistoric sites. Parts of the former provinces of Jehol and Chahar were incorporated into Hebei in 1956. Called Jizhou in ancient times. Formerly called Zhili, Chihli, or Chili. The name sometimes appears as Hopeh or Hopei.

Heber (HEE-buhr), unincorporated town, NAVAJO county, E central ARIZONA, 40 mi/64 km from WINSLOW, in Apache-Sitgreaves National Forest, in BLACK CANYON; elevation 6,439 ft/1,963 m. Timber. Fort Apache Indian Reservation and MOGOLLON Rim escarpment to S.

Heber, unincorporated town (2000 population 2,988), IMPERIAL county, S CALIFORNIA, 4 mi/6.4 km SE of EL CENTRO, in irrigated IMPERIAL VALLEY, 6 mi/9.7 km N of Mexican border; 32°44′N 115°31′W. Vegetables, tomatoes, sugar beets, melons, dates, corn, wheat, alfalfa; cotton; cattle, sheep.

Heber City (HEE-buhr), city (2006 population 9,775), ⊙ WASATCH county, N central UTAH, 23 mi/37 km NE of PROVO, near PROVO RIVER; 40°30′N 111°24′W. In mountain region; elevation 5,595 ft/1,705 m. Trade center and cattle-shipping point; alfalfa, barley; dairying; sheep; manufacturing (cheese, dehydrated foods; publishing). Lead, silver, zinc mines in vicinity; sand and gravel. Jordanelle Reservoir and State Park to N; parts of Uinta National Forest to W, S, and E. Heber Valley railroad, scenic railroad to Provo Canyon (SW). City settled 1859 by Mormons. Hot-water pools of extinct geysers nearby.

Heber Springs (HEE-buhr), town (2000 population 6,432), ⊙ CLEBURNE county, N central ARKANSAS, c.55 mi/89 km NNE of LITTLE ROCK, in the OZARK Mountains; 35°30′N 92°02′W. Agriculture; lumbering; manufacturing (wood products, machinery, building materials, leather products, medical equipment, consumer goods, tools). Dam and fish hatchery to NE. Laid out 1881. GREERS FERRY LAKE (reservoir) to N.

Hébertville (ai-ber-VEEL), village (□ 102 sq mi/265.2 sq km), SAGUENAY—Lac-Saint-Jean region, central QUEBEC, E CANADA, 23 mi/37 km W of JONQUIÈRE; 48°24′N 71°41′W. Lumbering; dairying.

Hébertville-Station (ai-ber-VEEL-STAI-shuhn), village (□ 13 sq mi/33.8 sq km; 2006 population 1,284), SAGUENAY—Lac-Saint-Jean region, S central QUEBEC, E CANADA, 7 mi/11 km from ALMA; 48°27′N 71°40′W.

Hebgen Lake (HEB-guhn), reservoir, GALLATIN county, SW MONTANA, at S end of MADISON RANGE, in Gallatin National Forest, 19 mi/31 km NW of WEST YELLOWSTONE; c.20 mi/32 km long, maximum 6 mi/9.7 km wide; 44°52′N 111°18′W. Formed by Hebgen Dam, built shortly after August 17, 1959, earthquake, to regulate water flow. Earthquake (QUAKE) Lake, natural reservoir formed by landslide during 1959 earthquake, is below dam. IDAHO to SW, WYOMING to E.

Hebi (HUH-BEE), city (□ 198 sq mi/514.8 sq km; 2000 population 377,346), N HENAN province, NE CHINA, at foot of the TAIHANG, near the Beijing-Wuhan railroad, 15 mi/24 km SW of ANYANG; 35°57′N 114°08′E. Heavy industry. Crop growing; animal husbandry. Grains, oil crops, cotton, vegetables, fruits, eggs; hogs. Manufacturing (textiles, chemicals, machinery, electronics). Coal mining.

Hebibchevo, Bulgaria: see LYUBIMETS.

Hebrides, New: see VANUATU.

Hebrides, Sea of the (HE-bri-deez) or **Gulf of the Hebrides**, sea (c.30 mi/48 km wide), NW Scotland, arm of the ATLANTIC OCEAN between the INNER HEBRIDES and the S OUTER HEBRIDES, opening N on the Little Minch; 57°07′N 06°55′W.

Hebrides, The (HE-bri-deez) or **Western Isles**, group of over 500 islands, W and NW Scotland, in EILEAN SIAR, HIGHLAND, and ARGYLL AND BUTE. The Outer Hebrides (sometimes also referred to as the Long Island) are separated from the mainland and from the Inner Hebrides by the straits of MINCH and Little Minch and by the SEA OF THE HEBRIDES; they extend for 130 mi/209 km from the Butt of Lewis on LEWIS AND HARRIS to BERNERAY. Other islands are NORTH UIST, BENBECULA, SOUTH UIST, BARRA, the FLANNAN ISLES (Seven Hunters), and ST. KILDA. The Inner Hebrides include the islands of SKYE, RAASAY, RUM, EIGG, COLL, TIREE, STAFFA, IONA, MULL, SCARBA, COLONSAY, ORONSAY, Jura, CANNA, MUCK, GIGHA, and ISLAY. Fishing; farming; sheep grazing; quarrying (slate); and catering to tourists are the chief means of livelihood. Previously manufacturing of tweeds. The original Celtic inhabitants, converted to Christianity by St. Columba (6th century), were conquered by the Norwegians (starting in the 8th century). They held the Southern Islands, as they called them, until 1266. From that time the islands were ruled by various Scottish chiefs, until the MacDonalds established themselves (1346) as lords of the isles. The Hebrides have belonged to the crown of Scotland since the 16th century. The tales of Sir Walter Scott did much to make the islands famous. The climate is mild, the scenery is beautiful, and there are prehistoric and ancient historical remains and geological structures. Emigration from the overpopulated islands has occurred in the 20th century, especially to Canada. Also known as Hebudae, Hebudai, Hebudes, and Ebudae.

Hebri, Jbel (HE-bree, zhe-BEL), extinct volcanic cone (6,902 ft/2,104 m), MOROCCO, central Middle ATLAS mountains, 9 mi/15km S of AZROU; 33°22′N 05°09′W. Cedar forest; ski slope.

Hebron, Arabic *Al-Khalil*, city, E ISRAEL, in the WEST BANK, 22 mi/35 km S of JERUSALEM; elevation 3,000 ft/914 m; 31°32′N 35°06′E. One of Judaism's four holy cities, it is situated in a region where grapes, cereal grains, and vegetables are grown. Tanning, food processing, glassblowing, and the manufacturing of sheepskin coats are the major industries. Road junction. Jewish population nearly 6,000. The site of ancient Hebron, which antedates the biblical record, has not been precisely determined. The Bible first mentions Hebron in connection with Abraham. The cave of Machpelah (now enclosed by the Haram, an important Muslim mosque) is the traditional burial place of the Patriarchs and Matriarchs. It is shared at alternate times as a place of worship by Jews and Arabs, and was site of 1991 massacre of Arab worshippers by an American-born Jewish fanatic that, perversely, may have hastened the Israel–PLO peace agreement. David ruled the Hebrews from here for seven years before moving his capital to Jerusalem. Absalom began his revolt here. The city has figured in every war in the region. It was taken (2nd century B.C.E.) by Judas Maccabeus and temporarily destroyed by the Romans. In 636 it was conquered by the Arabs, who named it El Khalil, and made it an important place of pilgrimage, later to be seized (1099) by the Crusaders and renamed St. Abraham, and retaken (1187) by Saladin. It later became (16th century) part of the Ottoman Empire, under British administration (1918–1948), and under Jordan (1948–1967). Hebron is a sacred place for Muslims and Jews. It usually had a significant Jewish population prior to an Arab massacre in 1929, when most of the Jews left. Some returned in 1931 but were forced to leave five years later. Jewish settlement revived after the 1967 Arab-Israeli War. A new Jewish quarter, Kiryat Arba, was established on the NW outskirts and a small Jewish enclave (developed around religious students), is in the center of town adjoining the main Arab market. The security of this enclave was a major issue in Israel's delay in handing over Hebron to the Palestinian Authority, in keeping with the 1995 territorial agreement between the two parties; the transfer to the Palestinian Authority took place in 1997. As one of the major towns in the Israeli-occupied West Bank, Hebron has been a focus of Jewish-Arab tensions. The emergence of the Intifada in the late 1980s was accompanied by an escalation of violence. El Arub Palestinian refugee camp (1995 population 8,300) 7.5 mi/12 km to the N; El Fawar refugee camp (1995 population 5,200) 3.1 mi/5 km to the S.

Hebron (HEE-bruhn), town, TOLLAND county, E central CONNECTICUT, on SALMON RIVER, and 18 mi/29 km SE of HARTFORD; 41°38′N 72°23′W. In agricultural area. Includes Gilead and Amston villages and Amston Lake (c.1 mi/1.6 km long). Reverand Samuel Peters, Tory author of exaggerated account of Connecticut "blue laws," lived here. Has 18th-century houses. Settled 1704, incorporated 1708.

Hebron, town (2000 population 3,596), PORTER county, NW INDIANA, 13 mi/21 km SW of VALPARAISO; 41°19′N 87°12′W. Agricultural area.

Hebron, unincorporated town, BOONE county, N KENTUCKY, suburb 14 mi/23 km WSW of CINCINNATI (OHIO), and 12 mi/19 km W of COVINGTON. Agriculture to W and SW (tobacco; livestock, poultry; dairying). Manufacturing (sand and gravel processing, tool and die, chemicals, machinery, plastic products, fabricated metal products, transportation equipment). International airport to SE.

Hebron, town, OXFORD county, W MAINE, just E of SOUTH PARIS; 44°12′N 70°23′W. Orchard center.

Hebron, town (2000 population 807), WICOMICO county, SE MARYLAND, 6 mi/9.7 km NW of SALISBURY; 38°25′N 75°42′W. In vegetable farm and timber area; lumber mill, clothing factories. A stone placed here in 1760 marks the town as exactly halfway between the Atlantic coast and CHESAPEAKE BAY.

Hebron, town (2006 population 1,379), ⊙ THAYER county, SE NEBRASKA, 65 mi/105 km SW of LINCOLN, and on LITTLE BLUE RIVER; 40°10′N 97°35′W. Railroad terminus. Grain; manufacturing (cheese products; publishing). On old OREGON TRAIL. Founded 1869.

Hebron (HE-bruhn), town, GRAFTON county, central NEW HAMPSHIRE, 18 mi/29 km NNW of FRANKLIN; 43°41′N 71°47′W. Drained by Cockermouth River, which flows into N end of NEWFOUND LAKE, on S border. Agriculture (poultry, cattle; apples, vegetables; dairying; nursery crops); tourism.

Hebron (HEE-bruhn), town (2006 population 749), MORTON CO., SW central NORTH DAKOTA, 55 mi/89 km W of MANDAN; 46°53′N 102°02′W. Dairy produce; livestock; wheat, corn, and barley. Known as the "Brick City" for the brick manufacturer that was founded here in 1904. Town was founded in 1885 and named for HEBRON, PALESTINE.

Hebron (HE-brawn), town (2006 population 158), DENTON county, N TEXAS, small residential suburb 17 mi/27 km NNW of downtown DALLAS, in urban fringe area; 33°02′N 96°54′W.

Hebron (HEE-bruhn), village, NE NEWFOUNDLAND AND LABRADOR, NE CANADA, on N side of entrance of Hebron Fiord (30 mi/48 km-long inlet of the ATLANTIC Ocean); 58°12′N 62°37′W. Established in 1831 as a Moravian mission for the Inuit; abandoned in 1959.

Hebron, village (2000 population 1,038), MCHENRY county, NE ILLINOIS, near WISCONSIN state line, 11 mi/18 km N of WOODSTOCK; 42°28′N 88°25′W. In agricultural area (corn; dairying); manufacturing (plastic products, garden tools).

Hebron (HEE-brahn), village (□ 3 sq mi/7.8 sq km; 2006 population 2,147), LICKING county, central OHIO, 8 mi/13 km SW of NEWARK; 39°58′N 82°29′W. Dairy products.

Hebros, Greece: see ÉVROS.

Hebros River, Bulgaria and Greece: see MARITSA RIVER.

Hebudae, Scotland: see HEBRIDES, THE.

Hebudai, Scotland: see HEBRIDES, THE.

Hebudes, Scotland: see HEBRIDES, THE.

Heby (HE-BEE), town, VÄSTMANLAND county, central SWEDEN, 8 mi/12.9 km E of SALA; 59°57′N 16°51′E. Railroad junction.

Hecate Island, CANADA: see CALVERT ISLAND.

Hecate Strait (HE-kuht), W BRITISH COLUMBIA, W CANADA, separates QUEEN CHARLOTTE ISLANDS from mainland; joins DIXON ENTRANCE (NNW) and QUEEN CHARLOTTE SOUND (SSE); 160 mi/257 km long, 40 mi/64 km–80 mi/129 km wide; 53°30′N 131°10′W. Salmon and halibut fisheries.

Hecelchakán (e-sel-chah-KAHN), city and township, ⊙ Hecelchakán municipio, CAMPECHE, SE MEXICO, on NW YUCATÁN peninsula, on railroad, and 65 mi/105 km SW of MÉRIDA; 20°10′N 90°09′W. Agricultural center (corn, sugarcane, henequen, tobacco, tropical fruit, livestock).

Heceta Beach (huh-KEE-tuh), unincorporated village, W OREGON, on the PACIFIC OCEAN, 4 mi/6.4 km N of FLORENCE, just N of mouth of SIUSLAW RIVER. HECETA HEAD Cape and lighthouse 7 mi/11.3 km to N. Siuslaw National Forest to N and E; Washburns Memorial and Devils Elbow state parks to N.

Heceta Head (huh-KEE-tuh), LANE county, W OREGON, coastal promontory, c.20 mi/32 km S of WALDPORT. Lighthouse.

Hechi (HUH-CHEE), city (□ 903 sq mi/2,339 sq km; 1994 estimated non-agrarian population 97,200; estimated total population 295,200), NW GUANGXI ZHUANG, S CHINA, 90 mi/145 km NW of LIUZHOU; 24°39′N 108°39′E. Heavy industry and commerce are the largest sectors of the city's economy. Crop growing; animal husbandry. Grain; hogs; manufacturing (textiles, utilities, chemicals, machinery).

Hechingen (HE-khing-uhn), town, Baden-Württemberg, in Swabian Jura Mountains, 12 mi/19 km SSW of Tübingen; 49°22′N 08°58′E. Manufacturing of textiles, machinery; woodworking. Has neoclassiccastle and church. Site of research institute for biology and physics. Residence of dukes of Hohenzollern-Hechingen until 1849, when it passedto Prussia. Just N, towering above town, on top of Hohenzollern(haw-e-TSUHL-lern) mt. (2,805 ft/855 m), isthe castle of Hohezollern (destroyed 1423; rebuilt in 1850s).

Hecho (AI-cho), town, HUESCA province, NE SPAIN, 16 mi/26 km NW of JACA, on S slopes of the PYRENEES MOUNTAINS; 42°44′N 00°45′W. Summer resort; cereals, alfalfa, potatoes; livestock; sawmills. Because of its proximity to French border, it was long a place of importance.

Hechtel-Eksel (HEKH-tuhl–EK-suhl), commune (2006 population 11,459), Maaseik district, LIMBURG province, NE BELGIUM, 7 mi/11 km SSE of LOMMEL.

Hechuan (HUH-CHWAHN), town, ⊙ Hechuan county, S central SICHUAN province, SW CHINA, on JIALING RIVER, at mouth of FU RIVER, and 35 mi/56 km NW of CHONGQING; 30°02′N 106°15′E. Rice, oilseeds, sugarcane; cotton textiles; food processing.

Hecker, village (2000 population 475), MONROE county, SW ILLINOIS, 9 mi/14.5 km ESE of WATERLOO; 38°17′N 89°59′W. In agricultural area.

Hecklingen (HEK-kling-uhn), town, SAXONY-ANHALT, central GERMANY, 3 mi/4.8 km WSW of STASSFURT; 51°52′N 10°32′E. Potash mining.

Heckmondwike (HEK-muhnd-weik), town (2004 population 16,500), WEST YORKSHIRE, N central ENGLAND, 2 mi/3.2 km WNW of DEWSBURY; 53°42′N 01°40′W. Carpet weaving; textile machinery, machine tools, shoes, asbestos, soap.

Heckscher State Park (HEK-shuhr), recreational area (□ 2 sq mi/5.2 sq km) on S LONG ISLAND, SE NEW YORK, on peninsula extending into GREAT SOUTH Bay SE of ISLIP; 40°42′N 73°09′W. Swimming, picnicking, hiking, horseback riding.

Hecla, village (2006 population 297), BROWN county, N SOUTH DAKOTA, 33 mi/53 km NNE of ABERDEEN, near JAMES RIVER; 45°52′N 98°09′W. Manufacturing (machinery). Sand Lake National Wildlife Refuge to W; Mud Lake Reservoir to SW.

Hecla and Griper Bay, inlet of the ARCTIC OCEAN, N MELVILLE Island, NORTHWEST TERRITORIES, N CANADA; 85 mi/137 km long, 25 mi/40 km–60 mi/97 km wide; 76°00′N 113°00′W.

Hecla, Cape (HE-kluh), NE ELLESMERE ISLAND, BAFFIN region, NUNAVUT territory, N CANADA, on LINCOLN SEA of the ARCTIC OCEAN; 82°54′N 64°40′W.

Hecla, Mount (HEK-lah), peak (1,988 ft/606 m) on E coast of SOUTH UIST island, OUTER HEBRIDES, EILEAN SIAR, NW Scotland.

Hector, town (2000 population 1,166), RENVILLE county, S central MINNESOTA, 14 mi/23 km E of OLIVIA; 44°44′N 94°42′W. Grain, sugar beets, soybeans, beans; livestock, poultry; dairying. Manufacturing (electrical equipment, tools, consumer goods; meat processing).

Hector (HEK-tuhr), unincorporated village, SE BRITISH COLUMBIA, E CANADA, near ALBERTA border, in ROCKY MOUNTAINS, in YOHO NATIONAL PARK, 40 mi/64 km NW of BANFF (Alberta), and in COLUMBIA SHUSWAP regional district; elevation 5,213 ft/1,589 m; 51°26′N 116°20′W. Between this point and FIELD, the Canadian Pacific railroad passes through the SPIRAL TUNNELS.

Hector Lake (HEK-tuhr), SW ALBERTA, W CANADA, near BRITISH COLUMBIA border, in ROCKY MOUNTAINS, in BANFF NATIONAL PARK, at foot of Mount HECTOR, 45 mi/72 km NW of BANFF; 4 mi/6 km long, 1 mi/2 km wide; elevation 5,704 ft/1,739 m; 51°35′N 116°22′W. Drains SE into BOW RIVER.

Hector, Mount (HEK-tuhr) (11,135 ft/3,394 m), SW ALBERTA, W CANADA, near BRITISH COLUMBIA border, in ROCKY MOUNTAINS, in BANFF NATIONAL PARK, 40 mi/64 km NW of BANFF; 51°39′N 116°16′W. At its foot is HECTOR LAKE.

Heda (he-DAH), village, Tagata district, SHIZUOKA prefecture, S HONSHU, E central JAPAN, 22 mi/35 km E of SHIZUOKA; 34°58′N 138°46′E. Crabs. Western-style sailing in Japan began here.

Hedaru (he-DAH-roo), village, KILIMANJARO region, NE TANZANIA, 85 mi/137 km SSE of MOSHI, at E edge of MASAI STEPPE, SW of PARE MOUNTAINS; 04°30′S 37°57′E. Cattle, sheep, goats; corn, wheat; timber.

Hedd, BAHRAIN: see HADD.

Heddalsvatn (HED-dahls-VAH-tuhn), lake, TELEMARK county, S NORWAY, extending 11 mi/18 km S from NOTODDEN (at mouth of TINNE and Heddøla rivers). Sauerelv River, its outlet, flows S to NORSJØ Lake.

Heddesheim (HED-des-heim), suburb of MANNHEIM, BADEN-WÜRTTEMBERG, GERMANY, 6 mi/9.7 km ENE of city center; 49°30′N 08°37′E. Tobacco growing.

Hédé (ai-dai), agricultural commune (□ 8 sq km), ILLE-ET-VILAINE department, in BRITTANY, NW FRANCE, 14 mi/23 km NNW of RENNES; 48°18′N 01°48′W. In nearby castle of Montmuran, Bertrand du Guesclin, constable of France, was knighted in 1354.

Area is shown by the symbol □, and capital city or county seat by ⊙.

Hedefors, SWEDEN: see LERUM.

Hedehusene (HAI-duh-hoo-suh-nuh), town (2000 population 11,271), Copenhagen county, SJÆLLAND, DENMARK, 16 mi/26 km W of COPENHAGEN; 55°39′N 12°12′E. Manufacturing (cement, stationery, nonmetallic mineral products); agriculture (dairy products).

Hedel (HAI-duhl), village, GELDERLAND province, central NETHERLANDS, on MAAS RIVER (bridge), and 4 mi/6.4 km NNW of 's-HERTOGENBOSCH; 51°44′N 05°15′E. Dairying; livestock; agriculture (vegetables, sugar beets); manufacturing (electronic components). Castle to W.

Hedemora (HE-de-MOO-rah), town, KOPPARBERG county, central SWEDEN, on DALÄLVEN RIVER, 25 mi/40 km SE of FALUN; 60°17′N 15°59′E. Manufacturing (ironworks). In former iron-mining region. Has fifteenth-century church; eighteenth-century Town Hall; museum. Former site of important annual trade fairs. Chartered 1459.

Hedenstad (HAI-duhn-stah), village, BUSKERUD county, SE NORWAY, 4 mi/6.4 km S of KONGSBERG. In hilly region. Agriculture; lumbering and sawmilling. Silver mined in region. Skrim, mountain, 7 mi/11.3 km S, rises 2,851 ft/869 m.

Hedensted (HAI-duhn-sted), town, VEJLE county, E JUTLAND, DENMARK, 7 mi/11.3 km NE of VEJLE; 55°45′N 09°20′E. Dairying; manufacturing (baked goods, margarine, cement).

Hédervár (HAI-dahr-vahr), Hungarian *Hédervár*, village, Györ-Sopron county, NW HUNGARY, on Mosoni branch of DANUBE River, on Szigetköz island, and 13 mi/21 km NW of Györ; 47°50′N 17°28′E. Castle here.

He Devil Mountain, highest peak (9,393 ft/2,863 m) in SEVEN DEVILS MOUNTAINS, SW IDAHO county, W IDAHO, in Nez Perce National Forest, 35 mi/56 km NW of MCCALL. Rises above HELLS CANYON on the SNAKE RIVER (W).

Hedge End (HEDJ), suburb (2001 population 18,696), S HAMPSHIRE, S ENGLAND, 4 mi/6.4 km E of SOUTHAMPTON; 50°54′N 01°18′W. Former agricultural market.

Hedgesville (HE-jez-vil), village (2006 population 244), BERKELEY county, NE WEST VIRGINIA, in EASTERN PANHANDLE, 7 mi/11.3 km NNW of MARTINSBURG; 39°32′N 77°59′W. Agriculture (grain, apples); livestock. Sleepy Creek Wildlife Management Area.

Hedjaz, SAUDI ARABIA: see WESTERN REGION.

Hedley (HED-lee), unincorporated village, S BRITISH COLUMBIA, W CANADA, in CASCADE Mountains, on SIMILKAMEEN RIVER, 26 mi/42 km WSW of PENTICTON, in OKANAGAN-Similkameen regional district; elevation 4,500 ft/1,372 m; 49°21′N 120°04′W. Gold and silver mining.

Hedley (HED-lee), village (2006 population 373), DONLEY county, extreme N TEXAS, in the PANHANDLE, 15 mi/24 km SE of CLARENDON; 34°52′N 100°39′W. In agricultural area (cotton, peanuts; cattle).

Hedmark (HED-mahrk), county (□ 10,575 sq mi/27,389 sq km; 2007 estimated population 188,664), SE NORWAY, bordering on SWEDEN in the E; ⊙ HAMAR. It is the chief forest area of Norway; production is especially important in the upper GLOMMA River valley. Productive farms.

Hedo, Cape (HE-do), Japanese *Hedo-misaki* (he-DO-mee-SAH-kee), northernmost point of OKINAWA island, OKINAWA prefecture, in RYUKYU ISLANDS, SW JAPAN, in EAST CHINA SEA; 26°52′N 128°16′E.

Hedon (HE-duhn), town (2006 population 7,600), East Riding of Yorkshire, NE ENGLAND, near HUMBER RIVER, 6 mi/9.7 km E of HULL; 53°44′N 00°11′W. Formerly a chemical center (industrial solvents). Has church dating back to 1180. Town was an important trade center, later displaced by the port of Hull.

Hedrick, town (2000 population 837), KEOKUK county, SE IOWA, 12 mi/19 km SW of SIGOURNEY; 41°10′N 92°18′W. Livestock; grain.

Hedrum (HED-room), village, VESTFOLD county, SE NORWAY, on LÅGEN River, and 5 mi/8 km N of LAR-

VIK. Market gardening; lumbering; tanning; stone cutting.

Hedwig Village, town (2006 population 2,319), HARRIS county, SE TEXAS, residential suburb 9 mi/14.5 km WNW of downtown HOUSTON, surrounded by city of Houston; 29°46′N 95°31′W.

Heeg (HAIKH), village, FRIESLAND province, N NETHERLANDS, on NE end of FLUESSEN LAKE, on Heegemeer Bay, 5 mi/8 km S of SNEEK; 52°58′N 05°37′E. Dairying; cattle; grain, vegetables. Lake fishing, water sports.

Heek (HAIK), village, NORTH RHINE–WESTPHALIA, NW GERMANY, near Dutch border, 25 mi/40 km NW of MÜNSTER; 52°07′N 07°07′E.

Heemskerk (HAIMS-kerk), city, NORTH HOLLAND province, NW NETHERLANDS, 9 mi/14.5 km NNE of HAARLEM, and 2 mi/3.2 km NE of BEVERWIJK; 52°31′N 04°41′E. NORTH SEA 4 mi/6.4 km to W (dunes); railroad junction to E. Dairying; cattle, sheep; flowers, flower bulbs, nursery stock, fruit, vegetables.

Heemstede (HAIM-stai-duh), city, NORTH HOLLAND province, W NETHERLANDS, 3 mi/4.8 km S of HAARLEM; 52°22′N 04°38′E. RINGVAART canal passes to SE; NORTH SEA 5 mi/8 km to W; National Park De Kennemerduinen to NW. Cattle; flowers, vegetables, fruit; manufacturing (concrete).

Heer (HER), suburb of MAASTRICHT, LIMBURG province, SE NETHERLANDS, 2 mi/3.2 km E of city center; 50°50′N 05°44′E. Dairying; cattle, hogs; vegetables, grain, fruit; light manufacturing.

Heer Arendskerke, 's, NETHERLANDS: see 's-HEER ARENDSKERKE.

Heerde (HER-duh), village, GELDERLAND province, N central NETHERLANDS, 12 mi/19 km NNE of APELDOORN; 52°23′N 06°03′E. Apeldoorns Canal passes to E, IJSSEL RIVER 2.5 mi/4 km to E. Dairying; cattle, poultry; grain, vegetables, fruit; manufacturing (detergent). Castle to NE.

Heerenberg, 's, NETHERLANDS: see 's-HEERENBERG.

Heerenveen (HER-uhn-vain), city (2001 population 28,691), FRIESLAND province, N NETHERLANDS, 18 mi/29 km ESE of LEEUWARDEN; 52°57′N 05°56′E. Tjonger Canal to SE, TJEUKEMEER Lake 5 mi/8 km to SW. Dairying; cattle, sheep; vegetables, grain, seeds; manufacturing (filters, bicycles; food processing). Museum (established seventeenth century), ice rink.

Heerhugowaard (her-HUH-khaw-vahrt), village (2001 population 59,209), NORTH HOLLAND province, NW NETHERLANDS, 5 mi/8 km NE of ALKMAAR; 52°40′N 04°50′E. Railroad junction. Cattle, sheep raising; dairying; flowers, vegetables, fruit; manufacturing (consumer goods, machinery).

Heerlen (HER-luhn), city (2001 population 184,899), LIMBURG province, SE NETHERLANDS, 14 mi/23 km ENE of MAASTRICHT, and 10 mi/16 km NW of AACHEN (GERMANY), 4 mi/6.4 km SW, W, and NW from German border; 50°53′N 05°59′E. Railroad junction. Industrial, office, and transportation center. Dairying; cattle, hogs; grain, vegetables, fruit; manufacturing (beverages, chemicals; food processing). Former coalmining center. Large archaeological museum. Site of (part of) headquarters of the Netherlands Central Bureau of Statistics. Seat of Open University (1984).

Heerlerheide (HER-luhr-hei-duh), suburb of HEERLEN, Limburg province, SE NETHERLANDS, 2 mi/3.2 km NNW of city center; 50°55′N 05°57′E. Dairying; cattle; vegetables, fruit. Former coal-mining center.

Heers (HERS), commune (2006 population 6,799), Tongeren district, LIMBURG province, NE BELGIUM, 7 mi/11 km SE of SINT-TRUIDEN; 50°45′N 05°17′E.

Heerwegen, POLAND: see POLKOWICE.

Heeslingen (HAIS-ling-uhn), village, LOWER SAXONY, N GERMANY, on right bank of OSTE RIVER, 28 mi/45 km NE of BREMEN; 53°18′N 09°20′E. In peat region.

Heeze (HAI-zuh), town, NORTH BRABANT province, SE NETHERLANDS, 6 mi/9.7 km SE of EINDHOVEN, on KLEINE DOMMEL RIVER; 51°23′N 05°35′E. Belgian

border 7 mi/11.3 km to S. Dairying; agriculture (grain, vegetables); cattle, hog raising.

Hefei (HUH-FAI), city (□ 177 sq mi/458 sq km; 1994 estimated non-agrarian population 866,800; estimated total population 1,126,600), ⊙ ANHUI province, E CHINA, c.110 mi/175 km WSW of NANJING; 31°51′N 117°17′E. A rapidly growing industrial city, it has textile mills, iron- and steelworks, chemical- and food-processing plants, and a variety of other manufacturing. Transportation hub, with railroad links to major cities and industrial centers. Convenient transportation conditions contribute to its position as a regional market for industrial consumer products and labor market. Well known for its city gardens, green belts, and its clean environment. Since the 1990s, the city has established a high-tech industrial park and an economic- and technological-development district to stimulate overseas investment. Seat of several higher learning institutes, including Anhui University, University of Science and Technology of China, a medical college, and agricultural and mining institutes. Formerly called Luzhou (Luchow). Sometimes appears as Hofei.

Hefeng (HUH-FUNG), town, ⊙ Hefeng county, SW HUBEI province, central CHINA, near HUNAN province border, 35 mi/56 km SE of ENSHI; 29°54′N 110°02′E. Rice, tobacco, medicinal herbs.

Heflin, town (2000 population 3,002), ⊙ Cleburne co., E Alabama, near Tallapoosa River, 15 mi/24 km E of Anniston. Lumber milling, clothing manufacturing, poultry processing. Settled 1883, incorporated 1892. Named for Wilson L. Heflin, an early settler and the father of U.S. senator J. Thomas Heflin. Became the county seat in 1906.

Heflin (HEF-len), village (2000 population 245), WEBSTER parish, NW LOUISIANA, 10 mi/16 km S of MINDEN; 32°28′N 93°16′W. In agricultural area (cotton, vegetables; cattle). Lake Bistinau reservoir to W.

Heftsi Bah or **Hefziba** (khef-tsee BAH), village, W ISRAEL, in PLAIN OF SHARON, 2 mi/3.2 km NW of HADERA; 32°31′N 35°25′E. Agricultural plantation that has become urbanized.

Heftsi Bah or **Heftziba** (khef-tsee BAH), kibbutz, NE ISRAEL, at SE end of PLAIN OF JEZREEL, at N foot of MT. GILBOA, 5 mi/8 km WNW of BEIT SHEAN; 32°28′N 34°54′E. Elevation 108 ft/32 m. Plastic products, computerized sprinkler systems; mixed farming, citriculture, livestock raising. Marble quarry. Agriculture school nearby. Founded 1922.

Hegang (HUH-GAHNG), city (□ 1,757 sq mi/4,551 sq km; 1994 estimated non-agrarian population 569,200; estimated total population 688,100), NE HEILONGJIANG province, NE CHINA, 40 mi/64 km N of JIAMUSI; 47°24′N 130°22′E. Railroad terminus and major coal-mining center. Mines extend along railroad spur to Heli, 20 mi/32 km N.

Hegeler, unincorporated town, VERMILION county, E central ILLINOIS, suburb 4 mi/6.4 km S of DANVILLE, near INDIANA state line; 40°04′N 87°38′W.

Hegenheim (ai-zhun-EM), German (HAI-guhn-eim), commune (2 sq mi/7.8 sq km), HAUT-RHIN department, in ALSACE, E FRANCE at Swiss border, 3 mi/4.8 km W of BASEL; 47°34′N 07°32′E. Precision instruments.

Heggadadevankote (heg-guh-DAI-de-vahn-ko-te), town, MYSORE district, KARNATAKA state, S INDIA, 25 mi/40 km SW of MYSORE; 12°06′N 76°20′E. Rice, millet, tobacco. Bamboo, sandalwood in nearby forests. Formerly known for tiger hunting. Also spelled Heggaddevankote.

Hegiri (HE-gee-ree), town, Ikoma district, NARA prefecture, S HONSHU, W central JAPAN, 8 mi/13 km S of NARA; 34°37′N 135°42′E. Flowers (roses, chrysanthemums). Ancient tombs in the area.

Hegyalja (HAHD-yahl-yah), mountainous region in NE HUNGARY, embracing S slopes of Zemplén Mountains, small spur of the CARPATHIANS; extends

22 mi/35 km between SZERENCS and Sárospatak; elevation around 3,500 ft/1,067 m; 46°59'N 17°56'E. Well-known Tokay wine produced here.

Hegyeshalom (HAH-dyahsh-hah-lom), village, Győr-Sopron county, NW HUNGARY, on LAJTA RIVER, and 24 mi/39 km NW of Győr; 47°55'N 17°10'E. Largest road and railroad crossing point to AUSTRIA; stone and gravel mining.

Heho (HAI-ho), village, YAWNGHWE township, SHAN STATE, MYANMAR, 15 mi/24 km WSW of TOUNGOO, on road and railroad to THAZI.

Heian-kyo, JAPAN: see KYOTO.

Heiban (hai-BAHN), village, South KURDOFAN state, central SUDAN, in NUBA MOUNTAINS, on road, and 50 mi/80 km S of DELAMI; 11°13'N 30°31'E. Gum arabic, sesame, and dura; livestock.

Heicheng (HAI-CHUNG) [Chinese=black city], ruined town, in W INNER MONGOLIA AUTONOMOUS REGION, N CHINA, in Naliu River (Etsui Gol or Dong River); 41°45'N 101°24'E. A trade center after c.1000 in the TANGUT kingdom, it was reported (13th century) by Marco Polo as Etzina, and was destroyed in the 14th century. Russian explorer Kozlov discovered the remains in 1909. Also known as Karakhoto.

Heidar, Jebel, ISRAEL: see HAARI, HAR.

Heide (HEI-de), town, SCHLESWIG-HOLSTEIN, N central GERMANY, in the center of the formerly important DITHMARSCHEN oil fields; 54°12'N 09°05'E. Manufacturing (machinery). A trade center, it has one of the largest market squares in Germany, especially noted for cattle trade. Was capital of the peasant republic of Dithmarschen (1447–1559), when it was destroyed by fire.

Heideck (HEI-dek), town, BAVARIA, S GERMANY, in MIDDLE FRANCONIA, 22 mi/35 km S of NUREMBERG; 49°08'N 11°07'E. Metalworking; plastics processing; manufacturing of medical equipment. Has two Gothic churches from 15th century. Town developed in 13th century around a castle (no longer standing).

Heidelberg (HEI-del-berg), city (2005 population 142,993), BADEN-WÜRTTEMBERG, SW GERMANY, on the NECKAR RIVER; 49°25'N 08°42'E. Manufacturing includes machinery, precision instruments, leather goods, and tobacco and wood products. Most important to Heidelberg, however, is the tourism industry. Primarily focused on the Heidelberg Castle, the trade brings in several million visitors per year. City first mentioned in the 12th century; in 1225 acquired by the count palatine of the RHINE and until 1720 was the residence of the electors palatine. The University of Heidelberg (Ruprecht-Karl-Universität) was founded in 1386 by Elector Rupert I and is the third-oldest German-speaking university in the world (after those in PRAGUE and VIENNA). It became a bulwark of the Reformation in the 16th century, declined after the Thirty Years War (1618–1648), and, recovering after the French Revolutionary Wars, became the leading university of 19th-century Germany. The city is an important research center, including the German Cancer Research Institute and the European Institute for Molecular Biology (EMBL). The university's professors have included noted theologians, the chemist River W. Bunsen, and the sociologist Max Weber. Since 1952 the city has been the headquarters of the U.S. Army in EUROPE. Heidelberg, the setting of the *Student Prince*, is famous for its ruined castle (built mainly in the 16th and early 17th century), which was largely destroyed by French troops in the late 17th century. In the castle's cellar is the Heidelberg Tun, a gigantic wine cask with a capacity of c.58,080 gallons/219,850 liters. Other points of interest in Heidelberg include the city hall (1701–1703) and the *Philosophenweg* (Philosopher's Way), a path overlooking the city.

Heidelberg, town, WESTERN CAPE province, SOUTH AFRICA, at foot of the LANGEBERG range, on Duiwenhoks River, on N2 highway, and 65 mi/105 km W of MOSSEL BAY; elevation 673 ft/205 m; 34°06'S

20°58'E. Agriculture center (wool, grain, tobacco, aloes). Established 1855 as trading, railroad, and administrative center for farming community.

Heidelberg, town, GAUTENG province, SOUTH AFRICA, on Blesbokspruit River, on major road R42 SE of WITWATERSRAND, and 30 mi/48 km SE of JOHANNESBURG; elevation 5,026 ft/1,532 m; 26°30'S 28°22'E. Gold-mining center from 1890 onwards; leather manufacturing. Nearby was headquarters of Zulu chief Moselikatse. Was capital of South African Republic during Transvaal revolt, 1880–1881. Site of military training base. Suikerbosrand Nature Reserve just W of town. Established 1865.

Heidelberg (HEI-duhl-buhrg), town (2000 population 840), JASPER county, E central MISSISSIPPI, 16 mi/26 km NNE of LAUREL; 31°53'N 88°59'W. Agriculture (corn, cotton; poultry, cattle; dairying; timber); manufacturing (automotive armatures); oil refinery. Oil field nearby.

Heidelberg (HEI-duhl-buhrg), unincorporated village, WATERLOO region, S ONTARIO, E central CANADA, 10 mi/17 km from from KITCHENER, and included in WELLESLEY township; 43°31'N 80°37'W.

Heidelberg, village (2000 population 72), LE SUEUR county, S MINNESOTA, 8 mi/12.9 km NE of LE CENTER; 44°29'N 93°37'W. Grain; livestock, poultry; dairying.

Heidelberg (HEI-duhl-buhrg), borough (2006 population 1,158), ALLEGHENY county, SW PENNSYLVANIA, residential suburb 6 mi/9.7 km SW of downtown PITTSBURGH, on Chartiers Creek. Woodville State Hospital to S.

Heidelberg (HEI-duhl-buhrg), suburb 11 mi/18 km NE of MELBOURNE, S VICTORIA, SE AUSTRALIA; 37°45'S 145°04'E. La Trobe University nearby. Railway station.

Heiden (HAH-duhn), commune, APPENZELL Ausser Rhoden half-canton, NE SWITZERLAND, 7 mi/11.3 km E of ST. GALLEN. Elevation 2,648 ft/807 m. Mountain health resort; textiles, embroideries.

Heiden (HEI-den), village, NORTH RHINE–WESTPHALIA, NW GERMANY, 18 mi/29 km NE of WESEL; 51°49'N 06°55'E.

Heidenau (HEI-de-nou), town, SAXONY, E central GERMANY, in SAXONIAN Switzerland, port on the ELBE RIVER, 8 mi/12.9 km SE of DRESDEN, near Czech border; 50°58'N 13°52'E. Manufacturing of printing presses, rubber products; paper milling.

Heidenheim an der Brenz (HEI-duhn-heim ahn der BRENTS), city, BADEN-WÜRTTEMBERG, SW GERMANY, 20 mi/32 km NNE of ULM; 48°41'N 10°10'E. Industrial center; manufacturing includes machinery, steel, electrical goods. Has Renaissance castle, ruined fortress. Built on the site of a Roman city (excavations nearby). Chartered 14th century.

Heidenreichstein (HEI-den-reich-shtein), town, NW LOWER AUSTRIA, 9 mi/14.5 km NE of Gmünd; 48°52'N 15°07'E. Railroad terminus; manufacturing of textiles, metals. Large moated castle.

Heidesheim am Rhein (HEI-des-heim ahm REIN), village, RHINELAND-PALATINATE, W Germany, near the Rhine River, 6 mi/9.7 km W of Mainz; 50°00'N 08°07'E. Manufacturing of plastics.

Heidrick (HED-rik), village, KNOX county, SE KENTUCKY, 2 mi/3.2 km NE of BARBOURVILLE, in the CUMBERLAND MOUNTAINS, near CUMBERLAND RIVER. Agriculture (tobacco); coal.

Heighington and Washingborough (HAI-ing-tuhn and WAHSH-ing-buh-ruh), villages (2001 population 6,613), LINCOLNSHIRE, E ENGLAND, on WITHAM RIVER, 4 mi/6.4 km E of LINCOLN; 51°13'N 00°27'W. Previously light engineering.

Heigun-jima (HAI-GUN-jee-mah), island (□ 7 sq mi/18.2 sq km), YAMAGUCHI prefecture, W JAPAN, in IYO SEA, just off SW coast of HONSHU, S of YANAI; 6 mi/9.7 km long, 2 mi/3.2 km wide. Mountainous, forested; rice, wheat, raw silk. Fishing.

Heihe (HAI-HUH), city (□ 5,571 sq mi/14,429 sq km; 1994 estimated non-agrarian population 97,400; esti-

mated total population 166,400), N HEILONGJIANG province, NE CHINA, on right bank of the AMUR RIVER (on Russian border), opposite BLAGOVESHCHENSK (Russia), 310 mi/499 km N of HARBIN; 50°16'N 127°25'E. Became a commercial exchange town in the mid-1800s. Designated as an open river port, it is currently a vibrant river trade center with RUSSIA. The largest joint Russia-China open air market is on Daheihe Island in the Amur River. Agriculture is the main source of income; products include grains and milk. Main industries include timber processing and food. Well known for its gold mining.

Heilbron, town, NE FREE STATE province, SOUTH AFRICA, 40 mi/64 km N of VEREENIGING; elevation 5,150 ft/1,570 m; 27°17'S 27°58'E. Agriculture center (corn; cattle, sheep, horses, poultry; dairying); grain elevator, cold-storage plant. Site of Voortrekker Museum (1934). Was ORANGE FREE STATE capital under president Martinus Theunis Steyn. Nearby Voortrekkers defeated (1836) force of tribal chief Msilikazi (Moselikatse).

Heilbronn (heil-BRUHN), city, BADEN-WÜRTTEMBERG, S GERMANY, a port on the NECKAR River; 49°09'N 09°13'E. A commercial and industrial center; manufacturing includes metal products, machinery, automotive parts, tools, electronics, paper, and wine. Salt is mined nearby. Site (early 9th century) of a Carolingian palace and in the 14th century became a free imperial city. Although it suffered in the wars of the 16th century, particularly in the Peasants' War, the city rose to great commercial prosperity in the late 16th and early 17th century. Passed to Württemberg 1802. Acquired industrial importance mid- to late-19th century. In World War II (especially 1944) much of the city was destroyed, but many of its historic buildings have been reconstructed. Points of interest include the church of St. Kilian (13th–15th century) and the Götzenturm, a tower built in 1392, which is mentioned in Goethe's drama *Götz von Berlichingen* (1772).

Heiligenbeil, RUSSIA: see MAMONOVO.

Heiligenblut (HEI-li-gen-bloot) [Ger.=holy blood], village, NW CARINTHIA province, S AUSTRIA, at the foot of the GROSSGLOCKNER Mountain; 47°02'N 12°51'E. Winter sports and mountain climbing center. Famous place of pilgrimage; in its Gothic church (late 15th century) is a vial that, according to tradition, contains some of Christ's blood. Southern starting point of the Grossglocknerstrasse. Nearby is Hohe Tauern, a mountain range and national park.

Heiligendamm, GERMANY: see BAD DOBERAN.

Heiligengrabe, GERMANY: see PRITZWALK.

Heiligenhafen (hei-lee-guhn-HAH-fuhn), town, SCHLESWIG-HOLSTEIN, NW GERMANY, port and seaside resort on the BALTIC SEA, 22 mi/35 km NE of EUTIN, opposite FEHMARN island; 54°22'N 10°58'E. Fishing; shipbuilding; tourism. Has church from 13th century. Chartered 1305.

Heiligenhaus (HEI-lee-guhn-hous), town, NORTH RHINE–WESTPHALIA, W GERMANY, 3 mi/4.8 km W of VELBERT; 51°19'N 06°58'E. Electronics and technology industry; foundries; manufacturing of machinery and plastics. Chartered 1947.

Heiligenkreuz (HEI-li-gen-KROITS), village, E LOWER AUSTRIA, 6 mi/9.7 km NNW of BADEN, in the WIENERWALD; 48°03'N 16°08'E. Has ancient abbey with 12th century Romanesque church, large library; tombs of Babenberg family. Gypsum quarries nearby (Preinsfeld).

Heiligenstadt (HEI-li-gen-SHTAHT), E section of Döbling district of VIENNA, AUSTRIA, 3 mi/4.8 km N of city center; 46°39'N 13°57'E. Was the long-time residence of Beethoven. The largest interwar apartment building in Vienna is here (Karl-Marx-Hof).

Heiligenstadt (HEI-lee-guhn-SHTAHT), town, THURINGIA, central GERMANY, on the LEINE RIVER, 18 mi/29 km NW of MÜHLHAUSEN, 10 mi/16 km E of

Eichenberg; 51°23′N 10°08′E. Manufacturing of textiles and paper. Health resort since 1950. Has several 14th-century churches; 18th-century castle. Was (1540–1802) capital of principality of Eichsfeld, property of the electors of Mainz.

Heiligenstadt in Oberfranken (HEI-lee-guhn-SHTAHT een O-buhr-frahng-kuhn), village, BAVARIA, central GERMANY, in UPPER FRANCONIA, 13 mi/21 km ESE of BAMBERG; 49°51′N 11°10′E. Has 17th-century church.

Heillecourt (ei-yuh-KOOR), town (☐ 1 sq mi/2.6 sq km), MEURTHE-ET-MOSELLE department, in LORRAINE, E FRANCE, a SE suburb of NANCY; 48°39′N 06°12′E. Electronic publishing industry.

Heilongjiang (HAI-lung-JYAHNG) [Chinese=black dragon river (the AMUR)] or **Hei** (HAI), province (☐ 179,000 sq mi/465,400 sq km; 2000 population 36,237,576), NE CHINA; ☉ HARBIN; 48°00′N 128°00′E. Constitutes the N part of the region known as NORTHEAST (formerly Manchuria) and is separated from RUSSIA by the Amur (Heilongjiang) River in the N and the USSURI RIVER in the E, and is bordered on the W by the INNER MONGOLIA AUTONOMOUS REGION. Heilongjiang's most important resources are coal, gold, crude oil, timber, and arable land. Coal is widespread. Major coal-mining centers are JIXI, HEGANG, and SHUANGYASHAN. Gold is widely mined along the Amur River, with HEIHE and HUMA as major mining centers. Both the GREATER HINGGAN and Lesser Hinggan mountain ranges traverse the province; their heavily forested slopes contain some of the finest timber in China. Lumbering is a major industry. The NW (the area W of the Greater Hinggan) was formerly (before 1969) a part of Inner Mongolia; it has a large Mongol population and is predominantly pastureland, with related industries (dairy products, leather). HAILAR is the principal city of this region. The S, which contains the agricultural, industrial, and economic base of the province, is watered by the SONGHUA, the Nen, the Hulan, and the MUDAN rivers, and is known as the Northeast plain or the Songhua-Nen plain. Rich crude-oil reserves have been found since the late 1950s in the plain. DAQING has become a major oil and petrochemical center. The Songhua-Nen plain is also a great wheat area—known as the Bread Basket of the North. Millet, kaoliang, soybeans, sugar beets, and flax are also grown. The Songhua-Nen plain has fertile agricultural land. Heilongjiang has a total of 34,484 sq mi/89,314 sq km of arable land, the largest in the nation. Its per capital arable land is three times the national average. The great potential for agriculture gives the province the name "the Northern Reserve." Farming here is highly mechanized, and vast reclamation projects have been instituted under the Communist government. The Northeast China Railroad crosses S Heilongjiang and has many branches to the N; Harbin is the junction point with the South Northeast railroad system. Heilongjiang, which produces almost half of China's oil, contains the great Daqing oil field, first worked in 1959. Major coal mines are in Jixi and Hegang. Iron and magnesite are also mined, and aluminum is produced. Gold is extracted in the Greater and Lesser Hinggan. Harbin is one of the country's leading industrial centers, known especially for its heavy machinery. QIQIHAR, JIAMUSI, and MUDANJIANG are also industrial cities, with manufacturing ranging from processed foods to locomotives. The boundaries of Heilongjiang have been changed several times. The former provinces of HINGGAN and NENJIANG were added to it in 1950, and SONGJIANG was incorporated in 1954. The NW section became part of Inner Mongolia in 1949 and was returned to Heilongjiang in the 1969–1970 redistricting. The name sometimes appears as Heilungkiang.

Heiloo (HEI-law), town (2001 population 20,510), NORTH HOLLAND province, NW NETHERLANDS, 3 mi/ 4.8 km SW of ALKMAAR; 52°36′N 04°43′E. NORTH SEA 4 mi/6.4 km to W (dunes); NORTH HOLLAND CANAL passes to E. Dairying; cattle, sheep; flowers (especially tulips), nursery stock, vegetables, fruit; manufacturing (precision parts).

Heilsberg, POLAND: see LIDZBARK WARMINSKI.

Heilsbronn (heilz-BRUHN), town, MIDDLE FRANCONIA, W BAVARIA, GERMANY, 10 mi/16 km ENE of ANSBACH; 49°20′N 10°47′E. Meat processing; manufacturing of toys; metalworking. Former Cistercian abbey (1132–1555) has mid-12th-century church.

Heilungkiang, CHINA: see HEILONGJIANG.

Heimaey, ICELAND: see VESTMANNAEYJAR.

Heimbach (HEIM-bahkh), town, NORTH RHINE–WESTPHALIA, W GERMANY, on RUHR RIVER, 13 mi/21 km S of DÜREN; 50°38′N 06°29′E. Has a Cistercian abbey from 15th century and a ruined castle from 13th century. To W is dam on Ruhr River with hydroelectric plant.

Heimburg (HEIM-boorg), commune, BERN canton, W SWITZERLAND, 2 mi/3.2 km NNW of THUN; 46°48′N 07°36′E.

Heimefront Range, mountain range, in NEW SCHWABENLAND, EAST ANTARCTICA, located in QUEEN MAUD LAND, 50 mi/81 km WSW of the KIRWAN ESCARPMENT, 65 mi/105 long; 74°35′S 11°00′W. Composed of three groups.

Heimer, Oued El-, MOROCCO: see SIDI BOUBKER.

Heimiswil (HEI-mis-vil), commune, BERN canton, NW SWITZERLAND, 13 mi/21 km NE of BERN.

Heimsheim (HEIMZ-heim), town, BADEN-WÜRTTEMBERG, S GERMANY, 9 mi/14.5 km SE of PFORZHEIM; 48°49′N 08°52′E. Has 15th-century castle.

Heinersreuth (HEI-nerz-roit), village, BAVARIA, central GERMANY, in UPPER FRANCONIA, 3 mi/4.8 km NW of BAYREUTH; 49°58′N 11°32′E.

Heinola (HAI-no-lah), town, MIKKELIN province, S FINLAND, 20 mi/32 km NE of LAHTI; elevation 198 ft/ 60 m; 61°13′N 26°02′E. On PÄIJÄNNE lake system. Textile mills; woodworking; vacation and health resort. Site of teachers' seminary for women. Incorporated 1839, it was capital (1776–1843) of former Kyminkartano county.

Heinrichswalde, RUSSIA: see SLAVSK.

Heinsberg (HEINS-berg), city, NORTH RHINE–WESTPHALIA, W GERMANY, 7 mi/11.3 km N of GEILENKIRCHEN, near Dutch border; 51°04′N 06°06′E. Manufacturing includes chemicals, electronics, machinery, and various small industries. Developed around a former castle. Has 15th-century church. Chartered 1255.

Heinzenberg (HEIN-tsuhn-berg), district, W central GRISONS canton, E SWITZERLAND. Main town is THUSIS; population is German-speaking and Protestant, but many Roman Catholics in Kreis Domleschg.

Hei River (HAI), Mandarin *Hei He* (HAI HUH), over 300 mi/483 km long, NW GANSU province, NW CHINA, rises in the NAN SHAN (mountains); flows NW, past Zhangye, joining the Beita River at Ding Xin to form Ruo Shui.

Heirnkut (HERN-koot), village, Leshi township, SAGAING division, MYANMAR, 27 mi/43 km NNW of HOMALIN. Once known as Fort Keary.

Heis (HAH-ees), village, N SOMALIA, on Gulf of ADEN, 35 mi/56 km NW of ERIGAVO. Important watering place for livestock. Sometimes HAIS.

Heis, YEMEN: see HAIS.

Heisdorf (HEIS-dorf), hamlet, STEINSEL commune, S central LUXEMBOURG, 4 mi/6.4 km N of LUXEMBOURG city, near ALZETTE RIVER. Rose-growing center.

Heise, unincorporated village, JEFFERSON county, E IDAHO, 20 mi/32 km NE of IDAHO FALLS, on SNAKE RIVER irrigated agricultural area, 43°38′N 111°41′W. Targhee National Forest and Kelly Canyon Ski Area to E.

Heishan (HAI-SHAHN), town, ☉ Heishan county, central LIAONING province, NE CHINA, on railroad, and 60 mi/97 km NE of JINZHOU; 41°40′N 122°03′E. Grain, jute, sugar beets, oilseeds. Building materials, machinery; food processing, papermaking; coal mining.

Heisker (HEIS-kuhr) or **Monach Isles**, group of small islands of the OUTER HEBRIDES, EILEAN SIAR, NW Scotland; 57°31′N 07°40′W. Separated from W coast of NORTH UIST island by 8-mi/12.9-km-wide. Sound of Monach, 12 mi/19 km S of HASKEIR (or Hyskier) island. Westernmost island, Shillay, is site of lighthouse (57°31′N 07°43′W). No permanent population.

Heisler (HEIZ-luhr), village (2001 population 183), E central ALBERTA, W CANADA, E of WETASKIWIN, in FLAGSTAFF County; 52°41′N 112°13′W. Incorporated 1920.

Heist-aan-Zee, BELGIUM: see KNOKKE-HEIST.

Heist-op-den-Berg (HEIST–op–duhn–BERG), agricultural commune (2006 population 38,537), Mechelen district, ANTWERPEN province, N BELGIUM, near GROTE NETE RIVER, 8 mi/12.9 km ESE of LIER; 51°05′N 04°43′E. Formerly spelled Heyst-op-den-Berg.

Heitersheim (HEI-ters-heim), town, BADEN-WÜRTTEMBERG, SW GERMANY, 12 mi/19 km SW of FREIBURG; 47°53′N 07°40′E. Residential community serving both Freiburg and BASEL (Switzerland). Has 16th-century castle.

Heithuizen, NETHERLANDS: see HEYTHUYSEN.

Heiwa (HAI-WAH), town, Nakashima district, AICHI prefecture, S central HONSHU, central JAPAN, 9 mi/15 km W of NAGOYA; 35°12′N 136°44′E.

Hejaz, SAUDI ARABIA: see WESTERN REGION.

Hejaz Railroad, connecting Al Madinah in SAUDI ARABIA to the Al Sham railroad line in SYRIA; 497 mi/ 800 km long. Built by the Ottomans in 1908 to strengthen the empire and ensure safe passage for pilgrims to HEJAZ. Extremely narrow (41 in/105 cm wide). It was the only transportation line in all Arabia for many decades. Destroyed during World War I and never reconstructed, although in the 1970s Syria, JORDAN, and Saudi Arabia discussed plans to construct a new line.

Hejian (HUH-JYEN), city (☐ 515 sq mi/1,334 sq km; 1994 estimated non-agrarian population 42,400; estimated total population 734,700), central HEBEI province, NE CHINA, 435 mi/700 km SE of BAODING; 38°27′N 116°02′E. Agriculture (especially crop growing) is the largest economic sector. Grains, fruits, cotton; hogs; eggs. Manufacturing (textiles and electrical equipment).

Hejiang, former province (☐ 52,300 sq mi/135,980 sq km), NE CHINA, in NORTHEAST (formerly Manchuria); its capital was JIAMUSI (Kiamusze). Created in 1945, largely out of the current province of JILIN, Hejiang was one of nine provinces established in Manchuria by the Chinese Nationalist government after World War II. In 1949, Hejiang was included in the province of SONGJIANG, which became part of HEILONGJIANG province in 1954. The name sometimes appears as Ho-chiang or Hokiang.

Hejin (HUH-JEEN), town, ☉ Hejin county, SW SHANXI province, NE CHINA, 55 mi/89 km NNE of YONGJI, and on FEN RIVER just above its confluence with HUANG HE (Yellow River); 35°35′N 110°42′E. Grain, cotton, oilseeds. Building materials, chemicals; food processing; non-ferrous metal smelting; coal mining.

Hejing (HUH-JEENG), town, ☉ Hejing county, central XINJIANG UYGUR AUTONOMOUS REGION, NW CHINA, 20 mi/32 km NNW of YANQI; 42°18′N 86°18′E. Livestock.

Hejnice (HAI-ni-TSE), German *Haindorf*, town, SEVEROCESKY province, N BOHEMIA, CZECH REPUBLIC, at SW foot of SMRK peak of the JIZERA MOUNTAINS, on railroad, 10 mi/16 km NNE of LIBEREC; 50°53′N 15°12′E. Manufacturing (machinery; jewelry). Health resort of LAZNE LIBVERDA (Czech *Lázně Libverda*, German *Liebwerda*), with carbonated springs, is just NNE.

Hekelgem (HAI-kul-khem), agricultural village in commune of Aalst, Hal-Vilvoorde district, BRABANT province, central BELGIUM, 4 mi/6.4 km SE of AALST; 50°54′N 04°06′E.

Heki (HE-kee), town, Otsu district, YAMAGUCHI prefecture, SW HONSHU, W JAPAN, 25 mi/40 km N of YAMAGUCHI; 34°22′N 131°06′E.

Hekimhan, township, E central TURKEY, on Malatya-Sivas railroad, 38 mi/61 km NW of MALATYA; 38°50′N 37°56′E. Wheat, sugar beets.

Hekinan (he-KEE-nahn), city, AICHI prefecture, S central HONSHU, central JAPAN, near CHITA BAY, 22 mi/35 km S of NAGOYA; 34°52′N 136°59′E. Vegetables, figs; tile.

Hekla (HEK-lah), volcano, c.4,900 ft/1,490 m, SW ICELAND. One of the highest in EUROPE. Since the early 11th century over twenty eruptions have been recorded; the worst occurred in 1766 and the most recent in 1947. Hekla emits steam and has several crater lakes. In medieval Icelandic folklore Hekla was believed to be one of the gates to purgatory; it is also a legendary gathering place for witches.

Hekou (HUH-ko), town, ⊙ Hekou county, SE YUNNAN province, SW CHINA, on Red River (Vietnam border), opposite LAO CAI (Vietnam), 65 mi/105 km SE of MENGZI, and on Kunming-Hanoi (Vietnam) railroad; 22°36′N 103°58′E. Commercial center. Opened to foreign trade in 1896.

Hel, POLAND: see HEL PENINSULA.

Hela, POLAND: see HEL PENINSULA.

Helagsfjället (HE-lahgs-FYEL-let), mountain (5,892 ft/1,796 m), JÄMTLAND county, W SWEDEN, near Norwegian border, 70 mi/113 km WSW of ÖSTERSUND; 62°56′N 12°20′E.

Helambu, region, central NEPAL, S of LANGTANG, on the INDRAWATI RIVER. Home of Sherpas and other ethnic groups supported by tourism, herding, and agriculture.

Helan (HUH-LAHN), town, ⊙ Helan county, NE NINGXIA HUI AUTONOMOUS REGION, N central CHINA, on railroad, and 20 mi/32 km NNE of YINCHUAN, near HUANG HE (Yellow River), at E foot of HELAN MOUNTAINS; 38°35′N 106°16′E. Rice, wheat, sugar beets, oilseeds.

Helan Mountains (HUH-LAHN), SE NINGXIA HUI AUTONOMOUS REGION, N central CHINA, extend parallel to HUANG HE (Yellow River), separating Ningxia agricultural district (E) from Alashan Desert (W); 39°00′N 106°00′E. Rise to over 10,000 ft/3,048 m at a point 40 mi/64 km NW of YINCHUAN. Cliff paintings from prehistoric times have been found in nearly thirty locations in the mountains. Sometimes called Alakshan or Alashan mountains.

Helbra (HEL-brah), town, SAXONY-ANHALT, central GERMANY, at E foot of the lower HARZ, 3 mi/4.8 km NW of EISLEBEN; 51°33′N 11°30′E. Copper-smelting center.

Helbronner, Pointe (el-bro-NER, PWAHNT), peak (11,356 ft/3,461 m) in the MONT BLANC massif, SE FRANCE, on Italian border, and 5 mi/8 km SE of CHAMONIX. A favorite destination of mountain climbers, it is also reached by aerial tramway from both Chamonix and COURMAYEUR (Italy).

Helchteren, BELGIUM: see HOUTHALEN-HELCHTEREN.

Helden (HEL-duhn), village, LIMBURG province, SE NETHERLANDS, 8 mi/12.9 km SW of VENLO; 51°39′N 06°00′E. Dairying; cattle; agriculture (sugar beets, vegetables, grain). Recreational center to E.

Helderbergs, the, N-facing limestone escarpment (c.1,700 ft/518 m) of the CATSKILL MOUNTAINS (NE edge of ALLEGHENY PLATEAU), E NEW YORK, along S side of Mohawk valley W of ALBANY; 42°39′N 74°01′W. Also known as the Helderberg Mountains.

Helechal (ai-lai-CHAHL), village, BADAJOZ province, W SPAIN, 9 mi/14.5 km ESE of CASTUERA, near CÓRDOBA province border. Olive oil industry. Castle and sanctuary are NE.

Helechosa (ai-lai-CHO-sah) or **Helechosa de los Montes** (ai-lai-CHO-sah dai los MON-tes), town, BADAJOZ province, W SPAIN, near the upper GUADIANA RIVER, 13 mi/21 km NE of HERRERA DEL DUQUE; 39°19′N 04°54′W. Olives, cereals; timber; livestock.

Hélécine (ai-lai-SEEN), commune (2006 population 3,078), Nivelles district, BRABANT province, central BELGIUM, 5 mi/8 km SSE of TIENEN, on Petite Gette branch of GETTE RIVER.

Helecine Provincial Domain (ail-SEEN), recreational park near Opheylissem, BRABANT province, central BELGIUM, 4 mi/6.4 km SSE of TIENEN.

Helen, village (2000 population 430), WHITE county, NE GEORGIA, 22 mi/35 km WNW of TOCCOA, near CHATTAHOOCHEE RIVER; 34°42′N 83°43′W. Resort community with a Bavarian alpine village theme. Also light manufacturing and retail discount outlet malls.

Helena (HE-luh-nuh), city (2000 population 10,296), Shelby co., central Alabama, near Cahaba River, suburb 15 mi/24 km S of Birmingham. First named 'Cove,' and then 'Hillsboro,' it was changed to Helena in 1864 in honor of Helen Lee Boyle, wife of Peter Boyle, an engineer making surveys for the Louisville and Nashville RR. Inc. in 1877.

Helena, city (2000 population 25,780), ⊙ MONTANA and LEWIS AND CLARK county, W central MONTANA, 48 mi/77 km NNE of BUTTE, on E slope of the ROCKY MOUNTAINS, just E of the CONTINENTAL DIVIDE; 46°36′N 112°01′W. Commercial, trading, and shipping center in a ranching and mining area. Manufacturing includes concrete and sheet metal products, dairy products, consumer goods; printing and publishing. The state corporate offices of major electronics, engineering, communications, and healthcare organizations, as well as the Federal Reserve bank, are here. Increased tourism and expanding artists' colonies have strengthened Helena's economic base. Agricultural area, including cattle, sheep; wheat, barley, oats. Founded after the discovery of gold (1864) in Last Chance Gulch (which has become Helena's main street) and grew rapidly. In 1875 a general election ratified the choice of Helena to replace VIRGINIA CITY as territorial capital. In the 1890s it maintained its position as state capital against the rivalry of ANACONDA. Original mining was principally for gold and silver. As these minerals were depleted, stores of copper, lead, and zinc were discovered and exploited. Seat of Carroll College. Has numerous museum, landmarks, and monuments. Original Governors Mansion (1888), Holter Museum of Art, Museum of Gold, Masonic Grand Lodge, Reeder's Alley, Helena Civic Center, Archie Bray Foundation (school of pottery), Poindexter Collection of Abstract Art, F. Jay Haynes Photo Collection, Montana Homeland Exhibit, Mackay Gallery, Montana Historical Society Museum, Myrna Loy Center for the Performing Arts. The capitol building has noteworthy historical and artistic collections. The city is surrounded by scenic mountains and is the headquarters of Helena National Forest. Lake HELENA formed on sidestream of MISSOURI River to NE. HAUSER Lake and Black Sandy State Park to NE, and Canyon Ferry Lake and State Park to E, both formed by dams on Missouri River, units of Helena National Forest to W, SE, and NE. Gates of the Mountains Wilderness to NE. Great Divide Ski Area to NW, Fort Harrison Military Reservation to W. Originally called Last Chance Gulch. Incorporated 1870.

Helena (HE-luh-nuh), town (2000 population 6,323), ⊙ PHILLIPS county, E central ARKANSAS, on the MISSISSIPPI RIVER, and at the S end of CROWLEY'S RIDGE; 34°31′N 90°35′W. Larger city of West Helena adjoins to W. Railroad center and river port with an economy based on cotton, lumber, and agricultural processing. Manufacturing (soybean meal, cottonseed products; printing). Occupied by Union troops in the Civil War; they were attacked unsuccessfully by Confederates in the Battle of Helena (July 4, 1863). Seat of Phillips County Community College. Incorporated 1833.

Helena (he-LEE-nuh), town (2000 population 2,307), TELFAIR county, S central GEORGIA, 32 mi/51 km S of DUBLIN, and on LITTLE OCMULGEE RIVER, adjacent to MCRAE (E); 32°05′N 82°55′W. Manufacturing includes vegetable processing; machinery.

Helena (HE-luh-nuh), town (2006 population 1,383), ALFALFA county, N OKLAHOMA, 24 mi/39 km WNW of ENID; 36°32′N 98°16′W. In grain and livestock area; manufacturing (concrete).

Helena (HE-le-nah), village, DEMERARA-MAHAICA district, N GUYANA, 10 mi/16 km SE of GEORGETOWN; 06°41′N 57°55′W.

Helena (HE-luh-nuh), village (2006 population 235), SANDUSKY county, N OHIO, 9 mi/14 km W of FREMONT; 41°20′N 83°17′W. In agricultural area.

Helena (huh-LAI-nuh), unincorporated village, NEWBERRY county, NW central SOUTH CAROLINA, 1 mi/1.6 km W of NEWBERRY.

Helena, Greece: see MAKRONISOS.

Helena Island, QUEEN ELIZABETH ISLANDS, NUNAVUT territory, N CANADA, just off N BATHURST ISLAND; 76°46′N 101°30′W. 23 mi/37 km long, 8 mi/13 km wide.

Helena Island, HONDURAS: see SANTA ELENA ISLAND.

Helena, Lake (HE-luh-nuh), reservoir, LEWIS AND CLARK county, W central MONTANA, on MISSOURI River (to NE), on backwater from HAUSER Dam (Lake Hauser); linked widened Silver Creek, 7 mi/11.3 km NNE of HELENA; 3 mi/4.8 km long, 2 mi/3.2 km wide; 46°41′N 111°55′W. Receives PRICKLY PEAR and Silver creeks.

Helenapolis, TURKEY: see YALOVA.

Helenaveen (he-LAI-nah-VAIN), village, NORTH BRABANT province, SE NETHERLANDS, 20 mi/32 km ENE of EINDHOVEN, at center of DE PEEL region; 51°23′N 05°55′E. Marshy area crossed by branch of ZUIDWILLEMSVAART canal. Dairying; livestock; sugar beets, vegetables.

Helene, Greece: see MAKRONISOS.

Helen Glacier, flows into the sea on the QUEEN MARY COAST, WILKES LAND, EAST ANTARCTICA; 66°40′S 93°55′E.

Helen Glacier Tongue, the floating extension of HELEN GLACIER, on the QUEEN MARY COAST of WILKES LAND, EAST ANTARCTICA; 66°33′S 94°00′E.

Helen Mine (HE-len MEIN) on **Helen** (HE-len), unincorporated village, central ONTARIO, E central CANADA, near Wawa Lake, 110 mi/177 km NNW of SAULT SAINTE MARIE, and included in MICHIPICOTEN township; 48°01′N 84°44′W. Iron-mining center.

Helenopolis, TURKEY: see HARAN.

Helensburgh (HE-linz-buh-ruh), town, E NEW SOUTH WALES, SE AUSTRALIA, 26 mi/42 km S of SYDNEY, and 21 mi/34 km N of WOLLONGONG, and at S end of ROYAL NATIONAL PARK; 34°11′S 151°00′E. Hindu temple. Waterfalls at nearby Kelly's Falls. Originally called Camp Creek.

Helensburgh (HEL-uhnz-buhr), town (2001 population 14,626), ARGYLL AND BUTE, W Scotland, on the CLYDE, at S end of GARE LOCH, and 8 mi/12.9 km NW of DUMBARTON; 56°01′N 04°44′W. Resort and port for Clyde and Gare Loch steamers. Birthplace of John Logie Baird, pioneer of television. Formerly in STRATHCLYDE, abolished 1996.

Helge å (HEL-ye-O), river, c. 200 mi/322 km long, S SWEDEN; rises NE of LJUNGBY; flows generally S, through LAKE MÖCKELN, past KRISTIANSTAD, to HANÖ BAY of BALTIC SEA at ÅHUS.

Helgeland (HEL-guh-lahn), region in S part of NORDLAND county, N central NORWAY; extends N-S between TRONDHEIM and RANA fjords; includes BØRGEFJELL mountains, SVARTISEN glacier, VEFSNA valley, and off-coast islands of HESTMONA, DØNNA, and Vega. Agriculture; lumber; manufacturing; hydroelectric power plants; fishing; iron mining (in Rana, N part of

Helgeland). Old Norse name *Hálogaland* covered area of entire present Nordland county.

Helgenæs, DENMARK: see MOLS.

Helgoland (HEL-go-lahnd), island (□ 0.25 sq mi/0.6 sq km; 1994 population 1,730), SCHLESWIG-HOLSTEIN, NW GERMANY, in the NORTH SEA; 54°11′N 07°53′E. Formed of red sandstone, it rises to c.200 ft/60 m above the sea and is largely covered with grazing land. Popular tourist resort and a center for scientific, especially ornithological research. Strategically located near the mouths of the WESER and the ELBE rivers, Helgoland was captured by the Danes in 1714, was occupied by the English in 1807, and was formally ceded to ENGLAND by DENMARK in 1814. In exchange for rights in AFRICA, England gave the island to Germany in 1890. The Germans installed fortifications, which were razed after World War I according to the terms of the Treaty of Versailles. However, Germany refortified Helgoland in 1936 and used it as a naval base in World War II. In 1947, British occupation authorities, after evacuating the islanders (mostly fishermen), blew up the fortifications and part of the island in one of the largest known nonatomic blasts. The island was largely rebuilt after British occupation forces returned it to West Germany in 1952. Sometimes spelled HELIGOLAND.

Helicon, Greece: see HELIKON.

Heliconia (ai-lee-KO-nee-ah), town, ☉ Heliconia municipio, ANTIOQUIA department, NW central COLOMBIA, on slopes of Cordillera CENTRAL, 12 mi/19 km WSW of MEDELLÍN; 06°12′N 75°44′W. Elevation 4,724 ft/1,440 m. In agricultural region (coffee, plantains, corn). Salt and coal mines nearby.

Heligoland, GERMANY: see HELGOLAND.

Helikon (e-lee-KON), Greek *Elikón*, mountain group, c.20 mi/32 km long, in BOEOTIA prefecture, CENTRAL GREECE department; rises to 5,736 ft/1,748 m; 38°20′N 22°50′E. Formed part of the border between ancient Boeotia and PHOCIS. In Greek legend the abode of the Muses and sacred to Apollo. The fountains of Hippocrene and Aganippe are on its slopes. The temple of the Muses was situated in the E part of the mountain, at the foot of which were Thespiae and Ascra, home of the poet Hesiod. Also Helicon.

Helinge'er, CHINA: see HORINGER.

Helıópolis (AI-lee-O-po-lees), city (2007 population 14,152), NE BAHIA, BRAZIL, near SERGIPE border, 25 mi/41 km NNE of Ribeira do Pombal; 10°41′S 38°17′W.

Héliopolis (ai-lyo-po-LEES), town, GUELMA wilaya, NE ALGERIA. Has hot springs and a Roman bath, restored for use today. Houari Boumeddiene, head of state 1965–1978, born here. Also called Hamam Berda.

Heliopolis (hel-YO-PO-lis) [Greek=city of the sun], residential suburb of CAIRO, Lower EGYPT, c.5 mi/8 km NE of Cairo city center. Largely middle and upper class. Cairo International Airport to E. Named after the ancient city whose site is NW. It was noted as the center of sun worship, and its god Ra (or Re) was the state deity until THEBES became capital (c.2100 B.C.E.). The god Amon was then joined with Ra as Amon-Ra or Amon Re. Under the New Empire (c.1570 B.C.E.–c.1085 B.C.E.), Heliopolis was capital of N Egypt. The obelisks called Cleopatra's Needles were erected here. Its schools of philosophy and astronomy declined after the founding of ALEXANDRIA in 332 B.C.E., but the city never wholly lost importance until the Christian Era. The Egyptian name was On; little of the ancient city remains.

Heliopolis, LEBANON: see BAALBEK.

Helix (HEE-liks), village (2006 population 182), UMATILLA county, NE OREGON, 15 mi/24 km NNE of PENDLETON; 45°51′N 118°39′W. Fruit, vegetables; cattle.

Hell, village, NORD-TRØNDELAG county, central NORWAY, on Strindafjorden (a bay in TRONDHEIMSFJORDEN), 16 mi/26 km E of TRONDHEIM. Junction for NORDLAND railroad and line to STORLIEN and Ös-

TERSUND (SWEDEN). Stone carvings from 4,000 B.C.E.–2,000 B.C.E.

Hell, village, LIVINGSTON county, SE MICHIGAN, 15 mi/24 km NW of ANN ARBOR; 42°26′N 83°59′W. In small lakes area. Resort; agriculture (grain, apples; cattle). Pinckney State Recreation Area to S.

Hellada River, Greece: see SPERCHIOS RIVER.

Hella, El, EGYPT: see HILLA, EL.

Hellam, Pennsylvania: see HALLAM.

Hellas: see GREECE.

Hell-Bourg (el-BOOR), village, central RÉUNION island, an overseas dependency of FRANCE in the INDIAN OCEAN; 11 mi/18 km SW of SAINT-ANDRÉ; elevation c.3,000 ft/915 m; 21°03′S 55°32′E. Health resort in SALAZIE cirque; Roman Catholic mission school, sanatorium.

Hellebæk (HE-luh-bek), town (2000 population 5,200), FREDERIKSBORG county, SJÆLLAND, DENMARK, on the ØRESUND, and 13 mi/21 km NE of HILLERØD; 56°04′N 12°34′E. Recreation area here; beaches.

Hellendoorn (HE-luhn-dawrn), town, OVERIJSSEL province, E NETHERLANDS, 9 mi/14.5 km WNW of ALMELO; 52°23′N 06°27′E. Regge River to E. Dairying; cattle, hogs; grain, vegetables. Manufacturing (food processing).

Hellenic League, Greece: see PELOPONNESIAN LEAGUE.

Hellenthal (HEL-len-tahl), village, NORTH RHINE-WESTPHALIA, W GERMANY, near Belgian border, 22 mi/35 km S of DÜREN; 50°30′N 06°26′E. Climatic health resort; metalworking; plastics processing. Reservoir nearby.

Hellertown (HE-luhr-toun), borough (2006 population 5,617), NORTHAMPTON county, E PENNSYLVANIA, suburb 3 mi/4.8 km SSE of downtown BETHLEHEM, on Saucon Creek. Manufacturing (apparel, bricks); sand quarry. Agriculture (corn, hay, apples, soybeans; dairying). Campus of Lehigh University to NW. Lost River Caverns to SE. Settled c.1740, incorporated 1872.

Helles, Cape, (Turkish=*Teke*) southernmost point of the GELIBOLU PENINSULA, NW TURKEY; 40°02′N 26°12′E. It commands the entrance to the DARDANELLES.

Hellesdon (HELZ-duhn), suburb (2001 population 18,980), NORFOLK, E ENGLAND, on WENSUM RIVER, 2 mi/3.2 km NE of NORWICH; 52°39′N 01°15′W. Agriculture.

Hellespont, TURKEY: see DARDANELLES.

Hellevoetsluis (HE-luh-voot-SLOIS), town (2001 population 36,416), SOUTH HOLLAND province, SW NETHERLANDS, in S part of VOORNE region, and 17 mi/27 km WSW of ROTTERDAM; 51°50′N 04°08′E. On N shore of HARINGVLIET estuary, at SW entrance of Voorns Canal. NORTH SEA 6 mi/9.7 km to W, Haringvliet Dam 3 mi/4.8 km to W. Dairying; cattle, hogs; sugar beets, vegetables; manufacturing (tires). Former navy yard; embarkation point (1688) for William III on his expedition against James II of England.

Hell Gate, narrow channel of the EAST RIVER, SE NEW YORK, between WARDS ISLAND and ASTORIA, QUEENS, NEW YORK city; 40°47′N 73°56′W. Named Hellegat by the Dutch navigator Adriaen Block, who passed through it into LONG ISLAND SOUND in 1614, it was dangerous to ships because of its strong tidal currents and rocks. Cleared of all obstacles, it allows ocean-going vessels to sail between NEW YORK HARBOR and Long Island Sound. It is crossed by the TRIBOROUGH highway bridge and by the Hell Gate railroad bridge.

Hell Hole Reservoir (□ 2 sq mi/5.2 sq km), PLACER county, E central CALIFORNIA, on RUBICON River, in Eldorado National Forest, 12 mi/19 km W of LAKE TAHOE; 39°03′N 120°07′W. Maximum capacity 209,000 acre-ft. Formed by Hell Hole Reservoir Dam (410 ft/125 m high), built (1966) by the Bureau of Reclamation for irrigation; also used for power generation, water supply, and recreation. Desolation Valley Wilderness Area nearby.

Helliar Holm, Scotland: see SHAPINSAY.

Hellier (HEL-uhr), unincorporated village, PIKE county, E KENTUCKY, near VIRGINIA state line, in CUMBERLAND MOUNTAINS, 13 mi/21 km SE of PIKEVILLE. Bituminous-coal mining. Jefferson National Forest to SE.

Hellifield (HEL-i-FEELD), agricultural village (2001 population 1,502), NORTH YORKSHIRE, N ENGLAND, near RIBBLE RIVER, 9 mi/14.5 km WNW of SKIPTON; 54°00′N 02°13′W. Previously an important railway junction.

Hellín (el-YEEN), city, ALBACETE province, SE SPAIN, in CASTILE-LA MANCHA; 38°31′N 01°41′W. An important marketing and distributing center, it is noted for its sulfur mines, worked since Roman times. Clay and gypsum are also quarried, and footwear, chemicals, and textiles are manufactured.

Hellingly (HEL-ing-LYE), village (2006 population 1,551), East SUSSEX, SE ENGLAND, on Cuckmere River, and 8 mi/12.9 km N of EASTBOURNE; 50°53′N 00°16′E. Former agricultural market with flour mills. Has 12th-century church.

Hellisandur (HET-lis-SAHN-duhr), village, Snæfellsnes county, W ICELAND, NW Snæfellsnes Peninsula, on BREIĐAFJÖRĐUR; 64°55′N 23°53′W. Cod and haddock fisheries; shrimp, lobster. Formerly called Sandur.

Hells Canyon, S IDAHO, extends c.125 mi/201 km N along the OREGON state line, reaches a maximum depth of c.7,900 ft/2,408 m. Sometimes called Grand Canyon of the SNAKE RIVER; the greatest of the Snake's many gorges and one of the deepest in the world.

Hells Canyon Reservoir, NE OREGON and W IDAHO, on SNAKE RIVER, at upper (S) end of HELLS CANYON, 60 mi/97 km E of LA GRANDE (Oregon); c.25 mi/40 km long; 45°10′N 116°16′W. Maximum capacity 200,000 acre-ft. Oxbow Dam at SW tip. Formed by Hells Canyon Dam (218 ft/66 m high), built by the Idaho Power Company, for power generation. HELLS CANYON National Recreation Area N of dam.

Hell's Gate National Park, RIFT VALLEY province, KENYA, in GREAT RIFT VALLEY S of Lake NAIVASHA. Features scenic cliffs, major gorge; buffalo, antelope, baboon.

Hellshire Hills (□ 45 sq mi/72 sq km), hills (800 ft/245 m) and KINGSTON suburb, SAINT CATHERINE parish, S JAMAICA, 6 mi/9.7 km WSW of greater PORTMORE section of twin city of Kingston; 17°53′N 76°58′W. These rolling limestone hills occupy the S promontory of the parish. A varied coastline of white sandy beach and rugged cliffs borders the S and large sugarcane estates of Innswood and Bernard Lodge separate it by 5 mi/8 km from SPANISH TOWN on the N. Further NE Portmore and Port Henderson housing estates link the area to Kingston via a causeway. West Point along the coastline at the mouth of Salt Island Creek emerges into Salt Island Lagoon at Galleon Harbor, and Old Harbor is to the W. Identified as an ideal development area with MANATEE BAY and Central Highlands.

Hell's Kitchen, New York: see CLINTON.

Hell-Ville, Réunion: see SALAZIE.

Hellyor Holm, Scotland: see SHAPINSAY.

Helmand (hel-MAHND), province (2005 population 767,300), SW AFGHANISTAN, on PAKISTAN (S) border; ☉ LASHKARGAH. Bordered by the provinces of GHOR and DAYKUNDI (NW), ORUZGAN (NE), FARAH and NIMRUZ (W), and KANDAHAR (E). Sand desert (Registan) in S and mountainous in N; major mountains include the Baghran, Naozad, Khan Neshin, Malik Dokan, and Kushtagan. The Helmand valley is a major area of opium poppy cultivation. Population largely Pashtun, with Hazara in N, Baluch in S. Agriculture (barley, cotton, and wheat; variety of fruit); sheep, goats, cattle, camels, horses, donkeys. Irrigated by HELMAND River; hydroelectric station at GIRISHK.

Helmand (hel-MAHND), c.700 mi/1,127 km long, longest river in AFGHANISTAN; rises near the Unai Kotal, c.50 mi/80 km W of KABUL, NE Afghanistan; flows generally SW to the Seistan Basin, SW Afghanistan, where it dissipates in the Hamun-e Helmand (31°12′N 61°34′E), a marshy lake that extends into IRAN. Drains 100,386 sq mi/260,000 sq km. Major tributaries include the Kaj Rud, Tirin, Arghandab, Tarnak, and Arghastan. The Helmand's ancient irrigation and river-control system was destroyed by Jenghiz Khan (13th century) and Tamerlane (14th century); the modern irrigation works are vital to both Iranians and Afghans, and in times of drought there are disputes over water rights. The Helmand Valley Authority extensively developed the region, improving irrigation and flood control. Multipurpose projects include storage and diversion dams and electric power generating (hydroelectric plant at GIRISHK) for industrial centers of KANDAHAR. Also spelled Helmund, Hilmand.

Helmarshausen (hel-muhrs-HOU-suhn), district of BAD KARLSHAFEN, central GERMANY, on the DIEMEL RIVER, 10 mi/16 km NNE of HOFGEISMAR; 51°38′N 09°29′E. Has ruined church and abbey from 11th century. The monastery was a jewelry and book-writing center until 15th century.

Helmbrechts (HELM-brekhts), town, UPPER FRANCONIA, BAVARIA, S GERMANY, in the FRANCONIAN FOREST, 10 mi/16 km SW of HOF; 50°14′N 11°43′E. Cloth-making and plastics industry. Limestone quarries in area. Chartered 1449. Completely destroyed in 1839 fire.

Helme River (HEL-me), c.55 mi/89 km long, central Germany; rises near S foot of the lower HARZ 5 mi/8 km SW of BAD SACHSA; 51°32′N 10°28′E; flows generally ESE, through the valley GOLDENE AUE, to the UNSTRUT RIVER. 3 mi/4.8 km SE of ARTERN.

Helmetta, borough (2006 population 2,023), MIDDLESEX county, E NEW JERSEY, 8 mi/12.9 km S of NEW BRUNSWICK; 40°22′N 74°25′W. Former snuff manufacturing center in 19th century. In rapidly suburbanizing area.

Helmond (HEL-mawnt), city (2001 population 74,740), NORTH BRABANT province, SE NETHERLANDS, on the AA RIVER, and the ZUID-WILLEMSVAART canal; 51°28′N 05°40′E. Dairying; cattle, hogs, poultry; grain, vegetables; manufacturing (fabricated metal products, sinks, machinery, consumer goods, tools, textiles). Its fifteenth-century castle serves as the town hall; castle to NW.

Helmsdale (HELMZ-dail), town, HIGHLAND, N Scotland, on MORAY FIRTH, at mouth of HELMSDALE RIVER, and 21 mi/34 km NE of DORNOCH; 58°07′N 03°40′W. Fishing port. Has ruins of ancient castle of dukes of SUTHERLAND.

Helmsdale River (HELMZ-dail), 20 mi/32 km long, N HIGHLAND, N Scotland; formed by two short headstreams 7 mi/11.3 km NNW of Kildonan; flows SE, past Kildonan, to MORAY FIRTH at HELMSDALE.

Helmsley (HELMZ-lee), town (2001 population 3,111), NORTH YORKSHIRE, N ENGLAND, on Rye River, and 13 mi/21 km NW of MALTON; 54°15′N 01°04′W. Stone quarrying; agricultural market. Has ruins of 12th-century castle.

Helmstedt (HELM-stet), city, LOWER SAXONY, N central GERMANY; 52°14′N 11°00′E. Manufacturing includes bricks, lignite, machinery, and yarn. Founded in the 9th century and later (15th–16th century) was a member of the HANSEATIC LEAGUE. From 1576 to 1810 it was the seat of a noted university, whose Renaissance-style buildings still stand. From 1945 until German reunification in 1990, Helmstedt was a major frontier post between EAST and WEST GERMANY.

Helmund, AFGHANISTAN: see HELMAND.

Helmville, village, POWELL county, W MONTANA, on Nevada Creek, near confluence with BLACKFOOT RIVER, and 50 mi/80 km WNW of HELENA. In agricultural and ranching region. Site of well-known Labor Day Rodeo. Helena National Forest to E, Nevada Lake Reservoir to SE, Garnet Range to S.

Helmyaziv (HEL-myah-zif) (Russian *Gelmyazov*), village, NE CHERKASY oblast, UKRAINE, 14 mi/23 km NW of ZOLOTONOSHA, on the Supiy River, left bank tributary of the DNIEPER River; 49°49′N 31°21′E. Elevation 459 ft/139 m. Formerly a town.

Helong (HUH-LUNG), town, ⊙ Helong county, E JILIN province, NE CHINA, 20 mi/32 km S of YANJI, near Tumen River (North Korea border); 42°31′N 128°59′E. Tobacco, jute; lumbering; coal mining.

Helotes (ai-LO-tes), town (2006 population 6,460), BEXAR county, S central TEXAS, suburb 14 mi/23 km NW of downtown SAN ANTONIO; 29°34′N 98°41′W. Agricultural area (cattle; wheat, corn, peanuts). Manufacturing (nut processing; plastic products, marble products, building materials, fabricated metal products).

Helpa (hel-yuh-PAH), Slovak *Hel'pa*, village, STREDOSLOVENSKY province, central SLOVAKIA, on HRON RIVER, on railroad, and 38 mi/61 km ENE of BANSKÁ BYSTRICA; 48°52′N 19°58′E. Lumbering; wheat and barley growing, cattle breeding. Folk costume and architecture (log cabins), annual festival of folk song and dance.

Hel Peninsula, Polish *Półwysep Hel* (poon-VEE-sep HEL) or *Mierzeja Pucka* (mee-ZEE-yah POOT-skah), German *Putziger Nehrung*, Gdansk province, N POLAND, extends SE into BALTIC SEA; separates PUCK BAY (W) from the Baltic Sea; 21 mi/34 km long, up to 2 mi/3.2 km wide. Hel, German *Hela*, village on S tip of peninsula, is railroad-spur terminus, fishing port, health resort; lighthouse. Jastarnia, village on middle of peninsula, is also a fishing port and health resort. In 17th century, present peninsula was still a series of islands; the peninsula was partially inundated in 1904 and 1914 by high seas. Population formerly largely Kashubs, with Germans only on SE tip (yielded to DANZIG in 1453); present population is Polish.

Helper, town (2006 population 1,886), CARBON county, E central UTAH, in canyon of PRICE RIVER, 6 mi/9.7 km NNW of PRICE; elevation 5,840 ft/1,780 m; 39°41′N 110°51′W. Shipping center for coal-mining and agriculture (fruit, hay, sugar beets) area. Known as Pratts Siding until 1892, renamed for locomotives called "helpers" that were stored here. SCOFIELD DAM, nearby on Price River, is used for irrigation. Price Canyon Recreation Area to NW; part of Ashley National Forest to NE. Uinta National Forest to NW; Manti–La Sal National Forest to W. Scofield Reservoir and State Park to NW. Western Mining and Railroad Museum. Junction of railroad spur to Wattis. Settled 1883, incorporated 1907.

Helpfau-Uttendorf (HEL-pfou–UT-ten-dorf), township, W UPPER AUSTRIA, in the INNVIERTEL, 8 mi/12.9 km SSE of BRAUNAU; 48°09′N 13°07′E. Brickworks, brewery; dairy farming.

Helsa (HEL-sah), village, HESSE, central GERMANY, 10 mi/16 km SE of KASSEL; 51°16′N 09°31′E. Has 16th-century church.

Helsby (HELZ-bee), village (2001 population 4,701), NW CHESHIRE, NW ENGLAND, near MERSEY RIVER, 8 mi/12.9 km NE of CHESTER; 53°16′N 02°46′W. Previously manufactured electric cables. Wood products. Residential.

Helsingborg (HEL-seeng-BOR-yuh), town, SKÅNE county, S SWEDEN, seaport on ÖRESUND, connected by ferry with HELSINGØR (DENMARK); 56°03′N 12°43′E. Manufacturing (processed chemicals, medicine, electrical goods, mineral water). Trade center and stronghold since the ninth century; destroyed during Danish-Swedish conflicts (seventeenth century), passing ultimately to Sweden. Rebuilt after unsuccessful Danish attack (1710). Modern industrial development dates from mid-nineteenth century. Well-preserved castle (twelfth–fifteenth century), Church of St. Mary (thirteenth–fifteenth century), numerous half-timber houses. Airport.

Helsinge (HEL-sing-uh), city (2000 population 6,339), FREDERIKSBORG county, SJÆLLAND, DENMARK, 7 mi/11.3 km NW of HILLERØD; 56°03′N 12°09′E.

Helsingfors, FINLAND: see HELSINKI.

Helsingør (HEL-sing-UHR) or **Elsinore** (EL-si-NOR), city, FREDERIKSBORG county, E DENMARK, on the ÈRESUND, opposite Hälsingborg (SWEDEN); 56°03′N 12°30′E. Industrial center, fishing port, and summer resort. Manufacturing include ships, rubber, machinery, beer, and textiles. Main ferry crossing point to Halsingborg. Known since the thirteenth century, Helsingør experienced its greatest prosperity from the fifteenth century to 1857, when it served as the port where the Danish kings collected tolls from ships passing through the Øresund. It is the site of Kronborg castle (1754–1785; completely restored 1925–1937), which, although the strongest fortress in Denmark at the time, was taken by the Swedes in 1660. The castle is now a maritime museum and is also used for performances of Shakespeare's *Hamlet,* which is set here. Also of note in Helsingør is the Church of Saint Mary (fifteenth century).

Helsinki (HEL-sing-kee), Swedish *Helsingfors,* city (1999 population 548,720), ⊙ Finland and Uudenmaan province, S Finland on a peninsula on the Gulf of Finland; elevation 17 ft/5 m; 60°10′N 25°00′E. Sheltered by islands, the city is a natural seaport (blocked by ice January–May) and the commercial, administrative, and intellectual center of Finland. It has machine shops, shipyards, food-processing plants, textile mills, clothing and china factories, printing plants, and electronics manufacturing. Also Finland's main tourist city, it is linked by ferry with Tallinn (Estonia), Stockholm (Sweden), and St. Petersburg (Russia). Founded in 1550 by Gustavus I of Sweden, it was devastated by a great fire in 1808; rebuilt as a well-planned, spacious metropolis, which includes the architecture of Alvar Aalto. Grew rapidly after Alexander I of Russia moved capital from Turku in 1812. When the University of Helsinki (est. 1640) was moved from Turku in 1828, Helsinki became the center of Finnish nationalism. The construction of the 1st Finnish railroad (1860), connecting Helsinki and Hämeenlinna, led to renewed prosperity. The older part of the city houses the state council building, the residence of the president, the University of Helsinki, the Church of St. Nicholas, the national art gallery, and the impressive railroad station (designed by Eliel Saarinen). Other cultural landmarks include an opera house, the house of representatives building, the technical university (1879), and the sports stadium (scene of the 1952 Olympics). SUOMENLINNA (Swed. *Sveaborg*) fortress, 2 mi/3.2 km SE, is tourist attraction. International airport at Vaanta to N. Distribution point for Vyborg-Finland natural-gas pipeline.

Helston (HEL-stun), town (2001 population 10,233), CORNWALL, SW ENGLAND, on Cober River, and 10 mi/16 km WSW of FALMOUTH; 50°06′N 05°17′W. Agricultural market; tanning industry; tourist site. Known for its Furry (or Flora) Dance, old flower and music festival in May.

Heltau, ROMANIA: see CISNĂDIE.

Helvecia (el-VAI-see-ah), town (1991 population 4,670), ⊙ Garay department (□ 1,555 sq mi/4,043 sq km), E central SANTA FE province, ARGENTINA, on SAN JAVIER RIVER, and 55 mi/89 km NE of SANTA FE; 31°06′S 60°05′W. In agricultural area (livestock; grain). Site of 19th-century Swiss colony.

Helvecia (el-VEE-see-ah), village, SE BAHIA, BRAZIL, on Rio Peruipe, 30 mi/49 km W of NOVA VIÇOSA; 17°51′S 39°39′W.

Helvellyn (hel-VEL-in), mountain (3,113 ft/949 m), in the Lake District, CUMBRIA, NW ENGLAND, 5 m/8 km SE of KESWICK; 54°32′N 03°02′W.

Helvetia (hel-VEE-zhah), unincorporated village, Brady township, CLEARFIELD county, W central PENNSYLVANIA, 5 mi/8 km S of DU BOIS, on Stump Creek; 41°02′N 78°46′W. Bituminous-coal region.

Helvetia (hel-VAI-tsee-ah), region of central EUROPE, occupying the plateau between the ALPS and the JURA range. The name is derived from the Roman term for its inhabitants, the predominantly Celtic Helvetii, who were defeated (58 B.C.E.) at BIBRACTE by Julius Caesar in the Gallic Wars. The Helvetii later prospered under Roman rule; their achievements are evidenced by the remains at AVENCHES. Helvetia corresponds roughly to the W part of modern SWITZERLAND, and the name is still used in poetic reference to that country and on its postage stamps.

Helvetic Republic, former Swiss state established under French auspices, 1798–1803. The survival of the Helvetic Republic until 1803 was largely due to the presence of French troops, since the Swiss were hostile to centralization. In February 1803, Napoleon, imposing the Act of Mediation, established a confederation of 19 cantons, with a federal diet subservient to FRANCE.

Helvick Head (HEL-vik), Gaelic *Ceann Heilbhic*, cape, S WATERFORD county, S IRELAND, at entrance to Dungarvan Bay; 52°03′N 07°32′W. Elevation 230 ft/70 m. Known for its views.

Helvoirt (HEL-vawrt), village, NORTH HOLLAND province, S NETHERLANDS, 5 mi/8 km SW of 's-HERTOGENBOSCH; 51°38′N 05°14′E. Loonse en Drunense Duinen National Park to W; De Ijzeren Man Recreational Park to N; Castle Zwijnsb to NW. Dairying; cattle, hogs; fruit, grain, vegetables; manufacturing (wheelbarrows).

Helwan (hel-WAHN), town, N EGYPT, on the E bank of the NILE River, opposite the ruins of MEMPHIS, suburb of CAIRO. Manufacturing includes iron and steel, cement, and textiles; there is a food-processing and motor vehicle industry. Has a metallurgical research center and an astronomical observatory. The town, the site of ancient settlements, is also a health resort, long known for its hot sulfur springs. An ancient burial chamber, one of the largest in the world, was discovered here in 1946. Also spelled Hilwan.

Hem, town (□ 4 sq mi/10.4 sq km), NORD department, NORD-PAS-DE-CALAIS region, N FRANCE, a suburb of both LILLE (10 mi/16 km W) and ROUBAIX (7 mi/11.3 km N), in the urban agglomeration of greater Lille; 50°39′N 03°12′E. Textile industry. Has modern church (completed 1958) with noteworthy translucent walls.

Hematheia, Greece: see EMATHEIA.

Hemau (HE-mou), town, UPPER PALATINATE, BAVARIA, S GERMANY, 14 mi/23 km WNW of REGENSBURG; 49°03′N 11°47′E. Woodworking; brewing.

Hemavati River (hai-MAH-vuh-tee), c.130 mi/209 km long, in KARNATAKA state, S INDIA; rises in WESTERN GHATS W of MUDIGERE; flows SSE past SAKLESHPUR, E past HOLE NARSIPUR, and S to Krishnaraja Sagara (reservoir on KAVERI River). Feeds several irrigation channels.

Hemby Bridge (HEM-bee), unincorporated town (□ 1 sq mi/2.6 sq km; 2006 population 1,782), UNION county, S NORTH CAROLINA, residential suburb 13 mi/21 km SE of downtown CHARLOTTE; 35°07′N 80°37′W. Service industries; retail trade.

Hemel Hempstead (HEM-uhl), town (2001 population 81,143), HERTFORDSHIRE, SE ENGLAND, 7 mi/11.3 km NW of WATFORD; 51°45′N 00°28′W. Designated one of the New Towns in 1946 to alleviate overpopulation in London. Market town and London suburb. Manufacturing includes paper, electrical and machine products, office machinery, and photographic apparatus.

Hemer (HE-mer), town, NORTH RHINE–WESTPHALIA, W GERMANY, 3 mi/4.8 km E of ISERLOHN; 51°23′N 07°46′E. Manufacturing of machinery, wire, paper; metalworking. Chartered 1936.

Hemet, city (2000 population 58,812), RIVERSIDE county, S CALIFORNIA; suburb 28 mi/45 km SE of RIVERSIDE, in San Jacinto valley; 33°44′N 117°00′W. Citrus, avocados, vegetables, grain; nursery stock; poultry, eggs; dairying. Manufacturing (carbon products, furniture, motor vehicles). In a county marked by fast growth during the 1970s and 1980s as a result of increased local agribusiness and the development of the aircraft industry in the area, much work has been provided, and Hemet is one of the many cities that has benefited from these developments. Seat of Mount San Jacinto College (two-year). A special tourist attraction is the Ramona Outdoor Play, staged annually by residents of the twin cities of Hemet and SAN JACINTO. SAN DIEGO AQUEDUCT passes N-S to W, branches from COLORADO RIVER AQUEDUCT to N; Soboba Indian Reservation to E, part of San Bernardino National Forest to E. Incorporated 1910.

Hemiksem (HAI-mik-sem), commune (2006 population 9,709), Antwerp district, ANTWERPEN province, N BELGIUM, on SCHELDT RIVER, and 5 mi/8 km SW of ANTWERP; 51°09′N 04°21′E. Manufacturing. Also spelled Hemixem.

Hemingford, village (2006 population 895), BOX BUTTE county, NW NEBRASKA, 18 mi/29 km NW of ALLIANCE; elevation 4,259 ft/1,298 m; 42°19′N 103°04′W. Poultry products, grain, sugar beets, potatoes; livestock. Fishing lures. Box Butte Reservoir State Recreation Area to N (Dawnes county).

Hemingway, town (2006 population 520), WILLIAMSBURG county, E SOUTH CAROLINA, 33 mi/53 km SSE of FLORENCE; 33°45′N 79°26′W. Manufacturing includes wire, consumer goods, foods, textiles, apparel, plastic products; poultry processing. Agriculture includes poultry, livestock; tobacco, grain, soybeans, cotton.

Hemis Gompa, INDIA: see HIMIS GOMPA.

Hemixem, BELGIUM: see HEMIKSEM.

Hemlingby (HEM-leeng-BEE), residential village, GÄVLEBORG county, E SWEDEN, just SSE of GÄVLE; 60°39′N 17°10′E.

Hemlock, village (2000 population 1,585), SAGINAW county, E central MICHIGAN, 14 mi/23 km W of SAGINAW; 43°25′N 84°13′W. Farm area. Manufacturing (medical equipment, building materials).

Hemlock (HEM-lahk), village (2006 population 147), PERRY county, central OHIO, 9 mi/14 km S of NEW LEXINGTON; 39°35′N 82°09′W.

Hemlock Lake, one of the FINGER LAKES, LIVINGSTON county, W central NEW YORK, 28 mi/45 km S of ROCHESTER; 7 mi/11.3 km long; 42°44′N 77°37′W. In agricultural area. Supplies water to Rochester.

Hemmingen (HE-ming-uhn), suburb of HANOVER, LOWER SAXONY, N GERMANY, on S edge of city; 52°20′N 09°44′E. Printing.

Hemmingford (HE-meeng-fuhrd), canton (□ 60 sq mi/156 sq km; 2006 population 1,753), Montérégie region, S QUEBEC, E CANADA, 29 mi/46 km from HUNTINGDON; 45°05′N 73°35′W.

Hemmingford (HE-meeng-fuhrd), village, Montérégie region, S QUEBEC, E CANADA, 32 mi/51 km S of MONTREAL, near U.S. (NEW YORK) border; 45°03′N 73°36′W. Agriculture; apple growing. Established in 1877.

Hemmingstedt (HEM-ming-stet), village, SCHLESWIG-HOLSTEIN, NW GERMANY, in the S DITHMARSCHEN, 3 mi/4.8 km S of HEIDE; 54°09′N 09°04′E. In oil-producing region; has refinery connected by pipeline with BRUNSBÜTTEL. Has monument (built 1900) commemorating defeat (1500) of Danes by peasants of the Dithmarschen.

Hemmoor (HEM-mawr), town, LOWER SAXONY, N GERMANY, 24 mi/39 km NE of BREMERHAVEN; 53°43′N 09°08′E. Metalworking; food processing; printing. Chartered 1982.

Hemnesberget (HEM-nais-BAR-guh), village, NORDLAND county, N central NORWAY, on RANA, 16 mi/26 km WSW of Missouri. Port with boatyards; lumber mills. Heavily damaged (1940) in World War II.

Hemp, NORTH CAROLINA: see ROBBINS.

Hemphill, county (□ 912 sq mi/2,371.2 sq km; 2006 population 3,412), extreme N TEXAS; ⊙ CANADIAN; elevation 2,500 ft/762 m–3,000 ft/914 m; 35°49′N 100°16′W. On high plains of the PANHANDLE, and bounded E by OKLAHOMA state line. Drained by CANADIAN and WASHITA rivers. Cattle-ranching region in N, producing also horses; some agriculture in S (wheat, grain sorghum; hay); oil and natural gas. Hunting, fishing in LAKE MARVIN and Canadian River regions. Black Kettle National Grassland in NE, includes small Lake Marvin, on small tributary of Canadian River. Formed 1876. Acquired parts of ELLIS and ROGER MILLS counties (Oklahoma), in resurvey of 100th meridian (1930).

Hemphill, town (2006 population 1,074), ⊙ SABINE county, E TEXAS, near LOUISIANA state line, c.50 mi/80 km E of LUFKIN; elevation 267 ft/81 m; 31°20′N 93°50′W. Trade point in timber; cattle, poultry; vegetables, fruit area; manufacturing (lumber). Recreation area. Located on W edge of Sabine National Forest; large TOLEDO BEND RESERVOIR (SABINE RIVER, Louisiana state line) to E; SAM RAYBURN RESERVOIR to SW. Incorporated as city 1939.

Hempstead, county (□ 741 sq mi/1,926.6 sq km; 2006 population 23,347), SW ARKANSAS; ⊙ HOPE; 33°43′N 93°39′W. Bounded SW by RED and LITTLE rivers and N by LITTLE MISSOURI RIVER and Hickory Creek. Agricultural area (soybeans; cattle, hogs, chickens); timber. Manufacturing at Hope. Part of MILLWOOD LAKE Reservoir (Little River) in W; Hope Wildlife Management Area at center; Bais D'Arc Wildlife Management Area to SW; Old Washington Historic State Park at WASHINGTON at center of county. Formed 1818.

Hempstead, town (□ 144 sq mi/374.4 sq km), NASSAU county, SE NEW YORK, on W LONG ISLAND; 40°39′N 73°36′W. Nation's most populous town. Occupies the S half of Nassau county. The Village of Hempstead (1990 population 49,453; 2000 population 56,554), located in the N central part of Hempstead township and having only 1% of the county's area, has 7% of the county's population. It is a retail center for the area. Diversified light manufacturing and extensive commercial services. Village population is predominantly African-American, but Latinos and whites make up a substantial proporation. The town grew significantly in the 1970s with the construction of nearby freeways, large retail outlets, and the expansion of regional suburban industries. Settled in 1644 by English colonists who named it for their old home in England, Hemel-Hempstead. Hofstra and Adelphi universities, and Nassau Community College. Has many colonial houses and monuments. Founded as a village, incorporated 1853.

Hempstead (hemp-STED), town (2006 population 6,837), ⊙ WALLER county, SE TEXAS, near BRAZOS RIVER, c.50 mi/80 km WNW of HOUSTON; 30°05′N 96°04′W. In agriculture, oil-producing, cattle area; light manufacturing. Nearby is Liendo, plantation of sculptor Elisabeth Ney (built 1853).

Hempstead Harbor, inlet of LONG ISLAND SOUND indenting N shore of W LONG ISLAND, SE NEW YORK. E of MANHASSET NECK, and c.5 mi/8 km N of HEMPSTEAD; 4 mi/6.4 km wide at entrance between MATINICOCK (E) and PROSPECT (W) points; 5 mi/8 km long. ROSLYN is at its head; GLEN COVE, SEA CLIFF, GLENWOOD LANDING, and GLEN HEAD are on or near its shores.

Hems, SYRIA: see HOMS.

Hemsbach (HEMS-bahkh), town, BADEN-WÜRTTEMBERG, SW GERMANY, 3 mi/4.8 km N of WEINHEIM; 49°35′N 08°43′E. Manufacturing (plastics); agriculture (fruit; vineyards).

Cross-references are shown in SMALL CAPITALS. The pronunciation guide is shown on page xix. The sources of population figures are shown on page xvii.

Hemse (HEM-SE), village, GOTLAND county, SE SWE-DEN, in S part of Gotland island, 25 mi/40 km S of VISBY; 57°14′N 18°22′E. Agriculture and trade center. Has thirteenth-century church.

Hemsedal (HEM-suh-dahl), village, BUSKERUD county, S NORWAY, in the Hemsedal (tributary valley of the upper HALLINGDAL), 65 mi/105 km WSW of LILLE-HAMMER. Agriculture; livestock raising; lumbering. Surrounded by the HEMSEDALSFJELL, and with several waterfalls, this is a favorite winter spot for tourists.

Hemsedalsfjell (HEM-suh-dahls-FYEL), mountain plateau in BUSKERUD, OPPLAND, and SOGN OG FJOR-DANE counties, S central NORWAY, between the upper HALLINGDAL and the HALLINGSKARVET (S), SOG-NEFJORDEN (W), JOTUNHEIMEN Mountains (N), and VALDRES region (E); c.70 mi/113 km long, 30 mi/48 km wide. Rises to 6,303 ft/1,921 m in the Jukleggi.

Hemsworth (HEMZ-wuhth), town (2001 population 13,965), WEST YORKSHIRE, N ENGLAND, 7 mi/11.3 km SE of WAKEFIELD; 53°36′N 01°21′W. Has grammar school founded 1546 and hospital founded 1555.

Hen (HAIN), village, BUSKERUD county, SE NORWAY, on BEGNA River, and 4 mi/6.4 km N of HØNEFOSS. Railroad junction. In agricultural and lumbering region. Lumber and wood-pulp mills; hydroelectric plant.

Henager (HE-nuh-guhr), town (2000 population 2,400), DE KALB county, NE ALABAMA, 13 mi/21 km NNW of FORT PAYNE; 34°38′N 85°44′W. Trucking business; textile manufacturing. Sequoyah Caverns located nearby.

Henan (HUH-nahn) [Chinese=south of the river] or **Yu** (YOO), province (□ 65,000 sq mi/169,000 sq km; 2000 population 91,236,854), NE CHINA; ☉ ZHENGZ-HOU; 34°00′N 114°00′E. It is sparsely settled in the mountainous W region but densely populated and cultivated in the E, which is the SW portion of the NORTH CHINA PLAIN. Although the climate is dry, the loess provides fertile soil irrigated by rich under-ground water. Henan is a major wheat- and cotton-producing province; other agricultural products include kaoliang, rice, millet, sweet potatoes, tobacco, fruit, oakleaf silk, and oilseed crops (sesame, peanuts). The province is well-watered, with the HUANG HE (Yellow River) flowing through the N section and the HUAI RIVER in the E; both are generally navigable only for small rivercraft. Floods and droughts, long suf-fered here, have been alleviated by the building (1960s) of both the Sanmen Dam on the Huang He in the SW and the People's Victory Canal (which diverts water from the Huang He to the Jin River). Foresta-tion efforts, as well as other irrigation and drainage programs, also have helped to combat droughts. In addition to its waterways, Henan has many good highways and a fine railroad system; the principal N-S and E-W railroad lines of China cross the province, intersecting at Zhengzhou. Coal, abundantly found here, and hydroelectric power from the Sanmen project supply burgeoning industries in Zhengzhou, LUOYANG, and KAIFENG. Has a growing variety of heavy industries, such as chemical works and tractor plants, and light industries, such as the production of textiles, appliances, and electronic equipment. An aluminum plant is located at the Sanmen gorge. Petroleum and natural gas are extracted at the Zhongyuan oil fields. In addition to coal, iron is mined, and lead and pottery clay are found. Stone Age remains have been discovered here, and the region was a center of Chinese civilization from c.2000 B.C.; ANYANG, Luoyang, and Kaifeng are historic cities. The province is well known for the SONGSHAN MOUNTAIN in the W, one of the five sacred mountain peaks in China. The Songyang Academy and Shaolin Temple are located here. Sometimes may appear as Honan.

Hen and Chickens, adjacent to Bream Bay and WHANGAREI HARBOUR, E of NORTHLAND region, NEW ZEALAND. A cluster of islands with a larger one (Taranga) beside smaller ones, now part of HAURAKI GULF MARITIME PARK, but preserved since 1925 as a reserve for rare animal and plant life, lacking main-land predators. Tuatara, Maori rat (kiore), giant kauri snail, and preservation of saddleback bird; plants linked to New Caledonian plants.

Hen and Chickens (HEN, CHI-kenz), S ONTARIO, E central CANADA, group of islets at W end of Lake ERIE, 9 mi/14 km W of PELEE ISLAND.

Henanfu, CHINA: see LUOYANG.

Henarejos (ai-nah-RAI-hos), village, CUENCA province, E central SPAIN, in the Serranía de Cuenca, 106 mi/171 km ESE of CUENCA. Cereals, grapes, saffron, honey; livestock. Wine production. Coal mining. Iron, gold, silver deposits.

Henares River (ai-NAH-res), c.90 mi/145 km long, CASTILE-LA MANCHA, central SPAIN; rises NE of SI-GÜENZA in GUADALAJARA province; flows SW, past Sigüenza, Guadalajara, and ALCALÁ DE HENARES, to JARAMA RIVER (affluent of the TAGUS RIVER) 9 mi/14.5 km E of MADRID.

Henbury, ENGLAND: see BRISTOL.

Henbury Meteorites Conservation Reserve (HEN-buh-ree), S central NORTHERN TERRITORY, N central AUSTRALIA, 100 mi/161 km SW of ALICE SPRINGS; extends more than 0.1 sq mi/0.3 sq km (39 acres/16 ha). Twelve meteorite craters from meteor that broke up prior to impact some 4,700 years ago. Largest crater 600 ft/183 m across. Event may have been wit-nessed by Aboriginal people. Habitat of bearded dragon lizard. Kangaroo, cockatoo, emu. Camping, picnicking.

Hendaye (ahn-DEI), Spanish *Hendaya* (en-DAH-yah), border town (□ 3 sq mi/7.8 sq km), PYRÉNÉES-ATLANTIQUES department, AQUITAINE region, SW FRANCE, port on Bay of BISCAY, at mouth of BIDASSOA RIVER (Spanish border), opposite FUENTERRABÍA, and 17 mi/27 km SW of BAYONNE in BASQUE COUNTRY; 43°22′N 01°47′W. International railroad station. Manufacturing of small arms, liqueur, and berets; sardine canning. Hendaye-Plage (1 mi/1.6 km N) is popular swimming resort with beach (2 mi/3.2 km long), casino, and hotels amidst subtropical vegeta-tion.

Hendek, town (2000 population 28,537), NW TURKEY, 18 mi/29 km E of ADAPAZARI; 40°47′N 30°45′E. Grain, tobacco.

Henderson, county (□ 395 sq mi/1,027 sq km; 2006 population 7,819), W ILLINOIS; ☉ OQUAWKA; 40°49′N 90°54′W. Bounded W by the MISSISSIPPI RIVER; drained by HENDERSON CREEK. Agriculture (cattle, hogs, poultry; corn, soybeans); limestone. Beaches, campsites along the Mississippi. Formed 1841.

Henderson, county (□ 467 sq mi/1,214.2 sq km; 2006 population 45,666), NW KENTUCKY; ☉ HENDERSON; elevation 401 ft/122 m; 37°47′N 87°34′W. Bounded N by the OHIO RIVER (INDIANA state line), SW by Highland Creek, E and SE by GREEN RIVER; drained by Green River (NE). Rolling plateau agricultural area (dark and burley tobacco, corn, wheat, soybeans, al-falfa, hay; hogs, cattle); bituminous-coal mines, oil wells; clay, sand and gravel. Manufacturing at Hen-derson. Includes DIAMOND ISLAND in Ohio River, in NW; JOHN JAMES AUDUBON STATE PARK in N; Sauerheber Wildlife Management Area in NW; Newburgh Lock and Dam, on Ohio River, in NE. Kentucky, on Indiana side of Ohio River, 4 mi/6.4 km S of EVANSVILLE (Indiana), isolated by changing river course, includes Ellis Park Racecourse. Formed 1798.

Henderson (HEN-duhr-suhn), county (□ 375 sq mi/975 sq km; 2006 population 99,033), SW NORTH CAR-OLINA; ☉ HENDERSONVILLE; 35°20′N 82°28′W. E part in the BLUE RIDGE MOUNTAINS; bounded S by SOUTH CAROLINA; drained by upper FRENCH BROAD, BROAD, and Green rivers. Resort area. Manufacturing (tex-tiles, stone quarrying) at FLETCHER and Henderson-ville; service industries; agriculture (hay, apples, corn; dairying; cattle). Holmes Educational State Forest in W; CARL SANDBURG HOME National Historical Site in S, at FLAT ROCK; Lake Summit reservoir (Green River) in S; part of Pisgah National Forest in NW. Formed 1838 from Buncombe County. Named for Leonard Henderson (1773–1833), Chief Justice of state Supreme Court.

Henderson, county (□ 515 sq mi/1,339 sq km; 2006 population 26,750), W TENNESSEE; ☉ LEXINGTON; 35°39′N 88°23′W. Drained by BIG SANDY RIVER. Once an agricultural area; now relies on diverse manufac-turing. Includes Natchez Trace Forest and State Park. Formed 1821.

Henderson, county (□ 949 sq mi/2,467.4 sq km; 2006 population 80,222), E TEXAS; ☉ ATHENS; 32°12′N 95°50′W. Bounded W by TRINITY RIVER, E by NECHES RIVER (forms LAKE PALESTINE); partly wooded (tim-ber). Agriculture (especially vegetables, melons, nursery crops, black-eyed peas, legumes); cattle, hogs, horses, ratites (emus, ostriches, rheas). Oil, natural gas; lignite, sulfur; sand and gravel. Hunting; fishing. Purtis Creek State Park in N; CEDAR CREEK RESER-VOIR in W; Lake Athens reservoir near center. Formed 1846.

Henderson, city (2000 population 27,373), ☉ HENDER-SON county, NW KENTUCKY, 9 mi/14.5 km S of EVANSVILLE (INDIANA), on the OHIO RIVER; 37°50′N 87°34′W. Railroad junction. Agriculture (dark and burley tobacco, corn; livestock); oil, coal. Manu-facturing (transportation equipment, plastic prod-ucts, furniture, consumer goods, chemicals, fabricated metal products, building materials, paper products, lumber, machinery, flour; printing and publishing, denim processing, aluminum smelting). Henderson City-County Airport to W. Naturalist and painter John J. Audubon lived here 1810–1819. JOHN JAMES AUDUBON STATE PARK, with a museum and a bird sanctuary to NE. Ellis Park Racecourse (annual thoroughbred racing) 5 mi/8 km to NNE in section of Kentucky on Indiana side of Ohio River, adjacent to Evansville. A branch of the University of Kentucky is here. Founded 1797, incorporated as a city 1867.

Henderson (HEN-duhr-suhn), city (2006 population 240,614), CLARK county, including suburb 12 mi/19 km of LAS VEGAS, SE NEVADA, in a desert area overlooking Las Vegas and surrounded by mountains; elevation 1,881 ft/573 m; 36°01′N 115°00′W. Limestone; manu-facturing (plastic products, fabricated metal products, lime products, foods, transportation equipment, con-sumer goods, chemicals, building materials; printing and publishing); tourism, recreation. Center for de-fense-related industries, specializing in large-volume chemical manufacturing. Hydroelectric power is supplied by HOOVER DAM. Founded (1942) to provide houses for employees of a magnesium plant. Southern Nevada Museum and Heritage Museum are here. LAKE MEAD Recreational Area adjoins city to NE. Sunrise Mountain Natural Area to N. Highland Range, crucial bighorn habitat area to S. Incorporated 1953.

Henderson (HEN-duhr-suhn), city (□ 8 sq mi/20.8 sq km; 2006 population 16,204), ☉ VANCE county, N NORTH CAROLINA, 41 mi/66 km NNW of RALEIGH; 36°19′N 78°24′W. Railroad junction. Manufacturing (apparel, textiles, furniture, fabricated metal prod-ucts, industrial minerals, building materials, mobile homes, consumer goods, machinery, foods; printing); service industries; agriculture (grain, soybeans, to-bacco; livestock). Seat of Vance-Granville Commu-nity College. Kerr Lake State Recreation Area to N; large S arm of Kerr Reservoir (ROANOKE RIVER) to N. Settled c.1811, incorporated 1841.

Henderson, city (2006 population 6,235), ☉ CHESTER county, SW TENNESSEE, 16 mi/26 km SE of JACKSON; 35°26′N 88°38′W. In timber and farm area; manu-facturing. Seat of Freed-Hardeman College. Chicka-saw State Park, with Native American burial mounds, is nearby.

Henderson, city (2006 population 11,584), ⊙ RUSK county, NE TEXAS, 25 mi/40 km S of LONGVIEW; 32°09′N 94°47′W. Elevation 505 ft/154 m. Prosperous oil and natural-gas city. Railroad terminus; dairying; cattle, horses; vegetables, watermelons; nursery crops; timber. Manufacturing (furniture, machinery, building materials, wood products; meat processing). Originally a pinewood lumbering town, then a cotton center, the city was transformed in 1930 when C. M. Joiner struck the first gusher of the fabulously rich East Texas Oil Field nearby. Site of an Old Shawnee village in the area. Seat of Texas Baptist Institute. Incorporated 1877.

Henderson (AIN-der-son), town, W central BUENOS AIRES province, ARGENTINA, 35 mi/56 km SSE of PEHUAJÓ; 36°18′S 61°43′W. Grain and livestock center; flour milling, dairying.

Henderson, unincorporated town, ADAMS county, N central COLORADO, suburb 15 mi/24 km NE of downtown DENVER, on SOUTH PLATTE RIVER; elevation 5,020 ft/1,530 m. Manufacturing (machinery, transportation equipment, fabricated metal products, building materials).

Henderson, town (2000 population 171), MILLS county, SW IOWA, near WEST NISHNABOTNA RIVER, 25 mi/40 km ESE of COUNCIL BLUFFS; 41°08′N 95°25′W. In grain and livestock region.

Henderson (HEN-duhr-suhn), town (2000 population 1,531), SAINT MARTIN parish, LOUISIANA, 12 mi/19 km E of LAFAYETTE; 30°18′N 91°47′W. Located at the edge of Atchafalaya Swamp. Crawfish ponds in area. Known for its seafood restaurants. Incorporated 1971.

Henderson, town (2000 population 118), CAROLINE county, E MARYLAND, near DELAWARE state line, 14 mi/23 km WSW of DOVER, Delaware; 39°04′N 75°46′W. Originally named Meredith's Crossing or River Bridges, but in 1868, when the railroad arrived, it was named for one of the BALTIMORE AND OHIO RAILROAD directors.

Henderson, town (2000 population 910), SIBLEY county, S MINNESOTA, 45 mi/72 km SW of MINNEAPOLIS, on MINNESOTA RIVER, at mouth of Rush River; 44°31′N 93°54′W. Grain; livestock, poultry; dairying; manufacturing (displays). Rush River State Park to W.

Henderson, village (2000 population 319), KNOX county, NW central ILLINOIS, near HENDERSON CREEK, 4 mi/6.4 km N of GALESBURG; 41°01′N 90°20′W. In agricultural area (cattle, hogs; corn, soybeans, sorghum; dairying).

Henderson, village (2006 population 1,000), YORK county, SE central NEBRASKA, 12 mi/19 km SW of YORK; 40°46′N 97°48′W. Irrigation equipment; printing, light manufacturing.

Henderson, village (2006 population 313), MASON county, W WEST VIRGINIA, 4 mi/6.4 km E of GALLIPOLIS (OHIO), and 2 mi/3.2 km S of POINT PLEASANT, on OHIO River (bridged to Galliopolis), at mouth of KANAWHA River (bridged to Point Pleasant); 38°49′N 82°08′W. Grain; poultry, livestock. Chief Cornstalk Wildlife Management Area to SE.

Henderson Creek, c.75 mi/121 km long, W ILLINOIS; rises in branches near GALESBURG; flows W and SW, to the MISSISSIPPI RIVER 5 mi/8 km above BURLINGTON (IOWA); 41°01′N 90°18′W.

Henderson Field, SOLOMON ISLANDS: see GUADALCANAL.

Henderson Harbor, resort village, JEFFERSON county, N NEW YORK, on Henderson Bay (an inlet of Lake ONTARIO), 16 mi/26 km SW of WATERTOWN; 43°52′N 76°11′W. Fishing.

Henderson Island, ANTARCTICA, 9 mi/14.5 km SE of MASSON ISLAND, in the SHACKLETON ICE SHELF; 9 mi/14.5 km long; 66°22′S 97°10′E. Discovered 1912.

Henderson Island, South Pacific, NEW ZEALAND: see PITCAIRN ISLAND.

Henderson Lake, reservoir, ESSEX county, NE NEW YORK, on HUDSON RIVER, in the ADIRONDACK MOUNTAINS, in Adirondack Park, 17 mi/27 km S of Saranac; c.1.5 mi/2.4 km long; 44°05′N 74°04′W. Drains S into SANFORD LAKE. Formed by dam 5 mi/8 km S of source of Hudson River.

Hendersons Knob, AUSTRALIA: see TAMBORINE MOUNTAIN.

Hendersonville, city (2006 population 46,218), SUMNER county, N central TENNESSEE, on OLD HICKORY Reservoir (CUMBERLAND RIVER), 14 mi/23 km NE of NASHVILLE, 36°18′N 86°37′W. Experiencing unprecedented growth in retail and industrial development. Home to a number of country music singers. Incorporated 1968.

Hendersonville (HEN-duhr-suhn-vil), town (□ c.6 sq mi/15.5 sq km; 1990 population 7,284; 2000 population 10,420), ⊙ HENDERSON county, SW NORTH CAROLINA, 20 mi/32 km SSE of ASHEVILLE, in the BLUE RIDGE MOUNTAINS; 35°19′N 82°27′W. Railroad junction. Resort center; manufacturing (clothing, electrical equipment, consumer goods; textiles, ceramic and foam filters, crushed stone; sheet metal fabricating); service industries; agriculture (corn, tobacco, soybeans; dairying; livestock). CARL SANDBURG HOME National Historical Site to S at FLAT ROCK. Pisgah National Forest to W; Holmes Educational State Forest to SW. Incorporated 1847.

Hendian, IRAN: see HENDIJAN.

Hendijan (hin-DEE-jahn), town, Khuzestān province, SW IRAN, 85 mi/137 km E of ABADAN, and on Zahreh River (here called Hendijan River) just above its mouth on PERSIAN GULF; 30°14′N 49°42′E. Livestock raising; grain, cotton. Also called HENDIAN, Hindijan.

Hendley, village (2006 population 36), FURNAS county, S NEBRASKA, 7 mi/11.3 km W of BEAVER CITY, and on BEAVER CREEK; 40°07′N 99°58′W.

Hendon, ENGLAND: see BARNET.

Hendourabi (hin-DOOR-ah-bee), island (□ 8 sq mi/20.8 sq km), Hormozgān province, PERSIAN GULF, Iran, S of LAVAN Island; 26°41′N 53°37′E.

Hendricks, county (□ 408 sq mi/1,060.8 sq km; 2006 population 131,204), central INDIANA, ⊙ DANVILLE; 39°46′N 86°31′W. Drained by Big Walnut, MILL, and WHITELICK creeks. Part of INDIANAPOLIS metro area; includes satellite communities of BROWNSBURG and PLAINFIELD. Corn, soybeans, fruit; hogs, cattle, sheep; flour milling, processing of dairy products and lumber; timber. Formed 1823.

Hendricks, village (2000 population 725), LINCOLN county, SW MINNESOTA, near SOUTH DAKOTA state line, and 9 mi/14.5 km WNW of IVANHOE, at E end of Lake Hendricks (extends into South Dakota), at exit of LAC QUI PARLE RIVER; 44°30′N 96°25′W. Grain; poultry, livestock; dairying.

Hendricks, village (2006 population 310), TUCKER co., NE WEST VIRGINIA, 3 mi/4.8 km ESE of PARSONS, on Laurel Fork River, at mouth of Blackwater River; 39°03′N 79°37′W. Coal-mining and agricultural area. Manufacturing (wood products). In Monongahela National Forest; Fernow Experimental Forest to SW.

Hendrik-Ido-Ambacht (HEN-drik—EE-daw—AHM-bahkht), suburb of DORDRECHT, S. HOLLAND province, W NETHERLANDS, 2 mi/3.2 km NNE of city center; 51°51′N 04°38′E. On Beneden (LOWER MERWEDE River) channel, which divides into NOORD and OUDE MAAS channels to W. Dairying; cattle; vegetables, potted plants, fruit, sugar beets; light manufacturing.

Hendrik-Kapelle, BELGIUM: see HENRI-CHAPELLE.

Hendrik Verwoerd Dam, FREE STATE province, SOUTH AFRICA: see GARIEP DAM.

Hendrix, village (2006 population 80), BRYAN county, S OKLAHOMA, 15 mi/24 km S of DURANT, and on RED RIVER; 33°46′N 96°24′W.

Hendrum (HEN-druhm), village (2000 population 315), NORMAN county, W MINNESOTA, on RED RIVER, near mouth of WILD RICE RIVER, 26 mi/42 km N of FARGO (NORTH DAKOTA); 47°15′N 96°48′W. Wheat, potatoes, sunflowers, sugar beets, soybeans.

Hendry (HEN-dree), county (□ 1,189 sq mi/3,091.4 sq km; 2006 population 40,459), S central FLORIDA; ⊙ LA BELLE; 28°54′N 82°22′W. Everglades sugarcane-growing and cattle-raising area, crossed in NW corner by CALOOSAHATCHEE RIVER; touches LAKE OKEECHOBEE (NE). Has Seminole Indian Reservation in SE. Formed 1923.

Henecán, HONDURAS: see SAN LORENZO.

Henefer, village (2006 population 722), SUMMIT county, N UTAH, 27 mi/43 km SE of OGDEN, and 25 mi/40 km NE of SALT LAKE CITY, on WEBER RIVER; elevation 5,280 ft/1,609 m; 41°01′N 111°29′W. Alfalfa; dairying; sheep, cattle. Wasatch National Forest to N and W; East Canyon Reservoir and State Park to SW. Lost Creek Reservoir and State Park to NE. Echo Reservoir to SE. Settled 1860s.

Henegouwen, BELGIUM: see HAINAUT.

Henfield (HEN-feeld), town (2001 population 4,810), West SUSSEX, S ENGLAND, near ADUR RIVER, 9 mi/14.5 km NW of BRIGHTON; 50°55′N 00°16′W. Agricultural market. Has 13th-century church.

Heng, GERMANY: see POSTBAUER-HENG.

Hengam (hin-GAHM), PERSIAN GULF island, of SE IRAN, in STRAIT of Hormuz, just off S coast of QESHM island; 6 mi/9.7 km long, 4 mi/6.4 km wide; 26°38′N 55°55′E. Site of British naval base during World War I; after 1934, Iranian naval and quarantine station. Also spelled HANGAM, HANJAM.

Henganofi (hahn-gaH-NO-fee), village, EASTERN HIGHLANDS province, NE NEW GUINEA island, N central PAPUA NEW GUINEA, in BISMARCK MOUNTAINS, 20 mi/32 km SE of GOROKA. Road access. Coffee, tea, sweet potatoes; cattle; timber.

Hengchun (HUNG-CHWUN), town, southernmost TAIWAN, 50 mi/80 km SE of PINGTUNG; 22°00′N 120°44′E. Sugar milling; rice, corn, onion, soybeans, peanuts. Limestone and oil deposits nearby.

Hengdaohezi (HUNG-DO-HUH-ZUH), town, SE HEILONGJIANG province, NE CHINA, on Chinese Eastern railroad, 33 mi/53 km WNW of MUDANJIANG; 44°46′N 129°03′E. Lumbering center.

Hengelo (HENG-uh-law), city (2001 population 56,369), OVERIJSSEL province, E NETHERLANDS, 36 mi/58 km ENE of APELDOORN, and 5 mi/8 km NW of ENSCHEDE; 52°03′N 06°19′E. TWENTE CANAL passes to S; railroad junction; Twenthe Airport to E. Salt mining. Dairying; cattle, hogs, poultry; vegetables, flowers, grain; manufacturing (electronic equipment, office supplies and machines, consumer goods; food processing).

Hengelo (HENG-uh-law), village, GELDERLAND province, E NETHERLANDS, 8 mi/12.9 km SE of ZUTPHEN; 52°16′N 06°48′E. Dairying; livestock; grain, vegetables.

Hengersberg (HENG-gers-berg), village, BAVARIA, S GERMANY, in LOWER BAVARIA, at SW foot of the BOHEMIAN FOREST, near the DANUBE River, 6 mi/9.7 km SE of DEGGENDORF; 48°47′N 13°03′E. Has a mid-13th-century and a late-16th-century church.

Hengfeng (HUNG-FUNG), town, ⊙ Hengfeng county, NE JIANGXI province, SE CHINA, 24 mi/39 km WSW of SHANGRAO, and on railroad; 28°25′N 117°35′E. Rice, oilseeds; timber processing; food and beverages. Coal mining.

Hengistbury Head (HENG-gist-buh-ree), promontory of DORSET, SW ENGLAND, on the CHANNEL, 2 mi/3.2 km SE of CHRISTCHURCH; 50°42′N 01°45′W. Tourist site.

Hengoed (HENG-goid), village (2001 population 5,044), ⊙ CAERPHILLY, SE Wales, on RHYMNEY RIVER, near GELLIGAER, and 5 mi/8 km N of Caerphilly; 51°39′N 03°14′W. Formerly in MID GLAMORGAN, abolished 1996.

Hengshan (HUNG-SHAHN), town, ⊙ Hengshan county, E HUNAN province, S central CHINA, on XIANG River, and 26 mi/42 km NNE of HENGYANG; 27°14′N

112°52′E. Rice, oilseeds; building materials, chemicals; timber and food processing. Non-ferrous ore mining. Heng Shan (mountain) is 10 mi/16 km W.

Hengshan (HUNG-SHAHN), town, ⊙ Hengshan county, N SHAANXI province, NW central CHINA, 40 mi/64 km SW of YULIN, near GREAT WALL; 37°58′N 109°17′E. Grain, oilseeds; livestock; food processing; building materials. Coal mining.

Hengshan Mountain (HUNG-SHAHN), NE SHANXI province, NE CHINA. Highest point is Xuanwu or the Northern Peak (6,617 ft/2,017 m), one of five sacred mountain peaks in China. Watershed for the SANG-GAN and HUTUO rivers, and known for its "eighteen scenic sites," the most famous of which is the Xuan-kong Temple (or the "Temple Hung in Midair"). Built during the N Wei period (450–589), this temple contains over forty halls and pavilions struck into the vertical walls of the cliff, making them look as if they are hanging in the air.

Hengshan Mountains (HUNG-SHAHN), mountain range, E HUNAN province, S central CHINA; has seventy-two peaks, including the Hengshan peak (4,232 ft/1,290 m), also called the Southern Peak (Chinese *Nanyue*), 30 mi/48 km NE of HENGYANG, which is one of five sacred mountain peaks in China. Hengshan Peak is the site of numerous Buddhist temples. An annual fair is held in the Nanyue Temple from July–September.

Hengshui (HUNG-SHWAI), district, HEBEI province, NE CHINA; 37°43′N 115°42′E. The district administers three cities and eight counties within its territory. Coincides approximately with the area that included the states of Yan and Zhao during the Warring States Period (481 B.C.–221 B.C.E.). Dong Zhongshu, a Confucian who lived during the Han dynasty (206 B.C.–A.D. 220), was born here, and poet Wang Zhi-huan held an official post here. First stop on the BEIJING-JIULONG RAILROAD S of BEIJING. Agriculture (wheat, corn, cotton).

Hengshui (HUNG-SHWAI), city (□ 228 sq mi/592.8 sq km; 2000 population 329,781), S HEBEI province, NE CHINA, on railroad, 35 mi/56 km NW of DEZHOU; 37°45′N 115°44′E. Light and heavy industry and agriculture (especially crop growing) are the major sources of income. Agriculture (grains, vegetables; hogs, poultry); manufacturing (textiles and chemicals).

Hengsteysee, GERMANY: see HERDECKE.

Heng Xian (HUNG-SHYEN), town, ⊙ Heng Xian county, S GUANGXI ZHUANG AUTONOMOUS REGION, S CHINA, 65 mi/105 km ESE of NANNING, and on left bank of Yu River; 22°41′N 109°16′E. Rice, oilseeds, sugarcane; sugar refining; iron smelting, electrical power generation.

Hengyang (HUNG-YAHNG), city (□ 216 sq mi/561.6 sq km; 2000 population 1,814,936), central HUNAN province, S central CHINA, on the XIANG RIVER, at the mouth of the LEI RIVER; 26°54′N 112°36′E. Leading transportation center of Hunan, linking water, railroad, and highway routes. Manufacturing includes chemicals, agricultural and mining equipment, textiles, paper, and processed foods. Lead, zinc, coal, and tin mines nearby. Its former name was Hengzhou (Hengchow).

Hengzhou, CHINA: see HENGYANG.

Heniches′k (he-NEE-chesk) (Russian *Genichesk*), city, SE KHERSON oblast, UKRAINE, port on Sea of AZOV, on railroad spur, 55 mi/89 km SW of MELITOPOL′; 46°10′N 34°48′E. Raion center; flour milling, fish canning, manufacturing (cheese, food flavoring, textiles, building materials, machines); saltworks. Seat of 2 professional-technical schools, medical school; heritage museum. Established in 1784, city since 1938.

Heniches′k Strait, UKRAINE: see ARABAT TONGUE.

Henik Lakes (HE-nik), NUNAVUT territory, N CANADA, group of two lakes S of YATHKYED LAKE. South Henik Lake (50 mi/80 km long, 1 mi/2 km–12 mi/19 km wide) is at 61°30′N 97°35′W; just N is North Henik Lake (18

mi/29 km long, 10 mi/16 km wide). Both drain E into HUDSON BAY through KAMINAK LAKE.

Hénin-Beaumont (ai-NAN–bo-MON), town (□ 8 sq mi/20.8 sq km), PAS-DE-CALAIS department, NORD-PAS-DE-CALAIS region, N FRANCE, 5 mi/8 km E of LENS, on a major highway interchange; 50°25′N 02°56′E. Metalworking center in former coal-mining district. Ready-to-wear clothing is also made here.

Henlawson (hen-LAW-suhn), unincorporated town, LOGAN county, SW WEST VIRGINIA, 4 mi/6.4 km NNE of LOGAN, near GUYANDOTTE RIVER. Coal-mining and agricultural area. Chief Logan State Park to W.

Henley, unincorporated town, COLE county, central MISSOURI, on OSAGE RIVER, and 18 mi/29 km SSW of JEFFERSON CITY. Former railroad bridge over Osage River.

Henley and Grange (HEN-lee, GRAINJ), town, SE SOUTH AUSTRALIA, on Gulf SAINT VINCENT; consists of Henley, a suburb of ADELAIDE, and Grange, 1.8 mi/2.8 km N of Henley. Bathing beaches.

Henley Harbour, village, E NEWFOUNDLAND AND LABRADOR, NE CANADA, on inlet of the ATLANTIC Ocean, 20 mi/32 km SW of BATTLE HARBOUR; 52°01′N 55°51′W. Fishing port; lumbering.

Henley-in-Arden (HEN-lee–in–AH-duhn), village (2001 population 4,176), W WARWICKSHIRE, central ENGLAND, on ALNE RIVER, and 8 mi/12.9 km W of WARWICK; 52°17′N 01°46′W. Produces cables and transportation equipment. Has many old houses (including an inn dating from 1358), guildhall (1448), and church dating from 15th century. Just S is agricultural village of WOOTTON WAWEN with church dating from Saxon times.

Henley-on-Thames (HEN-lee–on–TEMZ), town (2001 population 10,646), SE OXFORDSHIRE, S central ENGLAND, on the THAMES RIVER, and 7 mi/11.3 km NNE of READING, at foot of CHILTERN HILLS; 51°32′N 00°55′W. Chiefly known as scene of annual July rowing regatta, instituted 1839. Has 14th-century church, grammar school founded 1605, and five-arched stone bridge built 1786.

Henllys (HEN-hlis), village (2001 population 2,695), TORFAEN, SE Wales, 1 mi/1.6 km NE of RISCA; 51°38′N 03°05′W. Cwmcarn Forest Drive 5 mi/8 km to NW. Formerly in GWENT, abolished 1996.

Henlopen Acres, resort town (2000 population 139), SUSSEX county, S DELAWARE, on ATLANTIC OCEAN, 1 mi/1.6 km N of REHOBOTH BEACH; 38°44′N 75°05′W. Elevation 3 ft/0.9 m.

Henlopen, Cape (hen-LO-pen), E SUSSEX county, SE DELAWARE, at S side of entrance of DELAWARE BAY, opposite CAPE MAY (N.J.), 12 mi/19 km to NNE; 38°48′N 75°06′W. Ferry carries railroad and U.S. Highway 9 to Cape May. Site of U.S. Fort Miles, now Cape Henlopen State Park. LEWES AND REHOBOTH CANAL parallels coast SW of cape.

Henlow (HEN-lo), village (2003 population 3,230), E BEDFORDSHIRE, central ENGLAND, on IVEL RIVER, and 4 mi/6.4 km S of BIGGLESWADE; 52°02′N 00°18′W. Former agricultural market. Has 15th-century church.

Hennaya (hen-nah-YAH), village, TLEMCEN wilaya, NW ALGERIA, on railroad, and 6 mi/9.7 km NNW of TLEMCEN. Vineyards; olive-oil processing. Formerly named Eugéne Étienne Hennaya.

Henne, GERMANY: see MESCHEDE.

Henneberg (HEN-ne-berg), village, THURINGIA, central GERMANY, 6 mi/9.7 km SSW of MEININGEN; 50°30′N 10°22′E. Site of ruins of ancestral castle (destroyed 1525) of counts of HENNEBERG, who played important part in Thuringian history prior to 1353, when their property passed to house of WETTIN.

Hennebont (en-nuh-BON), town (□ 7 sq mi/18.2 sq km), MORBIHAN department, in BRITTANY, W FRANCE, seaport at head of BLAVET RIVER estuary, and 5 mi/8 km NNE of LORIENT; 43°40′N 03°30′E. Metallurgical industry; fish canning, distilling; horse breeding; stables. Town suffered heavy damage in World War II

(due to Allied bombing of Lorient submarine bunkers), including the massive 15th-century gate, 16th-century church, and many old houses.

Hennef (HEN-nef), town, NORTH RHINE–WESTPHALIA, W GERMANY, on the SIEG RIVER, 3 mi/4.8 km SE of SIEGBURG; 50°46′N 07°16′E. Many residents commute to COLOGNE (linked by railroad). Industry includes the manufacturing of agriculture and construction machinery, measuring instruments, and office equipment; crafts made with precious metals. Town first mentioned in 1075; fell to PRUSSIA in 1815. Incorporated a number of neighboring villages in 1935; chartered in 1981. There is a ruined castle here called Blanhenberg, around which there are many half-timbered medieval houses.

Hennepin (HE-ne-pin), county (□ 606 sq mi/1,575.6 sq km; 2006 population 1,122,093), E MINNESOTA; ⊙ MINNEAPOLIS, twin city to ST. PAUL (RAMSEY Co.) to E; 45°00′N 93°28′W. Urbanized area in Minneapolis–St. Paul metro region. Bounded NE and SE by MISSISSIPPI RIVER, NW by CROW RIVER, S by MINNESOTA RIVER. Manufacturing and commerce at Minneapolis, also BLOOMINGTON, MINNETONKA, PLYMOUTH, MAPLE GROVE, and other municipalities; agriculture (corn, soybeans, oats, alfalfa; cattle, sheep, poultry; dairying; nursery stock). Minneapolis–St. Paul International Airport in SE. Minnesota Valley National Wildlife Refuge along S border; FORT SNELLING National Cemetery and State Park in SE; MINNEHAHA falls, on Mississippi River, in E; Lake MINNETONKA in SW; numerous natural lakes in county. Formed 1852.

Hennepin, village (2000 population 707), ⊙ PUTNAM county, N central ILLINOIS, on ILLINOIS RIVER (bridged) near its great bend, and 8 mi/12.9 km SW of SPRING VALLEY; 41°15′N 89°19′W. Grain elevator. Fishing in Senachwine and Sawmill lakes. Oldest functional courthouse in Illinois.

Henne River, BELGIUM: see HAINE RIVER.

Hennersdorf or **Katholisch-Hennersdorf**, village, in LOWER SILESIA, Wroclaw province, SW POLAND, 12 mi/19 km E of Görlitz. Here Frederick the Great defeated (November 1745) Austrians and Saxons. Part of Poland since 1945.

Hennersdorf, CZECH REPUBLIC: see JINDRICHOV.

Hennessey, town (2006 population 2,048), KINGFISHER county, central OKLAHOMA, 21 mi/34 km S of ENID, near TURKEY CREEK; 36°06′N 97°54′W. In grain and livestock area; manufacturing (metal buildings). Laid out 1889.

Henniker (HE-ni-kuhr), town, MERRIMACK county, S central NEW HAMPSHIRE, 14 mi/23 km W of CONCORD; 43°10′N 71°49′W. Drained by CONTOOCOOK RIVER and Amey Brook. Agriculture (dairying; livestock, poultry; apples, vegetables; timber); manufacturing (medical supplies, lumber, building materials); summer resort. Seat of New England College. Craney Hill (1,402 ft/427 m) and Pat's Peak Ski Area in S; covered bridge. Settled 1763–1764, incorporated 1768.

Henning, town (2006 population 1,299), LAUDERDALE county, W TENNESSEE, 27 mi/43 km SSW of DYERSBURG; 35°41′N 89°34′W. In cotton-growing area. Alex Haley Museum here.

Henning, village (2000 population 241), VERMILION county, E ILLINOIS, 12 mi/19 km NNW of DANVILLE; 40°18′N 87°42′W. In agricultural and bituminous-coal area.

Henning, village (2000 population 719), OTTER TAIL county, W MINNESOTA, 30 mi/48 km E of FERGUS FALLS; 46°19′N 95°26′W. Grain, sunflowers; livestock, poultry; dairying. Resort area. East BATTLE LAKE to W, OTTER TAIL LAKE to NW; Inspiration Peak State Park to SW.

Henningsdorf bei Berlin (HEN-nings-dorf bei ber-LEEN), town, BRANDENBURG, E GERMANY, on the HAVEL River, 12 mi/19 km NW of BERLIN; 52°38′N 13°12′E. Manufacturing (steel products, locomotives, industrial ceramics). Chartered 1962. Developed industrially after 1900.

Henningsvær (HEN-nings-var), village, NORDLAND county, N NORWAY, on islet of same name in the VESTERÅLEN group, just S of AUSTVÅGØY, 11 mi/18 km SW of SVOLVÆR. Fishing center; summer resort.

Hennock (HEN-uhk), village, S central DEVON, SW ENGLAND, 9 mi/14.5 km SW of EXETER; 50°37′N 03°38′W. Previously granite quarrying. Has 15th-century church.

Henri-Chapelle (awn-REE–shah-PEL), Flemish *Hendrik-Kapelle*, village in commune of THIMISTER-CLERMONT, Verviers district, LIÈGE province, E BELGIUM, 6 mi/9.7 km NW of EUPEN; 50°40′N 05°56′E. U.S. World War II military cemetery.

Henrichemont (awn-reesh-MON), commune (□ 9 sq mi/23.4 sq km), CHER department, CENTRE administrative region, central FRANCE, 16 mi/26 km NNE of BOURGES; 47°18′N 02°32′E. Tanneries, pottery works. Planned and founded by Sully (finance minister under Henry IV) in 1608.

Henrichenburg (HEN-rikh-uhn-boorg), part of CASTROP-RAUXEL, NORTH RHINE–WESTPHALIA, W GERMANY, in the RUHR industrial district, 5 mi/8 km ESE of RECKLINGHAUSEN. Site of ship elevator joining DORTMUND-EMS and RHINE-HERNE canals (level difference: 46 ft/14 m); 51°33′N 07°19′E. The old elevator from 1918 is now a museum; new elevator built in 1962.

Henrico (hen-REI-ko), county (□ 243 sq mi/631.8 sq km; 2006 population 284,399), E central VIRGINIA; seat is RICHMOND (independent city separate from adjacent counties); 37°32′N 77°24′W. W part of county is in PIEDMONT region, E part on coastal plain; bounded S in part by JAMES RIVER N and NE by the CHICKAHOMINY RIVER. Agriculture (hay, barley, wheat, corn, soybeans; poultry, beef cattle). Most of county is urbanized, especially in center, around Richmond. Civil War battlefields in county are included in section of RICHMOND NATIONAL BATTLEFIELD PARK in S. Glendale National Cemetery in SE. First public school vocational education program in the U.S. established here by African-American educator and humanitarian Virginia Randolph. Formed 1634.

Henri Coanda Internation Airport, OTOPENI, ROMANIA; 10 mi/16 km N of BUCHAREST (highway and rail links); elevation 310 ft/95 m; 44°34′N 26°05′E. Two runways, one passenger terminal (3 million passengers annually); cargo facilities (49,000 tons/44,500 tonnes annually). Site of German military base in World War II; used by Romanian military until 1965 when a new civilian airport was authorized. Further expansion began in 1986, adding a second runway. Has largely supplanted Bucharest's BANSEA airport. Officially, Bucharest Henri Coanda Internation Airport (Romanian, *Aeroportul International Bucuresti Henri Coanda;* named in 2004 after the Romanian aviation pioneer); formerly called Bucharest Otopeni International Airport. Airport Code OTP.

Henrietta, city (2000 population 457), RAY county, NW MISSOURI, near MISSOURI River, 4 mi/6.4 km SE of RICHMOND; 39°14′N 93°56′W. Soybeans, corn; livestock. Manufacturing (aluminum granules). Damaged during 1993 flood.

Henrietta, town (2006 population 3,267), ⊙ CLAY county, N TEXAS, 18 mi/29 km ESE of WICHITA FALLS, and on LITTLE WICHITA RIVER; elevation 915 ft/279 m; 33°49′N 98°11′W. In cattle, cotton, oil and gas area; agricultural business; dairying; manufacturing (sheet metal fabrication, trophies, transportation equipment). Lake Arrowhead State Park (formed by Wichita River) to SW. Founded 1857, resettled 1873, incorporated 1882.

Henrietta (HEN-ree-et-ah), unincorporated village, RUTHERFORD county, SW NORTH CAROLINA, 15 mi/24 km W of SHELBY; 35°15′N 81°47′W. Manufacturing (apparel); agriculture (grain, soybeans; poultry, livestock).

Henrietta Island, Russian *Ostrov Genriyetta*, northernmost of DE LONG ISLANDS, in the EAST SIBERIAN SEA, 355 mi/571 km off N SAKHA REPUBLIC, RUSSIAN FAR EAST; 77°00′N 157°15′E. Arctic observation post.

Henrietta Maria, Cape (hen-ree-E-tuh muh-REE-uh), NE ONTARIO, E central CANADA, on HUDSON BAY, on W side of entrance of JAMES BAY; 55°09′N 82°20′W.

Henriette (HEN-ree-YET), village (2000 population 101), PINE county, E MINNESOTA, 8 mi/12.9 km WNW of PINE CITY; 45°52′N 93°07′W. Grain; livestock; dairying.

Henrieville, village (2006 population 145), GARFIELD county, S UTAH, 30 mi/48 km SE of PANGUITCH; 37°33′N 111°59′W. Alfalfa, barley; cattle. Kodachrome Basin State Park to S. Old schoolhouse (1881) serves as town hall. Settled 1870s.

Henri Pittier, Parque Nacional, national park (□ 416 sq mi/1,081.6 sq km), between ARAGUA and CARABOBO states, N VENEZUELA; 10°26′N 67°37′W. Adjoins SAN ESTEBAN national park to W. Covers much of the CORDILLERA DE LA COSTA, preserving dry-to-humid tropical forest and wide variety of animal and bird life. Created 1937 as the country's first national park. Until 1953 called Parque Nacional Rancho Grande. Named for Swiss biologist, Henri Pittier.

Henry, county (□ 568 sq mi/1,476.8 sq km; 2006 population 16,706), SE Alabama; ⊙ Abbeville; 31°30′N 85°15′W. Coastal plain bounded on E by Chattahoochee River (Ga. state line), drained (W) by Choctawhatchee River Peanuts, corn; hogs; bauxite; lumber. Formed 1819. Named for the American patriot Patrick Henry. County seats have been Richmond, 1819–1824; Columbia, 1824–1830; and the present one, Abbeville.

Henry, county (□ 331 sq mi/860.6 sq km; 2006 population 178,033), N central GEORGIA; ⊙ MCDONOUGH; 33°28′N 84°01′W. Rapidly urbanizing county on ATLANTA's S side. Bisected by I-75 which has become a major force in economic growth. Georgia National Golf Club in McDonough. Formed 1821.

Henry, county (□ 825 sq mi/2,145 sq km; 2006 population 50,339), NW ILLINOIS; ⊙ CAMBRIDGE; 41°21′N 90°09′W. Bounded NW by ROCK RIVER; drained by GREEN and EDWARDS rivers; old Illinois and Mississippi Canal crosses county. Agriculture (corn, soybeans, vegetables; cattle, hogs; dairy products). Manufacturing (food products, machinery; metal industries). Formed 1825.

Henry, county (□ 394 sq mi/1,024.4 sq km; 2006 population 46,947), E INDIANA; ⊙ NEW CASTLE; 39°56′N 85°24′W. Corn, soybeans; hogs, poultry. Manufacturing (especially transportation equipment) at New Castle and KNIGHTSTOWN. Drained by BIG BLUE RIVER, FLATROCK RIVER, and small Fall Creek. Wilbur Wright State Fish and Wildlife Area N of New Castle. Wilbur Wright Birthplace State Memorial in E. Formed 1821.

Henry, county (□ 436 sq mi/1,133.6 sq km; 2006 population 20,405), SE IOWA; ⊙ MOUNT PLEASANT; 40°59′N 91°33′W. Prairie agricultural area (sheep, cattle; corn, soybeans) drained by SKUNK RIVER; limestone quarries. Oakland Mills State Park in SW; Geode State Park in SE, known for its quartzite geodes. Flooding along Skunk River in 1993. Formed 1836.

Henry, county (□ 291 sq mi/756.6 sq km; 2006 population 16,025), N KENTUCKY; ⊙ NEW CASTLE, 38°27′N 85°09′W. Bounded E by KENTUCKY RIVER; drained by LITTLE KENTUCKY RIVER and FLOYDS FORK. Gently rolling upland agricultural area (burley tobacco, corn, hay, alfalfa, soybeans, wheat; cattle, hogs, horses, poultry; dairying; timber), in BLUEGRASS REGION. Primary metals industries, electronics. Zinc and lead mines in S were productive during World Wars I and II. Formed 1798.

Henry, county (□ 737 sq mi/1,916.2 sq km; 2006 population 22,719), W central MISSOURI; ⊙ CLINTON; 38°23′N 93°47′W. Drained by South GRAND RIVER.

Corn, wheat, soybeans; sorghum, sweet corn; cattle. Strip coal mines in SW. Some manufacturing at Clinton and WINDSOR. S Grand and Tebo Arms of Truman Lake enter county from E. Fishing, boating. Montrose Conservation Area in SW. Formed 1834.

Henry (HEN-ree), county (□ 416 sq mi/1,081.6 sq km; 2006 population 29,520), NW OHIO; ⊙ NAPOLEON; 41°21′N 84°06′W. Intersected by MAUMEE RIVER. In the Lake Plains physiographic region. Diversified farming (corn, wheat, oats, sugar beets, vegetables); manufacturing (preserved fruits and vegetables, fabricated metal products, cutlery and tools, transportation equipment). Formed 1824.

Henry, county (□ 599 sq mi/1,557.4 sq km; 2006 population 31,837), NW TENNESSEE; ⊙ PARIS; 36°20′N 88°18′W. Bounded N by KENTUCKY, E by KENTUCKY Reservoir (TENNESSEE RIVER; here receives the BIG SANDY); drained by East Fork CLARKS RIVER and forks of the OBION. Agricultural area; some manufacturing at Paris; tourism. Formed 1821.

Henry (HEN-ree), county (□ 384 sq mi/998.4 sq km; 2006 population 56,208), S VIRGINIA; ⊙ MARTINSVILLE (independent city, separate from county); 36°40′N 79°52′W. In PIEDMONT region, bounded S by NORTH CAROLINA state line; drained by SMITH and North and South MAYO rivers. Manufacturing at BASSETT (noted for its furniture products), FIELDALE, COLLINSVILLE, and RIDGEWAY; agriculture (especially tobacco; also hay; cattle, poultry; honey); timber. Part of FAIRY STONE STATE PARK, Philpott Reservoir in NW. Formed 1777. Named for Patrick Henry, first governor of Virginia after independence.

Henry, city (2000 population 2,540), MARSHALL county, N central ILLINOIS, on ILLINOIS RIVER (bridged here), and 7 mi/11.3 km N of LACON; 41°06′N 89°21′W. In agricultural area (corn, soybeans, fruit; cattle); ships grain; manufacturing (chemicals). Nurseries. Nearby is Senachwine Lake. Founded in early 1840s; incorporated 1854.

Henry (HEN-ree), former township, S central ONTARIO, E central CANADA; 46°34′N 80°21′W. Amalgamated into the town of MARKSTAY-WARREN in 1999.

Henry, town (2006 population 542), HENRY county, NW TENNESSEE, 8 mi/13 km SW of PARIS; 36°12′N 88°25′W.

Henry, village (2006 population 161), SCOTTS BLUFF county, extreme W NEBRASKA, 20 mi/32 km WNW of SCOTTSBLUFF, and on NORTH PLATTE RIVER at WYOMING state line; 42°00′N 104°02′W. OREGON TRAIL follows S side of river.

Henry, village (2006 population 258), CODINGTON county, NE SOUTH DAKOTA, 18 mi/29 km W of WATERTOWN; 44°52′N 97°27′W. Potatoes, corn; pheasant hunting.

Henry, Cape (HEN-ree), VIRGINIA BEACH (independent city), SE VIRGINIA, promontory at entrance to CHESAPEAKE BAY, 18 mi/29 km ENE of NORFOLK; 36°55′N 76°01′W. Cape Henry Memorial marks the approximate first landing site of English settlers of JAMESTOWN in 1607. In 1939 the site became a separate unit of COLONIAL NATIONAL HISTORICAL PARK (which also includes historic towns of YORKTOWN and Jamestown). Old Cape Henry Lighthouse, Fort Story Military Reservation, SEASHORE STATE PARK. Atlantic University to S.

Henryetta, town (2006 population 6,100), OKMULGEE county, E central OKLAHOMA, 12 mi/19 km S of OKMULGEE; 35°26′N 95°58′W. In diversified agricultural area; manufacturing (glass products, chemicals, consumer goods, machinery; sawmill). Coal mines; oil and natural-gas wells. Lake EUFAULA reservoir to E. Founded c.1900.

Henry, Fort (HEN-ree), provincial historic park, fortification, SE ONTARIO, E central CANADA, on the SAINT LAWRENCE RIVER, near LAKE ONTARIO, at SW end of RIDEAU CANAL, overlooking KINGSTON harbor; 44°14′N 76°28′W. Built 1812, original fort was demol-

Cross-references are shown in SMALL CAPITALS. The pronunciation guide is shown on page xix. The sources of population figures are shown on page xvii.

ished 1832 and replaced by present structure. Later a museum, it was a camp for prisoners of war in World Wars I and II. Commonly referred to as Old Fort Henry. Canadian Forces Base Kingston is 1 mi/2 km E.

Henry, Fort, TENNESSEE: see FORT HENRY.

Henry Hudson Bridge, NEW YORK CITY, SE NEW YORK, over Spuyten Duyvil (see HARLEM RIVER), connects N tip of MANHATTAN with the RIVERDALE section of the Bronx; 40°53′N 73°56′W. A double-deck vehicular structure, it is 2,000 ft/610 m long overall and 142 ft/43 m above the water. Lower level opened 1936, upper 1938.

Henry Ice Rise, in the SW RONNE ICE SHELF, WEST ANTARCTICA, between KORFF ICE RISE and BERKNER ISLAND, 70 mi/110 km long; 80°35′S 62°00′W.

Henry Island (HEN-ree), islet, in GULF OF ST. LAWRENCE, NE NOVA SCOTIA, E CANADA, off W CAPE BRETON ISLAND, 5 mi/8 km SW of PORT HOOD; 45°58′N 61°36′W.

Henry Kater, Cape, E BAFFIN ISLAND, BAFFIN region, NUNAVUT territory, N CANADA, on DAVIS STRAIT, E extremity of Henry Kater Peninsula (55 mi/89 km long, 7 mi/11 km–23 mi/37 km wide), forming N shore of HOME BAY; 69°04′N 66°46′W.

Henry Mountains, in E GARFIELD county (extends into WAYNE county), S UTAH, W of DIRTY DEVIL RIVER, E of ESCALANTE. Chief peaks are Mountains HILLERS (10,723 ft/3,268 m), PENNELL (11,371 ft/3,466 m), and ELLEN (11,522 ft/3,512 m). Laccolith intrusive formations. Recreation area.

Henrys Fork, river, c.110 mi/177 km long, SE IDAHO; rises in Henrys Lake (□ 8 sq mi/20.8 sq km), near MONTANA state line and CONTINENTAL DIVIDE, N FREMONT county, in Targhee National Forest; flows S through ISLAND PARK RESERVOIR, over UPPER MESA and Lower Mesa Falls (114 ft/35 m) and SW, past SAINT ANTHONY, to SNAKE RIVER 10 mi/16 km SW of REXBURG. Island Park Dam (91 ft/28 m high, 9,448 ft/2,880 m long; completed 1938) is in upper course near ISLAND PARK. Used in irrigation of upper Snake River valley; impounds Island Park Lake (12 mi/19 km long).

Henryville (HEN-ree-vil), village (□ 25 sq mi/65 sq km), Montérégie region, S QUEBEC, E CANADA, 12 mi/19 km SSE of SAINT-JEAN-SUR-RICHELIEU; 45°08′N 73°11′W. Dairying.

Henryville, village, CLARK county, S INDIANA, 7 mi/11.3 km NW of CHARLESTOWN. Clark State Forest to W. Cattle; corn.

Hensall (HEN-sol), former village (□ 1 sq mi/2.6 sq km), S ONTARIO, E central CANADA, 24 mi/39 km SSE of GODERICH; 43°26′N 81°30′W. Dairying, mixed farming. Amalgamated into the town of BLUEWATER in 2001.

Hensel, village, PEMBINA CO., NE NORTH DAKOTA, 8 mi/12.9 km SSW of CAVALIER; 48°41′N 97°39′W. Formerly called Canton. Founded in 1887 and named for Hensel, CANADA.

Henshaw, Lake, California: see SAN LUIS REY RIVER.

Hensies (awn-SEE), commune (2006 population 6,694), Mons district, HAINAUT province, SW BELGIUM, 12 mi/19 km W of MONS, near French border; 50°26′N 03°40′E.

Hensley, village, PULASKI county, central ARKANSAS, 16 mi/26 km SSE of LITTLE ROCK.

Henstedt-Ulzburg (HEN-stet–ULTS-boorg), commune, SCHLESWIG-HOLSTEIN, N GERMANY, 16 mi/26 km N of HAMBURG; 53°47′N 09°59′E. Residential commune; emerged from the unification of several villages.

Hentey, MONGOLIA: see HENTIYN NURUU.

Hentiy (KHEN-tai), province (□ 31,700 sq mi/82,420 sq km), E central MONGOLIA; ⊙ ÖNDÖRHAAN; 48°00′N 110°30′E. Bounded N by CHITA oblast of RUSSIAN FEDERATION, it extends from the wooded HENTIYN NURUU (Kentei Mountains) (NW) to the steppe (SE) and is traversed by KERULEN River.

Hentiy, MONGOLIA: see HENTIYN NURUU.

Hentiyn Nuruu (KHEN-tain NOO-roo), range, in N MONGOLIA, forming part of Transbaikalian mountain system; extends 150 mi/240 km NE from ULAANBAATAR; rises to 9,187 ft/2,800 m at the peak Asralt Khairkhan-Uul, 45 mi/70 km NE of Ulaanbaatar; 107°00′–110°00′E–48°00′–49°00′N. Densely forested, the range forms a major watershed of the YÖRÖÖ, HARAA, and TUUL rivers of the SELENGA-BAYKAL Basin, draining to the ARCTIC OCEAN, and the ONON and KERULEN rivers of the AMUR Basin, draining to the PACIFIC OCEAN. Also spelled Hentey, Kentei Mountains.

Henty (HEN-tee), town, S NEW SOUTH WALES, AUSTRALIA, 120 mi/193 km WSW of CANBERRA; 35°30′S 147°03′E. Railroad junction. Sheep; agriculture center (grain). Nature reserve (former dam) containing various aquatic species, birdlife.

Henzada (HEN-zah-DAH), Burmese *Hinthada*, formerly northernmost district (□ 2,807 sq mi/7,298.2 sq km) of AYEYARWADY division, MYANMAR; ⊙ HENZADA. Between ARAKAN YOMA and AYEYARWADY RIVER. Protected by embankments along AYEYARWADY RIVER, it is an intensive rice-growing area. Forests at foot of ARAKAN YOMA.

Henzada (HEN-zah-DAH), township, ⊙ AYEYARWADY division, MYANMAR, on right bank of AYEYARWADY RIVER opposite THARRAWAW (RR ferry), and 75 mi/121 km NE of PATHEIN. River port and railroad terminus of lines to KYANGIN and PATHEIN; wood carving.

Heorhiyivka (he-OR-hee-yif-kah) (Russian *Georgiyevka*), town, S LUHANS'K oblast, UKRAINE, in the DONBAS, 10 mi/16 km SSW of LUHANS'K; 48°25′N 39°15′E. Elevation 367 ft/111 m. Manufacturing (building materials); site of former coal mines. Known since the mid-18th century as Konoplyanka; renamed Heorhiyivka in 1917; town since 1938.

Hepburn, town (2000 population 39), PAGE county, SW IOWA, near NODAWAY RIVER, 8 mi/12.9 km N of CLARINDA; 40°51′N 95°01′W.

Hepburn (HEP-buhrn), village (2006 population 530), central SASKATCHEWAN, W CANADA, 28 mi/45 km N of SASKATOON; 52°43′N 106°43′W. Mixed farming, dairying.

Hepburn Springs (HEP-buhrn SPREENGZ), resort village, S central VICTORIA, AUSTRALIA, 55 mi/89 km NW of MELBOURNE, near DAYLESFORD, on spur of GREAT DIVIDING RANGE; 37°19′S 144°09′E. Health resort with mineral springs.

Hephzibah (HEP-zuh-buh), town (2000 population 3,880), RICHMOND county, E GEORGIA, 12 mi/19 km SSW of AUGUSTA; 33°17′N 82°07′W. Clay and wood products.

Heping (HUH-PEENG), town, ⊙ Heping county, NE GUANGDONG province, SE CHINA, at S foot of JIULIAN MOUNTAINS, 30 mi/48 km ENE of LIANPING, near JIANGXI province border; 24°26′N 114°56′E. Rice, sugarcane; pharmaceuticals, chemicals; logging.

Hepler, village (2000 population 154), CRAWFORD county, SE KANSAS, 23 mi/37 km NW of PITTSBURG; 37°39′N 94°58′W. Agriculture.

Heppenheim (HEP-pen-heim), town, S HESSE, central GERMANY, on the BERGSTRASSE, at W foot of the ODENWALD, 3 mi/4.8 km S of BENSHEIM; 49°38′N 08°39′E. Manufacturing of machinery, electrotechnical goods; food processing; agriculture (fruit, vineyards). Tourism; resort noted for its mild climate. The church of St. Peter, often called "The Cathedral of the Bergstrasse," was rebuilt 1900–1904. Town also has 13th-century castle, a former residence of archbishops of Mainz; 16th-century, half-timbered town hall. Ruins of 12th-century castle Starkenburg on hill just NE. A royal villa, Heppenheim came to LORSCH abbey in 8th century, passed to MAINZ in 1232, and was chartered 1318. Sometimes called Heppenheim an der Bergstrasse.

Heppner, town (2006 population 1,445), ⊙ MORROW county, N OREGON, on WILLOW CREEK (forms Willow Creek Reservoir to E), c.41 mi/66 km SW of PENDLETON; elevation 1,955 ft/596 m; 45°21′N 119°32′W. Railroad terminus. Potatoes, onions; sheep, cattle. Umatilla National Forest to SE.

Heptarchy [Gr.=seven kingdoms] (HEP-tar-chee), name applied to seven kingdoms of Anglo-Saxon ENGLAND: NORTHUMBRIA, EAST ANGLIA, MERCIA, ESSEX, SUSSEX, WESSEX, and KENT. Term probably first used in 16th century by writers who believed England was divided into just seven kingdoms founded by Angles and Saxons. In fact, political and geographical divisions of early England were more complex, and were fluid and shifting for centuries.

Hepu (HUH-POO), town, ⊙ Hepu county, S GUANGXI ZHUANG AUTONOMOUS REGION, S CHINA; port in Lim River delta on GULF OF TONKIN, 85 mi/137 km WNW of ZHANJIANG; 21°41′N 109°09′E. Rice, oilseeds, sugarcane, jute; food processing; crafts, textiles. Salt mining. Opened to foreign trade in 1877; later superseded by Beihai.

Hepworth (HEP-wuhrth), former village (□ 1 sq mi/2.6 sq km; 2001 population 485), SW ONTARIO, E central CANADA, 12 mi/19 km NW of OWEN SOUND; 44°38′N 81°08′W. Dairying, mixed farming. Amalgamated into town of SOUTH BRUCE Peninsula in 1999.

Hepworth, ENGLAND: see NEW MILL.

Heqing (HUH-CHEENG), town, ⊙ Heqing county, NW YUNNAN province, SW CHINA, near CHANG JIANG (Yangzi River), 60 mi/97 km N of Dali; 26°34′N 100°12′E. Elevation 6,831 ft/2,082 m. Rice, sugarcane; chemicals; food processing. Iron and lead-zinc mining; smelters.

Hequ (HUH-CHYOO), town, ⊙ Hequ county, northwesternmost SHANXI province, NE CHINA, at the GREAT WALL, 65 mi/105 km NW of NINGWU, and on HUANG HE (Yellow River); 39°23′N 111°08′E. Grain, oilseeds; chemicals, rubber and plastic products, building materials; coal mining.

Heraclea (he-ruh-KLEE-uh), ancient Greek city, in LUCANIA, S ITALY, near the Gulf of Tarentum (TARANTO). Here, Pyrrhus defeated the Romans in 280 B.C.E. Bronze tablets giving Roman municipal laws were found nearby.

Heraclea, TURKEY: see AYVALIK.

Heraclea Pontica, ancient Greek city, a port on the S shore of the BLACK SEA. Founded in the 6th century B.C.E. by colonists from MÉGARA and BOEOTIA, it rose to a position of great prominence, controlling much of the coast and sending out colonies. It was at its peak in the 4th century B.C.E. but was hindered by the rise of BITHYNIA (also in ASIA MINOR). Destroyed by the Romans in the wars against Mithridates VI of PONTUS. Modern EREĞLI (TURKEY), is on the site.

Heracleopolis (her-AKH-li-YO-PO-lis), ancient city, N EGYPT, just S of El FAIYUM. One of the oldest Egyptian cities, it was in existence before 3000 B.C.E. and was capital (c.2155 B.C.E.–c.2050 B.C.E.) of the IX and X dynasties. The temple of the local god was enlarged in the XII dynasty and again by Ramses II.

Heracleotica, Chersonesus, UKRAINE: see SEVASTOPOL'.

Heraclia, Greece: see IRAKLIA.

Herakleion, Greece: see IRÁKLION.

Herakol Daği, (Turkish=*Herakol Daği*) range, SE TURKEY, between TIGRIS (W) and Hazil (E) rivers, S of BOTAN RIVER valley, 10 mi/16 km S of PERVARI. Rises to 9,655 ft/2,943 m. Sometimes spelled Herekol Dagi.

Herald Harbor, summer resort (2000 population 2,313), ANNE ARUNDEL county, central MARYLAND, on SEVERN RIVER, 7 mi/11.3 km NW of ANNAPOLIS; 39°03′N 76°34′W. The name of this beach resort is attributed to promotional efforts made by the *Washington Herald* in 1924. One article was headlined, "Herald Harbor Plans Announced." One thousand lots were presumably sold on one day.

Herald Island, Russian *Ostrov Gerald, Ostrov Geral'da* (□ 4 sq mi/10.4 sq km), in the W CHUKCHI SEA, SAKHA

Area is shown by the symbol □, and capital city or county seat by ⊙.

REPUBLIC, RUSSIAN FAR EAST, 37 mi/60 km E of WRANGEL ISLAND; 71°20′N 175°40′W. Rises to 650 ft/198 m. Discovered in 1849 by the British navigator Henry Kellett.

Herald Square, in Midtown section of MANHATTAN borough of NEW YORK CITY, SE NEW YORK, intersection of 6th Avenue, 34th Street, and BROADWAY. Major commercial center; formerly called Greeley Square, it was renamed after the New York *Herald*. In 1902, Macy's built what was then the world's largest department store, and other retailers soon followed, including Gimbel's, Ohrbach, and Saks Fifth Avenue. Although the latter three stores closed in the 1980s, Macy's remains, and other shops also ring the area. Pennsylvania railroad station to SE, EMPIRE STATE BUILDING to SW.

Héraösvötn (HYE-rah[th]s-VUH-tuhn), river, 70 mi/113 km long, NW ICELAND; rises in several headstreams on HOFSJOKULL; flows N to SKAGAFJÖRDUR.

Heras, Las, ARGENTINA: see COLONIA LAS HERAS.

Heras, Las, ARGENTINA: see LAS HERAS.

Herat (he-RAHT), province (□ 50,000 sq mi/130,000 sq km; 2005 population 1,515,400), NW AFGHANISTAN; ⊙ HERAT; 34°30′N 62°00′E. Bounded by IRAN (W), TURKMENISTAN (N), BADGHIS province (N and NE), GHOR province (E), and FARAH province (S). It is penetrated E by the wild outliers of the HINDU KUSH (BAND-I-TURKESTAN and Paropamisus mountains in NE; Safed Koh and Siah Koh in SE) and forms desert lowland (*dasht*) along the W and N borders. Drained by the HARUT RUD R., which disappears into the desert, and the HARI RUD R. Population and agriculture is concentrated in a few oases producing grain (wheat, barley, millet, corn, rice), beans, oilseed, opium, cotton, fruit (peaches, pomegranates, melons), and wine. The Herat oasis along the HARI RUD River, extending from OBEH to the Iranian border, contains 70% of the province's population, and is the center of the most significant agricultural production. Petroleum is produced at TIRPUL; there are coal, silver-lead, iron, and rock-crystal deposits. Handicrafts and a few minor industries are concentrated in Herat, the chief commercial center, handling trade with Iran (via ISLAM QALA) and Turkmenistan (via TORGHONDI). Population is ethnically diverse, including Heratis (related to the Tajiks; NW) and Durani Afghans (SW) in oases, and seminomadic Persian-speaking peoples (Jamshidis, Firoz-Kohis, Taimanis, Hazaras) in mt. and desert areas. Also spelled Harat.

Herat (he-RAHT), city, ⊙ HERAT province, NW AFGHANISTAN, on the HARI RUD River, 400 mi/644 km W of KABUL, and c. 55 mi/89 km from the TURKMENISTAN and IRAN borders; 34°20′N 62°12′E. Elev. 3,030 ft/924 m. Located in the largest oasis of the country, Herat is a major center of Afghan trade with neighboring countries and is noted for its bazaars, market (wool, carpets, dried fruits, and nuts), and its highly decorated gharries (horse-drawn cabs). The fertile Hari Rud River valley is renowned for its fruits, especially grapes. Textile weaving and carpet industries; radio station. Oil fields. Paved roads lead to the Turkmenistan border; an E road is unpaved and passable only with four-wheel vehicles in summer months. Inhabitants are mainly Tajiks. The city walls are gone, but the great earthwork of a citadel remains. Mile-long walls on the massive earthwork are crossed by five gates. The old city is divided into quarters by main bazaar streets that meet in a domed central square known as Charsu. Landmarks include the Great Mosque (12th century) and several exquisite minarets; Masjid-i-Jami (Friday mosque) still preserves the brilliant blue of old tiles. Herat, an ancient city, is strategically located on the trade route from PERSIA to INDIA and on the caravan road from China and central Asia to Europe, which has long made Herat an object of contention. Although taken by various conquerors—the Mongols under Jenghiz Khan devastated

Herat in 1221; the Uzbeks took Herat in the early 16th century; and later it was disputed between the Persians and the rulers of Mogul India—it remained under the Persian empire for several century. In the mid-19th century, British pressure checked Persian claims to Herat, which in 1881 was taken by Abd ar-Rahman and finally confirmed as part of a united Afghanistan. During the 1979–1989 Soviet invasion and occupation, it was a military command center for Soviet forces. In the civil war following Soviet withdrawal, Herat was captured by the Taliban forces in September 1996.

Herau, IRAN: see KHALKHAL.

Hérault (ai-RO), department (□ 2,356 sq mi/6,125.6 sq km), in LANGUEDOC, S FRANCE; ⊙ MONTPELLIER; 43°40′N 03°30′E. Traversed SW-NE by southernmost ranges (CÉVENNES mountains) of the MASSIF CENTRAL, including the arid GARRIGUES range NW of Montpellier. Part of narrow coastal plain is occupied by lagoons near MEDITERRANEAN shoreline. Drained by short streams (ORB, HÉRAULT) that descend from the Cévennes to the NW. One of France's leading large-scale wine-producing departments, it also produces a great variety of fruits (strawberries, apricots, almonds) and vegetables on irrigated land. Olive and citrus trees thrive in the Mediterranean climate. There are bauxite mines at VILLEVEYRAC and some coal is mined; has marble quarries and saltworks. Most major industries (distilling and manufacturing of chemical fertilizer, glass bottles, casks) are connected to viticulture. Some textile mills still operate at LODÈVE and BÉDARIEUX. Chief towns are Montpellier (commercial and cultural city with impressive urban renewal), BÉZIERS (wine-shipping center), and SÈTE (important seaport). High-tech industries are concentrated in the Montpellier urban area, which accounts for one-third of the department's population. Here, too, are the major service businesses. Tourism has been of growing economic importance with the development of such seaside resorts as La GRANDE-MOTTE and beaches SE of Montpellier and along Étang de Thau (a coastal lagoon) near Sète. The department is part of the administrative region of LANGUEDOC-ROUSSILLON.

Hérault River (ai-RO), 100 mi/161 km long, in GARD and HÉRAULT departments, LANGUEDOC-ROUSSILLON administrative region, S FRANCE; rises at foot of Mont AIGOUAL, in the CÉVENNES mountains; flows S across LANGUEDOC region, past PÉZENAS, to the Gulf of LION (MEDITERRANEAN SEA) 3 mi/4.8 km below AGDE; 43°17′N 03°26′E. Vineyards in lower valley.

Herbert (HUHR-buhrt), town, S SASKATCHEWAN, W CANADA, near Rush Lake, 27 mi/43 km ENE of SWIFT CURRENT; 50°26′N 107°13′W. Grain elevators, lumber and flour mills.

Herbert G. West, Lake, reservoir (□ 10 sq mi/26 sq km), FRANKLIN county, SE WASHINGTON, on SNAKE RIVER, 33 mi/53 km NE of PASCO; 46°34′N 118°32′W. Maximum capacity 432,000 acre-ft. Formed by Lower Monumental Dam (226 ft/69 m high) built (1969) by Army Corps of Engineers for navigation; also used for power generation, flood control, and recreation and as a fish and wildlife pond. Lake Sacajewea downstream.

Herbert Hoover National Historic Site, E IOWA; extends over 187 acres/76 ha. Birthplace, childhood home, and gravesite of President Herbert Hoover; library, museum. Authorized 1965.

Herbertingen (HER-ber-ting-uhn), village, BADEN-WÜRTTEMBERG, S GERMANY, 12 mi/19 km E of SIGMARINGEN; 48°04′N 09°27′E.

Herbert, Mount, mountain peak (14,000 ft/4,627m) in the BISMARCK MOUNTAINS, PAPUA NEW GUINEA, 5 mi/8 km N of Mount WILHELM; 05°43′S 145°00′E

Herberton (HUHR-buhr-tuhn), town, NE QUEENSLAND, NE AUSTRALIA, 40 mi/64 km SSW of CAIRNS, on ATHERTON TABLELAND (Cairns Highlands); 17°24′S 145°23′E. Old mining center (tin, copper, silver, lead); citrus fruit.

Herbertshöhe, PAPUA NEW GUINEA: see KOKOPO.

Herbeumont (er-buh-MAW), commune (2006 population 1,507), Neufchâteau district, LUXEMBOURG province, SE BELGIUM, on SEMOIS RIVER, and 10 mi/16 km SW of NEUFCHÂTEAU, in the ARDENNES; 49°47′N 05°14′E. Quarrying.

Herbiers, Les (er-byai, laiz), town (□ 34 sq mi/88.4 sq km), VENDÉE department, PAYS DE LA LOIRE region, W FRANCE, 25 mi/40 km NE of La ROCHE-SUR-YON; 46°52′N 01°00′W. Hog market; manufacturing of footwear, furniture, and recreational boats. A historic steam railroad connects Les Herbiers with MORTAGNE-SUR-SÈVRES across fields and hedgerows typical of the Vendée. The Mont des Alouettes, an isolated hill (740 ft/226 m; 1 mi/1.6 km N), provides distant views as far N as NANTES and the LOIRE valley.

Herbignac (er-bee-NYAHK), town (□ 27 sq mi/70.2 sq km), LOIRE-ATLANTIQUE department, PAYS DE LA LOIRE region, W FRANCE, 11 mi/18 km NNW of SAINT-NAZAIRE, at N edge of the GRANDE-BRIÈRE bog; 47°27′N 02°19′W. Fishing and hunting in the regional park encompassing the Brière wetlands.

Herb Lake (UHRB), unincorporated village, N central MANITOBA, W central CANADA, on WEKUSKO LAKE, 85 mi/137 km E of FLIN FLON; 54°47′N 99°47′W. Former mining area.

Herblay (er-BLAI), residential town (□ 4 sq mi/10.4 sq km), VAL-D'OISE department, ÎLE-DE-FRANCE region, N central FRANCE, an outer NW suburb of PARIS, 13 mi/21 km from Notre Dame Cathedral, on right bank of the SEINE RIVER, opposite the Saint-Germain forest, and on main highway to HAUTE-NORMANDIE (Upper Normandy); 49°00′N 02°10′E. Formerly surrounded by vineyards and orchards.

Herbolzheim (HER-bohlts-heim), town, BADEN-WÜRTTEMBERG, SW GERMANY, at W foot of BLACK FOREST, 9 mi/14.5 km SW of LAHR; 48°14′N 07°47′E. Manufacturing of machinery, construction materials; metalworking.

Herborn (HER-born), town, HESSE, central GERMANY, in the WESTERWALD, on the DILL RIVER, 13 mi/21 km NW of WETZLAR; 50°41′N 08°18′E. Machinery manufacturing; metalworking. Has 13th-century castle, now housing noted Protestant theological seminary; and 16th-century town hall. Chartered 1251.

Herbrechtingen (HER-brech-ting-uhn), town, BADEN-WÜRTTEMBERG, S GERMANY, 4 mi/6.4 km S of HEIDENHEIM; 48°37′N 10°10′E. Largely residential; woodworking. Has church from 13th century.

Herbstein (HERB-shtein), town, HESSE, central GERMANY, 14 mi/23 km W of FULDA; 50°34′N 09°20′E. Health resort with thermal springs. Was enlarged in 1972 through incorporation of several villages; has an 18th-century palace in a large baroque park.

Herceg Novi (HER-tseg NO-vee), Italian *Castelnuovo* (kahs-tel-NWO-vo), city (2003 population 12,739), W MONTENEGRO, port on Bay of TOPLA (NW inlet of Gulf of KOTOR), on narrow-gauge railroad, and 37 mi/60 km W of PODGORICA, near CROATIA (DALMATIA, W) and BOSNIA-HERZEGOVINA (NW) borders; 42°27′N 18°32′E. Tourist center (chiefly in summer) in forested hills (vineyards); noted for mild climate; sea bathing. Ruined Saracen and Spanish castles here. Savina, a medieval monastery, stands on shore E of town. Dates from 1382; in Dalmatia until 1921. Sometimes spelled Ercegnovi; also Khertseg Novi.

Herculândia (ER-koo-lahn-zhee-ah), town (2007 population 8,573), W SÃO PAULO, BRAZIL, on railroad, and 30 mi/48 km WNW of MARÍLIA; 22°01′S 50°22′W. Coffee, cotton. Until 1944, called Herculânia.

Herculaneum (HUHR-kyoo-LAI-nee-uhm), ancient city of CAMPANIA, S ITALY, 5 mi/8 km SE of NAPLES, on the GULF of Naples at the foot of MOUNT VESUVIUS. Damaged by an earthquake in C.E. 63, it was completely buried, along with POMPEII, by the volcanic eruption of Mount Vesuvius in C.E. 79. Before the earthquake, it was a popular Roman resort and residential town with fine villas. The first discovery of ruins was made in 1709, and excavations have continued since. Important early finds were the sumptuous so-called Villa of the Papyri (with a large library, and bronze and marble statues), a basilica with fine murals, and a theater. The modern towns of ERCOLANO and PORTICI are on the site.

Herculaneum (HUHR-kyoo-LAI-nee-uhm), town (2000 population 2,805), JEFFERSON county, E MISSOURI, on MISSISSIPPI RIVER, satellite town 25 mi/ 40 km S of ST. LOUIS; 38°15′N 90°23′W. Limestone quarries; largest lead smelter in the U.S. Historic lead shot tower of 1809. Plotted 1808.

Herculânia, Brazil: see COXIM.

Herculânia, Brazil: see HERCULÂNDIA.

Hercules, city (2000 population 19,488), CONTRA COSTA county, W CALIFORNIA; suburb 13 mi/21 km N of downtown OAKLAND, on SAN PABLO BAY, and at mouth of Refugio Creek; 38°01′N 122°18′W. Manufacturing (computer equipment, laboratory instruments; petroleum refining).

Hercules, residential and industrial suburb, GAUTENG, SOUTH AFRICA, 3 mi/4.8 km NNW of TSHWANE; elevation 4,330 ft/1,320 m; 25°43′S 28°08′E.

Hércules, MEXICO: see CAYETANO RUBIO.

Hercules Dome, local ice dome, E ANTARCTIC ice sheet surface S of HORLICK MOUNTAINS; 86°00′S 105°00′W.

Hércules, Punta (ER-koo-les, POON-tah), cape in CHUBUT province, ARGENTINA, on ATLANTIC coast, easternmost point of VALDÉS PENINSULA; 42°37′S 65°36′W.

Herdecke (HER-dek-ke), town, NORTH RHINE–WESTPHALIA, W GERMANY, on the RUHR RIVER, just N of HAGEN; 51°24′N 07°26′E. Ironworking; manufacturing of chemicals and plastics. Site of first German private university founded 1982. Nearby is the HENGSTEYSEE, a dam on the Ruhr River. Has church from 13th century.

Herdorf (HER-dorf), village, RHINELAND-PALATINATE, W GERMANY, 7 mi/11.3 km SSW of SIEGEN; 50°47′N 07°57′E. Metalworking. Chartered 1981.

Heredia (eh-RAI-dee-ah), province (□ 1,027 sq mi/ 2,670.2 sq km; 2000 population 354,926), N COSTA RICA; ☉ HEREDIA; 10°20′N 84°20′W. Bounded N by SAN JUAN River (NICARAGUA border), it lies largely in N lowlands drained by SARAPIQUÍ RIVER. Extends S across BARVA volcano section of the Central CORDILLERA onto central plateau, where agriculture and population are concentrated. Coffee, sugarcane, potatoes, grain, vegetables; dairying. Bananas and livestock are raised in Sarapiquí lowlands. Plateau section (served by railroads and Inter-American Highway) is site of principal centers: Heredia, SANTO DOMINGO, SAN ISIDRO, and BARVA.

Heredia (eh-RAI-dee-ah), city, ☉ Heredia canton and HEREDIA province, central COSTA RICA; 09°59′N 84°07′W. On the central plateau, it is a center of the coffee industry and, with its colonial architecture, a tourist attraction. Largely a bedroom community for nearby SAN JOSÉ, to which it is linked by bus. Part of the GREATER METROPOLITAN AREA. Seat of the National University. Founded in the 1570s.

Hereford (HE-ruh-fuhd), city (2001 population 50,154) and district, Herefordshire, W central ENGLAND, 24 mi/38.4 km NW of GLOUCESTER; 52°03′N 02°43′W. Cattle-market town from which the cattle breed takes its name. Industries include food processing, brewing, cider making, and light manufacturing. At its cathedral, which probably dates from the 11th century, the Festival of the Three Choirs is held every third year. (In the other years, it is held at Gloucester or Worcester.) The nearby White Cross commemorates the termination of the great plague in the mid-14th century. A grammar school was founded here in the 14th century. Birthplace of Nell Gwyn, mistress of Charles II, and the stage actor David Garrick.

Hereford, city (2006 population 14,531), ☉ DEAF SMITH county, N TEXAS, 40 mi/64 km SW of AMARILLO, in the PANHANDLE; elevation 3,806 ft/1,160 m; 34°49′N 102°24′W. Livestock; cattle feeding is an important industry, along with meatpacking and sugar refining. Vegetables (especially onions), sugar beets, sunflowers, and grains are grown on irrigated farms in the semiarid plains. Former site of National Cowgirl Hall of Fame (1975–1996). Buffalo Lake National Wildlife Refuge to E. Incorporated 1906.

Hereford (HE-re-ford), village, BALTIMORE county, N MARYLAND, 20 mi/32 km N of BALTIMORE. Takes its name from the abandoned Hereford Farmhouse built (1714) by the Merryman family of HEREFORD, ENGLAND. Site of cross-country races for Grand National Steeplechase and My Lady's Manor race. Eighteen fences and two water jumps make the course one of the most dangerous in AMERICA.

Hereford and Worcester (HE-ruh-fuhd and WUStuh), former district, W central ENGLAND; 52°09′N 02°30′W. Composed of the county borough of Worcester and most of the former counties of Herefordshire and Worcestershire. Consisting of the Severn valley and the vale of EVESHAM and their hilly margins, it is noted for orchards, soft fruit, vegetables, and hops. Raising cattle and sheep dominates agriculture here. The former district contains the cathedral cities of Hereford and Worcester along with BROMSGROVE, Evesham, Kidderminster, Malvern, and Redditch. In addition to the SEVERN, it is drained by the Wye, Avon, Lugg, Monnow, and Teme rivers.

Hereford Beacon, ENGLAND: see MALVERN HILLS.

Hereford Inlet (HIR-fuhrd), S NEW JERSEY, passage (c.2 mi/3.2 km wide) connecting the ATLANTIC OCEAN and INTRACOASTAL WATERWAY channel, between NORTH WILDWOOD and S tip of SEVEN MILE BEACH.

Herefordshire (HE-ruh-fuhd-shir), county (□ 838 sq mi/2,178.8 sq km; 2001 population 174,871), SW ENGLAND; ☉ HEREFORD; 52°09′N 02°30′W. Created in 1998 when Hereford and Worcester was split up into unitary authorities of Herefordshire and Worcestershire. Fruit and potato growing.

Herefoss (HER-uh-faws), village, AUST-AGDER county, S NORWAY, on Herefoss Lake (8-mi/12.9-km-long widening of Tovdalelva River), on railroad, and 16 mi/ 26 km WNW of ARENDAL.

Hereke, township, NW TURKEY, on ISTANBUL-IZMIT railroad, and 15 mi/24 km W of Izmit; 40°49′N 29°38′E. Cotton and woolen goods.

Herekol Dagi, TURKEY: see HERAKOL DAGI.

Herelen River, MONGOLIA: see KERULEN.

Hérémence (ai-rai-MAWNS), commune, VALAIS canton, SW SWITZERLAND, in VAL D'HÉRÉMENCE, S of SION, which extends from Lake Dix (French *Lac des Dix*) to junction of Hérémence River with Hérens River. Dam and hydroelectric plant of GRANDE-DIXENCE lake are here.

Hérémence, Val d' (dai-rai-MAWNS, VAHL), tributary valley of RHÔNE RIVER in W VALAIS canton, S SWITZERLAND. Watered by Dixence River, which issues from Lake Dix.

Herencia (ai-REN-syah), town, CIUDAD REAL province, S central SPAIN, near TOLEDO province border, 40 mi/ 64 km NE of CIUDAD REAL; 39°21′N 03°22′W. Agricultural center on LA MANCHA plain (cereals, grapes, olives, saffron; livestock). Olive oil pressing, wine production, alcohol distilling, cheese making, textile mills, tanneries; soap and pharmaceuticals manufacturing; gypsum quarrying. Also exports wool. Notable church and convent here.

Herencias, Las (ai-REN-syahs, lahs), town, TOLEDO province, central SPAIN, on the TAGUS RIVER, and 7 mi/11.3 km SW of TALAVERA DE LA REINA. Grain, olives, grapes, cherries, melons; livestock.

Herencsvölgy, SLOVAKIA: see HRINOVA.

Herend (HA-rand), village, VESZPRÉM county, NW central HUNGARY, on Séd River, and 8 mi/13 km ENE of VESZPRÉM; 47°08′N 17°45′E. Noted porcelain and faïence factories. Founded 1837.

Herendeen Bay (HE-ren-deen), village, SW ALASKA, on ALASKA PENINSULA, at head of Herendeen Bay, inlet (20 mi/32 km long) of BRISTOL BAY of BERING SEA; 55°46′N 160°43′W. Fishing. Surface coal veins in vicinity.

Hérens (ai-RAWNS), district, S central VALAIS canton, S SWITZERLAND, S of SION. Population is French-speaking and Roman Catholic.

Hérens (ai-RAWNS), valley, VALAIS canton, S SWITZERLAND, watered by LA BORGNE RIVER. District includes commune of AYENT.

Hérens, Dend d', SWITZERLAND: see DENT D'HÉRENS.

Herent (HAI-ruhnt), commune (2006 population 19,279), Leuven district, BRABANT province, central BELGIUM, 2 mi/3.2 km NNW of LEUVEN; 50°54′N 04°40′E. Ceramics.

Herentals (HAI-ruhn-tahls), commune (2006 population 26,125), Turnhout district, ANTWERPEN province, N BELGIUM, on ALBERT CANAL, and 18 mi/29 km E of ANTWERP; 51°11′N 04°50′E. Has fifteenth-century church, sixteenth-century town hall. Het Netepark Municipal Recreational Area.

Herenthout (HAI-ruhnt-hout), commune (2006 population 8,410), Turnhout district, ANTWERPEN province, NE BELGIUM, 15 mi/24 km NE of MECHELEN; 51°09′N 04°46′E. Agricultural market.

Herero, town (2007 population 5,858), OROMIYA state, S central ETHIOPIA, 50 mi/80 km W of GOBA; 06°59′N 39°18′E.

Heretaunga Plains, HAWKE'S BAY, E NORTH ISLAND, NEW ZEALAND. A structural depression in-filled with alluvial soil from the Ngaruroro, Tukituki, and Tutaekuri rivers, with warm sunny climate utilized for horticulture in New Zealand's largest production area. Agriculture (apples, pears, peaches, vegetable crops for processing, wine grapes). HASTINGS focus of processing, NAPIER of exporting.

Herfølge (HER-ful-guh), town, ROSKILDE county, SJÆLLAND, DENMARK, 21 mi/34 km N of PRÆSTO; 55°25′N 12°10′E. Dairying; grain; cement.

Herford (HER-ford), city, NORTH RHINE–WESTPHALIA, NW GERMANY, on the WERRE RIVER; 52°07′N 08°40′E. Manufacturing includes cigars, textiles, chocolate, beer, carpets, machinery, and metal products. Also a major German producer of furniture. Developed around a 13th-century church (still standing) that formerly had been a Benedictine convent (founded in the 9th century). Was a member of the HANSEATIC LEAGUE and passed to BRANDENBURG in 1647.

Hergiswil (HER-gees-vil), commune, Nidwalden half-canton, central SWITZERLAND, on W shore of LAKE LUCERNE, at foot of the PILATUS, 4 mi/6.4 km S of LUCERNE. Elevation 1,473 ft/449 m. Resort.

Hergiswil bei Willisau (HER-gees-vil bei VIL-LI-sou), agricultural commune, LUCERNE canton, central SWITZERLAND, on Enzwigger River (affluent of WIGGER RIVER), and 3 mi/4.8 km SSW of WILLISAU.

Hergnies (er-NYEE), commune (□ 4 sq mi/10.4 sq km), NORD department, NORD-PAS-DE-CALAIS region, N FRANCE, near Belgian border, on Escaut (SCHELDT) River, and 8 mi/13 km N of VALENCIENNES; 50°28′N 03°31′E.

Herguijuela (er-gee-HWAI-lah), village, CÁCERES province, W SPAIN, 10 mi/16 km SE of TRUJILLO. Olive oil processing; cereals; livestock.

Héricourt (ai-ree-KOOR), town (□ 7 sq mi/18.2 sq km), HAUTE-SAÔNE department, FRANCHE-COMTÉ region, E FRANCE, in BELFORT GAP, 7 mi/11.3 km SW of BELFORT, and 6 mi/9.7 km NW of MONTBÉLIARD; 47°35′N 06°45′E. Metalworking center; cotton milling. Scene of battle between Charles the Bold of BURGUNDY and the Swiss (1474), and of Franco-German military action in 1871.

Hérimoncourt (ai-ree-mon-KOOR), commune (□ 2 sq mi/5.2 sq km), DOUBS department, FRANCHE-COMTÉ region, E FRANCE, 6 mi/9.7 km SE of MONTBÉLIARD, near Swiss border; 47°26′N 06°53′E. Metalworking (heavy metals) district.

Hérin (ai-RAN), residential town (□ 1 sq mi/2.6 sq km), NORD department, NORD-PAS-DE-CALAIS region, N FRANCE, 3 mi/4.8 km W of VALENCIENNES; 50°21′N 03°27′E. In old coal-mining district.

Heringen (HE-ring-uhn), town, central GERMANY, on the WERRA RIVER, and 12 mi/19 km E of HERSFELD; 50°54′N 10°E. Potash-mining; metal- and woodworking.

Heringsdorf (HE-rings-dorf) or **Seebad Heringsdorf**, village, MECKLENBURG–WESTERN POMERANIA, E GERMANY, on NE shore of USEDOM island, 5 mi/8 km NW of SWINEMÜNDE; 53°28′N 14°10′E. Popular seaside and health resort.

Herington (HER-ing-tuhn), town (2000 population 2,563), DICKINSON county, central KANSAS, 22 mi/35 km SE of ABILENE; 38°40′N 96°57′W. Railroad junction. Shipping and trade center, with railroad repair shops, for livestock and grain area. Manufacturing (industrial machinery). Monument to the missionary Juan de Padilla is here. Incorporated 1887.

Heri Rud, AFGHANISTAN, IRAN, and TURKMENISTAN: see HARI RUD.

Herisau (HER-re-sou), town (2000 population 15,882), ⊙ APPENZELL Ausser Rhoden half canton, NE SWITZERLAND. Elevation 2,530 ft/771 m. Manufacturing includes textiles, paper, wood products, machinery, and other metal goods. Also a cattle market and a popular tourist resort. Among its historic monuments is a church with elements dating back to the 10th century.

Héristal, BELGIUM: see HERSTAL.

Heritage Hills, incorporated village within the town of SOUTHBURY, NEW HAVEN county, SW CONNECTICUT Retirement center with 2,500 homes.

Heritage Range, in WEST ANTARCTICA, S of MINNESOTA GLACIER, forming the S half of the ELLSWORTH MOUNTAINS; 100 mi/160 km long and 30 mi/50 km wide; 79°45′S 83°00′W.

Herk-de-Stad (HERK–duh–STAHD), commune (2006 population 11,811), Hasselt district, LIMBURG province, NE BELGIUM, 8 mi/13 km W of HASSELT; 50°56′N 05°05′E.

Herkimer (HUHR-ki-muhr), county (□ 1,458 sq mi/3,790.8 sq km; 2006 population 63,332), N central NEW YORK; ⊙ HERKIMER; 43°24′N 74°57′W. Long, narrow area, with N part extending into the ADIRONDACK MOUNTAINS, and S part in fertile Mohawk valley, which contains virtually all of the county's population. Drained by MOHAWK RIVER, and by UNADILLA, BLACK, and MOOSE rivers. Manufacturing, farming, and dairying in S. Many mountain and lake resorts in N. Formed 1791.

Herkimer (HUHR-ki-muhr), industrial village (□ 2 sq mi/5.2 sq km; 2006 population 7,162), ⊙ HERKIMER county, central NEW YORK, on MOHAWK RIVER, at mouth of WEST CANADA CREEK, and 13 mi/21 km ESE of UTICA; 43°01′N 74°59′W. Formerly shipping, commercial, and trade center for surrounding Mohawk valley agricultural and industrial area that stretched W through village of MOHAWK to ILION and FRANKFORT. Today, however, the industrial center of the Mohawk Valley has greatly declined. Manufacturing includes machinery, consumer goods, wood products, and firearms. Herkimer County Historical Society has important documents and exhibits here. World-renowned "Herkimer diamonds" (actually a clear quartz) are to be found along West Canada Creek Valley N of the village. Settled c.1725, incorporated 1807.

Herkulesbad, ROMANIA: see BĂILE HERCULANE.

Herkulesfürdö, ROMANIA: see BĂILE HERCULANE.

Herlandsfoss (HER-lahns-faws), waterfall on small river, NW OSTERØY, HORDALAND county, SW NORWAY, 15 mi/24 km NNE of BERGEN. Site of hydroelectric station.

Herlany (her-LYAH-ni), Slovak *Herl'any*, Hungarian *Ránkfüred*, village, VYCHODOSLOVENSKY province, SE SLOVAKIA, in PRESOV MOUNTAINS, 11 mi/18 km NE of KOŠICE; 48°48′N 21°29′E. Summer resort featuring the only Slovak geyser, which erupts every thirty-three hours (130 ft/40 m).

Herlen Gol, MONGOLIA and CHINA: see KERULEN.

Herlen He, MONGOLIA and CHINA: see KERULEN.

Herlev (HER-lev), city, Copenhagen county, SJÆLLAND, DENMARK, 5 mi/8 km W of COPENHAGEN; 55°44′N 12°27′E.

Herlisheim, FRANCE: see HERRLISHEIM.

Herm, United Kingdom: see CHANNEL ISLANDS.

Hermagor (HER-muh-gor), town, CARINTHIA, S AUSTRIA, near GAIL RIVER, 22 mi/35 km W of VILLACH. Market center; manufacturing of refrigerators, wooden structural elements. Summer tourism; large winter sports area 9 mi/14.5 km SW; 46°38′N 13°22′E.

Herman, village (2000 population 452), GRANT county, W MINNESOTA, 15 mi/24 km SW of ELBOW LAKE town; 45°48′N 96°08′W. Grain; livestock; poultry; dairying; manufacturing (wire harnesses; light manufacturing). Several small lakes to E and SE.

Herman, village (2006 population 304), WASHINGTON county, E NEBRASKA, 10 mi/16 km NNW of BLAIR, near MISSOURI RIVER; 41°40′N 96°13′W. Grain; livestock. Manufacturing (portable grain augers).

Hermann (HUHR-muhn), city (2000 population 2,674), E central MISSOURI, ⊙ GASCONADE county, on MISSOURI River, 41 mi/66 km E of JEFFERSON CITY; 38°42′N 91°26′W. Dairying; cattle, poultry, hogs; manufacturing (plastic products, transportation equipment, consumer goods, wine; meat processing); clay. Wineries; tourism. Has annual Maifest and Oktoberfest celebrations. Settled 1837 by German immigrants. Deutschherm State Historic Site. Over 100 structures are on the Register of Historic Places. Incorporated 1839.

Hermannsbad, POLAND: see CIECHOCINEK.

Hermannsburg (HER-mahns-boorg), village, LOWER SAXONY, NW GERMANY, 14 mi/23 km N of CELLE; 53°50′N 10°05′E. Manufacturing of furniture; tourism.

Hermannsburg (HUHR-muhnz-buhrg), settlement, S central NORTHERN TERRITORY, N central AUSTRALIA, 75 mi/121 km WSW of ALICE SPRINGS, in MACDONNELL RANGES; 23°58′S 132°37′E. German mission for Aboriginal people; tourist center. Sheep. Meteorite craters nearby. Established by Lutherans 1877. Sometimes spelled Hermansburg.

Hermannseifen, CZECH REPUBLIC: see RUDNIK.

Hermannshöhle (HER-mahns-HUH-le), large stalactite cave, central GERMANY, in the lower HARZ, S of RÜBELAND; 51°46′N 10°51′E. Discovered 1866; has significant fossil remains.

Hermannstadt, CZECH REPUBLIC: see HERMANUV MESTEC.

Hermannstadt, ROMANIA: see SIBIU.

Hermanovy Sejfy, CZECH REPUBLIC: see RUDNIK.

Hermannsburg, AUSTRALIA: see HERMANNSBURG.

Hermansverk (HUHR-mahns-verk), village, SOGN OG FJORDANE county, W NORWAY, on N shore of SOGNEFJORDEN, 10 mi/16 km WSW of SOGNDAL. Canned fruit, vegetables.

Hermansville, village, MENOMINEE county, SW UPPER PENINSULA, N MICHIGAN, 24 mi/39 km SE of IRON MOUNTAIN city, and on Little Cedar River; 45°42′N 87°36′W. Lumber milling.

Hermantown, town (2000 population 7,448), ST. LOUIS county, NE MINNESOTA, residential suburb 6 mi/9.7 km NW of downtown DULUTH, on Midway River; 46°48′N 92°14′W. Manufacturing (sheet metal fabrication); agriculture (dairying; poultry; hay). Duluth Municipal Airport to NE.

Hermanus, town, WESTERN CAPE province, SOUTH AFRICA, on WALKER BAY of the ATLANTIC OCEAN, 55 mi/89 km SE of CAPE TOWN; elevation 91 ft/27 m; 34°54′S 19°07′E. Picturesque fishing and seaside resort. Fernkloof Nature Reserve has 10 mi/16 km of footpaths to view wide variety of fauna and flora.

Hermanuv Mestec (HERZH-mah-NOOF MNYES-tets), Czech *Heřmanův Městec*, German *Hermannstadt*, town, VYCHODOCESKY province, E BOHEMIA, CZECH REPUBLIC, and 7 mi/11.3 km SSE of PARDUBICE; 49°57′N 15°40′E. Railroad junction. Manufacturing (shoes); flowers and seeds. Has an 18th-century castle with park.

Hermanville, village, CLAIBORNE county, SW MISSISSIPPI, 8 mi/12.9 km E of PORT GIBSON. Agriculture (cattle; corn, cotton; timber); manufacturing (lumber).

Hermanville-sur-Mer (er-mahn-veel–syur-MER), commune (□ 3 sq mi/7.8 sq km), swimming resort in CALVADOS department, BASSE-NORMANDIE region, NW FRANCE, on ENGLISH CHANNEL, near mouth of the ORNE RIVER, 7 mi/11.3 km N of CAEN; 49°17′N 00°19′W. Its beach extends E to Ouistreham-Riva-Bella, a yachting resort. British troops landed here (June 1944) in NORMANDY invasion of World War II.

Hermel, LEBANON: see HERMIL.

Hermenegildo Galeana, MEXICO: see BIENVENIDO.

Hermeskeil (HER-mes-keil), village, RHINELAND-PALATINATE, W GERMANY, 16 mi/26 km SE of TRIER; 49°40′N 06°57′E. Manufacturing includes machinery, paper, ceramics; printing. Chartered 1970.

Hermigua (er-MEE-gwah), town, SANTA CRUZ DE TENERIFE province, GOMERA, CANARY ISLANDS, 7 mi/11.3 km NW of SAN SEBASTIÁN; 28°10′N 17°10′W. Cereals, bananas, tomatoes, grapes, sugarcane; sheep. Wine making.

Hermil (HER-mil), township, N LEBANON, 8 mi/13 km from SYRIAN border, 60 mi/97 km NE of BEIRUT. Cereals, fruits. Also spelled Hirmal, Hirmil, and Hermel.

Herminie (HER-mi-nee), unincorporated town, WESTMORELAND county, SW PENNSYLVANIA, 19 mi/31 km SE of PITTSBURGH, on Little Sewickley Creek; 40°15′N 74°43′W. Agricultural (corn, hay; dairying). Light manufacturing Shuster Cellars Winery to N.

Hermione, Greece: see ERMIONI.

Hermiston (HUHR-muhs-stuhn), city (2006 population 14,891), UMATILLA county, N OREGON, 27 mi/43 km NW of PENDLETON, near UMATILLA RIVER, 5 mi/8 km S of its confluence with the Columbia River; 45°49′N 119°16′W. Railroad junction to S. Manufacturing (frozen vegetables, mobile homes, building materials; printing and publishing). Grains; cattle. Umatilla Ordnance Depot (chemical weapons) to W. Cold Springs National Wildlife Refuge and Hot Rock State Park to NE. MCNARY DAM (Columbia River) to N. Incorporated 1910.

Hermitage (HUHR-mah-tuhj), city (2006 population 16,530), MERCER county, W PENNSYLVANIA, suburb 3 mi/4.8 km E of SHARON; 41°13′N 80°26′W. Manufacturing of fabricated metal products, machinery, food products, chemicals. Agriculture includes dairying; livestock; grain, potatoes, soybeans. Hermitage and Sharon airports in S. SHENANGO RIVER LAKE reservoir to N. Formerly a township.

Hermitage (HUHR-mi-tij), town (2000 population 406), ⊙ HICKORY county, central MISSOURI, in the OZARK MOUNTAINS, on POMME DE TERRE RIVER, and 50 mi/80 km N of SPRINGFIELD; 37°56′N 93°19′W. Wheat, soybeans; dairying; cattle; manufacturing (apparel). Tourism and lake region. Pomme de Terre Lake to S; extension of Harry S. Truman Lake to N; Pomme de Terre State Park. Plotted 1847.

Hermitage (HUHR-mi-tij), town, DAVIDSON county, central TENNESSEE, 10 mi/16 km E of NASHVILLE, 36°11′N 86°37′W. The Hermitage, home of Andrew Jackson, is just N.

Cross-references are shown in SMALL CAPITALS. The pronunciation guide is shown on page xix. The sources of population figures are shown on page xvii.

Hermitage, village (□ 18 sq mi/46.8 sq km; 2001 population 602), S NEWFOUNDLAND AND LABRADOR, E CANADA, on SE shore of Hermitage Bay, 80 mi/129 km E of BURGEO; 47°33′N 55°56′W. Fishing.

Hermitage (HERM-i-taj), village (2001 population 2,601), BERKSHIRE, S central ENGLAND, 5 mi/8 km NW of NEWBURY; 51°27′N 01°16′W. Farming.

Hermitage (HUHR-muh-tij), village (2000 population 769), BRADLEY county, S ARKANSAS, 32 mi/51 km ENE of EL DORADO in UNION county; 33°27′N 92°10′W. Railroad junction to N. Known for production of tomatoes.

Hermitage (HUHR-mi-tej), historic site, home of President Andrew Jackson, DAVIDSON county, central TENNESSEE, 12 mi/19 km E of NASHVILLE. The house, in a fine formal garden, was built 1819–1831; a church on the grounds was built 1823. Jackson and his wife are buried in the plantation graveyard.

Hermitage Bay, inlet of the ATLANTIC OCEAN, S NEWFOUNDLAND AND LABRADOR, E CANADA; 30 mi/48 km long, 10 mi/16 km wide at entrance; 47°35′N 55°50′W. N arm of Hermitage Bay is the Bay d'ESPOIR. Bay contains several islands.

Hermitage Castle, Scotland: see NEWCASTLETON.

Hermitage Island, S of RODRIGUEZ Island, dependency of MAURITIUS, 2 mi/4 km from Pointe Rafu; 19°45′S 63°26′E. Small basaltic island (7.4 acres/3 ha) with prominent rock faces and beaches; the only basalt island E of the cluster of coralline islands. Mixed and degraded vegetation with high invasion of weed species.

Hermit Island (HUHR-mit), largest (13 mi/21 km long, 5 mi/8 km wide) of small group of 4 main islands and many rocks, in TIERRA DEL FUEGO, CHILE, c.10 mi/16 km NW of CAPE HORN; Cape West is at 55°54′S 67°55′W.

Hermit Islands, small coral island group, BISMARCK ARCHIPELAGO, MANUS province, N PAPUA NEW GUINEA, SW PACIFIC OCEAN, 95 mi/153 km WNW of ADMIRALTY ISLANDS; circa 5 islands on reef 12 mi/19 km long, 10 mi/16 km wide; main island (Luf) is in lagoon; coconut plantations. BISMARCK SEA to S.

Hermit Kingdom: see NORTH KOREA.

Hermleigh (HUHR-mah-lee), unincorporated town, SCURRY county, NW central TEXAS, 10 mi/16 km SE of SNYDER; 32°38′N 100°45′W. In agricultural and cattle-ranching area.

Hermon (HUHR-muhn), town, PENOBSCOT county, S MAINE, 7 mi/11.3 km W of BANGOR; 44°49′N 68°55′W. In agricultural area. Settled c.1790.

Hermon, village (2006 population 386), ST. LAWRENCE county, N NEW YORK, 20 mi/32 km SE of OGDENSBURG; 44°28′N 75°13′W. Settled 1816, incorporated 1887.

Hermon, Mount (hehr-MON), Arabic *Jabal Ash Shaykh* [mountain of the chief] and *Jebel-eth-Thelj* [snowy mountain], on the Israeli, Lebanese, and Syrian borders; 33°24′N 35°50′E. The S part of the ANTI-LEBANON range. The highest of its three peaks (all of which are snow-covered in winter and spring) rises to 9,232 ft/2,814 m. Its seasonal snow melt is important to the headwater flow of the JORDAN River. Mount Hermon, a sacred ancient landmark, is mentioned often in the Bible as Hermon, Sion, Senir, and Sirion. The name Baal-Hermon records the reverence in which it was held by the worshipers of Baal. The Romans also revered it, as did the Druse (there is a Druse shrine near HASBEYA). The ancient city of CAESAREA PHILIPPI was at its foot. Mount Hermon is traditionally designated as the scene of the Transfiguration. Israel captured Mount Hermon's S and W slopes during the 1967 Six Day War, establishing military observational posts there, as well as developing it for recreational use, such as skiing.

Hermonthis (her-MON-this), ancient city, N EGYPT, 8 mi/13 km S of THEBES. Founded in prehistoric times and prominent during the period of Roman supremacy. Originally the shrine of Month, a hawk-headed deity, Hermonthis has a temple built c.1500 B.C.E. and reconstructed by the Ptolemies. Modern ARMANT is on the site.

Hermopolis, Greece: see ERMOUPOLIS.

Hermopolis Magna (her-MO-PO-lis MAG-nah), ancient city, central EGYPT, on the NILE River, and near the modern ASHMUNEIN, 22 mi/35 km S of El MINYA. It was the center for the worship of Thoth. At the modern Tunneh el Gebel, 7 mi/11 km SW of the site, is a Greco-Egyptian cemetery.

Hermopolis Parva, EGYPT: see DAMANHUR.

Hermosa (huhr-MO-sah), town, Bataan province, S LUZON, PHILIPPINES, at base of BATAAN PENINSULA, 36 mi/58 km WNW of MANILA; 14°51′N 120°29′E. Sugarcane, rice.

Hermosa, village (2006 population 354), CUSTER county, SW SOUTH DAKOTA, 17 mi/27 km S of RAPID CITY, and on Battle Creek; 43°50′N 103°11′W. Near E entrance to Custer State Park. Tourism at E edge of BLACK HILLS.

Hermosa Beach, city (2000 population 18,566), LOS ANGELES county, S CALIFORNIA; residential suburb 14 mi/23 km SW of downtown LOS ANGELES, on Santa Monica Bay; 33°56′N 118°25′W. Beach resort and recreation area. Nearby are Manhattan (N) and Redondo (S) state beaches. Incorporated 1907.

Hermosillo (er-mo-SEE-yo), city (2005 population 641,791) and township, ⊙ SONORA, NW MEXICO, on Mexico Highway 15; 29°15′N 110°59′W. At the entrance to the gorge of the SONORA RIVER. Hermosillo is a transportation, manufacturing (motor vehicles), and agricultural center in an irrigated area where cereals and cotton are grown and cattle are raised. Established 1700 as a Native American town with a Jesuit missionary, the city was later renamed in honor of the Spanish general José María González de Hermosillo. Airport.

Hermoso, Cerro (er-MO-so, SER-ro), Andean peak (15,216 ft/4,638 m), central ECUADOR, 27 mi/43 km ENE of AMBATO; 01°13′S 78°16′W.

Hermoupolis, Greece: see ERMOUPOLIS.

Hermsdorf (HERMS-dorf), town, THURINGIA, central GERMANY, 10 mi/16 km W of GERA; 50°54′N 11°51′E. Manufacturing of china and construction materials; optics industry.

Hermsdorf, POLAND: see SOBIECIN.

Hermupolis, Greece: see ERMOUPOLIS.

Hermus, TURKEY: see GEDIZ.

Hermus River, TURKEY: see GEDIZ RIVER.

Hernad River (hor-NAHT), Hungarian *Hernád*, Slovak *Hornád*, 165 mi/266 km long, S SLOVAKIA and NE HUNGARY; rises in Slovakia on N slope of the LOW TATRAS, 10 mi/16 km SW of POPRAD; flows E, past SPIŠSKA NOVÁ VES, and S, past KOŠICE, briefly along Slovak-Hungarian border, into Hungary, to SAJO RIVER, SE of MISKOLC. Length in Slovakia: 116 mi/187 km. Ružin Dam, Slovak *Ružín* (ru-ZHEEN) (965 acres/391 ha), is 11 mi/18 km NW of Košice.

Hernandarias (er-nahn-dah-REE-ahs), town, (2002 population 47,266), E PARAGUAY, near upper Paraná River (BRAZIL border), 190 mi/306 km E of Asunción; 25°23′S 54°42′W. Maté, lumber; livestock. Formerly called Tacurupucú. Former Alto Parana department capital.

Hernandarias, ARGENTINA: see VILLA HERNANDARIAS.

Hernández (er-NAHN-dez), village, W ENTRE RÍOS province, ARGENTINA, on railroad, and 50 mi/80 km SE of PARANÁ; 32°21′S 60°20′W. Flax, wheat, corn; livestock; food processing.

Hernando (her-NAN-do), county (□ 589 sq mi/1,531.4 sq km; 2006 population 165,409), W central FLORIDA, between WITHLACOOCHEE RIVER (E) and GULF OF MEXICO (W); ⊙ BROOKSVILLE; 28°33′N 82°28′W. Lowland area with marshy coast and small scattered lakes. Agricultural (poultry, cattle, hogs; corn, citrus fruit, peanuts); sawmilling; limestone quarrying. Formed 1843.

Hernando (er-NAHN-do), town, central CÓRDOBA province, ARGENTINA, 30 mi/48 km W of VILLA MARÍA; 32°25′S 63°44′W. Wheat, flax, corn, peanuts, soybeans, sunflowers, fruit; flour milling.

Hernando (her-NAN-do), town (□ 3 sq mi/7.8 sq km; 2000 population 8,253), CITRUS county, W central FLORIDA, 25 mi/40 km SW of OCALA, on TSALA APOPKA LAKE; 28°54′N 82°22′W. Fishing.

Hernando, town (2000 population 6,812), ⊙ DE SOTO county, extreme NW MISSISSIPPI, 22 mi/35 km S of MEMPHIS (TENNESSEE); 34°49′N 89°59′W. Trade center for agricultural region (soybeans, cotton, corn; cattle; dairying); manufacturing (aluminum and paper products, crushed stone, electrical goods, wood products; beef processing). Arkabutla Lake reservoir to SW; Hernando de Soto Memorial Trail. Incorporated 1837.

Hernando Siles (er-NAHN-do SEE-les), province (□ 2,113 sq mi/5,493.8 sq km), CHUQUISACA department, SE BOLIVIA; 19°49′S 63°59′W; ⊙ MONTEAGUDO (formerly Sauces). Terrain in the subandean mountains. Created in 1840 by Andrés de Santa Cruz. Formerly known as Azero.

Hernani, city, GUIPÚZCOA province, N SPAIN, on URUMEA RIVER, and 4 mi/6.4 km S of SAN SEBASTIÁN; 43°16′N 01°58′W. Manufacturing (paper, chemicals, cement, textiles). Lignite mines nearby. Has Renaissance church. Played notable role in Carlist Wars (19th century).

Hernani, town, EASTERN SAMAR province, SE SAMAR island, PHILIPPINES, on PHILIPPINE SEA, 60 mi/97 km SE of CATBALOGAN; 11°20′N 125°35′E. Iron mining since 1934.

Hernaudo de Magallanes (er-NOU-do dai mah-gah-YAH-nais), national park, in MAGALLANES Y LA ANTARTICA CHILENA region, includes chain of islands ranging from 150 mi/241 km W to 100 mi/161 km SW of PUNTA ARENAS.

Herndon (HERN-duhn), town (2006 population 21,877), FAIRFAX county, N VIRGINIA, suburb 20 mi/32 km WNW of WASHINGTON, D.C.; 38°58′N 77°23′W. Manufacturing (printing and publishing; electronic equipment, machinery, computer equipment, biological solutions, text encryption devices); some agriculture to NW (dairying; poultry, cattle; nursery stock). DULLES INTERNATIONAL AIRPORT to W. National Weather Service to W. Incorporated 1874; rechartered 1938.

Herndon (HUHRN-duhn), village (2000 population 149), RAWLINS county, NW KANSAS, on Beaver Creek, and 16 mi/26 km ENE of ATWOOD, near NEBRASKA state line; 39°54′N 100°47′W. Grain; livestock.

Herndon, borough (2006 population 357), NORTHUMBERLAND county, E central PENNSYLVANIA, 11 mi/18 km SSW of SUNBURY, on SUSQUEHANNA RIVER. Manufacturing (wood products, crushed limestone); agriculture (corn, hay; poultry; dairying).

Herne (HER-NUH), commune (2006 population 6,448), Halle-Vilvoorde district, BRABANT province, central BELGIUM, 7 mi/11 km S of NINOVE; 50°43′N 04°02′E.

Herne (HER-ne), city, NORTH RHINE–WESTPHALIA, W GERMANY, a port on the RHINE-HERNE CANAL; 51°32′N 07°13′E. Important railroad and road junction. Industrial center of the RUHR district; foundries; other manufacturing includes textiles, chemicals, and machinery. Also a spa with a thermal spring. Site of the medical campus of Bochum University. Grew steadily after 1850 with increase in coal mining and the development of the steel industry. In 1975, city incorporated Wanne (first mentioned 890), Eickel (first mentioned 1150), and another village. Coal mining ceased in 1978.

Herne Bay (HUHN), town (2006 population 55,000), KENT, SE ENGLAND, 7 mi/11.3 km N of CANTERBURY; 51°22′N 01°08′E. Resort with 7 mi/11.3 km of coast and 5 mi/8 km of promenades. The town developed after a railroad was built in 1833.

Area is shown by the symbol □, and capital city or county seat by ⊙.

Herne Hill (HUHRN HIL), residential suburb 2 mi/3 km WNW of GEELONG, VICTORIA, SE AUSTRALIA. A campus of Western Heights Secondary College here.

Herne Hill, ENGLAND: see LAMBETH.

Herning (HER-ning), city, RINGKJØBING county, W JUTLAND, DENMARK; 56°10′N 09°00′E. Important manufacturing center with textile mills and machine shops. An annual textiles fair is held here.

Herod (HER-uhd), town, TERRELL county, SW GEORGIA, 6 mi/9.7 km S of DAWSON; 31°41′N 84°26′W.

Herodion, mountain, in JUDEAN HIGHLANDS, 3.7 mi/6 km SE of BETHLEHEM; elevation 2,487 ft/758 m. The peak is conic shaped and its top looks like a crater. Named after Herod who built a fortress at the summit and a settlement at its foot.

Heroica Caborca (e-ro-EE-kah kah-BOR-kah), city and township, SONORA, NW MEXICO, on MAGDALENA RIVER (irrigation), and 80 mi/129 km SW of NOGALES; 30°42′N 112°10′W. Agriculture center (wheat, corn, cotton, beans). Silver, copper deposits nearby.

Heroica Ciudad de Tlaxiaco (he-RO-ee-kah see-OO-dahd dai tlahk-see-AH-ko), town, W central OAXACA, MEXICO, 65 mi/105 km WNW of OAXACA DE JUÁREZ, and on Mexico Highway 125, between PUTLA DE GUERRERO and SAN PEDRO Y SAN PABLO TEPOSCOLULA; elevation 6,555 ft/1,998 m; 17°15′N 97°40′W. On the banks of the Tlaxiaco River. The Los Tejocotes Antimony Mining is a short distance W of Tlaxiaco in San Juan Mixtepec municipio. Agriculture (corn, beans); cattle raising; forestry (pine, oak). An important commercial center that serves the surrounding rural population. Formerly SANTA MARÍA ASUNCIÓN TLAXIACO.

Heroica Guaymas (e-RO-ee-kah goo-AI-mas), city (2005 population 184,816) and township, SONORA, NW MEXICO, on the Bay of Guaymas, on Mexico Highway 15; 27°59′N 110°54′W. A port on the Gulf of CALIFORNIA, it is the outlet for HERMOSILLO. On a scenic inlet girt by desert mountains. Its fine beaches, excellent deep-sea fishing, and transportation facilities have made it a popular tourist resort. In addition to its role as a commercial center for the surrounding region, Guaymas has a substantial fishing industry. Although the surrounding area was explored as early as 1539, the city was not established until the early eighteenth century by Jesuit missionaries. UNITED STATES forces occupied Guaymas in 1846, during the Mexican War, and it was held by the French in 1865–1866.

Heroica Matamoros (e-RO-ee-kah mah-tah-MO-ros), city (2005 population 422,711) and township, ⊙ Matamoros municipio, TAMAULIPAS, NE MEXICO, near the mouth of the RIO GRANDE, opposite BROWNSVILLE (TEXAS), and on Mexico Highways 2 and 180; 25°53′N 97°30′W. Linked by railroad and highway with the UNITED STATES; international trading center and point of entry. Center of an agricultural region that has become increasingly industrialized. Foreign-owned manufacturing plants (motor vehicle parts), known as *maquiladoras*, proliferated in the area and constitute a dominant part of the economy. Founded in 1700 as San Juan de los Esteros, the city was renamed in 1851 in honor of the leader for Mexican independence Mariano Matamoros.

Heroica Nogales (e-RO-ee-kah no-GAH-les), city and township, SONORA, NW MEXICO, on UNITED STATES border, adjoining NOGALES (ARIZONA), 160 mi/257 km N of HERMOSILLO, on Mexico Highway 15; 31°20′N 111°00′W. Railroad junction at terminus of Mexico's W coast railroad; international trading point in cattle-raising and mining area (graphite, manganese, silver, gold, lead, antimony). Exports winter vegetables of S Sonora and SINALOA.

Heróica Puebla de Zaragoza (e-RO-ee-kah poo-EB-lah dai sah-rah-GO-sah) or **Puebla** (poo-EB-lah), city, ⊙ PUEBLA, E central MEXICO; elevation 7,093 ft/2,162 m; 19°03′N 98°10′W. Named Heróica Puebla de Zaragoza, in honor of General Ignacio Zaragoza, who defeated the French forces here in 1862. Located in a highland valley, it is an important agricultural, commercial, and manufacturing center, as well as a popular tourist spot. The site of Mexico's first textile-producing factory, Puebla has cotton mills, a motor vehicle factory, onyx quarries, and pottery and food industries. Noted for the colored tiles that decorate its buildings and numerous churches, as well as those of nearby CHOLULA. The cathedral, built 1552–1649, is one of the finest in Mexico; the theater, constructed in 1790, is said to be the oldest on the continent. Founded c.1535 as Puebla de los Ángeles, the city was historically a link between the coast and MEXICO CITY. Taken (1847) by UNITED STATES General Winfield Scott during the Mexican War. French troops captured Puebla in 1863 but were ousted by Porfirio Díaz in 1867. Puebla was the center of a large earthquake in 1973 that caused intense damage to the city and its surrounding region.

Heroica Zitácuaro (e-RO-ee-kah see-TAH-kwah-ro), city and township, ⊙ Zitácuaro municipio, MICHOACÁN, central MEXICO, in valley of central plateau, on railroad, and 50 mi/80 km SE of MORELIA, on Mexico Highway 15; 19°28′N 100°21′W. Lumbering and agricultural center (cereals, vegetables, fruit livestock); sawmilling, tanning, soapmaking, vegetable oil extracting, resin processing. Tourism. The spa of PURÚA or San José Purúa, with mineral waters, is 9 mi/14.5 km WNW.

Heroldsberg (HE-rohlds-berg), village, MIDDLE FRANCONIA, N central BAVARIA, GERMANY, 7 mi/11.3 km NE of NUREMBERG; 49°32′N 11°09′E.

Héron (ai-RO), commune (2006 population 4,532), Huy district, LIÈGE province, E BELGIUM, 7 mi/11 km WNW of HUY; 50°33′N 05°05′E.

Heron Island (HE-ruhn), largest of CAPRICORN ISLANDS, in CORAL SEA, 40 mi/64 km off CAPE CAPRICORN, on SE coast of QUEENSLAND, AUSTRALIA, 45 mi/72 km NE of GLADSTONE; 5 mi/8 km long, 2 mi/3 km wide; 23°27′S 151°55′E. Most southerly tourist resort within GREAT BARRIER REEF; wooded; birdlife; coral gardens. Marine research station.

Heron Lake, village (2000 population 768), JACKSON county, SW MINNESOTA, NW of HERON LAKE, near Jack Creek, 21 mi/34 km NW of JACKSON; 43°47′N 95°19′W. Grain, soybeans; livestock.

Heron Lake (□ 13 sq mi/34 sq km), JACKSON county, SW MINNESOTA, 20 mi/32 km NE of WORTHINGTON; 12 mi/19 km long (including South Heron Lake), 3 mi/4.8 km wide; elevation 1,400 ft/427 m. Has N outlet into DES MOINES RIVER; Jack Creek enters lake from W. Short channel connects lake with South Heron Lake, to SE. Abundance of game here.

Heron Lake (HE-ruhn), reservoir, RIO ARRIBA county, N NEW MEXICO, on Willow Creek, in Heron Lake State Park, 7 mi/11.3 km W of Tierra Amarilla; 3 mi/4.8 km long; 36°40′N 106°40′W. Maximum capacity 430,000 acre-ft. Formed by Heron Dam (254 ft/77 m high), built by the U.S. government for irrigation, water supply, and recreation.

Hérouville-Saint-Clair (ai-roo-veel–san–KLER), N suburb (□ 4 sq mi/10.4 sq km) of CAEN, CALVADOS department, BASSE-NORMANDIE administrative region, NW central FRANCE, on ORNE RIVER, and on main E-W highway across Lower NORMANDY; 49°12′N 00°19′W. Forms part of Caen's industrial complex.

Hérouxville (ai-roo-VEEL), parish (□ 21 sq mi/54.6 sq km; 2006 population 1,339), MAURICIE region, S QUEBEC, E CANADA, 5 mi/8 km from SAINT-TITE; 46°40′N 72°37′W.

Herowabad, IRAN: see KHALKHAL.

Herøya (HER-uh-yah), village, TELEMARK county, S NORWAY, on SKIENSELVA River, just SW of PORSGRUNN. Port facilities. Major manufacturing area; electrochemical works produce basic chemicals, magnesium, plastics.

Herradura (er-rah-DOO-rah), village (1991 population 1,488), ⊙ Laishi department, E FORMOSA province, ARGENTINA, on PARAGUAY RIVER, at mouth of the Rio Salado, and 23 mi/37 km SSW of FORMOSA; 26°29′S 58°18′W. Agriculture (corn, rice, cotton) and livestock center; lumbering.

Herradura, Isla (eh-rah-DOO-rah, EE-slah), on the PACIFIC OCEAN, in W COSTA RICA, at entrance to GULF OF NICOYA (c.1 mi/1.6 km long, ¼ mi/2/5 km wide); 09°38′N 84°41′W.

Herradura, La, CHILE: see LA HERRADURA.

Herradura, La, EL SALVADOR: see LA HERRADURA.

Herrán (er-RAHN), town, ⊙ Herrán municipio, NORTE DE SANTANDER department, N COLOMBIA, in the TÁCHIRA RIVER valley, 24 mi/39 km S of CÚCUTA; 07°30′N 72°30′W. Elevation 6,414 ft/1,955 m. Coffee, plantains; livestock.

Herräng (HER-ENG), village, STOCKHOLM county, E SWEDEN, on GULF OF BOTHNIA, 40 mi/64 km ENE of UPPSALA; 60°08′N 18°38′E. Site of one of Sweden's oldest iron mines (1584); no longer in operation.

Herreid, village (2006 population 396), CAMPBELL county, N SOUTH DAKOTA, 28 mi/45 km NE of MOBRIDGE, and on Spring Creek; 45°50′N 100°04′W. Pocasse National Wildlife Refuge to W.

Herrenalb, Bad, GERMANY: see BAD HERRENALB.

Herrenberg (HER-ren-berg), town, BADEN-WÜRTTEMBERG, S GERMANY, 10 mi/16 km NW of TÜBINGEN; 48°35′N 08°52′E. Metalworking; gypsum quarrying; manufacturing of pharmaceuticals. Has 14th–15th-century church.

Herrenbreitungen, GERMANY: see BREITUNGEN AN DER WERRA.

Herrera (e-RAI-rah), province (□ 568 sq mi/1,476.8 sq km; 2000 population 102,465), S central PANAMA, on the PACIFIC OCEAN; ⊙ CHITRÉ. Occupies NE part of AZUERO PENINSULA. Bounded (N) by SANTA MARIA RIVER. Important livestock region; agriculture (corn, rice, beans, vegetables). Crossed by INTER-AMERICAN HIGHWAY. Main centers are Chitré, OCÚ, and PARITA. Originally formed (1855) out of old Azuero province; was combined with LOS SANTOS province (1864–1915, 1941–1945).

Herrera (he-RER-rah), town, SEVILLE province, SW SPAIN, near GRANADA province border, 18 mi/29 km ENE of OSUNA. Agriculture center (olives, olive oil, cereals); liquor distilling, manufacturing of plaster. Hydroelectric plant nearby.

Herrera (er-RAI-rah), village (1991 population 1,405), ⊙ Avellaneda department, central SANTIAGO DEL ESTERO province, ARGENTINA, near RÍO SALADO, on railroad, and 13 mi/21 km W of AÑATUYA. Agricultural center (cotton); livestock.

Herrera de Alcántara (er-RER-rah dai ahl-KAHN-tah-rah), village, CÁCERES province, W SPAIN, near TAGUS RIVER, and Portuguese border, 15 mi/24 km SSE of CASTELO BRANCO (PORTUGAL). Cereals; livestock.

Herrera del Duque (er-RER-rah del DOO-kai), town, BADAJOZ province, W SPAIN, 75 mi/121 km ENE of MÉRIDA; 39°10′N 05°03′W. Agriculture center (cereals, olives, grapes, cork; livestock). Apiculture. Mineral springs.

Herrera de Pisuerga (er-RER-rah dai pee-SWER-gah), town, PALENCIA province, N central SPAIN, near PISUERGA RIVER, 43 mi/69 km NNE of PALENCIA; 42°36′N 04°20′W. Tanning, flour- and sawmilling, manufacturing of ceramics and burlap; ships cereals, potatoes, fruit.

Herreras, Los, MEXICO: see LOS HERRERAS.

Herreruela de Oropesa (e-rai-RWAI-lah dai o-ro-PAI-sah), village, TOLEDO province, central SPAIN, 22 mi/35 km W of TALAVERA DE LA REINA. Cereals, grapes; sheep, hogs, cattle.

Herrick (HE-rik), village, NE TASMANIA, AUSTRALIA, 45 mi/72 km NE of LAUNCESTON; 41°05′S 147°53′E. Railroad terminus.

Cross-references are shown in SMALL CAPITALS. The pronunciation guide is shown on page xix. The sources of population figures are shown on page xvii.

Herrick, village (2000 population 524), SHELBY county, central ILLINOIS, 17 mi/27 km SSW of SHELBYVILLE; 39°13′N 88°59′W. In agricultural area.

Herrick, village (2006 population 95), GREGORY county, S SOUTH DAKOTA, 7 mi/11.3 km SE of BURKE; 43°07′N 99°11′W. Small trading point for agricultural area. Burke Lake State Recreation Area to NW.

Herrieden (her-REE-den), town, MIDDLE FRANCONIA, W BAVARIA, GERMANY, on the ALTMÜHL RIVER, 5 mi/8 km SW of ANSBACH; 49°14′N 10°30′E.

Herrin, city (2000 population 11,298), WILLIAMSON county, S ILLINOIS, 8 mi/12.9 km NNW of MARION; 37°47′N 89°01′W. Trade center of an extensive coal-mining area. Manufacturing includes consumer goods, electrical goods, and clothing. The city was the site of the Herrin Massacre; in 1922 during a county-wide coal strike, clashes between unionized strikers and nonunion miners, who had been imported by the coal company, resulted in nearly 25 deaths. Settled 1818, incorporated 1898.

Herrings, village (2006 population 130), JEFFERSON county, N NEW YORK, near BLACK RIVER, 13 mi/21 km E of WATERTOWN; 44°01′N 39°75′W.

Herrington Lake, reservoir (□ 5 sq mi/13 sq km), MERCER county, central KENTUCKY, on DIX RIVER, 22 mi/35 km SSW of LEXINGTON; 37°47′N 84°43′W. Maximum capacity 300,000 acre-ft. Formed by Dix Dam (287 ft/87 m high), built (1925) for power generation.

Herrliberg (HER-lee-berg), commune, ZÜRICH canton, N SWITZERLAND, on NE shore of LAKE ZÜRICH, and 7 mi/11.3 km SSE of ZÜRICH.

Herrlisheim (er-lees-EM), German, *Herlisheim* (HER-lis-heim), residential town (□ 6 sq mi/15.6 sq km), BAS-RHIN department, ALSACE, E FRANCE, 8 mi/12.9 km SE of HAGUENAU, near RHINE RIVER (German border); 48°44′N 07°54′E. In agricultural lowland. Large oil refinery near DRUSENHEIM (2 mi/3.2 km NE).

Herrljunga (HER-YOONG-ah), town, ÄLVSBORG county, SW SWEDEN, 20 mi/32 km ENE of ALINGSÅS; 58°05′N 13°02′E. Railroad junction.

Herrnhut (HERN-hoot), town, SAXONY, SE GERMANY, 11.5 mi/18.5 km SW of GÖRLITZ; 51°01′N 14°45′E. Founded (1722) by Graf von Zinzendorf as a colony of Moravian Brethren and is today a Moravian center with archives, a publishing house, and a museum.

Herrnstadt, POLAND: see WASOSZ.

Herronton (HE-ruhn-tuhn), hamlet, S ALBERTA, W CANADA, 18 mi/28 km from VULCAN, in VULCAN County; 50°38′N 113°25′W.

Herrsching am Ammersee (HER-shing ahm UHM-mer-zai), village, BAVARIA, S GERMANY, in UPPER BAVARIA, on E shore of the AMMERSEE, 7.5 mi/12.1 km W of STARNBERG; 48°00′N 11°11′E. Manufacturing includes shipbuilding. Tourism; summer resort. S (1.5 mi/2.4 km) is ANDECHS, with former Benedictine priory and 15th-century pilgrimage church.

Herrumblar, El (e-room-BLAHR, el), town, CUENCA province, E central SPAIN, 31 mi/50 km NNE of ALBACETE. Cereals, saffron, grapes, vetch; sheep, goats; lumbering.

Hersbruck (HERS-bruk), town, MIDDLE FRANCONIA, N central BAVARIA, GERMANY, on the PEGNITZ RIVER, 7 mi/11.3 km E of LAUF; 49°30′N 11°26′E. Manufacturing of chemicals, plastics, precision instruments, metal products; hops-growing area and brewing center since 18th century. Has many half-timbered houses and remains of town wall. Chartered c.1235.

Herscheid (HER-sheid), village, NORTH RHINE–WEST-PHALIA, W GERMANY, 14 mi/23 km S of ISERLOHN; 51°10′N 07°45′E.

Herschel, district, EASTERN CAPE province, SOUTH AFRICA, 5 mi/8 km S of S LESOTHO border, in WIT-TEBERGE mountains (9,088 ft/2,770 m), 30 mi/48 km E of ALIWAL NORTH, 15 mi/24 km NNW of BARKLY EAST; elevation 4,922 ft/1,500 m; 30°36′S 27°05′E. Cattle raising is chief occupation.

Herschel Island, N YUKON, CANADA, in BEAUFORT Sea of the ARCTIC Ocean, on W side of entrance of MACKENZIE BAY; 11 mi/18 km long, 2 mi/3 km–7 mi/11 km wide; 69°35′N 139°05′W. Herschel village, on E coast, had Royal Canadian Mounted Police post; island now abandoned.

Herscher (HUHR-schuhr), village (2000 population 1,523), KANKAKEE county, NE ILLINOIS, 13 mi/21 km WSW of KANKAKEE; 41°02′N 88°05′W. In agricultural area.

Herseaux (er-SO), Flemish *Herzeeuw*, village, in commune of MOUSCRON, Mouscron district, HAINAUT province, W BELGIUM, 2 mi/3.2 km SE of Mouscron, near French border; 50°43′N 03°13′E. Textiles. Detached from WEST FLANDERS province as a result of 1960 linguistic census.

Herselt (HER-sult), commune (2006 population 13,870), Turnhout district, ANTWERPEN province, NE BELGIUM, 15 mi/24 km NE of LOUVAIN; 51°03′N 04°53′E. Agricultural market.

Herserange (er-suh-RAHNZH), town (□ 1 sq mi/2.6 sq km), E suburb of LONGWY, MEURTHE-ET-MOSELLE department, in LORRAINE, NE FRANCE, near LUX-EMBOURG border, in iron-mining and steel-milling district of Longwy; 49°31′N 05°47′E.

Hersey, town, AROOSTOOK county, E MAINE, 28 mi/45 km W of HOULTON; 46°04′N 68°22′W.

Hersey, village (2000 population 374), OSCEOLA county, central MICHIGAN, 3 mi/4.8 km SE of REED CITY, and on MUSKEGON RIVER; 43°51′N 85°26′W. In lake and agricultural area.

Hersfeld, Bad, GERMANY: see BAD HERSFELD.

Hershey, unincorporated city (□ 2 sq mi/5.2 sq km), Derry township, DAUPHIN county, S central PENN-SYLVANIA, 12 mi/19 km E of HARRISBURG. Company town owned by Hershey Corporation. Manufacturing (tool and die, machinery, explosives, food products). Agricultural area (apples, soybeans, grain; poultry, livestock; dairying). Hershey Park (□ 2 sq mi/5.2 sq km), theme park; Hershey's Chocolate World; Hershey Museum; ZooAmerica (North American wildlife); Hershey Medical Center; Indian Echo Caverns to SW. Founded 1903.

Hershey, village (2006 population 578), LINCOLN county, SW central NEBRASKA, 12 mi/19 km W of NORTH PLATTE city, between NORTH and SOUTH PLATTE rivers; 41°09′N 101°00′W. Livestock; grain; manufacturing (light aircraft). Railroad junction to W.

Hersin, IRAN: see HARSIN.

Hersin-Coupigny (er-san–koo-pee-NYEE), town (□ 4 sq mi/10.4 sq km), PAS-DE-CALAIS department, NORD-PAS-DE-CALAIS region, N FRANCE, 6 mi/9.7 km S of BÉTHUNE; 50°27′N 02°39′E. In old coal-mining area. Cementworks.

Hers River (ER), 75 mi/121 km long, in ARIÈGE department, MIDI-PYRÉNÉES region, S FRANCE; rises in the PYRENEES foothills E of FOIX; 43°18′N 01°33′E. Flows NNW, past MIREPOIX, to the ARIÈGE RIVER above CINTEGABELLE. Formerly called L'Hers.

Herstal (ER-stahl), commune (2006 population 37,398), Liège district, LIÈGE province, E BELGIUM, 4 mi/6.4 km N of LIÈGE, on the MEUSE RIVER, an industrial suburb of Liège; 50°40′N 05°38′E. Center of Belgium's armaments industry. Other manufacturing includes motor vehicles, aircraft engines, and electrical equipment. Once the residence of the early Carolingian mayors of the palace, including Pepin II Héristal, who was born here.

Herstappe (ER-STAHP), commune (2006 population 84), Tongeren district, LIMBURG province, NE BELGIUM, 4 mi/6 km SSW of TONGEREN; 50°44′N 05°25′E.

Herstmonceux (HUHST-muhn-SOO), village (2001 population 2,532), East SUSSEX, SE ENGLAND, near the CHANNEL, 8 mi/12.9 km N of EASTBOURNE; 50°53′N 00°20′E. Site of 1446 castle, restored 1907. After World War II, it was decided (1946) to move government meteorological station and Royal Observatory time clocks to the castle grounds from GREENWICH.

Herta, UKRAINE: see HERTSA.

Herten (HER-ten), city, North NORTH RHINE-WEST-PHALIA, W GERMANY, in the RUHR industrial district, just W of RECKLINGHAUSEN; 51°35′N 07°09′E. Manufacturing of mining and agricultural machinery, paper products; food processing (meat products, preserves). Coal mining. Has 16th-century castle. Chartered 1936.

Hertford (HUHRT-fuhrd), county (□ 360 sq mi/936 sq km; 2006 population 23,581), NE North Carolina; ⊙ WINTON; 36°21′N 76°58′W. Bounded N by VIRGINIA state line, E by CHOWAN RIVER, drained by MEHERRIN RIVER, forms NW boundary; coastal plain (W) and tidewater (E). Service industries; manufacturing (metal, plastic, and wood products; feeds; machinery); agriculture (peanuts, tobacco, cotton, corn, wheat, soybeans; poultry, hogs); timber (pine, gum, oak). Roanoke Chowan Community College, Chowan College. Formed 1759 from Bertie, Chowan, and Northampton counties. Named for the Marquis of Hertford, Francis Seymour Conway.

Hertford (HAHT-fuhd), town (2001 population 24,850), ⊙ HERTFORDSHIRE, E central ENGLAND, on the LEA RIVER; 51°48′N 00°04′W. Agricultural market with light industries. Manufacturing of leather goods and stationery. Nearby is Haileybury College, founded in 1862. The school merged with the Imperial Service College in 1942.

Hertford (HUHRT-fuhd), town (□ 2 sq mi/5.2 sq km; 2006 population 2,130), ⊙ PERQUIMANS county, NE NORTH CAROLINA, 16 mi/26 km SW of ELIZABETH CITY, and on PERQUIMANS RIVER; 36°10′N 76°28′W. Fish, crabs. Service industries; manufacturing (apparel, dental equipment and supplies); agriculture (peanuts, cotton, grain, soybeans; poultry, livestock). Newberg-White House (c. 1730). Settled before 1700; incorporated 1758.

Hertfordshire (HAHT-fuhd-shir) or **Herts**, county (□ 631 sq mi/1,640.6 sq km; 2001 population 1,033,977), E central ENGLAND; ⊙ HERTFORD, but Watford, Hemel Hempstead, Stevenage, Ware, Welwyn Garden City, Bishop's Stortford, Hitchin, Letchworth, and St. Albans are more important urban centers; 51°45′N 00°20′W. The S part of the county is part of London's suburbs. The terrain is level except for an extension of the CHILTERN HILLS in the NW. The chief streams are the Colne, the Lea, and the Stort, all of which drain into the THAMES RIVER. Contains four of the eight New Towns planned around London since 1947: Hatfield, Hemel Hempstead, Stevenage, and Welwyn. Hertfordshire has a large agricultural sector, producing wheat and hay as well as dairy products, vegetables, and flowers for the nearby London market. Diverse industries, such as brickmaking, printing, brewing (especially in Watford), papermaking, and engineering. Contains Neolithic, Bronze Age, and Roman remains, as well as a Norman castle.

Hertin, CZECH REPUBLIC: see RTYNE V PODKRKONOSI.

Hertogenbosch, 's, NETHERLANDS: see 's-HERTOGEN-BOSCH.

Herts, ENGLAND: see HERTFORDSHIRE.

Hertsa (HER-tsah), (Russian *Gertsa*), (Romanian *Herta*), city, S CHERNIVTSI oblast, UKRAINE, in N BUKOVYNA, near PRUT RIVER, 17 mi/27 km SE of CHERNIVTSI, on Romanian border; 48°09′N 26°15′E. Elevation 528 ft/160 m. Raion center; sewing; brickworks. Has an 18th-century church, 19th-century park. Known since 1408; in principality of Moldavia since 1437; annexed by Austria in 1774, by Romania in 1918, and by USSR in 1940. Since 1991, part of independent Ukraine. Most inhabitants are Romanian.

Herut or **Heruth** (khe-ROOT), moshav, W ISRAEL, in PLAIN OF SHARON, 3 mi/5.5 km from KFAR SABA;

32°14′N 34°54′E. Elevation 275 ft/83 m. Citriculture, mixed farming, flower growing. Founded 1930.

Herval, Brazil: see ERVAL.

Herval d'Oeste (ER-vahl DO-es-che), city, central SANTA CATARINA state, BRAZIL, adjacent to JOAÇABA; 27°11′S 57°29′W. Wheat, corn, rice; livestock.

Herval, Serra do, Brazil: see ERVAL, SERRA DO.

Hervás (er-VAHS), town, CÁCERES province, W SPAIN, 10 mi/16 km SW of BÉJAR; 40°16′N 05°51′W. Meat-processing center (sausage, cured hams); manufacturing of wool textiles, furniture, brandy; sawmilling. Livestock; wine, fruit in area. Mineral springs.

Herve (HER-vuh), commune (2006 population 16,774), Verviers district, LIÈGE province, E BELGIUM, 9 mi/14.5 km E of LIÈGE; 50°38′N 05°48′E. Market center for fruit-growing, dairying region.

Herveo (er-VAI-o), town, ⊙ Herveo municipio, TOLIMA province, W central COLOMBIA, 39 mi/63 km N of IBAGUÉ, at the foot of the Nevado del RUIZ; 05°04′N 75°10′W. Coffee, corn; livestock.

Hervey Bay (HAHR-vee), inlet of PACIFIC OCEAN, SE QUEENSLAND, AUSTRALIA, between Sandy Cape of FRASER Island (forming E shore) and Burnett Head; 50 mi/80 km long, 35 mi/56 km wide; 25°00′S 153°00′E. Receives BURNETT and MARY rivers. Tropical fish breeding for export. Town of Hervey on shore; tourism.

Hervey Islands, COOK ISLANDS: see MANUAE.

Herwen (HER-vuhn), village, GELDERLAND province, E NETHERLANDS, 11 mi/18 km SE of ARNHEM; 51°53′N 06°06′E. German border 1 mi/1.6 km to E. Dairying; livestock; agriculture (grain, fruit, vegetables).

Herxheim bei Landau (HERKS-heim bei LAHN-dou), village, RHINELAND-PALATINATE, W GERMANY, 6 mi/9.7 km SE of LANDAU; 49°09′N 08°12′E. Manufacturing of furniture; wine.

Herzberg am Harz (HERTS-berg ahm HAHRTS), town, LOWER SAXONY, W GERMANY, at S foot of the upper HARZ, 6 mi/9.7 km SE of OSTERODE; 51°39′N 10°20′E. Manufacturing (textiles, furniture, metal and wood products, paper). Has 16th-century castle. Was seat (1286–1596) of princes of Brunswick-Grubenhagen.

Herzberg an der Elster (HERTS-berg ahn der EL-ster), town, BRANDENBURG, central GERMANY, on island formed by two arms of the BLACK ELSTER RIVER, 13 mi/21 km NE of TORGAU; 51°32′N 13°14′E. Manufacturing includes metalworking; furniture.

Herzebrock-Charholz (HER-tse-bruk–KAHK-huhlts), town, NORTH RHINE–WESTPHALIA, NW GERMANY, 6 mi/9.7 km WSW of GÜTERSLOH; 51°53′N 08°12′E. Manufacturing of furniture, plastics, machinery; meat processing. Developed around two abbeys—Herzebroch (860) and Charholz (1133). Has late-Gothic church and monastery.

Herzeeuw, BELGIUM: see HERSEAUX.

Herzegovina (her-tse-go-VEE-nah), principal region, S BOSNIA AND HERZEGOVINA; 43°00′N 18°00′E. Covers approximately 20% of the total area of the country. Commonly divided into upper (Visoka) and lower (Niska) Herzegovina.

Herzele (HER-zai-luh), commune (2006 population 16,785), Aalst district, EAST FLANDERS province, W central BELGIUM, 7 mi/11 km SW of AALST; 50°53′N 03°53′E.

Herzlake (herts-LAH-ke), village, LOWER SAXONY, N GERMANY, on HASE RIVER, 13 mi/21 km E of MEPPEN; 52°31′N 07°37′E. In peat region.

Herzlia, ISRAEL: see HERZLIYA.

Herzliya (her-tsi-LEE-yah), coastal city, central ISRAEL, just N of TEL AVIV on the MEDITERRANEAN SEA; 32°09′N 34°50′E. Elevation 216 ft/65 m. Resort with large hotels and fine beaches. Has variety of industries, one of the main centers of the Israeli electronics industry. Founded in 1924 and named for Theodor Herzl, the founder of modern Zionism. Israel's largest film company based here. Also spelled Herzlia.

Herzogenaurach (her-tso-guhn-OU-rahkh), town, UPPER FRANCONIA, N BAVARIA, GERMANY, 6 mi/9.7 km WSW of ERLANGEN; 49°34′N 10°52′E. Manufacturing of sporting equipment, tools. Has Gothic church.

Herzogenbuchsee (HER-tzo-guhn-bookh-sai), commune, BERN canton, NW SWITZERLAND, 8 mi/12.9 km E of SOLOTHURN. Elevation 1,558 ft/475 m. Shoes, textiles; metalworking; canning.

Herzogenburg (HERTS-o-gen-boorg), town, central LOWER AUSTRIA, on the TRAISEN RIVER, and 7 mi/11.3 km NNE of Sankt Pölten; 48°17′N 15°42′E. Railroad junction; manufacturing includes a foundry with production of motor parts. Old abbey with large library, founded 1244, has frescoes by B. Altomonte.

Herzogenhorn, GERMANY: see BELCHEN.

Herzogenrath (HER-tso-guhn-raht), city, NORTH RHINE–WESTPHALIA, W GERMANY, 6 mi/9.7 km N of AACHEN, opposite KERKRADE (Netherlands); 50°52′N 06°05′E. Railroad junction. Glassworks; manufacturing of needles. Coal mined here from 1113 until 1972.

Hes, YEMEN: see HAIS.

Hesarak (he-sah-RAHK), village, Tehrān province, N central IRAN, 30 mi/48 km WNW of TEHRAN, just off highway to QAZVIN. Has noted medical institute; serum production.

Hesban, JORDAN: see HISBAN.

Hesbaye (es-BAI), low karstic plateau, N of the MEUSE RIVER, in NAMUR and LIÈGE provinces, BELGIUM.

Hesdin (ai-DAN), commune, PAS-DE-CALAIS department, NORD-PAS-DE-CALAIS region, N FRANCE, on the CANCHE RIVER, and 13 mi/21 km ESE of MONTREUIL, in the ARTOIS district; 50°22′N 02°02′E. Road junction; agricultural and market center; woodworking, hosiery manufacturing. Founded by Charles V in 1554; has 16th-century Gothic church and 17th-century town hall. Abbé Prévost, author of *Manon Lescaut*, born here, 1697.

Heshan (HUH-SHAHN), city (□ 135 sq mi/350 sq km; 1994 estimated non-agrarian population 61,700; estimated total population 140,300), central GUANGXI ZHUANG, S CHINA; 23°47′N 108°52′E. Heavy industry and agriculture. Crop growing; animal husbandry. Coal mining.

Heshbon (HESH-buhn), ancient city, in what is now JORDAN. It was an Amorite capital, located at the crossroads of the E-W road to JERICHO and the N-S road paralleling the JORDAN River. In the Hebrew Bible it was first allotted to Reuben, later to Gad. The modern village of Hisban is believed to be on or near the site of Heshbon. Wadi Hisban (river), running through the village, is known also as Nahal Heshban.

Heshui (HUH-SHWAI), town, ⊙ Heshui county, SE GANSU province, NW CHINA, 55 mi/89 km NE of JINGCHUAN; 35°49′N 108°02′E. Tobacco and food processing; grain, oilseeds.

Heshun (HUH-SHWUN), town, ⊙ Heshun county, E SHANXI province, NE CHINA, 30 mi/48 km S of XIYANG; 37°20′N 113°34′E. Grain. Chemicals; coal mining, iron smelting.

Hesketh (HES-keth), hamlet, S ALBERTA, W CANADA, 15 mi/23 km from DRUMHELLER, in KNEEHILL County; 51°28′N 112°58′W.

Hespeler (HES-puh-luhr), former village, WATERLOO region, S ONTARIO, E central CANADA, 8 mi/13 km from KITCHENER; 43°25′N 80°18′W. Amalgamated into city of CAMBRIDGE in 1973.

Hesperange (hes-pai-RAWNZH), town, Hesperange commune, S LUXEMBOURG, on ALZETTE river, and 3 mi/4.8 km SSE of LUXEMBOURG city; 49°34′N 06°09′E. Castle.

Hesperia, city (2000 population 62,582), SAN BERNARDINO county, S CALIFORNIA, 21 mi/34 km N of SAN BERNARDINO, near MOJAVE RIVER, in SW part of MOJAVE DESERT; 34°25′N 117°18′W. Mojave River Forks Reservoir to S; SAN BERNARDINO MOUNTAINS and San Bernardino National Forest to S; Sherwood Lake State Recreation Area to S. Cattle; fruit, alfalfa, grain. Manufacturing (fabricated metal products, trusses, concrete products, electronic equipment, wood products; printing and publishing). Limestone quarrying in area.

Hesperia (he-SPIR-ee-uh), village (2000 population 954), OCEANA and NEWAYGO counties, W MICHIGAN, 26 mi/42 km NNE of MUSKEGON, and on WHITE RIVER; 43°34′N 86°02′W. In resort and fruit-growing area. Manistee National Forest to W and N.

Hesperus, village, LA PLATA county, SW COLORADO, on LA PLATA RIVER, and 12 mi/19 km W of DURANGO; elevation 8,110 ft/2,472 m.

Hesperus Mountain (13,232 ft/4,033 m), in LA PLATA MOUNTAINS, SW COLORADO, 16 mi/26 km NW of DURANGO.

Hesse (HES-se), German *Hessen*, state (□ 8,150 sq mi/21,190 sq km; 2005 population 6,092,354), central GERMANY; ⊙ WIESBADEN; 49°24′N 07°48′E–51°37′N 10°17′E. Bounded S by BADEN-WÜRTTEMBERG and BAVARIA, W by RHINELAND-PALATINATE, N by NORTH RHINE–WESTPHALIA and LOWER SAXONY, and E by THURINGIA. Formed in 1945 through the consolidation of HESSE-NASSAU, a former Prussian province, and most of Hesse-Darmstadt, a former grand duchy. Nearly all of Hesse is a hilly, agricultural land, heavily forested in parts. Has the ODENWALD hills and the TAUNUS range and is drained by the RHINE, MAIN, LAHN, EDER, and FULDA rivers. Grain, potatoes, and fruit are grown, and cattle are raised. Along the Rhine valley some of the finest German wines are produced. Industry is centered in the FRANKFURT area and at KASSEL, Wiesbaden, and DARMSTADT. The chief manufacturing includes textiles, chemicals, machinery, and metal goods; others are electrical products and scientific instruments. In recent years E European immigrants have sparked a number of small industries, including glass, toy, and musical-instrument manufacturing. Wiesbaden, BAD HOMBURG, and BAD NAUHEIM are among the state's numerous health resorts. Frankfurt, MARBURG, GIESSEN, and Darmstadt have noted universities. A fiefdom first of the dukes of FRANCONIA, and later of the counts of THURINGIA, Hesse emerged in 1247 as a landgraviate immediately subject to the emperor under a branch of the house of BRABANT. Landgrave Philip the Magnanimous, a leading figure in the German Reformation, was responsible for reuniting a territory that had been torn by border disputes with neighboring areas. At his death (1567) Philip's lands were divided among his four sons, with Kassel, Marburg, RHEINFELS, and Darmstadt their respective capitals. Upon the demise, shortly afterward, of the Rheinfels (1583) and Marburg (1648) lines, the whole territory was held by the two remaining lines—Hesse-Kassel (known as Kurhessen 1803–1866) and Hesse-Darmstadt. The Congress of Vienna (1814–1815) awarded Hesse-Kassel and Hesse-Darmstadt substantial territorial gains. Electoral Hesse, the free city of Frankfurt, and Nassau—all three sided with AUSTRIA in the Austro-Prussian War (1866)—were annexed by PRUSSIA and were merged (1868) into the province of Hesse-Nassau, of which Kassel became the capital. The former state of WALDECK was incorporated 1929 into Hesse-Nassau. The grand duchy of Hesse-Darmstadt also had sided against Prussia. It ceded Hesse-Homburg (which it had just acquired through the extinction of that line), but it continued under its own dynasty until the German revolution of 1918. In 1871, Hesse-Darmstadt joined the newly founded German Empire.

Hesse-Darmstadt, GERMANY: see HESSE.

Hesse-Darmstadt, Germany: see HESSE.

Hesse-Homburg, Germany: see HESSE.

Hesse-Kassel, GERMANY: see HESSE KASSEL.

Hessel, village, MACKINAC county, SE UPPER PENINSULA, N MICHIGAN, 17 mi/27 km NE of SAINT

IGNACE, ON LAKE HURON; 46°00′N 84°25′W. Airport. LES CHENEAUX ISLANDS to SE. Hiawatha National Forest to W.

Hesselberg (HES-sel-berg), highest elevation (2,260 ft/ 689 m) of the SWABIAN JURA MOUNTAINS, BAVARIA, GERMANY, 8 mi/12.9 km E of DINKELSBÜHL.

Hessen, GERMANY: see HESSE.

Hesse-Nassau, GERMANY: see HESSE.

Hessen-Marburg, GERMANY: see MARBURG AN DER LAHN.

Hessen-Rheinfels, GERMANY: see SANKT GOAR.

Hessisch Lichtenau (HES-sish LIKH-tuh-nou), town, central GERMANY, 12 mi/19 km SE of KASSEL; 51°12′N 09°43′E. Manufacturing of textiles, plastics; woodworking; brickworks. Has remains of town wall and many half-timbered houses from 17th–19th century.

Hessisch Oldendorf (HES-sish OL-duhn-dorf), town, LOWER SAXONY, N GERMANY, at S foot of WESER MOUNTAINS, near right bank of the WESER RIVER, 6 mi/9.7 km NW of HAMELN; 52°10′N 09°15′E. Manufacturing of furniture and carpets. Belonged to former Prussian province of HESSE-NASSAU until 1932.

Hessle (HE-suhl), town (2001 population 14,767), East Riding of Yorkshire, NE ENGLAND, on HUMBER RIVER, and 4 mi/6.4 km WSW of HULL; 53°43′N 00°26′W. Has 13th–15th-century church. Former shipbuilding site.

Hessmer (HE-se-muhr), village (2000 population 642), AVOYELLES parish, E central LOUISIANA, 21 mi/34 km SE of ALEXANDRIA, near RED RIVER; 31°03′N 92°07′W. Grand Cote National Wildlife Refuge to NW. In agricultural area (cotton, rice, sugarcane); logging; catfish; manufacturing (canned vegetables, cypress lumber).

Hess, Mount (11,940 ft/3,639 m), E ALASKA, in ALASKA RANGE, 80 mi/129 km S of FAIRBANKS; 63°37′N 147°06′W.

Hesston, town (2000 population 3,509), HARVEY county, S central KANSAS, 8 mi/12.9 km NW of NEWTON; 38°08′N 97°25′W. In wheat region; farm equipment. Bethel College 6 mi/9.7 km SE near NORTH NEWTON.

Hestan Island (HES-tuhn), islet in SOLWAY FIRTH, DUMFRIES AND GALLOWAY, S Scotland, 7 mi/11.3 km S of DALBEATTIE; 54°50′N 03°48′W. Lighthouse.

Hester, village, GREER county, SW OKLAHOMA, 6 mi/9.7 km SSE of MANGUM, near NORTH FORK OF RED RIVER.

Hestmona (HEST-maw-nah) [Norwegian=horseman island], island (□ 4 sq mi/10.4 sq km) in NORTH SEA, on ARCTIC CIRCLE, Nordland co., N central NORWAY, 40 mi/64 km WNW of Missouri Rises to 1,863 ft/568 m in Hestmona mt.; associated with folk legends.

Heston and Isleworth, ENGLAND: see HOUNSLOW.

Hestur, Danish *Hestø*, island (□ 2 sq mi/5.2 sq km), of the N FAEROE ISLANDS, DENMARK. Separated from SW STREYMOY by Hestfjørður. Highest point 1,401 ft/ 427 m. Fishing; sheep raising.

Heswall (HEZ-uhl), town (2001 population 16,012), MERSEYSIDE, NW ENGLAND, on DEE RIVER, on Wirral Peninsula, 6 mi/9.7 km S of BIRKENHEAD; 53°19′N 03°06′W. Built on highest point of peninsula.

Hetao (HUH-TOU), agricultural district, W INNER MONGOLIA AUTONOMOUS REGION, N CHINA, on N bank of HUANG HE (Yellow River), W of BAO-TOU; chief town is WUYUAN. Long colonized by Chinese, the oasis has an elaborate irrigation system, permitting intensive agriculture. Grain, oilseeds, sugar beets. The rich land gives rise to the nickname "the land of rice and fish," comparable to the fertile, rich lower CHANG JIANG valley in S CHINA.

Hetauda (hai-TOU-dah), town, ☉ MAKWANPUR district, central NEPAL; elevation 1,529 ft/466 m; 27°25′N 85°03′E. Important crossroad on the E-W MAHENDRA RAJMARG and the N-S TRIBHUVAN RAJPATH. Large cement factory.

Het Bildt (uht BILT), region, FRIESLAND province, N NETHERLANDS, NW of LEEUWARDEN; reclaimed from NORTH SEA in 1508. Includes villages of Sint Anna-

Parochie, SINT JACOBI-PAROCHIE, and Vrouwen-Parochie. Dairying; cattle; vegetables, sugar beets. Also called 't-Bilt.

Hetch Hetchy Aqueduct, 156 mi/251 km long, TUO-LUMNE county, central CALIFORNIA; runs from HETCH HETCHY RESERVOIR W, roughly paralleling TUO-LUMNE RIVER, runs past Groveland, Oakdale and Modesto, crosses SAN JOAQUIN RIVER, 10 mi/16 km W of Modesto, continues across DIABLO RANGE, past Fremont and Newark, crosses lower end of SAN FRANCISCO BAY S of Dumbarton Bridge, continues past East Palo Alto, North Fair Oaks, and Redwood City, ending up at Upper Crystal Springs Reservoir (in San Andreas Rift Zone), 20 mi/32 km S of downtown SAN FRANCISCO, in San Mateo county.

Hetch Hetchy Reservoir (□ 3 sq mi/7.8 sq km), TUO-LUMNE county, E central CALIFORNIA, on TUOLUMNE RIVER, in N part of YOSEMITE NATIONAL PARK, 65 mi/ 105 km ESE of MERCED; 37°57′N 119°47′W. Maximum capacity 372,000 acre-ft. Formed by O'Shaughnessy Dam (430 ft/131 m), built (1923) for power generation and water supply for SAN FRANCISCO county. Enlarged 1938.

Hetch Hetchy Valley, TUOLUMNE county, central CA-LIFORNIA, on TUOLUMNE RIVER, in YOSEMITE NA-TIONAL PARK, 12 mi/19 km N of Yosemite Village. O'Shaughnessy Dam (forms HETCH HETCHY RE-SERVOIR in lower 10 mi/16 km of valley) turned the valley into a lake 10 mi/16 km long, which is used for generating power and for supplying water to SAN FRANCISCO county by HETCH HETCHY AQUEDUCT. Also called the Grand Canyon of the Tuolumne.

Heteren (HAI-tuh-ruhn), village, GELDERLAND prov-ince, central NETHERLANDS, 7 mi/11.3 km WSW of ARNHEM; 51°57′N 05°46′E. LOWER RHINE RIVER passes to N. Dairying; cattle; poultry; fruit, grain, vegetable growing.

Héthárs, SLOVAKIA: see LIPANY.

Hethersett (HE&hardTH;uh-set), village (1991 popu-lation 4,635; 2001 population 5,441), central NORFOLK, E ENGLAND, 5 mi/8 km WSW of NORWICH; 52°36′N 01°11′E. Previously an agricultural market, with farm-implement works. Has 14th-century church.

Hetian (HUH-TYEN), oasis city (□ 73 sq mi/189 sq km; 1994 estimated non-agrarian population 75,900; esti-mated total population 139,200), SW XINJIANG UYGUR, NW CHINA, near the headstream of the He-tian River; 37°07′N 79°57′E. Center of an area growing cotton, corn, wheat, rice, and fruit. Silk, cotton tex-tiles, and carpets are manufactured, and jewelry is made from the great quantity of jade in the area. On the S part of the SILK ROAD, Hetian was an early center for the spread of Buddhism from INDIA into China. It fell to the Arabs in the 8th century, and grew wealthy on the proceeds of the caravan trade that traveled the route between China and the West. Its prosperity ended with the conquest of Hetian by Jenghiz Khan. After many political changes, the region became (1878) permanently part of China. Connected by road with KASHI and URUMQI. The name sometimes ap-pears as Ho-t'ien or Hotan.

Het IJ (uht EI), inlet of the MARKERMEER lake (S section of IJSSELMEER), NORTH HOLLAND province, W NETHERLANDS, separated from lake by dike at its E end. Harbor for city of AMSTERDAM, joined from W by NOORDZEE Canal, shipping channel to NORTH SEA CANAL, 18 mi/29 km to W. Also joined from NW by NORTH HOLLAND CANAL from SE by AMSTERDAM-RHINE CANAL from LEK and WAAL rivers.

Hetland, village (2006 population 40), KINGSBURY county, E SOUTH DAKOTA, 5 mi/8 km W of ARLING-TON; 44°22′N 97°13′W.

Het Netepark (HET NAI-tuh-pahrk), municipal rec-reation area near HERENTALS, ANTWERPEN province, N BELGIUM, 18 mi/29 km E of ANTWERP.

Het Scheur (uht SKHUHR), channel, c.7 mi/11.3 km long, SOUTH HOLLAND province, SW NETHERLANDS;

formed by junction of NEW MAAS RIVER and OLD MAAS RIVER 7 mi/11.3 km WSW of ROTTERDAM; flows WNW past Maasluis; forms E part of the NEW WA-TERWAY and has been superceded by the new shipping channel. Navigable.

Het Sloe (uht SLOO), ZEELAND province, SW NETH-ERLANDS, former inlet of WESTERN SCHELDT estuary, in FLANDERS mainland, between WALCHEREN and SOUTH BEVELAND regions. Both former islands now joined by reclaimed land just E of VLISSINGEN.

Hettange-Grande (e-TAWNZH–GRAWND), German *Gross Hettingen*, town (□ 6 sq mi/15.6 sq km), MO-SELLE department, LORRAINE region, NE FRANCE, 3 mi/4.8 km N of THIONVILLE, near highway to LUX-EMBOURG; 49°24′N 06°09′E. In declining iron-mining district. Metalworking. The nuclear power plant of CATTENOM is located in the Forest of Garche, 3 mi/4.8 km E, on the MOSELLE RIVER.

Hettick, village (2000 population 182), MACOUPIN county, SW ILLINOIS, 10 mi/16 km WNW of CAR-LINVILLE; 39°21′N 90°02′W. In agricultural and bitu-minous-coal area.

Hettinger (HE-ting-guhr), county (□ 1,133 sq mi/ 2,945.8 sq km; 2006 population 2,564), SW NORTH DAKOTA; ☉ MOTT; 46°25′N 102°27′W. Agricultural area drained by CANNONBALL RIVER and Thirtymile Creek. Dairy products; wheat, rye; cattle. Formed 1883 and organized in 1907. Named for Mathias K. Het-tinger (1810–1890).

Hettinger (HE-ting-guhr), town (2006 population 1,185), ☉ ADAMS CO., SW NORTH DAKOTA, 62 mi/100 km SSE of DICKINSON, near SOUTH DAKOTA state line; 46°00′N 102°37′W. Agriculture; dairying. Grand River National Grassland to S. Founded 1907.

Hetton-le-Hole (HE-tun-luh–HOHL), town (2001 population 11,222), TYNE AND WEAR, NE ENGLAND, 6 mi/9.7 km NE of DURHAM; 54°49′N 01°27′W. Hetton Downs, just N, and EASINGTON LANE, just S. Some sand quarrying. Previously a major coal-mining site.

Hettstedt (HET-stet), town, SAXONY-ANHALT, central GERMANY, at E foot of the lower HARZ, on the WIPPER RIVER, 8 mi/12.9 km SSE of ASCHERSLEBEN; 51°39′N 11°30′E. Copper-slate mining and smelting center; cop-per and brass milling; manufacturing of light metals and chemicals. Has remains of Gothic castle. Just E is scene of battle (1115) in which Emperor Henry V and Count Mansfeld were defeated by Saxons.

Het Vinne Provincial Park (HET VI-nuh), recreational park, near ZOUTLEEUW, BRABANT province, central BELGIUM, 8 mi/12.9 km E of TIENEN.

Het Zomerhuis, recreational park near STEKENE, EAST FLANDERS province, NW BELGIUM, 5 mi/8 km NW of SINT-NIKLAAS.

Het Zoute, BELGIUM: see KNOKKE-HEIST.

Het Zwin (HET ZVIN), nature reserve, near KNOKKE-HEIST, WEST FLANDERS province, W BELGIUM, on NORTH SEA, 10 mi/16 km NNE of BRUGES.

Heubach (HOI-bahkh), town, BADEN-WÜRTTEMBERG, GERMANY, 6 mi/9.7 km ESE of SCHWÄBISCH GMÜND; 48°48′N 09°56′E. Manufacturing of textiles; metal-working.

Heuchelheim (HOI-khel-heim), village, HESSE, central GERMANY, 2 mi/3.2 km W of GIESSEN; 50°34′N 08°38′E. Largely residential.

Heukelum (HUH-kuh-lum), village, SOUTH HOLLAND province, W NETHERLANDS, 5 mi/8 km NE of GOR-INCHEM, on Linge River; 51°52′N 05°05′E. Dairying; livestock; vegetables, sugar beets.

Heule (HUH-luh), village in Kortrijk commune, Kor-trijk district, WEST FLANDERS province, W BELGIUM, 2 mi/3.2 km NW of KORTRIJK; 50°50′N 03°14′E. Has sixteenth-century church.

Heusay, BELGIUM: see BEYNE-HEUSAY.

Heuscheuer Mountains, German *Heuscheuer Gebirge*, Polish *Góry Stołowe* (GO-ree STO-wov), range of the Sudetes in LOWER SILESIA, SW POLAND, near CZECH border, W of KLODZKO (Glatz); extend c.10 mi/16 km

NW-SE; highest point (3,018 ft/920 m) is 7 mi/11.3 km NW of POLANICA ZDROJ. Separated from REICHENSTEIN MOUNTAINS (S) by the GLATZER NEISSE. NW slope extends toward TEPLICE (Czech Republic). In Poland since 1945.

Heusden (HUHZ-duhn), town, NORTH BRABANT province, S central NETHERLANDS, on MAAS RIVER, and 7 mi/11.3 km WNW of 's-HERTOGENBOSCH; 51°44′N 05°08′E. BERGSE MAAS channel divides 1 mi/1.6 km W. Dairying; cattle, hogs, poultry; fruit, vegetables, grain, sugar beets; manufacturing (shipbuilding; processed foods).

Heusden (HUHZ-duhn), agricultural village in commune of DESTELBERGEN, Ghent district, EAST FLANDERS province, NW BELGIUM, on SCHELDT RIVER, and 4 mi/6.4 km SE of GHENT. Cistercian monastery (1247–1578).

Heusden-Zolder (HUHZ-duhn–ZOL-duhr), commune (2006 population 30,889), Hasselt district, LIMBURG province, NE BELGIUM, 8 mi/12.9 km NNW of HASSELT. Agriculture; cattle raising.

Heusenstamm (HOI-sen-shtahm), town, HESSE, central GERMANY, 3 mi/4.8 km SSE of OFFENBACH; 50°03′N 08°48′E. Industry includes plastics; leather- and metalworking. Has moated castle from 12th–16th century.

Heusweiler (hois-VEI-luhr), town, SAARLAND, W GERMANY, 8 mi/12.9 km NNW of SAARBRÜCKEN; 49°20′N 06°56′E. Part of Saarbrücken conurbation. Electronics and technology industry; woodworking; brewing.

Heusy (uh-ZEE), town in commune of VERVIERS, Verviers district, LIÈGE province, E BELGIUM, S residential suburb of Verviers; 50°35′N 05°51′E.

Heuvelland (HUH-vuhl-lahnd), commune (2006 population 8,223), Ypres district, WEST FLANDERS province, W BELGIUM, 6 mi/10 km SSW of YPRES, near French border.

Heuvelton (HUH-vuhl-tuhn), village (2006 population 774), ST. LAWRENCE county, N NEW YORK, on OSWEGATCHIE RIVER, and 6 mi/9.7 km SE of OGDENSBURG; 44°37′N 75°24′W.

Hève, La (EV, lah), cape on ENGLISH CHANNEL, in SEINE-MARITIME department, HAUTE-NORMANDIE region, N FRANCE, 3 mi/4.8 km NW of Le Havre, at mouth of SEINE RIVER estuary; 49°30′N 00°04′E. Chalk cliff rises c.350 ft/107 m above the beach at suburb of SAINTE-ADRESSE (French merchant marine academy). The cape's lighthouse is nearby.

Hevenk, TURKEY: see BOZOVA.

Héverlé, BELGIUM: see HEVERLEE.

Heverlee (HAI-vuhr-lai), French *Héverlé* (AI-ver-lai), residential town in LEUVEN commune, Leuven district, BRABANT province, central BELGIUM, on DIJLE RIVER, S suburb of Leuven; 50°52′N 04°42′E. Ancient castle, once residence of chamberlains of Brabant, now owned by University of Leuven. Premonstratensian abbey was founded c.1130; most extant buildings date from fifteenth–seventeenth century.

Heves (HA-vash), county (1,492 sq mi/3,864 sq km), N HUNGARY; ⊙ EGER. Includes MÁTRA MOUNTAINS (N) and part of the Alföld (SE); drained by TARNA, ZAGYVA, EGER rivers; 47°50′N 20°15′E. Agriculture (grain, watermelons, cherries, peaches, caraway seed); livestock region (cattle, pigs, sheep, poultry). N part heavily forested. Industry at EGER, Gyöngyös, and HATVAN; largest lignite mine and coal-fired power plant in Hungary at Visonta.

Heves (HE-vash), city, HEVES county, N HUNGARY, 30 mi/29 km S of EGER; 47°36′N 20°17′E. Watermelons, peaches; vineyards; manufacturing (instruments); lumber. Lignite deposits nearby.

Héviz (HAI-viz), Hungarian *Héviz*, city, ZALA county, W HUNGARY, 17 mi/27 km ESE of ZALAEGERSZEG; 46°47′N 17°11′E. Vegetables, rapeseed; wine. On nearby small lake, fed by hot springs, is Hévizfürdö (Hungarian *Hévizfürdö*), a health resort.

Hevros River, Bulgaria and Greece: see MARITSA RIVER.

Hewanorra International Airport, SAINT LUCIA: see VIEUX FORT.

Hewitt, village (2000 population 267), TODD county, W central MINNESOTA, 8 mi/12.9 km SSE of WADENA, on Wing River; 46°19′N 95°05′W. Poultry, livestock; grain, potatoes; dairying; manufacturing (wood products).

Hewitt, village (2006 population 740), WOOD county, central WISCONSIN, 4 mi/6.4 km SE of MARSHFIELD; 44°38′N 90°06′W. Dairying; livestock; general farming. Agricultural research station to W.

Hewlett, village (2000 population 7,060), NASSAU county, SE NEW YORK, near S shore of W LONG ISLAND, 6 mi/9.7 km SE of JAMAICA; 40°38′N 73°41′W. It is one of "Five Towns" of Long Island; affluent residential area, with small light industry.

Hewlett Bay Park, residential village (2006 population 484), NASSAU county, SE NEW YORK, on S shore of W LONG ISLAND, on small Hewlett Bay, just S of VALLEY STREAM; 40°38′N 73°42′W.

Hewlett Neck, residential village (2006 population 517), NASSAU county, SE NEW YORK, on S shore of W LONG ISLAND, just S of VALLEY STREAM; 40°37′N 73°42′W.

Hewlett Point, NEW YORK: see MANHASSET BAY.

Hewlitt, town, MCLENNAN county, E central TEXAS, suburb 9 mi/14.5 km S of downtown WACO. Agricultural area (cattle; dairying; cotton). Manufacturing (cultured marble, wood and gypsum products; steel fabricating, diversified light manufacturing).

Hexham (HEK-suhm), town (2001 population 11,446), S NORTHUMBERLAND, NE ENGLAND, on the TYNE RIVER, and 19 mi/31 km W of NEWCASTLE UPON TYNE; 54°58′N 02°06′W. Agricultural market, with agriculture-machinery works. Remains of a monastery founded 674 by St. Wilfrid are now part of 13th-century abbey. Has 15th-century Moot Hall and 14th-century prison. Battle of Hexham (1464), in which Edward IV defeated the Lancastrians, was fought 2 mi/3.2 km SE. Racecourse 2 mi/3.2 km SW.

Hexi (HUH-SHEE), town, SE central YUNNAN province, SW CHINA, 60 mi/97 km S of KUNMING, on small lake. Rice, tobacco, sugarcane.

He Xian (HUH-SHYEN), town, ⊙ He Xian county, E ANHUI province, E CHINA, near JIANGSU province border, 35 mi/56 km SW of NANJING, and on CHANG JIANG (Yangzi River); 31°43′N 118°21′E. Rice, cotton, oilseeds; food processing. Site of Loushi (Simple House), built by Liu Yuxi in 824; famous from Liu's prose of the same title. The original building no longer exists; current building built in 1987 based on the original design.

He Xian, town, ⊙ He Xian county, E GUANGXI ZHUANG AUTONOMOUS REGION, S CHINA, 60 mi/97 km N of WUZHOU, near GUANGDONG province border; 24°25′N 111°31′E. Rice, tobacco, sugarcane; chemicals and pharmaceuticals; food processing; non-ferrous ore mining and smelting.

Hexi Corridor (HUH-SHEE), Mandarin *Hexizoulang* (HUH-SHEE-ZO-LAHNG), [corridor of the river], gorge, central GANSU province, NW CHINA; a long and narrow piedmont lowland area at the foot of the QILIAN MOUNTAINS. NW-SE trend; 6,000 mi/9,656 km long, 6 mi/9.7 km–30 mi/48 km wide; elevation c.4,900 ft/1,494 m. Between the high Qilian (S) and the Mazong, Heli, and Longshou (N) mountains. The corridor forms a portion of the SILK ROAD. Furnishes abundant evidence of earlier Chinese trade with S Asia and the Middle East, especially in Dunhuang caves. Historically this gorge is the W frontier of the Chinese empires. Various segments of the GREAT WALL were built along the N end of the corridor. The westernmost station on the Great Wall is in the corridor (Jiayuguan). The corridor is a path through which major roads and railroads connect Xinjiang with the E part of China. Important centers are Yumen, Jiayuguan, Wuwei, Zhangye, Jiuquan, and Dunhuang. The

Hexi Corridor is part of the NORTHERN PROTECTION BELT.

Hexington Qi (HUH-SHEENG-TUN CHEE), town, ⊙ Hexingten Qi county, central INNER MONGOLIA AUTONOMOUS REGION, N CHINA, 100 mi/161 km NW of CHIFENG; 43°15′N 117°31′E. Grain, oilseeds; livestock; food processing. Also known as Jingfeng. Also appears as Keshiketeng Qi.

Hexizoulang, CHINA: see HEXI CORRIDOR.

Hex River Mountains, WESTERN CAPE province, SOUTH AFRICA, extend 30 mi/48 km ENE from upper BREEDE RIVER valley N of WORCESTER; 33°25′S 19°30′E. Rise to 7,386 ft/2,251 m on MATROOSBERG, 20 mi/32 km E of CERES. In Fold Mountains, formed by meeting of N-S and E-W Trending Fold Mountains of Cape Boland.

Heyang (HUH-YAHNG), town, ⊙ Heyang county, E SHAANXI province, NW central CHINA, 95 mi/153 km NE of XI'AN, near HUANG HE (Yellow River); 35°14′N 110°08′E. Cotton, grain, tobacco, oilseeds; food processing, manufacturing of machinery and building materials. Coal mining.

Heybeli Island, one of KIZIL ADALAR (Princes Islands), in Sea of MARMARA, NW TURKEY, 12 mi/19 km SE of ISTANBUL, of which it is part; 1 mi/1.6 km long. Formerly Khalki or Copper Island.

Heybridge (HAI-brij), village (2001 population 7,627), E central ESSEX, SE ENGLAND, on BLACKWATER RIVER, opposite MALDON; 51°44′N 00°41′E. Has Norman church.

Heyburn, town (2000 population 2,899), MINIDOKA county, S IDAHO, 2 mi/3.2 km NE of BURLEY, and on SNAKE RIVER; elevation 4,342 ft/1,323 m; 42°34′N 113°46′W. Ships grain, sheep, cattle; dairying; potatoes, beans, sugar beets.

Heyburn, Lake, reservoir, CREEK county, E central OKLAHOMA, on Polecat Creek, 23 mi/37 km SW of TULSA; c.11 mi/18 km long; 35°56′N 96°17′W. Formed by Heyburn Dam. Heyburn State Park is here.

Heydebreck, POLAND: see KEDZIERZYN-KOŹLE.

Heyfield (HAI-feeld), town, VICTORIA, SE AUSTRALIA, 126 mi/203 km E of MELBOURNE, at S end of SNOWY MOUNTAINS; 37°57′S 146°47′E. Dairying; timber (treated hardwood). Glenmaggie Reservoir to N.

Heyrieux (ai-RYU), commune (□ 5 sq mi/13 sq km), ISÈRE department, RHÔNE-ALPES region, SE FRANCE, 14 mi/23 km SE of LYON; 45°38′N 05°03′E. The regional airport of Lyon-Satolas is 6 mi/9.7 km N and the new town of L'ISLE D'ABEAU is 4 mi/6.4 km E on Lyon-Grenoble highway.

Heys, YEMEN: see HAIS.

Heysham, ENGLAND: see MORECAMBE AND HEYSHAM.

Heyst-op-den-Berg, BELGIUM: see HEIST-OP-DEN-BERG.

Heythuisen, NETHERLANDS: see HEYTHUYSEN.

Heythuysen (HEI-toi-zuhn), village, LIMBURG province, SE NETHERLANDS, 7 mi/11.3 km NW of ROERMOND; 51°15′N 05°54′E. Tungelroijse Beek River to S. Dairying; cattle, hogs; grain, vegetables. Also spelled Heithuizen or Heythusisen.

Heyuan (HUH-yuh-WAHN), city (□ 1,704 sq mi/4,413 sq km; 1994 estimated non-agrarian population 127,500; estimated total population 182,800), E central GUANGDONG province, SE CHINA, on EAST RIVER, and 50 mi/80 km NNE of HUIZHOU; 22°41′N 114°45′E. Agriculture (crop growing; animal husbandry; commercial agriculture); heavy industry; forestry. Grain, sugarcane, oil crops, vegetables, fruits; hogs, poultry, eggs; manufacturing (apparel, utilities, pharmaceuticals).

Heywood (HAI-wud), town, SW VICTORIA, SE AUSTRALIA, 185 mi/298 km WSW of MELBOURNE; 38°08′S 141°38′E. Railroad junction; cattle center; cheese. Lake Condah Aboriginal mission 16 mi/25 km NE.

Heywood (HAI-wuhd), town (2001 population 29,240), GREATER MANCHESTER, NW ENGLAND, 8 mi/12.8 km N of MANCHESTER; 53°35′N 02°13′W. Light industry. Previously cotton milling.

Heywoods Resort, large luxury hotel complex set amid extensive gardens, just N of SPEIGHTSTOWN, BARBADOS.

Heyworth, village (2000 population 2,431), MCLEAN county, central ILLINOIS, 10 mi/16 km S of BLOOMINGTON; 40°18′N 88°58′W. Trade center in agricultural area (corn, soybeans; livestock).

Hezar Masjed Range, IRAN: see HAZAR MASJID RANGE.

Heze (HUH-ZUH), city (□ 540 sq mi/1,399 sq km; 1994 estimated non-agrarian population 237,000; estimated total population 1,189,000), SW SHANDONG province, NE CHINA; 35°16′N 115°27′E. The area is the birthplace of numerous historic figures, including Zuo Quming, Yin Yi, Cun Bin, Zin Ke, Wu Qi, Huang Chao, and Song Jiang. The city got its name from the He Lake and He Mountains, which are nearby; several other lakes are also in the vicinity, such as the Dayeze, Juyeze, and Lize. Rich in coal and petroleum, Heze's economy had been retarded by a lack of adequate transportation prior to the construction of the E-W Xinxiang-Shijiucuo railroad in the 1980s. The N-S BEIJING-JIULONG RAILROAD was completed in 1995. Agriculture and light industry are currently the largest sectors of the city's economy. Agriculture (grain, oil crops, cotton, vegetables, fruits, eggs; hogs, beef, lamb); manufacturing (food processing, tobacco, textiles, machinery). Well known for its peony raising and exporting.

Hezheng (HUH-JUNG), town, ⊙ Hezheng county, SE GANSU province, NW CHINA, 50 mi/80 km SSW of LANZHOU; 35°27′N 103°20′E. Livestock; grain.

Hezlev, UKRAINE: see YEVPATORIYA.

Hhohho (oo-HWAH-hwah), district (□ 1,378 sq mi/3,582.8 sq km), NW SWAZILAND, bounded N and W by SOUTH AFRICA, MBABANE (administrative ⊙ of Swaziland) in S; ⊙ PIGGS PEAK; 26°00′S 31°30′E. Drained from W to E by Lomati, Komati, and Black Mbuluzi (forms part of S border) rivers. Malolotja Nature Reserve and Millwane Game Reserve in SW; Phophonyane Nature Reserve in N. Timber; corn, cotton, vegetables, citrus, tea; cattle, goats, sheep, hogs. Asbestos mining in NW. Name derived from barking cry of baboons; pronounced at gutteral *ghou-ghou*.

Hialeah (hei-uh-LEE-uh), city (□ 19 sq mi/49.4 sq km; 2000 population 226,419), MIAMI-DADE county, SE FLORIDA, 10 mi/16 km NW of MIAMI; 25°51′N 80°17′W. Printing; manufacturing of metal and plastic goods. Nearby Miami International Airport is a major employer. A vibrant Cuban community dominates the city's population and adds to Hialeah's work force and housing developments. Incorporated 1925.

Hialeah Gardens (hei-uh-LEE-uh), town (□ 2 sq mi/5.2 sq km; 2000 population 19,297), MIAMI-DADE county, SE FLORIDA, 9 mi/14.5 km WNW of MIAMI; 25°52′N 80°20′W. Manufacturing includes figurines, wood products, building materials, apparel, marble products; textile printing.

Hian, town, UPPER WEST region, GHANA, 50 mi/80 km N of Wa, at crossroad W of Kulpawn River; 10°41′N 02°27′W. Livestock; groundnuts (peanuts), shea-nut butter. Also spelled Han.

Hiawassee (hei-uh-WAH-see), village (2000 population 808), ⊙ TOWNS county, NE GEORGIA, near NORTH CAROLINA state line, 34 mi/55 km NW of TOCCOA, on CHATUGE Reservoir; 34°57′N 83°45′W. Retirement and recreation area along Lake Chatuge in N Georgia Mountains. Annual Georgia Mountain fair near BRASSTOWN BALD. Annual bluegrass music festival held here. Manufacturing includes clothing, machinery, boat docks; fish processing.

Hiawatha (HEI-uh-WAH-thuh), city, LINN county, E central IOWA, suburb 4 mi/6.4 km N of CEDAR RAPIDS; 42°02′N 91°40′W. Manufacturing (concrete, fabricated metal products). Agriculture to N and E (corn; cattle, hogs).

Hiawatha (hei-uh-WAH-thuh), town (2000 population 3,417), ⊙ BROWN county, NE KANSAS, 37 mi/60 km W of SAINT JOSEPH (Mo.); 39°51′N 95°32′W. Railroad junction. Trade center for livestock-raising and agricultural area (grain, apples); dairying. Manufacturing (lumber, hand tools, feeds). Davis Memorial, with 11 statues of John Davis and wife, are in cemetery. Brown State Fishing Lake to E; Iowa Sac and Fox Indian Reservation to NE. Incorporated 1859.

Hiawatha, village, CARBON county, E central UTAH, 15 mi/24 km SW of PRICE; 39°30′N 111°01′W. Cattle; coal mining. Railroad spur terminus at Wattis to N. Manti–La Sal National Forest to W.

Hibar, SAUDI ARABIA: see TURAIF.

Hibbing (HIB-ing), city (2000 population 17,071), ST. LOUIS county, NE MINNESOTA, 58 mi/93 km NW of DULUTH, in the MESABI IRON RANGE, 90 mi/145 km S of the CANADA (ONTARIO) border; elevation 1,489 ft/454 m; 47°23′N 92°57′W. Iron mining, formerly the major industry, has declined. Manufacturing (paper and metal products, electronic equipment, chemicals, candy, storage tanks, mining equipment, explosives); poultry; alfalfa, oats; dairying; timber; tourism, recreation. In 1917, Hibbing was moved 2 mi/3.2 km S to make room for one of the world's largest open-pit iron mines. Mine viewing point located in densely forested region rich with wildlife. Hunting and camping are popular activities in the area. Hibbing Airport 5 mi/8 km to SE, Palucci Space Theater (astronomy and space programs), St. Louis County Fairgrounds. Seat of Hibbing Area Technical and Hibbing Community colleges. In area are George Washington (NW) and Superior (NE) national forests; McCarthy Beach State Park to N; numerous small natural lakes, especially to W. Incorporated 1893.

Hibernia (hei-BUHR-nee-uh), village, MORRIS county, N NEW JERSEY, 10 mi/16 km N of MORRISTOWN; 40°56′N 74°29′W. Largely residential. Until c.1912 was thriving iron-mining town; Hibernia Furnace here furnished munitions in American Revolution.

Hibernia: see IRELAND, island.

Hibiki Sea (HEE-bee-kee), Japanese *Hibiki-nada* (hee-BEE-kee–NAH-dah), S arm of SEA OF JAPAN, between SW coast of YAMAGUCHI prefecture, SW HONSHU (E) and small islands (W) which form E boundary of Tsushima Strait; connected with SUO SEA (SE) by SHIMONOSEKI STRAIT; merges with GENKAI SEA (SW). Wakamatsu on SE shore.

Hicacos Peninsula (hee-KAH-kos), narrow spit, MATANZAS province, NW CUBA, 20 mi/32 km ENE of MATANZAS, flanking NW Cárdenas Bay; 11 mi/18 km long NE-SW; 23°12′N 81°12′W. Terminates in Hicacos Cape. Has saltworks. VARADERO beach resort along N shore is a major international tourist destination.

Hichiso (hee-CHEE-so), town, Kamo district, GIFU prefecture, central HONSHU, central JAPAN, 25 mi/40 km N of GIFU; 35°32′N 137°07′E.

Hichuya Chico (ee-CHOO-yah CHEE-ko), town and canton, AROMA province, LA PAZ department, W BOLIVIA, N of DESAGUADERO RIVER; elevation 12,851 ft/3,917 m; 17°22′S 67°44′W. Gas resources in area. Copper, gold-bearing lode; clay, limestone, and gypsum deposits. Agriculture (potatoes, yucca, bananas, rye); cattle.

Hickam Field, U.S. Air Force base, S OAHU, HAWAII, 12 mi/19 km NW of HONOLULU, near PEARL HARBOR; 21°20′N 157°57′W. Completed 1935, it was bombed (December 7, 1941) by the Japanese.

Hickman, county (□ 252 sq mi/655.2 sq km; 2006 population 4,974), W KENTUCKY; ⊙ CLINTON, 36°40′N 88°59′W. Bounded W by the MISSISSIPPI RIVER (MISSOURI state line), SE corner by TENNESSEE state line; drained by OBION CREEK and BAYOU DE CHIEN. Gently rolling agricultural area (dark and burley tobacco, soybeans, wheat, corn, hay; hogs, cattle, poultry; dairying); manufacturing at Clinton. Includes COLUMBUS BELMONT BATTLEFIELD STATE PARK in NW; Obion Creek Wildlife Management Area in NE. Includes Wolf Island on Missouri side of Mississippi River, road access from Missouri across old channel of river. Formed 1821.

Hickman, county (□ 613 sq mi/1,593.8 sq km; 2006 population 23,812), central TENNESSEE; ⊙ CENTERVILLE; 35°48′N 87°28′W. Drained by DUCK RIVER and tributaries. Agriculture; lumbering; manufacturing. Formed 1807.

Hickman, city (2000 population 2,560), ⊙ FULTON county, extreme W KENTUCKY, 36 mi/58 km WSW of MAYFIELD; 36°33′N 89°11′W. Trade and shipping center for agricultural area (cotton, tobacco, grain, hay); timber; manufacturing (fabricated metal products; electronic equipment). Flood wall (1934). REELFOOT LAKE is 10 mi/16 km SW, in TENNESSEE. Reelfoot National Wildlife Refuge in Kentucky and Tennessee. Settled 1819; incorporated 1834.

Hickman, town (2006 population 1,404), LANCASTER county, SE NEBRASKA, 12 mi/19 km S of LINCOLN; 40°37′N 96°37′W. Dairy and poultry products, grain; livestock. Satellite community of Lincoln. In area are Wagon Train Lake (E) and Stagecoach Lake (S) state recreation areas.

Hickman, Mount (HIK-muhn), (9,700 ft/2,957 m), NW BRITISH COLUMBIA, W CANADA, near U.S. (ALASKA) border, in COAST MOUNTAINS, 70 mi/113 km NE of Wrangell (Alaska); 57°16′N 131°07′W.

Hickory, county (□ 410 sq mi/1,066 sq km; 2006 population 9,243), central MISSOURI, in the OZARK MOUNTAINS; ⊙ Hermitage; 37°56′N 93°19′W. Drained by POMME DE TERRE and LITTLE NIANGUA rivers. Corn, wheat, soybeans, hay; dairying; cattle. Recreation around lakes. Pomme de Terre State Park, POMME DE TERRE LAKE in S, arm of Harry S. Truman Lake in N. Formed 1845.

Hickory (HI-kuh-ree), city (□ 28 sq mi/72.8 sq km; 2006 population 40,583), BURKE, CALDWELL, and CATAWBA counties, W NORTH CAROLINA, 50 mi/80 km NW of CHARLOTTE, near CATAWBA RIVER, forms Lake Hickory reservoir to N, Lake Rhodhiss reservoir to NW, at E edge of the BLUE RIDGE MOUNTAINS; 35°44′N 81°19′W. Railroad junction. Processing and trade center for an abundant agricultural region (grain, soybeans; poultry, cattle; dairying), with related industries. Manufacturing (textiles and apparel; stone and plastic products and construction materials; electric and electronic equipment, furniture, optical fibres, fabricated metal products, consumer goods; printing); service industries. Hickory's location in the Blue Ridge Mountains has led to developing tourism. Seat of Lenoir Rhyne College Arts center of Catawba Valley; includes Hickory Museum of Art and Catawba Science Center. Catawba Valley Community College to SE, between Hickory and NEWTON. Hickory Motor Speedway to E. Incorporated 1870.

Hickory, unincorporated town, WASHINGTON county, SW PENNSYLVANIA, 9 mi/14.5 km W of WASHINGTON; 40°17′N 80°18′W. Manufacturing of lumber, machinery. Agriculture includes dairying; livestock; corn, hay.

Hickory, town, YORK county, N SOUTH CAROLINA, 10 mi/16 km W of YORK.

Hickory, village, GRAVES county, W KENTUCKY, on MAYFIELD CREEK, and 5 mi/8 km N of MAYFIELD. Agriculture (tobacco, grain; livestock, poultry); clay; manufacturing (lumber; clay processing, poultry processing). Also known as Hickory Grove.

Hickory, village (2000 population 499), NEWTON county, E central MISSISSIPPI, 19 mi/31 km W of MERIDIAN, on Potterchitto Creek; 32°19′N 89°01′W. Agriculture (cotton, corn; poultry, cattle; dairying).

Hickory, village (2006 population 89), MURRAY county, S OKLAHOMA, 17 mi/27 km SSW of ADA; 34°33′N 96°51′W. In agricultural area.

Hickory Flat, village (2000 population 565), BENTON county, N MISSISSIPPI, 13 mi/21 km NW of NEW AL-

BANY, in Holly Springs National Forest; 34°37′N 89°11′W. Agriculture (cattle; timber); manufacturing (furniture, wood products).

Hickory Grove, village, (2006 population 392), YORK county, SE SOUTH CAROLINA, 10 mi/16 km W of YORK; 34°58′N 81°25′W. Manufacturing includes meat processing; chemicals, apparel. Agriculture includes cotton, grain, soybeans, peaches; livestock, poultry; dairying.

Hickory Hills, city (2000 population 13,926), COOK county, NE ILLINOIS, suburb 15 mi/24 km SW of downtown CHICAGO; 41°43′N 87°49′W. Manufacturing (fabricated metal products, machinery, electronic goods); sheet metal fabricating, diverse light manufacturing.

Hickory Lake, NORTH CAROLINA: see CATAWBA RIVER.

Hickory Ridge, village (2000 population 384), CROSS county, E ARKANSAS, 27 mi/43 km NW of WYNNE, near Bayou DeView; 35°23′N 90°59′W. Rice processing.

Hickox (HI-kahks), town, BRANTLEY county, SE GEORGIA, 22 mi/35 km E of WAYCROSS; 31°09′N 81°59′W.

Hicksville, unincorporated suburban town (□ 6 sq mi/ 15.6 sq km; 2000 population 41,260), NASSAU county, SE NEW YORK, on LONG ISLAND; 40°45′N 73°31′W. It is chiefly residential, with electronic and metal products manufacturing and some nearby vegetable farming. Site of the Broadway Mall, one of the largest shopping centers in U.S. Founded 1648.

Hicksville (HIKS-vil), village (□ 3 sq mi/7.8 sq km; 2006 population 3,501), DEFIANCE county, NW OHIO, near INDIANA state line, 20 mi/32 km W of DEFIANCE; 41°17′N 84°46′W. Shipping and processing center in farming and dairying area; manufacturing (wood products, food products). Founded 1836.

Hico (HI-ko), town (2006 population 1,350), HAMILTON county, central TEXAS, on BOSQUE RIVER, and c.60 mi/ 97 km NW of WACO; 31°58′N 98°01′W. In agricultural area (dairying; cattle; grain); manufacturing (machining).

Hida, former province in central HONSHU, JAPAN; now part of GIFU prefecture.

Hida (HEE-dah), Hungarian *Hidalmás*, village, SĂLAJ county, W central ROMANIA, 24 mi/39 km NW of CLUJ-NAPOCA; 47°04′N 23°17′E. Agricultural center; coal mining. Under Hungarian rule, 1940–1945.

Hidaka (HEE-dah-kah), town, Hidaka district, HOK-KAIDO prefecture, N JAPAN, 56 mi/90 km E of SAP-PORO; 42°52′N 142°26′E. Lumber.

Hidaka (HEE-dah-kah), or **Hitaka**, town, Kinosaki district, HYOGO prefecture, S HONSHU, W central JAPAN, 58 mi/94 km N of KOBE; 35°28′N 134°46′E.

Hidaka (HEE-dah-kah), town, Hidaka district, WA-KAYAMA prefecture, S HONSHU, W central JAPAN, 22 mi/35 km S of WAKAYAMA; 33°55′N 135°08′E.

Hidaka (HEE-dah-kah), village, Takaoka district, KOCHI prefecture, S SHIKOKU, W JAPAN, 9 mi/15 km W of KOCHI; 33°31′N 133°22′E. Tea.

Hidalgo (ee-DAHL-go), state (□ 8,058 sq mi/20,950.8 sq km), central MEXICO; ⊙ PACHUCA DE SOTO; 20°30′N 99°00′W. Crossed by the SIERRA MADRE ORIENTAL, the state is extremely mountainous; in the S and W areas, however, are plains and fertile valleys lying within Mexico's central plateau. The climate is warm in the lower valleys, temperate on the plateau, and cold in the mountains. One of Hidalgo's chief crops is maguey, grown on the central plateau. Alfalfa, corn, sugarcane, and coffee are also cultivated. The state's main industry is mining (especially around Pachuca de Soto), and Hidalgo is a leading national producer of silver, gold, copper, lead, iron, and sulfur. Cement, textile, motor vehicle manufacturing and especially oil refining are other major industries. The territory was occupied successively by the Toltec (whose capital was Tollán—now TULA) and the Aztecs. Conquered by the Spanish in 1530, it was part of

the province and state of MEXICO until it became the separate state of Hidalgo in 1869.

Hidalgo (hi-DAHL-go), county (□ 3,447 sq mi/8,962.2 sq km; 2006 population 5,087), extreme SW NEW MEXICO; ⊙ LORDSBURG; 31°55′N 108°42′W. Watered N by GILA River; bounded by ARIZONA (W) and MEXICO (SONORA and CHIHUAHUA states; S, SE). Cattle, some sheep, chilies, cotton, hay, alfalfa, some sorghum, wheat, oats, barley, Christmas trees. Mining (clay, gold, silver, silica) near LORDSBURG. Includes parts of Coronado (SW) and Gila (NE) national forests. CONTINENTAL DIVIDE in S passes through Animas Mountains and part of Pyramid Mountains. Formed 1919.

Hidalgo, county (□ 1,582 sq mi/4,113.2 sq km; 2006 population 700,634), extreme S TEXAS; ⊙ EDINBURG; 26°23′N 98°10′W. S part is in rich irrigated valley of the RIO GRANDE (Mexican border), producing large part of Texas citrus crop, huge vegetable crops, sugarcane, grain; cotton; N part has ranches (beef and dairy cattle). Agribusiness important; oil, natural gas production and refining; stone, sand and gravel. Winter resort area. Santa Ana National Wildlife Refuge and Bentsen–Rio Grande Valley State Park in S, both on Rio Grande. Formed 1852.

Hidalgo (ee-DAHL-go), town, COAHUILA, N MEXICO, on the RIO GRANDE (TEXAS border), and 75 mi/121 km SE of PIEDRAS NEGRAS; 27°49′N 99°50′W. Cattle grazing. Also VILLA HIDALGO.

Hidalgo (ee-DAHL-go), town, TAMAULIPAS, N MEX-ICO, at E foot of SIERRA MADRE ORIENTAL, near NUEVO LEÓN border, 40 mi/64 km NNW of CIUDAD VICTORIA; 24°16′N 99°28′W. Sugarcane, beans, livestock.

Hidalgo, town (2006 population 11,357), HIDALGO county, extreme S TEXAS, port of entry on the RIO GRANDE (Mexican border; bridged) opposite REY-NOSA (MEXICO), and c.7 mi/11.3 km S of MCALLEN; 26°06′N 98°15′W. In irrigated agricultural area (citrus, vegetables; cotton); manufacturing (water filters). Santa Ana National Wildlife Refuge to E, Bentsen–Rio Grande Valley State Park to NW.

Hidalgo, village (2000 population 123), JASPER county, SE ILLINOIS, 13 mi/21 km N of NEWTON; 39°09′N 88°09′W. In agricultural area.

Hidalgo, MEXICO: see CIUDAD HIDALGO.

Hidalgo, MEXICO: see MESONES HIDALGO.

Hidalgo del Parral (ee-DAHL-go del pah-RAHL) or **Parral** (pah-RAHL), city and township, CHIHUAHUA, N MEXICO, on the Parral River, on Mexico Highway 45; 26°58′N 105°40′W. Railroad and highway junction. One of Mexico's large mining centers, especially for silver, which has been mined in the region since the sixteenth century. From 1640 to 1731, the city was the capital of the colonial province of NUEVA VIZCAYA. One of the first cities to take up arms during Francisco Madero's revolution of 1917, it was later (1923) the site of the assassination of Pancho Villa.

Hidalgo, Salina de, ARGENTINA: see SALINA GRANDE.

Hidalgotitlán (ee-dahl-go-teet-LAHN), town, VER-ACRUZ, SE MEXICO, on Isthmus of TEHUANTEPEC, on COATZACOALCOS River, and 17 mi/27 km SSW of MINATITLÁN; 17°46′N 94°39′W. Fruit; livestock.

Hidalgo, Villa, MEXICO: see VILLA HIDALGO.

Hidalgo Yalalag, MEXICO: see VILLA HIDALGO.

Hidalmás, ROMANIA: see HIDA.

Hidas, village, BARANYA county, S HUNGARY, 12 mi/19 km SW of SZEKSZÁRD; 46°16′N 18°30′E. Enamelware.

Hidasnémeti (HE-dahsh-nai-ma-ti), Hungarian *Hidasnémeti*, village, BORSOD-ABAÚJ-ZEMPLÉN county, NE HUNGARY, on HERNAD RIVER, and 33 mi/53 km NE of MISKOLC, near Slovak border; 48°30′N 21°14′E. Railroad and road crossing point.

Hidayatpur, INDIA: see GURGAON.

Hidden Hills, city (2000 population 1,875), LOS AN-GELES county, S CALIFORNIA; residential suburb 24 mi/39 km WNW of downtown LOS ANGELES, at ex-

treme W end of SAN FERNANDO VALLEY; 34°10′N 118°40′W. Santa Monica Mountains National Recreation Area to S.

Hidden Inlet, fishing village, extreme SE ALASKA, on PEARSE CANAL, 60 mi/97 km SE of KETCHIKAN; 54°59′N 130°21′W.

Hiddenite (HID-neit), unincorporated town, ALEX-ANDER county, W central NORTH CAROLINA, 6 mi/9.7 km ESE of TAYLORSVILLE; 35°54′N 81°05′W. Manufacturing (furniture, lumber, polyurethane foam, machining, yarn); agriculture (tobacco, grain; poultry, livestock; dairying). Named for gem hiddenite discovered here (c.1879) and mined for a time.

Hidden Meadows, unincorporated town (2000 population 3,463), SAN DIEGO county, S CALIFORNIA; residential suburb 3 mi/4.8 km NW of ESCONDIDO. Merriam Mountains to N. Citrus, avocados, nursery stock.

Hidden Peak, PAKISTAN: see GASHERBRUM I.

Hidden River Cave, KENTUCKY: see HORSE CAVE.

Hiddensee (HID-duhn-zai), island (□ 7 sq mi/ 18.2 sq km), NE GERMANY, just W of RÜGEN island, 2 mi/3.2 km off MECKLENBURG–WESTERN POMERANIAN coast; 11 mi/18 km long, 1 mi/1.6 km–2 mi/3.2 km wide; 54°28′N 13°06′E–54°37′N 13°06′E. Low and level, rising toward N, with even coastline (W) and slightly indented E coast. Frequented as seaside resort; fishing is chief occupation. Main village, KLOSTER; nearby are remains of 13th-century Cistercian monastery. Viking gold utensils have been found here. Site of ornithological station and of biological research station of Greifswald University. Island was separated (1308) from Rügen island by major storm wave. Also HID-DENSOE.

Hiddensoe, GERMANY: see HIDDENSEE.

Hidden Valley, Colorado: see ROCKY MOUNTAIN NA-TIONAL PARK.

Hidden Valley Lake, unincorporated town (2000 population 3,777), Lake county, NW CALIFORNIA, 19 mi/31 km NE of HEALDSBURG, on PUTAH CREEK, in Coyote Valley. Residential community.

Hiddeqel River, IRAQ and TURKEY: see TIGRIS.

Hidra (HEE-drah), island (8 sq mi/20.7 sq km) in NORTH SEA, VEST-AGDER county, S NORWAY, at mouth of a fjord near FLEKKEFJORD. Hidra is the largest island in Vest-Agder county, and has fishing, shipping, and quarrying. Stone Age findings made nearby. Fortified during Danish-English war (1807–1814).

Hidrolândia (EE-dro-LAHN-zhee-ah), city (2007 population 18,531), W central CEARÁ, BRAZIL, 19 mi/31 km SW of SANTA QUITÉRIA; 04°25′S 40°25′W.

Hidrolândia, city (2007 population 14,015), central GOIÁS, BRAZIL, 62 mi/100 km SW of ANÁPOLIS; 17°00′S 49°17′W.

Hieflau (HEEF-lou), village, N STYRIA, central AUSTRIA, on the ENNS RIVER, on the E end of Gesäuse, and 8 mi/ 12.9 km NW of EISENERZ; 47°36′N 14°45′E. Railroad junction; hydroelectric station.

Hiendelaencina (YEN-dai-lah-en-THEE-nah), village, GUADALAJARA province, central SPAIN, 33 mi/53 km NNE of GUADALAJARA. Grain growing; sheep raising. Has abandoned silver mines.

Hienghène (YEN-GEN), village, NEW CALEDONIA, a French territory in the S PACIFIC OCEAN, on E coast, 145 mi/233 km NW of NOUMÉA; 20°40′S 164°54′E.

Hiep Hoa (HYEP HWAH), village, LONG AN province, S VIETNAM, 25 mi/40 km NW of HO CHI MINH CITY; 10°56′N 106°20′E. Sugar-growing center; sugar mill, distillery. Formerly Hiephoa.

Hierapetra, Greece: see IERAPETRA.

Hierapolis, ancient city of PHRYGIA, W ASIA MINOR, 7 mi/11.3 km N of Laodicea, NW of present-day DENI-ZLI, and on a plateau 500 ft/152 m above the BÜYÜK MENDERES, in what is now known as Pamukkale, TURKEY. Devoted to the worship of Leto in ancient times, it became an early seat of Christianity

(Colossians 4.13). The Romans greatly enlarged and improved the city, building a large theater and numerous baths about the hot springs for which the site is famous. Near the city was a deep chasm called the Plutonium, which the ancients thought led to the nether regions; the fissure no longer exists. Extensive ruins survive from the Roman and early Christian periods.

Hierissos, Greece: see IERISSOS.

Hieroconpolis, EGYPT: see KOM EL AHMAR, EL.

Hierro (YE-ro) or **Ferro** (FE-ro), smallest and westernmost island (□ 107 sq mi/278.2 sq km) of the CANARY ISLANDS, SPAIN, 40 mi/64 km SW of GOMERA, and 40 mi/64 km S of PALMA, in SANTA CRUZ DE TENERIFE province; 18 mi/29 km long NE-SW; 27°45′N 18°00′W. Chief town and port, VALVERDE. Has an abrupt coastline and rises in volcanic interior to 4,330 ft/1,320 m (ALTO DE MAL PASO). Large tracts are wooded. Because of scarcity of water, there are only a few fertile valleys, yielding cereals, grapes, potatoes, tomatoes, figs, almonds. Also exports wine and cheese. Sheep, cattle, and goats are raised. Its iron deposits are not exploited. Hierro was anciently thought to be the end of the world, and the longitude of Hierro's westernmost point, CAPE ORCHILLA, was first used by Mercator, noted 16th-century geographer, as the prime meridian and a convenient dividing line between the Eastern and Western hemispheres. Confirmed 1634 as the zero meridian by a geographical congress in PARIS, it remained in use until the adoption (1884) of the Greenwich meridian. The Hierro meridian was originally defined as 20°00′W of Paris (i.e., 17°40′W of GREENWICH), but the actual longitude of Cape Orchilla is 18°20′W of Greenwich.

Hietzing (HEE-tsing), outer WSW district (□ 6 sq mi/15.6 sq km) of VIENNA, AUSTRIA. Has palace and park of Schönbrunn; Lainz zoological garden.

Hiezu (hee-EZ), village, Saihaku district, TOTTORI prefecture, S HONSHU, W JAPAN, 49 mi/79 km W of TOTTORI; 35°26′N 133°23′E.

Higashi (HEE-gah-shee), village, Kunigami district, OKINAWA prefecture, SW JAPAN, 40 mi/65 km N of NAHA; 26°37′N 128°09′E. Pineapples.

Higashiawakura (hee-GAH-shee-ah-WAHK-rah), village, Aida district, OKAYAMA prefecture, SW HONSHU, W JAPAN, 43 mi/70 km N of OKAYAMA; 35°08′N 134°21′E.

Higashichichibu (hee-GAH-shee-CHEE-chee-boo), village, Chichibu district, SAITAMA prefecture, E central HONSHU, E central JAPAN, 28 mi/45 km N of URAWA; 36°03′N 139°11′E. Traditional papermaking.

Higashidori (hee-GAH-shee-DO-ree), village, Shimokita district, Aomori prefecture, N HONSHU, N JAPAN, 43 mi/70 km N of AOMORI; 41°17′N 141°12′E.

Higashihiroshima (hee-GAH-shee-hee-RO-shi-mah), city (2005 population 184,430), HIROSHIMA prefecture, SW HONSHU, W JAPAN, 16 mi/25 km E of HIROSHIMA; 34°25′N 132°44′E. Stereos; sake.

Higashiichiki (hee-GAH-shee-EE-chee-kee), town, Hioki district, KAGOSHIMA prefecture, S KYUSHU, SW JAPAN, on NW SATSUMA PENINSULA, 12 mi/20 km N of KAGOSHIMA; 31°39′N 130°21′E. Satsuma pottery originated here. Hot springs, FUKIAGE sand dune nearby. *Yakko so* plants.

Higashiiyayama (hee-GAH-shee-ee-yah-YAH-mah), village, Miyoshi district, TOKUSHIMA prefecture, SE SHIKOKU, W JAPAN, 40 mi/65 km S of TOKUSHIMA; 33°51′N 133°54′E.

Higashiizu (hee-GAH-shee-EEZ), town, Kamo district, SHIZUOKA prefecture, S HONSHU, E central JAPAN, 40 mi/65 km E of SHIZUOKA; 34°46′N 139°02′E.

Higashiizumo (hee-GAH-shee-EEZ-mo), town, Yatsuka district, SHIMANE prefecture, SW HONSHU, W JAPAN, 6 mi/10 km S of MATSUE; 35°25′N 133°09′E. Agricultural machinery; fruit and fish processing.

Higashikagura (hee-GAH-shee-KAH-goo-rah), town, Kamikawa district, HOKKAIDO prefecture, N JAPAN, 71 mi/115 km N of SAPPORO; 43°41′N 142°27′E.

Higashikawa (hee-GAH-shee-KAH-wah), town, Kamikawa district, HOKKAIDO prefecture, N JAPAN, 74 mi/120 km N of SAPPORO; 43°41′N 142°30′E. Rice; furniture, folkcrafts. Skiing area.

Higashikurume (hee-GAH-shee-KOO-roo-me), city, Tokyo prefecture, E central HONSHU, E central JAPAN, 19 mi/30 km W of SHINJUKU; 35°45′N 139°31′E. Residential suburb of TOKYO.

Higashikushira (hee-GAH-shee-KOO-shee-rah), town, Kimotsuki district, KAGOSHIMA prefecture, S KYUSHU, SW JAPAN, on E OSUMI PENINSULA, 28 mi/45 km S of KAGOSHIMA; 31°22′N 130°58′E.

Higashimatsuyama (hee-GAH-shee-mahts-YAH-mah), city, SAITAMA prefecture, E central HONSHU, E central JAPAN, 19 mi/30 km N of URAWA; 36°02′N 139°24′E. Suburb of TOKYO. Automotive parts.

Higashimokoto (hee-GAH-shee-mo-ko-to), village, Abashiri district, HOKKAIDO prefecture, N JAPAN, 158 mi/255 km E of SAPPORO; 43°50′N 144°12′E. Potatoes, beans, barley; dairy farming (cheese); ham, sausage.

Higashimura (hee-GAH-shee-MOO-rah), village, West Shirakawa district, FUKUSHIMA prefecture, N central HONSHU, NE JAPAN, 43 mi/70 km S of FUKUSHIMA city; 37°05′N 140°21′E.

Higashimurayama (hee-GAH-shee-moo-rah-YAH-mah), city, TOKYO prefecture, E central HONSHU, E central JAPAN, 12 mi/20 km W of SHINJUKU; 35°45′N 139°28′E.

Higashi-naibuchi, RUSSIA: see UGLEZAVODSK.

Higashinaruse (hee-GAH-shee-NAH-roo-se), village, Ogachi district, Akita prefecture, N HONSHU, NE JAPAN, near Mt. Kurikoma, 47 mi/75 km S of AKITA city; 39°10′N 140°39′E.

Higashine (hee-GAH-shee-ne), city, YAMAGATA prefecture, N HONSHU, NE JAPAN, 12 mi/20 km N of YAMAGATA city; 38°25′N 140°23′E. Video parts; fusuma. Fruits (cherries, apples, pears, grapes), tobacco.

Higashino (hee-GAH-shee-no), town, Toyota district, HIROSHIMA prefecture, SW HONSHU, W JAPAN, 28 mi/45 km E of HIROSHIMA; 34°15′N 132°55′E.

Higashiosaka (hee-GAH-shee-O-sah-kah), city (2005 population 513,821), OSAKA prefecture, S HONSHU, W central JAPAN, on the Onii River, 5.6 mi/9 km E of OSAKA; 34°39′N 135°36′E. Residential and industrial suburb of Osaka, manufacturing vehicles, iron wire, wire netting, tools. Hiraoka shrine is here. Kongo Ikoma quasi-national park nearby.

Higashisefuri (hee-GAH-shee-SE-foo-ree), village, Kanzaki district, SAGA prefecture, N KYUSHU, SW JAPAN, 9 mi/15 km N of SAGA; 33°20′N 130°24′E.

Higashishirakawa (hee-GAH-shee-shee-RAH-kah-wah), village, Kamo district, GIFU prefecture, central HONSHU, central JAPAN, 37 mi/60 km N of GIFU; 35°38′N 137°19′E.

Higashisonogi (hee-GAH-shee-SO-no-gee), town, East Sonogi district, NAGASAKI prefecture, NW KYUSHU, SW JAPAN, 22 mi/35 km N of NAGASAKI; 33°02′N 129°55′E.

Higashitsuno (hee-GAH-shee-TSOO-no), village, Takaoka district, KOCHI prefecture, S SHIKOKU, W JAPAN, 31 mi/50 km S of KOCHI; 33°23′N 133°01′E. Karst zone.

Higashiura (hee-GAH-shee-OO-rah), town, Chita district, AICHI prefecture, S central HONSHU, central JAPAN, 16 mi/25 km S of NAGOYA; 34°58′N 136°58′E.

Higashiura (hee-GAH-shee-OO-rah), town, on NE coast of AWAJI-SHIMA island, Tsuna district, HYOGO prefecture, W central JAPAN, 16 mi/25 km S of KOBE; 34°31′N 134°59′E. Carnations. World Peace Great Kannon Statue (328 ft/100m high).

Higashiyama (hee-GAH-shee-YAH-mah), town, East Iwai district, IWATE prefecture, N HONSHU, NE JAPAN, 50 mi/80 km S of MORIOKA; 38°59′N 141°15′E. Geibi gorge is nearby.

Higashiyamato (hee-GAH-shee-YAH-mah-to), city, TOKYO prefecture, E central HONSHU, E central JAPAN, 6 mi/10 km W of SHINJUKU; 35°44′N 139°25′E.

Higashiyoka (hee-GAH-shee-YO-kah), town, Saga district, SAGA prefecture, N KYUSHU, SW JAPAN, 3.1 mi/5 km S of SAGA; 33°12′N 130°17′E.

Higashiyoshino (hee-GAH-shee-YO-shee-no), village, Yoshino district, NARA prefecture, S HONSHU, W central JAPAN, 22 mi/35 km S of NARA; 34°24′N 135°58′E. Logs.

Higashiyuri (hee-GAH-shee-YOO-ree), town, Yuri district, Akita prefecture, N HONSHU, NE JAPAN, 31 mi/50 km S of AKITA city; 39°18′N 140°17′E. Cryptomeria.

Higaturu (hee-gah-too-ROO), town, NORTHERN (ORO) province, SE PAPUA NEW GUINEA, SE NEW GUINEA island, 30 mi/48 km ENE of KOKODA. Palm oil; coconuts. Once the government center of the district, the capital was removed to POPONDETTA in 1951 after the eruption of MOUNT LAMINGTON destroyed Higaturu, killing 3,000.

Higaza (he-GAH-zuh), village, QENA province, Upper EGYPT, 7 mi/11.3 km SE of QUS; 25°51′N 32°50′E. Pottery making, sugar refining; cereals, dates. Also spelled Hijazah.

Higbee, city (2000 population 623), RANDOLPH county, N central MISSOURI, 9 mi/14.5 km SSW of MOBERLY; 39°18′N 92°30′W. Corn, wheat, soybeans; cattle; manufacturing (wine barrels). Former coal-mining area.

Higden, village (2000 population 101), CLEBURNE county, N central ARKANSAS, 11 mi/18 km WNW of HEBER SPRINGS, in the OZARK Mountains, on GREERS FERRY LAKE Reservoir (LITTLE RED RIVER); 35°34′N 92°12′W.

Higganum, village, MIDDLESEX county, S CONNECTICUT, just S of MIDDLETOWN; 41°29′N 72°33′W. Postal section of the town of HADDAM that lies entirely to the W of the CONNECTICUT River with part of Haddam. Publishing.

Higgins, village (2006 population 439), LIPSCOMB county, extreme N TEXAS, in the PANHANDLE, c.40 mi/64 km SW of WOODWARD (OKLAHOMA); 36°07′N 100°01′W. In wheat and cattle region. Black Kettle National Grassland to SW.

Higgins Bay, hamlet, HAMILTON county, E central NEW YORK, in the ADIRONDACK MOUNTAINS, on PISECO LAKE, c.40 mi/64 km NE of UTICA; 43°25′N 74°32′W.

Higgins Lake, ROSCOMMON county, N central MICHIGAN, 5 mi/8 km W of ROSCOMMON; 0.7 mi/11.3 km long, 4 mi/6.4 km wide; 44°43′N 85°40′W. Touches CRAWFORD county border on N. Resorts; fishing. South Higgins Lake State Park at SE end; North Higgins Lake State Park on N end. Joined to HOUGHTON LAKE (S) by small passage.

Higginson, village (2000 population 378), WHITE county, central ARKANSAS, 4 mi/6.4 km SSE of SEARCY; 35°12′N 91°42′W.

Higginsport (HIG-uhnz-sport), village (2006 population 295), BROWN county, SW OHIO, on OHIO RIVER, 6 mi/10 km SSW of GEORGETOWN; 38°48′N 83°58′W.

Higginsville, city (2000 population 4,682), LAFAYETTE county, W central MISSOURI, near MISSOURI River, 12 mi/19 km SE of LEXINGTON; 39°03′N 93°43′W. Wheat, corn, soybeans; cattle, hogs. Limestone quarry. Manufacturing (electronic equipment, plastic products). Laid out 1869. Confederate Memorial State Historic Site with cemetery.

Higgston, town (2000 population 316), MONTGOMERY county, E central GEORGIA, 3 mi/4.8 km W of VIDALIA; 32°13′N 82°28′W.

Higham Ferrers (HEI-uhm FE-rez), market town (2001 population 6,689), E NORTHAMPTONSHIRE, central ENGLAND, on NENE RIVER, and 4 mi/6.4 km E of WELLINGBOROUGH; 52°18′N 00°35′W. Leather and shoe manufacturing. Has 13th–14th-century church.

High Atlas, MOROCCO: see ATLAS MOUNTAINS.

Highbank, township, ASHBURTON district, CANTERBURY region, SOUTH ISLAND, NEW ZEALAND, 45 mi/72 km W of CHRISTCHURCH, and on S bank of RAKAIA

RIVER. Agricultural center. Hydroelectric plant and irrigation water.

High Bentham (HI BEN-thuhm), village (2001 population 3,513), NORTH YORKSHIRE, N ENGLAND, 7 mi/11.3 km SE of KIRKBY LONSDALE; 54°07′N 02°30′W. Previously a weaving center.

High Beskyd, UKRAINE: see SKOLE BESKYDS.

High Blantyre, Scotland: see BLANTYRE.

High Bridge, unincorporated village, JESSAMINE county, central KENTUCKY, 8 mi/12.9 km SW of NICHOLASVILLE, on KENTUCKY RIVER (Palisades of the Kentucky gorge), at DIX RIVER mouth, and 20 mi/32 km SW of LEXINGTON, in BLUEGRASS REGION. Limestone quarrying. Historic Shaker settlement of PLEASANT HILL (or Shakertown) to SW. High Bridge, a 317-ft/97-m-high railroad bridge across Kentucky River, is here. Dix Dam (HERRINGTON LAKE reservoir) on Dix River, to S.

High Bridge, borough (2006 population 3,763), HUNTERDON county, W NEW JERSEY, on South Branch of RARITAN River, and 15 mi/24 km E of PHILLIPSBURG; 40°40′N 74°54′W. Site of former iron works; agriculture. State park nearby. Settled before 1750, incorporated 1898.

Highbridge, a residential section of SW BRONX borough of NEW YORK city, SE NEW YORK; 40°50′N 73°56′W. Formerly an Irish stronghold, now predominantly African-American and Dominican.

Highbridge, ENGLAND: see BURNHAM-ON-SEA.

High Commission Territories, former group of landlocked territories, S AFRICA, including BECHUANALAND Protectorate (now BOTSWANA), Basutoland (now LESOTHO), and SWAZILAND. Bechuanaland Protectorate came under the British crown in September 1885 after the discovery of gold at TATI (1867).

Higher Walton (WAL-tun), town (2006 population 6,900), W central LANCASHIRE, N ENGLAND, 3 mi/4.8 km SE of PRESTON; 53°44′N 02°39′W. Nearby WALTON-LE-DALE has textile, electrical equipment, and machinery manufacturing.

Highett (HEI-uht), suburb 10 mi/16 km SE of MELBOURNE, VICTORIA, SE AUSTRALIA; 37°57′S 145°03′E.

High Falls, hamlet, ULSTER county, SE NEW YORK, on RONDOUT CREEK, and 10 mi/16 km SW of KINGSTON; 41°50′N 74°08′W. In resort and agricultural area.

Highfield, unincorporated village, Butler township, BUTLER county, W PENNSYLVANIA, suburb 1 mi/1.6 km W of BUTLER; 40°51′N 79°55′W.

Highgate, town, SAINT MARY parish, N JAMAICA, in uplands, on railroad, and 23 mi/37 km NNW of KINGSTON; 18°19′N 76°52′W. Fruit growing; livestock grazing. Busy agricultural and commercial center.

Highgate, town, FRANKLIN CO., NW VERMONT, on Lake CHAMPLAIN, at Canadian (QUEBEC) border, at mouth of Missisquoi River, and 8 mi/12.9 km N of St. Albans; 44°57′N 73°02′W. Metal products; lime, lumber. Port of entry. Highgate Springs is family-style resort village.

Highgate (HEI-gait), unincorporated village (2001 population 386), S ONTARIO, E central CANADA, included in city of CHATHAM-KENT; 42°30′N 81°48′W. Dairying; mixed farming; lumbering.

Highgate (HEI-gayt), residential area (2001 population 10,310), within CAMDEN, ISLINGTON, and HARINGEY boroughs, GREATER LONDON, SE ENGLAND; 51°34′N 00°08′W. The house where Francis Bacon died is in Highgate, and Herbert Spencer, George Eliot, and Karl Marx are buried in Highgate cemetery in Camden. Highgate School, a public school founded in 1565, is here.

Highgrove, unincorporated town (2000 population 3,445), RIVERSIDE county, S CALIFORNIA; residential suburb 5 mi/8 km NNE of RIVERSIDE, 8 mi/12.9 km SSW of SAN BERNARDINO. Citrus, nursery stock; dairying.

High Hill, town, MONTGOMERY county, E central MISSOURI, 9 mi/14.5 km SE of MONTGOMERY CITY; 38°52′N 91°22′W. Fire-clay.

High Island, unincorporated village, GALVESTON county, S TEXAS, 32 mi/51 km NE of GALVESTON, between Gulf coast (S) and GULF INTRACOASTAL WATERWAY, at base of BOLIVAR PENINSULA; 29°34′N 94°23′W. Oil field center, on salt dome in marshland. Resort area. Nearby are Anahuac (NW) and McFadden (NE) national wildlife refuges.

High Island, MICHIGAN: see BEAVER ISLANDS.

Highland (HI-luhnd), county (□ 10,085 sq mi/26,221 sq km; 2001 population 208,914) covering N half of Scotland; ⊙ INVERNESS; 57°35′N 05°00′W. Largest (in terms of area) of the thirty-two unitary authorities in Scotland. Borders MORAY, ABERDEENSHIRE, PERTH AND KINROSS, and ARGYLL AND BUTE. The MONADHLIATH MOUNTAINS are located in the SE, and LOCH NESS is in the central part of the county, just SW of Inverness. Highland is separated from LEWIS AND HARRIS island by THE MINCH. Consists of former counties of CAITHNESS, SUTHERLAND, Ross and Cromarty, Nairn, Inverness (including INNER HEBRIDES), and small parts of Argyll and Moray. A region of rugged beauty, the land is unsuitable for farming and since the 18th century has suffered from a steady decline of population—partly caused, initially, by the aftermath of the Jacobite rebellions. Crofting and fishing are the main occupations; in recent years the tourist trade has been an important source of income. Since the 1970s, North Sea oil has had a major impact on the economy of areas surrounding the MORAY FIRTH. Government-sponsored forestry and hydroelectric developments in the Highlands were part of a comprehensive effort to stem the flow of emigration and to relieve poverty and chronic depression. Gaelic is still spoken in parts of the W and on the islands. The distinctive marks of Highland, the dress and the clan system, were products of the late Middle Ages. The dress, including the kilt, tartan, sporran, tam, and dirk, was outlawed by the British government in the 18th century, when it became alarmed at the area's continued interest in the Jacobites. In the 19th century, as the language and sectional feeling declined, the government allowed the revival of clan dress and the use of bagpipes, long the national musical instrument of Scotland. In the remote areas, old customs survive more than anywhere else in the British Isles, and many of the Highlanders have remained Roman Catholic despite the vigor of the Scottish Reformation.

Highland (HEI-luhnd), county (□ 554 sq mi/1,440.4 sq km; 2006 population 42,833), SW OHIO; ⊙ HILLSBORO; 39°11′N 83°37′W. Drained by East Fork of Little Miami River and PAINT, White Oak, and Rattlesnake creeks. In the Till Plains physiographic region, except for the SE portion, which is in the Lexington Plain region. Agriculture (nursery and greenhouse crops, tobacco, grains; dairying; hogs); manufacturing at Hillsboro and GREENFIELD (textiles, plastic products, transportation equipment); limestone quarries. Formed 1805.

Highland (HEI-luhnd), county (□ 416 sq mi/1,081.6 sq km; 2006 population 2,510), NW VIRGINIA; ⊙ MONTEREY; 38°21′N 79°33′W. In ALLEGHENY MOUNTAINS, bounded (W, N) by WEST VIRGINIA state line; drained by JACKSON, BULLPASTURE, COWPASTURE, South Branch POTOMAC rivers and Back Creek. Some agriculture (hay, alfalfa, potatoes, sugar maples; cattle- and sheep-raising area); timber; freshwater fish (trout). Formed 1847.

Highland, city (2000 population 44,605), SAN BERNARDINO county, SE CALIFORNIA; suburb 4 mi/6.4 km E of SAN BERNARDINO, 59 mi/95 km E of downtown LOS ANGELES. Drained S by SANTA ANA River. In a citrus-grove area at the foot of the SAN BERNARDINO MOUNTAINS; also light manufacturing. Developed along with the Southern California area in the growth of agribusiness and aircraft industries. Patton State Hospital is here, and Norton Air Force Base to S. San Bernardino Mountains, in San Bernardino National Forest, to N. San Manuel Indian Reservation to N.

Highland, city (2000 population 8,438), MADISON county, SW ILLINOIS, 16 mi/26 km ESE of EDWARDSVILLE; 38°44′N 89°40′W. In dairying and agricultural area (corn, wheat; poultry, livestock); manufacturing (food products, machinery, transportation equipment). Founded 1831 by Swiss; incorporated 1863.

Highland, town (2000 population 23,546), Lake county, extreme NW INDIANA, in the CHICAGO metropolitan area; 41°33′N 87°28′W. Manufacturing (dairy products, mineral granules for sandblasting). Settled 1883 as Clough Postal Station, name changed to Highland in 1888. Incorporated 1910.

Highland, town (2006 population 13,889), UTAH county, N central UTAH, suburb 15 mi/24 km NNW of PROVO, and 25 mi/40 km SSE of SALT LAKE CITY, on American Fork River; 40°25′N 111°47′W. Dairying; cattle; fruit, vegetables, sugar beets. TIMPANOGOS CAVE National Forest, in WASATCH RANGE, to E. Settled 1875.

Highland, village (2000 population 976), DONIPHAN county, extreme NE KANSAS, 10 mi/16 km WNW of TROY; 39°51′N 95°16′W. In agricultural region (apples, grain; livestock, poultry). Seat of Highland Community College. Brown State Fishing Lake to W.

Highland, village (□ 4 sq mi/10.4 sq km; 2000 population 5,060), ULSTER county, SE NEW YORK, on W bank of the HUDSON RIVER (here crossed by Mid-Hudson Bridge), opposite POUGHKEEPSIE; 41°43′N 73°58′W. Summer residential area; liight manufacturing. Center of mid-Hudson apple-growing region, largest in state after WAYNE county region.

Highland (HEI-luhnd), village (2006 population 297), HIGHLAND county, SW OHIO, 10 mi/16 km N of HILLSBORO, on small Rattlesnake Creek; 39°20′N 83°35′W.

Highland, unincorporated village, MCCANDLESS township, ALLEGHENY county, SW PENNSYLVANIA, residential suburb 8 mi/12.9 km NNW of downtown PITTSBURGH; 40°28′N 79°55′W–40°33′N 80°02′W.

Highland, village (2006 population 840), IOWA county, SW WISCONSIN, 14 mi/23 km NW of DODGEVILLE; 43°02′N 90°22′W. Makes cheese; timber; stoneware pottery; winery. Black Hawk Lake reservoir to SE; on Pecatonica State Trail.

Highland, plantation, SOMERSET county, central MAINE, 10 mi/16 km WNW of BINGHAM; 45°05′N 70°04′W.

Highland Acres, unincorporated town (2000 population 3,379), KENT county, central DELAWARE, 2 mi/3.2 km S of DOVER; 39°07′N 75°31′W. Elevation 19 ft/5 m. Largely a residential suburb of Dover.

Highland Beach, town (2000 population 109), ANNE ARUNDEL county, central MARYLAND, 3 mi/4.8 km SE of ANNAPOLIS, on CHESAPEAKE BAY; 38°56′N 76°28′W. In shore-resort area. Charles Douglass, son of Frederick Douglass, the African-American abolitionist, bought land here in 1892 and built the first house in 1894 as part of a summer community for African-American intellectuals. It is still a private community with private beaches and homes. Incorporated 1922.

Highland City (HEI-land), town (2000 population 2,051), POLK county, central FLORIDA, 8 mi/12.9 km SE of LAKELAND; 27°57′N 81°52′W. Manufacturing includes metal fabrication, fresh fruit processing; mailboxes, powder coatings.

Highland Falls, summer residential-recreational village (□ 1 sq mi/2.6 sq km; 2006 population 3,736), ORANGE county, SE NEW YORK, on W bank of the HUDSON RIVER, and 10 mi/16 km S of NEWBURGH, just S of WEST POINT military academy; 41°21′N 73°58′W. Settled 1800, incorporated 1906.

Highland Heights (HEI-luhnd HEITZ), city (□ 5 sq mi/13 sq km; 2006 population 8,620), CUYAHOGA county, N OHIO; E suburb of CLEVELAND; 41°32′N 81°28′W.

Highland Heights, town (2000 population 6,554), CAMPBELL county, N KENTUCKY, a suburb 5 mi/8 km SSE of downtown CINCINNATI (OHIO), near OHIO RIVER; 39°02′N 84°27′W. Light manufacturing. Seat of Northern Kentucky University.

Highland Hills, unincorporated village, DU PAGE county, NE ILLINOIS, suburb 19 mi/31 km W of downtown CHICAGO, S of LOMBARD; 41°50′N 88°00′W.

Highland Lake, village, CUMBERLAND county, MAINE, suburb of PORTLAND.

Highland Lake, hamlet, SULLIVAN county, SE NEW YORK, on small Highland Lake, 13 mi/21 km SW of MONTICELLO; 41°32′N 74°51′W. In outdoor-recreation area.

Highland Lake, reservoir, WINCHESTER town, LITCH-FIELD county, NW CONNECTICUT, on small branch of Mad Still River, 9 mi/14.5 km NNE of TORRINGTON; 2 mi/3.2 km long; 41°55′N 73°05′W. Plat Hill State Park on SW shore.

Highland Lake, reservoir and resort lake, STODDARD and WASHINGTON towns, CHESHIRE and SULLIVAN counties, SW NEW HAMPSHIRE, on North Branch CONTOOCOOK River, 16 mi/26 km NNE of KEENE; 6 mi/9.7 km long; 43°06′N 72°04′W.

Highland Mills, village (□ 1 sq mi/2.6 sq km; 2000 population 3,468), ORANGE county, SE NEW YORK, 8 mi/12.9 km W of HIGHLAND FALLS; 41°21′N 74°07′W.

Highland Park, city (2000 population 31,365), Lake county, NE ILLINOIS, a suburb of CHICAGO on Lake MICHIGAN; 42°10′N 87°48′W. Retail and medical center for the North Shore area. Nearby Ravinia Park is the summer home to the Chicago Symphony Orchestra; the park also hosts a well-known music festival. Sheridan Reserve Center (formerly Fort Sheridan) is adjacent to the city. Incorporated 1869.

Highland Park, city (2000 population 16,746), WAYNE county, SE MICHIGAN, suburb 6 mi/9.7 km NNW of and completely surrounded by DETROIT; 42°23′N 83°05′W. SE corner touches city of HAMTRAMCK. Tractor and motor vehicle assembly are the city's main industries; manufacturing (coffee processing, printing; decorative chrome plates). Former headquarters (designed by Minoru Yamasaki) of the Chrysler Corporation. Grew mainly after Henry Ford established his first Model T auto plant here in 1910; it was here that assembly-line production was first based. Seat of Wayne County Community College. Laid out 1818, incorporated as a city 1917.

Highland Park, unincorporated town (2000 population 1,446), DERRY township, MIFFLIN county, central PENNSYLVANIA, residential suburb 2 mi/3.2 km N of LEWISTOWN, on Kishacoquillas Creek; 40°37′N 77°34′W.

Highland Park, unincorporated town, CUMBERLAND county, S PENNSYLVANIA, residential suburb 2 mi/3.2 km SW of HARRISBURG; 40°13′N 76°54′W.

Highland Park, town (2006 population 9,035), DALLAS county, N TEXAS, suburb 18 mi/6.4 km N of downtown DALLAS; 32°49′N 96°47′W. Surrounded by city of Dallas, as is neighboring UNIVERSITY PARK (N). Dallas Love Field airport to W. In 1916 V. C. Prather set aside land to build one of the first planned shopping centers; construction actually began in 1931 (Spanish architectural style). Settled 1907, incorporated 1913.

Highland Park, village, SULLIVAN county, NE TEN-NESSEE, suburb just E of KINGSPORT.

Highland Park, residential borough (2006 population 14,175), MIDDLESEX county, N central NEW JERSEY, on the RARITAN River, opposite NEW BRUNSWICK; 40°29′N 74°25′W. Incorporated 1905.

Highland Park, N residential section of LOS ANGELES, LOS ANGELES county, S CALIFORNIA, 4 mi/6.4 km NE of downtown Los Angeles, and SW of PASADENA. Founded 1887; annexed 1895 by Los Angeles.

Highland Park, Pennsylvania: see UPPER DARBY.

Highland Peak (10,935 ft/3,333 m), ALPINE county, E CALIFORNIA, in the SIERRA NEVADA, c.30 mi/48 km SSE of LAKE TAHOE, in Eldorado National Forest.

Highlands (HEI-luhndz), district municipality (□ 15 sq mi/39 sq km; 2001 population 1,674), SW BRITISH COLUMBIA, W CANADA, 8 mi/14 km NW of VICTORIA, in CAPITAL REGIONAL District; 48°31′N 123°30′W. Primarily residential. One-third of the area is protected parkland.

Highlands (HEI-landz), county (□ 1,106 sq mi/2,875.6 sq km; 2006 population 97,987), central FLORIDA, bounded E by KISSIMMEE RIVER; ⊙ SEBRING; 27°20′N 81°20′W. Rolling terrain with many lakes, notably Lake ISTOKPOGA; SE corner of county in EVERGLADES. Citrus-fruit and cattle area; also vegetable and poultry farming. Formed 1921.

Highlands, unincorporated town, SAN MATEO county, W CALIFORNIA; residential suburb 17 mi/27 km S of downtown SAN FRANCISCO, on San Mateo Creek. LOWER CRYSTAL SPRINGS reservoir, in San Francisco State Fish and Game Refuge, to SW.

Highlands (HEI-luhndz), town (5.6 sq mi/14.5 sq km; 1990 population 948; 2000 population 909), JACKSON and MACON counties, W NORTH CAROLINA, 13 mi/21 km SE of FRANKLIN, near GEORGIA state line; 35°02′N 83°12′W. Mountain resort in Nantahala National Forest; Chattahoochee National Forest (Georgia) to S, Sumter National Forest (S.C.) to SE. Common corner of Georgia, North Carolina, and SOUTH CAROLINA 7 mi/11.3 km to SE. Retail trade; manufacturing (printing and publishing); agriculture (tobacco, corn; cattle). Bridal Veil Falls (120 ft/37 m) to NW. Lake Sequayah reservoir to NW. Scaly Mountain Ski Area to W.

Highlands, unincorporated town (2000 population 7,089), HARRIS county, S TEXAS, suburb 18 mi/29 km E of downtown HOUSTON, and 7 mi/11.3 km NW of BAYTOWN, on SAN JACINTO RIVER (Winters Lake); 29°48′N 95°03′W. In agricultural area; manufacturing (chemicals, plastic products; sheet metal fabrication, sand and clay processing). Highlands Reservoir to E.

Highlands, resort borough (2006 population 4,987), MONMOUTH county, E NEW JERSEY, between NAVE-SINK RIVER and SANDY HOOK BAY, 5 mi/8 km NE of RED BANK; 40°23′N 73°59′W. Seafood. Largely residential. On nearby NAVESINK HIGHLANDS is the Twin Towers, one of most powerful lighthouse systems in U.S. First U.S. Navy wireless station (1903) nearby, and U.S. Army reservation just S. Incorporated 1900.

Highlands East (HEI-luhndz EEST), township (□ 271 sq mi/704.6 sq km; 2001 population 3,022), S ON-TARIO, E central CANADA, 20 mi/31 km from MINDEN; 44°58′N 78°16′W. Forestry, mining, tourism. Formed in 2001 from the former BICROFT, CARDIFF, GLA-MORGAN, and MONMOUTH townships.

Highlands Forge Lake, ESSEX county, NE NEW YORK, in ADIRONDACK MOUNTAINS, 3 mi/4.8 km N of WILLSBORO; 1.5 mi/2.4 km long; 44°30′N 73°26′W. Receives drainage from Long Pond, located immediately S. Drains NE to Lake CHAMPLAIN's Willsboro Bay. Dam at N end. Surface elevation 567 ft/173 m above mean sea level. Rattlesnake Mountain (elevation 1,316 ft/401 m) immediately SE.

Highlands of Navesink, NEW JERSEY: see NAVESINK HIGHLANDS.

Highland Springs (HEI-luhnd), unincorporated city (2000 population 15,137), HENRICO county, E central VIRGINIA, residential suburb 6 mi/10 km E of downtown RICHMOND, near CHICKAHOMINY RIVER; 37°32′N 77°19′W. Manufacturing (wood products, awards). SEVEN PINES National Cemetery to SE; Richmond International Airport, Virginia Aviation Museum to S.

Highlandville, town, CHRISTIAN county, SW MISSOURI, in the OZARK MOUNTAINS, near JAMES RIVER, 6 mi/9.7 km SSW of OZARK.

High Lane (HI LAIN), locality (2006 population 5,700), GREATER MANCHESTER, NW ENGLAND, on Macclesfield Canal, 5 mi/8 km SE of STOCKPORT; 53°21′N 02°05′W. Lyme Park (national trust) to S. Farming.

High Level (HEI LE-vuhl), town (□ 12 sq mi/31.2 sq km; 2005 population 3,849), NW ALBERTA, W CANADA, 150 mi/241 km N of PEACE River, on MACKENZIE HIGHWAY, on railroad to HAY RIVER (NORTHWEST TERRITORIES), and included in specialized municipality of MACKENZIE No.23; 58°31′N 117°08′W. Established 1940s; grew in 1960s with regional oil boom. Has strong agricultural economy; grain elevators, sawmill; timber, including distribution. Scheduled air service.

Highley (HEI-lee), village (2001 population 3,298), SE SHROPSHIRE, W ENGLAND, on SEVERN RIVER, and 7 mi/11.3 km NW of KIDDERMINSTER; 52°27′N 02°23′W. Former coal-mining site.

Highmore, city (2006 population 777), ⊙ HYDE county, central SOUTH DAKOTA, 45 mi/72 km ENE of PIERRE; 44°31′N 99°26′W. Trading center for farming region; dairy products; livestock; grain. State experiment farm nearby.

Highmount, hamlet, ULSTER county, SE NEW YORK, in the CATSKILL MOUNTAINS, NW of KINGSTON, and near Belle Ayr Mountain (hiking; skiing); 42°08′N 74°29′W.

High Peak, highest peak in ZAMBALES province, central LUZON, PHILIPPINES, c.30 mi/48 km NW of TARLAC; elevation 6,683 ft/2,037 m; 15°29′N 120°07′E.

High Peak, ENGLAND: see PEAK DISTRICT.

High Plains, Texas and New Mexico: see LLANO ES-TACADO.

High Point (HEI POINT), city (□ 50 sq mi/130 sq km; 2006 population 97,796), DAVIDSON, FORSYTH, GUIL-FORD, and RANDOLPH counties, N NORTH CAROLINA, suburb 16 mi/26 km SW of GREENSBORO; 35°58′N 80°00′W. DEEP RIVER passes to NE, in a heavily forested section of PIEDMONT region. Railroad junction. Along with GREENSBORO and WINSTON-SALEM, one of three cities comprising the Piedmont Triad region. Trade, industrial, and commercial center for an agricultural area, noted for the production of furniture and hosiery. Manufacturing (printing; wood, plastic, metal, and rubber products; furniture, chemicals, sandpaper, electronic equipment, apparel, textiles, embroidery, automobile parts, consumer goods); service industries. Annual furniture expositions held here. Of interest is the restored home of a blacksmith (1786). Seat of High Point University and North Carolina School of the Arts. International Home Furnishings Center; Furniture Discovery Center, museum on furniture manufacturing. High Point Museum and Historical Park. Angela Peterson Doll Museum Emerald Point theme park. Settled before 1750, incorporated 1859.

High Point, summit (1,803 ft/550 m) of KITTATINNY MOUNTAIN ridge, extreme NW NEW JERSEY; highest point in state, with view of three states. Has war memorial tower (225 ft/69 m high). Surrounding region included in High Point State Park (3 sq mi/7.8 sq km), year-round forest recreational area.

High Point, summit (3,075 ft/937 m) in the CATSKILL MOUNTAINS, ULSTER county, SE NEW YORK, 15 mi/24 km W of KINGSTON; 41°56′N 74°17′W.

High Prairie (HEI PRE-ree), town (□ 2 sq mi/5.2 sq km; 2001 population 2,737), W ALBERTA, W CANADA, 65 mi/105 km SE of PEACE River, and 15 mi/24 km W of LESSER SLAVE LAKE, in big lakes municipal district; 55°26′N 116°29′W. Lumbering; mixed farming; wheat, oats. Airport for light aircraft. Incorporated as a village in 1945, and as a town in 1950.

High Rhine, German *Hochrhein* (HOKH-rein), region (□ 748 sq mi/1,944.8 sq km), BADEN-WÜRTTEMBERG, S GERMANY. Bordered on S by SWITZERLAND; chief town is WALDSHUT-TIENGEN.

Area is shown by the symbol □, and capital city or county seat by ⊙.

High Ridge, unincorporated city, JEFFERSON county, E MISSOURI, 20 mi/32 km SW of ST. LOUIS; 38°27′N 90°31′W. Residential and manufacturing suburb. Manufacturing (fabricated metal products, electronic equipment, machinery).

High River (HEI RI-vuhr), town (□ 4 sq mi/10.4 sq km; 2001 population 9,345), S ALBERTA, SW CANADA, at foot of ROCKY MOUNTAINS, 35 mi/56 km SSE of CALGARY, in FOOTHILLS No. 31 municipal district; 50°35′N 113°52′W. Coal mining; grain, wheat; cattle; dairying. In oil-producing region. Formed as a village in 1901; became a town in 1906.

High Rock Lake (HEI RAHK LAIK), reservoir (□ 24 sq mi/62.4 sq km), on border between ROWAN and DAVIDSON counties, central NORTH CAROLINA, on Yadkin (PEE DEE) River, 32 mi/51 km S of WINSTON-SALEM; c.15 mi/24 km long; 35°36′N 80°14′W. Maximum capacity 248,800 acre-ft. Has several long arms. Formed by High Rock Dam (70 ft/21 m high), built (1927) by Army Corps of Engineers for power generation. TUCKERTOWN LAKE downstream.

High Shoals, town, OCONEE, WALTON, and MORGAN counties, NE central GEORGIA, 12 mi/19 km SW of ATHENS, and on APALACHEE RIVER; 33°49′N 83°30′W.

High Shoals (HEI SHOLZ), village (□ 2 sq mi/5.2 sq km; 2006 population 762), GASTON county, S NORTH CAROLINA, 9 mi/14.5 km N of GASTONIA, on South Fork CATAWBA RIVER; 35°23′N 81°12′W. Manufacturing; retail trade; agricultural area (grain; poultry, livestock; dairying).

High Sierra, California: see SIERRA NEVADA.

Highspire (HEI-SPEI-uhr), borough (2006 population 2,601), DAUPHIN county, S PENNSYLVANIA, residential suburb 6 mi/9.7 km SE of HARRISBURG, on Susquehanna River (bridged). Manufacturing (food products). Some agriculture in area (corn; poultry; dairying). Harrisburg International Airport to SE, on SUSQUEHANNA RIVER. Settled 1775, laid out 1814, incorporated 1867.

Highsplint, KENTUCKY, see LEJUNIOR.

High Springs (HEI), town (□ 9 sq mi/23.4 sq km; 2000 population 3,863), ALACHUA county, N central FLORIDA, near SANTA FE RIVER, 21 mi/34 km NW of GAINESVILLE; 29°49′N 82°35′W. Agricultural trade and shipping point; railroad shops. Founded c.1885.

High Tatras, SLOVAK Vysoké Tatry (vi-SO-kai TAHT-ri), POLISH Tatry Wysokie (TAH-tree wi-SO-kee), GERMAN Hohe Tatra (HUH-uh TAH-trah), HUNGARIAN Magas Tatra, highest group of the CARPATHIAN mountain system, in E central EUROPE; extends c.40 mi/60 km along the Polish-Slovak border; its highest peak, GERLACHOVSKÝ (8,711 ft/2,655 m), is in Slovakia. The extensively glaciated mountains have numerous lakes, moraines, and hanging valleys. High Tatras National Park (established in 1948) extends on both sides of the international border. The region's scenic beauty and excellent ski slopes have made it a year-round resort area. POPRAD (Slovakia) and ZAKOPANE (Poland) are the chief resort centers.

Highton (HEI-tuhn), suburb 2 mi/4 km from GEELONG, VICTORIA, SE AUSTRALIA. Church of England Girls' Grammar School.

Hightstown, borough (2000 population 5,216), MERCER county, central NEW JERSEY, near MILLSTONE RIVER, 12 mi/19 km ENE of TRENTON; 40°16′N 74°31′W. Trade center in agricultural area; light manufacturing; in process of suburbanizing. Seat of Peddie Preparatory School. Nearby (4 mi/6.4 km SE) is former JERSEY HOMESTEADS, now ROOSEVELT borough. Settled 1721, incorporated 1853.

High Valleyfield (VAH-lee-feeld), village (2001 population 2,940), FIFE, E Scotland, 1 mi/1.6 km N of Firth of Forth, and 6 mi/9.7 km W of DUNFERMLINE; 56°03′N 03°36′W. Adjacent to village of Low Valleyfield.

Highveld, E part of the central plateau above escarpment, extending through NORTHERN CAPE, FREE STATE and GAUTENG provinces, SOUTH AFRICA. Montane and warm temperate grassland; summer rainfall region, known for its thunderstorms and hail. Covers 20% of South African land surface from MALUTI MOUNTAINS in E, to Transvaal Drakensberg in NE. MAGALIESBERG MOUNTAINS in N slopes S and W drained by VAAL, Baledos and ORANGE rivers, falls 5,000 ft/1,524 m–6,000 ft/1,829 m in E and NE to 2,000 ft/610 m–3,000 ft/914 m in Ward SW. Rich in minerals, such as gold, iron ore, coal, and a range of nonferrous metals; major agricultural area with extensive livestock raising and arable farming.

Highview, unincorporated city (2000 population 15,161), JEFFERSON county, N KENTUCKY, residential suburb 9 mi/14.5 km SE of dowtown LOUISVILLE; 38°08′N 85°38′W. Agriculture to S (tobacco, grain; livestock).

Highway City, unincorporated town, FRESNO county, S central CALIFORNIA; suburb 8 mi/12.9 km NW of downtown FRESNO, near King River. Citrus, grain, cotton; dairying; cattle, poultry.

High Willhays, ENGLAND: see DARTMOOR.

Highwood, city (2000 population 4,143), Lake county, NE ILLINOIS, suburb 24 mi/39 km NNW of CHICAGO, on Lake MICHIGAN, 12 mi/19 km NNW of EVANSTON; 42°12′N 87°48′W. Site of Sheridan U.S. Army Reserve Center. Fort Sheridan decommissioned in 1990s. Incorporated 1886.

Highwoods, village, MARION county, central INDIANA, NW suburb of INDIANAPOLIS. Part of Indianapolis.

Highworth (HEI-wuhth), town (2001 population 8,347), N Swindon, S central ENGLAND, 6 mi/9.7 km NE of SWINDON; 51°37′N 01°42′W. Agricultural market in dairying region. Has 15th-century church.

High Wycombe (WI-kuhm), town (2006 population 81,117), BUCKINGHAMSHIRE, S ENGLAND, on the WYE RIVER, in a Chiltern valley; 51°38′N 00°46′W. Previously well known for its furniture industry; also has paper mills, saw mills, electrical equipment, and machinery works. Other industries include printing and the manufacturing of precision instruments and clothing. Ancient British and Roman remains nearby. The parish church dates from the 13th century. Wycombe Abbey, a mansion built in 1795, is an all girls public school. Includes WEST WYCOMBE and WYCOMBE MARSH.

Higley, unincorporated town, MARICOPA county, central ARIZONA, residential suburb 22 mi/35 km SE of downtown PHOENIX, surrounded by extensions of GILBERT city limits, which adjoins it to NW. Williams Air Force Base to E; Arizona Boys Ranch to SE; CHANDLER Municipal Airport to SW.

Higo, former province in W KYUSHU, JAPAN; now part of KUMAMOTO prefecture.

Higuera de Arjona (ee-GAI-rah dai ahr-HO-nah), town, JAÉN province, S SPAIN, 18 mi/29 km NW of JAÉN; 37°58′N 03°59′W. Agricultural trade (olive oil, cereals, vegetables).

Higuera de Calatrava (ee-GAI-rah dai kah-lah-TRAH-vah), town, JAÉN province, S SPAIN, 21 mi/34 km WNW of JAÉN; 37°48′N 04°10′W. Olive oil, cereals.

Higuera de las Dueñas (ee-GAI-rah dai lahs DWAI-nyahs), village, ÁVILA province, central SPAIN, on SE slopes of SIERRA DE GREDOS, 21 mi/34 km NE of TALAVERA DE LA REINA. Cereals, fruit; livestock.

Higuera de la Serena (ee-GAI-rah dai lah sai-RAI-nah), town, BADAJOZ province, W SPAIN, in LA SERENA region, 11 mi/18 km SW of CASTUERA; 38°39′N 05°44′W. Cereals, chickpeas, olives, grapes, tubers; livestock.

Higuera de la Sierra (ee-GAI-rah dai lah SYE-rah), or **Higuera junto a Aracena**, town, HUELVA province, SW SPAIN, 7 mi/11.3 km SE of ARACENA; 37°50′N 06°27′W. Lumbering and agriculture (cereals, olives, fruit, acorns, chestnuts, cork, chickpeas); charcoal burning. Antimony and silver-bearing lead deposits. Its railroad station is 4.5 mi/7.2 km NE.

Higuera de Llerena (ee-GAI-ra dai yai-RAI-nah), town, BADAJOZ province, W SPAIN, 11 mi/18 km N of LLERENA; 38°22′N 06°00′W. Cereals, chickpeas; livestock.

Higuera de Vargas (ee-GAI-rah dai VAHR-gahs), town, BADAJOZ province, W SPAIN, 30 mi/48 km S of BADAJOZ; 38°27′N 06°58′W. Agricultural center (flour, olive oil; timber; charcoal, acorns, cork; livestock).

Higuera, La, CHILE: see LA HIGUERA.

Higuera la Real (ee-GAI-rah lah rai-AHL), town, BADAJOZ province, W SPAIN, in the SIERRA MORENA, near HUELVA province border, 13 mi/21 km SSE of JEREZ DE LOS CABALLEROS; 38°08′N 06°41′W. Flour-milling center in agricultural region (cereals, olives; livestock).

Higueras (ee-GWE-rahs), town, NUEVO LEÓN, N MEXICO, 27 mi/43 km NE of MONTERREY; 25°58′N 100°01′W. Grain, cactus fibers; livestock.

Higuer, Cape (ee-GER), French Figuier (fee-GYAI), headland on BAY OF BISCAY, GUIPÚZCOA province, N SPAIN, near French border, 2 mi/3.2 km N of FUEN-TERRABÍA; 43°25′N 01°47′W. Lighthouse.

Higueritas (ee-ge-REE-tahs), village, COQUIMBO region, N central CHILE, 14 mi/23 km N of OVALLE; 29°40′S 70°52′W. Railroad junction in copper-mining area; fruit growing.

Higüero, Point (hee-GWE-ro), on NW coast of PUERTO RICO, 9 mi/14.5 km SW of AGUADILLA; 18°22′N 67°16′W. Former Bonus Thermoelectric Nuclear Plant nearby.

Higuerote (hee-gai-RO-tai), town, ⊙ Brión municipio, MIRANDA state, N VENEZUELA, port on the CARIBBEAN SEA, and 55 mi/89 km E of CARACAS; 10°29′N 66°06′W. Linked by railroad with RÍO CHICO and EL GUAPO. Port in cacao-growing region; exports hides, coffee, fruit.

Higueruela (ee-gai-RWAI-lah), town, ALBACETE province, SE central SPAIN, 23 mi/37 km E of ALBACETE. Wine, saffron, cereals.

Higuito River, HONDURAS: see JICATUYO RIVER.

Hihya (hee-HYAH), township, SHARQIYA province, Lower EGYPT, on the BAHR MUWEIS (stream), and 8 mi/12.9 km NE of ZAGAZIG. Cotton.

Hiipina, RUSSIA: see KIROVSK, MURMANSK oblast.

Hiisijärvi (HEE-si-YAR-vee), village, OULUN province, E central FINLAND, 30 mi/48 km ENE of KAJAANI; elevation 578 ft/175 m; 64°22′N 28°39′E. Iron deposits.

Hiiumaa (HEE-oo-muh), German Dagden, Russian Dago, Swedish Dagö, (□ 373 sq mi/969.8 sq km), 2nd-largest island (□ 373 sq mi/966 sq km) of ESTONIA, in BALTIC SEA, N of SAAREMAA island (separated by SOELA SOUND), 14 mi/23 km off mainland; 37 mi/60 km long, 28 mi/45 km wide. Level terrain; poor sandy soils on limestone base. Some agricultural, livestock raising (cattle, sheep); fisheries. Cotton-textile manufacturing at KARDLA, its main town and port. Ruled after 1200 by Livonian Knights; passed 1561 to SWEDEN; occupied 1710 by RUSSIA.

Híjar (EE-hahr), town, TERUEL province, E SPAIN, 18 mi/29 km WNW of ALCAÑIZ. Olive oil processing, manufacturing of candy and chocolate. Fruit, wheat; sheep in area. Irrigation reservoir nearby. Coal mines.

Hijazah, EGYPT: see HIGAZA.

Hiji (hee-JEE), town, Hayami district, OITA prefecture, NE KYUSHU, SW JAPAN, on SW KUNISAKI PENINSULA, on N shore of BEPPU BAY, 9 mi/15 km N of OITA; 33°21′N 131°32′E. Mandarin oranges; flatfish. Distilled spirits (shochu).

Hijikawa (hee-JEE-kah-wah), town, Kita district, EHIME prefecture, NW SHIKOKU, W JAPAN, 28 mi/45 km S of MATSUYAMA; 33°27′N 132°41′E. Shiitake mushrooms, chestnuts. Hot springs, Kanogawa lake nearby.

Hijiri, Mount (HEE-jee-ree), Japanese Hijiri-dake (hee-JEE-ree–DAH-ke) (9,936 ft/3,028 m), central HONSHU, central JAPAN, on border between SHIZUOKA and NAGANO prefectures, 30 mi/48 km SW of KOFU.

Hijuelas (ee-HWAI-lahs), town, ⊙ Hijuelas comuna, QUILLOTA province, VALPARAISO region, central CHILE, on ACONCAGUA RIVER, on PAN-AMERICAN HIGHWAY, 33 mi/53 km NE of VALPARAISO; 32°48′S 71°10′W. Agricultural center (fruit, grapes, cereals; livestock).

Hikami (hee-KAH-mee), town, Hikami district, HYOGO prefecture, S HONSHU, W central JAPAN, 34 mi/55 km N of KOBE; 35°10′N 135°02′E.

Hikari (hee-KAH-ree), city, YAMAGUCHI prefecture, SW HONSHU, W JAPAN, port on SUO SEA, 31 mi/50 km E of YAMAGUCHI; 33°57′N 131°56′E. Medicines and vitamins; steel manufacturing. Formed in early 1940s by combining adjacent towns of Hikari and Murozumi.

Hikari (hee-KAH-ree), town, Sosa district, CHIBA prefecture, E central HONSHU, E central JAPAN, 16 mi/25 km E of CHIBA; 35°39′N 140°30′E.

Hikata (hee-KAH-tah), town, Katori district, CHIBA prefecture, E central HONSHU, E central JAPAN, 22 mi/35 km E of CHIBA; 35°45′N 140°37′E. Pigs; vegetables, rice. One of the world's first agricultural cooperatives was set up here in 1838.

Hikawa (hee-KAH-wah), town, Hikawa district, SHIMANE prefecture, SW HONSHU, W JAPAN, 12 mi/20 km S of MATSUE; 35°23′N 132°50′E.

Hiketa (hee-KE-tah), town, Okawa district, KAGAWA prefecture, NE SHIKOKU, W JAPAN, on HARIMA SEA, 22 mi/35 km S of TAKAMATSU; 34°13′N 134°24′E. Fishing port (young yellowtail).

Hikigawa (hee-kee-GAH-wah), town, West Muro district, WAKAYAMA prefecture, S HONSHU, W central JAPAN, 48 mi/78 km S of WAKAYAMA; 33°33′N 135°26′E.

Hikimi (HEE-kee-mee), town, Mino district, SHIMANE prefecture, SW HONSHU, W JAPAN, 86 mi/138 km S of MATSUE; 34°34′N 132°01′E. Daikon; vegetable processing.

Hikkaduwa (HIK-kah-DOO-wuh), town, SOUTHERN PROVINCE, SRI LANKA, on SW coast, 10 mi/16 km NNW of GALLE; 06°08′N 80°06′E. Major tourist resort with surfing and a view of Sri Lanka's best-known coral gardens. Trades in batik, masks, jewelry; vegetables, rice.

Hikone (hee-KO-ne), city, SHIGA prefecture, S HONSHU, central JAPAN, on E shore of BIWA LAKE, 31 mi/50 km N of OTSU; 35°16′N 136°15′E. Buddhist altars; valves; cosmetics. Ruins of Hikone Castle.

Hiko-shima, island, JAPAN: see SHIMONOSEKI STRAIT.

Hikueru (hee-koo-E-roo), atoll, central TUAMOTU ARCHIPELAGO, FRENCH POLYNESIA, S PACIFIC; 17°36′S 142°40′W. Pearl shell. Formerly Melville Island.

Hiland Park (HEI-luhnd), town (□ 5 sq mi/13 sq km; 2000 population 999), BAY county, NW FLORIDA, 5 mi/8 km NE of PANAMA CITY; 30°11′N 85°37′W.

Hilbert, town (2006 population 1,071), CALUMET county, E WISCONSIN, 14 mi/23 km SE of APPLETON; 44°08′N 88°09′W. Manufacturing (cheese, animal feed). Railroad junction.

Hilchenbach (HIL-khen-bahkh), town, NORTH RHINE–WESTPHALIA, W GERMANY, 9 mi/14.5 km NE of SIEGEN; 51°00′N 08°07′E. Manufacturing of motors, plastics; iron- and leatherworking.

Hilda, village (2006 population 429), BARNWELL county, W SOUTH CAROLINA, c.7 mi/11.3 km ENE of BARNWELL; 33°16′N 81°15′W. Agriculture includes livestock; dairying; grain, cotton, peanuts.

Hilda (HIL-duh), hamlet, SE ALBERTA, W CANADA, 42 mi/67 km from REDCLIFF, in CYPRESS County; 50°28′N 110°03′W.

Hildale, town (2006 population 1,950), WASHINGTON county, SW UTAH, 32 mi/51 km ESE of SAINT GEORGE, on ARIZONA state line adjacent to COLORADO CITY (Arizona); 37°00′N 112°58′W. Manufacturing (cabinets). Cattle. ZION NATIONAL PARK to N. Cottonwood Point Wilderness Area, Kaibab Indian Reservation, and PIPE SPRING NATIONAL MONUMENT (all in Arizona) to SE.

Hildasay (HIL-dah-sai), uninhabited island (1 mi/1.6 km long) of the SHETLAND ISLANDS, extreme N Scotland, off SW coast of Mainland island, in the Deeps (bay), and 3 mi/4.8 km W of SCALLOWAY; 60°09′N 01°19′W. Previously granite quarrying.

Hildburghausen (hild-boorg-HOU-suhn), town, THURINGIA, central GERMANY, on the WERRA RIVER, and 15 mi/24 km NW of COBURG; 50°26′N 10°44′E. Printing, wood- and metalworking, food processing. Has town hall, rebuilt 1572; former ducal palace (17th century), now museum; 18th-century church. Chartered 1324, town was capital of duchy of Saxe-Hildburghausen (1683–1826), then passed to Saxe-Meiningen. Bibliographic Institute, founded here 1828, was moved to LEIPZIG in 1874.

Hildebran (HIL-duh-bran), town (□ 2 sq mi/5.2 sq km; 2006 population 1,852), BURKE county, W central NORTH CAROLINA, 4 mi/6.4 km WSW of HICKORY; 35°43′N 81°25′W. Manufacturing (apparel, lumber, furniture, wood products, fabricated metal products); agriculture (soybeans; poultry).

Hilden (HIL-den), city, NORTH RHINE–WESTPHALIA, W GERMANY, 7 mi/11.3 km SE of DÜSSELDORF (linked by tramway); 51°10′N 06°55′E. SOLINGEN steelworks; manufacturing of machinery, chemicals, plastics. Has 12th-century late-Romanesque church. Chartered 1861.

Hildenborough, ENGLAND: see TONBRIDGE AND MALLING.

Hilders (HIL-derz), village, HESSE, central GERMANY, 17 mi/27 km E of FULDA; 50°35′N 10°00′E. Ruined castle and many half-timbered houses.

Hildesheim (HIL-des-heim), city, LOWER SAXONY, N central GERMANY; 52°09′N 09°58′E. Industrial and transportation center; manufacturing includes stoves, radio and television sets, and agricultural and dairy machinery. In 815, Emperor Louis I made Hildesheim the seat of a bishopric; Hildesheim's bishops later became territorial princes of the Holy Roman Empire. Received a charter in 1249; soon afterwards joined the HANSEATIC LEAGUE. The bishopric was secularized at the beginning of the 19th century; passed (1813) to HANOVER, and passed (1866), with Hanover, to PRUSSIA. Almost all of Hildesheim's old buildings were badly damaged in World War II, but many have been restored. Noted buildings, all Romanesque in style, are the cathedral (11th century), the Church of St. Michael (11th–12th century), and the Church of St. Godehard (12th century). Site of museum of ancient Egyptian and Greco-Roman artifacts.

Hildreth, village (2006 population 344), FRANKLIN county, S NEBRASKA, 23 mi/37 km S of KEARNEY; 40°20′N 99°02′W. Railroad terminus. Dairying; grain; livestock.

Hilgard, Mount, peak (11,533 ft/3,515 m) in FISH LAKE PLATEAU, SEVIER county, S central UTAH, 30 mi/48 km ESE of RICHFIELD, in Fishlake National Forest.

Hilham (HI-luhm), town, OVERTON county, N TENNESSEE, 30 mi/48 km N of COOKEVILLE; 36°24′N 85°26′W.

Hili, INDIA: see HILLI.

Hilibandar, INDIA: see HILLI.

Hill, county (2,916 sq mi/7,552 sq km; 1990 population 17,654; 2000 population16,673), N MONTANA, on Canadian (SASKATCHEWAN and ALBERTA) border; ⊙ HAVRE; 48°38′N 110°07′W. Agricultural area drained by MILK River, forms FRESNO RESERVOIR in center of county. Drained also by Lodge, Big Sandy, and Sage creeks. Cattle, sheep, hogs; wheat, barley, oats, and hay. Part of Rocky Boy's Indian Reservation in SE. Formed 1912.

Hill, county (□ 985 sq mi/2,561 sq km; 2006 population 35,806), N central TEXAS; ⊙ HILLSBORO; 31°59′N 97°07′W. Rich blackland prairies; bounded W by BRAZOS RIVER; drained by Aquilla and Nolan creeks. Agriculture (especially cotton; also corn, wheat, sorghum; hay); cattle, horses, hogs; limestone; oil and gas. Manufacturing at Hillsboro. In W is Whitney Dam in Brazos River. Aquilla Lake reservoir in S center; Lake Whitney reservoir and State Park on W border (Brazos River). Formed 1853.

Hill, town, MERRIMACK county, S central NEW HAMPSHIRE, 5 mi/8 km NW of FRANKLIN; 43°31′N 71°46′W. Bounded E by PEMIGEWASSET RIVER (FRANKLIN FALLS RESERVOIR). Agriculture (poultry, livestock; apples, vegetables; dairying; nursery crops); manufacturing (machining). Flood-control dam on Pemigewasset River caused village to be moved to new site and rebuilt as model town, 1940–1941. Original town settled 1768.

Hill 112, CALVADOS department, BASSE-NORMANDIE region, NW FRANCE, 6 mi/9.7 km SW of CAEN, between ODON (W) and ORNE (E) rivers. Scene of heavy fighting in battle for Caen (July 1944), in NORMANDY campaign, World War II. Monument.

Hill 204, AISNE department, PICARDIE region, N FRANCE, 2 mi/3.2 km W of CHÂTEAU-THIERRY. Site of Aisne-Marne American Memorial commemorating successful Franco-American defense (June 1918) of road to PARIS.

Hill 609, BIZERTE province, N TUNISIA, 13 mi/21 km SW of MATIR. During World War II, it was hotly contested in last phase of the Tunisian campaign, and finally captured by the Americans on May 1, 1943. Also called Jabal Tahent.

Hill 70, height between LENS and LOOS, PAS-DE-CALAIS department, NORD PAS-DE-CALAIS region, N FRANCE. During World War I, it was attacked with heavy losses by British in 1915 battle of Loos, and carried by them (1917) in battle of Lens.

Hilla (HEE-luh), town, ⊙ BABYLON province, central IRAQ, on the SHATT HILLA (canal branching off the EUPHRATES RIVER), on railroad, and 60 mi/97 km S of BAGHDAD. Dates. Largely built of bricks from the ruins of Babylon, just NNE.

Hilla, SAUDI ARABIA: see HAWTAH.

Hillaby, Mount, hill (1,104 ft/336 m), highest in BARBADOS, 7 mi/11.3 km NNE of BRIDGETOWN. Potteries are nearby.

Hilla, El (HIL-luh, el), village, QENA province, Upper EGYPT, on E bank of the NILE River, opposite ISNA. Pottery making; cereals, sugarcane, dates. Sometimes spelled El Hella. On site of ancient Contra Latopolis.

Hill Air Force Base, UTAH: see OGDEN.

Hillary Coast, on the W edge of the ROSS ICE SHELF, ANTARCTICA, between MINNA BLUFF and Cape Selborne; 79°20′S 161°00′E.

Hilla, Shatt (HEE-luh, SHAHT), canal, c.120 mi/193 km long, central IRAQ, using largely, ancient deserted bed of the EUPHRATES RIVER; branches off the Euphrates below AL MUSAIYIB; flows SE past HILLA, DIWANIYA, and RUMAITHA, to SAMAWA, where it unites with the SHATT HINDIYA branch to reform the Euphrates. Irrigates BABYLON and QADISSIYA provinces. The Hindiya barrage diverts water from the main course of the Euphrates into the Hilla canal.

Hillburn, village (□ 2 sq mi/5.2 sq km; 2006 population 900), ROCKLAND county, SE NEW YORK, in the RAMAPOS, at NEW JERSEY state line, just W of SUFFERN; 41°07′N 74°10′W. In resort area. Incorporated 1893.

Hill City, town (2000 population 1,604), ⊙ GRAHAM county, NW KANSAS, on South Fork SOLOMON RIVER, and 45 mi/72 km NW of HAYS; 39°22′N 99°50′W. Wheat. Oil museum. Laid out 1880, incorporated 1888.

Hill City, village (2000 population 479), AITKIN county, N central MINNESOTA, 32 mi/51 km N of AITKIN, on W shore of Hill Lake, near Hill River; 46°59′N 93°35′W. Alfalfa, hay, wild rice; manufacturing (machinery, building materials). Hill River State Forest to W, S, and E, including Moose-Willow Wildlife Area to SE.

Hill City, village (2006 population 871), PENNINGTON county, SW SOUTH DAKOTA, 20 mi/32 km SW of RAPID CITY, in Black Hills National Forest; 43°55′N

103°34′W. Elevation 4,976 ft/1,517 m. Agriculture; gold, tungsten mines; timber. Nearby Sheridan Lake used for recreation. HARNEY PEAK (7,242 ft/2,207 m) to S.

Hill Country Village, town (2006 population 1,076), BEXAR county, S central TEXAS, residential suburb 10 mi/16 km N of downtown SAN ANTONIO; 29°34′N 98°29′W.

Hillcrest, village (2000 population 1,158), OGLE county, N central ILLINOIS, 2 mi/3.2 km N of ROCHELLE; 41°57′N 89°04′W. Agricultural area.

Hillcrest, unincorporated residential suburb (□ 1 sq mi/ 2.6 sq km; 2000 population 7,106), ROCKLAND county, S NEW YORK; 41°07′N 74°02′W. Part of dense cluster of residential commuter communities off PALISADES INTERSTATE PARKWAY at E foot of RAMAPO MOUNTAINS.

Hillcrest Center, residential suburb, KERN county, central CALIFORNIA, 2 mi/3.2 km E of downtown BAKERSFIELD. In irrigated agricultural area.

Hillcrest Heights, unincorporated residential town (2000 population 16,359), PRINCE GEORGES county, W central MARYLAND, suburb of WASHINGTON, D.C.; 38°50′N 76°58′W. In a fast-growing, moderate-income area.

Hillcrest Village, village (2006 population 720), BRAZORIA county, SE TEXAS, residential suburb 24 mi/39 km SSE of HOUSTON, and 2 mi/3.2 km SSE of ALVIN, on Mustang Bayou; 29°23′N 95°13′W.

Hille (HIL-le), town, NORTH RHINE–WESTPHALIA, N GERMANY, 8 mi/12.9 km WNW of MINDEN; 52°21′N 08°45′E. A port on MIDLAND CANAL. Meat processing; also spa with sulfur springs.

Hille (HIL-le), residential village, GÄVLEBORG county, E SWEDEN, 4 mi/6.4 km NNE of GÄVLE; 60°44′N 17°10′E.

Hillegersberg (HI-li-khuhrz-BERKH), district of ROTTERDAM, SOUTH HOLLAND province, W NETHERLANDS, 3 mi/4.8 km N of city center, on Rotte River; 51°57′N 04°32′E.

Hillegom (HI-luh-khawm), town (2001 population 18,419), SOUTH HOLLAND province, W NETHERLANDS, 7 mi/11.3 km SSW of HAARLEM, 5 mi/8 km E of NORTH SEA, and on the RINGVAART canal; 52°17′N 04°35′E. Important flower-bulb center; dairying; cattle raising; vegetables, fruit; manufacturing (medical systems).

Hill End (HIL END), town, E central NEW SOUTH WALES, AUSTRALIA, 120 mi/193 km NW of SYDNEY; 33°02′S 149°25′E. Old gold-mining center.

Hill End State Historic Site (HIL END), E central NEW SOUTH WALES, SE AUSTRALIA, 30 mi/48 km NNW of BATHURST; covers 0.5 sq mi/1.3 sq km (319 acres/129 ha). The SOFALA goldfields brought c.60,000 people to area in 1870s. Many restored buildings. Visitor center at old hospital building. Gold fossicking (hunting); tours, camping.

Hiller, unincorporated town (2000 population 1,234), Redstone township, FAYETTE county, SW PENNSYLVANIA, 30 mi/48 km S of PITTSBURGH, and 1 mi/1.6 km SW of BROWNSVILLE, near MONONGAHELA RIVER. Agricultural area (corn, hay; dairying).

Hillerød (HIL-uh-rudth), city (2000 population 27,675), ⊙ FREDERIKSBORG county, E DENMARK; 55°56′N 12°18′E. Industrial and tourist center. The city developed around the famous Renaissance-style FREDERIKSBORG CASTLE (1602–1620), which was once a royal residence and today houses the national museum. From 1640 to 1840, Danish kings were crowned in the castle church.

Hillers, Mount, 10,723 ft/3,268 m, E GARFIELD county, S UTAH, in HENRY MOUNTAINS, 32 mi/51 km S of HANKSVILLE.

Hillesøy, NORWAY: see KVALØYA.

Hillgrove (HIL-grov), town, NEW SOUTH WALES, SE AUSTRALIA, 346 mi/557 km NNE of SYDNEY, 20 mi/32 km E of ARMIDALE; 30°34′S 151°54′E. Old gold-mining town.

Hilli (HIL-lee), or **Hili**, town, WEST DINAJPUR district, N WEST BENGAL state, E INDIA, on JAMUNA RIVER (left

bank tributary of the ATRAI RIVER), and 60 mi/97 km ENE of Ingraj Bazar. Major rice-milling center; rice, jute, sugarcane, rape, and mustard. Also called HILIBANDAR.

Hilliard (HIL-yuhrd), city (□ 11 sq mi/28.6 sq km; 2006 population 26,812), FRANKLIN county, central OHIO; suburb 10 mi/16 km WNW of COLUMBUS; 40°02′N 83°08′W. In agricultural area.

Hilliard (HIL-yuhrd), township (□ 35 sq mi/91 sq km; 2001 population 241), E central ONTARIO, E central CANADA, 14 mi/23 km from NEW LISKEARD; 47°42′N 79°41′W.

Hilliard (HIL-yuhrd), town (□ 3 sq mi/7.8 sq km; 2000 population 2,702), NASSAU county, extreme NE FLORIDA, 28 mi/45 km NW of JACKSONVILLE; 30°41′N 81°55′W.

Hilliard (HIL-yuhrd), hamlet, E central ALBERTA, W CANADA, 21 mi/33 km from VEGREVILLE, in LAMONT County; 53°39′N 112°29′W.

Hillingdon (HIL-ing-DUN), outer borough (□ 45 sq mi/117 sq km; 2001 population 243,006) of GREATER LONDON, SE ENGLAND; 51°30′N 00°25′W. Printing, motion-picture production, and manufacturing (aircraft, food products, and electrical and musical instruments). Hillingdon contains a portion of London (HEATHROW) Airport and part of the Grand Union Canal. Within the borough, UXBRIDGE was an ancient market town on the COLNE RIVER. Charles I and Parliament representatives negotiated unsuccessfully in the "Treaty House" in Uxbridge in 1645. Uxbridge houses Brunel University. The parish church in Hayes was used by St. Anselm, archbishop of Canterbury. The other districts of the borough include Cowley, Eastcote, Northwood, Harefield, Harlington, Harmondsworth, Ruislip, West Drayton, Ickenham, Yiewsley, and Uxbridge.

Hillion (eel-YON), commune (□ 9 sq mi/23.4 sq km), CÔTES-D'ARMOR department, in BRITTANY, W FRANCE, 4 mi/ 6 km E of SAINT-BRIEUC, on the Bay of SAINT-BRIEUC; 48°31′N 02°41′W.

Hill Island (HIL), SE ONTARIO, E central CANADA, on U.S. (NEW YORK) border, in the SAINT LAWRENCE RIVER, in THOUSAND ISLANDS, 25 mi/40 km ENE of KINGSTON; connected with WELLESLEY ISLAND (New York), and with Ontario mainland by THOUSAND ISLANDS INTERNATIONAL BRIDGE; 44°21′N 75°57′W.

Hillman, village (2000 population 685), MONTMORENCY county, N MICHIGAN, 22 mi/35 km W of ALPENA, and on THUNDER BAY RIVER; 45°04′N 83°54′W. Manufacturing (wire cloth and filters).

Hillman, village (2000 population 29), MORRISON county, central MINNESOTA, 24 mi/39 km E of LITTLE FALLS, near Hillman Creek; 46°00′N 93°53′W. Dairying. Incorporated 1938.

Hill Military Reservation, Virginia: see FREDERICKSBURG.

Hillrose, village (2000 population 254), MORGAN county, NE COLORADO, on SOUTH PLATTE RIVER, at mouth of BEAVER CREEK, and 16 mi/26 km ENE of FORT MORGAN; elevation c.4,165 ft/1,269 m; 40°19′N 103°31′W. Railroad junction to N. Sand Hills nearby. Prewitt Reservoir to NE.

Hills, town (2000 population 679), JOHNSON county, E IOWA, 7 mi/11.3 km S of IOWA CITY, near IOWA RIVER; 41°34′N 91°32′W.

Hills, village (2000 population 565), ROCK county, extreme SW MINNESOTA, on IOWA (S) and SOUTH DAKOTA (W) state lines, 19 mi/31 km W of SIOUX FALLS (SOUTH DAKOTA), and 12 mi/19 km SW of LUVERNE on Mud Creek; 43°31′N 96°21′W. Corn, oats, soybeans; hogs, cattle; dairying; manufacturing (fertilizer). Railroad junction to W.

Hills and Dales, village (2006 population 265), STARK county, E central OHIO, 4 mi/6 km NW of CANTON; 40°49′N 81°26′W.

Hillsboro, city (2000 population 4,359), ⊙ MONTGOMERY county, S central ILLINOIS, 8 mi/12.9 km E

of LITCHFIELD; 39°09′N 89°28′W. In agricultural area (soybeans, corn, wheat, hay; cattle); manufacturing (zinc oxide, glass products). Incorporated 1855. Several recreational reservoirs in area include Lake Glenn Shoals to N, Lake Lou Yaeger to W, and Coffeen Lake to SE.

Hillsboro (HILZ-buh-ruh), city (□ 5 sq mi/13 sq km; 2006 population 6,694), ⊙ HIGHLAND county, SW OHIO, 34 mi/55 km WSW of CHILLICOTHE; 39°12′N 83°37′W. Trade center for livestock-raising, farming, and limestone-quarrying area; manufacturing. Women's Temperance Crusade was founded here 1873. Several mound builders' forts nearby. Plotted 1807.

Hillsboro, city (2006 population 87,732), ⊙ WASHINGTON county, NW OREGON, suburb 14 mi/23 km W of PORTLAND, on the TUALATIN RIVER, at mouth of Dairy Creek; 45°31′N 122°56′W. Railroad junction. Manufacturing (foods, wood products, furniture, plastics, sheet metal, electronic goods, computer equipment, harnesses, quartz crucibles, medical equipment, consumer goods; meat processing, printing, publishing). Agriculture (fruit, nuts, vegetables, grain; poultry, cattle); dairy products; wineries. Points of interest include pioneer museum and cemetery, notable old Scottish church. Portland Hillsboro Airport to NE. Bald Peak State Park to S. Settled c. 1845.

Hillsboro, town (2000 population 608), Lawrence co., NW Alabama, near Wheeler Reservoir (in Tennessee River), 12 mi/19 km WNW of Decatur. Inc. in 1899.

Hillsboro, town (2000 population 489), FOUNTAIN county, W INDIANA, 13 mi/21 km ESE of COVINGTON; 40°07′N 87°10′W. In agricultural area.

Hillsboro, town (2000 population 205), HENRY county, SE IOWA, 12 mi/19 km SW of MOUNT PLEASANT; 40°50′N 91°42′W. Limestone quarries.

Hillsboro, town (2000 population 2,854), MARION county, central KANSAS, 10 mi/16 km W of MARION, near COTTONWOOD RIVER; 38°21′N 97°12′W. In agricultural area (winter wheat, corn; poultry, livestock); bookbinding. Manufacturing (dairy products, feeds, transportation equipment). Seat of Tabor College. Marion Lake Reservoir to NE. Mennonite Heritage Center to SW at Goessel. Founded 1879, incorporated 1884.

Hillsboro, town (2000 population 163), CAROLINE county, E MARYLAND, on TUCKAHOE CREEK, and 6 mi/ 9.7 km WNW of DENTON; 38°55′N 75°56′W. Named for Lord Hillsboro, a relative of Lord Baltimore, it was previously called Tuckahoe Bridge because of the span over the river here. Charles Wilson Peale, a noted Maryland artist, used to live here. A stone monument marks the site of an early Methodist chapel.

Hillsboro, town (2000 population 1,675), ⊙ JEFFERSON county, E MISSOURI, 35 mi/56 km SW of ST. LOUIS; 38°13′N 90°34′W. Light manufacturing Semi-urban development in area; satellite community of St. Louis. Seat of Jefferson College. Plotted 1839.

Hillsboro, or **Hillsborough**, town, HILLSBOROUGH county, S NEW HAMPSHIRE, 18 mi/29 km WSW of CONCORD. Drained by CONTOOCOOK RIVER and Beards and Sand brooks. Manufacturing (transportation equipment, grinding wheels; printing and publishing); agriculture (vegetables, apples, nursery crops; livestock, poultry; dairying). President Franklin Pierce Homestead, in SW, was built 1804 by president's father, restored 1925. Fox Forest in E; part of Low State Forest in N; part of Franklin Pierce Lake in SW. Settled 1741, incorporated 1772.

Hillsboro, town (2006 population 1,508), ⊙ TRAILL CO., E NORTH DAKOTA, 39 mi/63 km N of FARGO, and on GOOSE RIVER; 47°23′N 97°03′W. Grain, potatoes; livestock. Founded in 1880 as Comstock and then renamed Hill City in honor of James J. Hill, president of the Great Northern Railroad and then named Hillsboro. Became the county seat in 1890.

Hillsboro, town (2006 population 9,064), ⊙ HILL county, N central TEXAS, 32 mi/51 km N of WACO; elevation 634 ft/193 m; 32°00'N 97°07'W. Railroad junction in rich blackland agriculture area (cotton, corn, sorghum; also cattle, horses); manufacturing (apparel, fabricated metal products, furniture, building materials); gas and oil. Seat of Hill Junior College Aquilla Lake reservoir to SW; Confederate Research Center and Gun Museum. Incorporated 1853.

Hillsboro (HILZ-buh-ro), town (2006 population 130), LOUDOUN county, N VIRGINIA, 10 mi/16 km NW of LEESBURG; 39°12'N 77°43'W. Agriculture (livestock; grain, apples).

Hillsboro, town (2006 population 1,286), VERNON county, SW WISCONSIN, on tributary of BARABOO RIVER, and 45 mi/72 km ESE of LA CROSSE; 43°38'N 90°40'W. In timber and farm area. Dairy products, flour, lumber; manufacturing (textiles, footwear). Hospital and airfield here. Settled 1854; incorporated as village in 1885, as city in 1939.

Hillsboro, unincorporated village, FLEMING county, NE KENTUCKY, 13 mi/21 km NW of Moreland. In agricultural area (tobacco, grain; livestock). Hillsboro Covered Bridge is here.

Hillsboro, village, SIERRA county, SW NEW MEXICO, in SE foothills of MIMBRES MOUNTAINS, extension of BLACK RANGE, near RIO GRANDE, 24 mi/39 km SW of TRUTH OR CONSEQUENCES; elevation 5,236 ft/1,596 m. Cattle, sheep. Once a busy silver-mining point with population (1886) of 5,000. Gila National Forest to W.

Hillsboro, village (2006 population 231), POCAHONTAS county, E WEST VIRGINIA, 9 mi/14.5 km SW of MARLINTON, near GREENBRIER RIVER; 38°08'N 80°12'W. DROOP MOUNTAIN BATTLEFIELD STATE PARK is 3 mi/4.8 km SW, at Droop village. Pearl Buck Museum (author, 1892–1973). Part of Monongahela National Forest to NW and SE; Watoga State Park to SE; Greenbrier River State Trail passes to E.

Hillsboro Beach (HILZ-buh-ruh), town (□ 1 sq mi/2.6 sq km; 2000 population 2,163), BROWARD county, SE FLORIDA, 2 mi/3.2 km N of POMPANO BEACH; 26°16'N 80°04'W.

Hillsboro River (HILZ-buh-ruh), c.55 mi/89 km long, central and W central FLORIDA; rises N of LAKELAND; flows NW and then SW to Hillsboro Bay (an arm of TAMPA BAY) at TAMPA.

Hillsborough (HILZ-buh-ruh), county (□ 1,266 sq mi/3,291.6 sq km; 2006 population 1,157,738), W central FLORIDA, on Gulf coast and bounded S and partly W by TAMPA BAY; ⊙ TAMPA; 27°54'N 82°20'W. Rolling and level terrain with many small lakes; drained by HILLSBORO RIVER. Agriculture (citrus fruit, tomatoes, strawberries, peanuts, corn); dairying, poultry raising; fishing. Manufacturing (food, tobacco, and wood products). Massive urban growth since 1975 as metropolitan TAMPA has expanded rapidly. Formed 1834.

Hillsborough (HILZ-buh-ro), county (□ 892 sq mi/2,319.2 sq km; 2006 population 402,789), S NEW HAMPSHIRE, on MASSACHUSETTS (S) state line; ⊙ NASHUA; 42°54'N 71°43'W. Manufacturing center of New Hampshire, one of leading industrial counties of U.S., with manufacturing at MANCHESTER, Nashua, BEDFORD, HUDSON, MERRIMACK, and MILFORD. Granite quarrying; timber; agriculture (vegetables, corn, beans, hay, apples; cattle, hogs, poultry; dairying). Resorts. Drained by CONTOOCOOK, PISCATAQUOG, SOUHEGAN, MERRIMACK, and NASHUA rivers. Franklin Pierce Homestead in NW, at HILLSBOROUGH; SILVER LAKE State Park in SE; Miller and Greenfield state parks and Sheiling Forest in W center; Clough State Park in NE; Fox Forest and part of Low State Park in NW. Formed 1769.

Hillsborough, city (2000 population 10,825), SAN MATEO county, W CALIFORNIA; residential suburb 14 mi/23 km S of downtown SAN FRANCISCO, on SAN FRANCISCO BAY. Software development. San Francisco International Airport to NW. Coyote Point to E; San Francisco State Fish and Game Refuge to SW, including SAN ANDREAS FAULT (Lower Crystal Springs Reservoir). Incorporated 1910.

Hillsborough (HILZ-buh-ruh), town (2001 population 3,311), N DOWN, SE Northern Ireland, 11 mi/18 km SW of BELFAST; 54°28'N 06°06'W. Market. Hillsborough Castle is official residence of governor of Northern Ireland. Fort dates from 17th century and church was built 1774.

Hillsborough, township, SOMERSET county, N NEW JERSEY, 5 mi/8 km S of SOMERVILLE; 40°30'N 74°40'W. Brick manufacturing. Incorporated 1798.

Hillsborough (HILZ-buhr-o), town (□ 4 sq mi/10.4 sq km; 2006 population 5,403), ⊙ ORANGE county, N central NORTH CAROLINA, 11 mi/18 km NW of DURHAM, on Eno River; 36°04'N 79°05'W. Service industries; manufacturing (machinery, plastic products, medical equipment, paper goods, packaging, salon equipment, power supplies); agriculture (tobacco, grain, soybeans; livestock; dairying). Eno River State Park to E. An early provincial capital of North Carolina; in 1768, scene of disturbances by the Regulators. The artist, Thomas Hart Benton born here. Settled before 1700; plotted 1754.

Hillsborough, town, W CARRIACOU island, GRENADINES, GRENADA, WEST INDIES, 35 mi/56 km NE of ST. GEORGE'S, Grenada; 12°28'N 61°27'W. Administrative and commercial center. Has fine harbor. Cotton, limes.

Hillsborough (HILZ-bro), village (2001 population 1,288), SE NEW BRUNSWICK, E CANADA, on PETITCODIAC RIVER estuary, and 13 mi/21 km SE of MONCTON; 45°56'N 64°39'W. Ships oil shale, albertite, manganese, coal, gypsum from Albert Mines. Fishing (cod, lobster, clams, oysters). Settled 1765.

Hillsborough Bay (HILZ-bro), inlet of NORTHUMBERLAND STRAIT, S central PRINCE EDWARD ISLAND, E CANADA; 10 mi/16 km long, 10 mi/16 km wide at entrance. Receives HILLSBOROUGH River at its head at CHARLOTTETOWN. SAINT PETERS Island is in entrance of bay.

Hillsborough Dam, lake, TOBAGO, TRINIDAD AND TOBAGO, in S central area. Supplies drinking water for the island

Hillsborough River (HILZ-buh-ruh), narrow lagoon, c.18 mi/29 km long, in VOLUSIA county, E central FLORIDA, sheltered from the ATLANTIC OCEAN by barrier beach; extends S from PONCE DE LEON INLET (at S end of HALIFAX RIVER lagoon) to N end of MOSQUITO LAGOON. Followed by INTRACOASTAL WATERWAY.

Hillsborough River (HILZ-bro), 30 mi/48 km long, central PRINCE EDWARD Island, E CANADA; rises 18 mi/29 km ENE of CHARLOTTETOWN; flows W and SW, past Charlottetown, to HILLSBOROUGH BAY at Charlottetown.

Hillsburg (HILZ-buhrg), rural municipality (□ 254 sq mi/660.4 sq km; 2001 population 470), SW MANITOBA, W central CANADA, NW of BRANDON; 51°20'N 101°00'W. Agriculture. Includes the communities of SHORTDALE, SHEVLIN, and BIELD. Incorporated 1913.

Hillsburgh (HILZ-buhrg), unincorporated village, S ONTARIO, E central CANADA, 18 mi/29 km NNE of GUELPH, and included in town of ERIN; 43°47'N 80°08'W. Dairying, mixed farming.

Hills Creek Lake, reservoir (□ 4 sq mi/10.4 sq km), LANE county, W central OREGON, on Middle Fork of WILLAMETTE RIVER, in Willamette National Forest, on W slope of CASCADE RANGE, 40 mi/64 km SE of EUGENE; 43°25'N 122°28'W. Maximum capacity 356,000 acre-ft. Formed by Hills Creek Dam (341 ft/104 m high), built (1962) by Army Corps of Engineers and U.S. Department of Agriculture Forest Service for flood control; also used for power generation, irrigation, recreation, and navigation.

Hillsdale, county (□ 607 sq mi/1,578.2 sq km; 2006 population 47,206), S MICHIGAN, on OHIO (S) and INDIANA (SW) state lines; ⊙ HILLSDALE; 41°53'N 84°35'W. Drained by headstreams of the KALAMAZOO RIVER and by SAINT JOSEPH RIVER. Agriculture (forage crops, corn, wheat, oats, apples, soybeans; cattle, hogs, sheep; dairy products). Manufacturing (consumer goods, plastic products, bakery products, playground equipment, transportation equipment; machining) at Hillsdale, JONESVILLE, and READING. Many small lakes. Organized 1835.

Hillsdale, town (2000 population 8,233), ⊙ HILLSDALE county, S MICHIGAN, 25 mi/40 km SSW of JACKSON, and on SAINT JOSEPH RIVER; 41°55'N 84°38'W. Trade and manufacturing center (transportation equipment, textiles; food processing). Seat of Hillsdale College, with 60-acre/24-ha Slayton Arboretum. Native American mounds nearby. Settled 1834; incorporated as village 1847, as city 1869.

Hillsdale, town (2000 population 1,477), SAINT LOUIS county, E MISSOURI, residential suburb 6 mi/9.7 km NW of downtown ST. LOUIS; 38°41'N 90°17'W.

Hillsdale (HILZ-dail), unincorporated village, S ONTARIO, E central CANADA, 17 mi/27 km W of ORILLIA, and included in SPRINGWATER; 44°34'N 79°45'W. Dairying, mixed farming.

Hillsdale, village (2000 population 588), ROCK ISLAND county, NW ILLINOIS, 19 mi/31 km ENE of MOLINE, on ROCK RIVER; 41°36'N 90°10'W. Corn, soybeans; cattle, hogs; dairying; manufacturing (firearms, wood products), meatpacking. Ancient MISSISSIPPI RIVER flowed across this site, toward SE.

Hillsdale, village, VERMILLION county, W INDIANA, 8 mi/12.9 km N of CLINTON, on WABASH RIVER. Corn, wheat; cattle.

Hillsdale, village, Miami county, E KANSAS, 7 mi/11.3 km N of PAOLA. Cattle raising, farming. Hillsdale Lake Reservation to W.

Hillsdale, village (2006 population 96), GARFIELD county, N OKLAHOMA, 13 mi/21 km NW of ENID; 36°33'N 97°59'W. In agricultural area.

Hillsdale, borough (2006 population 10,053), BERGEN county, NE NEW JERSEY, 8 mi/12.9 km N of HACKENSACK; 41°00'N 74°02'W. Primarily residential. Incorporated 1923.

Hillsdale, hamlet, COLUMBIA county, SE NEW YORK, near MASSACHUSETTS state line, 8 mi/12.9 km W of GREAT BARRINGTON (Massachesetts); 42°13'N 73°32'W. Skiing at nearby Catamount Mountain in the BERKSHIRES.

Hillsgrove, neighborhood, in WARWICK city, KENT county, E central RHODE ISLAND, 7 mi/11.3 km SSW of PROVIDENCE. Theodore F. Green State Airport (Providence) here.

Hillshire Village, village, HARRIS county, SE TEXAS, residential suburb 8 mi/12.9 km WNW of downtown HOUSTON, surrounded by city of Houston.

Hillside, township (2000 population 21,747), UNION county, NE NEW JERSEY, 2 mi/3.2 km SW of NEWARK; 40°42'N 74°13'W. Mixed industry. Incorporated 1913.

Hillside, village (2000 population 8,155), COOK county, NE ILLINOIS, W suburb of CHICAGO, 10 mi/16 km E of WHEATON; 41°52'N 87°54'W. Computer services. Incorporated 1905.

Hillside, village, PRINCE GEORGES county., central MARYLAND, ESE suburb of WASHINGTON, D.C.

Hillside Gardens, village and unincorporated suburb, JACKSON county, S MICHIGAN, 5 mi/8 km NW of JACKSON.

Hillspring (HIL-spreeng), village (2001 population 193), SW ALBERTA, W CANADA, in CARDSTON County; 49°17'N 113°38'W. Established 1961. Also written Hill Spring.

Hill States, INDIA: see PUNJAB HILL STATES.

Hillston (HIL-stuhn), town, S central NEW SOUTH WALES, SE AUSTRALIA, on LACHLAN RIVER, and 260 mi/418 km ESE of BROKEN HILL; 33°29'S 145°32'E. Sheep and agriculture center.

Hillsview, village (2006 population 3), MCPHERSON county, N SOUTH DAKOTA, 55 mi/89 km WNW of ABERDEEN; 45°40′N 99°33′W.

Hillsville (HILZ-vil), town (2006 population 2,709). ☉ CARROLL county, SW VIRGINIA, in the BLUE RIDGE, 12 mi/19 km NE of GALAX; 36°46′N 80°44′W. Manufacturing (machinery, clothing and textiles, lumber, crushed stone); agriculture (grain, cabbage, apples; cattle); dairying.

Hillswick (HILZ-wik), village, on NW coast of Mainland island, SHETLAND ISLANDS, extreme N Scotland, on N shore of ST. MAGNUS BAY, and 25 mi/40 km NNW of LERWICK; 60°28′N 01°30′W. Fishing port and tourist resort.

Hilltonia (hil-TO-nee-uh), city, SCREVEN county, E GEORGIA, 10 mi/16 km NNW of SYLVANIA; 32°53′N 81°40′W.

Hilltop (HIL-tahp), community, S MANITOBA, W central CANADA, 44 mi/70 km from BRANDON, in CLANWILLIAM rural municipality; 50°28′N 99°48′W.

Hilltop, town (2000 population 766), ANOKA county, E MINNESOTA, residential suburb 5 mi/8 km N of downtown MINNEAPOLIS, E of MISSISSIPPI RIVER, surrounded by COLUMBIA HEIGHTS city; 45°03′N 93°14′W.

Hilltop, unincorporated village, FAYETTE county, S central WEST VIRGINIA, 11 mi/18 km N of BECKLEY, near Dunloup Creek. In coal region. New River Gorge National River to E.

Hillview, town (2000 population 7,037), BULLITT and JEFFERSON counties, NW KENTUCKY, residential suburb 13 mi/21 km SSE of downtown LOUISVILLE; 38°04′N 85°41′W. Agricultural area (burley tobacco, grain; livestock, poultry; dairying).

Hillview, village, (2000 population 179), GREENE county, W ILLINOIS, near ILLINOIS RIVER, 13 mi/21 km NW of CARROLLTON; 39°27′N 90°32′W. In agricultural area.

Hillview Reservoir, in YONKERS, WESTCHESTER county, SE NEW YORK; covers 90 acres/36 ha; 40°55′N 73°53′W. Part of NEW YORK city water-supply system; completed 1915. Terminus of DELAWARE and Catskill aqueducts, and head of tunnel system delivering water to boroughs of New York city. See also ASHOKAN RESERVOIR.

Hillwood (HIL-wud), unincorporated town, FAIRFAX county, NE VIRGINIA, residential suburb 7 mi/11 km W of WASHINGTON, D.C.; 38°52′N 77°10′W.

Hilmar-Irwin, unincorporated town (2000 population 4,807), MERCED county, central CALIFORNIA, 4 mi/6.4 km S of TURLOCK, near MERCED RIVER. Dairying; poultry; fruit, sweet potatoes, beans, tomatoes, grain, cotton, almonds. Manufacturing (cheese).

Hilo (HEE-lo), city (2000 population 40,759), ☉ HAWAII county, HAWAII, on HILO BAY of HAWAII island (the Big Island), 200 mi/322 km SE of HONOLULU, at mouth of Wailuku River; 19°41′N 155°05′W. The second largest city of Hawaii, a port of entry, and the only metropolitan area on the island, Hilo is the trade and shipping center for an orchid, papaya, and macadamia nut region; fish, prawns, and until 1996, sugar. Manufacturing (chemicals, machinery, dairy products, machinery, food and beverages; printing and publishing). With the demise of sugar, the economy became based heavily on tourism, which was spurred by the inauguration in 1967 of direct air service to the U.S. mainland; Hilo International Airport (General Lyman Field) on E side of city. Among Hilo's points of interest are the peaks of MAUNA KEA and MAUNA LOA, which rise behind the city; waterfalls on the Wailuku River; the Lyman Mission House (c.1839) and Museum; East Hawaii Cultural Center; Wailoa River State Park includes Waiakea Pond, near center of city, near bay front; and an island-park in Hilo Bay. Seat of Hilo College, the University of Hawaii at Hilo, and Hilo Community College. Badly damaged by tidal waves in 1946 and 1960; after the

latter tidal wave, the lowland area was drained and a hill 26 ft/8 m above sea level was constructed. Akaka Falls State Park to N; Wailuku River State Parks to W (includes Rainbow Falls); Keaukaha Military Reserve to SE. S of airport (National Guard); Bayfront Beach Park, includes Lilioukalani Gardens, COCONUT (linked by footbridge) and Kaulainaiwi Islands, E of city center; Hilo Breakwater, on E side of bay, creates harbor and protects ships from Blonde Reef, immediately to E. Panaewa and Waiakea Forest Reserves to S, including Panaewa Zoo; Hilo Forest Reserve to W; Kaumana Caves to SW. Starting point of Hawaii Belt Road, which encircles the Big Island. Settled by missionaries c.1822, incorporated as a city 1911.

Hilo Bay, crescent-shaped indention, PACIFIC OCEAN, only anchorage on E coast, HAWAII county, HAWAII Island, HAWAII. Exposed to NE trade wind, vulnerable to tidal waves, protected on E by Hilo Breakwater, Blonde Reef lies beyond it. COCONUT ISLAND, connected to mainland by footbridge, and Kaulainaiwi Island, in the bay, both part of Bayfront Beach Park.

Hilongos (hee-LONG-gos), town, LEYTE province, W LEYTE, PHILIPPINES, on CANIGAO CHANNEL; 10°23′N 124°49′E. Agricultural center (rice, coconuts, corn). Airstrip.

Hilpoltstein (HIL-pohlt-shtein), town, MIDDLE FRANCONIA, central BAVARIA, GERMANY, 18 mi/29 km SSE of NUREMBERG; 49°12′N 11°11′E. Manufacturing of metal products, paper, pumps, chemicals; printing, lumber milling, brewing. Has ruins of castle and medieval fortifications. Chartered in 13th century.

Hilter am Teutoburger Wald (HIL-ter ahm TOI-tuh-boor-ger VAHLT), village, LOWER SAXONY, NW GERMANY, in TEUTOBURG FOREST, 10 mi/16 km SE of OSNABRÜCK; 52°08′N 08°09′E. Food processing.

Hilterfingen, commune, BERN canton, central SWITZERLAND, on LAKE THUN, and 2 mi/3.2 km SE of THUN. Elevation 1,900 ft/579 m.

Hilton (HIL-tuhn), community, SW MANITOBA, W central CANADA, 30 mi/49 km from BRANDON, and in STRATHCONA rural municipality; 49°30′N 99°31′W.

Hilton (HIL-tuhn), township (☐ 45 sq mi/117 sq km; 2001 population 258), S ONTARIO, E central CANADA, 33 mi/53 km from SAULT SAINTE MARIE; 46°12′N 83°52′W.

Hilton, village (☐ 1 sq mi/2.6 sq km; 2006 population 6,091), MONROE county, W NEW YORK 13 mi/21 km NW of ROCHESTER; 43°17′N 77°47′W. Light manufacturing; farming. Incorporated 1885.

Hilton Beach (HIL-tuhn BEECH), village (☐ 1 sq mi/2.6 sq km; 2001 population 174), S central ONTARIO, E central CANADA, on NE shore of SAINT JOSEPH ISLAND, on Saint Joseph Channel of Lake HURON, and 28 mi/45 km SE of SAULT SAINTE MARIE; 46°15′N 83°53′W. Lumbering. Also written Hiltonbeach.

Hiltonbeach, CANADA: see HILTON BEACH.

Hilton Head, city, BEAUFORT county, 17 mi/27 km ENE of SAVANNAH (GEORGIA), on HILTON HEAD ISLAND; 32°12′N 80°45′W. Tourism is the main industry.

Hilton Head Island, S SOUTH CAROLINA, one of the SEA ISLANDS, SW of PORT ROYAL SOUND, 17 mi/27 km ENE of SAVANNAH (GEORGIA): c.12 mi/19 km long, 1 mi/1.6 km–5 mi/8 km wide. Hilton Head village on W shore. Major resort and convention center. Airport.

Hiltrup (HIL-trup), suburb of MÜNSTER, NW GERMANY, NORTH RHINE–WESTPHALIA, 3 mi/4.8 km S of city center; 51°54′N 07°38′E. Independent village until 1975.

Hilvan, village, S TURKEY, 32 mi/51 km NNE of SANLIURFA; 37°35′N 38°58′E. Wheat, barley, sesame, lentils. Also called Karacurun.

Hilvarenbeek (HIL-vuh-ruhn-baik), town, NORTH BRABANT province, S NETHERLANDS, 6 mi/9.7 km SSE of TILBURG; 51°29′N 05°09′E. Belgian border 3 mi/4.8 km to SW. WILHELMINA CANAL passes to NE. Dairying; cattle, hogs; grain, vegetables; manufacturing (springs). Castle to E.

Hilversum (HIL-vuhr-sum), city (2001 population 77,494), NORTH HOLLAND province, central NETHERLANDS, 15 mi/24 km SE of AMSTERDAM, and 10 mi/16 km N of UTRECHT; 52°13′N 05°11′E. GOOIMEER channel 7 mi/11.3 km to N. Railroad junction; airport to S. Dairying; cattle, hogs, poultry; fruit, vegetables; manufacturing (fabricated metal products, chemicals, telecommunications equipment, building materials). Castle to W; nature reserve to S.

Hilwan, Egypt: see HELWAN.

Hilzingen (HIL-tsing-uhn), village, BADEN-WÜRTTEMBERG, SW GERMANY, near Swiss border, 8 mi/12.9 km NE of SCHAFFHAUSEN (SWITZERLAND); 47°46′N 08°46′E.

Hima (HEI-muh), unincorporated village, CLAY county, SE KENTUCKY, 17 mi/27 km E of LONDON, and 2 mi/3.2 km S of MANCHESTER, near Gross Creek, in CUMBERLAND foothills, in Daniel Boone National Forest. Coal, timber, and agriculture (tobacco; livestock).

Himachal Pradesh (hi-MAH-chuhl pruh-DAISH), state (☐ 21,495 sq mi/55,887 sq km), N INDIA, in the W HIMALAYAS; ☉ SHIMLA. Bordered E by CHINA (TIBET). The state is covered with forested mountains, and the valleys are extensively cultivated. The forests supplied large quantities of timber and wood products, but severe deforestation has led to a tree-planting program and the cessation of live tree-cutting. The emphasis in recent years is toward the protection of watersheds for future hydro-electric development. Potatoes, wheat, and maize are grown, as well as most of INDIA's apples, peaches, and apricots. Salt is mined and handicrafts are made. Tourism, timber, fruit, and horticulture are the mainstays of the economy. SHIMLA is a hill station and major tourist center. Pahari-speaking Hindus inhabit the lower hill area; peoples of Tibetan origin live in the high mountain regions. Principal languages are Hindi and Pahari. Formed as a union territory in 1948 by the merger of 30 former Punjabi princely hill states. The small state of BILASPUR was merged with it in 1954. In 1966, five more districts and parts of two others from PUNJAB state were added to the territory. Became a state in 1971. Governed by a chief minister and cabinet responsible to an elected unicameral legislature and by a governor appointed by the president of India.

Himalaya, [Sanskrit = *abode of snow*], great Asian mountain system, extending c.1,500 mi/2,414 km E from NANGA PARBAT on INDUS RIVER in PAKISTAN through N INDIA, TIBET (CHINA), NEPAL, E INDIA, and BHUTAN to NAMJAGBARWA (25,446 ft/7,756 m) on the BRAHMAPUTRA RIVER in SE TIBET. For most of its length, the Himalayas comprise two nearly parallel ranges separated by a wide valley in which the INDUS River flows W and the BRAHMAPUTRA in INDIA flows E. The N range is called the TRANS-HIMALAYAS. The S range has three parallel zones: the GREAT HIMALAYA, the perpetually snow-covered main range in which the highest peaks are found (average elevation 20,000 ft/6,100 m); the LESSER HIMALAYA with 7,000 ft/2,100 m–15,000 ft/4,600 m elevations; and the southernmost OUTER HIMALAYA, 2,000 ft/610 m–5,000 ft/1,500 m high. A relatively young and still-growing system subject to severe earthquakes, the main axis of the Himalayas was formed c.25 million–70 million years ago as the earth's crust folded against the N-moving Indian subcontinent. Current theories suggest the formation of the Himalayas may be responsible for climatic changes which caused the last Ice Age. Some thirty Himalayan peaks rise to over 25,000 ft/7,600 m, including Mount EVEREST (29,028 ft/8,848 m), K2 (28,251 ft/8,611 m), and KANCHENJUNGA (28,208 ft/8,598 m), respectively, the world's highest, second, and third highest peaks. All fourteen of the world's peaks over 26,247 ft/8,000 m are found in these ranges. Himalayan peaks have long been the goal of mountaineers. The towering ranges are penetrated by many roads and tracks with air flights linking remote towns to GILGIT, Sakardu (both in PAKISTAN), LEH, SRINAGAR (both in INDIA), KATHMANDU (NEPAL), LHASA (TIBET), and THIMPHU (BHUTAN). Railroads reach only the S foothills from where the main route follows footpaths across bridges and high mountain passes. Major metalled roads run between ISLAMABAD (PAKISTAN) and CHINA, Srinigar and LEH, KATHMANDU and LHASA. With the exception of major towns in NEPAL and BHUTAN almost all Himalayan villages are reached by four-wheel-drive vehicles. The aridity of the TIBETAN PLATEAU and the TARIM Basin of W CHINA results from the interception of the moisture-laden monsoon by the Himalayan S face. Consequently, the N slopes receive relatively light snowfall and have little drainage, while the snow-covered and extensively glaciated S slopes give rise to the Indian subcontinent's major rivers, including the INDUS, SUTLEJ, GANGA, and Tsangpo-Brahmaputra. Little of the high Himalayan region is inhabitable or of great current economic value. Numerous protected areas are of benefit for recreation and tourism. The S piedmont plains of TERAI and DUARS were formerly malarial jungle and swamps but have now been converted to agriculture, with many wild animals in nature reserves. Grazing is possible on high-elevation pastures and extensive farming is carried on in the valleys and adjacent slopes. Lumbering is in the extensive forests found below 12,000 ft/3,660 m. Military roads and trucks have completely transformed the frontier areas of PAKISTAN, CHINA, and INDIA since the 1960s. Migration of the Himalayan population to the urban plains is very common. The Himalayan rivers, with their perennial but highly seasonal flow, offer much scope for hydroelectric power and irrigation development, though seismic hazards and potential social and environmental impacts pose serious constraints. Major recreation hill resorts such as MURREE (PAKISTAN), SIMLA, NAINI TAL, MUSSOORIE, and DARJILING (all in INDIA) are popular summer retreats from the heat of the Pakistani and Indian plains. Extensive tourism is common throughout the Himalayas. The Himalayas are associated with many legends in Asian mythology. On isolated slopes are found the retreats of rishis (holy sages), gurus (teachers), and Tibetan monks. Mass tourism and pilgrimages to Hindu shrines have contributed to environmental degradation.

Himalchuli (hee-mahl-CHOO-lee), mountain peak (25,900 ft/7,894 m), central NEPAL; 28°26′N 84°39′E. First ascent by Japanese climbers in 1960.

Himamaylan (hee-mah-MEI-lan), town (2000 population 88,684), NEGROS OCCIDENTAL province, W NEGROS island, PHILIPPINES, on PANAY GULF, 7 mi/11.3 km S of BINALBAGAN; 10°04′N 122°55′E. Agricultural center (rice, coconuts).

Himarë (hee-MAHR), town, S ALBANIA, 27 mi/43 km SE of VLORË, near IONIAN SEA coast; 40°07′N 19°44′E. Situated at the food of the mountain, Keravnia, its name comes from the Greek "Himarros" meaning torrent. Tourist center. Citrus-growing center; oranges, lemons. Also called Himara.

Himatnagar (hi-MUHT-nuh-guhr), town, ⊙ SABAR KANTHA district, GUJARAT state, W INDIA, 45 mi/72 km NNE of AHMADABAD; 23°36′N 72°57′E. Market center for grain, oilseeds, cloth fabrics; cotton ginning, hand-loom weaving, oilseed milling, and manufacturing of matches, pottery, glass. Sandstone quarries nearby. Was capital of former W India state of Idar. Sometimes spelled Himmatnagar; formerly called Ahmadnagar.

Himberg (HIM-berg), township, SE LOWER AUSTRIA, 9 mi/14.5 km SSE of VIENNA; 48°05′N 16°26′E. Recycling industry.

Himbirti, town, central ERITREA, 12 mi/19 km SW of ASMARA, near the base of Mount TACARA; 15°16′N 38°44′E.

Himedo (HEE-me-do), town, on E coast of Kamishima island, AMAKUSA ISLANDS, Amakusa district, KUMAMOTO prefecture, SW JAPAN, 31 mi/50 km S of KUMAMOTO; 32°26′N 130°24′E.

Himeji (hee-ME-jee), city (2005 population 536,232), HYOGO prefecture, S HONSHU, W central JAPAN, 28 mi/45 km W of KOBE; 34°48′N 134°41′E. Railroad and market center. Manufacturing includes iron (supplying the OSAKA-KOBE industrial area), oil, computer components; white leather. The ruins of 14th-century Himeji Castle have been completely restored. Shosha Engyo temple is here.

Himera (HI-muh-ruh), ancient city on the N coast of SICILY. Founded by Greeks in the 7th century B.C.E. Here in 480 B.C.E. (a traditional date) forces led by Gelon routed the Carthaginians led by Hamilcar. Years later the Carthaginians destroyed (409 B.C.E.) the city. The citizens moved to nearby Thermae (modern TERMINI). The poet Stesichorus was born here.

Himeshima (hee-ME-shee-mah), village, East Kunisaki district, OITA prefecture, E KYUSHU, SW JAPAN, 34 mi/55 km N of OITA; 33°43′N 131°38′E. Prawns. Seto Naikai National Park nearby.

Himi (hee-MEE), city, TOYAMA prefecture, central HONSHU, central JAPAN, on TOYAMA BAY, 16 mi/25 km N of TOYAMA; 36°51′N 136°59′E. Fishery.

Himis Gompa (HEE-mees GOM-pah), noted Buddhist monastery, LEH district, JAMMU AND KASHMIR state, N INDIA, in E central KASHMIR, in ZASKAR Range, near the INDUS River, 20 mi/32 km SSE of LEH. Founded seventeenth century. Large annual (June) festival with devil dance. Also spelled Hemis Gompa.

Himlerville (HEIM-luhr-vil) or **Beauty**, locality, MARTIN county, E KENTUCKY, 3 mi/4.8 km ESE of INEZ, in CUMBERLAND MOUNTAINS. Historic Hungarian camp of immigrant coal miners. Has thirteen-room mansion of Marton Himler.

Himmelberg (HIM-mel-berg), village, CARINTHIA, S AUSTRIA, 15 mi/24 km NW of KLAGENFURT; 46°45′N 14°02′E. Summer resort.

Himmelbjærget (HI-muhl-byer-khuht), group of hills, E JUTLAND, DENMARK, rising to 531 ft/162 m. Largely wooded, with several lakes (largest, JULSØ) traversed by GUDENÅ RIVER.

Himmelpforten (HIM-mel-pfor-tuhn), village, NW GERMANY, 7 mi/11.3 km W of STADE; 53°37′N 09°18′E.

Himmerland (HI-muhr-lan), hilly region, N JUTLAND, DENMARK, along the KATTEGAT strait, and between Lim and MARIAGER fjords; highest point, 374 ft/114 m. Unfertile; largely moor and forest area. Sometimes Himmersyssel.

Himo (HEE-mo), village, KILIMANJARO region, N TANZANIA, 12 mi/19 km ESE of MOSHI, near KENYA border; 03°25′S 37°30′E. Border crossing 8 mi/12.9 km to E. Highway junction. KILIMANJARO Mountain and MOUNT KILIMANJARO NATIONAL PARK to N. Corn, wheat, pyrethrum; goats, sheep, cattle; timber.

Hims, SYRIA: see HOMS.

Himsworth South (HIMZ-wuhrth), former village, SE central ONTARIO, E central CANADA. Amalgamated into POWASSAN in 2001.

Hinai (hee-NAH-ee), town, North Akita district, Akita prefecture, N HONSHU, NE JAPAN, 43 mi/70 km N of AKITA city; 40°13′N 140°34′E. Chicken (*Hinai* chicken is known for its flavor); weeds used for brooms.

Hinase (HEE-nah-se), or **Hinashi**, town, Wake district, OKAYAMA prefecture, SW HONSHU, W JAPAN, on HARIMA SEA, 19 mi/30 km E of OKAYAMA; 36°51′N 136°59′E. Oysters.

Hinatuan (hee-nah-TOO-ahn), town, SURIGAO DEL SUR province, PHILIPPINES, on NE coast of MINDANAO; 08°24′N 126°19′E. Sawmill nearby.

Hin Boun (HIN BOON), village, KHAMMUAN province, central LAOS, 18 mi/29 km NW of Thakhek, on left bank of MEKONG River (THAILAND border), at mouth of the NAM HIN BOUN (river). Tin-shipping center for

Nam Patene tin-mining district (mines at BONENG and PHONTIOU). Founded 1895. Also spelled Hin Bun and sometimes called PAK HIN BAUN.

Hinche (ANGSH), town (2003 population 23,599), ⊙ CENTRE department, E central HAITI, 45 mi/72 km NE of PORT-AU-PRINCE; 19°01′N 72°01′W. On fertile plain. Coffee, sugarcane, sisal, cotton, fruit; cattle; beekeeping; cotton processing.

Hinchinbrook, Cape, S ALASKA, S tip of HINCHINBROOK ISLAND; 60°16′N 146°37′W.

Hinchinbrooke (HIN-chin-bruk), canton (□ 58 sq mi/150.8 sq km; 2006 population 2,410), Montérégie region, S QUEBEC, E CANADA, 4 mi/6 km from HUNTINGDON; 45°03′N 74°06′W.

Hinchinbrooke, ENGLAND: see HUNTINGDON.

Hinchinbrook Entrance (HIN-chin-bruk), S ALASKA, between HINCHINBROOK (E) and MONTAGUE islands (W), joins Gulf of ALASKA and PRINCE WILLIAM SOUND.

Hinchinbrook Island, S ALASKA, in GULF OF ALASKA, at mouth of PRINCE WILLIAM SOUND, 20 mi/32 km SW of CORDOVA; 22 mi/35 km long, 4 mi/6.4 km–13 mi/21 km wide; 60°23′N 146°25′W. NUCHEK village is in W.

Hinchinbrook Island (HIN-chin-bruk) (□ 152 sq mi/395.2 sq km), in CORAL SEA, just off E coast of QUEENSLAND, NE AUSTRALIA, within GREAT BARRIER REEF, N of LUCINDA POINT; 20 mi/32 km long, 9 mi/14 km wide. Mountainous, rising to 3,650 ft/1,113 m. Cape Sandwich, its N point (18°14′S 146°19′E), forms S side of entrance to ROCKINGHAM BAY; national park.

Hinchinbrook Island National Park (HIN-chin-bruk), (□ 154 sq mi/400.4 sq km), NE QUEENSLAND, NE AUSTRALIA, 85 mi/137 km NW of TOWNSVILLE; 28 mi/45 km long, 12 mi/19 km wide; 18°20′S 146°20′E. A mountainous continental island, separated from mainland by Hinchinbrook Channel, a submerged river course. Waterfall, secluded beaches, mangroves, rain forest, eucalyptus woodlands. Saltwater crocodile, dugong, sea turtle; sea birds. Highest elevation, Mount Bowen (3,671 ft/1,119 m). Resort at Cape Richards on N. Camping, picnicking, hiking. Established 1932.

Hinckley, town (2000 population 1,291), PINE county, E MINNESOTA, on Grindstone River, 14 mi/23 km N of PINE CITY; 46°00′N 92°56′W. Railroad junction. Oats; cattle, sheep, poultry; dairying. Manufacturing (machinery; meat processing); timber. Nearly destroyed in Great Fire of 1894, later rebuilt and became center for lumbering. Chengwatana State Forest to SE; Sandstone National Wildlife Refuge to NE; large St. Croix State Park to E.

Hinckley, village (2000 population 1,994), DE KALB county, N ILLINOIS, 16 mi/26 km S of SYCAMORE; 41°46′N 88°38′W. In rich agricultural area.

Hinckley, village (2006 population 734), MILLARD county, W UTAH, at S margin of Sevier desert, 5 mi/8 km SW of DELTA, near SEVIER RIVER; elevation 4,600 ft/1,402 m; 39°19′N 112°40′W. Alfalfa, wheat, barley, sugar beets; dairying; cattle. Precious metals mining. Fort Deseret State Historical Park to S. Gunnison Massacre Monument to SW. Topaz War Relocation Center Historical Site to NW.

Hinckley and Bosworth (HINK-lee and BOZ-wurth), district (□ 115 sq mi/299 sq km; 2001 population 100,141), LEICESTERSHIRE, central ENGLAND; 52°32′N 01°22′W. Hosiery and shoes are the chief manufacturing products, along with electrical equipment and machinery.

Hinckley Reservoir, New York: see WEST CANADA CREEK.

Hindaun (hin-DOUN), town, SAWAI MADHOPUR district, E RAJASTHAN state, NW INDIA, 75 mi/121 km ESE of JAIPUR; 26°43′N 77°01′E. Markets cotton, wheat, barley, millet, oilseeds. Sandstone quarried nearby. Cotton ginning at railroad station, 1 mi/1.6 km N.

Hindelang (HIN-de-lahng), village, BAVARIA, S GERMANY, in SWABIA, in ALLGÄU ALPS, 4 mi/6.4 km E of

SONTHOFEN; 47°30′N 10°22′E. Health resort (elevation 2,706 ft/825 m) in both summer and winter.

Hindeloopen (HIN-duh-law-puhn), village, FRIESLAND province, N NETHERLANDS, on the IJSSELMEER, 13 mi/21 km WSW of SNEEK; 52°56′N 05°24′E. Fishing; summer resort. Dairying; cattle, sheep; grain, vegetables. Has seventeenth-century church and town hall. Once famous for making painted furniture.

Hindenburg, POLAND: see ZABRZE.

Hindenburgdamm (HIN-duhn-boorg-duhm), dike, on NORTH SEA coast of SCHLESWIG-HOLSTEIN, NW GERMANY, connecting SYLT island and mainland; 8 mi/12.9 km long, 26 ft/8 m high, c.160 ft/49 m wide; 54°53′N 08°29′E–54°53′N 08°36′E. Built 1923–1927.

Hindenburg Line (HIN-den-buhrg), fortified German sector of the Western Front, between ARRAS and SOISSONS (NE FRANCE) during World War I; 93 mi/150 km long. First mass use of tanks in warfare failed to break line at CAMBRAI, November 1917. Launching area for German assaults which nearly broke Allied lines, March 1918. Line fell to Allied counterattack against worn German forces, September-October 1918. Name is British usage, from Imperial German Army Chief of Staff Paul von Hindenburg. Correct German name is Siegfried-Stellung [German=Position Siegfried].

Hinderwell (HIN-duh-wel), village (2005 population 2,315), NORTH YORKSHIRE, N ENGLAND, near NORTH SEA, 8 mi/12.9 km WNW of WHITBY; 54°32′N 00°46′W. Former agricultural market. Just NNW, on a small creek flowing into the North Sea, is resort and fishing village of STAITHES. James Cook born here.

Hindhead, ENGLAND: see FRENSHAM.

Hindijan, IRAN: see HENDIJAN.

Hindiya (hind-EE-yeh), town, KERBELA province, central IRAQ, on the W bank of the EUPHRATES RIVER, and 12 mi/19 km WNW of HILLA. Dates, corn, millet. Floating bridge. The HINDIYA barrage is 10 mi/16 km NNE at the bifurcation of the HILLA and HINDIYA branch canals of the Euphrates. Built in 1911 and repaired in 1917 mainly to feed the Hilla branch which had dried out, the barrage feeds numerous irrigation canals. Also called Tuwairij, Tuwairiq, Tuwayriq, or Tuwayrij.

Hindiya, Shatt (hind-EE-yeh, SHAHT), name given to part of the course of the EUPHRATES RIVER between HINDIYA and SAMAWA, central IRAQ. The river had changed its course repeatedly in this area.

Hindley (HEIND-lee), town (2001 population 13,677), GREATER MANCHESTER, W central ENGLAND, 3 mi/4.8 km SE of WIGAN; 53°32′N 02°35′W. Light industry. Previously site of coal mining and cotton manufacturing. Has 17th-century church, rebuilt in 18th century.

Hindman (HEIND-muhn), town (2000 population 787), ⊙ KNOTT county, E KENTUCKY, in CUMBERLAND foothills, 13 mi/21 km ENE of HAZARD; 37°19′N 82°58′W. In coal-mining and agricultural area; light manufacturing. Seat of Hindman Settlement School (founded 1902). Carr Fork Lake reservoir to S.

Hindmarsh (HEIND-mahrsh), NW suburb of ADELAIDE, SE SOUTH AUSTRALIA, on N shore of TORRENS RIVER; 34°54′S 138°34′E.

Hindmarsh, Lake (HEIND-mahrsh), salt lake (□ 47 sq mi/122.2 sq km), W central VICTORIA, AUSTRALIA, 200 mi/322 km NW of MELBOURNE; 12 mi/19 km long, 5 mi/8 km wide; 36°03′S 141°55′E. Frequently dry. JEPARIT village near S shore.

Hindola (hin-do-LAH), town, tahsil headquarters, DHENKANAL district, E central ORISSA state, E INDIA, 45 mi/72 km WNW of CUTTACK. Exports timber. Was capital of former princely state of Hindol in ORISSA STATES; state incorporated 1949 into newly-created DHENKANAL district.

Hindoli (hin-do-LEE), town, tahsil headquarters, SE RAJASTHAN state, NW INDIA, 13 mi/21 km NW of BUNDI; 25°35′N 75°30′E. Millet, gram.

Hinds (HEINDZ), county (877 mi/1,411 km; 1990 population 254,441; 2000 population 250,800), W MISSISSIPPI; ⊙ JACKSON (⊙ and RAYMOND; 32°15′N 90°26′W. Bounded E by PEARL RIVER, NW and partly W by BIG BLACK RIVER. Agriculture (cotton, corn, soybeans, vegetables, potatoes; cattle, poultry; timber); natural gas, paper and pulp. Manufacturing at Jackson and CLINTON. County is urbanized in E, including Jackson, state's largest city. LeFleur's Bluff State Park in E, in city of Jackson; Lake Dockery State Lake in E, S of Jackson. NATCHEZ TRACE PARKWAY, crosses county includes incomplete section. Formed 1821.

Hindsboro, village (2000 population 361), Douglas county, E central ILLINOIS, 10 mi/16 km SE of TUSCOLA; 39°40′N 88°07′W. In agricultural area.

Hindsdale (HEINDS-dail), village, VALLEY county, NE MONTANA, 24 mi/39 km NW of GLASGOW, on MILK River at its confluence with Rock Creek. Wheat, barley, oats; cattle, sheep, hogs. Larb Hills to SW.

Hind's Hill (2,158 ft/658 m), SW NEWFOUNDLAND AND LABRADOR, E CANADA, 45 mi/72 km E of CORNER BROOK, overlooking HIND'S LAKE.

Hind's Lake (□ 14 sq mi/36 sq km), SW NEWFOUNDLAND and LABRADOR, E CANADA, 40 mi/64 km E of CORNER BROOK, at foot of HIND'S HILL; 7 mi/11 km long, 3 mi/5 km wide.

Hindubagh, PAKISTAN: see MUSLIMBAGH.

Hindu Kush (HIN-doo-koo-shah) [=mountains of India], a high mountain system, extending c.500 mi/805 km W from the Pamir Knot, N PAKISTAN, into NE AFGHANISTAN; rising to 25,236 ft/7,692 m in TIRICH MIR, on the AFGHANISTAN/PAKISTAN border, where the HINDUKUSH divides S ASIA from central ASIA. Glaciated and receiving heavy snowfall, the mountains have permanently snow-covered peaks and little vegetation. Meltwater feeds the headstreams of the Amudarya (river) and the INDUS RIVER. Irrigated valleys are heavily populated. Extensive lumbering has greatly reduced forests on S slopes. Transhumant livestock graze mountain pastures with wheat-growing dominating agriculture. A major cultural contact zone where the Indian subcontinent meets central ASIA. The system is crossed by many high-elevation passes; once followed by Alexander the Great, Jenghis Khan, and Timur and Babur in their invasions of INDIA, they are now trade routes. An all-weather road links KABUL (AFGHANISTAN) with TAJIKISTAN through the Salang Tunnel, while a seasonal road over the DORAH PASS links PAKISTAN with central ASIA. The HINDUKUSH were called the Caucasus Indicus by the ancient Greeks.

Hindupur (HIN-duh-poor), city, ANANTAPUR district, ANDHRA PRADESH state, S INDIA, near PENNER RIVER, 60 mi/97 km S of ANANTAPUR; 13°49′N 77°29′E. Road junction; trade center; cloth goods, jaggery; peanut milling, hand-loom woolen weaving; silk growing. Corundum mines nearby. Notable sixteenth-century Shivaite temple 8 mi/12.9 km E, at village of Lepakshi.

Hindur, INDIA: see NALAGARH.

Hi-Nella (HEI-NE-luh), residential borough (2006 population 1,007), CAMDEN county, SW NEW JERSEY, 8 mi/12.9 km SE of CAMDEN; 39°50′N 75°01′W.

Hines, town (2006 population 1,488), HARNEY county, E central OREGON, 2 mi/3.2 km S of BURNS, near SILVIES RIVER; elevation 4,155 ft/1,266 m; 43°34′N 119°04′W. Lumber; millwork.

Hinesburg (HINS-buhrg), town, CHITTENDEN co., NW VERMONT, 12 mi/19 km SSE of BURLINGTON. Settled just after American Revolution. Chartered 1762. Named for the first town clerk, Abel Hine.

Hines Creek (HEINZ), village (□ 2 sq mi/5.2 sq km; 2001 population 437), W ALBERTA, W CANADA, near PEACE River, 50 mi/80 km W of PEACE RIVER town, and in CLEAR HILLS County; 56°14′N 118°36′W. Lumbering; mixed farming; wheat. Established 1952.

Hinesville (HEINZ-vil), city (2000 population 30,392), ⊙ LIBERTY county, SE GEORGIA, 30 mi/53 km WSW of SAVANNAH; 31°50′N 81°36′W. In farm area. FORT STEWART nearby is responsible for much of the city's growth. Manufacturing includes printing and publishing; concrete.

Hinganghat (HIN-guhn-ghaht), town, WARDHA district, MAHARASHTRA state, W INDIA, on tributary of WARDHA RIVER, and 45 mi/72 km SSW of NAGPUR; 20°34′N 78°50′E. Major cotton-textile center in important Indian cotton tract; millet, wheat, oilseeds, rice, turmeric.

Hinggan (SEENG-GAHN), former province (□ 107,605 sq mi/279,773 sq km) of the NORTHEAST (Manchuria), CHINA. Capital was HAILAR. It was one of nine Manchurian provinces under the Chinese Nationalist government after World War II. Incorporated 1949 into INNER MONGOLIA AUTONOMOUS REGION.

Hinggan, Greater (SEENG-AHN), mountain range, INNER MONGOLIA AUTONOMOUS REGION and NORTHEAST region, N CHINA; extends c.750 mi/1,207 km from the AMUR RIVER S to the LIAO RIVER; highest point is 5,657 ft/1,724 m. The heavily forested range forms the E edge of the Mongolian Plateau and has some of China's richest timber resources. The Xiao Hinggan Ling (Lesser Hinggan) in HEILONGJIANG province is a continuation of the BUREYA range in the Siberian Russian. The Lesser Hinggan extends c.400 mi/644 km NE from the SONGHUA JIANG RIVER and is linked to the Great Hinggan by the YILEHULI range, N Heilongjiang province. Also spelled Xing'an, or Khingan.

Hinggan Meng (SEENG-AHN MUNG), Mongolian league in NE INNER MONGOLIA AUTONOMOUS REGION, N CHINA, on SE slopes of the GREATER HINGGAN range; 46°05′N 122°05′E. Main town, Horqin Youyi Qianqi (Ulan Hot).

Hingham (HEENG-uhm), town, PLYMOUTH county, E MASSACHUSETTS, S of BOSTON, on the S shore of Hingham Bay; 42°13′N 70°54′W. HINGHAM is primarily residential with some diverse light industry (metals, scientific instruments). Once a fishing and shipbuilding center, its bay shore draws annual visitors. The Old Ship Church (1681), a fine example of American Gothic architecture, is one of the oldest houses of worship in the U.S. and has been in continuous use since it was built. Wompatuch State Park is here. Includes village of South Hingham. Incorporated 1635.

Hingham (HING-uhm), village (2000 population 157), HILL county, N MONTANA, 33 mi/53 km W of HAVRE; 48°34′N 110°25′W. Storage and shipping center for grain and livestock.

Hingoli (hin-GO-lee), town, tahsil headquarters, PARBHANI district, MAHARASHTRA state, W INDIA, 40 mi/64 km NNW of NANDED. Railroad terminus; cotton-trade center in agricultural area (millet, wheat); cotton ginning. Temple at Aundah is noted example of medieval Hindu architecture (14 mi/23 km SSW).

Hingol River (hin-go-lah), c.350 mi/563 km long, BALUCHISTAN province, SW PAKISTAN; rises on W slope of CENTRAL BRAHUI RANGE, c.15 mi/24 km SSW of KALAT; flows S, through barren hill ranges, cutting through E end of MAKRAN COAST RANGE (gorge used by Alexander the Great in 325 B.C.E.), to ARABIAN SEA 55 mi/89 km E of ORMARA. Called Nal in middle course. Main tributary, Mashkai River (right).

Hinigaran (hee-nee-GAH-rahn), town, NEGROS OCCIDENTAL province, W NEGROS island, PHILIPPINES, on GUIMARAS STRAIT, 5 mi/8 km N of BINALBAGAN; 10°16′N 122°54′E. Agricultural center (rice, sugarcane).

Hinis, township (2000 population 27,504), E TURKEY, 45 mi/72 km SE of ERZURUM; 39°22′N 41°44′E. Grain.

Hinlopen Strait (HIN-law-puhn), Norwegian *Hinlopenstredet*, passage of ARCTIC OCEAN, extending NW-SE between West Spitsbergen (W) and Northeast Land (E) of SVALBARD; 100 mi/161 km long, 7 mi/11.3 km–30 mi/48 km wide.

Hinna, ETHIOPIA: see IMI.

Hinnom (hee-NOM), valley, SSW of JERUSALEM, IS-RAEL. Its ill repute in the Hebrew Bible emanated from the worship there of foreign gods, including child sacrifice to Molech at Tophet. In later Jewish literature it was called Ge-Hinnom [Hebrew=valley of Hinnom] and in the Greek of the New Testament, Gehenna. A place for burning refuse in later Israelite times, it provided imagery for a fiery Hell in the Books of Isaiah and the New Testament. It appears as Jahannam in the Qur'an.

Hinnøya (HIN-uh-yah), island (□ 849 sq mi/2,207.4 sq km) in NORTH SEA, in NORDLAND and TROMS counties, N NORWAY, 55 mi/89 km long (N-S), up to 30 mi/48 km wide; largest of the VESTERÅLEN group and of Norway. Rises to 4,153 ft/1,266 m in MØYSALEN peak (SW). Separated from mainland by narrow strait, crossed by Tjelsund Bridge. Fishing is chief occupation; some agriculture. Main town: HARSTAD (N).

Hino (hee-NO), city (2005 population 176,538), Tokyo prefecture, E central HONSHU, E central JAPAN, on the Asa River, 12 mi/19 km WSW of SHINJUKU; 35°40′N 139°23′E. Residential suburb of TOKYO and part of the KEIHIN INDUSTRIAL ZONE. Manufacturing includes communications, electrical equipment, and vehicles.

Hino (hee-NO), town, Gamou district, SHIGA prefecture, S HONSHU, central JAPAN, 22 mi/35 km E of OTSU; 35°00′N 136°14′E. Pickles. Known for its Hino Festival.

Hino, town, Hino district, TOTTORI prefecture, S HONSHU, W JAPAN, 48 mi/78 km S of TOTTORI; 35°14′N 133°26′E.

Hinode (hee-NO-de), town, West Tama district, TOKYO prefecture, E central HONSHU, E central JAPAN, 22 mi/35 km W of SHINJUKU; 35°44′N 139°14′E. Pagodas, coffins.

Hinoemata (hee-NO-e-MAH-tah), village, South Aidzu district, FUKUSHIMA prefecture, N central HONSHU, NE JAPAN, 78 mi/125 km S of FUKUSHIMA city; 37°01′N 139°23′E. NIKKO NATIONAL PARK and Oze area nearby.

Hinohara (hee-NO-hah-rah), village, West Tama district, TOKYO prefecture, E central HONSHU, E central JAPAN, 16 mi/25 km W of SHINJUKU; 35°43′N 139°09′E.

Hinojal (ee-no-HAHL), village, CÁCERES province, W SPAIN, 17 mi/27 km NNE of CÁCERES. Livestock, cereals.

Hinojales (ee-no-HAH-les), village, HUELVA province, SW SPAIN, in the SIERRA MORENA, 8 mi/12.9 km N of ARACENA. Olives, cereals, honey, cork, timber.

Hinojos (ee-NO-hos), town, HUELVA province, SW SPAIN, 22 mi/35 km WSW of SEVILLE; 37°17′N 06°22′W. On the irrigated GUADALQUIVIR plain: vegetables, fruit, timber; viticulture and apiculture. Olive oil extracting, flour- and sawmilling, limestone quarries and kilns.

Hinojosa del Duero (ee-no-HO-sah dhel DHWAI-ro), town, SALAMANCA province, W SPAIN, 31 mi/50 km NNW of CIUDAD RODRIGO; 40°59′N 06°47′W. Cheese and meat processing, flour milling; agriculture (livestock; olive oil, fruit, wine). Dominated by hill with ruins of ancient castle. Topaz quarries nearby.

Hinojosa del Duque (ee-no-HO-sah dhel DOO-kai), town, CÓRDOBA province, S SPAIN, in the SIERRA MORENA, 50 mi/80 km NW of CÓRDOBA; 38°30′N 05°09′W. Agricultural trade center; ships wool. Manufacturing (olive oil processing, sawmilling, tanning); agriculture (cereals, wine; lumber; livestock). Lead deposits nearby.

Hinojosa del Valle (ee-no-HO-sah dhel VAH-yai), town, BADAJOZ province, W SPAIN, 18 mi/29 km SE of ALMENDRALEJO; 38°29′N 06°11′W. Cereals, acorns; livestock.

Hinojosa de San Vicente (ee-no-HO-sah dhai SAHN vee-SEN-tai), village, TOLEDO province, central SPAIN, 10 mi/16 km NE of TALAVERA DE LA REINA. Olives, grapes, potatoes, cherries, onions, figs.

Hinojosas or **Hinojosas de Calatrava**, town, CIUDAD REAL province, S central SPAIN, 27 mi/43 km SSW of CIUDAD REAL. Grain, olives, livestock; lead mining.

Hinojosos, Los (ee-no-HO-sos, los), town, CUENCA province, E central SPAIN, 50 mi/80 km SW of CUENCA; 39°36′N 02°50′W. Olives, cereals, grapes; livestock.

Hinokage (hee-NO-kah-GE), town, West Usuki district, MIYAZAKI prefecture, SE KYUSHU, SW JAPAN, 50 mi/80 km N of MIYAZAKI; 32°39′N 131°23′E. Lumber.

Hinomi Point (HEE-no-mee), Japanese *Hinomi-misaki* (hee-NO-mee-mee-SAH-kee), promontory in SHIMANE prefecture, SW HONSHU, W JAPAN, in Sea of JAPAN, just NW of TAISHA; 35°26′N 132° 32′E. Lighthouse.

Hinsdale, county (□ 1,123 sq mi/2,919.8 sq km; 2006 population 819), SW central COLORADO; ⊙ LAKE CITY; 37°49′N 107°16′W. Tourist and timber area, some livestock business. CONTINENTAL DIVIDE crosses twice, heading SW across N (also forming small E-W section of E boundary), turns to SE in neighboring SAN JUAN county, to W, and crosses S part of county, about 80% of county is national forest. S of DIVIDE, source of RIO GRANDE immediately W in San Juan county, headwaters flow E across center of county, through RIO GRANDE RESERVOIR; N part drained by Lake Fork of Gunnison and Cobolla Creek; S drained by Los Pinos (source) and Pindra rivers, both branches of San Juan. Includes parts of San Juan Mountains. Part of Uncompahgre National Forest in NW corner; parts of Gunnison National Forest N of Divide in NE and N center; Rio Grande National Forest in center, between Divide sections; San Juan National Forest in S. Williams Creek Reservoir in SE; Lake San Cristobal Reservoir in N center. Formed 1874.

Hinsdale, town, BERKSHIRE county, W MASSACHUSETTS, in the BERKSHIRES, 7 mi/11.3 km E of PITTSFIELD; 42°26′N 73°07′W. Resort; agriculture. Settled 1763; incorporated 1804. Peru State Forest nearby.

Hinsdale, town, CHESHIRE county, SW NEW HAMPSHIRE, 14 mi/23 km SE of KEENE; 42°48′N 72°30′W. Bounded S (for 1 mi/1.6 km) by MASSACHUSETTS state line, W by CONNECTICUT RIVER (Vermont state line); drained by ASHUELOT RIVER. Agriculture (dairying; cattle, poultry; apples, vegetables; timber); manufacturing (wire rope and cable, flashlights, paper products). Part of Pisgah State Park in NE, part of Wantastiquet Mountain State Forest in NW. Settled c.1742; incorporated 1753.

Hinsdale, village (2000 population 17,349), COOK and DU PAGE counties, NE ILLINOIS, part of the greater CHICAGO metropolitan area; 41°48′N 87°56′W. Computer systems software. Incorporated 1873.

Hintalo, village (2007 population 1,234), TIGRAY state, N ETHIOPIA, 15 mi/24 km S of MEK'ELĒ; 13°19′N 39°28′E. Cereals, honey, legumes.

Hinte (HIN-te), village, LOWER SAXONY, NW GERMANY, in EAST FRIESLAND, 3 mi/4.8 km N of EMDEN; 53°25′N 07°10′E.

Hinterland (HEEN-tuhr-lahnd), district W APPENZELL Ausser Rhoden half-canton, NE SWITZERLAND. Main town is HERISAU; population is German-speaking and Protestant.

Hinterpommern, Europe, GERMANY: see POMERANIA.

Hinterrhein (HIN-tuhr-rhein), district, SW GRISONS canton, E SWITZERLAND. Population is German-speaking and Protestant, but Romansch-speaking in Kreis Schams.

Hinterrhein River (HIN-tuhr-rhine), French *Rhin Postérieur*, 35 mi/56 km headstream of the RHINE, E SWITZERLAND; rises in ADULA GROUP of LEPONTINE ALPS, N of SAN BERNARDINO PASS; flows NE, through RHEINWALD valley, continues N, past THUSIS; joins the Adula River then continues on, joining the VORDERRHEIN RIVER W of CHUR to form the Rhine proper. Drains 654 sq mi/1,694 sq km.

Hinterrugg, SWITZERLAND: see CHURFIRSTEN.

Hinton (HIN-tuhn), town, NEW SOUTH WALES, SE AUSTRALIA, 25 mi/40 km NW of NEWCASTLE, 2 mi/3 km N of MORPETH, on HUNTER RIVER; 32°43′S 151°39′E. Historic buildings. Established 1840.

Hinton (HIN-tuhn), town (□ 10 sq mi/26 sq km; 2001 population 9,405), ALBERTA, W CANADA, in ALBERTA foothills, on E edge of JASPER NATIONAL PARK, 177 mi/285 km W of EDMONTON, and in YELLOWHEAD County; 53°24′N 117°35′W. Coal mining service center; winter recreation. Established 1956.

Hinton, town (2000 population 808), PLYMOUTH county, NW IOWA, 13 mi/21 km SSE of LE MARS; 42°37′N 96°17′W. Livestock; grain.

Hinton, town (2006 population 2,172), CADDO county, W central OKLAHOMA, 22 mi/35 km WSW of EL RENO; 35°30′N 98°21′W. In agricultural area (cotton, wheat, corn; cattle). Red Rock Canyon State Park to S.

Hinton, town (2006 population 2,651), ⊙ SUMMERS county, S WEST VIRGINIA, 18 mi/29 km SE of BECKLEY, on NEW RIVER, 1 mi/1.6 km N of mouth of GREENBRIER RIVER, and 4 mi/6.4 km NNE of mouth of BLUESTONE RIVER; 37°39′N 80°52′W. Railroad shops; shipping point for agricultural area (grain; livestock; dairy products); processes lumber, beverages. Manufacturing (concrete; printing and publishing). Hinton Visitor Center is here. Bluestone Dam (New River), in Bluestone State Park and Wildlife Management Area, to S; New River Gorge National River to N; Pipestem State Park and Bluestone National Scenic River to SW. Settled 1831; incorporated 1880.

Hinton (HIN-tuhn), unincorporated village, ROCKINGHAM county, NW VIRGINIA, 4 mi/6 km W of HARRISONBURG; 38°27′N 78°58′W. Manufacturing (furniture, poultry processing); agriculture (grain; poultry, livestock; dairying).

Hinunangan, town, SOUTHERN LEYTE province, SE LEYTE, PHILIPPINES, on small inlet of SURIGAO STRAIT; 10°24′N 125°08′E. Agricultural center (corn, hemp).

Hinwil (HIN-vil), commune, ZÜRICH canton, N SWITZERLAND, 15 mi/24 km ESE of ZÜRICH. Textiles; metalworking.

Hinwil (HEEN-veel), district, SE ZÜRICH canton, N SWITZERLAND. Main town is WETZIKON; population is German-speaking and Protestant.

Hinzir Dagi, (Turkish=*Hinzir Dağı*) mountain range, central TURKEY, 45 mi/72 km ENE of KAYSERI, W of the upper part of ZAMANTI RIVER. Rises to 9,022 ft/2,750 m.

Hippenus, ISRAEL: see ASHDOD.

Hipperholme (HIP-ruhm), village (2001 population 10,035), WEST YORKSHIRE, N central ENGLAND, 2 mi/3.2 km E of HALIFAX; 53°43′N 01°48′W. Light industry.

Hippo, ALGERIA: see ANNABA, city.

Hippone, ALGERIA: see ANNABA, city.

Hippo Regius, ALGERIA: see ANNABA, city.

Hippos, ancient city of JUDEA, ISRAEL, just E of SEA OF GALILEE, on the GOLAN HEIGHTS, above EIN GEV. One of the DECAPOLIS. The present ruins at Sussita are believed to be its site.

Hippo Valley, town, MASVINGO province, SE central ZIMBABWE, 10 mi/ 16 km SW of CHIREDZI, edge of RUNDE RIVER; 21°10′S 31°35′E. GONAREZHOU NATIONAL PARK to SE. Sugarcane, corn, wheat; cattle, sheep, goats.

Hirado (hee-RAH-do), city, NAGASAKI prefecture, on HIRADO island, off NW KYUSHU, SW JAPAN, 47 mi/75 km N of NAGASAKI; 33°21′N 129°33′E. The Portuguese traded at its port c.1550, and Dutch and English factories were established in the early 17th century. There was also trade with CHINA and KOREA. The Konyo Temple is one of the town's many historical sites.

Hirado-shima (hee-rah-DO-shee-mah), island (□ 66 sq mi/171.6 sq km), NAGASAKI prefecture, SW JAPAN, in EAST CHINA SEA, just off NW coast of HIZEN PENINSULA, NW KYUSHU; 20 mi/32 km long, 5 mi/8 km wide. Hilly, with rugged coast; fertile, forested.

Area is shown by the symbol □, and capital city or county seat by ⊙.

Hira-izumi (hee-RAH-ee-ZOO-mee), town, W Iwai district, IWATE prefecture, N HONSHU, NE JAPAN, 50 mi/80 km S of MORIOKA; 38°59′N 141°07′E. Site of Chusonji and Motsu temples.

Hiraka (hee-RAH-kah), town, Hiraka district, Akita prefecture, N HONSHU, NE JAPAN, 37 mi/60 km S of AKITA city; 39°15′N 140°29′E. Pickles.

Hiraka, town, S Tsugaru district, Aomori prefecture, N HONSHU, N JAPAN, 19 mi/30 km S of AOMORI; 40°34′N 140°34′E. Apples. *Oguni* sand (for garden work). Many historic ruins (160 sites).

Hirakata (hee-RAH-kah-tah), city (2005 population 404,044), OSAKA prefecture, S HONSHU, W central JAPAN, 12 mi/20 km N of OSAKA; 34°48′N 135°39′E. Tractors; noodles, local sake. Since early 1940s has included former town of Tonoyama.

Hirakud, INDIA: see SAMBALPUR, city.

Hiram (HEI-ruhm), town (2000 population 1,361), PAULDING county, NW GEORGIA, 23 mi/37 km WNW of ATLANTA; 33°52′N 84°46′W. Manufacturing (apparel, concrete); printing and publishing.

Hiram (HEI-ruhm), town, OXFORD county, W MAINE, on SACO RIVER and 27 mi/43 km SW of SOUTH PARIS; 43°52′N 70°49′W. Wood products.

Hiram (HEI-ruhm), unincorporated village, HARLAN county, SE KENTUCKY, 3 mi/4.8 km WSW of CUMBERLAND, on POOR FORK of CUMBERLAND RIVER. Bituminous coal; timber. Daniel Boone National Forest to NW.

Hiram (HIR-uhm), village (2006 population 1,187), PORTAGE county, NE OHIO, 23 mi/37 km NE of AKRON; 41°19′N 81°08′W. Seat of Hiram College.

Hiran, region (□ 13,000 sq mi/33,800 sq km), S central SOMALIA; ⊙ BELEDUEINE. Borders ETHIOPIA to NW. Scattered grazing, some crop cultivation (sorghum, maize, cassava, millet).

Hiranai (hee-RAH-nah-ee), town, East Tsugaru district, Aomori prefecture, N HONSHU, N JAPAN, 12 mi/20 km N of AOMORI; 40°55′N 140°57′E. Scallops; camellias (N limit).

Hirao (hee-ROU), town, Kumage district, YAMAGUCHI prefecture, SW HONSHU, W JAPAN, 40 mi/65 km E of YAMAGUCHI; 33°56′N 132°04′E. Marine product processing; mandarin oranges. Setonaikai National Park nearby.

Hirapur, INDIA: see BURNPUR.

Hirara (HEE-rah-rah), or **Taira**, city, port on W coast of MIYAKO-JIMA island of SAKISHIMA ISLANDS, in the RYUKYU ISLANDS, OKINAWA prefecture, extreme SW JAPAN, 180 mi/290 km S of NAHA, in EAST CHINA SEA; 24°48′N 125°16′E.

Hirata (hee-RAH-tah), city, SHIMANE prefecture, SW HONSHU, W JAPAN, 14 mi/22 km W of MATSUE across LAKE SHINJI; 35°25′N 132°49′E.

Hirata (hee-RAH-tah), town, Kaizu district, GIFU prefecture, central HONSHU, central JAPAN, 9 mi/15 km S of GIFU; 35°15′N 136°38′E.

Hirata (hee-RAH-tah), town, Akumi district, YAMAGATA prefecture, N HONSHU, NE JAPAN, 50 mi/80 km N of YAMAGATA city; 38°53′N 139°55′E. Ham, sausage; miso, soy sauce. Manufacturing (fusuma).

Hirata (hee-RAH-tah), village, Ishikawa district, FUKUSHIMA prefecture, N central HONSHU, NE JAPAN, 37 mi/60 km S of FUKUSHIMA city; 37°12′N 140°34′E. Tobacco.

Hiratori (hee-RAH-to-ree), town, Hidaka district, HOKKAIDO prefecture, N JAPAN, 50 mi/80 km S of SAPPORO; 42°34′N 142°07′E. Agriculture (vegetables; beef); lilies of the valley; lumber; *Yamabe* trout. Traditional Ainu wood carving. Hot springs and Mt. Horoshiridake nearby.

Hiratsuka (hee-RAHTS-kah), city (2005 population 258,958), KANAGAWA prefecture, E central HONSHU, E central JAPAN, on SAGAMI BAY and the Sagami River, 31 mi/50 km W of YOKOHAMA; 35°19′N 139°21′E. Commercial and industrial center. Tire manufactur-

ing; agriculture (dairying). Known for Tanabata Festival.

Hiraya (HEE-rah-yah), village, Shimoina district, NAGANO prefecture, central HONSHU, central JAPAN, 96 mi/155 km S of NAGANO; 35°19′N 137°37′E.

Hirekerur (hi-RAI-kai-roor), town, tahsil headquarters, DHARWAD district, KARNATAKA state, INDIA, 75 mi/121 km SSE of DHARWAD; 14°28′N 75°23′E. Chili, rice, sugarcane, betel leaf.

Hire Vadvatti (HEE-rai VUHD-vuht-tee), town, DHARWAD district, KARNATAKA state, INDIA, 55 mi/89 km ESE of DHARWAD. Local market for cotton, peanuts, wheat, millet. Sometimes spelled Hirewadwadi.

Hirghigo, Eritrea: see HIRGIGO.

Hirgigo (huhr-GEE-go), Italian *Archico*, village, SEMENAWI KAYIH BAHRI region, central ERITREA, fishing port on RED SEA, 8 mi/12.9 km S of MASSAWA; 15°32′N 39°27′E. Cereals, sesame, vegetables; thermal power station under construction. Formerly spelled Harkiko; also spelled Hirghigo.

Hirgis Nuur, MONGOLIA: see HYARGAS NUUR.

Hiri (hee-REE), volcanic island, N MALUKU, INDONESIA, in MALUKU SEA, just N of TERNATE; 2 mi/3.2 km in diameter; 00°53′N 127°19′E. Mountainous, rising to 2,067 ft/630 m. Fishing.

Hiriyur (hi-ri-YOOR), town, CHITRADURGA district, KARNATAKA state, INDIA, on Vedavati (HAGARI) River and 24 mi/39 km SE of CHITRADURGA; 13°58′N 76°36′E. Road center; coconuts, rice, sugarcane; cotton ginning.

Hîrlău (huhr-LUH-oo), town, IAŞI county, NE ROMANIA, in Moldavia, 24 mi/39 km SE of BOTOŞANI; 47°26′N 26°54′E. Railroad terminus in viticultural region. Glass and furniture manufacturing; distilling; flour milling; dairy products. Has 16th-century church, ruins of 17th-century palace.

Hirmal, LEBANON: see HERMIL.

Hirmil, LEBANON: see HERMIL.

Hirna, village (2007 population 17,524), OROMIYA state, E central ETHIOPIA, in CHERCHER highlands, 20 mi/32 km ENE of ASEBE TEFERI; 09°13′N 41°06′E. Coffee-collecting center.

Hirnyk, (hir-NIK), (Russian *Gornyak*), city, W Central DONETS'K oblast, UKRAINE, in the DONBAS, 16 mi/26 km WNW of DONETS'K and 6 mi/10 km SSE of SELYDOVE (Russian *Selidovo*); subordinated to the Selydove city council; 48°04′N 37°22′E. Elevation 662 ft/201 m. Two bituminous coal mines, prefabrication of building materials from reinforced concrete, bakery. Professional-technical school. Established in 1938; called Sotshorodok (Russian *Sotsgorodok*) until 1958; renamed Hirnyk (Ukrainian=miner).

Hirnyk (hir-NIK), (Russian *Gornyak*), (Ukrainian= miner), town, N L'VIV oblast, UKRAINE, on left bank of Rata River, left tributary of Western BUH RIVER, on road and on railroad 5 mi/8 km SSW of CHERVONOHRAD; 50°19′N 24°12′E. Elevation 574 ft/174 m. Coal mine, sewing factory. Established in 1954, town since 1956.

Hirnyts'ke (hir-NITS-ke), (Russian *Gornyatskoye*), town, E DONETS'K oblast, UKRAINE, on road 4.5 mi/7 km ESE of SNIZHNE; 48°00′N 38°47′E. Elevation 921 ft/280 m. Former coal mine. Established in 1912; town since 1938.

Hirochi, RUSSIA: see PRAVDA.

Hirod, INDIA: see FRENCH ROCKS.

Hirokami (hee-RO-kah-mee), village, North Uonuma district, NIIGATA prefecture, central HONSHU, N central JAPAN, 43 mi/70 km S of NIIGATA; 37°15′N 138°58′E.

Hirokawa (hee-RO-kah-wah), town, Yame district, FUKUOKA prefecture, N KYUSHU, SW JAPAN, 25 mi/40 km S of FUKUOKA; 33°14′N 130°33′E. Grapes.

Hirokawa (hee-RO-kah-wah), town, Arida district, WAKAYAMA prefecture, S HONSHU, W central JAPAN, 15mi/24 km S of WAKAYAMA; 34°01′N 135°10′E.

Hiromi (HEE-ro-mee), town, North Uwa district, EHIME prefecture, NW SHIKOKU, W JAPAN, 40 mi/65 km S of MATSUYAMA; 33°15′N 132°41′E.

Hirono (HEE-ro-no), town, Futaba district, FUKUSHIMA prefecture, central HONSHU, NE JAPAN, on the PACIFIC OCEAN, 47 mi/75 km S of FUKUSHIMA city; 37°12′N 140°59′E.

Hiroo (hee-RO), town, S HOKKAIDO prefecture, N JAPAN, fishing port on the PACIFIC OCEAN, 28 mi/45 km S of OBIHIRO. Agriculture, lumbering; fish canning.

Hiro-ochi (HEE-ro-O-chee), town, Tokachi district, HOKKAIDO prefecture, N JAPAN, 112 mi/180 km S of SAPPORO; 42°17′N 143°16′E. Dairying; pickles.

Hirosaki (hee-RO-sah-kee), city (2005 population 189,043), Aomori prefecture, N HONSHU, N JAPAN, 19 mi/35 km S of AOMORI; 40°36′N 140°28′E. Apples. Former castle town (17th-century ruins). Hirosaki Nebuta Festival held here.

Hirose (HEE-ro-se), town, Nogi district, SHIMANE prefecture, SW HONSHU, W JAPAN, 11 mi/17 km SE of MATSUE; 35°21′N 133°10′E. Beef cattle.

Hiroshima (hir-ro-SHEE-mah), prefecture [Jap. *ken*] (□ 3,258 sq mi/8,438 sq km; 1990 population 2,861,699), SW HONSHU, W JAPAN, on HIROSHIMA BAY (S) and across HIUCHI SEA (central section of INLAND SEA) from SHIKOKU; ⊙ HIROSHIMA. Includes offshore islands of NOMI-SHIMA, KURAHASHI-JIMA, ITSUKU-SHIMA, KAMI-KAMAKARI-JIMA, OSAKI-KAMI-SHIMA, IKUCHI-JIMA, MUKAI-SHIMA, OSAKI-SHIMO-SHIMA, SAKI-SHIMA, TA-JIMA, and several smaller islets. Bordered by Shimane and Tottori prefectures, E by Okayama prefecture, and W by Yamaguchi prefecture. Generally mountainous, with fertile valleys. Small streams drain agricultural and livestock-raising areas. Hot springs at Shobara in interior. Rice and oranges are grown extensively, cattle are raised, textiles are manufactured, and shipyards are plentiful. Mazda Motor Co. headquarters. Manufacturing of writing brushes, fountain pens, insecticide (made from chrysanthemums), dyes, floor mats, soy sauce. Exports sake, fountain pens, lumber, charcoal, textiles. Principal centers include Hiroshima, KURE (naval base), ONOMICHI, FUKUYAMA, and MIHARA.

Hiroshima (hir-ro-SHEE-mah), city (2005 population 1,154,391), ⊙ HIROSHIMA prefecture, SW HONSHU, W JAPAN, on HIROSHIMA BAY; 34°22′N 132°27′E. Important commercial and industrial center; manufacturing (machinery, motor vehicles, furniture, consumer goods, and Buddhist altars). Also a market for agricultural and marine products (oysters). Founded c.1594 as a castle city on the OTA RIVER delta, Hiroshima is divided by the river's seven mouths into six islands. After 1868, Hiroshima's port, Ujina, was enlarged, and railroad lines were built to link it with KOBE and SHIMONOSEKI. Target (Aug. 6, 1945) of the first atomic bomb ever dropped on a populated area; almost 130,000 people were killed, injured, or missing, and 90 percent of the city was leveled. Much of the city has been reconstructed, but a gutted section (around the Genbaku Dome, the center of the blast) has been set aside as a "Peace City" to illustrate the effect of an atomic bomb. Since 1955, an annual world conference against nuclear weapons has met here. Peace Memorial Park is located here; site of the Peace Memorial Ceremony. Habitat of *o sanshouo* salamander.

Hiroshima (hir-ro-SHEE-mah), city, Ishikari county, Hokkaido prefecture, N JAPAN, 20 mi/32 km S of SAPPORO; 42°59′N 141°34′E. Cold-climate rice planting originated here.

Hiroshima Bay (hir-ro-SHEE-mah), Japanese *Hiroshima-wan* (hee-RO-shi-mah-WAHN), inlet of INLAND SEA, SW HONSHU, W JAPAN; KURAHASHI-JIMA forms E side of entrance; 20 mi/32 km N-S, 15 mi/24 km E-W. Contains the islands of NOMI-SHIMA,

---

Cross-references are shown in SMALL CAPITALS. The pronunciation guide is shown on page xix. The sources of population figures are shown on page xvii.

ITSUKU-SHIMA, and many smaller islands HIROSHIMA is at head of bay, KURE is on E shore.

Hirota (HEE-ro-tah), village, Iyo district, EHIME prefecture, NW SHIKOKU, W JAPAN, 16 mi/25 km S of MATSUYAMA; 33°37'N 132°47'E. Clay for pottery.

Hirsau (HIR-sou), suburb of CALW, BADEN-WÜRTTEMBERG, S GERMANY, in BLACK FOREST, on the NAGOLD, 1.5 mi/2.4 km N city center; 48°43'N 08°45'E. Health resort. Has ruins of noted Benedictine monastery (destroyed 1692).

Hirschaid (HIR-sheit), village, BAVARIA, S GERMANY, in UPPER FRANCONIA, on RHINE-MAIN-DANUBE CANAL, near Regnitz River, and 7 mi/11.3 km SE of BAMBERG; 49°49'N 11°00'E. Hydroelectric station.

Hirschau (HIRSH-ou), town, BAVARIA, S GERMANY, in UPPER PALATINATE, 8 mi/12.9 km NNE of AMBERG; 49°33'N 11°57'E. Porcelain manufacturing, brewing. Has late-Gothic church and mid-16th-century town hall.

Hirschberg (HIRSH-berg), town, NORTH RHINE–WESTPHALIA, W GERMANY, 6 mi/9.7 km N of MESCHEDE; 51°26'N 08°16'E.

Hirschberg, CZECH REPUBLIC: see DOKSY.

Hirschberg, POLAND: see JELENIA GÓRA.

Hirschhorn am Neckar (HIRSH-horn ahm NEK-kahr), town, HESSE, central GERMANY, at S foot of the Odenwald, on the NECKAR, 10 mi/16 km ENE of HEIDELBERG; 49°27'N 08°53'E. Manufacturing of electrical goods and other small industries. Has 13th-century castle. Chartered 1391.

Hirsilä (HUH-si-lah), Swedish *Voitila*, village, HÄMEEN province, SW FINLAND, 25 mi/40 km NE of TAMPERE; 61°44'N 24°20'E. Elevation 264 ft/80 m. Shoe manufacturing; historic church.

Hirsingue (eer-SAN-guh), German, *Hirsingen* (HEER-sin-guhn), commune (□ 4 sq mi/10.4 sq km), HAUT-RHIN department, in ALSACE, E FRANCE, on upper ILL RIVER and 3 mi/4.8 km S of ALTKIRCH; 47°35'N 07°15'E. Textiles.

Hirske (HEERS-ke), (Russian *Gorskoye*), town, LUHANS'K oblast, UKRAINE, on road, on railroad, 39 mi/63 km W of LUHANS'K; 48°44'N 38°29'E. Elevation 620 ft/188 m. Poultry plant. Ukraine Institute of International Management is here; also Heritage museum.

Hirson (eer-son), town (□ 13 sq mi/33.8 sq km), AISNE department, PICARDIE region, N FRANCE, on the upper OISE RIVER and 35 mi/56 km NE of SAINT-QUENTIN, near Belgian border; 49°55'N 04°05'E. Railroad junction and metalworking center (die-casting mill); glassworks. Just E is the forest of SAINT-MICHEL, a recreation area adjacent to a 10th-century abbey, rebuilt in 1970s.

Hîrşova (HUHR-sho-vah), ancient *Carsium*, town, CONSTANŢA county, SE ROMANIA, in DOBRUJA, on DANUBE River, and 50 mi/80 km NW of CONSTANŢA. Small inland port. Manufacturing of tiles and bricks; food processing; distilling; stone and sand quarrying. Has notable remains of a Roman city and a Turkish fortress.

Hirta, Scotland: see ST. KILDA.

Hirtshals (HIRTS-hahls), city and port, NORDJYLLAND county, N JUTLAND, DENMARK, on the SKAGERRAK and 9 mi/14.5 km N of HJØRRING; 57°35'N 10°06'E. Railroad and ferry terminus; ferry to KRISTIANSAND, NORWAY.

Hirtzenberg (HIR-tzuhn-berk), hilltop (1,525 ft/465 m) of the ARDENNES, LUXEMBOURG province, SE BELGIUM, 2 mi/3.2 km SW of ARLON. Highest point in LUXEMBOURG Ardennes.

Hirukawa (hee-ROO-kah-wah), village, Ena district, GIFU prefecture, central HONSHU, central JAPAN, 37 mi/60 km N of GIFU; 35°31'N 137°23'E. Minerals. Oi Dam (powers Japan's first hydroelectric plant) and Ena Gorge nearby. Rare *yukino hana* [=snow flower] flower is native to this area.

Hirwaun, town (2001 population 4,851), RHONDDA CYNON TAFF, S Wales, adjoining ABERDARE, and 4 mi/

6.4 km SW of MERTHYR TYDFIL; 51°45'N 03°30'W. Site of Tower Colliery, the last deep-pit anthracite mine operating in S Wales. Formerly in MID GLAMORGAN, abolished 1996.

Hisaga-shima (hee-sah-GAH-shee-mah) or **Hisaka**, island (□ 15 sq mi/39 sq km) of island group GOTO-RETTO, NAGASAKI prefecture, SW JAPAN, in EAST CHINA SEA, 50 mi/80 km W of KYUSHU; 5.5 mi/8.9 km long, 4 mi/6.4 km wide. Large bay indents N coast; hilly.

Hisai (hee-SAH-ee), city, MIE prefecture, S HONSHU, central JAPAN, 5 mi/8 km S of TSU; 34°40'N 136°28'E. Hot springs.

Hisar (hi-SAHR), district (□ 2,424 sq mi/6,302.4 sq km), HARYANA state, N INDIA; ⊙ HISAR. On NE edge of THAR DESERT; bordered N by PUNJAB state, S by BHIWANI district, W by SIRSA district, SW by RAJASTHAN state, and SE by JIND and ROHTAK districts. Irrigated by branches of Bhakra, SIRHIND, and WESTERN YAMUNA canals. Agriculture (millet, gram, cotton, oilseeds); hand-loom weaving. Chief towns include Hisar, HANSI, FATEHABAD, and TOHANA. Several outbreaks occurred here during Sepoy Rebellion of 1857. Original district enlarged (1948) by the incorporation of former Punjab state of Loharu.

Hisar (hi-SAHR), city (2001 population 263,186), ⊙ HISAR district, HARYANA state, N INDIA, on the WESTERN YAMUNA CANAL; 29°10'N 75°43'E. A district administrative center in a well-irrigated area. Road and railroad center; market for cotton, grain, and oilseeds. An agricultural experimental farm is on the outskirts, one of N INDIA's largest livestock-breeding centers. Also seat of Agriculture University at HISAR. Cotton and silk fabrics are made by hand loom in the town. Founded 1356; important under the MOGUL EMPIRE. De-populated by famine in 1783; occupied by the British in 1803. A religious and historical site.

Hisarak, IRAN: see HESARAK.

Hisaronu, Gulf of, (Turkish=*Hisarönü*) inlet of SW coast of TURKEY, NW of RHODES; 18 mi/29 km long. Sombeki island at its mouth.

Hisarya (hee-SAHR-yah), city, PLOVDIV oblast, ⊙ Hisarya obshtina, central BULGARIA, at the S foot of the central SREDNA GORA, 11 mi/18 km SSW of KARLOVO; 42°31'N 24°42'E. Railroad terminus; health resort (thermal springs). Rye, aromatic oil crops, vineyards, fruit; livestock. Site of ancient Roman ruins of a city (Diokletianopol). Village formed by the amalgamation of Hisar and Momina Banya in 1942.

Hisayama (hee-SAH-yah-mah), town, Kasuya district, FUKUOKA prefecture, N KYUSHU, SW JAPAN, 6 mi/10 km N of FUKUOKA; 33°38'N 130°30'E.

Hisban (his-BAHN), village, central JORDAN, 12 mi/19 km SW of Amman; 31°48'N 35°48'E. Grain (wheat, barley), fruit. It was the biblical Heshbon. Also spelled Hesban.

Hiseville (HEIZ-vil), village (2000 population 224), BARREN county, S KENTUCKY, 10 mi/16 km NE of GLASGOW; 37°06'N 85°49'W. Agriculture (burley tobacco, grain; livestock. Caves in area.

Hishikari (hee-shee-KAH-ree), town, Isa district, KAGOSHIMA prefecture, S KYUSHU, SW JAPAN, 28 mi/45 km N of KAGOSHIMA; 32°00'N 130°38'E. Gold (Hishikari ore) mining. Alkaline hot springs nearby.

Hisnumansur, TURKEY: see ADIYAMAN.

Hisøy (HEES-uh-u), island (□ 3.5 sq mi/9.1 sq km) in the SKAGERRAK, AUST-AGDER county, S NORWAY, 1 mi/1.6 km off coast S of ARENDAL. Pilot station, formerly fortified. Fishing, sailing, light manufacturing Marine biological research center in Flødevigen. Part of the local Arendal economy. Stone Age findings made here.

Hispalis, SPAIN: see SEVILLE, city.

Hispania (his-PAI-nee-yuh), Latin name for IBERIAN PENINSULA including present-day SPAIN and PORTUGAL.

Hispanic America: see SPANISH AMERICA.

Hispaniola (ees-pah-nee-O-lah), Spanish *Española*, second-largest (□ 29,530 sq mi/76,483 sq km) island of the WEST INDIES, between CUBA and PUERTO RICO; 19°00'N 71°00'W. HAITI occupies the W one-third of the island and the DOMINICAN REPUBLIC the remainder. The largest cities include PORT-AU-PRINCE, Haiti, and SANTO DOMINGO, the Dominican Republic. Visited by Columbus in 1492, the island was called Española. The later French colony was called Saint-Domingue, after Santo Domingo, the Spanish colony in the E part of the island. The terrain, dominated by the Cordillera Central, is high and rugged; Pico Duarte (10,417 ft/3,175 m) is the tallest peak. Extending far W, like the claws of a crab, two mountain ranges form the scenic Gulf of Gonâve. The climate is subtropical, and agriculture (coffee, cocoa, sugarcane, and tobacco) flourishes in the abundant rainfall. In some areas of the island (especially Haiti), increased population has caused significant deforestation for cultivation.

Hissar, TAJIKISTAN: see GISSAR.

Hissarlik, TURKEY: see TROY.

Hissmofors, SWEDEN: see KROKOM.

Histiaia, Greece: see ISTIAIA.

Histon, ENGLAND: see CAMBRIDGE.

Historic Camden National Affiliated Area, KERSHAW county, N central SOUTH CAROLINA, authorized 1982. Colonial village. Established 1730; occupied by the British 1780–1781.

Histria, ROMANIA: see ISTRIA.

Hisua (HIS-wuh), town (2001 population 25,045), NAWADA district, S central BIHAR state, E INDIA, 26 mi/42 km E of GAYA; 24°50'N 85°25'E. Pottery manufacturing; rice, gram, wheat, barley. Hindu cave sculpture and inscriptions 9 mi/14.5 km NNW, at RAJPIND CAVE. In 2002, Hisua was declared India's first child labor free block (an administrative subdivision).

Hiswa (HIS-wuh), village, YEMEN, on mainland near ADEN, 4 mi/6.4 km SW of SHEIKH OTHMAN, on ADEN BAY at mouth of the Wadi Kebir (arm of the Wadi TIBAN); irrigated agriculture.

Hisya (HIS-yah), township, HOMS district, W SYRIA, 22 mi/35 km S of HOMS; 34°25'N 37°45'E. Cereals. Also spelled HASIYE.

Hit, township, ANBAR province, central IRAQ, on the W bank of the EUPHRATES and 32 mi/51 km WNW of AR RAMADI; 33°38'N 42°49'E. Dates, sesame, millet, corn. Site of ancient Mesopotamian city of *Is*, whose bitumen wells were utilized in the construction of the walls of BABYLON.

Hita (hee-TAH), city, OITA prefecture, N central KYUSHU, SW JAPAN, on the CHIKUGO RIVER and the HITA plain, 37 mi/60 km W of OITA; 33°19'N 130°56'E. Agriculture (shiitake mushrooms, pears) and industrial center; also produces lumber, videotapes, and pottery. Popular resort town known for its scenic beauty. Site of Kangi-En, a private school of the late Edo period.

Hita (EE-tah), village, GUADALAJARA province, central SPAIN, 15 mi/24 km NNE of GUADALAJARA. Cereals, olives, vegetables.

Hitachi, former province in central HONSHU, JAPAN; now IBARAKI prefecture.

Hitachi (HEE-tah-chee), city (2005 population 199,218), IBARAKI prefecture, central HONSHU, E central JAPAN, on the Kashima Sea, 19 mi/30 km N of MITO; 36°35'N 140°39'E. Leading producer of Japan's electrical equipment and the largest commercial port in the prefecture; manufacturing includes cables, computers, machinery, and nuclear and hydraulic power equipment.

Hitachiota (HEE-tah-chee-O-tah), city, IBARAKI prefecture, central HONSHU, E central JAPAN, 12 mi/20 km N of MITO; 36°32'N 140°32'E.

Hitchcock, county (□ 718 sq mi/1,866.8 sq km; 2006 population 2,926), S NEBRASKA; ⊙ TRENTON; 40°10'N

101°02′W. Agricultural area bounded S by KANSAS; drained by REPUBLICAN RIVER and Fredmen Creek. Corn, wheat, sorghum, sunflower seed; cattle, hogs. Petroleum. Central/Mountain time zone boundary follows W boundary. SWANSON RESERVOIR and State Recreation Area in W center; Massacre Canyon Monument in E center. Formed 1873.

Hitchcock, town (2006 population 7,265), GALVESTON county, S TEXAS, suburb 7 mi/11.3 km SW of TEXAS CITY, on Highland Bayou; 29°19′N 95°01′W. Oil and gas; cattle; rice, soybeans; manufacturing (industrial chemicals).

Hitchcock, village (2006 population 139), BLAINE county, W central OKLAHOMA, 9 mi/14.5 km NNE of WATONGA; 35°58′N 98°20′W. In agricultural area. Roman Nose State Park to SW.

Hitchcock, village (2006 population 98), BEADLE county, E central SOUTH DAKOTA, 20 mi/32 km NNW of HURON; 44°37′N 98°28′W.

Hitchin (HICH-in), town (2001 population 30,851), HERTFORDSHIRE, SE ENGLAND, 8 mi/12.8 km NE of LUTON; 51°57′N 00°16′W. Hitchin was the site of a monastery in Offa's time and appears in the *Domesday Book* as a royal manor named Hiz. Grain and cattle market. Engineering, tanning, parchment making, medicinal distilling, and rose growing. Henry Bessemer, the inventor of modern steel making, born here.

Hitchins (HICH-uhns), unincorporated village, CARTER county, NE KENTUCKY, 21 mi/34 km SW of ASHLAND. In agricultural and clay-producing area; agriculture (tobacco, corn; cattle). GRAYSON LAKE reservoir and State Park to SW.

Hitchita (hich-EET-uh), village (2006 population 112), MCINTOSH county, E OKLAHOMA, 14 mi/23 km ESE of OKMULGEE, on DEEP FORK (CANADIAN River) arm of EUFAULA LAKE (to S); 35°31′N 95°45′W. In agricultural area. Eufaula Lake nearby.

Hither Pomerania, EUROPE: see POMERANIA.

Hiti, TUAMOTU ARCHIPELAGO: see RAEVSKI ISLANDS.

Hito, El (el EE-to), village, CUENCA province, E central SPAIN, 35 mi/56 km WSW of CUENCA. Grain growing and wine producing; stock raising.

Hitokappu Bay, RUSSIA: see ITURUP ISLAND.

Hitoy-Cerere Biological Reserve (ee-TO-ee seh-REH-reh), in CORDILLERA DE TALAMANCA, COSTA RICA, 25 mi/40 km S of LIMÓN. Evergreen rainforest receives up to 140 in/356 cm of precipitation per year. Many rapids and falls characterize the broken topography (23,000 acres/9,308 ha).

Hitoyoshi (hee-TO-yo-shee), city, KUMAMOTO prefecture, W central KYUSHU, SW JAPAN, on the KUMA RIVER, 40 mi/65 km S of KUMAMOTO; 32°12′N 130°45′E. Agricultural center (shiitake mushrooms, tea), railroad junction, and popular resort area noted for its hot springs. Distilled liquor (*shochu*). Former Edo-era castle town noted for its Aoi Asho (Shinto) shrine built in 1611 A.D.

Hitra (EE-trah), island (□ 218 sq mi/566.8 sq km) in NORTH SEA, SØR-TRØNDELAG county, central NORWAY, separated from mainland by TRONDHEIM Channel, 37 mi/60 km W of TRONDHEIM; 29 mi/47 km long, 12 mi/19 km wide. Animal husbandry; herring fishing; fish farming; boatbuilding. Deposits of iron ore and galena, not mined since late nineteenth century.

Hitrino, Bulgaria: see GARA HITRINO.

Hitterdal (HI-tuhr-dawl), village (2000 population 201), CLAY county, W MINNESOTA, 26 mi/42 km ENE of FARGO, NORTH DAKOTA; 46°58′N 96°15′W. Grain, beans, sugar beets; dairying. White Earth Indian Reservation to E; small lakes to E and SE.

Hittin, Horns of (khee-TEEN), mountain (1,069 ft/326 m), LOWER GALILEE, NE ISRAEL, 3.7 mi/6 km W of TIBERIAS, situated in the center of an ancient volcano. Elevation 1,070 ft/326 m. Important fortress. Scene (1187) of decisive defeat of Crusaders by Saladin.

Hitzacker (HITS-ahk-ker), town, LOWER SAXONY, N GERMANY, on left bank of the ELBE, 26 mi/42 km ESE

of LÜNEBURG; 53°09′N 11°02′E. Mainly agriculture with some small industry. Mineral baths.

Hitzendorf (HIT-tsen-dorf), township, SW STYRIA, SE AUSTRIA, 6 mi/9.7 km WSW of GRAZ; 47°02′N 15°18′E. Corn, cattle.

Hiu, VANUATU: see TORRES ISLANDS.

Hiuchi Sea (hee-OO-chee), central section of INLAND SEA, W JAPAN, between S coast of HONSHU and N coast of SHIKOKU; merges E with HARIMA SEA, W with IYO SEA; c.60 mi/97 km E-W, c.35 mi/56 km N-S. Contains many islands; largest is OMI-SHIMA. IMABARI is on SW shore.

Hiva Oa (HEE-vah O-ah), volcanic island (□ 154 sq mi/400.4 sq km), South PACIFIC, second-largest and the most important of the MARQUESAS ISLANDS, FRENCH POLYNESIA; 09°45′S 139°00′W. Hiva Oa contains ATUONA, between 1904 and the 1940s the administrative center of the Marquesas. The Bay of Traitors, protected by ridges reaching 3,904 ft/1,190 m high, provides a good harbor. Copra is the chief export.

Hiw (HOO), ancient *Diospolis Parva*, village, QENA province, Upper EGYPT, 3 mi/4.8 km SE of NAG HAMMADI; 26°01′N 32°16′E. Cereals, sugarcane, dates.

Hiwa (hee-WAH), town, Hiba district, HIROSHIMA prefecture, SW HONSHU, W JAPAN, 50 mi/80 km N of HIROSHIMA; 34°58′N 132°59′E.

Hiwaki (HEE-wah-kee), town, KAGOSHIMA prefecture, SW KYUSHU, SW JAPAN, 19 mi/30 km N of KAGOSHIMA; 31°49′N 130°23′E. Distilled alcoholic beverage (*shochu*). Hot springs nearby.

Hiwasa (hee-WAH-sah), town, Kaifu district, TOKUSHIMA prefecture, E SHIKOKU, W JAPAN, on PHILIPPINE SEA, 22 mi/35 km S of TOKUSHIMA; 33°43′N 134°32′E. Site of Yakuo temple. Breeding ground for the red sea turtle on the nearby Ohama coast.

Hiwassee Lake (hei-WAW-see), reservoir (□ 3 sq mi/7.8 sq km), CHEROKEE county, extreme W NORTH CAROLINA, in Nantahala National Forest, on HIWASSEE RIVER, 70 m/112 km SW of WAYNESVILLE and near GEORGIA and TENNESSEE borders; 35°09′N 84°11′W. Formed by Hiwassee Dam (307 ft/97 m high), built (1940) by Tennessee Valley Authority for flood control.

Hiwassee River, c.140 mi/225 km long, GEORGIA, NORTH CAROLINA, and TENNESSEE; rises in the BLUE RIDGE in TOWNS county, N Georgia; flows NW through SW corner of North Carolina, past HAYESVILLE, and NW, past MURPHY, into Cherokee National Forest, into SE Tennessee, joins TENNESSEE RIVER (CHICKAMAUGA Reservoir) 13 mi/21 km N of CLEVELAND, Tennessee, in POLK county; 35°24′N 85°00′W. Drains 2,700 sq mi/6,993 sq km, including entire drainage basin of the OCOEE RIVER, a tributary. In North Carolina, Apalachia, HIWASSEE, and CHATUGE dams are major Tennessee Valley Authority units. Hiwassee Dam (307 ft/94 m high, 1,287 ft/392 m long; completed 1940) is 10 mi/16 km WNW of MURPHY; concrete, straight gravity, overflow construction; used for hydroelectric power and flood control. Impounds Hiwassee Reservoir (c. 10 sq mi/26 sq km; 22 mi/35 km long, capacity 438,000 acre-ft) in CHEROKEE county, North Carolina. Designated a Tennessee Scenic River in 1968. Whitewater rafting, canoeing.

Hixton, village (2006 population 445), JACKSON county, W central WISCONSIN, on TREMPEALEAU RIVER and 38 mi/61 km SE of EAU CLAIRE; 44°22′N 91°00′W. In dairying region. Timber.

Hiyoshi (hee-YO-shee), town, Hioki district, KAGOSHIMA prefecture, SW KYUSHU, SW JAPAN, 12 mi/20 km W of KAGOSHIMA; 31°35′N 130°20′E. Tile.

Hiyoshi (hee-YO-shee), town, Funai district, KYOTO prefecture, S HONSHU, W central JAPAN, 19 mi/30 km N of KYOTO; 35°09′N 135°30′E. *Matsutake* mushrooms.

Hiyoshi (hee-YO-shee), village, North Uwa district, EHIME prefecture, NW SHIKOKU, W JAPAN, 34 mi/55 km S of MATSUYAMA; 33°19′N 132°47′E.

Hiyoshi (hee-YO-shee), village, Kiso district, NAGANO prefecture, central HONSHU, central JAPAN, 59 mi/95 km S of NAGANO; 35°52′N 137°45′E.

Hizan, village, SE TURKEY, 25 mi/40 km SE of BITLIS; 38°10′N 42°26′E. Wheat, millet, barley. Formerly Karasufla or Karasusufla.

Hizen, former province in NW KYUSHU, JAPAN; now NAGASAKI and SAGA prefectures.

Hizen Peninsula (HEE-zen), Japanese *Hizen-hanto* (hee-ZEN-HAHN-to), W KYUSHU, SW JAPAN, in EAST CHINA SEA. Branching into three sprawling peninsulas: SONOGI PENINSULA (SW), NOMO PENINSULA (S), SHIMABARA PENINSULA (SE). Indented by OMURA BAY (W), TACHIBANA BAY (S), Ariakenoumi and SHIMABARA BAY (E); joined to Kyushu mainland by neck c.30 mi/48 km wide; major portion of peninsula is c.50 mi/80 km long. Contains NAGASAKI prefecture and part of SAGA prefecture. NAGASAKI city in S, SASEBO in W, KARATSU in N. Formerly all of HIZEN province.

Hizma, Arab village, Jerusalem district, 3.1 mi/4 km NE of JERUSALEM, on the W fringes of the Judaean Wilderness, WEST BANK; 31°50′N 35°16′E. Believed to be the biblical site of Beit Azmaweth. Well known for its cereals and milk products in the 19th century.

Hjallerup (YAH-luh-roop), town, NORDJYLLAND county, N JUTLAND, DENMARK, 21 mi/34 km SSE of HJØRRING; 57°10′N 10°09′E.

Hjälmaren (YEL-mahr-en), lake (□ 190 sq mi/494 sq km), in ÖREBRO and SÖDERMANLAND counties, S central SWEDEN. Drained into MÄLAREN LAKE by Hjälmaren. Canal connects lake with ARBOGAÅN (or Arboga) E.

Hjarnø (YAHRN-u), low island (□ 1 sq mi/2.6 sq km), DENMARK, at mouth of HORSENS FJORD, off E JUTLAND; 55°49′N 10°04′E. Farming.

Hjo (YOO), town, SKARABORG county, S SWEDEN, on W shore of LAKE VÄTTERN, 4 mi/6.4 km N of JÖNKÖPING; 58°19′N 14°17′E. Lake port; health and tourist resort. Manufacturing (metalworking). Europe Nostra Award for the preservation of its old wooden buildings (1990). First mentioned as town 1413.

Hjørring (YUHR-ing), Danish *Hjørring*, DENMARK'S most N county (□ 1,102 sq mi/2,865.2 sq km), N JUTLAND, N of Lim Fjord; ⊙ HJØRRING. Includes islands LÆSØ and Hirsholm in the KATTEGAT, Gjol in Lim Fjord. Drained by Rye and Uggerby rivers. Hilly in E and SE; soil rather poor. Agriculture (dairy farming); manufacturing (textiles; carpeting; wood and furniture; metal and machinery; feed; computer software). The chief cities are FREDERIKSHAVN, BRØNDERSLEV, Hjørring.

Hjørring (YUHR-ing), city, NORDJYLLAND county, N DENMARK; 57°25′N 10°00′E. The center of an agricultural region, it has food-processing plants, textile mills, machine shops, and shipyards. Hjørring dates from the twelfth century and has retained several medieval churches.

Hjørungavåg (YUH-roong-ah-VAWG), village, MØRE OG ROMSDAL county, W NORWAY, on E shore of HAREID island, 9 mi/14.5 km SSW of Ålesund. Fishing; shipyards. Scene of naval battle in 986 between Viking kings.

Hjulsbro (YOOLS-BROO), village, ÖSTERGÖTLAND county, SE SWEDEN, on STÅNGÅN River, 4 mi/6.4 km SE of LINKÖPING.

Hkakabo Razi (kah-KAH-bo RAH-zee), peak (19,296 ft/5,881 m), N MYANMAR, on an outlier of the Himalayan system. It is the highest point in MYANMAR.

Hkamti (KAHM-tee), township and village (□ 981 sq mi/2,550.6 sq km), in Naga Hills, MYANMAR, on left bank of CHINDWIN RIVER (ferry crossing) and 90 mi/145 km NE of HOMALIN; 26°00′N 95°42′E. Was (until 1947) the capital of SHAN STATE of Singkaling Hkamti, Shan *Kanti*.

Hkamti Long (KAHM-tee), former group of 8 Shan states (□ 296 sq mi/769.6 sq km) of MYITKYINA dist.,

KACHIN STATE, Myanmar, in Putao region. After 1947 inc. into Kachin State.

Hladir, NORWAY: see LADE.

Hladkyy, UKRAINE: see VOLODARS'KE.

Hlaingbwe (huh-LEING-bwai), village, MON STATE, MYANMAR, on Hlaingbwe River (headstream of GYA-ING RIVER) and 45 mi/72 km NNE of MAWLAMYINE. Rice, teak forests (government reserves).

Hlaing River, MYANMAR: see MYITMAKA RIVER.

Hlane Royal National Park (huh-LAHNE), (□ 151 sq mi/392.6 sq km), LUBOMBO district, NE SWAZILAND, c.40 mi/64 km ENE of MBABANE, bounded N by Mbuluzi River; 26°14'S 31°51'E. More than 10,000 animals; lions (reintroduced 1994), giraffe, wildebeest, cheetah, antelope, elephants, white rhinoceros. Established during 1940s as a preserve but only became a national sanctuary in 1967.

Hlatikulu (huh-LAH-ti-koo-loo), town, ⊙ SHISELWENI district, S SWAZILAND, 50 mi/80 km SSE of MBABANE; 26°58'S 31°19'E. Goats, sheep, cattle; corn, tobacco, cotton, vegetables. Near peak of the same name (4,134 ft/1,260 m).

Hlebarovo, Bulgaria: see TSAR KALOYAN.

Hlebine, village, central CROATIA, in PODRAVINA region, 6 mi/9.7 km SE of KOPRIVNICA. Since 1930 the center of Hlebine School of painting related to the primitive art movement; painters Ivan and Josip Generalić born and worked here.

Hlegu (huh-LE-goo), township, YANGON division, MYANMAR, 22 mi/35 km NNE of YANGON. Road center.

Hlevakha (hle-VAH-khah), (Russian *Glevakha*), town (2004 population 15,250), central KIEV (Ukrainian *Kyyiv*) oblast, UKRAINE, 14 mi/23 km SSW of KIEV city center; 50°17'N 30°17'E. Elevation 613 ft/186 m. Agricultural research station of the Academy of Sciences of Ukraine; formerly All-Union research institute for renovating worn-out machine parts, Ukrainian research institute for the mechanization and electrification of agriculture. Known since the second half of the 16th century; town since 1973.

Hleven (HLEV-en), peak (8,678 ft/2,645 m), PIRIN MOUNTAINS, SW BULGARIA; 41°40'N 23°32'E.

Hliboka, UKRAINE: see HLYBOKA.

Hlinik nad Hronom (hli-NYEEK NAHD hro-NOM), Slovak *Hliník nad Hronom*, Hungarian *Garamgeletnek*, village, STREDOSLOVENSKY province, SW central SLOVAKIA, on HRON RIVER, on railroad and 4 mi/6.4 km SW of ZIAR NAD HRONOM; 48°33'N 18°47'E. Manufacturing of mining machines. Gothic church. Health resort of Sklene Teplice (elev. 1,181 ft/360 m), Slovak *Sklené Teplice*, is 3 mi/4.8 km ESE.

Hlinik nad Vahom, SLOVAKIA: see BYTCA.

Hlinsko (HLIN-sko), town, VYCHODOCESKY province, E BOHEMIA, CZECH REPUBLIC, on CHRUDIMKA RIVER, on railroad and 20 mi/32 km SSE of PARDUBICE; 49°46'N 15°54'E. In sugar beet and potato district; fur-processing center; also has cotton and jute mills. Bathing place of HAMRY (HAHM-ri) is 2 mi/3.2 km SSE. Smokers' pipes made at village of Proseč (PRO-sech), 8 mi/12.9 ENE.

Hlobyne (HLO-bi-ne), (Russian *Globino*), city, SW POLTAVA oblast, UKRAINE, 21 mi/34 km NNW of KREMENCHUK; 49°23'N 33°17'E. Elevation 324 ft/98 m. Raion center; railroad station; grain elevator; sugar mill, food processing (butter, fruit and vegetable canning, flour). Heritage museum. Established at beginning of the 18th century; city since 1976.

Hlohovec (hlo-HO-vets), Hungarian *Galgóc*, city (2001 population 23,729), ZAPADOSLOVENSKY province, W SLOVAKIA, on VÁH RIVER, on railroad; 48°26'N 17°48'E. Agricultural area (wheat, corn, sugar beets, barley); manufacturing (medicine, clothing, wires; brewing, woodworking). Formerly also known as Frastak (frish-TAHK), Slovak *Frašták*, German *Freystadl*. Has 13th-century church and castle with park and theatre; also, museum and observatory. Military base.

Hlotse, town (2004 population 23,400), ⊙ LERIBE district, N LESOTHO, 47 mi/76 km NE of MASERU, on Hlotse River, 1 mi/1.6 km E of its mouth on CALEDON RIVER, on main N-S road, airfield to NE; 28°52'S 28°03'E. Leribe Plateau to E. Manufacturing (mohair products, handicrafts). Goats, sheep, cattle; corn, vegetables, sorghum, wheat. Once called Leribe.

Hlubocepy (HLU-bo-CHE-pi), Czech *Hlubočepy*, SSW district of PRAGUE, PRAGUE-CITY province, BOHEMIA, CZECH REPUBLIC, on left bank of VLTAVA RIVER, 4 mi/6.4 km from city center.

Hlubocky (HLU-boch-ki), Czech *Hlubočky*, German *Hombok*, village, SEVEROMORAVSKY province, N central MORAVIA, CZECH REPUBLIC, on railroad and 7 mi/11.3 km ENE of OLOMOUC; 49°38'N 17°24'E. Manufacturing (machinery, household equipment); foundry; wheat and hops growing.

Hluboczek Wielki, UKRAINE: see VELYKYY HLYBOCHOK.

Hlucin (HLU-cheen), Czech *Hlučín*, German *Hultschin*, town, SEVEROMORAVSKY province, NE SILESIA, CZECH REPUBLIC, on railroad and 6 mi/9.7 km NW of OSTRAVA; 49°54'N 18°11'E. Agriculture (sugar beets, oats); manufacturing (machines, food, wood); gravel pit. Has 16th-century castle and Gothic church.

Hluhluwe Game Reserve, SOUTH AFRICA: see also KWAZULU-NATAL. Famous for white rhino as well as a wide range of other game which inhabit this 98 sq mi/250 sq km sanctuary; wilderness part is situated 30 mi/48 km in NW of ST. LUCIA.

Hluhluwe-Umfolozi Park (□ 372.2 sq mi/964 sq km), KWAZULU-NATAL, SOUTH AFRICA, 35 mi/56 km N of RICHARD'S BAY on the confluence of White and Black MFOLOZI rivers. One of the oldest game sanctuaries in the republic (established 1895) and before that a hunting preserve for the Zulu kings. Home to the world's largest concentrations of rhino – white rhino (1,600) and black rhino (350) – as well as elephant, buffalo, lion, leopard, giraffe, hippo, wild dog, cheetah, crocodile, and many species of antelope. Birdlife is also prolific. Popular wilderness trails.

Hluk (HLUK), German *Hulken*, town, JIHOMORAVSKY province, SE MORAVIA, CZECH REPUBLIC, 6 mi/9.7 km SSE of UHERSKÉ Hradiště; 48°59'N 17°32'E. Manufacturing (machinery); woodworks; tanning and food processing; old administrative center of region (11th–12th centuries); 16th-century Renaissance stronghold.

Hlukhiv (HLOO-khif), (Russian *Glukhov*), city, NE SUMY oblast, UKRAINE, 65 mi/105 km NNW of SUMY; 51°41'N 33°55'E. Elevation 583 ft/177 m. Raion center; electronic machine building, hemp processing; dress making; food processing (dairying, fruit canning, flour milling), feed milling. Teachers college, agricultural technical school, two professional technical schools; heritage museum. Known as a city since 1152 in Chernihiv principality; in 13th–14th centuries, capital of an appanage principality; under Lithuania since 1350; annexed by Muscovy in 1503, by Poland in 1618; part of hetman state from 1648 and company town of the Nizhen regiment; after destruction of Baturyn (capital of Left-Bank Ukraine and seat of the hetman), Hlukhiv became capital of hetmans subservient to Russia (1708–1722, 1727–1734) and the Little Russian Collegium (1722–1727, 1764–1786). In 1918, Hlukhiv became part of the Ukrainian National Republic; in 1919, it was occupied by the Soviet forces; part of independent Ukraine since 1991. Architectural monuments: Triumphal Arch (or Kiev Gate, built in 1744), St. Nicholas church (1696), Transfiguration Church (1765) and St. Anastasia's Church (1884–1893).

Hlukhivtsi (HLOO-khif-tsee), (Russian *Glukhovtsy*), town, N VINNYTSYA oblast, UKRAINE, on railroad 6 mi/10 km NW of KOZYATYN; 49°46'N 28°44'E. Elevation 885 ft/269 m. Kaolin plant, reinforced concrete fabrication; mining pit S of the settlement. Known since 1595; town since 1981. The Kaolin, mined here since 1901, was used in the cellulose-paper industry of the USSR.

Hluti (huh-LOO-ti), town, SHISELWENI district, S SWAZILAND, 67 mi/108 km SSE of MBABANE, at source of Sitilo River; 27°14'S 31°35'E on N outsides of Mhlosheni Hills. Border with South Africa. 6 mi/9.7 km to S. Cattle, goats, sheep, hogs; corn, sugarcane, pineapples, citrus fruit.

Hlyboka (hli-BO-kah), (Russian *Glybokaya*), (German *Hiboka*), (Romanian *Adîncata*), town, S CHERNIVTSI oblast, UKRAINE, in N BUKOVYNA, 13 mi/21 km S of CHERNIVTSI, near Romanian border; 48°05'N 25°55'E. Elevation 1,154 ft/351 m. Raion center; major railroad junction (Adynkata station); manufacturing (glass, bricks, asphalt), food processing (butter, flour, food flavoring), grain milling; technical school, heritage museum, water healing sanitarium and park, established in 1860. Known since 1438 as part of Moldavia; passed to Austria (1775); to Romania (1919); to USSR (1940); town since 1956; part of independent Ukraine since 1991. Known as Adincata when part of Romania (1919–1940).

Hlyns'k (HLINSK), (Russian *Glinsk*), village, SW SUMY oblast, UKRAINE, near SULA river, 12 mi/19 km SW of ROMNY; 50°37'N 33°19'E. Elevation 416 ft/126 m. Wheat. Town in the 15th century, patrimony of the Hlyns'kyy noble family.

Hlynyany (hli-NYAH-nee), (Russian *Glinyany*), (Polish *Gliniany*), city, central L'VIV oblast, UKRAINE, 20 mi/32 km E of L'VIV; 49°49'N 24°31'E. Elevation 715 ft/217 m. Food processing (flour, butter), manufacturing (carpet, bricks). Has ruins of an old castle. Known since 1379, town since 1940, city since 1992.

H. Neely Henry Lake, CALHOUN county, NE ALABAMA, 17 mi/27 km SSW of GADSDEN, on the COOSA RIVER; 33°47'N 86°03'E. Maximum capacity 109,000 acre-ft. Formed by H. Neely Henry Dam (57 ft/17 m high), built (1966) by the Alabama Power Company for hydroelectric power and recreation; crossed by State Highway 144. Also called Neely Henry Lake and Dam; commonly called Henry Neely Lake.

Hnífsdalur (huh-NEEFS-DAH-luhr), fishing village, NORDUR-ISAFJARDARSYSLA county, NW ICELAND, on VESTFJARDA Peninsula, on ÝSAFJARÐARDJÚP, 3 mi/5 km N of ÝSAFJÖRÐUR.

Hnivan' (HNEE-vahn), (Russian *Gnivan'*), city, W VINNYTSYA oblast, UKRAINE, 11 mi/18 km SW of VINNYTSYA; 49°06'N 28°20'E. Elevation 807 ft/245 m. Railroad station; sugar refinery; ball bearing repair, tire repair depot, granite quarry, reinforced concrete construction materials. Known since 1629, city since 1981.

Hnizdychiv (hneez-DI-chif), (Russian *Gnezdychev*), (Polish *Hnizdyczów*), town, E central L'VIV oblast, UKRAINE, on right back of STRYY RIVER, on road and on railroad 4 mi/6 km SSW of ZHYDACHIV; 49°21'N 24°07'E. Elevation 833 ft/253 m. Feed mill, potato processing, gas equipment, manufacturing, paper manufacturing. Known since the 13th century in Halych Principality; passed to Hungary in 1378, to Poland in 1387, to Austria in 1772; part of West Ukrainian Republic (1918–1919); reverted to Poland in 1919; occupied by Soviets (1939–1941), by Germans (1941–1944); ceded to USSR (Ukrainian SSR) in 1945; town since 1957; part of independent Ukraine since 1991.

Hnizdyczów, UKRAINE: see HNIZDYCHIV.

Hnusta (hnoosh-TYAH), Slovak *Hnúšťa*, Hungarian *Nyustya*, town, STREDOSLOVENSKY province, S central SLOVAKIA, in SLOVAK ORE MOUNTAINS, on railroad and 38 mi/61 km ESE of BANSKÁ BYSTRICA; 48°35'N 19°58'E. Magnesite mining in vicinity; manufacturing (chemicals, building materials).

Hnyla Lypa River (hni-LAH LI-pah), 50 mi/80 km long, (Polish *Gnila Lipa*), (Russian *Gnilaya Lipa*), in W UKRAINE; rises in the HOLOHORY, ENE of PEREMYSHLYANY; flows SSE, past Peremyshlyany, ROHATYN, BURSHTYN, and BIL'SHIVTSI, to DNIESTER (Ukrainian *Dnister*) River, 2 mi/3.2 km N of HALYCH.

Hnyle More, UKRAINE: see SYVASH SEA.

Area is shown by the symbol □, and capital city or county seat by ⊙.

Hnylopyat' River (hni-lo-PYAT), river, mostly in ZHYTOMYR oblast, in NW UKRAINE; a tributary of the DNIEPER. Its eastern shore is the site of the oldest Catholic monastery in Ukraine - the "Barefoot Carmelites" monastery (established in 1627) - at BERDYCHIV.

Hnylyy Yelanets', UKRAINE: see YELANETS'.

Ho, town, ⊙ VOLTA region, GHANA, 35 mi/56 km NE of AKOSOMBO DAM in Toyo Mountains; 06°36′N 00°28′E. It is the largest commercial and administrative center of Volta region. Road junction; cacao, palm oil and kernels, cotton.

Hoa Binh, province (□ 1,780 sq mi/4,628 sq km), N VIETNAM, in N mountain and midland region, N border with VINH PHU province, NE border with HA TAY province, E border with NAM HA province, SE border with NINH BINH province, S border with THANH HOA province, W border with SON LA province; ⊙ HOA BINH; 20°30′N 105°30′E. Drained by the Song Da (BLACK RIVER), province embraces a variety of environmental regimes and localized economies. Its once-forested mountains and hills have undergone much deforestation in recent decades. Site of recently built HOA BINH Dam, which provides electricity for the entire country. Diverse soils and mineral resources (gold, granite, limestone, mineral water), shifting cultivation, wet rice farming in riverine niches, forest products (bamboo, gums, resins, medicinals, exotic plants), agroforestry, and commercial agriculture (coffee, tea, vegetables, fruits). Hydroelectricity, mining, handicrafts, cement production, brick and tile, light manufacturing, food processing, domestic tourism. Kinh population with Thai, Muong, H'mong, Dao, and other minorities.

Hoa Binh (HO-uh BIN), city, ⊙ HOA BINH province, N VIETNAM, on Song Da (BLACK RIVER) and 35 mi/56 km SW of HANOI, in forested region; 20°47′N 105°20′E. Shifting cultivation; gums and resin. Site of newly-completed HOA BINH Dam, which furnishes electricity to entire nation. Transportation hub, market center, and administrative complex; light manufacturing; food processing. Prehistoric caves. Formerly Hoabinh.

Hoa Binh (HO-uh BIN), dam, HOA BINH province, N VIETNAM, built in cooperation with the U.S.S.R. on the Da River (Song Da), creates the largest reservoir in the country; 20°49′N 105°20′E. Originally constructed to provide power for NORTH VIETNAM, a 500-kilovolt line was extended to the S in 1994 at enormous economic and human relocation costs.

Hoadley (HOD-lee), hamlet, S central ALBERTA, W CANADA, 35 mi/55 km from PONOKA, in PONOKA COUNTY; 52°51′N 114°32′W.

Hoagland, village, ALLEN county, NE INDIANA, 11 mi/18 km SE of FORT WAYNE. Corn, oats, soybeans. Manufacturing (wooden cabinets). Laid out 1872.

Hoai Nhon, town, BINH DINH province, central VIETNAM, on railroad and 45 mi/72 km NNW of QUY NHON; 14°26′N 109°01′E. Market and transportation center. Coconut-processing center, coconut-oil extraction; copra production; rope making; silk spinning, distilling. Formerly BONGSON or Bong Son.

Hoang Lien (HO-ahng lee-EN), mountains, YEN BAI and LAO CAI provinces, N VIETNAM; 20°18′N 105°45′E. Contains one of the country's major forest reserves. Tourism potential. Lolo, Meo, Tai, and other minorities. Illegal logging has become a major threat.

Hoang Long (HO-uhng LOUNG) or **nho quang**, town, NINH BINH province, N VIETNAM, 50 mi/80 km S of HANOI; 20°18′N 105°46′E. Coal mining, timber trading. Transportation and market hub. Formerly PHUNHOQUAN.

Hoback River (HO-bak), c.30 mi/48 km long, W WYOMING; rises in NW SUBLETTE county; flows NW, through picturesque Hoback Canyon, past N WYOMING RANGE, to SNAKE RIVER 10 mi/16 km S of JACKSON.

Hobara (HO-bah-rah), town, Date district, FUKUSHIMA prefecture, N central HONSHU, NE JAPAN, 6 mi/10 km N of FUKUSHIMA; 37°49′N 140°33′E. Fruits; apparel.

Hobart (HO-bahrt), city (2001 population 126,048), ⊙ and principal port of TASMANIA, SE AUSTRALIA, at the foot of Mount Wellington (4,166 ft/1,270 m high); 42°55′S 147°20′E. Hobart's harbor is one of the finest in the world. The city has diverse industries, including meat-packing, food processing, and the making of textiles, chemicals, and glass; zinc metal, alloys, fertilizer. It was founded in 1804 and named for Robert Hobart, the British colonial secretary. Hobart is the seat of the University of Tasmania (1890) and an important commercial and service center. The Hobart Theatre Royal (1836) is the oldest major theater in Australia.

Hobart, city (2000 population 25,363), Lake county, extreme NW INDIANA; 41°32′N 87°16′W. Metal products, electric coils and transformers, herbal products, tools and castings, and food products are made in Hobart. Part of the CALUMET region. Laid out c.1849, incorporated 1921.

Hobart (HO-buhrt), town (2006 population 3,771), ⊙ KIOWA county, SW OKLAHOMA, c.45 mi/72 km NW of LAWTON; 35°01′N 99°05′W. Elevation 1,547 ft/472 m. Trade center for rich agricultural area (cotton, grain, alfalfa, sorghum; cattle, sheep); manufacturing (oil seals, flexible aluminum packaging); cold-storage plants. ALTUS LAKE to SW. Incorporated 1901.

Hobart, village (2006 population 369), DELAWARE county, S NEW YORK, in the Catskills, on West Branch of DELAWARE RIVER and 20 mi/32 km ESE of ONEONTA; 42°22′N 74°40′W. Feed and pharmaceuticals.

Hoba West, NAMIBIA: see GROOTFONTEIN.

Hobbema (HAH-buh-muh), hamlet, S central ALBERTA, W CANADA, 11 mi/17 km from PONOKA, in PONOKA County; 52°50′N 113°27′W.

Hobbs, city (2006 population 29,292), LEA county, SE NEW MEXICO, 95 mi/153 km SE of ROSWELL; 32°43′N 103°09′W. Elevation 3,625 ft/1,105 m. Incorporated 1929. With the discovery (c.1928) of oil and natural gas in the area, Hobbs became one of the last great oil-boom towns in the U.S. It still remains a major shipping and trading center for oil-well supplies. Chemical production is of increasing importance, as are feedlots for livestock (sheep, cattle, dairying) and the raising of thoroughbred horses. Cotton, wheat, peanuts, vegetables, and melons are grown on irrigated farms in the area. Beef cattle have long been important in Hobbs; dairy farming is growing. Manufacturing (concrete, machinery, fabricated metal products, chemicals); mining. New Mexican Junior College and the College of the Southwest are in the city. Lea County Hobbs Airport to SW. Harry McAdams State Park is here.

Hobbs, village, TIPTON county, central INDIANA, 4 mi/6.4 km E of TIPTON. Agricultural area. Manufacturing (canned tomatoes and tomato juice). Established 1876.

Hobbs Coast, ANTARCTICA, along coast of MARIE BYRD LAND, extends from 127°–137°W. Discovered 1940 by U.S. Antarctic Service expedition.

Hobdo, MONGOLIA: see HOVD, city, or HOVD, province.

Hobdo Gol, MONGOLIA: see HOVD GOL.

Hobe Sound (HOB), resort town (□ 5 sq mi/13 sq km; 2000 population 11,376), MARTIN county, E central FLORIDA, 12 mi/19 km W of W. PALM BEACH, near JUPITER ISLAND; 27°04′N 80°08′W.

Hobetsu (HO-bets), town, Iburi district, HOKKAIDO prefecture, N JAPAN, 47 mi/75 km S of SAPPORO; 42°45′N 142°08′E.

Hobgood (HAHB-gud), village (□ 1 sq mi/2.6 sq km; 2006 population 385), HALIFAX county, NE NORTH CAROLINA, 12 mi/19 km NE of TARBORO near FISHING CREEK; 36°01′N 77°24′W. Manufacturing; agriculture (tobacco, cotton, soybeans, grain; poultry, hogs). Incorporated 1891.

Hobio, SOMALIA: see OBBIA.

Hobkirks Hill, SOUTH CAROLINA: see CAMDEN.

Hobo (O-bo), town, ⊙ Hobo municipio, HUILA department, S central COLOMBIA, in upper MAGDALENA valley, 27 mi/43 km SSW of NEIVA; 02°34′N 75°27′W. Elevation 2,175 ft/662 m. Rice, coffee, cotton; livestock.

Hoboken (HO-bo-kuhn), commune, Antwerp district, ANTWERPEN province, N BELGIUM, on the SCHELDT RIVER; 51°10′N 04°21′E. An industrial suburb of ANTWERP; annexed to Antwerp in 1983.

Hoboken (HO-BO-kin), city (2006 population 39,853), HUDSON county, NE NEW JERSEY, on the HUDSON RIVER adjoining JERSEY CITY and opposite MANHATTAN; 40°44′N 74°01′W. Settled by the Dutch c.1640, incorporated as a city 1855. It is a port of entry and a railroad terminal. The city has food-processing industries and factories that make electronic, chemical, and metal products. The site changed title many times before John Stevens gained possession in 1784. He built his home at Castle Point (an unusual rock formation overlooking the river) and laid out the town in 1804. Stevens built (c.1825) and ran on his estate the first locomotive to pull a train on tracks in the U.S. Hoboken became an important industrial and commercial center in the late 19th century with a major port, shipyards, and warehouses. In the 1970s and 1980s, professionals, artists, and students flocked to the city for its affordable, renovated housing and easy access to NEW YORK city. Hoboken's reputation has grown accordingly, and it has become a cultural community with art galleries, musical events, entertainment, and developing businesses. Higher rents brought by the arrival of these new Hobokenites have made it harder for poorer residents to remain in the community, though. John Jacob Astor lived here; his home was a gathering place for authors, including Fitz-Greene Halleck, Washington Irving, and William Cullen Bryant. Hoboken is the seat of the Stevens Institute of Technology. Frank Sinatra born here.

Hoboken (HO-bo-kihn), village (2000 population 463), BRANTLEY county, SE GEORGIA, 13 mi/21 km E of WAYCROSS; 31°11′N 82°08′W. Manufacturing of lumber.

Hobro (HO-bro), city (2000 population 10,837), NORDJYLLAND county, E JUTLAND, DENMARK, at head of MARIAGER FJORD, 36 mi/58 km NNW of ÅRHUS; 56°35′N 09°50′E. Agriculture (dairying, cattle, hogs); manufacturing (freezers); meat cannery, brewery.

Höbsögöl, MONGOLIA: see HÖVSGÖL.

Hobson, village (2000 population 244), JUDITH BASIN county, central MONTANA, on JUDITH RIVER and 21 mi/34 km W of LEWISTOWN; 47°00′N 109°52′W. Wheat, barley, alfalfa, cattle, sheep; sapphires. Sapphire Village and Mine 19 mi/31 km to WSW. Ackley Lake State Park to SW.

Hobson City, residential suburb (2000 population 878), Calhoun co., E ALABAMA, a town just SW of Anniston. Sportswear manufacturing. Named for Richmond Pearson Hobson, Spanish-American War naval hero born in AL.

Hobson Lake (HAHB-suhn), E BRITISH COLUMBIA, W CANADA, in CARIBOO MOUNTAINS, in WELLS GRAY PROVINCIAL PARK, 120 mi/193 km N of KAMLOOPS; 20 mi/32 km long, 1 mi/2 km–2 mi/3 km wide; 52°30′N 120°20′W. Elevation 2,735 ft/834 m. Drains S into North Thompson River through CLEARWATER LAKE.

Hobson's Bay (HAHB-suhnz BAI), S VICTORIA, AUSTRALIA, N arm of PORT PHILLIP BAY; 12 mi/19 km long, 7 mi/11 km wide. Receives YARRA RIVER. PORT MELBOURNE on N shore, WILLIAMSTOWN on SW shore.

Hocabá (o-kah-BAH), town, ⊙ Hocabá municipio, YUCATÁN, SE MEXICO, 26 mi/42 km ESE of MÉRIDA; 20°49′N 89°29′W. On railroad. Henequen, tropical fruit, corn, beans, livestock.

Hochalmspitze (HOKH-ahlm-shpi-tse), peak (10,241 ft/3,121 m) of the HOHE TAUERN mountain range, S

AUSTRIA, SE of the ANKOGEL mountain, in NW CARINTHIA; 47°01′N 13°19′E. Part of National Park Hohe Tauern.

Höchberg (HOKH-berg), village, BAVARIA, central GERMANY, in LOWER FRANCONIA, just E of WÜRZ-BURG; 49°47′N 09°52′E.

Hochburg-Ach (HOKH-burg-AHKH), village W UPPER AUSTRIA, on the SALZACH River, opposite BURGHAUSEN, GERMANY, and 12 mi/19 km SW of BRAUNAU; 48°08′N 12°53′E. Dairy farming.

Hochdorf (HO-dorf), commune, LUCERNE canton, central SWITZERLAND, 8 mi/12.9 km N of LUCERNE. Cement, chemicals, milk.

Hochdorf (HO-dorf), district, SW LUCERNE canton, central SWITZERLAND. Main town is EMMEN; population is German-speaking and Roman Catholic.

Hochelaga, CANADA: see MERCIER–HOCHELAGA-MAISONNEUVE.

Hochelaga, CANADA: see MONTREAL.

Höchenschwand (HO-khen-shvahnd), village, BADEN-WÜRTTEMBERG, SW GERMANY, in BLACK FOREST, 8 mi/12.9 km NNW of WALDSHUT; 47°44′N 08°10′E. Health resort (elevation 3,307 ft/1,008 m).

Hochfeiler (hokh-FEI-luhr), Italian *Gran Pilastro*, highest peak (10,698 ft/3,261 m) of ZILLERTAL ALPS, on Austro-Italian border, and 12 mi/19 km E of BRENNER PASS; 46°58′N 11°44′E.

Hochfeld (HOKH-felt), SW industrial district of DUISBURG, GERMANY, on right bank of the RHINE; 51°36′N 08°21′E. Deep-water port.

Hochfelden (ok-fel-DEN), German (HAHK-fel-duhn), commune (□ 5 sq mi/13 sq km), BAS-RHIN department, in ALSACE, E FRANCE, on MARNE-RHINE CANAL, and 15 mi/24 km NW of STRASBOURG; 48°45′N 07°34′E. Agricultural trade center on main PARIS-Strasbourg expressway; brewery.

Hochfilzen (hokh-FIL-tsen), village, TYROL, W AUS-TRIA, 11 mi/18 km E of Kitzbühel; just E is GRIESSEN PASS (elevation 2,935 ft/895 m); magnesite mining and manufacturing; 47°28′N 12°37′E.

Hochgolling (hokh-GAWL-ling), highest peak (8,726 ft/2,660 m) of the NIEDERE TAUERN, central AUSTRIA, on STYRIA-SALZBURG border; 47°16′N 13°46′E.

Hoch Hagi, SLOVAKIA: see VYSNE HAGY.

Hochheim am Main (HOKH-heim ahm MEIN), town, HESSE, W GERMANY, on right bank of the Main, 16 mi/26 km WSW of FRANKFURT; 50°01′N 08°21′E. Noted for its sparkling Hochheimer ("hock") wine. Manufacturing of packaging materials and machinery.

Ho Chi Minh City (1997 population 5,250,000), on the right bank of the SAIGON RIVER, a tributary of the DONG NAI, S VIETNAM; 10°46′N 106°43′E. HO CHI MINH CITY (formerly known as SAIGON) is the largest city, the greatest port, and the commercial and industrial center of Vietnam. It has an international airport and is the S hub of the country's highways and railroad, and the functional center of MEKONG delta waterways. The administrative boundaries of Thanh Pho Ho Chi Minh (Ho Chi Minh urban region) embrace Saigon center, CHOLON, several suburbs, and a number of nearby rural districts, including the former town of GIA DINH. Together they comprise a metropolitan realm destined for coordinated development. An ancient Khmer settlement, Saigon passed (17th century) to the Vietnamese during the S migration of the Kinh (Vietnamese) people. It was captured by the French in 1859 and ceded to FRANCE in 1862. A small village at the time of the French conquest, Saigon became a modern city under French rule, and was sometimes called the "Pearl of the Orient." Laid out by colonial administrators in rectilinear fashion with wide, tree-lined avenues and parks, it soon developed a reputation for tropical beauty and cosmopolitan atmosphere. It was the capital of COCHIN CHINA and from 1887 to 1902 served as the capital of the Union of INDOCHINA. For administrative purposes Saigon and Cholon, on oppo-

site banks of the Saigon River, were merged in 1932; in 1956 the two cities were included in the new prefecture of Saigon. Saigon functioned as the capital the newly created state of SOUTH VIETNAM 1954–1975. In the Vietnam War it served as military headquarters for U.S. and South Vietnamese forces, and it suffered considerable damage during the 1968 Tet offensive. During the 1960s and early 1970s, nearly 2 million refugees from rural areas poured into the city, creating serious housing problems and overcrowding. In 1975, after Saigon surrendered and Vietnam was reunited under the prevailing Communist government, the city lost its status as the capital, experienced some depopulation, and was renamed after the late North Vietnamese president. The local economy of Ho Chi Minh City was seriously disrupted during the early years of the new regime, which curtailed foreign investment and promoted collectivization. Beginning in the early 1980s, the national government unofficially liberalized some of its policies in response to continued poor economic activity. Under the current economic reform program (Doi Moi) of the 1990s, Ho Chi Minh City has undergone a considerable transformation. It is the fastest growing part of the country and a key tourist site. In addition to upgrading existing factories, a number of joint-venture projects are underway with Taiwanese, Korean, French, and other firms. Major industries include textiles, tobacco, fertilizer, building materials, and fish and food processing. The city center is expected to move to the S of its current location early in the 21st century as part of a major development scheme. A number of infrastructural improvements are planned to facilitate the process. These include a bridge (Thu Thiem) across the Saigon River and new highways to connect the current city center with the growing number of industrial zones, such as the Hiep Phuoc, Binh Hoa, Cat Lai, and Tan Thuan Zones. The city center still suffers from serious housing shortages, especially low-income housing, as rural migrants and former city residents flood the region. The city also has a number of key historical and cultural sites that merit preservation and renovation, including the Giac Lam Pagoda, Cholon Mosque, War Crimes Museum, Revolutionary Museum, Hall of Reunification. Ho Chi Minh City is one of the foremost educational centers of Vietnam, with colleges, institutes, training programs, and scattered facilities of Ho Chi Minh University.

Hochkirch (HOKH-kirkh), village, SAXONY, E GER-MANY, 16 mi/26 km W of GÖRLITZ; 51°09′N 14°34′E. At HOCHKIRCH in 1758 the Austrians under Daun defeated Frederick II of PRUSSIA. In 1813, Napoleon I defeated a Prussian-Russian army near the village.

Hochkönig (hawkh-KO-nik), mountain, massif and peak (9,639 ft/2,938 m) of the Salzburg–UPPER AUS-TRIAN LIMESTONE ALPS, W central AUSTRIA; 47°25′N 13°04′E. Glacier Übergossene. Almhe Manndlwend, a small range S of Hochkönig, offers ample opportunities for climbing.

Hoch-Königsburg, FRANCE: see HAUT-KOENIGSBOURG.

Hochlantsch (hokh-LAHNCH), highest peak (5,242 ft/1,598 m) of Graz Mountains (Grazer Bergland), in STYRIA, SE central AUSTRIA, 8 mi/12.9 km ESE of BRUCK AN DER MUR; 47°22′N 15°25′E. Has interesting limestone formations (gorges, caves).

Hochobir (HOKH-uh-bir), Slovenian *Ojstrc*, peak (6,520 ft/1,987 m) of the Obir (mountain in the KARAWANKEN mountain range), CARINTHIA, S AUS-TRIA, 12 mi/19 km SE of KLAGENFURT, between the Drau and Slovenian border; 46°30′N 14°29′E. Cave.

Hochrhein, GERMANY: see HIGH RHINE.

Hochschwab (hokh-SHVAHB), mountain, massif and peak (6,940 ft/2,115 m) in STYRIA, central AUSTRIA, in the Styrian Limestone Alps, 15 mi/24 km NNW of BRUCK AN DER MUR. Its springs supply drinking water for VIENNA. Many stalactite and ice caves; tourist destination.

Hochspeyer (HOKH-spei-er), village, RHINELAND-PA-LATINATE, W GERMANY, 5 mi/8 km E of KAI-SERSLAUTERN; 49°27′N 07°54′E.

Höchst (HOKHST), industrial district of FRANKFURT, in HESSE, central GERMANY; 50°07′N 08°40′E. A leading center of the German chemical industry and was formerly the site of the I. G. Farben chemical and dye works. Chartered in 1355 and incorporated 1928 into Frankfurt. In the Thirty Years War (1618–1648), Tilly defeated (1622) Christian of Brunswick here.

Höchst (HOKHST), village, VORARLBERG, W AUSTRIA, on the left bank of RHINE RIVER, near LAKE CON-STANCE and Swiss border, 6 mi/9.7 km SW of BRE-GENZ; elevation 1,228 ft/374 m; 47°28′N 09°38′E. Border station; manufacturing of metals, embroidery, plastics.

Hochstadt, CZECH REPUBLIC: see VYSOKE NAD JIZEROU.

Höchstadt an der Aisch (HOKH-shtat ahn der EISH), town, BAVARIA, S GERMANY, on the AISCH RIVER, 12 mi/19 km NW of ERLANGEN; 49°42′N 10°49′E. Manufacturing of leather and metal products; carp hatcheries. Has early-15th-century church and a chapel (1513).

Höchstädt an der Donau (HOKH-shtat ahn der DO-nou), town, BAVARIA, in SWABIA, on the DANUBE River, 4 mi/6.4 km NE of DILLINGEN; 48°37′N 10°34′E. Manufacturing includes toys; metalworking; lumber milling. Has several Gothic churches; late-16th-century castle with watchtower dating from 1292. Founded by the Hohenstaufen, it was chartered c.1280. Two battles of War of the Spanish Succession took place nearby, in 1703 and 1704.

Hochstetter Foreland (HOKH-ste-tuhr), Danish *Hochstetter Forland*, peninsula, NE GREENLAND, on GREENLAND SEA; 75 mi/121 km long, 20 mi/32 km–55 mi/89 km wide; 75°35′N 20°50′W. Generally ice-free, mountainous surface rises to over 4,000 ft/1,219 m. Meteorological station and hunting outpost at SE tip (near 75°08′N 19°37′W) operated by NORWAY until 1941, then dismantled; rebuilt after 1945; now abandoned.

Höchst im Odenwald (HOKHST eem O-duhn-vahlt), village, HESSE, central GERMANY, in the ODENWALD, on the MÜMLING, 8 mi/12.9 km N of MICHELSTADT; 49°48′N 09°00′E. Woodworking.

Hochstuhl (HOKH-shtool), Slovenian *Stol*, mountain peak (6,821 ft/2,079 m), highest of the KARAWANKEN ALPS, on Austro-Slovenian border, 15 mi/24 km SW of KLAGENFURT; 46°26′N 14°10′E.

Hochtaunus, GERMANY: see TAUNUS.

Hochtor, AUSTRIA: see ENNSTAL ALPS.

Hockanum River, c.25 mi/40 km long, central CON-NECTICUT; rises in SHENIPSIT LAKE; flows SW, past ROCKVILLE, to the CONNECTICUT at EAST HARTFORD. Many early mill sites.

Hockenheim (HOKH-kehn-heim), town, BADEN-WÜRTTEMBERG, W GERMANY, on the KRAICHBACH, 5 mi/8 km E of SPEYER; 49°19′N 08°33′E. In tobacco-growing area; manufacturing includes food processing, metalworking; agriculture (asparagus). Nearby is Hockenheim race track.

Hocking (HAHK-ing), county (□ 421 sq mi/1,094.6 sq km; 2006 population 28,973), S central OHIO; ⊙ LOGAN; 39°31′N 82°28′W. Intersected by HOCKING RIVER and small Rush, Salt, and Monday creeks. Primarily in the Unglaciated Plain physiographic region. Agriculture (nursery and greenhouse crops; beef cows, poultry; corn); manufacturing at Logan (rubber products, plastics, electrical goods); clay, sand, and gravel pits. Some coal mining. Formed 1818.

Hocking (HAHK-ing), village, ATHENS county, SE OHIO, on HOCKING RIVER, just N of ATHENS; 39°44′N 82°28′W.

Hocking River (HAHK-ing), c.100 mi/161 km long, SE OHIO; rises in FAIRFIELD county; flows SW, past LANCASTER, LOGAN, NELSONVILLE, and ATHENS, to the OHIO RIVER 22 mi/35 km SW of MARIETTA; 39°11′N 81°45′W.

Hockley, county (□ 908 sq mi/2,360.8 sq km; 2006 population 22,609), NW TEXAS; ⊙ LEVELLAND; 33°36′N 102°20′W. Drained in N by intermittent Yellow House Draw creek. Rich agricultural region, with extensive irrigated areas; cotton, grain sorghum; livestock (beef cattle, hogs; some sheep, horses, mules). Large oil and gas production. Formed 1876.

Hockley (HAHK-lee), village (2001 population 8,909), SE ESSEX, SE ENGLAND, 5 mi/8 km NNW of SOUTH-END-ON-SEA; 51°35′N 00°39′E. Former agricultural market. Has Norman church.

Hockley, unincorporated village, HARRIS county, SE TEXAS, 35 mi/56 km NW of downtown HOUSTON. Agricultural area (cattle, dairying, horses, rice, vegetables). Oil and natural gas, salt mines. Manufacturing (aggregated salt).

Hoc Mon (HOK MON), district, HO CHI MINH urban region, S VIETNAM, 10 mi/16 km NW of Ho Chi Minh City center; 10°53′N 106°36′E. As part of the suburban belt surrounding Ho Chi Minh City, the area has seen much change in recent years from primarily a rice-growing area to crop diversification and food processing. Population densities are increasing, and the town center has marketing, light manufacturing, and considerable infrastructural investment. Formerly Hocmon.

Hoctún (ok-TOON), town, YUCATÁN, SE MEXICO, 28 mi/45 km ESE of MÉRIDA; 20°48′N 89°14′W. Henequen, sugarcane, corn.

Hodaka, Mount, JAPAN: see HOTAKA, MOUNT.

Hodal (HO-dahl), town, FARIDABAD district, HARYANA state, N INDIA, 45 mi/72 km SSE of GURGAON; 27°54′N 77°22′E. Wheat, millet, cotton, sugarcane; cotton ginning.

Hoddesdon (HAHD-uhs-tuhn), town (2001 population 36,883), HERTFORDSHIRE, SE ENGLAND, 4 mi/6.4 km N of CHESHUNT; 51°45′N 00°01′W. A residential suburb of London, Hoddesdon has light industries and horticultural works. Broxbourne Woods attracts many London visitors. Several old inns and houses still stand. Izaak Walton, author of *The Compleat Angler*, fished in nearby LEA RIVER.

Hoddom Castle, Scotland: see ECCLEFECHAN.

Hodeida (ho-DAI-duh), city, W YEMEN, on the RED SEA. An important port, it exports coffee, cotton, dates, and hides. One of Yemen's main industrial centers; various metal products, textiles, food products, building materials. Has large thermal power station. Airport. It was developed as a seaport in the mid-19th century by the Turks. After a disastrous fire in January 1961 destroyed much of the city, it was rebuilt, particularly the port facilities, with Soviet aid. A highway from Hodeida to SANA, the capital, was completed in 1961. Hodeida was the site of a Soviet naval base in the 1970s and 1980s. Also spelled Al Hudaydah.

Hodgdon (HAHJ-duhn), town, AROOSTOOK county, E MAINE, 5 mi/8 km S of HOULTON. In agricultural, lumbering area. Incorporated 1832.

Hodge, village (2000 population 492), JACKSON parish, N central LOUISIANA, on DUGDEMONA RIVER, and 40 mi/64 km WSW of MONROE; 32°16′N 92°44′W. Railroad junction; in agricultural area; manufacturing (paper grocery bags, kraft paper). Caney Creek reservoir to E. Jackson Bienville State Wildlife Area to N.

Hodgeman (HAHJ-muhn), county (□ 860 sq mi/2,236 sq km; 2006 population 2,071), SW central KANSAS; ⊙ JETMORE; 38°05′N 99°54′W. Rolling prairie region, watered by PAWNEE RIVER and Buckner Creek. Wheat, sorghum, cattle. Hodgeman State Fishing Lake at center. Formed 1879.

Hodgenville (HAHJ-uhn-vil), town (2000 population 2,874), ⊙ LARUE county, central KENTUCKY, on NOLIN RIVER and 45 mi/72 km S of LOUISVILLE; 37°34′N 85°44′W. In agricultural area (burley tobacco, corn; livestock), limestone quarries, and timber area;

manufacturing (furniture, apparel, wooden products, lumber). ABRAHAM LINCOLN BIRTHPLACE NATIONAL HISTORIC SITE to S; Lincoln's boyhood home (Knob Creek Farm) to NE; The Lincoln Museum, Lincoln Statue on Lincoln Square. S boundary is Central/Eastern time zone boundary, county is in Eastern. Settled c.1789; established 1818.

Hodges (HAH-juhs), town (2000 population 261), Franklin co., NW Alabama, 17 mi/27 km SSW of Russellville. Named for a Stephen Hodge, a local resident. Inc. in 1913.

Hodges, village (2006 population 166), GREENWOOD county, W SOUTH CAROLINA, 8 mi/12.9 km NW of GREENWOOD; 34°17′N 82°15′W. Manufacturing of windows and door stock, textiles.

Hodges Hill (1,868 ft/569 m), S NEWFOUNDLAND AND LABRADOR, CANADA, 14 mi/23 km NW of GRAND FALLS.

Hodgeville (HAHJ-vil), village (2006 population 142), S SASKATCHEWAN, CANADA, on Wiwa Creek and 40 mi/64 km ESE of SWIFT CURRENT; 50°07′N 106°59′W. Grain elevators, dairying.

Hodgkins, village (2000 population 2,134), COOK county, NE ILLINOIS, W suburb of CHICAGO; 41°46′N 87°51′W. On the SANITARY AND SHIP CANAL.

Hod Hasharon (HOD hah-shah-RON), city (2006 population 44,200), S of KFAR SAVA, central ISRAEL, combining several townships that began as agricultural settlements; 32°09′N 34°53′E. Elevation 137 ft/41 m. Though agriculture (especially citrus fruit) is still important, now considered a TEL AVIV suburb.

Hodh Ech Charghy, administrative region, MAURITANIA: see HODH ECH CHARGUI.

Hodh Ech Chargui, administrative region (2000 population 281,600), E MAURITANIA; ⊙ NÉMA. Bordered by Hodh El Gharbi (SW), TAGANT (W), and Adrar (NWN) administrative regions) and MALI (NNE and S). In SAHARA DESERT. Dhar Oualâta hills in S central; part of El Mreyyé geographic region in N central. NOUAKCHOTT-NEMA HIGHWAY runs W out of Néma, through Timbedgha, and across Mauritania to NOUAKCHOTT; main road also runs S out of Néma into Mali; secondary roads run through S part of the region. Airports at TIMBÉDRA and near Néma. Also spelled Hodh Ech Charghy.

Hodh El Gharbi, administrative region (2000 population 212,156), SE MAURITANIA; ⊙ AIOUN EL ATROUS. Bordered by Assaba (W), TAGANT (N), and Hodh Ech Chargui (NEE) administrative regions and MALI (S). SAHARA DESERT extends into the region. Cheggué watercourse in SE. NOUAKCHOTT-NEMA HIGHWAY runs W-E across center of region, through Aioun el Atrous and TINTÂNE; secondary roads also run through the region. Airport at Aioun el Atrous. Also spelled Hodh El Gharby.

Hodh El Gharby, administrative region, MAURITANIA: see HODH EL GHARBI.

Hodkovice nad Mohelkou (HOD-ko-VI-tse NAHD MO-hel-KOU), German *Liebenau an der Mohelka*, town, SEVEROCESKY province, N BOHEMIA, CZECH REPUBLIC, on railroad and 6 mi/9.7 km SSE of LIBEREC; 50°40′N 15°05′E. Manufacturing (machinery, furniture, and costume jewelery). Preserved 18th-century wooden peasants' and stone burgess' houses. SVIJANY (SVI-yah-NI), brewery established in 1564, is 6 mi/9.7 km SSW.

Hódmezővásárhely (HOD-ma-zo-vah-shahr-hai), city (2001 population 49,382) with county rank, CSONGRÁD county, SE HUNGARY, near the TISZA RIVER; 46°26′N 20°22′E. Agricultural center and road hub; manufacturing (machinery, textiles, fireproof materials, tiles, clothing, footwear, meat products, pottery; grain milling, aluminum processing). Secondary education center in county, with technical college; has many churches.

Hodna (HAHD-nuh), interior drainage basin in M'SILA and BATNA wilaya, NE ALGERIA, at N edge of the

SAHARA, forming a break in the Saharan ATLAS (OULAD NAÏL MOUNTAINS, SW; AURÈS massif, E) and separated from the TELL (N) by the HODNA MOUNTAINS. The bottom of the arid depression (elev. 1,280 ft/390 m) is occupied by the Chott El HODNA, a playa lake of varying size (c.50 mi/80 km long, 10 mi/16 km wide). Only settlements, near basin's edges, are BOU SAÂDA (SW) and M'SILA (N).

Hodna Mountains, range of the TELL ATLAS, in NE ALGERIA, extending c.80 mi/129 km ESE from SOUR EL GHOZLANE, and bounding the HODNA depression on N. Rise to 6,000 ft/1,829 m. Oak and cedar forests. Phosphate mines on N slope.

Hodonín (HO-do-NYEEN), Czech *Hodonín*, German *Göding*, city (2001 population 27,361), JIHOMORAVSKY province, in MORAVIA, CZECH REPUBLIC, on MORAVA RIVER; 48°52′N 17°08′E. Railroad junction. Main agricultural center of MORAVIAN SLOVAKIA; tobacco processing, sugar refining; large brick kilns and brickyards. Nearby petroleum deposits have been exploited commercially since 1933. Has noted ethnographic museum. Thomas G. Masaryk born here.

Hodrusa-Hamre (ho-DRU-shah hah-MRE), Slovak *Hoduša-Hámre*, Hungarian *Hordusbríya*, village, STREDOSLOVENSKY province, SW central SLOVAKIA, 9 mi/14.5 km SSW of ZIAR NAD HRONOM; 48°28′N 18°49′E. Manufacturing of metal dishes. Has Gothic church (14th century).

Hodslavice, CZECH REPUBLIC: see NOVÝ JIČÍN.

Hodsock (HAHD-zuhk), village (2001 population 2,533), N NOTTINGHAMSHIRE, central ENGLAND, 4 mi/6.4 km NNE of WORKSOP; 53°21′N 01°04′W. Previous site of coal mining. The village is also referred to as Langold.

Hodur, SOMALIA: see HUDDUR.

Hoedeng-Goegnies, village in commune of LA LOUVIÈRE, Soignies district, HAINAUT province, SW BELGIUM, 13 mi/21 km WNW of CHARLEROI.

Hoëdic (o-e-deek), small island, in Bay of BISCAY, off BRITTANY coast (MORBIHAN department), W FRANCE, 12 mi/19 km SE of tip of QUIBERON PENINSULA and 5 mi/8 km E of BELLE-ÎLE; 2 mi/3.2 km long, 0.5 mi/1.3 km wide; 47°21′N 02°53′W.

Hoedjies Bay, SOUTH AFRICA: see SALDANHA.

Hoegaarden (HOO-gahr-duhn), French *hougarde* (oo-guard), commune (2006 population 6,247), Leuven district, BRABANT province, central BELGIUM, 3 mi/4.8 km SW of TIENEN; 50°47′N 04°54′E.

Hoehne, village, LAS ANIMAS county, S COLORADO, on PURGATOIRE RIVER and 10 mi/16 km NE of TRINIDAD. Elevation c.5,728 ft/1,746 m. In irrigated agricultural region; wheat, hay, sorghum, cattle. Rail junction to S.

Hoei, BELGIUM: see HUY.

Hoeilaart (HOO-ee-lahrt), commune (2006 population 10,099), Halle-Vilvoorde district, BRABANT province, central BELGIUM, 8 mi/12.9 km SE of BRUSSELS; 50°46′N 04°28′E.

Hoek van Holland, NETHERLANDS: see HOOK OF HOLLAND.

Hoendiep (HOON-deep), canal, GRONINGEN province, N NETHERLANDS; extends in circuitous route 16 mi/26 km E-W, between GRONINGEN and the VAN STARKENBORGH CANAL at ZUIDHORN, 9 mi/14.5 km W of GRONINGEN.

Hoengsong (HOING-SUHNG), county (□ 389 sq mi/1,011.4 sq km), SW KANGWON province, SOUTH KOREA, adjacent to HONGCHON on N, Yongwon and wonju on S, PYONGCHANG on E, and KYONGGI province on W. Mountainous. Som River forms Hoengsong Basin in center of county. Some agriculture (potatoes, vegetables, hop, fruits, ginseng); sericulture. YONGDONG Expressway. Chiak National Park.

Hoensbroek (HOONZ-brook), city, LIMBURG province, SE NETHERLANDS, 3 mi/4.8 km WNW of

HEERLEN; 50°55′N 05°56′E. Former coal mining center. Dairying; cattle; grain, vegetables, fruit; manufacturing (detergent).

Hoeryong (HUAE-RYOUNG), county, NORTH HAMGYONG province, NORTH KOREA, on TUMAN RIVER (MANCHURIA boundary) and 32 mi/51 km NW of NAJIN. Commercial center in agricultural area. Coal nearby.

Hoeselt (HOO-suhlt), commune (2006 population 9,321) Tongeren district of LIMBURG province, NE BELGIUM, 5 mi/8 km N of TONGEREN; 50°51′N 05°29′E. Agricultural market (fruit, potatoes, tobacco).

Hof, city, BAVARIA, E central GERMANY, in UPPER FRANCONIA, on the SAALE RIVER, near the Czech border; 50°20′N 11°55′E. Manufacturing includes textiles, machinery; brewing. City first mentioned in the early 13th century; in 1373 it passed to the Hohenzollern burgraviate of NUREMBERG. Went to Bavaria in 1810 and was included in Upper Franconia. Fire destroyed most of the city's historical buildings in 1823, but many have since been reconstructed. Among its points of interest are two churches, the Lorenzkirche (11th century) and the Michaelskirche (c.1230). From 1945 until German reunification in 1990, Hof was a significant border checkpoint between EAST and WEST GERMANY.

Hof, CZECH REPUBLIC: see DVORCE.

Höfe (HUH-fai), district, NW SCHWYZ canton, central SWITZERLAND. Main town is FREIENBACH; population is German-speaking and Roman Catholic.

Hofei, CHINA: see HEFEI.

Hoffman, village (2000 population 460), CLINTON county, SW central ILLINOIS, 8 mi/12.9 km W of CENTRALIA; 38°32′N 89°15′W. Sorghum, wheat; dairying; coal; manufacturing of mobile home units, metal stamping. Centralia Correctional Center to E.

Hoffman, village (2000 population 672), GRANT county, W MINNESOTA, 14 mi/23 km SE of ELBOW LAKE town, between POMME DE TERRE RIVER (W) and CHIPPEWA RIVER (E); 45°49′N 95°47′W. Agricultural area (grain, sunflowers, sugar beets; livestock, poultry; dairying; manufacturing (aseptic food products).

Hoffman (HAHF-muhn), village (□ 3 sq mi/7.8 sq km; 2006 population 616), RICHMOND county, S NORTH CAROLINA, 13 mi/21 km NE of ROCKINGHAM near Drowning Creek (LUMBER RIVER); 35°01′N 79°32′W. Service industries; manufacturing (sand and gravel processing); agriculture (grain, tobacco, cotton, soybeans; poultry, livestock). SANDHILLS Recreation Area to SW.

Hoffman, village (2006 population 149), OKMULGEE county, E central OKLAHOMA, 6 mi/9.7 km ENE of HENRYETTA, on DEEP FORK of Canadian River, arm of EUFAULA LAKE reservoir to E; 35°29′N 95°50′W. In agricultural area.

Hoffman Estates, village (2000 population 49,495), COOK county, NE ILLINOIS, suburb 30 mi/48 km NW of downtown CHICAGO, 6 mi/9.7 km E of ELGIN; 42°03′N 88°08′W. Residential area. Poplar Creek Theatre in W part.

Hoffman Island, artificial island, part of Richmond borough (Staten Island) of NEW YORK city, SE NEW YORK, in Lower NEW YORK BAY, just off E STATEN ISLAND; 40°35′N 74°03′W. Covers an area of c.10 acres/ 4 ha. Created 1872, it was formerly a quarantine station. Now part of Gateway National Recreation Area.

Hoffman Mountain (elevation 3,715 ft/1,132 m), ESSEX county, NE NEW YORK, in the High Peaks section of the ADIRONDACKS, 15 mi/24 km SSE of Mount MARCY and just NW of SCHROON LAKE village; 43°55′N 73°50′W.

Hofgastein, AUSTRIA: see BAD HOFGASTEIN.

Hofgeismar (hof-GEIS-mahr), town, HESSE, central GERMANY, 13 mi/21 km NNW of KASSEL; 51°29′N 09°23′E. Manufacturing of automobile parts; foundries; plastics industry. Nearby is Sababurg park. Has Gothic church and town hall from 13th century, also many half-timbered houses from 16th–19th centuries.

Hofheim am Taunus (HOF-heim ahm TOU-nus), town, HESSE, central GERMANY, at S foot of the TAUNUS, 10 mi/16 km W of FRANKFURT; 50°05′N 08°27′E. Largely residential; manufacturing of machinery, glass; woodworking; wine. Was Roman fort; excavations, both Roman and prehistoric, are nearby.

Hofheim in Unterfranken (HOF-heim in UN-tuhr-frahng-kuhn), town, BAVARIA, central GERMANY, in LOWER FRANCONIA, 14 mi/23 km NE of SCHWEINFURT; 50°06′N 10°31′E. Manufacturing of automobile parts; ceramics. Chartered 1576.

Hofit (kho-FEET), town, 4 mi/6.4 km N of NETANYA on the MEDITERRANEAN coast, ISRAEL; 32°23′N 34°52′E. Elevation 59 ft/17 m. Founded in 1955.

Hofmeyr, town, EASTERN CAPE, SOUTH AFRICA, 35 mi/56 km NNE of CRADOCK; 31°38′S 25°49′E. Elevation 4,659 ft/1,420 km. Railroad terminus; sheep, grain, feed crops. Salt pans in region. Airfield.

Höfn (HUH-puhn) or **Höfn i Hornafirdi**, fishing village (2000 population 1,769), AUSTUR-SKAFTAFELLSSYSLA county, SE ICELAND, on Hornafjörður; 64°15′N 15°11′W.

Hofnudstal, UKRAINE: see TSEBRYKOVE.

Hofors (HOO-FORSH), town, GÄVLEBORG county, E SWEDEN, 17 mi/27 km WSW of SANDVIKEN; 60°33′N 16°18′E. Manufacturing (iron- and steelworks).

Hofsjökull (HAWFS-YUH-kul), extensive glacier, central ICELAND. Rises to 5,581 ft/1,701 m at 64°46′N 18°43′W.

Hofstade (HOF-stah-duh), village, Aalst district, EAST FLANDERS province, N central BELGIUM, on DENDER RIVER and 2 mi/3.2 km N of AALST. BLOSO-DOMAIN Recreational Park in Hofstade-Strand.

Hofu (HO-foo), city, YAMAGUCHI prefecture, SW HONSHU, W JAPAN, on the SUO SEA, 9 mi/15 km S of YAMAGUCHI; 34°02′N 131°33′E. Medicine, motor vehicles; prawns; nori.

Hofuf (HUHF-uhl), town (2004 population 287,841) and oasis, E SAUDI ARABIA; 25°22′N 39°34′E. Has large resources of good water, much of which is used to irrigate the surroundings of the town. Textile manufacturing, food processing, and Arabian horse breeding are important economic activities. It is also a trade center for dates, wheat, and fruit, and has a large mosque. Originally called Hasa, it was the center of the Karmathian movement in the 10th century. The oil fields of Shimanya and Mubarraz oasis nearby. Also spelled Huhuf.

Höganäs (HUH-gah-NES), town, SKÅNE county, SW SWEDEN, on W coast of KULLEN peninsula, on KATTEGAT strait, at N end of ÖRESUND, 12 mi/19 km NNW of HELSINGBORG; 56°12′N 12°34′E. Manufacturing (pottery and ceramics, furniture; publishing).

Hogansburg, village, FRANKLIN county, N NEW YORK, on SAINT REGIS RIVER near its mouth on the SAINT LAWRENCE, and 11 mi/18 km ENE of MASSENA; 44°58′N 74°41′W. Akwesasne (Mohawk Indian) Museum here. Nearby is St. Regis Indian Reservation (partly in Canada), which has a gambling casino.

Hogansville, town (2000 population 2,774), TROUP county, W GEORGIA, 11 mi/18 km NE of LA GRANGE; 33°10′N 84°54′W. Manufacturing includes coated mesh fabrics, clothing, fabric chemicals, maps. Incorporated 1870.

Hogatza River (ho-GAT-suh), 130 mi/209 km long, W central ALASKA; rises in S foothills of BROOKS RANGE near 66°55′N 153°50′W; flows SSW to KOYUKUK RIVER at 66°N 155°23′W.

Hogback Mountain, peak (3,240 ft/988 m) in the BLUE RIDGE, GREENVILLE county, NW SOUTH CAROLINA, 6 mi/9.7 km W of LANDRUM, near NORTH CAROLINA state line.

Hogback Mountain, MONTANA: see SNOWCREST MOUNTAINS.

Hog Bay, AUSTRALIA: see PENNESHAW.

Högdalen (HUHG-DAHL-en), suburb, STOCKHOLM county, S central SWEDEN, 5 mi/8 km S of STOCKHOLM city center.

Hoge Vaart (HAW-khuh VAHRT), canal, EASTERN and SOUTHERN Flevoland polders, FLEVOLAND province, central NETHERLANDS; connects Kettlemeer, N of DRONTEN, with Oostvaardersdijk, c.5 mi/8 km N of ALMERE, crossing SE part of both polders. Parallels Loge Vaart canal for much of its course.

Högfors (HUHG-FORSH), village, VÄSTMANLAND county, central SWEDEN, in Bergslagen region, 7 mi/11.3 km NE of FAGERSTA; 60°02′N 16°01′E. Manufacturing (wood industries). In former iron-mining region.

Högfors, FINLAND: see KARKKILA.

Hoggar, AFRICA: see SAHARA.

Hoggar, ALGERIA: see AHAGGAR MOUNTAINS.

Hog Hammock, village, MCINTOSH county, SE GEORGIA on SAPELO ISLAND, 8 mi/12.9 km ESE of Eulonia; 31°23′N 81°16′W. Hog Hammock is home to the decendents of former slaves, who developed their own way of speaking, called Gullah.

Hog Harbour, VANUATU: see ESPÍRITU SANTO.

Hog Island (□ 22 sq mi/57 sq km), ESSEQUIBO ISLANDS, WEST DEMERARA district, in GUYANA, in ESSEQUIBO River estuary, SSW of WAKENAAM and LEGUAN islands, 25 mi/40 km W of GEORGETOWN; 06°49′N 58°32′W. Rice growing.

Hog Island, in city of PHILADELPHIA, PHILADELPHIA county, on W shore of DELAWARE RIVER, at W side of mouth of SCHUYLKILL RIVER. Separated from mainland by Mingo Creek, merged with TINICUM ISLAND to W by Philadelphia International Airport. In World War I, important government shipyard here; now an industrial and shipping area. Fort Mifflin Historical Site on Delaware River, in S.

Hog Island (HAHG), NORTHAMPTON county, E VIRGINIA, barrier island between the ATLANTIC OCEAN (E) and Hog Island Bay (W), 18 mi/29 km NE of CAPE CHARLES city; 9 mi/15 km long N–S, 6 mi/10 km wide; 37°24′N 75°41′W. Great Machipongo Inlet, at S end, links bay and ocean; Little Machipongo Inlet is at N end; PARRAMORE ISLAND to N, COBB ISLAND to S. Lighthouse.

Hog Island, BAHAMAS: see PARADISE ISLAND.

Hog Island, MICHIGAN: see BEAVER ISLANDS.

Hog Islands, Spanish *Cayos Cochinos* (KAH-yos ko-CHEE-nos) or *Islas Cochinos* (EES-lahs ko-CHEE-nos), archipelago (□ 1 sq mi/2.6 sq km) in ISLAS DE LA BAHÍA department, N HONDURAS, in CARIBBEAN SEA, c.8 mi/12.9 km off N Honduras coast, 22 mi/35 km ENE of LA CEIBA; 15°58′N 86°28′W. Consist of two larger islands (Big Hog Island, Little Hog Island) and thirteen islets. Coconuts. The entire archipelago lies within the CAYOS COCHINOS MARINE SANCTUARY.

Hogland, RUSSIA: see SURSAARI.

Hogolu, MICRONESIA: see CHUUK ISLANDS.

Hogoro (ho-GO-ro), village, DODOMA region, central TANZANIA, 50 mi/80 km ENE of DODOMA; 05°57′S 36°25′E. Corn, wheat; goats, sheep. This area was part of the failed groundnut, or peanut, growing scheme of the late 1940s and early 1950s.

Hog Point, easternmost point of BORNEO, MALAYSIA, in SULU SEA, at tip of wide NE peninsula; 05°19′N 119°16′E. Also called Tanjong Hog.

Hog Point, INDONESIA: see TUA, TANJUNG.

Högsby (HUHGS-BEE), town, KALMAR county, SE SWEDEN, on EMÅN River, 17 mi/27 km WSW of OSKARSHAMN; 57°10′N 16°02′E.

Høgsfjorden (HUHGS-fyawr), inlet, SE arm of BOKNAFJORDEN, ROGALAND county, SW NORWAY, extends c.17 mi/27 km to DIRDAL in RYFYLKE region.

Area is shown by the symbol □, and capital city or county seat by ⊙.

Name Høgsfjorden is sometimes applied to Høle village on W shore.

Hog's Head, Gaelic *Ceann na Muice*, cape, SW KERRY county, SW IRELAND, on E side of entrance of Ballinskelligs Bay, 4 mi/6.4 km SE of BALLINSKELLIGS; 51°47′N 10°14′W.

Högsjö (HUHG-SHUH), village, SÖDERMANLAND county, E SWEDEN, 19 mi/31 km W of KATRINEHOLM; 59°02′N 15°40′E.

Hogsty Reef, central BAHAMAS, 40 mi/64 km NW of GREAT INAGUA ISLAND, 310 mi/499 km SE of NASSAU; 21°40′N 73°50′W. The reef is c.6 mi/9.7 km long (W-E), 3 mi/4.8 km wide. Off it are 2 small cays. Called Les Etoiles on early French charts.

Högyész (HUH-dyais), Hungarian *Högyész*, village, TOLNA county, SW central HUNGARY, 16 mi/26 km NW of SZEKSZÁRD; 46°29′N 18°25′E. Corn, wheat, peas; hogs; vineyards.

Hoh, TURKEY: see SIVRICE.

Hohe Acht (HO-e AKHT), highest peak (2,447 ft/746 m) of the EIFEL, W GERMANY, 2 mi/3.2 km E of ADENAU; 50°24′N 07°00′E.

Hohe Eule, Polish *Góra Sowia* (goo-rah so-vee-ah), highest peak (3,327 ft/1,014 m) of the Eulengebrige, in SILESIA, after 1945 in SW POLAND, 11 mi/18 km S of SWIDNICA (Schweidntz).

Hohenau, town, Itapúa department, SE PARAGUAY, 20 mi/32 km NE of Encarnación; 27°04′S 55°45′W. Agricultural center (oranges, maté; livestock); sawmilling; winemaking. A German colony founded 1898, it stretches SE to the Paranâ. Sometimes Colonia Hohenau.

Hohenau an der March (HO-en-ou ahn der MAHRKH), town, NE LOWER AUSTRIA, near CZECH border, 35 mi/56 km NNE of VIENNA. Railroad junction and railroad border station opposite Czech Republic; 48°37′N 16°54′E. Sugar refinery.

Hohenberg (HO-en-berg), township, S LOWER AUSTRIA, on Unrechttraisen River, and 20 mi/32 km S of Sankt Pölten; 47°55′N 15°37′E. Machinery manufacturing.

Hohenbruck, CZECH REPUBLIC: see TREBECHOVICE POD OREBEM.

Hohenbrunn (ho-uhn-BRUN), suburb of MUNICH, BAVARIA, S GERMANY, in UPPER BAVARIA, 9 mi/14.5 km SE of city center; 48°03′N 11°43′E.

Hoheneck (HO-en-ek), N suburb of LUDWIGSBURG, GERMANY, on the NECKAR; 48°55′N 09°13′E. Resort with hot mineral springs.

Hohenelbe, CZECH REPUBLIC: see VRCHLABI.

Hohenems (HO-en-ems), town VORARLBERG, W AUSTRIA, in the Rhine Valley 4 mi/6.4 km SSW of DORNBIRN. Ski production, textiles, sulphur baths. Renaissance palace (built 1562–1567) where two manuscripts of the Nibelungen epic were discovered (1755 and 1779); remnants of a Jewish quarter; ruins of fortress on nearby mountain.

Hohenfriedeberg, Polish *Dobromierz* (do-bro-meez), town, Wałbrzych province, SW POLAND. In 1745 it was the site of the victory of Frederick II of PRUSSIA over the Austrian and Saxon forces in the War of the Austrian Succession. HOHENFRIEDEBERG was ceded to Poland after World War II.

Hohenfurth, CZECH REPUBLIC: see VYSSI BROD.

Hohenhameln (ho-en-HAHM-meln), village, LOWER SAXONY, N GERMANY, 9 mi/14.5 km NE of HILDESHEIM; 52°16′N 10°05′E.

Hohenheim (HO-en-heim), S suburb of STUTTGART, GERMANY, 5 mi/8 km S of city center; 48°46′N 09°10′E. Site of noted agricultural college, housed in 18th-century castle; large botanical garden.

Höhenkirchen-Siegertsbrunn (haw-en-KIR-khuhn–SEE-gerts-brun), suburb of MUNICH, BAVARIA, S GERMANY, in UPPER BAVARIA, 10 mi/16 km SE of city center; 48°01′N 11°43′E.

Hohenlimburg (ho-en-LIM-burg), part of HAGEN, NORTH RHINE–WESTPHALIA, W GERMANY, on the LENNE and 4 mi/6.4 km E of city center; 51°20′N 07°34′E. Steel and wire mills. Noted castle Hohenlimburg (1230) towers above district. Until 1879, called Limburg; independent town until 1975.

Hohenlockstedt (ho-en-LOHK-shtet), village, SCHLESWIG-HOLSTEIN, N GERMANY, 30 mi/48 km NW of HAMBURG; 53°58′N 09°38′E.

Hohenmauth, CZECH REPUBLIC: see VYSOKE MYTO.

Hohenmölsen (ho-en-MOL-sen), town, SAXONY-ANHALT, central GERMANY, 7 mi/11.3 km ESE of WEISSENFELS; 51°10′N 12°06′E. Lignite mining. Scene (1080) of battle in which King (later Emperor) Henry IV was defeated by his rival King Rudolf of SWABIA.

Hohen Neuendorf (HO-uhn NOI-en-dorf), town, BRANDENBURG, E GERMANY, 12 mi/19 km NW of BERLIN; 52°40′N 13°16′E.

Hohenpeissenberg (ho-en-PEIS-sen-berg), village, BAVARIA, S GERMANY, in UPPER BAVARIA, 32 mi/51 km SW of MUNICH; 47°48′N 11°01′E. 17th-century church.

Hohenrain (HO-en-rine), agricultural commune, LUCERNE canton, central SWITZERLAND, 9 mi/14.5 km N of LUCERNE.

Hohensalza, POLAND: see INOWROCLAW.

Hohensalzburg, RUSSIA: see LUNINO, KALININGRAD oblast.

Hohenschwangau, GERMANY: see SCHWANGAU.

Hohenstadt, CZECH REPUBLIC: see ZABREH.

Hohenstaufen (ho-en-SHTOU-fen), historic peak (2,244 ft/684 m) in the SWABIAN JURA, GERMANY, 4 mi/6.4 km NE of GÖPPINGEN; 48°45′N 09°43′E. Has ruins of ancestral Hohenstaufen castle, destroyed 1525.

Hohenstein, POLAND: see OLSZTYNEK.

Hohenstein-Ernstthal (HO-en-shtein–ERNST-tahl), town, SAXONY, E central GERMANY, 10 mi/16 km W of CHEMNITZ; 50°49′N 12°42′E. Textile center; metalworking. Author Karl May born here.

Hohentengen (ho-en-TENG-guhn), village, BADEN-WÜRTTEMBERG, SW GERMANY, at S foot of BLACK FOREST, on the RHINE (Swiss border; bridge), and 11 mi/18 km ESE of WALDSHUT; 47°34′N 08°26′E.

Hohentwiel (ho-en-TVEEL), mountain (2,260 ft/689 m), BADEN-WÜRTTEMBERG, SW GERMANY, just NE of SINGEN; 47°30′N 08°49′E. On the summit of Hohentwiel are the ruins of an ancient castle that was the seat of the dukes of SWABIA in the 10th century.

Hohenwald (HO-uhn-wohld), city (2006 population 3,831), ⊙ LEWIS county, central TENNESSEE, 30 mi/48 km W of COLUMBIA; 35°33′N 87°33′W. In forest and farm region; manufacturing. Meriwether Lewis National Monument is SE. Elephant Sanctuary nearby.

Hohenwart (HO-en-vahrt), village, BAVARIA, S GERMANY, in UPPER BAVARIA, on PAAR RIVER, and 12 mi/19 km S of INGOLSTADT; 48°35′N 11°23′E.

Hohenwartetalsperre (HO-en-vahr-te-TAHL-shperre), irrigation dam and reservoir, THURINGIA, central GERMANY, on the SAALE, 5 mi/8 km SE of SAALFELD; 50°37′N 11°30′E–50°37′N 11°38′E. Hydroelectric power station.

Hohenwarthe (HO-en-vahr-te), village, SAXONY-ANHALT, central GERMANY, on the ELBE, at SW end of Ihle Canal (ship elevator); 52°15′N 11°41′E. Connected here with the WESER-ELBE CANAL by aqueduct across the Elbe.

Hohenwestedt (ho-en-VES-stet), village, SCHLESWIG-HOLSTEIN, NW GERMANY, 13 mi/21 km W of NEUMÜNSTER; 54°05′N 09°39′E.

Hohenzollern Canal, GERMANY: see ODER-HAVEL CANAL.

Hoher Bogen (HO-er BO-guhn), Czech *Osek*, peak (3,517 ft/1,072 m) of BOHEMIAN FOREST, in BAVARIA, SE GERMANY, near German-Czech border, 5 mi/8 km NNE of KÖTZTING; 49°14′N 12°57′E.

Hoher Göll (HO-er GOEL), peak (7,687 ft/2,343 m) in Berchtesgaden Alps on Austro-German border, 8 mi/12.9 km SW of HALLEIN; 47°36′N 13°04′E.

Hohe Rhön (HAW-e ROEN), main range (SE) of the RHÖN MOUNTAINS, W central GERMANY; 50°10′N 09°40′E–50°50′N 10°10′E. Rises to 3,117 ft/950 m in the WASSERKUPPE. GERSFELD is the most frequented resort; mineral springs at BRÜCKENAU (SW foot).

Hoher Kasten (HO-uhr-KAS-tuhn), peak (5,889 ft/1,795 m) in the Appenzeller Alps, NE SWITZERLAND, 5 mi/8 km SE of Appenzell. Reached by overhead tramway. Views of Rhine valley and Appenzeller Alps.

Hoher Lindkogel, AUSTRIA: see WIENERWALD.

Hoher Nock, AUSTRIA: see SENGSENGEBIRGE.

Hoher Taunus, GERMANY: see TAUNUS.

Hohes Venn (HO-uhr VEN), French *Hautes Fagnes*, northernmost and highest section of the ARDENNES, E BELGIUM, on German border, bordered by the AMBLÈVE RIVER on the S and the VESDRE RIVER on the N, continues E as the EIFEL in GERMANY. Rises to 2,277 ft (694 m) in the BOTRANGE; highest point in Belgium. Sparsely populated region; extensive moors. Mainly in districts of Eupen and Malmédy, returned to Belgium after both World Wars.

Hohe Tatra, SLOVAKIA and POLAND: see HIGH TATRAS.

Hohe Tauern (HO-e TOU-ern), highest range of the Central Alps, S AUSTRIA, extending c.70 mi/113 km E from the Birnlücke at the Italian border. It rises to 11,576 ft/3,528 m in the GROSSGLOCKNER mountain. The Tauern railroad (built 1901–1908) traverses the range through a tunnel (5.3 mi/8.6 km long). It is a tourist spot for skiers and mountain climbers. The National Park Hohe Tauern was the first in Austria since 1983. It encompasses 690 sq mi/1,787 sq km in SALZBURG, CARINTHIA, and TYROL, and includes 304 peaks above 9,844 ft/3,000 m; 246 glaciers and several high alpine valleys with a traditional cultural landscape. Two roads cross the range: Felber-Tauern-Strasse between MITTERSILL and Matrei in EAST TYROL (Felber-Tauern-Tunnel 3.2 mi/5.1 km long; elevation 4,938 ft/1,505 m) and Grossglocknerstrasse (Hochtor-Tunnel. 0.2 mi/0.32 km long; elevation 7,632 ft/2,326 m) between Fusch and HEILIGENBLUT. The transalpine oil pipeline from Trieste to Ingolstadt crosses the HOHE TAUERN mountain range at Felbertauern. Subsidiary ranges are VENEDIGERGRUPPE, Granatspitzgruppe, Glocknergruppe, Sonnblickgruppe (or Goldberggruppe), and Ankogelgruppe. Hohe Tauern is the main hydropower resource of Austria; largest hydroelectric power stations in the STUBACHTAL valley, Kaprunertal (both in Salzburg), Fragent and Maultatal (Carinthia). The term Tauern signifies the high mountain passes which were used by shepherds in the past. Old tradition of mining, (especially gold), today only tungsten.

Hohe Warte (HO-e VAHR-te), Italian *Monte Coglians*, highest peak (8,473 ft/2,583 m) of CARNIC ALPS, on Austrian-Italian border, 17 mi/27 km SSE of LIENZ, AUSTRIA; 46°36′N 12°53′E.

Hohezollern, GERMANY: see HECHINGEN.

Hohhot (HU-HUH-HO-TUH) [Mongolian=Black City], city (□ 793 sq mi/2,054 sq km; 1994 estimated urban population 683,200; 1994 total estimated population 928,200), ⊙ INNER MONGOLIA AUTONOMOUS REGION, N CHINA; 40°48′N 111°37′E. The terminus of caravan routes to XINJIANG and to the Mongolian People's Republic, Hohhot is also connected by railroad with BEIJING and is a trade center for NW China. Manufacturing includes chemicals, textiles, fertilizers, agricultural machinery, construction materials, and beet sugar and other processed foods. Hohhot consists of two sections. The old town is a Mongolian political and religious center dating from the 9th century. It was the seat of the Lamaist Living Buddha until his removal (1664) to Urga (see ULAANBAATAR). The newer Chinese section, which grew around the railroad station after 1921, is the administrative center. Hohhot is the seat of Inner Mongolian University, a medical college, and several technical institutes. The city was called Guisui (Kweisui) until 1954.

Cross-references are shown in SMALL CAPITALS. The pronunciation guide is shown on page xix. The sources of population figures are shown on page xvii.

Hohndorf (HAWN-dorf), village, SAXONY, E central GERMANY, 8 mi/12.9 km E of ZWICKAU; 50°42′N 12°48′E.

Hohneck (ho-nek), summit (4,465 ft/1,361 m) of the central VOSGES MOUNTAINS, E FRANCE, on VOSGES–HAUT-RHIN department border, 6 mi/9.7 km W of MUNSTER, overlooking Col de la SCHLUCHT pass (N); 48°02′N 07°01′E. Road spur to summit. Petit Hohneck [French=little Hohneck], 4,225 ft/1,288 m, is 1 mi/1.6 km SE on trail in regional park of the Vosges summits.

Hohoe, town (2000 population 26,281), VOLTA region, GHANA, 38 mi/61 km N of Ho, on short tributary feeding Lake VOLTA; 07°09′N 00°28′E. Local council headquarters. Cacao, cotton, cassava, corn.

Hohokam Pima National Monument (□ 2 sq mi/5.2 sq km), PINAL county, S central ARIZONA, 22 mi/35 km SSE of PHOENIX, in Gila River Indian Reservation, NW of COOLIDGE. Archaeological remains of a large Hohokam Indian village of Sackstown culture, 300 B.C.–A.D. 1100. Authorized 1972; not yet established and not open to public as of 1996.

Hohoku (HO-ho-koo), town, Toyoura district, YAMAGUCHI prefecture, SW HONSHU, W JAPAN, 31 mi/50 km W of YAMAGUCHI; 34°17′N 130°57′E.

Ho-Ho-Kus (HO-HO-kuhs), borough (2006 population 4,095), BERGEN county, NE NEW JERSEY, on small Hohokus River, 7 mi/11.3 km NNW of HACKENSACK; 41°00′N 74°05′W. Largely residential. Has two notable 18th-century houses. Incorporated 1908.

Hoholeve (HO-ho-le-ve), (Russian *Gogolevo*), town, central POLTAVA oblast, UKRAINE, on road and on railroad, 11 mi/17 km ESE of MYRHOROD; 49°47′N 34°13′E. Elevation 528 ft/160 m. Oil and gas extraction, oil extraction equipment repair depot. Town since 1957.

Höhr-Grenzhausen (HUHR–grents-HOU-zuhn), town, RHINELAND-PALATINATE, W GERMANY, 6 mi/9.7 km NE of KOBLENZ; 50°26′N 07°40′E. Pottery works, school, and museum.

Hoh River (HO), c.30 mi/48 km long, mainly JEFFERSON county, NW WASHINGTON; formed at W end of OLYMPIC NATIONAL PARK by N and S forks (S fork c.20 mi/32 km long, N fork c.30 mi/48 km long) rising N and S of MOUNT OLYMPUS; main stream flows SW from junction to PACIFIC OCEAN at Hoh Indian Reservation, 15 mi/24 km S of Forks.

Höhronen (HER-o-nuhn), mountain (4,032 ft/1,229 m) in the ALPS, N central SWITZERLAND, 8 mi/12.9 km E of ZUG, on border between ZÜRICH and ZUG cantons.

Hohuanshan (HUH-HUAN-SHAN), [in Mandarin means "Harmonious Happiness Mountain"], ski resort, NANTOU county, W central TAIWAN, 24°09′N 121°16′E. Elevation 11,207 ft/3,416 m. On paved Tayuling-Wushe Highway, the highest road in Taiwan, reaching an elevation of 10,745 ft/3,275 m at Wuling Pass.

Hohwald, Le (ho-vahlt, luh), village (□ 8 sq mi/20.8 sq km), BAS-RHIN department, E FRANCE, in the E VOSGES of ALSACE, 8 mi/12.9 km SW of OBERNAI; 48°24′N 07°16′E. Resort in densely wooded area. Sawmills. Once a health resort for respiratory ailments. Winter sports in Hohwald region; vineyards in valleys.

Hoi An (HOI AHN) or **faifo**, city, QUANG NAM-DA NANG province, central VIETNAM, located 19 mi/31 km S of DA NANG city; 15°52′N 108°19′E. During the 17th–19th century, the area, known as Faifo to Western traders, was a major regional and international trading port. During period of French colonialism, Hoi An functioned as a key administrative center. Hoi An is currently experiencing extensive restoration work of earlier structures (e.g., Japanese covered bridge, Chinese communal buildings, and temples). City is also a major trading center for surrounding agricultural region (cinnamon, rice, corn, sugarcane, tea); great tourism potential.

Hoie, BELGIUM: see HUY.

Hoihow, CHINA: see HAIKOU.

Hoima (HOI-muh), administrative district (□ 2,230 sq mi/5,798 sq km; 2005 population 398,200), WESTERN region, W UGANDA, along shore of LAKE ALBERT (to W and N, with DEMOCRATIC REPUBLIC OF THE CONGO on opposite shore); ⊙ HOIMA; 01°25′N 31°05′E. Elevation averages 1,968 ft/600 m–3,281 ft/1,000 m. As of Uganda's division into eighty districts, borders BULIISA (NE), MASINDI (NEE), KIBOGA (E), and KIBAALE (S) districts. Vegetation is primarily savannah grasslands. Runyoro is primary spoken language. Largely agricultural area (including cocoa, cassava, sweet potatoes, bananas, maize, beans, rice, finger millet, and sorghum, cash crops include coffee, cotton, and tea). Livestock (primarily cattle). Several roads run through Hoima town, including one SE to KAMPALA city. In 1991 S portion of district carved out to form Kibaale district.

Hoima (HOI-muh), town (2002 population 27,934), ⊙ HOIMA district, WESTERN region, W UGANDA, 30 mi/48 km SW of MASINDI. Agricultural trade center (cotton, tobacco, coffee, bananas, corn). Inhabited by the Nyoro people. Was part of former WESTERN province.

Hoisington (HOI-zing-tuhn), town (2000 population 2,975), BARTON county, central KANSAS, 11 mi/18 km N of GREAT BEND; 38°31′N 98°46′W. In wheat, livestock, and poultry area; dairying; railroad maintenance. Manufacturing (wire drawing, feed mixers). Cheyenne Bottoms Lake to SE. Incorporated 1886.

Hoi Xuan (HOI SWAHN), village, THANH HOA province, N central VIETNAM, 65 mi/105 km SW of HANOI; 20°22′N 105°07′E. Forest products, agro-forestry. Muong, Thai, and other minority peoples. Formerly Hoixuan.

Hojancha (o-HAN-cha), town and ⊙ Hojancha canton, GUANACASTE province, NW COSTA RICA, on NICOYA PENINSULA, 6 mi/9.7 km SSE of NICOYA; 09°58′N 85°25′W. Livestock; rice, corn, beans, coffee.

Højer (HUH-yuhr), German *Hoyer*, town and port, SØNDERJYLLAND county, SW Jutland, DENMARK, on NORTH SEA, and 7 mi/11.3 km WNW of TØNDER; 54°58′N 08°44′E. Fisheries; serves as port for Tønder. Belonged to GERMANY before 1920 plebiscite. Also spelled Høyer.

Hojo (HO-jo), city, EHIME prefecture, N SHIKOKU, W JAPAN, on IYO SEA, 9 mi/15 km N of MATSUYAMA; 33°58′N 132°46′E.

Hojo (HO-jo), town, Tagawa district, FUKUOKA prefecture, N KYUSHU, SW JAPAN, 25 mi/40 km N of FUKUOKA; 33°41′N 130°47′E.

Hojo (HO-jo), town, Tohaku district, TOTTORI prefecture, S HONSHU, W JAPAN, 25 mi/40 km W of TOTTORI; 35°28′N 133°49′E.

Hokah (HO-kah), village (2000 population 614), HOUSTON county, extreme SE MINNESOTA, 8 mi/12.9 km WSW of LA CROSSE, WISCONSIN, on ROOT RIVER, W of its entrance to MISSISSIPPI RIVER; 43°45′N 91°20′W. Grain, soybeans; livestock, poultry; dairying; timber; manufacturing (lumber). Richard J. Dorer Memorial Hardwood State Forest to W, S, and E.

Hoke (HOK), county (□ 392 sq mi/1,019.2 sq km; 2006 population 42,303), S central NORTH CAROLINA; ⊙ RAEFORD; 35°01′N 79°13′W. Bounded W by Drowning Creek (LUMBER RIVER), N by Little River; source of Big Swamp River in SE. Forested sandhills. Manufacturing (textiles, poultry processing); service industries; agriculture (cotton, tobacco, corn, peaches, wheat, oats, soybeans, hay; cattle, poultry); timber. Part of large FORT BRAGG MILITARY RESERVATION in N. Formed 1911 from Cumberland and Robeson counties. Named for Robert F. Hoke (1837–1912), a general in the Confederate army.

Hokendauqua (HO-ken-DAW-kwah), unincorporated town (2000 population 3,411), LEHIGH county, E PENNSYLVANIA, residential suburb 4 mi/6.4 km N of downtown ALLENTOWN, on Lehigh River, opposite NORTH CATASAUQUA.

Hokes Bluff, city (2000 population 4,149), Etowah co., NE Alabama, a suburb 8 mi/12.9 km E of Gadsden. Inc. in 1946. Named in 1854 for Daniel Hoke, Jr., at the suggestion of his business partner, William B. Wynne.

Hoki, former province in SW HONSHU, JAPAN; now part of TOTTORI prefecture.

Hokiang, CHINA: see HEJIANG.

Hokianga Harbour (ho-ki-ANG-uh), ramifying inlet formed by submerged ancient river valley, W coast NORTHLAND Region, NORTH ISLAND, NEW ZEALAND, 120 mi/193 km NW of AUCKLAND. Outlet for early New Zealand kauri timber and gum trade. Small village of Omapere at entrance, RAWENE on S shore.

Hokitika (HO-ke-tee-ka), town and port, ⊙ WESTLAND district (□ 4,587 sq mi/11,926.2 sq km), in WEST COAST region of SOUTH ISLAND, NEW ZEALAND, at mouth of Hokitika River, 75 mi/121 km SSW of WESTPORT. Gold mining, sawmills. Tourism associated with forests, glaciers, and nearby Lake Kaniere.

Hokkaido (hok-KAH-ee-do) [Japan=island of the N sea], island (□ 30,130 sq mi/78,338 sq km), N JAPAN, coterminous with Hokkaido prefecture, separated from HONSHU island by the TSUGARU STRAIT and from SAKHALIN (Siberian RUSSIA), by the SOYA STRAIT. It is the second-largest, northernmost, and most sparsely populated of the major islands of Japan. Once called Yezo, or Ezo, it received its modern name in 1869. Its rugged interior with many volcanic peaks rises to 7,511 ft/2,289 m in ASAHI-DAKE. The ISHIKARI RIVER, second-longest in Japan, traverses W Hokkaido; its valley is an important urban and industrial region. Has a humid continental climate and receives much snow. Forests, covering most of the island, are a source of lumber, pulp, and paper (milled in Hokkaido). Coal, iron, and manganese are mined; the ISHIKARI coal field produces a major part of Japan's supply. Although large areas of the island are unsuited to farming, agriculture is an important occupation. One of the world's major fishing centers. Chief winter resort and sports area in Japan; the 1972 Winter Olympics were held at SAPPORO. Hokkaido's scenic beauty is preserved in several national parks. The population is concentrated largely in the W and SW. Sapporo, HAKODATE, and OTARU are the chief cities. KUSHIRO is the main port for E Hokkaido. Originally inhabited by Ainu, aborigines of uncertain ancestry. Until 1800 the Ainu outnumbered the Japanese, who had begun (16th century) to settle the SW peninsula; there are now c.16,000 Ainu remaining here. With the Meiji restoration (1868) Japan began the first serious effort to people the island as a means of strengthening the N frontier. Under a government-sponsored plan to develop the island, various American advisors were brought here, e.g., Horace Capron, an American agriculturalist, who introduced (1872–1876) scientific methods of farming. In 1885, Hokkaido was made an administrative unit and was granted a central government. The growth of the railroads helped speed settlement, but despite subsidies, the severe winters discouraged emigration from S Japan. Parts of the island, particularly in the N, are still largely uninhabited. The completion of the Seikan Tunnel (1988), which carries a railroad line connecting Hokkaido and HONSHU, has further decreased the isolation of Hokkaido.

Hokota (ho-KO-tah), town, Kashima district, IBARAKI prefecture, central HONSHU, E central JAPAN, on N shore of Kita-ura lagoon, 16 mi/25 km S of MITO; 36°09′N 140°31′E. Melons, sweet potato, peanuts; pigs.

Hokubo (ho-koo-BO), town, Jobo district, OKAYAMA prefecture, SW HONSHU, W JAPAN, 28 mi/45 km N of OKAYAMA; 34°57′N 133°38′E. Cucumbers (also pickled). Nearby Bicchu Kanachi limestone cave was the first to be discovered in Japan.

Area is shown by the symbol □, and capital city or county seat by ⊙.

Hokuse (hok-SE), town, Inabe district, MIE prefecture, S HONSHU, central JAPAN, 28 mi/45 km N of TSU; 35°08'N 136°31'E.

Hokutan (ho-koo-TAHN), town, on NW coast of AWAJI-SHIMA island, Tsuna district, HYOGO prefecture, W central JAPAN, 17 mi/28 km S of KOBE; 34°32'N 134°55'E. Nori.

Hol (HAWL), village, BUSKERUD county, S NORWAY, at E foot of HALLINGSKARVET mountains, in the HALLINGDAL, on railroad, and 70 mi/113 km NW of HØNEFOSS. Agriculture, livestock raising; lumbering; fishing; hunting; tourism. There are three hydroelectric power plants on the Holselva River.

Hölak, SLOVAKIA: see TRENCIANSKA TEPLA.

Holalkere (HO-lahl-ke-re), town, CHITRADURGA district, KARNATAKA state, INDIA, 20 mi/32 km SW of CHITRADURGA; 14°02'N 76°11'E. Agricultural market (millet, oilseeds, cotton); cotton ginning, hand-loom weaving. Manganese mines 5 mi/8 km NE.

Holandsfjorden (HAW-lahns-fyawr-uhn), village, NORDLAND county, N NORWAY, on small HOLANDSFJORDEN, at foot of SVARTISEN glacier, 30 mi/48 km NW of Missouri. Tourist center.

Hola Prystan' (HO-lah PRIS-tahn), (Russian Golaya Pristan'), city, SW KHERSON oblast, UKRAINE, on DNIEPER (Ukrainian Dnipro) River (landing) and 8 mi/12.9 km SSW of KHERSON; 46°31'N 32°31'E. Raion center; manufacturing (boats, furniture), food processing (butter, food flavoring). Mud bath sanatorium and health resort; museum, laboratories and offices of the BLACK SEA COASTAL BIOSPHERE NATURE PRESERVE. Known since 1786; city since 1958.

Holbæk (HOL-bek), city (2000 population 23,426), VESTSJÆLLAND county, E DENMARK, a seaport on HOLBÆK FJORD, an arm of the Isefjord; 55°42'N 11°42'E. Chartered 1288. It is a commercial and industrial center.

Holbæk Fjord, DENMARK: see ISE FJORD.

Holbeach (HAHL-beech), town (2001 population 10,263), LINCOLNSHIRE, E ENGLAND, 7 mi/11.3 km E of SPALDING; 52°48'N 00°01'E. Market town (chartered in 1252) in bulb- and fruit-growing region. Has 14th-century Gothic church.

Holbeck, ENGLAND: see LEEDS.

Holberg (HOL-buhrg), unincorporated village, SW BRITISH COLUMBIA, W CANADA, on N VANCOUVER ISLAND, in Mount WADDINGTON regional district, 9 mi/15 km N of WINTER HARBOUR; 50°39'N 128°01'W.

Holberg Inlet (HOL-buhrg), SW BRITISH COLUMBIA, W CANADA, N arm of QUATSINO SOUND, N VANCOUVER ISLAND, in lumbering, fishing area; 21 mi/34 km long, 1 mi/2 km wide; 50°36'N 127°44'W. HOLBERG village is at W end.

Holborn, ENGLAND: see CAMDEN.

Holborn Head, Scotland: see SCRABSTER.

Holbrook, town (2000 population 4,917), ⊙ NAVAJO county, E central ARIZONA, on LITTLE COLORADO RIVER, near mouth of Siver Creek and c.90 mi/145 km ESE of FLAGSTAFF; 34°54'N 110°09'W. Trade and tourist center in agricultural area (cattle, sheep, hogs and poultry raising; corn, alfalfa, hay); manufacturing (printing and publishing); junction of logging railroad spur to Apache-Sitgreaves National Forest to S. PETRIFIED FOREST NATIONAL PARK is 15 mi/24 km E; Hopi and Navajo Indian reservations are N. Northland Pioneer college (two year). Settled in 1870s, incorporated 1917.

Holbrook, town (2000 population 10,785), NORFOLK county, E MASSACHUSETTS. Settled 1710, set off from RANDOLPH and incorporated 1872; 42°09'N 71°01'W. It has both agriculture and manufacturing (paper, chemicals, metal products).

Holbrook (HAHL-brook), village, S NEW SOUTH WALES, AUSTRALIA, 105 mi/169 km WSW of CANBERRA; 35°44'S 147°19'E.

Holbrook, village (2006 population 216), FURNAS county, S NEBRASKA, 33 mi/53 km ENE of MCCOOK

and on REPUBLICAN RIVER; 40°17'N 100°00'W. Grain, livestock.

Holbrook, unincorporated village (□ 6 sq mi/15.6 sq km; 2000 population 27,512), SUFFOLK county, SE NEW YORK; abuts Long Island-MacArthur Islip Airport on N; 40°47'N 73°04'W. Diversified light manufacturing and commercial services.

Hol Chan Marine Reserve (□ 4.7 sq mi/12.2 sq km), BELIZE district, BELIZE, S of SAN PEDRO, at S end of AMBERGRIS CAY; 17°52'N 88°00'W. Important spawning area for marine organisms with important coral reef ecology.

Holcomb (HOL-kuhm), town (2000 population 2,026), FINNEY county, SW KANSAS, 7 mi/11.3 km WNW of GARDEN CITY, on ARKANSAS RIVER; 37°59'N 100°59'W. Wheat, cattle; manufacturing (beef processing).

Holcomb (HAW-kuhm), town, DUNKLIN county, in boot heel of extreme SE MISSOURI, near SAINT FRANCIS River, 12 mi/19 km N of KENNETT; 36°23'N 90°01'W. Cotton, soybeans.

Holcomb (HOL-kuhm), village, GRENADA county, N central MISSISSIPPI, 20 mi/32 km NNE of GREENWOOD, near YALOBUSHA RIVER. Agriculture (cotton, corn; cattle).

Holcomb, village, ONTARIO county, W central NEW YORK, 20 mi/32 km SE of ROCHESTER, 7 mi/11.3 km W of CANANDAIGUA; 42°53'N 77°25'W. In farming area; light manufacturing. In 1990, village was consolidated with East Bloomfield village to form BLOOMFIELD village; part of town of EAST BLOOMFIELD.

Holden, city (2000 population 2,510), JOHNSON county, W central MISSOURI, 14 mi/23 km W of WARRENSBURG; 38°42'N 93°59'W. Soybeans, corn; cattle; manufacturing (agricultural equipment). Laid out 1857.

Holden, town, PENOBSCOT county, S MAINE, just SE of BANGOR; 44°45'N 68°40'W. National Grange of Patrons of Husbandry founded here 1867.

Holden, town, WORCESTER county, central MASSACHUSETTS, a residential suburb 6 mi/9.7 km NNW of WORCESTER; settled 1723, set off and incorporated 1741; 42°22'N 71°52'W. Manufacturing (electric and metal products, plastics, and machinery). Includes village of Jefferson.

Holden, unincorporated town (2000 population 1,105), LOGAN county, SW WEST VIRGINIA, 5 mi/8 km WSW of LOGAN; 37°48'N 82°04'W. Coal-mining area. Manufacturing (coal processing, mining machinery).

Holden (HOL-den), village (□ 1 sq mi/2.6 sq km; 2001 population 374), central ALBERTA, W CANADA, 30 mi/48 km NE of CAMROSE, and in BEAVER COUNTY; 53°14'N 112°14'W. Dairying, mixed farming (wheat, oats, barley, canola). Incorporated 1909.

Holden, village (2006 population 388), MILLARD county, W central UTAH, 10 mi/16 km NNE of FILLMORE; 39°06'N 112°16'W. Elevation 5,115 ft/1,559 m. Alfalfa; dairying; cattle. PAVANT Range to the E. Parts of Fishlake National Forest to N and E. Pavant Range to E; Pavant Butte (5,771 ft/1,759 m) to W.

Holden Beach (HOL-duhn BEECH), village (□ 3.4 sq mi/8.8 sq km; 1990 population 626; 2000 population 787), BRUNSWICK county, SE NORTH CAROLINA, 30 mi/48 km SW of WILMINGTON, on ATLANTIC OCEAN; 33°54'N 78°18'W. Entrance to Lockwood Folly Inlet to E; INTRACOASTAL WATERWAY canal passes to N. Beach resort area. Service industries.

Holdenville, town (2006 population 5,564), ⊙ HUGHES county, central OKLAHOMA, 25 mi/40 km NE of ADA; 35°04'N 96°24'W. Elevation 903 ft/275 m. Railroad junction. Trade center for oil and agricultural area (corn, peanuts, watermelons); grain elevators, oil wells; manufacturing (catfish processing, sweaters, hardware); catfish farming. Lake Holdenville (c.3 mi/4.8 km long; fishing), with fish hatchery to SE, and site of old Fort Holmes are nearby. Incorporated 1898.

Holderness (HOL-duhr-ness), town, GRAFTON county, central NEW HAMPSHIRE, 14 mi/23 km NNW of LA-

CONIA; 43°44'N 71°35'W. Bounded in NW by PEMIGEWASSET RIVER; drained by Owl Brook. Agriculture (poultry, cattle; vegetables; nursery crops; timber); resort area. Science Center of New Hampshire is here. Part of SQUAM LAKE in E, Little Squam Lake in center; Mount Prospect (2,072 ft/632 m) in N.

Holderness (HAHL-duh-NES), low, fertile peninsula in East Riding of Yorkshire, NE ENGLAND, between NORTH SEA and HUMBER RIVER estuary; 53°40'N 00°16'W. Its S extremity is SPURN HEAD, a narrow, 4-mi/6.4-km promontory at mouth of Humber River opposite GRIMSBY, with two lighthouses and a nearby lightship.

Holdfast (HOLD-fast), village (2006 population 173), S central SASKATCHEWAN, CANADA, near LAST MOUNTAIN LAKE, 50 mi/80 km NW of REGINA; 50°58'N 105°28'W. Wheat.

Holdingford, village (2000 population 736), STEARNS county, central MINNESOTA, 18 mi/29 km NW of ST. CLOUD, on South Branch TWO RIVERS; 45°43'N 94°28'W. Grain; livestock, poultry; dairying; light manufacturing.

Holdorf (HUHL-dorf), village, LOWER SAXONY, NW GERMANY, 13 mi/21 km SW of VECHTA; 52°35'N 08°06'E.

Holdrege (HUHL-drej), city (2006 population 5,325), ⊙ PHELPS county, S NEBRASKA, located in PLATTE RIVER valley, 24 mi/39 km SW of KEARNEY; 40°26'N 99°22'W. Railroad junction. Flour; livestock; grain, dairy products. Manufacturing (feed and fertilizer, woven labels, diposable syringes, excavating equipment, agricultural equipment). County historical museum here. Settled 1883, incorporated 1884.

Hold with Hope, Cape, wide headland, E GREENLAND, on GREENLAND SEA; 73°22'–73°51'N 20°20'W. First sighted and named (1607) by Hudson. Forms N side of entrance to FRANZ JOSEF FJORD.

Hole (HOO-le), village, KOPPARBERG county, central SWEDEN, on VÄSTERDALÄLVEN River, close to MALUNG. Includes Idbäck (EED-BEEK), village.

Hole in the Wall, BAHAMA Islands: see GREAT ABACO ISLAND.

Holeischen, CZECH REPUBLIC: see HOLYSOV.

Hole Narsipur (HO-lai NUHR-si-poor), town, HASSAN district, KARNATAKA state, INDIA, on HEMAVATI RIVER, and 17 mi/27 km SSE of HASSAN; 12°47'N 76°15'E. Handicraft wickerwork; grain, rice. Chromite and asbestos workings nearby. Also written Hole-Narsipur or Holenarsipur.

Holesov (HO-le-SHOF), Czech Holešov, German Holleschau, town, JIHOMORAVSKY province, E central MORAVIA, CZECH REPUBLIC, on railroad, and 8 mi/12.9 km NNW of ZLÍN; 49°20'N 17°35'E. Manufacturing (textiles, furniture); sugar refinery; agricultural center (sugar beets, wheat, barley, oats). Neolithic-era archaeological site, baroque castle, synagogue and old Jewish cemetery.

Holesovice (HO-le-SHO-vi-TSE), Czech Holešovice, NNE district of PRAGUE, PRAGUE-CITY province, CZECH REPUBLIC, on left bank of VLTAVA RIVER, 2 mi/3.2 km from city center. Main river port, with dockyards, machines, chemicals, pharmaceuticals manufacturing, brewery.

Holeta Genet (ho-LAI-tah GEN-et), village (2007 population 31,438), OROMIYA state, central ETHIOPIA; 17 mi/27 km W of ADDIS ABABA; 09°04'N 38°30'E. Agricultural settlement (cereals, fruit, vegetables) founded here by Italians (1937); mills (flour, lumber). Formerly called Oletta. Also known as Holetta and Genet.

Holetown, town, W BARBADOS, BRITISH WEST INDIES, 6 mi/9.7 km N of BRIDGETOWN. Surrounded by hotels and vacation cottages. Has an old fort. Site where English made their first landing, in 1625, is commemorated by a column. Formerly called St. James Town.

Holetta, ETHIOPIA: see HOLETA GENET.

Cross-references are shown in SMALL CAPITALS. The pronunciation guide is shown on page xix. The sources of population figures are shown on page xvii.

Holgate (HOL-gait), village (□ 1 sq mi/2.6 sq km; 2006 population 1,152), HENRY county, NW OHIO, 10 mi/16 km S of NAPOLEON; 41°14′N 84°08′W. In dairying, poultry-raising, and grain-producing district.

Holguera (ol-GAI-rah), village, CÁCERES province, W SPAIN, 17 mi/27 km SW of PLASENCIA. Cereals, olive oil, pepper, tobacco.

Holguín (ol-GEEN), province (□ 3,440 sq mi/8,944 sq km; 2002 population 1,021,321), E CUBA; ⊙ HOLGUÍN. Bordered N by ATLANTIC OCEAN, SE by GUANTÁNAMO province, SW by GRANMA province, and W by LAS TUNAS province (□ 3,440 sq mi/8,910 sq km). One of 5 provinces created in 1976 jurisdictional reorganization, from old Oriente province. Important cities: HOLGUÍN (1994 estimated population 246,000), BANES (1994 estimated population 35,000). Drained by Nipe, MAYARÍ, Tacajó, GIBARA, and Moa rivers. Plantains, sugarcane, and other vegetables are grown in lowlands, whereas uplands and mountains constitute important ecological reserves and are covered with thick, tropical forests and pines. Important nickel concentrations are found in these mountainous areas and elsewhere in province, forming a key export mineral for the nation. By 1994, 10 sugar mills operated in province, processing cane produced on the nearly 700,000 acres/285,000 ha devoted to sugarcane. Tubers, greens, and plantains make it a major vegetable producer.

Holguín (ol-GEEN), city (2002 population 269,618), ⊙ HOLGUÍN province, E CUBA; 20°53′N 76°46′W. Cuba's fourth-largest city, it is a prosperous commercial center and transportation hub in a fertile region of diversified agriculture and the center of nickel production. Often called "Cuba's granary," Holguín is located in a region where corn, beans, sugarcane, tobacco, and cattle are raised. Manufacturing includes agricultural machinery. Most exports are handled by its port, GIBARA, c.16 mi/25 km N (though BANES and NIPE bays, c.25 mi/40 km E, are among Cuba's finest ports). Calixto García International Airport is 10 mi/16 km SSW. The city, founded in 1523, was named for Garcia Holguín, a 16th-century conquistador in Mexico. It was moved to its present site in the 18th century.

Hol Hol (HOL HOL), town, ALI SABIEH district, DJIBOUTI, on railroad to ADDIS ABABA (ETHIOPIA), 22 mi/35 km SW of DJIBOUTI; 11°19′N 42°54′E. Elevation 1,310 ft/399 m.

Holic (ho-LEECH), Slovak *Holič*, Hungarian *Holics*, town, ZAPADOSLOVENSKY province, W SLOVAKIA, 13 mi/21 km NW of SENICA; 48°48′N 17°10′E. Manufacturing (textiles, leather; food processing; distillery). Railroad junction. Has 11th-century castle, Baroque church, and Paleolithic-era archaeological site.

Holice (HO-li-TSE), German *Holitz*, town, VYCHODOCESKY province, E BOHEMIA, CZECH REPUBLIC, on railroad, and 11 mi/18 km ENE of PARDUBICE; 50°05′N 16°00′E. In sugar-beet and potato region; manufacturing of hand-worked footwear, clothing. Has baroque church, museum.

Holiday (HAHL-i-dai), city (□ 5 sq mi/13 sq km; 2000 population 21,904), PASCO county, W central FLORIDA, 25 mi/40 km NW of TAMPA; 28°11′N 82°44′W. Manufacturing includes women's swimwear, kitchen cabinets, commercial printing, and industrial tools.

Holiday Hills, village (2000 population 831), MCHENRY county, NE ILLINOIS, residential suburb 5 mi/8 km SE of MCHENRY, on FOX RIVER; 42°17′N 88°13′W. Moraine Hills State Park to N.

Holiday Lakes, town (2006 population 1,108), BRAZORIA county, SE TEXAS, residential suburb 7 mi/11.3 km NW of ANGLETON, in BRAZOSPORT area, on OYSTER CREEK. Agricultural area.

Holiday Shores, unincorporated village, MADISON county, SW ILLINOIS, suburb 25 mi/40 km NNE of downtown SAINT LOUIS, MISSOURI, 6 mi/9.7 km ENE of Behalto; 38°55′N 89°56′W. Centered on Lake Holiday.

Holikachuk (HO-lee-KA-chuhk), Indian village, W ALASKA, on INNOKO RIVER, and 50 mi/80 km N of HOLY CROSS. Salmon fishing; fur-trading post. Formerly spelled Holocachaket or Hologachaket.

Holin, MONGOLIA: see KARAKORUM.

Holin Hall Village, unincorporated town, residential suburb 5 mi/8 km S of ALEXANDRIA, 12 mi/19 km S of WASHINGTON, D.C., near POTOMAC River.

Holitz, CZECH REPUBLIC: see HOLICE.

Holker (HUK-uh), village (2001 population 1,808), CUMBRIA, NW ENGLAND, N of Cark; 54°11′N 02°58′W. Holker Hall has large gardens with motor museum.

Holkham Bay, ENGLAND: see WELLS.

Hollabrunn (hahl-lah-BRUN), town, N LOWER AUSTRIA, in the WEINVIERTEL, 13 mi/21 km NNW of STOCKERAU; 48°34′N 16°05′E. Market center of W Weiniertel; lumber mill; corn, cattle, vineyards; wine fair. Romanesque church (13th century) 3 mi/4.8 km N.

Hollam's Bird Island, islet, W NAMIBIA, in the ATLANTIC OCEAN, off Black Reef; 24°38′S 14°31′E. Guano. Breeding site for sea birds.

Holland, city (2000 population 35,048), OTTAWA and ALLEGAN counties, SW MICHIGAN, near LAKE MICHIGAN, on Lake Macatawa, in a dairy and poultry area; 42°46′N 86°05′W. Elevation 610 ft/186 m. Manufacturing (food and beverage, printing, machining, fabricated metal products, electronic equipment, furniture); 2 wooden shoe factories; delftware (earthenware) factory, only delft factory in U.S. Furnaces have been made there since 1906. Other products include chemicals and boats. Railroad junction. Tulip growing is an important industry, and the city's many Dutch descendants hold a week-long tulip festival each spring. Dutch Village, to N, has full scale buildings of Dutch architecture, windmills, canals, gardens. The Dutch Reformed Church operates Hope College and Western Theological Seminary. Coast guard station is on Lake Macatawa; Holland State Park to W on Lake Michigan; Saugatuck State Park to SW, also on Lake Michigan; Municipal airport to S. Baker Furniture Museum. The city is a popular summer resort. Founded 1847 by Dutch settlers, incorporated 1867.

Holland (HAH-luhnd), former township (□ 111 sq mi/288.6 sq km; 2001 population 2,988), S ONTARIO, E central CANADA, 13 mi/21 km from OWEN SOUND; 44°24′N 80°46′W. Amalgamated into the township of CHATSWORTH in 2000.

Holland, town (2000 population 695), DUBOIS county, SW INDIANA, 10 mi/16 km SSW of JASPER. In agricultural area. Manufacturing (ice cream).

Holland, town (2000 population 250), GRUNDY county, central IOWA, 4 mi/6.4 km NNW of GRUNDY CENTER; 42°23′N 92°47′W. In agricultural area.

Holland, agriculture town (2000 population 2,525), HAMPDEN county, S MASSACHUSETTS, on headstream of QUINEBAUG RIVER and 22 mi/35 km E of SPRINGFIELD. Hamilton Reservoir.

Holland, agricultural town (2000 population 246), PEMISCOT county, in bootheel of extreme SE MISSOURI, in MISSISSIPPI alluvial plain, 15 mi/24 km SW of CARUTHERSVILLE; 36°03′N 89°52′W. Cotton, rice, soybeans; manufacturing (typewriter ribbons).

Holland, unincorporated town, BUCKS county, SE PENNSYLVANIA, suburb 17 mi/27 km NNE of PHILADELPHIA and 12 mi/19 km WSW of TRENTON, NEW JERSEY on Mill Creek; 40°10′N 74°59′W. Manufacturing includes metal castings. Springfield Lake reservoir in NW.

Holland, town (2006 population 1,081), BELL county, central TEXAS, 22 mi/35 km W of TEMPLE, near LITTLE RIVER; 30°52′N 97°24′W. In cotton; corn, wheat, sorghum; cattle area; manufacturing (fertilizers).

Holland (HAW-lind), town, ORLEANS co., N VERMONT, on QUEBEC (CANADA) line, 10 mi/16 km E of Newport; 44°57′N 72°00′W.

Holland (HAH-luhnd), unincorporated village (2006 population 400), S MANITOBA, W central CANADA, 50 mi/80 km ESE of BRANDON, and in VICTORIA rural municipality; 49°35′N 98°52′W. Mixed farming, lumbering, grain elevators.

Holland, village (2000 population 215), PIPESTONE county, SW MINNESOTA, 9 mi/14.5 km NE of PIPESTONE, on ROCK RIVER; 44°05′N 96°11′W. Agricultural area (grain, soybeans, peas, potatoes; poultry, livestock; dairying).

Holland, residential village (□ 3 sq mi/7.8 sq km; 2000 population 1,261), ERIE county, W NEW YORK, on East Branch of Cazenovia Creek and 25 mi/40 km SE of BUFFALO; 42°38′N 78°32′W. In farm area. Also called Holland Village.

Holland (HAH-luhnd), village (2006 population 1,270), LUCAS county, NW OHIO, 9 mi/14 km W of TOLEDO; 41°37′N 83°42′W.

Holland, region, W NETHERLANDS; specifically refers to the Dutch provinces of NORTH and SOUTH Holland, which have been divided since 1840. Bounded E by MARKERMEER and IJSSELMEER (sections of former ZUIDER ZEE), N by WADDENZEE, W by NORTH SEA, and S by the estuaries of the MAAS and WAAL rivers. Includes the cities of AMSTERDAM, ROTTERDAM, THE HAGUE, and HAARLEM, as well as the towns of EDAM and GOUDA (after which varieties of cheese have been named). The economic and traditional heart of the Netherlands, it has most of the country's noted flower industry and its manufacturing and shipping facilities. The name is also used to refer to all of the Netherlands. The region was created in the early tenth century and originally controlled not only present North and South Holland, but also ZEELAND and part of medieval FRIESLAND. William II was elected (1247) German king, but was unable to exert his authority; he died in 1255 during a campaign against the independence-minded W Frisians. In 1299, John of Avesnes, count of Hainaut, seized Holland, which came (1345) into the hands of the Bavarian house of Wittelsbach through marriage. The house of Wittelsbach retained possession of Holland until 1433, when Philip the Good, duke of Burgundy, wrested it from Jacqueline, countess of Hainaut, Holland, Zeeland, and Friesland. Civil strife followed over the next fifty years. The cloth industry and commerce of Holland, though they developed later than those of FLANDERS and BRABANT, began to rival those of BRUGES and ANTWERP in the fifteenth century. The ports of Holland were closely linked with the Hanseatic League and later, after the Netherlands had gained independence, they became major entrepôts and shipbuilding centers. Holland led in the struggle (sixteenth–seventeenth centuries) for Dutch independence, and because it dominated the States-General, its subsequent history became virtually identical with that of the Netherlands.

Hollandale (HAHL-uhn-dail), town (2000 population 3,437), WASHINGTON county, W MISSISSIPPI, 20 mi/32 km SSE of GREENVILLE, and on Deer Creek; 33°10′N 90°50′W. In agricultural area (cotton, grain, soybeans, sorghum; cattle, catfish; timber); manufacturing (catfish processing, metal fabrication).

Hollandale, village (2000 population 292), FREEBORN county, S MINNESOTA, 12 mi/19 km NE of ALBERT LEA; 43°45′N 93°12′W. Grain, soybeans; livestock, poultry; dairying; light manufacturing.

Hollandale, village (2006 population 273), IOWA county, S WISCONSIN, 11 mi/18 km SE of DODGEVILLE; 42°52′N 89°55′W. In agricultural area. Septic tanks. Yellowstone Lake State Park to S.

Holland Bay, small inlet of the CARIBBEAN SEA in SAINT THOMAS parish, on E JAMAICA coast, on JAMAICA CHANNEL, 38 mi/61 km E of KINGSTON; 17°56′N 76°13′W. Receives plantain GARDEN River. Adjoining is village of Holland. Tourist attraction.

Hollandia, INDONESIA: see JAYAPURA.

Holland Landing (HAH-luhnd LAN-deeng), unincorporated village, YORK region, S ONTARIO, E central CANADA, 25 mi/41 km N of TORONTO, and included in the town of East GWILLIMBURY; 44°05′N 79°29′W.

Holland Patent, village (2006 population 458), ONEIDA county, central NEW YORK, 10 mi/16 km N of UTICA; 43°14′N 75°15′W. Named after the 1797 land grant establishing the area.

Hollandsche IJssel, NETHERLANDS: see HOLLANDSE IJSSEL RIVER.

Hollandsche Yssel, NETHERLANDS: see HOLLANDSE IJSSEL RIVER.

Hollands Diep (HAW-lahnts DEEP), Dutch *Hollandsediep*, river channel, 13 mi/21 km long, W NETHERLANDS, part of Maas (MEUSE) River system; formed by joining of AMER RIVER and NEW MERWEDE RIVER channels 10 mi/16 km NNW of BREDA; flows W through Volkerak Dam, past WILLEMSDORP and WILLEMSTAD. Divides into the VOLKERAK channel (SW, which continues as the KRAMMER channel) and the HARINGVLIET channel (W), 12 mi/19 km SW of DORDRECHT, joined from N by DORDTSE KIL channel, opposite MOERDIJK. Spanned by MOERDIJK BRIDGES (railroad and highway) in upper course, by Haringvliet Bridge at mouth, which joins Volkerak Dam at S end (at entrance to Volkerak). Forms part of boundary between SOUTH HOLLAND and NORTH BRABANT provinces.

Hollandse IJssel River (HAW-lahnt-suh EI-suhl), 48 mi/77 km long, UTRECHT and SOUTH HOLLAND provinces, W NETHERLANDS. Branches from LEK RIVER near VREESWIJK (5 mi/8 km S of UTRECHT); flows NNW past IJSSELSTEIN, and WSW past MONTFOORT, GOUDA, NIEUWERKERK, and KRIMPEN AAN DEN IJSSEL, to NEW MAAS RIVER (3 mi/4.8 km E of ROTTERDAM). Navigable by ship to Gouda, by shallow draft boats beyond Gouda. Formerly spelled Hollandsche IJssel and Hollandsche Yssel.

Holland Tunnel, vehicular tunnel (two tubes) under the HUDSON RIVER between MANHATTAN borough of NEW YORK city, SE NEW YORK, and JERSEY CITY, NEW JERSEY; 40°43′N 74°01′W. Completed in 1927, it is 9,250 ft/2,819 m long and serves approximately 35 million commuters annually. Designated a National Historic Landmark in 1993.

Holland Village, town, SW Singapore island, SINGAPORE, suburb 4 mi/6.4 km WNW of downtown SINGAPORE; 01°18′N 103°47′E. Junction of railroad spur to Jurong industrial area. On Malayan railroad line. Population is mainly Chinese. Known among tourists for its numerous inexpensive shops, for arts and crafts of SE Asia. Named for Holland Road which was named in 1907 for early resident, Hugh Holland.

Hollansburg (HAH-luhnz-buhrg), village (2006 population 211), DARKE county, W OHIO, 11 mi/18 km SW of GREENVILLE, at INDIANA state line; 39°59′N 84°48′W. Cheese.

Hollenberg (HAHL-en-buhrg), village (2000 population 31), WASHINGTON county, NE KANSAS, on LITTLE BLUE RIVER, near NEBRASKA line, and 12 mi/19 km NNE of WASHINGTON; 39°58′N 96°59′W. Grain, cattle. Washington State Fishing Lake to SW. Incorporated 1937.

Höllengebirge (HOL-len-ge-BIR-ge), mountain range in the SALZKAMMERGUT, in S UPPER AUSTRIA, between ATTERSEE and Lake Traun; rises to 5,675 ft/1,730 m in the Höllenkogel; bleak rough slopes.

Hollenstein an der Ybbs (HAHL-len-shtein ahn der IBS), village, SW LOWER AUSTRIA, in the EISENWURZEN region on YBBS RIVER, and 11 mi/18 km S of WAIDHOFEN AN DER YBBS. Dairy farming, recreation; 44°14′N 14°46′E.

Höllental (HOL-len-tahl), SE LOWER AUSTRIA, valley of SCHWARZA RIVER, between the RAX and the SCHNEEBERG mountains; beautiful scenery.

Holleschau, CZECH REPUBLIC: see HOLESOV.

Holley, village (□ 1 sq mi/2.6 sq km; 2006 population 1,738), ORLEANS county, W NEW YORK, on the BARGE CANAL, and 22 mi/35 km WNW of ROCHESTER; 43°13′N 78°01′W. Light manufacturing; agricultural commerce. Incorporated 1867.

Hollfeld (HUHL-felt), town, UPPER FRANCONIA, N BAVARIA, GERMANY, on the Wiesent, and 13 mi/21 km W of BAYREUTH; 49°56′N 11°17′E. Manufacturing of furniture and fishing equipment; lumber milling. Has late-18th-century church.

Hollick-Kenyon Peninsula, ice-covered spur projecting over 40 mi/65 km out from the mountains of WILKINS COAST, PALMER LAND, Antarctic Peninsula; 68°35′S 63°50′W. Cape Agassiz is at its tip.

Hollick-Kenyon Plateau, a large, featureless ice plateau in central WEST ANTARCTICA, bounded by the ELLSWORTH MOUNTAINS on the E and the CRARY MOUNTAINS on the W; it overlies the deep ice of the BYRD SUBGLACIAL BASIN and the BENTLEY SUBGLACIAL TRENCH; 78°00′S 105°00′W.

Holliday, town (2000 population 129), MONROE county, NE central MISSOURI, 6 mi/9.7 km W of PARIS; 39°29′N 92°07′W.

Holliday, town (2006 population 1,810), ARCHER county, N TEXAS, 12 mi/19 km SSW of WICHITA FALLS; 33°48′N 98°41′W. In oil-producing and irrigated farm area (cattle; dairying; wheat). LAKE WICHITA is just E.

Hollidaysburg, borough (2006 population 5,538), ⊙ BLAIR county, S central PENNSYLVANIA, 6 mi/9.7 km S of ALTOONA, near Frankstown Branch of the JUNIATA RIVER. Manufacturing (crystal oscillators, metal coil toys, steel fabricating); agr. (apples, corn, hay; dairying). Blue Knob Valley Airport to SW. Holidaysburg State Hospital to NW. ALLEGHENY PORTAGE RAILROAD NATIONAL HISTORIC SITE to W, carried canal boats over ALLEGHENY MOUNTAINS to JOHNSTOWN in mid-19th century; Canoe Creek State Park to NE. Settled 1768, laid out 1820.

Hollinger Mines, CANADA: see PORCUPINE.

Hollins (HAH-linz), unincorporated city, ROANOKE county, SW VIRGINIA, residential suburb 5 mi/8 km NNE of downtown ROANOKE; 37°20′N 79°57′W. Manufacturing (metal tubes). Hollins University (women). Carvin Cove Reservoir to NW.

Hollis, town, HILLSBOROUGH county, S NEW HAMPSHIRE, 6 mi/9.7 km W of NASHUA; 42°45′N 71°34′W. Bounded S by MASSACHUSETTS state line. Manufacturing (machining, commercial printing, mechanical equipment, computer software); agriculture (pumpkins, fruit, vegetables, nursery crops, corn; poultry, cattle, hogs; dairying). SILVER LAKE State Park in center.

Hollis, town (2006 population 2,095), ⊙ HARMON county, extreme SW OKLAHOMA, 33 mi/53 km W of ALTUS, near TEXAS state line (to S and W); 34°41′N 99°55′W. Elevation 1,615 ft/492 m. In livestock, cotton, and wheat area; peanuts, vegetables; manufacturing (meat processing); mesquite. Incorporated as town 1905, as city 1929.

Hollis, village, PRINCE OF WALES ISLAND, SE ALASKA, 40 mi/64 km W across CLARENCE STRAIT from KETCHIKAN, on E side of island; 55°29′N 132°43′W. Ferry to Ketchikan; island's road system connects village with CRAIG and other communities. A mining town in early 1900s, a logging camp revived the town during 1950–1962. Nearly a ghost town, residential development has brought a second rebirth to area.

Hollis, a residential section of E central QUEENS borough of NEW YORK city, ESE NEW YORK; 40°43′N 73°46′W. Population predominantly African-American.

Hollis Center, town, YORK county, SW MAINE, 10 mi/16 km NNE of ALFRED and on the SACO. Wood products. Incorporated 1798.

Hollister, city (2000 population 34,413), ⊙ SAN BENITO county, W CALIFORNIA, 38 mi/61 km SE of SAN JOSE, in San Benito valley, on SAN BENITO RIVER, 18 mi/29 km E of MONTEREY BAY. Elevation 291 ft/89 m. Nursery farming, flower-seed growing; vegetables; poultry, eggs; grapes, grain, nuts. Manufacturing (nut processing, metal stampings, wet-process systems). Has a community college. Holds annual rodeo. DIABLO RANGE to E; Hollister Hills State Park to S, San Juan Bautista State Historical Park to W. Settled 1868, incorporated 1874.

Hollister, resort city (2000 population 3,867), TANEY county, S MISSOURI, in the OZARKS, on Lake TANEYCOMO (formed by WHITE RIVER), across the lake from BRANSON; 36°37′N 93°13′W. Downtown business district built in Tudor style in early twentieth century. Tourist and recreation area.

Hollister, village (2000 population 237), TWIN FALLS county, S IDAHO, 15 mi/24 km SW of TWIN FALLS, near Demp Creek; 42°21′N 114°35′W. Elevation 4,500 ft/1,372 m. MUD LAKE to NW; Magic Mountain Ski Area to SE; part of Sawtooth National Forest to SE.

Hollister, village (2006 population 56), TILLMAN county, SW OKLAHOMA, 9 mi/14.5 km ESE of FREDERICK; 34°20′N 98°52′W. In cotton and grain area.

Holliston, town, MIDDLESEX county, E MASSACHUSETTS, residential suburb 20 mi/32 km SW of BOSTON; 42°12′N 71°27′W. Settled c.1659, incorporated 1724. Manufacturing include plastics, glass, wood, paper products, and computer-related products

Hölloch, SWITZERLAND: see MUOTATHAL.

Hollókö (HOL-lo-ko), village, NÓGRÁD county, HUNGARY, 12 mi/19 km SW of SALGÓTARJÁN; 48°00′N 19°36′E. Silage corn; cattle; poor soil and low-productivity agriculture. Manufacturing (apparel, handicrafts). Authentic village representing the regional Palóc culture and people. Preserved as a natural treasure.

Holloman Air Force Base, NEW MEXICO: see ALAMOGORDO.

Holloway, village (2000 population 112), SWIFT county, W central MINNESOTA, near POMME DE TERRE RIVER and Cottonwood Creek, 16 mi/26 km WSW of BENSON; 45°15′N 95°54′W. Grain; livestock; dairying; manufacturing (fertilizers).

Holloway (HAH-lo-wai), village (2006 population 336), BELMONT county, E OHIO, 11 mi/18 km SW of CADIZ, near PIEDMONT LAKE reservoir; 40°10′N 81°07′W. In former coal-mining area.

Hollow Creek, town (2000 population 815), JEFFERSON county, N KENTUCKY, residential suburb 8 mi/12.9 km SE of downtown LOUISVILLE, near FERN CREEK; 38°08′N 85°37′W. General Electric Appliance park (industrial) to NW.

Hollow Rock, town (2006 population 945), CARROLL county, NW TENNESSEE, 9 mi/14 km ENE of HUNTINGDON; 36°02′N 88°16′W.

Hollsopple (hol-SAH-puhl), unincorporated village, SOMERSET county, SW PENNSYLVANIA, 8 mi/12.9 km S of JOHNSTOWN on STONYCREEK RIVER; 40°12′N 78°55′W. Manufacturing includes steel ingots and billets, wood products QUEMAHONING RESERVOIR on Quemahoning Creek.

Hollum (HAW-luhm), village, FRIESLAND province, N NETHERLANDS, on AMELAND island, 6 mi/9.7 km W of Nes and 18 mi/29 km NNW of LEEUWARDEN; 53°26′N 05°39′E. Tourism; dairying; cattle, sheep; vegetables. Ferry terminal at Nes River; lighthouse on BORNDIEP strait 1 mi/1.6 km to W. Airport to NE, recreational area to N.

Holly, town (2000 population 1,048), PROWERS county, SE COLORADO, on ARKANSAS RIVER, at mouth of Wild Horses Creek, near KANSAS state line, and 24 mi/39 km E of LAMAR; 38°03′N 102°07′W. Elevation 3,367 ft/1,026 m. In wheat, cattle, corn, barley, sorghum region; manufacturing (dog and cat food).

Holly, town (2000 population 6,135) OAKLAND county, SE MICHIGAN, 15 mi/24 km S of FLINT and 20 mi/32 km NW of PONTIAC and on SHIAWASSEE RIVER, in an area of many lakes; 42°47′N 83°37′W. Manufacturing (auto parts, winery, foundry, machinery, automotive tubing, apple cider). Seven Lakes State Park to W; Holly State Recreation Area to NE; Mount Holly and Pine Knob Ski Areas to E. Settled 1836, incorporated 1865.

Holly Beach, LOUISIANA: see CAMERON.

Holly Grove, village (2000 population 722), MONROE county, E central ARKANSAS, 21 mi/34 km ENE of STUTTGART; 34°36′N 91°12′W. In agricultural area.

Holly Hill (HAH-lee), town (□ 4 sq mi/10.4 sq km; 2000 population 12,119), VOLUSIA county, E central FLORIDA, 3 mi/4.8 km N of DAYTONA BEACH on the ATLANTIC OCEAN; 29°14′N 81°02′W. Area noted for its flowers; coquina quarries.

Holly Hill, town (2006 population 1,355), ORANGEBURG county, S central SOUTH CAROLINA, 27 mi/43 km ESE of ORANGEBURG; 33°19′N 80°24′W. Manufacturing includes wood chairs, fiberboard, lumber, cement, printing and publishing, sportcoats, toys. Agriculture includes watermelons, tobacco and grains, livestock, cotton area. Santee-Cooper hydroelectric and navigation development is E. Incorporated 1887.

Hollymead (HAH-lee-meed), unincorporated town, ALBEMARLE county, central VIRGINIA, residential suburb 7 mi/16 km NNE of CHARLOTTESVILLE; 38°07′N 78°26′W.

Holly Pond, town (2000 population 645), Cullman co., N ALABAMA, 13 mi/21 km E of Cullman, near Mulberry Fork. First settled in 1875. Named after a pond that no longer exists. Inc. in 1912.

Holly Ridge (HAHL-ee RIJ), town (□ 1 sq mi/2.6 sq km; 2006 population 738), ONSLOW county, E NORTH CAROLINA, 20 mi/32 km SSW of JACKSONVILLE; 34°29′N 77°33′W. Manufacturing; agriculture (tobacco, cotton, corn, soybeans; poultry, livestock). ATLANTIC OCEAN and INTRACOASTAL WATERWAY separated by TOPSAIL ISLAND (sand-barrier island) to SE. Incorporated after 1940.

Holly River State Park (□ 13 sq mi/34 sq km), WEBSTER county, central WEST VIRGINIA, in the ALLEGHENY MOUNTAINS, 21 mi/34 km S of BUCKHANNON, drained by Holly River Wooded recreational area; facilities for fishing, swimming, other sports. Established 1938.

Holly Shelter Swamp (HAHL-ee SHEL-tuhr SWAHMP) (□ 100 sq mi/260 sq km), SE NORTH CAROLINA, NE of WILMINGTON, and bordering Northeast CAPE FEAR RIVER on E. Includes Holly Shelter Game Refuge (c.30,000 acres/12,141 ha), E of BURGAW. Angola Swamp is N.

Holly Springs, town (2000 population 3,195), CHEROKEE county, NW GEORGIA, 4 mi/6.4 km S of CANTON; 34°10′N 84°30′W. Manufacturing of hydrogen gas producing systems.

Holly Springs, town (2000 population 7,957), ⊙ MARSHALL county, N MISSISSIPPI, 41 mi/66 km SE of MEMPHIS, TENNESSEE; 34°46′N 89°26′W. Railroad junction. Trade and market center for cotton-growing and dairying region; manufacturing (wall paneling, steel and alloy fabrication, industrial machinery, printing and publishing, kitchen appliances, bricks, metal stampings); clay deposits. Rust College is here. Has many fine antebellum homes. Montrose Mansion (1858). In Civil War, town was captured (1862) by Confederates, thus delaying Grant's advance against Vicksburg. Clark Art Gallery; Marshall County Historical Museum, Wall Doxey State Park to S; Holly Springs National Forest to E. Incorporated 1837.

Holly Springs (HAHL-ee SPREENGZ), town (□ 7 sq mi/18.2 sq km; 2006 population 17,425), WAKE county, central NORTH CAROLINA, 14 mi/23 km SW of RALEIGH; 35°39′N 78°50′W. Service industries; manufacturing (electronic equipment, synthetic yarns, upholstered furniture); agriculture (tobacco, grain; livestock). Harris Lake reservoir to SW. Incorporated 1877.

Hollyville, village, JEFFERSON county, N KENTUCKY, residential suburb 12 mi/19 km S of downtown

LOUISVILLE; 38°05′N 85°45′W. Agricultural area (tobacco, grain; livestock).

Hollywood, unincorporated community, DUNKLIN county, in boot heel of extreme SE MISSOURI, near SAINT FRANCIS River, 15 mi/24 km SSW of KENNETT.

Hollywood (HAH-lee-wood), city (□ 30 sq mi/78 sq km; 2000 population 139,357), extreme SE BROWARD county, SE FLORIDA, on the ATLANTIC OCEAN; 26°01′N 80°09′W. A popular retirement center and part of the MIAMI/FORT LAUDERDALE metropolitan area, Hollywood produces electronic equipment and building materials and has a number of office parks. Most of Port Everglades, the county's largest port with an extensive warehouse complex, is within the city limits. Incorporated 1925.

Hollywood, town (2000 population 950), Jackson co., NE Alabama, near Tennessee River, 5 mi/8 km NE of Scottsboro; 34°43′N 85°58′W. First known as 'Bellefonte Station' and then 'Samples,' it was renamed around 1883.

Hollywood, town (2006 population 4,298), CHARLESTON county, SE SOUTH CAROLINA, 17 mi/27 km W of CHARLESTON; 32°45′N 80°12′W. Manufacturing of meat products and septic tanks.

Hollywood, village, SAINT MARYS county, S MARYLAND, 5 mi/8 km NE of LEONARDTOWN, near Patuxent River Resorts nearby (Sam Abell Cove, Clarks Landing; swimming, fishing, duck hunting, boating). Sotterly, begun in 1717, is still a working plantation with a house built in 1730. Nearby are St. Mary's Airport and Greenwell State Park.

Hollywood, suburban section of LOS ANGELES, LOS ANGELES county, S CALIFORNIA, on the E slopes of the SANTA MONICA MOUNTAINS. Often referred to as "Tinsel Town" for its glitter and high lifestyle (now fading). Manufacturing (broadcasting equipment, photographic equipment, electric light bulbs, motion-picture industry). Noted for its major film and television studios and their executive offices, many have relocated to nearby suburbs such as BURBANK and GLENDALE due to deterioration of community and increasing crime. Although many films are shot on location in cities and countries throughout the world, Hollywood remains the symbolic center of the U.S. motion-picture industry. Since the first film was made there c.1911, the community has come to signify the film industry in general—its morals, manners, and characteristics. Hollywood attracts large numbers of tourists. Points of interest include Hollywood Boulevard, Hollywood Bowl, and Grauman's Chinese Theatre and its Walk of Fame. LA BREA TAR PITS and Los Angeles County Art Museum to W. In surrounding hills is Griffith Park (with an observatory and planetarium) to NE, and the homes of film celebrities in Beverly Hills and Malibu, to W. The University of Judaism and a two-year college are in Hollywood. Incorporated 1903, consolidated with Los Angeles 1910.

Hollywood Beach, MARYLAND: see CHESAPEAKE CITY.

Hollywood by the Sea, village, VENTURA county, S CALIFORNIA.

Hollywood Park, city (2006 population 3,238), BEXAR county, S central TEXAS, residential suburb 11 mi/18 km N of downtown SAN ANTONIO; 29°36′N 98°28′W.

Holman Island (HOL-muhn), trading post, W VICTORIA ISLAND, NORTHWEST TERRITORIES, CANADA, on AMUNDSEN GULF at entrance of PRINCE ALBERT SOUND; 70°44′N 117°45′W. Radio station; site of Roman Catholic mission.

Hólmavík (HAWL-mah-VEEK), fishing village ⊙ STRANDASYSLA county, NW ICELAND, on VESTFJARÐA Peninsula, 45 mi/72 km SE of ÍSAFJÖRÐUR, on Steingrimsfjörður, 15 mi/24 km–long W arm of HUNAFLOI.

Holmberg, village, SW CÓRDOBA province, ARGENTINA, 5 mi/8 km SW of RÍO CUARTO. Grain, soybeans, flax; livestock.

Holmdel (HOLM-del), township, MONMOUTH county, E NEW JERSEY, 7 mi/11.3 km NE of FREEHOLD, in suburban area; 40°22′N 74°10′W. Site of AT&T Bell Laboratories was here.

Holmen (HOL-men), town (2006 population 7,342), LA CROSSE county, W WISCONSIN, 10 mi/16 km N of LA CROSSE, near MISSISSIPPI RIVER; 43°57′N 91°15′W. In farm and dairy region. Manufacturing (custom casework, silk screening, feather processing, concrete products). On Great River State Trail.

Holmenkollen (HAWL-muhn-KAWL-luhn), suburb of OSLO, SE NORWAY, in hilly region, 8 mi/12.9 km NW of city center. On electric railroad. A popular ski resort, it is the scene of the annual Holmenkollen Ski Festival, which includes famous ski-jump competition. National Ski Museum, located in the ski-jump, has exhibits including the equipment used by the polar explorers Nansen and Amundsen. Until 1948, in AKERSHUS county.

Holmes (HOMES), county (□ 483 sq mi/1,255.8 sq km; 2006 population 19,285), NW FLORIDA, on ALABAMA line (N) and bounded E by HOLMES CREEK; ⊙ BONIFAY; 30°52′N 85°48′W. Rolling agriculture area (corn, peanuts, cotton, vegetables, livestock) drained by CHOCTAWHATCHEE RIVER; has forest industries. Formed 1848.

Holmes (HOMZ), county (□ 764 sq mi/1,986.4 sq km; 2006 population 20,866), central MISSISSIPPI, ⊙ LEXINGTON; 33°07′N 90°05′W. Bounded E by BIG BLACK RIVER, NW and SW corners by YAZOO RIVER; drained by Tchula Lake slough, forms part of W boundary. Agriculture (cotton, corn, soybeans, sorghum; cattle; timber); clay deposits. Bee Lake, oxbow lake of Yazoo River, in SW corner. Morgan Brake National Wildlife Refuge in W, Hillside National Wildlife Refuge in SW; Holmes County State Park in E. Formed 1833.

Holmes (HOMZ), county (□ 424 sq mi/1,102.4 sq km; 2006 population 41,574), central OHIO, ⊙ MILLERSBURG; 40°33′N 81°54′W. Intersected by KILLBUCK CREEK and WALHONDING RIVER. In the Glaciated Plain physiographic region. Agriculture (livestock; dairy products; grain, nursery crops). Coal mining, sandstone quarries, and gravel pits. Manufacturing at Millersburg (rubber, plastic, and wood products). Formed 1825.

Holmes Beach (HOMZ), town (□ 1 sq mi/2.6 sq km; 2000 population 4,966), MANATEE county, W central FLORIDA, located on Anna Maria Key, 11 mi/18 km W of BRADENTON; 27°30′N 82°43′W.

Holmes Chapel (HOMZ), village (2001 population 5,669), central CHESHIRE, W central ENGLAND, on DANE RIVER and 8 mi/12.9 km ESE of NORTHWICH; 53°12′N 02°22′W. Site of agricultural and horticultural college affiliated with Manchester University. Has 14th-century church.

Holmes Creek (HOMZ), c.60 mi/97 km long, NW FLORIDA; rises in SE ALABAMA; flows SW into FLORIDA, to CHOCTAWHATCHEE RIVER 21 mi/34 km SE of DE FUNIAK SPRINGS; navigable below Vernon.

Holmes Educational State Forest, c.8.5 mi/13.7 km NW of HENDERSONVILLE, HENDERSON county, SW NORTH CAROLINA, in the GREAT SMOKY MOUNTAINS. Mountain Hardwoods, Rhododendron, Flame Azaleas, wildflowers grow here. One of 6 forests managed by the state as living environmental education centers; ranger-conducted programs, trails marked by displays and exhibits, natural ampitheaters.

Holmesglen (HOLMZ-glen), suburb of MELBOURNE, VICTORIA, SE AUSTRALIA, where Ashburton and Chadstone meet; 37°52′S 145°07′E. Its railway station is 9 mi/14 km SE of Melbourne. TAFE (technical and further education) college here.

Holmestrand (HAWL-muh-strahn), city, VESTFOLD county, SE NORWAY, on W shore of OSLOFJORDEN, on railroad, and 35 mi/56 km SSW of OSLO; 59°29′N 10°18′E. Aluminum manufacturing; dairy plant; tourist resort. The heavily trafficked highway E-18 has been rebuilt in tunnels under the city. Incorporated 1752.

Holmesville (HOMZ-vil), village (2006 population 412), HOLMES county, central OHIO, 33 mi/53 km ESE of MANSFIELD, on KILLBUCK CREEK; 40°37′N 81°55′W.

Holmfield (HOLM-feeld), unincorporated village, SW MANITOBA, W central CANADA, on WHITEMUD RIVER, 9 mi/14 km ESE of KILLARNEY, and in TURTLE MOUNTAIN rural municipality; 49°08′N 99°28′W. Grain, stock.

Holmfirth (HOM-fuhth), town (2001 population 25,047), WEST YORKSHIRE, N ENGLAND, 5 mi/8 km S of HUDDERSFIELD; 53°33′N 01°46′W. Light industry. Nearby are stone quarries and the suburbs of Cartworth (S), Wooldale (NE), and Netherthong (NNW).

Hol'mivs'kyy (HOL-meev-skee) (Russian *Gol'movsky*), town, central DONETS'K oblast, UKRAINE, in the DONBAS, 6 mi/9.7 km NNE of HORLIVKA city center; 48°24′N 38°05′E. Elevation 643 ft/195 m. Subordinated to the Mykyta raion council of the city of Horlivka. Dolomite quarry and dolomite processing. Established in 1875, town since 1938.

Holmsbu (HAWLMS-boo), village, BUSKERUD county, SE Norway, on E side of DRAMMENSFJORDEN near its mouth on OSLOFJORDEN, 14 mi/23 km SE of DRAMMEN. Fruitgrowing; fishing. Dates from Middle Ages. On SW Hurum peninsula (15 mi/24 km long, 4 mi/6.4 km to 8 mi/12.9 km wide), which is between Drammensfjorden (W) and Oslofjorden (E).

Holmslands Klit, DENMARK: see RINGKØBING FJORD.

Holmsund (HOLM-SUND), town, VÄSTERBOTTEN county, N SWEDEN, on GULF OF BOTHNIA, at mouth of UMEÄLVEN RIVER, 10 mi/16 km SSE of UMEÅ, for which it serves as port; 63°42′N 20°23′E.

Holne (HOLN), village (2001 population 273), S DEVON, SW ENGLAND, on DART RIVER in DARTMOOR, 8 mi/12.9 km NW of TOTNES; 50°30′N 03°49′W. Has medieval stone bridge and 15th-century church. Charles Kingsley born here.

Holoby (ho-LO-bi) (Russian *Goloby*) (Polish *Holoby*), town, central VOLYN' oblast, UKRAINE, 16 mi/26 km SE of KOVEL'; 51°05′N 25°01′E. Elevation 662 ft/201 m. Railroad station; flour milling, fruit canning; chemicals. Heritage museum. Known since the mid-16th century, town since 1957.

Holocachaket, Alaska: see HOLIKACHUK.

Hologachaket, Alaska: see HOLIKACHUK.

Holohory (ho-lo-HO-ri) (Russian and Polish *Gologory*), low mountains forming the NW high rim of the PODOLIAN UPLAND, W UKRAINE; extend approximately 35 mi/56 km from L'VIV (W) to ZOLOCHIV (E). Average elevation 1,500 ft/457 m; rises in SW to Mt. Kamula (1,544 ft/471 m).

Holon (kho-LON), city (2006 population 167,100), S satellite of TEL AVIV, W central ISRAEL; 32°00′N 34°46′E. Elevation 78 ft/23 m. Country's second-largest industrial zone. Manufacturing include textiles, metal and leather goods, processed foods, furniture, glassware, plastics, and construction materials. Holon was founded in mid-20th century with the merger of several small outer suburbs of Tel Aviv.

Holot Hadera (kho-LOT kha-DEH-rah), an area of sand dunes SW and W of HADERA, ISRAEL.

Holot Halutza (kho-LOT kha-loo-TZAH), a large area of sand dunes at the W boundary of the NEGEV, ISRAEL, SE of the GAZA STRIP.

Holoubkau, CZECH REPUBLIC: see HOLOUBKOV.

Holoubkov (HO-loup-KOF), village, German *Holoubkau*, ZAPADOCESKY province, SW BOHEMIA, CZECH REPUBLIC, on railroad, and 14 mi/23 km ENE of Plzeň; 49°47′N 13°42′E. Manufacturing (machine-tools). Summer resort.

Holovanivsk (ho-lo-vah-NEEVSK) (Russian *Golovanevsk*), town, W KIROVOHRAD oblast, UKRAINE, 83 mi/134 km W of KIROVOHRAD and 12 mi/19 km ENE of UL'YANOVKA; 48°23′N 30°28′E. Elevation 580 ft/176 m. Raion center; grain milling, feed milling, butter

milling; bricks. Professional technical school. Known since 1764, town since 1957. Jewish community since the 19th century, numbering 4,320 at its peak (1897); pogroms in 1905 and 1918 reduced it to 3,474 in 1939; completely exterminated by the Nazis between 1941 and 1943.

Holovne (HO-lov-ne) (Russian *Golovno*) (Polish *Holowno*), town, W VOLYN' oblast, UKRAINE, near source of PRIPET (Ukrainian *Prypyat*') River, 28 mi/45 km WNW of KOVEL'; 51°20′N 24°05′E. Elevation 666 ft/202 m. Vegetable drying plant. Known since 1564, town since 1957.

Holovyne (ho-lo-vi-NE) (Russian *Golovino*), town, E central ZHYTOMYR oblast, UKRAINE, 16 mi/26 km NNE of ZHYTOMYR, and 7 mi/11 km E of CHERNYAKHIV; 50°28′N 28°50′E. Elevation 698 ft/212 m. Labradorite quarry; vocational school. The labradorite, mined here since 1928, has been used as fascia in the Mausoleum of Lenin (Moscow) and other buildings in Moscow, Leningrad (now St. Petersburg), and Kiev.

Holowno, UKRAINE: see HOLOVNE.

Holroyd (HAHL-roid), W residential suburb of SYDNEY, E NEW SOUTH WALES, SE AUSTRALIA; 33°50′S 150°57′E. Manufacturing (electrical equipment, generators).

Holsbeek (HOLS-baik), commune (2006 population 9,280), Leuven district, BRABANT province, central BELGIUM, 4 mi/6 km NE of LEUVEN; 50°55′N 04°45′E.

Holsnøy (HAWL-suhn-uh-u), island (□ 34 sq mi/88.4 sq km) in NORTH SEA, HORDALAND county, SW NORWAY, 9 mi/14.5 km NNW of BERGEN; 13 mi/21 km long, 5 mi/8 km wide. Covered with moors and lakes.

Holstebro (HOL-stuh-BRO), city (2000 population 32,572), RINGKØBING county, W central DENMARK, on the STORÅ RIVER; 56°20′N 08°35′E. It is a commercial and industrial center and a railroad junction, producing foodstuffs, tobacco, and beer.

Holstebro River, DENMARK: see STORÅ.

Holsted (HAIL-stedh), town, RIBE county, S central JUTLAND, DENMARK, 18 mi/29 km ENE of ESBJERG; 55°38′N 08°55′E.

Holstein (HOL-shtein), former duchy, N central GERMANY, roughly comprised of the part of SCHLESWIG-HOLSTEIN S of the EIDER River; KIEL and RENDSBURG were the chief cities; 53°30′N 09°00′E–54°20′N 11°00′E. For a time part of the duchy of SAXONY, Holstein was created (1111) a county of the Holy Roman Empire and was bestowed on Adolf of Schauenburg. In 1459, Holstein's Christian I, a Danish heir to the throne, established a personal union with DENMARK, to the great displeasure of the German majority. In 1474, Emperor Frederick III raised Holstein to a duchy under the immediate suzerainty of the Holy Roman Empire (as distinct from Schleswig, which was outside the imperial jurisdiction).

Holstein, town (2000 population 1,470), IDA county, W IOWA, 10 mi/16 km NNW of IDA GROVE; 42°29′N 95°32′W. In livestock and grain area. Settled 1882.

Holstein, village (2006 population 249), ADAMS county, S NEBRASKA, 15 mi/24 km SW of HASTINGS and on branch of LITTLE BLUE RIVER; 40°27′N 98°39′W.

Holsteiner Schweiz (HOL-shtei-ner SHVEITS), hilly region with many lakes in SCHLESWIG-HOLSTEIN, N GERMANY; 54°10′N 10°25′E. The center is PLÖN.

Holsteinsborg (HOL-stains-BOR), Greenlandic *Sisimiut*, town in SISIMIUT (Holsteinsborg) commune, W GREENLAND; 66°55′N 53°40′W. The second-largest town in Greenland, it is a fishing center with a modern canning factory and a shipyard.

Holston, river, c.120 mi/193 km long, NE TENNESSEE; formed by the uniting of its N and S forks; flowing SW through the GREAT APPALACHIAN VALLEY, joining the FRENCH BROAD RIVER at KNOXVILLE to form the TENNESSEE RIVER; 35°57′N 83°51′W. Settlement along the Holston began before the American Revolution, and it was a major route of westward migration. On the river is Cherokee Dam, a flood control unit of the

Tennessee Valley Authority that impounds CHEROKEE LAKE; there are several smaller dams on the Holston's S fork.

Holston Mountain (HOL-stuhn), ridge (2,500 ft/762 m–4,000 ft/1,219 m) of the APPALACHIAN mountains between South Fork HOLSTON River (N) and IRON MOUNTAINS (SE), NE TENNESSEE and SW VIRGINIA, from ELIZABETHTON, Tennessee, extends c.30 mi/48 km NE to DAMASCUS, Virginia. Highest point (c.4,300 ft/1,311 m) is 7 mi/11 km NE of Elizabethton. Included in Cherokee and Jefferson national forests. Sometimes considered a range of UNAKA MOUNTAINS. The APPALACHIAN NATIONAL SCENIC TRAIL runs along part of its length. Summit, 36°30′N 82°38′W.

Holsworthy (HAHLZ-wuh-[th]ee), town (1991 population 1,890; 2001 population 3,435), W DEVON, SW ENGLAND, 9 mi/14.5 km E of BUDE; 50°49′N 04°21′W. Agricultural market. Scene of St. Peter's Fair, an annual horse festival. Has 13th-century church.

Holt, county (□ 456 sq mi/1,185.6 sq km; 2006 population 4,997), NW MISSOURI; ⊙ OREGON; 40°05′N 95°12′W. Between MISSOURI (W and S) and NODAWAY (E) rivers; drained by TARKIO RIVER. Corn, wheat, apples; cattle, hogs. Squaw Creek National Wildlife Refuge, Big Lake State Park. Areas along river were flooded in 1993. Formed 1841.

Holt, county (□ 2,417 sq mi/6,284.2 sq km; 2006 population 10,610), N NEBRASKA; ⊙ O'NEILL; 42°27′N 98°46′W. Grazing region with well irrigation, partly in Sand Hills bounded N by NIOBRARA RIVER; drained by ELKHORN RIVER. Cattle, hogs, corn, alfalfa, wild hay, soybeans, dairy and poultry products. Atkinson Lake State Recreation Area in W; Small Swan Lake in SW; Goose Lake in S. Formed 1876.

Holt (HOHLT), town (2001 population 3,550), N NORFOLK, E ENGLAND, 11 mi/18 km ENE of FAKENHAM; 52°54′N 01°05′E. Agricultural market. Has Gresham's School (16th-century) and 14th-century church.

Holt, town (2000 population 11,315), INGHAM county, S central MICHIGAN, suburb 7 mi/11.3 km S of LANSING; 42°38′N 84°31′W. In agricultural area (livestock; grain, vegetables, apples; dairy products); manufacturing (food processing, hydraulic hose fittings, fiberglass products).

Holt, town (2000 population 405), CLAY and CLINTON counties, W MISSOURI, on FISHING RIVER and 15 mi/24 km NNE of Liberty; 39°27′N 94°20′W. Corn, cattle.

Holt (HOHLT), agricultural village (2001 population 2,354), W WILTSHIRE, S central ENGLAND, 3 mi/4.8 km ENE of BRADFORD-ON-AVON; 51°21′N 02°12′W. Has 15th-century church and 18th-century house, The Courts.

Holt, village (2000 population 4,103), TUSCALOOSA county, W ALABAMA, on BLACK WARRIOR RIVER and 5 mi/8 km NE of TUSCALOOSA; near HOLT LOCK AND DAM.

Holt, village (2000 population 89), MARSHALL county, NW MINNESOTA, 12 mi/19 km N of THIEF RIVER FALLS; 48°17′N 96°11′W. Agricultural area (grain, beans, potatoes; livestock). Agassiz National Wildlife Refuge to E.

Holt (HOLT), village (2001 population 1,762), WREXHAM, NE Wales, on the DEE RIVER (14th-century bridge), 5 mi/8 km NE of Wrexham; 53°05′N 02°54′W. Formerly in CLWYD, abolished 1996.

Holt, ARGENTINA: see IBICUY.

Holte (HOL-tuh), city, Copenhagen county, DENMARK, 8 mi/12.9 km N of COPENHAGEN; 55°49′N 12°28′E. Vegetables, produce, fruit; light manufacturing. Includes communities of SØLLERØD (1992 population 30,609), ØVERØD, and DRONNINGGÅRD.

Holtemme River (HOL-tem-me), 28 mi/45 km long, central GERMANY; rises in the upper HARZ at foot of the BROCKEN (51°48′N 10°37′E); flows ENE, past WERNIGERODE and HALBERSTADT, to the BODE 2 mi/3.2 km NE of GRÖNINGEN.

Holten (HAWL-tuhn), village, OVERIJSSEL province, E NETHERLANDS, 12 mi/19 km E of DEVENTER; 52°17′N 06°26′E. Nature area (with Woodland Animal World Museum) to N. Dairying; cattle raising; vegetables, grain; manufacturing (furniture, snowplows, food processing).

Holter Lake (HOL-tuhr), reservoir, LEWIS AND CLARK county, W central MONTANA, on MISSOURI River, 30 mi/48 km N of HELENA; c.30 mi/48 km long; 47°00′N 112°00′W. Maximum capacity 265,000 acre-ft. HAUSER DAM at S end. Formed by Holter Dam (gravity; 125 ft/38 m high), built (1918) by the Montana Power Company for power generation. Upper Holter Lake separated from main reservoir by narrow section in Gates of the Rocky Mountains canyon.

Holte-Stukenbrock, GERMANY: see SCHLOSS HOLTE-STUKENBROCK.

Holt Lock and Dam, TUSCALOOSA county, W ALABAMA, on BLACK WARRIOR RIVER 7 mi/11.3 km ENE of TUSCALOOSA and 2 mi/3.2 km E of HOLT; 33°15′N 87°26′W. Dam (108 ft/33 m high) built by the Army Corps of Engineers (1968) for navigation and hydroelectric generation; maximum capacity c.117,990 acre-ft.

Holt Mine, KENTUCKY: see BRECKINRIDGE.

Holton, town (2000 population 407), RIPLEY county, SE INDIANA, 14 mi/23 km ENE of NORTH VERNON, near Otter Creek; 39°05′N 85°23′W. Versailles State Park to E; Jefferson Proving Ground to S. Cattle, poultry, corn.

Holton (HOL-tuhn), town (2000 population 3,353), ⊙ JACKSON county, NE KANSAS, 29 mi/47 km N of TOPEKA; 39°28′N 95°43′W. In livestock and grain region. Manufacturing (rotary mowers, feed, sausage and prepared meat). Potawatomi Indian Reservation is nearby. Laid out 1857 by Free Staters; incorporated 1870.

Holton, village, E NEWFOUNDLAND AND LABRADOR, CANADA, on coast of LABRADOR SEA, 5 mi/8 km N of Man of War Point. Fishing village in area.

Holts Landing State Park, DELAWARE, UNITED STATES: see MILLVILLE.

Holts Summit, town (2000 population 2,935), CALLAWAY county, central MISSOURI, 5 mi/8 km NE of JEFFERSON CITY; 38°38′N 92°07′W. Residential suburb of Jefferson City. Manufacturing (utility trailers, light manufacturing).

Holtville, city (2000 population 5,612), IMPERIAL county, S CALIFORNIA, 10 mi/16 km E of EL CENTRO, in irrigated IMPERIAL VALLEY, on Alamo River. Railroad terminus. Tomatoes, vegetables, melons, dates, wheat, alfalfa, corn, sugar beets; cattle, sheep. It is 10 mi/16 km N of Mexican border. Incorporated 1908.

Holtz Bay, Alaska: see ATTU Island.

Holualoa (HO-LOO-ah-LO-ah), town (2000 population 6,107), W HAWAII Island, HAWAII county, HAWAII, near KAILUA BAY, in the KONA district, 53 mi/85 km W of HILO, 2 mi/3.2 km inland from Kona (W) coast, on Mamalahoa Highway; 19°38′N 155°55′W. Produces coffee, but this production has been declining since the early 1980s. Waiaka Forest Reserve to E, Kakaluu Forest Reserve to SE.

Holubivka, UKRAINE: see KOMISARIVKA.

Holubivs'kyy Rudnyk, UKRAINE: see KIROVS'K, Luhans'k oblast.

Holwerd (HAWL-vuhrt), village, FRIESLAND province, N NETHERLANDS, 1 mi/1.6 km SE of the WADDENZEE, 12 mi/19 km NNE of LEEUWARDEN; 53°22′N 05°54′E. Car ferry to AMELAND island 2.5 mi/4 km to NW (at end of 1.5 mi/2.4 km jetty). Dairying; cattle, sheep; vegetables.

Holy Cross, town (2000 population 339), DUBUQUE county, E IOWA, 17 mi/27 km WNW of DUBUQUE; 42°36′N 91°00′W. Limestone quarry.

Holy Cross, village (2000 population 227), W ALASKA, on YUKON River at mouth of INNOKO RIVER and Reindeer River and 120 mi/193 km NE of BETHEL; 62°12′N 159°47′W. Established 1887.

Holycross, Gaelic *Mainistir na Croise Naomhtha*, agr. town, central Tipperary co., Ireland, on Suir R. and 4 mi/6.4 km SW of Thurles; 52°30′N 08°32′W. Site of ruins of abbey founded 1182 by Donal O'Brien, King of Limerick.

Holy Cross, Mount of the, peak (14,005 ft/4,269 m) in SAWATCH MOUNTAINS, EAGLE county, W central COLORADO, 18 mi/29 km NW of LEADVILLE. Near its summit, snow-filled crevices c.50 ft/15 m wide form a huge cross more than 1,000 ft/305 m long, with 750-ft/229-m-long arms. Formerly (1929–1950) Holy Cross National Monument; site now administered by U.S. Forest Service (White River National Forest).

Holyhead (HO-lee-hed), Welsh *Caergybi*, town (2001 population 11,237), ISLE OF ANGLESEY, NW Wales, at N end of HOLY ISLAND, on HOLYHEAD BAY, and 19 mi/31 km NW of CAERNARVON; 53°18′N 04°38′W. Chief port on mail and passenger route to DUBLIN, with direct railroad connection with mainland of Wales and England. Rocky promontory of North Stack, 2 mi/3.2 km WNW, has coves and sea inlets; promontory of South Stack, site of lighthouse, is 3 mi/4.8 km W.

Holyhead Bay (HO-lee-hed), arm of the IRISH SEA, ISLE OF ANGLESEY, NW Wales, on W coast of Anglesey island; 53°22′N 04°40′W. From North Stack in S to CARMEL HEAD in N. HOLYHEAD harbor on W side. Ferry routes to Dublin.

Holyhead Island, Wales: see HOLY ISLAND.

Holy Island (HO-lee) or **Lindisfarne**, island (2001 population 162) off the coast of NORTHUMBERLAND, NE ENGLAND; 55°41′N 01°47′W. At low tide the island is connected with the mainland by a stretch of sand. It is partly cultivated, and tourism and fishing are important. A church and monastery, built in 635 under St. Aidan, represented the first establishment of Celtic Christianity in England. The bishopric was maintained for 112 years here and moved to DURHAM in 995. A Benedictine priory was set up on the island in 1083 by monks from Durham. The Lindisfarne Gospels or Book of Durham was written at Holy Island before 700.

Holy Island or **Iniscaltra**, Gaelic *Inis Cealtra*, islet in LOUGH DERG, CLARE county, W IRELAND, 4 mi/6.4 km E of SCARIFF; 52°55′N 08°27′W. Site of 10th-century round tower and of remains of several churches and monastic buildings. Established in 7th century, destroyed by the Danes, and rebuilt by Brian Boru.

Holy Island (HO-lee), island, NORTH AYRSHIRE, SW Scotland, in Firth of Clyde just off E coast of ARRAN island, sheltering harbor of LAMLASH; 55°32′N 05°05′W. Rises to 1,030 ft/314 m; lighthouses at S end and on E coast. Formerly in STRATHCLYDE, abolished 1996.

Holy Island (HO-lee) or **Holyhead Island**, ISLE OF ANGLESEY, NW Wales, in IRISH SEA, just off W coast of Anglesey island; 7 mi/11.3 km long, 4 mi/6.4 km wide; 53°16′N 04°39′W. Chief town, HOLYHEAD. Linked by rail and road bridges.

Holy Land, MIDDLE EAST. Its boundaries, never constant, always included at least the land between the MEDITERRANEAN SEA and JORDAN river. Referred to in the Book of Zechariah, Chapter 2. It is called thus by Jews because it was promised to them by God; by Christians because it was the scene of Jesus' life; and by Muslims because they consider Islam to be the heir of Judaism and Christianity. Shrines of several religions cluster most numerously about JERUSALEM, BETHLEHEM, NAZARETH, and HEBRON. The term was used in its Latin form–Terra Sancta–by the Crusaders, in addition to Terra Israel. Holy Land was still the usual Western name for the land in the nineteenth century, although it alternated in vernacular with JUDEA, the Land of ISRAEL, Zion, PALESTINE, and Land of the Bible.

Holyoke (HOL-yok), city (2000 population 39,838), HAMPDEN county, S central MASSACHUSETTS, on the CONNECTICUT RIVER; 42°13′N 72°38′W. The city has varied manufacturing including printed materials, medical supplies, consumer goods, metal products, electronic goods, and chemicals. Holyoke Community College is here. Mount Holyoke College and a U.S. air force base are nearby. Ski resort, amusement park. Holyoke Mall at Ingelside, one of the largest shopping centers in U.S. Largest planned industrial city of U.S. Industrial Revolution. Largest dam in world in 1848 across Connecticut River. Dinosaur Footprints Park and Hampton Pond State Park. Settled 1745, incorporated 1873.

Holyoke, town (2000 population 2,261), ⊙ PHILLIPS county, NE COLORADO, on FRENCHMAN CREEK, near NEBRASKA line, and 45 mi/72 km E of STERLING; 40°34′N 102°17′W. Elevation 3,746 ft/1,142 m. Cattle, sunflowers, beans; light manufacturing. Incorporated 1888.

Holyoke, Mount, Massachusetts: see HOLYOKE RANGE.

Holyoke Range (HOL-yok), W central MASSACHUSETTS, E-W range just N of South HADLEY; c.8 mi/12.9 km long. Rises to 1,106 ft/337 m in Mount Norwottock. Mount Holyoke (878 ft/268 m), at W end, has road to summit. Volcanic basalt ridges within CONNECTICUT RIVER valley.

Holyrood (HAHL-ee-rood), village (2000 population 464), ELLSWORTH county, central KANSAS, 10 mi/16 km SW of ELLSWORTH; 38°35′N 98°24′W. Shipping point in wheat region.

Holyrood Palace (HO-lee-ROOD) [=holy cross], royal residence, EDINBURGH, SE Scotland; 55°57′N 03°10′W. In 1128, David I founded Holyrood Abbey on this site, where, according to legend, he was saved from an infuriated stag by the miraculous interception of a cross. The abbey's Chapel Royal, still standing, contains the remains of David II, James II, James V, Lord Darnley, and others. James IV began the present building c.1500. The palace, partially destroyed by the English in 1544, was the scene of the murder of David Rizzio in 1566. It was almost completely destroyed by fire in 1650. Charles II had the palace rebuilt (1671–1679) according to plans by William Bruce.

Holysov (HO-lee-SHOF), Czech *Holýšov*, German *Holeischen*, town, Zapadocesky province, SW Bohemia, on Radbuza River, on railroad, and 15 mi/24 km SW of Plzeň; 49°36′N 13°06′E. Manufacturing of machinery (parts fortrucks). Has 14th-cent. church, monument of anti-fascist revolt.

Holytown (HO-lee-toun), town (2001 population 5,483), NORTH LANARKSHIRE, central Scotland, 3 mi/4.8 km N of MOTHERWELL; 55°49′N 03°57′W. Previously steel and engineering. Formerly in STRATHCLYDE, abolished 1996.

Holywell (HO-lee-wel), town (2001 population 8,715), FLINTSHIRE, NE Wales, near the DEE RIVER, and 4 mi/6.4 km WNW of FLINT; 53°17′N 03°13′W. Previously woolen milling, paper manufacturing. Named for holy well sacred to the name of St. Winefride. On the Dee, 2 mi/3.2 km NE, is small port of Greenfield. Formerly in CLWYD, abolished 1996.

Holywell, ENGLAND: see EARSDON.

Holywood (HO-lee-wud), Gaelic *Ard Mhic Nasca*, town (2001 population 12,037), NE DOWN, Northern Ireland, on SE coast of BELFAST LOUGH, 6 mi/9.7 km NE of BELFAST. Port and seaside resort. Famous golf course. Ruins of 16th-century friary on site of 7th-century monastery. Nearby is seaside resort of Marino. Ulster Folk Museum is 2 mi/3.2 km NE at Cultra Manor.

Holzgerlingen (hohlts-GER-ling-uhn), town, BADEN-WÜRTTEMBERG, S GERMANY, 3 mi/4.8 km S of BÖBLINGEN; 48°38′N 09°00′E. Primarily residential.

Holzhausen (hohlts-HOU-suhn), village, SAXONY, E central GERMANY, 4 mi/6.4 km ESE of LEIPZIG city center; 51°20′N 12°28′E. Primarily residential.

Holzkirchen (hohlts-KIR-khuhn), town, BAVARIA, GERMANY, 10 mi/16 km NE of Bad BAD TÖLZ; 47°53′N

11°42′E. Manufacturing of machinery, tools; paper processing; brewing. Elevation 2,262 ft/689 m.

Holzminden (HOHLTS-min-den), town, LOWER SAX-ONY, central GERMANY, port on right bank of the WESER, 34 mi/55 km N of KASSEL; 51°50′N 09°27′E. Manufacturing of essential oils, machinery, construction materials, glass; brewing; tourism. Has architectural institute (founded 1831). Chartered 1245. Plundered and completely burned (1640) by imperial troops.

Holzwickede (hohlts-TSVIK-ke-de), town, NORTH RHINE-WESTPHALIA, W GERMANY, in the RUHR industrial district, 3 mi/4.8 km W of Unna; 51°30′N 07°37′E. Within commuting distance from DORT-MUND, the town is largely residential; manufacturing includes metalworking; food processing.

Homa Bay (HO-mah), town (1999 population 32,174), NYANZA province, W KENYA, on KAVIRONDO GULF of Lake VICTORIA, 36 mi/58 km SW of KISUMU; 00°31′S 34°30′E. Apatite deposits.

Homalin (HO-mah-lin), township, SAGAING division, MYANMAR, on left bank of CHINDWIN RIVER (head of navigation) and 120 mi/193 km NNE of KALEWA. Tea plantations; jungle-covered area.

Homathko River (ho-MATH-ko), 80 mi/129 km long, SW BRITISH COLUMBIA, W CANADA; rises in COAST MOUNTAINS near 51°50′N 124°45′W; flows generally S to head of BUTE INLET.

Homberg (HOHM-berg), town, HESSE, central GER-MANY, 19 mi/31 km SSW of KASSEL; 51°03′N 09°24′E. Manufacturing of motors, automobile parts; foundry; brickworks. In 1526, at a synod held in the Gothic church, the introduction of the Reformation into Hesse was decided. Has ruined castle with watch-tower.

Homberg (HOHM-berg), town, HESSE, central GER-MANY, on the OHM, 17 mi/27 km NE of GIESSEN; 50°43′N 09°00′E. Metal- and woodworking. Town is characterized by many half-timbered houses and church from 13th century.

Homberg (HOHM-berg), suburb of DUISBURG, NORTH RHINE–WESTPHALIA, W GERMANY, in the RUHR, port on left bank of the RHINE, nearly opposite of city center; 51°27′N 06°42′E. Manufacturing of machinery, dyes. Coal harbor. Incorporated into Duisburg 1975.

Hombok, CZECH REPUBLIC: see HLUBOCKY.

Hombori, township, FIFTH REGION/MOPTI, E central MALI, in hilly region (Hombori mountains), 120 mi/193 km WSW of GAO; 15°17′N 01°42′W. Rice, millet; livestock.

Hombourg-Haut (on-boorg–ho), German *Oberhom-burg*, town (□ 4 sq mi/10.4 sq km), MOSELLE department, LORRAINE region, NE FRANCE, 4 mi/6.4 km NE of SAINT-AVOLD, in old coal-mining district; 49°08′N 06°46′E. Boilerworks; manufacturing of auto springs. Has ancient fort of the bishops of Metz.

Hombrechtikon (HOHM-brech-tee-kon), commune, ZÜRICH canton, N SWITZERLAND, 13 mi/21 km SE of ZÜRICH. Metalworking.

Hombre Muerto, Salar de (OM-bre MWER-to, sah-LAHR dai), salt desert (□ 241 sq mi/626.6 sq km) in PUNA DE ATACAMA, N CATAMARCA province, AR-GENTINA, extends c.20 mi/32 km N-S and E-W on SALTA province border. Contains borax and sodium chloride. Sierra del Hombre Muerto, 80 mi/129 km SE, is a 40-mi/64-km subandean range rising to 16,500 ft/5,029 m.

Homburg (HOHM-burg), city, SAARLAND, SW GER-MANY, NE of SAARBRÜCKEN; 49°20′N 07°19′E. Iron, metal, and steel industry; manufacturing of vehicles; brewing. City is seat of the medical faculty of Saarland University; under French government in 17th century it was developed into a fortress; it was transferred to PRUSSIA in 1815. In 1920 it became part of the Saar region.

Homburg von der Höhe, Bad, GERMANY: see BAD HOMBURG VOR DER HÖHE.

Homeacre (HOM-ai-kuhr), unincorporated town, BUTLER county, W PENNSYLVANIA, residential suburb 2 mi/3.2 km W of BUTLER; 40°51′N 79°56′W.

Home Bay, E BAFFIN ISLAND, BAFFIN region, NUNAVUT territory, CANADA, inlet of DAVIS STRAIT; 40 mi/64 km long, 50 mi/80 km wide at mouth; 68°40′N 67°30′W. Bounded N by Henry Kater Peninsula.

Homebush (HOM-bush), suburb, E NEW SOUTH WALES, AUSTRALIA, 7 mi/11 km W of SYDNEY; 33°52′S 151°05′E. In metropolitan area; government-owned abattoirs and brickworks; dock facilties; manufac-turing (radios, phonograph records). Sydney Sports Complex is here. Included 1947 in Strathfield.

Home Corner, village, GRANT county, NE central IN-DIANA, suburb of MARION (SE).

Homécourt (o-mai-koor), town (□ 1 sq mi/2.6 sq km), MEURTHE-ET-MOSELLE department, LORRAINE region, NE FRANCE, in industrial ORNE RIVER valley, 10 mi/16 km NW of METZ; 49°14′N 05°59′E. Heavy metal-lurgical works. Has institute of metalworking technology.

Homecroft, town (2000 population 751), MARION county, central INDIANA; 39°40′N 86°08′W.

Homedale, town (2000 population 2,528), OWYHEE county, SW IDAHO, on SNAKE RIVER, at mouths of Jump Creek and Succor Creek, and 15 mi/24 km W of CALDWELL, near OREGON line; 43°37′N 116°57′W. In Idaho section of Owyhee irrigation project. Cattle; alfalfa, oats, grain; potatoes, sugar beets; manufac-turing. Succor Creek State Recreation Area to SW (Oregon).

Home Garden, unincorporated town (2000 population 1,702), KINGS county, central CALIFORNIA; residential suburb 2 mi/3.2 km SSE of HANFORD. Hanford Mu-nicipal Airport to N. Fruit, nuts, melons, tomatoes, grain, dairying, cattle, poultry.

Home Gardens, unincorporated town (2000 popula-tion 9,461), RIVERSIDE county, S CALIFORNIA; resi-dential suburb 11 mi/18 km WSW of RIVERSIDE. LAKE MATHEWS reservoir to SE. Fruit, grain, nursery products, grain, poultry, cattle, dairying.

Home Hill (HOM HIL), town, E QUEENSLAND, AUS-TRALIA, on BURDEKIN RIVER, and 50 mi/80 km SE of TOWNSVILLE, and a suburb of AYR; 19°40′S 147°25′E. Irrigation; sugar center.

Homei (HUH-MAI), town, W central TAIWAN, 4 mi/6.4 km NW of CHANG-HUA; 24°07′N 120°29′E. Sugar re-fining; rice, sweet potatoes, sugarcane.

Home Island, AUSTRALIA: see COCOS ISLANDS.

Homel, BELARUS: see GOMEL, city.

Homeland, city (2000 population 765), CHARLTON county, SE GEORGIA, 2 mi/3.2 km NNW of FOLKSTON; 30°51′N 82°01′W.

Homeland, unincorporated town (2000 population 3,710), RIVERSIDE county, S CALIFORNIA; residential suburb 21 mi/34 km SE of RIVERSIDE. Lakeview Mountains to N; Double Butte to S. SAN DIEGO AQUEDUCT passes to E. Cattle, dairying, poultry, nursery products, grain.

Homer, town (2000 population 950), ⊙ BANKS county, NE GEORGIA, 18 mi/29 km E of GAINESVILLE; 34°20′N 83°30′W. Manufacturing of clothing; lumber.

Homer, town (2000 population 3,788), ⊙ CLAIBORNE parish, N LOUISIANA, 45 mi/72 km NE of SHREVE-PORT; 32°48′N 93°04′W. Trade center in oil and nat-ural gas, timber area; lumber milling; manufacturing (plastic molding, wood products, packaging products, wood paneling); dairying. Unit of Kisatchie National Forest to NE, Lake CLAIBORNE reservoir to SE with Lake Claiborne State Park at SE end of lake. Hosts Louisiana Wildlife Festival. Greek Revival Court-house. Prior to Civil War a newspaper called the Homer Iliad was published here. Settled 1830, incor-porated 1850.

Homer, town (2000 population 1,851), CALHOUN county, S MICHIGAN, 7 mi/11.3 km SSW of ALBION and on a branch of KALAMAZOO RIVER; 42°08′N 84°48′W.

Agriculture (wheat, corn, and hay); poultry; manu-facturing (iron foundry, plastic foam molding, ma-chining). Settled 1832, incorporated 1871.

Homer, village (2000 population 3,946), S ALASKA, on KACHEMAK BAY, SW KENAI PENINSULA, 75 mi/121 km WSW of SEWARD; 59°36′N 151°25′W. Connected by highway with Seward, KENAI, Soldatna, and ANCHO-RAGE. Tourism; fishing, fish processing. Site of old Eskimo and Russian settlements.

Homer, village (2000 population 1,200), CHAMPAIGN county, E ILLINOIS, 16 mi/26 km ESE of CHAMPAIGN; 40°01′N 87°57′W. In agricultural area; corn, wheat, soybeans, livestock. Village formerly on SALT CREEK to N but moved in nineteenth century to be on rail-road.

Homer, village (2006 population 611), DAKOTA county, NE NEBRASKA, 12 mi/19 km S of SIOUX CITY, IOWA, near MISSOURI RIVER; 42°19′N 96°29′W. Farm trading center. Site of 18th-century Omaha Indian village nearby. Winnebago and Omaha Indian Reservations to S (both in THURSTON county).

Homer, village (□ 1 sq mi/2.6 sq km; 2006 population 3,280), CORTLAND county, central NEW YORK, in the TIOUGHNIOGA river valley, just N of CORTLAND; 42°38′N 76°10′W. Agricultural commerce; light man-ufacturing; quarrying; services. Old Homer Village Historic District here. Settled 1791, incorporated 1835.

Homer City, borough (2006 population 1,745), INDIANA county, SW central PENNSYLVANIA, 5 mi/8 km S of INDIANA, on Two Lick Creek. Manufacturing (bulk materials handling equipment); surface bituminous coal; agriculture (soybeans, grain; livestock, dairying). Incorporated 1872.

Homerville, city (2000 population 2,803), ⊙ CLINCH county, S GEORGIA, 25 mi/40 km WSW of WAYCROSS, near OKEFENOKEE SWAMP; 31°02′N 82°45′W. Manu-facturing of metal containers, modular buildings, clothing, plastics; pulpwood, lumber, wood products. Founded on the railroad 1859, incorporated as town 1869, as city 1931.

Homes Run Acres (HOMZ RUHN AI-kuhrz), unin-corporated town, FAIRFAX county, NE VIRGINIA, residential suburb 10 mi/16 km W of WASHINGTON, D.C.

Homestead (HOM-sted), city (□ 11 sq mi/28.6 sq km; 2000 population 31,909), MIAMI-DADE county, SE FLORIDA; incorporated 1913. A satellite town 25 mi/40 km S of MIAMI; 25°27′N 80°27′W. Homestead is a trade center for the Redland district to its N, known for its many varieties of citrus and other fruits and vegetables. Nearby Homestead Air Force Base, though sharply cut back during the 1990s, is still important to the economy. The city is adjacent to Everglades Na-tional Park and lies astride the gateway to the FLORIDA KEYS. Local attractions include several tropical gar-dens, a pioneer museum, and a castlelike building furnished with coral items. A state subtropical ex-periment station is there, and a nuclear power plant is nearby. In 1992, Hurricane Andrew swept through the city, leveling much of it. Homestead Air Force Base suffered serious hail damage. A massive, na-tionwide relief effort followed, and the city was rebuilt during the mid-1990s. The Turkey Point nuclear power facility is located on the shore of BISCAYNE BAY c.3 mi/4.8 km E.

Homestead, village, IOWA county, E IOWA, 7 AMANA COLONY villages, located S of IOWA RIVER. Agriculture (corn; cattle, hogs); winery; home industries.

Homestead, village, PORTSMOUTH town, PRUDENCE ISLAND, RHODE ISLAND; 41°37′N 71°18′W. Summer resort.

Homestead, borough (2006 population 3,486), ALLE-GHENY county, SW PENNSYLVANIA, suburb 5 mi/8 km SE of downtown PITTSBURGH, on the MONONGAHELA RIVER (bridged); 40°24′N 79°54′W. Manufacturing (babbitt bearings, carbide machinery, cleaning che-micals, cylindrical grinders). Once a foremost U.S.

steel producer. In 1892 the famous outbreak of the Homestead Strike, one of the most bitterly fought industrial disputes in U.S. labor history, occurred here. Incorporated 1880.

Homestead Monument, GAGE county, SE NEBRASKA, 5 mi/8 km SW of BEATRICE. Site of the first farm claimed under the Homestead Act of 1862. Freeman Schoolhouse; over 100 acres of tall grass prairie and wooded stream. Authorized 1936.

Homestead Valley, unincorporated town, MARIN county, W CALIFORNIA; residential suburb 10 mi/16 km NW of downtown SAN FRANCISCO. To SE is Mount Tamalpais Game Refuge, in Richardson Bay, arm of SAN FRANCISCO BAY; Mount Tamalpais State Park and MUIR WOODS NATIONAL MONUMENT to W. Statistically reported as Tamalpais-Homestead Valley (1990 population 9,601). TAMALPAIS VALLEY (unincorporated town) is 2 mi/3.2 km S.

Hometown, city (2000 population 4,467), COOK county, NE ILLINOIS, 14 mi/23 km SW of downtown CHICAGO, NE of OAK LAWN; 41°43′N 87°43′W. Residential suburb.

Hometown, unincorporated town (2000 population 1,399), SCHUYLKILL county, E central PENNSYLVANIA, 9 mi/14.5 km S of HAZLETON; 40°49′N 75°59′W. Anthracite coal mining. Tuscarora Lake and State Park to W.

Homewood, city (2000 population 25,043), Jefferson co., N central Alabama, a residential suburb S of Birmingham. Shades Mountain and Oak Mountain state parks nearby. Inc. 1921. Originally known as 'Edgewood,' it was renamed in 1926.

Homewood, village (2000 population 19,543), COOK county, NE ILLINOIS, a residential suburb S of CHICAGO; 41°33′N 87°39′W. Plotted 1852, incorporated 1893.

Homewood, borough (2006 population 142), BEAVER county, W PENNSYLVANIA, 4 mi/6.4 km N of BEAVER FALLS, near BEAVER RIVER. Corn, hay, dairying.

Homewood, suburb of ANNAPOLIS, ANNE ARUNDEL county, central MARYLAND.

Hominabad, INDIA: see HOMNABAD.

Hominy (HAHM-in-ee), town (2006 population 3,713), OSAGE county, N OKLAHOMA, 29 mi/47 km NW of TULSA; 36°25′N 96°23′W. In agricultural and oil- and natural gas-producing area; manufacturing (oil field equipment, T-shirts, lingerie). In Osage Indian Reservation. SKIATOOK LAKE reservoir to E. Established as an Indian subagency in 1874; laid out 1905, incorporated 1908.

Hommelvik (HAWM-muhl-veek), village, SØR-TRØN-DELAG county, central NORWAY, on S shore of TRONDHEIMSFJORDEN, on railroad and 12 mi/19 km E of TRONDHEIM. Iron casting; woodworking; exports lumber. Seaplane use and deep-water port.

Homnabad (HOM-nah-bahd), town, BIDAR district, KARNATAKA state, INDIA, 23 mi/37 km WSW of BIDAR; 17°46′N 77°08′E. Millet, cotton, rice. Road junction is 2 mi/3.2 km W. Sometimes spelled Hominabad.

Homochitto River (ho-muh-CHIT-uh), c.90 mi/145 km long, SW MISSISSIPPI; rises in SW COPIAH county; flows SW and W, through Homochitto National Forest, past BUDE, enters the MISSISSIPPI RIVER through two main channels, one 17 mi/27 km SSW of NATCHEZ, the other 25 mi/40 km SSW of Natchez, through OLD RIVER Lake.

Homoíne (o-MOIN), village, INHAMBANE province, SE MOZAMBIQUE, on road and 15 mi/24 km W of IN-HAMBANE. Coffee, cotton, mafura.

Homoine District, MOZAMBIQUE: see INHAMBANE.

Homonhon Island (ho-mon-HON) (□ 40 sq mi/104 sq km), EASTERN SAMAR province, PHILIPPINES, between LEYTE GULF and PHILIPPINE SEA, 14 mi/23 km S of narrow SE peninsula of SAMAR island; 12 mi/19 km long, 4 mi/6.4 km wide; 10°44′N 125°43′E. Deeply indented in E by Casogoran Bay. Hilly, rising to 1,120 ft/341 m. Rice growing; fishing. Chromite and iron

deposits. May have been Magellan's first landing site in Philippines in 1521. In mid-October 1944, the island was the site of a preliminary strike by U.S. liberation forces in World War II in preparation for their invasion of LEYTE a few days later.

Homonna, SLOVAKIA: see HUMENNÉ.

Homorod (ho-mo-ROD), Hungarian *Homoród*, village, BRAŞOV county, central ROMANIA, on railroad, and 25 mi/40 km SE of SIGHIŞOARA; lumbering center; stud farm. Has noted 15th-century fortified church.

Homosassa (ho-mo-SAS-uh), fishing port (□ 5 sq mi/13 sq km; 2000 population 2,294), CITRUS county, W central FLORIDA, 18 mi/29 km W of INVERNESS, on Gulf coast; 28°46′N 82°37′W. The Homosassa Islands, a group of many small mangrove islands, are just offshore. HOMOSASSA SPRINGS, a resort, is 2 mi/3.2 km ENE.

Homosassa Springs (ho-mo-SAS-uh), town (□ 13 sq mi/33.8 sq km; 2000 population 12,458), CITRUS county, W central FLORIDA, 25 mi/40 km NW of BROOKSVILLE; 28°48′N 82°32′W.

Hom'r, El (HOR, el), Saharan outpost, ADRAR wilaya, central ALGERIA, at S edge of the Great Western ERG, on El GOLÉA-ADRAR auto track and 95 mi/153 km SW of El GOLÉA; 29°46′N 01°37′E. Formerly Fort Mac-Mahon.

Homs (HOMS), district, central SYRIA; ⊙ HOMS. Extends from NE tip of LEBANON E and SE to borders of IRAQ and JORDAN. Mainly desert. The W part of the district is semi-arid with some areas irrigated by the ORONTES River and Lake HOMS; millet, wheat, corn, cotton are grown. Crossed E-W by the Iraq-TRIPOLI (LEBANON) and Iraq-BANIAS (Syria) oil pipeline, which passes through PALMYRA. Railroad N to HAMA and ALEPPO, and S to BAALBEK, BEIRUT, and DA-MASCUS. Important ruins, especially of the Roman period, are at Homs, PALMYRA, and Er RASTAN. Also spelled HEMS, HUMS, and HIMS.

Homs (HOMS), city, ⊙ HOMS district, W central SYRIA, on the ORONTES River; 34°43′N 36°43′E. It is a commercial center located in a fertile plain where wheat, grapes, fruit, and vegetables are grown. Manufacturing includes refined petroleum, flour, fertilizer, processed foods, handicrafts, and silk, cotton, and woolen textiles. The city is a road and railroad junction and is the home of Syria's main oil refinery. It refines oil both from the NE Syrian fields, and from IRAQ. The pipeline from Iraq to the Syrian and Lebanese coast (unoperational during the UN embargo on IRAQ) passes through the outskirts of HOMS. Many modern industries. Important thermal electric power station. In ancient times Homs, then called EMESA, was the site of a great temple to Baal (or Helios-Baal), the sun god. Emesa came into startling prominence in the early 3rd century C.E. when a priest of the temple became Roman emperor as Heliogabalus, or Elagabalus. Aurelian defeated the forces of Zenobia of PALMYRA there in 272. The Arabs took the town in 636, renaming it Homs. The Arab soldier Khalid died there in 642; a shrine and mosque in his honor were erected in 1908. Homs was part of the OTTOMAN Empire from the 16th century until after World War I, when it became part of the French League of Nations mandate. The city has a university. Also called HIMS.

Homs, LIBYA: see KHUMS, AL.

Homs, Lake, irrigation reservoir (□ 23 sq mi/59.8 sq km) on the ORONTES River (dammed), HOMS district, W SYRIA, 7 mi/11.3 km SW of HOMS, near Lebanese border; 9 mi/14.5 km long, c.2 mi/3.2 km wide; 34°37′N 36°33′E. QATTINE is near NE shore. Sometimes called Lake Qattine.

Homún (o-MOON), town, YUCATÁN, SE MEXICO, 27 mi/43 km SE of MÉRIDA. Henequen, tropical fruit, sugarcane, corn.

Hon, LIBYA: see HUN.

Honai (HO-nah-ee), town, West Uwa district, EHIME prefecture, NW SHIKOKU, W JAPAN, 34 mi/55 km S of MATSUYAMA; 33°28′N 132°24′E. Mandarin oranges.

Honaker (HON-aik-uhr), town (2006 population 913), RUSSELL county, SW VIRGINIA, in ALLEGHENY MOUNTAINS, near CLINCH RIVER, 10 mi/16 km NE of LEBANON; 37°01′N 81°58′W. Manufacturing (lumber, clothing); agriculture (corn, soybeans; livestock; dairying); timber.

Honalo (HO-nah-LO), town (2000 population 1,987), W HAWAII island, HAWAII county, HAWAII, 52 mi/84 km WSW of HILO, 3 mi/4.8 km N of CAPTAIN COOK, 2 mi/3.2 km inland from KONA (W) coast; 19°34′N 155°54′W. Coffee, fruit. Daifukuji Buddhist Temple. Kahaluu Forest Reserve to NE.

Honami (HO-nah-mee), town, Kaho district, FUKUOKA prefecture, N KYUSHU, SW JAPAN, 16 mi/25 km E of FUKUOKA; 33°36′N 130°41′E.

Honan, CHINA: see HENAN.

Honanfu, CHINA: see LUOYANG.

Honaunau (HO-NOU-NOU), town, W HAWAII island, HAWAII county, HAWAII, 53 mi/4.8 km WSW of HILO, 3 mi/4.8 km S of CAPTAIN COOK. Tourism. Coffee, cattle; fish. Honaunau Forest Reserve to E; PU'UHO-NUA O HONAUNAU NATIONAL HISTORIC PARK to W on Kiilae and Honaunau bays. Honaunau Rodeo Arena to W.

Honavar (HO-nah-vuhr), town, UTTAR KANNAD district, KARNATAKA state, INDIA, port on ARABIAN SEA, at mouth of SHARAVATI River, 42 mi/68 km SE of KARWAR; 14°17′N 74°27′E. Agriculture market (rice, coconuts, spices); fish-supplying center (mackerel, sardines, catfish, seerfish); sandalwood and ivory carving, fruit canning, manufacturing of coir products; exports matchwood. Handicraft school. Casuarina plantations nearby. Fortified 1505 by Portuguese; seized 1783 from Hyder Ali by British.

Honaz Dagi (Turkish=*Honaz Daği*) peak (8,435 ft/2,571 m), SW TURKEY, 12 mi/19 km SE of DENIZLI.

Honbetsu (HON-bets), town, Tokachi district, S central HOKKAIDO prefecture, N JAPAN, 115 mi/185 km E of SAPPORO; 43°07′N 143°36′E. Kidney and adzuki beans; dairying; lumber.

Honcharivs'ke (hon-chah-RIF-ske) (Russian *Gonch-arovskoye*), town, NW CHERNIHIV oblast, UKRAINE, on right bank of the DESNA River, on crossroads, 21 mi/33 km SW of CHERNIHIV city center; 51°17′N 31°00′E. Elevation 337 ft/102 m. Town since 1990.

Hon Chong (HON CHONG), town, S VIETNAM, on Gulf of THAILAND, 17 mi/27 km SE of HA TIEN; 10°10′N 104°37′E. Fishing center; pepper farms. Tourism potential. Nearby are offshore islands with white-sand beaches. Vietnamese and Cambodian (Khmer) people in area. Formerly Hongchong.

Honctö, ROMANIA: see GURAHONŢ.

Honda (ON-dah), city (2005 population 25,991), ⊙ Honda municipio, TOLIMA department, W central COLOMBIA, port on MAGDALENA River at Honda falls, and 60 mi/97 km NW of BOGOTÁ; 05°11′N 74°46′W. An early river-railroad transshipping point between lower and upper Magdalena River. Trading and processing center in agricultural region (coffee, rice, sorghum; livestock); food processing; consumer goods.

Honda Bay (HON-dah), on CARIBBEAN, N coast of LA GUAJIRA peninsula, N COLOMBIA, SW of Punta GAL-LINAS; c.6.0 mi/9.7 km wide; 12°20′N 71°45′W. Saltworks.

Honda Bay (ON-dah), sheltered inlet, PINAR DEL RÍO prov., NW CUBA, just N of town of BAHÍA HONDA, 50 mi/80 km WSW of HAVANA; 22°57′N 83°11′W. Near shark, lobster, and red snapper fishing grounds. Also called Bahía Honda.

Honda, Ensenada, PUERTO RICO: see ENSENADA HONDA.

Hondagua, PHILIPPINES: see LOPEZ.

Honddu River (HON-[th]ee), 10 mi/16 km long, POWYS, E WALES; rises on MYNYDD EPPYNT; flows SE to the USK RIVER at Brecon.

Hondo (HON-do), city, on E coast of SHIMO-JIMA island, of AMAKUSA ISLANDS, KUMAMOTO prefecture, SW JAPAN, 37 mi/60 km S of KUMAMOTO; 32°27′N 130°11′E. Mandarin oranges; prawns, *asari* clams; stone for pottery. Amakusa Shiro battle flag originated here.

Hondo, town (2006 population 8,933), ⊙ MEDINA county, SW TEXAS, W of SAN ANTONIO; 29°21′N 99°09′W. Elevation 905 ft/276 m. In ranching and irrigated farm area (cattle, sheep, goats; grain, peanuts, cotton, vegetables); manufacturing (grain processing, rebuilt aircraft engines, feeds, bathroom fixtures). MEDINA LAKE (irrigation, recreation) is 17 mi/27 km NE. Incorporated after 1940.

Hondo (ON-do), unincorporated village, LINCOLN county, central NEW MEXICO, 47 mi/76 km W of ROSWELL, at confluence of Rio Ruidoso and Rio Bonito, which form RIO HONDO. Cattle, sheep, alfalfa. Timber. Parts of Lincoln National Forest to N and W; Mescalero Apache Indian Reservation to SW. SACRAMENTO MOUNTAINS to W. Lincoln State Monument to NW.

Hondón de las Nieves (on-DON dhai lahs NYAI-ves), town, ALICANTE province, E SPAIN, 9 mi/14.5 km NW of ELCHE; 38°18′N 00°51′W. Footwear manufacturing, olive oil processing.

Hondo, Río (ON-do, REE-o), short stream on TUCUMÁN-SANTIAGO DEL ESTERO province border, NW ARGENTINA; formed by union of the RÍO CHICO and MARAPA RIVER, flows c.5 mi/8 km to the Sali River (upper course of the RIO DULCE) at Rio Hondo town. The name Rio Hondo is sometimes applied to the Santiago del Estero section of the Rio Dulce.

Hondo, Río, or **Hondo River** (ON-do), c.130 mi/209 km long, YUCATÁN Peninsula, SE MEXICO (CAMPECHE and QUINTANA ROO states), NE GUATEMALA, and N BELIZE, on Belize-Mexico border; rises as Río Azul NE of UAXACTÚN (Guatemala); flows NE, along international border, to CHETUMAL BAY (CARIBBEAN SEA) at CHETUMAL, Mexico. Receives Booth's River at Dos Bocas, Belize, and from there until it exits Belize, it is called Blue Creek.

Hondo, Río (ON-do), 106 mi/171 km long, 11th-longest river in CUBA; flows through PINAR DEL RÍO province, emptying into CARIBBEAN SEA. Drains 578 sq mi/1,497 sq km.

Hondschoote (HOND-shot), border town (□ 9 sq mi/23.4 sq km; 2004 population 3,749), NORD department, NORD-PAS-DE-CALAIS region, N FRANCE, at Belgian border, 10 mi/16 km SE of DUNKERQUE; 50°59′N 02°35′E. Agriculture market; brewing. Has 16th-century church and town hall. Flemish is spoken here.

Hondsrug (HAWNZ-ruhkh), region, in DRENTHE and GRONINGEN provinces, N central NETHERLANDS; extends 30 mi/48 km from Emmens (NW) to GRONINGEN. Highest elevation 100 ft/30 m.

Honduras (hon-DOO-rahs), republic (□ 43,277 sq mi/112,088 sq km; 2004 estimated population 6,823,568; 2007 estimated population 7,483,763), CENTRAL AMERICA; ⊙ TEGUCIGALPA; 15°00′N 86°30′W.

Geography

Tegucigalpa is the chief commercial center. The second largest of the Central American countries, Honduras is bounded on the N by the CARIBBEAN SEA, on the E and S by NICARAGUA, on the SW by EL SALVADOR, and on the W by GUATEMALA. The short stretch of S coast on the Gulf of FONSECA, with a small port, SAN LORENZO (Henecán), is the sole Pacific outlet. Honduras has a tropical, rainy climate. Over 80% of the land is mountainous; ranges extend from E to W at elevations of 5,000–9,000 ft/1,520–2,740 m and limit heavy rainfall to the N. In the E are the swamps and forests of the MOSQUITO COAST. Several river systems, of which the PATUCA and the ULÚA are most important, drain most of the N.

Population

The population is a mix of mestizo (mixed Amerindian and European), 90%, Amerindian, 7%, black,

2%, and white, 1%. About 40% of the population is illiterate. Hondurans are overwhelmingly Roman Catholic (97%).

Economy

The economy is based on agriculture; bananas and coffee are the most important exports. The vast banana plantations, established by U.S. companies, are mainly along the N coast; the United Fruit Company and the Standard Fruit Company, fiercely resented by many Latin Americans as exploitative monopolies, had much social and political influence in Honduras. Timber, minerals (silver, lead, zinc), beef, and seafood are also exported. Other important food crops include corn, beans, rice, and sugarcane. Honduras has rich forest resources and deposits of silver, lead, zinc, gold, cadmium, antimony, and copper, but exploitation is hampered by inadequate road and railroad systems, and the country remains underdeveloped. Its only railroads link the banana plantations in the N to SAN PEDRO SULA and the principal ports, LA CEIBA, PUERTO CORTÉS, and TELA; they do not penetrate more than 75 mi/121 km inland. Air transportation, however, has opened up remote areas. Industry, concentrated chiefly in San Pedro Sula, is small and consumer-oriented, including the production of processed food, cement, lumber, and chemicals. Clothing manufacture, based around low-wage sweatshops developed by South Korean and U.S. companies, is the country's fastest-growing industry; clothing is now the third-largest export.

History: to 1948

The restored Mayan ruins of COPÁN in the W, first discovered by the Spaniards in 1576 and rediscovered in 1839, reflect the great Mayan culture that arose in the 4th century. It had declined when Columbus sighted the region in 1502, naming it Honduras (meaning "depths") for the deep water off the coast. Hernán Cortés arrived in 1524 and ordered Pedro de Alvarado to found settlements along the coast. COMAYAGUA and Tegucigalpa developed as early mining centers. Honduras gained independence from SPAIN in 1821 and became part of Iturbide's Mexican Empire; it was a member of the Central American Federation from 1825 until the organization was dissolved in 1838. Great Britain long controlled the Mosquito Coast and the ISLAS DE LA BAHÍA; William Walker attempted a "liberation" in 1860 but was unsuccessful. Foreign capital, plantation life, and conservative politics constituted a trio of dominant forces that held sway in Honduras from the late 19th century to the end of the regime (1933–1948) of Tiburcio Carías Andino.

History: 1948 to Present

The illegal immigration of several hundred thousand Salvadorans across the ill-defined El Salvador–Honduras border and the expulsion of thousands of the immigrants by Honduras led to a war with El Salvador in July 1969. Although the war lasted only five days, its effects were serious, including the country's withdrawal from and the subsequent collapse of the Central American Common Market as well as continued border incidents. A 3.7-mi/6-km demilitarized zone was set up in 1970. After several more military incidents in 1976, negotiations began between the two sides. A peace agreement reached in 1980 demarcated two-thirds of the 213-mi/343-km border. The Caribbean coast of Honduras was devastated by a hurricane in late 1974. During the 1980s Honduras served as a base for insurgent activity against the government of Nicaragua by the Contra rebels. The country's economy became heavily dependent on aid from the U.S., which supported the rebel bases. Since 1990 Honduras has benefited from regional peace and cooperation to stabilize its economy and establish economic viability independent of the U.S. In 1992 Honduras signed an agreement with El Salvador, largely settling the border controversy between the two countries; the last disputed section of the border was demarcated in 2006.

In 1998 the country was once again devastated by a hurricane, which killed some 5,600 people and caused approximately $2 billion in damage.

Government

Honduras is governed under the constitution of 1982 as amended. The president, who is both head of state and head of government, is popularly elected for a four-year term. The unicameral legislature, the National Congress, has 128 members, also elected for four years. Administratively, the country is divided into eighteen departments. The current head of state, President Manuel Zelaya Rosales, has been in office since January 2006.

Honduras, Cape (hon-DOO-rahs), northernmost point of HONDURAS mainland, on CARIBBEAN SEA, 7 mi/11.3 km NNW of TRUJILLO; 16°01′N 86°01′W. Site of PUERTO CASTILLA. Also called Punta Castilla (POON-tah kah-STEE-yah).

Honduras, Gulf of (hon-DOO-rahs), wide inlet of CARIBBEAN SEA on coasts of HONDURAS, GUATEMALA, and BELIZE; mostly between 16°10′N 87°50′W. Includes BAY OF AMATIQUE. Receives MOTAGUA RIVER. It contains the S cays of Belize (N).

Honea Path (HUH-nee), town (2006 population 3,615), ANDERSON and ABBEVILLE counties, NW SOUTH CAROLINA, 16 mi/26 km ESE of ANDERSON; 34°27′N 82°23′W. Manufacturing of cotton and synthetic textiles, clothing, rubber gloves; agr. includes poultry, dairying, grain, soybeans, livestock.

Hønefoss (HUH-nuh-faws), town, BUSKERUD county, SE NORWAY, center of RINGERIKE region, on BEGNA River, at mouth of short Randselva River, on the main OSLO-BERGEN highway, N of TYRIFJORDEN, and 25 mi/40 km NW of Oslo. Railroad center. Paper milling, woodworking, brewing, food packing; manufacturing of truck and bus bodies. Falls. Hydroelectric power plant.

Honeoye (HUHN-ee-oi), resort village, ONTARIO county, W central NEW YORK, at N end of HONEOYE LAKE, 27 mi/43 km S of ROCHESTER; 42°47′N 77°31′W.

Honeoye Creek, New York: see HONEOYE LAKE.

Honeoye Falls (HUHN-ee-oi), residential village (□ 2 sq mi/5.2 sq km; 2006 population 2,556), MONROE county, W NEW YORK, on Honeoye Creek and 15 mi/24 km S of ROCHESTER; 42°57′N 77°35′W. Light manufacturing; agricultural products; services. Incorporated 1838. HONEOYE village is 12 mi/19 km S, on HONEOYE LAKE.

Honeoye Lake (HUHN-ee-oi), one of the FINGER LAKES, ONTARIO county, W central NEW YORK, W of CANANDAIGUA LAKE and 25 mi/40 km S of ROCHESTER; 4 mi/6.4 km long, up to 0.75 mi/1.21 km wide; 42°45′N 77°31′W. Resorts. Drained by Honeoye Creek, which flows c.35 mi/56 km NW and W, past Honeoye Falls, to GENESEE River 5 mi/8 km N of AVON.

Honesdale, borough (2006 population 4,806), ⊙ WAYNE county, NE PENNSYLVANIA, 24 mi/39 km ENE of SCRANTON, on LACKAWAXEN RIVER at mouth of Dyberry Creek. Manufacturing (furniture, food processing, sand and gravel processing, printing and publishing, concrete). First trial run of a locomotive in U.S. made here (1829). Was W terminus of DELAWARE AND HUDSON CANAL. Cherry Ridge Airport to S; Dorflinger Glass Museum; Stourbridge Line Rail Excursion; Wayne County Historical Society Museum; Cadjaw Pond to SW; Prompton Lake reservoir to NW. Settled 1803, incorporated 1831.

Honey (ON-ei), town, ⊙ Chila Honey municipio, PUEBLA, central MEXICO, in SIERRA MADRE ORIENTAL, on HIDALGO border, 12 mi/19 km WNW of HUAUCHINANGO; 20°16′N 98°11′W. Railroad terminus.

Honey Brook, borough (2006 population 1,439), CHESTER county, SE PENNSYLVANIA, 22 mi/35 km E of LANCASTER, near West Branch Brandywine Creek (source 1 mi/1.6 km to N). Manufacturing (apparel, food processing, metal fabricated products, truck trailers); rock quarries. Agricultural area (apples, grain; livestock, poultry, dairying). Struble Lake reservoir to NE.

Honey Grove, town (2006 population 1,838), FANNIN county, NE TEXAS, 21 mi/34 km WSW of PARIS; 33°34′N 95°54′W. Elevation 668 ft/204 m. In cotton, sorghum, soybeans; cattle; manufacturing (fertilizers, burial vaults). Nearby is large game and reforestation preserve; Caddo National Grassland to N. Inc. 1872.

Honey Island, unincorporated village, HARDIN county, SE TEXAS, 30 mi/48 km NW of BEAUMONT; 30°23′N 94°26′W. In agricultural area. Units of BIG THICKET NATIONAL PRESERVE to S and N.

Honey Lake, LASSEN county, NE CALIFORNIA, at E base of the SIERRA NEVADA, near NEVADA state line; 5 mi/8 km–10 mi/16 km wide. Elevation 3,949 ft/1,204 m. Intermittently dry bed. Receives SUSAN RIVER from NW, and Honey Valley Creek from SE. Its valley has irrigated agriculture. Honey Lake State Wildlife Area on N shore; Plumas National Forest to SW; Sierra Army Depot to SE.

Honeyville, town (2006 population 1,316), BOX ELDER county, N UTAH, 10 mi/16 km NNW of BRIGHAM CITY; 41°37′N 112°04′W. Fruit, sugar beets, barley, wheat, alfalfa; dairying; cattle, sheep. WASATCH RANGE and National Forest, including Wellsville Mountain Wilderness Area, to E. Crystal Hot Springs here. Elevation 4,268 ft/1,301 m. Settled 1862.

Honfleur (on-fluhr), town (□ 4 sq mi/10.4 sq km; 2004 population 8,139), CALVADOS department, BASSE-NORMANDIE region, NW FRANCE, historic port, on S shore of SEINE RIVER estuary on ENGLISH CHANNEL, 7 mi/11.3 km SE of Le HAVRE; 49°25′N 00°14′E. Production of audio-visual materials. Exports cider, dairy produce, and fruits (chiefly to England). A flotilla of fishing vessels supplies PARIS with crustaceans and ocean fishes. Boatbuilding. Has 15th-century wooden church, old wooden houses, museum of ethnography and Norman art, and a picturesque small 17th-century harbor from which navigators of 16th–17th century set sail to explore the Atlantic coast of N America. Jacques Cartier and Samuel de Champlain explored Canadian territory from here. The new Normandy bridge links Honfleur with Le Havre across the wide Seine estuary. Painters of the Impressionist school frequented Honfleur.

Honfleur (on-FLUR), village (□ 20 sq mi/52 sq km; 2006 population 866), Chaudière-APPALACHES region, S QUEBEC, E CANADA, 4 mi/7 km from SAINT-ANSELME; 46°39′N 70°53′W.

Høng (HUNG), town, VESTSJÆLLAND county, W SJÆLLAND, DENMARK, 22 mi/35 km SW of HOLBÆK; 55°30′N 11°20′E. Machinery manufacturing.

Honga, MARYLAND: see HOOPER ISLANDS.

Hongal, INDIA: see BAIL HONGAL.

Hong'an (HUNG-AN), town, ⊙ Hong'an county, E HUBEI province, central CHINA, near HENAN border, 60 mi/97 km NNE of WUHAN; 31°18′N 114°33′E. Rice, oilseeds, tobacco, cotton; tobacco processing. Also spelled Huang'an.

Honga River (HON-ga), E MARYLAND, wide arm of CHESAPEAKE BAY, between HOOPER ISLANDS (W) and the EASTERN SHORE (E), in DORCHESTER county; c.15 mi/24 km long.

Hongawa (HONG-gah-wah), village, Tosa district, KOCHI prefecture, S SHIKOKU, W JAPAN, 16 mi/25 km N of KOCHI; 33°43′N 133°18′E.

Hongay, VIETNAM: see HONG GAI.

Hongchon (HONG-CHUHN), county (□ 690 sq mi/1,794 sq km), W central KANGWON province, SOUTH KOREA, bordering CHUNCHON and INJE on N, YANGYANG and MYONGJU on E, PYONGCHANG and HOENGSONG on S, and KYONGGI province on W. Mountainous area. Hongchon River, a tributary of Pukhan River, in center of county. Some agriculture (tobacco, hop, pine nuts); sericulture; animal husbandry; logging and charcoal production. Suta Buddhist Temple, Yongso Valley.

Hong Don (HONG DON) or **phuoc long,** town, SOC TRANG province, S VIETNAM, 10 mi/16 km N of BAC LIEU; 09°26′N 105°44′E. Rice. Formerly PHULOC.

Hongdong, CHINA: see HONGTONG.

Hong Gai (HONG GEI), city, ⊙ QUANG NINH province, N VIETNAM, on HA LONG BAY, 80 mi/129 km E of HANOI; 20°57′N 107°05′E. Extensive open-cast anthracite mines, coal distributing and exporting center; coke-, briquette-, and coal-treating plant. Transportation and market center; light manufacturing. Fisheries; tourism. Called Port Courbet in the French period, it was sometimes the scene of labor unrest when Vietnamese miners protested intolerable working conditions under their colonial overseers. Formerly HONGAY.

Honghai Bay (HUNG-HEI), shallow inlet of SOUTH CHINA SEA, SE GUANGDONG province, S CHINA, 10 mi/16 km SW of HAIFENG; 25 mi/40 km wide, 10 mi/16 km long. Exposed to winds. Port of SHANWEI on E shore.

Honghu (HUNG-HU), city (□ 976 sq mi/2,528 sq km; 1994 estimated urban population 202,500; 1994 estimated total population 856,600), SE HUBEI province, central CHINA, on CHANG JIANG (Yangzi River), and 7 mi/11.3 km E of lake Hong Hu, on HUNAN border, and 50 mi/80 km NE of YUEYANG; 29°51′N 113°27′E. Agriculture and light industry are the largest sectors of the city's economy. Crop growing, fishing, and animal husbandry. Grain, oil crops, cotton, vegetables, hogs, eggs; manufacturing (food processing, textiles, apparel).

Hongjiang (HUNG-JIANG), city (□ 83 sq mi/215.8 sq km; 2000 population 89,451), SW HUNAN province, S central CHINA; port on the upper YUAN RIVER; 27°08′N 109°51′E. Light industry is the largest source of income. Crop growing, animal husbandry. Main industries include textiles and paper.

Hong Kong, Mandarin *Xiang-gang* [Mandarin= fragrant harbor], city and special administrative zone (□ 399 sq mi/1,037.4 sq km; 2007 population 6,980,000), S CHINA, adjacent to Guangdong province, on the estuary of the PEARL RIVER, 40 mi/64 km E of MACAO, and 90 mi/145 km SE of GUANGZHOU. The zone comprises Hong Kong island (□ 29 sq mi/75 sq km), ceded by China in 1842 under the Treaty of Nanjing; Kowloon (Mandarin *Jiulong*) peninsula (□ 3.5 sq mi/9 sq km), ceded (with STONECUTTERS ISLAND) in 1860 under the Beijing Convention; and the New Territories (□ 366 sq mi/948 sq km), a mountainous mainland area adjoining Kowloon, which (with DEEP BAY on the W and Dapeng Bay on the E and some 235 offshore islands), was leased to Britain from China (under British pressure) in 1898 for ninety-nine years. China regained sovereignty of Hong Kong on July 1, 1997.

Located in a subtropical zone, Hong Kong has a foggy spring, a hot and rainy summer with typhoons, a dry and warm autumn, and a cool and dry winter. The traditional core built-up area, officially named Victoria but commonly called Hong Kong, is on the NW shore of Hong Kong island, at the foot of Victoria Peak (elevation 1,805 ft/550 m), the center of an extensively quarried granite range covering much of the island. Hong Kong has many natural harbors, that of Victoria (□ c.17 sq mi/44 sq km) being among the world's finest. In the past century, part of Victoria Harbor has been reclaimed from the sea, mainly on the Kowloon side. Additional reclamation is planned to create space for highways, buildings, and parks. The colony grew around this sheltered deep water port, and an estimated 75% of the population is concentrated here and on the Kowloon peninsula across the harbor. Hong Kong is a free port, a bustling trade center, and a shipping and banking emporium—a principal trading and transshipment center of E Asia and the PACIFIC RIM (half of U.S.-China trade passes through it). Shipping continues to be a major

function, and Hong Kong is often cited as the world's largest container port.

After 1950, when much of its trade with the People's Republic of China was halted because of UN and U.S. embargoes, Hong Kong began to industrialize. Overcoming such handicaps as a scarcity of minerals, power sources, usable land, and fresh water, and utilizing its abundant supply of cheap labor, Hong Kong became a leading light-manufacturing center. Now, as a result of high per-capita GDP (c.$38,700 in 2005), and high wages and living costs, much of Hong Kong's industry has migrated to nearby mainland China, especially SHENZHEN. It is also a leading financial center and one of the largest gold markets in the world. The textile and garment industry, despite recent declines, remains important, and manufacturing of plastics and electrical and electronic equipment has grown in recent years. Other industries include the manufacturing of appliances, metal products, printed materials, rubber products, chemicals, ceramics, furniture, jewelry, industrial machinery, and toys. Manufacturing in general is in decline in Hong Kong because of the high cost of labor and the transfer of many industries to adjacent GUANGDONG province, where labor costs are cheaper. Tourism is a significant source of revenue which continues to grow; other major sources of income are motion-picture production, insurance, publishing, and overseas capital investments.

Because of the mountainous and rocky terrain, only about 8% of the land is arable; farming is carried on principally in the New Territories; the Yuanlong valley has the best farmland. Rice and a variety of vegetables, including cabbage, eggplant, maize, red pepper, leek, and watercress are grown. The yield cannot meet the needs of the overcrowded city, and much of its food comes from the mainland. Fishing is a common occupation (though both farming and fishing are becoming less important economically as Hong Kong continues to industrialize); Deep Bay is known for its oyster beds. About 98% of the population is Chinese, most of whom speak Cantonese. Buddhism is widely practiced, and Buddhist temples are widespread. Hong Kong has small British and American communities. The University of Hong Kong (1912) is a coeducational institution under government control, organized on the model of British universities. The merger of three existing colleges in 1963 created Chinese University. There are also now more than a dozen institutions of higher education, including a new science and technical university.

The region of Hong Kong, which had long been barren, rocky, and sparsely settled—its many islands and inlets a haven for coastal pirates—was occupied by the British during the Opium War (1839–1842). The area prospered as an E-W trading center, the commercial gateway to, and distribution center for, S China. It was efficiently governed, and its banking, insurance, and shipping services quickly became known as the most reliable in SE Asia. The British agreed to limit the fortifications of Hong Kong in 1921, and this decision contributed to Japan's easy conquest (December 25, 1941) of the area. Hong Kong was reoccupied by the British on September 16, 1945. Since 1949, when the Communists took control of mainland China, thousands of refugees have crossed the border, making Hong Kong's urban areas some of the most densely populated in the world. Immigration from China has been stimulated by the 1997 handover, and the territory's population continues to grow.

The city's long-standing water problem has been eased by the construction of an elaborate system of giant reservoirs, as well as by the piping in of a considerable amount of water from mainland China. Serious problems of housing, health, drug addiction, and crime have been the target of aggressive government programs. Increased public housing, preventive medicine programs, and limits on immigration have helped to moderate the impact of these social problems. In May 1967, Hong Kong was struck by a wave of riots and strikes inspired by China's Cultural Revolution. The government reacted firmly and, although the Chinese retaliated by briefly stopping the piping of water and by attacking British representatives in BEIJING, relations between Hong Kong and mainland China soon resumed the surface harmony that has existed since about 1959. For the Chinese government, Hong Kong remains a major source of foreign exchange and an important commercial link with the outside world. After several years of negotiations, Britain and the People's Republic of China agreed in a Joint Declaration on December 19, 1984, that Hong Kong (comprising Hong Kong island, Kowloon, and the New Territories) would become a special administrative region of China in July 1997, when Britain's lease expired. Declaring a policy of "One Country, Two Systems," China agreed to give Hong Kong considerable autonomy, allowing its existing social and economic systems to remain unchanged for a period of fifty years.

Hong Kong's railroad link with the mainland is by the Kowloon-Guangzhou railroad. Kowloon is connected with Hong Kong island, 1 mi/1.6 km away, by ferry, subway, and a vehicular tunnel (opened 1972). The Tsing Ma double-decker suspension bridge, one of the longest in the world, connects Lantau Island to the New Territories, crossing Tsing Yi and Ma Wan islands. Hong Kong has shipping connections with all major world ports and is an international air hub; the airport at Kai Tak (opened 1958) closed after the new CHEK LAP KOK international airport (E of Kowloon, on Lantau Island, W of Victoria) opened in 1998. The world's largest public infrastructure project, involving railroads, bridges, highways, a tunnel, land reclamation, and a new town, is underway at a cost of $21 billion.

Hongkou (HUNG-KO), N industrial district of SHANGHAI, E CHINA, on HUANGPU RIVER, in former U.S. concession of International Settlement. Machinery, textiles and clothing, chemicals, rubber products, eletronics, pharmaceuticals; non-ferrous metal smelting.

Hongo (HONG-go), town, Toyota district, HIROSHIMA prefecture, SW HONSHU, W JAPAN, 31 mi/50 km E of HIROSHIMA; 34°24′N 132°59′E.

Hongo (HONG-go), village, Kuga district, YAMAGUCHI prefecture, SW HONSHU, W JAPAN, 34 mi/55 km E of YAMAGUCHI; 34°17′N 132°02′E.

Hongqiao (HUNG-CHIOU), town, E CHINA, 11 mi/18 km SW of SHANGHAI. Major airport of Shanghai.

Hongshui River (HUNG-SHAI), 900 mi/1,448 km long, main left headstream of West River (Xi River), S CHINA; rises in GUIZHOU-YUNNAN border region, E edge of the YUNNAN AND GUIZHOU PLATEAU in two branches (Beipan and Nanpan rivers) which join at Guizhou-GUANGXI ZHUANG border; traverses Guangxi Zhuang in deep sandstone gorges, obstructed by rapids, and joins Yu River at Guiping to form West River.

Hongsong (HONG-SUHNG), town, SOUTH CHUNG-CHONG province, SOUTH KOREA, 40 mi/64 km N of KUNSAN. Rice, soybeans, cotton, tobacco; lumber.

Hongtong (HUNG-TUNG), town, ⊙ Hongtong county, S SHANXI province, NE CHINA, on FEN RIVER, on railroad, and 15 mi/24 km NNE of LINFEN; 36°16′N 111°41′E. Grain, cotton. Also appears as Hongdong.

Hongu (HONG-goo), resort town, East Muro district, WAKAYAMA prefecture, S HONSHU, W central JAPAN, on S KII PENINSULA, on KUMANO RIVER, 45 mi/72 km S of WAKAYAMA; 33°50′N 135°46′E.

Hongwon, county, SOUTH HAMGYONG province, NORTH KOREA, on E KOREA BAY, 22 mi/35 km NE of HUNG-NAM. Fishing center; chemicals.

Hongya (HUNG-YAH), town, ⊙ Hongya county, central SICHUAN province, SW CHINA, 30 mi/48 km NW of LESHAN; 29°56′N 103°25′E. Rice, oilseeds, medicinal herbs.

Hongze (HUNG-ZUH), lake (□ 757 sq mi/1,968.2 sq km), E CHINA, on border of ANHUI and JIANGSU provinces. It receives the HUAI RIVER and is connected with the GRAND CANAL and the Gaoyu Lake. The Sanhe dam, with the largest hydraulic works along the Huai River, is at the outlet of the lake. Main towns on shore include Shuanggou (W), Hongze (E), Xuyi (S), and Jiangba (SE). Agricultural center; rice, oilseeds, cotton, tobacco; fisheries. The name sometimes appears as Hungtse.

Honiara (ho-nee-AH-rah), town (1999 population 49,107), GUADALCANAL, SOLOMON ISLANDS, SW PACIFIC, 1,600 mi/2,575 km NE of SYDNEY and 1,000 mi/1,609 km E of PORT MORESBY. Was rebuilt to replace TULAGHI as the capital of the Solomon Islands at the end of World War II, and occupies the site of an important American campaign against the Japanese. Rebuilt from the 1950s with an expanded harbor, reconstructed Government House, a National Museum, and educational institutions.

Honington (HUHN-ing-tuhn), village (2001 population 1,247), SUFFOLK, E ENGLAND, 8 mi/12.9 km NW of BURY SAINT EDMUNDS; 52°59′N 00°36′W. Norman church.

Honiton (HUH-ni-tuhn), town (2001 population 11,213), E DEVON, ENGLAND, on OTTER RIVER and 15 mi/24 km ENE of EXETER; 50°48′N 03°12′W. Agricultural market in farming and dairying region. Town was previously well-known for its lace-manufacturing center. The 14th-century St. Margaret's almshouse was formerly a leper hospital. Parish church founded 1482.

Honjo (HON-jo), city, Akita prefecture, N HONSHU, NE JAPAN, 22 mi/35 km N of AKITA city; 39°22′N 140°03′E. Rice; traditional handicrafts.

Honjo (HON-jo), city, SAITAMA prefecture, E central HONSHU, E central JAPAN, on the TONE RIVER, 37 mi/60 km N of URAWA; 36°14′N 139°11′E. Communication equipment. Spring onions, cucumbers.

Honjo (HON-jo), village, East Chikuma district, NAGANO prefecture, central HONSHU, central JAPAN, 19 mi/30 km S of NAGANO; 36°23′N 138°00′E.

Honjo, village, South Amabe district, OITA prefecture, E KYUSHU, SW JAPAN, 22 mi/35 km S of OITA; 32°56′N 131°47′E. Ayu; distilled alcoholic beverage (*shochu*), miso. Known for "kombu roll" (fish wrapped in nori). Onagara limestone cave nearby.

Honkawane (hon-KAH-wah-ne), town, Haibara district, SHIZUOKA prefecture, central HONSHU, E central JAPAN, 16 mi/25 km N of SHIZUOKA; 35°06′N 138°08′E. Tea, wasabi, shiitake mushrooms; traditional confections. Wood and bamboo products. Hot springs and Settsu Gorge nearby.

Honley (HAHN-lee), town (2001 population 5,897), WEST YORKSHIRE, N ENGLAND, 3 mi/4.8 km S of HUDDERSFIELD; 53°36′N 01°48′W. Light industry.

Honnali (ho-NAH-lee), town, SHIMOGA district, KARNATAKA state, INDIA, on TUNGABHADRA RIVER and 21 mi/34 km N of SHIMOGA; 14°15′N 75°40′E. Rice milling; handicraft wickerwork. Granite quarrying nearby.

Honnedaga Lake, HERKIMER county, N central NEW YORK, in the ADIRONDACK MOUNTAINS, 35 mi/56 km NE of UTICA; c.4 mi/6.4 km long; 43°31′N 74°49′W. Recreational fishing.

Honnef (HOHN-nef) or **Bad Honnef**, town, NORTH RHINE-WESTPHALIA, W GERMANY, on right bank of the RHINE (landing), and 9 mi/14.5 km SE of BONN; 50°39′N 07°12′E. Resort with alkaline salt spring. Manufacturing of machinery, chemicals, and pharmaceuticals; wine. Has late-Gothic church.

Honnelles (aw-NEL), commune (2006 population 4,982), Mons district, HAINAUT province, SW BELGIUM, 13 mi/21 km SW of MONS, near French border.

Hönningen (HUHN-ning-guhn) or **Bad Hönningen**, village, RHINELAND-PALATINATE, W GERMANY, on right bank of the RHINE RIVER, 9 mi/14.5 km NW of NEUWIED; 50°33′N 07°17′E. Resort with thermal springs; manufacturing of chemicals; wine.

Honningsvåg (HAWN-nings-vawg), village, on SE MAGERØY, FINNMARK county, N NORWAY, on PORSANGEN Fjord of BARENTS SEA, 55 mi/89 km ENE of HAMMERFEST. Fishing center.

Hönö (HUHN-UH), town (□ 2 sq mi/5.2 sq km), GÖTEBORG OCH BOHUS county, SW SWEDEN, on island (□ 2 sq mi/5.2 sq km) of Hönö in KATTEGAT strait, 10 mi/16 km W of GÖTEBORG; 57°41′N 11°39′E.

Honokaa (HO-no-KAH-ah), town (2000 population 2,233), NE HAWAII island, HAWAII county, HAWAII, 35 mi/56 km NW of HILO, centered 1 mi/1.6 km from Hamakua Coast, on elevated coast; 20°04′N 155°28′W. Macadamia nuts; Macadamia nuts factory to N, near coast; manufacturing; rodeo. Parts of Hamakua Forest Reserve to W, SW, and SE; Kalopa State Recreational Area to S.

Honokahua (HO-no-kah-HOO-ah), village, W MAUI, MAUI county, HAWAII, 14 mi/23 km NW of KAHULUI, and 8 mi/12.9 km N of LAHAINA, at mouth of Honokahua Stream. Cattle; resorts. Honolua Bay Marine Conservation District to NE; D. T. Fleming Beach Park is here; Honokahua Bay (NE); Oneloa Bay (NW); West Maui Forest Reserve to SE.

Honokowai (HO-no-ko-WEI), town, MAUI island, MAUI county, HAWAII, 14 mi/23 km WNW of KAHULUI, 5 mi/8 km N of LAHAINA. Pineapples. Honokowai Beach Park here; Kaanapali Airport to S; gateway to W Maui resort areas.

Honolulu, county (□ 2,126 sq mi/5,506 sq km; 1990 population 836,231; 2000 population 876,156), central HAWAII; ⊙ HONOLULU, 23°34′N 164°41′W. Most populous county in Hawaii; includes OAHU island and all NW Hawaiian islands W of KAULA and NIIHAU, except for MIDWAY, uninhabited; Midway administered by U.S. Navy. Honolulu city is on S coast of Oahu island. Locally administered by 7 districts.

Honolulu (HO-no-LOO-loo), city (2000 population 371,657), ⊙ the state of HAWAII and ⊙ HONOLULU county, on the S coast of the island of OAHU; 21°19′N 157°48′W. With cruise ship and air connections to the U.S. mainland, ASIA, AUSTRALIA, and NEW ZEALAND, Honolulu is the crossroads of the PACIFIC, as well as the economic center and principal port of the Hawaiian Islands. The city is famous for its beauty and the variety of its ethnic groups. It lies on a narrow plain between the sea and the KOOLAU RANGE and climbs the slopes of PUNCHBOWL. Bypassed by Captain James Cook when he explored the islands in 1778, Honolulu's harbor was entered and praised in 1794 by William Brown, an English captain. Honolulu's history from 1820, when missionaries arrived on the isles, is much the same as that of Hawaii. Growing from a settlement of thatched grass huts into the main residence of Hawaiian royalty and later of foreign consuls, Honolulu became the permanent capital of the kingdom of Hawaii in 1845. In the 19th century, American and European whalers and sandalwood traders visited its port. It remained Hawaii's capital when the islands were annexed by the UNITED STATES in 1898 and achieved statehood in 1959. The Japanese bombed PEARL HARBOR, the naval base W of Honolulu, on December 7, 1941, and during World War II the port became a strategic naval base and a staging area for U.S. forces in the Pacific. Since the war, a rise in tourism, diversification of industry, and construction of luxury hotels and housing developments have made Honolulu the business and population center of Hawaii. Sugar processing and pineapple canning are no longer Honolulu's major industries. Increased peacetime defense activity at the many military installations in the area (Pearl Harbor Naval Shipyard, Schofield Barracks, Fort Shafter, Camp H. M. Smith, Hickam Field), expansion of harbor facilities, and the completion of an international airport further aided the city's growth. Honolulu Harbor near downtown; Pearl Harbor 5 mi/8 km NW of downtown. Manufacturing (jewelry, printing and publishing, apparel, food and beverages, rubber products, construction materials, consumer goods, electronics and computer equipment, machinery, metal products). The largest of Honolulu's parks is Kapiolani, containing a ZOO, an aquarium, and Waikiki Shell, where the Honolulu Symphony gives concerts. Also in Honolulu is the USS ARIZONA MEMORIAL for the 1,100 who died during the bombing of Pearl Harbor. Notable institutions are the University of Hawaii at Manoa; Kapiolani Community College; Honolulu Community College; Chaminade University; Hawaii Pacific University; Hawaiian Baptitst Academy; the Bishop Museum, noted for its studies of POLYNESIA; the Honolulu Academy of Arts; and Kawaiahao Church (1841), where funerals for Hawaiian monarchs and nobility were held. Iolani Palace, the former home of Hawaii's kings, is the only royal palace in the U.S. Neal Blaisdell Concert Hall and Convention Center; Foreign Trade Zone; Quarantine Station at Honolulu Harbor; Ala Moana Center, one of the largest shopping centers in the U.S.; WAIKIKI Beach, especially noted for bathing and surfing, and famous DIAMOND HEAD crater are both in E part of the city; National Cemetery of the Pacific (at Punchbowl Crater) N of Downtown; Honolulu Watershed Forest Reserve to NE; Kewalo Basin State Park on waterfront.

Honomu (HO-no-MOO), village (2000 population 541), E HAWAII island, HAWAII county, HAWAII, on the Hamakua Coast (NE) at Kohola Point, 10 mi/16 km N of HILO, near mouth of Kolekole Stream; 19°52′N 155°06′W. Hilo Forest Reserve and Akaka Falls State Park to W; Kolekole Beach Park to N.

Honor, village (2000 population 299), BENZIE county, NW MICHIGAN, 4 mi/6.4 km NE of BEULAH, on Platte River, SE of PLATTE LAKE; 44°40′N 86°01′W.

Honora, hamlet, ONTARIO, E central CANADA, included in the town of NORTHEASTERN MANITOULIN and The Islands; 45°54′N 82°06′W.

Honrubia (on-ROO-vyah), town, CUENCA province, E central SPAIN, 32 mi/51 km SSW of CUENCA; 39°37′N 02°16′W. Grain-growing center.

Honshu (HON-shoo), island (□ 89,000 sq mi/231,400 sq km), central Japan; c.800 mi/1,287 km long, c.30 mi/48 km–150 mi/241 km wide. Largest and most important island of Japan, including most of its area and population. Separated from Hokkaido (N) by the Tsugaru Strait, from Kyushu (SW) by Kanmon Strait, and from Shikoku (S) by the Inland Sea. Predominantly mountainous, rising to 12,389 ft/3,776 m at Fuji-san (the highest peak of Japan), and has many volcanoes. Has valuable forest, but a limited amount of arable land. Oil, zinc, and copper are found here. The Shinano, the longest river of Japan, traverses central Honshu. Most of the rivers of the island are short and swift, feeding many small hydroelectric plants. Earthquakes are common, especially around Fuji-san. The climate has a wide range from the N with its snowywinters to the sub tropical S. Agr. is varied; rice, other grains, fruits, and vegetables are grown. The bulk of Japan's tea and silk comes from Honshu. The population is concentrated in lowland areas. Most important of these is the Kanto Plain (c.5,000 sq mi/12,950 sq km) in the central part of the island; it contains the Tokyo-Yokohama industrial belt. Other large industrial regions include Osaka-Kobe (in the Kinkidistrict), and Nagoya (on the Nobi Plain). Most of Japan's great ports are on the Pacific side of Honshu. Kyoto, formerly the capital of Japan, is an ancient seat of culture and also the chief handicraft center of Honshu. Electronics, metallurgical,

chemical, and textile industries are very important on the island, although the larger cities have diverse industries. Politically the island is divided into 34 prefectures. Japan has steadily increased the number of bridges and tunnels connecting Honshu with the other islands Three new bridge systems have been built across the Inland Sea between Honshu and Shikoku, and the Seikan Tunnel (completed 1988) now connects Honshu with Hokkaido.

Hontalbilla (on-tahl-VEE-lyah), town, SEGOVIA province, central SPAIN, 28 mi/45 km N of SEGOVIA; 41°21′N 04°07′W. Cereals, grapes, chicory, resins.

Hontanaya (on-tah-NEI-ah), town, CUENCA province, E central SPAIN, 45 mi/72 km ESE of ARANJUEZ; 39°44′N 02°50′W. Cereals, olives, anise, vegetables, timber, livestock; apiculture. Limestone quarrying.

Honto, RUSSIA: see NEVEL'SK.

Hontoria del Pinar (on-TO-ryah dhel pee-NAHR), town, BURGOS province, N SPAIN, on BURGOS-SORIA railroad, on highway, and 37 mi/60 km W of Soria; 41°51′N 03°10′W. Cereals, vegetables, livestock; timber, naval stores. Flour milling, sawmilling. Iron and coal deposits.

Honyabakei (HON-yah-BAH-kah-ee), town, Shimoge district, OITA prefecture, E KYUSHU, SW JAPAN, 31 mi/50 km N of OITA; 33°29′N 131°10′E.

Hoo (HOO), village (2001 population 7,356), Medway, SE ENGLAND, on MEDWAY RIVER and 4 mi/6.4 km NNE of CHATHAM; 51°26′N 00°34′E. Previously known for its brickworks. Just S, in Medway River, is extensive salt marsh. Has church dating back to 8th-century.

Hood, county (□ 436 sq mi/1,133.6 sq km; 2006 population 49,238), N central TEXAS; ⊙ GRANBURY; 32°25′N 97°49′W. Drained by BRAZOS RIVER. Agricultural (mainly peanuts, hay, pecans); cattle. Oil and gas. Part of Squaw Creek Lake reservoir in S; Lake Granbury reservoir and Acton State Historic Site in E. Formed 1866.

Hood Bay, SE ALASKA, inlet of CHATHAM STRAIT, on W coast of ADMIRALTY island, just S of ANGOON; 13 mi/21 km long, 5 mi/8 km wide at mouth. Hood Bay or Killisnoo fishing village is on N shore.

Hood Canal, inlet, W WASHINGTON narrow arm extending SSW from entrance of PUGET SOUND channel hooks NE at S end, c.15 mi/24 km, coming within 1 mi/1.6 km of Case Inlet of Puget Sound, creating narrow isthmus of Kitsap Peninsula; on ADMIRALTY INLET c.50 mi/80 km along E side of OLYMPIC PENINSULA. Dabob Bay, arm in N part of inlet; highway bridge crosses inlet near its mouth.

Hood Island, GALÁPAGOS, ECUADOR: see ESPAÑOLA, ISLA.

Hood, Mount (11,235 ft/3,424 m), on Hood River and CLACKAMAS county lines, NW OREGON, in the CASCADE RANGE, E of PORTLAND. Highest point in the state and the center of Mount Hood National Forest and Mount Hood Wilderness Area. A symmetrical, cone-shaped, dormant volcano with glaciers and forested lower slopes, it rises high above the surrounding range and is a favorite mountain-climbing and skiing center. Timberline Lodge at S boundary. Several ski areas to S and SE.

Hood River, county (□ 533 sq mi/1,385.8 sq km; 2006 population 21,533), N OREGON; ⊙ HOOD RIVER; 45°31′N 121°38′W. MOUNT HOOD (highest point in state) is in CASCADE RANGE on SW boundary. COLUMBIA RIVER (Bonneville Reservoir) forms N boundary. Agriculture (apples, cherries, pears, grapes, peaches; cattle); timber; wineries. Recreation. Mount Hood National Forest occupies most of county, in W, S, and SE. Eight state parks in W, all on Columbia River; they include Viento, Wygant, and Starvation Creek State Parks. Parts of Columbia and Mount Hood Wilderness Areas in W and S. Ski area in S. Steady winds through Hood River Gorge make it the "Sailboard Capital of the World." Formed 1908.

Hood River, town (2006 population 6,673), ⊙ HOOD RIVER county, N OREGON, on COLUMBIA RIVER (Bonneville Reservoir) at mouth of Hood River, c.55 mi/89 km ENE of PORTLAND; 45°42′N 121°31′W. Elevation 54 ft/16 m. Highway bridge to WHITE SALMON, WASHINGTON 2 mi/3.2 km to NE. Trade, packing, shipping center for irrigated orchards (pears, apples, cherries, grapes) of Hood River Valley, stretching to base of MOUNT HOOD, 25 mi/40 km S, whose glaciers feed the river. Manufacturing (beer, bakery goods, lumber, millwork, concrete, utility substations). Wineries. Tourism. Annual international sailboarding events and music festival. Settled 1854, incorporated 1895.

Hoodsport, unincorporated town, MASON county, W WASHINGTON, 13 mi/21 km N of SHELTON, on HOOD CANAL inlet, N of the Great Bend of inlet. Fish hatchery. Winery. Junction of road to SE corner of OLYMPIC NATIONAL PARK (park is NW). Potlatch State Park and Skokomich Indian Reservation to S; Olympic National Forest and Lake Cushman reservoir to W.

Hoofddorp (HAWF-dawrp), town (2001 population 56,492), SOUTH HOLLAND province, W NETHERLANDS, 10 mi/16 km SW of AMSTERDAM, in HAARLEMMERMEER polder, surrounded by RINGVAART canal; 52°18′N 04°42′E. Dairying; cattle, poultry; flowers, nursery stock, vegetables, potatoes; manufacturing (beer, separation equipment, trenching machines, electronic controls, food processing).

Hooge (HOO-ge), NORTH SEA island (□ 3 sq mi/7.8 sq km) of N FRISIAN group, NW GERMANY, in HALLIG ISLANDS, 14 mi/23 km off SCHLESWIG-HOLSTEIN coast; 54°35′N 08°30′E. Grazing; tourism.

Hoogeveen, city (2001 population 37,156), DRENTHE province, NE central NETHERLANDS, 20 mi/32 km S of ASSEN, at juncture of Hoogevensevaart and Verlengde Hoogevensevaart canals; 52°43′N 06°29′E. Railroad junction, airport to E. Dairying; cattle, sheep; grain, vegetables; manufacturing (rubber items, motor vehicles, wrapping paper). National Park Het Dwingelderveld to N; recreational center to E.

Hoogezand-Sappemeer, city (2001 population 27,435), GRONINGEN province, N NETHERLANDS, on the WINSCHOTERDIEP canal and 9 mi/14.5 km ESE of GRONINGEN; 53°10′N 06°46′E. Dairying; cattle, sheep; fruit, grain, vegetables; manufacturing (tools, valves, binder's board).

Hoogkarspel (hawk-KAHR-spuhl), town, NORTH HOLLAND province, NW NETHERLANDS, 7 mi/11.3 km NE of HOORN; 52°42′N 05°13′E. At center of small peninsula between IJSSELMEER (NE) and MARKERMEER (SE). Dairy farming; sheep, cattle; flowers, flower bulbs, apples, vegetables.

Hoogkerk (HAWKH-kerk), suburb of GRONINGEN, GRONINGEN province, N NETHERLANDS, on the HOENDIEP canal and 3 mi/4.8 km W of city center; 53°13′N 06°30′E. Dairying; cattle, poultry; grain, vegetables.

Hooglede (HOKH-lai-duh), agricultural commune (2006 population 9,841), in Roeselare district, WEST FLANDERS province, W BELGIUM, 3 mi/4.8 km NW of ROESELARE; 50°59′N 03°05′E. Scene of Austrian defeat (1794) by French in French revolutionary wars.

Hoogstraten (HOKH-strah-tuhn), agricultural commune (2006 population 18,733), Turnhout district, ANTWERPEN province, N BELGIUM, near NETHERLANDS border, 20 mi/32 km NE of ANTWERP; 51°24′N 04°46′E. Has sixteenth-century Gothic, seventeenth-century baroque churches.

Hoogvliet (HAWKH-vleet), town (2001 population 36,529), SOUTH HOLLAND province, W NETHERLANDS, 8 mi/12.9 km SW of ROTTERDAM; 51°52′N 04°21′E. OUDE MAAS River to S and W (BOTLEK Tunnel and Bridge to W). Dairying; cattle; vegetables, potatoes, sugar beets, fruit. Oil refineries, on NEW MAAS RIVER, to N.

Hook (HUK), village (2001 population 7,378), NE HAMPSHIRE, S ENGLAND, 6 mi/9.7 km E of BASINGSTOKE; 51°16′N 00°57′W. Former agricultural market. Has late-Norman church.

Hooker, county (□ 721 sq mi/1,874.6 sq km; 2006 population 756), central NEBRASKA; ⊙ MULLEN; 41°53′N 101°08′W. Located in the Sand Hills region. Agricultural area drained by MIDDLE LOUP and DISMAL rivers. Cattle, hogs. Central/Mountain time zone boundary follows E and S boundaries. Formed 1889.

Hooker, town (2006 population 1,733), TEXAS county, central OKLAHOMA. Panhandle, on high plains, 20 mi/32 km NE of GUYMON; 36°51′N 101°12′W. Shipping and trading point in wheat-growing area; cattle; manufacturing (liquid hydro-carbons); natural gas. Incorporated 1907.

Hooker Island, in S FRANZ JOSEF LAND, ARCHANGEL oblast, extreme N European Russia, in the ARCTIC OCEAN; 20 mi/32 km long, 15 mi/24 km wide; 80°15′N 53°00′E. Rises to 1,483 ft/452 m. Government observation station on the TIKHAYA Bay of the NW coast. Discovered in 1880 by the British explorer Leigh Smith.

Hooker, Mount, WYOMING: see WIND RIVER RANGE.

Hookerton (HUK-uhr-tuhn), village (2006 population 490), GREENE county, E central NORTH CAROLINA, 11 mi/18 km N of KINSTON, on Contentnea Creek; 35°25′N 77°35′W. Manufacturing; in agricultural area (tobacco, cotton, grain; poultry, livestock).

Hook Head or **Hook Point**, Gaelic *Rinn Dúain*, cape, SW WEXFORD county, SE IRELAND, at entrance to WATERFORD HARBOUR, 12 mi/19 km SE of WATERFORD. Lighthouse (52°07′N 06°56′W).

Hook Mountain Park, New York: see NYACK.

Hook Norton (HUK NOR-tun), village (2001 population 2,495), W OXFORDSHIRE, S ENGLAND, 8 mi/12.9 km SW of BANBURY; 51°59′N 01°29′W. Brewery. Former site of quarrying. Church dates from Norman times.

Hook of Holland, Dutch *Hoek van Holland*, town, SOUTH HOLLAND province, SW NETHERLANDS, 10 mi/16 km SW of The HAGUE; 51°59′N 04°08′E. On NEW WATERWAY, its NORTH SEA entrance 2 mi/3.2 km to NW. Fishing; dairying; cattle; flowers, vegetables; manufacturing (resins). Minor port; resort area; car ferry to HARWICH (ENGLAND). Large EUROPOORT facility on S side of waterway, opposite town. Lighthouse.

Hook Point, IRELAND: see HOOK HEAD.

Hooks, town (2006 population 2,938), BOWIE county, NE TEXAS, 15 mi/24 km W of TEXARKANA; 33°28′N 94°16′W. In agricultural area (vegetables, rice; cotton; cattle; dairying). Oil and gas; manufacturing (jams and jellies). Incorporated after 1940.

Hooksett (HUK-suht), town, MERRIMACK county, S NEW HAMPSHIRE, at falls of the MERRIMACK, 6 mi/9.7 km N of MANCHESTER, on both sides of Merrimack River; 43°03′N 71°26′W. Manufacturing (machinery, electronic assembly, food products, construction materials; printing); agriculture (apples, vegetables, nursery crops; poultry, livestock; dairying). Part of Bear Brook State Park in NE. Includes village of SOUTH HOOKSETT. Incorporated 1822.

Hookstown, borough (2006 population 142), BEAVER county, W PENNSYLVANIA 5 mi/8 km ESE of EAST LIVERPOOL, OHIO, on Mill Creek. Manufacturing (metal fabricating); agriculture (corn, hay; dairying). Raccoon Creek State Park to S.

Hoolehua (HO-o-lai-HOO-ah), village, W central MOLOKAI, MAUI county, HAWAII, 6 mi/9.7 km NW of KAUNAKAKAI, 1 mi/1.6 km N of Maunaloa Highway, near N coast. Hoolehua Airport to SW.

Hoonah, village (2000 population 860), SE ALASKA, on N shore of CHICHAGOF ISLAND, on ICY STRAIT 40 mi/64 km WSW of JUNEAU; 58°06′N 135°25′W. Fishing, fish processing. Harbor is called PORT FREDERICK.

Hoopa, unincorporated village, HUMBOLDT county, NW CALIFORNIA, on TRINITY RIVER, 30 mi/48 km NE of EUREKA. Headquarters of Hoopa Indian Reservation in S part. Sheep, cattle, lambs; timber. Six Rivers National Forest to S, E and N.

Hooper, village (2000 population 123), ALAMOSA county, S COLORADO in SAN LUIS VALLEY, 20 mi/32 km N of ALAMOSA; 37°45′N 105°52′W. Elevation 7,553 ft/2,302 m. GREAT SAND DUNES NATIONAL MONUMENT to E; San Luis Lakes State Park to SE.

Hooper, village (2006 population 796), DODGE county, E NEBRASKA, 12 mi/19 km NNW of FREMONT, and on ELKHORN RIVER; 41°36′N 96°32′W. Grain, livestock, poultry and dairy products. Manufacturing (classic car parts).

Hooper, New York: see ENDWELL.

Hooper Bald, NORTH CAROLINA: see UNICOI MOUNTAINS.

Hooper Bay, Inuit village (2000 population 1,014), W ALASKA, on HOOPER BAY (15 mi/24 km long) of BERING SEA, 4 mi/6.4 km ESE of Point DALL; 61°33′N 165°48′W.

Hooper Islands, DORCHESTER county, E MARYLAND, three low marshy islands extending c.12 mi/19 km N-S in CHESAPEAKE BAY, N of HOOPER STRAIT; separated from mainland by HONGA RIVER. Bridged to mainland (N). Named after Henry Hooper, who owned much of the land on Upper Hooper Island. The four islands which form a chain 14 mi/23 km long were settled around 1660 by colonists from St. Mary's across the bay. A church built in 1872 stands on the site of St. Mary, Star of the Sea (R.C.). One of the greatest concentrated areas of migrating geese and ducks in America, iron bars are placed over larger windows to prevent wild fowl, attracted by lights, from breaking panes. Bridge connects N and middle islands. Fisheries (fish, crabs, oysters) seafood packing houses, vegetable canneries. Excellent sport fishing, duck and goose hunting. Villages are Honga and Fishing Creek on N Island; Hoopersville on middle island. Lower island is uninhabited hunting grounds. John Smith, when caught in a storm here in 1608, called the area "Limbo."

Hooper Strait, E MARYLAND, narrow channel of CHESAPEAKE BAY, Maryland, between BLOODSWORTH ISLAND (S) and DORCHESTER county shore and HOOPER ISLANDS (N).

Hoopersville, MARYLAND: see HOOPER ISLANDS.

Hoopeston (HUPE-stun), city (2000 population 5,965), VERMILION county, E ILLINOIS, 23 mi/37 km N of DANVILLE; 40°28′N 87°40′W. Canning center; manufacturing of packaging machinery, castings; agriculture (vegetables, livestock). Plotted 1871, incorporated 1877.

Hoople (HOOP-uhl), village (2006 population 264), WALSH CO., NE NORTH DAKOTA, 13 mi/21 km NW of GRAFTON and on N branch of Park River; 48°32′N 97°38′W. Founded in 1890 and incorporated in 1898. Named for Allan Hoople (1849–1923) who owned the townsite.

Hooppole, village (2000 population 162), HENRY county, NW ILLINOIS, 22 mi/35 km NNE of CAMBRIDGE; 41°31′N 89°54′W. In agricultural area.

Höör (HUH-uhr), town, SKÅNE county, S SWEDEN, 20 mi/32 km NE of LUND; 55°56′N 13°33′E.

Hoorn (HAWRN), city (2001 population 65,151), NORTH HOLLAND province, NW NETHERLANDS, on MARKERMEER, S section of IJSSELMEER, 20 mi/32 km NNE of AMSTERDAM; 52°39′N 05°04′E. Markerwaardijk barrier dam 9 mi/14.5 km to ENE; railroad junction. Dairying; cattle, sheep; flowers; apples, vegetables, potatoes; fishing; printing, textiles; manufacturing (binders, aluminum sheeting, processed food, boats). In the seventeenth century many Dutch explorers embarked from the city, such as Willem Schouten, who was the first to round (and who also named) Cape Hoorn (later Horn); A. J. Tasman, who discovered NEW ZEALAND and TASMANIA; and J. P. Coen, founder of Batavia (now Djakarta), INDONESIA. City founded 1311.

Hoorn, Cape, CHILE: see HORN, CAPE.

Hoorne Islands, FRENCH POLYNESIA: see HORNE ISLANDS.

Hoosac Range (HOO-zuhk), S continuation of the GREEN MOUNTAINS, NW MASSACHUSETTS and SW VERMONT, running from N to S. Rises to c.3,000 ft/910 m. The Hoosac railroad tunnel, c.5 mi/8 km long, built from 1852 to 1873, at the cost of nearly 200 lives, cuts beneath the range from E to W.

Hoosac Tunnel (HOO-zuhk), village in FLORIDA town, BERKSHIRE county, NW MASSACHUSETTS, 6 mi/9.7 km E of NORTH ADAMS, at E end of Hoosac Tunnel (c.25,000 ft/7,620 m long; completed 1873), which carries railroad under Hoosac Range.

Hoosick Falls, industrial village (□ 1 sq mi/2.6 sq km; 2006 population 3,306), RENSSELAER county, E NEW YORK, on HOOSIC RIVER (water power from falls), near VERMONT border, and 27 mi/43 km NE of ALBANY; 42°53′N 73°20′W. In agricultural area; light manufacturing. Bennington Battlefield Park is NE. Small lakes (resorts) are nearby. Hoosick Falls Historic District. Incorporated 1827.

Hoosic River (HOO-zuhk), c.70 mi/113 km long, MASSACHUSETTS and NEW YORK; rises in HOOSAC RANGE in NW MASSACHUSETTS; flows N, NW, and W, past ADAMS, NORTH ADAMS, and WILLIAMSTOWN, MASSACHUSETTS, across SW corner of VERMONT, and past HOOSICK FALLS, New York (water power), to the HUDSON RIVER c.14 mi/23 km above TROY.

Hoosier Pass (11,539 ft/3,517 m), central COLORADO, in ROCKY MOUNTAINS, on Park and SUMMIT county line. Crossed by State Highway 9. Crosses CONTINENTAL DIVIDE N-S. Arapaho National Forest (N); Pike National Forest (S). MOUNT LINCOLN is nearby.

Hoosier State, The: see INDIANA.

Hoover, city (2000 population 62,742), Jefferson and Shelby counties, N central Alabama, residential suburb on S of Birmingham; 33°45′N 86°49′W. Named for William H. Hoover. Inc. in 1967.

Hoover Dam (HOO-vuhr), on the COLORADO RIVER between CLARK county, S Nevada and MOHAVE county, NW ARIZONA; 726 ft/221 m high, 1,244 ft/379 m long; 36°01′N 114°46′W. One of the world's largest dams. Built 1931–1936 by the U.S. Bureau of Reclamation; named for President Herbert Hoover; known as Boulder Dam 1933–1947. A key unit on the Colorado River, the dam is a major supplier of hydroelectric power and provides for flood control, river regulation, and improved navigation. Impounds Lake Mead, the largest man-made reservoir in the U.S. at 32,501 cu yd/24,850 cu m (but does not rank among top twenty-five in world). Dam is focal point of LAKE MEAD NATIONAL RECREATION AREA, which extends N and E around lake, and S around LAKE MOHAVE (DAVIS DAM). BOULDER CITY, NEVADA, was built to house workers on the project.

Hoover Memorial Reservoir (□ 5 sq mi/13 sq km), FRANKLIN and DELAWARE counties, central OHIO, on Big Walnut Creek, 5 mi/8 km NE of COLUMBUS; 40°06′N 82°53′W. Maximum capacity 144,079 acre-ft. Formed by Hoover Dam (67 ft/20 m high), built for water supply; also used for recreation.

Hooverson Heights, unincorporated town (2000 population 2,909), BROOKE county, N WEST VIRGINIA, residential community 6 mi/9.7 km S of WEIRTON and 1 mi/1.6 km E of FOLLANSBEE, near OHIO River; 40°19′N 80°34′W.

Hooversville, borough (2006 population 725), SOMERSET county, SW PENNSYLVANIA, 10 mi/16 km S of JOHNSTOWN, on STONYCREEK RIVER. Agriculture includes corn, oats, potatoes; livestock; dairying. QUEMAHONING RESERVOIR to W. Incorporated 1896.

Hooverville, Pennsylvania: see SEWARD.

Hop (HAWP), village, HORDALAND county, SW NORWAY, 4 mi/6.4 km S of BERGEN. Knitting mill. Troldhaugen, the home of composer Edvard Grieg is here.

Cross-references are shown in SMALL CAPITALS. The pronunciation guide is shown on page xix. The sources of population figures are shown on page xvii.

Hopa, village, NE TURKEY, port on BLACK SEA, 27 mi/43 km NW of ARTVIN, 20 mi/32 km SSW of BATUMI, Georgia; 41°26′N 41°22′E. Manganese, lead deposits nearby.

Hopatcong (ho-PAT-kuhng), borough (2006 population 15,884), SUSSEX county, N NEW JERSEY, 10 mi/16 km SE of NEWTON and on LAKE HOPATCONG (c.7 mi/11.3 km long), in hilly region NW of DOVER; 40°57′N 74°39′W. Resorts include LANDING, MOUNT ARLINGTON, Lake Hopatcong.

Hopatcong, Lake, reservoir (□ 4 sq mi/10.4 sq km), MORRIS county, N NEW JERSEY, on Muscontcong River, 12 mi/19 km WNW of PARSIPPANY; 40°55′N 74°40′W. Maximum capacity 48,209 acre-ft. Formed by Lake Hopatcong Dam (17 ft/5 m high), built for flood control.

Hop Bottom, borough (2006 population 315), SUSQUEHANNA county, NE PENNSYLVANIA, 11 mi/18 km SSE of MONTROSE, on Martins Creek. Resort area. Agriculture (corn, hay; dairying).

Hope (HOP), canton (□ 27 sq mi/70.2 sq km; 2006 population 772), Gaspésie—Îles-de-la-MADELEINE region, E QUEBEC, E CANADA, 13 mi/21 km from BONAVENTURE; 48°10′N 65°10′W.

Hope (HOP), district municipality (□ 16 sq mi/41.6 sq km; 2001 population 6,184), SW BRITISH COLUMBIA, W CANADA, at the intersection of FRASER, Coquihalla rivers, in FRASER VALLEY regional district; 49°23′N 121°26′W. Coquihalla Provincial Park. Formed in 1992 from the town of HOPE and adjacent rural electoral areas.

Hope, city (2000 population 10,616), ☉ HEMPSTEAD county, SW ARKANSAS; 33°40′N 93°35′W. Railroad junction. Commercial center. Agriculture (watermelons); manufacturing (food products and processing, machinery, apparel, printing, bakery products). The city is the birthplace and boyhood home of President Bill Clinton. Hope Wildlife Management Area to N; Bais D'Arc Wildlife Management Area to SW.

Hope (HOP), former town (□ 102 sq mi/265.2 sq km; 2001 population 3,887), SW BRITISH COLUMBIA, W CANADA, on FRASER River, at mouth of Coquihalla River, 27 mi/43 km NE of CHILLIWACK; 49°23′N 121°26′W. In mining (gold, silver) and lumbering region. Joined adjacent rural electoral areas to form the district of HOPE in 1992.

Hope (HOP), former township (□ 102 sq mi/265.2 sq km; 2001 population 3,887), SE ONTARIO, E central CANADA, and included in municipality of PORT HOPE and Hope; 43°53′N 79°31′W.

Hope, town (2000 population 2,140), BARTHOLOMEW county, S central INDIANA, 11 mi/18 km NE of COLUMBUS; 39°18′N 85°46′W. In agricultural area. Manufacturing (meat processing, special machinery, plastics, lumber, steel fabricating).

Hope, town, KNOX county, S MAINE, 10 mi/16 km N of ROCKLAND; 44°15′N 69°11′W. In agricultural, resort region. Wood products.

Hope (HOP), agricultural village (2006 population 1,100), N DERBYSHIRE, central ENGLAND, just E of CASTLETON; 53°20′N 01°45′W. Cement works. Has 14th-century church and remains of 10th-century Saxon cross. Nearby, site of excavation of a Roman settlement.

Hope, village, S ALASKA, on N shore of KENAI PENINSULA, on TURNAGAIN ARM, 23 mi/37 km SSE of ANCHORAGE, on highway from Anchorage. Tourism. Scene (1896) of gold rush.

Hope, village (2000 population 79), BONNER county, N IDAHO, on E shore of PEND OREILLE LAKE, 14 mi/23 km E of SANDPOINT. Site of Thompson's Trading Post (1809) to SE; 48°15′N 116°19′W. Kaniksu National Forest to NE.

Hope, village (2000 population 372), DICKINSON county, central KANSAS, 17 mi/27 km SSE of ABILENE; 38°41′N 97°04′W. Railroad junction. In wheat, cattle, sheep, and poultry area; dairy products.

Hope, village and township, WARREN county, NW NEW JERSEY, on small Beaver Brook, near JENNY JUMP MOUNTAIN, and 8 mi/12.9 km NE of BELVIDERE; 40°54′N 74°58′W. Has 18th-century buildings, including stone mill built 1768 by Moravian settlers. Amusement park.

Hope, village (2006 population 107), EDDY county, SE NEW MEXICO, on Rio Peñasco and 57 mi/92 km NW of CARLSBAD; 32°49′N 104°44′W. Corn, alfalfa; cattle, sheep.

Hope, village (2006 population 265), STEELE CO., E NORTH DAKOTA, 15 mi/24 km SSE of FINLEY; 47°19′N 97°43′W. Dairy products, grain. Founded in 1881 and incorporated in 1890. Named for Hope A. Hubbard Steele, wife of E. H. Steele, namesake of the county.

Hope (HOP), village (2001 population 4,172), FLINTSHIRE, NE Wales, on ALYN RIVER, and 8 mi/12.9 km SW of FLINT; 53°06′N 03°02′W. Just SSW is village of Caergwrle. Formerly in CLWYD, abolished 1996.

Hope, agricultural station, SAINT ANDREW parish, SE JAMAICA, on HOPE RIVER, and 3 mi/4.8 km NE of KINGSTON; 18°02′N 76°45′W. Comprises Jamaica School of Agriculture, research laboratories, and botanical gardens. Also a tourist site. Was once part of large Hope sugar estate, which prospered in eighteenth century. Bought by government in 1913. Also called Hope Gardens.

Hope, Rhode Island: see SCITUATE.

Hope Bay, town, PORTLAND parish, NE JAMAICA, on the coast, on railroad, and 7 mi/11.3 km W of PORT ANTONIO; 18°12′N 76°34′W. In fruit-growing region (bananas, coconuts, cacao).

Hope Creek Nuclear Power Plant, NEW JERSEY: see SALEM county.

Hopedale, town, including Hopedale village, WORCESTER county, S MASSACHUSETTS, 17 mi/27 km SE of WORCESTER; 42°08′N 71°32′W. Once a Christian communist community (1841–c.1857); later developed as textile "company town." Textile loom factory now closed but was once the largest loom factory in U.S. Settled 1660, incorporated 1886.

Hopedale, village (□ 1 sq mi/2.6 sq km; 2006 population 530), E NEWFOUNDLAND AND LABRADOR, E CANADA, on the ATLANTIC; 55°28′N 60°12′W. Fishing port and seaplane anchorage.

Hopedale, village (2000 population 929), TAZEWELL county, central ILLINOIS, 14 mi/23 km SE of PEKIN; 40°25′N 89°25′W. In agricultural area.

Hopedale (HOP-dail), village (2006 population 987), HARRISON county, E OHIO, 6 mi/10 km NE of CADIZ; 40°19′N 80°54′W. In former coal-mining area.

Hopefield, town, WESTERN CAPE province, SOUTH AFRICA, on the South River, a tributary of GREAT BERG RIVER, 35 mi/56 km NW of MALMESBURY and 22 mi/35 km W of SALDANHA; 33°03′S 18°20′E. Elevation 213 ft/65 km. Grain-growing center. Road and railroad links.

Hope Gardens, JAMAICA: see HOPE.

Hopeh, CHINA: see HEBEI.

Hopei, CHINA: see HEBEI.

Hopelawn, NEW JERSEY: see Woodbridge.

Hopelchén (o-pel-CHEN), city and township, CAMPECHE, SE MEXICO, on YUCATÁN peninsula, 45 mi/72 km E of CAMPECHE; 19°46′N 89°50′W. Timber, sugar, chicle, fruit, henequen.

Hopeless Reach, AUSTRALIA: see SHARK BAY.

Hopeman (HOP-man), fishing village (2001 population 1,624), ABERDEENSHIRE, NE Scotland, on MORAY FIRTH, and 2 mi/3.2 km ENE of BURGHEAD; 57°42′N 03°25′W. Formerly in Grampian, abolished 1996.

Hope Mills (HOP MILZ), town (□ 6 sq mi/15.6 sq km; 2006 population 12,630), CUMBERLAND county, S central NORTH CAROLINA, suburb 7 mi/11.3 km SSW of FAYETTEVILLE, on Rockfish Creek; 34°58′N 78°57′W. FORT BRAGG Army base and POPE AIR FORCE BASE are adjacent. Service industries; manufacturing (yarn, embroidered fabric, gift items, machining); agriculture (cotton, tobacco, grain, peanuts, soybeans, sweet

potatoes; livestock; poultry). Fayetteville Regional Airport to E.

Hope, Mount, Rhode Island: see BRISTOL.

Hopen (HAW-puhn), island (□ 23 sq mi/59.8 sq km) of the Norwegian possession SVALBARD, in BARENTS SEA of ARCTIC OCEAN, SE of Spitsbergen group; 76°35′N 25°30′E. Island is 20 mi/32 km long (SW-NE), 1 mi/1.6 km wide; rises to 1,198 ft/365 m. Site of meteorological radio station.

Hope, Point, headland, NW ALASKA, on CHUKCHI SEA; 68°20′N 166°45′W. Site of Point Hope Inuit village. Whaling.

Hope River, c.15 mi/24 km long, SAINT ANDREW parish, SE JAMAICA; flows S, entering the CARIBBEAN SEA 5 mi/8 km ESE of KINGSTON; 18°02′N 76°44′W. Used for water supply of Kingston and suburbs.

Hopes Advance, Cape (HOPS ad-VANS), N QUEBEC, E CANADA, on HUDSON STRAIT, on W side of entrance of UNGAVA BAY; 61°02′N 69°30′W.

Hopetoun, town, VICTORIA, SE AUSTRALIA, 251 mi/404 km NW of MELBOURNE; 35°43′S 142°20′E. Wheat, barley, oats.

Hopetoun, small town and port, S WESTERN AUSTRALIA state, W AUSTRALIA, on INDIAN OCEAN, and 230 mi/370 km SSW of KALGOORLIE; 33°52′S 120°08′E. Head of railroad to RAVENSTHORPE; wheat; gold export; tourism.

Hope Town, town, N Bahamas, on cay just off E central GREAT ABACO ISLAND, 115 mi/185 km NNE of NASSAU; 26°32′N 76°58′W. Fishing and trading. Lighthouse.

Hopetown, town, NORTHERN CAPE province, SOUTH AFRICA, on ORANGE RIVER (bridge), on N12 highway to KIMBERLEY and 75 mi/121 km SW of Kimberley; 29°36′S 24°05′E. Elevation 3,822 ft/1,165 m. Diamond mining, stock raising, dairying. Salt pans in region. Airfield. First diamonds in South Africa discovered here 1867. Airfield.

Hope Town (HOP TOUN), village (□ 19 sq mi/49.4 sq km; 2006 population 330), Gaspésie—Îles-de-la-MADELEINE region, E QUEBEC, E CANADA, 15 mi/24 km from BONAVENTURE; 48°03′N 65°10′W.

Hope Valley, village (population 1,445) in HOPKINTON town, WASHINGTON county, SW RHODE ISLAND, on WOOD RIVER (bridged here) and 11 mi/18 km NNE of WESTERLY; 41°31′N 71°43′W. Abolitionist and teacher Prudence Crandall born here, 1803.

Hopeville, town, CLARKE county, S IOWA, 13 mi/21 km WSW of OSCEOLA.

Hopewell (HOP-wel), independent city (□ 11 sq mi/28.6 sq km; 2006 population 22,731), SE VIRGINIA, 18 mi/29 km SSE of RICHMOND, on JAMES RIVER, at mouth of APPOMATTOX RIVER; 37°17′N 77°17′W. Railroad junction, a deep-water port; manufacturing (chemicals, polyester fibers, printing and publishing, paperboard). Founded as a munitions center. In 1926, Hopewell annexed City Point, U.S. General Grant's base of operations in 1864–1865, now part of Petersburg National Battlefield. Historic Weston Plantation on Appomattox River. Benjamin Harrison Bridge (James River) to E. Fort Lee Military Reservation to W. Founded 1613, incorporated 1916.

Hopewell, town, HANOVER parish, NW JAMAICA, on coast, 6 mi/9.7 km W of MONTEGO BAY; 18°15′N 78°10′W. Rice, bananas, yams.

Hopewell (hop-wel) village, NE NOVA SCOTIA, CANADA, 8 mi/13 km SSW of NEW GLASGOW; 45°28′N 62°42′W. Elevation 180 ft/54 m. Dairying, farming.

Hopewell, village (2000 population 396), MARSHALL county, central ILLINOIS, 20 mi/32 km NNE of PEORIA, on ILLINOIS RIVER; 40°58′N 89°27′W. Corn, soybeans.

Hopewell, village (2000 population 1,815), BRADLEY county, SE TENNESSEE, 4 mi/6 km N of CLEVELAND; 35°14′N 84°53′W.

Hopewell, borough (2006 population 2,022), MERCER county, W NEW JERSEY, 11 mi/18 km N of TRENTON; 40°23′N 74°45′W. In agricultural region; manufacturing with headquarters for several important corporations nearby. Has Baptist Church (1748); monument (1865) to John Hart, who lived here; historical museum State Children's Home—the former Lindbergh estate deeded (1941) to state. Settled before 1700, incorporated 1891.

Hopewell, borough (2006 population 212), BEDFORD county, S PENNSYLVANIA, 15 mi/24 km NE of Bedford, on Raystown Branch of JUNIATA RIVER.

Hopewell Cape, village, ⊙ ALBERT county, SE NEW BRUNSWICK, CANADA, at head of SHEPODY BAY, at mouth of PETITCODIAC River, and 19 mi/31 km SE of MONCTON; 45°48′N 64°35′W. Tourism.

Hopewell Culture National Historical Park, prehistoric mounds of Hopewell Native Americans, ROSS county, S OHIO, just N of CHILLICOTHE; 39°22′N 83°00′W. Authorized 1923. Formerly called Mound City National Group Monument.

Hopewell Furnace National Historic Site, BERKS and CHESTER counties, SE PENNSYLVANIA, 6 mi/9.7 km SW of POTTSTOWN; 40°12′N 75°46′W. Iron-making site from 19th century with reconstructed blast furnace, ironmaster's mansion, auxilliary structures. FRENCH CREEK State Park adjoins it to NW. Authorized 1938.

Hopin (HO-pin), village, MOHNYIN township, KACHIN STATE, MYANMAR, on railroad and 65 mi/105 km SW of MYITKYINA.

Hôpital, L' (lo-pee-TAHL), German *Spittel*, town (□ 1 sq mi/2.6 sq km), MOSELLE department, LORRAINE region, NE FRANCE, on German border, 8 mi/13 km WSW of FORBACH, and just E of CARLING. In coalmining district. A large nearby coal-fired power plant (capacity 1200 MW) added (1991) a fluidized bed facility, among the largest in Europe.

Hopkins, county (□ 554 sq mi/1,440.4 sq km; 2006 population 46,830), W KENTUCKY, ⊙ MADISONVILLE; 37°18′N 87°32′W. Bounded E by POND RIVER, W by TRADEWATER RIVER. Rolling agricultural area (burley and dark tobacco, hay, alfalfa, soybeans, wheat, corn; hogs, cattle); important bituminous coal mines; oil wells, hardwood timber; some manufacturing. Tradewater Wildlife Management Area in SW, White City Wildlife Management Area in E. Formed 1806.

Hopkins, county (□ 792 sq mi/2,059.2 sq km; 2006 population 33,496), NE TEXAS; ⊙ SULPHUR SPRINGS. Bounded N by South Fork of SULPHUR RIVER (forms COOPER LAKE in NW); drained by WHITE OAK and Lake Fork creeks; 33°08′N 95°33′W. Prairies in W; hilly in E. Dairying (a leading Texas county); agriculture (hay, silage; wheat, corn, rice, soybeans); cattle; timber. Oil and natural gas wells; clay mining; lignite. Milk processing. Manufacturing, processing of farm products at Sulphur Springs. Formed 1846.

Hopkins, city (2000 population 17,145), HENNEPIN county, SE MINNESOTA, a suburb 7 mi/11.3 km WSW of downtown MINNEAPOLIS; 44°55′N 93°24′W. Railroad junction. Manufacturing (machinery; computer and electronic parts, printing and publishing, steel siding, wall murals, air pollution equipment, gun drilling, packaging, labels, opthamalic lenses, tools, jellies and candy, lumber, bakery products, software). An annual raspberry festival is held in Hopkins. Small lakes in area, especially SW. Incorporated as West Minneapolis 1893, name changed 1928.

Hopkins, city (2000 population 579), NODAWAY county, NW MISSOURI, near ONE HUNDRED AND TWO RIVER, 14 mi/23 km N of MARYVILLE, near Iowa State University; 40°32′N 94°49′W. Corn, wheat, soybeans; hogs, cattle; manufacturing (corrugated metal pipe).

Hopkins, village (2000 population 592), ALLEGAN county, SW MICHIGAN, 8 mi/12.9 km NNE of ALLEGAN; 42°37′N 85°45′W. In farm area. Manufacturing (trash compactors).

Hopkins Landing (HAHP-kinz), unincorporated village, SW BRITISH COLUMBIA, W CANADA, on THORNBROUGH CHANNEL of HOWE SOUND, 20 mi/32 km NW of VANCOUVER, in SUNSHINE COAST re-

gional district; 49°26′N 123°29′W. Lumber-shipping port.

Hopkins Park, village (2000 population 711), KANKAKEE county, E central ILLINOIS, 14 mi/23 km ESE of KANKAKEE, near INDIANA state line; 41°04′N 87°36′W. Corn, soybeans; manufacturing liquid food supplements.

Hopkins River (HAHP-kinz), 135 mi/217 km long, S VICTORIA, AUSTRALIA; rises in GREAT DIVIDING RANGE SW of ARARAT; flows S, past ALLANSFORD, to INDIAN OCEAN just E of WARRNAMBOOL; 38°24′S 142°31′E.

Hopkinsville, city (2000 population 30,089), ⊙ CHRISTIAN county, SW KENTUCKY, 60 mi/97 km WSW of BOWLING GREEN, on SOUTH FORK LITTLE RIVER; 36°51′N 87°29′W. Elevation 548 ft/167 m. Railroad junction. Fertile agricultural lands surround Hopkinsville, which is a leading dark and burley tobacco and livestock market; also grain; poultry; dairying; heavy manufacturing. Hopkinsville Community College, part of the University of Kentucky, is in the city. Hopkinsville–Christian County Airport to E. Fort Campbell Military Reservation (Kentucky/TENNESSEE) to SW. Pennyrile Forest State Resort Park to NW. Incorporated 1804.

Hopkinton, town, Delaware co., E Iowa, on Maquoketa River, and 15 mi/24 km SE of Manchester; 42°20′N 91°15′W.

Hopkinton, town, including Hopkinton village, MIDDLESEX county, E central MASSACHUSETTS, 26 mi/42 km WSW of BOSTON; 42°14′N 71°32′W. Manufacturing (electronic and computer equipment, machinery). Annual Boston Marathon starts here. State Park. Settled c.1715, incorporated 1744.

Hopkinton, town, MERRIMACK county, S NEW HAMPSHIRE, 6 mi/9.7 km W of CONCORD; 43°11′N 71°41′W. Drained by CONTOOCOOK and WARNER rivers. Manufacturing (construction materials, sawmill machinery); agriculture (nursery crops, vegetables, apples; livestock, poultry; dairying). Legislature met here occasionally, 1798–1807. Congregational church (1789) has Revere bell; Long Memorial Library has New Hampshire Antiquarian Society collection. Settled 1736, incorporated 1765. Includes CONTOOCOOK village (1990 population 1,334; covered bridge is here) on Contoocook River, in N.

Hopkinton, town (2000 population 7,836), WASHINGTON county, SW RHODE ISLAND, 30 mi/48 km SSW of PROVIDENCE; 41°29′N 71°45′W. In agricultural area; manufacturing. Includes villages of ASHAWAY, CANONCHET, HOPE VALLEY, HOPKINTON, and ROCKVILLE, and part of POTTER HILL village. Set off from WESTERLY and incorporated 1757. Named in 1757 for Governor Stephen Hopkins.

Hopland, unincorporated town, MENDOCINO county, NW CALIFORNIA, 11 mi/18 km S of UKIAH, on RUSSIAN RIVER. Fruit, grapes, hops, beans, nursery products; dairying; cattle. Manufacturing (beer, winery).

Hopland, village, MENDOCINO county, NW CALIFORNIA, in RUSSIAN RIVER valley, 13 mi/21 km S of UKIAH.

Hopong (HO-pong), township (□ 212 sq mi/551.2 sq km), SHAN STATE, MYANMAR, ⊙ Hopong, village on Thazi-Kengtung road, 10 mi/16 km E of TAUNGGYI, at head of road (S) to LOIKAW.

Hoppenhof, LATVIA: see APE.

Hoppers Crossing (HAH-puhrz KRAH-seeng), suburb 16 mi/25 km WSW of MELBOURNE, VICTORIA, SE AUSTRALIA, and just E of WERRIBEE. Recreational activities at Skeleton Creek.

Hop River, small stream, c.15 mi/24 km long, E central CONNECTICUT; formed by several branches near ANDOVER; flows SE and E to WILLIMANTIC RIVER just NW of WILLIMANTIC. Site of Andover Dam (for flood control) and Lake, 8 mi/12.9 km W of Willimantic.

Hop River, CONNECTICUT: see COLUMBIA.

Hopsten (HOP-stehn), village, NORTH RHINE-WESTPHALIA, NW GERMANY, 20 mi/32 km WNW of OSNABRÜCK; 52°23′N 07°38′E.

Hopwood, unincorporated town (2000 population 2,006), South Union and North Union townships, FAYETTE county, SW PENNSYLVANIA, suburb 2 mi/3.2 km SE of UNIONTOWN. Manufacturing (concrete blocks, machinery). CHESTNUT RIDGE and Forbes State Forest to SE. Hutchinson reservoirs (Numbers one, two, and three) to S. Founded 1791.

Hoquiam (HO-kwee-uhm), town (2006 population 9,061), GRAYS HARBOR county, W WASHINGTON, 4 mi/6.4 km W of ABERDEEN, at mouth of HOQUIAM River, on GRAYS HARBOR; 46°59′N 123°54′W. Railroad junction. With its twin city, Aberdeen, it has fishing (including shellfish), lumbering, paper, cranberry, and tourist industries; manufacturing (cedar products, lumber, veneer, plywood, paper industry machinery). Bowerman Airport on peninsula in harbor, to SW. OLYMPIC NATIONAL PARK and Olympic National Forest are to the N. Incorporated 1890.

Hor (HUHR), town, Gīlān province, IRAN, N of Gazvin; 30°25′N 50°32′E. Rice and tea cultivation.

Hora Abyata, ETHIOPIA: see ABIYATA.

Horace, village (2000 population 143), GREELEY county, W KANSAS, 2 mi/3.2 km W of TRIBUNE; 38°28′N 101°47′W. In agricultural and cattle area.

Horace, village (2006 population 1,464), CASS co., E NORTH DAKOTA, 10 mi/16 km SW of FARGO; 46°45′N 96°54′W. Founded in 1875 and incorporated in 1942. Named for Horace Greeley (1811–1872), losing candidate for U.S. Presidency in 1872, who had strong local support.

Horacko (HO-rahts-KO), Czech *Horácko*, historical region, JIHOCESKY, JIHOMORAVSKY, and VYCHODOCESKY provinces, E BOHEMIA and W MORAVIA, CZECH REPUBLIC, in the center of the BOHEMIAN-MORAVIAN HEIGHTS. Its limits marked by HUMPOLEC (W), PRIBYSLAV and ŽD'ÁR NAD SÁZAVOU (N), NOVE MESTO NA MORAVE (E), and JIHLAVA (S). Covered by large fir forests (formerly very dense, thus settled relatively late). Original country estates with their enclosed courts and wooden cottages are still preserved in the N; replaced (already in 19th century) in the S by brick or stone houses. Many artists, especially painters, worked here.

Horado (ho-RAH-do), village, Mugi district, GIFU prefecture, central HONSHU, central JAPAN, 16 mi/25 km N of GIFU; 35°36′N 136°49′E.

Horai (HO-RAH-ee), town, South Shitara district, AICHI prefecture, S central HONSHU, central JAPAN, 40 mi/65 km S of NAGOYA; 34°55′N 137°34′E. Nagashino Castle ruins.

Horaidha, YEMEN: see HUREIDHA.

Hor al Hammar, IRAQ: see HAMMAR, HOR AL.

Horana, town, WESTERN PROVINCE, SRI LANKA, 20 mi/32 km SSE of COLOMBO; 06°42′N 80°03′E. Extensive rubber and coconut plantations. Trades in vegetables. Ancient Buddhist monastery and temple nearby.

Hora Svateho Sebestiana (HO-rah SVAH-te-HO SHE-bes-TI-yah-NAH), Czech *Hora Svatého Šebestiána*, German *Sebastiansberg*, village, SEVEROCESKY province, NW BOHEMIA, CZECH REPUBLIC, in the ORE MOUNTAINS, on railroad, and 17 mi/27 km W of MOST; 50°31′N 13°16′E. Mountain resort. Peat extraction. Mining settlement. 15th–16th centuries.

Hora Svate Kateriny (HO-rah SVAH-te KAH-te-RZHI-ni), Czech *Hora Svaté Kateřiny*, German *Katharinaberg*, village, SEVEROCESKY province, NW BOHEMIA, CZECH REPUBLIC, in the ORE MOUNTAINS, 10 mi/16 km NW of MOST, near German border; 50°36′N 13°27′E. Old mining settlement.

Horatio (hor-AISH-ee-o), town (2000 population 997), SEVIER county, SW ARKANSAS, 7 mi/11.3 km S of DE QUEEN, near LITTLE RIVER; 33°56′N 94°21′W. Railroad junction to SW. Manufacturing (lumber).

Horatio, village, SUMTER county, central SOUTH CAROLINA, 17 mi/27 km NE of SUMTER near WATEREE

River. Agriculture includes cotton, soybeans, corn; broilers, cattle.

Horazdovice (HO-rahzh-DYO-vi-TSE), Czech *Horažd'ovice*, German *Horazdowitz*, town, ZAPADOCESKY province, S BOHEMIA, CZECH REPUBLIC, on OTAVA RIVER, 10 mi/16 km NW of STRAKONICE; 49°19′N 13°43′E. Railroad junction. Manufacturing (knitwear [hosiery, gloves], furniture, starch). Has Renaissance castle, museum.

Horazdowitz, CZECH REPUBLIC: see HORAZDOVICE.

Horb am Neckar (HORP ahm NEK-ahr), town, BADEN-WÜRTTEMBERG, SW GERMANY, in BLACK FOREST, on the NECKAR, and 13 mi/21 km E of FREUDENSTADT; 48°33′N 08°42′E. Industry includes textile manufacturing, wood- and metalworking; food processing. Has Gothic church.

Horbourg (or-boor), residential suburb (□ 4 sq mi/10.4 sq km) of COLMAR, HAUT-RHIN department, E FRANCE, on the ILL RIVER of ALSACE; 48°05′N 07°23′E. Biscuit manufacturing, vegetable (especially asparagus) growing and preserving. Village of Wihr forms part of township.

Horbury (HAW-buh-ree), town (2001 population 14,983), WEST YORKSHIRE, N ENGLAND, on CALDER RIVER and 3 mi/4.8 km SW of WAKEFIELD; 53°40′N 01°33′W. Light manufacturing.

Hörby (HUHR-BEE), town, SKÅNE county, S SWEDEN, near LAKE RINGSJÖN, 20 mi/32 km ENE of LUND; 55°51′N 13°40′E. Manufacturing (light industries); agriculture market (grain, potatoes; livestock). Commuting town to Lund, ESLÖV (21 mi/34 km) and MALMÖ (29 mi/47 km).

Horcajada (or-kah-HAH-dhah), town, ÁVILA province, central SPAIN, 40 mi/64 km WSW of ÁVILA. Cereals, legumes, livestock.

Horcajo de las Torres (or-KAH-ho dhai lahs TO-res), town, ÁVILA province, central SPAIN, 30 mi/48 km E of SALAMANCA; 41°04′N 05°05′W. Cereals; livestock.

Horcajo de los Montes (or-KAH-ho dhai los MON-tes), town, CIUDAD REAL province, S central SPAIN, 45 mi/72 kmNW of CIUDAD REAL; 39°19′N 04°39′W. Cereals, cork, honey; livestock; apiculture. Lumbering, olive and rock extracting.

Horcajo de Santiago (sahn-TYAH-go), town, CUENCA province, E central SPAIN, 34 mi/55 km ESE of ARANJUEZ; 39°50′N 03°00′W. Agricultural center (grapes, cereals, chickpeas, anise, cumin).

Horcasitas, MEXICO: see SAN MIGUEL DE HORCASITAS.

Horche (OR-chai), town, GUADALAJARA province, central SPAIN, 7 mi/11.3 km SE of GUADALAJARA; 40°34′N 03°04′W. Wheat, olives, livestock. Flour milling, olive oil pressing; manufacturing of soft drinks, soap, plaster.

Horconcitos (or-kon-SEE-tos), town, ⊙ San Lorenzo District, CHIRIQUÍ province, W PANAMA, near the PACIFIC, 20 mi/32 km ESE of DAVID. Leatherworking center; livestock.

Horcones Valley (or-KO-nes), NW MENDOZA province, ARGENTINA, at S part of ACONCAGUA massif, near CHILE border, along Horcones River, a headstream of MENDOZA RIVER.

Hordaland (HAWR-dah-lahn), county (□ c.6,036 sq mi/15,634 sq km; 2007 estimated population 456,495), SW NORWAY, bordering on the NORTH SEA in the W; ⊙ BERGEN. Hordaland includes the HARDANGERFJORD region and numerous islands and is a favorite tourist area. Fishing, farming, and manufacturing (including chemicals and metal goods) are the chief occupations.

Hords Creek Lake, reservoir, COLEMAN county, central TEXAS, on Hords Creek (tributary of PECAN BAYOU), 10 mi/16 km W of COLEMAN; 7 mi/11.3 km long; 31°50′N 99°34′W. Maximum capacity, c.49,000 acre-ft. Formed by dam (91 ft/28 m high), a unit of COLORADO RIVER flood-control project.

Hordville, village (2006 population 148), HAMILTON county, SE central NEBRASKA, 16 mi/26 km NNE of AURORA, and near PLATTE RIVER; 41°04′N 97°53′W.

Horebeke (HO-ruh-bai-kuh), commune (2006 population 2,013), Oudenaarde district, EAST FLANDERS province, W central BELGIUM, 4 mi/6 km E of OUDENAARDE.

Horefto, Greece: see KHOREFTO.

Horezu (ho-RE-zoo), town, VÎLCEA county, S central ROMANIA, in WALACHIA, 16 mi/26 km WNW of RÎMNICU VÎLCEA. Lignite mining; shoe manufacturing, flour milling, lumbering. The 17th-century Horezu monastery, founded by Constantine Brancovan, and often visited by Carmen Sylva, is 2 mi/3.2 km N. Nearby are 15th-century Bistriţa monastery, restored in 19th century, and 17th-century church of former Arnota monastery. Sometimes spelled Hurezu.

Horfield, ENGLAND: see BRISTOL.

Horgen (HOR-guhn), district, S ZÜRICH canton, N SWITZERLAND. Main towns are ADLISWIL, THALWIL, and WÄDENSWIL; population is German-speaking and Protestant.

Horgen (HOR-guhn), town (2000 population 17,432), ZÜRICH canton, N SWITZERLAND, on LAKE ZÜRICH, 8 mi/12.9 km SSE of ZÜRICH. Elevation 1,339 ft/408 m. Manufacturing of electronic equipment. A textile center in the 17th century. There is a noteworthy 18th-century church in the town.

Horgheim (HAWRG-haim), village, MØRE OG ROMSDAL county, W NORWAY, on RAUMA River, on railroad, and 9 mi/14.5 km SSE of ÅNDALSNES. Tourist center in the valley ROMSDAL.

Horgoš (HOR-gosh), village, VOJVODINA, N SERBIA, near TISZA RIVER, 14 mi/23 km E of SUBOTICA, in the BACKA region, near Hungarian border. Railroad junction. Also called KHORGOSH.

Horice (HO-rzhi-TSE), Czech *Hořice*, German *Horitz*, town, VYCHODOCESKY province, NE BOHEMIA, CZECH REPUBLIC, on railroad, and 9 mi/14.5 km SW of DVUR KRALOVE; 50°22′N 15°38′E. In sugar-beet and potato region; manufacturing (machinery, textiles); biscuits. Has large sculptors' and stonecutters' school (sandstone quarries nearby). Has a 16th-century castle and baroque church.

Horice na Sumave (HO-rzhi-TSE NAH shu-MAH-vye), Czech *Hořice na Šumavě*, German *Höritz*, village, JIHOCESKY province, S BOHEMIA, CZECH REPUBLIC, in BOHEMIAN FOREST, on railroad, and 7 mi/11.3 km SW of CESKY KRUMLOV; 48°46′N 14°11′E. Summer resort noted for annual "Passion Play" performances.

Horicon (HOR-i-kahn), town (2006 population 3,656), DODGE county, S central WISCONSIN, on ROCK RIVER, and 10 mi/16 km E of BEAVER DAM; 43°26′N 88°38′W. In farming area. Agriculture (dairy; livestock, poultry; grain); manufacturing (plastics, machinery, furniture, wood, iron and steel products, foods). Railroad junction. Horicon Marsh Wildlife Area and Horicon National Wildlife Refuge to N; SINISSIPPI Lake to S. Incorporated 1897.

Horicon, New York: see BRANT LAKE.

Horigane (ho-REE-gah-ne), village, South Azumi district, NAGANO prefecture, central HONSHU, central JAPAN, 31 mi/50 km S of NAGANO; 36°17′N 137°52′E.

Horinger (HUH-LIN-GUH), town, ⊙ Horinger county, central INNER MONGOLIA AUTONOMOUS REGION, N CHINA, 33 mi/53 km S of HOHHOT; 40°20′N 111°50′E. Engineering; grain, sugar beets, cattle raising. Also appears as Helinge'er.

Horinouchi (ho-ree-no-OO-chee), town, North Uonuma district, NIIGATA prefecture, central HONSHU, N central JAPAN, 43 mi/70 km S of NIIGATA; 37°14′N 138°55′E.

Höritz, CZECH REPUBLIC: see HORICE NA SUMAVE.

Horitz, CZECH REPUBLIC: see HORICE.

Horizon City, town (2006 population 10,709), EL PASO county, extreme W TEXAS, residential suburb 17 mi/27 km ESE of downtown EL PASO, in irrigated agricultural area; 31°40′N 106°11′W.

Horizonte (O-ree-SON-chee), city (2007 population 49,067), E central CEARÁ, BRAZIL, 27 mi/43 km S of FORTALEZA; 04°12′S 38°25′W.

Horizontina (O-ree-son-chee-nah), city (2007 population 18,305), NW RIO GRANDE DO SUL state, BRAZIL, 17 mi/27 km NE of SANTA ROSA; 27°37′S 54°19′W. Wheat, soybeans, corn; livestock.

Horley (HAW-lee), town (2001 population 21,232), SE SURREY, SE ENGLAND, on MOLE RIVER and 5 mi/8 km SSE of REIGATE; 1 mi/1.6 km N of Gatwick airport; 51°10′N 00°10′W. Former agricultural market. Commuter town to London. Has 15th-century church and old inn.

Horlick Ice Stream, ice stream, flows along the N side of the HORLICK MOUNTAINS into the lower part of REEDY GLACIER, WEST ANTARCTICA; 83°23′S 121°00′W.

Horlick Mountains, range in the TRANSANTARCTIC MOUNTAINS, EAST ANTARCTICA, located E of REEDY GLACIER; 85°23′S 121°00′W.

Horlivka (HOR-leev-kah) (Russian *Gorlovka*), city (2001 population 292,250), E central DONETS'K oblast, UKRAINE, in the DONETS BASIN; 48°18′N 38°03′E. Elevation 774 ft/235 m. A major coal-mining and industrial center, with a major coke-chemical complex and coal mining machinery plant. Other industries include food processing, knitting, sewing, building materials, wood working. Institutions include a foreign pedagogical institute, a branch of the Donets'k Polytechnical Institute, and many other professional technical schools; three museums. Founded in 1867, it is dotted with slag and mine waste heaps. Absorbed adjoining Kalinins'k after 1955.

Horme, L' (lor-muh), industrial town (□ 1 sq mi/2.6 sq km), LOIRE department, RHÔNE-ALPES region, SE central FRANCE, on GIER RIVER, and 8 mi/12.9 km ENE of SAINT-ÉTIENNE. Metalworking, manufacturing of precision tools.

Hormigas de Afuera Islands (or-MEE-gahs dai ah-FWAI-rah), two small islands, 40 mi/64 km off coast of W central PERU, W of LIMA; 11°58′S 77°47′W. Guano deposits.

Hormigueros (or-mee-GAI-ros), town (2006 population 17,414), W PUERTO RICO, 4 mi/6.4 km S of Mayagüez. Satellite of Mayagüez, which has practically taken up Hormigueros; most industry oriented toward Mayagüez. Industrial and commercial area; light manufacturing. Famed for its shrine of Our Lady of Montserrate (1775).

Mount Hor, ISRAEL: see EDOM.

Hormoz (huhr-MUHZ), island, Hormozgän province, S IRAN, in the Strait of HORMUZ, between the PERSIAN Gulf and the Gulf of OMAN; 5 mi/8.1 km long and 3.5 mi/5.6 km wide; 27°04′N 56°25′E. Salt and red ochre are produced. The town of Hormoz, originally built on the mainland, was moved (c.1300) to the island after repeated attacks by marauding raiders. The new port prospered and served as a center of trade with INDIA and CHINA. It was attacked by the Portuguese under Alfonso de Albuquerque in 1507 and was captured by them in 1514. Its recapture in 1622 by Shah Abbas I with the aid of an English fleet marked the end of the island's prosperity; the shah abandoned Hormoz for the new mainland port of BANDAR ABBAS. The town's proximity to the PERSIAN GULF has lent its strategic importance in the 20th century. The name is also spelled Hormuz.

Hormozgän, province (□ 25,819 sq mi/67,129.4 sq km), S IRAN, bordering the provinces of Färs and Kermän to N, and Sïstän va Balüchestän to E, extending 450 mi/724 km along the PERSIAN, Strait of HORMUZ, and Gulf of OMAN; ⊙ BANDAR ABBAS; 27°45′N 56°00′E. A coastal province, it includes many small islands in the Strait of Hormuz. Iron deposits; salt production; cereal crops and citris fruits. The port of Bandar Abbas is the only large city. Natural gas fields N and W of the city.

Area is shown by the symbol □, and capital city or county seat by ⊙.

Hormuz, Strait of (hahr-MOOZ), channel connecting the PERSIAN Gulf with the Gulf of OMAN of ARABIAN SEA, and separating IRAN and ARABIA; 40 mi/64 km–60 mi/97 km wide. Of great strategic importance, it contains several islands including: Qishm, Hormuz, and HANGAM. Also spelled Strait of ORMUZ.

Horn (HORN), town, N LOWER AUSTRIA, in the WALDVIERTEL 17 mi/27 km N of Krems; 48°40′N 15°40′E. Market center of E Waldviertel; large printery; agriculture includes corn, vineyards. Old town (founded c.1150), medieval fortifications well preserved; center of Protestantism before Counter-Reformation. Several Renaissance, baroque, and Biedermeier buildings; castle. Benedictine monastery Altenburg; Rosenburg castle and pilgrim church Maria Dreieichen nearby.

Horn (HAWRN), village, LIMBURG province, SE NETHERLANDS, 3 mi/4.8 km WNW of ROERMOND; 51°12′N 05°57′E. Dairying; cattle, hog raising; agriculture (grain, vegetables). Has fifteenth-century Horn Castle to S.

Hornachos (or-NAH-chos), town, BADAJOZ province, W SPAIN, in SIERRA DE HORNACHOS, 20 mi/32 km ESE of ALMENDRALEJO; 38°33′N 06°04′W. Spa and agriculture center (cereals, oranges, cork, olives; livestock).

Hornachos, Sierra de (or-NAH-chos, SYE-rah dhai), low W spur of the SIERRA MORENA, BADAJOZ province, W SPAIN, extending c.25 mi/40 km NW. Rises to 1,870 ft/570 m.

Hornachuelos (or-nahch-WAI-los), town, CÓRDOBA province, S SPAIN, 27 mi/43 km W of CÓRDOBA; 37°50′N 05°14′W. Olive oil processing. Agricultural trade (cereals, cork, livestock); lumber.

Hornad River, SLOVAKIA and HUNGARY: see HERNAD RIVER.

Horna Suca (hor-NAH soo-CHAH), Slovak *Horná Súča*, Hungarian *Felsöszúcs*, village, ZAPADOSLOVENSKY province, W SLOVAKIA, on S slope of Mount Javornik, and 6 mi/9.7 km NW of TRENCIN; 48°58′N 17°59′E. Food processing, distilling (plum brandy); wheat, barley, fruit (plums, apples).

Hornavan (HOORN-AH-vahn), lake (□ 97 sq mi/252.2 sq km), expansion of Skellefteälven River, NORRBOTTEN county, Lappland, N SWEDEN, 100 mi/161 km W of BODEN; 50 mi/80 km long, 1 mi/1.6 km–2 mi/3.2 km wide. Connects SE with LAKE UDDJAURE.

Horn–Bad Meinberg (HORN-BAHD MEIN-berg), town, NORTH RHINE-WESTPHALIA, N GERMANY, on N slope of TEUTOBURG FOREST, 5 mi/8 km SSE of DETMOLD; 51°53′N 08°57′E. Woodworking center; manufacturing of furniture, machinery. Horn was chartered 1248; retains its medieval character with many half-timbered houses. BAD MEINBERG is a health resort with calcium and sulphur springs (since 1767). In 1970 both towns, along with fourteen other villages, were unified as Horn–Bad Meinberg.

Hornbæk (HORN-bek), town, FREDERIKSBORG county, SJÆLLAND, DENMARK, on the ØRESUND and 12 mi/19 km NNE of HILLERØD; 56°05′N 12°28′E. Fisheries.

Hornbeak, town (2006 population 426), OBION county, NW TENNESSEE, 17 mi/27 km SW of UNION CITY; 36°19′N 89°17′W.

Hornbeck (HOORN-bek), town (2000 population 435), VERNON parish, W LOUISIANA, 56 mi/90 km W of ALEXANDRIA, near TEXAS state line; 31°20′N 93°24′W. Agriculture; manufacturing of wood products. State Wildlife Area to E.

Hornberg (HORN-berg), town, BADEN-WÜRTTEMBERG, SW GERMANY, in BLACK FOREST, 22 mi/35 km SE of OFFENBURG; 48°13′N 08°14′E. Health resort (elevation 1,180 ft/360 m); manufacturing of ceramics and electronics.

Hornburg (HORN-burg), town, LOWER SAXONY, N GERMANY, 9 mi/14.5 km SSE of WOLFENBÜTTEL; 52°02′N 10°37′E. Tourism. Has many 16th-century, half-timbered houses. Area served as a hops-growing center (15th–19th centuries).

Hornby Island (HORN-bee) (□ 11 sq mi/28.6 sq km; 2006 population 966), SW BRITISH COLUMBIA, W CANADA, in Strait of GEORGIA off VANCOUVER ISLAND, just E of DENMAN ISLAND and 16 mi/26 km SE of COURTENAY; 6 mi/10 km long, 4 mi/6 km wide; 49°31′N 124°40′W. Fishing, logging. Ferry from S end of Denman Island. Part of ISLANDS TRUST regional district.

Horn, Cape (HORN), Spanish, *Cabo de Hornos* (KAH-bo de HOR-nos), headland, 1,391 ft/424 m high, S CHILE, southernmost point of SOUTH AMERICA, in the archipelago of TIERRA DEL FUEGO. It was discovered and first rounded by Willem Schouten, the Dutch navigator, on January 29, 1616, and named for HOORN in the NETHERLANDS. Lashing storms and strong currents made "rounding the Horn" one of the great hazards of ship-sailing days. With its cold and windy climate, it is still a formidable challenge to navigation.

Horncastle (HAWN-kah-suhl), town (2001 population 6,340), LINCOLNSHIRE, E ENGLAND, on Bain River and 18 mi/29 km E of LINCOLN; 53°12′N 00°07′W. Leather works; fruit-growing area. Former site of noted horse fair. Has 13th-century church and remains of wall of the Roman station *Banovallum*.

Hornchurch, ENGLAND: see HAVERING.

Horndal (HOORN-DAHL), village, KOPPARBERG county, central SWEDEN, on Lake Rossen, (3 mi/4.8 km long), 12 mi/19 km NE of AVESTA; 60°18′N 16°25′E. In former iron-mining region.

Horndean (HAWN-deen), town (2001 population 12,639), HAMPSHIRE, S ENGLAND, 2 mi/3.2 km N of WATERLOOVILLE; 50°55′N 00°54′W. Residential and farming.

Horneburg (HOR-ne-berg), village, LOWER SAXONY, NW GERMANY, 7 mi/11.3 km SSE of STADE; 53°31′N 09°35′E.

Hörnefors (HUHR-ne-FORSH), town, VÄSTERBOTTEN county, N SWEDEN, on GULF OF BOTHNIA, 19 mi/31 km SW of UMEÅ; 63°37′N 19°55′E. Old church. Site of battle between Swedes and Russians (1809).

Horne Islands, French *Îles de Horne* (eel de orn), Polynesian island group, in WALLIS AND FUTUNA, overseas territory of FRANCE, SW PACIFIC, 120 mi/193 km SW of WALLIS ISLANDS; 14°15′S 178°05′W. They include two volcanic islands (FUTUNA and uninhabited 'ALOFI) and some coral islets. Also called Futuna Islands, sometimes Hoorn Islands.

Hornelen, NORWAY: see BREMANGER.

Hornell, city (□ 2 sq mi/5.2 sq km; 2006 population 8,705), STEUBEN county, SW NEW YORK, on the CANISTEO RIVER; 42°19′N 77°39′W. Light manufacturing and commercial services. Formerly a major rail transfer point. Nicknamed "Maple City" for trees there and in surrounding area. Settled 1790, incorporated 1906.

Hornepayne (HORN=pain), town (□ 79 sq km; 2001 population 1,362), N central ONTARIO, E central CANADA, 200 mi/322 km N of SAULT SAINTE MARIE; 49°15′N 84°46′W. Elevation 1,074 ft/327 m. Gold mining, lumbering.

Hornersville, town (2000 population 686), DUNKLIN county, extreme SE MISSOURI, on LITTLE RIVER drainage channel in MISSISSIPPI alluvial plain, 14 mi/23 km S of KENNETT; 36°02′N 90°06′W. Soybeans, cotton, corn. Manufacturing (barbecue sauce).

Horne Saliby (hor-NE sah-LI-bi), Slovak *Horné Saliby*, Hungarian *Felsöszeli*, village, ZAPADOSLOVENSKY province, SW SLOVAKIA, 18 mi/29 km SSE of TRNAVA; 48°07′N 17°45′E. Wheat, corn, sugar beets, fruit; food processing. Summer resort. Strong Hungarian minority. Under HUNGARY between 1938–1945.

Horne Srnie (hor-NE suhr-NYE), Slovak *Horné Srnie*, Hungarian *Alsószernye*, village, ZAPADOSLOVENSKY province, W SLOVAKIA, in WHITE CARPATHIAN MOUNTAINS, on railroad and 8 mi/12.9 km NNE of TRENCÍN; 48°59′N 18°06′E. Large cement works.

Horn Head, Gaelic *Corrán Binne*, promontory, N DONEGAL county, N IRELAND, 3 mi/4.8 km N of DUNFANAGHY; 53°14′N 07°59′W.

Horni Becva River, CZECH REPUBLIC: see BECVA RIVER.

Horni Benesov (HOR-nyee BE-ne-SHOF), Czech *Horní Benešov*, German *Bennisch*, village, SEVEROMORAVSKY province, central SILESIA, CZECH REPUBLIC, 30 mi/48 km NNE of OLOMOUC; 49°58′N 17°36′E. Railroad terminus; metallurgy; textile manufacturing.

Horni Blatna (HOR-nyee BLAHT-nah), Czech *Horní Blatná*, German *Platten*, village, ZAPADOCESKY province, W BOHEMIA, CZECH REPUBLIC, in the ORE MOUNTAINS, on railroad, and 5 mi/8 km NNE of NEJDEK, near German border; 50°23′N 12°46′E. Elevation 2,913 ft/888 m. Winter skiing resort. Textile manufacturing. Was mining settlement in 16th century. Has a baroque church.

Horni Briza (HOR-nyee BRZHEE-zah), Czech *Horní Bříza*, German *Ober-Briz*, village, ZAPADOCESKY province, W BOHEMIA, CZECH REPUBLIC, on railroad, 7 mi/11.3 km NNW of Plzeň; 49°50′N 13°22′E. Kaolin and fire-clay quarrying; ceramics manufacturing.

Hornick, town (2000 population 253), WOODBURY county, W IOWA, on West Fork Little Sioux River, and 25 mi/40 km SE of SIOUX CITY; 42°13′N 96°05′W. In agricultural area.

Horni Dvoriste (HOR-nyee DVO-rzhish-TYE), Czech *Horní Dvořiště*, German *Oberhaid*, village, JIHOCESKY province, S BOHEMIA, CZECH REPUBLIC, on Austrian border, on railroad, 26 mi/42 km S of ČESKÉ BUDEJOVICE; 48°36′N 14°25′E. Customs station. Has a 16th-century Gothic church.

Horni Litvinov, CZECH REPUBLIC: see LITVÍNOV.

Hornillo, El (or-NEE-lyo, el), village, ÁVILA province, central SPAIN, in the SIERRA DE GREDOS, 35 mi/56 km SW of ÁVILA. Olives, grapes, beans, apricots, walnuts; livestock. Lumbering; olive oil pressing.

Horni Marsov (HOR-nyee MAHR-shof), Czech *Horní Maršov*, German *Obermarschendorf*, village, VYCHODOCESKY province, NE BOHEMIA, CZECH REPUBLIC, in the GIANT MOUNTAINS, at E foot of Cerna hora, 34 mi/55 km ESE of LIBEREC. Elevation 1,968 ft/600 m. Manufacturing (textiles), wood processing. Winter ski resort.

Hornindalsvatnet (HAWR-nin-dahls-VAHT-nuht), lake (□ 20 sq mi/52 sq km) in SOGN OG FJORDANE county, W NORWAY, 50 mi/80 km ENE of FLORØ; 16 mi/26 km long. Deepest lake (1,600 ft/488 m deep) in Europe. Drains into NORDFJORD at NORDFJORDEID. Hornindal village is on NE shore.

Horning's Mills (HOR-neengz MILZ), hamlet of MELANCTHON, S ONTARIO, E central CANADA; 44°09′N 80°12′W.

Horni Plana (HOR-nyee PLAH-nah), Czech *Horní Planá*, German *Oberplan*, town, JIHOCESKY province, S BOHEMIA, CZECH REPUBLIC, in BOHEMIAN FOREST, near Austrian border, on bank of LIPNO DAM, on railroad, and 25 mi/40 km SW of ČESKÉ BUDEJOVICE; 48°46′N 14°02′E. Woodworking, food processing (milk); graphite mining in vicinity. Summer resort. Has 13th-century Gothic church.

Horni Poustevna, CZECH REPUBLIC: see DOLNI POUSTEVNA.

Hornisgrinde (HOR-nis-grin-de), mountain (3,819 ft/1,164 m) in the BLACK FOREST, S GERMANY, 10 mi/16 km S of BADEN-BADEN; 48°38′N 08°13′E.

Horn Island (HORN) (□ 19 sq mi/49.4 sq km), in TORRES STRAIT, 20 mi/32 km N (across ENDEAVOUR STRAIT) of CAPE YORK PENINSULA, N QUEENSLAND, AUSTRALIA, near PRINCE OF WALES ISLAND. Circular, 15 mi/24 km in circumference; rises to 376 ft/115 m; 10°37′S 142°17′E. Wooded, sandy; uninhabited.

Horn Island, JACKSON county, SE MISSISSIPPI, part of coastal island chain in the Gulf of MEXICO, partly sheltering MISSISSIPPI SOUND (N), 10 mi/16 km SSW of PASCAGOULA; c.14 mi/23 km long. Lighthouse

Cross-references are shown in SMALL CAPITALS. The pronunciation guide is shown on page xix. The sources of population figures are shown on page xvii.

(30°14′N 88°29′W). To E is deepwater channel (Horn Island Pass) leading from the Gulf to Pascagoula Bay. Part of GULF ISLANDS National Seashore. PETIT BOIS Island to E, SHIP ISLAND to W.

Horn Island, CHILE: see HORN, CAPE.

Horni Slavkov (HOR-nyee SLAHF-kof), Czech *Horní Slavkov*, German *Schlaggenwald*, town, ZAPADOCESKY province, W BOHEMIA, CZECH REPUBLIC, on railroad, and 20 mi/32 km ENE of CHEB; 50°09′N 12°48′E. Porcelain (since 1792) and machinery manufacturing; tanning. Former mining settlement. Notable 16th-century buildings.

Horni Sucha (HOR-nyee SU-khah), Czech *Horní Suchá*, German *Ober Suchau*, village, SEVEROMORAVSKY province, E central SILESIA, CZECH REPUBLIC, on railroad, and 5 mi/8 km SW of KARVINÁ; 49°48′N 18°30′E. Coal-mining area.

Hornitos, unincorporated village, MARIPOSA county, central CALIFORNIA, 18 mi/29 km NE of MERCED. Cattle. Has buildings dating from gold-rush era. New Exchequer Dam (Lake McClure) to N.

Horn Lake, town (2000 population 14,099), DE SOTO county, NW MISSISSIPPI, suburb 12 mi/19 km S of downtown Memphis, TENNESSEE, and 2 mi/3.2 km S of Southampton; 34°57′N 90°02′W. Manufacturing (cast metal products, elevators, lumber, food seasonings, sheet metal products). Horn Lake to W.

Hornnes (HAWRN-nais), village, AUST-AGDER county, S NORWAY, on OTRA RIVER, and 30 mi/48 km N of KRISTIANSAND.

Horn of Africa: see AFRICA, HORN OF.

Hornomoravsky Lowland, CZECH REPUBLIC: see SEVEROMORAVSKY.

Hornopiren Volcano (or-no-PEE-ren), Andean peak (5,480 ft/1,670 m) in LLANQUIHUE province, LOS LAGOS region, S central CHILE, 40 mi/64 km SE of PUERTO MONTT.

Hornos, Cabo de, CHILE: see HORN, CAPE.

Hornos Islands (OR-nos), group of small rocky islets in the RÍO DE LA PLATA, COLONIA department, SW URUGUAY, 4 mi/6.4 km NW of Colonia city, 25 mi/40 km NE of BUENOS AIRES; 34°25′S 57°55′W.

Hornostayivka (hor-no-STAH-yeev-kah) (Russian *Gornostayevka*), town, central KHERSON oblast, UKRAINE, on the E shore of the KAKHOVKA RESERVOIR, 55 mi/89 km NE of KHERSON; 47°00′N 33°44′E. Elevation 216 ft/65 m. Raion center; butter, food flavoring, sewing. Established at the end of the 18th century; town since 1956.

Horn Peak, Colorado: see SANGRE DE CRISTO MOUNTAINS.

Hornsberg (HOORNS-BERY), village, JÄMTLAND county, N central SWEDEN, on FRÖSÖN island in STORSJÖN LAKE, just W of ÖSTERSUND (bridge). Tourist area.

Hornsby (HORNZ-bee), town, E NEW SOUTH WALES, AUSTRALIA, suburb 13 mi/21 km NW of SYDNEY; 33°42′S 151°06′E. Railroad junction; coal-mining center; manufacturing (baked goods).

Hornsby, town (2006 population 299), HARDEMAN county, SW TENNESSEE, 30 mi/48 km S of JACKSON; 35°14′N 88°50′W. In farm area; lumber.

Hornsea (HAWN-zee), town (2006 population 8,200), East Riding of Yorkshire, NE ENGLAND, on NORTH SEA, 14 mi/23 km NE of HULL; 53°54′N 00°10′W. Seaside resort. Pottery factory. Has 14th–15th-century church. Just inland from the town is the lake Hornsea Mere.

Hornsey, ENGLAND: see HARINGEY.

Hornslet (HORNS-let), town (2000 population 4,702), ÅRHUS county, E JUTLAND, DENMARK, 14 mi/23 km SE of RANDERS; 56°19′N 10°20′E. Agriculture (dairying; hogs; barley, rye); manufacturing (furniture, woodworking).

Hornstein (HORN-shtein), Croatian *Hornštajn*, township, N BURGENLAND, E AUSTRIA, on W slope of LEITHA MOUNTAINS, and 4 mi/6.4 km NW of EISEN-

STADT. Near the Lower Austrian border; 47°53′N 16°27′E. Manufacturing of fittings; vineyards.

Hornstrandir Landscape Reserve (HORNS-trahn-dir) (□ 224 sq mi/582.4 sq km), NORDUR-ISAFJARDARSYSLA county, NW ICELAND, 20 mi/32 km NNE of Isafjordur. Roadless wilderness of deep fjords and steep sea cliffs. Rivers with arctic char flow through U-shaped valleys called "viks," origin of the word "Vikings." Occupies large peninsula between DENMARK STRAIT and Jokulfirdir fjord. Lighthouse at the Horn, 7 mi/11.3 km from ARCTIC CIRCLE. Razorbills, puffins, flumars. Drift ice in winter sometimes bring polar bears to area.

Hornsundtind (HAWRN-soon-TIN), mountain (4,692 ft/1,430 m), S West SPITSBERGEN, NORWAY, Spitsbergen group, near S shore of Horn Sound (Norweigan *Hornsund*; 15 mi/24 km-long inlet of ARCTIC OCEAN), 90 mi/145 km S of LONGYEAR City; 76°55′N 16°10′E. Mountain first climbed (1897) by Sir Martin Conway.

Hornu (awr-NOO), town in Bossu commune, Mons district, HAINAUT province, SW BELGIUM, 6 mi/9.7 km WSW of MONS.

Horochow, UKRAINE: see HOROKHIV.

Horodenka (ho-ro-DEN-kah) (Russian *Gorodenka*) (Polish *Horodenka*), city, IVANO-FRANKIVS'K oblast, UKRAINE, on right tributary of the DNIESTER (Ukrainian *Dnister*), and 23 mi/37 km NE of KOLOMYYA; 48°40′N 25°30′E. Elevation 895 ft/272 m. Raion center; food processing (grain, dairy, hops, fruit, vegetables), sugar refining; distilling, tobacco curing; stone quarrying. Has ruins of a 15th-century castle. Known since the end of the 12th century as part of Halych principality; city since 1939.

Horodnya (ho-ROD-nyah) (Russian *Gorodnya*), city, NW CHERNIHIV oblast, UKRAINE, 31 mi/50 km NNE of CHERNIHIV; 51°53′N 31°36′E. Elevation 413 ft/125 m. Raion center; flax, flour and grain milling, dairying; peat cutting. Historical museum. Orphanage school for 310 children, receiving Western charitable donations. First mentioned in 1552; Polish, then Cossack company center (1635–1654), Cossack company town (1705–1782), company town (1782–1917). Site of one of the first sugar refineries in Ukraine (1848). Sizeable Jewish community since the first half of the 19th century, close to 10% of the total population at its peak (1926); liquidated by the Nazis during World War II; fewer than 100 Jews remaining in 2005. City since 1957.

Horodnytsya (ho-ROD-ni-tsyah) (Russian *Gorodnitsa*), town, W ZHYTOMYR oblast, UKRAINE, on SLUCH River, and 20 mi/32 km NW of NOVOHRAD-VOLYNS'KYY; 50°48′N 27°19′E. Elevation 600 ft/182 m. Railroad terminus; ceramics (feldspar quarries); sawmill. Has a 19th-century park. Established in the 8th century, town since 1938.

Horodok (ho-ro-DOK) (Russian *Gorodok*), city, W KHMEL'NYTS'KYY oblast, UKRAINE, 25 mi/40 km SW of KHMEL'NYTS'KYY; 49°10′N 26°34′E. Elevation 810 ft/246 m. Raion center; sugar refining, machine and tool manufacturing, fruit and milk canning. Professional technical school; regional historical museum. First mentioned in 1392; annexed by Russia in 1793; site of battles between armies of the Ukrainian National Republic and the Bolsheviks (1919); raion center of Proskuriv okruha (1923–1930), Vinnytsya oblast (1932–1937) and Kamyanets'-Podil's'kyy oblast (1937–1954).

Horodok (ho-ro-DOK) (Russian *Gorodok*) (Polish *Gródek Jagielloński*), city, W central L'VIV oblast, UKRAINE, 16 mi/26 km WSW of L'VIV; 49°47′N 23°39′E. Elevation 869 ft/264 m. Raion center; agricultural processing (grain, dairy, vegetables), clothing and construction industries. Fisheries along small Vereshytsya Lake (N). Has a monument to Bohdan Khmel'nyts'kyy, St. Nicholas's Church (1510), wooden church of St. John the Baptist (1670) and a Roman Catholic Church (15th–18th century). First

mentioned as Solyanyy Horodok in 1213, and then an important fortified town of Halych-Volyn Principality. Site of a cossack victory over Polish troops (1655), razed by the Tatars (1672), site of a Russian defeat of Austrian army (1914) and Polish-Ukrainian battles (1919). One of the earliest Jewish communities in Eastern Europe (since 1420), numbering approximately 3,300 at its peak (1931); exterminated by the Nazis during World War II.

Horodyshche (ho-ro-DI-shche) (Russian *Gorodishche*), city, central CHERKASY oblast, UKRAINE, on the Vil'shanka River, 30 mi/48 km WSW of CHERKASY; 49°17′N 31°27′E. Elevation 318 ft/96 m. Raion center; on the Kiev-Dnipropetrovs'k highway and railroad line; sugar refinery, dairy, brewery, grain mill, feed mill, building material fabrication, mining and crafting of Labradorite; professional-technical school of sugar industry. Museum in honor of Ukrainian composer and singer S. S. Hulak-Artemovskyy; 1771 church of St. Mary the Protectress. Known since the 16th century; city since 1956. Called (1929–1933) Horodyshche-Shevchenkivs'ke (Russian *Gorodishche-Shevchenkovskoye*), and in 1933–1944, Imeny H. I. Petrovs'koho (Russian *Imeni G. I. Petrovskogo*).

Horodyshche (ho-ro-DI-shche) (Russian *Gorodishche*), town, SW central LUHANS'K oblast, UKRAINE, in the Donets Upland, 12 mi/20 km SW of ALCHEVS'K and 5 mi/8 km E of CHORNUKHYNE; 49°25′N 39°15′E. Elevation 587 ft/178 m. Broiler farm. Established at the beginning of the 18th century; town since 1964.

Horodyshche, UKRAINE: see MARHANETS'.

Horokanai (ho-RO-kah-NAH-ee), town, Sorachi district, HOKKAIDO prefecture, N JAPAN, 78 mi/125 km N of SAPPORO; 44°00′N 142°09′E. Noodles.

Horokhiv (ho-RO-kheev) (Russian *Gorokhov*) (Polish *Horochów*), city, S VOLYN' oblast, UKRAINE, on left tributary of STYR River and 30 mi/48 km SW of LUTS'K; 50°30′N 24°46′E. Elevation 744 ft/226 m. Raion center; food processing; brick, machine manufacturing. Has an old church, ruins of a medieval palace. First mentioned in 1240 as part of Halych-Volyn' principality; passed to Lithuania in the 14th century, to Poland in the 16th century, to Russia in 1795; reverted back to Poland in 1921; ceded to USSR in 1945; part of independent Ukraine since 1991. Jewish community since the 15th century, numbering 3,500 at its peak (1939); eliminated completely by the Nazis in 1941.

Horonai-kawa, RUSSIA: see PORONAY RIVER.

Horonobe (ho-ro-NO-be), town, Rumoi district, HOKKAIDO prefecture, N JAPAN, 136 mi/220 km N of SAPPORO; 45°00′N 141°51′E. Butter.

Hororata (Ho-ro-rat-a), township, SELWYN district (□ 2,531 sq mi/6,580.6 sq km), CANTERBURY region, SOUTH ISLAND, NEW ZEALAND, 35 mi/56 km W of CHRISTCHURCH. Agriculture.

Horoshky, UKRAINE: see VOLODARS'K-VOLYNS'KYY.

Horovice (HO-rzho-VI-tse), Czech *Hořovice*, German *Horowitz*, town, STREDOCESKY province, W central BOHEMIA, CZECH REPUBLIC, on railroad, and 29 mi/47 km SW of PRAGUE; 49°50′N 13°55′E. In potato- and barley-growing district; iron foundries; manufacturing (metal and wooden furniture, enamelware, musical instruments). Has an 18th-century baroque castle.

Horowhenua (hor-or-FEN-new-ah), district (□ 411 sq mi/1,068.6 sq km), NEW ZEALAND. Includes LEVIN, between WELLINGTON and PALMERSTON NORTH cities, where productive alluvial soil, extending S from the MANAWATU LOWLANDS is compressed between the sea and sand dunes of COOK STRAIT to the W and the hard rock TARARUA RANGE to the E. It channels N-S traffic by road and railroad, with intensive urbanization, agriculture with dairying and market gardening, and recreational development.

Horowitz, CZECH REPUBLIC: see HOROVICE.

Area is shown by the symbol □, and capital city or county seat by ⊙.

Horqin Youyi Qianqi, CHINA: see WULANHAOTE.

Horqueta (or-KE-tah), town, (2002 population 9,946), Concepción department, central PARAGUAY, a transportation hub 25 mi/40 km E of Concepción (connected by railroad line); 23°20′S 57°02′W. Cattle-raising and lumbering center; tanneries.

Horqueta, Cerro (or-KAI-tah), peak (7,530 ft/2,295 m) in CONTINENTAL DIVIDE of W PANAMA, 29 mi/47 km N of DAVID.

Horra, La (lah O-rah), village, BURGOS province, N SPAIN, 11 mi/18 km WNW of ARANDA DE DUERO. Grain growing and wine producing.

Horrocks, resort village, WESTERN AUSTRALIA state, W AUSTRALIA, 308 mi/496 km N of PERTH, 14 mi/22 km W of NORTHAMPTON; 28°23′S 114°25′E. Fishing.

Horry (OR-ree), county (□ 1,255 sq mi/3,250 sq km; 1990 population 144,053; 2000 population 196,629), E SOUTH CAROLINA; ⊙ CONWAY; 33°54′N 78°58′W. Bounded W and NW by LITTLE PEE DEE RIVER, SE by the ATLANTIC OCEAN, NE by NORTH CAROLINA state line, and SW by the GREAT PEE DEE RIVER; drained by WACCAMAW RIVER. INTRACOASTAL WATERWAY canal passes near coast. Summer resort area; includes MYRTLE BEACH (state park here). Agricultural interests include timber, tobacco; also cotton, corn, hogs, cattle, soybeans, oats, wheat. Hunting and fishing attract tourists. Formed 1785 as Kingston, renamed 1801.

Hörsching (HYOOR-shing), township, central UPPER AUSTRIA, 7 mi/11.3 km SW of LINZ; 48°14′N 14°11′E. Transportation and retail enterprises. Linz airport here.

Horse Cave, town (2000 population 2,252), HART county, central KENTUCKY, 32 mi/51 km ENE of BOWLING GREEN; 37°10′N 85°54′W. Tourist resort for Kentucky limestone cave region and an E gateway to MAMMOTH CAVE NATIONAL PARK. Agriculture (livestock, poultry; dairying; burley tobacco, corn, wheat); manufacturing (benches, lumber, commercial printing, plastic cups, crushed stone, concrete, truck seats). Hidden River Cave, with an underground river containing blind fish, is here, also Kentucky Cave and Mammoth Onyx Cave. Horse Cave Theatre (1911).

Horse Creek, 136 mi/219 km long, in SE WYOMING and W NEBRASKA; rises in LARAMIE MOUNTAINS near LARAMIE, Wyoming, SW ALBANY county; flows E, N, and E, past LA GRANGE, to North Platte River near MORRILL, Nebraska, just E of Nebraska state line. Hawk Springs Reservoir formed on small tributary 6 mi/9.7 km N of La Grange.

Horse Creek Reservoir, SE COLORADO, on side channel of Horse Creek, on BENT-OTERO county border, 11 mi/18 km NW of LAS ANIMAS; 3 mi/4.8 km long; 38°09′N 103°23′W. Extends N. Maximum capacity of 43,125 acre-ft. formed by Horse Creek Dam (31 ft/9 m high), built (1971) for irrigation. Formerly Timber Lake.

Horsefly Lake (HORS-flei), S central BRITISH COLUMBIA, W CANADA, in CARIBOO MOUNTAINS, 120 mi/193 km SE of PRINCE GEORGE, S of QUESNEL LAKE; 52°25′N 121°00′W. Drained W into FRASER River by Quesnel River; 28 mi/45 km long, 1 km–2 km/1 m/3 km wide.

Horsehead Lake, KIDDER CO., central NORTH DAKOTA, 14 mi/23 km NNE of STEELE; 6 mi/9.7 km long; 47°02′N 99°47′W.

Horseheads, village (□ 3 sq mi/7.8 sq km; 2006 population 6,293), CHEMUNG county, S NEW YORK, 5 mi/8 km N of ELMIRA; 42°10′N 76°49′W. Light manufacturing; quarrying; agriculture; services. Settled 1789, incorporated 1837.

Horse Island, Scotland: see ARDROSSAN.

Horse Islands, CANADA: see ST. BARBE ISLANDS.

Horse Mesa Dam, ARIZONA: see APACHE LAKE.

Horseneck Beach, Massachusetts: see WESTPORT.

Horsens (HOR-suhns), city (2000 population 48,730), VEJLE county, central DENMARK, a port at the head of the HORSENS FJORD, an inlet of the KATTEGATSTRAIT; 55°52′N 09°52′E. It is a commercial and industrial center. Horsens was a fortified town in the Middle Ages. It retains a noteworthy thirteenth-century monastery and church.

Horsens Fjord (HOR-suhns) c.13 mi/21 km long, E JUTLAND, DENMARK; inlet of the KATTEGATSTRAIT. At mouth are HJARNØ and ALRØ islands; HORSENS city at head.

Horse Pasture (HORS pas-chuhr), unincorporated town, HENRY county, S VIRGINIA, 6 mi/10 km SW of MARTINSVILLE; 36°37′N 79°57′W. Agriculture (tobacco; cattle; poultry).

Horseshoe Bay (HORS-shoo), summer village (2001 population 52), E ALBERTA, W CANADA, in St. PAUL COUNTY No. 19; 54°07′N 111°22′W.

Horseshoe Bay (HOR-shoo BAI), unincorporated village, SW BRITISH COLUMBIA, W CANADA, and included in WEST VANCOUVER; 49°22′N 123°16′W.

Horseshoe Bend, village (2000 population 770), BOISE county, W IDAHO, 21 mi/34 km N of BOISE, in bend of PAYETTE RIVER; BLACK CANYON DAM downstream (W); 43°55′N 1116°11′W. Logging, sawmill products; cattle; alfalfa. Bogus Basin Ski Area to S.

Horseshoe Bend, a turn on the TALLAPOOSA RIVER, near DADEVILLE, E central ALABAMA. Site of a battle on March 27, 1814, in which the Creeks, led by chief William Weatherford, were significantly defeated by a militia under the command of Andrew Jackson. HORSESHOE BEND MILITARY PARK is here.

Horseshoe Bend Military Park (□ 3 sq mi/7.8 sq km), E central ALABAMA, 13 mi/21 km E of ALEXANDER CITY. Authorized 1956.

Horseshoe Lake Conservation Area, ALEXANDER county, extreme S ILLINOIS, 11 mi/19 km NW of CAIRO; c.7 mi/11.3 km long; 37°08′N 89°21′W. State game preserve. The lake is a remnant of the MISSISSIPPI RIVER.

Horseshoe Mountain, peak (13,898 ft/4,236 m) in ROCKY MOUNTAINS, between Park and Lake counties, central COLORADO, 7 mi/11.3 km ESE of LEADVILLE, and S of MOSQUITO PASS.

Horsetooth, reservoir (□ 3 sq mi/7.8 sq km), LARIMER county, N central COLORADO, on CACHE LA POUDRE RIVER, 3 mi/5 km W of FORT COLLINS; 40°36′N 105°10′W. Formed by Horsetooth Dam (155 ft/47 m high), built (1949) by the Bureau of Reclamation for irrigation. Roosevelt National Forest just W.

Horsforth, ENGLAND: see LEEDS.

Horsham (HOR-shuhm), municipality (2001 population 12,591), W central VICTORIA, AUSTRALIA, on WIMMERA RIVER, and 170 mi/274 km WNW of MELBOURNE; 36°43′S 142°12′E. Railroad and commercial center for wheat-raising area; flour mills. Longerenong Agricultural College nearby.

Horsham (HOR-shuhm), unincorporated city (2000 population 14,779), MONTGOMERY county, SE PENNSYLVANIA, 12 mi/19 km NE of NORRISTOWN. Manufacturing (electronic equipment, apparel, computers, chemicals, construction equipment, level controls, consumer goods). U.S. Naval Air Station (WILLOW GROVE) to NW.

Horsham (HAW-shuhm), town (2001 population 47,804) and district, West SUSSEX, SE ENGLAND, 8 mi/12.9 km SW of CRAWLEY; 51°03′N 00°20′W. The town is known primarily for its agricultural and merchandising activities, but it also serves as an engineering center. The Causeway, an old cobbled street, has Tudor and Stuart houses and a 13th-century parish church. Christ's Hospital, a public school whose pupils included Charles Lamb and Samuel Taylor Coleridge, was moved from London to Horsham in 1902. Includes the district of NEWTON SAINT FAITH.

Horsham St. Faith (HAW-shuhm), agricultural village (2001 population 1,642), central NORFOLK, E ENGLAND, 4 mi/6.4 km N of NORWICH; 52°41′N 01°16′E. There are remains of 12th-century priory; church dates from 15th century. To N is agricultural village of NEWTON ST. FAITH. Norwich airport to S.

Horsovsky Tyn (HOR-shof-SKEE TEEN), Czech Horšovský Týn, German Bischofteinitz, town, ZAPADOCESKY province, SW BOHEMIA, CZECH REPUBLIC, on RADBUZA RIVER, on railroad, and 24 mi/39 km SW of Plzeň; 49°32′N 12°57′E. In woolen- and linen-spinning district; food processing, distilling, plastic manufacturing. Established in the 12th century as a bishop's residence. Has many historical buildings.

Horst (HAWRST), town, LIMBURG province, E NETHERLANDS, 8 mi/12.9 km NW of VENLO; 51°27′N 06°04′E. Dairying; cattle, hogs; grain, vegetables, mushrooms; manufacturing (food processing, pumps). Recreational center to N.

Horst (HORST), village, SCHLESWIG-HOLSTEIN, NW GERMANY, 4 mi/6.4 km NNW of ELMSHORN; 53°48′N 09°37′E.

Hörstel (HUHR-stel), town, NORTH RHINE-WESTPHALIA, NW GERMANY, 20 mi/32 km W of OSNABRÜCK; 52°18′N 07°35′E. Manufacturing of machinery, textiles, and building materials. Built around a Cistercian abbey in 1256 (secularized 1808).

Horstmar (HORST-mahr), town, NORTH RHINE-WESTPHALIA, NW GERMANY, 5 mi/8 km SSW of BURGSTEINFURT; 52°05′N 07°20′E. Manufacturing of lamps. Has 14th-century church and 15th-century castle.

Horst Wessel Stadt, GERMANY: see FRIEDRICHSHAIN.

Horta (ORT-ah), district (□ 295 sq mi/767 sq km; 2001 population 34,289) of the AZORES, PORTUGAL, covering FAIAL, PICO, FLORES, and CORVO islands; ⊙ HORTA (on Faial island).

Horta (ORT-ah), town (2001 population 9,563), ⊙ HORTA district, in the AZORES, PORTUGAL, on FAIAL island. Has an excellent harbor with shipyards and an international airport that currently serves both military and commercial flights.

Horta or **Horta de San Juan** (OR-tah dhai SAHN HWAHN), town, TARRAGONA province, NE SPAIN, 9 mi/14.5 km SW of GANDESA. Olive oil processing, brandy distilling, flour milling; wheat, wine, almonds and other nuts.

Hortaleza (or-tah-LAI-thah), town, MADRID province, central SPAIN, 4 mi/6.4 km NNE of MADRID; 40°28′N 03°38′W.

Horten (HAWR-tuhn), town (2007 population 25,011), VESTFOLD county, SE NORWAY, a port on the OSLOFJORD (an arm of the SKAGERRAK). It is a commercial and industrial center. Horten was the main Norwegian naval base until after World War II, when its facilities were moved to BERGEN. Chartered 1907.

Hortensias, CHILE: see LAS HORTENSIAS.

Hortiatis, Greece: see KHORTIATIS.

Hortobágy (HOR-to-bahdy), Hungarian Hortobágy, section (□ 104 sq mi/270.4 sq km) of the Alföld, E HUNGARY; 47°35′N 21°10′E. Characteristics of the puszta (pasturelands of the Alföld) are preserved here for tourist trade; large herds of horses. In July or August a mirage frequently is visible. New irrigation development; rice and waterfowl on lake.

Hortobágy River (HOR-to-bah-dyuh), Hungarian Hortobágy, 70 mi/113 km long, E HUNGARY; rises 15 mi/24 km WNW of Hajduböszörmény; flows SE, draining the HORTOBÁGY; joins Berettyó Canal 9 mi/14 km E of KISÚJSZÁLLÁS to form another, shorter, BERETTYÓ RIVER.

Horton (HOR-tuhn), township (□ 61 sq mi/158.6 sq km; 2001 population 2,567), SE ONTARIO, E central CANADA, 32 mi/51 km from PEMBROKE, and on the OTTAWA River; 45°30′N 76°37′W. Tourism.

Horton, town (2000 population 1,967), BROWN county, NE KANSAS, 23 mi/37 km WNW of ATCHISON; 39°39′N 95°31′W. Trading point in grain, livestock, and fruit area; dairy products. Manufacturing (industrial machinery). Mission Lake, just NE, is source of city's water supply. Kickapoo Indian Reservation to W. Incorporated 1887.

Horton (HAW-tuhn), agricultural village (2006 population 900), S BUCKINGHAMSHIRE, central ENGLAND, 3 mi/4.8 km E of WINDSOR; 51°51′N 00°39′W. Residence of John Milton for six years; his mother is buried here.

Horton, village, JACKSON county, S MICHIGAN, 9 mi/14.5 km SW of JACKSON; 42°09′N 84°31′W. In farm and lake-resort area; manufacturing of specialty machinery.

Horton, resort village, DELAWARE county, S NEW YORK, in the Catskills, on BEAVER KILL, and c.50 mi/80 km W of KINGSTON; 41°58′N 75°01′W.

Horton, river, c.275 mi/440 km long, CANADA; rising in a lake N of GREAT BEAR Lake, INUVIK Region, NORTHWEST TERRITORIES; flowing NW to FRANKLIN BAY, a part of the Beaufort Sea.

Hortonia, Lake, in towns of SUDBURY and HUBBARDTON, RUTLAND CO., W VERMONT, on Hubbardton River, 15 mi/24 km NW of RUTLAND; c.2 mi/3.2 km long; 43°43′N 73°12′W. River drains SW. Resort.

Horton Kirby (HAW-tuhn KUH-bee), village (2001 population 2,942), NW KENT, SE ENGLAND, on DARENT RIVER and 4 mi/6.4 km SSE of DARTFORD; 51°24′N 00°15′E. Paper milling, although this industry has declined since the mid 20th-century. Has 13th-century church.

Horton Plains, tableland in SRI LANKA HILL COUNTRY, SE Central prov., S central SRI LANKA; 06°48′N 80°48′E. Elev. 7,000 ft/2,134 m; rise to 7,857 ft/2,395 m in KIRIGALPOTTA, to 7,741 ft/2,359 m in Totapola peaks.

Hortonville, town (2006 population 2,723), OUTAGAMIE county, E WISCONSIN, 11 mi/18 km NW of APPLETON; 44°20′N 88°37′W. In dairying and fruit-growing area. Manufacturing (toys, juvenile furniture, vegetable canning, concrete pavers, nails, and wire products).

Horusicky Pond (HO-ru-SITS-kee), Czech *Horusický rybník* (HO-ru-SITS-kee RIB-nyeek) (□ 161 sq mi/ 418.6 sq km), JIHOCESKY province, S BOHEMIA, CZECH REPUBLIC, on LUZNICE RIVER, and 14 mi/23 km NE of České Budějovice; maximum depth 20 ft/6 m; 49°09′N 14°40′E. Elevation 1,365 ft/416 m. Built in 1500. Fishing; peat area with valuable flora on SE bank, summer resort on N bank.

Hörvik (HUHR-VEEK), fishing village, Blekinge county, S SWEDEN, on E shore of LISTERLANDET peninsula, on BALTIC SEA, 6 mi/9.7 km E of SÖLVESBORG; 56°02′N 14°46′E.

Horw (HORV), town (2000 population 12,648), LUCERNE canton, central SWITZERLAND, near LAKE LUCERNE, 2 mi/3.2 km S of LUCERNE. Chemicals, cement.

Horwich (HAH-rich), town (2001 population 18,017), GREATER MANCHESTER, W central ENGLAND, 5 mi/8 km WNW of BOLTON; 53°36′N 02°33′W. Locomotive building, paper manufacturing. Site of Liverpool reservoir.

Horyn' (ho-RIN) (Russian *Goryn'*) (Polish *Horyń*), river, 410 mi/660 km long, in W UKRAINE; rises 65 mi/ 105 km5 E of L'VIV; flows N into the PRIPET (Ukrainian *Pripyat'*) River.

Horyniec (hoh-REE-nee-ts), Russian *Gorynets*, village, Przemyśl province, SE POLAND, 12 mi/19 km ENE of LUBACZOW. Health resort (sulphur springs); sawmilling.

Horyu-ji, JAPAN: see NARA.

Hosadurga, INDIA: see HOSDURGA.

Hosa'ina (ho-SAI-in-uh), town (2007 population 60,183), SOUTHERN NATIONS state, S central ETHIOPIA; 07°35′N 37°53′E. 70 mi/113 km E of JIMMA. Trade center (coffee, beeswax, hides).

Hosanagara (HO-suh-nuh-guh-ruh), town, SHIMOGA district, KARNATAKA state, INDIA, on SHARAVATI RIVER and 35 mi/56 km W of SHIMOGA; 13°55′N 75°04′E. Rice, betel palms and vines; cinnamon oil extracting, handicraft wickerwork. Called Kallurkotte until 1893, when it succeeded NAGAR (7 mi/11.3 km S) as a subdivisional administrative headquarters of district. Also spelled Hosanagar.

Hosanger (HAW-sahn-guhr), village, HORDALAND county, SW NORWAY, on NW shore of OSTERØY, on OSTERFJORDEN, 14 mi/23 km NNE of BERGEN. Agriculture, cattle raising, fishing; lumber mills. Former copper and nickel mines.

Hösbach (HUHS-bahkh), town, BAVARIA, central GERMANY, in LOWER FRANCONIA, 3 mi/4.8 km NE of ASCHAFFENBURG; 50°00′N 09°11′E.

Hoschton (HUHSH-tuhn), village (2000 population 1,070), JACKSON county, NE central GEORGIA, 22 mi/35 km WNW of ATHENS; 34°05′N 83°46′W. Manufacturing includes industrial threads, nursery bottles, sports car assembly.

Hosdurga (HOS-duhr-guh), town, CHITRADURGA district, KARNATAKA state, INDIA, 30 mi/48 km SSW of CHITRADURGA; 13°48′N 76°16′E. Coir products (rope, sacks); hand-loom cotton and woolen weaving, handicraft cap-making; coconuts, oilseeds. Livestock grazing in NW hills (manganese deposits). Hosdurga Road, railroad station, is 10 mi/16 km WNW. Also spelled Hosadurga.

Hoshang, CHINA: see EMIN.

Hoshangabad (ho-shahn-gah-BAHD), district (□ 2,080 sq mi/5,408 sq km; 2001 population 886,449), S MADHYA PRADESH state, INDIA, on DECCAN PLATEAU; ⊙ HOSHANGABAD. Bordered N by NARMADA RIVER and BHOPAL district; mainly in level, fertile wheat tract of NARMADA valley, with densely forested spurs of SATPURA RANGE (S) and outliers of VINDHYA RANGE (NW). Wheat, millet, cotton, oilseeds (chiefly sesame) in alluvial valley; mahua, date palms, mangoes; betel farms. Teak, sal, salai, bamboo, and khair trees on S hills; lac growing. Flour, oilseed, and dal milling, cotton ginning, sawmilling, cutch processing; sandstone quarrying. Hoshangabad is an agriculture trade center. PACHMARHI (climatic health resort and wildlife sanctuary) is summer headquarters of MADHYA PRADESH government. District enlarged in early 1930s by merger of former district of NARSIMHAPUR (E) and in 1948 by incorporation of former CENTRAL INDIA state of MAKRAI. NARSIMHAPUR has since become a separate district again. HARDA district separated from Hoshangabad and formed a new district in 1998.

Hoshangabad (ho-shahn-gah-BAHD), town (2001 population 97,357), ⊙ HOSHANGABAD district, S MADHYA PRADESH state, INDIA, on NARMADA RIVER, and 140 mi/225 km NW of NAGPUR; 22°44′N 77°45′E. Agriculture trade center (wheat, cotton, millet, oilseeds); roofing tiles (red Vindhyan sandstone quarries nearby), brass ware, bamboo canes. Livestock raising. Railroad junction of ITARSI is 10 mi/16 km SSE.

Hoshcha (HO-shchah) (Russian *Goshcha*) (Polish *Hoszcza*), town, SE RIVNE oblast, UKRAINE, on HORYN' River, and 16 mi/26 km E of RIVNE; 50°36′N 26°40′E. Elevation 574 ft/174 m. Raion center; food processing (flour, butter, vegetables). St. Michael's Church (1639); park. Known since the 14th century; in the 16th century, center of Socinians (a Protestant sect) in VOLHYNIA; Orthodox monastery (1638); origin of miraculous icon of Mother of God in Pochaiv Monastery; town since 1959. Small Jewish community since the 18th century, numbering 811 in 1939; eliminated during World War II.

Hoshiarpur (HOSH-yahr-poor), district (□ 1,478 sq mi/3,842.8 sq km), N central PUNJAB state, N INDIA; ⊙ HOSHIARPUR. Largely in BIST JALANDHAR DOAB. Bordered E by foothills of E PUNJAB HIMALAYAS, parallel to (and W of) which runs W SHIWALIK RANGE, NW by BEAS RIVER, W by JALANDHAR and KAPURTHALA districts, and S by SUTLEJ RIVER. Seasonal torrents flow SW. Agriculture (wheat, gram, corn, sugarcane, cotton); hand-loom weaving. Chief towns are HOSHIARPUR, TANDA-URMAR, DASUYA.

Hoshiarpur (HOSH-yahr-poor), town, ⊙ HOSHIARPUR district, N central PUNJAB state, N INDIA, 24 mi/39 km NE of JALANDHAR; 31°32′N 75°54′E. Railroad spur terminus; trades in wheat, gram, maize, mangoes, sugar; manufacturing of shellac, rosin, turpentine,

aerated water; beeswax refining, hand-loom silk weaving, oilseed pressing; handicrafts (ivory and brass inlay work, fibre products, leather goods). Has a college.

Hoshino (HO-shee-no), village, Yame district, FUKUOKA prefecture, N KYUSHU, SW JAPAN, 31 mi/50 km S of FUKUOKA; 33°14′N 130°45′E. Cryptomeria; tea.

Hoshiuyama (ho-SHYOO-yah-mah), village, Asakura district, FUKUOKA prefecture, N KYUSHU, SW JAPAN, 31 mi/50 km S of FUKUOKA; 33°23′N 130°52′E.

Hoshiv (HO-sheev) (Russian *Goshev*), village, W IVANO-FRANKIVS'K oblast, UKRAINE, on SVICHA RIVER, right tributary of the DNIESTER, on road and near railroad, 2 mi/3 km SSE of BOLEKHIV and 32 mi/ 51.5 km W of IVANO-FRANKIVS'K; 49°02′N 23°53′E. Elevation 1,220 ft/371 m. Has a Basilian monastery, with its miracle-working icon of the Weeping Mother of God, since the 18th century. Popularly known as Queen of the Carpathians; site of pilgrimages from Galicia and Bucovina; closed by the Soviet authorities in 1951, who converted the buildings into a hostel for indigent children. Re-opened following Ukraine's independence (1991).

Hoskins (HAHS-kins), town, ⊙ WEST NEW BRITAIN province, N central NEW BRITAIN island, E PAPUA NEW GUINEA, 145 mi/233 km SW of RABAUL. Located on ASTROLABE (KIMBE) Bay of SOLOMON SEA. Airport; coastal road. Cocoa, copra, palm oil, prawns, tuna.

Hoskins, village (2006 population 261), WAYNE county, NE NEBRASKA, 16 mi/26 km SW of WAYNE, and on branch of ELKHORN RIVER; 42°06′N 97°17′W. Machinery.

Hoskote (HOS-ko-te), town, BANGALORE district, KARNATAKA state, INDIA, 15 mi/24 km NE of BANGALORE; 13°04′N 77°48′E. Rice milling, tobacco curing; handicrafts (silk cloth, lacquerware, coir mats).

Hosmer, village (2006 population 256), EDMUNDS county, N SOUTH DAKOTA, 23 mi/37 km WNW of IPSWICH; 45°34′N 99°28′W. In agricultural area.

Hosoe (ho-SO-e), town, Inasa district, SHIZUOKA prefecture, central HONSHU, E central JAPAN, 43 mi/70 km S of SHIZUOKA; 34°48′N 137°39′E. Oranges.

Hosoiri (ho-SO-ee-ree), village, Nei district, TOYAMA prefecture, central HONSHU, central JAPAN, 12 mi/20 km S of TOYAMA; 36°31′N 137°13′E.

Hospers, town (2000 population 672), SIOUX county, NW IOWA, 9 mi/14.5 km SSW of SHELDON; 43°04′N 95°54′W. Agriculture (livestock; grain; beef products); plastic injection moldings.

Hospet (HOS-pet), city, BELLARY district, KARNATAKA state, S INDIA, 37 mi/60 km WNW of BELLARY. Railroad junction; agriculture trade center; rice milling, jaggery. Manufacturing (non-metallic, metal, and food products). Famous ruins of ancient city of Vijayanagar 6 mi/9.7 km NE, at HAMPI. Construction of dam and power plant on the TUNGABHADRA RIVER, 4 mi/6.4 km W, at Mallapuram (or Malapuram) completed by 1956; part of irrigation and hydroelectric project. Railroad spur to Gunda Road, railroad junction 7 mi/11.3 km S, with spurs to SWAMIHALLI (manganese ore) and KOTTURU.

Hospital, Gaelic *Óspidéal Ghleann Áine*, town (2006 population 628), E LIMERICK county, SW IRELAND, 12 mi/19 km W of TIPPERARY; 52°28′N 08°26′W. Agricultural market. Remnants of Knights Hospitaller establishment. Nearby is Scarteen, home of the Black and Tans pack of hounds.

Hospital (os-PEE-tahl), village, SANTIAGO province, METROPOLITANA DE SANTIAGO region, central CHILE, on railroad and 30 mi/48 km S of SANTIAGO; 33°52′S 70°45′W. Resort in agricultural area.

Hospital, Cuchilla del (os-pee-TAHL, koo-CHEE-yah del), hill range, RIVERA department, NE URUGUAY, branches off from the CUCHILLA DE SANTA ANA NE of Yaguarí, extending 20 mi/32 km SSW, forming watershed between the ARROYO YAGUARÍ (NE) and the RÍO NEGRO (SW); 31°40′S 54°53′W.

Hospital de Órbigo (os-pee-TAHL dhai OR-vee-go), town, LEÓN province, NW SPAIN, 9 mi/14.5 km E of ASTORGA; 42°28'N 05°53'W. Beans, potatoes, sugar beets, tobacco.

Hospitalet (o-spee-tah-LET) or **L'Hospitalet de Llobregat** (lo-spee-tah-LET dhai lyo-vrai-GAHT), satellite city (2001 population 239,019) of BARCELONA, BARCELONA province, NE SPAIN, in CATALONIA, 4 mi/6.4 km WSW of city center, on LLOBREGAT coastal plain; 40°59'N 00°56'E. Large industrial zone. Steel mills; manufacturing also includes chemicals, electrical equipment, paper, wood, textiles; food processing. Secondary and Professional Teaching Institutes here.

Hospitalet-près-l'Andorre, L' (lo-pee-tah-lai-prai-lahn-dor), village (□ 10 sq mi/26 sq km), ARIÈGE department, MIDI-PYRÉNÉES region, S FRANCE, near ANDORRA and Spanish border, on the upper ARIÈGE RIVER, and 27 mi/43 km SSE of FOIX; elevation 4,711 ft/1,436 m. Located at N end of trans-Pyrenean railroad tunnel under PUYMORENS pass. Winter sports. Hydroelectric power plant.

Hossegor, FRANCE: see CAPBRETON.

Hosston (HAHS-tuhn), village (2000 population 387), CADDO parish, extreme NW LOUISIANA, near RED RIVER, 25 mi/40 km N of SHREVEPORT, at E end of BLACK BAYOU Lake reservoir; 32°53'N 93°53'W. Oil and natural-gas field in vicinity. Black Bayou State Game and Fish Preserve.

Hosszúmező, ROMANIA: see CÎMPULUNG TISA.

Hostalrich (o-stahl-REECH), town, GERONA province, NE SPAIN, 19 mi/31 km SSW of GERONA; 41°45'N 02°38'E. Cork, lumber, fruit.

Hostau, CZECH REPUBLIC: see HOSTOUN.

Hoste Island (O-stai), TIERRA DEL FUEGO, CHILE, one of the larger islands (c.90 mi/145 km long) of the archipelago, just W of NAVARINO ISLAND; 54°55'–55°44'S 68°–70°W. It has several irregular, large peninsulas, notably Hardy Peninsula, which extends SE, terminating in FALSE CAPE HORN.

Hostinne (HOS-tyi-NE), Czech *Hostinné*, German *Arnau*, town, VYCHODOCESKY province, NE BOHEMIA, CZECH REPUBLIC, on ELBE River, on railroad and 34 mi/55 km SE of LIBEREC; 50°33'N 15°44'E. Large paper mills; textiles. Dam across Elbe River, c.2 mi/3.2 km downstream.

Hostivar (HOS-tyi-VARZH), Czech *Hostivař*, manufacturing district of SE PRAGUE, PRAGUE province, CZECH REPUBLIC, 5 mi/8 km from city center. Manufacturing (chemicals, machinery, electrical products); foundries.

Hostivice (HOS-tyi-VI-tse), German *Hostiwice*, town, STREDOCESKY province, central BOHEMIA, CZECH REPUBLIC, 8 mi/12.9 km W of PRAGUE; 50°05'N 14°16'E. Railroad junction. Manufacturing (machinery, bricks); food processing. Has a baroque castle, 13th century; Gothic church, 14th century; Gothic stronghold.

Hostiwice, CZECH REPUBLIC: see HOSTIVICE.

Hostomel' (HOS-to-mel) (Russian *Gostomel'*), town, N central KIEV oblast, UKRAINE, 15 mi/24 km NW of KIEV, 2 mi/3.2 km N of BUCHA; 50°35'N 30°16'E. Elevation 298 ft/90 m. Manufacturing (glass, starch). Cargo terminal of Antonov Airport. Known since 1496, town since 1938.

Hostomice (HOS-to-MI-tse), German *Hostomitz*, town, STREDOCESKY province, SW central BOHEMIA, CZECH REPUBLIC, on railroad, and 10 mi/16 km S of BEROUN; 49°50'N 14°03'E. Manufacturing (gloves, building materials). Has an 18th-century church. OSOV (18th century baroque castle) is 2 mi/3.2 km NE.

Hostomitz, CZECH REPUBLIC: see HOSTOMICE.

Hostos (OS-tos), village, DUARTE province, E central DOMINICAN REPUBLIC, in fertile LA VEGA REAL valley, on railroad and 33 mi/53 km E of LA VEGA. Cacao, rice. Until 1928, LA CEIBA.

Hostotipaquillo (os-to-tee-pah-KEE-yo), town, ⊙ Hostotipaquillo municipio, JALISCO, W MEXICO, 55 mi/89 km NW of GUADALAJARA; 21°04'N 104°04'W. Elevation 3,507 ft/1,069 m. Silver, gold, and copper mining; agriculture (beans, corn, wheat, alfalfa, livestock); commerce.

Hostoun (HOS-toun-yuh), Czech *Hostouní*, German *Hostau*, village, ZAPADOCESKY province, SW BOHEMIA, CZECH REPUBLIC, on railroad, and 30 mi/48 km SW of Plzeň; 49°34'N 12°45'E. In lumbering region. Has a baroque church.

Hostyn, CZECH REPUBLIC: see BYSTRICE POD HOSTYNEM.

Hosur (HO-soor), town, DHARMAPURI district, TAMIL NADU state, S INDIA, near SE border of KARNATAKA state, 23 mi/37 km SE of BANGALORE. Railroad spur terminus; manufacturing of tiles, sandalwood oil; silk growing. Livestock research center; remount depot. C. Rajagopalachari, first Indian governor-general of India, born in nearby village.

Hoszcza, UKRAINE: see HOSHCHA.

Hotaka (HO-tah-kah), town, S Azumi district, NAGANO prefecture, central HONSHU, central JAPAN, 28 mi/45 km S of NAGANO; 36°20'N 137°53'E. Trout; wasabi.

Hotaka, Mount (HO-tah-kah), Japanese *Hotaka-dake* (ho-TAH-kah-DAH-ke), peak (10,527 ft/3,209 m), central HONSHU, central JAPAN, on GIFU-NAGANO prefecture border, 20 mi/32 km WNW of MATSUMOTO; highest peak in CHUBU-SANGAKU NATIONAL PARK. Sometimes called Hodaka or Okuhotaka.

Hotan, CHINA: see HETIAN.

Hotan River, 400 mi/644 km long, S XINJIANG UYGUR AUTONOMOUS REGION, NW CHINA; rises in KUNLUN mountains in two branches—the Karakax (left) and the Yurungkax (right), which form the HOTAN RIVER at Koxlax; flows N, through TAKLIMAKAN DESERT, but rarely reaching the TARIM.

Hotazel, town, NORTHERN CAPE province, SOUTH AFRICA, on tributary MOLOPO RIVER, near border with NORTH-WEST province, 78 mi/125 km N of POSTMASBURG. Center of iron and manganese mining. Named by surveyors completing work on its railroad link around 1912.

Hotchkiss, town (2000 population 968), DELTA county, W COLORADO, on North Fork of GUNNISON RIVER, at mouth of Leroux Creek, just W of West ELK MOUNTAINS, and 20 mi/32 km ENE of DELTA; 38°47'N 107°43'W. Elevation 5,351 ft/1,631 m. In fruit-growing region. Hotchkiss National Fish Hatchery to SW. Gunnison National Forest to E.

Hotevilla (ho-tay-villa), unincorporated town, NE ARIZONA, atop a mesa c.65 mi/105 km N of WINSLOW, in Hopi Indian Reservation. Elevation c.5,900 ft/1,798 m. Sheep, cattle, hogs; crafts. Founded 1906 by dissenting residents of ORAIBI.

Hotham, Cape (HAH-thuhm), NW NORTHERN TERRITORY, AUSTRALIA, between CLARENCE STRAIT and VAN DIEMEN GULF of TIMOR SEA; 12°03'S 131°18'E.

Hotham Inlet, NW ALASKA, arm of KOTZEBUE SOUND, extending SE from KOTZEBUE, forms E side of BALDWIN PENINSULA; 50 mi/80 km long, 5 mi/8 km–20 mi/32 km wide. Receives KOBUK River. At head of inlet is SELAWIK Lake.

Hotham, Mount (HAH-thuhm) (6,100 ft/1,859 m), E central VICTORIA, SE AUSTRALIA, in AUSTRALIAN ALPS, 130 mi/209 km ENE of MELBOURNE; 36°58'S 147°07'E. Winter sports (May–September).

Ho That Ba (HO TAHT BAH), major reservoir, YEN BAI province, N VIETNAM, on SONG CHAY River; 21°54'N 104°54'W.

Hoti Mardan, PAKISTAN: see MARDAN.

Hotin, UKRAINE: see KHOTYN.

Hoton Nuur, MONGOLIA: see HOVD.

Hot Spring, county (□ 622 sq mi/1,617.2 sq km; 2006 population 31,730), central ARKANSAS; ⊙ MALVERN; 34°19'N 92°57'W. Drained by OUACHITA RIVER and small CADDO RIVER. Agriculture (cattle, hogs). Manufacturing at Malvern and JONES MILL. Barite and rutile clay mining; bentonite, sand, gravel deposits; timber. Remmel Dam forms Lake CATHERINE in N county; Lake Catherine State Park is on S shore; part of DEGRAY LAKE reservoir (Caddo River) on S boundary; small part of Ouachita National Forest in far NW corner. Formed 1829.

Hot Springs, county (□ 2,006 sq mi/5,215.6 sq km; 2006 population 4,588), N central WYOMING; ⊙ THERMOPOLIS; 43°42'N 108°26'W. Mining, natural gas and oil, and some agriculture in N (alfalfa, sugar beets; some cattle); drained by BIGHORN RIVER. Coal. Part of ABSAROKA RANGE in W; small part of Shoshone National Forest in extreme W. Hot Springs State Park in E center. Part of Wind River Indian Reservation in S; OWL CREEK MOUNTAINS in S. Formed 1911.

Hot Springs, city (2006 population 4,102), ⊙ FALL RIVER county, SW SOUTH DAKOTA, 45 mi/72 km SSW of RAPID CITY, just S of BLACK HILLS, and on Fall River; 43°25'N 103°28'W. Health resort; sulphur hot springs; dairy products, timber, sandstone, alfalfa seed. WIND CAVE NATIONAL PARK and CUSTER STATE PARK to N. Manufacturing (boots and saddles). Black Hills National Forest to W and S. Cold Brook Reservoir Recreation Area to S. First settled 1879, incorporated 1882.

Hot Springs, town, MANICALAND province, E ZIMBABWE, 47 mi/76 km SSW of MUTARE, on ODZI RIVER; 19°39'S 32°28'E. Odzi Gorge to S; CHIMANIMANI MOUNTAINS to E. Cattle, sheep, goats; corn, wheat, soybeans.

Hot Springs, village (2000 population 531), SANDERS county, NW MONTANA, 46 mi/74 km SSW of KALISPELL, in W part of Flathead Indian Reservation; 47°37'N 114°40'W. Mineral waters resort. Lolo National Forest to W.

Hot Springs (HAHT SPREENGZ), village (□ 7 sq mi/18.2 sq km; 2006 population 642), MADISON county, W NORTH CAROLINA, 25 mi/40 km NW of ASHEVILLE, on FRENCH BROAD RIVER, near TENNESSEE state line, in Pisgah National Forest; 35°53'N 82°49'W. Service industries; manufacturing (nylon carrying cases); agriculture (tobacco, corn; cattle). APPALACHIAN TRAIL (Appalachian National Scenic Trail) passes NE-SW through village.

Hot Springs (HAHT SPREENGZ), unincorporated village, BATH county, W VIRGINIA, in ALLEGHENY MOUNTAINS, 17 mi/27 km NE of COVINGTON, in George Washington National Forest; 37°59'N 79°49'W. Tourism. Mineral springs. Homestead Resort and ski area. Douthat State Park to SE, LAKE MOOMAW Recreation Area to W.

Hot Springs, NEW MEXICO: see TRUTH OR CONSEQUENCES.

Hot Springs National Park or **Hot Springs**, city, ⊙ GARLAND county, W central ARKANSAS; 47 mi/76 km WSW of LITTLE ROCK; 34°29'N 93°02'W. Settled 1807, incorporated 1876. The city's N side is nearly surrounded by HOT SPRINGS NATIONAL PARK, noted for its hot mineral springs that have made the city a famous health resort. Central Avenue, which runs N-S between sections of the park, is known as Bath House Row. Situated in the OUACHITA MOUNTAINS, the city is at the center of the reservoir system of the OUACHITA RIVER, Lake CATHERINE to E, Lake HAMILTON in S, and Lake OUACHITA to NW. Diversified manufacturing. Fish hatchery to S on Lake Hamilton. The area was visited by the Spanish explorer Hernando De Soto in 1541. The properties of the waters were investigated in 1804 under the authorization of President Thomas Jefferson. Garland County Community College here. Ouachita National Forest to W.

Hot Springs National Park (□ 9.12 sq mi/23.62 sq km), W central ARKANSAS. Dominates the N part of the city of HOT SPRINGS NATIONAL PARK (Hot Springs). Visited by Spanish explorer Hernando De Soto in 1541. The springs, long used by Native Americans for medicinal purposes, became a Federal Reservation in

1832. From 750,000 gals/2,838,975 liters to 950,000 gals/3,596,035 liters of water per day, with an average temperature of 143°F/62°C, flow from forty-seven springs. The National Park Service collects, cools, and supplies water to bathhouses in and out of the park. Camping, hiking. Established 1921.

Hot Springs Region, NEW ZEALAND: see THERMAL REGION.

Hot Sulphur Springs, village (2000 population 521), ⊙ GRAND county, N COLORADO, on COLORADO RIVER, in W foothills of FRONT RANGE, and 65 mi/105 km WNW of DENVER; 40°04′N 106°05′W. Elevation 7,670 ft/2,338 m. Resort; headquarters of Arapaho National Forest. WILLIAMS FORK RESERVOIR to SW; Windy Gap Reservoir to E; parts of forest located to N and S, and Williams Fork game preserve. Mineral springs here. County museum here with preserved buildings and Native American artifacts.

Hottah Lake (□ 377 sq mi/976 sq km), NORTHWEST TERRITORIES, CANADA, 60 mi/97 km S of PORT RADIUM; 40 mi/64 km long, 1 mi/2 km–20 mi/32 km wide; 65°5′N 118°30′W. Drained N into GREAT BEAR LAKE by Camsell River (50 mi/80 km long).

Hotte, Massif de la (mah-SEEF duh lah OT), mountain range in SW HAITI, stretches through JACMEL PENINSULA for c.130 mi/209 km to JACMEL; 18°24′N 74°02′W. Rises to over 7,500 ft/2,286 m. Lignite, bauxite, and manganese deposits.

Hottentots Holland Mountains, WESTERN CAPE, SOUTH AFRICA, extending circa 20 mi/32 km in a semicircle NE and then N from E side of FALSE BAY at Roman Point rising to 5,217 ft/1,590 m on Sneeuwkop at its N end, 8 mi/12.9 km ENE of SOMERSET WEST. Crossed by railroad and road on Sir Lowry's Pass (1,530 ft/466 m), 6 mi/9.7 km SE of Somerset West. In early days of CAPE COLONY range represented E limit of Dutch rule. Nature reserve to N of range and Stumbas Dam to S of range.

Hotton (aw-TAWN), commune (2006 population 5,053), March-en-Famenne district, LUXEMBOURG province, SE BELGIUM, 6 mi/10 km N of MARCHE-EN-FAMENNE.

Hotval'd, UKRAINE: see ZMIYIV.

Hotzenplotz, CZECH REPUBLIC: see OSOBLAHA.

Hötzing, ROMANIA: see HAȚEG.

Houaïlou (wah-ee-LOO), village, NEW CALEDONIA, a territory of FRANCE, on E coast, 85 mi/137 km NW of NOUMÉA; 21°18′S 165°33′E. E coast terminal of cross-island road, well located at E exit of Col de Rousettes pass across New Caledonia from BOURAIL; agricultural products, livestock.

Houat (oo-ah), island (□ 1 sq mi/2.6 sq km), in Bay of BISCAY off BRITTANY coast (MORBIHAN department), W FRANCE, 7 mi/11.3 km SE of tip of QUIBERON PENINSULA; 3 mi/4.8 km long, 1 mi/1.6 km wide; 47°23′N 02°58′W. New port of Houat (on N shore) is home of fishing boats. Crustaceans are shipped to mainland. Several times island was occupied by British in 17th and 18th century.

Houches, Les (OOSH, laiz), commune (□ 16 sq mi/41.6 sq km), winter resort in HAUTE-SAVOIE department, RHÔNE-ALPES region, SE FRANCE, resort at SW end of Chamonix valley, 4 mi/6.4 km SW of CHAMONIX; 45°53′N 06°49′E. Elevation 3,304 ft/1,007 m. Winter sports on N slope of Dôme du GOÛTER peak, NW neighbor of MONT BLANC. Aerial tramway to Bellevue (elevation 5,945 ft/1,812 m; hotel).

Houck, unincorporated town, APACHE county, E ARIZONA, 30 mi/48 km SW of GALLUP, NEW MEXICO, on PUERCO River (Spanish, *Rio Puerco*), in S part of Navajo Indian Reservation. Sheep, cattle, crafts. Zuni Indian Reservation (New Mexico) to SE.

Houdain (oo-dan), industrial town (□ 2 sq mi/5.2 sq km; 2004 population 7,636), PAS-DE-CALAIS department, NORD-PAS-DE-CALAIS region, N FRANCE, 7 mi/11.3 km SW of BÉTHUNE, in coal-mining district; 50°27′N 02°32′E.

Houdan (oo-dahn), commune (□ 4 sq mi/10.4 sq km; 2004 population 3,140), YVELINES department, N central FRANCE, in the ÎLE-DE-FRANCE region, 14 mi/23 km NW of RAMBOUILLET; 48°47′N 01°36′E. Traditional market for poultry and specialized poultry pâté. Town has picturesque 15th–16th-century wooden houses and a large 12th-century keep.

Houdeng-Aimeries (oo-DAW–aim-REE), village, contiguous with **Houdeng-Goegnies** in commune of LA LOUVIÈRE, in Charleroi district of HAINAUT province, S BELGIUM, 10 mi/16 km E of MONS. Manufacturing.

Houègbo, town, BENIN, 35 mi/56 km N of COTONOU; 06°58′N 02°26′E. Oil palm, coffee.

Houei Sai (HOO-wai SEI), town, ⊙ BOKEO province, NW LAOS, on left bank of MEKONG River (head of navigation), opposite CHIANG KHONG (THAILAND), and 230 mi/370 km NW of VIENTIANE; 21°13′N 102°48′E. Trading center (cattle); shifting cultivation. Ruby mines nearby. Teak pagoda. Formerly called Fort Carnot. Lamet, Thai, and various minority peoples.

Houet, province (□ 4,468 sq mi/11,616.8 sq km; 2005 population 865,862), HAUTS-BASSINS region, SW BURKINA FASO; ⊙ BOBO-DIOULASSO; 11°20′N 04°15′W. Borders BANWA (N), MOUHOUN (NE), TUY (E), BOUGOURIBA (SE), COMOÉ (S), and KÉNÉDOUGOU (W) provinces. Drained by MOUHOUN, BOUGOURIBA, and COMOÉ rivers. Major railroad runs through BOBO-DIOULASSO to CÔTE D'IVOIRE. Agriculture (groundnuts, cotton, sorghum, rice); cotton ginning. Manganese deposits. A portion of this province was excised in 1997 when fifteen additional provinces were formed.

Houffalize (oo-fah-LEEZ), commune (2006 population 4,790), Bastogne district, LUXEMBOURG province, SE BELGIUM, on headstream of OURTHE RIVER and 9 mi/14.5 km NNE of BASTOGNE, in the ARDENNES. Tourist resort; cattle and pig market. In World War II suffered much destruction in Battle of the Bulge (1944–1945).

Hougang (HOU-gahng), town, NE central Singapore island, SINGAPORE, suburb 7 mi/11.3 km NNE of downtown SINGAPORE at N end of Tampines Expressway; 01°20′N 103°53′E. Includes village of Teck Hock. Manufacturing (printing and publishing, apparel, clothing, computer equipment). Paya Lebar Airport to SE.

Hougarde, BELGIUM: see HOEGAARDEN.

Houghton, county (□ 1,501 sq mi/3,902.6 sq km; 2006 population 35,334), ⊙ HOUGHTON, NW UPPER PENINSULA, MICHIGAN, includes S part of Keweenaw Peninsula, extending into LAKE SUPERIOR. Partly bounded SE by KEWEENAW BAY; drained by ONTONAGON and STURGEON rivers. Intersected NE by KEWEENAW WATERWAY and traversed by Copper Range; 46°58′N 88°39′W. Some manufacturing at HANCOCK and HOUGHTON. Cattle, poultry; agriculture (forage crops, oats), dairying, lumbering; tourism, resorts. F. J. McLain State Park in N; Mount Ripley Ski Area NW of Houghton; Twin Lakes State Park at center; several lakes are in county; S ⅓ of county is in Ottawa National Forest; S boundary coincides with Central/Eastern time zone boundary (county is in Eastern). Region settled by Finnish immigrants. Organized 1846.

Houghton, city (2000 population 7,010), ⊙ HOUGHTON county, NW UPPER PENINSULA, MICHIGAN, opposite HANCOCK, on KEWEENAW WATERWAY (port facilities); 47°06′N 88°33′W. Shipping, distribution, and industrial center for Keweenaw Peninsula. Manufacturing (wood products, publishing), lumbering; tourism; farming. Seat of Michigan Technological University; mineralogical museum; has mainland headquarters of ISLE ROYALE NATIONAL PARK; passenger ferry to Rock Harbor lodge, Isle Royale; vertical lift bridge to Hancock and rest of Keweenaw Peninsula. Mount

Ripley Ski Area to NW. Settled 1851, incorporated 1867.

Houghton (HO-tuhn), town, KING county, W WASHINGTON, on E shore of LAKE WASHINGTON, just S of KIRKLAND.

Houghton, village (2000 population 130), LEE county, SE IOWA, 19 mi/31 km NW of FORT MADISON; 40°46′N 91°36′W. Manufacturing (prefabricated metal buildings, farm machinery). Corn, oats; cattle, hogs.

Houghton (HOT-uhn), village (□ 2 sq mi/5.2 sq km; 2000 population 1,748), ALLEGANY county, W NEW YORK, on GENESEE River, and 22 mi/35 km S of WARSAW; 42°25′N 78°09′W. Farming. Seat of Houghton College (1923).

Houghton, NNE suburb of JOHANNESBURG, GAUTENG province, SOUTH AFRICA, part of metropolitan Johannesburg; 26°10′S 28°04′E. Elevation 5513 ft/1,680 m. High income and well established.

Houghton Heights or **Houghton Lake Heights**, town, Roscommon county, N central MICHIGAN, residential and resort community on SW shore of HOUGHTON LAKE; 44°19′N 84°46′W.

Houghton Lake, ROSCOMMON county, N central MICHIGAN, c.20 mi/32 km WNW of WEST BRANCH; 44°21′N 84°43′W. Largest lake of state (c.16 mi/26 km long, 7 mi/11.3 km. wide). Source of MUSKEGON RIVER. Houghton Lake, resort village, is on SW shore. Light manufacturing. Year-round fishing; winter sports; waterfowl feeding grounds. Villages of Prudenville, Houghton Lake, and HOUGHTON HEIGHTS on S shore.

Houghton-le-Spring (HOU-tuhn–luh–SPUHRING), town (2001 population 36,746), TYNE AND WEAR, NE ENGLAND, 6 mi/9.7 km SW of SUNDERLAND; 54°51′N 01°28′W. Formerly a coal-mining center, Houghton-le-Spring is now a market town.

Houghton, Port (HOO-tuhn), SE ALASKA, arm of STEPHENS PASSAGE, 75 mi/121 km SE of JUNEAU; 20 mi/32 km long, 57°19′N 133°27′W.

Houghton Regis (HOU-tuhn REE-jis), suburb (2007 population 17,000), S BEDFORDSHIRE, central ENGLAND, just N of DUNSTABLE; 51°03′N 00°20′W. Previously the site of limestone quarries and cement works. Has 14th-century church. Nearby (WSW) is quarrying village of SEWELL.

Houilles (oo-yuh), town (□ 1 sq mi/2.6 sq km), YVELINES department, ÎLE-DE-FRANCE region, N central FRANCE, an outer NW suburb of PARIS, 9 mi/14.5 km from Notre Dame Cathedral, within a S loop of the SEINE; 48°55′N 02°12′E. Diverse manufacturing (pharmaceuticals, biscuits, cottons).

Houlgate (ool-gaht), commune (□ 1 sq mi/2.6 sq km; 2004 population 1,908), CALVADOS department, BASSE-NORMANDIE region, NW FRANCE, bathing resort on ENGLISH CHANNEL just E of mouth of DIVES RIVER, 15 mi/24 km NE of CAEN and 7 mi/11.3 km SW of DEAUVILLE; 49°18′N 00°04′W. Fine sandy beach. Dark eroded cliffs overlook the sea 1 mi/1.6 km E.

Houli (HO-LEE), town, W central TAIWAN, 4 mi/6.4 km N of FENGYUAN and on railroad; 24°19′N 120°43′E. Sugar milling; rice, grapes.

Houlka, Mississippi: see NEW HOULKA.

Houlme, Le (ool-muh, luh), town (□ 1 sq mi/2.6 sq km), SEINE-MARITIME department, HAUTE-NORMANDIE region, N FRANCE, a suburb 5 mi/8 km NNW of ROUEN; 49°30′N 01°02′E. Linoleum manufacturing.

Houlton (HOLT-uhn), town, including Houlton village, ⊙ AROOSTOOK county, E MAINE, on MEDUXNEKEAG RIVER, and 100 mi/161 km NNE of BANGOR, near NEW BRUNSWICK border; 46°08′N 67°50′W. Port of entry. Trade, railroad, shipping center for large potato-growing area; commercial center for tourist region (hunting, fishing, canoeing). Light manufacturing. Airport is E of town.

Houlung (HO-LUNG), town, NW TAIWAN, minor port on W coast, on railroad, and 17 mi/27 km SW of HSINCHU. Rice, sweet potatoes, peanuts, watermelons; fish products (mullet, sardines).

Houma (HO-MA), city (□ 106 sq mi/275.6 sq km; 2000 population 178,480), S SHANXI province, NE CHINA, on the FEN RIVER; 35°36′N 111°15′E. Industry and commerce are the main sources of income. Crop growing accounts for most agricultural activity (vegetables, grains, swine, eggs). Manufacturing (machinery, electrical equipment).

Houma (HO-muh), city (2000 population 32,393), ⊙ TERREBONNE parish, SE LOUISIANA, 44 mi/71 km SW of NEW ORLEANS, on Bayou TERREBONNE; 29°35′N 90°43′W. Port on the INTRACOASTAL WATERWAY; junction with HOUMA NAVIGATION Canal to Gulf of MEXICO. Pontoon bridge crosses canal; also fifty-five other bridges in area. Leading industries include manufacturing (motor vehicles, oil field equipment and machinery, pneumatic relays, chemical tanks, valve parts, process control systems, slings, wire rope, drilling tools, sign brackets, tugboats); shipbuilding; printing and publishing; shrimp processing. Hosts Freedom Festival. Many fine antebellum homes. South Louisiana Trade School is here; nearby is a U.S. sugarcane experiment station. Founded in 1834; incorporated 1848.

Houma Navigation Channel (HO-muh), canal, c. 30 mi/48 km long, TERREBONNE parish, SE LOUISIANA, S extension of GULF INTRACOASTAL WATERWAY. Intersects WATERWAY S of HOUMA, goes S and SSE to TERREBONNE BAY, Gulf of MEXICO. Provides direct barge transport of offshore oil and natural gas to processing facilities in the HOUMA area and other parts of LOUISIANA.

Hound (HOWND), village (2001 population 6,846), S HAMPSHIRE, S ENGLAND, near HAMBLE RIVER estuary, 5 mi/8 km ESE of SOUTHAMPTON; 51°52′N 01°20′W. Agricultural market. Has 13th-century church.

Houndé (HOON-dai), town, ⊙ TUY province, HAUTS-BASSINS region, W BURKINA FASO, 65 mi/105 km ENE of BOBO-DIOULASSO; 11°34′N 03°31′W. On main E-W road to OUAGADOUGOU. Peanuts, shea nuts, subsistence crops; livestock. Manganese deposits nearby.

Hounslow (HOUNZ-lo), outer borough (□ 22 sq mi/ 57.2 sq km; 2001 population 212,341) of GREATER LONDON, SE ENGLAND, on the THAMES RIVER; 51°30′N 00°20′W. Manufacturing includes razor blades, soap, tires, biscuits, precision instruments, pharmaceuticals, and heating equipment. The town of FELTHAM has grown rapidly in the latter half of the 20th century due to its connection to HEATHROW AIRPORT. In 1016, Edmund Ironside defeated the Danes at Brentford. The Hounslow district of Heston and Isleworth was the site of the first stop on an important coach route to SOUTHAMPTON and BATH. The former Hounslow Heath, the location of a Roman camp, was a refuge for highwaymen; the area has become a military installation. The artist William Hogarth is buried in Chiswick; his house is a tourist attraction. Other districts of the borough include Isleworth, Gunnersbury, Hanworth, and Osterly Park.

Houpi (HO-PEE), town, W central TAIWAN, on railroad and 9 mi/14.5 km SSW of CHIAI. Railroad junction for KUANTZULING hot springs resort.

Houplines (oo-pleen), NE suburb (□ 4 sq mi/10.4 sq km) of ARMENTIÈRES, NORD department, NORD-PAS-DE-CALAIS region, N FRANCE, on the LYS (Belgian border); 50°42′N 02°55′E. Textile-weaving center.

Hourtin (oor-tan) or **Hourtins**, commune, GIRONDE department, AQUITAINE region, SW FRANCE, 33 mi/53 km NW of BORDEAUX, in the N LANDES district; 45°12′N 01°04′W. Fisheries. Bathing resort on Hourtin Lake (10 mi/16 km long, 2.5 mi/4 km wide), separated from Bay of BISCAY by a dune belt 2 mi/3.2 km wide.

Housatonic, village, Massachusetts: see GREAT BARRINGTON.

Housatonic (HOO-zuh-TAW-nik), river, c.130 mi/210 km long, W MASSACHUSETTS; rising in the BERKSHIRES; flows generally S through W CONNECTICUT into LONG ISLAND SOUND at STRATFORD. The river has long

been used as a source of power, with various hydroelectric plants in Connecticut.

Housay, Scotland: see OUT SKERRIES.

House, unincorporated village (2006 population 63), QUAY county, E NEW MEXICO, 44 mi/71 km SSW of Tucumcari; 34°38′N 103°54′W. Livestock, cotton, grain, alfalfa, vegetables.

House Harbour (HOUS) or **Havre-aux-Maisons** (AH-vr-o-mai-ZON), former village, E QUEBEC, E CANADA, on ALRIGHT ISLAND, one of the MAGDALEN ISLANDS; 47°24′N 61°49′W. Fishing port. Forms part of the Îles-de-la-MADELEINE agglomeration.

House Island, SW MAINE, small island off SOUTH PORTLAND; one of first settlements in CASCO BAY area; fortified since seventeenth century. Fort Scammel built 1808; rebuilt 1862; now abandoned.

House Springs, unincorporated town, JEFFERSON county, E MISSOURI, suburb 24 mi/39 km SW of downtown ST. LOUIS, on BIG RIVER. Light manufacturing.

Houston, district municipality, BRITISH COLUMBIA, CANADA, on YELLOWHEAD HIGHWAY, at confluence of BULKLEY and Morice rivers. Lumber; mining; cattle ranching.

Houston (HYOO-stuhn), district municipality (□ 28 sq mi/72.8 sq km; 2001 population 3,577), NW BRITISH COLUMBIA, W CANADA, 290 mi/465 km E of PRINCE RUPERT, 199 mi/320 km W of PRINCE GEORGE, and in BULKLEY-Nechako regional district; 54°24′N 126°40′W.

Houston (HYOO-stuhn), county (□ 581 sq mi/1,510.6 sq km; 2006 population 95,660), extreme SE Alabama; ⊙ Dothan, 31°09′N 85°19′W. Bounded E by Chattahoochee River and Georgia, S by Florida Rich agr. region (peanuts, corn, soybeans; hogs, beef cattle). Nuclear power plants Farley 1 (initial criticality August 9, 1977; max. dependable capacity of 814 MWe) and Farley 2 (initial criticality May 8, 1981; max. dependable capacity of 824 MWe) are 18 mi/29 km SE of Dothan. Uses cooling water from the Chatahoochee River. Formed 1903. Named for George Smith Houston, governor of AL, 1874–78.

Houston (HOUS-tuhn), county (□ 379 sq mi/985.4 sq km; 2006 population 127,530), central GEORGIA; ⊙ PERRY; 32°28′N 83°40′W. Bounded E by OCMULGEE River. Coastal plain agriculture (cotton, corn, melons, soybeans, wheat, peanuts, pecans, peaches; cattle, hogs, poultry); timber area. Formed 1821.

Houston (HYOOS-tuhn), county (□ 568 sq mi/1,476.8 sq km; 2006 population 19,832), extreme SE MINNESOTA; ⊙ CALEDONIA; 43°40′N 91°30′W. Drained by ROOT RIVER; bounded E by MISSISSIPPI RIVER (forms WISCONSIN state line), S by IOWA state boundary. Agricultural area (corn, oats, soybeans, hay, alfalfa; hogs, cattle, poultry; dairying); timber; sand and gravel; limestone. Beaver Creek Valley State Park in center; part of Richard J. Dorer Memorial Hardwood State Forest covers NW, center, and SE parts of county; parts of Upper Mississippi River National Wildlife Refuge in NE and SE. Winnebago Indian Reservation in NE. Formed 1854.

Houston, county (□ 207 sq mi/538.2 sq km; 2006 population 8,076), NW TENNESSEE; ⊙ ERIN; 36°17′N 87°43′W. Bounded W by KENTUCKY Reservoir (TENNESSEE RIVER). Timber; manufacturing; recreation is a growth industry. Formed 1871.

Houston, county (□ 1,236 sq mi/3,213.6 sq km; 2006 population 23,044), E TEXAS; ⊙ CROCKETT; 31°19′N 95°25′W. Bounded W by TRINITY RIVER, E by NECHES RIVER; includes part of Davy Crockett National Forest. Rolling wooded area (much timber); agriculture (hay; cotton; peanuts, pecans, grains, watermelons); cattle, hogs, horses. Some oil, natural gas; sand and gravel. Lumber milling, oil refining, manufacturing, produce processing. Houston County Lake reservoir to NW; Davy Crockett National Forest in E; Mission Tejas State Historical Park in NE. Formed 1837.

Houston, city (2000 population 1,992), S central MISSOURI, ⊙ TEXAS county, near Big Piney River in the OZARKS, 32 mi/51 km SW of SALEM; 37°19′N 91°57′W. Cattle, recreational center for the region; lumber and stave factory. Manufacturing (apparel).

Houston (HYOOS-tun), city (2006 population 2,144,491), ⊙ HARRIS county, SE TEXAS, corporate limits extend SW into FORT BEND county and N into MONTGOMERY county; 29°46′N 95°23′W. Elevation 55 ft/17 m. A deepwater port on the HOUSTON SHIP CHANNEL, it is the fourth-largest city in the nation, largest city in Texas and the largest city in the entire South and Southwest, U.S. port of entry; a great industrial, commercial, and financial hub; one of the world's major oil centers; and the third busiest tonnage-handling port in the U.S. (after NEW YORK and S LOUISIANA). Numerous space and science research firms; electronics plants; giant oil refineries; high-tech industries; computer technology; one of the world's greatest concentrations of petrochemical works; steel and paper mills; shipyards; breweries; meatpacking houses; and manufacturing (oil drilling equipment, clothing, glass, household items, seismic instruments). Major center of finance with a large number of banks, many of them foreign. The Texas Medical Center is the world's largest hospital complex and a leading medical research facility. Founded in 1836 by J. K. and A. C. Allen, it was named for Sam Houston, and served (1837–1839) as capital of the Texas Republic. In the course of the 19th century it grew to a prosperous railroad center. The digging (1912–1914) of the Houston Ship Channel made it a deepwater port and led to its expansion. Coastal oil fields, natural gas, sulfur, salt, and limestone deposits, and shipbuilding during World War II, poured quick wealth into the city. NASA Manned Spacecraft Center (1961; renamed the Lyndon B. Johnson Space Center in 1973) in SE brought the aerospace industry. It is the seat of Rice University, Texas Southern University, the University of Houston, the University of St. Thomas, Dominican College, Houston Baptist University, Sam Houston College, and two year community colleges. Its many parks include the large Hermann Park, which has a zoo, a museum of natural science, and a planetarium. Houston has several notable art museums, an arboretum, and a botanical garden. The civic center includes the Sam Houston Coliseum and Music Hall and the massive George R. Brown Convention Center (1987), one of the nation's largest; the Jesse H. Jones Hall for the Performing Arts, home of the symphony orchestra; and a convention and exhibit center, featuring the National Space Hall of Fame. Other tourist attractions include the Galleria and the Greenspoint Mall, two of the largest shopping centers in the U.S.; Old Market Square; Sam Houston Historical Park, which contains restored homes (built 1824–1868) and reconstructed buildings; the Astrodome (opened 1965, officially the Harris County Domed Stadium, part of convention center complex, and includes Astrohall and Astroarena) and its adjacent "Astroworld," an amusement center. San Jacinto Battleground is in nearby PASADENA. LAKE HOUSTON to NE; large Addicks and Barker flood control reservoirs in W part of city, are generally dry except during heavy rains and hurricanes. Sheldon State Wildlife Management Area to NE. Served by Houston Intercontinental Airport (N), Ellington Field and William P. Hobby Airport., both in SE part of city. Incorporated 1837.

Houston (HYOOS-tuhn), town (2000 population 1,202), S ALASKA, 30 mi/48 km N of ANCHORAGE, 25 mi/40 km W of PALMER, on Little Sustina River; 61°37′N 149°46′W. Town is located on GEORGE PARKS HIGHWAY (Route 3) and Alaska railroad, 12 mi/19 km NW of KNIK ARM of COOK INLET. Salmon fishing center.

Houston, town (2000 population 1,020), HOUSTON county, extreme SE MINNESOTA, 17 mi/27 km W of LA

CROSSE, WISCONSIN, on ROOT RIVER, W of confluence of South Fork Root River, in Richard J. Dorer Memorial Hardwood State Forest; 43°45′N 91°34′W. Grain, soybeans; livestock, poultry; dairying; manufacturing (button parts and button machines, feeds).

Houston, town (2000 population 4,079), a ⊙ CHICKA-SAW county (seat shared with OKOLONA), NE central MISSISSIPPI, 32 mi/51 km SSW of TUPELO; 33°53′N 89°00′W. Railroad terminus. In agricultural (cotton, corn; dairying; timber) area; manufacturing (fiber optics, furniture, apparel, cotton batting, paper products, carpet padding, lumber). Nearby are Indian mounds and Geology Hill. Part of Tombigbee National Forest in N and center. Incorporated 1837.

Houston (HOUS-tuhn), village (2001 population 6,610), RENFREWSHIRE, W Scotland, 6 mi/9.7 km W of RENFREW; 55°52′N 04°33′W. Commuters to PAISLEY and GLASGOW. Formerly in STRATHCLYDE, abolished 1996.

Houston (YOO-stuhn), village (2000 population 159), PERRY county, central ARKANSAS, 15 mi/24 km WSW of CONWAY, near ARKANSAS RIVER (N) and FOURCHE LA FAVE RIVER (S); 35°01′N 92°41′W.

Houston, village (2000 population 430), KENT county, central DELAWARE, 16 mi/26 km S of DOVER and 4 mi/6.4 km W of MILFORD; 38°55′N 75°30′W. Elevation 19 ft/5 m. In agricultural area.

Houston (HYOOS-tuhn), borough (2006 population 1,264), WASHINGTON county, SW PENNSYLVANIA, suburb 2 mi/3.2 km SW of CANONSBURG and 5 mi/8 km N of WASHINGTON, on Chartiers Creek. Manufacturing (cement blocks, plastic products, steel). Laid out 1871, incorporated 1901.

Houston Acres (YOO-stuhn), village (2000 population 491), JEFFERSON county, N KENTUCKY, residential suburb 7 mi/11.3 km ESE of downtown LOUISVILLE; 38°12′N 85°36′W.

Houstonia, town (2000 population 275), PETTIS county, central MISSOURI, 15 mi/24 km NNW of SEDALIA; 38°53′N 93°21′W.

Houston Lake, village (2000 population 284), PLATTE county, E MISSOURI, residential suburb 7 mi/11.3 km NNW of downtown KANSAS CITY, just N of RIVER-SIDE; 39°11′N 94°37′W.

Houston, Lake, reservoir (□ 19 sq mi/49.4 sq km), HARRIS county, SE TEXAS, on SAN JACINTO RIVER, 25 mi/40 km ENE of HOUSTON; 29°55′N 95°08′W. Maximum capacity 281,800 acre-ft. Formed by Lake Houston Dam (66 ft/20 m high), built (1964) for water supply; also used for irrigation and recreation. Lake Houston State Park at N end of reservoir.

Houston Ship Channel, S TEXAS, dredged deepwater channel c.50 mi/80 km long, connecting port of HOUSTON with the GULF OF MEXICO via BUFFALO BAYOU, SAN JACINTO RIVER, and GALVESTON BAY. Vehicular tunnels under Buffalo Bayou between PA-SADENA (just E of Houston) and Salena Park and under San Jacinto River section, between BAYTOWN and LA PORTE. Development began 1912; channel has since been deepened and widened to accommodate large vessels. San Jacinto Monument and Battleship Texas on S shore.

Houtbaai, river, SOUTH AFRICA: see HOUT BAY.

Hout Bay, Afrikaans *Houtbaai,* town, on the Houtbaai River, WESTERN CAPE province, SOUTH AFRICA, on Hout Bay (3 mi/4.8 km long, 8 mi/12.9 km wide) of the Atlantic, 10 mi/16 km SSW of CAPE TOWN; 34°02′S 18°21′E. Rock-lobster and fishing harbor; fruit. Nearby fort, dating from 1796 during Dutch period, is now national monument. Popular resort town with boating and launch trips to TABLE BAY.

Houten (HOU-tuhn), town (2001 population 32,243), UTRECHT province, W central NETHERLANDS, 5 mi/8 km SSE of UTRECHT; 52°02′N 05°10′E. AMSTERDAM-RIJN Canal passes to S. Dairying; cattle, hogs, poultry; vegetables, fruit.

Houth, YEMEN: see HUTH.

Houthalen-Helchteren (HOUT-hah-luhn–HEL-tuh-ruhn), commune (2006 population 29,992), in Maaseik district, LIMBURG province, NE BELGIUM, 6 mi/9.7 km N of HASSELT. Kelchterhof Domein recreational park nearby.

Houthulst (HOUT-hulst), commune (2006 population 9,056), Diksmuide district, WEST FLANDERS province, W BELGIUM, 8 mi/13 km WNW of ROESELARE.

Houtman Abrolhos (HOOT-muhn, uh-BROL-hahs), small archipelago in INDIAN OCEAN, 35 mi/56 km off W coast of WESTERN AUSTRALIA state, W AUSTRALIA; 28°18′S 113°36′E–29°21′S 114°11′E. Extends 50 mi/80 km N-S. Comprises 3 coral groups: WALLABI ISLANDS (N), EASTER ISLANDS (central), PELSART ISLANDS (S). Low, sandy; mangrove forests. Tourist resorts. Sometimes called Houtman Rocks and, popularly, Abrolhos Islands.

Hou Tsang, CHINA: see ZANG.

Houtskär (HOT-sher), Swedish *Houtskari,* island, TURUN JA PORIN province, SW FINLAND; 7 mi/11.3 km long, 1 mi/1.6 km–3 mi/4.8 km wide; 60°12′N 21°22′E. Elevation 17 ft/5 m. In strait between BALTIC SEA and GULF OF BOTHNIA, 35 mi/56 km WSW of TURKU. Houtskär fishing village is on N coast. Population is mostly Swedish-speaking.

Houtzdale (HOUTS-dail), borough (2006 population 895), CLEARFIELD county, central PENNSYLVANIA, 22 mi/35 km N of ALTOONA. Logging; surface bituminous coal mining, clay. Laid out 1870, incorporated 1872.

Houyet (oo-YE), commune (2006 population 4,506), Dinant district, NAMUR province, S central BELGIUM, 7 mi/11 km SE of DINANT, in the ARDENNES.

Hou Zang, CHINA: see ZANG.

Hova (HOO-vah), village, SKARABORG county, S SWE-DEN, 18 mi/29 km NE of MARIESTAD; 58°51′N 14°13′E.

Hovd, province (□ 29,300 sq mi/76,180 sq km), W MONGOLIA; ⊙ HOVD; 47°00′N 92°30′E. Bounded S by CHINA'S XINJIANG UYGUR AUTONOMOUS REGION, it is traversed by the MONGOLIAN ALTAY, sloping N into the basin of HAR US NUUR (lake) and S to Dzungarian Gobi (desert). Population (largely West Mongol) engages in stock grazing and some irrigated agriculture. May also be expressed as Kobdo or Khobdo.

Hovd, [Mongolian=happy (city)], town, ⊙ HOVD province, W MONGOLIA, at N foot of the MONGOLIAN ALTAY, 700 mi/1,100 km W of ULAANBAATAR, W of HAR US NUUR (lake), and on small Buyantu River (one on the lake's tributaries); 48°01′N 91°38′E. Elevation 4,613 ft/1,406 m. Main trading center of W Mongolia; processing of livestock products (hides, wool). Linked with RUSSIA by Chuya highway. In appearance like the cities of Central ASIA, Hovd was founded in 1731 as a Manchu fort on HOVD GOL (N), but was moved in 1763 to present site because of floods. It developed as a center of trade with Russia and CHINA. May also be expressed as Kobdo or Khobdo. Since 1928 officially Dzhirgalantu or Jirgalanta.

Hovden, NORWAY: see BØ.

Hovd Gol, river, 320 mi/516 km long, W MONGOLIA; rises near boundaries of Mongolia, RUSSIA, CHINA and KAZAKHSTAN in TAVAN BOGD UUL (lake) in the MONGOLIAN ALTAY; flows past town of ÖLGIY and SE to HAR US NUUR (lake) near town of HOVD. It is the longest river in the Mongolian Altay and is the longest river within Mongolia, after the DZAVHAN GOL, without an outlet to the sea. Its interior drainage basin is c.20,000 sq mi/50,000 sq km. Its upper reaches flow thorough the HOVD (Kobdo) Lakes. Its lower courses divide into three main branches and form a marshy delta in the Basin of the Great Lakes. Since it is fed by high mountain snow, its high water comes in May and June. May also be expressed as Kobdo River or Khobdo River.

Hovd Lakes or **Kobdo Lakes,** MONGOLIA, freshwater mountain lakes on the N slopes of the MONGOLIAN ALTAY on the upper HOVD GOL (Kobdo River), c.75 mi/120 km SW of ÖLGIY; 48°37′N 88°29′E. The higher

HOTON NUUR (Choton Lake; □ 23 sq mi/60 sq km) is 14 mi/23 km long, c.3 mi/5 km wide, and 120 ft/37 m deep; the lower Horgon Nuur (Khurgan Nur or Lake Chorgon; □ 30 sq mi/78 sq km) is 14 mi/22 km long, up to 4 mi/6 km wide, and up to 56 ft/17 m deep. Both lakes are rich in fish.

Hove (HO-vuh), commune (2006 population 8,310), Antwerp district, ANTWERPEN province, N BELGIUM, 5 mi/8 km SSE of ANTWERP.

Hove, ENGLAND: see BRIGHTON and Hove.

Hovedøya (HAW-vuhd-uh-yah), islet in OSLOFJORD, just S of OSLO city center; 59°52′N 10°44′E. Opposite AKERSHUS fortress. Seaside resort. Has remains of Cistercian abbey, founded 1147 by English monks and destroyed 1532.

Hövel, GERMANY: see BOCKUM-HÖVEL.

Hövelhof (HUH-fel-huhf), town, NORTH RHINE-WESTPHALIA, N GERMANY, 8 mi/12.9 km NW of PADERBORN; 51°49′N 08°39′E. Woodworking; manufacturing of paper.

Hoven, village (2006 population 431), POTTER county, N central SOUTH DAKOTA, 18 mi/29 km NNE of GET-TYSBURG; 45°14′N 99°46′W. Manufacturing (cheese and ice cream).

Hovenäset (HOO-ve-NES-et), village, GÖTEBORG OCH BOHUS county, SW SWEDEN, on SKAGERRAK, 7 mi/11.3 km NW of LYSEKIL; 58°22′N 11°17′E. Seaside resort.

Hovenweep National Monument (□ 1 sq mi/2.6 sq km), SAN JUAN county and MONTEZUMA county, SW COLORADO. Six groups of pre-Columbian cliff dwellings, towns and pueblos. Utah contains 440 acres/178 ha of the cliffs, comprising one unit; Colorado contains the other unit, with an area 345 acres/140 ha. Authorized 1923.

Hoverla (ho-VER-lah), (Polish *Howerla),* (Russian *Goverla),* highest peak (6,760 ft/2,060 m) in CHOR-NOHORA section of the CARPATHIANS and in all of Ukrainian Carpathians, SW UKRAINE, 12 mi/19 km W of VERKHOVYNA. PRUT RIVER rises at E foot.

Hovey Lake (HUH-vee), POSEY county, extreme SW INDIANA, in Hovey Lake State Fish and Wildlife Area, 9 mi/14.5 km SSW of MOUNT VERNON; 37°48′N 87°56′W. Natural backwater lake of OHIO River (1 mi/1.6 km E and S).

Høvik (HUH-veek), village, AKERSHUS county, SE NORWAY, at head of OSLOFJORDEN, on railroad, and 6 mi/9.7 km WSW of OSLO city center. Metalworking; glass manufacturing; seaside resort. The Henie-Onstad cultural center is located at Høvikodden.

Hovmantorp (HOOV-mahn-TORP), town, Kronoberg county, S SWEDEN, on N side of Lake Rottnen (7 mi/11.3 km long, 1 mi/1.6 km–4 mi/6.4 km wide), 14 mi/23 km ESE of VÄXJÖ; 56°47′N 15°09′E.

Hövsgöl, province, NW MONGOLIA; ⊙ MÖRÖN; 50°00′N 100°00′E. Bounded N and NW by TUVA and BURYAT republics of RUSSIAN FEDERATION, the generally mountainous province contains HÖVSGÖL NUUR (Lake Khubsugul) in forested section (N) and a wooded steppe section (S), drained by SELENGA River. Population includes Tuvinians. Agriculture (S), hunting and yak raising (N). Graphite, gold, silver-lead, and copper deposits. Sometimes spelled Hubsugul, Khubsugul, or Kosogol. Also Hösbögöl.

Hövsgöl Nuur or **Khubsugul, Lake,** second-largest lake (□ 1,010 sq mi/2,626 sq km) of MONGOLIA (after UVS NUUR), and its largest freshwater body, near RUSSIA border, 130 mi/210 km from SW end of Lake BAIKAL, and 350 mi/560 km NW of ULAANBAATAR; 780 ft/238 m deep, 80 mi/134 km long (N-S) and 20 mi/35 km wide; 51°00′N 100°30′E. Elevation 5,400 ft/1,645 m. Situated amid picturesque, steep, wooded mountains, at S foot of EAST SAYAN RANGE; has clear water and an abundance of fish. Frozen December-May; its outlet is the EGIYN GOL (river). Steamer navigation between TURTU (N) and HATGAL (S) forms part of a Russia-Mongolia trade route. Sometimes called Kosogol. Also spelled Lake Höbsögöl or Lake Hubsugul.

Hövüün or **Noyan**, village, Ömnögovĭ province, S MONGOLIA, at foot of Noyon Uul (7,884 ft/2,403 m), 125 mi/200 km WSW of DALANDZADGAD, in GOBI DESERT; 43°09′N 102°07′E.

Howaizeh (huh-wah-yee-ZEH), town, Khuzestān province, SW IRAN, 40 mi/64 km WNW of AHVAZ, in Arab tribal area, near border with IRAQ. Grain, cotton, rice, dates. Also called HOWAIZEH or Hoveyzeh.

Howald, town, HESPERANGE commune, S LUXEMBOURG, 2 mi/3.2 km SE of LUXEMBOURG city; 49°35′N 06°09′E. Luxembourg city limits. Some industry.

Howard, county (□ 595 sq mi/1,547 sq km; 2006 population 14,415), SW ARKANSAS; ☉ NASHVILLE; 34°05′N 93°59′W. Drained by SALINE (forms most of W boundary) and Cassatot rivers. Agriculture (cattle, hogs, chickens). Cotton ginning, sawmilling; manufacturing of wood and cement products. Timber; cinnabar mines. Part of MILLWOOD LAKE Reservoir in extreme SW corner; part of Lake Greerson Wildlife Management Area in NE; Dierks Lake Reservoir (Saline River) on W boundary; Gillham Lake Reservoir (Cassatat River) in NW; Howard County Wildlife Management Area in NW. Formed 1873.

Howard, county (□ 293 sq mi/761.8 sq km; 2006 population 84,500), central INDIANA; ☉ KOKOMO; 40°29′N 86°07′W. Rich agricultural area (corn, soybeans, wheat; hogs, cattle, poultry); diversified manufacturing. Drained by WILDCAT CREEK. Formed 1844.

Howard, county (□ 473 sq mi/1,229.8 sq km; 2006 population 9,677), NE IOWA, on MINNESOTA line (N); ☉ CRESCO, located at E end of county; 43°21′N 92°19′W. Rolling prairie agricultural area (hogs, cattle, poultry; corn, oats; dairying) drained by Upper IOWA, WAPSIPINICON, and TURKEY rivers. Limestone quarries, sand and gravel pits. Hayden Prairie State Preserve in NW; Lidtke Mill in N, near Lime Springs. Formed 1851.

Howard, county (□ 253 sq mi/657.8 sq km; 2006 population 272,452), central MARYLAND; ☉ ELLICOTT CITY; 39°15′N 76°56′W. Bounded NE by PATAPSCO River, W and SW by PATUXENT River. Mostly rolling piedmont, with SE part in coastal plain. Agricultural produce (dairy products, poultry, vegetables, apples, some grain) marketed in metropolitan district of BALTIMORE (E). Also manufacturing, furniture, chemicals, fabricated metals, electronics. Named for College John Edgar Howard, Revolutionary War hero and governor of Maryland from 1788 to 1791. Until the appearance of the planned community of COLUMBIA, the county was almost entirely agricultural. It is the only Maryland county that borders near the CHESAPEAKE BAY for the state boundaries. Located entirely on the Piedmont Plateau. Formed in 1851.

Howard, county (□ 469 sq mi/1,219.4 sq km; 2006 population 9,949), central MISSOURI; ☉ FAYETTE; 39°08′N 92°42′W. Borders MISSOURI River on W and S. Corn, wheat, apples, soybeans; cattle, hogs, poultry; manufacturing at GLASGOW and Fayette. Floods of 1993 caused damage in Glasgow and NEW FRANKLIN. Formed 1816. Known as "Mother of Missouri counties" for its original large size and for so many counties having been formed from it. Center of the highest Boonslick region. Boonslick State Park in SW. Original E terminus of the SANTA FE TRAIL in the 1820s.

Howard, county (□ 575 sq mi/1,495 sq km; 2006 population 6,736), E central NEBRASKA; ☉ ST. PAUL; 41°13′N 98°31′W. Agricultural region drained by North Loup and Middle Loup rivers. Cattle, hogs, corn, alfalfa, soybeans, dairying, poultry products. North Loup State Wayside Area at center of county. Formed 1871.

Howard, county (□ 904 sq mi/2,350.4 sq km; 2006 population 32,463), NW TEXAS; ☉ BIG SPRING; 32°18′N 101°26′W. Rolling plains, drained by BEALS CREEK, Morgan Creek, and Mustang Draw. Elevation

2,200 ft/671 m–2,800 ft/853 m. Livestock (beef cattle); agriculture in E (cotton, wheat, vegetables, black-eyed peas, sesame). Oil, natural gas fields; oil and gas refining, stone, sand and gravel. Includes Big Spring State Park. Formed 1876.

Howard, city (2006 population 936), ☉ MINER county, SE central SOUTH DAKOTA, 45 mi/72 km NW of SIOUX FALLS; 44°00′N 97°31′W. Near West Fork of VERMILLION RIVER. Plotted 1881.

Howard, city (2006 population 16,219), BROWN county, E WISCONSIN, suburb 4 mi/6.4 km NW of GREEN BAY city, on GREEN BAY, arm of LAKE MICHIGAN; 44°34′N 88°04′W. Railroad junction. Oneida Indian Reservation to SW.

Howard, town (2000 population 808), ☉ ELK county, SE KANSAS, on headstream of ELK RIVER and 55 mi/89 km ESE of WICHITA; 37°28′N 96°15′W. Elevation 1,000 ft/305 m. Trade center for livestock region. Manufacturing (lumber and wood products). County museum Founded 1870, incorporated 1877.

Howard (HOU-uhrd), village, QUEENSLAND, AUSTRALIA, 150 mi/241 km N of BRISBANE, 23 mi/37 km W of HERVEY BAY; 25°19′S 152°34′E. Sugar plantations; orchards (citrus fruits).

Howard, borough (2006 population 662), CENTRE county, central PENNSYLVANIA, 9 mi/14.5 km NE of BELLEFONTE, on SE shore of Sayer (Blanchard) Lake reservoir (Bald Eagle Creek). Manufacturing (apparel). Bald Eagle State Park to NE and NW; Howard State Nursery to SW.

Howard A. Hanson Reservoir (□ 31 sq mi/80 sq km), KING county, W central WASHINGTON, on GREEN RIVER, in CASCADE MOUNTAINS, in Mount Baker–Snoqualmie National Forest, 35 mi/56 km SE of SEATTLE; 47°17′N 121°47′W. Maximum capacity 136,700 acre-ft. Formed by Howard A. Hanson Dam (235 ft/72 m high), built (1962) by Army Corps of Engineers for flood control.

Howard Beach, a residential neighborhood of S QUEENS borough of NEW YORK city, SE NEW YORK, on N shore of JAMAICA BAY; 40°39′N 73°51′W. Many houses here are located along the water. Population mostly Italian. In 1986, an assault by white teenagers upon three passing African-Americans (one of whom died) resulted in what became known as the Howard Beach case; the teens were convicted of manslaughter.

Howard City, town (2000 population 1,585), MONTCALM county, central MICHIGAN, 32 mi/51 km NNE of GRAND RAPIDS and on short Tamarack River; 43°23′N 85°28′W. In agricultural and lake-resort area. Agriculture (grain, potatoes, beans, vegetables, apples); livestock; dairy products; manufacturing (paper roller headers).

Howard City, Nebraska: see BOELUS.

Howard Field, military reservation, at SE foot of Cerro Galera, adjoining PANAMA city. FUERTE KOBBE adjoins E. Will come under Panamanian control after December 31, 1999.

Howard Lake, town (2000 population 1,853), WRIGHT county, S central MINNESOTA, 19 mi/19 km SW of BUFFALO, at S end of Howard Lake; 45°03′N 94°04′W. Diversified-farming area (poultry; grain, soybeans; dairying); manufacturing (feeds, kitchen cabinets, pewter awards, egg processing, wooden wagon wheels). Dutch Lake to SE.

Howard Mountain, Colorado: see NEVER SUMMER MOUNTAINS.

Howard Prairie Lake, reservoir, JACKSON county, SW OREGON, on Jenny Creek, in Howard Prairie Recreation Area, 25 mi/40 km ESE of MEDFORD; 5 mi/8 km long; 42°12′N 122°22′W. Maximum capacity 76,700 acre-ft. Formed by Howard Prairie Dam (82 ft/25 m high), built (1958) for irrigation, power generation, and flood control.

Howards Grove, town (2006 population 3,071), SHEBOYGAN county, E WISCONSIN, 8 mi/12.9 km NW of

SHEBOYGAN; 43°49′N 87°49′W. Dairying, grain, fruit, vegetables. Manufacturing (wood pallets).

Howardville, village, NEW MADRID county, SE MISSOURI, 3 mi/4.8 km WSW of NEW MADRID in the MISSISSIPPI alluvial plain; 36°34′N 89°35′W. Residential and commercial area.

Howden (HOU-duhn), town (2001 population 4,454), East Riding of Yorkshire, NE ENGLAND, near OUSE RIVER, 3 mi/4.8 km N of GOOLE; 53°44′N 00°52′W. Agricultural market. Has church dating mainly from 14th century.

Howe, town (2006 population 2,717), GRAYSON county, N TEXAS, 9 mi/14.5 km S of SHERMAN; 33°30′N 96°36′W. In agricultural area (cattle; peanuts); manufacturing (sheet metal fabrication).

Howe, village, LAGRANGE county, NE INDIANA, 5 mi/8 km N of LAGRANGE, on PIGEON RIVER. Pigeon River State Fish and Wildlife Area to E. Manufacturing (metal stamping, food products, mobile homes). Cattle, poultry; corn, wheat.

Howe, village (2006 population 721), LE FLORE county, E OKLAHOMA, suburb 7 mi/11.3 km S of POTEAU; 34°57′N 94°38′W. Railroad junction. In agricultural area.

Howe, Cape (HOU), southeasternmost point of AUSTRALIA, on boundary between NEW SOUTH WALES and VICTORIA, Australia; forms S end of entrance to DISASTER BAY of TASMAN SEA; 37°31′S 149°58′E.

Howe Caverns, New York: see HOWES CAVE.

Howe Island (HOU), (□ 12 sq mi/31.2 sq km), SE ONTARIO, E central CANADA, one of the THOUSAND ISLANDS, in the SAINT LAWRENCE RIVER, near its outlet from Lake ONTARIO, 8 mi/13 km E of KINGSTON; 8 mi/13 km long, 3 mi/5 km wide; 44°16′N 76°17′W. Separated from mainland by narrow Bateau Channel.

Howell, county (□ 920 sq mi/2,392 sq km; 2006 population 38,734), S MISSOURI, ☉ WEST PLAINS; 36°46′N 91°53′W. In the OZARKS; drained by ELEVEN POINT RIVER. Livestock, cattle, goats, horses, and agricultural region (corn, hay); oak, cedar, pine timber; stone quarries. Manufacturing at West Plains, WILLOW SPRINGS, and MOUNTAIN VIEW. Mark Twain National Forest in NW part. Tourism, canoeing. Formed 1857.

Howell, town, ECHOLS county, S GEORGIA, 13 mi/21 km E of VALDOSTA, near ALAPAHA RIVER; 30°49′N 83°03′W.

Howell, town (2000 population 9,232), ☉ LIVINGSTON county, SE MICHIGAN, 33 mi/53 km ESE of LANSING, on small Thompson Lake; 42°36′N 83°56′W. Railroad junction. In agricultural and dairying area. Manufacturing (electronic equipment, chemicals, metal products, lubricants and hydarulic fluids, metal plating, plastic molding, transformers, soft drinks, baking containers, electric breakers, hospital supplies, aluminum wheels); summer resort. Airport. Numerous lakes to SE; Mount Brighton Ski Area to NW. Settled 1834; incorporated as village 1863, as city 1915.

Howell, township, MONMOUTH county, NE NEW JERSEY, 8 mi/12.9 km W of ASBURY PARK; 40°10′N 74°11′W. Light industry. Incorporated 1801.

Howell, village (2006 population 229), BOX ELDER county, NW UTAH, 30 mi/48 km NW of BRIGHAM CITY; 41°46′N 112°26′W. Alfalfa, wheat, barley; dairying; cattle, sheep. GOLDEN SPIKE National Historical Site 15 mi/24 km SSW.

Howells, village (2006 population 617), COLFAX county, E NEBRASKA, 20 mi/32 km NNE of SCHUYLER, and on branch of ELKHORN RIVER; 41°43′N 97°00′W. Grain; cheese.

Howe of the Mearns (HOU uhv the MERNZ), fertile lowland area in SE ABERDEENSHIRE, NE Scotland, centered on LAURENCEKIRK, extending E to NORTH SEA coast. The Mearns is ancient name of Kincardine. Formerly in Grampian, abolished 1996.

Howerla, UKRAINE: see HOVERLA.

Howes Cave, village, SCHOHARIE county, E central NEW YORK, 32 mi/51 km W of ALBANY; 42°41′N 74°23′W.

Tourist trade attracted by Howe Caverns, among largest in NE U.S., with underground stream and lake. Nearby are Secret Caverns.

Howe Sound (HOU), inlet of Strait of GEORGIA, SW BRITISH COLUMBIA, W CANADA, 10 mi/km NW of VANCOUVER; 26 mi/42 km long, 1 mi/2 km–10 mi/16 km wide; 49°22′N 123°18′W. Receives Squamish River at head; contains GAMBIER, BOWEN, ANVIL, and KEATS islands. Fishing and lumbering area. Copper mining at BRITANNIA Beach on E shore, now closed. Recreation all season. PORT MELLON, LANGDALE, GIBSONS on W shore; SQUAMISH, Brittania Beach, and HORSESHOE Bay on E shore.

Howey-in-the-Hills (HOU-ee), village (□ 1 sq mi/2.6 sq km; 2000 population 956), Lake county, central FLORIDA, 10 mi/16 km SE of LEESBURG, on LAKE HARRIS; 28°43′N 81°46′W. Citrus-fruit packing and canning. Conference center.

Howick (HOU-ik), township (□ 111 sq mi/288.6 sq km; 2001 population 3,779), S ONTARIO, E central CANADA, 34 mi/54 km from GODERICH; 43°53′N 81°04′W. Primarily rural. Established 1854. Includes the villages of FORDWICH, GORRIE, and WROXETER, and the hamlets of Belmore and Lakelet.

Howick, town, central KWAZULU-NATAL province, SOUTH AFRICA, on UMGENI RIVER and 12 mi/19 km NW of POLOKWANE (PIETERMARITZBURG); 29°30′S 30°06′E. Elevation 4,232 ft/1,290 m. Lumbering industry; resort. Just E are karkloobfalls (364 ft/111 m high) of UMGENI RIVER; hydroelectric power. Midmar Dam 1.9 mi/3 km to SW is popular resort area.

Howick (HOU-ik), village, Montérégie region, SW QUEBEC, E CANADA, near Châteauguay River, 15 mi/24 km ESE of SALABERRY-DE-VALLEYFIELD; 45°11′N 73°51′W. Dairying.

Howkan Island, Alaska: see LONG ISLAND.

Howland (HOW-lund), town, PENOBSCOT county, central MAINE, 30 mi/48 km N of BANGOR and on PENOBSCOT RIVER, at mouth of the PISCATAQUIS; 45°15′N 68°42′W. Plywood, paper mills. Incorporated 1826.

Howland (HOU-luhnd), township (□ 18 sq mi/46.8 sq km), TRUMBULL county, NE OHIO, 3 mi/5 km E of WARREN; 41°15′N 80°45′W. Often called Howland Corners or Howland Center.

Howland Island, uninhabited island (□ 73 sq mi/189.8 sq km), central PACIFIC near the equator, c.1,620 mi/ 2,600 km SW of HONOLULU; 00°48′N 176°38′W. The island was discovered by American traders and was claimed by the U.S. in 1856, along with JARVIS and BAKER islands. The 3 islands were worked for guano deposits by British and American companies during the 19th century. The guano industry declined, and the islands were forgotten until they became a stop on the air route to AUSTRALIA. American colonists were brought from Hawaii in 1935 in order to establish U.S. control against British claims, but the colony was disbanded at the outbreak of World War II. While en route to Howland Island in 1937, the aviator Amelia Earhart was lost in the Pacific. Howland Island is under the U.S. Department of the Interior's Fish & Wildlife Service. It is also a haven for migratory birds.

Howlong (HOU-lahng), town, NEW SOUTH WALES, SE AUSTRALIA, 371 mi/597 km SW of SYDNEY, on New South Wales–VICTORIA border; 35°58′S 146°38′E.

Howrah, INDIA: see HAORA.

Howson Peak (HOU-suhn) (9,000 ft/2,743 m), W central BRITISH COLUMBIA, W CANADA, in COAST MOUNTAINS, 30 mi/48 km SW of SMITHERS; 54°25′N 127°45′W.

Howth (HOTH), Gaelic *Binn Éadair*, suburb (2006 population 8,196) of DUBLIN, E DUBLIN county, E central IRELAND, on the IRISH SEA, 9 mi/14.5 km ENE of Dublin, on a peninsula at foot of HILL OF HOWTH; 53°23′N 06°04′W. Fishing port and seaside resort. Has 16th-century castle and 14th-century church. Until 1830, when superseded by Dún Laoghaire (Kings-

town), Howth was terminal of Dublin mail steamers from England. Part of Dublin since 1940. Guns were landed here in 1914 for Irish Volunteers by Erskine Childers, author of "Riddle of the Sands."

Howth, Hill of (HOTH), Gaelic *Binn Éadair*, hill on a peninsula (5 mi/8 km long, 2 mi/3.2 km wide) in the IRISH SEA, E DUBLIN county, E central IRELAND, forming N shore of DUBLIN BAY; rises to 560 ft/171 m; 53°23′N 06°04′W. At E extremity is Lion's Head promontory, with fishing village of Baily or Bailey; site of lighthouse. Built 1814.

Hoxie (HAHK-see), town (2000 population 2,817), LAWRENCE county, NE ARKANSAS, 20 mi/32 km NW of JONESBORO, between BLACK and CACHE rivers; 36°02′N 90°58′W. Railroad junction; manufacturing (machinery).

Hoxie, town (2000 population 1,244), ⊙ SHERIDAN county, NW KANSAS, 30 mi/48 km E of COLBY; 39°21′N 100°26′W. Shipping and trading point in grain and livestock region.

Hoxsie, Rhode Island: see WARWICK.

Höxter (HUHKS-ter), town, NORTH RHINE-WESTPHALIA, central GERMANY, on left bank of the WESER RIVER, 26 mi/42 km E of PADERBORN; 51°47′N 09°23′E. Manufacturing of chemicals, paper; cement works; wood- and metalworking. Has 13th-century church, 14th-century town hall. Was member of HANSEATIC LEAGUE. Just E are baroque buildings of noted former Benedictine abbey of CORVEY, or KORVEY (founded 822; secularized 1803), with Romanesque church. Hoffmann von Fallersleben buried here. Architecture department of the University of Paderborn is here.

Hoxtolgay (HUH-SHI-TO-LUH-GEI), town, N XINJIANG UYGUR, NW CHINA, 140 mi/225 km E of TACHENG, and on highway; 46°34′N 86°00′E. Coal-mining center; cattle raising; agricultural products.

Hoxton, ENGLAND: see HACKNEY.

Hoy (HOI), island (□ 55 sq mi/143 sq km), off N Scotland, second largest of the ORKNEY ISLANDS; 58°51′N 03°16′W. At the SW side of the SCAPA FLOW anchorage. Ward Hill (1,565 ft/477 m) is one of many hills on the island; magnificent cliffs line the shore. Farms in NE; midland is barren moor. The Old Man of Hoy, a sandstone pinnacle 450 ft/137 m high, is a famous landmark for sailors. The Dwarfie Stane, a huge sandstone block with hollowed rooms inside, is a Viking relic.

Hoya (HO-yah), city, Tokyo prefecture, E central HONSHU, E central JAPAN, on the Shakoji River, 5.6 mi/9 km NW of SHINJUKU; 35°44′N 139°33′E. Residential suburb of TOKYO.

Hoya (HOI-ah), town, LOWER SAXONY, N GERMANY, on the WESER, 8 mi/12.9 km SSW of VERDEN; 52°48′N 09°08′E. Manufacturing of furniture, paper, pharmaceuticals.

Hoya-Gonzalo (OI-ah-gon-THAH-lo), town, ALBACETE province, SE central SPAIN, 17 mi/27 km ESE of ALBACETE; 38°57′N 01°33′W. Livestock; wine, saffron, cereals.

Hoyales de Roa (oi-AH-les dhai RO-ah), village, BURGOS province, N SPAIN, 9 mi/14.5 km W of ARANDA de DUERO. Cereals, vegetables, grapes.

Høyanger (HUH-uh-ahn-guhr), village, SOGN OG FJORDANE county, W NORWAY, landing on N shore of SOGNEFJORDEN, 65 mi/105 km NNE of BERGEN. Large hydroelectric works; aluminium-smelting center. Tourist center. Has county hospital.

Høyer, Denmark: see HØJER.

Hoyerswerda (hoi-ers-VER-dah), city, SAXONY, SE GERMANY, on the BLACK ELSTER RIVER; 51°27′N 14°15′E. Located in a lignite-mining area, it is an industrial city; manufacturing of glass, bricks, and other products. Industry has declined since the 1990 German unification. Chartered 1371.

Hoylake and West Kirby (HOI-laik and WEST KUH-bee), resorts (2001 population 19,300), MERSEYSIDE, NW ENGLAND, on DEE RIVER at NW end of WIRRAL

Peninsula; 53°23′N 03°11′W. Championship golf course at Hoylake. Hilbre Islands, off West Kirby, are a nature reserve and accessible at low tide.

Hoyland Nether (HOI-land, NE-[th]uh), town (1991 population 18,083; 2001 population 15,497), SOUTH YORKSHIRE, N ENGLAND, 8 mi/12.9 km N of SHEFFIELD; 53°30′N 01°27′W. Former coal-mining center. Nearby (E) is village of SAINT HELENS.

Hoyleton, village (2000 population 520), WASHINGTON county, SW ILLINOIS, 13 mi/21 km SSE of CARLYLE; 38°27′N 89°16′W. In agricultural area.

Hoym (HOIM), town, SAXONY-ANHALT, central GERMANY, at N foot of the lower HARZ, 7 mi/11.3 km E of QUEDLINBURG; 51°47′N 11°19′E. In lignite-mining region. Has 13th-century palace.

Hoyocasero (oi-o-kah-SAI-ro), town, ÁVILA province, central SPAIN, in fertile ALBERCHE valley, at N foot of the SIERRA DE GREDOS, 21 mi/34 km SW of ÁVILA; 40°24′N 04°58′W. Cereals, vegetables.

Hoyo de Manzanares (OI-o dai mahn-thah-NAH-res), town, MADRID province, central SPAIN, in E hills of the SIERRA DE GUADARRAMA, 18 mi/29 km NW of MADRID; 40°38′N 03°53′W. Summer resort, dairying.

Hoyo de Pinares, El (OI-o dhai pee-NAH-res, el), town, ÁVILA province, central SPAIN, 18 mi/29 km SE of ÁVILA. Surrounding pine forests (timber, pine cones) are its mainstay; also grapes, cereals, fruit, livestock. Olive oil pressing, flour milling, dairying; ceramics manufacturing.

Hoyos (OI-os), town, CÁCERES province, W SPAIN, 50 mi/80 km NNW of CÁCERES. Olive oil processing; livestock raising; citrus and other fruit.

Hoyo Strait (HO-yo), Japanese *Hoyo-kaikyo* (HO-yo-KAH-ee-kyo), SW JAPAN, between KYUSHU (W) and SHIKOKU (E); connects IYO SEA (SW section of INLAND SEA) with PHILIPPINE SEA; c.60 mi/97 km long, c.35 mi/56 km wide. BEPPU BAY is largest inlet. Sometimes called Bungo Strait.

Hoyran, Lake, TURKEY: see EGRIDIR, LAKE.

Hoyt, village (2000 population 571), JACKSON county, NE KANSAS, 13 mi/21 km N of TOPEKA; 39°15′N 95°42′W. In livestock and grain region.

Hoyt Lakes, town (2000 population 2,082), ST. LOUIS county, NE MINNESOTA, 17 mi/27 km E of VIRGINIA, on S shore of Colby Lake, formed on Partridge River, in part of Superior National Forest; 47°33′N 92°07′W. Railroad junction to NE. Manufacturing of hardwood products; agriculture (timber, dairying; cattle; oats, alfalfa). Recreation area.

Hoytville (HOIT-vil), village (2006 population 295), Wood county, NW OHIO, 14 mi/23 km SSW of BOWLING GREEN; 41°11′N 83°47′W.

Hozat, village, E central TURKEY, 30 mi/48 km N of ELAZIG; 39°09′N 39°13′E. Grain. Formerly Dersim.

Hozumi (HO-zoo-mee), town, Motosu district, GIFU prefecture, central HONSHU, central JAPAN, 1.9 mi/3 km W of GIFU; 35°23′N 136°41′E.

Hpa-an, town and township, ⊙ KAYIN STATE, MYANMAR, on left bank of THANLWIN RIVER and 27 mi/43 km N of MAWLAMYINE. As the headquarters of a state so long in opposition to the national government, little modernization has taken place; it remains a local trading center in an important rice-producing area.

Hpimaw (PEE-maw), village, Chipwi township, KACHIN STATE, MYANMAR, 20 mi/32 km ENE of HTAWGAW, on CHINESE (YUNNAN province) border.

Hpungan Pass (poong-GAHN) (elevation 10,000 ft/ 3,048 m), on MYANMAR-INDIA border, 30 mi/48 km WNW of PUTAO; 27°30′N 96°55′E. Difficult route, rarely used.

Hrabusice (hrah-BU-shi-TSE), Slovak *Hrabušice*, Hungarian *Káposztafalva*, village, VYCHODOSLOVENSKY province, NE SLOVAKIA, on Hernad River, on railroad, and 7 mi/11.3 km SE of POPRAD; 48°59′N 20°25′E. Woodworking industry. Has 13th-century church.

Hracholusky Dam, CZECH REPUBLIC: see MZE RIVER.

Area is shown by the symbol □, and capital city or county seat by ⊙.

Hradcany (HRAHT-chah-NI), Czech *Hradčany*, district of PRAGUE, PRAGUE province, CZECH REPUBLIC, on left bank of VLTAVA RIVER, on Hradcany hill; 50°09′N 15°16′E. Site of famous Hradcany Castle, revered in Czech history. The castle, of legendary foundation (probably 9th century), became the residence of various dynasties; surrounding fortified walls are mostly of 12th century; neglected during Reformation and greatly damaged by fire in 1541; later restored and under Rudolph II (16th century) it became an outstanding center of science and art. Second Defenestration of Prague took place from castle windows in 1618, precipitating Thirty Years War. Twice occupied by PRUSSIA under Maria Theresa; temporary asylum for Charles X in 1832. Residence of Czech president since 1918, except for German occupation (1939–1945). Present form dates mostly from end of 18th century. Notable features include St. Vitus Cathedral (926 C.E.), originally a rotunda erected by St. Wenceslaus (St. Nicholas and St. John of Nepomuk are buried here); St. George Basilica (925); royal palace; with 15th-century Vladislav Hall, 16th-century Spanish Hall, 12th-century Black Tower.

Hradec Králové (HRAH-dets KRAH-lo-VE), German *Königgrätz*, city (2001 population 97,155), ⊙ VYCHODOCESKY province, in BOHEMIA, CZECH REPUBLIC, on the ELBE (Labe) River; 50°13′N 15°50′E. Industrial center known for manufacturing (musical instruments; chemicals, textiles, electrical manufacturing; brewery). Founded in the 10th century, it was a leading town of medieval BOHEMIA. It suffered heavily in the Hussite and Thirty Years wars. Became a Roman Catholic bishopric in 1653. Has a 14th-century Gothic cathedral, a 14th-century town hall, a 17th-century baroque palace, and two huge marketplaces dating back to medieval times. Seat of a medical institute (founded in 1946). The battle of SADOVA, or KONIGGRATZ (1866), was fought in the vicinity. Military base.

Hradec nad Moravici (HRA-dets NAD MO-ra-vi-tsee), town, SEVEROMORAVSKY province, central SILESIA, CZECH REPUBLIC, on MORAVICE RIVER, and 2 mi/3 km S of OPAVA, railroad terminus. Manufacturing (machinery); lumbering; paper mill. Has a 16th-century Renaissance church; annual violin festival called Beethoven's Hradec.

Hradek (HRAH-dek), Czech *Hrádek*, town, ZAPADOCESKY province, SW central BOHEMIA, CZECH REPUBLIC, on railroad, and 9 mi/14 km ESE of PLZEŇ. Ironworks (special steels, rolled materials).

Hradek nad Nisou (HRAH-dek NAHD nyi-SOU), Czech *Hrádek nad Nisou*, German *Grottau an der Niesse*, town, SEVEROCESKY province, N BOHEMIA, CZECH REPUBLIC, on LUSATIAN NEISSE RIVER, on railroad, and 11 mi/18 km NW of LIBEREC, near German-Polish border across from ZITTAU, GERMANY; 50°51′N 14°51′E. Rubber products.

Hradyz'k (HRAH-dizk), (Russian *Gradizhsk*), town, SW POLTAVA oblast, UKRAINE, on the NE shore of the KREMENCHUK RESERVOIR, 17 mi/27 km WNW of KREMENCHUK; 49°13′N 33°07′E. Elevation 351 ft/106 m. Food processing; clothing industries. Site of a medieval fortified settlement, resettled in the 16th century, with Pyvohorsky Monastery built nearby; supported uprisings against the Poles in the 17th and 18th centuries; company center of Katerynoslav viceregency (late 18th century); town again since 1957.

Hranice, German *Mährisch Weisskirchen*, town (2001 population 19,670), SEVEROMORAVSKY province, N MORAVIA, CZECH, on right bank of BECVA RIVER, and 21 mi/34 km ESE of OLOMOUC; 49°33′N 17°44′E. Railroad junction; manufacturing of machinery (notably motors, pumps) and textiles, food processing; cement works. Has 16th-century castle, museum. Just S, in karst area, are health resort of Teplice nad Becvou, Czech *Teplice nad Bečvou* (TEP-li-TSE NAHD bech-VOU), with mineral baths, and Zbrasov Caves,

Czech *Zbrašovské jeskyně* (ZBRAH shof-SKE YES ki-NYE).

Hranice (HRAH-nyi-TSE), German *Rossbach*, town, ZAPADOCESKY province, W BOHEMIA, CZECH REPUBLIC, in the ORE MOUNTAINS, on border opposite ADORF, GERMANY, 6 mi/9.7 km N of AS; 50°44′N 15°26′E. Railroad terminus; frontier station. Manufacturing (glass, textiles).

Hranovnica (hrah-NAWF-nyi-TSAH), Hungarian *Grénic*, village, VYCHODOSLOVENSKY province, N SLOVAKIA, 5 mi/8 km S of POPRAD; 49°22′N 20°43′E. Woodworking (notably construction materials).

Hrasnica (HRAHS-neet-sah), town, central BOSNIA, BOSNIA AND HERZEGOVINA; 43°47′N 18°18′E.

Hrastnik (HRAHST-neek), town, central SLOVENIA, near Sava River, 29 mi/47 km ENE of LJUBLJANA; 46°08′N 15°05′E. Manufacturing of hollow glass, paints, varnishes, superphosphates, sulphuric acid. Lignite mines at nearby hamlets of Dol and Ojstro (Oye-stro).

Hrazdan, city (2001 population 52,808), administrative center of Hrazdan region, ARMENIA; 40°29′N 44°46′E. Mining and chemical combine engaged in processing local nepheline syenites; hydroelectric power plant; large-panel housing combine, and a refrigeration combine; clothing factory, milk plant; brewery. Until 1959, Akhta. An alternate spelling is Razdan.

Hrazdan or **Zanga**, river, ARMENIA, left tributary of the ARAS RIVER (KURA RIVER basin); 88 mi/141 km long; rises in LAKE SEVAN; flows through mountain valley; lower Hrazdan flows across Ararat Plain. A series of six hydroelectric power plants have been constructed on the river. Used for irrigation. Cities of SEVAN, HRAZDAN, CHARENTSAVAN, ARZNI, and YEREVAN are on river. Basin 988 sq mi/ 2,560 sq km (including basin of Lake Sevan, 2,822 sq mi/7,310 sq km). An alternate spelling is Razdan.

Hrcava, CZECH REPUBLIC: see MOSTY U JABLUNKOVA.

Hrebinka (hre-BEEN-kah) (Russian *Grebenka*), city, NW POLTAVA OBLAST, UKRAINE, 25 mi/40 km WNW of LUBNY. Raion center; railroad junction, railroad car and locomotive depot with; metalworking, brickworks, food processing, food flavoring plant. Also called Hrebinkivskyy (Russian *Grebenkovskiy*), a city named after the Ukrainian 19th-century writer, born nearby, Ye-Hrebinka. Established 1895, city since 1959. Site of battles between army of Ukrainian National Republic and Bolsheviks (1919).

Hrebinka (hre-BIN-kah) (Russian *Grebenka*), city, NW POLTAVA oblast, UKRAINE, on the Hnyla Orzhytsya River, 25 mi/40 km WNW of LUBNY; 50°07′N 32°26′E. Elevation 354 ft/107 m. Raion center; railroad junction on the Kiev-Poltava line; rolling stock and locomotive depot, brick making, food flavouring factory. Also called Hrebinkivskyy (Russian *Grebenkovskiy*), a city named after the 19th-century Ukrainian writer, born nearby, Ye Herebinka. Known for battles between armies of the Ukrainian National Republic and the Bolsheviks, 1918–1919, and especially the victory of the Sich Rifleman over the Bolsheviks in January 1919. Established in 1895; city since 1959.

Hrebinkivs'kyy, UKRAINE: see HREBINKA.

Hrebinky (hre-BEEN-kee) (Russian *Grebenki*), town, SW central KIEV oblast, UKRAINE, 35 mi/56 km SSW of KIEV and 11 mi/18 km N of BILA TSERKVA; 49°57′N 30°12′E. Elevation 557 ft/169 m. Sugar refinery, sugar refinery equipment manufacturing, feed milling, dairy. Sugar beet experimental farm. Established at beginning of the 17th century, town since 1958.

Hremyach (hrem-YAHCH) (Russian *Gremyach*), village, NE CHERNIHIV oblast, UKRAINE, on right bank of Sudost' River, a tributary near its confluence with the DESNA River, 22 mi/35 km N of NOVHOROD-SIVERS'KYY, and on the border with Russia (Bryansk oblast); 52°20′N 33°17′E. Elevation 390 ft/118 m. Hemp; potatoes. Customs office.

Hrensko (HRZHEN-sko), Czech *Hřensko*, village, SEVEROCESKY province, N BOHEMIA, CZECH REPUBLIC, on right bank of the ELBE RIVER, opposite Schöna, GERMANY, and 6 mi/9.7 km N of DĚČÍN; 50°52′N 14°14′E. Popular excursion center. The lowest point of the Czech Republic; elevation 377 ft/ 115 m.

Hresivs'kyy (HRE-sif-skiy) (Russian *Gresovskiy*), town, Republic of CRIMEA, UKRAINE, on the SALHYR RIVER, on road and near railroad, 5 mi/8 km NW of and subordinated to SIMFEROPOL'; 45°02′N 34°01′E. Elevation 721 ft/219 m. Thermal electric power station, 2 reinforced concrete fabrication plants. Town since 1962.

Hricov Dam, SLOVAKIA: see VELKE ROVNE.

Hrinova (hri-NYO-vah), Slovak *Hriňová*, Hungarian *Herencsvölgy*, town, STREDOSLOVENSKY province, S central SLOVAKIA, 17 mi/27 km E of ZVOLEN; 48°35′N 19°32′E. Manufacturing of machinery (building machines) and textiles.

Hrísey (huh-REE-sai), fishing village, EYJAFJARÐARSÝSLA county, N ICELAND, on Hrisey Island (30 mi/ 48 km long) in EYJAFJÖRÐUR, 20 mi/32 km NNW of AKUREYRI.

Hrob (HROP), German *Klostergrab*, town, SEVEROCESKY province, NW BOHEMIA, CZECH REPUBLIC, on railroad, and 5 mi/8 km WNW of TEPLICE; 50°40′N 13°43′E. Manufacturing (textiles, glass, chemicals); lignite mining in vicinity. Renaissance church.

Hrochow Teinitz, CZECH REPUBLIC: see HROCHUV TYNEC.

Hrochuv Tynec (HRO-khoof TEE-nets), Czech *Hrochův Tynec*, German *Hrochow Teinitz*, village, VYCHODOCESKY province, E BOHEMIA, CZECH REPUBLIC, on railroad, and 11 mi/18 km SE of PARDUBICE; 49°58′N 15°55′E. Large sugar refinery; sugar beets; poultry. Baroque church, children's home.

Hrodivka (HRO-deev-kah) (Russian *Grodovka*), town, central DONETS'K oblast, UKRAINE, in the DONBAS, on railroad spur, 9 mi/14.5 km ESE of KRASNOARMIYS'K and 4 mi/6.4 km NE of NOVOHRODIVKA; 48°15′N 37°23′E. Elevation 580 ft/176 m. Brickworks. Established in the 1770s, town since 1938. Site of minor tank engagements between the Soviet and German forces during the winter 1943–1944.

Hronov (HRO-nof), German *Hronow*, town, VYCHODOCESKY province, NE BOHEMIA, CZECH REPUBLIC, in the SUDETES, on Metuje River, on railroad, and 16 mi/ 26 km ENE of DVUR; 50°29′N 16°11′E. Textile manufacturing (notably linen). Has a 14th-century church. Chalybeate springs. Writer Alois Jirasek was born here in 1851; annual dramatic festival named after him held here.

Hronow, CZECH REPUBLIC: see HRONOV.

Hron River (HRON), GERMAN *Gran*, Hungarian *Garam*, c.176 mi/283 km long, S SLOVAKIA; rises on SE slope of KRALOVA HOLA; flows W, between the LOW TATRAS and SLOVAK ORE MOUNTAINS, past BANSKÁ BYSTRICA, and S, past ZVOLEN, to the DANUBE RIVER opposite ESZTERGOM (HUNGARY).

Hronsky Benadik, SLOVAKIA: see TLMACE.

Hrosulove, UKRAINE: see VELYKA MYKHAYLIVKA.

Hrotovice (HRO-to-VI-tse), German *Hrottowitz*, town, JIHOMORAVSKY province, S MORAVIA, CZECH REPUBLIC, 26 mi/42 km WSW of BRNO; 49°08′N 16°03′E. Agriculture (barley, oats); manufacturing (leather). Has a 16th-century Renaissance castle.

Hrottowitz, CZECH REPUBLIC: see HROTOVICE.

Hroznetin (HROZ-nye-TYEEN), Czech *Hroznětín*, German *Hroznetyn*, village, ZAPADOCESKY province, W BOHEMIA, CZECH REPUBLIC, on railroad, and 6 mi/ 9.7 km N of KARLOVY VARY; 50°18′N 12°52′E. Manufacturing (food processing, ceramics); kaolin mining in vicinity. Has a Gothic church and 15th–16th century Jewish cemetery. Paper mill in MERKLIN, Czech *Merklín* (MERK-leen), just N.

Hroznetyn, CZECH REPUBLIC: see HROZNETIN.

Hrpelje-Kozina (her-PE-lye ko-ZEE-nah), railroad junction, SW SLOVENIA, 8 mi/12.9 km ESE of TRIESTE (ITALY); 45°36′N 13°56′E. Part of Italy (1919–1947).

Hrubieszow (hroo-BEE-shov), Polish *Hrubieszów*, Russian *Grubeshov* or *Grubeshov*, town, Lublin province, E POLAND, on railroad and 28 mi/45 km E of ZAMOŚĆ, near BUG RIVER (UKRAINE). Manufacturing of candy, soap; chicory drying, flour milling; brickworks. Before World War II, population 50% Jewish.

Hrun' (HROON), (Russian *Grun'*), village, SE SUMY oblast, UKRAINE, 14 mi/23 km WSW of OKHTYRKA; 50°14′N 34°36′E. Elevation 492 ft/149 m. Wheat.

Hruschau, CZECH REPUBLIC: see HRUSOV.

Hrushka (HROOSH-kah), (Russian *Grushka*), village, W KIROVOHRAD oblast, UKRAINE, 4 mi/6.4 km NE of UL'YANOVKA; 48°21′N 30°17′E. Elevation 518 ft/157 m. Large sugar refinery.

Hrushka, UKRAINE: see UL'YANOVKA.

Hrušica (HROO-shee-tsah), plateau, in DINARIC ALPS, SW SLOVENIA, between NANOS and TRNOVSKI GOZD mountains, 8.7 mi/14 km W of POSTOJNA. Old Logatec-Ajdovščina road passes through here; modern highway via Postojna at lower elevation.

Hrusov (HRU-shof), Czech *Hrusov*, German *Hruschau*, town, SEVEROMORAVSKY province, E SILESIA, CZECH REPUBLIC, on right bank of OSTRAVICE RIVER, at its influx into the ODER RIVER, on railroad, and just N of OSTRAVA; 49°52′N 18°18′E. Manufacturing of chemicals (notably soda and drugs) and earthenware. Part of industrial complex of Greater OSTRAVA.

Hrusov, SLOVAKIA: see GABCIKOVO DAM.

Hrusovany nad Jevisovkou (HRU-sho-VAH-ni NAHD YE-vi-SHOF-kou), Czech *Hrušovany nad Jevišovkou*, German *Grusbach*, village, JIHOMORAVSKY province, S MORAVIA, CZECH REPUBLIC, 27 mi/43 km SSW of BRNO; 48°50′N 16°24′E. Railroad junction. Agricultural center (barley, oats); sugar refining, tanning. Baroque castle.

Hrusovany u Brna (HRU-sho-VAH-ni U buhr-NAH), Czech *Hrušovany u Brna*, German *Rohrbach bei Brünn*, village, JIHOMORAVSKY province, S MORAVIA, CZECH REPUBLIC, on railroad, and 11 mi/18 km S of BRNO; 49°03′N 16°36′E. Manufacturing (footwear); stone quarry in vicinity; agriculture (vineyards; wheat, sugar beets, fruit).

Hruz'ko-Zoryans'ke (HROOZ-ko–ZOR-yahn-ske) (Russian *Gruzsko-Zoryanskoye*), town, central DONETS'K oblast, UKRAINE, in the DONBAS, 6 mi/9.7 km SE of MAKIYIVKA, subordinated to the Makiyivka city council; 47°56′N 38°06′E. Elevation 469 ft/142 m. On railroad (Ryasne station); coal mines. Established in 1938.

Hrvatska, Czech *Hrusovany nad Jevišovkou*, German *Grusbach*, village, S MORAVIA, CZECH REPUBLIC, 27 mi/43 km SSW of BRNO. Agricultural center (barley, oats); sugar refining.

Hrvatska Dubica (HUHR-vah-tskah DOO-bee-tsah), village, central CROATIA, on UNA RIVER, opposite BOSANSKA DUBICA (BOSNIA and HERZEGOVINA), on railroad. Local trade center. Formerly called Dubica.

Hrvatska Kostajnica (HUHR-vah-tskah KO-stah-nee-tsah), village, central CROATIA, on UNA RIVER, on railroad and 20 mi/32 km SSE of SISAK, in BANIJA, on BOSNIA and HERZEGOVINA border; 45°15′N 16°34′E. Trade center in plum-growing region. First mentioned in 13th century Village of Bosanska Kostajnica lies across the Una. Occupied by Serbs 1991–1995.

Hrvatsko Zagorje, region, central CROATIA, N of ZAGREB, bounded N by DRAVA RIVER, W by SLOVENIA border, and ESE by MEDVEDNICA Mountain. Hilly. One of the most populated regions in the country. Agriculture (small fragmented fields) includes wheat, corn, beans, beets, potatoes; fruit., with small-scale manufacturing (textiles, leather goods, processed foods, metals). Many residents commute daily to

Zagreb for work. Health spas (Stubičke, Krapinske, and Varaždinske Toplice) and recreational tourism represent a growing economic trend. Varaždin is the chief city; other centers include KRAPINA, IVANEC, Zabok, and DONJA STUBICA. Numerous medieval castles; old buildings currently being turned into museums, galleries, and hotels. Received many immigrants fleeing Ottoman advances during the 16th and 17th centuries. Together with Zagreb, formed the core area of reduced Croatia, then otherwise divided among VENICE, AUSTRIA, and TURKEY.

Hryhorivka, UKRAINE: see VERKHN'ODNIPROVS'K.

Hrymayliv (hri-MEI-lif) (Russian *Grimailov*) (Polish *Grzymałów*), town, E TERNOPIL' oblast, UKRAINE, on right tributary of ZBRUCH RIVER and 13 mi/21 km ENE of TEREBOVLYA; 49°20′N 26°02′E. Elevation 987 ft/300 m. Railroad spur terminus; flour milling, brick manufacturing, canning. Has old palace. Known since 1600, town since 1956. Sizeable Jewish community since the 18th century, numbering close to 1,500 in 1939; wiped out by the Nazis during World War II.

Hryshkivtsi (HRISH-kif-tsee) (Russian *Grishkovtsy*), town, S ZHYTOMYR oblast, UKRAINE, 2 mi/3.2 km NNE of BERDYCHIV; 49°56′N 28°36′E. Elevation 849 ft/258 m. Cannery. Known since 1775, town since 1938.

Hryshyne, UKRAINE: see KRASNOARMIYS'K.

Hrytsiv (HRI-tsif) (Russian *Gritsev*), town, NE KHMEL'NYTS'KYY oblast, UKRAINE, 15 mi/24 km SSE of SHEPETIVKA; 49°58′N 27°13′E. Elevation 826 ft/251 m. Foundry press manufacturing, beverage cannery. Established in the 11th century, received Magdeburg Rights in 1675; town since 1959. Large Jewish community since the 18th century, accounting for almost half of the total population by 1939; eliminated by the Nazis in 1942—a marked mass grave is 6 mi/9.6 km E of the town.

Hsahtung (shah-TOONG), township (□ 471 sq mi/1,224.6 sq km) of SHAN STATE, MYANMAR, on the NAM PAWN; ⊙ Hsihseng.

Hsamonghkam (shah-MUHNG-kahm), township (□ 479 sq mi/1,245.4 sq km), SHAN STATE, MYANMAR; ⊙ Hsamonghkam. Village of Hsamonghkam on Thazi-Shwenyaung railroad and 20 mi/32 km WSW of TAUNGGYI.

Hsawnghsup (SHAWNG-shoop), Burmese *Thaungdut*, former SHAN STATE (□ 567 sq mi/1,469 sq km), SAGAING division, MYANMAR; ⊙ THAUNGDUT. In MANIPUR HILLS, between MANIPUR (India) border and CHINDWIN RIVER; teak, wood-oil trees. Shan (Thai) population influenced by nearby Nagas and Manipuris.

Hsenwi (shen-WEE), township, SHAN STATE, MYANMAR, on BURMA ROAD, and 29 mi/47 km NE of LASHIO and 12 mi/19 km S of KUTKAI; head of road (E) to KUNLONG. Ruins of old capital nearby.

Hsenwi, MYANMAR: see NORTH HSENWI, SOUTH HSENWI.

Hsia-men, CHINA: see XIAMEN.

Hsi-an, CHINA: see XI'AN.

Hsiang, CHINA: see XIANG.

Hsiang-t'an, CHINA: see XIANGTAN.

Hsichih, TAIWAN: see SHIHCHIH.

Hsichou (CHI-JO) or **Kichow**, town, W central TAIWAN, 23 mi/37 km SSW of TAICHUNG near CHOSHUI RIVER; sugar milling; rice, sugarcane, jute, sesame, turnips.

Hsi-ch'uan, CHINA: see SICHUAN.

Hsihu (SHEE-HOO), town, W central TAIWAN, 9 mi/14.5 km SSW of CHANG-HUA. Sugar milling; rice, sweet potatoes, fruit; livestock.

Hsi Hu, CHINA: see WEST LAKE.

Hsi-kang, CHINA: see XIKANG.

Hsilo (SHEE-LUH), town, W central TAIWAN, 20 mi/32 km N of CHIAI and on CHOSHUI RIVER; 23°48′N 120°27′E. Rice; watermelon, sweet potatoes, vegetables. Also spelled Silo.

Hsincheng (SHIN-CHENG), village, E central TAIWAN, on E coast, 10 mi/16 km N of HUALIEN; 24°08′N 121°39′E. Corn. Sometimes spelled Sincheng.

Hsinchu, Hsin-chu or **Xinzhu** (all: SHIN-JOO), city, NW TAIWAN, on W coast, 38 mi/61 km SW of TAIPEI and on railroad; 24°48′N 120°58′E. Industrial center; petroleum refining, manufacturing of cement, fertilizers, synthetic oil, textiles, paper, wood products, and glass; an important computer and electronic engineering and manufacturing center. Agriculture, noted for tea, rice, and oranges. Iron ore, coal gold, and silver mining. Has Confucian and Buddhist temples. Site of Ching-hua University, a major technical university. Immigrants from mainland CHINA formed a colony at Hsinchu in the early 1700s. One of the oldest cities on Taiwan, it was called Chuchien or Teukchan until 1875. Since the 19th century the city has been a thriving commercial center. Called Chuchien or Teukchan until 1875. The name Hsinchu was also applied (1945–1950) to TAOYUAN, 25 mi/40 km NE.

Hsin-chuang (SHIN-CHUANG), residential suburb of TAIPEI, N TAIWAN, 4 mi/6.4 km S of Taipei city center, and across TANSHUI RIVER. Rice milling, woodworking, confectionery. Dated from 1732. Originally called Haishankow. Sometimes spelled Sinchwang.

Hsingan, CHINA: see HINGGAN MENG.

Hsinhua (SHIN-HUAH), town, W central TAIWAN, 7 mi/11.3 km ENE of TAINAN; 23°46′N 120°17′E. Tiles, incense, soy sauce; petroleum nearby. Sometimes spelled Sinhwa.

Hsin-hui, CHINA: see XINHUI.

Hsi-ning, CHINA: see XINING.

Hsinkang (SHIN-GANG), town, W central TAIWAN, near W coast, 10 mi/16 km WNW of CHIAI. Sometimes spelled Sinkang.

Hsinpo (SHIN-BU-uh), town, NW TAIWAN, 7 mi/11.3 km E of HSINCHU; 25°01′N 121°08′E. Rice milling; bricks and tiles, wooden articles; agriculture (rice, oranges, tea, peanuts).

Hsintien (SHIN-DIAN), residential suburb of TAIPEI, N TAIWAN, 5 mi/8 km S of city center and on Hsintien River (tributary of TANSHUI RIVER); black-tea center; industry. Sometimes spelled Sintien.

Hsinying (SHIN-YING), town, W central TAIWAN, on railroad, and 22 mi/35 km NNE of TAINAN; 23°18′N 120°18′E. Sugar milling, distilling; tropical vegetables, rice. Sometimes spelled Sinying.

Hsipaw (see-BAW), township and former sawbwaship (□ 4,591 sq mi/11,936.6 sq km), SHAN STATE, MYANMAR; ⊙ HSIPAW. Central plain along MYITNGE RIVER, with hills (6,000 ft/1,829 m) NW and SE. Rice, cotton, teak, ginger. Served by Mandalay-Lashio railroad.

Hsipaw (see-BAW), town, ⊙ HSIPAW township, SHAN STATE, MYANMAR, on MYITNGE RIVER and Mandalay-Lashio railroad, and 35 mi/56 km SW of LASHIO; road S to LOILEM. Trading center; tea and tung plantations; brine wells.

Hsitou (SHEE-TOW), park and forest reserve, SW of JIHYÜEH TAN, in mountain area. Elevation 3,773 ft/1,150 m. Established by the Japanese during their occupation; now run by the forestry department of National Taiwan University in TAIPEI. In TAIWAN's prime tea-growing region; bamboo forests.

Hsiyu (SHEE-YOOI), village, on Yuwengtao island, in the PENGHU ISLANDS, off SW TAIWAN; 23°36′N 119°30′E. Known for its many hidden coves and Hsitai Fort at the S tip of the island, built in 1883 under the Qing Dynasty. Connected to PAISHA ISLAND by the Kuahai Bridge, over 3.1 mi/5 km in length (the longest in Taiwan). Also connected to the Hsiaomen islet by a narrow bridge.

Hsu-chou, CHINA: see XUZHOU.

Hsueh Shan (SHUAI SHAN), second-highest peak (12,897 ft/3,931 m) on TAIWAN, in N central range, 33 mi/53 km SE of HSINCHU. Was first climbed in 1935 by Japanese. Formerly called Mount Sylvia.

Area is shown by the symbol □, and capital city or county seat by ⊙.

Htawgaw (TAW-gaw), village, Chipwi township, KA-CHIN STATE, MYANMAR, between NMAI RIVER and CHINA border, 70 mi/113 km NE of MYITKYINA.

Htugyi (TOO-jee), village, Lugapu township, AYEYAR-WADY division, MYANMAR, on railroad and 25 mi/40 km NW of HENZADA.

Hua'an (HUAH-AN), town, ⊙ Hua'an county, S FU-JIAN province, SE CHINA, 50 mi/80 km NNW of ZHANGZHOU, and on JINLONG RIVER; 25°01′N 117°33′E. Rice, sugarcane; food processing, papermaking, hydroelectric power generation.

Hua Bei Ping Yuan, CHINA: see NORTH CHINA PLAIN.

Huacachi (hwah-KAH-chee), town, HUARI province, ANCASH region, W central PERU, on E slopes of Cordillera BLANCA, 40 mi/64 km ENE of HUARÁZ; 09°16′S 76°54′W. Cereals, potatoes; livestock.

Huacachina (hwah-kah-CHEE-nah), village, ICA region, SW PERU, 3 mi/5 km WSW of ICA; 14°04′S 75°45′W. Resort; thermal baths.

Huacana, La, MEXICO: see LA HUACANA.

Huacaya, town and canton, LUIS CALVO province, CHUQUISACA department, SE BOLIVIA, on HUACAYA River (left affluent of the PILCOMAYO) and 37 mi/60 km NNW of VILLA MONTES; 20°45′S 63°43′W. Corn, fruit.

Huacaybamba (hwah-kei-BAHM-bah), province, HUÁNUCO region, central PERU; ⊙ HUACAYBAMBA. On E slopes of the Cordillera CENTRAL. Drained by MARAÑÓN RIVER. Sugarcane, cereals, coffee; livestock.

Huacaybamba (hwah-kei-BAHM-bah), town, ⊙ HUA-CAYBAMBA province, HUÁNUCO region, central PERU, on W slopes of Cordillera CENTRAL, near MARAÑÓN RIVER, and 34 mi/55 km SE of HUACRACHUCO; 09°05′S 76°50′W. Sugarcane, corn, coffee, cereals.

Huachacalla (wah-chah-KAH-yah), town and canton, ⊙ Litoral province, ORURO department, W BOLIVIA, near confluence of TURCO and LAUCA rivers, in the ALTIPLANO, 95 mi/153 km SW of ORURO; 18°46′S 68°17′W. Elevation 12,448 ft/3,794 m.

Huachinera (wah-chee-NE-rah), town, NW SONORA, MEXICO, 85 mi/137 km SSE of AGUA PRIETA; 30°12′N 108°57′W. Elevation 3,320 ft/1,012 m. Hot dry climate. On Babidanchic River, tributary of BAVISPE River, in outlier ranges of SIERRA MADRE OCCIDENTAL. On unpaved road.

Huachipato, CHILE: see SAN VICENTE.

Huacho (HWAH-cho), city (2005 population 52,776), ⊙ HUAURA province, LIMA region, W central PERU; PACIFIC port on Huacho Bay (2 mi/3 km wide, 1 mi/2 km long), on PAN-AMERICAN HIGHWAY, and 75 mi/121 km NW of LIMA; 11°07′S 77°37′W. Cotton ginning; cottonseed milling; manufacturing; rice milling. Shipping point for products of surrounding region (cotton, rice, sugarcane); fisheries. Salt is mined nearby. Huacho port is an adjoining settlement (tramway connection).

Huacho Bay, PERU: see HUACHO.

Huachón (hwah-CHON), town, PASCO province, PASCO region, central PERU, in Cordillera ORIENTAL, 21 mi/34 km E of CERRO DE PASCO; 10°40′S 75°57′W. Cereals, potatoes.

Huachuan (HUAH-CHUAN), town, ⊙ Huachuan county, NE HEILONGJIANG province, NE CHINA, 25 mi/40 km NE of JIAMUSI, and on right bank of SON-GHUA RIVER; 47°01′N 130°43′E. Grain, tobacco, sugar beets. Also known as Yuelai.

Huachuca City, town (2000 population 1,751), COCHISE county, SE ARIZONA, suburb 5 mi/8 km N of SIERRA VISTA; 31°37′N 110°20′W. Cattle; cotton, grain, alfalfa; manufacturing (concrete). San Pedro Riparian National Conservation Area to E.

Huachuca Mountains, in SW COCHISE county, SE AR-IZONA, in SW part of SIERRA VISTA city, near Mexican border. MILLER PEAK (9,466 ft/2,885 m) is highest point. Huachuca Peak (8,406 ft/2,562 m), also in

range, is 5 mi/8 km SW of Sierra Vista. S part lies in Coronado National Forest.

Huaco (HWAH-ko), village, N central SAN JUAN province, ARGENTINA, 18 mi/29 km NE of JACHAL; 30°09′S 68°31′W. Corn, barley; livestock. Warm sulphur springs.

Huacrachuco (hwah-krah-CHOO-ko), town, ⊙ MAR-AÑÓN province, HUÁNUCO region, central PERU, on E slopes of Cordillera CENTRAL of the ANDES, 105 mi/169 km NW of HUÁNUCO; 08°38′S 76°53′W. Elevation 11,975 ft/3,650 m. Agricultural products (cereals, potatoes); livestock.

Huacullani (hwah-koo-YAH-nee), town and canton, INGAVI province, LA PAZ department, W BOLIVIA, near LA PAZ–TIQUINA road, on AYGACHI Bay E of Lake TITICACA; 16°22′S 68°38′W. Elevation 12,641 ft/3,853 m. Experimental agriculture and livestock program.

Huade (HUAH-DUH), town, ⊙ Huade county, central INNER MONGOLIA AUTONOMOUS REGION, N CHINA, 85 mi/137 km NNW of ZHANGJIAKOU, near HEBEI border; 41°57′N 114°04′E. Cattle raising; grain, oilseeds.

Huadian (HUAH-DIAN), city (□ 2,413 sq mi/6,250 sq km; 1994 estimated urban population 188,300; estimated total population 438,500), S central JILIN province, NE CHINA, 60 mi/97 km S of JILIN, on the SONGHUA RIVER; 43°01′N 126°45′E. Heavy industry is the most important source of income; non-ferrous mineral mining, logging, utilities, and rubber.

Huadquirca, PERU: see HUAQUIRCA.

Huafo Island, CHILE: see GUAFO ISLAND.

Hua Hin (HU-uh HIN), village (2000 population 41,953), PRACHUAB KHIRI KHAN province, S THAI-LAND, on GULF OF THAILAND, on railroad and 85 mi/137 km SSW of BANGKOK. Largest seaside resort of Thailand; site of royal residence, built by King Prachadipok (1925–1935).

Hua Hin (HU-uh HIN) [Thai=head of the rock], beach resort, NW PRACHUAB KHIRI KHAN province, THAI-LAND; 12°34′N 99°58′E. Site of a summer residence built by Rama VII still used by the royal family. Tourism, fishing.

Huahine (HOO-ah-HEE-nai), volcanic island (□ 28 sq mi/72.8 sq km), Leeward group, SOCIETY ISLANDS, FRENCH POLYNESIA, S PACIFIC, 25 mi/40 km E of RAI'ATEA. Consists of two islands joined by isthmus, with fish-rich Lake Mahiva; circumference 20 mi/32 km; 16°43′S 151°00′W. Fertile and mountainous. Mount Turi (elevation 2,230 ft/680 m) highest peak. Exports copra, fish. Tourism. Chief town, Fare-nui-atea, usually called Fare.

Huahua River, NICARAGUA: see WAWA RIVER.

Huai (HUEI), river, c.680 mi/1,094 km long, E CHINA; rising in the Tongbai mountains, HENAN province; flowing E across ANHUI province, through HONGZE LAKE, to the EAST CHINA SEA. The Huai, together with the Qingling Mountains, marks the boundary between the NORTH CHINA PLAIN and the CHANG JIANG delta. More than two-thirds of the fertile Huai basin is under cultivation; wheat, millet, and kaolin are the main crops. An irrigation canal branches from the river to the sea. Receiving many tributaries, the Huai floods more frequently and over a larger area than any river in N China; extensive flood-control facilities, including eight dams, have been built since 1950.

Huai'an (HUEI-AN), city (□ 602 sq mi/1,559 sq km; 1994 estimated urban population 144,000; estimated total population 1,175,700), N central JIANGSU prov-ince, E CHINA, on the GRAND CANAL, 10 mi/16 km SE of HUAIYIN; 33°30′N 119°09′E. Agriculture and light industry are the largest sources of income. Main in-dustries include food processing, textiles, chemicals, and machinery.

Huaibei (HUEI-BAI), city (□ 114 sq mi/296.4 sq km; 2000 population 535,823), N ANHUI province, E CHINA; 34°00′N 116°48′E. Industrial center. Crop growing, animal husbandry. Agriculture (grains, oil

crops, cotton, vegetables, fruits; hogs; eggs); manu-facturing (utilities, coal mining).

Huaidezhen (HUEI-DUH-JEN), town, W JILIN prov-ince, NE CHINA, 28 mi/45 km W of CHANGCHUN; 43°48′N 124°45′E. Grain, sugar beets, oilseeds.

Huaihua (HUEI-HAH), city (□ 844 sq mi/2,194.4 sq km; 2000 population 488,343), W HUNAN province, S central CHINA, 32 mi/51 km ENE of ZHIJIANG; 27°27′N 109°50′E. Agriculture is the largest source of income. Crop growing. Grain, oilseeds, hogs; manufacturing (food processing, textiles, chemicals, electronics). Also called Yushuhan.

Huaiji (HUEI-JEE), town, ⊙ Huaiji county, W GUANGDONG province, S CHINA, 25 mi/40 km NE of WUZHOU; 23°55′N 112°10′E. Rice, sugarcane; logging, chemicals, food industry, iron-ore mining.

Huailai (HUEI-LEI), town, ⊙ Huailai county, NE HEBEI province, NE CHINA, near GREAT WALL, 50 mi/80 km SE of ZHANGJIAKOU, and on railroad; 40°25′N 115°27′E. Grain, fruit, sesame oil; timber, food and beverages, chemicals; coal mining.

Huailas, PERU: see HUAYLAS.

Huailas, Callejón de, PERU: see HUAYLAS, CALLEJÓN DE.

Huaillati (hwei-YAH-tee), town, GRAU province, APURÍMAC region, S central PERU, in ANDEAN valley, near VILCABAMBA RIVER, 15 mi/24 km S of COTA-BAMBAS; 14°05′S 72°31′W. Cereals; livestock.

Huainan (HUEI-NAN), city (□ 421 sq mi/1,091 sq km; 1994 estimated urban population 769,200; estimated total population 1,279,800), N central ANHUI prov-ince, E CHINA, on the Hua River, on railroad, and 40 mi/64 km SW of BENGBU; 32°38′N 116°59′E. Established after 1949 as the center of China's chief coal-mining region, it is the site of a major colliery; nitrogenous fertilizer, papermaking, food processing, thermal power generation; grain, oilseeds, cotton, jute. Sometimes may appear as Hwainan.

Huaining (HUEI-NING), town, ⊙ Huaining county, SW ANHUI province, E CHINA, 25 mi/40 km WSW of ANQING; 30°23′N 116°44′E. Textiles, food processing, tanning; rice, medicinal herbs.

Huairen (HUEI-REN), town, ⊙ Huairen county, N SHANXI province, NE CHINA, 20 mi/32 km SSW of DATONG, and on railroad; 39°50′N 113°06′E. Grain, sugar beets, oilseeds; coal mining.

Huairou (HUEI-RO), town, ⊙ Huairou county, NE CHINA, 30 mi/48 km NNE of BEIJING, on railroad; an administrative unit of Beijing municipality; 40°20′N 116°37′E. Grapes, pears, peaches, grain, oilseeds; food industry, building materials, engineering, textiles and clothing, motor-vehicle parts. Center for farming and fishing.

Huaitiquina Pass (hwei-tee-KEE-nah) (14,025 ft/4,275 m), in the ANDES, on ARGENTINA-CHILE border; 23°44′S 67°13′W.

Huaiyang (HUEI-YANG), town, ⊙ Huaiyang county, E HENAN province, NE CHINA, 85 mi/137 km SSE of KAIFENG; 33°44′N 114°53′E. Sesame, watermelons, wheat, beans, cotton, jute, tobacco. Commercial center of ZHOUKOU is 15 mi/24 km SW. Formerly called Chenchow.

Huaiyang Mountains, CHINA: see DABIE MOUNTAINS.

Huaiyin (HUEI-YIN), city (□ 134 sq mi/348.4 sq km; 2000 population 2,613,804), N JIANGSU province, E CHINA; 33°35′N 119°02′E. The city is a center for light industry. Crop growing, animal husbandry, fishing, commercial agriculture, and forestry. Grains, oil crops, cotton, vegetables, fruits, hogs, aquatic prod-ucts, eggs, poultry, beef, lamb. Manufacturing (food, tobacco, textiles, utilities, chemicals, pharmaceuti-cals, synthetic fibers, rubber, iron and steel, machin-ery, transportation equipment).

Huaiyin (HUEI-YIN), town, ⊙ Huaiyin county, N JIANGSU province, E CHINA, 5 mi/8 km N of HUAIYIN city, and on GRAND CANAL, E of HONGZE LAKE; 33°40′N 119°07′E. Oilseeds, rice.

Huai Yot (HU-ei YAWD), village and district center, TRANG province, S THAILAND, on MALAY PENINSULA, 20 mi/32 km N of TRANG, and on railroad; 07°45′N 99°37′E. Highway to port of KRABI on W coast. Sometimes spelled HUEY YOT.

Huaiyuan (HUEI-YUAN), town, ⊙ Huaiyuan county, N ANHUI province, E CHINA, on HUAI RIVER at mouth of GUO RIVER, and 10 mi/16 km W of BENGBU; 32°57′N 117°12′E. Rice, wheat, cotton, beans, tobacco, jute.

Huajicori (wah-hee-KO-ree), town, NAYARIT, W MEXICO, on ACAPONETA RIVER and 12 mi/19 km N of ACAPONETA. Corn, sugarcane, beans; cattle; silver and gold deposits.

Huajlaya (wahzh-LAH-yah), town and canton, NOR CINTI province, CHUQUISACA department, SE BOLIVIA; 20°36′S 64°33′W. Elevation 7,894 ft/2,406 m. Lead, zinc, silver mines to NW; limestone deposits. Agriculture (potatoes, yucca, bananas, corn, wheat, oats, rye, peanuts); cattle.

Huajriri (wahzh-REE-ree), town and canton, SAJAMA province, ORURO department, W central BOLIVIA, 38 mi/60 km SW of TURCO, E of the LAUCA RIVER; 18°12′S 68°25′W. Elevation 12,789 ft/3,898 m. Some gas resources in area. Copper, clay, limestone and gypsum deposits. Agriculture (potatoes, yucca, bananas); cattle.

Huajuapam de León (wah-hoo-ah-PAHM de le-ON), city and township, ⊙ Huajuapan de Leon, OAXACA, S MEXICO, in SIERRA MADRE DEL SUR, on INTER-AMERICAN HIGHWAY, and 105 mi/169 km NW of OAXACA DE JUÁREZ; 17°48′N 97°46′W. Elevation 5,249 ft/1,600 m. Agricultural center (cereals, coffee, sugarcane, fruit); manufacturing straw hats. Formerly HUAJUAPAN DE LEÓN.

Huajuapan de León, MEXICO: see HUAJUAPAM DE LEÓN.

Hualahuises (wah-lah-WEE-ses), town, ⊙ Hualahuises municipio, NUEVO LEÓN, N MEXICO, in foothills of SIERRA MADRE ORIENTAL, 8 mi/12.9 km W of LINARES; 24°56′N 99°42′W. Grain; livestock.

Hualaihué (wah-lei-WAI), village, LLANQUIHUE province, LOS LAGOS region, S central CHILE, on a headland in GULF OF ANCUD, 40 mi/64 km SSE of PUERTO MONTT; 42°01′S 72°41′W. Dairying, lumbering. Sometimes Gualaihué.

Hualalai (HOO-ah-LAH-LEI), mountain (8,275 ft/ 2,522 m), near W coast, HAWAII island, HAWAII county, HAWAII.

Hualañé (wah-lah-NYAI), village, ⊙ Hualañé comuna, CURICÓ province, MAULE region, central CHILE, on MATAQUITO RIVER and railroad, 32 mi/51 km W of CURICÓ; 34°59′S 71°49′W. Agricultural center (cereals, vegetables, grapes; livestock).

Hualapai Mountains (hua-la-pi), range in S central MOHAVE county, W ARIZONA, extends c.50 mi/80 km S from point near KINGMAN. Hualapai Peak (8,417 ft/ 2,566 m), 12 mi/19 km SE of Kingman, is highest point. NE extension, Peacock Mountains, rise to 6,292 ft/ 1,918 m in Peacock Peak, 17 mi/27 km ENE of Kingman. Wabayuma Peak Wilderness Area in W center of Hualapai range.

Hualgayoc (hwahl-gei-YOK), province, CAJAMARCA region, NW PERU; ⊙ HUALGAYOC. Located between the provinces of SANTA CRUZ (W) and CELENDÍN (E).

Hualgayoc (hwahl-gah-YOK), city (1993 population 17,182), HUALGAYOC province, CAJAMARCA region, NW PERU, in CORDILLERA OCCIDENTAL, 28 mi/45 km NNW of CAJAMARCA, on road from SAN MIGUEL DE PALLAQUES to BAMBAMARCA; 06°46′S 78°30′W. Elevation 11,548 ft/3,520 m. Silver-mining center; agricultural products (corn, cereals, potatoes).

Hualien (HUAH-LIAN), city, E central TAIWAN, 50 mi/ 80 km N of ILAN; 23°59′N 121°36′E. Located in a mountainous and volcanic region, susceptible to earthquakes and typhoons. The largest city on the E

coast; an international port, limited by its isolated location and small hinterland; a market center for local agriculture and marble products; marble and limestone mining. Site of the Retired Servicemen's Engineering Agency that quarries marble from the nearby TAROKO GORGE area and also runs a plant producing over 100 marble items. The Ami aboriginal tribe lives in the area; dance performances. Also spelled Hwalien.

Huallaga (hwah-YAH-gah), province (☐ 4,839 sq mi/ 12,581.4 sq km), SAN MARTÍN region, N central PERU; ⊙ SAPOSOA; 06°55′S 77°00′W. N province bordering on AMAZONAS and LA LIBERTAD regions.

Huallaga River (hwah-YAH-gah), c.700 mi/1,127 km long, central and N Peru; rises in the ANDES S of CERRO DE PASCO (PASCO region) in CORDILLERA OCCIDENTAL; flows N, past HUÁNUCO and TINGO MARÍA (HUÁNUCO region), then NE, past PICOTA, SHAPAJA, Chasuta, and YURIMAGUAS (SAN MARTÍN and LORETO regions), entering the AMAZON basin to join MARAÑÓN RIVER at 05°10′S 75°33′W. Navigable for small craft up to Tingo María (385 mi/620 km), for large vessels to LAGUNAS (30 mi/48 km).

Huallanca (hwah-YAHN-kah), city, HUÁNUCO region, central PERU, in the ANDES, 12 mi/19 km SW of LA UNIÓN; 09°51′S 76°56′W. Cereals, potatoes, alfalfa; sheep raising. Silver mining nearby.

Huallanca (hwah-YAHN-kah), town, BOLOGNESI province, ANCASH region, W central PERU, on SANTA RIVER and 5 mi/8 km NE of HUAYLAS; 08°49′S 77°52′W. Corn, wheat. At steep gorge of Santa River (Cañón del Pato) nearby is a large government hydroelectric project (161,160 kwh produced here).

Huallatiri (hwah-yah-TEE-ree), town and canton, PANTALEÓN DALENCE province, ORURO department, W central BOLIVIA, on the ORURO-POTOSÍ road; 18°17′S 66°51′W. Elevation 12,900 ft/3,932 m. Lead-bearing lode, tin mining at Mina HUANUNI, SANTA FE, Japo, and MOROCOCALA; lead, zinc, silver mining, copper mining to the N; clay limestone, and gypsum deposits. Agriculture (potatoes, bananas, yucca); cattle.

Huallay Grande (hwah-YEI GRAHN-dai), town, ANGARAES province, HUANCAVELICA region, S central PERU, in CORDILLERA OCCIDENTAL, 3 mi/5 km NNE of LIRCAY; 12°55′S 74°42′W. Elevation 13,123 ft/3,999 m. Cereals, alfalfa; livestock.

Hualong (HUAH-LUNG), town, ⊙ Hualong county, E QINGHAI province, W CHINA, 45 mi/72 km NE of XINING; 36°08′N 102°16′E. Livestock; food industry, non-ferrous and mining.

Hualqui (WAHL-kee), town, ⊙ Hualqui comuna, CONCEPCIÓN province, BÍO-BÍO region, S central CHILE, on railroad, on BÍO-BÍO RIVER and 13 mi/21 km SE of CONCEPCIÓN; 36°58′S 72°56′W. Agricultural center (cereals, vegetables; livestock); food processing.

Huamachuco (hwah-mah-CHOO-ko), city (2005 population 23,766), ⊙ SÁNCHEZ CARRIÓN province, LA LIBERTAD region, NW PERU, in CORDILLERA OCCIDENTAL of the ANDES, 70 mi/113 km ENE of TRUJILLO; 07°48′S 78°04′W. Elevation 10,859 ft/3,310 m. Cereals, corn, potatoes.

Huamalíes (hwah-mah-LEE-es), province (☐ 1,220 sq mi/3,172 sq km), HUÁNUCO region, central PERU; ⊙ LLATA; 09°15′S 76°30′W. Bordered by ANCASH region on the W. Cereals, potatoes; livestock.

Huamanga (hwah-MAHN-gah), province (☐ 8,699 sq mi/22,617.4 sq km), S PERU; ⊙ AYACUCHO; 13°12′S 74°15′W. A highland province, rich in archaeological ruins; mines; agriculture (coffee, cereals, potatoes; livestock), and site of the SANTUARIO HISTÓRICO PAMPAS DE AYACUCHO.

Huamanga, Peru: see AYACUCHO.

Huamanguilla (hwah-mahn-GEE-yah), town, HUANTA province, AYACUCHO region, S central Peru, in the ANDES, 12 mi/19 km N of AYACUCHO; 13°00′S 74°10′W.

Sugarcane, coffee, coca; vineyards. Near SANTUARIO HISTÓRICO PAMPAS DE AYACUCHO.

Huamantanga (hwah-mahn-TAHN-gah), town, CANTA province, LIMA region, W central Peru, in CORDILLERA OCCIDENTAL, 5 mi/8 km WSW of CANTA; 11°30′S 76°47′W. Cereals, potatoes; livestock.

Huamantla (wah-MAHN-tlah), city and township, TLAXCALA, central MEXICO, at NE foot of MALINCHE volcano, on railroad, and 25 mi/40 km NE of PUEBLA; 19°18′N 97°55′W. Elevation 8,376 ft/2,553 m. Agricultural center (corn, wheat, barley, alfalfa, beans, maguey; livestock); flour milling, pulque distilling. Many churches. Also known as Heroica Ciudad de Huamantla.

Huambo (WAHM-bo), province (☐ 13,230 sq mi/34,398 sq km), W central ANGOLA; ⊙ HUAMBO. Bordered N by CUANZA SUL, E by Rio Cutalo and BIÉ province, S by Rio Cunene and HUILA province, W by BENGUELA province. Drained by Rio Cunene. BIE PLATEAU is located in the W central section. Agriculture includes corn, beans, potatoes, wheat, citrus, manioc, rice. Minerals include iron ore, tungsten, barium monoxide, gold magnesium, uranium. Main centers are Huambo, WAMA, CUIMA, CAALA, BAILUNDO, UKUMA.

Huambo (WAHM-bo), town (2004 population 173,600), W central ANGOLA. The chief town of inland Angola, Huambo stands on a high plateau and serves as a road, railroad, and air transport hub, and as a commercial and shipping center for a rich agricultural region. Its railroad repair shops are among the largest in Africa. Huambo exports grain, rice, hides, skins, and fruit. Milling and the production of lime are carried on in the city. Since independence in 1976, civil war has devastated the city's economy. Huambo was founded in 1912. Also known as Nova Lisboa.

Hua Muong (HOO-wah MOO-uhng), town, HUA PHAN province, NE LAOS, NW of SAM; 20°13′N 103°44′E. Administrative center.

Huamuxtitlán (wah-moosh-tee-TLAHN), city and township, ⊙ Huamuxtitlán municipio, GUERRERO, SW MEXICO, in SIERRA MADRE DEL SUR, 50 mi/80 km ENE of CHILAPA DE ÁLVAREZ; 17°48′N 98°34′W. Elevation 3,691 ft/1,125 m. Cereals, sugarcane, fruit.

Huanacache (hwah-nah-KAH-che) or **Guanacache**, swamp and lake district of ARGENTINA, on borders of SAN JUAN, MENDOZA, and SAN LUIS provinces, formed by SAN JUAN, MENDOZA, and BERMEJO rivers; extends c.90 mi/145 km SE and S. It is drained by the DESAGUADERO, part of the Río Salado. Popular fishing area.

Huañacota (hwahn-yah-KO-tah), town and canton, CAPINOTA province, COCHABAMBA department, central BOLIVIA, S of COCHABAMBA; 17°34′S 66°10′W. Elevation 7,815 ft/2,382 m. Lead-bearing lode; phosphates, clay, limestone and gypsum deposits. Agriculture (potatoes, yucca, bananas, corn, rye, sweet potatoes, soy, coffee); cattle for meat and dairy products.

Huañamarca, Lake, Peru and Bolivia: see HUYÑAYMARKA, LAKE.

Huananhuata, Laguna, Peru: see ARICOTA, LAGUNA.

Huancabamba (hwahn-kah-BAHM-bah), province (☐ 2,532 sq mi/6,583.2 sq km), PIURA region, NW Peru; ⊙ HUANCABAMBA; 05°30′S 79°35′W. Easternmost province of Piura region, bordering on CAJAMARCA region.

Huancabamba (hwahn-kah-BAHM-bah), city, ⊙ HUANCABAMBA province, PIURA region, NW Peru, in CORDILLERA OCCIDENTAL of the ANDES, on HUANCABAMBA RIVER, and 80 mi/129 km E of PIURA; 05°14′S 79°28′W. Elevation 6,420 ft/1,957 m. Tobacco, coffee, cereals, potatoes, cacao, hides; livestock.

Huancabamba (hwahn-kah-BAHM-bah), town, OXAPAMPA province, PASCO region, central Peru, on E slopes of Cordillera ORIENTAL, 6 mi/10 km NW of OXAPAMPA; 10°21′S 75°32′W. Coffee, cacao, fruit, cereals. Near PARQUE NACIONAL YANACHAGA-CHEMILLEN.

Huancabamba River (hwahn-kah-BAHM-bah), 120 mi/193 km long, NW Peru; rises in CORDILLERA OCCIDENTAL 15 mi/24 km N of HUANCABAMBA (PIURA region); flows S and NE, past Huancabamba, to MARAÑÓN RIVER 9 mi/15 km SE of JAÉN (CAJAMARCA region); 06°03′S 79°08′W. Also called Chamaya River in lower course.

Huancacuni Grande (hwahn-kah-KOO-nee GRAHN-dai), town and canton, CAMPERO province, COCHABAMBA department, central BOLIVIA, 9 mi/15 km W of OCONI. Elevation 7,356 ft/2,242 m. Limestone deposits. Agriculture (potatoes, yucca, bananas, corn, rye, soy); cattle raised for meat and dairy products.

Huancanapi (hwahn-kah-NAH-pee), town and canton, SAN PEDRO DE TOTORA province, ORURO department, W central BOLIVIA, 19 mi/30 km S of TOTORA; 17°50′S 68°05′W. Elevation 12,795 ft/3,900 m. Gas resources in area. Copper mining and gypsum deposits. Agriculture (potatoes, yucca, bananas); cattle.

Huancané (hwahn-kah-NAI), province (□ 3,855 sq mi/10,023 sq km), PUNO region, SE Peru; ⊙ HUANCANÉ; 16°16′S 69°19′W. E province of Puno region, bordering Lake TITICACA. Site of Sector Ramis of RESERVA NACIONAL TITICACA.

Huancané (hwahn-kah-NE), town and canton, SUD YUNGAS province, LA PAZ department, W BOLIVIA, 12 mi/20 km NE of CHULUMANI; 16°22′S 67°32′W. Elevation 5,686 ft/1,733 m. Antimony-bearing lode. Tungsten mining at Mina Reconquistada, Mina CHOJLLA, Mina Bolsa Negra; clay, limestone, gypsum deposits. Agriculture (potatoes, yucca, bananas, rye); cattle.

Huancané (wahng-kah-NAI), town, CHALLAPATA (AVAROA) province, ORURO department, W BOLIVIA, in the ALTIPLANO, 55 mi/89 km SSE of ORURO, on FCRR Rio Mulatos–Oruro; 16°22′S 67°32′W. Elevation 12,185 ft/3,714 m. Corn, potatoes.

Huancané (hwahn-kah-NAI), town, ⊙ HUANCANÉ province, PUNO region, SE Peru, on the ALTIPLANO, near N shore of Lake TITICACA, 48 mi/77 km NNE of PUNO; 15°12′S 69°46′W. Trade center; potatoes, cereals; livestock.

Huancapi (hwahn-KAH-pee), town, ⊙ VÍCTOR FAJARDO province, AYACUCHO region, S central Peru, on E slopes of CORDILLERA OCCIDENTAL, 40 mi/64 km SSE of AYACUCHO, on road from CANGALLO to QUEROBAMBA; 13°41′S 74°04′W. Elevation 10,607 ft/3,233 m. Cereals, alfalfa; livestock.

Huancapón (hwahn-kah-PON), town, LIMA region, W central Peru, in CORDILLERA OCCIDENTAL, 17 mi/27 km SW of CAJATAMBO; 10°34′S 77°06′W. Cereals; livestock.

Huancarama (hwahn-kah-rah-MAH), town, APURÍMAC region, S central Peru, in ANDEAN spur, 13 mi/21 km W of ABANCAY; 13°38′S 73°07′W. Gold, copper, iron and coal mining; potatoes, cereals; livestock.

Huancarhuas (hwahn-kah-RO-mah), mountain peak (20,531 ft/6,258 m), in the ANDES, ANCASH department, PERU; 08°54′S 77°42′W.

Huancaroma (hwahn-kah-RO-mah), town and canton, TOMÁS BARRÓN province, ORURO department, W central BOLIVIA, on the border with LA PAZ department, 12 mi/20 km N of LA JOYA; 17°35′S 67°31′W. Elevation 12,795 ft/3,900 m. Gas resources in area. Copper, clay, and gypsum deposits. Agriculture (potatoes, yucca, bananas); cattle.

Huanca Sancos (HWAHN-kah SAHN-kos), province, AYACUCHO region, S Peru; ⊙ HUANCA SANCOS; 15°07′S 74°09′W. Agriculture and livestock.

Huanca Sancos (HWAHN-kah SAHN-kos), city, ⊙ Huanca Sancos province, AYACUCHO region, S central Peru, in CORDILLERA OCCIDENTAL, 45 mi/72 km SSW of AYACUCHO; 13°48′S 74°16′W. Livestock-raising center (cattle, sheep, llamas, vicuñas, alpacas); cereals, alfalfa. Also Sancos.

Huancavelica (hwahn-kah-vai-LEE-kah), region (□ 8,545 sq mi/22,217 sq km), S central Peru; ⊙ HUANCAVELICA; 13°00′S 75°00′W. Mainly mountainous, drained by MANTARO RIVER, which separates

Cordillera CENTRAL from CORDILLERA OCCIDENTAL. Wheat, barley, corn, alfalfa; cattle and sheep raising in mountains. Mining and smelting (silver, lead, mercury) in area of CASTROVIRREYNA, HUANCAVELICA, and LIRCAY. Main centers: HUANCAVELICA, ACOBAMBA, PAMPAS, CHURCAMPA, Lircay, Castrovirreyna, and HUAYTARÁ. The region is served by HUANCAYO-Huancavelica railroad.

Huancavelica (hwahn-kah-vai-LEE-kah), province (□ 1,936 sq mi/5,033.6 sq km), HUANCAVELICA region, S central Peru; ⊙ HUANCAVELICA. High ANDEAN mountains and valleys. Served by HUANCAYO-Huancavelica city railroad. Cereals, potatoes; livestock. Mining.

Huancavelica (hwahn-kah-vai-LEE-kah), city (2005 population 33,144), ⊙ HUANCAVELICA province and HUANCAVELICA region, S central Peru, on Huancavelica River (right affluent of the MANTARO) and 50 mi/80 km SSE of HUANCAYO (JUNÍN region), 140 mi/225 km ESE of LIMA (connected by railroad); 12°46′S 74°58′W. Elevation 12,401 ft/3,780 m. Mining and smelting center (mercury, silver, lead); flour milling; exports wool. Famous for native crafts, including leather and alpaca products. Flourished in colonial period following opening (16th century) of mercury and silver mines. It declined in 19th century, but resumed its mining activity after construction (1926) of railroad from Huancayo. The San Cristóbal thermal springs are in NE outskirts.

Huancavelica River (hwahn-kah-vai-LEE-kah), affluent of MANTARO RIVER, HUANCAVELICA region, S central Peru; rises in the ANDES; flows NE past HUANCAVELICA and then N to join the Mantaro; 12°29′S 74°56′W.

Huancayo (hwahn-KAH-yo), province (□ 1,388 sq mi/3,608.8 sq km), JUNÍN region, central Peru; ⊙ HUANCAYO. Southernmost province of Junín region, bordered by LIMA and HUANCAVELICA regions.

Huancayo (hwahn-KAH-yo), city (2005 population 306,954), ⊙ HUANCAYO province and JUNÍN region, S central Peru, in the MANTARO RIVER valley; 12°03′S 75°11′W. Elevation 10,731 ft/3,271m. One of Peru's major commercial and agricultural centers, it markets and ships the wheat, maize, potatoes, and barley grown in the surrounding area. Silver and copper are mined in the region. Sunday craft market for wood carvings, textiles. Railroad.

Huanchaca (wahn-CHAH-kah), town, DANIEL CAMPOS province, POTOSÍ department, SW BOLIVIA, on W slopes of Cordillera de CHICHAS and 15 mi/24 km NE of UYUNI, 3 mi/4.8 km E of COLCHANI on the Salar de UYUNI, on LA PAZ–RÍO MULATO-Uyuni-VILLAZÓN railroad; 20°19′S 66°41′W. Elevation 14,895 ft/4,540 m.

Huanchaca (hwahn-CHAH-kah), town and canton, SE ANTONIO QUIJARRO province, POTOSÍ department, W central BOLIVIA, 19 mi/30 km NE of UYUNI and 19 mi/30 km E of the Uyuni-ORURO railroad, in the Chichas mountain range; 20°20′S 66°39′W. Elevation 14,895 ft/4,540 m. Gas deposits in the area. Tungsten-bearing lodes, antimony mining, limestone and some gypsum deposits. Agriculture (potatoes, yucca, bananas); cattle.

Huanchaca (hwahn-CHAH-kah), national park, VELASCO province, SANTA CRUZ department, E central BOLIVIA, in the HUANCHACA mountain range; 14°30′S 60°39′W.

Huanchaco (hwahn-CHAH-ko), town (2005 population 36,550), TRUJILLO province, LA LIBERTAD region, NW Peru, PACIFIC port and beach resort, on railroad and 7 mi/11 km WNW of TRUJILLO; 07°46′S 78°29′W. Fisheries. Near CHAN CHAN archaeological site.

Huandacareo (wahn-dah-KAR-ee-o), town, MICHOACÁN, central Mexico, on NW shore of Lake CUITZEO, 22 mi/35 km NW of MORELIA; 19°59′N 101°17′W. Elevation 6,047 ft/1,843 m. Agricultural center (cereals, fruit; livestock).

Huando (HWAHN-do), town, HUANCAVELICA region, S central Peru, in CORDILLERA OCCIDENTAL, 15 mi/24 km NNE of HUANCAVELICA; 12°29′S 74°58′W. Cereals, corn, alfalfa; livestock.

Huandoy, mountain (20,981 ft/6,395 m), in Cordillera BLANCA, ANDES, ANCASH department, PERU; 09°01′S 77°40′W. Has several peaks, N peak is highest.

Huang, CHINA: see HUANG HE.

Huang'an, CHINA: see HONG'AN.

Huangbu (HUANG-BU), city, S GUANGDONG province, SE CHINA, on an island in the PEARL RIVER; 23°08′N 113°31′E. It is c.9 mi/14.5 km SE of GUANGZHOU, of which it is an outer port; Huangbu has been enlarged and modernized since 1952 and now accommodates large oceangoing vessels. An important industrial city, Huangbu is economically, and administratively, a part of Guangzhou. The Huangbu Military Academy, founded here in 1924 as a Kuomintang training center, was organized by Chiang Kai-shek (Jiang Jie-shi). Several of its officers, notably Zhou Enlai, later became leaders of the Chinese Communist army.

Huangcaoba, CHINA: see XINGYI.

Huangchuan (HUANG-CHUAN), town, ⊙ Huangchuan county, SE HENAN province, NE CHINA, 55 mi/89 km E of XINYANG; 32°07′N 115°02′E. Rice, oilseeds, jute.

Huanggang, CHINA: see HUANGZHOU.

Huang Hai, CHINA: see YELLOW SEA.

Huanghai, CHINA: see YELLOW SEA.

Huang-hai, CHINA: see YELLOW SEA.

Huang He (HUANG HUH) or **Yellow River**, c.3,000 mi/4,828 km long, N CHINA; rising in the KUNLUN mountains, NW QINGHAI province; flowing generally E into the "great northern bend" (around the ORDOS DESERT), then E again to the BOHAI, an arm of the YELLOW SEA. The turbulent upper Huang He meanders E through a series of gorges to the fertile Lanzhou valley. Hydroelectric power dams in the Liujia gorge and at LANZHOU, the largest city on the river, generate electricity and impound irrigation water. Past Lanzhou, the Huang He becomes a wide, slow-moving stream as it begins its bend around the Ordos and separates the N uplands from the desert and loess lands of the S and W. It is navigable in places for small vessels. The W end of the "great northern bend" passes through the heavily populated NINGXIA agricultural region, an oasis c.60 mi/97 km long, where cereals and fruits are raised. At the NW corner of the "great northern bend," the Huang He divides into numerous branches, watering a fertile area (HETAO) where ancient irrigation canals have been repaired and are now in use. At the NE corner lies the most fertile land outside the GREAT WALL; it was farmed without irrigation until 1929. Turning S, the Huang He passes through the Great Wall and enters the fertile loess region (where rich coal deposits are also found on both sides of the river). Cutting deep into the loess, the river receives most of the yellow silt from which its name is derived. After receiving the WEI and FEN river, its chief tributaries, the Huang He turns E and flows through San-men gorge, site of the huge Sanmen dam (completed 1962), which is used for power production, flood control, and navigation.

Silt dropped at the Huang He's basin over the millennia has created a great alluvial plain called the NORTH CHINA PLAIN, which extends over much of HENAN, HEBEI, and SHANDONG provinces, and which merges with the Yangzi delta in N JIANGSU and N ANHUI provinces; the delta is constantly expanding E into the sea. The Huang He meanders over the fertile, densely populated plain to reach the Bohai. The plain is one of China's most important agricultural regions, producing corn, kaoliang, winter wheat, vegetables, and cotton. However, floods, insufficient rainfall, and overcrowding cause frequent famine. During the winter dry season the Huang He is slow-moving and silt-laden, and occupies only part of its huge bed; with the summer rains, it can become a raging torrent. Since the 2nd century B.C.E., the lower Huang He has inundated the surrounding region some 1,500 times and has made nine major changes in its course. In an attempt to halt the Japanese invasion of China in 1938, the Nationalist government bombed the dike at Huayuankou to divert the Huang He S, flooding more than 20,000 sq mi/51,800 sq km and killing some 900,000 people; it was returned to its present course in 1947. The Chinese have long sought to control the Huang He by building dikes and over-flow channels. Silt deposition, the principal cause of flooding, continually elevates the riverbed; in places the river flows 60 ft/18 m to 70 ft/21 m above the surrounding plains. During the summer high-water period, water pressure against the dikes frequently breaks through the retaining walls to cause devastating floods, which have led the Huang He to be called "China's Sorrow." The Chinese initiated a fifty-year construction plan for control of the river in 1955. Dikes are being repaired and reinforced, and a series of silt-retaining dams, when completed, will control the upper river, produce 23 million kw of electricity, and provide water for 31,250 sq mi/80,937 sq km. The People's Victory Canal, a 40-mi/64-km-long diversion and irrigation channel, connects the Huang He with the Wei.

Huanghua (HUANG-HUAH), city (□ 597 sq mi/1,552.2 sq km; 2000 population 353,330), E HEBEI province, NE CHINA, near GULF OF BOHAI, 50 mi/80 km SSE of TIANJIN; 38°26′N 117°23′E. Agriculture (especially crop growing, fishing) and light industry are the most important economic activities. Grains; aquatic products. Manufacturing (food processing, chemicals, metals, and machinery). Connected to the interior of China via railroad, the city is one of the coal-export ports on the Bohai. It was designated as a provincial special economic zone in 1993.

Huangling (HUANG-LING), [Mandarin=Tomb of Huangdi], town, ⊙ Huangling county, N central SHAANXI province, NW central CHINA, 70 mi/113 km S of YAN'AN, in mountain region; 35°35′N 109°15′E. Grain, oilseeds, tobacco. Site of tomb of Haung Di (2697–2597 B.C.E.), legendary founder of Chinese Empire. Annual worship and commemoration of Huang Di is held here and participated in by Chinese from around the world.

Huanglong (HUANG-LUNG), town, ⊙ Huanglong county, central SHAANXI province, NW central CHINA, 85 mi/137 km SSE of YAN'AN; 35°35′N 109°50′E. Grain, tobacco; logging. Also called Shipu.

Huangmei (HUANG-MAI), town, ⊙ Huangmei county, SE HUBEI province, central CHINA, near ANHUI border, 100 mi/161 km ESE of WUHAN; 30°04′N 115°56′E. Rice, cotton, oilseeds; textiles and clothing, food processing, chemicals, engineering. Has noted monastery.

Huangmian (HUANG-MIAN), town, NE central GUANGXI ZHUANG, S CHINA, 35 mi/56 km ENE of LIUZHOU, near railroad; 24°45′N 109°39′E. Rice, wheat, corn, beans, sugarcane.

Huang Mountains (HUANG), Chinese *Huang Shan* (HUANG SHAN), range (average elevation 8,694 ft/2,650 m), S ANHUI province, E CHINA; 30°00′N 118°00′E. Famous for their scenic and tourist spots. Legend says that Xuanyuan Huangdi, the Han Chinese ancestor, became enlightened at these mountains (originally called the Yi Mountains, or Yi Shan). The current name was adopted during the Tang dynasty (618–907). The mountains have seventy-two peaks, including Lotus Flower and Heavenly Capital. The high peaks, vertical cliffs, ancient pines, rapid streams, hot springs, and mist give these mountains their reputation for natural beauty.

Huangpi (HUANG-PEE), town, ⊙ Huangpi county, E HUBEI province, central CHINA, on railroad spur, and

25 mi/40 km NNE of WUHAN; 30°58′N 114°23′E. Rice, cotton, tobacco, oilseeds.

Huangping (HUANG-PING), town, ⊙ Huangping county, E GUIZHOU province, S central CHINA, 80 mi/129 km ENE of GUIYANG, and on main road to HUNAN; 26°54′N 107°53′E. Rice, tobacco; tobacco processing. Also known as Siping.

Huangpu (HUANG-PU), river, 60 mi/97 km long, SHANGHAI municipality, E CHINA; rising in the lake district; flowing NE past Shanghai into the CHANG JIANG (Yangzi River) estuary at WUSONG. It is a major navigational route. Its dredged channel, lined with wharves, warehouses, and industrial plants, provides access to Shanghai for oceangoing vessels. Pollution has been a major problem in recent years. The name may sometimes appear as Hwangpoo, Whangpoo, or Hwangpu River.

Huangqiao (HUANG-CHIOU), town, S central JIANGSU province, E CHINA, 25 mi/40 km SE of TAIZHOU. Rice, oilseeds.

Huangshan (HUANG-SHAN), city (□ 909 sq mi/2,363.4 sq km; 2000 population 376,521), S ANHUI province, E CHINA; 30°04′N 118°11′E. Agriculture and light industry are the largest sectors of the city's economy. Crop growing, animal husbandry, commercial agriculture, and forestry. Agriculture (grains, oil crops, vegetables, fruits, hogs); manufacturing (food and beverages, textiles, chemicals, plastics, nonmetallic materials, machinery). In major tea-growing region. Formerly known as Tunxi (or Tunki).

Huang Shan, CHINA: see HUANG MOUNTAINS.

Huangshi (HUANG-SHI), city (□ 69 sq mi/179.4 sq km; 2000 population 546,290), E HUBEI province, central CHINA, on CHANG JIANG (Yangzi River); 30°13′N 115°05′E. Site of a bridge across the river. The city also has the oldest modern cement-manufacturing factory in China. Now an industrial center, with a giant iron and steel complex supplied with iron from the mines at Daye and with coking coal from PINGXIANG. The Daye iron-ore mine and iron smelter, established during the late Qing dynasty (1644–1911), were the forerunner of today's Daye iron and steel company. Crop growing, animal husbandry. Grains, oil crops, vegetables, hogs; manufacturing (food, textiles, apparel, utilities, pharmaceuticals, rubber, iron and steel, non-ferrous metals, machinery, electrical equipment). An ancient metallic center; metal-smelting sites here date from the Western Zhou dynasty (1126–771 B.C.E.) to the Western Han dynasty (206 B.C.–A.D. 9). The name may appear as Hwangshih.

Huang Shui, CHINA: see XINING RIVER.

Huanguelén (hwahn-ge-LAIN), town, W central BUENOS AIRES province, ARGENTINA, 28 mi/45 km E of GUAMINÍ; 37°02′S 61°57′W. Railroad junction in grain-growing and livestock area.

Huang Xian (HUANG SIAN), town, NE SHANDONG province, E CHINA, on road, and 12 mi/19 km E of LONGKOU, near N coast; 37°40′N 120°30′E. Grain, oilseeds.

Huangyan (HUANG-YAN), city (□ 487 sq mi/1,266.2 sq km; 2000 population 888,631), SE ZHEJIANG province, SE CHINA, near Taizhou Bay of EAST CHINA SEA; 28°39′N 121°19′E. Light industry and agriculture are the largest sectors of the city's economy. Crop growing, animal husbandry. Agriculture (grains, fruits, hogs, eggs); manufacturing (food processing, crafts, chemicals, plastics, machinery, electrical equipment).

Huangyuan (HUANG-YUAN), town, ⊙ Huangyuan county, NE QINGHAI province, W CHINA, on XINING RIVER, 33 mi/53 km WNW of XINING, and on route to TIBET and Jyekundo; 36°40′N 101°12′E. Oilseeds, grain, livestock; food processing, fur processing, building materials. Called Tangar, Tenkar, or Donkyr until 1912.

Huangzhong (HUANG-JUNG), town, ⊙ Huangzhong county, NE QINGHAI province, W CHINA, 12 mi/19 km SW of XINING; 36°31′N 101°40′E. Grain, oilseeds; food industry, iron smelting.

Huangzhou (HUANG-JO), city (□ 463 sq mi/1,203.8 sq km), E HUBEI province, central CHINA, on left bank of CHANG JIANG (Yangzi River), opposite EZHOU, and 35 mi/56 km SE of WUHAN; 30°27′N 114°53′E. Agriculture is the largest sector of the city's economy; light and heavy industries are also important. Crop growing, animal husbandry, and fishing. Grains, oil crops, vegetables, hogs; manufacturing (textiles). Also known as Huanggang.

Huanímaro (wah-NEE-mah-ro), city, ⊙ Huanímaro municipio, GUANAJUATO, central MEXICO, on central plateau, 22 mi/35 km SSW of IRAPUATO; 20°22′N 101°29′W. Elevation 8,068 ft/2,459 m. Cereals, alfalfa, beans, sugarcane, fruit. Also Guanímaro.

Huaning (HUAN-NING), town, ⊙ Huaning county, SE central YUNNAN province, SW CHINA, 60 mi/97 km SSE of KUNMING, near railroad; 24°12′N 102°55′E. Food processing, building materials; rice, sugarcane, tobacco; coal mining.

Huañipaya (hwahn-yee-PAH-yah), town and canton, OROPEZA province, CHUQUISACA department, SE BOLIVIA, N of the HUATA gully near the Chico River, 6 mi/10 km SE of SUCRE, on the Sucre-POTOSÍ railroad and highway. Elevation 8,212 ft/2,503 m. Antimony-bearing lodes, iron mining at Mina Okekhasa and Mina Virgén de COPACABANA; clay, limestone, and gypsum deposits. Agriculture (potatoes, yucca, bananas, corn, barley, oats, rye, peanuts); cattle.

Huaniqueo de Morales (wah-nee-KAI-o de mo-RAH-les), town, ⊙ Huaniqueo municipio, MICHOACÁN, central MEXICO, 30 mi/48 km NW of MORELIA; 19°54′N 101°30′W. Elevation 6,775 ft/2,065 m. Cereals, fruit; livestock.

Huanjiang (HUAN-JIANG), town, ⊙ Huanjiang county, N GUANGXI ZHUANG AUTONOMOUS REGION, S CHINA, 70 mi/113 km NW of LIUZHOU, near railroad; 24°49′N 108°12′E. Iron smelting; rice, beans; coal mining, iron-ore mining.

Huan Mountains (HUAN), low hill range in N central ANHUI province, E CHINA, forming divide between HUAI and CHANG JIANG (Yangzi River) systems. Also called Huan Shan.

Huañoma (hwahn-YO-mah), town and canton, N OROPEZA province, CHUQUISACA department, SE BOLIVIA, 9 mi/14 km E of SACACA, NE of the ORURO-COCHABAMBA road; 18°26′S 65°28′W. Elevation 8,212 ft/2,503 m. Lead-bearing lodes and iron and manganese mining; clay, limestone, and gypsum deposits. Agriculture (potatoes, yucca, bananas, corn, barley, oats, peanuts); cattle.

Huanoquite (hwah-no-KEE-tai), town, PARURO province, CUSCO region, S Peru, 33 mi/53 km SSW of CUSCO; 13°45′S 72°01′W. Elevation 10,190 ft/3,105 m. Cereals, potatoes; livestock.

Huanren (HUAN-REN), town, ⊙ Huanren county, SE LIAONING province, NE CHINA, 40 mi/64 km SW of TONGHUA; 41°16′N 125°21′E. Road center; medicinal herbs; hydroelectric power.

Huan River (HUAN), over 150 mi/241 km long, S GANSU province, NW CHINA; rises in SHAANXI-Gansu-NINGXIA border area; flows S, past Huan Xian, Qingyang, and Ning Xian, to JING RIVER on Shaanxi border, 90 mi/145 km NW of XI'AN.

Huan Shan, CHINA: see HUAN MOUNTAINS.

Huanta (HWAHN-tah), province (□ 2,373 sq mi/6,169.8 sq km), AYACUCHO region, S Peru; ⊙ HUANTA; 12°30′S 74°10′W. Northernmost province of Ayacucho. Drained by MANTARO RIVER. Sugarcane, potatoes, cereals. Mining.

Huanta (HWAHN-tah), city (2005 population 27,814), ⊙ HUANTA province, AYACUCHO region, S central Peru, in the ANDES, 15 mi/24 km N of AYACUCHO, on

Ayacucho to PAMPAS road; 12°56′S 74°15′W. Elevation 10,552 ft/3,216 m. Sugarcane, potatoes, coca, coffee, cereals. Gold, silver, lead, copper and zinc mining nearby.

Huantai (HUAN-TEI), town, ⊙ Huantai county, central SHANDONG province, NE CHINA, 10 mi/16 km NNE of ZIBO; 36°57′N 118°05′E. Grain, cotton; textiles, food industry, papermaking, engineering. Also known as Suozhen.

Huantán (hwahn-TAHN), town, YAUYOS province, LIMA region, W central Peru, in CORDILLERA OCCIDENTAL, 6 mi/10 km E of YAUYOS; 12°27′S 75°49′W. Potatoes, cereals; livestock.

Huantar (hwahn-TAHR), town, HUARI province, ANCASH region, W central Peru, on E slopes of Cordillera BLANCA, 22 mi/35 km ENE of HUARÁZ; 09°04′S 78°10′W. Cereals, potatoes. Near PARQUE NACIONAL HUASCARÁN.

Huantraicó, Sierra de (hwahn-trai-KO, see-ER-rah dai), subandean range in N NEUQUÉN province, ARGENTINA, 25 mi/40 km E of CHOS MALAL; c.20 mi/32 km long. Rises to c.6,000 ft/1,829 m. Has coal deposits.

Huáncuo (HWAH-noo-ko), province, HUÁNUCO region, central Peru; ⊙ HUÁNUCO.

Huánuco (HWAH-noo-ko), region (□ 14,242 sq mi/37,029.2 sq km), central Peru; ⊙ HUÁNUCO; 09°30′S 75°50′W. Bordered by middle HUALLAGA RIVER (NE), upper MARAÑÓN RIVER (NW), and Cordillera BLANCA of the ANDES (SW). Crossed N-S by Cordillera CENTRAL (W of Huallaga River) and E by Cordillera ORIENTAL. Corn, cereals, vegetables, potatoes; sheep and cattle raising in mountains; coca, cacao, plantain, yucca, and coffee in subtropical valleys on slopes of mountain ranges. Rubber exploitation; lumbering in E tropical regions toward PACHITEA RIVER. Oil is found at Agua Caliente on the Pachitea. Deposits of copper, lead, zinc, silver, iron, mercery, antimony, gold. Region is served by highway from CERRO DE PASCO (SW) to PUCALLPA (NE). Main centers: Huánuco, AMBO, LA UNIÓN, PUERTO INCA, TINGO MARÍA, LLATA, HUACAYBAMBA, HUACRACHUCO. Tourist sites include Huanuco Viejo, Tantamayo, Temple of Kotosh, and PARQUE NACIONAL TINGO MARÍA.

Huánuco (HWAH-noo-ko), city (□ 3,868 sq mi/10,056.8 sq km; 2005 population 161,005), ⊙ Huánuco province and HUÁNUCO region (□ 3,868 sq mi/10,018 sq km), central Peru, on E slopes of Cordillera CENTRAL, on highway, and 50 mi/80 km N of CERRO DE PASCO, on road to TINGO MARÍA, and 155 mi/249 km NNE of LIMA; 09°55′S 76°12′W. Elevation 6,273 ft/1,912 m. Local trade center. Timber, agricultural products (sugarcane, cotton, coffee, cocoa, fruit). Tourist resort, near Temple of Kotosh archaeological site. Cotton ginning and sugar milling nearby. Bishopric. Colonial style churches and Museum of Natural History. Founded by the Spaniards in 1539, it became an important center in colonial times, declining afterward. Airport.

Huánuco Viejo (HWAH-noo-ko vee-AI-ho), [Spanish=Old Huánuco], former place, archaeological site, HUÁNUCO region, central Peru. The settlement is the original location of present day HUÁNUCO.

Huanuni, town (2001 population 15,106) and canton, ⊙ PANTALEÓN DALENCE province, ORURO department, W BOLIVIA, on W slope of Cordillera de AZANAQUES and 28 mi/45 km SSE of ORURO; 18°17′S 66°51′W. Elevation 12,897 ft/3,931 m. On LA PAZ–Río MULATO–UYUNI–VILLAZÓN railroad. Major tin-mining center; Mina HUANUNI, biggest in country, is here.

Huanusco (wah-NOOS-ko), town, ZACATECAS, N central MEXICO, 22 mi/35 km E of TLALTENANGO DE SÁNCHEZ ROMÁN, on Mexico Highway 54. Also San Pedro Huanusco.

Huan Xian (HUAN SIAN), town, ⊙ Huan Xian county, SE GANSU province, NW CHINA, on HUAN RIVER, and 32 mi/51 km NNW of QINGYANG; 36°36′N 107°06′E. Grain, oilseeds; livestock; food processing, building materials.

Huanzo, Cordillera de (HWAHN-so, kor-dee-YAI-rah dai), mountain range, S Peru, in CORDILLERA OCCIDENTAL of the ANDES; extends c.34 mi/57 km in a NW direction; 14°30′S 73°20′W. Drained by OCOÑA and APURÍMAC rivers.

Huapango, Lake (wah-PAHN-go), artificial lake in MEXICO, central MEXICO, 51 mi/82 km NW of MEXICO CITY; c.12 mi/19 km long.

Hua Phan (HOO-wah PAHN), province (2005 population 280,780), NE LAOS, NW border with LUANG PHABANG province, SW border with XIENG KHUANG province, other borders with VIETNAM, E side of TRUONG SON RANGE; ⊙ SAM NEUA; 20°30′N 104°00′E. Border crossing to VIETNAM in Sop Hao district to THANH HOA province, VIETNAM. Shifting cultivation. Meo, Thai, and other minority peoples.

Huapí Mountains (hwah-PEE) or **Wapi Mountains**, S central NICARAGUA, E spur of main continental divide, extending from area of CAMOAPA c.100 mi/161 km E toward PEARL LAGOON. Rise to c.4,000 ft/1,219 m. Form watershed between RÍO GRANDE (N) and SIQUIA RIVER (S).

Huaping (HUAH-PING), town, ⊙ Huaping county, N YUNNAN province, SW CHINA, 60 mi/97 km ENE of Jinjiang, in mountain region, near SICHUAN border; 26°37′N 101°13′E. Timber, rice; chemicals, food industry, coal mining. Old town of Huaping is 25 mi/40 km SSE.

Huaquechula (wah-kai-CHOO-lah), town, PUEBLA, central MEXICO, 12 mi/19 km SW of ATLIXCO. Cereals, sugarcane, vegetables, livestock.

Huaquirca (hwah-KEER-kah), town, ANTABAMBA province, APURÍMAC region, S central Peru, on affluent of PACHACHACA RIVER, and 2 mi/3 km N of ANTABAMBA; 14°20′S 72°53′W. Cereals; livestock. Variant spelling: Huadquirca.

Huara (WAH-rah), town, ⊙ Huara comuna, ARICA province, TARAPACÁ region, N CHILE, on railroad and 30 mi/48 km NE of IQUIQUE; 19°59′S 69°47′W. Nitrate mining and refining, practically abandoned today.

Huaral (hwah-RAHL), province, LIMA region, W central Peru; ⊙ HUARAL; 11°15′S 76°55′W. Extending from the PACIFIC OCEAN on the W to PASCO and JUNÍN regions on the E.

Huaral (hwah-RAHL), town (2005 population 75,455), ⊙ HUARAL province, LIMA region, W central Peru, on coastal plain, 35 mi/56 km NNW of LIMA; 11°30′S 77°12′W. Cotton, rice.

Huarás, Peru: see HUARÁZ.

Huaráz (hwah-RAHS), province, ANCASH region, W central Peru; ⊙ HUARÁZ; 09°25′S 77°40′W. Straddles the Callejón de HUAYLAS of the SANTA RIVER. Crossed by PARQUE NACIONAL HUASCARÁN. Cereals, potatoes; livestock. Variant spelling: Huarás.

Huaráz (hwah-RAHS), city (2005 population 94,443), ⊙ HUARÁZ province and ANCASH region, W central Peru, on road from LIMA to CARÁZ, 25 mi/40 km ESE of CARHUÁZ; 09°32′S 77°32′W. It is in a high valley at an elevation of 9,931 ft/3,027 m, and has a predominantly Native American population. Huaráz is the center of an agricultural district raising cereals, potatoes, and maize. Some minerals (silver, cinnabar, and coal) are mined. Tourist resort. Near the PARQUE NACIONAL HUASCARÁN; many archaeological sites. HUASCARÁN peak forms a superb backdrop to the city. In 1941 a mudslide wiped out a large section of the city and killed some 6,000 people. In 1970 the city was struck by an earthquake that left nearly 10,000 dead. Suffers from recurrent earthquakes and alluvions. Variant spelling: Huarás.

Huari (HWAH-ree), province (□ 2,096 sq mi/5,449.6 sq km), ANCASH region, W central Peru; ⊙ HUARI; 09°20′S 77°10′W. Easternmost province of Ancash. Crossed by PARQUE NACIONAL HUASCARÁN and the Cordillera BLANCA of the ANDES. Cereals, potatoes; livestock.

Huari (HWAH-ree), city, ⊙ HUARI province, ANCASH region, W central Peru, on E slopes of Cordillera BLANCA of the ANDES, 24 mi/39 km ENE of HUARÁZ; 09°22′S 77°14′W. Agricultural products (cereals, corn, potatoes). Near PARQUE NACIONAL HUASCARÁN.

Huari (WAH-ree), town and canton, CHALLAPATA (AVAROA) province, ORURO department, W BOLIVIA, in the ALTIPLANO, 75 mi/121 km SSE of ORURO, on Oruro-uyuni railroad; 18°46′S 65°41′W. Elevation 12,165 ft/3,739 m. Annual fair of cattle and agriculture products (potatoes).

Huariaca (hwah-ree-AH-kah), town, PASCO region, central Peru, on HUALLAGA RIVER and 17 mi/27 km NNE of CERRO DE PASCO; 10°27′S 76°07′W. Cereals, potatoes.

Huaricaya, town and canton, Panata province, COCHABAMBA department, central BOLIVIA.

Huari Huari (HWAH-ree HWAH-ree), town and canton, TOMÁS FRÍAS province, POTOSÍ department, W central BOLIVIA, 15 mi/25 km NE of POTOSÍ, on the Potosí-SUCRE railroad and 6 mi/10 km N of Don Diego; 19°25′S 65°38′W. Elevation 15,092 ft/4,600 m. Antimony-bearing lode; zinc and lead mining at Mina HUARI HUARI; bismuth, silver, limestone, and gypsum deposits. Agriculture (potatoes, yucca, bananas); cattle.

Huarina (wah-REE-nah), town and canton, OMASUYOS province, LA PAZ department, W BOLIVIA, port on Lake TITICACA and 46 mi/74 km W of LA PAZ; 16°12′S 68°38′W. Elevation 12,530 ft/3,819 m. Potatoes, sheep.

Huaritolo (hwah-ree-TO-lo), town and canton, INQUISIVI province, LA PAZ department, W BOLIVIA, 2 mi/3 km S of CAJUATA, on the INQUISIVI–LA PAZ road; 16°50′S 67°20′W. Elevation 8,986 ft/2,739 m. Tin-bearing lode, tin mining (Mina CARACOLES, Mina COLQUIRI), lead, zinc, and silver mining (Mina Colquiri), copper mining (to W), tungsten mining (Mina Chambilaya, Mina Chicote Grande); clay, limestone, and gypsum deposits. Agriculture (potatoes, yucca, bananas, rye); cattle.

Huarmaca (hwahr-MAH-kah), town, PIURA region, HUANCABAMBA province, NW Peru, on W slopes of CORDILLERA OCCIDENTAL, 23 mi/37 km SSW of HUANCABAMBA; 05°34′S 79°32′W. Corn, cereals, potatoes.

Huarmaco, Peru: see HUARMACA.

Huarmechi (hwar-ME-chee), town and canton, MÉNDEZ province, TARIJA department, S central BOLIVIA. Elevation 6,555 ft/1998 m.

Huarmei, Peru: see HUARMEY.

Huarmey (hwahr-MAI), province, ANCASH region, W central PERU; ⊙ HUARMEY; 09°55′S 78°00′W. Between the PACIFIC OCEAN on the W and the Cordillera NEGRA of the ANDES on the E. Sugarcane, rice.

Huarmey (hwar-MAI), town, ⊙ HUARMEY province, ANCASH region, W central PERU, on coastal plain, near mouth of Huarmey River, on PAN-AMERICAN HIGHWAY, and 40 mi/64 km SSE of CASMA; 10°04′S 78°10′W. Rice, corn, fruit. Its port, Puerto Huarmey, on the Pacific, is 3 mi/5 km SW. Also spelled Huarmei.

Huarmey River (HWAHR-mai), ANCASH region, W central PERU; rises in the ANDES; flows W to the PACIFIC OCEAN near HUARMEY; 10°05′S 78°10′W.

Huarochirí (hwah-ro-chee-REE), province (□ 2,002 sq mi/5,205.2 sq km), LIMA region, W central PERU; ⊙ MATUCANA; 11°55′S 76°25′W. Central province of Lima region, bordering JUNÍN province on the E.

Huarochirí (hwah-ro-chee-REE), city, LIMA region, W central PERU, on MALA RIVER, just opposite SAN LORENZO DE QUINTI, and 55 mi/89 km ESE of LIMA; 12°07′S 76°13′W. Cereals, corn.

Huarocondo (hwah-ro-KON-do), town, ANTA province, CUSCO region, S central PERU, in the ANDES, on

CUSCO-Machupicchu railroad, 15 mi/24 km WNW of CUSCO; 13°24′S 72°12′W. Elevation 11,072 ft/3,374 m. Cereals, potatoes.

Huarón (hwah-RON), town, PASCO region, central PERU; 11°00′W 76°25′W. Mining center.

Huarong (HUAH-RUNG), town, ⊙ Huarong county, NE HUNAN province, S central CHINA, on N shore of DONGTING LAKE, 32 mi/51 km WNW of YUEYANG, on HUBEI border; 29°34′N 112°34′E. Rice, cotton, oilseeds, sugarcane; pulp and paper making, food processing.

Huarte or **Uarte** (WAHR-tai), town, NAVARRE province, N SPAIN, 3 mi/4.8 km NE of PAMPLONA; 42°50′N 01°35′W.

Huásabas (WAH-sah-bas), town, SONORA, NW MEXICO, on BAVISPE River and 115 mi/185 km NE of HERMOSILLO; 29°47′N 109°18′W. Livestock, wheat.

Huasacalle (hwah-sah-CAH-yai), town and canton, Germán JORDAN province, COCHABAMBA department, central BOLIVIA. Elevation 8,878 ft/2,706 m.

Huasanó, Colombia: see RIOFRÍO.

Huasa Rancho (HWAH-sah RAHN-cho), town and canton, ESTEBAN ARCE province, COCHABAMBA department, central BOLIVIA 15 mi/25 km SE of TARATA, on the MIZQUE-COCHABAMBA railroad. Elevation 9,026 ft/2,751 m. Copper-bearing lodes; clay, limestone, and gypsum deposits. Agriculture (corn, potatoes, yucca, bananas, rye, coffee); cattle raising for meat and dairy products.

Huasca de Ocampo (WAS-kah de o-KAM-po), town, HIDALGO, central MEXICO, 12 mi/19 km NE of PACHUCA DE SOTO. Corn, maguey, beans, livestock.

Huascarán (hwahs-kah-RAHN), mountain peak and extinct volcano (22,205 ft/6,768 m), ANCASH region, W central PERU, in Cordillera BLANCA of CORDILLERA OCCIDENTAL, near HUARÁZ; 09°07′S 77°37′W. The highest mountain in Peru and one of the highest in the ANDES, Huascarán and other nearby peaks form an impressive snow-capped rampart. Avalanches in 1962 and 1970 swept down its slopes and buried villages below. Also known as Nevado [=snow-covered peak] Huascarán.

Huasco (WAHS-ko), southernmost province of ATACAMA region, N central CHILE, ⊙ VALLENAR; 28°30′S 70°30′W. Irrigated agriculture (fruit, vegetables, grapes); mining (especially copper, silver, gold).

Huasco (WAHS-ko), town, ⊙ Huasco comuna, HUASCO province, ATACAMA region, N central CHILE, PACIFIC port near mouth (2 mi/3 km SW) of HUASCO RIVER, 30 mi/48 km WNW of VALLENAR, 95 mi/153 km SW of COPIAPÓ; 28°30′S 70°30′W. Trading and manufacturing center on Huasco River. Ships copper, cobalt, nickel. Grapes, currants, alfalfa, fish. Copper smelters, wineries. Deposits of high-grade iron ores at ALGARROBAL.

Huasco Bajo (WAHS-ko BAH-ho), village, ATACAMA region, N central CHILE, on HUASCO RIVER, on railroad and 3 mi/5 km E of HUASCO; 28°28′S 71°11′W. In agricultural area (alfalfa, corn, wine; cattle, goats).

Huasco River (WAHS-ko), c.140 mi/225 km long, ATACAMA region, N CHILE; formed 20 mi/32 km SE of VALLENAR by confluence at NW foot of SIERRA DE TATUL of 2 headstreams; flows c.60 mi/97 km WNW, past Vallenar and HUASCO BAJO, to the PACIFIC 2 mi/3 km NE of HUASCO. Irrigates agricultural area where alfalfa, corn, oranges, and grapes are grown and cattle and goats are raised. Rich mining area. Sometime spelled Guasco River.

Hua Shan Mountain (HUAH-SHAN) (7,218 ft/2,200 m), S SHAANXI province, NW central CHINA, 10 mi/16 km S of HUAYIN, near HENAN border; 34°29′N 110°04′E. The mountain is one of the five sacred mountain peaks in China. Hua Shan (Buddhist shrine) is in the mountain. Also called the Western Peak.

Huaspuc River, NICARAGUA: see WASPUK RIVER.

Huasteca, La (was-TEK-ah), GULF lowlands in NE MEXICO, comprise central and lower PÁNUCO River

basin of VERACRUZ and TAMAULIPAS; rise near slopes of SIERRA MADRE ORIENTAL in HIDALGO and SAN LUIS POTOSÍ. Fertile region with abundant rain, tropical in the plains. Excellent pastures for livestock raising; large-scale sugarcane production, cereals, fruit. Formerly center of Mexico's petroleum industry. Named for Huastec Indians, some 90,000 of whom still live in the W margins of the region.

Huata (HWAH-tah), canton, OROPEZA province, CHUQUISACA department, SE BOLIVIA, 12 mi/20 km N of SUCRE, on the edges of HUATA gorge; 18°59′S 65°16′W. Elevation 8,212 ft/2,503 m. Iron mining at Mina Okekhasa and Mina Virgen de Copacabana; clay and limestone deposits. Agriculture (potatoes, yucca, bananas, corn, barley, oats, rye, peanuts); cattle. Hot springs.

Huata, BOLIVIA: see SANTIAGO DE HUATA.

Huatabampo (wah-tah-BAHM-po), city and township, SONORA, NW MEXICO, on the Gulf of CALIFORNIA, near MAYO RIVER mouth (Río Mayo Irrigation District), 21 mi/34 km SW of NAVOJOA; 26°49′N 109°40′W. Agricultural center (chickpeas, cereals, vegetables, fruit). Many agricultural products for export.

Huatajata (hwah-tah-HAH-tah), town and canton, OMASUYOS province, LA PAZ department, W BOLIVIA, on the E coast of Lake TITICACA, 54 mi/87 km NW of LA PAZ; 16°10′S 68°44′W. Elevation 12,543 ft/3,823 m. Port; tourist ships anchor here. Copper-bearing lode and clay, limestone, and gypsum deposits. Agriculture (potatoes, yucca, bananas, rye); cattle.

Huata Peninsula, BOLIVIA: see ACHACACHI PENINSULA.

Huating (HUAH-TING), town, ⊙ Huating county, SE GANSU province, NW CHINA, 20 mi/32 km SW of PINGLIANG; 35°09′N 106°38′E. Grain, jute, oilseeds, medicinal herbs; coal mining, construction materials.

Huatlatlauca (wah-tla-tla-OO-kah), town, PUEBLA, central MEXICO, near ATOYAC RIVER, 27 mi/43 km SSE of PUEBLA. Corn, sugarcane, fruit, livestock.

Huatulco (wah-TOOL-ko), village, OAXACA, MEXICO. Fashionable resort village surrounded by coffee plantations along the PACIFIC coast.

Huatulco Bays (wah-TOOL-ko), region, OAXACA, MEXICO. A series of nine small bays on the PACIFIC coast of Oaxaca, 24 mi/40 km E of PUERTO ANGEL. Developed in the early 1990s as a new luxury beach resort area. Served by Mexico Highway 200 and commercial air service.

Huatusco, MEXICO: see HUATUSCO DE CHICUELLAR.

Huatusco de Chicuellar (wah-TOOS-ko de chee-KWAI-yar), city and township, ⊙ Huatusco municipio, VERACRUZ, E MEXICO, in SIERRA MADRE ORIENTAL, 50 mi/80 km W of VERACRUZ on Mexico Highway 125; 19°12′N 96°08′W. Agricultural center (bananas, corn, coffee, sugarcane).

Huauchinango (wou-chee-NAN-go), city and township, PUEBLA, central MEXICO, in SE SIERRA MADRE ORIENTAL, 45 mi/72 km E of PACHUCA DE SOTO on Mexico Highway 130; 20°11′N 98°04′W. Processing and agricultural center (corn, coffee, sugarcane, tobacco, vegetables, fruit); floriculture; tanning; lumbering; shoe manufacturing. Necaxa dam and hydroelectric plant are nearby. Also known as Huauchinango de Degollado.

Huaura (HWOU-rah), province (□ 2,651 sq mi/6,892.6 sq km), LIMA region, W central PERU; ⊙ HUACHO. N province of Lima region. Extends from PACIFIC OCEAN on the W to PASCO region on the E.

Huaura (HWOU-rah), city (2005 population 26,044), LIMA region, W central PERU, on coastal plain, on HUAURA RIVER, and 2 mi/3 km NNE of HUACHO; 11°04′S 77°36′W. In important rice, cotton, and sugarcane area. San Martín proclaimed Peru's independence here.

Huaura River (HWOU-rah), 90 mi/145 km long, LIMA region, W central PERU; rises in CORDILLERA OCCIDENTAL of the ANDES 17 mi/27 km W of YANA-

HUANCA; flows SW and W, past SAYÁN and HUAURA, to the PACIFIC; 11°06′S 77°39′W. Irrigation in lower course.

Huautepec (WOU-te-pek), town, in N central OAXACA, MEXICO, 17 mi/27 km E of TEOTITLÁN DE FLORES MAGÓN. Elevation 8,104 ft/2,470 m. Steep terrain in the Papaloapan River basin. Cold to temperate climate. Agriculture (corn, beans, wheat, potatoes, fruits). Connected by unpaved road to HUAUTLA DE JIMÉNEZ. Also Santa María Asunción Huautepec.

Huautla (WOU-tlah), town, HIDALGO, central MEXICO, in SIERRA MADRE ORIENTAL foothills, 12 mi/19 km SE of HUEJUTLA; 21°02′N 98°17′W. Agricultural center (corn, rice, sugarcane, tobacco, fruit, livestock).

Huautla de Jiménez (WOU-tlah de hee-ME-nes), city and township, OAXACA, S MEXICO, in Sierra MAZATECA 15 mi/24 km E of TEOTITLÁN DE FLORES MAGÓN in Petlapa River drainage (to Papaloapan River); 18°10′N 96°51′W. Elevation 7,218 ft/2,200 m. Agricultural center (cereals, fruit). In Mazatec-speaking area. Also Huautla.

Hua Xian (HUAH SIAN), town, ⊙ Hua Xian county, central GUANGDONG province, S CHINA, 25 mi/40 km N of GUANGZHOU, near Wuhan-Guangzhou railroad; 23°22′N 113°12′E. Rice, sugarcane.

Hua Xian (HUAH-SIAN), town, ⊙ Hua Xian county, NE HENAN province, NE CHINA, on main road, and 40 mi/64 km NE of XINXIANG, near railroad (DAOKOU station); grain, cotton, oilseeds.

Hua Xian (HUAH SIAN), town, ⊙ Hua Xian county, E SHAANXI province, NW central CHINA, in WEI RIVER valley, 50 mi/80 km ENE of XI'AN, and on Longhai railroad; 34°31′N 109°46′E. Grain, cotton, oilseeds; metal smelting.

Huayabamba, PERU: see HUAYLLABAMBA.

Huayacocotla (wah-yah-ko-KO-tlah), town, VERACRUZ, E MEXICO, in SIERRA MADRE ORIENTAL, 33 mi/53 km NNE of PACHUCA DE SOTO (HIDALGO); 20°34′N 98°27′W. Cereals, sugarcane, coffee.

Huayapacha (hwah-yah-PAH-chah), town and canton, CARRASCO province, COCHABAMBA department, central BOLIVIA, 15 mi/25 km WNW of Monte Puncu, on the border of ARANI province; 17°29′S 65°28′W. Elevation 9,150 ft/2,789 m. Clay and limestone deposits. Agriculture (potatoes, yucca, bananas, corn, rice, wheat, rye, soy, sweet potatoes, coffee); cattle raising for meat and dairy products.

Huaychaca River (hwei-CHAH-kah), LA LIBERTAD region, NW PERU; 08°19′S 78°08′W.

Huayco (HWAI-koe), town and canton, BURNET O'Conner province, TARIJA department, S central BOLIVIA, 6 mi/10 km WSW of SALADILLO; 21°43′S 64°14′W. Elevation 4,035 ft/1,230 m. Extensive gas deposits in area. Clay and limestone deposits. Agriculture (potatoes, yucca, bananas, corn, barley, sweet potatoes, tobacco); cattle.

Huaycoma (hwai-KO-mah), town and canton, CHAYANTA province, POTOSÍ department, W central BOLIVIA, N of ocuri on POTOSÍ-ORURO road; 18°37′S 65°34′W. Elevation 13,642 ft/4,158 m. Copper-bearing lode; iron (insignificant), clay, limestone, and gypsum deposits. Agriculture (barley, potatoes, yucca, bananas).

Huayculi (hwai-KOO-lee), town and canton, ESTEBAN ARCE province, COCHABAMBA department, central BOLIVIA, SE of TARATA; 17°40′S 65°59′W. Elevation 9,026 ft/2,751 m. Clay, limestone, and gypsum deposits. Agriculture (corn, potatoes, yucca, bananas, barley, oats, rye, soy, coffee); cattle.

Huayhuash, Cordillera (hwei-HWASH, kor-dee-YAI-rah), mountain range, in HUANÚCO, ANCASH, and LIMA regions, PERU, in CORDILLERA OCCIDENTAL of the ANDES; 10°30′S 76°45′W. Rises to its highest point at Cerro YERUPAJA (21,758 ft/6,632 m). Variant spelling: Huayashuash.

Huayin (HUAH-YIN), city (□ 315 sq mi/819 sq km), E SHAANXI province, NW central CHINA; 34°36′N

110°09′E. Heavy industry is the largest source of income. Crop growing. Manufacturing includes utilities, pharmaceuticals, and machinery.

Huaying (HUAH-YING), city (□ 166 sq mi/431.6 sq km; 2000 population 339,371), E SICHUAN province, SW CHINA, in the WEI RIVER valley, near HUANG HE (Yellow River) bend, at foot of the HUA SHAN, 70 mi/113 km ENE of XI'AN; 30°23′N 106°45′E. Agriculture and heavy industry are the largest sources of income. Crop growing, animal husbandry. Grain, hogs; manufacturing (machinery, coal mining).

Huaylas (HWEI-lahs), province (□ 1,331 sq mi/3,460.6 sq km), ANCASH region, W central PERU; ⊙ CARÁZ; 08°55′S 77°50′W. In the Callejón de HUAYLAS of the SANTA RIVER. Crossed by the PARQUE NACIONAL HUASCARÁN. Cereals, potatoes; livestock. Also spelled Huailas.

Huaylas (HWEI-lahs), town, HUAYLAS province, ANCASH region, W central PERU, in the Callejón de HUAYLAS, 10 mi/16 km NNW of CARÁZ; 08°52′S 77°54′W. Cereals, potatoes, corn. Variant names: Huailas; Villa Sucre.

Huaylas, Callejón de (HWEI-lahs, kah-yai-HON dai), valley of upper SANTA RIVER in CORDILLERA OCCIDENTAL of the ANDES, ANCASH region, W central PERU, between Cordillera NEGRA (W) and Cordillera BLANCA (E). Extends 100 mi/161 km SSE from area of HUALLANCA to area of CHIQUIÁN; 09°10′S 77°45′W. Silver, lead, and copper mines in RECUAY, CARHUÁZ, and CARÁZ. Agriculture (cereals, corn, fruit, potatoes). Main centers (HUARÁZ, Caráz, YUNGAY, Carhuáz).

Huayllabamba (hwei-yah-BAHM-bah), town, SIHUAS province, CUSCO region, S central PERU, on URUBAMBA RIVER, 5 mi/8 km SE of URUBAMBA; 13°20′S 72°04′W. Elevation 11,095 ft/3,381 m. Cereals, potatoes. Variant spelling: Huayabamba.

Huayllamarca (hwai-yah-MAR-kah), town and canton, INQUISIVI province, LA PAZ department, W BOLIVIA. Elevation 8,986 ft/2,739 m. Copper-bearing lode; tin mining at Mina CARACOLES, Mina COLQUIRI; tungsten mining at Mina Chambilaya, Mina Chicote Grande. Agriculture (potatoes, yucca, bananas, rye); cattle.

Huayllamarca, town and canton, N CARANGAS province, ORURO department, W central BOLIVIA, 10 mi/15 km SE of TOTORA and NE of CHUQUICAMBI; 17°51′S 68°00′W. Elevation 12,448 ft/3,794 m. Gas resources in area. Lead-bearing lode, clay and gypsum deposits. Agriculture (potatoes, yucca, bananas); cattle.

Huayllas (HWAI-yahs), town and canton, OROPEZA province, W CHUQUISACA department, SE BOLIVIA, on the border of POTOSÍ department; 18°44′S 65°39′W. Elevation 8,212 ft/2,503 m. Iron mining at Mina Okekhasa and Mina Virgén de COPACABANA, manganese mining at Mina Lourdes; clay, limestone, and gypsum deposits. Agriculture (bananas, potatoes, yucca); cattle.

Huayllas, town and canton, LITORAL province, ORURO department, W central BOLIVIA. Elevation 12,451 ft/3,795 m. Gas resources in area. Lead-bearing lode; clay, limestone, and gypsum deposits. Agriculture (potatoes, yucca, bananas); cattle.

Huayllay (hwei-YEI), town, PASCO region, central PERU, in CORDILLERA OCCIDENTAL of the ANDES, 24 mi/39 km SSW of CERRO DE PASCO; 11°01′S 76°21′W. Elevation c.14,000 ft/4,267 m. The Bosque de Rocas [Spanish=rock forest], an unusually eroded rock formation, is nearby. Near SANTUARIO NACIONAL HUAYLLAY.

Huayñacota (hwai-nyah-KO-tah), canton, INQUISIVI province, LA PAZ department, W BOLIVIA, SW of QUIME; 17°15′S 67°10′W. Elevation 8,986 ft/2,739 m. Tin mining at Mina CARACOLES, Mina COLQUIRI; tungsten mining at Mina Chambilaya, Mina Chicote Grande; clay, limestone, and gypsum deposits. Agriculture (potatoes, yucca, coca, bananas, rye); cattle.

Huayña Pasto Grande (HWAI-nyah PAHS-to GRAHN-dai), canton, CERCADO province, ORURO department, W central BOLIVIA. Elevation 12,372 ft/3,771 m. Lead, zinc, and silver mining at San Jose; clay deposits; dolomite mining at Conde Auque. Agriculture (potatoes, yucca, bananas, barley, oats); cattle.

Huayna Potosí (HWAI-nah po-to-SEE), canton, LOS ANDES province, LA PAZ department, W BOLIVIA; 16°16′S 68°11′W. Elevation 12,625 ft/3,848 m. Gas resources in the area. Lead-bearing lode; clay, limestone, and gypsum deposits. Agriculture (potatoes, yucca, bananas, rye); cattle.

Huayna Potosí (WEI-nah po-to-SEE), peak (20,328 ft/6,196 m) in Cordillera de LA PAZ, Los Andes Nevado de la Cordillera Real province, LA PAZ department, W BOLIVIA, 17 mi/27 km NNW of LA PAZ; 16°16′S 68°11′W. Tin mines in region. Formerly also Huaina Potosí.

Huaytará (hwei-tah-RAH), province, HUANCAVELICA region, S central PERU; ⊙ HUAYTARÁ. Largest and southernmost province of Huancavelica. Bordered by ICA and AYACUCHO regions. Cereals, potatoes; livestock. Mining.

Huaytará (hwei-tah-RAH), town, ⊙ HUAYTARÁ province, HUANCAVELICA region, S central PERU, on Huaytará River (left affluent of the PISCO), on road from PISCO to AYACUCHO; 13°36′S 75°22′W. Elevation 9,613 ft/2,930 m. Cereals, corn; livestock. Variant spelling: Huaitará.

Huayuan (HUAH-YUAN), town, E HUBEI province, central CHINA, 50 mi/80 km NW of WUHAN, and on Beijing-Wuhan railroad; 31°17′N 114°01′E. Commercial center; rice, cotton, oilseeds.

Huazalingo (wah-sah-LEEN-go), town, HIDALGO, central MEXICO, 14 mi/23 km SW of HUEJUTLA DE REYES. Corn, rice, sugarcane, tobacco, livestock.

Huazhou (HUAH-JO), town, ⊙ Huazhou county, SW GUANGDONG province, S CHINA, on Foshan River, and 30 mi/48 km NE of ZHANJIANG; 21°38′N 110°35′E. Rice, oilseeds; food processing, building materials, textiles.

Hubbard, county (□ 999 sq mi/2,597.4 sq km; 2006 population 18,890), NW central MINNESOTA; ⊙ PARK RAPIDS; 47°06′N 94°54′W. MISSISSIPPI RIVER drains NW corner (source at Lake ITASCA to W of county). Alfalfa, oats, barley, rye, potatoes, beans; timber; peat deposits. Parts of Paul Bunyan State Forest in N and center; Badoura State Forest in SE; part of Lake Itasca State Park in W; numerous small natural lakes throughout county, especially in S. Formed 1883.

Hubbard, city (□ 4 sq mi/10.4 sq km; 2006 population 7,921), TRUMBULL county, NE OHIO, 5 mi/8 km NE of YOUNGSTOWN, near PENNSYLVANIA state line; 41°09′N 80°34′W. Incorporated 1868.

Hubbard, town (2000 population 885), HARDIN county, central IOWA, 11 mi/18 km WSW of ELDORA; 42°18′N 93°17′W. Feed, concrete blocks.

Hubbard, town (2006 population 2,606), MARION county, NW OREGON, 20 mi/32 km NW of SALEM and 25 mi/40 km SSW of PORTLAND, near Pudding River; 45°10′N 122°48′W. Manufacturing (apparel, metal forgings). Nuts, fruits. Dairy products; poultry.

Hubbard, town (2006 population 1,692), HILL county, N central TEXAS, 29 mi/47 km NE of WACO; 31°51′N 96°47′W. In cotton-growing area; oil and natural gas. Navarro Hills Lake reservoir to NE.

Hubbard, village (2006 population 251), DAKOTA county, NE NEBRASKA, 9 mi/14.5 km W of DAKOTA CITY; 42°23′N 96°35′W.

Hubbard Creek Lake, reservoir (□ 24 sq mi/62.4 sq km), STEPHENS county, N central TEXAS, on Hubbard Creek, 55 mi/88 km NE of ABILENE; 32°50′N 98°58′W. Maximum capacity 720,000 acre-ft. Formed by Hubbard Creek Lake Dam (112 ft/34 m high), built (1962) for water supply.

Hubbard Glacier, SE ALASKA, on DISENCHANTMENT BAY at head of YAKUTAT BAY; 50 mi/80 km long; 60°02′N 139°25′W. Part of SAINT ELIAS MOUNTAINS glacier system.

Hubbard Lake, ALCONA county, NE MICHIGAN, 15 mi/24 km SSW of ALPENA; c.7 mi/11.3 km long, 3 mi/4.8 km wide; 44°48′N 83°33′W. Fishing. Source of a branch of THUNDER BAY RIVER. HUBBARD LAKE, resort village, is near N end in ALPENA county (light manufacturing).

Hubbard, Mount (14,950 ft/4,557 m), on YUKON-ALASKA border in ST. ELIAS MOUNTAINS, CANADA, 140 mi/225 km W of WHITEHORSE, 60 mi/97 km NNE of YAKUTAT; 60°19′N 139°04′W.

Hubbardston, agricultural town, WORCESTER county, N central MASSACHUSETTS, 18 mi/29 km NNW of WORCESTER, near WACHUSETT MOUNTAIN; 42°29′N 72°00′W. Settled 1737, incorporated 1767. Manufacturing (backhoe attachments).

Hubbardston, village (2000 population 394), on IONIA-CLINTON county line, S central MICHIGAN, 14 mi/23 km NE of IONIA and on FISH CREEK; 43°05′N 84°50′W. In farming area.

Hubbardton, town, RUTLAND CO., W VERMONT, 12 mi/19 km NW of RUTLAND, at N end of Lake BOMOSEEN; 43°42′N 73°10′W. Named for Thomas Hubbard. The only military battle of the American Revolutionary War to be fought in what is now Vermont was the Battle of Hubbardton, July 17,1777.

Hubbell, town (2000 population 1,105), HOUGHTON county, NW UPPER PENINSULA, MICHIGAN, 8 mi/12.9 km NE of HOUGHTON, on TORCH LAKE (connected by deepwater channel to Keweenaw Waterway); 47°10′N 88°26′W. Copper mining area; copper refining; manufacturing (copper oxides). Houghton Airport to W.

Hubbell, village (2006 population 65), THAYER county, SE NEBRASKA, 12 mi/19 km SSE of HEBRON, at KANSAS state line, and on branch of LITTLE BLUE RIVER; 40°00′N 97°30′W.

Hubbell Trading Post National Historic Site, APACHE county, NE ARIZONA, in Navajo Indian Reservation, at GANADO, 47 mi/76 km WNW of GALLUP, NEW MEXICO; 160 acres/65 ha. Example of a late-19th-century trading post in the Southwest, still active. Authorized 1965.

Hubei (HU-BAI), short name **E** (UH), province (□ c.72,000 sq mi/186,480 sq km; 1994 estimated population 56,560,000; 2000 population 59,508,870), central CHINA; ⊙ WUHAN; 31°00′N 112°00′E. In this province the CHANG JIANG (Yangzi River), flowing through the S, is joined by the HAN RIVER, coming from the NW. At their junction lies Wuhan, a city comprising three former cities, HANKOU, HANYANG, and WUCHANG; Wuhan is a transportation hub and the major industrial and commercial center of central China. The central part of Hubei was once a huge lake (the Yunmengze Lake), and is now a basin (called Chiang-Han plain), at or below sea level, formed from silt deposited by Chang Jiang (Yangzi River) and Han rivers. Hubei's lakes and many rivers provide excellent irrigation facilities, and the warm climate, adequate rainfall, and rich soil make the province one of the most productive in China. Wheat, barley, rapeseed, and beans are raised in the winter, and rice, cotton, tea, soybeans, and corn in the summer. Rice production has increased significantly as a result of water conservation, modern fertilizer, better seed, and double-cropping; the province produces a surplus, which is sent to N China. Wheat is raised in the drier areas. Commercial crops include sesame, peanuts, and ramie. The minerals in the province are mostly nonferrous, although there are two huge steel complexes, one at Wuhan and one at HUANGSHI. Coal, copper, and gypsum are also mined. Chief manufacturing includes finished steel, copper, motor vehicles, machine tools, cotton yarn and cotton cloth, power generation, chemicals, and construction materials. Library, museum, and sports facilities have

Cross-references are shown in SMALL CAPITALS. The pronunciation guide is shown on page xix. The sources of population figures are shown on page xvii.

been opened in many of Hubei's cities. Mountains W of the province contain various points of interest: Shennunjia, a well-preserved ecosystem; the sacred Daoist peak, the Wudan Mountain; and the CHANG JIANG GORGES (the Three Gorges). Sometimes spelled Hupeh or Hupei.

Huberdeau (yoo-ber-DO), village (□ 22 sq mi/57.2 sq km; 2006 population 954), LAURENTIDES region, S QUEBEC, E CANADA, 23 mi/36 km from LABELLE; 45°58′N 74°38′W.

Huber Ridge (HYOO-buhr RIJ), unincorporated village (□ 1 sq mi/2.6 sq km; 2000 population 4,883), FRANKLIN county, central OHIO; NE suburb of COLUMBUS; 40°05′N 82°55′W.

Hubertusburg (HOO-ber-tus-burg), hunting lodge, SAXONY, E central GERMANY, 10 mi/16 km ENE of GRIMMA, 24 mi/39 km E of LEIPZIG; 51°18′N 12°56′E. Peace treaty between PRUSSIA, SAXONY, and AUSTRIA, at conclusion of Seven Years War, was signed here in February 1763. Built 1721–1733 for electors of Saxony.

Hubkiv, UKRAINE: see SOSNOVE.

Hubli-Dharwad, twin city (2001 population 786,195); ⊙ DHARWAD district, KARNATAKA (formerly Mysore) state, S INDIA, on the main MUMBAI (BOMBAY)-BANGALORE railroad and highway; 15°21′N 75°10′E. The cities of HUBLI AND DHARWAD, 13 mi/21 km apart, were incorporated 1961 as a single city. DHARWAD is a rice- and cotton-growing area. Hubli is a trade and transportation center, with cotton and silk factories, railroad workshops, and a major newspaper industry. Built around an 11th-century Hindu stone temple. Dharwad grew up around a fort thought to have been built in 1405 by an officer of the Hindu king of VIJAYANAGAR. Captured by the Muslims in 1685 and by the Marathas in 1753. Hyder Ali, ruler of MYSORE, occupied Dharwad in 1778. It was ceded to the British in 1818. Seat of many colleges and a growing university in the metropolitan area, making it an education center for S INDIA. Known for its cool weather, lush vegetation, and a growing economy.

Hubsugul, MONGOLIA: see HÖVSGÖL.

Hubynykha, (hoo-BI-ni-khah), (Russian *Gubinikha*), town, N DNIPROPETROVS'K oblast, UKRAINE, on road and on railroad 11 mi/8 km N of NOVOMOSKOVS'K; 48°49′N 35°15′E. Elevation 416 ft/126 m. Sugar refinery, feed mill, grain elevator. Established in 1704; town since 1964.

Hucal, ARGENTINA: see BERNASCONI.

Hucclecote, ENGLAND: see GLOUCESTER.

Hückelhoven (hyook-kel-HO-fuhn), city, NORTH RHINE-WESTPHALIA, W GERMANY, near the RUHR, 20 mi/32 km NNE of AACHEN; 51°04′N 06°13′E. Enlarged through incorporation 1935 of Hilfarth and Ratheim; chartered in 1969; enlarged again through the incorporated 1972 of other villages.

Hückeswagen (hyoo-kes-VAH-gen), town, NORTH RHINE-WESTPHALIA, W GERMANY, 5 mi/8 km E of WERMELSKIRCHEN; 51°09′N 07°20′E. Manufacturing of precision instruments and power tools. Beaver dam and reservoir is 1.5 mi/2.4 km E.

Hucknall or **Hucknall Torkard** (HUK-nul), town (2001 population 29,188), NOTTINGHAMSHIRE, central ENGLAND, 6 mi/9.6 km NW of NOTTINGHAM; 53°02′N 01°11′W. Previously the site of coal mines and manufacturing of hosiery. Lord George Byron born in the parish church.

Hudaida, YEMEN: see HODEIDA.

Hudaydah, Al, YEMEN: see HODEIDA.

Huddersfield (HUHD-uhz-feeld), town (2001 population 146,234), WEST YORKSHIRE, N central ENGLAND, on the COLNE RIVER; 53°39′N 01°47′W. Textile industry including cotton, woolen, and rayon goods, plays an important role in the town's economic success. Other manufacturing includes machinery, fabricated metal products, and chemicals. Districts of the town include Almondbury, Birkby, Bradley, Dalton, Deighton, Lindley, Marsh, Moldgreen, and Paddock.

Huddinge (HUD-deeng-e), suburb of STOCKHOLM, STOCKHOLM county, E SWEDEN, 7 mi/11.3 km SW of Stockholm city center; 59°15′N 17°53′E.

Huddur (HOOD-door), Italian *oddur*, town, W central SOMALIA, 170 mi/274 km NW of MOGADISHO, on plateau (elevation 1,500 ft/457 m), between Webē Shebelē and JUBBA rivers; 04°08′N 43°54′E. Road junction. Trade (hides, ivory, myrrh, frankincense; gum arabic); wood carving. Fort, airfield. Formerly HODUR.

Hude (HOO-de), town, LOWER SAXONY, NW GERMANY, 8 mi/12.9 km NW of DELMENHORST; 53°07′N 08°28′E. Manufacturing of agricultural machinery, scales; metalworking. Has ruined 13th-century monastery.

Hudiksvall (HOO-deeks-VAHL), town, GÄVLEBORG county, E SWEDEN, on Hudiksvallsfjärden, 15-mi/24-km-long inlet of GULF OF BOTHNIA, 45 mi/72 km S of SUNDSVALL; 61°44′N 17°6′E. Railroad junction; manufacturing (metalworking; chemicals). Airport. Archaeological museum. Several winter sports resorts nearby. Chartered as town in sixteenth century.

Hudin (HOOD-een), village, N SOMALIA, on road, and 50 mi/80 km N of LAS ANOD. Water well.

Hudlitz, CZECH REPUBLIC: see HUDLICE.

Hudson, county (□ 62 sq mi/161.2 sq km; 2006 population 601,146), NE New Jersey, bounded by PASSAIC River and NEWARK BAY (W), and HUDSON RIVER and Upper New York Bay (E); ⊙ JERSEY CITY; 40°43′N 74°04′W. Heavily industrialized, with varied manufacturing, oil refining, shipbuilding, railroad and ocean shipping. Jersey City is a commercial and transportation center of NEW YORK city metropolitan area. Site of Liberty State Park which includes the New Jersey State Science Center. Drained by HACKENSACK River. Formed 1840.

Hudson, city, HILLSBOROUGH county, S NEW HAMPSHIRE. Bounded S by MASSACHUSETTS state line; W by MERRIMACK RIVER opposite (1 mi/1.6 km E of) NASHUA; 42°46′N 71°24′W. Manufacturing (apparel, computer and electronic equipment, textiles, printing, sheet metal fabrication); agriculture (fruit, vegetables, corn, nursery crops; poultry, livestock; dairying). The city's main growth is due to the establishment of high-technology computer industries and added housing developments in the area. Established 1673 as part of DUNSTABLE, MASSACHUSETTS, included in NEW HAMPSHIRE as Nottingham West in 1746; name changed to Hudson in 1830.

Hudson, city (□ 2 sq mi/5.2 sq km; 2006 population 6,985), COLUMBIA county, SE NEW YORK on the HUDSON RIVER; 42°15′N 73°47′W. The city was a whaling and trading port until 1812. Its industries included textiles, furniture, cement, and metal products, but these are largely gone. Many colonial and Revolutionary era homes are in the area. Olana, estate of Frederic E. Church, 2.5 mi/4 km S of city. Settled c.1622 by the Dutch and later in 1783 by English whalers; incorporated 1785.

Hudson (HUHD-suhn), city (□ 26 sq mi/67.6 sq km; 2006 population 23,154), SUMMIT county, NE OHIO, 11 mi/18 km NE of AKRON; 41°14′N 81°27′W. Photographic equipment; manufacturing (fireworks). Western Reserve Academy (1826) here. Settled 1799, incorporated 1837.

Hudson (HUHD-suhn), township (□ 35 sq mi/91 sq km; 2001 population 490), E central ONTARIO, E central CANADA, 7 mi/11 km from NEW LISKEARD, in the CANADIAN SHIELD; 47°32′N 79°49′W. The Twin Lakes, Pike Lake are within Hudson.

Hudson (HUHD-suhn), former town, NW ONTARIO, E central CANADA, on inlet of Lake SEUL, in gold mining, dairying, grain-growing region; 50°05′N 92°10′W. Amalgamated into SIOUX LOOKOUT in 1998.

Hudson (HUHD-suhn), town (□ 9 sq mi/23.4 sq km), Montérégie region, S QUEBEC, E CANADA, on Lake of the TWO MOUNTAINS, 8 mi/13 km E of RIGAUD, and 25 mi/40 km from MONTREAL; 45°27′N 74°09′W. Dairy-

ing, potato growing; resort. Part of the METROPOLITAN COMMUNITY OF MONTREAL (*Communauté Metropolitaine de Montréal*).

Hudson, town (2000 population 1,565), WELD county, N COLORADO, 30 mi/48 km NE of DENVER; 40°04′N 104°38′W. Elevation 5,024 ft/1,531 m. In agricultural region: fruit, sugar beets, beans, wheat, cattle, horses; manufacturing (plastic injection molding).

Hudson (HUHD-suhn), town (□ 3 sq mi/7.8 sq km; 2000 population 12,765), PASCO county, W central FLORIDA, 30 mi/48 km NW of TAMPA; 28°21′N 82°42′W. Manufacturing includes construction materials, furniture, lime rock processing.

Hudson, town (2000 population 596), STEUBEN county, NE INDIANA, 9 mi/14.5 km SSW of ANGOLA; 41°32′N 85°05′W. Potatoes, onions; manufacturing (metal furniture, extruded plastics, metal stampings).

Hudson, town (2000 population 2,117), BLACK HAWK county, E central IOWA, 8 mi/12.9 km SW of WATERLOO; 42°25′N 92°27′W. In agricultural area.

Hudson, town, PENOBSCOT county, S central MAINE, 15 mi/24 km NNW of BANGOR; 45°00′N 68°52′W.

Hudson, industrial and residential town, MIDDLESEX county, E central MASSACHUSETTS, on the ASSABET RIVER; 42°23′N 71°33′W. In an apple-growing region. Manufacturing includes communications equipment, locks, chemicals, plastics, electronic and metal products, and semiconductors. Settled c.1699, incorporated 1866.

Hudson, town (2000 population 2,499), LENAWEE county, SE MICHIGAN, on TIFFIN RIVER and 16 mi/26 km WSW of ADRIAN; 41°51′N 84°20′W. Shipping point for rich farm area; manufacturing (wire forms, aluminum and sand castings, screw machine products). Native American mounds nearby. Lake Hudson State Recreation Area to E. Settled 1834; incorporated as village 1853, as city 1893.

Hudson (HUHD-suhn), town (□ 3 sq mi/7.8 sq km; 2006 population 3,081), CALDWELL county, W central NORTH CAROLINA; shares a border with LENOIR; 35°51′N 81°29′W. Manufacturing (leather products, industrial packaging, curved plywood, lumber handling systems, furniture, clothing, woven cloth, yarn); timber; agriculture (tobacco, grain; poultry, livestock). Lake Rhodhiss reservoir to S. Incorporated 1905.

Hudson, town (2006 population 4,149), ANGELINA county, E TEXAS, suburb 2 mi/3.2 km WSW of LUFKIN; 31°19′N 94°47′W. Timber area. Davy Crocket National Forest to SW.

Hudson, town (2006 population 11,913), ⊙ ST. CROIX county, W WISCONSIN, on ST. CROIX RIVER 15 mi/24 km E of SAINT PAUL, MINNESOTA; 44°58′N 92°44′W. In dairying and grain-growing area. Railroad workshops; manufacturing (dairy products, furniture, refrigerators). A nearby park has Indian mounds. LOWER ST. CROIX NATIONAL SCENIC RIVERWAY; Willow River State Park to NE. Incorporated 1856.

Hudson, village (2000 population 1,510), MCLEAN county, central ILLINOIS, 8 mi/12.9 km N of BLOOMINGTON; 40°36′N 88°59′W. In rich agricultural area. Lake Bloomington and Evergreen Lake are nearby.

Hudson, village (2000 population 133), STAFFORD county, S central KANSAS, 19 mi/31 km SSE of GREAT BEND, near RATTLESNAKE CREEK; 38°06′N 98°39′W. In wheat area. Quivira National Wildlife Refuge, with extensive salt marshes, is to E.

Hudson, village (2006 population 395), LINCOLN county, SE SOUTH DAKOTA, 14 mi/23 km SSE of CANTON, and on BIG SIOUX RIVER, near IOWA state line; 43°07′N 96°27′W. Honey; meat processing.

Hudson, village (2006 population 421), FREMONT county, W central WYOMING, on POPO AGIE RIVER and 9 mi/14.5 km NE of LANDER, on S boundary of Wind River Indian Reservation; 42°53′N 108°34′W. Elevation 5,094 ft/1,553 m. Trading and cattle-shipping point in vegetable (corn, beans, sugar beets) region; coal mines.

Hudson Bay (HUHD-suhn), village (2006 population 1,646), E SASKATCHEWAN, CANADA, 80 mi/129 km SW of THE PAS; 52°51′N 102°23′W. Railroad junction for the Pas–Flin Flon mining region; lumbering; clay quarrying, mixed farming.

Hudson Bay, inland sea (□ 475,000 sq mi/1,235,000 sq km), of NORTH AMERICA, E central CANADA; c.850 mi/1,370 km long and c.650 mi/1,050 km wide. Hudson Bay and JAMES BAY (its S extension) and all their islands border NUNAVUT, MANITOBA, ONTARIO, and QUEBEC. HUDSON STRAIT (c.450 mi/720 km long) connects Hudson Bay with the ATLANTIC OCEAN, and FOXE CHANNEL leads to the ARCTIC OCEAN. MANSEL, COATS, and SOUTHAMPTON islands are at the N end of the bay. Hudson Bay occupies the southernmost portion of the Hudson Bay Lowlands, a depression in the CANADIAN SHIELD formed during the Pleistocene epoch by the weight of the continental ice sheet. As the ice retreated, the region was flooded by the sea, and sediments were deposited in it. With the burden of ice removed, the floor of the lowlands has been slowly rising and the bay is gradually becoming shallower. The W shores are generally low and marshy and covered by tundra, while the E coast is barren and rocky, with the OTTAWA and BELCHER island groups offshore. Many rivers, including the CHURCHILL and NELSON, drain into the bay. Hudson Bay moderates the local climate; it is ice-free and open to navigation from mid-July to October. The bay was explored and named (1610) by Henry Hudson in his search for the NORTHWEST PASSAGE. The surrounding region was a rich source of furs, and FRANCE and England struggled for its possession until 1713, when France ceded its claim by the Peace of UTRECHT. Hudson's Bay Company set up many trading posts there, especially at river mouths; some of the posts have operated continuously since 1670. The Hudson Bay railroad (opened 1929) links the prairie provinces with CHURCHILL, Manitoba, a port for oceangoing freighters.

Hudson Falls, village (□ 1 sq mi/2.6 sq km; 2006 population 6,772), ⊙ WASHINGTON county, E NEW YORK, on E bank of the HUDSON and 3 mi/4.8 km E of GLENS FALLS; 43°17′N 73°34′W. In diversified farming area; lumbering; some manufacturing. Hudson Falls Historic District. Settled 1761, incorporated 1810.

Hudson Heights (HUHD-suhn HEITS), unincorporated village, S QUEBEC, E CANADA, on Lake of the TWO MOUNTAINS, 7 mi/11 km E of RIGAUD, and included in HUDSON; 45°27′N 74°10′W. Dairying, potato growing.

Hudson Island, GILBERT ISLANDS, KIRIBATI: see NANUMANGA.

Hudson Lake, village, LA PORTE county, NW INDIANA, 12 mi/19 km NE of LA PORTE, at E end of Hudson Lake. Railroad junction. Dairying, fruit. Laid out c.1831.

Hudson Mountains, a group of low mountains and nunataks on the WALGREEN COAST, WEST ANTARCTICA, located just E of CRANTON and PINE ISLAND Bays, at the E edge of the AMUNDSEN SEA; 70 mi/113 km long; 74°25′S 99°30′W.

Hudson Oaks, village (2006 population 1,905), PARKER county, N central TEXAS, 20 mi/32 km W of FORT WORTH, on SW shore of Lake Weatherford reservoir; 32°45′N 97°42′W. Residential satellite community W of DALLAS–Fort Worth urbanized area. Agricultural area.

Hudson River, c.315 mi/507 km long, NEW YORK; rising in Lake TEAR OF THE CLOUDS, on Mount MARCY in the ADIRONDACK MOUNTAINS, NE New York; flowing generally S to Upper NEW YORK BAY at NEW YORK city; the MOHAWK RIVER is its chief tributary. The Hudson is navigable by ocean vessels to Albany and by smaller vessels to TROY; leisure boats and self-propelled barges

use the canalized section between Troy and FORT EDWARD, the head of navigation. Divisions of the NEW YORK STATE BARGE CANAL connect the Hudson with the GREAT LAKES and with Lake CHAMPLAIN and the SAINT LAWRENCE River. The Hudson is tidal to Troy (c.150 mi/240 km upstream); this section is considered to be an estuary. The main headstream of the Hudson is Feldspar Creek–Opalescent River. The upper course of the river has many waterfalls and rapids. The middle course, between Albany and NEWBURGH, is noted for the CATSKILL and SHAWANGUNK mountains on the W and by the large estates (the Roosevelt home at HYDE PARK is the most famous) on the E bank. From Newburgh to PEEKSKILL the river crosses the mountainous and forested Hudson Highlands in a deep, scenic gorge. WEST POINT Military Academy overlooks the river there, and BEAR MOUNTAIN Bridge spans this section. Near TARRYTOWN the river widens to form the NEW YORK STATE THRUWAY and the TAPPAN ZEE Bridge; from there to its mouth the Hudson is flanked on the W by the sheer cliffs of the PALISADES. At the mouth are the ports of New York and New Jersey. The Hudson forms part of the New York–New Jersey border, and the two states are linked by the GEORGE WASHINGTON BRIDGE, the HOLLAND and LINCOLN vehicular tunnels, and railroad tubes. Sighted first by Verrazano in 1524, the river was explored by Henry Hudson in 1609. It was a major route for Native Americans and later for the Dutch and English traders and settlers. During the American Revolution both sides fought for control of the Hudson; many battles were fought along its banks. In 1825 the ERIE CANAL linked the river with the Great Lakes, providing the first all-water trans-Appalachian route. Many industries are located on the Hudson's banks, and pollution by raw sewage and industrial wastes became a serious problem in the 1900s; anti-pollution legislation passed in 1965 has sought to protect the river from further contamination. Although pollution continued throughout the remainder of the 20th century and into the 21st, the state and municipal governments and environmental groups have contributed a significant clean-up effort, including more anti-pollution regulation. A major sewage treatment plant was constructed in the 1980s along the river. The Hudson is featured in the legend of Rip Van Winkle and other stories by Washington Irving.

Hudson River Greenway, park, E NEW YORK, extending eventually along E shore of HUDSON RIVER from NEW YORK city line to the TROY dam in RENSSELAER county, and from the NEW JERSEY state line to the mouth of the MOHAWK RIVER, N of ALBANY, along the W shore. Length of c.130 mi/209 km on each side. Purpose of plan is to link suburban, rural, historical, cultural, and recreational resources of the Hudson Valley, as well as to preserve open land and develop Hudson Valley trails. Idea proposed in early 1980s; State Greenway Act of 1991 began implementation. (Ongoing as of 2007.)

Hudson's Hope (HUHD-suhnz HOP), district municipality (□ 358 sq mi/930.8 sq km; 2001 population 1,039), E BRITISH COLUMBIA, W CANADA, on banks of Peace River, in PEACE RIVER regional district; 56°00′N 122°00′W. Hydroelectric power, forestry, agriculture. Has one million-year-old dinosaur tracks. Nearby is a large earthen dam (Bennett Dam, completed 1967, 600 ft/183 m high). Established as a trading post in 1805.

Hudson Strait, between N QUEBEC and S BAFFIN ISLAND, NUNAVUT territory, CANADA. Arm of the ATLANTIC, extending from N extremity of Labrador to HUDSON BAY at NW extremity of UNGAVA PENINSULA, opposite SOUTHAMPTON ISLAND; 450 mi/724 km long, 40 mi/64 km–150 mi/241 km wide; 62°00′N 70°00′W. UNGAVA BAY, N Quebec, is S arm. At E entrance of strait are KILLINEK and RESOLUTION islands; at W

entrance are SALISBURY and NOTTINGHAM islands. FOXE CHANNEL connects strait with FOXE BASIN and thence with other arms of the ARCTIC OCEAN. Trading posts on strait include LAKE HARBOUR on Baffin Island and SUGLUK on Ungava Peninsula. Ice-free from mid-July until October, strait is navigable with ice breakers during greater part of the year. Reputedly entered by Sebastian Cabot, 1498, E end of Hudson Strait was explored by Sir Martin Frobisher, 1576–1578, and by John Davis, 1585–1587; in 1610, Henry Hudson first navigated its full length. It later became main route of Hudson's Bay Company vessels and, since 1931, of grain ships from CHURCHILL.

Hudsonville, town (2000 population 7,160), OTTAWA county, SW MICHIGAN, suburb 11 mi/18 km SW of downtown GRAND RAPIDS; 42°52′N 85°51′W. In orchard, dairy, and farming area. Manufacturing (routers and welders, plastic molding, screw machine products, wood products). Incorporated in 1926.

Hudspeth, county (□ 4,572 sq mi/11,887.2 sq km; 2006 population 3,320), extreme W TEXAS; ⊙ SIERRA BLANCA; 31°26′N 105°22′W. Bounded N by NEW MEXICO line, S by the RIO GRANDE (Mexican border); third largest county in state. High plateau region (elevation c.3,500 ft/1,067 m) with mountains (up to c.7,500 ft/2,286 m) surrounding a central bolson with intermittent drainage into large playas (saltworks; figured in Salt War, 1877) in NE. Irrigated agriculture (water from Elephant Butte Reservoir, New Mexico) in Rio Grande valley (cotton, alfalfa; vegetables); cattle, hogs; minerals (talc, gypsum). Part of SIERRA DIABLO is in E, EAGLE PEAK in SE, QUITMAN MOUNTAINS in S, SIERRA BLANCA MOUNTAIN in S center, FINLAY MOUNTAINS in W center, HUECO MOUNTAINS in NW, scattered mountains in N; part of GUADALUPE MOUNTAINS NATIONAL PARK in NE corner; small part of FORT BLISS MILITARY RESERVATION in NW corner; one of two W Texas counties wholly in Mountain time zone (also EL PASO county and part of CULBERSON county). Formed 1917.

Hue (HWAI), city, ⊙ THUA THIEN-HUE province, former ⊙ the historic region of ANNAM, central VIETNAM, in a rich farming area on the Hue River near the SOUTH CHINA SEA; 16°28′N 107°36′E. Probably founded in the 3rd century C.E., Hue was occupied in turn by the Chams and the Annamese. After the 16th century, it served as the seat of a dynasty that extended its power over S Annam, modern COCHIN CHINA, and parts of CAMBODIA and LAOS. The first king of Vietnam, Nguyen Anh, was crowned here in 1802, and shortly thereafter Hue became the capital of the new kingdom, emerging as an artistic and literary center. The French occupied the city in 1883. During World War II, the Japanese mined iron ore in the area. In the Vietnam War, Hue was the scene of the longest and heaviest fighting of the Tet offensive (January-February 1968); some 4,000 civilians were killed and most of the city, including the palaces and some of the tombs of the former Annamese kings, was destroyed. Much of the city has been rebuilt. In the late 20th century, it underwent considerable commercial development, especially in textile and fishing industries. Hue is an important market and education center, administrative complex, and key tourist site. Major attractions include the Citadel, Thai Hoa Palace, Imperial Museum, Royal Tombs, Forbidden Purple City, and Nam Giao (Temple of Heaven). Many historic sites are scheduled for restoration. Hue University is one of the foremost institutions of higher learning in Vietnam.

Huechucuicui Point (hwai-choo-koo-ee-KOO-ee), PACIFIC cape at NW tip of CHILOÉ ISLAND, S CHILE, 11 mi/18 km NW of ANCUD.

Huechulafquén, Lake (hwai-choo-lahf-KAIN) (□ 32 sq mi/83.2 sq km), in the ANDES, SW NEUQUÉN province, ARGENTINA, S of LANÍN VOLCANO, and extending c.20 mi/32 km E from CHILE border, drained by an affluent of COLLÓN CURÁ RIVER. Elevation 3,180 ft/969 m. Also spelled Huechulaufquén and Huechulauquén.

Huechupín (wai-choo-PEEN), village, ÑUBLE province, BÍO-BÍO region, S central CHILE, 14 mi/23 km WSW of CHILLÁN; 36°37′S 72°22′W. In agricultural area (grain, wine, potatoes, vegetables; livestock).

Hueco Mountains, extreme W TEXAS and S NEW MEXICO, N-S range c.55 mi/89 km long, NE of EL PASO. Rise to 6,767 ft. in Cerro Alto Peak, Texas. At Hueco Tanks (c.25 mi/40 km ENE of El Paso) are caves, natural rock reservoirs, and pictographs left by tribes whose stronghold this was; camp grounds here. DIABLO BOLSON is E.

Huecú, El, ARGENTINA: see EL HUECÚ.

Huedin (HWE-deen), Hungarian *Bánffyhunyad*, town, CLUJ county, W central ROMANIA, in TRANSYLVANIA, on Crişul Rapede River and 28 mi/45 km WNW of CLUJ-NAPOCA; 46°52′N 23°03′E. Railroad junction and trading center, notably for lumber; flour milling, manufacturing of consruction materials, textiles. Has 16th-century church. Large Hungarian minority. In HUNGARY, 1940-1945.

Huehuetán (we-we-TAHN), town, CHIAPAS, S MEXICO, in PACIFIC coastal lowland, near highway, on railroad, and 10 mi/16 km SE of HUIXTLA; 15°01′N 92°25′W. Coffee, sugarcane, fruit, livestock.

Huehuetenango (hwe-hue-ten-NAHN-go), department (□ 2,857 sq mi/7,428.2 sq km), W GUATEMALA; ⊙ HUEHUETENANGO; 15°40′N 91°35′W. On MEXICO border; drained by headstreams of GRIJALVA river system (W) and LACANTÚN RIVER (N and E); contains highest (W) section of CUCHUMATANES MOUNTAINS. Wheat, corn, beans, sheep raising on slopes; coffee, sugarcane, tropical fruit raising on mid-level slopes; coffee, sugarcane, tropical fruit in lower part. Local industries: textile weaving, pottery making. Lead mining near CHIANTLA and SAN MIGUEL ACATÁN. Main commercial center, Huehuetenango.

Huehuetenango (hwe-hwe-te-NAHN-go), city (2002 population 26,600), ⊙ HUEHUETENANGO department, W GUATEMALA, near headstreams of SELEGUA River, 80 mi/129 km NW of GUATEMALA city; 15°19′N 91°28′W. Elevation 6,240/1,902 m. Commercial center of NW highlands and CUCHUMATANES MOUNTAINS; flour milling, tanning, wool processing and milling; trade (pottery, leather, textiles). Restored ruins of Indian capital ZACULEU, 1 mi/1.6 km W.

Huehuetla (we-WE-tlah), town, ⊙ Huehuetla municipio, HIDALGO, central MEXICO, in E foothills of SIERRA MADRE ORIENTAL, 50 mi/80 km NE of PACHUCA DE SOTO; 20°35′N 98°04′W. Elevation 984 ft/300 m. Corn, sugarcane, coffee, fruit, beans, livestock.

Huehuetla, town, PUEBLA, central MEXICO, in SE foothills of SIERRA MADRE ORIENTAL, 28 mi/45 km ESE of HUAUCHINANGO. Sugarcane, coffee, tobacco, fruit. In Totonac Indian area.

Huehuetlán (we-we-TLAHN), town, SE SAN LUIS POTOSÍ, MEXICO, 30 mi/49 km S of CIUDAD VALLES in the Tancanhuitz Sierra. Water sources are from the Esperanza and Huichihuan rivers. Agriculture (sugarcane, coffee), cattle.

Huehuetlán, MEXICO: see SANTO DOMINGO HUEHUETLÁN.

Huehuetlán El Chico (we-we-TLAHN el CHEE-ko), town, PUEBLA, central MEXICO, 21 mi/34 km SW of IZÚCAR DE MATAMOROS; 18°22′N 98°41′W. Agricultural center (corn, rice, sugarcane, fruit; livestock).

Huehuetlán el Grande, MEXICO: see SANTO DOMINGO HUEHUETLÁN.

Huehuetoca (we-we-TO-kah), town, ⊙ Huehuetoca municipio, MEXICO, central MEXICO, on railroad and 28 mi/45 km N of MEXICO CITY, within the ZONA METROPOLITANA DE LA CIUDAD DE MÉXICO; 19°49′N 99°09′W. Elevation 7,536 ft/2,297 m. Corn, maguey; livestock.

Huejotitán (we-ho-tee-TAHN), town, CHIHUAHUA, N MEXICO, 30 mi/48 km WNW of HIDALGO DEL PARRAL. Corn, cotton, beans, sugarcane, tobacco.

Huejotzingo (we-hot-SEEN-go), city and township, PUEBLA, central MEXICO, on central plateau, 15 mi/24 km NW of PUEBLA on Mexico Highway 190; 19°10′N 98°23′W. Railroad terminus; fruit-growing center (apples, pears, plums, figs, nuts); cider manufacturing; handwoven serapes. Has seventeenth-century church and a monastery of San Francisco.

Huejúcar (we-HOO-kar), town, ⊙ Huejúcar municipio, JALISCO, N central MEXICO, on ZACATECAS border, 19 mi/31 km N of COLOTLÁN; 22°21′N 103°13′W. Cereals, vegetables, livestock.

Huejuquilla el Alto (we-hoo-KEE-yah el AHL-to), town, JALISCO, W MEXICO, near ZACATECAS border, 50 mi/80 km NW of COLOTLÁN; 22°40′N 103°52′W. Elevation 5,577 ft/1,700 m. Cereals, alfalfa, beans, livestock.

Huejutla de Reyes (we-HOO-tlah de RAI-yes), city and township, ⊙ Huejutla de Reyes municipio, HIDALGO, central MEXICO, in E foothills of SIERRA MADRE ORIENTAL, near VERACRUZ border, 75 mi/121 km NNE of PACHUCA DE SOTO; 21°08′N 98°24′W. Elevation 564 ft/172 m. Agricultural center (corn, rice, sugarcane, tobacco, coffee, fruit, cattle); cigars.

Huelgoat (wel-GWAH), commune (□ 6 sq mi/15.6 sq km), FINISTÈRE department, W FRANCE, in BRITTANY, 30 mi/48 km E of BREST; 48°22′N 03°45′W. Summer resort in a scenic setting surrounded by a recreational forest and next to a small lake. The regional park of Armorica extends westward to the sea, encompassing the ARRÉE MOUNTAINS (ARMORICAN MASSIF), Brittany's main hill range.

Huellas de Acahualinca (HWAI-yahs dai ah-kah-hwah-LEEN-ka), archaeological site within the city of MANAGUA, NICARAGUA, 7,000-9,000 year old human footprints are preserved in volcanic ash.

Huellelhue (we-YEL-wai), village, VALDIVIA province, LOS LAGOS region, S central CHILE, on CALLE-CALLE RIVER, on railroad, and 7 mi/11 km ENE of VALDIVIA, in lumbering and agricultural area (cereals, potatoes, peas, apples, livestock). Also spelled Hueyelhue.

Huelma (WEL-mah), town, JAÉN province, S SPAIN, in mountainous district, 20 mi/32 km SE of JAÉN; 37°39′N 03°27′W. Manufacturing of esparto rope, soap; olive oil and cheese processing. Stock raising, lumbering. Iron and lead deposits.

Huelmó (wel-MO), village, LLANQUIHUE province, LOS LAGOS region, S central CHILE, on RELONCAVÍ SOUND, and 13 mi/21 km SSW of PUERTO MONTT, in agricultural area (wheat, flax, potatoes, livestock); dairying. Small island of Huelmó is just off the coast.

Huelva (WEL-vah), province (□ 3,894 sq mi/10,124.4 sq km), SW SPAIN, in ANDALUSIA; ⊙ HUELVA. Borders S on the ATLANTIC, bounded W by PORTUGAL along GUADIANA and CHANZA rivers; BADAJOZ province is N, SEVILLE province E. Watered by the RÍO TINTO and ODIEL RIVER, which join S of Huelva city to form a navigable estuary. Has two principal regions: mountainous area (N) is formed by wooded spurs of the SIERRA MORENA, chiefly SIERRA DE ARACENA; in S are fertile, low plains (called La Campiña) adjoined SE by the marshy, scantily populated, alluvial LAS MARISMAS along the lower GUADALQUIVIR. Climate is temperate to subtropical, cooler in the sierras. Huelva ranks among the leading mining areas in Spain and was renowned as such since Phoenician days; the RÍO TINTO district is its most productive area. Copper is chief mineral; also iron, copper-iron pyrite, argentiferous lead, manganese, antimony, sulphur, graphite, coal, peat, limestone, jasper, marble. Agricultural products include wheat, corn, barley, olives and olive oil, acorns, chestnuts, cork, chick peas, almonds, peaches, strawberries, oranges, figs (dried). Noted for its wines (LA PALMA, NIEBLA, BOLLULLOS PAR DEL CONDADO), and brandies. Considerable

livestock raising is carried on, mainly of hogs that are fed acorns. Next to mining, the Atlantic fisheries of tuna and sardines, based at AYAMONTE, ISLA-CRISTINA, and Huelva (with salting and canning plants), are a major source of province's income. The hilly ranges yield a variety of timber (oak, walnut, poplar, beech, pine). Huelva city, its trading and processing center, is an important ore-shipping port, linked by narrow-gauge railroad with the interior mines (Ríotinto, ZALAMEA LA REAL, NERVA, THARSIS).

Huelva (WEL-vah), city (2001 population 142,284), ⊙ HUELVA province, SW SPAIN, in ANDALUSIA, on the ODIEL RIVER above its junction with the RÍO TINTO; 37°15′N 06°57′W. A busy port with copper, sulfur, and cork exports, it also has fishing, shipbuilding, oil refining, and summer resort industries. A Roman aqueduct supplies the city with water. Nearby LA RÁBIDA monastery, where Columbus made his plans, is a summer university.

Huéneja (WAI-nai-hah), town, GRANADA province, S SPAIN, 13 mi/21 km SE of GUADIX; 37°10′N 02°58′W. Olive oil processing; livestock raising; lumbering; cereals, vegetables, sugar beets. Mineral springs. Rich iron mines nearby.

Hueneme, California: see PORT HUENEME.

Huépac (WE-pak), town, ⊙ Huépac municipio, SONORA, NW MEXICO, on SONORA RIVER (irrigation) and 75 mi/121 km NE of HERMOSILLO; 29°54′N 110°10′W. Elevation 1,588 ft/484 m. Farming where irrigation water available. Livestock; wheat, corn, vegetables, sugarcane.

Huepil (WAI-peel), town, ⊙ Tucapei comuna, BÍO-BÍO province, BÍO-BÍO region, S central CHILE, 27 mi/43 km NE of LOS ÁNGELES; 37°14′S 71°56′W. On railroad; fruit, vegetables, livestock.

Huequén (wai-KEN), village, MALLECO province, ARAUCANIA region, S central CHILE, 3 mi/5 km SE of ANGOL; 37°49′S 72°40′W. In fruit-growing area (apples).

Huércal de Almería (WER-kahl dhai ahl-mai-REE-ah), town and N suburb of ALMERÍA, ALMERÍA province, S SPAIN; 36°53′N 02°26′W. Grapes. Mineral springs. Sulphur, iron, and calamine mines.

Huércal-Overa (WER-kahl-o-VAI-rah), town, ALMERÍA province, S SPAIN, near ALMANZORA RIVER, 48 mi/77 km NE of ALMERÍA; 37°23′N 01°57′W. Grapes, cereals, figs; aviculture; olive oil processing, fodder milling. Gypsum quarries.

Huércanos (WER-kah-nos), town, LA RIOJA province, N SPAIN, 12 mi/19 km WSW of LOGROÑO; 42°26′N 02°41′W. Cereals, potatoes, sugar beets, wine.

Huerfano, county (□ 1,593 sq mi/4,141.8 sq km; 2006 population 7,808), S COLORADO; ⊙ WALSENBURG; 37°41′N 104°57′W. Coal-mining and livestock grazing area, bounded W by SANGRE DE CRISTO MOUNTAINS; drained by CUCHARAS (forms Cucharas Reservoir in NE) and HUERFANO rivers. Part of San Isabel National Forest in NW, W, and S; S end of WET MOUNTAINS in NW. Orlando Reservoir in NE; Lathrop State Park at center. Cucharas Ski Area in S. Formed 1861.

Huerfano River, 99 mi/159 km long, S COLORADO, in SIERRA BLANCA, in W HUERFANO county, at GARDNER; confluence of Huerfano, Muddy, and (in downstream) Williams creeks; flows E and NE to ARKANSAS RIVER E of PUEBLO. Source Lily Lake.

Huerquehue (hwer-KAI-hwai), national park, in ARAUCANIA region, central CHILE, 65 mi/105 km ESE of TEMUCO, on LAKE CABURGUA. Trail system. Park protects Nothofagus forest.

Huerta del Rey (WER-tah dhel RAI), town, BURGOS province, N SPAIN, 40 mi/64 km SE of BURGOS; 41°50′N 03°20′W. Cereals, vegetables, resins, livestock. Lumbering; flour milling; plaster and tile manufacturing.

Huerta de Valdecarábanos (WER-tah dhai VAHL-dai-kah-RAH-vah-nos), town, TOLEDO province, central SPAIN, 12 mi/19 km S of ARANJUEZ; 39°52'N 03°37'W. Cereals, hemp, grapes, sugar beets; livestock.

Huerta Grande (HWAIR-tah GRAHN-de), town (1991 population including LA FALDA, VALLE HERMOSO, and Villa Giardino 27,292), NW CÓRDOBA province, ARGENTINA, 30 mi/48 km NW of CÓRDOBA; 31°04'S 64°30'W. Tourist resort in N Córdoba hills; lime and marble quarries; livestock raising.

Huerta, La, ARGENTINA: see LA HUERTA.

Huerta, Sierra de la (HWAIR-tah, see-YER-rah dai lah), pampean range in E SAN JUAN province, ARGENTINA, 60 mi/97 km NE of SAN JUAN; extends c.50 mi/80 km N-S. Rises to c.8,000 ft/2,438 m. Has gold mines, and silver, lead, arsenic, and sulphur deposits.

Huertgen, GERMANY: see HÜRTGENWALD.

Huerva River (WER-vah), in TERUEL and ZARAGOZA provinces, NE SPAIN, rises at NE edge of the central plateau 10 mi/16 km NW of MONTALBÁN, flows 90 mi/145 km generally N to the EBRO at ZARAGOZA. Irrigation reservoirs. Olive groves.

Huesa (WAI-sah), town, JAÉN province, S SPAIN, 40 mi/64 km E of JAÉN; 37°46'N 03°04'W. Olive oil processing, flour milling; cereals, esparto, lumber. Gypsum quarries. Mineral springs nearby.

Huesca (WE-skah), province (□ 6,054 sq mi/15,740.4 sq km), NE SPAIN, in ARAGÓN; ⊙ HUESCA. Bounded N by crest of the central PYRENEES (FRANCE border), which here rise to 11,168 ft/3,404 m in the PICO DE ANETO (highest in Pyrenees) and slope S toward the EBRO plain. Much outmigration in recent past. In N are ancient districts of SOBRARBE and RIBAGORZA. Spanish province second in hydroelectric power. Watered by CINCA and GÁLLEGO rivers, which feed network of irrigation canals. Lead, manganese, bauxite deposits. Essentially agricultural: wine, olive oil; livestock; cereals, sugar beets. Chief towns: Huesca, BARBASTRO, JACA.

Huesca (WAI-skah), town, ⊙ HUESCA province, NE SPAIN, in ARAGÓN, at the foot of the PYRENEES; 42°08'N 00°25'W. It is a farm center. In this ancient town Sertorius founded a school in 77 B.C.E. After Peter I of Aragón liberated it (1096) from the Moors, Huesca was the residence of the kings of ARAGÓN until 1118. A university, later discontinued, was founded there in 1354. The 13th-century Gothic cathedral, the early Romanesque Church of San Pedro, and the royal palace of the Aragonese monarchs are notable landmarks.

Huéscar (WE-skahr), town, GRANADA province, S SPAIN, 26 mi/42 km NE of BAZA; 37°49'N 02°32'W. Agricultural trade center. Olive pressing, flour milling, wood turning, chocolate manufacturing. Wine, cereals, esparto, produce; livestock raising, lumbering. Has 16th century church.

Huetamo de Núñez (we-TAH-mo de NOON-yez), city, ⊙ Huetamo municipio, MICHOACÁN, central MEXICO, in Río BALSAS valley, 24 mi/39 km NW of CIUDAD ALTAMIRANO (GUERRERO); 18°36'N 100°54'W. Agricultural center (sugar, coffee, fruit, cereals); tanning.

Huete (WAI-tai), town, CUENCA province, E central SPAIN, on railroad to MADRID and 30 mi/48 km W of CUENCA; 40°08'N 02°41'W. Agricultural center in irrigated region (grapes, honey, cereals, livestock). Manufacturing of flour, chocolate, jute bags. Flourished in Middle Ages.

Huétor-Santillán (WAI-tor-sahn-tee-YAHN), town, GRANADA province, S SPAIN, 6 mi/9.7 km NE of GRANADA; 37°13'N 03°31'W. Olive oil, cereals, produce. Sand pits nearby.

Huétor-Tájar (WAI-tor-TAH-hahr), town, GRANADA province, S SPAIN, near GENIL RIVER, 6 mi/9.7 km ENE of Loja; 37°12'N 04°02'W. Olive oil processing, flour milling. Cereals, sugar beets, fruit, lumber.

Huétor-Vega (WAI-tor-VAI-gah), SE suburb of GRANADA, GRANADA province, S SPAIN; 37°09'N 03°34'W. Olive oil, sugar beets, produce, cereals.

Huetter, village (2000 population 96), KOOTENAI county, N IDAHO, 6 mi/9.7 km W of COEUR D'ALENE and 2 mi/3.2 km E of POST FALLS, on SPOKANE RIVER; 47°42'N 116°51'W.

Huévar (WAI-vahr), town, SEVILLE province, SW SPAIN, 15 mi/24 km W of SEVILLE (linked by railroad); 37°22'N 06°16'W. Olives, cereals, grapes; livestock.

Huevos Island (HWAI-vos), off NW TRINIDAD, TRINIDAD AND TOBAGO, in the DRAGON'S MOUTH, between CHACACHACARE islet (W) and MONOS ISLAND (E); 253 acres/102 ha; 10°42'N 61°43'W. Elev. 680 ft/207 m. Bathing resort.

Huexotla (we-HO-tlah), village, MEXICO, central MEXICO, 4 mi/6.4 km SSW of TEXCOCO DE MORA; 19°30'N 98°50'W. Site of many archaeological remains (temples, pyramids, etc.). Also San Luis Huexora.

Huey, village (2000 population 196), CLINTON county, SW ILLINOIS, 4 mi/6.4 km E of CARLYLE; 38°36'N 89°17'W. In agricultural and oil-producing area. Near REND LAKE.

Hueyapan (we-YAH-pahn), town, Hueyapan municipio, PUEBLA, central MEXICO, in SIERRA MADRE ORIENTAL, 6 mi/9.7 km SE of TULANCINGO. Cereals, sugarcane, vegetables, coffee; wood, resins.

Hueyapan de Ocampo (we-YAH-pahn de o-KAM-po), town, VERACRUZ, SE MEXICO, in GULF lowland, 17 mi/27 km NW of ACAYUCAN; 18°07'N 95°09'W. Sugarcane, fruit.

Hueyelhue, CHILE: see HUELLELHUE.

Hueyotlipan (wai-o-TLEE-pahn), town, TLAXCALA, central MEXICO, 13 mi/21 km NW of TLAXCALA. Maguey, corn, wheat, beans; livestock.

Hueyotlipan, MEXICO: see SANTO TOMÁS HUEYOTLIPAN.

Hueypoxtla (wai-POSH-tlah), town, MEXICO, central MEXICO, 33 mi/53 km N of MEXICO CITY. Grain, maguey, stock.

Hueytamalco (wai-tah-MAHL-ko), town, PUEBLA, E MEXICO, in foothills of SIERRA MADRE ORIENTAL, 10 mi/16 km NE of TEZIUTLÁN. Sugarcane, fruit.

Hueytepec, Sierra de (WAI-te-pek), range in CHIAPAS, S MEXICO, a N spur of SIERRA MADRE, E of TUXTLA GUTIÉRREZ; extends c.60 mi/97 km NW-SE, forming E watershed of upper GRIJALVA River. The Cerro Hueytepec rises to 8,946 ft/2,727 m. SAN CRISTÓBAL DE LAS CASAS is its center. Sometimes spelled Huitepec; also known as Sierra Los Altos de Chiapas.

Hueytlalpan (wai-TLAL-pahn), town, PUEBLA, central MEXICO, in foothills of SIERRA MADRE ORIENTAL, 25 mi/40 km ESE of HUAUCHINANGO. Coffee, tobacco, sugarcane, fruit.

Hueytown (HYOO-ee-TOUN), city (2000 population 15,364), JEFFERSON county, N central ALABAMA, just SW of BIRMINGHAM. Fishing lures manufactured here.

Huey Yot, THAILAND: see HUAI YOT.

Huez, FRANCE: see ALPE-D'HUEZ L'.

Huffman Dam (HUHF-muhn), 73 ft/22 m high, GREENE county, SW OHIO, on MAD RIVER, 5 mi/8 km ENE of downtown DAYTON; 39°48'N 84°05'W. Built (1922) for flood control. Reservoir is dry except during flood periods; maximum potential capacity 297,000 acre-ft; extends into MONTGOMERY and CLARK counties.

Hüfingen (HYOO-fing-guhn), town, BADEN-WÜRTTEMBERG, SW GERMANY, in BLACK FOREST, on the BREG RIVER, 9 mi/14.5 km S of VILLINGEN; 47°56'N 08°34'E. Built on the site of a Roman fortress. Remains of Roman bath.

Hufuf, SAUDI ARABIA: see HOFUF.

Hugh Butler Lake, reservoir (c.10 mi/16 km long) in FRONTIER county and on RED WILLOW county border, S NEBRASKA, on Red Willow Creek; 40°22'N 100°39'W. Maximum capacity 163,500 acre-ft. Formed by Red Willow Dam (117 ft/36 m high), built (1962) by the Bureau of Reclamation for irrigation, flood control and recreation.

Hughenden (HYOO-uhn-duhn), town, central QUEENSLAND, AUSTRALIA, on FLINDERS RIVER and 190

mi/306 km WSW of TOWNSVILLE; 20°51'S 144°12'E. Railroad junction; fruit and livestock center. Home of the Muttaburrasaurus, the first complete fossil to be found in Australia.

Hughenden (HYOO-en-den), village (2001 population 235), E ALBERTA, W CANADA, near small Hughenden Lake, 23 mi/37 km S of WAINWRIGHT, in PROVOST No. 52 municipal district; 52°31'N 110°59'W. Grain elevators, lumbering, mixed farming. Incorporated 1917.

Hughenden Valley (HYOO-wuhn-duhn), village (2001 population 2,297), S BUCKINGHAMSHIRE, central ENGLAND, just N of HIGH WYCOMBE; 51°39'N 00°45'W. Parish includes Hughenden Manor, seat of former British Prime Minister Benjamin Disraeli, who is buried here.

Hughes, county (□ 814 sq mi/2,116.4 sq km; 2006 population 13,893), central OKLAHOMA; ⊙ HOLDENVILLE; 35°02'N 96°15'W. Intersected by CANADIAN and NORTH CANADIAN, and Little rivers; includes Lake Holdenville in SW center. Agriculture (corn, peanuts, hay, watermelons, peanuts; cattle, hogs). Some manufacturing at HOLDENVILLE. Oil and natural gas wells; catfish farming; manufacturing of apparel. Formed 1907.

Hughes, county (□ 800 sq mi/2,080 sq km; 2006 population 16,946), central SOUTH DAKOTA; ⊙ PIERRE; 44°23'N 99°59'W. Agricultural region bounded S and SW by MISSOURI RIVER, and watered by Medicine Knoll Creek and other intermittent streams (LAKE OAHE reservoir formed by Oahe Dam, upstream from Pierre). Part of Crow Creek Indian Reservation in SE. Wheat, corn, barley; hogs, cattle. Formed 1873.

Hughes, town, S SANTA FE province, ARGENTINA, 70 mi/113 km SW of ROSARIO; 33°48'S 61°20'W. Agriculture center (corn, soybeans, flax, wheat; livestock, poultry).

Hughes, town (2000 population 1,867), SAINT FRANCIS county, E ARKANSAS, 27 mi/43 km WSW of MEMPHIS (TENNESSEE), near MISSISSIPPI RIVER; 34°57'N 90°28'W. Mud Lake to SW. In agricultural area (cotton, rice, soybeans); lumber milling. Manufacturing (picture frames and bulletin boards). Hunting, fishing. Founded 1913.

Hughes, Indian village (2000 population 78), central ALASKA; on KOYUKUK RIVER and 80 mi/129 km NW of TANANA; 66°03'N 154°13'W. Airfield.

Hughesdale (HYOOZ-dail), residential suburb 9 mi/14 km SE of MELBOURNE, VICTORIA, SE AUSTRALIA; 37°54'S 145°05'E.

Hughes, Port (HYOOZ), inlet of SPENCER GULF, SOUTH AUSTRALIA, AUSTRALIA, on W YORKE PENINSULA; 10 mi/16 km long, 3 mi/5 km wide. MOONTA town near E shore.

Hughes River (HYOOS), 18 mi/29 km long, NW WEST VIRGINIA; formed in E RITCHIE county by junction of North Fork (c.50 mi/80 km long) and South Fork (c.40 mi/64 km long; flows generally W); flows W, through Hughes River Wildlife Management Area, to LITTLE KANAWHA RIVER, 12 mi/19 km SE of PARKERSBURG.

Hughes Springs, town (2006 population 1,860), CASS county, NE TEXAS, 35 mi/56 km NNW of MARSHALL. Resort, with mineral springs; lumber milling; nursery; manufacturing (oil field pipe couplings, consumer displays); 33°00'N 94°37'W. Daingerfield State Park to W.

Hugheston (HYOOS-tuhn), unincorporated town, KANAWHA county, W central WEST VIRGINIA, on KANAWHA River, c.18 mi/29 km SE of CHARLESTON. Manufacturing (mining equipment).

Hughestown (HYOOS-toun), borough (2006 population 1,463), LUZERNE county, NE central PENNSYLVANIA, residential suburb 8 mi/12.9 km NE of WILKES-BARRE and 8 mi/12.9 km SW of SCRANTON, near SUSQUEHANNA RIVER. Incorporated 1879.

Hughesville, village (2000 population 1,537), CHARLES county, S MARYLAND, 28 mi/45 km SSE of WA-

SHINGTON, D.C.; 38°32′N 76°47′W. Tobacco market and farm trade center.

Hughesville, borough (2006 population 2,083), LYCOMING county, N central PENNSYLVANIA, on Muncy Creek, 17 mi/27 km E of WILLIAMSPORT. Light manufacturing. Agricultural area (potatoes, soybeans, corn, hay, dairying). Part of Tiadaghton State Forest to N. Laid out 1816, incorporated 1852.

Hughson, city (2000 population 3,980), STANISLAUS county, central CALIFORNIA, in San Joaquin Valley, 7 mi/11.3 km SE of MODESTO, near TUOLUMNE RIVER. Dairying; polutry; fruit growing; irrigated farming (nuts, vegetables, melons, pumpkins); manufacturing (almond processing, cabinets, machining).

Hughsonville, village, DUTCHESS county, SE NEW YORK, near the HUDSON RIVER, 13 mi/21 km S of POUGHKEEPSIE; 41°35′N 73°45′W.

Hugh Town (HYOO TOUN), town (2006 population 1,100), on SW side of Isle of St. Mary's; ⊙ SCILLY ISLES; 49°55′N 06°19′W. Fishing center, tourist resort. Star Castle (1593) is now a hotel here.

Hugli (HOO-glee), district (□ 1,216 sq mi/3,161.6 sq km), S central WEST BENGAL state, E INDIA; ⊙ CHUNCHURA, or Chinsurah. Bounded E by HUGLI RIVER; drained by DAMODAR and RUPNARAYAN rivers. Swampy alluvial tract, rising in NW; rice, jute, potatoes, pulses, sugarcane. Highly industrialized section in area W of HUGLI RIVER; jute milling (center at CHAMPDANI), rice and cotton milling; chemical and glass manufacturing, jute pressing at KONNAGAR. Hugli College at CHUNCHURA; large annual mela (Hindu religious festival) near Shrirampur. SATGAON was mercantile capital of lower BENGAL for 1,500 years. Oldest Muslim buildings in BENGAL near HUGLI. Former European settlements include HUGLI (Port.), Shrirampur (Dan.), Chunchura (Du.), and CHANDANNAGAR (Fr.).

Hugli (HOO-glee), or **Hoogly**, or **Hooghly**, river, a distributary of the GANGA RIVER, c.160 mi/257 km long, WEST BENGAL state, E INDIA; formed by the confluence of the BHAGIRATHI, JALANGI, and MATABHANGA rivers, and flowing S to the BAY OF BENGAL; navigable for small ocean-going vessels to CALCUTTA, c.80 mi/129 km downstream. Despite its sandbars, it is the major shipping artery through the important Hugliside industrial area (jute and other factories) N of CALCUTTA. Its headwaters are important for inland traffic. Farakka Barrage aids its flow. A Portuguese trading post, established 1537 at Hugli town, was of commercial importance in the 16th and 17th centuries. The gigantic cantilever Haora Bridge across the river used to connect CALCUTTA with HAORA (for its railroad network); now a second Hugli Bridge downstream connects the two cities.

Hugli-Chinsura, town, WEST BENGAL state, E INDIA, on the HUGLI RIVER. A road and railroad junction; many large rice mills and manufacturing of rubber goods.

Hugo, town (2000 population 6,363), WASHINGTON county, E MINNESOTA, residential suburb 16 mi/26 km NNE of downtown ST. PAUL; 45°09′N 92°57′W. Manufacturing (boring equipment, wire forms, plastic molds). White Beard Lake to S, numerous small natural lakes in Hugo and vicinity.

Hugo, town (2006 population 5,573), ⊙ CHOCTAW county, SE OKLAHOMA, c.45 mi/72 km E of DURANT, near RED RIVER; 34°00′N 95°31′W. Railroad junction; trade center for agricultural area (grain, livestock, peanuts). Manufacturing (vegetable processing, onions, sportswear); railroad shops. HUGO LAKE reservoir is NE. Incorporated 1908.

Hugo, village (2000 population 885), ⊙ LINCOLN county, E COLORADO, on BIG SANDY CREEK and 90 mi/145 km SE of DENVER; 39°07′N 103°28′W. Elevation 5,046 ft/1,538 m. Cattle, wheat, sunflowers.

Hugo Lake, reservoir (□ 21 sq mi/54.6 sq km), CHOCTAW county, SE OKLAHOMA, on KIAMICHI RIVER, 5 mi/8 km ENE of HUGO; 34°01′N 95°23′W. Maximum

capacity 1,249,800 acre-ft. Formed by Hugo Dam (101 ft/31 m high), built (1974) by Army Corps of Engineers for flood control; also used for water supply and recreation. Indian National Turnpike just W.

Hugo Napoleão (OO-go NAH-po-lai-oun), town (2007 population 3,443), N PIAUÍ state, NE BRAZIL, 67 mi/108 km SE of TERESINA; 05°59′S 42°28′W.

Hugoton (YOO-go-tuhn), town (2000 population 3,708), ⊙ STEVENS county, SW KANSAS, 22 mi/35 km WNW of LIBERAL; 37°10′N 101°20′W. Elevation 3,107 ft/947 m. Shipping point in GREAT PLAINS wheat area; center of a major natural-gas field, with pipelines to NE. U.S. Gas distribution. County museum. Founded 1885, incorporated 1910.

Huguan (HU-GUAN), town, ⊙ Huguan county, SE SHANXI province, NE CHINA, 8 mi/12.9 km SE of CHANGZHI; 36°07′N 113°12′E. Medicinal herbs, engineering, coal and iron-ore mining.

Huguang (HU-GUANG), ancient province of S central CHINA, centered at DONGTING LAKE. Divided in 1660s into HUBEI (N) and HUNAN (S) provinces.

Huguenot, a section of STATEN ISLAND borough of NEW YORK CITY, SE NEW YORK; 40°32′N 74°12′W. Has experienced considerable growth in recent decades. Also called Huguenot Park.

Huhí (oo-EE), town, YUCATÁN, SE MEXICO, 33 mi/53 km SE of MÉRIDA. Henequen, tropical fruit, sugarcane, corn.

Hühnerstock (HYOO-nuhr-shtok), peak (10,850 ft/3,307 m) in BERNESE ALPS, BERN canton, S central SWITZERLAND, 10 mi/16 km S of MEIRINGEN.

Hui, CHINA: see ANHUI.

Hui'an (HUAI-AN), town, ⊙ Hui'an county, SE FUJIAN province, SE CHINA, 35 mi/56 km SSW of PUTIAN, on EAST CHINA SEA coast; 25°02′N 118°48′E. Oilseeds, sugarcane.

Huichang (HUAI-CHANG), town, ⊙ Huichang county, S JIANGXI province, SE CHINA, 56 mi/90 km ESE of GANZHOU, and on GONG RIVER (right headstream of GAN RIVER); 25°32′N 115°45′E. Rice, sugarcane; food industry, logging, engineering, non-ferrous ore mining.

Huichapan (wee-CHA-pahn), city and township, ⊙ Huichapan municipio, HIDALGO, central MEXICO, on central plateau, 60 mi/97 km WNW of PACHUCA DE SOTO on Mexico Highway 45; 20°22′N 99°38′W. Elevation 6,896 ft/2,102 m. Wine producing. Thermal springs nearby.

Huicheng, CHINA: see SHE XIAN.

Huichon (HEE-CHUHN), city, CHAGANG province, NORTH KOREA, on CHONGCHON RIVER, 85 mi/137 km NNE of PYONGYANG, in stock-raising and agricultural area (rice, soybeans, and cotton); 38°37′N 128°21′E. Produces textiles (silk, hemp), vegetable oil, woodwork, and paper. Connects primary railroads.

Huichuan (HUAI-CHUAN), town, SE GANSU province, NW CHINA, 70 mi/113 km S of LANZHOU. Grain, medicinal herbs.

Huicungo (hwee-KOON-go), town, MARISCAL CÁCERES province, SAN MARTÍN region, N central PERU, on affluent of HUALLAGA RIVER, and 28 mi/45 km S of TINGO DE SAPOSOA; 07°17′S 76°48′W. Coca, tobacco, yucca, plantains.

Huide, CHINA: see GONGZHULING.

Huigra (WEE-grah), village, CHIMBORAZO province, S central ECUADOR, on CHANCHÁN RIVER gorge and 10 mi/16 km WSW of ALAUSÍ. Elev. c.4,000 ft/1,219 m; 02°17′S 78°59′W. Trading post; cereals, potatoes, sheep.

Huíla (WEE-luh), province (□ 28,951 sq mi/75,272.6 sq km), SW ANGOLA; ⊙ LUBANGO. Bordered NW by BENGUELA, NE by Rio Cunene and HUAMBO province, E by Rio Cutalo and CUANDO CUBANGO province, S CUNENE province, W by Serra da Chela and NAMIBE province. Drained by Rio Cunene, Rio Catape, Rio Cubango. Includes Bikuar National Park (S) and Huila Plateau (N). Agriculture. Minerals include

iron ore, granite, marble. Main centers are LUBANGO, MATALA, CHIANGE, CACULA, KUVANGO, Techamutete.

Huila (HWEE-lah), department (□ 7,992 sq mi/20,779.2 sq km), S central COLOMBIA; ⊙ NEIVA; 02°30′N 75°45′W. Occupies the MAGDALENA valley flanked by Cordillera CENTRAL and Cordillera ORIENTAL. Climate is hot and wet in Magdalena valley, cooler in uplands. Mineral resources include silver and gold (AIPE, LA PLATA, CAMPOALEGRE), asphalt (GARZÓN). Main agricultural crops include rice, cotton, coffee. Considerable livestock raising (cattle, horses, mules). Neiva, its only important city, manufactures food products, consumer goods. Consists of c.37 municipios, and most people live in municipio centers and main cities, which include PITALITO, La Plata, GIGANTE, TIMANÁ, and SAN AGUSTÍN, as well as the capital, Neiva.

Huilai (HUAI-LEI), town, ⊙ Huilai county, SE GUANGDONG province, S CHINA, near coast, 32 mi/51 km SW of SHANTOU; 23°03′N 116°17′E. Rice, sugarcane, oilseeds; food processing, plastics.

Huila, Nevado del (HWEE-lah, nai-VAH-do del), snow-capped ANDEAN volcanic peak (18,865 ft/5,750 m), S central COLOMBIA, on HUILA-TOLIMA-CAUCA department border, highest in Cordillera CENTRAL, 50 mi/80 km SE of CALI; 03°00′N 76°00′W.

Huili (HUAI-LEE), town, ⊙ Huili county, SE SICHUAN province, SW CHINA, near CHANG JIANG (Yangzi River), 85 mi/137 km S of XICHANG, on highway; 26°41′N 102°15′E. Rice, tobacco, sugarcane; iron-ore mining, lead-zinc ore mining, coal mining.

Huiliches, ARGENTINA: see JUNÍN DE LOS ANDES.

Huillapima (hwee-yah-PEE-mah), town, SE CATAMARCA province, ARGENTINA, on LA RIOJA-CATAMARCA railroad and 22 mi/35 km SW of Catamarca; 28°44′S 65°59′W.

Huiloapan, MEXICO: see HUILOAPAN DE CUAUHTÉMOC.

Huiloapan de Cuauhtémoc (wee-LO-ah-pahn dai kwou-TE-mok), town, W VERACRUZ, MEXICO, 3.1 mi/5 km SW of ORIZABA. Elevation 4,265 ft/1,300 m. Temperate climate. Agriculture (corn, beans, chile, coffee, fruits), woods, cattle and poultry raising.

Huimanguillo (wee-mahn-GWEE-yo), city and township, ⊙ Huimanguillo municipio, TABASCO, SE MEXICO, on GRIJALVA RIVER (CHIAPAS border) and 33 mi/53 km WSW of VILLAHERMOSA; 17°50′N 93°23′W. Elevation 98 ft/30 m. Important oil production. Agricultural center (bananas, tobacco, mangoes, rice, coffee, beans). LA VENTA archaeological site is nearby.

Huimilpan (wee-MEEL-pahn), town, QUERÉTARO, central MEXICO, 15 mi/24 km SSE of QUERÉTARO. Elevation 7,382 ft/2,250 m. Grain, sugarcane, alfalfa, vegetables; livestock.

Huimin (HUAI-MIN), town, ⊙ Huimin county, NW SHANDONG province, NE CHINA, on road, and 60 mi/97 km NNE of JINAN; 37°29′N 117°29′E. Grain, cotton; textiles.

Huinan (HUAI-NAN), town, ⊙ Huinan county, S central JILIN province, NE CHINA, 65 mi/105 km NNE of TONGHUA, on Songhua reservoir; 42°40′N 126°01′E. Rice, wheat; engineering.

Huinan, CHINA: see CHAOYANG.

Huinca Renancó (HWEEN-kah re-nahn-KO), town, S CÓRDOBA province, ARGENTINA, 120 mi/193 km S of RÍO CUARTO; 34°50′S 64°23′W. Railroad junction, connected with BAHÍA BLANCA and SAN RAFAEL (MENDOZA province). Agricultural center (cereals, soybeans, flax, alfalfa; livestock).

Huining (HUAI-NING), town, ⊙ Huining county, SE GANSU province, NW CHINA, 75 mi/121 km ESE of LANZHOU; 35°42′N 105°06′E. Grain, oilseeds; food processing, chemicals, textiles and clothing, coal mining.

Huiramba (hoo-ee-RAHM-bah), town, in central MICHOACÁN, MEXICO, 16 mi/25 km from PÁTZCUARO. Mountainous with cold climate. Inhabitants are laborers dedicated to forestry and some local crafts. Poor roads.

Hui River (HUAI), N ANHUI province, E CHINA, rises in E HENAN province near Shangqiu; flows 150 mi/241 km SE past Linhuanji and Guzhen to HUAI RIVER near Wuhe.

Huishui (HUAI-SHUAI), town, ⊙ Huishui county, S GUIZHOU province, S central CHINA, 30 mi/48 km S of GUIYANG; 26°08′N 106°36′E. Rice, oilseeds; machinery, papermaking, food industry.

Huîsne River (WEEN), 80 mi/129 km long, in SARTHE department, PAYS DE LA LOIRE region, NW central FRANCE; rises 6 mi/9.7 km SW of MORTAGNE-AU-PERCHE (ORNE department); flows generally SW, past NOGENT-LE-ROTROU and La FERTÉ-BERNARD, to the SARTHE 1 mi/1.6 km SW of Le MANS; 48°00′N 00°11′E. Not navigable.

Huissen (HOI-suhn), town, GELDERLAND province, E NETHERLANDS, 4 mi/6.4 km SSE of ARNHEM, on Pannerdens Canal (NEDER RIJN River); 51°56′N 05°57′E. Dairying; cattle, hogs, poultry; grain, vegetables, potatoes; manufacturing (dairy products).

Huistán, MEXICO: see HUIXTÁN.

Huitan (hwee-TAHN), town, QUEZALTENANGO department, GUATEMALA, 16 mi/26 km NW of QUEZALTENANGO; 15°06′N 91°37′W. Elevation 8,507 ft/2,593 m. Mam-speaking population; subsistence farming (wheat, corn, potatoes; livestock).

Huité (hwee-TAY), town, ZACAPA department, GUATEMALA, 12 mi/19 km WSW of ZACAPA; 14°56′N 89°43′W. Elevation 1,306 ft/398 m. Subsistence agriculture (livestock).

Huitiupan (wee-tee-OU-pahn), town, CHIAPAS, S MEXICO, in N outliers of SIERRA MADRE, 8 mi/12.9 km NE of SIMOJOVEL; 17°13′N 92°39′W. A Tzotzil Indian community. Corn, fruit.

Huitong (HUAI-TUNG), town, ⊙ Huitong county, SW HUNAN province, S central CHINA, near GUIZHOU border, 30 mi/48 km S of ZHIJIANG; 26°52′N 109°43′E. Rice, oilseeds; logging, furniture.

Huitzilac (weet-SEE-lak), town, MORELOS, central MEXICO, 8 mi/12.9 km N of CUERNAVACA. Wheat, fruit; livestock.

Huitzilan (weet-SEE-lahn), town, ⊙ Huitzilan de Serdán municipio, PUEBLA, central MEXICO, in foothills of SIERRA MADRE ORIENTAL, 17 mi/27 km E of ZACATLÁN; 19°58′N 97°41′W. Corn, tobacco, sugarcane, fruit. In Totonac Indian area.

Huitzilan de Serdan, MEXICO: see HUITZILAN.

Huitziltepec, MEXICO: see SANTA CLARA HUITZILTEPEC.

Huitzuco de los Figueroa, MEXICO: see CIUDAD HUITZUCO.

Huixian (HUAI-SIAN), city (□ 775 sq mi/2,015 sq km; 2000 population 717,448), N HENAN province, NE CHINA, 10 mi/16 km N of XINXIANG; 35°26′N 113°51′E. Agriculture and heavy industry are the largest sectors of the city's economy. Cotton, grain; manufacturing (chemicals, food processing, coal mining).

Hui Xian (HUAI SIAN), town, ⊙ Hui Xian county, SE GANSU province, NW CHINA, 50 mi/80 km SSE of TIANSHUI; 33°46′N 106°06′E. Grain, livestock; beverages, building materials.

Huixquilucan de Degollado (weesh-kee-LOO-kahn de dai-go-YAH-do), town, ⊙ Huixquilucan municipio, MEXICO, central MEXICO, 14 mi/23 km WSW of MEXICO CITY, and in the ZONA METROPOLITANA DE LA CIUDAD DE MÉXICO.

Huixtán (weesh-TAN), town, CHIAPAS, S MEXICO, in Sierra Madre of HUEYTEPEC, 11 mi/18 km E of SAN CRISTÓBAL DE LAS CASAS; 16°43′N 92°03′W. Elevation 6,562 ft/2,000 m. Wheat, fruit. In Tzotzil Maya-speaking area. Formerly HUISTÁN.

Huixtla (WEESH-tlah), city and township, CHIAPAS, S MEXICO, in PACIFIC lowland, on railroad and 23 mi/37 km NW of TAPACHULA; 15°09′N 92°30′W. At junction of Mexico Highways 190 and 200. Trading, processing, and agricultural center (coffee, sugarcane, cacao,

cotton, fruit, livestock); manufacturing (furniture, shoes).

Huizcolotla, MEXICO: see SAN SALVADOR HUIXCOLOTLA.

Huize (HUAI-ZUH), town, ⊙ Huize county, NE YUNNAN province, SW CHINA, 100 mi/161 km NE of KUNMING; 26°21′N 103°25′E. Elevation 7,251 ft/2,210 m. Rice; chemical industries, smelting; major mining center for non-ferrous metals (copper, lead, zinc).

Huizen (HOI-zuhn), town (2001 population 45,292), NORTH HOLLAND province, W central NETHERLANDS, 15 mi/24 km ESE of AMSTERDAM, on GOOIMEER channel; 52°18′N 05°15′E. SOUTHERN FLEVOLAND polder to N. Dairying; cattle, hogs, poultry; vegetables, fruit, nursery stock; manufacturing (food processing). Recreational center to NW.

Huizhou (HUAI-JO), city (□ 162 sq mi/421.2 sq km; 2000 population 274,689), E GUANGDONG province, S CHINA; port on EAST RIVER, and 75 mi/121 km E of GUANGZHOU; 23°08′N 114°28′E. A center for light industry and commerce; crop growing. Grain, oil crops, vegetables, fruits, hogs, poultry, eggs; manufacturing (apparel, textiles, beverages, leather, chemicals, plastics, electrical equipment, electronics). Opened to foreign trade in 1902. Sometimes spelled Waichow.

Huizhou, CHINA: see SHE XIAN.

Huizingen Provincial Park (HWAH-zin-guhn), recreation park near Huizingen-BEERSEL, BRABANT province, central BELGIUM, 8 mi/12.9 km SSW of BRUSSELS.

Huizucar (WEE-ZOO-KAR), municipality and town, LA LIBERTAD department, EL SALVADOR, S of SAN SALVADOR.

Hukawng Valley (HOO-koun), circular basin (□ 2,000 sq mi/5,200 sq km) of KACHIN STATE, MYANMAR. Agriculture (rice, tobacco, cotton, opium); amber mines on S edge. Drained by headwaters of CHINDWIN RIVER; main village, TANAI. Surrounded by mountains (KUMON RANGE; E), it connects S with MOGAUNG valley. Population is largely Kachin. Scene of heavy fighting during World War II, when the Ledo Road was laid through the valley.

Hukeri (hoo-KAI-ree), town, BELGAUM district, KARNATAKA state, INDIA, 28 mi/45 km NNE of BELGAUM; 16°14′N 74°36′E. Tobacco, millet, peanuts, chili, sugarcane. Muslim ruins (sixteenth century) nearby.

Hukou (HU-KO), [Chinese=lake mouth], town, ⊙ Hukou county, northernmost JIANGXI province, SE CHINA, 20 mi/32 km E of JIUJIANG, and on right bank of CHANG JIANG (Yangzi River), at N end of 30-mi/48-km-long Hukou Canal linking Poyang Lake and the Chang Jiang; 29°44′N 116°13′E. River-navigation hub and transshipment point; rice, oilseeds, cotton; engineering, textiles, building materials, food industry.

Hukow (HOO-KO), town, NW TAIWAN, on railroad, and 7 mi/11.3 km NE of HSINCHU. Oil field; agriculture (rice, sweet potatoes, vegetables, tea); livestock. Site of army base.

Hukuntsi (hoo-koon-TSEE), village (2001 population 3,807), N KGALAGADI District, SW BOTSWANA, in KALAHARI DESERT, 250 mi/402 km WNW of GABORONE; 24°39′S 25°54′E. Important road junction.

Hukvaldy, CZECH REPUBLIC: see BRUSPERK.

Hula, ETHIOPIA: see HĀGERE SELAM.

Hula, Lake, (KHU-lah) Arabic *Bahr al Hulah*, formed by a natural dam of basalt-the HULA VALLEY-across the course of the JORDAN RIVER, NE ISRAEL. Elevation 230 ft/70 m. The JORDAN River exits from its S end. Between 1950 and 1958, c.12,350 acres/4,998 ha of the lake and its swampy shore were drained. The land has been irrigated by the Jordan and numerous springs, and is among the most fertile regions in Israel. Crops include grains, fruit, vegetables, and cotton; the lake is used for fishing. A small part of the lake was left as a nature reserve; since 1994 it has been expanded for ecological purposes as formerly reclaimed areas are being reflooded due to pollution from the intensive farming that has affected the water

table and drainage into Sea of GALILEE. The area is rich in flora and fauna and an important site for tourists and nature lovers. It is an important fly-way for aquatic birds moving from central Eurasia to Africa and returning. Sometimes spelled Lake Huleh.

Hulan (HU-LAN), town, ⊙ Hulan county, S HEILONGJIANG province, NE CHINA, on railroad, 16 mi/26 km N of HARBIN, and on HULAN RIVER near its mouth on the SONGHUA JIANG; 45°59′N 126°36′E. Grain, sugar beets, jute, tobacco; linen textiles; ore mining and dressing of precious metal.

Hulan Ergi (HU-LAN-UHR-GI), town, SW HEILONGJIANG province, NE CHINA, 15 mi/24 km NW of QIQIHAR, at crossing of Nen River and on North railroad; 47°14′N 123°36′E. Transshipment point.

Hulan River (HU-LAN), c.250 mi/402 km long, central HEILONGJIANG province, NE CHINA; rises in S outlier of the Lesser Hinggan Mountains; flows SW and S, through densely populated agricultural district (sugar beets, wheat, soybeans), to SONGHUA RIVER below HULAN. Navigable in lower course; frozen November–April.

Hulata (khoo-LAH-tah), kibbutz, 8 mi/12.9 km NE of ZEFAT (Safed) in HULA VALLEY, N ISRAEL; 33°03′N 35°36′E. Elevation 278 ft/84 m. Mixed farming and a shoe factory. A nature museum is on the kibbutz grounds. Founded in 1936, its name comes from the swamps of the HULA LAKE that were drained to eliminate malaria. Was on the Syrian front lines in Israel's 1948 War of Independence and again in the 1967 Six Day War.

Hula Valley (HOO-lah), region, GOLAN HEIGHTS, N ISRAEL, N GREAT RIFT VALLEY, 110 sq mi/180 sq km (16 mi/25 km long by 4 mi/7 km wide). Elevation 230 ft/70 m. Basaltic hills in S intercept the JORDAN RIVER, restricting water drainage downstream into the SEA OF GALILEE, forming LAKE HULA and surrounding wetlands. Lake Hula was formed approximately 20,000 years ago. During the 14,000 years prior to the formation of Lake Hula, the valley was swampland. More than 60 in/152 cm of precipitation falls on the Hermon mountain range, feeding underground springs, including the sources of the Jordan, and giving rise to much of the water flowing through the valley. At one time, the valley was a resting place for birds migrating from Europe to Africa and back. In addition, many species of rare fish and plants lived here. Following Israel's establishment in 1948, the government drained most of the swamps and ponds and converted them into agricultural fields, but set aside 800 acres/324 ha of wetlands for a nature reserve. Officially declared in 1964, the Hula Valley Nature Reserve was the first nature reserve in Israel. Tens of thousand of birds, including cranes, storks, pelicans, cormorants, and several types of herons, live in the reserve. More than 200 species of water fowl. Also rare water plants such as the yellow iris. Buffalo were placed in selected parts of the reserve because their grazing helps preserve the open meadow. In the spring of 1994, 250 acres/100 ha of wasteland just N of the reserve were intentionally flooded, improving the quality of the water in the Sea of Galilee

Hulbert, village (2006 population 532), CHEROKEE county, E OKLAHOMA, 10 mi/16 km W of TAHLEQUAH, near NEOSHO River (Ft. GIBSON LAKE to W); 35°55′N 95°08′W. Manufacturing (oil field parts). Sequoyah State Park to W.

Hulda (KHOOL-dah), kibbutz, W ISRAEL, on W foothills of the JUDAEAN HIGHLANDS, 6 mi/9.7 km SE of REHOVOT; 31°50′N 34°53′E. Elevation 488 ft/148 m. Production of electric transformers; grain, fruit, olives; dairying; poultry, sheep raising, beekeeping. Winery. Home of author Amos Oz. Founded 1909.

Huldenberg (HUL-duhn-berk), commune (2006 population 9,177), Leuven district, BRABANT province, central BELGIUM, 11 mi/18 km SE of BRUSSELS.

Huleh, Lake, ISRAEL: see HULA, LAKE.

Cross-references are shown in SMALL CAPITALS. The pronunciation guide is shown on page xix. The sources of population figures are shown on page xvii.

Hulett (HYOO-let), village (2006 population 442), CROOK county, NE WYOMING, 22 mi/35 km NW of SUNDANCE, on BELLE FOURCHE RIVER; 44°41'N 104°35'W. Elevation 3,755 ft/1,145 m. Cattle, sheep; sugar beets, wheat; timber; manufacturing (lumber). Bear Lodge Mountains and unit of Black Hills National Forest to SE; DEVILS TOWER NATIONAL MONUMENT to S; Missouri Buttes to SW.

Hulett's Landing, resort village, WASHINGTON county, E NEW YORK, on Lake GEORGE, 8 mi/12.9 km NW of WHITEHALL; 43°39'N 73°30'W.

Hulha Negra (OOL-yah NE-grah), city, S RIO GRANDE DO SUL state, BRAZIL, 13 mi/21 km SE of BAGÉ, on railroad; 31°24'S 53°53'W. Sheep.

Hulikal Durg, INDIA: see COONOOR.

Hulin (HU-LIN), town, ⊙ Hulin county, E HEILONGJIANG province, NE CHINA; river port on USSURI RIVER (Russian border), opposite Nevskoye, and 220 mi/354 km ENE of MUDANJIANG; 45°48'N 132°59'E. Railroad terminus. Tobacco, sugar beets; logging; food processing. Zhenbao Island in Ussuri River was site of Sino-Soviet military clash in the late 1960s.

Hulin (HU-leen), Czech *Hulín*, German *Hullein*, town, JIHOMORAVSKY province, central MORAVIA, CZECH REPUBLIC, 21 mi/34 km SSE of OLOMOUC; 49°19'N 17°28'E. Railroad junction. Agriculture (wheat, sugar beets); manufacturing (machinery); wood processing; sugar refining. Has a 13th-century castle.

Hulin Rocks, Northern Ireland: see MAIDENS, THE.

Hulken, CZECH REPUBLIC: see HLUK.

Hull (HUHL), former county (□ 139 sq mi/361.4 sq km), SW QUEBEC, E CANADA, on ONTARIO border, on OTTAWA River; 45°40'N 75°35'W. Its county seat was HULL.

Hull (HUHL), district, former city, SW QUEBEC, E CANADA, at the confluence of the OTTAWA and GATINEAU rivers, opposite the city of OTTAWA; 45°26'N 75°44'W. Hydroelectric power station; paper, pulp, textile, steel, and lumber mills; iron foundries; cement and meatpacking plants. Center for service industries and federal government offices, with civil servants forming its largest bloc of workers. French-speaking part of National Capital Region. Canadian Museum of Civilization. Nearby is Gatineau Park, a large recreation area. Incorporated 1875; amalgamated into the city of GATINEAU in 2002.

Hull or **Kingston upon Hull** (HUL), town and county (□ 27 sq mi/70.2 sq km; 2001 population 243,589), Kingston upon Hull, NE ENGLAND, on the N shore of the HUMBER estuary at the influx of the small HULL RIVER; 53°45'N 00°19'W. Its port is one of the chief outlets for the surrounding area, which is also accessible by railroad. Imports include oilseed, wood, foodstuffs, wool, metal ores, and petroleum; exports include coal, coke, machinery, motor vehicles, tractors, iron and steel products, and textiles. Hull is also one of the world's largest fishing ports. Manufacturing includes processed foods, chemicals, iron and steel products, and machinery. Hull was founded late in the 13th century by Edward I, and the construction of docks, which extend for miles along the Humber, was begun c.1775. In July 1981, the Western Humber Bridge was opened; communication with other cities thus improved, and Hull's economic value increased. The Wilberforce House, Municipal Museum, and Ferens Art Galleries are noteworthy. The grammar school, founded in 1486, was attended by Andrew Marvell and William Wilberforce, born in Hull. Schools include the University of Hull and University of Humberside. Trinity House, established in 1369 to aid sailors, has been Trinity House Navigation School since 1787. Hull's annual fair is one of the largest in England. N industrial section of town is called SCULCOATES. District includes HALTEMPRICE, 4 mi/6.4 km NW.

Hull, town (2000 population 160), MADISON county, NE GEORGIA, 6 mi/9.7 km NE of ATHENS; 34°01'N 83°17'W.

Hull, town (2000 population 1,960), SIOUX county, NW IOWA, 27 mi/43 km N of LE MARS; 43°11'N 96°07'W. Livestock, grain. Incorporated 1888.

Hull, town (2000 population 11,050), PLYMOUTH county, E MASSACHUSETTS, on narrow Nantasket Peninsula in MASSACHUSETTS BAY, and 10 mi/16 km ESE of Boston; 42°18'N 70°53'W. Summer resort. Settled 1624, incorporated 1644. Resort villages include Allerton, Kenberma, NANTASKET BEACH.

Hull, unincorporated town, LIBERTY county, SE TEXAS, 32 mi/51 km W of BEAUMONT; 30°08'N 94°38'W. In oil, timber, farm area (cattle; rice, soybeans). BIG THICKET NATIONAL PRESERVE to N and NE.

Hullavington (huh-LAV-ing-tuhn), village (2001 population 1,247), WILTSHIRE, S central ENGLAND, 4 mi/6.4 km SW of MALMESBURY; 51°33'N 02°09'W.

Hullbridge (HUL-brij), residential suburb (2001 population 6,445), ESSEX, SE ENGLAND, on CROUCH RIVER, 3 mi/4.8 km N of RAYLEIGH; 51°37'N 00°38'E.

Hullein, CZECH REPUBLIC: see HULIN.

Hullet (HUH-let), former township (□ 84 sq mi/218.4 sq km; 2001 population 1,796), S ONTARIO, E central CANADA; 43°40'N 81°27'W. Amalgamated into CENTRAL HURON township in 2001.

Hullet, NORWAY: see KONGSVOLL.

Hüllhorst (HUL-horst), village, NORTH RHINE–WESTPHALIA, N GERMANY, 12 mi/19 km N of HERFORD; 52°18'N 08°04'E. Resort with sulphuric springs; manufacturing of furniture; metalworking; brickworks.

Hull Island, AUSTRAL ISLANDS: see MARIA ISLAND.

Hull Island, PHOENIX ISLANDS, KIRIBATI: see ORONA.

Hull Mountain (6,873 ft/2,095 m), Lake/MENDOCINO county line, NW CALIFORNIA, 34 mi/55 km N of LAKEPORT, in the COAST RANGES. In Mendocino National Forest.

Hull River (HUL), 23 mi/37 km long, East Riding of Yorkshire, NE ENGLAND; rises near GREAT DRIFFIELD, flows S to HUMBER RIVER at HULL.

Hulmeville (HUHLM-vil), borough (2006 population 875), BUCKS county, SE PENNSYLVANIA, residential suburb 17 mi/27 km NE of downtown PHILADELPHIA and 3 mi/4.8 km WSW of LEVITTOWN, on NESHAMINY CREEK. Light manufacturing

Hulong, AUSTRALIA: see WHITTON.

Hulpe, la (UHL-puh, lah), Flemish *Terhulpen* (terhool-pen), commune, Nivelles district, BRABANT province, central BELGIUM, 9 mi/14.5 km SE of BRUSSELS. Paper manufacturing. Castle.

Hulsout (HULS-hout), commune, Turnhout district, ANTWERPEN province, N BELGIUM, 3 mi/5 km E of HEIST-OP-DEN-BERG.

Hulst (HUHLST), town, ZEELAND province, SW NETHERLANDS, on FLANDERS mainland, 11 mi/18 km ESE of TERNEUZEN, 2 mi/3.2 km SE of Belgian border; 51°17'N 04°04'E. WESTERN SCHELDT estuary nearby. Dairying; cattle, hogs; grain, vegetables, sugar beets. Picturesque fortressed town; site of fifteenth-century church.

Hultschin, CZECH REPUBLIC: see HLUCIN.

Hultsfred (HULTS-FRED), town, KALMAR county, SE SWEDEN, 30 mi/48 km NW of OSKARSHAMN; 57°30'N 15°51'E. Railroad junction.

Huludao (HU-LU-DOU), city (□ 46 sq mi/119.6 sq km), SW LIAONING province, NE CHINA, on the Hulu peninsula of the N BOHAI BAY, on railroad, and 27 mi/43 km S of JINZHOU; 40°47'N 121°00'E. The S half of the city's territory is hilly land and the N half is plain. There are long coastlines with many beaches, and the climate is mild, making the city a popular tourist destination. The Beijing-Shenyang, Beijing-Harbin railroads, and the Beijing-Shenyang express highway converge here. In addition, the city is a commercial port. Major industries include petroleum chemicals, non-ferrous metals, machinery, shipbuilding, and construction materials. Other important economic activities are port and ocean transportation, trade and commerce, and real estate. The city has numerous industrial resources, including zinc (zinc output here accounts for 40% of the country's total), lead, coal, and crude oil. It is also rich in ancient architecture, especially a segment of town wall in the sea and the ruin of a palace built by the first Chinese emperor Qin Shi Huangdi more than 2,000 years ago. Many monuments and buildings from more modern times are here as well. The city was designated a provincial open economic zone in 1993, and facilities have been established to encourage foreign investment. Huludao annexed the former city of JINXI.

Hulun, CHINA: see HAILAR.

Hulunbuir, CHINA: see HULUN BUIR MENG.

Hulun Buir Meng (HUH-LUN-BER-MENG), Mongolian league of Inner Mongolia, in N INNER MONGOLIA AUTONOMOUS REGION, N CHINA. Bounded W by RUSSIA (along ERGUN RIVER), S by Mongolian People's Republic, and E by the GREATER HINGGAN MOUNTAINS, the area is essentially a steppe plateau in S (average elevation 2,000 ft/610 m) and mountainous taiga in N; 49°12'N 119°42'E. The league contains the lakes HULUN NUR and Buir (Bor) Nur, for which it is sometimes called Hulunbuir. The Mongol population is engaged in livestock raising (mainly sheep, horses, cattle), with wool and hides the chief export products. Main centers at HAILAR, MANZHOULI, YAKESHI, BUGT, and YALU.

Hulun Nur (HUH-LUN-NUHR), largest lake of INNER MONGOLIA AUTONOMOUS REGION, N CHINA, 100 mi/161 km W of HAILAR; 35 mi/56 km long, 10 mi/16 km wide, 3 ft/0.9 m–5 ft/1.5 m deep; 49°00'N 117°27'E. Of fluctuating level, it is the flood reservoir of the ERGUN RIVER, with which it is connected by a channel (N). Receives Herlen River (SW) and Orxon River (E; outlet of Buyr Nur).

Hulwan, EGYPT: see HELWAN.

Hulyaypole (hoo-LYEI-po-le), (Russian *Gulyaypole* or *Gulyay-Pole*), city, NE ZAPORIZHZHYA oblast, UKRAINE, 50 mi/80 km ESE of ZAPORIZHZHYA; 47°40'N 36°15'E. Elevation 360 ft/109 m. Raion center; manufacturing (agricultural machines, enamel paint, building materials, footwear, leather), food processing (food flavoring, cheese). Professional-technical school; historical museum. Established in 1785 as a military settlement; city since 1938. Small Jewish community since the second half of the 19th century (approximately 1,200 by 1939); exterminated during World War II—fewer than 100 Jews remaining in 2003.

Huma (HU-MAH), town, ⊙ Huma county, N HEILONGJIANG province, NE CHINA, 110 mi/177 km NNW of HEIHE, and on right bank of AMUR RIVER (Russia border); 51°42'N 126°39'E. Logging, fur processing, ore mining of precious metal. Well known for its gold mining.

Humacao (oo-mah-KOU), town (2006 population 60,569), E PUERTO RICO, 28 mi/45 km SE of SAN JUAN, near the coast. Its port, PLAYA DE HUMACAO or PUNTA SANTIAGO is 5 mi/8 km ENE; 18°09'N 65°49'W. Port of entry. Commercial, industrial, tourism center. Sugarcane; manufacturing (plastic products, chemicals, electronics). Resort (Palmas del Mar). Has district and municipal courts. Small airport. Museum and cultural center here. University of Puerto Rico–Humacao located here.

Humahuaca (oo-mah-HWAH-kah), town (1991 population 6,170), ⊙ Humahuaca department (□ 1,525 sq mi/3,965 sq km), N central JUJUY province, ARGENTINA, in the QUEBRADA DE HUMAHUACA VALLEY (upper RÍO GRANDE DE JUJUY valley), on railroad and 70 mi/113 km N of JUJUY; 23°12'S 65°21'W. Health resort; mining, lumbering, agricultural center. Zinc, silver, copper deposits. Potatoes, alfalfa, corn, wheat, fruit; livestock.

Humahuaca, Quebrada de, ARGENTINA: see GRANDE DE JUJUY, RÍO.

Area is shown by the symbol □, and capital city or county seat by ⊙.

Humaitá (oo-mei-TAH), city (2007 population 38,559), S AMAZONAS, BRAZIL, steamer and hydroplane landing on left bank of Rio MADEIRA, and 100 mi/161 km NE of PÔRTO VELHO; 00°28′N 67°28′W. Rubber, Brazil nuts, cacao, hides. Formerly also spelled Humaytá.

Humaitá (oo-mei-TAH), town and canton, ABUNA province, PANDO department, N BOLIVIA, on ORTON RIVER and 24 mi/39 km WNW of RIBERALTA; 10°50′S 66°24′W.

Humaitá (OO-mah-ee-tah), town (2007 population 4,923), NW RIO GRANDE DO SUL state, BRAZIL, 48 mi/77 km NW of PALMEIRA DAS MISSÕES; 27°34′S 53°58′W. Wheat, soybeans, corn; livestock.

Humaitá (oo-mei-TAH), town, Ñeembucú department, S PARAGUAY, on Paraguay River (ARGENTINA border) above its confluence with the Paraná, and 19 mi/31 km SW of PILAR; 27°02′S 58°33′W. Minor port and agricultural center (cotton, corn, sugarcane, oranges; cattle). Here are ruins of fort and of the church of San Carlos, bombarded 1867 by allied fleet in War of the Triple Alliance and now revered as national shrine.

Human', UKRAINE: see UMAN'.

Humanes (oo-MAH-nes), officially **Humanes de Mohernando**, town, GUADALAJARA province, central SPAIN, near HENARES RIVER, 14 mi/23 km N of GUADALAJARA; 40°49′N 03°09′W. Cereals, olives, grapes, sheep, goats. Flour milling, meat-product manufacturing.

Humansdorp, town, EASTERN CAPE province, SOUTH AFRICA, on Deekeoi River near CAPE ST. FRANCIS on the INDIAN OCEAN, 50 mi/80 km W of PORT ELIZABETH (NELSON MANDELA METROPOLE). Elevation 804 ft/245 m. Lies at entrance of prosperous long Kloof on N2 highway and railroad. Agricultural center (livestock, wheat, fruit, vegetables). Resort and magisterial town.

Humansville, city (2000 population 946), POLK county, SW central MISSOURI, in the OZARKS. between SAC RIVER and POMME DE TERRE RIVER, 15 mi/24 km NW of BOLIVAR; 37°47′N 93°34′W. Corn, soybeans; cattle; manufacturing (boxes).

Humarock, Massachusetts: see MARSHFIELD.

Humaviza (oo-mah-VEE-zah), town and canton, General Bernardin BILBAO province, extreme N POTOSÍ department, W central BOLIVIA; 17°55′S 66°04′W. Lead-bearing lode, limestone and gypsum deposits. Agriculture (potatoes, yucca, bananas); cattle.

Humay (hoo-MAI), town, ICA region, SW PERU, on PISCO RIVER and 22 mi/35 km E of PISCO; 13°43′S 75°54′W. Cotton, rice, grapes, vegetables.

Humaya (hoo-MAH-yah), village, LIMA region, W central PERU, on HUAURA RIVER, on HUAURA–SAYÁN railroad and 15 mi/24 km ENE of HUACHO; 11°06′S 77°25′W. Cotton- and rice-growing center.

Humayrah, Al-, SYRIA: see HAMRA, EL.

Humaytá, Brazil: see HUMAITÁ.

Humbe (HOOM-be), town, CUNENE province, S ANGOLA, near Cunene River, on road, and 56 mi/90 km WNW of ONDJIVA. Cattle raising, dairying.

Humber, river, c.75 mi/120 km long, SW NEWFOUNDLAND AND LABRADOR, CANADA, rising in the Long Mountains; flowing SE then SW, through DEER LAKE, to the Bay of Islands at CORNER BROOK.

Humbermouth (HUHM-buhr-muhth), neighborhood of CORNER BROOK, SW NEWFOUNDLAND AND LABRADOR, CANADA, on Humber River estuary; 48°58′N 57°55′W. Also known as Corner Brook East. On railroad.

Humber River (HUHM-buh), navigable estuary of the TRENT and OUSE rivers, c.40 mi/64 km long, NE ENGLAND, in East Riding of Yorkshire. It has a width of 1 mi/1.6 km–8 mi/12.9 km. SPURN HEAD, with a lighthouse, is at the mouth of the Humber. The shores are generally low, and shoals obstruct shipping in parts. Encroachment of the sea has destroyed former ports, notably Ravenspur. In early English history the Humber was significant as a means of ingress. HULL and Great Grimsby (see GRIMSBY, GREAT) are chief cities and major fishing ports. The Humber Bridge (4,580 ft/1,396 m), linking Hull with the estuary's S shore, opened in July 1981, and is one of the longest suspension bridges in the world.

Humberside (HUHM-buh-seid), former county (□ 1,356 sq mi/3,525.6 sq km), NE ENGLAND; 53°52′N 00°40′W. A tidal channel reaches over 20 mi/32 km into the area. A deep-water channel from the HUMBER RIVER also comprises part of this territory. The former county consists mainly of lowlands, marshes, and chalk cliffs. It is currently comprised of East Riding of Yorkshire, Kingston upon Hull, North Lincolnshire, and North East Lincolnshire. The economy in these districts is based on the wool industry and agriculture. Major towns in this former district include Great Grimsby, Immingham, Hull, Beverley, and Scunthorpe. Many Middle Age parish churches and minsters still stand in this area.

Humberston (HUHM-buh-stuhn), village (2001 population 5,375), North East Lincolnshire, NE ENGLAND, on estuary of HUMBER RIVER, 3 mi/4.8 km SE of CLEETHORPES; 53°32′N 00°02′W. Fine 18th-century brick church with tower.

Humberstone (HUHM-buhr-stuhn), unincorporated village, S ONTARIO, E central CANADA, on Lake ERIE, at S end of WELLAND SHIP CANAL, and included in PORT COLBORNE; 42°53′N 79°15′W. Humberstone Lock, on the canal, is one of world's largest lift locks.

Humberstone, ENGLAND: see LEICESTER.

Humberto de Campos (oom-ber-to dee kahm-pos), city, (2007 population 24,337), N MARANHÃO, BRAZIL, 55 mi/89 km ESE of SÃO LUÍS; 02°30′S 43°30′W. Saltworks; manioc meal, corn. Formerly called Miritiba.

Humberto Primo (oom-BER-to PREE-mo) or **Humberto I**, town, central SANTA FE province, ARGENTINA, 65 mi/105 km NW of SANTA FE; 33°02′S 61°02′W. Agricultural center (alfalfa, flax, soybeans, wheat); dairying, tanning.

Humbird, village, CLARK county, central WISCONSIN, 36 mi/58 km SE of EAU CLAIRE, in dairying and farming area. Hay, cheese.

Humble, city (2006 population 14,927), HARRIS county, SE TEXAS, suburb 17 mi/27 km N of downtown HOUSTON; 29°59′N 95°15′W. Elevation 96 ft/29 m. Bounded N by West Fork SAN JACINTO RIVER, nearly surrounded by extended Houston city limits. In oil and gas field; manufacturing (rubber and plastic components, oil field parts, golf equipment, diversified light manufacturing). Houston Intercontinental Airport to W; LAKE HOUSTON reservoir (San Jacinto River) to E. Founded 1888, incorporated 1933. Had oil boom, 1904.

Humboldt, county (□ 3,573 sq mi/9,289.8 sq km; 2006 population 128,330), NW CALIFORNIA, on PACIFIC OCEAN; ⊙ EUREKA; 40°42′N 123°50′W. HUMBOLDT BAY indents central coastline, parallels coast for 18 mi/29 km. CAPE MENDOCINO is westernmost point of California. Drained by KLAMATH, TRINITY, MAD, EEL, and MATTOLE rivers. Mainly in COAST RANGES; includes part of KLAMATH MOUNTAINS in E and NE, King Mountain Range in SW, Rainbow Ridge in W. Six Rivers National Forest in E and NE; Hoopa Valley Indian Reservation in N; Humboldt Bay National Wildlife Refuge in W; part of REDWOOD NATIONAL PARK in NW; PRAIRIE CREEK REDWOODS, Dry Lagoon, and Patrick's Point state parks and Trinidad State Beach in NW; Grizzly Creek Redwoods and Humboldt Redwoods state parks in S center; Benbow Lake State Recreation Area in S. World's tallest tree (367 ft/112 m) is near DYERVILLE. Logging (redwood, Douglas fir, cedar, spruce); timber; dairying, cheese making; cattle, lambs, sheep raising (for Merino wool). Noted recreational area (fishing, camping, hiking, bathing). Salmon; crabs. Some quarrying and mining (sand, gravel, clay, gold, silver). Formed 1853.

Humboldt, county (□ 435 sq mi/1,131 sq km; 2006 population 9,975), N central IOWA; ⊙ DAKOTA CITY; 42°46′N 94°12′W. Prairie agricultural area (hogs, cattle, corn, oats, soybeans) drained by DES MOINES and East Des Moines rivers. Bituminous-coal deposits, limestone quarries. Winter World ski area to E. Widespread flooding occurred in 1993. Formed 1857.

Humboldt (HUHM-buhlt), county (□ 9,658 sq mi/25,110.8 sq km; 2006 population 17,446), NW NEVADA; ⊙ WINNEMUCCA; 41°24′N 118°07′W. Ranching and mining area watered by QUINN, LITTLE HUMBOLDT, and HUMBOLDT rivers and bordering on OREGON, on N; borders IDAHO for 1 mi/1.6 km in NE corner. Cattle, sheep; hay, potatoes; silver, copper, gold; sand and gravel, clay. SANTA ROSA RANGE is in NE, in part of Humboldt National Forest. Summit Lake Indian Reservation and Lahontan Cutthroat Trout Natural Area are in W part of BLACK ROCK DESERT in SW. Formed 1861. Part of Fort Mcdermitt Indian Reservation on Oregon boundary in NE; smaller section in N center of county. Part of Sheldon National Wildlife Refuge in NW. Chimney Dam Reservoir in E.

Humboldt, city (2000 population 4,452), HUMBOLDT county, N central IOWA, on DES MOINES RIVER, near mouth of the East Des Moines, and 15 mi/24 km N of FORT DODGE; 42°43′N 94°13′W. Manufacturing (belt conveyors, heating equipment, meat products, steel and plastic fabrication, concrete and wood products, beverages). Limestone quarries, sand pits nearby. Settled 1863, incorporated 1869.

Humboldt (HUHM-bolt), city (2006 population 9,244), GIBSON county, W central TENNESSEE, 75 mi/121 km NW of MEMPHIS; 35°50′N 88°58′W. Agriculture; diverse manufacturing; retail. Humboldt also has a nearby state fish hatchery. Incorporated 1865.

Humboldt (HUHM-bolt), town (2006 population 4,998), central SASKATCHEWAN, CANADA, 65 mi/105 km E of SASKATOON; 52°12′N 105°07′W. Grain elevators. Service center for region. Resort with medicinal springs.

Humboldt (HUHM-bolt), town (2000 population 1,999), ALLEN county, SE KANSAS, on Neosho River, and 9 mi/14.5 km N of CHANUTE; 37°48′N 95°26′W. Railroad junction. Shipping point in oil and grain region; manufacturing of cement products; special machinery. Laid out 1857; incorporated as village 1866, as city 1870.

Humboldt, town (2006 population 845), RICHARDSON county, extreme SE NEBRASKA, 19 mi/31 km WNW of FALLS CITY, and on North Fork of BIG NEMAHA RIVER; 40°10′N 95°56′W. Grain; manufacturing (apparel, flour, feed). Settled c.1856.

Humboldt, village (2000 population 458), COLES county, E central ILLINOIS, 12 mi/19 km NW of CHARLESTON; 39°36′N 88°19′W. In rich agricultural area.

Humboldt, village, KITTSON CO., MINNESOTA, 12 mi/19 km NW of HALLOCK, 5 mi/8 km S of CANADA (MANITOBA) border, RED RIVER (NORTH DAKOTA state line) 5 mi/8 km to W; 48°55′N 97°05′W. Manufacturing (feeds).

Humboldt, village (2006 population 563), MINNEHAHA county, E SOUTH DAKOTA, 20 mi/32 km WNW of SIOUX FALLS; 43°38′N 97°04′W. Chemicals; grain; hogs. Lake Vermillion State Recreation Area to SW.

Humboldt (HUHM-buhlt), river, c.300 mi/483 km long, W U.S.; begins at confluences of Mary's and Bishop's Creek, c.15 mi/24 km W of WELLS, NE NEVADA; flows generally WSW, receives North Fork Humboldt from N, flows past ELKO, receives South Fork Humboldt from S, then flows past CARLIN, BATTLE MOUNTAIN, receives LITTLE HUMBOLDT RIVER from NE before passing WINNEMUCCA, continues through RYE PATCH RESERVOIR, finally flowing past LOVELOCK before entering HUMBOLDT SINK; intermittent Humboldt Lake in sink; no ocean outlet. Along with its tributaries, the Humboldt drains most of N Nevada. Known to early explorers and named by

J. C. Frémont, the river was an important route followed by many of the emigrants from SALT LAKE CITY to central CALIFORNIA. Its course supplied wagon trains with water and grass. Its length varies with the season, and its volume decreases downstream. It is the longest river in GREAT BASIN, and served to open the way for the 1849 California gold rush. Most of the towns of N Nevada are located on the river in a valley used by Union Pacific and Southern Pacific railroads and interstate highway 80 (replaced U.S. highway 40 in 1970s) as an E-W route. Near Lovelock the Humboldt project of the U.S. Bureau of Reclamation is served by the RYE PATCH DAM (completed 1936), which impounds water for irrigation. Forage crops are raised along the river. Upper course, to North Fork, sometimes called East Fork Humboldt. North Fork Humboldt River rises in N ELKO county, joins Humboldt 14 mi/23 km ENE of ELKO, flows c.70 mi/113 km SSE. South Fork Humboldt River rises in NW WHITE PINE county, flows N c.65 mi/105 km through South Fork Reservoir, joins Humboldt 7 mi/11.3 km WSW of Elko.

Humboldt Bay, INDONESIA: see YOS SUDARSO BAY.

Humboldt Current, Pacific Ocean: see PERU CURRENT.

Humboldt Glacier, Danish *Humboldt Gletscher,* NW GREENLAND. The largest known glacier of the Northern Hemisphere, it debouches into KANE BASIN along a front c.60 mi/100 km wide and 300 ft/91 m high. U.S. explorer E. K. Kane discovered it on his expedition of 1853–1855.

Humboldt Hill, unincorporated town (2000 population 3,246), HUMBOLDT county, NW CALIFORNIA; residential suburb 6 mi/9.7 km SSW of EUREKA, at Humboldt Hill, 1 mi/1.6 km E of South Bay of HUMBOLDT BAY. College of the Redwoods to S.

Humboldt House, a national park, in TAXCO DE ALARCÓN, GUERRERO, MEXICO, on Juan Ruíz de Alarcón. Dating from the 16th century. While on a scientific journey to SOUTH AMERICA and CUBA, Baron von Humboldt spent the night here in April 1803. The Moorish style house has been a convent, hospital, and Taxco's first movie theater. It is now the Museo de Arte Virreinal.

Humboldt Peak (14,064 ft/4,287 m), in SANGRE DE CRISTO MOUNTAINS, CUSTER county, S COLORADO.

Humboldt, Pico, VENEZUELA: see LA CORONA.

Humboldt Range (HUHM-buhlt), NW NEVADA, in PERSHING county, extending generally N-S along E HUMBOLDT River, forms RYE PATCH RESERVOIR to NW. Rises to 9,834 ft/2,997 m in Star Peak, at N end. HUMBOLDT SINK is to SW. Gold, silver, diatomite.

Humboldt Salt Marsh (HUHM-buhlt), in N NEVADA, in CHURCHILL county, 45 mi/72 km NE of FALLON; 15 mi/24 km long, 6 mi/9.7 km wide. Fed by Spring Creek from NE and intermittent affluents from CLAN ALPINE MOUNTAINS (E). CARSON SINK to W. Separated by Stillwater Range.

Humboldt Sink, PERSHING and CHURCHILL counties, W NEVADA, N of CARSON SINK, c.30 mi/48 km N of FALLON; 11 mi/18 km long, maximum width 4 mi/6.4 km. Intermittently dry lake bed fed by HUMBOLDT River from NE; has no outlet. Humboldt Lake is body of water at center of sink. TRINITY RANGE to NW. Area protected by Humboldt Wildlife Management Area.

Humbolt, unincorporated town, YAVAPAI county, W central ARIZONA, 13 mi/21 km ESE of PRESCOTT, on headstream of Aqua Fria River. Elevation 4,980 ft/1,518 m. Cattle, sheep, hay, alfalfa; manufacturing (pesticides). Parts of Prescott National Forest to E and W.

Hume, town (2000 population 337), BATES county, W MISSOURI, 20 mi/32 km W of BUTLER; 38°05′N 94°34′W. Agriculture.

Hume, village (2000 population 382), EDGAR county, E ILLINOIS, 15 mi/24 km NNW of PARIS; 39°47′N 87°52′W. In agricultural area; ships grain.

Hume Dam, AUSTRALIA: see HUME RESERVOIR.

Humenné (hu-MEN-yai), German *Homenau,* Hungarian *Homonna,* town, VYCHODOSLOVENSKY province, E SLOVAKIA, on LABOREC RIVER; 48°56′N 21°55′E. Has railroad junction; manufacturing (chemicals, building materials; food processing). Has 17th-century church, museum, 18th-century synagogue. Exposition of folk architecture. Military base. Founded in 14th century.

Humera, town (2007 population 26,620), TIGRAY state, NW ETHIOPIA, 7 mi/11.3 km E of TEKEZE RIVER, near ERITREA and SUDAN; 14°17′N 36°36′E. Has airfield.

Hume Reservoir (HYOOM) (□ 70 sq mi/182 sq km), on the MURRAY River, near ALBURY-WODONGA, on the VICTORIA-NEW SOUTH WALES border, AUSTRALIA; 36°00′S 147°20′E. Impounded by Hume Dam (completed 1937), the reservoir regulates the Murray River. It receives additional water from the SNOWY MOUNTAINS Hydroelectric Scheme.

Humeston, town (2000 population 543), WAYNE county, S IOWA, 11 mi/18 km NW of CORYDON; 40°51′N 93°30′W. Livestock, grain.

Humewood, S residential suburb of PORT ELIZABETH (NELSON MANDELA METROPOLE), EASTERN CAPE, SOUTH AFRICA; 33°59′S 25°38′E; elevation 3 ft/1 m. Site of DRIFTSANDS airport; campus of University of Port Elizabeth; oceanarium, snake park, and museum. Resort and recreational area.

Humilladero (oo-mee-yah-DHAI-ro), town, MÁLAGA province, S SPAIN, 10 mi/16 km NW of ANTEQUERA; 37°07′N 04°42′W. Olive oil industry; cereals.

Humla (HOOM-lah), district, NW NEPAL, in KARNALI zone; ⊙ SIMIKOT.

Humlebæk (HOOM-luh-bek), city (2000 population 8,571) and port, FREDERIKSBORG county SJÆLLAND, DENMARK, on the ØRESUND and 9 mi/14.5 km ENE of HILLERØD; 55°58′N 12°33′E. Resorts; fisheries; fruit (apples). Louisiana Museum of Modern Art (founded 1958) in park overlooking Øresund.

Hummelo en Keppel (HUH-muh-law uhn KE-puhl), village, GELDERLAND province, E NETHERLANDS, 4 mi/6.4 km NW of DOETINCHEM; 52°01′N 06°14′E. Keppel Castle to S. Dairying; cattle, hogs; grain, vegetables, fruit.

Hummelstadt, POLAND: see LEWIN KLODZKI.

Hummelstown (HUH-muhls-toun), borough (2006 population 4,382), DAUPHIN county, S central PENNSYLVANIA, 9 mi/14.5 km E of HARRISBURG and 3 mi/4.8 km W of HERSHEY, on SWATARA CREEK. Manufacturing (food products, asphalt, machinery). Agricultural area (apples, soybeans, grain; poultry, livestock, dairying). Indian Echo Caverns to S. Founded c.1740, laid out 1762, incorporated 1874.

Hummels Wharf, unincorporated town (2000 population 641), SNYDER county, central PENNSYLVANIA, residential suburb 3 mi/4.8 km SW of SUNBURY on SUSQUEHANNA River; 40°49′N 76°50′W.

Humnoke (HUHM-nok), village (2000 population 280), LONOKE county, central ARKANSAS, 12 mi/19 km WNW of STUTTGART; 34°32′N 91°45′W.

Humocaro Alto (hoo-mo-KAH-ro AHL-to), town, LARA state, NW VENEZUELA, in N ANDEAN spur, 55 mi/89 km SW of BARQUISIMETO; 09°36′N 69°58′W. Elevation 3,589 ft/1,094 m. Coffee, cacao, sugarcane, cereals; livestock.

Humocaro Bajo (hoo-mo-KAH-ro BAH-ho), town, LARA state, NW VENEZUELA, in N ANDEAN spur, 50 mi/80 km SW of BARQUISIMETO; 09°40′N 69°58′W. Elevation 3,668 ft/1,118 m. Coffee, sugarcane, cereals, fruit; livestock.

Humorului, ROMANIA: see GURA HUMORULUI.

Humos, Cape (OO-mos), PACIFIC headland in MAULE region, S central CHILE, 6 mi/19 km SW of CONSTITUCIÓN.

Humos Island, CHILE: see CHONOS ARCHIPELAGO.

Humpata (HOOM-pah-tuh), town, HUÍLA province, SW ANGOLA, 10 mi/16 km SW of LUBANGO; elevation 6,150 ft/1,875 m. Dairying, wool clipping, skin curing, corn.

Settled 1876 by Boers from TRANSVAAL. Majority of Boer population resettled in South-West Africa between 1928–1929. Formerly called São Januaria.

Humphrey (HUHM-free), former township (□ 78 sq mi/202.8 sq km; 2001 population 1,169), S ONTARIO, E central CANADA, 14 mi/23 km from PARRY SOUND; 45°16′N 79°45′W. Amalgamated into SEGUIN township in 1998.

Humphrey, town (2000 population 806), JEFFERSON and ARKANSAS counties, central ARKANSAS, 21 mi/34 km NE of PINE BLUFF; 34°25′N 91°42′W. Bayou Meto Wildlife Management Area to S.

Humphrey, village (2006 population 788), PLATTE county, E NEBRASKA, 19 mi/31 km NNW of COLUMBUS, and on branch of ELKHORN RIVER; 41°41′N 97°29′W. Grain.

Humphrey Island, COOK ISLANDS: see MANIHIKI.

Humphreys (HUHM-freez), county (□ 431 sq mi/1,120.6 sq km; 2006 population 10,393), W MISSISSIPPI; ⊙ BELZONI; 33°07′N 90°31′W. Bounded partly W in part by SUNFLOWER RIVER, Tchula Lake Creek forms part of E boundary; drained by YAZOO RIVER (forms part of boundary in NE and SE) and tributaries. Agriculture (cotton, corn, rice, soybeans, wheat; cattle, catfish; timber). Formed 1918.

Humphreys, county (□ 555 sq mi/1,443 sq km; 2006 population 18,394), central TENNESSEE; ⊙ WAVERLY; 36°02′N 87°46′W. Bounded W by TENNESSEE RIVER; drained by DUCK and BUFFALO rivers. Includes part of KENTUCKY Reservoir. Manufacturing; agriculture; recreation. Formed 1809.

Humphreys, town (2000 population 164), SULLIVAN county, N MISSOURI, 12 mi/19 km SW of MILAN; 40°03′N 93°19′W. Cattle; corn, soybeans.

Humphreys, Fort, Virginia: see BELVOIR, FORT.

Humphreys, Mount (13,986 ft/4,263 m), E CALIFORNIA, in the SIERRA NEVADA, on FRESNO-INYO county line, 17 mi/27 km WSW of BISHOP.

Humphreys Peak (12,643 ft/3,854 m), COCONINO county, N ARIZONA, 10 mi/16 km N of FLAGSTAFF, in SAN FRANCISCO MOUNTAINS, in Coconino National Forest. On rim of eroded volcano; highest point in Arizona. Arizona Snowbowl Ski Area to SW.

Humphreyville, AUSTRALIA: see BUSHY PARK.

Humpolec (HUM-po-LETS), German *Gumpolds,* town, JIHOCESKY province, SE BOHEMIA, CZECH REPUBLIC, 11 mi/18 km SW of HAVLÍČKŮV BROD; 49°33′N 15°22′E. In potato, barley, rye, and timber region. Railroad terminus; manufacturing (woolen and linen textiles). Has 13th-century St. Nicholas church and a 13th-century synagogue, museum. Brewery (established 1551).

Humptulips River (huhmp-TOO-lips), SW WASHINGTON; formed 18 mi/29 km N of HOQUIAM by W (c.30 mi/48 km long) and E forks (c.20 mi/32 km long) both rising in Olympic National Forest; flows c.20 mi/32 km SW to GRAYS HARBOR, 8 mi/12.9 km NW of Hoquiam.

Humpty Doo (HUHM-tee DOO), town, NORTHERN TERRITORY, N central AUSTRALIA, 29 mi/47 km from DARWIN, on Arnhem Highway; 12°38′S 131°15′E. Agricultural products; tourism. In 1954 a U.S.-Australian rice company was established here, but wild animals and other problems caused the paddies' demise. Bird sanctuary. Open-air church.

Hums, SYRIA: see HOMS, district.

Humurgan, TURKEY: see SURMENE.

Humuya (hoo-MOO-yah), town, COMAYAGUA department, W central HONDURAS, on COMAYAGUA RIVER (here called Humaya River), and 14 mi/23 km SSW of COMAYAGUA; 14°15′N 87°40′W. Grain, sugarcane.

Humuya River, HONDURAS: see COMAYAGUA RIVER.

Hun (HUHN), oasis, TRIPOLITANIA region, central LIBYA, in Al JUFRAH group of oases, 150 mi/241 km SSW of SURT; 29°07′N 15°56′E. Road junction; trade center (fruit, grain; livestock, hides). It was the capital of the S military territory (Libyan SAHARA) under

Italian administration (c.1911–1943). Was capital of former Al Jufrah province. Also spelled Hon.

Húnaflói (HOO-nah-FLAW-ee), inlet of the GREEN-LAND SEA, c.60 mi/100 km long and 30 mi/50 km wide, NW ICELAND, between the VESTFJARÐA and SKA-GAFJARÐA peninsulas. It has several fishing ports.

Hunan (HU-NAN) [Mandarin=south of the lake], short name, Xiang (SIANG), province (□ 80,000 sq mi/208,000 sq km; 2000 population 63,274,173), S central CHINA, S of DONGTING LAKE; ⊙ CHANGSHA; 28°00′N 112°00′E. Largely hilly in the S and W, Hunan becomes an alluvial lowland in the Dongting basin in the NE; the XIANG RIVER, which traverses the province from N to S, and the lesser YUAN and ZI rivers, drain into Dongting Lake. The mountainous uplands include the NANLING range and HENGSHAN MOUNTAIN. Rice is the outstanding crop, particularly in the "rice bowl" of Dongting Lake; corn, sweet potatoes, barley, potatoes, buckwheat, rapeseed, fruits, and tea are also produced. Although much of the province's forested land has been cleared due to excessive cutting, many stands of cedar, pine, fir, oak, camphor, bamboo, and tung wood are found in the SW hills. Fishing and livestock-raising are important rural activities. Pulp and paper mills are found along the upper Yuan and Zi rivers. Hunan abounds in minerals such as iron ore, lead, zinc, antimony, tungsten, manganese, coal, mercury, gold, tin, and sulfur. Although agriculture is still its main industry, Hunan has a variety of heavy and light industries, such as food processing, aluminum smelting, iron, steel, and textile mills, as well as the manufacturing of machine tools along with traditional handicrafts.

Hunan's population, concentrated mainly in the Xiang and lower Yuan valleys and along the Wuhan-Guangzhou railroad, is overwhelmingly Chinese and speaks a variety of Mandarin. There are aboriginal Miao and Yao peoples in the hills of the S and W; several autonomous reserves have been established for these minorities since 1952. Under Chinese rule since the 3rd century B.C.E., the region was traditionally called Xiang for its main river. It belonged to the Kingdom of Wu at the time of the Three Kingdoms (C.E. 220–280) and later became part of the Chu Kingdom of the Five Dynasties (907–960). Its present name, first used (12th century) under the Song dynasty (960–1279), was revived in the 17th century by the Manchus when the historic province of HUGUANG was divided into the present provinces of HUBEI and Hunan. Hunan, traditionally the home of fighting men, supplied the troops that saved the Qing dynasty (1644–1911) from the Taiping rebels. Largely unoccupied by the Japanese in World War II, it passed to Communist rule in 1949. Mao Zedong was born in Hunan.

Hunchun (HUN-CHUN), city (□ 1,899 sq mi/4,937.4 sq km), E JILIN province, NE CHINA, 45 mi/72 km E of YANJI, near Russian and Korean borders; 42°55′N 130°28′E. Agriculture comprises the largest sector of the city's economy. Crop growing (mainly grain); animal husbandry. Major industries are lumber, coal mining, precious-metal mining, food processing, and utilities.

Hundested (HOO-nuhs-tedh), city (2000 population 8,384), FREDERIKSBORG county, SJÆLLAND, DENMARK, at mouth of ISE FJORD, 17 mi/27 km WNW of HIL-LERØD; 55°58′N 11°54′E. Fisheries; fish processing. There is a ferry to GRENA.

Hundred, village (2006 population 328), WETZEL county, N WEST VIRGINIA, 22 mi/35 km E of NEW MARTINSVILLE, near PENNSYLVANIA state line; 39°40′N 80°27′W. Oil, natural gas, and agricultural area. Fish Creek Covered Bridge (1881).

Hundred and Two River, IOWA and MISSOURI: see ONE HUNDRED AND TWO RIVER.

Hundwil (HOOND-ville), commune, APPENZELL Ausser Rhoden half-canton, NE SWITZERLAND, 2 mi/3.2 km SE of HERISAU. The Ausser-Rhoden Landesgemeinde (yearly outdoor assembly of male voters) meets here in odd-numbered years; even-numbered years in TROGEN.

Hunedoara (HOO-ne-DWAH-rah), county, W central ROMANIA, in TRANSYLVANIA; ⊙ DEVA; 45°45′N 23°00′E. Ranges from hilly to mountainous; MUREŞ RIVER runs through county. Mining, farming, timber, and industry.

Hunedoara (HOO-ne-DWAH-rah), Hungarian *Vajdahunyad*, German *Eisenmarkt*, city, HUNEDOARA county, W central ROMANIA, in TRANSYLVANIA, 62 mi/100 km W of SIBIU; 45°45′N 22°54′E. A major industrial center, it has extensive ironworks and steelworks. Iron ore and coal are mined nearby. The city is noted for its historic Hunyadi Castle, built in the 15th century on the site of an old citadel.

Hünenberg (HYOO-nen-buhrg), commune, ZUG canton, central SWITZERLAND, 5 mi/8 km W of ZUG; 47°10′N 08°25′E.

Hünfeld (HYOON-felt), town, HESSE, central GERMANY, 9 mi/14.5 km NNE of FULDA; 50°41′N 09°46′E. Manufacturing of cosmetics, textiles; food processing; wood- and metalworking. Chartered 1310.

Hünfelden (HYOON-fel-ten), village, HESSE, central GERMANY, 20 mi/32 km N of WIESBADEN; 50°21′N 08°06′E. Ruined castle and many half-timbered houses; chartered 1355.

Hungary, Hungarian *Magyarország*, republic (2007 population 9,956,108), central EUROPE; ⊙ BUDAPEST. Hungary borders on SLOVAKIA in the N, on UKRAINE in the NE, on ROMANIA in the E, on SLOVENIA, CROATIA, and SERBIA in the S, and on AUSTRIA in the W. The DANUBE RIVER forms the Slovak-Hungarian border from near BRATISLAVA to near ESZTERGOM, then turns sharply S and bisects the country. To the E of the Danube River the Great Hungarian Plain (Hungarian *Alföld*) extends beyond the Hungarian boundaries to the CARPATHIANS and the BIHOR Mountains. To the W of the Danube River is TRANSDANUBIA, which includes the LITTLE ALFÖLD. The BAKONY and VÉRTES mountains run through the middle of Transdanubia in a SW-NE direction. The MÁTRA Mountains in the N reach an elevation of 3,330 ft/1,015 m at KÉKES, the highest peak in Hungary. Lake BALATON, the largest lake in Hungary and in central Europe, is a leading resort area. Hungary has cold winters and hot summers; springs and falls are short. It has long been an agricultural country, but since World War II has become heavily industrialized. Nearly 55% of Hungary is arable; agriculture occupies about 18% of the employable population and accounts for around 15% of the gross national product. In the late 1980s, when a privatization movement began, about two-thirds of all agricultural products came from collectivized farms. With highly diversified crop and livestock production, Hungary is self-sufficient in food and in 1989 made 25% of its export earnings from agriculture. Corn, wheat, and barley are the major crops, followed by sugar beets, potatoes, and grapes. Pigs, cattle, and sheep are raised. Hungary has been an important producer of bauxite, and deposits of natural gas, coal, oil, and uranium have been exploited as well. All these minerals, however, are poor in quality and/or geologically complex. Apart from the modest reserves of natural gas and oil, it is not cost-effective to exploit them at world prices. As Hungary moved to a market economy during the 1990s, their mining had been drastically curtailed or completely stopped. The gradual decline of gas and oil production is due to the exhaustion of reserves. Industry, which is well-diversified, provides over 40% of the gross national product and employs about 30% of the work force. Many of Hungary's industries (engineering, food processing, textiles, clothing), however, produced goods chiefly for export to the former USSR;

the Soviet successor states no longer take them. Energy and metallurgical branches produced basic materials for these ailing processing industries. Hungary's economy in the 1990s was undergoing difficult readjustment because exporting to the West was much more difficult. Average unemployment in 1993–1995 rose to c.12%. In some counties, it has reached 20%. About one-third of Hungarian industry is located in or near Budapest. Other industrial centers are GYÖR, MISKOLC, PÉCS, DEBRECEN, SZEGED, and other county centers. Industry was largely nationally owned, and two-thirds of agricultural output also came from collective and state farms. A privatization process began in the early 1990s. By the end of 1995, almost all retail trade had been privatized and less than half of economic output as a whole originated from state-owned entrprises. At the end of the 1980's, major products included machinery and transport equipment, chemicals, pharmaceuticals, clothing, footwear, and processed food. The bulk of output was shipped to the USSR (primarily RUSSIA and Ukraine) from where Hungary imported half of the energy it consumed. Russia is still the supplier of all the imported gas and most of the imported oil, but GERMANY is now Hungary's largest trading partner overall. Foriegn investment is on the rise, especially in alcohol, food, and hotel industries and, since 1995, in electric power, natural gas, and refinery product distribution. Tourism is an important source of foreign capital, although the number of tourists has declined in the 1990s. Most economic activity takes place in the W section of the country; the E section and the region bordering on the former YUGOSLAVIA being much less developed. Situated on a plain near the geographic center of Europe, Hungary has been the meeting place and battleground of many peoples, and its heterogeneous population within the pre-World War I boundaries created significant ethnic tension. However, as a result of the separation of non-Hungarian territories after World War I, the great slaughter of the Jews in World War II, and the exchange after the war of Slavic and Romanian minorities for their Magyar counterparts, Hungary is essentially homogeneous. The Magyars constitute about 92% of the population. Hungary, however, still has the largest Jewish population in central and Eastern Europe (c.80,000–100,000). Hungarian is the official language. About two-thirds of the people are Roman Catholic, but there is a large Calvinist minority. Lutherans and Greek Catholics are also significant. The Roman provinces of PANNONIA and DACIA, conquered under Tiberius and Trajan (1st century A.C.E.), embraced part of what was to become Hungary. The Huns and later the Ostrogoths and the Avars settled there for brief periods. In the late 9th century the Magyars, a Finno-Ugric people from beyond the URALS, conquered all or most of Hungary and TRANSYLVANIA. The semilegendary leader, Árpád, founded their first dynasty. The Magyars apparently merged with the earlier mostly Slavic settlers, but they also continued to press W until defeated by Otto I of the Holy Roman Empire at the LECHFELD (955). Halted in its expansion, the Hungarian state began to solidify. Its first king, St. Stephen (reigned 1001–1038), completed the Christianization of the Magyars and built the authority of his crown—which for 945 years remained the symbol of national existence—on the strength of the Roman Catholic Church. Under Béla III (reigned 1172–1196), Hungary came into close contact with Western Europe, particularly French, culture. Through the favor of succeeding kings, a few very powerful nobles—the magnates—won ever-widening privileges at the expense of the lesser nobles, the peasants, and the towns. In 1222 the lesser nobles forced the extravagant Andrew II to grant the Golden Bull (the "Magna Carta of Hungary"), which limited the king's power to alienate the authority of the nobility, and established the beginnings of a parliament. Under Andrew's son, Béla IV, the kingdom barely escaped annihilation: Mongol invaders, crushing Béla at Muhi (1241), occupied the country for a year, and Ottocar II of BOHEMIA also defeated Béla, who was further threatened by his own rebellious son, Stephen V. Under Stephen's son, Ladislaus IV, Hungary fell into anarchy, and when the royal line of Árpád died out (1301) with Andrew III, the magnates seized the opportunity to increase their authority. In 1308, Charles Robert of Anjou was elected king of Hungary as Charles I, the first of the Angevin line. Through battle and autocratic rule, he checked the magnates somewhat and furthered the growth of the towns. Under his son, Louis I (Louis the Great), Hungary reached its greatest territorial extension, with power extending into Dalmatia, the BALKANS, and, through the joint kingship of Louis, into POLAND. After the death of Louis I, a series of foreign rulers succeeded. During their reigns the Turks began to advance through the Balkans, destroying the Serb state at KOSOVO (1389), defeating the Hungarians and their allies at NIKOPOL (1396) and VARNA (1444), where Uladislaus I fell in battle. The era of Matthias Corvinus, elected king in 1458, was a glorious period in Hungarian history. Matthias maintained a splendid court at Buda, kept the magnates subject to royal authority, and improved the central administration. Suleyman the Magnificent's defeat of Hungarian forces marked the beginning of Ottoman domination over Hungary, lasting in areas until 1686. After a long period of religious, social, and political strife, Hungarian nobles recognized the Austrian (Hapsburg) claim to the Hungarian throne. By the Peace of Karlowitz (1699), TURKEY ceded to Austria most of Hungary proper and Transylvania. Transylvania and most of the rest of Hungary continued to fight the Hapsburgs, but in 1711, with the defeat of Francis II Rákóczy, Austrian control was definitely established. In 1718 the Hapsburgs took the BANAT from Turkey. The Hapsburgs brought in Germans and Slavs to settle the newly freed territory, greatly reducing Hungary's ethnic homogeneity. In the second quarter of the 19th century a movement that combined Hungarian nationalism with constitutional liberalism gained strength. Inspired by the French Revolution of 1848, the Hungarian diet passed the March Laws (1848), which established a liberal constitutional monarchy for Hungary under the Hapsburgs. But the reforms did not deal with the national minorities problem. Several minority groups revolted, and, after Francis Joseph replaced Ferdinand VII as emperor, the Austrians waged war against Hungary (December 1848). The *Ausgleich* (compromise) of 1867 set up the Austro-Hungarian Monarchy, in which Austria and Hungary were nearly equal partners. Emperor Francis Joseph was crowned (1867) king of Hungary, which at that time also included Transylvania, Slovakia, RUTHENIA, Croatia and Slovenia, and the Bánát; minority problem persisted. During this period industrialization began in Hungary, while the condition of the peasantry deteriorated to the profit of landowners. Until World War I, when republican and socialist agitation began to threaten the established order, Hungary was one of the most aristocratic countries in Europe. As the military position of Austria-Hungary in World War I deteriorated, the situation in Hungary grew more unstable. Hungarian nationalists wanted independence and withdrawal from the war; the political left was inspired by the 1917 revolutions in Russia; and the minorities were receptive to the Allies' promises of self-determination. In November of 1918 the emperor abdicated, and the Dual Monarchy collapsed: Hungary was proclaimed an independent republic. The Treaty of Trianon, signed in 1920, reduced the size and population of

Hungary by about two-thirds, depriving Hungary of valuable natural resources and removing virtually all non-Magyar areas, but leaving more than 2.5 million Hungarians beyond the new borders. Budapest and Transdanubia retained a large German-speaking population. The next twenty-five years saw continual attempts by the Magyar government to recover the lost territories. During World War II, Hungary regained territories from CZECHOSLOVAKIA, the former Yugoslavia, and Romania with the help of the Axis powers. It declared war on the USSR (June 1941) and on the U.S. (December 1941). When the Hungarian government took steps to withdraw from the war, German troops occupied the country (March 1944). Most Jews were sent to concentration camps. The Germans were driven out by Soviet forces (October 1944–April 1945), but the campaign caused much devastation. More than half a million Hungarians, including much of its Jewish population, lost their lives in WWII, over 5% of the population. The peace treaty signed at PARIS in 1947 restored the Trianon boundaries and required Hungary to pay $300 million in reparations to the USSR, Czechoslovakia, and Yugoslavia. Early in 1948 the Communist party, through its control of the ministry of the interior, arrested leading politicians, forced the resignation of Premier Ferenc Nagy, and gained full control of the state. Hungary was proclaimed a People's Republic in 1949, after parliamentary elections in which there was only a single slate of candidates. Radical purges in the national Communist Party made it thoroughly subservient to that of the USSR. Industry was nationalized, and collectivization of land was ruthlessly pressed. By 1953 continuous purges of Communist leaders, constant economic difficulties, and peasant resentment of collectivization had led to profound crisis in Hungary. In 1955, Hungary joined the WARSAW Treaty Organization and was admitted to the UN. On Oct. 23, 1956, a popular anti-Communist revolution, centered in Budapest, broke out in Hungary. A new coalition government under Imre Nagy declared Hungary neutral, withdrew it from the Warsaw Treaty, and appealed to the UN for aid. In severe and brutal fighting Soviet forces suppressed the revolution, and some 190,000 refugees fled the country. In 1968 economic reforms, known as the New Economic Mechanism (NEM), were introduced to bring a measure of decentralization to the economy and to allow for supply and demand factors. Consumer goods production rose sharply, and Hungary achieved substantial improvements in its standard of living. Due to Soviet criticism, many of the NEM reforms were subverted during the mid-1970s only to be reinstituted at the end of the decade after the inefficiency of the centralized Soviet-style economy became all too apparent. During the 1980s, Hungary began to increasingly turn to the West for trade and assistance in the modernization of its economic system. The economy continued to decline and the high foreign debt became unpayable. Prime minister Károly Grósz was ousted in 1988, and in 1989 the Communist party congress voted to dissolve itself. That same year Hungary opened its borders with Austria, allowing thousands of East Germans to cross to the West. By 1990, a multi-party political system with free elections had been established; legislation was passed granting new political and economic reforms such as a free press, freedom of assembly, and the right to own a private business. The new prime minister, Jozsef Antall, elected in 1990, continued the drive toward a free market economy. The Soviet military occupation of Hungary ended in the summer of 1991 with the departure of the final Soviet troops. The president of Hungary as well as the presidential council are elected by the 368-member legislative body known as the national assembly, whose members are elected directly for four-year terms. The leading po-

litical parties are the Hungarian Democratic Forum, the Alliance of Free Democrats, the Young Democrats, the Independent Smallholder's party, and the Socialist Party. Antall died in early 1994, and Gyula Horn of the socialist party became Prime Minister and formed a coalition government with the Free Democrats to broaden his support. After a hesitant start, and under pressure from the International Monetary Fund and other foreign financial institutions, the government undertook severe measures to rein in the soaring budget deficit (9% of GDP) and reduce the current account deficit, which reached $4 billion in 1994. The overall economic difficulties and widening income differentials are compounded by growing geographical differentiation; the Budapest region and NW Transdanubia are much more prosperous than the rest of the country, and have been far more successful in restructuring their economic base.

Hungen (HUNG-uhn), town, HESSE, central GERMANY, 12 mi/19 km SE of GIESSEN; 50°28′N 08°54′E. Dairying; milk processing; metalworking; manufacturing of construction materials. Has 15th–16th-century castle.

Hungerburg, ESTONIA: see NARVA-JÕESUU.

Hungerford (HUHN-guhr-fuhrd), former township (☐ 153 sq mi/397.8 sq km; 2001 population 3,365), SE ONTARIO, E central CANADA, 21 mi/34 km from BELLEVILLE; 44°32′N 77°11′W. Amalgamated into TWEED township in 1998.

Hungerford (HUHNG-guh-fuhd), town (2001 population 5,559), West Berkshire, S central ENGLAND, on KENNET RIVER and 9 mi/14.5 km W of NEWBURY; 51°25′N 01°31′W. Agricultural market. Has 14th-century church.

Hungnam (HUNG-NAHM), city, SOUTH HAMGYONG province, NORTH KOREA, at mouth of Tongsongchon R, SE of HAMHUNG; 39°49′N 127°37′E. Industrial center (gold refinery; aluminum, chemical, and nitrogen-fertilizer plants); fisheries. Coal mined nearby. Small fishing village until late 1920s. Heavily bombed in Korean War (1950).

Hungry Hill, mountain (2,251 ft/686 m), SW CORK county, SW IRELAND, 6 mi/9.7 km ENE of CASTLETOWN BERE; highest point of CAHA MOUNTAINS, with a waterfall on its slopes; 51°41′N 09°47′W.

Hungry Horse, town, FLATHEAD county, NW MONTANA, on FLATHEAD River at mouth of South Fork Flathead River. Tourism, manufacturing (jams and jellies). HUNGRY HORSE Dam and Reservoir to SE (on South Fork), and 17 mi/27 km NE of KALISPELL.

Hungry Horse Reservoir, FLATHEAD county, NW MONTANA, on S Fork FLATHEAD River, in Flathead National Forest, 20 mi/32 km NE of KALISPELL; c.35 mi/56 km long; 48°21′N 114°01′W. Formed by Hungry Horse Dam (564 ft/172 m high, 2,115 ft/645 m long), built (1948–1953) as unit in Columbia River Basin project for power generation, flood control, and irrigation.

Hungtse, China: see HONGZE.

Hung-tse, CHINA: see HONGZE.

Hungund (HUHN-guhnd), town, BIJAPUR district, KARNATAKA state, INDIA, 60 mi/97 km SSE of BIJAPUR; 16°04′N 76°03′E. Cotton ginning; cotton, peanuts, millet, wheat.

Hung Yen (HOONG YEN), city, ⊙ HAI HUNG province, N VIETNAM, near SONG COI (RED RIVER), 30 mi/48 km SE of HANOI; 20°39′N 106°04′E. Timber center; market and administrative hub; rice growing; woodcarving; service facilities. Important European trading center in 17th century. Declined following changes in river course. Formerly Hungyen.

Hung Yen, VIETNAM: see HAI HUNG.

Huningue (oo-nan-guh), German *Hüningen*, industrial town (☐ 1 sq mi/2.6 sq km), HAUT-RHIN department, ALSACE, E FRANCE river port on left bank of the RHINE (at junction of Huningue branch of RHÔNE-RHINE CANAL), 3 mi/4.8 km N of BASEL, Switzerland; 47°36′N 07°35′E. Manufacturing (plastics, chemicals). Nearby is a fish hatchery. Purchased and fortified by Louis XIV; besieged by Austrians in 1815, and captured by Germans in 1871. The Basel-MULHOUSE airport is just W.

Huni Valley (HOO-nee), town, WESTERN REGION, GHANA, 13 mi/21 km NNE of TARKWA; 05°28′N 01°55′W. Railroad junction in gold-mining region; cacao, cassava, corn; copper, platinum, nickel deposits.

Hunjiang (HUN-JIANG), city (1991 estimated urban population 489,800; total population 727,500; in an area that includes surrounding counties, urban population 750,900; total population 1,250,300), S JILIN province, NE CHINA; 41°54′N 126°23′E. Heavy industry is the largest sector of the city's economy. Crop growing, commercial agriculture, animal husbandry. Agriculture (grains, vegetables, hogs, eggs); manufacturing (logging, food, textiles, utilities, plastics, and non-ferrous metal processing, coal mining).

Hunker, borough (2006 population 320), WESTMORE-LAND county, SW central PENNSYLVANIA, 8 mi/12.9 km SSW of GREENSBURG, on Sewickley Creek. Agriculture (apples, corn, dairying).

Hunnebostrand (HUN-ne-BOO-STRAHND), village, GÖTEBORG OCH BOHUS county, SW SWEDEN, on SKAGERRAK, 12 mi/19 km NNW of LYSEKIL; 58°27′N 11°18′E. Seaside resort. A fourth-century burial mound nearby.

Hunnewell (HUHN-nee-wel), city (2000 population 227), SHELBY county, NE MISSOURI, near SALT RIVER, 7 mi/11.3 km W of MONROE CITY; 39°40′N 91°51′W. Soybeans, corn; cattle, hogs. Hunnewell Lake to NW.

Hunnewell (HUHN-ee-wel), village (2000 population 83), SUMNER county, S KANSAS, at OKLAHOMA state line, 16 mi/26 km S of WELLINGTON; 37°00′N 97°24′W. In wheat area.

Hunnur (HUHN-nuhr), town, BIJAPUR district, KAR-NATAKA state, INDIA, 2 mi/3.2 km W of JAMKHANDI. Cotton, millet, peanuts.

Hünsdorf (HYOONS-dorf), town, LORENTZWEILER commune, S central LUXEMBOURG, on ALZETTE RIVER, and 6 mi/9.7 km N of LUXEMBOURG city; 49°42′N 06°08′E. Manufacturing; metal casting and stamping.

Hunsfoss (HOONS-faws) or **Hundsfoss**, waterfall (46 ft/14 m) on OTRA RIVER, VEST-AGDER county, S NORWAY, 8 mi/12.9 km N of KRISTIANSAND. Hydroelectric plant powers pulp and paper industry.

Hunslet, ENGLAND: see LEEDS.

Hunsrück (HUNS-rooik), mountain region, W GER-MANY, between RHINE (E), Mosel (N), SAAR (W), and NAHE (S) rivers; 49°40′N 07°00′E–50°10′N 07°50′E. Rises to 2,677 ft/816 m in the ERBESKOPF. Densely forested; wine grown on N and S rim. Main town is SIMMERN. Geologically it is considered part of RHENISH SLATE MOUNTAINS. Mountain range includes SCHANZER KOPF (SHAHN-tser KOHPF) (2,120 ft/646 m).

Hunstanton (huhn-STAN-tuhn), resort (2001 population 4,961), NORFOLK, E ENGLAND, on The WASH, 14 mi/23 km NE of KING'S LYNN; 52°57′N 00°30′E. Continues into NEW HUNSTANTON.

Hunsur (HUHN-soor), town, MYSORE district, KAR-NATAKA state, INDIA, on LAKSHMANTIRTHA RIVER and 25 mi/40 km W of MYSORE; 12°18′N 76°17′E. Road center; depot for timber from mountains of former COORG state, now KODAGU district (W); processing of coffee, tobacco, spices (pepper, cardamom). Livestock farm breeds well-known Mysore bullocks, sheep.

Hunt, county (□ 882 sq mi/2,293.2 sq km; 2006 population 83,338), NE TEXAS; ⊙ GREENVILLE; 33°07′N 96°05′W. Rich blackland prairie in W and NW; timbered in E; drained by SABINE RIVER (forms LAKE TAWAKONI in SE corner) and S Fork of SULPHUR RIVER. Agriculture (especially cotton; hay, nursery crops; wheat); dairying; cattle, horses. Some oil and gas production; sand and rock. Manufacturing, pro-

cessing at Greenville and COMMERCE. Part of Lake Tawakoni State Park on S boundary. Formed 1846.

Hunt, unincorporated village, KERR county, SW TEXAS, 14 mi/23 km WNW of KERRVILLE, on GUADALUPE RIVER (source to W); 30°04′N 99°20′W. Ranching area (cattle, sheep, goats); wheat, pecans.

Hunter, unincorporated community, CARTER county, S MISSOURI, in the OZARKS, near CURRENT RIVER, c.12 mi/19 km SE of VAN BUREN. OZARK NATIONAL SCENIC RIVERWAYS canoe access to W.

Hunter, village (2000 population 152), WOODRUFF county, E central ARKANSAS, 20 mi/32 km W of FORREST CITY; 35°02′N 91°07′W.

Hunter, village (2000 population 77), MITCHELL county, N central KANSAS, 18 mi/29 km SW of BELOIT; 39°14′N 98°24′W. Home of Veras. Grain, livestock. Oil.

Hunter, resort village (□ 1 sq mi/2.6 sq km; 2006 population 491), GREENE county, SE NEW YORK, in the CATSKILL MOUNTAINS, on SCHOHARIE CREEK and 17 mi/27 km W of CATSKILL; 42°12′N 74°13′W. At HUNTER MOUNTAIN.

Hunter, village (2006 population 170), GARFIELD county, N OKLAHOMA, 16 mi/26 km NE of ENID; 36°33′N 97°39′W. In grain and livestock area.

Hunter Army Airfield, GEORGIA, in SAVANNAH, CHATHAM county. Home of National Guard air unit, adjacent to the Savannah airport.

Hunter Army Airfield, GEORGIA: see FORT STEWART MILITARY RESERVATION.

Hunterdon, county (□ 437 sq mi/1,136.2 sq km; 2006 population 130,783), W NEW JERSEY, bounded W by DELAWARE River; ⊙ FLEMINGTON; 40°34′N 74°55′W. Agricultural area that is rapidly suburbanizing as an outer suburb; manufacturing. Includes Voorhees State Park. Drained by the MUSCONETCONG River and by South Branch of RARITAN River. MUSCONETCONG MOUNTAIN is in W. County formed 1714.

Hunter Island (HUHN-tuhr) (□ 129 sq mi/335.4 sq km), SW BRITISH COLUMBIA, W CANADA, in NE part of QUEEN CHARLOTTE SOUND, 6 mi/10 km S of BELLA BELLA; 21 mi/34 km long, 3 mi/5 km–10 mi/16 km wide; 51°55′N 128°05′W. Rises to 2,950 ft/899 m.

Hunter Island, island, SW PACIFIC OCEAN, S of ANATOM island, VANUATU. Uninhabited outcrop claimed by Vanuatu and by FRANCE on behalf of NEW CALEDONIA.

Hunter Island (HUHN-tuhr), tract of land in NW ONTARIO, E central CANADA, extending N from U.S. (MINNESOTA) border, 100 mi/161 km W of FORT WILLIAM; 50 mi/80 km long, 30 mi/48 km wide; 48°16′N 91°29′W. Region of small lakes and streams, flowing into Rainy River.

Hunter Island, MARSHALL ISLANDS: see KILI.

Hunter Islands (HUHN-tuhr), island group in BASS STRAIT, 3 mi/5 km off NW coast of TASMANIA, AUSTRALIA; 40°35′S 144°52′E. Comprise Hunter Island, Three Hummock Island, and several small islets; mountainous. Hunter Island (□ 33 sq mi/85 sq km; 14 mi/23 km long, 4 mi/6 km wide) is largest; formerly called Barren Island.

Hunte River (HUN-te), c.120 mi/193 km long, NW GERMANY; rises in the WIEHEN MOUNTAINS 4 mi/6.4 km N of MELLE; flows generally N, past WILD-ESHAUSEN and OLDENBURG, to the WESER at ELSFLETH. Canalized lower course forms part of EMS-HUNTE CANAL, which connects EMS and Weser rivers. Source at 52°15′N 08°20′E.

Hunter Liggett Military Reservation, California: see JOLON.

Hunter, Mount (14,573 ft/4,442 m), S central ALASKA, in ALASKA RANGE, in MOUNT MCKINLEY NATIONAL PARK, 130 mi/209 km NNW of ANCHORAGE; 62°57′N 151°05′W.

Hunter Mountain (4,025 ft/1,227 m), GREENE county, SE NEW YORK, in CATSKILL MOUNTAINS, 19 mi/31 km W of CATSKILL; 42°10′N 74°14′W. Skiing resort.

Hunter, Port (HUHN-tuhr) or **Newcastle Harbour** (NOO-ka-suhl), estuary of the HUNTER RIVER, NEW SOUTH WALES, AUSTRALIA; 3 mi/4.8 km long and 2 mi/3.2 km wide; 32°56′S 151°47′E. The coal-loading port of Newcastle, one of the largest ports in the state, is on the S shore near the entrance. Also spelled Newcastel Harbour.

Hunter River, village (2001 population 351), central PRINCE EDWARD Island, CANADA, on Hunter River and 15 mi/24 km WNW of CHARLOTTETOWN; 46°21′N 63°21′W. Mixed farming, dairying; potatoes.

Hunter River (HUHN-tuhr), 287 mi/462 km long, E NEW SOUTH WALES, AUSTRALIA; rises in LIVERPOOL RANGE; flows SW, past MUSWELLBROOK and DENMAN, and SE, past SINGLETON, MAITLAND, MORPETH, and RAYMOND TERRACE, to PACIFIC OCEAN at Port HUNTER (site of NEWCASTLE); 32°50′S 151°42′E. Navigable by steamer for 23 mi/37 km below Morpeth. Coal mines in river valley; frequent floods. GOULBURN RIVER, main tributary.

Hunter's Creek Village, town, HARRIS county, SE TEXAS, residential suburb 8 mi/12.9 km W of downtown HOUSTON, bounded S by BUFFALO BAYOU. Surrounded by city of Houston.

Hunters Hill (HUHN-tuhrz HIL), suburb NW of SYDNEY, E NEW SOUTH WALES, AUSTRALIA, on N shore of Parramatta River; 33°50′S 151°09′E. In metropolitan area; shipyards. Linked by bridge with DRUMMOYNE, on S shore of river.

Hunter's Island, New York: see PELHAM BAY PARK.

Hunter's Quay (huhn-TUHRZ QUAI), resort, ARGYLL AND BUTE, W Scotland, on Firth of Clyde, and at landing stage at entrance to Holy Loch, 2 mi/3.2 km N of DUNOON; 55°58′N 04°54′W. Formerly in STRATH-CLYDE, abolished 1996.

Hunter's Road, town, MIDLANDS province, central ZIMBABWE, 20 mi/32 km N of GWERU, near KWEKWE River, on railroad; 19°09′S 29°48′E. Elevation 4,218 ft/1,286 m. Gold mining. Tobacco, corn, soybeans, citrus fruit; cattle, sheep, goats.

Huntersville (HUHN-tuhrz-vil), town (□ 31 sq mi/80.6 sq km; 2006 population 38,796), MECKLENBURG county, S NORTH CAROLINA, 12 mi/19 km N of CHARLOTTE; 35°25′N 80°50′W. Service industries; manufacturing (materials handling, surgical, and electronic equipment; robotic machinery, hardwood products); agriculture (grain, soybeans; livestock). Energy Explorium, hands-on museum at dam. Cowans Ford Dam, on CATAWBA RIVER, to W, forms Lake NORMAN reservoir to NW, Mountain Island Lake reservoir (Catawba River) to SW. Incorporated 1873.

Huntertown, town (2006 population 2,144), ALLEN county, NE INDIANA, 9 mi/14.5 km N of FORT WAYNE, near source of EEL RIVER; 41°14′N 85°10′W. Manufacturing (machinery, asphalt). Corn, soybeans. Settled 1830s.

Hunter Valley (HUHN-tuhr), region of NEW SOUTH WALES, SE AUSTRALIA. The HUNTER RIVER and its tributaries occupy this valley S of the Mount Royal Range. The land in the upper valley is used for livestock grazing, dairying and agriculture. The lower valley is an important center for coal mining and wineries.

Huntingburg, city (2000 population 5,598), DUBOIS county, SW INDIANA, 6 mi/9.7 km S of JASPER; 38°18′N 86°58′W. Agricultural area (grain, poultry, strawberries; livestock; dairy products); manufacturing (carbide cutting tools, furniture and wood products, packed meat, turkey products); clay and limestone. Railroad junction. Founded 1839.

Huntingdale (HUHN-teeng-dail), residential and industrial suburb 11 mi/17 km SE of MELBOURNE, VICTORIA, SE AUSTRALIA, between OAKLEIGH and CLAYTON. Light industry. Railway station.

Huntingdon (HUHN-teeng-duhn), former county (□ 361 sq mi/938.6 sq km), S QUEBEC, E CANADA, on ONTARIO and U.S. (NEW YORK) borders, on the SAINT

LAWRENCE RIVER; 45°02′N 74°05′W. Its county seat was HUNTINGDON.

Huntingdon, county (□ 890 sq mi/2,314 sq km; 2006 population 45,771), S central PENNSYLVANIA; ⊙ HUNTINGDON. Hilly region, drained by JUNIATA RIVER and Raystown Branch of Juniata River (forms large RAYSTOWN LAKE reservoir in SW). Crossed by several ridges (NNE-SSW angles): BALD EAGLE MOUNTAIN (NW boundary); TUSSEY MOUNTAIN (SW boundary); JACKS MOUNTAIN (partly on NE boundary); BLACKLOG MOUNTAIN, SHADE MOUNTAIN, TUSCARORA MOUNTAIN (SE boundary). Clay, bituminous coal, glass sand; manufacturing at HUNTINGDON and MOUNT UNION. Agriculture (corn, wheat, oats, hay, alfalfa; poultry, hogs, cattle, dairying). Parts of Rothrock State Forest in NE, N, and center; Lincoln Caverns in W; Trough Creek State Park in SW; Greenwood Furnace and Whipple Dam State Parks in NE. Formed 1787.

Huntingdon (HUHN-teeng-duhn), former township (□ 85 sq mi/221 sq km; 2001 population 2,851), SE ONTARIO, E central CANADA, 12 mi/19 km from BELLEVILLE; 44°24′N 77°28′W. Amalgamated into CENTRE HASTINGS.

Huntingdon (HUHN-teeng-duhn), town, ⊙ Le Haut-Saint-LAURENT county, Montérégie region, S QUEBEC, E CANADA, on Châteauguay River and 40 mi/64 km SW of MONTREAL, near U.S. (NEW YORK) border; 45°05′N 74°11′W. Milling, lumbering, dairying.

Huntingdon (huhn-TING-duhn), town (2001 population 20,600), CAMBRIDGESHIRE, E central ENGLAND, on OUSE RIVER and 60 mi/97 km N of London, on the ancient ERMINE STREET; 52°20′N 00°11′W. Some industry. Oliver Cromwell, born here, attended the grammar school where Samuel Pepys later studied. Just W is large estate of HINCHINBROOKE, bought from the Cromwells by Earl of Sandwich.

Huntingdon, town (2006 population 4,186), ⊙ CARROLL county, NW TENNESSEE, 34 mi/55 km NE of JACKSON; 36°00′N 88°24′W. In agricultural area; manufacturing. Settled and incorporated 1821.

Huntingdon (HUHN-teeng-duhn), unincorporated village, SW BRITISH COLUMBIA, W CANADA, on U.S. (WASHINGTON) border, and included in ABBOTSFORD, and 10 mi/16 km S of MISSION; 49°00′N 122°16′W. Dairying; cattle, fruit, hops.

Huntingdon, borough (2006 population 6,827), ⊙ HUNTINGDON county, S central PENNSYLVANIA, 21 mi/34 km E of ALTOONA, on JUNIATA RIVER. Manufacturing (fabricated metal products, electronic connectors, printing and publishing, wood products, vinyl wallets, glass fiber yarns, canvas footwear); glass sand; timber. Juniata College; Lincoln Caverns are W; Trough Creek State Park to S; large Raystown Reservoir to S, on Raystown Branch of Juniata River; parts of Rothrock State Forest to N and S, including Pennsylvania State University Experimental Forest to NE; Whipple Dam and Greenwood Furnace state parks to NE. Settled c.1755, laid out 1767, incorporated 1796.

Huntingdon Valley, unincorporated village, MONTGOMERY county, SE PENNSYLVANIA, industrial suburb 12 mi/19 km NNE of PHILADELPHIA; 40°07′N 75°03′W. Manufacturing includes industrial equipment, machinery, chemicals, tool and die, apparel, commercial printing.

Hunting Island, BEAUFORT county, S SOUTH CAROLINA, one of SEA ISLANDS, just E of SAINT HELENA ISLAND, to which it is connected by highway bridge; c.5 mi/8 km long. Huntington Island State Park (c.5,000 acres/2,024 ha), resort colony, and wildlife sanctuary here.

Huntington, county (□ 387 sq mi/1,006.2 sq km; 2006 population 38,026), NE central INDIANA; ⊙ HUNTINGTON. Agricultural area (corn, soybeans, vegetables; poultry, hogs, cattle; dairy products). Diversified manufacturing at Huntington. Limestone quarrying; timber. Drained by WABASH, SALAMONIE, and LITTLE WABASH rivers, and by small Clear Creek. HUNTINGTON Reservoir and Little Turtle State Recreation Area SE of Huntington; SALAMONIE Reservoir, with Lost Bridge State Recreation Area in SE. Formed 1832.

Huntington, city (2000 population 17,450), ⊙ HUNTINGTON county, NE INDIANA; 40°53′N 85°31′W. It is a farm trade center and an industrial city. Manufacturing (automotive parts, machinery, construction material, food and beverages, ceaning agents, fireplaces, electrical equipment, rubber, plastic; printing, packaging). The city is the seat of Huntington College. The WABASH RIVER and its forks were a Native American gathering place and early trade center. Settled 1831, incorporated 1848.

Huntington, city (2006 population 2,061), EMERY county, central UTAH, 20 mi/32 km SSW of PRICE, on Huntington Creek; 39°19′N 110°57′W. Elevation 5,791 ft/1,765 m. Alfalfa; cattle; coal mining. Power plant. Huntington Lake (not on creek) and State Park to NE. Manti–La Sal National Forest to W; Electric Lake to NW. Settled 1878, incorporated 1891.

Huntington, city (2006 population 49,007), ⊙ CABELL county, in Cabell and WAYNE counties, W WEST VIRGINIA, 45 mi/72 km W of CHARLESTON, 10 mi/16 km ESE of ASHLAND, KENTUCKY, on the OHIO River (bridged to OHIO), at mouth of GUYANDOTTE RIVER; 38°24′N 82°25′W. Elevation c.550 ft/168 m. The second-largest city in state (pop. over 70,000 in 1970, city has lost population at a faster rate than Charleston, falling to second place). Commercial (wholesale and market) and manufacturing center for deep and surface bituminous coal; oil; natural gas. Farm (especially tobacco, fruit) region. Important railroad city (huge repair shops); river port (ships coal). Tri-State Airport (Walker Long Field) to W. Industries include reduction and fabrication of nickel alloys; manufacturing (transportation equipment, glass, furniture, wood products, chemicals (dyes), plastics, lubricating oils, electrical goods, metal goods, industrial machine parts, flour and other foods, beverages, packaging, printing and publishing). Marshall University. State institutions here include industrial school for girls, hospital for the insane, children's home. Railroad museum; Antique Radio Museum; Huntington Museum of Art; Camden Park, turn-of-century amusement park with wooden roller coaster, to W. Along river here is 11 mi/18 km flood wall. Wayne National Forest (Ohio) to N; Beech Fork State Park and Beech Fork Lake Wildlife Management Area to S. City founded 1871 as W terminus of Chesapeake & Ohio railroad.

Huntington, town (□ 137 sq mi/356.2 sq km), SUFFOLK county, SE NEW YORK, on N shore of LONG ISLAND; 40°53′N 73°22′W. Chiefly residential heart of township containing eighteen contiguous communities. Numerous harbors and boatyards; major retailing center. Seminary of the Immaculate Conception. Settled 1653.

Huntington, town (2006 population 2,099), ANGELINA county, E TEXAS, 10 mi/16 km ESE of LUFKIN, near ANGELINA RIVER; 31°16′N 94°34′W. In pine timber; poultry, cattle area; recreation. Incorporated 1938.

Huntington, town, CHITTENDEN CO., NW VERMONT, on small Huntington River and 17 mi/27 km SE of BURLINGTON; woodworking; dairy products; 44°17′N 72°57′W. Includes part of Camels Hump State Forest and home to CAMELS HUMP, the third tallest peak in the state. Cross-country ski centers. Originally named New Huntington; the new was dropped in 1795. Named for Josiah, Charles, and Marmaduke Hunt.

Huntington (HUHNT-eeng-tuhn), unincorporated town, FAIRFAX county, NE VIRGINIA, residential suburb, 1 mi/2 km SW of ALEXANDRIA, 7 mi/11 km S of WASHINGTON, D.C., on Cameron Run creek at its entrance to POTOMAC RIVER; 38°47′N 77°04′W. Woodrow Wilson Memorial Bridge to E. S terminus of Washington Metro (subway) yellow line.

Huntington, village (2000 population 688), SEBASTIAN county, W ARKANSAS, 22 mi/35 km SSE of FORT SMITH; 35°04′N 94°16′W. In agricultural area; manufacturing (concrete septic tanks).

Huntington, village (2006 population 479), BAKER county, E OREGON, 25 mi/40 km N of VALE, on BURNT RIVER, near its confluence with the SNAKE RIVER to E; 44°21′N 117°16′W. Elevation 2,113 ft/644 m. Agriculture (wheat, barley, oats, potatoes; sheep, cattle); dairy products. Farewell Bend State Park to SE.

Huntington Bay, residential village, SUFFOLK county (□ 2 sq mi/5.2 sq km; 2006 population 1,475), SE NEW YORK, on N shore of LONG ISLAND, on inlet of HUNTINGTON BAY, just NE of HUNTINGTON; 40°54′N 73°25′W. Site of capture of Nathan Hale by British forces during American Revolution. By early 20th century it had developed into a resort area with casino and summer estates. Some large properties broken into smaller residences following World War II. Incorporated 1924.

Huntington Bay, an arm of LONG ISLAND SOUND, SE NEW YORK, indenting N shore of LONG ISLAND N of HUNTINGTON, which is at head of Huntington Harbor (c.2 mi/3.2 km long), bay's S arm; 40°55′N 73°25′W. It is c.4.5 mi/7.2 km wide at entrance between LLOYD POINT (W) and EATONS NECK Point (E); 3.5 mi/5.6 km long. NORTHPORT BAY and Centerport Harbor connect with Huntington Bay on SE, Lloyd Harbor adjoins on W.

Huntington Beach (HUHNT-ing-tuhn), city (□ 27 sq mi/70.2 sq km; 2000 population 189,594), ORANGE county, S CALIFORNIA; suburb 26 mi/42 km SSE of downtown LOS ANGELES, 10 mi/16 km ESE of LONG BEACH, and on the PACIFIC OCEAN coast; 33°39′N 118°00′W. In an oil-producing area. It has aerospace vehicles, aircraft parts, optical instruments, and heat-transfer equipment industries. The city's population skyrocketed between 1950–1990, but has stabilized since. Such growth was due to migration from the surrounding area and to the major high-technology, aircraft, and oil industries that developed in the area. The city has long been known for its fine beaches and has become one of the surfing capitals of Southern California. Seat of Golden West College (two-year). John Wayne Airport to E. To W is Sunset Bay (Huntington Harbor), with marina with man-made residential islands and coves (former marsh); Sunset Beach between bay and ocean; Bolsa Chica State Beach in SW, with Bolsa Bay marsh behind beach; Huntington State Beach in Costa Mesa to S. Seal beach. U.S. Naval Weapons Station to NW, Los Alamitos Naval Air Station to N. Incorporated 1909.

Huntington Harbor, New York: see HUNTINGTON BAY, inlet.

Huntington Lake, reservoir (6 mi/9.7 km long), FRESNO county, E central CALIFORNIA, on Big Creek, in the SIERRA NEVADA, 45 mi/72 km NE of FRESNO; 37°14′N 119°11′W. Formed by Huntington Lake Dam (165 ft/50 m high, 1,310 ft/399 m long), built (1917) for power generation. Receives water through tunnel from FLORENCE LAKE reservoir (12 mi/19 km E); tunnel diverts water to SHAVER LAKE reservoir (7 mi/11.3 km SSW). LAKESHORE (unincorporated village) on NE shore.

Huntington Lake, HUNTINGTON county, NE central INDIANA, on WABASH RIVER, in Little Turtle State Recreation Area, 3 mi/4.8 km SE of HUNTINGTON; 6 mi/9.7 km long; 40°50′N 85°28′W. Maximum capacity 153,100 acre-ft. Formed by Huntington Dam (89 ft/27 m high), built (1968) by the Army Corps of Engineers for flood control.

Huntington Park, city (2000 population 61,348), LOS ANGELES county, S CALIFORNIA; residential and industrial suburb 4 mi/6.4 km S of downtown LOS ANGELES. Varied manufacturing includes metal fabrication and processing, glass, rubber products, and industrial equipment. LOS ANGELES RIVER to E and N. Founded 1856, incorporated 1906.

Huntington Station, unincorporated town (□ 5 sq mi/ 13 sq km; 2000 population 29,910), SUFFOLK county, SE NEW YORK, on N shore of LONG ISLAND; 40°50′N 73°24′W. Diverse manufacturing base. Popular vacation and fishing area on Long Island's N shore nearby. Walt Whitman born here.

Huntington Woods, city (2000 population 6,151), OAKLAND county, SE MICHIGAN, residential suburb 12 mi/19 km NW of downtown DETROIT; 42°28′N 83°10′W. Detroit Zoological Park in SE part of city. Incorporated as village 1926, as city 1932.

Hunting Valley (HUHN-ting VAL-ee), village (□ 8 sq mi/20.8 sq km; 2000 population 704), on CUYAHOGA-GEAUGA county line, N OHIO; E suburb of CLEVELAND; 41°28′N 81°24′W. Wealthy area.

Huntland, town (2006 population 886), FRANKLIN county, S TENNESSEE, near ALABAMA line, 13 mi/21 km, SW of WINCHESTER; 35°03′N 86°16′W.

Huntleigh (HUHNT-lee), village (2000 population 528), SAINT LOUIS county, E MISSOURI, residential suburb 12 mi/19 km W of downtown ST. LOUIS; 38°36′N 90°24′W.

Huntley, village (2000 population 5,730), MCHENRY and KANE counties, NE ILLINOIS, 11 mi/18 km NW of ELGIN; 42°10′N 88°25′W. In dairying area. Summer resorts nearby.

Huntley, village, Yellowstone co., S Montana, on Yellowstone R at mouth of Pryor Creek, and 12 mi/19 km NE of Billings. Railroad junction. Huntley Irrigation Project was the first one completed in Montana, 1907. Huntley Project Museum of Irrigated Agr. Agr. research center to E.

Huntley, village (2006 population 62), HARLAN county, S NEBRASKA, 9 mi/14.5 km NNE of ALMA, and on branch of REPUBLICAN RIVER; 40°12′N 99°17′W.

Huntly, town (2001 population 6,822), WAIKATO district, NORTH ISLAND, NEW ZEALAND, on WAIKATO river and 53 mi/86 km S of AUCKLAND. Railroad and road junction; coal mines and thermal power plant (mothballed in 1991). School of mines.

Huntly (HUNT-lee), town (2001 population 4,412), ABERDEENSHIRE, NE Scotland, on DEVERON RIVER, and 33 mi/53 km NW of ABERDEEN; 57°27′N 02°47′W. Former agricultural market. Previously woolen and hosiery milling, bacon and ham curing, and manufacturing of agricultural machinery. Summer and angling resort. Nearby, the 14th-century Huntly Castle (formerly Strathbogie Castle) was seat of earls of Huntly. George MacDonald born here. Formerly in Grampian, abolished 1996.

Hunt, Mount (9,000 ft/2,743 m), SE YUKON, CANADA, near NORTHWEST TERRITORIES border, in MACKENZIE MOUNTAINS; 61°28′N 129°14′W.

Hunt River, c.8 mi/12.9 km long, S central RHODE ISLAND; rises NE of EXETER; flows SE, then NE, between EAST GREENWICH and NORTH KINGSTOWN towns, to NARRAGANSETT BAY SE of East Greenwich village. Called Potowomut River in lower course, where it is S boundary of POTOWOMUT PENINSULA.

Huntsdale, unincorporated town, BOONE county, central MISSOURI, on MISSOURI River and 8 mi/12.9 km W of COLUMBIA.

Hunts Peak (13,067 ft/3,983 m), S central COLORADO, in N tip of SANGRE DE CRISTO MOUNTAINS, between FREMONT and SAGUACHE counties.

Huntspill (HUHNT-spil), village (2001 population 2,564), N central SOMERSET, SW ENGLAND, near PARRETT RIVER estuary, 2 mi/3.2 km S of BURNHAM-ON-SEA; 51°13′N 03°00′W. On flat area known as Huntspill Level. Has 15th-century church.

Hunt's Point, a residential, wholesale, and industrial section of S BRONX borough of NEW YORK city, SE NEW YORK, at confluence of BRONX and EAST RIVERS; 40°49′N 73°54′W. Site of New York city's major produce market, including the new Fulton Fish Market, which relocated here (from Manhattan) in 2005.

Huntsville, city (2000 population 158,216), seat of Madison co., N Alabama. A major center for U.S. space research, Huntsville is the site of the Redstone Arsenal, the U.S. army's control and procurement center for guided missiles and rockets, NASA's George C. Marshall Space Flight Center (est. 1960), and the U.S. Space and Rocket Center (home of Space Camp). Although Huntsville's economy centers around the aerospace and high-technology industries, tires, glass, machinery, electrical, copper tubing, and computer equipment are also produced. The constitutional convention of the Alabama Territory was held in 1819 in Huntsville, where the 1st state legislature met. Numerous antebellum buildings remain. Huntsville is the seat of Oakwood College, Alabama Agricultural and Mechanical University, and the University of Alabama at Huntsville. Monte Sano State Park nearby. Inc. 1811. Named for John Hunt, a revolutionary war veteran who built his cabin nearby in 1802.

Huntsville, city (2000 population 1,553), ⊙ RANDOLPH county, N central MISSOURI, 5 mi/8 km W of MOBERLY; 39°26′N 92°32′W. Corn, soybeans; crushed stone; limestone quarries; former extensive coal mines. Founded c.1830.

Huntsville, city (2006 population 37,537), ⊙ WALKER county, E central TEXAS; 30°42′N 95°32′W. Elevation 401 ft/122 m. Located in a pine area, with sawmills. Farming along with agriculture and livestock trading add to the city's economic base; manufacturing (mirrors, signs, printing, oil and gas field equipment, lumber). Huntsville, the home of Samuel Houston, contains his grave (with an impressive monument; tallest free-standing statue in UNITED STATES), his restored home, and other memorials. Also in the city are Sam Houston State University and the Texas Department of Criminal Justice, Institutional Division; headquarters and prison (the most active execution chamber in the U.S.). An annual rodeo held by the prisoners draws many spectators. Sam Houston National Forest to S and E; Huntsville State Park to S. Incorporated 1845.

Huntsville (HUHNTS-vil), town (□ 271 sq mi/704.6 sq km; 2001 population 17,338), SE ONTARIO, E central CANADA, on the MUSKOKA RIVER; 45°20′N 79°13′W. Lumber mills and a woodworking plant; main economic base is year-round tourist trade. ALGONQUIN PROVINCIAL PARK is nearby.

Huntsville, town (2000 population 1,931), ⊙ MADISON county, NW ARKANSAS, 24 mi/39 km E of FAYETTE-VILLE, in the OZARKS; 34°42′N 86°37′W. Manufacturing (processed turkey, printed circuit assemblies). Withrow State Park and Madison County Wildlife Management Area to N.

Huntsville, town (2006 population 1,033), ⊙ SCOTT county, N TENNESSEE, 45 mi/72 km NW of KNOX-VILLE; 36°25′N 84°29′W. Lumbering; oil deposits.

Huntsville (HUHNTZ-vil), village (2006 population 412), LOGAN county, W central OHIO, 6 mi/10 km NNW of BELLEFONTAINE; 40°26′N 83°48′W. In agricultural area.

Huntsville, village (2006 population 650), WEBER county, N UTAH, 9 mi/14.5 km E of OGDEN, in WASATCH RANGE; 41°15′N 111°46′W. Elevation 4,920 ft/1,500 m. Ogden River Canyon to W. Wasatch National Forest surrounds area except NW. Located on East Fork Arm of Pineview Reservoir. Ski resort area: Snow Basin to SW, Nordic Valley to NW, Powder Mountain to N. Trappist Monastery to SE. Tourism.

Hunucmá (oo-nook-MAH), town, ⊙ Hunucmá municipio, YUCATÁN, SE MEXICO, 16 mi/26 km W of MÉRIDA; elevation 16 ft/5 m; 21°01′N 89°52′W. Henequen, tropical fruit, citrus, corn, beans; livestock.

Hunugama, village, Southern province, Sri Lanka. Fishing. Trades in coconuts, rice. Tile manufacturing.

Hünxe (HYUNK-se), town, NORTH RHINE–WESTPHALIA, W GERMANY, on the WESEL-DATTELN CANAL, 7 mi/11.3 km E of WESEL; 51°39′N 08°46′E. Sand and clay quarries.

Hunyani Range, MASHONALAND WEST province, N central ZIMBABWE, extends from CHINHOYI in S to DOMA in N; c.40 mi/64 km long and 10 mi/16 km wide. Highest point is Kasikana (4,825 ft/1,471 m) at center of range. CHINHOYI CAVES RECREATIONAL PARK at S end.

Hunyani River (oon-YAH-nee), in N ZIMBABWE and W MOZAMBIQUE; rises 5 mi/8 km W of MARANDELLAS; flows 260 mi/418 km W and N, past Sinoia, to the ZAMBEZI in Mozambique. Prince Edward Dam on its upper course, 10 mi/16 km S of SALISBURY, supplies that city with water. Receives ANGWA RIVER (left).

Hunyani River, ZIMBABWE: see MANYAME RIVER.

Hunyuan (HUN-YUAN), town, ⊙ Hunyuan county, NE SHANXI province, NE CHINA, 35 mi/56 km SE of DATONG; 39°42′N 113°41′E. Grain; engineering, pharmaceuticals, food industry, clothing, coal mining.

Hunza (huhn-zah), division (□ 3,900 sq mi/10,140 sq km), GILGIT district, comprising HUNZA, Nager, and Gojal tehsils (former princely states; incorporated 1976 into PAKISTAN), NORTHERN AREAS, extreme NE PAKISTAN. Major agricultural products (wheat, dried fruit, seed potatoes); local trekking and mountaineering (tourism). Greatly transformed by completion of KARAKORAM HIGHWAY in 1979. Population consists of Ismaili Muslims, Nager, Shia Muslims. Mother tongue is Beershaski in S and Wahi in Gojal. Entry to KHUNJERAB NATIONAL PARK on Pakistan-China border.

Hunza River (huhn-zah), c.120 mi/193 km long, in NORTHERN AREAS, extreme NE PAKISTAN, in N KASHMIR region; rises in N KARAKORAM mountain system, in headstreams joining 10 mi/16 km E of MISGAR; flows S, W past Baltit and NAGAR, and S to GILGIT RIVER just E of GILGIT.

Huocheng (HU-uh-CHENG), town, ⊙ Huocheng county, W XINJIANG UYGUR AUTONOMOUS REGION, NW CHINA, 30 mi/48 km WNW of YINING, near KAZAKHSTAN border and near ILI RIVER; 44°03′N 80°49′E. Oilseeds, sugar beets; livestock; grain; food processing, coal mining. Opened to foreign trade in 1850. Also known as Shuiding.

Huojia (HAW-JIAH), town, ⊙ Huojia county, NW HENAN province, NE CHINA, on WEI RIVER, and 10 mi/16 km WSW of XINXIANG; on railroad; 35°16′N 113°39′E. Grain, cotton.

Huolin Gol (HU-uh-LIN GO-LUH), city (□ 227 sq mi/590.2 sq km), E INNER MONGOLIA AUTONOMOUS REGION, N CHINA, on railroad to TONGLIAO, 185 mi/298 km NW of Tongliao. Crop growing and animal husbandry account for most agricultural activity. Grain; livestock; coal mining.

Huolu (HU-uh-LU), town, ⊙ Huolu county, SW HEBEI province, NE CHINA, 10 mi/16 km WNW of SHI-JIAZHUANG, near railroad; 38°05′N 114°18′E. Grain, cotton, oilseeds; building materials, engineering, chemicals.

Huo Mountains, CHINA: see TAIYUE MOUNTAINS.

Huong Giang (HU-uhng YAHNG), river, THUA THIEN-HUE province, N central VIETNAM, runs through HUE city; 16°33′N 107°38′E. Also spelled Huong Zang and Huong Yang.

Huong Khe (HU-uhng KAI), town, HA TINH province, N central VIETNAM, on railroad, and 15 mi/24 km WSW of HA TINH; 18°12′N 105°41′E. Betel nuts, rubber, tangerines; forestry. Coal and iron deposits nearby. Formerly Huongkhe.

Huon Gulf (HOO-uhn), NE NEW GUINEA, bounded by HUON PENINSULA (N) and coast of MOROBE district (S); 80 mi/129 km wide, 65 mi/105 km deep. LAE and SALAMAUA are its harbors.

Huon Islands (YOO-uhn), usually uninhabited coral group forming two separate lagoons, islets cover 160 acres/65 ha, SW PACIFIC OCEAN, 170 mi/274 km NW of NEW CALEDONIA, a territory of FRANCE, with which it is associated; 18°01′S 162°55′E. Includes four islands.

Cross-references are shown in SMALL CAPITALS. The pronunciation guide is shown on page xix. The sources of population figures are shown on page xvii.

Huon Peninsula (HOO-uhn), NE PAPUA NEW GUINEA, between ASTROLABE BAY and HUON GULF; 60 mi/97 km wide. FINSCHHAFEN in E, LAE in S.

Huon River (HYOO-uhn), 105 mi/169 km long, S TASMANIA, SE AUSTRALIA; rises in mountains 50 mi/80 km W of HOBART; flows E, past HUONVILLE and FRANKLIN, to D'ENTRECASTEAUX CHANNEL, forming estuary 2.5 mi/4 km wide; 43°16′S 147°07′E. CYGNET is at head of N inlet of estuary.

Huonta, NICARAGUA: see WOUNTA.

Huonville (HYOO-uhn-vil), town, SE TASMANIA, AUSTRALIA, 17 mi/27 km SW of HOBART and on HUON RIVER; 43°02′S 147°02′E. Hops; apple production; softwood timber. Population figure includes RANELAGH.

Huoqiu (HU-uh-CHYOOI), town, ⊙ Huoqiu county, N ANHUI province, E CHINA, near HENAN border, 70 mi/113 km NW of HEFEI, in lake area; 32°20′N 116°15′E. Rice, cotton, oilseeds, jute.

Huoshan (HU-uh-SHAN), town, ⊙ Huoshan county, N ANHUI province, E CHINA, on PI RIVER, and 65 mi/105 km WSW of HEFEI, at foot of the peak Huo Shan, in DABIE MOUNTAINS; 31°24′N 116°20′E. Rice, oilseeds, jute; engineering, food industry.

Huo Shan, CHINA: see TAIYUE MOUNTAINS.

Huounta Lagoon, NICARAGUA: see WOUNTA LAGOON.

Huozhou (HU-uh-JO), city (□ 295 sq mi/767 sq km; 2000 population 248,163), S SHANXI province, NE CHINA, on FEN RIVER, at W foot of the TAIYUE MOUNTAINS, on railroad, and 35 mi/56 km N of LINFEN; 36°37′N 111°44′E. Heavy industry is the most important economic activity. Wheat growing accounts for most of city's agriculture. Main industries include coal mining and utilities. Formerly called Huo Xian.

Hupeh, CHINA: see HUBEI.

Hupei, CHINA: see HUBEI.

Hur, TURKEY: see MUTKI.

Huraidha, YEMEN: see HUREIDHA.

Hur al-Hammar, IRAQ: see HAMMAR, HOR AL.

Huraydah, YEMEN: see HUREIDHA.

Hurayn, EGYPT: see HUREIN.

Hurbanovo (huhr-BAH-no-VO), town, ZAPADOSLOVENSKY province, S SLOVAKIA, on railroad and 8 mi/12.9 km NNE of KOMÁRNO; 47°52′N 18°12′E. Agricultural center (wheat, corn, tobacco); noted brewery; food processing. Has meteorological observatory. Paleolithic-era archaeological site. Until 1948, called STARA DALA, (Slovak *Stará Ďala*, HUNGARIAN *Ógyalla*). Under Hungarian rule from 1938–1945.

Hurbat Sufa, ancient settlement, ISRAEL: see HATZERIM.

Hurdes, las or **Las Jurdes**, comarca (local region), N CÁCERES province, W SPAIN, bordering the SALAMANCA S of Batuecas.

Hurdio (HOOR-dyoo), town, NE SOMALIA, 100 mi/161 km SSE of ALULA, on INDIAN Ocean, on bay enclosed by HAFUN peninsula; 10°34′N 51°08′E. Extensive saltworks.

Hurdland, town (2000 population 239), KNOX county, NE MISSOURI, near North Fork of SALT RIVER, 7 mi/11.3 km W of EDINA; 40°08′N 92°17′W. Agriculture.

Hurdsfield, village (2006 population 81), WELLS CO., central NORTH DAKOTA, 20 mi/32 km SW of FESSENDEN; 47°27′N 99°55′W. Founded in 1903 and incorporated in 1926. Named for Warren W. Hurd, prominent local farmer.

Hureidha (hu-RAI-duh), township, S YEMEN, 35 mi/56 km SW of SHIBAM, chief settlement of the Wadi 'Amd, place of religious teaching; mosques. Airfield. Also spelled Horaidha, Huraidha, or Huraydah.

Hurein (hoo-RAIN), village, GHARBIYA province, Lower EGYPT, 6 mi/9.7 km S of El Santa; 30°39′N 31°08′E. Cotton. Sometimes Hurin.

Hurepoix (oo-ruh-pwah), agricultural region of ÎLE-DE-FRANCE, N central FRANCE, situated between the wheat-growing BEAUCE and the cheese-making BRIE districts, generally SW of PARIS and E of RAMBOUILLET; 48°37′N 02°07′E. Traversed by valleys of the Orge,

ESSONNE, and Yvette rivers S of densely settled suburban communities of the Paris region.

Hurezu, ROMANIA: see HOREZU.

Hurfeish (huhr-FAISH) or **Churfeish**, predominantly Druze village, 9 mi/15 km NW of ZEFAT (Safed) in UPPER GALILEE, ISRAEL; 33°00′N 35°20′E. Elevation 2,053 ft/625 m. Farming (olives, goats, fruit orchards and tobacco); some textile production. Ruins from Roman times through 10th century.

Hurghada, Arabic, *El Ghardaqa* or *Al-Ghardaqah* (el–GHAR-da-kah), township, ⊙ RED SEA PROVINCE, E EGYPT, on RED SEA at entrance to Gulf of SUEZ,100 mi/161 km NE of QENA. Producer of crude oil, which is refined in SUEZ. Airport. Resort area along 37 mi/60 km costal stretch to SAFAGA. Naval base.

Huriel (uh-ryel), commune (□ 13 sq mi/33.8 sq km), ALLIER department, AUVERGNE region, central FRANCE, 7 mi/11.3 km WNW of MONTLUÇON; 46°22′N 02°29′E. Brick and tile manufacturing, dairying. Has a 12th-century keep and museum of wine-making.

Hurin, EGYPT: see HUREIN.

Hurlets (huhr-LETS), village, MONTANA oblast, KOZLODUI obshtina, BULGARIA; 43°44′N 23°51′E.

Hurley, town (2006 population 1,377), GRANT county, SW NEW MEXICO, in W foothills of PINOS ALTOS MOUNTAINS, 15 mi/24 km SE of SILVER CITY; 32°42′N 108°07′W. Elevation 5,720 ft/1,743 m. Cattle, some sheep, alfalfa, grain. Silver City–Grant Airport to S; Gila National Forest is N; City of Rocks State Park to SE; Santa Rita Open Pit Copper Mine to NE.

Hurley, town (2006 population 1,646), ⊙ IRON county, N WISCONSIN, on MONTREAL RIVER, opposite IRONWOOD, MICHIGAN, in GOGEBIC RANGE; 46°26′N 90°12′W. Manufacturing (carbonated beverages). It was a boom town (mining, lumbering) until c.1910. Founded 1885, incorporated 1918.

Hurley (HUH-lee), village (2006 population 2,500), Windsor and Maidenhead, S central ENGLAND, on the THAMES, 4 mi/6.4 km WNW of MAIDENHEAD; 51°33′N 00°49′W. Former agricultural market. Has Norman church and remains of Norman monastery.

Hurley, village (□ 5 sq mi/13 sq km; 2000 population 3,561), ULSTER county, SE NEW YORK, just W of KINGSTON; 41°54′N 74°03′W. In agricultural and summer-resort area.

Hurley, village (2006 population 398), TURNER county, SE SOUTH DAKOTA, 8 mi/12.9 km S of PARKER; 43°16′N 97°05′W.

Hurleyville, hamlet, SULLIVAN county, SE NEW YORK, 6 mi/9.7 km N of MONTICELLO; 41°45′N 74°40′W. Summer resort area.

Hurlford (HUHRL-fuhd), town (2001 population 4,968), EAST AYRSHIRE, S Scotland, 2 mi/3.2 km ESE of KILMARNOCK; 55°35′N 04°27′W. "Burns Country." Formerly in STRATHCLYDE, abolished 1996.

Hurlingham, town, in Greater BUENOS AIRES, ARGENTINA, 15 mi/24 km WNW of BUENOS AIRES; 34°36′S 58°38′W. Residential and fashionable sport center.

Hurlock (HER-lok), town (2000 population 1,874), DORCHESTER county, E MARYLAND, on DELMARVA Peninsula, 13 mi/21 km ENE of CAMBRIDGE; 38°38′N 75°52′W. Vegetable processing center, but also has tin can and shirt factories. Named after John M. Hurlock, who erected the first store in 1869 and first home in 1872. Hurlock is said to have won the right to name the town after himself in a tree-felling contest with another land owner.

Huron (HYU-ruhn), county (□ 1,316 sq mi/3,421.6 sq km; 2001 population 59,701), S ONTARIO, E central CANADA, on Lake HURON; ⊙ GODERICH; 43°40′N 81°30′W. Includes the municipalities of Blue Water, Central HURON, HURON East, Morris-TURNBERRY, South HURON; the town of GODERICH; and the townships of Ashfield-Colborne-WAWANOSH, HOWICK, and North HURON.

Huron, county (□ 2,136 sq mi/5,553.6 sq km; 2006 population 34,143), E MICHIGAN; ⊙ BAD AXE; 43°57′N

82°50′W; at tip of the "thumb," in S LAKE HURON. Bounded E and N by Lake Huron; W by SAGINAW BAY (numerous islands off Saginaw Bay shores). Drained by headwaters of the CASS and by small Pigeon and Willow rivers. Poultry; cattle, hogs; dairy products; agriculture (beans, sugar beets, corn, wheat); manufacturing (plastics products, metal products, industrial machinery); resorts on Lake Anson. Albert E. Sleeper State Park and Port Crescent State Park in N. Organized 1859.

Huron (HYUHR-ahn), county (□ 497 sq mi/1,292.2 sq km; 2006 population 60,313), N OHIO; ⊙ NORWALK; 41°09′N 82°33′W. Drained by HURON and VERMILION rivers. In the Lake and Till Plains physiographic regions. Agricultural area (nursery and greenhouse crops, sheep, corn, vegetables); manufacturing at BELLEVUE, Norwalk, and NEW LONDON (bakery products, household furniture, farm and garden machinery). Gravel pits. Formed 1809.

Huron, city (2000 population 6,306), FRESNO county, central CALIFORNIA, 40 mi/64 km SSW of FRESNO, in San Joaquin Valley. KETTLEMAN HILLS to S; Black Mountain to W. California Aqueduct passes to E. Irrigated agricultural area (cotton, grain, fruit, vegetables, dairying, cattle). Lemoore Naval Air Station to E.

Huron (HYUHR-ahn), city (□ 8 sq mi/20.8 sq km; 2006 population 7,459), ERIE county, N OHIO, on harbor on LAKE ERIE, at mouth of HURON RIVER, and 10 mi/16 km ESE of SANDUSKY; 41°23′N 82°34′W. Fishing; tourist resort. Previously coal and iron-ore transshipping. Makes cement blocks, boats, sauerkraut, and pickles. Settled c.1805.

Huron, city (2006 population 10,909), ⊙ BEADLE county, E central SOUTH DAKOTA, on the JAMES RIVER. A shipping and trade center for a large livestock and grain area, it has meatpacking, lumbering, and tourism industries. It is also the administrative center for a number of state and Federal agencies. Huron was the hometown of Hubert Humphrey. The city is the seat of Si Tanka Huron University. The South Dakota State Fair is held annually in Huron. Manufacturing (asphalt, pork processing, mining equipment, concrete, capacitor banks, security doors). James River and Lake Byron State Lakeside Use Areas to N; state fairgrounds; Pioneer Museum; railroad junction. Incorporated 1883.

Huron (HYOR-ahn), village (2000 population 87), ATCHISON county, extreme NE KANSAS, 13 mi/21 km WNW of ATCHISON, in corn belt; 39°38′N 95°20′W. Railroad junction to N. Atchison State Fishing Lake to E.

Huron Bay, MICHIGAN, narrow inlet of Lake SUPERIOR indenting N shore of UPPER PENINSULA, c.20 mi/32 km SE of HOUGHTON and just E of KEWEENAW BAY, from which it is separated by narrow peninsula (c.18 mi/29 km long) terminating in Point ABBAYE; 46°53′N 88°13′W.

Huron East (HYU-ruhn EEST), town (□ 258 sq mi/670.8 sq km; 2001 population 9,680), S ONTARIO, E central CANADA; 43°37′N 81°17′W. In area, it is the largest municipality in HURON county. Formed in 2001 from the merger of SEAFORTH, MCKILLOP, BRUSSELS, GREY, and TUCKERSMITH.

Huron-Kinloss (HYU-ron–KIN-lahs), township (□ 183 sq mi/475.8 sq km; 2001 population 6,224), SW ONTARIO, E central CANADA; 44°03′N 81°32′W. Agriculture. Created in 1999 from the amalgamation of LUCKNOW, Ripley-Huron, and KINLOSS.

Huron, Lake (□ 23,010 sq mi/59,596 sq km), between ONTARIO, CANADA, and MICHIGAN; 206 mi/332 km long and 183 mi/295 km at its greatest width; second-largest of the GREAT LAKES. It has a surface elevation of 580 ft/177 m (177 m) above sea level and a maximum depth of 750 ft/229 m. Centrally located between the upper and lower Great Lakes, Lake Huron receives the waters of Lake SUPERIOR through the ST. MARYS

RIVER and those of Lake MICHIGAN through the Straits of MACKINAC; it drains into Lake ERIE through the ST. CLAIR RIVER–Lake ST. CLAIR–DETROIT River system. Large tributaries flowing into the lake include the MISSISSAGI, Wanapitei, SPANISH, and FRENCH rivers from Ontario, and the AU SABLE and SAGINAW rivers from Michigan.

The N shoreline is irregular, with many bays and inlets; the largest are GEORGIAN BAY and NORTH CHANNEL, which indent the Ontario shore and are nearly landlocked by MANITOULIN Island and the BRUCE Peninsula. SAGINAW BAY is the principal indentation on the S shores. Lake Huron is part of the ST. LAWRENCE–GREAT LAKES WATERWAY system and is navigated by oceangoing and lake vessels that carry cargoes of iron ore, grain, coal, limestone, and other goods. Navigation is impeded by ice in the shallower sections from mid-December to early April. The lake is subject to occasional violent storms. The principal lakeshore cities are PORT HURON, Michigan, and SARNIA, Ontario, at the lake's outlet; OWEN SOUND, MIDLAND, and PARRY SOUND, Ontario; and BAY CITY, ALPENA, and CHEBOYGAN, Michigan.

The waters of the lake are relatively unpolluted; commercial and sport fishing is important, and several resorts are located along the lake shore. Major salt deposits are worked at the S end of the lake. Georgian Bay is a popular resort area, and recreational facilities are provided at Georgian Bay Islands National Park (Canada), on the island in Mackinac Strait, and at numerous state and provincial parks along the lake's scenic shores. Samuel de Champlain visited Lake Huron in 1615.

Huron Mountains (c.1,500 ft/457 m–1,800 ft/549 m), granitic range in NW MARQUETTE and NE BARAGA counties, UPPER PENINSULA, MICHIGAN, extending c.20 mi/32 km NW-SE near S shore of LAKE SUPERIOR; 46°50′N 87°55′W. Wilderness recreational region, with lakes (INDEPENDENCE; Ives, Mountain). Mount Avron (1,979 ft/603 m), highest point in Michigan, in W in Avron Hills.

Huron River, c.97 mi/156 km long, SE MICHIGAN; rises in small lakes in OAKLAND and LIVINGSTON counties; flows SW and S to DEXTER, then SE, past ANN ARBOR and YPSILANTI, through Ford and Belleville lakes, then past BELLEVILLE and FLAT ROCK to LAKE ERIE, SE of ROCKWOOD; 46°51′N 88°04′W. Utilized for power. In SW part of DETROIT metropolitan area.

Huron River (HYUHR-ahn), c.11 mi/18 km long, N OHIO, formed by East and West branches in S ERIE county; flows N to LAKE ERIE at HURON. East Branch rises in HURON county, flows c.32 mi/51 km N. West Branch rises near SHILOH, flows c.38 mi/61 km N, past MONROEVILLE to union with East Branch.

Huron Shores (HYOO-ruhn), municipality (□ 176 sq mi/457.6 sq km; 2001 population 1,794), ONTARIO, E central CANADA; 46°17′N 83°22′W. Tourism. Composed of IRON BRIDGE, and sections of THESSALON, Thompson, Day and Bright.

Hurricane (HUHR-i-kuhn), town (2006 population 12,084), WASHINGTON county, SW UTAH, near VIRGIN RIVER, 17 mi/27 km ENE of SAINT GEORGE; 37°09′N 113°20′W. In fruit area; manufacturing (apparel). Copper mining. Settled 1906. Dixie National Forest to NW; Quail Creek Reservoir and State Park to W; ZION NATIONAL PARK to E.

Hurricane, town (2006 population 6,071), PUTNAM county, W central WEST VIRGINIA, 22 mi/35 km W of CHARLESTON; 38°26′N 82°01′W. Coal-mining area. Agriculture (corn, tobacco); livestock; poultry. Manufacturing (truck equipment parts, clothing, machining). Incorporated 1888.

Hurricane Creek, ARKANSAS: see BAUXITE.

Hurricane Island, KNOX county, S MAINE, in PENOBSCOT BAY, 3 mi/4.8 km WSW of VINALHAVEN village. Vinalhaven Island; 0.7 mi/1.2 km long.

Hurricane Mountain (3,687 ft/1,124 m), ESSEX county, NE NEW YORK, in High Peaks section of ADIRONDACK MOUNTAINS, 12 mi/19 km NE of Mount MARCY and 13 mi/21 km ESE of LAKE PLACID village; 44°14′N 73°44′W.

Hursovo (huhr-SO-vo), village, SOFIA oblast, SANDANSKI obshtina, BULGARIA. Wine production; 41°27′N 23°23′E. Sometimes spelled Harsovo.

Hurst, city (2000 population 805), WILLIAMSON county, S ILLINOIS, 14 mi/23 km NW of MARION; 37°50′N 89°08′W. In bituminous coal-mining and agricultural area.

Hurst, city (2006 population 38,182), TARRANT county, N TEXAS, suburb 8 mi/12.9 km NE of downtown FORT WORTH; 32°50′N 97°10′W. Manufacturing (helicopters, helicopter components, aviation parts, draperies and bedspreads). Seat of Tarrant County Junior College NE Campus. Dallas–Fort Worth International Airport to NE.

Hurstbourne (HUHRST-born), town (2000 population 3,884), JEFFERSON county, N KENTUCKY, residential suburb 9 mi/14.5 km E of downtown LOUISVILLE.

Hurstmonceux (HUHST-mahn-SOO) or **Herstmonceux**, village and parish (2001 population 2,532), East SUSSEX, SE ENGLAND, near the CHANNEL, 8 mi/12.9 km N of EASTBOURNE; 50°53′N 00°20′E. Site of 1446 castle, restored 1907. After World War II, it was decided (1946) to move government meteorological station and time clocks to the castle grounds from Greenwich, the move to be completed by 1953. GREENWICH, however, continues to be the point from which longitude is reckoned.

Hurstpierpoint (HUHST-pir-POINT), town (2001 population 8,124), West SUSSEX, SE ENGLAND, 7 mi/11.3 km N of HOVE; 50°55′N 00°10′W. Site of Hurstpierpoint College, a public school.

Hurstville, town, JACKSON county, E IOWA, 2 mi/3.2 km N of MAQUOKETA. Hybrid seed corn.

Hurstville (HUHRST-vil), residential suburb, E NEW SOUTH WALES, AUSTRALIA, 9 mi/14 km SSW of SYDNEY; 33°58′S 151°06′E. In metropolitan area. Primarily residential with retail, light industry. Sydney International Airport nearby.

Hurt (HUHRT), town, PITTSYLVANIA county, S VIRGINIA, 22 mi/35 km S of LYNCHBURG, 1 mi/2 km SW of ALTAVISTA, on ROANOKE RIVER; 37°05′N 79°17′W. Railroad junction. Manufacturing (textile dyeing and finishing, millwork, machining); agriculture (tobacco, grain, soybeans; cattle, dairying). LEESVILLE LAKE reservoir to SW.

Hürtgenwald (HYOORT-guhn-vahlt), village, NORTH RHINE–WESTPHALIA, W GERMANY, 8 mi/12.9 km SW of DÜREN; 50°44′N 06°40′E. Nearby Hürtgen Forest was scene of heavy fighting in November 1944. Sometimes spelled HUERTGEN.

Hürth (HYOORT), town, NORTH RHINE–WESTPHALIA, W GERMANY, 5 mi/8 km SW of COLOGNE; 50°52′N 06°53′E. A former lignite-mining center; manufacturing of machinery, steel, electrical goods. Lignite-fed power plant. Also an agricultural area with vegetables being the chief products. Chartered 1978.

Hurtsboro, town (2000 population 592), Russell co., E Alabama, 44 mi/45 km SW of Phenix City. Lumber. Originally named 'Hurtsville' for Joel Hurt, the founder, the confusion with 'Huntsville' caused the name to be changed. Inc. in 1872.

Hurumu, town (2007 population 6,376), OROMIYA state, SW ETHIOPIA, 15 mi/24 km NE of GORE; 08°20′N 35°41′E.

Hurup (HOO-roop), town, Voborg county, N JUTLAND, DENMARK, 17 mi/27 km SW of THISTED; 56°45′N 08°25′E. Limestone and chalk quarries; cattle.

Huruta (hoo-ROO-tah), town (2007 population 17,729), OROMIYA state, central ETHIOPIA, 10 mi/16 km NE of ASELA; 08°11′N 39°21′E.

Hurzuf (hoor-ZOOF) (Russian *Gurzuf*), town (2004 population 50,000), S Republic of CRIMEA, UKRAINE,

port on the BLACK SEA, 7 mi/11.3 km NE and jurisdiction of YALTA, to which it is subordinate; 44°33′N 34°17′E. Climatic and beach resort. Sanatoriums in shore park, over 1 mi/1.6 km long. Ruins of a 6th-century fortress, built under Justinian I and later used by the Genoese. Site of Artek, the former renowned All-Soviet Young Pioneer Camp—now an international health resort for children.

Husain, INDIA: see HASAYAN.

Husan, Arab village, BETHLEHEM district, 6 mi/10 km SW of JERUSALEM, in the Judaean Highlands, WEST BANK; 31°43′N 35°08′E. Some archaeological remains from the Israelite through medieval periods. It is believed that Husan relates to the biblical site of Husha. Cereal; vineyards.

Húsavík (HOO-sah-veek), town (2000 population 2,423), ⊙ SUDUR-THINGEYJARSYSLA county, N ICELAND, on Skjalfandi Bay, 30 mi/48 km NE of AKUREYRI. Fishing port. Sulphur formerly shipped from here.

Huseyinabat, TURKEY: see ALACA.

Hushan, CHINA: see CIXI.

Hushu (HU-SHU), town, SW JIANGSU province, E CHINA, 22 mi/35 km SE of NANJING. Rice, oilseeds, jute.

Huși (HOOSH), town, VASLUI county, E ROMANIA, in Moldavia, 40 mi/64 km SE of IAȘI. Railroad terminus and commercial center noted for its wine; also manufacturing of furniture, candles, soap, dyes, bricks, and knitwear; flour milling, tanning. Extensive vineyards in vicinity. Founded in 15th century. Has 15th-century cathedral built by Stephen the Great, valuable library Orthodox bishopric. Treaty of Prut (Pruth) was signed here (1711) between RUSSIA and TURKEY.

Husinec, CZECH REPUBLIC: see PRACHATICE.

Huskisson (HUHS-ki-suhn), township, NEW SOUTH WALES, SE AUSTRALIA, 122 mi/179 km S of SYDNEY, 15 mi/24 km from NOWRA, and on JERVIS BAY shores; 35°03′S 150°40′E. Fishing port; holiday resort.

Huskvarna (HOOS-KVAHR-nah), town, JÖNKÖPING county, S SWEDEN, at S end of LAKE VÄTTERN, agglomerated with JÖNKÖPING; 57°47′N 14°17′E. Manufacturing (sewing machines, foundry products). Trade center in Middle Ages. Formerly spelled Husqvarna. Chartered 1911.

Husn al 'Abr (HUHS-in el AH-ber), desert outpost of the former Quaiti state, E YEMEN, 90 mi/145 km WNW of Shibam, and on route to Yemen; 16°05′N 47°15′E. Airfield. Hq. of Sei'ar tribal country. Also called Al 'Abr.

Husn, El (HUHS-in, el), township, N JORDAN, 6 mi/9.7 km SSE of IRBID. Road junction near oil pipeline; grain, vineyards, olives. About spelling Husun.

Husnumansur, TURKEY: see ADIYAMAN.

Husqvarna, SWEDEN: see HUSKVARNA.

Hussar (hoo-ZAHR), village (2001 population 181), S ALBERTA, W CANADA, 30 mi/48 km S of DRUMHELLER, in WHEATLAND County; 51°02′N 112°41′W. Wheat. Incorporated 1928.

Hussein Dey (yoo-SANG DAI), industrial SE suburb of ALGIERS, ALGIERS wilaya, N central ALGERIA, on Algiers Bay. Metalworks (founding and stamping, boilermaking; structural shapes, agricultural machinery), sawmills (barrels, crates); wool washing, manufacturing of corks, electrical equipment, paint, varnish, lubricating oil, rubber goods, flour products. Here Emperor Charles V (in 1541) and O'Reilly (in 1775), leading Spanish troops, made their unsuccessful landings on Algerian coast. Named after Algiers' last Ottoman prince (Dey).

Hussigny-Godbrange (u-see-nyee–god-brahnzh), commune (□ 5 sq mi/13 sq km; 2004 population 3,142), MEURTHE-ET-MOSELLE department, LORRAINE region, NE FRANCE, on LUXEMBOURG border, 5 mi/8 km ESE of LONGWY; 49°29′N 05°52′E. Metalworks in old iron-mining district.

Hustad (HOOS-tah), village, MØRE OG ROMSDAL county, W NORWAY, on an inlet NORTH SEA, 22 mi/35 km SW of KRISTIANSUND. Fisheries; lime pit.

Hustav, UKRAINE: see LOTYKOVE.

Huste, UKRAINE: see KHUST.

Hustisford (HUHS-tis-fuhrd), town (2006 population 1,152), DODGE county, S central WISCONSIN, 14 mi/23 km SE of BEAVER DAM and on ROCK RIVER; 43°21′N 88°35′W. Here dammed to form Sinissippi Lake to N (c.3 mi/4.8 km long). Dairy products; manufacturing (air compressors, hardware).

Hustler, village (2006 population 114), JUNEAU county, central WISCONSIN, c.46 mi/74 km E of LA CROSSE; 43°52′N 90°16′W. In farming area. Butter and cheese.

Hustonville (YOO-stuhn-vil), village (2000 population 347), LINCOLN county, central KENTUCKY, 13 mi/21 km S of DANVILLE; 37°28′N 84°49′W. In agricultural area (burley tobacco, grain; livestock; dairying); manufacturing (farm equipment, raw lumber). Isaac Shelby State Historic Site to N.

Hustopece (HUS-to-PE-tche), Czech *Hustopeče*, German *Auspitz*, town, JIHOMORAVSKY province, S MORAVIA, CZECH REPUBLIC, on railroad, and 18 mi/29 km SSE of BRNO; 48°56′N 16°44′E. Wine; agriculture (barley, wheat growing; exports fruit and vegetables); machinery manufacturing. President Thomas G. Masaryk was a schoolboy here. DOLNI VESTONICE, Czech *Dolní Vestonice* (DOL-nye VYE-sto-NYI-tse), German *unter wisternitz*, site of important prehistoric finds, is 5 mi/8 km SW.

Husum (HOO-sum), city, SCHLESWIG-HOLSTEIN, NW GERMANY, a port on the NORTH SEA; 54°28′N 09°03′E. It is a fishing center and major cattle market; food processing; tourism. Husum, first mentioned in the 13th century, was chartered at the beginning of the 17th century, and soon became a prosperous commercial city, though it later declined. Many fine patrician houses remain. Theodor Storm born and buried here.

Husum (HOO-SOOM), town, VÄSTERNORRLAND county, NE SWEDEN, on GULF OF BOTHNIA, 14 mi/23 km NE of ÖRNSKÖLDSVIK; 63°20′N 19°10′E.

Husun, JORDAN: see HUSN, EL.

Husyatyn (hoo-SYAH-tin) (Russian *Gusyatin*), (Polish *Husiatyn*), town (2004 population 6,435), E TERNOPIL′ oblast, UKRAINE, on ZBRUCH RIVER and 18 mi/29 km ENE of CHORTKIV; 49°04′N 26°13′E. Elevation 964 ft/293 m. Raion center; railroad station; machine building and repair, furniture making, flour milling. Gas compressor stations on the "Soyuz" and the "Urengoy-Uzhhorod" pipelines. Politechnic; sanatorium; heritage museum. Has a 16th-century Bernardine monastery and two 16th-century churches, ruins of a 17th-century castle and 17th-century town hall and synagogue. First mentioned 1559; thrived as Austrian-Russian border town (1772–1918); site of battles between Ukrainian Galician Army and Bolshevik forces, and Polish Army (1919); Soviet-Polish frontier station (1921–1939). Small Jewish community since 1577, eliminated by the Nazis in World War II.

Huszt, UKRAINE: see KHUST.

Hutchins, town (2006 population 3,009), DALLAS county, N TEXAS, suburb 10 mi/16 km SSE of downtown DALLAS, on TRINITY RIVER; 32°38′N 96°42′W. Manufacturing (plastic moldings, business forms).

Hutchinson, county (□ 814 sq mi/2,116.4 sq km; 2006 population 7,426), SE SOUTH DAKOTA; ⊙ OLIVET; 43°20′N 97°45′W. Agricultural and cattle-raising area drained by JAMES RIVER and Wolf Creek. Dairy products; corn, wheat, barley, soybeans, oats; cattle, hogs, poultry. Formed 1862.

Hutchinson, county (□ 895 sq mi/2,327 sq km; 2006 population 22,460), extreme N TEXAS; ⊙ STINNETT; 35°50′N 101°21′W. On high treeless plains of the Texas PANHANDLE, here broken by gorge of CANADIAN RIVER; elevation 3,000 ft/914 m–3,500 ft/1,067 m. Much of county underlaid by huge Panhandle natural gas and oil field; here is one of world's largest natural gas pumping stations, and large carbon black and oil refining industries; manufacturing of petro-

leum products; some irrigated agriculture (corn, wheat, sorghum); livestock (beef cattle). Part of LAKE MEREDITH NATIONAL RECREATION AREA in SW corner, including SANFORD DAM which forms the lake in the Canadian River. Formed 1876.

Hutchinson (HUHCH-in-suhn), city (2000 population 40,787), ⊙ RENO county, S central KANSAS, on the ARKANSAS RIVER; 38°04′N 97°54′W. Elevation 1,538 ft/469 m. Railroad junction. It is a commercial and industrial center in a grain (especially wheat), livestock, and oil region. Manufacturing (grain milling, vehicle parts, fuel tanks, bakery products, textile bags, industrial valves, welding supplies, signs, food products, asphalt, ambulances). Its many facilities include a giant grain elevator, over ½ a mile long. Salt is extracted from great beds beneath the city. Hutchinson Community College, a planetarium, and the KANSAS state fairgrounds here. Kansas Cosmosphere and Space Center. Incorporated 1872.

Hutchinson, city (2000 population 13,080), MCLEOD county, S central MINNESOTA, 55 mi/89 km W of MINNEAPOLIS, on South Fork CROW RIVER; 44°53′N 94°22′W. Manufacturing (agricultural equipment, concrete blocks, wood furniture, fertilizer, sheet metal, dry yeast, wood boxes, magnetic tapes, machining); grain, soybeans, peas; livestock, poultry; dairying. Part of settlement burned in Sioux uprising of 1862. Several small lakes in area, Otter Lake to W. Founded 1855, incorporated 1881.

Hutchinson River, 5 mi/8 km long, small stream in SE NEW YORK; rises in S WESTCHESTER county, just E of SCARSDALE; flows generally S through the BRONX to EASTCHESTER BAY in PELHAM BAY PARK. Impounded to created three reservoirs along its modes and length. Paralleled by landscaped Hutchinson River Parkway, known colloquially as "The Hutch." Named for Anne Hutchinson.

Hutch Mountain, peak (8,532 ft/2,601 m) in high plateau (c.7,000 ft/2,134 m), COCONINO county, central ARIZONA, c.30 mi/48 km SE of FLAGSTAFF, SE of MORMON LAKE, in Coconino National Forest.

Huth (HUTH), township, N central YEMEN, on central plateau, 45 mi/72 km SE of SA′DA, on the main road to SANA; 16°14′N 43°58′E. Sometimes spelled Houth.

Hutou, CHINA: see PEARL RIVER.

Hut Point Peninsula, projects SW from MOUNT EREBUS on ROSS ISLAND in WEST ANTARCTICA, 15 mi/25 km long and 2 mi–3 km/3 mi–5 km wide; 77°46′S 166°51′E. Previously known as Winterquarters Peninsula and Cape Armitage Promontory.

Hutsonville, village (2000 population 568), CRAWFORD county, SE ILLINOIS, on the WABASH (bridged here) and 8 mi/12.9 km NNE of ROBINSON; 39°06′N 87°39′W. In agricultural area (wheat, corn, livestock, hay).

Hutsul Beskyd, UKRAINE: see POKUTIAN-BUKOVINIAN CARPATHIANS.

Hutsul's′kyy Beskyd, UKRAINE: see POKUTIAN-BUKOVINIAN CARPATHIANS.

Hutt City (□ 436 sq mi/1,130 sq km; 2001 population 95,022), NORTH ISLAND, NEW ZEALAND, extends around the E shore of Wellington harbor (PORT NICHOLSON) to W PALLISER BAY; 41°13′S 174°55′E. Incorporates heavily industrialized and residential Lower Hutt Valley bounded on the W by the straight line of the WELLINGTON FAULT, the residential, dormitory, and recreational areas of EASTBOURNE and Wainuiomata, and the forested S extensions of the Rimutaka Range, which terminates in COOK STRAIT. Good railroad and road links to WELLINGTON. Hutt Valley industries include metalworking, engineering, textiles, and the full range of consumer industries, although there have been a number of losses in production facilities in all of the above 1965–1984, as well as in footwear and clothing. An auto assembly plant, an electrical assembly plant, and a meat-works have also closed. Scientific research institutes. Formerly known as Lower Hutt.

Hüttenberg (HYOOT-ten-berg), town, CARINTHIA, S AUSTRIA, in the Görtschitz valley, 15 mi/24 km NE of Sankt Veit; 46°56′N 14°33′E. Summer resort; railroad terminus. Local iron mines (Austria's second largest) worked here from ancient times to 1907. Museum of historic iron mining and ironworks.

Hüttenheim (oo-tuhn-EM), German, *Hüttenheim* (HOO-tuhn-heim), commune (□ 5 sq mi/13 sq km; 2004 population 2,220), BAS-RHIN department, ALSACE, E FRANCE, on the Ill, and 9 mi/14.5 km NE of SÉLESTAT; 48°21′N 07°35′E. Cotton milling.

Hutti (HUHT-tee), town, RAICHUR district, KARNATAKA state, S INDIA, 45 mi/72 km W of RAICHUR; 16°09′N 76°39′E. Gold mines, abandoned in 1920, but now in operation since reopening (1949).

Huttig (HUHT-ig), town (2000 population 731), UNION county, S ARKANSAS, 30 mi/48 km ESE of EL DORADO, near LOUISIANA line; 33°02′N 92°10′W. Manufacturing (softwood lumber); woodworking. Lower Ouachita Wildlife Management Area to E; Felsenthal National Wildlife Refuge and Felsenthal Lock and Dam (impounds Lake JACK LEE) to NE OUACHITA RIVER. Incorporated 1904.

Hüttlingen (HYUT-ling-guhn), village, BADEN-WÜRTTEMBERG, S GERMANY, on N slope of SWABIAN JURA, 4 mi/6.4 km N of AALEN; 48°53′N 10°06′E.

Hutto, village (2006 population 9,572), WILLIAMSON county, central TEXAS, 22 mi/35 km NNE of AUSTIN, on Bushy Creek; 30°33′N 97°32′W. Cotton, corn, wheat; cattle; manufacturing (machine parts).

Hutton (HUHT-tuhn), village (2006 population 2,500), W LANCASHIRE, N ENGLAND, 3 mi/4.8 km SW of PRESTON; 53°44′N 02°46′W. Garden produce.

Hutton (HUHT-tuhn), village (2001 population 3,695), North Somerset, SW ENGLAND, 2 mi/3.2 km SE of WESTON-SUPER-MARE; 51°19′N 02°55′W. Has church dating back to 15th century.

Hutton-le-Hole (HUHT-tuhn–luh–HOL), village, NORTH YORKSHIRE, N ENGLAND; 54°17′N 00°46′W. Ryedale Folk Museum has exhibits illustrating daily life in region, with reconstructions of 16th-century manor and 18th-century cottage. Picturesque village on border of Moors.

Huttonsville, village (2006 population 214), RANDOLPH county, E central WEST VIRGINIA, 16 mi/26 km SSW of ELKINS, on Tygart River; 38°42′N 79°58′W. Monongahela National Forest to E; Kumbrabow State Forest to SW.

Huttwil (HOOT-veel), commune, BERN canton, NW central SWITZERLAND, on Langetedel River, and 22 mi/35 km NE of BERN. Elevation 2,093 ft/638 m. Shoes, knit goods, pastry; tanning, canning, woodworking.

Hutubi (HU-TU-BEE), town, ⊙ Hutubi county, central XINJIANG UYGUR AUTONOMOUS REGION, NW CHINA, 40 mi/64 km WNW of URUMQI, on highway N of the TIANSHAN; 44°07′N 86°57′E. Cattle raising; wheat, rice, fruit; coal mining. Sometimes spelled Hutupi.

Hutuo River (HU-TU-uh), over 400 mi/644 km long, in SHANXI and HEBEI provinces, NE CHINA; rises on N slope of WUTAI MOUNTAINS near Hebei-Shanxi-INNER MONGOLIA border; flows S and E, into Shanxi, past Dai Xian and Dingxiang, and into Hebei, past Zhengding and Jin Xian, to area of Xian Xian, where it joins Fuyang River. Called Ziya River in lower course.

Huty (HOO-tee) (Russian *Guty*), town (2004 population 4,384), NW KHARKIV oblast, UKRAINE, 8 mi/12.9 km WSW of BOHODUKHIV; 50°07′N 35°20′E. Elevation 475 ft/144 m. Sugar refining; forestry. Established in the mid-17th century, town since 1938.

Hutyn Ridge, UKRAINE: see VOLCANIC UKRAINIAN CARPATHIANS.

Huvadhu Atoll, S group of MALDIVES, in INDIAN OCEAN, between 00°11′N 73°56′E and 00°44′N 73°31′E. Has commercial airport and fish-refrigeration plants.

Huvek, TURKEY: see BOZOVA.

Hüvösvölgy, HUNGARY: see HÁRS, MOUNT.

Huvvinahadagalli, INDIA: see HADAGALLI.

Huwairib (hu-WAI-rib), village, SW YEMEN, 65 mi/105 km W of ADEN, and on the small Wadi Timnan 10 mi/16 km from its mouth on Gulf of ADEN. Former main center of the Atifi sheikdom. Also spelled Huwayrib, Hawarib, and Hawerib.

Huwei (HU-WAI), town, W central TAIWAN, 15 mi/24 km N of CHIAI; 23°42′N 120°25′E. Sugar-milling center; rice, sweet potatoes, peanuts.

Hu Xian (HU SIAN), town, ⊙ Hu Xian county, S central SHAANXI province, NW central CHINA, 25 mi/40 km WSW of XI'AN; 34°07′N 108°36′E. Grain, animal husbandry.

Huxley, town, STORY county, central IOWA, 20 mi/32 km N of DES MOINES, 8 mi/12.9 km S of AMES; 41°53′N 93°35′W. Livestock; grain; apparel.

Huxley (HUHKS-lee), unincorporated village, S central ALBERTA, W CANADA, 70 mi/113 km NE of CALGARY, in KNEEHILL County; 51°56′N 113°14′W. Coal mining; oil and gas; wheat, oats.

Huxley, Mount (12,216 ft/3,723 m), SE ALASKA, in SAINT ELIAS MOUNTAINS, 20 mi/32 km N of ICY BAY; 60°20′N 141°10′W.

Huy (WEE), Flemish *Hoie*, commune (□ 255 sq mi/663 sq km; 2006 population 20,102), ⊙ of Huy district, LIÈGE province, E BELGIUM, on the MEUSE RIVER. Founded in the ninth century. Manufacturing. Below the nineteenth-century citadel, which dominates the town, is a Gothic abbatial church (fourteenth–fifteenth century). Mont Mosan Recreational Park.

Huyamampa (hoo-yah-MAHM-pah), village, W central SANTIAGO DEL ESTERO province, ARGENTINA, on railroad and 25 mi/40 km N of SANTIAGO DEL ESTERO; 27°23′S 64°18′W. In cotton, corn, and livestock area; lumbering. Salt deposits nearby.

Huyñaymarka, Lake (hwee-nei-MAHR-kah), SE part of Lake TITICACA, in SE PERU and W BOLIVIA (LA PAZ department); 16°20′S 68°55′W. Also spelled Huañamarca, Wuiñaimarca, Wiñaymarca, Winamarca.

Huyton-with-Roby (HEIT-uhn–WITH–RO-bee), town (2001 population 56,500), MERSEYSIDE, NW ENGLAND; 53°24′N 02°51′W. Includes residential and granite-quarrying town of Huyton, 6 mi/9.7 km E of LIVERPOOL, having 15th-century church with 14th-century arches and a font of Saxon or Norman origin. Just E is residential town of Roby.

Huzhou (HU-JO), city (□ 587 sq mi/1,520 sq km; 1994 estimated urban population 238,500; estimated total population 1,045,400), N ZHEJIANG province, SE CHINA, on S shore of TAI LAKE, 40 mi/64 km N of HANGZHOU; 30°56′N 120°04′E. The city is a center for light industry and agriculture (grain, oil crops, vegetables, hogs, eggs, and poultry). Also manufacturing (food processing, textiles, apparel, leather, chemicals, iron and steel, non-ferrous metals, machinery, electrical equipment). Last stronghold of Taiping Rebellion (1864). Also called Wuxing.

Huzhu (HU-JU), town, ⊙ Huzhu county, NE QINGHAI province, W CHINA, 20 mi/32 km NE of XINING, on GANSU border; 36°51′N 102°03′E. Livestock; food processing, non-ferrous ore mining, coal mining.

Huzurabad (huh-ZOOR-ah-bahd), town, KARIMNAGAR district, ANDHRA PRADESH state, INDIA, 23 mi/37 km SE of KARIMNAGAR; 18°12′N 79°25′E. Cotton ginning, rice milling.

Huzurnagar (huh-ZOOR-nuh-guhr), town, tahsil headquarters, NALGONDA district, ANDHRA PRADESH state, India, 42 mi/68 km SE of NALGONDA; 16°54′N 79°53′E. Rice milling, cotton ginning, castor oil extraction.

Hvaler (VAH-ler), group of 550 islands, SE NORWAY, in the SKAGERRAK, near Swedish border, along E shore of mouth of OSLOFJORDEN. Largest island is Kirkøy (11.5 sq mi/29.8 sq km), 10 mi/16 km S of FREDRIKSTAD, Norway, and 8 mi/12.9 km NW of STRÖMSTAD, SWEDEN. All major islands are connected to the mainland by bridges and tunnel. Fishing and fish canning are chief occupations; seaside resorts.

Hvalpsund, DENMARK: see SKIVE FJORD.

Hvalsund, GREENLAND: see WHALE SOUND.

Hvannadalshnúkur, peak, ICELAND: see ÖRÖFAJÖKULL.

Hvar (HVAHR), Italian *Lesina*, town, S CROATIA, port on W coast of HVAR Island, 22 mi/35 km S of SPLIT; 43°10′N 16°40′E. Sea resort. Roman Catholic bishopric. Once a leading center of Croatian culture; called "museum of architecture and art." Has 12th-century cathedral, arsenal (built 1300), Renaissance loggia (1515–1517), 16th-century Venetian fort, Franciscan church and monastery.

Hvar (HVAHR), Greek *Pharos*, Italian *Lesina*, island (□ 112 sq mi/291.2 sq km), S CROATIA, in the ADRIATIC SEA off the Dalmatian coast. Fruit (grapes) and olive growing and fishing are the chief occupations. Leading tourist center. Chief town, HVAR.

Hvardiys'k, UKRAINE: see GVARDIYS'KE.

Hverageri (HWE-rah-GER-[th]ee), town, ARNESSÝSLA county, SW ICELAND, 25 mi/40 km ESE of REYKJAVIK. Noted for its fruit, vegetables (tomatoes, cucumbers, potatoes, turnips, carrots), and flowers grown in hothouses. Sheep, cattle, horses; hay. Nearby are several natural hot springs.

Hvidtland, RUSSIA: see BELAYA ZEMLYA.

Hvítá (HWEE-tou), river, 80 mi/129 km long, SW ICELAND; rises in the HVITARVATN, a lake at S edge of LANGJOKULL; flows SW to confluence with SOG River 10 mi/16 km NE of EYRARBAKKI, forming Ölfusá River. Receives several streams from HOFSJOKULL. On upper course is GULLFOSS, large waterfall.

Hvítá (HWEE-tou), river, 50 mi/80 km long, W ICELAND; rises on EIRIKSJOKULL; flows WSW to Borgarfjörður, arm of FAXAFLOI. Noted for its salmon.

Hvítárvatn (HWEE-tour-VAH-tuhn), lake (□ 13.5 sq mi/35 sq km), W ICELAND, at SE edge of LANGJOKULL, 70 mi/113 km ENE of REYKJAVIK; 7 mi/11 km long, 1 mi/1.6 km–2 mi/3 km wide. Drained S by HVITA River.

Hvitsten (VEET-stain), village, AKERSHUS county, SE NORWAY, on E shore of OSLOFJORD narrows, 4 mi/6.4 km SSE of DRØBAK. Summer resort.

Hvittingfoss (VIT-ting-faws), village, BUSKERUD county, SE NORWAY, on LÅGEN River (falls), and 20 mi/32 km SSW of DRAMMEN. Paper- and sawmilling. Power station supplies TØNSBERG. Rock carvings from Bronze Age found S of waterfalls.

Hvizdets (hveez-DETS), (Russian *Gvozdets*), (Polish *Gwozdziec*), town, E IVANO-FRANKIVS'K oblast, UKRAINE, on left tributary of the PRUT, and 11 mi/18 km ENE of KOLOMYYA. Railroad station; flour milling, brick manufacturing. Has old monastery and church.

Hvoynati Vruh (khvoi-NAH-tee VROOKH), peak (8,645 ft/2,635 m) in the PIRIN MOUNTAINS, SW BULGARIA; 41°53′N 24°44′E.

Hwachon (HWA-CHUHN), county (□ 369 sq mi/959.4 sq km), NW KANGWON province, SOUTH KOREA, adjacent to YANGGU on E, CHUNCHON on S, CHORWON on N and W, and KYONGGI province on W. Part of NORTH KOREA before Korean War. Hilly in NE, downslopes to SW. Hwachon dam forms Paroho artificial lake on Pukhan River in E. Some agriculture (maize, potatoes, beans, apples, pears, medical herbs); sericulture; animal husbandry. Hwachon hydroelectric power plant.

Hwai, CHINA: see HUAI.

Hwai Ho, CHINA: see HUAI.

Hwainan, CHINA: see HUAINAN.

Hwaining, CHINA: see ANQING.

Hwalien, TAIWAN: see Hualien.

Hwange, city, MATABELELAND NORTH province, W ZIMBABWE, 180 mi/290 km NW of BULAWAYO, near DEKA RIVER; 18°22′S 26°23′E. The city was founded in 1903 and named for a local chief. Airport to E. HWANGE NATIONAL PARK to S, ZAMBEZI and Victoria Falls national parks to NW, KAZUMA PAN NATIONAL PARK to W; Deka Safari Area to S. Goba and Chawato suphur springs to SW. Coal mining, with substantial reserve. Livestock; grain. Tourism. Formerly called Wankie.

Hwange National Park (□ 2,289 sq mi/5,929 sq km), MATABELELAND NORTH province, W ZIMBABWE, S of HWANGE. Bounded on SW by BOTSWANA, N by Mateti and Deka safari areas. Drained in NW by DEKA RIVER. Numerous salt pans, especially in NW, provide water holes for wide variety of wildlife in this semiarid grassland, including giraffe, lion, eland, zebra, kudu, waterbuck, elephant, and buffalo. Lodges at MAIN CAMP, SINAMATELA CAMP, and ROBINS CAMP, all in NW; airport at Main Camp. Largest national park in Zimbabwe.

Hwang Ho, CHINA: see HUANG HE.

Hwangju (HWAHNG-JOO), county, SOUTH HWANGHAE province, central NORTH KOREA, 24 mi/39 km S of PYONGYANG. Iron mining.

Hwangpoo, CHINA: see HUANGPU.

Hwangshih, CHINA: see HUANGSHI.

Hwasong (HWA-SUHNG), county (□ 282 sq mi/733.2 sq km), KYONGGI province, SOUTH KOREA, on SW coast of YELLOW SEA. Hilly in N and E; S and W form a wide field. Agriculture (rice, barley, beans, fruit); salt works and shell cultivation on coast. Kyongbu expressway and railroad; Su-in railroad.

Hwasun (HWA-SOON), county (□ 302 sq mi/785.2 sq km), central SOUTH CHOLLA province, SOUTH KOREA, contiguous to TAMYANG (N), KOKSONG and SUNGJU (E), POSONG and CHANGHUNG (S) and NAJU city (W). County in Sobaek Mountains; HWASUN, Nungju, and Chisok rivers in W form Nungju plain, an agricultural center (rice, barley, beans, sweet potatoes, potatos, maize). Horticulture due to close proximity to KWANGJU. Natural resources included limestone, agalmatolite, silica and coal (third-largest producer in Korea). Kyongjon railroad; Hwasun railroad (coal delivery); Sangbong and Unju Buddhist Temples.

Hwlffordd, Wales: see HAVERFORDWEST.

Hyannis (hei-YAN-uhs), resort town, BARNSTABLE county, SE MASSACHUSETTS, on CAPE COD, seat of town of BARNSTABLE offices; 41°39′N 70°18′W. It is the business center and shipping point of the area; major industries are tourism and home construction; large retail malls. Other industry includes recreational products, candles, foams (polyethylene). Hyannis provides ferry transportation to MARTHA'S VINEYARD and NANTUCKET island Barnstable Municipal Airport. A community college and a conservatory of music and arts are located there. Nearby Hyannisport is famous as the site of a compound of houses owned by the Kennedy family. Incorporated 1639.

Hyannis (hei-AN-is), village (2006 population 253), ⊙ GRANT county, W central NEBRASKA, in Sand Hills, 62 mi/100 km N of OGALLALA; 42°00′N 101°45′W. Livestock; wild hay. Numerous small natural lakes including Raymond and Collins to NE, George Lake to SW, Home and Big Buckboard to S.

Hyargas Nuur, salt lake (□ 545 sq mi/1,411.8 sq km) in NW MONGOLIA, 80 mi/130 km SE of ULAANGOM; 52 mi/84 km long, 20 mi/32 km wide; 49°12′N 93°24′E. Elevation 3,373 ft/1,028 m. Situated in a semi-desert depression called the Basin of Great Lakes; the terminus of a system of lakes and rivers in an interior drainage basin. It receives the water of the DZAVHAN GOL (river) and is the last in a series of lakes. It contains sodium sulphate, sodium carbonate and magnesium chloride, but has abundant fish. Also spelled Kirgis Nor or Khirgis Nur.

Hyatt Regency Resort, tourist complex, St. Kitts, Federation of ST. KITTS AND NEVIS, WEST INDIES. A 250-room all-inclusive resort at South Friar's Bay on SE peninsula; opened in 1995 as part of tourism development program.

Hyattstown (HI-ats-town), town, MONTGOMERY county, central MARYLAND, 17 mi/27 km NW of ROCKVILLE. A commuter suburb of WASHINGTON, D.C., the first house here was built by Seth Hyatt about 1800. A relative, Jesse (1763–1813) ran a hotel behind the Methodist Church. Union and Confederate forces skirmished here in 1862. An abandoned grist mill is on Hyattsville Mill Road.

Hyattsville, city (2000 population 14,733), PRINCE GEORGES county, W central MARYLAND, a suburb of WASHINGTON, D.C.; 38°58′N 76°57′W. A residential community with some light industry and commercial activity, Hyattsville is named after Christopher Hyatt. Christopher, who settled here in 1860, was a close relative of Seth Hyatt, the founder of HYATTSTOWN. It is located in an area of major housing development and service industries, particularly for middle-income families who work in Washington, D.C. The Marquis de Lafayette stayed at Bothwick Hall, and the inventor, James Harrison Rogers, was a resident. Since it is closer to the centers of population than the county seat, UPPER MARLBORO, the County Library and many county offices are located here. Incorporated 1886.

Hyattville (HEI-uht-vil), village, Big Horn co., N Wyoming, on Paintrock Creek branch of Big Horn River, located in W foothills of Bighorn Mountains and 24 mi/39 km ESE of Basin. Elev. c.4,457 ft/1,358 m. In bean and sugar beet growing area. Medicine Lodge State Archaeological Site is here. Bighorn National Forest to E.

Hybla Valley (HEI-bluh), unincorporated city, FAIRFAX county, NE VIRGINIA, residential suburb 4 mi/6 km SSW of ALEXANDRIA, 11 mi/18 km SSW of WASHINGTON, D.C.; 38°45′N 77°04′W. Huntley Meadows Park to W.

Hybo (HEE-BOO), village, GÄVLEBORG county, E SWEDEN, near LJUSNAN RIVER, 4 mi/6.4 km SE of LJUSDAL; 61°48′N 16°12′E.

Hydaburg (HEI-duh-buhrg), town (2000 population 382), SE ALASKA, on W coast of PRINCE OF WALES ISLAND, 23 mi/37 km SE of CRAIG; 55°12′N 132°49′W. Fishing, fish processing, some timber. Connected by road network to N parts of island Totem Park. Founded 1911, combined 3 Haida settlements: Sukkwan, Howkan, and Klinkwan. Incorporated 1927.

Hydaspes, INDIA and PAKISTAN: see JHELUM RIVER.

Hyde (HEID), county (□ 1,424 sq mi/3,702.4 sq km; 2006 population 5,341), E NORTH CAROLINA; ⊙ SWAN QUARTER; 35°24′N 76°09′W. Forested and swampy tidewater area; bounded SE by ATLANTIC OCEAN, bounded N in part (NE) by ALLIGATOR RIVER estuary, W in part by PUNGO RIVER estuary, E by Long Shoal River estuary; crossed by Alligator-Pungo Canal; includes Ocracoke Island sand barrier island, part of OUTER BANKS and CAPE HATTERAS NATIONAL SEASHORE 20 mi/32 km SE of mainland across PAMLICO SOUND (ferry from SWAN QUARTER to OCRACOKE village). Agriculture (cotton, corn, wheat, soybeans); timber; fish, crabs, oysters; service industries. Lake Mattamuskeet National Wildlife Refuge in center; Swanquarter National Wildlife Refuge in SW; Alligator Lake in N; part of Pungo Lake in NW; large Lake MATTAMUSKEET in center (natural lakes). Formed 1705 as Wickham Precinct of Bath County. Named Hyde around 1712 for Governor Edward Hyde who died in 1712.

Hyde, county (□ 866 sq mi/2,251.6 sq km; 2006 population 1,551), central SOUTH DAKOTA; ⊙ HIGHMORE; 44°32′N 99°28′W. Agricultural area drained by Medicine Knoll and Wolf creeks; MISSOURI RIVER and Crow Creek Indian Reservation in SW corner. Wheat; dairy products; cattle. The SW corner borders MISSISSIPPI RIVER. Formed 1873.

Hyde (HEID), town (2001 population 31,253), GREATER MANCHESTER, central ENGLAND, 7 mi/11.3 km ESE of MANCHESTER; 53°27′N 02°05′W. Machine works, iron foundries; heavy manufacturing. To the S is leather-manufacturing suburb of Gee Cross.

Hyde, unincorporated town (2000 population 1,491), CLEARFIELD county, central PENNSYLVANIA, residential suburb 2 mi/3.2 km SW of CLEARFIELD on West Branch of SUSQUEHANNA RIVER; 41°00′N 78°28′W. Agriculture, includes dairying.

Hyde Creek (HEID KREEK), unincorporated village, SW BRITISH COLUMBIA, W CANADA, 4 mi/6 km from ALERT BAY, in Mount WADDINGTON regional district; 50°35′N 127°00′W.

Hyden, town, WESTERN AUSTRALIA state, W AUSTRALIA, 209 mi/336 km SE of PERTH; 32°30′S 118°51′E. In wheat-growing region. Tourism. Just E of town is Wave Rock, (part of the larger Hyden Rock complex), a collapsing-wave-shaped granite outcrop c.50 ft/15 m high and more than 328 ft/100 m long.

Hyden (HEID-uhn), village (2000 population 204), ⊙ LESLIE county, SE KENTUCKY, 12 mi/19 km WSW of HAZARD, in CUMBERLAND MOUNTAINS, on Middle Fork KENTUCKY RIVER, surrounded by Daniel Boone National Forest; 37°09′N 83°22′W. Light manufacturing. Seat of Frontier Nursing Service Hospital, including Wendover Big House (1925), to W. Buckhorn Lake reservoir and State Resort Park to N. Established 1882.

Hyde Park, town (□ 39 sq mi/101.4 sq km), DUTCHESS county, SE NEW YORK, on HUDSON RIVER; 41°47′N 73°56′W. It is famous as the site of the Roosevelt estate, part of the 264-acre/107-ha FDR National Historic Site where President Franklin D. Roosevelt was born and is buried. The Roosevelt Library (1941) contains historical material dating from 1910 until Roosevelt's death in 1945. The adjacent 180-acre/73-ha Eleanor Roosevelt National Historic Site (Valkill), an estate used by Mrs. Roosevelt as her personal retreat, was built for her by her husband, in 1925. Hyde Park is also the site of the Frederick W. Vanderbilt mansion and two state parks. All three homes are national historic sites. Seat of Culinary Institute of America. Settled c.1740.

Hyde Park, unincorporated town, BERKS county, SE central PENNSYLVANIA, residential suburb 3 mi/4.8 km N of READING; 40°22′N 75°55′W. Railroad junction.

Hyde Park, town (2006 population 2,864), CACHE county, N UTAH, 8 mi/12.9 km N of LOGAN; 41°47′N 111°48′W. Wasatch National Forest and Mount Naomi Wilderness Area to E. Elevation 4,560 ft/1,390 m. Settled 1860s.

Hyde Park, town (2006 population 472), including Hyde Park village, ⊙ LAMOILLE CO., N central VERMONT, 23 mi/37 km N of MONTPELIER. Lumber; wood products; 44°37′N 72°33′W. Settled 1787. North Hyde Park village is a woodworking center. Named forf Jedediah Hyde.

Hyde Park, village, DEMERARA-MAHAICA district, N GUYANA; 06°30′N 58°16′W. At landing on right bank of DEMERARA RIVER and 23 mi/37 km SSW of GEORGETOWN, in agricultural region. Just SE a U.S. army base (with ATKINSON AIRFIELD) was established 1940.

Hyde Park, borough (2006 population 491), WESTMORELAND county, SW central PENNSYLVANIA, on KISKIMINETAS RIVER 1 mi/1.6 km E of LEECHBURG. Manufacturing (cast iron rolls); agriculture (corn, hay; livestock, dairying).

Hyde Park (HEID), Westminster borough (□ 1 sq mi/2.6 sq km), LONDON, ENGLAND; 51°30′N 00°10′W. Once the manor of Hyde, a part of the old WESTMINSTER ABBEY property, it became a deer park under Henry VIII. Races were held here in the 17th century. In 1730, Queen Caroline had the artificial lake, the Serpentine, constructed. It curves diagonally through Hyde Park; in Kensington Gardens the lake is called the Long Water. Distinctive features of the park are Hyde Park Corner (near the Marble Arch), the meeting place of soapbox orators, and Rotten Row, a famous bridle path.

Hyder (HEI-duhr), village, SE ALASKA, on BRITISH COLUMBIA border, at head of PORTLAND CANAL, 2 mi/3.2 km S of STEWART, British Columbia. Supply point; port of entry; small-scale mining (gold, silver, lead, tungsten). Originally named Portland City; renamed for geologist F.B. Hyder. Town destroyed by great fire of 1948. Most mining ceased 1956. Granduc copper mine closed 1984; Westmin gold and silver mine continues to operate. Air service, road connection through CANADA, ferry connections. Popular for tourist nightlife (2 bars). "Friendliest Ghost Town in Alaska." In Misty Fiords National Monument.

Hyderabad (HEI-duhr-ah-bahd), former princely state, S central INDIA. The former princedom of Hyderabad is now divided among the states of KARNATAKA, MAHARASHTRA, and ANDHRA PRADESH. Situated almost entirely within the DECCAN PLATEAU, it was without coastal frontage. Abundant crops of cotton and rice grew in the N, while grains were grown in the heavily irrigated S area. Cotton spinning, weaving, and food processing were the principal industries. The monuments of AJANTA and ELLORA were the chief relics of the region's ancient Hindu civilization. The Mogul empire conquered Hyderabad in the late seventeenth century. In 1724 the viceroy Nizam-al-Mulk, founder of the last royal line, became its independent ruler. Later nizams (rulers) sought to maintain their independence, but the dynasty was forced to accede to British protection in 1798. In 1903, Berar, then the northernmost section of the state, was transferred to British administration. When India was partitioned (1947), the nizam, one of India's most important Muslim princes, wished to remain independent. Some 80% of Hyderabad's inhabitants were Hindu, however. After a series of religious battles, allegedly staged by India, the Indian army invaded Hyderabad in 1948. The population, in a plebiscite, endorsed accession to India. Hyderabad became a state in 1950 but was partitioned among neighboring states in 1956. The nizam, forced to renounce nearly all of his fortune, was removed from power.

Hyderabad (HEI-duhr-ah-bahd), district (□ 84 sq mi/218.4 sq km), central ANDHRA PRADESH state, S central INDIA, on DECCAN PLATEAU; ⊙ HYDERABAD. In hilly area (livestock), drained by MUSI RIVER. Sandy red soil; millet, rice, oilseeds (chiefly peanuts). Surrounds Hyderabad, its only important city and also the twin city of SECUNDERABAD; noted GOLCONDA fort is just W of Hyderabad. Formerly the personal estate of the nizam; nationalized by Indian government in 1949. Population is majority Hindu, but there is a sizable minority Muslim population, especially in the old city. Formerly known as Atraf-i-Balda. The area has grown at an incredibly rapid pace in the 1990s and 2000s, with much multinational and local investment in new information technology and biotechnology ventures; due to this boom, Hyderabad is growing into one of the most important regions of twenty-first century India.

Hyderabad (HEI-duhr-ah-bahd), walled city, (2001 population 5,742,036), ⊙ HYDERABAD (Urban) and RANGAREDDI districts, and ANDHRA PRADESH state, S central INDIA; 17°23′N 78°28′E. An administrative and commercial center and a transportation hub, the city has fine historic structures, notably the Charminar (1591), a 150-ft/46-m-high civic monument, and the Old Bridge (1593). Handicrafts include metal inlay work, carpets, silk, and gold and pearl jewelry (known as "Pearl City of the South"). Seat of Osmania University (founded 1918) and of Central University of Hyderabad. Several scientific technical institutes have also been set up, including the National Geophysical Institute and the Remote Sensing Agency, making Hyderabad a new center of science and technology. The former British cantonment of SECUNDERABAD is now a twin city. Airport at Begampett, c.7 mi/12 km N. Other attractions include an artificial lake, mosques, tombs, museum, and an open-air zoo. Original settlement on MUSI RIVER established 1589 as Bhagyanagar, the capital of the Golconda kingdom. Muslim invaders changed name to Hyderabad. It was the capital of Hyderabad princely state. In the 1990s and 2000s, Hyderabad emerged as an information technology hub in South India, on par with BANGA-LORE in terms of sheer job and wealth creation; the city's interesting architectural mix reflects its past and current economic importance, with gleaming new office parks and high-end housing next to nizam-era buildings. Overall Hyderabad is a wealthy and developed region, both in economic and historic terms: courtly mannerisms, finely crafted jewelry (especially pearls), luxurious textiles, and rich cuisine are considered hallmarks of Hyderabadi culture.

Hyderabad (hai-dah-rah-bahd), city, ⊙ HYDERABAD district, SIND province, SE PAKISTAN, near left bank of INDUS River, 90 mi/145 km NE of KARACHI; 25°22′N 68°22′E. Pakistan's fourth-largest city. Railroad and road junction. Important trade center and market of millet, rice, wheat, cotton, and fruits. Long been noted for its embroideries, precious-metal goods, and cutlery. Manufacturing includes food processing; textiles, hosiery, cement, cigarettes, glass, soap, paper, leather goods, and plastics. Airport. Founded in 1768 by Ghulam Shah Kalhora, Hyderabad was laid out by his son, Sarfaraz Khan, in 1782 and was the capital of the emirs of SIND. The British East India Company occupied Hyderabad when the Sind became a British protectorate in 1839. Was the capital of Old Sind until its surrender to the British in 1843, when Sir Charles Napier defeated the emirs of SIND. Became the capital of SIND again after 1950, replacing KARACHI. Seat of University of Sind, over 30 affiliated colleges, commercial and agriculture colleges, medical school. UMARKOT, birthplace of the great Mogul emperor Akbar, is nearby. Sometimes spelled HAIDARABAD.

Hyderabad Deccan, INDIA: see HYDERABAD.

Hydernagar, INDIA: see NAGAR, village.

Hydesville, unincorporated town (2000 population 1,209), HUMBOLDT county, NW CALIFORNIA, 17 mi/27 km SSE of EUREKA, near Van Duzen River. Humboldt Redwoods State Park to S; Grizzly Creek Redwoods State Park to E. Timber, cattle, sheep.

Hydetown (HEID-toun), borough (2006 population 580), CRAWFORD county, NW PENNSYLVANIA, 3 mi/4.8 km NW of TITUSVILLE, on Thompson Creek, near its mouth on OIL CREEK, to SW. Dairying.

Hydeville, VERMONT: see CASTLETON.

Hydra (EE-dra), island (□ 21 sq mi/54.6 sq km), ATTICA prefecture, ATTICA department, SE GREECE, in the AEGEAN SEA, off the ARGOLIS PENINSULA of the PE-LOPONNESUS; 37°20′N 23°30′E. Mostly barren and rocky. Hydra town is the center of population. Sponge fishing, shipbuilding, textile manufacturing, and tourism are the main industries. Settled from ancient times; present population and culture derive from the various Greeks who, beginning in the 15th century, fled here escaping Turkish persecution on the mainland. Left largely alone by the Ottoman rulers. A shipbuilding and commercial center by the 17th century. Declined after the Greek War of Independence (1821–1829), in which its seafaring people played an important role. Notable architecture. No vehicular traffic owing to topography. Also spelled Idhra or Hydrea; formerly spelled Idhra or Ydra.

Hydro, town (2006 population 1,037), CADDO county, W central OKLAHOMA, 8 mi/12.9 km E of WEATHERFORD, on Deer Creek; 35°32′N 98°34′W. In agricultural area (cotton, grain); manufacturing (peanut processing, peanut butter, sand and gravel).

Hydrovia, proposed waterway, SOUTH AMERICA. Plans included straightening and deepening the PARANÁ RIVER system, improving PARAGUAY's and BOLIVIA's outlet to the sea and also serving BRAZIL's MATO GROSSO, URUGUAY, and ARGENTINA. Involves 2,000 mi/3,219 km of waterways from CÁCERES, Brazil on UPPER PARAGUAY RIVER to NUEVA PALMIRA in Uruguay at head of RÍO DE LA PLATA. This will permit year-round deep-draft vessels to serve ASUNCIÓN.

Shallows and rapids will be eliminated N of Asunción, eliminating need to unload and reload vessels upstream; improvements to thirty-one river ports and parallel highways are scheduled. The project was implemented in 1994 and it is expected that work will take over a decade. Opposition to project comes from environmentalists concerned that re-engineering the Upper Paraná will greatly damage EL PANTANAL, the 50,000 sq mi/129,500 sq km floodplain of the Upper Paraná that is the world's most biologically diverse wetland.

Hyères (YER), city, resort in VAR department, SE FRANCE, in PROVENCE, on the MEDITERRANEAN and 11 mi/18 km E of TOULON; 43°07′N 06°07′E. It is the oldest of the French RIVIERA's resorts, well sheltered from N winds by the MAURES range. Has mild climate (especially in winter) and rich subtropical vegetation. The old town (2 mi/3.2 km inland) is picturesque, with a central square named for the preacher Massillon (1663–1742), many Renaissance gates, and the 12th–17th-century collegiate church of St. Paul. In addition to its development for tourism (Hyères is a favorite destination for British tourists), the community is surrounded by intensive cultivation of early vegetables and fruit (especially strawberries, peaches), vineyards, and ornamental plants. The streets of the modern town are lined with palm trees; there is a casino. The Toulon-Hyéres airport is just E of town center. The beach, small harbor, and hippodrome are located on a sandspit (2.5 mi/4 km long) connecting Hyères with the rocky peninsula of GIENS, which in fact is a former island (c. 3 mi/4.8 km long, E-W) linked to the mainland by 2 sandbars, which enclose a salt lagoon (saltworks). Hyères was built atop Greek ruins. It served as a point of departure for several crusades (13th century). The town passed to the counts of Provence, but beginning in 16th century it was overshadowed by nearby Toulon. Offshore is a group of islands known as the Îles d'HYÈRES.

Hyères, Îles d' (eel-dyer), island group (□ 7 sq mi/18.2 sq km) in the MEDITERRANEAN just off S coast of FRANCE at HYÈRES, administratively in VAR department, PROVENCE-ALPES-CÔTE D'AZUR region. Chief islands are, E–W, LEVANT, PORT-CROS (which is designated as a national park), and PORQUEROLLES (lighthouse); 43°00′N 06°20′E. They protect the Rade d'Hyères, a sheltered bay (10 mi/16 km wide, 6 mi/9.7 km deep) on the RIVIERA, which extends from GIENS PENINSULA (SW) to Cap BÉNAT (E).

Hyesan (HYE-SAHN), city; ⊙ YANGGANG province, NE NORTH KOREA, 110 mi/177 km NNE of HAMHUNG on YALU RIVER (MANCHURIA line), opposite CHANGPAI; 41°20′N 128°12′E. Railroad terminus. Reached in 1950 by U.S. troops in Korean War. Connects primary railroads. Major lumber and paper manufacturing center. Coldest temperatures in Korea.

Hygiene, unincorporated village, BOULDER county, N central COLORADO, 4 mi/6.4 km W of LONGMONT, on SAINT VRAIN CREEK. Elevation 5,090 ft/1,551 m. Cattle; fruit, vegetables, wheat; manufacturing of feeds.

Hylike, Lake, Greece: see ILIKI, LAKE.

Hylo (HEI-lo), hamlet, E central ALBERTA, W CANADA, 12 mi/19 km from LAC LA BICHE town, in LAKELAND County; 54°41′N 112°13′W.

Hyltebruk (HIL-te-BROOK), village, JÖNKÖPING county, S SWEDEN, on NISSAN RIVER, 30 mi/48 km WSW of VÄRNAMO; 57°00′N 13°14′E.

Hymera, town (2000 population 833), SULLIVAN county, SW INDIANA, 9 mi/14.5 km NNE of SULLIVAN; 39°11′N 87°18′W. In agricultural area; bituminous-coal mining. Shakamak State Park is nearby. Plotted 1870.

Hyndman (HIND-muhn), borough (2006 population 967), BEDFORD county, S PENNSYLVANIA, on Wills Creek, 22 mi/35 km SW of BEDFORD. Manufacturing (sawmill, lumber); Agriculture (corn, oats, hay; live-

stock; dairying). Leap Airport to N. Little Allegheny Mountain ridge to W; WILLS MOUNTAIN ridge to E. Severely damaged by fire, 1949. Laid out 1840, incorporated 1877.

Hyndman, Peak, highest (12,078 ft/3,681 m) of PIONEER MOUNTAINS, on CUSTER-BLAINE county line, central IDAHO, 11 mi/18 km ENE of SUN VALLEY. On boundary of Challis (NE) and Sawtooth (SW) national forests.

Hyogo (HYO-go), prefecture (□ 3,213 sq mi/8,322 sq km; 1990 population 5,405,090), S HONSHU, W central JAPAN, on SEA OF JAPAN (N) and HARIMA SEA and OSAKA BAY (S); ⊙ KOBE. Bordered E by KYOTO and OSAKA prefectures, W by SHIGA and MIE prefectures. Iron, steel, textiles, food processing, and lumbering are the main industries. Industrial centers at Kobe, AKASHI, AMAGASAKI, HIMEJI, and NISHINOMIYA.

Hypate, Greece: see IPATI.

Hyphasis, INDIA: see BEAS RIVER.

Hyrcania, IRAN: see GORGAN.

Hyrum, town (2006 population 5,971), CACHE county, N UTAH, on LITTLE BEAR RIVER, 7 mi/11.3 km S of LOGAN, and near WASATCH RANGE; 41°37′N 111°50′W. Vegetables, wheat, barley; dairying; cattle; manufacturing (meat packing). Elevation 4,750 ft/1,448 m. HYRUM RESERVOIR and Hyrum State Park on SW end of town, impounds water for irrigation of surrounding region. Settled 1860.

Hyrum Reservoir, at HYRUM, CACHE county, N UTAH, on LITTLE BEAR RIVER, in Hyman State Park, 11 mi/18 km NE of BRIGHAM CITY; 2 mi/3.2 km long; 41°29′N 111°53′W.

Hysham (HI-shuhm), village (2000 population 330), ⊙ TREASURE county, S central MONTANA, on YELLOWSTONE RIVER, 74 mi/119 km NE of BILLINGS; 46°17′N 107°13′W. Railroad junction to E; shipping point for cattle, corn, hay, sugar beets.

Hyskier, Scotland: see HASKEIR.

Hythe (HEI[TH]), town (1991 population 13,118; 2001 population 20,467), KENT, SE ENGLAND, 10 mi/16 km SW of DOVER; 51°05′N 01°05′E. A summer resort and market town, it was the Roman Portus Lemanis, one of the CINQUE PORTS, until shingle drift lowered its value. Nearby is Saltwood Castle, once the property of the archbishops of Canterbury. The School of Small Arms is here.

Hythe (HEI[TH]), village (□ 1 sq mi/3 sq km; 2001 population 582), NW ALBERTA, W CANADA, near BRITISH COLUMBIA border, on Beaverlodge River, 32 mi/51 km WNW of GRANDE PRAIRIE, in GRANDE PRAIRIE County No. 1; 55°19′N 119°34′W. Lumbering, mixed farming. Established 1929.

Hythe (HEI[TH]), suburb (1991 population 7,923; 2001 population 19,599), HAMPSHIRE, S central ENGLAND, on W bank of SOUTHAMPTON WATER, S of SOUTHAMPTON; 50°52′N 01°24′W.

Hyttefossen (HUT-tuh-faws), waterfall (171 ft/52 m) on NIDELVA River, in SØR-TRØNDELAG county, central NORWAY, just W of SELBUSJØEN Lake, 13 mi/21 km S of TRONDHEIM. Hydroelectric plant.

Hyuga (HYOO-gah), city, MIYAZAKI prefecture, E KYUSHU, SW JAPAN, on the PACIFIC OCEAN, 37 mi/60 km N of MIYAZAKI; 32°25′N 131°37′E. Citrons; clams. Traditional-game pieces and tables. National railroad research center.

Hyuga Sea (HYOO-gah), Japanese *Hyuga-nada* (HYOO-gah-NAH-dah), N arm of PHILIPPINE SEA, forms wide bight in MIYAZAKI prefecture, E KYUSHU, SW JAPAN, between Cape Sen (N) and POINT TOZAKI (S); c.75 mi/121 km wide.

Hyvinkää (HUH-vin-KAH), Swedish *Hyvinge*, town, UUDENMAAN province, S FINLAND, 30 mi/48 km N of HELSINKI; 60°38′N 24°52′E. Elevation 330 ft/100 m. Commuter suburb of Helsinki; railroad junction. Mostly residential; some light industry.

I

Ia, Greece: see OIA.

Iacanga (EE-ah-kahn-gah), town (2007 population 9,074), central SÃO PAULO, BRAZIL, near TIETÊ RIVER, 30 mi/48 km N of BAURU; 21°54′S 49°01′W. Coffee processing, pottery manufacturing; cattle raising.

Iaciara (EE-ah-SEE-ah-rah), city (2007 population 12,755), E central GOIÁS, BRAZIL, 31 mi/50 km W of POSSE; 14°10′S 46°37′W.

Iacobeni (yah-ko-BEN), village, SUCEAVA county, N ROMANIA, on BISTRIŢA RIVER, on railroad, and 7 mi/11.3 km NNW of VATRA DORNEI, in the E CARPATHIANS. Health resort (elev. 2,789 ft/850 m). Also manganese-mining center; limestone quarrying; manufacturing of agricultural tools. Established by German colonists.

Iacri (EE-ah-kree), town (2007 population 6,605), W SÃO PAULO state, BRAZIL, 53 mi/85 km NW of MARÍLIA, on railroad; 21°51′S 51°39′W.

Iaçu (ee-ah-SOO), city (2007 population 27,740), E central BAHIA, BRAZIL, 19 mi/30 km SE of ITABERABA, on SALVADOR–BELO HORIZONTE railroad, on RIO PARAGUAÇÚ; 12°46′S 40°13′W.

Ia Drang (EE-uh DRANG), valley, GIA LAI province, central VIETNAM; 13°36′N 107°29′E. Site of major Vietnam War battle. Jarai, Krung, and other minorities.

Iaeger (EI-guhr), town (2006 population 309), MC-DOWELL county, S WEST VIRGINIA, on Tug Fork River, 13 mi/21 km W of WELCH; 37°27′N 81°48′W. Semibituminous-coal field. Manufacturing (coal processing). Railroad junction. Panther State Forest to W.

Ialomiţa (yah-lo-MEE-tsah), county, SE ROMANIA, in WALACHIA; ⊙ SLOBOZIA; 44°40′N 26°50′E. Flat plain. Agriculture.

Ialomiţa River, 200 mi/322 km long, S central and S ROMANIA; rises in BUCEGI MOUNTAINS 5 mi/8 km NW of SINAIA; flows S past PUCIOASA and TÎRGOVIŞTE, and E past SLOBOZIA, to the lower DANUBE River, 3 mi/4.8 km W of HÎRŞOVA. Receives PRAHOVA RIVER (left). Upper valley is noted for its stalactite caverns, formerly site of ancient hermitage.

Ialpug, UKRAINE: see YALPUH LAGOON.

Ialysus, Greece: see TRIANDA.

Ianca (YAHN-kah), town, BRĂILA county, SE ROMANIA, on railroad, and 24 mi/39 km SW of BRĂILA. Agricultural center.

Ianstown, Scotland: see BUCKIE.

Iara (YAH-rah), Hungarian *Jára*, village, CLUJ county, NW central ROMANIA, 12 mi/19 km W of TURDA; 46°33′N 23°31′E.

Iaşi (YAHSH), county, E ROMANIA, in MOLDOVA, on border with MOLDOVA; ⊙ IAŞI; 47°15′N 27°15′E. Terrain ranges from flat to hilly; SIRET RIVER runs through county. Agriculture; industry.

Iaşi (YAHSH), city; ⊙ IAŞI county, E ROMANIA, in Moldova, 62 mi/100 km ENE of PIATRA NEAMŢ, near MOLDOVA; 47°10′N 27°36′E. Fifth-largest city in Romania and the administrative and commercial center of a fertile agricultural region. Chemicals, pharmaceuticals, plastics, machinery, metals, and textiles are produced. Airport, thermoelectric power plant. In 1565, Iaşi succeeded SUCEAVA as capital of the Romanian principality of Moldavia, a position it held until Moldavia and WALACHIA were united in 1859. Repeatedly burned and sacked by Tatars, Turks, and Russians. A treaty signed in 1792 ended the 2nd of the Russo-Turk. Wars of Catherine II. Iaşi was long an important cultural center; here, the 1st book in the

Romanian language was printed (1643) and the national theater was founded (1849). During World War I the city served as Romania's temporary capital while German forces occupied Walachia. The oldest Jewish community in Moldavia (from 2nd half of 15th century); Jews constituted $\frac{1}{2}$ the population into the 1920s, and $\frac{1}{3}$ until World War II. The town's Jewish residents were the leading commercial force here from late 18th century. Scene of many pogroms against the Jews, including one of the worst in history, when most of the Jewish population was massacred by the Nazis during World War II. Soviet troops took the city in 1944. See of an Orthodox archbishop; has a university (founded 1860) and other institutions of higher education. Landmarks include the 17th-century cathedral, the Church of the Three Hierarchs (17th century), and the Church of St. Nicholas (15th century), all outstanding examples of the Moldavian adaptation of Byzantine architecture. Formerly spelled Jassy.

Iasmos (e-AHS-mos), village, RODOPI prefecture, EAST MACEDONIA AND THRACE department, NE GREECE, on railroad, and 10 mi/16 km W of KOMOTINÍ; 41°08′N 25°11′E. Tobacco, wheat, vegetables.

Iatt, Lake (EI-uht), reservoir, GRANT parish, central LOUISIANA, on Iatt Creek, 3 mi/5 km NE of COLFAX; c. 7 mi/11 km long; 31°33′N 92°39′W. Natural marshy lake now impounded by dam. Lake Iatt State Game and Fish Preserve here. Kisatchie National Forest to E.

Iauaretê (EE-ou-AH-re-che), town, NW AMAZONAS, BRAZIL, post on COLOMBIA border, on Rio UAUPÉS at mouth of Rio PAPURY, and 150 mi/241 km WNW of Uaupés; 50°00′N 69°03′W. Formerly spelled Iuaretê, Jauaretê, or Yauareté.

Iba (EE-bah), town, ⊙ ZAMBALES province, central LUZON, PHILIPPINES, near W coast, 45 mi/72 km WSW of TARLAC; 15°22′N 120°02′E. Trade center for rice-growing area. Chromite shipping center (the world's richest refractory chromite deposit lies 9 mi/14.5 km inland at Coto). Mineral pigments. Agriculture. Airstrip. Birthplace of President Ramon Magsaysay. Founded 1611.

Ibach (EE-bahkh), village, SCHWYZ canton, central SWITZERLAND, 2 mi/3.2 km S of SCHWYZ. Manufacturing of renowned "Swiss Army" knives.

Ibadan (EE-bah-DAHN), city (2004 population 3,200,000), ⊙ OYO state, SW NIGERIA. The second-largest city in Nigeria, it is a major commercial center. Manufacturing includes metal products, furniture, soap, and handicrafts. It also is an important market for cacao, which, along with cotton, is produced in the region. Founded in the 1830s as a military camp during the YORUBA civil wars and developed into the most powerful Yoruba city-state. EGBA people lived in area, but after Yoruba civil war, migrated to areas near LAGOS in the 1830s. In 1840, Ibadan forces defeated Fulani invaders from the N at the battle of Oshogbo, thus protecting S Yorubaland from attack. Came under British protection in 1893. Has some mosques. Seat of the University of Ibadan (1962). The city also has numerous parks as well as botanical and zoological gardens.

Ibagué (ee-bah-GAI), city, ⊙ TOLIMA department, W central COLOMBIA; 04°26′N 75°13′W. Elevation 4,300 ft/1,311 m. Major commercial center for the MAGDALENA and CAUCA valleys. Food processing; consumer goods. Noted for its conservatory of music, and sometimes called the music capital of Colombia. Airport.

Ibahernando (ee-vah-er-NAHN-do), town, CÁCERES province, W SPAIN, 10 mi/16 km SSW of TRUJILLO; 39°19′N 05°55′W. Cereals, olive oil, wine; sheep, poultry.

Ibaiti (EE-bei-chee), city (2007 population 28,050), NE PARANÁ, BRAZIL, on railroad, and 10 mi/16 km WSW of TOMAZINA; 23°50′S 50°10′W. Coffee and rice processing; coal deposits. Until 1944, called Barra Bonita.

Ibajay (EE-BAH-hei), town, AKLAN province, NW PANAY island, PHILIPPINES, on SIBUYAN SEA, 45 mi/72

km WNW of ROXAS; 11°46′N 122°09′E. Agricultural center (tobacco, rice).

Ibalaghan (EE-bel-ah-gen), town, SEVENTH REGION/GAO, MALI, 81 mi/135 km NNE of MÉNAKA. Archaeological site. Also spelled Ibelaghene.

Ibanda, administrative district, WESTERN region, SW UGANDA. As of Uganda's division into eighty districts, borders KAMWENGE (W and N), KIRUHURA (E), MBARARA (S), and BUSHENYI (SWW) districts. Agricultural area. Formed in 2005 from NW portion of former MBARARA district (current Mbarara district formed from W portion, ISINGIRO district from S portion, and Kiruhura district from N, E, and central portions).

Ibapah (ei-BUH-pah), unincorporated village, TOOELE county, W UTAH, 48 mi/77 km S of WENDOVER, near NEVADA state line. Cattle; gold mining. Goshute Indian Reservation to S and SW; Deseret Military Test Center to E. Deep Creek Range to E. Highway access to rest of Utah via Nevada.

Ibara (EE-bah-rah), or **Ihara**, city, OKAYAMA prefecture, SW HONSHU, W central JAPAN, 28 mi/45 km W of OKAYAMA; 34°35′N 133°27′E. Clothing. Grapes. *Chugoku*-style lullaby originated here.

Ibaraki (ee-BAH-rah-kee), prefecture (□ 2,352 sq mi/6,092 sq km; 1990 population 2,845,411), central HONSHU, E central JAPAN, on PACIFIC OCEAN (E); ⊙ MITO. Bordered N by FUKUSHIMA, S by CHIBA, and W by TOCHIGI prefectures. Yields tobacco, cereals; coal, copper; petrochemicals, electric machinery.

Ibaraki (ee-BAH-rah-kee), city, OSAKA prefecture, S HONSHU, W central JAPAN, 9 mi/15 km N of OSAKA; 34°48′N 135°34′E. Appliances.

Ibaraki (ee-BAH-rah-kee), town (2005 population 267,961), East Ibaraki district, IBARAKI prefecture, central HONSHU, E central JAPAN, near Mt. Tsukuba; 6 mi/10 km S of MITO; 36°17′N 140°25′E. Peanuts, chestnuts, melons.

Ibaretama (EE-bah-re-TAH-mah), city (2007 population 12,729), E central CEARÁ, BRAZIL, 25 mi/40 km NE of QUIXADÁ; 04°45′S 38°45′W.

Ibarlucea (ee-bahr-loo-SAI-ah), village, S SANTA FE province, ARGENTINA, 10 mi/16 km. NW of ROSARIO; 32°51′S 60°48′W. Agriculture (corn, flax, wheat; livestock).

Ibarra (ee-BAHR-rah), city (2001 population 108,535), ⊙ IMBABURA province, N ECUADOR, on PAN-AMERICAN HIGHWAY, on railroad to SAN LORENZO, and 50 mi/80 km NE of QUITO, in a high Andean valley, near Ecuadorian lake district, surrounded by the peaks COTACACHI (W) and IMBABURA (SW); 00°21′N 78°07′W. Elev. 7,300 ft/2,225 m. Trading center in fertile agricultural region (coffee, sugarcane, cotton, cereals, potatoes, fruit, vegetables; livestock); manufacturing of native woolen goods, cotton products, fine silverwork, wood carvings, *aguardiente*, furniture, panama hats; saltworks. An old Native American town with colonial appearance.

Ibarreta (ee-bahr-RAI-tah), town, central FORMOSA province, ARGENTINA, on railroad, and 125 mi/201 km NW of FORMOSA; 25°13′S 59°51′W. Agricultural center (corn, wheat, soybeans; livestock).

Ibar River (EE-bahr), c.150 mi/241 km long, MONTENEGRO and SERBIA; rises in MOKRA PLANINA c.10 mi/16 km ESE of IVANGRAD, flows E, past ROZAJ and KOSOVSKA MITROVICA, and N, past RASKA, to WESTERN MORAVA RIVER. just below KRALJEVO. Navigable for c.130 mi/209 km. Receives SITNICA RIVER (right). KOPAONIK mountain range lies E of its middle course; paralleled by railroad below Kosovska Mitrovica.

Ibárs de Urgel (ee-VAHRS dai oor-HEL), village, LÉRIDA province, NE SPAIN, on URGEL CANAL, and 19 mi/31 km ENE of LÉRIDA. In well-irrigated agricultural area (cereals, olive oil, wine, almonds).

Ibaté (EE-bah-tai), city (2007 population 28,037), central SÃO PAULO state, BRAZIL, 9 mi/14.5 km NW of SÃO CARLOS, on railroad; 21°57′S 48°00′W.

Ibateguara (EE-bah-te-GWAH-rah), city (2007 population 15,359), NE ALAGOAS state, BRAZIL, 7 mi/12 km E of SÃO JOSÉ DA LAJE, near PERNAMBUCO border; 08°58′S 35°55′W.

Ibatiba (ee-bah-CHEE-bah), city (2007 population 19,645), W central ESPÍRITO SANTO, BRAZIL, near border with MINAS GERAIS; 20°24′S 41°28′W.

Ibatuba, Brazil: see SOLEDADE DE MINAS.

Ibb (IB), town, S central YEMEN, on plateau, 95 mi/153 km S of SANA, and on main route to TAIZ; 13°58′N 44°11′E. Elev. 6,700 ft/2,042 m. Trading center; tanning, manufacturing of saddles and harnesses, clothing, food processing; livestock raising. Thermal power station. In a mountainous area, it is a walled town of lofty stone-built houses, entered by castle-like gate.

Ibbenbüren (ib-buhn-BYOO-ren), city, NORTH RHINE–WESTPHALIA, NW GERMANY, at foot of TEUTOBURG FOREST, just W of OSNABRÜCK; 52°17′N 07°43′E. A coal-mining and industrial city with the deepest coal mine in the world (4,656 ft/1,419 m); coal plant nearby. Manufacturing includes steel, machinery, leather products, and chemicals. Limestone quarrying. A mission center in N Germany during in Carolingian era; fell to TECKLENBURG in 15th century, and to PRUSSIA in 1702; chartered in 1721. Has a late-Gothic church.

Ibdes (EEV-des), town, ZARAGOZA province, NE SPAIN, 14 mi/23 km SW of CALATAYUD; 41°13′N 01°50′W. Cereals, fruit.

Ibefun (ee-bai-FAWNG), town, OGUN state, SW NIGERIA, 12 mi/19 km SW of IJEBU Ode; 06°43′N 03°47′E. Cacao industry; cotton weaving, indigo dyeing; palm oil and kernels. Sometimes spelled Ibefon.

Ibekwe (EE-be-kwai), town, EDO state, SW NIGERIA, on road, 20 mi/32 km N of BENIN. Market town. Cassava, plantains, yams.

Ibembo (ee-BEM-bo), village, ORIENTALE province, N CONGO, on ITIMBIRI RIVER, 50 mi/80 km W of BUTA; 02°38′N 23°37′E. Elev. 1,381 ft/420 m. Steamboat landing and trading center; rice, cotton. Hospital. Across ITIMBIRI was the Roman Catholic mission of Tongerloo-Saint-Norbert with teachers' school and small seminary.

Iberá, Esteros del (ee-be-RAH, es-TAI-ros del), region of swamps and lakes, in CORRIENTES province, ARGENTINA, extending c.150 mi/241 km NE-SW between PARANÁ and URUGUAY rivers. Includes the lakes IBERÁ, LUNA, and ITATÍ. Subtropical forests cover most of the area.

Iberá, Lake (ee-be-RAH) (□ 140 sq mi/364 sq km), in the Esteros del Iberá (swamps), N central CORRIENTES province, ARGENTINA, 70 mi/113 km NE of MERCEDES; c.20 mi/32 km long.

Iberia (ee-be-REE-ah), ancient country of TRANSCAUCASIA, roughly the E part of present-day republic of GEORGIA. It was inhabited in earliest times by various tribes, collectively called Iberians by ancient historians, although Herodotus called them Saspirams. Bet. the 6th and 4th centuries B.C.E. the kingdoms of COLCHIS (now W Georgia) and Iberia were founded. Iberia was allied to the Romans, ruled by the Sassanids of Persia, and became (C.E. 6th century) a Byzantine province. Its later history is that of Georgia.

Iberia (EI-bir-ee-uh), parish (□ 588 sq mi/1,528.8 sq km; 2006 population 75,509), S LOUISIANA, on Gulf of MEXICO; ⊙ NEW IBERIA; 30°01′N 91°49′W. Bounded SW by VERMILION BAY, E by Belle River; intersected by Bayou TECHE, Big Bayou Pigeon, and Atchafalaya Main Channel. Includes AVERY, JEFFERSON, and WEEKS islands (salt domes). Oil and gas wells, huge salt mines, sulphur mines, fisheries (crawfish, catfish, shrimp, crabs, finfish). Rich agricultural area (sugarcane, soybeans, rice, vegetables, nursery crops; home gardens; cattle; dairying). Varied manufacturing includes food (especially pepper) products, apparel, metal products, industrial machinery; shipbuilding.

Hunting, fur trapping. Crossed by INTRACOASTAL WATERWAY. First settled by Spanish colonists. Named after the IBERIAN PENINSULA. Lake PEIGNEUR in W. Part of Allakanas State Wildlife Area in E. Lake Fausse Pointe, including state park, in E center. Part of Cypremort Point State Park in S. Beyond VERMILION BAY, to S, lies large MARSH ISLAND (entirely in Russell Sage State Wildlife Refuge); 3 mi/5 km S of MARSH ISLAND, in GULF, are the Shell Keys (National Wildlife Refuge). Both MARSH ISLAND and Shell Keys are parts of Iberia parish. Formed 1868.

Iberia, town (2000 population 605), MILLER county, central MISSOURI, 14 mi/23 km SE of TUSCUMBIA; 38°05′N 92°17′W. Hay; cattle, turkeys; manufacturing (apparel, wood products).

Iberian Mountains (ei-BIR-ee-uhn), NE SPAIN, mountain system, extending c.250 mi/400 km, along the NE edge of the MESETA (central plateau). Moncayo (7,590 ft/2,313 m high) is the highest peak in the system. The DUERO RIVER and several tributaries of the EBRO RIVER rise here.

Iberian Peninsula (□ c.230,400 sq mi/596,740 sq km), SPAIN and PORTUGAL, SW EUROPE, separated from the rest of Europe by the PYRENEES MOUNTAINS. Washed on the N and W by the ATLANTIC OCEAN and on the S and E by the MEDITERRANEAN SEA; the STRAIT OF GIBRALTAR separates it from AFRICA. Dominated by the Meseta (central plateau), a great uplifted fault block (average elevation 2,000 ft/610 m) ringed and crossed by mountain ranges. It covers about two thirds of the peninsula. Coastal lowlands, the site of the major industrial cities, surround the primarily agrarian-oriented Meseta. Climatically, the Iberian Peninsula has hot summers, cold winters, and limited precipitation. Drained by five major rivers.

Ibérica, Cordillera (kor-dee-LYAI-rah ee-VAI-ree-kah) or **Iberian Mountains** (ei-BIR-ee-uhn), mountain system on E edge of the great central plateau (MESETA) of SPAIN, extends in a wide arc (c.250 mi/402 km long) from SIERRA DE LA DEMANDA S to the JÚCAR RIVER basin, roughly separating CASTILE-LA MANCHA and CASTILE-LEÓN from ARAGON and VALENCIA. Rises in SIERRA DEL MONCAYO to c.7,590 ft/2,313 m. Among other subranges are SIERRA CEBOLLERA and Serranía de Cuenca. Sometimes called Sistema Ibérico.

Iberus, SPAIN: see EBRO.

Iberville (ee-ber-VEEL), former county (□ 198 sq mi/514.8 sq km), S QUEBEC, E CANADA, near U.S. (NEW YORK) border, on RICHELIEU River; 45°15′N 73°10′W. The town of IBERVILLE was its county seat.

Iberville (IB-uhr-vil), parish (□ 611 sq mi/1,588.6 sq km; 2006 population 32,974), SE central LOUISIANA, ⊙ PLAQUEMINE; 30°17′N 91°14′W. Bounded W by ATCHAFALAYA RIVER; intersected by MISSISSIPPI RIVER in E. Agriculture (corn, hay, soybeans, nursery crops, sugarcane; home gardens; cattle, horses, exotic fowl; crawfish, catfish, alligators; logging; manufacturing (chemicals, plastics, lumber, metal products, industrial machinery); oil and gas fields. Drained by GRAND RIVER; intersected by Bayou Maringouin and Bayou des Glaises. Locks connect MISSISSIPPI RIVER to the Plaquemine branch of PORT ALLEN-MORGAN CITY canal. Part of Atchafalaya National Wildlife Refuge in NW corner. Named after Pierre Le Moyne, sieur d'Iberville, brother of sieur de Bienville. Occupied by Union forces during Civil War. Formed 1807.

Iberville (ee-ber-VEEL), former town, S QUEBEC, E CANADA, on RICHELIEU River, and 22 mi/35 km SE of MONTREAL; 45°19′N 73°14′W. Manufacturing of chemicals, pottery; iron founding, woodworking; market in dairying region. Site of American mother-house of the Marist Brothers. Former seat of historic IBERVILLE county. Amalgamated into Saint-Jean-sur-RICHELIEU in 2001.

Ibex (EI-beks), peak (8,263 ft/2,519 m) in SW Palni Hills, TAMIL NADU state, S INDIA, 10 mi/16 km NW of PERIYAKULAM.

Ibi (EE-vee), city, ALICANTE province, E SPAIN, 8 mi/12.9 km SW of ALCOY; 38°38′N 00°34′W. Ice cream and biscuit factories; toy manufacturing, olive-oil processing. Wine.

Ibi (ee-BEE), town, TARABA state, E central NIGERIA, port on BENUE River, 25 mi/40 km N of WUKARI. Salt mining; cassava, durra, yams. British commercial station in 19th century.

Ibiá (ee-bah-AH), city (2007 population 22,060), W MINAS GERAIS, BRAZIL, in the SERRA DA CANASTRA, 90 mi/145 km ENE of UBERABA; 19°30′S 46°32′W. Railroad junction. Cotton, coffee; cattle; dairy products. Spa of ARAXÁ is 25 mi/40 km WSW.

Ibiaçá (EE-bee-ah-kah), town (2007 population 4,681), N RIO GRANDE DO SUL state, BRAZIL, 31 mi/50 km NE of PASSO FUNDO. Wheat, corn, potatoes; livestock.

Ibiaí (ee-bee-ah-EE), city (2007 population 7,571), N central MINAS GERAIS, BRAZIL, on Rio SÃO FRANCISCO, 16 mi/26 km N of PIRAPORA; 16°47′S 44°32′W.

Ibiajara (ee-BEE-ah-ZHAH-rah), village, W central BAHIA, BRAZIL, 37 mi/60 km N of PARAMIRÍM; 13°00′S 42°14′W.

Ibiapaba, Serra, Brazil: see GRANDE, SERRA.

Ibiapina (ee-bee-ah-pee-nah), city (2007 population 23,049), NW CEARÁ, BRAZIL, in the SERRA GRANDE, 40 mi/64 km SW of SOBRAL, on PIAUÍ border; 03°54′S 40°49′W. Ships coffee, tobacco, lumber; distilling.

Ibiapinópolis, Brazil: see SOLEDADE.

Ibiara (EE-bee-ah-rah), town (2007 population 6,106), SW PARAÍBA, BRAZIL, 8 mi/12.9 km NE of Conceção; 07°31′S 38°24′W.

Ibiassucé (ee-BEE-ah-soo-SAI), city (2007 population 9,507), S central BAHIA, BRAZIL, 28 mi /45 km SE of CAETITÉ; 14°17′S 42°17′W. Also spelled Ibiaçucé.

Ibibo (ee-BEE-bo), village, MOROGORO region, E central TANZANIA, 65 mi/105 km NW of MOROGORO; 06°03′S 37°08′E. Corn, wheat; goats, sheep; timber, sisal.

Ibicaraí (EE-bee-kah-rah-EE), city (2007 population 24,742), SE BAHIA, BRAZIL, 19 mi/31 km W of ITABUNA; 14°52′S 39°35′W.

Ibicora (EE-bee-ko-RAH), town, BAHIA, BRAZIL.

Ibicuí (ee-BEE-koo-EE), city (2007 population 15,802), SE BAHIA, BRAZIL, 60 mi/96 km WNW of ITABUNA; 14°51′S 39°58′W.

Ibicuí River (EE-bee-koo-ee), c.300 mi/483 km, SW RIO GRANDE DO SUL, BRAZIL; rises N of SANTA MARIA (29°25′S 56°47′W); flows W to URUGUAY RIVER (ARGENTINA border) 30 mi/48 km NE of URUGUAIANA. Navigable. Old spelling, Ibicuhy.

Ibicuitinga (EE-bee-koo-ee-CHEEN-gah), town (2007 population 10,997), E central CEARÁ, BRAZIL, 31 mi/50 km E of Quixadá; 04°50′S 38°40′W.

Ibicuy (ee-bee-KWEE), town, SE ENTRE RÍOS province, ARGENTINA, port on PARANÁ IBICUY RIVER, and 45 mi/72 km S of GUALEGUAY; 33°44′S 59°10′W. Railroad terminus; livestock-raising center; ships cereals.

Ibi Gamin, INDIA: see KAMET.

Ibigawa (ee-BEE-gah-wah), town, Ibi district, GIFU prefecture, central HONSHU, central JAPAN, 12 mi/20 km N of GIFU; 35°29′N 136°34′E.

I'billin (i-bil-LEEN), Arab township, NE of HAIFA in W GALILEE, N ISRAEL; 32°49′N 35°11′E. Elevation 728 ft/221 m. Population 60% Christian; 40% Muslim. Remains of Crusader fort. In the local mosque are buried family members of Dahar el-Umar who ruled the Galilee in the 18th century. Name may derive from ancient Jewish town 'Ablim (ruins nearby), through which the Roman road passed between AKKO (Acre) and the UPPER GALILEE and Jordan Valley.

Ibillo (ee-BEEL-lo), town, EDO state, SW NIGERIA, on road, 80 mi/129 km NE of BENIN; 07°26′N 06°05′E. Market town. Yams, cassava, rice.

Ibipitanga (EE-bee-pee-tahn-gah), city (2007 population 13,610), W BAHIA, BRAZIL, on the RIO PRÊTO (navigable), and 80 mi/129 km W of BARRA; 11°01′S 44°30′W. Hides, sugar, manioc. Also known as Santa Rita de Cássia.

Ibiporã (EE-bee-po-RUH), city (2007 population 45,162), N PARANÁ, BRAZIL, on railroad, and 7 mi/11.3 km ENE of LONDRINA; 23°17′S 51°03′W. In coffee-growing zone; also grows cotton, rice, potatoes, corn, kidney beans; sawmilling, brick manufacturing.

Ibiquera (EE-bee-KWAI-rah), town (2007 population 5,037), central BAHIA, BRAZIL, 58 mi/93 km W SW of ITABERABA; 13°24′S 41°17′W.

Ibirá (EE-bee-RAH), town, N SÃO PAULO, BRAZIL, 20 mi/32 km SSE of SÃO JOSÉ DO RIO PRÊTO; 21°06′S 49°17′W. Resort with mineral springs (established 1928); pottery manufacturing, corn milling.

Ibiraba (EE-bee-RAH-bah), village, N central BAHIA, BRAZIL, on Rio SÃO FRANCISCO, across river from XIQUE-XIQUE; 10°46′S 42°50′W.

Ibiraci (ee-bee-ruh-see), town (2007 population 10,927), SW MINAS GERAIS, BRAZIL, near SÃO PAULO border, 40 mi/64 km NW of PASSOS; 20°28′S 47°00′W. Coffee, sugar; livestock. Formerly spelled Ibiracy.

Ibiraçu (ee-bee-rah-soo), town (2007 population 10,301), central ESPÍRITO SANTO, BRAZIL, on railroad, and 35 mi/56 km N of VITÓRIA; 19°47′S 40°25′W. Coffee, bananas, manioc. Until 1944, called Pau Gigante.

Ibiraiaras (EE-bee-rei-ah-rahs), town (2007 population 7,094), E RIO GRANDE DO SUL state, BRAZIL, 16 mi/26 km SSW of LAGOA VERMELHA; 28°22′S 51°39′W. Grapes, wheat, corn, potatoes; livestock.

Ibirajuba (EE-bee-rah-zhoo-bah), town (2007 population 7,482), E PERNAMBUCO state, BRAZIL, 28 mi/45 km SW of CARUARU; 08°35′S 36°11′W. Sugar, manioc, corn.

Ibirama (EE-bee-rah-mah), city (2007 population 16,716), E central SANTA CATARINA, BRAZIL, in ITAJAÍ AÇU river valley, 30 mi/48 km WSW of BLUMENAU; 27°05′S 49°32′W. Agricultural colony settled in 19th century by Germans and Slavs. Until 1944, called Hamônia (old spelling, Hammonia).

Ibiranhem (EE-bee-rahn-YEN), village, SE BAHIA, BRAZIL, near ESPÍRITO SANTO border; 17°53′S 40°09′W.

Ibirapitanga (EE-bee-RAH-pee-TAHN-gah), city (2007 population 23,297), E central BAHIA, BRAZIL, 13 mi/21 km N of UBAITABA; 14°11′S 39°22′W.

Ibirapuã (EE-bee-rah-poo-AHN), town (2007 population 7,553), SE BAHIA, BRAZIL, near MINAS GERAIS border, on Rio do Meio; 17°43′S 40°08′W.

Ibirapuitã (EE-bee-rah-poo-ee-tah), city, N RIO GRANDE DO SUL state, BRAZIL, 22 mi/35 km NW of PASSO FUNDO, on Ernestina Reservoir; 28°34′S 52°33′W. Wheat, corn, potatoes, manioc; livestock.

Ibirarema (EE-bee-rah-re-mah), town (2007 population 6,617), W SÃO PAULO, BRAZIL, on railroad, and 13 mi/21 km NW of OURINHOS; 22°49′S 50°06′W. Rice, corn, coffee; poultry. Until 1944, Pau d'Alho.

Ibirgarsama (ee-bir-gahr-SAH-mah), canton, CARRASCO province, COCHABAMBA department, central BOLIVIA; 17°42′S 65°09′W. Elevation 9,150 ft/2,789 m. Limestone and gypsum deposits. Agriculture (potatoes, yucca, bananas, corn, rice, wheat, rye, sweet potatoes, soy, coca, coffee); cattle raising for meat and dairy products.

Ibiri (ee-bee-REE), village, TABORA region, NW central TANZANIA, 18 mi/29 km NW of TABORA; 04°56′S 32°33′E. Timber; tobacco, corn, wheat, millet; goats, sheep.

Ibirité (EE-bee-ree-TAI), city (2007 population 148,075), central MINAS GERAIS, BRAZIL, 7 mi/11.3 km SW of BELO HORIZONTE; 20°06′S 44°11′W. Large industrial suburb.

Ibi River, JAPAN: see KISO RIVER.

Ibirtataia (EE-bir-tah-TEI-ah), city, E central BAHIA, BRAZIL, 10 mi/16 km E of IPAÚ; 14°06′S 39°38′W.

Ibirubá (EE-bee-roo-bah), city (2007 population 18,690), central RIO GRANDE DO SUL state, BRAZIL, 23 mi/37 km E of CRUZ ALTA; 28°38′S 53°06′W. Wheat, corn, potatoes; livestock.

Ibitiá (EE-bee-chee-AH), city, N central BAHIA, BRAZIL, 24 mi/39 km SW of IRECÉ; 11°34′S 41°58′W.

Ibitiara (EE-bee-CHEE-ah-rah), city (2007 population 16,389), central BAHIA, BRAZIL, in Serra das MANGABEIRA, 37 mi/60 km NE of MACAUBAS; 12°40′S 42°14′W.

Ibitinga (EE-bee-cheen-gah), city (2007 population 50,076), central SÃO PAULO, BRAZIL, near TIETÊ RIVER, 75 mi/121 km WSW of RIBEIRÃO PRÊTO; 21°45′S 48°49′W. Agricultural-processing center (manioc flour, sausages, soft drinks); cotton, coffee, and rice processing. Oil-shale deposits.

Ibitirama (ee-bee-CHEE-rah-mah), town (2007 population 8,994), SW ESPÍRITO SANTO, BRAZIL, on Rio Braço do Norte, 19 mi/31 km SW of IÚNA; 20°38′S 41°40′W.

Ibitunane (EE-bee-TOO-nah-ne), village, W central BAHIA, BRAZIL, near Rio SÃO FRANCISCO, 25 mi/40 km SW of GENTIO DO OURO; 11°34′S 42°44′W.

Ibitupa (EE-bee-TOO-pah), village, E central BAHIA, BRAZIL, 9 mi/15 km SE of JAGUAQUARA; 14°31′S 39°52′W.

Ibiúna (EE-bee-oo-nah), city (2007 population 64,930), SE SÃO PAULO, BRAZIL, 40 mi/64 km W of SÃO PAULO, near Ituperagranga Reservoir; 23°45′S 47°13′W. Grain, citrus fruit. Until 1944, called Una.

Ibiza (ee-VEE-thah) or **Iviza**, Catalan *Eivissa*, chief city of IBIZA island, BALEARIC ISLANDS, SPAIN, port on SE coast, 80 mi/129 km SW of PALMA (MAJORCA), 105 mi/169 km ESE of VALENCIA; 38°55′N 01°25′E. Bishopric. Ships chiefly salt, wool, fruit. Diverse industry includes construction; transformers; metals, and lumber. Lobster fisheries. The picturesque city is a favorite avant-garde tourist site. Whitewashed houses are grouped around old castle which once protected city against pirate attacks. Has fine collegiate church (former cathedral) and renowned archaeological museum, best of its kind in Spain for Phoenician and Carthaginian remains. In vicinity are several remarkable Phoenician necropolises.

Ibiza (ee-VEE-thah), Catalan *Eivissa*, island (□ 221 sq mi/574.6 sq km), Baleares province, SPAIN, third largest of the BALEARIC ISLANDS, in the W MEDITERRANEAN SEA; ⊙ IBIZA; 39°00′N 01°25′E. Fisheries, saltworks, and subsistence farming have been the traditional occupations. Now largely devoted to tourists and artists. Picturesque island with Roman, Phoenician, and Carthaginian remains.

Iblei, Monti (ee-BLAI, MON-tee), mountain range, SE SICILY, ITALY; extends 25 mi/40 km NW from CASSIBILE River to CALTAGIRONE; rises to 3,231 ft/985 m in MONTE LAURO; 37°10′N 14°55′E.

Ibnat, ETHIOPIA: see EBENAT.

Ibn Ziad (EE-buhn zee-AHD), village, CONSTANTINE wilaya, NE ALGERIA, 8 mi/12.9 km W of CONSTANTINE. Wheat. Formerly Rouffach.

Ibó (EE-bo), city, S PERNAMBUCO state, BRAZIL, on SÃO FRANCISCO River (BAHIA state border), 39 mi/63 km S of Saigueiro; 08°37′S 39°14′W.

Ibo (EE-BO), town (□ 17 sq mi/44.2 sq km), CABO DELGADO province, N MOZAMBIQUE, on Ibo Island (□ 17 sq mi/44 sq km) in Mozambique Channel just off mainland, 45 mi/72 km N of PEMBA; 12°20′S 40°38′E. Oilseed processing. Ships copra, ivory, wax, cashew nuts, ebony. Founded 17th century. Became a slave-trading center. Has three old forts.

Ibo (EE-bo), village, N BAHIA, BRAZIL, on Rio SÃO FRANCISCO, and on border with PERNAMBUCO; 08°38′S 39°17′W.

Ibo District, MOZAMBIQUE: see CABO DELGADO.

Ibogawa (ee-BO-gah-wah), town, Ibo district, HYOGO prefecture, S HONSHU, W central JAPAN, 37 mi/59 km W of KOBE; 34°49′N 134°31′E.

Iboro (ee-BO-ro), town, OYO state, SW NIGERIA, on road, 40 mi/64 km WNW of IBADAN. Market town. Yams, cassava, millet.

Iboti, BRAZIL: see NEVES PAULISTA.

Ibotirama (EE-bo-CHEE-rah-mah), city (2007 population 25,465), W central BAHIA, BRAZIL, on Rio SÃO FRANCISCO; 12°12′S 43°12′W. Port.

Iboundji (ee-BOON-jee), town, OGOOUÉ-LOLO province, central GABON, 45 mi/72 km W of KOULAMOUTOU, near MOUNT IBOUNDJI; 01°12′S 11°50′E.

Ibra (IB-re), chief town (2003 population 24,473) interior OMAN, SE Arabian Peninsula, 65 mi/105 km S of MUSCAT, across EASTERN HAJAR hill country; 22°43′N 58°32′E. Market town for agricultural area.

Ibrahimiya Canal (eb-rah-HEE-MAI-yuh), Upper EGYPT, runs parallel (W) to the NILE River, extending c.200 mi/322 km from ASYUT to GIZA province; navigable in some parts. The BAHR YUSUF branches off at DAIRUT. Passes BENI SUEF. With its numerous tributaries, it is an important irrigation canal.

Ibrahimiya, El (eb-rah-HEE-MAI-yuh, el), town, SHARQIYA province, Lower EGYPT, 10 mi/16 km NNE of ZAGAZIG; 30°57′N 30°35′E. Cotton, cereals.

Ibrahimpatan (eeb-RAH-heem-puh-tuhn), town, RANGAREDDI district, ANDHRA PRADESH state, S INDIA, 17 mi/27 km SE of HYDERABAD. Rice, oilseeds. Also written Ibrahim Patan.

Ibrány (EB-rah-nyuh), village, SZABOLCS-SZATMÁR county, NE HUNGARY, 12 mi/19 km N of NYIREGYHÁZA; 48°08′N 21°43′E. Wheat, potatoes, apples, tobacco; cattle, hogs, sheep. Rubber industry.

Ibred' (EE-bryet), village, central RYAZAN oblast, central European Russia, on the Para River (tributary of the OKA River), on railroad and near highway, 3 mi/5 km S of SHILOVO, to which it is administratively subordinate; 54°15′N 40°52′E. Elevation 390 ft/118 m. In agricultural area; starch making, sugar processing.

Ibresi (ee-BRYE-see), town (2005 population 9,085), central CHUVASH REPUBLIC, central European Russia, on road and railroad, 35 mi/56 km NNE of ALATYR; 55°18′N 47°02′E. Elevation 649 ft/197 m. Woodworking; furniture. Shale oil deposits in the vicinity.

Ibri (IB-ree), town (2003 population 24,473) A'DAHAHIRAH region of interior OMAN, at foot of WESTERN HAJAR hill country, 140 mi/225 km WSW of MUSCAT; 23°14′N 56°30′E. Fruit-growing center; dates, limes, mangoes, and pomegranates. Large mosque.

Ibros (EE-vros), town, JAÉN province, S SPAIN, 9 mi/14.5 km ESE of LINARES; 38°01′N 03°30′W. Soap, essential-oil manufacturing, olive-oil processing; wine and cereals.

Ibshawai, township, FAIYUM province, Upper EGYPT, on railroad, and 10 mi/16 km WNW of FAIYUM; 29°22′N 30°41′E. Cotton, cereals, sugar, fruits.

Ibstock (IB-stahk), village (2001 population 5,621), NW LEICESTERSHIRE, central ENGLAND, 12 mi/19 km WNW of LEICESTER; 52°40′N 01°24′W. Former coal-mining site. Has 14th-century church.

Ibuki (EE-boo-kee), town, Sakata district, SHIGA prefecture, S HONSHU, central JAPAN, 40 mi/65 km N of OTSU; 35°23′N 136°22′E.

Ibur (EEB-uhr), peak (8,747 ft/2,666 m) in the E RILA MOUNTAINS, W BULGARIA, 12 mi/19 km SE of SAMOKOV; 42°12′N 23°45′E.

Iburg, Bad, GERMANY: see BAD IBURG.

Ibusuki (ee-BOOS-kee), city, KAGOSHIMA prefecture, S KYUSHU, SW JAPAN, on SE SATSUMA PENINSULA, 25 mi/40 km S of KAGOSHIMA; 31°14′N 130°38′E. Hot springs resort.

Ibyar (ib-YAHR), village, GHARBIYA province, Lower EGYPT, 3 mi/4.8 km ENE of KAFR EZ ZAIYAT. Cotton, cereals, rice, fruits.

Ica (EE-kah), region (□ 8,235 sq mi/21,411 sq km), SW PERU, between the PACIFIC OCEAN and CORDILLERA OCCIDENTAL of the ANDES Mountains; ⊙ ICA; 14°20′S 75°30′W. Includes CHINCHA and Viejas islands. Situated chiefly on coastal plain, it is irrigated by the SAN JUAN, PISCO, and ICA and Grande rivers. Large cotton-producing center, important also for viticulture and fruit growing. Alcohol distilling; produces Pisco brandy. Fishing at PISCO and SAN ANDRÉS; fish canning at San Andrés. Gold placers near NAZCA; iron deposits at Marcona; thermal baths at HUACACHINA. Famous for the Reserva Nacional Paracas (pre-Inca

Paracas culture; wildlife; bird colonies) and the Nazca lines (geometrical designs created on the desert floor). Crossed by PAN-AMERICAN HIGHWAY.

Ica (EE-kah), province (□ 6,198 sq mi/16,114.8 sq km), ICA region, SW PERU, ⊙ ICA; 14°20′S 75°40′W. Central and largest province of Ica region. Largely desert. Reserva Nacional Paracas is on W edge. This coastal province was one of the hardest hit areas devastated by floods in 1998.

Ica (EE-kah), city (2005 population 189,620), ⊙ ICA province and ICA region, SW PERU, on the PAN-AMERICAN HIGHWAY; 14°04′S 75°42′W. Commercial center for the cotton, wool, and wine produced in the region. Distilling of wine and brandy. There are several summer resorts nearby. Also the archaeological name of the Chincha empire of ancient Peru, which had one of its major centers in the adjacent valley. The empire fell to the Incas in the 15th century. The Spanish settled the city in 1563. The shrine of El Señor de Luren, the site of colorful pilgrimages, is here. Leveled twice by earthquakes. Has a university and a regional museum with artifacts from Inca, Nazca, and Paracas cultures.

Icacos Point (ee-KAH-kos), SW TRINIDAD, TRINIDAD AND TOBAGO, W section of the SW peninsula, on the SERPENT'S MOUTH, and 10 mi/16 km off VENEZUELA coast, terminating in Icacos Point headland at 10°02′N 61°55′W.

Icacos, Punta, NICARAGUA: see CORINTO.

Içana (ee-SAH-nah), town, NW AMAZONAS, BRAZIL, on right bank of the RIO NEGRO just below influx of Rio Içana, and 40 mi/64 km NNW of Uaupés; 28°00′N 67°28′W. Until 1944, called São Felippe.

Içana, town, NW AMAZONAS, BRAZIL, on RIO NEGRO, near Rio Negro Forest Reserve; 00°22′S 67°21′W. Airport. Formerly known as Sáo Felipe.

Icaño (ee-KAH-nyo), village, SE CATAMARCA province, ARGENTINA, 31 mi/50 km SE of CATAMARCA; 28°54′S 65°19′W.

Içara (EE-kah-rah), city (2007 population 54,107), SE SANTA CATARINA state, BRAZIL, 5 mi/8 km SE of Ciriúma; 28°42′S 49°18′W.

Icaraima (EE-kah-rah-ee-mah), city (2007 population 9,172), W PARANÁ state, BRAZIL, near PARANÁ River; 23°23′S 53°41′W. Coffee, rice, corn, cotton; hogs.

Icard (EI-kuhrd), unincorporated town (□ 4 sq mi/10.4 sq km; 2000 population 2,734), BURKE county, W central NORTH CAROLINA, 7 mi/11.3 km W of Hickory Lake; 35°43′N 81°27′W. Rhodhiss reservoir (CATAWBA RIVER) to N. Manufacturing (packaging, apparel, fabricated metal products); agriculture (soybeans; poultry).

Icari (ee-KAH-ree), town and canton, AYOPAYA province, COCHABAMBA department, central BOLIVIA; 17°07′S 66°23′W. Elevation 8,602 ft/2,622 m. Tungsten mining at Mina KAMI, 5 mi/8 km S of Chicote Grande; clay, limestone, and gypsum deposits. Agriculture (potatoes, yucca, bananas, corn, barley, rye, coffee); cattle raising for meat and dairy products.

Icaria, Greece: see IKARÍA.

Icaria, Lake, reservoir, ADAMS county, SW IOWA, on Kemp Creek, 7 mi/11.3 km N of CORNING; 5 mi/8 km long; 41°04′N 94°44′W. Max. capacity 25,310 acre-ft. Formed by dam (49 ft/15 m high), built (1974) by the Soil Conservation District for flood control.

Icarian Sea, Greece: see IKARIAN SEA.

Ica River (EE-kah), HUANCAVELICA and ICA region, SW PERU; rises in CORDILLERA OCCIDENTAL E of HUAYTARÁ; flows SW and then S past ICA, to the PACIFIC OCEAN, NW of SAN NICOLÁS; 14°51′S 75°31′W. Irrigates agricultural region (cotton, rice, grapes).

Içá River, Brazil: see PUTUMAYO, RÍO.

Icatu (ee-kah-too), city (2007 population 24,324), N MARANHÃO, BRAZIL, on Rio Monim near its mouth on SÃO JOSÉ BAY of the Atlantic Ocean, and 23 mi/37 km ESE of SÃO LUÍS; 02°40′S 44°04′W. Alcohol distilling; corn, manioc, sugar growing.

Icaturama, Brazil: see SANTA ROSA DO VITERBO.

Ice Age National Reserve (□ 50 sq mi/130 sq km), WISCONSIN. Contains kettles, drumlins, eskers. First national scientific reserve in Wisconsin. Scattered across Wisconsin; nine units include Horicon Marsh Wildlife Area, Interstate Park, Kettle Moraine State Park (N unit), Mill Bluff and Devil's Lake state parks. Established 1971.

Ice Bay, ANTARCTICA: see AMUNDSEN BAY.

Ice Caves Mountain, national natural landmark (□ 6 sq mi/15.6 sq km), ULSTER county, SE NEW YORK, 26 mi/42 km SE of KINGSTON, and 20 mi/32 km NW of NEWBURGH; 41°40′N 74°22′W. Access to the site by road is via New York Route 52E from ELLENVILLE (4 mi/6.4 km). Features include Sam's Point (elevation 2,255 ft/687 m), glacially scoured Lake Maratanza, and two large ice caves, with enormous fissures and cracks where drifts of snow often linger even in summer. In the bottom "galleries" (c.100 ft/30 m below the surface), there are walls of ice and huge icicles as well.

Ice Fjord, NORWAY: see ISFJORDEN.

Ice House Reservoir, 3 mi/4.8 km long, EL DORADO county, central CALIFORNIA, on South Fork of Silver Creek, in Eldorado National Forest, 22 mi/35 km WSW of SOUTH LAKE TAHOE; 38°52′N 120°22′W. Elevation 5,433 ft/1,656 m. Extends E. Maximum capacity 45,960 acre-ft. Formed by Ice House Dam (132 ft/40 m high), built (1959) for water supply of SACRAMENTO.

Içel, TURKEY: see MERSIN.

Iceland, Icelandic *Ísland*, republic (□ 9,698 sq mi/102,819 sq km; 2004 estimated population 293,866; 2007 estimated population 301,931), the westernmost state of EUROPE, occupying an island in the ATLANTIC OCEAN just S of the ARCTIC CIRCLE, c.600 mi/970 km W of NORWAY, and c.180 mi/290 km SE of GREENLAND; ⊙ REYKJAVÍK.

Geography

Iceland includes several small islands, notably the VESTMANNAEYJAR off the S coast. Deep fjords indent the coasts, particularly in the N and W. The island itself is a geologically young basalt plateau, averaging 2,000 ft/610 m in elevation (Hvannadalshnúkur, newly buried by ÖRAEFAJÖKULL, c.6,950 ft/2,120 m high, is the highest point) and culminating in vast icefields, of which the VATNAJÖKULL, in the SE, is the largest. There are about 200 volcanoes, many of them still active; the highest is Mount HEKLA (c.4,900 ft/1,490 m). Hot springs abound and are used for inexpensive heating; the great GEYSIR is particularly famous. The watershed of Iceland runs roughly E-W; the chief river, the JÖKULSÁ, flows N into the AXARFJÖRUR (there are several other rivers of the same name). Most energy is obtained from hydroelectric power (large dams on SOG and Thonsa rivers; geothermal energy in SW and NE). Only about one-quarter of the island is habitable, and practically all the larger inhabited places are located on the coast; they are Reykjavík, AKUREYRÍ, HAFNARFJÖRUR, SIGLUFJÖRUR, AKRANES, and ISAFJÖRUR. The climate is relatively mild and humid (especially in the W and S), owing to the proximity of the North Atlantic Drift (GULF STREAM); however, N and E Iceland have a polar, tundra-like climate. Grasses predominate; timber is virtually absent, and much of the land is barren.

Population

96% of the population is a homogeneous mixture of the descendants of Norse and Celtic settlers. The Lutheran Church is the sole established church, but there is complete religious freedom. The official language is Icelandic (Old Norse), and virtually all Icelanders are literate. There is a university (established 1911) at Reykjavík.

Economy

About 15% of the land is potentially productive, but agriculture, cultivating mainly hay, potatoes, and

turnips, is restricted to 0.5% of the total area. Fruits and vegetables are raised in greenhouses. There are extensive grazing lands, used mainly for sheep raising, but also for horses and cattle. Fishing is very important, accounting for 12% of the GNP and 70% of exports. Increased emphasis on wool industry, cattle, and tourism. There was an expansion of the aluminum industry in the late 1990s designed to make the country Europe's largest aluminum producer based on the use of its vast untapped hydroelectric resources. Aside from aluminum smelting, Iceland has little heavy industry and relies on imports for many of the necessities and luxuries of life. Over half of Iceland's GNP comes from the communications, publishing (most books per capita in the world), trade, and service industries. Tourism is also an important industry (waterfalls, geysers, and hot springs, volcanoes, glaciers; sport fishing). Most trade is with the U.S. and Europe. In 1990 Iceland's per capita national income was higher than the average for Europe.

History: to 1264

Iceland may be the ULTIMA THULE of the ancients. Irish monks visited here before the 9th century but abandoned it on the arrival (c.850–875) of Norse settlers, many of whom had fled from the domination of Harold I. The Norse settlements also contained Scottish and Irish slaves. In 930 a general assembly, the Althing, was established near Reykjavík at THINGVELLIR, and Christianity was introduced c.1000 by the Norwegian Olaf I, although paganism seems to have survived for a time. These events are preserved in the literature of 13th-century Iceland, where old Norse literature reached its greatest flowering. (Modern Icelandic is virtually the same language as that of the sagas). Politically, Iceland became a feudal state, and the bloody civil wars of rival chieftains facilitated Norwegian intervention. The attempt of Snorri Sturluson (1179–1241) to establish the full control of King Haakon IV of Norway over Iceland was a failure; however, Haakon incorporated Iceland into the archdiocese of TRONDHEIM and between 1261 and 1264 obtained acknowledgment of his suzerainty by the Icelanders.

History: 1264 to 1854

Norwegian rule brought order, but high taxes and an imposed judicial system caused much discontent. When, with Norway, Iceland passed (1380) under the Danish crown, the Danes showed even less concern for Iceland's welfare; a national decline (1400–1550) set in. Lutheranism was imposed by force (1539–1551) over the opposition of Bishop Jon Aresson; the Reformation brought new intellectual activity. The 17th and 18th centuries were, in many ways, disastrous for Iceland. English, Spanish, and Algerian pirates raided the coasts and ruined trade; epidemics and volcanic eruptions killed a large part of the population; and the creation (1602) of a private trading company at COPENHAGEN, with exclusive rights to the Iceland trade, caused economic ruin. The private trade monopoly was at last revoked in 1771 and transferred to the Danish crown, and in 1786 trade with Iceland was opened to all Danish and Norwegian merchants.

History: 1854 to 1961

The exclusion of foreign traders was lifted in 1854. The 19th-century rebirth of Icelandic culture and desire for independence was led by Jón Sigursson. The Althing, abolished in 1800, was reestablished in 1843; a constitution and limited home rule were granted in 1874; and Iceland became a sovereign state in personal union with DENMARK in 1918. The German occupation (1940) of Denmark in World War II gave the Althing autonomy over the country's affairs. Iceland was defended from German attack by Great Britain (1940–1941), then by U.S. forces (1941–1945). Union with Denmark was terminated in 1944 by majority vote; the kingdom of Iceland was proclaimed an independent republic on June 17, 1944, with Sveinn

Area is shown by the symbol □, and capital city or county seat by ⊙.

Björrnsson as its first president. Iceland was admitted to the UN in 1946; it joined in the Marshall Plan and NATO. In 1946, Iceland granted the U.S. the right to use the American-built airport at KEFLAVÍK for military as well as commercial planes. Under a 1951 defense pact, U.S. troops were stationed here. Björnsson was succeeded by Ásgeir Ásgeirsson. To protect its vital fishing industry, Iceland extended the limits of its territorial waters (1958) from 4 mi/6.4 km to 12 mi/19 km, resulting in a conflict with Great Britain, which at times led to exchanges of fire between Iceland's coast guard vessels and British destroyers, until 1961 when Great Britain accepted the new limits.

History: 1961 to Present

SURTSEY Island, 8 mi/12.9 km SE of VESTMANNAEYJAR, was born November 16, 1963, with the eruption of a submerged volcano. Iceland joined EFTA in 1970. In 1971 elections the Independence–Social Democratic coalition government, which had governed for twelve years, lost its majority. A leftist coalition, composed of the Progressive party, the Communist-led Labor Alliance party, and the Liberal Left party, came to power. In 1972, the territorial waters were further extended to 50 mi/80 km, renewing the dispute with Britain over fishing rights. An interim agreement in October 1973, limited the British catch and restricted the areas and types of vessels they could use. In January, 1973, the Helgafell volcano on HEIMAEY Island erupted, damaging the town of Vestmannaeyjar. Later in the year Iceland and the U.S. began revising the 1951 defense pact, with a view toward ending the U.S. military presence, and U.S. forces ultimately left the NATO base at Keflavík Airport in 2006. In May 1974, the Althing was dissolved following a split in the ruling coalition over economic policies. In the June elections the Independence party won a large plurality and formed a new government. Iceland extended its fishing limits to 200 mi/320 km in 1975, which, after more skirmishes with Great Britain, was finally recognized in 1976. Vigdís Finnbogadóttir (Progressive Party) was elected president in 1980, and was reelected in 1984, 1988, and 1992. David Oddson (Independence Party) became prime minister in April 1991; his center-right coalition was returned to office in 1995, 1999, and, narrowly, 2003. The economy stabilized in 1990s with government attempts to expand hydroelectric and geothermal energy resources, reducing dependency on oil imports, and emphasis on diversifying exports. A rural electrification program in 1980s brought electric power to most of Iceland's farms and small settlements. In 1994, the "cod wars" resumed, this time with Iceland's closest ally, Norway, in waters surrounding SVALBARD Islands (SPITSBERGEN). Though the latter is under Norwegian sovereignty, forty signatories of the 1920 Paris Treaty are allowed access to its resources. In 1996, Ólafur Ragnar Grímsson was elected to succeed Finnbogadóttir, who retired as president. The highly popular Grímsson was reappointed to the post by parliament without an election in 2000; he was reelected in 2004. Oddsson resigned and exchanged posts with coalition partner and foreign minister Halldór Ásgrímsson, of the Progressive party, in September 2004. In June 2006, after the Progressive party suffered losses in local elections, Ásgrímsson resigned as prime minister; he was succeeded in the post by Geir Hilmar Haarde, a member of the Independence party. The next year, after the May 2007 parliamentary elections, the Independence party formed a new coalition with the Social Democrats; Haarde remained prime minister.

Government

Iceland is governed under the constitution of 1944 as amended. The president, who is the head of state, a largely ceremonial post, is popularly elected to a four-year term; there are no term limits. The head of government is the prime minister. The legislature is the unicameral Althing, whose sixty-three members are popularly elected to four-year terms. Administratively, Iceland is divided into eight regions. The current head of state is President Ólafur Ragnar Grímsson (since 1996); the current head of government is Prime Minister Geir Hilmar Haarde (since 2006).

Icém (EE-sen), town (2007 population 6,451), N SÃO PAULO state, BRAZIL, 39 mi/63 km NW of BARRETOS, on RIO GRANDE (MINAS GERAIS state border); 20°21'S 49°12'W. Coffee growing.

Icemorelee (EIS-mor-lee), town, UNION COUNTY, S NORTH CAROLINA, just NW of MONROE. In large annexation area of Monroe.

Ice Mountain, WEST VIRGINIA: see ROMNEY.

Ichalkaranji (ee-chuhl-KUHR-uhn-jee), town, KOLHAPUR district, MAHARASHTRA state, W INDIA, 16 mi/26 km E of KOLHAPUR. Trade center for millet, sugarcane, tobacco, cotton, wheat; cotton milling, hand-loom weaving (noted for its saris); machining workshop. Annual fair.

Ichalki (EE-chahl-kee), settlement, S NIZHEGOROD oblast, central European Russia, on the P'YANA RIVER, on highway, 62 mi/100 km SSE of NIZHNIY NOVGOROD, and 10 mi/16 km S of PEREVOZ, to which it is administratively subordinate; 55°26'N 44°30'E. Elevation 357 ft/108 m. Gravel and sand quarrying.

Ichamati River, BANGLADESH: see JAMUNA RIVER.

Ichang, CHINA: see YICHANG.

Ichapur (ICH-ah-puhr), village, NORTH 24-PARGANAS district, SE WEST BENGAL state, E INDIA, on HUGLI RIVER, and 15.5 mi/24.9 km N of CALCUTTA city center. Ordnance factory and depot of the Defense Ministry. Sometimes called ICHAPUR-NAWABGANJ.

Ichapur, INDIA: see ICHCHAPURAM.

Ichawaynochaway Creek (i-chuh-wai-NAH-chuh-wai), c.65 mi/105 km long, SW GEORGIA; rises NE of CUTHBERT; flows SSE, past MORGAN, to FLINT RIVER 13 mi/21 km SW of NEWTON; 31°58'N 84°37'W.

Ichchapuram (ICH-ah-puhr-uhm), town, SRIKAKULAM district, NE ANDHRA PRADESH state, S INDIA, near the ORISSA state border, 80 mi/129 km NE of SRIKAKULAM; 16°42'N 74°28'E. Saltworks (large salt pans on BAY OF BENGAL; E). Muslim pilgrimage center. Sometimes called Ichapur.

Ichenhausen (ikh-uhn-HOU-suhn), town, BAVARIA, S GERMANY, in SWABIA, on the GÜNZ, 7 mi/11.3 km S of GÜNZBURG; 48°22'N 10°18'E. Manufacturing of machinery, tools, plastic, bricks. Chartered 1406, and again in 1913.

Icheu (ee-CHAI-oo), town, KOGI state, S central NIGERIA, on NIGER River, 25 mi/40 km S of LOKOJA; 07°42'N 06°46'E. Market and fishing town.

Ichhawar (EE-chah-wuhr), town (2001 population 12,688), SEHORE district, MADHYA PRADESH state, central INDIA, 29 mi/47 km W of BHOPAL; 23°01'N 77°01'E. Agriculture: gram, wheat.

Ichiba (EE-chee-bah), town, Awa district, TOKUSHIMA prefecture, E SHIKOKU, W JAPAN, 16 mi/25 km W of TOKUSHIMA; 34°05'N 134°17'E.

Ichiban Point, point, SOUTH AFRICA: see DASSEN ISLAND.

Ichibusa-yama (ee-CHEE-boo-sah-YAH-mah) or **Ichifusa-yama**, peak (5,650 ft/1,722 m), highest point of MIYAZAKI prefecture, in center of KYUSHU (second-highest peak), SW JAPAN, 38 mi/61 km WSW of NOBEOKA.

Ichihara (ee-CHEE-hah-rah), city (2005 population 280,255), CHIBA prefecture, E central HONSHU, E central JAPAN, on TOKYO BAY, 6 mi/10 km S of CHIBA; 35°29'N 140°07'E. Part of the KEIHIN INDUSTRIAL ZONE; home of Keiyo industrial complex. Petrochemicals, cables; shipbuilding, oil manufacturing; peanuts. Its traditional agriculture and fishing industries have declined, in part due to air pollution.

Ichihasama (ee-chee-HAH-sah-mah), town, Kurihara district, MIYAGI prefecture, N HONSHU, NE JAPAN, 34 mi/55 km N of SENDAI; 38°44'N 140°57'E.

Ichijima (ee-CHEE-jee-mah), town, Hikami district, HYOGO prefecture, S HONSHU, W central JAPAN, 35 mi/57 km N of KOBE; 35°12'N 135°08'E.

Ichikai (ee-chee-KAH-ee), town, Haga district, TOCHIGI prefecture, central HONSHU, N central JAPAN, 12 mi/20 km E of UTSUNOMIYA; 36°32'N 140°06'E. Traditional paint.

Ichikawa (ee-chee-KAH-wah), city (2005 population 466,608), CHIBA prefecture, E central HONSHU, E central JAPAN, on the Edo River, 9 mi/15 km N of CHIBA; 35°43'N 139°56'E. Timepieces. Pears.

Ichikawa (ee-chee-KAH-wah), town, Kanzaki district, HYOGO prefecture, S HONSHU, W central JAPAN, 30 mi/48 mi N of KOBE; 34°59'N 134°45'E.

Ichikawadaimon (ee-chee-KAH-wah-DAH-ee-mon), town, West Yatsushiro district, YAMANASHI prefecture, central HONSHU, central JAPAN, 9 mi/15 km S of KOFU; 35°33'N 138°50'E.

Ichiki (ee-chee-KEE), town, Hioki district, KAGOSHIMA prefecture, S KYUSHU, SW JAPAN, on NW SATSUMA PENINSULA, 16 mi/25 km N of KAGOSHIMA; 31°41'N 130°17'E. Mandarin oranges (ponkan variety); distilled alcoholic beverage (shochu). FUKIAGE Beach (sand dunes) nearby. Has feudal castle.

Ichilo, province, SANTA CRUZ department, BOLIVIA; ⊙ BUENA VISTA; 17°00'S 64°00'W.

Ichilo River (ee-CHEE-lo), river, 170 mi/274 km long, on COCHABAMBA–SANTA CRUZ department border, central BOLIVIA; rises at NE end of Cordillera de COCHABAMBA NNW of COMARAPA; flows N, mostly through tropical lowlands, joining the CHAPARÉ RIVER 80 mi/129 km S of TRINIDAD to form MAMORÉ RIVER; 15°57'S 64°42'W. Only navigable sites on ICHILO-MAMORÉ are Puerto Grethea–Puerto Villarovel and Puerto Villarovel–Boca Río Chapas. Receives the CHIMORÉ RIVER (left).

Ichinohe (ee-CHEE-no-HE), town, Ninohe district, IWATE prefecture, N HONSHU, NE JAPAN, 37 mi/60 km N of MORIOKA; 40°12'N 141°17'E. Lettuce.

Ichinomiya (ee-chee-NO-mee-yah), city (2005 population 371,687), AICHI prefecture, S central HONSHU, central JAPAN, 9 mi/15 km N of NAGOYA; 35°18'N 136°48'E. Industrial satellite of Nagoya with a large woolen-textile industry; televisions. Ichinomiya Tanabata [=weaver star] Festival is held here.

Ichinomiya (ee-chee-NO-mee-yah), town, Hoi district, AICHI prefecture, S central HONSHU, central JAPAN, 37 mi/60 km S of NAGOYA; 34°51'N 137°25'E.

Ichinomiya (ee-CHEE-no-mee-yah), town, Chose district, CHIBA prefecture, E central HONSHU, E central JAPAN, on E BOSO PENINSULA, 16 mi/25 km S of CHIBA; 35°22'N 140°22'E. Horticulture; vegetables, fruits.

Ichinomiya (ee-chee-NO-mee-yah), town, Shiso district, HYOGO prefecture, S HONSHU, W central JAPAN, 42 mi/68 km N of KOBE; 35°05'N 134°35'E. Nori.

Ichinomiya, town, on W coast of AWAJI-SHIMA island, Tsuna district, HYOGO prefecture, W central JAPAN, 23 mi/37 km S of KOBE; 34°28'N 134°50'E. Joss sticks.

Ichinomiya (ee-chee-no-MEE-yah), town, Aso district, KUMAMOTO prefecture, SW KYUSHU, SW JAPAN, 25 mi/40 km N of KUMAMOTO; 32°56'N 131°07'E. Aso mountains and Aso Shrine nearby.

Ichinomiya (ee-chee-NO-mee-yah), town, East Yatsushiro district, YAMANASHI prefecture, central HONSHU, central JAPAN, 6 mi/10 km E of KOFU; 35°38'N 138°41'E. Peaches.

Ichinoseki (ee-CHEE-no-SE-kee) or **Itinoseki**, city, IWATE prefecture, N HONSHU, NE JAPAN, 53 mi/85 km S of MORIOKA, near Mt. Kurikoma; 38°55'N 141°07'E. Computers. Sukawa hot spring and Genbi Gorge nearby.

Ichinskaya Sopka (EE-cheen-skuh-yah–SOP-kuh), inactive volcano (11,834 ft/3,607 m) in central range of the KAMCHATKA PENINSULA, KAMCHATKA oblast, extreme E SIBERIA, RUSSIAN FAR EAST, 185 mi/298 km NNW of PETROPAVLOVSK-KAMCHATSKIY.

Ichishi (ee-chee-SHEE), town, Ichishi district, MIE prefecture, S HONSHU, central JAPAN, 9 mi/15 km S of TSU; 34°39′N 136°26′E.

Ichiu (EE-chee-oo), village, Mima district, TOKUSHIMA prefecture, SE SHIKOKU, W JAPAN, 31 mi/50 km W of TOKUSHIMA; 33°57′N 134°04′E.

Ichky, UKRAINE: see SOVYETS′KYY.

Ichnya (EECH-nyah), city, SE CHERNIHIV oblast, UKRAINE, 24 mi/39 km SE of NIZHYN; 50°52′N 32°24′E. Elevation 518 ft/157 m. Raion center; railroad station. Distilling, flour and feed milling, canning, food processing (butter, powdered milk), manufacturing (bricks, leather goods). Known since the 14th century as Malaya Ichnya, given city status and renamed in 1957. Small but active Jewish community since the second half of the 19th century, destroyed during World War II—fewer than 100 Jews remaining in 2005.

Ichoa River, BOLIVIA: see ISIBORO RIVER.

Ichoca (ee-CHO-kah), town and canton, Inquivisi province, LA PAZ department, W BOLIVIA, at SE end of Cordillera de TRES CRUCES, 18 mi/29 km S of IN-QUISIVI; 17°12′S 67°17′W. Elevation 11,811 ft/3,600 m.

Ichon (EE-CHUHN), county, KYONGGI province, central SOUTH KOREA, 50 mi/80 km SE of SEOUL. Rice, soybeans, tobacco, hemp, cotton, fruits and vegetables, horticulture. Ceramics.

Ichtegem (IKH-tuh-khum), commune (2006 population 13,425), Ostend district, WEST FLANDERS province, NW BELGIUM, 13 mi/21 km SW of BRUGES; 51°06′N 03°00′E. Agricultural market. Partly Romanesque church. Center of pottery industry in Middle Ages.

Ichtiman, Bulgaria: see IHTIMAN.

Ichu (EE-choo), town (2007 population 5,891), BAHIA, BRAZIL.

Ichuraya Grande (ee-choo-RAH-yah GRAHN-dai), canton, INGAVI province, LA PAZ department, W BOLIVIA; 16°39′S 68°18′W. Elevation 12,641 ft/3,853 m. Gas resources in area. Clay, limestone, gypsum deposits. Agriculture (potatoes, yucca, bananas, rye); cattle.

Ičići, resort village, W CROATIA, in KVARNER region, in OPATIJA Riviera, 2.5 mi/4 km S of Opatija. Former fishing village, now a popular seaside tourist center.

Ickenham, ENGLAND: see HILLINGDON.

Icklesham (IK-uhl-shuhm), agricultural village (2006 population 2,500), East SUSSEX, SE ENGLAND, 6 mi/9.7 km NE of HASTINGS; 50°53′N 00°40′E. Has Norman church.

Icla (EE-klah), town and canton, ZUDANEZ province, CHUQUISACA department, S central BOLIVIA, 14 mi/23 km SE of TARABUCO; 19°19′S 64°17′W. Corn, vegetables, fruit.

Icó (ee-KO), city (2007 population 63,219), SW CEARÁ, BRAZIL, near JAGUARIBE River, 32 mi/51 km E of IGUATU; 06°20′S 38°48′W. Irrigated agricultur (cotton, sugar); cattle raising.

Icod de los Vinos (ee-KOD dai los VEE-nos), city, SANTA CRUZ DE TENERIFE, CANARY ISLANDS, SPAIN, near ATLANTIC coast, served by adjoining port of Puerto de San Marcos, or Calela de San Marcos, 30 mi/48 km WSW of SANTA CRUZ DE TENERIFE; 28°21′N 16°42′W. Agricultural center (cereals, bananas, tomatoes, grapes, potatoes). Aviculture; fishing. Wine production, cornmeal mills. Tourism; beaches. A resort with benign climate. Has a tree over 3,000 years old.

Iconha (ee-kon-yah), city (2007 population 11,516), S ESPÍRITO SANTO, BRAZIL, 22 mi/35 km NE of CA-CHOEIRO DE ITAPEMIRIM; 20°47′S 40°46′W. Coffee; monazitic-sand quarries.

Iconi (ee-KO-nee), town, Njazidja island and district, NW Comoros Republic, 3 mi/4.8 km SW of Moroni, on SW coast of island, on Mozambique Channel, Indian Ocean; 11°44′S 43°14′E. Fish; livestock; vanilla, ylang-ylang, coconuts, bananas. Oldest settlement on island,

predates 11th century Has fort; palace. Iconi Airport to NNE, at edge of Moroni.

Iconium, ancient city of ASIA MINOR, the modern KONYA, TURKEY. In ancient times it was variously part of in PHRYGIA, LYCAONIA, CAPPADOCIA, and the Roman province of GALATIA. Visited by Paul, who converted part of the Greek and Jewish population and established an important church here. In the 3rd century C.E., Iconium became an active Christian colony.

Iconoclast Mountain (ei-KAHN-uh-klast), (10,630 ft/3,240 m), SE BRITISH COLUMBIA, W CANADA, in SELKIRK MOUNTAINS, on E edge of HAMBER PROVINCIAL PARK, 35 mi/56 km NE of REVELSTOKE; 51°27′N 117°45′W.

Iconozo (ee-kon-ON-zo), town, ⊙ Iconozo municipio, TOLIMA department, W central COLOMBIA, on W slopes of Cordillera ORIENTAL, 40 mi/64 km SW of BOGOTÁ; 04°11′N 74°32′W. Elevation 4,338 ft/1,322 m. Coffee-growing center; sugarcane, corn, sorghum; livestock.

Icoya (ee-KOI-ah), town and canton, AYOPAYA province, COCHABAMBA department, W central BOLIVIA, in Cordillera de COCHABAMBA, 45 mi/72 km W of COCHABAMBA; 17°26′S 66°50′W. Barley, potatoes. Tungsten and tin deposits at Kami, just NW.

Icpic, Alaska: see IKPEK.

Icuna (ee-KOO-nah), canton, CARRASCO province, COCHABAMBA department, central BOLIVIA; 17°10′S 65°13′W. Elevation 9,150 ft/2,789 m. Clay and limestone deposits. Agriculture (potatoes, yucca, bananas, corn, rice, wheat, rye, sweet potatoes, soy, coffee); cattle raising for meat and dairy products.

Icy Bay, SE ALASKA, at head of the Panhandle, on Gulf of ALASKA, 70 mi/113 km WNW of YAKUTAT; 12 mi/19 km long, 8 mi/12.9 km wide; 59°55′N 141°33′W; receives GUYOT and MALASPINA glaciers.

Icy Cape, in S ALASKA, on GULF OF ALASKA at W entrance to ICY BAY, 75 mi/121 km WNW of YAKUTAT; 59°55′N 141°38′W.

Icy Cape, in NW ALASKA, on CHUKCHI SEA, at W edge of National Petroleum Reserve–Alaska; 70°20′N 161°50′W. Akeonik settlement here.

Icy Strait, SE ALASKA, between CHICHAGOF ISLAND and the mainland, extends 40 mi/64 km NW from CHATHAM STRAIT (58°07′N 135°00′W) to Glacier Bay (58°22′N 136°00′W) and CROSS SOUND.

Ida, county (□ 432 sq mi/1,123.2 sq km; 2006 population 7,180), W IOWA; ⊙ IDA GROVE; 42°22′N 95°30′W. Prairie agricultural area (cattle, hogs, poultry; corn, oats, soybeans) drained by MAPLE and SOLDIER rivers; sand and gravel pits (N), bituminous-coal deposits (S). General flooding here in 1993. Formed 1851.

Ida (EE-dah), city, NAGANO prefecture, central HONSHU, central JAPAN, on the TENRYU RIVER, 81 mi/130 km S of NAGANO; 35°30′N 137°49′E. Pears, Japanese plums; *koya* tofu. *Mizuhike* thread. Tenryu Gorge is nearby.

Ida (EI-duh), village (2000 population 258), CADDO parish, NW LOUISIANA, 32 mi/51 km NNW of SHREVEPORT, at ARKANSAS state line; 33°00′N 93°54′W. In agricultural and timber area. Oil and natural-gas production.

Ida, village, MONROE county, extreme SE MICHIGAN, 9 mi/14.5 km W of MONROE; 41°54′N 83°34′W. In farm area. Feed and fertilizer.

Idabel (EI-duh-bel), town (2006 population 6,918), ⊙ MCCURTAIN county, extreme SE OKLAHOMA, 38 mi/61 km ESE of HUGO, between LITTLE (N) and RED (SW) rivers; 33°53′N 94°49′W. In farming and lumbering region; sawmills; manufacturing (apparel, lumber, transportation equipment, millwork, paper products; meat processing). Red River Museum here. Unit of Ouachita National Forest to E. Incorporated 1906.

Ida Grove, city (2000 population 2,350), ⊙ IDA county, W IOWA, on MAPLE RIVER, and 50 mi/80 km ESE of

SIOUX CITY; 42°20′N 95°28′W. In livestock and grain area. Manufacturing (consumer goods, construction materials, transportation equipment). Settled 1856, incorporated 1887.

Idah (ee-DAH), town, KOGI state, S central NIGERIA, on NIGER RIVER, 50 mi/80 km S of LOKOJA; 07°06′N 06°44′E. Agricultural trade center; shea-nut processing, cotton weaving; palm oil and kernels, durra, corn, plantains, yams. Limestone deposits. British commercial station in 19th century.

Idaho, state (□ 83,573 sq mi/216,456 sq km; 2000 population 1,293,953; 1995 estimated population 1,163,261), NW UNITED STATES, one of the Rocky Mountain states, admitted as the forty-third state of the Union in 1890; ⊙ and largest city BOISE; 44°14 114°17′W. Along with Boise, the most important cities are POCATELLO, and IDAHO FALLS. Idaho is known as the "Gem State" due to the erroneous understanding that Idaho was a Shoshone word meaning 'gem of the mountains.'

Geography

Bounded N by CANADA (BRITISH COLUMBIA), NE by MONTANA, E by WYOMING, S by UTAH and NEVADA, and W by OREGON and WASHINGTON. From the N Panhandle, where Idaho is about 45 mi/72 km wide, the state broadens S of the BITTERROOT RANGE to 310 mi/499 km in width. Much of Idaho has a primitive and unspoiled natural beauty, with rugged slopes and towering peaks, a vast expanse of timberland, scenic lakes, wild rivers, cascades, and spectacular gorges. HELLS CANYON (Grand Canyon of the SNAKE RIVER) on the W border, which at one point is 7,900 ft/2,408 m below the mountaintops, is the deepest gorge (5,500 ft/1,676 m deep) in NORTH AMERICA. The climate of the state ranges from hot summers in the arid volcanic plains of S to cold, snowy winters in the high wilderness areas of central and N Idaho. The Snake River flows in a great arc across S Idaho; with its tributaries the river has been harnessed to produce hydroelectric power and to irrigate vast areas of dry, fertile volcanic soil. To the N of the Snake River valley, in central and N central Idaho, are the massive SAWTOOTH MOUNTAINS and the SALMON RIVER MOUNTAINS, which shelter some of the most magnificent wilderness remaining in the U.S., including the Selway-Bitterroot Wilderness Area and the Frank Church–River of No Return Wilderness Area. In the central and N central regions and in the Panhandle there are tremendous expanses of national forests covering approximately 40% of the state and constituting one of the largest gross areas of national forests in the nation, notably, in the N, Kamiksu, St. Joe, Coeur d'Alene, Clearwater and Nez Perce national forests; in the S center, Payette, Salmon, Challis, and Boise national forests; in the E and SE, Targhee, Caribou, Cache, and Sawtooth national forests.

Economy

Agriculture is still the most important sector of the state's economy. Cattle and calves are among the leading agricultural products; dairy products are also important. The irrigated Snake Valley produces nearly all of Idaho's famous potato crop, with greatest concentration in the upper valley (E), around Pocatello; also produces most of Idaho's other crops. Idaho's chief crops are potatoes (for which the state is nationally famous and by far the nation's largest producer), hay, wheat, fruits and vegetables and sugar beets. Food processing is the chief industry; lumber and wood products, chemicals, and electronic components are other major manufactured items. Mining, once a major source of income, and still important, has been surpassed by agriculture, manufacturing, and tourism in annual income earned. Science and technology industries have become a major part of the state's economy. Silver, antimony, phosphate rock, gold, lead, and zinc are the principal minerals produced.

Area is shown by the symbol □, and capital city or county seat by ⊙.

Tourism

The state's jagged granite peaks include Mount BORAH, in Custer county, in the S center of state, which reaches an elevation of 12,662 ft/3,859 m. Rushing rivers such as the Salmon and the CLEARWATER, and many lakes, notably Lake PEND OREILLE, COEUR D'ALENE LAKE (often described as one of the world's loveliest), and PRIEST LAKE, as well as the state's mountain areas, make Idaho a superb fish and game preserve and vacation land. The state is noted for white-water rafting, especially on the Salmon River, also known as the River of No Return (so named by Lewis and Clark expedition), and is especially inviting to campers, anglers, and hunters (Idaho has one of the largest elk herds in the nation). The growth of the winter sports industry has also helped make Idaho a leading tourist state.

History: to 1811

Probably the first non–Native Americans to enter the area that is now Idaho were members of the Lewis and Clark expedition in 1805. They were not far ahead of the fur traders who came to the region shortly thereafter. Canadian trader David Thompson of the North West Company, established the first trading post here in 1809. The next year traders from SAINT LOUIS penetrated the mountains, and Andrew Henry of the Missouri Fur Company established a post near present-day REXBURG, the first American trading post in the area. In this period the fortunes of the Idaho region were wrapped up with those of the COLUMBIA River region, and the area encompassed by what is now the state of Idaho was part of Oregon country, held jointly by the U.S. and Great Britain 1818–1846. Fur traders in an expedition sent out by John Jacob Astor came to the Snake River region to trap for furs after having established (1811) a trading post at ASTORIA on the Columbia River.

History: 1811 to 1863

In 1821, two British trading companies operating in the Idaho region, the North West and the Hudson's Bay companies, were joined together as the Hudson's Bay Company which, after 1824, came into competition with American mountain men also trapping in the area. By the 1840s the two groups had severely depleted the region's fur supply. In 1846 the U.S. gained sole claim to Oregon country S of the 49th parallel by the Oregon Treaty with Great Britain. The area was established as a territory in 1848. Idaho still had no permanent settlement when Oregon Territory became a state in 1859 and the E part of Idaho was added to Washington Territory. A Mormon outpost founded at FRANKLIN in 1860 is considered the first permanent settlement, but it was not until the discovery of gold that settlers poured into Idaho. Gold was discovered on the Clearwater River in 1860, on the Salmon in 1861, in the BOISE RIVER basin in 1862, and gold and silver were found in the OWYHEE RIVER country in 1863. The usual rush of settlers followed, along with the spectacular but ephemeral growth of towns. Most of these settlements are only ghost towns now, but the many settlers who poured in during the gold rush—mainly from Washington, Oregon, and California, with smaller numbers from the E—formed a population large enough to demand a new government administration, and Idaho Territory was set up in 1863.

History: 1863 to 1902

Native Americans, mostly Kootenai, Nez Percé, Western Shoshone, Bannock, Coeur d'Alene, and Pend d'Oreille, became upset by the incursion of settlers and some resisted violently. The Federal government had subdued many of these ethnic groups by 1858, placing them on Native American reservations. The Bannock were defeated in 1863 and again in 1878. In 1876–1877 the Nez Percé, led by Chief Joseph, made their heroic but unsuccessful attempt to flee to Canada while being pursued by U.S. troops. The late 19th century also witnessed the growth of cattle and sheep

ranching, along with the strife that developed between the two groups of ranchers over grazing areas. A new mining boom started in 1882 with the discovery of gold in the COEUR D'ALENE MOUNTAINS, and although the gold strike ended in disappointment, it prefaced the discovery there of some of the richest silver mines in the world. WALLACE and KELLOGG became notable mining centers, and the Bunker Hill and Sullivan (a lead mine) became one of the most famous of mines. Severe labor troubles there at the end of the century led to political uprisings. Frank Steunenberg, who as governor, had used Federal troops to put down the uprisings, was assassinated in 1905. The trial of William Haywood and others accused of involvement in the murder drew national attention and marked the beginning of the long career of William E. Borah (who had prosecuted the mine leaders) as an outstanding Republican party leader in the state and nation. The coming of the railroads (notably the Northern Pacific) through here in the 1880s–1890s brought new settlers and aided in the founding of such cities as Idaho Falls, Pocatello, and AMERICAN FALLS. Farming expanded in the state and private interests developed irrigation projects. Some of these aroused public opposition, which led to establishment of state irrigation districts under the Carey Land Act of 1894.

History: 1902 to 1949

The Reclamation Act of 1902 brought direct Federal aid, and furthered reclamation work in Idaho. Notable among public reclamation projects are the Boise and Minidoka projects. The projects, both public and private, have also helped to increase the development of Idaho's enormous potential of hydroelectric power. Three new private hydroelectric projects along the Snake River were put into operation between 1959 and 1968. The unspoiled quality of much of Idaho's land has nourished one of the newest and most profitable of Idaho's businesses—the tourist trade. SUN VALLEY, one of the nation's notable year-round vacation spots, is an example of the development of resorts in Idaho. The state also contains the CRATERS OF THE MOON NATIONAL MONUMENT in the SE, a volcanic area, dormant only 2,100 years, a major contributor, along with Yellowstone, to the basaltic layers that underlie the SNAKE RIVER PLAIN; a small section of YELLOWSTONE NATIONAL PARK on Wyoming state line; Sawtooth National Recreation Area in the S center. Indian Reservations include Coeur d'Alene and Nez Perce in the N, Fort Hall in the SE, part of Duck Valley is on NEVADA state line in SW.

History: 1949 to Present

In 1949 a large Atomic Energy Commission project was begun here. The National Reactor Testing Station is situated near ARCO, the first American town to be lighted by electricity obtained from atomic-power plants. EBR-1 (Experimental Breeding Reactor), the first nuclear reactor in the world, is now a National Historic Landmark, on the grounds of Idaho (Lost River) National Engineering Laboratory, U.S. Department of Energy, SE Idaho. Idaho suffered during the recession of the early 1980s, largely due to the drop in energy prices. The state, however, was able to rebound later in the decade through the attraction of new business, including high-technology firms, notably at Boise and Coeur d'Alene. The state's universities include University of Idaho, at MOSCOW; Idaho State University, at Pocatello; and Boise State University, at Boise. The E-W section of the Salmon River in the N central Idaho serves as the time-zone boundary splitting the state between Mountain (S) and Pacific (N) time.

Government

Idaho's constitution was adopted in 1889 and became effective in 1890 upon statehood. The state's chief executive is a governor elected for a term of four years. The current governor is C. L. "Butch" Otter. The legislature consists of an eighty-four-member house of

representatives and a forty-two-member senate. State representatives and senators are elected every two years. The state also elects two Representatives and two Senators to the U.S. Congress and has four electoral votes.

Idaho has forty-four counties: ADA, ADAMS, BANNOCK, BEAR LAKE, BENEWAH, BINGHAM, BLAINE, BOISE, BONNER, BONNEVILLE, BOUNDARY, BUTTE, CAMAS, CANYON, CARIBOU, CASSIA, CLARK, CLEARWATER, CUSTER, ELMORE, FRANKLIN, FREMONT, GEM, GOODING, IDAHO, JEFFERSON, JEROME, KOOTENAI, LATAH, LEMHI, LEWIS, LINCOLN, MADISON, MINIDOKA, NEZ PERCE, ONEIDA, OWYHEE, PAYETTE, POWER, SHOSHONE, TETON, TWIN FALLS, VALLEY, and WASHINGTON.

Idaho, county (□ 8,502 sq mi/22,105.2 sq km; 2006 population 15,762), central IDAHO; ⊙ GRANGEVILLE; 45°51′N 115°28′W. Agricultural and mining area bounded E by BITTERROOT RANGE and MONTANA, W by HELLS CANYON of the SNAKE RIVER and OREGON; drained by the South and Middle forks of CLEARWATER and SALMON rivers. Wheat, barley; alfalfa, hay; sheep, cattle; copper, gold, silver, lead; manufacturing; forest products. In Bitterroot Mountain Region. Large part of Nez Perce National Forest throughout county, except NW; part of Salmon River primitive area. In area are SEVEN DEVILS (W), Salmon River (SE), and Clearwater (NE) mountains; part of Clearwater National Forest in N; part of Nez Perce Indian Reservation in NW; small part of Hells Canyon National Recreation Area in SW corner; units of Nez Perce National Historical Park in NW at East Kamish and Grangerville; Gospel Hump Wilderness Area in center; part of Selway-Bitterroot Wilderness Area in E, part of Frank Church–River of No Return Wilderness Area in SE. Pacific/Mountain time zone border follows Snake River, N on W side, then follows Salmon River back S and E, across county, then N on E border, putting S and far W in Mountain time zone, N center and E in Pacific time zone. Largest county in land area in Idaho, 17th-largest in U.S. Formed 1861.

Idaho City, village (2000 population 458), ⊙ BOISE county, SW IDAHO, on Morse Creek, and 24 mi/39 km NE of BOISE; 43°51′N 115°51′W. Elevation 3,906 ft/1,191 m. In mountain area. Lumber milling, placer mining for gold; tourism. Important gold-mining center in 1860s (estimated population 30,000); classic gold-mining town with saloon and Boot Hill cemetery (of 200 people buried here, only 28 are said to have died of natural causes). Boise Basin Museum in Boise National Forest.

Idaho Falls, city (2000 population 50,730), ⊙ BONNEVILLE county, SE IDAHO, 46 mi/74 km NNE of POCATELLO, traversed by the SNAKE RIVER; 43°29′N 112°02′W. Elevation 4,710 ft/1,436 m. The chief city of the extensively irrigated upper Snake valley, Idaho Falls is the prosperous commercial and processing center of a cattle, dairy, and farm region that produces potatoes, wheat, sugar beets, and alfalfa. Manufacturing includes building materials, consumer goods, food products, leather goods, electronic equipment, satellite antennas. Tourism is important since the city lies near several national parks and major recreational areas. Idaho National Laboratory, including EBR-1 National Historic Landmark, first nuclear reactor in world, is 45 mi/72 km W. Originally a miner's fording point over the Snake River; first settled by Mormons. Municipal Airport to NW. The impressive Idaho Falls Mormon Temple (opened 1944) is a prominent landmark. Idaho Vietnam Memorial is at Freeman Park. Seat of Eastern Idaho Technical College. Ririe Reservoir to E. Several annual rodeos are held here. Incorporated 1900.

Idaho Springs, town (2000 population 1,889), CLEAR CREEK county, N central COLORADO, on CLEAR CREEK, in FRONT RANGE, and 26 mi/42 km W of DENVER; 39°44′N 105°30′W. Elevation 7,540 ft/2,298 m. Trade center, resort with hot mineral springs. Manufacturing. Gold, silver, lead, and copper mines in vicinity.

Nearby are site of first important gold strike (1859) in Colorado and Edgar mine, operated by Colorado School of Mines for instruction of students. Parts of Arapaho National Forest to S and NW; Mt. Evans State Wildlife Area to S. Settled 1859–1860, incorporated 1885.

Idahue (ee-DAH-wai), town, LIBERTADOR GENERAL BERNARDO O'HIGGINS region, central CHILE, on CACHAPOAL RIVER, and 25 mi/40 km SW of RANCAGUA; 34°18′S 71°10′W. Agricultural center (wheat, alfalfa, potatoes, beans, fruit; livestock); flour milling; dairying.

Idaian Cave, Greece: see IDHEAN CAVE.

Ida, Lake, DOUGLAS county, W MINNESOTA, 5 mi/8 km N of ALEXANDRIA; 5 mi/8 km long, 2 mi/3.2 km wide; 45°59′N 95°24′W. Fed N from Lake MILTONA; drained from S. Resort area. Lake CARLOS to E.

Idalion (i-dah-lee-YON), archaeological site, S central CYPRUS, 13 mi/21 km S of NICOSIA, and 3 mi/4.8 km SSW of DHALI. Ancient city had temple to Aphrodite of Apollo. Was the center of the strongest kingdom of CYPRUS; fortification ruins date to 5th century B.C.E.

Idalou, town (2006 population 2,065), LUBBOCK county, NW TEXAS, on the LLANO ESTACADO, 11 mi/18 km NE of LUBBOCK; 33°39′N 101°40′W. Manufacturing (small agricultural equipment).

Ida, Mount (EI-duh), (10,472 ft/3,192 m), E BRITISH COLUMBIA, W CANADA, in ROCKY MOUNTAINS, 100 mi/161 km E of PRINCE GEORGE; 54°03′N 120°20′W.

Ida, Mount, Greece: see PSILORITIS, MOUNT.

Ida Mountains, TURKEY: see KAZ DAĞI.

Idanha (ee-DAN-uh), village (2006 population 230), MARION county, NW OREGON, on North SANTIAM RIVER, c.50 mi/80 km ESE of SALEM, surrounded by Willamette National Forest; 44°42′N 122°04′W. Timber. Fish hatchery to SE. DETROIT LAKE Reservoir and State Park to W; Mount JEFFERSON Wilderness Area to E.

Idanha-a-Nova (ee-DAHN-yah–ah–NO-vah), town, CASTELO BRANCO district, central PORTUGAL, 16 mi/26 km ENE of CASTELO BRANCO; 39°55′N 07°14′W. Agriculture trade center (wheat, corn, beans, olives, wine; livestock). Has remains of medieval fortifications.

Idanre (ee-DAHN-i-ree), town, OGUN state, S NIGERIA, 18 mi/29 km E of ONDO. Cacao, palm oil and kernels, timber, rubber, cotton.

Idappadi, town, SALEM district, TAMIL NADU state, S INDIA, on left bank tributary of KAVERI RIVER, and 24 mi/39 km WSW of SALEM; 11°35′N 77°51′E. Trade center in tobacco area; castor- and peanut-oil extraction. Limestone, saltpeter, and mica workings nearby. Also written EDAPPADI.

Idar, former princely state in RAJPUTANA STATES, in what is now W INDIA; capital was HIMATNAGAR. Rulers were Rajputs. In WESTERN INDIA STATES agency from 1924 to early 1940s, when transferred to RAJPUTANA STATES; incorporated 1949 into SABAR KANTHA and MAHESANA districts of BOMBAY, later GUJARAT, state.

Idar (ID-uhr), town, SABAR KANTHA district, GUJARAT state, W INDIA, 17 mi/27 km N of HIMATNAGAR; 23°50′N 73°00′E. Market center for grain, oilseeds, sugarcane; pottery and wooden toy manufacturing; distillery.

Idar-Oberstein (EE-dahr–O-buhr-shtein), city, RHINELAND-PALATINATE, W GERMANY, 30 mi/48 km E of TRIER; 49°43′N 07°19′E. A center of German precious- and semiprecious-stone polishing industry; jewelry manufacturing; metalworking. Has two ruined castles. Research institute for precious stones. Formed 1933 through incorporation of Idar, OBERSTEIN (both chartered 1865), ALGENRODT, and TIEFENSTEIN.

Idaville, village, WHITE county, N central INDIANA, 6 mi/9.7 km E of MONTICELLO. Dairying; soybeans, corn; hogs.

Idawgaw, NIGERIA: see IDOGO.

Idbäck, SWEDEN: see HOLE.

Iddefjord, NORWAY: see HALDEN.

Iddo (EE-do), town, LAGOS township, LAGOS state, SW NIGERIA, on IDDO Island (1 mi/1.6 km long, 1.2 mi/2 km wide) in LAGOS LAGOON, between LAGOS Island (connected by Carter Bridge) and mainland. Railroad terminus; linked with EBUTE METTA (on mainland) by railroad and road causeway. Sawmilling.

Ide (EE-de), town, Tsuzuki district, KYOTO prefecture, S HONSHU, W central JAPAN, 9 mi/15 km S of KYOTO; 34°47′N 135°48′E.

Ideal, town (2000 population 518), MACON county, central GEORGIA, 9 mi/14.5 km NW of OGLETHORPE; 32°22′N 84°11′W.

Idel' (EE-dyel), town, E central Republic of KARELIA, on the Murmansk railroad, on the WHITE SEA-BALTIC CANAL, 28 mi/45 km SSW of BELOMORSK; 64°08′N 34°14′E. Elevation 209 ft/63 m. Transportation establishments. Arose in 1933 in conjunction with construction of the canal.

Ideles (eed-LES), Saharan oasis, TAMANRASSET wilaya, S ALGERIA, on N slope of AHAGGAR MOUNTAINS, 80 mi/129 km N of TAMANRASSET; 23°50′N 05°55′E.

Idenburg River, INDONESIA: see TARITATU RIVER.

Idensalmi, FINLAND: see IISALMI.

Ider Gol (EE-der GOL), river, 280 mi/450 km long, NW MONGOLIA; rises in the HANGAYN NURUU (Khangai Mountains), 50 mi/80 km ENE of ULIASTAY; flows ENE, joining the DELGER Mörön (Muren River) to form the SELENGA River, 35 mi/55 km SE of MÖRÖN at 49°16′N 100°41′E. Drains an area of c.9,500 sq mi/24,600 sq km. Also called Ider River, Iderin Gol.

Iderin Gol, MONGOLIA: see IDER GOL.

Ider River, MONGOLIA: see IDER GOL.

Idete (ee-DE-te), village, MOROGORO region, S central TANZANIA, 75 mi/121 km SE of IRINGA, and W of Mount Chikweta (4,974 ft/1,516 m); 08°40′S 36°23′E. Cattle, sheep, goats; corn, pyrethrum, manioc, sorghum, beans.

Idfa (id-FAH), village, SOHAG province, central Upper EGYPT, 4 mi/6.4 km NW of SOHAG; 26°34′N 31°38′E. Cotton, cereals, dates, sugar. Sometimes Edfa.

Idfina (id-FEE-nuh), village, BEHEIRA province, Lower EGYPT, on RASHID branch of the NILE RIVER, 9 mi/14.5 km SE of ROSETTA. Cotton, rice, cereals.

Idfu (id-FOO), town, S central EGYPT, on the W bank of the NILE River; 24°58′N 32°52′E. Agricultural trade center. Has paper mills and a sugar refinery. Predynastic upper Egyptian kingdom capital that flourished c.3400 B.C.E. and worshipped Horus. Later, a large sandstone temple of Horus was built here by Ptolemy III and Ptolemy IV. It is one of the finest extant examples of Egyptian temple architecture. Excavations have yielded a field of mastabas dating from the Old Kingdom, a Roman necropolis, and Coptic and Byzantine remains. The town was known to the Greeks, who identified Horus with Apollo, as Apollinopolis Magna.

Idhean Cave (EE-[th]e-ahn), IRÁKLION prefecture, central CRETE department, GREECE, on E slope of PSILORITI (Ida) Range, and above the NIDHA PLAIN; elevation 5,576 ft/1,700 m. Excavations since the late 19th century have revealed the cave to have been a sacred locale from the Neolithic period to the classical age. One of several caves claimed as the birthplace of Zeus. Also Idaian Cave.

Idhomeni, Greece: see IDOMENI.

Ídhra, Greece: see HYDRA.

Idi (EEd), village, DEBUBAWI KAYIH BAHRI region, SE ERITREA, fishing port on RED SEA, 75 mi/121 km NW of ASEB; 13°56′N 41°42′E. Has mosque; hot springs nearby. Also spelled Edd.

Idice River (ee-DEE-che), 45 mi/72 km long, N central ITALY; rises in ETRUSCAN APENNINES 6 mi/10 km NNW of FIRENZUOLA; flows N and E to RENO RIVER 2 mi/3 km S of ARGENTA. Receives Savena River (left). Canalized in lower course; used for irrigation.

Idil, village, SE TURKEY, 65 mi/105 km E of MARDIN; 41°50′N 28°01′E. Cereals, lentils. Also called Hazak.

Idiofa (eed-YO-fah), village, BANDUNDU province, SW CONGO, 50 mi/80 km ENE of KIKWIT; 05°02′S 19°36′E. Elev. 1,820 ft/554 m. Trade (palm products, fibers). Jesuit mission. Also spelled IDIOFO.

Idiofo (eed-YO-fo), CONGO: see IDIOFA.

Iditarod (ei-DI-tah-rawd), village, W ALASKA, on IDITAROD RIVER, and 118 mi/190 km SE of UNALAKLEET; 62°28′N 158°02′W. Placer gold mining. Scene (1908) of gold rush.

Iditarod (ei-DI-tah-rawd), affiliated area and former ALASKA Gold Rush trail, Alaska. extending 1,049 mi/1,688 km from ANCHORAGE to NOME. Annual 1,049-mi/1,688-km sled-dog race in late February starts at Mulcahy Park, Anchorage, follows GLENN HIGHWAY through KNIK before entering wilderness and ending at Nome. Dog Mushers' Hall of Fame at Knik. Authorized 1980.

Iditarod River (ei-DI-tah-rawd), 150 mi/241 km long, W ALASKA; rises N of Chuathbaluk; flows in an arc NE, N, and finally W, to INNOKO RIVER at 63°02′N 158°45′W. Placer gold mining in valley.

Idjil, MAURITANIA: see FDERIK.

Idjwi Island (EEJ-wee) (□ 96 sq mi/249.6 sq km), SUD-KIVU province, E CONGO, in LAKE KIVU; 25 mi/40 km long, 5 mi/8 km–8 m/12.9 km wide; 02°09′S 29°04′E. Mountainous; forests (N), banana groves (S). Sometimes called KWIDJWI.

Idkerberget (EED-sher-BER-yuh-et), village, KOPPARBERG county, central SWEDEN, in BERGSLAGEN region, 10 mi/16 km SW of BORLÄNGE; 60°23′N 15°14′E. Former iron-mining area.

Idku (id-KOO), town, BEHEIRA province, Lower EGYPT, near N shore of Lake IDKU, 26 mi/42 km ENE of ALEXANDRIA; 31°18′N 30°18′E. Rice milling, silk weaving; fisheries.

Idku, Lake (id-KOO), salt lagoon (□ 57 sq mi/148.2 sq km), Lower EGYPT, 14 mi/23 km E of ALEXANDRIA; 15 mi/24 km long, 3 mi/4.8 km wide. Fisheries. Separated from MEDITERRANEAN SEA (ABUKIR BAY) to the N by narrow sandbank and connected with the bay by the short El Ma'adiya canal, said to be old Canopic branch of the Nile River.

Idlewild Airport, NEW YORK: see JFK (JOHN F. KENNEDY) INTERNATIONAL AIRPORT.

Idlib, district, N SYRIA; ⊙ IDLIB. Bordered by ALEPPO (NE), HAMA (S), and RAQQA (W) districts Fertile agricultural region. Main crops include grains, cotton, fruit, vegetables, olives.

Idlib (id-LEEB), town, ⊙ IDLIB district, NW SYRIA; 35°55′N 35°38′E. Important market center for a fertile agricultural region where grains, grapes, olives, sesame, and cotton are grown. Idlib's chief industries are textile manufacturing, olive pressing, and fig drying.

Idmiston (ID-mi-stuhn), agricultural village (2001 population 2,025), SE WILTSHIRE, S central ENGLAND, 6 mi/9.7 km NE of SALISBURY; 51°08′N 01°43′W. Has 14th–15th-century church.

Idna, Arab township, HEBRON district, 6.3 mi/10 km NW of Hebron, WEST BANK; 31°34′N 34°59′E. Olives and vineyards.

Ido (EE-do), town, TARABA state, SE NIGERIA, on road, 140 mi/225 km SW of JALINGO. Market town. Rice, millet, maize.

Idodi (ee-DO-dee), village, IRINGA region, central TANZANIA, 33 mi/53 km W of IRINGA; 07°47′S 35°11′E. Road junction. Cattle, goats, sheep; tobacco, pyrethrum, corn, sorghum; timber.

Idogo (ee-DAW-gaw), town, OGUN state, extreme SW NIGERIA, 8 mi/12.9 km WSW of ILARO. Railroad spur terminus; cotton weaving, indigo dyeing; cacao, palm oil and kernels. Phosphate deposits. Also Idawgaw.

Idomeni (ee-[th]o-ME-nee), village, KILKIS prefecture, CENTRAL MACEDONIA department, NE GREECE, on railroad, and 21 mi/34 km WNW of KILKIS, on border of former Yugoslavian MACEDONIA, just S of DJEVDJELIJA; 41°07′N 22°31′E. Also spelled Eidomene or Idhomeni; formerly Sechovon (Sekhovon).

Area is shown by the symbol □, and capital city or county seat by ⊙.

Idrija (EED-ree-yah), town, W SLOVENIA, 23 mi/37 km W of LJUBLJANA; 46°00′N 14°03′E. Major mercury-mining and lace-making center. Part of Italy (1919–1947).

Idrinskoye (EE-dreen-skuh-ye), village (2005 population 5,470), SW KRASNOYARSK TERRITORY, SE SIBERIA, Russia, in the YENISEY RIVER basin, on road, 45 mi/72 km NNE of MINUSINSK, and 46 mi/74 km N of ABAKAN; 54°22′N 92°08′E. Elevation 1,040 ft/316 m. Highway junction; local transshipment point.

Idris Dagi (Turkish=Idris Daği) peak (6,512 ft/1,985 m), central TURKEY, 24 mi/39 km ENE of ANKARA.

Idritsa (EE-dree-tsah), town (2006 population 5,570), SW PSKOV oblast, W European Russia, on the Idritsa River, surrounded by small lakes, on road and railroad, 60 mi/97 km W of VELIKIYE LUKI; 56°21′N 28°53′E. Elevation 419 ft/127 m. Railroad junction; woodworking, hemp retting.

Idro, Lago d' (DEE-dro, LAH-go) or Eridio (er-EE-dyo), lake (□ 4 sq mi/10.4 sq km), in BRESCIA province, LOMBARDY, N ITALY, between lakes GARDA (E) and ISEO (W), 18 mi/29 km NE of BRESCIA; 6 mi/10 km long, up to 1.25 mi/2.01 km wide; elevation 1,207 ft/368 m; maximum depth 400 ft/122 m; 45°47′N 10°30′E. Traversed by CHIESE RIVER. Used for irrigation and hydroelectric power.

Idron, FRANCE: see IDRON-OUSSE-SENDETS.

Idron-Ousse-Sendets (ee-DRON–oos–sahn-DE), commune (□ 8 sq mi/20.8 sq km), PYRÉNÉES-ATLANTIQUES department, AQUITAINE region, SW FRANCE, suburb 5 mi/ 8 km E of PAU; 43°17′N 00°19′W. Also called simply Idron.

Idstedt (EED-stet), village, SCHLESWIG-HOLSTEIN, N GERMANY, 5 mi/8 km N of SCHLESWIG, in the ANGELN; 54°35′N 09°30′E. Scene (1850) of decisive victory of Danes over Schleswig-Holsteiners.

Idstein (EED-shtein), town, HESSE, W central GERMANY, in the TAUNUS, 9 mi/14.5 km NNE of WIESBADEN; 50°13′N 08°16′E. Manufacturing of tools, plastics, leather goods; tourism. Has 15th-century castle and many half-timbered houses from 16th–18th century.

Idua Oron, NIGERIA: see ORON.

Idukki, district (□ 1,938 sq mi/5,038.8 sq km), KARNATAKA state, S INDIA; ⊙ PAINAVU.

Idumaea, region, ISRAEL: see EDOM.

Idumania, ENGLAND: see BLACKWATER RIVER.

Idumea, ISRAEL: see EDOM.

Idutywa, town (□ 448 sq mi/1,164.8 sq km), EASTERN CAPE province, SOUTH AFRICA, on Mputi River, on N2 highway, 45 mi/72 km SW of UMTATA; 32°06′S 32°24′E. Elevation 2,730 ft/832 m. Founded in 1884; was originally center of Idutywa Reserve (□ 448 sq mi/1,160 sq km) of TRANSKEI district of the Transkeian Territories. Livestock; dairying; grain. Trading and communication center.

Idutywa Reserve, SOUTH AFRICA: see IDUTYWA

'Idwa, El (ed-WAH, el), village, FAIYUM province, Upper EGYPT, 6 mi/9.7 km NE of FAIYUM; 29°21′N 30°55′E. Cotton ginning; cotton, cereals, sugar cane, fruits.

Idyllwild, CALIFORNIA: see SAN JACINTO MOUNTAINS.

Idyllwild–Pine Cove, unincorporated town, RIVERSIDE county, S CALIFORNIA; residential suburb 10 mi/16 km SW of PALM SPRINGS, in SAN JACINTO MOUNTAINS. Mount San Jacinto State Park to N. Idyllwild section in Strawberry Valley; Pine Cove extends 2 mi/3.2 km NW. Manufacturing (wire drawing).

Idylwood (EI-duhl-wud), unincorporated city, FAIRFAX county, NE VIRGINIA, residential suburb 9 mi/14.5 km W of WASHINGTON, D.C., 2 mi/3 km W of FALLS CHURCH; 38°53′N 77°12′W.

Idzhevan, ARMENIA: see IJEVAN.

Ie (EE-e), village, Kunigami district, OKINAWA prefecture, SW JAPAN, 37 mi/60 km N of NAHA; 26°42′N 127°48′E.

Iecava (YE-tsuh-vuh), city, ZEMGALE district, S LATVIA, 30 mi/48 km S of RĪGA, on N-S via BALTICA HIGH-

WAY; 56°36′N 24°12′E. Serviced by E-W railroad line. Agriculture. Market center.

Ie-jima (ee-E–jee-MAH) or Ie-shima, volcanic island (□ 9 sq mi/23.4 sq km) of OKINAWA ISLANDS, in the RYUKYU ISLANDS, Okinawa prefecture, SW JAPAN, in EAST CHINA SEA, just off W coast of OKINAWA; 5.5 mi/8.9 km long, 2.5 mi/4 km wide; surrounded by coral reef. Part of battle of Okinawa fought here in World War II. American journalist Ernie Pyle was killed during the battle and is buried here.

Iepê (EE-e-pai), town (2007 population 7,490), W SÃO PAULO, brazil, near PARANAPANEMA RIVER, 45 mi/72 km SE of PRESIDENTE PRUDENTE; 22°40′S 51°05′W. Coffee, cotton, grain.

Iepenburg (I-puhn-buhrg), recreational park near SCHOTEN, ANTWERPEN province, N BELGIUM, 5 mi/8 km NE of ANTWERP.

Ieper, BELGIUM: see YPRES.

Ierapetra (ee-e-RAH-pe-trah), town, LASITHI prefecture, E CRETE department, GREECE; port and largest town on S coast, 17 mi/27 km S of AYIOS NIKOLAOS; 35°01′N 25°45′E. Trades in carob, raisins; olive oil. Hothouse tomatoes produced for export. Also appears as Hierapetra.

Ierissos (ee-e-ree-SOS), town, KHALKIDHIKÍ prefecture, CENTRAL MACEDONIA department, NE GREECE, port on Gulf of IERISSOS (Akanthus) of AEGEAN SEA, 23 mi/37 km E of POLYGYROS; 40°24′N 23°53′E. Olive oil, wine; timber. Nearby was ancient AKANTHOS. Also Hierissos.

Iernut (yer-NOOTS), Hungarian Radnót, town, MUREŞ county, central ROMANIA, on MUREŞ RIVER, on railroad, and 18 mi/29 km SW of TÎRGU MURES; 46°27′N 24°15′E. Agriculture center; flour milling, reed weaving.

Ieshima (ee-E-shee-mah), town, Shikama district, HYOGO prefecture, S HONSHU, W central JAPAN, 36 mi/58 km W of KOBE; 34°40′N 134°32′E. Nori; seafood (sea bream, oysters).

Ie-shima, JAPAN: see IE-JIMA.

Iesolo (YAI-so-lo), town, VENEZIA province, VENETO, N ITALY, on N shore of Lagoon of Venice, 16 mi/26 km NE of VENICE; 45°32′N 12°38′E. Swimming resort. Has ruined Romanesque church (11th century). Called Cavazuccherina until c.1930.

Iet, ETHIOPIA: see YET.

If (EEF), rocky islet off S coast of FRANCE, in the MEDITERRANEAN SEA, 2 mi/3.2 km SW of harbor of MARSEILLE, BOUCHES-DU-RHÔNE department, PROVENCE-ALPES-CÔTE D'AZUR region; 43°17′N 05°20′E. Chateau d'If (on RATONNEAU), a fort built here 1524 by Francis I, was long used as a state prison and has been made famous by Alexandre Dumas' The Count of Monte Cristo.

Ifach, SPAIN: see CALPE.

Ifag, ETHIOPIA: see YIFAG.

Ifakara (ee-fah-KAH-rah), town, MOROGORO region, central TANZANIA, 110 mi/177 km SW of MOROGORO, near Klombero River (c. 70 mi/113 km long); 08°10′S 36°39′E. Road junction. SELOUS GAME RESERVE to E. Sheep, goats; grain; timber.

Ifalik (EE-fah-leek), atoll (□ 1 sq mi/2.6 sq km), State of YAP, W CAROLINE ISLANDS, Federated States of MICRONESIA, SM, W PACIFIC, 40 mi/64 km SE of WOLEAI; 2 mi/3.2 km long, 1.6 mi/2.6 km wide; 5 wooded islets on reef. Also Ifaluk.

Ifanadiana (ee-fah-nah-DEE-nah), town, FIANARANTSOA province, E MADAGASCAR, on highway, and 40 mi/64 km NE of FIANARANTSOA; 21°18′S 47°38′E. Trading center; coffee plantations. Gold and iron mined nearby. Hospital.

Ife (EE-fai), city, OSUN state, SW NIGERIA, 25 mi/40 km S of OSHOGBO. Main road center. Cocoa, oil palms, yams, cassava. In a farm region, the city is an important center for marketing and shipping cacao. According to tradition, Ife is the oldest YORUBA town (founded c. 1300). All Yoruba chiefs trace their descent from the first mythological ruler of Ife, Odu-

duwa, and they regard the reigning oni (king) of Ife as their ritual superior. Ife was the most powerful Yoruba kingdom until the late 17th century, when OYO surpassed it. Terra-cotta and naturalistic bronze sculptures made in the area as early as the 12th century are considered among the finest works of W African art; some are displayed in the Ife Museum. Seat of the University of Ife. Sometimes called Ile-Ife [=old Ife].

Iferouâne (EE-fer-oo-ahn), town, AGADEZ province, N central NIGER, on trans-SAHARA route, 149 mi/240 km NNE of AGADEZ; 19°04′N 08°24′E.

Iffezheim (IF-fets-heim), village, BADEN-WÜRTTEMBERG, SW GERMANY, 3.5 mi/5.6 km SW of RASTATT; 48°50′N 08°08′E. Gravel quarrying. Hydroelectric plant on the RHINE RIVER.

Iffley, ENGLAND: see OXFORD.

Ifield, ENGLAND: see CRAWLEY.

Ifinga (ee-FEEN-gah), village, RUVUMA region, S central TANZANIA, 80 mi/129 km N of SONGEA, near Ruhudji River; 09°30′S 35°31′E. Corn, sorghum, sweet potatoes, beans.

Ifni (EEF-nee), former Spanish possession (□ 580 sq mi/1,508 sq km), SW MOROCCO, on the ATLANTIC OCEAN. The main industry is fishing. Ceded by Morocco to SPAIN in 1860, but Spanish administration was nominal until 1934; from then until 1958 the capital, SIDI IFNI, was the residence of the governor-general of Spanish Sahara (later WESTERN SAHARA). Border clashes between Spanish and Moroccan troops occurred in 1957. Spain returned Ifni to Morocco in 1969.

Ifni (EEF-nee), lake, MOROCCO, in High ATLAS mountains, S of Jbel TOUBKAL, 43 mi/70 km S of MARRAKECH; 31°02′N 07°53′W.

Ifo (EE-fo), town, OGUN state, extreme SW NIGERIA, 12 mi/19 km ESE of ILARO; 06°49′N 03°12′E. Railroad and road junction. Cotton weaving, indigo dyeing; cacao, palm oil and kernels, cotton, yams, corn, cassava. Phosphate deposits.

Ifôghas, Adrar des, MALI: see IFORAS, ADRAR DES.

Ifon (EE-fong), town, ONDO state, SW NIGERIA, on road, 45 mi/72 km S of AKURE. Market town. Kola nuts, maize, cocoyams.

Iforas, Adrar des, highland of the SAHARA, EIGHTH REGION/KIDAL, MALI, along Mali-ALGERIA border, c.300 mi/483 km NE of TIMBUKTU (Mali). A SW extension of the AHAGGAR massif; rises over 2,000 ft/610 m. Sometimes called Adrar des Ifôghas.

Ifrane (EEF-rahn), city, Ifrane province, Meknès-Tafilalet administrative region, N central MOROCCO, on N spur of the Middle ATLAS mountains, 35 mi/56 km SE of MEKNES; 33°12′N 05°25′W. Elevation 5,413 ft/1,650 m. Summer resort and one of royal residences; situated amidst national forest of cedars and oaks. Hotels, casino, summer homes. Built by French. Airfield. Nearby is Mischliffen ski station.

Ifri (ee-FREE), village, W BEJAÏA wilaya, on MEDITERRANEAN coast of N central ALGERIA; 36°33′N 04°33′E. Site of the Congress of the Soumam on August 20, 1956, the first congress of the independence movement, the FLN. A new organizational strategy was set up, most notably the division of Algeria into war zones (wilaya).

Ifs (EEFS), town, CALVADOS department, BASSE-NORMANDIE region, NW FRANCE, a S suburb of CAEN; 49°08′N 00°21′W.

Ifugao (ee-foo-GOU), province (□ 972 sq mi/2,527.2 sq km), in CORDILLERA ADMINISTRATIVE REGION, N central LUZON, PHILIPPINES; ⊙ LAGAWE; 16°50′N 121°10′E. Population almost entirely rural; in 1991 80% of urban and 58% of villages had electricity. Mountainous region noted for terraced rice growing. In 1980s popular resistance to environmentally damaging development projects that benefited lowland residents was partly responsible for the creation of the Cordillera Administrative Region here in the Luzon highlands.

Iga, former province in S HONSHU, JAPAN; now part of MIE prefecture.

Iga (EE-gah), town, Ayama district, MIE prefecture, S HONSHU, central JAPAN, 19 mi/30 km S of TSU; 34°49′N 136°13′E.

Igaci (EE-gah-SEE), city (2007 population 25,119), E central ALAGOAS state, BRAZIL, 11 mi/18 km S of Paleira dos Indios; 09°37′S 36°38′W.

Igaliku (i-GAH-li-ko), settlement, Julianehaab district, SW GREENLAND, at head of IGALIKU FJORD, 25 mi/40 km NE of Julianehaab; 60°58′N 45°24′W. Sheep raising. The old Norse *Gardar*, it was seat (c.tenth century) of the *Gardar Thing*, parliament of Norse Eastern Settlement; after 1126 it was seat of Greenland bishopric. Region first settled (c.985) by Eric the Red who had his farm at BRATTAHLID, Greenlandic *Qassiarsuk*, 12 mi/20 km N of Igaliku, near NARSARSUAQ. Older spelling, Igaliko.

Igaliku Fjord, inlet of the ATLANTIC OCEAN, SW GREENLAND; 35 mi/56 km long, 1mi/1.6 km–3 mi/4.8 km wide. Near mouth is Qaqortoq (JULIANEHÅB).

Igalukilo (ee-gah-loo-kee-LO), village, RUKWA region, W TANZANIA, 115 mi/185 km SE of KIGOMA; 05°58′S 30°41′E. Timber; goats, sheep; corn, wheat.

Igalula (ee-gah-LOO-lah), name of two villages, TABORA region, W central TANZANIA. One is 20 mi/32 km SE of TABORA, on railroad; 05°15′S 32°59′E. Livestock; subsistence crops. The other is 42 mi/68 km SSW of Tabora; 05°38′S 32°37′E.

Igamba (ee-GAHM-bah), village, MBEYA region, SW TANZANIA, 35 mi/56 km S of MBEYA, near MALAWI border; 09°33′S 32°53′E. Coffee, tea, corn, wheat, pyrethrum; cattle. There is another Igamba 52 mi/84 km to SE.

Igamba (i-GAHM-bah), village, MBEYA region, SW TANZANIA, 38 mi/61 km W of MBEYA; 09°33′S 32°53′E. Road junction. Cattle, goats, sheep; coffee, tea, corn, pyrethrum. There is another Igamba 52 mi/84 km to NW.

Igana (ee-GAH-nah), town, OYO state, SW NIGERIA, on Ifiki River and 60 mi/97 km NW of IBADAN. Market town. Yams, cassava, millet; fish.

Igandu (i-GAHN-doo), village, DODOMA region, central TANZANIA, 30 mi/48 km ESE of DODOMA, on railroad; 06°25′S 36°21′E. Cattle, sheep, goats; peanuts, corn, sorghum.

Iganga, former administrative district (□ 980 sq mi/2,548 sq km; 2005 population 780,800), EASTERN region, SE UGANDA; capital was IGANGA; 00°45′N 33°35′E. As of Uganda's division into fifty-six districts, was bordered by PALLISA (NE), TORORO (E), BUGIRI (SE), MAYUGE (S), JINJA (SW), and KAMULI (W and N) districts. Primary languages spoken were English, Lusiki, and Lusoga. Towns included Iganga and BUSEMBATIA. Majority of population was agricultural (coffee, cotton, maize, rice, and sugercane). Created in 2000 from N portion of former IGANGA district (Bugiri district was created from SE portion in 1997 and Mayuge district from SW portion in 2000). In 2006 NE portion of district was carved out to form NAMUTUMBA district and remainder of district was formed into current IGANGA district.

Iganga, administrative district, EASTERN region, SE UGANDA; ⊙ IGANGA. As of Uganda's division into eighty districts, borders KALIRO (N), NAMUTUMBA (NEE), BUGIRI (ESE), MAYUGE (S), JINJA (SW), and KAMULI (W) districts. Agricultural area (including coffee and cotton). Railroad from KASESE town (W Uganda) and MOMBASA (SE KENYA) runs across district. Roads connect district to surrounding Uganda as well as Kenya. Formed in 2006 from all but NE portion of former IGANGA district created in 2000 (Namutumba district formed from NE portion).

Iganga (EE-gahn-guh), former administrative district (□ 5,063 sq mi/13,163.8 sq km), SE UGANDA, along N shore of LAKE VICTORIA; capital was IGANGA; 00°30′N 33°40′E. As of Uganda's division into thirty-nine districts, was bordered by MUKONO (SW), JINJA (W), KAMULI (N), PALLISA (NE), and TORORO (E) districts

and Lake Victoria (S). Largest towns were Iganga, BUSEMBATIA. Cotton, bananas, millet, groundnuts were grown. In 1997 SE portion of district was carved out to form BUGIRI district; in 2000 SW portion was carved out to form MAYUGE district and N portion was formed into (now former) new IGANGA district.

Iganga (EE-gahn-guh), town (2002 population 39,472), ⊙ IGANGA district, EASTERN region, SE UGANDA, 22 mi/35 km NE of JINJA. Cotton, tobacco, coffee, bananas, corn. Was part of former BUSOGA province.

Igangan (ee-GAHN-gahn), town, OYO state, SW NIGERIA, on road, 50 mi/80 km WNW of IBADAN; 07°41′N 03°11′E. Market town. Cassava, yams.

Igaporã (EE-gah-po-RAH), city (2007 population 14,980), W central BAHIA, BRAZIL, 32 mi/52 km NW of CAETITÉ; 13°44′S 42°43′W.

Igara (EE-gah-RAH), village, N central BAHIA, BRAZIL, 3.7 mi/6 km NE of SENHOR DO BONFIM; 10°24′S 40°07′W.

Igaraçu (EE-gah-rah-soo), city, E PERNAMBUCO state, NE BRAZIL, on inlet of the ATLANTIC OCEAN, and 15 mi/24 km NNW of RECIFE; 07°50′S 34°54′W. Sugar, manioc, coconuts. One of oldest Portuguese settlements (16th century) in Brazil, with several 16th-century churches, and convent of São Francisco noted for its Dutch tiles and sculpture. Also spelled Iguarassú.

Igaraçu do Tietê (EE-gah-rah-soo do chee-E-tai), city, central São Paulo state,19 mi/31 km NE of Lençois Paulista, on Tietê River; 22°31′S 48°34′W.

Igara Paraná River (ee-GAH-rah pah-rah-NAH), c.250 mi/402 km long, AMAZONAS department, SE COLOMBIA; rises near CAQUETÁ River; flows SE, through densely forested tropical lowlands, to Río PUTUMAYO (PERU border); 02°10′S 71°46′W.

Igarapava (EE-gah-rah-pah-vah), city (2007 population 26,853), NE SÃO PAULO state, BRAZIL, near the RIO GRANDE, on railroad, and 22 mi/35 km SSE of UBERABA (MINAS GERAIS state); 20°03′S 47°47′W. Sugar milling, distilling (rum, alcohol), cotton and rice processing.

Igarapé (EE-gah-rah-PAI), city (2007 population 31,135), MINAS GERAIS state, BRAZIL, 12 mi/20 km SW of BETIM; 20°12′S 44°26′W.

Igarapé Açu (ee-gah-rah-PAI ah-soo), city, E PARÁ state, BRAZIL, on BELÉM-BRAGANÇA railroad, and 65 mi/105 km NE of BELÉM; 01°00′S 47°36′W. Brazil nuts, rubber, cacao, corn.

Igarapé Grande (ee-gah-rah-PAI grahn-zhee), city (2007 population 9,756), N central MARANHÃO state, BRAZIL, 9 mi/15 km N of POÇÃO DE PEDRAS, on RIO MEARIM; 04°34′S 44°50′W.

Igarapé Miri (EE-gah-rah-PAI mee-ree), city, E PARÁ state, BRAZIL, 50 mi/80 km SW of BELÉM; 10°58′S 48°55′W.

Igaratá (EE-gah-rah-tah), town (2007 population 8,537), SE SÃO PAULO state, BRAZIL, 16 mi/26 km W of SÃO JOSÉ DOS CAMPOS; 23°12′S 46°07′W.

Igarite (EE-gah-REE-te), village, N central BAHIA state, BRAZIL, 19 mi/30 km E of GENTIO DO OURO; 11°36′S 43°24′W.

Igarka (ee-GAHR-kah), city (2005 population 7,110), N KRASNOYARSK TERRITORY, N central SIBERIA, RUSSIA, on the lower YENISEY RIVER, on road, 1,100 mi/1,770 km N of KRASNOYARSK; 67°28′N 86°35′E. Major lumber port accessible to seagoing vessels. Sawmills; fishing. Founded in 1928, city since 1931.

Igatpuri (i-GAHT-poo-ree), town, NASHIK district, MAHARASHTRA state, W INDIA, in WESTERN GHATS, on railroad (workshops), and 27 mi/43 km SW of NASHIK; 19°42′N 73°33′E. Health resort (sanitarium); millet, wheat, gur. Electric locomotives were attached to all long-distance trains to and from Mumbai (Bombay) at this junction.

Igbaja (ee-BAH-jah), town, KWARA state, SW central NIGERIA, 22 mi/35 km ESE of ILORIN; 08°23′N 04°53′E. Shea-nut processing, cotton weaving; cassava, yams, corn; cattle, skins.

Igbaras (eeg-bah-RAHS), town, ILOILO province, S PANAY island, PHILIPPINES, 20 mi/32 km W of ILOILO; 10°46′N 122°15′E. Rice-growing center.

Igbetti (ee-be-TEE), town, OYO state, W NIGERIA, on road, 35 mi/56 km NW of ILORIN; 08°45′N 04°08′E. Market center. Cassava, millet, tobacco.

Igbo-Ora (EE-bo-O-rah), town, OYO state, SW NIGERIA, on road, 40 mi/64 km WNW of IBADAN; 07°26′N 03°17′E. Market town. Yams, cassava, millet.

Igdir, (Turkish=*Iğdir*) town (2000 population 59,880), NE TURKEY, between Mount ARARAT and ARAS RIVER, 70 mi/113 km SE of KARS, 7 mi/11.3 km from Armenian border; 40°17′N 35°37′E. Cotton gin.

Igea (ee-HAI-ah), town, LA RIOJA province, N SPAIN, 16 mi/26 km S of CALAHORRA; 42°04′N 02°01′W. Olive-oil processing; wine, fruit, vegetables, honey.

Igede (ee-GAI-dai), town, NIGER state, W central NIGERIA, on road, and 50 mi/80 km WNW of MINNA; 10°03′N 05°56′E. Market town. Yams, cassava; livestock.

Igel (EE-gel), village, RHINELAND-PALATINATE, W GERMANY, on the Mosel River, 5 mi/8 km SW of TRIER; 49°43′N 06°33′E. The sandstone Igel obelisk (75 ft/23 m high) is one of the most remarkable Roman remains N of the ALPS.

Igelsta (EEG-el-stah), village, STOCKHOLM county, E SWEDEN, just E of SÖDERTALJE.

Igersheim (EE-gers-heim), village, BADEN-WÜRTTEMBERG, central GERMANY, 2 mi/3.2 km E of BAD MERGENTHEIM; 49°30′N 09°49′E.

Iggesund (IG-ge-SUND), town, GÄVLEBORG county, E SWEDEN, on small inlet of GULF OF BOTHNIA, 7 mi/11.3 km S of HUDIKSVALL; 61°39′N 17°05′E. Manufacturing (wood and sulphite works).

Igharghar, Oued (ee-gahr-GAHR, WED), intermittent stream, E ALGERIA, E of the Mzah-Tadenait plateau; flows S-N and contains ancient alluvial deposits and dunes of the Great Eastern ERG. During the Ice Age the Oued IGHARGHAR crossed the SAHARA from the AHAGGAR region to the foot of the Saharan ATLAS MOUNTAINS. It profoundly denuded the Tassili escarpments of the Ahaggan, cutting a deep trench. The oued disappears beneath the dunes. Near OUARGLA, the Oued Mya joins it. The eroded channel by which the Igharghar spreads out towards the Tunisian CHOTTS is known as the Oued R'HIR. This low-lying region was once filled with thriving palm-tree oases, but now the oases are drying up due to overexploitation of the artesian water resources here.

Igherm (eeg-RAHM), village, Taroudannt province, Souss-Massa-Draâ administrative region, SW MOROCCO, in the Anti-ATLAS mountains, 36 mi/58 km SE of TAROUDANNT; 30°05′N 08°27′W. Elevation 6,476 ft/1,974 m. Also spelled Irherm.

Ighil M'Goun, MOROCCO: see M'GOUN, JBEL.

Ightham (EI-tuhm), village (2001 population 1,940), W central KENT, SE ENGLAND, 10 mi/16 km W of MAIDSTONE; 51°17′N 00°17′W. Previously stone quarrying. Has 14th-century moated house and Norman church. Pre-Roman earthworks nearby. Just ENE is village of BOROUGH GREEN.

Iginniarfik, fishing settlement, KANGAATSIAQ commune, W GREENLAND, on DAVIS STRAIT, 40 mi/64 km SSW of EGEDESMINDE (Aasiaat); 68°09′N 53°10′W.

Igis (EE-gis), commune, GRISONS canton, E SWITZERLAND, at confluence of RHINE and LANDQUART rivers, 6 mi/9.7 km N of CHUR. Includes large village of LANDQUART.

Igiugig (i-GYOO-gik), village (2000 population 53), S ALASKA, on KVICHAK RIVER, at SW end of ILIAMNA Lake, 40 mi/64 km SW of NEWHALEN; 59°20′N 155°54′W. Fishing; supply point for sportsmen.

Iglas (ig-LAHS), town, tahsil headquarters, ALIGARH district, W UTTAR PRADESH state, N central INDIA, 11 mi/18 km NNW of HATHRAS; 27°43′N 77°56′E. Wheat, barley, millet, gram.

Iglau, CZECH REPUBLIC: see JIHLAVA.

Iglawa River, CZECH REPUBLIC: see JIHLAVA RIVER.

Iglesia (ee-GLAI-see-ah), town, central SAN JUAN province, ARGENTINA, in irrigated valley, 36 mi/58 km WSW of JACHAL; 30°24'S 69°13'W. Elevation 6,500 ft/ 1,981 m. Alfalfa, wheat, wine; sheep; flour milling.

Iglesia, ARGENTINA: see RODEO.

Iglesias (ee-GLAI-zyahs), town, CAGLIARI province, SW SARDINIA, 31 mi/50 km WNW of CAGLIARI; 39°19'N 08°32'E. Center of SW mining district (Iglesiente). Lead, zinc, and silver mines at nearby MONTEPONI, San Giovanni, Nebida, Masua, Malacalzetta, San Benedetto; lignite and iron; wine, olive oil, furniture; gasworks, foundry, tanneries. Bishopric. Has cathedral (built 1285–1288), castle (built 1325; now a factory), 16th-century church of San Francisco.

Iglesuela, La (ee-glai-SWAI-lah, lah), village, TOLEDO province, central SPAIN, 18 mi/29 km NNE of TALAVERA DE LA REINA. Cereals, grapes, olives, acorns; livestock. Lumber; flour milling, olive oil pressing.

Igli (ee-GLEE), Saharan outpost and oasis, BÉCHAR wilaya, W ALGERIA, on the Oued SAOURA, 75 mi/121 km S of BÉCHAR, at NW edge of the Great Western ERG; 30°28'N 02°19'W.

Iglino (ee-glee-NO), town (2005 population 13,970), E BASHKORTOSTAN Republic, E European Russia, on the W bank of the UFA RIVER, on road and railroad, 30 mi/48 km NE of UFA; 54°50'N 56°26'E. Elevation 367 ft/111 m. Highway junction. Fruit and vegetable canning. Industrial scale plant.

Igloo, village, NW ALASKA, on SEWARD PENINSULA, 45 mi/72 km N of NOME. Also called Marys Igloo.

Igloolik, village, BAFFIN region, NUNAVUT territory, N CANADA, on IGLOOLIK ISLAND, in FOXE BASIN, NE of MELVILLE PENINSULA; 69°24'N 81°48'W. Site has been continuously occupied by Inuit people since 2000 B.C.E. Governmental anthropological research center. Co-op manufacturing of handicrafts, clothing. Hunting, fishing, sealing. Radio station; Royal Canadian Mounted Police post. Scheduled air service.

Igloolik Island (ig-LOO-lik), BAFFIN region, NUNAVUT territory, N CANADA, in FOXE BASIN, just off NE MELVILLE PENINSULA; 10 mi/16 km long, 1 mi/2 km–7 mi/11 km wide; 69°24'N 81°49'W. Formerly site of trading post, radio station, and Roman Catholic mission.

Igls (EE-guhls), village, TYROL, W AUSTRIA, at foot of the PATSCHERKOFEL mountain (cable car), 2 mi/3.2 km S of INNSBRUCK. Summer resort; winter sports. Site of 1976 winter Olympic toboggan and bobsled runs.

Igman (EEG-mahn), mountain, central BOSNIA, BOSNIA AND HERZEGOVINA, overlooking SARAJEVO; 43°46'N 18°15'E. Elevation over 4,921 ft/1,500 m. A major battleground in the conflict between Muslims and Serbs during the 1992–1995 civil war. Now in the Muslim part of the Federation of Bosnia and Herzegovina.

Ignace (IG-nuhs), township (□ 28 sq mi/72.8 sq km; 2001 population 1,709), W ONTARIO, E central CANADA, on Agimak Lake (5 mi/8 km long), 50 mi/80 km SSE of SIOUX LOOKOUT; 49°25'N 91°40'W. In goldmining, lumbering region. Incorporated 1908.

Ignacio, village (2000 population 669), LA PLATA county, SW COLORADO, on LOS PINOS RIVER, in foothills of SAN JUAN MOUNTAINS, and 17 mi/27 km SE of DURANGO; 37°07'N 107°37'W. Elevation c.6,432 ft/ 1,960 m. Cattle, sheep; hay, oats. Headquarters of Southern Ute Indian Reservation, located to E and SW. U.S. Southern Ute Agency school and hospital for Native Americans here. Navajo State Park to SE; San Juan National Forest to NE.

Ignacio Allende, MEXICO: see ATLEQUIZAYAN.

Ignacio de la Llave (eeg-NAH-see-o dai lah YAH-vai), town, central E VERACRUZ, MEXICO, on the plains of SOTAVENTO, at edge of Papaloapan Delta, 12 mi/20 km S of VERACRUZ; 18°43'N 95°59'W. Hot climate. Agriculture (corn, beans, sesame, fruits, sugarcane). Sugar and alcohol industry; cattle and poultry industries.

Ignacio Uchoa, Brazil: see UCHÔA.

Ignacio Zaragoza (eeg-NAH-see-o zah-rah-GO-sah), town, in N central CHIHUAHUA, MEXICO, on Piedras Verdes River, 25 mi/40 km W of NUEVO CASAS GRANDES; elevation 4,921 ft/1,500 m. Partly mountainous near SIERRA MADRE OCCIDENTAL. Temperate climate. Agriculture (corn, wheat, fruit); cattle raising; wood products. Road connects with Mexico Highway 2 and JUÁREZ.

Ignalina (ig-NUH-lee-nuh), city and administrative territory, E LITHUANIA, 62 mi/100 km NE of VILNIUS, on E end of AUKSTAITIJOS NATIONAL PARK; 55°21'N 26°10'E. Construction of 6,000 MW nuclear power plant (four reactors, 1,500 MW each) began in 1975; first commissioned in 1983, 25 mi/40 km NE of town. Two reactors are of the same design as those located in CHERNOBYL. Food processing.

Ignaluk, Alaska: see LITTLE DIOMEDE ISLAND.

Ignatievo (eeg-NAH-tee-e-vo), village, VARNA oblast, AKSAKOVO obshtina, BULGARIA; 43°15'N 27°45'E.

Ignatitsa (eeg-NAH-tee-tsah), village, MONTANA oblast, MEZDRA obshtina, BULGARIA; 43°03'N 23°38'E.

Ignatovka (eeg-NAH-tuhf-kah), town (2006 population 2,845), central ULYANOVSK oblast, E central European Russia, on the Gushcha River (VOLGA RIVER basin), on crossroads, 40 mi/64 km SW of ULYANOVSK; 53°57'N 47°39'E. Elevation 692 ft/210 m. Wool milling.

Ignatovo, RUSSIA: see BOL'SHOYE IGNATOVO.

Igneada, (Turkish=Iğneada) village, European TURKEY, near BLACK SEA, 7 mi/11.3 km from Bulgarian border, 40 mi/64 km ENE of KIRKLARELI; 41°52'N 28°03'E. Just E, across a small inlet, is Cape Igneada (sometimes spelled Iniada; ancient *Thynias*).

Ignusi, Greece: see OINOUSSA.

Igny (ee-NYEE), town (2004 population 9,868), ESSONNE department, ÎLE-DE-FRANCE region, N central FRANCE, on small Bièvre River, and 10 mi/16 km SW of PARIS; 48°44'N 02°14'E. Has a school of horticulture. The forest of Verrières (just N) is a large park.

Igodovo (EE-guh-duh-vuh), village, S central KOSTROMA oblast, central European Russia, on road, 25 mi/40 km S of GALICH; 58°01'N 42°21'E. Elevation 541 ft/164 m. In agricultural area.

Igoma (ee-GO-mah), village, MBEYA region, SW central TANZANIA, 66 mi/106 km N of MBEYA, near LUPA RIVER; 07°37'S 33°20'E. Sheep, goats; corn, wheat, sorghum.

Igorevskaya (EE-guh-ryef-skah-yah), village, N SMOLENSK oblast, W European Russia, on railroad, 8 mi/13 km WSW of (and administratively subordinate to) KHOLM-ZHIRKOVSKIY; 55°28'N 33°17'E. Elevation 774 ft/235 m. Sawmilling, woodworking.

Igoumenitsa (ee-yoo-me-NEET-sah), town, THESPROTIA prefecture, S EPIRUS dept., NW GREECE, port on inlet of Channel of CORFU, and 33 mi/53 km WSW of IOÁNNINA; 39°30'N 20°16'E. Timber, barley, corn; olive oil; fisheries. Ferries to ITALY. Also spelled Egoumenitsa.

Igra (ee-GRAH), town (2006 population 21,955), central UDMURT REPUBLIC, E European Russia, on the Loza River (short left tributary of the CHEPTSA RIVER), on railroad, 48 mi/77 km N of IZHEVSK; 57°33'N 53°03'E. Elevation 498 ft/151 m. Highway junction; auto depots. Woodworking, food processing (meat, dairy).

Igrapiúna (EE-grah-pee-OO-nah), city (2007 population 13,236), BAHIA state, E central coast of BRAZIL, 9 mi/15 km NW of CAMAMU; 14°50'S 39°09'W.

Igreja Nova (EE-grai-zhah NO-vah), city, (2007 population 23,072), E ALAGOAS state, NE BRAZIL, 12 mi/19 km NNW of PENEDO; 10°09'S 36°40'W. Sugar, rice; livestock.

Igrejinha (EE-gre-zheen-yah), city (2007 population 31,113), E RIO GRANDE DO SUL state, BRAZIL, 19 mi/31 km NE of NOVO HAMBURGO; 29°34'S 50°48'W. Grapes, corn; livestock.

Igrim (EE-greem), town (2005 population 9,615), NW KHANTY-MANSI AUTONOMOUS OKRUG, W central SIBERIA, RUSSIA, at a confluence of the Malaya Sosva and Severnaya Sosva (left tributary of the OB' RIVER) rivers, approximately 220 mi/354 km NW of KHANTY-MANSIYSK, and 80 mi/129 km (by air) NW of the nearest railroad (at Priob'ye on the Ob' River, Sergino); 63°11'N 64°25'E. In oil and gas fields; agricultural and fish products. Made town in 1964.

Iguaçu (EE-gwah-soo), former federal territory (□ 25,427 sq mi/66,110.2 sq km), S BRAZIL, on PARAGUAY and ARGENTINA border. Created as a frontier defense zone in 1943, it was dissolved in 1946, and area was reincorporated into PARANÁ and SANTA CATARINA. Seat of government was at Iguaçu (now called LARANJEIRAS DO SUL).

Iguaçu, Brazil: see ITAETÉ.

Iguaçu, Brazil: see LARANJEIRAS DO SUL.

Iguaçu Falls (ee-gwah-KOO), in the IGUAÇU RIVER, on the ARGENTINA-BRAZIL border near the PARAGUAY border. Has two main sections composed of hundreds of waterfalls separated from each other by rocky islands along a 3-mi/4.8-km escarpment. The highest fall is 210 ft/64 m high; most of the falls are 100 ft/30 m–130 ft/40 m high. Argentina and Brazil maintain national parks on each side of the falls. The surroundings, in the midst of beautiful scenery, abound in begonias, orchids, brilliant-hued birds, and myriads of butterflies. The ASUNCIÓN-PARANAGUÁ highway passes near the falls. Their potential as a source of hydroelectric power is considerable. Also known as Iguassú Falls.

Iguaçu River, Brazil and Argentina: see IGUASSÚ RIVER.

Iguaí (EE-gwah-EE), city (2007 population 27,687), SE BAHIA state, BRAZIL, 71 mi/115 km WNW of ITABUNA; 14°45'S 40°05'W.

Iguak, Alaska: see OHOGAMIUT.

Igualada (ee-gwah-LAH-dah), city, BARCELONA province, NE SPAIN, in CATALONIA, 30 mi/48 km WNW of BARCELONA; 41°35'N 01°38'E. Industrial center (leather, knitwear, synthetic and cotton textiles). Notable church.

Iguala de la Independencia (ee-GWAH-lah dai lah een-dai-pen-DEN-see-ah), city and township, GUERRERO, S MEXICO, on the Cocula River (BALSAS River tributary); 18°21'N 99°31'W. Communications, distribution, and processing center of the surrounding mining and agricultural region. Famous historically as the place where Agustín de Iturbide proclaimed the Plan of Iguala, which contained the guarantees of independence, on February 24, 1821.

Igualapa (ee-gwah-LAH-pah), town, GUERRERO, SW MEXICO, in PACIFIC lowland, 8 mi/12.9 km WNW of OMETEPEC. Fruit, sugarcane, livestock.

Igualeja (ee-gwah-LAI-hah), town, MÁLAGA province, S SPAIN, 7 mi/11.3 km S of RONDA; 36°38'N 05°07'W. Grapes, fruit, vegetables, corn, chestnuts; timber. Marble and graphite quarrying, iron mining.

Iguape (EE-gwah-pai), city (2007 population 28,963), S SÃO PAULO, BRAZIL, 95 mi/153 km SW of SANTOS; 24°43'S 47°33'W. Fishing port on the RIBEIRA DE IGUAPE near its mouth on a tidal inlet protected from the open ATLANTIC Ocean by a long, narrow island. Fish processing, rice milling, lumbering; caustic-soda manufacturing. Large apatite deposits discovered in area (1947). Center of lowland region settled largely by Japanese immigrants.

Iguaraci (EE-gwah-rah-see), city (2007 population 11,927), N central PERNAMBUCO state, BRAZIL, 47 mi/ 76 km NW of ARCOVERDE, on railroad. Cotton, corn; livestock.

Iguarassú, Brazil: see IGARAÇU.

Iguaratinga, Brazil: see SÃO FRANCISCO DO MARANHÃO.

Iguassú Falls, ARGENTINA and BRAZIL: see IGUAÇU FALLS.

Iguassú River (EE-gwah-soo), Spanish *Iguazú*, in PARANÁ state, SE BRAZIL, c.820 mi/1,320 km long; rises in the Serra do MAR just E of CURITIBA; flows W in a meandering course to the PARANÁ at point where

Cross-references are shown in SMALL CAPITALS. The pronunciation guide is shown on page xix. The sources of population figures are shown on page xvii.

ARGENTINA, Brazil, and PARAGUAY meet, 3 mi/4.8 km below FOZ DO IGUAÇU (Brazil). In last 75 mi/121 km of course it forms Brazil-Argentina border; the famous IGUASSÚ FALLS are 14 mi/23 km above its mouth. Navigable only between PÔRTO AMAZONAS and UNIÃO DA VITÓRIA in Brazil. Receives the RIO NEGRO (left). Also spelled Iguaçu River

Iguatama (ee-gwah-tah-mah), town (2007 population 7,628), SW central MINAS GERAIS state, BRAZIL, on railroad, and 50 mi/80 km W of DIVINÓPOLIS; 20°13′S 45°44′W. Diamond washings in headwaters of Rio são francisco (W). Until 1944, Pôrto Real.

Iguatemi (EE-gwah-TE-mee), city (2007 population 14,624), S central Mato Grosso do Sul state, BRAZIL, 22 mi/35 km WNW of ELDORADO; 23°41′S 54°34′W. Airport.

Iguatu (ee-gwah-too), city, S CEARÁ state, BRAZIL, on FORTALEZA-CRATO railroad, and 190 mi/306 km SSW of Fortaleza; 06°25′S 39°19′W. Trade center of irrigated agriculture region shipping cotton, tobacco, carnauba, manioc, cattle. Magnesite mining. Airfield. Called Telha until 1874. Dam at ORÓS (28 mi/45 km ENE) impounds waters of RIO JAGUARIBE for irrigation and flood control.

Iguatú (EE-gwah-TOO), city (2004 population 90,728), S central CEARÁ, BRAZIL, on the FORTALEZA-CARIUS railroad; 06°25′S 39°19′W. Airport.

Iguazú, ARGENTINA: see PUERTO IGUAZÚ.

Iguazú, Brazil and Argentina: see IGUASSÚ RIVER.

Igüembe (ee-GEM-be), canton, LUIS CALVO province, CHUQUISACA department, SE BOLIVIA; 20°34′S 63°48′W. Elevation 3,885 ft/1,184 m. Abundant gas, petroleum and some unproductive drilled wells. Agriculture (potatoes, yucca, bananas, corn, barley, rye, sweet potatoes, peanuts, tobacco); cattle.

Igueña, SPAIN: see TREMOR.

Iguidi, Erg, ALGERIA: see ERG.

Igula (ee-GOO-lah), village, IRINGA region, central TANZANIA, 80 mi/129 km WNW of IRINGA, near NJOMBE RIVER, in W part of RUAHA NATIONAL PARK; 06°41′S 34°53′E. RUNGWA GAME RESERVE to W. Tourism.

Igumale (i-goo-MAH-le), town, BENUE state, S central NIGERIA, on railroad, 33 mi/53 km SSW of OTURKPO; 06°48′N 07°58′E. Shea nuts, cassava, durra, yams. Sometimes spelled Ogumali.

Igushik (I-goo-shik), village, SW ALASKA, on W shore of NUSHAGAK BAY, at mouth of small Igushik River, 30 mi/48 km SW of DILLINGHAM.

Iharana (ee-HAHRN), town, ANTSIRANANA province, NE MADAGASCAR, on paved highway, 95 mi/153 km N of SAMBAVA; 13°21′S 50°01′E. Seaport and center of vanilla production. Also known as Vohémar.

Ih Bogd Uul or **Ikhe Bogdo**, highest peak (12,987 ft/3,957 m) of GOBI ALTAY Mountains (*Govd Altayn Nuruu*), in S central MONGOLIA, 250 mi/400 km SE of ULIASTAY; 44°55′N 100°20′E. Also called Barun Bogdo or Baruun Bogdo.

Ihden (I-den), village, N LEBANON, 12 mi/19 km SE of TRIPOLI; 34°17′N 35°58′E. Elevation 4,800 ft/1,463 m. Summer resort; cereals, fruit. Medieval church, ruins.

Ihema, Lake (ee-HE-mah), RWANDA, W of KIGALI on border of TANZANIA, near Akagera River and Akagera Hotel.

Iheme (ee-HAI-mai), village, IRINGA region, central TANZANIA, 20 mi/32 km SW of IRINGA; 08°02′S 35°19′E. Cattle, sheep, goats; tobacco, pyrethrum, corn, wheat.

Iheya (ee-HAI-yah), village, Iheya island, IHEYA-SHOTO island group, Shimajiri district, OKINAWA prefecture, SW JAPAN, 59 mi/95 km N of NAHA; 27°02′N 127°58′E.

Iheya-shoto (ee-HAI-yah–SHO-TO), volcanic island group (□ 17 sq mi/44.2 sq km) of OKINAWA ISLANDS, in the RYUKYU ISLANDS, Okinawa prefecture, SW JAPAN, in the PHILIPPINE SEA, 18 mi/29 km NW of OKINAWA. Comprises Iheya-jima island (or Iheya-ushiro-jima; the largest, at 8 mi/12.9 km long, 1.5 mi/

2.4 km wide), Izena-shima island (also called Iheya-mae-shima), and several smaller islets. Hilly; surrounded by coral reef. Produces sugarcane, sweet potatoes, rice.

Ihlen (EE-lahn), village (2000 population 107), PIPESTONE county, SE MINNESOTA, near SOUTH DAKOTA state line, 7 mi/11.3 km SSW of PIPESTONE; 43°54′N 96°22′W. Grain; livestock. Split Rock Creek State Park, on Split Rock Lake reservoir to SE.

Ihna River, POLAND: see INA RIVER.

Ihnasya el Madina (i-NAHS-yuh el mah-DEE-nuh), township, BENI SUEF province, Upper EGYPT, 9 mi/14.5 km W of BENI SUEF; 29°05′N 30°56′E. Cotton, cereals, sugarcane.

Ihosy (EE-oosh), town, FIANARANTSOA province, SE MADAGASCAR, on important highway junction, and 90 mi/145 km SW of FIANARANTSOA; 22°25′S 46°08′E. Cattle market; rice, cassava. Garnets and mica are mined nearby. Hospital.

Ihringen (EE-ring-uhn), village, BADEN-WÜRTTEMBERG, SW GERMANY, at S foot of the KAISERSTUHL, 10 mi/16 km WNW of FREIBURG; 48°03′N 07°39′E. Noted for its wine. Warmest location in Germany.

Ihrlerstein (IR-ler-shtein), village, BAVARIA, S GERMANY, in LOWER BAVARIA, on the DANUBE River, 11 mi/18 km NW of REGENSBURG; 48°56′N 11°52′E.

Ihtiman (eeh-tee-MAHN), city (1993 population 12,898), SOFIA oblast, ⊙ Ihtiman obshtina (□ 45 sq mi/117 sq km), W central BULGARIA, at the S foot of Ikhtimanska SREDNA GORA, on a right tributary of the TOPOLNITSA RIVER, 30 mi/48 km SE of SOFIA; 42°26′N 23°49′E. Agricultural center of the Ihtiman Basin (□ 45 sq mi/117 sq km; average elevation 3,000 ft/914 m); hardy grain, potatoes, flax; livestock. Manufacturing includes shoes and cast-iron electrodes, wood processing. Asbestos quarried nearby. Sometimes spelled Ikhtiman or Ichtiman.

Ihu (ee-hyoo), village, GULF province, SE NEW GUINEA island, S central PAPUA NEW GUINEA, 90 mi/145 km SE of KIKORI, on GULF OF PAPUA, at mouth of Vailala River, 160 mi/257 km NW of PORT MORESBY. Boat access. Sago; fish; cattle; timber.

Ii (EE), Swedish *Ijo* (EE-o), town, OULUN province, W FINLAND, on GULF OF BOTHNIA, at mouth of IIJOKI (river), 20 mi/32 km N of OULU; 65°20′N 25°25′E. Road and railroad bridges; timber-shipping port.

Iide (EE-DE), town, West Okita district, YAMAGATA prefecture, N HONSHU, NE JAPAN, 25 mi/40 km S of YAMAGATA city, near Iide mountain chain; 38°02′N 139°59′E. *Yonezawa* breed of cows originated here.

Iijoki (EE-YO-kee), Swedish *Ijoälv*, river, 150 mi/241 km long, N central FINLAND; rises in lake region near 66°00′N 28°00′E; flows in winding course generally SW, past PUDASJÄRVI, to GULF OF BOTHNIA at II.

Iisalmi (EE-sahl-mee), Swedish *Idensalmi*, town, KUOPION province, S central FINLAND, 50 mi/80 km NNW of KUOPIO; 63°34′N 27°11′E. Elevation 264 ft/80 m. In lake region. Railroad junction; lumber and flour mills, major brewery; some light industry.

Iittala (EE-tah-lah), village, HÄMEEN province, S FINLAND, 12 mi/19 km NW of HÄMEENLINNA; 61°04′N 24°10′E. Elevation 297 ft/90 m. In lake region. Glassworks. Has museum.

Iitti (EET-tee), Swedish *Iittis*, village, KYMEN province, SE FINLAND, 25 mi/40 km E of LAHTI; 60°56′N 26°25′E. Elevation 248 ft/75 m. In lake region and lumbering area; part of UUDENMAAN province until 1949.

Ijaci (EE-zhah-see), town (2007 population 5,687), S central MINAS GERAIS state, BRAZIL, on railroad, 7 mi/11.3 km NE of LAVRAS; 21°14′S 44°55′W.

Ijâfene (ee-jah-fe-NAI), dunes, SE MAURITANIA, range of dune formation running NE c.186 mi/300 km NE of TIDJIKJA; c.20°30′N 09°00′W.

Ijebu Igbo (ee-jai-BOO ee-BOO), town, OGUN state, SW NIGERIA, 12 mi/19 km NNE of IJEBU-ODE, and 45 mi/72 km ESE of ABEOKUTA; 06°58′N 04°00′E. Road

center and market town. In cocoa belt; hardwood, rubber, palm oil and kernels, kola nuts.

Ijebu-Ode (ee-jai-BOO-o-DAI), Local Government Area, OGUN state, SW NIGERIA; ⊙ IJEBU-ODE. Mainly in rain forest zone, with some freshwater and mangrove-swamp forests (S). Main products include cacao, palm oil and kernels, rubber, kola nuts. Food crops are rice, corn, yams, cassava, plantains. Hardwood lumbering. Population is largely YORUBA. One of three such districts created from former Ijebu province.

Ijebu-Ode (ee-jai-BOO-o-DAI), town, ⊙ IJEBU-ODE Local Govt. Area, OGUN state, SW NIGERIA; 06°49′N 03°56′E. Commercial town and a collection point for cacao, kola nuts, and palm products. Manufacturing includes textiles, metal and clay products, processed timber and plywood, canned fruit and juice, and milled rice. Ijebu-Ode was capital of the YORUBA Ijebu kingdom that was founded by the 15th century. Long opposed to foreign contacts, the Ijebu kingdom remained closed to Europeans until 1892, when the British seized it in retaliation for the Ijebu's closing of the trade routes to the N during the Yoruba civil wars. Seat of the Ogun State College of Education and teachers colleges.

Ijero (ee-JAI-ro), town, ONDO state, SW NIGERIA, on road, 45 mi/72 km N of AKURE. Market town. Cocoyams, maize, rice, beans.

IJerseke, NETHERLANDS: see YERSEKE.

Ijevan, city (2001 population 20,223), NE ARMENIA, in the valley of the Agstev River, tributary of the KURA RIVER, on the Yerevan-Tbilisi Highway, and 26 mi/41 km SW of AKSTAFA (AZERBAIJAN); 40°52′N 45°08′E. Railroad station on the Tbilisi-Baku railroad line. Manufacturing includes carpets, wood products, construction materials, timber machinery; brewery; cheese processing; timber transactions. Community college of industrial technology. Formerly Karavansarai; also spelled Idzhevan.

Ijil, MAURITANIA: see FDERIK.

Ijima (EE-jee-mah), town, Kamina district, NAGANO prefecture, central HONSHU, central JAPAN, 68 mi/110 km S of NAGANO; 35°40′N 137°55′E. Mugwort.

Ijira (ee-JEE-rah), village, Yamagata district, GIFU prefecture, central HONSHU, central JAPAN, 9 mi/15 km N of GIFU; 35°31′N 136°43′E.

IJlst (EILST), village, FRIESLAND province, N NETHERLANDS, 2 mi/3.2 km SW of SNEEK; 53°01′N 05°38′E. Dairying; cattle, sheep; vegetables, grain; manufacturing (tools). Formerly spelled Ylst.

IJmeer (EI-mer), inlet, FLEVOLAND and NORTH HOLLAND provinces, W central NETHERLANDS; lies between SOUTHERN FLEVOLAND polder and AMSTERDAM, at S end of MARKERMEER, SW IJSSELMEER.

IJmuiden (EI-MOI-duhn), city (2001 population 107,368), NORTH HOLLAND province, W NETHERLANDS, on the NORTH SEA, 14 mi/23 km WNW of AMSTERDAM; 52°28′N 04°37′E. On S side of entrance to NOORDZEE Canal (lighthouse); Velsertunnel (highway and railroad) to E; National Park De Kennemerduinen to S. Dairying; fishing; cattle; flowers, nursery stock, vegetables, fruit; manufacturing (machine parts, lifeboats; food processing. Seat of Institute for Fisheries Research.

Ijo älv, FINLAND: see IIJOKI.

Ijofin (ee-JAW-feen), town, OGUN state, extreme SW NIGERIA, on BENIN border, 16 mi/26 km WSW of ADO; 06°30′N 02°43′E. Cotton weaving, indigo dyeing; cacao, palm oil and kernels. Phosphate deposits.

IJsselmeer (EI-suhl-mer), shallow lake, NORTH HOLLAND (W), FLEVOLAND (SE), and FRIESLAND (NE) provinces, NW NETHERLANDS. Formerly the ZUIDER ZEE, arm of NORTH SEA, enclosed in 1932 by 19-mi/31-km AFSLUITDIJK, at entrance in NW; enclosure has transformed the IJSSELMEER from salt to fresh water. Since 1932, large parts in E and SE have been reclaimed to form dry land: the NORTHEAST polder in E (formerly part of OVERIJSSEL province) and the Oostelijk-Flevoland and Zuidelijk-Flevoland polders in SE. The three polders now form the Flevoland province. The Markerwaarddijk barrier dam (built in 1980s) encloses the S half of the remaining IJsselmeer, which is set to become the Markerwaard polder and another polder of Flevoland. ENKHUIZEN, LELYSTAD, and LEMMER are minor fishing ports on IJsselmeer and are joined to the North Sea via locks. The lake serves as a recreational area.

IJsselmonde (EI-suhl-mawn-duh), island, SOUTH HOLLAND province, SW NETHERLANDS, between NOORD (E), OLD MAAS (S and SW), and NEW MAAS (N) rivers; 15 mi/24 km long, 5 mi/8 km wide. S part of ROTTERDAM city and most of its port facilities and oil refineries are in N; other cities are Bardendrecht (center), HOOGVLIET (W), and RIDDERKERK (NE). Agriculture (strawberries, vegetables, sugar beets); dairying.

IJssel River (EI-suhl), 72 mi/116 km long, GELDERLAND and OVERIJSSEL provinces, central NETHERLANDS; branches from the LOWER RHINE River, 3 mi/4.8 km SE of ARNHEM; flows N, past DOESBURG, ZUTPHEN, and DEVENTER, passes SW of ZWOLLE and flows past KAMPEN to the KETELMEER (8-mi/12.9-km channel of IJSSELMEER between Oostelijk and NORTHEAST polders), 4 mi/6.4 km WNW of Kampen. Apeldooorns Canal parallels river up to 8 mi/12.9 km to W.

IJsselstein (EI-suhl-stein), town (2001 population 30,196), UTRECHT province, W central NETHERLANDS, on HOLLANDSE IJSSEL RIVER, and 6 mi/9.7 km SSW of UTRECHT; 52°01′N 05°03′E. LEK RIVER to SE, Lek Canal to E. Agriculture (fruit, vegetables); cattle, hogs, poultry raising; dairying; manufacturing (motor vehicles, furniture; food processing). Formerly spelled Ysselstein.

Ijuí (EE-zhoo-ee), city (2007 population 76,761), NW RIO GRANDE DO SUL state, BRAZIL, on railroad, and 26 mi/42 km NW of CRUZ ALTA; 28°23′S 53°55′W. Center of agricultural colony; cattle and hog raising; processing of meat, grain, and maté. Mineral springs nearby. Airfield. Formerly spelled Ijuhy.

Ijuin (ee-JOO-in) or **Izyuin**, town, Hioki district, KAGOSHIMA prefecture, S KYUSHU, SW JAPAN, on N SATSUMA PENINSULA, 9 mi/15 km W of KAGOSHIMA; 31°37′N 130°24′E. Railroad junction. Computer components. Known for Satsuma porcelain ware. Inhabited by descendants of colony of potters brought from KOREA in 16th century.

Ijuí River (EE-zhoo-ee), 225 mi/362 km, N RIO GRANDE DO SUL state, BRAZIL; rises in the Serra GERAL (27°58′S 55°20′W); flows W to URUGUAY RIVER E of CONCEPCIÓN DE LA SIERRA (ARGENTINA). Not navigable. Formerly spelled Ijuhy.

IJzendijke (EI-zuhn-DEI-kah), town, ZEELAND province, SW NETHERLANDS, on FLANDERS mainland, and 9 mi/14.5 km W of TERNEUZEN; 51°19′N 03°37′E. Belgian border 2 mi/3.2 km to S. Light manufacturing; dairying; cattle, hogs; potatoes, vegetables, sugar beets, grain. Was first mentioned in tenth century; trade center in Middle Ages. Formerly Yzendyke.

Ikaalinen (I-kah-LI-nen), Swedish *Ikalis*, town, HÄMEEN province, SW FINLAND, 30 mi/48 km NW of TAMPERE; 61°46′N 23°03′E. Elevation 248 ft/75 m. In lake region; health resort and spa.

Ikaho (ee-KAH-ho), town, North Gumma district, GUMMA prefecture, central HONSHU, N central JAPAN, 12 mi/20 km N of MAEBASHI; 36°29′N 138°55′E. Hot springs.

Ikali (ee-KAH-lee), village, Équateur province, NW CONGO, on LUILAKA RIVER, and 130 mi/209 km SSE of BOENDE; 02°02′S 21°02′E. Elev. 1,377 ft/419 m. Terminus of steam navigation in palm-growing region.

Ikalis, FINLAND: see IKAALINEN.

Ikamba (ee-KAHM-bah), village, RUKWA region, W TANZANIA, 95 mi/153 km NW of SUMBAWANGA, at W edge of KATAVI PLAINS Game Reserve; 06°50′S 30°47′E. Sheep, goats; wheat, corn; timber.

Ikamiut (i-KAHM-yoot), fishing settlement, Qasigiannguit (CHRISTIANSHÅB) commune, W GREEN-

LAND, on inlet of DISKO BAY, 25 mi/40 km ESE of EGEDESMINDE (Aasiaat); 68°38′N 51°50′W.

Ikang (ee-KAHNG), town, CROSS RIVER state, extreme SE NIGERIA, on inlet of Bight of BIAFRA (CAMEROON border), 20 mi/32 km SE of CALABAR; 04°48′N 08°32′E. Minor port of entry; fishing industry; coconuts, plantains, bananas.

Ikara (EE-kah-rah), town, KADUNA state, N central NIGERIA, 35 mi/56 km ENE of ZARIA; 11°11′N 08°14′E. Agricultural trade center (cotton, peanuts, locust beans). Surrounding district freed of tsetse flies in connection with population resettlement scheme. Sometimes spelled Ikare.

Ikaría (ee-kah-REE-ah), mountainous island (□ 100 sq mi/260 sq km), in the SPORADES group, SÁMOS prefecture, at the S edge of NORTH AEGEAN department, GREECE, near W TURKEY, and 13 mi/21 km SW of SÁMOS; 37°35′N 26°10′E. Iron-ore deposits and sulfur springs. Airport. According to Greek mythology, Icarus fell into the sea near here. Also Icaria.

Ikarian Sea (ee-KAH-ree-ahn), the section of the E AEGEAN SEA to the W of the N DODECANESE islands of IKARÍA, FOURNI, and PÁTMOS islands, off NORTH AEGEAN and SOUTH AEGEAN departments, GREECE. Named for the mythological Icarus who, while flying from CRETE with his father Daedalus using wings made with wax, fell into the water here and drowned. Also spelled Icarian Sea.

Ikarigaseki (ee-KAH-ree-gah-SE-kee), village, South Tsugaru district, Aomori prefecture, N HONSHU, N JAPAN, 25 mi/40 km S of AOMORI; 40°28′N 140°37′E.

Ikaruga (ee-KAH-roo-gah), town, Ikoma district, NARA prefecture, HONSHU, W central JAPAN, 8 mi/13 km S of NARA; 34°36′N 135°43′E. Area temples (Horyu, Horin, Hoki); ancient tomb nearby.

Ikast (EE-kahst), city (2000 population 14,151), RINGKOBING county, central JUTLAND, DENMARK, 35 mi/56 km E of Ringkøbing; 56°10′N 09°15′E. Manufacturing (apparel; wood moldings); cattle.

Ikata (ee-KAH-tah), town, West Uwa district, EHIME prefecture, NW SHIKOKU, W JAPAN, 34 mi/55 km S of MATSUYAMA; 33°29′N 132°21′E. Mandarin oranges.

Ikatan, village, on SE UNIMAK Island, SW ALASKA, on Ikatan Bay; 54°45′N 163°19′W.

Ikau (ee-KAH-oo), village, Équateur province, NW CONGO, on left bank of LULONGA RIVER, and 120 mi/193 km NE of MBANDAKA; 01°15′N 19°45′E. Elev. 1,184 ft/360 m. Christian mission.

Ikawa (ee-KAH-wah), town, South Akita district, Akita prefecture, N HONSHU, NE JAPAN, 12 mi/20 km N of AKITA city; 39°54′N 140°05′E.

Ikawa (EE-kah-wah), town, Miyoshi district, TOKUSHIMA prefecture, SE SHIKOKU, W JAPAN, 40 mi/65 km W of TOKUSHIMA; 34°01′N 133°52′E.

Ikazaki (ee-KAH-zah-kee), agricultural town, Kita district, EHIME prefecture, W SHIKOKU, W JAPAN, 22 mi/35 km S of MATSUYAMA; 33°31′N 132°29′E. In mountainous area. Traditional papermaking; wooden clogs.

Ikeda (ee-KE-dah), city, OSAKA prefecture, S HONSHU, W central JAPAN, on the Ina River, 12 mi/20 km N of OSAKA; 34°49′N 135°25′E. Industrial and residential suburb of Osaka. Motor vehicle manufacturing; plant nurseries.

Ikeda (ee-KE-dah), town, Imadate district, FUKUI prefecture, central HONSHU, W central JAPAN, 16 mi/25 km S of FUKUI; 35°53′N 136°20′E.

Ikeda, town, Ibi district, GIFU prefecture, central HONSHU, central JAPAN, 9 mi/15 km N of GIFU; 35°26′N 136°34′E.

Ikeda, town, Tokachi district, S central HOKKAIDO prefecture, N JAPAN, on TOKACHI RIVER, and 105 mi/170 km E of SAPPORO; 42°55′N 143°27′E. Railroad junction. Adzuki beans, sweet bean paste (wine-based *yokan*); dairying, wine production.

Ikeda, town, on S coast of SHODO-SHIMA island, Syozu district, KAGAWA prefecture, off NE SHIKOKU, W

JAPAN, 12 mi/20 km N of TAKAMATSU; 34°28′N 134°14′E. Flowers; noodles.

Ikeda, town, North Azumi district, NAGANO prefecture, central HONSHU, central JAPAN, 22 mi/35 km S of NAGANO; 36°25′N 137°52′E. Sericulture. Rice; asparagus.

Ikeda, town, Miyoshi district, TOKUSHIMA prefecture, NE central SHIKOKU, W JAPAN, on YOSHINO RIVER, and 43 mi/70 km W of TOKUSHIMA; 34°28′N 134°14′E. Koboke Gorge is nearby.

Ikegawa (ee-KE-gah-wah), town, Agawa district, KOCHI prefecture, central SHIKOKU, W JAPAN, 22 mi/35 km N of KOCHI; 33°36′N 133°10′E.

Ikeja (ee-KAI-jah), town, ⊙ LAGOS state, extreme SW NIGERIA, on railroad, 9 mi/14.5 km NNW of LAGOS Island. High concentration of industries. Site of Murtala Muhammed airport.

Ikela (ee-KE-lah), village, Équateur province, central CONGO, on TSHUAPA RIVER, and 175 mi/282 km ESE of BOENDE. Steamboat landing and trade center in palm-growing region; rubber plantations in vicinity.

Ikelemba River (ee-ke-LEM-bah), c.200 mi/322 km long, W CONGO; rises 25 mi/40 km SW of BEFALE; flows NW and SW, past BALANGALA and BOMBIMBA, to CONGO RIVER opposite MBANDAKA. Navigable for steamboats for 85 mi/137 km below BOMBIMBA, for barges below BALANGALA.

Ikengo (ee-KEN-go), village, Équateur province, NW CONGO, on the left bank of the CONGO RIVER, and 15 mi/24 km S of MBANDAKA; 00°08′S 18°08′E. Elev. 990 ft/301 m. Fishing.

Ikeram (ee-KER-ahm), town, ONDO state, SW NIGERIA, on road, 50 mi/80 km NE of AKURE; 07°37′N 05°51′E. Market town. Cassava, rice, beans.

Ikerasak (i-KE-rah-sahk), fishing and hunting settlement, UUMMANNAQ commune, W GREENLAND, on small IKERASAK ISLAND in QARAJAQ ICE FJORD, 20 mi/32 km SE of UUMMANNAQ; 70°30′N 51°19′W.

Ikerre (ee-KYI-re), town, ONDO state, SW NIGERIA, on road, 25 mi/40 km NNE of AKURE; 07°30′N 05°14′E. Market town. Cassava, maize, kola nuts, cacao industry; palm oil and kernels, rubber, timber.

Ikey (ee-KYAI), village, SW IRKUTSK oblast, S central SIBERIA, Russia, on road, 30 mi/48 km SW of TULUN; 54°11′N 100°05′E. Elevation 1,699 ft/517 m. In agricultural area (wheat, barley, oats; livestock); forestry services.

Ikey, RUSSIA: see IKEI.

Ikhe Aral Nor, MONGOLIA: see HAR US NUUR.

Ikhe Bogdo, MONGOLIA: see IH BOGD UUL.

Ikhtiman, Bulgaria: see IHTIMAN.

Iki-Burul (EE-kee-boo-ROOL), settlement (2005 population 3,675), SW Republic of KALMYKIA-KHALMGTANGEH, S European Russia, on road, 35 mi/56 km SSE of ELISTA; 45°49′N 44°39′E. Elevation 426 ft/129 m. In agricultural area. Has a Buddhist temple. Formerly known as Chonin-Sala.

Ikimba, Lake (ee-KEEM-bah), KAGERA region, NW TANZANIA, 20 mi/32 km SW of BUKOBA, and 14 mi/23 km W of Lake VICTORIA; 14 mi/23 km long, 10 mi/16 km wide; 01°30′S 31°30′E. Elevation 3,848 ft/1,173 m.

Ikina (ee-KEE-nah), village, Ochi district, EHIME prefecture, NW SHIKOKU, W JAPAN, 37 mi/60 km N of MATSUYAMA; 34°15′N 133°11′E.

Ikire, town, OYO state, SW NIGERIA, 20 mi/32 km E of IBADAN; 07°21′N 04°11′E. Road center. Cotton weaving; cacao, palm oil and kernels, cotton. Sawmilling nearby.

Ikirun (ee-KI-roon), town, OSUN state, SW NIGERIA, on railroad and road, 65 mi/105 km NE of IBADAN, and 10 mi/16 km NE of OSHOGBO; 07°55′N 04°40′E. Market town. Cotton weaving, shea-nut processing, cacao industry; tobacco, yams, maize.

Iki-shima (ee-KEE–shee-mah) or **Iki**, island (□ 53 sq mi/137.8 sq km), NAGASAKI prefecture, SW JAPAN, in Tsushima Strait, 8 mi/13 km N of NW coast of KYUSHU; 8.5 mi/13.7 km long, 5 mi/8 km wide. Hilly, fertile; agriculture. Chief town, GONOURA. Ruins of

16th-century castle built by Hideyoshi Toyotomi, ruler of Japan.

Ikitsuki (ee-KEETS-kee), town, North Matsuura district, NAGASAKI prefecture, NW KYUSHU, SW JAPAN, 50 mi/80 km N of NAGASAKI; 33°23′N 129°26′E.

Ikitsuki-shima (ee-KEETS-kee–SHEE-mah), island (□ 6 sq mi/15.6 sq km), NAGASAKI prefecture, SW JAPAN, in EAST CHINA SEA, just NW of HIRADO-SHIMA island, 8 mi/12.9 km W of HIZEN PENINSULA, NW KYUSHU; 6 mi/9.7 km long, 2 mi/3.2 km wide; hilly. Rice, sweet potatoes; fish. Chief town, IKITSUKI, is on E coast.

Ikiz Tepe, peak (5,580 ft/1,701 m), AMANOS MOUNTAINS, S TURKEY, 16 mi/26 km WNW of ANTAKYA, 8 mi/12.9 km from MEDITERRANEAN coast.

Ikizu (ee-KEE-zoo), village, MARA region, NW TANZANIA, 33 mi/53 km SSE of MUSOMA; 01°58′S 34°01′E. SERENGETI PLAIN and SERENGETI NATIONAL PARK to SE. Cattle, sheep, goats; cotton, corn, wheat, millet.

Ikkatteq (I-kah-tek), former meteorological station and landing field, SE GREENLAND, on inlet of DENMARK STRAIT, 35 mi/56 km NE of AMMASSALIK; 65°56′N 36°33′W. Erected by U.S. military July 1942 as "Bluie East-2"; closed 1959 when the civilian air traffic was moved to Kulusuk (65°34′N 37°12′W), only 13 mi/21 km E of the main town of SE Greenland, Ammassalik.

Ikkatteq, hunting settlement, Tasiilaq (or Ammassalik) commune, 9 mi/15 km WNW of Ammassalik; 65°38′N 37°56′W.

Iklád, village, PEST county, HUNGARY, 14 mi/23 km NE of BUDAPEST; 47°40′N 19°27′E. Aluminum foundry; reduced production in 1990s.

Ikola (ee-KO-lah), village, RUKWA region, W TANZANIA, 110 mi/177 km NW of SUMBAWANGA, and 5 mi/8 km N of KAREMA, on Lake TANGANYIKA; 06°45′S 30°23′E. Fish; goats, sheep; corn, wheat, millet.

Ikole (EE-kaw-lai), town, ONDO state, S NIGERIA, 40 mi/64 km N of OWO; 07°48′N 05°31′E. Cacao industry; palm oil and kernels, kola nuts.

Ikom (ee-KAWM), town, CROSS RIVER state, extreme SE NIGERIA, near CAMEROON border, port on CROSS RIVER (head of high-water navigation), and 28 mi/45 km ESE of OBUBRA; 05°58′N 08°42′E. Trade center; hardwood and rubber; palm oil and kernels, cacao, kola nuts.

Ikoma (ee-KO-mah), city, NARA prefecture, S HONSHU, W central JAPAN, 7 mi/11 km W of NARA; 34°41′N 135°42′E. Railroad junction; mountain resort. Buddhist temple on nearby Mt. Ikoma (2,120 ft/646 m).

Ikoma (ee-KO-mah), village, MARA region, N TANZANIA, 65 mi/105 km SE of MUSOMA, on Grumeti River; 02°04′S 34°37′E. SERENGETI NATIONAL PARK to E and S. Cattle, sheep, goats; corn, wheat, millet.

Ikombe (ee-KOM-bai), village, MBEYA region, SW TANZANIA, 55 mi/89 km SE of MBEYA, on Lake NYASA, W of LIVINGSTONE MOUNTAINS; 09°30′S 34°04′E. Road terminus, lake port. Cattle; corn, pyrethrum, beans; timber.

Ikonde (ee-KON-dai), village, KIGOMA region, W TANZANIA, 105 mi/169 km SSE of KIGOMA; 06°18′S 30°23′E. Sheep, goats; corn, wheat. Also called Nkonde.

Ikongo (ee-KOONG-goo), town, FIANARANTSOA province, SE MADAGASCAR, 40 mi/64 km NW of MANAKARA; 21°52′S 47°28′E. Market center; coffee plantations; bananas. Hospital. Formerly Fort-Carnot.

Ikon-Khalk (EE-kuhn–HAHLK), village (2005 population 4,230), N KARACHEVO-CHERKESS REPUBLIC, S European RUSSIA, in NW CAUCASUS MOUNTAINS, on the LITTLE ZELENCHUK RIVER, on road, 8 mi/13 km WNW of CHERKESSK; 44°18′N 41°55′E. Elevation 1,528 ft/465 m. In agricultural area.

Ikonnikovo, RUSSIA: see GOR′KOVSKOYE.

Ikopa River (ee-KOOP), 250 mi/402 km long, central and NW MADAGASCAR, main tributary of the BETSIBOKA RIVER; rises in ANTANANARIVO province 25 mi/40 km SSE of ANTANANARIVO; flows NNW to BETSI-

BOKA RIVER just below MAEVATANANA (MAHAJANGA province). MANTASOA (mahn-tah-SOO) reservoir (completed 1939), the largest of Madagascar, is at its source; used for irrigation (notably of BETSIMITATRA rice fields) and fishing. Several barrages regulate the Ikopa near Antananarivo. Valley has gold placers.

Ikornnes (EE-kawrn-nais), village, MØRE OG ROMSDAL county, W NORWAY, on a S inlet of STORFJORD, 14 mi/23 km ESE of ÅLESUND. Furniture.

Ikorodu (ee-ko-ro-DOO), town, LAGOS state, SW NIGERIA, 15 mi/24 km NNE of LAGOS (across lagoon), in Lagos metropolis; 06°37′N 03°31′E. Port and market; cacao, palm oil and kernels, rice.

Ikot Ekpene (ee-KAWT ep-PE-ne), town, AKWA IBOM state, SE NIGERIA, on headstream of KWA IBO River, and 25 mi/40 km E of Aba. Road center. Toys, batteries; raffia industry. Palm oil and kernels, kola nuts. Has hospital.

Ikovka (EE-kuhf-kah), village (2005 population 5,205), central KURGAN oblast, SW SIBERIA, RUSSIA, on road and railroad, 18 mi/29 km NW of KURGAN; 55°36′N 64°56′E. Elevation 419 ft/127 m. Agricultural products; timber.

Ikoyi (ee-KAW-yee), area developed originally by the British, in LAGOS township, LAGOS state, SW NIGERIA, on E end of Lagos Island, seperated from crowded native city to W by channel. Has Roman Catholic mission, hospital. Federal Secretariat is here.

Ikpek, village, W ALASKA, on Nome Bay. Sometimes spelled Icpic.

Ikramovo, UZBEKISTAN: see DJUMA.

Ik River (EEK), 326 mi/525 km long, in BASHKORTOSTAN and TATARSTAN, E European Russia; rises in the W foothills of the S URAL Mountains, 10 mi/16 km S of Belebei (BASHKORTOSTAN); flows S, W, NNW, past Oktyabrskiy and MUSLYUMOVO (Tatarstan), and WNW to the KAMA River, 9 mi/14.5 km E of BONDYUZHSKIY. Receives Usen (right) and Menzelya (left) rivers. Forms part of BASHKORTOSTAN borders with ORENBURG oblast and TATARSTAN.

Ikryanoye (ee-kryah-NAW-ye), village (2005 population 9,870), E ASTRAKHAN oblast, SE European Russia, on the BAKHTEMIR arm of the VOLGA RIVER delta mouth, on road, 22 mi/35 km SSW of ASTRAKHAN; 46°05′N 47°44′E. Below sea level. Caviar; fisheries; forestry.

Iksal (ik-SAHL), Arab township, 1.2 mi/2 km SE of NAZARETH in LOWER GALILEE, ISRAEL; 32°40′N 35°19′E. Elevation 433 ft/131 m. Founded in the 18th century, population predominantly Muslim. Ruins include an ancient cemetery, remains of buildings and water wells.

Iksan (EEK-SAHN), city (□ 261 sq mi/678.6 sq km; 2005 population 308,144), NW NORTH CHOLLA province, SOUTH KOREA; 35°59′N 127°03′E. Bordered N by South Chungchon province (across KUM RIVER), E by WANJU province, W by OKKU province, and S by KIMJE province (across Mankyong River). Alluvial valley on lower Mankyong River forms center of Honam plain, where food grains, fruits, and vegetables are grown. Stone and carved stoneworks. Many railroad run through city; Honam Expressway. IRI, an agricultural center and transportation hub, was absorbed into the city.

Iksha (EEK-shah), town (2006 population 3,700), N central MOSCOW oblast, central European Russia, on the MOSCOW CANAL, on road and railroad, 35 mi/56 km N of MOSCOW, and 12 mi/19 km S of DMITROV, to which it is administratively subordinate; 56°10′N 37°31′E. Elevation 620 ft/188 m. Wire, nails, felt boots; veterinary station.

Īkšķile (EEK-shkee-lai), German *Uxküll*, village, central LATVIA, in VIDZEME, on right bank of the DVINA (DAUGAVA) River, and 17 mi/27 km SE of Rīga. Summer resort.

Ikuchi-jima (eek-CHEE-JEE-mah), island (□ 12 sq mi/31.2 sq km), HIROSHIMA prefecture, W JAPAN, in

HIUCHI SEA (central section of INLAND SEA), 6 mi/9.7 km S of MIHARA on SW HONSHU, between OMI-SHIMA (W) and Inno-shima (E) islands; 5.5 mi/8.9 km long, 3 mi/4.8 km wide. Mountainous. Fishing, raw-silk production. SETODA is chief town.

Ikungi (ee-KOON-gee), village, SINGIDA region, N central TANZANIA, 16 mi/26 km S of SINGIDA; 05°06′S 34°45′E. Cattle, sheep, goats; corn, wheat.

Ikungu (ee-KOON-goo), village, TABORA region, central TANZANIA, 75 mi/121 km ESE of TABORA, near Chona River, on railroad; 05°31′S 33°50′E. Road junction. Timber; sheep, goats; corn, wheat.

Ikuno (ee-KOO-no), town, Asago district, HYOGO prefecture, S HONSHU, W central JAPAN, 38 mi/61 km N of KOBE; 35°09′N 134°47′E. Mining center (Ikuno silver ore).

Ikusaka (ee-KOO-sah-kah), village, East Chikuma district, NAGANO prefecture, central HONSHU, central JAPAN, 20 mi/35 km N of NAGANO; 36°25′N 137°55′E.

Ikutahara (eek-TAH-hah-rah), town, Abashiri district, HOKKAIDO prefecture, N JAPAN, 127 mi/205 km N of SAPPORO; 43°55′N 143°32′E.

Ikutha (ee-KOO-dhah), town, Kitui district, EASTERN province, KENYA, on road, and 50 mi/80 km SSW of KITUI; 02°34′S 38°11′E. Market and trading center.

Ikva River (eek-VAH), (Polish *Ikwa*), 100 mi/161 km long, W UKRAINE; rises SE of BRODY in PODOLIAN UPLAND; flows E, N past DUBNO, and WNW to STYR RIVER S of LUTS′K.

Ikwa River, UKRAINE: see IKVA RIVER.

Ila (EE-lah), town, OSUN state, SW NIGERIA, on road, 25 mi/40 km NE of OSHOGBO. Oil palms, yams, cassava.

Ila (EI-luh), town (2000 population 328), MADISON county, NE GEORGIA, 15 mi/24 km NNE of ATHENS; 34°10′N 83°17′W.

Ilabaya (ee-lah-BAH-yah), town, JORGE BASADRE province, TACNA region, S PERU, in ANDEAN foothills, on affluent of LOCUMBA RIVER, and 32 mi/51 km SW of CANDARAVE; 17°25′S 70°30′W. Elevation 4,862 ft/1,481 m. Grapes, fruit; livestock; wine and liquor distilling. Silver deposits nearby.

Ilacaon Point (ee-lah-KAH-wahn), northernmost point of NEGROS island, PHILIPPINES, in GUIMARAS STRAIT, near its entrance; 11°00′N 123°11′E.

Ilagan (ee-LAH-gahn), town, ⊙ ISABELA province, N LUZON, PHILIPPINES, 100 mi/161 km NE of BAGUIO; 17°10′N 121°54′E. Near confluence of CAGAYAN RIVER and four feeder streams; important market center (rice, corn). Cigar making. Fuyot Springs National Park (wildlife preserve) nearby.

Īlām, province (□ 7,353 sq mi/19,044 sq km; 1991 population 440,693), W IRAN; ⊙ Īlām. Bordering IRAQ to W, and the provinces of Kermānshāhān (N), Loerstan (E), and Khuzestān (S). Kabir Kuh Mountains run NW-SE, dominating the region. Largely pasturelands, but crops of wheat and barley are grown. A quarter of the population lives in Īlām, the only large urban center.

Ilam (EE-lahm), district, SE NEPAL, in MECHI zone; ⊙ ILAM.

Īlām (EE-lahm), town (2006 population 160,355), ⊙ Īlām province, W IRAN, 60 mi/97 km SW of KERMANSHAH; 33°10′N 47°00′E. Main town of the PUSHT Kuh Mountains; sheep raising; grain. Sometimes spelled ELAM, EILAM, or EYLAM.

Ilam (EE-lahm), town (2001 population 16,237), ⊙ ILAM district, E NEPAL, 33 mi/53 km WSW of DARJILING (INDIA); 26°54′N 87°56′E. Elevation 4,125 ft/1,257 m. Tea plantations.

Ilama (ee-LAH-mah), town, SANTA BÁRBARA department, W HONDURAS, on ULÚA RIVER, and 10 mi/16 km N of SANTA BÁRBARA; 15°04′N 88°13′W. Commercial center; palm-hat manufacturing, mat weaving, rope-making; coffee, sugarcane.

Ilamatlán (ee-lah-mah-TLAN), town, VERACRUZ, E MEXICO, in SIERRA MADRE ORIENTAL, on PUEBLA border, 22 mi/35 km SW of CHICONTEPEC DE TEJEDA. Corn, sugarcane, coffee.

Ilan (EE-lahn), city, NE TAIWAN; 24°46′N 121°45′E. In an agricultural area, with one of the largest rice markets in Taiwan. Manufacturing (fertilizers, wood and paper products).

Ilaniya (ee-lah-nee-YAH) or **Sejerah**, moshav, LOWER GALILEE, N ISRAEL, 7 mi/11.3 km ENE of NAZARETH; 32°45′N 35°24′E. Elevation 639 ft/194 m. Mixed farming. Founded as training farm in 1899 by Jewish Colonization Association (J.C.A.). In 1907 Ilaniya became training farm for Galilee agricultural workers; many early Zionist Labor leaders, including David Ben-Gurion, trained here. Hashomer group founded here in 1909.

Ilanskiy (ee-LAHN-skeeye), city (2005 population 16,540), SE KRASNOYARSK TERRITORY, SE SIBERIA, RUSSIA, on road, the TRANS-SIBERIAN RAILROAD, 173 mi/278 km E of KRASNOYARSK, and 11 mi/18 km E of KANSK; 56°14′N 96°03′E. Elevation 921 ft/280 m. Highway junction. Railroad shops; wood industries, concrete products, clothing factory. Agriculture (wheat, rye, oats, barley). Lignite mining in the vicinity. Founded in 1645 and developed as an important trade route in 1733, when the Siberia-Moscow road ran through it. Railroad station was added in 1899; city status granted in 1939.

Ilanz (EE-lahnts), Romansh *Glion*, commune, GRISONS canton, E central SWITZERLAND, on the VORDERRHEIN RIVER where Glogn River and Valserrhein join it, and 18 mi/29 km WSW of CHUR.

Ilarionove (ee-lah-ree-O-no-ve), (Russian *Illarionovo*), town, central DNIPROPETROVS'K oblast, UKRAINE, 7 mi/11 km ESE of DNIPROPETROVS'K, across the DNIEPER (Ukrainian *Dnipro*) River; 48°24′N 35°16′E. Elevation 488 ft/148 m. Railroad station; asphalt manufacturing. Established in 1875, town since 1938.

Ilaro (ee-LAH-ro), town, OGUN state, extreme SW NIGERIA, on railroad spur, 30 mi/48 km SW of ABEOKUTA. Market town. Major cacao-producing center; cotton weaving, indigo dyeing; palm oil and kernels, kola nuts; livestock. Phosphate deposits.

Ilaskhan-Yurt (ee-lahs-HAHN–YOORT), village (2005 population 4,500), E CHECHEN REPUBLIC, S European Russia, on road, 15 mi/24 km NE of GROZNY, and 5 mi/8 km S of GUDERMES; 43°17′N 46°06′E. Elevation 456 ft/138 m. Has a mosque, school. Formerly known as Belorechye (1944–1959).

Ilava (i-LYAH-vah), GERMAN *Illau*, HUNGARIAN *Illava*, town, ZAPADOSLOVENSKY province, W SLOVAKIA, on VÁH RIVER, on railroad, and 12 mi/19 km NE of TRENČÍN; 49°00′N 18°14′E. Old trading center; brewery; furniture manufacturing. Has large penitentiary; power plant.

Ilave (ee-LAH-vai), town, ⊙ EL COLLAO province, PUNO region, SE PERU, near W shore of Lake TITICACA, near mouth of small Ilave River (16°00′S 69°27′W), 30 mi/48 km ESE of PUNO, on Puno-JULI road; 16°05′S 69°40′W. Elevation 13,097 ft/3,992 m. Cereals; potatoes; livestock.

Ilave River, PERU: see ILAVE.

Iława (ee-WAH-vah), Polish *Iława*, German *Deutsch Eylau*, town (2002 population 32,544), in OLSZTYN province, NE POLAND, at S end of Lake JEZIORAK, 40 mi/64 km SE of MALBORK (Marienburg). Railroad junction; grain and cattle market. Until 1939, German frontier station in EAST PRUSSIA, near Polish border. Its German population left after World War II, when the town passed to Poland.

Ilayangudi (i-luh-YAHNG-guh-dee), town, Pasumpon Muthuramalinga Thevar district, TAMIL NADU state, S INDIA, 39 mi/63 km SE of MADURAI; 09°38′N 78°38′E. Betel farms. Also ILAYANKUDI.

Ilayankudi, INDIA: see ILAYANGUDI.

Ilchester (IL-ches-tuh), agricultural village (2006 population 2,000), S SOMERSET, SW ENGLAND, on YEO RIVER, and 5 mi/8 km NW of YEOVIL; 51°01′N 02°41′W. Has 13th-century church. Roger Bacon born here. Site of important Roman station where FOSSE WAY crossed the Yeo.

Ilchester (IL-ches-tuh), village, HOWARD county, central MARYLAND, on PATAPSCO River, and 9 mi/14.5 km WSW of downtown BALTIMORE. The Thistle Factory, turning out cotton printed cloth, was established here by George and William Morris; now turns out recycled paper products in a big stone factory surrounded by stone homes of the workers. St. Mary's College (c.1868), on 170 acres/69 ha of ground belonging to the Redemptorist Fathers, has been vacant since 1972.

Ilcínea (EEL-SEE-nai-ah), town, SW MINAS GERAIS state, BRAZIL, 23 mi/37 km NW of BOA ESPERANÇA; 20°49′S 45°42′W.

Ileanda (ee-LAN-dah), Hungarian *Nagyilonda*, village, SĂLAJ county, N ROMANIA, near SOMEŞ RIVER, on railroad, and 18 mi/29 km NW of DEJ; 47°20′N 23°38′E. Agriculture center. Sulphurous springs nearby. Under Hungarian rule, 1940–1945.

Ile aux Aigrettes (EEL o e-GRET), islet, 1.6 mi/2.5 km off Mahebourg coast, MAURITIUS; 20°26′S 57°43′E. Tourism. Local fauna include rat, lizard; birds.

Ile aux Bénitiers (EEL o bai-nee-TYAI), islet, 1.2 mi/2 km from Le More Brabank, MAURITIUS, in lagoon at Rivière Noire; 20°27′S 57°21′E. Covers 437 acres/177 ha. Charcoal production; coconuts, vegetables. Rare plants.

Ile aux Certs (EEL o SER), islet, 0.6 mi/1 km of E coast of MAURITIUS; 20°16′S 57°48′E. Tourism; vast beaches fringed with palm trees; water sports.

Ile aux Cocos (EEL-o ko-KO), island, E of RODRIGUEZ Island, dependency of MAURITIUS, 2.5 mi/4 km from Pointe Mapou; 19°43′S 63°18′E. Coralline island; covers 35 acres/14 ha; surrounded by beach and shallow water; tourist attraction for Rodriguez. Abundant bird life.

Ile aux Fouquets (EEL o foo-KAI), islet, 4.7 mi/7.5 km off MAHEBOURG coast, MAURITIUS; 20°23′S 57°47′E. Tourism; rare birds and vegetation.

Ile aux Fous, MAURITIUS: see BOOBY ISLAND.

Île-aux-Moines, L' (leel-o-MWAN-uh), largest of several islands in the Gulf of MORBIHAN, SW BRITTANY, W FRANCE, 6 mi/9.7 km SW of VANNES. About 4 mi/6.4 km long, island is a quiet resort area with lush vegetation, including citrus trees. There are a few megalithic monuments.

Île-aux-Noix (eel-o-nwah), island (☐ 0.3 sq mi/0.8 sq km), in the RICHELIEU River near Saint Jean, S QUEBEC, E CANADA; 45°08′N 73°17′W. During the French and Indian War (1759), the French built a fort here to delay the British advance on MONTREAL but were forced to surrender it in 1760. Named Fort Lennox and occupied by a British garrison, the island fell (1775) to American forces and was used as a base by the American generals Schuyler and Montgomery for attacks on Montreal and Quebec until abandoned in 1776. The British then used the island to supply their operations against the American fleet on Lake CHAMPLAIN. The present Fort Lennox dates from the 1820s, when the old fortifications were repaired and additions were built. It was a military post until 1870. Site of Fort Lennox National Historic Park (established 1921).

Ile aux Sables (EEL o SAHBL), island, NE of RODRIGUEZ, dependency of MAURITIUS, 3.1 mi/5 km from Pointe Mapou; 19°42′S 63°18′E. Identical to ILE AUX COCOS but smaller (19.8 acres/8 ha); separated by a shallow sea channel 0.6 mi/1 km across. Retains a number of seabirds.

Ilebo (ee-LE-bo), town, KASAI-OCCIDENTAL province, central CONGO, on right bank of Kasai River, just above the influx of SANKURU RIVER, and 90 mi/145 km NNW of LUEBO; 04°19′S 20°35′E. Elev. 1,053 ft/320 m. As river port and terminus of railroad from BUKAMA, it is an important transshipment point for copper ores of SHABA. Has Roman Catholic mission, hospital, airport. Formerly known as PORT-FRANCQUI.

Île-Bouchard, L' (leel-BOO-shahr), commune, INDRE-ET-LOIRE department, CENTRE region, W central FRANCE, on VIENNE RIVER, and 9 mi/14.5 km ESE of CHINON; 47°07′N 00°25′E. Vineyards, livestock raising. Has four old churches.

Ile D'Ambre (eel DAHM-bruh), islet, 0.3 mi/0.5 km from N coast of MAURITIUS; 20°02′S 57°41′E. Famous for sinking of the St. Géran. Name given in 17th century for grey amber on shores.

Île de or **Île d'** (eel duh), in French names beginning thus: see under following part of the name.

Île-de-France (eel-duh-frahn-suh), administrative region (☐ 4,637 sq mi/12,056.2 sq km), historic region, and former province of N central FRANCE; 48°30′N 02°30′E. The term, referring to the bounding of the region by rivers, applies to 3 entities of very different sizes. The vernacular region (French *pays*) is the smallest, scarcely larger than the urban agglomeration of modern PARIS. The historic region, the patrimony of the counts of Paris, was almost as small, but the royal domain in the Île-de-France soon included several neighboring *pays* (BEAUCE, BRIE, VEXIN français, Beauvaisis, VALOIS), and the Orléanis region as well. The modern administrative region is smaller than the historic province, but it contains about 20% of the population of France; it is composed of the department of SEINE-ET-MARNE, and the new departments (created from the previous SEINE and SEINE-ET-OISE departments) of PARIS, YVELINES, ESSONNE, HAUTS-DE-SEINE, SEINE-SAINT-DENIS, VAL-DE-MARNE, and VAL-D'OISE. The physical region is a productive lowland, bordered by hills and gentle outward facing escarpments, drained by the SEINE and its tributaries. Agriculture (wheat, sugar beets; dairy products) is still significant, though diminishing due to encroaching residential suburbs of Paris. Many private and public corporate headquarters here. Radial highways and express trains (R.E.R.) make the region well connected, though growth has caused congestion. Places of historic or cultural importance include SAINT-DENIS, SAINT-GERMAIN-EN-LAYE, and VERSAILLES among the outer suburbs of Paris, as well as BEAUVAIS, COMPIEGNE, FONTAINEBLEAU, and RAMBOUILLET lying beyond the urban area. Île-de-France was the cradle of the French monarchy and thus the nuclear core of the French state. Today it is the geopolitical heartland of France.

Île de France: see MAURITIUS.

Île de la Camargue, FRANCE: see CAMARGUE.

Ile de La Passe (EEL duh lah PAHS), islet, MAURITIUS, 4.1 mi/6.5 km SE of GRAND PORT; 20°24′S 57°46′E. Islet covers 12.4 acres/5 ha. Tourism.

Ile de L'Est (EEL duh LEST), islet, 0.6 mi/1 km off E coast of MAURITIUS; 20°16′S 57°48′E. Tourist resort; manicured beaches. Also called Ile aux Margenie.

Île de l'Est (EEL duh LEST), in Gulf of SAINT LAWRENCE, E QUEBEC, E CANADA, one of MAGDALEN ISLANDS, between GROSSE and COFFIN islands, 85 mi/137 km NNE of PRINCE EDWARD ISLAND; 5 mi/8 km long, 2 mi/3 km wide; 47°37′N 61°27′W.

Île d'Entrée (EEL dahn-TRAI), island, in Gulf of SAINT LAWRENCE, E QUEBEC, E CANADA, one of MAGDALEN ISLANDS, 60 mi/97 km NNE of PRINCE EDWARD ISLAND; 2 mi/3 km long, 2 mi/3 km wide; 47°17′N 61°42′W.

Ile des Chênes (eel dai SHEN), unincorporated village, SE MANITOBA, W central CANADA, 14 mi/22 km from WINNIPEG, in RITCHOT rural municipality; 49°42′N 96°59′W.

Île de Tonti, CANADA: see AMHERST ISLAND.

Ile District, MOZAMBIQUE: see ZAMBÉZIA.

Île du Cap aux Meules (eel dyoo kahp o MUL), in the Gulf of SAINT LAWRENCE, E QUEBEC, E CANADA, one of the MAGDALEN ISLANDS, 60 mi/97 km N of PRINCE EDWARD ISLAND; 5 mi/8 km long, 4 mi/6 km wide; 47°23′N 61°55′W. ÉTANG DU NORD (W) is chief town. Fisheries. WOLFE ISLAND, narrow spit of land, extends NE to GROSSE island.

Île du Havre aux Maisons (eel dyoo AH-vruh o mai-ZON), in Gulf of SAINT LAWRENCE, E QUEBEC, E

CANADA, one of MAGDALEN ISLANDS, 65 mi/105 km NNE of PRINCE EDWARD ISLAND; 8 mi/13 km long, 3 mi/5 km wide; 47°26'N 61°46'W. HOUSE HARBOUR (SW) is chief settlement. Fisheries. Variant names: Saunders Island, Alright Island.

Île d'Yeu, FRANCE: see YEU, ÎLE D'.

Île Jésus, CANADA: see LAVAL.

Ilek (ee-LYEK), village (2006 population 9,885), SW ORENBURG oblast, SE European Russia, on the right bank of the URAL River (landing), at the mouth of the ILEK RIVER, less than 3 mi/5 km N of the KA-ZAKHSTAN border, on road, 70 mi/113 km WSW of ORENBURG; 51°31'N 53°22'E. Elevation 190 ft/57 m. Customs station. Before World War I, called Iletskiy Gorodok.

Ilek River (YI-lyek), 330 mi/531 km long, NW KA-ZAKHSTAN and RUSSIA; rises in Kazakhstan, in MU-GODZHAR HILLS NE of OKTYABRSK; flows N, past ALGA and Akhtyube (Aktyubinsk), and generally WNW across S ORENBURG oblast (Russia) to URAL River at ILEK. Not navigable. Forms part of Kazakhstan-Russia border E of Ilek.

Île-Maligne (eel–mah-LEEN-yuh), former town, central QUEBEC, E CANADA, on Île d'ALMA, on SAGUENAY River. Aluminum mill; hydroelectric plant. Amalga-mated into city of ALMA.

Ile Marianne (EEL mah-ree-YAHN), islet, 5 mi/8 km from GRAND PORT, MAURITIUS; 20°22'S 57°48'E. Harsh topography provides ideal conditions for studying reintroduction strategies for lizards.

Ilemera (ee-lai-MAI-rah), village, KAGERA region, NW TANZANIA, 40 mi/64 km S of BUKOBA, near Ruiga Bay, Lake VICTORIA; 01°56'S 31°38'E. Fish; timber; coffee, corn, cotton.

Ilerda, SPAIN: see LÉRIDA.

Île-Rousse, L' (leel-roos), commune (□ 1 sq mi/2.6 sq km), NW CORSICA, HAUTE-CORSE department, FRANCE, port on the MEDITERRANEAN, 10 mi/16 km NE of CALVI; 42°38'N 08°56'E. Shipping service to NICE and MARSEILLE; tourist resort; produces cheese, olive oil; lobster fishing. Just N is Pietra peninsula (formerly an island) with lighthouse. Founded 1758 as rival to Genoese stronghold of Calvi. Oceanographic museum.

Île Royale, CANADA: see CAPE BRETON ISLAND.

Iles (EE-les), town, ⊙ Iles municipio, NARIÑO depart-ment, SW COLOMBIA, 21 mi/34 km SW of PASTO; 00°58'N 77°31'W. Elevation 8,838 ft/2,693 m. Wheat, sugarcane; livestock.

Île-Saint-Denis, L' (LEEL-san-de-NEE), town (□ 0.7 sq mi/1.8 sq km); 1993 estimated pop. 7,429), SEINE-SAINT-DENIS department, ÎLE-DE-FRANCE region, N central FRANCE, a N suburb of PARIS, 6 mi/9.7 km from Notre Dame Cathedral, on narrow island (3 mi/4.8 km long) in SEINE River, opposite SAINT-DENIS; 48°57'N 02°23'E. Chemical works; warehouses for Paris department stores.

Île Saint Jean, CANADA: see PRINCE EDWARD ISLAND.

Îles d'Hyères, FRANCE: see HYÈRES, ÎLES D'.

Îles du, in French names: see under following part of the name; e.g., for Îles du Salut, see SALUT, ÎLES DU.

Ilesha (ee-LAI-shah), city, OSUN state, SW NIGERIA, on road, 20 mi/32 km SE of OSHOGBO. Formerly a caravan trade center, Ilesha today is an agricultural and com-mercial city. Cacao, kola nuts, oil palm, and yams are shipped from here. There is a sawmill, and alluvial gold is found. Was capital of the YORUBA Ilesha kingdom of the Oyo empire. After Oyo's collapse in the early 19th century, Ilesha became subject to IBADAN. Taken by the British in 1893.

Ilet', RUSSIA: see KRASNOGORSKIY, MARI EL Republic.

Ilet Fourneau (ee-LAI foor-NO), islet, 1.9 mi/3 km from Pointe Sud Est, MAURITIUS. Covers 69 acres/28 ha. Indigenous tree species.

Ilet' River (ee-LYET), 120 mi/193 km long, MARI EL REPUBLIC, NE European Russia; rises 7 mi/11.3 km E of PARANGA; flows generally SW, past KRASNOGORSKIY,

to the VOLGA RIVER 3 mi/4.8 km WNW of VOLZHSK. Logging.

Iletsk, RUSSIA: see SOL'-ILETSK.

Iletskaya Zashchita, RUSSIA: see SOL'-ILETSK.

Iletski Gorodok, RUSSIA: see ILEK.

Ilfeld (IL-feld), village, THURINGIA, central GERMANY, at S foot of the lower HARZ, 6 mi/9.7 km N of NORDHAUSEN; 51°34'N 10°47'E. Health resort. Site of former Premonstratensian monastery (founded 1196), converted into school in 1546.

Ilford, ENGLAND: see REDBRIDGE.

Ilfov Agricultural Sector (eel-FOV), county, S ROMA-NIA, in WALACHIA; ⊙ BUCHAREST; 44°30'N 26°06'E. Originally the food-producing hinterland of Buchar-est, the Ilfov Agricultural (or Farming) Sector has county status and encompasses the BUCHAREST MU-NICIPALITY. The sector has mainly rural villages, but some industry is located here. Domestic airport at BĂNEASA; international airport at OTOPENI.

Ilfracombe (IL-fruh-kuhm), township, QUEENSLAND, NE AUSTRALIA, 17 mi/27 km E of LONGREACH; 23°30'S 144°30'E. In sheep, cattle area. Folk museum.

Ilfracombe (IL-fruh-kuhm), town (2001 population 10,840), N DEVON, SW ENGLAND, at mouth of BRISTOL CHANNEL, 10 mi/16 km NNW of BARNSTAPLE; 51°13'N 04°07'W. Seaside resort, fishing port. Light engi-neering; former agricultural market. Has Norman manor, 14th-century church, and 16th-century light-house (51°12'N 04°08'W). Noted for mild climate.

Ilgaz, village, N central TURKEY, on DEVREZ RIVER, and 22 mi/35 km N of CANKIRI; 40°55'N 33°37'E. Grain, potatoes, vetch; mohair goats. Formerly Kochisarbala.

Ilgaz Mountains, N TURKEY, extend 100 mi/161 km S of KASTAMONU, between the Göksu and ARAC rivers in N, KIZIL IRMAK and DEVREZ rivers in S; rise to 8,415 ft/2,565 m in ILGAZ MOUNTAINS. Towns of KARGI and TOSYA on S slope.

Ilgin, township, W central TURKEY, near S end of Lake CAVUSCU, on railroad, and 40 mi/64 km NW of KONYA; 38°16'N 31°57'E. Wheat.

Ilhabela (EEL-yah-be-lah), city (2007 population 23,902), SE SÃO PAULO, BRAZIL, port on NW coast of SÃO SEBASTIÃO ISLAND, 60 mi/97 km ENE of SANTOS; 23°47'S 45°21'W. Fishing; coffee trade. Until 1944, called Formosa; originally called Villa Bella.

Ilha das Flores (EEL-yah dahs FLO-res), town (2007 population 8,598), N SERGIPE state, BRAZIL, 48 mi/77 km NE of ARACAJU, on SÃO FRANCISCO River (ALA-GOAS state border); 10°27'S 36°33'W. Rice growing.

Ilha dos Pombos (EEL-yah dos POM-bos), island, in PARAÍBA River, on RIO DE JANEIRO–MINAS GERAIS state line, BRAZIL, 5 mi/8 km NNE of CARMO; site of hydroelectric station (built 1920–1923) supplying RIO DE JANEIRO city.

Ilha Grande (EEL-yah GRAHN-zhee), town, NW AMAZONAS state, BRAZIL, head of navigation on left bank of the RIO NEGRO, and 130 mi/209 km E of SÃO GABRIEL DA CACHOEIRA; 00°30'S 65°06'W. Brazil nuts; rubber. Previously known as Tapuruquá and Santa Isabel.

Ilha Grande Bay (EEL-yah GRAHN-zhe), on SE coast of RIO DE JANEIRO state, BRAZIL, 70 mi/113 km W of RIO; 30 mi/48 km wide, 15 mi/24 km long; heavily indented by spurs of the Serra do MAR reaching the coast. On it are towns of ANGRA DOS REIS and PARATI. Across the SE entrance lies Ilha Grande, a moun-tainous island (elevation 3,200 ft/975 m); c.15 mi/24 km long.

Ílhavo (EEL-yah-voo), town (2001 population 35,688), AVEIRO district, N central PORTUGAL, near Aveiro Lagoon (inlet of the ATLANTIC OCEAN), 3 mi/4.8 km SSW of AVEIRO; 40°36'N 08°40'W. Equips high-sea fishing fleets; saltworks. Marine museum.

Ilhéus (eel-YAI-oos), city (2007 population 220,144), BAHIA state, E BRAZIL, a port on Ilhéus Bay, an inlet of the ATLANTIC OCEAN; 14°45'S 39°04'W. Founded in the mid-16th century, it became the world's chief

cacao port through the early 20th century; today it remains the cacao center of Brazil. Cacao was first planted in 1746 and became the basis for great pros-perity. However, crop dropped 50% in the 1990s due to drought, disease, and international competition from W Africa and Indonesia. Half of the cacao processing plants closed. Once a wealthy, glori-ous city, much now in decay. Exports rubber, timber, chemicals, and manioc. A hydroelectric project is lo-cated to N.

Ilhota (EEL-yo-tah), city (2007 population 11,561), E SANTA CATARINA state, BRAZIL, 11 mi/18 km W of ITAJAÍ; 26°24'S 48°52'W. Sugar, fruit, rice, manioc; livestock.

Ili (EE-LEE), district, W XINJIANG UYGUR AUTONO-MOUS REGION, NW CHINA, in the W Dzungaria, on KAZAKHSTAN border; 43°54'N 81°21'E. Main town, YINING (Gulja). Situated in region of upper reaches of ILI RIVER and enclosed by the TIANSHAN MOUNTAINS (S) and the BOROHORO RANGE (N), it is oriented W toward the lower Ili River valley in Kazakhstan. Its chief centers are YINING and HUOCHENG (Shuiding), all of which once had the name Ili.

Ilia (ee-LEE-ah), prefecture, WESTERN GREECE depart-ment, in W PELOPONNESUS, on Gulf of KIPARISSIA, part of IONIAN SEA, opposite ZÁKINTHOS island; ⊙ PÍRGOS; 37°45'N 21°35'E. Bordered N by AKHAIA prefecture, E and S by PELOPONNESE department (ERIMANTHOS and Alpheus rivers, borders). Includes historic sites of OLYMPIA and ELIS.

Ilia (EEL-yah), Hungarian *Marosillye*, village, HUNE-DOARA county, W central ROMANIA, on MUREŞ RIVER, 12 mi/19 km WNW of DEVA; 45°56'N 22°39'E. Railroad junction; gold production. Former fortress against the Turks.

Iliamna (i-lee-YAM-nuh), village (2000 population 102), S ALASKA, on N shore of ILIAMNA Lake, 120 mi/193 km W of HOMER; 59°45'N 153°53'W. Sport fishing on Iliamna Lake. NEWHALEN village (1990 population 160) is 3 mi/4.8 km WSW.

Iliamna (i-lee-YAM-nuh), lake (□ 1,000 sq mi/2,600 sq km), SW ALASKA, at the base of the ALASKA PE-NINSULA; 75 mi/120 km long and up to 22 mi/35 km wide. Largest lake in Alaska and the second-largest freshwater lake wholly within the UNITED STATES. Fed by many lakes and streams; the KVICHAK RIVER drains it SW into BRISTOL BAY. Noted for sport fishing. ILIAMNA, NEWHALEN, and KAKHONAK are the chief lakeside villages.

Iliamna Volcano (10,016 ft/3,053 m), active volcano, S ALASKA, W of COOK INLET, 150 mi/241 km SW of ANCHORAGE; 60°02'N 153°06'W.

Ilic, (Turkish=*Iliç*) village, E central TURKEY, on E bank of the EUPHRATES RIVER, on Erzincan-Sivas railroad, and 55 mi/89 km WSW of ERZINCAN; 39°27'N 38°34'E. Grain.

Ilich (EEL-yich), town, S SOUTH KAZAKHSTAN region, KAZAKHSTAN, on spur of TRANS-CASPIAN RAILROAD, 50 mi/80 km SW of TOSHKENT (Uzbekistan); 40°50'N 68°29'E. Textile center in BETPAK DALA irrigation area. Nearby state cotton farms were among the largest in the former USSR.

Ilich Bay (eel-YEECH), inlet of CASPIAN SEA, AZER-BAIJAN, on S shore of APSHERON Peninsula, 4 mi/6 km SSW of BAKY. Oil fields.

Ilichevsk, city, ANDIJAN wiloyat, UZBEKISTAN; 40°44'N 72°53'E. Also Il'ichevsk.

Il'ichevsk, UKRAINE: see ILLICHIVS'K.

Il'icheysk (eel-yeech-YESK), town, NW NAKHICHEVAN AUTONOMOUS REPUBLIC, AZERBAIJAN, on railroad, on the Eastern ARPA–Chai River near its confluence with ARAS River, and 30 mi/48 km NW of NAKHICHEVAN. Cotton ginning. Formerly Norashen.

Ilidža (ee-LEE-zhah), part of SARAJEVO city, central BOSNIA, BOSNIA AND HERZEGOVINA, on railroad, and 7 mi/11.3 km SW of Sarajevo city center. Health re-sort; sulphur springs and baths. Known since Roman period; popular under Turkish rule. A Serbian stronghold during the civil war. In 1995, by the terms of the Dayton Accord, incorporated into the Federation of Bosnia and Herzegovina and placed under Muslim control. Also spelled Ilidzha.

Iliff, village (2000 population 213), LOGAN COUNTY, NE COLORADO, on SOUTH PLATTE RIVER, near NEBRASKA state line, and 11 mi/18 km NE of STERLING; 40°45'N 103°04'W. Elevation 3,833 ft/1,168 m. In sugar beet, sunflowers region.

Iligan (ee-LEE-gahn), city (□ 282 sq mi/733.2 sq km; 2000 population 285,061), ⊙ LANAO DEL NORTE province, W central MINDANAO, the PHILIPPINES, port on ILIGAN BAY; 08°15'N 124°24'E. Former Spanish missionary and military center, it is now the site of a growing heavy industry complex, powered by hydroelectricity from the MARIA CRISTINA FALLS plant on the AGUS RIVER. The nation's first steel mill was established here in 1964. The city also has chemical, cement, and fertilizer plants. Airport, located on road to MANILA 11 mi/17 km S, is sometimes closed for security reasons.

Iligan Bay (ee-LEE-gahn), inlet of MINDANAO SEA, PHILIPPINES, in N central MINDANAO; 30 mi/48 km long, 30 mi/48 km wide. On it are KOLAMBUGAN, OROQUIETA, and Ozamiz (MISAMIS). PANGUIL BAY is inlet's SW arm.

Iliki, Lake (ee-LEE-kee), Latin *Hylice* (□ 8.5 sq mi/22 sq km), in BOEOTIA prefecture, E CENTRAL GREECE department, on lower KIFISSOS RIVER, N of THEBES; 6 mi/9.7 km long, 3 mi/4.8 km wide; 38°25'N 23°15'E. Formerly called Likeri. Also Lake Hylike.

Ilimpeia River (ee-LEEM-pye-yah), 380 mi/611 km long, EVENKI AUTONOMOUS OKRUG, KRASNOYARSK TERRITORY, NW Siberian Russia, left tributary of the NIZHNYAYA TUNGUSKA, in the CENTRAL SIBERIAN PLATEAU; basin area, 6,718 sq mi/17,400 sq km. Not navigable because of rapids.

Ilim Range (ee-LEEM), central IRKUTSK oblast, RUSSIA; forms a divide between upper LENA and ILIM rivers; highest point 3,300 ft/1,006 m. Angara-Ilim iron-ore basin is NW.

Ilim River (ee-LEEM), 225 mi/362 km long, IRKUTSK oblast, RUSSIA; rises in the ILIM RANGE; flows N, past Ilimsk and NIZHNE-ILIMSK, through Angara-Ilim iron-ore basin, to the ANGARA RIVER at a recently built city and hydroelectric plant at UST'-ILIMSK. Navigable for 125 mi/201 km below Ilimsk.

Ilin Island (EE-leen) (□ 30 sq mi/78 sq km), MINDORO OCCIDENTAL province, PHILIPPINES, in MINDORO STRAIT, just off S coast of MINDORO island, 4 mi/6.4 km S of SAN JOSE; 11 mi/18 km long, 4 mi/6.4 km wide; 12°14'N 121°05'E. Hilly, rises to 666 ft/203 m; rice.

Iliniza, Cerro (ee-lee-NEE-zah, SER-ro), Andean peak (17,405 ft/5,305 m), in COTOPAXI province, central ECUADOR, 18 mi/29 km W of COTOPAXI volcano; 00°40'S 78°42'W.

Il'inka (eel-YEEN-kah), village (2005 population 4,315), central BURYAT REPUBLIC, S Siberian Russia, on the SELENGA River, on road and the TRANS-SIBERIAN RAILROAD, 19 mi/31 km NW of ULAN-UDE; 52°07'N 107°16'E. Elevation 1,669 ft/508 m. Timbering.

Il'inka (eel-YEEN-kah), settlement (2005 population 4,600), S ASTRAKHAN oblast, S European Russia, on the VOLGA RIVER, on road, 5 mi/8 km SE of AS-TRAKHAN; 46°14'N 47°54'E. Below sea level. Geophysical observation and scientific station.

Il'ino (EEL-yee-nuh), settlement, W NIZHEGOROD oblast, central European Russia, on railroad and near highway, 10 mi/16 km W, and under administrative jurisdiction, of VOLODARSK; 56°14'N 42°55'E. Elevation 295 ft/89 m. Sawmilling, woodworking.

Il'inogorsk (eel-yee-nuh-GORSK), town (2006 population 8,245), W NIZHEGOROD oblast, central European Russia, just N of the KLYAZ'MA RIVER where it forms the natural border with VLADIMIR oblast, on railroad, 7 mi/11 km W of (and administratively

subordinate to) VOLODARSK; 56°13'N 42°59'E. Elevation 291 ft/88 m. In agricultural area; livestock feed plant.

Il'ino-Zaborskoye (eel-YEE-nuh–zah-BOR-skuh-ye), village, NW NIZHEGOROD oblast, central European Russia, on road junction, 32 mi/51 km N of SEMENOV; 57°14'N 44°24'E. Elevation 360 ft/109 m. In agricultural area.

Il'inskaya (eel-YEEN-skah-yah), village (2005 population 4,935), E KRASNODAR TERRITORY, S European Russia, on road, 20 mi/32 km NNE of KROPOTKIN; 45°43'N 40°43'E. Elevation 239 ft/72 m. Flour mill, metalworks; wheat, sunflowers, castor beans.

Il'inskiy (eel-YEEN-skeeyee), town (2006 population 835), SW SAKHALIN oblast, on the W coast of S SA-KHALIN Island, RUSSIAN FAR EAST, on the TATAR STRAIT, junction of the coastal and cross-island highways, 17 mi/27 km NNE of TOMARI; 48°00'N 142°12'E. Elevation 269 ft/81 m. N terminus of the W coast railroad; fishing and fish processing. Under Japanese rule (1905–1945), called Kushunnai.

Il'inskiy (eel-YEEN-skeeyee), town (2006 population 6,290), central PERM oblast, W URAL MOUNTAINS, E European Russia, on the Obva River, near the mouth in the KAMA reservoir of the KAMA River, on road, 40 mi/64 km N of PERM, and 19 mi/31 km SW of CHER-MOZ; 58°34'N 55°41'E. Elevation 561 ft/170 m. Woodworking and food processing. In oil-producing area.

Il'inskiy (eel-yeen-skeeyee), village (2005 population 3,090), SE Republic of KARELIA, NW European Russia, on the SW shore of Lake LADOGA, on road and near railroad, 9 mi/14 km W, and under adiministrative jurisdiction, of OLONETS; 61°01'N 32°40'E. Popular weekend getaway spot for residents of PET-ROZAVODSK and nearby towns.

Il'insko-Podomskoye (ee-LYEEN-skuh–puh-DOM-skuh-ye), village, S ARCHANGEL oblast, N European Russia, on a left affluent of the VYCHEGDA River, near highway terminus, 45 mi/72 km ESE of KOTLAS; 61°07'N 47°58'E. Elevation 334 ft/101 m. Flax processing. Formerly called Il'inskoye.

Il'inskoye (eel-YEEN-skuh-ye), village, E central YAR-OSLAVL oblast, central European Russia, near road and railroad, 11 mi/18 km N of YAROSLAVL; 57°45'N 40°13'E. Elevation 341 ft/103 m. In agricultural area.

Il'inskoye (eel-YEEN-skuh-ye), settlement, SW KOS-TROMA oblast, central European Russia, near the VOLGA RIVER, on highway, 9 mi/14 km S of KOS-TROMA, to which it is administratively subordinate; 57°38'N 40°55'E. Elevation 426 ft/129 m. Brick works.

Il'inskoye, RUSSIA: see IL'INSKO-PODOMSKOYE.

Il'inskoye-Khovanskoye (eel-YEEN-skuh-ye–huh-VAHN-skuh-ye), town (2005 population 3,600), W IVANOVO oblast, central European Russia, 45 mi/72 km W of IVANOVO; 56°58'N 39°46'E. Elevation 482 ft/146 m. Highway junction. In agricultural area.

Ilintsy, UKRAINE: see ILLINTSI.

Iliodromia, Greece: see ALONNISSOS.

Ilion (IL-ee-uhn), industrial village (□ 2 sq mi/5.2 sq km; 2006 population 8,237), HERKIMER county, central NEW YORK, on MOHAWK RIVER and BARGE CANAL, 11 mi/18 km SE of UTICA; 43°00'N 75°02'W. Diverse manufacturing (including small arms); agricultural products; services. Part of the former HERKIMER (village). In MOHAWK-Ilion-Frankfort industrial region. Remington Arms Museum. Incorporated 1852.

Ilion, TURKEY: see TROY.

Ilipa (IL-i-puh), ancient town of SPAIN, near the modern SEVILLE. Here Scipio Africanus Major defeated (206 B.C.E.) the Carthaginian forces after Hasdrubal had fled to GAUL. The overthrow of Carthaginian power in Spain paved the way for the defeat of Hannibal at Zama (202 B.C.E.).

Ili River (EE-LEE), 590 mi/949 km long, in CHINA and KAZAKHSTAN; formed in W XINJIANG UYGUR AU-TONOMOUS REGION, NW CHINA (65 mi/105 km E of YINING) by junction of two branches—the Künes (140

mi/225 km long) rising in E TIANSHAN MOUNTAINS, NW China, and the Tekes (270 mi/435 km long), rising in central Tianshan, Kazakhstan. Flows W, past Yining, between the Junggarian Alatau and the Trans-Ili Alatau, into Kazakhstan and, passing Kapchagay, turns NW through desert area and into SW end of Lake Balkhash through a delta. Important for irrigation; navigable in middle course to Bakanas. Receives CHILIK and Kaskelen rivers.

Ilirska Bistrica (ee-LIR-skah BEES-tree-tsah), town, SW SLOVENIA, on the Reka River, on railroad, and 17 mi/27 km NNW of RIJEKA (CROATIA), at W foot of Snežnik mountain; 45°34'N 14°14'E. Wood-processing and chemical industries. Part of Italy (1919–1947).

Ilişeşti (ee-lee-SHESHT), village, SUCEAVA county, N ROMANIA, 10 mi/16 km WSW of SUCEAVA. Leather tanning.

Ilissos River, Greece: see EILISSOS RIVER.

Ilium, TURKEY: see TROY.

Iliyantsi (ee-LEE-yahn-tsee), former village, now incorporated into SOFIA city, SOFIA oblast, W BULGARIA, 4 mi/6 km N of Sofia; 42°45'N 23°18'E. Railroad junction. Manufacturing. Cemetery.

Il'ka (EEL-kah), village, central BURYAT REPUBLIC, S SIBERIA, RUSSIA, on road and the TRANS-SIBERIAN RAILROAD, 41 mi/66 km WSW of ULAN-UDE; 51°43'N 108°32'E. Elevation 2,066 ft/629 m. Woodworking machinery and equipment.

Ilkal (IL-kahl), town, BIJAPUR district, KARNATAKA state, S INDIA, 65 mi/105 km SE of BIJAPUR; 15°58'N 76°08'E. Market center for cotton, peanuts, millet, wheat; handicraft cloth weaving and dyeing.

Ilkeston (ILK-stuhn), town (2001 population 37,270), DERBYSHIRE, central ENGLAND, 8 mi/12.8 km W of NOTTINGHAM; 52°58'N 01°18'W. Iron and coal mines lie to the S, but are not as active as in previous years. Some manufacturing, such as lace and hosiery. Mentioned in the *Domesday Book*. EASTWOOD, in a former coal-mining region, is the birthplace of D. H. Lawrence, and many of his novels are set in the area.

Ilkley (ILK-lee), ancient *Olicana*, town (2001 population 13,828), WEST YORKSHIRE, N ENGLAND, on WHARFE RIVER, and 10 mi/16 km NNW of BRADFORD; 53°55'N 01°49'W. Elevation 700 ft/213.4 m. Resort with mineral springs. Has remains of Roman fort. The churchyard contains three sculptured Saxon crosses. Ilkley Moor to S is famed in song.

Ill, FRANCE: see ILL RIVER.

Illampú (ee-yahm-POO), peak (elevation 20,938 ft/6,382 m), in the CORDILLERA Real de la Paz of the Bolivian ANDES, SE of LARECAJA province, E BOLIVIA; 15°50'S 68°34'W. Although lower than the adjacent peak, ANCOHUMA (elevation 21,489 ft/6,550 m), Illampú is the name usually given to the whole mountain. It is sometimes called SORATA, from the village high on its slopes. Permanently capped with snow, Illampú dominates the mountain scenery visible from LA PAZ. Link to colonial-era goldfields of Tipuami River. Tin mines nearby.

Illana (ee-LYAH-nah), town, GUADALAJARA province, central SPAIN, 45 mi/72 km ESE of MADRID; 40°11'N 02°54'W. Cereals, olives, grapes; livestock; olive-oil pressing.

Illana Bay (eel-YAH-nah), inlet of MORO GULF in SW coast of MINDANAO, PHILIPPINES; 50 mi/80 km wide at mouth, 40 mi/64 km long. PAGADIAN is on NW shore, COTABATO on SE shore. BONGO ISLAND in bay.

Illapel (eel-YAH-pel), town, ⊙ Illapel comuna (2002 population 21,826) and CHOAPA province, COQUIMBO region, N central CHILE, on Illapel River (affluent of the CHOAPA), on railroad, and 120 mi/193 km S of LA SERENA; 31°37'S 71°09'W. Agricultural center, with mild climate; cereals, fruit; livestock. In colonial times noted for its gold deposits. Airport.

I'llar, Arab town, TULKARM district, 6 mi/9.5 km NE of Tulkarm, in the Samarian Highlands, WEST BANK; 32°22'N 35°06'E. Agriculture (fruits, olives). Believed to be the biblical town of A'nnar. Ancient burial caves nearby.

Illarionovo, UKRAINE: see ILARIONOVE.

Illaunamid (i-LAW-nuh-mid), rocky islet in the AT-LANTIC OCEAN, W GALWAY county, W IRELAND, 10 mi/16 km SW of CLIFDEN; 53°24′N 10°14′W. Slyne Head lighthouse. Also spelled Illaunimmul.

Illava, SLOVAKIA: see ILAVA.

Illawarra District (I-luh-WA-ruh), NEW SOUTH WALES, AUSTRALIA, on E coast, extends 50 mi/80 km S from HELENSBURGH to NOWRA and lower SHOALHAVEN RIVER. Contains ILLAWARRA LAKE. Known for coastal scenery; many seaside resorts. Produces dairy foods, coal. WOLLONGONG, industrial and coal-mining center, chief town.

Illawarra Lake (I-luh-WA-ruh), lagoon (□ 13 sq mi/33.8 sq km), E NEW SOUTH WALES, AUSTRALIA, on coast, near WOLLONGONG, in ILLAWARRA DISTRICT, 45 mi/72 km S of SYDNEY; 4 mi/6 km long, 3 mi/5 km wide; 34°30′S 150°50′E.

Illawarra North (I-luh-WA-ruh), municipality, E NEW SOUTH WALES, SE AUSTRALIA, on the PACIFIC OCEAN, 40 mi/64 km SW of SYDNEY. In coal-mining area.

Illecillewaet (il-uh-SIL-uh-wet), mountain stream, c.50 mi/80 km long, SE BRITISH COLUMBIA, W CANADA; rises in Illecillewaet glacier, on the W slope of the SELKIRK MOUNTAINS; flows SW in a mountain valley to join the COLUMBIA River near REVELSTOKE; 50°59′N 118°10′W. For almost its entire distance, it is followed by the Canadian Pacific railroad and is well known to travelers for its exceptional beauty.

Ille-et-Vilaine (eel–ai–vee-LEN), department (□ 2,616 sq mi/6,801.6 sq km), in BRITTANY, W FRANCE, on ENGLISH CHANNEL; ⊙ RENNES; 48°10′N 01°30′W. Mostly level terrain drained by the VILAINE, Illet, RANCE, and COUESNON rivers. A leading cider-making region, with extensive apple and pear orchards. Other crops include wheat, barley, oats, potatoes, legumes. One of France's chief dairy-farming, hog-raising, and beekeeping areas. Has some iron deposits (in S) and granite quarries. Principal urban centers are Rennes (diversified manufacturing), FOUGÈRES (shoes), CANCALE (noted for its oysters). Coastal tourist resorts are DINARD and SAINT-MALO (chief port). On the Rance River estuary there is a unique tidal electric power plant. Administratively part of the *Bretagne* (Brittany) region (Rennes is capital).

Illéla (EEL-ai-luh), town, TAHOUA province, NIGER, 30 mi/48 km S of TAHOUA; 14°28′N 05°15′E. Livestock market. Administrative center.

Ille-Rance Canal (EE-luh–RAWNS), CÔTES-D'ARMOR and ILLE-ET-VILAINE departments, BRITTANY, W FRANCE, connects DINAN (on the RANCE RIVER) with RENNES (on the VILAINE RIVER at the mouth of the Illet); c.40 mi/64 km long.

Ille River, FRANCE: see ILLET RIVER.

Iller River (IL-ler), 91 mi/146 km long, GERMANY; rises in ALLGÄU ALPS near Austrian border; flows N, past KEMPTEN, to the DANUBE River at NEU-ULM; source at 47°16′N 10°12′E. In middle and upper course it forms border between BAVARIA and BADEN-WÜRTTEMBERG.

Illertissen (IL-ler-tis-suhn), town, BAVARIA, S GERMANY, near the ILLER RIVER, 13 mi/21 km SSE of ULM; 48°13′N 10°06′E. Manufacturing of chemicals, pharmaceuticals, ceramics; metal- and woodworking; food processing. Has Renaissance castle. Evidence of prehistoric settlement. Chartered 1430.

Illesca, Cerro, PERU: see ILLESCAS, CERRO.

Illescas (ee-LYE-skahs), town, TOLEDO province, central SPAIN, on railroad, and 21 mi/34 km SSW of MADRID; 40°07′N 03°50′W. Wheat, carobs, olives, potatoes; cattle, hogs. Olive-oil pressing, sawmilling, plaster and soap manufacturing. Has Caridad church and hospital Francis I of France lived here.

Illescas (ee-YE-skahs), village, FLORIDA department, SE central URUGUAY, in the CUCHILLA GRANDE PRINCIPAL, on railroad and road, and 13 mi/21 km SW

of JOSÉ BATLLE Y ORDÓÑEZ, on LAVALLEJA department border; 33°36′S 55°20′W. Agricultural settlement; wheat, corn, oats; cattle.

Illescas, Cerro (ee-YES-kahs, SER-ro), small hilly range on PACIFIC coast of PIURA region, NW PERU, extends c.20 mi/32 km S from AGUJA POINT; 06°00′S 81°03′W. Elevation 1,696 ft/517 m. At its S foot are the now abandoned Reventazón sulphur mines. Variant names include: Monte Illescas; Cerro Illesca.

Illescas, Monte, PERU: see ILLESCAS, CERRO.

Ille-sur-Têt (EEL-syur–TET), town (□ 12 sq mi/31.2 sq km), PYRÉNÉES-ORIENTALES department, LANGUE-DOC-ROUSSILLON region, S FRANCE, on the TÊT RIVER, and 14 mi/23 km W of PERPIGNAN; 42°40′N 02°37′E. Fruit preserving and shipping. Noted for its orchards and olive groves.

Illet River (eel-e), 28 mi/45 km long, ILLE-ET-VILAINE department, in BRITTANY, W FRANCE; rises 5 mi/8 km SE of COMBOURG; flows S, joining the VILAINE at RENNES. Its stream channel is used by ILLE-RANCE CANAL, which connects DINAN with Rennes; 48°12′N 01°39′W. Also called Ille.

Illichivs'k (eel-lee-CHEEVSK), (Russian *Il'ichevsk*), city (2001 population 54,151), central ODESSA oblast, UKRAINE, 14 mi/23 km SSW of ODESSA city center, on the BLACK SEA coast, at the mouth of the Sukhyy Lyman (river); 46°18′N 30°40′E. Seaport; passenger port to VARNA (Bulgaria), base for Antarctic whaling fleet; fish processing, ship repair, container port and container repair. Seat of branch of Odessa marine school, two professional-technical schools. Known since the end of the 18th century as Buhovi Khutory; city status and renamed Illichivs'k, since 1973.

Illiers-Combray (eel-YAI–kon-BRAI), commune (□ 13 sq mi/33.8 sq km), EURE-ET-LOIR department, CENTRE administrative region, NW central FRANCE, on the LOIR RIVER, and 15 mi/24 km SW of CHARTRES; 48°18′N 01°15′E. Agriculture market in wheat-growing and horse-raising area. The commune (notably Combray) is celebrated in Marcel Proust's *In Search of Lost Time*.

Illilovette Fall, California: see YOSEMITE NATIONAL PARK.

Illimani (ee-yee-MAH-nee), mountain (21,184 ft/6,457 m), E BOLIVIA; 16°39′S 67°48′W. One of the highest peaks of the CORDILLERA. Real of the Bolivian ANDES, it is permanently snow-capped. First climbed by Baron Conway of Allington in 1898.

Illimo (ee-YEE-mo), town, LAMBAYEQUE province, LAMBAYEQUE region, NW PERU, on coastal plain, on LA LECHE RIVER, and 15 mi/24 km N of LAMBAYEQUE, on PAN-AMERICAN HIGHWAY; 06°28′S 79°51′W. Cereals, corn, fruit.

Illingen (IL-ling-uhn), village, BADEN-WÜRTTEMBERG, S GERMANY, 10 mi/16 km NE of PFORZHEIM; 48°57′N 08°56′E.

Illingen (IL-ling-uhn), suburb of SAARBRÜCKEN, SAARLAND, SW GERMANY, 9 mi/14.5 km N of city center; 49°23′N 07°03′E. Meat processing.

Illinois, state (□ 57,918 sq mi/150,007 sq km; 2000 population 12,419,293; 2005 estimated population 12,765,427), N central UNITED STATES, in the MIDWEST, admitted as the twenty-first state of the Union in 1818; ⊙ SPRINGFIELD; 40°07′N 89°21′W. CHICAGO, ROCK-FORD, and PEORIA are the three largest cities.

Geography

Bounded N by WISCONSIN, E by INDIANA, SE and S by KENTUCKY (where the OHIO RIVER forms the state line), and W by MISSOURI and IOWA (where the MIS-SISSIPPI RIVER forms the state line). LAKE MICHIGAN is on the NE. The broad level lands that gave Illinois the nickname Prairie State were fashioned by late Cenozoic glaciation, which leveled rugged ridges and filled valleys in over 90% of the state. There is a cluster of about sixty small glacial lakes in NE, especially in LAKE, MCHENRY, and COOK counties. The fertile prairies are drained by over 275 rivers, most of which flow to the

Mississippi-Ohio systems; the ILLINOIS is the state's largest river. These rivers provided early explorers a way SW from Lake Michigan into the interior of the continent and later, in the days of canal building, played a big part in hastening settlement of the prairies. The completion of the ERIE CANAL linked Illinois, through the GREAT LAKES, to the E seaboard of the U.S. The ILLINOIS WATERWAY, which includes the CHICAGO SANITARY AND SHIP CANAL deepened for the waterway, links Chicago to the Mississippi basin as the old Chicago and ILLINOIS AND MICHIGAN canals once did, and the SAINT LAWRENCE SEAWAY provides access for oceangoing vessels. The waterways are but a part of a transportation complex that includes railroads, airlines, and a very extensive modern highway system.

Economy

Although the area's climate varies, with extreme temperatures in parts of the state, the rich land, adequate rainfall (32 in/81 cm–46 in/116 cm annually), and a long growing season make Illinois an important agricultural state. It consistently ranks near the top in the production of corn and soybeans. Hogs and cattle are also principal sources of farm income. Other major crops include hay and wheat. Beneath the fertile topsoil lies mineral wealth, and the state is a leading producer of fluorspar. Bituminous-coal fields and oil deposits make S Illinois a major source of fuel; Illinois ranks high among the states in the production of coal, and its reserves are greater than any other state E of the ROCKY MOUNTAINS. These agricultural and mineral resources encouraged the establishment of abundant industries along the state's excellent lines of communication and transportation, and by 1880 income from industry was almost double that from agriculture. Major industries include the manufacture of electrical and nonelectrical machinery, food products, fabricated and primary metal products, and chemicals; and printing and publishing. Metropolitan Chicago, the country's leading railroad center, is also a major industrial center, famous for its huge grain mills and elevators. Outside Chicago is the ARGONNE NATIONAL LABORATORY, a major research and development installation of the Atomic Energy Commission. Suburbs of Chicago such as SCHAUMBURG and OAK BROOK have become important business centers. Scattered across the N half of the state are cities with specialized industries—ELGIN, Peoria, ROCK ISLAND, MOLINE, and Rockford. Industrially important cities in central Illinois include JOLIET, SPRINGFIELD, and DECATUR.

History: to 1766

At the end of the eighteenth century the Illinois, Sac, Fox, and other Native American groups were living in the river forests, where many centuries before them the prehistoric Mound Builders had dwelt. French explorers and missionaries came to the region early. Father Marquette and Louis Joliet, on their return from a trip down the Mississippi, paddled up the Illinois in 1673, and two years later Marquette returned to establish a mission in the Illinois country. In 1679 the French explorer Robert Cavelier, sieur de La Salle, went from Lake Michigan to the Illinois, where he founded (1680) Fort CREVE COEUR and with his lieutenant, Henri de Tonti, completed (1682–1683) Fort Saint Louis on STARVED ROCK cliff. French occupation of the area was sparse, but the settlements of CAHOKIA and KASKASKIA achieved a minor importance in the eighteenth century, and the area was valued for fur trading. By the Treaty of PARIS of 1763, ending the French and Indian Wars, FRANCE ceded all of the Illinois country to GREAT BRITAIN. However, the British did not take possession until resistance, led by the Ottawa chief, Pontiac, was quelled (1766).

History: 1766 to 1847

In the American Revolution, George Rogers Clark and his expedition captured (1778) the British posts of Cahokia and Kaskaskia before going on to take VIN-

CENNES. The Illinois region was an integral part of the Old Northwest that came within U.S. borders by the 1783 Treaty of Paris ending the Revolution. Under the Ordinance of 1787 the area became the NORTHWEST TERRITORY. Made part of Indiana Territory in 1800, Illinois became a separate territory in 1809. The fur trade was still flourishing throughout most of Illinois when it became a state in 1818, but already settlers were pouring down the Ohio River by flatboat and barge and across the Genesee wagon road. In 1820 the capital was moved from Kaskaskia to VANDALIA. The Black Hawk War (1832) practically ended the tenure of the Native Americans here and drove them W of the Mississippi. In the 1830s there was heavy and uncontrolled land speculation. Mob fury broke out with the murder (1837) of the abolitionist Elijah P. Lovejoy at ALTON and in the lynching (1844) of the Mormon leader Joseph Smith and his brother Hyrum at CARTHAGE.

History: 1847 to 1871

Industrial development came with the opening of an agriculture implements factory by Cyrus H. McCormick at Chicago and John Deere at Moline in 1847 and the building of the railroads in the 1850s. During this period the career of Abraham Lincoln began. In the state legislature, Lincoln and his colleagues from SANGAMON county had worked hard and successfully to bring the capital to Springfield in 1839. As Illinois moved toward a wider role in the country's affairs, Lincoln and another Illinois lawyer, Stephen A. Douglas, won national attention with their debates on the slavery issue in the senatorial race of 1858. In 1861, Lincoln became President and fought to preserve the Union in the face of the South's secession. During the Civil War, Illinois supported the Union, but there was much proslavery sentiment in the S part of the state. By the 1860s industry was well established, and many immigrants from Europe had already settled here, foreshadowing the influx still to come. Immediately after the war, industry expanded to tremendous proportions, and the Illinois legislature, by setting aside acreage for stockyards, prepared the way for the development of the meatpacking industry. Economic development had outrun the construction of facilities, and Chicago, which had grown dramatically since the 1830s, was a mass of flimsy wooden structures when the fire of 1871 destroyed most of the city.

History: 1871 to World War II

In the latter part of the nineteenth century farmers in the state revolted against exorbitant freight rates, tariff discrimination, and the high price of manufactured goods. Illinois farmers enthusiastically joined the Granger movement. Laborers in factories, railroads, and mines also became restive, and from 1870 to 1900 Illinois was the scene of such violent labor incidents as the Haymarket Square riot of 1886 and the Pullman strike of 1894. In the twentieth-century labor conditions improved, but violent labor disputes persisted, notably the massacre at HERRIN in 1922 during a coalminers' strike and the bloody riot during a steel strike at Chicago in 1937. State politics became divided by the conflicting forces of farmers, laborers, and corporations, and opposing political machines came into being downstate and upstate. In 1937 new oil fields were discovered in S Illinois, further enhancing the state's industrial development.

History: World War II to 1980

During World War II the nation's first controlled nuclear reaction was executed at the University of Chicago, paving the way for development of nuclear weapons during the war. World War II spurred the growth of the Chicago metropolitan area. Adlai E. Stevenson, governor of Illinois (1949–1953), achieved national prominence in winning the Democratic presidential nomination in 1952 and 1956. Also during the 1950s the "gateway amendment" to the Illinois constitution simplified the state's constitutional amendment process. S Illinois experienced population declines in the 1950s and 1960s as farms in the S became more mechanized, providing fewer jobs in the area. Civil rights demonstrations were held in CICERO in 1967; Democratic convention riots occurred in Grant Park, Chicago in 1968.

History: 1980 to Present

The area was hard hit again in the 1980s as farm prices fell and farm machinery, the major industrial product of S Illinois, was no longer in high demand. The N portion of the state saw a major decline in manufacturing in the 1970s and 1980s, which was partially offset by an increase in the service and trade industry and Chicago's continued strength as a financial center. In the 1990s decline leveled off with opening of new foreign markets, such as E EUROPE. The flood of 1993 affected all counties bordering the Mississippi River, especially MONROE and JERSEY counties. The Illinois (below BEARDSTOWN) and the WABASH rivers (in separate flooding) were also affected. In 1970, Illinois adopted a new state constitution that, among other reforms, banned discrimination in employment and housing.

Educational Institutions and Places of Historic Interest

Institutions of higher learning in Illinois include the University of Illinois, at URBANA-CHAMPAIGN, Chicago, and Springfield; DePaul University, at Chicago; Northwestern University, at EVANSTON; the University of Chicago and the Illinois Institute of Technology, in Chicago; Illinois State University, at NORMAL; Southern Illinois University, at CARBONDALE and EDWARDSVILLE; Eastern Illinois University at CHARLESTON; Western Illinois University at MACOMB; and Governors State University at UNIVERSITY PARK. Among the state's many tourist attractions are Shawnee National Forest, with recreational facilities; the CAHOKIA MOUNDS; and many state parks and historical sites, including NEW SALEM and Lincoln's home and burial place in Springfield. An additional summer attraction is the Illinois State Fair in Springfield and the DU QUOIN State Fair.

Government

The governor of Illinois is elected for a term of four years and is currently Rod Blagojevich. The state legislature, called the general assembly, consists of a house of representatives with 118 members elected to serve for two years and a senate with fifty-nine members elected for two or four years. Illinois elects twenty Representatives and two Senators to the U.S. Congress and has twenty-two electoral votes.

Illinois has 102 counties: ADAMS, ALEXANDER, BOND, BOONE, BROWN, BUREAU, CALHOUN, CARROLL, CASS, CHAMPAIGN, CHRISTIAN, CLARK, CLAY, CLINTON, COLES, COOK, CRAWFORD, CUMBERLAND, DE KALB, DE WITT, DOUGLAS, DU PAGE, EDGAR, EDWARDS, EFFINGHAM, FAYETTE, FORD, FRANKLIN, FULTON, GALLATIN, GREENE, GRUNDY, HAMILTON, HANCOCK, HARDIN, HENDERSON, HENRY, IROQUOIS, JACKSON, JEFFERSON, JERSEY, JO DAVIESS, JOHNSON, KANE, KANKAKEE, KENDALL, KNOX, LAKE, LA SALLE, LAWRENCE, LEE, LIVINGSTON, LOGAN, MCDONOUGH, MCHENRY, MCLEAN, MACON, MACOUPIN, MADISON, MARION, MARSHALL, MASON, MASSAC, MENARD, MERCER, MONROE, MONTGOMERY, MORGAN, MOULTRIE, OGLE, PEORIA, PERRY, PIATT, PIKE, POPE, PULASKI, PUTNAM, RANDOLPH, RICHLAND, ROCK ISLAND, SAINT CLAIR, SALINE, SANGAMON, SCHUYLER, SCOTT, SHELBY, STARK, STEPHENSON, TAZEWELL, UNION, VERMILION, WABASH, WARREN, WASHINGTON, WAYNE, WHITE, WHITESIDE, WILL, WILLIAMSON, WINNEBAGO, and WOODFORD.

Illinois and Michigan Canal, ILLINOIS: see ILLINOIS WATERWAY.

Illinois and Michigan Canal National Heritage Corridor, Affiliated Area (□ 503 sq mi/1,307.8 sq km), NE Illinois Canal vital to W expansion and growth of CHICAGO. Corridor authorized 1984.

Illinois and Mississippi Canal, NW ILLINOIS, abandoned waterway (75 mi/121 km long) between the MISSISSIPPI RIVER at ROCK ISLAND and ILLINOIS RIVER near HENNEPIN; 41°28′N 90°11′W. Opened 1907; soon abandoned because of railroad competition. Resurrected in late twentieth century as a recreational waterway. Often called Hennepin Canal.

Illinois River, 273 mi/439 km long, NE ILLINOIS; formed by the confluence of the Des Plaines and KANKAKEE rivers; flows SW to the MISSISSIPPI RIVER at GRAFTON; 41°23′N 88°15′W. Important commercial and recreational waterway. Forms the greater part of the ILLINOIS WATERWAY, which links the GREAT LAKES with the Mississippi. The chief city on the river is PEORIA. Water quality has improved after decline in early twentieth century.

Illinois Waterway, 336 mi/541 km long, linking LAKE MICHIGAN with the MISSISSIPPI RIVER, N ILLINOIS. An important part of the waterway connecting the GREAT LAKES with the GULF OF MEXICO. The ILLINOIS WATERWAY extends from the mouth of the CHICAGO RIVER, on Lake Michigan, following the CHICAGO SANITARY AND SHIP CANAL, the lower Des Plaines River, and the ILLINOIS RIVER to the Mississippi at GRAFTON. The CALUMET channels branch SE from the waterway and link it with the CALUMET industrial region along the Illinois-INDIANA state line. Principal cargoes, carried chiefly by barges, are coal, petroleum, and grain products. Recreational areas, including the Illinois and MICHIGAN CANAL NATIONAL HERITAGE CORRIDOR, have been developed along the waterway. Also known as the ILLINOIS AND MICHIGAN CANAL.

Illintsi (eel-LEEN-tsee), (Russian *Ilintsy*), city (2004 population 11,810), E VINNYTSYA oblast, UKRAINE, 33 mi/53 km ESE of VINNYTSYA; 48°26′N 25°17′E. Elevation 826 ft/251 m. Raion center; sugar refining, food processing (including dairy and canning), manufacturing (furniture); brickworks. Heritage museum. Known since the mid-15th century as Lintsi until the late 19th century, when the name was changed to Illintsi. Site of a castle (14th–17th century). Town since 1925, city since 1987.

Illiopolis (il-lee-AW-po-lis), village (2000 population 916), SANGAMON county, central ILLINOIS, 15 mi/24 km W of DECATUR; 39°51′N 89°15′W. In agricultural area; ships grain.

Illizi (ee-lee-ZEE), wilaya, in the SAHARA Desert, E ALGERIA; ⊙ ILLIZI; 26°50′N 08°10′E. Stretches along Libyan frontier, NE of TAMANRASSET wilaya and SE of OUARGLA wilaya and includes some important oil and gas fields, notably at EDJELLEH and AÏN AMENAS. Important settlements at Illizi, BORDJ OMAR DRISS, AÏN AMENAS, and DJANET. Created from 4 communes carved out of Ouargla wilaya in 1984.

Illizi (ee-lee-ZEE), town, ⊙ ILLIZI wilaya, SE ALGERIA, at the foot of the Tassili; 26°32′N 08°33′E. Elev. 1,995 ft/608 m. Became a wilaya capital in 1984. Airport. Founded in 1904 by French Commandant Laperrine. Formerly Fort Polignac.

Illkirch-Graffenstaden (eel-KIRSH–grah-fen-stah-DEN), German (IL-kirk-GRAH-fuhn-shtah-duhn), town (□ 9 sq mi/23.4 sq km), industrial suburb of STRASBOURG, in BAS-RHIN department, ALSACE, E FRANCE, on the ILL River, on RHÔNE-RHINE CANAL, and 5 mi/8 km SSW of Strasbourg; 48°32′N 07°43′E. Metalworks (locomotives, machine tools, woodworking machinery) and glassworks; center for computer technology development.

Illmitz (IL-mits), township, N BURGENLAND, E AUSTRIA, in the SEEWINKEL, 11 mi/18 km NE of SOPRON (HUNGARY), across LAKE NEUSIEDL. Vineyards. Several small salt lakes, biosphere reserves and National Park Lake Neusiedl-Seewinkel; 47°46′N 16°48′E.

Illmo, MISSOURI: see SCOTT CITY.

Illnau (EEL-nou), town (2000 population 14,491), ZÜRICH canton, N SWITZERLAND, on Kempt River (affluent of TÖSS RIVER), and 9 mi/14.5 km ENE of ZÜRICH. Metalworking; cotton textiles, flour.

Illo (EE-lo), town, KEBBI state, NW NIGERIA, on NIGER River, near BENIN border, 80 mi/129 km SSW of BIRNIN KEBBI; 11°33′N 03°42′E. Market town. Beans, millet, maize.

Illogan (IL-uh-guhn), village (2001 population 5,585), W CORNWALL, SW ENGLAND, 2 mi/3.2 km NW of REDRUTH; 50°13′N 05°16′W. Former tin- and copper-mining community.

Íllora (EE-lyo-rah), town, GRANADA province, S SPAIN, 17 mi/27 km NW of GRANADA. Olive-oil processing, brandy distilling, flour milling. Cereals, sugar beets, wine, fruit; lumbering.

Illorsuit, fishing and hunting settlement, Uummannaq commune, W GREENLAND, on E Ubekendt Island, on Illorsuit Sound, 50 mi/80 km NW of UUMMANNAQ; 71°14′N 53°32′W.

Ill River, c.45 mi/72 km long, VORARLBERG, W AUSTRIA; rises in glacier of PIZ BUIN of SILVRETTA Group (mountain range); flows NW, through MONTAFON valley, past BLUDENZ and FELDKIRCH, to the RHINE RIVER, 5 mi/8 km NW of Feldkirch. Hydroelectric works in upper course.

Ill River (EEL), 127 mi/204 km long, in HAUT-RHIN and BAS-RHIN departments, ALSACE, E FRANCE; rises on N slope of JURA mountains (near Swiss border, SW of BASEL); flows NNE in a course almost parallel to the RHINE RIVER, draining Alsatian lowland, past MUL-HOUSE, SÉLESTAT and STRASBOURG, to the Rhine below Strasbourg. Navigable in Strasbourg area as part of that city's port area. Receives THUR, FECHT, and BRUCHE rivers (left), all rising in the VOSGES MOUNTAINS.

Illueca (ee-lyoo-AI-kah), town, ZARAGOZA province, NE SPAIN, 13 mi/21 km N of CALATAYUD; 41°32′N 01°37′W. Woolen mills, shoe factory; olive oil, wine, pears.

Illuxt, LATVIA: see ILŪKSTE.

Illyria (i-LIR-ree-yah), ancient region of the BALKAN PENINSULA. In prehistoric times, a group of tribes speaking dialects of an Indo-European language swept down to the N and E shores of the ADRIATIC SEA and established themselves here. The region that they occupied came to be known as Illyria, and therefore the name has vague limits. Among the Illyrian peoples were the tribes later called the Dalmatians and the Pannonians; therefore Illyria is sometimes taken in the widest sense to include the whole area occupied by the Pannonians, and thus to reach from Epirus in the S to the DANUBE River in the N. More usually Illyria is used to mean only the ADRIATIC coast N of central ALBANIA and W of the DINARIC ALPS. The Illyrians were much influenced by the Celts and mingled freely with them; the inhabitants of the later Rhaetia were a compound of Illyrians and Celts. The Illyrians were warlike and frequently engaged in piracy. The mines of the region, located inland, attracted the Greeks, but the terrain was too difficult. Greek cities were established on the coast in the 6th century B.C.E., but they did not flourish, and generally the Greeks left the Illyrians alone. Philip II of Macedon warred against them, but without permanent results. An Illyrian kingdom was set up in the 3rd century B.C.E. with capital at Scodra (present-day SHKODËR, Albania), but trouble over Illyrian piracy led the Romans to conduct 2 victorious wars against Scodra (229 B.C.–228 B.C.E., 219 B.C.E.). After the Dalmatians had split from the kingdom, the Romans conquered Genthius, king of Scodra, and established (168 B.C.–167 B.C.E.) one of the earliest Roman colonies as Illyricum. The colony was enlarged by the total conquest of Dalmatia in several wars (notably 156 B.C.E., 119 B.C.E., and 78 B.C.–77 B.C.E.). The Southern Illyrians were finally conquered (35 B.C.–34 B.C.E.) by Augustus—a conquest confirmed by the campaigns of 29 B.C.–27 B.C.E. Illyricum was expanded by conquests (12 B.C.– 11 B.C.E.) of the Pannonians. At the time of the

stubborn revolt of the Illyrians (C.E. 6–A.D. 9), the territory was split into the provinces of Dalmatia and Pannonia, but the term *Illyricum* was still used. It was later given to one of the great prefectures of the late Roman Empire. Illyricum then included much of the region N of the Adriatic as well as a large part of the Balkan Peninsula. When Napoleon revived (1809) the name for the Illyrian provinces of his empire, he included much of the region N of the Adriatic and what is today the E part of the former Yugoslavia. Roughly the same region was included in the administrative district of Austria called (1816–1849) the Illyrian kingdom.

Illzach (eel-ZAHK), German (ILT-sahk), N suburb (□ 3 sq mi/7.8 sq km) of MULHOUSE, HAUT-RHIN department, in ALSACE, E FRANCE, on the ILL RIVER; 47°47′N 07°20′E. Textile bleaching; manufacturing of chemicals, fertilizer, and automotive parts.

Il'men' (eel-MYEN), shallow lake (□ 212 sq mi/551.2 sq km), NOVGOROD oblast, NW European Russia; 58°20′N 31°25′E. Maximum depth approximately 14 ft/ 4.4 m. It empties through the VOLKHOV RIVER into Lake LADOGA. Freezes over between November and April. NOVGOROD and STARAYA RUSSA are nearby.

Ilmenau (IL-muhn-ou), town, THURINGIA, central GERMANY, in THURINGIAN FOREST, on ILM RIVER, 20 mi/32 km SSW of ERFURT; 50°42′N 10°55′E. In fluor-spar-mining region; china- and glass-manufacturing center; also manufacturing of optical and precision instruments, electric lamps, machinery. Health resort. Has Gothic church, rebuilt 1609; palace (17th century); 18th-century town hall. Town first mentioned in 10th century; it was silver- and copper-mining center in 17th and early 18th century. Frequented by Goethe, whose cottage on KICKELHAHN (KIK-el-hahn) mountain (2,825 ft/861 m), 2 mi/3.2 km SW, was rebuilt in 1874 after fire.

Ilmenau, POLAND: see JORDANOW.

Ilmenau, POLAND: see LIMANOWA.

Ilmenau River (IL-muhn-ou), c.60 mi/97 km long, NW GERMANY; formed below UELZEN; flows N and NW, past LÜNEBURG (head of navigation), to the ELBE RIVER 2 mi/3.2 km N of Winsen; source at 52°48′N 10°32′E.

Il'menskiy Preserve (eel-MEN-skeeye), wildlife refuge and natural museum (□ 119 sq mi/309.4 sq km), CHELYABINSK oblast, RUSSIA, on the E slopes of the S URAL Mountains. Established in 1920. Contains 30 lakes and 50 archeological sites containing the remains of prehistoric humans. More than 80% of the preserve's territory is forested. Contains 32 species of plant life that are considered rare or endangered.

Ilminster (IL-min-stuh), town (2001 population 4,753), S SOMERSET, SW ENGLAND, on Isle River (short tributary of PARRETT RIVER), and 10 mi/16 km SE of TAUNTON; 50°56′N 02°55′W. Agricultural market; manufacturing (lace, cement). Has 15th-century church and 16th-century grammar school.

Ilm River (ILM), 75 mi/121 km long, central GERMANY; rises in THURINGIAN FOREST SW of ILMENAU; flows NE, past Ilmenau and WEIMAR, to the THURINGIAN SAALE 7 mi/11.3 km WSW of NAUMBURG; source at 50°27′N 10°53′E.

Ilnica, UKRAINE: see ILNYTSYA.

Ilnitsa, UKRAINE: see ILNYTSYA.

Ilnytsya (EEL-ni-tsyah), (Russian *Ilnitsa*), (Czech *Il-nica*), (Hungarian *Ilonca*), town (2004 population 10,260), central TRANSCARPATHIAN oblast, UKRAINE, on railroad spur, and 15 mi/24 km NW of KHUST; 48°21′N 23°05′E. Elevation 702 ft/213 m. Lignite mining. Welding shops, lubricating equipment; forestry. Known since 1450, town since 1971.

Ilo (EE-lo), province, MOQUEGUA region, S PERU; ⊙ ILO; 17°30′S 71°10′W. Southernmost province of Moquegua region, bordering TACNA region.

Ilo (EE-lo), town (2005 population 57,393), ⊙ ILO province, MOQUEGUA region, S PERU; PACIFIC port at

mouth of Río Osmore, on railroad to and 37 mi/60 km SW of MOQUEGUA; 17°37′S 71°19′W. With a well-protected harbor, Ilo is shipping center for fertile irrigation area (olives, wine, rice, cotton); copper. Fishing, fruit canning, and olive-oil extracting. Airport.

Ilobasco (ee-lo-BAHS-ko), city and municipality, CA-BAÑAS department, N central EL SALVADOR, 25 mi/40 km ENE of SAN SALVADOR; 13°51′N 88°51′W. Major pottery center; clays quarried nearby. Agriculture, livestock raising.

Ilobu (ee-LAW-boo), town, OSUN state, SW NIGERIA, 7 mi/11.3 km NW of OSHOGBO. Cotton weaving; cacao, palm oil and kernels, cotton, yams, corn, cassava, plantains.

Ilocos (ee-LO-kos), region (□ 4,958 sq mi/12,890.8 sq km), NW LUZON, PHILIPPINES, consisting of ILOCOS NORTE, ILOCOS SUR, LA UNION, and PANGASINAN provinces. Population 37.8% urban, 62.2% rural. Created 1975. Agricultural area (rice grown as food; tobacco as cash crop). Mining (gold, copper). Regional center, SAN FERNANDO. ABRA, MOUNTAIN, and BENGUET provinces, formerly part of region, were assigned to the CORDILLERA ADMINISTRATIVE REGION in 1987. Also known as Region 1.

Ilocos Norte (ee-LO-kos NOR-te), province (□ 1,312 sq mi/3,411.2 sq km), in ILOCOS region, NW LUZON, PHILIPPINES, on SOUTH CHINA SEA (W and N); ⊙ LAOAG (which is also the chief port); 18°10′N 120°45′E. Population 28.4% urban, 71.6% rural; in 1991, 100% of urban and 98% of rural settlements had electricity. Largely mountainous, with fertile coastal strip; drained by many small streams. Rice is grown extensively. Has chrome-ore deposits and manganese mines.

Ilocos Sur (ee-LO-kos SOOR), province (□ 996 sq mi/ 2,589.6 sq km), N LUZON, PHILIPPINES, on SOUTH CHINA SEA (W); ⊙ VIGAN; 17°20′N 120°35′E. Population 23.7% urban, 76.3% rural; in 1991, 88% of urban and 89% of rural settlements had electricity. Generally level, except in E part. Rice is grown in coastal areas. One of the country's chief weaving centers.

Ilog (EE-log), town, NEGROS OCCIDENTAL province, W NEGROS island, PHILIPPINES, near PANAY GULF, 13 mi/ 21 km SSW of BINALBAGAN; 09°47′N 122°44′E. Agricultural center (rice; sugarcane).

Iloilo (EE-lo-EE-lo), province (□ 2,048 sq mi/5,324.8 sq km), WESTERN VISAYAS region, E PANAY island, PHILIPPINES, bounded E by Visayan Sea and GUIMARAS STRAIT, SE by PANAY GULF; ⊙ ILOILO; 11°00′N 122°40′E. Includes GUIMARAS ISLAND. Mountainous, with fertile valleys. Agriculture (rice, sugarcane, coconuts).

Iloilo (EE-lo-EE-lo), city (□ 35 sq mi/91 sq km; 2000 population 365,820), ⊙ ILOILO province, SE PANAY, the PHILIPPINES, on Iloilo Strait of PANAY GULF; 10°45′N 122°33′E. With a fine harbor sheltered by GUIMARAS ISLAND, it is the principal port on Panay, with both inter-island and overseas shipping. Also a busy commercial center, with some manufacturing. Known for its delicate, handwoven fabrics, made from silk and pineapple leaves. Seat of Central Philippine University, the University of San Augustín, and a branch of the University of the Philippines.

Iloilo Strait, PHILIPPINES: see GUIMARAS STRAIT.

Ilok (EE-lok), Hungarian *Újlak*, town, E CROATIA, on the Danube River, 28 mi/45 km E of VINKOVCI; the easternmost town in Croatia. In wine-growing region. Has castle noted for its wine cellars. Occupied by Serbs 1991–1997.

Ilolo (ee-LO-lo), village, IRINGA region, central TAN-ZANIA, 40 mi/64 km NW of IRINGA, N of Lufugwa River, at NE boundary of RUAHA NATIONAL PARK; 07°13′S 35°22′E. Cattle, goats, sheep; corn, wheat.

Ilonca, UKRAINE: see ILNYTSYA.

Ilop (EE-lahp), village, SANDAUN (West Sepik) province, N central NEW GUINEA island, NW PAPUA NEW

GUINEA, 20 mi/32 km SSW of VANIMO and PACIFIC OCEAN. 20 mi/32 km W of Indonesian (IRIAN JAYA) border. Coconuts, bananas, palm oil. Road access.

Ilopango (ee-lo-PAHN-go), town and municipality, SAN SALVADOR department, S central EL SALVADOR, on railroad, and INTER-AMERICAN HIGHWAY, 5 mi/8 km E of SAN SALVADOR, near W shore of LAKE ILOPANGO. Pleasure resort; agriculture (tobacco, sugarcane). Site of SAN SALVADOR airport, formerly the country's main international airport (commercial and military). Part of San Salvador metropolitan area. International Airport of El Salvador is 25 mi/40 km S of city of San Salvador. The Old Aeropuerto de Ilopango on the E outskirts of San Salvador serves only domestic flights and military traffic.

Ilopango, Lake (□ 30 sq mi/78 sq km), on border of SAN SALVADOR, LA PAZ, and CUSCATLÁN departments, S central EL SALVADOR, 6 mi/9.7 km E of SAN SALVADOR; 8 mi/12.9 km long, 5 mi/8 km wide; elevation 1,411 ft/430 m. Occupies old volcanic crater; sulphurous water. Popular weekend resort, with center at ASINO, small port on W shore, near ILOPANGO town. Ilopango volcano (150 ft/46 m high, 500 ft/152 m across) appeared 1880 in center of lake. Outlet (E) drains into JIBOA RIVER.

Ilópolis (EE-lo-po-lees), town (2007 population 4,202), E RIO GRANDE DO SUL state, BRAZIL, 27 mi/43 km N of LAJEADO; 28°56′S 52°07′W. Grapes, wheat, manioc, potatoes; livestock.

Ilorin (ee-LAW-reen), former province, one of the former Northern Provinces, W NIGERIA; capital was Ilorin, between BENIN (then Dahomey) and the NIGER River. E section of former BORGU kingdom formed a division within the province. Changed to KWARA state in 1967; subdivided into Kwara and KOGI states in 1991.

Ilorin (ee-LAW-reen), city (2004 population 805,800), ⊙ KWARA state, SW NIGERIA. It is an industrial city and the market (especially for cattle, poultry, palm products, and yams) and transport center for a wide region. Manufacturing includes cigarettes, matches, and sugar. Traditional artisans make woven goods, tin products, wood carvings, and pottery. University town. Ilorin was capital of a Yoruba kingdom that, with the assistance of the Fulani, successfully rebelled against the Oyo empire in 1817 but soon thereafter was incorporated into the Fulani state of Sokoto. Through warfare against Oyo and Ibadan in the later 19th century, Ilorin considerably increased its territory. In 1897 it was conquered by troops of the British-chartered Royal Niger Co. led by Sir George Goldie.

Ilosva, UKRAINE: see IRSHAVA.

Ilova River (EE-lo-vah), c.50 mi/80 km long, central CROATIA, between MOSLAVINA and SLAVONIA; rises in BILOGORA 5 mi/8 km S of VIROVITICA; flows SW to LONJA RIVER 5 mi/8 km S of KUTINA.

Ilovatka (ee-luh-VAHT-kah), village, NE VOLGOGRAD oblast, SE European Russia, on the left bank of the VOLGA RIVER, on road, 36 mi/58 km NE of KAMYSHIN; 50°31′N 45°51′E. In agricultural area; grain storage, flour mill. Formerly known as Llovatni-Yerik.

Ilovay-Brigadirskoye (ee-luh-VEI-bree-gah-DEER-skuh-ye), village, NW TAMBOV oblast, S central European Russia, on road and railroad, 10 mi/16 km S of (and administratively subordinate to) PERVOMAYSKIY; 53°06′N 40°20′E. Elevation 475 ft/144 m. In agricultural area; poultry factory.

Ilovays'k (ee-lo-VEISK), (Russian *Ilovaysk*), city, central DONETS'K oblast, UKRAINE, in the DONBAS, 13 mi/21 km SE of MAKIYIVKA; 47°55′N 38°11′E. Elevation 675 ft/205 m. Railroad junction; railroad servicing. Metalworks; flour mill. Established in 1869 as a railroad station, city since 1938.

Ilovaysk, UKRAINE: see ILOVAYS'K.

Ilovlinskaya, RUSSIA: see ILOVLYA.

Ilovlya (ee-luh-VLYAH), town (2006 population 11,945), central VOLGOGRAD oblast, SE European Russia, on the ILOVLYA RIVER near its confluence with the DON River, on road and railroad, 45 mi/72 km NW of VOLGOGRAD; 49°18′N 43°59′E. Elevation 144 ft/43 m. Asphalt manufacturing; dairying, meat processing. Railroad junction of the Volga lateral railroad (built during World War II), connecting with the main Moscow-Volgograd line in the town's vicinity. Formerly known as Ilovlinskaya.

Ilovlya River (EE-luhv-lyah), approximately 150 mi/241 km long, in S European Russia, rises W of ZOLOTOYE (SARATOV oblast); flows SSW, past SOLODCHA, to the DON River below ILOVLINSKAYA.

Ilpela Pass (eel-PAI-lah) (1,520 ft/463 m), in the ANDES, on ARGENTINA-CHILE border, W of LAKE LACAR; 40°10′S 71°50′W.

Il'pyrskiy (EEL-pir-skeeyee), town, NE KORYAK AUTONOMOUS OKRUG, RUSSIAN FAR EAST, on Cape Il'pyr, on the BERING SEA, 150 mi/241 km NE of PALANA; 59°57′N 164°10′E. Fish cannery.

Ilsede (IL-se-de), suburb of Peine, LOWER SAXONY, N GERMANY, 4 mi/6.4 km S of city center; 52°16′N 10°15′E. Former mining and steel-producing center.

Ilsenburg (IL-sen-boorg), town, SAXONY-ANHALT, central GERMANY, at N foot of the upper HARZ, 5 mi/8 km WNW of WERNIGERODE; 51°53′N 10°40′E. Manufacturing includes steel milling (foundry established 1540), copper smelting and refining. Health resort. The first German country boarding school founded here in 1898. Town grew around castle first mentioned in 10th century and converted 1003 into Benedictine monastery. Of the monastery, the Romanesque church and buildings (converted 1609 into palace) are extant.

Ilsfeld (ILS-felt), village, BADEN-WÜRTTEMBERG, SW GERMANY, 10 mi/16 km S of HEILBRONN; 49°03′N 09°15′E.

Ilshofen (ILS-ho-fuhn), town, BADEN-WÜRTTEMBERG, S GERMANY, 9 mi/14.5 km NE of SCHWÄBISCH HALL; 49°10′N 09°55′E. Manufacturing of machinery and transformers. Has ruined 13th-century castle.

Ilsington (IL-sing-tuhn), agricultural village (2001 population 2,444), central DEVON, SW ENGLAND, 6 mi/9.7 km NW of NEWTON ABBOT, at E end of DARTMOOR; 50°34′N 04°30′W. Has 14th-century church, old thatched cottages.

Il'skiy (EEL-skeeyee), town (2005 population 22,720), S central KRASNODAR TERRITORY, S European Russia, on road and railroad, 21 mi/34 km SW of KRASNODAR; 44°50′N 38°34′E. Elevation 291 ft/88 m. Equipment for drilling oil and natural-gas wells; oil field in the vicinity.

Ilubabor, former province (□ 19,000 sq mi/49,400 sq km), SW ETHIOPIA, now divided between OROMIYA and GAMBELA states; capital was GORE; 07°45′N 35°00′E. Much of the province was situated between BARO and AKOBO rivers and bordered on SUDAN. Trade centers, GORE and BURÉ. METU is sometimes cited as the capital. Annexed 1887 by Ethiopia.

Iluka (ei-LOO-kuh), resort village, NEW SOUTH WALES, SE AUSTRALIA, 431 mi/694 km N of SYDNEY; 29°25′S 153°21′E. Holiday destination; fishing port.

Ilūkste (EE-look-stai), German *illuxt*, city, SE LATVIA, in ZEMGALE, 11 mi/18 km NW of DAUGAVPILS, near left bank of the DVINA (Daugava) River Agricultural market (rye, fodder); wool processing, flour milling.

Ilula (ee-LOO-lah), village, MWANZA region, NW TANZANIA, 55 mi/89 km SSE of MWANZA, on Moame River; 02°56′S 33°19′E. Cattle; cotton, corn, wheat; timber.

Ilumán (ee-loo-MAHN), village, IMBABURA province, N ECUADOR, in the ANDES Mountains, 9 mi/14.5 km WSW of IBARRA. In agricultural region (known as San Juan de Ilumán (cereals, fruit, coffee; livestock); weaving of native ponchos, tapestries, felt hats, embroidery. Famous for traditional healers (*curanderos*).

Ilunde (ee-LOON-dai), village, KIGOMA region, W TANZANIA, 65 mi/105 km ESE of KIGOMA, on railroad; 05°06′S 30°35′E. Cattle; rice, corn.

Ilunde, village, RUKWA region, W TANZANIA, 110 mi/177 km SE of KIGOMA; 05°51′S 30°54′E. Timber; goats, sheep; rice, corn.

Ilunde, village, RUKWA region, W central TANZANIA, 95 mi/153 km NE of SUMBAWANGA; 06°48′S 32°03′E. Timber; cattle, goats, sheep; wheat, corn.

Ilut (i-LOOT), Arab township, 2.5 mi/4 km NW of NAZARETH, in LOWER GALILEE, N ISRAEL; 32°42′N 35°15′E. Elevation 1400 ft/426 m. In Roman period was a Jewish settlement, although there are signs of habitation extending further back in time.

Ilvesheim (IL-ves-heim), village, BADEN-WÜRTTEMBERG, SW GERMANY, on the NECKAR RIVER, and 4 mi/6.4 km E of MANNHEIM; 49°29′N 08°34′E.

Ilwaco (il-WAH-ko), town (2006 population 997), PACIFIC county, SW WASHINGTON, at mouth of the COLUMBIA, 12 mi/19 km NW of ASTORIA (OREGON); 46°19′N 124°02′W. Cranberries; salmon, halibut, oysters; manufacturing (shipbuilding). Tourism, recreation. Willapa National Wildlife Refuge to NE; CAPE DISAPPOINTMENT (N side of entrance to Columbia River estuary) and Fort Canby State Park to S; Fort Columbia State Park to SE.

Ilya (il-YAH), Polish *Ilja*, village, MINSK oblast, BELARUS, on Ilya River (left tributary of Viliya River), and 19 mi/31 km NE of MOLODECHNO. Rye, oats, flax.

Ilyas Dag, (Turkish=*Ilyas Daği*) peak (1,991 ft/607 m), on MARMARA Island, NW TURKEY, in Sea of MARMARA.

Il'yatino (eel-YAH-tee-nuh), settlement, NW TVER oblast, W European Russia, near highway, 7 mi/11 km SW of BOLOGOYE, to which it is administratively subordinate; 57°49′N 33°43′E. Elevation 610 ft/185 m. Furs factory.

Ilych River (ee-LICH), approximately 200 mi/322 km long, in KOMI REPUBLIC, NE European Russia; rises in the N URAL Mountains; 62°30′N 56°30′E–63°15′N 59°00′E; flows S and W to the PECHORA River at Ust'-Ilych, 20 mi/32 km SE of TROITSKO-PECHORSK. Forms N border of the Pechora-Ilych game reserve.

Il'yinskoye, RUSSIA: see KRASNYY STEKLOVAR.

Ilyov Vrukh (EEL-yov VRUHK), peak (5,915 ft/1,803 m), in the VLAHINA MOUNTAINS, on the Makedonian-Bulgarian border, 17 mi/27 km SSW of BLAGOEVGRAD (BULGARIA); 41°46′N 23°01′E. Also called Dzhama or Dzhema.

Ilza (EE-wuh-zhah), Polish *Iłża*, Russian *Ilzha*, town, Radom province, E central POLAND, 17 mi/27 km SSE of RADOM. Flour milling, tanning, brewing; stone quarrying. Castle ruins nearby.

Ilz River (ILTS), 36 mi/58 km long, BAVARIA, GERMANY; rises in the BOHEMIAN FOREST near Czech border; flows S to the DANUBE at PASSAU; source at 48°59′N 13°30′E.

Imabari (ee-mah-BAH-ree), city (2005 population 173,983), EHIME prefecture, N SHIKOKU, S JAPAN, on the HIUCHI SEA and Kurushima Strait, 19 mi/30 km N of MATSUYAMA; 34°03′N 133°00′E. Commercial and fishing port and a manufacturing center with industries producing cotton textiles, automotive items, food products, kitchenware, lacquerware, arcade shutters; shipbuilding. Mandarin oranges; pickles.

Imabetsu (ee-mah-BETS), town, Easta Tsugaru district, Aomori prefecture, N HONSHU, N JAPAN, 28 mi/45 km N of AOMORI; 41°10′N 140°29′E.

Imaculada (EE-mah-koo-lah-dah), city (2007 population 11,445), S central PARAÍBA state, BRAZIL, on PERNAMBUCO state border, 26 mi/42 km SW of TEIXEIRA; 07°23′S 37°31′W.

Imad (ee-MAHD), village, S YEMEN, near ADEN, 5 mi/8 km NE of SHEIKH OTHMAN; 14°23′N 47°10′E. Formerly at W limit of FADHLI sultanate.

Imadate (ee-MAH-dah-te), town, Imadate district, FUKUI prefecture, central HONSHU, W central JAPAN, 9 mi/15 km S of FUKUI; 35°54′N 136°14′E. Traditional papermaking (*Echizen washi*).

Imaichi (ee-MAH-ee-chee), city, TOCHIGI prefecture, central HONSHU, N central JAPAN, 12 mi/20 km N of

UTSUNOMIYA; 36°42′N 139°42′E. Agricultural market and tourist center near NIKKO NATIONAL PARK.

Imajo (ee-MAH-jo), town, Nanjo district, FUKUI prefecture, central HONSHU, W central JAPAN, 22 mi/35 km S of FUKUI; 35°46′N 136°12′E. Persimmons.

Imakane (ee-MAH-kah-ne), town, Hiyama district, Hokkaido prefecture, N JAPAN, 81 mi/130 km S of SAPPORO; 42°25′N 140°00′E. Potatoes.

Imaklit Island, RUSSIA: see RATMANOV ISLAND.

Imam el Hamza, IRAQ: see HAMZA.

Iman, RUSSIA: see DAL′NERECHENSK.

Imandra (ee-MUHN-druh), lake (□ 338 sq mi/878.8 sq km), MURMANSK oblast, NW European Russia, on the KOLA PENINSULA, S of MURMANSK; 67°33′N 33°00′E. Average depth 43 ft/13 m. Deeply indented, it contains around 140 islands and receives about 20 large rivers. It empties into the KANDALAKSHA BAY of the WHITE SEA through the NIVA RIVER. Usually remains frozen between November and May.

Imani (ee-MAH-nee), town, KOGI state, S central NIGERIA, on road, 80 mi/129 km SE of LOKOJA; 07°17′N 07°42′E. Market town. Yams, rice, maize.

Iman-Nezer (ee-MAHN–ne-ZER), village, SE LEBAP weloyat, SE TURKMENISTAN, on Afghan border, 39 mi/63 km S of KERKI. Customs station; trade with ANDKHUI (AFGHANISTAN).

Iman River (ee-MUHN), 220 mi/354 km long, W MARITIME TERRITORY, RUSSIAN FAR EAST, SE Siberian Russia; rises in S SIKHOTE-ALIN RANGE; flows N and W to the USSURI River at DAL′NERECHENSK. Gold along its upper course. Timber floating.

Imari (EE-mah-ree), city, SAGA prefecture, NW KYUSHU, SW JAPAN, on Imari Bay, 25 mi/40 km W of SAGA; 33°15′N 129°52′E. Fishing and commercial port that produces such products as *Imari yaki* porcelain, ships, and wine. Pears, grapes. Imari Bay is the habitat and breeding ground of the king crab.

Imaruí (EE-mah-roo-EE), city (2007 population 11,675), SE SANTA CATARINA state, BRAZIL, on inlet of the ATLANTIC OCEAN, 10 mi/16 km N of LAGUNA; 28°21′S 48°49′W. Lumber. Formerly spelled Imaruhy.

Imataca, Brazo (ee-mah-TAH-kah, BRAH-so), southernmost major arm of ORINOCO RIVER delta, c.60 mi/97 km long, DELTA AMACURO state, NE VENEZUELA; branches off from the main arm, the Río GRANDE; flows E to the ATLANTIC OCEAN in the Boca GRANDE (or Boca de Navíos) at CURIAPO; 08°35′N 60°58′W. Variant names include Caño Mataca, Brazo Imataco.

Imataca, Serranía de (ee-mah-TAH-kah, ser-rah-NEE-ah dai), low mountain range in E VENEZUELA, in DELTA AMACURO and BOLÍVAR states; outlier of GUIANA HIGHLANDS, S of lower ORINOCO RIVER and N of UPATA; c.90 mi/145 km long W-E; rises to c.2,700 ft/823 m; 07°11′N 60°37′W. Rich iron deposits. EL PAO is a mining center. Also known as Sierra Imataca.

Imataca, Sierra, VENEZUELA: see IMATACA, SERRANÍA DE.

Imataco, Brazo, VENEZUELA: see IMATACA, BRAZO.

Imathia, Greece: see EMATHEIA.

Imatong Mountains (ee-MAH-tahng), SE SUDAN, near UGANDA border, SE of JUBA; rises to 10,456 ft/3,187 m in Mount KINYETI, highest mountain in SUDAN.

Imatra (I-mah-trah), falls and rapids, in the VUOKSI-JOKI (river), SE FINLAND, between lakes SAIMAA (Finland) and LADOGA (RUSSIA). The river descends 60 ft/18 m in a series of rapids c.0.5 mi/0.8 km long. The hydroelectric station there supplies power to S Finland. Tourist attraction. Wood processing.

Imazu (ee-MAHZ), town, Takashima district, SHIGA prefecture, S HONSHU, central JAPAN, on NW shore of BIWA LAKE, 28 mi/45 km N of OTSU; 35°24′N 136°01′E. Railroad terminus. Traditional candles; persimmons (also dried); buckwheat. Carp sushi. Skiing area.

Imbabura (eem-bah-BOO-rah), province (□ 1,760 sq mi/4,558 sq km; 2001 population 344,044), N ECUADOR,

in the ANDES Mountains, just N of the equator; ⊙ IBARRA. A mountainous region including many volcanic peaks (IMBABURA, COTACACHI, Yanaurcu) and several picturesque crater lakes (San Pablo, Yaguarcocha, CUICOCHA), because of which it is often called the Lake District of Ecuador. The climate is semitropical to temperate in the settled, fertile valleys, frigid on the mountaintops. Among mineral resources are manganese, iron, and salt; but livestock raising (cattle, sheep, llamas) and agriculture predominate. Main crops are cereals, potatoes, sugarcane, vegetables, cotton, fruit. The towns of Ibarra, OTAVALO, COTACACHI, and ATUNTAQUI have leather, textile, and food industries. The large native population is engaged in artisan activities, including weaving native cotton and woolen goods sold in world-famous indigenous markets.

Imbabura, Cerro (eem-bah-BOO-rah, SER-ro), extinct Andean volcano (15,026 ft/4,580 m), IMBABURA province, N ECUADOR, 7 mi/11.3 km SW of IBARRA; 00°16′N 78°10′W.

Imbaú (EEM-bah-oo), town (2007 population 11,112), central PARANÁ state, BRAZIL, 59 mi/95 km NW of PONTA GROSSA; 24°23′S 50°39′W. Coffee, cotton, corn, rice.

Imbert (eem-BERT), town, PUERTO PLATA province, N DOMINICAN REPUBLIC, 10 mi/16 km W of PUERTO PLATA; 19°45′N 70°52′W. Agricultural center (cacao, coffee, sugar.) Sugar mill nearby. Until 1925, called BAJABONICO.

Imbetiba, Brazil: see MACAÉ.

Imbituba (EEM-bee-too-bah), city (2007 population 36,169), SE SANTA CATARINA state, BRAZIL, port on the ATLANTIC OCEAN, 45 mi/72 km S of FLORIANÓPOLIS; 28°14′S 48°40′W. Railroad terminus. Ships coal mined in TUBARÃO region to steel mill at VOLTA REDONDA (RIO DE JANEIRO state).

Imbituva (EEM-bee-too-vah), city (2007 population 27,052), S central PARANÁ, BRAZIL, on road, and 28 mi/45 km WSW of PONTA GROSSA; 25°12′S 50°35′W. Woodworking, linen milling; maté, tobacco, grain, corn; cattle. Formerly Santo Antônio de Imbituva.

Imbler (IM-bluhr), village (2006 population 279), UNION county, NE OREGON, 12 mi/19 km NE of LA GRANDE, on GRANDE RONDE RIVER; 45°27′N 117°57′W. Grain, potatoes; cattle. In area are Umatilla (W) and Wallowa-Whitman (E) national forests.

Imbléville (an-blai-veel), NW residential suburb (□ 2 sq mi/5.2 sq km) of Le HAVRE, SEINE-MARITIME department, HAUTE-NORMANDIE region, N FRANCE, near Cape La Hève, on ENGLISH CHANNEL; c.49°43′N 00°57′E. Le Havre airport is just N.

Imboden (EEM-bo-duhn), district, W central GRISONS canton, SWITZERLAND. Main town is DOMAT; population is German-speaking and Roman Catholic in Kreis Rhäzüns (S of RHINE RIVER), Romansch-speaking and Protestant in Kreis Trins (N of Rhine).

Imboden (im-BO-duhn), village (2000 population 684), LAWRENCE county, NE ARKANSAS, 13 mi/21 km W of POCAHONTAS, and on SPRING RIVER; 36°12′N 91°10′W. Manufacturing (silk trees and florals).

Imbros, TURKEY: see GÖKÇEADE.

Imbroz, TURKEY: see GÖKÇEADE.

Imbuial, Brazil: see BOCAIÚVA DO SUL.

Imefout Dam, MOROCCO: see OUM ER RBIA, OUED.

Imeni Gor′kogo (EE-mee-nee GOR-kuh-vuh) [Russian=named after Gorkiy], settlement, SW VLADIMIR oblast, central European Russia, 12 mi/19 km WNW of POKROV; 56°28′N 40°59′E. Elevation 334 ft/101 m. Textile mill. Named after Maxim Gorkiy, a famous Russian writer.

Imeni H. I. Petrovs′koho: see HORODYSHCHE, CHERKASY OBLAST, UKRAINE.

Imeni Kaganovicha, RUSSIA: see MIRSKOY.

Imeni Karla Libknekhta, RUSSIA: see KARL LIBKNEKHT.

Imeni Karla Libknekhta, UKRAINE: see SOLEDAR.

Imeni Kirova, UKRAINE: see KIROVS′K, DONETS′K oblast.

Imeni Molotova, RUSSIA: see OKTYABR′SKIY, Nizhegorod oblast.

Imeni Odinnadtsati Let Oktyabrya (EE-mye-nee uh-DEE-nuh-tsah-tee LYET uhk-tyah-BRYAH) [Russian=in the name of the 11th Anniversary of October (Revolution)], former town, N CHITA oblast, SE SIBERIA, RUSSIA, 215 mi/346 km NW of SKOVORODINO; 55°54′N 119°36′E. Elevation 3,536 ft/1,077 m. A gold-mining town (1930-1942).

Imeni Pervogo Maya, RUSSIA: see PERVOYE MAYA.

Imeni Sverdlova, UKRAINE: see SVERDLOVS′K.

Imeni Tovarishcha Katayevicha, UKRAINE: see SYNEL′NYKOVE.

Imeni Voronina, RUSSIA: see SUMSKIY POSAD.

Imeni Yunykh Kommunarov, UKRAINE: see YUNOKOMUNARIVS′K.

Imeny Sverdlova, UKRAINE: see SVERDLOVS′K.

Imeny Tovarysha Katayevycha, UKRAINE: see SYNEL′NYKOVE.

Imeny Yunykh Komunariv, UKRAINE: see YUNOKOMUNARIVS′K.

Imere (ee-ME-rai), village, Équateur province, NW CONGO, along the UBANGI RIVER, and on the REPUBLIC OF CONGO border; 02°07′N 18°06′E.

Imeretian Range, Georgia: see ADZHAR-IMERETIAN RANGE.

Imerimandroso (ee-MAI-ree-mahn-DROOS), town, TOAMASINA province, E MADAGASCAR, on Lake ALAOTRA, and 30 mi/48 km NNE of Ambatodrazaka; 17°25′S 48°35′E. Market center; rice, cassava. W terminus of "smuggler's path" through rain forest from VAVATENINA. Legendary point of arrival of Merina tribe in highlands from E.

Imerina (ee-MERN) or **Emyrna**, French *Imérina* or *Emyrne*, mountainous region in heart of MADAGASCAR, ANTANANARIVO province, embracing twelve hills and centered on ANTANANARIVO; rises to 4,835 ft/1,474 m; 18°55′S 47°30′E. Once a mosaic of forest, brush, and grass, it is now chiefly agricultural, with rice, cassava, pulses, corn, and market vegetables as main crops. Cattle raising. Has primarily a historic significance as original home of Malagasy kings of Merina tribe. Formerly noted as center of slave traffic. First European explorers penetrated here in 1770. Fell to FRANCE 1895.

Imerissoq, GREENLAND: see KRONPRINSEN ISLAND.

Imeritia (ee-me-re-TEE-ah), geographic and historic region, GEORGIA, in the upper RIONI River basin; historic ⊙ KUTAISI. CHIATURA is the other main city. Agricultural region, noted for its mulberry trees and vineyards. There are also manganese deposits. The Imeritians, now numbering c. 500,000, speak a Georgian dialect and probably represent a very early branch of the Caucasians. Imeritia has been known since 1442, when the Georgian ruler Alexander I divided his kingdom into three parts among his sons; one part was Imeritia. From 1510 it was often invaded by the Turks, to whom it was forced to pay tribute. It was an independent kingdom 16th–18th century. In 1804, Russia forcibly obtained an oath of allegiance from Imeritia, which, however, continued to fight until its annexation to the Russian Empire in 1810.

Imías (ee-MEE-ahs), small town, in GUANTÁNAMO province, SE CUBA, on lee side of El Plurial range, 2 mi/3 km from CARIBBEAN SEA, on river of same name; 20°04′N 74°32′W. Small beach resort.

Imilchil (i-mil-SHEEL), Berber village, Errachidia province, Meknès-Tafilalet administrative region, MOROCCO, 75 mi/120 km SW of MIDELT, at end of Jbel AYACHI range; 32°10′N 05°40′W. High in the mountains. Goats, sheep. Annual marriage fair held by Ait Haddaidou tribe, where marriages are arranged for young members. Center for access to Plateau of the Lakes and new national park in High ATLAS mountains.

Imini, MOROCCO: see OUARZAZATE.

Imintanoute (i-mee-uhn-tah-NOOT), town, Chichaoua province, Marrakech-Tensift-Al Haouz administrative region, SW MOROCCO, near SW extremity of the High ATLAS mountains, 60 mi/97 km SW of MARRAKECH, on road; 31°10′N 08°51′W. Argan and chestnut trees. Also spelled Imi n'Tanout.

Imishli (eem-ish-LEE), town and administrative center of Imishli region, S AZERBAIJAN, on left bank of ARAS RIVER, on railroad, and 25 mi/40 km WSW of SABIRABAD; 39°55′N 48°00′E. Cotton-ginning center in cotton district; developed in late 1930s. Manufacturing (butter and cheese, railroad equipment, reinforced concrete, bricks, asphalt); hydroelectric power plant.

Imlay (IM-lai), unincorporated town, PERSHING county, W central NEVADA, 30 mi/48 km SW of WINNEMUCCA, near HUMBOLDT River (forms RYE PATCH RESERVOIR; Rye Patch State Recreation Area on its shores, to W and SW). Tungsten. Cattle, sheep, barley, alfalfa. HUMBOLDT Range to S.

Imlay City (im-LAI), town (2000 population 3,869), LAPEER county, E MICHIGAN, near source of BELLE RIVER, 11 mi/18 km ESE of LAPEER; 43°01′N 83°04′W. Agriculture (potatoes, carrots, beans grain); manufacturing (motor vehicles and transportation equipment, tool and die, plastic products, potting soil, pickles). Native American mounds nearby. Incorporated 1873.

Immanu'el (ee-MAH-noo-EL), Jewish settlement, 9 mi/15 km SW of NABLUS, in the WEST BANK. Founded in 1981, residents are primarily ultra-orthodox Jews.

Immatin, Arab village, NABLUS district, 6.2 mi/10 km W of Nablus, in the Samarian Highlands, WEST BANK; 32°11′N 35°09′E. Agriculture (various fruits). Some archaeological remains from Israelite and Byzantine eras found here.

Immendingen (IM-muhn-ding-uhn), village, BADEN-WÜRTTEMBERG, SW GERMANY, in BLACK FOREST, on the DANUBE River (which here diminishes in volume through seepage), and 5 mi/8 km SW of TUTTLINGEN; 47°57′N 08°44′E. Foundry; manufacturing of machinery. Summer resort (elevation 2,160 ft/658 m).

Immenhausen (im-muhn-HOU-suhn), town, HESSE, central GERMANY, 7 mi/11.3 km N of KASSEL; 51°26′N 09°28′E. Manufacturing of glass and precision instruments. Has remains of town wall and half-timbered town hall from 17th century; also, 15th-century church.

Immenstaad am Bodensee (IM-muhn-shtaht ahm BO-den-sai), village, BADEN-WÜRTTEMBERG, S GERMANY, on N shore of LAKE CONSTANCE, 6 mi/9.7 km W of FRIEDRICHSHAFEN; 47°40′N 09°22′E. Tourism. Airplanes.

Immenstadt (IM-muhn-shtaht), town, BAVARIA, S GERMANY, in SWABIA, in ALLGÄU ALPS, 12 mi/19 km SSW of KEMPTEN; 47°33′N 10°13′E. Railroad junction. Tourism is the principal industry; summer and winter resort (elevation 2,300 ft/701 m). Metalworking. Has early-17th-century castle and mid-17th-century town hall. Chartered c.1360. Small, picturesque Alp Lake is 1 mi/1.6 km NW.

Immingham (IM-ming-uhm), town (2001 population 11,804), North East Lincolnshire, NE ENGLAND, on HUMBER RIVER, and 7 mi/11.3 km NW of GRIMSBY; 53°36′N 00°13′W. Light industry. Has church dating from 13th century. To NE on Humber River is port of Immingham Docks, a satellite of the port of Grimsby; and (N) on Humber River is small port of South Killingholme Haven, where the Pilgrim Fathers embarked for Holland (1608).

Immokalee (i-MOK-uh-lee), town (□ 7 sq mi/18.2 sq km; 2000 population 19,763), COLLIER county, SW FLORIDA, 32 mi/51 km ESE of FORT MYERS, in the EVERGLADES; 26°25′N 81°25′W. Vegetable farming.

Immola (IM-mo-lah), village, KYMEN province, SE FINLAND, near Russian border, at E end of Lake SAIMAA, near mouth of VUOKSIJOKI (river), 25 mi/40 km

ENE of LAPPEENRANTA; 61°14′N 28°51′E. Elevation 248 ft/75 m. Cellulose milling.

Imo (EE-mo), state (□ 2,135 sq mi/5,551 sq km; 2006 population 3,934,899), SE NIGERIA; ⊙ Oweri; 05°30′N 07°10′E. Bordered N by ANAMBRA, E by ABIA, and S and W by RIVERS states. Tropical rain forest zone, with Isuochi uplands and Okigwe rolling hills. Drained by IMO, Njaba, and Nwaorie rivers. OGUTA (resort) and Abadaba lakes are here. Agriculture includes cocoa, oil palms, rubber, soybeans, sugarcane, cashews, bananas, yams, cassava, rice, cocoyams, maize, and fruits. Manufacturing (aluminum, shoes, resin and paint, clay products; breweries). Natural resources include crude oil, natural gas, limestone, lead zinc, granite, kaolin. Main centers are Owerri, Aboh Mbaise, Nekede, OKIGWI, ORLU, Egbema, Izombe, Nkalagu, Usu Mbano, and Urula.

Imogene, town (2000 population 66), FREMONT county, SW IOWA, 8 mi/12.9 km NNW of SHENANDOAH; 40°52′N 95°25′W. Livestock; grain.

Imola (EE-mo-lah), city (2001 population 64,348), BOLOGNA province, EMILIA-ROMAGNA, N central ITALY, on the AEMILIAN WAY; 44°21′N 11°42′E. Agricultural and market center, known for its ceramics; manufacturing non-metallics, fabricated metals, machinery, clothing. A Roman town (*Forum Cornelii*), it later (11th century) became a free commune. Subsequently ruled by tyrants (including the Visconti and the Sforza) until passing to the papacy in the early 16th century. Landmarks include a Gothic cathedral, several Renaissance palaces, and the "Rocca," a large fortress (14th century).

Imón (ee-MON), town, GUADALAJARA province, central SPAIN, 8 mi/12.9 km NW of SIGÜENZA; 41°10′N 02°44′W. Saltworking center. Lumbering; grain growing; sheep raising.

Imori (EE-mo-ree), town, North Takaki district, NAGASAKI prefecture, NW KYUSHU, SW JAPAN, 9 mi/15 km E of NAGASAKI, on TACHIBANA BAY; 32°46′N 130°01′E.

Imo River (EE-mo), 150 mi/241 km long, S NIGERIA, in forest belt; rises W of OKIGWI; flows S, through rich oil-palm region, to Gulf of GUINEA below OPOBO (AKWA IBOM state). The stretch shortly before the mouth sometimes is referred to as the Opobo River.

Imotski (EE-mots-kee), medieval *Emotha* or *Imotha*, town, S CROATIA, 38 mi/61 km E of SPLIT, near BOSNIA-HERZEGOVINA border, in DALMATIA; 43°26′N 17°10′E. Local trade center; manufacturing (bricks, textiles, apparel); tobacco. Roman ruins nearby. Passed 1718 from Turkish rule into Venetian Dalmatia. Imotska Jezera Nature Monument here. Older names were Imoschi or Imoski.

Imotski Jezera, [=Imotski Lakes], pair of lakes, S CROATIA, in DALMATIA, near IMOTSKI. Consists of 2 lakes: Modro [=blue] Lake, 656-ft/200-m-deep depression collecting rainwater, with water level reaching 250 ft/76 m, named for its clear blue water; and Crveno [=red] Lake, a steep-walled well 1,640 ft/500 m deep and 656 ft/200 m wide. Preserved as a nature monument. Swimming in Blue Lake.

Imotsko Polje (EE-mots-ko PO-lye) or **Imotski Plain**, plain and historical region in CROATIA (S DALMATIA) and BOSNIA and HERZEGOVINA (SW Lower Herzegovina). Karst topography. Principal town is IMOTSKI (Croatia).

Imouzzer Kandar (EE-mooz-zer KAHN-dahr), town, Sefrou province, Fes-Boulemane administrative region, MOROCCO, 25 mi/40 km S of FES; 33°44′N 05°01′W. Summer resort town in the N Middle ATLAS mountains.

Imperatriz (eem-pe-rah-trees), city (2007 population 229,629), W MARANHÃO state, BRAZIL, on RIO TOCANTINS (TOCANTINS state border), and 125 mi/201 km N of CAROLINA; 05°30′S 47°30′W. Ships manioc meal, babassu and copaiba oil, cotton. Airport.

Imperia (eem-PAI-ree-ah), province (□ 457 sq mi/ 1,188.2 sq km), LIGURIA, NW ITALY; ⊙ IMPERIA; 43°58′N 07°47′E. Comprises W Riviera di Ponente, enclosed by LIGURIAN ALPS; drained by ROYA and TAGGIA rivers. Mountainous terrain, rising to 7,218 ft/ 2,200 m in MOUNT SACCARELLO (N), covers 82% of area. Bordered W by French department of ALPES-MARITIMES. A leading province of Liguria for olive oil, flowers, and wood; economy based on agriculture (olives, grapes, flowers, citrus fruit, peaches) and tourist trade (San Remo, Bordighera) of the Riviera and alpine foothills bordering it. Industry limited to Imperia. From 1860 to 1923, called Porto Maurizio.

Imperia (eem-PE-ree-ah), city, ⊙ IMPERIA province, LIGURIA, NW ITALY, on the LIGURIAN SEA, on the ITALIAN; 43°53′N 08°03′E. Port and winter resort; small varied industrial facilities; food processing. The cathedral (1780–1832) dominates the modern city. Andrea Doria, the admiral and statesman, was born here (1468).

Imperial, county (□ 4,175 sq mi/10,855 sq km; 2006 population 160,301), S CALIFORNIA, on Mexican (BAJA CALIFORNIA NORTE state) border; ⊙ EL CENTRO; 33°02′N 115°21′W. Bordered E by COLORADO RIVER (ARIZONA state line). In COLORADO DESERT; desert ranges (low SUPERSTITION MOUNTAINS, W; CHOCOLATE MOUNTAINS, E), enclose IMPERIAL VALLEY. Drainage channels (New and Alamo rivers) carry wastewater to SALTON SEA (salt lake), to NW. Winter garden agricultural region irrigated by ALL-AMERICAN CANAL, which closely parallels international border; asparagus, broccoli, cauliflower, carrots, tomatoes, onions, melons, dates, corn, sugar beets, wheat; cotton; alfalfa, hay; cattle, sheep. Quarrying, mining (gypsum, sand, gravel). Part of PALO VERDE VALLEY in NE. Fort Yuma Indian Reservation in SE; part of Torres Martinez Indian Reservation in NW corner; parts of large ANZA-BORREGO DESERT STATE PARK in W; Salton Sea extends N into RIVERSIDE county; part of Salton Sea State Recreation Area in N, on NE shore; Salton Sea National Wildlife Refuge in S part of lake; part of Cibola National Wildlife Refuge in E; Picacho State Recreation Area in E, on Colorado River. Chief communities are BRAWLEY, CALEXICO, CALIPATRIA, and El Centro. Formed 1866; irrigation development was begun in 1900.

Imperial, city (2000 population 7,560), IMPERIAL county, S CALIFORNIA, 3 mi/4.8 km N of EL CENTRO. Oldest community in the irrigated IMPERIAL VALLEY. Headquarters of Imperial Irrigation District and seat of county fair. Stockyards; vegetables, tomatoes, melons, sugar beets, wheat, corn, alfalfa; cattle, sheep. Seat of Imperial Valley College (two-year). Founded 1902, incorporated 1904.

Imperial, unincorporated city, JEFFERSON county, E MISSOURI, manufacturing, commercial, and residential suburb 21 mi/34 km SSW of downtown ST. LOUIS, on MISSISSIPPI RIVER; 38°22′N 90°22′W. Mastodon State Park. Manufacturing (chemicals, transportation equipment).

Imperial, city (2006 population 1,849), ⊙ CHASE county, SW NEBRASKA, 32 mi/51 km S of OGALLALA, N of FRENCHMAN CREEK, in GREAT PLAINS region; 40°31′N 101°38′W. Grain, potatoes, beans, sunflower seeds, popcorn, potatoes; livestock; poultry products. Champion Mill State Historical Park and Champion Lake State Recreation Area to SW; Enders Reservoir State Recreation Area to SE. Settled c.1885.

Imperial (im-PIR-ee-uhl) town (2006 population 321), S central SASKATCHEWAN, W CANADA, 70 mi/113 km NNW of REGINA; 51°21′N 105°28′W. Wheat; livestock.

Imperial (eem-pai-ree-AHL), town (2005 population 30,009), CAÑETE province, LIMA region, W central PERU, on irrigated coastal plain, on LIMA-SAN VICENTE DE CAÑETE highway, and 5 mi/8 km ENE of San Vicente de Cañete; 13°04′S 76°21′W. Road junction. Sugarcane plantations in surrounding area.

Imperial, unincorporated town, ALLEGHENY COUNTY, W PENNSYLVANIA, suburb 13 mi/21 km W of downtown PITTSBURGH; 40°26′N 80°14′W. Manufacturing (paper products, metal coatings; light manufacturing). Agriculture (dairying) to S. Pittsburgh International Airport immediately to N.

Imperial, unincorporated village, PECOS COUNTY, extreme W TEXAS, 29 mi/47 km NNE of FORT STOCKTON; 31°16′N 102°41′W. Irrigated area of Pecos valley. Cattle, sheep, goats; cotton; vegetables, pecans.

Imperial Beach, city (2000 population 26,992), SAN DIEGO COUNTY, S CALIFORNIA; residential suburb 9 mi/14.5 km S of downtown SAN DIEGO, on the Mexican (BAJA CALIFORNIA NORTE state) border, 6 mi/9.7 km NW of TIJUANA (Mexico), on PACIFIC OCEAN, and at base of Coronado Peninsula. SAN DIEGO BAY to NE. Drained by TIJUANA and OTAY rivers. Manufacturing (cabinets). Southwesternmost city in the continental U.S. Coronado Naval Air Station and Amphibious Base to N, Imperial Beach Naval Radio Station in N, Ream Field Auxiliary Naval Air Station in S. Border Field State Park in SW; Silver Strand State Beach to N. Incorporated 1956.

Imperial Canal (eem-pai-ree-AHL), 60 mi/97 km long, NAVARRE and ZARAGOZA, NE SPAIN; starts from the EBRO below TUDELA; flows SE along right bank of river, which it reenters 12 mi/19 km SE of ZARAGOZA. Begun by Emperor Charles V (16th century) for navigation, it is now used for irrigation.

Imperial Dam, on COLORADO RIVER, on ARIZONA-CALIFORNIA state line), 15 mi/24 km NE of YUMA (Arizona); 3,475 ft/1,059 m long, including diversion structures and dike; 31 ft/9 m high; 32°52′N 114°26′W. Completed 1938 by Bureau of Reclamation; water diverted into ALL-AMERICAN CANAL to IMPERIAL VALLEY. Highly silted IMPERIAL RESERVOIR includes two backwater lakes 7 mi/11.3 km N of dam: Martinez (Arizona) and Ferguson (California) lakes.

Imperial Reservoir (□ 15 sq mi/39 sq km), SE CALIFORNIA, on COLORADO RIVER, between SE California (IMPERIAL COUNTY) and SW ARIZONA (YUMA COUNTY), in Imperial National Reserve, 9 mi/14.5 km NE of YUMA (Arizona); 32°53′N 114°28′W. Maximum capacity 160,000 acre-ft. Formed in Arizona by Imperial Diversion (also known as Imperial Dam; 85 ft/26 m high), built (1938) by the Bureau of Reclamation for irrigation. Also known as Ferguson and Martinez lakes.

Imperial River (eem-pai-ree-AHL), c.35 mi/56 km long, CAUTÍN province, ARAUCANIA region, S central CHILE; formed by union of CAUTÍN RIVER and the lesser Quepe River 3 mi/5 km SE of NUEVA IMPERIAL; flows W, past Nueva Imperial, CARAHUE, and PUERTO SAAVEDRA, to the PACIFIC OCEAN. Navigable c.20 mi/32 km for small craft. Sometimes considered to include the Cautín; total length c.135 mi/217 km.

Imperial Valley, fertile region, IMPERIAL and RIVERSIDE counties, SE CALIFORNIA, in the COLORADO DESERT, and extending S into BAJA CALIFORNIA NORTE state, NW MEXICO. At various times it has either been part of the GULF OF CALIFORNIA or part of a river system that flowed into the gulf, now having no outlet; most of the region is below sea level; its lowest point is -235 ft/-72 m at the S shore of the SALTON SEA. Receiving only c.3 in/7.6 cm of rain annually, the valley experiences extremely high temperatures (maximum temperature 115°F/46°C) and has a great daily temperature range. Having one of the longest growing seasons in the U.S. (over 300 days), the Imperial Valleay supports two crops a year with extensive irrigation, with the COLORADO RIVER being the primary source of water; it was first irrigated in 1901. Several disastrous floods on the Colorado River in 1905–1906 inundated the area; not until 1935, with the completion of HOOVER DAM, was the valley safe from floods. Approximately 1,563 sq mi/4,048 sq km have been irrigated, chiefly by the ALL-AMERICAN CANAL,

but also the Coachella and Westside Main canals. The valley is an important source of winter fruits and vegetables for the N areas of the U.S.; cotton, dates, grains, and poultry and dairy products are also important. BRAWLEY, CALEXICO, and EL CENTRO (all in California) are the main U.S. cities in the valley; COACHELLA and INDIO are at NW end of Salton Sea; MEXICALI (Mexico), also in the valley, is the center of Mexico's important cotton-growing district.

Imperoyal (im-puh-ROI-uhl), locality, S NOVA SCOTIA, E CANADA, on HALIFAX HARBOUR, near its mouth on the ATLANTIC OCEAN, 6 mi/10 km SE of HALIFAX.

Impfondo (eemp-fon-DO), town, ⊙ LIKOUALA region, NE Congo Republic, on right bank of UBANGI RIVER, and 440 mi/708 km NNE of BRAZZAVILLE; 01°38′S 18°04′E. Steamboat landing and market center for palm products, kola nuts, copal; cacao plantations. Meteorological station; airport. Formerly called Desbordesville.

Imphal (IM-pahl), district (□ 474 sq mi/1,232.4 sq km), MANIPUR state, NE INDIA; ⊙ IMPHAL. Agriculture: rice. Industry: handloom goods, handicrafts. Distinctive classical dance.

Imphal (IM-pahl), city (2001 population 250,234), ⊙ IMPHAL district and MANIPUR state, NE INDIA, in the MANIPUR RIVER valley; 24°49′N 93°57′E. Elevation 2,500 ft/762 m above sea level. Industries include weaving and manufacturing of metalware; important trade market. Until 1813, when MANIPUR was conquered by the Burmese, Imphal was the seat of the Manipuri kings. The inhabitants, of Tibeto-Burman origin, are famous for their music and dance. Seat of several colleges, a university and a technical institute. Site a major victory (1944) for British troops against the Japanese invading from Burma during World War II. Tourist center.

Imphy (an-FEE), industrial town (□ 6 sq mi/15.6 sq km; 2004 population 3,850), NIÈVRE department, BURGUNDY, central FRANCE, on right bank of LOIRE RIVER, and 6 mi/9.7 km SE of NEVERS; 46°56′N 03°15′E. Special steel manufacturing.

Impilakhti (EEM-pee-luhkh-tee), Finnish *Impilahti*, town, SW Republic of KARELIA, NW European Russia, on Lake LADOGA, on road and railroad, 14 mi/23 km ESE of SORTAVALA; 61°40′N 31°10′E. Elevation 164 ft/49 m. Fishing; timbering and sawmilling. Pegmatite minerals mining in the vicinity.

Impington (IM-ing-tuhn), agricultural village (2001 population 4,028), S CAMBRIDGESHIRE, E ENGLAND, 3 mi/4.8 km N of CAMBRIDGE; 52°15′N 00°07′E. Has 14th–15th-century church. Associated with Samuel Pepys.

Imp Mountain, NEW HAMPSHIRE: see CARTER-MORIAH RANGE.

Impora (eem-PO-rah), canton, SUD CINTI province, CHUQUISACA department, SE BOLIVIA, at a fork in the VILLAZÓN-UYUNI highway S of SAN JOSE, 44 mi/70 km W of SAN LORENZO; 21°28′S 65°21′W. Elevation 7,575 ft/2,309 m. Zinc, lead, silver and copper mining in area; limestone deposits. Agriculture (potatoes, yucca, bananas, corn, wheat, barley, oats, rye, peanuts, coffee); cattle.

Improvement District No. 12 (Jasper National Park), (□ 3,931 sq mi/10,220.6 sq km; 2001 population 49), W ALBERTA, W CANADA; 52°45′N 117°53′W. Formed 1945.

Improvement District No. 13 (Elk Island National Park), (□ 65 sq mi/169 sq km; 2001 population 27), E central ALBERTA, W CANADA, 25 mi/39 km from EDMONTON; 53°35′N 112°52′W. Established 1958.

Improvement District No. 24 (Wood Buffalo National Park), (□ 12,905 sq mi/33,553 sq km; 2001 population 369), N ALBERTA, W CANADA; 59°02′N 113°19′W.

Improvement District No. 25 (Wilmore Wilderness), (□ 1,778 sq mi/4,622.8 sq km; 2001 population 0), W ALBERTA, W CANADA; 53°40′N 119°08′W.

Improvement District No. 4 (Waterton Lakes National Park), (□ 186 sq mi/483.6 sq km; 2001 population 155), far SW ALBERTA, W CANADA; 49°05′N 113°53′W. Established 1944.

Improvement District No. 9 (Banff National Park), (□ 2,619 sq mi/6,809.4 sq km; 2001 population 1,497), SW ALBERTA, W CANADA; 51°34′N 116°04′W. Includes the populated places BANFF and LAKE LOUISE. Established 1945.

Impruneta (eem-proo-NAI-tah), town, FIRENZE province, TUSCANY, central ITALY, 6 mi/10 km S of FLORENCE; 43°41′N 11°15′E. Manufacturing includes fabricated metals, leather, clothing. Has basilica of Santa Maria dell'Impruneta (rebuilt 15th century; heavy war damage repaired). Partly destroyed by bombing (1944) in World War II.

Imroz, TURKEY: see GÖKÇEADE.

Imrun, TURKEY: see PUTURGE.

Imshil (EEM-SIL), county (□ 231 sq mi/600.6 sq km), S central NORTH CHOLLA province, SOUTH KOREA. Bordered N by WANJU and CHINAN, E by CHANGSU and NAMWON, S by SUNCHANG, and W by CHONGUP counties. Noryong Mountains; deep valley in SE. Agriculture (rice, barley, vegetables, tobacco; dairy products); sericulture. Cholla railroad; state roads. Sasondae (up SOMJIN RIVER) and Okchong Lake (formed by Somjin multipurpose dam) are resorts.

Imst, town, TYROL, W AUSTRIA, near the INN River, and 31 mi/50 km W of INNSBRUCK; 47°15′N 10°44′E. Road junction. Lumber mills; manufacturing of metals and plastics; hydropower station. Remarkable Gothic church. Old mining center. In 1949, the first SOS village was founded here by Hermann Gmeiner for orphans. Procession of Ghosts Carnival held here every five years.

Imués (eem-WAIS), town, ⊙ Imués municipio, NARIÑO department, SW COLOMBIA, in the ANDES Mountains, 36 mi/58 km SW of PASTO; 01°03′N 77°30′W. Elevation 8,008 ft/2,440 m. Wheat, sugarcane; livestock.

Imuris (ee-MOO-rees), town, SONORA, NW MEXICO, on MAGDALENA RIVER (irrigation), on railroad, and 40 mi/64 km S of NOGALES at intersection of Mexico Highways 2 and 15; 30°48′N 110°52′W. Wheat, corn, cotton, alfalfa, sugarcane.

Imus (EE-moos), town (2000 population 195,482), ⊙ CAVITE province, S LUZON, PHILIPPINES, 12 mi/19 km SSW of MANILA, near BACOOR BAY; 14°24′N 120°56′E. Agricultural center (rice, fruit, coconuts).

'Imwas, WEST BANK: see EMMAUS.

In, RUSSIA: see SMIDOVICH.

Ina (EE-nah), city, NAGANO prefecture, central HONSHU, central JAPAN, on the TENRYU RIVER, 56 mi/90 km S of NAGANO; 35°50′N 137°57′E. Agricultural and industrial center, producing microscopes and electronic products (resistors, condensers).

Ina (EE-nah), town, Tsukuba district, IBARAKI prefecture, central HONSHU, E central JAPAN, 37 mi/60 km S of MITO; 35°57′N 140°02′E. Spring onions.

Ina, town, North Adachi district, SAITAMA prefecture, E central HONSHU, E central JAPAN, 9 mi/15 km N of URAWA; 35°59′N 139°37′E. Pears, grapes.

Ina (EE-nah), village, South Aidzu district, FUKUSHIMA prefecture, N central HONSHU, NE JAPAN, 65 mi/105 km S of FUKUSHIMA city; 37°10′N 139°31′E. Gentian; tomatoes, daikon. Ayu.

Ina (EI-nah), village (2000 population 2,455), JEFFERSON county, S ILLINOIS, 12 mi/19 km S of MOUNT VERNON; 38°08′N 88°54′W. Agriculture. Big Muddy River Correctional Center and REND LAKE nearby.

Inaba, former province in SW HONSHU, JAPAN; now part of TOTTORI prefecture.

Inabe (EE-nah-be), town, Inabe district, MIE prefecture, S HONSHU, central JAPAN, 28 mi/45 km N of TSU; 35°06′N 136°33′E. Formed in early 1940s by combining former villages of Kasada, Oizumihara, and Oizumi.

Inabu (ee-NAHB), town, North Shitara district, AICHI prefecture, S central HONSHU, central JAPAN, 34 mi/55 km E of NAGOYA; 35°12′N 137°30′E. Lumber.

Inaccessible Island, uninhabited rocky islet, in S ATLANTIC, 12 mi/19 km SW of TRISTAN DA CUNHA, with which it is a dependency (since 1938) of ST. HELENA; 2 mi/3.2 km long, 0.75 mi/1.2 km wide; 37°17′N 12°40′W.

Inaccessible Islands, westernmost of SOUTH ORKNEY ISLANDS (UNITED KINGDOM), in the South ATLANTIC, 27 mi/43 km W of CORONATION ISLAND; 60°35′S 46°43′W. Steep pinnacle rocks (400 ft/122 m–700 ft/213 m). Discovered in 1821.

Inachos River, Greece: see INAKHOS RIVER.

Inaciolândia (ee-NAH-see-o-LAHN-zhee-ah), town (2007 population 5,627), S central GOIÁS, BRAZIL, 62 mi/100 km W of ITUMBIARA; 18°25′S 50°00′W.

Inácio Martins (ee-NAH-see-o MAHR-cheens), city (2007 population 11,098), S PARANÁ state, BRAZIL, 66 mi/106 km SW of PONTA GROSSA; 25°31′S 51°08′W. Corn, wheat; livestock.

Inácio Uchôa, Brazil: see UCHÔA.

Inagaki (ee-nah-GAH-kee), village, West Tsugaru district, Aomori prefecture, N HONSHU, N JAPAN, 19 mi/30 km E of AOMORI; 40°52′N 140°23′E. Rice straw processing.

Inagawa (ee-NAH-gah-wah), town, Kawabe district, HYOGO prefecture, S HONSHU, W central JAPAN, 17 mi/28 km N of KOBE; 34°53′N 135°22′E. *Tada ginzan* (silver ore).

Inagi (EE-nah-gee), city, TOKYO municipality, E central HONSHU, E central JAPAN, 31 mi/50 km SW of SHINJUKU; 35°38′N 139°30′E. Computers.

Inagua, island group of the BAHAMAS (2000 population 969). A virtually isolated cluster at the S end of the archipelago; includes GREAT INAGUA ISLAND, LITTLE INAGUA ISLAND, and some islets; 21°22′N 73°24′W. MATTHEW TOWN is the chief settlement. Salt production is the primary economic activity. Also known for its flamingos.

Inajá (EE-nah-zhah), city (2007 population 15,227), S PERNAMBUCO state, BRAZIL, on ALAGOAS state border; 08°54′S 37°49′W.

Inajaroba, Brazil: see SANTA LUZIA DO ITANHI.

Inakadate (ee-NAH-kah-DAH-te), village, South Tsugaru district, Aomori prefecture, N HONSHU, N JAPAN, 16 mi/25 km S of AOMORI; 40°37′N 140°33′E. Rice.

Inakawa (ee-NAH-kah-wah), town, Ogachi district, Akita prefecture, N HONSHU, NE JAPAN, 47 mi/75 km S of AKITA city; 39°08′N 140°34′E. Livestock (cattle); apples. Lacquerware, Buddhist altars, barrels, noodles; traditional *kokeshi* dolls. Skiing area.

Inakhos River (ee-nah-KHOS), 15 mi/24 km long, in ARGOLIS prefecture, NE PELOPONNESE department, S mainland GREECE; rises in ARTEMISION Mountains; flows SSE to Gulf of ARGOLIS 4 mi/6.4 km W of NÁVPLION. Also Inachos River

Inambari (ee-nahm-BAH-ree), village, TAMBOPATA province, MADRE DE DIOS region, SE PERU; 12°42′S 69°43′W. Landing at junction of INAMBARI and MADRE DE DIOS rivers, 38 mi/61 km W of PUERTO MALDONADO. In oil, lumbering, rubber region.

Inambari River (ee-nahm-BAH-ree), 210 mi/338 km long, SE PERU; rises in several branches in the Nudo de Apolobamba of the ANDES Mountains near SANDIA; flows NE and NW, past SANTO DOMINGO and PUERTO LEGUÍA, to MADRE DE DIOS RIVER at INAMBARI; 12°41′S 69°44′W.

Inami (ee-NAH-mee), town, Kako district, HYOGO prefecture, S HONSHU, W central JAPAN, 16 mi/25 km W of KOBE; 34°44′N 134°54′E.

Inami (EE-nah-mee), town, East Tonami district, TOYAMA prefecture, central HONSHU, central JAPAN, 16 mi/25 km S of TOYAMA; 36°33′N 136°58′E. Woodcarving.

Inami, town, Hidaka district, WAKAYAMA prefecture, S HONSHU, W central JAPAN, on PHILIPPINE SEA, on SW

coast of KII PENINSULA, 29 mi/46 km S of WAKAYAMA; 33°48′N 135°13′E. Field peas.

Inan (EE-nahn), town, Inan district, in the center of MIE prefecture, S HONSHU, central JAPAN, 19 mi/30 km S of TSU; 34°26′N 136°23′E.

Inangahua Junction, township, NEW ZEALAND, at confluence of W-flowing BULLER RIVER with Inangahua tributary river from S. A railroad and road connecting point. Sometimes called simply Inangahua.

Inango, town (2007 population 6,701), OROMIYA state, ETHIOPIA, near GIMBI; 09°10′N 35°41′E.

Inaouene, Oued (ee-nah-WAIN, wahd), stream, c.75 mi/121 km long, in N MOROCCO; rises on S slope of RIF mountains N of TAZA; flows W to the OUED SEBOU river 13 mi/21 km N of FES. Idriss, first dam and hydroelectric plant, near Fes, built 1973, irrigates region.

Iñapari (een-yah-PAH-ree), town, ⊙ TAHUAMANU province, MADRE DE DIOS region, SE PERU, landing on ACRE RIVER, at border of BRAZIL and BOLIVIA, opposite BOLPEBRA (Bolivia), and 55 mi/89 km W of COBIJA (Bolivia); 12°03′S 69°24′W. In oil, lumbering, rubber region. Airport.

Inarajan (in-uhr-AH-juhn), town and municipality, S GUAM, on coast. Watermelons; livestock. Site of agricultural experiment station.

Inari (I-nah-ree), Swedish *Enare*, village, LAPIN province, N FINLAND, at SW end of Lake INARI, 90 mi/145 km SW of KIRKENES; 68°54′N 27°01′E. Elevation 495 ft/150 m. Lapp trading center; fishing. On ROVANIEMI-PECHENGA branch of Arctic Highway. After World War II, Lapps evacuated from Pechenga region to resettle here. Site of Lapp cultural museum and folk high school.

Inari (I-nah-ree), Swedish *Enare*, lake (□ 500 sq mi/1,300 sq km), N FINLAND. Fed by the Ivalojoki (river) and empties into the ARCTIC OCEAN through the Paatsjoki (lake). Contains over 3,000 islands and is a tourist attraction.

Ina River (EE-nah), German *Ihna*, 70 mi/113 km long, in NW POLAND; rises 13 mi/21 km N of RECZ; flows S past Recz, WNW, NNW past STARGARD SZCZECINSKI, and GOLENIOW, and W to N end of DAMM LAKE opposite POLICE. Formerly in POMERANIA (under German administration) until 1945, when the area it waters passed to Poland.

Inaruwa (e-nuhr-WAH) town (2001 population 23,200), ⊙ SUNSARI district, E NEPAL; 26°34′N 87°09′E.

Inasa (EE-nah-sah), town, Inasa district, SHIZUOKA prefecture, central HONSHU, E central JAPAN, 43 mi/70 km S of SHIZUOKA; 34°49′N 137°40′E. Flowers.

Inatsuki (ee-NAHTS-kee), town, Kaho district, FUKUOKA prefecture, N KYUSHU, SW JAPAN, 19 mi/30 km E of FUKUOKA; 33°35′N 130°43′E.

Inawashiro (ee-nah-WAH-shee-ro), town, Yama district, FUKUSHIMA prefecture, N central HONSHU, NE JAPAN, near LAKE INAWASHIRO and MT. BANDAI, 25 mi/40 km S of FUKUSHIMA city; 37°33′N 140°06′E. Skiing area. Bacteriologist Noguchi Hideo born here, 1876.

Inawashiro, Lake (ee-nah-WAH-shee-ro), Japanese *Inawashiro-ko* (ee-nah-WAH-shee-ro–KO) (□ 40 sq mi/104 sq km), N central HONSHU, N JAPAN, 4 mi/6.4 km E of WAKAMATSU; roughly circular, 7 mi/11.3 km in diameter.

'Inaza, JORDAN: see 'UNEIZA.

'Inazah, JORDAN: see 'UNEIZA.

Inazawa (ee-NAH-zah-wah), city, AICHI prefecture, S central HONSHU, central JAPAN, 9 mi/15 km N of NAGOYA; 35°14′N 136°46′E. Residential and industrial suburb of Nagoya. Manufacturing includes elevators, television tubes. Also horticulture (garden plants, saplings, bonsai plants).

Inba (EEN-bah), village, Inba district, CHIBA prefecture, E central HONSHU, E central JAPAN, 9 mi/15 km N of Chiba; 35°46′N 140°12′E.

Inca (EEN-kah), former planning region (□ 68,715 sq mi/178,659 sq km), SE PERU. Consisted of Apurímac,

Cusco, and Madre de Dios departments (today APURÍMAC, CUSCO, and MADRE DE DIOS regions) and their twenty-three provinces. Agriculture (corn, barley, quinine, yucca, rice). Mineral resources are copper, iron, silver, gold. Meat and meat products (fowl, sheep, pork, beef); dairying; wool. Created as part of Peru's 1988 regionalization program. These regions never caught on and were abandoned.

Inca (EENG-kah), city, MAJORCA, BALEARIC ISLANDS, SPAIN, 17 mi/27 km NE of PALMA (linked by railroad and highway); 39°43′N 02°54′E. Fertile inland region (cereals, almonds, grapes, olives, figs, carob beans; livestock). Leather, shoe, and textile manufacturing, food processing (oil, wine, pastries, cookies, sausages). Venerable city with magnificent Santa María la Mayor church (begun in 13th century).

Inca de Oro (EEN-kah de O-ro), town, ATACAMA region, N CHILE, on railroad, and 45 mi/72 km NNE of COPIAPÓ; 26°45′S 69°54′W. Copper and gold mining.

Incahuara de Ckullu Kuchu (een-kah-HWAH-rah dai KOO-loo KOO-choo), canton, NOR YUNGAS province, LA PAZ department, W BOLIVIA; 15°23′S 67°41′W. Elevation 5,676 ft/1,730 m. Copper mining in area; phosphate deposits at CARANAVI. Agriculture (potatoes, yucca, bananas, rye, soy, tobacco, coffee, tea, citrus fruit); cattle.

Incahuasi (een-kah-HWAH-see), canton, NOR CINTI province, CHUQUISACA department, SE BOLIVIA; 20°35′S 64°59′W. Elevation 7,894 ft/2,406 m. Gas resources to E; limestone deposits. Agriculture (potatoes, yucca, bananas, corn, wheat, oats, rye, peanuts); cattle.

Incahuasi, Cerro (een-kah-HWAH-see, SER-ro), volcano and mountain peak (21,720 ft/6,621 m), on Argentina-Chile border, 90 mi/145 km SW of ANTOFAGASTA (Chile); 27°02′S 68°17′W. Also spelled Inca Huasi.

Inca, Lake (EEN-kah), in the ANDES Mountains, SAN FELIPE DE ACONCAGUA region, VALPARAISO region, central CHILE, just N of PORTILLO, near ARGENTINA border; 3 mi/5 km long. Tourists.

Inca, Paso del (EEN-kah, PAH-so del) (15,520 ft/4,730 m), pass in the ANDES, on ARGENTINA-CHILE border, on road between SAN JUAN (Argentina) and TRÁNSITO (Chile); 28°40′S 69°48′W.

Inca Road, the highway system of the Incan Empire, in what are now ARGENTINA, BOLIVIA, ECUADOR, and PERU, consisting of two main N-S roads, one coastal and one inland through the ANDES mountains, with connecting transverse roads and connection to every village in the empire. The coastal road ran from TUMBES to AREQUIPA (Peru), with a little-used extension into what is now N Chile. The highland road ran from QUITO (Ecuador) S to CUZCO and Ayavire, then to TUCUMAN (Argentina) and on to SANTIAGO (Chile). The Inca had no wheeled vehicles, using human and pack animal transport, so roads could be as narrow as 3 ft/1 m and use steps on inclines. Important roads were paved with stone slabs. Extensive use of bridges and bridging techniques. Accommodations provided at regular intervals. Most transport by large trains of men and animals, but messages sent by a very efficient twenty-four-hour relay runner system that could approach a steady rate of 10 mph/16 kph. Roads maintained and the relay system manned by a labor tax on the population. Had no known formal name.

Inca Trail, a walking route, traversing the mountains above URUBAMBA RIVER; CUZCO, S PERU. It was formerly a roadway used for travel to the ceremonial center of the Incas (today the ruins of MACHU PICCHU).

Ince, Cape (in-JE), (Turkish=*Inceburun*) (in-je-boo-ROON), on BLACK SEA, N TURKEY, 12 mi/19 km WNW of SINOP, northernmost point of ANATOLIA; 42°06′N 34°58′E. Sometimes spelled Inje and Indje.

Ince-in-Makerfield (INS–in–MAIK-uhr-FEELD), town (2001 population 10,184), GREATER MANCHESTER, central ENGLAND, on Leeds and Liverpool

Canal, just SSE of WIGAN; 53°32′N 02°37′W. Machinery, electric equipment, and glass bottle manufacturing.

Incesu, village, central TURKEY, on railroad, 19 mi/31 km WSW of KAYSERI; 38°39′N 35°12′E. Wheat, rye, barley.

Inch, formerly a large island (4 mi/6.4 km long, 3 mi/4.8 km wide) in SE part of LOUGH SWILLY, E DONEGAL county, N IRELAND; 7 mi/11.3 km NW of LONDONDERRY; 55°04′N 07°24′W. Now joined to the mainland as a result of drainage work.

Inchagoill (inch-uh-GOIL), island in LOUGH CORRIB, NW GALWAY county, W IRELAND, 4 mi/6.4 km SW of CONG; 53°29′N 09°19′W. Site of two ancient church ruins.

Inchasi (een-CHAH-see), canton, JOSÉ MARÍA LINARES province, POTOSÍ department, W central BOLIVIA; 19°49′S 66°31′W. Elevation 10,748 ft/3,276 m. Phosphate deposits at Miculpaya and small gypsum deposits in area. Agriculture (potatoes, yucca, bananas, wheat); cattle.

Inchcape Rock, Scotland: see BELL ROCK.

Inchcolm (INCH-kolm), Gaelic *Innis Choluim* [=island of Columba], island, in Firth of Forth, off coast of FIFE, E Scotland, 2 mi/3.2 km S of ABERDOUR; 56°01′N 03°18′W. Has noted ruins of abbey of St. Columba, founded 1123. Just SE, in Firth of Forth, is lighthouse.

Inchelium (in-chuh-LEE-uhm), unincorporated village (2000 population 389), FERRY county, NE WASHINGTON, 40 mi/64 km NE of COULEE DAM, on COLUMBIA River (FRANKLIN D. ROOSEVELT LAKE reservoir; Coulee Dam National Recreation Area follows shore), in E part of Colville Indian Reservation; 48°20′N 118°15′W. Gifford Ferry, Columbia River, 4 mi/6.4 km to SE. Manufacturing (wood preserving).

Inchgarvie (INCH-gahr-vee), islet in the Firth of Forth, EDINBURGH, E Scotland, just NNW of QUEENSFERRY; 56°00′N 03°25′W. It is a support of the Forth Bridge. Lighthouse.

Inchicore, IRELAND: see DUBLIN, city.

Inchinnan (in-SHI-nuhn), town (2001 population 1,574), Renfrewshire, S central Scotland, on WHITE CART WATER (site of drawbridge), and just WNW of RENFREW; 55°53′N 04°25′W. Previously motor vehicle tire manufacturing.

Inchiri, administrative region (2000 population 11,500), W MAURITANIA; ⊙ AKJOUJT. Bordered by Dakhlet Nouâdhibou (W and N), Adrar (E), and Trarza (S) administrative regions and ATLANTIC OCEAN (SW tip). Khatt et Toueïrja watercourse in S; part of Sebkhet Te-n-Ioubrar salt marsh in SW (extends into Trarza administrative region). Part of AKCHÂR dunes here (extend into Adrar administrative region); Tijirit geographic region in N central; part of Azzeffâl hills in N (extend N into WESTERN SAHARA). Part of Banc d'Arguin National Park extends into SW (from Dakhlet Nouâdhibou administrative region). Main road between NOUAKCHOTT and ATAR runs SW-NE across SE of region, through Akjoujt. Airport at Akjoujt.

Inchkeith (INCH-keeth), fortified island (0.6 mi/1 km long) in Firth of Forth, off S coast of FIFE, E Scotland, 3 mi/4.8 km SE of KINGHORN; 56°02′N 03°08′W. A monastery was established here c.700. At N end is lighthouse built 1803.

Inchmahome, Scotland: see LAKE OF MENTEITH.

Inchmarnock (inch-MAHR-nuhk), island (2.5 mi/4 km long, 1 mi/1.6 km wide), Argyll and Bute, W Scotland, in the Sound of Bute, just off W coast of BUTE island; 55°47′N 05°09′W.

Inchnadamph, Scotland: see LOCH ASSYNT.

Inchon (IN-CHUHN), city (2005 population 2,531,280), KYONGGI province, SOUTH KOREA, on the YELLOW SEA; 37°28′N 126°38′E. Korea's fourth-largest city, it has the special status of province as well. The country's second-largest port, Inchon has an ice-free harbor (protected by a tidal basin) and is the port and commercial center for SEOUL. Inchon's economy was heavily dependent on shipping and the transshipment of goods. Since the mid-1960s, the city has experienced rapid urbanization, and it is one of the most industrialized cities in Korea. It will be the site of the new Supersonic Transport Airport for NE ASIA, as well as the terminus of the new W coast highway of South Korea. One of South Korea's major industrial centers; iron, steel, coke, light metals, plate-glass, textiles, chemicals, and lumber are among its manufacturing. Fishing is also an important industry. Large salt fields have been developed in the tidal flats offshore. Increasing urbanization and subway and expressway links with Seoul have made Inchon and Seoul into one large urban region. Opened to foreign trade in 1883. It was called Jinsen by the Japanese, who ruled Korea 1905–1945. During the Korean War, U.S. troops landed here (Sept. 15, 1950) to relieve pressure on the PUSAN perimeter and to launch the subsequent UN drive N. Seat of several universities, including Inha University. Formerly called CHEMULPO.

Incio (EEN-thyo), village, LUGO province, NW SPAIN, 12 mi/19 km NE of MONFORTE. In iron-mining area.

Incirli, TURKEY: see KARASU.

Incline Village, unincorporated town (1999 population 7,119), WASHOE county, W NEVADA, 11 mi/18 km WNW of CARSON CITY, 2 mi/3.2 km E of CALIFORNIA state line, on CRYSTAL BAY, NE end of Lake TAHOE, in SIERRA NEVADA; 39°16′N 119°57′W. Gambling resort; tourism; manufacturing (wood products). Toiyabe National Forest to N; Lake Tahoe Nevada State Park to SE. DIAMOND PEAK, Mount Rose, and Slide Mountain ski areas to NE. Ponderosa Ranch, site of "Bonanza" TV series, is here.

Inconfidencia, Brazil: see CORAÇÃO DE JESUS.

Incoronata, CROATIA: see KORNAT ISLAND.

Incourt (eng-KOOR), commune (2006 population 4,586), Nivelles district, BRABANT province, central BELGIUM, 14 mi/23 km SSE of LEUVEN; 50°42′N 04°47′E.

Inčukalns (IN-choo-kuhlns), city, 25 mi/40 km NE of RĪGA, on railroad to TALLINN (ESTONIA); 57°06′N 24°41′E. Site of large underground gas storage facility.

Indabaguna, ETHIOPIA: see ENDABAGUNA.

Indaial (EEN-dei-ahl), city (2007 population 47,686), NE SANTA CATARINA state, BRAZIL, on ITAJAÍ AÇU RIVER, on railroad, and 10 mi/16 km W of BLUMENAU; 28°22′S 48°54′W. Agricultural colony. Formerly spelled Indayal.

Indaiatuba (EEN-dei-ah-too-bah), city (2007 population 173,508), SE central SÃO PAULO state, BRAZIL, on railroad, and 16 mi/26 km SW of CAMPINAS; 23°05′S 47°14′W. Woodworking; sugar, cotton, coffee.

Indalsälven (IN-DAHLS-ELV-en), river, 260 mi/418 km long, N central SWEDEN; rises in Norwegian border mountains; flows in winding course generally E, over the TÄNNFORSEN, through STORSJÖN, to KRÅNGEDE (falls; power station); thence flows SE, past HAMMARSTRAND, RAGUNDA, and BISPGÅRDEN, to GULF OF BOTHNIA 10 mi/16 km NE of SUNDSVALL.

Inda Medhanī Alem (IN-dah med-HAH-nee AH-lem), village, TIGRAY state, N ETHIOPIA, on the road between ADDIS ABABA and MEK'ELĒ, and 34 mi/55 km S of MEK'ELĒ; 13°04′N 39°31′E. Cereals; livestock. Formerly Enda Medani Alem.

Indang (een-DAHNG), town, CAVITE province, S LUZON, PHILIPPINES, 29 mi/47 km SSW of MANILA; 14°12′N 120°53′E. Agricultural center (rice, fruit, coconuts).

Indanza (een-DAHN-zah), village, MORONA-SANTIAGO province, SE ECUADOR, on E slopes of the ANDES Mountains, 40 mi/64 km SE of CUENCA; 03°04′S 78°30′W. Livestock; fruit; timber.

Indaparapeo (een-dah-pah-RAH-pee-o), town, MICHOACÁN, central MEXICO, 15 mi/24 km NE of MORELIA; 19°47′N 100°58′W. Cereals, vegetables, fruit; livestock.

Indapur (in-DAH-poor), town, PUNE district, MAHARASHTRA state, W INDIA, 80 mi/129 km SE of PUNE. Road junction; market center (sugarcane, millet); handicraft cloth weaving.

Indargarh (IN-duhr-guhr), town, BUNDI district, RAJASTHAN state, NW INDIA, 45 mi/72 km NNE of KOTA. Wheat, barley; handicraft pottery and lacquered woodwork.

Inda Sillase, ETHIOPIA: see ENDASELASIE.

Indaw (IN-daw), township, SAGAING division, MYANMAR, on railroad, and 12 mi/19 km W of KATHA.

Indaw (IN-daw), village, INDAW township, SAGAING division, MYANMAR, 25 mi/40 km E of MAWLAIK. Oil field (opened 1918).

Indawgyi Lake (in-dou-JEE), largest lake (□ 80 sq mi/208 sq km) in MYANMAR. Located in KACHIN STATE, 65 mi/105 km WSW of MYITKYINA. Surrounded by thickly wooded hills; theme of many Burmese legends.

Indé (en-DAI), town, DURANGO, N MEXICO, 60 mi/97 km NNE of SANTIAGO PAPASQUIARO; 25°53′N 105°10′W. Elevation 6,102 ft/1,860 m. Silver, gold, lead, copper deposits, mined in the past.

Indefatigable Island, ECUADOR: see SANTA CRUZ ISLAND.

Inden (IN-den), village, NORTH RHINE–WESTPHALIA, W GERMANY, 13 mi/21 km NE of AACHEN; 50°52′N 06°22′E. Lignite mining.

Indented Head (in-DEN-tuhd HED), township, VICTORIA, SE AUSTRALIA, 27 mi/43 km SSW of MELBOURNE across bay, at end of Bellarine Peninsula, overlooking CORIO BAY, PORT PHILLIP BAY; 38°08′S 144°43′E.

Independence, county (□ 771 sq mi/2,004.6 sq km; 2006 population 34,909), NE central ARKANSAS; ⊙ BATESVILLE; 35°44′N 91°34′W. Bounded E by BLACK RIVER; drained by WHITE RIVER and Departee Creek. Agriculture (rice, soybeans, wheat; cattle, hogs). Manufacturing at Batesville; black marble and limestone quarries; timber. Part of OZARK Mountains in W. Founded 1820.

Independence, city (2000 population 6,014), ⊙ BUCHANAN county, E IOWA, on WAPSIPINICON RIVER, and 23 mi/37 km E of WATERLOO; 42°28′N 91°53′W. Manufacturing (sheet-metal fabrication; corn milling; dressed poultry, feed, dairy and metal products). Limestone quarries, sand and gravel pits nearby. Mental health institute W of town (1873).

Independence, city (2000 population 9,846), ⊙ MONTGOMERY county, SE KANSAS, on the VERDIGRIS RIVER, near the OKLAHOMA state line; 37°13′N 95°42′W. In an important oil-producing area where corn and wheat are also grown. Manufacturing (machinery, transportation equipment, fabricated metal products; natural-gas distribution. Founded (1869) on a former Osage reservation. Boomed with the discovery of natural gas in 1881 and oil in 1903. Seat of Independence Community College. Hometown of playwright William Inge and oil magnate Harry Sinclair.

Independence, city (2000 population 14,982), ⊙ KENTON county, N KENTUCKY, a residential suburb 11 mi/18 km S of CINCINNATI, OHIO, and 9 mi/14.5 km S of COVINGTON; 38°57′N 84°32′W. Agricultural area (burley tobacco; cattle; dairying); manufacturing (sheetmetal fabricating; machine tools).

Independence, city (2000 population 113,288), ⊙ JACKSON county, W MISSOURI, suburb 7 mi/11.3 km E of downtown KANSAS CITY; 39°05′N 94°20′W. Considered by residents as sister city to Kansas City (predates Kansas City). Agriculture to E (soybeans, corn, sorghum; dairying); manufacturing (machinery, building materials, apparel, foods, paper products, ordnance; printing). Natural gas in the area contributes to the city's industries and economy. In the 1830s and 1840s, Independence was the starting point for expeditions over the SANTA FE, OREGON, and California trails. A group of Mormons settled here in 1831 after the Mormon leader Joseph Smith declared

Jackson county as the original Garden of Eden. World headquarters of the Reorganized Church of Jesus Christ of Latter Day Saints (a break away from the Mormon Church) and a new temple was built in 1994. A major museum of the Church of Jesus Christ of Latter Day Saints (based in UTAH) is also located here. Home of President Harry S. Truman (preserved as National Historical Site) and seat of the Harry S. Truman Library and Museum, on whose grounds the former president is buried. Other points of interest include the old county jail and museum (1859; restored); the old county courthouse (1825; restored); and nearby Fort Osage on the MISSOURI RIVER (1808; reconstructed). Central Missouri State College has a residence center here. Incorporated 1849.

Independence (IN-duh-pen-duhns), city (□ 10 sq mi/ 26 sq km; 2006 population 6,789), CUYAHOGA county, NE OHIO; S suburb of CLEVELAND, on State Highway 21; 41°23′N 81°38′W. Chemicals manufacturing headquarters.

Independence, town (2000 population 1,724), TANGIPAHOA parish, SE LOUISIANA, 13 mi/21 km N of PONCHATOULA, on TANGIPAHOA RIVER; 30°39′N 90°31′W. Strawberries, vegetables; catfish, crawfish; manufacturing (furniture, chemicals). Incorporated 1903.

Independence, town (2000 population 3,236), HENNEPIN county, E MINNESOTA, residential suburb 19 mi/31 km W of downtown MINNEAPOLIS; 45°01′N 93°42′W. Lake Independence on E border, Sarah Lake on N border, in NE; Lake MINNETONKA to SE. Morris T. Baker Park Reserve to E.

Independence, town (2006 population 8,764), POLK county, NW OREGON, on WILLAMETTE RIVER, 8 mi/ 12.9 km SE of DALLAS; 44°51′N 123°11′W. Railroad junction. Agriculture (grapes, hops, corn, beans, grain; poultry, hogs, sheep, cattle); dairy products; nurseries. Ankeny National Wildlife Refuge to SE; Helnick State Park to SW. Incorporated 1874.

Independence (in-duh-PEN-dens), town (2006 population 909), ⊙ GRAYSON county, SW VIRGINIA, in the BLUE RIDGE, 13 mi/21 km WSW of GALAX, near NORTH CAROLINA state line; 36°37′N 81°09′W. Manufacturing (machinery, clothing, bottled water, consumer goods, fabricated metal products, electronic equipment; printing and publishing); agriculture (tobacco, corn; livestock; dairying). Point Lookout Mountain (4,554 ft/1,388 m) to N.

Independence, town (2006 population 1,244), TREMPEALEAU county, W WISCONSIN, on TREMPEALEAU RIVER, and 23 mi/37 km NE of WINONA (MINNESOTA); 44°21′N 91°25′W. Dairy and farm area (grain); dairy products; manufacturing (wood products, concrete products). Settled 1856; incorporated as village in 1876, as city in 1942.

Independence, unincorporated village (2000 population 574), ⊙ INYO county, E CALIFORNIA, in Owens Valley, 35 mi/56 km S of BISHOP; 36°48′N 118°12′W. Elevation 3,930 ft/1,198 m. Mining (gold, salt, lead, mercury); livestock raising (cattle). Winter sports center. Eastern California Museum. MOUNT WHITNEY to S. Large fish hatchery to NW. Old Fort Independence nearby. Fort Independence Indian Reservation to N. Nearby are KINGS CANYON (W) and SEQUOIA (SW) national parks; part of Inyo National Forest to E and W. Owens River to E, Los Angeles Aqueduct passes to E (branches form OWENS RIVER 15 mi/24 km to N).

Independence, village, TATE county, NW MISSISSIPPI, 28 mi/45 km SSE of MEMPHIS (TENNESSEE). Agriculture (cotton, grain; cattle).

Independence Fjord, inlet of GREENLAND SEA, NE GREENLAND, 80 mi/129 km long, 8 mi/12.9 km–15 mi/ 24 km wide; 81°40′–82°16′N 21°50′–33°50′W. Extends W to edge of inland ice, where it receives large ACADEMY GLACIER. Forms SE border of PEARY LAND.

Independence, Fort, Massachusetts: see CASTLE ISLAND.

Independence Island, KIRIBATI: see MALDEN ISLAND.

Independence, Lake, MARQUETTE county, NW UPPER PENINSULA, N MICHIGAN, 22 mi/35 km NW of MARQUETTE, between HURON MOUNTAINS (W and SW) and LAKE SUPERIOR (1 mi/1.6 km away); c.2 mi/3.2 km long, 1.5 mi/2.4 km wide. Village of BIG BAY on W end; 46°48′N 87°42′W.

Independence, Mount, W VERMONT, a rugged promontory 200 feet above Lake CHAMPLAIN, and directly opposite Fort TICONDEROGA in New York. Fort was built here in 1776 to defend against the British during the American Revolutionary War and named Mount Independence after the Declaration of Independence. Designated a National Historic Landmark. Current site covered by hiking trails and archaelogical remains of fort complex.

Independence Mountains, NE NEVADA, in ELKO county; extends N from HUMBOLDT River, NW of ELKO. Rises to McAfee (10,438 ft/3,182 m) and Jacks (10,198 ft/3,108 m) peaks. N point in Humboldt National Forest.

Independence National Historical Park, city of PHILADELPHIA, PHILADELPHIA county, SE PENNSYLVANIA, in E central Center City; covers 45 acres/18 ha; 39°56′N 75°08′W. Historic points of interest include Independence Hall, Congress Hall, Old City Hall, Franklin Court. Independence Hall is the site of the signing of the Declaration of Independence. Authorized 1948.

Independence Pass, Colorado: see SAWATCH MOUNTAINS.

Independence Rock, historic site, also known as "The Great Register of the Desert," located in central WYOMING, S of CASPER, N of RAWLINS. A well-known landmark on the Sweetwater River, Independence Rock was a favorite resting place for emigrants traveling along the OREGON TRAIL. The granite outcropping is 136 ft high, 1,900 ft long, and 700 ft wide. The origin of the name varies, but popular legend says that it was named such because emigrants needed to reach the rock before July 4th in order to beat the heavy snows in the mountains to the west. Another version credits a party of fur trappers who celebrated at the spot on July 4, 1830. While camping at the site, more than 5,000 emigrants carved their names in the sturdy granite. Thousands of names and dates can still be seen on the rock despite erosion and lichen growth.

Independencia (een-dai-pen-DEN-see-ah), province (2002 population 50,833), SW DOMINICAN REPUBLIC, between Lake ENRIQUILLO and HAITI border; ⊙ JIMANÍ; 18°15′N 71°30′W. Formed 1949 out of W BAHORUCO province.

Independência (een-dai-pen-DEN-see-ah), city (2007 population 25,387), W CEARÁ province, BRAZIL, 28 mi/ 45 km SE of CRATEÚS; 05°25′S 40°25′W. Cheese manufacturing; cotton, carnauba, palms.

Independencia (een-dai-pain-DAIN-syah), town and canton, ⊙ AYOPAYA province, COCHABAMBA department, W central BOLIVIA, in N outliers of Cordillera de COCHABAMBA, 50 mi/80 km WNW of COCHABAMBA; 17°05′S 66°49′W. Elevation 8,600 ft/2,621 m. Potatoes.

Independência (EEN-dai-pen-den-see-ah), town (2007 population 6,679), N RIO GRANDE DO SUL state, BRAZIL, 39 mi/63 km NW of IJUÍ; 28°20′S 52°17′W. Wheat, corn, soybeans; livestock.

Independencia, town, Guairá department, S PARAGUAY, 10 mi/16 km E of VILLARICA; 25°42′S 56°15′W. Cotton, corn, peanuts, tobacco; lumbering; vineyards. German colony. Also called Colonia Independencia

Independencia (in-dai-pen-DEN-see-ah), town, TÁCHIRA state, W VENEZUELA, in ANDEAN spur, 5 mi/8 km W of SAN CRISTÓBAL; 07°49′N 72°18′W. Coffee, cereals; livestock.

Independencia, small village, in FLORIDA department, 30 mi/48 km N of Veinticinco de AGOSTO, URUGUAY, on railroad and trail; 34°18′S 56°24′W.

Independencia, ARGENTINA: see PATQUÍA.

Independencia, ARGENTINA: see VILLA INDEPENDENCIA.

Independência, Brazil: see GUARABIRA.

Independência, Brazil: see PENDÊNCIAS.

Independencia Bay (in-dai-pen-DEN-see-ah), Spanish *Bahía de Independencia*, inlet of the PACIFIC OCEAN, ICA region, SW PERU, 25 mi/40 km S of PISCO; 17 mi/27 km wide, 5 mi/8 km long; 14°15′S 76°10′W. Guano deposits on small islands, including Viejas Island. Surrounded by Reserva Nacional Paracas.

Independencia, La, MEXICO: see LA INDEPENDENCIA.

Independenţa (een-de-pen-DEN-tsah), village, GALAŢI county, E ROMANIA, on railroad, and 14 mi/23 km W of GALAŢI.

Inderagiri River, INDONESIA: see INDRAGIRI RIVER.

Inderborski (in-der-BOR-skee), city, N ATYRAU region, KAZAKHSTAN, on URAL River, near Inder Lake, 95 mi/ 153 km N of ATYRAU; 48°32′N 51°44′E. Tertiary-level administrative center. Borax-extracting center. Also called Inderbor.

Inder Lake, KAZAKHSTAN: see INDERBORSKI.

Indersdorf, GERMANY: see MARKT INDERSDORF.

Indetu, village, OROMIYA state, S central ETHIOPIA, near the upper WABĒ SHEBELĒ RIVER (here called Wabe), 65 mi/105 km NW of GINIR; 07°34′N 38°54′E.

Index, village (2006 population 165), SNOHOMISH county, NW WASHINGTON, 32 mi/51 km ESE of EVERETT, and on SKYKOMISH RIVER, at junction of North and South forks, in CASCADE RANGE; 47°49′N 121°33′W. Gold, silver; granite quarries; salmon; timber. Area surrounded by mountains Mount Baker-Snoqualmie National Forest, including Henry M. Jackson Wilderness Area to NE and Alpine Lakes Wilderness to S. Sunset Falls (on South Fork) to S; Mount Index (5,979 ft/1,822 m) to S.

Indi (IN-dee), town, BIJAPUR district, KARNATAKA state, S INDIA, 29 mi/47 km NE of BIJAPUR; 17°10′N 75°58′E. Millet, peanuts, cotton, wheat.

India, republic (□ 1,256,958 sq mi/3,268,090 sq km; 2004 estimated population 1,065,070,607; 2007 estimated population 1,129,866,154), S ASIA; ⊙ NEW DELHI; 20°00′N 80°00′E. The second-most populous country in the world, it has also been referred to in its constitution as Bharat, its ancient name.

Geography

India's land frontier (c.9,500 mi/15,290 km long) stretches from the ARABIAN SEA on the W to the Bay of Bengal on the E and touches (W-E) PAKISTAN, CHINA (TIBET), NEPAL, BHUTAN, BANGLADESH, and MYANMAR. Bhutan is advised in foreign affairs by India. The S half of India is a largely upland area that thrusts a triangular peninsula (c.1,300 mi/2,090 km wide at the N) into the INDIAN OCEAN between the Bay of Bengal (E) and the Arabian Sea (W) and has a coastline c.3,500 mi/5,630 km long; at its S tip is CAPE COMORIN (*Kanya Kumari*). In the N, towering above peninsular India, is the Himalayan mountain wall, where rise the three great rivers of the Indian subcontinent—the INDUS, the GANGA (Ganges), and the BRAHMAPUTRA. The Ganga (Gangetic) alluvial plain, which has much of India's arable land, lies between the Himalayas and the dissected plateau occupying most of peninsular India. The ARAVALLI range, a ragged hill belt, extends from the borders of GUJARAT state (SE) to the fringes of DELHI state (NE). The plain is limited W by the THAR (Great Indian) Desert of RAJASTHAN state, which merges with the swampy Rann of CUTCH (S). The S boundary of the plain lies close to the YAMUNA and Ganga rivers, where the broken hills of the CHAMBAL, BETWA, and SON rivers rise to the low plateaus of MALWA (W) and CHOTA NAGPUR (E). The NARMADA River, S of the VINDHYA hills, marks the beginning of the DECCAN; it is one of only two rivers (TAPI RIVER is the other) of the Deccan that flows W to the Arabian Sea. The triangular plateau, scarped by the mountains of the EASTERN and

WESTERN GHATS, is drained by the GODAVARI, KRISHNA, and KAVERI rivers; they break through the Eastern Ghats and, flowing E into the Bay of Bengal, form broad deltas on the wide COROMANDEL COAST. Further N, the MAHANADI River drains into the Bay of Bengal. The much narrower W coast of peninsular India, comprising chiefly the MALABAR COAST and the fertile Gujarat plain, bends around the Gulf of Khambhat in the N to the Kathiawar and KACHCHH peninsulas. The coastal plains of peninsular India have a tropical, humid climate. The Deccan interior is partly semiarid on the W and wet on the E. The Indo-Ganga plain is subtropical, with the W interior areas experiencing frost in winter and very hot summers. India's rainfall, which depends upon the monsoon, is variable; it is heavy in ASSAM and WEST BENGAL states and along the S coasts, moderate in the inland peninsular regions, and scanty in the arid NW, especially in Rajasthan and PUNJAB states.

Population
The racial composition of modern India is exceedingly complex due to centuries of immigration. Demographers generally refer, however, to four main groups: Caucasoid, Mongoloid, Australoid, and Negroid. India is a land of great cultural diversity, as is evidenced by the enormous number of languages spoken throughout the country. Although Hindi and English are used officially, over 1,500 local dialects are also spoken. The Indian constitution recognizes fifteen regional languages (Assamese, Bengali, Gujarati, Hindi, Kannada, Kashmiri, Malayalam, Marathi, Oriya, Punjabi, Sanskrit, Sindhi, Tamil, Telugu, and Urdu). Ten of the major states of India are generally organized along linguistic lines. The constitution forbids the practice of "untouchability," and legislation has been used to reserve quotas for former untouchables, tribal peoples, and other traditionally disadvantaged groups in the legislatures, in education, and in the public services. All these measures and the social awakening of more than a century has considerably relaxed India's caste system. About 81% of the poplation is Hindu, and about 12% is Muslim. Other significant religious groups include Christians, Sikhs, Buddhists, and Jains. There is no state religion and India is declared to be a secular state in the constitution.

Holy Cities and Shrines
The holy cities of India attract pilgrims from throughout the East: VARANASI (formerly Benares), ALLAHABAD, PURI, and NASHIK are religious centers for the Hindus; AMRITSAR is the holy city of the Sikhs; and Satrunjaya Hill near PALITANA is sacred to the Jains. With its long and rich history, India retains many outstanding archaeological landmarks; preeminent of these are the Buddhist remains at SARNATH, SANCHI, and BODH GAYA; the cave temples at AJANTA, ELLORA, and ELEPHANTA; and the temple sites at MADURAI, THANJAVUR, ABU, BHUBANESHWAR, KONARAK, and Mamallapuram.

Economy: Arrgriculture
Agriculture supports about 60% of the Indian people. Vast quantities of rice are grown wherever the land is level and water plentiful; other crops are wheat, pulses, sugarcane, jowar (sorghum), and bajra (a cereal), and corn. Cotton, tobacco, oilseeds, and jute are the principal non-food crops. There are large tea plantations in ASSAM, WEST BENGAL, KARNATAKA, KERALA, and TAMIL NADU states. In recent years India's food output and distribution have been adequate for the needs of its enormous population, maintaining an annual 2% increase in food production that keeps pace with population growth. Fragmentation of holdings, outmoded methods of crop production, and delays in acceptance of newer, high-yielding grains have been characteristic of Indian agriculture in the past, although in the last few decades, since the Green Revolution, significant progress has been made in these areas. The subsistence-level existence of village

India, ever threatened by drought, flood, famine, and disease, has been largely alleviated in recent years by government agricultural modernization efforts and reclamation and irrigation projects. Goats and sheep are raised in the arid regions of the W and NW.

Economy: Natural Resources
India has forested mountain slopes, with stands of oak, pine, sal, teak, ebony, palms, and bamboo, and the cutting of timber is a major rural occupation. Aside from mica, manganese, and ilmenite, in which the country ranks among the world's highest, India's mineral resources, although large and varied, are not as yet fully exploited. There are very large and rich iron-ore deposits. The CHOTA NAGPUR PLATEAU of S Bihar state, the hill areas of SW West Bengal, N ORISSA, and E MADHYA PRADESH states are the most important mining areas; they are the source of coal, iron, mica, and copper. There are workings of magnesite, gold (in the KOLAR goldfields in Karnataka state), bauxite, chromite, salt, and gypsum. Despite oil fields in Assam and Gujarat states and the spectacular output (since the 1970s) of Bombay High offshore oil field, India is deficient in petroleum.

Economy: Industry
Industry in India, traditionally limited to agricultural processing and light manufacturing, especially of cotton, woolen, and silk textiles, jute, and leather products, has been greatly expanded and diversified in recent years but still employs less than 17% of the workforce. There are large textile works at MUMBAI (Bombay) and AHMADABAD and a huge iron and steel complex (mainly controlled by the Tata family) at JAMSHEDPUR. The public-sector Steel Authority of India runs steel plants at ROURKELA, BHILAINAGAR, DURGAPUR, and BOKARO. BANGALORE has electronics, machine-tool, aeronautics, and armaments industries. India's large motion picture industry is concentrated in Mumbai, other centers being CALCUTTA and CHENNAI (Madras). The government has departed from its traditional policy of self-reliant industrial activity and development and is working to deregulate Indian industry and attract foreign investment. With educational levels rising, India is capitalzing on its large numbers of well-educated people skilled in English to become a major player in software services.

Economy: Transportation
Most towns are connected by state-owned railroad systems, one of the most extensive networks in the world. The railroad system is made mainly of broad-gauge track (5.6 ft/1.7 m), but includes a variety of railroad gauges, which makes frequent transshipment necessary. This multiplicity, however, is to end soon, due to a rapidly progressing scheme of converting all tracks to broad gauge. Transportation by road is increasing, with the introduction of ordinary and luxury bus service on long-distance routes on improved highways, but in rural India the bullock cart is still an important means of short-haul transportation. There are international airports at New Delhi, Calcutta, Mumbai, and Chennai, and a few others are being built or planned. The leading ports are Mumbai, Chennai, Calcutta, KOCHI, and VISHAKHAPATNAM, with many minor ports growing in importance.

Economy: Exports
The leading exports are iron ore, iron and steel, cotton goods, tea, and jute, along with engineering goods, machine tools, watches, apparel, and handicraft goods. The chief imports are machinery, petroleum, fertilizers, cotton, steel, and rice. India's major trade partners are the U.S., China, Great Britain, Hong Kong, and Germany.

Urbanization
In 1991, India had twenty-three cities with urban areas of over one million people: Ahmadabad, Bangalore, BHOPAL, Calcutta, Chennai, COIMBATORE, DELHI, HYDERABAD, INDORE, JAIPUR, KANPUR, Kochi, LUCK-

NOW, LUDHIANA, MADURAI, Mumbai, NAGPUR, PATNA, PUNE, SURAT, VADODARA, VARANASI, and VISHAKHAPATNAM.

History: Early Civilization to Alexander the Great
One of the world's oldest civilizations, and the earliest on the Indian subcontinent, was the Indus Valley civilization, which flourished c.2500 B.C.–c.1700 B.C.E. It was an extensive and highly sophisticated culture, its chief urban centers being MOHENJO-DARO and HARAPPA. While the causes of the decline of the Indus Valley civilization are unclear, it is possible that the periodic shifts in the courses of the major rivers of the valley may have deprived the cities of floodwaters necessary for their surrounding agricultural lands. The cities thus became more vulnerable to raiding activity. At the same time, Indo-Aryan peoples were migrating into the Indian subcontinent through the NW mountain passes, settling in the Punjab and the Ganga River valley. Over the next 2,000 years the Indo-Aryans developed a Brahmanic civilization out of which Hinduism evolved. From Punjab they spread E over the Ganga plain and by c.800 B.C.E. were established as far E as Bihar and Bengal. The first important Aryan kingdom was MAGADHA, with its capital near present-day Patna; it was there, during the reign of Bimbisara (540 B.C.–490 B.C.E.), that the founders of Jainism and Buddhism preached. KOSALA was another kingdom of the period.

History: Alexander the Great to the Rise of Islam
In 327 B.C.–325 B.C.E., Alexander the Great invaded the provinces of Gandhara and Punjab in NW India that had been a part of the Persian empire. The Greek invaders were eventually driven out by Chandragupta of Magadha, founder of the Mauryan empire. The Mauryan emperor Asoka (d. 232 B.C.E.), Chandragupta's grandson, perhaps the greatest ruler of the ancient period, unified all of India except the S tip. Under Asoka, Buddhism was widely propagated, and spread to Ceylon (now SRI LANKA) and SE Asia. During the 200 years of disorder and invasions that followed the collapse of the Mauryan state (c.185 B.C.E.), Buddhism in India declined. S India enjoyed greater prosperity than the N, despite almost incessant warfare; among the Tamil-speaking kingdoms of the S were the Pandya and Chola states, which maintained an overseas trade with the Roman Empire. Indian culture was spread through the MALAY ARCHIPELAGO to CAMBODIA and INDONESIA by traders from the S Indian kingdoms. Meanwhile, Greeks following Alexander had settled in BACTRIA (in the area of present-day Afghanistan) and established an Indo-Greek kingdom. After the collapse (1st century B.C.E.) of Bactrian power, the Scythians, Parthians, Afghans, and Kushans swept into NW India. There, small states arose and disappeared in quick succession; among the most famous of these kingdoms was that of the Kushans, which, under its sovereign Kanishka, enjoyed (2nd century A.D.) great prosperity. In the 4th and 5th centuries C.E., N India experienced a golden age under the Gupta dynasty, when Indian art and literature reached a high level. Gupta splendor rose again under the emperor Harsha of Kanauj (c.606–647), and N India enjoyed a renaissance of art, letters, and theology. While the Guptas ruled the N, in this, the classical period of Indian history, the Pallava kings of Kanchi held sway in the S, and the Chalukyas controlled the Deccan.

History: Rise of Islam to the Mogul Empire
During the medieval period (8th–13th centuries) several independent kingdoms, notably the Palas and the Sens of Bihar and Bengal, the Ahoms of Assam, a later Chola empire at Thanjavur, and a second Chalukya dynasty in the Deccan, waxed powerful. In NW India, beyond the reach of the medieval dynasties, the Rajputs had grown strong and were able to resist the rising forces of Islam. Islam was first brought to Sind, W India, in the 8th century by seafaring Arab traders;

by the 10th century Muslim armies from the N were raiding India. From 999 to 1026, Mahmud of Ghazni several times breached Rajput defenses and plundered India. In the 11th and 12th centuries Ghaznavid power waned, to be replaced c.1150 by that of the Turkic principality of Ghor. In 1192 the legions of Ghor defeated the forces of Prithivi Raj, and the DELHI SULTANATE, the first Muslim kingdom in India, was established. The sultanate eventually reduced to vassalage almost every independent kingdom on the subcontinent, except that of Kashmir and the remote kingdoms of the S. The task of ruling such a vast territory proved impossible; difficulties in the S with the state of VIJAYANAGAR, the great Hindu kingdom, and the capture (1398) of the city of Delhi by Tamerlane finally brought the sultanate to an end. The Muslim kingdoms that succeeded it were defeated by a Turkic invader from Afghanistan, Babur, a remote descendant of Tamerlane, who, after the battle of Panipat in 1526, founded the Mogul empire.

History: Mogul Empire to 1757

The empire was consolidated by Akbar and reached its greatest territorial extent, the control of almost all of India, under Aurangzeb (ruled 1659–1707). Under the Delhi Sultanate and the Mogul empire a large Muslim following grew and a new culture evolved in India; Islam, however, never supplanted Hinduism as the faith of the majority. In 1498, only a few years before Babur's triumph, Vasco da Gama had landed at Calicut (now KOZHIKODE) and the Portuguese conquered GOA (1510). The splendor and wealth of the Mogul empire (from it comes some of India's greatest architecture, including the TAJ MAHAL) attracted British, Dutch, and French competition for the trade that Portugal had at first monopolized. The British East India Company, which established trading stations at Surat (1613), Bombay (1661), and Calcutta (1691), soon became dominant and with its command of the sea drove off the traders of Portugal and Holland. While the Mogul empire remained strong, only peaceful trade relations with it were sought; but in the 18th century, when an Afghan invasion, dynastic struggles, and incessant revolts of Hindu elements, especially the Marathas, were rending the empire, Great Britain and France seized the opportunity to increase trade and capture Indian wealth, each attempting to oust the other. From 1746 to 1763 India was a battleground for the forces of the two powers, each attaching to itself as many native rulers as possible in the struggle.

History: 1757 to the Indian Mutiny

Clive's defeat of the Nawab of Bengal at PLASSEY in 1757 traditionally marks the beginning of the British Empire in India (recognized in the Treaty of Paris of 1763). Warren Hastings, Clive's successor and the first governor-general of the company's domains to be appointed by Parliament, did much to consolidate Clive's conquests. By 1818 the British controlled nearly all of India S of the SUTLEJ River and had reduced to vassalage their most powerful Indian enemies, the state of Mysore and the Marathas. Only Sind and Punjab (the Sikh territory) remained completely independent. The East India Company, overseen by the government's India Office, administered the rich areas with the populous cities; the rest of India remained under Indian princes, but with British residents in effective control. Great Britain regarded India as an agricultural reservoir and a market for British goods. However, the export of cotton goods from India suffered because of the Industrial Revolution and the production of cloth by machine. The British also initiated projects to improve transportation and irrigation. British control was extended over Sind in 1843 and Punjab in 1849. Social unrest, added to the apprehension of several important native rulers about the aggrandizing policies of Governor-General Dalhousie, led to the bloody Indian Mutiny of 1857

(today, known in India as the "First Battle for Independence"). It was suppressed, and Great Britain, determined to prevent a recurrence, initiated long-needed reforms.

History: Indian Mutiny to World War I

Control passed from the East India Company to the crown. The common soldiers in the British army in India were drawn more and more from among the Indians, and these troops were later also used overseas. Sikhs and Gurkhas became famous as British soldiers. Native rulers were guaranteed the integrity of their domains as long as they recognized the British as paramount. In 1861 the first step was taken toward self-government in British India with the appointment of Indian councillors to advise the viceroy and the establishment of provincial councils with Indian members. But the power of Britain was symbolized and reinforced when Queen Victoria was crowned Empress of India in 1877. With the setting up of a few universities, an Indian middle class began to emerge and to advocate further reform. Popular nationalist sentiment was perhaps most strongly aroused when, for administrative reasons, Viceroy Curzon partitioned (1905) Bengal into two presidencies; newly created East Bengal had a Muslim majority. Because of a mass movement against it, the partition had to be ended in 1911. In the early 1900s the British had widened Indian participation in legislative councils (the Morley-Minto reforms). Separate Muslim constituencies, introduced for the first time, were to be a major factor in the growing split between the two communities. Muslim nationalist sentiment was expressed by Sayyid Ahmad Khan, Iqbal, and Muhammad Ali Jinnah.

History: World War I to World War II

At the outbreak of World War I all elements in India were firmly united behind Britain, but discontent arose as the war dragged on. The British, in the Montagu declaration (1917) and later in the Montagu-Chelmsford report (1918), held out the promise of eventual self-government. Crop failures and an influenza epidemic that killed millions plagued India in 1918–1919. Britain passed the Rowlatt Acts (1919), which enabled authorities to dispense with juries, and even trials, in dealing with agitators. In response Mohandas K. Gandhi organized the first of his many passive resistance campaigns. The massacre of innocent Indians by British troops at Jalianwala Bagh in Amritsar further inflamed the situation. The Government of India Act (late 1919) set up provincial legislatures with "dyarchy," which meant that elected Indian ministers, responsible to the legislatures, had to share power with appointed British governors and ministers. Although the act also provided for periodic revisions, Gandhi felt too little progress had been made, and he organized new protests. Imperial conferences (Round Table talks) concerning the status of India were held in 1930, 1931, and 1932. They led to the Government of India Act of 1935, which provided for the election of entirely Indian provincial governments and a federal legislature in Delhi that was to be largely elected. In the first elections (1937) held under the act, the Congress, led by Gandhi and Jawaharlal Nehru, won well over half the seats, mostly in general constituencies, and formed governments in seven of the eleven provinces. The Muslim League, led by Muhammad Ali Jinnah, won 109 of the 485 Muslim seats and formed governments in three of the remaining provinces. Fearing Hindu domination in a future independent India, Muslim nationalism in India began to argue for special safeguards for Muslims.

History: World War II to the Creation of Pakistan

World War II found India by no means unified behind Great Britain. There was even an "Indian national army" of anti-British soldiers of various religious groups, formed and led by Subhas Bose in SE Asia. In 1942, to procure India's more wholehearted

support, Sir Stafford Cripps, on behalf of the British cabinet, had presented a proposal for establishing an Indian interim government, in which Great Britain would maintain control only over defense and foreign policy, to be followed by full self-government after the war. The Congress adamantly demanded that the British leave India and, when the demand was refused, initiated civil disobedience and the Quit India movement in August 1942. Great Britain's response was to outlaw the Congress and jail Gandhi and other leaders. Jinnah gave conditional support to the war, but used it to build up the Muslim League. The British Labour government of Prime Minister Clement Attlee in 1946 offered self-government to India, but it warned that if no agreement was reached between the Congress and the Muslim League, Great Britain, on withdrawing in June 1948, would have to determine the apportionment of power between the two groups.

History: Creation of Pakistan to Death of Gandhi

Reluctantly the Congress agreed to the creation of Pakistan, and in August 1947, British India was divided into the dominions of India and Pakistan. No general referendum was involved, however. The princely states were nominally free to determine their own status, but realistically they were unable to stand alone. By a mixture of persuasion coercion, they joined one or the other of the new dominions. Hyderabad, in S central India, with a Muslim ruler and Hindu population, held out to the last and was finally incorporated (1948) into the Indian union by force. The future of Kashmir was never resolved. Nehru became prime minister of India, and Jinnah governor-general of Pakistan. Partition left large minorities of Hindus and Sikhs in Pakistan and Muslims in India. Most Hindus eventually came over to India within a short period, and large numbers of Muslims fled to Pakistan, although Muslim migration out of India has not been on the same scale. A sizeable number of Hindu minorities have stayed in East Pakistan (now Bangladesh) in spite of successive waves of refugees at different times. Widespread hostilities erupted between the communities at the time of the partition and continued while large numbers of people—about 16 million in all—fled across the borders seeking safety. Over 500,000 people died in the disorders (late 1947).

Histroy: Death of Gandhi to 1957

Gandhi was shot dead in a prayer meeting at Delhi by a Hindu fanatic in January 1948. The hostility between India and Pakistan was aggravated when warfare broke out (October 1947) over their conflicting claims to jurisdiction over the princely state of JAMMU AND KASHMIR. India became a sovereign democratic republic under a constitution adopted in 1949 and which came into force on January 26, 1950. In addition to staggering problems of partition, overpopulation, economic underdevelopment, and inadequate social services, India had to achieve the integration of the former princely states into the union and the creation of national unity from diverse cultural and linguistic groups. The states of the republic were reorganized several times, finally in 1956 along linguistic lines. India consolidated its territory by incorporating the former French settlements in 1956 and by forcibly annexing the Portuguese enclaves of GOA, DAMAN AND DIU in December 1961. In 1987 Goa became a separate state and Daman and Diu became a union territory. In world politics, India has been a leading exponent of nonalignment. Besides the ongoing difficulties with Pakistan, the republic's major foreign problem has involved a border dispute with China that first surfaced in 1957.

History: 1957 to 1967

Sino-Indian skirmishes began in 1959 and the controversy climaxed on October 20, 1962, when the Chinese launched a massive offensive against LADAKH in Kashmir and in other areas on the NE Indian border, including the Brahmaputra River valley in Assam state. The Chinese announced a cease-fire on November 21 after gaining some territory claimed by India. In the late 1960s there was friction with Nepal, which accused India of harboring Nepalese politicians hostile to the country's monarchy. In August 1965, fighting between India and Pakistan broke out in the Rann of Kachchh (CUTCH) frontier area and in Kashmir. The UN proclaimed a cease-fire in September, but clashes continued. Shri Lal Bahadur Shastri, who succeeded Nehru as prime minister after the latter's death in 1964, and Ayub Khan, president of Pakistan, met (1966) under Soviet auspices in TOSHKENT (USSR, now in UZBEKISTAN) to negotiate the Kashmir problem. They agreed on mutual troop withdrawals to the lines held before August 1965. Shastri died in Toshkent and was succeeded, after bitter debate within the Congress Party, by Indira Gandhi, Nehru's daughter.

History: 1967 to 1971
The Congress Party suffered a setback in the elections of 1967; its parliamentary majority was sharply reduced and it lost control of several state governments. In 1969 the party split: Gandhi and her followers formed the New Congress Party, and her opponents on the right remained in the Old Congress Party. In the elections of March 1971, the New Congress won an overwhelming victory. Acts of terrorism by Maoists, known as Naxalites, flared in 1970 and 1971. The situation was particularly serious in West Bengal and ANDHRA PRADESH states, where certain "liberated areas" were proclaimed by armed Naxalite groups. In Pakistan, attempts by the government (dominated by West Pakistanis) to suppress a Bengali uprising in East Pakistan led in 1971 to the exodus of millions of Bengali refugees (mostly Hindus) from East Pakistan into India. Caring for the refugees imposed a severe drain on India's slender resources. India supported the Awami League–led demand for the autonomy of East Bengal (East Pakistan) and when the Pakistani army used excessive force to repress this movement, India provided support for the Bengalis.

History: 1971 to 1974
As a result, war broke out in December 1971 between India and Pakistan in East Pakistan and in Kashmir. Indian forces rapidly advanced into East Pakistan; the war ended in two weeks with the creation of independent Bangladesh to replace East Pakistan, and approximately 900,000 East Bengali refugees returned from India. India's relations with the U.S. were strained because of the American-declared "tilt" toward Pakistan. In mid-1973, India and Pakistan signed an agreement (to which Bangladesh was a party but not a signatory) providing for the release of prisoners of war captured in 1971 and calling for peace and friendship on the Indian subcontinent. Also in 1973, India's ties with the USSR were strengthened by a new Indo-Soviet treaty that considerably increased Soviet economic and military aid; at the same time, relations with the U.S. improved somewhat. In May 1974, India became the world's 6th nuclear power by "imploding" an underground nuclear device in the THAR DESERT in RAJASTHAN state.

History: 1974 to 1989
In 1974, Indira Gandhi came under intense pressure from opponents who criticized her government for abuse of power and in June 1975, her 1971 election to the Lok Sabha was struck down by a High Court order. Despite the declaration of a state of emergency and the initiation of several relatively popular public-policy programs, the opposition campaign and the growing power of her son Sanjay Gandhi contributed to a 1977 election defeat for Gandhi and the New Congress Party at the hands of a coalition known as the Janata (People's) Party. The Janata Party soon became fractured, however, and in January 1980, In-

dira Gandhi and her Congress Party won another resounding election victory. Less than six months later Sanjay Gandhi, expected by many to be his mother's successor, was killed in a plane crash. In 1982 Sikh militants began a terrorism campaign intended to pressure the government to create an autonomous Sikh state in the PUNJAB. Government response escalated until in June 1984, army troops stormed the Golden Temple in Amritsar, the Sikh's holiest shrine and the center of the independence movement. Sikh protests across India added to the political tension, and in October Indira Gandhi was assassinated by two Sikh members of her personal guard. The resulting anti-Sikh riots prompted the government to appoint Indira's eldest son, Rajiv, prime minister. Rajiv Gandhi moved quickly to end the rioting and thereafter pursued a domestic policy emphasizing conciliation among India's various conflicting ethnic and religious groups.

History: 1989 to Present
In 1989 he was defeated by the Janata Dal Party under the leadership of Vishwanath Pratap Singh. While India's economic performance was generally stable in the 1980s, it experienced continuing problems politically, including border and immigration disputes with Bangladesh, internal agitation by Tamil separatists, violent conflicts in Assam state, strife caused by the Sikh question, and continued antagonism between Hindus and Muslims. A notable occurrence in 1984 was the BHOPAL environmental disaster, the worst industrial accident in history, in which toxic fumes escaped from a Union Carbide insecticide plant in Bhopal, killing 2,500 and injuring tens of thousands. From 1987 to 1990, the Indian military occupied the N area of Sri Lanka in an unsuccessful attempt to quell the Tamil separatist insurgency. In May 1991 Rajiv Gandhi was assassinated during an election rally. The Congress Party won the ensuing election and P.V. Narashimha Rao became prime minister. He immediately instituted sweeping economic reforms, moving away from the centralized planning that had characterized India's economic policy since Nehru. The Congress suffered a defeat in the 1996 elections. The Bharatiya Janata Party (BJP) formed the government with Atal Behari Vajpayee as the prime minister. As the BJP failed to attain a majority in Parliament within the period stipulated in the constitution, Vajpayee stepped down after thirteen days and a non-Congress coalition government was formed by H.D. Devegowda, who became the new prime minister on July 13, 1996. Less than a year later, in April 1997, the leadership changed hands again, and I.K. Gujaral became prime minister. India and Pakistan entered into negotiations on a non-aggression pact in 1997. In elections held in 1998, the BJP and its allies won the most seats and Vajpayee was named prime minister. His government fell in April 1999, but following elections held in September he formed a new coalition government. In May 1999, India launched a military campaign against Islamic guerrillas who had occupied strategic positions in the Indian-held part of Kashmir (and who were generally regarded as sponsored by Pakistan); most of the rebels had withdrawn by the end of July. A July 2001 summit between Vajpayee and Pakistan's leader ended without progress, and relations verged toward war after Muslim militants attacked (December 2001) the Indian parliamnent. Denouncing that and other attacks as Pakistani-sponsored, India mobilized its forces. Tensions eased somewhat by January 2002, when Pakistan announced that it would not tolerate any group engaging in terrorism. War again threatened in May, after more Kashmir-related violence, but the crisis again eased after Pakistan government-supported infiltration across the line of control in Kashmir and relations subsequently improved. Congress defeated the BJP in the 2004 elections and formed a coalition government.

Manmohan Singh, a former finance minister, became prime minister after Congress's leader, the foreign-born Sonia Gandhi (Rajiv's widow) declined the office. The December 2004 Indian Ocean tsunami devastated India's SE coast and Andaman and Nicobar Islands, killing more than 16,000.

Government
India is a federal republic with a parliamentary system of the Westminster type of government. It is governed under the 1949 constitution. The president of India is elected for a five-year term by the elected members of the bicameral Indian parliament and the state assemblies. Theoretically the president possesses full executive power, but actually that power is exercised by the prime minister (leader of the majority party in the federal parliament) and council of ministers (which includes the cabinet), who are appointed by the president on the recommendation of the prime minister. The ministers are responsible to the lower house of parliament (Lok Sabha), which has 545 members, and must be elected members of parliament. The upper house, the Council of States (Rajya Sabha), consists of a maximum of 250 members; the great majority are apportioned by state—each state's delegates are elected by its elected assembly—and twelve members are nominated by the president from among distinguished citizens in the realm of arts, sciences, and letters. State governors are appointed by the president for five-year terms. States have jurisdiction over police and public order, agriculture, education, public health, and local government. The Union Government at New Delhi has jurisdiction over any matter not specifically reserved to the states. The Congress Party has generally dominated Indian politics since independence. Other major parties include the Janata Dal, the Bharatiya Janata Party, the Communist Party of India, and the Communist Party of India (Marxist). There are also significant regional parties. India is a member of the UN and of the Commonwealth of Nations.The current head of state is President Pratibha PATIL. Manmohan Singh has been prime minister since May 2004. India is divided into twenty-eight states: ANDHRA PRADESH, ARUNACHAL PRADESH, ASSAM, BIHAR, Chhattisgarh, GOA, GUJARAT, HARYANA, HIMACHAL PRADESH, JAMMU AND KASHMIR, Jharkhand, KARNATAKA, KERALA, MADHYA PRADESH, MAHARASHTRA, MANIPUR, MEGHALAYA, MIZORAM, NAGALAND, ORISSA, PUNJAB, RAJASTHAN, SIKKIM, TAMIL NADU, TRIPURA, UTTAR PRADESH, Uttaranchal, and WEST BENGAL. There are also seven union territories, administered by the union government: the ANDAMAN AND NICOBAR ISLANDS, CHANDIGARH, DADRA AND NAGAR-HAVELI, DAMAN AND DIU, DELHI, LAKSHADWEEP, and PUDUCHERRY. Jammu and Kashmir's accession to India is disputed by Pakistan.

Indiahoma (in-dee-uh-HO-muh), village (2006 population 349), COMANCHE county, SW OKLAHOMA, 20 mi/32 km W of LAWTON, on WEST CACHE CREEK, S of the WICHITA MOUNTAINS; 34°37′N 98°45′W. Trade center for agricultural area. Fort Sill Military Reservation to NE; Wichita Mountains National Wildlife Refuge to N.

India Hook, unincorporated town (2000 population 1,614), YORK county, N SOUTH CAROLINA, residential suburb 4 mi/6.4 km NW of ROCK HILL; 35°00′N 81°02′W.

Indialantic (in-dee-uh-LAN-tik), town (□ 1 sq mi/2.6 sq km; 2000 population 2,944), BREVARD county, E central FLORIDA, 3 mi/4.8 km E of MELBOURNE; 28°05′N 80°34′W. Manufacturing includes medical equipment, baked goods.

Indiana, state (□ 36,420 sq mi/94,328 sq km; 1995 estimated population 5,803,471; 2000 population 6,080,485), N central U.S., in the MIDWEST, admitted as the nineteenth state of the Union in 1816; 39°47′N 86°08′W; ⊙ INDIANAPOLIS, in the central part of the state. Indianapolis is the largest city; other major cities are FORT WAYNE, EVANSVILLE, and GARY.

Geography

Indiana is bounded on the N by MICHIGAN and Lake MICHIGAN, on the E by OHIO, on the S by KENTUCKY (from which it is separated by the OHIO River), and on the W by ILLINOIS. Northern Indiana is a glaciated lake area, separated by the WABASH RIVER from the central agricultural plain, which is rich with deep glacial drift. The S portion of the state is a succession of bottomlands interspersed with knolls and ridges, gorges and valleys. Limestone caves, such as the big WYANDOTTE CAVE, and mineral springs, such as at FRENCH LICK and WEST BADEN SPRINGS, are found there.

Economy

The unglaciated soil is shallow in S Indiana, and the cutting of timber has caused erosion, but there is still extensive farming. Although Indiana as a whole is a manufacturing state, about three-quarters of the land is utilized for agriculture. With a growing season of about 170 days and an average rainfall of 40 in/102 cm per year, Indiana farms have rich yields. Grain crops, mainly corn and wheat, are important and also support the livestock and dairying industries. Indiana is the nation's leading producer of popcorn, and soybeans and hay are also principal crops. Vegetables and fruits are produced in great quantity and variety as well. Livestock, especially hogs, cattle, and poultry, is another major agricultural product. Meat packing is chief among the many industries related to agriculture. Although the urban population exceeds the rural, many towns are primarily service centers for agricultural communities. There are, however, cities with varied, heavy industries; prominent, besides Indianapolis, Evansville, Fort Wayne, and Gary, are KOKOMO, SOUTH BEND, ELKHART, and TERRE HAUTE. These cities were among the highest in the nation in unemployment during the recession of the early 1980s. In the CALUMET region along Indiana's Lake Michigan shoreline, marshy wastelands were drained and transformed into an area supporting a complex of factories and refineries. In the mid-1990s, Indiana led the nation in the production of steel. Other leading manufacturing includes electrical equipment, transportation equipment, mobile homes, nonelectrical machinery, chemicals, food products, and fabricated metals. Rich mineral deposits of coal and stone (the S central Indiana area is the nation's leading producer of building limestone) have encouraged construction and industry. Throughout the state the products of farms and factories are transported by truck and by railroad. Indiana calls itself the crossroads of America, and its extreme NW corner—where transportation lines head E after converging on nearby CHICAGO from all directions—is one of the most heavily traveled areas in the world in terms of railroad, road, and air traffic. Waterborne traffic is also important to Indiana. Improvements on the Ohio River and the opening (1959) of the ST. LAWRENCE SEAWAY, linking the GREAT LAKES with the ATLANTIC OCEAN, have benefited the state. With the opening in 1970 of the Burns Waterway Harbor on Lake Michigan, Indiana gained its first public port and enhanced its shipping facilities.

History: to 1787

The Mound Builders were some of Indiana's earliest known inhabitants; their cultural remains have been found along Indiana's rivers and bottomlands. The region was first explored by Europeans, notably the French, in the late 17th century. The leading French explorer was Robert Cavalier, sieur de La Salle, who came to the area in 1679. At the time of exploration, the area was occupied mainly by Native American groups of the Miami, Delaware, Potawato, and Shawnee descents. VINCENNES, the first permanent settlement, was fortified in 1732, but for the early 1700s, most of the settlers in the area were Jesuit missionaries

or fur traders. By the Treaty of PARIS of 1763 ending the French and Indian Wars, Indiana, then part of the area known as the OLD NORTHWEST, passed from French to British control. Along with the rest of the Old Northwest, Indiana was united with CANADA under the QUEBEC Act of 1774. During the American Revolution an expedition led by George Rogers Clark captured, lost, and then recaptured Vincennes from the British. By the Treaty of Paris of 1783 ending the Revolutionary War, Great Britain ceded the Old Northwest to the U.S. Indiana was still largely unsettled when the NORTHWEST TERRITORY, of which it formed a part, was established in 1787.

History: 1787 to 1860

Native Americans in the territory resisted settlement, but General Anthony Wayne's victory at FALLEN TIMBERS in 1794 effectively ended Native American resistance in the Old Northwest. U.S. forces led by General William Henry Harrison also defeated the Native American forces in the battle of TIPPECANOE RIVER (1811) in the Wabash country. In 1800, Indiana Territory was formed and included the states of Indiana, Illinois, and WISCONSIN, and parts of Michigan and MINNESOTA. Vincennes was made the capital, which in 1813 was moved to CORYDON. A constitutional convention met in 1816, and Indiana achieved statehood. Jonathan Jennings, an opponent of slavery, was elected governor. Indianapolis was laid out as the state capital, and the government moved there in 1824–1825. Indiana was the site of several experimental communities in the early 19th century, notably the Rappite (1815) and Owenite (1825) settlements at NEW HARMONY. In the 1840s the Wabash and Erie Canal opened between LAFAYETTE and TOLEDO (Ohio), giving Indiana a water route via Lake ERIE to markets in the E. Also in the 1840s the state's first railroad line was completed between Indianapolis and MADISON. The Hoosier spirit of simplicity and forthrightness that developed during Indiana's early years of statehood figured in the writings of Edward Eggleston in The Hoosier Schoolmaster and was represented much later in works by James Whitcomb Riley, George Ade, and Gene Stratton Porter. This also led to the state's nickname as the "Hoosier State."

History: 1860 to World War I

The Civil War brought great changes in the state. In the elections of 1860, Indiana voted for Lincoln, who had spent his boyhood here. Although there was some proslavery sentiment in the state, represented by the Knights of the Golden Circle, Oliver P. Morton, governor during the war, held the state unswervingly to the Union cause even after the constitutional government broke down in 1862. Confederate general John Hunt Morgan led a raid into Indiana in 1863, but otherwise little action occurred in the state. Manufacturing, which had been stimulated in Indiana by the needs of the war, developed rapidly after the war. Factories sprang up, and the old rustic pattern was broken. However, Indiana's farmers continued to be an important force in the state, and in the hard times following the Panic of 1873 indebted farmers expressed their discontent by supporting the Granger movement and later the Greenback party in 1876 and the Populist party in the 1890s. Industrial development came to the Calumet region in the late 19th century with the establishment of an oil refinery at WHITING. As the 19th century drew to a close, industry continued to expand and the growing numbers of industrial workers in the state sought to organize through labor unions. Eugene V. Debs, one of the great early labor leaders, was from Indiana, and the labor movement at Gary in the Calumet area figured prominently in the nationwide steel strike just after World War I.

History: World War I to Present

Indiana was an early leader in the production of automobiles. Before Detroit took control of the industry

in the 1920s, Indiana boasted over 300 automobile companies. Indiana industries contributed heavily to the war effort. Indiana society in the early 20th century has been described in a number of studies and books. The classic sociological study by Robert S. Lynd and Helen M. Lynd of an American manufacturing town, Middletown (1929), was based on data from MUNCIE, Indiana. In the 1920s religious and racial intolerance was exploited in Indiana by the Ku Klux Klan. In the 1930s and 1940s, Wendell Willkie and Ernie Pyle, both natives of Indiana, became nationally prominent figures in politics and journalism, respectively. In the 1980s, Indianapolis experienced significant growth with a diversified economy, something the N industrial portion of the state has been unable to achieve.

Academic Institutions and Places of Interest

Concern for education has long been manifest in the state. Robert Dale Owen, son of the English reformer Robert Owen (who founded an idealistic community at New Harmony), promoted tax support of public schools, and this policy was incorporated into the state constitution of 1851. Among the institutions of higher learning in Indiana are Indiana University, at BLOOMINGTON; Purdue University, at WEST LAFAYETTE; the University of Notre Dame, at SOUTH BEND; Indiana University/Purdue University at Indianapolis (IUPUI); Indiana State University, at Terre Haute; DePauw University, at GREENCASTLE; Butler University, at Indianapolis; Valparaiso University, at VALPARAISO; Wabash College, at CRAWFORDSVILLE; Earlham College, at RICHMOND; and Goshen College, at GOSHEN. In 1962, the U.S. Congress authorized the establishment of the LINCOLN BOYHOOD NATIONAL MEMORIAL in S Indiana. INDIANA DUNES National Lakeshore, with a 3-mi/4.8-km frontage on Lake Michigan, is noted for its beautiful shifting sand dunes. Formerly a state park, the area was made a National Lakeshore in 1966. The Indianapolis Motor Speedway is the site of the famous annual 500-mi/805-km auto race.

Government

Indiana's constitution dates from 1851 and provides for an elected executive and legislature. A governor serves as the chief executive for a term of four years. The current governor is Mitch Daniels. The legislature, called the General Assembly, has a senate with fifty members elected for four years and a house of representatives with 100 members elected for two years. Indiana elects ten Representatives and two Senators to the U.S. Congress and has twelve electoral votes. Although Indiana in the late 19th century was regarded as a "swing state" electorally, in the 20th century its voting pattern has been generally conservative and Republican. Republican J. Danforth Quayle, elected to the U.S. Senate in 1980 and 1986, was elected Vice-President of the U.S. in 1988. However, Democrats have had some successes in gubernatorial and congressional elections.

Indiana has ninety-two counties: ADAMS, ALLEN, BARTHOLOMEW, BENTON, BLACKFORD, BOONE, BROWN, CARROLL, CASS, CLARK, CLAY, CLINTON, CRAWFORD, DAVIESS, DEARBORN, DECATUR, DEKALB, DELAWARE, DUBOIS, ELKHART, FAYETTE, FLOYD, FOUNTAIN, FRANKLIN, FULTON, GIBSON, GRANT, GREENE, HAMILTON, HANCOCK, HARRISON, HENDRICKS, HENRY, HOWARD, HUNTINGTON, JACKSON, JASPER, JAY, JEFFERSON, JENNINGS, JOHNSON, KNOX, KOSCIUSKO, LAGRANGE, LAKE, LA PORTE, LAWRENCE, MADISON, MARION, MARSHALL, MARTIN, MIAMI, MONROE, MONTGOMERY, MORGAN, NEWTON, NOBLE, OHIO, ORANGE, OWEN, PARKE, PERRY, PIKE, PORTER, POSEY, PULASKI, PUTNAM, RANDOLPH, RIPLEY, RUSH, SAINT JOSEPH, SCOTT, SHELBY, SPENCER, STARKE, STEUBEN, SULLIVAN, SWITZERLAND, TIPPECANOE, TIPTON, UNION, VANDERBURGH, VERMILLION, VIGO, WABASH, WARREN, WARRICK, WASHINGTON, WAYNE, WELLS, WHITE, and WHITLEY.

Indiana, county (□ 834 sq mi/2,168.4 sq km; 2006 population 88,234), W central PENNSYLVANIA; ⊙ INDIANA. Bounded S by CONEMAUGH RIVER; drained by Blacklick, Two Lick, MAHONING, and Little Mahoning creeks. Coal-mining, manufacturing, agricultural region. Bituminous coal, limestone; manufacturing at Indiana and BLAIRSVILLE. Agriculture (corn, wheat, oats, hay, alfalfa, potatoes; poultry, sheep, hogs, cattle; dairying). Small part of Gallitzin State Forest in SE corner; Yellow Creek State Park in E center; YELLOW CREEK LAKE and Two Lick reservoirs in E center. Formed 1803.

Indiana (EEN-zhee-ah-nah), town (2007 population 4,685), W SÃO PAULO state, BRAZIL, 9 mi/14.5 km E of PRESIDENTE PRUDENTE; 22°12′S 51°14′W.

Indiana, borough (2006 population 14,817), ⊙ INDIANA county, W central PENNSYLVANIA, railroad terminus and principal supply and trading center for a bituminous-coal mining area in the ALLEGHENY MOUNTAINS; 40°37′N 79°09′W. Manufacturing (diesel engines, medical products, food, printing and publishing, lab equipment, rubber products); surface coal mining. Agricultural area (corn, hay; livestock, dairying). Actor Jimmy Stewart born here 1908. Seat of Indiana University of Pennsylvania. Jimmy Stewart Field (public airport) to E; Jimmy Stewart Museum; County Historical Society; Two Lick Reservoir to E; YELLOW CREEK Reservoir and State Park to SE. Incorporated 1816.

Indiana Dunes (□ 20 sq mi/52 sq km), NW INDIANA, 200-ft/60-m sand dunes, beaches, and marshes along the S shore of Lake MICHIGAN. National Lakeshore, authorized 1966.

Indiana Harbor, INDIANA: see EAST CHICAGO.

Indianapolis, city (2006 population 785,597), ⊙ INDIANA and MARION county, central Indiana, on the WHITE RIVER; 39°47′N 86°09′W. Selected 1820 as the site of the state capital (which was moved here 1824–1825). The largest city in Indiana, it is the chief processing point in a rich agricultural region and is a major grain market. It is also the commercial, transportation, and industrial center for a large area and is Indiana's leading manufacturing city (printing and publishing, flour milling, construction equipment, clay products, electronics, paper products, chemicals, auto parts, food products, feeds and fertilizers, lumber products, agricultural equipment, crushed limestone, dairy products, apparel, pharmaceuticals). The site for the city was deliberately located at a point equidistant from the four corners of Indiana; largest metropolitan area in U.S. not situated on a navigable river. On January 1, 1970, Indianapolis consolidated with all of Marion county, except for the municipalities of BEECH GROVE, SOUTHPORT, SPEEDWAY, and LAWRENCE. The city is the seat of Butler University, Marian College, University of Indianapolis, Christian Theological Seminary, and Indiana University–Purdue University at Indianapolis (IUPUI), with many units, including the Medical Center and the Herron School of Art. The American Legion has its national headquarters here in a building erected as a war memorial. Landmarks are the state capitol (1878–1888); the state library and historical building; the home and burial place of James Whitcomb Riley; the home and burial place of Benjamin Harrison (twenty-third president of the U.S.); a Carmelite monastery; the Soldiers and Sailors Monument (1902); the Bank One Tower, tallest building in the state; and the Indianapolis Motor Speedway, site of the world-famous annual 500-mi/805-km automobile race (Indy 500). In the city's downtown section is the RCA Dome (formerly Hoosierdome), a massive indoor sports facility and convention center. In the 1980s, Indianapolis acquired a National Football League team, the Colts, and Market Square Arena is the home of the National Basketball Association's Indiana Pacers. The city hosts numerous cultural events and has

noteworthy museum, a symphony orchestra, and a zoo. Indianapolis International Airport is on the W edge of the city. Fort Benjamin Harrison (to the NE) has been closed; part of its area has been converted to a state park, and the rest is being developed as residential and commercial property. The Naval Air Warfare Center (NAWC) was privatized in 1996 and is now devoted to the design and production of advanced electronics; this was the largest privatization of an American military base in history. Incorporated 1847.

Indian Arm (IN-dee-uhn AHRM), N arm of BURRARD INLET, SW BRITISH COLUMBIA, CANADA, 8 mi/13 km NE of VANCOUVER; 13 mi/21 km long, 1 mi/2 km wide; 49°22′N 122°53′W. Receives small Indian River at head. On both shores mountains rise to 3,000 ft/914 m–5,000 ft/1,524 m.

Indian Beach (IN-dee-uhn BEECH), village (□ 1 sq mi/2.6 sq km; 2006 population 96), CARTERET county, E NORTH CAROLINA, 12 mi/19 km WSW of MOREHEAD CITY, on ATLANTIC OCEAN, on BOGUE ISLAND; 34°41′N 76°53′W. Beach resort area. Theodore Roosevelt State Natural Area and North Carolina Aquarium to E.

Indian Creek, village (2000 population 194), Lake county, NE ILLINOIS, residential suburb 27 mi/43 km NNW of CHICAGO, 8 mi/12.9 km W of LAKE FOREST; 42°13′N 87°58′W.

Indian Creek, c.60 mi/97 km long, S INDIANA; rises in SW CLARK county, flows SW to the OHIO River 11 mi/18 km SW of CORYDON.

Indian Desert, INDIA: see THAR DESERT.

Indian Empire, those parts of the INDIAN subcontinent which were formerly directly or indirectly under British rule or protection; comprised BRITISH INDIA and INDIAN (or Native) States. Until 1937 included BURMA and ADEN. In accordance with Indian Independence Act of 1947, INDIAN EMPIRE was partitioned into the two self-governing dominions of INDIA and PAKISTAN.

Indian Harbour Beach (IN-dee-uhn), town (□ 2 sq mi/5.2 sq km; 2000 population 8,152), BREVARD county, E central FLORIDA, 20 mi/32 km S of CAPE CANAVERAL; 28°08′N 80°35′W. Manufacturing includes water and waste systems, communications equipment, and cosmetics.

Indian Head (IN-DEE-uhn), town (2006 population 1,634), SE SASKATCHEWAN, CANADA, E of REGINA; 50°32′N 103°40′W. In wheat-growing region. Flour mills and grain elevators. Agricultural.

Indian Head, town (2000 population 3,422), CHARLES county, S MARYLAND, on the POTOMAC c.28 mi/45 km below WASHINGTON, D.C.; 38°36′N 77°10′W. The U.S. Naval Ordnance Station, sometimes called the U.S. Naval Propellant Plant, a facility covering 2,072 acres/839 ha and with 1,347 buildings, has been located here since 1890. Producing solid rocket fuel, it is the largest employer in CARROLL county, with 2,200 civilian workers. Original houses built by the Navy as residences for its workers have been privately sold.

Indian Head Park, village (2000 population 3,685), COOK county, NE ILLINOIS, residential suburb 15 mi/24 km WSW of CHICAGO, near Des Plaines River; 41°46′N 87°54′W.

Indian Hills, town, JEFFERSON county, N KENTUCKY, residential suburb 5 mi/8 km ENE of LOUISVILLE, near OHIO RIVER. Zachary Taylor National Cemetery to NE.

Indian Hills Cherokee Section (CHER-uh-kee), town, JEFFERSON county, N KENTUCKY, residential suburb 7 mi/11.3 km ENE of LOUISVILLE. Zachary Taylor National Cemetery is here, including grave of President Taylor.

Indian House Lake (IN-dyuhn HOUS), (□ 125 sq mi/325 sq km), NE QUEBEC, E CANADA, on GEORGE River; 56°00′N 64°30′W; 35 mi/56 km long, 2 mi/3 km wide. On E shore hills rise to c.1,800 ft/549 m.

Indian Island, KNOX county, S MAINE, small lighthouse island off harbor of ROCKPORT.

Indian Islands, CANADA: see EAST INDIAN ISLAND and WEST INDIAN ISLAND.

Indian Lake, resort village (□ 7 sq mi/18.2 sq km), HAMILTON county, NE central NEW YORK, in ADIRONDACK MOUNTAINS, near N end of Indian Lake (□ c.7 sq mi/18.1 sq km; c.7 mi/11.3 km long), c.45 mi/72 km NW of GLENS FALLS; 43°47′N 74°18′W. Lumber and wood products.

Indian Lake, village (2006 population 569), CAMERON county, extreme S TEXAS, residential community 12 mi/19 km N of BROWNSVILLE, W of LOS FRESNOS; 26°05′N 97°30′W. Irrigated agriculture area of Rio Grande Valley.

Indian Lake, borough (2006 population 451), SOMERSET county, SW PENNSYLVANIA, residential community 11 mi/18 km E of SOMERSET; 40°02′N 78°51′W. Town surrounds Indian Lake reservoir. Lake Stonycreek reservoir immediately to S.

Indian Lake (IN-dee-uhn), c.4 mi/6 km wide, LOGAN county, W central OHIO, at source of GREAT MIAMI RIVER, 10 mi/16 km NW of BELLEFONTAINE; 40°28′N 83°53′W. Formed by Indian Lake Dam. Indian Lake State Park on S shore.

Indian Lake, MICHIGAN: see INDIAN RIVER.

Indian Lake, NEW JERSEY: see Denville.

Indian Lorette, CANADA: see LORETTEVILLE.

Indian Ocean, third-largest ocean (□ c.28,350,000 sq mi/73,426,500 sq km), extending from S ASIA to ANTARCTICA and from E AFRICA to SE AUSTRALIA; c.4,000 mi/6,437 km wide at the equator; has an average depth of c.11,000 ft/3,353 m. Constitutes about 20% of the world's total ocean area.

Relationship to Other Bodies of Water

Connected with the PACIFIC OCEAN by passages through the MALAY ARCHIPELAGO and between Australia and Antarctica; with the ATLANTIC OCEAN by the expanse between Africa and Antarctica, and with the MEDITERRANEAN SEA via the SUEZ CANAL. Its chief arms are the ARABIAN SEA (with the RED SEA, the Gulf of ADEN, and the PERSIAN GULF), the Bay of BENGAL, and the ANDAMAN SEA.

Continental Shelf and Volcanic Cones

The continental shelf of the Indian Ocean is narrow. MADAGASCAR and SRI LANKA, its largest islands, are structurally parts of their respective continents as are SOCOTRA, the ANDAMAN Islands, and the NICOBAR Islands; the SEYCHELLES and the KERGUELEN Islands are exposed tops of submerged ridges. The LAKSHADWEEP (Laccadives), the MALDIVES, and the CHAGOS are low coral islands, and MAURITIUS and RÉUNION are high volcanic cones.

Mid-Oceanic Ridge and Deep-sea Basins

The Mid-Oceanic Ridge, a broad submarine mountain range extending from Asia to Antarctica, divides the Indian Ocean into three major sections—the African, Antardis, and Australasian. The ridge rises to an average elevation of c.10,000 ft/3,048 m, and a few peaks emerge as islands. A large rift, an extension of the E branch of the GREAT RIFT VALLEY that runs through the Gulf of Aden, extends along most of its length. The Mid-Oceanic Ridge, along with other submarine ridges, encloses a series of deep-sea basins (abyssal plains). The greatest depth (25,344 ft/7,725 m) is in the JAVA TRENCH, S of JAVA (INDONESIA).

Waters That Flow Into the Ocean

Receives the waters of the ZAMBEZI, TIGRIS and EUPHRATES, INDUS, GANGA-BRAHMAPUTRA, and AYEYARWADY (Irawadi) rivers.

Temperature and Circulation Systems

The surface waters of the ocean are generally warm, although close to Antarctica pack ice and icebergs are found. The ocean has two water circulation systems—a regular counterclockwise S system (SOUTH EQUATORIAL CURRENT, MOZAMBIQUE CURRENT, WEST WIND DRIFT, WEST AUSTRALIAN CURRENT) and a N system, the Monsoon Drift, whose currents are directly related to the seasonal shift of monsoon winds.

Area is shown by the symbol □, and capital city or county seat by ⊙.

The SW monsoon draws moisture from the Indian Ocean and drops heavy rainfall on the Indian subcontinent and SE Asia.

Economic Activity

Fishing is comparatively less developed in the Indian Ocean than in the Atlantic and PACIFIC oceans. Fishing activity takes place in the Arabian Sea and Bay of Bengal. The Indian Ocean has long served as a major highway for domestic and foreign commerce, sailed by Arabs, Indians, and Chinese for centuries.

Indianola (in-dee-uh-NO-luh), city (2000 population 12,998), ⊙ WARREN county, S central IOWA, 16 mi/26 km S of DES MOINES, N of SOUTH RIVER; 41°21′N 93°34′W. Manufacturing (building equipment, agricultural machinery, plastic and metal products, consumer goods, animal feed). Seat of Simpson College (1860). Lake Ahquabi State Park to S. Incorporated 1863.

Indianola, city (2000 population 12,066), ⊙ SUNFLOWER county, W MISSISSIPPI, 24 mi/39 km E of GREENVILLE, near SUNFLOWER RIVER; 33°27′N 90°38′W. In rich agricultural area (cotton, corn, alfalfa, rice, soybeans, pecans; catfish); manufacturing (catfish, meat, and pecan processing; paper products). Settled in mid-19th century; incorporated 1886.

Indianola (IN-dee-ah-NO-lah), unincorporated town, ALLEGHENY county, W PENNSYLVANIA, on Deer Creek, suburb 11 mi/18 km NW of PITTSBURGH; 40°34′N 79°51′W. Manufacturing of medical equipment and plastic products.

Indianola, unincorporated town (2000 population 3,026), KITSAP county, NW WASHINGTON, 7 mi/11.3 km SW of EDMONDS, on Port Madison bay, arm of PUGET SOUND; 47°45′N 122°31′W. Manufacturing (consumer goods). Port Madison Indian Reservation to E.

Indianola (en-de-an-OH-la), village (2000 population 207), VERMILION county, E ILLINOIS, 14 mi/23 km SSW of DANVILLE; 39°55′N 87°44′W. In agricultural and bituminous-coal area. Near LITTLE VERMILION RIVER.

Indianola, village (2006 population 600), RED WILLOW county, S NEBRASKA, 10 mi/16 km E of MCCOOK, on REPUBLICAN RIVER; 40°13′N 100°25′W. Grain, livestock; butchering.

Indianola, village (2006 population 194), PITTSBURG county, SE OKLAHOMA, 16 mi/26 km N of MCALESTER, near CANADIAN River (EUFAULA LAKE); 35°10′N 95°46′W.

Indianola, unincorporated village, CALHOUN county, S TEXAS, on MATAGORDA BAY, c.11 mi/18 km SE of PORT LAVACA. Founded 1844, it was once most active port in state and port of entry for many immigrants; destroyed by hurricanes of 1875 and 1886. Myrtle Foester Whitmire Division, Aransas National Wildlife Refuge to W.

Indianópolis (EEN-shee-ah-NAH-po-lees), city, W PARANÁ state, BRAZIL, 50 mi/80 km WSW of MARINGÁ; 23°25′S 52°44′W. Coffee, corn, manioc; rice; livestock.

Indianópolis (EEN-zhee-ah-NO-po-les), town (2007 population 6,212), MINAS GERAIS, BRAZIL, on Rio Araquari, 31 mi/50 km SSE of UBERLÂNDIA; 19°01′S 47°56′W.

Indian Orchard, Massachusetts: see SPRINGFIELD.

Indian Pass, gorge in ADIRONDACK MOUNTAINS, ESSEX county, NE NEW YORK, between WALLFACE MOUNTAIN (W) and Mount MACINTYRE (E), c.6 mi/9.7 km WNW of Mount MARCY, c.1,300 ft/396 m deep, c.1 mi/1.6 km long; 44°08′N 74°02′W. Hiking trails.

Indian Peak (IN-dee-uhn PEEK), (9,817 ft/2,992 m), SE BRITISH COLUMBIA, W CANADA, near ALBERTA border, in ROCKY MOUNTAINS, near SE side of KOOTENAY NATIONAL PARK, 20 mi/32 km SSW of BANFF (Alberta); 50°55′N 115°45′W.

Indian Point Nuclear Power Plants, two nuclear power–generating stations, WESTCHESTER county, at

BUCHANAN on E side of HUDSON RIVER, 35 mi/56 km N of NEW YORK city; 41°16′N 73°56′W. Entergy Nuclear Northeast owns and operates both units. Indian Point Unit 2, a 975-MW reactor, began operating in 1974. Indian Point Unit 3, a 960-MW generator, began operating in 1976. Unit 1, opened in 1962, halted operations in 1974; its nuclear fuel was removed and its facilities now support the operation of Unit 2. About 1,500 people work at Indian Point. The facility provides up to 30% of the electricity used by the New York metro area. Unit 3 was shut down from February 1993–July 1995 due to acute operational and management problems. Indian Point's nuclear facilities are surrounded by more people than any other nuclear power site in the country: 10% of U.S. population lives within 60 mi/97 km of the site.

Indian Pond, SOMERSET county, central MAINE, in SAINT ALBANS town, 18 mi/29 km NE of SKOWHEGAN; 3.5 mi/5.6 km long.

Indian Pond, reservoir, SOMERSET county, W central MAINE, on KENNEBEC RIVER, 15 mi/24 km W of GREENVILLE; 8 mi/12.9 km long; 45°30′N 69°50′W. Receives old and new channels of Kennebec River from MOOSEHEAD LAKE reservoir (NE).

Indian River (IN-dee-uhn), county (616 sq mi/1,601.6 sq km; 2006 population 130,100), E central FLORIDA, on the ATLANTIC (E); ⊙ VERO BEACH; 27°41′N 80°34′W. Coastal lowland bordered by barrier beach enclosing INDIAN RIVER lagoon; interior is a marshy peat area containing Lake Wilmington. County forms part of Indian River district noted for its citrus fruit, especially oranges; also a farming and tourist region; sugarcane grown around FELLSMERE. Formed 1925.

Indian River, village, E ALASKA, on COPPER RIVER, 120 mi/193 km NE of VALDEZ, on TOK CUT-OFF.

Indian River, village, CHEBOYGAN county, N MICHIGAN, 18 mi/29 km SSW of CHEBOYGAN, on SE shore of BURT LAKE; 45°24′N 84°36′W. In resort and forest area. Light manufacturing. Burt Lake State Park to W.

Indian River (IN-dee-uhn), lagoon, c.100 mi/161 km long, E central FLORIDA; parallel to the ATLANTIC coast from N of TITUSVILLE S to STUART. Along the lagoon a variety of citrus and vegetable products are grown and transported by small boats to towns on its waterway and those farther inland. The river's coasts have been marked by housing developments, especially for retired communities and for vacationers. Notable resort towns along its shores include Titusville, VERO BEACH, SEBASTIAN, and FORT PIERCE.

Indian River, 10 mi/16 km long, SUSSEX county, SE DELAWARE, tidal estuary formed by small streams just W of MILLSBORO (dam here); flows E, widens into INDIAN RIVER BAY before entering ATLANTIC ocean through narrow passage in Delaware Seashore State Park 5 mi/8 km N of BETHANY BEACH.

Indian River, c.40 mi/64 km long, in S UPPER PENINSULA, MICHIGAN; rises in NW SCHOOLCRAFT county, flows SE to INDIAN LAKE (c.6 mi/9.7 km long, 4 mi/6.4 km wide) just NW of MANISTIQUE, then short distance E to MANISTIQUE RIVER; 45°59′N 86°17′W. Palms Book State Park on NW end; two units of Indian Lake State Park on E and W sides; Big Spring on W side. Sometimes called BIG INDIAN RIVER.

Indian River, c.80 mi/129 km long, in N NEW YORK; rises in N LEWIS county, flows NW to ANTWERP, and SW, past PHILADELPHIA village, then generally N, past THERESA, to S end of BLACK LAKE in SAINT LAWRENCE county. At NATURAL BRIDGE, river has cut limestone bridge and caverns.

Indian River Bay, estuary, SE DELAWARE, widening of INDIAN RIVER, which enters from West REHOBOTH BAY and forms large N arm of bay, c.6 mi/9.7 km long, 4 mi/6.4 km wide; 38°36′N 75°06′W. Barrier beach (Delaware Seashore State Park), cut by dredged passage, protects bay from the ATLANTIC.

Indian River Shores (IN-dee-uhn), town (□ 7 sq mi/18.2 sq km; 2000 population 3,448), INDIAN RIVER

county, E central FLORIDA, 5 mi/8 km NE of VERO BEACH; 27°42′N 80°22′W.

Indian Rocks Beach (IN-dee-uhn), town (2000 population 5,072), PINELLAS county, W central FLORIDA, 10 mi/16 km S of CLEARWATER; 27°53′N 82°50′W.

Indian Shores (IN-dee-uhn), town (2000 population 1,705), PINELLAS county, W central FLORIDA, 12 mi/19 km S of CLEARWATER; 27°51′N 82°50′W.

Indian Springs, unincorporated village (2000 population 1,302), CLARK county, S NEVADA, 40 mi/64 km NW of LAS VEGAS; 36°34′N 115°40′W. Indian Springs Air Force Base is here (N). Headquarters for Nellis Air Force Bombing and Gunnery Range (□ c.5,000 sq mi/12,950 sq km), which includes Nevada Test Site (U.S. Atomic Energy Commission) to NW, and FRENCHMAN FLAT, desert basin which was site (Jan. 1951) of experimental atomic explosions. Part of Toiyabe National Forest, including Mount Charleston to S. Desert National Wildlife Range to N. Desert View Natl Area to SE.

Indian Stream, c.25 mi/40 km long, COOS county, N NEW HAMPSHIRE; rises near QUEBEC border, flows S to the CONNECTICUT RIVER below PITTSBURG.

Indian Stream Republic, short-lived independent territory at headwaters of CONNECTICUT River, over which neither UNITED STATES nor Canadian jurisdiction was established. Set up 1832 by local inhabitants; annexed 1835 by NEW HAMPSHIRE; awarded to United States by Webster-Ashburton Treaty of 1842.

Indian Subcontinent, region, S central ASIA, comprising the countries of AFGHANISTAN, PAKISTAN, INDIA, and BANGLADESH and the Himalayan states of NEPAL and BHUTAN. SRI LANKA, an island off the SE tip of the Indian peninsula, is also considered a part of the subcontinent.

Indian Territory, in U.S. history, name applied to the region in U.S. GREAT PLAINS set aside for Native Americans by the Indian Intercourse Act (1834). In the 1820s, the Federal government began moving the Five Civilized Tribes of the Southeast (Cherokee, Creek, Seminole, Choctaw, and Chickasaw) to lands W of the MISSISSIPPI RIVER. The Indian Removal Act of 1830 gave the president authority to designate specific lands for them, and in 1834 Congress formally approved the choice. The Indian Territory included present-day OKLAHOMA N and E of the RED RIVER, as well as parts of KANSAS and NEBRASKA; the lands were delimited in 1854, however, by the creation of the Kansas and Nebraska territories. Tribes other than the original five also moved there, but each one maintained its own government. As white settlers continued to move W, pressure to abolish the Indian Territory mounted. With the opening of W Oklahoma to whites in 1889 the way was prepared for the extinction of the territory, achieved in 1907 with the entrance of Oklahoma into the Union. Oklahoma Territory est. 1890; opened the region to white settlement. The Indian Removal Act of 1830 exchanged Indian lands in SE U.S. for land W of Mississippi. The five civilized tribes functioned as nations. The relocation of the tribes, known as the Trail of Tears, in the winter of 1838–1839 caused many deaths. See OKLAHOMA.

Indian Tibet, INDIA and PAKISTAN: see LADAKH.

Indiantown (IN-dee-uhn), town (□ 5 sq mi/13 sq km; 2000 population 5,588), MARTIN county, E central FLORIDA, 35 mi/56 km NW of WEST PALM BEACH; 27°02′N 80°28′W. Manufacturing includes flour, concrete, beverages, and building equipment.

Indiantown Gap Military Reservation, Pennsylvania: see ANNVILLE.

Indian Trail (IN-dee-uhn TRAI-uhl), town (□ 15 sq mi/39 sq km; 2006 population 17,491), UNION county, S NORTH CAROLINA, suburb 14 mi/23 km SE of CHARLOTTE; 35°04′N 80°40′W. Retail trade; service industries; manufacturing (transformers, paper products, metal fabrication); agriculture (cotton, grain; livestock; dairying).

Indian Village, town (2000 population 144), SAINT JOSEPH county, N INDIANA, suburb of SOUTH BEND; 41°43′N 86°14′W.

Indian Wells, affluent city (2000 population 3,816), RIVERSIDE county, S CALIFORNIA, c.15 mi/24 km ESE of PALM SPRINGS; 33°43′N 116°18′W.

Indiaporã (EEN-zhee-AH-po-ruh), city, extreme NW São Paulo state, 22 mi/35 kmN of Ferdinanópolis, on Agua Vermelha Reservoir; 19°57′S 50°17′W.

Indiara (EE-zhee-AH-rah), city (2007 population 12,753), central GOIÁS, BRAZIL, 62 mi/100 km SW of GOIÂNIA; 17°10′S 50°02′W.

Indiaroba (EEN-zhee-ah-ro-bah), city (2007 population 17,043), S SERGIPE, NE BRAZIL, on BAHIA border, 16 mi/26 km S of ESTÂNCIA; 11°32′S 37°31′W. Manioc. Until 1944, called Espírito Santo.

Indiavaí (EEN-dee-ah-vah-EE), town (2007 population 2,506), MATO GROSSO, BRAZIL, c.215 mi/346 km W of CUIABÁ; 15°30′S 58°34′W.

Indies: see EAST INDIES and WEST INDIES.

Indiga (EEN-dee-guh), settlement, W NENETS AUTONOMOUS OKRUG, ARCHANGEL oblast, N European Russia, port on the BARENTS SEA, 110 mi/177 km W of NARYAN-MAR; 67°39′N 49°02′E. Fish cannery.

Indigirka (een-dee-GEER-kah), river, approximately 1,100 mi/1,770 km long, N RUSSIAN FAR EAST, in the NE SAKHA REPUBLIC. It rises in the OYMYAKON Plateau and flows N into the ARCTIC OCEAN, cutting through the Cherski Range and through tundra zone, past Khomu (head of navigation), DRUZHINA, and CHOKURDAKH, to the EAST SIBERIAN SEA, forming large delta mouth. Main tributaries are Selennyakh (left) and Moma (right) rivers. It is navigable (June–September) from its confluence with the Moma River to the Arctic Ocean.

Indigo Crossing, AUSTRALIA: see BARNAWATHA.

Indin Lake, NORTHWEST TERRITORIES, CANADA, 120 mi/193 km NNW of YELLOWKNIFE, 25 mi/40 km long, 1 mi/2 km–8 mi/13 km wide; 64°15′N 115°15′W. Drains S into GREAT SLAVE LAKE by Snare River. Gold deposits discovered here 1945.

Indio (IN-dee-o), city (□ 27 sq mi/70.2 sq km; 2000 population 49,116), RIVERSIDE county, SE CALIFORNIA, 65 mi/105 km ESE of RIVERSIDE, 23 mi/37 km ESE of PALM SPRINGS, in the COACHELLA VALLEY of the COLORADO DESERT; 33°43′N 116°14′W. The city is 22 ft/7 m below sea level. It is the trade and administrative center for a citrus, grape, and date area; cotton; grain; poultry. Indio was once the center of one of the largest date-producing areas in the U.S., but that industry has gradually moved S of the city. Manufacturing (machinery). The National Date Festival is held on the Indio county fairgrounds. The area has also benefited from the regional growth of it becoming a resort and retirement community. Date Gardens; JOSHUA TREE NATIONAL MONUMENT to NE; San Bernardino National Forest and SANTA ROSA MOUNTAINS to SW; SALTON SEA, inland saltwater lake, 15 mi/24 km to SE; Cabazon Indian Reservation to SE. Founded 1876, incorporated 1930.

Indios (EEN-zhee-os), city, S SANTA CATARINA state, BRAZIL, 8 mi/12.9 km E of Lages; 27°47′S 50°12′W. Corn, manioc; livestock.

Indira Gandhi Canal, an extension of the BHAKRA-NANGAL irrigation system (among the world's largest), in the W areas of GANGANAGAR, BIKANER, and JAISALMER districts, RAJASTHAN state, NW INDIA. Supplies water to parts of the THAR DESERT. Fed by a fully lined feeder canal from the Harike Barrage on the BEAS and SUTLEJ rivers in PUNJAB state. When fully completed, it will be c.400 mi/650 km long, irrigating over 3,860 sq mi/10,000 sq km of desert land. To accommodate the influx of new farmers, who will be provided with 14.8-acre/6-ha plots in the area, several new settlements are being planned. The current BIKANER CANAL draws water from FIROZPUR headworks further downstream and provides water mainly to GANGANAGAR district. Formerly known as the RAJASTHAN CANAL PROJECT; initially scheduled to be completed by 1999; as of 2007, still under construction.

Indispensable Strait, Solomon Islands, SW Pacific, separates Malaita (E) and islands of Guadalcanal and Santa Isabel (W); 40 mi/64 km wide.

Indje, Cape, TURKEY: see INCE, CAPE.

Indjija (een-JEE-yah), town (2002 population 49,609), SREM district, in SREM region, VOJVODINA, N SERBIA, 25 mi/40 km NW of BELGRADE; 45°03′N 020°04′E. Railroad junction. Also spelled Indyiya.

Indochina, French *Indochine*, former federation of colonial states, SE ASIA. It comprised the French colony of COCHIN CHINA and the French protectorates of TONKIN, ANNAM, LAOS, and CAMBODIA (Cochin China, Tonkin, and Annam were later united to form VIETNAM). The capital was HANOI. The federation formed the easternmost region of mainland SE Asia (which it shared with THAILAND and the British colonies of BURMA and MALAYA) and faced E on the SOUTH CHINA SEA. The cultures of Indochina were influenced by CHINA and INDIA. The centuries before European intervention saw the growth and decline of the KHMER EMPIRE in Cambodia, the rise and fall of CHAMPA, and the steady expansion of Annam. European penetration began in the 16th century; in the 19th-century race for a colonial empire, the French took (1862, 1867) Cochin China as a colony and gained protectorates over Cambodia (1863), Annam (1884), and Tonkin (1884). In 1887 they forged those four states into a union of Indochina, with a governor general at its head; LAOS was added to the union in 1893. In World War II, FRANCE was forced to accept Japanese intervention in N Indochina (1940); the subsequent Japanese move into S Indochina (July 1941) was viewed by the U.S. as a threat to the PHILIPPINES; it prompted the freezing of all Japanese assets in the U.S. and precipitated diplomatic exchanges which were cut short by the JAPANESE attack on PEARL HARBOR. Even before the end of the war, the French announced plans for a federation of Indochina within the French Union, with greater self-government for the various states. The federation was accepted in Cambodia and Laos. Vietnamese nationalists, however, demanded (1945) the complete independence of Annam, Tonkin, and Cochin China as Vietnam, and after December 1946, these regions were plunged into bitter fighting between the French and the extreme nationalists, oftentimes led by Communists. The war in Vietnam dragged on for years, culminating in the French defeat at DIEN BIEN PHU. The Geneva Conference in 1954 effectively ended French control of Indochina.

Indonesia, republic (□ c.735,000 sq mi/1,903,650 sq km; 2004 estimated population 238,452,952; 2007 estimated population 234,693,997), SE ASIA, in the MALAY ARCHIPELAGO (sometimes called Nusantara); ⊙ JAKARTA (largest city), on JAVA.

Geography

The fifth-most populous country in the world, Indonesia comprises more than 13,000 islands extending c.3,000 mi/4,830 km along the equator from the MALAYSIA mainland toward AUSTRALIA; the archipelago forms a natural barrier between the INDIAN and PACIFIC oceans. Consisting of the territory of the former Dutch (or Netherland) East Indies, Indonesia's main island groups are the Greater SUNDA ISLANDS, which include Java, SUMATRA, central and S BORNEO (Kalimantan), and SULAWESI (formerly Celebes); the Lesser SUNDA ISLANDS, consisting of BALI, FLORES, SUMBA, LOMBOK, and the W part of TIMOR; the MALUKU (Moluccas), with AMBON, BURU, SERAM, and HALMAHERA and the RIAU ARCHIPELAGO. IRIAN JAYA (West New Guinea), after years of dispute with the Dutch, was formally annexed by Indonesia in August 1969. The most important islands, culturally and economically, are Java, Sumatra, and Bali. All the larger islands have a central volcanic mountainous area flanked by coastal plains; there are more than 100 active volcanoes. Earthquakes are frequent and, occasionally severe. On December 26, 2004, the western islands of Indonesia were devastated by a massive tsunami, generated by a 9.0 magnitude earthquake off the W coast of Sumatra, which sustained the heaviest damage and the loss of tens of thousands of lives; particularly affected was the Aceh province and the city of BANDA ACEH. The animal life of Indonesia roughly forms a connecting link between the fauna of Asia and that of Australia. Elephants are found in Sumatra and Borneo, tigers as far S as Java and Bali, and marsupials in Timor and Irian Jaya. Crocodiles, snakes, and richly colored birds are everywhere. The tropical climate, abundant rainfall, and remarkably fertile volcanic soils permit a rich agricultural yield. Administratively, Indonesia is divided into thirty provinces (Sumatra, seven provinces; Riau Islands; Bangka-Belitung Islands; Java, four provinces; Bali; Kalimantan (Indonesian Borneo), four provinces; Sulawesi, six provinces; Lesser Sundas, two provinces; Maluku, two provinces; and Indonesian New Guinea, two provinces); two special regions (Aceh and Yogyakarta), and the capital district, and is further divided into districts (*kabupaten*) and municipalities (*kotamadya*). Indonesia is one of five countries claiming sovereignty over the Spratly Islands in the South China Sea.

Population

The population falls roughly into two groups, the Malays and the Papuans, with many of the inhabitants E of Bali representing a transition between the two types. Within each group are numerous subdivisions, and cultural development ranges from the modern Javanese and Balinese to ancient tribes in Borneo, Sumatra, and Irian Jaya. The complex ethnic structure is the result of several great migrations many centuries ago largely from Asia. More than three hundred languages are spoken in Indonesia, but an official language, Bahasa Indonesia (regarded as the purest Malay and originally spoken in the Rim Archipelago) has been adopted; it has spread rapidly and is now understood in all but the most remote villages. English is considered to be the country's second language. About 87% of the population is Muslim, some 9% is Christian, and about 2% is Hindu. Hindus are concentrated principally on Bali, which is known for its distinctive culture. Animism, sometimes combined with Islam, is common among some tribal groups. The Chinese constitute by far the greatest majority of the nonindigenous population; they number between two to three million and play an important role in the country's economic life. There are smaller minorities of Arabs and Indians.

Higher Education

Notable among the many state universities scattered throughout the islands are the University of Indonesia, at Jakarta; the Bandung Technology Institute, one of the country's oldest and most prestigious universities; Airlangga Surabaya University, at Surabaya; Gadjah Mada University, at Yogyakarta; and the University of North Sumatra, at Medan. Private schools include the Islamic University of Indonesia, at Yogyakarta, and National University, at Jakarta.

Economy

Crude oil is Indonesia's most valuable natural resource. Also, Indonesia is the world's leading supplier of liquified natural gas. Nearly all of the country's oil and gas deposits are located on Sumatra, though oil fields have recently been discovered in Kalimantan Timur province. Indonesia is one of the world's major rubber producers; other plantation and smallholder crops include sugarcane, coffee, tea, tobacco, palm oil, cinchona, cacao, sisal, coconuts, and spices. Despite plantation cultivation, Indonesia has a wide landholding base; the majority of the people are largely self-sufficient in food. Rice is the major crop; cassava, maize, yams, soybeans, peanuts, and fruit are also grown. Horses and cattle are raised on some of the Lesser Sunda Islands. Fish are abundant, both in the ocean and in inland ponds. In natural-resource potential, Indonesia is one of the wealthiest countries in the world. It has great timberlands; vast rain forests of giant trees (among the world's tallest) cover the mountain slopes, and teak, sandalwood, ironwood, camphor, and ebony are cut. Palm, rattan, and bamboo abound, and a great variety of forest products are produced. Indonesia is a major exporter of timber, accounting for nearly half of the world's tropical hardwood trade. However, the rapid deforestation of Indonesia's hardwoods, mainly due to its expanding population and growing timber-related industries, has caused concern among international environmental groups regarding the impact on global warming. Tin, bauxite, nickel, coal, manganese, salt, copper, gold, and silver are also mined. Salt is available in large quantities from shallow enclosed seashore lagoons, especially Madura Island and Java Timur province. Iron is believed to exist in great quantity, and uranium has been reported. Primarily a supplier of raw materials, the country has begun to industrialize. However, much of its working force is still engaged in agriculture. Industry is mainly limited to food, mineral and wood processing, a variety of light manufacturing, and cement production. However, Indonesia faces economic development concerns because of recent acts of terrorism, endemic corruption, and the weaknesses of its banking system.

History: to 16th Century
Early in the Christian era, Indonesia came under the influence of Indian civilization through the gradual influx of Indian traders and Buddhist and Hindu monks. By the 7th and 8th centuries, kingdoms closely connected with India had developed in Sumatra and Java; the Hindu temple, PRAMBANAN, and the spectacular Buddhist temples of BOROBUDUR date from this period. Sumatra was the seat (7th–13th centuries) of the important Buddhist kingdom of SRI VIJAYA, in the SE in the PALEMBANG and JAMBI areas. In the late 13th century the center of power shifted to Java, where the fabulous Hindu kingdom of Majapahit had arisen; for two centuries it held sway over Indonesia and large areas of the Malay Peninsula. A gradual infiltration of Islam began in the 14th and 15th centuries with the arrival of Indian Muslim and Arab traders, and by the end of the 16th century Islam had replaced Buddhism and Hinduism as the dominant religion. The once-powerful kingdoms broke into smaller Islamic states whose internecine strife made them vulnerable to European imperialism.

History: 16th Century to 1825
Early in the 16th century the Portuguese, in pursuit of the rich spice trade, began establishing trading posts in Indonesia, after taking the strategic commercial center of MELAKA (Malacca) in 1511 on the Malay Peninsula. The Dutch followed in 1596 and the English in 1600. By 1610 the Dutch had ousted the Portuguese, who were allowed to retain only the E part of Timor Island, but English competition remained strong, and it was only after a series of Anglo-Dutch conflicts (1610–1623) that the Dutch emerged as the dominant power in Indonesia. Throughout the 17th and 18th centuries the Dutch East India Company steadily expanded its control over the entire area. When the company was liquidated in 1799, the Dutch government assumed its holdings, which were thereafter known in English as the Netherlands (or Dutch) East Indies. Dutch rule was briefly broken (1811–1814) during the Napoleonic Wars when the islands were occupied by the British under T. Stamford Raffles. Bengkulu, the last British holding in SW Sumatra, was relinquished in 1824.

History: 1825 to 1957
The Dutch exploited the riches of the islands throughout the 19th century, but their rule did not go unchallenged by the Indonesians. In 1825, Prince Diponegoro of Java launched a long and bloody guerrilla war against the Dutch; in 1906 and again in 1908 the native rulers of Bali led their subjects in suicidal charges against Dutch fortifications; the Acehnese in N Sumatra fought Dutch rule from 1820 into the early 1900s. The Indonesian independence movement began early in the 20th century. The Indonesian Communist Party (PKI) was founded in 1920; in 1927 the Indonesian Nationalist Party (PNI) arose under the leadership of Sukarno. It received its impetus during World War II, when the Japanese drove out (1942) the Dutch and occupied the islands. In August 1945, immediately after the Japanese surrender, Sukarno and Muhammad Hatta, another nationalist leader, proclaimed Indonesia an independent republic. The Dutch bitterly resisted the nationalists, and four years of intermittent and sometimes heavy fighting followed. Under UN pressure, an agreement was finally reached (November 1949) for the creation of an independent republic of Indonesia. A new constitution provided for a parliamentary form of government. Sukarno was elected president, and Hatta became premier and then vice president. Although Sukarno had achieved a major accomplishment in uniting so many diverse peoples and regions under one government and one language, his administration was marked by inefficiency, injustice, corruption, and chaos.

History: 1957 to 1965
The rapid expropriation of Dutch property and the ousting of Dutch citizens (1957) severely dislocated the economy; the country's great wealth was not exploited, and soaring inflation and great economic hardship ensued. A popular revolt, stemming from a desire for greater autonomy, began on Sumatra early in 1958 and spread to Sulawesi (Celebes) and other islands; the disorders led to increasingly authoritarian rule by Sukarno, who dissolved (1960) the parliament and reinstated the constitution of 1945, which had provided for a strong, independent executive. The army, whose influence was strengthened by its role in quickly quelling the revolts, and the Communist Party, whose ranks were growing very rapidly, constituted two important power blocs in Indonesian politics, with Sukarno holding the balance of power between them. In early 1962, Sukarno dispatched paratroopers to Netherlands New Guinea—territory claimed by Indonesia but firmly held by the Dutch—forcing the Dutch to agree to transfer that area to the UN with the understanding that it would pass under Indonesian administration in May 1963, pending a referendum that was to be held by 1970. After the referendum, in August 1969, Netherland New Guinea was formally annexed by Indonesia, and its name was changed to Irian Barat (or West Irian), which was then renamed Irian Jaya. Sukarno began to lean increasingly toward the left, openly summoning Communist leaders for advice, exhibiting hostility toward the U.S., and cultivating the friendship of Communist China.

History: 1965 to 1970
In 1965, he withdrew Indonesia from the UN. An abortive Communist coup against the army began in September 1965 with the assassination of six high army officials. The coup was swiftly thwarted by strategic army forces under General Suharto, who gradually assumed power (although retaining Sukarno as symbolic leader). Thousands of alleged Communists were executed, and a widespread massacre ensued (October–December 1965). As many as 750,000 people may have been killed throughout Indonesia, including many ethnic Chinese; in E and central Java and in Bali entire villages were wiped out. The new government

steadily increased its power, aided by massive student demonstrations against Sukarno. General Suharto banned the PKI, reestablished close ties with the U.S., and reentered (1966) the UN. On March 12, 1967, the national assembly voted Sukarno out of power altogether and named General Suharto acting president. Indonesia was one of the founding countries of the Association of Southeast Asian Nations (ASEAN) in 1967. Suharto was elected president in 1968. The government re-instituted an earlier Dutch colonial policy of "transmigration," in which farmers from the overpopulated islands of Java and Bali were moved to underpopulated areas such as Sumatra, Kalimantan (Indoesian Borneo), and most recently, Irian Jaya.

History: 1970 to Present
The policy, which has continued through the 1990s, has had mixed results; though more than six million have moved Java and Bali continue to be heavily populated (especially Java, which has received large numbers of immigrants in recent years and whose population had nearly quadrupled in the 20th century). The economy grew rapidly during the 1970s and 1980s due to expanded oil, gas, timber, textiles, and coffee exporting, but corruption also remained a problem. After the financial collapse of the national oil corporation Pertamina in 1975, the government instituted a series of reforms to liberalize the economy and promote private enterprise. In the 1980s, government policies promoting export manufacturing was successful, and the country's industrial sector grew quickly, particularly in chemicals, electronic components, cement, rubber tires, paper, and textiles. During the 1990s, the Jakarta Stock Exchange grew rapidly, becoming one of the largest in the region. The country has also promoted tourism, and Bali has become a popular tourist attraction. In 1975–1976, Indonesia annexed East Timor (a former Portuguese colony), and incorporated it as the province of Timor Timur. Indonesia defended its actions by claiming that it was preventing growing Soviet influence, but the takeover was not recognized by the UN. Since the annexation, separatists resisted Indonesian control, suffering substantial loss of life, and in the 1980s and 1990s, Indonesia came under increasing criticism from the U.S. and international organizations for human-rights abuses in the area. In August 1999, the East Timorese voted for independence, leading to widespread violence by Indonesian militias backed by some elements of the army, and the UN est. a peacekeeping mission and interim administrative force in October 1999; EAST TIMOR officially became an independent country in May 2002. Meanwhile, In October 1997 the country was plunged into economic upheaval when its currency and stock market plummeted, but Suharto was reluctant to implement the reforms required by the International Monetary Fund (IMF) in return for aid. Protests and riots over rising prices led Suharto resigned in May 1998, and Vice President B.J. Habibie succeeded him. The event of 1999 in East Timor led to increased calls for independence in Aceh, in N Sumatra, and Papua. Laws granting those regions limited autonomy were passed in 2001, but military action continued in both. In the 1999 presidential election, Abdurrahman Wahid of the National Awakening party became the country's first democratically elected president. After a series of clashes with parliament, Wahid was removed (2001) from office; and Vice President Megawati Sukarnoputri was elected to succeed him. In 2004 Susilo Bambang Yudhoyono, a former general and security minister, defeated Megawati in Indonesia's first direct presidential election. In 2005 the Acehnese rebels signed a peace accord that led to their disarming in exchange for local self-government in Aceh.

Government
Indonesia is governed under the constitution of 1945 (which was restored in 1959) as amended. The

president, who is both head of state and head of government, is popularly elected for a five-year term and is eligible for a second term. The vice president is similarly elected. The unicameral legislature consists of the 550-seat House of Representatives (Dewan Perwakilan Rakyat; DPR), whose members are popularly elected (by proportional representation) from multimember constituences. This body plus 195 indirectly selected members make up the People's Consultative Assembly (Majelis Permusyawaratan Rakyat; MPR), which meets every five years to determine national policy and annually to consider constitutional amendments and other changes. Prior to 2004 the president and vice president were chosen by the MPR. For over 30 years, until 1999, the government was essentially controlled by the quasi-official Golkar party. The current head of state is President Susilo Bambang Yudhoyono (since October 2004).

Indonesian Borneo: see BORNEO.

Indor, ISRAEL: see EIN-DOR.

Indore (in-DOR), former native state, W central INDIA. Established c.1728 by Malhar Rao Holkar, a soldier in the service of the Marathas and the founder of the ruling dynasty. In 1818, Indore became tributary to the British. Territory is now part of SW MADHYA PRADESH state.

Indore (in-DOR), district (□ 1,505 sq mi/3,913 sq km; 2001 population 2,585,321), SW MADHYA PRADESH state, INDIA; ⊙ INDORE. Several universities, noted architectural sites, famous for shopping and rich cuisine, royal history.

Indore (in-DOR), city (2001 population 1,506,062), ⊙ INDORE district, MADHYA PRADESH state, W central INDIA, on the MALWA PLATEAU near the Vindhya escarpment; 22°43′N 75°50′E. Indore is a commercial and industrial center. Manufacturing includes cotton textiles, chemicals, tiles, cement, iron and steel, furniture, hosiery, sporting goods, and motor vehicles. Indore has several colleges, a medical school, and a university. Became important in the late 18th century when it was established as the capital city of the Holkar dynasty; Maharathas ruled Indore until Indian independence. Remnants of their rule are scattered around the city, as well as many other noted buildings: their imposing Holkar and Lalbagh Palaces, the Krishnapura Chhatris (cenotaphs of the Holkars), Mahatma Gandhi Hall, Gopal Mandir, Hindu god Ganesh temples Khajrana and Bada Ganapati (housing the largest Ganesh idol in the world), the Kanch Mandir (a glass Jain temple).

Indragiri River (in-dah-GIR-ee), 250 mi/402 km long, central SUMATRA, Indonesia; rises in PADANG HIGHLANDS, part of the W BARISAN MOUNTAIN range, near BUKITTINGGI; flows generally E past RENGAT, in RIAU province, thence through marshy area to BERHALA STRAIT 45 mi/72 km ENE of Rengat; 00°22′S 103°30′E. Also spelled Inderagiri. Formerly called Koeantan or Kuantan.

Indramayu (in-drah-MAH-yoo), town, ⊙ INDRAMAYU district, Java Barat province, INDONESIA, near the N coast, 30 mi/48 km NW of Ceribon, near INDRAMAYU POINT; 06°20′S 108°19′E. Rice-production center. An important trade center. Formerly spelled Indramaju or Indramajoe.

Indramayu Point (in-drah-MAH-yoo), promontory on JAVA SEA, NE Java Barat province, INDONESIA, just N of town of INDRAMAYU; 06°15′S 108°19′E.

Indraprastha, INDIA: see DELHI.

Indrapura (in-drah-POO-rah), town, Sumatra Barat province, INDONESIA, near INDIAN OCEAN, 90 mi/145 km SSE of PADANG; 02°04′S 100°55′E. Tea, pepper. Minor trading center in 1990s. The Dutch gained foothold here in 1668; area was under British rule (1685–1693). Formerly capital of sultanate of Indrapura (abolished 1792).

Indrapura (in-drah-POO-rah), village, NW SUMATRA, INDONESIA, 50 mi/80 km ESE of MEDAN; 03°18′N

99°23′E. In area producing rubber and palm oil. Formerly spelled Indrapoera.

Indrapura Peak, INDONESIA: see KERINCI, MOUNT.

Indravati River (in-DRAH-vuh-tee), 315 mi/507 km long, SW ORISSA and SE MADHYA PRADESH states, E INDIA; rises in EASTERN GHATS c.25 mi/40 km SSW of BHAWANIPATNA (ORISSA state), flows SSW, W, and SSW, past JAGDALPUR, to Godavari River at ANDHRA state border. Creates a gigantic waterfall at CHITRAKUT (BASTAR district, MADHYA PRADESH state).

Indrawati River (in-DRAH-wo-ti), central NEPAL; rises in LANGTANG NATIONAL PARK; flows through the HELAMBU region, joining the SUN KOSI RIVER at 27°39′N 85°42′E.

Indre (AN-druh), department (□ 2,622 sq mi/6,817.2 sq km), CENTRE administrative region, central FRANCE, formed from parts of historic BERRY, ORLÉANAIS, MARCHE, and TOURAINE provinces; ⊙ CHÂTEAUROUX; 46°50′N 01°40′E. Gently rolling terrain at S margin of PARIS basin. Drained by INDRE and CREUSE rivers which flank the marshy BRENNE district (W). Department produces wheat, oats, vegetables (mainly artichokes and beans), and some wine; cattle and sheep raising. Although chiefly agricultural, it has textile, leather, and paper manufacturing and handicraft industries. Chief towns are Châteauroux, ISSOUDUN, ARGENTON-SUR-CREUSE, Le BLANC. Dept. is losing population to PARIS region. It contains the regional park of the Brenne (LA BRENNE NATURAL REGIONAL PARK, a lake district N of Le Blanc).

Indre (AN-druh), W commune (□ 1 sq mi/2.6 sq km), suburb of NANTES, LOIRE-ATLANTIQUE department, PAYS DE LA LOIRE region, W FRANCE; 47°12′N 01°40′E. It includes Basse-Indre (heavy metallurgy), on right bank of LOIRE RIVER and 6 mi/9.7 km W of Nantes, and Indret island (in Loire River opposite Basse-Indre), with naval machine shops and arsenal.

Indre Arna (IN-druh AHR-nah), village, part of BERGEN city, HORDALAND county, SW NORWAY, on an inlet of SØRFJORDEN, on railroad, 5 mi/8 km E of Bergen (city center). Connected to the center of Bergen by the Ulriken-Tunnel.

Indre-et-Loire (AN-drai-LWAHR), department (□ 2,366 sq mi/6,151.6 sq km), in historic TOURAINE, CENTRE administrative region, W central FRANCE; ⊙ TOURS; 47°15′N 00°45′E. Traversed E-W by broad valley of the LOIRE, and drained by its large tributaries, the CHER, INDRE, and VIENNE. The department produces fruits and vegetables, small grains, and flowers. Its Loire Valley vineyards (VOUVRAY, BOURGUEIL) are as famous as its chain of châteaux (CHINON, CHENONCEAUX, AMBOISE, AZAY-LE-RIDEAU). Tourism is a major regional industry. Tours is the only major urban center with a diversified economic base and major railroad facilities.

Indre River (AN-druh), 165 mi/266 km long, in INDRE and INDRE-ET-LOIRE departments, CENTRE administrative region, central FRANCE; rises in N foothills of MASSIF CENTRAL 5 mi/8 km NW of Boussac, flows NW past La CHÂTRE, CHÂTEAUROUX, CHÂTILLON-SUR-INDRE, LOCHES, and AZAY-LE-RIDEAU, to the LOIRE 25 mi/40 km below TOURS; 47°15′N 00°11′E. Its lower course flows through the LOIRE CHÂTEAU country.

Indura (in-DOO-ruh), town, W GRODNO oblast, BELARUS, 15 mi/24 km S of GRODNO, near Polish border; 53°27′N 23°53′E. Distilling, brewing.

Indus (IN-duhs), chief river of PAKISTAN, c.1,900 mi/3,058 km long (longest of the Himalayan rivers); rising in the KAILAS RANGE at an elevation of 17,000 ft/5,182 m, near SENGE (SHICHUANGHE) village, in SW Tibetan Himalayas; flowing W across LADAKH region (N INDIA), then SW through PAKISTAN to the ARABIAN SEA SE of KARACHI. The upper Indus, fed by snow and glacial meltwater from the KARAKORUM, HINDU KUSH, and HIMALAYA mountains, flows through deep gorges and scenic valleys; its turbulence makes it unsuitable

for navigation. Receives the combined waters of the 5 rivers of PUNJAB (CHENAB, JHELUM, RAVI, BEAS, and SUTLEJ), its chief affluent supplemented by the KABUL RIVER. The Indus then flows onto the dry PUNJAB plains of PAKISTAN and becomes a broad, slow-moving, silt-laden river with many braided channels. Many fish live in the river. The irrigated plain of Indus valley is Pakistan's most densely populated region and its main agricultural area; wheat, maize, rice, cotton, vegetables, and fruits are the chief crops. In PAKISTAN the Indus is extensively used for irrigation and generation of hydroelectric power (especially the TARBELA DAM NW of ISLAMABAD). The Jinnah, Sukker, and KOTRI barrages feed the main Indus canals in W PUNJAB and SIND provinces. The use of the Indus and its tributaries has been a source of conflict between PAKISTAN and INDIA, although a treaty by which the waters were to be shared was signed in 1960. The lower Indus is navigable for small boats but is little used for transportation, at least since the development of railroads. The delta is a level muddy area with little cultivation. The river valley was the site of the prehistoric Indus valley civilization, which flourished c.2500 B.C.E.–c.1500 B.C.E. At its height, its geographical reach exceeded that of EGYPT or MESOPOTAMIA. Since 1921 this civilization has been revealed by spectacular finds at MOHENJO-DARO, an archaeological site in NW SIND province, and at HARAPPA, in central PUNJAB province near the RAVI River. These sites were once the chief cities of the Indus civilization. They had large and complex hill citadels, housing palaces, granaries, and baths that were probably used for sacred ablutions; the great bath at MOHENJO-DARO was c.40 ft/12 m long and 23 ft/7 m wide. The economy of the Indus civilization was largely based on agriculture, and the arts flourished here; many objects of copper, bronze, and pottery, including a large collection of terra-cotta toys, have been uncovered. Most notable, however, are the steatite seals, exquisitely engraved with animal figures and often bearing a line of pictographic script. On some seals are depicted a pipal tree or, as some authorities hold, a Babylonian tree of life, and others have as their central figure the god Shiva, who later became preeminent in the Hindu pantheon. The writing, long a riddle to archaeologists, has yet to be satisfactorily deciphered; the language appears to be structurally related to the Dravidian languages. The origin, rise, and decline of the Indus valley civilization remain a mystery, but it seems most probable that the civilization fell (c.1500 B.C.E.) to invading Aryans.

Industrial'nyy (een-doo-stree-AHL-niyee) [Russian=industrial], urban settlement (2005 population 3,940), central KRASNODAR TERRITORY, S European Russia, 5 mi/8 km NE, and under administrative jurisdiction, of KRASNODAR (connected by railroad); 45°06′N 39°06′E. Elevation 147 ft/44 m.

Industrial'nyy (een-doo-stree-AHL-niyee), former town, now a suburb of PETROPAVLOVSK-KAMCHATSKIY, E KAMCHATKA oblast, RUSSIAN FAR EAST; 16 mi/26 km SE of the city center; 52°57′N 158°44′E. Elevation 167 ft/50 m. Dry docks; shipbuilding.

Industry, town, FRANKLIN county, W central MAINE, 7 mi/11.3 km NE of FARMINGTON; 44°45′N 70°03′W. Contains villages of Allen Mills and West Mills.

Industry, village (2000 population 540), MCDONOUGH county, W ILLINOIS, 9 mi/14.5 km SSE of MACOMB; 40°19′N 90°36′W. Agriculture (corn, sorghum; cattle, hogs); coal processing.

Industry, unincorporated village (2006 population 336), AUSTIN county, S TEXAS, 68 mi/109 km WNW of Houston; 29°58′N 96°30′W. Oil and natural gas. Agriculture (livestock; cotton, peanuts). Timber. Manufacturing (lumber, concrete).

Industry, borough (2006 population 1,833), BEAVER county, W PENNSYLVANIA, 7 mi/11.3 km WSW of BEAVER, on OHIO RIVER; 40°39′N 80°24′W. Manufacturing includes steel beams. Agriculture includes dairying; livestock. Nuclear-power plant at SHIPPINGPORT across the river to S.

Industry, City of, city (2000 population 777), LOS ANGELES county, S CALIFORNIA; industrial suburb 17 mi/27 km E of LOS ANGELES, near San Jose Creek; 34°01'N 117°57'W. Manufacturing (consumer goods, dairy products, plastic products, carpeting, fabricated metal products, lumber, chemicals, industrial equipment, food, ground minerals, leather products, resins, rubber products, paper products, motors, printing). Created 1957, it was incorporated for the sole purpose of industrial development.

Indwe, town, EASTERN CAPE province, SOUTH AFRICA, in STORMBERG range, 40 mi/64 km NE of QUEENSTOWN, on railroad link at Mairiva en route to MACLEAR; 31°29'S 27°19'E. Elevation 4,954 ft/1,510 m. Agricultural center (dairying; grain). Former coal-mining center; mines were opened 1859, closed down 1918, except for shafts now worked to cover local needs. Airfield. Water from nearby Doring River Dam.

Indyiya, SERBIA: see INDJIJA.

Ine (EE-ne), town, Yosa district, KYOTO prefecture, S HONSHU, W central JAPAN, 53 mi/85 km N of KYOTO; 35°40'N 135°17'E.

Ine-bekchi, Bulgaria: see STRYAMA.

Ineboli, TURKEY: see INEBOLU.

Inebolu, (Turkish=Inebolu) village, N TURKEY, port on BLACK SEA, 40 mi/64 km N of KASTAMONU; 41°57'N 33°45'E. Shipbuilding; hemp, wheat, corn. Sometimes spelled Ineboli.

Inegol, (Turkish=Inegöl) town (2000 population 105,959), NW TURKEY, 25 mi/40 km ESE of BURSA; 40°06'N 29°31'E. Wheat, barley, corn.

'Ineiba, former village, Upper EGYPT. The site of village and its lands were flooded after the inauguration of the ASWAN Dam and are now covered by Lake NASSER. The population was evacuated and resettled N of the dam.

Ineli, Greece: see PALAIOMILOS.

Ineu (ee-NE-oo), Hungarian *Borosjenö*, town, ARAD county, W ROMANIA, on Crişul Alb River, 30 mi/48 km NE of ARAD. Railroad junction; mining and agricultural center; manufacturing of furniture, construction materials; wine production, food products.

Inevi, TURKEY: see CIHANBEYLI.

Inewarī, ETHIOPIA: see ENEWARI.

Inez, town (2000 population 1,787), VICTORIA county, S TEXAS, 13 mi/21 km NE of VICTORIA, near Arenosa Creek; 28°52'N 96°47'W. Oil and natural gas. Agriculture (cotton, rice; cattle; dairying).

Inez (ei-NEZ), village (2000 population 466), ⊙ MARTIN county, E KENTUCKY, in CUMBERLAND MOUNTAINS, 26 mi/42 km N of PIKEVILLE; 37°52'N 82°32'W. Coal; agriculture (livestock; tobacco); coal processing, light manufacturing. Big Sandy Regional Airport to N. Martin County Reservoir to N.

Inezgane (EENZ-gahn), city, Souss-Massa-Draâ administrative region, SW MOROCCO, on right bank of the Oued SOUSS river, 7 mi/11.3 km SE of AGADIR; 30°21'N 09°32'W. Trade center (olives, almonds, farming). Major international airport 3 mi/4.8 km NW serves Agadir and the region's tourist industry.

Infanta (een-FAHN-tah), town, QUEZON province, central LUZON, PHILIPPINES, near POLILLO STRAIT, 45 mi/72 km ENE of MANILA; 14°44'N 121°39'E. Fishing and agricultural center (coconuts, rice).

Infantes (een-FAHN-tes) or **Villanueva de Infantes** (vee-lyah-NWAI-vah dhai een-FAHN-tes), town, CIUDAD REAL province, S central SPAIN, in CASTILE-LA MANCHA, 50 mi/80 km ESE of CIUDAD REAL; 38°44'N 03°01'W. Agricultural center (olives, cereals, grapes, livestock); olive-oil pressing, flour milling, cheese processing; manufacturing of woolen goods, firearms, plaster. Had greater importance as center of ancient Campo de Montiel. The writer Francisco de Quevedo is buried here.

Infesta, PORTUGAL: see SÃO MAMEDE DE INFESTA.

Infiesto (eem-FYE-sto), town, OVIEDO province, NW SPAIN, 24 mi/39 km E of OVIEDO; 43°21'N 05°22'W. Agricultural center shipping almonds, nuts, apples. Vegetable canning, cider distilling; stock raising. Has fish hatchery. Mineral springs nearby.

Ingá (een-GAH), city (2007 population 18,168), E PARAÍBA, NE BRAZIL, on railroad, 18 mi/29 km E of CAMPINA GRANDE; 07°18'S 35°36'W. Cotton, sugar, coffee, tobacco, wool.

Ingá, Brazil: see ANDIRÁ.

Ingabu (IN-gah-boo), town, AYEYARWADY division, MYANMAR, on railroad, 16 mi/26 km NW of HENZADA.

In gall (EEN gahl), town, AGADEZ province, NIGER, 74 mi/119 km WSW of AGADEZ; 16°47'N 06°56'E. Oasis. Commercial center. Coal and salt production nearby. Sometimes spelled I-n-gall.

Ingalls, town (2000 population 1,168), MADISON county, E central INDIANA, on small Fall Creek, 24 mi/39 km NE of INDIANAPOLIS; 39°58'N 85°48'W. In agricultural area.

Ingalls (ING-guhls), village (2000 population 328), GRAY county, SW KANSAS, on ARKANSAS RIVER, 6 mi/9.7 km WNW of CIMARRON; 37°49'N 100°27'W. Grain; cattle.

Ingalls, Mount (8,377 ft/2,553 m), PLUMAS county, NE CALIFORNIA, in the SIERRA NEVADA, 17 mi/27 km E of QUINCY. In Plumas National Forest. Copper mine on its slopes.

Inganda (ing-GAHN-duh), CONGO: see BOENDE.

Ingapirca (een-gah-PEER-kah), the best-known Inca ruins in ECUADOR, N of CUENCA, near CAÑAR in CAÑAR province. Built in 1400s on Inca road from CUZCO to QUITO.

Ingatestone (ING-gait-ston), town (2001 population 4,508), S central ESSEX, SE ENGLAND, 6 mi/9.7 km SW of CHELMSFORD; 51°41'N 00°22'E. Church dating from c.1000. Also site of 16th-century Ingatestone Hall. Just W is agricultural village of Fryerning.

Ingavi, province, LA PAZ department, BOLIVIA, ⊙ VIACHA; 16°50'S 68°40'W.

Ingavi (eeng-GAH-vee), town and canton, ABUÑA province, PANDO department, NW BOLIVIA, on ORTON RIVER, 50 mi/80 km W of RIBERALTA; 11°31'S 67°34'W.

Ingelfingen (ING-gel-fing-uhn), town, BADEN-WÜRTTEMBERG, GERMANY, on the KOCHER, 14 mi/23 km SSW of MERGENTHEIM; 48°18'N 09°39'E. Manufacturing of electrical goods; jewelry; wine.

Ingelheim am Rhein (ING-gel-heim ahm REIN), town, RHINELAND-PALATINATE, W GERMANY, on left bank of the RHINE, 10 mi/16 km W of MAINZ; 49°59'N 08°04'E. Market center for wine; fruit and vegetable region. Chemical and pharmaceutical industry; manufacturing of electronics. Has 13th-century church. Formed 1939 through unification of NIEDER-INGELHEIM, OBER-INGELHEIM, and FREI-WEINHEIM.

Ingelmunster (ING-uhl-mun-stuhr), commune (2006 population 10,602), Roeselare district, WEST FLANDERS province, W BELGIUM, 7 mi/11.3 km N of KORTRIJK; 50°55'N 03°15'E.

Ingenbohl (EENG-uhn-bol), commune, SCHWYZ canton, central SWITZERLAND, near LAKE LUCERNE, 2 mi/3.2 km SW of SCHWYZ. Cementworks.

Ingende (ing-GEN-dai), village, Équateur province, W CONGO, on RUKI RIVER at confluence of MOMBOYO and BUSIRA RIVERS, 40 mi/64 km ESE of MBANDAKA; 00°15'S 18°57'E. Elev. 1,131 ft/344 m. Trading and agricultural center (palm products, bananas, manioc), steamboat landing.

Ingeniero Balloffet, ARGENTINA: see RINCÓN DEL ATUEL.

Ingeniero Boasi, ARGENTINA: see SARMIENTO, SANTA FE province.

Ingeniero Giagnoni (een-he-nee-AI-ro ee-ahg-NO-nee), town, N MENDOZA province, ARGENTINA, on railroad, 30 mi/48 km SE of MENDOZA; 33°07'S

68°25'W. Agricultural center; wine making, dried-fruit processing.

Ingeniero Huergo (een-he-nee-AI-ro HWER-go) or **Ingeniero Luis A. Huergo,** town, N RÍO NEGRO province, ARGENTINA, in RÍO NEGRO valley (irrigation area), on railroad, 17 mi/27 km E of FUERTE GENERAL ROCA. Agriculture center (alfalfa, fruit, wine, potatoes); wine making, lumbering.

Ingeniero Jacobacci (een-he-nee-AI-ro jah-ko-BAH-chee), town, S RÍO NEGRO province, ARGENTINA, 45 mi/72 km W of MAQUINCHAO; 41°18'S 69°35'W. Railroad junction; livestock raising (cattle, sheep).

Ingeniero Luiggi (een-he-nee-AI-ro loo-EE-jee), town, NE LA PAMPA province, ARGENTINA, 45 mi/72 km WNW of GENERAL PICO; 35°25'S 64°29'W. Railroad terminus; grain and livestock center; flour milling; dairying.

Ingeniero Luis A. Huergo, ARGENTINA: see INGENIERO HUERGO.

Ingeniero Montero Hoyos (mon-TAI-ro OI-os), town and canton, Andrez de IBAÑEZ province, SANTA CRUZ department, central BOLIVIA, on RÍO GRANDE, 35 mi/56 km E of Santa Cruz; 17°38'S 62°48'W. Corn.

Ingeniero White (een-he-nee-AI-ro WEIT) or **Puerto Ingeniero White,** S suburb and main port of BAHÍA BLANCA, SW BUENOS AIRES province, ARGENTINA, 350 mi/563 km SW of BUENOS AIRES. Located 14 mi/23 km inland. Modern harbor facilities (oil tanks, cranes, grain elevators); railroad terminus.

Ingenio (een-HAI-nyo), city, Grand Canary, CANARY ISLANDS, 13 mi/21 km S of LAS PALMAS; 27°55'N 15°26'W. Fishing; vegetables.

Ingenio San Pablo (een-he-nee-AI-ro sahn PAH-blo), town, Lules department, TUCUMÁN province, ARGENTINA, on railroad line, 10 mi/16 km SW of TUCUMÁN; 26°54'S 65°19'W. Sugar-producing region.

Ingermanland (EEN-gyer-muhn-lent) or **Ingria,** Finnish *Ingerinta,* historic region in what is now LENINGRAD oblast, NW European Russia, along the NEVA RIVER, on the E bank of the Gulf of FINLAND. Its name derives from the ancient Finnic inhabitants, the Ingers, some of whose descendants (about 93,000) still live in the SAINT PETERSBURG area and are called Ingrians or Finns. In medieval times, the region was subject to Great Novgorod, with which it passed in 1478 to the grand duchy of MOSCOW. Conquered in the early 17th century by SWEDEN, it remained Swedish until Peter I of RUSSIA captured it in 1702 and built his new capital of Saint Petersburg here. The area was formally ceded to Russia by the Treaty of Nystad (1721), which ended the Northern War between Russia and Sweden.

Ingerois, FINLAND: see INKEROINEN.

Ingersheim (an-zher-ZEM), German (ING-uhrsheim), town (☐ 2 sq mi/5.2 sq km), HAUT-RHIN department, E FRANCE, on the FECHT RIVER, 2 mi/3.2 km WNW of COLMAR, in the ALSACE vineyard area; 48°06'N 07°18'E.

Ingersoll (IN-guhr-sahl), town (☐ 5 sq mi/13 sq km; 2001 population 10,977), SW ONTARIO, E central CANADA, on the THAMES River, E of LONDON; 43°02'N 80°53'W. Large dairy-processing industry and manufacturing (paper boxes, wire products, motor vehicles; and automotive, machine, and tool parts).

Ingersoll (ING-uhr-sahl), village, ALFALFA county, N OKLAHOMA, 3 mi/4.8 km NW of CHEROKEE. In grain-growing area.

Ingham (ING-uhm), county (☐ 560 sq mi/1,456 sq km; 2006 population 276,898), S central MICHIGAN; 42°36'N 84°22'W; ⊙ MASON. LANSING, the state capital, is in NW corner. Drained by GRAND and RED CEDAR rivers, and small Sycamore Creek. Agriculture (apples, wheat, soybeans, corn, hay, beans, onions, cucumbers, carrots); hogs, cattle, sheep, poultry; dairy products. Manufacturing at Lansing. Oil and gas extraction. Also includes suburbs of E. Lansing and MERIDIAN TOWNSHIP. Formed 1838.

Cross-references are shown in SMALL CAPITALS. The pronunciation guide is shown on page xix. The sources of population figures are shown on page xvii.

Ingham (ING-uhm), town, E QUEENSLAND, AUSTRALIA, 55 mi/89 km NW of TOWNSVILLE, and on Herbert River; 18°39′S 146°10′E. Sugar-producing center; connected with LUCINDA POINT (its port) by electric railroad. Annual Australian-Italian festival.

Ingichka (in-GEECH-kah), town, W SAMARKAND wiloyat, UZBEKISTAN, c.15 mi/24 km S of AKTASH; 39°32′N 66°34′E. Mining. Also Ingichkapayan.

Ingierstrand Bad (ING-ei-er-strahn BAHD), village, AKERSHUS county, SE NORWAY, on E shore of Bundefjord (SE arm of OSLOFJORD), 7 mi/11.3 km S of OSLO. Seaside resort. Amundsen lived here; his residence is now a museum

Inginiyagala (ING-gi-ni-yah-GAH-luh), town, EASTERN PROVINCE, E SRI LANKA, on the GAL OYA River, 37 mi/60 km ENE of BADULLA; 07°13′N 81°32′E. Dam (3,600 ft/1,097 m long, 122 ft/37 m high, and 19,200 acres/7,770 ha in size), completed in the 1950s, forms a reservoir known as Senanayake Samudra, or Gal Oya reservoir. A 350-sq-mi/907-sq-km area has opened for settlement and irrigation. A sugar refinery and distillery is at Hingurana. Upstream is the 234-sq-mi/606-sq-km Gal Oya National Park, a wildlife preserve.

Ingleburn (ING-guhl-buhrn), outer residential suburb, E NEW SOUTH WALES, AUSTRALIA, 18 mi/29 km SW of SYDNEY; 34°00′S 150°52′E. In metropolitan area. Large public housing development as green-field suburb.

Ingleby Barwick (ING-guhl-bee BA-rik), village (2001 population 5,230), NORTH YORKSHIRE, N ENGLAND, 3 mi/4.8 km S of STOCKTON-ON-TEES; 54°31′N 01°19′W.

Inglefield Gulf, Danish *Inglefield Bredning*, inlet of N BAFFIN BAY, NW GREENLAND, 60 mi/97 km long, 10 mi/16 km–20 mi/32 km wide; 77°27′N 68°00′W. Extends to edge of inland ice; receives several glaciers. At mouth Northumberland Island (20 mi/32 km long, 7 mi/11.3 km wide), Herbert Island (19 mi/31 km long, 2 mi/3.2 km–6 mi/9.7 km wide), and HAKLUYT Island divide approaches to gulf into Murchison (N) and WHALE (S) sounds.

Inglefield Land, ice-free region, NW GREENLAND, on SMITH SOUND and KANE BASIN, between HUMBOLDT GLACIER (N) and PRUDHOE LAND (S), in N part of HAYES PENINSULA; 78°45′N 69°00′W.

Ingleses do Rio Vermelho (EEN-gle-ses do REE-o VER-mel-yo), city, E SANTA CATARINA state, Brazil, 9 mi/14.5 km N of FLORIANÓPOLIS on SANTA CATARINA ISLAND; 27°54′S 48°52′W. Beach community with upscale resorts, hotels, and condominiums.

Ingleside, town (2006 population 9,357), SAN PATRICIO county, S TEXAS, suburb 12 mi/19 km ENE of CORPUS CHRISTI, across CORPUS CHRISTI BAY; 27°52′N 97°12′W. Oil refining. Corpus Christi Bay to SW, Redfish Bay and INTRACOASTAL WATERWAY to E; Ingleside Naval Station at cape to S.

Inglestat, Alaska: see KOYUK.

Ingleton (ING-guhl-tuhn), village (2001 population 1,692), NORTH YORKSHIRE, N ENGLAND, 15 mi/24 km ENE of LANCASTER; 54°09′N 02°28′W. In limestone (karst) area, center for caving.

Inglewood (ING-guhl-wud), municipality, N central VICTORIA, SE AUSTRALIA, 105 mi/169 km NW of MELBOURNE, 28 mi/45 km NW of BENDIGO, near LODDON RIVER; 36°35′S 143°56′E. Railroad junction in old gold-mining area; eucalyptus oil.

Inglewood, city (2000 population 112,580), LOS ANGELES county, S CALIFORNIA; residential and industrial suburb 7 mi/11.3 km SW of LOS ANGELES; 33°57′N 118°21′W. The city grew substantially during the 1950s and 1960s. Bounded by Los Angeles on the E, N, and W, and by HAWTHORNE and EL SEGUNDO on the S. In an oil-producing area. Its manufacturing includes motor-vehicle parts, furniture, processed food, plastics products, and electronic equipment. The city has greatly benefited from the regional advancement, extension, and development of these industries. Inglewood's population has grown accordingly. LOS ANGELES INTERNATIONAL AIRPORT to SW. Attractions

include the Hollywood Park and the (Los Angeles) Great Western Forum, a massive arena that hosted Los Angeles Lakers basketball games until 1999 (the team now plays at Staples Center in downtown Los Angeles). Former site of Northrop University, closed in 1993. Dockweiler State Beach, on Pacific Ocean, 5 mi/8 km W. Founded 1873, incorporated 1908.

Inglewood, town, NEW PLYMOUTH district (□ 859 sq mi/2,233.4 sq km), TARANAKI region, NEW ZEALAND, at base of MOUNT EGMONT, 11 mi/18 km SE of NEW PLYMOUTH. Dairying; some petroleum nearby.

Inglewood, unincorporated town, KING county, W WASHINGTON, residential suburb 11 mi/18 km NNE of SEATTLE, on E shore of LAKE WASHINGTON, near its N end, at mouth of Sammamish River. St. Edward State Park to SW, on lake shore.

Inglewood (ING-guhl-wud), village, SE QUEENSLAND, NE AUSTRALIA, 135 mi/217 km WSW of BRISBANE; 28°25′S 151°02′E. Railroad junction in diversified agricultural area; formerly a tobacco-growing center. Recreational activities at Coolmunda Dam, 12 mi/20 km E. Originally called Brown's Inn.

Inglis (IN-glis), community, SW MANITOBA, W central CANADA, 44 mi/71 km W of RIDING MOUNTAIN NATIONAL PARK, in SHELLMOUTH-Boulton rural municipality; 50°56′N 101°15′W.

Inglis (ING-gliss), town (□ 3 sq mi/7.8 sq km; 2000 population 1,491), LEVY county, W central FLORIDA, 35 mi/56 km WSW of OCALA; 29°01′N 82°39′W. Manufacturing (machine parts).

Inglis Island, AUSTRALIA: see ENGLISH COMPANY'S ISLANDS.

Ingoda River (EEN-guh-duh), 500 mi/805 km long, S central CHITA oblast, RUSSIA; rises near SOKHONDO peak in the BORSHCHOVOCHNY RANGE; flows generally NE, past ULETY, CHITA, and KARYMSKOYE, joining the ONON RIVER to form the SHILKA River, 15 mi/24 km above SHILKA. Navigable below Chita. TRANS-SIBERIAN RAILROAD runs along its lower course.

Ingolstadt (ING-guhl-shtaht), city, BAVARIA, S central GERMANY, in UPPER BAVARIA, on the DANUBE River; 48°46′N 11°25′E. A commercial and industrial center; manufacturing includes engines, machinery, refined oil, and motor vehicles. Major oil pipelines link Ingolstadt to MARSEILLES, FRANCE, and the Italian cities of GENOA and TRIESTE. Chartered in 1250, Ingolstadt was besieged (1632) by Gustavus II of SWEDEN during the Thirty Years War. The University of Ingolstadt (founded 1472 and removed to LANDSHUT in 1802 and then to MUNICH in 1826) was a stronghold of the Counter-Reformation; Joseph von Eck taught at the university from 1510 to 1543. The city's noteworthy buildings include the Gothic Liebfrauenmünster (15th–16th centuries) and other churches; town gate from 1385 with seven towers.

Ingomar (ING-guh-mahr), village, UNION county, N MISSISSIPPI, 6 mi/9.7 km SSW of NEW ALBANY. In agricultural and dairying area.

Ingomar, village, ROSEBUD county, SE central MONTANA, 41 mi/66 km NW of FORSYTH, near source of East Fork Froze to Death Creek. Sheep and cattle. Black Sea Reservoir to NE. One of the largest sheep-shearing plants in the state.

Ingonish (ing-guh-NISH), village, NE NOVA SCOTIA, CANADA, on NE coast of CAPE BRETON ISLAND, 40 mi/64 km NNW of SYDNEY; 46°41′N 60°22′W. Elevation 0 ft/0 m. Fishing port, tourist resort. Extending W is Cape Breton Highland National Park.

Ingram, town (2006 population 1,868), KERR county, SW TEXAS, on GUADALUPE RIVER, c.60 mi/97 km NW of SAN ANTONIO; 30°04′N 99°14′W. Elevation 1,600 ft/488 m. Trading point in ranching area (cattle, sheep, goats; wheat, pecan, apples). Established 1883; new section built after flood of 1936.

Ingram, village (2006 population 77), RUSK county, N WISCONSIN, 14 mi/23 km ENE of LADYSMITH; 45°30′N 90°48′W. In dairying and stock-raising area.

Ingram (ING-rahm), borough (2006 population 3,427), ALLEGHENY county, SW PENNSYLVANIA, residential suburb 3 mi/4.8 km W of PITTSBURGH, near OHIO RIVER. Incorporated 1902.

Ingram Beach, village, HORRY county, E SOUTH CAROLINA, 14 mi/23 km NE of MYRTLE BEACH, on ATLANTIC OCEAN, in Grand Strand beach resort area.

Ingrandes (an-grahn-duh), commune (□ 13 sq mi/33.8 sq km), VIENNE department, POITOU-CHARENTES region, central FRANCE, on VIENNE RIVER, 5 mi/8 km N of CHÂTELLERAULT; 46°52′N 00°34′E. Makes rubber products. Coal deposits nearby but not mined. Also called Ingrandes-sur-Vienne.

Ingrandes-sur-Vienne, FRANCE: see INGRANDES.

Ingré (an-grai), town (□ 8 sq mi/20.8 sq km), LOIRET department, CENTRE administrative region, central FRANCE. WNW suburb of ORLÉANS; 47°55′N 01°49′E.

Ingria, RUSSIA: see INGERMANLAND.

Ingrid Christensen Coast, ANTARCTICA, on INDIAN OCEAN; extends from 72°30′E to 81°30′E. Discovered 1935 by Klarius Mikkelsen, Norwegian explorer.

Ingrowitz, CZECH REPUBLIC: see JIMRAMOV.

Ingulets, UKRAINE: see INHULETS', river.

Ingul River, UKRAINE: see INHUL River.

Inguri River (in-GOO-ree), 125 mi/201 km long, NW GEORGIA; rises in the Greater Caucasus, at S foot of peak DYKH-TAU; flows W and SW, through SVANETIA, to BLACK SEA at ANAKLIA. Forms border of ABKHAZ Autonomous Republic in lower course. Powers Zugdidi paper mill. Its basin covers 2,517 mi/4,060 km. Used for irrigation. Large hydroelectric power plant.

Ingushetia, RUSSIA: see INGUSH REPUBLIC.

Ingush Republic (een-GOOSH) or **Ingushetia** (een-goo-SHET-ee-yah), constituent republic (□ 144,788 sq mi/376,448.8 sq km), SW RUSSIA; ⊙ NAZRAN. The crest line of the Greater CAUCASUS mountain range forms its S boundary with GEORGIA; the CHECHEN REPUBLIC lies to the E, and the republic of Alania to the W and N. Main economic activities are farming (on the plains), cattle raising (primarily in the mountains), and horticulture. Population largely Sunnite Muslims who speak Inguish, an Ibero-Caucasian language that had no written form before the 1917 revolution. Was originally populated by separate tribal groups in the central part of the Greater Caucasus mountain range. The general name for the Inguish people came from the village name Anguish (or Inguish), which was one of the first Inguish settlements in the lowlands. The migration of the Inguish from the mountains to the lowlands began in the 16th and 17th centuries and was especially large from the 1830s to the 1860s. In 1943 and 1944, the Inguish and the Chechens, who then shared an autonomous republic, were among several ethnic groups of the Caucasus who were forcibly relocated to various regions of Middle Asia and KAZAKHSTAN for collaborating with the Germans in World War II. In addition, the Prigorny region, which had been part of the Chechen-Inguish Republic, was given to the NORTH OSSETIAN REPUBLIC, an action deeply resented by the Inguish. In 1957, the deportees were allowed to return to their native areas, but the Prigorny region was retained by the Ossetians. As a result, relations between the Ossetians and the Inguish remained strained. The relationship between the two groups became even more tense after the Chechen-Inguish Republic split in two parts in 1991, when the Chechens refused to ratify the treaty that created the RUSSIAN FEDERATION. For many months thereafter, the Ingush Republic lacked any legitimate governing authority because Moscow refused to recognize the Chechen Republic as a separate entity and was content to leave Ingushetia in legal limbo. As a result, Ingushetia was not included among those Russian regions that signed the Union Treaty (February 1992), which legally allowed for the co-existence of regions and republics of the Russian Federation. After the Chechen victory over the Russian army in 1995, the Chechen Republic was granted autonomy, and now the Ingush Republic is also considered separate. However, its status in the Russian Federation remains uncertain.

Ingwavuma, town, NE KWAZULU-NATAL, SOUTH AFRICA, on E SWAZILAND border, 75 mi/121 km E of PIET RETIEF, on Ngwavuma River, in LEBOMBO mountains; 27°07′S 32°00′E. Elevation 2,313 ft/705 m.

Ingwiller (ang-vee-LER), German, *Ingweiler* (ING-vei-luhr), commune (□ 7 sq mi/18.2 sq km), BAS-RHIN department, in ALSACE, E FRANCE, 11 mi/18 km NNE of SAVERNE at E edge of the regional park of the N Vosges (VOSGES DU NORD NATURAL REGIONAL PARK); 48°37′N 07°29′E. Brewing, diverse small manufacturing. Lichtenberg castle is 4 mi/6.4 km N on a peak of the N VOSGES MOUNTAINS.

Ingylchek, KYRGYZSTAN and CHINA: see ENGILCHEK.

Inhaca Island (en-YAH-kuh), off S coast of MOZAMBIQUE, guarding entrance to MAPUTO BAY, 20 mi/32 km E of MAPUTO city; c.5 mi/8 km long.

Inhambane (en-YAM-bahn), province (2004 population 1,401,216), MOZAMBIQUE; ⊙ INHAMBANE. Bounded in N by the provinces of SOFALA and MANICA, W and SW by province of GAZA, E and SE by INDIAN OCEAN. One of the longest coastlines among the coastal provinces. Commercial agriculture (sugar, coconuts); oil. Inhabited by the musically gifted Chope and Bitonga. Large population of Tswa speakers. There are 12 districts: FUNHALOURO, GOVURO, HOMOINE, INHASSOURO, INHARRIME, JANGAMO, MABOTE, MASSINGA, MORRUMBENE, VILANCULOS, PANDA, and ZAVALA; and approximately 101 habitats.

Inhambane (en-YAM-bahn), city, ⊙ INHAMBANE province, SE MOZAMBIQUE, on Inhambane Bay, an inlet of MOZAMBIQUE CHANNEL on the INDIAN OCEAN; 23°52′S 35°23′E. It is a port and the center of an important agricultural region. The bay was discovered in 1498 by Vasco da Gama, who claimed it for PORTUGAL. Inhambane developed as a trade center, notably for slaves and ivory. Sugar and coconut products are significant exports.

Inhambupe (een-yahn-BOO-pi), city (2007 population 33,325), NE BAHIA, BRAZIL, 25 mi/40 km N of ALAGOINHAS; 11°45′S 38°26′W. Tobacco, coffee, oranges.

Inhaminga (en-YAM-een-guh), village, central MOZAMBIQUE, on railroad, 100 mi/161 km N of BEIRA. Hardwood lumbering, pottery manufacturing; cotton.

Inhangapi (een-yah-gah-pee), town (2007 population 9,592), E PARÁ, BRAZIL, 40 mi/64 km E of BELÉM; 01°25′S 47°50′W.

Inhantehtaat (IN-hahn-te-taht), village, VAASAN province, W FINLAND, 55 mi/89 km WNW of JYVÄSKYLÄ; 62°31′N 24°10′E. Elevation 578 ft/175 m. In lake region; steel and lumber mills.

Inhapi (EEN-yah-pee), city (2007 population 17,456), W ALAGOAS state, BRAZIL, NE of DELMIRO GOUVEIA; 09°14′S 37°43′W.

Inharrime (en-YAH-reem-e), village, INHAMBANE province, SE MOZAMBIQUE, 50 mi/80 km SSW of INHAMBANE (linked by railroad). Cotton, coffee, manioc, mafura. Roman Catholic mission here.

Inharrime District, MOZAMBIQUE: see INHAMBANE, province.

Inhassouro District, MOZAMBIQUE: see INHAMBANE, province.

Inhassunge District, MOZAMBIQUE: see ZAMBÉZIA, province.

Inhaúma (EEN-yah-OO-mah), town (2007 population 5,332), central MINAS GERAIS, BRAZIL, 12 mi/20 km SW of SETE LAGOAS; 19°16′S 44°30′W.

Inhobim (EEN-yo-been), village, S central BAHIA, BRAZIL, near RIO PARDO and MINAS GERAIS border; 15°15′S 40°58′W.

Inhoca, town, CABINDA province, ANGOLA, on road 55 mi/89 km NE of CABINDA; 04°48′S 12°23′E. Market town.

Inhulets' (een-hoo-LETS), (Russian *Ingulets*), city, SW DNIPROPETROVS'K oblast, UKRAINE, on right bank of INHULETS' River, 14 mi/23 km SSW of KRYVYY RIH; 46°48′N 32°49′E. Iron ore benefication combine; highway building materials, flour mill, yeast factory. Ore-mining technical school, two professional-technical schools. Iron-ore mining pit S of the city. Village of Inhulets' (1990 estimated population 500) is across Inhulets' River, 2 mi/3.2 km NE. Established in the early 20th century as a mining town; city since 1956.

Inhulets' (een-hoo-LETS), (Russian *Ingulets*), river, approximately 340 mi/547 km long, S UKRAINE; rising 15 mi/24 km NNE of KIROVOHRAD, it flows E past OLEKSANDRIYA, and then S through the KRYVYY RIH iron district to join the DNIEPER (Ukrainian *Dnipro*) River above KHERSON.

Inhulets' Irrigation and Water Supply System (een-hoo-LETS), (Ukrainian *Inhulets'ka zroshuval'no-obvodnyuval'na systema*), irrigation and water supply system in SE MYKOLAYIV and W KHERSON oblasts, UKRAINE, on the BLACK SEA LOWLAND, between BUH LIMAN (W), lower INHULETS' River (E), lower DNIEPER River (SE), and DNIEPER LIMAN (SW). The system uses Dnieper water, diverted 53 mi/85 km up the Inhulets' by deepening its meandering channel to a point 4 mi/6 km S of SNIHURIVKA, then pumped up (200 ft/60 m) to the head of the Inhulets' Mainline Canal (Ukrainian *Inhulets'kyy mahistral'nyy kanal*), which transports the water WSW by gravity flow for 33 mi/53 km for distribution S via 13 major distributaries. Built in 1951–1963, the canal and its distributaries were reconstructed and lined with cement and polyethylene in the 1980s, to curtail seepage and reduce salinization and waterlogging. The system irrigates 155,000 acres/62,700 ha, supplies water to 432,000 acres/175,000 ha, and is served by drainage on 14,000 acres/5,600 ha. Water is applied by sprinklers to grow grain, vegetables, and feed crops.

Inhulets'ka zroshuval'no-obvodnyuval'na systema, UKRAINE: see INHULETS' IRRIGATION AND WATER SUPPLY SYSTEM.

Inhulets'kyy mahistral'nyy kanal, UKRAINE: see INHULETS' IRRIGATION AND WATER SUPPLY SYSTEM.

Inhulets' Mainline Canal, UKRAINE: see INHULETS' IRRIGATION AND WATER SUPPLY SYSTEM.

Inhul River (een-HOOL), (Russian *Ingul*), river, S UKRAINE; rising 14 mi/23 km NNW of KIROVOHRAD, it flows through the city and then S for approximately 210 mi/340 km to empty into the BUH estuary, an inlet of the BLACK SEA, at MYKOLAYIV.

Inhuma (EEN-yoo-mah), city (2007 population 14,953), central PIAUÍ state, BRAZIL, 133 mi/214 km SE of TERESINA; 06°40′S 41°42′W.

Inhumas (een-YOO-mahs), city (2007 population 41,406), S central GOIÁS, central BRAZIL, 20 mi/32 km NW of GOIÂNIA; 16°26′S 49°31′W.

Inhuporauga (EEN-yoo-po-RAH-gah), city, N central CEARÁ, BRAZIL, 37 mi/60 km SW of FORTALEZA; 04°06′S 39°07′W.

Iniada, TURKEY: see IGNEADA.

Inicua (ee-NEE-kwah), canton, NOR YUNGAS province, LA PAZ department, W BOLIVIA; 15°19′S 67°31′W. Elevation 5,676 ft/1,730 m. Clay, gypsum, and limestone deposits; phosphates at CARANAVI. Agriculture (potatoes, yucca, bananas, rye, soy, tobacco, coffee, tea, citrus fruits); cattle.

Iniesta (ee-NYE-stah), town, CUENCA province, E central SPAIN, 32 mi/51 km N of ALBACETE; 39°26′N 01°45′W. Agricultural center (saffron, cereals, grapes, sheep); lumbering; liquor distilling.

Inimutaba (EE-nee-moo-tah-bah), town (2007 population 6,420), N central MINAS GERAIS, BRAZIL, 3.1 mi/5 km NE of CURVELO; 18°40′S 44°25′W.

Ining, CHINA: see YINING.

Inini (ee-nee-NEE), former territory (□ c.30,300 sq mi/78,477 sq km), S FRENCH GUIANA. Bordering S on BRAZIL (TUMUC-HUMAC MOUNTAINS). Vast, densely forested hinterland, it included low outliers of the GUIANA HIGHLANDS and was intersected by many streams. Little exploited and unexplored, the region yielded some rosewood, fine cabinet wood, balata, and alluvial gold (SAINT-ÉLIE). The chief outposts were GRAND-PONT, MARIPA, and MARIPASOULA. Set up in 1930 as a separate unit, it was reunited in 1946 as a dependency of French Guiana.

Inini River (ee-nee-NEE), c.100 mi/161 km long, S FRENCH GUIANA; rises at S foot of CHAÎNE GRANITIQUE; flows W through tropical forests, to MARONI RIVER (here called LAWA RIVER) at 03°40′N 54°02′W. Interrupted by rapids. Gold placers near source.

Inírida River (een-EE-ree-dah), c.450 mi/724 km long, SE COLOMBIA; rises in GUAVIARE department; flows E (GUAINÍA department) in meandering course, forming side channels, oxbows, and islands, through wild rainforests, to GUAVIARE RIVER just before latter joins ORINOCO RIVER; 03°55′N 67°52′W. Cataracts make it unnavigable.

Iniscaltra, IRELAND: see HOLY ISLAND.

Inishark, IRELAND: see INISHSHARK.

Inishbofin (i-nish-BAH-fuhn), Gaelic *Inis Bó Finne*, island (□ 1 sq mi/2.6 sq km) off NW DONEGAL county, N IRELAND, 5 mi/8 km ENE of BLOODY FORELAND promontory.

Inishbofin, Gaelic *Inis Bó Finne*, island (□ 4 sq mi/10.4 sq km) in the ATLANTIC OCEAN, 4 mi/6.4 km NNW of NW coast of GALWAY county, W IRELAND; 55°10′N 08°10′W. Rises to 292 ft/89 m. Surrounded by several small rocky islets, largest of which is INISHSHARK.

Inisheer (i-ni-SHIR) or **Inishere**, Gaelic *Inis Oírr*, rocky island (□ 2 sq mi/5.2 sq km) of the ARAN ISLANDS, in GALWAY BAY off SW GALWAY county, W IRELAND, 24 mi/39 km SW of GALWAY; 53°03′N 09°31′W. Rises to 202 ft/62 m. Has prehistoric fort, remains of church of St. Kevin, and medieval stronghold of the O'Briens. On S coast of island is lighthouse.

Inishglora (i-nish-GLOR-ruh), Gaelic *Inis Glora*, island (1 mi/1.6 km long) just off MULLET PENINSULA, NW MAYO county, NW IRELAND, 3 mi/4.8 km W of BINGHAMSTOWN; 54°12′N 10°08′W. Site of remains of church built by St. Brendan.

Inishmaan (i-nish-MAN), Gaelic *Inis Meáin*, rocky island (□ 4 sq mi/10.4 sq km) of the ARAN ISLANDS, SW GALWAY county, W IRELAND, in GALWAY BAY, and 25 mi/40 km WSW of GALWAY; 53°05′N 09°35′W. Rises to 275 ft/84 m. Has prehistoric fort (Dun Chonchobair) and remains of ancient churches. Airport.

Inishmore (i-nish-MOR) or **Deer Island**, Gaelic *Inis Mór*, island (□ 1 sq mi/2.6 sq km) in Fergus River estuary, S CLARE county, W IRELAND, 7 mi/11.3 km S of ENNIS; 52°42′N 09°01′W.

Inishmore (i-nish-MOR), Gaelic *Árainn na Naomh* or *Árainn Mór*, rocky island (□ 12 sq mi/31.2 sq km) of the ARAN ISLANDS, SW GALWAY county, W IRELAND, in GALWAY BAY, 27 mi/43 km WSW of GALWAY; 53°07′N 09°43′W. Largest of the Aran Islands. Rises to 403 ft/123 m. Towns on island are KILRONAN, Killeany, and Onaght. Fishing. Numerous remains of 6th-century churches and monastic establishments; several prehistoric forts. Arkyn Castle held out against Cromwell for a year after capture of Galway. After fall of Galway in 1691 island was garrisoned for several years by William III's troops.

Inishmurray (i-nish-MUH-ree), Gaelic *Inis Muirigh*, island at entrance to DONEGAL BAY, N SLIGO county, NW IRELAND, 14 mi/23 km NW of SLIGO; 54°26′N 08°40′W. Contains large number of antiquities, including pagan stone fort (converted into monastery in 6th century by St. Molaise) and three ancient oratories.

Inishowen Head (i-ni-SHO-wuhn), Gaelic *Srúibh Brain*, promontory, NE DONEGAL county, N IRELAND, on W side of entrance to LOUGH FOYLE, 6 mi/9.7 km NE of MOVILLE. Lighthouse (55°13′N 06°56′W).

Inishshark (i-ni-SHAHRK) or **Inishark**, Gaelic *Inis Airc*, island (□ 1 sq mi/2.6 sq km) off NW GALWAY county, W IRELAND, just W of INISHBOFIN island; 53°37′N 10°17′W.

Inishturk (i-nish-TUHRK), Gaelic *Inis Toirc*, island, NW GALWAY county, W IRELAND, 15 mi/24 km NNW of CLIFDEN; 53°31′N 10°09′W. Rises to 629 ft/192 m. Remains of church built by St. Columcille.

Inistioge (i-nish-TEEG), Gaelic *Inis Téog*, town (2006 population 263), E KILKENNY county, SE IRELAND, on the NORE RIVER (bridged), 9 mi/14.5 km NW of NEW ROSS; 52°29′N 07°04′W. Agricultural market. Has remains of Austin priory (1210).

Initao (ee-NEE-tou), town, MISAMIS ORIENTAL province, N MINDANAO, PHILIPPINES, 23 mi/37 km W of CAGAYAN DE ORO; 08°30′N 124°20′E. Agricultural center (corn, coconuts). Chromite deposits. Initao National Park (forest preserve).

Inje (IN-JEI), county (□ 759 sq mi/1,973.4 sq km), NE KANGWON province, SOUTH KOREA, on W slope of TAEBAEK Mountains, facing cease-fire line on N; 38°04′N 128°10′E. Small rivers originating from mountains join Sobaek River. Mountain passes to E. Some agriculture (vegetables, peppers, garlic, honey, mushrooms); forestry; logging. SORAK MOUNTAIN NATIONAL PARK. Shinhung Buddhist Temple.

Inje, Cape, TURKEY: see INCE, CAPE.

Injibara, ETHIOPIA: see ENJEBARA.

Injune (IN-joon), town, E central QUEENSLAND, NE AUSTRALIA; 25°53′S 148°30′E. S entry to CARNARVON NATIONAL PARK.

Inkak, town, SEVENTH REGION/GAO, MALI, E of GAO.

Inkerman (EEN-ker-mahn), city (2004 population 51,200), S Republic of CRIMEA, UKRAINE, less than 3 mi/4.8 km E of SEVASTOPOL'; 44°36′N 33°36′E. Elevation 262 ft/79 m. Port on the BLACK SEA. Railroad station. Manufacturing (construction materials); electric-generating station. Site of a 9th-century B.C.E. Taurian settlement, and 4th-century C.E. Scythian burial ground. From the 6th to 15th centuries, site of fortified city Kalamita, with many dwellings and churches inside caves; captured in 1475 by the Turks and named Inkerman [Turkish=stone fortress]. The site of the Russian Imperial Army's defeat at the hands of the British and French forces in the Crimean War on November 5, 1854, despite the Russian's significant numerical superiority in troops and artillery. Made city in 1976 from what had become E suburb of Sevastopol', and given name of Bilokam'yans'k (Russian *Belokamensk*) [=white stone]. Renamed Inkerman in 1991.

Inkeroinen (ING-ke-ROI-nen), Swedish *Ingerois*, village, KYMEN province, SE FINLAND, on the KYMIJOKI (river), and 16 mi/26 km N of KOTKA; 60°42′N 26°51′E. Railroad junction; lumber, pulp, and paper mills.

Inkisi (ing-KEE-see), village, BAS-CONGO province, W CONGO, on INKISI RIVER, on railroad, 140 mi/225 km ENE of BOMA; 05°08′S 15°04′E. Elev. 1,354 ft/412 m. Trading center; sugarcane, coffee, staple foodstuffs.

Inkisi-Kisantu, CONGO: see KISANTU.

Inkisi River (ing-KEE-see), W CONGO; rises in NW ANGOLA, flows generally NNW, past LEMFU, INKISI, and KISANTU to CONGO RIVER 40 mi/64 km SW of KINSHASA. Has two hydroelectric power stations at SANGA and ZONGO.

Inkom (IN-kum), town (2000 population 738), BANNOCK county, SE IDAHO, 12 mi/19 km SE of POCATELLO, on PORT-NEUF RIVER; 42°48′N 112°15′W. Manufacturing (portland cement). Parts of Caribou National Forest to SW and E; Pebble Creek Ski Area to E; Fort Hall Indian Reservation to N (port of entry).

Inkomo, town, MASHONALAND EAST province, NW ZIMBABWE, 25 mi/42 km NW of HARARE; 17°40′S 30°44′E. Tobacco; dairying; cattle, sheep, goats. Location of army barracks.

Inks Lake, reservoir (□ 1 sq mi/2.6 sq km), BURNET-LLANO county line, S central TEXAS, on COLORADO RIVER, 3 mi/4.8 km below Buchanan Dam, c.10 mi/16 km WSW of BURNET; 30°43′N 98°22′W. Formed by Roy Inks Dam. National Fish Hatchery. Inks Lake State Park on E shore.

Inkster, city (2000 population 30,115), WAYNE county, SE MICHIGAN, suburb 14 mi/23 km WSW of DEARBORN, on the ROUGE RIVER; 42°17′N 83°19′W. Manufacturing (fabricated metal products, storage tanks, chemicals, gaskets and seals). Nearby auto plants suffer from the decline of the auto industry in the 1970s and 1980s. Drained by Lower River Rouge. Settled 1825 as Moulin Rouge, renamed 1863, incorporated as a city 1964.

Inkster, village (2006 population 94), GRAND FORKS CO., E NORTH DAKOTA, 33 mi/53 km NW of GRAND FORKS, near FOREST RIVER; 48°08′N 97°38′W. Founded in 1880 and named for local settler, George T. Inkster.

Inland Empire, name given to vast region of COLUMBIA River basin in E WASHINGTON, N OREGON E of the CASCADES, and N IDAHO, and sometimes including NW MONTANA; SPOKANE (Wash.) is chief center. Dry farming (wheat) and livestock grazing have long predominated, but large-scale irrigation is contemplated through the Columbia River project, so as to increase the variety of the region's agricultural products. Mining and lumbering are important in Idaho and Montana portions.

Inland Sea, Japanese *Seto-naikai* (se-to-NAH-ee-kah-ee), arm of the PACIFIC OCEAN (□ 255 sq mi/663 sq km), S JAPAN, between HONSHU, SHIKOKU, and KYUSHU islands. Linked to the SEA OF JAPAN by a narrow channel. The shallow sea is dotted with over nine hundred and fifty islands, the largest of which is AWAJI-SHIMA. The shores of the Inland Sea are heavily populated and are part of Japan's most important industrial belt. Many industrial cities line the sea from the Osaka-Kobe complex on the E to the N KYUSHU industrial complex on the W. Many of Japan's greatest ports, including OSAKA, KOBE, and HIROSHIMA, are here. Also famed for its scenic beauty and is the site of Inland Sea (Seto-naikai) National Park (□ 255 sq mi/ 660 sq km), established 1934, which includes some six hundred islands and coastal segments.

Inle Lake (IN-LAI), in YAWNGHWE township, SHAN STATE, MYANMAR, on SHAN PLATEAU, SW of TAUNGGYI; 12 mi/19 km long, 4 mi/6.4 km wide. Elevation c.3,000 ft/914 m. Fisheries; rice area. Source of NAM PILU, it is fed by small streams N and W and is bordered by floating islands (cultivated) of decayed vegetable matter.

Inlet, hamlet, HAMILTON county, NE central NEW YORK, in ADIRONDACK MOUNTAINS, between two lakes of FULTON CHAIN OF LAKES, c.50 mi/80 km NNE of UTICA; 43°44′N 74°44′W. It is the only significant community in the town of the same name. Recreation and resort area.

Inman, town (2006 population 1,935), SPARTANBURG county, NW SOUTH CAROLINA, 12 mi/19 km NW of SPARTANBURG; 35°02′N 82°05′W. Manufacturing (textiles, paper, chemicals, copper wire). Agriculture includes soybeans, apples; poultry, hogs; dairying. Painter Henry Inman and others in South Carolina created some of the earliest genre painting done in the UNITED STATES. (1825–1850).

Inman, village (2000 population 1,142), MCPHERSON county, S central KANSAS, 15 mi/24 km NE of HUTCHINSON; 38°13′N 97°46′W. In wheat region; flour milling. Manufacturing (lumber). Oil wells nearby.

Inman, village (2006 population 129), HOLT county, N NEBRASKA, 8 mi/12.9 km SE of O'NEILL, on ELKHORN RIVER; 42°22′N 98°31′W.

Inman (IN-muhn), unincorporated village, WISE county, SW VIRGINIA, 9 mi/15 km W of NORTON, 2 mi/ 3 km NW of APPALACHIA, in Jefferson National Forest; 36°54′N 82°48′W. Coal mining.

Inman Mills, unincorporated village (2000 population 1,151), SPARTANBURG county, NW SOUTH CAROLINA, 12 mi/19 km NW of SPARTANBURG, 1 mi/1.6 km SW of INMAN; 35°02′N 82°05′W. Textiles and paper mills in area.

Inn (EEN), district, E GRISONS canton, E SWITZERLAND. Population is Romansch-speaking and mainly Protestant, but commune of TARASP on Inn River is Roman Catholic, and commune of SAMNAUN on Austrian border is German-speaking and Roman Catholic.

Inn (EEN), ancient *Aenus*, river, c.320 mi/515 km long, SE SWITZERLAND; rises in GRISONS canton near LAKE SILS and near MALOJA pass, flows NE through the ENGADINE valley, then through W AUSTRIA, past INNSBRUCK and Solbad Hall (the head of navigation), and into S central GERMANY, where it is joined by the SALZACH RIVER. The Inn forms part of the German-Austrian border before entering the DANUBE RIVER at PASSAU. There are more than twenty hydroelectric power plants on the river's swift-flowing stream.

Innai (EEN-nah-ee), town, Usa district, OITA prefecture, E KYUSHU, SW JAPAN, 22 mi/35 km N of OITA; 33°25′N 131°19′E. Citron and citron processing.

Innamincka (I-nuh-MING-kuh), settlement, NE SOUTH AUSTRALIA, on BARCOO RIVER, 390 mi/628 km NNE of PORT PIRIE, near QUEENSLAND border; 27°45′S 140°44′E. Cattle. Established 1882. Became a ghost town after a 1956 flood, but discovery of underground gas fields revived interest in the area in the 1970s.

Innerer Landesteil (EEN-uhr-uhr LAHN-duhs-teil), main district in APPENZELL Inner Rhoden half-canton, NE SWITZERLAND. Main town is Appenzell; population is German-speaking and Roman Catholic.

Inner Hebrides, Scotland: see THE HEBRIDES.

Innerleithen (I-nuhr-LEI-thuhn), town (2001 population 2,586), Scottish Borders, SE Scotland, on the TWEED RIVER, and 5 mi/8 km ESE of PEEBLES; 55°37′N 03°04′W. Health resort, commuter town to EDINBURGH. Previously woolen (tweed) milling and hosiery knitting.

Inner Mongolia, CHINA: see INNER MONGOLIA AUTONOMOUS REGION.

Inner Mongolia Autonomous Region, Mandarin *Nei Menggu zizhi qu* (NEI-MENG-GUH ZI-CHI CHU), short name *Nei Menggu* or **Inner Mongolia**, autonomous region (□ c.455,000 sq mi/1,178,755 sq km; 1994 estimated population 22,170,000), N CHINA; ⊙ HOHHOT; 44°00′N 112°00′E. It is bounded on the N by the Mongolian People's Republic. Inner Mongolia is largely steppe country that becomes increasingly arid toward the GOBI DESERT in the W. The climate is continental, with cold, dry winters and hot summers. Livestock raising, mainly of sheep, goats, horses, and camels, is a major occupation; wool, hides, and skins are important exports. Rainfall is scanty, but irrigation makes agr. possible, and in recent years much grazing land has been converted to raising spring wheat. The main farming areas are in the bend of the HUANG HE (Yellow River) and in the Hohhot plains. Principal crops are wheat, kaoliang, millet, oats, corn, linseed, soybeans, sugar beets, and rice. There are valuable mineral deposits (coal, lignite, iron ore, lead, zinc, and gold) that are only partially exploited. The region's industries, centered at BAOTOU, include iron and steel mills, as well as plants producing fertilizer, cement, textiles, and machinery. A railroad built in 1958, linking RUSSIA (through the Mongolian People's Republic) with LANZHOU, passes through Hohhot and Baotou. The Beijing-Ulaanbaatar road traverses the region. Considerable additional road and railroad improvements have been made with the vigorous industrialization of Baotou.The Mongols of China are concentrated in the Inner Mongolia Autonomous Region, but there has been much Chinese immigration and the Mongols now comprise less than 20% of the population. The Chinese live mostly in the farming areas. The traditionally nomadic Mongols are beginning to settle as their pastoral economy is collectivized. Originally the S part of MONGOLIA, Inner Mongolia was settled chiefly by the Tumet and Chahar tribes. From 1530 to 1583, Inner Mongolia was held by Anda (Altan Khan), chief of the Tumets, who harried N China and once besieged Beijing. After his death, Likdan Khan of the Chahars became (c.1605) ruler, but in 1635 he was defeated by the Manchus, who soon annexed Inner Mongolia. Under Manchu rule S Mongolia became known as Inner Mongolia; N Mongolia, conquered by the Manchus at the end of the 17th century, became known as Outer Mongolia. Until 1911, Inner Mongolia was only under nominal Chinese rule; Chinese settlers in the region, however, soon forced the Mongol tribes into the steppe and arid parts of the region. After the Revolution of 1911, Inner Mongolia became an integral part of the Chinese Republic. Inner Mongolia was divided among the Chinese provinces of NINGXIA, SUIYUAN, and Chaha'er in 1928. After the outbreak (1937) of the Sino-Japanese War, the Mongols of Suiyuan and Chahar established the Japanese-controlled state of Mengkiang, or Mengjiang, with its capital at Guihua. The Chinese Communists, after their conquest of Inner Mongolia in 1945, supported the traditional aspirations of the Mongols for autonomy, and in May 1947, the Inner Mongolia Autonomous Region—with limited powers of self-government within the Communist state—was formally proclaimed. It was the first autonomous region established by the Communist government. From 1949 to 1956 the area of the region was expanded through the incorporation of the former province of Suiyuan and parts of the provinces of LIAOBEI, REHE, Chaha'er, and GANSU. Extensive boundary changes in 1969, however, reduced the size of the province considerably. The W Alashan Desert region was given to Gansu and Ningxia Hui Autonomous Region, and the NE corner, which bordered on Russia, was divided between the provinces of NE China. HEBEI province also received a section of Inner Mongolia. These border changes were reversed in 1979, and the region was restored to its former size. The primary administrative unit in the region is a *qi*, the equivalent of a county. A number of *qis* form a *meng*, the equivalent of a district. Hohhot has been the capital since 1952; the capital was at Ulan Hot from 1947 to 1950, and at ZHANGJIAKOU (Kalgan; now in Hebei province) from 1950 to 1952. Inner Mongolian University is in Hohhot.

Inner Pintades, Mauritius: see PINTADES ISLAND.

Inner Sound (in-UHR SOUND), strait (5 mi/8 km–8 mi/12.9 km wide) between mainland HIGHLAND, N Scotland, and RAASAY island; 57°25′N 05°56′W. Links the MINCH (N) and Loch Alsh (S).

Innerste River (IN-ner-ste), c.50 mi/80 km long, NW GERMANY; rises in the upper HARZ near CLAUSTHAL-ZELLERFELD; 51°46′N 10°21′E. It flows N and NW, past HILDESHEIM, to the LEINE 1 mi/1.6 km NW of SARSTEDT.

Innertkirchen (EEN-uhrt-KIR-hen), commune, BERN canton, S central SWITZERLAND, on AARE RIVER, at junction of HASLITAL and Gadmental, 2 mi/3.2 km SE of MEIRINGEN, in valley among peaks of BERNESE ALPS. Elevation 2,041 ft/622 m. Has hydroelectric plant.

Inner West (I-nuhr WEST), district comprising the inner-city suburbs adjacent and W of SYDNEY, NEW SOUTH WALES, SE AUSTRALIA. Suburbs include: STRATHFIELD, NEWTOWN, LEICHHARDT, GLEBE, BALMAIN. Mostly residential; university campuses; railway line.

Innes National Park, national park (□ 36 sq mi/93.6 sq km), SOUTH AUSTRALIA state, S central AUSTRALIA, on S tip of YORKE PENINSULA. Habitat types include heathlands, woodlands, grasslands, mallee, samphire flats, salt lakes, beaches, and rocky cliffs. Dedicated in 1970 to conserving the rare Western Whipbird (*Pso-

phodes nigrogularis). The Narungga Aboriginal people are the area's original landowners.

Innherad (IN-ha-rahd), region around inner TRONDHEIMSFJORDEN, NORD-TRØNDELAG county, central NORWAY. STEINKJER and LEVANGER are its centers.

Inning am Ammersee (IN-ning ahm AHM-mer-sai), village, BAVARIA, S GERMANY, in UPPER BAVARIA, on N shore of AMMERSEE Lake, 20 mi/32 km E of MUNICH; 48°04′N 11°10′E. Tourism.

Innisfail (I-nis-FAIL), town, NE QUEENSLAND, NE AUSTRALIA, 45 mi/72 km SSE of CAIRNS, on coast; 17°32′S 146°02′E. Sugar-producing center; sugar mill; bananas. Mean annual rainfall 144 in/366 cm. Crocodile farm 5 mi/8 km from town. Heavily damaged by a tropical cyclone in 2006.

Innisfail (I-nis-fail), town (□ 4 sq mi/10.4 sq km; 2001 population 6,928), S central ALBERTA, W CANADA, near RED DEER River, 18 mi/29 km SSW of RED DEER, and in RED DEER County; 52°02′N 113°57′W. Grain elevators; dairying; ranching. Established as a village in 1899, and as a town in 1903.

Innisfallen (i-nish-FAH-luhn) or **Inisfallen**, Gaelic *Inis Faithlenn*, islet in Lough Leane (one of the LAKES OF KILLARNEY), KERRY county, SW IRELAND, 3 mi/4.8 km SW of KILLARNEY; 52°02′N 09°33′W. Site of ruins of 6th-century abbey founded by St. Finian, where the *Annals of Innisfallen*, now in the Bodleian Library at OXFORD, were composed. Island is the "sweet Innisfallen" of Moore's poem.

Innisfil (I-nis-fil), town (□ 110 sq mi/286 sq km; 2001 population 28,666), S ONTARIO, E central CANADA, 7 mi/12 km from BARRIE; 44°16′N 79°37′W. First surveyed 1820.

Innisfree (I-nuhs-free), village (2001 population 219), E ALBERTA, W CANADA, 28 mi/45 km W of VERMILION, in MINBURN County No. 27; 53°22′N 111°32′W. Dairying; grain; stock.

Innisfree (i-nish-FREE), small island in LOUGH GILL, NE SLIGO county, NW IRELAND. Celebrated in Yeat's poem.

Innishcrone (i-nish-KRON), Gaelic *Inis Easgrach Abhann*, town (2006 population 829), NW SLIGO county, NW IRELAND, on KILLALA BAY, 5 mi/8 km E of KILLALA; 54°13′N 09°06′W. Fishing port and resort.

Innokentyevka (een-nuh-KYEN-tyeef-kuh), village, SE AMUR oblast, RUSSIAN FAR EAST, on the AMUR River (landing), 120 mi/193 km ESE of BLAGOVESHCHENSK; 49°58′N 129°11′E. Elevation 728 ft/221 m. In agricultural area. Originally established in 1858 as a Cossack *stanitsa* (village).

Innokentyevka (een-nah-KYEN-tyeef-kuh), settlement, SE KHABAROVSK TERRITORY, RUSSIAN FAR EAST, on the lower AMUR River, 150 mi/241 km NE of KHABAROVSK; 53°13′N 140°19′E. Town status received in 1949; revoked in 1992.

Innoko River, c.450 mi/724 km long, W ALASKA; rises near 63°00′N 156°30′W, flows first N, then in a winding course generally SW, to YUKON River opposite HOLY CROSS.

Innoshima (EEN-no-shee-mah), city, on Innoshima Island, HIROSHIMA prefecture, off SW HONSHU, W JAPAN, on the HIUCHI SEA, 43 mi/70 km SE of HIROSHIMA; 34°17′N 133°10′E. Shipbuilding. Known for its bridges (Innoshima, Iguchi). *Hassaku* variety of orange originated here.

Innsbruck (INS-bruk), city, TYROL province, W AUSTRIA, on INN River; 47°16′N 11°24′E. Famous summer and winter tourist center, also a commercial, transport, and industrial center; international airport to W. Episcopal see. Manufacturing of textiles, metal products, processed food, beer. Strategically located in Eastern Alps, Innsbruck grew to early prominence as transalpine trading post. Established as fortified town by 1180, received city rights in early 13th century. Supplanted Merano as capital of Tyrol in 1420. The Tyrolese peasants, led by Andreas Hofer, made their

heroic stand (1809) against French and Bavarian troops near INNSBRUCK; a monument on Bergisel commemorates the event. The Hofkirche (built 1553–1563), Franciscan church, is an architectural gem. Equally famous is Fürstenburg, a 15th century castle, which has a balcony with gilded copper roof (*Goldenes Dachl*). The Column of St. Anne (1706) is a landmark in Innsbruck's main thoroughfare, the Maria Theresienstrasse. The city has several museums, notably the Ferdinandeum; a botanical garden, which has a large collection of Alpine plants; several monasteries, the most notable of which is Wilten; and a university (founded 1677). The Winter Olympics were held in Innsbruck in 1964 and 1976.

Innsworth and Churchdown (INNZ-wuhrth and CHUHRCH-doun), suburbs (2001 population 12,998), GLOUCESTERSHIRE, central ENGLAND, 4 mi/6.4 km NE of GLOUCESTER; 51°53′N 02°10′W. Computer technology.

Innuitians (in-yoo-ISH-uhnz), mountain range, stretching c.800 mi/1,290 km through the ARCTIC ARCHIPELAGO, NORTHWEST TERRITORIES, N CANADA. Largely unexplored, the range runs NE from the Parry Islands to N ELLESMERE ISLAND. S along the island's E coast. It rises to c.9,000 ft/2,740 m on Ellesmere Island, where most of the range is covered by an ice cap. The Innuitians form the N rim of the CANADIAN SHIELD.

Innuit, Mount (4,554 ft/1,388 m), E NEWFOUNDLAND AND LABRADOR, CANADA, at head of NACHVAK Fiord; 59°02′N 64°09′W.

Innviertel (IN-feer-tel) or **Innkreis** (IN-kreiz), region of UPPER AUSTRIA between INN River and HAUSRUCK MOUNTAINS, W central AUSTRIA, 47°59′N 12°45′E– 48°35′N 13°44′E. Bounded on SW by FLACHGAU (SALZBURG), NW by BAVARIA (GERMANY), N by the Danube, E by Hausruck Mountains Fertile. Densely populated hilly region drained by tributaries of Inn River Dairy farming, cattle breeding, and industry with some outstanding industrial locations (aluminium production in RANSHOFEN, sports equipment in RIED IM INNKREIS), hydroelectric power generation on Inn River, large thermoelectric power station in SW (Riedersbach), gas and oil fields. Urban centers are Ried im Innkreis, Braunau am Inn, and Schärding. Railroad and motorway (LINZ-WELS-PASSAU) cross the region. The Innviertel is part of Austria with the strongest Bavarian character. It first joined Austria in 1704, finally in 1814.

Ino (EE-no), town, Date district, FUKUSHIMA prefecture, N central HONSHU, NE JAPAN, 6 mi/10 km SE of FUKUSHIMA city; 37°39′N 140°32′E.

Ino, town, Agawa district, KOCHI prefecture, S SHIKOKU, W JAPAN, on NIYODO RIVER, 5 mi/8 km W of KOCHI (linked by railroad); 33°32′N 133°25′E.

Inocência (EE-no-SEN-see-ah), town (2007 population 7,339), NE Mato Grosso do Sul, BRAZIL, 45 mi/72 km W of PARANAÍBA; 19°43′S 51°53′W.

Inokuchi (ee-NOK-chee), village, East Tonami district, TOYAMA prefecture, central HONSHU, central JAPAN, 19 mi/30 km S of TOYAMA; 36°32′N 136°56′E.

Inola (ei-NO-luh), town (2006 population 1,732), ROGERS county, NE OKLAHOMA, 25 mi/40 km E of TULSA. In livestock-raising and agricultural area; manufacturing (machine parts, figurines, lead, medical equipment).

Inongo (ee-NAWNG-go), town, BANDUNDU province, W CONGO, on E shore of LAKE MAI-NDOMBE, 270 mi/435 km NE of KINSHASA; 01°57′S 18°16′E. Elev. 1,102 ft/335 m. Trading center in region noted for palm products; steamboat landing. Has church, teachers and trade schools, hospital, and airport.

Inonu, (Turkish=Inönü) township, NW TURKEY, on Eskisehir-Bursan railroad, 20 mi/32 km WNW of ESKISEHIR; 39°49′N 30°07′E. Scene of Turkish victory over the Greeks after World War I; General Ismet Pasha, who commanded the Turks and later took his

---

Cross-references are shown in SMALL CAPITALS. The pronunciation guide is shown on page xix. The sources of population figures are shown on page xvii.

name from the town, succeeded (1938) Atatürk as president of Turkey.

Inovo (EEN-o-vo), village, MONTANA oblast, VIDIN obshtina, BULGARIA; 44°02′N 22°50′E.

Inowroclaw (ee-no-VRO-tsahv), Polish *Inowrocław*, German *Hohensalza*, city (2002 population 77,986), Bydgoszcz province, central POLAND, 21 mi/34 km SW of TORUŃ, near NOTEĆ River; 52°48′N 18°16′E. Railroad junction; trade center; a leading city in the KUJAWY; health resort, with saltwater springs, hot, salt, and mud baths; many hospitals. Manufacturing of machinery, pumps, bricks, glass; brass working; beet-sugar milling; brewing. Rock salt (c.500 ft/152 m underground) and gypsum (for art uses) worked in vicinity. Passed 1772 to PRUSSIA; reverted 1919 to Poland. Here, in 1909, a church collapsed when the dissolution of underground salt created a cave. During World War II the town suffered relatively little damage.

İnoz, TURKEY: see ENEZ.

Inozemtsevo (ee-nuh-ZYEM-tsi-vuh), town (2006 population 27,775), S STAVROPOL TERRITORY, N CAUCASUS, S European Russia, at the base of Mount Beshtau, on highway and the Mineral'nyye Vody-Kislovodsk railroad, 42 mi/68 km N of NAL'CHIK (KABARDINO-BALKAR REPUBLIC), and 7 mi/11 km S of MINERAL'NYYE VODY; 44°06′N 43°05′E. Elevation 1,469 ft/447 m. Wineries.

Inquisivi (een-kee-SEE-vee), province (□ 3,968 sq mi/10,316.8 sq km), LA PAZ department, W BOLIVIA; ⊙ INQUISIVI; 16°41′S 67°10′W. Elevation 8,986 ft/2,739 m. Agriculture (coca, potatoes, yucca, rye, bananas); cattle. Tin mining at Mina COLQUIRI; tungsten mining at Mina Chambilaya, Mina Chicote Grande; clay, limestone, and gypsum deposits.

Inquisivi (ing-kee-SEE-vee), town and canton, ⊙ INQUISIVI province, LA PAZ department, W BOLIVIA, on E slopes of Cordillera de TRES CRUCES, 70 mi/113 km ESE of LA PAZ; 16°41′S 67°10′W. Elevation 8,984 ft/2,738 m. Potatoes, oca, orchards, sugarcane, fruits, tea, coca.

Inrim, PAPUA NEW GUINEA: see LORENGAU.

Inriville, town, E CÓRDOBA province, ARGENTINA, on the RÍO TERCERO, 20 mi/32 km SW of MARCOS JUÁREZ; 32°56′S 62°14′W. Manufacturing of agricultural implements; agricultural center (corn, soybeans, wheat, flax, alfalfa, oats); horticulture; cattle raising.

Ins, French *Anet*, commune, BERN canton, W SWITZERLAND, 16 mi/26 km WNW of BERN, near Lakes NEUCHÂTEL, BIEL, and MORAT. Elevation 1,562 ft/476 m.

In Salah (EEN sah-LAH), town, TAMANRASSET wilaya, S ALGERIA, in the TIDIKELT, one of the oases in the TADEMAÏT PLATEAU, 230 mi/370 km S of El GOLÉA; 27°12′N 02°29′E. Once an important trading post for the exchange of tea, sugar, gold, and cloth, it is now a center for date cultivation. Surrounded by palm groves irrigated by artesian wells and *fogaras*, an underground irrigation system. The market trades in grains, foodstuffs, carpets and local crafts, and wool, cloth, and leather goods. Recently discovered natural gas deposits to be linked to HASSI R'MEL by pipeline. A meeting place for the Tuaregs of the region, it is an important desert junction. Airport. It has an average of 55 days of sandstorms annually and much effort is spent digging the town out of the sand. Also Aïn Salah.

Insar (een-SAHR), city (2006 population 8,800), S MORDVA REPUBLIC, central European Russia, on the Issa River (right branch of the MOKSHA River, OKA River basin), on crossroads, 50 mi/80 km SW of SARANSK; 53°52′N 44°22′E. Elevation 531 ft/161 m. Bast fiber processing, ribbons. Made city in 1958.

Insar River (een-SAHR), approximately 80 mi/129 km long, MORDVA REPUBLIC, E European Russia; rises just SE of KADOSHKINO, flows ENE, past RUZAYEVKA and N, past SARANSK, ROMODANOVO, and LADA, to the ALATYR′ River just NE of KEMLYA.

Inscription, Cape (in-SKRIP-shuhn), cape, DIRK HARTOG ISLAND, in INDIAN OCEAN, 1 mi/1.6 km off W coast of WESTERN AUSTRALIA, W AUSTRALIA; 25°29′S 112°59′E.

Inscription Point (in-SKRIP-shuhn), E NEW SOUTH WALES, AUSTRALIA, on S shore of BOTANY BAY; 30°00′S 151°13′E. Site of monument in honor of Captain Cook's landing (1770) and establishment of British claim.

Insein (IN-sain), township (□ 1,903 sq mi/4,947.8 sq km), YANGON division, MYANMAR, between AYEYARWADY RIVER and on S spurs of PEGU YOMA; ⊙ INSEIN. Part of the AYEYARWADY delta, it is drained by MYITMAKA RIVER. Rice fields and fisheries in plains; teak forests on PEGU YOMA. Served by Yangon-Mandalay railroad.

Insein, town, ⊙ INSEIN township, YANGON division, MYANMAR, port on MYITMAKA RIVER (here called HLAING RIVER) and Yangon-Mandalay railroad, 5 mi/8 km N of YANGON. Woolen mills; glass factory; railroad shops. Jail.

Inselsberg (IN-sels-berg), second-highest peak (3,005 ft/916 m) of the THURINGIAN FOREST, central GERMANY, 13 mi/21 km SW of GOTHA; 50°52′N 10°29′E.

Inseno, town (2007 population 10,585), SOUTHERN NATIONS state, central ETHIOPIA, 5 mi/8 km E of BUTAJIRA; 08°04′N 38°28′E. In livestock raising area.

Inside Passage (IN-seid), natural, protected waterway, c.950 mi/1,530 km long, threading through the ALEXANDER ARCHIPELAGO off the coast of BRITISH COLUMBIA (CANADA) and SE ALASKA. From SEATTLE, WASHINGTON, to SKAGWAY, ALASKA, or via CROSS SOUND to the Gulf of ALASKA, the route uses channels and straits between islands and the mainland that afford protection from the storms and open waters of the PACIFIC OCEAN. Snow-capped mountains, forests, waterfalls, glaciers, and deep, narrow channels give the Inside Passage great scenic beauty. It was known to Spanish, Russian, English, and American explorers. Important coastal route for Canadian shipping as well as the route generally used by ships sailing between the continental U.S. and Alaska.

Insiza, ZIMBABWE: see NSIZA.

Insjön (IN-SHUHN), town, KOPPARBERG county, central SWEDEN, on Österdalälven River, 17 mi/27 km NW of BORLÄNGE; 60°41′N 15°10′E.

Inskaya (EEN-skah-yah), settlement, SE NOVOSIBIRSK oblast, SW SIBERIA, Russia, on road, 52 mi/84 km S of NOVOSIBIRSK, and 52 mi/84 km N of BARNAUL; 54°13′N 83°13′E. Elevation 711 ft/216 m. Railroad junction on the TURKISTAN-SIBERIA RAILROAD (Turk-Sib); freight yards; sawmilling; metalworks.

Insko (EEN-sko), Polish *Ińsko*, German *Nörenberg*, town, Szczecin province, NW POLAND, on small lake, 40 mi/64 km E of SZCZECIN. Tourist resort. Before 1945, was in POMERANIA.

Inskoy (een-SKO-yee), town (2005 population 13,215), W central KEMEROVO oblast, S central SIBERIA, RUSSIA, on the INYA RIVER (tributary of the OB′ RIVER), 9 mi/14 km E of BELOVO, to which it is administratively subordinate; 54°26′N 86°25′E. Elevation 685 ft/208 m. Belovo regional electric power station is here.

Insnotú, VENEZUELA: see ISNOTÚ.

Insoemanai, INDONESIA: see WAKDE ISLANDS.

Insoemoar, INDONESIA: see WAKDE ISLANDS.

Inspiration, unincorporated village, GILA county, SE central ARIZONA, in PINAL MOUNTAINS, 2 mi/3.2 km N of MIAMI. Copper mining.

Insterburg, RUSSIA: see CHERNYAKHOVSK.

Inster River (EEN-styer), right headstream of the PREGEL River, 45 mi/72 km long, in KALININGRAD oblast, NW European Russia; rises NE of DOBROVOL′SK, flows W and SW, joining the Angerapp at CHERNYAKHOVSK to form the PREGEL River.

Instinción (een-steen-THYON), town, ALMERÍA province, S SPAIN, at foot of the SIERRA DE GÁDOR, 15 mi/24 km NW of ALMERÍA; 36°59′N 02°39′W.

Ships grapes; produces olive oil, cereals. Iron mines nearby.

Institute, unincorporated town, KANAWHA county, W central WEST VIRGINIA, suburb 8 mi/12.9 km WNW of CHARLESTON, on KANAWHA River. Manufacturing (asphalt products, industrial and agr. chemicals). West Virginia State College.

Institute Ice Stream, WEST ANTARCTICA; flows into the RONNE ICE SHELF SE of Hercules Inlet; 82°00′S 75°00′W.

Insu (IN-soo), town, WESTERN REGION, GHANA, on railroad, 19 mi/31 km NNE of TARKWA. Gold mining; cacao, cassava, corn.

Insulan, town, ⊙ SULTAN KUDARAT province, SW MINDANAO, PHILIPPINES, 75 mi/120 km SW of DAVAO; 06°38′N 124°26′E. Agricultural center (rice, corn; livestock). Road junction in Alah River valley.

Insull (IN-suhl), village, HARLAN county, SE KENTUCKY, in the CUMBERLAND MOUNTAINS, 16 mi/26 km NE of MIDDLESBORO, 3 mi/4.8 km NW of ALVA. Bituminous coal.

Insumanai, INDONESIA: see WAKDE ISLANDS.

Insumuar, INDONESIA: see WAKDE ISLANDS.

Insurăței (een-soo-ruh-TSAI), town, BRĂILA county, ROMANIA, 31 mi/50 km SSW of BRĂILA; 44°55′N 27°36′E.

Insurgente José María Morelos y Pavón (een-soor-HEN-tai ho-SAI mah-REE-ah mo-RE-los ee pah-VON), national park, in central MICHOACÁN, MEXICO, 16 mi/26 km E of MORELIA. Mexico Highway 15 leads to the park known for its scenic views.

Insurgente Miguel Hidalgo Y Costilla (een-soor-HEN-tai mee-GEL ee-DAHL-go ee ko-STEE-yah), national park (□ 7 sq mi/18.2 sq km), W of San Ángel (ÁLVARO OBREGÓN), MEXICO, suburb of MEXICO CITY on Mexico Highway 15. This park is locally called La Marquesa. The battle of Monte de las Cruces took place here in the nineteenth century. It was an important battle in the War of Independence. There are picnic sites, a trout hatchery, and a lake here.

Insuta, GHANA: see NSUTA.

Inta (een-TAH), city (2005 population 37,485), NE KOMI REPUBLIC, NE European Russia, on the Inta River (PECHORA River basin), on the North Pechora railroad, 377 mi/607 km NE of SYKTYVKAR, and 150 mi/241 km SW of VORKUTA; 66°05′N 60°08′E. Elevation 173 ft/52 m. Coal-mining center in the PECHORA basin. Vocational school (reindeer herding). Founded in 1940 as part of Stalin's Gulag prison camp system, and called Rudnik. Developed during World War II. Made city and renamed in 1954. Population decline is partially because of the closing of a former Soviet military air base in the vicinity.

Intendente Alvear (een-ten-DAIN-te ahl-vai-AHR), town (1991 population 5,688), ⊙ Chapaleufú department, NE LA PAMPA province, ARGENTINA, 33 mi/53 km NNE of GENERAL PICO; 35°14′S 63°35′W. Agriculture center (wheat, oats, corn, barley, alfalfa, soybeans; livestock).

Inter-American Highway, section of the PAN-AMERICAN HIGHWAY system from NUEVO LAREDO (MEXICO) to YAVIZA (PANAMA); c.3,400 mi/5,472 km long. The principal highway connecting the countries of CENTRAL AMERICA. In Mexico, N of MEXICO CITY, it is designated Highway 85; from Mexico City to GUATEMALA border, Highway 90. In Central America, it is usually called Highway C.A.1.

Interborough (Interboro) Parkway, QUEENS and BROOKLYN boroughs of NEW YORK city, SE NEW YORK, at extreme W end of LONG ISLAND. Short parkway (4.7 mi/7.6 km long) snakes its way from the EAST NEW YORK–Highland Park area of Brooklyn E to the Grand Central Parkway in KEW GARDENS, Queens. Completed August 1935, its lack of breakdown lanes, narrow, winding character, difficult lane changes, and low median barrier made it one of the most hazardous thoroughfares in the metropolitan area until 1973,

Area is shown by the symbol □, and capital city or county seat by ⊙.

when its safety was improved to meet modern standards. Renamed Jackie Robinson Parkway in 1997.

Intercourse, unincorporated town, LANCASTER county, SE PENNSYLVANIA, 10 mi/16 km E of LANCASTER; 40°02′N 76°06′W. In rich agricultural area. Manufacturing (furniture, food products). Agriculture (grain, potatoes, soybeans, apples; livestock; dairying). Pennsylvania Dutch region. People's Place Quilt Museum.

Interior, village (2006 population 78), JACKSON county, SW central SOUTH DAKOTA, 25 mi/40 km WSW of KADOKA, just E of BADLANDS NATIONAL PARK, and on WHITE RIVER; 43°43′N 101°58′W. Located S of park headquarters. Pine Ridge Indian Reservation S of river; Buffalo Gap National Grassland N of river. Center of tourist trade.

Interior and Labuan (LAH-boo-ahn), residency (□ 8,042 sq mi/20,829 sq km), E MALAYSIA; ⊙ BEAUFORT. Until 1946, when LABUAN island was included, it was called simply Interior residency. Labuan Federal Territory takes in Labuan Island; chief town is Victoria. It was used as military headquarters for Commonwealth forces sent to fight INDONESIA's attempt in mid-1960s to disrupt Malaysia's formation. It is a federal territory, a political unit. Separated from SABAH state in 1984 in attempt to establish a duty-free international trading zone.

Interior, Cadena del (in-tai-ree-OR, kah-DAI-nah del), mountain range, N VENEZUELA; 10°00′N 67°00′W. S of Lago de VALENCIA and CARACAS VALLEY, Cadena del Interior makes up the S portion of W CORDILLERA DE LA COSTA. Variant name: Cordillera del Interior.

Interior, Cordillera del, VENEZUELA: see INTERIOR, CADENA DEL.

Interior, Serranía del, VENEZUELA: see INTERIOR, CADENA DEL.

Interlachen (IN-tuhr-LAH-kuhn), town (□ 6 sq mi/15.6 sq km; 2000 population 1,475), PUTNAM county, N central FLORIDA, 15 mi/24 km W of PALATKA; 29°37′N 81°54′W.

Interlaken, commune, BERN canton, central SWITZERLAND, between Lake Brienz and Lake Thun. Elevation 1,870 ft/570 m. Its manufacturing includes textiles and watches. Interlaken is one of the largest tourist resorts (mainly summer) in the BERNESE ALPS, and its yearly visitors far outnumber its permanent inhabitants. The region is famous for its magnificent view of the JUNGFRAU.

Interlaken, district, S BERN canton, W central SWITZERLAND. Main town is INTERLAKEN; population is German-speaking and Protestant.

Interlaken, unincorporated town (2000 population 7,328), SANTA CRUZ county, W CALIFORNIA; residential suburb 2 mi/3.2 km NE of WATSONVILLE, in Pajaro Valley; 36°57′N 121°45′W. Between Kelly, Drew, and Tynan lakes. SANTA CRUZ MOUNTAINS to NE. Fruit, nuts, vegetables, grain, flowers, nursery products.

Interlaken, village, central TASMANIA, AUSTRALIA, 50 mi/80 km N of HOBART, between Lake SORELL and Lake CRESCENT; 42°09′S 147°11′E. Sheep.

Interlaken, village (2006 population 666), SENECA county, W central NEW YORK, in FINGER LAKES region, near CAYUGA LAKE, 18 mi/29 km NW of ITHACA; 42°37′N 76°43′W. In agricultural area. Incorporated 1904.

Interlaken (IN-tuhr-LAH-kuhn), borough (2006 population 881), MONMOUTH county, E NEW JERSEY, near coast, 13 mi/21 km E of FREEHOLD and just N of ASBURY PARK; 40°13′N 74°01′W.

Interlaken, Massachusetts: see STOCKBRIDGE.

Interlochen (IN-tur-LAH-kun), village, GRAND TRAVERSE county, NW MICHIGAN, 10 mi/16 km SSW of TRAVERSE CITY; 44°38′N 85°46′W. In fruit-growing and resort area. Headquarters for Interlochen Center for the Arts (formerly National Music Camp). The eight week Interlochen Arts festival takes place every summer, with performances of more than 360 concerts. Interlochen State Park to S between Duck and Green lakes.

International Amistad Reservoir, in TEXAS and COAHUILA (Mexico), on RIO GRANDE, 10 mi/16 km WNW of DEL RIO (Texas); c.60 mi/97 km long; 29°27′N 101°01′W. Maximum capacity 5,658,600 acre-ft. DEVIL'S RIVER forms 30-mi/48-km N arm. Formed by International Amistad Dam (247 ft/75 m high), built (1969) by the UNITED STATES and MEXICO for flood control and irrigation. AMISTAD NATIONAL RECREATION AREA (Texas) on NE shore, Parque Nacional los Novillos (Mexico) on SW shore.

International Falcon Reservoir (□ 180 sq mi/466 sq km), on S TEXAS–NW TAMAULIPAS (Mexico) border, on RIO GRANDE, 70 mi/113 km SSE of LAREDO (Texas); 25°34′N 99°10′W. Maximum capacity 3,177,000 acre-ft. Formed by International Falcon Lake Dam (also known as Falcon Dam; 175 ft/53 m high), built (1953) by International Boundary and Water Commission for flood control; also used for irrigation and power generation. Falcon State Park near dam.

International Falls, town (2000 population 6,703), ⊙ KOOCHICHING county, N MINNESOTA, 140 mi/225 km NW of DULUTH, on Rainy River below its outlet from RAINY LAKE, opposite FORT FRANCES, ONTARIO (CANADA); 48°35′N 93°24′W. Port of entry; toll bridge to Fort Frances. Manufacturing (pulpwood, wood products, paper, beverages, printing and publishing); timber; cattle; alfalfa; dairying. Tourism. Growth followed construction of paper mill (1904) at falls in Rainy River. Recognized on national weather reports for its frequent coldest daily temperature readings for U.S. (outside ALASKA); sometimes referred to as "Frostbite Falls" and the "Icebox of the Nation." VOYAGEURS NATIONAL PARK to E.

International Peace Garden, affiliated area (□ 4 sq mi/10.4 sq km) of U.S. and Canadian national park systems, N NORTH DAKOTA and SW MANITOBA (CANADA), where U.S. Highway 281 meets Manitoba Highway 10; 15 mi/24 km N of DUNSEITH (North Dakota), 16 mi/26 km S of BOISSEVAIN (Manitoba); 48°59′N 100°03′W. A total of 1.4 sq mi/3.6 sq km is in North Dakota; 2.6 sq mi/6.7 sq km is in Manitoba. Dedicated 1932, the garden commemorates peaceful relations between Canada and the U.S. It includes the 120-ft/37-m-tall Peace Tower. Camping facilities. One of the nicknames for North Dakota is "The Peace Tower State."

Interstate State Park (□ 2 sq mi/5.2 sq km), POLK county, W WISCONSIN; located SW of ST. CROIX FALLS (town on ST. CROIX RIVER). Its central feature is the scenic gorge called the Dalles of the St. Croix, with rock walls c.200 ft/61 m high; there are curious rock formations and potholes. There is also another Interstate State Park (□ 293 acres/119 ha) in Chicago county, MINNESOTA, across St. Croix River; administered separately.

Inthanon (IN-tuh-NON), peak (8,451 ft/2,576 m), in the THANON TONG CHAI range, NW THAILAND, 35 mi/56 km SW of CHIANG MAI; 18°35′N 98°30′E. It is the highest point in THAILAND. Usually referred to as DOI INTHANON.

Intibucá (een-tee-boo-KAH), department (□ 1,057 sq mi/2,748.2 sq km; 2001 population 179,862), SW HONDURAS, on EL SALVADOR border; ⊙ LA ESPERANZA; 14°20′N 88°10′W. Astride Continental Divide, sloping into valleys of Río Grande de Otoro (N) and Lempa River (S). Mainly agricultural (corn, beans, rice, tobacco, coffee); cattle, hogs. Lumbering and sawmilling. The making of mats, baskets, and rope is the local industry. Main centers are La Esperanza and adjoining Intibucá. Formed 1883.

Intibucá (een-tee-boo-KAH), town (2001 population 12,320), INTIBUCÁ department, SW HONDURAS, 2 mi/3.2 km E of LA ESPERANZA; 14°19′N 88°10′W. Elevation 6,529 ft/1,990 m. Mat weaving, basket making. Population largely Lenca Indian.

Intipucá (een-tee-poo-KAH), municipality and town, LA UNIÓN department, EL SALVADOR, to extreme S near coast.

Intisār (in-tee-SAHR), group of five oil fields, CYRENAICA region, N central LIBYA, in SURT BASIN, 30 mi/48 km W of AWJILAH.

Intorsătura Buzăului, ROMANIA: see ÎNTORSURA BUZĂULUI.

Întorsura Buzăului (uhn-tor-SUH-rah boo-ZUH-oo-loo-ee), Hungarian *Bodzaforduló*, town, COVASNA county, central ROMANIA, on BUZĂU RIVER, and 22 mi/35 km E of BRAŞOV; 45°41′N 26°02′E. Railroad terminus, agricultural and mining center; manufacturing (construction materials, tractor components). Also known as Intorsătura Buzăului.

Int'ot'o, central ETHIOPIA, on road, 3 mi/4.8 km N of ADDIS ABABA; 09°06′N 38°42′E. Former village, now part of federal district that includes ADDIS ABABA. Was capital of ETHIOPIA (c.1880–1889) until replaced by ADDIS ABABA. Has ruins of a palace and churches. Also spelled Entotto.

Intra (EEN-trah), town, NOVARA province, PIEDMONT, N ITALY, port on W shore of LAGO MAGGIORE, 18 mi/29 km SE of DOMODOSSOLA; 45°56′N 08°34′E. Commercial center; textiles (cotton, silk), machinery (textile, quarry), glass, hats, sausages. Small museum.

Intracoastal Waterway, partly natural, partly artificial waterway providing sheltered passage for commercial and leisure boats along the U.S. Atlantic coast from BOSTON (MASSACHUSETTS) S to KEY WEST (FLORIDA), where it wraps along the GULF OF MEXICO coast to BROWNSVILLE (TEXAS) at the mouth of the RIO GRANDE; c.3,000 mi/4,828 km long. The toll-free waterway, authorized by Congress in 1919, is maintained by the Army Corps of Engineers at a minimum depth of 12 ft/4 m for most of its length; some parts have 7 ft/2.1 m and 9 ft/2.7 m depths. Among some of the waterway's most often used canals along the ATLANTIC route are the Chesapeake and Delaware, and Albemarle; along the Gulf route the most used are the New Orleans Rigolets Cut, the Port Arthur–Corpus Christi Channel, and the Inner Harbor Navigational Canal at NEW ORLEANS. It runs along the entire coastline of Florida. The separate OKEECHOBEE WATERWAY in E central and SW Florida crosses the Florida peninsula. Plans to build a canal across N Florida to link the Atlantic and Gulf sections were blocked in 1971 by a presidential order to prevent potential environmental damage. Many miles of navigable waterways connect with the coastal system, including the HUDSON RIVER, NEW YORK STATE BARGE CANAL, CHESAPEAKE BAY, the SAVANNAH RIVER, the APALACHICOLA RIVER, and the entire MISSISSIPPI RIVER system. The Intercoastal Waterway has a good deal of commercial activity; barges haul petroleum products, foodstuffs, building materials, and manufactured goods.

Intundhla (ee-en-too-en-DLIH), village, MATABELELAND NORTH province, W central ZIMBABWE, 67 mi/108 km SE of HWANGE, on railroad, at NE boundary of HWANGE NATIONAL PARK; 18°58′S 27°18′E. Livestock.

Inúbia Paulista (EE-noo-bee-ah POU-lee-stah), town (2007 population 3,595), W SÃO PAULO state, BRAZIL, 5 mi/8 km SE of LUCÉLIA, on railroad; 21°45′S 50°58′W.

Inubo, Cape (EE-noo-bo), Japanese *Inubo-misaki* (ee-noo-BO-mee-SAH-kee), CHIBA prefecture, E central HONSHU, E central JAPAN, near CHOSHI, at NE tip of base of BOSO PENINSULA, in the PACIFIC OCEAN; 35°42′N 140°52′E. Important lighthouse here.

Inukai (ee-NOO-kah-ee), town, Ono district, OITA prefecture, NE KYUSHU, SW JAPAN, 12 mi/20 km S of OITA; 33°04′N 131°38′E.

Inukjuak or **Port-Harrison** (port–HA-ri-suhn), village, NORD-du-Québec region, NW QUEBEC, E CANADA, on E shore of HUDSON BAY, W side of UNGAVA PENINSULA, at mouth of Innuksuak River; 58°27′N 78°06′W. Populated by Inuit people. Scheduled air service. Hunting, fishing, trapping.

Cross-references are shown in SMALL CAPITALS. The pronunciation guide is shown on page xix. The sources of population figures are shown on page xvii.

Inútil, Bahía, CHILE: see USELESS BAY.

Inuvik (IN-oo-vik), NW region (□ 152,089 sq mi/ 395,431.4 sq km), NORTHWEST TERRITORIES, Canada, extending E from the YUKON to the Hornaday River and N from WRIGLEY to include BANKS ISLAND. The town of INUVIK, built in the 1950s and 1960s, is the regional headquarters. Inuvik is the largest community N of the Arctic Circle. The N MACKENZIE RIVER forms the heart of the region, which is steadily developing its oil and natural-gas resources. Native hunting, sealing, and crafts. The region is a mix of subarctic coniferous forest and arctic tundra.

Inuvik (IN-oo-vik), town (□ 19 sq mi/49.4 sq km; 2006 population 3,484), INUVIK region, NORTHWEST TERRITORIES, CANADA, on E channel of the MACKENZIE RIVER; 68°21′N 133°42′W. It was built (1954–1962) as a new town site for AKLAVIK and was the first model town in the Canadian Arctic. Airport.

Inuyama (ee-NOO-yah-mah), city, AICHI prefecture, S central HONSHU, central JAPAN, on the KISO RIVER, 12 mi/20 km N of NAGOYA; 35°22′N 136°56′E. Inuyama Castle has the oldest castle tower in Japan.

Inverallochy (IN-vuhr-ah-LAW-kee), fishing village (2001 population 1,197), Aberdeenshire, NE Scotland, on NORTH SEA, and 3 mi/4.8 km ESE of FRASERBURGH; 57°40′N 01°55′W. Nearby is Inverallochy Castle, ancient stronghold of the Comyns. Just SW of Inverallochy is fishing village of Cairnbulg.

Inveraray (IN-vuhr-RE-ree), town (2001 population 1,201), Argyll and Bute, W Scotland, on LOCH FYNE, and 45 mi/72 km NW of GLASGOW; 56°14′N 05°04′W. Inveraray Castle is seat of the duke of Argyll and thus the center of the Campbell clan.

Inverbervie (IN-vuhr-BER-vee) or **Bervie**, town (2001 population 2,094), Aberdeenshire, NE Scotland, on NORTH SEA at mouth of BERVIE WATER, and 9 mi/14.5 km SSW of STONEHAVEN; 56°51′N 02°16′W. Fishing port, fish curing, textile manufacturing. Just S are ruins of Hallgreen Castle.

Invercargill (in-vuhr-KAHR-gil), city (□ 379 sq mi/ 985.4 sq km; 2006 population 46,773), (cap.) SOUTHLAND region (□ 20,515 sq mi/53,132 sq km), SOUTH ISLAND, NEW ZEALAND, on the alluvial Southland Plain; 46°24′S 168°21′E. It is an agricultural center with timber, wool, and meat and milk processing industries. Bluff on FOVEAUX STRAIT is main port.

Inverell (IN-vuh-REL), municipality, NE NEW SOUTH WALES, SE AUSTRALIA, on MACINTYRE (Barwon) River, 220 mi/354 km NNW of NEWCASTLE; 29°47′S 151°07′E. Railroad terminus; mining center (silver, lead, tin; gemstones, especially sapphires, industrial diamonds); dairy foods, fruit, tobacco, grain; sheep, cattle.

Inveresk, Scotland: see MUSSELBURGH.

Invergarry (IN-vuhr-GE-ree), village, central HIGHLAND, N Scotland, in Glen Garry near LOCH OICH, and 7 mi/11.3 km SW of FORT AUGUSTUS; 57°02′N 04°47′W. Ruins of ancient Invergarry Castle, former seat of the MacDonnells of Glengarry, whose last chief was reputedly the prototype of Fergus MacIvor in Scott's *Waverley*.

Invergordon (IN-vuhr-gor-duhn), port town (2001 population 3,840), E HIGHLAND, N Scotland, on CROMARTY FIRTH, 15 mi/24 km NNE of INVERNESS; 57°41′N 04°10′W. Previously an important naval base. Aluminum works, distillery, and NORTH SEA oil service base.

Invergowrie, Scotland: see DUNDEE.

Inver Grove Heights, city (2000 population 29,751), DAKOTA county, SE MINNESOTA, suburb 7 mi/11.3 km S of ST. PAUL, on MISSISSIPPI RIVER (bridged); 44°49′N 93°03′W. It has benefited from the industrial and cultural growth of the greater MINNEAPOLIS–St. Paul area. Manufacturing (motor-vehicle and aircraft parts, asphalt, feeds, consumer goods, building materials, paper products, medical equipment). South St. Paul Airport to NE. Seat of Inver Hills Community College. Several small lakes in city.

Inverhuron (in-vuhr-HYU-ron), unincorporated village, SW ONTARIO, E central CANADA, included in KINCARDINE; 44°17′N 81°35′W. Beach.

Inverkeithing (IN-vuhr-KEE-theeng), town (2001 population 5,412), FIFE, E Scotland, near the Firth of Forth, and 4 mi/6.4 km SE of DUNFERMLINE; 56°02′N 03°23′W. Paper-milling center; stone quarries nearby. On the Firth of Forth is port, with shipyards. Has 14th-century Hospitium of Greyfriars and 16th-century market cross. Town was residence of David I. Cromwell defeated supporters of Charles II at battle of Inverkeithing (1651). On the Firth of Forth, 2 mi/3.2 km S, is town of North Queensferry, N terminal of the Forth Bridge. Just NW is ROSYTH.

Inverkip (IN-vuhr-kip), village (2001 population 1,548), Inverclyde, W Scotland, on Firth of Clyde, and 6 mi/ 9.7 km WSW of GREENOCK; 55°55′N 04°52′W. Just N is Ardgowan with whiskey distillery.

Inverleigh (IN-vuhr-lee), township, VICTORIA, SE AUSTRALIA, 64 mi/103 km from MELBOURNE, 17 mi/28 km W of GEELONG, and at junction of Leigh, Barwon rivers; 38°06′S 144°03′E. Notable church buildings.

Inverloch (IN-vuhr-lahk), village, S VICTORIA, SE AUSTRALIA. 70 mi/113 km SSE of MELBOURNE, near WONTHAGGI, on small inlet of BASS STRAIT; 38°38′S 145°43′E. Tourism; seaside resort. Dairying; wool. Once an important port. Name is Gaelic for "entrance to lake."

Inverlochy, Scotland: see FORT WILLIAM.

Invermay (IN-vuhr-mai), village (2006 population 262), SE SASKATCHEWAN, CANADA, near small Saline and Stonewall lakes, 33 mi/53 km WNW of CANORA; 51°47′N 103°09′W. Mixed farming.

Invermere (IN-vuhr-meer), resort village (□ 3 sq mi/7.8 sq km; 2001 population 2,858), SE BRITISH COLUMBIA, W CANADA, on slope of ROCKY MOUNTAINS, on WINDERMERE LAKE, 50 mi/80 km SSW of BANFF (ALBERTA), and in EAST KOOTENAY regional district; 50°30′N 116°02′W. Elevation 2,863 ft/859 m. Logging; tourism, ski-resort area. Near KOOTENAY NATIONAL PARK.

Inverness (IN-vuhr-NES), county (□ 1,409 sq mi/ 3,663.4 sq km; 2001 population 19,937), NE NOVA SCOTIA, CANADA, in E part of CAPE BRETON ISLAND, on GULF OF ST. LAWRENCE; ⊙ PORT HOOD. Split from CAPE BRETON county in 1835.

Inverness (IN-vuhr-NES), city (2001 population 40,949), ⊙ HIGHLAND, N Scotland, on the MORAY FIRTH at mouth of the Ness River, and 9 mi/14.4 km E of BEAULY; 57°28′N 04°14′W. It is a seaport and transportation center due to its proximity to the river and the CALEDONIAN CANAL. Light industries, including printing and food processing. Previously wool weaving, distilling, and shipbuilding. Electrical and mechanical products and motor-vehicle parts are also manufactured. The castle, reputedly built under Malcolm III (late 11th century) was involved in many wars and was blown up by the Jacobites in 1746. A new castle was built in 1835. Frequent invasions have destroyed most of the town's old buildings. Cromwell's Fort was demolished by Charles II during the Restoration. Inverness, a thriving tourist center, has a museum of Highland relics and hosts an annual Highland Gathering. Inverness Airport, a small air-transportation crossroads, is located at Dalcross. Located 2 mi/3.2 km to the W is Craig Phadrick with ancient vitrified fortifications.

Inverness (IN-vuhr-NES), town, NOVA SCOTIA, CANADA, on W coast of CAPE BRETON ISLAND, NORTHUMBERLAND STRAIT, 55 mi/89 km W of SYDNEY; 46°13′N 61°16′W. Elevation 144 ft/43 m. Formerly coal-mining center and coal-shipping port.

Inverness (IN-vuhr-NES), unincorporated town (2000 population 1,421), MARIN county, W CALIFORNIA, 20 mi/32 km NW of SAN RAFAEL, on W side of TOMALES BAY, on POINT REYES peninsula, bounded N, W, and S by POINT REYES NATIONAL SEASHORE. Tomales Bay

lies in SAN ANDREAS FAULT. Clams, oysters, mussels. Tourism.

Inverness (in-VUHR-ness), town (□ 8 sq mi/20.8 sq km; 2000 population 6,789), CITRUS county, W central FLORIDA, c.60 mi/97 km NNE of TAMPA, on TSALA APOPKA LAKE; 28°50′N 82°20′W. Vegetable packing; fishing.

Inverness (in-VUHR-nes), town (2000 population 1,153), SUNFLOWER county, W MISSISSIPPI, 8 mi/12.9 km SSE of INDIANOLA; 33°21′N 90°35′W. In rich agricultural area (cotton, corn, rice, soybeans; cattle); manufacturing (catfish processing, agriculture equipment).

Inverness (in-vuhr-NES), village (□ 68 sq mi/176.8 sq km), CENTRE-du-Québec region, S QUEBEC, E CANADA, 16 mi/26 km NW of THETFORD MINES; 46°16′N 71°31′W. Copper and magnesite mining; dairying; cattle and pig raising; lumbering. Was seat of historic MEGANTIC county.

Inverness (IN-vuhr-NES), village (2000 population 6,749), COOK county, NE ILLINOIS, residential suburb 30 mi/48 km NW of CHICAGO; 42°07′N 88°05′W. Some remnant agriculture. Incorporated 1962.

Inverness, village, HILL county, N MONTANA, 47 mi/76 km W of HAVRE. Cattle, sheep, hogs; wheat, barley, oats, hay.

Inverness (IN-vuhr-NES), residential suburb of Birmingham, Shelby co., N central ALABAMA; 33°24′N 86°43′W. Named for the Scotland home of the ancestors of John C. Graham, a local resident.

Inversnaid (IN-vuhrs-naid), village (2003 population 28), W STIRLING, central Scotland, on NE shore of LOCH LOMOND, and 30 mi/48 km WNW of Stirling; 56°15′N 04°41′W. Just N is Rob Roy's Cave, where Robert the Bruce hid after 1306 battle of Dalrigh.

Inveruno (een-ve-ROO-no), town, MILANO province, LOMBARDY, N ITALY, 17 mi/27 km WNW of MILAN; 45°31′N 08°29′E. Fabricated metal products; cotton mill.

Inverurie (IN-vuhr-RE-ree), town (2001 population 10,882), Aberdeenshire, NE Scotland, on URIE RIVER just above its mouth on DON RIVER, 14 mi/23 km NW of ABERDEEN; 57°16′N 02°22′W. Agricultural market, with paper milling and oil production. Robert the Bruce defeated John Comyn, Earl of Buchan, in a battle here 1308.

Investigator Islands (in-VES-ti-gai-tuhr), 40-mi/64-km island chain, in GREAT AUSTRALIAN BIGHT, 4 mi/6 km off W coast of EYRE PENINSULA, SOUTH AUSTRALIA; comprising Flinders Island, Pearson and Waldegrave islands; 33°45′S 134°30′E. Sandy, hilly.

Investigator Strait (in-VES-ti-gai-tuhr), channel of INDIAN OCEAN, forms SW entrance to Gulf SAINT VINCENT, SE SOUTH AUSTRALIA, between S coast of YORKE PENINSULA and N coast of KANGAROO ISLAND; merges with BACKSTAIRS PASSAGE (E); 60 mi/97 km long, 35 mi/56 km wide; 35°25′S 137°02′E.

Invisible Mountain, Idaho: see LOST RIVER RANGE.

Inwood (IN-wood), town (□ 2 sq mi/5.2 sq km; 2000 population 6,925), POLK county, central FLORIDA, 10 mi/16 km E of LAKELAND; 28°02′N 81°46′W.

Inwood, town (2000 population 875), LYON county, NW IOWA, 11 mi/24 km SW of ROCK RAPIDS; 43°17′N 96°26′W. In livestock and grain area.

Inwood, unincorporated town (2000 population 2,084), BERKELEY county, NE WEST VIRGINIA, 7 mi/11.3 km SSW of MARTINSBURG; 39°21′N 78°02′W. Agriculture (grain, apples); livestock; dairying. Limestone quarrying. Manufacturing (food and beverages, crushed limestone).

Inwood (IN-wud), unincorporated village, S ONTARIO, E central CANADA, 10 mi/16 km ESE of PETROLIA, and included in BROOKE-Alvinston township; 50°30′N 97°30′W. Dairying; mixed farming.

Inwood, residential village (□ 2 sq mi/5.2 sq km; 2000 population 9,325), NASSAU county, SE NEW YORK, on W LONG ISLAND, near E shore of JAMAICA BAY, one of the "Five Towns of Long Island," 9 mi/14.5 km SW of

HEMPSTEAD; 40°37′N 73°45′W. Affluent residential area; some light manufacturing.

Inwood, residential district, NW MANHATTAN borough, NEW YORK city, SE NEW YORK, along the HUDSON and HARLEM rivers; 40°52′N 73°56′W. Northernmost part of Manhattan island. Population is largely Hispanic and African-American. Contains Inwood Hill and Fort Tryon parks; site of the Cloisters, noted museum of medieval art. Also here is Dyckman House (built 1783), the last 18th-century farmhouse in Manhattan.

Inwood (IN-wud), hamlet, S MANITOBA, W central CANADA, 13 mi/21 km NW of TEULON, and in ARMSTRONG rural municipality; 50°30′N 97°29′W.

Inya (IN-yah), lake, MYANMAR, 2.5 mi/4 km N of Shwedagon Pagoda, YANGON. Artificial lake, very irregular in shape; maximum length and width are about 2 mi/3.2 km by 1.5 mi/2.4 km. Location of sailing clubs, hotel; major government agencies; several parks.

Inyanga, ZIMBABWE: see NYANGA.

Inyangani Mountains, ZIMBABWE: see NYANGA MOUNTAINS.

Inyan Kara Creek (IN-yuhn KAR-uh), 43 mi/69 km long, NE WYOMING; rises in NE corner of WESTON county in BLACK HILLS near INYAN KARA MOUNTAIN, flows NW to BELLE FOURCHE RIVER 16 mi/26 km NE of MOORCROFT.

Inyan Kara Mountain (IN-yuhn KAR-uh), peak (6,368 ft/1,941 m) in W part of BLACK HILLS, CROOK county, NE WYOMING, 14 mi/23 km S of SUNDANCE.

Inyantue, town, MATABELELAND NORTH province, W ZIMBABWE, 17 mi/27 km SE of HWANGE, on railroad, at N edge of HWANGE NATIONAL PARK; 18°34′S 26°40′E. Mtoa Ruins to S. Livestock; grain.

Inya River (EEN-yah), 331 mi/533 km long, RUSSIA; rises in central KUZNETSK BASIN (KEMEROVO oblast), flows generally WNW, past LENINSK-KUZNETSKIY, PROMYSHLENNAYA, and TOGUCHIN (Novosibirsk oblast), to the OB' RIVER just above NOVOSIBIRSK.

Inyati, town, MATABELELAND NORTH province, SW central ZIMBABWE, 36 mi/58 km NNE of BULAWAYO; 19°41′S 28°51′E. Cattle, sheep, goats; tobacco, corn, cotton, peanuts. Gold mining to S. Seat of Inyati Mission. Ndumba Hills (4,672 ft/1,424 m) to S.

Inylchek, KYRGYZSTAN and CHINA: see ENGILCHEK.

Inyo, county (□ 10,192 sq mi/26,499.2 sq km; 2006 population 17,980), E CALIFORNIA; ⊙ INDEPENDENCE; 36°36′N 117°30′W. Crest of SIERRA NEVADA (High Sierras) along W boundary, also forms E boundary of KINGS CANYON and SEQUOIA national parks. County is leading producer in state of lead, tungsten, and talc; also mining of molybdenum, zinc, silver; and extraction of borax, potash, salt, and soda. Some irrigated farming (in Owens Valley); stock raising; dairying. Camping, hunting, fishing, and winter sports in mountains; winter resorts in DEATH VALLEY. Includes MOUNT WHITNEY (14,494 ft/4,418 m), highest peak in U.S., outside Alaska, and nine other peaks over 14,000 ft/4,267 m. In E, bounded by NEVADA state line, includes large part of Death Valley National Monument, which has lowest point (282 ft/86 m below sea level) in Western Hemisphere. Between the Sierra Nevada and PANAMINT RANGE (W wall of Death Valley) are arid basins (notably Owens Valley), INYO MOUNTAINS, and other ranges. AMARGOSA RANGE is E of Death Valley. OWENS RIVER supplies water to LOS ANGELES AQUEDUCT (begins in NW part of county, runs S at base of Sierra Nevada, to Los Angeles); AMARGOSA RIVER, Furnace Creek vanish in Death Valley. Includes Big Pine, Fort Independence, and Lone Pine Indian reservations, all in NW (Owens Valley). Partly in Inyo National Forest. Owens Lake (dry) in W; part of large China Lake Naval Air Weapons Station in SW. Formed 1866.

Inyokern, unincorporated village (2000 population 984), KERN county, S central CALIFORNIA, in MOJAVE DESERT, 62 mi/100 km ENE of BAKERSFIELD. Railroad junction. China Lake Naval Air Weapons Station (one of two large sections to NE); China Lake (dry) to NE; Sequoia National Forest and Pacific Crest Trail to W; LOS ANGELES AQUEDUCT passes to W.

Inyo Mountains, range, INYO county, E CALIFORNIA, between Owens Valley (W) and Saline Valley (E), extending 70 mi/113 km SSE from S end of White Mountains, SE of Bishop, to point just SE of Owens Lake, on E side of SIERRA NEVADA. Rise to 11,123 ft/3,390 m at Waucoba Mountain, 18 mi/29 km SE of Big Pine. Crossed by Westgard Pass (State Highway 168; elevation 7,313 ft/2,229 m) just NE of Big Pine. N part of range in Inyo National Forest.

Inyonga (ee-NYON-gah), village, RUKWA region, W central TANZANIA, 90 mi/145 km NNE of SUMBAWANGA; 06°42′S 32°03′E. MLALA HILLS to W. Road junction, landing strip. Timber; cattle, sheep, goats; corn, wheat.

Inywa (IN-ywah), village, SAGAING division, MYANMAR, on left bank of AYEYARWADY RIVER, 20 mi/32 km S of KATHA, at mouth of SHWELI RIVER.

Inza (EEN-zah), city (2000 population 19,555), W ULYANOVSK oblast, E central European Russia, near the Inza River (right tributary of the SURA River), 104 mi/167 km WSW of ULYANOVSK; 53°51′N 46°21′E. Elevation 534 ft/162 m. Railroad and highway junction; woodworking, wood cracking; washing of wool, diatomaceous bricks; mining and processing of diatomaceous earth. Founded in 1897. Became city in 1946. Construction of a new gas pipeline, to run through the city when completed, was scheduled to begin in 2004.

Inzá (een-SAH), town, ⊙ Inzá municipio, CAUCA department, SW COLOMBIA, in Cordillera CENTRAL, 40 mi/64 km ENE of POPAYÁN; 02°33′N 76°04′W. Elevation 5,755 ft/1,754 m. Coffee, corn.

Inzago (een-ZAH-go), village, MILANO province, LOMBARDY, N ITALY, near ADDA RIVER, 15 mi/24 km ENE of MILAN; 45°32′N 09°29′E. Small varied industries including silk mill.

Inzai (EEN-zah-ee), town, Inba district, CHIBA prefecture, E central HONSHU, E central JAPAN, 9 mi/15 km N of CHIBA; 35°49′N 140°08′E. Rice crackers (sembei).

Inzell (IN-tsel), village, BAVARIA, SE GERMANY, in UPPER BAVARIA, in Bavarian Alps, 14 mi/23 km WSW of SALZBURG (AUSTRIA); 47°45′N 12°46′E. Elevation 2,272 ft/ 693 m. Climatic health resort and tourist center with winter sports.

Inzer (een-ZYER), town (2005 population 4,390), E BASHKORTOSTAN Republic, SW URALS, European Russia, on the INZER RIVER, on road, 40 mi/64 km NW of BELORETSK; 54°13′N 57°33′E. Elevation 853 ft/259 m. On a railroad spur from MAGNITOGORSK; mining center in Komarovo-Zigazinskiy iron district. Until 1946, called Inzerskiy Zavod.

Inzer River (een-ZYER), 185 mi/298 km long, BASHKORTOSTAN, RUSSIA; rises in the S URAL Mountains, WNW of TIRLYANSKIY, flows SSW, then NNW, past INZER, and W to the SIM RIVER just above its mouth.

Inzersdorf (IN-tsers-dorf), part of the LIESING district of VIENNA, AUSTRIA, 4 mi/6.4 km S of city center. Formerly known for its brickyards, it is known today as a motorway junction and for its market for fruits and vegetables. Trade and transportation enterprises.

Inzerskiy Zavod, RUSSIA: see INZER.

Inzhavino (een-ZHAH-vee-nuh), town (2006 population 10,175), E TAMBOV oblast, S central European Russia, on the VORONA RIVER, on crossroads and terminus of local railroad, 24 mi/39 km SSW of KIRSANOV; 52°19′N 42°29′E. Elevation 380 ft/115 m. In agricultural area; grain storage, flour mill, sunflower-oil press, poultry factory.

Inzinzac-Lochrist (an-zan-zahk–lo-KREEST), town (□ 17 sq mi/44.2 sq km), MORBIHAN department, in BRITTANY, FRANCE, near BLAVET RIVER, 8 mi/13 km NE of LORIENT; 47°51′N 03°16′E.

Ioakim Gruevo (yo-ah-KEEM GROO-ev-o), village, PLOVDIV oblast, Rodopi obshtina, BULGARIA; 42°07′N 24°35′E.

Ioánnina (ee-o-AH-nee-nah), prefecture (□ 1,947 sq mi/5,062.2 sq km), EPIRUS department, NW GREECE; ⊙ IOÁNNINA; 39°45′N 20°40′E. Bounded NW by ALBANIA, E by WEST MACEDONIA and THESSALY departments (near PINDOS mountains and TSOUMERKA massif), S by ÁRTA and PREVEZA prefectures, and W by THESPROTIA prefecture. Drained by Aoos (VIJOSË) (N) and Arakhthus (S) rivers and contains Lake IOÁNNINA. Major livestock-raising region, known for its cheese, milk, and skin exports. Served by major highway between Albania and Greece.

Ioánnina (ee-o-AH-nee-nah), city (2001 population 75,179), ⊙ IOÁNNINA prefecture, EPIRUS department, NW GREECE, on Lake IOÁNNINA; 39°40′N 20°50′E. The chief city of Epirus, it is the commercial center for an agricultural region. Manufacturing includes textiles and gold and silver products. International airport. A Byzantine city founded c.527 by Justinian, Ioánnina became an important center in the 11th century. Taken (1081) by the Normans, and in 1204, Michael I, despot of Epirus, made it his capital. Conquered by the Ottoman Turks in 1430. Ali Pasha made it (1788) the stronghold of his virtually independent state. Passed to Greece in 1913 as a result of the First Balkan War. Held for a short time by Italy at the start of World War II; later occupied by Germany. Long a center of Greek learning, it is today the seat of the University of Ioánnina. Also spelled Janina, Jannina, Yanina, or Yannina.

Ioánnina, Lake (ee-o-AH-nee-nah), ancient *Pambotis* (□ 8.5 sq mi/22 sq km), IOÁNNINA prefecture, S EPIRUS department, NW GREECE; 5 mi/8 km long, 3 mi/4.8 km wide; 39°40′N 20°53′E. IOÁNNINA city on W shore. Important fisheries.

Ioba, province (□ 1,255 sq mi/3,263 sq km; 2005 population 181,227), SUD-OUEST region, SW BURKINA FASO; ⊙ Dano; 11°05′N 03°05′W. Bordered N by TUY province, NE by BALÉ province, ENE by SISSILI province, ESE by GHANA, S by PONI province, and WSW by BOUGOURIBA province. Established in 1997 with fourteen other new provinces.

Ioco (ei-O-ko), unincorporated town, SW BRITISH COLUMBIA, W CANADA, at W end of BURRARD INLET of the Strait of GEORGIA, 10 mi/16 km ENE of VANCOUVER, included in PORT MOODY; 49°18′N 122°52′W. Oil refinery.

Iojima (ee-O-jee-mah), town, West Sonogi county, NAGASAKI prefecture, NW KYUSHU, SW JAPAN, 6 mi/10 km S of NAGASAKI; 32°42′N 129°46′E. Fish farming (flattfish, lobster, globefish, sea bream, horse mackerel). French-style lighthouse (second-oldest in Japan).

Ioka (EE-o-kah), town, Kaijo county, CHIBA prefecture, E central HONSHU, E central JAPAN, at NE base of BOSO, 35 mi/56 km E of CHIBA; 35°47′N 140°43′E. Summer resort; fishing center. Dried sweet salmon, dried whitebait.

Iokhannes, RUSSIA: see SOVETSKIY, LENINGRAD oblast.

Iola (ei-O-luh), city (2000 population 6,302), ⊙ ALLEN county, SE KANSAS, on Neosho River, 35 mi/56 km W of FORT SCOTT; 37°55′N 95°24′W. Elevation 1,040 ft/317 m. Railroad junction. Trade center for wheat, cattle, and hog region; dairying; manufacturing of cement, rubber products, clothing, building materials, consumer goods, motor-vehicle parts, honey. County fair takes place here annually in August. Allen County Community College. Founded 1859, incorporated 1870.

Iola (ei-O-luh), town (2006 population 1,266), WAUPACA county, central WISCONSIN, 22 mi/35 km E of STEVENS POINT; 44°30′N 89°07′W. In agricultural area (dairy products; oxen); manufacturing (book publishing).

Iola (ei-O-lah), village (2000 population 171), CLAY county, S central ILLINOIS, 8 mi/12.9 km NW of

LOUISVILLE; 38°49′N 88°37′W. In agricultural (wheat, corn, soybeans, cattle), oil, and natural-gas area. Incorporated 1914.

Iolotan (yo-lo-TAHN), city, central MARY weloyat, SE TURKMENISTAN, on MURGAB RIVER, on railroad, 35 mi/56 km SE of MARY; 37°18′N 62°21′E. Tertiary-level administrative center. Food processing; cotton ginning. Irrigation works of same name associated with Gindukush hydroelectric dam. Also spelled Yoloton.

Ioma (ei-O-muh), town, NORTHERN province, PAPUA NEW GUINEA, 90 mi/145 km NE of PORT MORESBY. Palm oil, tapa cloth, coconuts.

Iona (ei-O-nuh), town (□ 10 sq mi/26 sq km; 2000 population 11,756), WALTON county, NW FLORIDA, 5 mi/8 km E of DE FUNIAK SPRINGS; 26°30′N 81°57′W.

Iona, town (2000 population 1,201), BONNEVILLE county, SE IDAHO, near SNAKE RIVER, residential suburb 5 mi/8 km ENE of IDAHO FALLS; 43°32′N 111°56′W. Elevation 4,788 ft/1,459 m. Junction of railroad terminus to AMMON. In irrigated region (cattle, sheep; dairying; grains, alfalfa; potatoes). Ririe Reservoir to E.

Iona (ei-O-nuh), village (2000 population 173), MURRAY county, SW MINNESOTA, 5 mi/8 km SSW of SLAYTON; 43°54′N 95°47′W. Grain; livestock; dairying. Lakes to E.

Iona (ei-O-nuh), Gaelic *Ioua*, island (3.5 mi/5.6 km long, 1.5 mi/2.4 km wide), ARGYLL and BUTE, W Scotland, one of the INNER HEBRIDES; 56°34′N 06°04′W. Separated from the island of MULL by the Sound of Iona, the island is hilly, with shell beaches. Farming, livestock grazing, and fishing are done, but tourism is the main industry here. Iona is famous as the early center of Celtic Christianity. St. Columba, with his companions, landed here from Ireland in 563. They founded a monastery, which was burned by the Danes in the 8th or 9th century. Iona was a bishopric from 838 to 1098. A Benedictine monastery, of which there are remains, was established in 1203. The cathedral, formerly the Church of St. Mary, dates from the early 13th century. The cemetery of St. Oran's Church contains the graves of many monarchs of Scotland, Ireland, Norway, and France. The Iona Community, dedicated to reviving the spirit of Celtic Christianity, has restored many ancient buildings. Iona College (New Rochelle, New York state) is named after this island.

Iona National Park, NAMIBE province, SW ANGOLA, 100 mi/161 km SE of NAMIBE.

Ionava, LITHUANIA: see JONAVA.

Ione, city (2000 population 7,129), AMADOR county, central CALIFORNIA, 33 mi/53 km SE of SACRAMENTO, on Sutter Creek; 38°22′N 120°57′W. Grain, grapes, walnuts; cattle. Clay refractory. Preston School of Industry. Camanche Reservation to S; Pardee Reservoir to SE.

Ione (ei-ON), village (2006 population 334), MORROW county, N OREGON, 16 mi/26 km NW of HEPPNER, on WILLOW CREEK; 45°30′N 119°49′W. Agriculture (wheat, alfalfa, potatoes; sheep, cattle).

Ione, village (2006 population 508), PEND OREILLE county, NE WASHINGTON, 25 mi/40 km NE of COLVILLE, on PEND OREILLE RIVER; 48°45′N 117°25′W. Manufacturing (lumber). Parts of Colville National Forest to E and W; SELKIRK MOUNTAINS to W. Box Canyon Dam 2 mi/3.2 km N (downstream).

Ionia, ancient region of ASIA MINOR. It occupied a narrow coastal strip on the E MEDITERRANEAN (in present-day W TURKEY) as well as the neighboring AEGEAN islands, which now mainly belong to GREECE. The region was of the utmost importance in ancient times, for it was there that Greek settlers established colonies before 1000 B.C.E. These colonists were called Ionians, and tradition says that they fled to Asia Minor from the mainland of Greece to escape from the conquering Dorians. ATHENS claimed to be the mother city of all the Ionian colonists, but modern

scholars believe that the Ionians were actually a mixed group (mainly from ATTICA and BOEOTIA). There came to be twelve important cities—MILETUS, Myus, PRIENE, SÁMOS, EPHESUS, COLOPHON, Lebedos, TEOS, ERYTHRAE, KHÍOS, CLAZOMENAE, and PHOCAEA. A religious league (which reached its full power in the 8th century B.C.E.) was formed, with its center at the temple of Poseidon near Mycale. Smyrna (IZMIR), originally an Aeolian colony, later joined the league. The fertility of the region and its excellent harbors brought prosperity to the cities. Traders and colonists were sent into the Mediterranean as far west as SPAIN and up to the shores of the BLACK SEA. When Croesus was conquered (before 546 B.C.E.) by Cyrus the Great of PERSIA, the Greek cities came under Persian rule. That rule was not very exacting, but it was despotic in nature, and at the beginning of the 5th century B.C.E. the cities rose in revolt against Darius I, and were easily put down. Their fate continued to be subject to treaties with the Persians and changed as Persian fortunes waxed and waned. Alexander the Great easily took (c.335) all the Ionian cities in his power, and the Diadochi quarreled over them. The cities continued to be rich and important through the time of the Roman and Byzantine empires. It was only after the Turkish conquest in the 15th century that their culture was destroyed.

Ionia (ei-O-nee-uh), county (□ 580 sq mi/1,508 sq km; 2006 population 64,821), S central MICHIGAN; ⊙ IONIA; 42°56′N 85°04′W. Intersected by GRAND RIVER, and drained by FLAT, LOOKING GLASS, and MAPLE rivers. Poultry, cattle, hogs; forage crops, wheat, oats, barley, corn, green beans, beans, peas, dry strawberries, apples, peaches. Manufacturing at IONIA, BELDING, and PORTLAND; lake resorts. Ionia State Park in W center of county. Organized 1837.

Ionia, town (2000 population 277), CHICKASAW county, NE IOWA, 7 mi/11.3 km W of NEW HAMPTON; 43°02′N 92°27′W. In livestock and grain area.

Ionia, town (2000 population 10,569), ⊙ IONIA county, S central MICHIGAN, 30 mi/48 km E of GRAND RAPIDS, on GRAND RIVER; 42°59′N 85°03′W. Poultry; agriculture (grain, fruit); manufacturing (furniture, food processing, motor-vehicle parts, fabricated metal products, printing). Airport. Has a state reformatory and a state mental hospital. Ionia State Park to SW. Settled 1833; incorporated as village 1865, as city 1873.

Ionia (ei-ON-yah), town (2000 population 108), BENTON county, central MISSOURI, 15 mi/24 km S of SEDALIA; 38°30′N 93°19′W.

Ionian Islands, department (□ 891 sq mi/2,316.6 sq km; 2001 population 212,984), comprising and contiguous with the IONIAN ISLANDS off the W shore of GREECE; ⊙ KÉRKIRA. Divided into KEFALLINÍA, KÉRKIRA, LEFKAS, and ZÁKINTHOS prefectures. Important towns include Kérkira, ARGOSTOLI, LEFKAS, and ZÁKINTHOS.

Ionian Islands, chain of islands, coterminous with IONIAN ISLANDS department, off W coast of mainland GREECE, in the IONIAN SEA, along the coasts of EPIRUS and the PELOPONNESUS; 38°30′N 20°30′E. Made up of seven main islands—KÉRKIRA, PAXÍ, LEFKAS, KEFALLINÍA, ITHÁKI, ZÁKINTHOS, and KÍTHIRA—and numerous islets. Kíthira, off the SE coast of the Peloponnesus, is in ATTICA department. Largely mountainous, the islands reach their highest point at Mount AINOS (c.5,340 ft/1,630 m) on Kefallinía. Fruits (especially currants), grains, timber, olives, wine, and cotton are produced, and sheep, goats, and hogs are raised. Industries include fishing, shipping, and tourism. The islands had no unified history until the 10th century, when they were made a province of the BYZANTINE EMPIRE. Taken by VENICE in the 14th and 15th centuries and held until 1797, when the Treaty of Campo Formio, which ended the Venetian republic, gave the islands to France. Seized 1799 by a Russo-Turkic fleet and constituted a republic under Russian protection. Returned to France in 1807, by the Treaty

of Tilsit. From 1809 to 1814 the British navy occupied all the islands except Kérkira. Placed under British protection in 1815; known as the "United States of the Ionian Islands." Ceded by Britain to Greece in 1864 after considerable popular agitation on the islands. A series of earthquakes in 1953 caused extensive damage.

Ionian Sea (ei-O-nee-ahn), Latin *Mare Ionium,* part of the MEDITERRANEAN SEA, SE EUROPE, between W GREECE and S ITALY. Connected with the ADRIATIC SEA by the Strait of OTRANTO. The Gulf of TARANTO and the Gulf of CORINTH are its chief arms. The IONIAN ISLANDS lie in its E part. KÉRKIRA and PATRAS (Greece) and CATANIA and Taranto (Italy) are the chief ports.

Ionishkis, LITHUANIA: see JONIŠKIS.

Iori Plateau (ee-O-ree) or **Gare Kakheti Plateau**, between the KURA and Alazani rivers, GEORGIA and S AZERBAIJAN; stretches WNW to ESE, bisected longitudinally by the valley of the IORI RIVER. Elevation is 656 ft/200 m–2,953 ft/600 m. E part is called SHIRAKI STEPPE.

Iori River (ee-O-ree) or **Iora River**, 180 mi/290 km long, in GEORGIA and AZERBAIJAN; rises on S slopes of Glavnyi and VODORAZDEL′NYI RANGE, in the E Greater Caucasus, flows S past TIANETI, and SE through dry steppe in lower course, to MINGECHAUR RESERVOIR (formerly the Alazani River) on KURA River. Irrigates SAMGORA steppe. After construction of the Samgori irrigation system, some of the Iori's waters flowed through a canal to a reservoir E of TBILISI.

Ios (EE-os), island (□ 43 sq mi/111.8 sq km), CYCLADES prefecture, SOUTH AEGEAN department, GREECE, in AEGEAN SEA, SSW of NAXOS island; 12 mi/19 km long, 5 mi/8 km wide; 36°42′N 25°25′E. Rises to 2,410 ft/735 m. Produces olive oil, wine, cotton, barley. Main town, Ios, is on W coast. One legend says it is the burial place of Homer. Also spelled Nios.

Iosco, county (□ 1,890 sq mi/4,914 sq km; 2006 population 26,831), NE MICHIGAN; 44°16′N 83°20′W; ⊙ TAWAS CITY. Port on Tawas Bay. Bounded E by LAKE HURON; drained by AU SABLE and AU GRES rivers and small Tawas River. Cattle, hogs; agriculture (corn, wheat, oats, barley); manufacturing (metal forgings, industrial machinery, gypsum products, dairy products, lumber milling, woodworking); cement plants; resorts. Includes part of Huron National Forest, in N $\frac{1}{2}$ of county, also TAWAS and VAN ETTAN lakes; Tawas Point State Park in E; Paul B. Wurtsmith Air Force Base in NE. Organized 1857.

Ioshkar-Ola, RUSSIA: see YOSHKAR-OLA.

Iota (ei-O-tuh), town (2000 population 1,376), ACADIA parish, S LOUISIANA, 11 mi/18 km NW of CROWLEY, near Bayou DES CANNES; 30°20′N 92°30′W. In rice-growing area; manufacturing of plastic pipe. Incorporated 1902.

Iotry, Lake (ee-OOCH) (□ 45 sq mi/117 sq km), TOLIARY province, SW MADAGASCAR, 15 mi/24 km SE of MOROMBE; 10 mi/16 km long, 7 mi/11.3 km wide; 21°57′S 43°40′E.

Iowa, state (□ 56,276 sq mi/145,755 sq km; 2005 estimated population 2,965,524; 2000 population 2,926,324), N central UNITED STATES, in the MIDWEST, admitted to the Union in 1846 as the 29th state; ⊙ DES MOINES; 42°02′N 93°28′W. Des Moines is the largest city; other major cities are CEDAR RAPIDS, DAVENPORT, and SIOUX CITY.

Geography

Nicknamed the Hawkeye State, Iowa is bordered on two sides by rivers; the MISSISSIPPI separates it on the E from WISCONSIN and ILLINOIS, and the MISSOURI and the BIG SIOUX separate it on the W from NEBRASKA and SOUTH DAKOTA. The state is bounded on the N by MINNESOTA and on the S by MISSOURI; 22 mi/35 km of the border with Missouri is formed by the DES MOINES RIVER, in SE. Iowa is an area of rich, rolling plains, interrupted by many rivers. The terrain is low and gently sloping, except for the hills in the unglaciated area of NE Iowa, the steeply sloping bluffs on the banks of the Mississippi, and the moundlike bluffs on the banks of the Missouri. The rivers of the E two-thirds of Iowa flow to the Mississippi; those of the remaining one-third flow to the Missouri. The original woodlands, which included black walnut and hickory, were destroyed by lumbering and land clearing in the 19th century, and the present wooded sections are covered only with a second or third growth of timber. Typical of Iowa is the prairie. The Iowa lakes district, extending S from Minnesota, has over forty small natural lakes of glacial origin, most generally NE of SPENCER, and a smaller group generally W of MASON CITY; a regional recreation area. Covered a little more than a century ago with grass higher than the wheels of the pioneers' prairie schooners (covered wagons), the prairies are now covered with fields of corn and other grains. The wildflowers that once blossomed among the prairie grass still brighten the roadsides; however, few areas of the original grassland remain, and prairie grass preserves have been established. The cornfields have replaced the grasslands as the habitat of wild turkeys, prairie chickens, and quail. Iowa abounds with migratory geese and ducks and the imported ring-necked pheasant and European partridge, all of which are hunted in the autumn. The climate is continental: NW winds drive the mercury down to below 0°F/−18°C in winter, and in the summer hot air masses bring oppressive heat; violent thunderstorms, hail, and occasional droughts vex the farmer. The average annual rainfall is 31 in/78.7 cm, and, since most of the rain falls in summer, the soil is often washed away. Iowans have had to fight erosion with modern plowing and planting practices, control of water flow, and reforestation. In addition, floods have inflicted great loss of life and property damage on cities and countryside alike; therefore, flood-control projects are vitally important to Iowa. The state was at the center of the catastrophic floods of 1993, and suffered greatly as a result. Des Moines and Davenport were hit especially hard.

Economy

Yet Iowa has some of the most fertile agricultural land in the world. The deep, porous soil yields corn and other grains in tremendous quantities, and the corn-fed hogs and cattle are nationally known. Iowa has consistently led the nation in the production of corn and hogs, and ranks in the top 10 in the raising of cattle. In addition to corn, Iowa's other major crops are soybeans, hay, and oats. Iowans have used the rich earth and its bounty to gain the nation's second-highest total cash receipts from farm marketing. Agriculture in Iowa also benefits the state's chief industry, food processing, and in Sioux City and Cedar Rapids many factories process farm products. Machinery, tires, appliances, electronic equipment, and chemicals are among the other manufacturing. Cement is the most important mineral product; others are stone, sand, gravel, lead, zinc, and gypsum. Mineral production is small, however.

History: to 1832

In prehistoric times, the Mound Builders, a farming people, lived in the Iowa area. When Europeans first came to explore the region in the 17th century, various Native American groups, including the Iowa (reputedly the source of the state's name), occupied the land. The Sac and Fox also ranged over the land, but it was the combative Sioux who dominated the area. In 1673 the French explorers Father Jacques Marquette and Louis Jolliet traveled down the Mississippi River and touched upon the Iowa shores, as did Robert Cavelier, sieur de La Salle, in 1681–1682. The areas surrounding the Des Moines and Mississippi rivers were profitable for fur traders, and a number of Iowa towns developed from trading posts. Late in the 18th century a French Canadian, Julien Dubuque, leased land from Native Americans around the DUBUQUE

area and opened lead mines there. After his death they refused to permit others to work the mines, and U.S. troops under Lieutenant Jefferson Davis protected Native American rights to the land as late as 1830. However, their hold was doomed after the U.S. acquired Iowa as part of the Louisiana Purchase of 1803.

History: 1832 to 1846

In 1832 the Black Hawk War broke out as the Sac and Fox, led by their chief, Black Hawk, fought to regain their former lands in Illinois along the Mississippi River. They were defeated by U.S. troops and were forced to leave the Illinois lands and cede to the U.S. much of their land along the river on the Iowa side. Within two decades after the Black Hawk War, all Native American lands had been ceded to the U.S. Meanwhile, a great rush of frontiersmen came to settle the prairies and take the mines. Slavery was prohibited in Iowa under the Missouri Compromise of 1820, which excluded it from the lands of the Louisiana Purchase N of latitude 36°30′N. Part of Missouri Territory prior to 1821, Iowa was subsequently part of both the Michigan and Wisconsin territories. By 1838, Iowa Territory was organized, with BURLINGTON as the temporary capital. In the following year, IOWA CITY became the capital. The Iowans quickly built a rural civilization like that of NEW ENGLAND, where many of them had lived. Later, immigrants from EUROPE, notably Germans, Czechs, Dutch, and Scandinavians, brought their agricultural skills and their own customs to enrich Iowa's rural life, and a group of German Pietists established the AMANA CHURCH SOCIETY, a successful attempt at communal social organization. A system of public schools was set up in 1839, and successful efforts soon were made to establish colleges and universities.

History: 1846 to 1874

Iowa became a state in 1846, and Ansel Briggs was elected as the first governor. In 1857 the capital was moved from Iowa City to Des Moines. In that same year the state adopted its second constitution. Iowa prospered greatly with the beginning of railroad construction, and the rivalry between towns to get the lines was so fierce that the grant of big land tracts to railroad companies was curtailed by legislative act in 1857. In 1855 the state's first railroad line was completed between Davenport and MUSCATINE along the E border. Before and during the Civil War, Iowans, generally owners of small, independent farms, were naturally sympathetic to the antislavery side, and many fought for the Union. The Underground Railroad, which helped many fugitive slaves escape to free states, was active in Iowa, and the abolitionist John Brown made his headquarters there for a time. Iowa's farmers prospered after the Civil War, but during the hard times that afflicted the country in the 1870s they found themselves burdened with debts. Feeling oppressed by the currency system, corporations, and high railroad and grain-storage rates, many of Iowa's farmers supported the Granger movement, the Greenback Party, and the Populist Party. The reform movements had some success in the state.

History: 1874 to Present

Granger laws were enacted in 1874 and 1876 regulating railroad rates, but these laws were repealed in 1877 under pressure from the railroad companies. By the end of the 19th century, times improved, and the agricultural movements declined. Farm units grew larger, and mechanization brought great increases in productivity. Much of the state's society may still resemble that depicted in the paintings of Iowan artist Grant Wood, but the state's industrial economy as well as other elements of modernization has altered this image. The volatile nature of agriculture prices combined with a steady decline in manufacturing has made Iowa susceptible to economic recession. This was especially true in the 1980s, when Iowa was

second in the U.S. in outmigration with a 4.7% decline in population. On July 19, 1989, United Airlines DC-10 crashed at Sioux City, killing 111.

Famous Iowans, Tourist Attractions, and Academic Institutions

Among Iowa's colorful figures were Buffalo Bill, John Wayne, Bix Beiderbecke, Glenn Miller, and Billy Sunday. Other public figures associated with the state are James Wilson, U.S. Secretary of Agriculture for 16 years (1897–1913), and the noted members of the Wallace family—Henry Wallace, Henry Cantwell Wallace, and Henry Agard Wallace. Herbert C. Hoover and Harry Lake Hopkins were born in Iowa. HERBERT HOOVER NATIONAL HISTORIC SITE, which contains Hoover's birthplace, childhood home, and grave, and the Herbert Hoover Presidential Library are at WEST BRANCH. N of MARQUETTE is EFFIGY MOUNDS NATIONAL MONUMENT, site of Native American mounds built by the area's earliest inhabitants. Many state parks and forests provide recreational facilities. Among the educational institutions in Iowa are Iowa State University of Science and Technology, at AMES; the University of Iowa, at Iowa City; Grinnell College, at GRINNELL; Cornell College, at MOUNT VERNON; Drake University, at Des Moines; University of Northern Iowa, at CEDAR FALLS; and the University of Dubuque, Loras College, and Clarke College, at Dubuque.

Government

Iowa's constitution was adopted in 1857. The governor is elected for a term of four years and may be re-elected. The current governor is Chet Culver. The general assembly, or legislature, has a senate with fifty members elected for four-year terms and a house of representatives with one hundred members elected for two-year terms. Iowa is represented in the U.S. Congress by two senators and five representatives. The state has seven electoral votes.

Iowa has ninety-nine counties: ADAIR, ADAMS, ALLAMAKEE, APPANOOSE, AUDUBON, BENTON, BLACK HAWK, BOONE, BREMER, BUCHANAN, BUENA VISTA, BUTLER, CALHOUN, CARROLL, CASS, CEDAR, CERRO GORDO, CHEROKEE, CHICKASAW, CLARKE, CLAY, CLAYTON, CLINTON, CRAWFORD, DALLAS, DAVIS, DECATUR, DELAWARE, DES MOINES, DICKINSON, DUBUQUE, EMMET, FAYETTE, FLOYD, FRANKLIN, FREMONT, GREENE, GRUNDY, GUTHRIE, HAMILTON, HANCOCK, HARDIN, HARRISON, HENRY, HOWARD, HUMBOLDT, IDA, IOWA, JACKSON, JASPER, JEFFERSON, JOHNSON, JONES, KEOKUK, KOSSUTH, LEE, LINN, LOUISA, LUCAS, LYON, MADISON, MAHASKA, MARION, MARSHALL, MILLS, MITCHELL, MONONA, MONROE, MONTGOMERY, MUSCATINE, O'BRIEN, OSCEOLA, PAGE, PALO ALTO, PLYMOUTH, POCAHONTAS, POLK, POTTAWATTAMIE, POWESHIEK, RINGGOLD, SAC, SCOTT, SHELBY, SIOUX, STORY, TAMA, TAYLOR, UNION, VAN BUREN, WAPELLO, WARREN, WASHINGTON, WAYNE, WEBSTER, WINNEBAGO, WINNESHIEK, WOODBURY, WORTH, and WRIGHT.

Iowa, county (□ 587 sq mi/1,526.2 sq km; 2006 population 16,140), E central IOWA; ⊙ MARENGO; 41°41′N 92°04′W. Rolling prairie agricultural area (cattle, hogs, sheep, poultry; corn, oats) drained by IOWA and ENGLISH rivers. In NE are 7 villages of the AMANA COLONIES. Formed 1843.

Iowa, county (□ 768 sq mi/1,996.8 sq km; 2006 population 23,756), S WISCONSIN; ⊙ DODGEVILLE; 43°00′N 90°07′W. Dairy product processing is chief industry; agriculture (barley, oats, corn, soybeans; cattle, hogs, sheep, poultry); lead and zinc deposits. Bordered N by WISCONSIN RIVER; drained by PECATONICA and Blue rivers. Tower Hill State Park in N; Governor Dodge State Park in center; Wintergreen Ski Area in N; Timberline Ski Area in NE; W part of Military Ridge State Trail runs through center of county, terminating at Dodgeville. Frank Lloyd Wright–designed homes Taliesin and House on the Rock (both open to visitors) are in N. Formed 1829.

Iowa, town (2000 population 2,663), CALCASIEU parish, SW LOUISIANA, 11 mi/18 km E of LAKE CHARLES; 30°14′N 93°01′W. Railroad junction to E. Soybeans, sorghum; cattle, horses, hogs.

Iowa, river, 329 mi/529 km long, rising in the lakes of N IOWA and flowing SE to the MISSISSIPPI RIVER, SE Iowa; CEDAR RIVER (300 mi/483 km long) is its chief tributary. A power dam crosses the gorge at IOWA FALLS. The Iowa River has an extensive flood-control system; Coralville Dam and reservoir, N of IOWA CITY, is the largest unit.

Iowa City, city (2000 population 62,220), ⊙ JOHNSON county, E IOWA, on both sides of the IOWA RIVER; 41°39′N 91°32′W. Manufacturing (foam rubber, animal feed, paper, food products). Founded 1839 as the capital of Iowa Territory. The old stone capitol in the city was begun in 1840; the legislature sat there until the seat of government was moved to DES MOINES in 1857. With the arrival of the railroad (1855), Iowa City became an important outfitting center for the westward trails. The seat of the University of Iowa (1855), the city is a major center of medical treatment and research. The university sponsors the Iowa Writers' Workshop, one of the most prestigious of U.S. creative writing programs, attracting teachers and students from all over the world. The city's activities center greatly around the university. The library of the state historical society is in Iowa City. Nearby are the villages of the AMANA Society, Coralville dam and reservoir, and the Herbert Hoover Presidential Library, as well as his birthplace (in WEST BRANCH). Incorporated 1853.

Iowa Falls, city (2000 population 5,193), HARDIN county, central IOWA, on IOWA RIVER, 38 mi/61 km NNW of MARSHALLTOWN; 42°31′N 93°16′W. Railroad junction. Agriculture-processing center (packed poultry and eggs, dairy products; feeds; soybean meal and oil; tankage; manufacturing (concrete blocks, luggage, plastic products). Limestone quarries nearby. Has Ellsworth Community College and museum of pioneer relics. Annual state Baptist convention held here. Settled 1853, incorporated 1889.

Iowa Park, town (2006 population 6,142), WICHITA county, N TEXAS, suburb 11 mi/18 km W of WICHITA FALLS, near WICHITA RIVER; 33°57′N 98°40′W. In oil, agricultural area (cotton, cattle, wheat); manufacturing of oil-field supplies. Texas-Oklahoma Fair, Southwestern Oil Exposition held here. Seat of agriculture experiment station.

Ipaguazú (ee-pah-gwah-ZOO), canton, BURNET O'Conner province, TARIJA department, S central BOLIVIA; 21°19′S 63°56′W. Elevation 4,035 ft/1,230 m. Abundant oil and gas resources in area. Clay, limestone, and gypsum deposits. Agriculture (potatoes, yucca, bananas, corn, sweet potatoes, tobacco); cattle.

Ipala (ee-PAH-lah), town, CHIQUIMULA department, GUATEMALA, N of AGUA BLANCA; 14°37′N 89°37′W. Elevation 2,700 ft/823 m. Locally prominent cinder cone with a crater lake, Laguna de Ipala. Noted for harness-making, sisal goods (cordage). Agriculture is mostly subsistence farming (rice, corn, beans, sugarcane).

Ipala (ee-pah-LAH), village, TABORA region, NW central TANZANIA, 40 mi/64 km N of TABORA, on Igombe River; 04°30′S 30°41′E. Timber; corn; wheat.

Ipameri (ee-pah-me-ree), city (2007 population 23,207), SE GOIÁS, central BRAZIL, on railroad, 100 mi/161 km SE of GOIÂNIA; 17°43′S 48°12′W. Cattle-shipping center; meat processing, rice hulling. Rock crystals and rutile exploited in area. Airfield. Formerly spelled Ipamery.

Ipamu, CONGO: see MANGAI.

Ipanema (ee-pah-ne-mah), city (2007 population 17,128), E MINAS GERAIS, BRAZIL, 35 mi/56 km NNE of Manhuaçu; 19°40′S 41°40′W. Agricultural trade. Formerly called José Pedro.

Ipanguaçu (EE-pahn-gwah-soo), city (2007 population 13,441), central RIO GRANDE DO NORTE state, BRAZIL, 6 mi/9.7 km NE of Açu, on Açu River; 05°30′S 36°52′W.

Ipati (ee-PAH-tee), village, FTHIOTIDA prefecture, E CENTRAL GREECE department, 10 mi/16 km W of LAMIA; 38°52′N 22°14′E. Health resort; sulfur springs. Also spelled Hypate or Ypati.

Ipatinga (EE-pah-cheen-gah), city (2007 population 238,397), E central MINAS GERAIS, BRAZIL, 7 mi/11.3 km NE of CORONEL FABRICIANO, in RIO DOCE Valley, on VITÓRIA-Minas railroad; 19°30′S 42°30′W.

Ipatovo (ee-PAH-tuh-vuh), city (2006 population 29,240), N STAVROPOL TERRITORY, N CAUCASUS, S European Russia, on railroad, 75 mi/121 km NE of STAVROPOL; 45°43′N 42°54′E. Elevation 360 ft/109 m. Highway junction; local transshipment point. Flour milling; clothing industry; brewery. Until the 1930s, called Vinodel'noye. Made city in 1979.

Ipauçu (EE-pah-oo-soo), city, SW central SÃO PAULO, BRAZIL, on railroad, 16 mi/26 km E of OURINHOS; 23°03′S 49°37′W. Coffee processing, distilling, pottery manufacturing. Formerly spelled Ipaussú.

Ipaumirím (EE-pou-mee-REEN), city (2007 population 11,610), SE CEARÁ, BRAZIL, near border with PARAÍBA, 5 mi/8 km S of Baxio; 06°45′S 38°46′W.

Ipava (ei-PAI-vah), village (2000 population 506), FULTON county, W central ILLINOIS, 9 mi/14.5 km SW of Lewiston; 40°21′N 90°19′W. In agricultural (corn, wheat, sorghum, soybeans, cattle) and bituminous-coal-mining area; ships grain. Vestiges of Camp Ellis, a World War II prisoner-of-war camp, nearby.

Ipecaetá (EE-pe-kah-ee-TAH), city (2007 population 15,883), E central BAHIA, BRAZIL, 16 mi/25 km W of Feira de Santana; 12°19′S 39°18′W.

Ipel River (i-PEL), Slovak **Ipel'**, GERMAN **Eipel**, HUNGARIAN **Ipoly**, c.145 mi/233 km long, S SLOVAKIA; rises on S slope of SLOVAK ORE MOUNTAINS, 25 mi/40 km ENE of ZVOLEN; flows c.50 mi/80 km S, then generally SW, forming Slovak-Hungarian border, to the DANUBE RIVER E of ESZTERGOM (Hungary).

Iperó (EE-pe-ro), city (2007 population 24,239), S SÃO PAULO state, BRAZIL, 17 mi/27 km NW of SOROCABA, on railroad; 23°21′S 47°41′W.

Iphofen (IP-ho-fuhn), town, MIDDLE FRANCONIA, BAVARIA, S GERMANY, 16 mi/26 km ESE of WÜRZBURG; 49°42′N 10°15′E. Vineyards. Has Gothic church and baroque town hall.

Ipiaçu (EE-pee-ah-SOO), town (2007 population 4,187), W central MINAS GERAIS, BRAZIL, 25 mi/40 km W of CAPINÓPOLIS, in TRIÂNGULO MINEIRO on São Simão Reservoir; 18°45′S 50°01′W.

Ipiales (ee-pee-AH-les), town, ⊙ Ipiales municipio, NARIÑO department, SW COLOMBIA, in the ANDES, on Carchi River, on ECUADOR border opposite TULCÁN (linked by bridge and highway), 35 mi/56 km SW of PASTO; 00°50′N 77°37′W. Elevation 7,516 ft/2,291 m. Trading and textile manufacturing center in agricultural region (corn, wheat, sugarcane, coffee; livestock). In canyon of the Guaitará River nearby is the shrine of Nuestra Señora de las Lajas. Airport nearby.

Ipiaú (ee-pee-ah-OO), city (2007 population 42,590), BAHIA, BRAZIL, on the RIO DE CONTAS, 23 mi/37 km SE of JIQUIÉ; 14°11′S 39°45′W. Ships cacao and coffee. Until 1944, called Rio Novo.

Ipin, CHINA: see YIBIN.

Ipirá (ee-peer-AH), city (2007 population 60,343), E central BAHIA, BRAZIL, 60 mi/96 km NW of Feira de Santana; 12°09′S 39°44′W.

Ipiranga (EE-pee-rahn-gah), city, S central PARANÁ, BRAZIL, 26 mi/42 km WSW of PONTA GROSSA; 25°01′S 50°35′W. Coffee and lard processing; maté, rice, corn, kidney beans, grapes. Coal deposits. Formerly spelled Ypiranga.

Ipiranga (EE-pee-rahn-gah), town, E RIO GRANDE DO SUL, BRAZIL, 5 mi/8 km NNW of GRAVATAÍ, 15 mi/24 km NE of PÔRTO ALEGRE; 29°53′S 51°03′W. Oil refinery. Also called Ipiranga do Sul.

Ipiranga (EE-pee-RAHN-gah), stream flowing near the city of SÃO PAULO, BRAZIL. On its banks the regent Pedro (later Emperor Pedro I) issued the *Grito do Ipiranga*, the declaration of the independence of Brazil from Portugal, on September 7, 1822. The event is commemorated by a monument and a historical museum.

Ipita (ee-PEE-tah), canton, CORDILLERA province, SANTA CRUZ department, E central BOLIVIA, 15 mi/25 km N of LAGUNILLAS, on the YACUIBA–SANTA CRUZ highway; 19°20′S 63°32′W. Elevation 3,950 ft/1,204 m. Abundant gas and oil resources in area; limestone and gypsum deposits. Agriculture (potatoes, yucca, bananas, corn, cotton, peanuts, soy, coffee); cattle.

Ipixuna (EE-pee-SHOO-nah), city (2007 population 17,177), SW AMAZONAS, BRAZIL, 47 mi/75 km from ACRE border, on Rio JURUÁ; 07°03′S 71°42′W.

Ipixuna do Pará (EE-pee-shoo-nah do pah-RAH), city (2007 population 39,367), N central PARÁ, BRAZIL, 43 mi/69 km NE of GURUPÁ, on island in AMAZON delta; 00°44′S 50°59′W.

Ipoh (EE-po), city (2000 population 574,041), ⊙ PERAK state, MALAYSIA, central MALAY PENINSULA, in the KINTA RIVER valley; 04°35′N 101°05′E. A modern commercial town, it is a major tin-mining center. Nearby are rubber plantations and limestone quarries; has diversified into light industry. The mine laborers and the population are mainly Chinese. The city has noted Chinese rock temples.

Ipojuca (EE-po-zhoo-kah), city (2007 population 69,781), E PERNAMBUCO, NE BRAZIL, near the Atlantic, 27 mi/43 km SSW of RECIFE; 08°24′S 35°04′W. Sugar, manioc, coconuts. Clay deposits.

Ipole (ee-po-LAI), village, TABORA region, NW central TANZANIA, 50 mi/80 km S of TABORA; 05°46′S 32°44′E. Road junction. Timber; corn; wheat.

Ipoly River, SLOVAKIA and HUNGARY: see IPEL RIVER.

Iporá (EE-po-RAH), city (2007 population 31,005), W central GOIÁS, BRAZIL, 50 mi/80 km NE of PALESTINA; 16°03′S 51°08′W.

Iporã (EE-po-ru), city (2007 population 15,086), W PARANÁ state, BRAZIL, 34 mi/55 km ENE of GUAÍRA; 23°59′S 53°37′W. Coffee.

Iporanga (EE-po-rahn-gah), town (2007 population 4,507), S SÃO PAULO, BRAZIL, on RIBEIRA RIVER, 75 mi/121 km SSW of ITAPETININGA; 24°35′S 48°35′W. Lead, silver, vanadium, gold, copper deposits. Santana cavern nearby. Formerly spelled Yporanga.

Ipperwash (I-puhr-wahsh), military reservation, 25 mi/40 km NE of Lake HURON, E central ONTARIO, E central CANADA; 43°11′N 81°55′W. Land was given back to the Chippewa tribe, who seized part of the base in 1995 to demand its return.

Ipponmatsu (eep-PON-mahts), town, South Uwa district, EHIME prefecture, NW SHIKOKU, W JAPAN, 62 mi/100 km S of MATSUYAMA; 32°57′N 132°39′E.

Ippy (ee-PEE), village, OUAKA prefecture, central CENTRAL AFRICAN REPUBLIC, 45 mi/72 km NE of BAMBARI; 06°05′N 21°07′E. Cotton center; diamond mining.

Ipsala, ancient *Cypsela*, village, European TURKEY, near MARITSA RIVER (Greek border), 50 mi/80 km SSW of EDIRNE; 40°56′N 26°23′E. Wheat, rice, rye, sugar beets.

Ipsara, Greece: see PSARA.

Ipsus, small town in ancient PHRYGIA, ASIA MINOR (NW of modern AKSEHIR). Antigonus I, who had summoned his son Demetrius to his aid, was defeated and slain here by his rivals Seleucus and Lysimachus in 301 B.C.E. The battle of Ipsus resulted in the dissolution of Alexander's empire.

Ipswich (IP-swich), community, SW MANITOBA, W central CANADA, 4 mi/6 km from STRATHCLAIR village, and in STRATHCLAIR rural municipality; 50°25′N 100°28′W.

Ipswich (IP-swich), city, SE QUEENSLAND, AUSTRALIA, on Bremer River, 20 mi/32 km WSW of BRISBANE; 27°37′S 152°46′E. Woolen mills, sawmills, earthenware works, abattoirs, foundries; limestone.

Ipswich (IPS-wich), town (2001 population 117,069) and district, ⊙ SUFFOLK, E ENGLAND, on the ORWELL estuary 12 mi/19 km from its entry into the NORTH SEA; 52°05′N 01°10′E. A market and port, it exports barley, malt, and fertilizers; imports coal, petroleum, phosphates, grain, and timber. Agricultural machinery and construction vehicles are the chief manufacturing products of Ipswich, which also makes fertilizer, cigarettes, malting, milling, brewing, printing, and textile industries. The city reached the peak of its significance in the woolen trade in the 16th century. Vestiges of Roman habitation remain; twelve old churches, several 15th- and 16th-century houses, Christchurch mansion (1548), the public school (14th century), and Sparrowe's House (1567). Wolsey's Gate is the only remnant of the college founded in the early 16th century.

Ipswich, town, SAINT ELIZABETH parish, W JAMAICA, on Jamaica railroad, 21 mi/34 km SSE of MONTEGO BAY; 18°13′N 76°42′W. In agricultural region (corn, vegetables; livestock).

Ipswich, town, ESSEX county, NE MASSACHUSETTS, on the IPSWICH RIVER and IPSWICH BAY, 11 mi/18 km NNE of BEVERLY; 42°42′N 70°50′W. Ipswich clams are found here. Tourism and the production of electronic and wood products are important; also fishing and shellfish. Crane's Beach, one of the country's most beautiful beaches, is in Ipswich. Of interest are the many well-preserved colonial and historic buildings; Choate Bridge, the first stone bridge in the U.S. (1764); and the John Whipple House (c.1640), with the Ipswich Historical Society collection. An Air Force radar experimental station is also here. Plum Island State Park. Incorporated 1634.

Ipswich, town (2006 population 875), ⊙ EDMUNDS county, N SOUTH DAKOTA, 25 mi/40 km W of ABERDEEN; 45°26′N 99°01′W. In farming and cattle-raising region. Prayer Rock E of town; Mina State Recreational Area to E.

Ipswich Bay, bight (c.6 mi/9.7 km wide) of the ATLANTIC, E MASSACHUSETTS, E of IPSWICH; sheltered on S and E by Cape Ann.

Ipswich River, c.35 mi/56 km long, NE MASSACHUSETTS; rises in NE MIDDLESEX county, flows generally NE, past IPSWICH, to IPSWICH BAY. Includes wildlife sanctuary.

Ipu (ee-poo) town (2007 population 39,433), W CEARÁ, BRAZIL, on E slope of SERRA GRANDE, 50 mi/80 km SSW of SOBRAL, on CAMOCIM-CRATEÚS railroad; 04°05′S 40°40′W. Cotton, sugar, tobacco, skins. Gold, nitrate, iron deposits nearby.

Ipuã (EE-po-ah), city (2007 population 14,344), N SÃO PAULO state, BRAZIL, 36 mi/58 km NE of BARRETOS; 20°27′S 48°02′W. Coffee growing.

Ipubi (EE-poo-bee), city (2007 population 25,893), extreme NW PERNAMBUCO state, BRAZIL; 07°39′S 40°07′W.

Ipueiras (ee-pwe-rahs), city (2007 population 38,046), W CEARÁ, BRAZIL, on CAMOCIM-CRATEÚS railroad, 60 mi/97 km SSW of SOBRAL; 04°30′S 40°42′W. Cotton, sugar, tobacco.

Ipuiúna (EE-poo-e-OO-nah), town (2007 population 9,170), S central MINAS GERAIS, BRAZIL, 29 mi/47 km NW of POUSO ALEGRE; 22°12′S 46°25′W.

Ipun Island, CHILE: see CHONOS ARCHIPELAGO.

Ipupiara (EE-poo-pee-AH-rah), town (2007 population 8,931), W central BAHIA, BRAZIL, 18 mi/29 km N of BROTAS DE MACAUBAS; 11°49′S 42°37′W.

Iput' River (EE-poot), 295 mi/475 km long, BELARUS and W European RUSSIA; rises E of KLIMOVICHI (Belarus) in SMOLENSK-MOSCOW UPLAND, flows E into RUSSIA, S, and WSW in wide bend, past SURAZH (BRYANSK oblast, RUSSIA) and DOBRUSH (Belarus), to SOZH RIVER at GOMEL, Belarus. Navigable for 20 mi/32 km, to Dobrush. Used for floating timber.

Iqaluit (ee-KAH-loo-it), town (□ 20 sq mi/52 sq km; 2001 population 5,236), ⊙ NUNAVUT territory, CA-NADA, S BAFFIN ISLAND, BAFFIN region, at head of FROBISHER BAY; 63°44′N 68°30′W. In June, it receives twenty-four hours of daylight, but in December only six. One of twenty-eight settlements in Nunavut. Sir Martin Frobisher explored the area in 1576 and celebrated Thanksgiving at this site in 1577. Became part of Distant Early Warning (D.E.W.) Line system in 1953 (Crystal II). Administrative, transportation, commercial, and service center for Canadian Arctic. Trapping, hunting, fishing, sealing. Manufacturing (jewelry, apparel, crafts). Tourism. Inuit Cultural Museum Hospital, radio and TV stations, scheduled air service. Qaummaarviit Historic Park. Town established in 1942 as a U.S. Air Force base. In December 1995, Iqaluit was selected as the capital of the new territory Nunavut, established April 1, 1999; on April 19, 2001, Iqaluit received its Order of Official Status as a City. Formerly FROBISHER BAY (1942–1984).

Iqlid, IRAN: see EQLID.

Iqlit (ik-LEET), village, ASWAN province, S EGYPT, on the E bank of the NILE River, 30 mi/48 km N of ASWAN; 24°31′N 32°54′E. Cereals, dates.

Iquiaca de Umala (ee-kee-AH-kah dai oo-MAH-lah), canton, AROMA province, LA PAZ department, W BOLIVIA, on the ORURO–LA PAZ highway, 62 mi/100 km S of La Paz; 17°24′S 67°58′W. Elevation 12,625 ft/3,848 m. Airstrip and secondary airport here. Abundant gas resources in area; lead-bearing lode; clay, limestone and gypsum deposits. Agriculture (potatoes, yucca, bananas, rye); cattle.

Iquique (ee-KEE-kai), city (2002 population 164,396), ⊙ Iquique comuna and province, TARAPACÁ region, N CHILE; 20°13′S 70°10′W. Railroad. A port on the PACIFIC, it exports nitrates and ore from the ATACAMA DESERT. Fish-meal processing. Duty-free zone. The city, founded in the 16th century, was taken (1879) from PERU by Chile during the War of the Pacific. The city has fine beaches and excellent deep-sea fishing. Airport.

Iquira (ee-KEE-rah), town, ⊙ Iquira municipio, HUILA department, S central COLOMBIA, in the foothills of the Nevado del HUILA, 35 mi/56 km SW of NEIVA; 02°39′N 75°38′W. Elevation 4,015 m/1,223 m. Coffee, plantains.

Iquitos (ee-KEE-tos), city (2005 population 356,549), ⊙ MAYNAS province and LORETO region, NE PERU, on the AMAZON River, c.2,300 mi/3,701 km from the Amazon's mouth; 03°46′S 73°15′W. It is the farthest inland port of any considerable size in the world. Founded in 1750s by missionaries. With the boom in wild rubber at the beginning of the 20th century the city gained prominence, but it declined after the collapse of the market. Today coffee, tobacco, bananas, cotton, timber, balatá, and BRAZIL nuts, as well as rubber, are exported. Oil prospecting and producing center. There is launch service some distance up the MARAÑÓN and UCAYALI rivers, but the ANDES Mountains are so formidable a barrier to the transport of most commercial goods that Iquitos has been oriented toward the ATLANTIC rather than the PACIFIC. City founded 1863. Tourist center for jungle tours. Coronel Francisco Secada Vignetta International Airport.

Ira, town, RUTLAND co., W central VERMONT, 6 mi/9.7 km SW of RUTLAND; 43°32′N 73°04′W. Named for Ira Allen.

Iraan (I-ruh-an), town (2006 population 1,196), PECOS county, extreme W TEXAS, on PECOS RIVER, c.55 mi/89 km E of FORT STOCKTON; 30°54′N 101°54′W. Elevation 2,200 ft/671 m. Oil field; agriculture (livestock; cotton; vegetables, pecans, grain); manufacturing (gas processing). Archaeological museum.

Irabu (EE-rah-boo), town, IRABU-SHIMA island, Miyako district, OKINAWA prefecture, extreme SW JAPAN, 180 mi/290 km S of NAHA; 24°49′N 125°10′E. Dried bonito; pumpkins.

Irabu-shima (ee-rah-BOO–shee-mah), volcanic island (□ 15 sq mi/39 sq km) of SAKISHIMA ISLANDS, in the RYUKYU ISLANDS, OKINAWA prefecture, extreme SW JAPAN, between EAST CHINA (W) and PHILIPPINE (E) seas, 3.5 mi/5.6 km W of MIYAKO-JIMA; 5 mi/8 km long, 3 mi/4.8 km wide. Generally low elevation, fertile (sugarcane). Fishing. Just off its W coast is Shimoji-shima (3 mi/4.8 km long, 1.5 mi/2.4 km wide). Coral reef encircles both islands.

Iracema (EE-rah-SE-mah), city (2007 population 14,333), E central CEARÁ, BRAZIL, near Serra das Melancias and border with RIO GRANDE DO NORTE, 22 mi/35 km ENE of JAGUARIBÉ; 05°48′S 38°25′W.

Iracema (EE-rah-SE-mah), township, central ACRE, BRAZIL, on Rio Envira; 08°35′S 70°30′W.

Iracemápolis (EE-rah-se-MAH-po-lees), city (2007 population 18,026), central SÃO PAULO state, BRAZIL, 3.7 mi/6 km W of LIMEIRA; 22°35′S 47°32′W.

Iracoubo (ee-rah-koo-BO), town, N FRENCH GUIANA, near the coast, 70 mi/113 km WNW of CAYENNE; 05°29′N 53°13′W.

Irafale, Eritrea: see IRAFAYLE.

Irafayle (IR-ah-fah-lee), village, SEMENAWI KAYIH BAHRI region, central ERITREA, on coast, 45 mi/72 km SSE of MASSAWA; 15°05′N 39°46′E. Fishing port on the S shores of the Gulf of ZULA; trade center. Pozzolana deposits and hot springs nearby. Also spelled Irafale and Arafali.

Iraí (EE-rah-ee), city (2007 population 8,468), N RIO GRANDE DO SUL, BRAZIL, on URUGUAY RIVER (SANTA CATARINA border), 70 mi/113 km WNW of ERECHIM; 27°13′S 53°15′W. Mineral springs. Semiprecious stones nearby. Formerly spelled Irahy.

Iraí de Minas (ee-rah-EE dee MEE-nahs), town (2007 population 6,306), W central MINAS GERAIS, BRAZIL, 40 mi/64 km W of PATROCINIO; 18°55′S 47°30′W.

Irajuba (EE-rah-ZHOO-bah), town (2007 population 7,245), E central BAHIA, BRAZIL, 16 mi/25 km W of SANTA INÉS; 13°15′S 40°05′W.

Irak: see IRAQ.

Iraklia (ee-RAH-klee-ah), town, SÉRRAI prefecture, CENTRAL MACEDONIA department, NE GREECE, 15 mi/24 km NW of SÉRRAI; 41°10′N 23°16′E. Cotton, barley, corn. Formerly called Tzoumaia (also spelled Tzoumayia, Jumaya, and Dzhumaya) or Kato Tzoumagia [Greek=lower Tzoumagia], as opposed to Gorna Dzhumaya [Bulgarian=upper Dzhumaya], Bulgaria, which is now called BLAGOEVGRAD. Also spelled Herakleia and Heraclia.

Iraklia (ee-RAH-klee-ah), island (□ 7 sq mi/18.2 sq km), CYCLADES prefecture, SOUTH AEGEAN department, GREECE, S of NAXOS island; 5 mi/8 km long, 2.5 mi/4 km wide; 36°50′N 25°26′E. Also spelled Herakleia or Heraclia.

Iráklion (ee-RAH-klee-on), prefecture (□ 989 sq mi/2,571.4 sq km), E central CRETE department, GREECE, between Mount DIKTI (E) and Mount Psilioti (W); ⊙ IRÁKLION; 35°10′N 25°10′E. Largest and most populous prefecture on CRETE. Bordered E by LASITHI prefecture and W by RETHYMNON prefecture. Main port, Iráklion, on N shore. Most fertile area of Crete, producing raisins, carob, wheat, citrus fruits; olive oil; livestock raising (sheep, goats, cows); fisheries. Tourism. Includes historic sites of KNOSSOS, Phaistos, and Gortis.

Iráklion (ee-RAH-klee-on), ancient *Heracleum*, city (2001 population 144,642), ⊙ CRETE department and IRÁKLION prefecture, N CRETE, GREECE; 35°10′N 25°10′E. A port on the Sea of Crete. Largest city on Crete; ships wine, olive oil, raisins, and almonds. Tourism is especially important. Founded (9th century) by the Muslim Saracens. In 961 it was conquered by the Byzantine emperor Nicephorus II, and in the 13th century it became a Venetian colony. The Venetians, who named the city Candia, fortified it and improved its port. In 1669 it was captured by the Ottoman Turks after a two-year siege. Was capital of Crete until 1841, and in 1913 it passed to Greece. Severely damaged during German invasion (1941). Has a

---

Cross-references are shown in SMALL CAPITALS. The pronunciation guide is shown on page xix. The sources of population figures are shown on page xvii.

museum of Minoan antiquities that were excavated at the site of ancient KNOSSOS, just outside the city. Local historic monuments include a cathedral, several mosques, and remains of Venetian walls and fortifications. International airport to E. Also called Herakleion.

Iramaia (EE-rah-MEI-ah), city (2007 population 13,856), central BAHIA, BRAZIL, on SALVADOR–BELO HORIZONTE railroad, 35 mi/56 km S of ITAETÉ; 13°18′S 40°58′W.

Iramba Plateau (ee-RAHM-bah), N central TANZANIA, N of SINGIDA and S of SERENGETI PLAIN; grassland drained by Ndurumo River. Elevation c. 1,500 ft/457 m. Gold deposits. Gemstones.

Iran (ee-RAHN), Islamic republic (□ 1,648,195 sq mi/ 4,268,825 sq km; 2004 estimated population 69,018,924; 2007 estimated population 65,397,521), SW ASIA; ⊙ TEHRAN. Name was changed from PERSIA to Iran in 1935 by royal decree.

Geography

TEHRAN is the largest city of Iran and is the political, cultural, commercial, and industrial center of the nation. ESFAHAN, MASHHAD, TABRIZ, RASHT, HAMADAN, KERMANSHAH, ABADAN, SHIRAZ, and AHVAZ are other major cities. Bordered in N by ARMENIA, AZERBAIJAN, and TURKMENISTAN (all parts of former USSR) and the CASPIAN SEA; E by AFGHANISTAN and PAKISTAN; S by the PERSIAN GULF and the Gulf of OMAN; and W by TURKEY and IRAQ. The SHATT AL ARAB forms part of the Iran-IRAQ border. Physiographically, Iran lies within the Alpine-Himalayan mountain system and is composed of a vast central plateau rimmed by mountain ranges and limited lowland regions. Iran is subject to numerous and often severe earthquakes and volcanic eruptions. The Iranian Plateau (elev. c.4,000 ft/1,200 m), which extends beyond the low ranges of E Iran into AFGHANISTAN, is a region of interior drainage. It consists of a number of arid basins of salt and sand, including DASHT-E KAVIR and DASHT-E LUT, and some marshlands, such as the area along the Afghanistan border. The plateau is surrounded by high folded and volcanic mountain chains including the Kopet Mountains (KOPET DAG) in the NW, the ELBURZ Mountains (rising to 18,934 ft/5,771 m at Mount DAMAVAND, Iran's highest point) in the N, and the complex ZAGROS Mountains in the W. Lake URMIA, the country's largest inland body of water, is in the ZAGROS of NW Iran. Narrow coastal plains are found along the shores of the Persian Gulf, Gulf of OMAN, and the Caspian Sea; at the head of the Persian Gulf is the Iranian section of the Mesopotamian lowlands. Of the few perennial rivers in Iran, only the KARUN in the W is navigable for large craft; other rivers are the KARKHEH and the SEFID. Iran's climate is continental, with hot summers and cold, rainy winters; the mountain regions of the N and W have a subtropical climate. Temperature and precipitation vary with elevation, as winds bring heavy moisture from the Mediterranean and Persian Gulf. The Caspian region receives over 40 in/102 cm of rain annually. Precipitation occurs mainly in the winter and decreases from NW to SW; much of the precipitation in the mountains is in the form of snow. Snowmelt is vital for Iran's water supply. The central portion of the plateau and the S coastal plain (MAKRAN) receive less than 5 in/12.7 cm of rain annually. The N slopes of the ELBURZ Mountains are heavily wooded. Tree cutting is rigidly controlled by the government, which also has a reforestation program. Rivers toward the Caspian Sea have salmon, carp, trout, and pike; sturgeon are abundant in the sea. Of the variety of natural resources found in Iran, petroleum (discovered in 1908 in KHUZESTAN) and natural gas are by far the most important. The chief oil fields are in the central and SW parts of the Zagros Mountains with other fields in N Iran and in the offshore waters of the Persian Gulf.

Domestic oil and gas, along with hydroelectric facilities, provide the country with power.

Population

Iran's central position has made it a crossroads of migration; the population is not homogeneous, although it has a Persian core. The migrant ethnic groups of the mountains and highlands, including the Kurds, Lurs, Qashqai, and Bakhtiari, are of the least mixed descent of the original Iranians. In the N provinces, Turkic and Tatar influences are evident; Arab strains predominate in the SE. Islam entered the country in the 7th century and is now the official religion; about 95% of Iranians are Muslims, mainly of the Shiite section. The remainder, mostly Kurds and Arabs, are Sunnis. Colonies of Zoroastrians remain at YAZD, KERMAN, and other large towns. In addition to Armenian and Assyrian Christian sects, there are Jews, Protestants, and Roman Catholics. Extreme measures are in force to suppress Babism and its successor, Bahaism, in Iran. Other religious movements, such as Mithraism and Manichaeism, originated in Iran. The principal language of the country is Persian (Farsi), which is written in Arabic characters, but others include Iranian languages and dialects, Turkic dialects, Turkish, Kurdish, Arabic, and Armenian. Among the educated classes, English and French are spoken. Iran has a large rural population (more than 90% of people live in agricultural villages); there are also small groups of nomadic and seminomadic pastoralists throughout the country.

Economy

KHORRAMSHAHR, a port on the SHATT AL ARAB most accessible to large vessels, was badly damaged and put out of action during the Iran-Iraq war. BUSHEHR has since become a major port; BANDAR-E ANZALI is the chief Caspian Sea port. A network of roads links the villages with the larger cities; most of the principal routes are paved. The Trans-Iranian railroad links N Iran with the Persian Gulf; numerous branch lines connect with points E and W of the main line. About 8% of land in Iran is arable. The main food-producing areas are in the Caspian region and in the valleys of the NW. Wheat, the most important crop, is grown mainly in the W and NW; rice is the major crop in the Caspian region. Barley, corn, cotton, tea, hemp, tobacco, sugar beets, fruits (including citrus), nuts, and dates are also grown. Fish is abundant, and livestock is raised in most inhabited parts of the country. Forestry products, from areas in the Caspian region, are important. Cultivation of the opium poppy was prohibited in 1955. The principal obstacles to agricultural production are primitive farming methods, overworked and underfertilized soil, poor seed, and the scarcity of water. About one-third of the cultivated land is irrigated; the construction of multipurpose dams and reservoirs along the rivers in the Zagros and Elburz mountains have increased the amount of water available for irrigation. Agricultural programs of modernization, mechanization, and crop and livestock improvement, and programs for the redistribution of land are increasing agricultural production. The petroleum industry is Iran's economic mainstay; oil accounts for over 90% of Iran's export revenues. Finances are used to stimulate industrial growth and diversification as well as to provide for better social conditions and to lure private and foreign investments. Iran also has the world's second-largest reserves of natural gas. Petroleum production is concentrated in W Iran and major refineries are at ABADAN (site of Iran's first refinery, built in 1913), Bakhtaran, and Tehran. Pipelines move oil from the fields to the refineries and to exporting ports of Abadan, BANDAR MAHSHAHR, and KHARK Island, which during the Iran-Iraq war was a focal point of attacks after Abadan was destroyed. Iran is a member of the Organization of Petroleum Exporting Countries (OPEC). Textiles are Iran's second most im-

portant industrial product; Tehran and ESFAHAN are the chief textile-producing centers. Other major industries are sugar refining, food processing, and petrochemical production and machinery. There is an iron and steel plant at Esfahan and a fertilizer plant at SHIRAZ. Traditional handicrafts, such as carpet weaving, and the manufacturing of ceramics, silk, and jewelry are also important parts of Iran's economy. Besides crude and refined petroleum, Iran's chief exports are cotton, carpets, and fruit; its chief imports are food, metals, machinery, military supplies, and chemicals. Iran's chief trading partners are JAPAN, CHINA, ITALY, SOUTH KOREA, and the NETHERLANDS.

History: to 16th Century

Until 1979, Iran was a constitutional monarchy with a parliamentary form of government, which adhered to the constitution of 1906. Iran became a theocratic Islamic republic in 1979. A new constitution was written, incorporating a four-year presidential term, a prime minister, and a parliament. Many religious political parties exist, some of which are suppressed by the government. An important Islamic religious advisory board works in close conjunction with the government Iran has a long and rich history. Some of the world's most ancient settlements have been excavated in the Caspian region and on the Iranian plateau; village life began there c.4000 B.C.E. The Aryans came about 2000 B.C.E. and split into two main groups, the Medes and the Persians. The Persian Empire founded by Cyrus the Great (c.550 B.C.E.) was succeeded, after a period of Greek and Parthian rule, by the Sassanid in the early 3rd century C.E. This empire ruled until Arabs invaded, bringing Islam (641). For the next several centuries, the area was ruled by the Turks (10th century), the Mongols (13th century), and finally Tamerlane (14th century).

History: 16th Century to 1813

The Safavid dynasty (1502–1736), founded by Shah Ismail, restored internal order in Iran and established the Shiite sect of Islam as the state religion; it reached its height during the reign (1587–1629) of Shah Abbas I (Abbas the Great). He drove out the Portuguese, who had established colonies on the Persian Gulf early in the 16th century. Shah Abbas also established trade relations with GREAT BRITAIN and reorganized the army. Religious differences led to frequent wars with the Ottoman Turks, whose interest in Iran was to continue well into the 20th century. The fall of the Safavid dynasty was brought about by the Afghans, who overthrew the weak shah, Husein, in 1722 and ruled for a short time. Under the subsequent Afshar dynasty (1736), and the Zand dynasty (1750–1794), Iran enjoyed a period of wealth, peace, and prosperity. The capital was established at SHIRAZ. In the early years of the Qajar dynasty (1794–1925), Iran steadily lost territory to neighboring countries and fell under the increasing influence of European nations, particularly czarist Russia.

History: 1813 to 1919

The treaties of GULISTAN (1813) and TURKMANCHAI (1828) forced Iran to give up the Caucasian lands. HERAT, the rich city on the HARI RUD, which had been part of the ancient Persian Empire, was taken by the Afghans. A series of campaigns to reclaim it ended with the intervention of the British on behalf of AFGHANISTAN and resulted in the recognition of Afghan independence by Iran in 1857. The discovery of oil in the early 1900s intensified the rivalry of Great Britain and RUSSIA for power over the nation. Internally, the early 20th century saw the rise of the constitutional movement and a constitution establishing a parliament was accepted by the shah in 1906. Meanwhile, British-Russian rivalry continued and, in 1907, resulted in an Anglo-Russian agreement (annulled after World War I) that divided Iran into spheres of influence. The period preceding World War I was one of political and financial difficulty. During the war,

Iran was occupied by the British and Russians but remained neutral; after the war, Iran was admitted to the League of Nations as an original member.

History: 1919 to 1951

In 1919, Iran made a trade agreement with Great Britain in which Britain formally reaffirmed Iran's independence but actually attempted to establish a complete protectorate over it. After Iranian recognition of the USSR in a 1921 treaty, the SOVIET UNION renounced czarist imperialistic policies toward Iran, canceled all debts and concessions, and withdrew occupation forces from Iranian territory. In 1921, Reza Khan, an army officer, effected a coup d'etat and established a military dictatorship. He was subsequently (1925) elected hereditary shah, ending the Qajar dynasty and founding the Pahlavi dynasty. Reza Shah Pahlavi abolished the British treaty, reorganized the army, introduced many reforms, and encouraged the development of industry and education. British, Soviet, and U.S. troops occupied Iran after Germany's 1941 invasion of the USSR. In 1943 the Tehran Declaration was signed by these nations to guarantee the independence and territorial integrity of Iran. However, the USSR, dissatisfied with the refusal of the Iranian government to grant it oil concessions, fomented a revolt in the N which led to the establishment (December 1945) of the People's Republic of AZERBAIJAN and the Kurdish People's Republic, headed by Soviet-controlled leaders. The Soviets finally withdrew (May 1946) after receiving a promise of oil concessions from Iran subject to approval by the parliament. The Soviet-established governments in the N, lacking popular support, were deposed by Iranian troops late in 1946, and the parliament subsequently rejected the oil concessions.

History: 1951 to 1978

In 1951, the National Front movement nationalized the oil industry and formed the National Iranian Oil Company (NIOC). A British blockade led to the virtual collapse of the oil industry and serious internal economic troubles until, in 1954, Iran allowed an international consortium of British, U.S., French, and Dutch oil companies to operate its oil facilities, with profits shared equally between Iran and the consortium. Starting in the 1960s and continuing into the 1970s, the Iranian government, at the shah's initiative, undertook a broad program designed to improve economic and social conditions. Land reform was a major priority and within three years, 1.5 million former tenant farmers were landowners. Women received the right to vote in national elections in 1963. In April 1969, Iran voided the 1937 accord with Iraq on the control of the SHATT AL ARAB and demanded that the treaty, which had given Iraq virtual control of the river, be renegotiated. In 1971, although Iran renounced all claims to BAHRAIN in 1970, it took control (November 1971) of three small, Arab-owned islands at the mouth of the Persian Gulf. In 1975, the Iran-Iraq Border Agreement provided that the two countries would define their frontiers on the basis of the protocol of CONSTANTINOPLE of 1913, and the verbal agreement on frontiers of 1914, and that the Shatt al Arab frontier would be defined according to the Thalweg line. This treaty later became one of the key issues of the war with Iraq that broke out in September 1980.

History: 1978 to 1989

The shah's autocratic rule and his extensive use of the secret police led to widespread popular unrest throughout 1978. The religious-based protests were conservative in nature, directed against the shah's policies. The Ayatollah Khomeini, who had been expelled from Iraq in February 1978, called for the abdication of the shah. Martial law was declared in September for all major cities. As government controls faltered, the Shah Pahlavi fled Iran on January 16, 1979. Khomeini returned and led religious revolutionaries to the final overthrow of the shah's gov-

ernment on February 11th. The new government represented a major shift toward conservatism. It nationalized industries and banks and revived Islamic traditions. Western influence and music were banned, women were forced to return to traditional veiled dress, and Westernized elites fled the country. A new constitution was written allowing for a presidential system, but Khomeini remained at the executive helm. On November 4, 1979, Iranian militants seized the U.S. Embassy in Tehran, taking fifty-two American hostages. The hostage crisis lasted 444 days and was finally resolved on January 20, 1980. Nearly all Iranian conditions had been met, including the unfreezing of nearly $8 billion in Iranian assets. On September 22, Iraq invaded Iran, commencing an eight-year war primarily over the disputed Shatt al Arab waterway. Fighting crippled both nations, devastating Iran's military supply and oil industry and leading to an estimated 500,000 to one million casualties. Chemical weapons were used by both countries. In July 1988, Khomeini agreed to accept a UN ceasefire with Iraq, ending the eight-year war. Iran immediately began rebuilding the nation's economy, especially its oil industry.

History: 1989 to Present

Khomeini died of illness in June 1989 and was succeeded as supreme leader by Ayatollah Ali Hoseini Khamenei. The same year, Ali Akbar Hashemi Rafsanjani, who sought improved relations and financial aid with Western nations, was elected president. A major earthquake hit N Iran on June 21, 1990, killing nearly 40,000 people. Iran remained neutral during the Persian Gulf War, aside from harboring defecting Iraqi pilots and grounding their planes during the fighting. Countries of the EU have renewed economic ties with Iran; the U.S. has blocked more normalized relations until Iran stops its support of what the U.S. considers terrorist organizations. In May 1997, Mohammed Khatami, a moderately liberal Muslim cleric, was elected president, which was widely seen as a reaction against the country's repressive social policies and lack of economic progress. Reformers won about two thirds of the seats in the 2000 parliamentary elections, but Khamenei and other conservative elements in the government restricted their ability to pass reformist legislation. Khatami was overwhelmingly reelected in 2001. After the Anglo-American invasion of Iraq, U.S. unhappiness with Iranian support for Iraqi Shiite militias and U.S. charges of Iranian development of nuclear weapons (under the guise of nuclear power development) increased tensions internationally. An earthquake in SE Iran in December 2003 killed more than 26,000 people. In 2003 Iran agreed to stricter inspections of its nuclear sites, but other actions, such as its processing of nuclear fuel in September 2004, called into question its willingness to cooperate while not clearly indicating a program to develop weapons. In November 2004 Iran reached an agreement with the International Atomic Energy Agency (IAEA) to suspend uranium enrichment, but the issue remained contentious and unresolved in the following months. In the 2004 parliamentary elections conservatives won control of parliament, and in 2005 the hardline conservative mayor of Tehran, Mahmud Ahmadinejad, was elected president. In 2006 the IAEA reported Iran to the Security Council for nuclear nonproliferation treaty violations; Iran then announced that it was resuming enrichment. Iran has since remained defiant concerning enrichment and its right to develop nuclear power despite negotiations and Security Council sanctions.

Government

Iran is a theocratic Islamic republic governed under the constitution of 1979 as amended. Appointed, rather than elected, offices and bodies hold the real power in the government. The supreme leader, who is the head of state, is appointed for life by an Islamic

religious advisory board (the Assembly of Experts). The supreme leader oversees the military and judiciary and appoints members of the Guardian Council and the Expediency Discernment Council. The former, some of whose members are appointed by the judiciary and approved by parliament, works in close conjunction with the government and must approve both candidates for political office and legislation passed by parliament. The latter is a body responsible for resolving disputes between parliament and the Guardian Council over legislation. The president, who is popularly elected for a four-year term, serves as the head of government. The unicameral legislature consists of the 290-seat Islamic Consultative Assembly, whose members are elected by popular vote for four-year terms. Supreme Leader Ayatollah Ali Hoseini Khamenei has been head of state since 1989; President Mahmud Ahmadinejad has been head of government since 2005. Administratively, Iran is divided into thirty provinces: ARDABĪL, AZARBAYJAN-E GHARBI, Azarbayjan-e Sharqi, BUSHEHR, CHAHAR MAHALL VA BAKHTIARI, ESFAHAN, FĀRS, GILAN, Golestan, HAMADAN, HORMOZGAN, ILAM, KERMĀN, KERMANSHAH, KHORASAN-E JANUBI, KHORASAN-E RAZAVI, KHORASAN-E SHEMALI, KHUZESTĀN, KOHGILUYEH VA BUYER AHMAD, KORDESTAN, LORESTAN, MARKAZĪ, MĀZANDARĀN, QAZVIN, QOM, SEMNAN, SĪSTĀN VA BALŪCHESTĀN, TEHRĀN, YAZD, and ZANJĀN.

Iranduba (EE-rahn-DOO-bah), city (2007 population 30,472), E central AMAZONAS, BRAZIL, on Rio Solimões, across river, 5 mi/8 km from MANAUS; 03°10′S 60°27′W. Airport.

Iraniel (i-ruhn-YEL), town, KERALA state, S INDIA, 30 mi/48 km SE of THIRUVANANTHAPURAM. Coir rope and mats, copra, palmyra jaggery. Monazite workings nearby. Formerly spelled Eraniel. Also called Neyyur.

Iran Mountains (i-RAN), range in N central BORNEO, on border between SARAWAK to W and KALIMANTAN (Indonesian Borneo) to E, c.170 mi/274 km S of Brunei town. Extends c.50 mi/80 km N-S; rises to c.8,000 ft/2,438 m.

Iranshahr (EE-rahn-shahr), town (2006 population 100,642), Sīstān va Balūchestān province, SE IRAN, 15 mi/24 km E of old fortress of BAMPUR; 27°12′N 60°41′E. Largely superseded as an agricultural center; dates, corn, barley, cotton. Sometimes called FAHRAJ, it is commonly identified with ancient *Pura*, the capital of Gedrosia, where Alexander the Great passed in 325 B.C.E. on his return from INDIA. The modern town dates from 1892.

Irapa (ee-RAH-pah), town, ⊙ Mariño municipio, SUCRE state, NE VENEZUELA, port on S coast of PARIA PENINSULA, on Gulf of PARIA, 45 mi/72 km E of CARÚPANO; 10°34′N 62°35′W. Livestock; cassava, plantains.

Irapiranga, Brazil: see ITAPORANGA D'AJUDA.

Irapuã (EE-rah-poo-roo), town (2007 population 6,710), N central SÃO PAULO, BRAZIL, 17 mi/27 km NW of NOVO HORIZONTE; 21°18′S 49°24′W. Rice, coffee, sugar.

Irapuato (ee-ra-PWAH-to), city (2005 population 342,561) and township, ⊙ Irapuato municipio, GUANAJUATO, W central MEXICO, on the Irapuato River, on Mexico Highway 45-110; 20°40′N 101°20′W. Elevation 5,656 ft/1,724 m. It is the commercial and communications center of the surrounding mining and agricultural (cereals and cattle) region. The fruits and flowers of Irapuato's luxurious gardens are famous throughout Mexico.

Irapuru (EE-rah-poo-roo), town (2007 population 7,566), W SÃO PAULO state, BRAZIL, 8 mi/12.9 km S of DRACENA; 21°34′S 51°21′W. Coffee growing.

Iraq (i-RAHK), republic (□ 167,924 sq mi/434,924 sq km; 2004 estimated population 25,374,691; 2007 estimated population 27,499,638), SW ASIA; ⊙ BAGHDAD; 33°00′N 44°00′E. Iraq is sometimes spelled Irak.

Geography

Iraq is bordered on the S by KUWAIT, the PERSIAN GULF, and SAUDI ARABIA; on the W by JORDAN and SYRIA; on the N by TURKEY; and on the E by IRAN. Iraq formerly shared a neutral zone with Saudi Arabia that is now divided between the two countries. Iraq's only outlet to the sea is a short stretch of coast on the NW end of the Persian Gulf, including the SHATT AL ARAB waterway. BASRA and UMM QASR are the main ports. Iraq is approximately coextensive with ancient MESOPOTAMIA and occupies the structural depression between the plateaus of Iran and Saudi Arabia, extending from a short c.55-mi/89-km-long frontage at head of the Persian Gulf c.600 mi/966 km NW (maximum width c.400 mi/644 km) to the highlands of KURDISTAN, where it borders on Turkey. With Syria, Jordan, and Saudi Arabia, it shares the great SYRIAN DESERT. Iraq has three physiographic regions: the NE highlands (rising above 10,000 ft/3048 m), the rugged and sparsely wooded home of the Kurds, with some good pastures and fertile peneplanes (e.g., SULAIMANIYA); the vast Syrian Desert, contiguous with the ARABIAN DESERT and inhabited by a small population of nomadic and seminomadic tribes; and the all-important lowland between the mountains and the desert watered by the EUPHRATES and TIGRIS rivers, the lifeblood of the country, which come together in the Shatt al Arab and form a delta with marshes and lakes at the head of the Persian Gulf. After Iraq's defeat in the Gulf War (1991), Saddam Hussein drained many of the marshes in the S area of the country, leaving cracked desert land, and much of the wildlife, including fish and birds, is now gone. Between the two rivers are numerous arms and old canals. The rivers used to cause frequent destructive floods, though they occur less often as a result of flood-control projects undertaken in the last forty years. Iraq has excessively hot summers, while its winters are quite cool, considering the latitude. In Baghdad, the temperature in August frequently rises to 115°F/46°C, while the January mean is barely 40°F/4.5°C. The Kurdish uplands have a rigorous mountain climate with snowfall. Rains occur almost entirely during the winter (November–April). Rainfall increases toward the N and is highest in the mountains. The plains of Upper Iraq receive enough rain to permit dry farming, mainly of barley and wheat as winter crops.

Population

Nearly 80% of the population of Iraq is Arabic-speaking and Muslim (Sunni and Shiite) in religion. There are, among the Arabic-speaking people, twice as many Shiites as Sunnis, the latter sect being more numerous throughout the majority of Arab countries. The hilly uplands of NE Iraq are primarily inhabited by restive Kurds; other minorities include Turks, Armenians, and Assyrians (Nestorian Christians). Most of the country's once large Jewish population emigrated to ISRAEL in the early 1950s.

Economy

Iraq is not self-sufficient in food supply, though it is the world's largest exporter of dates. The oil industry dominates Iraq's economy, accounting for nearly 95% of the country's revenues. Oil is produced mainly by the Iraq Petroleum Company, which was owned by an international group of shareholders until it was nationalized by the Iraqi government in June 1972. The oil was piped to Turkey, TRIPOLI (LEBANON), BANIYAS (Syria), and the Persian Gulf before lines were closed by the Persian Gulf War. Aside from petroleum, Iraq has a small, diversified industrial sector, including the production of textiles, cement, food products, construction materials, leather goods, and machinery. It also has a large military industry. New industries have been started in electronics products, fertilizers, and refined sugar. Livestock is raised. The Baghdad railroad, long an important means of transport, is declining in importance in favor of travel by road and air. There are international airports at Baghdad and Basra, and a state-owned airline handles domestic and international traffic. However, because of the current war, much of the central economic structure was shutdown. Currently, the rebuilding of oil, electricity, and other production is proceeding with foreign support.

History: to World War I

Iraq is a veritable treasure house of antiquities, and recent archaeological excavations have greatly expanded the knowledge of ancient history. Prior to the Arab conquest in the 7th century, Iraq had been the site of a number of flourishing civilizations, including the Sumerians (who developed one of the earliest known writing systems), the Akkadians, the Babylonians, and the Assyrians. The capital of the Abbasid caliphate was established at Baghdad in the 8th century, and the city became a famous center for learning and the arts. Mesopotamia fell to the Ottoman Turks in the 16th century and passed under direct Ottoman administration in the 19th century, when it came to constitute the three Turkish provinces of Basra, Baghdad, and MOSUL. At this time the area became of great interest to the European powers, especially the Germans, who wanted to extend the Berlin-Baghdad railroad all the way to the port of Kuwait.

History: World War I to End of World War II

In World War I the British invaded Iraq in their war against the OTTOMAN EMPIRE; Britain declared then that it intended to return to Iraq some control of its own affairs. The Treaty of SÈVRES (1920) established Iraq as a mandate of the League of Nations under British administration, and in 1921 the country was made a kingdom headed by Faisal I. The British mandate was terminated in 1932, and Iraq was admitted to the League of Nations. In 1933, the small Christian Assyrian community revolted, culminating in a government military crackdown and heavy loss of life, setting a precedent for internal minority uprisings in Iraq. The first oil concession was granted in 1925, and in 1934 the export of oil began. Late in 1936, the country experienced the first of seven military coups that were to take place in the next five years. In April 1941, Rashid Ali al-Gaylani, leader of an anti-British and pro-Axis military group, seized power and ousted Emir Abd al-Ilah, the pro-British regent for the child king, Faisal II. The British reinforced their garrisons by landing troops at Basra, and in May, al-Gaylani, with some German and Italian support, opened hostilities. He was utterly defeated, and Emir Abd al-Ilah was recalled.

History: End of World War II to 1966

Iraq, with other members of the Arab League, participated in 1948 in the unsuccessful war against Israel. In 1955, under the leadership of Prime Minister Nuri-al Said, Iraq joined the Baghdad Pact council with Turkey, Pakistan, Iran, and Britain, the U.S. being an associate member. In February 1958, following announcement of the merger of Syria and Egypt into the UNITED ARAB REPUBLIC, Iraq and Jordan announced the federation of their countries into the Arab Union. In a swift coup d'etat in 1958, the army led by General Abd al-Karim Kassem seized control of Baghdad, withdrew from the Baghdad Pact, and proclaimed a republic, with Islam declared the national religion. King Faisal, Crown Prince Abd al-Ilah, and Nuri al-Said were killed, and the Arab Union was dissolved. In 1962 the chronic Kurdish problem flared up when tribes led by Mustafa al-Barzani revolted, demanded an autonomous KURDISTAN, and gained control of much of N Iraq. An agreement was reached in March 1970. In 1974, following Iraq's rejection of the Kurdish bid for autonomy, fighting erupted between the Kurds and Iraqi troops. In 1963, Colonel Abd al-Salam Aref led a coup that overthrew the Kassem regime. The new regime was dominated by members of the Iraqi Ba'ath party, a socialist group whose overall goal was Arab unity.

History: 1966 to 1979

In 1966, the president and two cabinet members died in a helicopter crash. The president's brother, General Abd al-Rahman Aref, assumed office; he was overthrown by a bloodless coup in 1968. Major General Ahmad Hasan al-Bakr of the Ba'ath party became president. Relations with the USSR improved, and in April 1972, a fifteen-year friendship treaty was signed. The Communist party in Iraq was legalized in 1972. Iraq took an active part in the 1973 Arab-Israeli War; it also participated in the oil boycott against nations supporting Israel. In 1975, the Kurds once again fought for their independence in N Iraq, but suffered heavily when Iran withdrew support. As the Islamic Revolution in neighboring Iran grew in the late 1970s, Iraqi leaders recognized its threat.

History: 1979 to 1990

In 1979, President Bakr resigned, and Saddam Hussein Takriti assumed control of the government. War between Iran and Iraq, primarily over the Shatt al Arab waterway, erupted full-scale in September 1980. The eight-year war became a series of mutual attacks and stalemates, as both countries' oil production fell drastically, the death toll rose, and great mutual destruction was inflicted. Poison gas was reportedly used by Iraq against Iran and on Kurdish villages as the Kurdish rebellion continued. Eventually, a cease-fire under the auspices of the UN led to the war's end in 1988. Throughout 1989 and into 1990, Hussein's repressive policies and continued arms buildup caused international criticism, particularly in the U.S., which had favored Iraq during the war with Iran. On August 2, 1990, some 120,000 Iraqi troops invaded Kuwait, and Hussein declared its annexation. The UN, led by intense U.S. pressure, established international trade sanctions against Iraq. Nevertheless, Hussein desisted from recalling his troops. U.S.-led coalition forces began air attacks on Iraq on January 16, 1991, which led to a ground invasion to retake Kuwait. During this time, Iraq launched Scud missiles against both Israel and Saudi Arabia. Iraqi forces quickly succumbed to coalition troops and were forced out of Kuwait. Hussein remained in power. The Kurdish rebellion continued despite heavy-handed Iraqi military attacks. Trade embargoes against Iraq remained in force, and the UN sent teams to investigate Iraq's existing military facilities. To restrict the movement of Iraqi military aircraft, two "No Fly Zones" were established in the N and S areas of Iraq. The S Zone (below 33°N) was patrolled by U.S. planes operating from Saudi Arabian airbases or from U.S. aircraft carriers. The N Zone (above 36°N) was patrolled by U.S. and coalition planes operating from NATO bases in Turkey. The war left huge amounts of wreckage in the country's major cities and ports, and created hundreds of thousands of Iraqi refugees.

History: 1990 to Present

During the 1990s, the Hussein government alternately thwarted and allowed UN weapons inspections, until those teams were withdrawn due to Iraqi non-cooperation on December 16, 1998, followed by four days of air strikes against weapons facilities by U.S. and British aircraft. Sporadic challenges of the "No Fly Zones" over the following years resulted in bombing by the U.S. and Britain. After the terrorist attacks in the U.S. on September 11, 2001, U.S. President Bush declared Iraq to be part of an "Axis of Evil" supporting terrorism and began a diplomatic process in the UN toward forcibly disarming the Hussein government. Though weapons inspections resumed in November 2002, the U.S. said they were ineffective and a coalition led by the U.S. and Britain launched an invasion of Iraq to overthrow the Hussein government on March 20, 2003. Coalition troops captured Baghdad, effectively toppling the Hussein government in April 2003. Amid continuing attacks by rebel insurgents, the Coalition Authority turned over sovereignty to an interim Iraqi government headed by Ayad Allawi as prime minister on June 28, 2004. On April 7, 2005, Iraq's newly elected lawmakers (January 30, 2005) named their first prime minister of the post-Saddam Hussein era, Ibrahim al-Jaafari. Hussein was put on trial for crimes against humanity in October 2005; he was convicted and sentenced to death a year later. The December 2005 National Assembly elections gave a near majority of the seats to the Shiite religious parties, but Kurdish and Sunni objections to Jaafari delayed a new government until April 2006, when Nuri Kamal al-Maliki became prime minister. By late 2006 increased sectarian violence, including some between rival Shiite militias, raised the specter of civil war. In December 2006, Saddam Hussein was hanged for crimes against humanity. In 2007 U.S. forces in Iraq were increased in an effort to bring security to Baghdad and end sectarian violence. Due to the fighting, some two million Iraqis were estimated to have fled to Jordan, Syria, and other nations by mid-2007.

Government

Iraq is a parliamentary democracy governed under a constitution that was ratified in 2005. The president, who is head of state, is elected by the Council of Representatives. The government is headed by the prime minister. The bicameral legislature consists of the 275-seat Council of Representatives, whose members are elected by proportional representation, and a Federation Council, whose membership had not been defined as of late 2007. The current head of state is President Jalal Talabani (since 2005); the current head of government is Prime Minister Nuri al-Maliki (since 2006). The country is divided into eighteen provinces, or governates: ANBAR, BABYLON, BAGHDAD, BASRA, DAHUK, DIYALA, ERBIL, KERBELA, MAYSAN, MUTHANNA, NAJAF, NINEVEH, QADISSIYA, SALAH ED-DIN, SULAIMANIYA, TA'MEEM, THI-GAR, and WASIT.

Iraquara (EE-rah-KWAH-rah), city (2007 population 22,601), central BAHIA, BRAZIL, 19 mi/31 km NNW of PALMEIRAS; 12°15′S 41°36′W.

Irará (ee-rah-RAH), city (2007 population 25,636), E BAHIA, BRAZIL, 5 mi/8 km WNW of ALAGOINHAS; 12°05′S 38°45′W. Tobacco, coffee, oranges, cotton.

Irasburg, town, ORLEANS CO., N VERMONT, on BLACK RIVER, 10 mi/16 km SSW of Newport; 44°49′N 72°16′W. Dairying. Settled 1798 on Ira Allen's 1781 grant.

Irati (EE-rah-chee), city (2007 population 54,141), S PARANÁ, BRAZIL, on railroad, 50 mi/80 km SW of PONTA GROSSA; 25°27′S 50°39′W. Sawmilling center; maté processing, grain milling. Formerly Iraty.

Irauçuba (EE-rou-SOO-bah), city (2007 population 21,858), N central CEARÁ, BRAZIL, 20 mi/32 km E of PATOS; 03°35′S 39°45′W.

Irawadi, MYANMAR: see AYEYARWADY, river.

Irazú (ee-rah-SOO), active volcano (c.11,260 ft/3,432 m), central COSTA RICA. It erupted in 1723, destroying CARTAGO, and in 1963, covering SAN JOSÉ with ash. Vapor constantly rises from one of its craters. From the summit, the PACIFIC OCEAN, the CARIBBEAN SEA, and LAKE NICARAGUA can be seen. Included within Irazú Volcano National Park (□ 5,900 acres/2,388 ha), which offers spectacular views from the crater rim, which is accessible by paved road via Cartago.

Irbe Strait (IR-bai), joins BALTIC SEA and GULF OF RIGA, separating CAPE KOLKA from SAAREMAA island (ESTONIA); 17 mi/27 km wide. Receives small Irbe River of KURZEME (LATVIA), on S shore.

Irbeyskoye (eer-BYAI-skuh-ye), village (2005 population 4,660), SE KRASNOYARSK TERRITORY, SE SIBERIA, RUSSIA, on the KAN RIVER, on road and railroad, 40 mi/64 km S of KANSK; 55°38′N 95°27′E. Elevation 859 ft/261 m. In agricultural area.

Irbid (ER-bid), city (2004 population 250,645), N JORDAN, 40 mi/64 km N of AMMAN; 32°33′N 35°51′E. Elevation 1,920 ft/585 m. Road and agricultural center;

grain (wheat, barley), olives. Jordan's third-largest city, it serves as the main administrative and commercial center in the N. Important university. Some industry. Husn (1996 estimated population 22,000) and Irbid (1996 estimated population 30,000) refugee camps located nearby.

Irbil, IRAQ: see ERBIL, town.

Irbit (eer-BEET), city (2006 population 41,100), SE central SVERDLOVSK oblast, E URALS, W Siberian Russia, on the NITSA RIVER (landing), at the mouth of the Irbit River (right tributary), in the OB' RIVER basin, on crossroads and railroad, 125 mi/201 km NE of YEKATERINBURG; 57°40′N 63°03′E. Elevation 242 ft/73 m. Manufacturing center (agricultural machinery, motorcycles, auto trailers, industrial glass, pharmaceuticals, building materials, furniture, garments); flour milling, distilling, meat preserving, dairying. Has old churches. Founded in 1643 as Irbitskaya Sloboda; chartered in 1776. Developed as timber-, livestock-, and grain-trading center. Prior to the 1930s, site of a noted annual fair. Heavy industrial pollution of the city's water supplies is among the main reasons for recent population decline.

Irbitskaya Sloboda, RUSSIA: see IRBIT.

Irbitskiye Vershiny, RUSSIA: see ALTYNAY.

Irbitskiy Zavod, RUSSIA: see KRASNOGVARDEYSKIY, SVERDLOVSK oblast.

Ircalaya (eer-kah-LAH-yah), canton, MÉNDEZ province, TARIJA department, S central BOLIVIA, near EL PUENTE; 21°12′S 65°17′W. Elevation 6,555 ft/1,998 m. Lead, silver, and zinc mining; copper mining in area; clay and limestone deposits. Agriculture (potatoes, yucca, bananas, corn, wheat, barley, sweet potatoes); cattle.

Irchester (UH-ches-tuh), village (2001 population 4,807), E NORTHAMPTONSHIRE, central ENGLAND, 2 mi/3.2 km SE of WELLINGBOROUGH; 52°16′N 00°38′W. Leather and shoe manufacturing. Has church dating from 13th century. Site of Saxon and Roman camps. Just NW, on NENE RIVER, is former ironstone-quarrying town of LITTLE IRCHESTER.

Irdyn' (eer-DIN), town (2004 population 5,753), E central CHERKASY oblast, UKRAINE, on road, 17 mi/28 km W of CHERKASY; 49°23′N 31°44′E. Elevation 374 ft/113 m. Peat-extraction and peat-bricketing works. Established in 1930, town since 1941.

Irebu (ee-RE-boo), village, ÉQUATEUR province, W CONGO, on left bank of CONGO RIVER opposite the influx of the UBANGI, at mouth of IREBU channel (outlet of LAKE TUMBA), 65 mi/105 km SSW of MBANDAKA; 00°37′S 17°45′E. Elev. 967 ft/294 m. Training center for troops; steamboat landing. Has Roman Catholic mission.

Irecê (ee-re-SE), city (2007 population 62,672), N central BAHIA, BRAZIL, 68 mi/109 km ESE of XIQUE-XIQUE; 11°18′S 41°15′W.

Irechek (EER-ee-chek), peak (9,357 ft/ 2,852 m) in the RILA MOUNTAINS, SW BULGARIA; 42°11′N 23°35′E.

Iredell (EIR-del), county (□ 597 sq mi/1,552.2 sq km; 2006 population 146,206), W central NORTH CAROLINA, ⊙ STATESVILLE; 35°48′N 80°52′W. In PIEDMONT region; bounded SW by CATAWBA RIVER (forms Lake NORMAN reservoir in SW corner); drained by South Fork of Yadkin (PEE DEE) River. Manufacturing (rubber, plastic, and metal products) at Statesboro, MOORESVILLE, TROUTMAN; timber; service industries; agriculture (tobacco, corn, wheat, hay, barley, soybeans). Duke Power State Park in SW on Lake Norman. Fort Dobbs State Historical Site in center. Formed 1788 from Rowan County. Named for James Iredell (1751–1799), Attorney General of state during the Revolution and delegate to the Constitutional Convention of 1788.

Iredell, village (2006 population 387), BOSQUE county, central TEXAS, on BOSQUE RIVER, c.50 mi/80 km NW of WACO; 31°59′N 97°52′W. In farm area. Manufacturing (apparel).

Ireland, village, DUBOIS county, SW INDIANA, 4 mi/6.4 km NW of JASPER. Manufacturing (luggage). Wheat, corn, soybeans; cattle, hogs. Flat land.

Ireland, Gaelic *Éire* (to it are related the poetic *Erin* and, perhaps, the Latin *Hibernia*), island (□ 32,598 sq mi/84,754.8 sq km), second-largest of the BRITISH ISLES. It lies W of the island of GREAT BRITAIN, from which it is separated by the narrow NORTH CHANNEL, the IRISH SEA (which attains a width of 130 mi/209 km), and ST. GEORGE'S CHANNEL. Slightly more than one-third the size of Britain, the island averages 225 mi/362 km in length and 140 mi/225 km in width. A large central plain extending to the Irish Sea between the MOURNE MOUNTAINS in the N and the Wicklow Mountains in the S is roughly enclosed by a highland rim. The highlands of the N, W, and S, which rise to more than 3,000 ft/914 m, are generally barren, but the central plain is fairly fertile and the climate is temperate and moist, warmed by SW winds. The rains, which are heaviest in the W (some areas have more than 80 in/203 cm annually), are responsible for the brilliant green grass of the "Emerald Isle," and for the large stretches of peat bog, a source of fuel. The coastline is irregular, affording many natural harbors. Off the W coast are numerous small islands, including the ARAN ISLANDS, the BLASKET ISLANDS, ACHILL ISLAND, and CLARE ISLAND. The interior is dotted with lakes (the most celebrated are the LAKES OF KILLARNEY) and wide stretches of river called loughs. The SHANNON RIVER, the longest Irish river, drains the W plain and widens into the beautiful loughs ALLEN, REE, and DERG. The LIFFEY RIVER empties into DUBLIN BAY, the LEE RIVER into CORK HARBOUR at CÓBH, the FOYLE RIVER into LOUGH FOYLE near LONDONDERRY, and the LAGAN RIVER into BELFAST LOUGH. The island is divided into two political units—NORTHERN IRELAND, which is joined with Great Britain in the United Kingdom, and the REPUBLIC OF IRELAND. Of the thirty-two counties of Ireland, twenty-six lie in the Republic, and of the four historic provinces (ULSTER, MUNSTER, CONNACHT, and LEINSTER), three and part of the fourth (Ulster) are in the Republic, while six counties and the rest of Ulster make up Northern Ireland. The earliest known people in Ireland belonged to the groups that inhabited all of the British Isles in prehistoric times. In the several centuries preceding the birth of Christ, a number of Celtic tribes invaded and conquered Ireland and established their distinctive culture, although they do not seem to have come in great numbers. Ancient Irish legend tells of four successive peoples who invaded the country—the Firbolgs, the Fomors, the Tuatha de Danann, and the Milesians. Oddly enough, the Romans, who occupied Britain for 400 years, never came to Ireland, and the Anglo-Saxon invaders of Britain, who largely replaced the Celtic population there, did not greatly affect Ireland. Until the raids of the Norsemen in the late 8th century, Ireland remained relatively untouched by foreign incursions and enjoyed the golden age of its culture. The people, Celtic and non-Celtic alike, were organized into clans, or tribes, which in the early period owed allegiance to one of five provincial kings—of Ulster, Munster, Connacht, Leinster, and MEATH (now the N part of Leinster). These kings nominally served the high king of all Ireland at TARA (in Meath). The clans fought constantly among themselves, but despite civil strife, literature and art were held in high respect. Each chief or king kept an official poet (Druid) who preserved the oral traditions of the people. The Gaelic language and culture were extended into SCOTLAND by Irish emigrants in the 5th and 6th centuries. Parts of Ireland had already been Christianized before the arrival of St. Patrick in the 5th century, but pagan tradition continued to appeal to the imagination of Irish poets even after the complete conversion of the country. The Celtic Christianity of Ireland produced many scholars and missionaries who traveled to ENGLAND and the European Continent, and the religion attracted students to Irish monasteries, perhaps the most brilliant in Europe until the 8th century. St. Columba and St. Columban were among the most famous of Ireland's missionaries. All the arts flourished; Irish-illuminated manuscripts were particularly noteworthy, and the Book of Kells is especially famous. The country did not develop a strong central government, however, and it was not united to meet the invasions of the Norsemen who settled on the shores of the island late in the 8th century, establishing trading towns (including DUBLIN, WATERFORD, and LIMERICK) and creating new petty kingdoms. Brian Boru, who had become high king by conquest in 1002, broke the strength of the Norse invaders at Clontarf in 1014. A period of 150 years followed during which Ireland was free from foreign interference but was torn by clan warfare. In the 12th century, Pope Adrian IV granted overlordship of Ireland to Henry II of England. Richard de Clare, second earl of Pembroke (known as Strongbow), began the English conquest of Ireland; he intervened in behalf of a claimant to the throne of Leinster. Henry himself went to Ireland in 1171, establishing his overlordship there temporarily. With this invasion commenced an Anglo-Irish struggle that continued for nearly 800 years. The English established themselves in Dublin. Roughly a century of warfare ensued as Ireland was divided into English shires ruled from Dublin (the domains of feudal magnates who acknowledged English sovereignty) and the independent Irish kingdoms. Many English people intermarried with the Irish and were assimilated into Irish society. The English introduced a parliament in Ireland in the late 13th century. Edward Bruce of Scotland invaded Ireland in 1315; many Irish kings joined him. Although Bruce was killed in 1318, the English authority in Ireland was weakening, becoming limited to a small district around Dublin known as the Pale; the rest of the country fell into a struggle for power among the ruling Anglo-Irish families and Irish chieftains. English attention was diverted by the Hundred Years War with France (1337–1453) and the Wars of the Roses (1455–1485). However, under Henry VII new interest in the island was aroused by Irish support for Lambert Simnel, a Yorkist pretender to the English throne. To crush this support, Henry sent to Ireland Sir Edward Poynings, who summoned an Irish Parliament at DROGHEDA and forced it to pass the legislation known as Poynings' Law (1495). These acts provided that future Irish Parliaments and legislation receive prior approval from the English Privy Council, thus rendering a free Irish Parliament impossible. The English Reformation under Henry VIII gave rise in England to increased fears of foreign, Catholic invasion; control of Ireland thus became even more imperative. Henry VIII put down a rebellion (1534–1537), abolished the monasteries, confiscated lands, and established a Protestant "Church of Ireland" (1537). But since the vast majority of Irish remained Roman Catholic, the seeds of bitter religious contention were added to the already rancorous Anglo-Irish relations. The Irish rebelled three times during the reign of Elizabeth I and were brutally suppressed. Under James I, Ulster was settled by Scottish and English Protestants, and many of the Catholic inhabitants were driven off their lands; thus two sharply antagonistic communities were established. Another Irish rebellion, begun in 1641 in reaction to the hated rule of Charles I's deputy, Thomas Wentworth, first earl of Strafford, was crushed (1649–1650) by Oliver Cromwell with the loss of hundreds of thousands of lives. More land was confiscated (and often given to absentee landlords), and more Protestants settled in Ireland. The intractable landlord-tenant problem that plagued Ireland in later centuries can be traced to the English confiscations of the 16th

and 17th centuries. Irish Catholics rallied to the cause of James II after his overthrow (1688) in England, while the Protestants in Ulster enthusiastically supported William III. In 1690, William defeated James and his French allies at the battle of the Boyne. The English-controlled Irish Parliament passed harsh penal laws designed to keep the Catholic Irish powerless; the parliament also denied political equality to Presbyterians. At the same time, English trade policy depressed the economy of Protestant Ireland, causing many so-called Scots-Irish to emigrate to America. A newly flourishing woolen industry was destroyed when export from Ireland was forbidden. During the American Revolution, fear of a French invasion of Ireland led Irish Protestants to form (1778–1782) the Protestant Volunteer Army. The Protestants, led by Henry Grattan, and even supported by some Catholics, used their military strength to extract concessions for Ireland from Britain. Trade concessions were granted in 1779, and, with the repeal of Poynings' Law (1782), the Irish Parliament had its independence restored. But the Parliament was still chosen undemocratically and Catholics continued to be denied the right to hold political office. Wolfe Tone, a Protestant who had formed the Society of United Irishmen and who accepted French aid in the uprising, staged another unsuccessful rebellion in 1798. The reliance on French assistance revived anti-Catholic feeling among the Irish Protestants, who remembered French support of the Jacobite restoration. The rebellion convinced the British prime minister, William Pitt, that the Irish problem could be solved by the adoption of three policies: abolition of the Irish Parliament, legislative union with Britain in a United Kingdom of Great Britain and Ireland, and Catholic emancipation. The first two goals were achieved in 1800, but the opposition of George III and British Protestants prevented the enactment of the Catholic Emancipation Act until 1829. The Irish representatives in the British Parliament attempted to maintain the Irish question as a major issue in British politics after 1829. Daniel O'Connell worked to repeal the union with Britain, which was felt to operate to Ireland's disadvantage, and to reform the government in Ireland. Towards the middle of the century, the Irish Land Question grew increasingly urgent. But the Great Potato Famine (1845–1849), one of the worst natural disasters in history, dwarfed political developments. During these years a blight ruined the potato crop, the staple food of the Irish population, and hundreds of thousands perished from hunger and disease. Many thousands of others emigrated; about 1.6 million people went to the U.S. between 1847 and 1854. The population dropped from an estimated 8.5 million in 1845 to 6.5 million in 1851. Irish emigrants in America formed the secret Fenian Movement, dedicated to Irish independence. In 1869, British Prime Minister William Gladstone sponsored an act disestablishing the Protestant "Church of Ireland" and thereby removed one Irish grievance. In the 1870s, Irish politicians renewed efforts to achieve home rule within the union, while in Britain, Gladstone and others attempted to solve the Irish problem through land legislation and home rule. Gladstone twice submitted home rule bills (1886 and 1893) that failed. The proposals alarmed Protestant Ulster, which began to organize against home rule. In 1905, Arthur Griffith founded Sinn Féin among Irish Catholics, but for the time being the dominant Irish nationalist group was the Home Rule Party of John Redmond. Home rule was finally enacted in 1914, with the provision that Ulster could remain in the union for six more years, but the act was suspended for the duration of World War I and never went into effect. Volunteer military groups were formed in both Ulster and Catholic Ireland. The Irish Republican Brotherhood, a descendent of the Fenians, organized a rebellion on Easter Sun-

day, 1916; although unsuccessful, the rising acquired great propaganda value when the British executed its leaders. Sinn Fein, linked in the Irish public's mind with the rising and aided by Britain's attempt to apply conscription to Ireland, scored a tremendous victory in the parliamentary elections of 1918. Its members refused to take their seats in Westminster, declared themselves the Dáil Éireann (Irish Assembly), and proclaimed an Irish Republic. The British outlawed both Sinn Féin and the Dáil, which went underground and engaged in guerrilla warfare (1919–1921) against local Irish authorities representing the union. The British sent troops, the Black and Tans, who inflamed the situation further. A new home rule bill was enacted in 1920, establishing separate parliaments for Ulster and Catholic Ireland. Ulster accepted this bill, thus creating Northern Ireland. The Dáil rejected the plan, but in autumn 1921, Prime Minister Lloyd George negotiated a treaty with Griffith and Michael Collins of the Dáil, granting Dominion status within the BRITISH EMPIRE to Catholic Ireland. The Irish Free State was established in January 1922. A new constitution was ratified in 1937 that terminated Great Britain's sovereignty. In 1948, all semblance of Commonwealth membership ended with the Republic of Ireland Act. (Also see IRELAND, REPUBLIC OF and IRELAND, NORTHERN).

Ireland Island, northwesternmost part of BERMUDA, at entrance to GREAT SOUND, W of HAMILTON; 1.5 mi/2.4 km long, 0.25 mi/0.4 km wide; 32°19′N 64°50′W. A narrow channel divides it into two parts. Has British naval base.

Ireland, Northern: see NORTHERN IRELAND.

Ireland, Republic of, country (□ 27,136 sq mi/70,553.6 sq km; 2007 population 4,109,086); ⊙ DUBLIN. (For physical geography, see IRELAND.) The country was known as the Irish Free State from 1922 to 1937, and as Éire from 1937 to 1949. Besides Dublin, other urban areas are LIMERICK, CORK, DÚN LAOGHAIRE, WATERFORD, GALWAY, and DUNDALK.

Economy
In the 1990s, Ireland had the fastest growing economy within the EU, largely based on the information technology and pharmaceutical industries. It provides a major share of Europe's business applications and PC software and most of Europe's teleservicing. Agriculture engages about 70% of the land, 8% of the work force, 5% of GNP, and 20% of total exports. The raising of dairy and beef cattle, sheep, pigs, and poultry is the chief agricultural enterprise. Among the leading crops are oats, wheat, turnips, potatoes, sugar beets, and barley. Of the 2003 GDP ($116.2 billion), 46% was accounted for by industry (including construction) and manufacturing, 5% by agriculture, and 49% by service (2002 estimate). The republic's industries produce such items as linen and laces (for which Ireland is famous), Waterford crystal, food products, textiles, ships, iron products, pharmaceuticals, software, and handicrafts. Dublin and Cork are the main ports. A new highway infrastructure has been constructed. Around the customs-free SHANNON AIRPORT are factories producing electronic equipment, chemicals, plastics, and textiles. The NAVAN lead-zinc mines, the largest in Europe, opened in 1977. Oil and natural gas are also produced offshore.

History: to World War II
After the establishment of the Irish Free State by treaty with GREAT BRITAIN (January 1922), civil war broke out between supporters of the treaty and opponents, who refused to accept the partition of Ireland and the retention of any ties with Britain. The antitreaty forces, embodied in the Irish Republican Army (IRA) and led by Eamon de Valera, were defeated, although the IRA continued as a secret terrorist organization. William Cosgrave became the first prime minister. De Valera and his followers, the Fianna Fáil Party, agreed to take the oath of allegiance to the British crown and

entered the Dáil Éireann in 1927. De Valera became prime minister in 1932, and, under his administration, a new constitution was promulgated (1937), establishing the sovereign nation of Ireland, Éire, within the British Commonwealth of Nations. De Valera's policies aimed at the political and economic independence and union of all of Ireland. The loyalty oath to the British crown was abolished, and certain economic provisions of the 1921 treaty with England were repudiated, leading to an "economic war" (1932–1938) with Britain.

History: World War II to 1973
During World War II, Éire remained neutral and vigorously protested Allied military activity in Northern Ireland. Ireland denied the British access to their ports, and allowed German and Japanese agents to operate in the country. However, great numbers of Irishmen volunteered to serve with the British armed forces. The people of Éire suffered relatively little hardship during the war and even profited from increased food exports. The postwar period brought a sharp rise in the cost of living and a decline in population, due in great part to steady emigration to NORTHERN IRELAND, Great Britain, and other countries. In 1948, Prime Minister John A. Costello demanded total independence from Great Britain and reunification with the six counties of Northern Ireland. The Republic of Ireland was proclaimed on April 18, 1949. The country withdrew from the Commonwealth and formally claimed jurisdiction over the ULSTER counties. Ireland was admitted to the UN in 1955. Nothing came of the claim to Ulster, and the republic and Northern Ireland improved their economic relations during the 1950s and 1960s. But the problem of Northern Ireland flared up again in the late 1960s with bitter fighting between the Protestant majority and Catholic minority there, providing an opportunity for the IRA to emerge again.

History: 1973 to Present
Erskine H. Childers succeeded de Valera as president of Ireland in 1973, and Liam Cosgrave, at the head of a Fine Gael–Labour coalition, replaced Jack Lynch, a member of Fianna Fáil, as prime minister. The republic joined the EEC the same year. Childers died in late 1974 and was succeeded by Cearbhall O. Dalaigh. Lynch led Fianna Fáil back into office in 1977; in 1979 fellow party member Charles Haughey replaced Lynch as prime minister. By 1981 a Fine Gael–Labour coalition headed by Garret FitzGerald took power on an economic platform. Although ousted in 1982, the coalition was back in power six months later. Throughout the 1980s the republic's political situation was more fluid than it had been; there were several general elections and a variety of party schisms. In 1987 Haughey again became prime minister. Ireland elected its first female president, Mary Robinson, in early 1991. In 1992, Albert REYNOLDS, of Fianna Fáil, replaced Haughey. The Reynolds government fell in 1994, and Fine Gael leader John BRUTON succeeded him, heading a Fine Gael–Labour coalition. In 1997 Bertie Ahern became prime minister, heading a Fianna Fáil–Progressive Democrat coalition. The next year Irish and Northern Irish voters endorsed the Belfast (or Good Friday) Agreement, a peace settlement for Northern Ireland; by May 2007, the sometime fitful peace process seemed finally to have led to stable home rule in the North. Ahern's coalition was returned to power in 2002 and 2007, with the addition of the Green party in 2007.

Government
The republic is governed under the constitution of 1937. The president, who is the head of state, is popularly elected to a seven-year term and is eligible for a second term. The prime minister, who is the head of government, is appointed by the president, as is the cabinet. There is a bicameral Parliament, the Oireachtas. The House of Representatives or Dáil Éireann is the more powerful chamber. Its 166 members are elected by popular vote on the basis of proportional representation. Members of the 60-seat Senate or Seanad Éireann are indirectly elected or appointed. All legislators serve five-year terms. The main political parties are the Fianna Fáil, the Fine Gael, and the Labour Party. Gaelic and English are the official languages, but English is used more. The current president is Mary McAleese (since November 1997). Bertie Ahern has been prime minister since June 1997.The republic is divided into twenty-six counties: MONAGHAN, CAVAN, and DONEGAL (constituting part of the historic province of ULSTER); LOUTH, MEATH, DUBLIN, KILDARE, WICKLOW, CARLOW, WEXFORD, KILKENNY, LAOIGHIS, OFFALY, WESTMEATH, and LONGFORD (comprising LEINSTER); TIPPERARY, WATERFORD, CORK, KERRY, LIMERICK, and CLARE (comprising MUNSTER); and LEITRIM, ROSCOMMON, GALWAY, MAYO, and SLIGO (comprising CONNACHT).

Ireland's Eye, Gaelic *Inis Mac Neasáin*, islet in the IRISH SEA, E DUBLIN county, E central IRELAND, just N of HOWTH; 53°24′N 06°03′W. Has remains of ancient chapel, built on site of 7th-century church of St. Nessan.

Ireland's Eye Island (□ 2 sq mi/5 sq km), SE NEWFOUNDLAND AND LABRADOR, CANADA, on NW side of Trinity Bay, 35 mi/56 km NNW of CARBONEAR; 48°13′N 53°30′W. Fishing.

Iremel' (ee-rye-MYEL'), peak (5,197 ft/1,584 m) in S URAL Mountains, CHELYABINSK oblast and BASHKORTOSTAN, RUSSIA, SW of ZLATOUST; 54°30′N 58°50′E. The BELAYA and AI rivers rise here.

Irene, village (2006 population 406), TURNER, CLAY, and YANKTON counties, SE SOUTH DAKOTA, 20 mi/32 km NE of YANKTON; 43°04′N 97°09′W.

Iretama (EE-re-tah-mah), city (2007 population 11,174), W central PARANÁ state, BRAZIL, 25 mi/40 km SE of CAMPO MOURÃO; 24°17′S 52°02′W. Coffee, corn; livestock.

Ireton, town (2000 population 585), SIOUX county, NW IOWA, 14 mi/23 km W of ORANGE CITY; 42°58′N 96°19′W. In livestock and grain area; feeds; meat processing.

Irgalem, ETHIOPIA: see YIRGA-ALEM.

Irgiz (ir-GIZ), town, SE AKTÖBE region, KAZAKHSTAN, on IRGIZ RIVER, 90 mi/145 km NE of SHALKAR; 48°36′N 61°14′E. On dry steppe. Tertiary-level (raion) administrative center. Dairying. Founded (1847) during Russian conquest of Kazakhstan.

Irgiz River (ir-GIZ), 300 mi/483 km long, E AKTÖBE region, KAZAKHSTAN; rises in the foothills of the S URAL Mountains, c.70 mi/113 km SE of ORSK; flows S, through salt steppe, past KARABUTAK, and SE, past IRGIZ, to lake SHALKAR TENGIZ.

Irgiz River (eer-GEEZ), two left affluents of the lower VOLGA RIVER, in SAMARA and SARATOV oblasts, SE European Russia. The Greater IRGIZ River (Russian *Bol'shoy Irgiz*) rises in the OBSHCHIY SYRT (SAMARA oblast) 50 mi/80 km SSW of BUZULUK, flows 355 mi/571 km generally WSW, past PUGACHEV (Saratov oblast) and SULAK, to the Volga River opposite VOLSK. Navigable at high water in the lower course. The LESSER IRGIZ River (Russian *Malyy Irgiz*) flows 110 mi/177 km intermittently WSW, parallel to and N of the Greater Irgiz, past IVANTEYEVKA (Saratov oblast) to an arm of the Volga near ALEKSEYEVKA.

Irherm, MOROCCO: see IGHERM.

Irhir, MOROCCO: see OUARZAZATE.

Iri, SOUTH KOREA: see IKSAN.

Irian Jaya (EE-ryahn JAH-yah), province (162,927 sq mi/421,981 sq km; 1990 population 1,641,430), INDONESIA, comprising the W half of the island of NEW GUINEA and about twelve offshore islands; ☉ JAYAPURA; 05°00′S 138°00′E. A rugged, densely forested region, with mountains rising to over 16,535 ft/5,029 m, it is inhabited chiefly by Papuans living in hundreds of tribes, each with its own language and customs. Many of the tribes have been influenced by aggressive activity by Christian missionaries. The coastal lowlands are swampy and cut by many rivers, including the DIGUL RIVER and the MAMBERAMO RIVER, Indonesia's longest. Subsistence farming is carried on (some of the highland tribes terrace and cultivate the mountains at 45-degree angles); taro, bananas, sugarcane, and sweet potatoes are the principal crops. Wild game is trapped, and there is fishing along the coast and the rivers. Copper in W mountains (Vogelkop) inland from SORONG, as well as in South Central, the world's largest copper mine, near Tembagapura; magnetite has been found in the Star Mountains, a region unexplored until 1959. Also major gold and silver deposits.There are deposits of oil in the W and nickel and cobalt on WAIGEO island. The Dutch first visited the W coast of the island in 1606; they extended their rule along the coastal areas in the eighteenth century and claimed possession of the coast W of the 141st meridian (1828) and of the N coast W of Humboldt Bay (1848), now YOS SUDARSO BAY. The Dutch claim to the W half of the island was recognized by GREAT BRITAIN and GERMANY in treaties of 1885 and 1895. In World War II, the N coastal areas and offshore islands were occupied (1942) by the Japanese but retaken (1944) by the Allies, after which Hollandia (now JAYAPURA) became a staging base for operations in the PHILIPPINES. Following Indonesian independence (1949), the Dutch retained control of West New Guinea. Years of dispute over the territory culminated in the landing (early 1962) of Indonesian guerrillas and paratroopers here. The conflict ended in late 1962 when the NETHERLANDS agreed to UN administration of West New Guinea and, after May 1, 1963, transfer of the territory to Indonesian control pending a plebiscite, which was held under UN supervision in August 1969. Tribal leaders, voting as representatives of their people, chose to remain under Indonesian rule, and Indonesia then formally annexed the territory. The province was officially renamed Irian Jaya in 1973. Nearly 60,000 families were displaced to Irian Jaya by the Indonesian government as part of its voluntary resettlement program initiated forty years ago. The province has since been targeted for increased settlement, but the policy is garnering protest from environmentalists concerned with overexploitation of timber and from those concerned about the alienation of the land of indigenous tribes living in the jungle. Formerly Netherlands (or Dutch) New Guinea. Also called Irian Barat, W Irian, or W New Guinea. See also NEW GUINEA.

Iriba (ee-ree-BAH), town, WADI FIRA administrative region, E CHAD, 100 mi/161 km NE of ABÉCHÉ; 15°07′N 22°15′E.

Iridere, Bulgaria: see ARDINO.

Irig (EER-eeg), town (2002 population 12,329), SREM district, in the SREM region, VOJVODINA, SERBIA, 11 mi/18 km S of NOVI SAD, at S foot of FRUŠKA GORA; 45°06′N 19°51′E.

Iriga (ee-ree-GAH), city (□ 46 sq mi/119.6 sq km; 2000 population 88,893), CAMARINES SUR province, SE LUZON, PHILIPPINES, on railroad, 29 mi/47 km NW of LEGASPI; 13°25′N 123°25′E. Agricultural center (rice, corn). Indigenous people are the Bicolano.

Irigny (ee-ree-nyee), town (□ 3 sq mi/7.8 sq km), RHÔNE department, RHÔNE-ALPES region, E central FRANCE, on right bank of the RHÔNE, 6 mi/9.7 km S of LYON; 45°40′N 04°49′E. Metalworking and meat packing.

Irigoyen, ARGENTINA: see BERNARDO DE IRIGOYEN.

Iriha, SYRIA: see ERIHA.

Irihirose (ee-REE-HEE-ro-se), village, North Uonuma county, NIIGATA prefecture, central HONSHU, N central JAPAN, 37 mi/60 km S of NIIGATA; 37°21′N 139°04′E.

Iriki (ee-REE-kee), town, Satsuma district, KAGOSHIMA prefecture, SW KYUSHU, SW JAPAN, 16 mi/25 km N of KAGOSHIMA; 31°48′N 130°25′E. Known for historic

---

Cross-references are shown in SMALL CAPITALS. The pronunciation guide is shown on page xix. The sources of population figures are shown on page xvii.

late-16th century streetscape, with samurai residences (*buke yashiki*).

Iriklinskiy (ee-ree-KLEEN-skeeye), town, NE OREN-BURG oblast, extreme SE European Russia, in the SE foothills of the S URALS, at a dam for the IR-IKLINSKOYE RESERVOIR on the URAL River, on road and railroad spur, 30 mi/48 km N of ORSK; 51°40′N 58°38′E. Elevation 849 ft/258 m. In agricultural area.

Iriklinskoye Reservoir (ee-REE-kleen-skuh-ye), Russian *Iriklinskoye Vodokhranilishche*, artificial lake (surface □ 97 sq mi/250 sq km) on the upper course of the URAL River, NE ORENBURG oblast, in the SE URALS, extreme SE European Russia. Completed in 1958 and used primarily for industrial purposes (electricity generation, oil and gas processing, metalworking). Plans to establish a natural reserve on its shores are in the works as of 2006.

Irimbo (ee-REEM-bo), town, MICHOACÁN, central MEXICO, 6 mi/9.7 km E of CIUDAD HIDALGO. Elevation 6,906 ft/2,105 m. Corn; livestock.

Iringa (ee-RING-gah), region (2006 population 1,618,000), S central TANZANIA, ⊙ IRINGA; 07°46′S 35°40′E. Bounded SW by Lake NYASA (NE shoreline forms MALAWI border), and by LIVINGSTONE MOUNTAINS rise above lake. Great RUAHA and KISIGO rivers form N boundary. Tobacco, pyrethrum, corn, wheat; cattle, sheep, goats; timber. Part of RUAHA NATIONAL PARK in NW. Part of former SOUTHERN HIGHLANDS province.

Iringa (ee-RING-gah), town, ⊙ IRINGA region, central TANZANIA, 110 mi/177 km S of DODOMA, on LITTLE RUAHA RIVER Airstrip; highway junction. Agricultural trade center (tobacco, corn, wheat, pyrethrum; cattle, sheep, goats; timber); manufacturing (tobacco processing). ISIMILA Archaelogical Site to S; RUAHA NATIONAL Park to NW.

Irinjalakuda (i-rin-JAH-luh-kuh-duh), city, THRISSUR district, KERALA state, S INDIA, 11 mi/18 km SSW of THRISSUR; 10°20′N 76°14′E. Coir products (rope, mats); betel-nut curing, rice milling. Also spelled Irinjalakkuda.

Iriomote-jima (ee-REE-o-MO-te-JEE-mah), volcanic island (□ 144 sq mi/374.4 sq km) of SAKISHIMA IS-LANDS, in the RYUKYU ISLANDS, OKINAWA prefecture, extreme SW JAPAN, between EAST CHINA (W) and PHILIPPINE (E) seas, 10 mi/16 km W of ISHIGAKI-JIMA; 18 mi/29 km long, 9 mi/14.5 km wide. Mountainous, rising to c.1,670 ft/509 m. Home of endangered Iriomote Mountain cat.

Irion, county (□ 1,051 sq mi/2,732.6 sq km; 2006 population 1,814), W TEXAS; ⊙ MERTZON; 31°18′N 100°58′W. Elevation c.2,000 ft/610 m–2,500 ft/762 m. Broken prairie, drained by Middle Concho River and spring-fed small streams. Ranching region (sheep, Angora goats, cattle), wool and mohair; some agriculture (milo, cotton). Oil and gas. Formed 1889.

Iriona (ee-ree-OE-nah), town, COLÓN department, N HONDURAS; port on CARIBBEAN SEA, in Mosquitia, 55 mi/89 km E of TRUJILLO; 15°57′N 85°11′W. Transit center between Trujillo and Mosquitia region of E Honduras; coconuts; livestock.

Iriondo, ARGENTINA: see CAÑADA DE GÓMEZ.

Iriri, MOROCCO: see OUARZAZATE.

Iri River, Greece: see EVROTÁS RIVER.

Irish Free State: see IRELAND and IRELAND, RE-PUBLIC OF.

Irish Sea, arm of the ATLANTIC OCEAN (□ 40,000 sq mi/104,000 sq km), 130 mi/209 km long and up to c.140 mi/225 km wide, lying between IRELAND and GREAT BRITAIN. It is connected with the Atlantic by the NORTH CHANNEL and (on the S) by ST. GEORGE'S CHANNEL. The REPUBLIC OF IRELAND and NORTHERN IRELAND are on its W shore; SCOTLAND, ENGLAND, and WALES on the E. The principal islands in the sea are the ISLE OF MAN, ANGLESEY, and HOLY ISLAND. The chief ports are DUBLIN, LIVERPOOL, MANCHESTER, FLEETWOOD, and DÚN LAOGHAIRE.

Irishtown (EI-rish-toun), village, NW TASMANIA, SE AUSTRALIA, 110 mi/177 km WNW of LAUNCESTON; 40°54′S 145°08′E. Sheep.

Iriston (ee-ree-STON), former village, now a suburb of BESLAN (2 mi/3.2 km SW of the city center), central NORTH OSSETIAN REPUBLIC, in the N CAUCASUS Mountains, RUSSIA, on railroad, on the TEREK RIVER, 13 mi/21 km NNW of VLADIKAVKAZ; 43°11′N 44°33′E. Elevation 1,673 ft/509 m. Until 1941, called Tulatovo.

Irituia (ee-ree-too-EE-ah), city (2007 population 29,771), E PARÁ, BRAZIL, 90 mi/145 km SE of BELÉM; 01°40′S 47°20′W. Brazil nuts, rubber.

Irka, YEMEN: see IRQA.

Irkleyev, UKRAINE: see IRKLIYIV.

Irkliyevskaya (eer-klee-YEFS-kah-yah), village (2005 population 3,085), central KRASNODAR TERRITORY, S European Russia, on the Beysuzhek River (tributary of the BEYSUG RIVER), on road, 53 mi/85 km NNE of KRASNODAR; 45°51′N 39°39′E. Elevation 150 ft/45 m. In agricultural area (grain, fruit, livestock).

Irkliyiv (ir-KLEE-yif) (Russian *Irkleyev*), village (2004 population 2,687), E CHERKASY oblast, UKRAINE, on the N shore of the KREMENCHUK RESERVOIR, 16 mi/26 km SE of ZOLOTONOSHA; 49°32′N 32°20′E. Elevation 331 ft/100 m. Dairy; sewing.

Irkut River (eer-KOOT), 270 mi/435 km long, SW BURYAT REPUBLIC and SE IRKUTSK oblast, RUSSIA; rises in the EASTERN SAYAN MOUNTAINS, flows NE to the ANGARA RIVER at IRKUTSK. Flows through a narrow canyon at its head, with numerous rapids; whitewater rafting tours are common in this area. In the vicinity of BAYKAL′SK, the river has been severely polluted by waste products from the wood pulp mill.

Irkutsk (eer-KOOTSK), city (2005 population 584,440), ⊙ IRKUTSK oblast, S SIBERIA, RUSSIA, at the confluence of the ANGARA and IRKUT rivers, near the SW outlet from Lake BAYKAL, 3,126 mi/5,031 km E of MOSCOW; 52°16′N 104°20′E. Elevation 1,627 ft/495 m. It is an industrial and cultural center, a port, the site of a hydroelectric dam, and a major stop on the TRANS-SIBERIAN RAILROAD. Manufacturing includes aircraft, automobiles, heavy machinery, machine tools, textiles, chemicals, food products, and metals. Founded as a Cossack fortress in 1652, Irkutsk became the capital of Eastern SIBERIA in 1822. It has been a place of exile since the 18th century. Developed and expanded rapidly with the discovery of gold deposits in the nearby LENA River basin in the late 19th century. It is an educational and research center, with nine vocational schools and thirty-six institutes of higher learning, including universities and research institutes of the Russian Academy of Sciences. Art establishments include drama, musical, and children's theaters, philarmonics, circus, and museums of arts, history, and regional heritage.

Irkutsk (eer-KOOTSK), oblast (□ 296,488 sq mi/770,868.8 sq km; 2004 population 2,795,000) in S Siberian Russia; ⊙ IRKUTSK. Bounded SE by Lake BAY-KAL, SW by the EASTERN SAYAN MOUNTAINS; extends N into the CENTRAL SIBERIAN PLATEAU along the upper LOWER TUNGUSKA RIVER, NE to the PATOM PLATEAU. Drained by upper LENA and ANGARA rivers. Population is almost 80% urban and includes Russians, Buryat-Mongols (in Ust-Ord, Buryat Autonomous Okrug), and Evenki (N and NE). Forests (mainly coniferous) cover 86% of region, and fur-bearing wildlife, such as ermine, sable, squirrel, and muskrat, proliferates, fueling the region's booming fur trade. Steppe and wooded steppe along the TRANS-SIBERIAN RAILROAD (S) contain major industrial centers. Coal (CHEREMKHOVO), salt (USOLYE); other mineral deposits include iron in Angara-Ilim area, gold (BOD-AYBO), mica (MAMA, SLYUDYANKA), and asbestos. Main exports are lumber, gold, furs. Harnessing of the ANGARA River water power has transformed the region into a major industrial area (electrometallurgy, aluminum, woodworking, oil refining, chemicals). Dams

and hydroelectric plants at BRATSK, UST′-IL′IMSK, and Irkutsk. Consists of twenty-two cities, fifty-nine towns, and 380 villages.

Irlam (UH-luhm), town (2001 population 10,209), GREATER MANCHESTER, central ENGLAND, on Manchester Ship Canal, 7 mi/11.3 km WSW of MANCHE-STER; 53°27′N 02°25′W. Some manufacturing. Nearby (SW) town of CADISHEAD, on Manchester Ship Canal, specializes in benzol refining, tar distilling, and paper milling.

Irlande (eer-LAHND), village (□ 43 sq mi/111.8 sq km; 2006 population 988), Chaudière-APPALACHES region, S QUEBEC, E CANADA, 9 mi/15 km from THET-FORD MINES; 46°04′N 71°29′W.

Irma (UHR-muh), village (2001 population 435), E ALBERTA, W CANADA, near SASKATCHEWAN border, 17 mi/27 km WNW of WAINWRIGHT, in WAINWRIGHT No. 61 municipal district; 52°55′N 111°14′W. Dairying; grain, mixed farming; oil; Wainwright Military Base. Established 1912.

Irmak, village and railroad junction, central TURKEY, 30 mi/48 km E of ANKARA.

Irmãos, Serra dos, Brazil: see DOIS IRMÃOS, SERRA DOS.

Irmine, UKRAINE: see TEPLOHIRS′K.

Irminger Sea, part of North ATLANTIC OCEAN, off SE GREENLAND, linked by DENMARK STRAIT with GREENLAND SEA of ARCTIC OCEAN. The Irminger Current, a branch of the North Atlantic Current, flows N and W, past S coast of ICELAND, and joins the East Greenland Current issuing from Denmark Strait.

Irminio River (eer-MEE-nyo), 28 mi/45 km long, SE SICILY; rises on MONTE LAURO, flows SSW, past RA-GUSA, to MEDITERRANEAN SEA 6 mi/10 km E of Cape Scaramia. Sometimes called Ragusa River.

Irmino, UKRAINE: see TEPLOHIRS′K.

Irminskiy Rudnik, UKRAINE: see TEPLOHIRS′K.

Irmins′kyy Rudnyk, UKRAINE: see TEPLOHIRS′K.

Irmo (UHR-mo), city (2006 population 11,338), LEX-INGTON county, central SOUTH CAROLINA, residential suburb 9 mi/14.5 km WNW of COLUMBIA, near Lake MURRAY; 34°05′N 81°11′W. Manufacturing of nylon filament and log homes.

Iro, Cape (EE-ro), Japanese *Iro-misaki* (ee-RO-mee-SAH-kee), SHIZUOKA prefecture, central HONSHU, central JAPAN, in PHILIPPINE SEA; southernmost point of IZU PENINSULA; 34°36′N 138°51′E. Lighthouse.

Irois, Cape (ee-RWAH), on westernmost headland of JACMEL PENINSULA, HAITI, 5 mi/8 km SSW of ANSE-D′HAINAULT; 18°26′N 74°28′W.

Irois, Les (ee-RWAH, laiz), town, GRANDE-ANSE department, HAITI, on the far W end of JACMEL PE-NINSULA; 18°24′N 74°27′W. Cocoa growing.

Iron, county (□ 1,211 sq mi/3,148.6 sq km; 2006 population 12,377), SW UPPER PENINSULA, MICHIGAN; ⊙ CRYSTAL FALLS; 46°12′N 88°30′W. Bounded S by WISCONSIN line. Drained by BRULE, MICHIGAMME, PAINT, and IRON rivers. Lumbering; cattle, sheep; agriculture (potatoes, oats, forage crops); dairy products; some manufacturing at Crystal Falls; lake resorts. Many small lakes and streams. Part of Ottawa National Forest in W; Bewabic State Park in S; Brule Mountain Ski Area in S. Includes part of ME-NOMINEE IRON RANGE. University of Michigan forestry school's summer camp is here. One of four counties in Michigan in Central time zone (N boundary and N part of E and W boundaries coincide with Eastern/Central time zone boundary). Formed and organized 1885.

Iron, county (□ 554 sq mi/1,440.4 sq km; 2006 population 10,279), SE central MISSOURI; ⊙ IRONTON; 37°32′N 90°45′W. In the OZARKS, in SAINT FRANCOIS MOUNTAINS; includes TAUM SAUK MOUNTAIN (1,772 ft/540 m), highest point in state. Agriculture (cattle, hogs, mixed farming), mining (manganese, lead, iron, granite, zinc); oak, pine timber; recreation; resorts. Units of Mark Twain National Forest in SE and W; Elephant Rocks State Park in NE. Formed 1857.

Iron, county (□ 3,302 sq mi/8,585.2 sq km; 2006 population 40,544), SW UTAH; ⊙ PAROWAN; 37°51′N 113°16′W. Mountain and plateau region bordering on NEVADA. Alfalfa, barley; dairying; cattle; iron ore in IRON MOUNTAINS (Columbia Mine, S). CEDAR BREAKS National Monument in SE; parts of Dixie National Forest in MARKAGUNT PLATEAU in S, SE and SW. KOLOB TERRACE S of CEDAR CITY; desert area in N. Iron Mission State Historical Park in E center. Small part of Fishlake National Forest in NE corner; small part of ZION NATIONAL PARK in S. Formed 1850.

Iron, county (□ 918 sq mi/2,386.8 sq km; 2006 population 6,502), N WISCONSIN; ⊙ HURLEY; 46°19′N 90°15′W. Near HURLEY, iron mining is principal industry. Bounded partly N by LAKE SUPERIOR and MONTREAL RIVER (here forming MICHIGAN border); drained by tributaries of BAD RIVER and by Montreal River GOGEBIC RANGE (valuable source of iron ore) extends across county. S half of Iron county is largely wooded, with many lakes, and forms a large resort area. Whitecap Mountain Ski Area in N; large TURTLE-FLAMBEAU FLOWAGE reservoir in S; FLAMBEAU RIVER State Forest extends downstream from dam; part of LAC DU FLAMBEAU Indian Reservation in SE corner; small part of Bad River Indian Reservation in NW corner; part of Northern Highlands State Forest in E. Formed 1893.

Iron, MINNESOTA: see IRON JUNCTION.

Iron Acton (EI-uhn AK-tuhn), agricultural village (2001 population 1,253), SW GLOUCESTERSHIRE, central ENGLAND, 9 mi/14.5 km NE of BRISTOL; 51°33′N 02°28′W. Has 14th-century church.

Ironbark, AUSTRALIA: see STUART TOWN.

Iron Baron (EI-uhrn BA-ruhn), town, SOUTH AUSTRALIA, S central AUSTRALIA, 271 mi/436 km from ADELAIDE, 77 mi/124 km S of PORT AUGUSTA, and on EYRE PENINSULA; 32°58′S 137°09′E. Iron ore mining; ore is transported E to WHYALLA for processing.

Iron Belt, village, IRON county, N WISCONSIN, 6 mi/9.7 km WSW of HURLEY. Manufacturing (paving products).

Ironbridge (EI-uhn-brij), town (2001 population 2,417), SHROPSHIRE, W ENGLAND, about 6 mi/9.7 km S of TELFORD; 52°38′N 02°29′W. On side of SEVERN RIVER gorge, spanned by first cast iron bridge in 1778. An important center of the Industrial Revolution. Museum of reconstructed "authentic" early Industrial-era buildings located here.

Iron Bridge (EI-uhrn BRIJ), unincorporated village (□ 13 sq mi/33.8 sq km; 2001 population 686), N ONTARIO, E central CANADA, 51 mi/81 km from SAULT SAINTE MARIE, and included in HURON SHORES municipality; 46°17′N 83°22′W.

Iron Bridge Dam, Texas: see TAWAKONI, LAKE.

Iron City, village (2000 population 321), SEMINOLE county, extreme SW GEORGIA, 16 mi/26 km WNW of BAINBRIDGE; 31°01′N 84°49′W.

Iron City, Pennsylvania: see PITTSBURGH.

Irondale (EI-ruh-dail), city (2000 population 9,813), Jefferson co., N central ALABAMA, suburb E of Birmingham. Inc. in 1887.

Irondale, unincorporated town, ADAMS county, N central COLORADO, near SOUTH PLATTE RIVER, residential suburb 7 mi/11.3 km NE of DENVER and NE of COMMERCE CITY, at NW boundary of Rocky Mountain Arsenal. Elevation c.5,115 ft/1,559 m.

Irondale, town (2000 population 437), WASHINGTON county, E central MISSOURI, in the SAINT FRANCIS MOUNTAINS, on BIG RIVER, and 10 mi/16 km SSE of POTOSI; 37°49′N 90°40′W.

Irondale (EI-uhrn-dail), village (□ 1 sq mi/2.6 sq km; 2006 population 408), JEFFERSON county, E OHIO, 15 mi/24 km NNW of STEUBENVILLE, near the OHIO RIVER; 40°34′N 80°43′W. Clay products, lumber.

Irondequoit (eer-AHN-duh-koit), town (□ 16 sq mi/41.6 sq km), MONROE county, W NEW YORK, on Lake ONTARIO and IRONDEQUOIT BAY, 5 mi/8 km NNE of downtown ROCHESTER; 43°12′N 77°34′W. Partly enclosed by the city. Seebreeze Amusement Park (fourth oldest in U.S.) is here. Settled 1791, organized 1839.

Irondequoit Bay (eer-AHN-duh-koit), MONROE county, W NEW YORK, inlet of Lake ONTARIO just NE of ROCHESTER; c.4 mi/6.4 km long, 0.5 mi/0.8 km–1 mi/1.6 km wide; 43°13′N 77°52′W. Receives small Irondequoit Creek from S.

Iron Gate (EI-uhrn GAIT), village (2006 population 382), ALLEGHANY county, NW VIRGINIA, 3 mi/5 km SE of CLIFTON FORGE, on JACKSON RIVER, in George Washington National Forest; 37°47′N 79°47′W. COWPASTURE RIVER joins Jackson River 2 mi/3 km SE to form JAMES RIVER. Manufacturing (molded rubber); agriculture (corn, alfalfa, apples; cattle).

Iron Gate, Romanian *Porţile de Fier*, Serbo-Croatian *Djerdap*, gorge of the DANUBE River, on the SERBIA-Romanian border between Orşova (ROMANIA) and Drobeta-Turnu Severin (Romania); c.2 mi/3.2 km long and c.550 ft/168 m wide. Here the river narrows and swiftly flows through a gap between the CARPATHIAN and BALKAN mountains. Iron Gate, formerly an obstacle to shipping, was cleared of rock obstructions in the 1860s; the SIP Canal (opened 1896) permitted large river crafts to pass through the gorge. Iron Gate is the site of one of Europe's largest hydroelectric power dams. The joint Yugoslav-Romanian project (opened 1971) improved river navigation by impounding a large lake and has a huge electricity generating capacity. Now small oceangoing vessels can reach BELGRADE port. On the Serbian side of the gorge is a national park (158,080 acres/64,000 ha).

Iron Islands, AUSTRALIA: see YAMBI SOUND.

Iron Junction, village (2000 population 93), ST. LOUIS county, NE MINNESOTA, just S of MESABI IRON RANGE, 8 mi/12.9 km SSW of VIRGINIA; 47°25′N 92°36′W. Railroad junction. Oats, alfalfa; dairying; manufacturing (electronic ice fishing equipment); iron mines in area.

Iron Knob (EI-uhrn NOB), village, S SOUTH AUSTRALIA, AUSTRALIA, in MIDDLEBACK RANGE, NE EYRE PENINSULA, 155 mi/249 km NE of PORT LINCOLN; 32°44′S 137°08′E. Head of railroad to WHYALLA. Iron mines.

Iron, Lough (EI-ruhn, LAHK), lake (2.5 mi/4 km long, 0.5 mi/0.8 km wide), NW WESTMEATH county, central IRELAND, 7 mi/11.3 km NW of MULLINGAR; 53°36′N 07°28′W.

Iron Mine Hill, village, MIDLANDS province, central ZIMBABWE, 33 mi/53 km ENE of GWERU, on railroad, 19°18′S 30°17′E. Elevation 4,752 ft/1,448 m. Iron mining.

Iron Monarch, AUSTRALIA: see MIDDLEBACK RANGE.

Iron Mountain, town (2000 population 8,154), ⊙ DICKINSON county, SW UPPER PENINSULA, MICHIGAN, c.50 mi/80 km W of ESCANABA, near WISCONSIN state line; 45°49′N 88°03′W. Distribution point for region. Manufacturing (wood products, auto parts, food processing, machining, wood furniture); livestock; hay; timber; dairying; resort (winter sports). Veterans' hospital nearby. Mining museum. Pine Mountain Ski Area to N; Ford Airport to SW; Iron Mountain Iron Mine (tourist attraction) to E, at NORWAY. Settled 1879, incorporated 1889.

Iron Mountains, IRON and WASHINGTON counties, SW UTAH. Rise to 7,831 ft/2,387 m in Iron Mountain, 17 mi/27 km W of CEDAR CITY. S end in Dixie National Forest. Iron mining.

Iron Mountains, range, in NE TENNESSEE and SW VIRGINIA, ridge (2,500 ft/762 m–4,500 ft/1,372 m) of the APPALACHIAN MOUNTAINS between HOLSTON (SW) and STONE (E) mountains, from DOE RIVER near ELIZABETHTON, Tennessee, extending c.80 mi/129 km NE to NEW RIVER SE of WYTHEVILLE, Virginia. A spur in Virginia, just NE of Tennessee-NORTH CAROLINA state line, includes Mount ROGERS (5,729 ft/1,746 m) and Whitetop Mountain (5,520 ft/1,682 m). Included in Cherokee and Jefferson national forests. Sometimes considered a range of UNAKA MOUNTAINS.

Iron Quadrilateral, mining zone, central MINAS GERAIS, BRAZIL, SE of BELO HORIZONTE. Rectangular area rich in iron ore and other minerals. Some of Brazil's largest iron mines and steel manufacturing plants are located here.

Iron Ridge, town (2006 population 991), DODGE county, S central WISCONSIN, near ROCK RIVER, 15 mi/24 km ESE of BEAVER DAM; 43°23′N 88°31′W. In dairying region. Manufacturing (custom tools). Railroad junction. Horicon Marsh Wildlife Area to NW.

Iron River, town (2000 population 1,929), IRON county, SW UPPER PENINSULA, MICHIGAN, 33 mi/53 km NW of Iron, and on IRON RIVER; 46°06′N 88°38′W. In lumbering region. Livestock; potatoes; dairy products; manufacturing (machining, naval equipment and cranes); lake resort. Ottawa National Forest to NW; Brule Mountain Ski Area to SW; Bewabic State Park to E. Settled by iron-ore prospectors c.1881; incorporated as village 1885, as city 1926.

Iron River, village, BAYFIELD county, extreme N WISCONSIN, 25 mi/40 km W of ASHLAND; 46°32′N 91°21′W. In lake-resort region. Lumbering, dairying, farming; manufacturing (logging machinery, precision machining). Chequamegon National Forest to E and SE; BRULE RIVER State Forest to W.

Iron River, c.25 mi/40 km long, SW UPPER PENINSULA, MICHIGAN; rises in small lakes in SW IRON county; flows SE, past STAMBAUGH, to BRULE RIVER 5 mi/8 km SE of CASPIAN; 46°07′N 88°43′W.

Iron Station (EI-uhrn STAI-shuhn), unincorporated village, LINCOLN county, W central NORTH CAROLINA, 6 mi/9.7 km SE of LINCOLNTON; 35°26′N 81°09′W. Agriculture (grain; poultry; livestock); manufacturing (industrial products).

Ironton (EI-uhrn-tuhng), city (2000 population 1,471), SE central MISSOURI, ⊙ IRON county, in SAINT FRANCIS MOUNTAINS, 17 mi/27 km SW of FARMINGTON; 37°36′N 90°38′W. Tourism, resorts, timber and wood products, mixed farming; former major iron mines. Historic court house (1858–1860). Civil War battle of Pilot Knob, fought nearby on September 27, 1864 (Fort Davidson State Historic Site), resulted in reverse for Confederates under Sterling Price. Mark Twain National Forest nearby; TAUM SAUK MOUNTAIN, highest point in Missouri, to W. Founded 1857.

Ironton (EI-uhrn-tuhn), city (□ 4 sq mi/10.4 sq km; 2006 population 11,416), ⊙ LAWRENCE county, S OHIO, on OHIO RIVER; 38°32′N 82°40′W. Manufacturing includes chemicals, metal pipes, plastics, and iron products; some coal mining. Ironton was a great iron-producing center during the Civil War. From 1900–1910, it had the largest blast furnace in the world, Big Etna, with a capacity of 100 tons per day. However, the development of the N iron-ore ranges and improved transportation by railroad and on the GREAT LAKES led to the decline of its iron industry by the early 20th century. The remains of many giant charcoal iron furnaces are local landmarks. Incorporated as a city 1865.

Ironton, village, OURAY county, SW COLORADO, S of OURAY on Red Mountain Creek, 12 mi/19 km N of SILVERTON. Elevation c.9,800 ft/2,987 m. In mining area.

Ironton, village (2000 population 498), CROW WING county, central MINNESOTA, 14 mi/23 km NE of BRAINERD and 1 mi/1.6 km W of CROSBY, in CUYUNA IRON RANGE; 46°28′N 94°00′W. Grain, oats, alfalfa; dairying; poultry; manufacturing (machining, signs); iron-mining district. In lake and forest area, Crow Wing State Forest to N.

Ironton, village (2006 population 233), SAUK county, S central WISCONSIN, on BARABOO RIVER and 21 mi/34 km WNW of BARABOO; 43°32′N 90°08′W. In dairy and livestock region.

Ironville, unincorporated village, BOYD county, NE KENTUCKY, residential suburb 3 mi/4.8 km SW of ASHLAND. Tobacco, alfalfa, cattle.

Ironville, ENGLAND: see ALFRETON.

Ironwood, town (2000 population 6,293), GOGEBIC county, W UPPER PENINSULA of MICHIGAN, on MONTREAL RIVER, 95 mi/153 km ESE of DULUTH, MINNESOTA, 12 mi/19 km SE of LAKE SUPERIOR shore, opposite HURLEY, WISCONSIN; 46°27′N 90°09′W. Elevation 1,503 ft/458 m. Trade center for Gogebic Range region; manufacturing (sportswear, publishing, plastic molding, concrete blocks, canvas products); lumbering, dairy and vegetable farming; resort (winter sports). Gogebic Community College. Ottawa National Forest to NE; Mount Zion Ski Area to NW, Big Powderhorn Ski Area to E; Gogebic Municipal Airport to N. Founded 1885, incorporated 1889.

Iroquois (EE-rah-kwoi), county (□ 1,118 sq mi/2,906.8 sq km; 2006 population 30,598), E ILLINOIS, on INDIANA line (E); ☉ WATSEKA; 40°44′N 87°49′W. Agriculture (corn, sorghum, soybeans, cattle, hogs; dairying). Manufacturing (food products, wood products, industrial machinery). Drained by IROQUOIS RIVER and small Sugar Creek. Formed 1833.

Iroquois (EE-ruh-kwoi), unincorporated village (□ 2 sq mi/5.2 sq km; 2001 population 1,228), SE ONTARIO, E central CANADA, on the SAINT LAWRENCE RIVER, 15 mi/24 km NE of PRESCOTT, and included in SOUTH DUNDAS township; 44°51′N 75°19′W. Milling, manufacturing, dairying.

Iroquois, village (2000 population 207), Iroquois co., E Illinois, on Iroquois River (bridged here), and 9 mi/14.5 km ENE of Watseka; 40°49′N 87°34′W. In agr. area (corn, sorghum, soybeans, cattle, hogs).

Iroquois, village (2006 population 253), KINGSBURY and BEADLE counties, E central SOUTH DAKOTA, 15 mi/24 km W of DE SMET; 44°22′N 97°50′W.

Iroquois Falls (I-ruh-kwah FAHLZ), town (□ 231 sq mi/600.6 sq km; 2001 population 5,217), E central ONTARIO, E central CANADA, on ABITIBI RIVER, and 27 mi/43 km SE of COCHRANE; 48°46′N 80°40′W. Pulp, paper, and sulphite milling center. Nearby are waterfalls and hydroelectric station.

Iroquois River, c.85 mi/137 km long, in NW INDIANA and NE ILLINOIS; rises in JASPER county, NW Indiana; flows WSW to WATSEKA, Illinois, then generally N to KANKAKEE RIVER c.4 mi/6.4 km above KANKAKEE.

Iro River, MONGOLIA: see YÖRÖÖ GOL.

Irosin (ee-RO-seen), town, SORSOGON province, extreme SE LUZON, PHILIPPINES, 19 mi/31 km S of SORSOGON; 12°45′N 124°02′E. Agricultural center (coconuts, rice).

Irpen', UKRAINE: see IRPIN'.

Irpin' (ir-PEEN) (Russian *Irpen'*), town, N central KIEV oblast, UKRAINE, on railroad, and 13 mi/21 km NW of KIEV; 50°31′N 30°15′E. Elevation 354 ft/107 m. Manufacturing (construction materials, furniture, leather, machines), peat cutting. Technical schools, sanatoria, youth summer camps. Established as a railroad junction in 1902; city since 1956.

Irpuma Irpa Grande (eer-POO-mah EER-pah GRAHN-dai), canton, INGAVI province, LA PAZ department, W BOLIVIA, 25 mi/45 km SW of LA PAZ, near La Paz–CHARAÑA railroad; 16°54′S 68°20′W. Elevation 12,641 ft/3,853 m. Gas resources in area. Lead-bearing lode; clay, limestone, and gypsum deposits. Agriculture (potatoes, yucca, bananas, rye, barley); cattle.

Irqa (ir-GUH), formerly small sheikdom of S YEMEN, on Gulf of ADEN, 130 mi/209 km WSW of MUKALLA; consists of Irqa village and environs. Submitted to British protection in 1902. Sometimes spelled Irka.

Irricana (i-ruh-KAH-nuh), town (□ 1 sq mi/2.6 sq km; 2001 population 1,038), S central ALBERTA, CANADA, near Rosebud River, 28 mi/45 km NE of CALGARY, in ROCKY VIEW No. 44 municipal district; 51°19′N 113°37′W. Railroad junction in irrigated area. Established as a village in 1911; became a town in 2005.

Irrigon (IR-uh-gahn), village (2006 population 1,829), MORROW county, N OREGON, 10 mi/16 km WNW of HERMISTON, on Columbia River (LAKE UMATILLA reservoir); 45°53′N 119°29′W. Railroad terminus. Corn, alfalfa, wheat, potatoes; cattle. Manufacturing (dried corn, alfalfa pellets). Umatilla National Wildlife Refuge to W. Umatilla Ordnance Depot to S.

Iršava, UKRAINE: see IRSHAVA.

Irsha (eer-SHAH), town (2005 population 1,430), SE KRASNOYARSK TERRITORY, SE SIBERIA, RUSSIA, on road and near railroad, 40 mi/64 km SW of KANSK, and 4 mi/6 km SSE, and under administrative jurisdiction of ZAOZËRNYY; 55°55′N 94°47′E. Elevation 1,000 ft/304 m. Lignite mines.

Irshans'k (eer-SHAHNSK) (Russian *Irshansk*), town, central ZHYTOMYR oblast, UKRAINE, on right bank of the Irsha River, tributary of the TETERIV RIVER, on road, 14 mi/23 km SSE of KOROSTEN'; 50°44′N 28°44′E. Elevation 570 ft/173 m. Mining beneficiation plant. Topaz deposits nearby. Established in 1953; town since 1960.

Irshansk, UKRAINE: see IRSHANS'K.

Irshava (ir-SHAH-vah) (Czech *Iršava*) (Hungarian *Ilosva*), city (2004 population 10,400), S central TRANSCARPATHIAN oblast, UKRAINE, 17 mi/27 km SE of MUKACHEVE; 48°19′N 23°03′E. Elevation 472 ft/143 m. On railroad. Manufacturing (abrasives, cotton textiles, furniture, wooden toys), food processing (including flour milling). Known since 1341, city since 1982.

Ir-Shemesh, ISRAEL: see BEIT SHEMESH.

Irsina (eer-SEE-nah), town, MATERA province, BASILICATA, S ITALY, near BRADANO river, 20 mi/32 km WNW of MATERA; 40°45′N 16°14′E. In cereal- and grape-growing region. Has cathedral and museum of antiquities.

Irthlingborough (UH-thling-buh-ruh), town (2001 population 7,033), E NORTHAMPTONSHIRE, central ENGLAND, on NENE RIVER (crossed by 14th-century bridge), and 4 mi/6.4 km ENE of WELLINGBOROUGH; 52°19′N 00°36′W. Light manufacturing. Has 13th-century church.

Irtysh (eer-TISH), river, approximately 2,650 mi/4,265 km long, W Siberian Russia and KAZAKHSTAN. It is the chief tributary of the OB' RIVER and one of the two major rivers of W SIBERIA. As the ERTIX River, it rises in XINJIANG UYGUR AUTONOMOUS REGION, extreme NW CHINA, in the Mongolian ALTAI Mountains; flows NW through Lake ZAYSAN (in Kazakhstan), where it is known as the Irtysh, through NE Kazakhstan, past SEMEY, and enters W Siberia (OMSK oblast). There it receives the Ishim and Tobol rivers, its chief tributaries. The Irtysh flows past OMSK and TOBOL'SK (TYUMEN oblast) and joins the Ob near KHANTY-MANSIYSK. Major hydroelectric stations are at UST'-KAMENOGORSK and Zhana Buktyrma, Kazakhstan (1959). The riverbanks were occupied by the Chinese, Kalmyks, and Mongols until the Russians arrived in the late 16th century. The Russian conquest of the basin was completed by the early 19th century. Also called Kara-Irtysh River in the upper course, above Lake Zaysan.

Irtyshsk (ir-TISHSK), city, N PAVLODAR region, KAZAKHSTAN, on IRTYSH River, 95 mi/153 km NNW of Pavlodar; 53°22′N 75°30′E. Tertiary-level (raion) administrative center. Wheat. Also called Ertys, Irtyshskoye.

Irtyshskiy (eer-TISH-skeeye), settlement (2006 population 4,425), central TYUMEN oblast, SW Siberian Russia, on the IRTYSH River, on highway, 9 mi/14 km NNW of TOBOL'SK, to which it is administratively subordinate; 58°20′N 68°08′E. Elevation 147 ft/44 m. Port facilities.

Irtyshskiy (eer-TISH-skeeye), settlement (2006 population 3,270), SE OMSK oblast, SW SIBERIA, RUSSIA, on the IRTYSH River, on highway, 14 mi/23 km SSE of OMSK, to which it is administratively subordinate;

54°48′N 73°35′E. Elevation 255 ft/77 m. In agricultural and woodworking region.

Irugwa Island (ee-ru-GWAH), MWANZA region, NW TANZANIA, in Lake VICTORIA, 35 mi/56 km WSW of MUSOMA, 4 mi/6.4 km N of mainland; 4 mi/6.4 km long, 2 mi/3.2 km wide; 01°42′S 33°21′E.

Iruma (EE-roo-mah), city, SAITAMA prefecture, E central HONSHU, E central JAPAN, on the Iruma River, 16 mi/25 km W of URAWA; 35°49′N 139°23′E. Residential and industrial suburb of TOKYO famous for the cultivation of SAYAMA green tea.

Irumu (ee-ROO-moo), town, ORIENTALE province, NE CONGO, on SHARI RIVER near its confluence with ITURI RIVER, and 315 mi/507 km ENE of KISANGANI; 01°27′N 29°52′E. Elev. 2,762 ft/841 m. Commercial center. Large hospitals, airport. Has decreased in importance since development of BUNIA and of mining centers in the KILO-MOTO gold fields.

Irún (ee-ROON), city, GUIPÚZCOA province, N SPAIN, in the BASQUE region near the French border, on the BIDASSOA RIVER near the BAY OF BISCAY; 43°21′N 01°47′W. It is a commercial and manufacturing center producing paper and leather goods. Lead and iron mines are nearby. Irún was staunchly defended by the Loyalists in the civil war of 1936–1939.

Irupana (ee-roo-PAH-nah), city and canton, SUD YUNGAS province, LA PAZ department, W BOLIVIA, in the yungas, c. 5 mi/8 km SE of CHULUMANI, on road; 16°28′S 67°28′W. Elevation 6,184 ft/1,885 m. In subtropical agriculture area (coffee, cacao, quinoa).

Irurita (ee-roo-REE-tah), village, NAVARRE province, N SPAIN, in the W PYRENEES, 22 mi/35 km NNE of PAMPLONA. Flour milling, chocolate manufacturing; lumber milling.

Iruya (ee-ROO-yah), village (1991 population 585), ☉ Iruya department (□ 1,360 sq mi/3,536 sq km), N SALTA province, ARGENTINA, 60 mi/97 km NW of ORÁN, in stock-raising area; 22°46′S 65°14′W.

Irvine (UHR-vein), city (□ 47 sq mi/122.2 sq km; 2000 population 143,072), ORANGE county, S CALIFORNIA; suburb 36 mi/58 km SE of downtown LOS ANGELES, and 10 mi/16 km SSE of ANAHEIM; 33°40′N 117°48′W. Its industries include the research and development of high-technology electronics, especially computer products, as well as service and retailing; also manufacturing of motor vehicles, pharmaceuticals, aerospace vehicles and aircraft parts, and medical instruments. However, Irvine is best known as an educational center and the seat of University of California, Irvine (established 1965). City was built in the 1970s as a planned community on farmland that was part of the Irvine Ranch (which had been carved out of three Spanish and Mexican land grants in 1876). It is one of the fastest-growing cities in the U.S., marked by a population increase of nearly 78% between 1980 and 1990. Irvine Valley College (two-year). Of interest are several old preserved buildings. John Wayne Airport (Orange county) to W. SANTA ANA MOUNTAINS and Cleveland National Forest to NE; Lion Country safari theme park in E; Corona del Mar State Beach in SW; San Joaquin Hills to S. Santa Ana Marine Corps Air Base (helicopter station) to N, El Toro Marine Corps Air Station to E. Incorporated 1971.

Irvine (EER-vin), former town (□ 1 sq mi/2.6 sq km; 2001 population 356), SE ALBERTA, W CANADA, near SASKATCHEWAN border, 19 mi/31 km ESE of MEDICINE HAT, in CYPRESS County; 49°57′N 110°16′W. Grain elevators. Coal and gas deposits nearby. Dissolved; amalgamated into Cypress County.

Irvine (UHR-vin), town (2001 population 33,090), ☉ North Ayrshire, SW Scotland, on the IRVINE RIVER estuary, and 6 mi/9.6 km W of KILMARNOCK; 55°36′N 04°40′W. Designated a New Town in 1965. Chemicals, electric goods, glass, plastics, and clothing. Previously iron and brass foundries. Once a major exporting point, Irvine now engages primarily in coastal trade.

Area is shown by the symbol □, and capital city or county seat by ☉.

Irvine (UHR-vuhn), town (2000 population 2,843), ⊙ ESTILL county, E central KENTUCKY, 38 mi/61 km SE of LEXINGTON, on KENTUCKY RIVER, N of Station Camp Creek mouth; 37°42′N 83°58′W. In agricultural area (burley tobacco, soybeans, corn; hogs, cattle; timber); manufacturing (overalls and jackets, printing and publishing, lumber). Mountain Mushroom Festival (April). Daniel Boone National Forest is to E and S; Lexington-Bluegrass Army Depot (MADISON county) to W.

Irvine, unincorporated village, WARREN county, NW PENNSYLVANIA, 7 mi/11.3 km W of WARREN on Brokenstraw Creek just W of its mouth on ALLEGHENY RIVER; 41°50′N 79°16′W. Manufacturing includes iron and steel forgings.

Irvinebank (UHR-vein-bank), QUEENSLAND, NE AUSTRALIA, 50 mi/80 km SW of CAIRNS; 17°26′S 145°13′E. In tin-mining area. Also spelled Irvine Bank.

Irvine Bank, AUSTRALIA: see IRVINEBANK.

Irvine River (UHR-vin), 29 mi/47 km long, South Lanarkshire, SW Scotland; rises in South Lanarkshire 3 mi/4.8 km ENE of DARVEL; flows W, past Darvel, Galston, Kilmarnock, Dreghorn, and IRVINE, to Firth of Clyde just WSW of Irvine.

Irvines Landing (UHR-veinz), unincorporated village, SW BRITISH COLUMBIA, W CANADA, on MALASPINA STRAIT of Strait of GEORGIA, at mouth of JERVIS INLET, 30 mi/48 km N of NANAIMO, and in SUNSHINE COAST regional district; 49°37′N 124°02′W. Lumber-shipping port.

Irvinestown (UHR-vuhnz-toun), Gaelic *Baile an Irbhinigh*, town (2001 population 2,100), N FERMANAGH, SW Northern Ireland, 9 mi/14.5 km N of ENNISKILLEN; 54°28′N 07°38′W. Market.

Irving, city (2006 population 196,084), DALLAS county, N TEXAS, a growing suburb 9 mi/14.5 km NW of downtown DALLAS and 21 mi/34 km ENE of downtown FORT WORTH; 32°51′N 96°58′W. Elevation 470 ft/143 m. Bounds city of Dallas (E) and extension of city of Fort Worth (W). Bounded on E by ELM FORK TRINITY RIVER and S in part by West Fork Trinity River. Manufacturing (building supplies, chemicals, electronic equipment, and airplane parts). The city has grown rapidly along with the expanding business community of the Dallas–Fort Worth metropolitan area, and the city population increased by more than 40% between 1980 and 1990. City is now hemmed in by neighboring municipalities. Irving has profited from the nearby oil, aerospace, electronic, engineering, and auto industries. The prosperous business center, Las Colinas, which is headquarters to the ExxonMobil, is nearby. The Texas Stadium, home of the Dallas Cowboys professional football team; the University of Dallas, North Lake College (two year); and the Dallas–Fort Worth Regional Airport (opened 1974) are partly within city on W, on Dallas-TARRANT county boundary. North Lake Park in N. Incorporated as a city 1952.

Irving, village (2000 population 2,484), MONTGOMERY county, S central ILLINOIS, 5 mi/8 km NE of HILLSBORO; 39°12′N 89°24′W. In agricultural and bituminous coal area.

Irvington, town (2000 population 1,257), BRECKINRIDGE county, NW KENTUCKY, 40 mi/64 km SW of LOUISVILLE; 37°52′N 86°16′W. Trade center in agricultural area (burley tobacco, corn, wheat, hay; livestock); manufacturing (crushed stone).

Irvington, town (2000 population 60,695), ESSEX county, NE NEW JERSEY, an industrial suburb just to the W of NEWARK; 40°43′N 74°13′W. Settled 1692 as Camptown, renamed 1852, incorporated 1898.

Irvington (UHR-veeng-tuhn), town (2006 population 651), LANCASTER county, E VIRGINIA, 18 mi/29 km NNE of GLOUCESTER, on RAPPAHANNOCK RIVER (bridged to SE), arm of CHESAPEAKE BAY; 37°39′N 76°25′W. Manufacturing (valves, sailboat masts, seafood packing and canning). Tides Inn (resort). Christ Church (1732) to N.

Irvington, unincorporated village, ALAMEDA county, W CALIFORNIA; suburb 25 mi/40 km SE of OAKLAND. In orchard and vineyard region. Manufacturing (electronic components, motor-vehicle parts, candy, lumber, wire products). HETCH HETCHY AQUEDUCT runs E-W to N. MISSION SAN JOSE to E.

Irvington, village (2000 population 736), WASHINGTON county, SW ILLINOIS, 9 mi/14.5 km NE of NASHVILLE; 38°26′N 89°09′W. In fruit-growing area.

Irvington, residential village (□ 4 sq mi/10.4 sq km; 2006 population 6,656), WESTCHESTER county, SE NEW YORK, on E bank of the HUDSON RIVER, between DOBBS FERRY (S) and TARRYTOWN; 41°02′N 73°52′W. Light manufacturing, research and development, and commercial services. Here at "Nevis," once the estate of Alexander Hamilton's son, are a Columbia University arboretum and a children's museum. Home for cardiac children here. Originally called Dearman; renamed (1857) for Washington Irving, who bought the estate "Sunnyside" (extant) here in 1835. Settled c.1655, incorporated 1872.

Irvington, neighborhood, MARION county, central INDIANA, c.4 mi/6.4 km ESE of downtown INDIANAPOLIS. Historic district, noteworthy architecture. Laid out 1870. Annexed to Indianapolis 1902.

Irvona (uhr-VO-nah), borough (2006 population 649), CLEARFIELD county, central PENNSYLVANIA, 20 mi/32 km NNW of ALTOONA, on Clearfield Creek. Manufacturing (refractory specialties); agriculture (dairying).

Irwell River (UH-wel), 40 mi/64 km long, LANCASHIRE and GREATER MANCHESTER, central ENGLAND; rises just N of Bacup; flows S, past Bacup, Bury, Radcliffe, and Manchester, to MERSEY river at Irlam. Receives ROCH river 2 mi/3.2 km S of Bury.

Irwin, county (□ 372 sq mi/967.2 sq km; 2006 population 10,403), S central GEORGIA; ⊙ OCILLA; 31°36′N 83°16′W. Coastal plain agriculture produces wheat, cotton, corn, tobacco, peanuts, peaches; cattle, hogs, poultry; and timber area drained by ALAPAHA and SATILLA rivers. Jefferson Davis Memorial State Park (W). Formed 1818.

Irwin, town (2000 population 372), SHELBY county, W IOWA, on WEST NISHNABOTNA RIVER, and 11 mi/18 km NE of HARLAN; 41°47′N 95°12′W. Manufacturing (steel grain boxes).

Irwin, unincorporated town (2000 population 1,343), LANCASTER county, N SOUTH CAROLINA, residential suburb 2 mi/3.2 km SW of LANCASTER; 34°41′N 80°49′W.

Irwin, village (2000 population 157), BONNEVILLE county, SE IDAHO, on SNAKE RIVER, and 38 mi/61 km E of IDAHO FALLS, near WYOMING. Caribou National Forest to SW, Targhee National Forest to NE, PALISADES RESERVOIR and Dam to SE (Snake River); 43°24′N 111°16′W.

Irwin, village (2000 population 92), KANKAKEE county, NE ILLINOIS, 7 mi/11.3 km SW of KANKAKEE; 41°02′N 87°58′W. In agricultural area (corn, soybeans; dairying).

Irwin, borough (2006 population 4,145), WESTMORELAND county, SW PENNSYLVANIA, suburb 16 mi/26 km SE of downtown PITTSBURGH. Coal, limestone; manufacturing (metal products, computer and electronic equipment, machinery, motor vehicle parts). Agriculture (grain; livestock, dairying). Laid out 1853, incorporated 1864.

Irwin Canal, irrigation channel in MANDYA district, KARNATAKA state, S INDIA; from dam on KAVERI RIVER at KRISHNARAJASAGARA extends 28 mi/45 km NE to 9,000-ft/2,743-m-long underground tunnel, from which subsidiary canals (total length, 180 mi/290 km) branch out, irrigating over 120,000 acres/48,564 ha.

Irwindale, city (2000 population 1,446), LOS ANGELES county, S CALIFORNIA; residential suburb 18 mi/29 km ENE of downtown LOS ANGELES, S of Azusa, near SAN GABRIEL RIVER; 34°07′N 117°58′W.

Irwinton, village (2000 population 587), ⊙ WILKINSON county, central GEORGIA, 27 mi/43 km E of Macon; 32°49′N 83°10′W. Manufacturing of lumber, wooden pallets. Major center for kaolin mining and processing.

Irymple (i-RIM-puhl), town, NW VICTORIA, AUSTRALIA, near MILDURA, 205 mi/330 km ENE of ADELAIDE; 34°15′S 142°10′E. In irrigated fruit-growing area; dried fruit.

Is (EES), town (2006 population 4,705), W SVERDLOVSK oblast, W Siberian Russia, in the E central URALS, on the Is River (left tributary of the TURA RIVER), on highway, 60 mi/97 km NNW of NIZHNIY TAGIL, and 11 mi/18 km N of NIZHNYAYA TURA, to which it is administratively subordinate; 58°47′N 59°43′E. Elevation 551 ft/167 m. Railroad junction on spur from Vyya; center of gold and platinum mining region. Developed in the 1930s; called Sverdlovskiy Priisk until 1933.

Is, IRAQ: see HIT.

Isa (EE-sah), town, SOKOTO state, NW NIGERIA, 45 mi/72 km NNW of KAURA NAMODA. Cotton, millet; cattle, skins.

Isaac Lake (EI-zuhk), E BRITISH COLUMBIA, W CANADA, in CARIBOO MOUNTAINS, 70 mi/113 km ENE of QUESNEL; 28 mi/45 km long, 1 mi/2 km–2 mi/3 km wide; 53°11′N 120°55′W. Elevation 3,200 ft/975 m. Drains SE through LANEZI LAKE and Cariboo River into Quesnel River.

Isaacs Harbour (EI-zaks), village, E NOVA SCOTIA, CANADA, on Isaacs Harbour River, near its mouth on the ATLANTIC, 17 mi/27 km SSW of GUYSBOROUGH; 45°11′N 61°40′W. Elevation 85 ft/25 m. Fishing. Gold mining region.

Isabel, town, LEYTE province, W LEYTE, PHILIPPINES, 43 mi/70 km NE of TACLOBAN; 10°57′N 124°27′E. Site of Leyte Industrial Estate (copper smelting, phosphate fertilizer plant), which uses geothermal power from Tongonan Geothermal Plant.

Isabel (IZ-uh-bel), village (2000 population 108), BARBER county, S KANSAS, 14 mi/23 km SE of PRATT; 37°28′N 98°32′W. In cattle and wheat region; feeds.

Isabel, village (2006 population 233), DEWEY county, N central SOUTH DAKOTA, 17 mi/27 km W of TIMBER LAKE village; 45°23′N 101°25′W. Lignite mines nearby; in N part of Cheyenne River Indian Reservation.

Isabela (ee-sah-BE-lah), province (□ 4,118 sq mi/10,706.8 sq km), in CAGAYAN VALLEY region, NE LUZON, the PHILIPPINES; ⊙ ILAGAN; 17°00′N 122°00′E. Population 23.4% urban, 76.6% rural; in 1991, 89% of urban and 68% of rural settlements had electricity. The fertile, densely populated CAGAYAN RIVER valley, which is in the central and E part of the province, is a leading tobacco- and cacao-producing region. The rugged, less populated SIERRA MADRE on the E is mineral rich; iron, nickel, and copper deposits, as well as the nation's major manganese mines, are here. Logging is important, as is corn and rice growing. MAGAT RIVER HIGH DAM (at Aguinaldo) is a major irrigation and hydroelectric power project. Airport at CAUAYAN, airstrip at PALANAN. Original inhabitants are the tribal Ybanag, Yogad, Gaddong; the Negrito, Ilocomo, and Tagalog peoples have emigrated here. Spanish settlement from 1598; province created 1856.

Isabela (ee-sah-BAI-lah), ruins of a town on the N shore of HISPANIOLA, in DOMINICAN REPUBLIC, at the base of Cape ISABELA. Believed to have been founded by Columbus (c. 1494), it was one of the first Spanish settlements in the New World.

Isabela (ee-sah-BAI-lah), town (2000 population 73,032), ⊙ BASILAN province, PHILIPPINES, on Basilan Strait opposite ZAMBOANGA; 06°38′N 121°58′E. Small port, shipping lumber. Fishing, agriculture (coffee, rubber, copra, coconuts).

Isabela, town, NEGROS OCCIDENTAL province, W NEGROS island, PHILIPPINES, 9 mi/14.5 km E of BINALBAGAN 10°11′N 123°02′E. Agricultural center (rice, sugarcane); sugar milling.

Isabela (ee-sah-BAE-lah), town (2006 population 47,301), NW PUERTO RICO, near the ATLANTIC, 9 mi/ 14.5 km NE of AGUADILLA. Manufacturing (shoes, women's clothing, electric transformers, water meters, tiles); horse raising, cattle. Fishing; tourism. Site of experimental agricultural station and branch of University of Puerto Rico-Mayaguez. Large sand dune reserves have been exploited for the construction industry.

Isabela, Cape (ee-sah-BAI-lah), headland on N coast of DOMINICAN REPUBLIC, 23 mi/37 km WNW of PUERTO PLATA; 19°58′N 71°W. The ruined town of ISABELA, reputedly the first town settled by Spanish in AMERICA, is nearby.

Isabela de Sagua (ee-sah-BAI-lah dai SAH-gwah) or **La Isabela**, town, VILLA CLARA province, central CUBA, port for SAGUA LA GRANDE, on NICHOLAS CHANNEL, 37 mi/60 km N of SANTA CLARA; 22°56′N 80°01′W. Railroad terminus; sugar-shipping and fishing (sharks, oysters, crabs) center. Also a seaside resort.

Isabela Island (ee-sah-BAI-lah), ECUADOR, largest island in GALÁPAGOS ISLANDS (□ 2,249 sq mi/5,825 sq km). There are ten visitor sites for eco-tourism; flightless cormorants and penguins, also flamingos and Galapagos tortoises. Has highest point in Galapagos, Volcán Wolf (5,597 ft/1,706 m). Sometimes called ALBEMARLE ISLAND.

Isabela, La, village, GUADALAJARA province, central SPAIN, 3 mi/4.8 km SSE of SACEDÓN, 28 mi/45 km ESE of GUADALAJARA. Thermal springs.

Isabela, La, CUBA: see ISABELA DE SAGUA.

Isabelia, Cordillera (ee-sah-BAI-lee-ah, kor-dee-YAI-rah), N central NICARAGUA, E spur of main continental divide, extending from area between CONDEGA and SAN RAFAEL DEL NORTE 150 mi/241 km generally ENE to BONANZA mining district. Forms watershed between COCO RIVER (N) and RÍO GRANDE (S); rises to over 6,500 ft/1,981 m in SASLAYA peak. Sometimes Cordillera Isabella.

Isabel II Canal or **Lozoya Canal** (lo-THOI-ah), c.40 mi/64 km long, MADRID province, central SPAIN; begins near confluence of LOZOYA and JARAMA rivers (reservoirs) E of PATONES; flows S to MADRID, which it supplies with water. Begun 1902.

Isabella (i-zuh-BE-luh), community, SW MANITOBA, W central CANADA, 11 mi/18 km NNE of MINIOTA village, and in MINIOTA rural municipality; 50°16′N 100°52′W.

Isabella, county (□ 577 sq mi/1,500.2 sq km; 2006 population 65,818), central MICHIGAN; ⊙ MOUNT PLEASANT; 43°38′N 84°50′W. Drained by CHIPPEWA and PINE rivers. Cattle, hogs, poultry, sheep; dairying; agriculture (sugar beets, beans, wheat, oats, soybeans, hay, asparagus); manufacturing at MOUNT PLEASANT; oil wells, refineries. The large Isabella Indian Reservation dominates center of county (c.100 sq mi/259 sq km), includes city of Mount Pleasant. Organized 1859.

Isabella, village, POLK county, SE TENNESSEE, near GEORGIA-NORTH CAROLINA-Tennessee border, 50 mi/ 80 km E of CHATTANOOGA; 35°01′N 84°21′W.

Isabella, Cape, SE ELLESMERE ISLAND, BAFFIN region, NUNAVUT territory, CANADA, on SMITH SOUND; 78°21′N 75°00′W. Named (1818) by John Ross after one of his expedition vessels.

Isabella Dam, California: see KERN RIVER.

Isabella River, 30 mi/48 km long, Lake county, NE MINNESOTA; rises in Lake Isabella (3 mi/4.8 km long, 2 mi/3.2 km wide; 47°47′N 91°31′W), fed by Perent River from E; flows NW, through Superior National Forest; receives Island River from SE below Isabella Lake; receives Snake River from S in BALD EAGLE LAKE where it turns NW, passes through Gabbro Lake reservoir before it enters South Fork KAWISHIWI RIVER.

Isabelle, Lake, reservoir (□ 18 sq mi/46.8 sq km), KERN county, S central CALIFORNIA, on KERN RIVER, in S part of Sequoia National Forest, 33 mi/53 km ENE of BAKERSFIELD; 35°37′N 118°28′W. Maximum capacity 568,000 acre-ft. Formed by Isabelle Dam (185 ft/56 m high), built (1953) by Army Corps of Engineers for flood control, irrigation, recreation, and power generation.

Isabel Segunda (ee-sah-BEL se-GOON-dah) or **Isabela II**, town, on N shore of VIEQUES Island, off E PUERTO RICO, 50 mi/80 km ESE of SAN JUAN; 18°09′N 65°26′W. Main town and landing of the island. Tourism. U.S. naval base. Has last fort built by SPAIN in New World. Lighthouse. Museum.

Isaccea (ee-SAH-chah), Latin *Noviodunum*, town, TULCEA county, SE ROMANIA, in DOBRUJA, on DANUBE River, and 26 mi/42 km W of Brăila; 45°16′N 28°28′E. Inland port trading in fish, lumber, grain. Manufacturing textiles and foodstuffs. Extensive vineyards and marble quarries in vicinity (S). Has 16th-century mosque.

Isachsen, Cape, NW extremity of ELLEF RINGNES ISLAND, NUNAVUT territory, CANADA, on the ARCTIC OCEAN; 79°25′N 105°30′W.

Isachsen Peninsula, CANADA: see ELLEF RINGNES ISLAND.

Isady, RUSSIA: see SEMIBRATOVO.

Ísafjarardjúp (EE-sah-FYAHR-[th]ahr-dyoop) or **Isafjardhardjup**, inlet (50 mi/80 km long, 1 mi/1.6 km–12 mi/19 km wide) of DENMARK STRAIT, NW ICELAND, on NW coast of VESTFJARA Peninsula; 66°05′N 22°45′W. Extends numerous arms (S). On a S arm, near its mouth, is Ísafjörur city.

Ísafjörur (EE-sah-FYUHR-[th]ur), town (2000 population 2,781), NW ICELAND, on the Ísafjarardjúp, an arm of the DENMARK STRAIT. It is a fishing port and has refrigeration plants, shrimp and fish-meal factories, shipyards, and machine workshops. Chartered 1866.

Isagarh (IS-ah-guhr), town (2001 population 10,347), GUNA district, MADHYA PRADESH state, INDIA, 38 mi/ 61 km ENE of GUNA; 24°50′N 77°53′E. Millet, wheat, gram.

Isahaya (ee-SAH-hah-yah), city, NAGASAKI prefecture, W KYUSHU, SW JAPAN, 12 mi/20 km N of NAGASAKI, on Ariake Sea (E), TACHIBANA BAY (S), and OMURA BAY (W); 32°50′N 130°03′E. An agricultural center (persimmons; barley) and railroad junction, Isahaya's economy is based on food processing (rice cakes; baked eel), electronics (computer components), manufacturing (underwater pumps, torpedoes), and tourism to UNZEN-AMAKUSA NATIONAL PARK. Mudskippers. Pavement stone.

Isaías Coelho (EE-sah-EE-ahs KO-el-yo), town (2007 population 7,753), SE PIAUÍ state, BRAZIL, 45 mi/72 km SSW of PICOS; 07°50′S 41°52′W.

Isaka (ee-SAH-kah), village, TABORA region, N central TANZANIA, 75 mi/121 km N of TABORA on railroad; 03°55′S 32°55′E. Timber, subsistence crops; livestock. Gold and limestone deposits.

Isa Khel (i-sah KHAIL), town, MIANWALI district, NW PUNJAB province, central PAKISTAN, 15 mi/24 km WNW of MIANWALI; 32°41′N 71°17′E. Local market. Also written Isakhel.

Isakly (ee-sah-KLI), village (2006 population 4,405), NE SAMARA oblast, E European Russia, on the SOK RIVER, on crossroads, 21 mi/34 km NNE of SERGIYEVSK; 54°08′N 51°32′E. Elevation 213 ft/64 m. In agricultural area; bakery, creamery.

Isakogorka (ee-SAH-kuh-GOR-kuh), S suburb of ARCHANGEL, ARCHANGEL oblast, N European Russia, on railroad, 7 mi/11.3 km S of city center; 64°26′N 40°37′E. Railroad junction of a branch W to SEVERODVINSK.

Isalo National Park (ee-SHAHL) (□ 315 sq mi/819 sq km), FIANARANTSOA province, SW MADAGASCAR, 120 mi/193 km NE of TOLIARY, in region of Isalo massif (elevation up to 4,160 ft/1,268 m); 22°30′S 45°15′E. Ruiniform sandstone formations, oases; lemurs.

Tourism facilities at nearby Ranohira town. Established 1962.

Isanapura, CAMBODIA: see SAMBOR PREI KUK.

Isandlwana, hill, KWAZULU-NATAL, SOUTH AFRICA, near BUFFALO RIVER, 30 mi/48 km ESE of DUNDEE; 28°22′S 30°37′E. Mission station was built nearby at RORKE'S DRIFT. During Zulu War (January 22, 1879), British force of 1,200 men was annihilated here by Cetshwayo's troops. Louis Napoleon, son of Napoleon III, serving with the British, was killed 20 mi/32 km NNE during the same campaign.

Isanga (ee-SAHN-gah), village, SINGIDA region, central TANZANIA, 90 mi/145 km SW of MANYONI, near Nkuhulu River, at NW edge of RUNGWA GAME RESERVE; 07°24′S 34°15′E. Timber; goats, sheep; corn, wheat.

Isangele (ee-SAHN-gai-lai), town, South-West province, CAMEROON, near NIGERIA border; 04°45′N 08°41′E. Offshore oil deposits nearby. Formerly called RIO-DEL-REY.

Isangi (ee-SAHNG-gee), village, ORIENTALE province, central CONGO, on left bank of CONGO RIVER at mouth of LOMAMI RIVER, and 70 mi/113 km WNW of KISANGANI; 00°46′N 24°15′E. Elev. 1,407 ft/428 m. Steamboat landing and trans-shipment point. Roman Catholic mission. Former Arab post, it still preserves picturesque ruins of old mosque. Local tribe, the Topoke, is noted for its inordinate tattooing.

Isangila (ee-sahng-GEE-lah), village, BAS-CONGO province, W CONGO, on right bank of CONGO RIVER, and 45 mi/72 km NE of BOMA; 05°17′S 13°36′E. Elev. 1,026 ft/312 m. Though this section of CONGO RIVER is part of the noted LIVINGSTONE FALLS, the river is navigable between here and MANYANGA. British explorer Capt. J. K. Tuckey reached Isangila from W in 1876.

Isanlu Makatu (ee-SHAHN-loo mah-kah-TOO), town, KWARA state, W central NIGERIA, on road, 38 mi/61 km NW of KABBA; 08°16′N 05°48′E. Market and tin-mining center; columbite and tantalite also mined. Shea-nut processing; millet, maize, kolanuts, tobacco.

Isanti (ei-SAN-tee), county (□ 451 sq mi/1,172.6 sq km; 2006 population 38,576), E MINNESOTA; ⊙ CAMBRIDGE; 45°33′N 93°17′W. Agricultural area drained by RUM RIVER. Corn, soybeans, oats, rye, alfalfa; hogs, sheep, poultry; dairying. Numerous small lakes throughout county; Green Lake in W, Lake Fannie in E. Formed 1849.

Isanti (ei-SAN-tee), town (2000 population 2,324), ISANTI county, E MINNESOTA, 5 mi/8 km S of CAMBRIDGE on RUM RIVER; 45°29′N 93°15′W. Poultry; grain, soybeans; dairying; manufacturing (machine products, food processing, metal fabrication, trusses, consumer goods).

Isarco River (ee-ZAHR-ko), German *Eisack*, 55 mi/89 km long, N ITALY; rises in the ALPS 1 mi/1.6 km W of BRENNER PASS; flows generally S, past BRESSANONE and BOLZANO, to ADIGE RIVER 4 mi/6 km SW of Bolzano.

Isar River (EE-sahr), 160 mi/257 km long, W AUSTRIA; rising in the TYROL; flowing NE through S central BAVARIA, past MUNICH, to the DANUBE River. There are more than twenty-five large hydroelectric plants below Munich.

Isaszeg (E-shah-sag), village, PEST county, N central HUNGARY, on Rákos River, and 15 mi/24 km E of BUDAPEST; 47°32′N 19°24′E. Hungarian victory here (1849) over Austrian forces.

Isata (ee-SAH-tah), canton, ESTEBAN ARCE province, COCHABAMBA department, central BOLIVIA, 19 mi/30 km NW of ANZALDO, S of TARATA; 17°46′S 66°03′W. Elevation 9,026 ft/2,751 m. Silver-bearing lode; important clay deposits, limestone, some gypsum deposits. Agriculture (potatoes, yucca, bananas, corn, rye, soy, coffee, grapes); cattle raising for meat and dairy products.

Area is shown by the symbol □, and capital city or county seat by ⊙.

Isauria, ancient district of S ASIA MINOR, on the borders of PISIDIA and CILICIA, N of the TAURUS range, in present S central TURKEY. It was a wild region inhabited by marauding bands. When the capital of Isaura or Isaura Vetus [old Isaura], a strongly fortified city at the foot of Mount Taurus, was besieged by the Macedonian regent Perdiccas in the 4th century B.C.E., the Isaurians destroyed the town by fire rather than submit to capture. The Isaurians were brought partially under control (76–75 B.C.E.) by the Romans, and again by the Byzantines under Justinian I, but were not completely subdued until the arrival in the 11th century C.E. of the Seljuk Turks. The site contains ruins of the town and its fortifications.

Isawa (ee-SAH-wah), town, Isawa district, IWATE prefecture, N HONSHU, NE JAPAN, 37 mi/60 km S of MORIOKA, near Mt. Yakeishi; 39°06′N 141°04′E. Green peppers and green pepper products; lacquering.

Isawa, town, East Yatsushiro district, YAMANASHI prefecture, central HONSHU, central JAPAN, 4.3 mi/7 km S of KOFU; 35°38′N 138°38′E. Hot springs.

Isayevo-Dedovo, RUSSIA: see OKTYABR′SKOYE, ORENBURG oblast.

Isbergues (ee-ber-guh), town (□ 5 sq mi/13 sq km), PAS-DE-CALAIS department, NORD-PAS-DE-CALAIS region, N FRANCE, 11 mi/18 km NW of BÉTHUNE; 50°37′N 02°27′E. Metalworks. Has 15th-century church.

Isca, Wales: see CAERLEON.

Isca Dumnoniorum, ENGLAND: see EXETER.

Iscar (EE-skahr), town, VALLADOLID province N central SPAIN, 22 mi/35 km SE of VALLADOLID; 41°22′N 04°32′W. Flour- and sawmills; grain, chicory, sheep.

Iscayachi (ees-kah-YAH-chee), town and canton, MÉNDEZ province, TARIJA department, S BOLIVIA, on TARIJA-VILLAZÓN road, 34 mi/55 km W of TARIJA. Vineyards, fruit (bananas, grapes, peaches, apples), grain (barley, corn, wheat), potatoes.

Ischgl (ISHGL), village, TYROL, W AUSTRIA, in PAZNAUN valley on TRISANNA RIVER, 16 mi/26 km SW of LANDECK; 47°01′N, 10°18′E. Elev. 4,194 ft/1,278 m. Wintersports center; funiculars to Idaalpe (7,035 ft/2,144 m) and Pardatscher Grat (7,980 ft/2,432 m).

Ischia (EES-kyah), volcanic island (□ 18 sq mi/46.8 sq km), CAMPANIA, S ITALY, in the TYRRHENIAN SEA between the GULF OF GAETA and the BAY OF NAPLES; 40°44′N 13°57′E. Known as the Emerald Isle, it is a health resort and a tourist center, celebrated for its warm mineral springs and for its scenery. Fishing and farming are also pursued, and wine (Epomeo), tiles, and pottery are made. Settled in the 8th century B.C.E., the island was abandoned several times because of volcanic eruptions (the last of which occurred in 1301). There was a severe earthquake in 1883. Monte Epomeo (2,585 ft/788 m) is the highest point. Ischia, the main town, has an imposing 15th-century castle, constructed on foundations built by the Greeks in the 5th century B.C.E.

Ischia di Castro (dee KAH-stro), village, VITERBO province, LATIUM, central ITALY, 20 mi/32 km WNW of VITERBO; 42°33′N 11°45′E. Agriculture. Medieval palace.

Ischilín, ARGENTINA: see DEÁN FUNES.

Ischilín, Sierra de (ees-chee-LEEN, see-YER-rah dai), pampean mountain range of SIERRA DE CÓRDOBA, NW CÓRDOBA province, ARGENTINA, extending c.12 mi/19 km S from DEÁN FUNES. Rises to c.3,000 ft/914 m.

Ischl, AUSTRIA: see BAD ISCHL.

Ischua (ISH-yoo-ai), hamlet, CATTARAUGUS county, W NEW YORK, on ISCHUA CREEK, 12 mi/19 km N of OLEAN; 42°13′N 78°23′W.

Ischua Creek (ISH-yoo-ai), c.30 mi/48 km long, W NEW YORK; rises W of MACHIAS; flows S to ALLEGHENY RIVER at OLEAN. Called Olean Creek below junction with small Oil Creek.

Iscia Baidoa, SOMALIA: see BAIDABO.

Iscuandé, Colombia: see SANTA BÁRBARA.

Ise (EE-se), city, MIE prefecture, S HONSHU, central JAPAN, on ISE BAY, 19 mi/30 km S of TSU; 34°29′N 136°42′E. Manufacturing of electrical machinery. The three Ise shrines form one of the most important Shinto religious centers. Set deep in a forest, they are said to have been built in 4 B.C.E. They exhibit an archaic style of architecture, completely without Chinese or Buddhist influence; until 1868 Buddhist priests and nuns were forbidden to enter the shrines. The Naigu, or Inner Shrine, is dedicated to Amaterasu-o-mikami, the "divine ancestress" of the imperial family, and still houses the Sacred Mirror, one of the three treasures that comprise the imperial regalia. Seat of a university and several museum of antiquities. Pilgrimages here support a steady tourism industry. Called Uji-yamada until 1955.

Isebania, town, NYANZA province, SW KENYA, on TANZANIA-Kenya border; 01°12′N 34°29′E. Agriculture (tobacco, maize, finger millet) and livestock. Important border-crossing point. Market and trade center. Formerly known as Nyabikaye.

Ise Bay (EE-se), Japanese *Ise-wan* (ee-SE-wahn), inlet of PHILIPPINE SEA, central HONSHU, between MIE (W) and AICHI (E) prefectures, S central JAPAN; bounded E by CHITA PENINSULA; 15 mi/24 km long, 5 mi/8 km–10 mi/16 km wide. Has two arms: CHITA BAY (NE) and ATSUMI BAY (E). NAGOYA is at head of bay, TSU and ISE on W shore. Formerly sometimes called Owari Bay.

Ise Fjord (EE-suh), 20 mi/32 km long, SJÆLLAND, DENMARK. Branches into ROSKILDE FJORD (25 mi/40 km long) on E side, NYKØBING BAY, LAMME FJORD, and HOLBÆK FJORD on W side. Orø island (□ 6 sq mi/15.5 sq km) is 11 mi/18 km S of mouth.

Isehara (ee-SE-hah-rah), city, KANAGAWA prefecture, E central HONSHU, E central JAPAN, 19 mi/30 km W of YOKOHAMA; 35°23′N 139°19′E.

Iseke (ee-SAI-kai), village, SINGIDA region, central TANZANIA, 50 mi/80 km WSW of DODOMA; KISIGO Game Reserve to SW; 06°27′S 35°01′E. Cattle, sheep, goats; corn, wheat.

Iselin Bank (IS-uh-luhn), Anarctica, submarine plateau, in central ROSS SEA, NE of PENNELL BANK; 72°30′S 179°W.

Iselin Seamount (IS-uh-luhn), ANTARCTICA, seamount in N ROSS SEA, an extension of ISELIN BANK, S PACIFIC OCEAN; 70°45′S 178°15′E.

Isel River (EE-sel), 30 mi/48 km long, river, E TYROL, S AUSTRIA; rises in the VENEDIGERGRUPPE mountains of the HOHE TAUERN; flows SE, through East Tyrol, to the Drau at LIENZ.

Iselsberg (EE-sels-berg), mountain pass between lower PUSTERTAL and Möll valley, on EAST TYROL-Carinthian border, SW AUSTRIA, 5 mi/8 km NE of LIENZ; 46°51′N, 12°52′E. Elevation 3,408 ft/1,039 m. Crossed by road between Lienz and WINKLERN.

Isen (EE-sen), town, S TOKUNO-SHIMA island, Oshima district, KAGOSHIMA prefecture, SW KYUSHU, SW JAPAN, 285 mi/460 km S of KAGOSHIMA; 27°40′N 128°56′E. Distilled alcoholic drinks (*shochu*).

Isen (EE-sen), village, BAVARIA, S GERMANY, in UPPER BAVARIA, on the ISEN, 10 mi/16 km SE of ERDING; 48°12′N 12°03′E. Has late-12th-century church. Chartered 1434.

Isenbayevo (ee-seen-BAH-ee-vuh), settlement, NE TATARSTAN Republic, E European Russia, on the administrative border with UDMURT REPUBLIC, 24 mi/39 km NE of MENZELINSK; 56°02′N 53°25′E. Elevation 396 ft/120 m. Brick works.

Isenbüttel (EE-sen-but-tel), village, LOWER SAXONY, N GERMANY, 10 mi/16 km E of WOLFSBURG; 52°22′N 10°36′E.

Isen River (EE-sen), 40 mi/64 km long, BAVARIA, GERMANY; rises 4 mi/6.4 km S of ISEN; flows E to the INN, 1.5 mi/2.4 km N of ALTÖTTING; source at 48°10′N 12°04′E.

Iseo (ee-ZAI-o), town, BRESCIA province, LOMBARDY, N ITALY, port on SE shore of LAGO D'ISEO, 12 mi/19 km NW of BRESCIA; 45°39′N 10°03′E. Resort; silk and flax mills, dyeworks, tannery, pottery factory; fishing.

Iseo, Lago d' (dee-ZAI-o, LAH-go) or **Sebino** (se-BEE-no), ancient *Lacus Sebinus*, lake (□ 24 sq mi/62.4 sq km) in LOMBARDY, N ITALY, in BRESCIA (E) and BERGAMO (W) provinces, 18 mi/29 km E of BERGAMO; 15.5 mi/24.9 km long, up to 3 mi/5 km wide; 45°43′N 10°04′E. Elevation 610 ft/186 m, maximum depth 820 ft/250 m. Fishing; fish hatcheries. Used for irrigation. In center of lake is Monte Isole or Montisola, largest freshwater island (□ 5 sq mi/13 sq km; 1991 population 1,745) in Italy; c.2 mi/3 km long, 1 mi/1.6 km wide; rises to 1,965 ft/599 m. Lake traversed by OGLIO RIVER. On its picturesque banks are ISEO, PISOGNE, and Lovere. Small steamers ply between them.

Iseran, Col de l' (lee-ze-rahn, kol duh), high Alpine pass (8,900 ft/2,713 m), SAVOIE department, RHÔNE-ALPES region, SE FRANCE, in E part of Massif de la VANOISE, connects upper ISÈRE RIVER valley (known as the TARENTAISE; N) with ARC RIVER valley (known as the MAURIENNE; S); 45°25′N 07°02′E. Scenic road (one of the highest in Europe) between BOURG-SAINT-MAURICE (20 mi/32 km NW) and LANSLEBOURG-MONT-CENIS (12 mi/19 km SE) completed 1937. Winter sports resort of VAL-D'ISÈRE is a few miles N of the pass amid spectacular ski slopes known as Space Killy. The pass lies at E edge of the VANOISE NATIONAL PARK.

Isère (ee-zer), department (□ 3,180 sq mi/8,268 sq km), in historic DAUPHINÉ region, SE FRANCE; ⊙ GRENOBLE; 45°10′N 05°50′E. Extending E from the RHÔNE RIVER valley (below LYON), it includes the outer Alpine massifs of CHARTREUSE, BELLEDONNE, VERCORS, and part of the rugged Massif des ÉCRINS (rising to 13,461 ft/4,103 m). Department is bisected by ISÈRE RIVER, which, between the CHAMBÉRY trough and Grenoble, flows in fertile GRÉSIVAUDAN valley. There are vineyards and orchards in the Rhône valley and in lower Isère valley, where tobacco and cereals also are grown. Cattle raising and timber operations in outer mountain ranges. Most mineral deposits are too remote for economic exploitation. Extensive hydroelectric development in the gorges of the ROMANCHE (CHAMBON DAM; power plants at LIVET-ET-GAVET), and of the DRAC (SAUTET DAM; power plants near La MURE, PONT-DE-CLAIX) activates electrometallurgical (aluminum, steel) and electrochemical works around Grenoble. Department also specializes in the production of kid gloves (at Grenoble), woolens (at VIENNE); it also has numerous paper mills and cement works. It exports Grande Chartreuse liqueur from VOIRON. Chief towns are Grenoble (in an Alpine basin at confluence of the Drac and Isère rivers) and Vienne (in the Rhône valley). The Grésivaudan valley is also densely populated. Tourism is a major contributor to the local economy. The department has some of the finest winter sport resorts in France and contains the W half of the ÉCRINS NATIONAL PARK as well as most of the VERCORS regional park. Administratively, Isère department lies within the RHÔNE-ALPES region, of which Lyon is the capital.

Isère River (ee-zer), ancient *Isara*, 180 mi/290 km long, RHÔNE-ALPES region, in SE FRANCE; rises in the high ALPS near Italian border above the resort of VAL-D'ISÈRE; flows generally W in 3 great arcs through Savoy Alps (ALPES FRANÇAISES) in deep TARENTAISE valley, then turns SW near ALBERTVILLE, cutting through the COMBE DE SAVOIE range; continuing S, it flows in the broad and fertile GRÉSIVAUDAN glacial valley to GRENOBLE, then it veers NW around the commanding CHARTREUSE range, winds around the N end of the VERCORS massif, and continues SW, to the RHÔNE 4 mi/6.4 km above VALENCE; 44°59′N 04°51′E. An important stream of French Alps. It is a principal tributary of the Rhône, collecting the waters of the ARLY (right), ARC, DRAC, and Bourne rivers (left). Upper valley has numerous hydroelectric plants which power chemical and metallurgical works. Above Albertville, the Isère valley provides access to some of France's finest winter sport resorts, many of which provided venues for the 1992 Winter Olympics.

Isergebirge, Czech *Jizerské Hory*, Polish *Góry Izerskie*, mountain range of the SUDETES, in N BOHEMIA, CZECH REPUBLIC, and LOWER SILESIA (after 1945, SW POLAND). Extends c.20 mi/32 km between CHRASTAVA, Czech Republic (W), and headwaters of KWISA (QUEIS) RIVER, Poland (E); rises to 3,681 ft/1,122 m in SMRK, to c.3,675 ft/1,120 m in JIZERA MOUNTAINS. Novy Svet Pass is at S foot.

Isergebirge, CZECH REPUBLIC: see JIZERA MOUNTAINS.

Ise River (EIZ), NORTHAMPTONSHIRE, central ENGLAND; rises just N of Naseby; flows 20 mi/32 km E and S, past Kettering, to NENE RIVER at Wellingborough.

Iserlohn (EE-ser-lon), city, NORTH RHINE–WESTPHALIA, W GERMANY; 51°22′N 07°42′E. Commercial and industrial center; manufacturing includes metal goods, machinery, chemicals, and rubber and leather goods. Iserlohn became an important town in the 13th century and was known for the manufacturing of armor, chains, and needles. Nearby is DECHENHÖHLE, a stalactite cave. Until the 19th century it was the largest city in WESTPHALIA.

Isernhagen (ee-sern-HAH-gen), suburb of HANOVER, LOWER SAXONY, N GERMANY, just N of city; 52°27′N 09°50′E.

Isernia (ee-ZER-nyah), town, Isernia province, MOLISE, S central ITALY, 23 mi/37 km W of CAMPOBASSO; 41°36′N 14°14′E. Woolen mills, foundry; lace manufacturing. Bishopric. Badly damaged (1943) in World War II.

Iser River, CZECH REPUBLIC: see JIZERA RIVER.

Isesaki (ee-SE-sah-kee), city (2005 population 202,447), GUMMA prefecture, central HONSHU, N central JAPAN, 12 mi/20 km S of MAEBASHI; 36°18′N 139°12′E. Manufacturing include textiles (with traditional splashed pattern), automotive parts. Edo-era castle town of the Sakai family.

Iset' (ee-SYET), village (2006 population 3,175), S SVERDLOVSK oblast, extreme W Siberian Russia, in the E outliers of the central URALS, on the S shore of Iset' Lake, on railroad, 10 mi/16 km NW of YEKATERINBURG, and 8 mi/13 km W of VERKHNYAYA PYSHMA, to which it is administratively subordinate; 56°58′N 60°22′E. Elevation 843 ft/256 m. Building materials; poultry factory.

Iset' River (ee-SYET), approximately 325 mi/523 km long, in SW Siberian Russia; rises in SVERDLOVSK oblast, in the central URAL Mountains, approximately 10 mi/16 km NE of BILIMBAY; flows SE, forming dammed Iset' Lake (□ 8 sq mi/21 sq km; SREDNEURAL'SK on the E shore), Verkhne-Iset pond (W of YEKATERINBURG city center), and Nizhne-Iset pond (S of Yekaterinburg city center), past Yekaterinburg and ARAMIL', and generally ESE, past KAMENSK-URALSKIY, KATAYSK (KURGAN oblast), DALMATOVO, and SHADRINSK, into S TYUMEN oblast, to the TOBOL River, 4 mi/6 km S of YALUTOROVSK. Timber floating. Part of the projected Kama-Irtysh waterway. Dammed at Yekaterinburg for industrial water supply. Receives Techa and Miass rivers.

Isetskoye (ee-SYET-skuh-ye), village (2006 population 7,155), SW TYUMEN oblast, W SIBERIA, RUSSIA, on the ISET' RIVER, on road, 42 mi/67 km S of TYUMEN, and 40 mi/64 km SW of YALUTOROVSK; 56°29′N 65°21′E. Elevation 203 ft/61 m. In agricultural area; produce processing, fertilizer manufacturing. Regional power station.

Iseum, EGYPT: see BEHBET EL HAGAR.

Iseyin (ee-SAI-ang), town, OYO state, SW NIGERIA; 07°58′N 03°36′E. The city, in a tobacco-growing region, has an important traditional textile industry. Iseyin was the capital of a small YORUBA kingdom under the Oyo empire. In 1893 it came under British control. Scene of an unsuccessful rebellion in 1916 against British-imposed taxes.

Isfahan, IRAN: see ESFAHAN.

Isfana (is-fuh-NAH), village, W OSH region, KYRGYZSTAN, on N slope of TURKESTAN RANGE and KARASU RANGE, 30 mi/48 km S of KHUDJAND (TAJIKISTAN); 39°48′N 69°34′E. Wheat. Tertiary-level administrative center.

Isfara (ees-fahr-AH), city (2000 population 37,000), E LENINOBOD viloyat, TAJIKISTAN, on railroad, and 15 mi/24 km SE of KANIBADAM; 40°07′N 70°38′W. In oil-producing area; fruit-canning (apricots), tobacco industry; cotton. Ozocerite quarries nearby.

Isfarainesfarayen, IRAN: see ESFARAYEN.

Isfendiyar Mountains, N TURKEY, part of the Pontus mountain system, extending 150 mi/241 km E-W along the BLACK SEA N of KASTAMONU, between ZAMANTI RIVER (W) and KIZIL IRMAK (E); rise to 6,514 ft/1,985 m in YARALIGOZ DAG. Coal, lignite in W, arsenic and mercury in center. Sometimes called Kure.

Isfjorden (EES-fyawr-duhn) [Norwegian=ice fjord], inlet of the GREENLAND SEA and largest fjord of SPITSBERGEN island, SVALBARD, NORWAY; 65 mi/105 km long, 8 mi/12.9 km–20 mi/32 km wide. Receives several glaciers. Mining towns of LONGYEARBYEN and BARENTSBURG on inlet.

Isfjorden, ANTARCTICA: see AMUNDSEN BAY.

Isha Baidoa, SOMALIA: see BAIDABO.

Ishcherskaya (EE-shcheer-skah-yah), village (2005 population 4,945), NW CHECHEN REPUBLIC, S European Russia, on the TEREK River, on road and the Moscow-Groznyy railroad, 36 mi/58 km NW of GROZNYY; 43°43′N 45°07′E. Elevation 318 ft/96 m. Originally a residential settlement for railroad constuction and servicing crews.

Isherim (ee-she-REEM), peak (4,367 ft/1,331 m) in the URAL Mountains, NE PERM oblast, RUSSIA, at a junction of the N and central Urals; 61°00′N 59°00′E. The Vishera and Lozva rivers rise nearby.

Isheyevka (ee-SHEH-eef-kah), town (2006 population 9,920), N ULYANOVSK oblast, E central European Russia, on the SVIYAGA RIVER, near highway and railroad, 9 mi/14 km NNW of ULYANOVSK; 54°25′N 48°16′E. Elevation 367 ft/111 m. Woolen milling.

Ishibashi (ee-shee-BAH-shee), town, Shimotsuga district, TOCHIGI prefecture, central HONSHU, N central JAPAN, 9 mi/15 km S of UTSUNOMIYA; 36°26′N 139°51′E. Gourd processing.

Ishibe (ee-SHEE-bai), town, Koka county, SHIGA prefecture, S HONSHU, central JAPAN, 12 mi/20 km E of OTSU; 36°03′N 139°51′E.

Ishida (ee-SHEE-dah), town, Iki district, NAGASAKI prefecture, NW KYUSHU, SW JAPAN, 71 mi/115 km N of NAGASAKI; 33°44′N 129°45′E.

Ishidoriya (ee-SHEE-do-REE-ah), town, Hienuki district, IWATE prefecture, N HONSHU, NE JAPAN, on KITAKAMI RIVER, 16 mi/13 km S of MORIOKA; 39°28′N 141°09′E. Sake-brewing equipment.

Ishigaki (ee-shee-GAH-kee), city, ISHIGAKI-JIMA island, SAKISHIMA ISLANDS, OKINAWA prefecture, RYUKYU ISLANDS, southernmost city of JAPAN, 273 mi/440 km SW of NAHA; 24°20′N 124°09′E. Iriomote National Park (fauna include Iriomote wildcat, *Kanmuri* eagle, *Kin* pigeon, *Semaruhako* turtle) is nearby. Lagoon in the vicinity has Japan's largest collection of coral.

Ishigaki-jima (ee-shee-GAH-kee–JEE-mah), largest island (□ 83 sq mi/215.8 sq km) of SAKISHIMA ISLANDS, Okinawa prefecture, extreme SW JAPAN, in the Ryukyus, between EAST CHINA (W) and PHILIPPINE (E) seas, 230 mi/370 km WSW of OKINAWA; 11 mi/18 km long (with long peninsula in N), 8.5 mi/13.7 km wide; 24°24′N 124°12′E. Mountainous, rising to 1,670 ft/509 m. Produces pottery, sake, raw silk, dried tuna. Agricultural products include sweet potatoes, sugarcane, rice. Formerly called Pachungsan. Chief city is ISHIGAKI.

Ishige (ee-shee-GE), town, Yuki district, IBARAKI prefecture, central HONSHU, E central JAPAN, 34 mi/55 km S of MITO; 36°06′N 139°58′E. Pongee.

Ishii (ee-SHEE), town, Myozai district, TOKUSHIMA prefecture, E SHIKOKU, W JAPAN, 6 mi/10 km W of TOKUSHIMA; 34°04′N 134°26′E.

Ishikari (ee-shee-KAH-ree), town, Ishikari district, SW HOKKAIDO prefecture, N JAPAN, fishing port on ISHIKARI BAY, at mouth of Ishikari River, 6 mi/10 km N of SAPPORO; 43°14′N 141°12′E. Salmon.

Ishikari (ee-shee-KAH-ree), c.225 mi/362 km long, Japan's second-longest river, Hokkaido prefecture, N JAPAN; rises in the mountainous interior of HOKKAIDO; flows generally SW to ISHIKARI BAY near OTARU. Drains an extensive coal area and waters the Ishikari lowland, a fertile agricultural region.

Ishikari Bay (ee-shee-KAH-ree), Japanese *Ishikari-wan* (ee-shee-KAH-ree-WAHN), or **Otaru Bay** (o-TAH-roo), inlet of Sea of JAPAN, in W Hokkaido prefecture, N JAPAN; 50 mi/80 km long, 30 mi/48 km wide. OTARU is on S shore. Formerly sometimes called Strogonov Bay.

Ishikawa (ee-shee-KAH-wah), prefecture (□ 1,619 sq mi/4,193 sq km; 1990 population 1,164,627), central HONSHU, W central JAPAN, on SEA OF JAPAN; ⊙ KANAZAWA, the largest city. Includes NOTO PENINSULA, NANATSU-JIMA and Hegura-jima islands. Bordered SE by TOYAMA and GIFU prefectures, SW by FUKUI prefecture. Wooded mountains in the interior and fertile plains along the coast. Rice, machinery, lumber, raw silk, and lacquerware are produced.

Ishikawa (ee-shee-KAH-wah), city, on E coast of central OKINAWA island, OKINAWA prefecture, SW JAPAN, in the RYUKYU ISLANDS, 12 mi/20 km N of NAHA; 26°25′N 127°49′E. Fishing port.

Ishikawa (ee-shee-KAH-wah), town, Ishikawa district, FUKUSHIMA prefecture, N central HONSHU, NE JAPAN, 40 mi/65 km S of FUKUSHIMA city; 37°08′N 140°27′E. Minerals.

Ishikoshi (ee-shee-KO-shee), town, Tome county, MIYAGI prefecture, N HONSHU, NE JAPAN, 37 mi/60 km N of SENDAI; 38°45′N 141°11′E.

Ishim (ee-SHIM), city (2006 population 67,785), SE TYUMEN oblast, W SIBERIA, RUSSIA, on the ISHIM River, on highway junction and the TRANS-SIBERIAN RAILROAD, 200 mi/322 km SE of TYUMEN; 56°09′N 69°27′E. Elevation 278 ft/84 m. An agricultural center, it produces agricultural machinery and chemicals, concrete products, furniture, garments, shoes, distilled liquers, and food products. Has a pedagogical institute. An old trading village known as Korkina, it was renamed and made a city in 1782. Underwent a period of rapid development and expansion during World War II, with many industries being relocated to the region from European Russia ahead of advancing German armies. Seat of the local Russian Orthodox diocese; has the Bogoyavlenskiy (Russian=God's manifestation) Cathedral (built in the early 18th century).

Ishim (ee-SHIM), river, 1,530 mi/2,450 km long, W Siberian Russia and KAZAKHSTAN; rises N of KARAGANDA (Kazakhstan), flows W past ATBASAR, N past PETROPAVLOVSK, and into RUSSIA (OMSK oblast), joining the IRTYSH River at UST'-ISHIM. Partially navigable. The Ishim space complex, specializing in launching small military and scientific satellites, is on its shores, in Kazakhstan.

Ishimbay (ee-shim-BEI), city (2005 population 70,500), S central BASHKORTOSTAN Republic, E European Russia, on the BELAYA RIVER, near highway, 100 mi/161 km S of UFA; 53°27′N 56°02′E. Elevation 544 ft/165 m. Founded in 1932 and made a city in 1940, Ishimbay developed around the first major oil field of the Volga-Ural region, formerly the leading oil area of Russia. Ishimbay's chief industries are oil refining, petrochemical production, and manufacturing of machinery for oil production and processing; catalysts;

clothing factory; winery and distillery. Center of a network of oil and gas pipelines.

Ishim Steppe (ee-SHIM), black-earth area of WEST SIBERIAN PLAIN, TYUMEN and OMSK oblasts, RUSSIA, between lower IRTYSH and TOBOL rivers; drained by the lower ISHIM River. Rich agricultural area (wheat, dairy farming).

Ishinomaki (ee-SHEE-no-MAH-kee), city (2005 population 167,324), Miyagi prefecture, N HONSHU, NE JAPAN, on ISHINOMAKI BAY, 25 mi/40 km N of SENDAI; 38°25′N 141°18′E. Commercial and fishing port and a center for marine product (oysters) processing and paper manufacturing.

Ishinomaki Bay (ee-SHEE-no-mah-kee), Japanese *Ishinomaki-wan* (ee-SHEE-no-MAH-kee-WAHN) or **Sendai Bay** (SEN-dah-ee), inlet of PACIFIC OCEAN, in MIYAGI prefecture, N HONSHU, NE JAPAN; sheltered E by OJIKA PENINSULA; 25 mi/40 km E-W, 10 mi/16 km N-S. Contains, near MATSUSHIMA, hundreds of pine-clad islets. ISHINOMAKI is on N, SENDAI on W shore.

Ishioka (ee-shee-O-kah), city, IBARAKI prefecture, central HONSHU, E central JAPAN, 16 mi/25 km S of MITO; 36°11′N 140°17′E.

Ishizuchi, Mount (ee-SHEE-zoo-chee), Japanese *Ishizuchiyama* (ee-shee-ZOO-chee-YAH-mah), highest peak (6,497 ft/1,980 m) of SHIKOKU, in NW part of island, W JAPAN, in EHIME prefecture, 11 mi/18 km SSW of SAIJO. Site of Shinto temple. Cherry and fir trees on slopes.

Ishkal, Lake (esh-KEL), BIZERTE province, N TUNISIA, at foot of the Jabal Ishkal (elevation 1,667 ft/508 m), 7 mi/11.3 km N of MATIR; 9 mi/14.5 km long, 4 mi/6.4 km wide. Marshy banks. Outlet (E) to LAKE BIZERTE. On the SE shore is BAHIRA ISHKAL NATIONAL PARK, a marshland nature preserve with wetlands on a bird migration route from EUROPE to AFRICA. Also spelled Ischkeul.

Ishkashim (ish-kah-SHEEM), village, SW BADAKHSHAN AUTONOMOUS VILOYAT, TAJIKISTAN, in the PAMIR, on PANJ RIVER (AFGHANISTAN border), and 55 mi/89 km S of KHORUGH; 36°44′N 71°37′E. Tertiary level administrative center; horses.

Ishkhoy-Yurt (EESH-hoyee-YOORT), village (2005 population 3,055), E CHECHEN REPUBLIC, S European Russia, in the NE foothills of the Greater CAUCASUS Mountains, 15 mi/24 km ESE of GUDERMES; 43°13′N 46°23′E. Elevation 767 ft/233 m. Meat and dairy livestock raising.

Ishkuman (ISH-koo-muhn), town, GILGIT district, NORTHERN AREAS, extreme NE PAKISTAN; 36°32′N 73°49′E. In fertile valley in HINDU KUSH mountains; traversed (N-S) by ISHKUMAN River (left tributary of GILGIT RIVER). Exports seed potatoes and fruit. Popular for tourism and trekking.

Ishley (eesh-LYAI), village (2005 population 3,150), N CHUVASH REPUBLIC, central European Russia, on road and near railroad, 10 mi/16 km SE of CHEBOKSARY; 56°01′N 47°03′E. Elevation 685 ft/208 m. High voltage electric devices manufacturing; brick factory; agricultural collective farm. Formerly called Ishley-Pokrovskoye.

Ishley-Pokrovskoye, RUSSIA: see ISHLEY.

Ishmant (ish-MAHNT), village, BENI SUEF province, Upper EGYPT, on IBRAHIMIYA CANAL, on railroad, and 15 mi/24 km NNE of BENI SUEF; 29°12′N 31°11′E. Cotton, cereals, sugarcane.

Ishm River (EESHM), coastal stream, c.25 mi/40 km long, ALBANIA; formed by 3 parallel mountain torrents rising in area of TIRANË; flows NW to DRIN GULF of the ADRIATIC SEA 12 mi/19 km S of LEZHË. Also Ishmi River

Ishnya (eesh-NYAH), town (2006 population 3,275), SE YAROSLAVL oblast, central European Russia, in the KOTOROSL' RIVER basin near NERO LAKE, on highway and near railroad junction, less than 2 mi/3.2 km SSW of (and administratively subordinate to) ROSTOV; 57°11′N 39°25′E. Elevation 295 ft/89 m.

Manufacturing (ceramic products), agricultural equipment servicing.

Ishpeming (ish-PEM-ing), town (2000 population 6,686), MARQUETTE county, NW UPPER PENINSULA, MICHIGAN, 14 mi/23 km SW of MARQUETTE, in MARQUETTE IRON RANGE; 46°29′N 87°39′W. Railroad junction. Lumbering; cattle; manufacturing (logging, construction, sand and gravel, hardwood parquet); iron mining. U.S. National Ski Hall of Fame; birthplace of skiing in America; ski tournaments held here since 1888. Incorporated as a village 1871, as city 1873.

Ishperu Arkari, PAKISTAN: see ARKARI.

Ishpushta (ish-POOSH-tah) or **Sabz Ishpishta**, village, SAMANGAN province, NE AFGHANISTAN, on N slopes of the Hindu Kush, 37 mi/60 km SW of DOSHI, near SURKHAB RIVER; 35°18′N 68°06′E. Site of one of Afghanistan's leading coal mines.

Ishtykhan (EESH-tee-kahn), city, W SAMARKAND wiloyat, UZBEKISTAN, c.20 mi/32 km E of KATTAKURGAN, between KARA DARYA and Ak Darya rivers; 39°58′N 66°29′E. Cotton.

Ishuatán, MEXICO: see IXHUATÁN.

Ishurdi (eesh-shor-dee), town (2001 population 62,617), PABNA district, W EAST BENGAL, BANGLADESH, 14 mi/23 km NW of PABNA; 24°11′N 88°59′E. Railroad junction (workshops). Manufacturing (rice milling, tobacco processing); agriculture (rice, jute, rape and mustard, tobacco).

Isiboro River (ee-see-BO-ro), 130 mi/209 km long, COCHABAMBA department, central BOLIVIA; rises in Serranía de MOSETENES NNE of COCHABAMBA; flows NE, through tropical lowlands, to the SECURE RIVER c. 20 mi/32 km W of LIMOQUIJE; 15°28′S 65°05′W. Receives ICHOA RIVER (left). Navigable for 110 mi/177 km.

Isiboro-Sécure (ee-see-BO-ro–SE-koo-rai), national park, CHAPARE province, COCHABAMBA department, and MOXOS province, BENI department, NE BOLIVIA, between the ISIBORO and SÉCURE rivers and the Serranías Sejerruma, Yanakaka, and MOSETENES. Covers 2.97 million acres/1.2 million ha. The more remote parts of this park still have a diverse population of Amazonic wildlife. Loggers and farmers have infiltrated much of the park. Established 1965.

Isidro Fabela (ee-SEED-ro fah-BE-lah), town, ⊙ Isidro Fabela municipio, MEXICO, 20 mi/33 km WNW of MEXICO CITY on E slopes of Cerro Monte Alto, included within the ZONA METROPOLITANA DE LA CIUDAD DE MÉXICO; 19°34′N 99°26′W. Elevation 9,186 ft/2,800 m. Cool climate, in wooded area on W slopes bordering the Basin of Mexico; agriculture (small farming). Town formerly Tlazala de Fabela, municipio formerly Iturbide.

Isigny-le-Buat (ee-zee-nyee–luh–boo-ah), commune (□ 28 sq mi/72.8 sq km), MANCHE department, BASSENORMANDIE region, NW FRANCE, 10 mi/16 km SE of AVRANCHES; 48°37′N 01°10′W. Dairying.

Isigny-sur-Mer (ee-zee-nyee–syur–mer), commune (□ 6 sq mi/15.6 sq km), CALVADOS department, BASSENORMANDIE region, NW FRANCE, small port at confluence of VIRE and AURE rivers, near ENGLISH CHANNEL, 14 mi/23 km N of SAINT-LÔ; 49°19′N 01°06′W. Well-known distributor of dairy products, (especially butter) and pastries. American troops landed at nearby UTAH BEACH (NW) and OMAHA BEACH (NE) on June 6, 1944, in NORMANDY invasion of World War II.

Isikizya (ee-see-KEE-zyah), village, TABORA region, NW central TANZANIA, 25 mi/40 km NE of TABORA; 04°54′S 32°48′E. Timber; goats, sheep; wheat, corn. Also called Pigawasi.

Isili (EE-zee-lee), village, NUORO province, S central SARDINIA, ITALY, 36 mi/58 km N of CAGLIARI; 39°44′N 09°06′E. Site of nuraghi, including one of the best-preserved and largest in Sardinia.

Isil Kol, RUSSIA: see ISIL'KUL'.

Isil'kul' (ee-seel-KOOL), city (2006 population 26,395) SW OMSK oblast, W SIBERIA, RUSSIA, 12 mi/19 km E of the Russia-KAZAKHSTAN border, on road junction and the TRANS-SIBERIAN RAILROAD, 90 mi/145 km W of OMSK; 54°55′N 71°16′E. Elevation 387 ft/117 m. Highway junction; local transshipment point. Knitwear, reinforced concrete; grain storage. Sometimes spelled Isil Kol.

Isimala (ee-see-MAH-lah), village, IRINGA region, W central TANZANIA, 100 mi/161 km WNW of IRINGA, on NJOMBE RIVER, in W edge of RUAHA NATIONAL PARK; RUNGWA GAME RESERVE to W; 07°25′S 34°11′E. Livestock, timber.

Isimila (ee-see-MEE-lah), archaeological site, IRINGA region, central TANZANIA, 5 mi/8 km SE of IRINGA, near LITTLE RUAHA RIVER; 07°50′S 35°43′E. Prehistoric tools, artifacts; fossilized hominids.

Isimlian, Bulgaria: see SMILYAN.

Isin (IS-in), the capital city of an ancient Semitic kingdom of N BABYLONIA. The city became important after the third dynasty of Ur fell to the Elamites and the Amorites (c.2025 B.C.E.). The phase from c.2025–c.1763 B.C.E. is sometimes called the Isin-Larsa period. Many city-states vied with one another, but Isin and LARSA were the most powerful of these. Excavations have brought to light the law code of King Lipit-Ishtar of Isin. This code is one of several codes that predate the stele of Hammurabi. Also spelled Issin.

Isingiro, administrative district, WESTERN region, SW UGANDA, on TANZANIA border (to S). As of Uganda's division into eighty districts, Isingiro's border districts include NTUNGAMO (W), MBARARA (NW), and KIRUHURA (N). Agricultural area. Formed in 2005 from S portion of former MBARARA district (current Mbarara district formed from W portion, IBANDA district from NW portion, and Kiruhura district from N, E, and central portions).

Isinglass River (EI-zin-glas), c.15 mi/24 km long, SE NEW HAMPSHIRE; rises at Bow Lake in W STRAFFORD county; flows E to the COCHECO below ROCHESTER.

Isiolo (ee-see-O-lo), town, district administrative center, EASTERN province, central KENYA, N of Mount KENYA, 120 mi/193 km NNE of NAIROBI; 00°25′N 37°37′E. Trade center; coffee, sisal, wheat, corn; livestock. Airfield; military barracks.

Isirio, CONGO: see ISIRO.

Isiro (ee-SEE-ro), town, ORIENTALE province, NE CONGO, on railroad, and 185 mi/298 km E of BUTA; 02°46′N 27°37′E. Elev. 2,398 ft/730 m. Commercial center in cotton-producing area. Cotton ginning; palm products; repair shops for automobiles. Also center of Mangbettu tribe. Has Dominican mission, hospital, airport. Formerly called PAULIS. GOSSAMULEZ-PAULIS cotton center is 2 mi/3.2 km E. Nala coffee plantations and Protestant mission are 6 mi/9.7 km N. Also spelled ISIRIO.

Isis, ENGLAND: see THAMES.

Isisford (EI-sis-fuhrd), township, QUEENSLAND, NE AUSTRALIA, 73 mi/117 km S of Ilfracombe; 24°17′S 144°30′E.

Iskandar (ees-kahn-DAHR), town, TOSHKENT wiloyat, NE UZBEKISTAN; 41°36′N 69°41′E.

Iskandariyah, Al, EGYPT: see ALEXANDRIA.

Iskander, town, NAMANGAN wiloyat, NE UZBEKISTAN, on KAZAKHSTAN border, on CHIRCHIK RIVER, 25 mi/40 km NE of TOSHKENT (linked by railroad); 41°34′N 69°14′E. Cotton; fruit.

Iskanderun, TURKEY: see ISKENDERUN.

Iskar, Bulgaria: see ISKUR.

Iskateley (ees-KAH-tee-lye-ee), urban settlement (2006 population 6,405), central NENETS AUTONOMOUS OKRUG, administratively subordinate to ARCHANGEL oblast, NE European Russia, on the right bank of the PECHORA River, on local road, 2 mi/3.2 km N of NAR'YAN-MAR; 67°40′N 53°01′E. Logging and lumbering. Has a hospital.

Iskele, CYPRUS: see TRIKOMO.

Iskelib, TURKEY: see ISKILIP.

Iskenderon, TURKEY: see ISKENDERUN.

Iskenderun, city (2000 population 159,149), S TURKEY, on the GULF OF ISKENDERUN, an inlet of the MEDITERRANEAN SEA; 36°37'N 36°08'E. The principal Turkish port on the Mediterranean, it has a large steel plant, produces chemicals, and is the terminus for an oil pipeline. The city was founded by Alexander the Great to commemorate his victory over the Persians at ISSUS in 333 B.C. Developed as an outlet for overland trade from Iran, India, and E Asia. In C.E. 1515 the Ottoman Empire under Selim I, its ruler, captured the city. Iskenderun was transferred (1920) to the French Syria League of Nations mandate as part of the sanjak of ALEXANDRETTA, but was returned to Turkey in 1939. Also spelled Iskenderun, Iskenderon; formerly called Alexandretta.

Iskenderun, Gulf of (Turkish=*Iskenderun Körfezi*) an inlet of the MEDITERRANEAN SEA, S TURKEY. Location of ISKENDERUN, Turkey's main port on the Mediterranean. The SEYHAN and CEYHAN rivers flow into the Gulf; AMANOS MOUNTAINS along its E shore. Also called Gulf of Alexandretta.

Iskilip (is-ki-LEEP), town, N central TURKEY, near the KIZIL IRMAK, 28 mi/45 km NW of ÇORUM; 40°45'N 34°28'E. Wheat, mohair goats. Formerly Iskelib.

Iskininskiy, KAZAKHSTAN: see ESKENE.

Iskitim (ees-kee-TEEM), city (2006 population 61,520), E NOVOSIBIRSK oblast, SW SIBERIA, RUSSIA, on the TURK-SIB railroad, near the mouth of the Berd River, on the NOVOSIBIRSK RESERVOIR of the OB' RIVER, 40 mi/64 km SSE of NOVOSIBIRSK; 54°38'N 83°18'E. Elevation 374 ft/113 m. Production of heating equipment, plumbing, cement; synthetic fibers; grain storage, food industries (bakery, meat cannery). Developed after 1935, and made city in 1938.

Isko (EES-ko), village, BANDUNDU province, W CONGO, S of LAKE MAI-NDOMBE.

Iskra (ees-KRAH), village, PLOVDIV oblast, PURVOMAI obshtina, S central BULGARIA, at the N foot of the E RODOPI Mountains, 12 mi/19 km SSW of PURVOMAI; 41°56'N 25°08'E. Tobacco, fruit, truck, grain, vineyards; manufacturing (shipbuilding parts, agricultural machinery parts, foodstuffs, machine parts, vegetable oil, construction materials). Formerly Karadzhilar (until 1906), later Popovo (until 1950).

Iskrets (ees-KRETS), village, SOFIA oblast, svoge obshtina, W BULGARIA, in the W BALKAN MOUNTAINS, on the Iskrets River (left tributary of the ISKUR River), 20 mi/32 km N of SOFIA; 42°59'N 23°15'E. Grain, sheep.

Iskur (EES-kuhr), river, c.250 mi/402 km long, W BULGARIA; rising in the RILA MOUNTAINS and flowing generally NE past SOFIA, and through the STARA PLANINA Mountains to the DANUBE River; 43°20'N 24°08'E. Bulgaria's largest reservoir is on the Iskur, supplying water to Sofia. The gorge of the Iskur is one of the chief passes through the Stara Planina Mountains.

Iskur Dam (EES-kuhr), in SW BULGARIA, on the ISKUR River, 20 mi/32 km SE of SOFIA. Construction project (begun 1951) includes three hydroelectric stations below the dam—Pasarel (at dam site), Kokalyane (12 mi/19 km SE of Sofia), and Sofia (just NE of the city); 42°28'N 23°55'E. A multipurpose project, the reservoir supplies Sofia's potable water and irrigates the semiarid Sofia plain. Largest reservoir in Bulgaria. Formerly called Pasarel Dam, later Stalin Dam.

Iskushuban (es-koo-SHOO-bahn), Italian *scusciuban*, village, NE SOMALIA, in GARGORE JAEIL Valley, 70 mi/113 km W of HAFUN. Water hole, livestock market. Formerly SKUSHUBAN.

Isla (EES-lah), town, SE VERACRUZ, MEXICO, 40 mi/58 km SSW of SAN ANDRÉS TUXTLA; on railroad and 3 mi/5 km S of Mexico Highway 145; 18°01'N 95°35'W.

Isla Angel de la Guarda National Park (EES-lah AHN-hel dai la GWAHR-dah), a biological reserve in BAJA CALIFORNIA, MEXICO. It is an island located in the Gulf of CALIFORNIA separated from the peninsula by the Canal of Whales (Canal Las BALLENAS). A chain of mountains reaching 3,281 ft/1,000 m cover the area. The W coast is inaccessible. The island is uninhabited.

Isla, Cabo del (EES-lah, KAH-bo del), island off CARIBBEAN coast of MARGARITA ISLAND (NE), NUEVA ESPARTA state, NE VENEZUELA, 10 mi/16 km N of LA ASUNCIÓN; 11°11'N 63°58'W.

Isla Cañas Wildlife Refuge, LOS SANTOS province, PANAMA, near mouth of Limón River; occupies a barrier island off S shore of AZUERO PENINSULA.

Isla Cancún, MEXICO: see CANCÚN.

Isla-Cristina (IS-lah–kree-STEE-nah), city, HUELVA province, SW SPAIN, in ANDALUSIA, on peninsula at the GUADIANA estuary, 4 mi/6.4 km ESE of AYAMONTE, 21 mi/34 km W of HUELVA; 37°12'N 07°19'W. Fishing (tuna, sardines) and canning center. Manufacturing of fertilizers; saltworks. Sometimes written Isla Cristina.

Isla de las Perlas National Park, Pearl Islands, PANAMA, in PANAMA BAY. Park under development.

Isla del Caño Biological Reserve (EE-sla dehl CAHN-yo), island of OSA PENINSULA, COSTA RICA. Small volcanic island preserves archaeological sites and wildlife. Uninhabited, access by launch.

Isla de León, SPAIN: see LEÓN ISLAND.

Isla de los Estados (EES-lah dai los es-TAH-dos), English *Staten Island*, rocky island (□ 209 sq mi/543.4 sq km) in the South Atlantic, 18 mi/29 km E of SE tip of main island of Tierra del Fuego, Argentina, across Le Maire Strait; c.45 mi/72 km long. Rises to 3,000 ft/914 m. Has cold, humid climate and is sparsely populated. Some goat and sheep raising, seal and nutria hunting. Has seismographic station. A small port, San Juan de Salvamento, is on E coast. Its E tip is Cape San Juan (54°42'S 63°43'W). Just N of Isla de los Estados are the small New Year's Islands.

Isla de Maipo (EES-lah dai MEI-po), town, ⊙ Isla de Maipo comuna, TALAGANTE province, METROPOLITANA de Santiago region, central CHILE, surrounded by arms of MAIPO RIVER, 25 mi/40 km SW of SANTIAGO; 33°45'S 70°54'W. Resort and agricultural center (cereals, alfalfa, fruit, livestock).

Isla de Pascua, CHILE: see EASTER ISLAND.

Isla de Pinos, CUBA: see JUVENTUD, ISLA DE LA.

Isla Grande Airport, PUERTO RICO: see FERNANDO LUIS RIBA DOMINICCI AIRPORT.

Isla Guadalupe Biosphere Reserve (EES-lah wah-dah-LOO-pe), biosphere reserve in BAJA CALIFORNIA, MEXICO, in the PACIFIC OCEAN, 168 mi/270 km from the W coast of Baja California; 28°45'N 118°10'W–29°15'N 118°30'W. It measures 21 mi/33 km from the N to the S and is only 6 mi/10 km wide. It is of volcanic origin and a mountain range in the N reaches 4,593 ft/1,400 m. The island is covered with desert vegetation and is home to herds of sea lions.

Isla Guambin (EES-lah gwahm-BEEN), national park, in Aisén de GENERAL CARLOS Ibañez del Campo region, an island 135 mi/217 km W of PUERTO CISNES.

Islahiye (Turkish=*Islâhiye*) township, S TURKEY, on ALEPPO-ADANA railroad, and 40 mi/64 km W of GAZIANTEP; 37°02'N 36°37'E. Wheat, barley, rice, lentils.

Isla Iguana Wildlife Refuge, LOS SANTOS province, PANAMA, N of PEDASÍ, in GULF OF PANAMA; 40 acres/16 ha. Refuge for birds and whales (in season).

Isla, La, MEXICO: see SAN ANTONIO LA ISLA.

Isla la Libertad (EES-lah lah LEE-ber-tahd), island, S URUGUAY, MONTEVIDEO department; 34°54'S 56°14'W. Small island in BAHIA DE MONTEVIDEO.

Islamabad, federal capital district, NE PAKISTAN; ⊙ ISLAMABAD. Created around national capital of ISLAMABAD, between N PUNJAB province and NORTHWEST FRONTIER PROVINCE.

Islamabad (IS-lah-mah-BAHD) [=city of Islam], city, ⊙ Pakistan and Islamabad federal capital district, NE Pakistan, just NE of RAWALPINDI, the former interim capital. Construction began in 1960, under direction of Constantine Doxiades and Edward Durrell Stone. A totally planned city, linked to the military garrison of Rawalpindi. Well-defined diplomatic residential areas and industrial-commercial zones. Rawalpindi being incorporated in the plan for Islamabad Internatl airport to SW, at Rawalpindi. Light manufacturing industries. Points of interest include Pakistan House, the home of the president; the national assembly building; the National and Islamic University; the Grand National Mosque; Supreme Court; and the botanical gardens. The nearby Margala and Murree Hills serve as recreation and summer resort destinations for many diplomatic missions. Near the city are the historical ruins of Taxila.

Isla Mayor (EE-slah mei-OR), sprawling village and island, SEVILLE province, SW SPAIN, formed by right arm of the lower GUADALQUIVIR 13 mi/21 km SSW of SEVILLE; c.25 mi/40 km long N-S, up to 9 mi/14.5 km wide. Largely marshland, used for stock raising.

Islamey (ees-lah-MYAI), village (2005 population 11,015), N central KABARDINO-BALKAR REPUBLIC, S European Russia, in the foothills of the N CAUCASUS Mountains, on highway, 12 mi/19 km WNW of NAL'-CHIK, and 4 mi/6 km W of BAKSAN; 43°40'N 43°27'E. Elevation 1,663 ft/506 m. In agricultural area (grain, vegetables; livestock); food processing. Formerly called Kyzburun Vtoroy (Russian=Kyzburun the second).

Islamnagar (is-LAHM-nuh-guhr), town, BUDAUN district, N central UTTAR PRADESH state, INDIA, 31 mi/50 km NW of BUDAUN. Trades in wheat, pearl millet, mustard, sugarcane.

Islamorada (ei-luh-muh-RUH-dah), town (□ 1 sq mi/2.6 sq km), MONROE county, upper FLORIDA KEYS, 20 mi/32 km SW of KEY LARGO; 24°55'N 80°38'W. Manufacturing includes printing and publishing; tourism, sport fishing.

Islampur (IS-lahm-poor), town (2001 population 29,855), NALANDA district, S central BIHAR state, E INDIA; 25°09'N 85°12'E.

Islampur, town, SANGLI district, MAHARASHTRA state, W central INDIA, 22 mi/35 km NW of SANGLI. Market center for millet, peanuts, wheat, rice; hand-loom weaving. Sometimes called Urun Islampur.

Islampur, BANGLADESH: see JAMALPUR.

Islam Qala (is-LAHM KAH-lah), frontier post, HERAT province, NW AFGHANISTAN, on Iranian border, 65 mi/105 km WNW of HERAT, near the Hari Rud River; 34°40'N 61°04'E. Formerly called Kafir Qala. Also spelled Islam Kala or Eslam Kala.

Islam-Terek, UKRAINE: see KIROVS'KE.

Isla Mujeres (EES-lah moo-HE-res), town (□ 1 sq mi/2.6 sq km), QUINTANA ROO, SE MEXICO, on small Mujeres Island (□ 1.3 sq mi/3.4 sq km), 5 mi/8 km off NE YUCATÁN Peninsula, 16 mi/25 km NE of CANCÚN. Tourist resort. Formerly belonging to great Maya federation, island has many archaeological remains.

Island, county (□ 517 sq mi/1,344.2 sq km; 2006 population 81,489), NW WASHINGTON; ⊙ COUPEVILLE, Whidbey Island county consists of WHIDBEY ISLAND and CAMANO ISLAND in Puget Sound, NW of EVERETT, also SMITH ISLAND in STRAIT OF JUAN DE FUCA (W); 48°09'N 122°35'W. Alfalfa, hay, berries, vegetables; fishing and summer resorts. County bounded by Skagit Bay, Davis Slough and Port Susan Bay (arm of Puget Sound) on E, by main channel of Puget Sound, on SW. Saratoga Passage separates N 2 islands. Main town is OAK HARBOR, on WHIDBEY ISLAND. Ebey's Landing National Historical Reserve at Coupville; Camano Island State Park in E; DECEPTION PASS and Joseph Whidbey state parks in N; South Whidbey State Park in S. Formed 1853.

Island, village (2000 population 435), MCLEAN county, W KENTUCKY, 22 mi/35 km S of OWENSBORO; 37°26′N 87°09′W. In agricultural area (tobacco, grain; livestock; timber); manufacturing (wooden pallets).

Island: see ICELAND.

Island Beach, borough and beach, OCEAN county, E NEW JERSEY, on lower 8 mi/12.9 km of peninsula between BARNEGAT BAY and the ATLANTIC OCEAN, S of SEASIDE PARK and N of Barnegat Inlet; 39°50′N 74°05′W. It is a state park, opened in 1959, noted for plant life and wildlife. Borough set off from BERKELEY township and incorporated 1933.

Island City, village (2006 population 902), UNION county, NE OREGON, on GRANDE RONDE RIVER, 4 mi/6.4 km E of LA GRANDE; 45°20′N 118°02′W. Elevation 2,750 ft/838 m. Agriculture (cherries, apples, grain, potatoes; cattle). Part of Wallowa-Whitman National Forest to NW.

Island Falls, town, AROOSTOOK county, central MAINE, 22 mi/35 km SW of HOULTON, and on Matawamkeag Lake; 45°59′N 68°14′W. Trade center for agricultural, lumbering, recreation (hunting, fishing) area. Settled 1843, incorporated 1872.

Island Falls (EI-luhnd FAHLZ), unincorporated village, NE ONTARIO, E central CANADA, on ABITIBI RIVER (66-ft/20-m falls), and included in unorganized N part of COCHRANE; 49°34′N 81°22′W. Hydroelectric power center, supplying Cochrane mining region.

Island Falls (AI-land), village, E SASKATCHEWAN, CANADA, on CHURCHILL RIVER (falls), and 55 mi/89 km NNW of FLIN FLON; 55°31′N 102°21′W. Hydroelectric power center, supplying Flin Flon and SHERRIDON mining region.

Island Heights, resort borough (2006 population 1,877), OCEAN county, E NEW JERSEY, on TOMS RIVER, near BARNEGAT BAY, just E of TOMS RIVER; 39°56′N 74°09′W.

Islandia, village (2006 population 3,086), SUFFOLK county, SE NEW YORK, 5 mi/8 km N of CENTRAL ISLIP; 40°49′N 73°10′W. Manufacturing.

Island Lake, city (2000 population 8,153), MCHENRY and Lake counties, NE ILLINOIS, suburb 34 mi/55 km NW of downtown CHICAGO, 6 mi/9.7 km SE of MC-HENRY; 42°16′N 88°12′W. Small lakes in area. Manufacturing (printing, oil field valves and fittings). Moraine Hills State Park to NW.

Island Lake (EI-luhnd) summer village (□ 1 sq mi/2.6 sq km; 2001 population 199), central ALBERTA, W CANADA, 14 mi/23 km from ATHABASCA, in Athabasca County No. 12; 54°51′N 113°33′W. Established 1958.

Island Lake (EI-luhnd) (□ 550 sq mi/1,430 sq km), E MANITOBA, W central CANADA, on ONTARIO border; 55 mi/89 km long, 20 mi/32 km wide; 53°45′N 94°30′W. Drains into HUDSON BAY.

Island Lake, reservoir (□ 18 sq mi/46.8 sq km), ST. LOUIS county, NE MINNESOTA, on CLOQUET RIVER, 20 mi/32 km N of DULUTH; 47°00′N 92°14′W. Maximum capacity 171,520 acre-ft. Formed by Island Lake Dam (45 ft/14 m high), built (1915) for power generation; also used for recreation. Cloquet Valley State Forest to N.

Island Lake South (EI-luhnd LAIK), summer village (2001 population 71), central ALBERTA, W CANADA, 13 mi/21 km from ATHABASCA, in ATHABASCA County No. 12; 54°50′N 113°32′W. Incorporated 1983.

Island No. 10, former island in the MISSISSIPPI RIVER, between NW TENNESSEE and SE MISSOURI. Site of an important campaign of the Civil War.

Island Park, village (2000 population 215), FREMONT county, E IDAHO, 35 mi/56 km NE of SAINT ANTHONY, on HENRYS FORK river; 44°34′N 111°20′W. In Targhee National Forest; ISLAND PARK DAM and RESERVOIR on Henrys Fork is 1 mi/1.6 km to W, Harriman State Park is on its S shore; YELLOWSTONE NATIONAL PARK to E; Henrys Lake reservoir and State Park to N.

Island Park, residential village (2006 population 4,707), NASSAU county, SE NEW YORK, on island off S shore of

W LONG ISLAND, 7 mi/11.3 km S of HEMPSTEAD; 40°36′N 73°39′W. Connected by causeways to LONG BEACH (S) and Long Island. Incorporated 1926.

Island Park, village, PORTSMOUTH town, NEWPORT county, SE RHODE ISLAND, on RHODE ISLAND, 2 mi/3.2 km E of Portsmouth village; 41°37′N 71°13′W. Yacht club, harbor, and numerous beaches located here.

Island Park, locality, HENNEPIN county, E MINNESOTA, residential section in S part of town of MOUND, on Phelps island in Lake MINNETONKA, 20 mi/32 km W of MINNEAPOLIS; 44°55′N 93°38′W. Cooks Bay to W.

Island Park Dam, Idaho: see HENRYS FORK.

Island Park Reservoir (□ 14 sq mi/36 sq km), FREMONT county, E IDAHO, on HENRYS FORK of the SNAKE RIVER, abutting Targhee National Forest; 45 mi/72 km NNE of REXBURG; 44°25′N 111°24′W. Maximum capacity 169,646 acre-ft. Formed by ISLAND PARK DAM (94 ft/29 m high), built (1938) by the Bureau of Reclamation for irrigation; also used for recreation.

Island Pond, NEW HAMPSHIRE: see HAMPSTEAD.

Island Pond, VERMONT: see BRIGHTON.

Islands, Bay of, inlet (20 mi/32 km long, 10 mi/16 km wide at entrance) of the Gulf of St. Lawrence, SW NEWFOUNDLAND AND LABRADOR, CANADA, 20 mi/32 km N of CORNER BROOK; 49°10′N 58°15′W. Contains TWEED, PEARL, Woods, GUERNSEY, and GOVERNORS islands. E shore is deeply indented by three arms; S arm receives HUMBER RIVER estuary. On shore are several fishing settlements.

Islands, Bay of, branching inlet of the PACIFIC, NE NORTHLAND region, NORTH ISLAND, NEW ZEALAND, c.100 mi/240 km N of AUCKLAND. Geologically a submerged river system, with c.150 islands. RUSSELL is on E shore and WAITANGI on the W shore. Site of many Maori-Pakeha (Westerner) contacts. Now popular historic and vacation resort with deep-sea fishing. Semi-tropical fruits including citrus. Bay of Islands Maritime and Historic Park formed 1978 to conserve historic and natural sites on shoreline and islands.

Islands Trust (EI-luhndz TRUHST), regional district (2006 population 23,000), BRITISH COLUMBIA, W CANADA. Federation of independent local governments working to protect and preserve the environment and culture of member communities; encompasses the land and water between the British Columbia mainland and S VANCOUVER ISLAND. Includes the municipality of BOWEN ISLAND; the islands of DENMAN, GABRIOLA, GALIANO, HORNBY, LAS-QUETI, MAYNE, NORTH PENDER, South PENDER, SATURNA, SALT SPRING, and THETIS; and the EXECUTIVE Islands.

Island View, village, KOOCHICHING county, N MINNESOTA, 9 mi/14.5 km E of INTERNATIONAL FALLS, on peninsula in RAINY LAKE, Black Bay to S, Black Bay Narrows to E; VOYAGEURS NATIONAL PARK to E; 48°36′N 93°11′W.

Isla Pucú (EES-lah poo-KOO), town, La Cordillera department, S central PARAGUAY, 45 mi/72 km E of ASUNCIÓN; 25°17′S 56°54′W. In agricultural area (fruit, tobacco, livestock).

Isla River (IS-lah), 46 mi/74 km long, Angus, central Scotland; rises in NW Angus, 10 mi/16 km SSE of BRAEMAR; flows S and SW, into Perth and Kinross and past COUPAR ANGUS, to the Tay River 4 mi/6.4 km WSW of Coupar Angus.

Isla River (IS-lah), 18 mi/29 km long, in Aberdeenshire, NE Scotland; rises 3 mi/4.8 km NE of DUFFTOWN; flows NE, past Dufftown and Keith, to DEVERON RIVER just N of Rothiemay.

Islas de Golfo de California, a special biological reserve in S BAJA CALIFORNIA SUR, MEXICO. This is a multi-island reserve including ÁNGEL DE LA GUARDA, SAN MARCOS, Coronado, CARMEN, Monserrate, Santa Catarina, SANTA CRUZ, SAN JOSÉ, ESPÍRITU SANTO, CERRALVO (in front of the E coast of the BAJA CALI-

FORNIA peninsula), and LA TIBURÓN, the largest island in the country.

Islas de la Bahía, HONDURAS: see BAHÍA, ISLAS DE LA.

Islas del Maiz, NICARAGUA: see CORN ISLANDS.

Islas Taboga y Urabá Wildlife Preserve, includes parts of TABOGA and Urabá islands in PANAMA BAY, PANAMA.

Isla Tiburón Biosphere Reserve (EES-lah tee-boo-RON), a biological reserve off the W central coast of SONORA, MEXICO. This is the largest island in the Gulf of California and it is a reserve for flora and fauna, such as cormorants, seagulls, pelicans, and sea lions. A permit is required to visit this island.

Isla Tigre (EES-lah TEE-grai), island, S URUGUAY, SAN JOSE department, 15 mi/24 km WNW of MONTEVIDEO, at mouth of SANTA LUCIA RIVER where it enters the Rio de la Plata; 34°47′S 56°23′W.

Isla Umbú (EES-lah oom-BOO), town, Ñeembucú department, S PARAGUAY, 13 mi/21 km SE of PILAR; 26°59′S 58°18′W. Located on road to PASO DE PATRIA. Stock-raising center; tanneries.

Isla Verde (EES-lah VER-de), town, SE CÓRDOBA province, ARGENTINA, 45 mi/72 km SSW of MARCOS JUÁREZ. Corn, soybeans, wheat, flax, alfalfa; cattle, hogs; flour milling, dairying.

Isla Verde (EES-lah VER-de), beach and residential sector, NE PUERTO RICO, 5.5 mi/8.9 km E of OLD SAN JUAN. Tourist hotels; residential condominiums. Luiz Muñoz Marin International Airport.

Isla Vista, unincorporated city (2000 population 18,344), SANTA BARBARA county, SW CALIFORNIA, 10 mi/16 km W of SANTA BARBARA, on Pacific Ocean; 34°25′N 119°52′W. University of California (Santa Barbara) to E; Santa Barbara Airport to NE. El Capitan State Park to W; Los Padres National Forest to N. Residential and resort community. Fruit, vegetables, avocados, grain, cattle.

Islay (EES-lei), province (□ 2,432 sq mi/6,323.2 sq km), AREQUIPA region, S PERU; ⊙ MOLLENDO; 17°00′S 71°50′W. A PACIFIC province. Drained and irrigated by Mollendo River, crossed by PAN-AMERICAN HIGHWAY. Rice, cotton, sugarcane.

Islay (EI-lai), hamlet, E central ALBERTA, W CANADA, 13 mi/20 km from VERMILION, in VERMILION RIVER County No. 24, near SASKATCHEWAN border; 53°24′N 110°33′W.

Islay (IS-lai), island (□ 240 sq mi/624 sq km; 2006 population 3,500), Argyll and Bute, W Scotland, southernmost of the INNER HEBRIDES; 55°46′N 06°11′W. Bowmore is the ancient capital, but PORT ELLEN is the main town. Other towns are Port Askaig, Port Charlotte, and Portnahaven. The land is fertile, with large livestock and dairy farms and vast fields of peat. Oats and potatoes are the main crops, and cheese is made. Distilling and tourism are important. Memorials to victims of the sinkings (1918) of the *Tuscania* and the *Otranto* are on the headland of the Mull of Oa in the S. This was the seat of the Lords of the Isles (the MacDonalds and Campbells). There is an airfield near Port Ellen and ferry services to Kintyre and Jura from Port Askaig.

Isla Zacatillo (EES-lah zah-kah-TE-yo), island, EL SALVADOR, in GULF OF FONSECA, 5 mi/8 km ESE of LA UNIÓN at entry to LA UNIÓN BAY. One of several islands in W Gulf of Fonseca belonging to El Salvador but disputed with HONDURAS.

Isle (EEL), town (□ 7 sq mi/18.2 sq km), HAUTE-VIENNE department, LIMOUSIN region, W central FRANCE; 3 mi/4.8 km SW of LIMOGES, on VIENNE RIVER; 45°49′N 01°13′E.

Isle (EIL), village (2000 population 707), MILLE LACS county, E MINNESOTA, 36 mi/58 km SE of BRAINERD, on SE shore of MILLE LACS LAKE on Isle Harbor; 46°08′N 93°27′W. Grain; livestock, poultry; dairying; manufacturing (fishing tackle, electroplating, printing and publishing); timber; sand and gravel. Granite quarries nearby. Mille Lacs Wildlife Area to SW; Father Hennepin State Park to NW, at Pope Point; Mille Lacs Lake; small part of Mille Lacs Indian Reservation to NE.

Isle, FRANCE: see ISLE RIVER.

Isle-Adam, L' (leel–ah-dahm), residential town (□ 5 sq mi/13 sq km), VAL-D'OISE department, ÎLE-DE-FRANCE region, N central FRANCE, on left bank of OISE RIVER, and 7 mi/11.3 km NE of PONTOISE; 49°07′N 02°14′E. Fashionable water sports resort (river beach); manufacturing of gardening implements. Its 16th-century Renaissance church has a 19th-century spire. Just SE is the forest of same name, laid out as a formal suburban park.

Isle au Haut (EI-luh HO), town (□ 9 sq mi/23.4 sq km), KNOX county, S MAINE, on Isle au Haut (□ c.9 sq mi/23.3 sq km) and adjacent islands, 25 mi/40 km ESE of ROCKLAND in Isle au Haut Bay; 44°00′N 68°34′W. Part of the island is in ACADIA NATIONAL PARK.

Isle-d'Abeau, L' (leel–dah-bo), new town (□ 4 sq mi/10.4 sq km), ISÈRE department, RHÔNE-ALPES region, SE FRANCE. Established as a computer research and operations center on the LYON-GRENOBLE highway, 25 mi/40 km ESE of Lyon, and just NW of BOURGOIN-JALLIEU, an older textile weaving and silk-screen production center.

Isle-d' Espagnac, L' (leel–des-pah-nyahk), town (□ 2 sq mi/5.2 sq km), NW suburb of ANGOULÊME, CHARENTE department, POITOU-CHARENTES region, W central FRANCE. Electronics.

Isle-en-Dodon, L' (leel–ahn-do-don), commune, HAUTE-GARONNE department, MIDI-PYRÉNÉES region, S FRANCE, on SAVE RIVER, and 20 mi/32 km NNE of SAINT-GAUDENS. Fruit growing, poultry and livestock raising on LANNEMEZAN PLATEAU.

Isleham (EI-luhm), agricultural village (2001 population 2,347), E CAMBRIDGESHIRE, E ENGLAND, 7 mi/11.3 km N of NEWMARKET; 52°22′N 00°25′E. Former site of limestone quarrying. Has 11th-century chapel and 14th-century church.

Isle-Jourdain, L' (leel–zhoor-dan), town (□ 27 sq mi/70.2 sq km), GERS department, MIDI-PYRÉNÉES region, SW FRANCE, on the SAVE, and 18 mi/29 km W of TOULOUSE. Agriculture market in fruit-growing valley.

Isle La Motte, island and town, 6 mi/9.7 km long and 2 mi/3.2 km wide, in Lake CHAMPLAIN, NW VERMONT; 44°52′N 73°19′W. The French chose the island as the site for Fort Sainte. Anne (built 1666), the first recorded European settlement in Vermont. Sainte Anne's shrine is here. Limestone quarries. A twenty-five foot cast iron lighthouse was built in 1880; after many uses of no use the lighthouse was reactivated in October 2002.

Isle La Ronde (EIL luh-ROON), islet, GRENADA, WEST INDIES, NNE of SAUTEURS. Home to a few fishing families, sea bird drops.

Isle Madame, CANADA: see MADAME ISLAND.

Isle of Hope, resort town, CHATHAM county, E GEORGIA, 6 mi/9.7 km SSE of SAVANNAH, near the ATLANTIC OCEAN; 31°58′N 81°03′W.

Isle of Palms, town (2006 population 4,643), CHARLESTON county, SE SOUTH CAROLINA, on ISLE OF PALMS island (5.5 mi/8.9 km long), 8 mi/12.9 km E of CHARLESTON; 32°47′N 79°45′W. Suburb of Charleston and resort center. Bridged to SULLIVANS ISLAND (SW). INTRACOASTAL WATERWAY passes to W. Destroyed by Hurricane Hugo (1989); since rebuilt.

Isle of Springs, island, LINCOLN county, S MAINE, in SHEEPSCOT RIVER just W of Boothbay Harbor town; c.0.5 mi/0.8 km in diameter. Formerly called Sweet Island.

Isle of Wight (EIL, WEIT), county (□ 362 sq mi/941.2 sq km), 2006 population 34,723), SE VIRGINIA, ☉ ISLE OF WIGHT; 36°54′N 76°42′W. In the Tidewater region; bounded W by BLACKWATER RIVER, NE by the JAMES estuary. Agriculture (sandy soils produce peanuts, corn, hay, soybeans, melons, barley, wheat, cotton;

hogs; cattle); some timber. Historic buildings include St. Luke's Church, one of the oldest churches in America, at BENNS CHURCH in NE. Lake Burnt Mills reservoir in NE, on Western Branch NANSEMOND RIVER. Formed 1634.

Isle of Wight (EIL, WEIT), unincorporated village, ☉ ISLE OF WIGHT county, SE VIRGINIA, 12 mi/19 km NNW of SUFFOLK; 36°54′N 76°42′W. Agriculture (grain, soybeans, melons, cotton, peanuts; livestock).

Isle Ornsay (EI-uhl ORN-sai) or **Isleornsay**, Gaelic *Eilean Iarmain*, village, Inner Hebrides, HIGHLAND, N Scotland, on E coast of Sleat peninsula; 57°09′N 05°47′W.

Isle River (EEL), 145 mi/233 km long, in DORDOGNE and GIRONDE departments, AQUITAINE region, SW central FRANCE; rises in LIMOUSIN hills; flows WSW through a fertile valley, past PÉRIGUEUX, to the DORDOGNE at LIBOURNE; 44°55′N 00°17′W. Receives the AUVÉZÈRE (left) and the DRONNE (right). Navigable in lower half of its course.

Isle Royale National Park (EIL roi-AL) (□ 210 sq mi/546 sq km), comprising Isle Royale and about 200 smaller islands, in LAKE SUPERIOR, NW MICHIGAN. Isle Royale—210 sq mi/544 sq km, 45 mi/72 km long, max. 8 mi/12.9 km wide—is the largest island in Lake Superior; Greenstone Ridge and Trail extend along its entire length. It includes Lake Siskiwit. Ryan Island, in Lake Siskiwit, is the largest island in the world's largest fresh-water lake. Glaciated, the island has about 50 lakes, also streams and inlets. It remains a roadless (does have trails and portage system), forested wilderness. Campgrounds located throughout inland; canoeing, fishing, hiking. Its abundant wildlife includes squirrels, beaver, fox, moose, wolves, and many birds. The French, lured by the fur trade, named the island in 1671. Isle Royale became U.S. territory in 1783 and was ceded to the U.S. by the Chippewa in 1843. It was mined for copper from 1843 to 1899; large areas of forest were burned to expose the ore and to build settlements. Has pre-Columbian copper-mine sites. In the early 1900s the island was a popular vacation retreat; still has a few private cabins. Season mid-May to Oct. 1. 40 mi/64 km from Michigan's UPPER PENINSULA, 15 mi/24 km from ONTARIO shore, and 20 mi/32 km from Minnesota Acreage includes park office and ranger ferry terminal at HOUGHTON. Accommodations and services at Rock Harbor Lodge area (NE); reached by ferries from COPPER HARBOR and Houghton. Ferry from GRAND PORTAGE, MINNESOTA, to Windigo ranger station at SW end; shuttle service encircles island. Established in 1940.

Islesboro (EILZ-buhr-o), island and town, WALDO county, S MAINE, in PENOBSCOT BAY SE of BELFAST; c.11.5 mi/19 km long, 0.5 mi/0.8 km–1.5 mi/2.4 km wide; 44°17′N 68°55′W. Includes resort villages of Dark Harbor, Pripet, North Islesboro. Sometimes Islesborough.

Isles Dernieres (EEL dern-YER), S LOUISIANA, uninhabited island chain (c. 18.5 mi/31 km long), 38 mi/61 km S of HOUMA, between CAILLOU BAY and Lake PELTO (N) and Gulf of MEXICO (S); 29°02′N 90°48′W. Once a continuous barrier beach; fashionable mid-nineteenth-century resorts here were destroyed by hurricane (1856), which took many lives.

Islesford, MAINE: see CRANBERRY ISLES.

Isles of Scilly, ENGLAND: see SCILLY ISLES.

Isles of Shoals, MAINE and NEW HAMPSHIRE, islands 10 mi/16 km SE of PORTSMOUTH, New Hampshire. Appledore, Cedar, Duck, and Smuttynose (or Haley's) islands are in Maine; Lunging, White (lighthouse), and Star islands are in New Hampshire. Resorts on Appledore and Star islands.

Isle St. George, Ohio: see BASS ISLANDS.

Isle-sur-la-Sorgue, L' (leel–syur–lah–sor-guh), town (□ 17 sq mi/44.2 sq km), VAUCLUSE department, PROVENCE-ALPES-CÔTE D'AZUR region, SE FRANCE, on the SORGUE DE VAUCLUSE RIVER, and 12 mi/19 km

E of AVIGNON; 43°55′N 05°03′E. Once an active industrial center manufacturing dyes and paper, and milling olive oil and grain, it now weaves rugs and wool blankets, and has a small chemical and plastics industry. Has beautifully decorated 14th–17th-century church, and a large hospital dating from 18th century.

Isle-sur-le-Doubs, L' (leel-syur-luh-doo), commune (□ 4 sq mi/10.4 sq km), DOUBS department, FRANCHE-COMTÉ region, E FRANCE, on DOUBS RIVER and RHÔNE-RHINE CANAL, and 11 mi/18 km SW of MONTBÉLIARD; 47°27′N 06°35′E. Metalworking, furniture manufacturing.

Isleta (iz-LET-ah), pueblo, BERNALILLO county, central NEW MEXICO, 12 mi/19 km S of downtown ALBUQUERQUE, on the E bank of the RIO GRANDE, in Isleta Indian Reservation (1990 population 2,915); 34°52′N 106°40′W. Confluence of RIO PUERCO and Rio San Juan in W part of reservation. It is a tourist attraction. According to many experts, the pueblo stands on the site it occupied when discovered in 1540. It was the seat of the Franciscan mission of San Antonio de Isleta from c.1621 until the Pueblo revolt of 1680. The Spanish captured the pueblo in 1681, and most of the captives were ultimately settled at YSLETA, TEXAS. In the early 18th century, when N Isleta was either rebuilt or repopulated, it became the mission of San Agustín de Isleta. The Pueblo in Isleta are mainly farmers; the language there is Tiwa.

Isleta (ees-LAI-tah), peninsula, NE Grand Canary, CANARY ISLANDS, 3 mi/4.8 km N of LAS PALMAS; c.3 mi/4.8 km long, 2 mi/3.2 km wide. On its isthmus is PUERTO DE LA LUZ.

Isleton, city (2000 population 828), SACRAMENTO county, central CALIFORNIA, 28 mi/45 km S of SACRAMENTO, 13 mi/21 km NE of ANTIOCH, on SACRAMENTO RIVER; 38°10′N 121°36′W. Vegetables, corn, tomatoes, beans, sugar beets, rice; cattle. Founded 1874, incorporated 1923.

Isleworth, ENGLAND: see HOUNSLOW.

Isli (EES-lee), lake, MOROCCO, in Plateau of the Lakes between the Middle and High ATLAS mountains, 50 mi/80 km E of BENI MELLAL; 32°13′N 05°32′W.

Islington, town, SAINT MARY parish, NE JAMAICA, 23 mi/37 km NNW of KINGSTON; 18°19′N 76°51′W. Road junction.

Islington (IZ-ling-tuhn), inner borough (□ 6 sq mi/15.6 sq km; 2001 population 175,797) of GREATER LONDON, SE ENGLAND; 51°35′N 00°05′W. Islington, in the N, is mostly residential, while Finsbury, in the S, is highly industrialized (special and electrical engineering, printing, food processing, brewing, clothing, furniture, and scientific, surgical, and optical instruments). Finsbury has an important wholesale trade in industrial equipment and supplies. Bunhill Fields in Finsbury contains the graves of William Blake, Daniel Defoe, John Bunyan, and Isaac Watts. John Wesley's chapel and house and the Sadler's Wells Theatre, former home of the Royal Ballet, are also in Finsbury. The City University and the University of North London (established 1992; formerly Polytechnic of North London) are located in the borough.

Islington, Massachusetts: see WESTWOOD.

Islip, suburban residential village (2000 population 20,575), SUFFOLK county, SE NEW YORK, on S shore of LONG ISLAND, on GREAT SOUTH BAY, just E of BAY SHORE; 40°43′N 73°11′W. Light manufacturing; in recreational and duck-farming area; horticultural crops. HECKSCHER STATE PARK is SE of Islip, and Seatuck National Wildlife Refuge is immediately S. S terminus of Long Island Greenbelt trail, which runs N from Great South Bay to Sunken Meadow State Park on N shore of Long Island, N of KINGS PARK, on SMITHTOWN BAY.

Islip Terrace, suburban residential village (□ 1 sq mi/2.6 sq km; 2000 population 5,641), SUFFOLK county, SE NEW YORK, on central LONG ISLAND, 2 mi/3.2 km NE of ISLIP; 40°45′N 73°11′W.

Isluga Volcano (ees-LOO-gah), Andean peak (elevation 18,145 ft/5,531 m), N CHILE, near BOLIVIA border; 19°10′S.

Isly, Oued (wahd EES-lee), short stream of NE MOROCCO, just W of OUJDA, near ALGERIA border. Here, in 1844, French general Bugeaud defeated Abd-el-Kader and a Moroccan army in a decisive battle.

Ismael Cortinas, town, FLORES department, SW URUGUAY, 33 mi/53 km SSW of TRINIDAD, on SORIANO-COLONIA department border; 33°58′S 57°06′W. Road and railroad service from SAN JOSE DE MAYO. Formerly called Arroyo Grande.

Ismail (ees-MEI), mining center, SKIKDA wilaya, NE ALGERIA, 11 mi/18 km NE of SKIKDA. Center for the exploitation and treatment of mercury and antimony. The mercury mine, the most important in Algeria, has extensive galleries and layers of rich ore. The plant treating the ore will make Algeria one of the world's largest suppliers.

Ismail, UKRAINE: see IZMAYIL.

Ismailia, province (2004 population 844,091), E EGYPT, bounded W by the SUEZ CANAL, N by PORT SAID province, S by SUEZ province, and W by SHARQIYA province; ⊙ ISMAILIA. Main urban centers include Ismailia, Nifisma, Qantara. Served by railroad from CAIRO and the ISMAILIA CANAL.

Ismailia (is-MAH-il-LAI-yuh), city, ⊙ ISMAILIA province, NE EGYPT; 30°35′N 32°16′E. It is the headquarters of the SUEZ CANAL administration. Extensive irrigation is used for growing fruits and vegetables; livestock is raised. Ismailia was founded in 1863 by Ferdinand de Lesseps, who used it as his base of operations during the construction of the canal. The city was named after Ismail, a khedive of Egypt. A nearby military base established by the British in World War I is now under Egyptian control. Part of the city's civilian population was evacuated after Israeli forces shelled Ismailia in the war of attrition which followed the 1967 Arab-Israeli War. In the 1973 war, Israeli ground forces pushed to within the city's outskirts.

Ismailia Canal (is-MAH-il-LAI-yuh), c.80 mi/129 km long, navigable fresh-water canal, E EGYPT, in Wadi TUMILAT, from CAIRO to ISMAILIA and Lake TIMSAH on the SUEZ CANAL. At NIFISHA, just W of Ismailia, it branches off S and at Ismailia it continues N as the El' Abbasiya Canal. It was constructed (c.1860) to supply the villages on the projected Suez Canal with drinking water. Used chiefly for irrigation.

Ismailly (ees-mei-EE-lee), city, E central AZERBAIJAN, on S slope of the Greater CAUCASUS, 55 mi/89 km ENE of YEVLAKH. Wheat, livestock; manufacturing (furniture, dairy products); lumbering. Administrative center.

Ismaning (is-MAH-ning), suburb of MUNICH, BAVARIA, S GERMANY, in UPPER BAVARIA, on ISAR RIVER, 7 mi/11.3 km NE of city center; 48°13′N 11°41′E. Chemical industry.

Ismaros (EES-mah-ros), hill range in EAST MACEDONIA AND THRACE department, NE GREECE, on AEGEAN SEA; 40°54′N 25°33′E. Rises to 2,224 ft/678 m, 17 mi/27 km SE of Komotiní.

Ismay (IS-mai), village (2000 population 26), CUSTER county, E MONTANA, on O'FALLON Creek at mouth of Sandstone Creek, and 52 mi/84 km E of MILES CITY; 46°30′N 104°48′W. Cattle, sheep, hay. Also called Joe.

Ismeli, RUSSIA: see OKTYABR'SKOYE, CHUVASH REPUBLIC.

Isna (is-NAH), town, central EGYPT, on the W bank of the NILE River; 25°18′N 32°33′W. It is the center for an agricultural area that is irrigated by the Nile. Manufacturing includes cotton fabrics and ceramics. The Ptolemaic temple (with Roman additions) to the ram-headed deity Khnum is the town's outstanding monument. Nearby is a Coptic Christian monastery, said to have been founded in the fourth century to commemorate those martyred by Diocletian, but now believed to date from the tenth or eleventh century.

Isnos (EES-nos), town, ⊙ Isnos municipio, HUILA department, S central COLOMBIA, 96 mi/154 km SW of NEIVA, in the Cordillera CENTRAL; 01°56′N 76°14′W. Elevation 6,811 ft/2,075 m. Coffee, plantains; livestock.

Isnotú (ees-no-TOO), town, TRUJILLO state, W VENEZUELA, in ANDEAN spur, 7.0 mi/11.3 km WNW of VALERA; 09°22′N 70°42′W. Elevation 2,867 ft/873 m. Coffee, cereals, sugarcane.

Isny (IS-nee), town, BADEN-WÜRTTEMBERG, GERMANY, in the ALLGÄU, 13 mi/21 km W of KEMPTEN; 47°42′N 10°02′E. Summer resort and winter sports center. Manufacturing of precision instruments. Site of technical school. Has 13th- and 17th-century churches. Was free imperial city.

Isobe (EE-so-be), town, Shima district, MIE prefecture, S HONSHU, central JAPAN, on Matoya Bay, 28 mi/45 km S of TSU; 34°22′N 136°48′E. Seafood (oysters, eels), nori; pearls. Pottery.

Isoka, township, NORTHERN province, NE ZAMBIA, 110 mi/177 km E of KASAMA; 10°08′S 32°38′E. Road junction. Corn; cattle; phosphate mining. NYIKA PLATEAU National Park to SE. Oil pipeline from NDOLA, Zambia, to TANZANIA passes to E.

Isoko (ee-SO-ko), village, MBEYA region, SW TANZANIA, 30 mi/48 km S of MBEYA; 09°27′S 33°31′E. Coffee, tea, corn; cattle, sheep, goats. Coal mining in area.

Isokyrö (I-so-kuh-ruh), Swedish *Storkyrö*, village, VAASAN province, W FINLAND, 20 mi/32 km ESE of VAASA; 63°00′N 22°19′E. Elevation 116 ft/35 m. In lumbering, grain-growing region. Has fourteenth-century church. Napue, scene of 1714 Russian victory over Finns, is just S.

Isola (ei-SO-luh), town (2000 population 768), HUMPHREYS county, W MISSISSIPPI, 8 mi/12.9 km NW of BELZONI; 33°15′N 90°35′W. In rich agricultural area (cotton, corn, rice; cattle); manufacturing (catfish processing, cattle feed).

Isola (EE-zo-lah), village, BRESCIA province, LOMBARDY, N ITALY, in VAL CAMONICA, on small branch of OGLIO RIVER and 9 mi/14 km SE of EDOLO. Has one of major Italian hydroelectric plants.

Isola 2000, modern winter sports resort, ALPES-MARITIMES department, PROVENCE-ALPES-CÔTE D'AZUR region, SE FRANCE, at edge of MERCANTOUR NATIONAL PARK, close to Italian border, and 33 mi/53 km NNW of NICE. Ski terrain at elevation 6,400 ft/1,951 m–8,300 ft/2,530 m; has sunny winter climate due to proximity of the MEDITERRANEAN SEA. Established 1972. Fine summer excursions to Mont Saint-Sauveur (elevation 8,700 ft/2,652 m) and across mountain passes to Italian ALPS.

Isola della Scala (EE-zo-lah DEL-lah SKAH-lah), town, VERONA province, VENETO, N ITALY, on TARTARO RIVER and 11 mi/18 km S of VERONA; 45°16′N 11°00′E. Railroad junction; manufacturing (fabricated metals, clothing, wood products).

Isola del Liri (del LEE-ree), town, FROSINONE province, LATIUM, S central ITALY, 12 mi/19 km ENE of FROSINONE; 41°41′N 13°34′E. On island in LIRI RIVER, which here forms several waterfalls utilized for industrial power. A major paper manufacturing center; felt and woolen mills, cement works, pasta factories; machinery.

Isola Vicentina (EE-zo-lah vee-chen-TEE-nah), town, VICENZA province, VENETO, N ITALY, 7 mi/11 km NW of VICENZA; 45°38′N 11°25′E. Fabricated metals, machinery, textiles.

Isonzo River, SLOVENIA and ITALY: see SOČA RIVER.

Iso Saimaa, FINLAND: see SAIMAA.

Ispahan, IRAN: see ESFAHAN.

Isparta, city (2000 population 148,496), ⊙ Isparta province, W central TURKEY; 37°46′N 30°32′E. Railroad terminus 105 mi/169 km W of KONYA. Cotton, carpets, attar of roses. Severely damaged by an earthquake in 1889. Sometimes called Sparta. Formerly Hamidabat.

Isperih (EES-pe-REEK), city (1993 population 10,631), RUSE oblast, ⊙ Isperih obshtina, NE BULGARIA, 20 mi/32 km NE of RAZGRAD; 43°42′N 26°50′E. Market center; agriculture (grain, sunflowers, beets, tobacco, vineyards, fruit); manufacturing (fodder, spare parts for forklifts and textile machinery); repair plant for construction machinery. Until 1934, called Kemanlar. City since 1948.

Isperikh, Bulgaria: see ISPERIH.

Ispica (EE-spee-kah), town, RAGUSA province, SE SICILY, ITALY, 10 mi/16 km SE of MODICA; 36°47′N 14°55′E. Wine. Called Spaccaforno until 1935. Cava d'Ispica is near Modica. Archaeological park nearby.

Ispir, village, NE TURKEY, on CORUH RIVER and 40 mi/64 km NNW of ERZURUM, in mountainous area; 40°29′N 41°02′E. Elevation c.6,500 ft/1,981 m. Barley, potatoes.

Ispiriz Daglari, mountain range, E TURKEY, 40 mi/64 km SE of VAN, near Iranian border. Rises to 11,604 ft/3,537 m.

Ispravnaya (ees-PRAHV-nah-yah), settlement (2005 population 4,615), central KARACHEVO-CHERKESS REPUBLIC, N CAUCASUS, S European Russia, on the GREAT ZELENCHUK RIVER, on road, 25 mi/40 km SW of CHERKESSK; 44°06′N 41°37′E. Elevation 2,312 ft/704 m. Mineral deposits in the vicinity.

Ispringen (I-shpring-uhn), suburb of PFORZHEIM, BADEN-WÜRTTEMBERG, SW GERMANY, just NW of city; 48°55′N 08°40′E.

Israel (IZ-rah-el), Hebrew Medinat Yisrael, Arabic *Dalwat Israil*, republic (□ 13,572 sq mi/21,843 sq km; 2004 estimated population 6,809,008; 2007 estimated population 6,426,679), SW ASIA, MIDDLE EAST, on the MEDITERRANEAN SEA; ⊙ JERUSALEM. Important cities include TEL AVIV, JAFFA, HAIFA, BEERSHEBA, NETANYA, and ZEFAT. The country is divided into six districts (Hebrew=*mehoz*, plural *mehozot*): Haifa, Jerusalem, Northern, Southern, and Tel Aviv.

Geography

The country is a narrow, irregularly shaped strip of land bounded on the N by LEBANON, on the E by SYRIA and JORDAN, on the W by the Mediterranean Sea, on the SW by EGYPT, and on the S by the GULF OF AQABA (an arm of the RED SEA). It has four principal regions: the plain along the Mediterranean coast; the mountains, which are E of this coastal plain; the NEGEV, which comprises the S half of the country; and the portion of Israel that forms part of the JORDAN VALLEY, in turn a part of the GREAT RIFT VALLEY. N of the Negev, Israel enjoys a Mediterranean climate, with long, hot, dry summers and short, cool, rainy winters. This N half of the country has a limited but adequate supply of water, except in times of drought. The N part of the Negev is semi-arid with an average annual rainfall of 8 in/20 cm–10 in/25 cm. The central and S parts are arid desert, most of which receives less than 4 in/10.2 cm of rainfall per year. Because Israel is plagued by a shortage of water, Israeli scientists have worked on the desalination of seawater; small desalination plants operate in the EILAT region. Israel uses more than 90% of its available water supply. Over 60% of the water is consumed by agriculture. Israel has two main aquifers—the coastal aquifer which is shallow and therefore becomes polluted; and the mountain aquifer (between the West Bank and the coastal plain), which is deep and clean. The most important river in Israel is the JORDAN. Other smaller rivers are the YARKON, the KISHON, and the YARMUK, a tributary of the Jordan. In the S part of the country are many wadis, or riverbeds, that are dry except in the brief rainy spells during the winter. Other bodies of water include the Sea of GALILEE, the main water reservoir for Israel's national water carrier system, and the DEAD SEA (part of which belongs to neighboring Jordan). Owing to interior drainage and an elevation below sea level, the waters of the Dead Sea have about eight times as much salt as the ocean. The draining in 1957 of Lake HULA, located in N Israel, served to increase both the farming area and the number of fish ponds in the region. Because of the unforeseen negative consequences to the environment, especially the polluting effects of chemicals in the runoff waters and damage to the natural ecology, a major effort is underway to restore part of the former lake area. The highest point in Israel is Mount MERON (3,960 ft/1,208 m) near Zefat. The lowest point is the surface of the Dead Sea, which is 1,321 ft/403 m below sea level and which is also the lowest point on land in the world. As the result of an intensive reforestation program, well over 100 million trees (20% of the entire cultivated area) have been planted since 1948, the year the state of Israel was established.

Population
Israel, the world's only Jewish state, is made up of about 80% Jews and about 20% Arabs, including Druze. The Arab population is primarily Muslim; a smaller proportion is Christian. Hebrew is the official language of Israel. Arabic is used officially within the Arab minority, and English is the most commonly used foreign language.

Economy: Agriculture
The economy of Israel is based on both state and private ownership and operation. Despite adverse conditions, agriculture in Israel has been developed to a degree that compares favorably with the agriculture of advanced countries. In 1948, Israel produced only 30% of the food it needed; by the early 1970s it produced enough fruits and vegetables, poultry and eggs, and milk and dairy products to meet all domestic needs, although most of the cereals and feed grains, as well as decreasing amounts of coffee, sugar, and beef, are still imported. The area of land under cultivation has been increased by over 250% since the founding of the state in 1948, and extensive irrigation (one-third of all field crops and half of total agricultural areas) has been provided to develop farmland and compensate for the shortage of rainfall. Greenhouses and hydroponics are other means through which agriculture has been intensified. While a major exporter of citrus, Israel's fresh non-citrus fruits and vegetables now exceed citrus in export value, and citrus dropped from representing two-thirds of Israel's agricultural exports in 1970 to 25% in 1991. Western Europe takes 85% of these exports, which include, in addition to citrus, flowers (such as roses, carnations, and gladioli), non-citrus fruits (such as avocados, melons, bananas, and peaches), and vegetables (such as eggplants and tomatoes). They are especially important as out-of-season crops in the winter. Other sizable crops are cotton, wheat, barley, peanuts, sunflowers, grapes, and olives. Poultry (chicken and turkeys) and livestock are raised. Agricultural production adds up to roughly 2.8% of Israel's gross national product.

The Israel Lands Authority leases land to kibbutzim, which are communal agricultural settlements; to moshavim, which are cooperative agricultural settlements; and to other agricultural or rural villages. On the kibbutz, members formerly received all the necessities of life (housing, food, clothing, medical care, education, recreation, vacations, and spending money) instead of wages in return for their labor, but changes to traditional kibbutz practices and privatization since the 1980s have reduced the communal and collective elements on many kibbutzim. Industry and tourism are of major importance to most kibbutzim. In the cooperative moshav each family unit has its own home and cultivates its own plot of land, but the members own farming equipment collectively and market their crops as a group. Many moshav dwellers now hold non-farming jobs in projects developed on the moshav or outside the village, and some have developed manufacturing or service enterprises. Some moshav lands are now being made available for private residences for outsiders.

Economy: Industry
Israeli industry has developed in an explosive manner since 1948, despite a comparative scarcity of raw materials, many of which have to be imported. The owners of industry in Israel include individuals, the government, the Histadrut union (which has been divested of some of its enterprises through privatization), and other public organizations. Israel encourages foreign investment in its industry by low rates of taxation and by permitting the withdrawal abroad of most of the profits. The major industries include the cutting and polishing of diamonds, manufacturing of chemical fertilizers from the potash obtained from the Dead Sea and from the phosphates found in the Negev, apparel manufacturing, and the increasing production of military and electronic equipment. High-technology industries are Israel's fastest-growing developments, with emphasis on computers, software, telecommunications, biotechnology, and medical electronics. These industries have matched diamonds in export importance. Several international research and development centers are located in Israel. The Dead Sea has other minerals of commercial value, such as magnesium, bromine, and salt. Building construction has become a very large industry, partly because of the need to provide homes for the over three million persons who have come to Israel since 1948, and particularly since 1989, as an influx of some 1,000,000 Jews from the former Soviet Union and Ethiopia were allowed to emigrate from their respective countries to Israel. A number of light industries also produce processed foods, precision instruments, shoes, clothing, and various plastic goods.

Economy: Exports/Imports
Diamonds and high-tech industrial products, including military hardware and weapons systems, are the major exports, followed by chemicals, pharmaceuticals, textiles, and apparel. The leading imports are military equipment, machinery, rough diamonds, crude oil, transport equipment, and wheat. Although Israel still imports more than it exports, the balance of trade is far more favorable now than it was in the early years of the state. Israel's chief trading partners are the EU, especially the U.K. and Belgium, and the U.S. Israel also trades with countries in Asia, Africa, and South America.

Economy: Nuclear Power
Israel has two nuclear reactors: one S of Tel Aviv, and another near DIMONA in the Negev, where scientists are conducting research on using atomic energy in the production of electricity and for the desalination of seawater. The plant at Dimona has been credited with nuclear-weapons capacities.

Economy: Tourism and GNP
Another major industry is tourism, which is one of Israel's largest sources of revenue. Terrorist attacks have had a negative impact on this, especially since the beginning of the second intifada (2000), though tourism began to increase again in 2003. The standard of living in Israel is high for a Middle Eastern nation and is comparable to that which prevails in Western Europe; its GNP is $19,800, but it has a growing trade deficit.

Economy: Privatization
The move toward privatization of state-run monopolies that began in the 1970s accelerated in the early 2000s under finance minister Benjamin Netanyahu, who oversaw sale of such concerns as the Israel Ports and Railways Authority, telephony Bezeq, and El Al Israel Airlines, and introduced pension reforms, dealing a blow to the powerful countrywide Histadrut union.

Education
Israel has major universities and technical and research institutes in Jerusalem, Tel Aviv, RAMAT GAN, Haifa, Beersheba, and REHOVOT, as well as many smaller colleges and other institutes of higher education located throughout the country. Muslim and Christian Arabs in Israel have their own schools in which the language of instruction is Arabic. Education at the primary level is free and compulsory.

History: Founding to 1949
The state of Israel is the culmination of nearly seventy years of activity in Zionism. Following World War I, Great Britain occupied (1917–1918) and later received (1922) Palestine, comprised of parts of present-day Israel, West Bank, Gaza Strip, Jordan, and Egypt, as a mandate from the League of Nations, called the Mandate of Palestine or, more popularly, the British Mandate. The struggle for a Jewish state here had begun in the mid-late 19th century and gained momentum on the eve of, during, and after World War I. As Jewish immigration to Palestine grew during the 1920s and 1930s, there were sporadic clashes between the Jewish settlers and Arabs. International pressure to allow more Jewish immigration after World War II, along with the desire of the British to end their administration of the region, lead to increased calls for a Jewish state in Palestine. The militant opposition of the Arabs to such a state and the inability of the British to solve the problem eventually led to a session of the General Assembly of the UN in April 1947, which established the UN Special Committee on Palestine (UNSCOP). UNSCOP reported a plan to divide the territory into a Jewish state, an Arab state, and a small internationally administered zone including Jerusalem. The General Assembly adopted the UNSCOP recommendations on November 29, 1947. The Jews accepted the plan, albeit with mixed feelings; the Arabs rejected it, leaving the meeting and asserting their intention to resist. On May 14, 1948, when the British high commissioner for the mandate departed, the state of Israel was proclaimed at Tel Aviv. On the same day it received the de facto recognition of the U.S. (on May 17 the USSR extended de jure recognition). The Arab states of Lebanon, Syria, Jordan, Egypt, and IRAQ invaded Israel with their regular armies on May 14, 1948. The Israelis were prepared, however, through their self-defense force known as the Haganah, and the flight of most Palestinian Arabs from Jewish territory facilitated defense.

History: 1949 to 1956
Not until the spring of 1949 were armistice agreements reached. In its War of Independence, Israel had increased its holdings by about half, although it lost about 6,000 of its 650,000 Jews. There seemed little likelihood of a new Arab state, for Jordan annexed the area adjoining its territory, which became known as the West Bank, and Egypt was occupying the SW coastal strip, known as the GAZA STRIP. In January 1949, elections were held for the Knesset, and the Mapai (moderate socialist) and the religious parties formed a government. David Ben-Gurion (Mapai) became prime minister, and Chaim Weizmann was elected president. The elected government received recognition from most European and American countries, Australia, New Zealand, and some Asian countries. On May 11, 1949, Israel was admitted to the UN. Jerusalem was declared the country's capital on December 14, 1949, though the city remained divided between Israel and Jordan. Following the Lausanne Conference of 1949, Israel allowed the return of 150,000 Arab refugees, mostly to reunite families. An ever-expanding Arab economic boycott hampered the new country's growth, and continuous attacks by Arab border marauders threatened its security. One major aim of the government was to gather in all Jews who wished to immigrate to Israel. This led to the 1950 Law of Return, which provided for free and automatic citizenship for all immigrant Jews. Border incidents with Egypt, Syria, and Jordan continued, and bloody attacks and reprisals were sharply condemned by the UN.

History: 1956 to 1963
In 1956, Egyptian president Gamal Abdel Nasser nationalized the SUEZ CANAL. On October 25, 1956,

Nasser, signed a tripartite agreement with Syria and Jordan, placing Nasser in command of all three armies. Threatened by this combined military and aggressive rhetoric, Israel, in cooperation with Great Britain and France, attacked Egypt on October 29. Within a few days, Israel conquered the Gaza Strip and the SINAI Peninsula, while Britain and France invaded the area of the Suez Canal. Israel eventually yielded to strong pressure from the U.S. and the USSR, and removed its troops from Sinai in November 1956, and from Gaza by March 1957, as UN forces were sent to the Sinai and Gaza to keep peace between Egypt and Israel. In May 1960, Nazi Adolf Eichmann was captured by Israeli agents in Argentina brought to Israel to stand trial. He was found guilty of crimes against humanity, and hanged in May 1962. By 1962, despite pressure from the USSR and Egypt, most of the new nations of Africa had signed aid agreements with Israel. In 1963, Ben-Gurion resigned as prime minister and was succeeded in that office by Levi Eshkol.

History: The Six-Day War, 1967

In May 1967, Nasser mobilized the Egyptian army in Sinai. He next demanded that the UN Emergency Force withdraw from the Israeli-Egyptian border, where it had been stationed since 1956. Nasser then blockaded the Eilat port (on the Gulf of AQABA) by closing the Straits of TIRAN. On June 5, 1967, Israel launched preemptive attacks against Egypt and Syria. It appealed to Jordan to remain neutral, but the latter launched an attack, which Israel countered. In just six days, Israel captured the GAZA STRIP and the SINAI peninsula of Egypt, the GOLAN HEIGHTS of Syria, and the West Bank and Arab sector of E Jerusalem (both under Jordanian rule), thereby giving the conflict the name of the Six-Day War. Israel unified the Arab and Israeli sectors of Jerusalem and annexed E Jerusalem.

History: 1967 to 1973

On November 22, 1967, the Security Council adopted Resolution 242, calling for the withdrawal of Israeli forces from Arab territories captured in the war, the right of all states to live in peace within secure and recognized boundaries, freedom of navigation through international waterways in the area, and a just settlement of the Arab refugee problem. After Eshkol's death on February 26, 1969, Golda Meir became prime minister. In 1968, Egypt began hostilities that grew into the "war of attrition." Using heavy artillery, new MiG aircraft, Soviet advisers, and an advanced Soviet-designed surface-to-air missile system, Egypt inflicted heavy losses on Israel. Meir practiced a policy of asymmetrical response, ordering massive air raids deep into Egypt. Israel was also beset by raids from Jordan, launched by the Palestine Liberation Organization (PLO). These attacks were often on civilian targets, and Israel soon declared the PLO as a terrorist organization and refused to negotiate with it.

The U.S. brokered a treaty in August 1970 between Israel, Egypt, and Jordan. This plan specified limits on the deployment of missiles. In Jordan, Hussein's acceptance of the cease-fire ignited fighting between his army and the PLO. As the battles intensified, the Syrian-backed Palestinian Liberation Army (PLA) sent tanks to aid the Palestinians, the violence peaking during what became known as Black September. Coordinated Israeli, American, and Jordanian military actions forced the withdrawal of the PLA forces and expelled the PLO, whose members fled mostly to Lebanon. In September 1970, Nasser died and was succeeded by Anwar Sadat. In 1972, Palestinian terrorists called the Black September group took hostage and killed 11 Israeli athletes during the Olympic games in Munich.

History: The Yom Kippur War, 1973

On October 6, 1973, on the Jewish holy day of Yom Kippur, Egypt and Syria launched a surprise attack on

Israel, with Egypt sending troops across the Suez Canal and Syria's army penetrating the Golan Heights. The Soviet Union, Iraq, Saudi Arabia, Kuwait, Libya, Algeria, Tunisia, Sudan, and Morocco sent troops, arms, and funding to aid in the attack. Jordan sent troops to fight from Syria, but did not attack Israel directly across the common border. After 18 days of fighting IDF troops advanced within 20 mi/32 km of Damascus, and crossed to the Suez Canal's W bank, encircling Egypt's Third Army on the E bank, and were within 65 mi/105 km of Cairo when UN efforts brought an end to the fighting.

History: 1973 to 1982

On December 21, 1973, the first Arab-Israeli peace conference opened in Geneva, Switzerland, under UN auspices. An agreement to disengage Israeli and Egyptian forces was reached in January 1974, largely through the "shuttle diplomacy" mediation of U.S. Secretary of State Henry Kissinger. Israeli troops withdrew several miles into the Sinai, a UN buffer zone was established, and Egyptian forces reoccupied the E bank of the Suez Canal and a small, adjoining strip of land in the Sinai. A similar agreement between Israel and Syria was achieved in May 1974, and a UN buffer zone was created. In July 1976, an Air France jetliner carrying over 100 Israeli citizens was hijacked by Palestinian terrorists and taken to ENTEBBE, UGANDA. Israeli forces undertook a dramatic rescue of 103 hostages; three hostages and one Israeli colonel were killed by the terrorists and Ugandan troops. On May 17, 1977, the Likud party under the leadership of Menachem Begin defeated the Labor party. As prime minister, Begin strongly supported the development of Jewish settlements in the Israeli-occupied territories. Egypt began peace initiatives with Israel in late 1977, when Sadat visited Jerusalem. A year later, with the help of U.S. president Jimmy Carter, terms of peace between Egypt and Israel were negotiated at CAMP DAVID, MARYLAND. It was here that self-rule for the Palestinians was also first discussed. A formal treaty, signed on March 26, 1979, in WASHINGTON, D.C., granted full recognition of Israel by Egypt, opened trade relations between the two countries, and limited Egyptian military buildup in the Sinai. Israel agreed to return the final portion of Sinai to Egypt; the transfer was completed in 1982. Later negotiations resulted in the return of TABA to Israel, which developed the land into a resort, then returned it to Egypt in 1988 following the ruling of an international arbitration panel.

History: 1982 to 1991

Begin ordered (June 1982) the bombing of the Osirak nuclear facility in Iraq, stunting Iraqi weapons capacity. On June 6, 1982, Israel initiated a full-scale invasion of S Lebanon in order to destroy PLO guerrilla bases; troops moved toward BEIRUT and surrounded the W part of the city, which housed PLO headquarters. Nearly 7,000 PLO members were forced to flee. Israeli troops began a gradual withdrawal from Lebanon starting in 1983, leaving only a small force to monitor, with Lebanese militia forces, a 6-mi/9.7-km-deep security zone. Begin was replaced in 1983 by Likud's Yitzhak Shamir. Large numbers of emigrants, primarily from ETHIOPIA and the USSR, increased Israel's population by some 10% in three years (1989–1992). Unemployment and lack of housing were major problems for the new immigrants. Shamir was replaced in 1984 by Labor's Shimon Peres, then regained the post of prime minister in 1986. In December 1987, an accident involving a military vehicle that killed four Palestinians in the Gaza Strip triggered a number of Palestinian protests in the West Bank and Gaza Strip, which marked the beginning of the *Intifada* [=uprising or awakening]. Israel was affected by the Persian Gulf War in early 1991 when it was bombed regularly by Iraqi missiles, particularly in the areas of Tel Aviv and Haifa.

History: 1991 to 1999

With U.S. involvement, peace talks began in August 1991, with Israel, Syria, Lebanon, and a joint Jordanian-Palestinian delegation. Rabin reentered the political scene in June 1992, becoming prime minister after Labor's defeat of the Likud party, and the establishment of a governing coalition. He began his tenure by halting Israeli settlements in the occupied territories and pursuing the Arab-Israeli peace talks. Agreements with the Palestinian Liberation Organization at Oslo and Taba in 1993 provided for the gradual transfer of six of the seven major towns in the West Bank and 30% of its land to Palestinian Authority control; the accord also effectively ended the violence of the Intifada. A peace treaty with Jordan was signed in 1994, establishing full political and economic relations between the countries. Terms of the agreement include lease of small tracts of land and water rights to Israel (NAHARAYIM area along the Jordan River in N, Zohar area in ARAVA). Peace negotiations with Syria, which were halted early in 1996, began in 1994 as well. Rabin was assassinated on November 4, 1995 by an Israeli extremist opposed to the Oslo agreements. Deputy prime minister Shimon Peres replaced him and called for early elections in mid-1996, while adamantly pursuing Rabin's Labor party goals. General elections in May 1996 brought the Likud party back into power when Benjamin Netanyahu was narrowly elected in the first direct election for the office of prime minister. Labor remained the largest party, but was substantially weakened as a result. It came back into power in 1999, with the election of Ehud Barak.

History: 2000 to Present

Israel completed its withdrawal from Lebanon security zone on May 24, 2000. On September 28, 2000, following a visit by Likud leader Ariel Sharon to the Temple Mount (Arabic *Haram esh-Sherif*), Arab riots broke out, marking the beginning of the second Intifada, also known as the Al-Aqsa Intifada. Barak resigned in December 2000, and was defeated in the February 2001 election by Ariel Sharon, who formed a national unity government. In 2002, U.S. President Bush laid out a "road map" for peace between Israel and the Palestinians, envisioning a two-state solution. However, ongoing violence undermined the agreement and prompted Sharon to reoccupy some West Bank towns and begin construction of a 400 mi/640 km "security fence," a move that has draw international criticism. Arafat died in 2005 and was succeeded by Mahmoud Abbas as head of the Palestinian Authority. Prime Minister Sharon received official backing in January 2005 for a unilateral disengagement plan calling for Israeli withdrawal from all settlements in Gaza and four in the West Bank. Implementation of the plan began in Gaza in August 2005, after a grace period for voluntary evacuation of Jewish settlers; Israeli troops forcibly removed unwilling settlers and the process was formally completed with the withdrawal of Israeli forces from Gaza on September 12, 2005. In November 2005, responding to pressure from within his party, Sharon left the Likud to form a new party, Kadima (Hebrew=forward). Many politicians left their parties—notably Shimon Peres and Dalia Itzik of Labor, Shaul Mofaz and Tzipi Livni of Likud, and Shinui founder Uriel Reichman—in order to back Sharon. In January 2006, Sharon suffered a massive stroke and Ehud Olmert became acting prime minister; Sharon never recovered. In the general elections in March 2006 Olmert and the Kadima party won the largest block of seats in the Knesset and assembled a coalition government. Olmert officially became prime minister in April 2006. Tensions between Israel and the Palestinian Authority grew after the January 2006 election victory of Hamas, a rejectionist Palestinian Islamist party. Later in 2006, in reaction to the capture of Israeli soldiers by Hamas in June and Hezbollah in

July, Israel launched air attacks against targets in Gaza and Lebanon, respectively, and also sent troops into Gaza and S Lebanon. There also were Hamas rockets attacks against S Israel, and more significant Hezbollah rocket attacks against N Israel, including Haifa. A cease-fire (August) led to an Israeli withdrawal (September–October) from Lebanon, while the situation in Gaza remained unsettled and marked by sporadic fighting. The failure of Israeli forces to dislodge or disarm Hezbollah hurt Olmert's government, and an independent report subsequently criticized his handling of the invasion.

Government

Israel has no constitution; it is governed under the 1948 Declaration of Establishment as well as parliamentary and citizenship laws. The government consists of a legislature (Knesset), a president, a prime minister, and the cabinet. The Knesset has a single chamber with 120 seats and is elected for four years by popular vote. The president, who is head of state, is elected for seven years by the Knesset. The prime minister, who is head of government, is the dominant figure in Israeli politics, and appoints a cabinet that must be approved by the Knesset; both the prime minister and the cabinet are responsible to the Knesset. The current head of state, Shimon Peres, has been president since July 2007. The current head of government is Prime Minister Ehud Olmert (April 2006).

Israelândia (EES-rah-e-LAHN-zhee-ah), town (2007 population 2,820), W central GOIÁS, BRAZIL, 22 mi/35 km NE of IPORÁ; 16°26′S 51°00′W.

Israel River, c.25 mi/40 km long, COOS county, N NEW HAMPSHIRE; rises in PRESIDENTIAL RANGE; flows NW to the CONNECTICUT RIVER near LANCASTER.

Issa (EES-sah), town (2005 population 5,500), N PENZA oblast, E European Russia, on the Issa River (right branch of the MOKSHA River), on road, 12 mi/19 km SSW of RUZAYEVKA; 53°51′N 44°51′E. Elevation 643 ft/ 195 m. In agricultural area; regional produce market. Electronics repair; construction materials.

Issa, CROATIA: see VIS.

Issano (i-SA-no), village, CUYUNI-MAZARUNI district, central GUYANA, a communication point on MAZARUNI River, 70 mi/113 km SW of BARTICA (connected by road); 05°49′N 59°25′W.

Issaquah (I-suh-kwah), town (2006 population 18,373), KING county, W central WASHINGTON, suburb 15 mi/ 24 km ESE of downtown SEATTLE, on Sammamish River; 47°32′N 122°02′W. Railroad terminus. Dairy products; poultry; timber; manufacturing (in vitro diagnostic substances, printed circuit boards, X-ray apparatus, metal products, electromedical apparatus, printing, wiring devices, refrigeration equipment, measurement equipment). LAKE SAMMAMISH STATE PARK to NW.

Issaquena (is-uh-KWEE-nuh), county (□ 441 sq mi/ 1,146.6 sq km; 2006 population 1,805), W MISSISSIPPI; ⊙ MAYERSVILLE; 32°44′N 90°59′W. Bounded W by the MISSISSIPPI RIVER (LOUISIANA state line), NW corner of county touches SW corner of ARKANSAS, YAZOO RIVER forms parts of S and E boundaries; state boundary follows old channel of Mississippi River, including Albemarle Lake and other oxbow lakes on either side of river. Agriculture (cotton, corn, oats, sorghum, soybeans, wheat; cattle; timber). Includes part of Delta National Forest in SE. Has oxbow lakes along the Mississippi. Mahannah Wildlife Management Area in S, Anderson-Tully Wildlife Management Area in center, Shipland Wildlife Management Area in W. Formed 1844.

Issarlès, Lac d' (dee-sahr-le, lahk), crater lake, ARDÈCHE department, RHÔNE-ALPES region, SE central FRANCE, in the VIVARAIS mountains of the MASSIF CENTRAL, 15 mi/24 km SSE of LE-PUY-EN-VELAY; 44°49′N 04°04′E. Lake has an area of 225 acres/91 ha, and is 1,444 ft/440 m deep. Noted for its scenic site

and deep blue waters. Lake acts as the upper reservoir (elevation 3,270 ft/997 m) of an interconnected hydroelectric power system on the upper LOIRE RIVER.

Issel, GERMANY and NETHERLANDS: see OLD IJSSEL RIVER.

Isselburg (IS-sel-burg), town, NORTH RHINE–WESTPHALIA, W GERMANY, 6 mi/9.7 km W of BOCHOLT, near Dutch border; 51°50′N 06°28′E. Manufacturing of machinery; ironworking. Moated castle nearby.

Issele Uku (EE-she-lai OO-koo), town, DELTA state, S NIGERIA, 20 mi/32 km WNW of ASABA; 06°19′N 06°28′E. Road center; palm oil and kernels, kola nuts, yams, cassava, corn, plantains. Lignite deposits.

Isser, Oued (ee-SER, WED), 87 mi/140 km long, stream in N central ALGERIA; rises in the TELL ATLAS S of Berrouaghia; flows NE, past LAKHDARIA and BORDJ MÉNAÏEL, to the MEDITERRANEAN SEA 35 mi/56 km ENE of ALGIERS.

Isshiki (EE-shee-kee), town, Hazu county, AICHI prefecture, S central HONSHU, central JAPAN, on CHITA BAY and 28 mi/45 km S of NAGOYA; 34°48′N 137°01′E. Nori; eels; carnations; shrimp crackers.

Issia (EE-syah), village, Haut-Sassandra region, W CÔTE D'IVOIRE, 30 mi/48 km SSW of DALOA; 06°29′N 06°35′W. Agriculture (coffee, cacao, palm kernels, bananas, rice, corn, kola nuts); timber. Lithium and berylium deposits.

Issin, IRAQ: see ISIN.

Issoire (ee-swahr), town (□ 7 sq mi/18.2 sq km), sub-prefecture of PUY-DE-DÔME department, central FRANCE, in the MASSIF CENTRAL, near the ALLIER, 18 mi/29 km SSE of CLERMONT-FERRAND; 45°33′N 03°15′E. An important industrial center with modern machine shops; also produces aluminum and titanium alloys for Europe's aeronautical industry. The 12th-century Benedictine abbey of Saint-Austremoine was restored in 19th century and is a tourist attraction. Issoire also has a 12th-century Romanesque church, one of the finest in the AUVERGNE region. Town is a renowned base for sport gliders.

Issoudun (ee-soo-duhn), ancient *Uxellodunum*, historic town (□ 14 sq mi/36.4 sq km), sub-prefecture of INDRE department, CENTRE administrative region, central FRANCE, on Théols River, and 18 mi/29 km NE of CHÂTEAUROUX; 46°57′N 02°00′E. Leading morocco-leather manufacturing center, with tanneries. Also makes electrical equipment and ready-to-wear clothing. The town is surrounded by produce gardens and vineyards. It was a medieval stronghold fought over by Richard the Lionhearted and Philip Augustus. It was a royal seat during the reign of Louis IX, and Louis XI made a pilgrimage to the castle's abbey. The museum of St. Roche occupies a 12th–16th-century hospital built over the river.

Issum (IS-sum), town, NORTH RHINE–WESTPHALIA, W GERMANY, 13 mi/21 km N of KREFELD; 51°32′N 06°25′E.

Issuna (ee-soo-NAH), village, SINGIDA region, central TANZANIA, 40 mi/64 km S of SINGIDA; 05°27′S 34°44′E. Highway junction. Cattle, sheep, goats; corn, wheat.

Is-sur-Tille (ee–syur–teel), town (□ 8 sq mi/20.8 sq km), CÔTE-D'OR department, in BURGUNDY, E central FRANCE, 14 mi/23 km N of DIJON; 47°31′N 05°06′E. Lumber market.

Issuru, Congo: see ISURU.

Issus (I-suhs), ancient town of SE ASIA MINOR, now in TURKEY, 5 mi/8 km NW of DÖRTYOL. Located near the head of a gulf (the modern Gulf of ISKENDERUN), Issus was on a narrow strip of land backed by mountains. Nearby, in 333 B.C.E., Alexander defeated the forces of Darius III of PERSIA. In C.E. 194, Septimius Severus conquered Pescennius Niger. In 622 the Byzantine emperor Heraclius won the first of a series of battles at Issus in which the West regained territory formerly lost to the Persians.

Issyk (is-SIK), city, S ALMATY region, KAZAKHSTAN, 20 mi/32 km E of ALMATY; 42°23′N 77°25′E. Irrigated

agriculture (wheat, tobacco, fruit, vegetables). Also spelled Esyk.

Issyk-Ata (uh-suhk-ah-TAH), village, S CHÜY region, KYRGYZSTAN, on N slope of KYRGYZ Range, 20 mi/32 km S of KANT; 42°36′N 74°52′E. Elevation 5,990 ft/ 1,826 m. Health resort; mineral water. Coats. Also called (locally) Arashan.

Issyk-Kol (us-suhk-KUL), region (□ 43,144 sq mi/ 112,174.4 sq km; 1999 population 413,149), E KYRGYZSTAN, on border of SE KAZAKHSTAN; ⊙ KARAKOL. Other centers are BALYKCHY and CHOLPON-ATA. Karkara River ("milk valley") at E end (Kazakhstan border). Dairying in E. Tourism at Lake ISSYK-KOL (beaches) and in TESKEY ALA-TOO (TERSKEI ALATAU) mountains (trekking). Formed in 1939; joined (1959) to NARYN oblast until independence (1991), when it was reestablished. Also spelled Ysyk-Köl.

Issyk-Kol (us-suhk-KUL) [=hot lake], lake (□ 2,408 sq mi/6,260.8 sq km), ISSYK-KOL region, KYRGYZSTAN, in the Kyrgyz Ala-Too (ALATAU) Mountains, between the KÜNGEI ALA-TOO (KUNGEI ALATAU, N) and TESKEY ALA-TOO (TERSKEI ALATAU, S) mountains; 105 mi/169 km long, 35 mi/56 km wide; 42°30′N 77°00′E. Maximum depth 2,191 ft/668 m. Elevation 5,268 ft/ 1,606 m. World's second-largest mountain lake, exceeded only by Lake TITICACA. Slightly saline; ice-free in winter; shipping route between BALYKCHY and PRISTAN PRZHEVALSK. Tourist area; health resorts along shores. Some fishing, though diminished since non-native species (trout) replaced native ones (carp, herring). Receives many streams, including TÜP RIVER (E); intermittent link with CHU RIVER (W). Ruins of an ARMENIA monastery, reputed to be the burial place of St. Matthew, were discovered on the E shore of the lake in 2005. Also spelled Issy-Kul, Yssk-Kol.

Issyk-Kol, KYRGYZSTAN: see BALYKCHY.

Issy-les-Moulineaux (ee-see–lai–moo-lee-no), SW suburb (□ 1 sq mi/2.6 sq km) of PARIS, HAUTS-DE-SEINE department, ÎLE-DE-FRANCE region, N central FRANCE, 4 mi/6.4 km from Notre-Dame Cathedral; 48°49′N 02°16′E. It is an industrial center (metals, aeronautical equipment, chemicals, cigarettes, and beer). The terminal at the Porte de Sèvres of the PARIS-BRUSSELS railroad and other short-distance helicopter services, as well as a theological seminary, are located in Issy-les-Moulineaux.

Istakhr (EES-tahk-her), old town, S IRAN. Built largely from the ruins of ancient PERSEPOLIS, 3 mi/4.8 km away, it was a capital of the Sassanid dynasty. Istakhr stubbornly resisted (640–649) the Arabs but soon afterward lost its importance to the city of SHIRAZ. The name also appears as STAKHR.

Istalif (is-tah-LEEF), town, KABUL province, E AFGHANISTAN, 22 mi/35 km NNW of KABUL, on slopes of PAGHMAN MOUNTAINS; 34°50′N 69°05′E. Handicrafts (famous for its blue-colored pottery); weaving; graphite mining. Summer resort. Stormed (1842) by British troops in first Afghan War.

Istállóskö, Mount (ESH-tal-losh-kuh), Hungarian *Istállóskö*, highest point (3,145 ft/959 m) of Bükk Mountains, NE HUNGARY; 48°04′N 20°26′E. Istállóskö Cave, on N slope, contains remains from Stone Age.

Istán (ees-TAHN), town, MÁLAGA province, S SPAIN, in coastal spur of the CORDILLERA PENIBÉTICA, 30 mi/48 km WSW of MÁLAGA; 36°35′N 04°57′W. Cork, corn, carob beans, oranges, figs, almonds, esparto, resins.

Istanbul, city (2000 population 8,803,468), NW TURKEY, on both sides of the BOSPORUS at its entrance into the Sea of MARMARA; 41°02′N 28°57′E. Its name, often distorted to Islambol ("Islam abounding") was officially changed from CONSTANTINOPLE to Istanbul in 1930; before C.E. 330 it was known as BYZANTIUM. Constantine I then founded a new city on the site to serve as the capital of the Byzantine Empire. One of the great historic cities of the world, Istanbul is the chief city and seaport of Turkey as well as its commercial, industrial, and financial center. Manufacturing (textiles, glass, pottery, shoes, motor vehicles, ships, chemicals, printing, food products, soap, cement). The city is visited by many tourists and is a popular resort. (For the history of the city, see BYZANTIUM and CONSTANTINOPLE.) Istanbul is the seat of Istanbul University (founded 1453 as a theological school; completely reorganized 1933), a technical university, University of the Bosporus (formerly Robert College), Marmara University, Mimar Sinan University, and Yildiz University. It is the see of the patriarch of the Greek Orthodox Church, of a Latin-rite patriarch of the Roman Catholic Church, and of a patriarch of the Armenian Church. The European part of Istanbul is the terminus of an international railroad service (formerly called the Orient Express), and at Haydarpaşa station, on the Asian side, begins the Baghdad railroad. About 75% of the people live on the European side. Ataturk (YESILKÖY) International Airport is on the European side to the SW. Istanbul is the only city in the world to straddle two continents, and the only one to have been a capital during two consecutive empires—Christian and Islamic. The part of Istanbul corresponding to historic Constantinople is situated entirely on the European side. It rises on both sides of the Golden Horn (the river Heliz), which empties into an inlet of the Bosporus and then into the straits of the Sea of Marmara. Like ROME, the city is built on seven hills. Several miles of its ancient moated and turreted walls are still standing. Outside the walls and N of the Golden Horn are the commercial quarter of Galata, originally a Genoese settlement; the quarter of BEYOĞLU (formerly Pera), which under the Ottoman sultans was reserved for foreigners and their embassies; and Hasköy, the Jewish quarter. The Golden Horn is crossed by two bridges, the famous Galata bridge (which was replaced by a modern bridge in the early 1990s) and the modern Atatürk bridge. The former leads into the historic quarter of Stambul, the ancient core of the city, abutting the Bosporus and the Sea of Marmara. The quarter of PHANAR (Fanar) in the NW, near the former site of the palace of Blachernae of the Byzantine emperors, contains the see of the Greek Orthodox Church and is inhabited mainly by Greeks. Some of the palace walls are still standing. The present administrative districts of Istanbul include FATIH and EMINÖNÜ on the European side and KADIKÖY (ancient CHALCEDON) and ÜSKÜDAR (Scutari) on the Asian side. The chief monument surviving from Byzantine times is the Hagia Sophia, one of the world's most renowned works of architecture, and a UNESCO World Heritage site. (Other World Heritage sites in Istanbul include the Archaeological Park, the Süleymaniye Mosque and its associated Conservation Area, the Zeyrek Mosque, or Pantocrator Church, and its associated Conservation Area, and the Land Walls of Istanbul.) Originally a church, the Hagia Sophia was converted into a mosque after the Ottoman conquest in 1453 and is now a museum. Excavations on the sites of the former Byzantine palaces have brought to light fine works of art, and Istanbul has many monuments of the Byzantine past. The city was destroyed (1509) by an earthquake and was rebuilt by Sultan Beyazid II. Turkish culture reached its height in the 16th century and from that period date most of its magnificent mosques, notably those of Beyazid II, Sulayman I, and Ahmed I. They all reflect the influence of the Hagia Sophia—yet are distinctly Turkish—and give the skyline of Istanbul its unique character, a succession of perfectly proportioned domes broken by minarets. The Topkapi palace and the Grand Bazaar, the world's largest covered market, were also built during the Ottoman period. In the gardens by the Bosporus stand the buildings of the Seraglio, the former palace of the Ottoman sultans, now a museum. The Seraglio, begun by Muhammad II in 1462, consists of many buildings and kiosks, grouped into three courts, the last of which

contained the treasury, the harem, and the private apartments of the ruler. In the 19th century the sultans shifted (1853) their residence to the Dolma Bahçe Palace and the Yildiz Kiosk, N of Beyoğlu on the Bosporus. The environs of Istanbul, particularly the villas, gardens, castles, and small communities along the Bosporus, are famed for their beauty. In 1973 the European and Asian sections of the city were linked by the opening of the Bosporus Bridge, one of the world's longest (3,524 ft/1,074 m) suspension bridges. This was followed by the Second Bosporous Bridge (3,322 ft/1,012 m), completed in 1988. The city experienced explosive population growth in the 1970s and 1980s (it tripled in size), with the Turkish Muslim majority increasing. In the 1908s, the city tore down factories and slums along the Golden Horn, and built new parks and playgrounds. Massive efforts have been made to keep up with the growth by modernizing the city's infrastructure and municipal services. The first section of a new subway system opened in September 2000. With the establishment of the Environment Ministry in 1991, Turkey moved forward in addressing some of its most pressing environmental problems, including significant reductions of air pollution in ANKARA and Istanbul. The city is governed by a mayor and an elected city council. In August 1999, a powerful earthquake that killed more than 17,000 people struck NW Turkey, including Istanbul.

Istanos, TURKEY: see KORKUTELI.

Istarske Toplice, spa and health resort village, W CROATIA, in ISTRIA, on MIRNA RIVER, 16.1 mi/25.9 km E of NOVIGRAD. Toplice is Croatian for spa. First mentioned in 17th century; in operation since 1817. Also called Sveti [=St.] Stjepan.

Istebne (is-TYEB-ne), Slovak Istebné, HUNGARIAN Isztebne, village, STREDOSLOVENSKY province, N SLOVAKIA, on railroad, on ORAVA RIVER, and 19 mi/31 km WNW of LIPTOVSKÝ MIKULÁŠ; 49°13′N 19°13′E. Metallurgy. Has 17th-century wooden church. SUTOVSKY WATERFALL (125 ft/38 m high), Slovak Šútovský (shu-TOU-skee), is 6 mi/9.7 km WSW in Little Fatra.

Ister, ancient name for DANUBE River.

Isthmia (ISTH-mee-ah), village, KORINTHIA prefecture, extreme NE corner of PELOPONNESE department, S mainland GREECE, port at SE end of CORINTH CANAL, 4 mi/6.4 km ESE of CORINTH; 37°55′N 23°00′E. Nearby took place the Isthmian games of Corinth.

Istiaia (ee-sti-AI-ah), Latin Histiaea, town, on NW ÉVVIA island, ÉVVIA prefecture, CENTRAL GREECE department, 43 mi/69 km NW of KHALKÍS; 38°57′N 23°09′E. Wheat, wines, olive oil, citrus fruit; livestock. Also spelled Istiea or Histiaia; formerly Xerochori or Xirokhori.

Istiea, Greece: see ISTIAIA.

Istisu (eest-yi-SOO), urban settlement, SW AZERBAIJAN, on KARABAKH Upland of the Lesser CAUCASUS, in the KURDISTAN, on TERTER RIVER, and 11 mi/18 km SSW of KELBADZHAR. Health resort.

Istmia, Colombia: see ISTMINA.

Istmina (eest-MEE-nah), town, ⊙ Istmina municipio, CHOCÓ department, W COLOMBIA, landing on SAN JUAN RIVER, and 38 mi/61 km S of QUIBDÓ; 05°09′N 76°40′W. Gold and platinum mining; plantains, cacao, yucca. The rich ANDAGOYA placers are 4.0 mi/6.4 km S. Founded 1784. Alternate spellings include Itsmina, Istmino, Istmia.

Istmino, Colombia: see ISTMINA.

Istobensk (ees-TO-byensk), village, W central KIROV oblast, E central European Russia, on the E shore of the VYATKA River, terminus of local highway branch and railroad spur, 9 mi/14 km W of (and administratively subordinate to) ORICHI; 58°25′N 48°47′E. Elevation 413 ft/125 m. Pedigree livestock breeding.

Istok (ees-TOK), village (2004 population 710), W central BURYAT REPUBLIC, S Siberian Russia, on the SE shore of Lake BAYKAL, on coastal road, 43 mi/69

km WSW of ULAN-UDE; 52°05′N 106°15′E. Elevation 1,525 ft/464 m. Fisheries.

Istok (EE-stok), village, SW SERBIA, 20 mi/32 km WSW of MITROVICA, in the METOHIJA valley, at S foot of MOKRA PLANINA.

Istok (ees-TOK), settlement (2006 population 5,640), S SVERDLOVSK oblast, W Siberian Russia, on railroad junction and near highway, 7 mi/11 km SE of YEKATERINBURG; 56°47′N 60°47′E. Elevation 790 ft/240 m. Region's main airport is 2 mi/3.2 km to the S.

Istokpoga, Lake (iz-stock-PO-guh), HIGHLANDS county, central FLORIDA, 25 mi/40 km NW of LAKE OKEECHOBEE; c.10 mi/16 km long, 5 mi/8 km wide; connected by channel (N) with Lakes Weohyakapka and KISSIMMEE; has short outlet (dredged) at E end to KISSIMMEE RIVER; 24°55′N 80°38′W.

Istra (EES-trah), city (2006 population 33,305), central MOSCOW oblast, central European Russia, on the ISTRA RIVER, on road and railroad, 36 mi/58 km WNW of MOSCOW; 55°55′N 36°52′E. Elevation 482 ft/146 m. Clothing, furniture, metalworking, gas pipeline equipment; food industries (baby food, gruels). Created through amalgamation of the village of Voskresenskoye and another three neighboring hamlets into a town by an Orthodox patriarch Nikon in 1656, which, until 1930, was called Voskresensk. Nearby is the famous Novo-Iyerusalimskiy (New Jerusalem) monastery (17th century), which served as a heritage museum under the Soviet rule. Chartered in 1781.

Istra, CROATIA: see ISTRIA.

Istranca Mountains, Bulgaria and Turkey: see STRANDZHA MOUNTAINS.

Istranja, Bulgaria: see STRANDZHA MOUNTAINS.

Istra River (EES-trah), approximately 50 mi/80 km long, MOSCOW oblast, European Russia; rises in the KLIN-DMITROV RIDGE NW of SOLNECHNOGORSK; flows SE, past ISTRA, to the MOSKVA River, 6 mi/10 km SSE of PAVLOVSKAYA SLOBODA; reservoir and hydroelectric station in the upper course.

Istres (ees-truh), town, sub-prefecture of BOUCHES-DU-RHÔNE department, PROVENCE-ALPES-CÔTE D'AZUR region, SE FRANCE, near W shore of Étang de BERRE, a lagoon of the MEDITERRANEAN SEA, 24 mi/39 km NW of MARSEILLE; 43°31′N 04°59′E. Aeronautical industry at large military air base.

Istria (EES-tree-ah), Croatian Istra, region and largest peninsula (□ c.1,500 sq mi/3,885 sq km) in ADRIATIC SEA, predominantly in CROATIA, projecting into the N ADRIATIC bet. the gulfs of TRIESTE and RIJEKA. A sect. of the NW portion, including the city of TRIESTE, belongs to ITALY. Mountainous in the NE and hilly in the center; open plain in the SW. The area is forested both along the coast and inland. Agr. (olives, grapes, vegetables; dairy cattle, poultry); fishing; mining (Raša); mfg. (shipbuilding in PULA, fish canning; electronics, building materials); tourism (POREC, OPATIJA, and MEDULIN Rivieras; spas). Chief city, Pula. The population is about 2/3 Croatian. Inhabited by Illyrian tribes when it passed (2nd cent. B.C.E.) to Rome. Remained under nominal Byzantine rule until the 8th cent. By that time, Slavs had settled in the rural areas and Romans in the cities. By the 15th cent. Austria and Venice had absorbed, respectively, the NE and SW parts of the region. The Treaty of Campo Formio (1797) and the Congress of Vienna (1815) added the Venetian part to Austria. In 1919 all Istria passed to Italy, but the Ital peace treaty of 1947 gave most of it to the former Yugoslavia. The NW sect. passed to Italy in 1954.

Istria (EES-tree-yah), village, Constanţa county, SE ROMANIA, in DOBRUJA, on W shore of Lake SINOE, 27 mi/43 km N of Constanţa; 44°34′N 28°43′E. Noted as site of the ruins of ancient city of HISTRIA, lying just SE, the object of extensive excavations. Histria was founded by Milesians in 7th century B.C.E. as a trading colony, then passed successively under Macedonian (339 B.C.E.) and Roman (72 B.C.E.) domination; destroyed by the Goths (238 C.E.), it was rebuilt by Constantine the Great, and destroyed again by the Barbarians.

Istrouma (is-TROO-muh), suburb, EAST BATON ROUGE parish, SE central LOUISIANA, part of BATON ROUGE.

Ist'ye (EEST-ye), settlement, central RYAZAN oblast, central European Russia, on the Ist'ya River (right tributary of the OKA River), on highway branch and railroad spur, 14 mi/23 km SW of SPASSK-RYA-ZANSKIY; 54°18′N 40°06′E. Elevation 380 ft/115 m. Machine-building plant.

Isumi (EES-mee), town, Isumi district, CHIBA prefecture, E central JAPAN, 19 mi/30 km S of CHIBA; 35°16′N 140°18′E.

Isuru (ee-SOO-roo), village, Équateur province, NE CONGO, 11 mi/18 km W of KILO-MINES. Gold mining and processing. Sometimes spelled ISSURU.

Isvoron, Greece: see STRATONIKI.

Iswaripur (eesh-sho-ree-poor), village, KHULNA district, SW EAST BENGAL, BANGLADESH, in the SUN-DARBANS, 45 mi/72 km SW of KHULNA; 22°22′N 89°02′E. Road terminus. Agriculture (rice, jute, oil-seeds). Was the capital of a 16th-century independent Muslim kingdom until defeated in 1576 by Akbar's Hindu general.

Isyangulovo (ee-syahn-GOO-luh-vuh), village (2005 population 7,460), S BASHKORTOSTAN Republic, on the SW slopes of the S URALS, RUSSIA, on the GREATER IK RIVER, on road, 70 mi/113 km NW of ORENBURG; 52°11′N 56°34′E. Elevation 672 ft/204 m. In protected old-growth forest region. Weather station. Dairy processing.

Iszkaszentgyörgy (ES-kah-sant-dyuhr-dyuh), mining settlement, FEJÉR county, N central HUNGARY, on SE slope of BAKONY MOUNTAINS, 6 mi/10 km NW of Szekesfehérvár; 47°14′N 18°18′E. Bauxite mine. Production was to close during the 1990s.

Isztebne, SLOVAKIA: see ISTEBNE.

Itá (ee-TAH), town (2002 population 17,469), CENTRAL department, S PARAGUAY, 26 mi/42 km SE of Asunción; 25°03′S 57°21′W. Processing and agricultural center (cotton, sugarcane, tobacco, fruit); vegetable oil manufacturing; liquor distilling; tanning. Founded 1536.

Itabaiana (ee-tah-bei-ah-nah), city (2007 population 24,644), E PARAÍBA, NE BRAZIL, on right bank of RIO PARAÍBA, and 35 mi/56 km SW of JOÃO PESSOA; 07°20′S 35°27′W. Road and railroad junction (spur to CAMPINA GRANDE); important trade center (cotton, livestock, leather, cereals). Called Tabaiana, 1944–1948. Formerly spelled Itabayana.

Itabaiana (EE-tah-bei-ah-nah), city (2007 population 83,167), central SERGIPE, NE BRAZIL, on W slope of Serra ITABAIANA, 30 mi/48 km NW of ARACAJU; 11°32′S 37°31′W. Agricultural trade center (corn, manioc, cereals, livestock). Has model farms.

Itabaiana, Serra (EE-tah-bei-ah-nah), hill range in E SERGIPE, NE BRAZIL, forming escarpment above coastal lowland c.25 mi/40 km NW of ARACAJU. Rises to 2,800 ft/853 m. Dense forest cover.

Itabaianinha (EE-tah-bei-ah-neen-yah), city (2007 population 37,431), S SERGIPE, NE BRAZIL, on railroad, and 22 mi/35 km W of ESTÂNCIA; 11°16′S 37°47′W. Manioc; livestock.

Itabapoana (EE-tah-bah-po-ah-nah), town, NE RIO DE JANEIRO state, BRAZIL, on the Atlantic at mouth of ITABAPOANA River (ESPÍRITO SANTO border), and 36 mi/58 km NNE of CAMPOS; 21°18′S 40°58′W. Sugarcane, fish.

Itabashi (ee-TAH-bah-shee), ward, NW TOKYO city, Tokyo prefecture, E central HONSHU, E central JAPAN, NW of central Tokyo. Bordered N and NW by SAI-TAMA prefecture, E by KITA ward, SE by TOSHIMA ward, and SW by NERIMA ward.

Itabayanna, Brazil: see ITABAIANA.

Itabela (EE-tah-BE-lah), city (2007 population 25,821), SE BAHIA, BRAZIL, 25 mi/41 km S of EUNÁPOLIS; 16°45′S 39°35′W.

Itaberá (EE-tah-be-RAH), city (2007 population 17,605), S SÃO PAULO, BRAZIL, 19 mi/31 km NW of ITAPEVA; 23°51′S 49°09′W. Coffee, cotton, tobacco, grain.

Itaberaba (ee-tah-be-rah-bah), city (2007 population 59,501), E central BAHIA, BRAZIL, on railroad spur, 100 mi/161 km W of CACHOEIRA; 12°30′S 40°20′W. Tobacco, coffee, manioc, livestock. Asbestos deposits.

Itaberaí (ee-tah-be-rah-EE), city (2007 population 30,621), S central GOIÁS, central BRAZIL, 50 mi/80 km NW of GOIÂNIA; 16°01′S 49°45′W. Distilling, cattle-shipping. Emeralds and beryls found in area. Formerly spelled Itaberahy.

Itabi (EE-tah-bee), town (2007 population 4,736), N SERGIPE state, BRAZIL, 50 mi/80 km N of ARACAJU; 10°08′S 37°06′W. Manioc.

Itabira (ee-tah-bee-rah), city (2007 population 105,199), E central MINAS GERAIS, BRAZIL, in an offshoot of the SERRA DO ESPINHAÇO, 50 mi/80 km NE of BELO HORIZONTE; 19°34′S 43°42′W. Leading iron-mining center of Brazil. Caué peak (4,281 ft/1,305 m), an actively worked iron mt. containing over 100,000,000 tons/90,700,000 metric tons of ore (67% pure hematite), and Conceição peak, with similar deposits, surround the city. Crushed ore is shipped 300 mi/483 km by narrow-gauge railroad through RIO DOCE valley to port of VITÓRIA-TUBARÃO (ESPÍRITO SANTO). Semi-precious stones and graphite also found in area. Originally called Itabira or Itabira de Matto Dentro; renamed Presidente Vargas, 1944; reverted back to Itabira, 1948.

Itabirinha de Montena (EE-tah-bee-reen-yah zhe man-TAI-nah), town (2007 population 10,343), E central MINAS GERAIS, BRAZIL, near border with ESPÍRITO SANTO, 56 mi/90 km NE of GOVERNADOR VALADARES; 18°29′S 41°14′W.

Itabirito (ee-tah-bee-ree-to), city (2007 population 41,523), S central MINAS GERAIS, BRAZIL, on railroad, and 25 mi/40 km SSE of BELO HORIZONTE; 20°21′S 43°45′W. Iron-mining center, second only to ITABIRA, working high-grade hematite ore of cone-shaped Itabirito peak; pig-iron plant. Ore is shipped by rail-road to RIO DE JANEIRO.

Itaboraí (EE-tah-bo-rah-ee), city (2007 population 215,792), S central RIO DE JANEIRO state, BRAZIL, on railroad, and 19 mi/31 km NE of NITERÓI; 22°45′S 42°52′W. Alcohol distilling; poultry, sugar, fruit. Formerly spelled Itaborahy.

Itabuna (ee-tah-boo-nah), city (2007 population 210,604), BAHIA state, E BRAZIL, on the Rio Itabuna; 14°46′S 39°16′W. A cacao-producing center, it also has a well-developed cattle industry and a chemical factory.

Itacajá (ee-tah-kah-zhah), town (2007 population 6,913), E TOCANTINS state, BRAZIL, 81 mi/130 km SE of PALMAS; 08°19′S 47°46′W.

Itacambira (EE-tah-kahm-bee-rah), town (2007 population 5,018), N central MINAS GERAIS, BRAZIL, 59 mi/95 km SE of MONTES CLAROS; 17°08′S 43°17′W.

Itacarambi (EE-tah-kah-rahn-bee), city (2007 population 17,626), N central MINAS GERAIS, BRAZIL, on Rio SÃO FRANCISCO, 37 mi/59 km NE of JANUÁRI; 05°10′S 44°18′W.

Itacaré (ee-tah-kah-RAI), city (2007 population 24,639), E BAHIA, BRAZIL, on the Atlantic at mouth of the RIO DE CONTAS, 35 mi/56 km N of ILHÉUS; 14°16′S 39°00′W. Ships cacao, piassava, manioc. Formerly called Barra do Rio de Contas.

Itacoatiara (EE-tah-ko-AH-chee-AH-rah), city (2007 population 84,671), E AMAZONAS, BRAZIL, steamer and hydroplane landing on left bank of the AMAZON below influx of the MADEIRA, and 110 mi/177 km E of MANAUS; 03°00′S 58°30′W. Ships rubber, Brazil nuts, cacao, guaraná, hardwood, fish, and copaiba. Important archaeological site; tinted petroglyphs found here. Old name, Serpa.

Itacolomi, Pico de, mountain (5,896 ft/1,797 m) in offshoot of the SERRA DO ESPINHAÇO, SE central MINAS GERAIS, BRAZIL, just SW of OURO PRÊTO.

Itacurubí de la Cordillera (ee-tah-koo-roo-BEE DE LA kor-dee-YE-rah), town, La Cordillera department, S central PARAGUAY, in Cordillera de los Altos, 55 mi/89 km ESE of Asunción; 25°26′S 56°51′W. Resort and agr. center (fruit, maté, tobacco; cattle); tanning, tile making.

Itacurubí del Rosario (ee-tah-koo-roo-BEE DEL ro-SAH-ree-o), town, San Pedro department, central PARAGUAY, 75 mi/121 km NE of Asunción; 24°30′S 56°40′W. Lumbering, agricultural (oranges, maté; livestock); tanneries, distilleries, sawmills.

Itadori (ee-TAH-do-ree), village, Mugi district, GIFU prefecture, central HONSHU, central JAPAN, 22 mi/35 km N of GIFU; 35°42′N 136°48′E.

Itaetê (ee-tah-e-TAI), city (2007 population 14,079), central BAHIA, BRAZIL, W terminus of railroad from SALVADOR, and 26 mi/42 km SE of ANDARAÍ; 13°00′S 40°58′W. Diamond shipping. Until 1944, called Iguaçu (formerly spelled Iguassú).

Itagawa (ee-TAH-gah-wah), town, South Akita district, Akita prefecture, N HONSHU, NE JAPAN, 12 mi/20 km N of AKITA city; 39°52′N 140°04′E.

Itaguaçu (EE-tah-gwah-soo), city (2007 population 13,876), central ESPÍRITO SANTO, BRAZIL, 45 mi/72 km NW of VITÓRIA; 19°43′S 40°48′W. Coffee, corn. Formerly spelled Itaguassú.

Itaguaí (EE-tah-gwah-ee), city (2007 population 95,468), SW RIO DE JANEIRO state, BRAZIL, on railroad and 32 mi/51 km W of RIO DE JANEIRO; 22°52′S 43°47′W. At nearby site called Kilometro 47, on Rio–SÃO PAULO highway, is Rural University (established 1943; operated by federal government).

Itaguatins (EE-tah-gwah-cheens), town (2007 population 6,074), TOCANTINS state, N central BRAZIL, on left bank of TOCANTINS River (rapids) and 15 mi/24 km S of IMPERATRIZ (MARANHÃO); 05°47′S 47°29′W. Babassu nuts, mangabeira rubber. Until 1944, called Santo Antônio da Cachoeira.

Itagüí (ee-tah-GWEE), town, ⊙ Itagüí municipio, AN-TIOQUIA department, NW central COLOMBIA, in Cordillera CENTRAL, on Río PORCE, 6.0 mi/9.7 km SSW of MEDELLÍN; 06°10′N 75°36′W. Elevation 5,331 ft/1,625 m. Agricultural region (sugarcane, coffee, corn, beans, potatoes); consumer goods. Resort.

Itahari (EE-tuh-huh-ree), village (2001 population 41,210), SE NEPAL; 26°38′N 87°17′E. Major crossroads on the E-W highway and the road from BIRATNAGAR to DHARAN and DHANKUTA.

Itaí (EE-tah-EE), city (2007 population 22,608), S SÃO PAULO, BRAZIL, 25 mi/40 km SSW of AVARÉ, near Armando Lavdner Reservoir; 23°24′S 49°06′W. Tanning, lumber shipping. Formerly Itahy.

Itá-Ibaté (ee-TAH–ee-bah-TAI), town, N CORRIENTES province, ARGENTINA, river port on PARANÁ RIVER (PARAGUAY border), and 95 mi/153 km E of COR-RIENTES; 27°26′S 57°20′W. Rice, citrus fruit; livestock.

Itaiçaba (EE-tai-SOO-bah), town (2007 population 7,433), NE CEARÁ, BRAZIL, on RIO JAGUARIBE, 9 mi/14.5 km SW of ARACATÍ; 04°40′S 37°45′W.

Itaiópolis (EE-tah-ee-O-po-lees), city (2007 population 19,748), N SANTA CATARINA, BRAZIL, 28 mi/45 km S of MAFRA; 26°20′S 49°56′W. Agricultural colony (grain, livestock, timber) founded 1891 by British and Slavic settlers. Formerly spelled Itayopolis.

Itaipú Dam, BRAZIL-PARAGUAY border, 8 mi/12.9 km N of CIUDAD DEL ESTE on PARANÁ River; 25°22′S 54°34′W. One of the world's largest hydroelectric dams (12,600 MW); started in 1979, finished in 1991. Most electricity exported to Brazil.

Itaipú Reservoir, BRAZIL-PARAGUAY border; 25°00′S 54°28′W. Created by ITAIPÚ DAM, it reaches nearly 100 mi/161 km upstream and flooded GUAIRÁ FALLS.

Itaituba (ee-tah-EE-too-bah), city (2007 population 118,403), W PARÁ, BRAZIL, head of navigation on left bank of TAPAJÓS, and 140 mi/225 km SSW of SAN-TARÉM; 04°17′S 56°02′W.

Itajahy do Su, Brazil: see RIO DO SUL.

Area is shown by the symbol □, and capital city or county seat by ⊙.

Itajaí (EE-tah-zhah-EE), city (2007 population 163,298), NE SANTA CATARINA, BRAZIL, Atlantic port at mouth of ITAJAÍ AÇU RIVER, and 50 mi/80 km N of FLORIANÓPOLIS; 26°53′S 48°39′W. Port for BLUMENAU (25 mi/40 km W; river navigation) and its fertile agricultural hinterland settled by German immigrants in mid-19th century. Exports timber (cedar and rosewood, mahogany), dairy products, rice, meat. Textile mills. Iron and molybdenum deposits in area. Airfield. Formerly spelled Itajahy.

Itajaí Açu River (EE-tah-zhah-EE AH-soo), c.125 mi/201 km long, E SANTA CATARINA, BRAZIL; rises in several headstreams in the Serra do MAR; flows E, through fertile, thickly settled lowland, past RIO DO SUL and BLUMENAU (head of navigation), to the ATLANTIC at ITAJAÍ.

Itajaí do Sul, Brazil: see RIO DO SUL.

Itajobi (EE-tah-zhah-bee), city (2007 population 14,182), N central SÃO PAULO, BRAZIL, 13 mi/21 km SSW of CATANDUVA; 21°19′S 49°04′W. Cotton ginning, wine making; cotton, coffee, cattle.

Itajubá (ee-tah-zhoo-BAH), city (2007 population 86,693), SW MINAS GERAIS, BRAZIL, on slope of SERRA DA MANTIQUEIRA, near SÃO PAULO border, 110 mi/177 km NE of SÃO PAULO city; 22°25′S 45°27′W. Ships coffee, sugar, tobacco, lard; manufacturing (textiles, matches, hats). Has electrotechnical institute (established 1913).

Itaka (EE-tah-kah), town, Inan district, MIE prefecture, S HONSHU, central JAPAN, 22 mi/35 km S of TSU; 34°25′N 136°20′E.

Itaka (ee-tah-KAH), town (2005 population 425), central CHITA oblast, S SIBERIA, RUSSIA, on the Itaka River, 37 mi/60 km N of KSENYEVKA; 53°53′N 118°42′E. Elevation 2,234 ft/680 m. Gold mining.

Itaka (ee-TAH-kah), village, MBEYA region, SW TANZANIA, 43 mi/69 km W of MBEYA; 08°52′S 32°48′E. Coffee, tea, corn, wheat; sheep, goats.

Itakhola (ee-tah-ko-lah), village, SYLHET district, E EAST BENGAL, BANGLADESH, 60 mi/97 km SSW of SYLHET; 24°51′N 90°48′E. Rice, tea, oilseeds; umbrella manufacturing. Tea processing nearby.

Itako (ee-TAH-ko), town, Namegata county, IBARAKI prefecture, central HONSHU, E central JAPAN, 28 mi/45 km S of MITO; 35°56′N 140°33′E. Summer resort. Also Idako.

Itakura (ee-TAHK-rah), town, Oura district, GUMMA prefecture, central HONSHU, N central JAPAN, 37 mi/60 km S of MAEBASHI; 36°13′N 139°36′E. Vegetables (cucumbers, eggplants).

Itakura, town, Nakakubiki county, NIIGATA prefecture, central HONSHU, N central JAPAN, 65 mi/105 km S of NIIGATA; 37°02′N 138°17′E. Rice.

Itakyry, town, E PARAGUAY, ALTO PARANA department, 50 mi/80 km NW of CIUDAD DEL ESTE; 25°02′S 55°05′W. Agricultural center (soybeans, cotton, corn).

Itala, SOMALIA: see ADALE.

Italaque (ee-tah-LAH-kai), town and canton, CAMACHO province, LA PAZ department, W BOLIVIA, in CORDILLERA Real, 15 mi/24 km NE of PUERTO ACOSTA; 15°24′S 69°04′W. Potatoes, oca.

Italian East Africa, former federation of the Italian colonies of ERITREA and ITALIAN SOMALILAND and the kingdom of ETHIOPIA. The federation was formed (1936) to consolidate the administration of the three areas. During the federation's existence, efforts were made to construct road systems and to establish new industries and agricultural plantations. Resistance to Italian rule was particularly strong in ETHIOPIA, and when British forces invaded the federation in January 1941 they received widespread support. By December 1942 the Italians had been totally defeated. Ethiopia was restored its independence; Eritrea was placed under Ethiopian control in 1952; and Italian Somaliland, after a period as a UN trusteeship, became part of SOMALIA in 1960.

Italian Mountain, peak (13,378 ft/4,078 m) in ROCKY MOUNTAINS, GUNNISON county, W central COLORADO, 13 mi/21 km ENE of CRESTED BUTTE, in Gunnison National Forest.

Italian Riviera, ITALY: see RIVIERA.

Italian Somaliland: see SOMALIA.

Itálica (ee-TAH-lee-kah), ruined Roman town, SEVILLE province, SW SPAIN, just NW of SANTIPONCE, 5 mi/8 km NW of SEVILLE. Founded c.205 B.C.E. by Scipio Africanus for retired soldiers. Noted as birthplace of Emperor Hadrian, and probably of Trajan and Theodosius. Has remains of amphitheater.

Italy, Italian *Italia*, republic (□ 116,303 sq mi/301,225 sq km; 2004 estimated population 58,057,477; 2007 estimated population 58,147,733), S EUROPE, bordering on FRANCE in the NW, the LIGURIAN SEA and the TYRRHENIAN SEA in the W, the IONIAN SEA in the S, the ADRIATIC SEA in the E, SLOVENIA in the NE, and AUSTRIA and SWITZERLAND in the N; ⊙ ROME; 43°00′N 13°00′E.

Geography

The country includes the large Mediterranean islands of SICILY and SARDINIA and several small islands, notably ELBA, CAPRI, ISCHIA, and the LIPARI ISLANDS. Vatican City (see under VATICAN) and SAN MARINO are two independent enclaves on the Italian mainland. Rome is Italy's largest city; other important cities include MILAN, NAPLES, TURIN, GENOA, BOLOGNA, FLORENCE, CATANIA, VENICE, BARI, TRIESTE, MESSINA, VERONA, PADUA, CAGLIARI, TARANTO, BRESCIA, and LIVORNO. The country is divided into twenty regions, which are subdivided into a total of ninety-four provinces. About 75% of Italy is mountainous or hilly, and roughly 20% of the country is forested. There are narrow strips of low-lying land along the Adriatic coast and parts of the Tyrrhenian coast. N Italy, made up largely of a vast plain that is contained by the ALPS in the N and drained by the PO RIVER and its tributaries, comprises the regions of LIGURIA, PIEDMONT, Valle d'Aosta (see AOSTA, VALLE D'), LOMBARDY, Trentino–lto Adige, VENETIA, FRIULI-VENEZIA GIULIA, and part of EMILIA-ROMAGNA(which extends into central Italy). It is the richest part of the country, with the best farmland, the chief port (GENOA), and the largest industrial centers. N Italy also has a flourishing tourist trade on the ITALIAN RIVIERA, in the Alps (including the DOLOMITES), on the shores of its beautiful lakes (LAKE MAGGIORE, LAKE COMO, and LAKE GARDA), and in Venice. GRAN PARADISO (13,323 ft/4,061 m), the highest peak wholly situated within Italy, rises in Valle d'Aosta. The Italian peninsula, bootlike in shape and traversed in its entire length by the APENNINES (which continue on into Sicily), comprises central Italy (MARCHE, TUSCANY, UMBRIA, and LATIUM regions) and S Italy (CAMPANIA, BASILICATA, ABRUZZI, MOLISE, CALABRIA, and APULIA regions). Central Italy contains great historic and cultural centers such as Rome, Florence, PISA, SIENA, PERUGIA, ASSISI, URBINO, Bologna, RAVENNA, RIMINI, FERRARA, and PARMA. The major cities of S Italy, generally the poorest and least developed part of the country, include Naples, Bari, BRINDISI, FOGGIA, and Taranto. Except for the Po and ADIGE, Italy has only short rivers, among which the ARNO and the TIBER are the best known. Most of Italy enjoys a Mediterranean climate; however, that of Sicily is subtropical, and in the Alps there are long and severe winters. The country has great scenic beauty—the majestic Alps in the N, the soft and undulating hills of Umbria and Tuscany, and the romantically rugged landscape of the S Apennines. The Bay of NAPLES, dominated by Mount VESUVIUS, is one of the world's most famous sights.

Population

The great majority of the population speaks Italian (including several dialects); there are small German-, French-, and Slavic-speaking minorities. Most Italians are Roman Catholic. There are numerous universities in Italy, including ones at Bari, Bologna, Genoa, Milan, Naples, Turin, Padua, Palermo, and Rome.

Economy

Italy has greatly improved its highway system in the post-war years, especially in the S. Italy began to industrialize late in comparison to other European nations, and until World War II was largely an agricultural country. However, after 1950 industry was developed rapidly so that by the early 1990s manufacturing contributed about 28% of the annual GDP and agriculture only about 2.2%. The principal farm products are wheat, sugar beets, maize, tomatoes, potatoes, citrus fruit, olives, and livestock (especially cattle, pigs, sheep, and goats). In addition, much wine is produced from grapes grown throughout the country. Industry is centered in the N, particularly in the "golden triangle" of Milan-Turin-Genoa. Italy's economy has been gradually diversifying, shifting from food and textiles and apparel, to engineering, steel, and chemical products. The chief manufacture of the country includes iron, steel and other metal products, refined petroleum, chemicals, electrical and nonelectrical machinery, motor vehicles, textiles and apparel, printed materials, and plastics. Although many of Italy's important industries are state-owned, the trend in recent years has been toward privatization. There is a small fishing industry. Italy has only limited mineral resources and has consistently increased its mineral imports; the chief minerals produced are petroleum (especially in Sicily), lignite, iron ore, iron pyrites, bauxite, sulfur, and mercury. There are also large deposits of natural gas (methane). Much hydroelectricity is generated, and there are several nuclear power stations in the country. Italy, however, is still greatly dependent on oil to meet its energy requirements. In order to further the economic development of the S, the *Cassa per il mezzogiorno* (Southern Italy Development Fund) was founded in 1950; it allocated considerable funds, especially for improving the economic infrastructure of the region, but it is now defunct. A plan was instituted in the 1980s by the government to cut consumption and reduce public spending; nevertheless, by 2003 the national debt was 106.4% of GNP. Italy has a large foreign trade, facilitated by its sizable commercial shipping fleet. The leading exports are machinery, textiles, chemicals, motor vehicles, and metals; the main imports are machinery, transport equipment, chemicals, food and food products, and minerals (especially petroleum). Tourism is a major source of foreign exchange. The chief trade partners are Germany, France, the U.S., Spain, and Great Britain. Italy is a member of the EU, NATO, and the UN.

History: to 5th Century

Little is known of Italian history before the 5th century B.C.E., except for the regions (S Italy and Sicily) where the Greeks had established colonies. The earliest known inhabitants seem to have been of Ligurian stock. The Etruscans, coming probably from ASIA MINOR, established themselves in central Italy before 800 B.C.E. They reduced the indigenous population to servile status and established a prosperous empire with a complex culture. In the 4th century B.C.E., the Celts (called Gauls by Roman historians) invaded Italy and drove the Etruscans from the Po valley. In the S, the Etruscan advance was checked about the same time by the Samnites, who had adapted the civilization of their Greek neighbors and who in the 4th century B.C.E. drove the Etruscans out of Campania. The Latins, living along the coast of Latium, had not been fully subjected to the Etruscans; they and their neighbors, the Sabines, were the ancestors of the Romans. The history of Italy from the 5th century B.C.E. to the 5th century C.E. is largely that of the growth of Rome and of the Roman Empire, of which Italy was the core. By the beginning of the Christian era (4th century), all of Italy had been thoroughly latinized, Roman citizenship had been extended to all free Italians, an excellent system of roads had been built, and Italy, made tax exempt, shared fully in the wealth of Rome. Never since has Italy known an equal degree of prosperity or as long a period of peace.

History: 5th Century to 962

Christianity spread rapidly. Like the rest of the Roman Empire, Italy in the early 5th century C.E. began to be invaded by successive waves of barbarian tribes—the Germanic Visigoths, the Huns, and the Germanic Heruli and Ostrogoths. The deposition (476) of Romulus Augustulus, the last Roman emperor of the West, and the assumption by Odoacer of the rule over Italy is commonly regarded as the end of the Roman Empire. The Eastern emperors, residing at CONSTANTINOPLE, never renounced their claim to Italy, and the Byzantine emperor Justinian I retook the land in 535. Except in the exarchate of Ravenna, the PENTAPOLIS, and the coast of S Italy, Byzantine rule was soon displaced by that of the Lombards, who under Alboin established (569) a new kingdom. During this time the papacy emerged as the chief bulwark of Latin civilization. The Lombards warded off Byzantine efforts at reconquest and in 751 took Ravenna; their advance on Rome resulted in the appeal of Pope Stephen II to Pepin the Short, ruler of the Franks, who expelled the Lombards from the exarchate of Ravenna and from the Pentapolis, which he donated (754) to the pope. Pepin's intervention was followed by that of his son Charlemagne, who defeated the Lombard king, Desiderius; was crowned king of the Lombards; confirmed his father's donation to the papacy; and in 800 was crowned emperor of the West at Rome. These events shaped much of the later history of Italy and of the papacy. Among the direct results were the claim of later emperors to Italy and the temporal power of the popes.

History: 962 to 15th Century

As Carolingian power waned and Italy fell victim to internal strife and foreign raiders, the German king Otto I invaded Italy at the request of the papacy. In 962, he was crowned emperor by the pope. This union of Italy and Germany marked the beginning of the Holy Roman Empire. Although the Alps had never prevented invaders from entering Italy, they did prevent the emperors from exercising effective control over it. Again and again the emperors and German kings crossed the Alps to assert their authority; each time their authority virtually vanished when they left Italy. This led to the rise of the Italian city, beginning in the 10th century. The rise was partly political in origin—the burghers were drawing together to protect themselves from the nobles—and partly economic—contact with the Muslim world was making the Italian merchants the middlemen and the Italian cities the entrepôts of Western Europe. To protect their commerce and their industries (particularly the wool industry), cities grouped together in leagues, which often were at war with each other. Rivalry among the cities, however, prevented the formation of any union strong enough to consolidate even a part of Italy. The most powerful princes and the most powerful republics tended to increase their territories at the expense of weaker neighbors. The cities in the PAPAL STATES passed under local tyrants during the Babylonian captivity of the popes at AVIGNON (1309–1378) and during the Great Schism (1378–1417).

History: 15th Century to 1701

By the end of the 15th century Italy had fallen into the following chief component parts: in the S, the kingdoms of Sicily and Naples, torn by the rival claims of the French Angevin dynasty and the Spanish house of ARAGÓN; in central Italy, the Papal States, the republics of Siena, Florence, and LUCCA, and the cities of Bologna, FORLÌ, RIMINI, and FAENZA (only nominally subject to the pope); in the N, the duchies of

FERRARA and MODENA, Mantua, Milan, and SAVOY. The two great merchant republics, Venice and Genoa, with their far-flung possessions, colonies, and outposts, were distinct in character and outlook from the rest of Italy. Constant warfare among these many states resulted in political turmoil, but did little to diminish their wealth or to hinder their cultural output. In fact, their prosperity facilitated the great cultural flowering of the Italian Renaissance, which permanently changed the civilization of Western Europe. The Renaissance reached its peak in the late 15th century, just as Italy's political independence was threatened by the growing nations of France, Spain, and Austria. Quarrels among Italian states invited foreign intervention. The invasion (1494) of Italy by Charles VIII of France marked the beginning of the Italian Wars, which ended in 1559 with most of Italy subjected to Spanish rule or influence. By the Treaty of Cateau-Cambrésis (1559), Spain gained the kingdoms of Sicily and Naples and the duchy of Milan.

History: 1701 to 1796

Foreign domination continued with the War of the Spanish Succession (1701–1714). By 1748, Naples, Sicily, and the duchies of PARMA and PIACENZA had passed to branches of the Spanish Bourbons, and the duchies of Milan, Mantua, Tuscany, and Modena to Austria. These centuries of political weakness were also a period of economic decline. The center of European trade shifted away from the Mediterranean, and commerce and industry suffered from the mercantilist policies of the European states. Nevertheless, Italy continued to have considerable influence on European culture, especially in architecture and music. Yet to subsequent generations in Italy (especially in the 19th century), preoccupied with the concepts of national independence and political power, the political condition of 18th-century Italy represented national degradation. The French Revolution rekindled Italian national aspirations, and the French Revolutionary Wars swept away the political institutions of 18th-century Italy.

History: 1796 to 1861

General Bonaparte (later Napoleon I), who defeated Sardinian and Austrian armies in his Italian campaign of 1796–1797, was at first acclaimed by most Italians. Napoleon redrew the Italian map several times. Extensive land reforms were carried out, especially in N Italy. But Napoleon's failure to unite Italy and to give it self-government disappointed Italian patriots, some of whom formed secret revolutionary societies such as the Carbonari, which later played a vital role in Italian unification. The Congress of VIENNA (1814–1815) generally restored the pre-Napoleonic status quo and the old ruling families. In 1848–1849, there were several short-lived revolutionary outbreaks, notably in Naples, Venice, Tuscany, Rome, and the kingdom of Sardinia (whose new liberal constitution survived). Unification was ultimately achieved under the house of Savoy, and Victor Emmanuel II became king of Italy in 1861. At that time, the kingdom of Italy did not include Venetia, Rome, and part of the Papal States. Relations between the Italian government and the papacy, which refused to concede the loss of its temporal power, remained a major problem until 1929, when the Lateran Treaty made the pope sovereign within VATICAN City.

History: 1861 to 1922

From 1861 until the Fascist dictatorship (1922–1943) of Benito Mussolini, Italy was governed under the liberal constitution adopted by Sardinia in 1848. The reigns of Victor Emmanuel II (1861–1878) and Humbert I (1878–1900), and the first half of the reign of Victor Emmanuel III (1900–1946), were marked by moderate social and political reforms and by some industrial expansion in N Italy (mainly in the 20th century). In the underdeveloped S, rapid population growth led to mass emigration, both to the industrial centers of N

Italy and to the Americas. In World War I, Italy at first remained neutral. After the Allies offered substantial territorial rewards, Italy denounced the Triple Alliance and entered (1915) the war on the Allied side. Although the Italians initially suffered serious reverses, they won (1918) a great victory at VITTORIO VENETO, which was followed by the surrender of Austria-Hungary. Within Italy, political and social unrest increased, furthering the growth of Fascism.

History: 1922 to 1941

The Fascist leader (Italian *Il Duce*) Mussolini, promising the restoration of social order and of political greatness, directed (October 27, 1922) a successful march on Rome and was made premier by the king. Granted dictatorial powers, Mussolini quashed opposition to the state (especially that of socialists and communists), regimented the press and the schools, imposed controls on industry and labor, and created a corporative state controlled by the Fascist party and the militia. Mussolini followed an aggressive foreign policy, and after 1935 he turned increasingly to militarist and imperialist solutions to Italy's problems. Italy conquered ETHIOPIA in 1935–1936, easily overcoming the ineffective sanctions imposed by the League of Nations (from which Italy withdrew in 1937). At the same time, Italy drew closer to Nazi Germany and to Japan; in 1936, Italy formed an entente with Germany. Italy intervened on the Insurgent side in the Spanish civil war (1936–1939), and in 1939 it seized ALBANIA. At the outbreak of World War II, Italy assumed a neutral stance friendly to Germany, but in June 1940 it declared war on collapsing France and on Great Britain.

History: 1941 to 1951

In 1941, Italy declared war on the USSR and on the U.S. Soon Italy suffered major reverses, and by July 1943 it had lost its African possessions, its army was shattered, Sicily was falling to U.S. troops, and Italian cities (especially ports) were being bombed by the Allies. In July 1943, discontent among Italians culminated in the rebellion of the Fascist grand council against Mussolini, Mussolini's dismissal by Victor Emmanuel III, the appointment of Pietro Badoglio as premier, and the dissolution of the Fascist party. In September 1943, Italy surrendered unconditionally to the Allies, while German forces quickly occupied N and central Italy. Aided by the Germans, Mussolini escaped from prison and established a puppet republic in N Italy. In April 1945, partisans captured and summarily executed Mussolini. In May 1945, the Germans surrendered. After the war, Italy's borders were established by the peace treaty of 1947, which assigned several small Alpine districts to France; the DODECANESE to GREECE; and TRIESTE, ISTRIA, most of VENEZIA GIULIA, and several Adriatic islands to the former YUGOSLAVIA and to the Free Territory of Trieste. In 1954, Trieste and its environs were returned to Italy. As a result of the war, Italy also lost its colonies of LIBYA, ERITREA, and ITALIAN SOMALILAND. The Christian Democrats, Communists, and Socialists emerged from the war as Italy's chief political parties; the political scene in post-war Italy, however, was characterized by ever-crumbling power structures, government scandal, and popular unrest. In international affairs, postwar Italy was firmly tied to the West, joining NATO at its inception in 1949.

History: 1951 to Present

Despite Italy's pervasive political instability, Italy's economy, particularly the industrial sector, expanded dramatically between 1950 and 1970. Beginning in the late 1960s, there was considerable industrial unrest in the country as workers demanded higher wages (to offset inflation), better social services, and increased opportunities for education. Italy was admitted to the UN in 1955. In 1971 a treaty between Austria and Italy, granting increased autonomy to the German-speaking province of BOLZANO in Trentino-Alto Adige, was signed and ratified. In mid-1974, Italy faced its worst economic crisis in thirty years. In 1978 former premier Aldo Moro, a Christian Democrat, was kidnapped and assassinated by the Red Brigade, a left-wing terrorist group. Center-left coalitions dominated by the Christian Democrats held power until 1983, when the republic's first socialist-led coalition took power under Premier Bettino Craxi. He led the government for four years, a relatively long time by Italian standards, until he resigned in 1987 and was replaced by Christian Democrat Giovanni Goria. In 1989, Giulio Andreotti became premier for the sixth time at the age of seventy. That same year, the Italian judicial system was significantly changed, allowing for cross-examination of witnesses and the assumption of innocence on the part of the defendant. In 1991 the Italian Communist party changed its name to the Democratic Party of the Left. In the 1992 elections the Christian Democrats barely maintained their coalition with the Socialists, the Liberals, and the Social Democrats. Giuliano Amato was named premier. Corruption probes, begun in 1992, led to the arrest of hundreds of business and political figures and the investigation of many others. Responding to a call for political change, Carlo Azeglio Ciampi, head of Italy's central bank, became premier in 1993. In new elections in 1994, a coalition of conservatives and neofascists won a majority of the seats in parliament, and Silvio BERLUSCONI became premier. The government splintered, however, and collapsed by year's end. Romano PRODI became premier in 1995, heading a center-left government that included the Democratic Party of the Left. Following a series of budget impasses, Prodi's government fell in 1998 and he was succeeded by Massimo D'Alema of the Democratic Party of the Left, as head of a multiparty coalition. Losses in regional elections led D'Alema to resign in 2000; Giuliano Amato succeeded him, heading a substantially unchanged center-left multiparty coalition. The 2001 parliamentary elections, however, ended six years of liberal rule and gave Berlusconi's conservative coalition a solid victory. In the 2006 elections, however, the conservatives narrowly lost to the center-left, led this time by Prodi.

Government
Italy is governed under the constitution of 1948 as amended. The president, who is the head of state, is elected by both houses of Parliament and 58 regional representatives for a seven-year term; there are no term limits. The prime minister, who is the head of government, is appointed by the president and approved by Parliament. The Council of Ministers, head by the prime minister, serves as the country's executive; it must have the confidence of parliament. The bicameral parliament consists of the 630-seat Chamber of Deputies, whose members are popularly elected, and the Senate, with 315 members elected by region, plus a few life members. All legislators serve five-year terms. In 1994, 1996, and 2001, most deputies and senators were directly elected, with approximately a quarter of the seats in both houses assigned on a proportional basis. Changes enacted in 2005 returned the country to a proportional system for electing national legislators except for those seats awarded to the winning coalition as a bonus. The country's regions also have parliaments and governments. The current head of state is President Giorgio Napolitano (May 2006); the current head of government is Prime Minister Romano Prodi (May 2006).

Italy, town (2006 population 2,108), ELLIS county, N central TEXAS, 13 mi/21 km SSW of WAXAHACHIE and 38 mi/61 km S of DALLAS, near Chambers creek; 32°10′N 96°52′W. In rich blackland agriculture area (cotton, corn, grain; cattle; dairying). Manufacturing (disposable protective clothing, chain-link fence hardware, industrial aprons).

Itamaracá Island (EE-tah-mah-rah-KAH), in the ATLANTIC, off coast of PERNAMBUCO, NE BRAZIL, 20 mi/ 32 km N of RECIFE, separated from mainland by narrow channel; c.9 mi/14.5 km long, 4 mi/6.4 km wide. Just S is city of IGARAÇU.

Itamarandiba (ee-tah-mah-rahn-zhee-bah), city (2007 population 32,064), NE central MINAS GERAIS, BRAZIL, in the SERRA DA PENHA, 60 mi/97 km NE of DIAMANTINA; 17°48′S 42°54′W. Tourmalines found here; also iron and kaolin deposits.

Itamaraty, Brazil: see PEDRO II.

Itambacuri (ee-tahm-bah-koo-ree), city (2007 population 22,507), NE MINAS GERAIS, BRAZIL, 20 mi/32 km SW of TEÓFILO OTONI; 12°04′S 41°43′W. Coffee, lard, alcohol; mica mines. Formerly spelled Itambacury.

Itambé (ee-tahm-BAI), city (2007 population 5,897), S BAHIA, BRAZIL, on RIO PARDO and 30 mi/48 km SSE of Vitória da Conquista; 15°18′S 40°35′W. Livestock, cotton, coffee.

Itambé, Brazil: see TAMBÉ.

Itambé, Pico do (EE-tahn-bai), peak (6,155 ft/1,876 m) in central MINAS GERAIS, BRAZIL, 20 mi/32 km SE of DIAMANTINA, in E outlier of the SERRA DO ESPINHAÇO. RIO JEQUITINHONHA rises here. Large iron deposits.

Itami (ee-TAH-mee), city (2005 population 192,250), HYOGO prefecture, S HONSHU, W central JAPAN, on the Muko River and Osaka Bay, 14 mi/23 km N of KOBE; 34°47′N 135°24′E. Residential suburb of OSAKA and the site of Osaka Airport. Bread; automotive parts.

Itamorotinga, Brazil: see SERRA BRANCA.

Itanagar, city (2001 population 34,970), ⊙ ARUNACHAL PRADESH state, extreme NE INDIA, near the ASSAM state border, 56 mi/90 km NE of TEZPUR (linked by road); 27°06′N 93°37′E. Tourism known for its beautiful scenery; local attractions include excavated ruins of Ita fort, a Buddhist monastery, Nehru Museum, and Ganga Lake (boating). Nearest commercial airport at Tezpur.

Itanhaém (EE-tahn-yah-em), city (2007 population 80,787), SE SÃO PAULO BRAZIL, unsheltered port on the ATLANTIC, on railroad and 35 mi/56 km SW of SANTOS; 24°11′S 46°47′W. Fishing. Church and convent buildings date from 1534. Sometimes spelled Itanhaen.

Itanhandu (ee-tahn-yahn-doo), city (2007 population 14,429), S MINAS GERAIS, BRAZIL, in the SERRA DA MANTIQUEIRA, on railroad and 20 mi/32 km N of CRUZEIRO (SÃO PAULO); 22°17′S 45°00′W. Tobacco-growing center.

Itanhomi (EE-tahn-yo-mee), city (2007 population 11,878), E central MINAS GERAIS, BRAZIL, 40 mi/65 km W of CONSELHEIRO PENA; 19°13′S 41°50′W.

Itano (ee-TAH-no), town, Itano county, TOKUSHIMA prefecture, SE SHIKOKU, W JAPAN, 6 mi/10 km N of TOKUSHIMA; 34°08′N 134°27′E.

Itany River (EE-tah-nee), Dutch *Litani*, c.100 mi/161 km long, in the GUIANAS; rises in TUMUC-HUMAC MOUNTAINS near BRAZIL border; flows through tropical forests along S border of FRENCH GUIANA and SURINAME to MARONI RIVER (of which it is also considered the upper course) at 03°18′N 54°05′W.

Itaobim (EE-tou-been), city (2007 population 21,023), NE MINAS GERAIS, BRAZIL, on RIO JEQUITINHONHA, 27 mi/44 km S of MEDINA; 16°35′S 41°30′W.

Itaocara (EE-tah-O-kah-rah), city (2007 population 22,068), NE RIO DE JANEIRO state, BRAZIL, on PARAÍBA River above influx of POMBA River, and 50 mi/80 km W of CAMPOS; 21°41′S 42°04′W. Coffee, sugar.

Itapaci (EE-tah-pah-SEE), city (2007 population 16,502), GOIÁS, BRAZIL; 14°57′S 49°34′W.

Itapagé (ee-tah-pah-ZHAI), city (2007 population 45,526), N CEARÁ, BRAZIL, in Uruburetama hills, 75 mi/121 km W of FORTALEZA; 03°40′S 39°34′W. Cotton, coffee, sugar. Until 1944, called São Francisco.

Itapagipe (EE-tah-pah-zhee-pai), city (2007 population 14,019), W central MINAS GERAIS, BRAZIL, in TRIÂNGULO MINEIRO near RIO GRANDE, 45 km/28 mi S of CAMPINA VERDE; 19°47′S 49°25′W.

Itaparica (ee-tah-pah-ree-kah), city (2007 population 19,897), E BAHIA, BRAZIL, fishing port at N tip of ITAPARICA ISLAND, 12 mi/19 km NW of SALVADOR; 12°50′S 38°42′W. Oil and natural-gas wells. Ships whale oil, salt, coconuts, fruit. Its fort (built 1711) was attacked in 1823 by Portuguese fleet.

Itaparica, Brazil: see PETROLÂNDIA.

Itaparica Island (ee-tah-pah-ree-kah), in TODOS OS SANTOS BAY of the ATLANTIC, E BAHIA, BRAZIL, 10 mi/16 km W of SALVADOR; 18 mi/29 km long, 5 mi/8 km wide. ITAPARICA city is at N tip. Petroleum, saltworks.

Itapaya (ee-tah-PAH-yah), canton, QUILLACOLLO province, COCHABAMBA department, central BOLIVIA; 17°34′S 66°21′W. Elevation 8,343 ft/2,543 m. Clay, limestone, and gypsum deposits. Agriculture (potatoes, yucca, bananas, corn, rice, rye, soy, coffee); cattle raising for meat and dairy products.

Itapé (EE-tah-PAI), city (2007 population 11,112), SE BAHIA, BRAZIL, on Rio Côlonia, 15 mi/24 km SW of ITABUNA; 14°55′S 39°26′W.

Itapé (ee-tah-PAI), town, Guairá department, S PARAGUAY, 12 mi/19 km SW of VILLARRICA. Liquor-distilling center in agricultural area (sugarcane, tobacco, maté; cattle); 25°50′S 56°37′W. Founded 1680. Sometimes spelled Ytapé.

Itapebi (EE-tah-PE-bee), city (2007 population 11,494), SE BAHIA, BRAZIL, on RIO JEQUITINHONHA; 15°59′S 39°32′W. Also spelled Itabebi.

Itapecerica (ee-tah-pe-ke-ree-kah), city (2007 population 20,491), S central MINAS GERAIS, BRAZIL, 25 mi/40 km SW of DIVINÓPOLIS; 20°28′S 45°15′W. Railroad terminus; agricultural trade (coffee, fruit, dairy products, sugar). Graphite deposits.

Itapecerica, Brazil: see ITAPECERICA DA SERRA.

Itapecerica da Serra (EE-tah-pe-se-ree-kah dah SE-rah), city (2007 population 148,567), SE SÃO PAULO, BRAZIL, 20 mi/32 km SW of SÃO PAULO; 23°43′S 46°50′W. Timber; kaolin quarries; mica, gold deposits. Until 1944, Itapecerica.

Itapecuru-Mirim (ee-tah-pe-koo-roo–mee-reen), city (2007 population 54,575), N MARANHÃO, BRAZIL, on lower RIO ITAPECURU, on SÃO LUÍS-TERESINA railroad and 60 mi/97 km S of São Luís; 03°25′S 44°45′W. Sugar, cotton, babassu nuts, rubber, tobacco. Road to BREJO.

Itapejara d'Oeste (EE-tah-pe-zhah-rah DO-es-che), town, SW PARANÁ state, BRAZIL, 22 mi/35 km NNW of PATO BRANCO; 25°58′S 52°49′W. Also called Itapejara.

Itapema (EE-tah-pe-mah), city (2007 population 33,766), E SANTA CATARINA, BRAZIL, on the Atlantic, and 15 mi/24 km S of ITAJAÍ; 27°06′S 48°37′W. Rice, bananas; fishing.

Itapemirim (e-tah-pe-mee-reen), city (2007 population 30,764), S ESPÍRITO SANTO, BRAZIL, on the Atlantic, near mouth of RIO ITAPEMIRIM, 23 mi/37 km ESE of CACHOEIRO DE ITAPEMIRIM (railroad link); 21°01′S 40°59′W. Sugar, coffee, rice, bananas.

Itaperuna (EE-tah-pe-roo-nah), city (2007 population 92,862), NE RIO DE JANEIRO state, BRAZIL, on MURIAÉ River, on railroad, and 55 mi/89 km NW of CAMPOS; 21°12′S 41°54′W. Coffee- and rice-processing center; dairying; alcohol distilling; cotton ginning; manufacturing (chemicals).

Itapetinga (EE-tah-pe-CHEEN-gah), city (2007 population 63,177), SE BAHIA, BRAZIL, on RIO PARDO, 87 mi/140 km SW of ITABUNA; 15°15′S 40°15′W.

Itapetininga (EE-tah-pe-chee-neen-gah), city (2007 population 138,791), S central SÃO PAULO, BRAZIL, on railroad, and 90 mi/145 km W of SÃO PAULO, in fertile cotton-growing valley; 23°36′S 48°03′W. Agricultural trade and processing center; cheese factories. Cottonseed-oil mills. Trade in cattle (for São Paulo slaughterhouses), cotton, grain. Founded at end of 18th century. Has fine old church.

Itapeva (EE-tah-pe-vah), city (2007 population 85,656), S SÃO PAULO, BRAZIL, on railroad, and 60 mi/97 km SW of ITAPETININGA; 23°58′S 48°52′W. Livestock center; cotton processing. Copper and marble deposits. Airfield. Until mid-1930s, called Faxina.

Itapipoca (ee-tah-pee-po-kah), city (2007 population 107,567), N CEARÁ, BRAZIL. Cotton-growing center on N slope of Uruburetama hills; 03°27′S 39°32′W. Coffee, carnauba wax.

Itapira (EE-tah-pee-rah), city (2007 population 68,131), E SÃO PAULO, BRAZIL, near MINAS GERAIS border, on railroad, and 35 mi/56 km NNE of CAMPINAS; 22°26′S 46°50′W. Manufacturing (shoes, hats, chairs, tile, bricks, sugar, alcohol); coffee processing. Limestone, marble deposits nearby.

Itapira, Brazil: see UBAITABA.

Itapiranga (ee-tah-pee-rahn-gah), town (2007 population 7,828), E AMAZONAS, BRAZIL, on left bank of the AMAZON, and 40 mi/64 km NE of ITACOATIARA; 02°50′S 58°15′W.

Itapoã (EE-tah-po-AH), city, E RIO GRANDE DO SUL state, BRAZIL, 22 mi/35 km SE of PÔRTO ALEGRE, on Lagoa dos PATOS; 30°16′S 51°01′W.

Itapocu (ee-tah-po-koo), city, NE SANTA CATARINA state, BRAZIL, 17 mi/27 km SE of JOINVILLE; 26°33′S 48°43′W. Fruit, rice, manioc.

Itápolis (ee-TAH-po-lees), city (2007 population 38,633), central SÃO PAULO, BRAZIL, 40 mi/64 km WNW of ARARAQUARA; 23°42′S 49°29′W. Pottery manufacturing; coffee, cotton, rice, and corn processing; distilling.

Itaporã (EE-tah-po-RUHN), city (2007 population 18,525), S central Mato Grosso do Sul, BRAZIL, 10 mi/ 16 km N of DOURADOS; 22°05′S 54°48′W. Agriculture, including soybeans.

Itaporanga (ee-tah-po-rahn-gah), city (2007 population 22,433), W PARAÍBA, NE BRAZIL, 40 mi/64 km SE of CAJÀZEIRAS; 07°15′S 38°13′W. Cotton, rice, sugar; goat cheese. Called Misericórdia until 1939; Itaporanga, 1939–1943; and again Misericórdia, 1944–1948.

Itaporanga (EE-tah-po-rahn-gah), city (2007 population 14,284), S SÃO PAULO, BRAZIL, near PARANÁ border, 55 mi/89 km SSW of AVARÉ; 23°42′S 49°29′W. Cotton, corn, beans, manioc, coffee.

Itaporanga, Brazil: see ITAPORANGA D'AJUDA.

Itaporanga d'Ajuda (EE-tah-po-rahn-gah DAH-zhoo-dah), city, E SERGIPE, NE BRAZIL, on lower VASA BARRIS River, on railroad, and 18 mi/29 km SW of ARACAJU; 10°59′S 37°18′W. Ships sugar, cattle; sugar milling. Until 1944 called Itaporanga; and Irapiranga, 1944–1948.

Itapúa (ee-tah-POO-ah), department (☐ 6,380 sq mi/16,588 sq km; 2002 population 453,692), SE PARAGUAY; ☉ ENCARNACIÓN; 26°50′S 55°50′W. Forested lowlands bounded S and E by upper PARANÁ RIVER (ARGENTINA border) and TEBICUARY RIVER (NW). Has subtropical, humid climate. Some copper deposits. A rich lumbering and agr. area producing maté, cotton, soybeans, corn, rice, tobacco, sugarcane, grapes; cattle. Sawmilling and processing industries concentrated at Encarnación, CARMEN DEL PARANÁ, HOHENAU, CORONEL BOGADO, SAN PEDRO DEL PARANÁ. Area was colonized by 18th-century Jesuit missions; later German, Russian, and Czech colonies founded.

Itapuí (ee-tah-poo-EE), town (2007 population 11,605), central SÃO PAULO, BRAZIL, on TIETÊ RIVER, 9 mi/14.5 km WNW of JAÚ; 22°14′S 48°41′W. Cotton, coffee, and rice processing. Formerly Bica de Pedra.

Itapuranga (EE-tah-poo-RAHN-gah), city (2007 population 24,790), N central GOIÁS, BRAZIL, 27 mi/43 km NNE of GOIÁS, BRAZIL; 15°30′S 49°56′W.

Itaquara (ee-tah-kwah-rah), town (2007 population 7,623), E BAHIA, BRAZIL, on railroad and 75 mi/121 km SW of NAZARÉ; 13°23′S 39°51′W. Coffee, fruit, manioc.

Itaquarai (EE-tah-KWAH-rah-EE), village, S central BAHIA, BRAZIL, 15 mi/24 km NNW of BRUMADO; 14°04′S 41°44′W.

Itaqui (ee-tah-kwee), city (2007 population 36,191), W RIO GRANDE DO SUL, BRAZIL, on URUGUAY RIVER (ARGENTINA border; connected by bridge) at influx of AGUAPEY RIVER, opposite ALVEAR (Argentina), on railroad and 55 mi/89 km NE of URUGUAIANA; 29°08′S

56°33′W. Exports oranges and bergamots (for essential oils) to Argentina; slaughterhouses; horse raising; flax growing. Airfield. Formerly spelled Itaquy.

Itaquiraí (EE-tah-kwee-rah-EE), city (2007 population 16,919), SE Mato Grosso do Sul, BRAZIL, 25 mi/40 km N of ELDORADO; 23°29′S 54°11′W. Airport.

Itarana (ee-tah-RAH-nah), city (2007 population 10,828), central ESPÍRITO SANTO, BRAZIL, 24 mi/39 km WNW of SANTA TERESA; 19°50′S 40°50′W.

Itarantim (EE-tah-rahn-CHEEN), city (2007 population 17,619), SE BAHIA, BRAZIL, near MINAS GERAIS border; 15°45′S 40°03′W.

Itararé (EE-tah-rah-RAI), city (2007 population 48,737), S SÃO PAULO, BRAZIL, near PARANÁ border, on railroad to JAGUARIAÍVA, and 90 mi/145 km N of CURITIBA; 24°07′S 49°20′W. Livestock center; grain, cotton, fruit; lumbering.

Itararé River (EE-tah-rah-RAI), S BRAZIL, left tributary of PARANAPANEMA River, forms SÃO PAULO-PARANÁ border throughout its N-S course of 120 mi/193 km.

Itarema (EE-tah-RE-mah), city (2007 population 34,410), W central coast of CEARÁ, BRAZIL, 15 mi/24 km E of ACARAÚ; 02°52′S 39°50′W.

Itaretama, Brazil: see LAJES.

Itarsi (i-TAHR-see), town (2001 population 93,783), HOSHANGABAD district, S MADHYA PRADESH state, INDIA, in fertile NARMADA valley, 10 mi/16 km SSE of HOSHANGABAD; 22°39′N 77°48′E. Important railroad junction (workshops); trade center; livestock market. Products of dense teak, sal, and khair forests (S) include lac and cutch.

Itarumã (EE-tah-roo-MAH), town (2007 population 5,338), SW GOIÁS, BRAZIL, 37 mi/60 km NE of Itajá; 18°50′S 51°30′W.

Itasca (ei-TAS-kuh), county (☐ 2,927 sq mi/7,610.2 sq km; 2006 population 44,729), N MINNESOTA; ☉ GRAND RAPIDS; 47°30′N 93°37′W. Drained by MISSISSIPPI RIVER (forms last of SW boundary, including Lake WINNIBIGOSHISH reservoir). Agricultural area (hay, alfalfa; cattle; some dairying); timber; peat deposits; iron mining in W part of MESABI IRON RANGE. Parts of Leech Lake Indian Reservation in SW; part of Chippewa National Forest in N; Big Fork and George Washington state forests in N; Golden Anniversary State Forest in S; Scenic State Park in N; Hill Annex Mine State Park in E; Schoolcraft State Park on SW boundary. Numerous lakes, including BOWSTRING LAKE, in W, and POKEGAMA LAKE. Formed 1849.

Itasca, town (2006 population 1,637), HILL county, N central TEXAS, 42 mi/68 km SSE of FORT WORTH; 32°09′N 97°09′W. In cotton, grain; cattle area; manufacturing (fertilizer). Settled 1882; incorporated as city 1910.

Itasca (i-TAS-kuh), village (2000 population 8,302), DU PAGE county, NE ILLINOIS, residential suburb 18 mi/ 29 km WNW of downtown CHICAGO and 15 mi/24 km ESE of ELGIN; 41°58′N 88°01′W. Some agriculture (corn, oats).

Itasca, Lake (ei-TAS-kuh) (☐ 2 sq mi/5.2 sq km), CLEARWATER county (SE arm extends into HUBBARD county), NW central MINNESOTA, shallow lake, in a pine-wooded swampy region; 47°13′N 95°12′W. It is the source of the MISSISSIPPI RIVER, which drains N from lake before turning E and S; stepping stones cross river at its exit point. Henry River Schoolcraft identified it (1832) as the source of the Mississippi. Although unarguably the source of the named river, the river's true physical course has been disputed by geographers. In 1891 the lake was included in Itasca State Park, which has a historical and natural-history museum. A school of forestry and a biological-research station are nearby.

Itaska Beach (i-TAS-kuh), summer village (2001 population 10), central ALBERTA, W CANADA, 26 mi/42 km from LEDUC, in LEDUC County; 53°04′N 114°05′W.

Itasy, Lake (ee-TAHSH) (☐ 35 sq mi/91 sq km), AN-TANANARIVO province, central MADAGASCAR, 28 mi/

45 km WSW of ANTANANARIVO; c.10 mi/16 km long, 10 mi/16 km wide; 19°05′S 46°48′E. Drains into TSIR-IBIHINA RIVER. Center of fishing and agricultural activities (rice, peanuts, tobacco, corn, pulses, aleurites). Landscape of volcanic cones and craters; geothermal resources, tourism.

Itata, CHILE: see QUIRIHUE.

Itata River (ee-TAH-tah), 110 mi/177 km long, S central CHILE, in BÍO-BÍO region; rises in ANDES foothills W of YUNGAY; flows NW, through the irrigated central valley, to the PACIFIC 16 mi/26 km NNE of TOMÉ. It is joined by its main tributary, the ÑUBLE, 20 mi/32 km W of CHILLÁN. The Itata-Ñuble is c.150 mi/241 km long. A swift, torrential stream, it is navigable only a short distance by small craft.

Itate (ee-tah-te), village, Soma county, FUKUSHIMA prefecture, N central HONSHU, NE JAPAN, 16 mi/25 km S of FUKUSHIMA; 37°41′N 140°44′E. Beef cattle; lumber; granite.

Itatí (ee-tah-TEE), village (1991 population 4,644), ☉ Itatí department, NW CORRIENTES province, ARGENTINA, port on PARANÁ RIVER (PARAGUAY border) and 40 mi/64 km ENE of CORRIENTES; 27°16′S 58°15′W. Agricultural area (rice, cotton, oranges); stock raising, fishing; forest industry; tanning. Has old church of Our Lady of Itatí. Founded 1516.

Itatiaia (EE-tah-chee-ei-ah), city, W RIO DE JANEIRO state, BRAZIL, on PARAÍBA River, on railroad, and 25 mi/40 km W of BARRA MANSA, near SÃO PAULO border; 22°30′S 44°34′W. Cattle; bauxite deposits. Starting point for Itatiaia National Park (N). Until 1943, called Campo Belo.

Itatiaia, mountain (9,145 ft/2,787 m) in the SERRA DA MANTIQUEIRA, SE BRAZIL, on MINAS GERAIS-RIO DE JANEIRO border, 30 mi/48 km WNW of BARRA MANSA. Surrounding it is a national park (☐ 46 sq mi/119 sq km), est. 1937. Peak also called Agulhas Negras [Port.=black needles].

Itatiba (EE-tah-chee-bah), city (2007 population 91,382), E SÃO PAULO, BRAZIL, 16 mi/26 km SE of CAMPINAS; 23°00′S 46°51′W. In coffee-growing region; textile milling, distilling.

Itatiba do Sul (EE-tah-chee-bah do sool), town (2007 population 4,574), N RIO GRANDE DO SUL state, BRAZIL, 19 mi/31 km NW of ERECHIM; 27°22′S 52°27′W. Wheat, soybeans, corn, manioc, potatoes; livestock. Also called Itatiba.

Itatí, Lake (ee-tah-TEE), W central CORRIENTES province, ARGENTINA, at SW end of ESTEROS DEL IBERÁ, 30 mi/48 km N of MERCEDES; 12 mi/19 km long, 4 mi/6.4 km wide. The CORRIENTES RIVER flows through part of it.

Itatim (EE-tah-CHEEN), city (2007 population 14,659), BAHIA, BRAZIL, 50 mi/80 km SE of Feira de Santana; 12°43′S 39°42′W.

Itatinga (EE-tah-cheen-gah), city (2007 population 17,570), S central SÃO PAULO, BRAZIL, 20 mi/32 km E of AVARÉ, on railroad; 23°07′S 48°36′W. Butter processing, flour milling, tanning.

Itatira (EE-tah-CHEE-rah), city (2007 population 17,608), central CEARÁ, BRAZIL, in Serra do Machado, 22 mi/35 km SW of CANINDE; 04°30′S 39°36′W.

Itatskiy (ee-TAHT-skeeyee), town (2005 population 4,200), NE KEMEROVO oblast, S central SIBERIA, RUSSIA, on the TRANS-SIBERIAN RAILROAD (Itatskaya station), 50 mi/80 km E of MARYINSK; 56°04′N 89°02′E. Elevation 810 ft/246 m. Coal enrichment plant. In agricultural area; food industries (dairying, vegetable drying).

Itatuba (EE-tah-too-bah), town (2007 population 9,841), SE PARAÍBA, BRAZIL, 7 mi/11.3 km SW of INGÁ; 07°22′S 35°44′W.

Itatupã (ee-tah-too-puh), town, NE PARÁ, BRAZIL, on N coast of GURUPÁ ISLAND in Amazon delta, 40 mi/64 km SSW of MACAPÁ; 00°42′S 51°06′W. Until 1943, Sacramento.

Area is shown by the symbol ☐, and capital city or county seat by ☉.

Itaú (EE-tah-oo), town (2007 population 5,755), extreme W RIO GRANDE DO NORTE state, BRAZIL, 66 mi/106 km SW of MOSSORÓ; 05°50′S 37°59′W.

Itaúba (EE-tah-OO-bah), town (2007 population 4,634), NC MATO GROSSO, BRAZIL, 60 mi/96 km N of SINOP; 11°00′S 55°10′W.

Itaubal (ee-TOU-bahl), town (2007 population 3,439), E central AMAPÁ, BRAZIL, 86 mi/138 km NE of MACAPÁ, on coastal marshes of AMAZON delta; 00°40′N 50°45′W.

Itauçu (EE-tah-oo-SOO), town (2007 population 8,682), central GOIÁS, BRAZIL, 50 mi/80 km NW of GOIÂNIA; 16°11′S 49°38′W.

Itaú de Minas (EE-tah-OO zhe MEE-nahs), city (2007 population 14,551), SW MINAS GERAIS, BRAZIL, 11 mi/17 km NW of PASSOS; 20°40′S 46°45′W.

Itaú de Minas, Brazil: see PASSOS.

Itaueira (EE-tah-oo-ai-rah), city (2007 population 10,578), SW PIAUÍ state, BRAZIL, 36 mi/58 km S of FLORIANO; 07°36′S 43°02′W.

Itaugua (ee-TOU-gwah), town (2002 population 45,577), CENTRAL department, S PARAGUAY, 23 mi/37 km ESE of Asunción; 25°22′S 57°20W. In agricultural area (fruit, tobacco, cotton, rice). Known for its fine lace. Tanneries. Notable 19th-century church.

Itaúna (ee-tah-oo-nah), city (2007 population 81,878), S central MINAS GERAIS, BRAZIL, on railroad, and 40 mi/64 km WSW of BELO HORIZONTE; 20°06′S 44°32′W. Ships coffee, sugarcane, cattle; textile milling.

Itaúna do Sul (EE-tah-oo-nah do SOOL), town (2007 population 3,686), far NW PARANÁ state, BRAZIL; 22°40′S 52°49′W. Coffee, cotton, rice, corn; livestock.

Itaú, Sierra del (ee-tah-TOO, see-YER-rah del), sub-Andean mountain range in NE SALTA province, ARGENTINA, and parallel to BOLIVIA-ARGENTINA border, NW of TARTAGAL, extending c.45 mi/72 km NNE-SSW. Rises to c.3,500 ft/1,067 m.

Itawamba (it-uh-WAHM-buh), county (□ 540 sq mi/1,404 sq km; 2006 population 23,352), NE MISSISSIPPI, borders E on ALABAMA; ⊙ FULTON; 34°16′N 88°21′W. Drained by East Fork of TOMBIGBEE River Agr. (cotton, corn, soybeans; poultry; cattle; timber). NATCHEZ TRACE PARKWAY passes through NW corner. Part of John Bell Williams Wildlife Management Area (canal section) in N. Formed 1836.

Itayanagi (ee-TAH-yah-NAH-gee), town, North Tsugaru district, Aomori prefecture, N HONSHU, N JAPAN, 19 mi/30 km S of AOMORI; 40°41′N 140°27′E. Apples.

Itayopolis, Brazil: see ITAIÓPOLIS.

Itbayat Island (eet-BAH-yaht), largest island (□ 33 sq mi/85.8 sq km) of BATAN ISLANDS, Batanes province, N PHILIPPINES, in LUZON STRAIT, 200 mi/322 km N of LUZON; 20°46′N 121°50′E; 11 mi/18 km long, 4 mi/6.4 km wide. Rocky coast with steep cliffs; no harbor, bay, or accessible beaches. Interior is generally sloping, coarse grassland. Highest points are Mount Santa Rosa (912 ft/278 m) and Mount Riposet (758 ft/231m). Chief settlement Mayan; airstrip near Raele.

Itchen River (ITCH-uhn), 25 mi/40 km long, HAMPSHIRE, S ENGLAND; rises just E of NEW ALRESFORD; flows W and SW, past New Alresford and Winchester, to SOUTHAMPTON WATER at Southampton. Navigable below WINCHESTER.

Itea (ee-TE-ah), town, PHOCIS prefecture, W CENTRAL GREECE department, GREECE, port on Bay of Crisa (or Itea) of Gulf of CORINTH, 7 mi/11.3 km S of AMPHISSA; 38°26′N 22°25′E. Wheat; olive oil; wine; fisheries.

Itebej (EE-te-bai), Hungarian *Ittebe*, village, VOJVODINA, NE SERBIA, near Romanian border, on Begej Canal, and 20 mi/32 km NE of ZRENJANIN, in the BANAT region; 45°34′N 20°42′E. Population is largely Serb. Until 1947, called Srpski Itebej, Hungarian *Felsőittebe*. Novi Itebej, Hungarian *Alsóittebe*, smaller village with Magyar population, is just SW. Also spelled Itebei, Itebey, or Ittebe.

Iten, town, ⊙ Keiyo district, RIFT VALLEY province, W KENYA, 23 mi/37 km NE of ELDORET. Agriculture

(coffee, maize) and livestock; flourspar mining, saw milling. Trade center.

Itende (ee-TAIN-dai), village, SINGIDA region, central TANZANIA, 75 mi/121 km SSW of MANYONI, on common boundary between RUNGWA (S) and KISIGO (N) game reserves; 06°46′S 34°23′E. Road junction. Livestock; corn, wheat.

Iténez, province, BENI department, NE BOLIVIA; ⊙ MAGDALENA; 13°25′S 63°30′W.

Iteya (ee-TAI-yah), town (2007 population 13,598), OROMIYA state, central ETHIOPIA, 10 mi/16 km NE of ASELA; 08°13′N 39°16′E.

Ithaca, city (□ 6 sq mi/15.6 sq km; 2006 population 29,829), ⊙ TOMPKINS county, S central NEW YORK, at S end of CAYUGA LAKE, in the FINGER LAKES region; 42°26′N 76°30′W. It is important chiefly as an educational center, the seat of Cornell University, Ithaca College, and State University of New York's Cornell campus. Light manufacturing, research and development, agricultural distribution, and commercial services. City has access to the NEW YORK STATE BARGE CANAL. Tourism in the Finger Lakes area is important to the city's economy. A state hospital is also in Ithaca. Settled 1789, incorporated as a city 1888.

Ithaca (I-thuh-kuh), town (2000 population 3,098), ⊙ GRATIOT county, central MICHIGAN, 34 mi/55 km WSW of SAGINAW; 43°17′N 84°35′W. In agricultural area (livestock; beans, sugar beets, dairy products); manufacturing (aircraft-engine components, die sets, printing, molded plastics. Incorporated 1869.

Ithaca, village (2006 population 164), SAUNDERS county, E NEBRASKA, 5 mi/8 km SE of WAHOO, and on branch of PLATTE RIVER; 41°09′N 96°32′W. Pioneer State Wayside Area to SW.

Ithaca (I-thuh-kuh), village (2006 population 101), DARKE county, W OHIO, 12 mi/19 km SSE of GREENVILLE; 39°56′N 84°33′W. In agricultural area.

Ithaca, Greece: see ITHÁKI.

Itháki (ee-THAH-kee), island (□ 36 sq mi/93.6 sq km), KEFALLINÍA prefecture, IONIAN ISLANDS department, W GREECE, one of the IONIAN ISLANDS, NE of KEFALLINÍA; 38°24′N 20°40′E. It is mountainous, rising to c.2,650 ft/808 m at Mount ANOYI, and has little arable land. Chief products are olive oil, currants, and wine. The main town is Itháki (1981 population 2,037), on the E coast. Traditionally celebrated as the home of Odysseus. Cyclopean walls and remains of a Corinthian colony (c.8th century B.C.E.) have been found. In 1953, Itháki was devastated by tidal waves. Occupied by Germans in 1941. Also Ithaca.

Ithmaniya (ith-mahn-EE-yeh), oil field and settlement in AL AHSA region, SAUDI ARABIA, 17 mi/27 km SW of HOFUF. Discovered 1950. Natural gas processing plant. Also known as Othmaniya and Uthmaniya.

Ithomi, Mount (ee-THO-mee) (2,617 ft/798 m), in MESSENIA prefecture, SW PELOPONNESE department, extreme SW GREECE, 14 mi/23 km NW of KALAMATA; 37°11′N 21°56′E. Ruins of ancient Messene (MEKENE) on W slopes; ruined monastery on summit. Also Mount Ithome.

Iti, Greece: see OITI.

Itigi (ee-TEE-gee), town, SINGIDA region, central TANZANIA, 22 mi/35 km W of MANYONI, on railroad; 05°44′S 34°50′E. Road junction. Source of KISIGO RIVER to NE. Livestock; grain.

Itil (EE-teel), former town in the VOLGA RIVER delta, SE European Russia, on site of modern ASTRAKHAN. Was the capital of Khazar state (8th–10th centuries).

Itilleq, fishing settlement, SISIMIUT (Holsteinsborg commune), SW GREENLAND, on islet in DAVIS STRAIT, at mouth of Itilleq Fjord (30 mi/48 km long), 25 mi/40 km SSE of HOLSTEINSBORG (Sisimiut); 66°34′N 53°31′W.

Itimbiri River (ee-tim-BEE-ree), c.350 mi/563 km long, right tributary of CONGO RIVER, N CONGO; rises as RUBI RIVER 3 mi/4.8 km W of NIAPU; flows c.165 mi/266 km W, past BUTA, to EKWANGATANA, where it

becomes the ITIMBIRI and flows c.180 mi/290 km SW, past AKETI and IBEMBO, to CONGO RIVER 25 mi/40 km SE of BUMBA. Navigable for 160 mi/257 km below AKETI. It is an important freight waterway, carrying agricultural produce of UELE region, notably cotton and rice.

Itinga (ee-cheen-gah), city (2007 population 14,592), NE MINAS GERAIS, BRAZIL, on left bank of JEQUITINHONHA River, and 20 mi/32 km NE of ARAÇUAÍ; 16°34′S 41°45′W. Rice, manioc.

Itinga do Maranhão (ee-cheen-gah do MAH-rahn-YOUN), city (2007 population 25,102), NW MARANHÃO state, BRAZIL, near PARÁ border; 04°30′S 47°34′W.

Itiquira (EE-chee-kwee-rah), town (2007 population 12,159), SE MATO GROSSO, BRAZIL, 68 mi/109 km W of ALTO ARAGUAIA; 17°12′S 54°10′W.

Itirapina (EE-chee-rah-pee-nah), city (2007 population 13,889), E central SÃO PAULO, BRAZIL, 17 mi/27 km SSE of SÃO CARLOS; 22°15′S 47°49′W. Railroad junction; coffee, cotton, grain, cattle.

Itirjuro (ee-teer-HOO-ro), canton, GRAN CHACO province, TARIJA department, S central BOLIVIA, 38 mi/60 km E of Capirenda; 21°20′S 62°36′W. Elevation 2,054 ft/626 m. Abundant oil and gas resources to W and SW. Agriculture (potatoes, yucca, bananas, corn); cattle.

Itiruçu (EE-chee-roo-SOO), city (2007 population 15,764), E central BAHIA, BRAZIL, 13 mi/21 km W of JAGUAQUARA; 13°31′S 40°08′W.

Itiuba (EE-chee-OO-bah), city (2007 population 35,134), NE BAHIA, BRAZIL, on SALVADOR-JUAZEIRO railroad, 43 mi/69 km SSE of SENHOR DO BONFIM; 10°41′S 39°51′W.

Itkillik River (IT-ki-lik), c.180 mi/290 km long, N ALASKA; rises in N BROOKS RANGE near 68°10′N 150°W; flows N to COLVILLE RIVER near its mouth at 70°03′N 151°02′W.

Itlidim (it-MEE-duh), village, ASYUT province, central Upper EGYPT, on W bank of the NILE River, on IBRAHIMIYA CANAL, on railroad, and 15 mi/24 km SSE of MINYA; 27°52′N 30°48′E. Cereals, dates, sugarcane.

Itmadpur, INDIA: see ETMADPUR.

Itmida, village, DAQAHLIYA province, Lower EGYPT, 6 mi/9.7 km NE of MIT GHAMR; 30°46′N 31°20′E. Cotton, cereals.

Ito (ee-TO), city, SHIZUOKA prefecture, central HONSHU, E central JAPAN, on the IZU PENINSULA and the SAGAMI SEA, 40 mi/65 km E of SHIZUOKA; 34°58′N 139°06′E. Important fishing port and hot-spring resort.

Itobi (EE-to-bee), town (2007 population 7,444), E SÃO PAULO state, BRAZIL, 9 mi/14.5 km SW of SÃO JOSÉ DO RIO PRETO; 21°44′S 46°58′W.

Itoda (ee-TO-dah), town, Tagawa district, FUKUOKA prefecture, N KYUSHU, SW JAPAN, 22 mi/35 km E of FUKUOKA; 33°39′N 130°46′E.

Itogon (ee-TO-gon), town, BENGUET province, N LUZON, PHILIPPINES, 6 mi/9.7 km ESE of BAGUIO; 16°19′N 120°44′E. Gold mining.

Itoigawa (ee-TO-ee-gah-wah), or **Itoikawa**, city, NIIGATA prefecture, central HONSHU, N central JAPAN, on TOYAMA BAY, 90 mi/145 km S of NIIGATA; 37°02′N 137°51′E. Marine-product processing, stone (jade) processing, cement manufacturing, sake brewing. Skiing area.

Itoman (ee-TO-MAHN), city, southernmost point of OKINAWA island, OKINAWA prefecture, SW JAPAN, in the RYUKYU ISLANDS, 5.6 mi/9 km S of NAHA; 26°07′N 127°40′E. Fishing port; sugarcane. Site of Peace Memorial Park and Himiyuri-no-to tower.

Itonamas, Lake (ee-to-NAH-mahs), BENI department, NE BOLIVIA, 50 mi/80 km SSE of MAGDALENA; 10 mi/16 km long, c. 5 mi/8 km wide. Inlet: SAN MIGUEL RIVER; outlet: ITONAMAS RIVER. Also called Lake SAN LUIS and Lake CARMEN.

Itonamas River, c. 120 mi/193 km long, BENI department, NE BOLIVIA; rises in Lake ITONAMAS 50 mi/80 km SSE of MAGDALENA; flows NNW, past Magdalena, to Guaporé River on BRAZIL border, 4 mi/6.4 km SE of FORTE PRÍNCIPE DA BEIRA (Brazil); 12°28′S 64°24′W. Receives MACHUPO RIVER (left). Navigable for c. 70 mi/113 km below Magdalena.

Iton River (ee-ton), c.70 mi/113 km long, in EURE department, HAUTE-NORMANDIE region, NW FRANCE; rises 6 mi/9.7 km N of Mortagne; flows in a lowland generally NE, past BRETEUIL and ÉVREUX, to the EURE above LOUVIERS; 49°09′N 01°12′E. It flows partially underground above Évreux, and feeds the PARIS water supply.

Itonuki (ee-TO-noo-kee), town, Motosu district, GIFU prefecture, central HONSHU, central JAPAN, 5 mi/8 km N of GIFU; 35°26′N 136°40′E. Persimmons.

Itororó (EE-to-ro-RO), city (2007 population 20,117), SE BAHIA, BRAZIL, in Serra da Ouricana, 66 mi/107 km SW of ITABUNA; 15°06′S 40°04′W.

Itororó do Paranapanema (EE-to-ro-ro do PAH-rah-nah-pah-ne-mah), city, W SÃO PAULO state, BRAZIL, on PARANAPANEMA River and 34 mi/55 km SW of PRESIDENTE PRUDENTE; 22°36′S 51°43′W.

Itoupava (EE-too-oo-pah-vah), city, NE SANTA CATARINA state, BRAZIL, 9 mi/14.5 km N of BLUMENAU; 26°48′S 49°06′W. Fruit, corn, manioc; livestock.

Ítrabo (EE-trah-vo), town, GRANADA province, S SPAIN, 8 mi/12.9 km WNW of MOTRIL. Olive-oil processing; raisins, almonds, wine.

Itsa (IT-sah), township, FAIYUM province, Upper EGYPT, on railroad, and 5 mi/8 km SW of FAIYUM; 29°15′N 30°48′E. Cotton, cereals, sugarcane, fruits.

Itsmina, Colombia: see ISTMINA.

Itsukaichi (eets-KAH-ee-chee), town, West Tama district, TOKYO prefecture, E central HONSHU, E central JAPAN, 25 mi/40 km W of SHINJUKU; 35°43′N 139°13′E. Akikawa Gorge is nearby.

Itsuki (eets-KEE), village, Kuma county, KUMAMOTO prefecture, SW KYUSHU, SW JAPAN, 28 mi/45 km S of KUMAMOTO; 32°23′N 130°49′E. Lumber.

Itsuku-shima (eets-KOO–shee-mah) or **Miya-jima** (mee-YAH-jee-mah) [Japanese=Shrine Isle], sacred island (□ 12 sq mi/31.2 sq km), in the INLAND SEA, off SW HONSHU, W JAPAN, SW of HIROSHIMA. Site of an ancient Shinto shrine, famous for its magical beauty. Also known for a 9th century Buddhist temple, a pagoda (built 1407), a 16th century hall built by military ruler Hideyoshi Toyotomi, and a huge torii (1875). One of the three most famous scenic areas in Japan.

Itsuwa (eets-WAH), town, Amakusa county, in NE corner of Amkusa Shimo-jima island, KUMAMOTO prefecture, SW KYUSHU, SW JAPAN, 37 mi/60 km S of KUMAMOTO; 32°30′N 130°11′E. Loquats; marine products (abalone, sea vegetables, smoked octopus).

Itta Bena (it-uh BEEN-uh), town (2000 population 2,208), LEFLORE county, W central MISSISSIPPI, 8 mi/12.9 km WSW of GREENWOOD; 33°30′N 90°19′W. In cotton-growing area, agriculture (cattle; corn, rice, sorghum, soybeans); manufacturing (catfish processing). Seat of Mississippi Valley State University (to N).

Ittebe, SERBIA: see ITEBEJ.

Itteville (eet-VEEL), town (□ 4 sq mi/10.4 sq km), ESSONNE department, ÎLE-DE-FRANCE region, N central FRANCE, on Juin River just upstream from its confluence with ESSONNE RIVER, and 9 mi/15 km NE of ÉTAMPES; 48°31′N 02°20′E.

Ittifak (eet-tee-FAHK), town, NAMANGAN viloyat, NE UZBEKISTAN.

Ittigen (IT-i-gen), commune (2000 population 10,991), BERN canton, N SWITZERLAND; NE suburb of BERN, created from BOLLINGEN commune in 1983, 7 mi/11.3 km ENE of ZÜRICH; 46°49′N 07°30′E.

Ittre (IT-truh), commune (2006 population 6,047), Nivelles district, BRABANT province, central BELGIUM, 5 mi/8 km NW of NIVELLES; 50°39′N 04°16′E.

Itu (EE-too), city (2007 population 147,260), S SÃO PAULO, BRAZIL, on railroad and 45 mi/72 km NW of SÃO PAULO; 23°16′S 47°19′W. Industrial center (cotton

textiles, chemicals, flour products, alcohol); metalworking; agricultural processing (rice, corn, cotton, oranges). Slate quarries. Hydroelectric plant on TIETÊ RIVER (just N). Founded 1657. Has two 17th-century churches, an 18th-century convent, and a historical museum. Formerly spelled Ytú.

Itu (EE-choo), town, AKWA IBOM state, SE NIGERIA, on CROSS RIVER (head of year-round navigation), 30 mi/48 km NW of CALABAR. Road center; palm oil and kernels, kola nuts. Ceramics. Has leper settlement.

Ituaçu (ee-twah-soo), city (2007 population 17,988), S central BAHIA, BRAZIL, on railroad and 70 mi/113 km S of ANDARAÍ; 13°45′S 41°25′W. Diamonds, gold, and semiprecious stones (aquamarines, amethysts) found in area.

Ituango (ee-too-AHN-go), town, ⊙ Ituango municipio, ANTIOQUIA department, NW central COLOMBIA, in CAUCA valley, 55 mi/89 km NNW of MEDELLÍN; 07°10′N 75°45′W. Elevation 4,921 ft/1,500 m. Corn, beans, coffee; livestock.

Ituberá (ee-too-be-RAH), city (2007 population 23,453), E BAHIA, BRAZIL, 45 mi/72 km SSW of NAZARÉ; 13°45′S 39°12′W. Vegetable-oil processing, coffee, and lumber shipping. Until 1944, called Santarém.

Ituiutaba (ee-too-ee-oo-tah-bah), city (2007 population 92,754), extreme W MINAS GERAIS, BRAZIL, in the TRIÁNGULO MINEIRO, 70 mi/113 km W of UBERLÂNDIA; 18°56′S 49°29′W. Cattle raising, sugar growing, soybeans. Formerly spelled Ituyutaba.

Itula (ee-TOO-lah), village, SUD-KIVU province, E CONGO, on ELILA RIVER and 95 mi/153 km SW of BUKAVU. Communications point; palm products.

Itumbiara (ee-toom-bee-ah-rah), city (2007 population 88,122), S GOIÁS, central BRAZIL, on right bank of RIO PARANAÍBA (MINAS GERAIS border) and 70 mi/113 km NW of UBERLÂNDIA; 18°25′S 49°14′W. Cattle-raising. Extensive nitrate deposits. Until 1944 called Santa Rita or Santa Rita do Paranaíba.

Itum-Kale (ee-TOOM–KAH-lye), village (2005 population 2,925), S CHECHEN REPUBLIC, N central CAUCASUS, S European Russia, approximately 7 mi/11 km N of the RUSSIA-GEORGIA border, on mountain highway, 35 mi/56 km S of GROZNYY; 42°44′N 45°34′E. Elevation 3,562 ft/1,085 m. Livestock. Relatively unaffected by the first Russian-Chechen conflict, but besieged and occupied by Russian troops in 1999, after hostilities resumed. Formerly known as Akhalkhevi.

Ituna (ei-TOO-nuh), town (2006 population 622), SE SASKATCHEWAN, CANADA, in the Beaver Hills, 45 mi/72 km W of YORKTON; 51°10′N 103°30′W.

Itungi (ee-TOON-gee) or **Itungi Port**, town, MBEYA region, SW TANZANIA, 55 mi/89 km SE of MBEYA, on Lake NYASA, 5 mi/8 km N of mouth of SONGWE RIVER (MALAWI border); 09°36′S 33°56′E. Lake port. Fish; cattle; corn, wheat. Coal mining to NW.

Ituni (ei-TOO-nee), village, UPPER DEMERARA–BERBICE district, N central GUYANA; 05°32′N 58°15′W. Located on ITUNI River, a small affluent of the BERBICE, on railroad, and 30 mi/48 km S of MACKENZIE, on watershed between DEMERARA and BERBICE rivers. Bauxite-mining area.

Itupeva (EE-too-pe-vah), city (2007 population 36,802), S SÃO PAULO state, BRAZIL, 13 mi/21 km W of JUNDIAÍ; 23°09′S 47°04′W.

Itupiranga (ee-too-pee-rahn-gah), city (2007 population 42,165), E PARÁ, BRAZIL, on left bank of TOCANTINS and 5 mi/8 km NW of MARABÁ; 05°10′S 49°20′W. Railroad around rapids of Tocantins.

Ituporanga (ee-too-po-rahn-gah), city (2007 population 20,577), E SANTA CATARINA state, BRAZIL, 30 mi/48 km SW of BLUMENAU; 29°25′S 49°36′W. Manioc, corn, rice, fruit; livestock.

Ituraea (ee-tur-yah), ancient country on the N border of Canaan, mainly in the Beqa'a Valley of LEBANON. Jetur, the son of Ishmael, was its founder. Ancient geographers are not agreed as to the exact limits of the country. The inhabitants were Arabians with their

capital at Chalchis and their religious center at Heliopolis (BAALBEK). Ituraea was conquered in 105 B.C.E. by Aristobulus, king of Judea, who annexed it to Judea and converted many of the inhabitants to Judaism. Later, after a brief period of independence, the country was subdued by Pompey. It remained thereafter chiefly in Roman hands, being united (C.E. c.50) to the Roman province of Syria. Many Ituraeans served in the armies of Rome and were renowned for their skill as horsemen and archers.

Iturama (EE-too-rah-mah), city (2007 population 31,495), W central MINAS GERAIS, BRAZIL, in TRIÁNGULO MINEIRO, 30 mi/49 km W of São Francisco de Sales; 19°45′S 50°18′W.

Iturbe (ee-TOOR-bai), town, Guairá department, S PARAGUAY, on branch of TEBICUARY RIVER, on railroad, and 20 mi/32 km S of VILLARRICA; 26°03′S 56°29′W. Agricultural center (fruit, sugarcane; livestock); alcohol distilling, winemaking, sugar refining. Road and railroad junction.

Iturbide (ee-tor-BEE-de), town, NUEVO LEÓN, N MEXICO, in SIERRA MADRE ORIENTAL, 23 mi/37 km WSW of LINARES on Mexico Highway 60; 24°45′N 99°53′W. Grain, livestock.

Iturbide, MEXICO: see VILLA HIDALGO, SAN LUIS POTOSÍ.

Ituri River, CONGO: see ARUWIMI RIVER.

Iturralde, province, LA PAZ department, W BOLIVIA, ⊙ IXIAMAS; 13°00′S 68°00′W.

Iturup Island (EE-too-roop), Japanese *Etorofu-to*, largest (□ 2,587 sq mi/6,726.2 sq km) and most important of KURIL ISLANDS, SAKHALIN oblast, extreme E Siberia, RUSSIAN FAR EAST, in the S section of the main chain; separated from URUP ISLAND (N) by FRIZ STRAIT, from KUNASHIR ISLAND (S) by YEKATERINA STRAIT; 126 mi/203 km long, 28 mi/45 km wide; 44°55′N 147°40′E. Largely mountainous; rises to 5,207 ft/1,587 m. Hunting (bear, red fox, rabbit) for fur; fishing, whaling, lumbering (pine forests), and sulphur mining are chief economic activities. Main centers: KURIL'SK, KUIBYSHEVO, KASATKA (chief port; on HITOKAPPU Bay). First visited (1643) and named Staten Island by Dutch navigator David Pietersz De Vries. Reached by the Russians in the 1760s; occupied (1800) by Japan; seized by Russia in 1945. As one of the southernmost Kuril Islands, Iturup is the subject of extensive negotiations and tension between Russia and Japan, with the latter claiming the islands as its Northern Territories.

Ituverava (EE-too-ve-rah-vah), city (2007 population 38,563), NE SÃO PAULO, BRAZIL, on railroad, and 60 mi/97 km N of RIBEIRÃO PRÊTO; 20°20′S 47°47′W. Livestock center.

Ituyutaba, Brazil: see ITUIUTABA.

Ituzaingó (ee-too-zein-GO), town, in Greater BUENOS AIRES, ARGENTINA, 17 mi/27 km WSW of BUENOS AIRES. Agricultural and industrial center. Alfalfa, grain; livestock; plant nurseries.

Ituzaingó, town (1991 population 17,136), ⊙ Ituzaingó department, N CORRIENTES province, ARGENTINA, port on PARANÁ RIVER (PARAGUAY border), 10 mi/16 km S of Apipé Rapids, and 50 mi/80 km WSW of POSADAS. Farming center (oranges, maté, tobacco, cotton; livestock). Near site of large hydroelectric dam.

Ituzaingó, village, SAN JOSÉ department, S URUGUAY, on railroad, and 3 mi/4.8 km NW of SANTA LUCÍA; 34°25′S 56°26′W. In agricultural region (cereals, livestock). Founded 1875.

Itwangi (ee-TWAN-gee), village, SHINYANGA region, NW central TANZANIA, 25 mi/40 km SW of SHINYANGA, near MANONGA RIVER; 03°52′S 33°05′E. Timber, cattle, goats, sheep; corn, wheat, millet.

Ityai el Barud (it-YEIL bah-ROOD), village, BEHEIRA province, Lower EGYPT, railroad junction 17 mi/27 km SE of DAMANHUR; 30°53′N 30°40′E. Cotton, rice, cereals. Important freight yard. Sometimes called Teh el Barud. The ruins of ancient NAUCRATIS are 4 mi/6.4 km W.

Area is shown by the symbol □, and capital city or county seat by ⊙.

Itzalco, EL SALVADOR: see IZALCO, city.

Itzapa, GUATEMALA: see SAN ANDRÉS ITZAPA.

Itzehoe (IT-tse-hoe), city, SCHLESWIG-HOLSTEIN, N central GERMANY, a port on the STÖR RIVER; 53°55′N 09°32′E. Commercial center; manufacturing includes cement and machinery. Founded c.810 by Charlemagne (one of the oldest cities in Schleswig-Holstein); passed to PRUSSIA in 1866.

Itzer (IT-zer), village, Khénifra province, Meknès-Tafilalet administrative region, central MOROCCO, in upper MOULOUYA river valley, on S slope of the Middle ATLAS mountains, 22 mi/35 km NW of MIDELT; 32°53′N 05°03′W. Periodic market. Oak forests.

Itz River (ITS), 40 mi/64 km long, GERMANY; rises N of Bavarian-Thuringian border; flows S, past COBURG, to the Main 2 mi/3.2 km S of Rattelsdorf; source at 50°26′N 11°00′E. Receives the RODACH (right).

Iuaretê, Brazil: see IAUARETÊ.

Iuiú (ee-OO-ee-OO), city (2007 population 11,528), BAHIA, BRAZIL.

Iuka (ei-YOO-kuh), town (2000 population 3,059), ⊙ TISHOMINGO county, extreme NE MISSISSIPPI, 20 mi/32 km ESE of CORINTH; 34°48′N 88°12′W. Agriculture (grain, soybeans; hogs); manufacturing (apparel, consumer goods, shoes, limestone); sandstone, clay deposits; mineral spring. A Civil War battle was fought here in 1862. Fine antebellum houses survive. Old Courthouse (1870). WOODALL MOUNTAIN (806 ft/246 m), highest point in Mississippi, 3 mi/4.8 km to W; PICKWICK LAKE reservoir (TENNESSEE RIVER) to NE, Bear River Arm to E; J. P. Coleman State Park, on SW shore of Pickwick Lake, to N. Inc. 1857.

Iuka (ei-YOO-kah), village (2000 population 598), MARION county, S ILLINOIS, 8 mi/12.9 km E of SALEM; 38°36′N 88°47′W. In agricultural (grain, cattle) and oil area.

Iuka (ei-YOO-kuh), village (2000 population 185), PRATT county, S KANSAS, 6 mi/9.7 km N of PRATT; 37°43′N 98°43′W. In wheat area.

Iulis, Greece: see KÉA.

Iúna (ee-OO-nah), city (2007 population 25,562), SW ESPÍRITO SANTO, BRAZIL, at foot of the Pico da Bandeira, 45 mi/72 km NW of CACHOEIRO DE ITAPEMIRIM; 20°30′S 41°35′W. Coffee, bananas; sawmill. Until 1944, called Rio Pardo.

Iuripick, MICRONESIA: see EAURIPIK.

Iva (EI-vuh), town (2006 population 1,184), ANDERSON county, NW SOUTH CAROLINA, 14 mi/23 km S of ANDERSON; 34°18′N 82°39′W. Manufacturing includes clothing, cotton and polyester cloth; agricultural area that produces poultry, cattle, hogs, dairying, grain, soybeans. Founded c.1885.

Ivahy River, Brazil: see IVAÍ RIVER.

Ivailo (ee-VEI-lo), village, PLOVDIV oblast, PAZARDZHIK obshtina; 42°13′N 24°20′E.

Ivailovgrad (ee-VEI-lof-grahd), city, HASKOVO oblast, ⊙ Ivailovgrad obshtina (1993 population 10,355), SE BULGARIA, on the E slope of the RODOPI Mountains, near the ARDA RIVER and Greek border, 40 mi/64 km SE of HASKOVO; 41°32′N 26°07′E. Sericulture center in agricultural area; exports tobacco, cotton, silk, sesame oil, sweet potatoes, almonds. Metal processing. Lead, marble, and copper deposits nearby (W). Sometimes spelled Ivaylovgrad. Until 1934, known as Orta-koi.

Ivailovgrad (ee-VEI-lof-grahd), reservoir, HASKOVO oblast, BULGARIA, on the ARDA RIVER; 41°39′N 25°59′E.

Ivaiporã (EE-vah-ee-po-ruh), city (2007 population 31,344), central PARANÁ state, BRAZIL, 55 mi/89 km SSE of MARINGÁ; 24°15′S 51°45′W. Coffee growing.

Ivaí River (EE-vah-ee), c.400 mi/644 km long, central and W PARANÁ state, BRAZIL; rises in the Serra de ESPERANÇA E of GUARAPUAVA; flows NW to the PARANÁ 80 mi/129 km above GUAÍRA FALLS. Navigable in lower course. Formerly spelled Ivahy.

Ivalo (I-vah-lo), village, LAPIN province, N FINLAND, 95 mi/153 km SW of KIRKENES; 68°39′N 27°36′E. Eleva-

tion 410 ft/125 m. Near mouth of Ivalojoki River (120 mi/193 km long), on S shore of Lake INARI, on ROVANIEMI-PECHENGA section of Arctic Highway. Administrative and commercial center. Gold panning in river; regional airport.

Ivančica (EE-vahn-chee-tsah), mountain (3,480 ft/1,061 m), central CROATIA, 13 mi/21 km SW of Varaždin, in HRVATSKO ZAGORJE.

Ivancice (I-vahn-CHI-tse), Czech *Ivančice*, German *Eibenschitz*, town, JIHOMORAVSKY province, S MORAVIA, CZECH REPUBLIC, on JIHLAVA RIVER, on railroad, and 12 mi/19 km SW of BRNO; 49°06′N 16°23′E. Agricultural center, noted for vegetables (especially horseradishes) and fruit; manufacturing (textile and machinery). Neolithic-era archaeological site.

Ivanec (EE-vah-nets), town, central CROATIA, on railroad, 12 mi/19 km WSW of Varaždin, at N foot of the Ivančica Mountain, in HRVATSKO ZAGORJE; 46°14′N 16°08′E. Local trade center; lignite mine.

Ivanestii-Noui, UKRAINE: see NOVA IVANIVKA.

Ivangorod (ee-vahn-GO-ruht), city (2005 population 11,030), W LENINGRAD oblast, NW European Russia, on the NARVA River (Estonian border) opposite NARVA, on road and railroad, 90 mi/145 km W of SAINT PETERSBURG; 59°22′N 28°13′E. Elevation 108 ft/32 m. Manufacturing of boiler apparatus and pipes, flax and jute processing. Narva hydroelectric station. Site of Ivangorod castle fortress (built 1492 under Ivan III). In ESTONIA as part of Narva city, 1920–1945.

Ivangrad (EE-vahn-grahd), town, E MONTENEGRO, on LIM RIVER, and 40 mi/64 km NE of PODGORICA. Road junction; trade center (cattle, corn). Dates from 1862. Sometimes referred to as BERANE, its official name until 1948. Monastery of Djurdjevi Stupovi nearby.

Ivanhoe, unincorporated town (2000 population 4,474), TULARE county, central CALIFORNIA, 5 mi/8 km NE of VISALIA; 36°23′N 119°13′W. Orchards, citrus groves, grapes, nuts, grain; cattle; nursery products.

Ivanhoe, town (2000 population 679), ⊙ LINCOLN county, SW MINNESOTA, near SOUTH DAKOTA state line, 23 mi/37 km W of MARSHALL; 44°27′N 96°15′W. Elevation 1,658 ft/505 m. Grain, soybeans, alfalfa; livestock, poultry; dairying; light manufacturing.

Ivanhoe (EI-vuhn-ho), village, W central NEW SOUTH WALES, SE AUSTRALIA, 180 mi/290 km ESE of BROKEN HILL; 32°54′S 144°18′E. Sheep center. Railhead.

Ivanhoe (EI-vuhn-ho), unincorporated village, WYTHE county, SW VIRGINIA, near NEW RIVER, 10 mi/16 km SSE of WYTHEVILLE; 36°50′N 80°58′W. Agriculture (livestock; grain, cotton, peanuts). New River Trail passes to E.

Ivan-Horod, UKRAINE: see RZHYSHCHIV.

Ivanić Grad (EE-vah-neech GRAHD), town, central CROATIA, on LONJA RIVER, on railroad, 22 mi/35 km ESE of ZAGREB; 45°42′N 16°24′E. Center of wine-growing region; oil fields nearby. Fortress in Turkish wars.

Ivanichi, UKRAINE: see IVANYCHI.

Ivanino (ee-VAH-nee-nuh), village (2005 population 2,315), central KURSK oblast, SW European Russia, near the confluence of the Reut and SEYM rivers, on road and railroad, 25 mi/40 km W of KURSK, and 11 mi/18 km E of L'GOV; 51°38′N 35°35′E. Elevation 643 ft/195 m. In agricultural area (wheat, oats, rye, sugar beets).

Ivanishchevskiy, RUSSIA: see IVANISHCHI.

Ivanishchi (ee-VAH-nee-shchee), town (2006 population 2,130), S central VLADIMIR oblast, central European Russia, on road and railroad spur, 25 mi/40 km S of VLADIMIR; 55°46′N 40°26′E. Elevation 429 ft/130 m. Glassworks. Established as Ivanishchevskiy settlement; renamed Ukrepleniye Kommunizma (Russian= Strengthening of Communism) in the early 1920s; town status and current name given in 1942.

Ivanitsa, UKRAINE: see IVANYTSYA.

Ivanivka (ee-VAH-nif-kah) (Russian *Ivanovka*), town, E KHERSON oblast, UKRAINE, 40 mi/64 km WSW of MELITOPOL'; 46°42′N 34°33′E. Raion center; feed and

flour mill, dairy, food flavoring factory. Professional technical school. Established in 1820, town since 1956.

Ivanivka (ee-VAH-neev-kah) (Russian *Ivanovka*), town (2004 population 2,700), S ODESSA oblast, UKRAINE, 35 mi/56 km NNW of ODESSA; 46°58′N 30°28′E. Raion center; grain elevator, flour mill, oil pressing, dairy. Founded in 1793 as Malobaranivka (Russian *Malobaranovka*); from 1858 to 1946, called Yanivka (Russian *Yanovka*).

Ivanivka (ee-VAH-neev-kah) (Russian *Ivanovka*), town, SW LUHANS'K oblast, UKRAINE, in the DONBAS, 7 mi/11.3 km N of KRASNYY LUCH. Machine and tool manufacturing, food processing, brewery. Formerly also called Mala Ivanivka (Russian *Malaya Ivanovka*).

Ivanivs'kyy Rudnyk, UKRAINE: see LOTYKOVE.

Ivanjica (ee-VAH-nyee-tsah), town (2002 population 35,445), MORAVA district, W SERBIA, on WESTERN MORAVA RIVER, and 25 mi/40 km SW of KRALJEVO; 43°34′N 20°14′E. Hydroelectric plant. Antimony mine (Lisanski Rudnik) and smelter 4 mi/6.4 km NE. Also spelled Ivanyitsa.

Ivanjska (ee-VAHN-skah), town, N BOSNIA, BOSNIA AND HERZEGOVINA, 11 mi/18 km NNW of Banja Luka. Also spelled Ivanska or Ivan'ska.

Ivanka pri Dunaji (i-VAHN-kah PRI-du-NAH-yi), village, ZAPADOSLOVENSKY province, SW SLOVAKIA, on railroad and 7 mi/11.3 km NE of BRATISLAVA; 48°11′N 17°16′E. Poultry; machinery. Airport. Paleolithic-era archaeological site. Until 1927, was called Bratislava Ivanka, Slovak *Bratislavsda Ivánka*, HUNGARIAN *Pozsonyivánka*.

Ivankiv (ee-VAHN-kif) (Russian *Ivankov*), town, N KIEV oblast, UKRAINE, on TETERIV RIVER (head of navigation), and 45 mi/72 km NW of KIEV; 50°56′N 29°54′E. Elevation 301 ft/91 m. Raion center; tools, food processing (vegetables, butter, fish), woodworking. Heritage museum. Established in 1589, town since 1940.

Ivankov, UKRAINE: see IVANKIV.

Ivankovo (ee-VAHN-kuh-vuh), former city, central European Russia, 12 mi/19 km SW of KIMRY and on right bank of the VOLGA RIVER, at the E end of the Ivankovo Reservoir, at the efflux of the MOSCOW CANAL; 55°15′N 35°52′E. Elevation 770 ft/234 m. Site of a dam and hydroelectric station (completed 1937); railroad spur terminus (Bol'shaya Volga station). Became part of KRYUKOVO city, MOSCOW oblast, in 1960, which was then in turn administratively incorporated into MOSCOW proper.

Ivan'kovo (ee-VAHN-kuh-vuh), village, NE TVER oblast, W central European Russia, 37 mi/60 km N of TVER; 57°24′N 36°13′E. Elevation 666 ft/202 m. In agricultural area (wheat, rye, oats, flax).

Ivan'kovo, RUSSIA: see IVAN'KOVSKIY.

Ivan'kovskiy (ee-VAHN-kuhf-skeeyee), settlement (2005 population 445), SW IVANOVO oblast, central European Russia, near highway, 18 mi/29 km SW of TEYKOVO; 56°39′N 40°05′E. Elevation 456 ft/138 m. In agricultural area (grain, sunflowers, sugarbeets; livestock). Also known as Ivan'kovo.

Ivan Mountains (ee-VAHN), in DINARIC ALPS, on Bosnia-Herzegovina line, BOSNIA AND HERZEGOVINA. Highest point (5,720 ft/1,743 m) is 8 mi/12.9 km N of KONJIC. SARAJEVO-Konjic railroad passes through 3,172-ft/967-m-high Ivan Saddle, which lies 8 mi/12.9 km NNE of Konjic.

Ivano-Alekseevka (ee-VAH-no–ah-lek-SAIV-kuh), village, TALAS region, KYRGYZSTAN, 2 mi/3.2 km NW of TALAS; 42°34′N 72°11′E. Wheat; horses. Tertiary-level administrative center. Also Ivanovo-Alekseyevka.

Ivano-Frankivs'k (ee-VAH-no–frahn-KIFSK) (Ukrainian *Ivano-Frankivs'ka*), (Russian *Ivano-Frankovskaya*), oblast (⊙ 5,378 sq mi/13,982.8 sq km; 2001 population 1,409,760), W UKRAINE, ⊙ IVANO-FRANKIVS'K. From N slopes of CARPATHIAN Mountains extends N into DNIESTER (Ukrainian *Dnister*) River valley; bounded SE by CHEREMOSH RIVER. Drained by right tributaries of the Dniester (BYSTRYTSYA, LIMNYTSYA, and SVICHA rivers), and by PRUT RIVER. Has dark prairie soils; loess (N); humid continental climate. Population mostly Ukrainian (95%), with Russian (4%), Polish (0.2%), Belorussian (0.2%), and Jewish (0.1%) minorities. Mining region: petroleum, natural gas, ozokerite, and salt in the Carpathian foothills (DOLYNA, BYTKIV, KOSMACH); potassium (KALUSH, Dolyna), lignite, and gypsum. Grain, potatoes (N), sugar beets, hogs, tobacco, fruit and vegetables (Pokutye upland), sheep (in mountains). Industries based on mining (petroleum refining, saltworks), agriculture (sugar refining, flour milling, tanning, distilling, brewing, vegetable-oil extracting), and timber (sawmilling, paper and plywood manufacturing). Metalworking and light industries in main urban centers (Ivano-Frankivs'k, KOLOMYYA, Kalush). Health resorts (YAREMCHA, VOROKHTA) in GORGANY and Hutsul Beskyd of the Carpathian Mountains. Formed in 1939 out of Polish Stanislawow province, following Soviet occupation, and named Stanislaviv (Russian *Stanislav*); held by Germany (1941–1944); ceded to USSR in 1945 as part of Ukrainian SSR. Renamed after Ukrainian poet, Ivan Franko, in 1962. In 1993 had 14 cities, 25 towns, and 14 rural raions.

Ivano-Frankivs'k (ee-VAH-no–frahn-KIFSK) (Russian *Ivano-Frankovsk*), city (2001 population 218,359), ⊙ IVANO-FRANKIVS'K oblast, W UKRAINE, on the BYSTRYTSYA RIVER; 48°55′N 24°42′E. Elevation 820 ft/249 m. It is a railroad junction and industrial and cultural center, situated in a fertile agricultural zone of the CARPATHIAN foothills. Diverse light manufacturing (woodworking, armature, locomotive repair, machinery, furniture making, textiles, food processing). It has institutes of medicine, education, petroleum gas, and many professional technical schools; two theaters; several museums. An old Ukrainian settlement, Stanislaviv, was chartered in 1662 on the site of the former village of Zabolotiv, as the Polish town of Stanisławów. Despite Tatar and Turkish raids, it flourished as a well-developed trade center in the 17th and 18th centuries. Merchants were principally Armenians and Jews. Stanislaviv passed to Austria in 1772, and ownership of the city from Count Potocki to the Austrian state in 1801. Ukrainian national movement gained force after 1848, and the city became the bishopric of the Ukrainian Catholic (Uniate) Church in 1885. In 1918 it was part of, and between December 1918 and May 1919 served as capital of, the West Ukrainian National Republic, but passed to Poland in 1919 and then to the USSR (Ukrainian SSR) in 1939. The city and oblast were renamed in 1962 in honor of the Ukrainian poet and writer Ivan Franko. Landmarks include a wooden church (1601), a Ukrainian Catholic cathedral, and an 18th-century palace.

Ivano-Frankivs'ka oblast, UKRAINE: see IVANO-FRANKIVS'K OBLAST.

Ivano-Frankove (ee-VAH-no–frahn-KO-ve) (Russian *Ivano-Frankovo*), town, central L'VIV oblast, UKRAINE, 12 mi/19 km WNW of L'VIV; 49°55′N 23°44′E. Elevation 869 ft/264 m. Lumber, sawmilling, furniture manufacturing, woodworking. Fisheries along a small lake (just N). Until 1945, called Yaniv (Russian *Yanov*, Polish *Janow*).

Ivano-Frankovo, UKRAINE: see IVANO-FRANKOVE.

Ivano-Frankovsk, city, UKRAINE: see IVANO-FRANKIVS'K.

Ivano-Frankovskaya oblast, UKRAINE: see IVANOFRANKIVS'K oblast.

Ivanopil' (ee-vah-NO-peel) (Russian *Ivanopol'*), town, S ZHYTOMYR oblast, UKRAINE, 16 mi/26 km W of BERDYCHIV; 49°51′N 28°13′E. Elevation 853 ft/259 m. Sugar refining, flour milling. Professional technical school. Known since 1714; town since 1924. Until 1946, Yanushpil' (Polish *Januszpol*, Russian *Yanushpol'*).

Ivanopol', UKRAINE: see IVANOPIL'.

Ivanovice na Hane (I-vah-NO-vi-TSE NAH HAH-ne), Czech *Ivanovice na Hané*, German *Eiwanowitz in der*, town, JIHOMORAVSKY province, S central MORAVIA, CZECH REPUBLIC, on railroad, and 5 mi/8 km NE of VYŠKOV; 49°19′N 17°06′E. Manufacturing (chemicals, textiles); woodworking; brick kiln; malt house. Has a Renaissance castle and a baroque church.

Ivanovka (ee-VAH-nov-kah)), city, W LUHANS'K oblast, UKRAINE, in the DONBAS, 10 mi/16 km NNE of POPASNA; 48°46′N 38°30′E. Elevation 469 ft/142 m. Coal-mining center, with coal mine, enrichment plant, and footwear factory. Established in 1898, city since 1938. Until about 1940, known as Hirsko-Ivanivske (Russian *Gorsko-Ivanovskoye*).

Ivanovka (ee-VAH-nuhv-kuh), town, N CHÜY region, KYRGYZSTAN, in CHU valley, on railroad, and 28 mi/45 km E of BISHKEK. Sugar beets. Tertiary-level administrative center. Also called Ak-Chey.

Ivanovka (ee-VAH-nuhf-kah), village, NW ORENBURG oblast, SW URALS, SE European Russia, in the TOK RIVER basin, on road, 44 mi/71 km NE of ORENBURG; 53°18′N 52°32′E. Elevation 669 ft/203 m. Vegetable growing; livestock.

Ivanovka (ee-VAH-nuhf-kah), village, N KIROV oblast, E central European Russia, on the MOLOMA River (tributary of the VYATKA River), on road, 53 mi/85 km N of KOTEL'NICH; 59°04′N 48°15′E. Elevation 452 ft/137 m. Sawmilling, logging, lumbering.

Ivanovka (ee-VAH-nuhf-kah), village (2005 population 6,400), SE AMUR oblast, RUSSIAN FAR EAST, 22 mi/35 km ENE of BLAGOVESHCHENSK; 50°22′N 128°00′E. Elevation 465 ft/141 m. In agricultural area.

Ivanovka (ee-VAH-nuhf-kah), village (2006 population 2,985), SW MARITIME TERRITORY, SE RUSSIAN FAR EAST, on crossroads, 36 mi/58 km N of VLADIVOSTOK; 43°58′N 132°29′E. Elevation 416 ft/126 m. In agricultural area (rice, soybeans).

Ivanovo (ee-VAH-nuh-vuh), oblast (□ 8,417 sq mi/ 21,884.2 sq km; 2004 population 1,266,000) in central European Russia; ⊙ IVANOVO. On a level plain between VOLGA and KLYAZMA rivers; drained by NERL, UVOD, and TEZA rivers (left affluents of the Klyazma); mixed forest zone (about 30% of the territory); marshes (SE). Mineral resources include peat (W of Ivanovo and in SE; peat-power station at KOMSOMOL'SK), phosphorites near KINESHMA, quartz sands. Principal textile-producing region of Russia; cotton-milling centers at Ivanovo, KOKHMA, SHUYA, TEYKOVO, RODNIKI, FURMANOV, and VICHUGA; linen milling (PRIVOLZHSK, PUCHEZH); also chemicals (acids, alkalis, dyes), flax fibers, starch. Handicraft industries (homespun goods, embroidery, weaving shuttles, carved wooden articles). Sawmilling along the Volga River (Kineshma, NOVAYA SLOBODKA). Textile and peatworking machine manufacturing (Ivanovo). Agriculture includes vegetables, potatoes, dairy products chicory and tobacco (W), wheat, flax, fodder crops (E). Part of the Gold Ring of Russia, a designation for the area containing a large number of historical and cultural monuments. Formed in 1929 out of Ivanovo-Voznesensk (formed in 1918) and Vladimir, Kostroma, and Yaroslavl; originally called Ivanovo Industrial Oblast.

Ivanovo (ee-VAH-nuh-vuh), city (2005 population 417,290), ⊙ IVANOVO oblast, central European Russia, in the Moscow industrial region, 198 mi/319 km NE of MOSCOW; 56°59′N 40°59′E. Elevation 413 ft/125 m. Railroad junction. The city is the historic center of Russia's cotton-milling industry (blended yarns, worsteds, cotton cloth). Other manufacturing includes machinery for textiles, peat, mining cranes, machine tools, synthetics, chemicals, wood, and food products. Known as Ivanovo-Voznesensk until 1932.

Ivanovo (ee-VUH-nah-voh), town, BREST oblast, BELARUS, 25 mi/40 km W of PINSK, 52°10′N 25°31′E. Poultry; cannery, bakery, powdered-milk plant. Until 1945 called Yanov, Polish *Janów* or *Janow Poleski*.

Ivanovo (ee-vah-NO-vo), village (1993 population 1,108), RUSE oblast, ⊙ Ivanovo obshtina, BULGARIA; 43°42′N 25°57′E. Grain, fruit; wine; produce warehousing.

Ivanovo-Voznesensk, RUSSIA: see IVANOVO, city.

Ivanovskaya (ee-VAH-nuhfs-kah-yah), village (2005 population 9,460), central KRASNODAR TERRITORY, S European Russia, in the KUBAN' River basin, on road, 32 mi/51 km WNW of KRASNODAR, and 16 mi/26 km E of SLAVYANSK-NA-KUBANI; 45°16′N 38°28′E. In oil- and gas-producing region.

Ivanovskiy Rudnik, UKRAINE: see LOTYKOVE.

Ivanovskoye, RUSSIA: see SMYCHKA.

Ivanska, BOSNIA AND HERZEGOVINA: see IVANJSKA.

Ivanski (ee-VAHN-skee), village, VARNA oblast, SHUMEN obshtina, E BULGARIA, on the GOLYAMA KAMCHIYA RIVER, 11 mi/18 km SSE of Shumen; 43°08′N 27°03′E. Vineyards, livestock. Formerly Kopryu-koi (until 1878), then Zlokuchen (until 1950).

Ivanteyevka (ee-vahn-TYE-eef-kah), city (2006 population 50,970), E central MOSCOW oblast, central European Russia, on railroad, on the UCHA RIVER, 25 mi/ 40 km NE of MOSCOW, and 4 mi/6 km SE of PUSHKINO; 55°58′N 37°55′E. Elevation 472 ft/143 m. Woolen- and cotton-milling center; knitwear; concrete structures; forest nursery; electric wiring; food industries (bakery, brewery). Known since 1586. Called Ivanteyevskiy after 1928, until city status granted in 1938.

Ivanteyevka (ee-vahn-TYE-eef-kah), village (2006 population 6,115), NE SARATOV oblast, SE European Russia, on the LESSER IRGIZ RIVER, on crossroads and railroad, 22 mi/35 km NNE of PUGACHËV; 52°16′N 49°06′E. Elevation 203 ft/61 m. Highway junction, railroad station; local transshipment point.

Ivanteyevskiy, RUSSIA: see IVANTEYEVKA, MOSCOW oblast.

Ivanychi (ee-VAH-ni-chee) (Russian *Ivanichi*) (Polish *Iwanicze*), town (2004 population 5,360), SW VOLYN' oblast, UKRAINE, 13 mi/21 km S of VOLODYMYR-VOLYNS'KYY; 50°39′N 24°21′E. Elevation 656 ft/199 m. Raion center. Railroad junction; sugar refinery, cannery, feed mill. Historical museum. Known since 1545, town since 1951.

Ivanyitsa, SERBIA: see IVANJICA.

Ivanytsya (ee-vah-NI-tsyah) (Russian *Ivanitsa*), village, central SUMY oblast, UKRAINE; 50°47′N 32°38′E. Elevation 465 ft/141 m. Dairying.

Ivashchenkovo, RUSSIA: see CHAPAYEVSK.

Ivato (ee-VAHT), town, ANTANANARIVO province, central MADAGASCAR, 10 mi/16 km NNW of ANTANANARIVO; 18°47′S 47°29′E. Site of internatioal airport and military base.

Ivatsevichi (ee-vuh-TSE-vee-chee), Polish *Iwacewicze*, town, NE BREST oblast, BELARUS, ⊙ IVATSEVICHI region, 8 mi/13 km ESE of KOSSOVO, 54°50′N 30°13′E. Railroad; manufacturing (linen, reinforced concrete products, baked goods, chemical timber).

Ivatuba (EE-vah-too-bah), town (2007 population 2,715), NW PARANÁ state, BRAZIL, 22 mi/35 km SW of MARINGÁ; 23°37′S 52°13′W. Coffee, cotton, rice, corn; livestock.

Ivaylovgrad, Bulgaria: see IVAILOVGRAD.

Ivdel' (EEV-dyel), city (2006 population 19,400), N SVERDLOVSK oblast, W Siberian Russia, in the E foothills of the central URALS, on the Ivdel' River (right tributary of the LOZVA RIVER, OB' RIVER basin), on road, 316 mi/509 km N of YEKATERINBURG, and 75 mi/121 km N of SEROV; 60°41′N 60°25′E. Elevation 269 ft/81 m. On the Serov-Priob'ye railroad and on a natural-gas pipeline; sawmilling and wood hydrolysis. Became city in 1943.

Ivel River (EI-vuhl), 30 mi/48 km long, in HERTFORDSHIRE and BEDFORDSHIRE, central ENGLAND, rises just N of Baldock, flows N past Biggleswade to the OUSE 3 mi/4.8 km N of Sandy. Navigable below BIGGLESWADE.

Area is shown by the symbol □, and capital city or county seat by ⊙.

Ivenets (ee-vah-NETS), Polish *Iwieniec*, town, NE BARANOVICHI oblast, BELARUS, 29 mi/47 km N of STOLBTSY, in woodland, 53°50′N 26°40′E. Cannery, creamery.

Iver (EI-vuh), village (2001 population 9,925), SE BUCKINGHAMSHIRE, central ENGLAND, 2 mi/3.2 km SW of UXBRIDGE; 51°31′N 00°30′W. Makes cables and tile. Has Saxon church with 13th–15th-century additions.

Ivesdale, village (2000 population 288), CHAMPAIGN county, E central ILLINOIS, 15 mi/24 km SW of CHAMPAIGN; 39°57′N 88°27′W. In agricultural area (corn, soybeans).

Iveşti (ee-VESHT), agricultural village, Galaţi county, E ROMANIA, on Birlad River, on railroad, and 14 mi/23 km SSE of TECUCI.

Ivey, town (2000 population 1,100), WILKINSON county, central GEORGIA, 11 mi/18 km SSW of MILLEDGEVILLE; 32°55′N 83°18′W. Manufacturing of steel tanks.

Ivi, Cape (ee-VEE), headland of MOSTAGANEM wilaya, NW ALGERIA, on the MEDITERRANEAN SEA, at E end of Gulf of ARZEW, 14 mi/23 km NNE of MOSTAGANEM; 36°07′N 00°14′E. Lighthouse.

Ivindo River (EE-vin-do), c.225 mi/362 km long, in S CAMEROON and N GABON, rises on Cameroon-Gabon border 20 mi/32 km ENE of MINVOUL. Flows S and SW, past MAKOKOU, Gabon, to Ogooué River 15 mi/24 km E of Booué. Its middle course is navigable intermittently for c.200 mi/322 km; rapids in lower course. Sometimes called LIVINDO River.

Ivinghoe (EI-ving-ho), village (2006 population 950), E BUCKINGHAMSHIRE, central ENGLAND, 8 mi/12.9 km E of AYLESBURY at foot of CHILTERN escarpment; 51°51′N 00°39′W. Agricultural market.

Ivins (EI-vihnz), town (2006 population 7,205), WASHINGTON county, SW UTAH, 5 mi/8 km NW of SAINT GEORGE, near SANTA CLARA RIVER; 37°10′N 113°40′W. Snow Canyon State Park to NE; Gunlock State Park to NW.

Ivittuut, town, SW GREENLAND, on the Arsuk Fjord. The world's largest known cryolite deposit was discovered here in 1806. Mined since 1858, the deposit has been recently exhausted; stockpiled cryolite has been exported since 1969. The town is now deserted.

Iviza, SPAIN: see IBIZA.

Ivnya (EEV-nyah), village (2005 population 7,860), NW BELGOROD oblast, S central European Russia, on road, 41 mi/66 km S of KURSK, 30 mi/48 km N of BELGOROD; 51°03′N 36°08′E. Elevation 656 ft/199 m. Sugar refinery; creamery.

Ivo (EE-vo), canton, LUIS CALVO province, CHUQUISACA department, SE BOLIVIA, 10 mi/15 km SW of BOYUIBE, on YACUIBA–SANTA CRUZ highway, 6 mi/10 km S of SANTA CRUZ department border; 20°27′S 63°26′W. Elevation 3,885 ft/1,184 m. Abundant gas and oil resources in area. Agriculture (potatoes, yucca, bananas, corn, rye, sweet potatoes, peanuts, tobacco); cattle. Also Ibo.

Ivohibe (ee-voo-ee-BAI), town, FIANARANTSOA province, SE central MADAGASCAR, 70 mi/113 km SSE of FIANARANTSOA; 22°28′S 46°53′E. Market center; rice, cassava, coffee; cattle.

Ivolga, RUSSIA: see IVOLGINSK.

Ivolginsk (EE-vuhl-geensk), town (2005 population 7,065), SE BURYAT REPUBLIC, S SIBERIA, RUSSIA, on road, 20 mi/32 km SW of ULAN-UDE; 51°45′N 107°14′E. Elevation 1,794 ft/546 m. In agricultural area; state farm. Has a Buddhist religious school. Formerly called Ivolga.

Ivoloina (ee-voo-LOO-en), village, TOAMASINA province, E MADAGASCAR, near coast, on the Canal des PANGALANES, and NNW of TOAMASINA; 18°03′S 49°23′E. Site of oldest (1898) agricultural station in MADAGASCAR, specializing in tropical cultures (bananas, peppers, vanilla, rice). Zoo.

Ivón (ee-VON), canton, VACA DIEZ province, BENI department, NE BOLIVIA, 28 mi/45 km S of RIBERALTA, on the BENI RIVER, on unpaved Highway 8 (GUAYARAMERÍN-RURRENABAQUE); 11°07′S 66°09′W. Elevation 443 ft/135 m. Clay deposits in area. Agriculture (yucca, bananas, rice, potatoes, cacao, coffee, cotton, quinine, rubber); cattle and horse raising.

Ivondro River (ee-VOON-druh), 100 mi/161 km long, in TOAMASINA province, E MADAGASCAR, rises 50 mi/80 km NNE of MORAMANGA (18°06′S 48°21′E), flows in a wide curve NE and E, cutting through Canal des Pangalanes to the INDIAN OCEAN 5 mi/8 km S of TOAMASINA. Used for water power.

Ivor (EI-vor), town (2000 population 316), SOUTHAMPTON county, SE VIRGINIA, 17 mi/27 km N of FRANKLIN, near BLACKWATER RIVER; 36°53′N 76°54′W. Manufacturing (meat packing, fertilizer).

Ivory Coast: see CÔTE D'IVOIRE.

Ivoryton, CONNECTICUT: see ESSEX.

Ivösjön (EEV-UH-SHUHN), lake, KRISTIANSTAD county, S SWEDEN, extends NW from BROMÖLLA; 16 mi/26 km long, 7 mi/11.3 km–10 mi/16 km wide. Drains SE into BALTIC SEA. Contains Ivön island (□ 4.9 sq mi/12.7 sq km).

Ivot (EE-vuht), town (2005 population 6,250), NE BRYANSK oblast, W central European Russia, on road and railroad, 7 mi/11 km NW of DYAT′KOVO; 53°40′N 34°11′E. Elevation 633 ft/192 m. Railroad junction. Glassworks.

Ivrea (eev-RE-ah), town, TORINO province, PIEDMONT, NW ITALY, on the DORA BALTEA RIVER; 45°28′N 07°52′E. It is a commercial and industrial center, and the headquarters of Olivetti. Manufacturing includes typewriters, computers, machinery, fabricated metals, synthetic fibers, and textiles. A Roman town (*Eporedia*), it was later the capital of a Lombard duchy and then the seat of a marquisate. Berengar II, one of its rulers, was briefly king of Italy (mid-10th century). Ivrea passed to the house of Savoy in the 14th century. The city is dominated by a picturesque castle (14th century), which has four red brick towers.

Ivrindi, village, NW TURKEY, 22 mi/35 km WSW of BALIKESIR; 39°33′N 27°27′E. Cereals.

Ivry-la-Bataille (ee-vree-lah-bah-tei), commune (□ 3 sq mi/7.8 sq km; 2004 population 2,653), EURE department, HAUTE-NORMANDIE region, NW central FRANCE, on left bank of EURE RIVER, and 17 mi/27 km SE of ÉVREUX; 48°53′N 01°28′E. Manufacturing of woodwind musical instruments at nearby La Couture-Boussey. At Ivry, king-to-be Henry IV won a decisive victory (1590) over the Catholic League led by the duke of Mayenne.

Ivry-sur-le-Lac (ee-VREE–syur-luh–LAHK), village (□ 12 sq mi/31.2 sq km; 2006 population 449), LAURENTIDES region, S QUEBEC, E CANADA; 46°04′N 74°20′W. An independent municipality, it forms part of the SAINTE-AGATHE-DES-MONTS agglomeration.

Ivry-sur-Seine (ee-vree–syur-sen), city (□ 2 sq mi/5.2 sq km), SSE industrial and commercial suburb of PARIS, VAL-DE-MARNE department, ÎLE-DE-FRANCE region, N central FRANCE, about 3 mi/4.8 km from Notre-Dame Cathedral; 48°49′N 02°23′E. Connected to Paris by subway. On railroad line to LYON and MARSEILLE. Its port on the SEINE RIVER, where it receives the MARNE RIVER, engages in wholesale trade in fuel, timber, barrels, and foodstuffs. Manufacturing (chemicals, refractory products, portland cement, pharmaceuticals, ball bearings, and food products). Has several old churches.

Ivujivik or **Notre-Dame-d'Ivugivic**, village (□ 14 sq mi/36.4 sq km), NORD-DU-QUÉBEC region, N QUEBEC, E CANADA, at NW tip of UNGAVA PENINSULA, where HUDSON STRAIT enters HUDSON BAY; 62°25′N 77°54′W. It is Quebec's northernmost village. Population is Inuit. Scheduled air service. Hunting, fishing, trapping. Large colony of thick-billed murre here. Name *Inujivik* means "place where ice accumulates due to strong currents."

Ivuna (ee-VOO-nah), village, MBEYA region, SW TANZANIA, 70 mi/113 km NW of MBEYA; 08°24′S 32°30′E. Cattle, goats, sheep; corn, wheat, coffee. Lake RUKWA to NE.

Ivy (EI-vee), unincorporated village, ALBEMARLE county, central VIRGINIA, 5 mi/8 km W of CHARLOTTESVILLE; 38°03′N 78°35′W. Manufacturing (clothing, wall clocks); agriculture (dairying; livestock; grain, apples). Meriwether Lewis born here, 1774. Also called Ivy Depot.

Ivybridge (EI-vee-BRIJ), town (2001 population 12,056), S DEVON, SW ENGLAND, 10 mi/16 km E of PLYMOUTH; 50°23′N 03°55′W. Former agricultural market; light manufacturing, but mainly residential. Has old bridge.

Ivy Depot, Virginia: see IVY.

Ivye (ee-VYE), Polish *Iwje*, town, GRODNO oblast, BELARUS, 19 mi/31 km ENE of LIDA. Railroad-spur terminus; tanning, flour milling, alcohol distilling, manufacturing of sweets.

Ivyland (EI-vee-luhnd), borough (2006 population 818), BUCKS county, SE PENNSYLVANIA, industrial suburb 16 mi/26 km NNE of PHILADELPHIA. Manufacturing (machinery, pressure vessels, laboratory equipment, electronic components, canvas products, windows, graphic-arts equipment, food preparations, flavors). Agriculture to NE (grain; livestock; dairying). U.S. Naval Air Development Center to S.

Ivywild, village, EL PASO county, E central COLORADO, 2 mi/3.2 km SSW of, and part of, COLORADO SPRINGS.

Iwade (ee-WAH-de), town, Naga county, WAKAYAMA prefecture, S HONSHU, W central JAPAN, on W KII PENINSULA, 9 mi/15 km E of WAKAYAMA; 34°15′N 135°18′E.

Iwadeyama (ee-WAH-de-YAH-mah), town, Tamatsukuri county, MIYAGI prefecture, N HONSHU, NE JAPAN, 28 mi/45 km N of SENDAI; 38°38′N 140°52′E. Handmade bamboo crafts. Processed soybeans (tofu). Sazaragi ruins.

Iwafune (ee-WAHF-ne), town, Shimotsuga county, TOCHIGI prefecture, central HONSHU, N central JAPAN, 22 mi/35 km S of UTSUNOMIYA; 36°18′N 139°39′E.

Iwagi (EE-wah-gee), village, Ochi county, EHIME prefecture, NW SHIKOKU, W Japan, 34 mi/55 km N of MATSUYAMA; 34°14′N 133°09′W.

Iwai (ee-WAH-ee), city, IBARAKI prefecture, central HONSHU, E central JAPAN, on small lake, 40 mi/65 km S of MITO; 39°50′N 141°48′E. Manufacturing (televisions); agriculture (spring onions, lettuce).

Iwaizumi (ee-WAH-ee-ZOO-mee), town, Shimohei county, IWATE prefecture, N HONSHU, NE JAPAN, 34 mi/55 km N of MORIOKA; 39°50′N 141°48′E. Lumber; wooden furniture. Nearby attractions include Ryusen and Akka caves and Rikuchu Kaigan [=seacoast] National Park.

Iwaki (EE-wah-kee), city (2005 population 354,492), FUKUSHIMA prefecture, N central HONSHU, NE JAPAN, on the Iwaki River, 53 mi/85 km S of FUKUSHIMA; 37°02′N 140°53′E. Formerly a major coal-mining center. Manufacturing includes electronics (stereos), lumber, medicines, and processed fish (*ita kamaboko*). Cucumbers. Hot springs in the vicinity. Ruins of Tokugawa era tollgate at Nakosono.

Iwaki (ee-wah-kee), town, Yuri county, AKITA prefecture, N HONSHU, NE JAPAN, 12 mi/20 km S of AKITA; 39°33′N 140°03′E. Lobster farming; plum wine; textile manufacturing.

Iwaki, town, Nakatsugaru county, Aomori prefecture, N HONSHU, N JAPAN, near Mount Iwaki, 22 mi/35 km S of AOMORI; 40°36′N 140°25′E. Apples; pickles, vegetable processing. Bamboo handicrafts.

Iwakuni (ee-WAH-koo-nee), city (2005 population 149,702), YAMAGUCHI prefecture, SW HONSHU, W JAPAN, on the Aki Sea, 43 mi/70 km E of YAMAGUCHI; 34°09′N 132°13′E. Synthetic fibers. The Kintai Bridge here, built in 1673, is famous for providing an escape route from floods.

Iwakura (ee-WAH-koo-rah), city, AICHI prefecture, S central HONSHU, central JAPAN, 6 mi/10 km N of NAGOYA; 35°16′N 136°52′E. Glass; traditional carp streamers.

Iwama (ee-WAH-mah), town, West Ibaraki county, IBARAKI prefecture, central HONSHU, E central JAPAN, 12 mi/20 km S of MITO; 36°17′N 140°16′E. Aikido martial art originated here.

Iwami, former province in SW HONSHU, JAPAN; now part of SHIMANE prefecture.

Iwami (ee-WAH-mee), town, Ochi county, SHIMANE prefecture, SW HONSHU, W JAPAN, 54 mi/87 km S of MATSUE; 34°53′N 132°26′E.

Iwami, town, Iwami county, TOTTORI prefecture, S HONSHU, W JAPAN, 7 mi/11 km N of TOTTORI; 35°34′N 134°20′E.

Iwamizawa (ee-WAH-mee-ZAH-wah), city, W central HOKKAIDO prefecture, N JAPAN, 22 mi/35 km E of SAPPORO; 43°11′N 141°43′E. Vegetables (onions, cabbages, lily roots). Trench diggers.

Iwamura (ee-WAH-moo-rah), town, Ena county, GIFU prefecture, central HONSHU, central JAPAN, 40 mi/65 km E of GIFU; 35°21′N 137°26′E. Baked goods, rice cakes, pickles, local sake. Iwamura Yamajiro [=mountain castle] is located here.

Iwamuro (ee-WAH-moo-ro), village, W Kanbara county, NIIGATA prefecture, central HONSHU, N central JAPAN, 12 mi/20 km S of NIIGATA; 37°43′N 138°52′E. Rice.

Iwanai (ee-WAH-nah-ee), town, Shiribeshi district, SW Hokkaido prefecture, N JAPAN, port on SEA OF JAPAN, 43 mi/70 km W of SAPPORO; 42°58′N 140°30′E. Food processing (cod roe, smoked trout, herring, herring roe). The first asparagus grown in Japan was planted here.

Iwanicze, UKRAINE: see IVANYCHI.

Iwanuma (ee-WAH-noo-mah), city, MIYAGI prefecture, N HONSHU, NE JAPAN, on ABUKUMA RIVER, 12 mi/20 km S of SENDAI; 38°06′N 140°32′E. Paper, tires. Gourds, cucumbers, watermelon; pickles, rice crackers (*sembei*), sake.

Iwasaki (ee-wah-SAH-kee), village, West Tsugaru county, Aomori prefecture, N HONSHU, N Japan, near Shiragami mountain range, 47 mi/75 km E of AOMORI; 40°34′N 139°55′E. Tsugaru park is nearby.

Iwase (EE-wah-se), town, West Ibaraki county, IBARAKI prefecture, central HONSHU, E central JAPAN, 22 mi/35 km W of MITO; 36°21′N 140°06′E. Granite tombstones.

Iwase (EE-wah-se), village, Iwase county, FUKUSHIMA prefecture, N central HONSHU, NE JAPAN, 31 mi/50 km S of FUKUSHIMA; 37°18′N 140°16′E.

Iwashiro, former province in N HONSHU, JAPAN; now part of FUKUSHIMA prefecture.

Iwashiro (ee-WAH-shee-ro), town, Adachi county, FUKUSHIMA prefecture, N central HONSHU, NE JAPAN, 12 mi/20 km S of FUKUSHIMA; 37°33′N 140°30′E. Shiitake mushrooms.

Iwata (ee-WAH-tah), city (2005 population 170,899), SHIZUOKA prefecture, central HONSHU, E central JAPAN, on the estuary of the TENRYU RIVER, 34 mi/55 km S of SHIZUOKA; 34°42′N 137°51′E. Agriculture (melons, white onions). Manufacturing includes motorcycles, sushi-making machines, velveteen, bearings. Site of Iwata Bunko, Japan's first library.

Iwataki (ee-WAH-tah-kee), town, Yosa county, KYOTO prefecture, S HONSHU, W central JAPAN, on inlet of WAKASA BAY, 50 mi/80 km N of KYOTO; 35°33′N 135°09′E. *Tango* crepe.

Iwate (EE-wah-te), prefecture (Japan *ken*) (□ 5,882 sq mi/15,234 sq km; 1990 population 1,428,646), N HONSHU, NE JAPAN; ⊙ MORIOKA. Bordered N by Aomori prefecture, S by MIYAGI prefecture, and W by Akita prefecture. Mountainous terrain; drained by KITAKAMI RIVER. Chief port, KAMA-ISHI. Major livestock-raising area. Extensive lumbering, fishing, raw-silk culture, agriculture (soybeans, rice, potatoes, tobac-

co). Iron, gold, silver, and copper mines. Manufacturing (silk textiles, lacquerware, chemicals, integrated circuits), sake brewing. Principal centers are MORIOKA, KAMA-ISHI, MIYAKO.

Iwate (EE-wah-te), town, Iwate county, IWATE prefecture, N HONSHU, NE JAPAN, 19 mi/30 km N of MORIOKA; 39°58′N 141°13′E. Tobacco.

Iwatsuki (ee-WAHTS-kee), city, SAITAMA prefecture, E central HONSHU, E central JAPAN, on the Edo River, 6 mi/10 km N of URAWA; 35°36′N 139°42′E. Traditional dolls.

Iwimbi, TANZANIA: see UWIMBI.

Iwo (EE-wo), city, OSUN state, SW NIGERIA. It is the trade center for a farm region specializing in cacao. A coffee plantation is nearby. Former capital of a YORUBA kingdom (founded in the 17th century) that grew rapidly in the 19th century by taking in refugees during the Yoruba civil wars.

Iwo Jima, JAPAN: see IWO TO.

Iwo To (EE-wo toh), Japanese *Io-to* (ee-O-toh), volcanic island (□ 8 sq mi/20.8 sq km), largest and most important of the VOLCANO ISLANDS, Tokyo prefecture, extreme SE JAPAN; 24°47′N 141°20′E. MOUNT SURIBACHI (546 ft/166 m), on the S side of the island, is an extinct volcano. The main industries are sulfur mining and sugar refining. Historically known as Iwo Jima, during World War II, the island, site of a Japanese air base, was taken (February–March 1945), at great cost to U.S. and Japanese forces. A photograph of the U.S. marines raising the flag over Mount Suribachi, which they called Meatgrinder Hill, is one of the most famous images of the war. Occupied by the U.S. until 1968, when it was returned to Japan. Name changed from Iwo Jima in June 2007, restoring name used by the original inhabitants prior to their evacuation during World War II. Currently occupied only by several hundred Japanese soldiers.

Ixcamilpa (eesh-kam-MEEL-pah), town, ⊙ IXCAMILPA DE GUERRERO municipio, PUEBLA, central MEXICO, 40 mi/64 km SSW of IZÚCAR DE MATAMOROS (on the GUERRERO border); 18°00′N 98°42′W. Corn, sugarcane, livestock.

Ixcamilpa de Guerrero, MEXICO: see IXCAMILPA.

Ixcán (eesh-KAHN), region, N HUEHUETENANGO department, GUATEMALA; 15°49′N 91°04′W. NE lowlands; site of a colonization project in 1970s, which was destroyed with much loss of life during civil war of 1980s. Also known as Ixcán Grande.

Ixcán, municipio, GUATEMALA: see CANTABAL.

Ixcán Grande, GUATEMALA: see IXCÁN.

Ixcapuzalco (eesh-kah-poo-SAHL-ko), town, ⊙ PEDRO ASCENCIO ALQUISIRAS municipio, GUERRERO, SW MEXICO, on S slope of central plateau, 21 mi/34 km WSW of TAXCO DE ALARCÓN. Cereals, sugarcane, fruit.

Ixcateopan de Cuautémoc (ish-kah-te-O-pan de kwou-TE-mok), town, GUERRERO, SW MEXICO, 17 mi/27 km WSW of TAXCO DE ALARCÓN. Cereals, sugarcane, fruit, timber.

Ixcatepec (eesh-KA-te-pek), town, VERACRUZ, E MEXICO, in SIERRA MADRE ORIENTAL foothills, 40 mi/64 km NW of TÚXPAM DE RODRÍGUEZ CANO. Cereals, sugarcane, coffee, fruit.

Ixchiguán (eesh-chee-GWAHN), town, SAN MARCOS department, SW GUATEMALA, in the Sierra Madre, 16 mi/26 km NNW of SAN MARCOS; 15°12′N 91°53′W. Elevation 8,858 ft/2,700 m. Corn, wheat, fodder grasses; livestock.

Ixelles (ik-SEL), Flemish *Elsene,* commune (2006 population 77,341), industrial suburb in capital district of BRUSSELS, BRABANT province, central BELGIUM; 50°50′N 04°22′E.

Ixhuacán de los Reyes (eesh-wah-KAN de los RE-yes), town, VERACRUZ, E MEXICO, in SIERRA MADRE ORIENTAL, 17 mi/27 km SW of XALAPA ENRÍQUEZ. Corn, fruit. Formerly Ixhuacán.

Ixhuatán (eesh-wah-TAHN), town, in NW CHIAPAS, MEXICO, 37 mi/60 km NW of SAN CRISTÓBAL DE LAS

CASAS, on Mexico Highway 195; 17°17′N 93°02′W. Elevation 1,640 ft/500 m. A Zoque Indian community. Formerly ISHUATÁN.

Ixhuatán, GUATEMALA: see SANTA MARÍA IXHUATÁN.

Ixhuatlancillo (eesh-wah-tlahn-SEE-yo), town, ⊙ Ixhuatlancillo municipio, VERACRUZ, E MEXICO, in SIERRA MADRE ORIENTAL, 4 mi/6.4 km NW of ORIZABA. Coffee, sugarcane, tobacco, fruit.

Ixhuatlán del Café (eex-waht-LAHN del ka-FE), town, VERACRUZ, E MEXICO, in SIERRA MADRE ORIENTAL, 12 mi/19 km N of CÓRDOBA; 20°42′N 98°00′W. Coffee, corn, fruit. Formerly Ixhuatlán.

Ixhuatlán del Sureste (eesh-wat-LAHN del soor-ES-te), town, VERACRUZ, SE MEXICO, on Isthmus of TEHUANTEPEC, 11 mi/18 km ENE of MINATITLÁN; 20°42′N 98°00′W. Tropical fruit, livestock. Petroleum production. Sometimes Chapopotla.

Ixhuatlán de Madero (eesh-wat-LAHN de ma-DAI-ro), town, VERACRUZ, E MEXICO, in SIERRA MADRE ORIENTAL foothills, 40 mi/64 km SE of TÚXPAM DE RODRÍGUEZ CANO; 20°42′N 98°00′W. Corn, sugarcane, fruit.

Ixiamas (ee-see-AH-mahs), town and canton, ITURRALDE province, LA PAZ department, NW BOLIVIA, 60 mi/97 km NNW of San Buenaventura; 13°45′S 68°09′W. Cacao, tobacco.

Ixil (eesh-EEL), town, YUCATÁN, SE MEXICO, 16 mi/26 km NE of MÉRIDA. Henequen.

Iximché (eeks-eem-CHE), archaeological site, CHIMALTENANGO department, GUATEMALA, 1.9 mi/3 km S of TECPÁN GUATEMALA; 14°44′N 90°59′W. Post-classic site; was the principal center of the Cakchiquel Indians at the time of conquest, served as capital of Spanish colony (1523–1524).

Ixmatlahuacán (eesh-mah-tlah-wah-KAHN), town, VERACRUZ, SE MEXICO, in SOTAVENTO lowlands, 6 mi/9.7 km NNW of COSAMALOAPAN. Sugarcane, bananas.

Ixmiquilpan (eesh-mee-KEEL-pahn), city and township, HIDALGO, central MEXICO, on TULA RIVER, on INTER-AMERICAN HIGHWAY, and 40 mi/64 km NW of PACHUCA DE SOTO; 20°29′N 99°14′W. Elevation 5,577 ft/1,700 m. Cereals, maguey, livestock, native textiles. Ancient Otomi Indian capital.

Ixonia (iks-O-nee-ah), village, JEFFERSON county, S WISCONSIN, 8 mi/12.9 km SE of WATERTOWN. Dairying, wheat. Transitional urban area. Manufacturing (tool and die, metal products, feeds, furniture).

Ixopo, town, KWAZULU-NATAL province, SOUTH AFRICA, on small Ixopo River, and 40 mi/64 km SW of MSUNDUZI (PIETERMARITZBURG); 30°10′S 30°04′E. Elevation 3,773 ft/1,150 m. Railroad and road junction; livestock; fruit. In heavily wooded area 40 mi/64 km from the coast.

Ixpantepec Nieves (eesh-PAHN-te-pek nee-E-ves), town, in far NW OAXACA, MEXICO, 19 mi/30 km SW of HUAJUAPAM DE LEÓN. Elevation 7,677 ft/2,340 m. Steep terrain with cold climate. Corn, wheat, beans; mescal; local straw textiles. On unpaved road with connections to Huajuapam de León and SILACAYOAPAM.

Ixtacamaxtitlán (eesh-tah-ka-mash-teet-LAHN), town, PUEBLA, central MEXICO, 27 mi/43 km SSE of ZACATLÁN; 19°37′N 97°49′W. Corn, maguey. Silver, gold, copper deposits nearby.

Ixtaccihuatl, MEXICO: see IZTACCIHUATL.

Ixtacomitán (eesh-tah-ko-mee-TAHN), town, CHIAPAS, S MEXICO, 45 mi/72 km N of TUXTLA GUTIÉRREZ, on Mexico Highway 195; 17°26′N 93°05′W. Rice, fruit. A Zoque Indian community.

Ixtacuixtla, MEXICO: see VILLA MARIANO MATAMOROS.

Ixtacuixtla de Mariano Matamoros, MEXICO: see VILLA MARIANO MATAMOROS.

Ixtaczoquitlán (eeks-tak-zo-keet-LAHN), town, VERACRUZ, E MEXICO, inSIERRA MADRE ORIENTAL, 3 mi/4.8 km NE of ORIZABA. Elevation 3,891 ft/1,186 m. Coffee, sugarcane, fruit.

Ixtahuacán (eeks-tah-hwah-KAHN), town, HUEHUETENANGO department, W GUATEMALA, near CUILCO

RIVER, 21 mi/34 km WNW of HUEHUETENANGO; 15°25′N 91°46′W. Elevation 4,521 ft/1,500 m. Coffee, sugarcane, fruit, vegetables. Also known as San Ildefonso Ixtahuacán (SAHN EEL-de-FON-so eeks-tah-hwah-KAHN).

Ixtahuacán, GUATEMALA: see SANTA CATARINA IXTAHUACÁN.

Ixtapa (eesh-TAH-pah), town, CHIAPAS, S MEXICO, in N spur of SIERRA MADRE, 13 mi/21 km E of TUXTLA GUTIÉRREZ. Elevation 3,625 ft/1,105 m. Cereals, fruit, livestock. A Tzotzil-Maya community.

Ixtapaluca (eeks-tah-pah-LOO-kah), town (2005 population 290,076), MEXICO, central MEXICO, 19 mi/31 km SE of MEXICO CITY, and part of the ZONA METROPOLITANA DE LA CIUDAD DE MÉXICO. Cereals, maguey, livestock.

Ixtapan de la Sal (eeks-TAH-pahn de lah SAHL), town, MEXICO, central MEXICO, 32 mi/51 km S of TOLUCA DE LERDO; 18°50′N 99°41′W. Sugarcane, coffee, cereals, fruit. Thermal springs.

Ixtapan del Oro (eeks-TAH-pahn del O-ro), town, MEXICO, central MEXICO, 40 mi/64 km W of TOLUCA DE LERDO. Grain, fruit; livestock.

Ixtapangajoya (eeks-tah-pahn-gah-HO-yah), town, CHIAPAS, S MEXICO, in GULF lowland, 37 mi/60 km S of VILLAHERMOSA; 17°30′N 92°02′W. Fruit. A Zoque Indian community.

Ixtapa Point (eesh-TAH-pah), cape on PACIFIC OCEAN coast of GUERRERO, SW MEXICO, 6 mi/10 km W of Zinuatanejo; 17°40′N 101°40′W.

Ixtenco or **San Juan Ixtenco** (eeks-TEN-ko), town, TLAXCALA, central MEXICO, at E foot of MALINCHE volcano, 25 mi/40 km NE of PUEBLA. Agricultural center (corn, wheat, barley, alfalfa, beans; livestock).

Ixtepec (EEKS-te-pek), town, ⊙ Ixtepec municipio, PUEBLA, central MEXICO, in E foothills of SIERRA MADRE ORIENTAL, 25 mi/40 km SE of HUAUCHINANGO; 18°26′N 100°09′W. Sugarcane, coffee, tobacco, fruit.

Ixtlahuaca de Rayón (eesh-tlah-WAH-kah de rah-YON), town, ⊙ Ixtlahuaca municipio, MEXICO, central MEXICO, on LERMA River, and 40 mi/64 km WNW of MEXICO CITY. Cereals, fruit; livestock. Silver deposits nearby.

Ixtlahuacán (eesh-tlah-wah-KAHN), town, COLIMA, W MEXICO, on coastal plain, 16 mi/26 km S of COLIMA; 19°00′N 103°40′W. Rice, corn, sugarcane, coffee, tobacco, fruit.

Ixtlahuacán de los Membrillos (eesh-tlah-wah-KAHN de los mem-BREE-yos), town, ⊙ Ixtlahuacán de los Membrillos municipio, JALISCO, central MEXICO, near Lake CHAPALA, on railroad, and 25 mi/40 km SSE of GUADALAJARA. Wheat-growing center.

Ixtlahuacán del Río (eesh-tlah-wah-KAHN del REE-o), town, JALISCO, central MEXICO, on affluent of SANTIAGO RIVER, and 14 mi/23 km NE of GUADALAJARA; 20°50′N 103°20′W. Grain, sugarcane, fruit, livestock.

Ixtlán de Juárez (eesh-TLAHN de HWAH-res), town, OAXACA, S MEXICO, surrounded by spurs of SIERRA MADRE DEL SUR, 25 mi/40 km NE of OAXACA DE JUÁREZ, on Mexico Highway 175; 17°22′N 96°20′W. Elevation 5,577 ft/1,700 m. Corn, beans, wheat, fruits; livestock; manufacturing serapes.

Ixtlán de los Hervores (eesh-TLAHN de los er-VO-res), town, ⊙ Ixtlán municipio, MICHOACÁN, central MEXICO, on central plateau, 18 mi/29 km NW of ZAMORA DE HIDALGO, on Mexico Highway 35; 20°11′N 102°24′W. Agricultural center (cereals, fruit, vegetables; livestock). Geysers and springs nearby. Formerly Ixtlán.

Ixtlán del Río (eesh-TLAHN del REE-o), city, NAYARIT, W MEXICO, amid W outliers of SIERRA MADRE OCCIDENTAL, 50 mi/80 km SE of TEPIC, on railroad, and on Mexico Highway 15; 21°02′N 104°22′W. Elevation 3,419 ft/1,042 m. Silver, gold, lead deposits. Agricultural center (corn, beans, sugarcane, bananas); sugar refineries, tanneries. Formerly Ixtlán.

Iyama (EE-yah-mah), city, NAGANO prefecture, central HONSHU, central JAPAN, 19 mi/30 km N of NAGANO; 36°50′N 138°22′E. Asparagus. Buddhist altars (well-known street of shops); traditional papermaking. Ski areas nearby.

Iya River (ee-YAH), 370 mi/595 km long, in SW IRKUTSK oblast, RUSSIA, rises in the EASTERN SAYAN MOUNTAINS, flows NE, past TULUN, to the OKA River (tributary of the ANGARA RIVER), 35 mi/56 km SSE of BRATSK.

Iylanly (yi-lahn-LEE), city, NE DASHHOWUZ weloyat, N TURKMENISTAN, on Khiva oasis, 17 mi/27 km W of DASHHOWUZ; 41°50′N 59°39′E. Cotton. Also spelled Yylanly.

Iyo, former province in W SHIKOKU, JAPAN; now EHIME prefecture.

Iyo (EE-yo), city, EHIME prefecture, NW SHIKOKU, W JAPAN, 5.6 mi/9 km S of MATSUYAMA; 33°45′N 132°42′E. Mandarin oranges; bonito.

Iyomishima (ee-yo-MEE-shee-mah), city, EHIME prefecture, NW SHIKOKU, W JAPAN, 43 mi/70 km E of MATSUYAMA; 33°58′N 133°33′E. Paper.

Iyo Sea (EE-yo), Japanese *Iyo-nada* (ee-YO–nah-dah), SW section of INLAND SEA, W JAPAN, between SW coast of HONSHU and NW coast of SHIKOKU; merges with HOYO STRAIT (S), SUO SEA (W), and HIUCHI SEA (E). Its largest inlet is HIROSHIMA BAY. Contains numerous islands, OSHIMA (in YAMAGUCHI prefecture) being the largest.

Iza (EE-sah), town, ⊙ Iza municipio, BOYACÁ department, central COLOMBIA, in Cordillera ORIENTAL of the ANDES, near Lago TOTA, 24 mi/39 km NE of TUNJA; 05°35′N 72°58′W. Elevation 8,766 ft/2,671 m. Coffee, corn, sugarcane; livestock.

Izabal (ee-sah-BAHL), department (□ 3,489 sq mi/9,071.4 sq km), E GUATEMALA; ⊙ PUERTO BARRIOS; 15°30′N 89°00′W. On Gulf of HONDURAS of CARIBBEAN SEA (forming Bay of AMATIQUE). Includes Lake IZABAL and its outlet, Río DULCE, separated from lower MOTAGUA River valley by the Sierra del MICO. Railroad serves Motagua River valley. Hot, humid climate. Agriculture (bananas, corn, beans, coconuts); lumbering in tropical forests. Main centers: Puerto Barrios (chief port and railroad terminus), LÍVINGSTON (port and capital until 1920).

Izabal (ee-sah-BAHL), village, IZABAL department, E GUATEMALA, port on S shore of Lake IZABAL, 40 mi/64 km WSW of PUERTO BARRIOS; 15°24′N 89°08′W. Grain, bananas, coconuts. Once a thriving port on water route between sea and highlands; declined after construction of GUATEMALA city–Puerto Barrios railroad. Subjected to pirate raids in 17th century.

Izabal (ee-sah-BAHL), lake (□ 228 sq mi/592.8 sq km), c.30 mi/48 km long and 15 mi/24 km wide, E GUATEMALA, largest lake in the country; 15°30′N 89°10′W. Known also as the Golfo Dulce, it drains to the CARIBBEAN SEA through El Golfete, a small adjacent lake, and the Río DULCE, a broad river. In Spanish colonial times, Lake Izabal carried lively trading between the seacoast and the highlands, and the small town of IZABAL on its S shore was a thriving port, constantly subjected to raids in the 17th century by English and Dutch buccaneers. Today shipping is negligible, although LÍVINGSTON, at the mouth of the Río Dulce, is of some importance. Nearby are many pre-Columbian ruins.

Izad Khrast (EE-zahd KRAST), town, Fārs province, central IRAN, 85 mi/137 km SE of ESFAHAN, along road between Esfahan and SHIRAZ; 31°31′N 52°08′E. Center of agricultural area; grain, fruit, cotton; rugmaking. Sometimes called Samirum.

Izalco (ee-ZAHL-ko), city and municipality, SONSONATE department, W EL SALVADOR, at SW foot of IZALCO volcano, 5 mi/8 km NE of SONSONATE; 13°45′N 89°40′W. Pottery and basket making; grain, fruit, sugarcane; livestock. Limestone quarries. City formed 1869 from merging adjacent Ladino town of DOLORES

and Indian settlement of Asunción. Sometimes spelled ITZALCO.

Izalco, volcano (7,828 ft/2,386 m), W EL SALVADOR; 13°43′N 89°40′W. It was sometimes called the Lighthouse of the Pacific because its eruption was visible to 19th-century navigators off the coast. Although it has not erupted since 1926, the volcano is still considered active. Terrain is lava covered and slippery. Two-hour walk to top and 1-hour walk around crater.

Izamal (ee-SAH-mal), city and township, YUCATÁN, SE MEXICO, 39 mi/63 km E of MÉRIDA; 20°56′N 89°01′W. On railroad. Agricultural center (henequen, sugarcane, corn). Site of ancient Maya town (believed to be older than CHICHÉN ITZÁ), an aboriginal pilgrimage site. Many religious remains, pyramids, mausoleum. Monastery and cathedral were erected 1553 on site of Maya temples.

Izard (IZ-uhrd), county (□ 584 sq mi/1,518.4 sq km; 2006 population 13,356), N ARKANSAS; ⊙ MELBOURNE; 36°05′N 91°54′W. Bounded SW by WHITE RIVER; drained by STRAWBERRY RIVER. Agriculture (cattle, hogs, chickens; hay; dairying); lumber milling, cotton ginning; glass sand, gravel pits. North Central Correctional Unit in W. Formed 1825.

Izberbash (eez-byer-BAHSH), city (2005 population 43,110), E DAGESTAN republic, N CAUCASUS, SE European Russia, on the Caspian coastal railroad, 35 mi/56 km SSE of MAKHACHKALA; 42°33′N 47°51′E. Below sea level. In an oil district. Manufacturing (equipment for electricity-generating stations). Arose in 1932 as an oil town, with the first major pipeline laid down in 1937. Developed mainly after World War II. Made city in 1949. Significant population increase largely due to influx of refugees from the nearby CHECHEN REPUBLIC.

Izbushechnaya, RUSSIA: see KHREBTOVAYA.

Izdebnyk, UKRAINE: see STEBNYK.

Izdeshkovo (eez-DYESH-kuh-vuh), town, central SMOLENSK oblast, W European Russia, on road and near railroad, 25 mi/40 km W of VYAZ'MA; 55°08′N 33°35′E. Elevation 685 ft/208 m. Limestone works; flax retting.

Izdihar Dam, TUNISIA: see SIDI ABDELLI.

Izeda (ee-ZAI-dah), village, BRAGANÇA district, N PORTUGAL, 17 mi/27 km S of BRAGANÇA. Wheat, rye, almonds.

Izegem (I-zuh-khem), commune (2006 population 26,526), Roeselare district, WEST FLANDERS province, W central BELGIUM, 5 mi/8 km ESE of ROESELARE; 50°55′N 03°12′E.

Izeh (EE-ze), town (2006 population 104,364), Khuzestān province, Iran, 80 mi/129 km NE of Ahvaz, on mtn. plain near upper Karun River. Center of Bakhtiari region; 31°50′N 49°52′E. Sheep raising; rugmaking. Situated on site of Sassanian city of Izeh (ruins, inscriptions), it was later known as Malamir, before regaining the ancient name.

Izena (ee-ZE-nah), village, on IZENA-SHIMA island, IHEYA-SHOTO island group, Shimajiri county, Okinawa prefecture, SW JAPAN, 50 mi/80 km N of NAHA; 26°55′N 127°56′E.

Izena-shima, JAPAN: see IHEYA-SHOTO.

Izera, CZECH REPUBLIC: see JIZERA.

Izerskie, Góry, POLAND and CZECH REPUBLIC: see JIZERA MOUNTAINS.

Izgrev (EEZ-graiv), village, SOFIA oblast, BLAGOEVGRAD obshtina, BULGARIA; 41°59′N 23°06′E.

Izgrev (EEZ-graiv), village, LOVECH oblast, LEVSKI obshtina, BULGARIA; 43°27′N 25°04′E.

Izhevsk (ee-ZHEFSK), city (2006 population 630,755), ⊙ UDMURT REPUBLIC, E European Russia, on the IZH RIVER, 702 mi/1,130 km E of MOSCOW; 56°51′N 53°14′E. Elevation 433 ft/131 m. Railroad junction. A major steel-milling, armament, machinery, and metallurgical center; also timber factory and furniture making. Has higher education institutes, several theaters, and a philharmonics. Founded in 1760 as Izhevskiy Zavod, producing iron bands and anchors for ships, expanding into weapons and tool production in the early 19th century. Made city in 1918. Briefly called Ustinov (1984–1987).

Izhevskiy Zavod, RUSSIA: see IZHEVSK.

Izhevskoye (ee-ZHEF-skuh-ye), village, central RYAZAN oblast, central European Russia, in the OKA River valley, on road, 40 mi/64 km E of RYAZAN; 54°33′N 40°52′E. Elevation 341 ft/103 m. In agricultural area (rye, oats, flax).

Izhma (EEZH-mah), village (2005 population 3,540), N central KOMI REPUBLIC, NE European Russia, on the IZHMA RIVER (landing), on road, 100 mi/161 km N of UKHTA; 65°04′N 53°55′E. Elevation 124 ft/37 m. Reindeer raising.

Izhma, RUSSIA: see SOSNOGORSK.

Izhora River (ee-ZHO-rah), 50 mi/80 km long, in LENINGRAD oblast, NW European Russia, rises W of GATCHINA, flows E and N, past KOLPINO (shipyards), to the NEVA RIVER at UST′-IZHORA.

Izh River (EEZH), 132 mi/212 km long, in E European Russia, rises in UDMURT REPUBLIC E of YAKSHUR-BODYA, flows generally S, past IZHEVSK, to the KAMA River E of BONDYUZHSKIY (TATARSTAN). Timber floating.

Iž Island (EEZH), Italian *Eso* (E-so), Dalmatian island, in ADRIATIC SEA, S CROATIA; 7 mi/11.3 km long, 1 mi/1.6 km wide; 1991 population 657. Chief village, Iž Veliki (Italian *Eso Grande*), is 7 mi/11.3 km WSW of ZADAR.

Izki (IZ-kee), township (2003 population 35,173), OMAN, 70 mi/113 km SW of MUSCAT, at S foot of the JABAL AKHDAR; 22°56′N 57°46′E. Center controlling Wadi Sama'il route from coast to interior of Arabian Peninsula. Sometimes spelled Azki, Ziki, and Zikki.

Izluchinsk (eez-LOO-cheensk), industrial settlement (2005 population 15,920), S central KHANTY-MANSI AUTONOMOUS OKRUG, central SIBERIA, RUSSIA, in the OB′ RIVER basin, on local road and near railroad, 16 mi/26 km ENE of NIZHNEVARTOVSK; 61°02′N 76°51′E. Power plant and industries supporting it (machinery and equipment manufacturing and repair).

Izmail, UKRAINE: see IZMAYIL, city.

Izmail oblast, UKRAINE: see ODESSA, oblast.

Izmalkovo (eez-MAHL-kuh-vuh), village (2006 population 4,060), W LIPETSK oblast, S central European Russia, on road and railroad, 21 mi/34 km WNW of YELETS; 52°41′N 37°58′E. Elevation 764 ft/232 m. In agricultural area (wheat, rye, oats); creamery.

Izmayil (eez-mah-YEEL) (Russian *Izmail*), city (2001 population 84,815), SW ODESSA oblast, UKRAINE, on an arm of the DANUBE delta and near the Romanian border; 45°21′N 28°50′E. Elevation 104 ft/31 m. Raion center. Railroad junction, river port, commercial center, and the naval base of the Danube fleet. Orchards and vineyards surround the city. Industries include food and fish processing, winemaking, and the manufacture of bricks and tiles. Settlement dates to 1st–2nd centuries C.E., known as Smil; from 1484 to 1812 and from 1856 to 1877 it belonged to the Turks, who gave it its current name and made it a Turkish fortress and capital of a Turkish sanjak. It was attacked and sacked by the Zaporozhian Cossacks in the 17th century. Russian forces took the city twice (1770, 1790) during the Russo-Turkish wars of Catherine II. Recaptured by the Russians in 1809, it was ceded to them by the Treaty of Bucharest (1812). At the Congress of Paris in 1856, Izmail was returned to TURKEY; but RUSSIA seized the city again in 1878 and held it until 1918, when Romania took it. Transferred to the USSR in 1940, it was reconquered by the Romanians the following year but restored to the USSR in 1947. Remains of the old Turkish fortress have been preserved. Small Jewish community since the 16th century (1,680 in 1939), one of the very few that was mostly evacuated and/or drafted into the Soviet Army rather than exterminated by the Nazis during World War II; the bulk of it returned following the end of the war—close to 1,000 Jews in Izmayil in 2005.

Izmayil oblast, UKRAINE: see ODESSA, oblast.

Izmaylovo (eez-MEI-luh-vuh), town (2006 population 3,065), central ULYANOVSK oblast, E central European Russia, on road, 7 mi/11 km NE of BARYSH; 53°43′N 47°14′E. Elevation 754 ft/229 m. Textile industry since 1845.

Izmaylovo (eez-MEI-luh-vuh), W suburb of MOSCOW, central European Russia, less than 6 mi/10 km WNW of the city center; 55°48′N 37°46′E. Elevation 495 ft/150 m. Experimental plant for the Russian academy of agricultural sciences. Was the 17th century summer residence of Russian tsars; has a 17th century cathedral and castle towers.

Izmir (iz-MEER), city (2000 population 2,232,265), W TURKEY, on the Gulf of IZMIR, an arm of the AEGEAN SEA; 38°25′N 27°10′E. The second-largest Turkish seaport (after ISTANBUL) and the country's third-largest city, its exports include cotton, tobacco, dried and fresh fruits and vegetables, carpets, chrome ore, and olive oil. An important commercial and industrial center; manufacturing includes processed food, textiles, tobacco, cement, petrochemicals. Tourism is increasingly important. The Birince Kordon promenade stretches the length of the city up to the Alsancak Ferry Terminal, and there are many cafés along the waterfront. It is a road and railroad transportation center, and an annual trade fair is held here. International airport to S. The city was settled during the Bronze Age (c.3000 B.C.E.). It was colonized (c.1000 B.C.E.) by Ionians and was destroyed (627 B.C.E.) by the Lydians. It was rebuilt on a different site in the early 4th century B.C.E. by Antigonus I, was enlarged and beautified by Lysimachus, and became one of the largest and most prosperous cities of ASIA MINOR. The city had a sizable Jewish colony, was an early center of Christianity, and was one of the Seven Churches in Asia (Rev. 2:8). It fell to the Seljuk Turks in the 11th century, was recaptured for BYZANTIUM during the First Crusade, and formed part of the empire of Nicaea from 1204 to 1261, when the Byzantine Empire was restored. Also in 1261 the Genoese obtained trading privileges here, which they retained until the city fell (c.1329) to the Seljuk Turks. The Knights Hospitalers captured the city in 1344, restored Genoese privileges, and held the city until 1402, when it was captured and sacked by Tamerlane. The Mongols were succeeded in 1424 by the Ottoman Turks. A Greek Orthodox archiepiscopal see, the city retained a large Greek population and remained a center of Greek culture and the chief MEDITERRANEAN port of Asia Minor. After the collapse of the Ottoman Empire in World War I, the city was occupied (1919) by Greek forces. Izmir fell to the Turks in September 1922, and a few days later was destroyed by fire. Thousands of Greek civilian refugees fled from the city. The Treaty of Lausanne (1923) restored Izmir to Turkey. A separate convention between Greece and Turkey provided for the exchange of their minorities, which was carried out under League of Nations supervision. Thus the population of Izmir became predominantly Turkish. The city suffered greatly from severe earthquakes in 1928 and 1939. It is a NATO command center for SE Europe and a base for the Turkish fleet. It is also the site of the Aegean University and several museum, and was possibly the birthplace of the poet Homer. Formerly known as Smyrna.

Izmir, Gulf of (iz-MEER), inlet of AEGEAN SEA in W TURKEY; 35 mi/56 km long, 14 mi/23 km wide. IZMIR is at its head, island of LESBOS opposite its mouth.

Izmit (iz-MEET), city (2000 population 195,699), NW TURKEY, a port on the Bay of Izmit, at the E end of the Sea of MARMARA; 40°47′N 29°55′E. A port, it is the center of a rich tobacco- and olive-growing region. Manufacturing includes paper, petrochemicals, textiles, beer, and cement. Founded c.712 B.C.E., the city became famous after Nicomedus I of BITHYNIA rebuilt it in 264 B.C.E. as his capital, NICOMEDIA. In 1999, Izmit was shattered by an earthquake that killed more than 17,000 people in NW Turkey.

Izmit, Gulf of (iz-MEET), long narrow inlet at E end of Sea of MARMARA, NW TURKEY, 25 mi/40 km SE of ISTANBUL; 45 mi/72 km long. Town of IZMIT at E end, GÖLCÜK on S shore.

Iznájar (eeth-NAH-hahr), town, CÓRDOBA province, S SPAIN, on GENIL RIVER, and 15 mi/24 km SE of LUCENA; 37°15′N 04°18′W. Olive-oil processing, flour milling. Gypsum quarries nearby.

Iznalloz (eeth-nah-LYOTH), town, GRANADA province, S SPAIN, 16 mi/26 km NNE of GRANADA; 37°23′N 03°31′W. Brandy distilling, flour milling. Olive oil, sugar beets, cereals, vegetables; livestock; lumber.

Iznate (eeth-NAH-tai), town, MÁLAGA province, S SPAIN, 14 mi/23 km ENE of MÁLAGA; 36°47′N 04°11′W. Grapes, raisins, olives and olive oil, cereals.

Iznatoraf (eeth-nah-to-RAHF), town, JAÉN province, S SPAIN, 4 mi/6.4 km NE of VILLACARRILLO; 38°09′N 03°02′W. Olive-oil processing; cereals, vegetables. Gypsum quarries. Has remains of ancient walls.

Iznik, city (2000 population 20,169), N TURKEY; 40°27′N 29°43′E. The ancient city of Nicaea, N ASIA MINOR, it was built in the 4th century B.C.E. by Antigonus I as Antigonia and renamed Nicaea by Lysimachus for his wife. It flourished under the Romans and was the scene of the ecumenical council called in C.E. 325 by Constantine I. Another council held in 787 sanctioned the devotional use of images. The city, captured by the Turks in 1078 and by the Crusaders in 1097, passed finally to the Turks in 1330. It is sometimes called Nice.

Iznik, Lake, ancient *Ascanius*, lake (□ 80 sq mi/208 sq km), NW TURKEY, 23 mi/37 km NE of BURSA; 20 mi/32 km long, 7 mi/11.3 km wide; elevation 260 ft/79 m.

Iznoski (eez-NOS-kee), village (2005 population 1,900), N KALUGA oblast, central European Russia, on road and railroad, 45 mi/72 km NW of KALUGA; 54°59′N 35°18′E. Elevation 679 ft/206 m. Peat works.

Izoard, Col d' (dee-zo-ahr, kol), high pass (7,700 ft/2,347 m) of the DAUPHINÉ ALPS, in HAUTES-ALPES department, SE FRANCE, on BRIANÇON-GUILLESTRE road known as the *Route des Grandes Alpes*, 7 mi/11.3 km SE of Briançon. Just E is the Pic de ROCHEBRUNE (10,700 ft/3,261 m). Numerous switchbacks on S slope. The Château-Queyras fortress is in the valley, 5 mi/8 km S.

Izobil'no-Tishchenskiy, RUSSIA: see IZOBIL'NYY.

Izobil'noye, RUSSIA: see IZOBIL'NYY.

Izobil'nyy (ee-zuh-BEEL-niyee), city (2006 population 40,545), NW STAVROPOL TERRITORY, N CAUCASUS, S European Russia, on crossroads and railroad, and 40 mi/64 km NNW of STAVROPOL; 45°22′N 41°42′E. Elevation 656 ft/199 m. In agricultural area; food industries (canning, sugar, meat packing, flour milling). Until the mid-1930s, called Izobil'no-Tishchenskiy, later Izobil'noye. City status and current name since 1965.

Izola, town (2002 population 10,152), SE SLOVENIA, fishing port on Gulf of TRIESTE, in KOPRSKO PRIMORJE (Koper Riviera), 9 mi/14.5 km SW of TRIESTE (ITALY), on small island linked to mainland; 45°32′N 13°40′E. Fish and vegetable canning, wine making, boatbuilding; tourism. Has 15th-century cathedral. Part of Italy (1919–1947), then passed to the former YUGOSLAVIA.

Izoplit (ee-zuh-PLEET), town (2006 population 4,255), SE TVER oblast, W central European Russia, near the VOLGA RIVER, on railroad spur, 19 mi/31 km SE of TVER; 56°38′N 36°12′E. Elevation 423 ft/128 m. Manufacturing of mineral cotton wool.

Izounar (EE-zoo-nahr), lake, MOROCCO, in central High ATLAS mountains, S of Jbel AZOURKI, 47 mi/75 km S of BENI MELLAL.

Izozog (ee-ZO-zog), town and canton, CORDILLERA province, SANTA CRUZ department, E central BOLIVIA, on the SANTA CRUZ–Abapo road (which is only paved to here), 13 mi/20 km W of the PARAPETI RIVER, and 13 mi/20 km S of Fortín Ortíz; 19°24′S 62°45′W. Elevation 3,950 ft/1,204 m. Potatoes, yucca, bananas, corn, cotton, peanuts, coffee; cattle. There is a secondary airstrip here and an airport at Fortín Ortíz. Gas deposits (unexploited) located to W.

Izozog, Bañados de (bah-NYAH-dos dai ee-so-SOK), marshy area in SANTA CRUZ department, SE BOLIVIA, 80 mi/129 km NE of CHARAGUA; 80 mi/129 km long, c. 30 mi/48 km wide; 18°48′S 62°10′W. Receives the PARAPETÍ RIVER; has several outlets, one of which flows to Lake CONCEPCIÓN (N). Also called Bañados del Parapetí.

Izra' (iz-RAH), ancient *Zoroa*, township, Der'a district, SW SYRIA, on railroad and highway from DAMASCUS to the Jordanian border, 19 mi/31 km NNE of Der'a; 32°50′N 36°14′E. Cereals, wheat. Has ruins dating from Roman era. Also called EZREA.

Izsák (E-zhahk), village, PEST county, central HUNGARY, 17 mi/27 km SW of KECSKEMÉT. Grain, tobacco, cherries, apricots; cattle. Minor road hub.

Iztacalco (eez-ta-KAHL-ko), city and delegación, Federal Distrito, S central MEXICO; 19°23′N 99°07′W. It is an industrial center, now a part of greater MEXICO CITY, SE of city center. Several historic landmarks have been preserved. La Calzada de la Viga built on site occupied by ancient National Canal de la Viga. It was a trade route to Mexico City for cargoes of flowers, vegetables, and fruits.

Iztaccihuatl or **Ixtaccihuatl** (eez-tah-SEE-watl) (Aztec=white woman), dormant volcano (17,342 ft/5,286 m), central MEXICO, on the border between PUEBLA and MEXICO states; 19°11′N 98°38′W. Irregular in outline, and snow-capped, it is also popularly known as the Sleeping Woman. Also Ixtacihuatl.

Iztaccihuatl-Popocatépetl (eez-tah-SEE-watl–po-po-kah-TE-petl), national park (□ 100 sq mi/260 sq km), in SE MEXICO, MEXICO. These are snow covered twin volcanoes, Popocatépetl (17,890 ft/5,453 m), called "smoking mountain" and Iztaccihuatl (17,346 ft/5,287 m), called "sleeping woman." The peak of Popocatépetl, 55 mi/88 km SE of MEXICO CITY, lies on the borders of the states of Mexico, MORELOS, and PUEBLA. It is the second highest mountain in Mexico after PICO DE ORIZABA. A saddle valley separates the two volcanoes, which can be reached by a paved road branching off Mexico Route 115, 1 mi/1.6 km S of the town of AMECAMECA. Cold climate. Rough terrain. It is a two-day climb to the summit of these volcanoes. Also known as Izta-Popo.

Iztapa (ees-TAH-pah), town, ESCUINTLA department, S GUATEMALA, on the PACIFIC OCEAN, at mouth of MICHATOYA RIVER, and 7 mi/11.3 km E of SAN JOSÉ; 13°56′N 90°43′W. Tropical resort. During the Spanish colonial era a ship-building center and minor port. Currently has no operational harbor. Sometimes called Puerto de Iztapa (PWER-to de ees-TAH-pah).

Iztapalapa (ees-tah-pah-LAH-pah), delegación, Federal Distrito, S central MEXICO. Part of greater MEXICO CITY, SE of city center. It is a commercial and industrial center; the city's main wholesale market is here. Founded on the site of an important pre-Columbian city.

Iztochno Shivachevo, Bulgaria: see SHIVACHEVO.

Izu, former province in central HONSHU, JAPAN; now part of SHIZUOKA prefecture. Also Idzu.

Izúcar de Matamoros (ee-SOO-kar de ma-tah-MO-ros), city and township, PUEBLA, central MEXICO, on railroad, on INTER-AMERICAN HIGHWAY (Mexico Highway 190), and 35 mi/56 km SW of PUEBLA; 18°38′N 98°30′W. Center for growing and refining sugar; rice, fruit; livestock. Site of battle in revolution against SPAIN. Sometimes Matamoros.

Izuhara (ee-ZOO-hah-rah), town, on SE coast of TSUSHIMA island, Shimoagata county, NAGASAKI prefecture, SW JAPAN, on Tsushima Strait, 109 mi/175 km N of NAGASAKI; 34°11′N 129°17′E. Chief town of island Inkstones. Shiitake mushrooms; yellowtail. Ruins of 16th century castle built by Hideyoshi Toyotomi. Sometimes spelled Idzuhara.

Izu Islands, JAPAN: see IZU-SHICHITO.

Izuka (EEZ-kah), city, FUKUOKA prefecture, N KYUSHU, SW JAPAN, on the Onga River, 19 mi/30 km E of FUKUOKA; 33°38′N 130°41′E. Formerly an important coal-mining center.

Izumi (EE-zoo-mee), city, KAGOSHIMA prefecture, W KYUSHU, SW JAPAN, 34 mi/55 km N of KAGOSHIMA; 32°05′N 130°21′E. Electronic equipment.

Izumi, city (2005 population 177,856), OSAKA prefecture, S HONSHU, W central JAPAN, 16 mi/25 km S of OSAKA; 34°28′N 135°25′E. Residential and commercial suburb of Osaka, with numerous textile mills. Artificial pearls, glass products.

Izumi (EE-zoo-mee), village, Ono county, FUKUI prefecture, central HONSHU, W central JAPAN, 28 mi/45 km S of FUKUI; 35°54′N 136°40′E.

Izumi (ee-ZOO-mee), village, Yatsushiro county, KUMAMOTO prefecture, W KYUSHU, SW JAPAN, 16 mi/25 km N of KUMAMOTO; 32°32′N 130°48′E. Tea, miso-pickled tofu.

Izumiotsu (ee-ZOO-mee-OTS), city, OSAKA prefecture, S HONSHU, W central JAPAN, on OSAKA BAY, 16 mi/25 km S of OSAKA; 34°30′N 135°24′E. Commercial port. Manufacturing (blankets).

Izumisano (ee-ZOO-mee-SAH-no), city, OSAKA prefecture, S HONSHU, W central JAPAN, on OSAKA BAY, 22 mi/35 km S of OSAKA; 34°24′N 135°19′E. Commercial port. Manufacturing (towels). Open-air market for fish and vegetables (onions). Gateway to Kanku (Kansai International Airport).

Izumizaki (ee-ZOO-mee-ZAH-kee), village, W. Shirakawa county, FUKUSHIMA prefecture, N central HONSHU, NE JAPAN, 40 mi/65 km S of FUKUSHIMA city; 37°09′N 140°17′E.

Izumo, former province in SW HONSHU, JAPAN; now part of SHIMANE prefecture. Also Idzumo.

Izumo (EEZ-mo), city, SHIMANE prefecture, SW HONSHU, W JAPAN, on the Hii River, 19 mi/30 km S of MATSUE; 35°21′N 132°45′E. An important railroad and road hub. Livestock-raising center.

Izumozaki (ee-ZOO-mo-ZAH-kee), town, Santo county, NIIGATA prefecture, central HONSHU, N central JAPAN, on SEA OF JAPAN, 31 mi/50 km S of NIIGATA; 37°31′N 138°42′E. Japanese plums, shiitake mushrooms; fish; milk. Balloon manufacturing Japanese petroleum industry founded here. *Okesa* folk songs and dances and *ryokan* -style traditional inns originated here.

Izumrud (ee-zoom-ROOT) [Russian=emerald], town (2006 population 1,300), S SVERDLOVSK oblast, E URALS, W Siberian Russia, on a left tributary of the PYSHMA RIVER, on road, 6 mi/10 km NW of ASBEST; 57°04′N 61°24′E. Elevation 711 ft/216 m. Noted emerald-mining center since the 1830s.

Izunagaoka (ee-ZOO-nah-gah-O-kah), town, Tagata county, SHIZUOKA prefecture, central HONSHU, E central JAPAN, on NW IZU PENINSULA, 31 mi/50 km E of SHIZUOKA; 35°01′N 138°55′E. Strawberries. Hot springs.

Izu Peninsula (ee-ZOO), Japanese *Izu-hanto* (ee-ZOO–HAHN-to), SHIZUOKA prefecture, central HONSHU, central JAPAN, between SURUGA BAY (W) and SAGAMI BAY (E); 40 mi/64 km long, 10 mi/16 km–20 mi/32 km wide. Mountainous, known for hot springs, seaside resorts. ATAMI is chief town.

Izushi (EEZ-shee), town, Izushi district, HYOGO prefecture, S HONSHU, W central JAPAN, 58 mi/94 km N of KOBE; 35°27′N 134°52′E. Pottery; noodles.

Izu-shichito (ee-ZOO–SHEE-chee-to) [=seven Izu islands], island group, extending c.300 mi/483 km S of TOKYO BAY, Tokyo prefecture, SE JAPAN. O-SHIMA is the largest of these volcanic islands, which are now tourist attractions. Known for their production of camellia oil. Formerly used for penal settlements.

Izvarino, UKRAINE: see IZVARYNE.

Izvarinskiy Rudnik, UKRAINE: see IZVARYNE.

Izvaryne (eez-VAH-ri-ne) (Russian *Izvarino*), town, SE LUHANS'K oblast, UKRAINE, in the DONBAS, 5 mi/8 km E of KRASNODON; 48°19′N 39°52′E. Elevation 554 ft/168 m. Coal mines. Formerly Izvaryns'kyy Rudnyk (Russian *Izvarinskiy Rudnik*).

Izvaryns'kyy Rudnyk, UKRAINE: see IZVARYNE.

Izvestiya TsIK Islands (eez-VYES-tee-yuh–TSIK), Russian *Ostrova Izvestiy Ts.I.K.* (short for Tsentral'nyy Ispolnitel'nyy Komitet [Central Executive Committee]), in the KARA SEA of the ARCTIC OCEAN, 100 mi/161 km off NW TAYMYR PENINSULA, in KRASNOYARSK TERRITORY, Russia; 75°50′N 82°30′E.

Izvestkovyy (eez-vyest-KO-viyee), town, NW JEWISH AUTONOMOUS OBLAST, RUSSIAN FAR EAST, on highway and the TRANS-SIBERIAN RAILROAD, 125 mi/201 km W of BIROBIDZHAN; 48°59′N 131°33′E. Elevation 839 ft/255 m. Junction for railroad north to Chegdomymo (to coal fields of the BUREYA Range) and the BAYKAL-AMUR MAINLINE. Railroad enterprises; limestone works.

Izvestkovyy (eez-vyest-KO-viyee), rural settlement (2005 population 2,135), N central JEWISH AUTONOMOUS OBLAST, S RUSSIAN FAR EAST, on highway and the TRANS-SIBERIAN RAILROAD, 19 mi/30 km E of OBLUCHYE, and 6 mi/10 km W of BIRAKAN; 48°59′N 131°33′E. Elevation 839 ft/255 m. In agricultural area (grain, soybeans). Clay quarries in the vicinity.

Izvor (eez-VOR), village, BURGAS oblast, SOZOPOL obshtina, BULGARIA; 42°22′N 27°27′E.

Izvorets (eez-VOR-ets), peak (9,065 ft/2,763 m), in the PIRIN MOUNTAINS, SW BULGARIA; 41°34′N 24°40′E.

Izvorul Muntelui-Bicaz (eez-VO-rool MOON-te-looi–bee-KAHZ), reservoir for BICAZ power station, NEAMŢ county, ROMANIA, just N of Bicaz; 47°03′N 26°00′E. Also called Bicaz Reservoir.

Izyaslav (eez-yahs-LAHF), city, N KHMELNYTS'KYY oblast, UKRAINE, on HORYN' River, and 11 mi/18 km WSW of SHEPETIVKA; 50°07′N 26°48′E. Elevation 715 ft/217 m. Raion center. Food processing, food industries, manufacturing (machines, furniture). Founded in 987 as Izyaslavl' in Kievan Rus', it became part of the Grand Duchy of Lithuania in the 14th century, and Poland in the 16th century, when it was renamed Zaslav (Polish *Zasław*); passed to Russia in 1793; renamed Zaslav (Ukrainian *Izaslav* (1910); part of Ukrainian state (1918–1920); passed to USSR (Ukrainian SSR), 1920–1991. In the 1930s, it was called Zaslav.

Izyayu (eez-YAH-yoo), settlement (2005 population 1,605), E central KOMI REPUBLIC, NE European Russia, on the W bank of the PECHORA River, 5 mi/8 km W, and under administrative jurisdiction of PECHORA; 65°06′N 56°59′E. Elevation 229 ft/69 m. Geological survey station in the vicinity.

Izyum (eez-YOOM), city (2001 population 56,114), E KHARKIV oblast, UKRAINE, on the DONETS River, and 70 mi/113 km SE of KHARKIV; 49°13′N 37°15′E. Elevation 223 ft/67 m. Raion center. Metalworking center; railroad shops; manufacturing (optical equipment, building materials, ceramics, furniture), sawmilling, food processing (dairy, beer, flour, food flavoring). Founded in 1639 by the Ukrainian Cossacks; fortified by Colonel Donets' against Tatar raids in 1681. Regimental capital of Slobidska Ukraina (1685–1765).

Izzan, ADEN: see AZZAN.

J

Ja'alan, OMAN: see JA'LAN.

Jääski, RUSSIA: see LESOGORSKIY.

Jaba', Arab township, JENIN district, NE of JERUSALEM, in the E Judaean Highlands, West Bank; 31°51′N 35°16′E. Agriculture (wheat, olives, vineyards). It is believed to be the settlement Geva' or Geva' Benjamin, mentioned in the Bible several times. An Arab suburb of Jerusalem now that the municipal boundaries have expanded. Contains village of Jaba' (1995 population 1,300).

Jabal-Abyad (JE-bel–ahb-YAHD), village, BAJAH province, N central TUNISIA, in the Majardah Mountains, 18 mi/29 km NNW of BAJAH. Iron, lead, and zinc mining.

Jabal Al-Barkal, archaeological site, Northern state, N SUDAN, on the right bank of the NILE RIVER W of KARIMA. Located near NAPATA, the old capital of the Cush Kingdom (or Meroetic Kingdom), it is the site of the tombs and temples of the Meroetic kings. The Cush Kingdom began c.2000 B.C.E. and extended into the 4th century C.E. The Cushites ruled over Egypt as the 25th dynasty.

Jabal Ali Free Zone, DUBAI emirate, UNITED ARAB EMIRATES, 35 mi/56 km S of city of Dubai. Established in 1985 to attract foreign companies. Served by Jabal Ali port. Offers incentives to international corporations.

Jabal Ash-Sha'nabi, TUNISIA: see ASH-SHA'NABI, JABAL.

Jabal as-Sardj (JE-bel e–SAHRJ), mountain (4,452 ft/1,357 m.), SILYANAH province, central TUNISIA, 20 mi/32 km ENE of MAKTHAR.

Jabal Awliya (JE-bel OU-li-yuh), Arabic *Jabal al Awliya,* township, KHARTOUM state, N central SUDAN, on the right bank of the WHITE NILE River, 15 mi/24 km S of KHARTOUM; 15°14′N 32°30′E. Nearby on the WHITE NILE, a large dam (completed in 1937) is used to control the flow of the NILE and that helps the Aswan Dam regulate the storage of water.

Jabal, Bahr Al, SUDAN: see BAHR AL-GABAL.

Jabal Bargū, TUNISIA: see BARGŪ, JABAL.

Jabal Dahar (JE-bel dah-HAHR), mountain range, QABIS and MADANIYINA provinces, SE TUNISIA. Runs SE from MATMATAH to FUM TATAWINAH; separates AL JIFARAH coastal plain from Saharan region of S TUNISIA. There are numerous Berber villages and troglodyte dwellings in the MATMATAH area of the mountains.

Jabal Diss, TUNISIA: see DISS, JABAL.

Jabal-Hallūf (JE-bel–hah-LOOF), village, BAJAH province, NW TUNISIA, 12 mi/19 km W of BAJAH; lead mining and smelting.

Jabalia, Arab town, 1.5 mi/2.5 km NE of GAZA, in the GAZA STRIP; 31°32′N 34°29′E. It was a large village (1945 population 2,500), and has grown nearly 10-fold since, mainly due to the influx of refugees in 1948. The refugee population of the adjacent Jabalia refugee camp (1995 population 75,500) is the dominant factor in the life of the town, and with it forms a continuous urban area.

Jabalí Island (hah-bah-LEE), (8 mi/12.9 km long, 3 mi/4.8 km wide), off ATLANTIC coast, SW BUENOS AIRES province, ARGENTINA, S of SAN BLAS BAY, 45 mi/72 km ENE of CARMEN DE PATAGONES.

Jabal-Jallūd (JE-bel–je-LOOD), SE industrial suburb of TUNIS, TUNIS province, N TUNISIA, on S shore of LAKE OF TUNIS; sulphur refinery, cement and metalworks, flour mills. Railroad yards.

Jabal Kassala (kah-SAH-lah), mountain (4,415 ft/1,346 m), just NE of KASALA, E SUDAN, near ERITREA border; 15°28′N 36°24′E.

Jabal Matmatah (JE-bel met-me-TAH), mountain range (average elevation 2,000 ft/610 m), extending c.70 mi/113 km NW-SE from Matmata (N). Barren upland inhabited by nomadic Berbers. During World War II, it formed SW anchor of the Marith Line.

Jabal Nafusah (JE-bel nah-FOO-sah), hilly desert plateau (1,500 ft/457 m–2,380 ft/725 m) in W TRIPOLITANIA region, NW LIBYA, between JIFARAH plain (N) and HAMMADA AL HAMRA desert (S); extends from NALUT (W), near Tunisian border, to AL QASABAT (NE), near MEDITERRANEAN SEA. Rainfall ranges from 2 in/5.1 cm (W) to 16 in/41 cm (center and NE); occurs mostly in winter. Agriculture (grain, esparto grass, olives, fruit; tobaaco at TIGRINNA; sheep, goats); quartz deposits. Contains towns of GHARYAN, TARHUNAH, YAFRAN, and JADU. Road runs almost entire length; connects with Tripoli-Mizdah (N-S) road.

Jabalón River, SPAIN: see JAVALÓN RIVER.

Jabalpur (juh-BAHL-poor), district (□ 3,923 sq mi/10,199.8 sq km; 2001 population 2,151,203), E central MADHYA PRADESH state, INDIA; ⊙ JABALPUR. Bordered N by KATNI district, SW by NARSINGHPUR district, S by Seori district, E by SHAHDOL district; S escarpment of central VINDHYA RANGE with DAMOH district in NW, spurs of SATPURA RANGE in MANDLA, DINDORI districts in SE; drained by NARMADA RIVER and its tributaries (S) and by tributaries of the SON RIVER (N). Flour, rice, and dal milling, sawmilling, chemical manufacturing. Wheat, gram, rice, millet, oilseeds in alluvial river valleys; tamarind, mahua, mangoes. Bamboo, sal, ebony, sunn hemp in forested hills (lac growing). KATNI and JABALPUR are railroad junctions and industrial centers (ordnance, cement) near extensive bauxite, limestone, steatite, clay, and ocher workings. Tropical Forest Research Institute, Mandla Road; State Forest Research Institute, Polipathar; Weed Research Institute, Marhatal located here. A rock edict of Asoka, 22 mi/35 km N of SIHORA, proves extension of his empire over N part of state in 3rd century B.C.E. District abounds in other Buddhist remains and in Hindu and Jain temple ruins (many dating from 5th century A.D.).

Jabalpur (juh-BAHL-poor), city (2001 population 1,098,000), ⊙ JABALPUR district, N MADHYA PRADESH state, central INDIA, on the NARMADA RIVER, 150 mi/241 km NNE of NAGPUR; 23°10′N 79°57′E. Important railroad junction and military post. Seat of MADHYA PRADESH State High Court and State Electricity Board; Bargi Dam Project, located here, was an ambitious project to provide electricity and water to surrounding areas, however, it has become a point of contention for its failure to live up to expectations and its detrimental effect on local populations. Industrial center in agricultural area (wheat, gram, rice, millet, oilseeds). Manufacturing includes ordnance for India's Defence Ministry, foods, textiles, electronic equipment, furniture, and wood products; major cement-, iron-, ceramics-, and glassworks; lapidary industry. Seat of Jabalpur University Experimental farm (silk raising; dairy products). Limestone, bauxite, clay, and steatite workings nearby. Nearby popular tourist attractions include the MARBLE ROCKS on NARMADA RIVER and the waterfall at BHERAGHAT.

Jabalquinto (hah-vahl-KEEN-to), town, JAÉN province, S SPAIN, near confluence of GUADALIMAR and GUADALQUIVIR rivers, 7 mi/11.3 km SW of LINARES; 38°01′N 03°43′W. Olive oil, cereals.

Jabal Sammama, TUNISIA: see SAMMAMA, JABAL.

Jabal Taburuq, TUNISIA: see TABURSUQ, JABAL.

Jabal Tahant, TUNISIA: see HILL 609.

Jabal-us-Siraj (jah-bal–as–SEE-rahj), industrial town, PARWAN province, E AFGHANISTAN, 40 mi/64 km N of KABUL, on Salang River (short left tributary of Ghorband River), at foot of the Hindu Kush; 35°07′N 69°14′E. Modern cotton mill; hydroelectric station. Iron deposits nearby. Strategic town on plain leading from Kabul to Salang tunnel.

Jabal Zaghouan, TUNISIA: see ZAGHWAN, JABAL.

Jabal Zaghwan, TUNISIA: see ZAGHWAN, JABAL.

Jabal Zaltān (JE-bel zuhl-TAN), mountain range, CYRENAICA region, N central LIBYA, in SURT BASIN, 140 mi/225 km SSW of AJDABIYAH, near TRIPOLITANIA border. Several rich oil fields at E foot, including NASSIR (Zaltān), Waha, and Defa.

Jabarhera, INDIA: see JHABRERA.

Jabba (je-BAH), village, BAJAH province, N central TUNISIA, 6 mi/9.7 km WNW of TABURSUQ; calamine and lead mining.

Jabbeke (YAH-bai-kuh), agricultural commune (2006 population 13,586), Bruges district, WEST FLANDERS province, NW BELGIUM, 6 mi/9.7 km WSW of BRUGES; 51°11′N 03°05′E.

Jabeit (je-BAIT), village, Red Sea state, NE SUDAN, on road, and 110 mi/177 km NNW of PORT SUDAN; 21°04′N 36°19′E. Gold-mining center, connected by road with its RED SEA port, MUHAMMAD GOL.

Jabel el-Marad, ISRAEL: see HANITA.

Jabinyanah (je-been-YAH-nah), village (2004 population 6,576), SAFAQIS province, E TUNISIA, 22 mi/35 km NNE of SAFAQIS; 35°02′N 10°55′E. Administrative center. Olive oil pressing.

Jabiri (JAH-bee-ree), town, BORNO state, extreme N NIGERIA, near current border with CAMEROON in area once part of N. (British) CAMEROON, 50 mi/80 km S of DIKWA. Peanuts, millet, cotton; cattle, skins.

Jabiru (JA-bi-ru), township, N central NORTHERN TERRITORY, AUSTRALIA, 180 mi/290 km E of DARWIN, near end of Arnhem Highway; 12°38′S 132°52′E. Named for species of stork common to area. Situated in KAKADU NATIONAL PARK. Residential and service center for Ranger Uranium Mine 9 mi/14 km SE. Airport 4 mi/6 km E. Mineral rights leased from Arnhem Land Aboriginal Trust. Park visitor center 2 mi/3 km SW. Tourism; Aboriginal arts and culture. Aboriginal rock paintings at Nourlangie and Ubirr.

Jabitacá (ZHAH-bee-tah-kah), city, N PERNAMBUCO state, BRAZIL, 44 mi/71 km NW of ARCOVERDE; 07°50′S 37°23′W. Cotton, corn.

Jablanica, Serbian *Jablanicki Okrug,* district (□ 1,069 sq mi/2,779.4 sq km; 2002 population 240,923), ⊙ LESKOVAC, S central SERBIA; 42°53′N 21°35′E. Includes municipalities (*opštinas*) of BOJNIK, Crna Trava, LEBANE, Leskovac, Medvedja, and Vlasotince. Manufacturing (pharmaceuticals, chemicals, textiles); meatpacking.

Jablanica (yah-blah-NEET-sah), village, upper HERZEGOVINA, BOSNIA AND HERZEGOVINA, on the NERETVA RIVER, on railroad, and 35 mi/56 km SSE of PROZOR. Climatic resort; tourist center for excursions into nearby mountains. Hydroelectric plant, large dam and reservoir. Also spelled Yablanitsa.

Jablanica (yah-blah-NEET-sah), mountain (7,403 ft/2,256 m) in the PINDUS system, on Macedonia-Albania border, between the BLACK DRIN and the Shkumbi rivers, 10 mi/16 km NW of STRUGA (Macedonia). Also spelled Yablanitsa.

Jablanica Lake (yah-blah-NEET-sah), Serbo-Croatian *Jablaničko jezero,* reservoir, upper HERZEGOVINA, BOSNIA AND HERZEGOVINA, on RAMA and NERETVA Rivers.

Jableh, SYRIA: see JEBLE.

Jablonec nad Jizerou (YAHB-lo-NETS NAHD YI-ze-ROU), German *Jablonetz an der Iser,* town, VYCHODOCESKY province, NE BOHEMIA, CZECH REPUBLIC, in GIANT MOUNTAINS, on JIZERA river, on railroad, and 11 mi/18 km ESE of JABLONEC NAD NISOU; 50°42′N 15°26′E. Elevation 1,476 ft/450 m. Manufacturing (textiles, machinery); winter skiing resort. Has an 18th-century Baroque church; folk architecture.

Jablonec nad Nisou (YAHB-lo-NETS NAHD NYI-sou), German *Gablonz an der Neisse,* city (2001 population 45,266), SEVEROCESKY province, CZECH REPUBLIC, N BOHEMIA, on the LAUSITZER NEISSE

RIVER; 50°43′N 15°11′E. A glassware center, it produces high-quality glassware, jewelry, ornaments, and vans. Has a 14th-century Gothic church and a museum of glass and jewelry.

Jablonetz in der Iser, CZECH REPUBLIC: see JABLONEC NAD JIZEROU.

Jablonica Pass, UKRAINE: see YABLUNYTSYA PASS.

Jablonka, POLAND: see ORAVA.

Jablonne nad Orlici (YAHB-lo-NE NAHD or-LI-tsee), Czech *Jablonné nad Orlicí*, German *Gabel an der Adler*, town, VYCHODOCESKY province, NE BOHEMIA, CZECH REPUBLIC, on railroad, and 17 mi/27 km WNW of ŠUMPERK; 50°02′N 16°36′E. Manufacturing (electronic equipment, furniture). Has a baroque church; preserved typical wooden, and 18th-century baroque-style houses.

Jablonne v Podjestedi (YAHB-lo-NE FPOD-yesh-TYE), Czech *Jablonné v Podještědí*, town, SEVEROCESKY province, N BOHEMIA, CZECH REPUBLIC, on railroad, 32 mi/51 km ENE of PRAGUE; 50°46′N 14°47′E. Precious and semi-precious stone-grinding. Formerly known as NEMECKE JABLONNE, Czech *Německé Jablonné* (NYE-mets-KE YAHB-lo-NE), German *Deutsch Gabel*.

Jablonów, UKRAINE: see YABLUNIV.

Jablunkau, CZECH REPUBLIC: see JABLUNKOV.

Jablunkov (YAHB-lun-KOF), German *Jablunkau*, town, SEVEROMORAVSKY province, SE SILESIA, CZECH REPUBLIC, in the BESKIDS, on OLSE RIVER and on railroad, and 23 mi/37 km SE of OSTRAVA, near Polish border; 49°35′N 18°46′E. Manufacturing (fabricated metal products, electronic equipment, wood products). Jablunkov Pass (elev. 1,807 ft/551 m), with railroad tunnel, is 4 mi/6.4 km S.

Jaboatão (ZHAH-bo-ah-TOUN), city (2000 population 568,474), PERNAMBUCO, NE BRAZIL, on railroad, 11 mi/18 km WSW of RECIFE; 08°07′S 35°01′W. Sugar, manioc, tobacco. Here were fought two battles in Dutch Wars (17th century).

Jaboatão, Brazil: see JAPOATÃ.

Jabón (hah-BON), town, LARA state, NW VENEZUELA, in SEGOVIA HIGHLANDS, 35 mi/56 km SW of CARORA. Vegetables; livestock.

Jaborá (ZHAH-bo-rah), town (2007 population 4,032), W SANTA CATARINA state, BRAZIL, 13 mi/21 km W of JOAÇABA; 27°11′S 51°44′W. Wheat, potatoes, fruit, corn; livestock.

Jaborandi (ZHAH-bo-rahn-ZHEE), city (2007 population 6,462), BAHIA, BRAZIL.

Jaboti (ZHAH-bo-chee), town (2007 population 5,019), N PARANÁ state, BRAZIL, 31 mi/50 km SSE of SANTO ANTÔNIO DA PLATINA; 23°46′S 50°04′W. Corn, rice, potatoes.

Jaboticabal (ZHAH-bo-chee-kah-bahl), city (2007 population 69,624), N central SÃO PAULO, BRAZIL, 33 mi/53 km WSW of RIBEIRÃO PRÊTO; 21°16′S 48°19′W. Manufacturing (machinery, food); sugar refining, alcohol distilling, cotton ginning. Trades in livestock, coffee, fruit. Has agricultural school with airfield.

Jaboticatubas (ZHAH-bo-CHEE-kah-too-bahs), city (2007 population 15,496), central MINAS GERAIS, BRAZIL, 38 km/24 mi N of SANTA LUZIA; 19°15′S 43°45′W.

Jabrin (JAB-rin), oasis and main village of oasis in S EASTERN PROVINCE, SAUDI ARABIA, 150 mi/241 km S of HOFUF. Agriculture, cereals, dates, vegetables.

Jabugo (hah-VOO-go), town, HUELVA province, SW SPAIN, in the SIERRA MORENA, 10 mi/16 km W of ARACENA; 37°55′N 06°44′W. Acorns, chestnuts, walnuts, olives, fruit, hogs; manufacturing of food products; sawmilling. Famous throughout Spain for its cured ham.

Jabuka Island (YAH-boo-kah), Italian *Pomo*, Dalmatian island (□ 1 sq mi/2.6 sq km) in ADRIATIC SEA, S CROATIA, c.55 mi/89 km SW of SPLIT. Designated as Jabuka Island Nature Monument (□ 1/10 sq mi/3/10 sq mi) because it is one of two (Brusnik is the other) volcanic islands on Croatia's Adriatic coast.

Jabukovac (YAH-oo-ko-vahts), village, E SERBIA, 11 mi/18 km NNW of NEGOTIN. Also spelled Yabukovats.

Jabwot (JAHB-wot), coral island, Ralik Chain, Kwajalein district, Marshall Islands, W central Pacific, 10 mi/16 km N of Ailinglapalap; 07°44′N 168°59′E. It is c.10 mi/16 km incircumference. Phosphorite deposit.

Jaca (HAH-kah), town (1990 10,874), HUESCA province, NE SPAIN, in BASQUE COUNTRY near the PYRENEES, near the French border on the ARAGÓN RIVER; 42°34′N 00°33′W. Elevation c.2,700 ft/820 m. A communications center and a processing center for lumber and for the farm products of the fertile Aragón valley. After its recapture from the Moors it was (11th century) the cradle of the Aragonese kingdom. HUESCA, taken in 1097, replaced it as the capital. Jaca has ancient walls and towers and a Romanesque cathedral (11th–15th century).

Jacala (hah-KAH-lah), town, HIDALGO, central MEXICO, in SIERRA MADRE ORIENTAL, on INTER-AMERICAN HIGHWAY (Mexico Highway 85), and 70 mi/113 km NNW of PACHUCA DE SOTO; ☉ Jacala de Ledesma municipio; 21°01′N 99°12′W. Small farming.

Jacaleapa (hah-kah-le-AH-pah), town, EL PARAÍSO department, S HONDURAS, 5 mi/8 km W of DANLÍ; 14°00′N 86°40′W. Tobacco, corn, beans.

Jacaltenango (hah-kahl-te-NAHN-go), town, HUEHUETENANGO department, W GUATEMALA, on W slopes of CUCHUMATANES MOUNTAINS, 30 mi/48 km NW of HUEHUETENANGO; 15°40′N 91°44′W. Elevation 4,718 ft/1,438 m. Manufacturing (textiles, food); agriculture (subsistence farming; livestock raising). Jacalteca-speaking population.

Jaçana (ZHAH-kah-nah), town (2007 population 7,763), S RIO GRANDE DO NORTE state, BRAZIL, on PARAÍBA border, 25 mi/40 km SE of CURRAIS NOVOS; 06°24′S 36°14′W. Cotton, corn, manioc. Also spelled Jaçanã.

Jacaraci (ZHAH-kah-rah-SEE), city (2007 population 14,625), S central BAHIA, BRAZIL, in SERRA DO ESPINHAÇO, near MINAS GERAIS border; 14°50′S 42°26′W.

Jacaraú (ZHAH-kah-rah-OO), city (2007 population 13,708), NE PARAÍBA, BRAZIL, 11 mi/18 km NE of DUAS ESTRADAS; 06°39′S 35°18′W.

Jacareacanga (ZHAH-kah-rai-ah-kahn-gah), city (2007 population 37,055), SW PARÁ, BRAZIL, on Rio TAPAJÓS near border with AMAZONAS; 06°13′S 57°46′W. Airport.

Jacaré dos Homens (ZHAH-kah-RAI dos O-mains), town (2007 population 5,724), W central ALAGOAS state, BRAZIL, 11 mi/18 km NW of BATALHA; 09°40′S 37°12′W.

Jacareí (ZHAH-kah-RAI-ee), city (2007 population 207,028), E SÃO PAULO, BRAZIL, on PARAÍBA River, on railroad, and 45 mi/72 km ENE of SÃO PAULO; 23°19′S 45°48′W. Textile milling; light manufacturing; dairying, distilling. Old spelling, Jacarehy.

Jacarèzinho (ZHAH-kah-rai-seen-yo), city (2007 population 39,327), NE PARANÁ, BRAZIL, on railroad, and 75 mi/121 km E of LONDRINA; 23°09′S 49°59′W. Furniture manufacturing, coffee and rice processing; cotton, tobacco, grain, cattle.

Jachacha (hah-CHAH-hah), canton, RAFAEL BUSTILLO province, POTOSÍ department, W central BOLIVIA; 18°28′S 66°36′W. Elevation 12,680 ft/3,865 m. Limestone and gypsum deposits. Agriculture (potatoes, yucca, bananas); cattle. Tin mining in the area formerly at Mina Llallagua Siglo XX (closed 1986).

Jachal (hah-CHAHL), town, central SAN JUAN province, ARGENTINA, on Jachal or ZANJÓN RIVER, and 90 mi/145 km N of SAN JUAN; ☉ Jachal department (□ c.8,500 sq mi/22,015 sq km; 1991 population 19,989); 30°30′S 68°30′W. Farming center (alfalfa, wheat, corn, olives; livestock). Lime deposits; sawmills. Irrigation dam and copper, lead, and zinc mines nearby.

Jachal River, ARGENTINA: see ZANJÓN RIVER.

Jachymov (YAH-khi-MOF), Czech *Jáchymov*, German *Joachimsthal*, town, ZAPADOCESKY province, W BOHEMIA, CZECH REPUBLIC, in the ORE MOUNTAINS; 50°22′N 12°55′E. Noted health resort, with thermal radioactive springs. It was the main center of silver mining in EUROPE after the 16th century. Uranium mining started in 1840s; once extensive deposits began to decline after World War II, until finally the mines were closed in 1960. The word *Thaler*, from which *dollar* is derived, is an abbreviation of *Joachimsthaler*, the name of a coin first struck there, in the 16th century. The world's first miners' school was established here in 1716.

Jaciara (ZHAH-see-ah-rah), city (2007 population 25,028), SE MatoGrosso, 38 mi/62 km NW of Rondonópolis; 15°50′S 55°00′W. Soybeans, sugarcane.

Jaciba (ZHAH-see-bah), city, central PARANÁ state, BRAZIL, 72 mi/116 km WNW of PONTA GROSSA; 24°56′S 51°12′W. Corn; livestock.

Jacinto (ZHAH-seen-to), city (2007 population 12,423), NE MINAS GERAIS, BRAZIL, on Rio JEQUITINHONHA, 33 mi/54 km ENE of ALMENARA; 16°12′S 40°17′W.

Jacinto Aráuz (hah-SEEN-to ah-ROOZ), town, SE LA PAMPA province, ARGENTINA, on railroad, and 85 mi/137 km SE of GENERAL ACHA; 38°04′S 63°26′W. Wheat and livestock center.

Jacinto City, town (2006 population 9,939), HARRIS county, SE TEXAS, residential suburb 6 mi/9.7 km E of downtown HOUSTON, on Hunting Bayou; 29°46′N 95°14′W. Incorporated since 1940.

Jacinto Machado (ZHAH-seen-to MAH-shah-do), city (2007 population 10,738), extreme SE SANTA CATARINA state, BRAZIL, 34 mi/55 km SW of Circiúma; 29°00′S 49°46′W. Corn, rice, potatoes, manioc; livestock.

Jaciparaná (ZHAH-see-pah-rah-NAH), town, W RORAIMA state, W BRAZIL, on Madeira-MAMORÉ railroad, and 50 mi/80 km SW of PÔRTO VELHO; 09°15′S 64°23′W. Manufacturing of rubber. Until 1944, called Generoso Ponce.

Jaciporã (ZHAH-see-po-ruh), city, W SÃO PAULO state, BRAZIL, 8 mi/12.9 km S of DRACENA, on PEIXE River; 21°36′S 51°35′W.

Jack, county (□ 920 sq mi/2,392 sq km; 2006 population 9,110), N TEXAS; ☉ JACKSBORO; 33°14′N 98°10′W. Drained by W Fork of TRINITY RIVER. Livestock (cattle, horses, ostriches); wheat, pecans; wool, mohair marketed. Oil and natural-gas wells; gravel; tourism; timber. Fort Richardson State Historical Park at Jacksboro (center of county). Formed 1856.

Jackfish (JAK-fish), unincorporated village, central ONTARIO, E central CANADA, on JACKFISH BAY of Lake SUPERIOR, 14 mi/23 km E of SCHREIBER, and included in TERRACE BAY township; 48°48′N 86°58′W. Gold mining.

Jackfish Bay (5 mi/8 km long, 1 mi/1.6 km wide), Lake county, NE MINNESOTA, near CANADA (ONTARIO) border, 10 mi/16 km NNE of ELY, SW extension of BASSWOOD LAKE (connected by narrow passage), in BOUNDARY WATERS Canoe Area of Superior National Forest; 48°02′N 91°44′W. PIPESTONE BAY extends SW from near entrance to Jackfish Bay. Also known as Jackfish Lake.

Jackfish Lake (JAK-fish), (10 mi/16 km long, 6 mi/10 km wide), W SASKATCHEWAN, CANADA, 17 mi/27 km N of NORTH BATTLEFORD; 53°05′N 108°20′W. Drains into NORTH SASKATCHEWAN RIVER. Just E is Murray Lake (6 mi/10 km long, 3 mi/5 km wide).

Jackie Robinson Parkway, NEW YORK: see INTERBOROUGH (INTERBORO) PARKWAY.

Jack Lee, Lake, reservoir (□ 36 sq mi/93.6 sq km), extreme S central ARKANSAS, near LOUISIANA border, on OUACHITA RIVER, in Felsenthal National Wildlife Refuge, 7 mi/11.3 km W of CROSSETT; 33°03′N 92°12′W. Maximum capacity 76,700 acre-ft. Fed by SALINE RIVER. Formed by Felsenthal Lock and Dam (105 ft/32 m high), built (1978) by Army Corps of Engineers for navigation and recreation. Dam also known as Lock and Dam #6; reservoir also known as Ouachita Reservoir.

Jackman, town, SOMERSET COUNTY, W MAINE, 25 mi/40 km W of MOOSEHEAD LAKE, 15 mi/24 km E of QUEBEC line, 33 mi/53 km NW of GREENVILLE; 45°36'N 70°12'W. Port of entry; center of wilderness hunting, fishing, camping region; lumbering.

Jackpot, unincorporated village, ELKO COUNTY, NE NEVADA, 60 mi/97 km NNE of WELLS and 40 mi/64 km SSW of TWIN FALLS, Idaho, on IDAHO state line, on SALMON FALLS CREEK. Jackpot has become one of several gambling resorts on that line that have been developed at the Nevada/Idaho border. Part of Humboldt National Forest to W. Established in 1959.

Jacksboro, coal-mining town (2006 population 2,034), NE TENNESSEE, 28 mi/45 km NW of KNOXVILLE; ☉ CAMPBELL COUNTY; 36°20'N 84°11'W.

Jacksboro, town (2006 population 4,639), N TEXAS, 50 mi/80 km SSE of WICHITA FALLS; ☉ JACK COUNTY; 33°13'N 98°09'W. Elevation 1,074 ft/327 m. In cattle ranching; oil and natural gas; timber area; light manufacturing. Fort Richardson State Historic Site (founded 1867; extant). Settled 1855; incorporated 1899.

Jacks Mountain, ridge (c.2,000 ft/610 m), central PENNSYLVANIA, 2 mi/3.2 km wide (SW-NE), runs c.70 mi/113 km from central part of HUNTINGDON county to NW part of SNYDER county, where it merges with Creek Mountain ridge; 40°37'N 77°37'W–40°51'N 77°06'W. JUNIATA RIVER flows through gap W of MOUNT UNION. Sandstone, silica.

Jackson, county (□ 1,126 sq mi/2,927.6 sq km; 2006 population 53,745), NE Alabama; ☉ Scottsboro; 34°46'N 86°00'W. Agr. region bordering on Georgia and Tennessee, drained by Tennessee and Paint Rock rivers and Guntersville Reservoir. Soybeans, hay, corn; cattle, poultry; timber; deposits of coal and limestone. Formed 1819. Named for Andrew Jackson of TN, commanding general in the Creek Indian War of 1813–1814, who was visiting Huntsville at the time the general assembly was meeting.

Jackson, county (□ 641 sq mi/1,666.6 sq km; 2006 population 17,426), NE ARKANSAS; ☉ NEWPORT, BLACK RIVER forms NW boundary; 35°36'N 91°12'W. Drained by WHITE and CACHE rivers, and by Departee Creek. Agriculture (wheat, sorghum, soybeans, rice, hogs). Manufacturing at Newport. Sand and gravel pits. Formed 1829. Jacksonport State Park in NW.

Jackson, county (□ 1,621 sq mi/4,214.6 sq km; 2006 population 1,406), N COLORADO; ☉ WALDEN; 40°40'N 106°20'W. CONTINENTAL DIVIDE forms S boundary. Agricultural area, bordering on WYOMING; drained by headwaters of North Platte River. Livestock, lumber. Walden Lake, Lake John and Delany Butte Lakes at county center. Most of Colorado State Forest in E. Includes part of Routt National Forest in W and SE, small part of Arapaho National Forest on S boundary. Part of PARK RANGE in W and MEDICINE BOW MOUNTAINS in E. Formed 1909.

Jackson (JAK-suhn), county (□ 942 sq mi/2,449.2 sq km; 2006 population 49,288), NW FLORIDA, on ALABAMA (N) and GEORGIA (E; CHATTAHOOCHEE RIVER) state lines; ☉ MARIANNA; 30°48'N 85°12'W. Rolling agriculture area with many small lakes; drained by CHIPOLA RIVER. Some manufacturing (food products; lumber). Formed 1822.

Jackson, county (□ 337 sq mi/876.2 sq km; 2006 population 55,778), NE central GEORGIA; ☉ JEFFERSON; 34°08'N 83°34'W. Piedmont area drained by OCONEE River. Agriculture (hay, sweet potatoes, apples, peaches; cattle, hogs, poultry); textile manufacturing. Formed 1796.

Jackson, county (□ 602 sq mi/1,565.2 sq km; 2006 population 57,778), SW ILLINOIS; ☉ MURPHYSBORO; 37°46'N 89°21'W. Bounded SW by MISSISSIPPI RIVER; drained by BIG MUDDY and LITTLE MUDDY rivers and BEAUCOUP CREEK. CARBONDALE (with Southern Illinois University) is dominant city, in E part. Agricultural area (wheat, sorghum, fruit, cattle; dairying),

with some manufacturing (wood products, paper products, fabricated metal products). Bituminous coal mining. Includes part of Shawnee National Forest and Lake Murphysboro and Grant City State Parks, also Lake Kincaid and Ceder Lake reservoirs. Formed 1816.

Jackson, county (□ 513 sq mi/1,333.8 sq km; 2006 population 42,404), S INDIANA; ☉ BROWNSTOWN; 38°55'N 86°02'W. Bounded S by MUSCATATUCK RIVER; drained by East Fork of WHITE RIVER and tributaries of the Muscatatuck River. Agricultural area (corn, wheat, vegetables, cattle, hogs, truck, poultry), with diversified manufacturing; timber. Hoosier National Forest in NW corner; Cypress Lake State Fishing Area in NE. Brownstown State Fishing Area W of Brownstown; MUSCATATUCK NATIONAL WILDLIFE REFUGE on E boundary. Starve Hollow State Beach and part of Jackson-Washington State Forest in S, includes Driftwood State Fish Hatchery. Formed 1816.

Jackson, county (□ 649 sq mi/1,687.4 sq km; 2006 population 20,290), E IOWA, on ILLINOIS line (E; formed here by MISSISSIPPI RIVER); ☉ MAQUOKETA; 42°10'N 90°35'W. Prairie agricultural area (hogs, cattle, poultry; corn, oats, soybeans) drained by MAQUOKETA and North Fork Maquoketa rivers; limestone quarries. Maquoketa Caves State Park in SW; Bellevue State Park in E on Mississippi River, below Lock and Dam No. 12. SAINT DONATUS, located in NE corner near Mississippi River, is a picturesque Luxembourger village, that was settled by LUXEMBOURG immigrants. General river flooding occurred in 1993. Formed 1837.

Jackson, county (□ 657 sq mi/1,708.2 sq km; 2006 population 13,500), NE KANSAS; ☉ HOLTON; 39°24'N 95°50'W. Rolling prairie, watered by DELAWARE RIVER. Wheat, sorghum, soybeans, cattle, hogs. Potawatomi Indian Reservation, dominant feature of county, is W of MAYETTA. Formed 1857.

Jackson, county (□ 346 sq mi/899.6 sq km; 2006 population 13,810), SE central KENTUCKY, in CUMBERLAND foothills; ☉ MCKEE; 37°25'N 84°00'W. Bounded SW by ROCKCASTLE RIVER; drained by Middle Fork Rockcastle River and by Station Camp Creek. Large part of county in Daniel Boone National Forest in NE, center, and SW. Mountain agricultural area (burley tobacco, hay, alfalfa, corn; cattle, poultry; dairying); timber; limestone. Formed 1858.

Jackson, county (□ 723 sq mi/1,879.8 sq km; 2006 population 163,851), S MICHIGAN; ☉ JACKSON; 42°15'N 84°25'W. Drained by GRAND and RAISIN rivers and headstreams of the KALAMAZOO. Agriculture (forage crops, corn, hay, apples, soybean, wheat, oats; cattle, hogs, sheep, dairy products); manufacturing at Jackson. Contains many small lakes (fishing, swimming), in E ½ of county. Organized 1832.

Jackson, county (□ 719 sq mi/1,869.4 sq km; 2006 population 11,150), SW MINNESOTA; ☉ JACKSON; 43°40'N 95°09'W. Bordering IOWA (S) and watered by DES MOINES RIVER and headwaters of Little Sioux River. Agricultural area (corn, oats, soybeans, alfalfa; sheep, hogs, cattle). Includes part of COTEAU DES PRAIRIES; Kilen Woods State Park in NE center; HERON LAKE and South Heron Lake are in NW center. Several small natural lakes, especially near S boundary. Formed 1857.

Jackson, county (1,043 sq mi/2,701 sq km; 1990 population 115,243; 2000 population 131,420), extreme SE MISSISSIPPI; ☉ PASCAGOULA; 30°27'N 88°37'W. Bounded E by ALABAMA state line, S on MISSISSIPPI SOUND, Gulf of MEXICO; drained by PASCAGOULA and ESCATAWPA rivers. Cotton, corn, pecans, honey; timber; fish, shrimp, crab; catfish farming. Includes part of De Soto National Forest in NW; Grand Bay National Wildlife Refuge in SE, Mississippi Sandhill Crane National Wildlife Refuge in SW, Shepard State Park in S, Gulf Marine State Park in SW, Ward Bayou Wildlife Management Area in center, part of Pascagoula River Wildlife Management Area in N, included

HORN and PETIT BOIS Islands in Gulf of Mexico, part of GULF ISLANDS National Seashore, visitors center on mainland in SW part of county. Formed 1812.

Jackson, county (□ 603 sq mi/1,567.8 sq km; 2006 population 664,078), W MISSOURI; ☉ INDEPENDENCE; 39°01'N 94°21'W. Bounded N by MISSOURI RIVER; KANSAS River enters the Missouri at NW corner. Has the downtown and major part of city of KANSAS CITY, (also goes in to CLAY and PLATTE counties). Other major cities include Independence, RAYTOWN, LEE'S Summit, BLUE SPRINGS, and GRANDVIEW. Agriculture (wheat, corn, soybeans, dairy products); manufacturing grain, and livestock industries centered at Kansas City. County was designated as the original Garden of Eden by Mormon prophet Joseph Smith. James A. Reed Wildlife Area in S; Burr Oak Woods Wildlife Area at Blue Springs; Lake Jacomo, Longview Lake, and Blue Springs Lake are Army Corps of Engineers reservoirs, Lake Sacamo Park at Lee's Summit. Professional football and baseball stadiums at intersection of I-70 and I-435. Formed 1826.

Jackson (JAK-suhn), county (□ 494 sq mi/1,284.4 sq km; 2006 population 35,562), W NORTH CAROLINA; ☉ SYLVA; 35°17'N 83°08'W. Partly in the BLUE RIDGE MOUNTAINS (SE); bounded S by SOUTH CAROLINA state line; BALSAM MOUNTAIN in E, COWEE MOUNTAINS in W; drained by TUCKASEGEE RIVER. All but N end of county in Nantahala National Forest (Glenville). Thorpe Lake reservoir in S center. Service industries; manufacturing at Sylva; agriculture (apples, tobacco, hay; cattle); timber; mica and talc mining; resort region. BLUE RIDGE (NATIONAL) PARKWAY follows NE county line. Part of Eastern Cherokee (Qualla Boundary) Indian Reservation in N. Formed 1851 from Haywood and Macon counties. Named for Andrew Jackson, president of the US from 1829 to 1837.

Jackson (JAK-suhn), county (□ 420 sq mi/1,092 sq km; 2006 population 33,543), S OHIO; ☉ JACKSON; 39°02'N 82°37'W. Drained by LITTLE SCIOTO and SYMMES rivers and Little Raccoon Creek. Includes Buckeye Furnace and Leo Petroglyph state parks, as well as Canter's Cave. In the Unglaciated Plain physiographic region. Agricultural area (poultry, corn); manufacturing at Jackson, OAK HILL, and WELLSTON; some coal mining. Formed 1816.

Jackson, county (□ 804 sq mi/2,090.4 sq km; 2006 population 26,042), extreme SW OKLAHOMA; ☉ ALTUS; 34°35'N 99°24'W. Bounded S by TEXAS (RED RIVER), on E by NORTH FORK OF RED RIVER; drained by SALT FORK OF RED RIVER. Agriculture (cotton, wheat, sorghum, alfalfa); cattle; light manufacturing. Greyhound breeding. Irrigation from Altus Dam to N (not in county). Formed 1907.

Jackson, county (□ 2,801 sq mi/7,282.6 sq km; 2006 population 197,071), SW OREGON; ☉ MEDFORD; 42°25'N 122°44'W. Mountainous area bordering CALIFORNIA on S, crossed by ROGUE RIVER. Agriculture (pears, plums, peaches, grapes, cherries, apples, wheat, oats, barley, hops, nuts; poultry, hogs, sheep, cattle); wineries, nurseries. Includes part of SISKIYOU MOUNTAINS in SW and parts of Rogue River National Forest, including part of Sky Lakes Wilderness Area on E boundary, in S, N, and E. Small part of Klamath National Forest in S; part of CRATER LAKE NATIONAL PARK on NE boundary; Stewart and Casey State Parks in NE; TouVelle State Park in center; Tub Springs State Wayside in SE; Ben Hur Lampman State Wayside and Valley of the Rogue State Park in W. Formed 1852.

Jackson, county (□ 1,871 sq mi/4,864.6 sq km; 2006 population 2,900), SW central SOUTH DAKOTA; ☉ KADOKA; 43°42'N 101°38'W. Agricultural area watered by intermittent streams; also Pass and Bear-in-the-Lodge creeks. Hay, soybeans; cattle. Extreme NE part of BADLANDS NATIONAL PARK in W (including park headquarters). Buffalo Gap National Grassland in NW. Area S of WHITE RIVER is former Washabaugh county. All of county S of White River (over half of county) is part of the large Pine Ridge Indian Reservation. Formed 1883 and later absorbed into other counties; reconstituted 1914.

Jackson, county (□ 327 sq mi/850.2 sq km; 2006 population 10,918), N central TENNESSEE; ⊙ GAINESBORO; 36°22′N 85°40′W. Crossed by CUMBERLAND RIVER. Agriculture; timber. Formed 1801.

Jackson, county (□ 857 sq mi/2,228.2 sq km; 2006 population 14,249), S TEXAS; ⊙ EDNA; 28°56′N 96°34′W. On GULF OF MEXICO coastal plain; indented by Lavaca Bay in S. Bounded SW by Arenosa Creek; drained by LAVACA and NAVIDAD rivers. Oil, natural-gas wells; cattle ranching; agriculture (cotton, rice, grain sorghum); livestock raising. LAKE TEXANA reservoir (Navidad River) and State Park in center of county. Formed as municipality 1835, as county 1836.

Jackson, county (□ 472 sq mi/1,227.2 sq km; 2006 population 28,451), W West Virginia; ⊙ RIPLEY; 38°49′N 81°40′W. Bounded NW by OHIO RIVER (OHIO state line); drained by MILL CREEK. Natural-gas and oil wells; some coal. Agriculture (corn, wheat, oats, tobacco, potatoes, alfalfa, hay, nursery crops); cattle; some dairying. Some manufacturing (metals, chemicals) at Ripley, RAVENSWOOD. Frozen Camp Wildlife Management Area in E; Woodrum Wildlife Management Area in S; Cedar Lake State Camp (FFA-FHA) in center. Formed 1831.

Jackson, county (□ 1,000 sq mi/2,600 sq km; 2006 population 19,853), W central WISCONSIN; ⊙ BLACK RIVER FALLS, 44°20′N 90°52′W. Dairying area. Agriculture (potatoes, barley, corn, soybeans, peas, brans, cranberries, cattle, hogs, sheep, poultry); some manufacturing. Intersected by BLACK, BUFFALO, and TREMPEALEAU rivers. Winnebago Indian Reservation at center; Arbutus Lake on N boundary; Black River State Forest in E center, crosses county N-S; smaller units of Black River Falls, includes Castle Mound. Formed 1853.

Jackson, parish (□ 583 sq mi/1,515.8 sq km; 2006 population 15,202), N central LOUISIANA; ⊙ JONESBORO; 32°14′N 92°43′W. Agriculture (hay, vegetables, cattle, poultry). Manufacturing of paper products; logging; timber. Drained by DUGDEMONA RIVER and CASTOR CREEK. Caney Creek reservoir in S. Part of Jackson-Bienville State Wildlife Area in NW. Named after President Andrew Jackson. Formed 1845.

Jackson (JAK-suhn), city (2000 population 3,989), ⊙ AMADOR county, central CALIFORNIA, 40 mi/64 km ESE of SACRAMENTO, on Jackson Creek; 38°21′N 120°46′W. Railroad terminus. Trade and mining center in 1849 California Gold Rush region. Gold mines, marble quarries, clay pits, farms (grapes, walnuts, grain; cattle), vineyards nearby; manufacturing (printing and publishing). Argonaut and Kennedy quartz mines, which began operating in early 1850s, are more than 1 mi/1.6 km deep; they were closed in the 1940s. PARDEE RESERVOIR to SW (MOKELUMNE RIVER); Jackson Butte (2,310 ft/704 m) to E. Founded during gold rush; made county seat in 1851; incorporated 1905.

Jackson, city (2000 population 36,316), S MICHIGAN, on the GRAND RIVER; ⊙ JACKSON county; 42°14′N 84°24′W. Elevation 960 ft/293 m. It is an industrial and commercial center in a farm region. Manufacturing (machinery, aerospace components, transportation equipment, food, fabricated metal products, electronic equipment, construction materials). Railroad junction. Several automobile models were pioneered in Jackson in the early 20th century. The first Republican convention was held in the city on July 6, 1854; a tablet marks the site. Jackson Community College to S; Michigan Space Center to S; Reynolds Field Airport to W. Nearby are Spring Arbor College and a state prison. Incorporated 1857.

Jackson, city (2000 population 184,256), HINDS and MADISON counties, W central MISSISSIPPI; ⊙ of Mississippi. and Hinds county(shares county seat with RAYMOND), on the PEARL RIVER (forms large ROSS BARNETT RESERVOIR to NE); 32°19′N 90°12′W. Railroad junction. It is the state's largest city and commercial center, with important railroad, warehouse, and distributing operations. Manufacturing (food, construction materials, glass, paper products, printing and publishing, lumber, machinery, consumer goods, furniture, concrete, fabricated metal products). The site of the city, a trading post known as Le Fleur's Bluff near the NATCHEZ TRACE, was chosen and laid out as the state capital in 1821 and named for Andrew Jackson. The first U.S. law giving property rights to married women was passed here in 1839. During the Civil War, Jackson was a military center for the Vicksburg Campaign and was largely destroyed by Sherman's forces in 1863. The old capitol (1839) is preserved as a museum; the new capitol was completed in 1903. Among the many points of interest are the governor's mansion (erected 1839); city hall, which was used as a hospital during the Civil War; a 220-acres/89-ha scale model of the Mississippi River flood control system; Mynelle's Gardens; a Jackson Zoological Park; Municipal Art Gallery; a notable Confederate monument; and many antebellum homes; Mississippi Agriculture and Forestry Museum; Mississippi Museum of Natural History; Mississippi State Historical Museum; Dizzy Dean Museum; Davis Planetarium/McNair Space Theater; Smith Robertson Museum (African-Amer. culture). Belhaven College, Jackson State University, Millsaps College, the University of Mississippi Medical Center, and several state institutions for the physically and mentally handicapped are here. Nearby are Tougaloo College (RIDGELAND) and Mississippi College (CLINTON). During the 1960s, Jackson was the scene of considerable racial unrest. In May 1970, demonstrations at the predominantly black Jackson State College resulted in the deaths of 2 students. Allen C. Thompson Airport to E; Hawkins Field Airport in NW. LeFleurs Bluff State Park in city, NE of downtown. Incorporated 1833.

Jackson, city (2000 population 11,947), SE MISSOURI, 10 mi/16 km NW of CAPE GIRARDEAU; ⊙ CAPE GIRARDEAU county; 37°22′N 89°39′W. Corn, lumber, cattle, dairy products; manufacturing (transportation equipment, machinery, fabricated metal products, construction materials). Founded c.1815.

Jackson (JAK-suhn), city (□ 8 sq mi/20.8 sq km; 2006 population 6,232), ⊙ JACKSON county, S OHIO, 28 mi/45 km NE of PORTSMOUTH; 39°03′N 82°37′W. Steel mills, foundries; also produces clay products, lumber. Clay, sand, gravel, and silica pits. Some coal mining. Buckeye Furnace and Leo Petroglyph state parks and Canter's Cave are nearby. Founded 1817.

Jackson, city (2006 population 62,711), W TENNESSEE, on the South Fork of the FORKED DEER RIVER, c.81 mi. ENE of MEMPHIS; ⊙ MADISON county, 35°37′N 88°50′W. Transportation and shipping hub; manufacturing; recreation. Jackson experienced development as a trucking center. It is the seat of Lane College, Lambuth College, Union University, and a community college. Nearby is a state park with Native American mounds. Home and burial site of Casey Jones; Casey Jones railroad museum is here. Founded by a nephew of Andrew Jackson, incorporated 1823.

Jackson, town (2000 population 5,419), Clarke co., SW ALABAMA, on Tombigbee River (RR bridge), and 16 mi/26 kmSSW of Grove Hill. Lumber, apparel. Artesian mineral wells and SaltSprings State Park nearby. Inc. 1816.

Jackson, town (2000 population 3,934), central GEORGIA, c.40 mi/64 km SE of ATLANTA; ⊙ BUTTS county; 33°17′N 83°58′W. Manufacturing includes fixtures, apparel, concrete, fabricated metal products. Indian Springs State Park nearby, where Creek nation and the U.S. government signed treaty ceding territory to the U.S. Lake Jackson and Dauset Trails Nature Center. Once a thriving resort area. Incorporated 1826.

Jackson, town (2000 population 2,490), E central KENTUCKY, in the CUMBERLAND MOUNTAINS, on North Fork KENTUCKY RIVER, and 24 mi/39 km NNW of HAZARD; ⊙ BREATHITT county; 37°33′N 83°22′W. Trade center for coal mining and agricultural (cattle, poultry; corn, fruit, potatoes, tobacco; honey) area; manufacturing (computer equipment). Juliann Carroll Airport to SE. Seat of Lees College (1864; two-year). County Museum. Parts of Robinson Forest, University of Kentucky, research tract, to SE; Pan Bowl Lake to N, natural oxbow lake. Established 1883.

Jackson, town (2000 population 4,130), EAST FELICIANA parish, SE central LOUISIANA, on Thompson Creek, 27 mi/43 km N of BATON ROUGE; 30°50′N 91°13′W. Rolling hill country. In agricultural area; manufacturing (concrete, lumber); winery. State mental hospital here. Audubon State Commemorative Area to SW, site of antebellum home of John James Audubon. Centenary State Commemorative Area to E, site of former Centenary College (1800s).

Jackson, town, WALDO county, S MAINE, 14 mi/23 km NNW of BELFAST; 44°36′N 69°09′W. Agriculture, lumbering.

Jackson, town (2000 population 3,501), SW MINNESOTA, 27 mi/43 km W of FAIRMONT, on DES MOINES RIVER, near IOWA state line; ⊙ JACKSON county; 43°37′N 94°59′W. Elevation 1,459 ft/445 m. Trade and shipping point; grain, soybeans, alfalfa; livestock; light manufacturing. Kilen Woods State Park to NW; SPIRIT LAKE (Iowa) to SW. Settled before 1857, scene of Sioux uprising 1862, incorporated 1881.

Jackson, town, CARROLL county, E NEW HAMPSHIRE, 22 mi/35 km S of BERLIN; 44°11′N 71°12′W. Drained by ELLIS RIVER. Agriculture (livestock; dairying); timber; resort area. Parts of White Mountain National Forest in W, N, and E; Black Mountain ski area and Nestlenook Farm cross-country ski area in center; covered bridge.

Jackson, town (2006 population 1,647), AIKEN county, W SOUTH CAROLINA, near SAVANNAH River, 18 mi/29 km SSW of AIKEN; 33°19′N 81°47′W. Grew with establishment nearby (1951) of SAVANNAH RIVER Nuclear Power Plant of Atomic Energy Commission. Redcliffe Plantation State Park N and E. Manufacturing of concrete, ordnance, transportation equipment. Agriculture includes cotton, peanuts, grains, livestock, poultry.

Jackson, town (2000 population 4,938), WASHINGTON county, E WISCONSIN, 23 mi/37 km. NNW of MILWAUKEE; 43°19′N 88°09′W. In dairying and farming area; manufacturing (fabricated metal products, pharmaceuticals).

Jackson, town (2006 population 9,215), NW WYOMING, on Gros Ventre River near its confluence with SNAKE RIVER, just S of GRAND TETON NATIONAL PARK, and 65 mi/105 km E of IDAHO FALLS, IDAHO; ⊙ TETON county; 43°28′N 110°45′W. Elevation c.6,209 ft/1,893 m. Resort and trading point in JACKSON HOLE; agriculture, livestock; manufacturing (food, apparel, construction materials, printing and publishing). A very popular tourist community and a haven for artists and writers. Headquarters of Bridger-Teton National Forest (to E and S). Rodeo takes place annually. Hunting and dude ranches in vicinity. Daily gunfight enactment takes place on town square during the summer. National Wildlife Art Museum (1994) is here. National Elk Refuge to NE; Targhee National Forest to W.

Jackson, village (2006 population 209), DAKOTA county, NE NEBRASKA, 7 mi/11.3 km W of DAKOTA CITY, near MISSOURI RIVER and SOUTH DAKOTA state line; 42°27′N 96°34′W.

Jackson (JAK-suhn), village (□ 1 sq mi/2.6 sq km; 2006 population 662), NE NORTH CAROLINA, 14 mi/23 km SE of ROANOKE RAPIDS; ⊙ NORTHAMPTON county;

Area is shown by the symbol □, and capital city or county seat by ⊙.

36°23'N 77°25'W. Service industries; agriculture (peanuts, cotton, tobacco, grain; poultry, livestock). Incorporated in 1823 and named after Andrew Jackson, seventh President of the United States, who was born on the North Carolina/South Carolina line.

Jackson, borough, OCEAN county, E NEW JERSEY; 40°06'N 74°21'W. Site of Six Flags Great Adventure Theme Park with looping roller coasters and drive-through animal safari park.

Jackson Bay, village, NEW ZEALAND, southernmost settlement in WEST COAST region of SOUTH ISLAND, 150 mi/241 km SW of HOKITIKA, on SW shore of Jackson Bay (13 mi/21 km wide). Some tourism.

Jacksonboro, village, COLLETON county, S SOUTH CAROLINA, near EDISTO RIVER, 15 mi/24 km SE of WALTERBORO. Site of State legislature in 1781 when the British occupied Charleston; timber.

Jacksonburg (JAK-suhn-buhrg), village (2006 population 69), BUTLER county, extreme SW OHIO, 7 mi/11 km W of MIDDLETOWN; 39°32'N 84°30'W. Sometimes spelled Jacksonburgh.

Jackson Center (JAK-suhn SEN-tuhr), village (□ 1 sq mi/2.6 sq km; 2006 population 1,453), SHELBY county, W OHIO, 12 mi/19 km NE of SIDNEY; 40°26'N 84°02'W.

Jackson Center, borough (2006 population 214), MERCER county, W PENNSYLVANIA, 6 mi/9.7 km NE of MERCER, on Yellow Creek. Light manufacturing; dairying. Lake Latonka reservoir to W.

Jackson Creek, AUSTRALIA: see KEILOR.

Jackson, Fort, fort in LOUISIANA: see TRIUMPH.

Jackson, Fort, fort in SOUTH CAROLINA: see COLUMBIA.

Jackson Gulch Dam, Colorado: see MANCOS RIVER.

Jackson Heights, section of borough of QUEENS, NEW YORK city, SE NEW YORK; 40°45'N 73°53'W. Just S of LAGUARDIA AIRPORT; Flushing Meadows to E, BROOKLYN-QUEENS EXPRESSWAY to W. Its main commercial thoroughfare is 37th Avenue. Once a mix of Italians, Jews, Irish, Poles, and later Colombians, the neighborhood saw a major influx of Asian Indians, Koreans, Thais, and Chinese in the early 1980s, which has given it a remarkable ethnic diversity. Begun in 1910s as a project by the Queensboro Corp., most of the apartment buildings and homes were built in Georgian and Tudor style. The purpose was to create a unified community modeled on the garden suburbs in England and Germany. With their architectural flourishes, luxurious gardens, and arboreal sidewalks, the structures here coined the term "garden apartments." The thirty-block historic district here is one of only two in Queens.

Jackson Hole, valley (c. 50 mi/80 km long and 6 mi/9.7 km to 8 mi/12.9 km wide), TETON county, NW WYOMING. GRAND TETON NATIONAL PARK protects all but S end of valley. The valley is hemmed in—hence the term "hole"—by mountains, the TETON RANGE (W), Snake River Range (SW), GROS VENTRE RANGE (SE), and lesser ranges (NE). JACKSON LAKE is in N, 39 sq mi/101 sq km, originally a natural lake on the SNAKE RIVER, was dammed in 1911 and 1916 to control the river's flow. The average altitude of the valley is over 6,500 ft. Descriptions of the valley and its features were recorded in the journals of John Colter, who had been a member of the Lewis and Clark Expedition. After returning to the ROCKY MOUNTAINS, Colter entered the region in 1806 in the vicinity of Togwotee Pass and became the first white American to see the valley. His reports of the valley, the Teton Range and of the Yellowstone region to the north were viewed by people of the day with skepticism. Jackson Hole was first settled in 1880s, but was popular with hunters and trappers from the time U.S. trapper David Jackson, for whom it was named, wintered there 1828–1829. In the 1890s, cattle ranching became the major focus of the area, and with cattle ranching came a larger and more permanent settlement. The valley is now a major tourist destination, with the Jackson Hole Mountain Resort and Grand Targhee ski areas attracting tourists year round, and

with the only airport in a U.S. national park. The town of JACKSON (often mistakenly called 'Jackson Hole') is the only incorporated municipality of the county. Teton Village is another community in Jackson Hole, and is the home of the Jackson Hole Mountain Resort. In addition to the ski resort, there is the aerial tram, specialty shops, restaurants, motels and condominiums. The National Elk Refuge, SE of Grand Teton National Park (1,500 acres/607 ha), is the winter home of the largest elk herd in N. America. Concern for wintering elk began early in Jackson Hole. The severe winter of 1908–1909 brought the concern to a head; thousands of elk were starving in the valley. The townspeople, with the help of the state of Wyoming, bought hay to help the animals through the winter, but the following winter was no better. The U.S. Biological Survey Elk Refuge was established in 1912 with an allotment of 1,000 acres. Today the National Elk Refuge, the direct descendant of the original refuge, feeds over 7,000 elk every winter. Bald eagles and the rare trumpeter swan inhabit the area.

Jackson Junction, town (2000 population 60), WINNESHIEK county, NE IOWA, 18 mi/29 km SW of DECORAH; 43°06'N 92°02'W. Limestone quarries.

Jackson Lake, reservoir (c.10 mi/16 km long, 1 mi/1.6 km wide), on BUTTS-JASPER county border, central GEORGIA, on OCMULGEE RIVER, 7 mi/11.3 km E of JACKSON; 33°19'N 83°50'W. Receives (N) ALCOVY, YELLOW, and south rivers, which form Ocmulgee River here. Formed by LLOYD SHOALS Dam (c.100 ft/30 m high, 500 ft/152 m long), built (1910) for power generation. Popular recreation area. Also called Lloyd Shoals Reservoir.

Jackson, Lake (JAK-suhn), reservoir (c.2 mi/3.2 km long), PRINCE WILLIAM county, NE VIRGINIA, at confluence of Cedar Run and Broad Run forming OCCOQUAN Creek, 4 mi/6.4 km SSE of MANASSAS; 38°42'N 77°24'W. Village of Lake Jackson near dam.

Jackson Lake (□ 40 sq mi/104 sq km; 18 mi/29 km long, average 4 mi/6.4 km wide), TETON county, NW WYOMING, on SNAKE River, in JACKSON HOLE valley, in GRAND TETON NATIONAL PARK, 28 mi/45 km NNE of JACKSON; 43°51'N 110°36'W. Elevation c. 6,750 ft/2,057 m. Natural lake with level raised by JACKSON LAKE Dam (70 ft/21 m high), at SE corner, built (1916) for irrigation and flood control. River enters from N. TETON RANGE Mountains bound lake on W.

Jackson, Mount (10,007 ft/3,050 m), ANTARCTICA, in S part of PALMER LAND; 71°23'S 63°22'W. Highest point on the ANTARCTIC PENINSULA. Also known as Mount Gruening.

Jackson, Mount, peak in MONTANA: see LEWIS RANGE.

Jackson, Mount, peak in NEW HAMPSHIRE: see PRESIDENTIAL RANGE.

Jackson Mountains, (8,923 ft/2,720 m at KING LEAR peak), NW NEVADA, in HUMBOLDT county, E of BLACK ROCK DESERT. QUINN RIVER to W.

Jackson, Port (JAK-suhn, PORT) or **Sydney Harbour** (SID-nee HAHR-buhr), inlet of the PACIFIC OCEAN, 22 sq mi/57 sq km, 12 mi/19 km long and 1.5 mi/2.4 km wide at its mouth, NEW SOUTH WALES, AUSTRALIA, forming Australia's finest harbor; 33°51'S 151°15'E. The Parramatta River forms its W arm. Sydney on the S shore is connected with its northern suburbs by Sydney Harbour Bridge (1932), the second-longest steel-arch bridge in the world, with an arch span of 1,650 ft/503 m. Variant name: Port Jackson Bay.

Jacksonport, village (2000 population 235), JACKSON county, NE ARKANSAS, 3 mi/4.8 km NW of NEWPORT, and on WHITE RIVER; 35°38'N 91°18'W. Jacksonport State Park is here.

Jacksonport, village, DOOR county, NE WISCONSIN, on DOOR PENINSULA, on LAKE MICHIGAN, 13 mi/21 km NE of STURGEON BAY. Formerly a lumber port. Whitefish Dunes State Park to S.

Jackson Reservoir (□ 4 sq mi/10.4 sq km), MORGAN county, NE COLORADO, on tributary of SOUTH PLATTE

RIVER, 30 mi/48 km E of GREELEY; 40°22'N 104°05'W. Maximum capacity 47,000 acre-ft. Formed by Jackson Lake Dam (38 ft/12 m high), built (1900) for irrigation. Jackson Lake State Park at N end of reservoir.

Jackson River (JAK-suhn), c.75 mi/1,221 km long, W VIRGINIA; rises in ALLEGHENY MOUNTAINS in central HIGHLAND county, flows SSW, through LAKE MOOMAW reservoir (formed by Gathright Dam), past COVINGTON, and ENE, past CLIFTON FORGE, joins COWPASTURE RIVER SE of IRON GATE to form JAMES RIVER.

Jackson Springs (JAK-suhn SPREENGZ), unincorporated village, MOORE county, central NORTH CAROLINA, 13 mi/21 km WNW of SOUTHERN PINES near Drowning Creek (LUMBER RIVER); 35°12'N 79°37'W. Agriculture (tobacco, grain; livestock).

Jacksonville, city (2000 population 8,404), Calhoun co., E Alabama, 12 mi/19 km NNE of Anniston. In dairying area; textiles, lumber, fabricated metal products, furniture. Jacksonville State University here. Fort McClellan nearby. First named 'Drayton' and then 'Madison,' it was named 'Jacksonville' for Andrew Jackson, whose troops under John Coffee had fought here during the Creek Indian War of 1813–1814. Inc. in 1836.

Jacksonville, city (2000 population 29,916), PULASKI county, central ARKANSAS; 34°52'N 92°07'W. Railroad junction. Manufacturing (printing and publishing, electronic equipment, ordnance, plastic products, fabricated metal products, fixtures). The nearby Little Rock Air Force Base to N, a tactical air command installation; defense-related industries and missile bases are also important to Jacksonville's economy. Incorporated 1941.

Jacksonville (JAK-suhn-vil), city (□ 918 sq mi/2,386.8 sq km), coextensive (since 1968) with ⊙ DUVAL county, extreme NE FLORIDA, on the ST. JOHNS RIVER near its mouth on the ATLANTIC OCEAN; 30°19'N 81°39'W. The largest city in the state (and in the contiguous U.S. in total area), it is a railroad, air, and highway focal point and a busy port of entry, with ship repair yards and extensive freight-handling facilities. Lumber, paper, and wood pulp are the principal exports; automobiles and coffee are among imports. The city has a large and diverse manufacturing base. Jacksonville is one of the most important Southern centers of commerce, finance, and insurance on the Atlantic coast. It is also a major East Coast center of U.S. navy operations; three important naval installations are in the area, including Jacksonville Naval Air Station and the large MAYPORT base at the mouth of the ST. JOHNS RIVER. Jacksonville is also a tourist resort, with ocean beaches, fishing and yachting facilities, and inland hunting areas. Educational facilities include the University of North Florida, Jacksonville University, Edward Walters College, Jones College, and a junior college. Incorporated 1832.

Jacksonville, city (2000 population 18,940), W central ILLINOIS, 31 mi/50 km W of SPRINGFIELD; ⊙ MORGAN county; 39°44'N 90°13'W. Its industries include bookbinding and manufacturing of plastics, fabricated metal products. It is the seat of Illinois College, MacMurray College, Jacksonville. Correctional Center, a state mental hospital, and schools for the deaf and blind. Stephen A. Douglas and William Jennings Bryan lived there. Jacksonville was a station on the Underground Railroad and on Illinois's first true railroad, the Northern Cross. Laid out 1825, incorporated 1840.

Jacksonville (JAK-suhn-vil), city (□ 45 sq mi/117 sq km; 2006 population 69,688), E NORTH CAROLINA, c.105 mi/169 km SE of RALEIGH on the NEW RIVER; ⊙ ONSLOW county; 34°45'N 77°24'W. Service industries; retail trade; manufacturing (food, printing and publishing, machinery). It is also a summer resort. CAMP LEJEUNE U.S. Marine Corps training base, is adjacent to the city, to S and SE; New River, a Marine Air Station, to S; both installations play a major role in Jacksonville's economy. A Coastal Carolina

Community College is here. Hoffmann Forest to NE; Hammocks Beach State Park to SE; TOPSAIL ISLAND beach resort to S. Settled c.1757.

Jacksonville, city (2006 population 14,402), CHEROKEE county, E TEXAS, 27 mi/43 km S of TYLER; 31°58′N 95°15′W. Elevation 516 ft/157 m. Railroad junction; tomato-shipping center in rich vegetable-growing region; canneries; timber, nursery plants; manufacturing (furniture, wood products, apparel); oil and gas. Baptist Missionary Theologogian College, Jacksonville College (two year), Lon Morris College (two year). Lake Jacksonville to SW; LAKE PALESTINE to NW. Founded nearby 1847; moved to present site 1872.

Jacksonville, town (2000 population 118), TELFAIR county, S GEORGIA, 22 mi/35 km NNW of DOUGLAS; 31°49′N 82°59′W.

Jacksonville, town (2000 population 163), RANDOLPH county, N central MISSOURI, 11 mi/18 km N of MOBERLY; 39°35′N 92°28′W. Corn, soybeans, cattle.

Jacksonville, city (2000 population 2,195), JACKSON county, SW OREGON, 5 mi/8 km WSW of MEDFORD; 42°18′N 122°58′W. Timber. Wineries. Historic gold rush town. Site of Museum of Southern Oregon History. Music festivals.

Jacksonville (JAK-suhn-vil), village (2006 population 560), ATHENS county, SE OHIO, 9 mi/14 km N of ATHENS, on small Sunday Creek; 39°28′N 82°05′W. In former coal-mining area.

Jacksonville, village (2006 population 224), WINDHAM CO., VERMONT; part of WHITINGHAM, town.

Jacksonville, borough, INDIANA county, SW central PENNSYLVANIA, 9 mi/14.5 km SW of INDIANA, on Aultsmans Run. Dairying.

Jacksonville Beach (JAK-suhn-vil), city (□ 21 sq mi/54.6 sq km; 2000 population 20,990), DUVAL county, extreme NE FLORIDA, 17 mi/27 km ESE of JACKSONVILLE, on the Atlantic; 30°16′N 81°22′W. Incorporated 1907.

Jacktown, town, SINOE county, SE LIBERIA, on SINOE RIVER and 14 mi/23 km NNE of GREENVILLE. Citrus fruit, palm oil and kernels, cacao, coffee.

Jack Wade, village, E ALASKA, near YUKON border, 65 mi/105 km W of DAWSON, on TAYLOR "Top of the World" Highway. Gold placers; tourism.

Jacmel (zhak-MEL), city, S HAITI, c.25 mi/40 km S of PORT-AU-PRINCE; ⊙ SUD-EST department; 18°14′N 72°32′W. Jacmel is an important fishing port on the CARIBBEAN SEA. Agriculture (cacao, cotton, tobacco, coffee growing and processing); essential oils and pectin distilleries, cotton-seed oil extraction, cigar manufacturing; bauxite, manganese deposits nearby. Also spelled Jaquemel.

Jacmel Peninsula (zhak-MEL), (140 mi/225 km long), SW HAITI, between GULF OF GONAÏVES (N) and the CARIBBEAN SEA (S); traversed by the MASSIF DE LA HOTTE. PORT-AU-PRINCE is at its base. Along its coast are a number of ports–JÉRÉMIE, LES CAYES, JACMEL, and PETIT-GOÂVE–which ship the products of the fertile region (sugarcane, coffee, cacao, cotton, tropical fruit, tobacco, and sisal) and are home to fishing fleets. Manganese, lignite, and bauxite deposits. Sometimes called TIBURON PENINSULA.

Jacó (hah-KO), town, PUNTARENAS province, COSTA RICA, on PACIFIC coast, and 35 mi/56 km SSE of PUNTARENAS; ⊙ Garabito canton (1995 estimated population 4,690); 09°37′N 84°38′W. International tourist resort; minor fishing; agriculture.

Jaco (jah-KO), uninhabited island (□ 5 sq mi/13 sq km) off E tip of TIMOR ISLAND, in TIMOR SEA; 08°26′S 127°20′E. Also spelled Yako; also called Nusa Besi.

Jacobabad (JAH-ko-ba-BAHD), district (□ 1,969 sq mi/5,119.4 sq km), N SIND province, SE PAKISTAN; ⊙ JACOBABAD. A hot, dry tract, irrigated by canals, including NORTH WESTERN CANAL (W) of SUKKUR BARRAGE system; bordered E by INDUS River. Before British occupation of N BALUCHISTAN in late 19th century, area constituted a military frontier, with

cantonment at JACOBABAD. Known to be the hottest place in the entire subcontinent during the summer. Formerly Upper Sind Frontier.

Jacobabad (JAH-ko-ba-BAHD), town, N SIND province, SE PAKISTAN, 255 mi/410 km NNE of KARACHI; ⊙ JACOBABAD district; 28°17′N 68°26′E. Railroad junction. Trade center (grain, dairy products, leather goods); markets millet, rice, wheat; rice milling, handicrafts (carpets, saddlery, palm mats); gas fields. Founded 1847 by General John Jacob; frontier post before British occupation of QUETTA. Noted for intense summer heat (known to reach 127°F/52.7°C in June); hottest point in the INDIAN SUBCONTINENT.

Jacobi Island, Alaska: see YAKOBI ISLAND.

Jacobina (zhah-ko-bee-nah), city, (2007 population 76,452), NE central BAHIA, BRAZIL, near head of RIO ITAPICURU, on railroad, and 55 mi/89 km SSW of SENHOR DO BONFIM; 11°15′S 40°30′W. Ships cattle and caroa fibers. Manganese mines in Serra da Jacobina (NW). Aquamarines and amethysts found in area. Old gold mining center.

Jacob Riis, a municipal park of NEW YORK city, SE NEW YORK, on ROCKAWAY PENINSULA in S QUEENS borough; 40°34′N 73°53′W. Ocean bathing; recreational facilities. Marine Parkway bridge across Rockaway Inlet is here.

Jacobshagen, POLAND: see DOBRZANY.

Jacobs Pillow, Massachusetts: see LEE.

Jacobus (JA-kuh-bus), borough (2006 population 1,199), YORK county, S PENNSYLVANIA, 6 mi/9.7 km S of YORK. Manufacturing (electronic products). Agricultural area (grain, soybeans, apples; poultry, livestock, dairying). Richard M. Nixon County Park to W. Lakes William and Redman (both on East Fork of Codorus Creek) to N.

Jacomino (hah-ko-MEE-no), town, CIUDAD DE LA HABANA province, W CUBA, 3 mi/5 km SE of HAVANA in suburbs.

Jacona de Plancarte (hah-KO-nah de plan-KAHR-tai), town, MICHOACÁN, central MEXICO, on central plateau, 2 mi/3.2 km SSW of ZAMORA DE HIDALGO on Mexico Highway 40; ⊙ Jacona municipio; 19°58′N 102°19′W. Agricultural center (cereals, sugarcane, fruit; livestock).

Jacou (zhah-koo), commune (□ 1 sq mi/2.6 sq km; 2004 population 5,044), HÉRAULT department, LANGUEDOC-ROUSSILLON region, S FRANCE; 43°40′N 03°54′E.

Jacque Peak (13,205 ft/4,025 m), in ROCKY MOUNTAINS, SUMMIT county, NW central COLORADO, 8 mi/12.9 km WSW of BRECKENRIDGE, in Arapaho National Forest. Also known as Eagle River Peak.

Jacques-Cartier (ZHAHK–kahr-TYAI), borough (French *arrondissement*) of SHERBROOKE, S QUEBEC, E CANADA; 45°24′N 71°55′W.

Jacques-Cartier, La (zhahk–kahr-TYAI, la), county (□ 1,278 sq mi/3,322.8 sq km; 2006 population 29,150), Capitale-NATIONALE region, QUEBEC, E Canada; ⊙ SHANNON; 47°01′N 71°34′W. Composed of ten municipalities. Formed in 1981.

Jacques Cartier, Mount (ZHAHK kahr-TYAI) or **Tabletop** (TAI-buhl-tahp), peak (4,160 ft/1,268 m), E QUEBEC, E CANADA, on N side of GASPÉ PENINSULA, 70 mi/113 km W of GASPÉ; 48°59′N 65°57′W. Highest peak of SHICKSHOCK MOUNTAINS, in Gaspesian Provincial Park.

Jacqueville (zhak-VEEL), town, Lagunes region, S coast of CÔTE D'IVOIRE, on Gulf of Guinea, 28 mi/45 km W of ABIDJAN; 05°12′N 04°25′W. Fishing port.

Jacuhy, Brazil: see SOBRADINHO.

Jacuhy River, Brazil: see JACUÍ RIVER.

Jacuí (ZHAH-koo-EE), town (2007 population 7,224), SW MINAS GERAIS, BRAZIL, 25 mi/40 km SE of SÃO SEBASTIÃO DO PARAÍSO; 21°02′S 46°40′W.

Jacuí, Brazil: see SOBRADINHO.

Jacuipe (zhah-koo-EE-pai), town (2007 population 6,680), NE ALAGOAS state, BRAZIL, on Rio Jacuipe and PERNAMBUCO border; 08°48′S 35°28′W.

Jacuí River (ZHAH-soo-ee), 280 mi/451 km long, central RIO GRANDE DO SUL, BRAZIL; rises in the Serra GERAL NE of CRUZ ALTA (30°02′S 51°15′W); flows S and E, past CACHOEIRA DO SUL and RIO PARDO, through SÃO JERÔNIMO coal basin, to PÔRTO ALEGRE, where it is joined by three short streams (CAÍ, SINOS, Gravataí) to form the GUAÍBA, a shallow, widening estuary at N end of the Lagoa dos PATOS. Navigable for river steamers beyond Cachoeira do Sul; used for coal shipments to seaports. Receives the TAQUARI (left). Its valley, settled by German immigrants after 1824, is a rich agricultural district. Rice is grown in flood plain. Formerly spelled Jacuhy.

Jacuizinho (ZHAH-koo-ee-seen-yo), city, central RIO GRANDE DO SUL state, BRAZIL, 42 mi/68 km SE of CRUZ ALTA; 29°02′S 53°04′W. Livestock.

Jacumba (ha-KOOM-bah) unincorporated village, SAN DIEGO county, S CALIFORNIA, at Mexican (BAJA CALIFORNIA NORTE state) border, opposite La Rumorosa, Mexico (no crossing), c.55 mi/89 km ESE of SAN DIEGO. Hot springs. CARRIZO GORGE nearby. ANZA-BORREGO DESERT STATE PARK to N.

Jacundá (ZHAH-koon-DAH), city (2007 population 51,811), E central PARÁ, BRAZIL, 71 mi/114 km N of MARABÁ; 04°30′S 49°27′W.

Jacupiranga (ZHAH-koo-pee-rahn-gah), city (2007 population 16,217), S SÃO PAULO, BRAZIL, in the RIBEIRA DE IGUAPE valley, 27 mi/43 km W of IGUAPE; 24°42′S 48°00′W. Apatite mined here for SANTO ANDRÉ superphosphate plant.

Jacura (hah-KOO-rah), town, ⊙ Jacura district, FALCÓN state, NW VENEZUELA, 40 mi/64 km NW of TUCACAS; 11°04′N 68°51′W. Elevation 1,305 ft/397 m. Corn, yucca, fruit.

Jacutinga (zhah-koo-CHEEN-gah), city (2007 population 20,393), SW MINAS GERAIS, BRAZIL, near SÃO PAULO border, on railroad, and 35 mi/56 km S of POÇOS DE CALDAS; 22°15′S 46°34′W. Coffee-growing center. Mica deposits.

Jacutinga, city, central PARANÁ state, BRAZIL; 25°04′S 51°51′W.

Jada (JAH-dah), town, ADAMAWA state, E NIGERIA, on road, 40 mi/64 km S of YOLA. Market center. Yams, groundnuts, maize, cassava.

Jadacaquiva (hah-dah-kah-KEE-vah), town, FALCÓN state, NW VENEZUELA, on PARAGUANÁ PENINSULA, 13 mi/21 km WSW of PUEBLO NUEVO; 11°54′N 70°05′W. Livestock.

Jadar (YAH-dahr), former county, now in MAČVA district, W SERBIA; ⊙ LOZNICA. Drained by JADAR RIVER (right affluent of the DRINA River). Also spelled Yadar.

Jadar River (YAH-dahr), c. 40 mi/64 km long, E BOSNIA, BOSNIA AND HERZEGOVINA; flows NW and N to Drinjaca River 7 mi/11.3 km S of zvornik. Another Jadar River in W SERBIA is a right affluent of DRINA River. Also spelled Yadar River.

Jade (YAH-de), village, LOWER SAXONY, NW GERMANY, 6 mi/9.7 km SE of VAREL; 53°21′N 08°14′E. In peat region.

Jade Bay (YAH-de), German *Jadebusen* (YAH-de-bus-en), NORTH SEA inlet (10 mi/16 km long, 10 mi/16 km wide), in LOWER SAXONY, NW GERMANY; 53°24′N 08°03′E–53°7′N 08°18′E. Formed by floods (1218, 1511). Receives small Jade River WILHELMSHAVEN is near its mouth (W).

Jadita, LEBANON: see JEDITA.

Jadotsville, CONGO: see LIKASI.

Jadotville, CONGO: see LIKASI.

Jädran (YE-drahn), river, 50 mi/80 km long, E SWEDEN, rises NNE of FALUN, flows generally SE to STORSJÖN at SANDVIKEN.

Jadraque (hah-DHRAH-kai), town, GUADALAJARA province, central SPAIN, on HENARES RIVER, on railroad, and 24 mi/39 km NNE of GUADALAJARA; 40°55′N 02°55′W. In horticultural region; also cereals, grapes, fruit, livestock. Gypsum quarrying; plaster manufacturing. Has ruins of old castle.

Jadu (JE-doo), village, TRIPOLITANIA region, NW LIBYA, on JABAL NAFUSAH plateau, 30 mi/48 km WSW of YAFRAN; 31°57′N 12°01′E. Elevation c. 2,160 ft/658 m. Road junction; manufacturing (textiles; terra-cotta vases); agriculture region (olives, grain; sheep, goats). Formerly Giado.

Jægerspris (YAI-yuhr-sprees), town, FREDERIKSBORG county, SJÆLLAND, DENMARK, 13 mi/21 km SW of HILLERØD; 55°50′N 12°00′E. Orchards.

Jægervasstind, NORWAY: see LYNGEN.

Ja-ela (JAH-A-luh), town (2001 population 30,910), WESTERN PROVINCE, SRI LANKA, 10 mi/16 km NNE of COLOMBO; 07°04′N 79°53′E. Textiles; coconuts, rice, cinnamon. Also written Jaela.

Jaén (hah-EN), province (□ 3,920 sq mi/10,192 sq km), CAJAMARCA province, NW PERU; ⊙ JAÉN; 05°25′S 79°00′W. Ranges E-W from MARAÑÓN RIVER to PIURA region, across CORDILLERA OCCIDENTAL. Agricultural center, with potatoes, cereals; livestock.

Jaén (hai-EN), province (□ 5,209 sq mi/13,543.4 sq km), S SPAIN, in NE ANDALUSIA; ⊙ JAÉN. Bounded N by the SIERRA MORENA; crossed by several mountain ranges; includes plain in central and NW section. Drained by GUADALQUIVIR RIVER and its tributaries and by the SEGURA, which rises here. Climate hot and dry in summer with winter rainfall. Has some dams and hydroelectric power plants. Lead mines in the Sierra Morena (LINARES, LA CAROLINA, SANTA ELENA districts) among richest in EUROPE; also some iron and copper. Lead widely exported. Gypsum, limestone, and marble quarries; some saltworks; mineral springs. Livestock raising (including fine breeds of horses); lumbering in mountain areas. Besides mining, agriculture is chief resource; Jaén is first province in Spain for olive oil production (one-third of total). Broad Guadalquivir valley has hills covered with olive groves, vineyards, and fruit orchards, and fertile plains yielding large crops of cereals. Also produces esparto, vegetables, some potatoes, and tobacco. Except for mining and metalworking (Linares, La Carolina), industries are mostly derived from agriculture: olive oil and esparto processing, brandy and liqueur distilling; flour; sawmills; tanning. Manufacturing of apparel, furniture, some chemicals. Chief cities: Jaén, Linares, ÚBEDA, ANDÚJAR, BAEZA.

Jaén (hah-EN), city (2005 population 63,464), ⊙ JAÉN province, CAJAMARCA region, NW PERU, in CORDILLERA OCCIDENTAL, 100 mi/161 km N of CAJAMARCA, on OLMOS to SAN IGNACIO road; 05°42′S 78°50′W. Elevation 2,428 ft/740 m. Agriculture (cereals, potatoes, coffee, sugarcane, tobacco; livestock).

Jaén (hah-AIN), city (2001 population 112,590), S SPAIN, in ANDALUSIA, ⊙ JAÉN province; 37°46′N 03°47′W. It is a marketing and distribution center for a fertile area producing olive oil and wine. Nearby lead mines are believed to be among the richest in EUROPE; iron and copper are also exploited. Once the seat of a small Moorish kingdom, Jaén was conquered by Ferdinand III of CASTILE in 1246. There are remains of a Moorish castle and walls; an imposing cathedral (16th–18th century); and several palaces.

Jaen (HAH-en), town, NUEVA ECIJA province, PHILIPPINES, on PAMPANGA RIVER, and 11 mi/18 km SSW of CABANATUAN; 15°22′N 120°54′E. Agricultural center (rice, corn).

Jæren (YAHR-uhn), lowland region in ROGALAND county, SW NORWAY, extending from EGERSUND c.25 mi/40 km N along NORTH SEA and c.10 mi/16 km inland. Unlike the rest of Norwegian coast, it is unprotected by offshore islands. Partly covered with lakes, woods, and moors, it has good soil and is Norway's best agricultural area. Exports food and dairy products via SANDNES and STAVANGER.

Jærstrendene (YAHR-stren-duh-nai), protected coastal landscape area, ROGALAND county, NORWAY. Covers 70 km of shoreline from Tungenes, 10 km NE of STAVANGER, S to Sirevåg, with the exclusion of a small

part N of Kolnes. All islands off the JÆREN coast included in the protected area, making it a total of 16 sq km. Parts of the area are subject to special protection of the rich bird life. An ornithological research station has been in operation at Reve since 1937. Established 1977.

Jafarabad (JAH-fuh-rah-BAHD), town, AMRELI district of former SAURASHTRA, now in GUJARAT state, INDIA; port on GULF OF KHAMBAT, on KATHIAWAR peninsula, 75 mi/121 km SE of Junagarh; 20°12′N 76°00′E. Trades in timber, ghee, oilseeds; fishing (chiefly Bombay duck) off coast. Lighthouse (SE). Founded c.1575. Was ⊙ former princely state of Jafarabad of WEST INDIA STATES agency; state merged 1948 with SAURASHTRA, later becoming part of Gujarat state. Also Jafrabad.

Jafarabad, town, JAUNPUR district, SE UTTAR PRADESH state, N central INDIA, on the GOMATI RIVER, 5 mi/8 km SE of JAUNPUR; 20°52′N 71°22′E. Railroad junction; barley, rice, corn, wheat. Muslim fort ruins, 14th-century mosque. Under KANAUJ kingdom, 11th–12th centuries; under Tughlaks in 14th century.

Ja'fariyah, Al-, EGYPT: see GA'FARIYA, EL.

Jaffa, Cape (JA-fuh), SE SOUTH AUSTRALIA, in INDIAN OCEAN, at SW end of LACEPEDE BAY; 36°57′S 139°40′E.

Jaffna, district (□ 359 sq mi/933.4 sq km; 2001 population 490,621), NORTHERN PROVINCE, N tip of SRI LANKA; ⊙ JAFFNA; 09°45′N 80°05′E.

Jaffna, town (2001 population 145,600), N tip of SRI LANKA, on SW JAFFNA Peninsula, on JAFFNA LAGOON, 190 mi/306 km N of COLOMBO; ⊙ JAFFNA district and NORTHERN PROVINCE; 09°39′N 80°00′E. Trade (grain, tobacco, coconut, chili, mangoes, yams); small port (fishing, chank, bêche-de-mer, and turtle fishing). Seat of University of Jaffna. Major saltern is 2 mi/3.2 km E, at CHIVVIYATERU. The city includes NALLUR, capital of the independent Tamil kingdom conquered by the Portuguese in 1617. Jaffna was the last Portuguese possession to fall to the Dutch (1658) and it was occupied by the British in 1795. The old Dutch fort (1680) and buildings were damaged by internal warfare in the 1990s.

Jaffna, peninsula, NORTHERN PROVINCE, northernmost part of SRI LANKA, separated from S INDIA by PALK STRAIT; 09°40′N 80°00′E. The peninsula is densely inhabited, almost entirely by Tamil-speaking people. Tobacco, rice, coconuts, palmyra palm, and vegetables are grown; fishing is an important occupation. The main industries are salt, cement, chemical, and tobacco production. The center of an independent kingdom from the 13th to the 15th centuries; occupied by the Portuguese (1617–1658) and the Dutch (1658–1795) until the British conquest. Since the 1970s it has been the center of violent Tamil resistance against the Sinhalese-dominated government of Sri Lanka. Militant groups, particularly the Liberation Tigers, controlled the area 1986–1988 and 1990–1996. Indian troops occupied the area 1988–1990 in an attempt to implement the Indo-Sri Lankan Agreement of 1987, which provided for some political devolution. Sri Lankan armed forces reoccupied the peninsula in early 1996.

Jaffna Lagoon (c.50 mi/80 km long, 10 mi/16 km wide), NORTHERN PROVINCE, N tip of SRI LANKA, between JAFFNA Peninsula (N) and Sri Lanka proper (S), separated from PALK STRAIT (NW) by MANDAITIVU, VELANAI, and KARAITIVU islands; 09°35′N 80°05′E. Chank and bêche-de-mer fishing. Saltern near ELEPHANT PASS. Army camp at POONERYN, near S shore. JAFFNA and CHAVAKACHCHERI are on N shore.

Jaffrey, town, CHESHIRE county, SW NEW HAMPSHIRE, 15 mi/24 km SE of KEENE; 42°49′N 72°03′W. Railroad terminus. Agriculture (cattle, sheep, poultry; vegetables; dairying; nursery crops); manufacturing (fabricated metal products, printing and publishing, electronic equipment). Thorndike Pond in N, Contoocook Lake on S boundary; MONADNOCK MOUN-

TAIN (3,165 ft/965 m), in Monadnock State Park, in NW. Settled c.1758, incorporated 1773.

Jafi (jah-FEE), town, IRIAN JAYA, INDONESIA, 61 mi/97 km SE of JAYAPURA, on PAPUA NEW GUINEA border; 03°20′S 140°55′E. Also spelled Yafi.

Jafr, El (JIF-er, el), desert post, E JORDAN, 25 mi/40 km ENE of Ma'an. Police post; sheep, goat raising. Provides some services to bedouins in the area.

Jafura (juh-FOO-ruh), sandy desert, N outlier of the RUB' AL KHALI, SAUDI ARABIA, W of QATAR peninsula.

Jagadalpur, INDIA: see JAGDALPUR.

Jagadhri (juh-GAH-dree), town, YAMUNANAGAR district, HARYANA state, N INDIA, 34 mi/55 km SE of AMBALA; 30°10′N 77°18′E. Trade center for grain, cotton, timber, borax, sugarcane; paper milling, cotton ginning, chemical manufacturing, handicraft cloth weaving and dyeing, sugar refining; fabricated metal products. Village of SUGH, 4 mi/6.4 km SE, has ruins of 7th-century Buddhist and Brahman seat of learning.

Jagalur (JUH-gah-loor), town, CHITRADURGA district, KARNATAKA state, INDIA, 20 mi/32 km N of CHITRADURGA; 14°32′N 76°21′E. Livestock grazing in nearby scrub forests.

Jagannathapuram, INDIA: see KAKINADA.

Jagannathganj, BANGLADESH: see JAMALPUR.

Jagatsinghpur (juh-guht-SING-poor), district (□ 679 sq mi/1,765.4 sq km), ORISSA state, E INDIA; ⊙ JAGATSINGHPUR.

Jagatsinghpur (juh-guht-SING-poor), town, ORISSA state, E INDIA; ⊙ JAGATSINGHPUR district; 20°16′N 86°10′E.

Jagdalak (JAHG-dah-lak) or **Jigdalik**, village, NANGARHAR province, AFGHANISTAN, 40 mi/64 km W of JALALABAD on highway; 34°26′N 69°46′E. Ruby mining. Silk mill.

Jagdalpur (JUHG-duhl-puhr), town (2001 population 73,687), SE BASTAR district, S CHHATTISGARH state, central INDIA, on INDRAVATI RIVER, and 155 mi/249 km S of RAIPUR; ⊙ BASTAR district; 19°04′N 82°05′E. Serves as trade center for a large, underdeveloped hinterland. Rice, millet, oilseeds. Surrounded by dense forests (sal, bamboo, myrobalan). Was capital of former princely state of BASTAR, one of CHHATTISGARH STATES; state capital moved here from BASTAR village, 11 mi/18 km NW, in 18th century. Sometimes spelled JAGADALPUR; sometimes called BASTAR.

Jagdispur (juhg-DEESH-poor), or **Jagdishpur**, town (2001 population 28,071), BHOJPUR district, W central BIHAR state, E INDIA, on SON CANALS branch, and 16 mi/26 km WSW of ARA. Sugar-processing center; rice, gram, wheat, oilseeds, barley, corn.

Jägerndorf, CZECH REPUBLIC: see KRNOV.

Jagersfontein, town, SW FREE STATE province, SOUTH AFRICA, 70 mi/113 km SW of MANGAUNG (BLOEMFONTEIN); 29°45′S 25°25′E. Elevation 4,618 ft/1,408 m. Diamond-mining center; gold deposits nearby. Airfield. First diamond discovered here 1870; mines closed 1932; reopened 1949.

Jaggayyapeta (juh-GEI-yuh-pet-tuh), town, KRISHNA district, ANDHRA PRADESH state, INDIA, on left tributary of KRISHNA RIVER, and 40 mi/64 km NW of VIJAYAWADA; 16°54′N 80°06′E. Cotton ginning, handloom silk weaving; rice, peanuts. Bamboo, myrobalan in nearby forests. Has Buddhist stupa. Also spelled Jaggayapet or Jagaiahpet.

Jaghatay (jahg-hat-AI), village, Khorāsān province, NE IRAN, 45 mi/72 km NW of SABZEVAR, at N foot of short Jaghatay range; 28°30′N 53°33′E. Grain, silk. Copper deposits nearby.

Jaghatu River, IRAN: see ZARINEH RIVER.

Jaghbub (jeg-BOOB), Italian *Giarabub*, oasis and township, CYRENAICA region, NE LIBYA, near Egyptian border, 140 mi/225 km S of BARDIYAH, in N part of LIBYAN DESERT; 29°45′N 24°31′E. Agriculture (fruit). Holy site and religious center of Senusi; sanctuary contains tomb of sect founder. Awarded to Italian Libya by Italo-Egyptian treaty (1925). Formerly Jarabub.

Jagnair (juhg-NEIR), town, AGRA district, W UTTAR PRADESH state, INDIA, 33 mi/53 km SW of AGRA; 26°52′N 77°36′E. Pearl millet, gram, wheat, barley, oilseeds. Also spelled JAGNER.

Jagner, INDIA: see JAGNAIR.

Jagniatkow, POLAND: see AGNETENDORF.

Jagodina (YAH-go-dee-nah), city, POMORAVLJE district, central SERBIA, on railroad, and 70 mi/113 km SSE of BELGRADE, near MORAVA R.; 43°59′N 21°15′E. Meat-packing, brewing. Wine growing in vicinity. Known as SVETOZAREVO under Tito, its pre-1947 name has been restored. Also spelled Yagodina.

Jagodzin, UKRAINE: see YAHODYN.

Jagraon (JUHG-roun), town, LUDHIANA district, PUNJAB state, N INDIA, 24 mi/39 km WSW of LUDHIANA; 30°47′N 75°29′E. Trades in wheat, gram, sugarcane, horses; hand-loom woolen weaving, cotton ginning, ivory carving; metalware.

Jagstfeld, GERMANY: see BAD FRIEDRICHSHALL.

Jagst River (YAHGST), over 100 mi/161 km long, in S GERMANY; rises 8 mi/12.9 km E of ELLWANGEN, meanders generally N and W, past CRAILSHEIM, to the NECKAR at BAD FRIEDRICHSHALL; source at 48°59′N 10°19′E.

Jagtial (JUHG-tyuhl), town, KARIMNAGAR district, ANDHRA PRADESH state, INDIA, 29 mi/47 km NNW of KARIMNAGAR; 18°48′N 78°56′E. Millet, rice, oilseeds, cotton. Nearby forests supply bamboo pulp to SIRPUR paper mill.

Jaguapitã (ZHAH-gwah-pee-tah), city, NW Paraná state, 31 mi/50 km NW of Londrina; 23°07′S 51°33′W. Coffee, corn, rice, cotton; livestock.

Jaguaquara (zhah-goo-ah-KWAH-rah), city, (2007 population 46,505), E BAHIA, BRAZIL, on railroad, and 20 mi/32 km NNE of JIQUIÉ; 13°32′S 39°58′W. Coffee, tobacco, livestock.

Jaguarão (zhah-gwah-ROUN), city (2007 population 27,944), S RIO GRANDE DO SUL, BRAZIL, on JAGUARÃO River (URUGUAY border; international bridge), opposite RÍO BRANCO, on railroad, and 75 mi/121 km SW of PELOTAS; 32°34′S 53°23′W. Livestock raising and shipping center; trades in cereals, wool, wine. Airfield; custom station.

Jaguarão River (zhah-gwah-ROUN), Sp. *Yaguarón*, 135 mi/217 km long; rises in BRAZIL in S RIO GRANDE DO SUL E of BAGÉ (32°39′S 53°12′W), flows SE to MIRIM LAKE, forming (through most of its course) Brazil–URUGUAY border. Crossed by international road and railroad bridge between JAGUARÃO (Brazil) and RÍO BRANCO (Uruguay).

Jaguarari (zhah-goo-ah-RAH-ree), city, (2007 population 29,128), N BAHIA, BRAZIL, on railroad to JUÀ-ZEIRO, and 13 mi/21 km N of SENHOR DO BONFIM; 10°15′S 40°15′W. Sugarcane, cotton, tobacco; rose-quartz and kaolin deposits. Formerly spelled Jaguarary.

Jaguaré (ZHAH-gwah-RAI), city (2007 population 21,923), NE ESPÍRITO SANTO, BRAZIL, 25 mi/40 km SW of SÃO MATEUS; 18°52′S 40°08′W.

Jaguaretama (ZHAH-gwah-re-TAH-mah), city (2007 population 17,854), E central CEARÁ, BRAZIL, near RIO JAGUARIBÉ, 45 mi/72 km SW of MORADA NOVA; 05°37′S 38°40′W.

Jaguari (ZHAH-gwah-ree), city (2007 population 11,626), W central RIO GRANDE DO SUL, BRAZIL, on railroad, and 55 mi/89 km WNW of SANTA MARIA; 29°30′S 54°41′W. Cattle raising. Formerly spelled Jaguary.

Jaguari, Brazil: see JAGUARIÚNA.

Jaguariaíva (ZHAH-gwah-ree-ah-EE-vah), city (2007 population 31,865), E PARANÁ, BRAZIL, 85 mi/137 km NNW of CURITIBA; 24°15′S 49°47′W. Important meat-packing center (hogs); manufacturing (food, paper, tannin); corn and rice milling. Railroad. Old spelling, Jaguariahyva.

Jaguaribara (ZHAH-gwah-ree-BAH-rah), town (2007 population 9,992), E central CEARÁ, BRAZIL, on RIO JAGUARIBE, 19 mi/31 km downriver from JAGUARIBE; 05°38′S 38°39′W.

Jaguaribe (zhah-gwah-REE-be), city, (2007 population 33,389), E CEARÁ, BRAZIL, on RIO JAGUARIBE, and 60 mi/97 km NE of IGUATU; 05°47′S 38°31′W. Cotton, sugar, cattle. Formerly called Jaguaribe Mirim.

Jaguaripe (zhah-goo-ah-REE-pai), city, (2007 population 16,205), E BAHIA, BRAZIL, on inlet of the Atlantic, 8 mi/12.9 km SE of NAZARÉ; 13°10′S 38°50′W. Ceramics; lumber, piassava.

Jaguariúna (ZHAH-gwah-ree-OO-nah), city (2007 population 36,801), E SÃO PAULO, BRAZIL, 15 mi/24 km NE of CAMPINAS; 22°41′S 46°59′W. Coffee. Until 1944, called Jaguari.

Jaguaruana (zhah-goo-ah-roo-AH-nah), city, (2007 population 30,843), NE CEARÁ, BRAZIL, on left bank of lower RIO JAGUARIBE, and 20 mi/32 km S of ARACATI; 04°47′S 37°45′W. Cotton, carnauba wax. Until 1944, called União.

Jaguaruna (zhah-gwah-roo-nah), city (2007 population 15,668), SE SANTA CATARINA, BRAZIL, near the Atlantic, on railroad, and 18 mi/29 km SW of LAGUNA; 28°36′S 49°02′W. Lumber.

Jaguary, Brazil: see JAGUARI, Rio Grande do Sul.

Jagüel, Sierra de (hah-GAIL, see-YER-rah dai), subandean mountain range rising to c.10,000 ft/3,048 m, in NW LA RIOJA province, ARGENTINA, N of VINCHINA. Extends c.30 mi/48 km SSW from CATAMARCA province border. Sometimes called Sierra de Vinchina.

Jaguey, Colombia: see REMOLINO.

Jagüey Grande (hah-GWAI GRAHN-dai), town (2002 population 27,248), MATANZAS province, W CUBA, on railroad and National Highway, and 45 mi/72 km SE of MATANZAS; 22°32′N 81°08′W. Agricultural center (sugarcane, honey, poultry, cattle); lumbering and charcoal burning. The Australia sugar central is 2 mi/3 km S.

Jaguito, El, PANAMA: see EL JAGUITO.

Jahanabad (juh-HAH-nuh-bahd), or **Jehanabad**, district (☐ 606 sq mi/1,575.6 sq km; 2001 population 1,511,406), BIHAR state, E INDIA; ☉ JAHANABAD. Rice, wheat, lentils, maize. Carved out of N GAYA district in 1986. Was site of widespread Naxalite violence in early 2000s.

Jahanabad (juh-HAH-nuh-bahd), or **Jehanabad**, town (2001 population 81,723), JAHANABAD district, W central BIHAR state, E INDIA, on GANGA plain, on tributary of the GANGA River, and 30 mi/48 km N of GAYA. Road junction; trades in rice, gram, oilseeds, wheat, barley. Saltpeter processing nearby.

Jahanabad, INDIA: see ARAMBAGH.

Jahangirabad (juh-hahng-GEE-rah-bahd), town, BULANDSHAHR district, W UTTAR PRADESH state, INDIA, 14 mi/23 km E of BULANDSHAHR. Trades in wheat, oilseeds, barley, jowar, sugarcane; hand-loom weaving. Founded 16th century.

Jahannam, ISRAEL: see HINNOM.

Jahazpur (juh-HAHZ-puhr), town, BHILWARA district, S central RAJASTHAN state, NW INDIA, 12 mi/19 km SSW of DEVLI; 25°37′N 75°17′E. Millet, wheat, oilseeds.

Jahorina (yah-ho-REE-nah), mountain (6,284 ft/1,915 m), in DINARIC ALPS, central BOSNIA, BOSNIA AND HERZEGOVINA; extends c. 10 mi/16 km NW-SE; 44°07′N 19°13′E. Highest point, Sjenište peak, is 12 mi/19 km SSE of SARAJEVO. Also spelled Yakhorina.

Jahra (ZHA-her-uh), town (2005 population 28,387), KUWAIT, near SW corner of Kuwait Bay, 20 mi/32 km W of Kuwait town; ☉ Al Jahra governorate; 29°19′N 47°40′E. Irrigated agricultural center; wheat, barley, lucerne, dates, melons.

Jahra, Al, governate (2005 population 272,373), NE KUWAIT, at head of PERSIAN GULF; 29°20′N 47°40′E. Famous Red Palace; battle of Al Jahra. Al Rawdhatain oil fields.

Jahrom (jah-RUHM), town (2006 population 105,285), FĀRS province, S IRAN, 100 mi/161 km SE of SHIRAZ. Dates, tobacco, grain, cotton.

Jahrum, IRAN: see JAHROM.

Jahú, Brazil: see JAÚ.

Jahuel (jah-WAIL), mountain resort (3,900 ft/1,189 m), SAN FELIPE DE ACONCAGUA province, central CHILE, in Andean foothills, 10 mi/16 km NE of SAN FELIPE; 32°43′S 70°39′W. Has noted thermal springs and mineral waters. Sometimes Baños de Jahuel or Termas de Jahuel.

Jahwarian, PAKISTAN: see JHAWARIAN.

Jaíba (zhah-EE-bah), city (2007 population 29,849), N central MINAS GERAIS, BRAZIL, 43 mi/69 km NW of JANAÚBA; 15°22′S 43°22′W. Center of large irrigation farming development along Rio Verde Grande. Purported to be one of the largest such projects in South America.

Jaicós (ZHAH-ee-kos), city (2007 population 16,827), E central PIAUÍ, BRAZIL, 65 mi/105 km SE of OEIRAS; 07°21′S 41°08′W. Carnauba wax, maniçoba rubber, hides. Airfield.

Jaihan, TURKEY: see CEYHAN RIVER.

Jailolo (JEI-lo-lo), town, HALMAHERA island; 01°05′N 127°30′E. Airport.

Jailolo, INDONESIA: see HALMAHERA.

Jaime Prats (HEI-me), village, SE central MENDOZA province, ARGENTINA, near ATUEL RIVER (irrigation), 36 mi/58 km SE of SAN RAFAEL; 34°54′S 67°48′W. Railroad terminus; agricultural center (wine, alfalfa, grain; livestock; apiculture). Formerly Atuel Sud.

Jainagar, INDIA: see JAYNAGAR.

Jainpur, INDIA: see JIANPUR.

Jainti, INDIA: see RAJABHAT KHAWA.

Jaintia Hills (JEINT-yuh), district (☐ 1,475 sq mi/3,835 sq km), MEGHALAYA state, NE INDIA; ☉ JOWAI.

Jaintia Hills (JEINT-yuh), range (c.130 mi/209 km long E-W; rises to c.4,430 ft/1,350 m), part of JAINTIA HILLS district, MEGHALAYA state, NE INDIA, forming part of W ASSAM RANGE, or Meghalaya Plateau, between BRAHMAPUTRA RIVER valley (N) and SHILLONG PLATEAU (S) which separates them from KHASI HILLS. The range lies to the E of GARO and Khasi Hills.

Jaipur (JEI-puhr), former native state, NW INDIA. Founded in the 12th century by the Kachwaha clan of the Rajputs. Became (c.1550) a feudatory of the Mogul empire. In 1818, GREAT BRITAIN exacted a treaty providing for an annual tribute. Now part of RAJASTHAN state.

Jaipur (JEI-puhr), district (☐ 1,114 sq mi/2,896.4 sq km), ORISSA state, E INDIA; ☉ Panikoili.

Jaipur (JEI-poor), district (☐ 5,420 sq mi/14,092 sq km; 2001 population 5,252,388), RAJASTHAN state, NW INDIA; ☉ JAIPUR.

Jaipur (JEI-puhr), city (2001 population 2,324,319), NW INDIA; ☉ JAIPUR district and RAJASTHAN state; 26°55′N 75°49′E. Founded (1728) by Maharaja Jai Singh II, who ordered it to be colored pink, it is therefore known as the "pink city" from the color of its houses. Transportation junction and commercial center. Enclosed by a crenellated wall 20 ft/6 m high. An unusual feature for an Indian city of this size are the wide, regular streets. The grounds of the former maharaja's palace occupy approximately one seventh of the municipal area. Among Jaipur's famed art products are jewelry, enamels, and muslins, as well as paintings and miniature work. Its industries include metalworking, engineering, and manufacturing of beer, glass, carpets, apparel, blankets, and chemicals. Also has a large banking industry. Seat of Rajasthan University. Popular tourist center. Has an open-air observatory (Jantar Mantar). The deserted city of AMBER, 5 mi/8 km away, was capital of JAIPUR state until 1728; has the palace of the Winds, or "Hawa Mahal," a fine example of Rajput architecture.

Jaipur Hāt (jei-poor-haht), town (2001 population 56,585), BOGRA district, N central EAST BENGAL, BANGLADESH, 30 mi/48 km NW of BOGRA, 23 miles N of SANTAHAR; 24°58′N 84°52′E. Agriculture (rice, jute, rape and mustard); rice milling. Also spelled Joypurhat or Jaypurhat; also written Jaipurhat.

Jaipur Residency, India: see RAJPUTANA STATES.

Jais (JEIS), town, RAE BARELI district, central UTTAR PRADESH state, INDIA, 20 mi/32 km W of RAE BARELI; 26°15′N 81°32′E. Trades in rice, wheat, barley, oilseeds, cotton cloth. Jama Masjid built from remains of Hindu temple.

Jaisalmer (JEI-suhl-mair), former principality, RAJASTHAN state, NW INDIA. Jaisalmer was brought under the Mogul empire by Akbar in 1570. It became a British protectorate in 1818. Incorporated 1949 into Rajasthan state.

Jaisalmer (JEI-suhl-mair), district (□ 2,675 sq mi/ 6,955 sq km), RAJASTHAN state, NW INDIA; ⊙ JAISALMER.

Jaisalmer (JEI-suhl-mair), town, RAJASTHAN state, NW INDIA; ⊙ JAISALMER district; 26°55′N 70°54′E. railroad terminus and strategic military post close to the PAKISTAN border. Nearby, in the THAR DESERT, is the future location of a 50-mw solar-powered plant proposed by an American company.

Jai Samand, INDIA: see DHEBAR LAKE.

Jaitaran (jei-TAH-ruhn), town, PALI district, RAJASTHAN state, NW INDIA, 55 mi/89 km E of JODHPUR; 26°12′N 73°56′E. Millet.

Jaito (JEI-too), town, formerly in PATIALA AND EAST PUNJAB STATES UNION, now in FARIDKOT district, PUNJAB state, N INDIA, 10 mi/16 km SSE of KOT KAPURA; 30°28′N 74°53′E. Agricultural market center (gram, wheat, millet, cotton); hand-loom weaving, oilseed milling. Important Sikh religious center. Annual livestock fair. Also Jaitu.

Jaivergi, INDIA: see JEVARGI.

Jajapur, town, CUTTACK district, E ORISSA state, E INDIA, on BAITARANI RIVER, and 38 mi/61 km NE of CUTTACK; 20°51′N 86°20′E. Hindu pilgrimage center (annual festival fair); rice growing. College. Was capital of Orissa (C.E. c.500–950) under Hindu dynasty.

Jajarkot (JAH-juhr-kot), distict, W NEPAL, in BHERI zone; ⊙ JAJARKOT.

Jajarkot (JAH-juhr-kot), village, W central NEPAL, 22 mi/35 km N of SALYAN; ⊙ JAJARKOT district; 28°42′N 82°14′E. Elevation 4,019 ft/1,225 m. Castle ruins. Absorbed by Nepal in late 18th century.

Jajarm, town, Khorāsān province, NE IRAN, on road, and 60 mi/97 km SW of BOJNURD; 36°57′N 56°22′E. Grain, gums.

Jajce (YAH-eet-se), town, W central BOSNIA, BOSNIA AND HERZEGOVINA, on VRBAS RIVER, 30 mi/48 km S of Banja Luka; 44°20′N 17°15′E. In Muslim part of the Federation of BOSNIA AND HERZEGOVINA. Hydroelectric plant at c. 100-ft-/30-m-high waterfall at mouth of PLIVA RIVER (tributary of the VRBAS RIVER); electrochemical (calcium carbide, ferrosilicon, caustic soda) and electrometallurgical (ferromanganese, spiegeleisen) industries; lumber works; light manufacturing. Scenic town. Ruined 13th century castle (captured by the Turks in 1528) and Franciscan church with tomb of last Bosnian king here; seat of Bosnian kings in early 15th century. Site of second session (1943) of provisional parliament which drafted post-World War II constitution of the former YUGOSLAVIA. Also spelled Yaitse or Yaytse.

Jají (hah-HEE), town, MÉRIDA state, W VENEZUELA, in the ANDES, c.18 mi/29 km E of MÉRIDA; 08°34′N 71°20′W. Elevation 5,928 ft/1806 m. Restored buildings. Tourism.

Jajibiriri (jah-jee-BEE-ree-ree), town, YOBE state, NE NIGERIA, on road, 110 mi/177 km NE of POTISKUM; 13°00′N 11°52′E. Market center. Millet, cotton, beans, groundnuts.

Jajja Abbasian (jah-JAH ah-BAHS-yuhn), town, RAHIMYARKHAN district, PUNJAB province, central PAKISTAN, 9 mi/14.5 km NW of KHANPUR; 28°45′N 70°34′E. On railroad spur.

Jajó (hah-HO), town, TRUJILLO state, W VENEZUELA, in ANDEAN spur, 17 mi/27 km S of VALERA. Elevation 5,892 ft/1,796 m. Wheat, corn, potatoes. Colonial architecture.

Jájome Alto, resort, SE central PUERTO RICO, in CORDILLERA CENTRAL, 6 mi/9.7 km NNW of GUAYAMA. Governor's summer residence. SE is a replica of the Grotto of Our Lady of Lourdes.

Jajorm, IRAN: see JAJARM.

Jaj River (JAHJ), c.80 mi/129 km long, N IRAN; rises in ELBURZ mountains 25 mi/40 km NNE of TEHRAN, flows S, past LATYAN (projected dam), and through VARAMIN plain, to NAMAK LAKE salt flats.

Jajura, town (2007 population 5,687), SOUTHERN NATIONS state, central ETHIOPIA, 10 mi/16 km SW of HOSA'INA; 07°28′N 37°47′E. In an agricultural (cereal crops) region.

Ják (YAHK), Hungarian *Ják*, village, VAS county, W HUNGARY, 7 mi/11 km SSW of SZOMBATHELY; 47°08′N 16°35′E. Has Hungary's largest and best-preserved Romanesque church, dating from 1256.

Jakar (JA-kahr), town, center for Bumthang district, central BHUTAN, on the Chamkhar Chu, 43 mi/69 km E of TRONGSA by road; 27°33′N 90°43′E. Area is well known for its woolen weavings and is the site of several industries, including apple growing, cheese making, and animal husbandry. Nearby is Kurjey Lakhang, a temple regarded as one of the holiest spots in Bhutan; here Guru Rimpoche, who first introduced Buddhism into Bhutan in the 8th century, meditated (he is said to have left the imprint of his body on a rock here). Tamshing Lakhang monastery (1505) also nearby. Also called Byakar.

Jakarta (jah-KAHR-tah), city (2000 population 8,389,443), ⊙ and largest city of Indonesia, NW Java, at the mouth of the canalized Ciliwung (or Tjiliwung) River, on Jakarta Bay, an inlet of the Java Sea; 06°10′S 106°48′E. It is the administrative, commercial, industrial, and transportation center of the country, comprising the DKI (Daerah Khusus Ibukoto), a Special Capital Area split into 5 divisions. Food-processing plants, ironworks, automobile and small aircraft assembly plants, textile mills, chemical factories, tanneries, saw mills, soap factories, and printing establishments. Its port, Tanjungpriok, is the largest in Indonesia, handling most of the country's export-import trade. Exports consist mainly of agr., forest, and mining products The city is divided into 3 sections: the old town in the N, with Javanese, Chinese, and Arab quarters; central Jakarta, with high-rise government buildings, hotels, and offices; and the modern residential garden suburb of Kebayoran in the S. With its many canals and drawbridges, N Jakarta somewhat resembles a Dutch town. Landmarks include the great architectural monuments built during President Sukarno's long rule (1945–1965)—freedom statues; a huge sports complex (financed by the Soviet Union); Taman Mini, SE of Jakarta, a theme park highlighting the 27 provinces and major peoples of Indonesia; and the Istiqlal Mosque (the country's largest). Jakarta is the seat of the University of Indonesia. There are notable museum and several 17th-cent. houses and churches. The Dutch founded (c.1619) the fort of Batavia near the Javanese settlement of Jakarta, repulsing English and native attempts to oust them. Batavia became the headquarters of the Dutch E. India Co. and was a major trade center in the 17th century. It declined in the 18th century, following rebellions against the Dutch, but prospered again with the introduction of plantation cultivation in the 19th century From 1811 to 1814, Jakarta was the center of British rule in Java. Batavia was renamed Jakarta in

December 1949, and was proclaimed ⊙ the newly independent Indonesia. Its international airport (Soekarno Hatta), opened 1985; an older, smaller airport at Kemayoran serves local air traffic. Also spelled Djakarta.

Jakhal (JAH-kuhl), town, HISAR district, HARYANA state, INDIA, 45 mi/72 km NNE of HISAR; 29°48′N 75°50′E. Railroad junction.

Jakhau (juh-KOU), town, KACHCHH district, GUJARAT state, INDIA, near ARABIAN SEA, 60 mi/97 km W of BHUJ; 23°13′N 68°43′E. Market center for wheat, fish, barley.

Jakin (JAI-kuhn), village (2000 population 157), EARLY county, SW GEORGIA, 7 mi/11.3 km WNW of DONALSONVILLE, near the CHATTAHOOCHEE RIVER; 31°05′N 84°59′W.

Jakiri (jah-KEE-ree), town, North-West province, CAMEROON, 35 mi/56 km NE of BAMENDA; 06°04′N 10°40′E. Coffee-producing center.

Jakkals, river, SOUTH AFRICA: see LAMBERT'S BAY.

Jakobsberg (YAH-kobs-BER-yuh), suburb of STOCKHOLM, STOCKHOLM county, E SWEDEN, 11 mi/18 km NW of Stockholm city center; 59°26′N 17°47′E.

Jakobshavn (YAH-kops-HOUN), [Greenlandic= Ilulissat], town, W GREENLAND, in DISKO BAY, at mouth of JAKOBSHAVN ICE FJORD; ⊙ Ilulissat (Jakobshavn) commune; 69°13′N 51°04′W. Fishing port and hunting base; meteorological station, hospital. Founded 1741.

Jakobshavn Ice Fjord, Danish *JaKobshavn Isfjord*, inlet (25 mi/40 km long, 3 mi/4.8 km–6 mi/9.7 km wide) of DISKO BAY, W GREENLAND; 69°10′N 50°35′W. At its head it receives large Jakobshavn Glacier, Danish *Jakobshavn Isbræ*, the fastest and most calf-ice producing glacier known—an estimated 30 cubic km annually.

Jakobstad (YAH-kop-STAHD), Finnish *Pietarsaari*, town, VAASAN province, W FINLAND, 50 mi/80 km NE of VAASA; 63°40′N 22°42′E. Elevation 33 ft/10 m. On GULF OF BOTHNIA. Seaport for shipping tobacco and wood products; tobacco-processing center; lumber and pulp mills, machine shops. Over half the population speaks Swedish. Has fourteenth-century church. Founded 1652. Finnish national poet, J.L. Runeberg, born here 1804.

Jakobstadt, LATVIA: see JĒKABPILS.

Jakpa (JAH-bah), town, EDO state, SW NIGERIA, on coast of Bight of BENIN, and 50 mi/80 km SW of BENIN. Market town. Yams, cassava, plantains.

Jakupica (yah-koo-PEET-sah), mountain (8,331 ft/2,539 m), N MACEDONIA, SOLUNSK GLAVA or MOKRA, 20 mi/32 km SSW of SKOPJE. Also spelled Jakubica, Yakubitsa, or Yakupitsa.

Jal, town (2006 population 2,040), LEA county, extreme SE NEW MEXICO, on LLANO ESTACADO, near SE corner of New Mexico, TEXAS state boundary to E and S; 40 mi/64 km S of HOBBS; 32°06′N 103°11′W. Oil fields; cattle, sheep, cotton, wheat, alfalfa. The name is derived from a cattle brand and the Jal Ranch. The origins of the initials are unclear. Settled c.1916, incorporated 1928. Developed with discovery of oil (1927) in vicinity.

Jala (HAH-lah), town, NAYARIT, W MEXICO, at E foot of CEBORUCO VOLCANO, 45 mi/72 km SE of TEPIC; 21°05′N 104°26′W. Corn, beans, sugarcane, fruit; cattle.

Jalacingo (hah-lah-SEEN-go), city and township, VERACRUZ, E MEXICO, in SIERRA MADRE ORIENTAL, 33 mi/53 km NW of XALAPA ENRÍQUEZ on Mexico Highway 129; 19°48′N 97°19′W. Agricultural center (corn, sugarcane, coffee, tobacco, fruit).

Jalai Nur (JAH-LEI-NOH), town, NE INNER MONGOLIA AUTONOMOUS REGION, N CHINA, on railroad, and 18 mi/29 km ESE of MANZHOULI, on channel linking HULUN NUR (lake) and ERGUN RIVER; 49°26′N 117°44′E. Livestock; coal mining.

Jalakandapuram (JAH-luhn-kuhn-kuh-puh-RUHM), or **Jalakandpuram**, or **Jalakantapuram**, town, SALEM district, TAMIL NADU state, S INDIA, 20 mi/32 km WNW of SALEM; 11°42′N 77°53′E. Cotton weaving; castor beans, peanuts, millet.

Jalal-Abad (jah-LAHL-ah-bahd), region (□ 12,991 sq mi/33,776.6 sq km; 1999 population 869,259), NW KYRGYZSTAN; ⊙ JALAL-ABAD. Mostly mountainous; CHATKAL RANGE (NW), the Talas Ala-Too (ALATAU; N), FERGANA RANGE (E); includes NE section of FERGANA VALLEY; drained by NARYN RIVER. Wheat, barley, and cotton grown in mountain valleys and lower irrigated areas; cattle and goat raising; walnut and almond woods; some sericulture. Coal mines at TASH-KÖMÜR and KÖK-JANGGAK, oil fields and sulphur and vanadium mines at MAYLUU-SUU. Manufacturing at Jalal-Abad. Formed in 1939.

Jalalabad (jah-lah-LAH-bahd), city, E AFGHANISTAN, 75 mi/121 km E of KABUL, on KABUL RIVER, just above the mouth of KUNAR RIVER; ⊙ NANGARHAR province; 34°26′N 70°28′E. Near the KHYBER PASS; elevation 1,950 ft/594 m. At center of a large irrigated plain, the city dominates the entrances to the Laghman and Kunar valleys and is a leading trading center with INDIA and PAKISTAN, handling 50% of Afghanistan's exports and imports on Kabul-Peshawar Route, via KHYBER PASS. Agriculture (oranges, rice, sugarcane). Manufacturing (food, paper). Military center; winter resort. Old walled town contains former royal winter residence (1892) in subtropical gardens; it has large bazaars (E), handicrafts shops, and modern factories in NE and residential section in NW. Garden suburbs extend outside walled town. Was the major city of the ancient Greco-Buddhist center of Gandhara. Modern city built c.1570 by Jalaluddin Akbar, and named after him. During the first Afghan War, British troops held (1842) Jalalabad against an Afghan siege. The Pashtuns constitute most of the population. The city has a university and medical school. Under Taliban control in the late 1990s.

Jalal-Abad (jah-LAHL-ah-bahd), city (1999 population 70,401), JALAL-ABAD region, KYRGYZSTAN, in E FERGANA VALLEY, on railroad, 34 mi/55 km ENE of Andijon (UZBEKISTAN), and 155 mi/249 km SSW of BISHKEK; 40°57′N 73°00′E. Industrial center; cotton ginning, food processing; tobacco products, apparel, paper. Health resort (hot springs) nearby. Until 1937 spelled Dzhalyal-Abad. Prior to industrialization in 1930s, known chiefly for its resort. Also spelled DZHALAL-ABAD.

Jalalabad (juh-LAH-lah-BAHD), town, FIROZPUR district, PUNJAB state, N INDIA, 31 mi/50 km SW of FIROZPUR. Hand-loom weaving; wheat, millet, cotton; livestock breeding.

Jalalabad, town, MUZAFFARNAGAR district, N UTTAR PRADESH state, INDIA, 18 mi/29 km NW of MUZAFFARNAGAR. Wheat, gram, sugarcane, oilseeds. Ruins of 18th-century Afghan fort (S).

Jalalabad, town, SHAHJAHANPUR district, central UTTAR PRADESH state, INDIA, 18 mi/29 km SW of SHAHJAHANPUR. Wheat, rice, gram, oilseeds, sugarcane. Founded 13th century.

Jalali (juh-LAH-lee), town, ALIGARH district, W UTTAR PRADESH state, N central INDIA, 11 mi/18 km E of ALIGARH; 27°52′N 78°16′E. Wheat, barley, gram, corn, pearl millet. Extensive imambarahs (mausoleums).

Jalalpur (juh-LAHL-puhr), town, VALSAD district, GUJARAT state, INDIA, on PURNA RIVER and 18 mi/29 km S of SURAT. Market center for timber, rice, fruit; millet, sugarcane, plantains; salt drying. Also spelled Jalapore.

Jalalpur, town, FAIZABAD district, E central UTTAR PRADESH state, N central INDIA, on TONS RIVER, and 15 mi/24 km SE of FAIZABAD. Rice, wheat, gram, sugarcane. Has 18th-century imambarah (mausoleum).

Jalalpur (ja-LAHL-puhr), village, JHELUM district, N PUNJAB province, central PAKISTAN, near JHELUM RIVER, 25 mi/40 km SW of JHELUM; 32°40′N 73°24′E. Identified by some as site from which Alexander the Great crossed the JHELUM to defeat Porus in battle of Hydaspes (326 B.C.E.), afterward erecting the town of BUCEPHALA, in memory of his famous horse. Others place site near town of JHELUM. Also called Jalalpur Kiknan.

Jalalpur Jattan (ja-LAHL-puhr jah-THUN), town, GUJRAT district, NE PUNJAB province, central PAKISTAN, 8 mi/12.9 km NE of GUJRAT; 32°38′N 74°12′E.

Jalalpur Pirwala (ja-LAHL-puhr pir-wah-lah), town, MULTAN district, S PUNJAB province, central PAKISTAN, 50 mi/80 km SSW of MULTAN; 29°30′N 71°13′E. Sometimes called JALALPUR.

Jalama, Arab village, JENIN district, 3.1 mi/5 km N of Jenin, in the S part of the Estraelon Valley, WEST BANK; 32°31′N 35°19′E. Fruits and cereals.

Ja'lan (zhah-LAN), sandy desert region of SE OMAN, on landward side of EASTERN HAJAR hill country and extending to ARABIAN SEA S of RAS AL HADD. Sometimes spelled Ja'alan.

Jalance (hah-LAHN-thai), town, VALENCIA province, E SPAIN, 22 mi/35 km S of REQUENA; 39°12′N 01°04′W. Flour milling, fruit canning, olive-oil processing; esparto.

Jalandhar (juh-luhn-DUHR), district (□ 1,313 sq mi/3,413.8 sq km), central PUNJAB state, N INDIA; ⊙ JALANDHAR. Bounded S by SUTLEJ River, W by KAPURTHALA district, E by HOSHIARPUR district and a pocket of Kapurthala district. Carpentry, hand-loom weaving; construction materials. Agriculture (wheat, corn, gram, cotton, sugarcane). Chief towns are Jalandhar, KARTARPUR, PHILLAUR, NAKODAR. Formerly spelled Jullundur.

Jalandhar (juh-luhn-DUHR), city (2001 population 714,077), PUNJAB state, N INDIA, on the intensively irrigated "doab" between the BEAS and SUTLEJ rivers; ⊙ JALANDHAR district; 23°54′N 78°26′E. Has major road and railroad connections; market for agricultural products. A major automotive-repair center; manufacturing includes textiles, leather goods, wood products. One of capitals of Punjab from India's independence (1947) until CHANDIGARH was built in 1953. Has several colleges.

Jalangi River (juh-LUHN-gee), c.120 mi/193 km long, E WEST BENGAL state, E INDIA; a main distributary of GANGA DELTA, leaves PADMA RIVER 14 mi/23 km SSE of RAJSHAHI, flows generally S past Krishnanagar, and W, joining BHAGIRATHI RIVER at NABADWIP to form HUGLI River. Main distributary, YAMUNA RIVER (left).

Jalán River (hah-LAHN), c.100 mi/161 km long, E central HONDURAS; rises NW of GUAIMACA, flows SE, past Guaimaca, E, and NE to GUAYAPE RIVER in CATACAMAS VALLEY, 4 mi/6.4 km SE of JUTICALPA; 15°43′N 86°30′W. Gold placers along upper course.

Jalapa (ha-LAH-pah), department (□ 797 sq mi/2,072.2 sq km), E central GUATEMALA, in E highlands, on continental divide; ⊙ JALAPA; 14°35′N 89°55′W. Drained by affluents of MOTAGUA River (N) and Ostúa River (inlet of Lake Güija; S); includes volcanoes Alzatate and JUMAY. Agricultural and livestock-raising region; corn, wheat, beans, tobacco, rice; dairying (cheese production); lumber. Main centers: Jalapa, SAN LUIS JILOTEPEQUE, MONJAS.

Jalapa (hah-LAH-pah), city (2002 population 30,500), ⊙ JALAPA department, E central GUATEMALA, in highlands, 36 mi/58 km E of GUATEMALA city; 14°38′N 89°59′W. Elevation 4,469 ft/1,362 m. Commercial center in agricultural area (corn, beans, wheat; cattle, hogs; dairying in N).

Jalapa (hah-LAH-pah), town, NUEVA SEGOVIA department, NW NICARAGUA, at E foot of CORDILLERA DE DIPILTO Y JALAPA, 32 mi/51 km NE of OCOTAL. Tobacco center; sugarcane, coffee, rice, corn.

Jalapa, MEXICO: see XALAPA ENRÍQUEZ.

Jalapa de Méndez (hah-LAH-pah dai MEN-dez), city and township, TABASCO, SE MEXICO, on affluent of GRIJALVA River, and 20 mi/32 km SSE of VILLAHERMOSA; 18°45′N 92°48′W. Rice, coffee, beans, fruit.

Jalapahar (jah-LAH-puh-hahr), town, DARJILING district, N WEST BENGAL state, E INDIA; S suburb of DARJILING; 27°01′N 88°16′E. Elevation c.7,870 ft/2,399 m. Former British cantonment.

Jalaput (juh-LAH-puht), village, KORAPUT district, SW ORISSA state, E INDIA, on Machkhund River, and 28 mi/45 km S of JAYPUR, near ANDHRA PRADESH state border. A demand reservoir constructed at Jalaput, along with installation of a hydroelectric plant.

Jalarpet, INDIA: see TIRUPPATTUR.

Jalatlaco (hah-lah-TLAH-ko), town, MEXICO, central MEXICO, 25 mi/40 km SW of MEXICO CITY. Agricultural center (cereals, vegetables, livestock; dairying). Also Xalatlaco.

Jalaun (jah-LOUN), district (□ 1,763 sq mi/4,583.8 sq km), S UTTAR PRADESH state, INDIA; ⊙ ORAI. Bounded NE by the YAMUNA River; irrigated by distributaries of BETWA Canal. Agriculture (gram, wheat, mustard, jowar, linseed, pearl millet, sesame); babul plantations near KALPI. Main towns: Kalpi, KUNCH, Orai. S area of district was formerly part of British BUNDELKHAND. District enlarged 1950 by incorporated of several former petty states.

Jalaun (jah-LOUN), town, JALAUN district, S UTTAR PRADESH state, INDIA, 12 mi/19 km NW of ORAI; 26°09′N 79°21′E. Trades in gram, wheat, oilseeds, jowar, pearl millet. Was ⊙ 18th-century Maratha state.

Jalbun, Arab village, JENIN district, 7 mi/11 km E of Jenin, the SE slopes of the GILBOA' range, WEST BANK. Agriculture (fruit, wheat). Believed to be on the site of the biblical settlement Gilboa'.

Jalcomulco (hahl-ko-MOOL-ko), town, VERACRUZ, E MEXICO, in SIERRA MADRE ORIENTAL, 17 mi/27 km SE of XALAPA ENRÍQUEZ; 19°20′N 96°33′W. Corn, coffee, fruit.

Jaldak (jal-DAHK), town, ZABUL province, SE AFGHANISTAN, on TARNAK RIVER, and 60 mi/97 km NE of KANDAHAR; 31°58′N 66°43′E. On highway to Kabul.

Jaldessa, ETHIOPIA: see JELDÉSA.

Jaldhaka River (c.145 mi/233 km long), in NE INDIA and BANGLADESH; rises in ASSAM HIMALAYAS in SE SIKKIM state, INDIA, 15 mi/24 km E of GANGTOK, flows S and SSE, past MATABHANGA (WEST BENGAL state) and KURIGRAM (Bangladesh) to the BRAHMAPUTRA RIVER 10 mi/16 km SSE of Kurigram. Receives Dharla River (arm of the TORSA RIVER) 18 mi/29 km S of KOCH BIHAR, West Bengal (below this confluence also known as Dharla).

Jales (ZHAH-les), city (2007 population 47,649), NW SÃO PAULO state, BRAZIL, 69 mi/111 km NNW of ARAÇATUBA, on railroad; 20°16′S 50°33′W. Coffee growing.

Jalesar (juh-LAI-suhr), town, ETAH district, W UTTAR PRADESH state, INDIA, 22 mi/35 km WSW of ETAH; 27°29′N 78°19′E. Hand-loom cotton weaving, saltpeter processing, manufacturing of glass bangles; wheat, pearl millet, barley, corn, oilseeds, cotton. Ruins of 15th-century fort.

Jaleshwar, town, BALESHWAR district, NE ORISSA state, E INDIA, on SUBARNAREKHA RIVER, and 28 mi/45 km NE of BALESHWAR; 21°49′N 87°13′E. Rice milling, hand-loom weaving.

Jaleswar (JUH-les-wor), town (2001 population 22,046), SE NEPAL, in the TERAI, near INDIA border, 10 mi/16 km SW of JANAKPUR; ⊙ MAHOTTARI district; 26°38′N 85°48′E. Trades in rice, corn, wheat, barley, oilseeds, jute. Border transit point between Nepal and India. Nepal's only railroad runs from here to Janakpur.

Jalgaon (JAHL-goun), district (□ 4,542 sq mi/11,809.2 sq km), MAHARASHTRA state, INDIA, on DECCAN PLATEAU; ⊙ JALGAON. Bordered N by SATPURA RANGE, S by AJANTA HILLS; drained by TAPTI and its tributary GIRNA rivers (JAMDA irrigation system begins 8 mi/12.9 km NNW of CHALISGAON). To the N are Asirgarh Hills. The major part of the district is a part of Khandesh Plain. Cotton ginning, handicraft cloth weaving, oilseed milling, dyeing, tanning; also food and paper products. Agriculture (sugarcane, cotton, peanuts, millet, wheat, mangoes); timber in forests at foot of Satpura Range (markets at YAWAL, FAIZPUR). AMALNER, Chalisgaon, Jalgaon, and BHUSAWAL are trade and manufacturing centers and cotton markets. Jalgaon is a major Indian manufacturing center. Annexed 1601 to Mughal empire by Akbar; under Marathas in late seventeenth century. Formerly joined with W Khandesh in one district called Khandesh; divided 1906. Boundaries altered 1950 by exchange of enclaves with HYDERABAD. A wildlife sanctuary N of Yawal.

Jalgaon (JAHL-goun), city (2001 population 368,618), MAHARASHTRA state, W central INDIA; ⊙ JALGAON district; 21°01′N 75°34′E. It is the center of a significant cotton- and banana-growing district. A growing educational center for N Maharashtra state.

Jalhay (zhah-LAI), commune (2006 population 7,981), Verviers district, LIÈGE province, E BELGIUM, 5 mi/8 km ESE of VERVIERS; 50°34′N 05°58′E. Agriculture; lumber.

Jalingo (jah-LEENG-go), town, E NIGERIA, 85 mi/137 km WSW of YOLA; ⊙ TARABA state. Cassava, durra, yams; cattle, skins.

Jalisco (hah-LEES-ko), state (31,152 sq mi/80,684 sq km; 1990 population 5,302,689), W MEXICO; ⊙ GUADALAJARA; 18°58′N 101°28′W. Bounded on the W by the PACIFIC OCEAN, Jalisco is dominated by the S end of the SIERRA MADRE OCCIDENTAL and the W extremity of the TRANSVERSE VOLCANIC AXIS, extending across central Mexico. The hot, tropical plains of the coast are broken by spurs of the Sierra, and most of the E part of the state lies within the central plateau. In the central part of Jalisco is an intermontane basin containing Lake CHAPALA, Mexico's largest lake; it is drained by the LERMA-SANTIAGO (RAMOS) system. Because of the variety of climate, landform, and elevation, nearly every kind of fruit and vegetable grows somewhere in Jalisco. Corn and wheat from the central plateau make it known as the "granary of Mexico"; rice and wheat are grown in the S; and the mountains yield timber and minerals (especially iron, silver, some gold, and precious stones). The raising of livestock and the processing of food products are also important. Although Jalisco was explored as early as 1522, a serious invasion of the area, later included in NUEVA GALICIA, was not undertaken until 1529 by Nuño de Guzmán. Shortly before the War of the Reform (1858–1861), Jalisco became a leading state in the great liberal revolution of Benito Juárez. It was occupied by the French in the wars of intervention but was recaptured in 1866. In 1884 the territory of NAYARIT was separated from Jalisco.

Jalisco, MEXICO: see XALISCO.

Jaljuliya, Arab village, RAMALLAH district, 9.3 mi/15 km of Ramallah, in the Samarian Highlands, on the road to NABLUS, WEST BANK. Agriculture (olives, wheat).

Jaljulye (jal-JUL-yah), Arab township, 2.5 mi/4 km SE of KFAR SAVA on SHARON plain, ISRAEL; 32°23′N 35°02′E. Elevation 124 ft/37 m. Citrus and vegetable growing. Signs of Bronze, Iron Ages, 8th century and Crusader habitation. Ruins include *khan* (inn) from medieval period and Mameluke-period mosque.

Jallieu, FRANCE: see BOURGOIN-JALLIEU.

Jallo (jahl-LO), town, LAHORE district, E PUNJAB province, central PAKISTAN, 9 mi/14.5 km E of LAHORE; 31°35′N 74°30′E. Manufacturing of apparel, chemicals. Dairy farm (S). Modern water park.

Jalna (JAHL-nah), district (□ 2,980 sq mi/7,748 sq km), MAHARASHTRA state, INDIA; ⊙ JALNA.

Jalna (JAHL-nah), city (2001 population 235,795), JALNA district, MAHARASHTRA state, INDIA, 37 mi/60 km E of AURANGABAD; 19°50′N 75°53′E. Road junction; agriculture trade center (chiefly cotton, millet, wheat, peanuts); cotton and oilseed milling, biri

manufacturing. Kadirabad suburb is sometimes spelled Qadirabad or Khadirabad.

Jalo, LIBYA: see JALU.

Jalón (hah-LON), town, ALICANTE province, E SPAIN, 26 mi/42 km ENE of ALCOY. Olive-oil processing; raisins, wine, almonds.

Jalón River, flows 145 mi/233 km generally NE, in SORIA and ZARAGOZA provinces, NE central SPAIN; rises at NE edge of the central plateau 6 mi/9.7 km S of MEDINACELI, past CALATAYUD, to the EBRO 2 mi/3.2 km W of ALAGÓN. Used for irrigation and hydroelectric power. Olive groves and vineyards along its course.

Jalor, district (□ 4,108 sq mi/10,680.8 sq km), RAJASTHAN state, NW INDIA; ⊙ JALOR.

Jalor, town, SW central RAJASTHAN state, NW INDIA, 70 mi/113 km SSW of JODHPUR; ⊙ JALOR district; 25°21′N 72°37′E. Local market for cotton, millet, oilseeds, wheat; light manufacturing. Noted hill fort (S) dates from c.11th century; a medieval Rajput stronghold. Sometimes spelled Jhalore.

Jalostotitlán (hah-lo-sto-tee-TLAHN), city and township, JALISCO, central MEXICO, on interior plateau, 65 mi/105 km NE of GUADALAJARA on Mexico Highway 80; ⊙ Jalostotitlán municipio; 21°11′N 102°29′W. Elevation 5,686 ft/1,733 m. Agricultural center (corn, wheat, beans, chickpeas, livestock).

Jalovec (YAH-lo-vets), peak (8,671 ft/2,643 m) in JULIAN ALPS, NW SLOVENIA, 8 mi/12.9 km WNW of Triglav peak. On Italo-Slovenian border from 1919 to 1947.

Jalpa (HAHL-pah), town, ZACATECAS, N central MEXICO, on JUCHIPILA River, and 27 mi/43 km SE of TLALTENANGO DE SÁNCHEZ ROMÁN, near junction of Mexico Highways 70 and 54; ⊙ Jalpa municipio; 21°40′N 103°00′W. Elevation 5,906 ft/1,800 m. Agricultural center (grain, beans, sugarcane; livestock).

Jalpa de Méndez (HAHL-pah dai MEN-dez), town, TABASCO, SE MEXICO, at W edge of GRIJALVA River delta, and 17 mi/27 km NW of VILLAHERMOSA; ⊙ Jalpa de Méndez municipio; 18°08′N 93°05′W. Elevation 131 ft/40 m. Corn, rice, beans, tobacco, fruit; livestock.

Jalpaiguri (juhl-PEI-guh-ree), district (□ 2,404 sq mi/6,250.4 sq km), N WEST BENGAL state, E INDIA; ⊙ JALPAIGURI. Bounded N by BHUTAN, E by ASSAM state, S by KOCH BIHAR district and BANGLADESH, W by DARJILING district; drained by upper ATRAI, TISTA, JALDHAKA, and TORSA rivers. Rice milling, sawmilling, tea processing; railroad workshops at Domohani. Alluvial soil; rice, tea, tobacco (major tea- and tobacco-growing district of West Bengal), rape and mustard, jute, sugarcane, potatoes, tamarind. Spurs of W ASSAM HIMALAYAN foothills in NE; extensive sal, sissoo, and khair tracts. WESTERN DUARS in E; sal, sissoo, and bamboo tracts. Minerals include coal deposits near BAGRAKOT, copper-ore deposits (BUXA DUAR, MATIALI); limestone quarries near Bhutan border. Noted for wild game (elephant, tiger, rhinoceros, buffalo). Part of ancient Kamarupa and Buddhist Pal kingdom. Ruins of city of Bhitargarh (built c.ninth century A.D.) in S. Under sixteenth-century Koch kingdom (absorbed 1639 into Mogul empire); passed to English in 1765. Original district was reduced 1947 by transfer of SW portion into Dinajpur district of E BENGAL, following creation of PAKISTAN.

Jalpaiguri (juhl-PEI-guh-ree), town, WEST BENGAL state, E INDIA, on the TISTA RIVER; ⊙ JALPAIGURI district; 26°31′N 88°44′E. On major road and railroad lines. Sawmilling and jute-pressing are major industries. Agriculture includes tea, rice, jute, tobacco, timber, and medicinal herbs; W. Bengal's primary agriculture market. An educational center with an engineering college.

Jalpan (HAHL-pahn), town, PUEBLA, central MEXICO, 22 mi/35 km NNE of HUAUCHINANGO. Sugarcane, coffee, fruit. In Totonac Indian area.

Jalpan de Serra (HAHL-pahn dai SE-rah), city and township, QUERÉTARO, central MEXICO, in a valley of SIERRA MADRE, 75 mi/121 km NE of QUERÉTARO on Mexico Highway 120. Elevation 2,526 ft/770 m. Grain, sugarcane, bananas, dates, pineapples, pomegranates, limes, sweet potatoes, coffee, maguey.

Jalpatagua (hahl-pah-TAH-gwah), town, JUTIAPA department, SE GUATEMALA, in highlands, 13 mi/21 km SW of JUTIAPA; 14°08′N 90°01′W. Elevation 1,827 ft/557 m. Corn, beans, livestock.

Jalsuri (hal-SOO-ree), canton, GUALBERTO VILLARROEL province, LA PAZ department, W BOLIVIA, 6 mi/10 km NE of TOTORA, on the LA PAZ-SAJAMA road; 17°50′S 68°04′W. Elevation 12,723 ft/3,878 m. Gas resources in area. Lead-bearing lode; clay, limestone, and gypsum deposits. Agriculture (potatoes, yucca, bananas, rye; cattle).

Jalta (jel-TAH), village, BIZERTE province, N TUNISIA, 23 mi/37 km SW of BIZERTE; cereals, livestock. Lead mining.

Jaltenango de la Paz (hahl-te-NAHN-go dai la pahz), town, CHIAPAS, SE MEXICO, in a valley of the Sierra de Chiapas, on Jaltenango River, 60 mi/96 km S of SAN CRISTOBAL DE LAS CASAS; ⊙ Jaltenango municipio; 15°51′N 92°43′W. Elevation 2,133 ft/650 m. Agriculture (small farming).

Jaltenco (hahl-TEN-ko), town, MEXICO, central MEXICO, 23 mi/37 km N of MEXICO CITY and in the ZONA METROPOLITANA DE LA CIUDAD DE MÉXICO. Cereals, fruit; livestock. Also San Andrés Jaltenco.

Jaltepeque Lagoon (hahl-te-PE-kai), (15 mi/24 km long, c.1 mi/1.6 km wide), Spanish *Estero Grande de Jaltepeque*, salt-marsh inlet of PACIFIC OCEAN, in LA PAZ department, S EL SALVADOR, 25 mi/40 km SE of SAN SALVADOR. Salt extraction.

Jáltipan, MEXICO: see JÁLTIPAN DE MORELOS.

Jáltipan de Morelos (HAHL-te-pahn dai mo-RE-los), city and township, VERACRUZ, SE MEXICO, on Isthmus of TEHUANTEPEC, 12 mi/19 km W of MINATITLÁN on Mexico Highway 180; 17°58′N 94°42′W. Agricultural center (rice, fruit, coffee; livestock).

Jaltocán (hahl-to-KAN), town, HIDALGO, N MEXICO, in foothills of SIERRA MADRE ORIENTAL, 8 mi/12.9 km W of HUEJUTLA DE REYES; 21°09′N 98°32′W. Rice, corn, sugarcane, tobacco, coffee, fruit.

Jalu (JAH-loo), township, main settlement of Jalu oasis, CYRENAICA region, NE LIBYA, 145 mi/233 km SE of AJDABIYAH; 29°02′N 21°33′E. In oasis (c. 9 mi/14.5 km long) at N end of LIBYAN DESERT. Agriculture (fruit, olives, grain; goats). Also name of group of oases, WAHAT JALU, in area with rich oil fields. Also spelled Jalo. Formerly Gialo.

Jalu (JAH-loo), major oil field, CYRENAICA region, NE LIBYA, in SURT BASIN, 25 mi/40 km S of AWJILAH.

Jaluit (JAH-loo-it), atoll (□ 4 sq mi/10.4 sq km; 1999 population 1,669), central PACIFIC, one of the RALIK CHAIN in the MARSHALL ISLANDS; 05°51′N 169°38′E. Comprised of some eighty-five islets, of which Jaluit Island (□ 4 sq mi/10.4 sq km) is the largest. In World War II it was the headquarters of the Japanese Admiralty for the Marshall Islands U.S. forces captured the atoll in 1944. Jaluit is a seaport and trade center for the Marshalls.

Jamaari (JUH-mah-ree), town, BAUCHI state, N NIGERIA, 20 mi/32 km W of AZARE; 11°40′N 09°56′E. Agricultural trade center; cotton, peanuts. durra, cattle, skins. Sometimes spelled Jamaare.

Jamaat Shain (ZHAM-aht sah-HAIN), town, Safi province, Doukkala-Abda administrative region, MOROCCO, 25 mi/40 km E of SAFI.

Jamaica, republic (4,411 sq mi/11,424 sq km; 2004 estimated population 2,713,130; 2007 estimated population 2,780,132), coextensive with the island of Jamaica, 146 mi/235 km long, 22 mi/35 km–51 mi/82 km wide, WEST INDIES, 90 mi/145 km S of Cuba and 100 mi/161 km W of HAITI; ⊙ KINGSTON; 17°43′N–18°32′N 76°05′W–78°26′W.

Geography

Jamaica is the third-largest island in the CARIBBEAN. Besides Kingston, other important cities are SPANISH TOWN and MONTEGO BAY. The Jamaica railroad connects Kingston and MONTEGO BAY and links inland settlements to PORT ANTONIO. The two international airports—the Norman Manley Palisadoes International Airport in Kingston and the Donald Sangster International Airport in Montego Bay—facilitate travel with the rest of world. Although largely a limestone plateau more than 3,000 ft/914 m above sea level, Jamaica has a mountainous backbone that extends across the island from the W and rises to the BLUE MOUNTAINS in the E; BLUE MOUNTAIN (7,402 ft/2,256 m) is the highest point. Rainfall is heavy in this region (where there are extensive timber reserves) but diminishes westward across the plateau, which is a rugged area deeply dissected by streams and underlain by subterranean rivers. The heart of the plateau, known as the COCKPIT COUNTRY or Cockpits, is used mostly for livestock grazing. A narrow plain along the N coast and several larger plains near the S shore are Jamaica's major agricultural zones. The N coast also has fine beaches and is the focus of tourism. The Rio Grande and the Black River are the country's chief waterways, but neither is navigable for long distances.

The country is divided into three counties and subdivided into fourteen parishes. On the E is SURREY county, with parishes of Kingston, SAINT ANDREW and SAINT THOMAS to the S and PORTLAND to the N. MIDDLESEX, the central county, has SAINT MARY and SAINT ANN parishes to the N, bordered by SAINT CATHERINE, CLARENDON, and MANCHESTER to the S. On the W is CORNWALL county, with TRELAWNY, SAINT JAMES and HANOVER to the N and bounded to the S by SAINT ELIZABETH and WESTMORELAND.

Economy

The coastal bands widened by broad river valleys, as well as the mountain slopes, support the bulk of Jamaica's export crops: the famed Blue Mountain coffee, sugarcane—from which rum and molasses are also made—bananas, ginger, citrus fruits, cocoa, pimento, and tobacco. Most of these crops are grown on large plantations. Small peasant farms produce some ginger, bananas, and sugarcane for export but mainly raise such subsistence crops as yams, breadfruit, and cassava. Mining is a major source of wealth; since large, easily accessible deposits of bauxite were discovered in 1942, Jamaica has become one of the world's leading suppliers of this ore. Along with the alumina made from it, bauxite accounts for almost half of Jamaica's foreign exchange. Tourism is the biggest earner of exchange. Among Jamaica's internationally known resort areas are Montego Bay, OCHO RIOS, and NEGRIL. Apparel constitutes the chief export item of the manufacturing sector. Jamaica's other industries (mainly concentrated in the Kingston area) include oil refining, tobacco processing, flour milling, and the production of cement, textiles, and processed foods. Since the late 1960s industry has generated a greater share of the national income than agriculture, which, however, still employs a larger percentage of the work force. The U.K., U.S., and Canada, Jamaica's top trading partners, also provide much needed capital for economic development.

Population

English is the official language, but many Jamaicans also speak a Jamaican creole dialect. The unit of currency is the Jamaican dollar of 100 centuries About half of the population is rural, but migration to the cities continues; the greatest urban concentration is around Kingston. Adequate health facilities and islandwide education from the early childhood to the university level are available to all Jamaicans. People of African descent predominate in Jamaica, making up approximately 90.9% of the population. A small upper class is largely of European descent. Afro-Europeans and such Middle Eastern and Asian groups as Lebanese, Syrians, Chinese, and Indians, make up the rest of the population. The chief religion is Protestantism, although there is considerable religious variety.

History: to 1865

Sighted by Christopher Columbus in 1494, Jamaica was conquered and settled in 1509 by Spaniards under a license from Columbus's son. Spanish exploitation decimated the native Arawaks. The island remained Spanish until 1655, when Admiral William Penn and Robert Venables captured it; it was formally ceded to England in 1670, but the local European population obtained a degree of autonomy. Jamaica prospered from the wealth brought by buccaneers, notably Sir Henry Morgan, to Port Royal, the capital; in 1692, however, much of the city sank into the sea during an earthquake, and SPANISH TOWN became the new capital. A huge, mostly African, slave population grew up around the sugarcane plantations in the 18th century, when Jamaica was a leading world sugar producer. Freed and escaped slaves, sometimes aided by the maroons (slaves who had escaped to remote areas after Spain lost control of Jamaica), succeeded in organizing frequent uprisings against the European landowners. The sugar industry declined in the 19th century, partly because of the abolition of slavery in 1833 (effective 1838) and partly because of the elimination in 1846 of the imperial preference tariff for colonial products entering the British market.

History: 1865 to 1944

Economic hardship was the prime motive behind the MORANT BAY rebellion by freedmen in 1865. The British ruthlessly quelled the uprising and also forced the frightened legislature to surrender its powers; Jamaica became a crown colony. Poverty and economic decline led many blacks to seek temporary work in neighboring Caribbean areas and in the United States; many left the island permanently, emigrating to England, Canada, and the U.S. Indians were imported to meet the labor shortage on the plantations after the slaves were freed, and agriculture was diversified to lessen dependence on sugar exports. A new constitution in 1884 marked the initial revival of local autonomy for Jamaica. Despite labor and other reforms, black riots recurred, notably those of 1938, which were caused mainly by unemployment and resentment against British racial policies. Jamaican blacks had been considerably influenced by the theories of black nationalism promulgated by the American expatriate Marcus Garvey. A royal commission investigating the 1938 riots recommended an increase of economic development funds and a faster restoration of representative government for Jamaica.

History: 1944 to 1976

Universal adult suffrage was introduced in 1944, and a new constitution provided for a popularly elected house of representatives. By 1958, Jamaica became a key member of the British-sponsored West Indies Federation. The fact that Jamaica received only one-third of the representation in the federation, despite its having more than half the land area and population of the grouping, bred resentment; a campaign by the nationalist labor leader Sir Alexander Bustamante led to a 1961 decision, by popular referendum, to withdraw from the federation. The following year Jamaica won complete independence from Great Britain. The country has a two-party system: the Jamaica Labor Party (JLP) favors private enterprise, while the People's National Party (PNP) advocates moderate socialism. Bustamante, leader of the JLP, became the first prime minister of independent Jamaica. The party continued in power until 1972, when the PNP won an impressive victory. Although the PNP administration worked effectively to promote civil liberties and reduce illiteracy, economic problems proved more difficult.

History: 1976 to Present

In 1976 the PNP won decisively after a violent election contest between the two parties. The PNP continued to promote socialist policies, nationalizing businesses and strengthening ties to Cuba. Lack of foreign investment and aid continued to hurt the economy. In 1980 the JLP returned to power, with the administration favoring privatization, distancing itself from Cuba, attracting foreign investment, stimulating tourism, and finding the U.S. willing to provide substantial aid. Nonetheless, two major hurricanes (1980, 1988) hit Jamaica, setting back prospects for substantial economic progress. In 1989 the PNP won control of Parliament with a more conservative program, and the party remained in power into the 21st century. Portia Simpson-Miller, who became prime minister in 2006, was the first woman to hold the post, but she lost the office in 2007 when the JLP narrowly won control of parliament. Jamaica is internationally known as the home of "reggae" music, whose best-known performer is Bob Marley.

Government

Jamaica is a parliamentary democracy governed under the constitution of 1962. It has a bicameral Parliament made up of a twenty-one–member Senate and a sixty-member House of Representatives. The prime minister is the head of government. The head of state is the British monarch, as represented by the governor general. The current head of state is Queen Elizabeth II, represented by Governor General Kenneth O. Hall (since 2006). The current head of government is Prime Minister Bruce Golding (since 2007).

Jamaica (hah-MEI-kah), town, GUANTÁNAMO province, E CUBA, on railroad, and 5 mi/8 km NE of GUANTÁNAMO; 20°11′N 75°08′W. Agricultural center (cacao, coffee, fruit, sugarcane). There are 4 sugar mills within a 10 mi/16 km radius.

Jamaica, town (2000 population 237), GUTHRIE county, W central IOWA, 15 mi/24 km NE of GUTHRIE CENTER; 41°51′N 94°18′W. In agricultural area.

Jamaica, town, WINDHAM CO., SE VERMONT, on West River, and 10 mi/16 km NW of NEWFANE; 43°06′N 72°47′W. Partly in Green Mountain National Forest. Name is derived from the native American Natick word for beaver.

Jamaica, a commercial, industrial, and residential center in central QUEENS borough of NEW YORK CITY, SE NEW YORK; 40°41′N 73°49′W. Chief transfer station of Long Island railroad, subway terminus, highway hub. Diversified manufacturing; extensive residential sections; has declined as retailing center. County courthouse and the northeastern regional office of the U.S. Social Security Administration. Points of interest include King Mansion (c.1750) and edifice (1813) of first Presbyterian Church, organized 1662. The main unit of the Queens Borough Public Library is here, as are York College (City University of New York) and a campus of St. John's University. Jamaica Race Track was a popular sports and gambling venue before its demise in the 1960s, when nearby Aqueduct Raceway (OZONE PARK) was rebuilt. Settled in mid–17th century, it was first capital of Queens county.

Jamaica Bay (□ c.20 sq mi/52 sq km), SW LONG ISLAND, SE NEW YORK, separated from the ATLANTIC OCEAN by ROCKAWAY PENINSULA; 40°37′N 73°51′W. The Rockaway Inlet links it to the sea. The shallow bay has many islands, and its shores are generally marshy. There is a minimum of water movement and pollution is a problem. Nearly all of the bay is in the boroughs of BROOKLYN and QUEENS in NEW YORK city; since 1950 much of the adjacent area has been reclaimed for housing. JFK INTERNATIONAL AIRPORT extends into the bay. Part of GATEWAY NATIONAL RECREATION AREA, the bay is used for boating and fishing and is a wildlife refuge.

Jamaica Beach, village (2006 population 1,120), GALVESTON county, SE TEXAS, on GALVESTON ISLAND, suburb 11 mi/18 km SW of downtown GALVESTON, on GULF OF MEXICO, surrounded by city of Galveston; 29°11′N 94°58′W. Residential and recreational beach community. Galveston Island State Park to SW.

Jamaica Channel, channel in the CARIBBEAN SEA, separating JAMAICA (W) from HISPANIOLA island (E) and forming SW continuation of WINDWARD PASSAGE, 120 mi/193 km wide between 18°N 74°30′W and 18°N 76°01′W. In the channel are NAVASSA ISLAND (N) and MORANT CAYS (at its S entrance).

Jamaica Estates, S section of E central QUEENS borough, NEW YORK city, SE NEW YORK, bounded on E by 188th Street, S by Hillside Avenue, W by Home Lawn Street, and N by Union Turnpike; 40°44′N 73°46′W. Affluent residential neighborhood. Began with purchase of 507-acre/205-ha tract by Jamaica Estates Co. in 1907; layout of residential park of one- and two-story houses, some occupying three lots, was designed to conform to the contour of the land. Today there are 1,700 houses on eighty blocks of tree-lined streets. When initial deed restrictions expired in 1929, Jamaica Estates Association was formed to preserve the neighborhood's character; the only place apartment buildings were permitted was along Hillside Avenue.

Jamaica Plain, Massachusetts: see BOSTON.

Jam, Al (JEM, el), ancient *Thysdrus* or *Tysdrus*, town (2004 population 18,302), MAHDIYA province, E TUNISIA, in the coastal region (*sahel*), on railroad, and 39 mi/63 km S of SUSAH; 35°18′N 10°43′E. Olive oil pressing; sheep and cattle raising; agricultural trade; administrative center. Has remains of large amphitheater (one of the finest Roman ruins in N Africa). Thysdrus, a 3rd-century Punic settlement, prospered under Roman rule (2nd century C.E.) from olive cultivation and olive oil trading.

Jamal (hah-MAHL), unincorporated town, SAN DIEGO county, S CALIFORNIA; residential suburb 17 mi/27 km E of downtown SAN DIEGO. Sequan Indian Reservation to N; Cleveland National Forest to E. Dairying, poultry, fruit, flowers.

Jamaliyah, Al-, EGYPT: see GAMALIYA, EL.

Jamalpur (jah-mahl-poor), city (2001 population 120,955), MYMENSINGH district, NE EAST BENGAL, BANGLADESH, on the old BRAHMAPUTRA RIVER, and 31 mi/50 km WNW of MYMENSINGH; 24°58′N 89°52′E. Railroad junction (Singjhani station), with spurs to JAGANNATHGANJ (16 mi/26 km SW) and BAHADURABAD GHAT (2.7 mi/4.3 km NW). Trade center (rice, jute, oilseeds, sugarcane, tobacco). Jute-pressing center 15 mi/24 km SW, at SARISHABARI; fabricated metal products 14 mi/23 km NW, at Islampur.

Jamalpur (juh-MAHL-poor), town (2001 population 96,659), MUNGER district, SE BIHAR state, E INDIA, 6 mi/9.7 km S of MUNGER. Railroad junction; major railroad workshops; locomotive manufacturing; iron and steel foundries. Slate quarries 7 mi/11.3 km SW, at Dharhara. Has two colleges affiliated with Bhagalpur University.

Jamame (jah-MAH-me), town, SE SOMALIA, on E bank of JUBBA River, 30 mi/48 km NE of KISMAYO; ⊙ JUBBADA HOOSE region; 00°04′N 42°45′E. Agricultural center; livestock market. Formerly MARGHERITA.

Jamapa (hah-MAH-pah), town, VERACRUZ, E MEXICO, in GULF lowland, 13 mi/21 km SW of VERACRUZ; 19°02′N 96°08′W. Fruit.

Jamasi (jah-MAH-see), town, ASHANTI region, GHANA, 30 mi/48 km NE of KUMASI, on Kumasi-Yeji road; 06°58′N 01°28′W. Cacao, coffee, timber.

Jamay (HAH-mai), town, JALISCO, central MEXICO, on E shore of Lake CHAPALA, 50 mi/80 km SE of GUADALAJARA; 20°20′N 102°41′W. Agricultural center (wheat, corn, oranges, beans, alfalfa, livestock).

Jambaló (hahm-bah-LO), town, ⊙ Jambaló municipio, CAUCA department, SW COLOMBIA, 28 mi/45 km NE of POPAYÁN; 02°47′N 76°19′W. Coffee, corn, sugarcane; livestock. In Jambaló and nearby TORIBÍO, over 90% of residents are Nasa indigenous people.

Jambeiro (ZHAHM-bai-roo), town, E SÃO PAULO, BRAZIL, 17 mi/27 km E of JACAREÍ; 23°16′S 45°41′W. Corn, rice

Jambelí (hahm-be-LEE), island archipelago, the northernmost of a group of mangrove islands off the coast of EL ORO province, W and SW of PUERTO BOLÍVAR; 03°20′S 80°04′W. Growing tourism (beaches; birdwatching along mangrove channels).

Jambelí Channel (hahm-be-LEE), Spanish *Canal de Jambelí*, c.50 mi/80 km long, 7 mi/11.3 km–20 mi/32 km wide, on coast of GUAYAS province, SW ECUADOR, E of PUNÁ Island, links Gulf of GUAYAQUIL with GUAYAS RIVER estuary.

Jamberoo (JAM-buh-ROO), municipality, E NEW SOUTH WALES, SE AUSTRALIA, near coast, 55 mi/89 km SSW of SYDNEY, and 4 mi/7 km W of KIAMA; 34°39′S 150°47′E. Cabbage-tree palms.

Jambes (ZHAHM-buh), town in commune of NAMUR, Namur district, NAMUR province, S central BELGIUM, near SAMBRE RIVER, just SE of Namur; 50°28′N 04°52′E.

Jambi (JAHM-bee), city (2000 population 417,568), SE SUMATRA, INDONESIA, a port at the head of navigation on the HARI RIVER; ⊙ Jambi province; 01°36′S 103°37′E. Shipping and commercial center for an area producing oil, rubber, and timber. Jambi Province University is here. Also spelled Djambi.

Jambi River, INDONESIA: see HARI RIVER.

Jambol, BULGARIA: see YAMBOL.

Jambuair, Cape (jahm-boo-EI-yuhr) or **Cape Tanjung**, Dutch *Diamantpunt*, promontory at NE tip of SUMATRA, ACEH province, INDONESIA, in INDIAN OCEAN, at entrance to Strait of MALACCA, 62 mi/100 km NW of LANGSA; 05°15′N 97°30′E. Lighthouse. Important oil fields nearby. Formerly called Diamond Point.

Jambughoda (juhm-buh-GO-duh), town, tahsil headquarters, Panch Mahalas district of former Bombay Presidency, now in GUJARAT state, INDIA, 29 mi/47 km SSE of GODHRA; 22°22′N 73°43′E. Agriculture market (oilseeds, cotton, grain); rice husking. Was capital of former princely state of Jambughoda, in Gujarat States, BOMBAY; state incorporated 1949 into PANCH MAHALS district.

Jambusar (JUHM-buh-suhr), town, BHARUCH district, in former Bombay Presidency, now in GUJARAT state, INDIA, 27 mi/43 km NNW of BHARUCH; 22°03′N 72°48′E. Railroad junction; markets millet, wheat, cotton; tanning, cotton ginning, calico printing; light manufacturing.

Jamda, INDIA: see GIRNA RIVER.

Jamdena, INDONESIA: see TANIMBAR ISLANDS.

James, former county, E TENNESSEE: see HAMILTON county.

James (JAIMZ), township (□ 33 sq mi/85.8 sq km; 2001 population 467), E central ONTARIO, E central CANADA, 33 mi/53 km from city of TEMISKAMING Shores, and located where the Makobe River flows into MONTREAL RIVER; 47°42′N 80°20′W.

Jamesabad (JAIM-sah-bahd), village, THAR PARKAR district, S SIND province, SE PAKISTAN, on JAMRAO CANAL, and 31 mi/50 km WSW of UMARKOT; 25°17′N 69°15′E.

James A. FitzPatrick Nuclear Power Plant, OSWEGO county, central NEW YORK, on S shore of Lake ONTARIO, 7 mi/11.3 km NE of OSWEGO; 43°31′N 76°25′W. Began commercial operation July 1975; has generating capacity of 800 MW. Lake water is used to cool the reactor. Owned and operated by the private firm Entergy. The plant, which is adjacent to NINE MILE POINT NUCLEAR POWER PLANTS, generates power for municipal electric systems, rural cooperatives, New York state's major private utilities, NEW YORK city's public agencies, WESTCHESTER county, government agencies, and industries. Closed by Nuclear Regulatory Commission due to mechanical and safety inadequacies in November 1991; reopened January 1993.

James A. Garfield National Historic Site (GAHRfeeld), MENTOR, Lake county, NE OHIO; home of the 20th President and site of first presidential memorial library; 41°40′N 81°20′W. Authorized 1980.

James Bay, shallow S arm of HUDSON BAY, c.300 mi/480 km long and 140 mi/230 km wide, E central CANADA, in NUNAVUT territory between ONTARIO and QUEBEC. Numerous rivers flow into the bay. Of its many islands, the largest is Akimiski (1,158 sq mi/3,000 sq km). The bay was discovered (1610) by Henry Hudson but was named for Captain Thomas James, an Englishman who explored much of it in 1631. An early fur-trading post established by Groseilliers and Radisson became (1670) RUPERT HOUSE, the first post established here by the Hudson's Bay Company. Other important posts on James Bay are FORT ALBANY, Fort George, and Eastmain. The shores of the bay and some of its islands are wildlife reserves. Akimiski Island Bird Sanctuary; game sanctuaries at TWIN ISLANDS, Trodely Island, CHARLTON ISLAND; bird sanctuaries also at Hannah Bay, Ontario and Boatswain Bay, Quebec. Entire bay is set aside as James Bay Preserve. The JAMES BAY PROJECT, a colossal hydroelectric development on the E coast of James Bay, has evoked a tremendous negative response from environmentalists and Cree Indians, who claim that the project is disrupting the lives of the natives and destroying the region. Rivers have been diverted, forests have been incinerated, and wilderness areas have been inundated. Phase I, finished in 1984, created the world's largest underground powerhouse, a tiered spillway on GRAND RIVER 3 times the height of NIAGARA FALLS, and five reservoirs that total half the volume of Lake ONTARIO. The completion of the project was threatened in 1992 when the New York State Power Authority refused to sign a purchase contract. Includes GREAT WHALE RIVER (Hudson Bay), La Grande River, and NOTTAWAY/Broadback river basins.

James Bay Project (JAIMZ BAI), French, *projet de la Baie James* (pro-JAI duh lah BAI JAIMZ), hydroelectric scheme, central QUEBEC, E Canada; partially completed project of Hydro-Quebec and Quebec government. Extends from Jarvis and HUDSON bays on W to NEWFOUNDLAND and LABRADOR border on E; from NE of VAL-D'OR on S to Nastapoca River on N. Since its inception in 1975, numerous dams, diversion channels, and power houses have been planned or built. There were four planned stages within the scheme: La Grande Phase One, completed 1985, involved construction of reservoirs on La Grande River; La Grande Phase Two, involving the redirection of flow from EASTMAIN, Laforge, and CANIAPISCAU rivers (the latter flows N to UNGAVA BAY) into La Grande River, was largely completed when work was suspended in 1994; the Great Whale Project, which called for the diversion of Little Whale and Nastapoca rivers to GREAT WHALE RIVER; and the NBR Project, involving the diversion of RUPERT and NOTTAWAY rivers to Broadback River basin. Phase One currently generates more power than all of Quebec's twenty-five coal and one nuclear power plants combined. In 1984, 10,000 caribou were drowned in flood waters on lower Caniapiscau River attributed to release from dam; Hydro-Quebec blamed heavy rains. General concern over environmental and social impact on Cree and Inuit people, including damage to wildlife and fish populations and possible effects on beluga whale populations off coast (deprived of ice-free channels at mouths of diverted rivers), led to tremendous opposition to the project. As a result, the Great Whale and NBR projects were suspended along with completion of La Grande Phase Two. In 2002, an agreement with the Cree allowed for finishing work on La Grande Phase Two and the diversion of the Rupert River into the La Grande. The NBR project was canceled.

Jamesburg, borough (2006 population 6,429), MIDDLESEX county, E NEW JERSEY, 10 mi/16 km S of NEW

BRUNSWICK; 40°21′N 74°26′W. Some light industry. Incorporated 1887.

James City (JAIMZ SIT-ee), county (□ 179 sq mi/465.4 sq km; 2006 population 59,741), SE VIRGINIA; ⊙ WILLIAMSBURG (independent city separated from adjoining James City, YORK counties); 37°18′N 76°46′W. In Tidewater region, bounded NE by YORK RIVER, W by CHICKAHOMINY RIVER, S by JAMES RIVER. Includes Jamestown Island, site of first permanent English settlement in America (1607); both part of COLONIAL NATIONAL HISTORICAL PARK and Colonial Parkway. Agriculture (hay, barley, wheat, soybeans, tobacco, corn, potatoes, fruit; cattle; dairying; fish, oysters, crabs. York River State Park in NE. Formed 1634.

James City (JAIMZ SIT-ee), unincorporated town (□ 14 sq mi/36.4 sq km; 2000 population 5,420), CRAVEN COUNTY, E NORTH CAROLINA, residential suburb 1 mi/1.6 km S of and opposite NEW BERN, on TRENT RIVER (bridged), at its entrance to NEUSE RIVER; 35°04′N 77°01′W. Croatan National Forest to S. Service industries; manufacturing.

James Craik or **James Craig**, town, central CÓRDOBA province, ARGENTINA, 70 mi/113 km SE of CÓRDOBA; 32°09′S 63°28′W. Railroad junction and agriculture center (alfalfa, wheat, soybeans, corn, flax; livestock, dairy products).

James Creek, DELAWARE, UNITED STATES: see BROAD CREEK.

James Creek, Pennsylvania: see MARKLESBURG.

James H. Turner Dam, California: see SAN ANTONIO RESERVOIR (Alameda county).

James Island, unincorporated town, CHARLESTON county, SE SOUTH CAROLINA, suburb 3 mi/4.8 km SW of downtown CHARLESTON, in N center of James Island. One of the SEA ISLANDS, 8 mi/12.9 km long and 6 mi/9.7 km wide.

James Island (JAIMZ), village, SW BRITISH COLUMBIA, W CANADA, on James Island (2 mi/3 km long, 1 mi/2 km wide), in HARO STRAIT, off SE VANCOUVER ISLAND, 14 mi/23 km N of VICTORIA; 48°37′N 123°22′W. Fishing, lumbering.

James Island, CHILE: see CHONOS ARCHIPELAGO.

James Island, GALÁPAGOS: see SAN SALVADOR, ISLA.

James Island, SOUTH CAROLINA: see CHARLESTON.

James, Lake, natural lake, in central STEUBEN county, NE INDIANA. Indiana's fourth largest natural lake. Glacially formed. Average depth 86 ft/26 m.

James, Lake (JAIMZ, LAIK), reservoir (□ 10 sq mi/26 sq km), BURKE county, W NORTH CAROLINA, on CATAWBA RIVER, 40 mi/64 km ENE of ASHEVILLE; 35°44′N 81°53′W. Maximum capacity 277,960 acre-ft. Formed by Bridgewater Dam (100 ft/30 m high), built (1919) for power generation. Lake James State Park on W end of reservoir.

Jameson (JAI-muh-suhn), town (2000 population 120), DAVIESS county, NW MISSOURI, near the GRAND RIVER, and 9 mi/14.5 km N of GALLATIN; 40°00′N 93°59′W. Mormon Shrine of Adam-ondi-Ahman to S.

Jameson Point, KNOX county, S MAINE, forms N side of ROCKLAND harbor. Has breakwater, lighthouse.

James Peak (13,294 ft/4,052 m), in FRONT RANGE, GRAND and BOULDER counties, N central COLORADO, c.35 mi/56 km WNW of DENVER, on CONTINENTAL DIVIDE, E of Winter Rock. MOFFAT TUNNEL passes through part of mountain.

Jamesport, city (2000 population 505), DAVIESS county, NW MISSOURI, 18 mi/29 km NW of CHILLICOTHE; 39°58′N 93°47′W. Corn, soybeans, wheat; cattle, hogs; concrete. Large Amish community established 1953. Numerous antique shops. Platted 1857.

Jamesport, village (□ 7 sq mi/18.2 sq km), SUFFOLK county, SE NEW YORK, on NE LONG ISLAND, near GREAT PECONIC BAY, 4 mi/6.4 km ENE of RIVERHEAD; 40°57′N 72°34′W. In summer resort area.

James River, flows c.80 mi/129 km SW, S central MISSOURI; rises in the OZARKS in WEBSTER county, to TABLE ROCK LAKE (WHITE RIVER) S of GALENA.

Fishing, canoeing. Lake Springfield on it on S side of SPRINGFIELD.

James River, 710 mi/1,143 km long, in NORTH DAKOTA and SOUTH DAKOTA; rises in WELLS CO. in central North Dakota; flows across South Dakota to the MISSOURI River at YANKTON, South Dakota; 47°28′N 99°51′W. Jamestown Dam, forming Jim Lake (1,500 acres/607 ha), on the river is an irrigation and flood control unit of the Missouri River Basin Project of the U.S. Bureau of Reclamation. New Rockford Canal connects upper James River with upper SHEYENNE RIVER in WELLS co. irrigation canal. In North Dakota, James River flows E past FESSENDEN and NEW ROCKFORD, then S through Arrowwood Lake and Jim Lake reservoirs, past LA MOURE and OAKES. In South Dakota, the river continues through Mud Lake and Columbia Road reservoirs, both of which are in Sand Lake National Wildlife Refuge, past REDFIELD, HURON, and MITCHELL. The James is also known as the Jim River or the Dakota River.

James River (JAIMZ), 340 mi/547 km long, formed in W central VIRGINIA by JACKSON and COWPASTURE rivers; 37°47′N 79°46′W. Flows generally E past GLASGOW, where it receives MAURY (North) River from N, cuts through BLUE RIDGE at Balcony Falls, past LYNCHBURG, GOOCHLAND, RICHMOND, overfalls and around island including Williams Island and Belle Isle, past HOPEWELL, where it becomes a tidal river, continues ESE through series of historic plantations, widens into estuary c.12 mi/19 km W of WILLIAMSBURG and receives CHICKAHOMINY RIVER from NW. Continues past JAMESTOWN and NEWPORT NEWS, forms HAMPTON ROADS, harbor for Newport News, HAMPTON, PORTSMOUTH, and NORFOLK, before entering CHESAPEAKE BAY, c.20 mi/32 km W of mouth of the bay.

James Ross Island (40 mi/64 km long), ANTARCTICA, in WEDDELL SEA off N tip of ANTARCTIC PENINSULA; 64°10′S 57°45′W. Discovered 1903 by Otto Nordenskjöld. Formerly known as Ross Island; the name James Ross Island was adopted to prevent confusion with ROSS ISLAND in the ROSS SEA.

James Ross Strait, arm (110 mi long, 30–40 mi wide) of the ARCTIC OCEAN, NUNAVUT territory, CANADA, between KING WILLIAM ISLAND (SW) and BOOTHIA PENINSULA (NE); 70°00′N 96°00′W. Connects N with sea area leading to MCCLINTOCK CHANNEL and FRANKLIN STRAIT.

Jamestown, city (□ 8 sq mi/20.8 sq km; 2000 population 31,730), CHAUTAUQUA county, W NEW YORK, on CHAUTAUQUA LAKE; 42°06′N 79°14′W. It is the commercial center of an agricultural area. Formerly it was a major producer of wood products and furniture; today only remnants of such industries survive. Seat of Jamestown Community College. Nearby are Allegany State Park and the Chautauqua Institute, a cultural and recreational center on the lake. Founded c.1806, incorporated as a city 1886. Supreme Court justice Robert H. Jackson attended school and practiced law here.

Jamestown, city (2000 population 15,527), seat of STUTSMAN CO., SE NORTH DAKOTA, on the JAMES RIVER, in a farm area, 85 mi/137 km W of FARGO; 46°54′N 98°42′W. It is the trade and processing center for an agricultural area where sunflowers, grain and flour are produced and livestock is raised, manufacturing (food, printing and publishing, ordnance, construction materials). Jamestown College, a state home for handicapped children, and the North Dakota State Hospital, the state mental hospital, are in the city. Fort Seward Historic Site to W, and a restored frontier village lie to S on the outskirts. Fish hatchery to NE. Municipal Airport in NE part of city. Jamestown Reservoir (Jim Lake) to N. Pipestone Lake to NW. Founded 1872 when Fort Seward was established to protect railroad workers. Named for JAMESTOWN, VIRGINIA.

Jamestown, city (2000 population 1,839), N TENNESSEE, 37 mi/60 km NE of COOKEVILLE; ⊙ FENTRESS county; 36°26′N 84°56′W. In hilly coal, oil, gas, and timber area; lumbering; manufacturing; agritourism. Alvin C. York State Historic Site and BIG SOUTH FORK NATIONAL RIVER AND RECREATION AREA nearby. Settled 1827; incorporated 1837.

Jamestown (JAIMZ-toun), town, S SOUTH AUSTRALIA, 34 mi/55 km E of PORT PIRIE, on Belalie River banks; 33°12′S 138°36′E. Wheat; wool, sheep, dairy products; wine, timber.

Jamestown, town, EASTERN CAPE, SOUTH AFRICA, 30 mi/48 km SSE of ALIWAL NORTH, on the Klipspruit River, in foothills of DRAKENSBERG RANGE; 31°06′S 26°48′E. Elevation 6,002 ft/1,830 m. Sheep, wool, grain. Masakhane siding in town is the terminus of railroad. On N6 highway.

Jamestown, town (2006 population 605), port, and ⊙ ST. HELENA, in S ATLANTIC, on the James Bay, on the island's NW shore; 15°56′S 05°43′W. Elevation 961 ft/292 m. Ships flax fibers (phormium tenax) and some lily bulbs. Once a busy coaling station on East India route, it lost its importance after the opening of the SUEZ CANAL. Still supplies drinking water to vessels. Adjoining are barracks and batteries. Plantation House (governor's residence) and the cathedral of St. Paul's are approximately 2 mi/3.2 km S in the uplands.

Jamestown, unincorporated town (2000 population 3,017), TUOLUMNE county, central CALIFORNIA, 3 mi/4.8 km SW of SONORA; 37°58′N 120°25′W. Timber; cattle; hay; apples. Gold-mining center in 1849 California Gold Rush region; received its nickname "Jimtown" in gold rush. TABLE MOUNTAIN is nearby.

Jamestown, town (2006 population 966), BOONE county, central INDIANA, 13 mi/21 km SW of LEBANON; 39°56′N 86°38′W. In agricultural area; lumber.

Jamestown, town (2000 population 1,624), S KENTUCKY, in CUMBERLAND foothills, 27 mi/43 km WSW of SOMERSET, WOLF CREEK DAM in CUMBERLAND RIVER is SW; ⊙ RUSSELL county; 36°59′N 85°04′W. Light manufacturing. Russell County Airport to NW. LAKE CUMBERLAND reservoir to S and SE; Lake Cumberland State Resort Park to S; Wolf Creek Dam National Fish Hatchery to SW. Established 1827.

Jamestown, town (2006 population 382), BIENVILLE parish, NW LOUISIANA, 4 mi/6.4 km NE of RINGGOLD; 32°10′N 93°12′W. Gas field. Named in honor of JAMESTOWN, VIRGINIA.

Jamestown, town (2000 population 382), MONITEAU county, central MISSOURI, near MISSOURI River, 11 mi/18 km NNE of CALIFORNIA; 38°46′N 92°28′W. Cattle, hogs, corn.

Jamestown (JAIMZ-toun), town (□ 2 sq mi/5.2 sq km; 2000 population 3,088), GUILFORD county, N central NORTH CAROLINA, suburb 5 mi/8 km NE of HIGH POINT and 11 mi/18 km SW of GREENSBORO, on DEEP RIVER; 36°00′N 79°55′W. Manufacturing (food, electronic equipment, textiles, chemicals); service industries. Incorporated 1947.

Jamestown, suburb and resort town (2000 population 5,622), including Jamestown village, NEWPORT county, S RHODE ISLAND, coextensive with CONANICUT ISLAND (c.9 mi/14.5 km long, 1 mi/1.6 km—2 mi/3.2 km wide), in NARRAGANSETT BAY W of NEWPORT; 41°31′N 71°22′W. Farming. Jamestown Bridge (1940) connected with NORTH KINGSTOWN (W), replaced by Jamestown-Verrazano Bridge, 1992. Pell Bridge (1969) connects island and Newport; longest New England bridge. Beavertail Light, at S tip of island, was established before 1750. Several pre-Revolutionary buildings remain. Named in honor of James II, Duke of York and Albany. Incorporated 1678.

Jamestown (JAIMZ-toun), village, West Dunbartonshire, W SCOTLAND, on the LEVEN RIVER just N of BONHILL; 55°59′N 04°35′W. Textiles.

Jamestown, village, INDEPENDENCE county, NE central ARKANSAS, 6 mi/9.7 km SW of BATESVILLE.

Area is shown by the symbol □, and capital city or county seat by ⊙.

Jamestown, village (2000 population 205), BOULDER county, N COLORADO, in foothills of FRONT RANGE, 8 mi/12.9 km NW of BOULDER; 40°07′N 105°23′W. Elevation c.6,920 ft/2,109 m. Supply point in gold-mining region. ROCKY MOUNTAIN NATIONAL PARK to NW. Surrounded by Roosevelt National Forest.

Jamestown, village (2000 population 399), CLOUD county, N KANSAS, 11 mi/18 km W of CONCORDIA; 39°36′N 97°51′W. Railroad junction. In wheat region.

Jamestown (JAIMZ-toun), village (□ 1 sq mi/2.6 sq km; 2000 population 1,917), GREENE county, SW central OHIO, 10 mi/16 km ESE of XENIA, on small Caesar Creek; 39°39′N 83°44′W. Settled 1806, laid out 1815.

Jamestown, village (2000 population 97), BERKELEY county, SE SOUTH CAROLINA, 35 mi/56 km NNE of CHARLESTON, near SANTEE RIVER, in Francis Marion National Forest; 33°17′N 79°42′W. Light manufacturing. Agriculture includes timber; livestock, poultry; cotton, grain.

Jamestown, borough (2006 population 596), MERCER county, NW PENNSYLVANIA, 22 mi/35 km SW of MEADVILLE, on SHENANGO RIVER (forms PYMATUNING RESERVOIR to NW). Light manufacturing; agriculture (grain, potatoes; dairying). Greenville Airport to SE.

Jamestown (JAIMZ-toun), locality, JAMES CITY county, SE VIRGINIA, 5 mi/8 km SW of WILLIAMSBURG, on Jamestown Island, on JAMES RIVER, part of COLONIAL NATIONAL HISTORICAL PARK, connected with Williamsburg and YORKTOWN by Colonial Parkway; 37°19′N 78°17′W. First permanent English settlement in America; established May 14, 1607, by the London Company on a marshy peninsula (now an island) in the James River and named for the reigning English monarch, James I. Disease, starvation, and Native American attacks wiped out most of the colony, but the London Company continually sent more men and supplies. John Rolfe cultivated the first tobacco here in 1612, introducing a successful source of livelihood; in 1619 the first representative government in the New World met at Jamestown, which remained the capital of Virginia throughout the 17th century. The village was almost entirely destroyed during Bacon's Rebellion; it was partially rebuilt but fell into decay with the removal of the capital to Williamsburg (1698–1700). Of the 17th-century settlement, only the old church tower (built c.1639) and a few gravestones remain. It is included in Colonial National Historical Park.

Jamestown Bay, district, SE ALASKA, on W shore of BARANOF ISLAND, 4 mi/6.4 km E of SITKA; 57°03′N 135°17′W. Fishing.

Jamestown National Historic Site (JAIMZ-toun), 21 acres/8 ha, JAMES CITY county, SE VIRGINIA, an affiliated area of National Park system, on upper part of Jamestown Island, 5 mi/8 km SW of WILLIAMSBURG; 37°19′N 78°10′W. Site of the first permanent English settlement in America. Authorized 1940. See JAMESTOWN, Virginia.

Jamesville (JAIMZ-vil), village (□ 1 sq mi/2.6 sq km; 2006 population 472), MARTIN county, NE NORTH CAROLINA, on ROANOKE RIVER, and 9 mi/14.5 km SE of WILLIAMSTON; 35°48′N 76°54′W. Manufacturing (plastic and wood products, fabrics); service industries; agriculture (tobacco, peanuts, cotton, grain; chickens, hogs). Fishing. Incorporated 1785.

Jamesville, hamlet, ONONDAGA county, central NEW YORK, 5 mi/8 km SE of SYRACUSE; 42°59′N 76°04′W. Light manufacturing. Part of DE WITT, New York, census area.

James W. Dalton Highway, dirt and gravel road (257 mi/414 km long), ALASKA, running parallel to Trans-Alaska pipeline from LIVENGOOD (84 mi/135 km N of FAIRBANKS) to Deadhorse, just S of PRUDHOE BAY. Cuts across 3 wilderness preserves and GATES OF THE ARCTIC NATIONAL PARK.

James W. Trimble Lock and Dam (39 ft/12 m high), on border of SEBASTIAN and CRAWFORD counties, W ARKANSAS, on ARKANSAS RIVER, 8 mi/12.9 km ESE of FORT SMITH, at city limits; 35°22′N 94°20′W. Built (1969) by the Army Corps of Engineers for navigation. The raised channel has a maximum capacity of 59,100 acre-ft.; extends W past downtown Fort Smith into OKLAHOMA. Originally called Lock and Dam #13.

Jamieson, village, VICTORIA, SE AUSTRALIA, 75 mi/120 km NE of MELBOURNE, 23 mi/37 km S of MANSFIELD, at junction of Jamieson, GOULBURN rivers; 37°19′S 146°08′E. Tourism; sawmilling; agriculture. Old gold mining town; amateur gold panning.

Jamieson River, river, VICTORIA, SE AUSTRALIA; 37°18′S 146°08′E. Meets the GOULBURN RIVER at Jamieson village.

Jamikunta (juhm-ee-KUHN-tah), town, KARIMNAGAR district, ANDHRA PRADESH state, INDIA, 23 mi/37 km ESE of KARIMNAGAR; 18°17′N 79°27′E. Rice milling. Sometimes spelled Jumekoonta, Jammikunta.

Jamilena (hah-mee-LAI-nah), town, JAÉN province, S SPAIN, 7 mi/11.3 km WSW of JAÉN; 37°45′N 03°55′W. Olive oil, cereals, livestock. Mineral springs.

Jam Jodhpur (JUHM JOO-puhr), town, JAMNAGAR district, formerly in W central SAURASHTRA, now in GUJARAT state, INDIA, 39 mi/63 km S of JAMNAGAR. Markets millet, cotton, oilseeds, gram; cotton ginning, flour and oilseed milling, hand-loom weaving.

Jam jo Tando, PAKISTAN: see TANDO JAM.

Jamkhandi (JUHM-kuhn-dee), town, BIJAPUR district, KARNATAKA state, INDIA, 35 mi/56 km SW of BIJAPUR; 16°31′N 75°18′E. Local trade center for cotton and silk fabrics, wheat, millet; cotton ginning, hand-loom weaving. Was capital of former princely state of Jamkhandi in DECCAN STATES, Bombay. State incorporated 1949 into districts of DHARWAD, Bijapur, BELGAUM, SOLAPUR, Sangli, and Ahmadnagar.

Jammain, large Arab village, NABLUS district, 7.5 mi/12 km SW of Nablus, in the Samarian Highlands, WEST BANK. Agriculture (olives, fruit, cereals).

Jammalamadugu (JUH-muh-luh-MUH-duh-goo), town, CUDDAPAH district, ANDHRA PRADESH state, INDIA, on PENNER RIVER, and 38 mi/61 km NW of CUDDAPAH; 14°50′N 78°24′E. Peanut milling; handicraft lacquerware. Agriculture (rice, cotton, turmeric). Extensive limestone quarrying nearby.

Jammikunta, INDIA: see JAMIKUNTA.

Jammu, former province, SW part of former KASHMIR state. Area was old kingdom of Dogra Rajputs, who maintained semi-independence during Mogul occupation (1586–1750) of Kashmir Valley. Came under Sikh control in 1819, and in 1820 Gulab Singh, a Dogra, was made Raja of Jammu for distinguished service in Ranjit Singh's army. Gulab Singh conquered LADAKH (1830s) and in 1846 was acknowledged (by British) ruler of most of what became KASHMIR state; died in 1857. Province continued under rule of Dogra maharajas of JAMMU AND KASHMIR until after 1948, when, following Indian-Pakistani hostilities in KASHMIR, districts of JAMMU, KATHUA, UDHAMPUR, RIASI (renamed RAJAORI), DODA, and dependencies of CHINENI and part of PUNCH remained under Indian administration; MIRPUR and most of PUNCH lay in Pakistan-held territory. Indian districts now part of JAMMU AND KASHMIR state. Formerly JUMMOO.

Jammu (JUHM-moo), district (□ 1,196 sq mi/3,109.6 sq km), JAMMU AND KASHMIR state, N INDIA; ⊙ JAMMU.

Jammu (JUHM-moo), city (2001 population 612,163), ⊙ JAMMU district and winter ⊙ JAMMU AND KASHMIR state, N INDIA, on the Tawi River, and in the HIMALAYAN foothills; 32°44′N 74°52′E. Strategically important as the S terminus of a highway linking the VALE OF KASHMIR with the N Indian plain. A railroad terminus and transportation node. Once the seat of a Rajput dynasty, Jammu became the nucleus of the dominions of Gulab Singh, founder of the last ruling house of KASHMIR. On one bank of the river is Jammu's old Fort of Bahu; on the other bank is the maharaja's palace. Seat of a university. Jammu has long

been the site of border disputes with PAKISTAN in the N and with CHINA in the NE.

Jammu and Kashmir (JUHM-moo KASH-mir), state (□ 85,805 sq mi/222,236 sq km; 1991 estimated population 7,718,700, excluding population of areas under Pakistani and Chinese occupation; 2001 population 10,143,700), N INDIA, portion of KASHMIR region; ⊙ SRINAGAR (summer), JAMMU (winter). Bordered N by PAKISTAN and CHINA, S by HIMACHAL PRADESH and PUNJAB states, E by China, and W by Pakistan (Pakistan-administered Kashmir). Principal languages are Urdu, Kashmiri, Dogri, Pahari, Balti, Ladakhi, Punjabi, Gujri, and Dadri. Main products include wool, fruit, maize, wheat, and rice. Noted for superior textiles and handicraft industries (luxury goods): shawls, carpets, papier mache boxes and decorative pieces, crewel embroidery, woodcarving (especially walnut), sericulture, copper and silverware; saffron farming. Traditionally a popular tourist destination, with much of the best hiking, skiing, golfing, water and mountain sports, fishing in India; however, both tourism and the handicraft industry, two major economic pillars of the area, have suffered due to troubled situation. Although the Maharaja of Kashmir opted to join the Republic of India when given the choice in 1947, the Pakistani government has long disputed his decision due to the fact that Kashmir is a Muslim-majority territory. Thus, claimed by both India and Pakistan after Partition, the Kashmir region was divided between the two countries, with India governing most of the area and Pakistan securing three western districts. The India-administered section became Jammu and Kashmir state by a vote of the assembly in Indian Kashmir in 1956. The only state with a Muslim majority, it has been the source of tensions between India, China (NE section), Pakistan, as well as local rebels. In the late 1980s, Muslim resistance (possibly with foreign assistance) to Indian rule escalated. Violence forced the suspension of the legislature and in, 1990, the imposition of direct presidential rule. Also in 1990, Muslim pogroms against nearly 400,000 Kashmiri Brahmins (referred to as Pandits) resulted in this Hindu population being internally displaced, and as of 2007, has yet to be resolved. Normally governed by a chief minister responsible to a bicameral legislature with one elected house and by a governor appointed by the president of India.

Jamna (jem-NAH), ancient *Zama*, village (2004 population 6,128), QABILI province, N central TUNISIA, 27 mi/43 km N of MAKTHAR; 33°34′N 09°01′E. Nearby, Hannibal was defeated (202 B.C.E.) by Scipio Africanus Major in one of history's decisive battles, ending the strength of CARTHAGE.

Jamnagar (JAHM-nuh-guhr), district (□ 5,454 sq mi/14,180.4 sq km), GUJARAT state, INDIA; ⊙ JAMNAGAR.

Jamnagar (JAHM-nuh-guhr), city (2001 population 556,956), ⊙ JAMNAGAR district, GUJARAT state, W central INDIA, port on the GULF OF KACHCHH (arm of the ARABIAN SEA); 22°28′N 70°04′E. Transportation hub, with road, railroad, and air transport. Known for its silk, embroidery, and marble. There are cotton textile mills and manufacturing of cement and pottery. Has naval and aeronautical schools and a radium institute.

Jamner, town, JALGAON district, MAHARASHTRA state, INDIA, 19 mi/31 km SE of JALGAON. Railroad terminus; market center for cotton, peanuts, millet; cotton ginning.

Jamnitz, CZECH REPUBLIC: see JEMNICE.

Jamoigne (zhahm-WAHN-ye), village in commune of TINTIGNY, VIRTON district, LUXEMBOURG province, SE BELGIUM, in the ARDENNES, on SEMOIS, and 19 mi/31 km W of ARLON; 49°42′N 05°25′E. Agriculture. Has 16th-century church.

Jampol, UKRAINE: see YAMPIL′, Khmel′nyts′kyy oblast.

Jampur (JAHM-puhr), town, DERA GHAZI KHAN district, SW PUNJAB province, central PAKISTAN, 28 mi/45 km S of DERA GHAZI KHAN; 29°39′N 70°36′E.

Jamrao Canal (JAHM-rou), irrigation canal, 125 mi/201 km long, E SIND province, SE PAKISTAN; from EASTERN NARA CANAL (17 mi/27 km NE of MOHATTANAGAR) runs S, past JAMESABAD, to point c.15 mi/24 km SW of NABISAR ROAD; 25°20′N 69°10′E. Has several branches. Opened 1899.

Jamrud (jum-ROO-duh), fort at E mouth of KHYBER PASS, Khyber centrally administered tribal region, W North-West Frontier prov., N Pakistan, 10 mi/16 km W of PESHAWAR; 34°00′N 71°22′E. Military post, also customs and immigration post. Fort expanded to large town noted for smuggling activity from Afghanistan. Home to many Afghan refugees. Built 1836 by Sikhs.

Jämsänkoski (YAM-san-KOS-kee), village, KESKI-SUOMEN province, S central FINLAND, 30 mi/48 km SW of JYVÄSKYLÄ; 61°55′N 25°11′E. Elevation 396 ft/120 m. In PÄIJÄNNE lake region. Pulp, cellulose, and paper mills.

Jamsar, village, BIKANER district, N RAJASTHAN state, NW INDIA, 17 mi/27 km NNE of BIKANER. Nearby gypsum deposits worked.

Jamshedpur (juhm-SHAID-puhr), city (2001 population 1,101,804), ⊙ EAST SINGHBHUM district, Jharkhand state, E central INDIA, at the confluence of the SUBARNAREKHA and Kharkai rivers; 22°48′N 86°11′E. A great iron- and steel-producing center known as "The Steel City" and the "Pittsburgh of India." Other manufacturing (transportation equipment, agriculture equipment). Built by private business in the early 20th century, it was named for Jamshedji Tata, founder of the Tata Iron and Steel Company (TISCO). Nearby are extensive coal and iron deposits. Seat of the National Metallurgical Laboratory.

Jamtara (juhm-THAHR-ah), district (□ 690 sq mi/1,794 sq km; 2001 population 652,354), NE Jharkhand state, E central INDIA; ⊙ JAMTARA. Underdeveloped; suffers from low literacy rates, underinvestment in education, little industry, high unemployment. A children's park, Parvat Vihar, is 5 km from the railroad station.

Jamtara (juhm-THAHR-ah), town (2001 population 22,426), ⊙ JAMTARA, NE Jharkhand state, E central INDIA; 23°57′N 86°48′E. 158 mi/252 km NE of RANCHI, 78 mi/125 km NE of JAMSHEDPUR, 220 mi/350 km SE of PATNA; 156 mi/250 km NW of KOLKATA. Railroad junction; one of three stops (Chittranjan, Jamtara, Vidyasagar) in Jamtara district along the Delhi-Howrah line.

Jämtland (YEMT-lahnd), county (□ 19,966 sq mi/51,911.6 sq km), NW SWEDEN; ⊙ ÖSTERSUND; 63°00′N 14°40′E. On NORWAY border, N part comprises JÄMTLAND province (□ 14,543 sq mi/37,666 sq km; 1995 population 119,400); S part comprises HÄRJEDALEN province (□ 4,961 sq mi/12,849 sq km; 1995 population 4,961). Agriculture (dairying); lumber; quarrying (limestone, marble, soapstone). Hilly, becoming mountainous near Norwegian border (HELAGSFJÄLLET, elevation 5,892 ft/1,796 m). Many lakes (STORSJÖN LAKE is largest) and rivers, including LJUSNAN and INDALSÄLVEN. Numerous health and winter sports resorts.

Jamui (juh-MOO-ee), district (□ 1,196 sq mi/3,109.6 sq km; 2001 population 1,397,474), SE BIHAR state, E INDIA; ⊙ JAMUI. Carved out of MUNGER district in 1991. Kshatriya Kund Gram, 9.5 mi/15 km S of Lachchuar is a sacred Jain site, where Lord Mahaveer is supposed to have been born.

Jamui (juh-MOO-ee), town (2001 population 66,752), ⊙ JAMUI district, SE BIHAR state, E INDIA, on tributary of the GANGA River, and 32 mi/51 km SSW of MUNGER; 24°55′N 86°13′E. Road and trade (rice, corn, gram, wheat, barley) center. Mica mines 15 mi/24 km SE, near Nawadih.

Jamuna River, c.100 mi/161 km long, in BANGLADESH and WEST BENGAL state, INDIA, left tributary of the ATRAI RIVER; rises 30 mi/48 km WNW of RANGPUR, flows S, past PHULBARI, HILLI (INDIA), and NAOGAON, to ATRAI RIVER 28 mi/45 km NE of RAJSHAHI.

Jamuna River (JUH-muh-nuh), c.1,800 mi/2,897 km long, in WEST BENGAL state, E INDIA, and BANGLADESH, a main distributary of the GANGA River; leaves JALANGI RIVER 19 mi/31 km SSE of RAJSHAHI (Bangladesh); flows 225 mi/362 km S via the Ganga-Brahmaputra delta, past CHUADANGA (Bangladesh), BANGAON, BASIRHAT, TAKI, DEBHATA, and KALIGANJ, through the SUNDARBANS (here partly forming India-Bangladesh border), to BAY OF BENGAL, forming estuary mouth. Agriculture in lower valley (tea, rice, sugarcane). In upper course, above BASIRHAT, called Ichamati; in parts of lower course, also called Kalindi and Raimangal. Navigable throughout course.

Jamuna River, INDIA: see YAMUNA RIVER, UTTAR PRADESH state.

Jamundí (hah-moon-DEE), town, ⊙ Jamundí municipio, VALLE DEL CAUCA department, W COLOMBIA, in CAUCA valley, on railroad, and 12 mi/19 km SSW of CALI; 03°15′N 76°32′W. Elevation 2,854 ft/869 m. Agriculture (sugarcane, coffee, bananas, corn, soybeans, sorghum; livestock.

Jamursba, Tanjung, INDONESIA: see YAMURSBA, TANJUNG.

Janakkala (YAH-nah-kah-lah), village, HÄMEEN province, FINLAND, 12 mi/20 km SE of HÄMEENLINNA; 60°54′N 24°36′E. Elevation 248 ft/75 m. Grain, potatoes; livestock; lumbering.

Janakpur (JUH-nuhk-poor), administrative zone (2001 population 2,557,004), E NEPAL. Includes the districts of DHANUSA, DOLAKHA, MAHOTTARI, RAMECHAP, SARLAHI, and SINDHULI.

Janakpur (JUH-nuhk-poor), city (2001 population 74,192), ⊙ DHANUSA district, S central NEPAL; 26°40′N 85°55′E. Elevation 256 ft/78 m. Site of an important temple dedicated to Hindu god Ram. Airport.

Jana, La (HAH-nah, lah), town, Castellón de la Plana province, E SPAIN, 10 mi/16 km WNW of VINAROZ. Olive-oil processing; cereals, wine.

Janaúba (zhahn-ah-OO-bah), city (2007 population 65,377), N MINAS GERAIS, BRAZIL, 60 mi/97 km NE of MONTES CLAROS; 15°45′S 43°25′W.

Jancacola (hahn-kah-CO-lah), canton, CARANGAS province, ORURO department, W central BOLIVIA; 18°21′S 67°42′W. Elevation 12,448 ft/3,794 m. Gas resources in area. Lead-bearing lode; extensive gypsum deposits. Agriculture (potatoes, yucca, bananas); cattle.

Jancko Amaya (ZHAN-ko ah-MAH-yah), canton, OMASUYOS province, LA PAZ department, W BOLIVIA; 16°12′S 68°49′W. Elevation 12,543 ft/3,823 m. Clay, limestone, and gypsum deposits. Agriculture (potatoes, yucca, bananas, rye, barley); cattle.

Jancko Marca Sirpa (ZHAN-ko MAR-kah SEER-pah), canton, PACAJES province, LA PAZ department, W BOLIVIA; 17°12′S 68°29′W. Elevation 12,989 ft/3,959 m. Gas deposits in area. Lead-bearing lode; former COROCORO copper mine (closed 1986); clay, limestone, and gypsum deposits. Agriculture (potatoes, yucca, bananas, rye); cattle.

Janco Marca (ZHAN-ko MAR-kah), canton, GUALBERTO VILLARROEL province, extreme S central LA PAZ department, on the border with ORURO department, W BOLIVIA; 17°40′S 68°03′W. Elevation 12,707 ft/3,873 m. Some gas resources in area. Lead-bearing lode, gypsum deposits. Agriculture (bananas, yucca, potatoes); cattle.

Jand (JAH-und), town, ATTOCK district, NW PUNJAB province, central PAKISTAN, 30 mi/48 km SW of CAMPBELLPUR; 33°26′N 72°01′E.

Jandaia do Sul (ZHAHN-dah-ee-ah do SOOL), city (2007 population 18,916), N PARANÁ state, BRAZIL, 23 mi/37 km SE of MARINGÁ; 23°36′S 51°39′W. Coffee, corn, rice, cotton; livestock. Also called Jandaia.

Jandaíra (ZHAHN-dai-rah), town (2007 population 6,451), NE RIO GRANDE DO NORTE state, BRAZIL, 41 mi/66 km SE of MACAU; 05°25′S 36°03′W.

Jandakot, town, SW WESTERN AUSTRALIA, SE suburb of FREMANTLE, in turn a suburb of PERTH. Wool scouring. Jandakot Airport, opened 1963, with an "air work" rather than passenger service function.

Janda, Laguna de (HAHN-dah, lah-GOO-nah dhai), shallow lake (8 mi/12.9 km long, up to 3 mi/4.8 km wide), CÁDIZ province, SW SPAIN, near ATLANTIC coast, 30 mi/48 km SE of CÁDIZ; 36°15′N 05°51′W. Through it flows BARBATE RIVER. Here, and not on the GUADALETE RIVER, was most likely fought the decisive "battle of Guadalete," in which the last Visigothic king, Roderick, was defeated (July 19, 711) by the Moors under Tarik.

Jandaq (JAHN-dahk), village, Esfahān province, central IRAN, 145 mi/233 km N of YAZD, in the desert DASHT-i-Kavir; 34°03′N 54°25′E. Barley, gums; camel breeding.

Jandiala (juhn-DYAH-luh), town, AMRITSAR district, W PUNJAB state, N INDIA, 10 mi/16 km ESE of AMRITSAR; 31°10′N 75°37′E. Local trade in wheat, cotton, gram, rice, oilseeds; hand-loom weaving (cotton, wool); copper and brass ware. Sometimes called Jandiala Guru.

Jandía Peninsula (hahn-DEE-ah), SW extremity of FUERTEVENTURA, CANARY ISLANDS, 30 mi/48 km SW of PUERTO DE CABRAS, ends in Jandía Point; 28°03′N 14°30′W. Rises to highest point (2,648 ft/807 m) of island.

Jandowae (jan-DOU-ee), town, QUEENSLAND, NE AUSTRALIA, 30 mi/48 km N of DALBY; 26°46′S 151°03′E. Agriculture (wheat, grains); timber; dairy, beef cattle; growing of native flowers.

Janduba (jen-DOO-bah), Arabic *Jundūbah*, province (□ 1,198 sq mi/3,114.8 sq km; 2006 population 419,100), NW TUNISIA; ⊙ JANDUBA; 36°40′N 08°45′E. Borders ALGERIA (to W). Also Jendouba.

Janduba (jen-DOO-bah), town (2004 population 43,997), ⊙ JANDUBA province, NE TUNISIA, 30 mi/48 km SE of BAJAH. Regional transportation center on the MAJARDAH RIVER.

Janduís (ZHAHN-doo-ees), town (2007 population 5,440), S RIO GRANDE DO NORTE state, BRAZIL, 59 mi/95 km S of MOSSORÓ; 06°01′S 36°25′W.

Jane Franklin, Cape, NW KING WILLIAM ISLAND, NUNAVUT territory, CANADA, on VICTORIA STRAIT; 69°36′N 98°20′W. Remains of camp, graves, and other relics of Franklin expedition, 1847–1848, were found here by McClintock (1859) and Hall (1861–1865).

Jane Lew, village (2006 population 407), LEWIS county, central WEST VIRGINIA, 12 mi/19 km S of CLARKSBURG; 39°06′N 80°24′W. Coal-mining. Manufacturing (hand-blown glassware, steel tubing, coal processing). Agriculture (corn, potatoes); livestock. Jacksons Mill State 4-H Camp to SW.

Janesville, city (2006 population 62,998), ⊙ ROCK county, S central WISCONSIN, on the ROCK RIVER, 30 mi/48 km SE of MADISON; twin city with BELOIT, 13 mi/21 km S; 42°40′N 89°01′W. Industrial and commercial center in a grain, dairy farm, and tobacco area. Manufacturing (agricultural equipment, machinery, consumer goods, metal products, feeds, printing and publishing, concrete, transportation equipment, plastic products, prepared food). Major railroad junction. Wisconsin School for visually handicapped, University of Wisconsin (Janesville campus), and Blackhawk Technical College (to S) are located here. Points of interest include the twenty-six room Tallman House, where Lincoln spent a weekend in 1859; the Stone House (1842), of Greek Revival style; and the Milton House (1844), which is connected to a log cabin by a tunnel used by runaway slaves as a stop on the Underground Railroad. Rock County Airport to S. Incorporated 1853.

Area is shown by the symbol □, and capital city or county seat by ⊙.

Janesville, town (2000 population 829), on BLACK HAWK-BREMER county line, E central IOWA, on CEDAR RIVER, and 12 mi/19 km NNW of WATERLOO; 42°38′N 92°27′W. Feed milling. Limestone quarries, sand and gravel pits nearby.

Janesville, town (2000 population 2,109), WASECA county, S MINNESOTA, 10 mi/16 km WNW of WASECA, and 14 mi/23 km E of MANKATO; 44°07′N 93°42′W. Agriculture (grain, soybeans; livestock, poultry; dairying). Manufacturing (feeds and fertilizers). Lake Elysian reservoir to N; Buffalo Lake to S. Plotted 1855, deserted in Sioux outbreak of 1862, incorporated 1870.

Janga (JAHNG-gah), town, W BALKAN weloyat, TURKMENISTAN, on TRANS-CASPIAN RAILROAD, on N Krasnovodsk Gulf, 7 mi/11.3 km E of TURKMENBASHI; 40°02′N 53°08′E. Tertiary-level administrative center. Fisheries, salt extraction. Also spelled Dzhanga.

Jangada (ZHAHN-gah-dah), town (2007 population 8,056), S central MATO GROSSO, BRAZIL, 43 mi/69 km NW of CUIABÁ; 15°12′S 56°38′W. Stock raising and agriculture.

Jangada do Sul (ZHAHN-gah-dah do SOOL), city, S PARANÁ state, BRAZIL, on SANTA CATARINA border; 26°22′S 51°15′W. Corn, potatoes, wheat, rice.

Jangamo District, MOZAMBIQUE: see INHAMBANE, province.

Jangaon (JUHN-goun), town, Warengal district, ANDHRA PRADESH state, INDIA, 24 mi/39 km NE of BHONGIR; 17°43′N 79°11′E. Rice and castor bean milling.

Jangipur (JUHNG-gi-puhr), town, MURSHIDABAD district, central WEST BENGAL state, E INDIA, on the BHAGIRATHI RIVER, and 28 mi/45 km NNW of BAHARAMPUR; 24°28′N 88°04′E. Cotton weaving; rice, gram, oilseeds, jute, barley. Reputedly founded by Jahangir. Silk-trade center in 18th century. Extensive silk production nearby.

Jangshahi, PAKISTAN: see JUNGSHAHI.

Jangy-Bazar (jahng–bah-ZUHR), village, W JALAL-ABAD region, KYRGYZSTAN, on CHATKAL RIVER; 41°33′N 70°40′E. Wheat; livestock. Also spelled DZANY-BAZAR.

Jangy-Jol (jahng-guh-JOL), village, JALAL-ABAD region, KYRGYZSTAN, near NARYN RIVER, on TOKTOGUL reservoir, 15 mi/24 km NNE of TASH-KÖMÜR; 41°45′N 72°06′E. Pastures. Until 1942, called Chong-Ak-Dzhol; also spelled Dzhangi-Dzol.

Janicho, MEXICO: see JANITZIO.

Jánico (HAH-nee-ko), officially Santo Tomás de Jánico, town, SANTIAGO province, N central DOMINICAN REPUBLIC, 11 mi/18 km SW of SANTIAGO; 19°24′N 70°48′W. In agricultural region (tobacco, cacao, coffee.)

Janikow, POLAND: see JANKAU.

Janina, Greece: see IOÁNNINA, city.

Janitza, Greece: see YIANNITSÁ.

Janitzio (hah-NEET-see-o), island, in Lake PÁTZCUARO, MICHOACÁN, central MEXICO, 7 mi/11.3 km NNW of PÁTZCUARO; c.1 mi/1.6 km long. Fishing village. Sometimes JANICHO.

Janiuay (hah-NEE-wei), town, ILOILO province, S central PANAY island, PHILIPPINES, 17 mi/27 km NNW of ILOILO; 11°00′N 122°26′E. Agricultural center (rice, sugarcane); sugar milling.

Janja (YAHN-yah), town, NE BOSNIA, BOSNIA AND HERZEGOVINA, near DRINA RIVER (SERBIA border), 6 mi/9.7 km S of BIJELJINA; 44°40′N 19°14′E. Also spelled Yanya.

Janjanbureh, city, THE GAMBIA: see GEORGETOWN city.

Janjevo (YAH-nye-vo), village, KOSOVO province, S SERBIA, 8 mi/12.9 km SSE of PRISTINA; 42°34′N 021°14′E. Handicraft (jewelry). Also spelled Yanyevo.

Janjgir (JAHNG-gir) or **Naila Janjgir**, town (2001 population 32,495), JANJGIR-CHAMPA district, E central Chhattisgarh state, INDIA, 28 mi/45 km ESE of BILASPUR; 22°01′N 82°34′E. Manufacturing (shellac);

rice and oilseed milling. Includes adjoining railroad settlement of Naila. Was part of BILASPUR district until 1998.

Janjgir-Champa (JAHNJ-gir CHUM-pah), district (□ 1,720 sq mi/4,472 sq km; 2001 population 1,316,140), ⊙ JANJGIR, E central Chhattisgarh state. Formed in 1998 out of BILASPUR district in Madhya Pradesh. Became a part of Chhattisgarh in 2000.

Janjira (juhn-JEE-ruh), former princely state in DECCAN STATES, BOMBAY, W INDIA. Incorporated 1949 into Kolaba district, and now included in GREATER MUMBAI (Bombay), MAHARASHTRA state.

Jan Juc, township, VICTORIA, SE AUSTRALIA, 13 mi/21 km S of GEELONG, just W of Torquay, and on BASS STRAIT. On coast, near surfing beaches.

Jankampet, INDIA: see BODHAN.

Jankau (YAHN-kah), Polish Janikow, village in LOWER SILESIA, after 1945 in Wrocław province, SW POLAND, 13 mi/21 km SE of WROCŁAW (Breslau). In Thirty Years War, scene (February 1645) of decisive victory of Swedes under Torstensson over imperial forces under Emperor Ferdinand and Hatzfeld. After 1937, called Grünaue; sometimes called Jankowitz.

Jankau, CZECH REPUBLIC: see JANKOV.

Jan Kempdorp, town, S NORTH-WEST province, N central SOUTH AFRICA, near borders with NORTHERN CAPE and FREE STATE provinces, c.60 mi/97 km N of KIMBERLEY; 27°55′S 24°51′E. On railroad from Kimberley, N of fourteen streams.

Jankho Jankho (ZHAN-ko ZHAN-ko), canton, ALONSO DE IBÁÑEZ province, POTOSÍ department, W central BOLIVIA; 18°29′S 66°27′W. Elevation 12,008 ft/3,660 m. Lead-bearing lode; clay and limestone deposits. Agriculture (potatoes, yucca, bananas); cattle.

Jankov (YAHN-kof), German Jankau, village, STREDOCESKY province, S central BOHEMIA, CZECH REPUBLIC, 9 mi/15 km ENE of BENESOV; 49°39′N 14°44′E. Bavarian and Imperialist armies were defeated here in 1645 by Swedish commander Torstensson, during the Thirty Years War.

Jankowitz, POLAND: see JANKAU.

Jan Mayen (YAHN MEI-uhn), island (□ 145 sq mi/377 sq km), in the ARCTIC OCEAN, c.300 mi/483 km E of SCORESBY SOUND, E GREENLAND. It was annexed by NORWAY in 1929. The island is barren tundra land rising abruptly to Håkon VII Toppen (c.7,450 ft/2,270 m) on Mt. Beerenberg, an extinct volcano. It was discovered (1607) by Henry Hudson and named for Jan Jacobsz May, a Dutch whaler who landed there in 1614.

Jannale (jahn-NAH-le), township, SE central SOMALIA, on the E bank of the Wabē Shebelē River (dammed here 1924–1926 for irrigation), and 8 mi/12.9 km NW of MARCA; 01°48′N 44°42′E. Center of major agricultural area (□ 104 sq mi/269 sq km) developed by Italians after 1912; cotton, corn, castor beans, tobacco, peanuts, sugarcane, tropical fruit; chief banana-producing region of Somalia (exports via Marca). Formerly GENALE.

Jano (HAH-no), town, OLANCHO department, N central HONDURAS, on unpaved road 31 mi/50 km N of SALAMÁ; 15°03′N 86°30′W. Elevation 3,281 ft/1,000 m. Small farming, corn, beans.

Janos (HAH-nos), town, CHIHUAHUA, N MEXICO, on affluent of CASAS GRANDES River, and 90 mi/145 km SE of DOUGLAS, ARIZONA, on Mexico Highway 2; 30°50′N 108°10′W. Elevation 4,452 ft/1,357 m. Cotton, cereals, cattle.

Jánoshalma (YAH-nosh-hahl-mah), Hungarian Jánoshalma, Serbo-Croatian Jankovać, city, BÁCS-KISKÚN county, S HUNGARY, 12 mi/19 km SW of KISKUNHALAS; 46°18′N 19°20′E. Wine production, food; flour mills, brickworks.

Jánosmorja (YAH-nosh-mor-yah), village, HUNGARY, 24 mi/39 km WNW of Győr; 47°48′N 17°07′E. Wheat, corn, silage corn, sugar beets; dairying; cattle. Large confectionery plant.

János, Mount (YAH-nosh), Hungarian Jánoshegy, highest peak (1,738 ft/530 m) in S range of BUDA MOUNTAINS, N central HUNGARY; 47°31′N 18°58′E. Residential districts on E slope.

Janov, LITHUANIA: see JONAVA.

Janow (YAH-noov), Polish Janów Lubelski, Russian Yanov Lyubelski or Lyubel'ski, province, E POLAND, 17 mi/27 km SSE of KRAŚNIK. Stone quarrying; sawmilling, tanning, brewing.

Janow, Polish Janów Podlaski, Russian Yanov Podlyaski, town, Biała Podlaska province, E POLAND, 12 mi/19 km NNE of Biala Podlaska, near BUG RIVER (BELARUS border). Railroad spur terminus; flour milling, tanning, tile manufacturing (clay pit); horse-breeding station. Sometimes called Janow nad Bugiem.

Janow, Belarus: see IVANOVO.

Janow, UKRAINE: see IVANO-FRANKOVE.

Janow, UKRAINE: see DOLYNA, TERNOPIL' oblast.

Janowiec Wielkopolski (yah-NO-vyets vyel-ko-POL-skee), German Janowitz, town, Bydgoszcz province, W central POLAND, on WEŁNA River, and 34 mi/55 km NE of POZNAN. Railroad junction. Manufacturing (cement, bricks); tanning, flour milling.

Janpol, UKRAINE: see YAMPIL'.

Janpur (JUHN-poor), town, BAHAWALPUR district, PUNJAB province, central PAKISTAN, 60 mi/97 km SW of BAHAWALPUR; 29°01′N 70°49′E. Dates.

Jansath, town, MUZAFFARNAGAR district, N UTTAR PRADESH state, INDIA, 14 mi/23 km SE of MUZAFFARNAGAR; 29°20′N 77°51′E. Wheat, gram, sugarcane, oilseeds. Home of Jansath Sayids, chief power in Delhi empire in early 18th century. Hydroelectric station 3 mi/4.8 km NNW, near village of Chitaura.

Jansen, village (2006 population 137), JEFFERSON county, SE NEBRASKA, 5 mi/8 km NE of FAIRBURY; 40°11′N 97°04′W.

Jansenville, town, EASTERN CAPE province, SOUTH AFRICA, on SUNDAYS RIVER, and 47 mi/75 km S of GRAAFF-REINET; 32°57′S 24°40′E. Elevation 1,312 ft/400 m. Livestock; grain, feed crops, fruit. Airfield.

Janske Lazne (YAHN-ske LAHZ-nye), Czech Janské Lázně, German Johannisbad, town, VYCHODOCESKY province, NE BOHEMIA, CZECH REPUBLIC, in the KRKONOŠE MOUNTAINS, on railroad, and 33 mi/53 km ESE of LIBEREC; 50°38′N 15°48′E. Elevation 1,703 ft/519 m. Health resort with thermal (82°F/28°C) and chalybeate springs, especially equipped, since 1935, for helping to cure poliomyelitis. Also noted as winter sports center with excellent skiing facilities on slopes of Cerna Mountain, Czech Černá hora, German Schwarzenberg; cable railroad leads to top.

Jan Smuts Airport, SOUTH AFRICA: see JOHANNESBURG.

Jantetelco (hahn-te-TEL-ko), town, MORELOS, central MEXICO, 13 mi/21 km SE of CUAUTLA; 18°42′N 98°45′W. Elevation 3,806 ft/1,160 m. Rice, sugarcane, fruit. Also known as SAN PEDRO.

Jan Tiel (yahn TEEL), village, S CURAÇAO, NETHERLANDS ANTILLES; beach resort on coastal lagoon, 3 mi/4.8 km ESE of WILLEMSTAD.

Jantra River, Bulgaria: see YANTRA RIVER.

Januária (zhahn-oo-AH-ree-ah), city (2007 population 64,983), N MINAS GERAIS, BRAZIL, river port on SÃO FRANCISCO, and 90 mi/145 km NNW of MONTES CLAROS (connected by new road); 15°31′S 44°26′W. Ships cotton, sugarcane, resins; zinc deposits. Hospital; airfield. First settled in late 17th century.

Januário Cicco (ZHAH-noo-ah-ree-o SEE-cho), town (2007 population 8,283), E RIO GRANDE DO NORTE state, BRAZIL, 34 mi/55 km SW of NATAL; 06°09′S 35°35′W. Cotton, corn, aloe; livestock.

Janub Darfur, state, SUDAN: see S DARFUR.

Janub Kurdufan, state, SUDAN: see S KURDOFAN.

Januszpol, UKRAINE: see YAMPIL'.

Janvier/Chard (ZHAHN-vyai, SHAHRD), hamlet (2006 population 143), NE ALBERTA, W CANADA, 62 mi/100 km S of FORT MCMURRAY, and included in

WOOD BUFFALO; 55°50′N 110°55′W. Oil sands, gas; forestry.

Janvrin Island, in the ATLANTIC, E NOVA SCOTIA, CANADA, off S CAPE BRETON ISLAND, just E of MADAME ISLAND, at entrance of the STRAIT OF CANSO; 3 mi/5 km long, 2 mi/3 km wide; 45°32′N 61°10′W.

Janwada, village, BIDAR district, KARNATAKA state, INDIA, 8 mi/12.9 km NW of BIDAR. Cotton, rice, sugarcane.

Janzé (zhahn-zai), town (□ 15 sq mi/39 sq km), ILLE-ET-VILAINE department, in BRITTANY, W FRANCE, 13 mi/21 km NE of RENNES; 47°58′N 01°30′W. Market center (poultry; cheese, cider).

Janzur, EGYPT: see GANZUR.

Jaora (JOU-rah), town (2001 population 63,736), RATLAM district, W MADHYA PRADESH state, INDIA, 22 mi/35 km NNE of RATLAM; 23°38′N 75°08′E. Trades in corn, millet, cotton, sugarcane, opium; sugar milling, cotton ginning, hand-loom weaving. Was capital of former princely state of Jaora of CENTRAL INDIA agency; state established early 19th century by Pathan chieftain, in 1948 merged with MADHYA BHARAT, and later with the enlarged Madhya Pradesh state.

Japan (jah-PAN), Japanese *Nihon* (nee-HON) or *Nippon* (neep-PON) [from Chinese=the place where the sun comes from, or Land of the Rising Sun], country (142,811 sq mi/369,881 sq km; 1991 population 124,017,137; 2004 estimated population 127,330,002; 2007 estimated population 127,433,494), occupying an archipelago off the coast of E ASIA; ⊙ TOKYO, which, along with neighboring YOKOHAMA, forms the world's most populous metropolitan region.

Geography

Japan proper has four main islands, which are, from N to S, HOKKAIDO, HONSHU (the largest island, where the capital and most major cities are located), SHIKOKU, and KYUSHU. There are also many smaller islands stretched in an arc between the Sea of JAPAN and the EAST CHINA SEA (W) and the PACIFIC OCEAN proper (E). Honshu, Shikoku, and Kyushu enclose the INLAND SEA. The general features of the four main islands are shapely mountains, sometimes snowcapped, the highest and most famous being the sacred FUJI-SAN; short rushing rivers; forested slopes; irregular and lovely lakes; and small, rich plains. Mountains, many of them volcanoes, cover two-thirds of Japan's surface, hampering transportation and limiting agriculture. On the arable land, which is only one-eighth of Japan's total land area, the population density is among the highest in the world. The climate ranges from chilly humid continental to humid subtropical. Rainfall is abundant, and typhoons and earthquakes are frequent. (For a more detailed description of geography, see separate articles on the individual islands.)

Population

The Japanese people are primarily the descendants of various peoples who migrated from Asia in prehistoric times; the dominant strain is N Asian or Mongoloid, with some Malay and Indonesian admixture. One of the earliest groups, the Ainu, who still persist to some extent in HOKKAIDO, are physically somewhat similar to Caucasians. Non-Japanese, mostly Koreans, make up less than 1% of the population.

Religion

Japan's principal religions are Shinto and Buddhism. While the development of Shinto was radically altered by the influence of Buddhism, which was brought from China in the 6th century, special varieties of Japanese Buddhism have developed in sects such as Jodo, Shingon, Nichiren, and Zen. Numerous cults formed after World War II and called the "New Religions" have attracted many members. One of these, the Sokagakkai, a Buddhist sect, built up a large following in the 1950s and 1960s, and became a strong social and political force. Less than 1% of the population are Christians. Confucianism has deeply affected Japanese thought and was part of the generally

significant influence that Chinese culture wielded on the formation of Japanese civilization.

Family

The family has long been the basic social unit in Japan. Family elders command much respect, and even in the 21st century many parents continue to select marriage partners for their children. The status of women improved after the end of World War II, when they received the right to vote, but social customs still tend to restrict their freedom. Because many young women have chosen to concentrate on their careers, the Japanese government has turned to actively promoting marriage and parenting in order to maintain an adequate level of population growth.

Education

The Japanese educational system, established during the Allied occupation after World War II, is one of the most comprehensive and effective in the world. Nine years of schooling is compulsory, although the great majority of citizens are in school much longer. The two leading national universities are at Tokyo and KYOTO.

Social Welfare

The standard of living improved dramatically between the 1950s and the early 1970s, and the Japanese have the highest per capita income of all Asians. Programs for social welfare and health insurance are fairly comprehensive. Since 1961 Japan has had a health-insurance system that covers all of its citizens. A major concern confronting policy planners is the large and growing portion of the population that is elderly. Traditional Japanese sports include judo, kendo (a kind of fencing), and sumo wrestling. Baseball and golf, though not native to Japan, are also very popular.

Economy

Mineral resources are meager, except for coal, which was an important source of industrial energy. The rapid streams supply hydroelectric power. Imported oil, however, is the major source of energy. One third of Japan's electricity comes from nuclear power, with c.50 reactors having been built since the 1960s and several additional ones being planned or under construction. The rivers are generally unsuited for navigation (only two, the ISHIKARI and the SHINANO, are over 200 mi/322 km long), and railroads and ships along the coast are the chief means of transportation. The "Shinkansen" bullet train, the second-fastest train system in the world after France's TGV, was inaugurated in 1964 between Tokyo and OSAKA, and then extended to OKAYAMA.

Japan's farming population has been declining steadily and was about 5.3% of the total population in 2003; agriculture accounted for only 1.3% of the GNP. Arable land is intensively cultivated; farmers use irrigation, terracing, and multiple cropping to coax crops from the overworked soil. Rice and other cereals are the main crops; some vegetables and industrial crops, such as mulberry trees (for feeding silkworms), are also grown, and livestock is raised.

Fishing is highly developed, and the annual catch is one of the biggest in the world. The decision by many nations to extend economic zones 200 mi/322 km offshore has forced Japan to concentrate on more efficiently exploiting its own coastal and inland waters. In the late 19th century Japan was rapidly and thoroughly industrialized.

Textiles were a leading item; vast quantities of light manufactures were also produced, and, in the 1920s and 1930s, heavy industries were greatly expanded, principally to support Japan's growing imperialistic ambitions.

Japan's economy collapsed after the defeat in World War II, and its merchant marine, one of the world's largest in the 1930s, was almost totally destroyed. In the late 1950s, however, the nation reemerged as a major industrial power. By the 1970s it had become the most industrialized country in Asia

and the second-greatest economic power in the world after the U.S. Japanese industry is concentrated mainly in S Honshu and N Kyushu, with centers at Tokyo, Yokohama, OSAKA, KOBE, and NAGOYA. In the 1950s and 1960s, textiles became less important in Japanese industry while the production of heavy machinery expanded.

Japanese industry depends heavily on imported raw materials, which make up a large share of the country's imports. Japan receives all it needs of bauxite, phosphate, steel scrap, and iron ore from imports, as well as virtually all of its crude oil and copper ore. Japan became one of the world's leading producers of motor vehicles and steel, and by the 1980s had become a leading exporter of high-technology goods, including electrical and electronic appliances. It has increasingly shifted some of its industries overseas through outsourcing, and has made massive capital investments abroad, especially in the U.S. and the PACIFIC RIM. Trade unions, organized by enterprise rather than by occupation, represent about one-third of all employed workers. The two largest unions are the General Council of Trade Unions and the Japan Confederation of Labor.

Since the late 1960s, its economy has been marked by a large trade surplus, with the U.S. and Europe accounting for more than half its exports. Japan has also become a global leader in financial services; as of 1990 it had seven of the world's ten largest banks.

History: to 9th Century

Japan's early history is lost in legend. The divine design of the empire—supposedly founded in 660 B.C.E. by the emperor Jimmu, a lineal descendant of the sun goddess and ancestor of the present emperor—was held as official dogma until 1945. Actually, reliable records date back only to about C.E. 400. In the 1st century C.E. the country was inhabited by numerous clans or tribal kingdoms ruled by priest-chiefs. Contacts with KOREA were close, and bronze and iron implements were probably introduced by invaders from Korea around the 1st century. By the 5th century the Yamato clan, whose original home was apparently in Kyushu, had settled in the vicinity of modern Kyoto and had established a loose control over the other clans of central and W Japan, laying the foundation of the Japanese state. From the 6th to the 8th centuries the rapidly developing society gained much in the arts of civilization under the strong cultural influence of CHINA, then flourishing in the splendor of the Tang dynasty. Buddhism was introduced, and the Japanese upper classes assiduously studied Chinese language, literature, philosophy, art, science, and government, creating their own forms adapted from Chinese models. A partially successful attempt was made to set up a centralized, bureaucratic government like that of imperial China. The Yamato priest-chief assumed the dignity of an emperor, and an imposing capital city, modeled on the Tang capital, was erected at NARA, to be succeeded by an equally imposing capital at Kyoto.

History: 9th Century to 1274

By the 9th century, however, the powerful Fujiwara family had established a firm control over the imperial court. The Fujiwara influence and the power of the Buddhist priesthood undermined the authority of the imperial government Provincial gentry—particularly the great clans that opposed the Fujiwara—evaded imperial taxes and grew strong. A feudal system developed. Civil warfare was almost continuous in the 12th century The Minamoto family defeated its rivals, the Taira, and became masters of Japan. Its great leader, Yoritomo, took the title of shogun, established his capital at KAMAKURA, and set up a military dictatorship. For the next 700 years Japan was ruled by warriors. The old civil administration was not abolished, but gradually decayed, and the imperial court at Kyoto fell into obscurity. The Minamoto soon gave way to the Hojo, who managed the Kamakura administration as regents for puppet shoguns, much as the Fujiwara had controlled the imperial court.

History: 1274 to 1542

In 1274 and again in 1281 the Mongols under Kublai Khan tried unsuccessfully to invade the country. In 1331 the emperor Daigo II attempted to restore imperial rule. He failed, but the revolt brought about the downfall of the Kamakura regime. The Ashikaga family took over the shogunate in 1338 and settled at Kyoto, but was unable to consolidate its power. The next 250 years were marked by civil wars, during which the feudal barons (the daimyo) and the Buddhist monasteries built up local domains and private armies. Nevertheless, in the midst of incessant wars there was a brisk development of manufacture and trade, typified by the rise of Sakai (later Osaka) as a free city not subject to feudal control. This period saw the birth of a middle class. Extensive maritime commerce was carried on with the continent and with SE Asia; Japanese traders and pirates dominated East Asian waters until the arrival of the Europeans in the 16th century.

History: 1542 to Early 19th Century

The 1st European contact with Japan was made by Portugese sailors in 1542. A small trade with the West developed. Christianity was introduced by St. Francis Xavier, who reached Japan in 1549. In the late 16th century three warriors, Oda Nobunaga, Hideyoshi Toyotomi, and Tokugawa Ieyasu, established military control over the whole country and succeeded one another in the dictatorship. Hideyoshi unsuccessfully invaded Korea in 1592 and 1596 in an effort to conquer China. After Hideyoshi's death, Ieyasu took the title of shogun, and his family ruled Japan for over 250 years. They set up at Yedo (later Tokyo) a centralized, efficient, but repressive system of feudal government. Stability and internal peace were secured, but social progress was stifled. Christianity was suppressed, and all intercourse with foreign countries was prohibited except for a Dutch trading post at NAGASAKI. Tokugawa society was rigidly divided into the daimyo, samurai, peasants, artisans, and merchants, in that order. The system was imbued with Confucian ideas of loyalty to superiors, and military virtues were cultivated by the ruling aristocracy. Oppression of the peasants led to many sporadic uprisings. Yet despite feudal restrictions, production and trade expanded, the use of money and credit increased, flourishing cities grew up, and the rising merchant class acquired great wealth and economic power. Japan was, in fact, moving toward a capitalist system.

History: Early 19th Century to 1871

By the middle of the 19th century the country was ripe for change. Most daimyo were in debt to the merchants, and discontent was rife among impoverished but ambitious samurai. The great clans of W Japan, notably Choshu and Satsuma, had long been impatient of Tokugawa control. In 1854 an American naval officer, Matthew C. Perry, forced the opening of trade with the West. Japan was compelled to admit foreign merchants and to sign unequal treaties. Attacks on foreigners were answered by the bombardment of KAGOSHIMA and SHIMONOSEKI. Threatened from within and without, the shogunate collapsed. In 1867 a conspiracy engineered by the W clans and imperial court nobles forced the shogun's resignation. After brief fighting, the boy emperor Meiji was "restored" to power in 1868, and the imperial capital was transferred from Kyoto to Tokyo. This was the Meiji Restoration. Although the Meiji Restoration was originally inspired by antiforeign sentiment, Japan's new rulers quickly realized the impossibility of expelling the foreigners. Instead, they strove to strengthen Japan by adopting the techniques of Western civilization. Under the leadership of an exceptionally able group of statesmen (who were chiefly samurai of the W clans) Japan was rapidly trans-

formed into a modern industrial state and a great military power. Feudalism was abolished in 1871.

History: 1877 to 1899

The defeat of the Satsuma rebellion in 1877 marked the end of opposition to the new regime. Emissaries were sent abroad to study Western military science, industrial technology, and political institutions. The administration was reorganized on Western lines. An efficient modern army and navy were created, and military conscription was introduced. Industrial development was actively fostered by the state, working in close cooperation with the great merchant houses. A new currency and banking system were established. New law codes were enacted. Primary education was made compulsory. In 1889 the emperor granted a constitution, modeled in part on that of PRUSSIA. Supreme authority was vested in the emperor, who in practice was largely a figurehead controlled by the clan oligarchy. Subordinate organs of government included a privy council, a cabinet, and a diet consisting of a partially elected house of peers and a fully elected house of representatives. Universal manhood suffrage was not granted until 1925. After the Meiji Restoration, nationalistic feeling ran high. The old myths of imperial and racial divinity, were revived, and the sentiment of loyalty to the emperor was actively propagated by the new government. Feudal glorification of the warrior and belief in the unique virtues of Japan's "Imperial Way" combined with the expansive drives of modern industrialism to produce a vigorous imperialism. At first concerned with defending Japanese independence against the Western powers, Japan soon joined them in the competition for an empire in the Orient.

History: 1899 to 1915

By 1899, Japan cast off the shackles of extraterritoriality, which allowed foreign powers to exempt themselves from Japanese law, thus avoiding taxes and tariffs. It was not until 1911 that full tariff autonomy was gained. The first Sino-Japanese War (1894–1895) marked the real emergence of imperial Japan, with acquisition of TAIWAN (then Formosa) and the PESCADORES islands. and also of the Liao-dong Peninsula in MANCHURIA, which the great powers forced it to relinquish. An alliance with Great Britain in 1902 increased Japanese prestige, which reached a peak as a result of the Russo-Japanese War in 1904–1905. Unexpectedly, the Japanese smashed the might of Russia with speed and efficiency. The treaty of PORTSMOUTH, ending the war, recognized Japan as a world power. A territorial foothold had been gained in Manchuria. In 1910, Japan was able to officially annex Korea, which it had controlled de facto since 1905. During World War I the Japanese secured the German interests in SHANDONG (later restored to China) and received the German-owned islands in the Pacific as mandates.

History: 1915 to 1929

In 1915, Japan presented the Twenty-One Demands, designed to reduce China to a protectorate. The other world powers opposed the demands giving Japan policy control in Chinese affairs and prevented their execution, but China accepted the rest of the demands. In 1918, Japan took the lead in Allied military intervention in SIBERIA, and Japanese troops remained there until 1922. These moves, together with an intensive program of naval armament, led to some friction with the U.S., which was temporarily adjusted by the WASHINGTON Conference of 1921–1922. During the next decade, the expansionist drive abated in Japan, and liberal and democratic forces gained ground. The power of the diet increased, party cabinets were formed, and despite police repression, labor and peasant unions attained some strength. Liberal and radical ideas became popular among students and intellectuals. Politics was dominated by zaibatsu (big business conglomerates), and businessmen were more interested in economic than in military expansion.

Trade and industry, stimulated by World War I, continued to expand, though interrupted by the earthquake of 1923, which destroyed much of Tokyo and Yokohama. Agriculture, in contrast, remained depressed. Japan pursued a moderate policy toward China, relying chiefly on economic penetration and diplomacy to advance Japanese interests. This and other foreign policies pursued by the government displeased more extreme militarist and nationalist elements developing in Japan, some of whom disliked capitalism and advocated state socialism. Chief among these groups were the Kwantung army in Manchuria, young army and navy officers, and various organizations such as the Amur River Society, which included many prominent men.

History: 1929 to 1940

Militarist propaganda was aided by the depression of 1929, which ruined Japan's silk trade. In 1931 the Kwantung army precipitated an incident at Mukden (now SHENYANG) and promptly overran all of Manchuria, which was detached from China and set up as the puppet state of Manchukuo. When the League of Nations condemned Japan's action, Japan withdrew from the organization. During the 1930s, the military party gradually extended its control over the government, brought about an increase in armaments, and reached a working agreement with the zaibatsu. Military extremists instigated the assassination of Prime Minister Inukai in 1932 and an attempted coup d'etat in 1936. At the same time, Japan was experiencing a great export boom, due largely to currency depreciation. From 1932 to 1937, Japan engaged in gradual economic and political penetration of N China. In 1937, after an incident at Peking (now BEIJING), Japanese troops invaded the N provinces. Chinese resistance led to a full-scale though undeclared war. A puppet Chinese government was installed at Nanking (now NANJING) in 1940. Meanwhile, relations with the Soviet Union were tense and worsened after Japan and Germany joined together against the Soviet Union in the Anti-Comintern Pact of 1936. In 1938 and 1939, armed clashes took place on the Manchurian border. Japan then stepped up an armament program, extended state control over industry through the National Mobilization Act (1938), and intensified police repression of dissident elements.

History: 1940 to 1945

In 1940, all political parties were dissolved and replaced by the state-sponsored Imperial Rule Assistance Association. After World War II erupted (1939) in Europe, Japan signed a military alliance with GERMANY and ITALY, sent troops to INDOCHINA (1940), and announced the intention of creating a "Greater East Asia Co-Prosperity Sphere" under Japan's leadership. In April 1941, a neutrality treaty with the SOVIET UNION was triumphantly concluded. In October 1941, the militarists achieved complete control in Japan, when General Hideki Tojo succeeded a civilian, Prince Fumimaro Konoye, as prime minister. Unable to neutralize U.S. opposition to its actions in SE Asia, Japan opened hostilities against the UNITED STATES and GREAT BRITAIN on December 7, 1941, by striking at PEARL HARBOR, SINGAPORE, and other Pacific possessions. The fortunes of war at first ran in favor of Japan, and by the end of 1942 the spread of Japanese military might over the Pacific to the doors of INDIA and of ALASKA was prodigious. Then the tide turned; territory was lost to the Allies island by island; warfare reached Japan itself with intensive bombing; and finally in 1945, following the explosion of atomic bombs by the U.S. over HIROSHIMA and NAGASAKI, Japan surrendered on August 14, the formal surrender being on the U.S. battleship *Missouri* in Tokyo Harbor on Septemberember 2, 1945.

History: 1945 to 1946

The surrender was unconditional, but the terms for Allied treatment of the conquered power had been laid down at the POTSDAM Conference. The empire was dissolved, and Japan was deprived of all territories it had seized by force. The Japanese Empire at its height had included the S half of SAKHALIN, the Kuril Islands, the RYUKYU ISLANDS, Taiwan, the Pescadores, Korea, the BONIN ISLANDS, the Kwantung protectorate in Manchuria, and the island groups held as mandates from the League of Nations (the CAROLINE ISLANDS, MARSHALL ISLANDS, and MARIANA ISLANDS). In the early years of the war, Japan had conquered vast new territories, including a large part of China, SE Asia, the PHILIPPINES, and the DUTCH EAST INDIES. With defeat, Japan was reduced to its size before the imperialist adventure began. The country was demilitarized, and steps were taken to bring forth "a peacefully inclined and responsible government" Industry was to be adequate for peacetime needs, but war-potential industries were forbidden. Until these conditions were fulfilled, Japan was to be under Allied military occupation. The occupation began immediately under the command of Douglas MacArthur. A Far Eastern Commission, representing eleven Allied nations and an Allied council in Tokyo, was to supervise general policy. The commission, however, suffered from the general rising tension between the USSR and the Western nations and did not function effectively, leaving the U.S. occupation forces in virtual control. The occupation force controlled Japan through the existing machinery of the Japanese government. Japan did not sign a peace treaty with the USSR because of a dispute over control of four islands off its N coast, including the two southernmost Kuril Islands that had formerly been held by Japan but occupied by the USSR after the war. The two countries did, however, sign (1956) a peace declaration and established fishing and trading agreements. In 1997 a Russo-Japanese accord was signed to settle the dispute over the islands and to sign a permanent peace agreement.

History: 1946 to 1954

A new constitution was adopted in 1946 and went into effect in 1947; the emperor publicly disclaimed his divinity. The general conservative trend in politics was tempered by the elections of 1947, which made the Social Democratic party headed by Tetsu Katayama the dominant force in a two-party coalition government. In 1948, the Social Democrats slipped to a secondary position in the coalition, and in 1949 they lost power completely when the conservatives took full charge under Shigeru Yoshida. An attempt was made to break up the *zaibatsu*. Many of the militarist leaders and generals were tried as war criminals, and in 1948 many were convicted and executed. Economic revival proceeded slowly with much unemployment and a low level of production, which improved only gradually. In 1949, however, MacArthur loosened the bonds of military government, and many responsibilities were restored to local authorities. At SAN FRANCISCO in September 1951, a peace treaty was signed between Japan and most of its opponents in World War II. India and Burma refused to attend the conference, and the USSR, Czechoslovakia, and Poland refused to sign the treaty. It nevertheless went into effect on April 28, 1952, and Japan again assumed full sovereignty. The elections in 1952 kept the conservative Liberal party and Premier Shigeru Yoshida in power.

History: 1954 to 1960

In November 1954, the Japan Democratic party was founded. This new group attacked governmental corruption and advocated stable relations with the USSR and Communist China. In December 1954, Yoshida resigned, and Ichiro Hatoyama, leader of the opposition, succeeded him. The Liberal and Japan Democratic parties merged in November 1955, to become the Liberal Democratic party (LDP). Hatoyama resigned because of illness in December 1956,

and was succeeded by Tanzan Ishibashi of the LDP. Ishibashi was also forced to resign because of illness and was followed by his fellow party member Nobusuke Kishi in February 1957. In the 1950s Japan signed peace treaties with Taiwan, India, Burma, the Philippines, and Indonesia. Reparations agreements were concluded with Burma, the Philippines, Indonesia, and South Vietnam, with reparations to be paid in the form of goods and services to stimulate Asian economic development. In 1951, Japan signed a security treaty with the U.S., providing for U.S. defense of Japan against external attack and allowing the U.S. to station troops in the country. New security treaties with the U.S. were negotiated in 1960 and 1970. Many Japanese felt that military ties with the U.S. would draw them into another war. Student groups and labor unions, often led by Communists, demonstrated during the 1950s and 1960s against military alliances and nuclear testing. One such demonstration (June 1960) forced U.S. President Eisenhower to cancel a scheduled trip to Japan.

History: 1960 to the Early 1970s

Prime Minister Kishi was forced to resign in 1960 following the diet's acceptance, under pressure, of the U.S.-Japanese security treaty. He was succeeded by Hayato Ikeda, also of the LDP. Ikeda led his party to two resounding victories in 1960 and 1963. He resigned in 1964 because of illness and was replaced by Eisaku Sato, also of the LDP. Sato overcame strong opposition to his policies and managed to keep himself and his party in firm control of the government. The LDP maintained its sizable strength in the diet in the 1967 and 1969 elections. Opposition to the government because of its U.S. ties abated somewhat in the early 1970s when the U.S. agreed to relinquish its control of the Ryukyu Islands, including Okinawa, which had come under U.S. administration after World War II. All of the Ryukyus formally reverted to Japanese control in 1972. In that same year, Sato resigned and was succeeded by Kakuei Tanaka, also a Liberal Democrat. In the early 1970s, however, U.S.-Japanese relations became strained after the U.S. pressured Japan to revalue the yen and again when it opened communications with Communist China without prior consultation with Japan. Partly in response, the Tanaka government established (1972) diplomatic relations with Communist China and announced plans for negotiation of a peace treaty. Relations also became strained with South Korea and Taiwan. For his efforts in opposing the development of nuclear weapons in Japan, Sato was awarded the Nobel Peace Prize in 1974.

History: The Mid-1970s to 1980s

In 1974, Tanaka resigned and was replaced as prime minister by Takeo Miki, another Liberal Democrat. Miki, who became embroiled in a scandal over his personal finances, was replaced by Takeo Fukuda. Beginning in late 1973, when the Arab nations began a cutback in oil exports, Japan faced a grave economic situation that threatened to reduce power and industrial production. In addition, a high annual inflation rate (19% in 1973), a price freeze, and the instability of the yen on the international money markets slowed Japan's economy as it entered the mid-1970s. Though Fukuda was considered to be an expert in economic policy, he had difficulty in combating the economic downturn of the late 1970s, but the continued growth of foreign markets brought Japan out of its slump. Fukuda was replaced by Masayoshi Ohira, who died in office in 1980 and was replaced by Zenko Suzuki. In 1982, the more outspoken Yasuhiro Nakasone took office. He argued for an increase in Japan's defensive capability, extended his second term by an extra year, and appointed his own successor, Noboru Takeshita. The terms of both Takeshita and his replacement, Sosuke Uno, were cut short by influence-peddling scandals that shook up

the LDP and caused a public outcry for governmental reform. In the general election of 1989, the LDP lost in the upper house of the parliament for the first time in thirty-five years. However, party president Toshiki Kaifu was still elected later that year. He stressed the need for honesty in government, and drew much criticism for pledging $9 million to the U.S. for military operations in the Persian Gulf. As the world's second-largest economy, Japan has struggled to define its role as a superpower. Although Article Nine of the constitution forbids the maintenance of armed forces, Japan has a sizable military capability for defensive warfare. The U.S. has put increasing pressure on Japan to assume a larger share of responsibility for the defense of its region, although with the dissolution of the Soviet Union, the threat has decreased somewhat. (However, Japanese troops are part of the coalition forces currently fighting in Iraq.)

History: 1980s to Present

In the 1980s Japan continued its strategy of investing heavily in other countries, and had a surplus with virtually every nation with which it traded. The high level of government involvement in banking and industry led many other countries to accuse Japan of protectionism. The U.S. in particular sought to reduce its huge trade deficit with Japan, which was mostly a result of the growing success of Japan's auto industry. Japan has also had to deal with growing economic competition within its own region from such countries as Korea and Taiwan. In addition to economic pressures, great political pressure was put on Japan to assume a larger role in world affairs. The Persian Gulf War caused great dissension in Japan. The government, which felt tremendous pressure to contribute to the UN effort in accordance with its economic power, also had to address the decidedly antimilitaristic bias of the Japanese people. This situation exemplifies Japan's growing need to take a larger role in world politics in order to promote the kind of stable, open markets that its economy requires. In 1991 Kiichi Miyazawa succeeded Kaifu as prime minister. Miyazawa's government fell in 1993 after the LDP split over political reforms. No party won a majority in the subsequent elections, but an opposition coalition formed a government, and Morihiro Hosokawa became prime minister. Hosokawa resigned in 1994. Hata Tsutomu Hata served briefly as prime minister but was forced to resign. An unlikely LDP-Socialist coalition then formed a government, with Socialist Tomiich Murayama as prime minister; in early 1996 he was succeeded by leader Ryutaroashimoto. An earthquake in the Kobe region in January 1995, the worst in Japan in more than 70 years, killed more than 6,000 people. In 1997 Japan suffered a major economic crisis resulting from the failure of stock brokerage firms and banks. The country's bad debt was estimated at $1 trillion when Keizo Obuchi was elected head of the LDP and succeeded Hashimoto as prime minister in mid-1998. By the time the recession showed signs of ending in 1999, the government had spent more than $1 trillion in a series of economic stimulus packages. In 1999 the LDP formed a coalition government with the Liberals and, later, New Komeito. The Liberals withdrew in April 2000, and shortly afterward Obuchi was incapacitated; Yoshiro Mori became prime minister. A series of political blunders undermined Mori, who was replaced by Junichiro Koizumi in April 2001; the New Conservative party joined the government the same month (and later merged with the LDP). The economy underwent another downturn in 2001, after a period of anemic growth, but improved beginning in 2002. Shinzo Abe succeeded the retiring Koizumi as prime minister in September 2006 but left office a year later after a series of government scandals and LDP electoral losses. Yasuo Fukuda replaced Abe as prime minister.

Government

Japan is governed under the constitution of 1947, drafted by the Allied occupation authorities and approved by the Japanese Diet. It declares that the emperor is the symbol of the state but that sovereignty rests with the people. Executive power is vested in a cabinet appointed and headed by the prime minister, who is elected by the Diet and is usually the leader of the majority party in that body. Japan's bicameral Diet has sole legislative power. The House of Representatives has 480 members, who are popularly elected for four-year terms; approximately three fifths of them are chosen by single-seat constituencies and the rest proportionally. The House of Councilors has 242 members; they elected for six-year terms. A supreme court heads an independent judiciary. The current head of state is Emperor Akihito. Prime Minister Yasuo Fukuda has been head of government since 2007.

Most political parties in Japan are small and do not have broad, mass memberships; their members are mainly professional politicians. The Liberal Democratic party (LDP), which supports close ties with the U.S. and a strong relationship between government and business, held the majority of seats in the Diet from 1955, when the party was formed, to 1993, when an opposition coalition formed a government; however, it was back in government in 1994. The Social Democratic party (SDP, formerly the Socialist party), was long the chief LDP rival; in 1994–1999, however, the party formed a governing coalition with the LDP. Other significant parties currently include the Democratic party of Japan, which is now Japan's largest opposition party, and New Komeito, a Buddhist-influenced party.

A popularly elected governor and a single-house legislature govern each of the country's forty-seven prefectures: Akita, Aichi, Aomori, Chiba, Ehime, Pukui, Pukuoka, Pukushima, Gifu, Gumma, Hiroshima, Hokkaido, Hyogo, Ibaraki, Ishikawa, Iwate, Kagawa, Kagoshima, Kanagawa, Kochi, Kumamoto, Kyoto, Mie, Miyagi, Miyazaki, Nagasaki, Nakano, Nara, Niigata, Oita, Okayama, Okinawa, Osaka, Saga, Saitama, Shiga, Shimane, Shizuoka, Tochigi, Tokushime, Toyama, Tottori, Toyama, Wakayama, Yamagata, Yamaguchi, Yamanashi. Cities, towns, and villages elect their own mayors and assemblies."

Japan Current, Japanese *Kuroshio* (koo-RO-shee-YO) [Japanese=black stream], warm ocean current of the PACIFIC OCEAN, off E ASIA. A N-flowing branch of the NORTH EQUATORIAL CURRENT, it runs E of TAIWAN and JAPAN; the TSUSHIMA CURRENT separates from the main current and flows into the Sea of JAPAN. At about latitude 35°N it divides to form an E branch flowing nearly to the HAWAIIAN Islands and a N branch that skirts the coast of Asia and merges with the waters of the cold OYASHIO Current to form the NORTH PACIFIC CURRENT. Dense fogs develop along the boundary between the Japan and Oyashio currents. Air moving over the warm Japan Current becomes more temperate and acts to moderate the climate of Taiwan and Japan.

Japanese Alps, a name sometimes given to the volcanic ranges extending N-S through central and widest portion of HONSHU, JAPAN. Highest peak, MOUNT HOTAKA (10,527 ft/3,209 m).

Japan, Sea of, enclosed arm of the PACIFIC OCEAN (□ c.405,000 sq mi/1,048,950 sq km), located between JAPAN and the ASIAN mainland, connecting with the EAST CHINA SEA, the PACIFIC OCEAN, and the Sea of OKHOTSK through several straits. The shallower N and S portions of the sea are important fishing areas. The sea has depths of over 10,000 ft/3,048 m. A branch of the warm JAPAN CURRENT flows NE through the sea, modifying the climate of the region; VLADIVOSTOK, the only ice-free port of eastern RUSSIA, is here. Also called, especially in NORTH KOREA and SOUTH KOREA, the East Sea.

Japan Trench, submarine depression in N Pacific Ocean, extending in a concave curve from BONIN ISLANDS along E coast of N HONSHU and HOKKAIDO to the KURIL ISLANDS. The TUSCARORA DEEP (N) was long considered the greatest ocean depth (27,929 ft/8,513 m) in the world. However, the depth of 30,954 ft/9,435 m was sounded in 1926 at 30°49′N 142°18′E by the Japanese ship *Manshu*, and the Ramapo Deep (first thought to be 34,626 ft/10,554 m, later corrected to 34,038 ft/10,375 m) was sounded during World War II at 30°43′N 142°28′E by the U.S.S. *Ramapo*. One of the deepest parts of the ocean is in the MINDANAO TRENCH.

Japara, INDONESIA: see JEPARA.

Japaratinga (ZHAH-pah-rah-CHEEN-gah), town (2007 population 7,463), ALAGOAS state, BRAZIL, on coast SW of MARAGOJI; 09°08′S 35°14′W.

Japaratuba (ZHAH-pah-rah-too-bah), city (2007 population 15,473), NE SERGIPE, NE BRAZIL, on railroad, 23 mi/37 km N of ARACAJU; 10°35′S 36°57′W. Ships rice, manioc, alcohol, fruit.

Japen Islands, INDONESIA: see YAPEN ISLANDS.

Japeri (ZHAH-pe-ree), city (2007 population 89,300), W RIO DE JANEIRO state, BRAZIL, 13 mi/21 km NW of NOVA IGUAÇU, on railroad; 22°39′S 43°40′W.

Japi (ZHAH-pee), town (2007 population 5,566), SE RIO GRANDE DO NORTE state, BRAZIL, 66 mi/106 km SW of NATAL, on PARAÍBA border; 06°27′S 35°56′W. Aloe, cotton, corn; livestock.

Japira (ZHAH-pee-rah), town (2007 population 4,694), N PARANÁ state, BRAZIL, 31 mi/50 km S of SANTO ANTÔNIO DA PLATINA; 23°48′S 50°09′W. Coffee, corn, rice; livestock.

Japla, INDIA: see DALTONGANJ.

Japlim (zhah-PLEEN), township, extreme W ACRE, BRAZIL, near terminum of BR 364 Highway at PERU border; 07°40′S 73°20′W.

Japoatã (ZHAH-po-ah-TUH), city (2007 population 13,583), NE SERGIPE, NE BRAZIL, 14 mi/23 km S of PROPRIÁ; 10°20′S 36°48′W. Manioc, rice. Kaolin quarries. Until 1944, spelled Jaboatão.

Japonski Island (juh-PAWN-skee), SE ALASKA, in ALEXANDER ARCHIPELAGO, in SITKA SOUND, just W of SITKA; 57°03′N 135°22′W. At Edgecumbe, site of U.S. naval base in World War II; installations now occupied by state boarding school. Russians had magnetic observatory here. Site of Sitka airport; connected to Sitka by bridge.

Jappen Islands, INDONESIA: see YAPEN ISLANDS.

Japurá (zhah-poo-RAH), town (2007 population 5,281), NW AMAZONAS, BRAZIL, 53 mi/85 km NW of FONTE BOA, on Rio Japurá; 01°51′S 66°35′W.

Japurá, town (2007 population 8,248), NW PARANÁ state, BRAZIL, 39 mi/63 km WSW of MARINGÁ; 23°25′S 52°34′W. Coffee, cotton, corn, rice; livestock.

Japurá, Colombia: see CAQUETÁ.

Jaqué (hah-KAI), village and minor civil division of Sambú district, DARIÉN province, E PANAMA, on PACIFIC OCEAN, and 60 mi/97 km S of LA PALMA. Stock raising, lumbering.

Jaquirana (ZHAH-kwee-rah-nah), town (2007 population 4,404), E RIO GRANDE DO SUL state, BRAZIL, 53 mi/85 km NE of CAXIAS DO SUL; 28°54′S 50°23′W. Livestock.

Jar (YAHR), village, AKERSHUS county, SE NORWAY, suburb of OSLO, on railroad, and 5 mi/8 km W of Oslo city center. The Øvrevoll racetrack is here.

Jara, town (2007 population 6,524), OROMIYA state, S central ETHIOPIA, 40 mi/64 km NE of GOBA; 07°21′N 40°31′E.

Jára, ROMANIA: see IARA.

Jarabacoa (hah-rah-bah-KO-ah), town (2002 population 27,370), LA VEGA province, central DOMINICAN REPUBLIC, in the Cordillera CENTRAL, on YAQUE DEL NORTE, and 10 mi/16 km SW of LA VEGA; 19°08′N 70°40′W. Resort in fruit-growing valley. Has fine mountain climate. Just E are JIMENOA falls, a tour-

ist site and hydroelectric project. Nickel deposit nearby.

Jarabub, LIBYA: see JAGHBUB.

Jarabulus, SYRIA: see JERABLUS.

Jara, Cerrito (HAH-rah, se-REE-to), hill on BOLIVIA (SANTA CRUZ department)-PARAGUAY border, 65 mi/105 km SW of PUERTO SUÁREZ (Bolivia), in the Chaco; 19°46′S 58°14′W. Boundary marker established 1938, in Chaco Peace Conference.

Jaraco (hah-RAH-ko), town, VALENCIA province, E SPAIN, near the MEDITERRANEAN, 7 mi/11.3 km NNW of GANDÍA; 39°02′N 00°13′W. Rice, oranges, vegetables; poultry.

Jarácuaro (ha-RAH-kwah-ro), small island and town, Erongaricuaro municipio, MICHOACÁN, central MEXICO, in Lake PÁTZCUARO, 6 mi/9.7 km NW of PÁTZCUARO. Fruit growing; fishing.

Jaradu, EGYPT: see GARADU.

Jarafuel (hah-rah-FWEL), town, VALENCIA province, E SPAIN, 25 mi/40 km S of REQUENA; 39°08′N 01°04′W. Manufacturing (walking canes); olive-oil processing, flour milling; lumbering. Honey.

Jaraguá (zhah-rah-goo-AH), city (2007 population 38,825), S central GOIÁS, central BRAZIL, on the Rio das Almas (headstream of RIO TOCANTINS), and 60 mi/97 km N of GOIÂNIA; 15°40′S 49°26′W. Tobacco; livestock; rare skins; tanning. Gold placers (abandoned).

Jaraguá, Brazil: see JARAGUÁ DO SUL.

Jaraguá do Sul (ZHAH-rah-goo-ah do SOOL), city (2007 population 130,060), NE SANTA CATARINA, BRAZIL, on railroad, and 18 mi/29 km SW of JOINVILLE; 26°29′S 49°04′W. Stock raising; rice, manioc. Textile mill. Until 1944, Jaraguá.

Jaraguari (zhan-rahg-wah-REE), town (2007 population 5,657), central Mato Grosso do Sul, BRAZIL, 32 mi/51 km N of CAMPO GRANDE; 20°10′S 54°12′W.

Jarahueca (hahr-ah-WAI-kah), village, SANCTI SPÍRITUS province, central CUBA, on railroad, and 23 mi/37 km SE of CAIBARIÉN; 22°14′N 79°21′W. In agricultural region (sugarcane; cattle). Has deposits of light oil.

Jaraicejo (hah-rei-THAI-ho), town, CÁCERES province, W SPAIN, 16 mi/26 km NNE of TRUJILLO; 39°40′N 05°49′W. Flour milling; stock raising; cereals, cork. Has picturesque parochial church.

Jarai Plateau, VIETNAM: see KON TUM PLATEAU.

Jaraíz (hah-rah-EETH), town, CÁCERES province, W SPAIN, 18 mi/29 km E of PLASENCIA. Pepper- and paprika-shipping center. Manufacturing (soap, baskets, candy); food processing, olive pressing, sawmilling. Agriculture (figs, fruit; wine; tobacco).

Jara, La (HAH-rah, lah), high region in W central SPAIN, cutting across CÁCERES and TOLEDO province border along left bank of the TAGUS. The MONTES DE TOLEDO range is S. Picturesque wooded upland; grain; livestock.

Jaral del Progreso (hah-RAHL del pro-GRE-so), city, in the S part of GUANAJUATO, MEXICO, 22 mi/35 km SE of SALAMANCA, NE of YURIRIA Lake, and on the LERMA River; 20°22′N 101°04′W. Elevation 6,204 ft/1,891 m. Terminus of branch railroad line. Agriculture (corn, beans, wheat, peaches); livestock.

Jaral, El, HONDURAS: see EL JARAL.

Jarales (huh-RAH-les), unincorporated village, VALENCIA county, W central NEW MEXICO, on RIO GRANDE, and 33 mi/53 km S of ALBUQUERQUE. Trading point in irrigated region. Cattle, sheep; dairying; corn, grain, alfalfa, fruit. MANZANO RANGE and part of Cibola National Forest E.

Jarama River (hah-RAH-mah), c.100 mi/161 km long, CASTILE-LA MANCHA, central SPAIN; rises in the SOMOSIERRA range near SEGOVIA province border, flows S, almost entirely through MADRID province, to the TAGUS 2 mi/3.2 km W of ARANJUEZ. Its lower course is accompanied by a canal. Waters used to supply Madrid's needs. Widely used for irrigation of central plateau (MESETA). Receives LOZOYA, MANZANARES,

HENARES, and TAJUÑA rivers. At its lower course, the Nationalist advance was held (February 1937) during Spanish civil war.

Jaramataia (ZHAH-rah-mah-TEI-ah), town (2007 population 6,096), C ALAGOAS state, BRAZIL, 8 mi/13 km E of BATALHA; 09°42′S 36°59′W.

Jaramillo (hah-rah-MEE-yo), village, SE COMODORO RIVADAVIA military zone, ARGENTINA, on railroad, near DESEADO RIVER (irrigation area), 70 mi/113 km NW of PUERTO DESEADO. Fruit-growing and sheep-raising center. Petrified forest nearby.

Jarandilla (hah-rahn-DEE-lyah), town, CÁCERES province, W SPAIN, 24 mi/39 km ENE of PLASENCIA. Agricultural trade center (livestock; cereals, wine, flax, honey); ships pepper. Manufacturing (footwear), meat processing, olive pressing, flour- and saw-milling. Has medieval castle, the residence (1556–1557) of Emperor Charles V before he moved to YUSTE.

Jaranwala (ju-RAHN-wah-luh), town, FAISALABAD district, E PUNJAB province, central PAKISTAN, 20 mi/32 km ESE of FAISALABAD; 31°20′N 73°26′E. Railroad junction. Manufacturing (jute, sugarcane; sulphuric acid).

Jarash, JORDAN: see GERASA.

Jara, Villa, MEXICO: see PASO DEL MACHO.

Jarbah (jer-BAH), ancient *Meninx*, island (□ 197 sq mi/512.2 sq km; 2004 population 139,517) in the central MEDITERRANEAN SEA, MADANIYINA province, just off S coast of TUNISIA (of which it forms an administrative district), at S entrance to the GULF OF QABIS, 40 mi/64 km E of QABIS; 17 mi/27 km long, 16 mi/26 km wide; ⊙ HAWMAT AS-SUQ (or Jarbah); 33°48′N 10°54′E. Island has no natural water supply (water is piped in from the mainland and collected in cisterns). Tourism is the main source of income; there are also vineyards, fig and pomegranate trees, date palms, and olive groves. Sponge and oyster fishing. Berber population is noted for handmade pottery, wool and silk cloth, jewelry, and esparto products Access to mainland is across narrow channels at AJIM (1 mi/1.6 km wide) and by road (once a Roman causeway) at Al Kantara (4.5 mi/7.2 km wide) on S shore. A small Jewish community, engaged in business dating back to ancient (and, according to tradition, King Solomon's) times remains, though most of the community emigrated to Israel after Tunisia's independence in 1956 and after the 6-Day War of June 1967. A legend makes Jarbah the land of the lotus-eaters in the *Odyssey*. Jarbah was first settled by Phoenicians in 6th century B.C.E. and was an important Phoenician trading center. Conquered by Arabs in 667, it was abandoned until the mid-13th century. Because of its location, it was a base for Spanish and Ottoman corsairs from the late 13th century on. Sometimes spelled Jerba or Djerba.

Jarbah, TUNISIA: see HAWMAT AS-SUQ.

Järbo (YER-BOO), village, GÄVLEBORG county, E SWEDEN, on JÄDRÅN RIVER, 9 mi/14.5 km NW of SANDVIKEN; 60°43′N 16°36′E.

Jarboesville, MARYLAND: see LEXINGTON PARK.

Jarbridge, unincorporated village, ELKO county, NE NEVADA, 57 mi/92 km NNW of WELLS, 8 mi/12.9 km S of IDAHO state line, in Humboldt National Forest. Gold.

Jardim (ZHAHR-zheen), city (2007 population 25,769), extreme S CEARÁ, BRAZIL, in the Serra do ARARIPE, near PERNAMBUCO border, 27 mi/43 km SSE of CRATO; 07°37′S 39°20′W. Sugar, cotton, cattle. Gypsum quarries; copper deposits.

Jardim, city (2004 population 24,193), SW Mato Grosso do Sul, BRAZIL, 62 mi/100 km WNW of MARACAJU; 21°29′S 56°10′W. Airport.

Jardim Alegre (ZHAHR-zheen AH-le-gre), city (2007 population 14,310), NW PARANÁ state, BRAZIL, 48 mi/77 km SSE of MARINGÁ; 24°09′S 51°43′W. Coffee; livestock.

Jardim de Angicos (ZHAHR-zheen zhe AHN-zhee-kos), town (2007 population 2,673), E RIO GRANDE DO NORTE state, BRAZIL, 55 mi/89 km NW of NATAL. Cotton; livestock.

Jardim de Piranhas (ZHAHR-zheen zhe PEE-rahn-yahs), town (2007 population 13,719), S RIO GRANDE DO NORTE, NE Brazil, near PARAÍBA border, 21 mi/34 km W of CAICÓ; 06°22′S 37°20′W. Livestock.

Jardim do Seridó (ZHAHR-zheen do SE-ree-do), city (2007 population 12,008), S RIO GRANDE DO NORTE, NE BRAZIL, on BORBOREMA PLATEAU, 24 mi/39 km SE of CAICÓ; 06°35′S 36°46′W. Mining of rare minerals (beryl, tantalite, columbite).

Jardín (hahr-DEEN), town, ⊙ Jardín municipio, ANTIOQUIA department, NW central COLOMBIA, in CORDILLERA OCCIDENTAL, 50 mi/80 km SSW of MEDELLÍN; 05°35′N 75°50′W. Elevation 5,928 ft/1,807 m. Coffee-growing; corn, beans, sugarcane, plantains. Manufacturing (consumer goods).

Jardín América (hahr-DEEN ah-MAI-ree-kah), town, San Ignacio department, SW MISIONES province, ARGENTINA, 45 mi/72 km NE of POSADAS city; 27°03′S 55°14′W. Cotton, mate, and fruit region.

Jardine (JAHR-deen), village, Park county, S MONTANA, on Bear Creek near YELLOWSTONE RIVER, just W of Buffalo Plateau, 43 mi/69 km S of LIVINGSTON. Elevation c.7,000 ft/2,134 m. Mining; ski basin. YELLOWSTONE NATIONAL PARK just S; ABSAROKA-BEARTOOTH Wilderness Area of Gallatin National Forest to W, N, and E.

Jardine River National Park (jahr-DEEN) (□ 915 sq mi/2,379 sq km), N QUEENSLAND, NE AUSTRALIA, 500 mi/805 km NW of CAIRNS, on CAPE YORK PENINSULA Developmental Road and PACIFIC OCEAN; 50 mi/80 km long, 30 mi/48 km wide; 11°20′S 142°40′E. Vast wilderness plateau at end of peninsula and N end of GREAT DIVIDING RANGE. Swamps and bogs in coastal areas, resulting in oppressive mosquito populations. Large variety of flora and fauna not fully recorded. Possums, tree kangaroos, cassowaries, crocodiles. Camping. Established 1977.

Jardines de la Reina (hahr-DEE-naiz dai lah REI-nah), archipelago of coral reefs, off CARIBBEAN coast of CIEGO DE ÁVILA province, E CUBA, 70 mi/113 km SW of CAMAGÜEY; c.85 mi/137 km long NW–SE; 21°03′N 79°17′W. More than 400 keys, consisting of Cayos de las DOCE LEGUAS (NW) and Laberinto de las DOCE LEGUAS (SE), separated by the Canal de CABALLONES; bounded by the Gran Banco de BUENA ESPERANZA (NE), sometimes considered to be a part of the Jardines de la Reina.

Jardines del Rey, CUBA: see CAMAGÜEY ARCHIPELAGO.

Jardinópolis (ZHAHR-zheen-NO-po-lees), city (2007 population 34,606), NE SÃO PAULO state, BRAZIL, on railroad junction, and 11 mi/18 km NNE of RIBEIRÃO PRÊTO; 21°02′S 47°46′W. In coffee-growing region. Pottery; hats.

Jardins-de-Napierville, Les (zhahr-DAN–duh–NAI-pyuhr-vil, lai), county (□ 308 sq mi/800.8 sq km), Montérégie region, S QUEBEC, S CANADA; ⊙ NAPIERVILLE; 45°10′N 73°31′W. Composed of eleven municipalities. Formed in 1982.

Jare, town (2007 population 5,434), OROMIYA state, W central ETHIOPIA, 35 mi/56 km ENE of NEKEMTE; 09°11′N 37°05′E. In coffee growing area.

Jaremcze, UKRAINE: see YAREMCHA.

Jarez (zhah-rez), small region in LOIRE department, RHÔNE-ALPES region, E central FRANCE, extending generally NE from SAINT-ÉTIENNE between the Monts du LYONNAIS (N) and Mont PILAT (S) along the valley of the GIER RIVER. Includes E section of Saint-Étienne coal field (no longer mined) and industrial district. Densely populated, it contains several small but active manufacturing communities with metalworking shops.

Jargeau (zhahr-zho), commune (□ 5 sq mi/13 sq km), LOIRET department, CENTRE administrative region, N central FRANCE, on left bank of the LOIRE, and 11 mi/18 km ESE of ORLÉANS; 47°52′N 02°07′E. Metalworking.

Here Joan of Arc was wounded in a victorious battle (1429) against English invaders.

Jari (ZHAH-ree), city, central RIO GRANDE DO SUL state, BRAZIL, 38 mi/61 km NW of SANTA MARIA; 29°17′S 54°13′W. Livestock.

Jaria Jhanjail, BANGLADESH: see SHAMGANJ.

Jarīd, TUNISIA: see BILAD AL JARĪD.

Jarīd, Shatt al (je-REED, SHAWT el), salt lake (□ 1,900 sq mi/4,940 sq km), W central TUNISIA. Salt marshes in S; causeway from TAWZAR to QABILI built in 1986; small villages and date palm oases at E end.

Jarinilla (hah-ree-NEE-yah), canton, LADISLAO CABRERA province, ORURO department, W central BOLIVIA. Elevation 12,218 ft/3,724 m. Gas resources in area. Lead-bearing lode; clay, limestone, gypsum, and large salt deposits. Agriculture (potatoes, yucca, bananas, rye); cattle.

Jari River (ZHAH-ree REE-o), c.350 mi/563 km long, N BRAZIL; rises on S slope of the Serra de Tumucumaque; flows SE, forming PARÁ-AMAPÁ border, to the AMAZON delta opposite GURUPÁ ISLAND. Navigable in lower course. Formerly spelled Jary.

Jarissah (je-REE-sah), village (2004 population 11,298), AL KAF province, W TUNISIA, 23 mi/37 km S of AL KAF. Important iron-mining and administrative center. Also called Djebel-Djérissa.

Jarjis (jer-JEES), town and oasis (2004 population 70,895), MADINIYINA province, SE TUNISIA, small port on the central MEDITERRANEAN SEA, 65 mi/105 km ESE of QABIS; 33°30′N 11°07′E. Olive oil processing; tuna and sponge fishing. Saltworks just S. Roman ruins nearby. Formerly called Zarzis.

Jarkent, KAZAKHSTAN: see ZHARKENT.

Jarmah (JER-muh), village, FAZZAN region, SW LIBYA, 70 mi/113 km NW of Murzuq, in oasis. Agriculture (fruit). Roman ruins (fort, walls, mausoleum). Formerly Germa.

Jarmeritz an der Rokytna, CZECH REPUBLIC: see JAROMERICE NAD ROKYTNOU.

Jarmo, IRAQ: see MESOPOTAMIA, region.

Järna (YER-nah), town, STOCKHOLM county, E SWEDEN, 7 mi/11.3 km SSW of SÖDERTÄLJE; 59°06′N 17°34′E. Railroad junction.

Jarnac (zhahr-nahk), town (□ 4 sq mi/10.4 sq km), CHARENTE department, POITOU-CHARENTES region, W FRANCE, on CHARENTE RIVER, and 16 mi/26 km W of ANGOULÊME; 45°41′N 00°10′W. In COGNAC district, with brandy-distilling and -distribution facilities. Here, in 1569, French Catholics under the duke of Anjou (later Henry III) defeated the Huguenots, whose leader, the Prince of Condé, was killed. Hometown and burial site of French President François Mitterrand (in office 1981–1996).

Järnforsen (YERN-FORSH-en), village, KALMAR county, SE SWEDEN, on EMÅN RIVER, 19 mi/31 km E of VETLANDA; 57°25′N 15°37′E.

Jarny (zhahr-nee), town (□ 6 sq mi/15.6 sq km), MEURTHE-ET-MOSELLE department, in LORRAINE, NE FRANCE, in BRIEY iron basin, 13 mi/21 km WNW of METZ; 49°09′N 05°54′E. Railroad yards, steel working.

Jaro (HAH-ro), town, LEYTE province, N central LEYTE, PHILIPPINES, 15 mi/24 km WSW of TACLOBAN in mountainous area; 11°10′N 124°46′E. Agricultural center (rice, coconuts).

Jarocin (yah-RO-cheen), German *Jarotschin*, town (2002 population 25,897), Poznań province, W POLAND, 38 mi/61 km SSE of POZNAŃ. Railroad junction. Manufacturing (cement and building materials, agricultural machinery and tools, furniture, agricultural products, leather). Castle ruins.

Jaromer (YAH-rom-YERZH), Czech *Jaroměř*, German *Jermer*, town, VYCHODOCESKY province, NE BOHEMIA, CZECH REPUBLIC, on ELBE RIVER, and 7 mi/11.3 km SE of DVUR KRALOVE; 50°22′N 15°55′E. Railroad junction of JOSEFOV is 2 mi/3.2 km downstream. Processing (hides, furs, jute, milk, poultry). Manufacturing (machinery, industrial leather goods). Has an old belfry, noted church. A stronghold of Hussites in 1421; later (1634–1645) occupied by Swedes. Health resort of VELICHOVKY (VE-li-KHOF-ki) is 3 mi/4.8 km W.

Jaromerice nad Rokytnou (YAH-rom-YER-zhi-TSE NAHD RO-kit-NOU), Czech *Jaroměřice nad Rokytnou*, German *Jarmeritz an der Rokytna*, town, JIHOMORAVSKY province, SW MORAVIA, CZECH REPUBLIC, on railroad, and 8 mi/12.9 km S of TŘEBÍČ; 49°06′N 15°54′E. Manufacturing (furniture, machinery, seeds); tanning. Has an 18th-century castle with park and open-air theater. Neolithic-era archaeological site.

Jaronú (hahr-o-NOO), village, CAMAGÜEY province, E CUBA, on railroad, and 29 mi/47 km N of CAMAGÜEY; 21°49′N 77°57′W. Adjacent to modern sugar-mill village of BRASIL, one of nation's largest.

Jaroslavice (YAH-ro-SLAH-vi-TSE), German *Joslowitz*, village, JIHOMORAVSKY province, S MORAVIA, CZECH REPUBLIC, 10 mi/16 km SW of ZNOJMO, near Austrian border; 48°45′N 16°14′E. Vineyards; barley. Traditional (white) wine-making center since the 12th century. Has a large castle. Paleolithic-era archaeological site (evidence of mammoth hunters).

Jarosław (yah-ROS-lahv), town (2002 population 40,235), SE POLAND, on the SAN RIVER. Natural-gas deposits nearby. The town was founded by Yaroslav the Wise, duke of KIEV, in the 11th century. It passed to Poland in 1382. Despite continuous Tatar raids, it developed as an important trade center in the 15th and 16th centuries. It passed to AUSTRIA in 1772 and was restored to Poland in 1919.

Jarosov, CZECH REPUBLIC: see STARE MESTO, JIHOMORAVSKY province.

Jarotschin, POLAND: see JAROCIN.

Jarovnice (yah-ROU-nyi-tse), HUNGARIAN *Jernye*, village, VYCHODOSLOVENSKY province, NE SLOVAKIA, 9 mi/14.5 km NW of PREŠOV; 49°03′N 21°04′E. Wheat, barley, potato, rape; food processing. The 17th-century manor house of FRICOVCE (fri-CHOU-tse), Slovak *Fričovce*, 4 mi/6.4 km SW, is a typical building of Slovak renaissance.

Järpås (YERP-OS), village, SKARABORG county, SW SWEDEN, 10 mi/16 km SW of LIDKÖPING; 58°23′N 12°58′E.

Järpen (YER-pen), village, JÄMTLAND county, NW SWEDEN, on upper INDALSÄLVEN RIVER at mouth of small Järpströmmen River, 40 mi/64 km WNW of ÖSTERSUND; 63°21′N 13°28′E.

Jarque (HAHR-kai), town, ZARAGOZA province, NE SPAIN, 14 mi/23 km NNW of CALATAYUD; 41°34′N 01°40′W. Flour milling; olive oil, potatoes, sugar beets, fruit.

Jarratt, town (2006 population 557), GREENSVILLE and SUSSEX counties, S VIRGINIA, 10 mi NNE of EMPORIA, near NOTTOWAY RIVER; 36°49′N 77°28′W. Manufacturing (food products, wood fiber sheathing). Agriculture (livestock, poultry; cotton, tobacco; peanuts, grain).

Jarrell, unincorporated village (2006 population 1,423), WILLIAMSON county, central TEXAS, 22 mi/35 km SSW of TEMPLE. Agriculture (cattle; cotton, corn, wheat); limestone; manufacturing (fertilizer, flagstone).

Jarrie (zhah-ree), town (□ 5 sq mi/13 sq km), outer S suburban township of GRENOBLE, ISÈRE department, RHÔNE-ALPES region, SE FRANCE, on mountain overlooking ROMANCHE RIVER, at its junction with the DRAC, 6 mi/9.7 km SSE of city center; 45°07′N 05°46′E. Electrochemical industry.

Jarrie, La (zhah-ree, lah), commune (□ 3 sq mi/7.8 sq km; 2004 population 3,915), CHARENTE-MARITIME department, POITOU-CHARENTES region, W FRANCE, 7 mi/11.3 km ESE of La ROCHELLE; 46°08′N 01°01′W. Flour milling, distilling.

Jarrow (JA-ro), town (2001 population 27,526), TYNE AND WEAR, NE ENGLAND, on the TYNE estuary; 54°58′N 01°28′W. Manufacturing (iron and steel

products, oil installations, shipbuilding and repairing). Light industry is found at the Bede Industrial Estate. St. Paul's Church and an adjacent Benedictine monastery (now in ruins) were both founded in the 7th century. The Venerable Bede lived, worked, and died in the monastery. Jarrow lent its name to the hunger marches that were made across England to London during the 1930s. In 1967 the Tyne Tunnel (beneath the Tyne River) was opened, connecting Jarrow with WILLINGTON.

Jaru (ZHAH-roo), city (2007 population 52,476), N RONDÔNIA state, BRAZIL, 154 mi/248 km SE of PÔRTO VELHO; 10°26′S 62°27′W.

Jaruco (hahr-OO-ko), town, LA HABANA province, W CUBA, on small Jaruco River, and 23 mi/37 km ESE of HAVANA; 23°04′N 82°01′W. Railroad junction and agricultural center (sugarcane, fruit, vegetables). Copper deposits nearby. Adjoining (W) are the Escaleras de Jaruco, a picturesque hilly range.

Jarud Qi (JAH-LU-TUH CHI), town, ⊙ Jarud Qi county, NE INNER MONGOLIA AUTONOMOUS REGION, N CHINA, 135 mi/217 km SW of TAONAN; 44°36′N 121°00′E. Food processing. Agriculture (grain, oilseeds; livestock). Also appears as Zalute Qi.

Järva (YER-vah), region with suburbs of STOCKHOLM, STOCKHOLM county, S central SWEDEN, 8 mi/12.9 km NW of Stockholm city center.

Järva-Jaani (YAR-vuh-YAH-nee) or **yarva-yani**, N central ESTONIA, 16 mi/26 km NE of PAIDE; 59°02′N 25°53′E. Agricultural area.

Järved (YER-VED), village, VÄSTERNORRLAND county, NE SWEDEN, on small inlet of GULF OF BOTHNIA, agglomerated with ÖRNSKÖLDSVIK.

Järvenpää (YAR-ven-PAH), Swedish *Träskända*, village, UUDENMAAN province, S FINLAND, 20 mi/32 km NNE of HELSINKI; 60°28′N 25°06′E. Elevation 330 ft/100 m. On Tuusulanjärvi (lake). Commuter suburb of Helsinki; lumber mills, rubber works; site of agricultural college and teachers' school. Houses of many authors and artists are on nearby lake. Famous composer, Sibelius, lived here 1904–1957.

Jarvie (JAHR-vee), hamlet, central ALBERTA, W CANADA, 21 mi/34 km N of WESTLOCK, in WESTLOCK County; 54°27′N 113°59′W.

Jarville-la-Malgrange (zhahr-veel–lah–mahl-grahnzh), SSE suburb of NANCY, MEURTHE-ET-MOSELLE department, in LORRAINE, NE FRANCE, on MEURTHE RIVER and MARNE-RHINE CANAL; 48°40′N 06°13′E. Railroad shops and classification yards; heavy metalworks. Museum traces the history of iron and its utilization.

Jarvis (JAHR-vis), unincorporated village, S ONTARIO, E central CANADA, 11 mi/18 km E of SIMCOE, and included in HALDIMAND; 42°53′N 80°06′W. Grist milling, dairying; bees; poultry.

Jarvis Bay (JAHR-vis), summer village (2001 population 124), ALBERTA, W CANADA, 12 mi/19 km from RED DEER, in RED DEER COUNTY; 52°19′N 114°04′W. Incorporated 1986.

Jarvis Island, coral island (□ 2 sq mi/5.2 sq km), central PACIFIC, geologically one of the LINE ISLANDS, just S of the equator and c.1,300 mi/2,000 km S of HONOLULU. Known to British and American mariners, it was claimed in 1856 by the U.S. along with HOWLAND ISLAND and BAKER ISLAND, but was annexed by GREAT BRITAIN in 1889. American colonists were brought to Jarvis in 1935; the following year the island was placed under the U.S. Department of the Interior's Fish and Wildlife Service. Jarvis is now uninhabited.

Jarvis Sound, S NEW JERSEY, inlet (1.5 mi/2.4 km long, 0.75 mi/1.2km wide) just N of CAPE MAY HARBOR, to which it is joined by INTRACOASTAL WATERWAY channel, which enters from RICHARDSON SOUND (N); 38°58′N 74°52′W.

Järvsö (YERVS-UH), village, GÄVLEBORG county, E central SWEDEN, on LJUSNAN RIVER, 7 mi/11.3 km SSE of LJUSDAL; 61°43′N 16°10′E. Vacation sports center.

Jaryczow Nowy, UKRAINE: see NOVVY YARYCHIV.

Jary River, Brazil: see JARI RIVER.

Jasamba, NEPAL: see PASSANG LHAMU CHULI.

Jaša Tomic (YAH-shah TO-meets), Hungarian *Módos*, village, VOJVODINA, NE SERBIA, on Tamis River, on railroad, and 23 mi/37 km E of ZRENJANIN, on ROMANIAN border, in the BANAT region; 45°26′N 020°51′E. Also spelled Yasha Tomich.

Jasdan (JUHS-duhn), town, RAJKOT district, in former SAURASHTRA, now in GUJARAT state, INDIA, 32 mi/51 km SE of RAJKOT; 22°02′N 71°12′E. Market center (millet, cotton, ghee); oilseed milling, match manufacturing, hand-loom weaving. Railroad spur terminus just E. Was capital of former West KATHIAWAR state of Jasdan of WESTERN INDIA STATES agency; state merged 1948 with Saurashtra and later with Gujarat state.

Jasenov, SLOVAKIA: see STRAZSKE.

Jasenovac (YAH-se-no-vahts), village, E CROATIA, on Sava River, opposite UNA RIVER mouth, 30 mi/48 km SE of SISAK, in W SLAVONIA. Trade center in plum-growing region; poultry raising. In World War II, site of a concentration camp, where Serbs, Jews, and Gypsies died. Occupied by Serb forces 1991–1994.

Jashpur (JUHSH-puhr), former princely state of CHHATTISGARH STATES, central INDIA. Incorporated 1948 into RAIGARH district of MADHYA PRADESH state; in 1998, became JASHPURNAGAR district; in 2000, joined fifteen other districts to form Chhattisgarh state.

Jashpurnagar (JUHSH-poor-nuh-GUHR), district (□ 2,390 sq mi/6,214 sq km; 2001 population 739,780), NE central Chhattisgarh state, E central INDIA; ⊙ JASHPURNAGAR. Hilly, forested territory populated mainly by tribals. Primarily agricultural area: rice, maize, oil seeds, lentila, potatoes, eggplant, tomatoes, garlic, onions, okra, mangoes, jackfruit, tamarind, blackberries, papaya, lychee. In 1948, became part of RAIGARH district in MADHYA PRADESH; formed separate district in 1998; joined fifteen other districts to form Chhattisgarh in 2000. Attractions include: Loroghat's Phoolon ki Ghati (Valley of Flowers); Ranidah Waterfall's Girma Valley, Fish Point, and Dudhdhara; Sougarha's Awadhuth ashram and temple.

Jashpurnagar (JUHSH-poor-nuh-GUHR), town, JASHPURNAGAR district, NE Chhattisgarh state, INDIA, 85 mi/137 km NNE of RAIGARH; 22°54′N 84°09′E. Rice, oilseeds. Lac in nearby sal forests. Was capital of former princely state of JASHPUR, one of CHHATTISGARH STATES; in 1948 became a town in RAIGARH district, MADHYA PRADESH state; in 1998, became capital of Jashpurnagar district; in 2000, became part of Chhattisgarh state.

Jasien (YAH-seen), Polish *Jasień*, German *Gassen*, town in BRANDENBURG, after 1945 in Zielona Góra province, W POLAND, 16 mi/26 km E of FORST. Railroad junction. Agricultural market (grain, potatoes; livestock).

Jasikan (jah-SEE-kan), town, VOLTA region, GHANA, 50 mi/80 km N of HO, and 17 mi/27 km N of HOHOE on road to N; 07°24′N 00°28′E. Market center; cacao. Also spelled GJASIKAN.

Jasina, UKRAINE: see YASINYA.

Jask (JUHSK), town, Hormozgān province, SE IRAN, port on Gulf of OMAN, 145 mi/233 km SE of BANDAR ABBAS; 25°39′N 57°43′E. Airport, wireless station, customs and police station. Fishing.

Jasliq, UZBEKISTAN: see ZHASLIK.

Jaslo (YAH-slo), Polish *Jasło*, town (2002 population 37,916), KROSNO province, SE POLAND, on Wisłoka River, and 31 mi/50 km SW of RZESZÓW. Railroad junction. Center of region producing petroleum and natural gas (gas pipe line to OSTROWIEC, RADOM, SANDOMIERZ); food processing, lumbering, tanning. Manufacturing (cement ware). Explosives factory planned here before World War II. Manganese and oil deposits nearby. During World War II, under German rule, called Jessel.

Cross-references are shown in SMALL CAPITALS. The pronunciation guide is shown on page xix. The sources of population figures are shown on page xvii.

Jaslovce, SLOVAKIA: see JASLOVSKE BOHUNICE.

Jaslovske Bohunice (yahs-LOU-ske bo-HU-nyi-TSE), SLOVAK *Jaslovské Bohunice*, village, ZAPADOSLO-VENSKY province, W Slovakia, 7 mi/11.3 km NNE of TRNAVA; 48°28′N 17°39′E. Manufacturing (textiles); nuclear power plant (four reactors of SOVIET design, total output 1,632 MW). Formed in 1960 by the uni-fication of JASLOVCE and BOHUNICE.

Jasmergarh (juhs-MER-guhr), village, KATHUA dis-trict, JAMMU AND KASHMIR state, N INDIA, in SW KASHMIR, 17 mi/27 km WNW of KATHUA. Wheat, rice, corn, bajra. Sometimes spelled Jasmirgarh.

Jasmund Lake, GERMANY: see RÜGEN.

Jasna Gora, POLAND: see CZĘSTOCHOWA.

Jaso (jah-SO), village, W central SATNA district, NE MADHYA PRADESH state, central INDIA, 25 mi/40 km SE of PANNA. Agriculture (millet, wheat, gram). Was capital of former petty state of Jaso of CENTRAL INDIA agency; since 1948, state merged with VINDHYA PRA-DESH and later with Madhya Pradesh state.

Jason Peninsula, juts out into the LARSEN ICE SHELF from the S of GRAHAM LAND's E coast, ANTARCTIC PENINSULA; 66°10′S 61°00′W. Cape Framnes is on its tip.

Jasonville, city (2000 population 2,490), GREENE county, SW INDIANA, 16 mi/26 km NW of BLOOM-FIELD; 39°10′N 87°12′W. In agricultural area (grain, fruit). Manufacturing (electrical equipment, telephone line coils, wood products); bituminous-coal mines. Shakamak State Park nearby to W. Laid out 1859.

Jasov (yah-SAWF), HUNGARIAN *Jászó*, village, VYCHO-DOSLOVENSKY province, S SLOVAKIA, on BODVA RIVER, on railroad, and 13 mi/21 km WSW of KOŠICE; 48°41′N 20°59′E. Noted for 18th-century monastery with large library, art collections, and botanical gar-dens. Has castle ruins, stalactite caverns. Numerous prehistoric (Paleolithic and Neolithic) remains. Under Hungarian rule from 1938–1945.

Jasper (JAS-puhr), specialized municipality, (□ 357 sq mi/928.2 sq km; 2005 population 4,511), W ALBERTA, W CANADA, near BRITISH COLUMBIA border, in ROCKY MOUNTAINS, on ATHABASCA RIVER, and 200 mi/322 km WSW of EDMONTON; 52°53′N 118°05′W. Elevation 3,470 ft/1,058 m. Tourist center in JASPER NATIONAL PARK. Overlooked by peaks over 10,000 ft/3,048 m high. Formerly designated an improvement district (1995); the municipality of Jasper was formed in 2001.

Jasper, county (□ 373 sq mi/969.8 sq km; 2006 popu-lation 13,624), central GEORGIA; ⊙ MONTICELLO; 33°19′N 83°41′W. Bounded W by OCMULGEE River (forms LLOYD SHOALS RESERVOIR here); drained by LITTLE RIVER. Piedmont agriculture; cattle, poultry; and timber area. Feldspar mining. Formed 1807.

Jasper, county (□ 498 sq mi/1,294.8 sq km; 2006 population 9,880), SE central ILLINOIS; ⊙ NEWTON; 39°00′N 88°09′W. Agriculture (soybeans, corn, sor-ghum, wheat; cattle, hogs; dairy products). Oil. Some manufacturing (auto parts). Drained by EMBARRAS RIVER. In SW of county is NEWTON LAKE; Sam Parr State Park at center. Formed 1831.

Jasper, county (□ 561 sq mi/1,458.6 sq km; 2006 pop-ulation 32,296), NW INDIANA; ⊙ RENSSELAER; 41°02′N 87°07′W. Bounded N by KANKAKEE River; drained by IROQUOIS RIVER. Corn, soybeans; cattle, hogs; dairying. Jasper-Pulaski State Fish and Wildlife Area and Nursery in NE. Formed 1835.

Jasper, county (□ 732 sq mi/1,903.2 sq km; 2006 pop-ulation 37,440), central IOWA; ⊙ NEWTON; 41°41′N 93°02′W. Prairie agricultural area (hogs, cattle, poul-try; corn, oats), drained by SKUNK and NORTH SKUNK rivers, and with bituminous-coal deposits. Rock Creek Lake and Rock Creek State Park in E. Wide-spread flooding in 1993. Formed 1846.

Jasper, county (□ 677 sq mi/1,760.2 sq km; 2006 population 18,197), E central MISSISSIPPI; ⊙ PAULDING and BAY SPRINGS; 32°01′N 89°07′W. Drained by

TALLAHALA CREEK and short Tallahoma and Souin-lovey creeks. Agriculture (corn, cotton; poultry, cattle; dairying); timber. Oil fields. Includes part of Bienville National Forest in NW. Lake Claude Bennett State Lake in NE. Formed 1833.

Jasper, county (□ 642 sq mi/1,669.2 sq km; 2006 pop-ulation 112,505), SW MISSOURI; ⊙ CARTHAGE; 37°12′N 94°20′W. Borders KANSAS on W; drained by SPRING RIVER. Agriculture (grain, soybeans; poultry, cattle, dairying). Manufacturing at JOPLIN, Carthage, WEBB CITY; former major lead, zinc mines; numerous abandoned surface and underground mines; lime-stone (marketed as marble) quarries; oak timber. Formed 1841.

Jasper, county (□ 685 sq mi/1,781 sq km; 2006 popu-lation 21,809), extreme S SOUTH CAROLINA; ⊙ RIDGE-LAND; 32°26′N 81°01′W. Bounded W by SAVANNAH River, NE by COOSAWHATCHIE RIVER, SE by NEW RIVER. Agricultural area (cattle, hogs; wheat, soy-beans, hay, corn). Formed 1912.

Jasper (JAS-puhr), county (□ 969 sq mi/2,519.4 sq km; 2006 population 35,293), E TEXAS; ⊙ JASPER; 30°44′N 94°01′W. Bounded W by NECHES RIVER (forms B. P. STEINHAGEN LAKE [State Park] in NW). Heavily wooded; lumbering chief industry. Diversified agri-culture (cattle, hogs, horses, poultry; vegetables, pe-cans, fruit); oil and gas. Part of BIG THICKET NATIONAL PRESERVE follows course of Neches River downstream from Steinhagen Lake; part of large SAM RAYBURN RESERVOIR (ANGELINA RIVER) on N boundary; part of Angelina National Forest in NW. Formed 1836.

Jasper, city (2000 population 14,052), ⊙ Walker co., NW central Alabama; 33°51′N87 °16′W. Jasper is a trade and processing center in a coal and timber area. Abundant agr., coal mining, andvaried light manu-facturing (sporting goods, furniture, bottling; poul-tryprocessing). Walker College of the University of Alabama at Birmingham is here.Inc. 1889.

Jasper, city (2000 population 12,100), ⊙ DUBOIS county, SW INDIANA, on PATOKA River, and 45 mi/72 km NE of EVANSVILLE; 38°23′N 86°56′W. Agricultural area (grain, strawberries; livestock, poultry). Manufactur-ing (wood products, machinery, rubber products, furniture, plastic products, electronic equipment); timber. Founded 1818, laid out 1830.

Jasper, city (2000 population 1,011), JASPER county, SW MISSOURI, 11 mi/18 km N of CARTHAGE; 37°19′N 94°17′W. Agriculture (wheat, soybeans; dairying; cattle), Manufacturing (popcorn, auto parts).

Jasper (JAS-puhr), city (2006 population 7,465), ⊙ JASPER county, E TEXAS, c.60 mi/97 km NNE of BEAUMONT; 30°55′N 94°00′W. Elevation 221 ft/67 m. In pine woods area; lumber milling, oil and gas. Agriculture (cattle, horses, hogs; vegetables, fruit, pecans). Diversified light manufacturing. Angelina National Forest to NW; Sabine National Forest to NE; SAM RAYBURN RESERVOIR to N; Martin Dies Jr. State Park, on STEINHAGEN LAKE, to W. Settled 1824, in-corporated 1926. In 1998, the town gained notoriety due to the case of James Byrd, Jr., a black man who was chained to a truck and dragged to his death by three white men.

Jasper (JAS-puhr), town (□ 1 sq mi/2.6 sq km; 2000 population 1,780), ⊙ HAMILTON county, NE FLORIDA, near GEORGIA line, 15 mi/24 km N of LIVE OAK; 30°31′N 82°57′W. Trade and processing center, in to-bacco and timber region. Settled c.1825.

Jasper, town (2000 population 2,167), ⊙ PICKENS county, N GEORGIA, c.50 mi/80 km N of ATLANTA; 34°28′N 84°26′W. Manufacturing (metal products, yarn dying, apparel, molded rubber products, print-ing and publishing). Several marble-clad, older buildings. Incorporated 1857.

Jasper, town (2006 population 3,101), ⊙ MARION county, SE TENNESSEE, near ALABAMA and GEORGIA

state lines, 18 mi/29 km W of CHATTANOOGA; 35°05′N 85°38′W. In fertile SEQUATCHIE River valley; manu-facturing; recreation.

Jasper, village (2000 population 498), ⊙ NEWTON county, NW ARKANSAS, 15 mi/24 km SSW of HARRI-SON, in the OZARKS, near Ozark National Forest and BUFFALO NATIONAL RIVER; 36°00′N 93°11′W. Gene Bush–Buffalo River Wildlife Management Area to SE; small unit of Ozark National Forest to N, main unit to S. Tourism.

Jasper, village (2000 population 597), PIPESTONE and ROCK counties, SW MINNESOTA, near SOUTH DAKOTA state line, 11 mi/18 km SSW of PIPESTONE, on Split Rock Creek; 43°51′N 96°24′W. Agricultural area (grain, soybeans; livestock, poultry; dairying). Man-ufacturing (feeds and fertilizers); silica quarries nearby. Split Rock Creek State Park to NE.

Jasper Lake (JAS-puhr), expansion of ATHABASCA RIVER, W ALBERTA, W CANADA, in ROCKY MOUN-TAINS, in JASPER NATIONAL PARK, 13 mi/21 km N of JASPER; 8 mi/13 km long, 1 mi/2 km wide; 53°06′N 118°01′W. N end of lake was last site of Jasper House, Hudson's Bay Company trading post, moved here 1801 from BRÛLÉ LAKE, abandoned 1875.

Jasper National Park (□ 4,200 sq mi/10,920 sq km), W ALBERTA, in the CANADIAN ROCKY MOUNTAINS. It is the second-largest of the Canadian scenic national parks and contains many high peaks, glaciers, lakes, hot springs, and streams. It is a game reserve and a popular recreation area, with mountain climbing and excellent fishing. The park was named for Jasper Hawes, agent of the North West Company fur-trading post established 1813 on the ATHABASCA RIVER. JAS-PER, a resort town, is the park headquarters and is a station on the Canadian National Railways system. Established 1907.

Jaspur (JUHS-puhr), town, Naini Tal district, N UTTAR PRADESH state, INDIA, 9 mi/14.5 km NW of KASHIPUR. Hand-loom cotton weaving; trades in rice, wheat, mustard, sugarcane, timber.

Jasrana (jahs-RAH-nah), town, MAINPURI district, W UTTAR PRADESH state, INDIA, 22 mi/35 km W of MAINPURI; 27°15′N 78°41′E. Wheat, gram, pearl millet, barley, jowar.

Jassans-Riottier (zhah-sahn–ree-o-tyai), town (□ 1 sq mi/2.6 sq km), AIN department, RHÔNE-ALPES region, E FRANCE, on left bank of the SAÔNE, and 16 mi/26 km N of LYON; 45°59′N 04°45′E. It forms part of VILLE-FRANCHE-SUR-SAÔNE (just W) industrial district (metallurgy, chemicals).

Jassy, ROMANIA: see IAŞI, city.

Jastrebarsko (YAH-stre-bahr-sko), town, central CROATIA, on railroad, 16 mi/26 km SW of ZAGREB; 45°40′N 15°40′E. Trade center in wine-growing re-gion; fishery. Castle. Formerly called Jaska.

Jastrowie (yah-STRO-vee), German *Jastrow*, town, in POMERANIA, Piła province, NW POLAND, 20 mi/32 km N of Piła (Schneidermühl). Agricultural market (grain, sugar beets, potatoes; livestock); textiles. Until 1938, in former Prussian province of Grenzmark Posen-Westpreussen.

Jaswantnagar (JUHS-wuhnt-nuh-guhr), town, ETA-WAH district, W UTTAR PRADESH state, INDIA, 10 mi/16 km NW of ETAWAH; 26°53′N 78°55′E. Road center. Manufacturing (ornamental brass ware). Trades in pearl millet, wheat, barley, oilseeds, yarn, livestock.

Jászapáti (YAHS-ah-pah-ti), Hungarian *Jászapáti*, city, SZOLNOK county, E central HUNGARY, 10 mi/16 km E of JÁSZBERÉNY; 47°31′N 20°09′E. Wheat, corn; dairy; vineyards; hogs, sheep. The city is almost perfectly round, a legacy of the Ottoman era and the subse-quent recolonization of the Great Plain.

Jászárokszállás (YAHS-ah-rok-sahl-lahsh), city, SZOLNOK county, N central HUNGARY, on short Gyöngyös River, and 10 mi/16 km SSE of Gyöngyös; 47°38′N 19°59′E. Wheat, corn, sugar beets; vineyards.

Jászberény (YAHS-ba-ra-nyuh), city (2001 population 28,203), central HUNGARY, on the ZAGYVA RIVER, a tributary of the TISZA; 47°30′N 19°55′E. Dairy; manufacturing (fabricated metal products, leather, apparel). Reputedly was the seat of Attila the Hun.

Jászladány (YAHS-lah-dah-nyuh), Hungarian *Jászladány*, village, SZOLNOK county, central HUNGARY, 13 mi/21 km N of SZOLNOK; 47°22′N 20°10′E. Corn, barley, green peppers, onions, wheat; cattle, hogs; manufacturing (apparel).

Jászó, SLOVAKIA: see JASOV.

Jászság (YAHS-shahg), Hungarian *Jászság*, region in NE central HUNGARY, E of BUDAPEST, on right bank of the TISZA RIVER. Chief city, JÁSZBERÉNY. Its name is related to the Jazyges, a tribe of Sarmatian origin, thought to have settled here in the early Middle Ages.

Jataí (zhah-tah-EE), city (2007 population 82,010), SW GOIÁS, central BRAZIL, 50 mi/80 km W of RIO VERDE; 17°47′S 51°45′W. Coffee, grapes; livestock. Formerly spelled Jatahy.

Jataizinho (ZHAH-tei-seen-yo), city (2007 population 11,245), N PARANÁ state, BRAZIL, 8 mi/12.9 km E of LONDRINA, on railroad; 23°13′S 50°58′W. Coffee, corn, rice; cotton; livestock.

Jataté River (hah-tah-TE), c.150 mi/241 km long, in CHIAPAS, S MEXICO; rises in Sierra de HUEYTEPEC S of OCOSINGO, flows SE to join LACANTÚN RIVER (USUMACINTA system) near GUATEMALA border; 16°15′N 91°17′W.

Jataúba (ZHAH-tah-oo-bah), city (2007 population 15,074), N PERNAMBUCO state, BRAZIL, 42 mi/68 km NW of CARUARU; 70°58′S 36°29′W. Corn, aloe.

Jateí (ZHAH-te-EE), town (2007 population 3,808), S central Mato Grosso do Sul, BRAZIL, 36 mi/58 km SE of DOURADOS; 22°29′S 54°18′W. Agriculture.

Jath, town, SANGLI district, MAHARASHTRA state, INDIA, 45 mi/72 km ENE of SANGLI; 17°03′N 75°13′E. Local market for millet, cotton, wheat, rice. Was capital of former princely state of JATH in DECCAN STATES, BOMBAY; state incorporated 1949 into SATARA SOUTH, SOLAPUR, and BELGAUM districts, Bombay, and later incorporated into Maharashtra state.

Jati (zhah-CHEE), town (2007 population 7,306), S CEARÁ, BRAZIL, 20 mi/32 km S of BREJO SANTO; 07°38′S 39°00′W.

Jati (JAH-tee), village, TATTA district, SW SIND province, SE PAKISTAN, 85 mi/137 km ESE of KARACHI; 24°21′N 68°16′E. Market center for salt, millet, rice. Also called MUGHALBHIN.

Jatibonico (hah-tee-bo-NEE-ko), town (2002 population 22,962), SANCTI SPÍRITUS province, E CUBA, on RÍO JATIBONICO DEL SUR, on Central Highway, on railroad, and 27 mi/43 km W of CIEGO DE ÁVILA; 21°56′N 79°10′W. In agricultural region (sugarcane, tobacco, livestock). Manufacturing (pottery and cigars). The sugar mill of URUGUAY is just SW.

Jatibonico del Norte, Río (hah-tee-bo-NEE-ko del NOR-te, REE-o), 43 mi/70 km long, central CUBA; rises in the Sierra de Jatibonico 8 mi/13 km SSE of CAIBARIÉN, flows along SANCTI SPÍRITUS–CIEGO DE ÁVILA province border to N coast; 21°15′N 78°58′W. Has irregular course, obstructed by cataracts, and flowing partly through subterranean trench (3 mi/4 km long).

Jatibonico del Sur, Río (hah-tee-bo-NEE-ko del soor), 73 mi/117 km long, seventh-longest river in CUBA; flows through SANCTI SPÍRITUS province. Drains 322 sq mi/835 sq km area and empties into CARIBBEAN SEA.

Jatiluhur, Bendung (jah-TEE-loo-hoor) or **Lake Jatiluhur**, hydroelectric and irrigation dam and reservoir, Java Barat province, INDONESIA, 31 mi/50 km NW of BANDUNG; 06°35′S 107°20′E. Multipurpose dam project completed in 1965.

Jatinã (ZHAH-chee-nah), city, W central PERNAMBUCO, NE BRAZIL, on left bank of SÃO FRANCISCO River (BAHIA border), and 60 mi/97 km SW of SERRA TALHADA; 08°47′S 38°49′W. Onions, cereals. Until 1944, called Belém.

Jatiúca (ZHAH-chee-oo-kah), city, N PERNAMBUCO state, BRAZIL, 9 mi/14.5 km N of Serra Taihada; 07°53′S 38°13′W. Cotton.

Játiva (HAH-tee-vah), Catalan *Xátiva*, city, VALENCIA province, E SPAIN, in Valencia, 20 mi/32 km W of GANDÍA; 38°59′N 00°31′W. The city is a processing and distribution center for farm products. Its famous linen industry dates back to Roman times. James I of Aragón liberated Játiva from the Moors in the 13th century. There are many fine public and private buildings, notably the well-preserved Spanish-Moorish castle, a former Mozarabic church, and the Gothic collegiate church (15th century). Játiva was long the residence of the Borgia, or Borja, family. Popes Calixtus III and Alexander VI were born here, as was the painter Jusepe Ribera.

Jatiwangi (jah-tee-WAHNG-gee), town, Java Barat province, INDONESIA, 30 mi/48 km W of Ceribon; 06°44′S 108°15′E. Trade center, with highway and railroad connections for agricultural area (rice and sugar); sugar and textile mills. Also spelled Djatiwangi.

Jatoba, Brazil: see PETROLÂNDIA.

Jatt (JAHT), Arab township, ISRAEL, ESE of HADERA, and 5.6 mi/9 km N of TULKARM in N Samarian highlands. Established in 19th century on foundations of ancient settlement that was alongside the Via Maris route. Bronze Age artifacts found. Also Druze village in W UPPER GALILEE, 8 mi/13 km NE of NAHARIA. Mixed farming, olive growing. The site was inhabited from the Canaanite to the Arab periods, with many remnants of ancient buildings.

Jatta Ismail Khel (JUT-tuh is-MEIL KAIL), village, KOHAT district, E NORTH-WEST FRONTIER PROVINCE, N PAKISTAN, 18 mi/29 km SSW of KOHAT; 33°20′N 71°17′E. Sometimes called Jatta.

Jaú (ZHAU-oo), city (2007 population 125,469), central SÃO PAULO, BRAZIL, near TIETÊ RIVER, 100 mi/161 km WNW of CAMPINAS; 22°18′S 48°33′W. Agricultural processing center at E edge of coffee-growing region. Manufacturing of machinery, furniture; manioc and cottonseed-oil processing, brewing, distilling. Trades in coffee, sugar, potatoes, rice, livestock. Has noteworthy church and town hall. Airfield. Formerly spelled Jahú.

Jauareté, Brazil: see IAUARETÊ.

Jauer, POLAND: see JAWOR.

Jauernig, CZECH REPUBLIC: see JAVORNIK.

Jauf (JOWF), town and oasis (2004 population 26,179), Jouf province, N of HEJAZ, northernmost SAUDI ARABIA, at N edge of the NAFUD, 250 mi/402 km E of Ma'an, near the head of the WADI SIRHAN; 29°58′N 39°34′E. Agricultural center (dates, wheat, barley, millet, corn, alfalfa, vegetables, fruit). Handicrafts (weaving, leatherworking). A major caravan center between SYRIAN DESERT and central ARABIAN PENINSULA, it forms a long, narrow belt of gardens and palm groves, dominated (SE) by a stone castle. Also called Jauf al Amir (or Jauf el Amr), the oasis was formerly considered part of JEBEL SHAMMAR. Also spelled Jawf.

Jauf (JOF), large oasis in YEMEN hinterland, at edge of the desert Rub' AL KHALI; 16°10′N 44°50′E. Here are the ruins of Ma'in, capital of Minaean kingdom (c.1200–650 B.C.E.), which was succeeded by the Sabaean kingdom of Marib. Also spelled Jawf.

Jaugarh, ruined fort in GANJAM district, SE ORISSA state, E INDIA, on RUSHIKULYA RIVER, and 14 mi/23 km NNE of BRAHMAPUR. Contains Asokan rock edicts.

Jauja (HOU-hah), province (□ 4,200 sq mi/10,920 sq km), W central JUNÍN region, central PERU; ⊙ JAUJA; 11°25′S 74°50′W. Traversed by MANTARO RIVER.

Jauja (HOU-hah), city, ⊙ JAUJA province, JUNÍN region, central PERU, on CERRO DE PASCO-HUANCAVELICA railroad, on MANTARO RIVER, 25 mi/40 km NW of HUANCAYO; 11°48′S 75°30′W. Elevation 11,187 ft/3,410 m. Agricultural products (cereals, potatoes); livestock. Pizarro's first capital in Peru. Indian ruins. Health resort.

Jaumave (hwah-MAH-ve), town, TAMAULIPAS, NE MEXICO, in a valley of E SIERRA MADRE OCCIDENTAL outliers, 28 mi/45 km SW of CIUDAD VICTORIA; 23°28′N 99°22′W. Small farming and ranching.

Jaunay-Clan (zho-nai–klahn), town (□ 10 sq mi/26 sq km), VIENNE department, POITOU-CHARENTES region, W FRANCE, 7 mi/11 km ENE of POITIERS on CLAIN RIVER; 46°41′N 00°22′E.

Jaungulbene, LATVIA: see GULBENE.

Jaunjelgava (YOUN-yel-gah-vuh) or **yaunyelgava**, German *friedrichstadt*, city, S central LATVIA, in ZEMGALE, on left bank of the DVINA (DAUGAVA) River, and 45 mi/72 km SE of RIGA; 56°37′N 25°05′E. Leather manufacturing, sawmilling. Has castle, botanic garden.

Jaunlantgale, RUSSIA: see PYTALOVO.

Jaunlatgale, RUSSIA: see PYTALOVO.

Jaun Pass (DJAW-uhn), French *La Jogne* (4,951 ft/1,509 m), in BERNESE ALPS, W central SWITZERLAND, N of ZWEISIMMEN; joins the SIMMENTAL valley in BERN to the LA GRUYÈRE country in FRIBOURG canton.

Jaunpur (JOUN-puhr), district (□ 1,559 sq mi/4,053.4 sq km), SE UTTAR PRADESH state, N INDIA, on GANGA PLAIN; ⊙ JAUNPUR. Drained by GOMATI RIVER. Agriculture (barley, rice, corn, wheat, sugarcane, gram, mustard, millets); mango, mahua, sissoo, and babul groves. Main towns are Jaunpur, MACHHLISHAHR, SHAHGANJ.

Jaunpur (JOUN-puhr), city, ⊙ JAUNPUR district, UTTAR PRADESH state, NE INDIA, on the GOMATI RIVER; 25°44′N 82°41′E. Market town where perfume is made. Also a religious site, Jaunpur was a well-known center of Muslim learning and architecture in the 15th century. Of the many buildings from this period, the great Atala Devi Masjid mosque (completed 1408) is the most notable.

Jauntal (YOUN-tahl), Slovenian *Podjuna*, region between DRAU RIVER and eastern KARAWANKEN mountains, S of Völkermarkt, CARINTHIA, S AUSTRIA; 46°32′–38′N, 14°30′–49′E. Flat or hilly, in the W several small lakes (KLOPEINER SEE, Turnersee). Main settlements are BLEIBURG, Kühnsdorf, EBERNDORF, Globasnitz (Slovenian *Globasnica*). Small industry, summer tourism; crossed by Jauntal railroad in E-W direction and by Seeberg road N-S. Slovenian minority.

Jáuregui (HWAH-re-gee), town, BUENOS AIRES province, ARGENTINA, 4 mi/6.4 km SW of LUJÁN; 34°36′S 59°10′W. Livestock center.

Jauru (ZHAH-oo-roo), city (2007 population 10,760), SW MATO GROSSO, BRAZIL, 73 mi/118 km NW of CÁCERES; 15°28′S 56°38′W. Stock raising and agriculture.

Java, village (2006 population 179), WALWORTH county, N central SOUTH DAKOTA, 8 mi/12.9 km E of SELBY; 45°30′N 99°52′W. In cattle-raising region. Lake Hiddenwood State Park to NW.

Java (JAH-vuh), unincorporated village, PITTSYLVANIA county, S VIRGINIA, 12 mi/32 km NNE of DANVILLE; 36°50′N 79°13′W. Manufacturing (lumber, hickory chips). Agriculture (dairying; cattle; grain, soybeans); timber.

Java (JAH-vah), island (□ 51,038 sq mi/132,188 sq km; 1990 population 107,525,520), INDONESIA, S of BORNEO, from which it is separated by the JAVA SEA, and SE of SUMATRA across SUNDA STRAIT; 07°30′S 110°00′E. Although Java is the fifth largest island of

Indonesia, constituting only a fraction of the country's total area, it contains a majority of the country's population; its population density (more than 2,000 people/sq mi) is one of the highest in the world. For centuries it has been the cultural, political, and economic center of the area. In Java are Indonesia's capital and largest city, JAKARTA (formerly called Batavia), and the second and third largest cities, SURABAYA and BANDUNG. TANJUNGPRIOK, serving Jakarta in the NW, is the chief port, and YOGYAKARTA and SURAKARTA are cultural centers. A chain of volcanic mountains—most of them densely forested with teak, palms, and other woods—traverses the length of the island E-W; Mount SEMERU rises to 12,060 ft/3,676 m. Although Java contains only about 3% of the country's forest land, it accounts for much of its teak production. There are almost 2,000,000 acres/809,400 ha of planted teak forests. The climate is warm and humid, and the volcanic soil is exceptionally fertile. There are elaborate irrigation systems, supplied by the island's numerous short, turbulent rivers. Agricultural productivity has grown in recent years and industrialization of the economy has boosted the island's capacity to support its people despite the overcrowded conditions. Most of Indonesia's sugarcane and kapok are grown in Java. Rubber, tea, coffee, tobacco, cacao, and cinchona are produced in highland plantations. Rice is the chief small-farm crop. Cattle are raised in the E. In the NE are important oil fields; gold, silver, manganese, phosphate, and sulphur are mined. Most of the country's manufacturing establishments are in Java. Industry is centered chiefly in Jakarta and Surabaya, but Bandung is a noted textile center. Found mostly in the interior are such animals as tigers, rhinoceros, and crocodiles; birds of brilliant plumage are numerous. Java was a home of early humans; here in 1891 were found the fossilized remains of the so-called Java man, or *Pithecanthropus erectus*. The typically Malayan inhabitants of the island comprise the Javanese (the most numerous), the Sundanese, and the Madurese. Numerous Chinese and Arabs live in the cities. Like Bali, Java is known for its highly developed arts. There is a rich literature, and the *wayang*, or shadow play, employing puppets and musical accompaniment, is an important dramatic form. Java has many state and private institutions of higher learning; most are in Jakarta, but Bandung, BOGOR, Yogyakarta, and Surabaya all have several universities. Early in the Christian era Indians began colonizing Java, and by the seventh century "Indianized" and "Hindu" kingdoms were dominant in both Java and Sumatra. The Sailendra dynasty (760–860 in Java) unified the Sumatran and Javan kingdoms and built in Java the magnificent Buddhist temple BOROBUDUR. From the tent century to the fifteenth century, E Java was the center of Hindu-Javanese culture. The high point of Javanese history was the rise of the powerful Hindu-Javanese state of Majapahit (founded 1293), which extended its rule over much of Indonesia and the MALAY PENINSULA. Islam, which had been introduced in the thirteenth century, peacefully spread its influence, and the new Muslim state of Mataram emerged in the sixteenth century. Following the Portuguese, the Dutch arrived in 1596, and in 1619 the Dutch East India Company established its chief post in Batavia (Jakarta), thence gradually absorbing the native states into which the once-powerful Javanese empire had disintegrated. Between 1811 and 1815, Java was briefly under British rule headed by Sir Thomas S. Raffles, who instituted many reforms. The Dutch ignored these when they returned to power, resorting to a system of enforced labor, which, along with harsh methods of exploitation, led to a native uprising (1825–1830) under Prince Diponegoro; the Dutch subsequently adopted a more humane approach. In the early phase of World War II in Asia, Java was left open to Japanese invasion after the disastrous Allied defeat in the battle of the Java Sea in February 1942; Java was occupied by the Japanese until the end of the war. After the war the island was the scene of much fighting between Dutch and Indonesian independence forces; in 1946 the Dutch occupied many of the key cities, and the republic's capital was moved to Yogyakarta. Java became part of Indonesia in December 1949, and now constitutes three provinces of Indonesia: Java Barat (W Java; twenty districts), Java Tengah (Central Java; twenty-nine districts), and Java Timur (E Java; twenty-nine districts), as well as the autonomous districts of Yogyakarta and Jakarta, which have equal standing to province administration. Overcrowding on Java led the government to make a policy approximately forty years ago of "transmigration." Since then, more than six million people have been relocated to less populated Indonesian islands. Nevertheless, Java continues to receive large numbers of immigrants, which complicates the overcrowding problem.

Javadi Hills (juh-VAH-dee), E outlying group of S EASTERN GHATS (separated by upper PALAR RIVER valley), in KOLAR district, KARNATAKA state, S INDIA; c.60 mi/97 km long, maximum c.35 mi/56 km wide; rises to over 1,000 ft/305 m. Chief products are sandalwood, tanbark, nux vomica, hemp narcotics, tamarind, beeswax. CHEYYAR River rises on central plateau. Historic rock fortresses at Karnaticgarh (SE; 3,180 ft/969 m) and Kailasagarh (NE; c.2,740 ft/835 m) peaks.

Java Head, INDONESIA: see GEDE POINT.

Javalón River or **Jabalón River** (both: hah-vah-LON), c.100 mi/161 km long, CIUDAD REAL province, S central SPAIN, in CASTILE-LA MANCHA; rises near ALBACETE province border SE of MONTIEL, flows W and NW, past VALDEPEÑAS, to the GUADIANA 11 mi/18 km SW of CIUDAD REAL.

Javanrud (jah-vah-ROOD), town, Kermānshāhān province, W IRAN, 45 mi/75 km NW of KERMANSHAH, 40 mi/65 km E of IRAQ border; 34°48'N 46°30'E.

Javari River (hah-VAH-ree), Spanish, *Yavarí*, c.500 mi/805 km long, E Peru; rising in the Cerro de Canchyuaya; flows NE, forming part of the border between BRAZIL and PERU, before entering the AMAZON near TABATINGA; 04°21'S 70°02'W. It is navigable for most of its length.

Java Sea (JAH-vah) (□ 167,000 sq mi/434,200 sq km), part of the PACIFIC OCEAN, between BORNEO (N) and JAVA (S), connected with SOUTH CHINA SEA by KARIMATA STRAIT, with CELEBES SEA by MAKASAR STRAIT, and with INDIAN OCEAN by SUNDA STRAIT; c.900 mi/1,448 km E-W, c.260 mi/418 km N-S; 05°00'S 110°00'E. In World War II, scene of disastrous Allied naval defeat (Feb. 27–March 1, 1942), which exposed Java to JAPANESE invasion. The sea has become an increasingly important shipping area as a result of the development of oil and timber resources in SUMATRA and Borneo (Kalimantan) and manufacturing on Java. The area off SE Borneo is thought to have oil reserves.

Java Trench, submarine depression in INDIAN OCEAN, off S coast of JAVA. Here the greatest depth (24,440 ft/7,449 m) of the Indian Ocean was obtained (1925–1928) at 10°21'S 110°6'E, E of CHRISTMAS ISLAND. Here the Eurasian Plate rises above the Australian Continental Plate.

Jávea (HAH-vai-ah), seaport, ALICANTE province, E SPAIN, in picturesque site on the MEDITERRANEAN, 22 mi/35 km SE of GANDÍA; 38°47'N 00°10'E. Exports raisins, wine, citrus from fertile hinterland. Olive-oil processing; light manufacturing. Has remains of ancient walls and medieval castle. Stalactite caves nearby.

Javier Island (hah-vee-ER), just off coast of AISÉN province, S CHILE, on inner GULF OF PEÑAS; 47°06'S 74°24'W.

Javorice (YAH-vo-RZHI-tse), Czech *Javořice*, second-highest mountain (2,746 ft/837 m) in BOHEMIAN-MORAVIAN HEIGHTS, JIHOMORAVSKY province, SW MORAVIA, CZECH REPUBLIC, 16 mi/26 km SW of JIHLAVA; 49°13'N 15°20'E. Granite quarries at SE foot.

Javorina (yah-VO-ri-NAH), village, VYCHODOSLOVENSKY province, N SLOVAKIA, on NE slope of the HIGH TATRAS, 16 mi/26 km NNW of POPRAD, near POLISH border; 49°16'N 20°09'E. Elevation 3,280 ft/1,000 m. Woodworking. Noted natural park and game reserve to SE. Along with surrounding area, incorporated 1938 into Poland after MUNICH Pact; returned to Slovakia after partition (1939) of Poland.

Javor Mountains, in DINARIC ALPS, E central BOSNIA, BOSNIA AND HERZEGOVINA; extend c. 15 mi/24 km NW-SE; highest peak (5,041 ft/1,536 m) is 4 mi/6.4 km ESE of HAN PIJESAK; 43°18'N 18°20'E. Also spelled Yavor Mountains.

Javornik (YAH-vor-NYEEK), Czech *Javorník*, German *Jauernig*, village, SEVEROMORAVSKY province, NW SILESIA, CZECH REPUBLIC, 14 mi/23 km NW of JESENIK, near Polish border; 50°24'N 17°01'E. Railroad terminus. Manufacturing (textiles); agriculture (oats). Has an old castle.

Javornik (yah-VOR-nik), mountain (3,483 ft/1,062 m), DINARIC ALPS, E BOSNIA, BOSNIA AND HERZEGOVINA, 7 mi/11.3 km NNE of KLADANJ. c. DRINJACA River flows along S and E foot. Also spelled Yavornik.

Javorniks, CZECH and SLOVAK *Javorníky* (Slovak: yah-VOR-nyee-KI; Czech: YAH-vor-NYEE-ki), HUNGARIAN FEHÉR KARPATOK, mountain range, STREDOSLOVENSKY province, NW SLOVAKIA and NE CZECH REPUBLIC (MORAVIA); extends c.40 mi/64 km NE-SW, between ČADCA (N) and LYSA PASS (S); rises to 3,514 ft/1,071 m in MOUNT VELKY JAVORNIK.

Jawad (JAH-wuhd), town (2001 population 16,143), formerly in W MADHYA BHARAT, now in NIMACH district, NW MADHYA PRADESH state, INDIA, 9 mi/14.5 km N of NIMACH; 24°36'N 74°51'E. Market center for cotton, millet, wheat, gram; oilseed milling, cotton ginning; handicrafts (cloth, bracelets).

Jawar (JUH-wahr), town (2001 population 7,131), formerly in W BHOPAL state, now in SEHORE district, central MADHYA PRADESH state, central INDIA, 60 mi/97 km WSW of BHOPAL. Hand-loom weaving. Agriculture: wheat, gram, cotton, millet.

Jawf, YEMEN: see JAUF.

Jawf, Al (JOUF, el), largest oasis in KUFRA group of oases, CYRENAICA region, SE LIBYA, 350 mi/563 km SSE of JALU; 24°12'N 23°18'E. Was capital of former Al Kufrah province. Formerly El Giof.

Jawf, Wadi al, seasonal watercourse, YEMEN, draining a large area of NE Yemen from Hajjah W to the RAMLAT AS SABATAYN desert.

Jawhar, village, THANE district, W MAHARASHTRA state, W INDIA, 50 mi/80 km NNE of Thane, on outlier of WESTERN GHATS. Agriculture market (millet, pulses). Teak in nearby forests. Was capital of former princely state of JAWHAR in Gujarat States, BOMBAY; state incorporated 1949 into Thane district.

Jawor (YAH-vor), German *Jauer*, town (2002 population 24,762) in LOWER SILESIA, Legnica province, SW POLAND, 11 mi/18 km S of LEGNICA (Liegnitz). Manufacturing (leather goods, chemicals, soap, stoves); textiles, metalworking; granite quarrying. Has late-Gothic church. In 14th century, capital of principality under branch of Polish Piast dynasty.

Jaworów, UKRAINE: see YAVORIV.

Jaworzno (yah-VOZ-no), town (2002 population 96,826), Katowice province, S POLAND, 12 mi/19 km E of KATOWICE. Manufacturing (chemicals, wood products, flour); coal mining. Abandoned galena mine.

Jaxartes, KAZAKHSTAN: see SYR DARYA.

Jay, county (□ 383 sq mi/995.8 sq km; 2006 population 21,605), E INDIANA, bounded E by OHIO state line; ⊙ PORTLAND; 40°26'N 85°01'W. Agricultural area (corn, oats, vegetables, soybeans; hogs, cattle, poultry). Diversified manufacturing at Portland; lumber milling. Natural-gas and oil wells; timber. Drained by SALAMONIE RIVER. Formed 1836.

Jay, town, FRANKLIN county, W central MAINE, on the ANDROSCOGGIN RIVER, 13 mi/21 km SSE of FARMINGTON. Village of Chisholm (1990 population 1,653) has pulp and paper mills; 44°31′N 70°13′W. Includes villages of Jay and North Jay. Incorporated 1795.

Jay, town (2006 population 2,947), ⊙ DELAWARE county, NE OKLAHOMA, near ARKANSAS state line, 25 mi/40 km SE of VINITA; 36°25′N 94°47′W. Elevation 1,032 ft/315 m. Trade center for agricultural and recreation area (fruit, berries; livestock, poultry). Manufacturing (concrete, electronic equipment, poultry processing). Lake Eucha (formerly Upper Spavinaw Lake) and Upper Spavinaw State Park to S; LAKE OF THE CHEROKEES to NW.

Jay, town, ORLEANS CO., N VERMONT, on QUEBEC (CANADA) border, 11 mi/18 km W of Newport; 44°58′N 72°28′W. Jay Peak is W, with ski resort. Named for American founding father, John Jay (1745–1829).

Jay, hamlet, ESSEX county, NE NEW YORK, in ADIRONDACK MOUNTAINS, on East Branch of AUSABLE RIVER, 26 mi/42 km SSW of PLATTSBURGH; 44°22′N 73°42′W. In resort area.

Jayabhum, Thailand: see CHAIYAPHUM, town.

Jayanca (hah-YAHN-kah), town, LAMBAYEQUE province, LAMBAYEQUE region, NW PERU, on coastal plain, on PAN-AMERICAN HIGHWAY, and 22 mi/35 km NNE of LAMBAYEQUE; 06°24′S 79°50′W. In irrigated Motupe valley (rice, corn, fruit); apiaries; vineyards.

Jayankondacholapuram (juh-YUHN-kon-duh-CHOluh-puh-ruhm), town, TIRUCHIRAPPALLI district, S TAMIL NADU state, S INDIA, 50 mi/80 km NE of Tiruchchirappalli; 11°13′N 79°22′E. Manufacturing (cotton weaving; brass vessels). Nearby village of Gangaikondapuram is site of large 11th-century Dravidian temple. Formerly spelled Jeyamkondacholapuram; also spelled Jayamkondacholapuram.

Jaya Peak (JAH-yah), Indonesian *Puncak Jayawijaya,* mountain peak (16,503 ft/5,031 m), the highest in the SUDIRMAN mountains, of NEW GUINEA, and INDONESIA, IRIAN JAYA province, Indonesia, on W New Guinea island; 04°05′S 137°11′E. The peaks are snow covered, but the lower slopes have lush tropical vegetation. It was formerly called Mount Carstensz and Mount Sukarno or Soekarno. Elevation sometimes cited at 16,024 ft/4,884 m.

Jayapura (jah-yah-POO-rah), town (2000 population 166,201), ⊙ IRIAN JAYA province, W NEW GUINEA, INDONESIA; 02°32′S 140°42′E. A regional trade center and seaport, it is on YOS SUDARSO BAY (formerly Humboldt Bay; an inlet of the PACIFIC OCEAN) near the border of PAPUA NEW GUINEA. Occupied by the Japanese in World War II, it was liberated by U.S. forces in April 1944 and served as General MacArthur's headquarters. Formerly called Hollandia. Sentani Airport.

Jayaque (hah-YAH-kai), city and municipality, LA LIBERTAD department, SW EL SALVADOR, in coastal range, 8 mi/12.9 km W of NUEVA SAN SALVADOR; 13°40′N 89°26′W. Coffee-growing center; coffee processing, light manufacturing. Gypsum and marble deposits nearby.

Jayena (hei-AI-nah), town, GRANADA province, S SPAIN, 20 mi/32 km SW of GRANADA; 36°57′N 03°49′W. Olive oil, cereals; lumber. Completely destroyed in 1884 by earthquake; was soon rebuilt.

Jaynagar (JEI-nah-guhr) or **Jainagar**, town (2001 population 19,493), N MADHUBANI district, N BIHAR, INDIA, 33 mi/53 km NNE of DARBHANGA, near NEPAL border. Railroad terminus. Agriculture (rice, corn, sugarcane, barley, oilseeds).

Jaynagar, town, South 24–Paraganas district, SE WEST BENGAL state, E INDIA, on railroad spur, and 28 mi/45 km SSE of CALCUTTA city center. Rice, jute, pulses. Also spelled Joynagar. Railroad-spur terminus 8 mi/12.9 km SW, at Lakshmikantapur.

Jaynagar, INDIA: see JAYNAGAR.

Jay Peak (3,861 ft/1,177 m), in GREEN MOUNTAINS, N Vermont, near QUEBEC (CANADA) border, 15 mi/24 km W of Newport. N terminus of Long Trail. Downhill ski area.

Jaypur, town, KORAPUT district, SW ORISSA state, INDIA, 240 mi/386 km SW of CUTTACK. Trades in rice, hides, forest products (sal, teak, bamboo, lac); rice milling, tanning; tile works, distillery (arrack). College.

Jayton, village (2006 population 443), KENT county, NW TEXAS, 40 mi/64 km NNE of SNYDER, near Salt Fork Brazos River; 33°15′N 100°34′W. Elevation 2,015 ft/614 m. In cattle and agricultural (cotton; wheat, sorghum) region.

Jayuma Llallagua (hah-YOO-mah yay-YAH-gwah), canton, PACAJES province, LA PAZ department, W BOLIVIA; 17°16′S 68°34′W. Elevation 12,992 ft/3,960 m. Gas resources. Lead-bearing lode, copper resources (although copper mine COROCORO closed in 1986); clay, limestone, gypsum deposits. Agriculture (bananas, yucca, potatoes, rye); cattle.

Jayus, Arab village, TULKARM district, 7.5 mi/12 km S of Tulkarm, in the W foothills of the Samarian Highlands, WEST BANK; 32°12′N 35°02′E. Agriculture (olives, grapes).

Jayuya (hah-YOO-yah), town (2006 population 18,194), central PUERTO RICO, in Toro Negro Forest, 14 mi/23 km N of PONCE. Coffee-growing and production center. Agriculture (tomatoes, citrus fruits, plantains); light manufacturing. Traditional woodcarving. Starting point for ascent of Tres Picachos peak, 3 mi/4.8 km E.

Jaza'ir Bin Ghalfan, OMAN: see KURIA MURIA ISLANDS.

Jazer (JEZ-er), ancient city E of the JORDAN River, probably about 10 mi/16 km N of HISBAN, JORDAN. In the Hebrew Bible it was assigned to Gad. Called Ya'azer in Hebrew.

Jazira, Al (jah-ZEE-rah, al), the region of MESOPOTAMIA between the TIGRIS River and the EUPHRATES River in NE SYRIA and NW IRAQ, NW of BAGHDAD. The Syrian JAZIRA has undergone extensive development and settlement since the 1960s, especially after the diversion of large quantities of water to irrigate extensive areas in S Syria. The region is a steppe, arid in the SE, but in the N and W receives sufficient rainfall for growing wheat and barley. KHABUR River, an affluent of the EUPHRATES RIVER, is used for irrigation. Sometimes spelled El JEZIREH or El Jezire.

Jazirat al-Malik, SUDAN: see URONARTI.

Jazirat ash-Sharik, TUNISIA: see BON, CAPE.

Jazirat Banzart, TUNISIA: see BIZERTE, CAPE.

Jazirat Carthage, TUNISIA: see CARTHAGE, CAPE.

Jazire-ye Forur, island, IRAN, in PERSIAN GULF, 30 mi/48 km SW of BANDAR-E-LENGEH. JAZIRE-YE Bani Forur island is 15 mi/24 km S.

Jazłowiec, UKRAINE: see YABLUNIVKA.

Jaz Murian (JUHZ MOO-ree-ahn), salt-lake depression in SE IRAN, 200 mi/322 km SE of KERMAN; 27°29′N 58°32′E. Receives BAMPUR (E) and HALIL (W) rivers.

Jazzin, LEBANON: see JEZZIN.

Jbel Aklim, MOROCCO: see AKLIM, ADRAR-N-.

Jbilet (zhe-bee-LET), low, E-W mountain range, MOROCCO, 12 mi/20 km N of MARRAKECH, separating HAOUZ basin from BAHIRA plain and plateaus to N; 31°50′N 08°00′W.

Jdeide (zh-DAI-dai), residential suburb of BEIRUT, central LEBANON, near MEDITERRANEAN SEA, 5 mi/8 km E of Beirut. Sericulture, cotton, tobacco; cereals, lemons. Also spelled Jedeideh and Judeide.

Jean, unincorporated village, CLARK county, S NEVADA, 25 mi/40 km SSW of LAS VEGAS, 11 mi/18 km NE of CALIFORNIA state line. Cattle. Manufacturing (plastics products). Toiyabe National Forest to NW.

Jeanerette (jen-uh-RET), city, IBERIA parish, S LOUISIANA, on navigable Bayou TECHE, and 10 mi/16 km SE of NEW IBERIA; 29°55′N 91°41′W. Market center for oil, natural gas. Agricultural area (sugarcane, rice, soybeans); crawfish, catfish. Manufacturing (boats, sugar-mill machinery, apparel); sugar milling. Jeanerette Museum features sugarcane industry. Antebellum homes in area. Lake Fausse Pointe (State Park on E shore to NE).

Jean Lafitte (ZHAWN lah-FEET), town (2000 population 2,137), elevation 2 ft/0.6 m, JEFFERSON parish, SE LOUISIANA, 14 mi/23 km S of NEW ORLEANS, on Dupre Cut-Off Canal; 29°45′N 90°07′W. Agriculture (home gardens, nursery crops). JEAN LAFITTE NATIONAL HISTORICAL PARK, Barataria Unit, to N; swamp tours. Village of LAFITTE 4 mi/6 km S.

Jean Lafitte National Historical Park and Preserve (ZHAWN lah-FEET), SE LOUISIANA. This park includes four units, total acreage 20,020 acres/8,108 ha. Part of the NEW ORLEANS' French Quarter (18.6 acres/7.5 ha) is in ORLEANS parish, in the French Quarter of downtown NEW ORLEANS. Depicts life and French heritage of LOUISIANA delta region. Includes park visitor center. The Chalmette Battlefield (143 acres/58 ha) is in SAINT BERNARD parish, 7 mi/11 km E of NEW ORLEANS. Established 1907; site of Battle of NEW ORLEANS (1812) and Chalmette National Cemetery. The BARATARIA unit (19,851 acres/8,040 ha) is in JEFFERSON and LAFOURCHE parishes, 15 mi/24 km S of NEW ORLEANS. Sample of MISSISSIPPI RIVER delta ecology, including cypress swamps, marshes, bayous. Smuggling site provided to U.S. forces by Jean Lafitte's band during War of 1812. Lafitte helped Andrew Jackson defeat British in 1815 Battle of New Orleans. Isleno or Acadian unit (7.4 acres/3 ha) includes three small sites in LAFAYETTE and SAINT LANDRY parishes. Headquarters and cultural resources center in LAFAYETTE.

Jeannette (juh-NET), city (2006 population 10,096), WESTMORELAND county, SW PENNSYLVANIA, suburb 20 mi/32 km SE of PITTSBURGH, on Brush Creek; 40°19′N 79°36′W. Located in a coal and natural-gas area. Manufacturing (machinery, plastic products, specialty glass, printing). Agriculture (corn; livestock; dairying). Its glassworks date from 1889. Bushy Run Battleground historic site to N. Laid out 1888, incorporated as a city 1937.

Jeannette Island, Russian *Ostrov Zhannetta,* easternmost of DE LONG ISLANDS, in EAST SIBERIAN SEA, 345 mi/555 km off N SAKHA REPUBLIC, RUSSIAN FAR EAST; 76°35′N 158°30′E.

Jeannette, Mount (11,700 ft/3,566 m), SW YUKON, CANADA, near ALASKA border, in ST. ELIAS MOUNTAINS, 200 mi/322 km W of WHITEHORSE; 60°31′N 140°57′W.

Jean-Rabel (ZHAWNG–rah-BEL), town, NORD-OUEST department, NW HAITI, near NW tip of HISPANIOLA island, 23 mi/37 km WSW of PORT-DE-PAIX; 19°52′N 73°11′W. Agricultural center (sugarcane, sisal, fruit); bee-keeping; sugar processing. Copper deposits nearby. Its port, BORD-DE-MER-JEAN-RABEL, is 4 mi/6.4 km NNW.

Jebail, LEBANON: see BYBLOS.

Jebba (JE-bah), town, KWARA state, W NIGERIA, the head of navigation on the NIGER River; 09°08′N 04°50′E. The second-largest town in the state, it is a port as well as a railroad and road center. Manufacturing (paper). Jebba was conquered by the British in 1897 and served as the temporary capital of the Protectorate of North Nigeria from 1900 to 1902. The railroad reached Jebba in 1909 and, in 1916, Nigeria's first bridge across the Niger was built here. The Niger Dams project, begun in the 1980s and designed to provide hydroelectric power to the country, is upstream.

Jebel (JE-bel), town, W BALKAN weloyat, W TURKMENISTAN, in the KARA KUM desert, at W base of GREATER BALKAN MOUNTAINS, on TRANS-CASPIAN RAILROAD, and 10 mi/16 km NW of NEBITDAG; 39°38′N 54°14′E. Railroad shops; salt extraction; health resort Molla-Kara (just SW). Also spelled Dzhebel.

Cross-references are shown in SMALL CAPITALS. The pronunciation guide is shown on page xix. The sources of population figures are shown on page xvii.

Jebel [Arabic=mountain], for names beginning thus and not found here: see under following part of the name.

Jebel-Bereket, TURKEY: see OSMANIYE.

Jebel Esh-Shifa, peak (7,828 ft/2,386 m.), Hejaz Mountains, SAUDI ARABIA, near TAIF; 21°31′N 39°55′E. Highest peak in HEJAZ.

Jebel Oda (Je-bel O-duh), peak (7,412 ft/2,259 m), in the Red Sea highlands bordering RED SEA, NE SUDAN, 70 mi/113 km NNW of PORT SUDAN; 20°21′N 36°39′E.

Jebel Shammar, SAUDI ARABIA: see SHAMMAR, JEBEL.

Jeberos (hai-BAI-ros), town, ALTO AMAZONAS province, LORETO region, N central PERU, in AMAZON basin, 45 mi/72 km N of YURIMAGUAS; 05°17′S 76°17′W. Agriculture (banana, plantains, yucca; tobacco). Founded 1640.

Jebja (ZHEB-zhuh), beach resort, Tanger-Tétouan administrative region, MOROCCO, on the MEDITERRANEAN SEA coast, 6 mi/10 km SE of the mouth of Oued LAOU river, and 22 mi/35 km NE of CHEFCHAOUEN.

Jebla, SYRIA: see JEBLE.

Jeble (JE-ble), town, LATTAKIA district, NW SYRIA, on the MEDITERRANEAN Sea, 14 mi/23 km SE of LATTAKIA; 35°21′N 35°55′E. Agriculture (cotton, tobacco; cereals). Site of a once-flourishing Phoenician town, also important in Byzantine times; extensive ruins remain. Ship anchorage. Sometimes spelled JABLEH or JEBLA.

Jech Doab, PAKISTAN: see CHAJ DOAB.

Jechica Island, CHILE: see CHONOS ARCHIPELAGO.

Jechnitz, CZECH REPUBLIC: see JESENICE.

Jedburgh (JED-buhrg), town (2001 population 4,090), Scottish Borders, SE Scotland, on the JED WATER, 10 mi/16 km NE of HAWICK; 55°28′N 02°34′W. Light manufacturing; electrical engineering. Previously corn mills. The red sandstone ruins of an abbey founded 1118 are notable.

Jedda, SAUDI ARABIA: see JIDDA.

Jeddo, borough (2006 population 136), LUZERNE county, E central PENNSYLVANIA, 5 mi/8 km NE of HAZLETON; 40°59′N 75°54′W. Former anthracite coal-mining area.

Jeddore Harbour (jeh-DOR), inlet (7 mi/11 km long, 3 mi/5 km wide) of the ATLANTIC, S NOVA SCOTIA, CANADA, 30 mi/48 km ENE of HALIFAX; 44°45′N 63°01′W. At head of bay is HEAD OF JEDDORE.

Jedeideh, LEBANON: see JDEIDE.

Jedeira and Qalandia, Arab village, Jerusalem district, 5 mi/8 km N of JERUSALEM, in the Judaean Highlands, WEST BANK. Agiculture (cereals, vineyards). Believed to be on the site of the biblical city Gdor.

Jedita (zhe-DEE-tuh), town, central LEBANON, 18 mi/29 km E of BEIRUT, on the Beirut-DAMASCUS (SYRIA) railroad; 33°49′N 35°50′E. Grapes, cereals, fruit. Summer resort. Also spelled Jadita.

Jedrzejow (yen-DZHE-yoov), Polish *Jędrzejów*, town, KIELCE province, S central POLAND, 21 mi/34 km SSW of KIELCE. Railroad junction. Manufacturing (bricks, metalware); brewing, flour milling, sawmilling, tanning; gypsum mining. In Russian Poland, 1815–1919, called Andreyev. Before World War II, population 40% Jewish.

Jedwabne (yed-VAH-bne), town, ŁOMŻ a province, NE POLAND, 12 mi/19 km NE of ŁOMŻ A.

Jed Water (JED), river, 21 mi/34 km long, Scottish Borders, SE Scotland; rises in CHEVIOT HILLS, flows N, past Jedburgh, to TEVIOT RIVER.

Jefara, LIBYA: see JIFARAH.

Jeff, village, PERRY county, SE KENTUCKY, 4 mi/6.4 km SE of HAZARD, in CUMBERLAND foothills, on North Fork KENTUCKY RIVER opposite mouth of Carr Fork Kentucky River. Bituminous coal. Agriculture (tobacco; livestock). Manufacturing (coal processing). Daniel Boone National Forest to SW.

Jeff Davis, county (□ 331 sq mi/860.6 sq km; 2006 population 13,278), SE central GEORGIA; ⊙ HAZLEHURST; 31°48′N 82°38′W. Bounded NW by OCMULGEE River, N by ALTAMAHA RIVER; drained by LITTLE SATILLA RIVER. Coastal plain agriculture (cotton, soybeans, tobacco, corn, sugarcane, peanuts, pecans; cattle, hogs, poultry). Textile manufacturing at Hazlehurst. Formed 1905.

Jeff Davis, county (□ 2,264 sq mi/5,886.4 sq km; 2006 population 2,315), extreme W TEXAS; ⊙ FORT DAVIS; 30°43′N 104°07′W. Primarily a high plateau (c.4,500 ft/1,372 m–8,382 ft/2,555 m), diamond-shaped county extending W, touching the RIO GRANDE (Mexican border); rises to scenic DAVIS MOUNTAINS in center of county, including MOUNT LIVERMORE (8,382 ft/2,555 m; second-highest peak in state) and MOUNT LOCKE, with McDonald Observatory in center. Part of SIERRA VIEJA MOUNTAINS is in W. Cattle-ranching area; dairying; goats, hogs; sorghum, cotton, melons, corn, hay, wheat; wine grapes. Davis Mountains State Park and FORT DAVIS NATIONAL HISTORIC SITE both in SE center. Formed 1887.

Jeffers, village (2000 population 396), COTTONWOOD county, SW MINNESOTA, 14 mi/23 km NNW of WINDOM; 44°03′N 95°12′W. Agriculture (grain, soybeans; livestock). Manufacturing (hydraulic cylinders, carts).

Jefferson, county (□ 1,123 sq mi/2,919.8 sq km; 2006 population 656,700), N central ALABAMA; ⊙ Birmingham; 33°35′N 86°52′W. Industrial area crossed by Locust Fork; and drained by the Black Warrior River (W) and the Cahaba River (E). Coal and iron mining, limestone quarrying, natural-gas production. The co., along with Walker co., once accounted for 60% of Alabama's coal production, but now the industry is declining. Iron and steel products are made at Birmingham, Bessemer, Fairfield, Tarrant, and Leeds. Formed 1819. Named for Thomas Jefferson. The county seats have been Carrollsville, (1819–1821), Elyton, (1821–1873), and Birmingham.

Jefferson, county (□ 913 sq mi/2,373.8 sq km; 2006 population 80,655), central ARKANSAS; ⊙ PINE BLUFF; 34°14′N 91°54′W. Intersected by ARKANSAS RIVER NW to SE; drained by Wabbaseka River in NE, by Bayou BARTHOLOMEW in SW; Bayou Meto forms extreme E boundary. Agriculture (cotton, hay, wheat, rice, soybeans; hogs, turkeys); timber. Manufacturing at Pine Bluff. LOCK AND DAM NUMBER 5 in NW, Emmett Sanders Lock and Dam near center. Pine Bluff Arsenal in NW; part of Bayou Meto Wildlife Management Area in E. Formed 1829.

Jefferson, county (□ 778 sq mi/2,022.8 sq km; 2006 population 526,994), central COLORADO; ⊙ GOLDEN; 39°38′N 105°16′W. Coal-mining and irrigated agricultural region, bounded by SOUTH PLATTE RIVER (forms CHATFIELD and CHEESMAN reservoirs); drained by CLEAR CREEK. Remnant agriculture in NE part, most agriculture replaced by urban growth from DENVER in NE and growth of LAKEWOOD, Golden, ARVADA, and other JEFFERSON county communities (sugar beets; beans; livestock). Fur farms. All but NE part of county is in FRONT RANGE of ROCKY MOUNTAINS and its foothills. Includes parts of Pike National Forest in W and S; part of Chatfield State Park on SE boundary; part of Golden Gate State Park in NW. Land area was 785 sq mi/2,033 sq km in 1960, reduced by the encroachment of DENVER county in NE, which adjusts its boundaries every decennial year to absorb city of Denver's annexations into neighboring counties. Formed 1861.

Jefferson (JEF-uhr-suhn), county (□ 598 sq mi/1,554.8 sq km; 2006 population 14,677), NW FLORIDA; ⊙ MONTICELLO; 30°25′N 83°54′W. Bounded by GEORGIA line (N), GULF OF MEXICO (S), and AUCILLA RIVER (E). Lowland area, partly swampy, with rolling terrain and LAKE MICCOSUKEE in N. Agriculture and some forestry. Formed 1827.

Jefferson, county (□ 532 sq mi/1,383.2 sq km; 2006 population 16,768), E GEORGIA; ⊙ LOUISVILLE; 33°03′N 82°25′W. Coastal plain agriculture (cotton, peanuts) and sawmilling area drained by OGEECHEE RIVER. Manufacturing (trusses, textiles, apparel, consumer goods); printing and publishing. Formed 1796.

Jefferson, county (□ 1,105 sq mi/2,873 sq km; 2006 population 22,350), E IDAHO; ⊙ RIGBY; 43°49′N 112°19′W. Drained and irrigated by SNAKE RIVER in SE, forms part of NE boundary. Livestock-raising and irrigated agricultural area in SNAKE RIVER PLAIN. Clover, legumes, sugar beets, potatoes, alfalfa, orchards; wheat, barley, oats; sheep, cattle; poultry. Part of Idaho National Laboratory (U.S. Department of Energy) on W boundary; Jefferson Rays Lake and Mud Lake reservoirs at center of county; Camas National Wildlife Refuge in N center. Formed 1913.

Jefferson, county (□ 583 sq mi/1,515.8 sq km; 2006 population 40,523), S ILLINOIS; ⊙ MOUNT VERNON; 38°18′N 88°56′W. Major railroad and road junction at Mount Vernon. Agriculture (cattle; sorghum, wheat). Manufacturing (RR cars, rubber products, machinery, electronic equipment). Bituminous-coal mining, oil. Drained by BIG MUDDY RIVER. Formed 1819. REND LAKE on S boundary.

Jefferson, county (□ 362 sq mi/941.2 sq km; 2006 population 32,668), SE INDIANA; ⊙ MADISON; 38°47′N 85°26′W. Bounded partly S by OHIO River (here forming KENTUCKY line); drained by BIG CREEK, Clifty Creek, and Indian-Kentucky Creek. Corn; cattle, poultry, hogs; diversified manufacturing; timber. Contains Clifty Falls State Park on W edge of Madison; Hardy Lake State Recreation Area on W county line. Part of Jefferson Proving Ground in N part. Formed 1811.

Jefferson, county (□ 436 sq mi/1,133.6 sq km; 2006 population 15,945), SE IOWA; ⊙ FAIRFIELD; 41°02′N 91°57′W. Prairie agricultural area (hogs, cattle, poultry; corn, soybeans, hay) drained by SKUNK RIVER; coal mines, limestone quarries. Skunk River flooded in 1993. Formed 1839.

Jefferson, county (□ 556 sq mi/1,445.6 sq km; 2006 population 18,848), NE KANSAS; ⊙ OSKALOOSA; 39°13′N 95°24′W. Hilly area, crossed by DELAWARE RIVER; bounded S by KANSAS River. Corn, hogs, cattle, sorghum, hay, wheat; dairying. Limestone mining. Perry Lake Reservoir at center of county, Perry State Park at dam. Formed 1855.

Jefferson, county (□ 398 sq mi/1,034.8 sq km; 2006 population 701,500), N KENTUCKY; ⊙ LOUISVILLE, state's largest city and major transportation, commerce, and manufacturing center for the South and Midwest; 38°11′N 85°39′W. Bounded W and N by OHIO RIVER (Indiana state line), drained by FLOYDS FORK river. Majority of the city is urbanized, while S and E margins remain agricultural. Agriculture (vegetables, burley tobacco, hay, alfalfa, soybeans, wheat, corn; some livestock); quarrying (limestone, clay, sand, gravel); manufacturing remains centered in Louisville. F.P. "Tom" Sawyer State Park in NE; also numerous other city and county parks. Formed in 1780 from old Kentucky county, VIRGINIA, becoming one of three counties of Kentucky district, then part of Virginia.

Jefferson, county (□ 527 sq mi/1,370.2 sq km; 2006 population 9,194), SW MISSISSIPPI; ⊙ FAYETTE; 31°44′N 91°02′W. Bounded W by the MISSISSIPPI RIVER, including Rodney Lake in NW, Oxbow lake and former channel of Mississippi River forms part of LOUISIANA-Mississippi state line. Includes part of Homochitto National Forest in E and SE. Agriculture (cotton, corn, soybeans; cattle); timber. NATCHEZ TRACE PARKWAY passes N-S through county. Formed 1802.

Jefferson, county (□ 667 sq mi/1,734.2 sq km; 2006 population 216,469), E MISSOURI; ⊙ HILLSBORO; 38°16′N 90°35′W. On MISSISSIPPI RIVER (E) and MERAMEC RIVER (NE and NW); drained by BIG RIVER. Agriculture (corn, hay; livestock); silica (sand), barite mines. Very hilly area experiencing urban growth, especially in N and E. Part of ST. LOUIS metropolitan area; major towns with manufacturing include AR-NOLD, PEVELY, HERCULANEUM (lead smelter), FESTUS–CRYSTAL CITY, and DE SOTO. Major unincorporated residential areas include IMPERIAL, BARNHART, HIGH RIDGE, HOUSE SPRINGS, FENTON (city of Fenton is in Saint Louis county). Mastodon State Park archaeological site and museum at Imperial. Sandy Creek Covered Bridge State Historic Site. Formed 1818.

Jefferson, county (1,658 sq mi/4,294 sq km; 1990 population 7,939; 2000 population 10,049), SW central MONTANA; ⊙ BOULDER; 46°11′N 112°07′W. Agricultural and mining region drained by BOULDER RIVER; bounded S by JEFFERSON RIVER, W by CONTINENTAL DIVIDE. Cattle, sheep, hogs; gold, silver, lead, zinc; hay, wheat, oats. LEWIS AND CLARK CAVERNS STATE PARK in SE; part of Deerlodge National Forest in W and E; part of Helena National Forest in NE and NW. Formed 1865.

Jefferson, county (□ 575 sq mi/1,495 sq km; 2006 population 7,874), SE NEBRASKA; ⊙ FAIRBURY; 40°10′N 97°09′W. Agricultural region bounded S by KANSAS state line; drained by LITTLE BLUE RIVER. Cattle, hogs, corn, wheat, soybeans, sorghum, alfalfa; dairying. Clay quarry. Old OREGON TRAIL crosses county from SE to NW with Historic Rock Creek Station SE of Fairbury. Alexandria Lakes State Recreation Area in W. Formed 1871.

Jefferson, county (□ 1,293 sq mi/3,361.8 sq km; 2006 population 114,264), N NEW YORK; ⊙ WATERTOWN; 43°59′N 76°02′W. Bounded W by Lake ONTARIO and NW by SAINT LAWRENCE River; drained by BLACK and INDIAN rivers (water power). Part of Tug Hill Plateau in southern portion. Manufacturing, especially at Watertown and CARTHAGE; dairying region; minerals mining. Includes FORT DRUM army base and other military installations. Named for Thomas Jefferson, who was president at time of county's creation. New Englanders, attracted by water power at Watertown, led movement to establish county from large tract from Macomb Purchase that had been bought by French nobleman Le Ray de Chaumont and refugees from the French Revolution, who returned to France unable to accept the rigorous difficulties of pioneer life. Winter and summer recreational areas and state parks on Lake Ontario and in THOUSAND ISLANDS region of Saint Lawrence River. Formed 1805.

Jefferson (JEF-uhr-suhn), county (□ 411 sq mi/1,068.6 sq km; 2006 population 70,125), E OHIO; ⊙ STEUBEN-VILLE; 40°22′N 80°45′W. Bounded E by OHIO RIVER, here forming WEST VIRGINIA state line; drained by small Yellow and Cross creeks. In the Unglaciated Plain physiographic region. Manufacturing (lumber, wood and metal products); agriculture (cattle, corn). Ceramics plants. Some coal mining. Formed 1797.

Jefferson, county (□ 773 sq mi/2,009.8 sq km; 2006 population 6,385), S OKLAHOMA; ⊙ WAURIKA; 34°06′N 97°50′W. Bounded S by RED RIVER, here forming TEXAS line; and drained by Beaver and Mud creeks. Agriculture (grain, corn); sheep, cattle. Manufacturing at Waurika. Oil. Part of WAURIKA LAKE in NW. Formed 1907.

Jefferson, county (□ 1,791 sq mi/4,656.6 sq km; 2006 population 20,352), central OREGON; ⊙ MADRAS; 44°37′N 121°10′W. MOUNT JEFFERSON, in CASCADE RANGE, on W boundary. Drained by DESCHUTES RIVER. Part of Warm Springs Indian Reservation in NW. Agriculture (wheat, barley, oats, potatoes; poultry, sheep, cattle); lumber milling; mercury mining. CROOKED RIVER National Grassland in S; part of Deschutes National Forest in SW; Corbett State Park in SW corner; Cove Palisades State Park on LAKE CHINOOK Reservoir in W center. Formed 1914.

Jefferson, county (□ 656 sq mi/1,705.6 sq km; 2006 population 45,725), W central PENNSYLVANIA; ⊙ BROOKVILLE; 41°07′N 79°00′W. Railroad junction. Bounded N by CLARION RIVER; drained by MAHONING

and RED BANK creeks. Agricultural and manufacturing area. Bituminous coal; manufacturing at PUNX-SUTAWNEY and BROOKVILLE; clay, sand, and gravel. Agriculture (corn, wheat, oats, barley, hay, alfalfa, soybeans; sheep, hogs, cattle, poultry; dairying). Clear Creek State Park and State Forest in N. Formed 1804.

Jefferson, county (□ 318 sq mi/826.8 sq km; 2006 population 49,372), E TENNESSEE; ⊙ DANDRIDGE; 36°03′N 83°27′W. In GREAT APPALACHIAN VALLEY; traversed by BAYS MOUNTAIN; bounded NW by HOLSTON River; drained by FRENCH BROAD RIVER. Includes parts of CHEROKEE and DOUGLAS reservoirs. Agriculture; some zinc mining. Formed 1792.

Jefferson, county (□ 1,111 sq mi/2,888.6 sq km; 2006 population 243,914), SE TEXAS; ⊙ BEAUMONT; 29°53′N 94°09′W. On Gulf coastal plain; bounded E by SABINE LAKE (here forming LOUISIANA state line); NE by NECHES RIVER, N by Pine Island Bayou, S by GULF OF MEXICO; crossed by GULF INTRACOASTAL WATERWAY (to parallel Gulf Coast, also NE along Texas shore of Sabine Lake). Sabine-Neches Waterway gives access from Gulf to deep-water ports of Beaumont and PORT ARTHUR, important oil-shipping, oil-refining, and industrial centers. Oil, natural-gas fields. Cattle raising, agriculture (rice, soybeans). Fishing, duck hunting. Small part of BIG THICKET NATIONAL PRESERVE in N; Texas Point National Wildlife Refuge, Sea Rim State Park, Sabine Pass Battleground State Historical Park in SE; McFadden National Wildlife Refuge in S, on coast. Formed 1836.

Jefferson, county (□ 2,177 sq mi/5,660.2 sq km; 2006 population 29,279), W WASHINGTON; ⊙ PORT TOWNSEND; 47°51′N 123°35′W. Bounded W by PACIFIC OCEAN, E by HOOD CANAL and PUGET SOUND, far NE by STRAIT OF JUAN DE FUCA; peaks of OLYMPIC MOUNTAINS in interior. Timber, wood pulp; fish; cattle; dairying. Includes Hoh Indian Reservation in W and part of Quinault Indian Reservation in SW. Central part of OLYMPIC NATIONAL PARK and Olympic Mountains, including Mount Olympus (7,965 ft/ 2,428 m), crosses county N-S in center, isolating W part of county, along Pacific Coast, from E part, on Puget Sound. Protection Island National Wildlife Refuge in NE; Quillayute National Wildlife Refuge off W coast; S part of coastal section of Olympic National Park in W; Dosewallips and Pleasant Harbor state parks in SE; ANDERSON LAKE, Fort Worden, Fort Flagler, Old Fort Townsend, and Mystery Bay state parks in NE; parts of Olympic National Forest in SW and E, including the Brothers Wilderness Area and part of Buckhorn Wilderness Area, both in E. Includes Marrowstone and INDIAN ISLANDS in NE (Puget Sound), latter is U.S. Naval Reservation. Formed 1852.

Jefferson, county (□ 211 sq mi/548.6 sq km; 2006 population 557), NE WEST VIRGINIA, at end of E PANHANDLE; ⊙ CHARLES TOWN, 39°18′N 77°51′W. In S part of GREAT APPALACHIAN VALLEY; BLUE RIDGE is along SE border. Bounded NE by POTOMAC RIVER (MARYLAND state line), SE and SW by VIRGINIA; drained by SHENANDOAH RIVER, which joins the Potomac River at HARPERS FERRY, and by short Opequon Creek. Scenic resort region. Agriculture (corn, wheat, oats, barley, soybeans, potatoes, alfalfa, hay, sorghum, apples; livestock; poultry; dairying); limestone and dolomite quarrying. Industry at RANSON, Charles Town. Part of Harpers Ferry National Historic Park in NE (1,102 acres/446 ha); James Rumsey State Historic Monument in N; Shannondale Springs Wildlife Management Area in SE. Formed 1801.

Jefferson, county (□ 582 sq mi/1,513.2 sq km; 2006 population 80,025), S WISCONSIN; ⊙ JEFFERSON; 43°02′N 88°46′W. Dairying is chief industry; agriculture (wheat, corn, soybeans, potatoes; cattle, hogs, sheep, poultry); some manufacturing at LAKE MILLS, WATERTOWN, Jefferson, and WHITEWATER. AZTALAN STATE PARK in W; larger LAKE KOSHKONONG in SW; Glacial Drumlin State Trail passes E-W through

center of county; small part of Kettle Moraine State Forest in SE corner. Drained by ROCK, BARK, and CRAWFISH rivers. Has lake resorts. Formed 1836.

Jefferson, city (2000 population 4,626), ⊙ GREENE county, central IOWA, on RACCOON RIVER, and 43 mi/ 69 km W of AMES; 42°01′N 94°22′W. Railroad junction. Agricultural trade center with dairy products, feed, tankage; manufacturing (consumer goods, fabricated metal products, paper products). Sand and gravel pits nearby. Mahaney Memorial Carillon Tower here; Spring Lake State Park to NE. Settled c.1854, incorporated 1871.

Jefferson, unincorporated city (2000 population 11,843), JEFFERSON parish, SE LOUISIANA, suburb 4 mi/ 6 km E of downtown NEW ORLEANS, on MISSISSIPPI RIVER; 29°58′N 90°10′W. Manufacturing (paper products, building materials, food processing, fabricated steel, spirits and cordials, lubricating oils, pulp-mill equipment); printing and publishing.

Jefferson, city (2006 population 592), UNION county, SE SOUTH DAKOTA, 10 mi/16 km SE of ELK POINT, 13 mi/21 km NW of SIOUX CITY, Iowa, and near BIG SIOUX RIVER; 42°36′N 96°33′W.

Jefferson (JE-fuhr-suhn), unincorporated city (2000 population 27,422), FAIRFAX county, N VIRGINIA, residential suburb 8 mi/13 km W of WASHINGTON, D.C., includes communities of JEFFERSON VILLAGE and HILLWOOD.

Jefferson, town (□ 15 sq mi/39 sq km; 2000 population 3,825), ⊙ JACKSON county, NE central GEORGIA, 15 mi/ 24 km NW of ATHENS; 34°08′N 83°36′W. Manufacturing (textiles, plastics, apparel; printing and publishing). Dr. Crawford W. Long performed an operation here in 1842 using ether as an anesthetic; commemorated in a museum. Many nineteenth-century homes. Incorporated 1806.

Jefferson, town, LINCOLN county, S MAINE, on DAMARISCOTTA LAKE, and 18 mi/29 km NE of WISCASSET; 44°11′N 69°30′W.

Jefferson, town, COOS county, N central NEW HAMPSHIRE, 14 mi/23 km WSW of BERLIN; 44°23′N 71°28′W. Drained by ISRAEL RIVER. Agriculture (cattle, poultry; dairying; timber); manufacturing (lumber, furniture, computer software). Santa's Village and Six Gun City theme parks are here. Parts of White Mountain National Forest in S and NE; Agnew State Forest in S.

Jefferson, township, MORRIS county, N NEW JERSEY, 10 mi/16 km W of PATERSON; 41°00′N 74°32′W. Incorporated 1809.

Jefferson (JEF-uhr-suhn), town (□ 2 sq mi/5.2 sq km; 2006 population 1,370), ⊙ ASHE county, NW NORTH CAROLINA, 20 mi/32 km NE of BOONE, in the BLUE RIDGE MOUNTAINS; 36°25′N 81°28′W. Service industries; manufacturing (lumber; printing and publishing); agriculture (tobacco; cattle). Mount Jefferson State Park to SE, New River State Park to E. BLUE RIDGE PARKWAY passes to SE.

Jefferson, town, MARION county, W OREGON, 8 mi/12.9 km NNE of ALBANY, on SANTIAM RIVER; 44°43′N 123°00′W. Agriculture (fruit, nuts, berries, hops; poultry; dairy products); fabrics. Ankeny National Wildlife Refuge to NW.

Jefferson, town (2006 population 698), CHESTERFIELD county, N SOUTH CAROLINA, 21 mi/34 km ESE of LANCASTER; 34°38′N 80°23′W. Manufacturing (construction materials, textiles, gold bullion); agriculture (livestock, poultry; grain, watermelons, tobacco, peaches).

Jefferson, town (2000 population 2,024), ⊙ MARION county, NE TEXAS, on BIG CYPRESS CREEK, and 14 mi/ 23 km N of MARSHALL; 32°45′N 94°20′W. Elevation 200 ft/61 m. Railroad junction. In oil, vegetables, timber (pine, cypress) area. Lumber milling; manufacturing (fabricated metal products, food processing, wood and flat-glass products). Tourism. CADDO LAKE (hunting, fishing) is to E, Caddo Lake State Park to

SE. Grew as a river port and lumbering center in area settled in 1830s; reached population of c.30,000 in 1875, later declined. LAKE O' THE PINES to W.

Jefferson, town (2006 population 7,721), ⊙ JEFFERSON county, S WISCONSIN on ROCK RIVER, at confluence of CRAWFISH RIVER, 30 mi/48 km ESE of MADISON; 43°00′N 88°48′W. In dairying and farming region. Manufacturing (furniture, shoes, textiles, wood products, food and meat products, vegetable processing; printing). Glacial Drumlin State Trail passes to N. Settled c.1836, incorporated 1878.

Jefferson, village, Park county, central COLORADO, on TARRYALL CREEK, in ROCKY MOUNTAINS, and 50 mi/80 km SW of DENVER; elevation c.9,500 ft/2,896 m. Shipping point for livestock and timber below KENOSHA PASS (to NE, 10,001 ft/3,048 m). Pike National Forest to W, N, and E. Jefferson Lake to NW. Tarryall State Wildlife Area to SE.

Jefferson (JEF-uhr-suhn), village (□ 2 sq mi/5.2 sq km; 2000 population 3,572), ⊙ ASHTABULA county, extreme NE OHIO, 9 mi/14 km S of ASHTABULA; 39°39′N 83°33′W. Livestock and dairying area. Founded c.1804.

Jefferson, village (2006 population 34), GRANT county, N OKLAHOMA, 8 mi/12.9 km SSW of MEDFORD; 36°43′N 97°47′W. In agricultural area.

Jefferson, borough, ALLEGHENY county, W PENNSYLVANIA, residential suburb 11 mi/18 km SSE of downtown PITTSBURGH, and 2 mi/3.2 km W of CLAIRTON on Peters Creek; 40°17′N 79°55′W.

Jefferson, borough (2000 population 337), GREENE county, SW PENNSYLVANIA, 7 mi/11.3 km ENE of WAYNESBURG, near South Fork of Tenmile Creek. Manufacturing (apparel, lumber, wood products); agriculture (sheep; dairying).

Jefferson, borough (2000 population 631), YORK county, S PENNSYLVANIA, 12 mi/19 km SSW of YORK. Agriculture (apples, soybeans, grain; livestock, poultry; dairying). Codorus State Park and Lake Marburg reservoir to SW.

Jefferson, parish (□ 409 sq mi/1,063.4 sq km; 2006 population 431,361), extreme SE LOUISIANA; ⊙ GRETNA; 29°55′N 90°03′W. Situated in the delta of the MISSISSIPPI RIVER, which intersects parish in N; bounded S by Gulf of MEXICO (BARATARIA BAY), N by Lake PONTCHARTRAIN, W by CATAOUATCHE, SALVADOR, and LITTLE lakes. Important industrial parish, adjoining (in N) NEW ORLEANS. Widely varied manufacturing in N. Home gardens, nursery crops, horses; oysters, shrimp, crabs, finfish, alligators, exotic fowl; oil and natural-gas wells. New Orleans International Airport in NW at KENNER. Traversed by INTRACOASTAL WATERWAY. GRAND ISLE, 15 mi/24 km S of mainland at entrance to BARATARIA BAY, is part of parish, connected by bridge to LAFOURCHE parish (W); includes Grand Isle State Park. City of NEW ORLEANS, to NE of parish, urbanized in N. Bayou Segnette State Park in N. Named after Thomas Jefferson. Formed 1825.

Jefferson, Massachusetts: see HOLDEN.

Jefferson Barracks, former military base, MISSOURI, 10 mi/16 km S of downtown ST. LOUIS, on MISSISSIPPI RIVER. Supply depot and training center for deployment of troops in the American West (established 1826), prominent in American Civil War; now a county park. Has national cemetery, veterans' hospital.

Jefferson City, city (2000 population 39,636), COLE and CALLAWAY counties, central MISSOURI, on the S bank of the MISSOURI RIVER, W of mouth of the OSAGE; ⊙ Missouri and ⊙ Cole co.; 38°34′N 92°11′W. The state government is the major employer, but the city, with railroad and river facilities, is also the commercial and processing center of an agricultural area. It was chosen (1821) for the state capital, the legislature moved from SAINT CHARLES in 1826. Because of divided loyalties and the difficulties of holding the state in the Union, Jefferson City was occupied by Federal troops during the Civil War. The Italian-

Renaissance capitol of CARTHAGE marble (completed 1917) contains murals by Thomas Hart Benton and N. C. Wyeth, and is the site of the Missouri State Museum. Manufacturing (machinery, construction materials, dairy products, consumer goods, building materials, furniture, transportation equipment, printing). In or near the city are Lincoln University (a historic black state university), the state penitentiary and three other facilities, and a national cemetery. Regional shopping mall on W side. Major highway and property damage in 1993 floods, especially in areas to N. Cedar City on N annexed in 1989. Incorporated 1825. Commonly referred to as Jeff City.

Jefferson City, city (2006 population 8,028), JEFFERSON county, E TENNESSEE, near Cherokee Dam (HOLSTON River), 26 mi/42 km ENE of KNOXVILLE; 36°07′N 83°30′W. Manufacturing. Seat of Carson-Newman College. Zinc mines nearby. Settled c.1810; incorporated 1900.

Jefferson City, village, JEFFERSON county, SW central MONTANA, on PRICKLY PEAR CREEK at mouth of Spring Creek, and 15 mi/24 km S of HELENA. Gold and silver mines nearby. Helena National Forest to E and W.

Jefferson Davis, county (□ 409 sq mi/1,063.4 sq km; 2006 population 13,184), S central MISSISSIPPI; ⊙ PRENTISS; 31°33′N 89°49′W. Drained by Bowie Creek (forms part of E boundary), other creeks. Agriculture (cotton, corn; poultry, cattle, hogs); timber. Formed 1906.

Jefferson Davis, parish (□ 658 sq mi/1,710.8 sq km; 2006 population 31,418), SW LOUISIANA; ⊙ JENNINGS; 30°16′N 92°49′W. Bounded E by Bayou NEZPIQUE, SE by MERMENTAU RIVER and Lake ARTHUR, and Bayou Lacassine forms part of W boundary. Drained in far NW by CALCASIEU RIVER. Oil, natural gas; agriculture (sorghum, cotton, rice, soybeans, sweet potatoes, sod production; cattle, exotic fowl); fishing (crawfish, catfish); manufacturing (apparel; shipbuilding; logging). First oil in LOUISIANA discovered here (1901). Includes Lake ARTHUR (recreation). Named after the President of the Confederacy. Formed 1910.

Jefferson Island, unincorporated village, on one of the FIVE ISLANDS, in IBERIA parish, S LOUISIANA, a salt dome rising from prairies just E of Lake PEIGNEUR, 9 mi/14 km W of NEW IBERIA; 29°58′N 91°58′W. Low-draft port; fishing (shrimp, crawfish, crabs, fish, catfish); oil and natural-gas deposits. Large rock-salt mine to SE. Scenic gardens to N.

Jefferson Memorial, DISTRICT OF COLUMBIA: see THOMAS JEFFERSON MEMORIAL.

Jefferson, Mount, peak (10,495 ft/3,199 m), at joining of JEFFERSON (E), MARION (NW), and LINN (SW) county boundaries, NW central OREGON, in CASCADE RANGE, 65 mi/105 km ESE of SALEM at center of MOUNT JEFFERSON Wilderness Area.

Jefferson, Mount, NEVADA: see TOQUIMA RANGE.

Jefferson, Mount, NEW HAMPSHIRE: see PRESIDENTIAL RANGE.

Jefferson National Expansion Memorial Park, on the riverfront in downtown ST. LOUIS, E MISSOURI. A national park commemorating westward exploration and settlement; includes Gateway Arch and Museum of Westward Expansion. Authorized 1935. See also St. Louis, Missouri.

Jefferson River, 207 mi/333 km long, SW MONTANA; rises in Centennial Mountains as Red Rock River near CONTINENTAL DIVIDE (Montana-IDAHO state line); flows W through Upper and Lower RED ROCK lakes, then NNW past LIMA, to CLARK CANYON RESERVOIR where it becomes Beaverhead River; then flows NNE past DILLON joined by BIG HOLE and RUBY rivers near TWIN BRIDGES; continues as Jefferson River N and E to point just NE of THREE FORKS, where it joins MADISON and GALLATIN rivers to form the MISSOURI River (28 mi/45 km WNW of Bozeman).

Jefferson Springs, resort village, RUTHERFORD county, central TENNESSEE, on STONES RIVER, and 24 mi/39 km SE of NASHVILLE; 36°00′N 86°27′W.

Jefferson, Territory of, in U.S. history, region that roughly encompassed the present-day state of COLORADO, although extending 2° farther S and 1° farther N, organized by its inhabitants (1859–1861), but never given congressional sanction. After a great increase in emigration in the 1850s, settlers in ARAPAHOE county, Kansas Territory, felt the need to be closer to the seat of government. They met in convention in DENVER on August 1, 1859, to discuss alternatives to the region's status. The 166 delegates present debated the benefits of reorganization as a state or as a territory and submitted the question on September 5 to the public, which voted overwhelmingly for territorial status. Subsequently, Beverly D. Williams was sent as a representative to Congress, which, however, refused his petition. Nevertheless, the constitution of the Territory of Jefferson was adopted on October 24, and the first session of its legislature met on November 7. Robert W. Steele was elected provisional governor. Although illegal, the new government coexisted peacefully with the official county institutions. Laws were passed regarding taxation, and the franchise was denied Native and African-Americans. On February 28, 1861, Congress passed the Organic Act, which created the Territory of Colorado. The provisional government quickly dismantled, and William Gilpin replaced Steele as governor.

Jeffersontown, city (2000 population 26,633), JEFFERSON county, N KENTUCKY, suburb 12 mi/19 km ESE of downtown LOUISVILLE; 38°12′N 85°34′W. Some agriculture (potatoes, corn, apples, peaches); nursery products; manufacturing (plastic products, consumer goods, food and pharmaceutical processing equipment, machinery, restaurant equipment).

Jefferson Village (JE-fuhr-suhn), unincorporated town, FAIRFAX county, NE VIRGINIA, residential suburb 8 mi/13 km W of WASHINGTON, D.C.; 38°52′N 77°10′W.

Jeffersonville, city (2000 population 27,362), ⊙ CLARK county, S INDIANA, at the falls of the OHIO River opposite LOUISVILLE, KENTUCKY; 38°18′N 85°44′W. Together with CLARKSVILLE and NEW ALBANY (Indiana) and Louisville (Kentucky), referred to as Falls City Area. Located in a rich agricultural area, the city is a shipping point for farm products. Manufacturing (chemicals, steel and wood products, oil lubricants, electronic and transportation equipment, textiles, construction materials, furniture, consumer goods, agricultural machinery); food processing. The city was founded (1802) on the site of Fort Steuben (formerly Fort Finney) by veterans of George Rogers Clark's NW expedition, who were given the land in gratitude for their services. The original town was built according to plans suggested by Thomas Jefferson, after whom it is named. A branch of Indiana Vocational Technical College (Ivy Tech) and a U.S. Census Bureau Mapping Center are located here. Incorporated 1817.

Jeffersonville, town (2000 population 1,209), ⊙ TWIGGS county, central GEORGIA, 19 mi/31 km SE of MACON; 32°41′N 83°20′W. Manufacturing includes lumber, kaolin-clay processing.

Jeffersonville, town (2000 population 1,804), MONTGOMERY county, NE central KENTUCKY, 7 mi/11.3 km SE of MOUNT STERLING, near Slate Creek; 37°58′N 83°49′W. Agricultural area (burley tobacco, corn; cattle, poultry; dairying). Manufacturing (lumber).

Jeffersonville, village (2000 population 366), WAYNE county, SE ILLINOIS, 5 mi/8 km NNW of FAIRFIELD; 38°26′N 88°24′W. In agricultural area. Also known as Geff.

Jeffersonville, hamlet (2006 population 404), SULLIVAN county, SE NEW YORK, 10 mi/16 km W of LIBERTY; 41°46′N 74°55′W. Some manufacturing. In resort area.

Jeffersonville (JEF-uhr-suhn-vil), village (□ 2 sq mi/ 5.2 sq km; 2006 population 1,239), FAYETTE county, S central OHIO, 10 mi/16 km NW of WASHINGTON COURT HOUSE; 39°39′N 83°33′W. Livestock raising and farming.

Jeffersonville, unincorporated village, WEST NORRITON township, MONTGOMERY county, SE PENNSYLVANIA, 16 mi/26 km NW of downtown PHILADELPHIA, and 2 mi/3.2 km NW of Norristown, near SCHUYLKILL RIVER; 40°07′N 75°22′W. Manufacturing (transportation equipment).

Jeffersonville, village (2006 population 565), LAMOILLE CO., VERMONT: 44°38′N 72°49′W. See CAMBRIDGE. Named for Thomas Jefferson (1743–1826), third president of the United States.

Jeffrey's Bay, town, EASTERN CAPE province, SOUTH AFRICA, ON ST. FRANCIS BAY on INDIAN OCEAN coast, 40 mi/64 km W of PORT ELIZABETH (NELSON MANDELA METROPOLE), and 15 mi/24 km E of HUMANSDORP; 34°03′S 24°55′E. Elevation 3 ft/1 m. Town grew out of a holiday resort, still famous for its amazing variety of seashells; known internationally for its surfing area. Geared to tourist trade.

Jefren, LIBYA: see YAFRAN.

Jega (JAI-gah), town, KEBBI state, NW NIGERIA, on ZAMFARA RIVER, 70 mi/113 km SW of SOKOTO; 12°13′N 04°23′E. Agricultural trade center (cotton, millet, rice; cattle; skins).

Jegenstorf (YEH-guhns-dohrf), commune, BERN canton, central SWITZERLAND, 8 mi/12.8 km NNE of BERN.

Jegindø (YAI-yin-DU), island (□ 3 sq mi/7.8 sq km) in W Lim Fjord, NW JUTLAND, DENMARK, 1 mi/1.6 km S of MORS island; 56°39′N 08°38′E. Highest point, 43 ft/13 m; flat and fertile.

Jehanabad, INDIA: see JAHANABAD.

Jehlam, INDIA and PAKISTAN: see JHELUM RIVER.

Jehlam River, INDIA and PAKISTAN: see JHELUM RIVER.

Jehol, CHINA: see REHE.

Jeida, ISRAEL: see RAMAT YISHAI.

Jeinemeni, Cerro (jai-ne-MAI-nee, SER-ro), Andean peak (8,530 ft/2,600 m) in AISÉN province, S CHILE, between LAKE BUENOS AIRES and Lake Cochrane (PUEYRREDÓN), 110 mi/177 km SSE of PUERTO AISÉN; 46°52′S 72°14′W.

Jejuí Guazú (he-HWEE gwah-SOO), c.150 mi/241 km long, river in central PARAGUAY; rises on W slopes of Cordillera de Mbaracayú near boundary with BRAZIL; flows W to Paraguay River 13 mi/21 km SW of San Pedro; 24°13′S 55°42′W. Village of same name at 24°13′S 55°36′W.

Jejuri (JAI-juh-ree), town, PUNE district, central MAHARASHTRA state, W INDIA, 25 mi/40 km SE of PUNE; 18°17′N 74°10′E. Hindu pilgrimage center.

Jekabmiests, LATVIA: see JĒKABPILS.

Jēkabpils (YAI-kuhb-peels) or **yekabpils**, German *Jakobstadt*, city (2000 population 27,871), S central LATVIA, in ZEMGALE, on left bank of the DVINA (DAUGAVA) River, and 75 mi/121 km ESE of RĪGA; 56°29′N 25°51′E. Manufacturing (leather, apparel, beer, liquor); dairying. Important trading center in 15th century Formerly also called JEKABMIESTS.

Jekyll Island, one of the SEA ISLANDS, in GLYNN county, SE GEORGIA, just off the S coast of BRUNSWICK, 2 mi/3.2 km to the SE (reached by crossing a causeway); c.7 mi/11.3 km long, 1 mi/1.6 km–2 mi/3.2 km wide; 31°04′N 81°24′W. Georgia's first brewery was once located here; hops, barley, and cotton once grown. The island residents own their homes but lease the land from the state. Made a state park in 1947, it was formerly a winter-resort colony of large estates, known as Millionaire's Village. The Jekyll Island club was once among the most exclusive clubs in the U.S. Famous luminaries living here included the Rockefellers, Pulitzers, Astors, Vanderbilts, Goulds, Morgans, and Jennings. First transcontinental telephone call made here in 1915. The Jekyll Island Authority now operates the 240-acre/97-ha historic district.

Jelalabad, AFGHANISTAN: see JALALABAD.

Jelbart Ice Shelf, an ice shelf on the PRINCESS MARTHA COAST of QUEEN MAUD LAND, EAST ANTARCTICA, lies between the EKSTRÖM and FIMBUL ice shelves, with which it is continuous; 70°30′S 04°30′W.

Jeldēsa (jel-DAI-sah), village, SOMALI state, E central ETHIOPIA, 19 mi/31 km ENE of DIRE DAWA; 09°42′N 42°06′E. Also spelled Jaldessa.

Jelebu (je-LE-boo), district (□ 528 sq mi/1,372.8 sq km) in N NEGERI Sembilansia, MALAYSIA; 02°55′N 102°05′E. Main town, Kuala Kelawang. One of the original NEGERI SEMBILAN states. Within the original tin-mining region that brought the original Malayan Federation into existence; declining in importance.

Jelec, CZECH REPUBLIC: see USTEK.

Jelenia Góra (ye-LE-nyah GO-rah), German *Hirschberg*, city (2002 population 89,339), SW POLAND, at N foot of the RIESENGEBIRGE, on BOBRAWA RIVER, and 60 mi/97 km WSW of WROCŁAW (Breslau); 50°54′N 15°44′E. Railroad junction. It is an industrial and commercial center known for its woolen textiles; paper milling; manufacturing of machinery, optical lenses, glass, castings chemicals, pharmaceuticals, building materials; wood-pulp processing; copper deposits nearby. Tourist center. First mentioned c.1280; chartered in 1312. Passed to BOHEMIA after 1368. A prosperous weaving center in the 15th and 16th centuries, the city was destroyed by the Thirty Years War (1618–1648) and, in 1640, by the plague. It was rebuilt and suffered again under Prussian rule in the 18th century. Became a linen-milling center in the 18th century and has one of six Churches of Grace allowed to Silesian Protestants (1707) under Treaty of Altranstädt.

Jelep La (JOU-LUHP-LAH), pass (c.14,000 ft/4,267 m) in SW ASSAM HIMALAYAS, on SIKKIM-TIBET border, 15 mi/24 km ENE of GANGTOK, INDIA. Heavily traveled road leads from Sikkim into CHUMBI VALLEY (Tibet) and connects with main India-Tibet trade route.

Jelgava (YEL-gah-vuh) or **Yelgava**, German *mitau*, city (2000 population 63,652), in LATVIA, 26 mi/42 km SE of RIGA, on the LIELUPE River; 56°39′N 23°42′E. It is a major railroad hub. Center for textiles, grain, timber, and minibus manufacturing The city grew around a fortress established by the Livonian Knights in the 13th century, but was destroyed by the Lithuanians in 1345. In 1561, Jelgava became the residence of the dukes of COURLAND; it passed to RUSSIA with the duchy in 1795. German troops held JELGAVA during World War I. In 1919, during the struggle for Latvian independence, the city was occupied in turn by Soviet forces, by German free corps, and by the Latvians. Part of independent Latvia from 1920 to 1940, Jelgava was then seized by the USSR, held by the Germans from 1941 to 1944, and taken by Soviet troops. City landmarks include the 16th-century Trinity Church and the 18th-century ducal palace.

Jelka (yel-KAH), HUNGARIAN *Jóka*, village, ZAPADOSLOVENSKY province, SW SLOVAKIA, on LITTLE DANUBE RIVER, and 17 mi/27 km E of BRATISLAVA; 48°09′N 17°31′E. Wheat, corn, and sugar beets; food processing. Strong Hungarian minority. Was under Hungary between 1938–1945.

Jellico (JEH-li-koh), city (2006 population 2,534), CAMPBELL county, NE TENNESSEE, at KENTUCKY line, 45 mi/72 km NNW of KNOXVILLE, in foothills of the CUMBERLAND MOUNTAINS; 36°35′N 84°08′W. Manufacturing; lumbering. U.S. mine-rescue station here. Settled 1795. Indian Mountain State Park just W.

Jellico Creek (JEL-li-ko), unincorporated village, WHITLEY county, SE KENTUCKY, in the CUMBERLAND MOUNTAINS, 6 mi/9.7 km SW of WILLIAMSBURG, on TENNESSEE state line, opposite JELLICO, Tennessee. In coal mining area.

Jellicoe (JE-li-ko), unincorporated village, N central ONTARIO, E central CANADA, near Lake NIPIGON, 110 mi/177 km NE of THUNDER BAY, and included in town of GREENSTONE; 49°41′N 87°31′W. Elevation 1,087 ft/331 m. Gold mining.

Jelling (YE-ling), town, VEJLE county, E JUTLAND, DENMARK, 5 mi/8 km NW of VEJLE; 55°45′N 09°25′E. Cement; fruit; dairying. Burial mounds of King Gorm the Old (died c.935) and wife (Thyra Danebod) here.

Jellison, Cape (JEL-uh-suhn), HANCOCK county, S MAINE, peninsula on W shore of PENOBSCOT BAY, and 8 mi/12.9 km NE of BELFAST. Fort Point has lighthouse.

Jelsa, Italian *Gelsa*, village, S CROATIA, port on N coast of HVAR Island, 13 mi/21 km E of HVAR, in DALMATIA. Seaside resort; center of wine-growing region.

Jelsava (yel-SHAH-vah), Slovak *Jelšava*, Hungarian *Jólsva*, town, VYCHODOSLOVENSKY province, S central SLOVAKIA, on railroad, and 47 mi/76 km WSW of KOŠICE; 48°38′N 20°14′E. Manufacturing (ceramics); magnesite mining in vicinity. Has noted 19th-century Empire-style castle. Military base. Under Hungarian rule between 1938–1945.

Jemaa (je-MAH), town, PLATEAU state, central NIGERIA, 10 mi/16 km SSE of KAFANCHAN. Tin-mining center.

Jemaja, INDONESIA: see ANAMBAS ISLANDS.

Jemaluang (je-MAH LOO-ahng), town, NE JOHOR, MALAYSIA, 10 mi/16 km S of MERSING, 6 mi/9.7 km from SOUTH CHINA SEA; 02°16′N 103°51′E. Road junction; coconuts, oil-palm; tin mining. Sometimes called Bandar Jemaluang.

Jemappes (zhuh-MAHP), town in commune of QUAREGNON, Mons district, HAINAUT province, S BELGIUM; 50°27′N 03°53′E. It is a center of the Borinage region, once concerned primarily with coal mining, but now undergoing economic reconversion; iron and steel manufacturing. At Jemappes in 1792 the French defeated the Austrians in one of the first important battles of the French Revolutionary Wars.

Jember (JUHM-buhr), city and ⊙ Jember district, Java Timur province, INDONESIA, 95 mi/153 km SE of SURABAYA, at foot of Mount ARGAPURA; 08°10′S 113°42′E. Trade center for agricultural area (sugar, tobacco, corn, peanuts); lumber mills, machine shops. Also spelled Djember.

Jemeppe (zhuh-MEP) or **Jemeppe-sur-Euse**, town in commune of FLÉMALLE, Liège district, LIÈGE province, E BELGIUM, on MEUSE RIVER, and 4 mi/6.4 km WSW of LIÈGE. Manufacturing (soda, machinery).

Jemeppe-sur-Sambre (zhuh-MEP–suhr–SAHM-BRUH), commune (2006 population 18,022), Namur district, NAMUR province, S central BELGIUM, on SAMBRE RIVER, and 10 mi/16 km W of NAMUR. Glass manufacturing. In nearby grotto are prehistoric relics.

Jemez (HAI-mez), pueblo, SANDOVAL county, central NEW MEXICO, in Jemez Indian Reservation (1990 population 1,750), 31 mi/50 km NW of BERNALILLO on the East Fork of the JEMEZ RIVER; 35°38′N 106°47′W. In the 16th century there were seven Jemez pueblos; by 1622 there were only two. One of the remaining pueblos was abandoned prior to the Pueblo revolt of 1680. The other took a prominent part in the revolt; the Jemez Native Americans attacked the Spanish repeatedly. In 1694 the pueblo was stormed and captured by the Spanish. Although the Jemez promised to remain at peace, they revolted in 1696, killed the missionaries there, and then fled into Navajo country, where they remained for several years. Some later returned to build (c.1700) the present village. The inhabitants are Pueblos of the Tanoan linguistic stock, speaking Towa.

Jemez Canyon Reservoir (HAI-mez), BERNALILLO county, N central NEW MEXICO, on JEMEZ RIVER, near its mouth on the RIO GRANDE, 20 mi/32 km N of ALBUQUERQUE; c.23 mi/37 km long; 35°22′N 106°29′W. Maximum capacity 118,818 acre-ft. Intermittent. Formed by Jemez Canyon Dam (131 ft/40 m), built (1953) by the federal government for flood and debris control. Extends into Santa Ana and Zia Indian reservations.

Jemez Mountain, NEW MEXICO: see VALLE GRANDE MOUNTAINS.

Jemez River (HAI-mez), SANDOVAL county, N central NEW MEXICO; rises in several branches near REDONDO PEAK; flows S, past JEMEZ SPRINGS village, through JEMEZ and ZIA Indian reservations, and SE to RIO GRANDE 5 mi/8 km N of BERNALILLO; c.60 mi/97 km long.

Jemez Springs (HAI-mez), village (2006 population 398), SANDOVAL county, N central NEW MEXICO, on JEMEZ RIVER, in VALLE GRANDE MOUNTAINS, and 45 mi/72 km N of ALBUQUERQUE; 35°46′N 106°41′W. Agriculture and livestock area. JEMEZ pueblo to S and parts of Indian Reservation to S and W; fish hatchery and Fenton Lake State Park to NW; Jemez State Monument is here.

Jeminay (JEE-MU-NEI), town, ⊙ Jeminay county, N XINJIANG UYGUR AUTONOMOUS REGION, NW CHINA, on KAZAKHSTAN border, 40 mi/64 km E of ZAISAN; 47°25′N 85°53′E. Livestock, grain; food processing, coal mining, building materials. Also appears as Ji-munai.

Jemison (JE-mi-suhn), town (2000 population 2,248), Chilton co., central Alabama, 11 mi/18 km NW of Clanton. Lumber; clothing. Originally called 'Langston Station' or 'Langstonville,' for Louise Langston, owner of much of the land occupied by the settlement. The name was changed to 'Jemison' for Robert Jemison, operator of a stage line that ran between Montgomery and Tuscaloosa.

Jemnice (YEM-nyi-TSE), German *Jamnitz*, town, JI-HOMORAVSKY province, SW MORAVIA, CZECH RE-PUBLIC, 26 mi/42 km S of JIHLAVA. Railroad terminus. Manufacturing (machinery and clothing); food processing. Former gold-mining center (beginning in the 13th century). Has a 12th-century Romanesque rotunda tower, and a 17th-century baroque castle.

Jemo (JE-mo), uninhabited coral island, RATAK CHAIN, MARSHALL Islands, W central PACIFIC, 20 mi/32 km NE of LIKIEP; 5 mi/8 km long; 10°07′N 169°33′E.

Jemseg (JEHM-sehg), village (2001 population 573), S NEW BRUNSWICK, CANADA, on short Jemseg stream (connecting GRAND LAKE with St. John River), and 40 mi/64 km N of SAINT JOHN; 45°49′N 66°03′W. Farmers market.

Jena (YAI-nuh), city, THURINGIA, E central GERMANY, on the SAALE RIVER; 50°56′N 11°35′E. Manufacturing of this industrial center includes pharmaceuticals, glass, and optical and precision instruments. The Zeiss works, one of the most prosperous and well-known camera and optical equipment manufacturers in the world, was the city's most important employer. In 1990, it merged with the Carl Zeiss Co., which had opened at OBERKOCHEN after World War II. Jena was known in the 9th century and was chartered in the 13th century, passed to the house of Wettin in the 14th century, and in 1485 to its Ernestine line. In 1806, Napoleon I decisively defeated the Prussians at Jena. The University of Jena was founded in 1557–1558 and reached its height in the late 18th and early 19th centuries. At that time the dramatist Friedrich von Schiller, the philosophers Hegel, Fichte, Schelling, Friedrich von Schlegel, and his poet brother August Wilhelm taught or lectured here. Schiller wrote the Wallenstein trilogy and Goethe wrote *Hermann und Dorothea* at Jena; Karl Marx was granted a doctorate in 1841. Noteworthy structures in the city include the Church of St. Michael (13th century), a 15th-century city hall, and parts of the city's medieval fortifications.

Jena (JEN-uh), town (2000 population 2,971), ⊙ LA SALLE parish, central LOUISIANA, 34 mi/55 km NE of ALEXANDRIA; 31°42′N 92°08′W. In agricultural area (cotton, soybeans; cattle, hogs); sawmills; metal industry; manufacturing (concrete, fabricated metal products, lumber, pulpwood); publishing; oil fields nearby. Incorporated 1927.

Jenai, TAIWAN: see Wushe.

Jenbach (YEN-bahkh), township, TYROL, W AUSTRIA, on INN River, and 13 mi/21 km ENE of INNSBRUCK; 47°24′N 10°46′E. Railroad junction; large hydroelectric plant. Manufacturing of wagons, compressors, plastics. Cogwheel railroad to the ACHENSEE Lake, whose water supplies hydroelectric plant.

Jendouba, province, TUNISIA: see JANDUBA, province.

Jeneponto (JEH-nai-POHN-to), town, ⊙ Jeneponto district, S tip of Sulawesi Selatan province, INDONESIA, 54 mi/87 km S of UJUNG PANDANG.

Jenera (JE-ni-ruh), village (2006 population 223), HANCOCK county, NW OHIO, 11 mi/18 km SSW of FINDLAY; 40°54′N 83°43′W.

Jenesano (hai-nai-SAH-no), town, ⊙ Jenesano municipio, BOYACÁ department, central COLOMBIA, 12 mi/19 km S of TUNJA; 05°23′N 73°22′W. Elevation 6,955 ft/2,119 m. Coffee, sugarcane, corn; livestock. Alternate names include Jenezano and Piranguata.

Jenezano, Colombia: see JENESANO.

Jengish Peak, KYRGYZSTAN and CHINA: see POBEDA PEAK.

Jenin, Arab town, Jenin district, at the S entrance to the JEZREEL VALLEY, at the N edge of the SAMARIAN HIGHLANDS, WEST BANK; 32°27′N 35°17′E. Elevation 820 ft/250 m. An important crossroads, urban center, and market town for N SAMARIA. Its economy is based on service, craft, and some industry; agriculture (wheat, barley, and vegetables). Jenin is believed to be on the site of the ancient city Ein-Ganam, mentioned in AMARNA Tablets, or the biblical town Ginnat. During World War I, Jenin was a military base for the Turkish-German army.

Jenkinjones (JENK-in-jonz), unincorporated village, MCDOWELL county, S WEST VIRGINIA, at VIRGINIA state line, 10 mi/16 km W of BLUEFIELD; 37°17′N 81°25′W. In bituminous-coal region.

Jenkins, county (□ 351 sq mi/912.6 sq km; 2006 population 8,725), E GEORGIA; ⊙ MILLEN; 32°47′N 81°58′W. Coastal plain agriculture (cotton, corn, soybeans, peanuts, tobacco, wheat); cattle, hogs; lumber and wood products, and timber area intersected by OGEECHEE RIVER. Formed 1905.

Jenkins, town (2000 population 2,401), LETCHER county, SE KENTUCKY, 21 mi/34 km SSW of PIKE-VILLE, in the CUMBERLANDS, at VIRGINIA state line; 37°10′N 82°37′W. Center of important bituminous-coal region. POUND GAP is 2 mi/3.2 km SW, pass through PINE MOUNTAIN ridge (2,380 ft/725 m). Manufacturing (concrete, crushed stone).

Jenkins, unincorporated town, BARRY county, SW MISSOURI, in the OZARKS, 14 mi/23 km NE of CASS-VILLE.

Jenkins, village (2000 population 287), CROW WING county, central MINNESOTA, 21 mi/34 km NNW of BRAINERD, in lakes and woods region; 46°38′N 94°19′W. Dairying; poultry; oats, alfalfa. Hay Lake to E, WHITEFISH Lake to NE.

Jenkinsburg, town (2000 population 203), BUTTS county, central GEORGIA, 4 mi/6.4 km WNW of JACKSON; 33°19′N 84°02′W.

Jenkintown (JAIN-kuhn-toun), borough (2006 population 4,350), MONTGOMERY county, SE PENNSYLVANIA, suburb 10 mi/16 km N of downtown PHILADELPHIA. Railroad junction. Manufacturing (machinery, commercial graphics, packaging, hardware). Seat of Beaver College to W (GLENSIDE). Only synagogue designed by Frank Lloyd Wright is here. Settled 1750, incorporated 1874.

Jenks, town (2006 population 14,123), TULSA county, NE OKLAHOMA, residential suburb 10 mi/16 km S of downtown TULSA, and on ARKANSAS RIVER; 36°00′N 95°58′W. Manufacturing (asphalt, paper products). Oral Roberts University to NE; Jones Airport to N.

Jenné, MALI: see DJENNÉ.

Jenner, unincorporated village, SONOMA county, W CALIFORNIA, at mouth of RUSSIAN RIVER on the PACIFIC OCEAN, 22 mi/35 km W of SANTA ROSA. Resort area. Dairying; fish, sheep, poultry; nursery products; fruit, grain. Sonoma Coast State Beach to S; Arm-strong Redwoods State Park and Austin Creek State Recreation Area to NE; Fort Ross and Salt Point state parks, on coast, to NW. Also called Jenner-by-the-Sea.

Jennersdorf (YEN-ners-dorf), town, S BURGENLAND, E AUSTRIA, near the RAAB and Hungarian-Slovenian border, 33 mi/53 km ESE of GRAZ; 46°56′N 16°09′E. Manufacturing of textiles; orchards (cherries, peaches), vineyards; poultry, pig breeding.

Jennerstown (JE-nuhrs-toun), borough (2006 population 701), SOMERSET county, SW PENNSYLVANIA, 12 mi/19 km SW of JOHNSTOWN. Agriculture (corn, hay; livestock; dairying). LAUREL MOUNTAIN village and ski resort to W of Stoughton Lake reservoir to NE.

Jennings, county (□ 378 sq mi/982.8 sq km; 2006 population 28,473), SE INDIANA; ⊙ VERNON; 39°00′N 85°38′W. Agricultural area (corn, wheat, tobacco; cattle, hogs). Manufacturing at NORTH VERNON. Timber; limestone quarries. Drained by small MUS-CATATUCK RIVER and by Vernon, Graham, and Sand creeks. Brush Creek State Fish and Wildlife Area, Purdue Southeast Agricultural Center, and Selmier State Forest NE of North Vernon. Crosley State Fish and Wildlife Area S of Vernon; part of MUSCATATUCK NATIONAL WILDLIFE REFUGE in W. Formed 1816.

Jennings (JEN-eengz), city (2000 population 10,986), ⊙ JEFFERSON DAVIS parish, SW LOUISIANA, 34 mi/55 km W of LAFAYETTE, near Bayou NEZPIQUE and its entrance to MERMENTAU RIVER; 30°13′N 92°40′W. In agricultural area (cotton, rice, and small crops); bottling plant; manufacturing (drugs, machinery, apparel, water-treatment systems, fabricated metal products); transportation equipment; oil field nearby. Barge port on MERMENTAU RIVER, 5 mi/8 km SE. Incorporated 1888.

Jennings, city (2000 population 15,469), SAINT LOUIS county, E MISSOURI, near MISSISSIPPI RIVER, a residential and industrial suburb 7 mi/11.3 km NW of downtown ST. LOUIS; 38°43′N 90°15′W. Manufacturing (signs, blowpipes); food products.

Jennings, village (2000 population 146), DECATUR county, NW KANSAS, on PRAIRIE DOG CREEK, and 23 mi/37 km SW of NORTON; 39°40′N 100°17′W. Agriculture and livestock raising.

Jennings, village (2006 population 381), PAWNEE county, N OKLAHOMA, 17 mi/27 km SE of PAWNEE; 36°10′N 96°34′W. In agricultural area; manufacturing (food processing, reclamation of lead products).

Jennings Lodge, unincorporated town (2000 population 7,036), CLACKAMAS county, NW OREGON, residential suburb 8 mi/12.9 km SSE of downtown PORTLAND, and 1 mi/1.6 km NW of GLADSTONE, on WILLAMETTE RIVER; 45°23′N 122°36′W.

Jenny Jump Mountain (c.1,100 ft/335 m), ridge of AP-PALACHIAN MOUNTAINS in NW NEW JERSEY, NW of BELVIDERE; 40°55′N 74°54′W. State forest here.

Jenny Lind Island or **Lind Island**, NUNAVUT territory, CANADA, in QUEEN MAUD GULF, at SW end of VIC-TORIA STRAIT, off SE VICTORIA ISLAND; 17 mi/27 km long, 10 mi/16 km wide; 68°52′N 101°30′W.

Jenolan Caves, AUSTRALIA: see OBERON.

Jensen, village, UINTAH county, NE UTAH, on GREEN RIVER, and 12 mi/19 km SE of VERNAL. Oil and natural gas. Headquarters for DINOSAUR NATIONAL MONU-MENT (N). Stewart Lake Waterfowl Management Area to S (small lake 1 mi/1.6 km W of Green River).

Jensen Beach (JEN-suhn), town (□ 5 sq mi/13 sq km; 2000 population 11,100), ST. LUCIE county, E central FLORIDA, 7 mi/11.3 km SE of PORT ST. LUCIE; 27°14′N 80°13′W. Manufacturing includes steel fabrication, stone processing, and metal doors.

Jens Munk Island, NUNAVUT territory, CANADA, at head of FOXE BASIN, off NW BAFFIN ISLAND; 45 mi/72 km long, 17 mi/27 km wide; 69°42′N 79°40′W.

Jepara (juh-PAH-rah), city, ⊙ Jepara district, Java Tengah province, INDONESIA, on JAVA SEA, 30 mi/48 km NE of SEMARANG; 06°35′S 110°39′E. Trade center for agricultural and forested area (sugar, rice, kapok, teak, cassava). Has mosque dating from sixteenth-seventeenth centuries, when town was seat of important Muslim sultanate. It was supplanted (early nineteenth century) by nearby KUDUS. Also spelled Djepara or Japara.

Jeparit (juh-PA-rit), village, W central VICTORIA, SE AUSTRALIA, on WIMMERA RIVER, and 205 mi/330 km WNW of MELBOURNE, and 22 mi/35 km N of DIMBOOLA, near S shore of Lake HINDMARSH; 36°09′S 141°59′E. Railroad junction; sheep (wool); wheat, barley, oats. Birthplace of Sir Robert Menzies, Australian prime minister 1939–1941, 1949–1966.

Jepelacio (hai-pai-LAH-see-o), town, MOYOBAMBA province, SAN MARTÍN region, N central PERU, in E outliers of the ANDES, 5 mi/8 km SSE of MOYOBAMBA; 06°07′S 76°57′W. Sugarcane, cereals, corn.

Jeppe, E suburb of JOHANNESBURG, E end of central business district, GAUTENG province, SOUTH AFRICA; 26°13′S 28°04′E. Elevation 5,740 ft/1,750 m.

Jequerí (ZHE-ker-EE), city (2007 population 12,963), SE MINAS GERAIS, BRAZIL, 28 mi/45 km E of PONTE NOVA; 20°28′S 44°31′W.

Jequetepeque (hai-kai-tai-PAI-kai), town, PACASMAYO province, LA LIBERTAD region, NW PERU, on coastal plain, near mouth of JEQUETEPEQUE RIVER, 7 mi/11 km NW of SAN PEDRO DE LLOC; 07°20′S 79°35′W. In irrigated agricultural area (rice, cotton, alfalfa).

Jequetepeque River (hai-kai-tai-PAI-kai), 100 mi/161 km long, NW PERU; rises in CORDILLERA OCCIDENTAL of the ANDES, 15 mi/24 km W of CAJAMARCA; flows W across CAJAMARCA and LA LIBERTAD regions, past CHILETE, to the PACIFIC 2 mi/3 km W of JEQUETEPEQUE; 07°21′S 79°36′W. Feeds numerous irrigation channels in lower course.

Jequitaí (ZHE-kee-tah-EE), town (2007 population 8,026), N central MINAS GERAIS, BRAZIL, on Rio Jequitaí, 52 mi/84 km NE of PIRAPORA; 17°18′S 44°31′W.

Jequitibá (ZHE-kee-chee-BAH), town (2007 population 5,496), central MINAS GERAIS, BRAZIL, on RIO DAS VELHAS, 26 mi/42 km NE of SETE LAGOAS; 19°15′S 44°03′W.

Jequitiba (ZHE-kwee-CHEE-bah), village, central BAHIA, BRAZIL, 11 mi/18 km S of MUNDO NOVO; 12°05′S 40°29′W.

Jequitinhonha (zhe-kwee-cheen-yon-yah), city (2007 population 23,966), NE MINAS GERAIS, BRAZIL, on navigable RIO JEQUITINHONHA, and 90 mi/145 km NNE of TEÓFILO OTONI; 16°28′S 41°01′W. Rock crystals and semiprecious stones found here.

Jerablus (je-RAH-BLUS), town, ALEPPO district, NW SYRIA, on W bank of the EUPHRATES River at Turkish border, and 65 mi/105 km NE of ALEPPO; 36°49′N 38°00′E. Cereals. Across the border in TURKEY is site of the ancient city of CARCHEMISH. Also spelled JARABULUS.

Jerai, Gunong, Malaysia: see KEDAH PEAK.

Jerantut (JER-ahn-TOOT), town (2000 population 24,737), central PAHANG, Malaysia, 28 mi/45 km SE of Kuala Lipis. Junction of highway and E coast railroad; important agr. and regional development settlement.

Jerash, JORDAN: see GERASA.

Jerauld, county (□ 532 sq mi/1,383.2 sq km; 2006 population 2,071), SE central SOUTH DAKOTA; ⊙ WESSINGTON SPRINGS; 44°03′N 98°37′W. Agricultural area watered by intermittent streams; drained by Firesteel, Sand, and Smith creeks. Corn, wheat; cattle. Formed 1883.

Jerba, TUNISIA: see HAWMAT AS-SUQ and JARBAH, island.

Jerdun, village, MOKA district, MAURITIUS, 11.5 mi/18.4 km SSE of PORT LOUIS. Sugarcane, vegetables; livestock.

Jerécuaro (he-RE-kwah-ro), city and township, GUANAJUATO, central MEXICO, on affluent of LERMA River, and 20 mi/33 km NE of ACÁMBARO, on railroad; 20°09′N 100°31′W. Grain, sugarcane, alfalfa, fruit, vegetables.

Jérémie (zhai-rai-MEE), town (2003 population 27,510), ⊙ GRANDE-ANSE department, SW HAITI, port on NW coast of JACMEL PENINSULA, on GULF OF GONAÏVES, 120 mi/193 km W of PORT-AU-PRINCE; 18°39′N 74°08′W. Port ships produce of fertile region (cacao, coffee, sugarcane, mangoes, logwood, hides); beekeeping. Fishing; manufacturing of soap, cigars; bauxite deposits nearby.

Jeremoabo (zhe-re-mo-ah-bo), city (2007 population 37,469), NE BAHIA, BRAZIL, on VASA BARRIS River, and 105 mi/169 km NW of ARACAJU (SERGIPE); 10°05′S 38°20′W. Nitrate deposits. Formerly spelled Geremoabo.

Jerer River (JER-uhr), c.180 mi/290 km long; SOMALI state, SE ETHIOPIA; rises in mountains at edge of the GREAT RIFT VALLEY, 20 mi/32 km NW of JIJIGA; seasonal stream flows intermittently SE, past JIJIGA and DEGEH BUR, through the OGADEN to FAFEN RIVER 20 mi/32 km S of SASABENEH. Also spelled Gerrer River.

Jeresa (hai-RAI-sah), village, VALENCIA province, E SPAIN, 4 mi/6.4 km NW of GANDÍA. Rice, oranges, wine, vegetables.

Jeres del Marquesado (HAI-res dhel mahr-kai-SAH-dho), town, GRANADA province, S SPAIN, 8 mi/12.9 km SSW of GUADIX; 37°11′N 03°09′W. Cereals, chestnuts; livestock. Copper deposits.

Jerez (he-RES), town, JUTIAPA department, SE GUATEMALA, in highlands, at SW foot of CHINGO volcano, near EL SALVADOR border, 16 mi/26 km SE of JUTIAPA; 14°06′N 89°45′W. Elevation 2,297 ft/700 m. Corn, beans; livestock.

Jérez, MEXICO: see JÉREZ DE GARCÍA SALINAS, ZACATECAS.

Jérez de García Salinas (HE-res de gar-SEE-ah sah-LEE-nahs), city, ZACATECAS, N central MEXICO, on interior plateau, 30 mi/48 km WSW of ZACATECAS; 22°39′N 103°00′W. Mining (tin, mercury) and agricultural center (cereals, vegetables, sugarcane, livestock); tanning.

Jerez de la Frontera (hai-raith dai lah fron-TAI-rah), city (2001 population 183,273), CÁDIZ province, SW SPAIN, in ANDALUSIA, 15 mi/24 km NE of CÁDIZ; 36°41′N 06°08′W. Important commercial center noted for its sherry and brandy. Most of its production is exported. World-famous horses of mixed Spanish, Arab, and English blood. City was captured by the Moors in 711 and later recovered by Alfonso X of Castile (1264). Of interest are its Gothic churches and an 11th-century Arabian alcazar.

Jerez de los Caballeros (dhai los kah-vah-LYAI-ros), town, BADAJOZ province, W SPAIN, in EXTREMADURA, in outliers of the SIERRA MORENA, near Portuguese border, 40 mi/64 km ESE of BADAJOZ; 38°19′N 06°46′W. Processing, lumbering, and agricultural center (cereals, olives, grapes; livestock; cork). Olive-oil pressing, liquor distilling, meat packing, sawmilling, wood turning, tanning; copper, tungsten, and iron mining. Its chief industry is manufacturing of bottle corks. Founded in 1229, it was formerly known as Jerez de Badajoz, but later was named Jerez de los Caballeros for the Knights Templars, to whom the city had been given. Balboa was born here.

Jergucat (yer-goo-TSAHT), village, S ALBANIA, near Greek border, 12 mi/19 km SE of GJIROKASTËR; 39°56′N 20°15′E. Road junction. Also called Jergucati.

Jérica (HAI-ree-kah), town, Castellón de la Plana province, E SPAIN, 30 mi/48 km WSW of CASTELLÓN DE LA PLANA; 39°55′N 00°34′W. Light manufacturing; olive oil, wine, fruit. Remains of medieval walls and castle.

Jericho [Heb.=fragrant, or city of the moon god], town (pop. 17,000), WEST BANK, N of the DEAD SEA; 31°52′N 35°27′E. The urban center for the W part of the Lower Jordan valley. Oldest known city. The modern Ariha lies near the ancient site. Jericho is an oasis watered by a number of springs, and the town is surrounded by orchards and intensive market gardening (dates, oranges, bananas, papayas). A large part of the population is engaged in agriculture, with Jericho supplying early fruit and vegetables and tropical fruit to both JERUSALEM and to ISRAEL. The 1st Arab town in the West Bank that became autonomous under the agreement between Israel and the Palestine Liberation Organization, in 1994. Jericho figures prominently in the Bible. According to the account in the Book of Joshua, it was captured from the Canaanites by Joshua and then destroyed, an event several times repeated in its history. Herod the Great (one of its many conquerors) sacked and rebuilt it. Later it was taken by the Muslims. Excavations of the mound of the original site, TELL ES SULTAN, were begun early in the 20th century and have revealed the oldest known settlement in the world, dating perhaps from c.8000 B.C.E. Because the town of Joshua was destroyed by erosion, scholars have been unable to fix the date of the conquest of Canaan but generally place it between 1400 B.C.E. and 1250 B.C.E. At the newer site of Herodian Jericho, 2 mi/3.2 km S of Tell es Sultan, a Hellenistic fortress and the palace of Herod have been excavated.

Jericho (JER-i-ko), unincorporated town (□ 3 sq mi/7.8 sq km; 2000 population 13,045), NASSAU county, SE NEW YORK, on LONG ISLAND; 40°47′N 73°32′W. Chiefly residential; some light manufacturing; commercial services.

Jericho, town, including Jericho village, CHITTENDEN CO., NW VERMONT, 10 mi/16 km E of Burlington; 44°28′N 72°57′W. Settled in late 18th century. Named for the biblical Jericho.

Jericho (JE-ri-ko), village, central QUEENSLAND, AUSTRALIA, 280 mi/451 km W of ROCKHAMPTON; 23°34′S 146°10′E. Railroad junction; sheep.

Jericho, unincorporated village, JUAB county, central UTAH, 20 mi/32 km WNW of NEPHI, near Tanner Creek. On railroad; sheep and cattle center for wide region. Little Sahara Recreation Area to SW.

Jericho Bay (JER-i-ko), HANCOCK county, S MAINE, bounded W, S, and E by Deer Island, ISLE AU HAUT, and SWANS ISLAND; opens NE into BLUE HILL BAY.

Jerichow (YE-rikh-ou), town, SAXONY-ANHALT, central GERMANY, near the ELBE, 10 mi/16 km SE of STENDAL; 52°30′N 12°01′E. Has noted 12th-century church of former Premonstratensian monastery (founded 1144; dissolved 1552).

Jericó (AHE-ree-KO), town (2007 population 7,825), NW PARAÍBA, BRAZIL, 18 mi/29 km N of POMBAL; 06°33′S 37°49′W.

Jericó (hai-ree-KO), town, ⊙ Jericó department, ANTIOQUIA department, NW central COLOMBIA, on E slopes of CORDILLERA OCCIDENTAL, near CAUCA River, 35 mi/56 km SSW of MEDELLÍN; 05°47′N 75°47′W. Elevation 6,453 ft/1,967 m. In fertile agricultural region (coffee, corn, sugarcane, beans, plantains; livestock). Light industry.

Jericó (hai-ree-KO), town, ⊙ Jericó municipio, BOYACÁ department, central COLOMBIA, in the ANDES, 70 mi/113 km NE of TUNJA; 06°08′N 72°33′W. Elevation 8,182 ft/2,493 m. Coffee, sugarcane, corn; livestock.

Jerico Springs, city (2000 population 259), CEDAR county, W MISSOURI, 24 mi/39 km SE of NEVADA; 37°37′N 94°00′W. Soybeans, corn, wheat; cattle. Former health springs resort. Plotted 1882.

Jerilderie (juh-RIL-duh-ree), village, S NEW SOUTH WALES, AUSTRALIA, 190 mi/306 km W of CANBERRA; 35°22′S 145°44′E. Wheat; sheep (merino stud region); rice, vegetables grown through irrigation; tomato-processing plants. Ned Kelly and his gang raided the bank and held c.300 hostages in 1879.

Jerimoth Hill (je-REI-moth) (812 ft/247 m), highest point in RHODE ISLAND, in FOSTER town, near CONNECTICUT state line, c.20 mi/32 km W of PROVIDENCE.

Jermaq, Jebel, ISRAEL: see MEIRON, HAR.

Jerma River, Bulgaria: see ERMA RIVER.

Jermer, CZECH REPUBLIC: see JAROMER.

Jermuk, village, E ARMENIA, in DARALAGEZ RANGE, 20 mi/32 km ENE of Mikoyan; 39°50′N 45°39′E. Health resort (elevation c.7,000 ft/2,134 m).

Jermyn (JUHR-min), borough (2006 population 2,237), LACKAWANNA county, NE PENNSYLVANIA, 10 mi/16 km NE of SCRANTON, on LACKAWANNA RIVER; 41°31′N 75°32′W. Former anthracite-coal center. Manufacturing (machinery, burial caskets). Includes community of Nebraska to E. Incorporated 1870.

Jernye, SLOVAKIA: see JAROVNICE.

Jerome, county (□ 601 sq mi/1,562.6 sq km; 2006 population 20,130), S IDAHO; ⊙ JEROME; 42°42′N 114°16′W. County is opposite TWIN FALLS. Livestock-raising and irrigated agricultural area (sheep, cattle; dairying; corn, barley, wheat; alfalfa, hay; potatoes, apples, sugar beets). Bounded on S by SNAKE RIVER and in SNAKE RIVER PLAIN. SHOSHONE FALLS to W (212 ft/65 m); Wilson Lake reservoir in E. Formed 1919.

Jerome, town (2000 population 7,780), ⊙ JEROME county, S IDAHO, near SNAKE RIVER, 12 mi/19 km NNW of TWIN FALLS; 42°44′N 114°31′W. Elevation 3,600 ft/1,097 m. In irrigated agricultural area (grain, vegetables, melons; poultry); dairying; manufacturing (paper products, concrete blocks, wooden trusses, fertilizers). On SNAKE RIVER PLAIN. Laid out 1907, incorporated 1909.

Jerome, unincorporated town (2000 population 1,068), SOMERSET county, SW PENNSYLVANIA, 9 mi/14.5 km SSW of JOHNSTOWN; 40°12′N 78°58′W. Agriculture includes dairying; livestock; corn, oats, hay, potatoes. QUEMAHONING RESERVOIR to SE; Laurel Ridge State Park to NW.

Jerome, village, YAVAPAI county, central ARIZONA, in BLACK HILLS, 25 mi/40 km NE of PRESCOTT; elevation 5,354 ft/1,632 m. Former copper-mining center; railroad terminus. Grew in 1880s after discovery of copper in 1870s; once had 13,000 people; incorporated 1899. Jerome State Historic Park is here; located in Prescott National Forest; MINGUS MOUNTAIN (7,743 ft/2,360 m) to S.

Jerome, village (2000 population 46), DREW county, SE ARKANSAS, 9 mi/14.5 km S of DERMOTT, near Bayou BARTHOLOMEW; 33°23′N 91°28′W. Cut-off Creek Wildlife Management Area to W.

Jerome, village (2000 population 1,414), SANGAMON county, central ILLINOIS, residential suburb 3 mi/4.8 km SW of downtown SPRINGFIELD; 39°46′N 89°40′W. In agricultural and oil area.

Jeromesville (je-ROMZ-vil), village (2006 population 485), ASHLAND county, N central OHIO, 8 mi/13 km SE of ASHLAND, on Jerome Fork of MOHICAN RIVER; 40°48′N 82°11′W. Formerly called Jeromeville.

Jerônimo Monteiro (zhe-RO-nee-mo MON-tai-ro), town (2007 population 10,722), SW ESPÍRITO SANTO, BRAZIL, 14 mi/23 km ESE of ALEGRE; 20°49′S 41°25′W.

Jerramungup (jer-uh-MUHN-guhp), town, WESTERN AUSTRALIA state, W AUSTRALIA, 267 mi/429 km SE of PERTH, 111 mi/179 km N of ALBANY. Wheat, sheep.

Jerry City (JER-ee), village (□ 1 sq mi/2.6 sq km; 2006 population 466), Wood county, NW OHIO, 8 mi/13 km SSE of BOWLING GREEN; 41°15′N 83°36′W. In agricultural area.

Jerrys Plains (JE-reez PLAINZ), village, NEW SOUTH WALES, SE AUSTRALIA, 153 mi/247 km NW of SYDNEY, in HUNTER RIVER valley; 32°30′S 150°55′E.

Jersey, county (□ 377 sq mi/980.2 sq km; 2006 population 22,628), W ILLINOIS; ⊙ JERSEYVILLE; 39°05′N 90°21′W. Bounded S by the MISSISSIPPI and W by ILLINOIS RIVER; drained by MACOUPIN CREEK. Agriculture (soybeans, sorghum, apples, corn, wheat; cattle, hogs; dairying); manufacturing. Resorts on Illinois River. Free ferry across Illinois River to CALHOUN county. Pere Marquette State Park; towns of ELSAH and GRAFTON, on Great River Road, popular tourist areas; Great River Road bicycle trail; limestone cliffs noted for bald eagle nesting. Formed 1839.

Jersey, town (2000 population 163), WALTON county, N central GEORGIA, 7 mi/11.3 km SW of MONROE; 33°43′N 83°48′W.

Jersey, island (□ 46 sq mi/119.6 sq km; 2004 population 87,700), in the ENGLISH CHANNEL, largest of the CHANNEL ISLANDS, which are dependencies of the British Crown. It is 15 mi/24 km from the Normandy coast of France and SE of GUERNSEY. SAINT HELIER, the capital, is on St. Aubin's Bay. The mild climate (plants requiring subtropical conditions grow without protection), the moderate rainfall (30 in/76 cm–35 in/89 cm), and the scenery have contributed to make Jersey, like other Channel Islands, a vacation resort. The soil is generally fertile, and large quantities of vegetables (especially potatoes, tomatoes, and broccoli) and fruits are grown. Cattle raising and dairying (Jersey cattle) are also important, as is fishing, light industry, and tourism. It also has become a financial center for investors and an important e-commerce hub. The inhabitants are mostly of Norman descent; English, French (the official language), and a Norman dialect are spoken. The Jersey Zoological Park was founded in 1959 to protect endangered animals.

Jersey City, city (2006 population 241,789), ⊙ HUDSON county, NE NEW JERSEY, a port on a peninsula formed by the HUDSON and HACKENSACK rivers and Upper New York Bay, opposite the lower area of MANHATTAN island; 40°42′N 74°03′W. Settled before 1650, incorporated as Jersey City 1836. The second largest city in the state and a commercial and industrial center surpassed only by NEWARK. It is a port of entry and a manufacturing center. With 11 mi/18 km of waterfront and significant railroad connections, Jersey City is an important transportation terminal point and distribution center. It has railroad shops, oil refineries, warehouses, and plants manufacturing a diverse assortment of products, such as chemicals, petroleum and electrical goods, newspapers, textiles, and cosmetics. The city has benefited from its position across from Manhattan, and many Jersey City companies are extensions of those originating in NEW YORK city. Further developments have included housing and shopping areas and marinas along the waterfront; other parts of the city, however, remain run-down after years of commercial inactivity. A large number of ethnic groups throughout U.S. history have settled in Jersey City before venturing out across the country. The area was acquired by Michiel Pauw c.1629. The Dutch soon set up the trading posts of PAULUS HOOK, Communipaw, and Horsimus. In 1674 the site fell permanently under British rule. The fort at Paulus Hook was captured by Light-Horse Harry Lee under Washington's plan, August 19, 1779. Nearby BERGEN was a stockaded Dutch village dating from before 1620 and had New Jersey's first municipal government, church (Dutch Reformed), and school (1662). Jersey City was consolidated with Bergen and Hudson City in 1869; the town of Greenville was added in 1873. The city's industrial growth began in the 1840s with the arrival of the railroad and the improvement of its water transport system. In 1916, Jersey City docks were the scene of the "BLACK TOM" explosion that caused widespread property damage. The city has a modern medical center and is the seat of Jersey City State College, Hudson County Community College, and St. Peter's College. Site of waterfront-renewal project. In Lincoln Park is a statue of Lincoln, built in 1929.

Jersey Homesteads, NEW JERSEY: see ROOSEVELT.

Jersey Shore, borough (2006 population 4,376), LYCOMING county, N central PENNSYLVANIA, 7 mi/11.3 km WSW of Williamsport, on West Branch of SUSQUEHANNA RIVER (bridged). Manufacturing (tool and die, seed hybridization, fabricated metal products, machinery, construction materials, corrugated containers, medical supplies, apparel); agriculture (grain, potatoes; livestock; dairying). Jersey Shore

Airport to E. Little Pine State Park to N; parts of Tiadaghton State Forest to N and S; PINE CREEK to W. Settled 1785, incorporated 1826.

Jersey Village, town (2006 population 7,138), HARRIS county, SE TEXAS, residential suburb 14 mi/23 km NW of downtown HOUSTON; 29°53′N 95°34′W. Drained by White Oak Bayou. Oil and natural gas; agriculture (cattle, horses; dairying; nurseries; vegetables).

Jerseyville, city (2000 population 7,984), ⊙ JERSEY county, W ILLINOIS, 17 mi/27 km NNW of ALTON; 39°07′N 90°19′W. Trade and shipping center in agricultural area (apples, corn, wheat; livestock). Plotted 1834, incorporated 1855.

Jerte (HER-tai), town, CÁCERES province, W SPAIN, 23 mi/37 km NE of PLASENCIA; 40°13′N 05°45′W. Olive-oil processing; fruit, wine, honey.

Jerumenha (ZHE-roo-men-yah), town (2007 population 4,371), W central PIAUÍ, BRAZIL, on GURGUEIA RIVER, and 40 mi/64 km SW of FLORIANO, in cattle region; 07°05′S 43°30′W. Carnauba wax, hides. Formerly spelled Jeromenha.

Jerusalem, Hebrew **Yerushalayim**, Arabic **Al Quds**, city (2005 population 704,900), ⊙ ISRAEL; 32°23′N 35°02′E, often referred to as the City of David. Elevation 1,906 ft/580 m. It is situated on a ridge 2,295 ft/700 m–2,738 ft/835 m above sea level, lying W of the DEAD SEA and the JORDAN RIVER. Jerusalem is an administrative, religious, educational, cultural, and market center. Tourism and the construction of houses and hotels are the city's major industries. Manufactures include cut and polished diamonds, plastics, apparel, and shoes; electronic printing and other high-technology, science-based industries have also developed in Jerusalem. The city is served by road, railroad, and air transport. Jerusalem is the holiest city for Jews and Christians since biblical times, and is more recently claimed by some as one of the holy centers of Islam. Sometimes under the name of Zion, it figures familiarly in Jewish and Christian literature, and is named in the Old and New Testaments more than 800 times. It is the site of much of Jesus's story. Jerusalem is mentioned in Jewish blessings on bread as well as Passover holiday prayers. Jews throughout the world pray daily in the direction of Jerusalem, and those in Jerusalem pray in the direction of the Western Wall. Jerusalem contains an ethnographic mix of people; while the majority are Jewish, about one-third (some 228,000) are not—mostly Arabs.

In the eastern part of Jerusalem is the Old City, a quadrangular area built on two hills and surrounded by a wall completed (1542) by the Ottoman sultan Suleiman I. The city within the walls is divided into four quarters. The Muslim quarter, in the E and NE, contains a sacred enclosure, the Haram esh-Sherif; within which, built on the old Mount Moriah, are the Dome of the Rock (completed 691; getting new gold covering as gift from Jordan's King Hussein), or Mosque of Omar, and the Mosque of al-Aksa. The wall of the Haram incorporates the only extant piece of the Second Temple; this, the Western Wall, or Wailing Wall, is a holy place for Jews. Nearby and SW of the Western Wall is the Jewish quarter, with several famous old synagogues. The synagogues were destroyed in the 1948 Arab-Israeli fighting, some reconstructed after 1967, the Old City was taken by the Jordanians in 1948 and recaptured in 1967 by the Israelis, who rebuilt and renovated the Jewish quarter. To the W of the Jewish quarter is the Armenian quarter, site of the Gulbenkian Library. The Christian quarter occupies the NW part of the Old City. Its greatest monument is the Church of the Holy Sepulcher. Through the area runs the Via Dolorosa, where Jesus is said to have carried his cross. The rest of the city, extending W, N, and NW of the Old City, has developed tremendously since the 19th century. It is the site of major educational institutions (led by the Hebrew University), as well as the Knesset (the Israeli parliament), the Supreme Court, the seat of the President, and other government buildings

To the E of the Old City is the Valley of the Kidron, across which lie the Garden of GETHSEMANE and the MOUNT OF OLIVES. To the N is MOUNT SCOPUS, which is the site of part of Hadassah Hospital, the Mormon University, and the Hebrew University. Another campus of Hebrew University is located on the W side of the city at Givat Ram, and it includes the Israel National Library. From 1948 to 1967, Mount Scopus was an Israeli exclave in Arab territory, although Jews were denied access to it. To the W and S of the Old City runs the valley of HINNOM; this meets the Kidron near the pool of SILOAM, which is next to the site of the original city of Jerusalem, now partly excavated and called the City of David. Jerusalem's churches and shrines are innumerable. The traditional identifications vary in reliability from certainty (such as Gethsemane) to pious supposition (such as the Tomb of the Virgin). Excavations have been made in Jerusalem since 1835, and after 1967, the Israelis increased this activity, uncovering remains of the Herodian period and ruins of a Muslim structure of the 7th or 8th century. Many of Jerusalem's original streets, including the main Cardo, have been excavated and preserved as tourist sites.

Despite the incomplete archaeological work, it is evident that Jerusalem was occupied as far back as the 4th millennium B.C.E. In the late Bronze Age (2000–1550 B.C.E.), it was a Jebusite (Canaanite) stronghold. David captured it (c.1000 B.C.E.) from the Jebusites and walled the city. After Solomon built the Temple on Mount Moriah in the 10th century B.C.E., Jerusalem became the spiritual and political capital of the Hebrews. In 586 B.C.E. it fell to the Babylonians, and the Temple was destroyed. The city was restored to Hebrew rule later in the 6th century B.C.E. by Cyrus the Great, king of PERSIA. The Temple was rebuilt (538–515 B.C.E.; known as the Second Temple) by Zerubbabel, a governor of Jerusalem under the Persians. The city was the capital of the Maccabees in the 2nd and 1st centuries B.C.E. After Jerusalem had been taken for the Romans by Pompey, it became the capital of the Herod dynasty, which ruled under the aegis of Rome. The Roman emperor Titus ruined the city and destroyed the Temple (C.E. 70) in order to punish and discourage the Jews. After the revolt of Bar Kokba (132–135), Hadrian rebuilt the city as a pagan shrine called Aelia Capitolina but forbade the Jews to live on the site.

With the imperial toleration of Christianity (from 313), Jerusalem underwent a revival, greatly aided by St. Helena, who sponsored much building in the early 4th century. Since that time Jerusalem has been a world pilgrimage spot. The Muslims, who believe that the city was visited by Muhammad, treated Jerusalem favorably after they captured it in 637, making it their chief shrine after MAKKA. From 688 to 691 the Dome of the Rock mosque was constructed. Jerusalem was conquered by the Crusaders in 1099 and for most of the 12th century was the capital of the Latin Kingdom of Jerusalem. In 1187, Muslims under Saladin recaptured the city. Thereafter, under Mamluk and then Ottoman rule, Jerusalem was rebuilt and restored; but by the late 16th century it was declining as a commercial and religious center. In the early 19th century, Jerusalem began to revive. The flow of Christian pilgrims increased, and churches, hospices, and other institutions were built. Jewish immigration accelerated, and by 1850, Jews made up the largest community in the city, expanding outside the Old City walls. In 1917, during World War I, Jerusalem was captured by British forces under Gen. Edmund Allenby. After the war it was made the capital of British Mandatory Palestine (1918–1948). The UN, in partitioning the

mandate into Arab and Jewish states, declared that Jerusalem and its environs (including Bethlehem) would be an internationally administered enclave in the projected Arab state. Even before the partition went into effect (May 14, 1948), fighting between Jews and Arabs broke out in the city (Dec. 1947). On May 28, the Jews in the Old City surrendered. The rest of the city remained in Jewish hands. The Old City and all areas held by the Arab Legion (the eastern section of Jerusalem) were annexed by Jordan in April 1949.

On December 14, 1949, Jerusalem was made the capital of Israel. In the Six Day War, Israeli forces took the Jordanian-held part, including the Old City. Late in June of that year the Israeli government formally annexed the Old City and its surroundings, increasing its area to 40 sq mi/104 sq km from its 1947 size of 12 sq mi/30 sq km. Arab Jerusalemites are citizens and vote in Israeli national elections. With increasing suburbanization and added housing developments in formerly Jordanian-held territory, Jerusalem has become Israel's largest city. Arts and education centers, parks, promenades, playgrounds, rehabilitated neighborhoods, and museum expansions have improved the city's social standards. A central issue of ongoing negotiations between Israel, the Palestinian Authority, and surrounding Arab nations is the status of Jerusalem. Jerusalem is the seat of Hebrew University, the Rubin Academy of Music, the Israel Academy of Sciences and Humanities, the Institute of Jerusalem Studies, Van Leer Institute, the Israel Museum, the British School of Archaeology, the Dominican Fathers' Convent of St. Étienne, with the attached Bible School and French Archaeological School, the American College, the Greek Catholic Seminary of St. Anne, the Pontifical Biblical Institute, the Swedish Theological Institute, the Near East School of Archaeology. Military cemetery, Bible Lands Museum, Rockefeller Museum of Archaeology, Teddy Kollek stadium, City of David Museum, Islamic Art Museum here.

Jerusalem (je-ROO-suh-luhm), village (2006 population 150), MONROE county, E OHIO, 11 mi/18 km SSE of BARNESVILLE; 39°51′N 81°05′W.

Jerusalem, village, NARRAGANSETT town, WASHINGTON county, S RHODE ISLAND; 41°22′N 71°31′W. Summer resort.

Jerusalem, Kingdom of or **Latin Kingdom of Jerusalem**, French kingdom, (1099–1291) established in the LEVANT by the First Crusade. Comprised present-day ISRAEL, southern LEBANON, and southwestern JORDAN. The kingdom was not wealthy, depending on trade with the Muslims, banking activities, and taxes on pilgrims to sustain itself. Much of the land was barren; in bad years grain had to be imported from SYRIA. Destroyed the first time by Saladin in 1187, it was re-established around AKKO and maintained until the capture of that city in 1291.

Jerusalem Mills, MARYLAND: see KINGSVILLE.

Jerusalén (he-roo-sah-LEN), municipality and town, LA PAZ department, EL SALVADOR, N of ZACATECOLUCA close to border with Cuscatlan department.

Jerusalén (hai-roo-sah-LAIN), town, ⊙ Jerusalén municipio, CUNDINAMARCA department, central COLOMBIA, 34 mi/55 km WSW of BOGOTÁ; 04°40′N 73°55′W. Elevation 1,171 ft/357 m. Coffee, corn, sugarcane; livestock.

Jervaulx (JUH-vis), location, NORTH YORKSHIRE, N ENGLAND, on URE RIVER, and 10 mi/16 km S of RICHMOND; 54°16′N 01°44′W. Nearby are remains of Jervaulx Abbey, a Cistercian abbey founded 1156 and dismantled 1537. The last abbot was executed after taking part in the Pilgrimage of Grace (1536).

Jervis Bay (JAHR-vis), sheltered inlet of the PACIFIC OCEAN, SE AUSTRALIA; 10 mi/16.1 km long and 6 mi/9.7 km wide; 35°05′S 150°45′E. In 1915 the harbor and part of the coast were transferred to the federal government by

NEW SOUTH WALES. Jervis Bay, connected by railroad with CANBERRA, 85 mi/137 km inland, then became the port of the landlocked Australian Capital Territory. The area around the bay is a popular summer resort. Associated with Royal Australian navy bases.

Jervis, Cape (JAHR-vis), SE SOUTH AUSTRALIA, at E side of entrance of Gulf SAINT VINCENT, 67 mi/108 km S of ADELAIDE; 35°38′S 138°06′E. Looks out onto BACKSTAIRS PASSAGE. Fishing. Access point for travelers to KANGAROO ISLAND.

Jervis Inlet (JAHR-vis), SW BRITISH COLUMBIA, W CANADA, NE arm of MALASPINA STRAIT of Strait of GEORGIA; 51 mi/82 km long, 1 mi/2 km–8 mi/13 km wide; mouth opposite TEXADA ISLAND, head near 50°13′N 123°58′W. At mouth is Nelson Island; 12 mi/19 km long, 5 mi/8 km wide.

Jervois (JUHR-vis), village, SE SOUTH AUSTRALIA, 55 mi/89 km ESE of ADELAIDE, on MURRAY RIVER, near Tailem Bend; 35°16′S 139°26′E. Dairy products; livestock.

Jeschken, CZECH REPUBLIC: see JESTED.

Jesenice, German *Jechnitz*, village, STREDOCESKY province, W BOHEMIA, CZECH REPUBLIC, on railroad, and 25 mi/40 km N of PLZEŇ; 50°06′N 13°29′E. Museum; 13th-century church.

Jesenice (ye-SE-nee-tse), town (2002 population 13,236), NW SLOVENIA, on SAVA DOLINKA RIVER, and 35 mi/56 km NNW of LJUBLJANA, near Austrian border, at S foot of the KARAWANKEN mountains; 46°27′N 14°04′E. Railroad junction, with lines to Ljubljana, TRIESTE (ITALY), VILLACH (AUSTRIA), and TARVISIO (ITALY). Industrial center (iron- and steelworks here and in vicinity). Skiing; mountain climbing.

Jesenik (YE-se-NYEEK), Czech *Jeseník*, town, SEVEROMORAVSKY province, NW SILESIA, CZECH REPUBLIC, on railroad, and 45 mi/72 km N of OLOMOUC; 50°14′N 17°12′E. Elevation 1,387 ft/423 m. Manufacturing (textiles, chiefly linen; furniture); granite quarrying. Noted health and winter-sports resort in JESENIKS mountains, with main spa facilities at LAZNE JESENIK (Czech *Lázně Jeseník*, also called *Gräfenberg*), just NW. Has a 15th-century castle, an 18th-century town hall, and several sanatoriums. Town plays host to a 4-mi/6.4-km bobsled run. Known for its healthful waters since 13th century. Until 1946, called FRYVALDOV, Czech *Frývaldov* (FREE-vahl-DOF), German *Freiwaldau*.

Jeseniks (YE-se-NYEEKS), Czech *Jeseníky*, German *Altvatergebirge*, mountain range of the SUDETES, SEVEROMORAVSKY province, W SILESIA and NW MORAVIA, CZECH REPUBLIC. From Czech-Polish border NE of STARE MESTO extends c.50 mi/80 km SE to MORAVIAN GATE; between KRALICKY SNEZNIK MOUNTAIN group (W) and ODER MOUNTAINS (SE). Rises to 4,892 ft/1,491 m at PRADED (mountain). Extensive forests and pastures; great power resources; scattered iron and graphite deposits on Moravian slopes; marble quarries and numerous mineral springs in N. Noted for health resorts (LAZNE JESENIK, KARLOVA) and textile industries.

Jeseníky, CZECH REPUBLIC: see JESENIKS.

Jesenske (ye-SEN-skai), Slovak *Jesenské*, town, STREDOSLOVENSKY province, S SLOVAKIA, 2 mi/3.2 km SE of LUCENEC; 48°18′N 20°05′E. Vegetables, sugar beets. Railroad junction. Until 1949, called FELDINCE (HUNGARIAN *Feled*). Under Hungarian rule between 1938–1945.

Jesi (YAI-zee), ancient *Aesis*, town, ANCONA province, THE MARCHES, central ITALY, near ESINO RIVER, 15 mi/24 km SW of ANCONA. Silk, woolen, and paper mills, foundry; manufacturing (agricultural machinery, furniture, pasta, consumer goods, liquor). Bishopric. Has cathedral, early Renaissance palace, remains of 14th-century town walls. Emperor Frederick II was born here. Formerly called Iesi.

Jessamine (JEZ-uh-muhn), county (□ 174 sq mi/452.4 sq km; 2006 population 44,790), central KENTUCKY;

⊙ NICHOLASVILLE; 37°52′N 84°34′W. Bounded SW, S, and SE by KENTUCKY RIVER; drained by Hickman Creek. Gently rolling upland agricultural area in BLUEGRASS region (burley tobacco, corn, hay, alfalfa, soybeans; cattle, horses, poultry; dairying). Palisades of Kentucky River (gorge) in SW; Jim Beam Nature Preserve; Daniel Boone's Cave and CAMP NELSON National Cemetery in S. Formed 1798.

Jessel, POLAND: see JASLO.

Jesselton, Malaysia: see KOTA KINABALU.

Jessen (YES-sen), town, SAXONY-ANHALT, central GERMANY, on the BLACK ELSTER, 14 mi/23 km ESE of WITTENBERG; 51°48′N 12°57′E. Manufacturing (furniture). Chartered 1350.

Jessie, village, MATABELELAND SOUTH province, S central ZIMBABWE, 4 mi/6.4 km WNW of WEST NICHOLSON, on railroad; 21°03′S 29°18′E. Gold mining. Livestock; corn, wheat, tobacco.

Jessnitz (YES-nits), town, SAXONY-ANHALT, central GERMANY, near the MULDE, 5 mi/8 km N of BITTERFELD; 51°32′N 12°17′E.

Jessore, administrative district, W EAST BENGAL, BANGLADESH; 22°47′N 88°53′E. Agriculture (paddy, pulses, oilseeds, sugarcane, and date-palm). One of the least industrialized districts of BANGLADESH.

Jessore (jaw-shor), city (□ 26 sq mi/67.6 sq km; 2001 population 176,655), SW BANGLADESH, on the Bhairab River; 23°12′N 89°08′E. Modern Jessore trades in rice and sugar; rice and oilseed milling; celluloid and plastics industries. Four affiliated counties of Rajshahi University are in the city. Airport.

Jessup (JESS-up), village (2000 population 7,865), HOWARD county, central MARYLAND, 14 mi/23 km SW of downtown BALTIMORE; 39°09′N 76°46′W. Originally named Jessup's Cut in the mid-18th century, shortened to Jessup in 1863. Original site of the Maryland House of Correction, medium-security prison established in 1878. Also site of the Maryland State Reformatory for Women and the Perkins Hospital for male prisoners. Famous for the Baltimore Produce Terminal, housing all the city's produce markets under one roof.

Jessup, borough (2006 population 4,583), LACKAWANNA county, NE PENNSYLVANIA, suburb 7 mi/11.3 km NE of SCRANTON on LACKAWANNA RIVER; 41°27′N 75°32′W. Manufacturing includes security printing, artificial Christmas trees, fabricated metal products, apparel, consumer goods. Moosic Mountains to SE; Archbald Pothole State Park to NE. Includes the community of WINTON.

Jessup, Lake (JES-uhp), SEMINOLE county, central FLORIDA, 13 mi/21 km NE of ORLANDO, N end connected with ST. JOHNS RIVER; c.10 mi/16 km long, 1 mi/1.6 km–3 mi/4.8 km wide; 28°43′N 81°01′W.

Jesteburg (YES-te-burg), village, LOWER SAXONY, N GERMANY, 17 mi/27 km S of HAMBURG; 53°48′N 09°56′E.

Jested (YESH-tyet), Czech *Ještěd*, German *Jeschken*, mountain, highest peak (3,320 ft/1,012 m) of LUSATIAN MOUNTAINS, SEVEROCESKY province, N BOHEMIA, CZECH REPUBLIC, 3 mi/4.8 km SW of LIBEREC; 50°44′N 14°59′E. Noted winter-sports resort; cable railroad to summit. Amethyst and agate deposits, stone quarries.

Jestetten (YE-shtet-tuhn), village, BADEN-WÜRTTEMBERG, SW GERMANY, near the RHINE (bridge), 4 mi/6.4 km SE of SCHAFFHAUSEN, near Swiss border; 47°39′N 08°35′E.

Jesup (JES-uhp), town (2000 population 9,279), ⊙ WAYNE county, SE GEORGIA, c.55 mi/89 km SW of SAVANNAH, near ALTAMAHA RIVER; 31°36′N 81°53′W. Trade and processing center for agriculture and timber area; manufacturing (apparel, furniture, lumber and pulp, machinery, consumer goods, food processing, plastics, printing and publishing); railroad responsible for town's growth. Incorporated 1870.

Jesup, town (2000 population 2,212), BUCHANAN county, E IOWA, 10 mi/16 km W of WATERLOO; 42°28′N 92°04′W.

Jesup, Fort, LOUISIANA: see MANY.

Jesús (hai-SOOS), town, Itapúa department, SE PARAGUAY, 22 mi/35 km NNE of Encarnación; 27°02′S 55°47′W. Maté center; viticulture. Former Jesuit mission, founded 1685. Just NE, at Tabarangué, is Jesuit church (1767).

Jesús (hai-SOOS), village, IBIZA, BALEARIC ISLANDS, 1 mi/1.6 km NNE of IBIZA. Cereals, fruit; livestock.

Jesús Carranza (hai-SOOS kah-RAHN-zah), town, SE VERACRUZ, MEXICO, 6 mi/9 km N of the CHIAPAS state border, 50 mi/80 km SW of MINATITLÁN, on railroad, and 4 mi/6 km E of Mexico Highway 185 (Trans-Isthmian Highway); 17°28′N 95°01′W. Agriculture (corn, beans, rice, sesame, chiles, fruits); petroleum resources, precious woods, construction lumber, and cattle. Formerly Santa Lucrecia.

Jesús de Cavinas (he-SOOS dai kah-VEE-nahs), canton, ITURRALDE province, LA PAZ department, W BOLIVIA; 12°36′S 67°04′W. Elevation 833 ft/254 m. Agriculture (potatoes, yucca, bananas, rye); cattle, sheep.

Jesús de Machaca (he-SOOS dai mah-CHAH-kah), town and canton, Ingava province, LA PAZ department, W BOLIVIA, in the ALTIPLANO, 11 mi/18 km SSE of GUAQUI; 16°45′S 68°50′W. Potatoes, barley; sheep.

Jesús de Otoro (he-SOOS de o-TO-ro), town (2001 population 10,703), INTIBUCÁ department, W central HONDURAS, in OTORO VALLEY, 20 mi/32 km NE of LA ESPERANZA, near Río Grande de Otoro; 14°29′N 87°59′W. Palm-hat manufacturing; livestock.

Jesus Island, CANADA: see LAVAL.

Jesús María (hai-SOOS mah-REE-ah), city and township, AGUASCALIENTES, N central MEXICO, on SAN PEDRO RIVER, and 11 mi/16 km NW of AGUASCALIENTES; 22°00′N 102°20′W. Cereals, fruit, vegetables, tobacco; livestock.

Jesús María (he-SOOS mah-REE-ah), town (1991 population including Colonia Caroya 31,485), ⊙ Colón department, N central CÓRDOBA province, ARGENTINA, 30 mi/48 km N of CÓRDOBA; elevation 1,749 ft/533 m. Tourism and farming center (vineyards, orchards, crops; livestock); ancillary industries. Formerly a Jesuit mission (founded 1618). Has 18th-century church and cloister, now Jesuit museum.

Jesús María (hai-SOOS mah-REE-ah), town, ⊙ Jesús María municipio, SANTANDER department, N central COLOMBIA, 90 mi/145 km SW of BUCARAMANGA; 05°52′N 73°47′W. Elevation 5,990 ft/1,825 m. Coffee, cacao, corn.

Jesús María (hai-SOOS mah-REE-ah), town, JALISCO, central MEXICO, 21 mi/32 km E of ATOTONILCO EL ALTO. Grain, beans, livestock.

Jesús María, town, ⊙ El Nayar municipio, in NW NAYARIT, MEXICO, 16 mi/25 km NW of Arteaga. In Cora Indian–occupied area. Formerly named Nayar.

Jesús Menéndez (hai-SOOS men-EN-des), sugar-mill town, LAS TUNAS province, E CUBA, near Chaparra Bay (N), 23 mi/37 km NW of HOLGUÍN; 21°10′N 76°28′W. Has Cuba's largest sugar central. Formerly called Chaparra.

Jet, village (2006 population 215), ALFALFA county, N OKLAHOMA, 25 mi/40 km NW of ENID; 36°40′N 98°10′W. In agricultural area (wheat; livestock, poultry). GREAT SALT PLAINS LAKE reservoir, including Great Salt Plains State Park and National Wildlife Refuge, to N.

Jetalsar (JAI-tahl-suhr), village, RAJKOT district, GUJARAT state, INDIA, 2 mi/3.2 km SW of JETPUR. Railroad junction. Formerly in old central SAURASHTRA state.

Jethou, United Kingdom: see CHANNEL ISLANDS.

Jeti-Ögüz (je-tee-uh-GUHZ), village, central ISSYK-KOL region, KYRGYZSTAN, on N slope of the TESKEY ALA-TOO (TERSKEI ALATAU) mountains, 10 mi/16 km SW of KARAKOL; 42°20′N 78°15′E. Health resort of Jeti-Ögüz Kurortu (7 mi/11.3 km S) nearby. Also spelled DZHETY-OGUZ.

Jetmore, town (2000 population 903), ⊙ HODGEMAN county, SW central KANSAS, on Buckner Creek of PAWNEE RIVER, and 24 mi/39 km NNE of DODGE CITY; 38°04′N 99°53′W. Grain; livestock. County museum here; state park nearby. Hodgeman State Fishing Lake to SE.

Jetpur (JAIT-puhr), town, RAJKOT district, GUJARAT state, INDIA, on BHADAR RIVER, and 40 mi/64 km SSW of RAJKOT. Agriculture market center (millet, oilseeds, wheat, cotton); handicraft cloth weaving. Was the capital of the former WEST KATHIAWAR state of Jetpur of WEST INDIA STATES agency; state merged 1948 with SAURASHTRA and later with Gujarat state.

Jette (ZHET), commune (2006 population 43,269), in ⊙ district of Brussels, BRABANT province, central BELGIUM, NW suburb of BRUSSELS; 50°52′N 04°20′E. Resort.

Jettingen (YET-ting-uhn), village, BADEN-WÜRTTEMBERG, SW GERMANY, 12 mi/19 km WNW of TÜBINGEN; 48°35′N 08°46′E.

Jettingen-Scheppach (YET-ting-uhn–SHEP-pahkh), village, BAVARIA, S GERMANY, in SWABIA, 20 mi/32 km W of AUGSBURG; 48°23′N 10°25′E.

Jeumont (zhu-mon), town (□ 3 sq mi/7.8 sq km), NORD department, N FRANCE, NORD-PAS-DE-CALAIS region, on Belgian border opposite ERQUELINNES, 6 mi/9.7 km ENE of MAUBEUGE, on the SAMBRE RIVER; 50°18′N 04°06′E. Railroad classification yards. Produces machine tools and electrical equipment; stone and marble quarries.

Jevany, CZECH REPUBLIC: see KOSTELEC NAD CERNYMI LESY.

Jevargi (JAI-vuhr-gee), town, tahsil headquarters, GULBARGA district, KARNATAKA state, S INDIA, 22 mi/35 km S of GULBARGA; 17°01′N 76°46′E. Millet, rice. Sheep raising nearby. Sometimes spelled Jaivergi.

Jever (YEH-ver), town, LOWER SAXONY, NW GERMANY, in EAST FRIESLAND, 10 mi/16 km WNW of WILHELMSHAVEN; 53°34′N 07°53′E. Main town of Friesland district; brewing, meat processing; horse and cattle markets. Has 17th-century town hall.

Jevicko (YE-veech-ko), Czech *Jevíčko*, German *Gewitsch*, town, VYCHODOCESKY province, W MORAVIA, CZECH REPUBLIC, on railroad, and 24 mi/39 km W of OLOMOUC; 49°38′N 16°44′E. Brewery (established 1896); health resort with sanatorium for patients suffering from tuberculosis. Has a 16th-century Renaissance tower.

Jewar (JAI-wuhr), town, BULANDSHAHR district, W UTTAR PRADESH state, INDIA, 20 mi/32 km WSW of KHURJA; 28°08′N 77°33′E. Hand-loom cotton weaving; wheat, oilseeds, barley, cotton, jowar, corn.

Jewe, ESTONIA: see JÕHVI.

Jewel Cave National Monument (□ 1 sq mi/2.6 sq km), limestone caves with chambers connected by narrow passages, CUSTER county, SW SOUTH DAKOTA, in the BLACK HILLS; 43°44′N 103°51′W. At 139 mi/222 km, Jewel Cave is the second longest cave in the world. Authorized 1908.

Jewell (JOO-uhl), county (□ 914 sq mi/2,376.4 sq km; 2006 population 3,324), N KANSAS; ⊙ MANKATO; 39°47′N 98°13′W. Rolling plain area, bordering N on NEBRASKA. Drained by REPUBLICAN RIVER in extreme NE and its tributaries; White Rock Creek in N. Wheat, corn, sorghum, alfalfa; hogs, sheep, cattle. Formed 1870.

Jewell, town, HAMILTON county, central IOWA, 19 mi/31 km N of AMES. In livestock and grain area; pork processing. Plotted 1880.

Jewell (JOO-uhl), village (2000 population 483), JEWELL county, N KANSAS, 9 mi/14.5 km SSE of MANKATO; 39°40′N 98°09′W. In grain and livestock area.

Jewett, village (2000 population 232), CUMBERLAND CO., SE central ILLINOIS, 5 mi/8 km S of TOLEDO; 39°12′N 88°14′W. In agricultural area.

Jewett, hamlet, GREENE county, SE NEW YORK, in CATSKILL MOUNTAINS, 23 mi/37 km E of CATSKILL; 42°15′N 74°13′W. Summer resort area.

Jewett (JYOO-uht), village (2006 population 788), HARRISON county, E OHIO, 6 mi/10 km N of CADIZ, on small Conotton Creek; 40°22′N 81°00′W. Sawmilling. Previously coal mining. Laid out 1851, incorporated 1886.

Jewett, village (2006 population 933), LEON county, E central TEXAS, c.60 mi/97 km ESE of WACO; 31°21′N 96°09′W. Railroad junction in agricultural area. Oil and gas; manufacturing (industrial glass, steel bars). LAKE LIMESTONE reservoir to W.

Jewett City, CONNECTICUT: see GRISWOLD.

Jewish Autonomous Oblast or **Birobidzhan**, autonomous region (□ 13,900 sq mi/36,140 sq km; 2005 population 210,000), KHABAROVSK TERRITORY, SE SIBERIA, RUSSIAN FAR EAST, in the basins of the Biro and BIDZHAN rivers (tributaries of the AMUR River); ☉ BIROBIDZHAN. Bounded S by northeasternmost section of CHINA (HEILONGJIANG province; Amur River is the border) and N by the BUREYA and Khingan mountains (gold, tin, iron ore, and graphite). Mostly swampy plains, with forested low hills in NE. Mining (tin at Khingansk), agriculture (chiefly carried on in the rich Amur plain), lumbering (along railroad), and light manufacturing are the major economic activities. Linked to the TRANS-SIBERIAN RAILROAD to the N. Formed in 1928 to give Soviet Jews a home territory, reduce emigration to British Mandate Palestine, and increase settlement along the sparsely populated and vulnerable borders of the Soviet Far East. The region was raised to the status of an autonomous region in 1934. The area never became a center for the Soviet Jewry, with fewer than 6,000 relocating there between 1928 and 1940, and today, non-Jewish Russians and Ukrainians heavily outnumber Jews (only 4.3% of the population, according to the 2001 census), while the harsh climate has discouraged settlement. Despite some Yiddish influences—including a Yiddish theater, newspaper, and radio station, almost all of them in the capital city of Birobidzhan—Jewish cultural activity has begun its slow revival only recently after decades of repressions that began with Stalin's anti-cosmopolitanism campaigns of the late 1940s, and almost two-thirds of the region's Jewish population emigrating abroad since the liberalization of the immigration process in the mid-1970s.

Jezerce (ye-ZERTS), peak (8,839 ft/2,694 m) in NORTH ALBANIAN ALPS, on ALBANIA-MONTENEGRO border, 3.7 mi/6 km W of VALBONË (Albania). One of the highest mountains in the Albanian Alps.

Jezerska Česma (YEZ-er-skah CHES-mah), peak (8,541 ft/2,603 m) in SAR Mountains, on SERBIA-MACEDONIA border, 10 mi/16 km NNE of TETOVO (MACEDONIA). Chromium mining. Also spelled Yezerska Chesma.

Jezerski Vrh, AUSTRIA and SLOVENIA: see SEEBERG PASS.

Jeziorak, Lake (ye-ZHO-rahk) or **Jezierzyce, Lake**, German *Geserich*, (□ 12 sq mi/31.2 sq km), N POLAND, N of IŁAWA; 18 mi/29 km long N-S, 1 mi/1.6 km–5 mi/8 km wide; drains SW into the VISTULA. In what was formerly EAST PRUSSIA (before 1945).

Jeziorany (ye-zho-RAH-nee), German *Seeburg*, town OLSZTYN province, NE POLAND, in Masurian Lakes region, 17 mi/27 km NNE of OLSZTYN (Allenstein). Grain and cattle market. In what was formerly EAST PRUSSIA (before 1945).

Jezireh, El, MESOPOTAMIA: see JAZIRA, AL.

Jezreel Valley (yeez-ri-EL) or **Plain of Jezreel**, Hebrew *'Emeq Yezreel* (□ c.200 sq mi/518 sq km), extending c.25 mi/40 km NW-SE between SE foot of Mount CARMEL and the JORDAN River valley near BEIT SHE'AN, SE of HAIFA, N ISRAEL. Separates LOWER GALILEE (N) from hills of SAMARIA (S). The plain is drained in the W by the KISHON River and in the E by the HAROD. AFULA is economic center of region. Once a swampy, malarial lowland, the Jezreel Valley has been drained and turned into one of Israel's most fertile and densely populated rural regions, with a high proportion of kibbutzim (collective farms) and moshavim (cooperative farms). Diverse crops are produced here in abundance. Since ancient times the plain has been a battleground, especially around MEGIDDO. Once known as Esdraelon (Greek=Jezreel). Sometimes spelled Yizre'el.

Jezupol, UKRAINE: see ZHOVTEN', IVANO-FRANKIVS'K oblast.

Jezzin (zhe-ZEEN), town, S LEBANON, 12 mi/19 km E of SAIDA; 33°32′N 35°34′E. Summer resort; cereals, oranges. Has 130-ft/40-m waterfall. The town has been under Israeli protection since the early 1980s. Although it is just N of the Israeli security zone in S Lebanon, the Israeli and S Lebanon armies extend their presence to Jezzin whenever necessary. Also spelled Jazzin.

JFK (John F. Kennedy) International Airport (□ 7.7 sq mi/19.9 sq km), S QUEENS borough of NEW YORK city, SE NEW YORK, on NE side of JAMAICA BAY; elevation 12.7 ft/3.9 m; 40°38′N 73°47W. One of three major airports serving metropolitan New York city (see also LAGUARDIA AIRPORT and NEWARK LIBERTY INTERNATIONAL AIRPORT), 15 mi/24 km E of MANHATTAN. Part of the PORT AUTHORITY OF NEW YORK AND NEW JERSEY's facilities and operations. Nine passenger terminals (over 50,000 passengers daily) and ten cargo facilities (averaging 1.6 million tons/1.4 million tonnes annually); two pair of parallel runways (four runways total); 30 mi/48 km of roadway; AirTrail light rail connection to city subways and Long Island Railroad opened December, 2003. Opened in 1948 as Idlewild Airport, much of it was built on filled marshland of Jamaica Bay. A major employer in the area, generating roughly 35,000 on-site and another 98,000 off-site jobs. Has largest U.S. Customs facility in nation. Together with the Port Authority's LaGuardia and Newark airports, it handles over 25% of the country's international air-cargo traffic. Currently undergoing a $10 billion redevelopment. Airport Code JFK.

Jhabrera, town, HARIDWAR district, N UTTAR PRADESH state, INDIA, 16 mi/26 km SE of SAHARANPUR. Wheat, rice, rape and mustard, gram, corn. Also spelled Jabarhera.

Jhabua (JAHB-oo-uh), district (□ 2,619 sq mi; 2001 population 1,394,000), SW MADHYA PRADESH state, INDIA; ☉ JHABUA. Over 80% of district is tribal (Bhils, Bhilalas) and extremely poor, with very low literacy rates. Soil erosion, poor quality soil, and disappearing forest cover plague most of district. Bounded on S by NARMADA RIVER. Handicrafts: bamboo products, beaded jewelry; industry: cotton ginning, dolomite, marble, copper, aluminum processing.

Jhabua (JAHB-oo-uh), town (2001 population 30,577), ☉ JHABUA district, MADHYA PRADESH state, INDIA, 40 mi/64 km SW of RATLAM; 22°46′N 74°36′E. Local market (corn, millet, timber); cotton ginning. Was capital of former princely state of Jhabua of CENTRAL INDIA AGENCY; a Rajput state, founded 16th century, since 1948 merged with MADHYA BHARAT and after Reorganisation of States in 1956, included within the new Madhya Pradesh state.

Jhagadiya, town, BHARUCH district, GUJARAT state, INDIA, 11 mi/18 km E of BHARUCH; 21°42′N 73°09′E. Railroad junction; trades in cotton, tobacco, millet, timber. Formerly in state of BOMBAY.

Jhajjar (JUHJ-juhr), town, ROHTAK district, HARYANA state, N INDIA, 20 mi/32 km SSE of ROHTAK; 28°37′N 76°39′E. Trade center for grain, salt, textiles; hand-loom weaving, embroidering, dyeing, pottery manufacturing; millet, gram, cotton.

Jhalakati (jah-law-kah-tee), town (2001 population 45,428), Bakarganj district, S EAST BENGAL, BANGLA-DESH, on BISHKHALI RIVER (distributary of Arial Khan River), and 12 mi/19 km WSW of BARISAL; 22°36′N 90°13′E. Trade center (rice, oilseeds, jute, sugarcane, sundari timber, betel nuts); rice and oilseed milling. Also spelled Jhalokati.

Jhalawan (jhah-lah-WAHN), region, comprising KHUZDAR district of E BALUCHISTAN province, SW PAKISTAN. Bordered E by KIRTHAR RANGE; hilly region, watered by HINGOL, PORALI, and HAB rivers.

Jhalawar (JAH-luh-wahr), former princely state in RAJPUTANA STATES, INDIA. Created 1838 from original KOTAH state. In 1948, merged with union of Rajasthan; now part of RAJASTHAN state.

Jhalawar (JAH-luh-wahr), district (□ 2,675 sq mi/6,955 sq km), SE RAJASTHAN state, NW INDIA; ☉ JHALAWAR.

Jhalawar (JAH-luh-wahr), town, SE RAJASTHAN state, NW INDIA, near KALI SINDH RIVER, 160 mi/257 km SSE of JAIPUR; ☉ JHALAWAR district; 24°36′N 76°09′E. Market center for cotton, millet, wheat; hand-loom weaving. Has college. Was capital of former Rajputana state of JHALAWAR. Formerly called Jhalrapatan Chhaoni. Town of JHALRAPATAN is 4 mi/6.4 km S.

Jhalida (JUHL-duh), town, Puruliya district, WEST BENGAL state, E INDIA, 25 mi/40 km W of Puruliya; 23°22′N 85°58′E. Manufacturing (fabricated metal products, chemicals, ordnance); rice, corn, oilseeds, bajra, sugarcane; lac growing. Also Jhalda.

Jhalod (JAH-lod), town, PANCH MAHALS district, GU-JARAT state, INDIA, 40 mi/64 km NE of GODHRA; 23°06′N 74°09′E. Market center for corn, rice, wheat, cotton cloth; pottery, lac bracelets. Flagstone quarried nearby.

Jhalokati, BANGLADESH: see JHALAKATI.

Jhalore, INDIA: see JALOR.

Jhalrapatan (jahl-ruh-PAH-tuhn), town, JHALAWAR district, SE RAJASTHAN state, NW INDIA, 4 mi/6.4 km S of JHALAWAR; 24°33′N 76°10′E. Trades in wheat, cotton, oilseeds, millet, livestock; hand-loom weaving, oilseed pressing, pottery making. Sometimes called Patan.

Jhalrapatan Chhaoni, INDIA: see JHALAWAR.

Jhalu (JAH-loo), town, BIJNOR district, N UTTAR PRADESH state, INDIA, 6 mi/9.7 km ESE of BIJNOR; 29°21′N 78°15′E. Rice, wheat, gram, sugarcane.

Jhang (jhah-ung), district (□ 3,415 sq mi/8,879 sq km), central PUNJAB province, central PAKISTAN; ☉ JHANG-MAGHIANA. Bounded S by RAVI River; includes parts of RECHNA DOAB (E), CHAJ DOAB (N), and SIND-SAGAR DOAB (W); drained by CHENAB and JHELUM rivers. Wheat, cotton, millet grown in large alluvial areas; textile manufacturing; cattle breeding. Chief towns are JHANG-MAGHIANA, CHINIOT. Formerly included present FAISALABAD district.

Jhang-Maghiana (jhah-ung-mug-ee-yah-nah), twin cities; ☉ JHANG district, PUNJAB province, central PAKISTAN, on the CHENAB River, 120 mi/193 km WSW of LAHORE; 31°16′N 72°19′E. Maghiana is above, in the highlands overlooking a valley; Jhang is below, separated by c.2 mi/3.2 km. Trades in grain, wool, cloth fabrics; cotton ginning, hand-loom weaving; light manufacturing. Two roads link the two cities, and a government college is halfway between them. Jhang has a government center that supplies blankets to the army and to hospitals Maghiana, where many refugee weavers from INDIA settled after Partition (1947), is an important wool-collection center. In the center of Jhang is the temple of Lal Nath, who founded the city in the late 17th century Formerly two towns, joined to form one municipality.

Jhansi (JHAN-see), district (□ 1,940 sq mi/5,044 sq km), S UTTAR PRADESH state, INDIA; ☉ JHANSI. Drained by BETWA RIVER; irrigated by Betwa Canal; foothills of VINDHYA RANGE in S. Agriculture (jowar, oilseeds, wheat, gram, barley, rice, corn). Main centers: Jhansi, MAU, BARWA SAGAR. Archaeological landmarks include remains near Mahonri. Formerly part of British BUNDELKHAND. Ruled by Chandel and Bundela Rajput dynasties until conquered (18th century) by Marathas. Last Maratha raja died without issue in 1853, and the state was escheated to GREAT

BRITAIN and made a district; his widow, Rani of Jhansi (Lakshmibai), actively collaborated with other native rulers and became an important military leader during Sepoy Rebellion of 1857. Original district enlarged 1950 by incorporated of several former petty states. After the Reorganisation of States in 1956, the district was reduced by loss of territory; e.g., LALITPUR is now a separate district which includes DEOGARH.

Jhansi (JHAN-see), city (2001 population 460,278), UTTAR PRADESH state, N central INDIA; ⊙ JHANSI district; 25°26′N 78°35′E. A road and railroad junction, agricultural market, and small industrial center, it has iron and steel mills and railroad workshops. Grew around a fort built in 1613 by the Rajputs and strengthened in 1742 by the Marathas. Queen Laxmibai of Jhansi fought the British unsuccessfully. Reverted to GREAT BRITAIN in 1853, when the ruling prince died without heirs. British residents in Jhansi were massacred during the Indian Mutiny (1857) at the still-existing 17th-century Mogul fort. Seat of a college and an agricultural research institute.

Jhapa (JAH-pah), district, SE NEPAL, in MECHI zone; ⊙ CHADRAGADHI.

Jharia (JAIR-ryuh) or **Jherria**, town (2001 population 81,979), DHANBAD district, E central Jharkhand state, E INDIA, near DAMODAR River, 28 mi/45 km N of Puruliya; 23°45′N 86°26′E. Mining center in Jharia coalfield, located in Damodar Valley, producing high-quality coal. Jharia and RANIGANJ fields produce most of India's coal.

Jharsuguda (JAHR-suh-guh-duh), district (□ 850 sq mi/2,210 sq km), ORISSA state, INDIA; ⊙ JHARSUGUDA.

Jharsuguda (JAHR-suh-guh-duh), town, NW ORISSA state, E INDIA, 27 mi/43 km N of SAMBALPUR, on tributary of MAHANADI River; JHARSUGUDA district; 21°51′N 84°02′E. Important railroad junction (workshops) on CALCUTTA-MUMBAI (Bombay) route. Trades in rice, timber, oilseeds, hides; biri manufacturing. Sometimes called Jharsugra. Light manufacturing at Brajrajnagar village. A major center of paper manufacturing has sprung up by the river.

Jhawarian (jhah-WAH-ree-uhn), town, SARGODHA district, central PUNJAB province, central PAKISTAN, 18 mi/29 km N of SARGODHA; 32°22′N 72°38′E. Sometimes spelled JAHWARIAN.

Jhelum (jhe-LUHM), district (□ 2,774 sq mi/7,212.4 sq km), N PUNJAB province, central PAKISTAN; ⊙ JHELUM. Bounded E and S by JHELUM RIVER; crossed S by E SALT RANGE. Plateau (N, center) and riverbank area produce wheat, millet, oilseeds. Rainfed as well as irrigated agriculture. Petroleum wells at BALKASSAR and JOYA MAIR; some coal and rock-salt deposits (works at KHEWRA and DANDOT) in SALT RANGE. Trade centers are JHELUM, PIND DADAN KHAN.

Jhelum (jhe-LUHM), town, PUNJAB province, NE PAKISTAN, 95 mi/153 km NNW of LAHORE, on the JHELUM RIVER; ⊙ of JHELUM district; 32°56′N 73°44′E. Located on the main railroad and road on Punjab piedmont plain. Important market for wheat, millet, and timber. Has sawmills; plywood, textile, cigarette, and glass industries; cement making; boatbuilding. Timber depot for logs floated down from N forests. An army-supply corps training center and two colleges. Also a Military College. The area's history dates back at least to the 3rd century B.C.E. Old Jhelum stood on the left bank of the river; boatmen crossed the river (c.1532) and founded the new town on the right bank.

Jhelum Canal, Lower (jhe-LUHM), runs 39 mi/63 km SW, irrigation channel in N PUNJAB province, central PAKISTAN; flows from left bank of JHELUM RIVER (headworks near RASUL, GUJRAT district). Here it divides into two main branches, with N branch flowing c.110 mi/177 km S, S branch flowing c.85 mi/137 km S; numerous distributaries. Irrigates extensive areas of GUJRAT, SHAHPUR, and JHANG districts in CHAJ DOAB. Opened 1901.

Jhelum Canal, Upper (jhe-LUHM), runs 85 mi/137 km S and SE, irrigation channel mainly in GUJRAT district, N PUNJAB province, central PAKISTAN; flows from left bank of JHELUM RIVER (headworks at MANGLA) to CHENAB River 5 mi/8 km W of WAZIRABAD. Used chiefly as feeder for Lower Chenab Canal. Opened 1915.

Jhelum River (JAI-luhm), 480 mi/772 km long, westernmost of the five rivers of PUNJAB, in N INDIA and NE PAKISTAN; rising in JAMMU AND KASHMIR state (India), in a spring near VERNAG, at the foot of Banchal Pan on N slope of Pir Pryal range; flows W through the VALE OF KASHMIR, S through foothills in AZAD KASHMIR (NE Pakistan), SW across Pakistani Punjab to the CHENAB River 124 mi/200 km W of LAHORE. The Lower Jhelum (opened 1901) and the Upper Jhelum (1915) canals irrigate extensive areas of Pakistani Punjab. The Mangla Dam and Reservoir (1960) has greatly improved irrigation along the river's lower course, as well as supplying extensive hydroelectric power on Lower Jhelum Canal (headwaters at Rasul). The Jhelum was crossed in 326 B.C.E. by Alexander the Great, who defeated the Indian king Porus. The river's ancient and Greek name was Hydaspes, known as Veth in Kashmir. Also spelled Jehlam.

Jhenaidah, BANGLADESH: see JHENIDA.

Jhenida (jee-nah-ee-dah), town (2001 population 86,919), JESSORE district, W EAST BENGAL, BANGLADESH, on distributary of the MADHUMATI River, and 26 mi/42 km N of JESSORE; 23°36′N 89°10′E. Trades in rice, jute, linseed, sugarcane, pepper; manufacturing (food). Also spelled Jhenaidah.

Jhind, INDIA: see JIND.

Jhinjhak (JIN-juhk), town, Kanpur district, S UTTAR PRADESH state, INDIA, on LOWER GANGA CANAL, and 38 mi/61 km WNW of KANPUR; 26°34′N 79°44′E. Gram, wheat, jowar, barley, mustard.

Jhinjhana (jin-JAH-nuh), town, MUZAFFARNAGAR district, N UTTAR PRADESH state, N INDIA, 29 mi/47 km W of MUZAFFARNAGAR; 29°31′N 77°13′E. Wheat, gram, sugarcane, oilseeds. Has tomb of Muslim saint (built 1495), 17th-century mosque.

Jhomo Lhari, BHUTAN: see JUMOLHARI.

Jhuldabhaj, INDIA: see BARAUNI.

Jhumra, PAKISTAN: see CHAK JHUMRA.

Jhunjhunun (juhn-JOO-nuhn), district (□ 2,289 sq mi/5,951.4 sq km), RAJASTHAN state, NW INDIA; ⊙ JHUNJHUNUN.

Jhusi (JOO-see), town, ALLAHABAD district, SE UTTAR PRADESH state, INDIA, at YAMUNA RIVER mouth, 4 mi/6.4 km E of ALLAHABAD city center; 28°08′N 75°24′E. Sugar processing. Gold coins dating from Gupta empire found here, also copperplate of a Pratihara Rajput king dated C.E. 1027. Area identified with *Pratisthan* or *Kesi* of the Puranic histories, when it was the capital of the first king of the Lunar dynasty.

Ji, CHINA: see HEBEI.

Ji, CHINA: see JILIN.

Jiading, CHINA: see LESHAN.

Jiahe (JIAH-HUH), town, ⊙ Jiahe county, S HUNAN province, S central CHINA, 50 mi/80 km WSW of CHEN XIAN; 25°33′N 112°15′E. Rice, tobacco.

Jiajiang (JIAH-JIANG), town, ⊙ Jiajiang county, S central SICHUAN province, SW CHINA, 15 mi/24 km NW of LESHAN; 29°45′N 103°35′E. Oilseeds, tobacco, jute. Site of "Thousand Buddha Cliff," a mountain cliff that has more than 2,700 Buddhas carved into it. The earliest carvings date from the Sui dynasty (589–618).

Jialing (JIAH-LING), river, c.450 mi/724 km long; rising in S GANSU province, central CHINA, and flowing S through SW SHAANXI and E SICHUAN provinces, past Guangyuan, Lang Zheng, and Nancheng to join the CHANG JIANG (Yangzi River) at CHONGQING; it receives the FU and QU rivers near Hechuan. One of the Chang Jiang's chief tributaries, the Jialing is navigable up through the Sichuan basin, an important

agricultural and industrial area. It is one of the four chief rivers in the Sichuan [=four rivers] basin that gives the province its name. The name sometimes appears as Chia-ling or Kialing.

Jiamusi (JIAH-MU-SI), city (□ 353 sq mi/913 sq km; 1994 estimated non-agrarian population 548,600; estimated total population 781,100), E HEILONGJIANG province, NE CHINA; 46°50′N 130°21′E. It is the chief port on the lower reaches of the SONGHUA RIVER; the city has coal, agricultural and mining equipment, lumber, chemicals, paper, plastics, textiles, and beet sugar-processing industries. There are railroad connections to HARBIN, NORTH KOREA, and RUSSIA. Nearby Sandaogang is site of huge state farm. Name sometimes appears as Chia-mu-ssu or Kiamusze.

Ji'an (JEE-AN), city (□ 197 sq mi/510 sq km; 1994 estimated non-agrarian population 163,800; estimated total population 317,400), central JIANGXI province, SE CHINA; 27°08′N 115°00′E. A major commercial port on the GAN RIVER and important road hub and market center. Light industry is the largest sector of the city's economy; agriculture (grains, hogs) and heavy industry. Manufacturing (food, electronic equipment). Known for its pagoda. Well-preserved residential complexes built during the Ming and Qing dynasties have been found in W suburb. Formerly called Luling. Sometimes spelled Chi'an or Kian.

Ji'an, city (□ 1,242 sq mi/3,217 sq km; 1994 estimated non-agrarian population 73,500; estimated total population 226,900), S JILIN province, NE CHINA; 41°06′N 126°10′E. Agriculture and heavy industry are the largest sectors of the city's economy. A major industry is pharmaceuticals.

Jianchuan (JIAN-CHUAN), town, ⊙ Jianchuan county, NW YUNNAN province, SW CHINA, 55 mi/89 km NNW of DALI; 26°28′N 99°52′E. Elevation 7,447 ft/2,270 m. Rice; timber; food, apparel; coal mining.

Jiande (JIAN-DUH), town, ⊙ Jiande county, NW central ZHEJIANG province, SE CHINA, 65 mi/105 km SW of HANGZHOU, and on Fuchun River; 29°29′N 119°16′E. Rice; chemicals, construction materials, food, hydroelectric power generation, iron smelting. Also known as Baisha.

Jiang'an (JIANG-AN), town, ⊙ Jiang'an county, S central SICHUAN province, SW CHINA, 23 mi/37 km WSW of LU XIAN, and on right bank of CHANG JIANG (Yangzi River); 28°44′N 105°04′E. Rice, jute; food, chemicals.

Jiangbei (JIANG-BAI), town, ⊙ Jiangbei county, SE central SICHUAN province, SW CHINA, on left bank of CHANG JIANG (Yangzi River), at mouth of JIALING RIVER, opposite CHONGQING; 29°43′N 106°37′E. Chemicals, pharmaceuticals, textiles; coal mining.

Jiangcheng (JIANG-CHENG), town, ⊙ Jiangcheng county, S YUNNAN province, SW CHINA, 55 mi/89 km SE of SIMAO, near LAOS border; 22°36′N 101°50′E. Elevation 4,150 ft/1,265 m. Rice; food and beverages; non-ferrous ore mining.

Jiangchuan (JIANG-CHUAN), town, ⊙ Jiangchuan county, SE central YUNNAN province, SW CHINA, 45 mi/72 km S of KUNMING, on W shore of LAKE FUXIAN, in mountain region; 24°19′N 102°47′E. Rice, tobacco; food, construction materials; coal mining.

Jiange (JIAN-UH), town, ⊙ Jiange county, N SICHUAN province, SW CHINA, 40 mi/64 km NW of LANGZHONG; 32°04′N 105°26′E. Food, textiles, apparel; grain, oilseeds, sugarcane; tobacco.

Jianghua (JIANG-HUAH), town, ⊙ Jianghua county, S HUNAN province, S central CHINA, near GUANGXI ZHUANG border, in NANLING mountains; 25°02′N 111°45′E. Rice; tobacco; logging.

Jiangjin (JIANG-JIN), town, ⊙ JIANGJIN county, S SICHUAN province, SW CHINA, 28 mi/45 km SW of CHONGQING, and on right bank of CHANG JIANG (Yangzi River); 29°17′N 106°15′E. Rice, tobacco, oilseeds, jute; textiles, food, construction materials, paper.

Jiangkou (JIANG-KO), town, ⊙ Jiangkou county, NE GUIZHOU province, S central CHINA, 35 mi/56 km SE of SINAN; 27°42′N 108°50′E. Rice, tobacco, oilseeds; food and beverages, furniture; logging.

Jiangle (JIANG-LUH), town, ⊙ Jiangle county, NW FUJIAN province, SE CHINA, 45 mi/72 km WNW of NANPING, and on tributary of MIN RIVER; 26°46′N 117°22′E. Rice; logging, coal mining.

Jiangling (JIANG-LING), town, ⊙ Jiangling county, S HUBEI province, central CHINA, 125 mi/201 km W of WUHAN, and on left bank of CHANG JIANG (Yangzi River); 30°21′N 112°11′E. Rice, sugarcane, oilseeds, cotton. An ancient walled city, one of the oldest of China, dating from c.1000 B.C.E. Jiangling flourished until the rise of the nearby treaty port of Shashi after 1876. Jiangling was called Jiangzhou.

Jiangmen (JIANG-MEN), city (□ 50 sq mi/130 sq km; 1994 estimated non-agrarian population 292,600; estimated total population 389,200; 2000 population 284,935), SW GUANGDONG province, S CHINA, 50 mi/80 km SW of GUANGZHOU, on the PEARL RIVER DELTA; 22°40′N 113°05′E. An industrial and commercial center; crop growing, animal husbandry, fishing. Grain, oil crops, vegetables, fruits, hogs, poultry, eggs; manufacturing (food, textiles, apparel, leather, paper, chemicals, plastics, machinery, electronic equipment).

Jiangnan (JIANG-NAN), former province of E CHINA during the Ming dynasty (1368–1644), on lower CHANG JIANG (Yangzi River) and HUAI RIVER. Capital was NANJING (Yingtianfu). Separated in 1667, in early Qing dynasty (1644–1911), into JIANGSU and ANHUI provinces. Also known as Nanjing.

Jiangning (JIANG-NING), town, ⊙ Jiangning county, S JIANGSU province, E CHINA, 7 mi/11.3 km SSE of NANJING, and on QINHAI RIVER; 31°55′N 118°50′E. Building materials; metal smelting; rice, wheat, beans, kaoliang, corn; metal-ore mining.

Jiangpu (JIANG-PU), town, ⊙ Jiangpu county, S JIANGSU province, E CHINA, 8 mi/12.9 km W of NANJING, across CHANG JIANG (Yangzi River); 32°03′N 118°37′E. Rice.

Jiangshan (JIANG-SHAN), city (□ 779 sq mi/2,018 sq km; 1994 estimated non-agrarian population 58,100; 1994 estimated total population 552,000; 2000 total population 519,929), SW ZHEJIANG province, SE CHINA, near JIANGXI border, on railroad, and 20 mi/32 km SW of QUZHOU; 28°44′N 118°38′E. Agriculture and heavy industry are the largest sectors of the city's economy. Rice, oilseeds; manufacturing (chemicals, electronic equipment). Also known as Qu Xian.

Jiangsu (JIANG-SU) or short name **Su** (SU), province (c.41,000 sq mi/106,190 sq km; 1994 estimated population 68,310,000; 2000 population 73,043,577), E CHINA, on the YELLOW SEA; ⊙ NANJING; 33°00′N 120°00′E. Jiangsu consists largely of the alluvial plain of the CHANG JIANG (Yangzi River) and includes much of its delta; in elevation it rarely rises above sea level, although there are hills in the SW. The fairly warm climate, moderate rainfall, and fertile soil make Jiangsu one of the richest agricultural regions of China and one of the most densely populated. The HUAI RIVER divides the province into Subei (or N Jiangsu), and Sunan (or S Jiangsu). The N Jiangsu belongs to the HUANG HE (Yellow River) and the Huai River basins, and S Jiangsu is in the Chang Jiang (Yangzi River) basin.

The province straddles two agricultural zones, with wheat, millet, kaoliang, corn, soybeans, and peanuts cultivated in the N and rice, tea, sugarcane, and barley raised in the S. Cotton is grown along the coast (N and S) in the saline soil, which is not suited for other crops. Tea is planted in the W hills, and some experimenting with oak trees for silk culture has been initiated. Intensive land reclamation has been accomplished, with extensive dikes and the use of the raised-field system. Fish are abundant in the many lakes (of which Tai is the most famous), in the streams

and canals, and off the Yangzi delta; Jiangsu, which is known to the Chinese as "the land of rice and fish," is rich in marine products. Industrial development has been helped greatly by productive agriculture which generates investment in rural and town industries. The enterprises account for half of the total industrial output of the province. It is also a major salt-producing area.

Jiangsu is bisected by the Yangzi, which can be navigated by steamers up to 15,000 tons, and by a portion of the GRAND CANAL. Its first-class roads and extensive railroad system, including the busiest railroad in China, along the SHANGHAI-NANJING border, make for excellent communications. As one of the most prosperous provinces in China, Jiangsu is deficient only in timber and minerals. A major part of China's foreign trade clears through the port of Shanghai into Jiangsu. Shanghai, one of the world's great seaports and the chief manufacturing center of China, is in Jiangsu province but is administered directly by the central government. Nanjing has been developed into an industrial center, producing petrochemicals, motor vehicles, machinery, and construction materials. SUZHOU, WUXI, and ZHENJIANG are known for their silk. Textile, food, cement, and fertilizer industries are found throughout the province.

Jiangsu is among the first provinces where the township enterprises developed. In many developed rural counties, township enterprises have become a most important economic component. In S Jiangsu, many developed rural counties have had more than 70% of the total labor forces working in manufacturing and commercial businesses. Jiangsu was originally part of the Wu kingdom, and the name Wu is still its traditional name. Jiangsu received its present name, derived from Jiangning (Nanjing) and Suzhou (Soochow), in 1667, when it was formed from the old JIANGNAN province. The gateway to central China, Jiangsu became the main scene of European commercial activity after the Treaty of Nanjing (1842). The capture of Jiangsu in 1937 was an important phase of Japan's effort to conquer all of China. Liberated by the Chinese Nationalists in 1945, Jiangsu fell to the Communists in 1949. For a time Jiangsu was administered as two regional units, North and South Jiangsu, but the province was reunited in 1952. Many archaeological sites have been excavated in Jiangsu since 1956. In 1984 the province was made a part of the Shanghai special economic zone, which has increased investment in, and exports from, port cities like NANTONG. The name sometimes spelled Kiangsu.

Jiangxi (JIANG-SEE) or short name **Gan** (GAN), province (c.66,000 sq mi/170,940 sq km; 1994 estimated population 38,930,000; 2000 population 40,397,598), SE CHINA; ⊙ NANCHANG; 28°00′N 116°00′E. The largely hilly and mountainous surface is drained by many rivers; the longest of which is the navigable GAN, which flows NE to Poyang Lake. Agriculture flourishes in Jiangxi's fertile soil and mild climate; the growing season is nine to eleven months long, and more than one-third of the area is cultivated. Jiangxi is one of China's leading rice producers; other food crops include wheat, sweet potatoes, barley, and corn. Commercial crops are cotton, oil-bearing plants (rapeseed, sesame, soybeans, and peanuts), ramie, tea, sugarcane, tobacco, and oranges. Ten percent of the province is forested, and a lumbering industry has developed. Tung and mulberry trees are grown; a large, integrated silk complex is at Nanchang. Livestock raising and fish culture are important. Jiangxi is an important source of tungsten; it also has high-grade coking coal (near PINGXIANG) and kaolin, which supplies the ancient porcelain industry of Jingdezhen. Manganese, tin, lead, zinc, iron, and antimony are also found. The province has a variety of heavy and light industries that produce petrochemicals, textiles, paper, machinery, transportation equip-

ment, food, printing and publishing. Cities, such as Nanchang, JIUJIANG, GANZHOU, and FUZHOU, are generally situated along the Gan River or on the province's two main railroads. The population in the N consists mainly of Chinese who speak the Gan (Jiangxi) variety of Mandarin, while in the S, adjoining GUANGDONG province, there is a large Kejia (Hakka) minority.

Jiangxi, linked with Guangdong by the MEILING PASS, has been China's main N-S corridor for migration and communication for centuries. Traditionally known as Gan, Jiangxi was ruled by the Zhou dynasty (722–481 B.C.E.); it received its present name only under the Southern Song dynasty (C.E. 1127–1280). The province, whose present boundaries date from the Ming dynasty (1368–1644), passed under Manchu rule in 1650. The Chinese Communist movement began (1927) in Jiangxi; the province was a stronghold for the Communists until they were dislodged in 1934. The famous Long March began from Jiangxi. Following World War II, during which Jiangxi was largely free of Japanese forces, the province passed (1949) to the Communists. The name sometimes appears as Chiang-hsi or Kiangsi.

Jiang Xian (JIANG SIAN), town, ⊙ Jiang Xian county, S SHAANXI province, NW central CHINA, 45 mi/72 km S of LINFEN; 35°29′N 111°33′E. Grain; textiles, chemicals; non-ferrous-ore mining.

Jiangyin (JIANG-YIN), city (□ 378 sq mi/979 sq km; 1994 estimated urban population 244,700; estimated total population 1,133,600; 2000 total population 1,108,406), S JIANGSU province, E CHINA, 50 mi/80 km ESE of ZHENJIANG, and on CHANG JIANG (Yangzi River); 31°54′N 120°16′E. In rice-growing region. Manufacturing center with both light and heavy industries; manufacturing of food, textiles, chemicals, rubber, plastics, fabricated metal products, machinery, transportation equipment, and electronic equipment.

Jiangyong (JIANG-RUNG), town, ⊙ Jiangyong county, S HUNAN province, S central CHINA, near GUANGXI ZHUANG border, 65 mi/105 km SSW of YONGZHOU; 27°20′N 111°20′E. Food processing; rice, wheat, corn, sugarcane, oilseeds, tobacco; logging, tin mining.

Jiangyou (JIANG-YO), city (□ 1,050 sq mi/2,720 sq km; 1994 estimated urban population 198,300; estimated total population 843,300; 2000 total population 825,521), N central SICHUAN province, SW CHINA, on the FU RIVER, and 55 mi/89 km NNW of SANTAI, in mountain region; 31°46′N 104°43′E. Heavy industry and agriculture are the largest sources of income. Main industries include oil, food processing, nonmetallic materials, iron and steel, and fabricated metal products. Main agricultural products are rice, medicinal herbs, oilseeds, tobacco, and jute.

Jianhe (JIAN-HUH), town, ⊙ Jianhe county, E GUIZHOU province, S central CHINA, 35 mi/56 km SSE of ZHENYUAN, on upper YUAN RIVER; 26°39′N 108°35′E. Rice; logging.

Jianjun, CHINA: see YONGSHOU.

Jianli (JIAN-LEE), town, ⊙ Jianli county, S HUBEI province, central CHINA, 55 mi/89 km SE of JIANGLING, and on CHANG JIANG (Yangzi River); 29°49′N 112°53′E. Rice, cotton; textiles, apparel, food, paper, construction materials.

Jianning (JIAN-NING), town, ⊙ Jianning county, W FUJIAN province, SE CHINA, near JIANGXI border, 80 mi/129 km WNW of NANPING, and on tributary of MIN RIVER; 26°48′N 116°50′E. Rice; logging, papermaking.

Jian'ou (JIAN-O), town, ⊙ Jian'ou county, N FUJIAN province, SE CHINA, 30 mi/48 km NNE of NANPING, and on JIAN RIVER (tributary of MIN RIVER); 27°03′N 118°19′E. Commercial center in major rice-growing region; logging.

Jianping (JIAN-PING), town, W LIAONING province, NE CHINA, 45 mi/72 km SE of CHIFENG; 41°27′N 119°37′E. Grain, sugar beets, oilseeds.

Jianping, town, ⊙ Jianping county, W LIAONING province, NE CHINA, 95 mi/153 km ENE of CHENGDE; 41°24′N 119°38′E. Railroad junction for CHIFENG. Grain, oilseeds, sugar beets. Also known as Yebaishou.

Jianpur (JEE-uhn-puhr), village, AZAMGARH district, E UTTAR PRADESH state, INDIA, 11 mi/18 km NE of AZAMGARH. Rice, barley, wheat, sugarcane. Also called Sagri and Jainpur.

Jianqiao (JIAN-CHIOU), town, N ZHEJIANG province, SE CHINA, on railroad to SHANGHAI, 5 mi/8 km NE of HANGZHOU, near estuary of Fuchun River. Rice, jute, oilseeds.

Jian River (JIAN), NW headstream (flows c.50 mi/80 km S) of MIN RIVER, in FUJIAN province, SE CHINA, formed by three headstreams (Chongyang from NW, Nanpu from N, and Song from NE) in vicinity of Jian'ou, to Min River at Nanping. Navigable for small vessels.

Jianshi (JIAN-SHI), town, ⊙ Jianshi county, SW HUBEI province, central CHINA, near SICHUAN border, 30 mi/48 km NNE of ENSHI; 30°37′N 109°38′E. Rice, tobacco, oilseeds; chemicals, food; coal mining.

Jianshui (JIAN-SHUAI), town, ⊙ Jianshui county, SE central YUNNAN province, SW CHINA, on railroad, and 100 mi/161 km S of KUNMING; 23°37′N 102°49′E. Elevation 4,836 ft/1,474 m. Rice, sugarcane, oilseeds; sugar refining.

Jianyang (JIAN-YANG), town, ⊙ Jianyang county, N FUJIAN province, SE CHINA, 50 mi/80 km N of NANPING, and on tributary of MIN RIVER; 27°20′N 118°07′E. Rice; logging; food, chemicals, pharmaceuticals, paper.

Jianyang, town, ⊙ Jianyang county, central SICHUAN province, SW CHINA, on railroad, 35 mi/56 km SE of CHENGDU, and on TUO RIVER (head of navigation); 30°24′N 104°33′E. Rice, sugarcane, oilseeds, cotton; chemicals, plastics, food, construction materials.

Jiaocheng (JIOU-CHENG), town, ⊙ Jiaocheng county, central SHANXI province, NE CHINA, 30 mi/48 km SW of TAIYUAN; 37°33′N 112°09′E. Grain, cotton; textiles, apparel, chemicals, construction materials, furniture; coal mining.

Jiaohe (JIOU-HUH), city (□ 2,336 sq mi/6,050 sq km; 1994 estimated urban population 169,200; estimated total population 463,000; 2000 total population 479,135), central JILIN province, NE CHINA, 40 mi/64 km ESE of JILIN, and on railroad; 43°36′N 127°23′E. Agriculture and heavy industry are the main sources of income. Grain, tobacco; food; logging, coal mining.

Jiaoling (JIOU-LING), town, ⊙ Jiaoling county, NE GUANGDONG province, S CHINA, on branch of MEI RIVER, and 24 mi/39 km N of Meixian; 24°40′N 116°10′E. Rice, sugarcane, oranges; logging; coal mining; construction materials, chemicals.

Jiaonan (JIOU-NAN), city (□ 744 sq mi/1,927 sq km; 1994 estimated urban population 166,900; 1994 estimated total population 850,100), E SHANDONG province, NE CHINA, 30 mi/48 km SW of QINGDAO across the JIAOZHOU BAY, on the YELLOW SEA; 35°55′N 119°58′E. Agriculture and light industry are the largest sectors of the city's economy; heavy industry is also important. Manufacturing includes food, textiles, chemicals, and machinery. Formerly called Wanggezhuang.

Jiaozhou (JIOU-JO), former German territory (□ c.200 sq mi/518 sq km), along the S coast of SHANDONG province, NE CHINA. Its administrative center was the city of QINGDAO. Germany leased Jiaozhou in 1898 for ninety-nine years, but Japan seized it in 1914. Jiaozhou was returned to China through agreements reached at the Washington Conference in 1922.

Jiaozhou (JIOU-JO), city (□ 467 sq mi/1,214.2 sq km), E SHANDONG province, NE CHINA, near JIAOZHOU BAY of the YELLOW SEA, 25 mi/40 km NW of QINGDAO, and on Qingdao-Jinan railroad; 36°20′N 120°00′E. Agriculture and light industry are the largest sectors of the city's economy; heavy industry is also important.

Main industries include food, textiles, apparel, chemicals, and machinery. Formerly called Jiao Xian.

Jiaozhou Bay (JIOU-JO), Mandarin *Jiaozhou Wan*, well-sheltered inlet (□ 200 sq mi/520 sq km; 2000 population 707,536) of the YELLOW SEA, in SHANDONG province, NE CHINA, 7 mi/11.3 km SE of JIAOZHOU; 20 mi/32 km long; 15 mi/24 km wide. QINGDAO (deepwater harbor) is at entrance. Occupied 1898–1914 by GERMANY as part of JIAOZHOU lease, and 1914–1922 by JAPAN.

Jiaozuo (JIOU-ZU-uh), city (□ 143 sq mi/370 sq km; 1994 estimated urban population 476,800; estimated total population 662,300; 2000 total population 1,578,461), NW HENAN province, NE CHINA; 35°14′N 113°13′E. A railroad hub; a center for heavy industry. Crop growing, commercial agriculture, animal husbandry, and forestry. Grains, oil crops, cotton, vegetables, fruits; hogs, eggs, beef, lamb. Manufacturing (textiles, chemicals, rubber, iron and steel, non-ferrous metals, machinery, electrical equipment); coal mining. Utilities.

Jiashan (JIAH-SHAN), town, ⊙ Jiashan county, NE ANHUI province, E CHINA, 65 mi/105 km NW of NANJING, and on Tianjin-Pukou railroad; 32°45′N 117°59′E. Rice, oilseeds; beverages.

Jiashan, town, ⊙ Jiashan county, NE ZHEJIANG province, SE CHINA, near JIANGSU border, 60 mi/97 km NE of HANGZHOU, and on railroad to SHANGHAI; 30°51′N 120°54′E. Rice, sugarcane, oilseeds; food processing, textiles and clothing; chemicals, plastics, electronics; engineering.

Jiashi (JIAH-SHI), town and oasis, W XINJIANG UYGUR AUTONOMOUS REGION, NW CHINA, 35 mi/56 km E of KASHI, on highway S of the TIANSHAN; 39°29′N 76°39′E. Grain, livestock; textiles, food processing. Sometimes called Payzawat.

Jiawang (JIAH-WANG), iron-mining settlement, NW JIANGSU province, E CHINA, 20 mi/32 km NE of XUZHOU, and on spur of Tianjin-Pukou railroad; 34°30′N 117°26′E.

Jia Xian (JIAH SIAN), town, ⊙ Jia Xian county, NW central HENAN province, NE CHINA, 45 mi/72 km SSW of ZHENGZHOU; 33°58′N 113°13′E. Grain, tobacco, oilseeds; coal mining.

Jia Xian, town, ⊙ Jia Xian county, N SHAANXI province, NW central CHINA, 40 mi/64 km ESE of YULIN, and on HUANG HE (Yellow River); 38°02′N 110°29′E. Grain, livestock.

Jiaxing (JIAH-SING), city (□ 374 sq mi/969 sq km; 1994 estimated urban population 227,600; estimated total population 760,500; 2000 total population 1,358,733), N ZHEJIANG province, SE CHINA, at the junction of the GRAND CANAL, the HUANGPU RIVER, and the Hangzhou-Shanghai railroad; 30°51′N 120°52′E. Center for light industry and agriculture. Fishing. Grain, oil crops, vegetables, hogs, eggs, poultry. Food processing. Manufacturing (textiles, apparel, paper, chemicals, iron and steel, machinery, transportation equipment, electrical equipment). The South Lake, or Nanhu, is a scenic point of interest. On Huxin Island is the Fog and Rain Pagoda, which is featured in an ancient Chinese poem. The city is known as one of the birthplaces of the Chinese Communist Party. Sometimes spelled Kashing or Chia-hsing.

Jiayin (JIAH-YIN), town, ⊙ Jiayin county, N HEILONGJIANG province, NE CHINA, 130 mi/209 km N of JIAMUSI, and on right bank of AMUR RIVER (Russian border); 48°52′N 130°25′E. Timber processing; grain, livestock; non-ferrous ore mining.

Jiayu (JIAH-YOOI), town, ⊙ Jiayu county, SE HUBEI province, central CHINA, 45 mi/72 km SW of WUHAN, and on CHANG JIANG (Yangzi River); 29°59′N 113°54′E. Grain, jute, cotton, fisheries.

Jiayuguan (JIAH-YOOI-GUAN), city (□ 1,133 sq mi/2,935 sq km; 1994 estimated urban population 96,800; estimated total population 119,700; 2000 total population 109,987), NW GANSU province, NW CHINA, on

SILK ROAD, on highway and railroad that connects Gansu and XINJIANG UYGUR, 13 mi/21 km WNW of JIUQUAN, across BEIDA RIVER, and at W end of the GREAT WALL; 39°47′N 98°14′E. As one of three major military forts on the Silk Road (the other two were Juyangguan and Yumenguan) in ancient times, Jiayuguan had a strategic location and a tall wall. Among all the walls, gates, and forts along the Great Wall, those in Jiayuguan are the best preserved. The Jiayuguan fort has three layers of walls and thirteen turrets. The area has a desert climate. Sandy land and soil erosion are the major threats to agriculture and ecology. The Northern Protection project, which stabilizes the soil and environment through reforestation, passes through the region. Jiayuguan is an important iron- and steel-industrial center, which has earned it the nickname "Steel City of the Northwest."

Jib, Arab village, Jerusalem district, 4.7 mi/7.5 km NW of JERUSALEM, in the JUDAEAN HIGHLANDS, WEST BANK. Agriculture (cereals, vegetables, grapes, fruit). Believed to be on the site of the biblical Giveon. Some archaeological remains from the Canaanite and the Israelite periods.

Jibacoa del Norte (hee-bah-KO-ah del NOR-tai), town, LA HABANA province, W CUBA, on railroad, and 30 mi/48 km E of HAVANA; 23°09′N 81°53′W. In sugar-growing region.

Jibhalanta, MONGOLIA: see ULIASTAY.

Jibiya (JI-bi-yah), town, KATSINA state, N NIGERIA, on upper KEBBI RIVER, 28 mi/45 km WNW of KATSINA, on NIGER border; 13°06′N 07°14′E. Peanuts, cotton; cattle, skins. Also spelled Jibia.

Jibla (JIB-luh), township, S YEMEN, 3 mi/4.8 km SW of IBB; 13°55′N 44°09′E.

Jibliya, KURIA MURIA ISLANDS: see QIBLIYA.

Jiboa River (hee-BO-ah), c.45 mi/72 km long, S EL SALVADOR; rises near SAN RAFAEL; flows SSW, through LA PAZ department, to the PACIFIC 17 mi/27 km ESE of LA LIBERTAD. Receives waters of LAKE ILOPANGO (right).

Jibondo Island (jee-BON-doh), PWANI region, E TANZANIA, in INDIAN OCEAN, 5 mi/8 km S of MAFIA ISLAND; JUANI ISLAND to NE; 2 mi/3.2 km long, 1 mi/1.6 km wide; 08°03′S 39°43′E.

Jibou (zhee-BO-oo), Hungarian *Zsibó*, town, SĂLAJ county, NW ROMANIA, near SOMEŞ RIVER, 10 mi/16 km NE of ZALĂU; 47°16′N 23°15′E. Railroad junction; iron ore and coal mining. Manufacturing (furniture, linen, and alcohol). In HUNGARY, 1940–1945.

Jicalapa (hee-kah-LAH-pah), municipality and town, EL SALVADOR, ESE of NUEVA SAN SALVADOR near the coast.

Jicaral, El, NICARAGUA: see EL JICARAL.

Jicarilla Mountains (hi-kuh-REE-uh), S central NEW MEXICO, N range of SACRAMENTO MOUNTAINS, in LINCOLN county, NE of CARRIZOZO; lies in part of Lincoln National Forest. Highest peak is CARRIZO PEAK (9,650 ft/2,941 m). Gold is mined.

Jícaro, GUATEMALA: see EL JÍCARO.

Jícaro, NICARAGUA: see CIUDAD SANDINO.

Jícaro, El, NICARAGUA: see CIUDAD SANDINO.

Jícaro Galán (hee-KAH-ro gah-LAHN), village, VALLE department, S HONDURAS, 3 mi/4.8 km ESE of NACAOME; 13°32′N 87°26′W. Major road junction, on INTER-AMERICAN HIGHWAY and road to TEGUCIGALPA.

Jícaro River, El (HEE-kah-ro), c.50 mi/80 km, NW NICARAGUA; rises in CORDILLERA DE DIPILTO Y JALAPA near JALAPA, flows S and SE past CIUDAD SANDINO and San Albino gold mines (hydroelectric station), to COCO RIVER below QUILALÍ.

Jicatuyo River (hee-kah-TOO-yo), c.100 mi/161 km long, W HONDURAS; rises as Alash River in continental divide, near EL SALVADOR border and S of San Marcos (Ocotepeque department); flows N, past CUCUYAGUA (here called Higuito River), and generally E, to ULÚA

RIVER 3 mi/4.8 km N of SANTA BÁRBARA; 14°59′N 88°16′W.

Jichu Drake, Bhutan: see JIWUCHUDRAKEY.

Jicin (YI-cheen), Czech *Jičín*, German *Gitschin*, town, VYCHODOCESKY province, N BOHEMIA, CZECH REPUBLIC, on Cidlina River, 26 mi/42 km SSE of LIBEREC; 50°26′N 15°22′E. Railroad junction. Manufacturing (machinery, motors, pianos, textiles, furniture). Has a 17th-century castle built by Wallenstein; remains of medieval fortifications (notably Walditz Gate). Emperor Francis of Austria held conference here in 1813 with Prussian and Russian representatives, before battle of Leipzig. Picturesque formations of sandstone rocks (PRACHOV ROCKS), part of so-called BOHEMIAN PARADISE, are 3 mi/4.8 km–6 mi/9.7 km NW.

Jico, MEXICO: see XICO.

Jidanyile (JEE-dahn-yee-lai), town, N REGION, GHANA, 20 mi/32 km SW of TAMALE; 09°05′N 00°33′W. Livestock; groundnuts, shea-nut butter.

Jidda (JED-duh), Arabic *Jiddah*, city (2004 population 2,801,481), SW HEJAZ, MAKKA province, W SAUDI ARABIA, on the RED SEA. Jidda is the port of Makka (c.45 mi/72 km E) and annually receives many pilgrims from all over the world, especially from central ASIA and the former Soviet Islamic states since the fall of communism. Unlike Makka, Jidda has always accepted visitors of all religions. The diverse local population includes a large mixture of Africans, Persians, Yemenis, and Indians. There are few exports, but many goods are imported to support the pilgrims. It is, to a great extent, the main economic center of the kingdom. Several government ministries are in the city. Jidda was ruled by the Turks until 1916, when it became part of the independent Hejaz. In 1925 it was conquered by Ibn Saud. Oil wealth brought expansion of the city and its seaport; the city walls came down in 1947, and a desalinization plant was built in the 1970s. The city has many industries (mainly of consumer goods), plus workshops and service stations of many foreign companies. Modern Jidda is not more than three centuries old, but Old Jidda, c.12 mi/19 km S of the city, was founded c.646 by the caliph Uthman. The city has a university and several technical and professional high schools. An international airport. Also called Jedda or Judda.

Jidda, small island of BAHRAIN, in PERSIAN GULF, 3 mi/5 km NW of Bahrain Island; 26°11′N 50°24′E. Known for palm trees, fruit, flowers, and birds. Prison for long-term prisoners. Sometimes spelled Jiddah.

Jiddat Al Harasis, wildlife reserve, AL WUSTA governate, central OMAN. Oryx protection area; seven oryx were imported from the U.S., and by the end of the 1990s the reserve plans to have 350.

Jieshi Bay (JIE-SHI), inlet of SOUTH CHINA SEA, in GUANGDONG province, S CHINA, 65 mi/105 km SW of SHANTOU; 15 mi/24 km wide, 10 mi/16 km long. Town and port of Jieshi on SE shore.

Jieshou (JIE-SHO), city (□ 257 sq mi/666 sq km; 1994 estimated urban population 80,000; estimated total population 690,400; 2000 total population 642,474), NW ANHUI province, E CHINA, on YING RIVER, and on HENAN province border. Crop growing, forestry. Agriculture (vegetables, hogs) is main source of income; manufacturing (plastics). Food and beverages.

Jieshou (JIE-SHO), town, W central JIANGSU province, E CHINA, on E shore of GAOYOU LAKE, 15 mi/24 km NNE of GAOYOU, and on GRAND CANAL; 33°15′N 115°21′E. Commercial center.

Jiexiu (JIE-SÜ), town, ⊙ Jiexiu county, S central SHANXI province, NE CHINA, on FEN RIVER, on railroad, and 20 mi/32 km SE of FENYANG; 37°02′N 111°55′E. Coal-mining center; grain, cotton, oilseeds; cotton textiles.

Jieyang (JIE-YANG), city (□ 70 sq mi/182 sq km), E GUANGDONG province, S CHINA; riverport on coastal stream, and 20 mi/32 km WSW of CHAOZHOU (Chao'an); 23°33′N 116°20′E. Rice, food process-

ing; engineering, chemicals, plastics, textiles, electronics.

Jifarah (ji-FAH-rah), arid, sandy plain in TRIPOLITANIA region, NW LIBYA, between fertile coastal strip (N) and plateau JABAL NAFUSAH (S). Arid torrid climate; annual rainfall 2 in/5.1 cm (W)–16 in/41 cm (E). Chief center AZIZIYAH. Predominantly pastoral (goats, sheep); agricultural settlements in E. Crossed (N-S) by roads. Sometimes Jefara. Formerly Gefara.

Jifarah, Al (el jee-FAH-rah), coastal plain, MADINIYINA province, SE TUNISIA. W extension of the Libyan AL JIFARAH to QABIS. Separated from Saharan TUNISIA by JABAL DAHAR mountain range.

Jifna, Arab village, Ramallah district, 3.8 mi/6 km N of RAMALLAH, in the Samarian Highlands, WEST BANK; 31°58′N 35°13′E. Agriculture (olives, fruit). Pop. largely Christian. Believed to be the biblical settlement of Gufna, an important urban center in the Roman period. Remains of several ancient churches; Byzantine and Crusader's fortress.

Jiga, town (2007 population 13,600), AMHARA state, NW ETHIOPIA; 10°40′N 37°24′E. Roadside service town 32 mi/51 km NW of DEBRE MARKOS.

Jigat, INDIA: see DWARKA.

Jigawa (ji-GAH-wah), state (□ 8,939 sq mi/23,241.4 sq km; 2006 population 4,348,649), N NIGERIA; ⊙ Dutse; 12°00′N 09°45′E. Bordered N by NIGER, NE by YOBE state, S and E by BAUCHI state, W by KANO state, and NW by KATSINA state. Tropical climate with savanna vegetation. Drained by Hadejia and Jamaari rivers; Kazaure rock ranges here. Agriculture includes groundnuts, cotton, wheat, maize, millet, and guinea corn. Traditional arts and crafts. Tin and columbite. Dutse, GUMEL, and Maigatari are the main centers, as well as the resorts of HADEJIA and KAZAURE.

Jigdalik, AFGHANISTAN: see JAGDALAK.

Jigjiga, ETHIOPIA: see JIJIGA.

Jigni (JIG-nee), village, HAMIRPUR district, S UTTAR PRADESH state, INDIA, on DHASAN RIVER, and 45 mi/72 km NW of MAHOBA. Was capital of former petty state of Jigni of CENTRAL INDIA agency; in 1948, state merged with VINDHYA PRADESH, in 1950, incorporated into Hamirpur district of Uttar Pradesh state.

Jiguaní (hee-gwah-NEE), town (2002 population 21,130), GRANMA province, E CUBA, on Central Highway, on railroad, and 13 mi/21 km E of BAYAMO; 20°22′N 76°26′W. Dairying center. Also produces sugarcane, fruit, coffee, cacao. Granite quarrying. In its picturesque surroundings are ruins of a colonial castle and the Pepú caves.

Jigüey Bay (hee-GWAI), shallow inlet of OLD BAHAMA CHANNEL, off CAMAGÜEY province, E CUBA, between CAYO ROMANO (island) and Cuba, 30 mi/48 km E of MORÓN; c.30 mi/48 km long NW-SE, 6 mi/10 km wide; 22°05′N 78°00′W. Receives CAONAO River and small Jigüey River.

Jih-ka-tse, CHINA: see XIGAZE.

Jihlava (YI-hlah-VAH), German *Iglau*, city (2001 population 50,702), JIHOMORAVSKY province, on the BOHEMIA-MORAVIA border, CZECH REPUBLIC, in BOHEMIAN-MORAVIAN HEIGHTS, on the JIHLAVA RIVER; 49°24′N 15°35′E. Railroad junction. KOSTELEC (KOSte-LETS), whose main industry is meat processing, lies 5 mi/8 km SW. Manufacturing (linen, woolen cloth, machinery, motors, furniture); brewery (established 1860). Chartered in 1227, it was the site of the signing of the Compactata in 1436—the Magna Carta of the Hussites. City has two medieval churches and a 16th-century town hall.

Jihlava River (YI-hlah-VAH), German *iglawa*, 115 mi/ 185 km long, JIHOMORAVSKY province, SW MORAVIA, CZECH REPUBLIC; rises in BOHEMIAN-MORAVIAN HEIGHTS. 3 mi/4.8 km NE of POCATKY; flows NNE, then E, past JIHLAVA, and SE, past TŘEBÍČ and IVANCICE, to SVRATKA RIVER 19 mi/31 km S of BRNO. DALESICE DAM, Czech *Dalešice*, capacity 1,186 acres/ 480 ha, ESE of Třebíč.

Jihocesky (YI-ho-CHES-kee), Czech *Jihočeský kraj*, province (□ 4,380 sq mi/11,388 sq km), S BOHEMIA and SE MORAVIA, CZECH REPUBLIC; ⊙ ČESKÉ BUDĚJOVICE. Bordered W by ZAPADOCESKY province, N by STREDOCESKY province, NE by VYCHODOCESKY province, E by JIHOMORAVSKY province, S by AUSTRIA, and SW by GERMANY. Consists of the following eight districts: ČESKÉ BUDĚJOVICE, CESKY KRUMLOV, JINDŘICHŮV HRADEC, PELHRIMOV, PISEK, PRACHATICE, STRAKONICE, and TÁBOR. Between BOHEMIAN-MORAVIAN HEIGHTS (E) and Sumava Mountains (SW) are BUDEJOVICKA BASIN (Czech *Budějovická*), and TREBONSKA BASIN (Czech *Třeboňská*), with noted ponds. Important centers include České Budějovice (machinery manufacturing; brewery), Tábor and Strakonice (machinery), Pisek (textiles), Cesky Krumlov (woodworking). Agriculture (wheat, potatoes, flax); cattle; fish; lumbering (fir). Province with the lowest population.

Jihomoravsky (YI-ho-MO-rahf-SKEE), Czech *Jihomoravský kraj*, province (□ 5,802 sq mi/15,085.2 sq km), S MORAVIA and SW BOHEMIA, CZECH REPUBLIC; ⊙ BRNO. Bordered W by Jihocesky province, N by Vychodocesky province, N and NE by Severomoravsky province, SE by Slovakia, and S by Austria. Consists of the following fourteen districts: Blansko, Breclav, Brno-City, Brno-Co., Hodonín, Jihlava, Kroměříž, Prostějov, Třebíč, Uherské Hradiště, Vyškov, Žďár nad Sázavou, Zlín, and Znojmo. Between Bohemian-Moravian Heights (W and NW) and White Carpathian Mountains (E) is a hilly country with Dyjsko-svratecky Lowland, Czech *Dyjsko-svratecký*, and Dolnomoravsky Lowland, Czech *Dolnomoravský*, along the rivers (Dyje, Svratka, Morava). Important centers include Brno (machinery; fair), Zlín and Třebíč (footwear), Prostějov (textiles), Kroměříž, Žďár nad Sázavou, Jihlava and Uherské Hradiště (machinery), and Znojmo (food processing). Wheat, barley, corn, potatoes, sugar beets, grapes, fruits, vegetables; cattle, hogs, horses, poultry.

Jihuacuta (hee-hwah-KOO-tah), canton, PACAJES province, LA PAZ department, W BOLIVIA; 17°12′S 68°29′W. Elevation 12,992 ft/3,960 m. Gas resources in area. Clay, limestone, gypsum, and some copper deposits. Agriculture (potatoes, yucca, bananas, rye); cattle.

Jihun River, TURKEY: see CEYHAN RIVER.

Jihÿüeh Tan (RI-YOOE TAN) or **Sun Moon Lake** (□ 2 sq mi/5.2 sq km) in central range of TAIWAN, 24 mi/39 km SE of TAICHUNG; 15 ft/5 m deep. Elevation 2,400 ft/732 m. Consists of 2 sections, Sun Lake (NE) and Moon Lake (SW). Tourism. Lake feeds one of Taiwan's first and largest hydroelectric plants (completed 1934, during the period of Japanese colonial control). Formed in 1920s by damming small streams (rather than a major river), during early years of Japanese occupation, forcing relocation of Ami aboriginal tribe.

Jijel (jee-JEL), wilaya, N central ALGERIA, in Lesser KABYLIA; ⊙ JIJEL; 36°45′N 06°00′E. One of the most wooded parts of the country, with predominantly cork oak trees. Industries include glassmaking at TAHER and ceramics at El MILIA.

Jijel (jee-JEL), town, ⊙ JIJEL wilaya, N central ALGERIA, port on the MEDITERRANEAN SEA, in Lesser KABYLIA; 36°50′N 05°43′E. Long cut off from the rest of the country by Kabylia uplands, it has grown into a major city, producing cork, skins, and hides. Training institutes here; recently endowed with a university, an airport (El Achouatt) and a new port, Djendjen, to E, which caters to larger vessels than those at the port of Jijel itself. Founded by Carthaginians, it became Roman colony of *Ighilgili* under Augustus. In 16th century, it was the Barbarossa brothers's capital. Repulsed French expedition in 1664. Formerly Djidjelli.

Jijiga (ji-JEE-gah), town (2007 population 102,489), SOMALI state, E central ETHIOPIA; 09°21′N 42°49′E. On plateau, near JERER RIVER, and 45 mi/72 km E of

---

Cross-references are shown in SMALL CAPITALS. The pronunciation guide is shown on page xix. The sources of population figures are shown on page xvii.

HARAR. Commercial center (coffee, hides, wax) at the junction of roads to ADDIS ABABA, BERBERA (NW SOMALIA), and Mogadishu (SE SOMALIA); also has an airfield. Occupied by Italians (1936) in the Italo-Ethiopian War and by British (1941) in World War II. Sometimes spelled Jigjiga.

Jijoca de Jericoacoara (ZHEE-zho-kah dee ZHE-ree-ko-AH-ko-ah-RAH), town (2007 population 15,442), CEARÁ, BRAZIL.

Jijona (hee-HO-nah), town, ALICANTE province, E SPAIN, 15 mi/24 km N of ALICANTE; 38°32′N 00°30′W. Candy- and conserve-manufacturing center, shipping nougats and raisins. Olive-oil processing, flour and sawmilling. Sheep raising; honey, cereals, vegetables in area. Has remains of Moorish castle.

Jilantaiyanchi (JEE-LAN-TEI-YAN-CHI), salt lake, W INNER MONGOLIA AUTONOMOUS REGION, N CHINA, 100 mi/161 km W of YINCHUAN, in Alashan Desert. Salt extraction.

Jilava (zhee-LAH-vah), village, ILFOV AGRICULTURAL SECTOR, S ROMANIA, 7 mi/11.3 km S of BUCHAREST; 44°20′N 26°05′E. Rubber processing; dairying.

Jilawiyah, Al-, EGYPT: see GILAWIA, EL.

Jilazun, Arab village, Ramallah district, 3.7 mi/6 km N of RAMALLAH, in the Samarian Highlands, WEST BANK. Agriculture (cereals, vegetables, fruit). A large refugee camp (1995 est. population 7,000) located here.

Jilib (JEE-lib), town, SE SOMALIA, on the E bank of JUBBA River and 60 mi/97 km NE of KISMAYO. Agricultural region (sesame, corn, sorghum). Formerly GELIB.

Jilin (JEE-LIN), short name: JI (JEE), province (□ 72,000 sq mi/187,200 sq km; 2000 population 26,802,191), NE CHINA; ⊙ CHANGCHUN; 43°00′N 126°00′E. It is bordered by HEILONGJIANG province and RUSSIA on the NE, by LIAONING province and NORTH KOREA on the S, and by the INNER MONGOLIA AUTONOMOUS REGION on the W. Jilin, crossed by the SONGHUA RIVER and forming part of the fertile alluvial NE plain, enjoys great agricultural prosperity; soybeans, wheat, upland rice, sweet potatoes, and beans are grown. Mountains in the E rise to more than 9,000 ft/2,743 m. Vast timberlands, among the best in China, are exploited; primary metals are extracted. Manufacturing includes chemicals, machine tools, automobiles, and metals; petrochemicals and automobiles are two pillars of Jilin's industry. Jilin has a good network of railroads, including the line between SHENYANG, Changchun, and HARBIN. The population, mainly Chinese, is concentrated in the industrial cities of Changchun, JILIN, SIPING, and LIAOYUAN. Near the North Korean border is Yanbian Korean autonomous region (established 1952), which has a large Korean population. Jilin University is in Changchun.

Jilin (JEE-LIN), city (□ 475 sq mi/1748 sq km; 1994 estimated urban population 1,117,800; 1994 estimated total population 1,367,700; 2000 total population 2,251,848), central JILIN province, NE CHINA, on the SONGHUA RIVER 60 mi/97 km E of CHANGCHUN; 43°51′N 126°33′E. It is a shipping port, railroad junction, and commercial and industrial center. Major manufacturing includes chemicals (organic/inorganic), artificial fiber, steel rolling and rare metal smelting; mining and petroleum equipment, transport equipment, coal gas, papermaking. Jilin was the capital of Jilin province until 1954. Formerly called Yung-ki, Jilin is one of the oldest cities in NE China. The name sometimes appears as Chi-lin, or Kirin.

Jiloca River or **Giloca River** (both: hee-LO-kah), c.80 mi/129 km long, in TERUEL and ZARAGOZA provinces, NE SPAIN; rises in the MONTES UNIVERSALES N of TERUEL; flows NNW of the JALÓN at CALATAYUD.

Jilotepec (hee-LO-te-pek), town, VERACRUZ, E MEXICO, in SIERRA MADRE ORIENTAL, 6 mi/9.7 km N of XALAPA ENRÍQUEZ; 19°55′N 99°30′W. Elevation 6,201 ft/1,890 m. Corn, coffee, fruit.

Jilotepec, MEXICO: see JILOTEPEC DE MOLINA ENRÍQUEZ.

Jilotepec de Abasolo, MEXICO: see JILOTEPEC DE MOLINA ENRÍQUEZ.

Jilotepec de Molina Enríquez (hee-LO-te-pek dai mo-LEE-nah en-REE-kez), city and township, ⊙ Jilotepec municipio, MEXICO, central MEXICO, 40 mi/64 km NW of MEXICO CITY. Cereals, fruit, livestock. Formerly Jilotepec de Abasolo.

Jilotlán de los Dolores (hee-lo-TLAWN dai los do-LO-res), town, JALISCO, central MEXICO, 45 mi/72 km E of COLIMA. Isolated small farming community in the TRANSVERSE VOLCANIC AXIS; 19°14′N 102°59′W. Sugarcane, corn, fruit.

Jilotzingo, MEXICO: see SANTA ANA JILOTZINGO.

Jilove (YEE-lo-VE), Czech *Jílové*, German *eulau*, town, SEVEROCESKY province, N BOHEMIA, CZECH REPUBLIC, on railroad, and 4 mi/6.4 km WSW of DĚČÍN; 50°39′N 15°06′E. Paper mill, fluorite mining in vicinity. Town was a 17th-century church; folk architecture.

Jilove u Prahy (YEE-lo-VE U-prah-HI), Czech *Jílové*, German *eule*, town, SEVEROCESKY province, S central BOHEMIA, CZECH REPUBLIC, on railroad, and 13 mi/21 km SSE of PRAGUE; 49°54′N 14°30′E. Excursion center. Woodworking; manufacturing (building materials). Former gold-mining settlement (has museum).

Jima, ETHIOPIA: see JIMMA.

Jimaguayú (hee-mah-gwah-YOO), small agricultural town, CAMAGÜEY province, E central CUBA, 2 mi/3 km from Central Highway, and 9 mi/15 km SE of CAMAGÜEY city; 21°18′N 77°52′W.

Jimaní (hee-mah-nee), town (2002 population 5,842), BAHORUCO province, SW DOMINICAN REPUBLIC, between Lake ENRIQUILLO and HAITI border, 30 mi/48 km W of NEIBA; 18°28′N 71°50′W. Coffee, fruit, timber. In 1949, became capital of newly formed INDEPENDENCIA province.

Jimbolia (zheem-BO-lyah), Hungarian *Zsombolya*, German *Hatzfeld*, town, TIMIŞ county, W ROMANIA, 25 mi/40 km W of TIMIŞOARA; 45°47′N 20°43′E. Railroad junction and frontier station on Yugoslav border. Manufacturing of agricultural machinery, clay products; tiles, felt hats, combs, buttons, and flour.

Jimbour, hamlet, QUEENSLAND, NE AUSTRALIA, 148 mi/238 km NW of BRISBANE, 17 mi/27 km N from DALBY; 26°58′S 151°13′E. Jimbour House, built 1874, was the property from which German immigrant and Australia explorer Ludwig Leichhardt began his journey across W Queensland to NE of DARWIN, NORTHERN TERRITORY.

Jim Chapman Lake, reservoir (□ 30 sq mi/78 sq km), HOPKINS and DELTA counties, NE TEXAS, on South Fork of SULPHUR RIVER, 20 mi/32 km S of PARIS; 33°20′N 95°26′W. Maximum capacity 797,300 acre-ft. Formed by Cooper Dam (79 ft/24 m high), built (1991) by Army Corps of Engineers for water supply; also used for flood control and recreation. Cooper Lake State Park near dam. Previously called Cooper Lake, the reservoir was renamed in 1998 in honor of politician Jim Chapman.

Jimena (hee-MAI-nah), town, JAÉN province, S SPAIN, 18 mi/29 km ENE of JAÉN; 37°50′N 03°28′W. Olive-oil processing, flour milling, soap manufacturing; cereals, fruit. Stone quarries.

Jimena de la Frontera (dhai lah fron-TAI-rah), town, CÁDIZ province, SW SPAIN, 21 mi/34 km N of ALGECIRAS; 36°26′N 05°27′W. Agricultural center (cereals, vegetables, fruit, tubers, livestock); cork, timber; apiculture; manufacturing of shoes; limekilns. At site of old Iberian settlement; has Moorish castle.

Jiménez, canton, COSTA RICA: see JUAN VIÑAS.

Jiménez (hee-ME-nes), city and township, CHIHUAHUA, N MEXICO, on plateau E of SIERRA MADRE OCCIDENTAL, on FLORIDO River, and 125 mi/201 km SE of CHIHUAHUA on Mexico Highway 49; 27°09′N 104°54′W. Elevation 4,531 ft/1,381 m. Mining center (fluorspar, mercury); silver, gold, lead, copper deposits; cotton gins. The region nearby is known for large number of meteorites, some of them discovered by the Spaniards in sixteenth and seventeenth century, and now exhibited in School of Mines in Mexico City. Also Ciudad Jimenez.

Jiménez (hee-ME-nes), town, COAHUILA, N MEXICO, near the RIO GRANDE (TEXAS border), 27 mi/43 km NW of PIEDRAS NEGRAS on Mexico Highway 2; 29°05′N 100°40′W. Elevation 755 ft/230 m. Wheat, bran, cattle, istle fibers, candelilla wax.

Jimenez (hee-ME-nes), town, MISAMIS OCCIDENTAL province, W MINDANAO, PHILIPPINES, port on PANGUIL BAY, 11 mi/18 km SSE of OROQUIETA; 08°19′N 123°42′E. Agricultural center (corn, coconuts); ships copra.

Jiménez (hee-MEH-nehs), village, LIMÓN province, E COSTA RICA, 3 mi/4.8 km E of GUÁPILES, on San José-Limón highway; 10°13′N 83°44′W. Bananas, corn, livestock.

Jiménez, ARGENTINA: see GRAMILLA.

Jiménez, MEXICO: see VILLA JIMÉNEZ.

Jiménez, MEXICO: see SANTANDER JIMÉNEZ.

Jiménez del Teul (hee-ME-nes del TE-ool), town, ZACATECAS, N central MEXICO, near DURANGO border, 59 mi/95 km W of FRESNILLO; 23°10′N 104°05′W. An isolated town S of Chalchihuites Mining District. Maguey, corn; livestock raising.

Jiménez, Villa, MEXICO: see VILLA JIMÉNEZ.

Jimenoa (hee-mai-NO-ah), falls and hydroelectric project, LA VEGA province, central DOMINICAN REPUBLIC, on affluent of the YAQUE DEL NORTE, just E of JARABACOA, 10 mi/16 km SW of LA VEGA.

Jimera de Libar (hee-MAI-rah dhai lee-VAHR), village, MÁLAGA province, S SPAIN, on GUADIARO RIVER, on railroad, and 9 mi/14.5 km SW of RONDA. Acorns, almonds, grapes, olives, wheat, vegetables; flour milling. Has mineral springs.

Jim Hogg, county (□ 1,136 sq mi/2,953.6 sq km; 2006 population 5,027), extreme S TEXAS; ⊙ HEBBRONVILLE; 27°03′N 98°40′W. Oil and natural gas; mainly cattle and ranching; some agriculture (sorghum). Formed 1913.

Jim Lake, reservoir (□ 27 sq mi/70.2 sq km), STUTSMAN co., E central NORTH DAKOTA, on JAMES RIVER, 3 mi/5 km N of Jamestown; 46°56′N 98°43′W. Max. capacity 379,636 acre-ft. Formed by dam (110 ft/33 m high), built (1953) by the Bureau of Reclamation for irrigation; also used for flood control, recreation, and water supply. N part of reservoir within Arrowroot National Wildlife Refuge.

Jimma (JI-mah), city (2007 population 166,592), OROMIYA state, SW ETHIOPIA; 07°40′N 36°50′E. Former ⊙ KEF (Kaffa) province. Commercial center for coffee producing region; food and clothing industry; airfield and road junction. Sometimes spelled Jima.

Jimmi Bagbo, village, SIERRA LEONE; 07°36′N 11°49′W. Chiefdom headquarters. Rice production. Has secondary school.

Jimmy Carter National Historic Site, SW GEORGIA, in PLAINS. Carter home and other buildings and exhibits associated with the thirty-ninth President's life are showcased here. Visitor center. Authorized 1987.

Jimo (JEE-MO), city (□ 666 sq mi/1,725 sq km; 1994 estimated urban population 165,800; estimated total population 1,044,800; 2000 population 1,018,709), E SHANDONG province, NE CHINA, 25 mi/40 km NNE of QINGDAO; 36°22′N 120°28′E. Agriculture, light industry are the largest sectors of the city's economy; heavy industry is also important. Grain, fruits, oil crops. Manufacturing (food and beverages, textiles, crafts, machinery).

Jimokuji (jee-MOK-jee), town, Ama county, AICHI prefecture, S central HONSHU, central JAPAN, 5 mi/8 km N of NAGOYA; 35°11′N 136°49′E.

Jimramov (YIM-rah-MOF), German *ingrowitz*, village, JIHOMORAVSKY province, W MORAVIA, CZECH REPUBLIC, in BOHEMIAN-MORAVIAN HEIGHTS, on

SVRATKA RIVER, 35 mi/56 km NNW of BRNO; 49°38′N 16°13′E. Excursion center; textile industry.

Jimsah, EGYPT: see GEMSA.

Jimsar (JIM-SAH), town, ⊙ Jimsar county, E central XINJIANG UYGUR AUTONOMOUS REGION, NW CHINA, 70 mi/113 km E of URUMQI, and on branch of SILK ROAD; 43°59′N 89°04′E. Grain, cotton, oilseeds; livestock. Also appears as Jimusa'er.

Jim Thorpe, borough (2006 population 4,886), ⊙ CARBON county, E PENNSYLVANIA, 22 mi/35 km NW of ALLENTOWN on LEHIGH RIVER; 40°52′N 75°44′W. Agriculture includes dairying. Borough created in the late 1950s by merging Mauch Chunk (the former county seat) and East Mauch Chunk boroughs. Borough named for Oklahoma football legend Jim Thorpe (1888–1953). State parks and reservoirs nearby.

Jimunai, CHINA: see JEMINAY.

Jimusa'er, CHINA: see JIMSAR.

Jim Wells, county (□ 868 sq mi/2,256.8 sq km; 2006 population 41,131), S TEXAS; ⊙ ALICE; 27°43′N 98°05′W. Bounded NE by NUECES RIVER (including dam of Lake Corpus Christi) and drained by small Los Olmos, AGUA DULCE and San Diego creeks. Large oil, natural gas production; caliche. Agriculture includes grain sorghum, wheat, corn, vegetables; cotton; cattle, hogs; dairying. Lipantitlan State Park in NE corner. Formed 1911.

Jim Woodruff Dam, GEORGIA and FLORIDA: see SEMINOLE, LAKE.

Jin, CHINA: see SHANXI.

Jin, CHINA: see TIANJIN.

Jinan (JEE-NAN), city (□ 818 sq mi/2,126.8 sq km; 2000 population 2,403,946), ⊙ Shandong province, E China, 3 mi/4.8 km S of the Huang He (Yellow River), and 30 mi/48 km NNW of the Lushan; 36°40′N 117°00′E. A railroad junction on the network linking Shanghai and Nanjing with Tianjin; it has connections to Qingdao and Yantai and at W end of the Jinan-Qingdao expressway. Jinan is an industrial center with textile, flour, paper, and iron and steel mills; food-processing, machine shops; production of trucks, agr. machinery, motor vehicles, chemicals, and fertilizer. It belongs to the E coastal open zone of China. In 1992, the city started direct international flights. An ancient walled city, Jinan was a provincial center as early asthe 12th century. It came under the control of the Communists in September 1948. Jinan is rich in springwater and lakes, which gives it the namethe "City of Springs." The city is well known for its Buddhist culture. Among them are China's earliest stone pagoda; the Lingyan-Temple which has China's largest forest of tomb stones; a Buddhist cave which contains a 26-ft/8-m-tall Buddha statue; a 2,100-sq-ft/195-sq-m cliff writing; and the thousand-Buddha cave which houses over 1,000 stone Buddha statuescarved through various dynasties. Jinan is the seat of Jinan Technical University, a medical college, and 14 other insts. of higher learning. The name sometimes appears as Chi-nan or Tsinan.

Jinchang (JIN-CHANG), city (□ 407 sq mi/1,058.2 sq km; 2000 population 156,907), central Gansu province, China; 38°28′N 102°10′E. Site of nuclear industry and a major military research and development laboratory. Heavy industry is the largest source of income for the city. Agr. (grain, oil crops), animal husbandry (hogs). Manufacturing (nonferrous metals).

Jincheng (JIN-CHENG), city (□ 838 sq mi/2,178.8 sq km; 2000 population 677,045), SE Shanxi province, China, near S end of Taihang Mountains, 45 mi/72 km S of Changzhi; 35°30′N 112°52′E. The city is a heavy industrial center. Crop growing and animal husbandry account for most of agr. activity (grains, fruits, vegetables; hogs, eggs, cattle, sheep). Manufacturing (iron andsteel, machinery); coal mining.

Jind (JEEND), former princely state of PUNJAB STATES, INDIA. Formed 1763 by Sikhs on breakup of Mogul empire. In 1948, merged with PATIALA AND EAST

PUNJAB STATES UNION; part of HARYANA state since 1966. Sometimes spelled Jhind.

Jind (JEEND), district (□ 1,056 sq mi/2,745.6 sq km), HARYANA state, N INDIA; ⊙ JIND.

Jind (JEEND), town, formerly in S central part of PATIALA AND EAST PUNJAB STATES UNION, now ⊙ JIND district, HARYANA state, N INDIA, 70 mi/113 km S of PATIALA; 29°19′N 76°19′E. Railroad junction; local trade in millet, gram, wheat; cotton ginning. Sometimes spelled Jhind.

Jindabyne (JIN-duh-bein), town, NEW SOUTH WALES, SE AUSTRALIA, 287 mi/462 km from SYDNEY, 38 mi/61 km from COOMA; 36°22′S 148°36′E. Near SNOWY MOUNTAINS ski resorts. Elevation 3,248 ft/990 m; occasional light snowfalls. Original town site drowned late 1960s by the Snowy Mountains Hydro-electric Authority to create Lake Jindabyne; trout fishing, water skiing.

Jindera, township, NEW SOUTH WALES, SE AUSTRALIA, 359 mi/578 km SW of SYDNEY, 10 mi/16 km N of ALBURY; 35°57′S 146°54′E. Pioneer museum of early German settlement.

Jindrichov (YIN-drzhi-KHOF), Czech Jindřichov, village, SEVEROMORAVSKY province, NW MORAVIA, CZECH REPUBLIC, on railroad, and 14 mi/23 km NNW of KRNOV; 50°15′N 17°31′E. Agriculture (oats); paper mills.

Jindrichov, Czech Jindřichov, German HENNERSDORF, village, SEVEROMORAVSKY province, NW MORAVIA, CZECH REPUBLIC, on railroad, and 9 mi/14.5 km N of ŠUMPERK; 50°45′N 15°11′E. Agriculture (flax); lumbering; paper mills. Has a 16th century castle with park, and a 17th-century church. Irregular stone blocks in the area are a residue of last Ice Age.

Jindřichův Hradec (YIN-drzhi-KHOOF HRAH-dets), German Neuhaus, city (2001 population 22,695), JIHOCESKY province, S BOHEMIA, CZECH REPUBLIC, in foothills of BOHEMIAN-MORAVIAN HEIGHTS; 49°09′N 15°00′E. Railroad junction. Manufacturing (cotton goods, machinery, mother-of-pearl buttons); food processing. Pond fishing in vicinity. Noted for castle museum (Gothic to Renaissance), a 13th century Hunger Tower, and a 15th-century Gothic church.

Jing'an (JING-AN), town, ⊙ Jing'an county, NW JIANGXI province, SE CHINA, 27 mi/43 km NW of NANCHANG; 28°52′N 115°22′E. Rice, oilseeds.

Jingbian (JING-BIAN), town, ⊙ Jingbian county, NW SHAANXI province, NW central CHINA, 70 mi/113 km NW of YAN'AN, at GREAT WALL; 37°36′N 108°48′E. Grain, oilseeds, livestock. Also called Zhangjiapan.

Jingchuan (JING-CHUAN), town, ⊙ Jingchuan county, SE GANSU province, NW CHINA, on JING RIVER, and 105 mi/169 km NE of TIANSHUI; 35°15′N 107°22′E. Grain, oilseeds, tobacco; food processing, clothing, engineering.

Jingde (JING-DUH), town, ⊙ Jingde co. (2000 population 369,995), SE Anhui province, China, 45 mi/72 km SSW of Xuancheng, in Huang Mountains; 30°18′N 118°32′E. Rice, jute; textiles, food processing, building materials.

Jingdezhen (JING-DUH-JEN), city (□ 157 sq mi/408.2 sq km), NE JIANGXI province, SE CHINA, on the CHANG RIVER; 29°17′N 117°12′E. It is an industrial and commercial center. Crop growing, animal husbandry. Grain, oil crops, vegetables, hogs, eggs; utilities. Manufacturing (coal, chemicals, transportation equipment, electric and electronic equipment). The city is well known for its fine porcelain, made since the Han dynasty (202 B.C.–A.D. 220) from kaolin, a white clay found near BOYANG LAKE to the W. Most of the antique porcelain objects found today were made during the Ming (1368–1644) and Qing (1644–1911) dynasties; the city also continues to export porcelain made during modern times. Sometimes appears as Ching-te-chen.

Jingdong (JING-DUNG), town, ⊙ Jingdong county, central YUNNAN province, SW CHINA, 120 mi/193 km

WSW of KUNMING; 24°28′N 100°54′E. Elevation 3,848 ft/1,173 m. Rice, wheat, millet, sugarcane, timber. Iron mines nearby.

Jingfeng, China: see HEXINGTEN QI.

Jinggangshan (JING-GANG-SHAN), city (□ 255 sq mi/663 sq km; 2000 population 45,585), Jiangxi province, China; 26°37′N 114°05′E. Named after the nearby Jinggang Mountains (highest elevation 4,718 ft/1,438 m); the city was one of the sites of the Communist base during the civil war in the 1930s. Thereare numerous museums and memorials dedicated to Communist leaders and solders. Agr. is the largest sector of the city's economy. Cropgrowing, forestry, and commercial agr. Main industries include foodprocessing, apparel, paper, and machinery.

Jinggu (JING-GU), town, ⊙ Jinggu county, SW YUNNAN province, SW CHINA, 35 mi/56 km NW of SIMAO; 23°28′N 100°42′E. Elevation 3,150 ft/960 m. Timber, tea, rice, sugarcane; sugar refining.

Jinghai (JING-HEI), town, ⊙ Jinghai county, NE CHINA, 20 mi/32 km SW of TIANJIN, an administrative unit of Tianjin municipality, on Tianjin-Pukou railroad and GRAND CANAL; 38°56′N 116°55′E. Grain; textiles and clothing, engineering, building materials, food industry.

Jinghong (ING-HUNG), town, ⊙ Jinghong county, southernmost YUNNAN province, SW CHINA, on MEKONG RIVER, on route to THAILAND, and 70 mi/113 km S of SIMAO; 21°58′N 100°50′E. Rice, millet, beans, sugarcane.

Jingjiang (JING-JIANG), town, ⊙ Jingjiang county, S JIANGSU province, E CHINA, 50 mi/80 km ESE of ZHENJIANG, across CHANG JIANG (Yangzi River); 32°02′N 120°16′E. Rice, cotton, jute; engineering, chemicals.

Jingle (JING-LUH), town, ⊙ Jingle county, N SHANXI province, NE CHINA, 50 mi/80 km NW of TAIYUAN; 38°22′N 111°56′E. Grain, oilseeds; building materials, textiles and clothing, food processing, coal mining.

Jingmen (JING-MEN), city (□ 1,703 sq mi/4,427.8 sq km; 2000 population 1,017,021), W central Hubei province, China, 30 mi/48 km WSW of Zhongxiang; 31°02′N 112°06′E. Road center; agr. and heavy industry are the largest sectors of the city's economy. Commerce also accounts for alarge share. Crop growing, animal husbandry, fishing. Agr. products include grains, oil crops, cotton, vegetables, hogs, and eggs. Industries include food processing, textiles, utilities, oil refining, and chemicals.

Jing Mountains (JING), NW HUBEI province, central CHINA, between HAN RIVER and CHANG JIANG (Yangzi River); rise to c.5,000 ft/1,524 m, 50 mi/80 km NW of JINGMEN; 31°30′N 111°36′E. Also called Jing Shan.

Jingning, town, ⊙ Jingning county, SE GANSU province, NW CHINA, 70 mi/113 km N of TIANSHUI; 35°30′N 105°45′E. Grain, oilseeds; chemicals, food processing, crafts, building materials, clothing.

Jingning (JIN-NING), town, ⊙ Jingning county, S ZHEJIANG province, SE CHINA, 60 mi/97 km W of WENZHOU, and on headstream of WU RIVER. Paper making; rice, wheat, corn, tea; logging.

Jingpo Lake (JING-po), natural reservoir on upper MUDAN RIVER, SE HEILONGJIANG province and NE JILIN province, NE CHINA, 50 mi/80 km SW of MUDANJIANG; formed by lava flow across river course; 25 mi/40 km long, 5 mi/8 km wide; 43°50′N 128°53′E. Jingpo is on N shore and Talazhan is near S shore. Hydroelectric station.

Jing River (JING), 200 mi/322 km long, NW central CHINA; rises in the LIUPAN MOUNTAINS of GANSU province, in two main headstreams, the Pu River (W) and the Huan-Malian River (E), which join to form the Jing River near Changwu; flows ESE, past Pingliang, Jingchuan, Qingyuan, and Ning Xian, and into SHAANXI province, past Changwu and Binxian, to WEI RIVER NE of Xi'an. Also called Jing Shui.

Jingshan (JING-SHAN), town, ☉ Jingshan county, central HUBEI province, central CHINA, 33 mi/53 km ESE of ZHONGXIANG; 31°02′N 113°03′E. Rice, oilseeds, cotton, sugarcane, jute; food processing, engineering, building materials.

Jing Shan, China: see JING MOUNTAINS.

Jing Shui, China: see JING RIVER.

Jingtai (JING-TEI), town, ☉ Jingtai county, central GANSU province, NW CHINA, near HUANG HE (Yellow River), 80 mi/129 km NNE of LANZHOU; 37°10′N 104°02′E. Wheat; cattle and sheep raising; building materials, food processing, coal mining.

Jingxi (JING-SEE), town, ☉ Jingxi county, SW GUANGXI ZHUANG AUTONOMOUS REGION, S CHINA, 50 mi/80 km S of BOSE, near VIETNAM border; 23°08′N 106°25′E. Rice, medicinal herbs, sugarcane; food and forage processing, chemicals, iron-ore mining.

Jing Xian (JING SIAN), town, ☉ Jing Xian county, S ANHUI province, E CHINA, 40 mi/64 km S of WUHU; 30°42′N 118°24′E. Rice, oilseeds, papermaking, engineering, food industry, building materials, iron smelting.

Jing Xian, town, ☉ Jing Xian county, S HEBEI province, NE CHINA, near SHANDONG border, 15 mi/24 km N of DEZHOU; 37°42′N 116°16′E. Cotton, grain, oilseeds; textiles.

Jing Xian, CHINA: see JINGZHOU.

Jingxing (JING-SING), town, ☉ Jingxing county, SW HEBEI province, NE CHINA, near SHANXI border, 25 mi/40 km W of SHIJIAZHUANG, on railroad spur; 38°02′N 114°08′E. Coal-mining center, thermal electric power generation.

Jingxing, town, SW HEILONGJIANG province, NE CHINA, 50 mi/80 km WSW of QIQIHAR; 47°00′N 123°05′E. Grain, soybeans, sugar beets, oilseeds.

Jingyan (JING-YAN), town, ☉ Jingyan county, S central SICHUAN province, SW CHINA, 20 mi/32 km NE of LESHAN; 29°43′N 104°04′E. Rice, tobacco, oilseeds, cotton.

Jingyang (JING-YANG), town, ☉ Jingyang county, central SHAANXI province, NW CHINA, on JING RIVER, and 15 mi/24 km N of XI'AN, and on railroad; 34°32′N 108°50′E. Grain, oilseeds; manufacturing of tractors.

Jingyu (JING-YOOI), town, ☉ Jingyu county, SE JILIN province, NE CHINA, 50 mi/80 km ESE of HAILONG; 42°21′N 126°49′E. Grain, soybeans, medicinal herbs; timber, chemicals, coal mining.

Jingyuan (JING-YUAN), town, ☉ Jingyuan county, SE GANSU province, NW CHINA, on HUANG RIVER, and 60 mi/97 km NE of LANZHOU; 36°35′N 104°02′E. Grain, oilseeds; food processing, engineering, building materials, coal mining.

Jingzhou (JING-JO), town, ☉ Jingzhou county, SW HUNAN province, S central CHINA, near GUIZHOU border, 50 mi/80 km S of ZHIJIANG; 26°35′N 109°41′E. Rice, oilseeds; industries include logging, papermaking, food. Also called Jing Xian.

Jinhua (JIN-HUAH), city (□ 116 sq mi/301.6 sq km; 2000 population 2,215,666), central Zhejiang province, China; 29°06′N 119°40′E. A transportation hub on the Zhejiang-Jiangxi railroad, the city has long been famous for producing hams. It is an industrial and commercial center. Crop growing, animal husbandry, and commercial agr. (grains, oil crops, vegetables, fruits, hogs, milk, eggs). Manufacturing (food processing, textiles, chemicals, pharmaceuticals, machinery). Sometimes appears as Kinhwa.

Jining (JEE-NING), city (□ 44 sq mi/114.4 sq km; 2000 population 193,085), central Inner Mongolian Autonomous Region, China; 40°58′N 113°01′E. A railroad center at the junction of the system connecting Beijing and Lanzhou, with the line traversing the Mongolian People's Republic to Russia. Industry and commerce are the main economic activities. Crop growing. Main industries include food processing and utilities. The name may appeares Chi'ning or Chi-ning.

Jining, city (□ 350 sq mi/910 sq km; 2000 population 1,465,656), W Shandong province, China, on railroad, and on the Grand Canal; 35°35′N 116°40′E. Agr. is the largest sector of the city's economy. Industry and commerce are also important. Crop growing, animal husbandry. Utilities. Agr. includes grains, oil crops, cotton, vegetables, fruits, hogs, beef, lamb, eggs, poultry, aquatic products. Manufacturing includes food and beverages; textiles, coal, chemicals, pharmaceuticals, plastics, machinery, electrical equipment. The city is famous for its Confucian culture. Nearby are Qufu, birthplace of Confucius, and Zoucheng, birthplace of Mencius. Liangshanco is well known for its thematic recreational park about the peasants who rebelled during the Song dynasty. The Han Chinese ancestor of the Huangdi is believed to have been born in this area as well. There are more than 400 historic sites of ancient buildings, ruins, and tombs. Also Chi-ning.

Jinja (JIN-juh), administrative district (□ 296 sq mi/769.6 sq km; 2005 population 448,700), EASTERN region, S central UGANDA, along E bank of VICTORIA NILE RIVER (forms W border); ☉ Jinja; 00°35′N 33°15′E. Elevation 3,707 ft/1,130 m. As of Uganda's division into eighty districts, borders MUKONO (SWW), KAYUNGA (NW), KAMULI (N), IGANGA (NEE), and MAYUGE (ESE) districts and LAKE VICTORIA (S). OWENS FALLS dam and hydroelectric station located here. Bananas, sugar, groundnuts are grown. Some industry around Jinja, Uganda's second largest city. Railroad between KASESE town (W Uganda) and MOMBASA (SE KENYA) travels W-E through Jinja town, connecting it to KAMPALA city to W. Several main roads also run through Jinja town, connecting it to Kampala city, surrounding Uganda, and Kenya.

Jinja-Kawempe (JIN-ja-kah-WEM-pe), town, WAKISO district, CENTRAL region, S UGANDA, 6 mi/10 km N of KAMPALA; 00°23′N 32°33′E. Suburb of Kampala. Sometimes called Kawempe.

Jinka, town (2007 population 23,548), SOUTHERN NATIONS state, SW ETHIOPIA; 05°38′N 36°39′E. 70 mi/113 km WSW of ARBA MINCH.

Jinkouhe (JIN-KO-HUH), town, SW SICHUAN province, SW CHINA, 5 mi/8 km WSW of EMEI, on S slope of Emei Shan. Summer resort.

Jinlong River (JIN-LUNG), FUJIAN province, SE CHINA; rises in WUYI MOUNTAINS N of LONGYAN; flows 120 mi/193 km SE, past ZHANGPING, HUA'AN, and CHANGTAI, forming common estuary with LONG RIVER on XIAMEN BAY of TAIWAN STRAIT.

Jinmen (JIN-MEN), island group, TAIWAN STRAIT, just off coast of FUJIAN province, CHINA, c.150 mi/241 km W of TAIWAN; 24°30′N 118°20′E. The group consists of the islands of Jinmen and Little Jinmen and 12 islets in the mouth of Xiamen Bay. The town of Jinmen, on Jinmen Island, is the chief population center. Farming is the main occupation; a large portion of the land is under cultivation. Crops include sweet potatoes, peanuts, sorghum, barley, wheat, soybeans, vegetables, and rice. Fishing is also important. Jinmen Island is heavily fortified and is honeycombed by tunnels and pillboxes. For many years it was subjected to periodic bombardment from mainland China, including a failed assault by Communist China in 1949. An incident in 1958 led to the deployment of the U.S. 7th Fleet, but an escalation of hostilities was avoided. During the height of tensions, the military garrison had 80,000 men; in 1997 the number was reduced to 8,000. The islands are no longer an important point of contention due to a lack of fresh water. The name is also spelled Kinmen or Chin-men. Also called Quemoy.

Jinning (JIN-NING), town, ☉ Jinning county, central YUNNAN province, SW CHINA, 25 mi/40 km SSW of KUNMING, on SW shore of DIAN CHI lake; 24°40′N

102°35′E. Elevation 6,552 ft/1,997 m. Chemicals, iron smelting; rice, tobacco. Also known as Kunyang.

Jinotega (hee-no-TAI-gah), department (□ 5,870 sq mi/15,262 sq km), N NICARAGUA, on HONDURAS border; ☉ JINOTEGA. COLÓN Mountains in NW, CORDILLERA ISABELIA in S; drained by COCO and BOCAY rivers. Agriculture concentrated in SW part, site of main populated centers. Coffee (YALÍ), sugarcane (LA CONCORDIA), wheat (SAN RAFAEL DEL NORTE), potatoes, grain, fodder crops. Lumber in undeveloped NE area is floated down Coco and Bocay rivers. Flour milling at Jinotega. Main centers linked by road with MATAGALPA.

Jinotega (hee-no-TAI-gah), city and township, (2005 population 41,134), ☉ JINOTEGA department, W central NICARAGUA, 70 mi/113 km NNE of MANAGUA; 13°06′N 86°00′W. Commercial center; coffee processing, flour milling, tanning, manufacturing (hats, mats). Scene of heavy fighting in revolution of 1978–1979, city was destroyed.

Jinotepe (hee-no-TAI-pe), city and township (2005 population 31,257), ☉ CARAZO department, SW NICARAGUA, on INTER-AMERICAN HIGHWAY, and 22 mi/35 km SSE of MANAGUA; 11°55′N 86°12′W. Agricultural processing and commercial center in coffee area; rice, sugarcane, sesame, livestock. Limestone, saltworks, lumber trade. Became city in 1883, capital department in 1891.

Jinping (JIN-PING), town, ☉ Jinping county, E GUIZHOU province, S central CHINA, near HUNAN border, 50 mi/80 km SE of ZHENYUAN, and on upper YUAN RIVER; 26°42′N 109°08′E. Rice; logging.

Jinping, town, ☉ Jinping county, S YUNNAN province, SW CHINA, 60 mi/97 km SW of KAIYUAN, near VIETNAM border; 22°46′N 103°15′E. Rice, food and beverages; building materials, textiles, papermaking, non-ferrous ore mining.

Jin River (JIN), 130 mi/209 km long, NW JIANGXI province, SE CHINA; rises in Jiuling Mountains near HUNAN border; flows ENE, past WANZAI, SHANGGAO, and GAO'AN, to GAN RIVER.

Jinsafut, Arab village, Nablus district, 8 mi/13 km SW of NABLUS, in the Samarian Highlands, WEST BANK. Agriculture (olives, cereal). Has remains from the Roman and Byzantine periods.

Jinseki (JEEN-se-kee), town, Jinseki county, HIROSHIMA prefecture, SW HONSHU, W JAPAN, 50 mi/80 km N of HIROSHIMA; 34°47′N 133°10′E.

Jinsha (JIN-SHAH), city (□ 1,139 sq mi/2,959 sq km; 1994 estimated urban population 683,000; estimated total population 1,434,500), S HUBEI province, central CHINA; port on CHANG JIANG (Yangzi River), connected to HUNAN province by canal; 30°16′N 112°20′E. An important trade center for the N DONGTING LAKE basin; site of a massive reservoir built to protect the central Hubei plains from floods; an industrial center. Crop growing, animal husbandry. Main industries include food processing, textiles, chemicals, plastics, machinery, and electrical equipment. Well known for its silk fabrics. Formerly Shashi.

Jinsha (JIN-SHAH), town, ☉ Jinsha county, NW GUIZHOU province, S central CHINA, 65 mi/105 km NNW of GUIYANG; 27°29′N 106°15′E. Grain, tobacco, oilseeds; tobacco processing, food and beverages, coal mining, iron-ore mining.

Jinsha, CHINA: see NANTONG.

Jinsha Jiang, China: see CHANG JIANG.

Jinshan (JIN-SHAN), town, ☉ Jinshan county, E CHINA, 32 mi/51 km SW of SHANGHAI, an administrative unit of Shanghai municipality, in canal district, near ZHEJIANG border; 30°53′N 121°09′E. Rice, oilseeds, jute; cement, major chemical-fiber facilities. Also known as Zhujing.

Jinshi (JIN-SHI), city (□ 212 sq mi/551.2 sq km), N HUNAN province, S central CHINA; port on LI RIVER, and 7 mi/11.3 km E of LI XIAN; 29°38′N 111°53′E.

Commercial center of Li River valley; textiles, transport equipment, food processing, chemicals, salt mining.

Jinta (JIN-TAH), town, ⊙ Jinta county, NW GANSU province, NW CHINA, on BEIDA RIVER, and 30 mi/48 km NE of JIUQUAN, near the GREAT WALL; 39°59′N 98°52′E. Elevation 4,199 ft/1,280 m. Grain, sugar beets, livestock.

Jintan (JIN-TAN), town, ⊙ Jintan county, S JIANGSU province, E CHINA, 32 mi/51 km SSE of ZHENJIANG; 31°44′N 119°31′E. Rice, cotton, jute, sugarcane; textiles, engineering, chemicals, food processing, iron smelting.

Jintang (JIN-TANG), town, ⊙ Jintang county, central SICHUAN province, SW CHINA, 22 mi/35 km NE of CHENGDU; 30°51′N 104°27′E. Rice, medicinal herbs, oilseeds. Sometimes appears as Zhao Zhou.

Jin, Tell (JIN, tel), village, ALEPPO district, NW SYRIA, on railroad, 26 mi/42 km SSW of ALEPPO; 35°50′N 37°02′E. Cotton, cereals.

Jintotolo Channel (heen-to-TO-lo), PHILIPPINES, between Masbate Island (N) and PANAY (S), connecting Visayan Sea with SIBUYAN SEA; 20 mi/32 km wide.

Jintsu River (JEENTS), Japanese *Jintsu-gawa* (jeen-TSOO-GAH-wah), 78 mi/126 km, TOYAMA prefecture, central HONSHU, central JAPAN; rises in mountains W of MOUNT HOTAKA; flows WNW, past TOYAMA, to TOYAMA BAY (inlet of SEA OF JAPAN). Hydroelectric plants on upper course.

Jintur (jin-TOOR), town, PARBHANI district, MAHARASHTRA state, INDIA, 24 mi/39 km NNW of PARBHANI; 19°37′N 76°42′E. Cotton, millet, wheat. Cotton ginning nearby.

Jinxi (JIN-SEE), former city, SW LIAONING province, NE CHINA. Annexed by HULUDAO.

Jinxi (JIN-SEE), town, ⊙ Jinxi county, E central JIANGXI province, SE CHINA, 30 mi/48 km SE of FUZHOU; 27°54′N 116°44′E. Rice, jute, fruit; timber, papermaking, crafts, food processing.

Jin Xian (JIN SIAN), town in LÜSHUN port district, SE LIAONING province, NE CHINA, at S tip of Liaodong peninsula, 15 mi/24 km NE of DALIAN, on inlet of BOHAI; 41°10′N 121°21′E. Railroad junction; oilseeds, fruits.

Jinxian (JIN-SIAN), town, ⊙ Jinxian county, N central JIANGXI province, SE CHINA, on S shore of BOYANG lake, on Zhejiang-Jiangxi railroad, and 30 mi/48 km SE of NANCHANG; 28°22′N 116°14′E. Rice, cotton, oilseeds, jute; food processing, papermaking, crafts, engineering.

Jin Xian, China: see JINZHOU.

Jinyuan (JIN-YUAN), town, central SHANXI province, NE CHINA, 20 mi/32 km SW of TAIYUAN. Grain, oilseeds.

Jinyun (JIN-YOOIN), town, ⊙ Jinyun county, S central ZHEJIANG province, SE CHINA, 40 mi/64 km SE of JINHUA; 28°39′N 120°03′E. Rice, medicinal herbs; textiles, engineering, electronics, chemicals, food industry.

Jinzhai (JIN-JEI), town, ⊙ Jinzhai county, W ANHUI province, E CHINA, near HENAN-HUBEI border, 45 mi/72 km WSW of LIU'AN, N of DABIE MOUNTAINS; 31°41′N 115°51′E. Rice, tea; hydroelectric power generation.

Jinzhou (JIN-JO), city (□ 239 sq mi/619 sq km; 1994 estimated urban population 31,500; estimated total population 499,600; 2000 total population 736,297), S HEBEI province, NE CHINA, on railroad, and 30 mi/48 km SE of SHIJIAZHUANG; 38°03′N 115°03′E. Agriculture (especially crop growing and animal husbandry) is the most important economic activity; others include light and heavy industry. Grain, fruits; hogs, eggs; manufacturing (textiles, chemicals, and machinery). Formerly called Jin Xian. Sometimes called Kinchow.

Jinzhou, city (□ 170 sq mi/442 sq km), W LIAONING province, NE CHINA, 130 mi/209 km SW of SHENYANG, and on railroad to BEIJING, near GULF OF LIAODONG; 41°07′N 121°06′E. Center of the Shanhaiguan corridor connecting NE and N China; railroad hub and industrial center. The major manufacturing includes electronics, petroleum, textiles, and papermaking. Coal mining at Dayaokou, 20 mi/32 km W of Jinzhou. An old city dating from the 12th century, Jinzhou developed mainly after construction of Beijing-Shenyang railroad. In 1992 the city established its economic and technological development zone at the coast 20 mi/32 km S of the old city. This satellite town is situated at the port of Jinzhou, 30 sq mi/78 sq km, connected with major railroad. The Jinzhou port started its operation in 1990. It is the northernmost first-class seaport in E coastal China. An all-season harbor, it has three berths with capacity above 10,000 tons. Between the satellite town and Jinzhou, there are expressways and pipelines that handle passenger and freight transportation.

Ji-Paraná (zhee–pah-rah-NAH), city (2007 population 107,638), central RONDÔNIA state, BRAZIL, 185 mi/298 km SE of PÔRTO VELHO; 08°03′S 62°52′W.

Jipe (JEE-pai), lake, COAST province, KENYA, 15 mi/24 km SW of SERENGETI plains, on Kenya-TANZANIA border; 10 mi/16 km long, 2 mi/3.2 km wide.

Jipijapa (hee-pee-HAH-pah), city (2001 population 36,078), MANABI province, W ECUADOR, on the equatorial lowlands; 01°20′S 80°35′W. On main road from GUAYAQUIL to MANTA, near Parque Nacional MACHALILLA. Jipijapa is famous for the manufacture of high-grade Panama hats, made from the Carludovica palmata plant. It is also the trade center for an agricultural region (coffee, cotton).

Jiquié (zhe-KYEE-ay), city, BAHIA state, E BRAZIL, on left bank of the RIO DE CONTAS, on RIO de JANEIRO–BAHIA highway, and 125 mi/201 km SW of SALVADOR; W terminus of railroad from SÃO ROQUE DO PARAGUAÇU and NAZARÉ; 13°50′S 40°09′W. Asbestos-mining center; also ships tobacco, cacao, sugar, coffee, livestock. Iron deposits near by. Also Jequié.

Jiquilillo (hee-kee-LEE-yo), beach resort, NICARAGUA, 25 mi/40 km E of CHINANDEGA. Minor beach resort.

Jiquilisco (hee-kee-LEES-ko), city and municipality, USULUTÁN department, SE EL SALVADOR, 9 mi/14.5 km W of USULUTÁN; 13°19′N 88°35′W. Commercial center at road junction for PUERTO EL TRIUNFO; grain, livestock raising.

Jiquilisco Bay, lagoonlike inlet of PACIFIC OCEAN, in USULUTÁN department, SE EL SALVADOR, S of USULUTÁN; 25 mi/40 km long, 1 mi/1.6 km–2 mi/3.2 km wide. Fisheries; salt extraction. PUERTO EL TRIUNFO on N shore.

Jiquilpan de Járez (hee-KEEL-pahn dai HWAH-res), city and township, MICHOACÁN, central MEXICO, on central plateau, 60 mi/97 km SE of GUADALAJARA; 19°57′N 102°42′W. Agricultural center (cereals, sugarcane, tobacco, beans, fruit, livestock); flour milling.

Jiquipilas (hee-kee-PEE-las), town, CHIAPAS, S MEXICO, on N slopes of SIERRA MADRE, 36 mi/58 km W of TUXTLA GUTIERREZ and just S of Mexico Highway 190 (INTER-AMERICAN HIGHWAY); 16°40′N 93°39′W. Corn, beans, sugarcane, fruit.

Jiquipilco (hee-kee-PEEL-ko), town, MEXICO, central MEXICO, 20 mi/32 km N of TOLUCA DE LERDO; 19°32′N 99°36′W. Agricultural center (cereals, fruit; livestock).

Jiquiriçá (ZHEE-kwee-ree-SAH), city (2007 population 13,306), E central BAHIA, BRAZIL, 19 mi/31 km E of SANTA INÉS; 13°16′S 39°35′W.

Jiquitaia, Brazil: see PRADO.

Jirays, EGYPT: see GIREIS.

Jirem Meng (JUH-LEE-MU MENG), Mongolian league in N INNER MONGOLIA AUTONOMOUS REGION, N CHINA, N of upper Liao River. Main centers are KAILU, TONGLIAO, and JARUD QI.

Jirgalanta, MONGOLIA: see HOVD.

Jiri (jee-REE), village, central NEPAL, at the end of a 69-mi/110-km-long Swiss-built road leading into the hills; 27°38′N 86°14′E. Elevation 6,000 ft/1,830 m. Airport.

Jirikov (YI-rzhee-KOF), Czech *Jiříkov*, German *Georgswalde*, town, SEVEROCESKY province, N BOHEMIA, CZECH REPUBLIC, 32 mi/51 km NE of ÚSTÍ NAD LABEM, on border opposite EBERSBACH and NEUGERSDORF, both in GERMANY); 50°59′N 14°35′E. Railroad junction. Manufacturing (pianos, chandeliers, textiles).

Jiring, VIETNAM: see DI LINH.

Jirira (hee-REE-rah), canton, LADISLAO CABRERA province, ORURO department, W central BOLIVIA, 6 mi/9.7 km E of SALINAS DE GARCI MENDOZA and NE of the Salar de Uyuni; 19°46′S 67°47′W. Elevation 12,218 ft/3,724 m. Gas resources in area. Lead-bearing lode; clay, limestone, gypsum, and large salt deposits. Agriculture (potatoes, yucca, bananas, rye); cattle.

Jirja, EGYPT: see GIRGA.

Jirkov (YIR-kof), German *Görkau*, town (2001 population 20,717), SEVEROCESKY province, NW BOHEMIA, CZECH REPUBLIC, on railroad, and 8 mi/12.9 km WSW of MOST; 50°30′N 13°27′E. Coal mining; manufacturing (textiles, glass). Has a 14th century Gothic church.

Jiroft (jee-RUHFT), town (2006 population 97,988), Kermān province, SE IRAN, on HALIL RIVER, and 120 mi/193 km SSE of KERMAN. Dates, cotton, rice, tobacco; rugmaking. Charcoal burning in nearby tamarisk woods. Formerly called Sabzevaran.

Jiroft, IRAN: see SABZEVARAN.

Jirriban, village, W central SOMALIA, 53 mi/85 km SE of EIL.

Jish (JEESH), Hebrew Gush-Halav, village, UPPER GALILEE, N ISRAEL, at NW foot of MOUNT MERON, 5 mi/8 km NW of ZEFAT; 33°01′N 35°26′E. Elevation 2,634 ft/802 m. A mostly Christian village with some Muslims. In Roman times, a fortified town, noted for its olive oil, it was home of Johanan of Gischala, leader of Jewish uprising (C.E. 66) against Romans. Built on site of ancient city of Gischala.

Jishan (JEE-SHAN), town, ⊙ Jishan county, SW SHANXI province, NE CHINA, on FEN RIVER, and 45 mi/72 km SW of LINFEN; 35°36′N 110°59′E. Grain, cotton.

Jishou (JI-SHO), city (□ 409 sq mi/1,059 sq km; 1994 estimated urban population 98,000; estimated total population 246,100; 2000 total population 230,621), NW HUNAN province, S central CHINA, on railroad, and near GUIZHOU border, on the Guiling River, 70 mi/113 km WNW of XUPU; 28°14′N 109°39′E. Agriculture and light industry are the largest sources of income. Crop growing, animal husbandry. Grain, hogs. Manufacturing (food and beverages, textiles, chemicals).

Jishui (JEE-SHUAI), town, ⊙ Jishui county, W central JIANGXI province, SE CHINA, 10 mi/16 km NE of JI'AN, and on GAN RIVER; 27°13′N 115°07′E. Oilseeds, jute.

Jisr el Majami (JIS-er el me-JA-mee), former frontier post, N JORDAN, 17 mi/27 km WNW of IRBID, below mouth of YARMUK RIVER. As a result of 1948 fighting between Israel and the Arabs, road and railroad bridge over JORDAN River not in use, railroad dismantled, and the hydroelectric station was totally destroyed.

Jisr esh Shughur (ZHIS-resh shuh-GOOR), town, IDLIB district, NW SYRIA, 8 mi/12.9 km from the Turkish border, 55 mi/89 km WSW of ALEPPO, on the left bank of the ORONTES River at the N end of the GHAB valley; 35°48′N 36°19′E. Cereals, tobacco. Bridge crossing on the Aleppo-LATTAKIA road. Surrounding marshy areas drained.

Jisr e-Zarka (JIS-RE ez–ZAHR-kah), Arab township, N of HADERA near TANINIM stream mouth, ISRAEL;

32°31′N 34°55′E. Elevation 104 ft/31 m. Remains of Roman aqueduct nearby, built to transport water from the sources of the Taninim to CAESAREA. Founded by Beduin in 1924.

Jissa (ZHEE-suh) or **Jissah**, inlet of GULF OF OMAN, in OMAN, 5 mi/8 km SE of MUSCAT; 23°33′N 58°39′E. Good natural harbor sheltered by rocky islet. Also called Bandar Jissah.

Jit, Arab village, Nablus district, 5.6 mi/9 km of NABLUS, in the Samarian Highlands, WEST BANK; 32°12′N 35°10′E. Agriculture (olives, cereals).

Jitaúna (ZHEE-tah-OO-nah), city (2007 population 16,880), E central BAHIA, BRAZIL, 14 mi/23 km SE of Jequié; 14°05′S 39°53′W.

Jitotol de Zaragoza (hee-TO-tol dai sah-rah-GO-sah), town, ⊙ Jitotol municipio, CHIAPAS, S MEXICO, in N spur of SIERRA MADRE DEL SUR, 27 mi/43 km NE of TUXTLA GUTIÉRREZ on Mexico Highway 195; 18°01′N 92°52′W. Cereals, tobacco, fruit. Zoque and Tzotzil Maya speakers in rural areas.

Jitra, (JIT-rah), town (2000 population 49,455), N KEDAH, MALAYSIA, 11 mi/18 km NNE of Alor Star.

Jiujiang (JIU-JIANG), city (□ 270 sq mi/699 sq km; 1994 estimated urban population 322,300; estimated total population 456,600; 2000 total population 791,224), N JIANGXI province, SE CHINA, on the CHANG JIANG (Yangzi River), and on HUBEI border, 40 mi/60 km N of the BOYANG LAKE; 29°41′N 116°03′E. A major river port, the city is the second-largest passenger port and the fourth-largest freight port on the Chang Jiang. The city was the capital of Jiujiang prefecture during the Qing dynasty (221–206 B.C.E.), and it opened to foreign trade in 1862. The earliest railroad (1904) and earliest highway (1910) in Jiangxi province passed through here. The city is an industrial and commercial center. In a major tea-growing area, Jiujiang is a large processing, marketing, and shipping point. Crop growing, animal husbandry, fishing, commercial agriculture, and forestry. Grains, oil crops, cotton, vegetables, fruits, hogs, eggs. Urban utilities. Manufacturing (food processing, textiles, oil refining, chemicals, machinery). The city has notable botanical gardens and an arboretum. Just S is the wooded Lu Shan, the location of GULING resort and of White Deer Cave, in which Zhu Xi (Chu Hsi), the 13th-century Confucian philosopher, lived and taught. Sometimes appears as Chiu-chiang.

Jiujiang (JIU-JIANG), town, S GUANGDONG province, S CHINA, on West River, and 15 mi/24 km SW of Nanhai. Rice, sugarcane, oilseeds. Sometimes spelled Kiukiang.

Jiulian Mountains (JIU-LIAN), section of the NANLING, on GUANGDONG-JIANGXI border, S CHINA, extending c.50 mi/80 km E-W along border in area of DINGNAN (Jiangxi province). Mount Jiulian, S of main range and 20 mi/32 km ENE of LIANPING, rises over 2,000 ft/610 m.

Jiulong (JIU-LUNG), town, ⊙ Jiulong county, SW SICHUAN province, SW CHINA, 80 mi/129 km SSW of KANGDING; 28°59′N 101°32′E. Logging.

Jiulong Mountains, China: see DABA SHAN.

Jiuquan (JU-CHUAN), city (□ 1,307 sq mi/3,385 sq km; 1994 estimated urban population 84,800; estimated total population 311,200; 2000 total population 294,080), W GANSU province, NW CHINA, on BEIDA RIVER, and 120 mi/193 km NW of ZHANGYE, near the GREAT WALL; 39°47′N 98°34′E. Historically, Jiuquan has been a commercial center for XINJIANG UYGUR. Today the city is the site of a major iron and steel complex. Agriculture and commerce are the largest sources of income. Crop growing, animal husbandry. Agricultural products include grain, vegetables, and hogs. Manufacturing includes food processing, iron and steel, and machinery. Opened to foreign trade in 1881. Formerly known as Suzhou.

Jiu River (ZHEE-oo), German Schyl, SW and S ROMANIA, in TRANSYLVANIA and OLTENIA, formed by two headstreams just S of PETROŞANI; cuts a deep gorge (SURDUC or VULCAN pass) through the TRANSYLVANIAN ALPS, past CRAIOVA; flows S to the DANUBE 8 mi/12.9 km W of ц; 135 mi/217 km long. Upper Jiu River valley is the leading hard coal-mining region of Romania.

Jiutai (JIU-TEI), city (□ 1,197 sq mi/3,100 sq km; 1994 estimated urban population 191,800; estimated total population 814,900; 2000 total population 788,550), W central JILIN province, NE CHINA, 35 mi/56 km NE of CHANGCHUN, and on railroad; 44°10′N 125°49′E. Agriculture is the largest sector of the city's economy. Industry and commerce are other economic activities. Agriculture (soybeans, grain); manufacturing (food processing, machinery).

Jiutepec (hee-DU-te-pek), town, MORELOS, central MEXICO, 5 mi/8 km SE of CUERNAVACA. Sugarcane, rice, fruit; livestock.

Jiutiao Mountains, China: see DABA SHAN.

Jiuzhaigou Valley (JIU-JEI-GO), nature reserve (□ 234 sq mi/608.4 sq km), N SICHUAN province, SW CHINA, in the MIN SHAN MOUNTAINS, near GANSU border, in Nanping county. Highest elevation 15,000 ft/4,572 m. An undisturbed area with the original forest cover, lakes, waterfalls, and a variety of karst landscapes. Jiuzhaigou is well known for its scenery and the giant panda. There are reports about unknown creatures living in the lakes. Residents are mainly Tibetans. This tourist destination was listed as a UN World Heritage Site in 1992.

Jiwani, Cape (ji-WAH-nee), SW BALUCHISTAN province, SW PAKISTAN, on E GWADAR Bay of ARABIAN SEA, near IRAN border; 25°01′N 61°44′E.

Jiwo Drashing (JO-wo DRAH-shing), mountain range, S spur of W HIMALAYAS, central BHUTAN, between PUNATSANG River (W) and MANGDE River (E), boundary between W and central Bhutan; highest point (16,404 ft/5,000 m), BLACK Mountain, is 17 mi/27 km SSW of TONGSA; 27°17′N 90°23′E. PELE LA pass is main crossing point. Known in English as Black Mountains

Jiwuchudrakey (JEE-CHOO-drah-kai), mountain peak (22,274 ft/6,789 m), W HIMALAYAS, BHUTAN, near border of TIBET, 31 mi/50 km N of PARO. Also spelled Jichu Drake.

Jixi (JEE-SEE), town, ⊙ Jixi county (2000 population 835,496), S ANHUI province, E CHINA, near ZHEJIANG border, 15 mi/24 km NE of Shexian, and on railroad; 30°05′N 118°36′E. Rice, oilseeds, jute; engineering, textiles, electronics.

Ji Xian (JEE SIAN), town, ⊙ Ji Xian county, SW HEBEI province, NE CHINA, 40 mi/64 km WNW of DEZHOU; 36°08′N 110°39′E. Grain, cotton; textiles, construction materials, chemicals.

Ji Xian, town, ⊙ Ji Xian county, SW SHANXI province, NE CHINA, 45 mi/72 km W of LINFEN, near HUANG HE (Yellow River). Grain; food, chemicals.

Ji Xian, town, ⊙ Ji Xian county, NE CHINA, 50 mi/80 km ENE of BEIJING, near GREAT WALL; 40°03′N 117°24′E. An administrative unit of TIANJIN municipality. Cotton, grain; textiles, building materials, food.

Jiyang (JEE-YANG), town, ⊙ Jiyang county, N SHANDONG province, NE CHINA, on HUANG HE (Yellow River), and 25 mi/40 km NNE of JINAN; 36°59′N 117°11′E. Peanuts, grain, cotton.

Jiyuan (JEE-YUAN), city (□ 746 sq mi/1,939.6 sq km; 2000 population 399,079), NW HENAN province, NE CHINA, 70 mi/113 km WSW of XINXIANG, near SHANXI border; 35°08′N 112°30′E. Agriculture and heavy industry are the largest sectors of the city's economy. Grain, tobacco, oilseeds; manufacturing (chemicals, non-ferrous metals).

Jizah, al, EGYPT: see GIZA.

Jizah, Al-, EGYPT: see GIZA.

Jizan (gee-ZAHN), town (2004 population 100,694), ASIR region, ⊙ Jizan province, SAUDI ARABIA, port on RED SEA, 80 mi/129 km S of ABHA. Chief export and shipping center of Asir, sheltered by FARASAN ISLANDS; dhow construction; fishing. Exports grain, sesame, dried fish, dates. Salt pans (S). Sometimes spelled Gizan.

Jizay, EGYPT: see GIZAI.

Jize (JEE-ZUH), town, ⊙ Jize county, SW HEBEI province, NE CHINA, 22 mi/35 km ESE of XINGTAI; 36°54′N 114°52′E. Grain, oilseeds, cotton.

Jizera (YI-ze-RAH), German sieghübel, Polish izera, mountain, second-highest peak (c.3,681 ft/1,122 m) of the JIZERA MOUNTAINS, in SEVEROCESKY province, N BOHEMIA, CZECH REPUBLIC, 10 mi/16 km NE of LIBEREC; 50°50′N 15°16′E.

Jizera Mountains (YI-ze-RAH), Czech jizerské hory, German isergebirge, Polish izerskie, góry, mountain range of the SUDETES, in SEVEROCESKY province, N BOHEMIA, CZECH REPUBLIC and LOWER SILESIA, SW POLAND (since 1945). Extends c.20 mi/32 km between CHRASTAVA (Czech Republic; W) and headwaters of KWISA (QUEIS) RIVER (Poland; E); rises to 3,688 ft/1,124 m in SMRK and to c.3,681 ft/1,122 m in Jizera peaks. NOVY SVET Pass is at SE foot.

Jizera River (YI-ze-RAH), German ISER, 103 mi/166 km long, N BOHEMIA, CZECH REPUBLIC; rises in the JIZERA MOUNTAINS, at S foot of SMRK MOUNTAIN, on POLAND border; flows SSE, along border for c.8 mi/12.9 km, and generally SSW, past TURNOV and MLADÁ BOLESLAV, to ELBE RIVER at LAZNE TOUSEN.

Jizerské hory, CZECH REPUBLIC: see JIZERA MOUNTAINS.

Jižní Město, CZECH REPUBLIC: see SOUTHERN TOWN.

Jizzakh (jee-ZAHK), wiloyat, UZBEKISTAN; ⊙ JIZZAKH. Bordered N by KAZAKHSTAN and S by TAJIKISTAN. TRANS-CASPIAN RAILROAD runs E-W across center of wiloyat. Aydarkul Lake in N. MIRZACHUL STEPPE in NE, NURATAU range in NW.

Jizzakh (jee-ZAHK), city, JIZZAKH wiloyat, UZBEKISTAN, on TRANS-CASPIAN RAILROAD, and 55 mi/89 km NE of SAMARKAND; 40°06′N 67°50′E. Metalworks, tobacco products; cotton. Hot climate. Until late 19th century, a major trade center at junction of caravan routes, but declined after the construction of railroad. Fell to Russians (1866). Also Dzhizak.

Joaçaba (ZHO-ah-kah-bah), city (2007 population 24,210), W SANTA CATARINA, BRAZIL, on PEIXE RIVER, on railroad, and 70 mi/113 km SSW of UNIÃO DA VITÓRIA (PARANÁ); 27°10′S 57°30′W. Livestock center (cattle, horses); meat preserving, fruit and maté shipping. Originally called Limeira; then Cruzeiro do Sul, c.1928–1938; and Cruzeiro, 1939–1943.

Joachimsthal, CZECH REPUBLIC: see JACHYMOV.

Joaíma (ZHO-ah-EE-mah), city (2007 population 14,846), NE MINAS GERAIS, BRAZIL, 17 mi/27 km S of JEQUITINHONHA; 16°40′S 41°00′W.

Joal (ZHO-ahl), town, THIÈS administrative region, W SENEGAL, on the ATLANTIC OCEAN, 55 mi/89 km SE of DAKAR; 14°10′N 16°51′W. Fishing and oyster-breeding center. High school, Roman Catholic church. Tourist attraction.

Joanna, unincorporated town (2000 population 1,609), LAURENS county, at SW edge of section of Sumter National Forest; 34°25′N 81°48′W. Manufacturing includes cotton insulation products. Agriculture includes timber, dairying, poultry, livestock, grain.

Joannès (zho-ah-NE), village, W QUEBEC, E CANADA, 15 mi/24 km E of ROUYN-NORANDA; 48°13′N 78°41′W. Gold mining.

Joanópolis (zho-ah-NO-po-lees), town (2007 population 10,677), E SÃO PAULO, BRAZIL, near MINAS GERAIS border, 50 mi/80 km NNE of SÃO PAULO; 22°56′S 46°17′W. Dairying.

João Alfredo (zho-oun AHL-fre-do), city (2007 population 28,460), E PERNAMBUCO, NE BRAZIL, on CAPIBERIBE RIVER and 10 mi/16 km WSW of LIMOEIRO; 07°52′S 35°35′W. Cotton, sugarcane, corn, kidney beans, fruit, manioc.

João Armaro (zho-OUN AH-mah-ro), village, central BAHIA, BRAZIL, on SALVADOR-BELO HORIZONTE railroad, 31 mi/50 km S of ITABERABA; 12°47′S 40°20′W.

João Belo, Vila de, MOZAMBIQUE: see XAIXAI.

João Câmara (ZHO-oun KAH-mah-rah), city (2007 population 30,333), NE RIO GRANDE DO NORTE state, BRAZIL, on railroad, and 44 mi/71 km NW of NATAL. Cotton; livestock.

João Câncio (zho-OUN KAHN-see-o), township, central ACRE, BRAZIL, on Rio ENVIRA; 08°35′S 70°30′W.

João Dias (ZHO-oun ZHEE-ahs), town (2007 population 2,739), extreme SW RIO GRANDE DO NORTE state, BRAZIL, 83 mi/134 km SSW of MOSSORÓ, on PARAÍBA border; 06°16′S 37°48′W. Cotton.

João Lisboa (ZHO-oun LEEZH-bo-ah), city (2007 population 19,938), W MARANHÃO state, BRAZIL, 7 mi/11 km NE of IMPERATRIZ; 06°26′S 47°25′W.

João Monlevade (zho-OUN mon-le-vah-zhee), city (2007 population 71,658), SE central MINAS GERAIS, BRAZIL, on PIRACICABA River, on railroad, and 50 mi/80 km E of BELO HORIZONTE; 19°47′S 43°13′W. Steel-milling center (blast and open-hearth furnaces; blooming, rolling, and wire mills) using ore from ITABIRA (18 mi/29 km NNW) and manganese mined in region.

João Neiva (zho-OUN NAI-vah), city (2007 population 14,435), central ESPÍRITO SANTO, BRAZIL, 6 mi/9.7 km N of IBIRAÇU; 19°41′S 40°25′W.

João Pessoa (zho-OUN pe-so-ah), city (2000 population 597,934), ⊙ PARAÍBA state, NE BRAZIL, at the confluence of the Sanhauá and PARAÍBA DO NORTE rivers; 07°10′S 34°49′W. Exports (cotton, sugar, minerals). Manufacturing (chemicals, plastics, metals, and electrical goods). The city was established in the late 16th century and named (1585) Filipea, in honor of Philip II of Spain and Portugal. During the brief Dutch occupation (17th century) it was called Frederickstadt, and, after its reconquest by the Portuguese, Paraíba. Its present name was acquired in 1930, in honor of the state governor who was assassinated in RECIFE during the Vargas revolution. João Pessoa is the site of a federal university; colonial architecture (especially Franciscan convent and church); beach resort areas nearby.

João Pessoa, Brazil: see EIRUNEPÉ.

João Pessoa, Brazil: see MIMOSO DO SUL.

João Pessoa, Brazil: see PÔRTO.

João Pinheiro (zho-OUN peen-yai-ro), city (2007 population 43,217), W MINAS GERAIS, BRAZIL, 55 mi/89 km SE of PARACATU; 17°44′S 46°13′W; cattle.

Joaquim Gomes (ZHWAH-keen GO-mes), city (2007 population 21,735), NE ALAGOAS state, BRAZIL, on Rio Camarajibe; 09°10′S 35°44′W.

Joaquim Murtinho (ZWHAH-keen MOOR-cheen-yo), city, E PARANÁ state, BRAZIL, 48 mi/77 km NNE of PONTA GROSSA; 24°24′S 49°52′W. Railroad junction.

Joaquim Nabuco (ZHWAH-keen NAH-boo-ko), city (2007 population 15,953), E PERNAMBUCO state, BRAZIL, 53 mi/85 km SW of RECIFE; 08°37′S 35°32′W. Sugar, aloe, corn, fruit.

Joaquim Távora (ZHWAH-keen TAH-vo-rah), town (2007 population 10,247), NE PARANÁ, BRAZIL, on railroad, and 23 mi/37 km S of JACAREZINHO; 23°30′S 49°58′W. Sawmilling; rice, corn, coffee, cotton.

Joaquin, town (2006 population 952), SHELBY county, E TEXAS, near the SABINE RIVER (LOUISIANA boundary), headwaters of TOLEDO BEND RESERVOIR, opposite LOGANSPORT, Louisiana (bridge), 45 mi/72 km SSE of MARSHALL; 31°58′N 94°02′W. Timber; oil and gas; poultry, cattle; vegetables. Sabine National Forest to SE.

Joaquín Suárez, URUGUAY: see SUÁREZ.

Joaquín V. González (hwah-KEEN gon-ZAH-lez), town, SE SALTA province, ARGENTINA, on Pasaje or Juramento River, 85 mi/137 km ESE of SALTA; 25°05′S 64°11′W. Railroad junction; agricultural center (corn,

alfalfa; livestock); sawmills. Formerly called Kilómetro 1082.

Joateca (hoe-ah-TE-kah), municipality and town, MORAZÁN department, EL SALVADOR, N of SAN FRANCISCO GOTERA, near HONDURAN border.

Joazeiro, Brazil: see JUAZEIRO DO NORTE.

Job (JAHB), unincorporated village, RANDOLPH county, E central WEST VIRGINIA, 15 mi/24 km ESE of ELKINS, on Gandy Creek, in Monongahela National Forest; 38°51′N 79°33′W.

Jobabo (ho-BAH-bo), town, LAS TUNAS province, E CUBA, near JOBABO RIVER, 22 mi/35 km WSW of Victoria de las Tunas; 20°55′N 77°17′W. Home of Peru sugar mill.

Jobabo River (ho-BAH-bo), 48 mi/77 km long, in LAS TUNAS province, E CUBA; flows S to the swamps at head of the GULF OF GUACANAYABO; 20°50′N 77°15′W.

Joban (JO-bahn), major coal field of NE JAPAN, in IBARAKI and FUKUSHIMA prefectures, central HONSHU, on the PACIFIC OCEAN, c.120 mi/193 km NNE of TOKYO. Provides fuel for industrial Tokyo-Yokohama area.

Jobat (JO-buht), former princely state of CENTRAL INDIA agency. In 1948, merged with MADHYA BHARAT and later included in JHABUA district, far west MADHYA PRADESH state.

Joboji (JO-BO-jee), town, Ninohe county, IWATE prefecture, N HONSHU, NE JAPAN, 34 mi/55 km N of MORIOKA; 40°10′N 141°09′E. Tobacco; lacquer and lacquerware.

Jobos (HO-bos), village, S PUERTO RICO, 3 mi/4.8 km SW of GUAYAMA. Port of entry. Former sugar milling village. PUERTO JOBOS is 1.5 mi/2.4 km W.

Jobourg, Nez de (zho-boor, nai duh), rocky headland of COTENTIN PENINSULA, MANCHE department, BASSE-NORMANDIE region, NW FRANCE, on Race of ALDERNEY of ENGLISH CHANNEL, 11 mi/18 km ESE of ALDERNEY; 49°41′N 01°57′W. The HAGUE uranium reprocessing facility is located here.

Jobson, ARGENTINA: see VERA.

Jocassee (jo-KAS-ee), village, OCONEE county, NW SOUTH CAROLINA, in BLUE RIDGE MOUNTAINS, 33 mi/53 km WNW of GREENVILLE. JOCASSEE Dam is E.

Jocassee, Lake, reservoir, on PICKENS-OCONEE county border, NW SOUTH CAROLINA, on SENECA (Keowee) River, 30 mi/48 km WNW of GREENVILLE; 8 mi/12.9 km long, 3 mi/4.8 km wide; 34°58′N 82°55′W. Max. capacity 1,315,670 acre-ft. Formed by Jocassee Dam (385 ft/117 m high), built (1973) by the Duke Power Company for power generation. Bounded W by Sumter National Forest; Devils Fork State Park on SW shore.

Jocelyn (JAHS-lin), township (□ 51 sq mi/132.6 sq km; 2001 population 298), on SAINT JOSEPH ISLAND, ONTARIO, E central CANADA. Fort Saint Joseph National Historic Site, woodland gardens.

Jochmus Lake, AUSTRALIA: see GALILEE, LAKE.

Jockgrimm (YUHK-grim), village, RHINELAND-PALATINATE, W GERMANY, near the RHINE RIVER, 8 mi/12.9 km NW of KARLSRUHE; 49°06′N 08°17′E.

Jockis, FINLAND: see JOKIOINEN.

Jocoaitique (ho-ko-ai-TEE-kai), city and municipality, MORAZÁN department, NE EL SALVADOR, near HONDURAS border, 17 mi/27 km N of SAN FRANCISCO GOTERA; 13°54′N 88°09′W. Henequen center; cordage manufacturing.

Jocón (ho-KON), town, YORO department, NW HONDURAS, on paved road, 38 mi/61 km NE of YORO, 3 mi/4.8 km W of Yaguala River; 15°17′N 86°58′W. Elevation 2,982 ft/909 m. Small farming.

Jocoro (ho-KO-ro), city and municipality, MORAZÁN department, E EL SALVADOR, on branch of INTER-AMERICAN HIGHWAY, and 13 mi/21 km NE of SAN MIGUEL; 13°37′N 88°01′W. In former gold- and silver-mining district; grain, sugarcane.

Jocotán (ho-ko-TAHN), town, CHIQUIMULA department, E GUATEMALA, on branch of RÍO GRANDE DE ZACAPA, and 11 mi/18 km E of CHIQUIMULA; 14°49′N

89°23′W. Elevation 1,532 ft/467 m. Corn, beans; livestock. Chortí-speaking population.

Jocotenango (ho-ko-te-NAHN-go), town (2002 population 17,000), SACATEPÉQUEZ department, S central GUATEMALA, 2 mi/3.2 km NW of ANTIGUA GUATEMALA; 14°35′N 90°44′W. Elevation 4,757 ft/1,450 m. Coffee. Founded 1542. Now effectively an extension of Antigua Guatemala.

Jocotepec (ho-KO-te-pek), town, JALISCO, W MEXICO, near W shore of Lake CHAPALA, 27 mi/43 km SSW of GUADALAJARA, and just off Mexico Highway 15; 20°18′N 103°26′W. Beans, grain, fruit, livestock.

Jocotitlán (ho-ko-teet-LAHN), town, MEXICO, central MEXICO, 30 mi/48 km N of TOLUCA DE LERDO; 19°42′N 99°48′W. Cereals; livestock.

Jódar (HO-dhar), town, JAÉN province, S SPAIN, 12 mi/19 km S of ÚBEDA; 37°50′N 03°21′W. Olive-oil and esparto processing; light manufacturing; brandy distilling, flour milling. Agricultural trade (cereals, wine, esparto, livestock). Marble and gypsum quarries.

Jo Daviess (jo-DAI-viss), county (□ 618 sq mi/1,606.6 sq km; 2006 population 22,594), extreme NW ILLINOIS, bounded N by WISCONSIN state line and W by the MISSISSIPPI (here forming IOWA state line); ⊙ GALENA; 42°21′N 90°12′W. Drained by APPLE, PLUM, and GALENA rivers. Agriculture (cattle, hogs, sheep, oats, alfalfa; dairying). Lead and zinc mines. Manufacturing: dairy products, metal products, fertilizer, iron. Includes hilly area near Wisconsin line; CHARLES MOUND (1,235 ft/376 m), highest point in Illinois, is here. Contains Apple River Canyon State Park, portion of Upper Mississippi River National Wildlife Refuge and U.S. Grant Home State Historical Site. Formed 1827. Savanna Army Depot in SW on Mississippi River. Tourism a major industry.

Jodhpur (JAHD-puhr), former princely state in RAJPUTANA STATES, INDIA. Established in early 13th century by Rathor Rajputs. Invaded 16th–17th century by Moguls, late 18th century by Marathas. Sought British protection by alliance in 1818. In 1949, joined union of Rajasthan, later included as a district within RAJASTHAN state. Also called Marwar.

Jodhpur (JAHD-puhr), district (□ 8,822 sq mi/22,937.2 sq km), RAJASTHAN state, NW INDIA; ⊙ JODHPUR.

Jodhpur (JAHD-puhr), city (2001 population 860,818), ⊙ JODHPUR district, RAJASTHAN state, NW INDIA; 26°17′N 73°02′E. Important marketplace for wool and agricultural products. Has a domestic airport. Manufacturing includes textiles, metal utensils, bicycles, ink, and sporting goods. Noted for diversified cottage industries (glass bangles, cutlery, carpets, and marble products). Surrounded by a wall nearly 6 mi/9.7 km long. Towering above the city on a rock 400 ft/122 m high is an old fortress housing several palaces and the treasury of the maharaja. The Indian Air Force maintains a training center here. Seat of university and an Arid Zone Research Institute. Center for equestrian sports; the close-fitting riding breeches, "jodi purs," take their name from here. Was capital of the former princely state of JODHPUR. Also Marwar. Founded in 1459.

Jodiya (JOD-yuh), town, in JAMNAGAR district, formerly in SAURASHTRA, now in GUJARAT state, INDIA, near GULF OF KACHCHH, 21 mi/34 km NE of JAMNAGAR; 22°42′N 70°18′E. From wharf (2 mi/3.2 km NW) exports cotton, wool, and oilseeds; oilseed pressing, cotton ginning, handicraft cloth weaving.

Jodoigne (zho-DWAHN-ye), Flemish *Geldenaken*, commune (2006 population 12,498), Nivelles district, BRABANT province, central BELGIUM, 6 mi/9.7 km SSW of TIENEN; 50°43′N 04°52′E. Agriculture market.

Jodrell Bank Observatory (JAH-druhl) observatory for radio astronomy located at Jodrell Bank, MACCLESFIELD, CHESHIRE, NW ENGLAND; 53°15′N 02°12′W. Administered by the University of Manchester. It is part of the MERLIN array (Multi-Element Radio-Linked Interferometer Network) that includes other radio telescopes throughout England. The Lovell Radio Telescope underwent extensive renovations here from 2001–2002.

Joe B. Hoggsett Dam, Texas: see CEDAR CREEK RESERVOIR.

Joensuu (YO-en-soo), city, ⊙ POHJOIS-KARJALAN province, SE FINLAND; 62°35′N 29°45′E. Trade center of NE Karelia forest region on the Pielsjoki River. Railroad junction and lake port; plywood mills, dairy processing. Chartered in 1848 as a copper-mining town. The city hall was designed by Eliel Saarinen in 1914. Joensum University established 1969. Airport.

Joetsu (JO-ETS), city (2005 population 208,082), NIIGATA prefecture, central HONSHU, N central JAPAN, 65 mi/105 km S of NIIGATA; 37°08′N 138°14′E. Textiles, agricultural machinery, wooden furniture, lace. Ruins of Kasugayama castle.

Joeuf (ZHUF), town (□ 1 sq mi/2.6 sq km; 2004 population 7,116), MEURTHE-ET-MOSELLE department, in LORRAINE, NE FRANCE, in briey iron basin, on ORNE RIVER, and 4 mi/6.4 km SE of Briey, adjoining HOMÉCOURT; 49°14′N 06°01′E. Metalworks. Iron mined nearby.

Joffre (JAH-fuhr), hamlet, S central ALBERTA, W CANADA, 13 mi/20 km from LACOMBE, in LACOMBE COUNTY; 52°20′N 113°32′W.

Joffre, Mount (JAH-fuhr), (11,316 ft/3,449 m), SE BRITISH COLUMBIA, W CANADA, on ALBERTA border, in ROCKY MOUNTAINS, 50 mi/80 km SSE of BANFF (Alberta); 50°31′N 115°14′W.

Joffreville (JO-frah-VEEL), village, ANTSIRANANA province, N MADAGASCAR, at foot of Mount Ambre, 20 mi/32 km S of ANTSIRANANA; 12°29′S 49°13′E. Elevation c.3,900 ft/1,189 m. Formerly called Camp d'Ambre [Malagasy=Ambohitra] (awm-BOO-eetch), former climatic resort, now gateway to AMBER MOUNTAIN NATIONAL PARK (4 mi/6.4 km S).

Jofra, LIBYA: see JUFRAH, AL.

Jog, settlement, SHIMOGA district, KARNATAKA state, S INDIA, on SHARAVATI RIVER, near railroad terminus at TALGUPPA (6 mi/9.7 km E), and 55 mi/89 km WNW of SHIMOGA. Headworks of Mahatma Gandhi Hydroelectric Works nearby consist of storage dam on the Sharavati River and hydroelectric plant, in operation since 1948. GERSOPPA Falls are downstream, 1.5 mi/2.4 km W. The system, with substations at Shimoga and other towns of W Karnataka state, is intended to aid industrial expansion in this mountainous area and to merge its power with that of Sivasamudram and Shimsha Falls, in SE part of state.

Jogbani (JOG-bah-nee), village (2001 population 29,962), ARARIA district, NE BIHAR state, INDIA, on NEPAL border, 45 mi/72 km NNW of PURNIA. Railroad terminus.

Joge (JO-ge), town, Konu county, HIROSHIMA prefecture, SW HONSHU, W JAPAN, 43 mi/70 km N of HIROSHIMA; 34°41′N 133°07′E. Has old town streetscape. Hot springs nearby.

Jogeshvari (jo-GAISH-vuh-ree), a growing settlement in MUMBAI (BOMBAY) SUBURBAN district, W MAHARASHTRA state, INDIA, on Salsette Island, 14 mi/23 km N of MUMBAI city center. Manufacturing (matches, chemicals, rubber goods); rice growing. A state-run college is on the hillock. On electric railroad. Has large Brahmanic cave temple of late 7th century. Sometimes spelled Jogeshwari or Jogeswar.

Jõgeva (YUH-ge-vah), German Laisholm, city, E ESTONIA, on railroad, and 28 mi/45 km NNW of TARTU; 58°45′N 26°24′E. Agricultural market (dairying, cereals); clay quarry, sawmill, leather works, wool.

Joggins (JAH-ginz), town, N NOVA SCOTIA, CANADA, on CHIGNECTO BAY, 16 mi/26 km SW of AMHERST; 45°42′N 64°26′W. Coal mining.

Jogindarnagar (JO-gin-duhr-nuh-guhr), town, MANDI district, N central HIMACHAL PRADESH state, INDIA, 20 mi/32 km NNW of MANDI, in S PUNJAB HIMALAYAS; 31°59′N 76°46′E. Railroad terminus. Hydroelectric plant (in operation since 1932) supplied by dam (3 mi/4.8 km NNE) across UHL RIVER; powers industries in large cities of Indian Punjab.

Jogipet (JO-gi-pait), town, MEDAK district, ANDHRA PRADESH state, INDIA, near MANJRA RIVER, 40 mi/64 km NNW of HYDERABAD; 17°51′N 78°04′E. Rice, sugarcane.

Jogjakarta, INDONESIA: see YOGYAKARTA.

Johana (JO-hah-nah), town, E. Tonami county, TOYAMA prefecture, central HONSHU, central JAPAN, 22 mi/35 km S of TOYAMA; 36°30′N 136°54′E. Railroad terminus. Radioactive mineral springs nearby.

Johanna Island, COMOROS: see NZWANI.

Johannesburg, Zuli eGoli [=city of gold], city (2001 population 3,225,810), ⊙ GAUTENG province, NE SOUTH AFRICA, on the S slopes of the WITWATERSRAND; 26°12′S 28°03′E. Elevation 5,750 ft/1,753 m. The largest city of South Africa, it is the center of its important gold-mining industry, its manufacturing and commercial center, and the hub of its transportation network. Gold mining is the sprawling city's chief industry. Manufacturing includes cut diamonds, industrial chemicals, plastics, cement, electrical and mining equipment, paper and paper products, glass, textiles, food products, and beer. The country's main stock exchange (founded 1887) is in the city. In accordance with apartheid law, racial groups were once restricted to separate residential areas; most blacks still live in SOWETO and TEMBISA (NE). Formerly a group of townships SW of the city, Soweto became an independent city in 1983. Greater metropolitan Johnnesburg incorporates the independent municipalities of Randburg, Sandton, EDENVALE, and Bedfordview.

Rand Afrikaans University (1966), the University of the Witwatersrand (1922), and Witwatersrand College for Advanced Technical Education (1925) are here; several museums, an art gallery, planetarium, zoo, bird sanctuary, and numerous parks. Jan Smuts House. Nearby is Kyalami Circuit, where international motor races are held. Johannesburg International Airport nearby to NE is in Kempton Park; the largest airport in South Africa, still known by former name Jan Smuts Airport. The city was founded as a mining settlement in 1886, when gold was found on the Witwatersrand; by 1900 it had a population of c.100,000. Johannesburg's large black population provided labor for the mines.

Johannesburg, unincorporated village, KERN county, S central CALIFORNIA, 50 mi/80 km NW of BARSTOW, in RAND MOUNTAINS (MOJAVE DESERT). Silver, tungsten, gold mines. Part of China Lake Naval Air Weapons Station to E; Cuddeback Lake (dry) to SE.

Johanngeorgenstadt (YO-hahn-ge-OR-gen-shtaht), town, SAXONY, E central GERMANY, in the ERZGEBIRGE, 11 mi/18 km S of AUE; frontier station on Czech border, opposite POTUCKY; 50°26′N 12°43′E. Winter sports center. Manufacturing of kid gloves; metal- and woodworking. Founded 1654 as silver-mining settlement by Bohemian Protestants.

Johannisbad, CZECH REPUBLIC: see JANSKE LAZNE.

Johannisburg, POLAND: see PISZ.

Johen (JO-HEN), town, South Uwa county, EHIME prefecture, NW SHIKOKU, W JAPAN, 62 mi/100 km S of MATSUYAMA; 32°57′N 132°35′E. Oranges; marine product processing (dried bonito). Sometimes spelled Zyohen.

Johi (jo-HEE), village, DADU district, W SIND province, SE PAKISTAN, 10 mi/16 km W of DADU; 26°41′N 67°37′E.

John Day, town (2006 population 1,582), GRANT county, NE central OREGON, on JOHN DAY RIVER, at mouth of Canyon Creek, 90 mi/145 km S of PENDLETON; 44°25′N 118°57′W. Chromite deposits nearby. Site of Kam Wah Chung Museum. Parts of Malheur National Forest to N and S, including Strawberry Mountain Wilderness Area to SE (STRAWBERRY RANGE). JOHN DAY FOSSIL BEDS NATIONAL MONUMENT (Sheep Rock Unit) 35 mi/56 km WNW. Clyde Holliday State Park to W.

John Day, river, 281 mi/452 km long, rising in E GRANT county, in BLUE MOUNTAINS, NE OREGON, flows W past the town of JOHN DAY, N to the Columbia River, past JOHN DAY FOSSIL BEDS NATIONAL MONUMENT, 27 mi/43 km ENE of THE DALLES, just upstream (E) of JOHN DAY Dam. The lower half of the river is in John Day River State Scenic Waterway. Unnavigable, the river is used to irrigate vegetable farms.

John Day Fossil Beds National Monument (□ 14,014 sq mi/36,296 sq km), GRANT and WHEELER counties, N central OREGON. Consists of Sheep Rock, Painted Hills, and Clarno units. Sheep Rock Unit on JOHN DAY RIVER, 35 mi/56 km WNW of JOHN DAY. Clarno Unit on John Day River, 10 mi/16 km SW of Fossil Beds. Painted Hills Unit on Ridge Creek, 65 mi/105 km WNW of John Day. Rich fossil remains extend over 5 geological epochs, Eocene through Pleistocene. Authorized 1974.

John Day Lock and Dam, Oregon/Washington: see UMATILLA, LAKE.

John D. Rockefeller, Jr. Memorial Parkway, TETON county, NW WYOMING, scenic 8.2 mi/13.2 km corridor between YELLOWSTONE (N) and GRAND TETON (S) national parks, commemorating Rockefeller's role in the creation of many national parks, including Grand Teton. Authorized 1972.

John Fitzgerald Kennedy Historic Site, BROOKLINE, E MASSACHUSETTS, birthplace and early boyhood home of President John F. Kennedy. Authorized 1967.

John F. Kennedy Center for the Performing Arts Memorial, WASHINGTON, D.C., opened 1971. Site of cultural performances in its theater, concert hall, and opera house.

John H. Kerr Reservoir (KUHR), (□ 78 sq mi/202.8 sq km), S central VIRGINIA (HALIFAX and MECKLENBURG counties) and N central NORTH CAROLINA (VANCE county), on ROANOKE RIVER, 64 mi/103 km SW of PETERSBURG (Virginia); 36°36′N 78°18′W. Maximum capacity 3,363,500 acre-ft. Has two main branches. Formed by John H. Kerr Dam (144 ft/44 m high), built (1953) by Army Corps of Engineers for power generation; also used for water supply, recreation, and as a fish and wildlife pond. Kerr Lake State Recreational Area (North Carolina) and two Virginia state parks here.

John James Audubon State Park (□ 1 sq mi/2.6 sq km), HENDERSON county, W KENTUCKY, near the OHIO RIVER, 3 mi/4.8 km NE of downtown HENDERSON, in city of Henderson. Memorial to famed naturalist and painter; includes migratory-bird refuge, museum, camping facilities, nature center.

John Martin Reservoir (□ 27.5 sq mi/71.2 sq km), BENT county, SE COLORADO, on ARKANSAS RIVER, in John Martin Reservoir State Wildlife Area, 15 mi/24 km E of LAS ANIMAS; 12 mi/19 km long, 2 mi/3.2 km wide; 39°35′N 104°54′W. Formed by John Martin Dam (153 ft/47 m high, 2.6 mi/4.2 km long), built (1948) of concrete and earthfill for flood control and irrigation. Formerly Caddoa Reservoir.

John Muir National Historic Site (□ 1 sq mi/2.6 sq km), CONTRA COSTA county, W CALIFORNIA. John Muir House and Martinez Adobe, 1 mi/1.6 km S of MARTINEZ, on Alhambra Avenue, commemorate contributions of John Muir to conservation and literature. Authorized 1964.

John Muir Trail, mountain footpath (c.200 mi/322 km long), E central CALIFORNIA, follows crest of the SIERRA NEVADA from YOSEMITE NATIONAL PARK (N) to MOUNT WHITNEY in SEQUOIA NATIONAL PARK (S). Now coincides with portion of Pacific Crest National Scenic Trail.

John o'Groats (JAHN o-GROTZ), locality, HIGHLAND, N Scotland, on the PENTLAND FIRTH, 14 mi/23 km N of WICK, near Duncansby Head; 58°38′N 03°05′W. It is often erroneously called the northernmost point of island of Britain; the phrase "from LAND'S END to John o'Groat's" is commonly used to denote greatest land distance in Britain (876 mi/1,401 km). In fact, DUNNET HEAD, 15 mi/24 km WNW, is the northernmost point of the island of Britain. The house, of which there are no remains, was, according to legend, built in octagonal shape by a Dutchman, John de Groot or John o'Groat, who settled in Scotland in 16th century.

John Redmond Lake, reservoir (□ 15 sq mi/39 sq km), COFFEY county, E central KANSAS, on Grand NEOSHO River, in Flint Hills National Wildlife Refuge, 23 mi/37 km SE of EMPORIA; 38°15′N 95°46′W. Maximum capacity 630,250 acre-ft. Formed by John Redmond Lake Dam (74 ft/23 m high), built (1964) by Army Corps of Engineers for flood control; also used for water supply and recreation. John Redmond State Park near dam. Also known as John Redmond Reservoir.

Johns, Alabama: see NORTH JOHNS. Named for Llewelyn Johns, a Welch mining engineer for the nearby coal and iron mines. Although it was inc. in 1912 as 'North Johns,' it is now commonly referred to as 'Johns.'

Johnsburg, village, MCHENRY county, NE ILLINOIS, suburb 40 mi/64 km NW of downtown CHICAGO, 4 mi/6.4 km NE of MCHENRY; 42°22′N 88°14′W. Located on FOX RIVER at its outflow from Lake PISTAKEE.

Johnsburg, hamlet, WARREN county, E NEW YORK, in ADIRONDACK MOUNTAINS, 26 mi/42 km NW of GLENS FALLS; 43°38′N 74°02′W. Resort. area. Lumbering.

Johnshaven (JAHNZ-hai-VUHN), fishing port and village (2001 population 646), Aberdeenshire, NE Scotland, on NORTH SEA, 6 mi/9.7 km ESE of LAURENCEKIRK; 56°47′N 02°20′W.

Johns Island, CHARLESTON county, S SOUTH CAROLINA, one of SEA ISLANDS, c.5 mi/8 km WSW of CHARLESTON; c.11 mi/18 km long, 5 mi/8 km–10 mi/16 km wide. Manufacturing of neon signs and agricultural chemicals; agriculture includes vegetables, sweet potatoes, watermelons.

Johnson (JAHN-suhn), community, central MANITOBA, W central CANADA, 19 mi/30 km from THOMPSON, in MYSTERY LAKE local government district; 55°32′N 97°31′W.

Johnson, county (□ 682 sq mi/1,773.2 sq km; 2006 population 24,453), NW ARKANSAS; ⊙ CLARKSVILLE; 35°34′N 93°27′W. Bounded S by ARKANSAS RIVER (Lake DARDANELLE in E part); drained by small MULBERRY RIVER and Piney Creek. Agriculture (soybeans; cattle, hogs, poultry [turkeys, chickens]). Coal mines; timber. Part of Ozark National Forest is in N half of county. Formed 1833.

Johnson, county (□ 313 sq mi/813.8 sq km; 2006 population 9,626), E central GEORGIA; ⊙ WRIGHTSVILLE; 32°42′N 82°40′W. Bounded W by OCONEE River; drained by OHOOPEE RIVER. Coastal plain agricultural area (cotton, corn, potatoes, soybeans, peanuts, fruit); manufacturing of apparel, textiles; lumber and millwork, wholesale trade. Formed 1858.

Johnson, county (□ 348 sq mi/904.8 sq km; 2006 population 13,360), S ILLINOIS; ⊙ VIENNA; 37°27′N 88°52′W. Agricultural area (fruit, sorghum, wheat; cattle; dairy products). Lumbering; wood products. Drained by CACHE RIVER; includes part of Shawnee National Forest (N); Ferne Clyffe State Park in NW; Lake Egypt Recreational Area in N. Formed 1812. One of 17 Illinois counties to retain Southern-style commission form of county government

Johnson, county (□ 321 sq mi/834.6 sq km; 2006 population 133,316), central INDIANA; ⊙ FRANKLIN; 39°29′N 86°04′W. Drained by West Fork of WHITE RIVER and tributaries of the East Fork of White River. Part of INDIANAPOLIS metropolitan area; urban growth in N, especially around GREENWOOD. Agriculture (wheat, corn, soybeans, vegetables; dairy products; hogs, cattle); manufacturing at Franklin, Greenwood, EDINBURGH. Atterbury State Fish and Wildlife Area and part of CAMP ATTERBURY military reservation in S. Formed 1822.

Johnson, county (□ 623 sq mi/1,619.8 sq km; 2006 population 118,038), E IOWA; ⊙ IOWA CITY; 41°40′N

91°32′W. Prairie agricultural area (corn; hogs, cattle, poultry) drained by IOWA RIVER; limestone quarries. Manufacturing at Iowa City and CORALVILLE. Includes Lake Macbride State Park in N; Coralville Reservoir (Iowa River), N of Iowa City. Formed 1839.

Johnson, county (□ 480 sq mi/1,248 sq km; 2006 population 516,731), E KANSAS; ⊙ Olathe; 38°52′N 94°52′W. Rolling plain area with low hills; bounded N by KANSAS River, E by MISSOURI. Agriculture in far S and W (wheat, soybeans, hay; cattle). Manufacturing (sand and gravel, oil and gas, paper products, printing, chemicals, plastics products, glass products, electronic equipment). Scattered oil and gas fields. Fast-growing suburban area in NE, adjacent to KANSAS CITY, KANSAS and KANSAS CITY, MISSOURI. Formed 1855.

Johnson, county (□ 263 sq mi/683.8 sq km; 2006 population 24,188), E KENTUCKY; ⊙ PAINTSVILLE, 37°51′N 82°51′W. Drained by LEVISA FORK. Agricultural area in foothills of the CUMBERLAND MOUNTAINS (burley tobacco, hay; some cattle and hogs). Coal mines, oil wells. Part of Paintsville Lake reservoir in W; Paintsville Lake State Park in W center. Formed 1843.

Johnson, county (□ 826 sq mi/2,147.6 sq km; 2006 population 50,646), W central MISSOURI; ⊙ WARRENSBURG; 38°45′N 93°48′W. Drained by BLACKWATER RIVER. Corn, wheat, sorghum, soybeans, hay, grapes; cattle, horses. Former coal mines, stone quarries, clay pits. Manufacturing at Warrensburg, HOLDEN, KINGSVILLE. Knob Noster State Park and WHITEMAN AIR FORCE BASE in E. Formed 1834.

Johnson, county (□ 376 sq mi/977.6 sq km; 2006 population 4,683), SE NEBRASKA; ⊙ TECUMSEH; 40°24′N 96°16′W. Agricultural area drained by branches of North Fork of BIG NEMAHA RIVER. Cattle, hogs, poultry; corn, sorghum, soybean, wheat; dairying; poultry products, feed. Formed 1856.

Johnson, county (□ 299 sq mi/777.4 sq km; 2006 population 18,043), extreme NE TENNESSEE; ⊙ MOUNTAIN CITY; 36°28′N 81°52′W. Bounded N by VIRGINIA, E and SE by NORTH CAROLINA; STONE MOUNTAINS lie along North Carolina state line; traversed by IRON MOUNTAINS; drained by WATAUGA River. Includes parts of Watauga Reservoir and Cherokee National Forest. Agriculture; manufacturing; natural resources; tourism. Formed 1836.

Johnson, county (□ 734 sq mi/1,908.4 sq km; 2006 population 149,016), N central TEXAS; ⊙ CLEBURNE; 32°22′N 97°21′W. Bounded SW by BRAZOS RIVER; drained by tributaries of the Brazos (in W), including Nolan Creek, and the TRINITY (in E). Shipping, processing center. Rich agricultural area: cotton, wheat, grain sorghum, corn, silage, hay; extensive dairying; cattle, hogs, horses. Limestone, sand and gravel. Manufacturing, processing at Cleburne. Cleburne State Park in SW; Lake Pat Cleburne in SW center. Formed 1854.

Johnson, county (□ 4,174 sq mi/10,852.4 sq km; 2006 population 8,014), N central WYOMING; ⊙ BUFFALO; 44°02′N 106°34′W. Agriculture and coal-mining region; watered by POWDER RIVER and Clear and Crazy Woman creeks. Sugar beets, hay, alfalfa; sheep, cattle; timber; sand and gravel; oil, uranium, coal). Part of Bighorn National Forest in NW, including part of Cloud Peak Wilderness Area; part of BIGHORN MOUNTAINS in W 0.3 of county. Small Lake De Smet Reservoir in N. Formed 1875.

Johnson (JAHN-suhn), township (□ 46 sq mi/119.6 sq km; 2001 population 658), ALGOMA district, ONTARIO, E central CANADA.

Johnson (JAWN-son), town (2006 population 1,402), including Johnson village, LAMOILLE CO., N central VERMONT, on LAMOILLE RIVER and 5 mi/8 km NW of Hyde Park; 44°38′N 72°40′W. Johnson State College here. Settled 1784. Named for landowner, William Samuel Johnson.

Johnson, village (2000 population 32), BIG STONE county, W MINNESOTA, 20 mi/32 km NNE of ORTONVILLE; 45°34′N 96°17′W. Grain; manufacturing (feeds).

Johnson, village (2006 population 259), NEMAHA county, SE NEBRASKA, 8 mi/12.9 km W of AUBURN; 40°24′N 96°00′W. In agricultural region.

Johnson, village (2006 population 232), POTTAWATOMIE county, central OKLAHOMA, 5 mi/8 km NE of SHAWNEE, on NORTH CANADIAN River; 35°24′N 96°50′W. Agricultural area.

Johnsonburg, village, WARREN county, NW NEW JERSEY, 9 mi/14.5 km SW of NEWTON, in hilly region; 40°57′N 74°52′W.

Johnsonburg, borough (2006 population 2,777), ELK county, N central PENNSYLVANIA, 32 mi/51 km S of BRADFORD, on CLARION RIVER; 41°20′N 78°40′W. Manufacturing (fabricated metal products, paper products); natural gas; agriculture (grain; livestock; dairying); timber. Allegheny National Forest to W; Bendingo State Park to NE. Settled 1810, laid out 1888.

Johnson City, city (2006 population 59,866), WASHINGTON and CARTER counties, extreme NE TENNESSEE, 20 mi/32 km SE of KINGSPORT, 36°20′N 82°23′W. Rich natural resources; diverse manufacturing. East Tennessee State University is here. Incorporated 1869.

Johnson City, town (2000 population 1,528), ⊙ STANTON county, SW KANSAS, 50 mi/80 km SW of GARDEN CITY; 37°34′N 101°45′W. Elevation 3,330 ft/1,015 m. In wheat area; feeds, fertilizers.

Johnson City, town (2006 population 1,537), ⊙ BLANCO county, central TEXAS, 40 mi/64 km W of AUSTIN, on PEDERNALES RIVER; 30°16′N 98°24′W. Elevation 1,197 ft/365 m. Manufacturing. The LYNDON B. JOHNSON NATIONAL HISTORIC SITE includes Johnson's boyhood home and an information center. His birthplace and the family cemetery where he is buried are 15 mi/24 km W at STONEWALL, at the LBJ ranch. Pedernales Falls State Park to E.

Johnson City, village (□ 4 sq mi/10.4 sq km; 2006 population 14,889), Broome county, S NEW YORK, in tri-city area including ENDICOTT and BINGHAMTON; 42°07′N 75°57′W. Light manufacturing and services. Formerly noted for its Endicott-Johnson shoes. Originally called Lestershire, the area remained rural until a shoe company built a factory here in 1890. The name was changed in 1916. Incorporated 1892.

Johnson City, village (2006 population 668), CLACKAMAS CO., NW OREGON, residential suburb, 8 mi/12.9 km SSE of downtown PORTLAND and 1 mi/1.6 km NE of GLADSTONE, on Kellogg Creek; 45°23′N 122°34′W.

Johnson Creek, town (2006 population 2,158), JEFFERSON county, S WISCONSIN, on small Johnson Creek, and 8 mi/12.9 km S of WATERTOWN; 43°04′N 88°46′W. In dairying region. Ships dairy products, eggs; manufacturing (grain processing, rubber products, furniture, consumer goods). Aztalan State Park to W; Glacial Drumlin State Trail passes to S.

Johnson Industrial Airport, industrial area, JOHNSON county, KANSAS, 4 mi/6.4 km SW of OLATHE, in urban-growth fringe of KANSAS CITY, Kansas, and KANSAS CITY, MISSOURI. Manufacturing (consumer goods, printing, food emulsifiers, wood products, communications equipment, farm machinery, fats and oils). Agriculture to S and W. Airport. Airport Code JCI.

Johnson Island, CHILE: see CHONOS ARCHIPELAGO.

Johnsonville, town, MONTSERRADO county, W LIBERIA, 10 mi/16 km ENE of MONROVIA, on road. Palm oil and kernels, rubber, coffee.

Johnsonville, town (2006 population 1,455), FLORENCE county, E SOUTH CAROLINA, 30 mi/48 km SSE of FLORENCE, near LYNCHES RIVER; 33°49′N 79°27′W. Hogs, chickens; grain, cotton, tobacco; timber.

Johnsonville, village (2000 population 69), WAYNE county, SE ILLINOIS, 14 mi/23 km NW of FAIRFIELD; 38°31′N 88°32′W. In agricultural area; oil wells.

Johnston (JAHN-stuhn), county (□ 795 sq mi/2,067 sq km; 2006 population 152,143), central NORTH CAROLINA; ⊙ SMITHFIELD; 35°31′N 78°22′W. On coastal plain; drained by NEUSE RIVER; source of SOUTH RIVER in SW. Service industries; manufacturing (cotton and tobacco processing, lumber milling) at Smithfield, SELMA and CLAYTON; agriculture (tobacco, cotton, wheat, oats, soybeans, hay, sweet potatoes; poultry, cattle, hogs); timber (pine, gum). Bentonville Battleground State Historical Site in S; Clemmons Educational State Forest in NW. Formed 1746 from Craven County. Named for Gabriel Johnson (1699–1752), governor of North Carolina from 1734–1752.

Johnston, county (□ 658 sq mi/1,710.8 sq km; 2006 population 10,436), S OKLAHOMA; ⊙ TISHOMINGO; 34°18′N 96°39′W. Bounded on S in part by Lake TEXOMA (WASHITA River arm); drained by BLUE and Washita rivers. Cattle raising; dairying; peanuts. Sand and gravel pits. Part of Tishamingo National Wildlife Refuge on arm of Lake Texoma in S. Formed 1907.

Johnston, town (2000 population 8,649), POLK county, central IOWA, near DES MOINES RIVER, suburb, 7 mi/11.3 km NW of downtown DES MOINES; 41°41′N 93°42′W. Manufacturing (machinery, concrete; sand and gravel processing). Saylorville Dam to N; Margo Frankel Woods State Park to E. U.S. Camp Dodge nearby was active in World War II.

Johnston, town (2000 population 28,195), PROVIDENCE county, N central RHODE ISLAND, a suburb of PROVIDENCE; 41°50′N 71°31′W. Johnston is the home of several insurance companies. Its many historic landmarks include the Clemence-Irons House (c.1680). Named for Augustus Johnston, Attorney General for the county from 1758–1776. Set off from Providence and incorporated 1759.

Johnston, town (2006 population 2,346), EDGEFIELD county, W SOUTH CAROLINA, 10 mi/16 km ENE of EDGEFIELD; 33°49′N 81°48′W. Manufacturing of cotton products; agriculture includes livestock; dairying; grain, cotton, soybeans.

Johnston City, city (2000 population 3,557), WILLIAMSON county, S ILLINOIS, 7 mi/11.3 km N of MARION; 37°49′N 88°55′W. In bituminous-coal mining and agricultural area (corn, wheat, hay). Incorporated 1896.

Johnstone (jahn-STON), town (2001 population 16,468), Renfrewshire, W Scotland, 4 mi/6.4 km W of PAISLEY; 55°50′N 04°30′W. Engineering works. Light manufacturing (chemicals, fabricated metal products, shoelaces). Previously cotton and flax mills.

Johnstone Strait (JAHN-ston), SW BRITISH COLUMBIA, W CANADA, joins QUEEN CHARLOTTE Strait with Strait of GEORGIA, via DISCOVERY PASSAGE, separating VANCOUVER ISLAND from mainland; 70 mi/113 km long, 2 mi/3 km–3 mi/5 km wide; 50°27′N 126°00′W. Mainland shore deeply indented. ALERT BAY is at W entrance.

Johnston Island, atoll, central PACIFIC, c.700 mi/1,100 km SW of HONOLULU; c.3,000 ft/900 m long and c.600 ft/183 m wide. It was discovered by Americans in 1796 (or the British in 1807) and claimed by the UNITED STATES in 1858. It was not, however, included in HAWAII statehood, but became an unincorporated territory of the U.S. The U.S. Navy took over the atoll in 1934 and used it as a seaplane and submarine base during World War II. Operational control was given to the Defense Nuclear Agency in 1958, and the U.S. conducted a series of high altitude nuclear tests here during the 1950s and 1960s. It is still designated as a standby site should the U.S. resume testing. The Johnston Atoll Chemical Agent Disposal System (JACADS), a facility for the stockpiling and incineration of chemical weapons, is located here. The U.S. Army has used the island as a site for the destruction

of chemical weapons since 1990. A bird reservation since 1923, the atoll was declared a Pacific Islands National Wildlife Refuge in 1974.

Johnstons Station or **Johnston Station**, village, PIKE county, SW MISSISSIPPI, on the BOGUE CHITTO, and 7 mi/11.3 km N of MCCOMB.

Johnston Station, IOWA: see JOHNSTON.

Johnstown, city (□ 4 sq mi/10.4 sq km; 2000 population 8,511), ⊙ FULTON county, E central NEW YORK; 43°00′N 74°22′W. Some light manufacturing. Its leather-glove industry dates back to 1800, and Johnstown and nearby GLOVERSVILLE are known as the "Glove Cities." Only token production exists today. Notable buildings include the county courthouse (1774) and Fort Johnstown (1771), the county jail. The last American Revolutionary battle in New York state was fought in Johnstown on October 25, 1781. Elizabeth Cady Stanton born here. Seat of Fulton Montgomery Community College. Founded 1772, incorporated 1895.

Johnstown, city (2006 population 22,269), CAMBRIA county, W central PENNSYLVANIA, 55 mi/89 km E of PITTSBURGH, at confluence of Little Conemaugh and STONYCREEK rivers, which form CONEMAUGH RIVER; 40°19′N 78°55′W. Formerly one of the great centers of U.S. heavy industry, using coal from nearby Conemaugh Valley mines. Manufacturing (fabricated metal products, apparel, dairy products, machinery, printing and publishing, consumer goods, construction materials, furniture, ice rinks). Branches of U.S. Steel and Bethlehem Steel were here before the decline of the steel industry in the 1970s and 1980s when thousands of jobs were lost. Part of abandoned Bethlehem Steel Plant used for Heritage Museum. The first Kelly pneumatic converter for the transformation of crude iron into steel was built here in 1862. On May 31, 1889, South Fork Dam with its large upriver reservoir c.12 mi/19 km above Johnstown broke as a result of heavy rains, and the city was flooded, with the devastating loss of nearly 2,200 lives; this was one of the greatest disasters of 19th-century industrialized America. The river was later channeled (completed 1943) for flood prevention, but the city continues to be subject to recurrent flooding. The University of Pittsburgh at Johnstown 6 mi/9.7 km to SE; a state rehabilitation center to SW at FERNDALE; Johnstown Flood National Memorial 8 mi/12.9 km to E; ALLEGHENY PORTAGE RAILROAD NATIONAL HISTORIC SITE 20 mi/32 km to NE; Johnstown Flood Museum; Johnstown-Cambria County Airport to E; Cambria County Arts Center, now a major tourist center; Johnstown Inclined Plane, Pennsylvania's steepest inclined railroad; parts of Gallitzin State Forest to NW and SE; Laurel Ridge State Park to SW; QUEMAHONING RESERVOIR to S. Settled 1770, incorporated as a city 1936.

Johnstown, Gaelic *Cúirt an Phúca*, town (2006 population 445), NW KILKENNY county, SE IRELAND, 13 mi/21 km WNW of KILKENNY; 52°45′N 07°33′W. Agricultural market. Once a Purcell stronghold. Reputed grave of Oisin nearby.

Johnstown, town (2000 population 3,827), WELD county, N COLORADO, on Little Thompson River, 13 mi/21 km WSW of GREELEY; 40°20′N 104°54′W. Elevation 4,818 ft/1,469 m. Grain, beans, and sugar beet region. Fertilizer.

Johnstown, village (2000 population 53), BROWN county, N NEBRASKA, 10 mi/16 km W of AINSWORTH; 42°34′N 100°03′W.

Johnstown (JAHNS-toun), village (□ 2 sq mi/5.2 sq km; 2000 population 3,440), LICKING county, central OHIO, 16 mi/26 km WNW of NEWARK, on RACCOON CREEK; 40°09′N 82°41′W. In diversified farming area.

Johnstown Flood National Memorial, CAMBRIA county, W central PENNSYLVANIA, 8 mi/12.9 km E of JOHNSTOWN, at South Fork Dam site, on South Fork of Little Conemaugh River; 40°20′N 78°46′W. Memorializes the Johnstown flood of May 31, 1889, in which 2,200 people lost their lives, after dam collapsed

during heavy rains. See JOHNSTOWN, Pennsylvania. Authorized 1964.

John W. Flannagan Reservoir (FLAN-uh-guhn), DICKENSON county, extreme SW VIRGINIA, in Pound River, 25 mi/40 km N of NORTON, near KENTUCKY state line, on boundary of Jefferson National Forest; 37°14′N 82°20′W. Dam is 216 ft/66 m high, built in 1971 by the Army Corps of Engineers for flood control. Reservoir used for recreational purposes; has a maximum water storage capacity of 145,700 acre-ft.

Johoku (JO-ho-koo), town, East Ibaraki county, IBARAKI prefecture, central HONSHU, E central JAPAN, 9 mi/15 km N of MITO; 36°28′N 140°22′E.

Johol (JO-hole), town, S Negri SEMBILAN, MALAYSIA, 23 mi/37 km ESE of SEREMBAN. Rice, rubber. Chief town of JOHOL, one of the original Negri Sembilan states.

Johor (JO-hoar), state (□ 7,360 sq mi/19,136 sq km; 2000 population 2,740,625), at the S extremity of the MALAY PENINSULA, MALAYSIA, opposite SINGAPORE; ⊙ JOHOR BARU, across the strait from Singapore. The principal rivers and communication routes are the MUAR and the JOHOR. The Chinese and the Malays are the two largest groups in the population, and there is a significant Indian minority. Johor has extensive rubber and oil-palm plantations; other agricultural products are rice, copra, pineapples, and gambier. Tin and bauxite are mined. The far S region of the state is becoming rapidly industrialized because of foreign investment and from industrial relocation from Singapore. It anchors the Malaysian end of the Singapore and RIAU (INDONESIA) growth triangle. After the fall of MALACCA (MELAKA) to the Portuguese (1511), the former sultan of Malacca continued to rule over Johor and the RIAU. In the 18th century the Bugis, a Malay people from CELEBES, became dominant in Johor. In 1819 a British-installed sultan granted the site of Singapore to the British East India Company and became for practical purposes an independent ruler. Thereafter relations with GREAT BRITAIN were friendly. Johor remained one of the most peaceful of the Malay states. In 1885, Johor and Great Britain established formal treaty relations, and in 1914 Johor became a British protectorate. Until 1948, when it entered the FEDERATION OF MALAYA, Johor was classified as one of the UNFEDERATED MALAY STATES. In the 1970s Johor Tenggara became the site of a major resettlement and agricultural development project. See MALAYSIA, FEDERATION OF. Sometimes spelled Jahore.

Johor Baru (JO-hoar BAH-roo), **Johore Bharu** or **Johore Bahru**, city (2000 population 630,603), ⊙ JOHOR, MALAYSIA, S MALAY PENINSULA, opposite SINGAPORE, 13 mi/21 km NNW of Singapore City; satellite city of Singapore; 01°27′N 103°45′E. The city is connected with Singapore by a stone causeway across the narrow JOHOR STRAIT. It is a trade center for rubber and tropical produce, and is an increasingly important industrial center as part of an emerging Singapore-Johor-RIAU "growth triangle." Has flour and feed mills. The seat of the sultan of Johor is in Johor Bahru; his residence, Bakit Serene, contains priceless art treasures. The city has a large Chinese population; Jaro Handicraft Center; Sultan Abu Bakar College; Istana Gardens.

Johor River, 40 mi/64 km long, JOHOR, S MALAYSIA; formed 15 mi/24 km NW of KOTA; flows SE, past KOTA TINGGI, to SINGAPORE STRAIT forming wide estuary in low mangrove coast; 01°40′N 103°57′E. Formerly spelled Johore River.

Johor Strait (jo-HOR) or **Selat Johor**, arm of the Straits of SINGAPORE, between Singapore Island (S) and JOHOR, MALAYSIA(N); c.40 mi/64 km long and 1 mi/1.6 km–3 mi/4.8 km wide; 01°28′N 103°47′E. The E part of the strait separates into two channels, Serangoon Harbour, a deep channel leading to part of Selatar, N Singapore, and Nanas Channel, between Malaysia and Pulau UBIN island, Singapore. Johor

Causeway (3,443 ft/1,049 m long; opened 1924) connects Johor Baharu, Malaysia, and Woodlands, Singapore railroad and highway). Large Pulau Tekong island, Singapore, at strait's E entrance.

Jõhvi (YUH-vee), German *Jewe*, city, NE ESTONIA, on railroad and 27 mi/43 km W of NARVA; 59°21′N 27°25′E. In oil-shale mining area; brewery.

Jóia (ZHOI-zh), town (2007 population 8,279), NW RIO GRANDE DO SUL state, BRAZIL, 20 mi/32 km SW of IJUÍ; 28°39′S 54°08′W. Wheat, soybeans, corn; livestock.

Joice, town (2000 population 231), WORTH county, N IOWA, 19 mi/31 km NW of MASON CITY; 43°21′N 93°27′W. Livestock-shipping point; animal feeds. Rice Lake State Park to NW.

Joides Basin (JOI-duhz), ANTARCTICA, basin beneath W ROSS SEA, between MAWSON and CRARY Banks and PENNELL Bank, S PACIFIC OCEAN; 74°30′S 174°00′E.

Joigny (zhwah-nyee), town (□ 17 sq mi/44.2 sq km), YONNE department, N central FRANCE, on the YONNE, and 15 mi/24 km NNW of AUXERRE; 47°59′N 03°24′E. Vineyards in N BURGUNDY; woodworking, fruit preserving. Has two 16th-century churches and many restored 15th- and 16th-century houses in old town center.

Joiner, village, MISSISSIPPI CO., NE ARKANSAS, 27 mi/43 km NNW of MEMPHIS, TENNESSEE; 35°30′N 90°09′W. Cotton, rice, soybeans. MISSISSIPPI RIVER to SE.

Joinerville, unincorporated village, RUSK county, E TEXAS, 5 mi/8 km W of HENDERSON. In East Texas oil field.

Joinville (ZHO-een-vee-lee), city (2007 population 487,003), NE SANTA CATARINA, BRAZIL, on an Atlantic inlet, opposite São Francisco Island, on railroad, and 90 mi/145 km NNW of FLORIANÓPOLIS; 26°18′S 48°50′W. State's chief industrial center, with textile mills, breweries, distilleries; manufacturing (maté processing, furniture, ships). Ships rice, sugar, tapioca, corn, tobacco, arrowroot, and dairy products via seaport of SÃO FRANCISCO DO SUL (12 mi/19 km NE). Airfield. Iron deposits in area. Founded c.1850 by German immigrants as center of agricultural colony (sometimes called Doña Francisca). Sometimes spelled Joinville.

Joinville (zhwan-veel), town (□ 7 sq mi/18.2 sq km), HAUTE-MARNE department, CHAMPAGNE-ARDENNE region, NE FRANCE, on MARNE RIVER and MARNE-SAÔNE CANAL, 16 mi/26 km SE of SAINT-DIZIER; 48°27′N 05°08′E. Paper and flour mills. Has ruined castle of seneschals of CHAMPAGNE (most famous was Jean de Joinville). Here treaty allying Spain with the Catholic League was signed in 1584.

Joinville Island, ANTARCTICA, off NE tip of GRAHAM LAND; 46 mi/74 km long, 14 mi/23 km wide; 63°15′S 55°45′W. Discovered 1838 by Dumont d'Urville, French navigator.

Joinville-le-Pont (zhwan-veel-luh–pon), residential town, VAL-DE-MARNE department, ÎLE-DE-FRANCE region, N central FRANCE, a SE suburb of PARIS, 6 mi/9.7 km from Notre Dame Cathedral, at narrow neck of MARNE RIVER meander and at SE edge of the forest of VINCENNES; 48°49′N 02°28′E. Film-making studios; diverse consumer goods manufacturing.

Jojima (JO-jee-mah), town, Mizuma county, FUKUOKA prefecture, NW KYUSHU, SW JAPAN, 22 mi/35 km S of FUKUOKA; 33°14′N 130°25′E. Site of feudal castle.

Jojutla de Juárez (ho-HOOT-lah dai HWAH-res), city and township (□ Jojutla municipio, MORELOS, central MEXICO, on S slope of central plateau, on railroad, 22 mi/35 km S of CUERNAVACA; 18°36′N 99°10′W. Elevation 2,920 ft/890 m. Agricultural center (rice, sugarcane, melons, tropical fruit; livestock). Lake TEQUESQUITENGO, popular fishing and hunting resort, is 4 mi/6.4 km W.

Jóka, SLOVAKIA: see JELKA.

Jokela (YO-ke-lah), village, Uudenmaan province, S Finland, 25 mi/40 km N of Helsinki; 60°33′N 24°59′E. Elev. 330 ft/100 m.

Jokioinen (YO-kee-OI-nen), Swedish *Jockis* or *Jokkis*, village, Hämeen province, SW Finland, 35 mi/56 km WSW of Hameenlinna; 60°49′N 23°28′E. Elev. 330 ft/100 m. Site of experimental farm and plant nursery.

Jokjakarta, INDONESIA: see YOGYAKARTA.

Jokkis, FINLAND: see JOKIOINEN.

Jokkmokk (YOK-mok), town, NORRBOTTEN county, N SWEDEN, within ARCTIC CIRCLE, on headstream of LULEÄLVEN RIVER, 75 mi/121 km NW of BODEN; 66°36′N 19°50′E. Road junction and trade center.

Jokneam, ISRAEL: see YOKNEAM.

Jökulsá (YUH-kul-soo), name of several Icelandic rivers formed by glaciers. The best known is the JOKULSA á Fjöllum, which rises on the N slope of the VATNAJOKULL in SE ICELAND and flows c.130 mi/210 km N into the AXARFJÖRÐUR, forming the DETTIFOSS c.30 mi/50 km from its mouth.

Jokulsargljufur National Park (□ 58 sq mi/150.8 sq km), NORDUR-THINGEYJARSYSLA county, NE ICELAND, 20 mi/32 km E of HUSAVIK. Canyon of Jokuls A Fjollum (river), begins at DETTIFOSS (waterfall; 148 ft/45 m high, 328 ft/100 m wide) in S, past 2 more falls. Rare and isolated plant communities. Birdlife includes meadow pipits, ptarmigans, gyrfalcons; nesting areas of great skua. Established 1973.

Jokyakarta, INDONESIA: see YOGYAKARTA.

Jolalpan (ho-LAHL-pan), town, PUEBLA, central MEXICO, 30 mi/48 km SW of IZÚCAR DE MATAMOROS. Elevation 2,690 ft/820 m. Corn, rice, fruit, sugar; livestock.

Jolburi, THAILAND: see CHON BURI.

Jolfa, IRAN: see JOLFA.

Joliet (JO-lee-EHT), city (2000 population 106,221), ⊙ WILL county, NE ILLINOIS, on the DES PLAINES River, satellite city of CHICAGO; 41°31′N 88°07′W. A river port and an industrial shipping center, with limestone quarries and coal mines in the area. Manufacturing (machinery, electronic equipment, chemicals, metal products, paper products, transportation equipment, plastic products, food products, consumer goods); oil refineries. Riverboat gambling now primary industry. Joliet is the seat of the College of St. Francis and Joliet Junior College. Joliet Army Ammunition Plant SW of city, now abandoned, to become a tall grass prairie, landfill, industrial park, and veterans' cemetery. New stockyards constructed early 1970s. Joliet Correctional Center here; Stateville Correctional Center, a maximum security facility, nearby. Fastest-growing city in Illinois. Incorporated 1845.

Joliet (jo-lee-ET), village (2000 population 575), CARBON county, S MONTANA, on Rock Creek, and 31 mi/50 km SW of BILLINGS; 45°29′N 108°58′W. In irrigated agricultural region: barley, oats, corn, sugar beets, beans, garden crops; cattle, sheep. Cooney Reservoir State Park on RED LODGE CREEK to WSW.

Joliette (zhol-YEHT), county (□ 162 sq mi/421.2 sq km; 2006 population 57), S QUEBEC, E CANADA, on the SAINT LAWRENCE RIVER; ⊙ JOLIETTE; 46°01′N 73°27′W. Composed of ten municipalities. Formed in 1982.

Joliette (zho-LYET), city (□ 8 sq mi/20.8 sq km), ⊙ JOLIETTE county, Lanaudière region, S QUEBEC, E CANADA, on L'ASSOMPTION RIVER, NE of MONTREAL; 46°02′N 73°26′W. Industries include steel, paper, textile, and ceramic manufacturing, tobacco processing, and limestone quarrying. The Séminaire de Joliette, affiliated with the University of Montreal, is here.

Jolimont (JAH-li-mahnt), locality, EAST MELBOURNE suburb, VICTORIA, SE AUSTRALIA. Railway station (37°49′S 144°59′E). Melbourne Cricket Ground. Residential, commercial buildings.

Jolivue, unincorporated town, AUGUSTA county, NW VIRGINIA, 3 mi/5 km S of STAUNTON; 38°07′N 79°04′W. Agriculture (apples, grain; livestock; dairying).

Jolley, town (2000 population 54), CALHOUN county, central IOWA, 6 mi/9.7 km NW of ROCKWELL CITY; 42°28′N 94°43′W. In agricultural area.

Jolly, village (2006 population 189), CLAY county, N TEXAS, 9 mi/14.5 km E of WICHITA FALLS; 33°52′N 98°20′W. Oil and natural gas. Cattle; dairying; cotton, peaches. ARROWHEAD LAKE reservoir and State Park to S.

Jollyville, unincorporated city (2000 population 15,813), WILLIAMSON county, S TEXAS, residential suburb 13 mi/21 km N of downtown AUSTIN; 30°26′N 97°45′W. Located in growing N fringe of smaller urbanized area.

Jolo (JO-lo), unincorporated town, MCDOWELL county, S WEST VIRGINIA, near DRY FORK, 14 mi/23 km SW of WELCH near VIRGINIA state line; 37°19′N 81°48′W. Bituminous-coal region. Panther State Forest to NW; Berwind Lake Wildlife Management Area to SE.

Jolo, island (□ 345 sq mi/897 sq km; 2000 population 87,998), SULU province, SULU ARCHIPELAGO, the PHILIPPINES; 05°58′N 121°06′E. The seaport city, Jolo (1990 population 53,055), on the NW coast of the island, is ⊙ Sulu province, the province's largest town, the trading and shipping hub of the archipelago (between islands and to SABAH and SINGAPORE), and a Muslim center. Marine products. An ancient walled city, it was once a pirate base and served as the residence of a sultan until the sultanate was abolished in 1940. The city was almost completely destroyed in 1974 when fighting erupted between government forces and Muslim insurgents who were seeking to establish a secessionist state. After the battle, the rebels withdrew into the island's interior to fight a war of attrition, which continued until 1990, when a detente was reached. Airport 0.3 mi/0.5 km from downtown.

Jolon (hoh-LON), village, MONTEREY county, W CALIFORNIA, in valley of COAST RANGES, 35 mi/56 km NW of PASO ROBLES. In E edge of HUNTER LIGGETT MILITARY RESERVATION; restored Mission San Antonio de Padua (founded 1771) to NW.

Jølstravatn (YUHL-strah-VAH-tuhn), lake (□ 15 sq mi/39 sq km) in SOGN OG FJORDANE county, W NORWAY, at W foot of the glacier JOSTEDALSBREEN, 40 mi/64 km E of FLORØ. Tourist attraction.

Jólsva, SLOVAKIA: see JELSAVA.

Joly (JO-lee), township (□ 75 sq mi/195 sq km; 2001 population 290), S ONTARIO, E central CANADA, 13 mi/20 km from KEARNEY; 45°47′N 79°14′W.

Joly (zho-LEE), unincorporated village, S QUEBEC, E CANADA, 30 mi/48 km SW of QUEBEC city, and included in Saint-JANVIER-de-Joly; 46°29′N 71°40′W. Garnet mining.

Joly, Mont (zho-LEE), summit (8,283 ft/2,525 m) of Savoy Alps (ALPES FRANÇAISES), HAUTE-SAVOIE department, RHÔNE-ALPES region, SE FRANCE 5 mi/8 km S of SAINT-GERVAIS-LES-BAINS, overlooking the valley of the Bon Nant torrent; 45°49′N 06°41′E. Splendid views of MONT BLANC (8 mi/12.9 km E).

Jomala (YO-mah-lah), village, ÅLAND province, SW FINLAND, in central part of Åland Island, 5 mi/8 km N of MAARIANHAMINA; 60°09′N 19°58′E. Elevation 17 ft/5 m. Grain.

Jo-Mary Lakes, central MAINE, 10 mi/16 km W of MILLINOCKET; chain of three lakes (Upper, Middle, and Lower), each c.3 mi/4.8 km long, joined to PEMADUMCOOK LAKE to N.

Jombang (JOM-bahng), town, ⊙ JOMBANG district, Java Timur province, INDONESIA, on BRANTAS RIVER, and 40 mi/64 km WSW of SURABAYA; 07°33′S 112°19′E. Trade center for agricultural area (rice, corn, cassava, peanuts); also highway and railroad hub. Also spelled Djombang.

Jombe, village, MIDLANDS province, central ZIMBABWE, 17 mi/27 km WNW of KWEKWE; 18°51′S 29°34′E. Livestock; corn, wheat, cotton, tobacco, soybeans.

Jomda (JIANG-DAH), town, E TIBET, SW CHINA, near Yangzi River, 80 mi/129 km NE of QAMDO; 31°30′N 98°16′E. Livestock; electric power generation. Also appears as Rangsum, Tungpu, or Chiang-ta.

Jomsburg, POLAND: see WOLIN, town.

Jomsom (JUHM-suhm), town, ⊙ MUSTANG district, central NEPAL; 28°46′N 83°43′E. Elevation 8,800 ft/ 2,682 m. Wheat, barley. Nepali Army high-altitude warfare training center. Tourism. Airport.

Jomu (jo-MOO), village, SHINYANGA region, NW central TANZANIA, 20 mi/32 km SW of SHINYANGA, near MANONGA RIVER; 03°55′S 33°11′E. Cattle, goats, sheep; corn, wheat; timber.

Jona, town (2000 population 16,947), ST. GALLEN canton, N SWITZERLAND, near RAPPERSWIL on LAKE ZÜRICH. Textiles; woodworking.

Jonacatepec (ho-nah-KAH-te-pek), city and township, MORELOS, central MEXICO, 32 mi/51 km ESE of CUERNAVACA; 18°41′N 98°48′W. Agricultural center (rice, coffee, sugarcane, limes, and other fruit).

Jonage (zho-nahzh), town (□ 4 sq mi/10.4 sq km; 2004 population 5,679), RHÔNE department, RHÔNE-ALPES region, E central FRANCE; 45°48′N 05°02′E.

Jonan (JO-NAHN), town, Shimomashiki county, KUMAMOTO prefecture, W KYUSHU, SW JAPAN, 5 mi/8 km S of KUMAMOTO; 32°42′N 130°43′E.

Jonava (YO-nuh-vuh), German *Janow*, Russian *Janov*, city (2001 population 34,954), central LITHUANIA, on VILIYA RIVER, and 20 mi/32 km NE of KAUNAS; 55°05′N 24°17′E. Furniture manufacturing center; chemicals, fertilizer, apparel, textiles, construction materials; sawmilling. Railroad junction. In Russian KOVNO government until 1920. Also spelled Ionava.

Jondaryan (jahn-DE-ree-uhn), hamlet, QUEENSLAND, NE AUSTRALIA, 107 mi/172 km W of BRISBANE, 27 mi/ 43 km from TOOWOOMBA; 27°22′S 151°36′E. Jondaryan Woolshed tourist complex. Annual Australian Heritage Festival. Variant spelling: Jondaryn.

Jondaryn, AUSTRALIA: see JONDARYAN.

Jone (JO-NI), town, ⊙ Jone county, SE GANSU province, NW CHINA, 100 mi/161 km S of LANZHOU, on TAO RIVER, in mountain region; 34°35′N 103°32′E. Livestock; building materials. Also appears as Zhuoni.

Jones, county (□ 402 sq mi/1,045.2 sq km; 2006 population 26,973), central GEORGIA, ⊙ GRAY; 33°02′N 83°34′W. Bounded SW by OCMULGEE River. Intersected by fall line. Agriculture (peaches, soybeans, pimientos; cattle, poultry, hogs); sawmilling. Part of CHATTAHOOCHEE National Forest in W. Formed 1807.

Jones, county (□ 576 sq mi/1,497.6 sq km; 2006 population 20,505), E IOWA; ⊙ ANAMOSA; 42°07′N 91°07′W. Prairie agricultural area (cattle, hogs, poultry; corn, oats) drained by WAPSIPINICON and South and North Forks of MAQUOKETA rivers. Limestone quarries, sand and gravel pits. Wapsipinicon State Park to W. Formed 1837.

Jones, county (□ 699 sq mi/1,817.4 sq km; 2006 population 66,715), SE MISSISSIPPI; ⊙ LAUREL and ELLISVILLE; 31°37′N 89°10′W. Drained by LEAF RIVER and TALLAHALA CREEK. Agriculture (cotton, corn, sweet potatoes, honey; poultry, cattle; dairying); timber. Oil and natural gas. Includes part of De Soto National Forest in SE. Lake Bogue Homa State Lake in NE. Formed 1826.

Jones (JONZ), county (□ 473 sq mi/1,229.8 sq km; 2006 population 10,204), E NORTH CAROLINA; ⊙ TRENTON; 35°00′N 77°22′W. Forested and swampy tidewater area. Bounded S in part by White Oak River. White Oak Swamp (SE) including Catfish Lake on E boundary; drained by TRENT RIVER. Service industries; manufacturing; agriculture (tobacco, corn, cotton, beans, wheat, soybeans, hay; hogs); timber (pine, gum). Stone quarrying. Includes part of Croatan National Forest (SE), Wolf Swamp (S). Formed 1778 from Craven County. Named for Willie Jones (1740–1801), Revolutionary leader and later opponent to the adoption of the Constitution of the U.S.

Jones, county (□ 971 sq mi/2,524.6 sq km; 2006 population 1,067), S central SOUTH DAKOTA; ⊙ MURDO; 43°57′N 100°41′W. Agricultural area drained by BAD RIVER and Dry and White Clay creeks; bounded S by WHITE RIVER. Wheat, cattle. Part of Fort Pierre National Grassland in NE. Mountain/Central time zone boundary splits county down the middle, then follows White River E; DRAPER is in Central (E of boundary), Murdo and Okaton in Mountain (W). Formed 1916.

Jones (JONS), county (□ 937 sq mi/2,436.2 sq km; 2006 population 19,645), W central TEXAS; ⊙ ANSON; 32°44′N 99°52′W. Rolling plains, drained by Clear Fork of BRAZOS RIVER. Rich agricultural county (cotton, wheat, milo, hay, watermelons, peanuts); livestock (cattle). Oil and gas wells; gypsum, sand and gravel, stone. Formed 1858.

Jones, town (2006 population 2,654), OKLAHOMA county, central OKLAHOMA, residential suburb 15 mi/ 24 km ENE of OKLAHOMA CITY, and on NORTH CANADIAN River; 35°34′N 97°17′W. In oil-producing and agricultural area.

Jones Beach, state park (□ 4 sq mi/10.4 sq km), on offshore bar, SW LONG ISLAND, NASSAU county, SE NEW YORK; 40°36′N 73°30′W. Noted for its wide, white sand beaches, outdoor marine theater, and varied recreational facilities. Established 1929 under Robert Moses's direction. Serves an estimated eight million visitors annually.

Jones Bluff Lake, reservoir (□ 19 sq mi/49.4 sq km), central ALABAMA, on ALABAMA RIVER, between AUTAUGA county (N) and MONTGOMERY and LOWNDES counties (S), 13 mi/21 km W of MONTGOMERY; 32°19′N 86°47′W. Maximum capacity 234,200 acre-ft. Formed by Robert F. Henry Lock and Dam (110 ft/ 34 m high), built (1971) by Army Corps of Engineers for navigation, power generation, and recreation. Extends along ALABAMA RIVER in general E-W direction. Also known as R. E. (Bob) Woodruff Lake and Jones Bluff Lake.

Jonesboro, city (2000 population 55,515), shares ⊙ functions with LAKE CITY, CRAIGHEAD county, NE ARKANSAS, on CROWLEY'S RIDGE; 35°49′N 90°41′W. Founded 1859, incorporated 1883. Railroad junction and center. The city services a rich agricultural area with many processing plants and much manufacturing (construction materials, plastic products, glass products, food products, transportation equipment, apparel, consumer goods, machinery, paper products). Arkansas State University is here, and a state park is nearby. Lake Frierson State Park to N.

Jonesboro, city (2000 population 1,853), ⊙ UNION county, S ILLINOIS, 30 mi/48 km N of CAIRO, in Illinois Ozarks; 37°27′N 89°16′W. Agriculture (fruit, wheat, sorghum, vegetables; dairying); limestone quarries; sawmill. A Lincoln-Douglas debate was held here in 1858. Shawnee National Forest nearby. Laid out 1816, incorporated 1857.

Jonesboro, city (2000 population 1,887), GRANT county, E central INDIANA, on MISSISSINEWA RIVER, and 6 mi/ 9.7 km S of MARION; 40°29′N 85°38′W. Farm trading center in agricultural area; manufacturing (fabricted metal products). Plotted 1837.

Jonesboro, town (2000 population 3,829), ⊙ CLAYTON county, N central GEORGIA, 17 mi/27 km S of ATLANTA; 33°31′N 84°21′W. Suburb of Atlanta; once a major clothing and textile manufacturing center. Historical museum nearby on site of Civil War battle of Jonesboro in General William Tecumseh Sherman's Atlanta campaign (1864). The ancestors of Margaret Mitchell, author of *Gone with the Wind*, lived here. Confederate cemetary. Clayton State College and University, a unit of the University system of Georgia, located nearby. Settled 1823, incorporated 1859.

Jonesboro (JONZ-buhr-uh), town (2000 population 3,914), ⊙ JACKSON parish, N central LOUISIANA, 40 mi/64 km WSW of MONROE; 32°14′N 92°43′W. Railroad junction to N. In agricultural area (sweet potatoes, vegetables; cattle, poultry); concrete manufacturing, publishing. Caney Creek reservoir to E, Jackson Bienville State Wildlife Area to N.

Jonesboro, town, WASHINGTON county, E MAINE, at mouth of Chandler River, 7 mi/11.3 km SW of MACHIAS; 44°40′N 67°34′W. Sometimes Jonesborough.

Jonesborough, formerly (until 1983) **Jonesboro**, town (2006 population 4,721), ⊙ WASHINGTON county, NE TENNESSEE, 6 mi/10 km WSW of JOHNSON CITY; 36°18′N 82°29′W. Laid out 1779; oldest town in Tennessee; was first capitol of State of Franklin. Andrew Jackson admitted to law practice here in 1788. Among many old buildings is an inn built c.1798. Annual storytelling festival here.

Jonesburg, town (2000 population 695), MONTGOMERY county, E central MISSOURI, 13 mi/21 km SE of MONTGOMERY CITY; 38°51′N 91°18′W. Manufacturing.

Jones Creek (JONS CREEK), town (2006 population 2,122), BRAZORIA county, SE TEXAS, residential suburb 7 mi/11.3 km WNW of FREEPORT, in BRAZOSPORT area; 28°58′N 95°28′W. San Bernard National Wildlife Refuge to SW.

Jones Mill, village, HOT SPRING county, central ARKANSAS, near Remmel Dam and Lake CATHERINE, 7 mi/11.3 km NW of MALVERN. Manufacturing (fabricated metal products).

Jones Mountains, an isolated group of mountains on the EIGHTS COAST of ELLSWORTH LAND, WEST ANTARCTICA, overlooking the BELLINGSHAUSEN SEA; 27 mi/44 km long; 73°32′S 94°00′W.

Jones Pass (12,451 ft/3,795 m), N central COLORADO, in FRONT RANGE, on CONTINENTAL DIVIDE, on GRAND-CLEAR CREEK county line. Jones Pass Tunnel, 0.5 mi/ 0.8 km S (c.3 mi/4.8 km long; finished 1939), unit in Denver sewage-disposal system. Henderson tunnel, 8 mi/12.9 km long, runs beneath Jones Pass. Hiking trail crosses pass.

Jones, Plaines des, VIETNAM: see PLAIN OF REEDS.

Jonesport, town, WASHINGTON county, E MAINE, on peninsula W of WOHOA BAY and 15 mi/24 km SW of MACHIAS; 44°32′N 67°30′W. Port of entry. Summer resort, fishing center. Settled 1763–1764; incorporated 1832; included BEALS until 1925.

Jones Sound, NUNAVUT territory, CANADA, arm of BAFFIN BAY, between ELLESMERE (N) and DEVON (S) islands; 250 mi/402 km long, 15 mi/24 km–60 mi/97 km wide; 76°00′N 85°00′W. At BAFFIN BAY end is COBOURG ISLAND. Discovered (1616) by William Baffin.

Jonestown, town (2000 population 1,701), COAHOMA county, NW MISSISSIPPI, 10 mi/16 km NE of CLARKSDALE, near COLDWATER RIVER; 34°19′N 90°27′W. railroad terminus. Agriculture (cotton, corn, wheat, rice, soybeans; cattle); manufacturing (cottonseed oil).

Jonestown, borough (2006 population 1,009), LEBANON county, SE central PENNSYLVANIA, 6 mi/9.7 km NW of LEBANON, on SWATARA CREEK; 40°24′N 76°28′W. Manufacturing (paper products, transportation equipment); agriculture (grain, soybeans, apples; poultry, livestock; dairying). APPALACHIAN TRAIL passes to N on BLUE MOUNTAIN ridge; Swatara State Park to N; Memorial Lake State Park to W; Fort Indiantown Military Reserve to NW; Bellgrove Airport to NW.

Jonesville, town (2000 population 220), BARTHOLOMEW county, S central INDIANA, 10 mi/16 km S of COLUMBUS; 39°04′N 85°53′W. In agricultural area. Manufacturing (furniture and wood products). Laid out 1851.

Jonesville (JONZ-vil), town, CATAHOULA parish, E LOUISIANA, on OUACHITA RIVER (BLACK RIVER), at influx of TENSAS and LITTLE rivers, and 24 mi/39 km W of NATCHEZ, MISSISSIPPI; 31°37′N 91°50′W. In agricultural area (corn, cotton); lumbering; manufacturing (nylon twine, nets); fisheries. Built on site of ancient Native American village. CATAHOULA National Wildlife Reserve to SW (LA SALLE parish), Bayou Cocodrie National Wildlife Refuge to SE (CONCORDIA parish). Incorporated 1904.

Jonesville, town (2000 population 2,337), HILLSDALE county, S MICHIGAN, 4 mi/6.4 km NNW of HILLS-

DALE, and on SAINT JOSEPH RIVER; 41°58′N 84°40′W. In diversified agricultural and manufacturing area. Settled 1828, incorporated 1855.

Jonesville (JONZ-vil), town (□ 2 sq mi/5.2 sq km; 2006 population 2,294), YADKIN county, N NORTH CAROLINA, on Yadkin (PEE DEE) River opposite (2 mi/3.2 km SE of) ELKIN; 36°13′N 80°50′W. Manufacturing (concrete, fabricated metal products, apparel); service industries; agriculture (tobacco, grain; poultry, livestock; dairying).

Jonesville, town (2006 population 922), UNION county, N SOUTH CAROLINA, 14 mi/23 km ESE of SPARTANBURG; 34°50′N 81°40′W. Manufacturing of cotton products; agriculture includes poultry; grain, soybeans, peaches, and apples. Settled 1808, incorporated 1876.

Jonesville (JONZ-vil), town (2006 population 982), ⊙ LEE county, extreme SW VIRGINIA, near POWELL RIVER, 32 mi/51 km SW of NORTON; 36°41′N 83°07′W. KENTUCKY state line to NW, TENNESSEE state line to S. Manufacturing; agriculture (tobacco, corn, alfalfa; cattle).

Jonesville, village, S ALASKA, in MATANUSKA VALLEY, 10 mi/16 km NE of PALMER; 61°48′N 148°51′W. Former coal mining center.

Jongka, China: see GYIRONG.

Jongkha, China: see GYIRONG.

Jonglei, Arabic *Junqali*, state, Upper Nile region, SE SUDAN; ⊙ BOR; 08°00′N 32°00′E. Bordered by East Equatoria (S), Central Equatoria (SW), Bohairat (W), Unity (WNW), and UPPER NILE (NNE) states and ETHIOPIA (E). Created in 1996 following a reorganization of administrative divisions. BAHR AL-GABAL River (part of WHITE NILE River) forms entire W border with Bohairat and Unity states, with swampy SUDD region along much of this; BAHR AL-ZARAF River in N. Lake No—where BAHR AL-GABAL and BAHR AL-GHAZAL rivers meet—is in NW on border with Unity state. The JONGLEI CANAL is here. Main road runs N-S through W of state, traveling S to JUBA (in Central Equatoria state) and N to MALAKAL (in UPPER NILE state) and eventually KHARTOUM. The area making up Jonglei state was part of former UPPER NILE province.

Jonglei (jahng-LAI), village, Jonglei state, S SUDAN, on right bank of the BAHR AL-GABAL (WHITE NILE), and 50 mi/80 km NNW of BOR. S terminus of the JONGLEI CANAL. Oil reserves in the area.

Jonglei Canal (jahng-LAI), 174 mi/280 km long, SUDD region, S SUDAN. Planned canal between the village of JONGLEI, from which it derives its name, N of BOR to the confluence of the BAHR AL-GABAL and SOBAT rivers, c.19 mi/31 km S of MALAKAL, essentially cutting a great W loop of the river. Canal is a joint project of EGYPT and SUDAN for the purpose of saving water usually lost through evaporation of the WHITE NILE; designed to carry a maximum amount of 880 million cu ft/25 million cu m per day. Digging of the canal started in 1974, but stopped in 1983 because of the civil war in the S after a length of 160 mi/260 km out of the total 225 mi/360 km had been dug.

Jong River (JAWNG), c.100 mi/161 km long, S SIERRA LEONE; formed 10 mi/16 km W of YELE by union of TEYE and PAMPANA rivers; flows S past MANO, to SHERBRO RIVER (inlet of the ATLANTIC) opposite BONTHE. Navigable in lower course. Sometimes called TEYE RIVER.

Jongsong La, NEPAL and INDIA: see SINGALILA RANGE.

Joniškis (YO-nish-kis), German *Janischken*, Russian *Yanishki*, city (2005 population 11,329), N LITHUANIA, 24 mi/39 km NNE of SIAULIAI, near Latvian border; 56°14′N 23°37′E. Manufacturing (shoes, furniture), flour milling, dairying, flax processing. Dates from 15th century In Russian Kovno government until 1920. Also spelled Ionishkis and Yonishkis.

Jönköping (YUHN-SHUH-peeng), county (□ 4,449 sq mi/11,567.4 sq km), S Sweden; ⊙ JÖNKÖPING; 57°30′N 14°30′E. Extends S from LAKE VÄTTERN; forms N part

of SMÅLAND province. Manufacturing (machinery, furniture; wood products, fabricated metal products); agriculture (grain; cattle). Marshy, undulating surface studded with many lakes; drained by EMÅN, LAGAN, and NISSAN rivers. Important towns include JÖNKÖPING, NÄSSJÖ (railroad center), HUSKVARNA, VÄRNAMO, VETLANDA, EKSJÖ, TRANÅS, and GRÄNNA.

Jönköping (YUHN-SHUH-peeng), town, capital of JÖNKÖPING county, S SWEDEN, at S end of LAKE VÄTTERN; 57°47′N 14°11′E. Commercial and administrative center; manufacturing (machinery). Airport. Site of invention of the safety match and former home to large match factories (founded 1844). Chartered by Magnus I (1284). Gustavus Adolphus awarded it special privileges (1620) after citizens burned city to thwart Danish attack. Modern prosperity began in nineteenth century with opening of nearby GÖTA CANAL and coming of railroad.

Jonotla (ho-NO-tlah), town, PUEBLA, central MEXICO, 28 mi/45 km E of HUAUCHINANGO. Sugar, coffee, tobacco, fruit. Also San Juan Jonotla.

Jonquière (zhohn-KYER), former city, borough (French *arrondissement*) of SAGUENAY, SAGUENAY—Lac-Saint-Jean region, S central QUEBEC, E CANADA, on the SAGUENAY River, W of CHICOUTIMI; 48°25′N 71°15′W. Its chief industries produce paper, pulp, and aluminum. The city was reincorporated in 1976, when it absorbed the surrounding cities of Arvida, established in 1925 as an Alcan company town around one of the then-largest aluminum smelters in the world, Kénogami, and the municipality of Saint-Dominique-de-Jonquière. Jonquière has a college and a school of technology.

Jonquières (zhon-KYER), commune (□ 9 sq mi/23.4 sq km), VAUCLUSE department, PROVENCE-ALPES-CÔTE D'AZUR region, SE FRANCE, 13 mi/20 km NNE of AVIGNON, and on OUVÈZE RIVER; 44°07′N 04°54′E.

Jonsered (YOON-ser-ED), village, GÖTEBORG OCH BOHUS county, SW SWEDEN, on SÄVEÅN river, 7 mi/11.3 km ENE of GÖTEBORG; 57°45′N 12°11′E.

Jonte River (zhon-tuh), 25 mi/40 km long, LOZÈRE and AVEYRON departments, S FRANCE; rises in the MASSIF CENTRAL at Mont AIGOUAL, flows W in a deep gorge past MEYRUEIS between Causse Méjan (N) and Causse Noir (S), both limestone tablelands to the TARN at Le Rozier; 44°11′N 03°13′E. The river's canyon is a tourist attraction.

Jonuta (ho-NOO-tah), city and township, TABASCO, SE MEXICO, on USUMACINTA River, near CAMPECHE border, and 45 mi/72 km SE of FRONTERA; 18°08′N 92°10′W. Rice, beans, tobacco, fruit.

Jonzac (zhon-zahk), town (□ 5 sq mi/13 sq km), subprefecture of CHARENTE-MARITIME department, POITOU-CHARENTES region, W FRANCE, on the SEUGNE RIVER, and 18 mi/29 km SSW of COGNAC; 45°27′N 00°26′W. Cognac-brandy distilling; ships butter and biscuits.

Jopala (ho-PAH-lah), town, PUEBLA, central MEXICO, 24 mi/39 km E of HUAUCHINANGO. Sugar, coffee, fruit. In Totonal Indian area.

Joplin, city (2000 population 45,504), JASPER and NEWTON counties, SW MISSOURI; 37°04′N 94°30′W. It is a railroad center, a major truck stop on I-44, a regional shopping center, the shipping and processing point of a grain and livestock region with dairy and fruit farms, and the industrial center of a former major lead and zinc area, the Tri-State Mining District. Mining has ceased in the Joplin area. Manufacturing (transportation equipment, plastic products, food products, fabricated metal products, machinery, construction materials, paper products, chemicals, leather products; machining, printing, meat packing). The city has a mineral museum for the historic lead and zinc industry. It is the seat of Missouri Southern State College and Ozark Bible College. The GEORGE WASHINGTON CARVER National Monument is nearby. Settled c.1839, incorporated 1873.

Joppa (JAH-pa), village (2000 population 409), MASSAC county, extreme S ILLINOIS, on OHIO River, and 8 mi/12.9 km WNW of METROPOLIS; 37°12′N 88°50′W. Steam power plant here for Atomic Energy Commission plant near PADUCAH, KENTUCKY.

Joppa, ISRAEL: see JAFFA.

Joppa, MARYLAND: see JOPPATOWNE.

Joppatowne, city (2000 population 11,391), HARFORD county, NE MARYLAND, near LITTLE GUNPOWDER FALLS (stream), 17 mi/27 km NE of BALTIMORE; 39°25′N 76°21′W. It is a large modern development, surrounding the mansion of Benjamin Rumsey (c.1773), which has been restored as the home of the Harford county executive. Nearby is site of Joppa hamlet, capital of old BALTIMORE county (1712–1768), which was a major American tobacco market until c.1750.

Joquicingo (ho-kee-SEEN-go), town, MEXICO, central MEXICO, 21 mi/34 km SE of TOLUCA DE LERDO; 19°03′N 99°28′W. Sugarcane, cereals; livestock.

Jora (JO-rah) or **Joura** or **Jora Alapur**, town (2001 population 25,514), SE MORENA district, N central MADHYA PRADESH state, central INDIA, 24 mi/39 km WNW of LASHKAR; 26°20′N 77°49′E. Gram, millet, wheat.

Jorat (zhoh-RAH), chain of heights in W SWITZERLAND, between BERNESE ALPS and the JURA; forms part of the watershed between NEUCHÂTEL and GENEVA lakes. Rises to 3,047 ft/929 m at Mount Jorat; LAUSANNE is on its slopes, 10 mi/16 km SSE.

Joravarnagar, INDIA: see SURENDRANAGAR, town.

Jordan, officially Hashemite Kingdom of Jordan, formerly TRANSJORDAN, kingdom (□ 31,189 sq mi/89,206 sq km; 2004 population 5,103,639; 2007 estimated population 6,053,193), SW ASIA, bordering ISRAEL on the W, SYRIA on the N, IRAQ on the NE, and SAUDI ARABIA on the E and S, ⊙ AMMAN; 31°00′N 36°00′E.

Geography

Major cities in Jordan include ZARQA, IRBID, AQABA, SALT, and Al Karak (see KARAK, AL). In the Arab-Israeli War of 1967, ISRAEL captured and occupied all of the territory of former BRITISH MANDATE PALESTINE, seized by Jordan during the Israeli War of Independence in 1948 and formally annexed to the kingdom in 1950. The 1994 peace treaty between Israel and Jordan sets Jordan's W boundary along the JORDAN RIVER, the DEAD SEA, and the Wadi ARABA except where the river enters the Dead Sea, and at the point where the river met the 1948 armistice line S of BEIT SHEAN. The final determination of the boundary is to be made when Israeli and Palestinian authorities arrive at a lasting peace agreement. This territory is located W of the Jordan River and NW of the Dead Sea (the area known collectively as the WEST BANK), which comprises about 2,165 sq mi/5,607 sq km and includes HEBRON, JERICHO, and NABLUS (for population and history, see West Bank). Annexation of the West Bank was recognized only by BRITAIN and PAKISTAN, but not by the Arab League. Jordan is made up of a section (average elevation 2,500 ft/760 m) of the Arabian Plateau that includes part of the SYRIAN DESERT in the NE. In the W part of the plateau are the Jordanian Highlands, and S of MA'AN near the Gulf of AQABA is a mountainous region including Jabal Ramm (elevation 5,755 ft/1,754 m), Jordan's loftiest point. Most of the uplands consists of barren, calcareous rock in advanced stages of soil erosion. Jordan's most distinctive feature is the deep Jordan Valley, generally called the GHOR, which together with the saline Dead Sea and the Araba (an arid valley) forms a continuous depression to the Gulf of Aqaba on the RED SEA. This depression, part of the GREAT RIFT VALLEY of AFRICA and the NEAR EAST, is 1,322 ft/403 m below sea level at the surface of the Dead Sea. Both sides are lined by steep cliffs. The E mountains rise to a height of 3,280 ft/1,000 m and then descend gently to merge with the SYRIAN DESERT. The majority

---

Cross-references are shown in SMALL CAPITALS. The pronunciation guide is shown on page xix. The sources of population figures are shown on page xvii.

of the country's population is concentrated in the N and central parts of this mountainous rim. The climate, influenced by the Mediterranean, has dry, hot summers; winters are mild and cool with occasional torrential rains.

Population

Until 1948, Jordan's population was largely of Bedouin ancestry and more than half was nomadic. As a result of the Arab-Israeli war, Palestinian refugees have changed the country's demographic character so that two-thirds of the populace is of Palestinian Arab origin. There are small minorities of Armenians and Circassians. Arabic, the official language, is spoken by virtually everyone. About 92% of the people are Sunni Muslims. Christians make up about 8% of the population, half of whom are Greek Orthodox.

Economy

Jordan's economy is largely agricultural. Less than 10% of the country's land is arable, although irrigation has been extended considerably, especially on the E side of the JORDAN VALLEY by the diversion of the YARMUK and other, smaller rivers. The principal crops of the valley are bananas, lentils, tomatoes, eggplants, and citrus fruits. They are grown under hothouse, drip-irrigation conditions. There is dry farming of wheat and barley on the plateau, and olives and grapes are also cultivated. Pastoralism (sheep and goat raising) has declined with the sedentation of the Bedouins. Manufacturing includes items such as food, beverages, clothing, construction materials (especially cement and glass), soap, dairy products, plastics, fertilizers, chemicals, and cigarettes. Most of the country's industry is based in the Amman-ZARQA area. Numerous artisans make items of leather, wood, and metal. Phosphate rock and potash are produced in significant quantities. Oil was discovered in 1982, and hence a small oil and natural-gas industry has been developed, including petroleum refining. Tourism is of growing importance, especially in the Gulf of Aqaba resorts and in ancient sites. In addition to the Jordan, other important rivers include the Yarmuk (which partly flows along Jordan's N border, abutting the Israeli and Syrian borders), WADI ZARQA, Wadi el MUJIB and Wadi Hasa. Jordan's transportation system consists of a small network of all-weather roads and a railroad (formerly part of the HEJAZ RAILROAD) that enters Jordan from Syria and runs S through Amman and has been extended to Aqaba (the country's only seaport). In a territorial agreement with Saudi Arabia (1963), Jordan added 6.2 mi/10 km of coast, almost double the length of its outlet to the Gulf of Aqaba. A new highway running from Aqaba to Iraq has made trade and transportation more efficient. The principal imports are foodstuffs, textiles, machinery, iron and steel, and chemicals; the main exports are phosphates, tomatoes, and bananas. Jordan's leading trade partners are the U.S., Iraq, INDIA, and Saudi Arabia.

History: to 1921

The region of present-day Jordan roughly corresponds to the biblical lands of (N-S) GILEAD, Ammon, MOAB, and EDOM. The area was conquered by the Seleucids in the 4th century B.C.E. and was part of the Nabatean empire, whose capital was PETRA, from the 1st century B.C.E. to the mid-1st century C.E., when it was captured by the Romans under Pompey. In the period between the 6th and 7th centuries it was the scene of considerable fighting between the BYZANTINE EMPIRE and PERSIA. In the early 7th century the region was invaded by the Muslim Arabs, and after the Crusaders captured Jerusalem in 1099, it became part of the Latin Kingdom of Jerusalem. In 1516 the Ottoman Turks gained control of what is now Jordan, and it remained part of the OTTOMAN EMPIRE until the 20th century. After the fall of the Ottoman Empire in World War I, the region came under (1919) the government of Faisal I, centered at DAMASCUS. When Faisal was ejected by French troops in July 1920,

TRANSJORDAN (as Jordan was then known) was made (1920) part of the British League of Nations mandate of Palestine.

History: 1921 to 1968

In 1921, Abdullah ibn Husain, a member of the Hashemite dynasty and the brother of Faisal, was made head of Transjordan, and later became king. By a treaty with Great Britain signed in 1946, it became (May 25) independent as the Hashemite Kingdom of Transjordan. The Arab League (of which it was a part) entered into military conflict with the newly created Israel in 1948, and as a result Transjordan acquired the area of W central British Mandate Palestine that the UN had designated as Arab territory. In April 1949, the country's name was changed to Jordan, reflecting its acquisition of Palestinian lands. In December 1949, Jordan concluded an armistice with Israel, and early in 1950, it formally annexed the West Bank. The annexation of the West Bank increased Jordan's population by about 550,000 persons, many of them homeless refugees from Israel. In 1951, Abdullah was assassinated in Jerusalem by a Palestinian and was succeeded the following year by his grandson, Hussein I. The next decade marked a period of tension and hostility not only with Israel, but also Jordan's Arab neighbors, causing short-lived alliances and eventual U.S. and British intervention. In the mid-1960s, Jordanian politics were calm, the economy expanded as its international trade increased, and the kingdom was on good terms with EGYPT (which had previously called for an overthrow of Jordan's government). Despite Israeli attempts to persuade Jordan to abstain from battle, the nations became embroiled in the 1967 Arab-Israeli War. As a result, Jordan was ejected from the West Bank area, which Israeli troops then occupied. A large number of Palestinian refugees fled to Jordan during and after the war.

History: 1968 to 1991

In 1968–1969 there were clashes along the frontier with Israel, but of greater significance was the growing hostility between the Jordanian government and the Palestinian guerrilla organizations operating in Jordan, culminating in a bloody ten-day civil war in September 1970. The Palestinians suffered heavy casualties, and many of them fled to LEBANON and Syria, which shifted the locus of the Palestinian refugee problem. In July 1971, the army carried out a successful offensive that destroyed the remaining guerrilla bases in Jordan. The kingdom played a minor role in the Arab-Israeli War of October 1973, sending a small number of troops to fight on the Syrian front. In 1974, Hussein complied with the Arab League's ruling that the PLO (Palestinian Liberation Organization) was to be the single legitimate representative of the Palestinians. Jordan sided with Iraq in the Iran-Iraq War, despite Syrian threats, and supplied the country with large amounts of war materials via the port of Aqaba. In 1988, Hussein formally relinquished claim to the West Bank in acknowledgement of Palestinian sovereignty. He approved the creation of an independent Palestinian state in the West Bank, and Arabs residing in that area lost their Jordanian citizenship. Plagued by serious economic problems since the mid-1980s, Jordan received increased U.S. economic aid in 1990.

History: 1991 to Present

However, the outbreak (1991) of the Persian Gulf War led to a repeal of U.S. aid to Jordan due to Hussein's support of Iraq. Jordan also suffered a loss of aid from Saudi Arabia and KUWAIT during the war. The country endured further economic hardship when approximately 700,000 Jordanian workers and refugees returned to Jordan as a result of the fighting in the PERSIAN GULF, causing housing and employment shortages. Jordanian trade relations with Iraq had been minimal due to international sanctions, which continued throughout 1992. Peace talks between Israel, Syria, LEBANON, and a joint Palestinian-Jorda-

nian delegation began in August 1991, and lasted through the late 1990s. In 1994, Israel and Jordan signed a peace treaty, opening a new chapter of economic and political relations between the two countries. Hussein continued to promote peace between Arabs and Israelis until his death in 1999; he was succeeded by his son Abdallah II.

Government

Jordan is governed under the 1952 constitution as amended. The most powerful political and military figure in the country is the king, who is head of state. The government is headed by the prime minister, who is appointed by the monarch. The bicameral parliament consists of the fifty-five-seat Senate, whose members are appointed by the king, and the 110-seat House of Deputies, whose members are popularly elected, with six seats reserved for women. All legislators serve four-year terms. The current head of state is King Abdallah II. The head of government is Prime Minister Marouf al-Bakhit (since November 2005). Administratively, the country is divided into eight governorates (*muhafazat*): AMMAN, Balqa, IRBID, KARAK, MA'AN, MAFRAQ, Tafilah, and ZARQA; the governorates are further divided into districts and sub-districts.

Jordán (hor-DAHN), town, ☉ Jordán municipio, SANTANDER department, N central COLOMBIA, 22 mi/ 35 km S of BUCARAMANGA in the CHICAMOCHA RIVER valley; 06°44′N 73°05′W. Elevation 1,824 ft/555 m. Sugarcane, corn, cassava; livestock. Sometimes called El Jordán.

Jordan, town, ☉ GUIMARAS province, NE coast GUIMARAS ISLAND, PHILIPPINES, across Iloilo Strait from ILOILO; 10°36′N 122°37′E. Ferry terminal. Sugar refining; sugar and molasses shipping center; mango processing.

Jordan, town (2000 population 3,833), SCOTT county, S MINNESOTA, near MINNESOTA RIVER, 28 mi/45 km SW of MINNEAPOLIS; 44°40′N 93°37′W. Trading point in agricultural area (grain; livestock, poultry; dairying); manufacturing (fabricated metal products, transportation equipment, food products, electronic equipment, machinery); sawmill; sand and gravel pits nearby. Thompson Ferry State Wayside, on Minnesota Valley State Trail, to N. Plotted 1854, incorporated as village 1872, as city 1891.

Jordan, village (2000 population 364), in Justun township, ☉ GARFIELD county, E central MONTANA, 80 mi/129 km NW of MILES CITY on Big Dry Creek; 47°19′N 106°55′W. Trading point in irrigated ranching region; wheat, barley, hay; sheep, cattle. Hell Creek State Park and Hell Creek Bay of FORT PECK Reservoir to N; Charles M. Russell National Wildlife Refuge to N. Site of FBI seizure of anti-government militants in 1996.

Jordan, village (□ 1 sq mi/2.6 sq km; 2006 population 1,335), ONONDAGA county, central NEW YORK, 17 mi/27 km W of SYRACUSE; 43°04′N 76°28′W. Some manufacturing and farming. Incorporated 1835.

Jordan, Hebrew *Yarden* river, c.200 mi/322 km long, formed in the HULA basin, N ISRAEL, by the confluence of three headwater streams (DAN, BANIYAS, and Hazbani) and meandering S through the SEA OF GALILEE to the DEAD SEA; 31°45′N 35°33′E. Israel's longest and most important river and the world's lowest river below sea level. It flows through the N section of the Jordan trough, a part of the GREAT RIFT VALLEY, between the Sea of Galilee and the Dead Sea (the Jordan valley is called the Ghor in Arabic). The Jordan is fed by many small streams, with headwaters in SYRIA, JORDAN, LEBANON, and ISRAEL. The YARMUK River is its largest tributary. Deep and turbulent during the rainy season, the Jordan is reduced to a sluggish, shallow stream during the summer. As it nears the Dead Sea, its salinity increases. The river is not navigable. Its waters are valuable for irrigation. The cultivation of fruits and vegetables has been made possible by Israel's National Water Carrier, which uses the Sea of Galilee as a reservoir, and Jordan's East Ghor project, which diverts water from the Yarmuk River. Because of these two projects, very little of the water reaches the Dead Sea. Since 1967 the Jordan River has served as the de facto border between Israel and Jordan, until formally recognized as such by the Israel-Jordan peace treaty of 1994.

Jordânia (zhor-DAH-nee-ah), city (2007 population 10,766), NE MINAS GERAIS, BRAZIL, on BAHIA border, 80 mi/129 km W of CANAVIEIRAS (BAHIA); 15°51'S 40°11'W. Agr. settlement. Until 1944, called Palestina.

Jordan, Lake (□ 7.7 sq mi/19.9 sq km), on border of Coosa and Chilton counties, central Alabama, on Coosa River, 17 mi/27 km NNE of Montgomery; 18 mi/29 km long; 32°36'N 86°15'W. Formed by Jordan Dam, privately built power dam (125 ft/38 m high, 2,066 ft/630 m long) completed 1929. New channel of Coosa River exits Lake Jordan S 1 mi/1.6 km WNW of dam and enters reservoir of Bouldin Dam. Given the maiden name of the mother of Reuben A. and Sidney Z. Mitchell, officers of Alabama Power Co.

Jordan Lake, NORTH CAROLINA: see B. EVERETT JORDAN LAKE.

Jordanow (yor-DAH-noov), Polish *Jordanów*, town, Nowy Sacz province, S POLAND, on SKAWA RIVER, and 29 mi/47 km S of CRACOW. Flour milling, sawmilling, tanning; stone quarrying. Tourist resort. During World War II, under German rule, called Ilmenau.

Jordan River, 60 mi/97 km long, draining UTAH LAKE N into GREAT SALT LAKE, N central UTAH, passing through LEHI, SANDY, WEST VALLEY CITY, and SALT LAKE CITY. Fed by numerous streams flowing off the WASATCH RANGE, the Jordan is used for irrigation and forms the heart of the Utah Oasis. Mormons settled along its banks in the mid-1800s. Named for Jordan River in Middle East to which it bears an uncanny resemblance, flowing from a freshwater lake to a salt water lake and providing sustenance to a dry climate region.

Jordan River, CANADA: see RIVER JORDAN.

Jordans, ENGLAND: see CHALFONT SAINT GILES.

Jordan Valley, village (2006 population 232), MALHEUR county, SE OREGON, on Jordan Creek, 70 mi/113 km S of VALE on IDAHO boundary; 42°58'N 117°02'W. Elevation 4,389 ft/1,338 m.

Jordan Valley, JORDAN: see GHOR.

Jordanville, hamlet, HERKIMER county, central NEW YORK, 20 mi/32 km SE of UTICA; 42°55'N 74°57'W. In agricultural area. Site of Holy Trinity Russian Orthodox Seminary (largest Eastern Orthodox monastery in North America), with traditional Russian architecture and gold-clad Byzantine domes that are visible for miles.

Jordão, Brazil: see FOZ DO JORDÃO.

Jorenko (ho-REN-ko), canton, BOLIVAR province, COCHABAMBA department, central BOLIVIA, 17°47'S 66°32'W. Elevation 13,287 ft/4,050 m. Tin-bearing lode; clay, limestone, and gypsum deposits. Agriculture (potatoes, yucca, bananas, corn, rye, soy, coffee); cattle for dairy products and meat.

Jorf-Lasfar (jorf las-FAHR), port, El Jadida province, Doukkala-Abda administrative region, MOROCCO, 12 mi/19 km S of EL JADIDA. Morocco's third busiest port; ships phosphates and phosphate products from KHOURIBGA and YOUSSOUFIA mines.

Jorge Basadre (HOR-hai bah-SAH-dre), province, TACNA region, S PERU; ⊙ LOCUMBA. W province of Tacna region, on PACIFIC OCEAN on W.

Jorge Island, CHILE: see CHONOS ARCHIPELAGO.

Jorge Montt Island (HOR-hai), off coast of S CHILE, 90 mi/145 km WNW of PUERTO NATALES, just NE of NELSON STRAIT; 28 mi/45 km long, 25 mi/40 km wide; 51°20'S 74°45'W.

Jorhat (JOR-haht), district (□ 1,101 sq mi/2,862.6 sq km), ASSAM state, NE INDIA; ⊙ JORHAT.

Jorhat (JOR-haht), town, ⊙ JORHAT district, E central ASSAM state, NE INDIA, in BRAHMAPUTRA RIVER val-

ley, on tributary of the Brahmaputra River, and 165 mi/266 km NE of SHILLONG; 26°45'N 94°13'E. Railroad junction, with spur to Kokilamukh (steamer service), 8 mi/12.9 km NW; road and trade center (tea, rice, rape and mustard, sugarcane, jute); has Assamese jewelry manufacturing center. Seat of Assam Agricultural College and a Regional Engineering College; a Tea Research Institute nearby. Was capital of Ahom (Shan) kingdom in late eighteenth century.

Jork (YORK), village, LOWER SAXONY, N GERMANY, on left bank of ELBE estuary, 14 mi/23 km W of HAMBURG; 53°32'N 09°41'E.

Jorm, AFGHANISTAN: see JURM.

Jörn (YUHRN), village, VÄSTERBOTTEN county, N SWEDEN, 35 mi/56 km NW of SKELLEFTEÅ; 65°04'N 20°02'E. Railroad junction.

Jornada del Muerte, New Mexico: see EL CAMINO REAL.

Jornada del Muerto (hor-NAH-duh del MWER-to), arid region of desert and lava beds in S central NEW MEXICO; extends N-S, E of RIO GRANDE and just W of SAN ANDRES MOUNTAINS, in SOCORRO, SIERRA, and DOÑA ANA counties. Once crossed by EL CAMINO REAL (early trade route); area was much feared by travelers because of robbers and arid conditions. Now crossed N-S by railroad; irrigated in places. Trinity Site, site of first atomic explosion, July 16, 1945, on E flank of region at foot of SIERRA OSCURA.

Joroinen (YO-roi-nen), Swedish *Jorois*, village, MIKKELIN province, SE FINLAND, 40 mi/64 km NW of SAVONLINNA; 62°11'N 27°50'E. Elevation 248 ft/75 m. Located on lake of SAIMAA system; agriculture center (grain, potatoes; livestock; dairy farming).

Jørpeland (YUHR-puh-lahn), village, ROGALAND county, SW NORWAY, on SE shore of BOKNAFJORDEN, 12 mi/19 km ENE of STAVANGER. Industrial center; manufacturing of fabricated metal products. Connected to Stavanger by ferry from Tau, 7 mi/11 km NW.

Jorullo (ho-ROO-yo), volcano (4,330 ft/1,320 m) in MICHOACÁN, central MEXICO, on SW slope of central plateau, 33 mi/53 km SE of URUAPAN; crater is 1.2 mi/1.9 km wide; 19°00'N 101°35'W. Elevation 10,167 ft/3,099 m. One of two Mexican volcanoes to develop during historic time (the other is PARÍCUTIN), it first erupted in 1759, destroying what had been a rich agricultural area. Has numerous craters and fumaroles; thermal waters. Last erupted 1958.

Jos (JAWS), city, ⊙ PLATEAU state, central NIGERIA, on the JOS PLATEAU; 09°55'N 08°55'E. It is a mining center for tin ore, which is processed here, and a collection point for hides and skins and for market-garden produce to be sent to LAGOS. It also is a resort. Jos was developed in the early 20th century by the British as an administrative center and mining town. The railroad reached here in 1927. The Jos Museum includes a collection of neolithic Nok terra-cotta figurines; the UNESCO School for Museum Technicians is attached. The University of Jos (1975), which has a teaching hospital, is here.

Jošanička Banja (YOSH-ahn-eech-ka BAH-nyah), village, S central SERBIA, on railroad 23 mi/37 km S of KRALJEVO; 43°23'N 020°45'E. Health resort. Also spelled Yoshanichka Banya.

José Agustín de Palacios (ho-SE ah-goos-TEEN de pah-LAH-see-os), canton, YACUMA province, BENI department, NE BOLIVIA, 28 mi/45 km NE of El Triunfo (airstrip), 28 mi/45 km E of REYES-RIBERALTA highway and 12 mi/19 km NW of YACUMA RIVER; 13°48'S 66°14'W. Elevation 476 ft/145 m. Agriculture (bananas, coffee, tobacco, cotton, cacao, peanuts, potatoes, cassava); cattle for dairy products and meat.

José Azueta, MEXICO: see ZIHUATANEJO.

José Azueta, MEXICO: see VILLA AZUETA.

José Batlle y Ordoñez (ho-SAI BAHT-yai ee or-DOnyes) or **Batlle y Ordóñez**, town, LAVALLEJA department, SE central URUGUAY, in the CUCHILLA GRANDE PRINCIPAL, on railroad (Nico Pérez station), and 45

mi/72 km WSW of TREINTA Y TRES, on FLORIDA department line; 33°28'S 55°07'W. Local trade center; railroad and road junction; airport. Agricultural products (wheat, corn, oats); cattle, sheep. Until 1907 called Nico Pérez, a name still applied to the town section extending W into FLORIDA department.

José Benifacio (zho-SAI BO-nee-fah-see-o), city, extreme E RONDÔNIA state, BRAZIL; 12°20'S 61°00'W.

José Bonifácio (ZHO-sai BO-nee-FAH-see-o), city (2007 population 30,639), N SÃO PAULO, BRAZIL, 25 mi/40 km SW of SÃO JOSÉ DO RIO PRÊTO; 21°03'S 49°41'W. Coffee, cotton, rice; macaroni processing; sawmilling.

José Bonifácio, Brazil: see ERECHIM.

José Cardel (ho-SAI kahr-DEL), town, ⊙ La Antigua township, VERACRUZ, E MEXICO, in GULF lowland, on railroad, on Mexico Highway 140, and 15 mi/24 km NW of VERACRUZ; 19°21'N 96°23'W. Coffee, fruit; livestock. Site of Mexico's only nuclear power plant, which began operation in 1989. Cortés landed here in April 1519. Also known as LA ANTIGUA; formerly San Francisco de las Peñas.

José Carlos Mariátegui (ho-SAI KAHR-los mah-ree-AH-tai-gee), former region (□ 39,339 sq mi/102,281.4 sq km), SE PERU. Consisted of Moquegua, Puno, and Tacna departments (today MOQUEGUA, PUNO, and TACNA regions), including seventeen provinces and 147 districts. Mineral resources include copper, silver, pirite, lead, clay, limestone, gold, carbon, and marble. Agriculture (olives, wheat, and corn). Fishing resources (anchovies, tuna, bass, sardines, snails). Created as part of Peru's 1988 regionalization program. These regions never quite caught on and were abandoned.

José C. Paz (ho-SAI sai PAHZ), town, BUENOS AIRES province, ARGENTINA, 2.5 mi/4 km NW of SAN MIGUEL; 34°30'S 58°45'W.

José de Alencar, São, Brazil: see ALENCAR.

José de Freitas (ZHO-sai zhe FRAI-tahs), city (2007 population 35,188), N PIAUÍ state, BRAZIL, 28 mi/45 km NNE of TERESINA; 04°54'S 42°35'W. Carnauba wax, maniçoba rubber. Until 1939, called Livramento.

José de Penha (zho-SAI zhe PEN-yah), town, southwesternmost RIO GRANDE DO NORTE state, BRAZIL, 102 mi/164 km SW of MOSSORÓ. Cotton.

José de San Martín (ho-SAI dai sahn mahr-TEEN), village (1991 population 1,371), ⊙ Tehuelches department, SW CHUBUT province, ARGENTINA, 85 mi/137 km SE of ESQUEL; 44°02'S 70°29'W. Resort, sheep-raising center. Formerly Colonia San Martín.

José Enrique Rodó, URUGUAY: see RODÓ.

Josefov (YO-ze-FOF), German JOSEFSTADT, town, VYCHODOCESKY province, NE BOHEMIA, CZECH REPUBLIC, on ELBE RIVER, and 8 mi/12.9 km SSE of DVUR KRALOVE; 50°20'N 15°55'E. Railroad junction. Manufacturing (clothing, underwear, embroidery). A fortress until 1890, it still preserves much of its fortifications. Town of JAROMER is 1 mi/1.6 km NNW.

Josefov (YO-ze-FOF), NNW district of PRAGUE, PRAGUE-CITY, central BOHEMIA, CZECH REPUBLIC, on right bank of VLTAVA RIVER. Site of a former Jewish ghetto, almost entirely demolished at end of 19th century. It still retains a remarkable 14th century synagogue, 18th century Jewish town hall, and a famous Jewish cemetery, which was in use 1439–1787.

Josefstadt (YO-sef-shtaht), district of VIENNA, AUSTRIA, just W of city center. Piaristen church constructed by Lukas von Hildebrandt with paintings by Franz Maulbertsch. Noted theater here (Theater in der Josefstadt).

Josefstadt, CZECH REPUBLIC: see JOSEFOV.

Josefstahl, CZECH REPUBLIC: see JOSEFUV DUL.

Josefuv Dul (YO-ze-FOOF DOOL), Czech *Josefův Důl*, German JOSEFSTAHL, village, SEVEROCESKY province, N BOHEMIA, CZECH REPUBLIC, in JIZERA MOUNTAINS, and 7 mi/11.3 km ENE of LIBEREC; elevation 2,051 ft/625 m. Railroad terminus; glass- and woodworks; winter skiing resort. Has a Gothic church.

José Gonçalves (zho-SAI gon-SAHL-ves), village, S central BAHIA, BRAZIL, in Serra Geral, 17 mi/28 km NE of VITÓRIA DA CONQUISTA; 14°41′S 40°43′W.

José Ignacio, Lake (ho-SAI eeg-NAH-syo), freshwater coastal lagoon, on the ATLANTIC OCEAN, MALDONADO department, S URUGUAY, 13 mi/21 km ENE of MALDONADO; 6 mi/9.7 km long, 2.5 mi/4 km wide; 34°51′S 54°43′W.

José Ignacio Point, headland on the ATLANTIC OCEAN, MALDONADO department, S URUGUAY, 19 mi/31 km E of MALDONADO city; 34°51′S 54°38′W.

José Ingenieros (ho-SAI dai een-he-nee-AI-ros), town in W Greater BUENOS AIRES, ARGENTINA. Residential center.

Joselândia (ZHO-sai-LAHN-zhee-ah), city (2007 population 15,595), central MARANHÃO state, BRAZIL, 48 km/30 mi WNW of PRESIDENTE DUTRA; 04°54′S 44°40′W.

José María Linares (ho-SAI mah-REE-ah lee-NAH-res), province (□ 1,983 sq mi/5,155.8 sq km), POTOSÍ department, W central BOLIVIA; ⊙ Puna a Villa Talavera; 19°49′S 66°31′W. Elevation 10,748 ft/3,276 m.

José María Morelos (ho-SAI mah-REE-ah mo-RAI-los), town, QUINTANA ROO, NE MEXICO, 25 mi/40 km SW of CANCÚN; 20°42′N 97°32′W. Small farming; livestock.

José María Morelos, town, N TLAXCALA, MEXICO, 14 mi/23 km NW of APIZACO, near railroad. Has good road connections. Temperate to cold climate. Agriculture (cereals, fruits); woods; cattle and poultry raising.

José María Morelos, MEXICO: see MAZATECOCHCO.

José Martí International Airport (ho-SE mahr-TEE), RANCHO BOYEROS, CUBA, 11 mi/18 km from HAVANA; 23°59′N 82°24′W. Single runway; three passenger terminals. Airport Code HAV.

José Néstor Lencinas, ARGENTINA: see LAS CATITAS.

Jøsenfjord (YUH-suhn-fyawr), inlet, NE branch of BOKNAFJORDEN, ROGALAND county, SW NORWAY, extends from OMBO 16 mi/26 km NE.

Joseni (zho-SEN), Hungarian *Gyergyóalfalu*, village, HARGHITA county, E central ROMANIA, near MUREȘ RIVER, 15 mi/24 km W of GHEORGHENI. Mineral springs. Under Hungarian rule, 1940–1945.

Jose Pañganiban (ho-SAI pahng-ah-NEE-bahn), town, CAMARINES NORTE province, SE LUZON, PHILIPPINES, port on small inlet of PHILIPPINE SEA, 105 mi/169 km NW of LEGASPI; 14°15′N 122°43′E. Trade center for iron-mining area; some small-scale gold mining; sawmilling. Exports iron ore. Formerly Mambulao.

José Pedro Varela (ho-SAI PAI-dro vah-RAI-lah), town LAVALLEJA department, SE URUGUAY, on railroad (CORRALES station) and highway junction, and c.16 mi/26 km SW of TREINTA Y TRES; 33°27′S 54°32′W. Road junction; commercial, distributing center; wheat, corn; cattle, sheep.

Joseph, town (2006 population 980), WALLOWA county, NE OREGON, 6 mi/9.7 km SE of ENTERPRISE, on Wallowa River; 45°21′N 117°13′W. Elevation 4,191 ft/1,277 m. Railroad terminus. Manufacturing (lumber, bronze and silver castings). Grain, potatoes; cattle. Part of Wallowa-Whitman National Forest, including EAGLE CAP Wilderness Area, to S; HELLS CANYON National Recreation Area (SNAKE RIVER) to E. Hells Canyon Dam 25 mi/40 km to E. WALLOWA LAKE to S. Wallowa Lake State Park at S end.

Joseph, village (2006 population 271), SEVIER county, SW central UTAH, 10 mi/16 km SW of RICHFIELD, and on SEVIER RIVER; 38°37′N 112°13′W. Elevation 5,435 ft/1,657 m. Alfalfa, barley; dairying; cattle. Parts of Fishlake National Forest to E and NW. Fremont Indian State Park to SW; Big Rock Candy Mountain to S. Old Farm Museum Settled 1864.

Joseph Bonaparte Gulf (JO-sef BO-nuh-pahrt), arm of TIMOR SEA, N AUSTRALIA, between Cape LONDONDERRY (W) and Point BLAZE (E); 225 mi/362 km E-W, 100 mi/161 km N-S. Divides into CAMBRIDGE GULF (W; site of WYNDHAM) and QUEENS CHANNEL (E).

Josephburg (JO-sef-buhrg), hamlet (2006 population 230), central ALBERTA, W CANADA, 20 mi/31 km from EDMONTON, and included in N of municipality of strathcona County; 53°43′N 113°05′W. Farming. First settled 1890s.

Joseph City, unincorporated village, NAVAJO county, E central ARIZONA, 10 mi/16 km WNW of HOLBROOK, on LITTLE COLORADO RIVER. Timber. Navajo Indian Reservation to N.

Joseph Henry, Cape, NE ELLESMERE ISLAND, BAFFIN region, NUNAVUT territory, N CANADA, on LINCOLN SEA of the ARCTIC OCEAN, NE extremity of FIELDEN PENINSULA; 82°49′N 63°35′W.

Josephine, county (□ 1,641 sq mi/4,266.6 sq km; 2006 population 81,688), SW OREGON; ⊙ GRANTS PASS; 42°21′N 123°33′W. Mountainous area bordering on CALIFORNIA (S) and crossed by ROGUE (N) and ILLINOIS (SW) rivers; State Scenic Waterways in W. Agriculture (wheat, oats, barley, apples, pears, plums, peaches, grapes; poultry, hogs, sheep, cattle); nurseries, wineries. Lumber. Gold. Parts of Siskiyou National Forest, including Kalmiopsis Wilderness Area, in W; part of Rogue River National Forest in SE corner; parts of SISKIYOU MOUNTAINS in W and S, including Oregon Caves National Monument in SE; Illinois River State Park in SW. Formed 1856.

Josephine (JOS-uh-feen), village (2006 population 825), COLLIN county, N TEXAS, 34 mi/55 km ENE of DALLAS; 33°03′N 96°19′W. Agricultural area just beyond urban fringe (cotton, sorghum, wheat; cattle, horses).

Josephinenhütte, POLAND: see SZKLARSKA POREBA.

Joseph, Lake (JO-suhf), S ONTARIO, E central CANADA, in MUSKOKA lake region, 16 mi/26 km SE of PARRY SOUND; 12 mi/19 km long, 4 mi/6 km wide; 45°10′N 79°43′W.

Josephstaal (JO-sef-stahl), village, MADANG province, NE NEW GUINEA island, N central PAPUA NEW GUINEA, 25 mi/40 km inland from BISMARCK SEA, SW of ADELBERT RANGE, and 60 mi/97 km NW of MADANG. Bananas, sugarcane; cattle.

José Santos Guardiola (ho-SAI SAHN-tos gwahr-dee-O-lah), town, on S coast of ROATÁN ISLAND, ISLAS DE LA BAHÍA department, N HONDURAS, 16 mi/26 km ENE of ROATÁN; 16°24′N 86°21′W. Shipbuilding; fishing; tourism. Formerly called Oak Ridge and still commonly known by that name in Islas de la Bahía.

José Sixto Verduzco, MEXICO: see PASTOR ORTIZ.

Joshimath (JO-shee-muht), town, tahsil headquarters, CHAMOLI district, N UTTAR PRADESH state, N central INDIA, at confluence of headstreams of the ALAKNANDA RIVER, 55 mi/89 km NE of PAURI; 30°34′N 79°34′E. Winter headquarters of chief priest of BADRINATH. Religious and tourist center.

Joshua (JAHSH-oo-wah), town (2006 population 5,574), JOHNSON county, N central TEXAS, 20 mi/32 km S of FORT WORTH; 32°27′N 97°22′W. Dairying; cotton; cattle; manufacturing (transportation equipment, machinery).

Joshua Tree, unincorporated town (2000 population 4,207), SAN BERNARDINO county, S CALIFORNIA, 54 mi/87 km E of SAN BERNARDINO, and 25 mi/40 km NE of PALM SPRINGS; 34°08′N 116°19′W. Twentynine Palms Marine Corps Base to NE, main entrance to E. JOSHUA TREE NATIONAL MONUMENT to SE. Cattle.

Joshua Tree National Monument (□ 875 sq mi/2,266 sq km), RIVERSIDE and SAN BERNARDINO counties, S CALIFORNIA. Main (S) entrance c.40 mi/64 km ESE of PALM SPRINGS; park road connects S entrance with TWENTYNINE PALMS to N. Rare Joshua trees (plentiful here), or "praying plant"; named by Mormons because of upstretched arms; includes wilderness area (1976). Variety of flora and fauna. Part of LITTLE SAN BERNARDINO MOUNTAINS in W; Twentynine Palms Indian Reservation on N. Authorized 1936.

Jo Shui, China: see RUO SHUI.

Joslowitz, CZECH REPUBLIC: see JAROSLAVICE.

Jos Plateau (JAWS), region (around □ 3,000 sq mi/7,770 sq km), central NIGERIA, W Africa. The plateau (elevation around 4,200 ft/1,280 m), composed mainly of granite, slopes gently to the N and is covered by grasslands; the GONGOLA RIVER rises here. The region has one of the higher population densities in Nigeria and is free from disease than the surrounding lowlands. Tin is mined and processed on the plateau, and farming and grazing are important. The city of Jos is the region's chief center. In the 19th century the plateau was a refuge for non-Islamic people fleeing the Islamic Fulani people.

Jössefors (YUHS-se-FORSH), village, VÄRMLAND county, W SWEDEN, on BYÄLVEN RIVER (falls), at N end of GLAFSFJORDEN LAKE, 4 mi/6.4 km WNW of ARVIKA; 59°40′N 12°30′E. Manufacturing (plastic products).

Josselin (zho-suh-lan), commune (□ 1 sq mi/2.6 sq km), MORBIHAN department, W FRANCE in BRITTANY, on OUST River (BREST-NANTES CANAL), and 7 mi/11.3 km W of PLOËRMEL; 47°57′N 02°33′W. Meat canning, carton manufacturing. Its fine 15th–16th-century castle has been beautifully restored with high towers, crenellated walls and a moat, surrounded by gardens.

Jossgrund (YUHS-grund), village, HESSE, central GERMANY, 21 mi/34 km NE of ASCHAFFENBURG; 50°11′N 09°29′E.

Jostedalsbreen (YAW-stuh-dahls-BRAI-uhn), largest glacier (□ 315 sq mi/819 sq km) of the European mainland, SOGN OG FJORDANE county, SW NORWAY, W of the JOTUNHEIMEN mts., between NORDFJORD and SOGNEFJORDEN; 60 mi/97 km long, 15 mi/24 km wide. Head is c.6,700 ft/2,042 m above sea level. It has many tributary glaciers.

Jost Van Dyke, islet, BRITISH VIRGIN ISLANDS, 4 mi/6.4 km W of TORTOLA; 18°27′N 64°45′W. Rugged, mountainous, rising to 1,070 ft/326 m. Dr. William Thornton, a designer of the Washington capitol, b. here. Little Jost Van Dyke island is just E. Port of entry, ferries to Tortola and Virgin Gorda (BVI) and ST. THOMAS and SAINT JOHN (U.S. VIRGIN ISLANDS).

Jósvafő (YOSH-vah-fuh), Hungarian *Jósvafő*, village, BORSOD-ABAÚJ-ZEMPLÉN county, NE HUNGARY, 29 mi/47 km NNW of MISKOLC; 48°29′N 20°34′E. Entrance to Jósvafő Cave, one of BARADLA CAVES.

Jotiba's Hill, INDIA: see PANHALA.

Jotsoma (jot-SO-muh), village, KOHIMA district, NAGALAND state, NE INDIA, in Naga Hills, 3 mi/4.8 km W of KOHIMA. Rice, cotton, oranges. Former stronghold of Naga tribes.

Jotunheimen (YAW-toon-haim), mountain group and national park, S central NORWAY; highest of SCANDINAVIA. It culminates in GALDHØPIGGEN (8,100 ft/2,469 m high) and GLITTERTIND (8,087 ft/2,465 m). The JOSTEDALSBREEN, a huge glacier, is to the W. Sparsely inhabited, the region is used for summer pasture. In Norse mythology, it was the home of the giants, the Jotuns.

Jouarre (zhoo-AHR), commune (□ 16 sq mi/41.6 sq km), SEINE-ET-MARNE department, ÎLE-DE-FRANCE region, N central FRANCE, 2 mi/3.2 km S of La FERTÉSOUS-JOUARRE, and 12 mi/19 km E of MEAUX; 48°56′N 03°08′E. Former seat of a Merovingian abbey. Behind the 15th-century parochial church is a crypt containing 7th–8th-century sculpted tombstones, among the oldest French religious monuments.

Jouars-Pontchartrain (zhoo-ahr–pon-shahr-TRAN), town (□ 3 sq mi/7.8 sq km; 2004 population 4,932), YVELINES department, ÎLE-DE-FRANCE region, N central FRANCE, 10 mi/16 km W of VERSAILLES; 48°47′N 01°54′E.

Joubert Mountains (JOO-ber), N extension (6,422 ft/1,784 m) of escarpment, KUNENE REGION, NW NAMIBIA; 18°30′S 14°00′E. Part of the watershed between Etosha drainage basin and coastal rivers.

Area is shown by the symbol □, and capital city or county seat by ⊙.

Joubert, Point (zhoo-BER, pwan), point, FRENCH GUIANA, at mouth of SINNAMARY RIVER, 50 mi/80 km NW of CAYENNE; 05°23′N 52°57′W.

Joué-lès-Tours (zhoo-AI–lai–TOOR), town (□ 12 sq mi/31.2 sq km), INDRE-ET-LOIRE department, CENTRE administrative region, W central FRANCE, a suburb 2 mi/3.2 km SW of TOURS, on the CHER RIVER above its junction with the LOIRE RIVER; 47°21′N 00°40′E. Has a modern tire factory, a technology park, and some workshops maintaining the area's velvet-making tradition.

Jougne (ZHOO-nyuh), commune (□ 10 sq mi/26 sq km), resort in DOUBS department, FRANCHE-COMTÉ region, E FRANCE, in JURA mountains on Swiss border, opposite VALLORBE, 10 mi/16 km S of PONTARLIER; 46°46′N 06°24′E. Elevation 3,280 ft/1,000 m. Winter sports.

Jouhar, town, ⊙ SHABEELLAHA DHEXE region, SW SOMALIA, on SHEBELI River, N of MOGADISHO. Irrigated land, producing sugarcane, bananas, and cotton.

Jouques (ZHOOK), town (□ 31 sq mi/80.6 sq km), BOUCHES-DU-RHÔNE department, PROVENCE-ALPES-CÔTE D'AZUR region, S FRANCE, near the DURANCE RIVER, and 12 mi/19 km NE of AIX-EN-PROVENCE; 43°38′N 05°38′E. Hydroelectric plant on canal parallel to the Durance River (3 mi/4.8 km N). The regional park of the LUBÉRON extends along N bank of Durance River valley. A nuclear research center conducting studies in controlled nuclear fusion and in agriculture and biotechnology applications is at CADARACHE, 6 mi/9.7 km NE.

Joura, India: see JORA.

Jourdanton (JOR-duhn-tun), town (2006 population 4,353), ⊙ ATASCOSA county, SW TEXAS, 35 mi/56 km S of SAN ANTONIO, near ATASCOSA RIVER; 28°54′N 98°32′W. In agricultural area (peanuts, corn); dairying; manufacturing (machinery).

Joure (YOO-ruh), town, FRIESLAND province, N NETHERLANDS, 8 mi/12.9 km SE of SNEEK; 52°58′N 05°48′E. TJEUKEMEER lake 4 mi/6.4 km to S, branch of PRINSES MARGRIET CANAL to SW. Dairying; cattle, sheep; vegetables, nursery stock; manufacturing (motor vehicles, clocks). Joure Museum.

Jourimain, Cape (JOOR-i-main), promontory on NORTHUMBERLAND STRAIT, SE NEW BRUNSWICK, E CANADA, 32 mi/51 km ENE of SACKVILLE; 46°10′N 63°49′W.

Joussard (zhoo-SAHRD), hamlet, central ALBERTA, W CANADA, 21 mi/34 km E of HIGH PRAIRIE, in BIG LAKES municipal district; 55°24′N 115°57′W.

Joutseno (YOT-se-no), village, KYMEN province, SE FINLAND, 12 mi/19 km ENE of LAPPEENRANTA, near RUSSIAN border, by S shore of LAKE SAIMAA; 61°06′N 28°30′E. Elevation 248 ft/75 m. Lumber and cellulose mills. Seat of University of Joutseno.

Joux, Fort de, FRANCE: see PONTARLIER.

Joux, Lac de (ZHOO, LAHK duh), lake (□ 4 sq mi/10.4 sq km), VAUD canton, W SWITZERLAND, in JURA mountains; 6 mi/9.7 km long; maximum depth 112 ft/34 m. Elevation 3,300 ft/1,006 m. Area (above) includes small Lac des Brenets. ORBE RIVER enters Lac de Joux from S and leaves Lac des Brenets in N. Vallée de Joux extends W of lake and river to Le Mont Risoux on French border.

Jouy-en-Josas (zhoo-ee–ahn–zho-ZAH), residential town (□ 4 sq mi/10.4 sq km), YVELINES department, ÎLE-DE-FRANCE region, N central FRANCE, 4 mi/6.4 km SE of VERSAILLES; 48°46′N 02°10′E. Surrounded by greenbelts, it is the site of several academic institutions, notably the College of Advanced Business Studies, known in France as H.E.C., and a national center for zoological research.

Jouy-le-Moutier (zhoo-ee–luh–moo-TYAI), town (□ 2 sq mi/5.2 sq km), VAL-D'OISE department, ÎLE-DE-FRANCE region, N central FRANCE, in the PARIS basin, 4 mi/6.4 km SSW of PONTOISE and NW of PARIS, near the inflow of the OISE RIVER into the SEINE RIVER;

49°01′N 02°02′E. The "new town" (planned community) of Cergy-Pontoise was developed nearby, just N of Oise River bend.

Jovellanos (ho-vai-YAHN-os), town (2002 population 26,726), MATANZAS province, W CUBA, on Central Highway, and 29 mi/47 km ESE of MATANZAS; 22°48′N 81°13′W. Railroad junction and commercial center in sugar-growing region. Has foundries, machine shops, tobacco factories. Julio Reyes Cairo sugar mill is nearby (E).

Jovero, DOMINICAN REPUBLIC: see MICHES.

Jovet, Mont (zho-VAI, MON), peak (8,378 ft/2,554 m) of Savoy Alps (ALPES FRANÇAISES), SAVOIE department, RHÔNE-ALPES region, SE FRANCE, between ISÈRE (N) and DORON DE BOZEL (S) river valleys, overlooking MOÛTIERS (5 mi/8 km W) and VANOISE massif; 45°30′N 06°39′E. The extensive La PLAGNE ski terrain is just E.

Joveymand, IRAN: see GONABAD.

Joviânia (zho-vee-AH-nee-ah), town (2007 population 6,798), S central GOIÁS state, BRAZIL, in Serra Campo Alegre, 40 mi/64 km N of BOM JESUS DE GOIÁS; 18°49′S 49°39′W.

Jovita, ARGENTINA: see SANTA MAGDALENA.

Jowai (JO-wei), town, ⊙ JAINTIA HILLS district, MEGHALAYA state, NE INDIA, in JAINTIA HILLS, 22 mi/35 km ESE of SHILLONG; 25°27′N 92°12′E. Rice, sesame, cotton.

Jowara (JOU-er-uh), town, NORTH BANK division, GAMBIA, 35 mi/56 km ENE of BANJUL; 13°34′N 16°05′W.

Joy, village (2000 population 373), MERCER county, NW ILLINOIS, 7 mi/11.3 km W of ALEDO; 41°12′N 90°52′W. In agricultural area.

Joyabaj (ho-yah-BAH), town, QUICHÉ department, W central GUATEMALA, on S slope of Sierra de CHUACÚS, 23 mi/37 km E of SANTA CRUZ DEL QUICHÉ; 14°59′N 90°48′W. Elevation 4,701 ft/1,433 m. Corn, beans; livestock. Quiché-speaking population.

Joya de Cerén (HOI-yah dai se-REN), archaeological site, EL SALVADOR, between SANTA ANA and SAN SALVADOR, off INTER-AMERICAN HIGHWAY. Has 1,500-year-old remains of Maya settlement preserved by a volcanic eruption; source of important information about Classic Mayan life.

Joya Mair, PAKISTAN: see BALKASSAR.

Joyce's Country (JOI-suhz), Gaelic *Dúiche Sheoigheach*, mountainous district of NW GALWAY county, W IRELAND, extending between KILLARY HARBOUR (W) and LOUGH MASK (E). Noted for wild scenery. Named for a Welsh family that came here in 12th century.

Joydevpur, BANGLADESH: see GAZIPUR.

Joynagar, INDIA: see JAYNAGAR.

Joyo (JO-YO), town, Yame county, FUKUOKA prefecture, N KYUSHU, SW JAPAN, 28 mi/45 km S of FUKUOKA; 33°14′N 130°38′E. Tea, shiitake mushrooms; forest management.

Joypurhat, BANGLADESH: see JAIPUR HĀT.

Jozjan (jooz-JAHN), province (2005 population 443,300), N central AFGHANISTAN, ⊙ SHIBARGHAN. Borders on TURKMENISTAN (N) and UZBEKISTAN (NE tip), BALKH province (E), SAR-I-PUL province (S), and FARYAB province (W). Flat sand desert in N; mountainous in S. Population largely Uzbek, with some Pashtun and Tajiks. Rich in mineral resources, oil, and natural gas; a pipeline, completed in 1968, transported gas to SOVIET UNION. Now some local use in production of fertilizer. Major exports include carpets and karakul skins; the province is famous for its melons. Domestic airports in AQCHAH and SHIBARGHAN. A militia raised from here supported the Marxist government of KABUL. In 1992, at the fall of the Kabul regime, General Dostum, an Uzbek, became ruler of N central Afghanistan, and an opponent of the Taliban government. Also spelled Jowzjan.

J. Percy Priest Lake, reservoir (c.20 mi/32 km long), DAVIDSON and RUTHERFORD counties, N TENNESSEE,

on STONES RIVER, 10 mi/16 km E of downtown NASHVILLE; 36°08′N 86°36′W. Maximum capacity 652,000 acre-ft. Formed by J. Percy Priest Dam (129 ft/39 m high), built (1967) by the Army Corps of Engineers for flood control and power generation. Long Hunter State Park on NE shore, near dam.

J. P. Koch Fjord (KOKH fyor), inlet (70 mi/113 km long, 2 mi/3.2 km–10 mi/16 km wide) of LINCOLN SEA of ARCTIC OCEAN, PEARY LAND region, N GREENLAND; 82°35′N 41°00′W. Receives large glacier. It is the W part of what was long thought to be PEARY CHANNEL.

Jrada (zhe-RAH-duh), town, Jerada province, Oriental administrative region, NE MOROCCO, 26 mi/42 km SSW of OUJDA; 34°18′N 02°09′W. Morocco's major coal-mining center, yielding high-grade anthracite. Mining operations, begun 1931, were greatly expanded during World War II. Coal was taken by cable car to Guenfouda (10 mi/16 km NNE, on railroad to Oujda), where a washing and briquette-manufacturing plant was built 1936. Formerly called Al Aouïnet. Formerly spelled Djerada; also spelled Jerada or Djerada.

Jreida (juh-re-EE-dah), hamlet, Trarza administrative region, MAURITANIA, on ATLANTIC OCEAN coast, c.17 mi/27 km N of NOUAKCHOTT; 18°27′N 16°01′W. Fishing. Formerly called Coppólani.

J. Strom Thurmond Lake or **Clark Hill Lake**, reservoir (36 mi/58 km long), GEORGIA and SOUTH CAROLINA, on SAVANNAH RIVER (Ga.-S.C. border), 20 mi/32 km NNW of AUGUSTA, Georgia; 33°39′N 82°11′W. Maximum capacity 2,900,000 acre-ft. Formed by J. Strom Thurmond Dam (or Clarks Hill Dam; 200 ft/61 m high, 5,660 ft/1,725 m long) for flood control, navigation, power generation, and recreation. Several state parks on both sides; Sumter National Forest (S.C.) on NE shore. Wealthy coastal residents established summer homes here in early 19th century to escape oppressive weather in S Georgia. One of the largest inland man-made lakes in the S.

Juab (JOO-ab), county (□ 3,406 sq mi/8,855.6 sq km; 2006 population 9,420), W UTAH, on NEVADA (W) state line; ⊙ NEPHI; 39°42′N 112°47′W. Mining and agriculture area watered in SE by SEVIER RIVER. Wheat, barley, alfalfa; cattle; silver. Part of Goshute Indian Reservation is in NW, on Nevada state line. Little Sahara Recreation Area (Bureau of Land Management) in SE, in White Sand Dunes; Fish Springs National Wildlife Refuge in N center in Fish Springs Flats. Tintic Mining District in NE, around EUREKA (gold, silver, lead). Part of Fishlake National Forest in SE, parts of Uinta National Forest in E, including part of Mount Nebo Wilderness Area in NE. N end of SEVIER BRIDGE RESERVOIR in SE corner, including Yuba (W shore) and Painted Rock (E shore) state parks, on SANPETE county border. Mona Reservoir in NE. Formed 1852.

Juabeso (jwah-bai-so), town (2000 population 3,639), WESTERN REGION, GHANA, 30 mi/48 km NW of WIAWSO; 06°20′N 02°50′W. Timber, cocoa, rubber.

Juana Azurduy de Padilla (HWAH-nah ah-soor-DOO-ee dai pah-DEE-yah), canton, TOMINA province, CHUQUISACA department, SE BOLIVIA, at the fork of the MONTEAGUDO-SUCRE and Monteagudo-SANTA CRUZ highways; 19°19′S 64°20′W. Elevation 6,824 ft/2,080 m. Airstrip. Limestone deposits. Agriculture (potatoes, yucca, bananas, corn, rye, sweet potatoes, peanuts); cattle.

Juanacatlán (hwah-nah-kah-TLAHN), town, JALISCO, central MEXICO, on Santiago (LERMA) River, and 16 mi/26 km SE of GUADALAJARA; 20°31′N 103°10′W. Wheat, vegetables; livestock. EL SALTO hydroelectric plant (capacity 2,975 kw). The famous falls here, second in NORTH AMERICA only to NIAGARA Falls, are much reduced because of lowered water level in Lake CHAPALA.

Juana Díaz (HWAH-nah DEE ahz), town (2006 population 52,770), S central PUERTO RICO, 8 mi/12.9 km ENE of PONCE. Some agriculture including hogs, poultry; large mango growths (for export). Varied manufacturing. Marble, manganese deposits nearby. U.S. Navy Fort Allen is c. 3 mi/4.8 km S.

Juan A. Escudero, MEXICO: see TIERRA COLORADA.

Juan Aldama (hwan ahl-DAH-mah), town, ⊙ Juan Aldama municipio, ZACATECAS, N central MEXICO, on interior plateau, near DURANGO border, 32 mi/51 km NW of NIEVES on Mexico Highway 49; 24°17′N 103°23′W. Agriculture (maguey, corn; livestock). Formerly San Juan de Mezquital.

Juan Antonio Artigas (HWAHN an-TO-nee-o ahr-TEE-gahs), city (2004 population 13,553) and suburb of MONTEVIDEO, S URUGUAY, CANELONES department, along the Pan-American Highway, 12 mi/19 km NE of Montevideo; 34°44′S 56°01′W.

Juan B. Arruabarrena, ARGENTINA: see SAN JAIME.

Juan Bautista Alberdi (HWAHN bou-TEES-tah ahl-BER-dee), town, SW TUCUMÁN province, ARGENTINA, on MARAPA RIVER, and 70 mi/113 km SW of TUCUMÁN. Dam and hydroelectric station. Formerly Escaba.

Juan B. Molina (HWAHN bai mo-LEE-nah), village, SE SANTA FE province, ARGENTINA, 38 mi/61 km S of ROSARIO; 33°29′S 60°30′W. Agricultural center (corn, soybeans, flax; livestock).

Juan C. Bonilla, MEXICO: see CUANALÁ.

Juancheng (RWAHN-CHUHNG), town, ⊙ Juancheng county, SW SHANDONG province, NE CHINA, 65 mi/105 km ESE of ANYANG; 35°34′N 115°30′E. Grain, oilseeds, cotton, medicinal herbs. Textiles, building materials. Manufacturing; food industry.

Juan Cumatzi, Contla de, MEXICO: see CONTLA.

Juan de Acosta (HWAHN dai ah-KOS-tuh), town, ⊙ Juan de Acosta municipio, ATLÁNTICO department, N COLOMBIA, in CARIBBEAN LOWLANDS, 20 mi/32 km SW of BARRANQUILLA; 10°50′N 75°02′W. Cotton growing; corn, yucca, sorghum; livestock.

Juan de Fuca, Strait of (HWAHN duh FYOO-kuh), inlet of the PACIFIC OCEAN, between VANCOUVER ISLAND, BRITISH COLUMBIA (SW CANADA; N) and OLYMPIC PENINSULA, WASHINGTON (U.S.; S), linking the Strait of GEORGIA through HARO and ROSARIO straits, and PUGET SOUND with the Pacific; forms part of the U.S.-Canada border; 100 mi/161 km long, 11 mi/18 km–20 mi/32 km wide; 48°20′N 124°00′W. VICTORIA (British Columbia), the strait's largest city, is located on N side; ferries connect it with the U.S. mainland. It is the main shipping channel to SEATTLE, TACOMA, VANCOUVER (British Columbia), Victoria, other parts. Cape FLATTERY (Washington), at S side of Pacific entrance to strait; SAN JUAN ISLANDS (Washington), to NE; WHIDBEY ISLAND at E end of strait. West Coast Trail Unit of PACIFIC RIM NATIONAL PARK on NW coast. PORT ANGELES (Washington), on S side. Discovered by the English captain Charles W. Barkley in 1787, the strait was named for a sailor, Juan de Fuca, who reputedly had explored it here for SPAIN in 1592.

Juan de Mena (HWAHN DAI ME-nah), town, La Cordillera department, S central PARAGUAY, 60 mi/97 km ENE of Asunción; 24°54′S 56°44′W. Fruit; livestock. Formerly called San Rafael.

Juan de Nova Island (HWAHN dai NO-vah), small island, dependency of MADAGASCAR, in MOZAMBIQUE CHANNEL of INDIAN OCEAN, 90 mi/145 km off W coast; 17°05′S 42°18′E. Guano deposits. Ceded by FRANCE in 1990. Also known as Saint-Christophe.

Juan Díaz (hwahn DEE-ahs), village, PANAMA province, central PANAMA, in PACIFIC lowland, 7 mi/11.3 km NE of PANAMA City. Largely absorbed by Panama City urban expansion.

Juan Dolio (HWAHN DO-lee-o), beach resort and residential community, DOMINICAN REPUBLIC, 20 mi/32 km E of SANTO DOMINGO airport.

Juan Emilio O'Leary (HWAHN ai-MEE-lyo O-LIR-ee), town, E PARAGUAY, ALTO PARANA department, 50 mi/80 km W of CIUDAD DEL ESTE; 25°25′S 55°23′W. Cotton- and soybean-producing center.

Juan Fernandez (HWAHN fer-NAHN-des), national park, consisting of 3 islands in the PACIFIC OCEAN W of CHILE, in the VALPARAISO region. Practically pristine natural environment.

Juan Fernández (HWAHN fer-NAHN-des), group of small islands, S PACIFIC, c.400 mi/644 km W of VALPARAISO, CHILE; 33°00′S 80°00′W. They belong to Chile and are administered as a part of VALPARAISO region. The 2 principal islands are Isla ROBINSON CRUSOE (formerly Más a Tierra) and Isla Alejandro Selkirk (formerly Más Afuera). Volcanic in origin, they have a pleasant climate and are rugged and heavily wooded. The village of San Juan Bautista, on Isla Robinson Crusoe, is the only settlement. Fishing is the principal occupation. Airport. Discovered by an obscure Spanish navigator in 1563, the islands achieved fame with the publication of Daniel Defoe's *Robinson Crusoe* (1719), generally acknowledged to have been inspired by the confinement on Más a Tierra (1704–1709) of Alexander Selkirk, a Scottish sailor. Occupied by the Spanish in 1750, the islands passed to Chile when it won independence. In the 19th century Robinson Crusoe island was used as a penal colony. The islands are now a national park and have been designated a World Biosphere Reserve by UNESCO.

Juan Galindo, MEXICO: see NUEVO NECAXA.

Juan Gallegos Island (hwahn gah-YAI-gos), in GATÚN LAKE, PANAMA CANAL ZONE, 9 mi/14.5 km SSE of COLÓN; 4 mi/6.4 km long, 1.5 mi/2.4 km wide. Covered by rain forest.

Juan Godoy (HWAHN go-DOI), village, ATACAMA region, N central CHILE, 32 mi/51 km S of COPIAPÓ. Railroad terminus in copper- and silver-mining area.

Juan Griego (HWAHN gree-AI-go), town, ⊙ Marcano municipio, NUEVA ESPARTA state, NE VENEZUELA, port on MARGARITA ISLAND, in the CARIBBEAN SEA, 7 mi/11 km NW of LA ASUNCIÓN; 11°50′N 63°58′W. Fishing (pearls; sardines, mackerel, tuna, herring).

Juan Guerra (HWAHN GAI-rah), town, SAN MARTÍN province, SAN MARTÍN region, N central PERU, landing on affluent of MAYO RIVER, and 8 mi/13 km S of TARAPOTO; 06°35′S 76°21′W. In agricultural region (tobacco, coffee, cotton, sugarcane).

Juan, Gulf of (ZHWAHN), on the MEDITERRANEAN SEA, SE FRANCE, on French RIVIERA, between Pointe de la CROISETTE (W) and Cap d'Antibes (E); 43°33′N 07°05′E. Good fleet anchorage sheltered by the small Îles de Lérins (S). On it are the fashionable resorts of JUAN-LES-PINS and GOLFE-JUAN.

Juani Island (joo-AH-nee), PWANI region, E TANZANIA, 85 mi/137 km SSE of DAR ES SALAAM, and 5 mi/8 km SE of MAFIA ISLAND, in INDIAN OCEAN; 6 mi/9.7 km long, 2 mi/3.2 km wide; 08°00′S 39°46′E.

Juanita (hwahn-EE-tuh), unincorporated town, KING county, W WASHINGTON, residential suburb 8 mi/12.9 km NE of downtown SEATTLE, on E shore of LAKE WASHINGTON. St. Edward State Park to NW.

Juan José Castelli (ho-SAI kahs-TAI-lee), town (pop. 13,206), General Güemes department, Chaco province, ARGENTINA, 145 mi/233 km NW of city of RESISTENCIA. Railroad terminus. Cotton and cattle region.

Juanjuí (hwahn-HWEE), city (2005 population 22,756), ⊙ MARISCAL CÁCERES province, SAN MARTÍN region, N central PERU, landing on HUALLAGA RIVER, and 80 mi/129 km SSE of MOYOBAMBA, on road from SAPOSOA to TOCACHE NUEVO; 07°11′S 76°45′W. Coca, bananas, yucca, plantains. Airport.

Juankoski (YOO-ahn-KOS-kee), Swedish *Strömsdal*, village, KUOPION province, S central FINLAND, in SAIMAA lake region, 20 mi/32 km NE of KUOPIO; elevation 297 ft/ 90 m; 63°04′N 28°21′E. Pulp and board mills.

Juan Lacaze (HWAHN lah-KAH-sai), city (2004 population 13,196), COLONIA department, SW URUGUAY, on N bank of the Río de la PLATA, and 23 mi/37 km E of Colonia; 34°26′S 57°27′W. The department's leading port, visited by oceangoing vessels; exports cereals, cattle, and building materials to BUENOS AIRES (55 mi/89 km W). Textile and paper mills. Quarries nearby (NW). Officially called Sauce until 1909; its port is still sometimes referred to as Puerto Sauce or Puerto del Sauce. Also Juan Lake Lacaze.

Juan Latino (hwahn lah-TEE-no), canton, WARNES province, SANTA CRUZ department, E central BOLIVIA, 31 mi/50 km N of SANTA CRUZ, on the Santa Cruz–COCHABAMBA highway; elevation 1,089 ft/332 m; 17°26′S 63°13′W. Abundant gas and oil resources in area and some limestone. Agriculture (bananas, sugarcane, rice, soybeans, sesame, peanuts, corn, potatoes, cotton, coffee, citrus fruits); cattle for meat and dairy products.

Juan-les-Pins (ZHWAHN–lai–PAN), noted resort on the French RIVIERA, ALPES-MARITIMES department, PROVENCE-ALPES-CÔTE D'AZUR region, SE FRANCE, 1 mi/1.6 km SSW of ANTIBES, on sheltered Gulf of JUAN; 43°34′N 07°06′E. It has an excellent beach, waterfront hotels, casino; it hosts an annual world jazz festival. Administratively, it forms part of the town of Antibes.

Juan N. Méndez, MEXICO: see ATENAYUCA.

Juan Ortíz (hwahn or-TEEZ), town, SE SANTA FE province, ARGENTINA, near PARANÁ RIVER, 10 mi/16 km NNW of ROSARIO. Agricultural center (flax, corn, wheat, alfalfa; livestock).

Juan Rodríguez Clara (hwahn ro-DREE-gez KLAH-rah), town, SE VERACRUZ, MEXICO, 34 mi/55 km from SAN ANDRÉS TUXTLA, on Mexico Highway 145, on railroad; 17°59′N 95°24′W. Located on the plains of the Papaloapan River basin. Hot climate. Agriculture (corn, beans, plantains, rice, mangoes); cattle; forestry. Cooking oil.

Juan Soldado, Cerro de (HWAHN sol-DAH-do, SER-ro dai), pre-Andean mountain (elevation 3,900 ft/ 1,189 m), COQUIMBO region, N central CHILE, near PACIFIC coast, N of LA SERENA; 29°41′S 71°16′W. Limestone and high-grade iron-ore deposits.

Juan Stuven Island (HWAHN STOO-ven), off coast of AISÉN province, AÍSEN DEL GENERAL CARLOS IBAÑEZ DEL CAMPO region, S CHILE, just S of GUAYANECO ISLANDS; c.20 mi/32 km long; 47°58′S 75°00′W. Uninhabited.

Juan Vicente (HWAHN vee-SAIN-tai), beach resort, HOLGUÍN province, E CUBA, on NIPE BAY (ATLANTIC OCEAN), 2 mi/3 km NW of MAYARÍ, 50 mi/80 km N of SANTIAGO DE CUBA; 20°42′N 75°44′W.

Juan Viñas (hwahn VEEN-yahs), town, ⊙ JIMÉNEZ canton, CARTAGO province, COSTA RICA, in REVENTAZÓN RIVER valley, on old San José–Limón Highway, and 12 mi/19 km ENE of CARTAGO; 09°54′N 83°45′W. Sugarcane center; sugar mill. Coffee; livestock.

Juan W. Gez (HWAHN DO-bluh-vai GAIZ), village, N central SAN LUIS province, ARGENTINA, at S foot of SIERRA DE SAN LUIS, on railroad, and 20 mi/32 km ESE of SAN LUIS; elevation 3,100 ft/945 m; 33°23′S 66°08′W. Resort in wheat and livestock area. Formerly La Cumbre.

Juara (ZHOO-ah-rah), city (2007 population 32,096), NW Mato Grossostate, near Rio Arinos; 11°12′S 57°29′W.

Juárez (HWAH-rez), city (2005 population 1,313,338) and township, CHIHUAHUA, N MEXICO, on the RIO GRANDE, opposite EL PASO (TEXAS); 31°44′N 106°29′W. Elevation 3,734 ft/1,138 m. Connected with the UNITED STATES by three international bridges, it is a shipping point and highway and railroad terminus. It is also the commercial and processing center for the surrounding cotton-growing area. Except for the river valley, under intense cultivation SE of the city, Juárez is surrounded by desert. It has experienced extremely rapid population growth and has been a favored location for *ma-*

quiladoras, foreign-owned manufacturing plants that finish goods, including computer and mechanical equipment parts for sale in the U.S. It was originally called El Paso del Norte and included settlements on both sides of the river, until they were split by the Treaty of Guadalupe Hidalgo (1848). In 1888 the name of the Mexican town was changed to honor Benito Juárez, who made it his capital when exiled from central Mexico. Sometimes called Ciudad Juárez.

Juárez (HWAH-rez), town, CHIAPAS, S MEXICO, in Gulf lowland, 27 mi/43 km SW of VILLAHERMOSA, on railroad, in Zoque Indian area; 17°40′N 93°10′W. Cacao, rice.

Juárez, town, COAHUILA, N MEXICO, near DON MARTÍN DAM, in irrigated plain, 26 mi/42 km SE of SABINAS; 27°39′N 100°43′W. Elevation 984 ft/300 m. Cereals, fruit, istle fibers, cattle.

Juárez, town, ⊙ JUÁREZ HIDALGO municipio, HIDALGO, central MEXICO, 50 mi/80 km NNW of PACHUCA DE SOTO. Corn, wheat, beans, fruit, livestock. Also known as Juárez Hidalgo.

Juárez (HWAH-rez), settlement, Nuevo Casas Grandes municipio, CHIHUAHUA, NW MEXICO, on Piedras Verdes River, 12 mi/20 km from NUEVO CASAS GRANDES, on Chihuahua State Highway 10; 30°18′N 108°04′W. Elevation 5,085 ft/1,550 m. Mixed farming and dairying. Agricultural colony established 1886 by Mormon immigrants from UNITED STATES. Was the principal center for the Mormon colonies, which still survive in the area. Also Colonia Juárez.

Juárez, ARGENTINA: see BENITO JUÁREZ.

Juárez, MEXICO: see BENITO JUÁREZ.

Juárez Celman, ARGENTINA: see LA CARLOTA.

Juárez Hidalgo, MEXICO: see JUÁREZ.

Juárez, Sierra (HWAH-rez, SYER-rah), range, in N BAJA CALIFORNIA, NW MEXICO, extends c.90 mi/145 km SE from UNITED STATES border, W of Laguna SALADA; elevation 2,953 ft/900 m–6,562 ft/2,000 m. Gold placers. Sometimes Sierra de Juárez.

Juárez, Sierra de, MEXICO: see OAXACA, SIERRA MADRE DE.

Juarez Távora (ZHOO-wah-res TAH-vo-rah), town (2007 population 7,618), E central PARAÍBA state, BRAZIL, 11 mi/18 km SE of ALAGOA GRANDE; 07°10′S 35°33′W.

Juárez, Villa, MEXICO: see CIUDAD MANTE.

Juaryen (joo-AHR-yen), town, LIBERIA, 30 mi/48 km E of GREENVILLE. Cocoa, oil palm.

Juaso (JOO-ah-so), town, local council headquarters, ASHANTI region, GHANA, on railroad, and 40 mi/64 km ESE of KUMASI; 06°35′N 01°07′W. Gold mining; cacao, cassava, corn. Also spelled Dwaaso.

Juatuba, city (2007 population 19,548), Minas Gerais state, Brazil.

Juayúa (hwah-YOO-ah), city and municipality, SONSONATE department, W EL SALVADOR, on Río Grande, and 8 mi/12.9 km N of SONSONATE; 13°51′N 89°45′W. Coffee-growing center; coffee processing.

Juazeirinho (ZHOO-wah-sai-reen-yo), city (2007 population 15,913), central PARAÍBA state, BRAZIL, on JOÃO PESSOA–FORTALEZA railroad, 21 mi/34 km W of SOLEDADE; 07°04′S 36°35′W.

Juazeiro (zhwah-SAI-ro), city (2007 population 230,538), N BAHIA state, BRAZIL, on right bank of Rio SÃO FRANCISCO, opposite PETROLINA (PERNAMBUCO), 280 mi/451 km N of SALVADOR; 09°28′S 40°30′W. Downstream end of river navigation from PIRAPORA (MINAS GERAIS state), and terminus of railroad from Salvador. Important center for shipping of caroa fibers. Railroad bridge to Petrolina built 1951. Rock crystals found in area. Has 18th-century church. Formerly spelled Joazeiro.

Juàzeiro (ZHWAH-sai-ro), city, N PIAUÍ state, BRAZIL, 70 mi/113 km E of TERESINA; 05°05′S 41°50′W.

Juazeiro do Norte (zwah-SAI-ro do nor-chee), city (2007 population 242,139), S CEARÁ state, BRAZIL, on

FORTALEZA-CRATO railroad, and 6 mi/9.7 km E of Crato; 07°15′S 39°20′W. Center of fertile agricultural area growing sugar, cotton, cereals; sugar distilling, cotton ginning. Large population increase in recent decades. Residence of Padre Cicero, important figure in the folk history of Nordeste (the northeastern region of Brazil). Until 1944, called Juázeiro; previously spelled Joazeiro.

Juazohn (joo-AH-zahn), town, SINOE county, LIBERIA, 30 mi/48 km NE of GREENVILLE, on road to ZWEDRU; 05°23′N 08°51′W.

Juba (JOO-bah), city, ⊙ Central Equatoria state, S SUDAN, a port on the WHITE NILE River; 04°51′N 31°37′E. S terminus of river traffic in the SUDAN; highway hub, with roads radiating into UGANDA, KENYA, and CONGO. A unification agreement arranged here in 1947 joined the N and S parts of SUDAN, which dashed British hopes of adding the S to UGANDA. Became the center of S resistance to alleged N dominance of the country. In 1955 a mutiny of S troops at Juba caused a Sudanese civil war, which was not settled until 1969. In 1983, civil war in the S broke out again and continues still, caused by S Sudanese opposition to the JONGLEI CANAL (which they believe will serve only EGYPT while the ecology of the SUDD region will change dramatically), and by president Numeri's violation of the conditions of the Addis Ababa agreement that brought peace to the S in 1972. Seat of the University of Juba (1975). Juba was the capital of former EQUATORIA province.

Juba Hill, affluent suburb along the coast, 5 mi/8 km SE of FREETOWN, SIERRA LEONE; 08°27′N 13°16′W. Recently built presidential residence called Cabassa Lodge here; military barracks nearby.

Jubail (joo-BAIL), town (2004 population 222,544), AL AHSA region, E SAUDI ARABIA, port on PERSIAN GULF, 55 mi/89 km NNW of DHAHRAN; 27°00′N 49°39′E. Trading center; pearling. It is a large modern industrial center and port. Oil refining. Was major Al Ahsa port in 1920s and 1930s until development of oil industry. Also spelled Jubayl. Also called Jubail al Bahri.

Jubaland, historic territory, SOMALIA, drained by JUBBA River, a seasonal stream. Area population by Sab tribes (of Bantu origin, some descendants of slaves), pastoralists, and sedentary farmers. Steppe vegetation on higher ground, jungle on the edges of the river. Territory was granted to Italy by Negus of Abyssinia in 1908. In 1924, Great Britain ceded lands S and W of the Jubba River (TRANS-JUBALAND) to Italy. Was divided into Upper and Lower Juba provinces during Italian colonial rule.

Juban Bey (JOO-BAHN BAI), village, ALEPPO district, NW SYRIA, near Turkish border, on railroad, 32 mi/51 km NNE of ALEPPO; 36°36′N 37°27′E. Cotton, cereals.

Jubany Station (JOO-buh-nee), ANTARCTICA, Argentinian station on KING GEORGE ISLAND; 62°14′S 58°40′W.

Jubayl, SAUDI ARABIA: see JUBAIL.

Jubbada Dhexe, region, SW SOMALIA; ⊙ Bu'aale. JUBBA River flows through the region. Some grazing, cultivation (sorghum, maize, cassava, millet).

Jubbada Hoose, region (□ 24,000 sq mi/62,400 sq km), SW SOMALIA, borders KENYA on the W; ⊙ KISMAYO. Sugar refining and food- and tobacco-processing industry.

Jubbal (JUHB-buhl), former princely state of PUNJAB HILL STATES, INDIA. Since 1948, merged in SHIMLA district, HIMACHAL PRADESH state. Jubbal-Kotkhai is a politically important constituency of the state. Two Chief Ministers elected from here.

Jubb al Jarrah, SYRIA: see JUBB EL JARRAH.

Jubba River, c.1,000 mi/1,609 km long, E Africa; rises in S ETHIOPIA by the confluence of the DAWA (Daua) and GENALE rivers; meanders S through SW SOMALIA to the INDIAN OCEAN near KISMAYO. Navigable for shallow craft to Bardero. The JUBBA has flood seasons

in both spring and autumn. The valley is part of SOMALIA's chief agricultural region, and the river is extensively used for irrigation. Agriculture includes bananas, sugarcane, sorghum, maize, cassava, and millet.

Jubb el Jarrah (JUB el JAHR-rah), township, HOMS district, W SYRIA, 36 mi/58 km ENE of HOMS; 34°50′N 37°19′E. Cereals. Also spelled JUBB al Jarrah.

Jubilee Lake, SE NEWFOUNDLAND AND LABRADOR, E CANADA, 30 mi/48 km N of FORTUNE BAY; 6 mi/10 km long, 4 mi/6 km wide; 48°03′N 55°10′W.

Jublains (zhoo-BLAN), small village (□ 13 sq mi/33.8 sq km), MAYENNE department, PAYS DE LA LOIRE region, W FRANCE, 6 mi/9.7 km SE of MAYENNE; 48°15′N 00°30′W. Has partially buried remains of 3rd-century Roman camp, enclosed by a double wall.

Jubones River (hoo-BO-nes), c.100 mi/161 km long, S ECUADOR; rises in the ANDES Mountains S of CUENCA; flows W, past PASAJE, to the JAMBELI CHANNEL on the PACIFIC OCEAN, 7 mi/11.3 km NNE of MACHALA.

Jubrique (hoo-VREE-kai), town, MÁLAGA province, S SPAIN, 12 mi/19 km S of RONDA; 36°34′N 05°13′W. Grapes, chestnuts, olives; liquor distilling, sawmilling.

Juby, Cap (ZHOO-bee, KAHP), prominent coastal headland, on ATLANTIC OCEAN, Laâyoune province, Laâyoune-Boujdour-Sakia El Hamra administrative region, SW MOROCCO, just N of WESTERN SAHARA territory, opposite CANARY ISLANDS (SPAIN), 150 mi/241 km E of Las PALMAS (Spain); 27°57′N 12°55′W. On it is town of TARFAYA (formerly Villa Bens). Also spelled Cap Yubi.

Juçara (zhoo-SAH-rah), city, N central BAHIA state, BRAZIL, 18 mi/29 km NW of IRECÉ; 11°02′S 41°58′W. Also spelled Jussara.

Júcaro (HOO-kah-ro), village, CIEGO DE ÁVILA province, E CUBA, landing on the CARIBBEAN Gulf of ANA MARÍA, 16 mi/26 km SSW of CIEGO DE ÁVILA (linked by railroad); 21°36′N 78°52′W. Timber.

Júcar River (HOO-kahr), 310 mi/499 km long, E SPAIN, in CUENCA, ALBACETE, and VALENCIA provinces; rises on W slopes of the MONTES UNIVERSALES, flows S, past Cuenca, turning E near La Roda, to the MEDITERRANEAN SEA at Cullera. River feeds several economically important irrigation canals in lower course.

Jucás (zhoo-KAHS), city (2007 population 22,874), S CEARÁ state, BRAZIL, on upper RIO JAGUARIBE, and 22 mi/35 km SW of IGUATU; 06°30′S 39°31′W. Coffee, cotton, tobacco. Magnesite deposits. Irrigation dam and reservoir under construction at Cariús, just SE. Until 1944, called São Mateus (formerly spelled São Matheus).

Jüchen (YOOI-khuhn), commune, NORTH RHINE–WESTPHALIA, W GERMANY, 5 mi/8 km SSE of RHEYDT; 51°04′N 06°29′E. Lignite mining; manufacturing of textiles. Has 17th-century castle and 14th-century monastery.

Juchipila (hoo-chee-PEE-lah), city and township, ZACATECAS, N central MEXICO, on JUCHIPILA RIVER, and 32 mi/51 km SSE of TLALTENANGO DE SÁNCHEZ ROMÁN on Mexico Highway 54; 21°25′N 103°07′W. Agricultural center (grain, sugarcane, tobacco, fruit; livestock).

Juchipila River (hoo-chee-PEE-lah), c.120 mi/193 km long, in ZACATECAS, N central MEXICO; rises on interior plateau N of VILLANUEVA; flows S, past Villanueva, JALPA, and JUCHIPILA, to Santiago (LERMA) River 28 mi/45 km NNW of GUADALAJARA.

Juchique de Ferrer (hoo-CHEE-kai dai fe-RER), town, VERACRUZ, E MEXICO, 27 mi/43 km NE of XALAPA ENRÍQUEZ; elevation 2,657 ft/810 m. Corn, coffee. Often Juchique.

Juchitán de Zaragoza (hoo-chee-TAHN dai zah-rah-GO-zah), city, in SE OAXACA, MEXICO, on S part of Isthmus of TEHUANTEPEC, on banks of Rio de los Perros (flows to MAR MUERTO SUPERIOR), on railroad and Mexico Highways 190 (INTER-AMERICAN HIGH-WAY) and 185 (Transisthmian Highway), 25 mi/40 km NE of the port of SALINA CRUZ; 16°25′N 95°01′W. Very hot climate. Agriculture (corn, beans); livestock. Important center for commerce and communications. Well known for straw-hat production. There is a creosote plant here. Export salt. Zapotec center; market and handicrafts attract tourists.

Juchitepec de Mariano Riva Palacio (hoo-CHEE-tai-pek dai mah-ree-AH-no REE-vah pah-LAH-see-o), town, ⊙ Juchitepec municipio, MEXICO state, central MEXICO, 29 mi/47 km SE of MEXICO CITY, and in the ZONA METROPOLITANA DE LA CIUDAD DE MEXICO. Agricultural center (cereals, vegetables; livestock). Formerly Xuchitepec and Juchitepec.

Juchitlán (hoo-chee-TLAHN), town, JALISCO, W MEXICO, 31 mi/50 km. S of AMECA, on Mexico Highway 80; 20°04′N 104°06′W. Agricultural center (grain, sugarcane, flax, fruit, tobacco).

Jucuapa (hoo-koo-AH-pah), city and municipality, USULUTÁN department, E EL SALVADOR, near INTER-AMERICAN HIGHWAY, 12 mi/19 km NNE of USULUTÁN, at NE foot of volcano TECAPA (elev. 5,230 ft/1,594 m); 13°30′N 88°23′W. Coffee, grain; livestock.

Jucuarán (hoo-kwah-RAHN), town and municipality, USULUTÁN department, SE EL SALVADOR, 13 mi/21 km SE of USULUTÁN, at W end of JUCUARÁN coastal range (rising to 3,113 ft/949 m). Grain; collecting of medicinal plants.

Jucuri (ZHOO-koo-ree), city, W RIO GRANDE DO NORTE state, BRAZIL, 13 mi/21 km W of MOSSORÓ; 05°13′S 37°29′W.

Jucuruçu (ZHOO-koo-roo-SOO), city (2007 population 10,633), SE BAHIA state, BRAZIL, on Rio Jucuruçu, near border with MINAS GERAIS state; 16°54′S 40°02′W.

Jucurutu (ZHOO-koo-roo-too), city (2007 population 17,534), central RIO GRANDE DO NORTE state, NE BRAZIL, on PIRANHAS River, 35 mi/56 km S of Açu; 06°02′S 37°01′W. Formerly called São Miguel de Jucurutu.

Jud, village (2006 population 70), LA MOURE CO., SE central NORTH DAKOTA, 32 mi/51 km WNW of LA MOURE; 46°31′N 98°54′W. Founded in 1906 and incorporated in 1909. Named for Judson LaMoure (1839–1918), politician and namesake of county.

Juda, village, GREEN county, S WISCONSIN, 7 mi/11.3 km E of MONROE. In livestock, dairy, and grain region. Manufacturing (cheese, whey products; meat processing).

Jud ad Daim (JED ed DAH-im), village, TRIPOLITANIA region, NW LIBYA, on MEDITERRANEAN SEA coast, 23 mi/37 km WSW of TRIPOLI. Agricultural settlement (grain, alfalfa, olives, nuts). Founded 1938–1939 by Italians, who left after World War II, replaced by Libyan population. Formerly Oliveti.

Judaea, ISRAEL: see JUDEA.

Judaea and Samaria, WEST BANK and Gaza: see JU-DAEA.

Judayyidah, Al (joo-dah-YEE-dah, el), village, TUNIS province, N TUNISIA, on the MAJARDAH RIVER, and 13 mi/21 km W of TUNIS; 36°51′N 09°56′E. Agricultural center. Ruins of Roman aqueduct nearby.

Judea (joo-DAI-ah), Hebrew Yehudah, region, Greco-Roman name for S ISRAEL and JORDAN. It varied in size in different periods. The origin of the name comes from the lands held by the tribe of Judah after the conquest of CANAAN by Joshua. These lands extended W to the coastal plain, E to the DEAD SEA, N to Jebul (JERUSALEM) and S to the NEGEV. David became king of the lands and under him (and Solomon) it became part of the united kingdom of Israel (922 B.C.E.–721 B.C.E.). Then a split occurred, and the region constantly warred with Israel and its other neighbors. While Israel (modern SAMARIA) fell to the Assyrians in 721 B.C.E., Judah remained as a kingdom under Assyrian domination until 597 B.C.E. when a

revolt was put down by the Assyrians, Jerusalem was destroyed, and its people sent into exile. In the time of Jesus Christ it was both part of the Roman province of Syria and a kingdom ruled by the Herods. It was the southernmost of the Roman divisions, the others being GALILEE, Samaria, and Peraea. Idumaea was to S. A strip of Samaria lay between Judea and the MEDITERRANEAN SEA. Changed to Syria Palaestina in 135 C.E. by Roman emperor Hadrian following the bloody Bar Kochba revolt. With Israel's capture of the WEST BANK during the 1967 Arab-Israeli war (or Six Day War) the term Judea (as well as Samaria, the N half of the West Bank) was revived by some Israel advocates. The phrase Judea and Samaria is used to refer to the West Bank, in an effort to link it to its biblical past. Also spelled Judaea.

Judean Highlands, an upfold with a broad crest, central ISRAEL. Sometimes referred to in general terms as the Judean Hills or the Judean Mountains. The upfold is divided into three subregions: the Judean Foothills (W), the Judean Hills (center), and the Judean Desert (E). The highest point exceeds 3,000 ft/1,000 m. The Hebron and Bethel Hills are domes of the upfold in the S and N, with the intermediate Jerusalem Hills acting as the connecting saddle. The Judaean Foothills are a narrow strip up to 8 mi/12 km wide, running 40 mi/65 km N-S. This gently rolling landscape adjoins the coastal plain that lies farther W. The Judaean Desert, which descends into the GREAT RIFT VALLEY, is dry and barren, cut by canyons and includes the MASADA rock fortress. The foothills include the valleys of Ayalon, Soreq, and Ela, and the towns of KIRYAT GAT, RAMLA, and BEIT SHEMESH. The Shefela, a larger highlands region 62 mi/100 km long and 12 mi/20 km–25 mi/40 km at its widest, lies between Samarian and Judean highlands and the coastal plain, and incorporates the Judean Highlands into its S reaches.

Judeida, Arab village, 8.8 mi/14 km S of JENIN, in the S Sanur Valley of the Samarian Highlands, WEST BANK; 32°20′N 35°18′E. Olives and cereals.

Judeide, LEBANON: see JDEIDE.

Judeide-Makr (JUH-daid–mahk-re), Arab township, E of AKKO (Acre), in WESTERN GALILEE, ISRAEL. Population mostly Muslim, with some Christians. The two villages of Judeide and el-Makr were combined in 1989; the former was established 16th century by settlers from Syria and the latter a century earlier, on the foundations of a Byzantine village, by Christians from Lebanon. Some Roman and Hellenistic ruins.

Judeita (joo-DEE-tuh), township, N JORDAN, 10 mi/16 km NW of ʿAJLUN. Vineyards; olives, fruit, grain. Also spelled Jdaitta.

Judenburg (YOO-den-boorg), town, central STYRIA province, S central AUSTRIA, on the MUR River; 47°10′N 14°40′E. Traffic node and industrial town and winter sports center; manufacturing of refined steel, metals. Originally a settlement along a Roman road, Judenburg was settled by Jewish merchants and moneylenders in the 11th century and became an important regional commercial and financial center (iron and salt trade with Venice). The municipal arms depict a head wearing a Jewish hat. In 1496 Jews were expelled (as they were from all of Styria) and not permitted to return until the second half of the 19th century when a very small community was re-established (only to disappear in 1939). Has a Romanesque church and a 15th century bell tower. Fortifications of the medieval town, its walls and towers are partly intact. Abandoned monasteries and castles nearby. In Strettweg, near JUDENBURG, an important relic of the Hallstatt culture (600 B.C.E.), the Strettweger-Kultwagen (Strettweg cult carriage), has been found.

Judendorf-Strassengel (YOO-den-dorf–SHTRAH-sen-gehl), township, STYRIA, SE AUSTRIA, 7 mi/11.3 km NW of GRAZ, at its urban fringe; 47°07′N 15°21′E. Gothic pilgrim church (1346–1355).

Judge and Clerk, AUSTRALIA: see MACQUARIE ISLAND.

Judge Daly Promontory (DAI-lee), NE ELLESMERE ISLAND, BAFFIN region, NUNAVUT territory, N CANADA, peninsula, on KENNEDY CHANNEL; 100 mi/161 km long, 10 mi/16 km–48 mi/77 km wide; 81°00′N 67°00′W. Terminates NE at Cape BAIRD. Rises to c.6,000 ft/1,829 m near base of peninsula.

Judiaí River (ZHOO-zhe-ah-ee), 50 mi/80 km long, RIO GRANDE DO NORTE state, NE BRAZIL; flows E to the POTENGI River just above NATAL. Navigable below MACAÍBA. Intermittent stream. Formerly spelled Jundiahy.

Judith Basin, county (□ 2,282 sq mi/5,910 sq km; 1990 population 2,282; 2000 population 2,329), central MONTANA; ⊙ STANFORD; 47°02′N 110°16′W. Agricultural region drained by JUDITH RIVER and Arrow and Wolf creeks. Wheat, barley, hay; cattle, sheep, poultry. Sapphires, silver, and other metals. Part of Lewis and Clark National Forest in SW and N, and LITTLE BELT MOUNTAINS in SW. Ackley Lake State Park in SE. Formed 1920.

Judith Gap, village (2000 population 164), WHEATLAND county, central MONTANA, on East Fork of Roberts Creek, 33 mi/53 km SSW of LEWISTOWN; 46°41′N 109°45′W. Grain, hay; livestock. Parts of Lewis and Clark National Forest to W and NE. LITTLE BELT MOUNTAINS to W. Big Snowy Mountains to NE.

Judith, Point, Rhode Island: see POINT JUDITH.

Judith River, 124 mi/200 km long, central MONTANA; rises in LITTLE BELT MOUNTAINS; flows NE, past UTICA and HOBSON, to MISSOURI River 18 mi/29 km NW of WINIFRED.

Judson, village, PARKE county, W INDIANA, 6 mi/9.7 km NNE of ROCKVILLE; 39°49′N 87°08′W. In agricultural area. Laid out 1872.

Judson, village (2000 population 2,456), GREENVILLE county, NW SOUTH CAROLINA, a residential suburb 2 mi/3.2 km W of downtown GREENVILLE; 34°49′N 82°25′W.

Judsonia (juhd-SON-ee-uh), town (2000 population 1,982), WHITE county, central ARKANSAS, 5 mi/8 km NE of SEARCY, and on LITTLE RED River; 35°16′N 91°38′W. In strawberry-producing area. Manufacturing (bronze casting; asphalt products).

Juegang, China: see RUDONG.

Juelsminde (YOOL-smi-nuh), town, VEJLE county, E JUTLAND, DENMARK, on the LILLE BÆLT, and 19 mi/31 km E of VEJLE; 55°45′N 09°54′E. Cement; fish processing (cod, herring, eels); dairying; hogs; barley.

Juffure (JOO-for-ai), village, NORTH BANK division, the GAMBIA, on N bank of GAMBIA River, and 15 mi/24 km SE of BANJUL; 13°20′N 16°23′W. Peanut-shipping center. Now a tourist center made famous as a setting in Alex Haley's book and film *Roots*.

Jufrah, Al (JUF-ruh, el), group of SAHARAN OASES in TRIPOLITANIA region, N central LIBYA, 150 mi/241 km SSW of Sirte, near N foot of JABAL as Sawda range; 29°10′N 16°00′E. Chief oases are HUN, SAWKNAH, and WADDAN. Agriculture (fruit). Linked by road with MISRATAH. Formerly Giofra or Jofra.

Jugenheim, GERMANY: see SEEHEIM-JUGENHEIM.

Jugoslavia: see YUGOSLAVIA.

Jugsalai, INDIA: see TATANAGAR.

Juhi (JOO-hee), suburban town, in KANPUR municipality, Kanpur district, S UTTAR PRADESH state, N central INDIA, 2.5 mi/4 km SW of Kanpur city center. Oilseed milling, soap manufacturing. Also called Juhi Bari.

Juhu (JOO-hoo), town, MUMBAI (Bombay) Suburban district, W MAHARASHTRA state, W INDIA, on SAL-SETTE ISLAND, 12 mi/19 km N of MUMBAI (Bombay) city center; 19°06′N 72°49′E. Popular swimming resort on ARABIAN SEA; fishing. Seat of a university (for women). Has an airport and several hotels.

Juifang (RWAI-FAHNG), town, N TAIWAN, on railroad, and 5 mi/8 km ESE of CHI-LUNG; 25°07′N 121°48′E. Gold- and coal-mining center; also silver and copper mining.

Juigalpa (hoo-ee-GAHL-pah), city and township (2005 population 42,763), ⊙ CHONTALES department, S NICARAGUA, on RAMA Highway, 65 mi/105 km E of MANAGUA (linked by road); 12°05′N 85°24′W. Agriculture center (sugarcane, coffee, grain); livestock. Beverage manufacturing, coffee processing, tanning. Port of LAKE NICARAGUA in PUERTO DÍAZ.

Juimand, IRAN: see GONABAD.

Juína (zhoo-EE-nah), city (2007 population 38,497), NW MATO GROSSO state, BRAZIL, near RIO JURUENA; 11°27′S 58°43′W. Airport.

Juist (YOO-IST), NORTH SEA island (□ 6 sq mi/15.6 sq km) of E FRISIAN group, GERMANY, off EMS River estuary, 7 mi/11.3 km NW of NORDDEICH (vehicular ferry connection); 9 mi/14.5 km long (E-W), c. 0.5 mi/0.8 km wide (N-S); 53°41′N 06°52′E–53°41′N 07°05′E. Nordseebad Juist (center) is resort.

Juiz de Fora (zhoo-ees dee fo-rah), city (2007 population 513,348), MINAS GERAIS state, SE BRAZIL, on the road to RIO DE JANEIRO; 21°44′S 43°25′W. Industrial and commercial city; manufacturing (textiles, foodstuffs, plastics); coffee cultivation (late 19th century). Established late 18th century; grew rapidly because of its strategic location on the highway. The first railroad in Brazil was constructed (1861) between here and PETRÓPOLIS. Seat of university and museum.

Jujutla (hoo-HOOT-lah), municipality and town, AHUACHAPÁN department, EL SALVADOR, S of AHUACHAPÁN city.

Jujuy (hoo-HWEE), province (□ 22,962 sq mi/59,701.2 sq km; 2001 population 611,888), NW ARGENTINA, ⊙ JUJUY. In the ANDES and the *Puna* region, bordering CHILE and BOLIVIA. Drained by the RÍO GRANDE DE JUJUY and SAN FRANCISCO RIVER. Its lower mountain ranges are covered with forests. Climate varies according to altitude, but the fertile inhabited valleys have a humid, semitropical climate. Among its rich mineral resources are lead (AGUILAR, HUMAHUACA, MINA PIRQUITAS), iron (SIERRA DE ZAPLA), tin (CERRO GALÁN, Rinconada department), copper (PURMAMARCA), kaolin, sodium and borax salts (SALAR DE CAURCHARI, SALINAS GRANDES). Agriculture mainly in irrigated valleys (chiefly sugarcane and tobacco; also sheep, goats, cattle); donkeys and llamas also raised in higher elevations. Fisheries in main rivers. Rural industries include mining, lumbering, spinning, dairying, cotton ginning. Manufacturing concentrated at Jujuy, LA ESPERANZA, SAN PEDRO, LEDESMA, LA MENDIETA, PALPALÁ. Major resorts, noted for mild winters, are REYES, Humahuaca, and Jujuy. Founded 1834, when it was separated from SALTA. When Los Andes territory was dissolved (1943), Susques department (SW) was incorporated into Jujuy province.

Jujuy (hoo-HWEE), city, ⊙ JUJUY province, NW ARGENTINA, on the BERMEJO RIVER, 800 mi/1,287 km NW of BUENOS AIRES, 45 mi/72 km NNE of SALTA; 24°12′S 65°17′W. Elevation 4,215 ft/1,285 m. In the scenic foothill region of the E ANDES MOUNTAINS, it is the center of an agriculture, mining, and cattle-raising area. Manufacturing (flour; sawmills; dairy industry). Trade in timber, minerals, and agricultural products (grain, fruit, livestock). Asphalt and lead deposits nearby. Hydroelectric station on the river. Fisheries. Has administrative buildings, government house, cathedral, 17th-century chapel, national college, historical museum, seismographic station. An old colonial city, founded 1593 near a Native village. Here, Manuel Belgrano, the patriot general, created the first Argentine national flag. Juan Lavalle, after a futile attempt to depose the caudillo Juan Manuel de Rosas, was killed here in 1841. There are interesting Native ruins nearby. Also known as San Salvador de Jujuy.

Jujuy River, ARGENTINA: see GRANDE DE JUJUY, RÍO.

Jukkasjärvi (YUK-kahs-YER-vee), village, NORRBOTTEN county, N SWEDEN, 12 mi/19 km E of KIRUNA; 67°51′N 20°37′E. Wooden eighteenth-century church with triptych by artist Bror Hjort.

Jukleggi, NORWAY: see HEMSEDALSFJELL.

Jukwa (JOO-kwuh), town, WESTERN REGION, GHANA, on ANKOBRA RIVER, 50 mi/80 km NNW of TARKWA; 05°57′N 02°11′W. Bauxite, rubber, caoca, palm oil, kola nuts; timber.

Julaca (hoo-LAH-kah), canton, NOR LÍPEZ province, POTOSÍ department, W central BOLIVIA, on the NE edge of the Salar CHIGUANA, on the UYUNI-ANTOFAGASTA (CHILE) railroad and highway; 20°57′S 67°33′W. Elevation 12,402 ft/3,780 m. Military checkpoint for passports. Tungsten-bearing lode; Mina Toldos mines zinc and lead; manganese mining at Mina Negra; clay, limestone. Agriculture (potatoes, yucca, bananas); cattle.

Julamerk, TURKEY: see HAKKARI.

Julcán (hool-KAHN), province, LA LIBERTAD region, NW PERU; ⊙ Julcán. Bordered N by OTUZCO, E by SANTIAGO DE CHUCO, and SW by Virú provinces.

Julcán (hool-KAHN), town, ⊙ Julcán province, LA LIBERTAD region, NW PERU; 08°03′S 78°30′W.

Julesburg, town (2000 population 1,467), ⊙ SEDGWICK county, extreme NE COLORADO, on SOUTH PLATTE RIVER, near NEBRASKA state line, and 55 mi/89 km NE of STERLING; 40°59′N 102°15′W. Elevation 3,477 ft/1,060 m. Trade center and railroad division point in sugar beet region; cattle, poultry; wheat, corn, sunflowers. Founded 1881, incorporated 1886.

Julfa (JOOL-fah), town, Āzerbāyjān-e SHARQI province, NW IRAN, 65 mi/105 km NNW of TABRIZ, on ARAS RIVER, and on border of AZERBAIJAN Republic, adjoining Dzhulfa. Road center and terminus of railroad from TABRIZ. Population largely Armenian. Sometimes spelled JOLFA or Djulfa.

Julfa (JOOL-fah), former suburb and now the S section of ISFAHAN, IRAN.

Juli (HOO-lee), town, ⊙ CHUCUÍTO province, PUNO region, SE PERU, on SW shore of Lake TITICACA, near mouth of small Juli River, 40 mi/64 km SE of PUNO, on the PUNO–LA PAZ road; 16°13′S 69°27′W. Elevation 12,697 ft/3,870 m. Potatoes, cereals; livestock.

Juliaca (hoo-lee-AH-kah), city (2005 population 208,553), ⊙ SAN ROMÁN province, PUNO region, SE PERU, on the ALTIPLANO, on CUSCO-AREQUIPA-PUNO railroad, 24 mi/39 km NNW of Puno; 15°30′S 70°08′W. Elevation 12,730 ft/3,880 m. Railroad junction; commercial center. Native woolen textiles; wool, hides; cereals, potatoes; livestock. Airport.

Julia Creek (JOO-lee-uh, JOO-lyuh), settlement, NW QUEENSLAND, NE AUSTRALIA, 393 mi/632 km W of TOWNSVILLE, 84 mi/135 km E of CLONCURRY; 20°40′S 141°40′E. Wool, beef production.

Juliaetta, village (2000 population 609), LATAH county, W IDAHO, 18 mi/29 km SE of MOSCOW, and on POTLATCH RIVER; 46°34′N 116°43′W. Grain, barley, oats; alfalfa; sheep, cattle; manufacturing (sawmill products, lumber). Nez Perce Indian Reservation to S.

Julian (zhoo-lyahn), commune, HAUTES-PYRÉNÉES department, MIDI-PYRÉNÉES region, SE FRANCE, 3 mi/5 km SW of TARBES.

Julian, unincorporated town (2000 population 1,621), SAN DIEGO county, S CALIFORNIA, 40 mi/64 km ENE of SAN DIEGO; 33°04′N 116°36′W. Nearby are ANZABORREGO DESERT (E) and Cuyama Rancho (S) state parks; Santa Ysabel (N) and Inaja-Cosmit (SW) Indian reservations also nearby. Pacific Coast Trail passes to E. Cattle, poultry; grain; dairying.

Julian, village (2006 population 74), OTOE and NEMAHA counties, SE NEBRASKA, 8 mi/12.9 km N of AUBURN, near MISSOURI RIVER; 40°31′N 95°52′W.

Juliâna (ZHOO-lee-ah-nah), city, W SÃO PAULO state, BRAZIL, 16 mi/26 km ENE of TUPÃ; 21°52′S 51°22′W.

Juliana Canal, extends 21 mi/34 km N-S, LIMBURG province, SE NETHERLANDS; lateral to non-navigable section of MAAS RIVER on E between Maasbracht (6 mi/9.7 km SW of ROERMOND) and Borgharen (1 mi/1.6 km N of MAASTRICHT).

Julian Alps, Italian *Alpe Julia*, Slovenian *Julijske Alpe*, mountain range, NE ITALY and NW SLOVENIA, between the CARNIC (N) and the DINARIC (S) Alps, rising to 9,396 ft/2,864 m at TRIGLAV, the highest peak in all of the former YUGOSLAVIA (now Slovenia). Mainly limestone and dolomite. The forested, glacier-scoured region is a popular resort area.

Julianehåb (yoo-LYAH-nuh-HOP) [Greenlandic= *Qaqortoq*], town in Qaqortoq (Julianehåb) commune, SW GREENLAND. Fishing port with canneries. Sheep raising, meatpacking, and sealing are other significant industries.

Juliasdale, town, MANICALAND province, E ZIMBABWE, 42 mi/68 km N of MUTARE, NYANGA MOUNTAINS; 18°22′S 32°41′E. Nearby are MTARAZI FALLS (SE) and Nyanga (NE) national parks. Manufacturing (fruit and nut processing). Citrus fruit, macadamia nuts, coffee, tea; dairying; livestock; timber.

Jülich (YOO-likh), town, NORTH RHINE–WESTPHALIA, W GERMANY, on ROER RIVER, 12 mi/19 km NE of AACHEN; 50°55′N 06°22′E. Leather, paper, and wire industries; sugar refining. Site of a nuclear research institute. Founded c.350 C.E. as a Roman fortress; has a notable Roman museum and remains of medieval town wall.

Juliénas (zhoo-lyai-nah), commune (□ 2 sq mi/5.2 sq km), RHÔNE department, RHÔNE-ALPES region, E central FRANCE, in BEAUJOLAIS district, 8 mi/12.9 km SW of MÂCON; 46°14′N 04°43′E. Noted for its red wines.

Julier, Piz, SWITZERLAND: see ALBULA ALPS.

Juliette, town, MONROE county, GEORGIA, on railroad, 9 mi/14.5 km NE of FORSYTH; 33°06′N 83°48′W. Manufacturing includes textile finishing; crushed stone. Nearly abandoned when it was chosen in the early 1990s as the site for the filming of the motion picture *Fried Green Tomatoes*. Now a bustling tourist town with antique and craft stores, and the Whistle Stop Cafe. Near the Jarrell Plantation, Lake Juliette, and the Rum Creek Wildlife Management Area.

Julimes (hoo-LEE-mes), town, CHIHUAHUA, N MEXICO, on CONCHOS RIVER (irrigation), and 40 mi/64 km ESE of CHIHUAHUA; 28°25′N 105°26′W. Elevation 3,839 ft/1,170 m. Grain, beans, cattle. Sulphur thermal springs.

Julin, POLAND: see WOLIN.

Julio, 18 de, URUGUAY: see DIEZ Y OCHO DE JULIO, DIECIOCHO DE JULIO.

Julio, 9 de, ARGENTINA: see SIERRA COLORADA.

Julio, 9 de, ARGENTINA: see NUEVE DE JULIO.

Júlio de Castilhos (ZHOO-lee-o zhe KAH-steel-yos), city, central RIO GRANDE DO SUL, BRAZIL, in the Serra GERAL, on railroad, and 30 mi/48 km N of SANTA MARIA; 27°49′S 52°10′W. Cattle; rice, tobacco.

Juli River, PERU: see JULI.

Julis (JOO-lis), Druze township, 6 mi/10 km NE of AKKO (Acre), in WESTERN GALILEE, ISRAEL; 32°56′N 35°11′E. Elevation 534 ft/162 m. Home of Druze spiritual leader Sheikh Amin Tarif, whose grandson Saleh Tarif was Israel's first Druze minister.

Jullundur, INDIA: see JALANDHAR.

Julo (HOO-lo), canton, ATAHUALLPA province, ORURO department, W central BOLIVIA, NW of the Poquentica mountains, 9 mi/15 km E of the Chilean border, and 22 mi/35 km W of BELLA VISTA; elevation 12,113 ft/3,692 m.

Julsø (YOOL-soo), largest (5 mi/8 km long, 1 mi/1.6 km wide) of the HIMMELBJÆRGET lakes, E JUTLAND, DENMARK, c.20 mi/32 km E of Århus. Surrounded by forests.

Julu (JOO-LOO), town, ⊙ Julu county, SW HEBEI province, NE CHINA, 30 mi/48 km ENE of XINGTAI; 37°13′N 115°01′E. Grain, cotton, oilseeds, medicinal herbs. Textiles, beverages, rubber products; engineering.

Jumaima, SAUDI ARABIA: see RAFHA.

Jumaisa, IRAQ: see SUWAIRA.

Jumaitepeque, GUATEMALA: see JUMAYTEPEQUE.

Jumay (hoo-MAI), extinct volcano (7,139 ft/2,176 m), JALAPA department, E central GUATEMALA, 3 mi/4.8 km N of JALAPA; 14°41′N 89°59′W.

Jumaya, Greece: see IRAKLIA.

Jumaytepeque (hoo-mai-te-PE-kai), extinct volcano (5,955 ft/1,815 m), SANTA ROSA department, S GUATEMALA, 5 mi/8 km NNE of CUILAPA; 14°20′N 90°16′W. On N slope is Jumaitepeque village. Sometimes spelled Jumaitepeque.

Jumbo (JOOM-bo), town, SE SOMALIA, near mouth of JUBBA River, on INDIAN Ocean opposite GOBUEN. Formerly GIUMBO.

Jumbo, town, MASHONALAND EAST province, N ZIMBABWE, 27 mi/43 km NW of HARARE, on railroad; 17°28′S 30°50′E. Citrus fruit, tobacco, corn; cattle, sheep, goats; dairying.

Jumbo Mountain (JUHM-bo) (11,217 ft/3,419 m), SE BRITISH COLUMBIA, W CANADA, in SELKIRK MOUNTAINS, 70 mi/113 km NNE of NELSON; 50°24′N 116°34′W.

Jumbor, THAILAND: see CHUMPHON.

Jumbunna, town, VICTORIA, SE AUSTRALIA, in S GIPPSLAND region; 38°28′S 145°46′E. In agricultural area. Old coal-mining center.

Jumekoonta, INDIA: see JAMIKUNTA.

Jumento Cays, BAHAMAS: see RAGGED ISLAND AND CAYS.

Jumento Cays, BAHAMA Islands: see RAGGED ISLAND AND CAYS.

Jumet (zhuh-MET), town, in commune of CHARLEROI, Charleroi district, HAINAUT province, S BELGIUM; 50°26′N 04°25′E. Manufacturing.

Jumgal River (joom-GAHL), c.50 mi/80 km long, N NARYN region, KYRGYZSTAN; rises N of SONG-KOL lake; flows W, past CHAEK, to Kokomeren River (left tributary of NARYN River); in fertile agricultural and grazing valley. Also spelled DZHUMGAL.

Jumièges (zhoo-MYEZH), commune (□ 7 sq mi/18.2 sq km), SEINE-MARITIME department, HAUTE-NORMANDIE region, N FRANCE, on SEINE RIVER, and 13 mi/21 km W of ROUEN; 49°26′N 00°49′E. Has interesting ruins of well-known Benedictine abbey (founded 7th century) with 11th-century Romanesque abbatial church; major tourist attraction of the lower Seine valley. The forest of Brotonne, near CAUDEBEC-EN-CAUX, is just W.

Jumilla (hoo-MEE-lyah), city, MURCIA province, SE SPAIN, 16 mi/26 km N of CIEZA; 38°29′N 01°17′W. Wine-producing center. Brandy and alcohol distilling, esparto processing, flour- and sawmilling, olive pressing; manufacturing of footwear, knit goods, furniture, baskets, candy, sausage. Ships saffron, almonds, and olives. Limestone and gypsum quarries, saltworks nearby. Has 15th-century church, medieval castle.

Jumla (JOOM-lah), district, NW NEPAL, in KARNALI zone; ⊙ JUMLA.

Jumla (JOOM-lah), town, ⊙ JUMLA district, W NEPAL, on Tila Khola River (E tributary of KARNALI, or GHAGHARA, River), and 95 mi/153 km NNE of NEPALGUNJ; 29°18′N 82°12′E. Elevation 7,700 ft/2347 m. Corn, millet, vegetables, fruit, rice. Airport.

Jumlikhalanga (JOOM-li-kuh-LAHNG-guh), town, ⊙ RUKUM district, W NEPAL; 28°37′N 82°15′E.

Jummoo, INDIA: see JAMMU.

Jumna, INDIA: see YAMUNA RIVER.

Jumneta, POLAND: see WOLIN.

Jumolhari (JOO-mo-LHAHR-ee), mountain peak (24,003 ft/7,316 m), BHUTAN, in the W HIMALAYA, near the TIBET (CHINA) border, 31 mi/50 km N of PARO; 27°50′N 89°16′E. Sacred to the Buddhists. Also spelled Jhomo Lhari.

Jumonji (JYOO-MON-jee), town, Hiraka county, Akita prefecture, N HONSHU, NE JAPAN, 40 mi/65 km S of AKITA city; 39°13′N 140°31′E. Cherries.

Jumpertown, village (2000 population 404), PRENTISS county, NE MISSISSIPPI, 17 mi/27 km SSW of CORINTH; 34°42′N 88°39′W. Cotton, corn, wheat, soybeans; cattle; dairying.

Jump River, c.30 mi/48 km long, N central WISCONSIN; formed by two forks rising in PRICE county. N Fork rises in central Price county and flows SW; S Fork, c.40 mi/64 km long, rises in E Price county, and flows SW, past PRENTICE, then W, where it joins N Fork. The Jump River flows another 35 mi/56 km SW through wooded region, to Holcombe Flowage (reservoir) of CHIPPEWA RIVER, 13 mi/21 km S of LADYSMITH. Main river is also called Big Jump River.

Junagadh (joo-NAH-gahd), former principality, KATHIAWAR PENINSULA, W INDIA, on the ARABIAN SEA. Wrested from the Mogul empire in the mid-eighteenth century by Sher Kahn Babi, a Muslim freebooter who established a dynasty later supported by the British. In 1947 the Muslim ruler ceded his state to PAKISTAN, although the population was overwhelmingly Hindu. He was forced to flee when Indian forces invaded and made Junagadh part of India, first part of SAURASHTRA then (1960) a district of Gujarat state. Also called Junagarh.

Junagadh (joo-NAH-gahd), district (□ 4,095 sq mi/10,647 sq km), GUJARAT state, W INDIA; ⊙ JUNAGADH.

Junagadh (joo-NAH-gahd), city (2001 population 252,108), ⊙ JUNAGADH district, GUJARAT state, W INDIA; 21°31′N 70°28′E. Market for gold and silver embroidery, perfume, and copper and brass vessels. Has ancient Buddhist caves and Rajput forts, as well as a modern college. Nearby is the Gir forest, a major wildlife sanctuary which is the only place in Asia where lions are found. Also called Junagarh.

Juncal (hoon-KAHL), village, S SANTA FE province, ARGENTINA, 55 mi/89 km SSW of ROSARIO; 33°43′S 61°03′W. In grain, soybean, and livestock area.

Juncal, Cerro (hoon-KAHL, SER-ro), ANDEAN peak (19,880 ft/6,059 m) on ARGENTINA-CHILE border, N of NEVADO DEL PLOMO massif; 33°06′S 70°04′W.

Juncal, Lake (hoon-KAHL) (□ 50 sq mi/130 sq km), E RÍO NEGRO province, ARGENTINA, 4 mi/6.4 km SE of VIEDMA, along the lower course of the RÍO NEGRO; 24 mi/39 km long, 2 mi/3.2 km–4 mi/6.4 km wide.

Junco (ZHOON-ko), village, N central BAHIA state, BRAZIL, 25 mi/40 km SW of JUAZEIRO; 09°39′S 40°35′W.

Junco de Seridó (ZHOON-ko zhe SE-ree-DO), town, central PARAÍBA state, BRAZIL, 25 mi/40 km SE of SANTA LUZIA; 06°59′S 36°43′W.

Juncos (HOON-kos), town (2006 population 40,129), E PUERTO RICO, 21 mi/34 km SE of SAN JUAN. Industrial and commercial area; varied manufacturing. Iron deposits nearby.

Juncosa (hoong-KO-sah), village, LÉRIDA province, NE SPAIN, 19 mi/31 km SSE of LÉRIDA. Sheep raising; olive oil, wine, almonds.

Junction (JUNK-shun), town (2006 population 2,645), ⊙ KIMBLE county, W central TEXAS, on EDWARDS PLATEAU, c.100 mi/161 km NW of SAN ANTONIO; 30°29′N 99°46′W. Elevation 1,710 ft/521 m. At junction of North Llano and South Llano rivers to form LLANO RIVER. Important market and shipping center for wool, mohair, cattle, sheep, goats; pecans, sorghum; manufacturing (building materials, wood products, cedar wood oil); sand and gravel; oil and natural gas. Scenery, hunting, fishing attract visitors. South Llano River State Park to SW. Settled 1876, incorporated 1928.

Junction, village (2000 population 139), GALLATIN county, SE ILLINOIS, 3 mi/4.8 km WNW of SHAWNEETOWN; 37°43′N 88°14′W. In agricultural area.

Junction, village (2006 population 164), ⊙ PIUTE county, S UTAH, 22 mi/35 km E of BEAVER, and on SEVIER RIVER, near junction of EAST FORK SEVIER RIVER (impounds water of Sevier River for irrigation); 38°14′N 112°13′W. Elevation 6,250 ft/1,905 m. Alfalfa, vegetables; dairying; cattle, sheep; gold, silver. Parts of Fishlake National Forest to W and NE; Dixie National Forest to SE. PIUTE RESERVOIR and State Park to N. County Courthouse (1903).

Junction, JAMAICA: see BULL SAVANNA/Junction.

Junction City, city (2000 population 18,886), ⊙ Geary county, NE Kansas, at the confluence of the REPUBLICAN and SMOKY HILL rivers, 22 mi/35 km E of ABILENE, 18 mi/29 km SW of MANHATTAN; 39°01′N 96°50′W. The railroad, trade, and processing center of an agricultural and dairy area, it grew as the service center for Fort Riley Military Reserve 8 mi/12.9 km N, which contributes to the city's economy. Manufacturing (transportation equipment; foundry; gas distribution). Limestone quarries nearby. Milford Lake Reservoir and Milford State Park to NW; Geary State Fishing Lake to SW. Incorporated 1859.

Junction City, town (2000 population 2,184), BOYLE county, central KENTUCKY, 4 mi/6.4 km S of DANVILLE, in BLUEGRASS REGION; 37°35′N 84°47′W. Agriculture (burley tobacco, grain; livestock, poultry; dairying); manufacturing (machinery, wood products). Isaac Shelby Cemetery State Historic Site to S.

Junction City, town (2006 population 5,362), LANE county, W OREGON, 12 mi/19 km NNW of EUGENE, near WILLAMETTE RIVER; 44°13′N 123°12′W. Shipping point for seeds and canned goods. Agriculture (fruit, sweet corn, grain; poultry); dairy products. Wood products, chemicals. Nearby are Washburne (NW) and Alderwood (W) state waysides. FERN RIDGE RESERVOIR to SW. Incorporated 1872.

Junction City, village (2000 population 721), in UNION county, S ARKANSAS, on LOUISIANA state line, 14 mi/23 km S of EL DORADO, and adjacent to JUNCTION CITY (Louisiana); 33°01′N 92°43′W. Manufacturing (wood products), lumber milling; poultry.

Junction City, village, TALBOT county, W GEORGIA, 32 mi/51 km ENE of COLUMBUS; 32°36′N 84°28′W. Furniture manufacturing; crushed stone, sand processing.

Junction City, village (2000 population 652), UNION and CLAIBORNE parishes, N LOUISIANA, on ARKANSAS state line, 4 mi/6 km S of EL DORADO (ARKANSAS), adjacent to JUNCTION CITY (ARKANSAS); 33°01′N 92°43′W. Lumber milling. Unit of Kisatchie National Forest (including CORNEY LAKE) to SW.

Junction City (JUHNK-shuhn), village (2006 population 854), PERRY county, central OHIO, 5 mi/8 km W of NEW LEXINGTON; 39°43′N 82°18′W.

Junction City, village (2006 population 407), PORTAGE county, central WISCONSIN, 14 mi/23 km NNE of WISCONSIN RAPIDS; 44°35′N 89°46′W. Makes cheese. Railroad junction.

Junction Gate, town, MANICALAND province, E ZIMBABWE, 12 mi/19 km E of CHIPINGE; 20°08′S 32°47′E. Coffee, tea, corn, citrus fruit; cattle, sheep, goats. CHIRUNDA FOREST/Botanical Reserve to S.

Jundiá (zhoon-zee-AH), city (2007 population 4,569), NE ALAGOAS state, BRAZIL, near PERNAMBUCO state border; 08°56′S 35°34′W.

Jundiaí (ZHOON-zhe-ah-EE), city (2007 population 342,983), SÃO PAULO state, S BRAZIL, on the JUNDIAÍ River; 23°11′S 46°52′W. Agricultural and industrial center; manufacturing (textiles, ceramics, furniture, wines, foodstuffs, chemicals, and agr. tools). Established 17th century. Airfield nearby.

Jundiaí do Sul (ZHOON-zhei-ee do SOOL), town (2007 population 3,654), N PARANÁ state, BRAZIL, 13 mi/21 km SW of SANTO ANTÔNIO DA PLATINA; 23°27′S 50°17′W. Coffee, corn, cotton, rice; livestock.

Jundubah or Jundûbah, province, TUNISIA: see JANDUBA, province.

Juneau (JOO-no), county (□ 804 sq mi/2,090.4 sq km), 2006 population 26,855), central WISCONSIN; ⊙ MAUSTON; 43°55′N 90°06′W. Predominantly agricultural area (dairy products, potatoes, beans, corn, soybeans; cattle, hogs, sheep, poultry). Processing of dairy products is principal industry. Bounded E by WISCONSIN RIVER; drained by YELLOW, LEMONWEIR, and BARABOO rivers. WISCONSIN DELLS (canyon area) are in SE. Includes part of Necedah National Wildlife

Refuge and central Wisconsin Conservation Area in N; Rocky Arbor State in SE corner; Buckhorn State Park in E; E end of La Crosse State Trail in SW corner. Formed 1856.

Juneau (JOO-no), city (2000 population 30,711), ⊙ ALASKA, in the Alaska Panhandle, in SE corner of state, at the foot of two spectacular peaks, Mountains Juneau and Roberts; 61°48′N 134°27′W. A port on GASTINEAU CHANNEL, Juneau is a trade center for the Panhandle area, with an ice-free harbor and an airport. Surrounding road system extends 40.5 mi/65.2 km NW; no road connection to outside. Ferry connections to other Panhandle communities; bridge to DOUGLAS ISLAND. The state and Federal government are the major employers. Salmon and halibut fishing, mining, and tourism are important economic activities. Joseph Juneau and a partner discovered gold nearby in 1880, and the city developed as a gold rush town. Officially designated as capital of the Territory of Alaska in 1900 but did not function as such until the government offices were moved from Sitka in 1906. In 1959 it became state capital with the admission of Alaska to the Union. Douglas Island, a part of the city, lies across the channel. The Alaska Historical Library and Museum and Alaska State Museum are here. University of Alaska Southeast Campus serves the area. In 1970 city and borough governments were united, including neighboring Douglas, making Juneau the largest city in area in the U.S., at 3,108 sq mi/8,050 sq km. GLACIER BAY NATIONAL PARK AND PRESERVE to the NW. Alaskans voted in 1976 to move capital to WILLOW, W of PALMER, but defeated funding for it, 1982. Incorporated 1900.

Juneau, town (2006 population 2,631), ⊙ DODGE county, S central WISCONSIN, 7 mi/11.3 km SE of BEAVER DAM; 43°24′N 88°42′W. In dairying region. Manufacturing (porcelain products, machinery; food and dairy processing). Horicon Marsh Wildlife Area to NE. Incorporated 1887.

Juneau (JOO-no), borough (□ 2,594 sq mi/6,744.4 sq km; 2006 population 30,737), SE ALASKA, coterminous with JUNEAU city, which includes the former town of Douglas and all of DOUGLAS ISLAND. Bounded on SW by STEPHENS PASSAGE, on W by Long Canal inlet, and on E by CANADIAN (BRITISH COLUMBIA) border. Part of COAST MOUNTAINS and TONGASS NATIONAL FOREST to E. Point Bridget State Park to W. Tourism, fishing, timber.

Juneda (hoo-NAI-dah), town, LÉRIDA province, NE SPAIN, 12 mi/19 km ESE of LÉRIDA; 41°32′N 00°49′E. Soap manufacturing, olive-oil and wine processing; agricultural trade (cereals, sugar beets, tobacco).

Junee (JOO-NEE), municipality, S NEW SOUTH WALES, SE AUSTRALIA, 95 mi/153 km WNW of CANBERRA; 34°51′S 147°40′E. Railroad junction; sheep, lambs, deer; agriculture center (canola, wheat, oats, barley, seeds, olives). Correctional facility.

June Lake, unincorporated village, MONO county, E CALIFORNIA, on June Lake (resort), in the SIERRA NEVADA, c.10 mi/16 km S of MONO LAKE, and 45 mi/72 km NW of BISHOP. In Inyo National Forest; YOSEMITE NATIONAL PARK to W; June Mountain Ski Area to W; Mono Craters to N. Trout fishing.

June Park (JOON), town (□ 3 sq mi/7.8 sq km; 2000 population 4,367), BREVARD county, E central FLORIDA, 4 mi/6.4 km W of MELBOURNE; 28°04′N 80°41′W.

Junga (JOON-gah), village, SHIMLA district, central HIMACHAL PRADESH state, N INDIA, 6 mi/9.7 km SSE of SHIMLA. Local market for corn, rice, timber. Was capital of former PUNJAB HILL state of KEONTHAL.

Jungapeo de Juárez (hoon-GAH-pai-o dai HWAHR-res), town, ⊙ Jungapeo municipio, MICHOACÁN, central MEXICO, 20 mi/32 km S of HIDALGO; 19°30′N 100°30′W. Cereals, fruit; livestock.

Jungaria, China: see JUNGGAR.

Jungbuch, CZECH REPUBLIC: see MLADE BUKY.

Jungbunzlau, CZECH REPUBLIC: see MLADÁ BOLESLAV.

Jungfrau, peak (13,642 ft/4,158 m), on border of BERN and VALAIS cantons, S central SWITZERLAND, in the BERNESE ALPS. First ascended by the Meyer brothers in 1811. ALETSCH GLACIER is on the S side. The Jungfraujoch is a mountain saddle (11,401 ft/3,475 m), the highest point in Europe reached by railroad. The joch is part scientific (avalanche and other mountain studies) and part tourist resort (outstanding views, summer skiing, dogsled rides). There is also a meteorological station on the belvedere atop nearby Sphinx summit (11,723 ft/3,573 m).

Jungfraujoch, SWITZERLAND: see JUNGFRAU.

Junggar (JUN-GAHR) or **Dzungaria**, basin (□ 77,220 sq mi/200,772 sq km), in N XINJIANG UYGUR, NW CHINA. The longer S edge is 530 mi/850 km long and the shorter E and W edges are each 240 mi/380 km long. The basin is largely a steppe and semi-desert area surrounded by the TIANSHAN (S) and ALTAI (N) mountain ranges. The Gurbantunggut desert is in the center. The basin, along with the surrounding mountains, forms most of N Xinjiang. Wheat, barley, oats, and sugar beets are grown; cattle, sheep, and horses are raised. The fields are watered by melted snow from the permanently white-capped mountains. URUMQI, KARAMAY, and YINING are the main cities; other smaller oases dot the piedmont area. These settlements are mostly on the N branch of the SILK ROAD.

The population is Uygur, Kazak, Kyrgyz, Mongol, and Han Chinese. Since 1953, there has been a massive influx of Han Chinese to work on water-conservation and industrial projects. The basin has deposits of coal, iron, and gold, as well as large oil fields. Dzungaria (named after the Dzungar, a Mongol tribe) was ruled by the confederation of Western Mongols that established (17th century) a large empire in central Asia. The region passed to the Chinese in the mid-18th century. The Dzungarian Alatau is a mountain chain that lies on the boundary of Xinjiang and KAZAKHSTAN. At the E end of the chain, on the Kazakhstan-China border, lies the Dzungaria Gate (Chinese *Alatou Shankou*), which was used for centuries as an invasion route by central Asian conquerors of China. The name also appears as Sungaria, Zungaria, or Jungaria.

Junggarian Alatau, China: see DZUNGARIAN ALA-TAU.

Junggarian Ala-Tau, China: see DZUNGARIAN ALA-TAU.

Junggarian Gate, China: see DZUNGARIAN GATE.

Junglinster (yung-LIN-stuhr), village, Junglingster commune (2001 population 5,753), E LUXEMBOURG, 9 mi/14.5 km NE of LUXEMBOURG city; 49°43′N 06°15′E. Sawmills; orchards. Radio-Television Luxembourg.

Jungshahi (jung-SHAH-hei), village, TATTA district, SW SIND province, SE PAKISTAN, 45 mi/72 km E of KARACHI; 24°51′N 67°46′E. Sometimes spelled JANGSHAHI.

Jungurudó, Serranía de (huhn-goo-roo-DO, sai-rah-NEE-an dai), mountain range, DARIÉN province, PANAMA, along PACIFIC coast, S of Serranía del Sapo. Max. elevation 5,597 ft/1,706 m.

Jungwoschitz, CZECH REPUBLIC: see MLADA VOZICE.

Juniata (JOO-nee-A-tah), county (□ 393 sq mi/1,021.8 sq km; 2006 population 23,512), central PENNSYLVANIA; ⊙ MIFFLINTOWN. Agricultural area, drained by JUNIATA RIVER and Tuscarora Creek. BLACKLOG MOUNTAIN lies along SW, SHADE MOUNTAIN along NW border; TUSCARORA MOUNTAIN ridge forms SE border; bounded extreme E by SUSQUEHANNA RIVER, NE by MAHANTANGO and West Branch Mahantango creeks. Manufacturing at Mifflintown. Agriculture (corn, wheat, oats, hay, alfalfa, apples; poultry, sheep, hogs, cattle; dairying); timber. Parts of Tuscarora State Forest along SE and NW borders. Formed 1831.

Juniata (JOO-nee-tuh), village (2006 population 789), ADAMS county, S NEBRASKA, 5 mi/8 km W of HASTINGS, and on branch of LITTLE BLUE RIVER; 40°35′N 98°30′W. Manufacturing (aluminum foundry; irrigation supplies).

Juniata River (JOO-nee-A-tah), scenic stream, 90 mi/145 km long, central PENNSYLVANIA; formed 3 mi/4.8 km SE of HUNTINGDON by junction of Little Juniata River (35 mi/56 km long) and Raystown Branch of Juniata River; flows generally E, past MOUNT UNION, LEWISTOWN, and NEWPORT to SUSQUEHANNA RIVER 1 mi/1.6 km NE of DUNCANNON; passes through gaps in JACKS MOUNTAIN ridge at Mount Union and through TUSCARORA MOUNTAIN ridge at MILLERSTOWN; 40°33′N 78°04′W. Raystown Branch rises in ALLEGHENY MOUNTAINS in E SOMERSET county, flows E past BEDFORD and EVERETT, and NNE past SAXTON and through large RAYSTOWN LAKE reservoir, joins Little Juniata River 3 mi/4.8 km SE of Huntingdon to form Juniata River Frankstown Branch; 105 mi/169 km long, rises in N edge of BEDFORD county, flows generally NE c.50 mi/80 km past CLAYSBURG and WILLIAMSBURG to Little Juniata River, 6 mi/9.7 km NW of Huntingdon.

Juniata Terrace (JOO-nee-A-tah), borough (2006 population 480), JUNIATA county, central PENNSYLVANIA, residential suburb 1 mi/1.6 km S and across the JUNIATA RIVER from LEWISTOWN; 40°34′N 77°34′W. BLUE RIDGE MOUTNAINS to SE.

Junín (hoo-NEEN), region (□ 17,147 sq mi/44,582.2 sq km), central PERU; ⊙ HUANCAYO; 11°30′S 75°00′W. Bordered by Ene and Tambo rivers (APURÍMAC RIVER) (E), CORDILLERA ORIENTAL (NE) and CORDILLERA OCCIDENTAL (W). Includes Lake JUNÍN; drained by MANTARO and PERENÉ rivers. Cordillera CENTRAL crosses region NW-SE. Important mining region, with centers in LA OROYA (metallurgy), MOROCOCHA (copper), and Cercapuquio (cadmium). Lead, zinc, silver, gold, molybdenum, tungsten are other important minerals. Hydroelectric plants at La Oroya and Pachachaca, fed by water reservoirs on Yauli and Mantaro rivers. Agriculture (cereals, potatoes); cattle and sheep raising in mountains; subtropical products (coffee, cacao, sugarcane) on E slopes of the ANDES Mountains; rubber and lumbering in E tropical lowlands toward the Perené. Thermal baths at LLOCLLAPAMPA and Yauli. Served by trans-Andean railroad from LIMA. Main centers are Huancayo, La Oroya, JAUJA, TARMA. Attractive for tourists because of its textiles, silverwork, and ceramics. Site of RESERVA NACIONAL JUNÍN.

Junín (hoo-NEEN), northwesternmost province of JUNÍN region, central PERU; ⊙ JUNÍN; 11°05′S 76°00′W. Borders on PASCO region. Contains RESERVA NACIONAL JUNÍN, protecting Lake JUNÍN (or Chinchaycocha) and its native birdlife.

Junín (hoo-NEEN), city, N BUENOS AIRES province, E ARGENTINA, on the Salado River, 150 mi/241 km of BUENOS AIRES; 34°30′S 61°00′W. Cattle; soybeans, grain; (manufacturing of ceramics, furniture, metal goods). Fish from nearby lagoons are basis of developing industry. Airport.

Junín (hoo-NEEN), city, ⊙ JUNÍN province, JUNÍN region, W central PERU, in the ANDES Mountains; 11°10′S 76°00′W. In the vicinity on August 6, 1824, Simón Bolívar, aided by Antonio José de Sucre, defeated the Spanish general, José Canterac, in the first important battle leading to Peruvian independence. The surrounding area is rich in mineral, hydroelectric, and agricultural resources. Near Lake JUNÍN and RESERVA NACIONAL JUNÍN.

Junín (hoo-NEEN), town (1991 population 4,492), ⊙ Junín department (□ 130 sq mi/338 sq km), N MENDOZA province, ARGENTINA, in MENDOZA RIVER valley (irrigation area), 27 mi/43 km SE of MENDOZA. Wine; fruit; apiculture; wine making, dried-fruit processing, sawmilling.

Junín (hoo-NEEN), town, ⊙ Junín municipio, CUNDINAMARCA department, central COLOMBIA, 23 mi/37 km ENE of BOGOTÁ; 04°47′N 73°38′W. Elevation 7,546 ft/2,300 m. Coffee, corn, sugarcane; livestock. Sometimes called Chipasaque.

Junín (hoo-NEEN), village, TARAPACÁ region, N CHILE, port on the PACIFIC OCEAN, 4 mi/6 km SSE of PISAGUA; 19°40′S 70°10′W. Railroad terminus, formerly shipping nitrate; almost abandoned in 1930s.

Junín, ARGENTINA: see SANTA ROSA DEL CONLARA.

Junín de los Andes (hoo-NEEN dai los AHN-des), town (1991 population 7,339), ⊙ Huiliches department, SW NEUQUÉN province, ARGENTINA, in Argentinian lake district, 90 mi/145 km SW of ZAPALA; 39°56′S 71°05′W. Resort; trout fishing. Horses, sheep, cattle.

Junín, Lake (hoo-NEEN), Spanish *Lago de Junín*, JUNÍN region, central PERU, in Cordillera CENTRAL of the ANDES Mountains, 8 mi/13 km NNW of JUNÍN; 14 mi/23 km long, 6 mi/10 km wide; 11°02′S 76°06′W. Elevation 13,560 ft/4,133 m. Outlet; MANTARO RIVER. Peru's second-largest lake (after Lake TITICACA). Wide variety of birdlife in surrounding RESERVA NACIONAL JUNÍN. Also called Chinchaycocha.

Junio, 4 de, ARGENTINA: see LA TOMA.

Junior, village (2006 population 448), BARBOUR county, E WEST VIRGINIA, on TYGART RIVER, 5 mi/8 km NW of ELKINS; 38°58′N 79°57′W. Agriculture (corn); livestock, poultry. Coal-mining area.

Junipero Serra Peak, California: see SANTA LUCIA RANGE.

Juniye (ZHOON-ye), town, central LEBANON, near the MEDITERRANEAN SEA, overlooking a little bay, 9 mi/15 km NE of BEIRUT; 33°58′N 35°37′E. Cotton, cereals, oranges. During the civil war (in the 1970s) and the Israeli military operations in Lebanon (early 1980s), when the port of Beirut was paralyzed, Juniye became the main port of the Christian-held areas of Lebanon and levied its own customs taxes. It developed into a town which in addition to its functions as a port became the center of numerous economic and social activities of parts of the Christian (Maronite) community and in effect is capital of Maronite Lebanon. Also called Djouni.

Junk, LIBERIA: see MARSHALL.

Junk Bay, Chinese *Tseung Kwan O*, town, HONG KONG, SE CHINA, in SE corner of New Territories; 22°20′N 114°15′E. Recently created; located SE of former Kai Tak Airport. Built partially on landfill and connected to the E leg of the cross-harbor tunnel.

Junkceylon, THAILAND: see PHUKET.

Junken, Cape (JUHN-kin), S ALASKA, on S KENAI PENINSULA, 30 mi/48 km ESE of SEWARD; 59°55′N 148°39′W.

Junlian (JWUN-LEE-yuhn), town, ⊙ Junlian county, S SICHUAN province, SW CHINA, 40 mi/64 km SSW of YIBIN, on YUNNAN province border; 28°08′N 104°29′E. Rice, wheat, sugarcane, oilseeds, tobacco, jute; food and beverages, textiles, building materials.

Junnar (JUHN-nuhr), town, PUNE district, MAHARASHTRA state, W INDIA, in WESTERN GHATS, 50 mi/80 km N of PUNE; 19°12′N 73°53′E. Market center for millet, wheat, rice. Ancient Buddhist center (c. C.E. 1st century–C.E. 3rd century). Nearby hill fort of Shivneri was birthplace (1627) of Shivaji, the great Maratha hero; later a Maratha stronghold.

Juno Beach (JOO-no), town (□ 1 sq mi/2.6 sq km; 2000 population 3,262), PALM BEACH county, SE FLORIDA, 13 mi/21 km N of WEST PALM BEACH; 26°52′N 80°03′W.

Junonis Promontorium, SPAIN: see TRAFALGAR, CAPE.

Junqali, state, SUDAN: see Jonglei.

Junqueiro (zhoon-KAI-ro), city (2007 population 24,429), E ALAGOAS state, BRAZIL, 26 mi/42 km N of PENEDO; 09°05′S 36°39′W. Sugar, hides.

Junqueirópolis (ZHOON-kai-RO-po-lees), city (2007 population 18,637), W SÃO PAULO state, BRAZIL, 6 mi/9.7 km E of DRACENA, on railroad; 21°32′S 51°26′W. Coffee growing.

Junquera, La (hoong-KAI-rah, lah), town, GERONA province, NE SPAIN, 11 mi/18 km NNW of FIGUERAS. Customs station near French border on road to PERTHUS PASS. Lumbering; cork, cereal, and wine processing.

Junta, La, COSTA RICA: see LA JUNTA.

Juntas (HOON-tahs), canton, Aviléz province, TARIJA department, S central BOLIVIA, 13 mi/20 km SW of TARIJA, on the Tarija–FORTÍN CAMPERO highway; 21°45′S 64°46′W. Elevation 5,600 ft/1,707 m. Clay deposits. Agriculture (potatoes, yucca, bananas, corn, barley, sweet potatoes); cattle.

Juntas (HOON-tahs), town) ⊙ Abangares canton, Guanacaste province, NW Costa Rica, on Abangares River, and 40 mi/64 km SE of Liberia. Formerly (1884–1930) a gold-mining area; trading center. Grain; livestock; lumbering. Also known as Las Juntas.

Juntas de San Antoñio (HOON-tahs dai sahn ahn-TO-nee-oh), canton, ANICETO ARCE province, TARIJA department, S central BOLIVIA, at the junction of the BERMEJO and Grande rivers; 21°54′S 64°20′W. Elevation 6,549 ft/1,996 m. Abundant gas deposits in area; clay, limestone, and gypsum deposits. Agriculture (potatoes, yucca, bananas, corn); cattle.

Juntavi (hoon-TAH-vee), canton, ALONSO DE IBÁÑEZ province, POTOSÍ department, W central BOLIVIA, 4 mi/6 km W of CARIPUYO, on the CHAYANTA RIVER; 18°15′S 66°31′W. Elevation 12,008 ft/3,660 m. Some lead, silver, and zinc in area; limestone and gypsum deposits. Agriculture (potatoes, yucca, bananas); cattle.

Juntura (juhn-TUHR-uh), unincorporated village, MALHEUR county, E OREGON, 45 mi/72 km WSW of VALE, on MALHEUR RIVER at mouth of North Fork.

Junturas, Las, ARGENTINA: see LAS JUNTURAS.

Juo (JYOO-O), town, Taga county, IBARAKI prefecture, central HONSHU, E central JAPAN, 22 mi/35 km N of MITO; 36°40′N 140°41′E.

Juodkrantė (yawd-KRAHN-tai), German *Schwarzort*, climatic and seaside resort, W LITHUANIA, on lagoon side of COURLAND Spit, 11 mi/18 km S of Klaipeda; 55°33′N 21°08′E. In Kursiu Nerijos National Park. Fisheries; amber diggings. In MEMEL TERRITORY, 1920–1939. Also spelled Yuodkrante.

Juoksengi (YOOK-seng-ee), village, NORRBOTTEN county, N SWEDEN, on TORNEÄLVEN RIVER (Finnish border), 55 mi/89 km NNW of HAPARANDA; 66°34′N 23°51′E.

Juozapinė (YO-zuh-pee-nai), E LITHUANIA, in the Medininkai Uplands. Highest point in Lithuania (965 ft/294 m).

Jupagua (ZHOO-pah-GWAH), village, W BAHIA state, BRAZIL, on RIO GRANDE, 68 mi/110 km NE of BARREIRAS; 11°48′S 44°20′W.

Juphal (JOO-pahl), village, W NEPAL; 28°58′N 82°49′E. Elevation 8,200 ft/2,499 m. Dolpo airport located here.

Jupi (ZHOO-pee), city (2007 population 13,234), E PERNAMBUCO state, BRAZIL, 42 mi/68 km SW of CARUARU; 08°43′S 36°25′W. Corn, sugar, aloe.

Jupille (zhuh-PEEL), town, in commune of Liège, Liège district, LIÈGE province, E BELGIUM, on MEUSE RIVER, and 3 mi/4.8 km E of LIÈGE. Favorite residence of Pepin of HÉRISTAL (died here) and of Charlemagne.

Jupiter (JOO-pi-tuhr), city (□ 14 sq mi/36.4 sq km; 2000 population 39,328), PALM BEACH county, SE FLORIDA, 16 mi/26 km N of WEST PALM BEACH, on Atlantic coast, near S end of JUPITER ISLAND; 26°55′N 80°05′W. Resort city experiencing much growth since 1970.

Jupiter (JOO-pi-tuhr), unincorporated village, BUNCOMBE county, W NORTH CAROLINA, 11 mi/18 km NNW of ASHEVILLE; 35°45′N 82°35′W. Agriculture (tobacco, corn; cattle).

Jupiter Island (JOO-pi-tuhr), (□ 3 sq mi/7.8 sq km), MARTIN county, E central FLORIDA, barrier beach on the ATLANTIC OCEAN, separated from mainland by narrow channel 15 mi/24 km long; 27°02′N 80°06′W. At N end is ST. LUCIE INLET, at S end Jupiter Inlet. Lighthouse at S tip.

Juprelle (zhuh-PREL), commune (2006 population 8,447), Liège district, LIÈGE province, E BELGIUM, 6 mi/10 km NW of LIÈGE; 50°43′N 05°32′E.

Juqueri, Brazil: see MAIRIPORÃ.

Juquiá (ZHOO-kee-AH), town (2007 population 19,340), S SÃO PAULO state, BRAZIL, head of navigation on JUQUIÁ River, and 25 mi/40 km N of IGUAPE; 24°19′S 47°38′W. On railroad from Santos to Registro, servicing Japanese settlements in Registro-Iguape area.

Juquiá River (ZHOO-kee-ah), 100 mi/161 km long, S SÃO PAULO state, BRAZIL; rises in the Serra do MAR 35 mi/56 km W of SANTOS; flows SW, past JUQUIÁ (head of navigation), joining RIBEIRA River above REGISTRO to form RIBEIRA DE IGUAPE River.

Juquilla, MEXICO: see SANTA CATARINA JUQUILA.

Juquitiba (ZHOO-kee-chee-bah), city (2007 population 27,717), SE SÃO PAULO state, BRAZIL, 38 mi/61 km SW of SÃO PAULO; 23°57′S 47°03′W.

Jura (zhuhr-ah), department (□ 1,930 sq mi/5,018 sq km), E FRANCE, bordering (SE) SWITZERLAND; ⊙ LONS-LE-SAUNIER; 46°50′N 05°50′E. The SE part of the department is in the JURA mountains; further W is the Jura Plateau with a few hills (REVERMONT in the SW). The NW portion is occupied by the BRESSE district and the upper part of the SAÔNE RIVER valley. Principal rivers are the DOUBS and the AIN, both of which flow to the RHÔNE. Forestry and agriculture provide resources for woodworking, paper, and food-processing industries. Specialities are poultry in the Bresse, blue cheese (Morbier), and small-scale industry, notably clock- and watchmaking, gemstone cutting, and the manufacturing of optical instruments, brier pipes, and plastic products. Chief towns are Lons-le-Saunier, DOLE (metalworking, chemical industry), SAINT-CLAUDE, and MOREZ (eyeglasses, clockworks). The tourist industry is especially active in the high valleys and the regional park of the High Jura (along the Swiss border). The department forms part of the administrative region of FRANCHE-COMTÉ (headquarters in BESANÇON).

Jura, canton (□ 323 sq mi/839.8 sq km), NW SWITZERLAND, in the JURA mountains; ⊙ DELÉMONT. Bordered by the Swiss cantons of BERN on the S and SOLOTHURN in the E and by FRANCE in the N and W. Its chief rivers are the DOUBS and BIRS. Agricultural products, horses, and cattle are the major economic concerns. The traditional watchmaking industry has long been important in the Jura region, but it is in decline; textiles and tobacco are also manufactured. Jura became Switzerland's newest canton in 1978. After 800 years as part of the independent Bishopric of Basel, the region was given to the canton of Bern in 1815. Frustration grew among the French-speaking Roman Catholics in the N region of the ex-Bishopric. They felt slighted and discriminated against the overwhelmingly German-speaking Protestant canton. An active separatist movement of twenty-five years culminated in the vote of 1974 that created the canton of Jura out of the N part of the former bishopric.

Jura (zhuhr-ah) or **Jura Mountains**, tertiary mountain range, extending in an arc of narrow, parallel ridges for some 60 mi/260 km from E FRANCE (FRANCHE-COMTÉ region) to NW Switzerland (VAUD, NEUCHÂTEL, JURA, SOLOTHURN, BASEL, AARGAU cantons); 46°50′N 05°50′E. The calcareous sediments that constitute this range, deposited c.200–150 million years ago and rich in fossils, gave their name to the Jurassic Period of the Mesozoic Era. The uplift of the sediments that created the range came later, in the Tertiary, as part of the great Alpine orogeny c.60 million years ago. The Jura consists of two parts, an E sector of folded ridges and valleys, higher in the S than in the N (reaching 5636 ft/1,718 m at the CRÊT DE LA NEIGE in the Pays de Gex (AIN department, France); and a W tabular plateau overlooking the Saene lowlands. The orientation and the altitude result in surprisingly cold winters (La BRÉVINE, in the Neuchâtel Jura, where Gide wrote his Symphonie Pastorale, has experienced temperatures of −40°F/−40°C); and considerable precipitation (c.60 in/1500 mm per year), favoring forests and pastures on the rounded crests and intervening parallel valleys where the principal economic activities involve forestry, dairying, and tourism. Watchmaking is concentrated in La CHAUX-DE-FONDS, Le LOCLE, and a series of small towns in the Bernese Jura; light industry involving both wood and metal is ubiquitous in the valleys. NW-SE passage across the grain of the land is difficult; there are few transverse gorges (called cluses) and these are filled by railroad and road. The main traffic is by tunnel; the Swiss expertise in piercing the Alps began with the construction of lower, but equally long, tunnels through the Jura (Weissenstein, Hauenstein, Kilchberg, Bözberg). There are few towns in the Jura, including PONTARLIER (France) and NEUCHÂTEL, BIEL (or Bienne), SOLOTHURN, and OLTEN (all in Switzerland). The Regional Natural Park of the High Jura (□ c.240 sq mi/620 sq km), was established in France (Jura department) along the Swiss border.

Jura, Scotland: see THE HEBRIDES.

Juradó (hoo-rah-DO), town, ⊙ Juradó municipio, CHOCÓ department, NW COLOMBIA, minor port on the PACIFIC OCEAN, near PANAMA border, 20 mi/32 km NNW of Cape MARZO; 07°07′N 77°46′W. Coconuts, bananas, plantains.

Jurançon (zhuhr-ahn-SON), town (□ 6 sq mi/15.6 sq km), SW suburb of PAU, PYRÉNÉES-ATLANTIQUES department, AQUITAINE region, SW FRANCE; 43°18′N 00°23′E. It gives its name to the high-quality vineyards, which produce Jurançon amber-colored wines. They are the finest of the BÉARN region's wines.

Juranda (ZHOO-rahn-dah), town (2007 population 7,684), W central PARANÁ state, BRAZIL, 39 mi/63 km SW of CAMPO MOURÃO; 24°26′S 52°50′W.

Jurba, Al (JUHR-buh, el), village, S YEMEN, 6 mi/9.7 km SW of MAHJABA; 13°47′N 45°10′E. Center of the former Maflahi sheikdom of Upper Yafa.

Jurbarkas (YUHR-buhr-kuhs), German Jurburg or Georgenburg, city (2001 population 13,797), SW LITHUANIA, on right bank of NEMAN River, and 45 mi/72 km WNW of KAUNAS; 55°04′N 22°46′E. Manufacturing (shoes, furniture); sawmilling, oil-seed pressing, brewing. Dates from 14th century, when Teutonic Knights built Georgenburg castle; passed (16th century) to Lithuania; in Russian KOVNO government until 1920. Also spelled as Yurbarkas.

Jurbise (zhuhr-BEEZ), Flemish Jurbeke (ZHOOR-bai-ke), commune (2006 population 9,587), Mons district, HAINAUT province, SW BELGIUM, on DENDER RIVER, and 6 mi/9.7 km NNW of MONS; 50°32′N 03°56′E. Agriculture; pottery.

Jurburg, LITHUANIA: see JURBARKAS.

Jurema (ZHOO-re-mah), city (2007 population 14,018), E central PERNAMBUCO state, NE BRAZIL, 28 mi/45 km NE of GARANHUNS; 08°23′S 36°11′W. Coffee, sugar, corn, kidney beans, fruit, manioc.

Juremal (ZHOO-re-mahl), village, N central BAHIA state, BRAZIL, 32 km/20 mi SE of JUAZEIRO; 09°36′S 40°22′W.

Jurf (JOORF), village, MADINIYINA province, SE TUNISIA, 43 mi/69 km NNE of MADINIYINA. Small harbor with ferry to AJIM on JARBAH ISLAND.

Jurf al-Darawish, JORDAN: see JURUF ED DARAWISH.

Jurien Bay (JOO-ree-uhn) or **Jurien**, town, WESTERN AUSTRALIA state, W AUSTRALIA, 165 mi/266 km N of PERTH, and at S end of Jurien Bay (6 mi/9 km long); 30°18′S 115°02′E. Tourism; resort. Recreational fishing, commercial crayfishing.

Jurien Bay (JOO-ree-uhn), bay, WESTERN AUSTRALIA state, W AUSTRALIA, c.165 mi/266 km N of PERTH; 30°17′S 115°00′E. Jurien Bay town at S end. Bay extends 6 mi/9 km from Island Point at S to North Head; a series of islands and reefs just off coast shelter bay waters. Important fishing location.

Jurilovca (zhoo-ree-LOU-kah), village, TULCEA county, SE ROMANIA, on LAKE GOLOVIȚA, 12 mi/19 km SE of

BABADAG; 44°46′N 28°52′E. Fishing port, center for production of black caviar.

Juripiranga (zhoo-ree-pee-rahn-gah), town (2007 population 10,235), E PARAÍBA state, NE BRAZIL, on PERNAMBUCO state border, 30 mi/48 km SW of JOÃO PESSOA; 07°22′S 35°26′W. Cotton, pineapples. Until 1944, called Serrinha.

Jurisdicciones, Cerro Las (hoo-rees-dee-see-O-nes, SER-ro LAHS), ANDEAN peak (12,631 ft/3,850 m) of Cordillera ORIENTAL, on NORTE DE SANTANDER-MAGDALENA department border, N COLOMBIA, 45 mi/72 km W of CÚCUTA; 07°51′N 73°14′W.

Jurish, Arab village, Nablus district, 8.1 mi/13 km SE of NABLUS, in the Samarian Mountains, WEST BANK; 32°06′N 35°19′E. Agriculture (olives, cereal).

Jurm, town, BADAKHSHAN province, NE AFGHANISTAN, on tributary of KOKCHA RIVER, 22 mi/35 km SE of FAIZABAD; 36°52′N 70°51′E.

Jūrmala (YOOR-muh-luh) or **yurmala** [Lettish=Rīga beach], BALTIC Sea seaside resort of LATVIA, located W of Rīga, on 10 mi/16 km-long pine-clad sand dunes, between lower LIELUPE RIVER and Gulf of RIGA; 56°58′N 23°34′E. Consists of six amalgamated resorts (linked to Rīga by railroad): BULDURI (German bilderlingshof), 10 mi/16 km WNW of Rīga city center; EDINBURG; MAJORI or MAYORI (German majorenhof); DUBULTI (German dubbeln); MELLUZI or Melluzhi (German karlsbad); ASARI (German assern), 17 mi/27 km W of Rīga city center. Agriculture (fruits and vegetables). Psychiatric hospitals, rest homes.

Jurong (JOO-RUNG), town, ⊙ Jurong county, SW JIANGSU province, E CHINA, 20 mi/32 km ESE of NANJING; 31°56′N 119°10′E. Rice, wheat, beans, cotton, oilseeds; textiles and clothing, building materials; engineering.

Jurong (JOO-RONG), town, SW Singapore island, SINGAPORE, suburb 11 mi/18 km W of downtown SINGAPORE, on Straits of SINGAPORE, at W entrance to Selat Johor strait, W of mouth of Jurong River; 01°17′N 103°41′E. Railroad spur terminus; W terminus of Mass Rail Transit (MRT) line and W terminus of Pan-Island Expressway. Large Jurong Industrial Estate and Pioneer Sector port and shipbuilding facilities at W end of island, surround former village of TUAS. Area developed since 1970s, rivals Singapore city's KEPPEL HARBOUR and older manufacturing areas. Manufacturing (computer peripherals, electron tubes, electronic parts, fabricated structural metals, soft drinks, beer, vegetable oils, small arms ammunition, batteries, telephone apparatus, motors and generators, baked goods, tools and jigs; shipbuilding, metal insulating, metal stamping, steel milling). Seat of Nanyang Technical Institute (formerly Nanyang University); Singapore Science Centre research facility; also Jurong Bird Park, Jurong Crocodile Paradise. Ferries to Pulau Pesek, Pulau Ayer Chawan, and Pulau Seraya islands, to S. Tengeh and Poyan reservoirs to NW, both are enclosed arms of JOHOR STRAIT. Much of W part of area is land reclaimed from marsh and shallow waters of strait.

Jurong Industrial Estate, SINGAPORE: see JURONG SINGAPORE.

Jur River (JOOR), over 350 mi/563 km long, SW SUDAN; rises on NILE-CONGO watershed SW of WAU; flows N and E, past WAU and GOGRIAL, joining TONJ RIVER in Lake AMBADI to form, together with the BAHR AL-ARAB, the BAHR AL-GHAZAL. Navigable (July–Oct.) below WAU.

Juru (ZHOO-roo), city (2007 population 10,222), SW PARAÍBA state, BRAZIL, 19 mi/31 km NE of Princessa Isabel; 07°32′S 37°49′W.

Juruá (joo-ruh-AH), town (2007 population 8,658), W central AMAZONAS state, BRAZIL, 56 mi/90 km S of FONTE BOA, on Rio Jururá; 03°30′S 66°04′W.

Juruá River (hoo-roo-AH), c.1,500 mi/2,414 km long, PERU and BRAZIL; rises in the Cerros de Canchyuaya, E Peru; flows in a winding course generally NE

through ACRE and AMAZONAS states (W BRAZIL), to the AMAZON River E of FONTE BOA; 03°00′S 66°00′W. One of the Amazon's longer tributaries, it is navigable along ⅓ of its course and was important for transport during the wild-rubber boom.

Juruena (ZHOO-roo-en-ah), town, MATO GROSSO state, BRAZIL, on Rio JURUENA, 65 mi/105 km NE of Castrahelta; 11°05′S 58°00′W.

Juruf ed Darawish (JOO-ruf ed dah-rah-WEESH), village, S central JORDAN, on HEJAZ RAILROAD, and 18 mi/29 km SE of TAFILA; 30°42′N 35°52′E. Camel raising. Sometimes spelled Juruf el Darawish or Jurf al-Darawish.

Juruf el Darawish, JORDAN: see JURUF ED DARAWISH.

Juruti (zhoo-roo-chee), city (2007 population 34,338), extreme W PARÁ state, BRAZIL, on right bank of the AMAZON River, near AMAZONAS state border, and 40 mi/64 km WSW of ÓBIDOS; 02°16′S 56°05′W. Rubber, Brazil nuts, jute.

Juscimeira (zhoo-see-MAI-rah), city (2007 population 11,999), SE MatoGrosso state, BRAZIL, 43 mi/70 km NW of Rondonópolis; 16°11′S 55°02′W. Soybeans, sugarcane.

Josepín (hoo-sai-PEEN), town, MONAGAS state, NE VENEZUELA, 22 mi/35 km W of MATURÍN; 09°45′N 63°33′W. Oil wells, linked by pipeline with CARIPITO. Vegetables, sugarcane.

Jushiyama (JOO-shee-YAH-mah), village, Ama county, AICHI prefecture, S central HONSHU, central JAPAN, 9 mi/15 km S of NAGOYA; 35°05′N 136°46′E.

Jussari (zhoo-SAH-ree), town (2007 population 6,750), SE BAHIA state, BRAZIL, 34 mi/55 km SSW of ITABUNA; 15°11′S 39°29′W. Also spelled Juçari.

Jussiape (ZHOO-see-AH-pai), city (2007 population 8,444), central BAHIA state, BRAZIL, on RIO DE CONTAS, 32 mi/52 km NE of LIVRAMENTO DO BRUMADO; 13°31′S 41°35′W. Also spelled Juciape.

Justice (JUHS-tis), community, SW MANITOBA, W central CANADA, 12 mi/20 km NNE of BRANDON, in ELTON rural municipality; 49°59′N 99°47′W.

Justice, village (2000 population 12,193), COOK county, NE ILLINOIS, W suburb of CHICAGO, on ILLINOIS WATERWAY; 41°45′N 87°50′W.

Justin (JUHST-in), town (2006 population 3,119), DENTON county, N TEXAS, 13 mi/21 km SW of DENTON; 33°05′N 97°17′W. Manufacturing (building materials).

Justiniano Posse (hoos-tee-nee-AH-no POS-sai), town, E CÓRDOBA province, ARGENTINA, 45 mi/72 km SE of VILLA MARÍA; 32°53′S 62°40′W. Wheat, flax, corn; livestock; dairying.

Justo Daract (HOOS-to dah-RAHKT), town, E SAN LUIS province, ARGENTINA, on railroad, on the RÍO QUINTO, and 22 mi/35 km SE of MERCEDES; 33°52′S 65°11′W. Agriculture center (alfalfa, soybeans, corn, wheat; poultry, livestock).

Justøya (YOOST-uh-yah), island (□ 3 sq mi/7.8 sq km) in the SKAGERRAK strait, AUST-AGDER county, S NORWAY, 3 mi/4.8 km S of LILLESAND.

Jutaí (zhoo-tah-EE), city, SW AMAZONAS state, BRAZIL, 93 mi/150 km NNE of EIRUNEPÉ, on Rio Jutaí; 05°31′S 69°09′W.

Jutaí (ZHOO-tah-ee), city, extreme W PERNAMBUCO state, BRAZIL, 56 mi/90 km NNE of PETROLINA; 08°39′S 40°14′W.

Jüterbog (YOOI-tuhr-bawg), town, BRANDENBURG, E GERMANY, 20 mi/32 km ENE of WITTENBERG; 52°00′N 13°05′E. Manufacturing of machinery and furniture. Has 12th- and 15th-century churches, several 15th-century town gates. Chartered 1174; was property of archbishops of MAGDEBURG; passed to SAXONY in 1648. Went to PRUSSIA in 1815. Formerly a military center with extensive training grounds.

Jutfaas, NETHERLANDS: see JUTPHAAS.

Juti (ZHOO-chee), town (2007 population 5,358), S central Mato Grosso do Sul state, BRAZIL, 17 mi/27 km SE of CAARAPÓ; 22°53′S 54°18′W. Airport. Also called Santa Luzia (SAHN-tah loo-SEE-ah).

Jutiapa (hoo-tee-AH-pah), department (□ 1,243 sq mi/ 3,231.8 sq km), SE GUATEMALA, on PACIFIC OCEAN and EL SALVADOR border; ⊙ JUTIAPA; 14°10′N 89°50′W. In SE highlands, sloping S to Pacific coastal plain; drained by the Río de la PAZ (which forms its SE border); includes SUCHITÁN and Moyuta volcanoes. CHINGO volcano and Lake Güija are on El Salvador border. Mainly agricultural (corn, beans, fodder grasses, rice); extensive livestock raising (hogs, cattle). Coffee and sugarcane are grown on lower slopes. Main centers are Jutiapa and ASUNCIÓN MITA, served by INTER-AMERICAN HIGHWAY.

Jutiapa (hoo-tee-AH-pah), city (2002 population 22,200), ⊙ JUTIAPA department, SE GUATEMALA, near INTER-AMERICAN HIGHWAY, 45 mi/72 km ESE of GUATEMALA city, in highlands, 14°17′N 89°53′W. Elevation 2,926 ft/892 m. Commercial center in dairying region; corn, beans; livestock.

Jutiapa (hoo-tee-AH-pah), municipality and town, CABAÑAS department, EL SALVADOR, W of SENSUN-TEPEQUE in W of department.

Jutiapa (hoo-tee-AH-pah), town, ATLÁNTIDA department, N central HONDURAS, on paved road and railroad, 37 mi/60 km SE of LA CEIBA; 15°46′N 86°34′W. Market center in banana-producing area.

Juticalpa (hoo-tee-KAHL-pah), city (2001 population 30,072), ⊙ OLANCHO department, central HONDURAS, in CATACAMAS VALLEY, on Juticalpa River (left affluent of GUAYAPE RIVER), and 75 mi/121 km NE of TE-GUCIGALPA (linked by paved road); 14°39′N 86°12′W. Elevation 5,423 ft/1,653 m. Commercial center in livestock area; dairying; coffee, tobacco, rice. Has government buildings. Airfield (W). Founded c.1611; became city in 1835. Flourished in 19th century through trade with Caribbean ports and placer mining in nearby Guayape River.

Jutland (JUT-luhnd), Danish *Jylland*, German *Jütland*, peninsula, N EUROPE, comprising continental DEN-MARK and N SCHLESWIG-HOLSTEIN state, GERMANY; c.250 mi/400 km long, up to 110 mi/177 km wide. Bounded N by the SKAGERRAK, W by the NORTH SEA, E by the KATTEGAT strait and LILLE BÆLT, and S by the EIDER RIVER. The term usually is applied only to the Danish territory. Danish Jutland, including adjacent islands, has an area of 11,441 sq mi/29,632 sq km and contains about half the population of Denmark. The LIMFJORD strait cuts across N Jutland. A glacial ridge extending through central Jutland divides the peninsula into two sections. W Jutland is windswept and sandy and has poor soil. Its coast is marshy, with many lagoons, and ESBJERG is the only good port. The E coast of Jutland is fertile and densely populated. Dairying and livestock raising are the main occupations of E Jutland; ÅRHUS and ÅLBORG are the chief ports. The peninsula has many lakes and is traversed by the GUDENÅ, Denmark's principal river. Yding Skovhøj, the highest point (568 ft/173 m) in Denmark,

is in E Jutland. SØNDERJYLLAND [=South Jutland] is the name applied in Denmark to the N part of the former duchy of Schleswig, including the towns of ÅBENRÅ, HADERSLEV, and SØNDERBORG. Jutland was known to the ancients as the Cimbric Peninsula (Lat. *Chersonesus Cimbrica*). In 1916, off the coast of W Jutland, British and German fleets engaged in the largest naval battle of World War I.

Jutogh (juh-TOG), town, SHIMLA district, central HI-MACHAL PRADESH state, N INDIA, cantonment 3 mi/ 4.8 km W of SHIMLA; 31°06′N 77°07′E. A geological series was named after Jutogh.

Jutphaas (YUHT-fahs), district of NIEUWEGEIN, UTRECHT province, central NETHERLANDS, 5 mi/8 km SSW of UTRECHT, on LEK Canal; 52°01′N 05°04′E. Also spelled Jutfaas.

Jutrosin (yoo-TRO-sheen), German *Jutroschin*, town, Leszno province, W POLAND, 11 mi/18 km WSW of KROTOSZYN. Manufacturing of cement goods, flour milling, sawmilling.

Juttah, ancient city mentioned in the Old Testament, near what is now Yattah, a large Arab village on the WEST BANK, S of HEBRON.

Jutulstraumen Glacier, 120 mi/200 km long, in EAST ANTARCTICA; flows into the FIMBUL ICE SHELF through the mountains of NEW SCHWABENLAND; 71°35′S 00°30′W.

Juvavum, AUSTRIA: see SALZBURG.

Juventino Rosas, MEXICO: see SANTA CRUZ DE JU-VENTINO ROSAS.

Juventud, Isla de la (hoo-vain-TOOD, EE-sluh dai luh), English, *Isle of Youth*, island (□ 1,180 sq mi/3,068 sq km) and Special Municipality, off SW CUBA, from which it is separated by the GULF OF BATABANÓ; ⊙ NUEVA GERONA. Until 1978 it was called Isla de Pinos (English, *Isle of Pines*). Pine forests cover much of the island, and there are many mineral springs. Marble is quarried from low ridges in the N part; the S quarter of the island is an elevated plain. The economy is based on fishing and agriculture (primarily citrus fruits, some vegetables). Until the break in U.S.-Cuban relations in the early 1960s, much of the land was owned by American citizens, and the climate and excellent fishing waters made the island an attractive resort. Bibijagua beach remains popular. Sighted by Columbus in 1494, Isla de la Juventud was later used as a penal colony and was a rendezvous for buccaneers. During the colonial period it was a summer resort and a rest area for the Spanish military. Ceded to the U.S. after the Spanish-American War (1898) and later claimed by both the U.S. and Cuba because its name was omitted from the Platt Amendment, which defined Cuba's borders. Finally, in 1907, the U.S. Supreme Court declared that the island did not belong to the U.S.; a treaty was later signed (1925) confirming the island as Cuban. Near Nueva Gerona is a large prison, often used for political prisoners. During the gov-

ernment of Fidel Castro, himself jailed here in 1953, the island has been extensively beautified, but political prisoners are incarcerated here. The name change also reflects large numbers of schools and training programs for young people. Isla de la Juventud has suffered frequent damage from hurricanes.

Juvisy-sur-Orge (zhoo-vee-zee–syur–ORZH), town, ESSONNE department, ÎLE-DE-FRANCE region, N central FRANCE, on left bank of the SEINE RIVER, and 11 mi/18 km S of PARIS; 48°41′N 02°23′E. Diversified industries and major railroad freight yards.

Juwaimisa, IRAQ: see SUWAIRA.

Ju Xian (JOO SHYEN), town, ⊙ Ju Xian county, S central SHANDONG province, NE CHINA, 45 mi/72 km NE of LINYI; 35°35′N 118°49′E. Grain, tobacco, oilseeds. Tobacco processing, manufacturing (food and beverages, textiles, building materials), papermaking; crafts.

Juxtlahuaca, MEXICO: see SANTIAGO JUXTLAHUACA.

Jūybār (JOO-yub-AR), town, Māzandarān province, N IRAN, 8 mi/13 km NW of SARI; 36°38′N 52°55′W.

Juye (JOO-YE), town ⊙ Juye county, SW SHANDONG province, NE CHINA, 38 mi/61 km ENE of HEZE; 35°23′N 116°06′E. Grain, cotton, peanuts, oilseeds, jute. Textiles.

Juymand, IRAN: see GONABAD.

Júzcar (HOOTH-kahr), village, MÁLAGA province, S SPAIN, 7 mi/11.3 km S of RONDA. Olives, grapes, cereals, oranges, nuts; liquor.

Južna Morava River, SERBIA: see SOUTHERN MORAVA RIVER.

Juzni Brod, MACEDONIA: see BROD.

Jwaneng (JWAI-neng), town (2001 population 15,179), SOUTHERN DISTRICT, BOTSWANA, 50 mi/80 km W of GABORONE; 24°35′S 24°43′E. Town Council administrative area. Major diamond-mining town. Hospital, airfield.

Jyderup (JUH-duh-roop), town, VESTSJÆLLAND county, SJÆLLAND, DENMARK, 12 mi/19 km WSW of HOLBÆK. Containers.

Jyekundo, China: see YUSHU.

Jyhun, TURKEY: see CEYHAN RIVER.

Jylland, DENMARK: see JUTLAND.

Jyrgalang (jer-guh-LAHN), town, E ISSYK-KOL region, KYRGYZSTAN, on TÜP RIVER, and 35 mi/56 km E of KARAKOL (linked by narrow-gauge railroad); 42°36′N 79°00′E. Coal mining. Also spelled DZHERGALAN.

Jyväskylä (YUH-vas-KUH-lah), city, ⊙ KESKI-SUOMEN province, S central FINLAND, on LAKE PÄIJÄNNE; 62°14′N 25°44′E. Elevation 462 ft/140 m. Important port, railroad, and road junction. Manufacturing includes paper and wood products, paper-making machinery, tractors, metals, wool, and foodstuffs. Chartered in 1837; site of the first Finnish-language secondary school (1858). Seat of University of Jyväskylä (1966), designed by architect Alvar Aalto. Arts festival held here in July.

K

K2, PAKISTAN: see GODWIN-AUSTEN, MOUNT.

K5, China: see MUZTAG.

Kaaawa (KAH-ah-AH-VAH), town (2000 population 1,324), OAHU island, HONOLULU county, HAWAII, 16 mi/26 km N of HONOLULU, on NE coast; 21°33′N 157°51′W. Pineapples, sugarcane; fish, prawns. Swanzy Beach Park to NW; Kaaawa Beach Park to SE; Waiahole Forest Reserve to SW; Kahana Valley State Park to W, including Crouching Lion rock formation.

Ka'abiye-Tabbash (KAH-AH-bi-ye-TAHB-BAHSH), Beduin village, 8 mi/12.9 km NW of NAZARETH, ISRAEL, in LOWER GALILEE.

Kaabong, administrative district, NORTHERN region, extreme NE UGANDA, on SUDAN (NWN) and KENYA (N and E) borders. As of Uganda's division into eighty districts, borders MOROTO (SE tip), KOTIDO (S), and KITGUM (W) districts. Includes MOUNT MORUNGOLE (9,020 ft/2,749 m) and KIDEPO NATIONAL PARK in N. Primary inhabitants are the Dodoth people. Agriculture; cattle raising is important. Formed in 2005 from N portion of former KOTIDO district (ABIM district formed from S portion in 2006 and central portion became current Kotido district).

Kaabong (ka-BONG), town (2002 population 30,728), KAABONG district, NORTHERN region, NE UGANDA, 15 mi/25 km W of KENYA border; 03°31′N 34°08′E. Commercial center in a cattle- and grain-producing area.

Kaaden, CZECH REPUBLIC: see KADAN.

Ka'aiti, YEMEN: see QUAITI.

Kaakamojoki (KAH-kah-mo-YO-kee), river, 40 mi/64 km long, NW FINLAND; rises N of TORNIO; flows generally S to GULF OF BOTHNIA 4 mi/6.4 km SW of KEMI. Boundary (from 1374) between bishoprics of UPPSALA and TURKU; later, until 1809, border between SWEDEN and Finland.

Kaa-Khem (KAH-ah-HYEM), industrial settlement (2006 population 10,350), central TUVA REPUBLIC, S SIBERIA, RUSSIA, on the Kyzyl-Khem River (tributary of the YENISEY RIVER), on road, 11 mi/18 km E of KYZYL; 51°41′N 94°44′E. Elevation 2,506 ft/763 m. Mining of gold and nonferrous metals; lumbering, woodworking. Ecotourism base.

Kaakhka (KAH-kah), city, SE AHAL weloyat, S central TURKMENISTAN, on TRANS-CASPIAN RAILROAD, 75 mi/121 km SE of ASHGABAT; 37°21′N 59°36′E. In orchard area. Tertiary-level administrative center. Called Ginsburg, c.1920–1927. Also spelled Qaqa.

Kaala (ka-AH-lah), peak (4,046 ft/1,233 m) of WAIANAE RANGE, W OAHU, HONOLULU county, HAWAII, 20 mi/32 km NW of HONOLULU. Highest point on Oahu.

Kaala-Gomen (kah-AH-lah-GO-men), village, W NEW CALEDONIA, 170 mi/274 km NW of NOUMÉA; 20°44′S 164°29′E. Agricultural produce, extensive livestock; meat cannery at nearby Ouaco.

Kaanapali (KAH-AH-nah-PAH-lee), village (2000 population 1,375), MAUI island, MAUI county, HAWAII, 3 mi/4.8 km N of LAHAINA, on W coast of West Maui Peninsula; 20°55′N 156°40′W. Tourism. Pineapples. Kaanapali Airport to N. N terminus of scenic train from Lahaina. Whalers Village Shopping Center and Museum. West Maui Forest Reserve to E.

Kaap Plateau, region in N part of the NORTHERN KAROO, NORTHERN CAPE, SOUTH AFRICA, in GRIQUALAND WEST, between ORANGE RIVER (S), ASBESTOS MOUNTAINS (W), KALAHARI DESERT (N), and HARTZ and VAAL rivers (E and SE). Elevation 3,700 ft/1,128 m–5,000 ft/1,524 m. Lower W portion of central plateau mainly sub-desert steppe.

Kaarlakhti, RUSSIA: see KUZNECHNOYE.

Kaarst (KAHRST), town, NORTH RHINE–WESTPHALIA, W GERMANY, 4 mi/6.4 km NW of NEUSS; 51°12′N 06°37′E. Manufacturing of machinery. Has been incorporated into Neuss.

Kaaterskill Clove, SE NEW YORK, in CATSKILL MOUNTAINS, scenic gorge of small Kaaterskill Creek, extends c.5 mi/8 km between HAINES FALLS village (W; noted waterfall here) and PALENVILLE (E); 42°11′N 74°04′W. Traversed by state highway.

Kaatsheuvel (KAHTS-huh-vuhl), village (2001 population 19,843), NORTH BRABANT province, S NETHERLANDS, 7 mi/11.3 km NNW of TILBURG; 51°39′N 05°02′E. WAAL RIVER 3 mi/4.8 km to N, De Efteling Sprookjespark (amusement park) to S. Dairying; cattle, hogs; grain, vegetables; light manufacturing.

Kaba (KAH-bah), village, HAJDU-BIHAR county, E HUNGARY, 22 mi/35 km SW of DEBRECEN; 47°21′N 21°17′E. Wheat, corn, sunflowers; cattle, hogs. Hungary's largest sugar refinery.

Kaba, China: see HABAHE.

Kaba Aye Pagoda (KAH-bah AH-yai) [Nor.=peace pagoda], religious site, YANGON, MYANMAR. Built by Prime Minister U Nu between 1950 and 1952, this pagoda is distinguished by having the same diameter and height: 112 ft/34 m. It was dedicated to the cause of world peace and was the site of the Sixth Buddhist Synod. An inner temple contains a Buddha image cast in solid silver.

Kabaena (kah-BEI-nah), island (□ 338 sq mi/878.8 sq km), INDONESIA, off SE coast of SULAWESI, at E side of entrance to Gulf of BONE; 05°15′S 121°55′E. Island is 30 mi/48 km long, 20 mi/32 km wide; partly mountainous, rising to c.5,400 ft/1,646 m in center of island. Buffalo raising, agriculture (sago, coconuts).

Kabáh (kah-BAH), historic site, SW YUCATÁN, MEXICO, 12 mi/19 km S of Mexico Highway 261 from UXMAL. This site is representative of Puuc architectural style and is dominated by a Palace of Masks, which is decorated with elaborate stone masks of Chac, the rain god. Across the road is a freestanding arch marking the spot where a Mayan road once entered KABAH from Uxmal.

Kabakovo (kah-bah-KO-vuh), village (2005 population 3,080), central BASHKORTOSTAN Republic, E European Russia, on the INZER RIVER, near railroad, 12 mi/19 km SE of UFA; 54°32′N 56°08′E. Elevation 305 ft/92 m. In oil-producing and -refining area.

Kabakovsk, RUSSIA: see SEROV.

Kabala (kah-BAH-lah), town, ⊙ KOINADUGU district, NORTHERN province, N SIERRA LEONE, on road, 60 mi/97 km NNE of MAKENI; 09°35′N 11°33′W. Trade center; peanuts, rice, livestock, hides, and skins.

Kabale (kuh-BAH-lai), administrative district (□ 705 sq mi/1,833 sq km; 2005 population 483,400), WESTERN region, SW UGANDA, along RWANDA border (to S); ⊙ KABALE; 01°15′S 30°00′E. Elevation ranges from 3,937 ft/1,200 m–9,842 ft/3,000 m. As of Uganda's division into eighty districts, borders KISORO (W), KANUNGU (NW), RUKUNGIRI (N), and NTUNGAMO (E) districts. LAKE BUNYONI located here. Mountainous region, also containing forests and wetlands. Vegetation ranges from high altitude forests to savannah and swamps. Primarily rural and agricultural area (producing bananas, beans, finger millet, groundnut, maize, potatoes, sorghum, sweet potatoes, and wheat, cash crops include tobacco; cattle, goats, sheep, pigs, and rabbits; some fishing). There is also quarrying of mineral reserves (including wolfram, irone ore, and lava ash). Kabale town is only urban area. Main roads connect it to surrounding Uganda and S into Rwanda.

Kabale (kuh-BAH-lai), town (2002 population 41,344), ⊙ KABALE district, WESTERN region, SW UGANDA, in mountain and lake region, 65 mi/105 km SW of MBARARA. Elevation c.6,000 ft/1,829 m. Agricultural trade center; coffee, corn, millet; cattle, sheep, goats.

Tourist center for surrounding mountains and LAKE BUNYONI (SW). Inhabited by Bantu-speaking Chiga people. Gold, tin, tungsten deposits. Kikungiri, 2 mi/3.2 km S, was Roman Catholic mission center. Was part of former SOUTHERN province.

Kabalega Falls (kah-buh-LAI-guh), rapids on lower VICTORIA NILE RIVER, on border between MASINDI and AMURU districts, NW UGANDA, in KABALEGA NATIONAL PARK, 22 mi/35 km E of LAKE ALBERT. Here the river (only 19 ft/6 m wide) flows through precipitously cleft rocks, dropping 400 ft/122 m in a series of three cascades to level of Lake Albert. Also called Murchison Falls.

Kabalega National Park (kah-buh-LAI-guh) (□ 1,483 sq mi/3,855.8 sq km), NW UGANDA, on border of NORTHERN and WESTERN regions and traversed by VICTORIA NILE RIVER. Includes KABALEGA FALLS. Established 1952. Also called Kabalega Falls National Park and Murchison Falls National Park.

Kabalo (kah-BAH-laoh), town, KATANGA province, E CONGO, on LUALABA RIVER, on railroad, 160 mi/257 km W of Kalamie; 06°03′S 26°55′E. Elev. 1,742 ft/530 m. Trans-shipment point for produce of middle LUALABA region and for imports from E Africa. Food staples (manioc, yams, sweet potatoes) are grown on a large scale in vicinity. Has Roman Catholic and Protestant missions. Airport.

Kabambare (kah-bahm-BAH-rai), village, SUD-KIVU province, central CONGO, 70 mi/113 km ESE of KASONGO; 04°42′S 27°43′E. Elev. 2,719 ft/828 m. Tin mining, cotton growing, cotton ginning; also tantalite mining. Local airfield.

Kabanjahe (kah-BAHN-jah-he), city, N SUMATRA, Indonesia, 46 mi/74 km SW of MEDAN; ⊙ Karo district; 03°06′N 98°30′E. Elevation 3,936 ft/1,200 m.Traditionalist Karo Batak group lives in area.

Kabankalan (kah-bahn-KAH-lahn), town (2000 population 149,769), NEGROS OCCIDENTAL province, W NEGROS island, PHILIPPINES, 15 mi/24 km S of BINALBAGAN; 09°53′N 122°50′E. Sugar mill, distillery; gypsum mine.

Kabansk (kah-BAHNSK), village (2005 population 6,220), central BURYAT REPUBLIC, S SIBERIA, RUSSIA, on the SELENGA River, on road and near the TRANS-SIBERIAN RAILROAD (Timlyuy station), 45 mi/72 km WNW of ULAN-UDE; 52°03′N 106°39′E. Elevation 1,581 ft/481 m. Dairy plant, fish canneries; forestry services. Oil and natural gas deposits in the vicinity.

Kabara, village, SIXTH REGION/TIMBUKTU, central MALI, port for TIMBUKTU (just N) on left bank of the NIGER RIVER.

Kabara (kahm-BAH-rah), composite limestone-volcanic island (□ 12 sq mi/31.2 sq km), Lau group, FIJI, SW PACIFIC OCEAN; 18°57′S 178°57′W. Island is 4.5 mi/7.2 km long, 3 mi/4.8 km wide; rises to 470 ft/143 m. Timber, copra. Sometimes spelled Kambara.

Kabardinka (kah-bahr-DEEN-kah), settlement (2005 population 7,430), S KRASNODAR TERRITORY, S European Russia, on the NE shore of the BLACK SEA, on coastal highway, 8 mi/13 km SE of NOVOROSSIYSK; 44°39′N 37°56′E. Elevation 141 ft/42 m. Seaside resort, with spa, sanatoria, and summer homes.

Kabardino-Balkar Republic (kah-bahr-DEE-nuh-bahl-KAHR), constituent republic (□ 4,826 sq mi/12,547.6 sq km; 2004 population 790,000), SE European Russia, in the northern part of the CAUCASUS mountains; ⊙ NAL'CHIK. The area is a largely unsettled, roadless mountain wilderness. The population—Kabards, Balkars, Russians, and Ukrainians—is concentrated in the narrow gorges of the streams flowing into the TEREK RIVER. The Kabards speak a Caucasian language and are Sunni Muslims; the Balkars speak a Turkic language. Kabards and Balkars make up 57% of the population; Russians make up 30%. Livestock and poultry are raised; and wheat, corn, hemp, and fruit are grown. Much of the republic's industry is related to agriculture processing. Lumbering and mining are also important, specifically iron ore and gold. The Kabards were known since the 9th century. They occupied the land in the foothills of the central Caucasus between the 13th and 15th centuries. It is not known when the Balkars settled. They have a mixed Black Bulgar, Alan, and Cuman heritage. The Kabard area became a Muscovite protectorate in 1557. Its annexation by Russia began with the treaty of Kuchuk Kainarji (1774) and was completed in 1827. The area was organized briefly as the Gorskiy (Russian= mountainous) autonomous socialist republic following the revolution of 1917; reorganized as an autonomous region in 1922, and became an autonomous republic in 1936. In 1943, the Balkars, accused of collaborating with the Germans, were deported, and their area, the upper Baksan valley, was ceded to the Georgian SSR. The area was then renamed Kabardinian Autonomous SSR. In 1956, the Balkars were returned, and in 1957 the area assumed its old name. Karbardino-Balkar became a full republic in 1991 and was a signatory to the March 31, 1992, treaty that created the Russian Federation. The republic has a 200-member parliament and is divided into eight districts, which include three cities, thirteen towns, and ninety-four villages.

Kabardino-Balkarskiy Preserve (kah-bahr-DEE-nuh– bahl-KAHR-skeeyee), wildlife refuge (□ 286 sq mi/ 743.6 sq km), KABARDINO-BALKAR REPUBLIC, SE European Russia, on the N slope of the central part of Glavnyy and BOKOVOY ranges. Glaciers and mountains account for half of the area. Local fauna include goat, lynx, wild boar, and leopard. Established in 1976.

Kabare (kah-BAH-rai), village, SUD-KIVU province, E CONGO, 7 mi/11.3 km WSW of BUKAVU; 02°28′S 28°49′E. Elev. 4,927 ft/1,501 m. Center of an area of European agricultural colonization (notably coffee plantations). Has large Roman Catholic mission.

Kabarnet (kah-bahr-NAIT), town (1999 population 9,583), ☉ Baringo district, RIFT VALLEY province, W KENYA, on road, 35 mi/56 km E of ELDORET; 00°33′N 35°46′E. Agriculture (coffee, wheat, corn, wattle) and livestock; trade and market center.

Kabarole (kah-buh-RO-lai), administrative district (□ 712 sq mi/1,851.2 sq km; 2005 population 373,300), WESTERN region, W UGANDA; ☉ FORT PORTAL; 00°35′N 30°15′E. Elevation ranges from 1,165 ft/355 m– 3,002 ft/915 m. As of Uganda's division into eighty districts, borders BUNDIBUGYO (W and N), KIBAALE (NE tip), KYENJOJO (E), KAMWENGE (SE), and KASESE (SSW) districts. Primary inhabitants are the Bakiga, Bakonjo, Bamba, and Batooro peoples; primary languages are Rukiga, Runyankole, and Rutooro. RUWENZORI mountain range in W. Agriculture is main economic activity (coffee and tea are important cash crops, also bananas, cassava, groundnuts, maize, millet, soy beans, sunflowers, sweet potatoes, and yams). Livestock also important to economy (mainly cattle, also poultry, goats, and pigs). Main road runs E out of Fort Portal to KAMPALA city and S to KASESE and NTUNGAMO towns. Formed in 2000 from NW portion of former KABAROLE district (Kyenjojo district was formed from N and E portions and Kamwenge district from S portion).

Kabarole (kah-buh-RO-lai), former administrative district (□ 3,228 sq mi/8,392.8 sq km), W UGANDA, NE of LAKE GEORGE and S of LAKE ALBERT; capital was FORT PORTAL; 00°30′N 30°30′E. As of Uganda's division into thirty-nine districts, was bordered by KIBAALE (N), MUBENDE (E), MASAKA (SE tip), MBARARA (S), BUSHENYI (SW tip), KASESE (W), and BUNDIBUGYO (NW) districts. Tea, coffee, cotton, bananas, sweet potatoes were grown around Fort Portal, the only city. In 2000 NW portion of district was formed into current KABAROLE district, S portion was formed into KAMWENGE district, and N and E portions were formed into KYENJOJO district.

Kabarole, town, W UGANDA: see FORT PORTAL.

Kabasalan (kah-bah-SAH-lahn), town, ZAMBOANGA DEL SUR province, W MINDANAO, PHILIPPINES, near head of SIBUGUEY BAY, 45 mi/72 km W of PAGADIAN; 07°50′N 122°47′E. Coconuts, corn, rice.

Kabaw Valley (kah-BAW), longitudinal lowland in SAGAING division, MYANMAR, on MANIPUR (INDIA) border, between MANIPUR and CHIN HILLS (W) and CHINDWIN RIVER; 100 mi/161 km long. It is sometimes defined as including additional 60 mi/97 km section of MYITTHA VALLEY (S). Valley is used by major Myanmar-India trade route via KALEMYO, TAMU, and IMPHAL.

Kabaya (KAH-bah-yah), village, RWANDA, 85 mi/137 km ENE of KIGALI; 01°45′S 29°32′E. Minor administrative center.

Kabba (KAH-BAH), one of the 13 former Northern provinces, SW central NIGERIA, in what now is KOGI state; capital was LOKOJA.

Kabba (KAH-BAH), town, KWARA state, S central NIGERIA, 45 mi/72 km W of LOKOJA; 07°50′N 06°04′E. Agricultural trade center; shea-nut processing, cotton weaving, sack making; cassava, corn, durra.

Kabbani River (KUH-buh-nee) c.120 mi/193 km long, KERALA and KARNATAKA states, S INDIA; rises in WAYANAD district, Kerala state in WESTERN GHATS; flows N and generally ENE past SARGUR and NANJANGUD to KAVERI RIVER at TIRUMAKUDAL NARSIPUR.

Kabe (KAH-be), town, HIROSHIMA prefecture, SW HONSHU, W Japan, 9 mi/14.5 km NNE of HIROSHIMA. Commercial center for agricultural area (rice, wheat); sake, soy sauce, ironware.

Kab, El (KAHB, el), Greek *Eileithyaspolis*, village, ASWAN province, S EGYPT, 12 mi/19 km NNW of IDFU. Site of ruins of ancient Nekheb or Nikhab, ancient capital of Upper Egypt.

Kabelvåg (KAH-buhl-vawg), village, NORDLAND county, N NORWAY, on S AUSTVÅGØY, LOFOTEN Islands, on VESTFJORDEN, 4 mi/6.4 km SW of SVOLVÆR. Light industries and services are main economic activities. Oldest fishing (cod) village on Lofoten Islands; summer resort. Church built by King Øystein (1120).

Kaberamaido, administrative district (□ 636 sq mi/ 1,653.6 sq km; 2005 population 135,000), EASTERN region, E central UGANDA, on N shore of LAKE KYOGA; ☉ KABERAMAIDO; 01°50′N 33°10′E. Elevation averages 3,281 ft/1,000 m–4,921 ft/1,500 m. As of Uganda's division into eighty districts, borders LIRA (N), AMURIA (NE), SOROTI (ESE), KAMULI (S, on opposite shore of Lake Kyoga), and AMOLATAR and DOKOLO (W) districts. Primary inhabitants are the Kumam and Iteso peoples. Wetlands and swamps near lake. Primarily rural and agricultural (including beans, cassava, groundnuts, maize, millet, potatoes, rice, simsim, and sorghum, cotton is primary cash crop; cattle). Formed in 2001 from SW portion of former SOROTI district ([now former] KATAKWI district formed from NE portion in 1997 and current Soroti district formed from S portion in 2001).

Kaberamaido (kah-ber-uh-MEI-do), village (2002 population 2,349), ☉ KABERAMAIDO district, EASTERN region, E central UGANDA, 30 mi/48 km W of SOROTI, near LAKE KYOGA. Cotton, peanuts, sesame. Was part of former EASTERN province.

Kabete (kah-BAI-tai), town, CENTRAL province, S central KENYA, on railroad, 7 mi/11.3 km W of NAIROBI; 01°10′N 36°44′E. Coffee center; corn; dairying. Veterinary research station (bacteriological institute), agriculture departments of university, and experimental farm.

Kabetogama, Lake (ka-buh-TO-guh-muh) (□ 31 sq mi/80.6 sq km), in ST. LOUIS county, extending W into KOOCHICHING county, NE MINNESOTA, 16 mi/26 km ESE of INTERNATIONAL FALLS, in VOYAGEURS NATIONAL PARK. Elevation 1,119 ft/341 m; 20 mi/32 km long, maximum width 5 mi/8 km. Connected by channel to NAMAKAN LAKE to E. Lake has numerous peninsulas, bays, and islands. Kettle Falls Dam at outlet (N) of Namakan Lake, has merged the lakes and evened their levels. Kabetogama State Forest to S; RAINY LAKE to N.

Kabgaye (kab-GAH-yee), town, central RWANDA, 23 mi/37 km WSW of KIGALI. Elevation 6,100 ft/1,859 m. Roman Catholic missionary center with two seminaries, trade and teachers' schools. Has noted museum of Rwandan life.

Kabilcoz, TURKEY: see SASON.

Kabin, TURKEY: see BESIRI.

Kabinakagami Lake (ka-bin-uh-KA-guh-mee), central ONTARIO, E central CANADA, 150 mi/241 km W of TIMMINS; 16 mi/26 km long, 10 mi/16 km wide. Drains N into KENOGAMI RIVER; 48°54′N 84°25′W.

Kabinburi (GUH-BIN-BU-REE), village and district center, PRACHIN BURI province, S THAILAND, on BANG PAKONG RIVER (high-water head of navigation), on BANGKOK railroad, 80 mi/129 km ENE of BANGKOK; 13°59′N 101°43′E. Rice; gold. Local name, KABIN. Also spelled KRABIN and Krabinburi.

Kabinda (kah-BEEN-dah), town, KABINDA district, KASAI-ORIENTAL province, SE CONGO, 110 mi/177 km SE of LUSAMBO; 05°28′S 28°15′E. Trading and cotton center, road communications hub; cotton ginning. Has hospital, Roman Catholic mission, several schools.

Kabinda, ANGOLA: see CABINDA.

Kabinyuk-syuyutlyu, Bulgaria: see VURBAK.

Kabirdham (kah-BEER-dhum), district (□ 1,710 sq mi/ 4,446 sq km), W central Chhattisgarh state, E central INDIA; ☉ KAWARDHA. Formed in 1998 in MADHYA PRADESH by combining the areas of Kawardha from Rajnandgaon district and of Pandariya from BILASPUR district. Became a district of Chhattisgarh in 2000.

Kabirdham (kah-BEER-dhum), INDIA: see KAWARDHA district.

Kabir Kuh (kah-BIR KOO), mountain range, SW IRAN, in the PUSHT (ZAGROS Mountain system), on right bank of upper KARKHEH River. Rises to 9,000 ft/2,743 m at 33°20′N 46°40′E.

Kabir, mountain, WEST BANK: see ELON MOREH.

Kableshkovo (kah-BLESH-ko-vo), city, BURGAS oblast, POMORIYE obshtina, BULGARIA; 43°39′N 27°34′E. Cattle; vineyards, fruit. Supplies nearby tourist areas.

Kabo (kah-BO), town, OUHAM prefecture, N CENTRAL AFRICAN REPUBLIC, near CHAD border, 35 mi/56 km NNE of BATANGAFO; 07°48′N 18°35′E. Market town; agriculture (cotton).

Kabo (kah-BO), village, YE-U township, SAGAING division, MYANMAR, on MU RIVER, 27 mi/43 km NNW of SHWEBO. Site of major irrigation headworks feeding Shwebo and Ye-u canals.

Kabodiion (kah-buh-dee-YON), village, SW KHATLON viloyat, TAJIKISTAN, on KAFIRNIGAN RIVER, 40 mi/64 km SSW of QURGHONTEPPA; 37°24′N 68°11′E. Long-staple cotton; karakul sheep. Until c.1935 called Kabadian, later Mikoyanad; also spelled Kabodiyon, Kabdien.

Kaboeroeang, INDONESIA: see KABURUANG.

Kabolela (kah-bo-LE-lah), village, KATANGA province, SE CONGO, 15 mi/24 km NW of LIKASI. Copper mining and concentrating; also cobalt mining.

Kabompo (kah-BO-em-po), town, NORTH-WESTERN province, NW ZAMBIA, 155 mi/249 km SW of SOLWEZI, on Kabompo River; 13°36′S 24°12′E. West Lunga National Park to NE. Honey, beeswax; cattle. Timber. Manganese deposits.

Kabongo (kah-BAWNG-go), village, KATANGA province, SE CONGO, 95 mi/153 km NNE of KAMINA; 07°25′S 25°38′E. Elev. 3,001 ft/914 m. Cotton ginning. Roman Catholic and Protestant missions.

Kabou (kah-BOO), town, KARA region, N central TOGO, 10 mi/16 km N of BASSAR, road junction on tributary of Môr; 09°27′N 00°49′E. Nearby are iron ore deposits.

Kabri (KAH-bree) or **Cabri**, kibbutz, NE of AKKO (Acre) and E of NAHARIYA in W GALILEE, ISRAEL;

33°01′N 35°08′E. Elevation 374 ft/113 m. Mixed farming and industry (non-ferrous mold factory, jewelry). Nearby are springs (with a mineral-water plant). The surrounding area has produced many artifacts from various eras, including mosaics from Roman and Byzantine periods and also remains of an Ottoman aqueduct built (1814) by the Turkish caliph Suleiman Faha to transport spring water to Akko. Also nearby is Tel Kabri, where ruins indicate signs of settlement in Chalcolithic and later times. Ruins include parts of a house and Bronze-Age pottery pieces, some brought from Cyprus in 17th century B.C.E. Founded in 1949.

Kabru (kah-BROO), peak (24,002 ft/7,316 m), in SIN-GALILA RANGE, on border with SIKKIM state, NE INDIA, 40 mi/64 km NNW of DARJILING (India); 27°36′N 88°07′E. Ascent in 1883 claimed by W. W. Graham.

Kabūdarāhang (kahb-OOD-ahr-AH-AHNG), town, Hamadān province, W IRAN, 32 mi/51 km NNE of HAMADAN; 35°11′N 48°43′E. Grain, opium, fruit; sheep raising.

Kabudi (kah-BOO-DEE), island (□ 12 sq mi/31.2 sq km) in Lake URMIA, IRAN. Highest peak, 5,249 ft/1,600 m. Forested conservation area with large populations of sheep and flamingoes. Also known as Jazire ye Ghoyun Dashi, or SHEEP ISLANDS.

Kabūdiyah, Ras (kah-boo-DEE-yah, RAHS), cape on the MEDITERRANEAN coast of E TUNISIA, 40 mi/64 km NNE of SAFAQIS; 35°13′N 11°10′E. Also spelled Ras Kapoudia.

Kabul (KAH-bool), province (2005 population 3,013,200), E central AFGHANISTAN; ⊙ KABUL; 34°30′N 69°25′E. Bordered by WARDAK province (W), PARWAN and KAPISA provinces (N), LAGHMAN and NANGAR-HAR provinces (E), and LOGAR province (S). Includes the rich economic areas of Kabul city in the E, where irrigated agriculture, industry, and population are concentrated. Largely mountainous including the prominent PAGHMAN chain in E. A network of paved roads connects Kabul province via SALANG PASS to Mazar-i Sharif and UZBEKISTAN to N, via Tang-i Gharu to PAKISTAN in the E, and via KANDAHAR and HERAT to IRAN in the W. Hydroelectric power stations in SAROBI, Mahipar, and Naghlu provide electricity for Kabul.

Kabul (KAH-bool), city (2002 population 2,678,000), ⊙ AFGHANISTAN and KABUL province, E Afghanistan, on the KABUL RIVER; 34°31′N 69°12′E. Elev. 5,895 ft/1,797 m. Afghanistan's largest city and economic and cultural center; strategically located in a high narrow valley, wedged between mountain ranges that command the main approaches to the KHYBER PASS. It is the nodal point of all major roads within the country; a paved road links Kabul, through the Salang tunnel (elev. 11,100 ft/3,383 m) under the HINDU KUSH mountains, with the TAJIKISTAN border. Located 20 mi/32 km N of the city is Hotki Pass, an important pass traversed by the "new road" that runs N across the Shamwali Plain to the S escarpment of the Hindu Kush mountains Another 30 mi/48 km farther N is Khotal-e-Salane, another pass of great strategic importance. Chief manufacturing is woolen and cotton cloth, beet sugar, ordnance, and furniture. Kabul's history dates back more than 3,000 years, although the city has been destroyed and rebuilt on several different sites. Called Kubha in the Rigveda (c.1,500 B.C.E.) and Kabura by Ptolemy (2nd century). Conquered by Arabs (7th century), Monguls (16th century), and Persians (18th century). It succeeded Kandahar as capital of Afghanistan in 1773. During two Anglo-Afghan wars (1839–1842 and 1878–1880), the British took Kabul. In 1842 the withdrawing British troops were ambushed and almost annihilated after the Afghans had promised them safe conduct; in retaliation another British force partly burned Kabul. On December 23, 1979, Soviet armed forces landed at Kabul airport to help bolster an increasingly unpopular

Communist government; ten years of devasting conflict ensued. Kabul became the Soviet command center, but was little damaged by the conflict. In February 1989, Soviet forces withdrew from the city, and in spring of 1992 Kabul fell to guerrilla armies. Destruction of the city increased as the coalition of guerrilla forces broke into rival warring factions. On September 26, 1996, the Taliban captured Kabul and assumed control of its administrative apparatus. The population grew from c.700,000 in the 1970s to 2 million in 1992. Due to the half decade of violence, however, the city's population is estimated to have fallen to c.1,500,000. Kabul's old section, with its narrow, crooked streets, contains extensive bazaars; the modern section has administrative and commercial buildings. An educational center, Kabul has a university (est. 1931), numerous colleges, and a museum, now largely plundered. Also in the city are Babur's tomb and garden; the mausoleum of Nadir Shah; the Minar-i-Istilal (column of independence), built in 1919 after the Third Afghan War; the tomb of Timur Shah (reigned 1773–1793); and several important mosques. The fort of Bala Hissar, destroyed by the British in 1879 to avenge the death of their envoy in Kabul, is now a military college. The royal palace and an ancient citadel are near the center of the city. Pul-i-Charki fortress, located outside the city, was transformed into a prison that is the most notorious in the country, both under the Communists and the Taliban. The Taliban government abandoned the city, which fell to Northern Alliance forces, supported by U.S. airstrikes, on November 13, 2001. An international airport connects Afghanistan with Europe and neighboring countries. Hotels.

Kabul (KAH-bool), Arab township, 9 mi/15 km SE of AKKO (Acre) in LOWER GALILEE, ISRAEL, on site of biblical Cabul; 32°52′N 35°12′E. Elevation 833 ft/253 m. King Saul presented a gift here to Hiram, King of Tyre. Known for its wise men in biblical times, it became a Jewish pilgrimage site. Ruins include a synagogue and cemetery from early Byzantine times.

Kabul River (KAH-bool), ancient *Cophus*, 320 mi/515 km long, in E AFGHANISTAN and PAKISTAN; 33°55′N 72°14′E; rises in PAGHMAN MOUNTAINS at foot of UNAI PASS 45 mi/72 km W of KABUL, flows E, past Kabul, JALALABAD, and Nowshera, to the INDUS just N of Attock. Receives PANJSHIR, KUNAR, and SWAT rivers (left), LOGAR RIVER (right). Irrigates Jalalabad valley in Afghanistan and PESHAWAR district in Pakistan (headworks at Warsak). Alexander the Great invaded INDIA (327 B.C.E.) via the Kabul River valley, which is traversed, since 1945, by Peshawar-Jalalabad-Kabul highway.

Kabundi (kuh-BOON-dee), village, KATANGA province, SE CONGO, close to MANIKA PLATEAU (UPEMBA NATIONAL PARK); 10°07′S 27°16′E. Elev. 2,883 ft/878 m.

Kaburuang (kah-BOO-roo-wuhng), island (□ 36 sq mi/93.6 sq km), TALAUD ISLANDS, N SULAWESI, INDONESIA, 10 mi/16 km S of KARAKELONG; 03°47′N 126°48′E. Island is 10 mi/16 km long, 5 mi/8 km wide; hilly, forested. Timber, copra, sago; nutmeg; fishing. Formerly spelled Kaboeroeang.

Kabushia (kuh-BOO-shi-yuh), township, River Nile state, SUDAN, on right bank of the NILE, on road and railroad, 25 mi/40 km NE of SHENDI; 16°53′N 33°42′E. Cotton, wheat, fruit, and durra; livestock. Nearby (N) site of ancient city of MARAWI.

Kabwe, city (□ 47,000 sq mi/122,200 sq km; 2000 population 176,758), ⊙ CENTRAL province, central ZAMBIA, 70 mi/113 km N of LUSAKA; 14°28′S 28°25′E. On railroad. Mining center (lime, lead, zinc, manganese, cadmium); sulphide-ore treating plant. Agriculture (cotton, tobacco). Manufacturing (transmission poles, lumber). Airport to South. Mulungushi Dam on branch of LUNSEMFWA RIVER to SE; Lunsemfwa (Mita Hills) Dam on Lunsemfwa River to NE; hydroelectric power station at both dams. Lu-

kanga swamps to W. Formerly called Broken Hill, which was capital of former Luangwa province. Broken Hill Mine township adjoins to W. Mine is noted for discovery (1921) of skull of prehistoric *Homo Rhodesiensis*.

Kabylia (kuh-BEEL-yuh), French *Kabylie*, mountainous coastal region of ALGERIA, in BOUMERDES, BEJAÏA, JIJEL, and TIZI OUZOU wilaya, extending from the MITIDJA lowland near ALGIERS (W) to the SKIKDA area (E), and reaching up to 30 mi/48 km inland. It is geographically divided by the SAHEL-SOUMMAM valley into 2 sections, Great KABYLIA (W) and Little Kabylia (E). The former has the more rugged terrain, including the DJURDJURA range, and has its main center at TIZI OUZOU. It is densely populated, grows olives on mountain slopes and wheat, tobacco, wine grapes, and figs in sheltered valleys. Cork is exploited in coastal strip between DELLYS and BEJAÏA (Kabylia's chief port). Little Kabylia, longitudinally traversed by the BABOR range, is Algeria's ranking cork-producing region. Its main port is JIJEL, but there are few inland towns. Scattered mineral deposits (only a few economically exploited) yield iron, lead, zinc, copper. The MEDITERRANEAN shore of Kabylia is followed (with interruptions) by a scenic highway. Peopled by fiercely independent Berbers, Kabylia offered desperate resistance to French occupation. Even after Abd El Kader's surrender, the Kabylians fought French expeditions throughout the 1850s, and were not pacified until the failure of the 1871 insurrection.

Kacabaş, TURKEY: see GRANICUS.

Kačanik (KAH-chah-neek), village, S SERBIA, on LE-PENAC RIVER, on railroad, 18 mi/29 km NNW of SKOPJE (MACEDONIA), near Macedonia border, in KOSOVO; 42°14′N 21°15′E. Kacanik defile (Serbian *Kač anič ka Klisura*), between SAR Mountains and CRNA GORA, extends from Kacanik SSE along Lepenac River into Macedonia. Also spelled Kachanik.

Kacēni, LATVIA: see KACHANOVO.

Kačerginė (kaht-chuhr-gee-nai), town, S central LI-THUANIA, on left bank of the Nemunas, 7 mi/11 km W of KAUNAS. Summer resort. Also spelled Kachergine.

Kacha (KAH-chah), town (2004 population 32,400), SW Republic of CRIMEA, UKRAINE, on BLACK SEA coast, N of KACHA RIVER mouth, 12 mi/19 km N and under jurisdiction of SEVASTOPOL'; 44°46′N 33°33′E. Seaside recreation; fisheries; fish processing. Established in 1912 as Aleksandro-Mikhaylovka; renamed and town status since 1938.

Kacha, NIGERIA: see KATCHA.

Kachanovo (kah-CHAH-nuh-vuh), Latvian *Kacēni*, village, W PSKOV oblast, W European Russia, near the border with LATVIA, terminus of local road, 32 mi/51 km SW of PSKOV; 57°28′N 27°46′E. Elevation 291 ft/88 m. Flax, grain. In Russian Pskov government until 1920; in Latvian Latgale province until 1945.

Kachar (kah-CHAHR), city, KOSTANAI region, KA-ZAKHSTAN, near RUDNYI, 33 mi/55 km WNW of KOSTANAI; 53°27′N 62°59′E. Gold mining; metallurgy. Agricultural products.

Kachar, INDIA: see CACHAR.

Kacha River (KAH-chah), approximately 40 mi/64 km long, S Republic of CRIMEA, UKRAINE; rises in CRIM-EAN MOUNTAINS WSW of ALUSHTA, flows W to BLACK SEA 7 mi/11.3 km N of SEVASTOPOL'.

Kachchh (kuh-CHUH), district (□ 17,626 sq mi/45,827.6 sq km), GUJARAT state, W central INDIA, on GULF OF KACHCHH (S); ⊙ BHUJ. Bounded SW by ARABIAN SEA, N by THAR DESERT and PAKISTAN. Mostly treeless and barren, with rocky hills (S), except for a fertile band along the Gulf of Kachchh, though its agricultural yield (chiefly wheat, barley, cotton) is small. MANDVI, Bhuj, and Kadla, a new port, are the chief towns. Grains, cotton and lentils are grown; limestone, marble, and gypsum are mined. Embroidering, handicraft work (cloth weaving, silver products). Fishing in SW. Formerly a princely state,

Kachchh was established in the fourteenth century by Rajputs and passed under British rule in 1815. Incorporated into Gujarat state in 1960. The RANN OF CUTCH is mainly in the N of the the district. Formerly spelled Cutch, Kutch; sometimes spelled Kachh or Kachchi.

Kachchh, Gulf of (kuh-CHUH) or **Cutch**, inlet of ARABIAN SEA on W coast of INDIA, between KACHCHH (N) and KATHIAWAR Peninsula (S), off GUJARAT state, near PAKISTAN border; 110 mi/177 km long, 10 mi/16 km—40 mi/64 km wide. Head and sides composed largely of mudflats. Ports at MANDVI, KANDLA, NAVLAKHI, BEDI, and OKHA. Has several islands (S), including Beyt.

Kachchhiahana, INDIA: see SAUSAR.

Kacheliba (kah-CHAI-lee-bah), town, RIFT VALLEY province, W KENYA, on right bank of Suam TURKWEL RIVER near UGANDA border, on road, 32 mi/51 km N of KITALE. Agriculture (sisal, coffee, wheat, corn).

Kachemak Bay, arm (40 mi/64 km long, 20 mi/32 km wide) of COOK INLET, S ALASKA, SW KENAI PENINSULA; 59°34′N 51°31′W. HALIBUT COVE, S; SELDOVIA village, S; HOMER village, N.

Kachergine, LITHUANIA: see KAČERGINĖ.

Kach Gandava, PAKISTAN: see KACHHI.

Kachh, INDIA: see CUTCH KACHCHH.

Kachhi (KUCH-ee), district, SIBI division, BALUCHISTAN province, SW PAKISTAN. A dry, low-lying plain, bounded W, N, and NE by CENTRAL BRAHUI RANGE; watered by seasonal mountain streams, including NARI and BOLAN rivers. Sometimes spelled CUTCHI; formerly called Kach (Cutch) GANDAVA.

Kachhwa (KUHCH-wah), town, MIRZAPUR district, SE UTTAR PRADESH state, INDIA, near the GANGA RIVER, 10 mi/16 km ENE of MIRZAPUR; 25°13′N 82°43′E. Rice, barley, gram, wheat, oilseeds.

Kachia (KAH-chah), town, KADUNA state, central NIGERIA, 60 mi/97 km SE of KADUNA; 09°52′N 07°57′E. Cotton, peanuts, ginger. Tin mining.

Kachin State (kah-CHEEN), state (□ 33,903 sq mi/88,147.8 sq km), extreme N MYANMAR; ⊙ MYITKYINA. Rice and sugarcane are grown, jade and amber mined, and timber and bamboo cut. It is a mountainous region, bounded on the NW by INDIA and on the N and E by CHINA, and traversed by tributaries of the AYEYARWADY RIVER. MYITKYINA and BHAMO are the chief towns. The state is sparsely populated; Jinghpaw-speaking Kachins constitute the largest group. They maintain the tribal forms of organization under chiefs, many practice shifting cultivation, and are mostly animists. The territory was never subject to the Burman kings, and after the establishment of British rule it was governed by the British directly, not as part of Burma. The territory was invaded by the Chinese between 1945 and 1947, but a border agreement was signed between BURMA and mainland CHINA in 1960. Antigovernment insurgents have been active in Kachin State since MYANMAR achieved independence in 1948. A cease-fire was signed between the Kachins and the government of Myanmar in February 1994.

Kachira (kah-CHEE-ruh), lake (□ 16 sq mi/41.6 sq km), WESTERN region, SW UGANDA. Elevation 4,048 ft/1,234 m, maximum depth 13 ft/4 m. One of five interconnected lakes that constitute the Koki lakes.

Kachiry (kah-chee-REE), city, N PAVLODAR region, KAZAKHSTAN, on IRTYSH River, 65 mi/105 km NNW of Pavlodar; 53°07′N 76°08′E. In agricultural area (wheat). Tertiary-level (raion) administrative center.

Kachise, town (2007 population 8,177), OROMIYA state, central ETHIOPIA, 70 mi/113 km NW of ADDIS ABABA; 09°38′N 37°51′E.

Kachkanar (kahch-kah-NAHR), city (2006 population 43,755), W SVERDLOVSK oblast, central URALS, extreme W Siberian Russia, in the OB' RIVER basin, just E of the administrative border with PERM oblast, terminus of a branch line of railroad, 182 mi/293 km N of YEKATERINBURG, and 62 mi/100 km N of NIZHNIY TAGIL;

58°42′N 59°29′E. Elevation 915 ft/278 m. Mining and concentrating of iron ore. Arose in 1958 with development of mines. Made city in 1968.

Kachreti (kahch-RE-tee), village, E GEORGIA, 45 mi/72 km E of TBILISI. Railroad junction for MIRZAANI oil fields.

Kachug (KAH-chook), town (2005 population 7,470), SE IRKUTSK oblast, S SIBERIA, Russia, port on the LENA RIVER (high-water head of navigation), on highway, 135 mi/217 km NNE of IRKUTSK; 53°57′N 105°53′E. Elevation 1,630 ft/496 m. Transit center for the Irkutsk–upper Lena River route; shipyard for river tankers, sawmilling, woodworking, dairy processing. Also known as Kachuga.

Kachuga, RUSSIA: see KACHUG.

Kachybey, UKRAINE: see ODESSA.

Kacina, CZECH REPUBLIC: see KUTNÁ HORA.

Kačkar Dagi (Turkish=Kaççkar Dağ), highest peak (12,917 ft/3,937 m) of RIZE MOUNTAINS, NE TURKEY, 38 mi/61 km ESE of RIZE.

Kaczawa (kah-CHAH-vah), river in SW POLAND; rises NE of JELENIA GÓRA (Hirschberg), flows N past ZLOTORYJA, and NE past LEGNICA (Liegnitz) and PROCHOWICE, to ODER River 3 mi/4.8 km NE of Prochowice. In battle of the Katzbach (August 1813), Blcher defeated French at Bremberg, 9 mi/14.5 km S of LIEGNITZ.

Kada-boaz Pass, Bulgaria and Serbia: see BELOGRADCHISHKI PASS.

Kadaiyam, INDIA: see KADAYAM.

Kadaiyanallur (kuh-DAH-yuh-nuhl-LOOR), town, Tirunelveli Kattabomman district, TAMIL NADU state, S INDIA, 9 mi/14.5 km NNE of TENKASI; 09°05′N 77°21′E. Major cotton-weaving center. Formerly spelled Kadayanallur.

Kadakoi, Bulgaria: see STRAZHITSA.

Kadamatt Island (kuh-DAH-maht), coral island of Amindivi group of Lakshadweep Islands, in LAKSHADWEEP Union Territory, INDIA, in ARABIAN SEA; 11°15′N 72°45′E. Centrally administered. Coconuts; coir manufacturing. Also spelled Kadmatt Island.

Kadam, Mount (KUH-dahm) (10,063 ft/3,067 m), NAKAPIRIPIRIT district, NORTHERN region, E UGANDA, N of MT. ELGON; 01°45′N 34°43′E. Also known as Dabesian or Tabasiak peak. Was in former Karamoga province.

Kadan, Czech Kadaň, German KAADEN, town (1991 population 17,796), NW BOHEMIA, CZECH REPUBLIC, on OHRE RIVER, on railroad, 8 mi/12.9 km SW of CHOMUTOV. Manufacturing (ceramics, clothing, machinery). Town has a 15th century town hall; Franciscan monastery with late-Gothic church just outside of town.

Kadarkút (KAH-dahr-kut), Hungarian Kadarkút, village, SOMOGY county, SW HUNGARY, 12 mi/19 km SW of KAPOSVÁR; 46°14′N 17°38′E. Wheat, corn, tomatoes; hogs; manufacturing (apparel).

Kadavu (kahn-DAH-voo), andesitic volcanic island (□ 159 sq mi/413.4 sq km), southwesternmost of FIJI islands and third-largest of Fijian group, SW PACIFIC OCEAN, 50 mi/80 km S of Viti Levu across Kadavu Passage; 35 mi/56 km long. Central mountain range, with dissected slopes, rises (SW) to Mt. Washington, Fijian Buke Levu (2,750 ft/838 m). Isthmus connects SW and NE parts of island Barrier reef, but there are some harbors, the best of which is Galoa Harbor, near S end of isthmus. Produces timber, copra, fruits. Sometimes spelled Kandavu.

Kadawatha, town, WESTERN PROVINCE, SRI LANKA; 07°10′N 79°57′E. Food processing. Trades in vegetables, rice. Residential and transport center.

Kadaya (kah-dah-YAH), village (2005 population 1,150), central CHITA oblast, S Siberian Russia, near the Nercha River, 38 mi/61 km NNW of NERCHINSK; 52°32′N 116°35′E. Elevation 2,404 ft/732 m. Seabuckthorn harvesting. Gold mines in the vicinity.

Kadayam (kuh-duh-YUHM), town, Tirunelveli Kattabomman district, TAMIL NADU state, S INDIA, 22 mi/

35 km WNW of TIRUNELVELI. Silk growing. Also spelled Kadaiyam.

Kadayanallur, INDIA: see KADAIYANALLUR.

Kadduwa, village, SOUTHERN PROVINCE, SRI LANKA; 06°02′N 80°31′E. Trades in coconuts, vegetables, and rice.

Kade (kah-DAI), town, EASTERN REGION, GHANA, 14 mi/23 km NE of ODA; 06°05′N 00°50′W. Railroad terminus; diamond-mining center. Located in densely populated central region.

Kadéï River (kah-DAI), main headstream of SANGA River, c.250 mi/402 km long, in E CAMEROON and W UBANGI-SHARI (Central Afr. Republic), rises on Cameroon–Central Afr. Republic border 15 mi/24 km NW of BABOUA. Flows in Cameroon S past BATOURI, and turns E and SE, crossing into Central Afr. Republic, joining another headstream at NOLA to form the Sanga.

Kadena (kah-DE-nah), town, S OKINAWA island, Nakagami county, OKINAWA prefecture, SW JAPAN, 12 mi/20 km N of NAHA; 26°21′N 127°45′E.

Kadesh (kah-DESH), ancient city of SYRIA, in the BEQA'A valley S of HOMS, on the ORONTES River; 34°35′N 36°31′E. Ramses II fought (c.1300 B.C.E.) the Hittites in a great battle here that ended in a truce. Tell Nebi Mend is believed to be the site of the ancient KADESH.

Kadesh (kah-DESH), oasis in the desert, E SINAI Peninsula, EGYPT, mentioned frequently in the Bible. Another biblical name is En-mishpat. Also known as Kadesh Barne'a.

Kadey or **Kadei**, department (2001 population 192,927), East province, CAMEROON; ⊙ BATOURI.

Kadhimain, Al (KED-hi-main, ahl), town, BAGHDAD province, central IRAQ, on W bank of the TIGRIS RIVER, on railroad, 5 mi/8 km NW of BAGHDAD. Dates, sesame, fruit; textile mill. Has notable mosques. Is now part of greater Baghdad. Sometimes spelled Al-Kazimayn.

Kadi (KUHD-ee), town, MAHESANA district, GUJARAT state, INDIA, 20 mi/32 km SSW of MAHESANA; 23°18′N 72°20′E. Trades in millet, cotton, cloth fabrics; cotton ginning and milling, handicraft cloth weaving, calico printing, oilseed milling; copper and brass products.

Kadiak, Alaska: see KODIAK.

Kadiam, INDIA: see DOWLAISWARAM.

Kadiitsa (kah-DEE-tsah), Bulgarian Ogreyak, highest peak (6,337 ft/1,932 m) of VLAHINA MOUNTAINS, on Macedonian-Bulgarian border, 8 mi/12.9 km NE of BEROVO (MACEDONIA); 41°47′N 22°58′E. Sometimes spelled Kadiyitsa; formerly called Kadijica.

Kadijica, Bulgaria: see KADIITSA.

Kadikoy (kah-DI-koi) (Turkish=Kadiköy), city, NW TURKEY in ASIA, at S entrance of the BOSPORUS on the Sea of MARMARA opposite ISTANBUL (of which it forms a part); 39°25′N 29°56′E. Railroad terminus and commercial center. On site of ancient CHALCEDON.

Kadima (kah-DEE-mah) or **Qadima**, township, W ISRAEL, in Plain of SHARON, 5 mi/8 km SE of NETANYA; 32°17′N 34°55′E. Elevation 173 ft/52 m. Mixed farming, citrus fruits. Founded 1933.

Kadina (kuh-DEI-nuh), town, S SOUTH AUSTRALIA state, S central AUSTRALIA, on W YORKE PENINSULA, 55 mi/89 km SSW of PORT PIRIE, near WALLAROO; 33°58′S 137°43′E. Railroad junction; wheat, barley; sheep, lambs; wool. Former copper-mining center. Manufacturing of farm machinery. Banking & Currency Museum.

Kadinhani, township, W central TURKEY, 29 mi/47 km NNW of KONYA; 38°15′N 32°14′E. Road junction. Wheat. Formerly called Saideli or Saiteli.

Kadiogo, province (□ 1,083 sq mi/2,815.8 sq km; 2005 population 1,291,966), CENTRE region, central BURKINA FASO; ⊙ OUAGADOUGOU; 12°20′N 01°30′W. Borders KOURWÉOGO (NWN), OUBRITENGA (N and E), GANZOURGOU (SE), BAZÉGA (S), and BOULKIEMDÉ (W) provinces. Intersection of major cross-country road network. Agriculture (groundnuts, cotton); manufacturing (cotton ginning).

Kadiolo (kah-JEE-o-lo), village, THIRD REGION/SIKASSO, MALI, 60 mi/100 km S of SIKASSO, near CÔTE D'IVOIRE border; 10°33'N 05°46'W.

Kadipur (KAH-dee-puhr), town, SULTANPUR district, E central UTTAR PRADESH state, INDIA, 20 mi/32 km ESE of SULTANPUR; 26°10'N 82°23'E. Rice, wheat, gram, barley, oilseeds.

Kadirabad, INDIA: see JALNA.

Kadiri (KUH-di-ree), town, ANANTAPUR district, ANDHRA PRADESH state, INDIA, 55 mi/89 km SE of ANANTAPUR; 14°07'N 78°10'E. Road center; grain trade; jaggery. Headquarters of regional malaria control. Sheep grazing in nearby forested hills (bamboo, satinwood).

Kadiri, INDONESIA: see KEDIRI.

Kadirli, township, S TURKEY, 50 mi/80 km NE of ADANA; 37°22'N 36°05'E. Grain. Formerly called Karszulkadrive.

Kadiyevka, UKRAINE: see STAKHANOV.

Kadiyitsa, Bulgaria: see KADIITSA.

Kadiyivka, UKRAINE: see STAKHANOV.

Kadjebi (kah-JAI-bee), town (2000 population 8,249), VOLTA region, GHANA, on route to NORTHERN REGION and upper Ghana, near TOGO border, 27 mi/43 km N of Hohoe; 07°07'N 00°19'E. Cacao market.

Kadmatt Island, INDIA: see KADAMATT ISLAND.

Kadnikov (KAHD-nee-kuhf), city (2006 population 5,295), central VOLOGDA oblast, N central European Russia, on road, 24 mi/39 km NNE of VOLOGDA; 59°30'N 40°20'E. Elevation 485 ft/147 m. In farming region. Flax processing. Chartered in 1780.

Kadogawa (kah-DO-gah-wah), town, Usuki county, MIYAZAKI prefecture, E KYUSHU, SW JAPAN, on HYUGA SEA, 40 mi/65 km N of MIYAZAKI; 32°27'N 131°39'E.

Kadoka, city (2006 population 677), ⊙ JACKSON county, SW central SOUTH DAKOTA, 70 mi/113 km SW of PIERRE, 25 mi/40 km ENE of BADLANDS NATIONAL PARK (park headquarters); at E end of Buffalo Gap National Grassland; 43°49'N 101°30'W. Petrified Gardens near town; Pine Ridge Indian Reservation to S; tourist center; agriculture (livestock).

Kadom (KAH-duhm), town (2006 population 5,975), E RYAZAN oblast, central European Russia, on the MOKSHA River, on road junction, 152 mi/245 km E of RYAZAN, and 27 mi/43 km NE of SASOVO; 54°33'N 42°28'E. Elevation 334 ft/101 m. Woodworking; starch and distilling works. An old Russian city, reduced to status of a village in 1926; made town in 1958.

Kadoma (kah-DO-mah), city, OSAKA prefecture, S HONSHU, W central JAPAN, on the Furu River, 5.6 mi/9 km N of OSAKA; 34°44'N 135°35'E. Industrial and residential suburb of Osaka. Headquarters of the Matsushita Electric Industrial Co., which produces electrical products.

Kadoma, city, MASHONALAND WEST province, central ZIMBABWE, 85 mi/137 km SW of HARARE, on railroad; 18°20'S 29°55'E. Elevation 3,814 ft/1,163 m. Airstrip. Gold mining at EIFFEL FLATS, 5 mi/8 km to ENE; Golden Valley Mine 12 mi/19 km to NW. Agriculture (tobacco, cotton, peanuts, citrus fruit, corn); cattle, sheep, goats. Manufacturing (textile machinery, cotton milling). Site of Cotton Research Institute. Formerly called Gatooma.

Kadoshkino (kah-DOSH-kee-nuh), town (2006 population 4,725), S central MORDVA REPUBLIC, central European Russia, on road and railroad, 20 mi/32 km W of RUZAYEVKA; 54°01'N 44°25'E. Elevation 761 ft/231 m. Electrical products manufacturing and repair; grain storage.

Kadugannawa (KAH-du-GAHN-nah-wuh), town, CENTRAL PROVINCE, SRI LANKA, on KANDY PLATEAU, 8 mi/12.9 km WSW of KANDY; 07°15'N 80°31'E. Tea, rice, vegetables. Ancient Buddhist temples nearby. A 125-ft/38-m tower is dedicated to British engineer

Dawson, who built the first paved road from Colombo to Kandy.

Kadugli (KAH-duhg-lee), township, ⊙ S KURDOFAN state, central SUDAN, in NUBA MOUNTAINS, on road, 70 mi/113 km S of DALANG; 11°01'N 29°43'E. Gum arabic; dura; livestock.

Kaduna (KAH-doo-NAH), state (□ 17,781 sq mi/46,230.6 sq km; 2006 population 6,066,562), N central NIGERIA; ⊙ KADUNA; 10°20'N 07°45'E. Bordered NW by ZAMFARA state, N by KATSINA state, NE by KANO state, E tip by BAUCHI state, SE by PLATEAU state, S by NASSARAWA state, SW by ABUJA Federal Capital Territory, and W by NIGER state. Kagoro Hills are here. Drained by KADUNA RIVER (Zaria Dam); Kaangimi Dam also is here. Agriculture includes groundnuts, cotton, tobacco, sugarcane, wheat, maize, millet, guinea corn, beans, yams, vegetables, and potatoes. Textile mills, motor vehicle assembly plants, and cigarette factories. Minerals include kaolin, columbite, sapphire, and gold. Main centers are KADUNA and ZARIA. Nok excavations are in this state, as are Matsirga waterfalls and the Nimbia and Afaka forest preserves.

Kaduna (KAH-doo-NAH), town (2004 population 1,560,000), ⊙ KADUNA state, N NIGERIA, on KADUNA RIVER; 10°31'N 07°26'E. A leading commercial and industrial center of N Nigeria, Kaduna has cotton textile, beverage, and furniture factories. It also is a railroad and road junction and the trade center for the surrounding agricultural area near Kaangimi Dam. A pipeline connects the city's oil refinery and petrochemical plant to oil fields in the NIGER River delta. Cotton, peanuts, sorghum, and ginger are shipped. The city was founded by the British in 1913 and became capital of Nigeria's Northern region in 1917. Training colleges for teachers, police, and the military, two universities (one military), and a technical institute are here.

Kaduna River (KAH-doo-NAH), around 325 mi/523 km long, N central NIGERIA; rises in KADUNA state, on BAUCHI PLATEAU SW of JOS, flows NW and SW, past KADUNA and into NIGER state, past ZUNGERU, to the NIGER River at MUREJI. Navigable (July–Oct.) below Zungeru for barges.

Kadur (kuh-DOOR), town, CHIKMAGALUR district, KARNATAKA state, S INDIA, 20 mi/32 km NE of CHIKMAGALUR; 13°34'N 76°01'E. Biri making; tobacco, cotton, millet.

Kadurupokuna, village, SOUTHERN PROVINCE, SRI LANKA; 06°01'N 80°46'E. Buddhist temple and pilgrim center. Known for wood carving. Trades in rice and coconuts.

Kaduy (KAH-dooyee), town (2006 population 11,495), SW VOLOGDA oblast, N central European Russia, on the Voron River near its confluence with the SUDA RIVER (VOLGA RIVER basin), on highway branch and railroad, 27 mi/43 km W of CHEREPOVETS; 59°12'N 37°09'E. Elevation 347 ft/105 m. Food processing, veneering.

Kaduy, RUSSIA: see KADUI.

Kadykchan (kah-dik-CHAHN), settlement (2006 population 790), NW MAGADAN oblast, E RUSSIAN FAR EAST, in the KOLYMA River basin, on road, 43 mi/69 km NW of SUSUMAN; 63°04'N 147°00'E. Elevation 2,664 ft/811 m. Formerly a tin-mining center, mostly abandoned after the mines' shutdown in the late 1990s.

Kadyy (kah-DI-yee), town (2005 population 3,750), S KOSTROMA oblast, central European Russia, on the Vottat River (VOLGA RIVER basin), on the Kostroma-Manturovo highway, 88 mi/142 km E of KOSTROMA, and 45 mi/72 km NE of KINESHMA; 57°47'N 43°11'E. Elevation 311 ft/94 m. Wood and food industries. Founded in 1673. Became city in 1778; reduced to status of a village in 1917; made town in 1971.

Kadzherom (kah-ji-ROM), settlement, N central KOMI REPUBLIC, NE European Russia, in the Kozhva River

(tributary of the IZHMA River) basin, on road and railroad, 42 mi/68 km SW of PECHORA; 64°41'N 55°55'E. Elevation 298 ft/90 m. Sawmilling, timbering.

Kadzhi-Sai, KYRGYZSTAN: see KAJI-SAI.

Kadzuno (KAHDZ-no), city, Akita prefecture, N HONSHU, NE JAPAN, 50 mi/80 km S of AKITA city; 40°12'N 140°47'E. Lumber; apples; cows; dyeing, soap manufacturing. Known for Towada wine and perila rolls. Habitat of the Akita dog. Towada-Hachimantai National Park (hot springs) is nearby.

Kaechon (GAI-CHUHN), county, SOUTH PYONGAN province, NORTH KOREA, 40 mi/64 km NNE of PYONGYANG. Mining area (coal, iron); graphite works.

Kaédi (kah-AI-dee), town (2000 population 34,227), ⊙ Gorgol administrative region, S MAURITANIA, on right bank of SÉNÉGAL RIVER at mouth of GORGOL RIVER, 195 mi/314 km E of SAINT-LOUIS (SENEGAL); 16°09'N 13°30'W. Produces gum, hides, skins, butter for export, subsistence crops (millet, potatoes); livestock. Airport.

Kaélé (kah-HAI-lai), town (2001 population 23,400), ⊙ Mayo-Kani department, Far-North province, CAMEROON, 45 mi/72 km SE of MAROUA, near CHAD border; 10°07'N 14°28'E. Native trade center; livestock raising; cotton oil.

Kaélé, CAMEROON: see MAYO-KANI DEPARTMENT.

Kaena Point (kah-AI-nah), NW tip of OAHU, HONOLULU county, HAWAII, 30 mi/48 km NW of HONOLULU; KAUAI CHANNEL to N. Kaena Military Reserve on N and SW shores of point.

Kaeng Krachan National Park (GANG GRUH-JAHN), Phetburi province, THAILAND. The largest park in THAILAND, it occupies nearly half the area of Phetburi province. Its wildlife includes elephants, tigers, bears, boars, deer, gaurs, and wild cattle.

Kaeo (KAI-oh), township, Far North district, NORTH ISLAND, NEW ZEALAND, 130 mi/209 km NNW of AUCKLAND, close to WHANGAROA Harbour. Center of fruit-growing area.

Kaersorsuaq, GREENLAND: see QAARSORSUAQ.

Kaersut, GREENLAND: see QAARSUT.

Kaesong (GAI-SUHNG), city (□ 505 sq mi/1,313 sq km), S NORTH KOREA; 38°02'N 126°33'E. It has had the status of a special city since the Korean War. A longtime commercial center, it is important chiefly for its exports of ginseng, a valuable medicinal root. There is also active trade in rice, barley, and wheat. Textiles are made in the city, and there is some heavy industry. In the 10th century Wang, founder of the Koryo dynasty, made Kaesong his capital; the city, then called Songdo, remained Korea's capital until 1392, when the Yi dynasty moved the capital to Seoul. Intersected by the 38th parallel, Kaesong served as the main contact point between North and SOUTH KOREA from 1945 to 1951 and passed from UN to North Korean forces several times during the Korean War. The armistice talks, first held at Kaesong, were later transferred to PANMUNJOM. Special economic zone established here in 2003 for South Korean industry. Historic landmarks include the tombs of several Korean kings, the old city walls, and the remains of a royal palace from the Koryo period.

Kaf (KAF), township and oasis, Qurayyat province, northwesternmost SAUDI ARABIA, in the WADI SIRHAN, 170 mi/274 km NW of JAUF, near JORDAN border, on the highway from the PERSIAN GULF to Jordan and to SYRIA; 31°25'N 37°22'E. Also spelled Qaf.

Kafa, ETHIOPIA: see KEFA.

Kaf, Al (KEF, el), province (□ 1,917 sq mi/4,984.2 sq km; 2006 population 257,900), NW TUNISIA; ⊙ AL KAF; 36°05'N 08°45'E. Borders on ALGERIA (to W). Also Kef, El Kef, Le Kef, and Al Kāf.

Kaf, Al (KEF, el), ancient *Sicca Veneria*, town (2004 population 45,191), ⊙ AL KAF province, NW TUNISIA, 104 mi/167 km SW of TUNIS; 36°11'N 08°43'E. Railroad-spur terminus and road junction. Agricultural market (grain, livestock, wool, camel hair); olive oil production and flour milling. Handicraft industries. Iron mines nearby. Carthiginian fortress town in 450 B.C.E. Roman colony (27 B.C.E.), destroyed by Arabs in 7th century. The French razed site to build modern town.

Kafanchan (kah-fang-SHANG), town, PLATEAU state, central NIGERIA, 45 mi/72 km SW of JOS; 09°35′N 08°18′E. Railroad junction; tin-mining center.

Kaffa, ETHIOPIA: see KEFA.

Kaffa, UKRAINE: see FEODOSIYA.

Kaffraria, former name for a region in the TRANSKEI, E SOUTH AFRICA. Founded in 1848 as the dependency of British KAFFRARIA, it was added to CAPE COLONY in 1865.

Kaffrine (kaf-REEN), town (2007 population 27,416), KAOLACK administrative region, W SENEGAL, on SALOUM RIVER, on Dakar-Niger railroad, 35 mi/56 km E of KAOLACK; 14°06′N 15°33′W. Peanut growing and cotton production.

Kafireos, Cape (kah-fee-RE-os), Latin *Caphereus*, SE extremity of ÉVVIA island, ÉVVIA prefecture, CENTRAL GREECE department, on AEGEAN SEA at NE entrance of KAFIREOS CHANNEL; 38°09′N 24°36′E. Also spelled Kaphereus; formerly called Cape d'Oro or Cape Doro.

Kafireos Channel (kah-fee-RE-os), in AEGEAN SEA, off E CENTRAL GREECE department, separates ÉVVIA (NW) and ÁNDROS islands (SE); 8 mi/12.9 km wide. Formerly called Doro Channel or Channel d'Oro. Also spelled Kaphereus.

Kafiristan, AFGHANISTAN: see NURISTAN.

Kafirnigan River (kah-FEER-nee-gahn), 220 mi/354 km long, SW TAJIKISTAN; rises in several branches in E GISSAR RANGE; flows WSW, past Ramit and ORDZ-HONIKIDZEABAD, and S, past SHAARTUZ, to the AMU DARYA. Rich cotton areas along middle and lower course. Receives (right) DYUSHAMBINKA RIVER, on which is DUSHANBE. Joined to the SURKHAN DARYA by Gissar Canal, at GISSAR.

Kafir Qala, AFGHANISTAN: see ISLAM QALA.

Kafr Bara (kah-FUHR BAH-rah), Arab village, NW ISRAEL, 4.3 mi/7 km NE of PETAH TIKVA on SHARON PLAIN; 32°08′N 34°58′E. Elevation 400 ft/121 m. Named for a Muslim saint, a-Nabi Bara, who is buried here. Roman ruins.

Kafr Behum (ka-FUHR be-HOOM), township, HAMA district, W SYRIA, on railroad, 5 mi/8 km SSW of HAMA; 35°03′N 36°42′E. Cotton, cereals. Also spelled Kafr BUHUM.

Kafr Biheida (ka-fuhr be-HAI-duh), village, DA-QAHLIYA province, Lower EGYPT, 5 mi/8 km NNE of MIT GHAMR. Cotton, cereals.

Kafr Buhum, SYRIA: see KAFR BEHUM.

Kafr Bulin (ka-fuhr boo-LEEN), village, BEHEIRA province, Lower EGYPT, 3 mi/4.8 km SE of KOM HAMADA; 30°44′N 30°44′E. Cotton. Has canal regulator.

Kafr Dan, large Arab village, JENIN district, WEST BANK, 3.1 mi/5 km NW of Jenin, next to the road that leads to HAIFA, on the NE slopes of the Samarian Highlands; 32°28′N 35°15′E. Elevation 525 ft/160 m. Cereal and olives.

Kafr Dawud (ka-fuhr dah-WOOD), village, BEHEIRA province, W Lower EGYPT. On RASHID branch of the NILE River, 22 mi/35 km SSE of KOM HAMADA; 30°28′N 30°49′E. Cotton.

Kafr ed Dawar (KAHF ed dahw-WAHR), township, BEHEIRA province, on railroad, Lower EGYPT, 15 mi/24 km ESE of ALEXANDRIA; 31°08′N 30°07′E. Cotton ginning, cigarette manufacturing; cotton, rice, cereals. Also spelled Kafr el Dauwar.

Kafr Ein, Arab village, Ramallah district, WEST BANK, 11.3 mi/18 km NW of RAMALLAH, on the W slopes of the Samarian Highlands; 32°02′N 35°07′E. Agriculture (olives, fruits, cereals).

Kafr el Battikh (KAHF el baht-TEEK), township, KAFR ESH SHEIKH province, Lower EGYPT, on railroad, near mouth of DUMYAT branch of the NILE, 5 mi/8 km W of Damietta; 31°24′N 31°44′E. Cotton. fruits.

Kafr esh Sheikh, province (2004 population 2,541,124), Lower EGYPT, bounded N by the MEDITERRANEAN SEA, NW by the RASHID branch of the NILE, and S by GHARBIYA province; ⊙ Kafr esh Sheikh; 31°15′N 30°50′E. Formerly a part of Gharbiya province. Rich agricultural area: cereals, rice, fruit, cotton, vegetables. Industries include cotton ginning, rice husking, textile milling. Served by railroad from CAIRO.

Kafr esh Sheikh (KAHF resh SHAIKH), town, ⊙ KAFR ESH SHEIKH province, Lower EGYPT, on railroad, 22 mi/35 km NNW of TANTA; 31°07′N 30°56′E. Cotton ginning, cigarette manufacturing; cotton, cereals, rice, fruits.

Kafr ez Zaiyat (KAHF rez zah-TAHT), agricultural and industrial town, GHARBIYA province, Lower EGYPT, on RASHID branch of the NILE, on CAIRO-ALEXANDRIA railroad, 11 mi/18 km WNW of TANTA; 30°49′N 30°49′E. Cotton ginning, wool spinning, cigarette manufacturing, important chemical industry (cottonseed oil, soap, phosphates, fatty acids). Agriculture (cotton, cereals, rice, fruits).

Kafr Jamal, Arab village, Tulkarm district, WEST BANK, 6.3 mi/10 km S of TULKARM, on the W slopes of the Samarian Highlands; 32°13′N 35°02′E. Agriculture (wheat, barley, vegetables, grapes, olives).

Kafr Kadum, large Arab village, Nablus district, WEST BANK, 4.4 mi/7 km NW of NABLUS, in the Samarian Highlands. Agriculture (olives, fruit, vegetables, wheat).

Kafr Kama (kah-FUHR KAH-mah), Circassian Muslim village, SW of TIBERIAS in LOWER GALILEE, ISRAEL; 32°43′N 35°26′E. Elevation 803 ft/244 m. Ruins of a church and monastery with mosaic floors. Founded in 1876 by Circassians fleeing Russian pursuit in the Caucasus.

Kafr Kanna, Arab township, N ISRAEL, 3 mi/4.8 km NNE of NAZARETH; 32°45′N 35°20′E. Believed to occupy the same site as the ancient town of CANA.

Kafr Kanna (kah-FUHR kahn-NAH), Arab township, 3.7 mi/6 km NE of NAZARETH in LOWER GALILEE, IS-RAEL; 32°45′N 35°20′E. Elevation 1,125 ft/342 m. Building materials; sewing shops. Population is 74% Muslim, the rest Christian. Ancient Kanna (also CANA) is referred to in 20th century B.C.E. and was one of the cities conquered by Tiglath-Pilesser. In Christian tradition, the place where Jesus turned water into wine (John 2:1-11). A Jewish settlement in biblical times and possibly until the 16th century; in the Middle Ages a station on the DAMASCUS-EGYPT route. There are two churches devoted to the water-wine miracle: a late-19th century Franciscan church, built on the foundations of a 6th-century church, which in turn had a mosaic floor with Aramaic writing (indicating it was a synagogue at the time) dating to the 4th century, and a Greek Orthodox church.

Kafr Kara' (kah-FUHR kah-RAH) or **Kafr Qara'**, Arab township NE of HADERA, ISRAEL; 32° 30′ N 35° 04′ E. Elevation 364 ft/110 m. Established by Beduin in early 18th century on remains of Roman/Byzantine settlement. Population is almost entirely Muslim.

Kafr Kassem (kah-FUHR kah-SEM) or **Kafr Qasem**, ISRAEL, Arab township, Israel, 5 mi/8 km NE of PETAH TIKVA, on the E edge of SHARON plain; 32°06′N 34°58′E. Elevation 498 ft/151 m. Site of ancient Samarian town mentioned in the Bible. The Arab village dates from 17th century. In the 1956 Sinai War, around fifty residents were killed by Israeli border patrol soldiers and their subsequent trial for the shootings led to the writing of a military code for the Israeli army. Current population is almost entirely Muslim.

Kafr Kila el Bab (ka-fuhr ke-lah el BAHB), village, GHARBIYA province, Lower EGYPT, 11 mi/18 km SE of TANTA; 30°41′N 31°09′E. Cotton.

Kafr Labad, Arab town, Tulkarm district, WEST BANK, 5 mi/8 km E of TULKARM, in the Samarian Highlands. Agriculture (olives, fruits, vegetables).

Kafr Maleq, large Arab village, Ramallah district, WEST BANK, 9.4 mi/15 km NE of RAMALLAH, in the Samarian

Highlands. Agriculture (olives, cereals, fruits). Believed to be the biblical city Beit Melek; 31°30′N 35°02′E. The tomb of A-Nebi Shma'il, a Muslim saint, located here.

Kafr Manda (kah-FUHR mahn-DAH), Arab township, NW of NAZARETH in LOWER GALILEE, ISRAEL; 32°48′N 35°15′E. Elevation 498 ft/151 m. Mixed farming, cereals, vegetables. Site of ancient Jewish settlement (biblical and later times) and burial place of several great rabbis.

Kafr Misr (kah-FUHR MIS-re), Arab village, NE of AFULA in LOWER GALILEE, ISRAEL; 32°38′N 35°25′E. Elevation 446 ft/135 m. Mixed farming. The name derives from the origin of the founders of the village—immigrants from Egypt (Misr) in the 19th century.

Kafr Ni'ma, large Arab village, Ramallah district, WEST BANK, 6.3 mi/10 km NW of RAMALLAH, in the Samarian Highlands; 31°56′N 35°06′E. Agriculture (olives, fruit, cereals, vegetables).

Kafr Qalil, Arab village, Nablus district, WEST BANK, 1.8 mi/3 km S of NABLUS, in the Samarian Highlands, at the S foot of MOUNT GERIZIM; 32°12′N 35°17′E. Agriculture (olives, fruit). Kafr Qalil was a Samaritan settlement until the 13th century.

Kafr Qud, Arab village, Jenin district, WEST BANK, 4 mi/ 6.5 km W of JENIN, in the N Samarian Highlands. Agriculture (wheat, barley, vegetables, fruit).

Kafr Saqr (ka-fuhr SAH-kur), village, SHARQIYA province, Lower EGYPT, on the BAHR MUWEIS, on railroad, 16 mi/26 km NNE of ZAGAZIG; 30°48′N 31°37′E. Cotton. Sometimes spelled Kafr Sakr.

Kafr Shibin (ka-fuhr shi-BEEN), village, QALYUBIYA province, Lower EGYPT, 18 mi/29 km N of CAIRO; 30°18′N 31°18′E. Cotton, flax, cereals, fruits.

Kafr Shukr (ka-fuhr SHU-kuhr), village, DAQAHLIYA province, Lower EGYPT, on DUMYAT branch of the NILE, 11 mi/18 km S of MIT GHAMR. Cotton, cereals.

Kafr Thulth, Arab village, Tulkarm district, WEST BANK, 5 mi/8 km SE of QALQILYA, in the Samarian Highlands; 32°08′N 35°02′E. Agriculture (olives, fruits).

Kafr Yasif (kah-FUHR YAH-sif), Arab township, 5 mi/ 8 km NE of AKKO (Acre) in W GALILEE, ISRAEL; 32°57′N 35°09′E. Elevation 134 ft/40 m. Site of Jewish settlement (possibly named for Josephus Flavius, Jewish historian and military leader during the anti-Roman revolt) from ancient times to 1840. Now populated by Christians and Muslims, with some Druze. Ruins include Jewish graves from biblical and later times and a well with a mosaic floor.

Kafr Zibad, Arab village, Tulkarm district, WEST BANK, 6.3 mi/10 km SE of TULKARM, in the Samarian Highlands. Agriculture (olives, cereals, grapes).

Kafubu (kah-FOO-boo), village, KATANGA province, SE CONGO, on railroad, 8 mi/12.9 km SE of LUBUMBASHI; 11°45′S 27°34′E. Elev. 3,923 ft/1,195 m. Agricultural center; large citrus-fruit plantations, vineyards, and cattle-raising farms. Has Roman Catholic mission, small seminary, and school of agr. Formerly SALESI-ENNE; also known as LA KAFUBU.

Kafue (kah-foo-ai), township, LUSAKA PROVINCE, S central ZAMBIA, 25 mi/40 km SSW of LUSAKA, on KAFUE RIVER (boundary of SOUTHERN province); 15°46′S 28°11′E. On railroad. Kafue River bridged here; Kafue Dam and Gorge to E (downstream); hydroelectric power station. Kafue Flats swamp to W, Chirundu Fossil Forest National Monument to SE. Agriculture (cotton, sugarcane, vegetables, corn, soybeans, flowers); poultry; cattle; fish. Manufacturing (chemicals, cloth, fishing nets).

Kafue (kah-foo-ai), river, c.600 mi/966 km long, ZAMBIA; rises in NE part of NORTH-WESTERN province, at the CONGO border, 30 mi/48 km W of LU-BUMBASHI, Congo; flows SE through COPPERBELT region, past CHINGOLA and KITWE, then SW through W edge of Lukanga Swamp and KAFUE NATIONAL

PARK (Itezhi-tezhi Dam at E border of park forms 143-sq-mi/370-sq-km lake); turns E, through Kafue Flats Swamp, past Kafue township and through Kafue Dam and Gorge; enters ZAMBEZI RIVER at CHIRUNDU 50 mi/80 km SE of LUSAKA. It provides water to Zambia's Copperbelt region. Agriculture in lower Kafue valley is sugarcane and cotton. Hydroelectric power plant at Kafue Gorge. Main tributaries are Lushwishi, Lunga, and Lufupa rivers, which enter upper Kafue from N.

Kafue National Park (kah-foo-ai) (□ 8,650 sq mi/22,490 sq km), NORTH-WESTERN, CENTRAL, and SOUTHERN provinces, ZAMBIA, 165 mi/266 km W of LUSAKA. Established 1950, it is a haven for the animal and bird life of a diverse region that includes grasslands ("dumbos"), forests ("miombos"), and marshes. Drained by KAFUE RIVER and its tributaries, the Lunga and Lufupa rivers. Busanga Plain in N is KAFUE's main wildlife area. Flora and fauna include elephant, buffalo, leopard, lion, antelope, cheetah, wildebeest, zebra, sable, hippo; crocodile, monitor lizards; fig trees, phoenix palms. Ngoma Lodge at SE edge of park.

Kafulafuta (kah-foo-lah-FOO-tah), township, COPPERBELT province, N central ZAMBIA, near CONGO border, 19 mi/31 km SSE of NDOLA. On railroad. Kafulafuta Mission is 28 mi/45 km W, on Kafulafuta River (enters KAFUE RIVER from E). Agriculture (grains, soybeans, vegetables, cotton, tobacco, flowers); cattle.

Kafura (kah-FOO-rah), village, KIGOMA region, NW TANZANIA, 12 mi/19 km S of KIBONDO; 03°45′S 30°41′E. Timber; goats, sheep; corn, wheat.

Kafu River (KAH-foo), c.150 mi/241 km long, W UGANDA; rises in marshes W of LAKE KYOGA, flows WSW to S end of LAKE ALBERT. Called Nkusi River in lower course.

Kafyr-Kumukh (kah-FIR–koo-MOOKH), village (2005 population 4,375), central DAGESTAN REPUBLIC, in the NE foothills of the Greater CAUCASUS Mountains, near road and railroad, 21 mi/34 km SW of MAKHACHKALA, and 2 mi/3.2 km E of BUYNAKSK, to which it is administratively subordinate; 42°50′N 47°09′E. Elevation 1,381 ft/420 m. In oil- and gas-producing region.

Kaga (KAH-gah), city, ISHIKAWA prefecture, central HONSHU, central JAPAN, 25 mi/40 km S of KANAZAWA; 36°17′N 136°19′E. Pears. Hot-spring resort.

Kaga Bandoro, village, ⊙ NANA-GRÉBIZI prefecture, central CENTRAL AFRICAN REPUBLIC, on GRIBINGUI RIVER, 90 mi/145 km W of SIBUT; 07°00′N 19°10′E. Cotton ginning. Chrome deposits nearby. Formerly called Fort-Crampel.

Kagal (KAH-guhl), town, KOLHAPUR district, MAHARASHTRA state, central INDIA, 10 mi/16 km SE of KOLHAPUR; 16°35′N 74°19′E. Market center for rice, sugarcane, tobacco, mangoes; handicraft cloth weaving. Was part of former princely state of Kolhapur.

Kagalaska Island (ka-guh-LAS-kah) (9 mi/14.5 km long, 7 mi/11.3 km wide), ANDREANOF ISLANDS, ALEUTIAN ISLANDS, SW ALASKA, just E of ADAK.

Kagal'nik (kah-GAHL-neek), village (2006 population 7,940), SW ROSTOV oblast, S European Russia, on the Taganrog Gulf of the Sea of AZOV, at the mouths of the KAGAL'NIK RIVER and the S arm of the DON River delta, on local highway branch, 23 mi/37 km W of ROSTOV-NA-DONU, and 5 mi/8 km SW of AZOV; 47°04′N 39°19′E. Fisheries.

Kagal'nik, RUSSIA: see KAGAL'NITSKAYA.

Kagal'nik River (kuh-GAHL-neek), approximately 70 mi/113 km long, S ROSTOV oblast, SE European Russia; rises N of MECHETINSKAYA; flows WNW, past KAGALNITSKAYA and SAMARSKOYE, to the Sea of AZOV, here joining the S arm of the DON River delta mouth at KAGAL'NIK.

Kagal'nitskaya (kah-gahl-NEETS-kah-yah), village (2006 population 6,835), S ROSTOV oblast, S European Russia, on the KAGAL'NIK RIVER, on road junction

and railroad, 33 mi/53 km SE of ROSTOV; 46°53′N 40°09′E. Elevation 124 ft/37 m. Flour milling; dairy plant. Formerly known as Kagal'nik.

Kagami (KAH-gah-mee), town, Kami county, KOCHI prefecture, S SHIKOKU, W JAPAN, 12 mi/20 km E of KOCHI; 33°33′N 133°44′E.

Kagami, town, Yatsushiro county, KUMAMOTO prefecture, W KYUSHU, SW JAPAN, on YATSUSHIRO BAY, 16 mi/25 km S of KUMAMOTO; 32°33′N 130°39′E. Rush.

Kagami (KAH-gah-mee), village, Tosa county, KOCHI prefecture, S SHIKOKU, W JAPAN, 4.3 mi/7 km N of KOCHI; 33°35′N 133°28′E.

Kagami-ishi (kah-gah-MEE–ee-shee), town, Iwase county, FUKUSHIMA prefecture, N central HONSHU, NE JAPAN, 34 mi/55 km S of FUKUSHIMA city; 37°14′N 140°20′E.

Kagamino (kah-GAH-mee-no), town, Tomata county, OKAYAMA prefecture, SW HONSHU, W JAPAN, 31 mi/50 km N of OKAYAMA; 35°05′N 133°56′E. Pickles; Japanese plums.

Kagan (kah-GAHN), city, S BUKHARA wiloyat, UZBEKISTAN, on TRANS-CASPIAN RAILROAD (junction for TERMEZ and DUSHANBE, TAJIKISTAN), 8 mi/12.9 km SE of BUKHARA; 39°43′N 64°33′E. Railroad center (repair shops); tobacco and wine processing, cotton ginning, cottonseed-oil extraction. Developed in late 19th century around Bukhara station; residence of Russian political agent to khan of Bukhara. Until c.1935, called Novaya Bukhara [Russian=new Bukhara].

Kaganovich, RUSSIA: see NOVOKASHIRSK.

Kaganovich, RUSSIA: see TOVARKOVSKIY.

Kaganovicha, Imeni, RUSSIA: see CHERNYSHEVSK.

Kaganovicha, Imeni L. M., UKRAINE: see POPASNA.

Kaganovichesk, town, E LEBAP weloyat, TURKMENISTAN, suburb 5 mi/8 km SSE of CHARJEW; 39°02′N 63°36′E. Cotton, orchards; metalworks. Site of ancient Amul, later known as Chardzhui. Following rise of new city of Chardzhui, it was called Stary Chardzhui [Russian=old Chardzhui] until 1937.

Kaganovichi Pervyye, UKRAINE: see POLIS'KE.

Kagarko (kah-GAHR-ko), town, KADUNA state, central NIGERIA; 70 mi/113 km SSE of KADUNA; 09°29′N 07°41′E. Tin, columbite, wolfram mining. Sometimes spelled Kagerko.

Kagarlyk, UKRAINE: see KAHARLYK.

Kagawa (kah-GAH-wah), prefecture (□ 724 sq mi/1,882.4 sq km), N SHIKOKU, W JAPAN, on HARIMA SEA (NE) and HIUCHI SEA (NW) sections of INLAND SEA; ⊙ TAKAMATSU. Bordered S by TOKUSHIMA prefecture, SW by EHIME prefecture. Includes TONOSHO and TOSHIMA islands. Agricultural region (rice, barley, wheat, mandarin oranges) with a mountainous and forested interior. The coast has fishing ports and salt-producing centers. Lacquerwork, gloves, and fans are also produced.

Kagawa (KAH-gah-wah), town, Kagawa county, KAGAWA prefecture, NW SHIKOKU, W JAPAN, 6 mi/10 km S of TAKAMATSU; 34°14′N 134°02′E.

Kagawong (KA-guh-wahng), unincorporated village, S central ONTARIO, E central CANADA, on N Manitoulin Island, on NORTH CHANNEL of Lake HURON, 17 mi/27 km WSW of LITTLE CURRENT, and included in BILLINGS township; 45°54′N 82°15′W. Fishing, lumbering.

Kagayan, PHILIPPINES: see CAGAYAN SULU ISLAND.

Kåge (KO-ge), town, VÄSTERBOTTEN county, N SWEDEN, on small inlet of GULF OF BOTHNIA, 5 mi/8 km N of SKELLEFTEÅ; 64°50′N 21°00′E.

Kagera (kah-GE-rah), region (2006 population 2,210,000), NW TANZANIA; ⊙ BUKOBA. Bounded W by BURUNDI and RWANDA, N by UGANDA, E by Lake VICTORIA; lakes IKIMBA and BURIGI in center, Lake Bisongu in W. Burigi Game Reserve in S center. Coffee, grain, tobacco; sheep, goats; fish. Tin mining in NW corner. Formerly West Lake (Ziwa Magharibi) province.

Kagera National Park, RWANDA: see AKAGERA NATIONAL PARK.

Kagera River (kah-GE-rah), c.275 mi/443 km long, in BURUNDI, RWANDA, TANZANIA, and UGANDA; formed on the Rwanda-Tanzania border, E central Africa, by the confluence of the NYABARONGO and Ruvuvu rivers. The Kagera's headwaters, which rise in the highlands of Rwanda and Burundi, are the remotest sources of the NILE. The Kagera flows N and E, forming part of Tanzania's borders with Rwanda and Uganda, before emptying into Lake VICTORIA. There is a small hydroelectric power plant at KIKAGATI (Uganda). Called Akagera River in Rwanda, where it forms the E boundary of AKAGERA NATIONAL PARK.

Kagerko, NIGERIA: see KAGARKO.

Kågeröd (KO-ge-RUHD), village, SKÅNE county, S SWEDEN, 16 mi/26 km ESE of HELSINGBORG; 55°60′N 13°05′E.

Kagitumba (kah-gee-TOOM-bah), village, NE RWANDA, on KAGERA RIVER, on UGANDA and TANZANIA borders, 65 mi/105 km NE of KIGALI. Customs station; coffee growing.

Kagizman (Turkish=Kağizman), town (2000 population 21,685), NE TURKEY, on S bank of ARAS RIVER, 32 mi/51 km S of KARS; 40°08′N 43°07′E. Arsenic, some gold.

Kagmari, BANGLADESH: see TANGAIL, town.

Kagna River (KUHG-nuh), c.150 mi/220 km long, S INDIA; rises in Balaghat Hills in ANDHRA PRADESH state, flows SSW, past TANDUR in RANGAREDDI district to BHIMA RIVER 7 mi/11.3 km S of SHAHABAD in GULBARGA district, KARNATAKA state.

Kagoro (KAH-go-ro), town, KADUNA state, central NIGERIA, on railroad, 5 mi/8 km E of KAFANCHAN; 09°36′N 08°23′E. Tin-mining center.

Kagoshima (kah-GO-shee-mah), prefecture [Japanese ken] (□ 3,515 sq mi/9,104 sq km; 1990 population 1,797,766), S KYUSHU, SW JAPAN; ⊙ KAGOSHIMA, its chief port. Mainland portion bounded W by EAST CHINA SEA, S by KAGOSHIMA BAY, OSUMI STRAIT, and ARIAKE BAY. Bordered N by KUMAMOTO prefecture, E by MIYAZAKI prefecture. Includes TANEGA-SHIMA, YAKU-SHIMA, NAGA-SHIMA, and SHISHI-JIMA of AMAKUSA ISLANDS, and KOSHIKI-RETTO. SATSUMA and OSUMI peninsulas comprise major part of prefecture. Mountainous; hot springs in Kirishima National Park. No large streams. Gold and silver mines in S, iron and copper in W. Subtropical vegetation, with gum, camphor, and litchi trees. Principal agricultural products are sweet potatoes, tobacco, soybeans; livestock breeding; sawmilling. Outlying islands produce sugarcane, sweet potatoes, fish. Kagoshima is chief manufacturing center.

Kagoshima (kah-GO-shee-mah), city (2005 population 604,367), ⊙ KAGOSHIMA prefecture, extreme S KYUSHU, SW JAPAN, on SATSUMA PENINSULA and KAGOSHIMA BAY; 31°35′N 130°33′E. Tinware, cryptomeria items, bamboo work, lacquerware, pottery; traditional toys. Tea processing, distilling of shochu (an alcoholic drink), bamboo-shoot canning. Agriculture includes rape seeds, daikon, mandarin oranges; pickles. Known for Satsuma cuisine, including local confections. Seat of two universities. Emperor Antoku mausoleum located here. Historically important as the castle town of the Shimazu family and as the birthplace of Takamori Saigo, Toshimichi Okubo, and Heihachiro Togo. The center of the Satsuma Rebellion, the city was destroyed in 1877. In 1914 it suffered damage from the eruption of Ontake volcano on SAKURAJIMA, an island in the bay now connected to the city. Bombed (1945) in World War II. St. Francis Xavier landed here in 1549. Hot springs nearby.

Kagoshima Bay (kah-GO-shee-mah), Japanese Kagoshima-wan (kah-GO-shee-mah–WAHN), inlet of EAST CHINA SEA, S KYUSHU, SW JAPAN, in KAGOSHIMA prefecture, between SATSUMA (W) and OSUMI (E) peninsulas; c.45 mi/72 km long, 13 mi/21 km wide. Sakurajima (NW projection of Osumi Peninsula) is opposite KAGOSHIMA on W shore of bay. Also called Kinko Bay.

Kagul, MOLDOVA: see CAHUL.

Kagul Lagoon, UKRAINE: see KAHUL LAGOON.

Kaguyak (ka-GEI-yak), fishing village, S ALASKA, S KODIAK ISLAND, 80 mi/129 km SW of KODIAK; 56°52′N 153°46′W.

Kaguyak, fishing village, S ALASKA, on SHELIKOF STRAIT, NE ALASKA PENINSULA, in KATMAI National Monument, 80 mi/129 km NW of KODIAK. Formerly spelled Kayuyak or Kayayak.

Kahaluu (KAH-hah-LOO-oo), town, W HAWAII island, HAWAII county, HAWAII, on Kahaluu Bay, KONA (W) Coast, 55 mi/89 km W of HILO, 4 mi/6.4 km S of KAILUA-KONA; 21°27′N 157°49′W. Tourism. White Sands (Disappearing Sands) Beach Park here; Farmer's Market.

Kahaluu, town (2000 population 2,935), OAHU island, HONOLULU county, HAWAII, on E coast, 10 mi/16 km N of HONOLULU, Kamehameha Highway. Laenani Beach Park here. Waiahole Forest Reserve and KOO-LAU RANGE to W; Kahaluu Beach Park to S; Kahaluu Forest Reserve to SE.

Kahama (ka-HA-mah), town, SHINYANGA region, NW central TANZANIA, 80 mi/129 km N of TABORA; Siga Hills to N; 03°49′S 32°33′E. Timber; cotton, subsistence crops.

Kaharlyk (kah-hahr-LIK) (Russian *Kagarlyk*), city, S KIEV oblast, UKRAINE, 40 mi/64 km SSE of KIEV; 49°51′N 30°50′E. Elevation 505 ft/153 m. Raion center; sugar-refining center; elevator; flour and feed mill, dairy, asphalt. Has a 19th-century park. Known since 1142, city since 1971. Jewish community since 1847, reaching its peak (1,414) in 1897; reduced during the pogroms of 1919 and completely eliminated during World War II.

Kahawa (kah-HA-wah), town, S central KENYA, on railroad, 10 mi/16 km NNE of NAIROBI; 01°09′S 36°55′E. Coffee center; sisal, wheat, corn. Military barracks. Kenyatta University.

Kahawatta (KUH-huh-wath-thuh), town, SABAR-AGAMUWA PROVINCE, SW central SRI LANKA, 15 mi/24 km ESE of RATNAPURA; 06°15′N 80°14′E. Rubber, tea, rice, vegetables; gem mining. Also spelled Kahawatte.

Kahe (KAH-hai), village, KILIMANJARO region, N TANZANIA, 12 mi/19 km SE of MOSHI; 03°30′S 37°26′E. Railroad junction. Sugarcane, sisal, corn, wheat, coffee, tea; cattle. Nyumba ya Mungu reservoir, in PAN-GANI RIVER, to S.

Kahemba (kah-HEM-bah), village, BANDUNDU province, SW CONGO, 140 mi/225 km S of KIKWIT, near ANGOLA border; 07°17′S 19°00′E. Elev. 3,225 ft/982 m. Customs station, trade center; copal.

Kahili (KAH-HEE-lee), peak (3,089 ft/942 m), NW KAUAI, HAWAII.

Kahla (KAH-LAH), town, THURINGIA, central GER-MANY, on the SAALE, 9 mi/14.5 km S of JENA; 50°49′N 11°35′E. Glass- and china-manufacturing center. Towered over by 12th-century Leuchtenburg castle. Has 15th-century church and remains of medieval town walls.

Kahl am Main (KAHL ahm MEIN), village, LOWER FRANCONIA, BAVARIA, central GERMANY, on small Kahl River near its junction with the Main, 9 mi/14.5 km NW of ASCHAFFENBURG; 50°04′N 09°00′E. Manufacturing of metal products, electrical goods, machinery, construction materials.

Kahle Asten (KAH-le AHS-ten), highest peak (2,759 ft/841 m) of the ROTHAARGEBIRGE, W GERMANY, 2 mi/3.2 km W of WINTERBERG; 51°11′N 08°29′E. LENNE RIVER rises here.

Kahlotus (kah-LO-tuhs), village (2006 population 230), FRANKLIN county, SE WASHINGTON, 35 mi/56 km NE of PASCO; 46°38′N 118°33′W. Railroad junction. In COLUMBIA basin agricultural region; wheat, vegetables, alfalfa, beans, potatoes. Palousa Falls and Lyons Ferry state parks to E; Lake Kahlotus reservoir to E.

Lower Monumental Dam, 6 mi/9.7 km S, forms Lake Herbery on SNAKE RIVER. Juniper Dunes Wilderness Area.

Kahmard (KAH-mah-erd), village, BAMIAN province, E AFGHANISTAN, at N foot of the Hindu Kush, 105 mi/169 km NW of KABUL, on Kamard River (tributary of the Surkhab); 35°20′N 67°30′E. Sulphur deposits in valley, chiefly at DASHT-I-SAFED (15 mi/24 km E). Sometimes spelled Kamard.

Kahnuj (kah-NOOJ), town, Kermān province, SE IRAN, 105 mi/169 km NE of BANDAR.

Kaho (KAH-ho), town, Kaho county, FUKUOKA prefecture, N KYUSHU, SW JAPAN, 19 mi/30 km E of FUKUOKA; 33°32′N 130°44′E.

Kahoka (kuh-HO-kuh), city (2000 population 2,241), ⊙ CLARK county, extreme NE MISSOURI, 48 mi/77 km ENE of KIRKSVILLE; 40°25′N 91°43′W. Corn, soybeans. hogs; manufacturing (candles). Laid out 1858. Battle of Athems State Park to N.

Kahoku (KAH-ho-koo), town, Kami county, KOCHI prefecture, S SHIKOKU, W JAPAN, 16 mi/25 km N of KOCHI; 33°38′N 133°46′E. Lumber.

Kahoku, town, Kamoto county, KUMAMOTO prefecture, W KYUSHU, SW JAPAN, 22 mi/35 km N of KUMAMOTO; 33°06′N 130°41′E.

Kahoku (kah-HO-koo), town, Mono county, MIYAGI prefecture, N HONSHU, NE JAPAN, 28 mi/45 km N of SENDAI; 30°38′N 141°19′E.

Kahoku (KAH-ho-koo), town, West Murayama county, YAMAGATA prefecture, N HONSHU, NE JAPAN, 12 mi/20 km N of YAMAGATA city; 38°25′N 140°19′E. Cherries, cinnamon vines, safflowers. Footwear manufacturing, dyeing.

Kahoolawe (kah-HO-o-LAH-vai), uninhabited island (□ 45 sq mi/117 sq km), MAUI county, central HAWAII, 110 mi/177 km SE of HONOLULU; separated from MAUI island to the NE by Alalakeiki Channel and from LANAI island to NW by KEALAIKAHIKI CHANNEL. The island, in the rain shadow of Haleakala, is low and unfertile; has served as a penal colony and as a U.S. military bombing range; now has been returned to the state for restoration. Kanapou Bay on E end; Kamohio Bay on S coast. Highest point is Lua Makika (1,477 ft/450 m) in E.

Kahramanmaraş (kah-rah-mahn-MAH-rahsh), city (2000 population 326,198), S TURKEY; 37°34′N 36°54′E. The city lies on a fertile plain at the foot of the TAURUS MOUNTAINS. A center for light industry and trade; chiefly spices, olive oil, and handicrafts. The area surrounding the city is noted for agriculture (cotton) and mining (especially silver and iron). The capital of a Hittite kingdom in the 12th century B.C.E., the city was later conquered by the Romans, and in Roman times was called Germanikeia. In the 12th century, the city came under the control of the Seljuq Turks. Kahramanmaraş has several medieval mosques. Large Shiite population.

Kahramanmaras, city (2000 population 326,198), S central TURKEY, S of the TAURUS MOUNTAINS. It is an agricultural trade center and a transportation hub. Ancient inscriptions found there indicate that Kahramanmarus was a Hittite city-state c.1000 B.C.E. The city was called Germanikeia in Roman times. It was captured by the Arabs in C.E. 638 and was annexed by the Ottoman Empire in the early 16th century

Kahrizak (kah-REE-zahk), village, Tehrān province, N IRAN, 10 mi/16 km S of TEHRAN and on road to QOM. Grain, cotton, sugar beets, other produce; sugar refinery.

Kahror (kah-hah-ro-RAH), town, MULTAN district, S PUNJAB province, central PAKISTAN, 45 mi/72 km SE of MULTAN; 29°37′N 71°55′E. Also called Kahror Pakka.

Kahta (Turkish=*Kâhta*), township, E central TURKEY, on W shore of ATATÜRK DAM and reservoir, 45 mi/72 km SSE of MALATYA; 37°48′N 38°35′E. Barley, wheat, chickpeas, lentils, tobacco. Also called Koluk.

Kahuku (kah-HOO-koo), village (2000 population 2,097), OAHU, HONOLULU county, HAWAII, about 1 mi/1.6 km off N coast, 26 mi/42 km NNW of HONOLULU; 21°40′N 157°56′W. Comsat Earth Station to W; Makahoa Point to E; KAHUKU POINT to NW.

Kahuku Point, the N tip of OAHU, HONOLULU county, HAWAII; 21°42′N 157°59′W.

Kahul Lagoon (kah-HOOL) (Romanian *Cahul*) (Russian *Kagul*), SW ODESSA oblast, UKRAINE, near DANUBE River, SE of RENI; 15 mi/24 km long, 6 mi/9.7 km wide (in S); 19 ft/6 m at its deepest. Receives minor Kahul River (N).

Kahului (KAH-hoo-LOO-ee), city (2000 population 20,146), N MAUI, MAUI county, HAWAII, 92 mi/148 km ESE of HONOLULU, on Kahului Harbor, principal Maui port; 20°52′N 156°27′W. Twin city to the smaller WAILUKU, to W; manufacturing (meat packing, canned fruit, printing, service industry machines, lumber). Has Maui Community College; fairgrounds. The second largest city (after HILO), in Hawaii, outside of OAHU.

Kahurangi National Park (KAR-or-ANG-ee), extreme NW of SOUTH ISLAND, NEW ZEALAND, in rugged, forested TASMAN MOUNTAIN region. The second largest national park (after Fiordland) in New Zealand, with lowland and alpine karst, including the three largest cave systems in New Zealand Vegetation ranging from wild shoreline to montane forest and grassland and rare birds, including great spotted kiwi. Diversity of giant snails. Portion of Heaphy Track.

Kahusi Range (kah-HOO-see), group of dormant volcanoes in E CONGO, extending c.60 mi/97 km parallel to W shore of LAKE KIVU. Rises to 10,738 ft/3,273 m; bamboo forests. Was first explored and mapped 1930–1932.

Kai, former province in central HONSHU, JAPAN; now part of YAMANASHI prefecture.

Kaiafa, Lake, Greece: see ZAKHARO.

Kaiama (kah-YAH-mah), town, Kaiama Local Government Area, KWARA state, W NIGERIA, 60 mi/97 km NW of JEBBA. Gold-mining center; shea-nut processing, cotton weaving; cassava, millet, durra; cattle, skins.

Kaiapha, Lake, Greece: see ZAKHARO.

Kaiapit (KEI-a-pit), PAPUA NEW GUINEA; see MUTSING.

Kaiapoi, town, CANTERBURY region, E SOUTH ISLAND, NEW ZEALAND, 10 mi/16 km N of CHRISTCHURCH, near mouth of WAIMAKARIRI RIVER. Mixed crop and livestock farming, dairying, vegetables, with processing; wool, fish.

Kaibab Plateau, high tableland largely in COCONINO county, N ARIZONA, extends N from GRAND CANYON, past MARBLE GORGE, into S UTAH. Elevation 7,500 ft/2,286 m–9,300 ft/2,835 m. Includes part of Kaibab National Forest and N part of Grand Canyon National Park; Kaibab Creek to W.

Kaibara (KAH-ee-bah-rah), town, Hikami district, HYOGO prefecture, S HONSHU, W central JAPAN, 31 mi/50 km N of KOBE; 35°07′N 135°04′E.

Kaiba-to, RUSSIA: see MONERON ISLAND.

Kaida (KAH-ee-dah), town, Aki county, HIROSHIMA prefecture, SW HONSHU, W JAPAN, 5 mi/8 km E of HIROSHIMA; 34°22′N 132°32′E.

Kaida (KAH-ee-dah), village, Kiso county, NAGANO prefecture, central HONSHU, central JAPAN, 59 mi/95 km S of NAGANO; 35°56′N 137°36′E. Pickles; spring water from nearby Mount Ontake.

Kaidanovo, BELARUS: see DZERZHINSK.

Kaieiewaho Channel, HAWAII: see KAUAI CHANNEL.

Kaien Island (KAIN), W BRITISH COLUMBIA, W CA-NADA, in CHATHAM SOUND, off Tsimpshian Peninsula, near mouth of SKEENA River; 6 mi/10 km long, 4 mi/6 km wide; 54°17′N 130°18′W. On NW coast is PRINCE RUPERT city. Island is linked with mainland (S) by railroad and road bridges.

Kaieteur Falls (kei-CHOOR), waterfall in POTARO RIVER, POTARO-SIPARUNI district, W GUYANA;

05°10′N 59°28′W. Elev. 741 ft/226 m. Plunging over an escarpment of the Guyana Highlands, it is a major tourist attraction in Guyana and one of the most impressive falls in SOUTH AMERICA. It is included in Kaieteur National Park.

Kaifeng (KEI-FENG), city (□ 140 sq mi/362 sq km; 1994 estimated urban population 535,300; 1994 estimated total population 724,200; 2000 total population 693,148), NE HENAN province, NE CHINA, on the Longhai railroad; 34°51′N 114°21′E. It is a commercial, agricultural, and industrial center. Manufacturing includes agricultural machinery, zinc, textiles, fertilizer, chemicals, and processed foods. The HUANG HE (Yellow River), just N, has frequently flooded the city. Kaifeng has often been a major center of Chinese political and cultural life. Founded in the 3rd century B.C.E., it was, as Bianliang, capital of Five Dynasties (906–959) and then capital of the Northern Song dynasty (960–1127). Zoroastrians worshipped here, and a Jewish colony was established in the 12th century. The city fell to the Mongols in the 13th century. Kaifeng was the provincial capital until superseded (1954) by ZHENGZHOU. The city is well known for its numerous historic sites, including an iron tower built during the Song dynasty.

Kaifu (KAH-eef), town, Kaifu county, TOKUSHIMA prefecture, SE SHIKOKU, W JAPAN, 34 mi/55 km S of TOKUSHIMA; 33°35′N 134°21′E.

Kaihua (KEI-HWAH), town, ⊙ Kaihua county, SW ZHEJIANG province, SE CHINA, near JIANGXI-ANHUI border, 32 mi/51 km WNW of JIANGSHAN; 29°09′N 118°24′E. Rice, wheat, oilseeds; logging, engineering, food and beverages.

Kaihyo-to, RUSSIA: see TYULENIY ISLAND.

Kaiingveld, SOUTH AFRICA: see GROBLERSHOOP.

Kai Islands (KEI), island group (c.572 sq mi/1,481 sq km), S MALUKU, E INDONESIA, SE of SERAM, in the BANDA SEA; 05°35′S 132°45′E. It is densely forested with valuable timber; the people are skilled boat builders. The chief island is NUHU CUT (or Great Kai) and KAI KECIL (Little Kai). The chief port and community is TUAL. The group is sometimes called the Kei Islands.

Kaijiang (KEI-JI-ANG), town, ⊙ Kaijiang county, E SICHUAN province, SW CHINA, 27 mi/43 km ESE of DACHUAN; 31°02′N 107°53′E. Rice, tobacco, oilseeds, jute, medicinal herbs; food and beverages, engineering, textiles, coal mining.

Kaikalur (KEI-kah-loor), town, KRISHNA district, ANDHRA PRADESH state, SE INDIA, in KRISHNA RIVER delta, 13 mi/21 km SSE of ELURU; 16°34′N 81°12′E. Rice, peanuts, tobacco.

Kai Kecil (KEI) or **Little Kai**, island, in the KAI ISLANDS, S MALUKU, INDONESIA, in BANDA SEA, just W of NUHU CUT, 175 mi/282 km SE of SERAM; 05°37′S 132°44′E. Island is 30 mi/48 km long, 10 mi/16 km wide. Produces coconuts, trepang.

Kaikohe, town, ⊙ Far North District (□ 2,897 sq mi/ 7,532.2 sq km), located between BAY OF ISLANDS and HOKIANGA Harbor, near Lake Omapere and thermal Ngawha Springs. Agricultural service center, with grassland and forestry research.

Kaikoura (kei-KOO-ruh), township, KAIKOURA district (□ 790 sq mi/2,054 sq km), E SOUTH ISLAND, NEW ZEALAND, 65 mi/105 km SSW of BLENHEIM, on small limestone peninsula. Livestock raising, fishing. Tourism (caverns, whale watching). Marine Research Center (University of Canterbury) is here.

Kaikoura Ranges (kei-KOO-ruh), Two fault-bounded mountain chains (Inland Kaikouras and Seaward Kaikouras), E SOUTH ISLAND, NEW ZEALAND, extending c. 62 mi/100 km parallel with E coast, flanking lower course of CLARENCE RIVER. Highest peak is TAPUAE-NUKU (9,469 ft/2,886 m); grassy and forested slopes.

Kailahun (kei-lah-HOON), town, Kailahun district, EASTERN province, E SIERRA LEONE, near MOA RIVER (GUINEA border), 85 mi/137 km ENE of BO; 08°17′N

10°34′W. Road and trade center; palm oil and kernels, cacao, rice, coffee. Has hospital.

Kailali (KEI-lah-lee), district, SW NEPAL, in SETI zone; ⊙ DHANGADHI.

Kailali (KEI-lah-lee), village (2001 population 33,260), SW NEPAL, in the TERAI, on small Kateni River, 16 mi/ 26 km NE of PALIA (INDIA), 2 mi/3.2 km NE of railroad spur terminus at CHANDAN CHAUKI (India); 28°34′N 80°47′E. Rice, jute, oilseeds, sabai grass, vegetables.

Kailas (KEI-LUS), peak (c.22,280 ft/6,791 m), SW TIBET, SW CHINA, highest point of the KAILAS RANGE, in the HIMALAYAS; near the sources of the SUTLEJ, INDUS, and BRAHMAPUTRA rivers. Hindus consider Kailas to be the dwelling place of the god Shiva, and it is the goal of pilgrimages. The pilgrim road that girdles the mountain reaches 18,000 ft/5,486 m. Tibetan Buddhists identify Kailas with Mount Sumeru, which they consider to be the cosmic center of the universe. Also Kangrinboqê.

Kailasagarh, INDIA: see JAVADI HILLS.

Kailashahar (KEI-lahsh-uh-huhr), town, ⊙ NORTH TRIPURA district, TRIPURA state, NE INDIA, on tributary of the KUSIYARA RIVER, 48 mi/77 km NE of AGARTALA; 24°20′N 92°01′E. Trades in rice, cotton, tea, mustard, jute. Religious center. Also spelled Kailasahar.

Kailas-Karakoram Range (kah-lahs–KAHR-kor-uhm), S lateral range of KARAKORAM mountain system, NORTHERN AREAS, extreme NE PAKISTAN, in N KASHMIR region; from HUNZA RIVER bend extends c.200 mi/322 km SE to main bend of SHYOK RIVER; parallels main KAILAS RANGE. Consists of four groups (NW-SE): RAKAPOSHI, HARAMOSH, MASHERBRUM, and SALTORO ranges. MASHERBRUM (25,660 ft/7,821 m) is highest peak; average elevation c.18,000 ft/5,486 m–19,700 ft/6,005 m. CHANG CHENMO and PANGONG ranges are SE extensions.

Kailas Range (KEI-LUS), mountain range of the SW Trans-Himalayas, S central TIBET, SW CHINA; from upper INDUS RIVER; c.20 mi/32 km wide; extends c.360 mi/579 km ESE, 31°00′N 80°00′E–31°00′N 85°00′E. Highest peaks are KAILAS (22,280 ft/6,791 m) in W and LOMBO KANGRA (23,165 ft/7,061 m) in E. Gives rise to Indus River (near Tibetan village of SHICHUANGHE) and to headstreams of the BRAHMAPUTRA RIVER.

Kaili (KEI-LEE), city (□ 504 sq mi/1,305 sq km; 1994 estimated urban population 128,300; 1994 estimated total population 395,300; 2000 total population 382,026), SE GUIZHOU province, S central CHINA; 26°35′N 107°55′E. Agriculture and heavy industry are the largest sources of income. Crop growing, animal husbandry. Grain, hogs; manufacturing (textiles, synthetic fibers, electronics).

Kailo (KEI-lo), village, MANIEMA province, E CONGO, 135 mi/217 km NNW of KASONGO; 02°38′S 26°06′E. Elev. 1,978 ft/602 m. Center of tin-mining area; tin concentrating; also wolfram mining.

Kailu (KEI-LOO), town, ⊙ Kailu county, SE INNER MONGOLIA AUTONOMOUS REGION, N CHINA, 50 mi/80 km W of TONGLIAO; 43°35′N 121°12′E. Grain, sugar beets, oilseeds; livestock; manufacturing (food and beverages).

Kailua (kei-LOO-ah), unincorporated city (2000 population 36,513), HONOLULU county, HAWAII, 10 mi/16 km NE of HONOLULU on the SE coast of OAHU, on KAILUA BAY; 21°23′N 157°44′W. Bellows Air Force Base (inactive) to SE; Kaneohe Bay Marine Corps Base to N on Mokapu Peninsula. Kailua Beach County Park is here, to E; Ulu Po Heiau State Monument (temple) is here.

Kailua Bay, W HAWAII island, HAWAII county, HAWAII. Historically, chief landing in KONA district for cattle shipment. The first missionaries in Hawaii landed here in 1820. Lighthouse at Kukailimoku Point. City of KAILUA-KONA on shore. Mooring for pleasure cruise ships.

Kailua Bay, E OAHU island, HONOLULU county, HAWAII, near MOKAPU POINT, 9 mi/14.5 km NE of HONOLULU. City of KAILUA on shore; state seabird sanctuary, coral reef, beach.

Kailua-Kona (kei-LOO-ah–KO-nah), town, W HAWAII island, HAWAII county, HAWAII, on KAILUA BAY, KONA (W) Coast, 56 mi/90 km W of HILO; 19°39′N 155°57′W. The second largest town on the Big Island (Hawaii). Tourism. Coffee; fish, prawns. Manufacturing (concrete products, printing and publishing, brewing). Officially named Kailua, Kona is added to distinguish it from other Hawaiian towns of same name. Hulihee Palace, University of the Nations here. Old Kona Airport State Park to NW. HONOKOHAU NATIONAL HISTORIC PARK to N.

Kailuan (KEI-LU-AN), coal-mining district, NE HEBEI province, NE CHINA, extending along Beijing-Shenyang railroad, from KAIPING to LUAN XIAN, for which it is named. Includes coal-mining centers of Tangshan, Kaiping, and Luan Xian. Sometimes spelled Kailwan.

Kailwan, China: see KAILUAN.

Kaimai Tunnel, NE NORTH ISLAND, NEW ZEALAND, W of TAURANGA. Completed 1978 through the Kaimai Range separating the WAIKATO and HAURAKI basins from the TAURANGA (Mount Manganui) harbor. The longest tunnel in New Zealand.

Kaimakcalan, Greece and Macedonia: see KAJMAKCALAN.

Kaimakli (kai-mahk-LEE), NE suburb of NICOSIA, LEFKOSIA district, N central CYPRUS, 1 mi/1.6 km NNE of city center; 35°10′N 33°23′E. Comprises former suburbs of KAIMAKLI BEUYUK and KAIMAKLI KUTCHUK. Attila Line marking Turkish zone divides community N-S in W; Turkish sector called OMORPHITA.

Kaimakli Beuyuk, CYPRUS: see KAIMAKLI.

Kaimakli Kutchuk, CYPRUS: see KAIMAKLI.

Kaimanawa Mountains (KEI-man-a-WAH), central NORTH ISLAND, NEW ZEALAND, forested portion of Main Range, extending about 37 mi/60 km along the watershed between LAKE TAUPO and HAWKE BAY; rising to 5,666 ft/1,727 m. The hard-rock ranges reach WELLINGTON FAULT on E, W slope sinks to volcanic plateau near TAUPO. Includes Kaimanawa State Forest Park.

Kaimganj (KEIM-guhnj), town, FARRUKHABAD district, central UTTAR PRADESH state, N central INDIA, 18 mi/29 km NW of FARRUKHABAD; 27°34′N 79°21′E. Wheat, gram, corn, tobacco. Founded 1713 by first nawab of Farrukhabad.

Kaimon (KAH-ee-mon), town, Ibusuki county, KAGOSHIMA prefecture, SW KYUSHU, SW JAPAN, 28 mi/ 45 km S of KAGOSHIMA; 31°12′N 130°32′E. Site of Tama-no, Japan's oldest well.

Kaimpur, PAKISTAN: see QAIMPUR.

Kaimur, INDIA: see BHABHUA.

Kaimur Hills (KEI-muhr), NE branch of VINDHYA RANGE, central INDIA; extend c.300 mi/483 km ENE from S JABALPUR district in MADHYA PRADESH state, across SHAHDOL and SIDHI districts in Madhya Pradesh and MIRZAPUR district of UTTAR PRADESH state, to just E of SASARAM (ROHTAS district, BIHAR state). Range is c.10 mi/16 km–35 mi/56 km wide; maximum elevation c.2,000 ft/610 m. Drained largely by SON RIVER.

Kain (KAN), village, Tournai district, HAINAUT province, SW BELGIUM, 2 mi/3.2 km N of TOURNAI, near SCHELDT RIVER; 50°38′N 03°22′E. Liquor distilling; agriculture. Site of former Cistercian nunnery.

Kain, IRAN: see QAEN.

Kainan (KAH-ee-nahn), city, WAKAYAMA prefecture, S HONSHU, W central JAPAN, on the Kii Sound, 7 mi/11 km S of WAKAYAMA; 34°09′N 135°12′E. Port, railroad junction, and industrial center; manufacturing (lacquerware, steel pipes, household goods, ropes, and nets).

Cross-references are shown in SMALL CAPITALS. The pronunciation guide is shown on page xix. The sources of population figures are shown on page xvii.

Kainan (KAH-ee-nahn), town, Kaifu county, TO-KUSHIMA prefecture, SE KYUSHU, W JAPAN, 34 mi/55 km S of TOKUSHIMA; 33°35′N 134°21′E.

Kainan Bay, indentation of the front of the ROSS ICE SHELF, ANTARCTICA, N of ROOSEVELT ISLAND; 78°07′S 162°30′W.

Kainantu (kei-NAN-tyoo), village (2000 population 6,723), EASTERN HIGHLANDS province, NE NEW GUINEA island, N central PAPUA NEW GUINEA, 85 mi/137 km NW of LAE. Road junction. Coffee, tea; cattle; timber.

Kainar (kei-NAHR), village, NW SEMEY region, KA-ZAKHSTAN, 165 mi/266 km WSW of SEMEY (Semi-palatinsk); 49°15′N 77°25′E. Cattle. Also spelled Kaynar.

Kainardzha (kei-yee-nahrd-ZHAH), Romanian *Cai-nargeana-Mica*, village, RUSE oblast, ⊙ Kainardzha obshtina (1993 population 6,140), NE BULGARIA, in the S DOBRUDZHANSKO plateau, 15 mi/24 km SE of SILISTRA; 43°59′N 27°29′E. Grain, sheep. A peace treaty was concluded here (1774), by which Turkey ceded to Russia the CRIMEA and S UKRAINE. Called Kuchuk Kainarji or Kutchuk Kainarji under Turkey. In Romania 1913–1940. Also spelled Kaynardzha.

Kainary, MOLDOVA: see CAINARI.

Kainda, KYRGYZSTAN: see KAYYNGDY.

Kaindorf (KEIN-dorf), village, E STYRIA, SE AUSTRIA, 25 mi/40 km NE of GRAZ; 47°14′N 15°55′E. Fruits.

Kaingaroa Forest, between LAKE TAUPO and BAY OF PLENTY, NEW ZEALAND. A level or undulating plain or plateau with underlying ignimbrite and a pumice surface, brought into production in the 1920s when soil seemed to preclude farming; therefore was transformed from scrub to Monterey pine. Often viewed as the world's most extensive man-made forest, containing planned timber milling towns, including Kawerau and Murupara.

Kainji Lake, reservoir, W NIGERIA; formed by Kainji Dam about 1 mi/1.6 km E of NEW BUSSA; part of the Niger Dams project designed to control floodwaters of the NIGER River. The dam is 200 ft/61 m high at a point where the river is confined to hills on both sides; provides hydroelectricity and water storage. Fishing on lake and agricultural development on floodplain S of dam.

Kainsk, RUSSIA: see KUYBYSHEV, NOVOSIBIRSK oblast.

Kaintira, INDIA: see ATHMALLIK.

Kaipara Harbour, N NORTH ISLAND, NEW ZEALAND, 25 mi/40 km NNW of AUCKLAND. Shallow, branching inlet originating in geological submergence of river system, 40 mi/65 km N-S, 5 mi/8 km E-W; connected with TASMAN SEA by unstable sandy passage 5 mi/8 km across. Receives WAIROA RIVER in N. Early kauri timber; long replaced by agricultural townships near harbor. Dairy and sheep products

Kaiparowits Plateau (kuh-PAHR-o-witz), S UTAH, mainly in KANE county, extends N into GARFIELD county. Elevation c.7,000 ft/2,134 m–8,000 ft/2438 m. Sparsely settled tableland W of COLORADO RIVER, between ESCALANTE (to NE) and PARIA (SW) rivers, SE end reaches Colorado River (L. Powell); bordered on NE by the Straight Cliffs. Sheep and cattle grazing.

Kaiping (KEI-PING), town, ⊙ Kaiping county, S GUANGDONG province, S CHINA, 21 mi/34 km NW of TAISHAN; 22°23′N 112°41′E. Rice, wheat, cotton, sugarcane, fruits; polyester fiber, knitwear. River port.

Kaiping, town, NE HEBEI province, NE CHINA, on railroad, 5 mi/8 km NE of TANGSHAN; 39°40′N 118°12′E. One of the major coal-mining centers of KAILUAN district. Spur leads to Matiakou mine, 3 mi/4.8 km N.

Kaipong Islands (KEI-PUNG), Mandarin *Jia-Peng* (JAH-PENG), group in SOUTH CHINA SEA, GUANG-DONG province, S CHINA, off PEARL RIVER estuary, extending 10 mi/16 km SW-NE at 21°50′N 114°00′E. Largest island (NE) is Pak Tsim (Mandarin *Pei Jian*), 3 mi/4.8 km long.

Kairakkum (kei-rahk-KOOM), city, LENINOBOD viloyat, N TAJIKISTAN, 9 mi/15 km E of KHUDJAND, adjacent to Kairakkum reservoir, W edge of FERGANA VALLEY; 40°16′N 69°49′W. Also spelled Kayrakkum.

Kairana (kei-RAH-nuh), town, MUZAFFARNAGAR district, N UTTAR PRADESH state, INDIA, 30 mi/48 km WSW of MUZAFFARNAGAR; 29°24′N 77°12′E. Trades in wheat, gram, sugarcane, oilseeds. Sugar processing, ornamental curtain manufacturing.

Kairiru (kei-ree-ROO), volcanic island, EAST SEPIK province, NW PAPUA NEW GUINEA, BISMARCK SEA, SW PACIFIC OCEAN, 20 mi/32 km NNW of WEWAK; MUSHU lies between KAIRIRU and NEW GUINEA. Island is circa 9 mi/14.5 km long; rises to 3,350 ft/1,021 m. Coconuts, fish, palm oil. Has active volcanoes, hot springs, and waterfalls.

Kairouan, TUNISIA: see QAYRAWAN, AL.

Kairuku, PAPUA NEW GUINEA: see YULE ISLAND.

Kaisarganj (KEI-suhr-guhnj), town, BAHRAICH district, N UTTAR PRADESH state, N central INDIA, 22 mi/35 km S of BAHRAICH; 27°15′N 81°33′E. Rice, wheat, corn, gram.

Kaisariani (ke-sah-ryah-NEE), ESE suburb of ATHENS, ATTICA prefecture, ATTICA department, E central GREECE, 2 mi/3.2 km from city center; 37°58′N 23°48′E. Also spelled Kaisariane, Kessariani.

Kaise (KAH-ee-SE), town, Ashigarakami county, KA-NAGAWA prefecture, E central HONSHU, E central JAPAN, 31 mi/50 km S of YOKOHAMA; 35°19′N 139°07′E.

Kaiseraugst, SWITZERLAND: see AUGST.

Kaiseregg, peak (7,169 ft/2,185 m) in BERNESE ALPS, on border of FRIBOURG and BERN cantons, W central SWITZERLAND, 7 mi/11.3 km NNE of ZWEISIMMEN.

Kaisergebirge (KEI-ser-ge-beer-ge), mountain group of TYROLEAN-BAVARIAN LIMESTONE ALPS, TYROL, W AUSTRIA, E of KUFSTEIN; subsidiary ranges are WILDER KAISER and Zahmer Kaiser; 47°36′N 12°20′E. Rises to 7,144 ft/2,177 m in the ELLMAUER HALT. Eldorado of mountain climbers. Summer and winter tourism.

Kaiser Peak (10,320 ft/3,146 m), FRESNO county, E central CALIFORNIA, in Kaiser Ridge, in the SIERRA NEVADA, 50 mi/80 km NE of FRESNO. Kaiser Pass (2,797 ft/853 m) to E.

Kaiserslautern (kei-sers-LOU-tuhern), city, RHINE-LAND-PALATINATE, W GERMANY, on the LAUTER RIVER; 49°27′N 07°47′E. Commercial, industrial, and cultural center, and a center for banking and railroad shipment. There are textile mills and machine and automobile factories. Charlemagne built a castle here that was later enlarged (1153–1158) by Emperor Frederick I (Barbarossa); some ruins of the castle remain today. The city was repeatedly devastated by warring armies, notably by the Spanish (1635) in the Thirty Years War. During the French Revolutionary Wars the Prussians defeated (1793) the French here. Has a noted early Gothic collegiate church (13th–14th century) and an art gallery. Formerly the seat of part of the University of Trier and Kaiserslautern (founded 1970), the Kaiserslautern campus became independent in 1975. Major U.S. and NATO military base here.

Kaiserstuhl (KEI-ser-shtool), volcanic massif on right bank of the RHINE, BADEN-WÜRTTEMBERG, SW GER-MANY, 8 mi/12.9 km NW of FREIBURG; 48°05′N 07°40′E. Rises to 1,826 ft/557 m in the TOTENKOPF. One of Germany's warmest regions. Noted for wine and fruit.

Kaiserswerth (KEI-sers-vert), district (since 1929) of DÜSSELDORF, W GERMANY, on right bank of the RHINE (landing), 6 mi/9.7 km N of city center; 51°13′N 06°47′E. Has early Romanesque church. Site of noted school for deaconesses, founded 1836.

Kaiser Wilhelm Canal, GERMANY: see KIEL CANAL.

Kaiser-Wilhelmsland: see PAPUA NEW GUINEA.

Kaishantun (KEI-SHAN-TUN), town, SE JILIN province, NE CHINA, on railroad, 20 mi/32 km SE of YANJI, on Tumen River (North Korea border). Chemical-fiber manufacturing, papermaking, engineering.

Kaisheim (KEIS-heim), village, SWABIA, BAVARIA, S GERMANY, 3 mi/4.8 km NNE of DONAUWÖRTH; 48°46′N 10°48′E. Former Cistercian monastery (founded c.1135), with late-Gothic church.

Kaišiadorys (KEISH-yuh-do-REES), Russian *Koshed-ary*, city (2001 population 10,002), S central LITHUA-NIA, 20 mi/32 km E of KAUNAS; 54°52′N 24°27′E. Railroad junction; manufacturing (shoes, lubricating oils, bone meal, bricks, flour). Hydroelectric power plant (1979). Has museum. In Russian VILNA government until 1920. Also spelled Kayshyadoris and Kaishyadorys.

Kaita (KAH-ee-TAH), town, Kaho county, FUKUOKA prefecture, N KYUSHU, SW JAPAN, 22 mi/35 km N of FUKUOKA; 33°40′N 130°43′E.

Kaitaia (kei-TEI-uh), township (2001 population 5,151), Far North District (□ 14,139 sq mi/36,761.4 sq km; 2001 population 5,151), N NORTH ISLAND, NEW ZEALAND, 145 mi/233 km NW of AUCKLAND. Agricultural center.

Kai Tak: see HONG KONG.

Kaitangata (kei-tan-GAT-uh), township, SE SOUTH ISLAND, NEW ZEALAND, on CLUTHA RIVER, 45 mi/72 km SW of DUNEDIN. Lignite mines.

Kaithal (KEI-tuhl), district (□ 1,081 sq mi/2,810.6 sq km), HARYANA state, N INDIA; ⊙ KAITHAL.

Kaithal (KEI-tuhl), town, ⊙ KAITHAL district, HAR-YANA state, N INDIA; 29°48′N 76°23′E. Market center for wheat, millet, cotton, rice; cotton ginning, salt-peter refining; metalwork, pottery, lacquered wood-work. Large sacred bathing tank, or *ghat*.

Kaiti (KA-ee-tee), village, N TANZANIA, 60 mi/97 km WSW of ARUSHA, at S end of Lake MANYARA; 03°47′S 35°51′E. Cattle, goats, sheep; corn, wheat. Phosphate mining.

Kaitong, China: see TONGYU.

Kaiwi Channel (kah-EE-vee), PACIFIC OCEAN, between MOLOKAI and OAHU islands, HAWAII, 26.5 mi/42.6 km wide.

Kai Xian (KEI SI-AN), town, ⊙ Kai Xian county, E SICHUAN province, SW CHINA, 25 mi/40 km N of WANXIAN; 31°12′N 108°25′E. Rice, oilseeds, tobacco, jute, medicinal herbs; food processing, coal mining, building materials.

Kaiyang (KEI-YANG), town, ⊙ Kaiyang county, central GUIZHOU province, S central CHINA, 35 mi/56 km NNE of GUIYANG; 27°04′N 106°58′E. Rice, tobacco, oilseeds; food processing, chemical-ore mining.

Kaiyuan (KEI-YOO-AN), city (□ 1,222 sq mi/3,165 sq km; 1994 estimated urban population 129,800; 1994 estimated total population 599,600; 2000 total population 587,530), NE LIAONING province, NE CHINA, 40 mi/64 km SSW of SIPING, on railroad; 42°39′N 124°04′E. Agriculture is the largest source of income. Main crops are soybean and grain; main industries include food and machinery. Has old walled city dating from the 13th century.

Kaiyuan, city (□ 753 sq mi/1,950 sq km; 1994 estimated urban population 100,600; 1994 estimated total population 249,900; 2000 total population 248,303), SE YUNNAN province, SW CHINA, on railroad, 95 mi/153 km SE of KUNMING, on highway leading E to GUIZHOU province; 23°42′N 103°14′E. Elevation 3,445 ft/1,050 m. Heavy industry is the largest sector within the city's economy; agriculture and commerce are also impor-tant. Crop growing, animal husbandry. Grain, hogs; manufacturing (food processing, utilities, chemicals, coal mining).

Kaizu (KAH-eedz), town, Kaizu county, GIFU prefec-ture, central HONSHU, central JAPAN, 12 mi/20 km S of GIFU; 35°13′N 136°38′E.

Kaizuka (KAH-EEZ-kah), city, OSAKA prefecture, S HONSHU, W central JAPAN, on OSAKA BAY, 19 mi/30 km S of OSAKA; 34°26′N 135°21′E. Commercial port and industrial center; manufacturing (fibers and textiles, combs, rope). Agriculture includes veg-etables, taro; mandarin oranges, strawberries; butter-burs.

Area is shown by the symbol □, and capital city or county seat by ⊙.

Kajaani (KAH-yah-nee), Swedish *Kajana*, city, OULUN province, central FINLAND; 64°14′N 27°41′E. On Kajaaninjoki (river). Manufacturing includes paper goods, woodworks, and sports equipment. Hydroelectric power station, transportation center. Kajaani was chartered in 1651; the Kajaneborg (Kajaaninlinna) fortress, around which the city grew, was taken by the Russians in 1716. Restored in 1937, the fortress is now a tourist attraction.

Kajabbi, settlement, W QUEENSLAND, NE AUSTRALIA, 73 mi/118 km NE of MOUNT ISA, 62 mi/100 km NW of CLONCURRY, on LEICHHARDT RIVER banks; 20°03′S 140°02′E. Lake Julius 16 mi/25 km W.

Kajakai (kah-JAH-kei), village, HELMAND province, S central AFGHANISTAN, on Helmand River, 45 mi/72 km NE of GIRISHK; 32°16′N 65°03′E. Storage dam built here (c.1950) regulates seasonal level changes of Helmand River Hydroelectric power plant and is Afghanistan's largest hydroelectric dam.

Kajang (KAH-jahng), city, SE SELANGOR, MALAYSIA, on railroad, 12 mi/19 km SSE of KUALA LUMPUR, on LANGAT RIVER. Rubber, rice; declining tin mining; cattle, hogs; manufacturing (handicrafts). Growing suburban settlement on the edge of Kuala Lumpur conurbation.

Kajani (kah-HAH-nee), canton, AROMA province, LA PAZ department, W BOLIVIA. Elevation 12,851 ft/3,917 m. Gas deposits in area. Lead-bearing lode; clay, limestone, and gypsum deposits. Agriculture (bananas, yucca, potatoes, rye); cattle.

Kajansi (kah-JAHN-see), town, WAKISO district, CENTRAL region, S UGANDA, 6 mi/10 km S of KAMPALA; 00°12′N 32°33′E. Trade center in a coffee- and tea-growing area.

K'ajaran, town, S ARMENIA, in ZANGEZUR RANGE (8,000 ft/2,438 m), 15 mi/24 km WSW of Kafan (railroad spur terminus); 39°09′N 46°07′E. Molybdenum- and copper-mining center; concentrating plant; building-materials combine. Developed after World War II.

Kajiado (kah-jee-AH-do), town (1999 population 9,128), ⊙ Kajiado district, RIFT VALLEY province, S KENYA, on railroad (at road crossing), 35 mi/56 km ENE of MAGADI; 01°51′S 36°49′E. Livestock raising; peanuts, corn. Marble quarrying nearby. Hospital, schools.

Kajikawa (kah-JEE-kah-wah), village, North Kanbara county, NIIGATA prefecture, central HONSHU, N central JAPAN, 19 mi/30 km N of NIIGATA; 37°58′N 139°21′E. Rice.

Kajikazawa (kah-jee-kah-ZAH-wah), town, South Koma county, YAMANASHI prefecture, central HONSHU, central JAPAN, 9 mi/15 km S of KOFU; 35°32′N 138°27′E. Trout.

Kajiki (KAH-jee-kee), town, Aira county, KAGOSHIMA prefecture, S KYUSHU, SW JAPAN, on N shore of KAGOSHIMA BAY, 12 mi/20 km N of KAGOSHIMA; 31°44′N 130°39′E. Dumplings; pottery.

Kajima (KAH-jee-mah), town, Kashima county, ISHIKAWA prefecture, HONSHU, central JAPAN, 31 mi/50 km N of KANAZAWA; 36°57′N 136°55′E. Textiles. Noto Hanto park is nearby.

Kaji-Say (kuh-jee-SEI), town, SW ISSYK-KOL region, KYRGYZSTAN, near S shore of ISSYK-KOL lake, on small stream of Kaji-Say, 70 mi/113 km WSW of KARAKOL; 42°09′N 77°11′E. Coal mining (since late 1940s). Also spelled KADZHI-SAI.

Kajmakcalan (kei-MAHK-chah-LAHN), Serbo-Croatian *Kajmakčalan*, Greek *Voras*, peak (8,280 ft/2,524 m) in NIDZE massif on border of N GREECE and MACEDONIA, E of PELAGONIJA valley; rises to its maximum elevation 15 mi/24 km NW of EDHESSA (Greece). The name Kajmakcalan is sometimes applied to the entire Nidze massif. Also spelled Kajmakchalan or Kaymakchalan.

Kajora, INDIA: see RANIGANJ.

Kaka (KA-kah), village, UPPER NILE state, S central SUDAN, on left bank of the WHITE NILE River, on road,

80 mi/129 km NNE of MALAKAL; 10°36′N 32°11′E. Shilluk tribes in area.

Kakabeka Falls (ka-kuh-BE-kuh), unincorporated village, W ONTARIO, E central CANADA, on KAMINISTIKWIA RIVER (with 130-ft/40-m falls), 17 mi/27 km W of FORT WILLIAM, and included in OLIVER PAIPOONGE; 48°25′N 89°37′W. Hydroelectric power center; dairying, grain growing.

Kakadu National Park (KAK-uh-doo) (☐ 7,646 sq mi/19,879.6 sq km), N central NORTHERN TERRITORY, N central AUSTRALIA, 130 mi/209 km E of DARWIN; 13°00′S 132°45′E. Largest national park in Australia, archaeological reserve, 100 mi/161 km N–S, 80 mi/129 km E–W. Declared UNESCO World Heritage Area 1981. Park area excludes mineral leases E and S of JABIRU. Coastal and estuarine wetlands N; grasslands, eucalypt woodlands in hinterland; ARNHEM LAND escarpment and plateau SE. Aboriginal rock art galleries at Nourlangie Rock, other sites. Jim Jim Falls (elevation 118 ft/36 m). Crocodiles, kangaroos, wallabies, cockatoos, magpie (pied), geese, jabirus. Information center SW of Jabiru. Services at Kakadu Village, Jabiru, Cooinda. Camping, picnicking, hiking, boating, fishing. Park partially closed during wet season (November–March). Administered by Australian National Parks and Wildlife Service. Established 1979.

Kakagi Lake, W ONTARIO, E central CANADA, 3 mi/5 km E of head of WHITEFISH BAY (Lake of the Woods), 40 mi/64 km SE of KENORA; 20 mi/32 km long, 6 mi/10 km wide; 49°12′N 93°50′W.

Kakaidy (kah-KEI-ti), oil town, S SURKHANDARYO wiloyat, UZBEKISTAN, 25 mi/40 km N of TERMEZ, on left bank of the SURKHAN DARYA River; 37°37′N 67°30′E. Also spelled Kakaydy, KOKAYTY, and Kokaity.

Kakaji (kah-KAH-jee), town, West Kunisaki county, OITA prefecture, NE KYUSHU, SW JAPAN, on SUO SEA, 31 mi/50 km N of OITA. Mandarin oranges.

Kakamas, town, NORTHERN CAPE province, SOUTH AFRICA, on ORANGE RIVER between Bushmanland and Gordonia, 45 mi/72 km WSW of UPINGTON; 28°45′S 20°38′E. Elevation 2,263 ft/689 m. Originally settled as a mission station in 1870, later in 1895 began to be developed as irrigation area for viticulture and fruit growing. Railroad terminus.

Kakamega (ka-kah-MAI-gah), town (1999 population 57,128), ⊙ WESTERN province, W KENYA, on road, 28 mi/45 km W of KISUMU; 00°17′N 34°50′E. Center of gold field (extending S to TANZANIA border), discovered 1930. Also pyrite mining. Asbestos-cement sheet manufacturing. Airfield.

Kakamigahara (kah-KAH-mee-gah-HAH-rah), city, GIFU prefecture, central HONSHU, central JAPAN, 6 mi/10 km E of GIFU; 35°23′N 136°51′E. Agriculture and commercial center. Airplane manufacturing. Hida Kiso River park.

Kakamoeka (kah-kah-mwe-KAH), village, KOUILOU region, SW Congo Republic, on KOUILOU RIVER, 50 mi/80 km NNE of POINTE-NOIRE; 04°07′S 12°04′E. Terminus of navigation; gold mining, plantations.

Kakanda (kah-KAHN-dah), village, KATANGA province, SE CONGO, 28 mi/45 km NW of LIKASI; 10°44′S 26°23′E. Elev. 4,232 ft/1,289 m. Copper mining.

Kakanj (KAH-kahn-yuh), town, central BOSNIA, BOSNIA AND HERZEGOVINA, on BOSNA RIVER, on railroad, 10 mi/16 km ESE of ZENICA; 44°08′N 18°07′E. In SARAJEVO coal area; lignite mine. Also spelled Kakan.

Kakarbhitta (kah-KUHD-bee-TAH), village, E NEPAL, near INDIA border; 26°38′N 88°10′E. Elevation 413 ft/126 m. Border crossing point between E Nepal and WEST BENGAL state, E India, with connections to Silaguri and DARJILING.

Kaka River (KAH-kah), 40 mi/64 km long, LA PAZ department, W BOLIVIA; formed by confluence of MAPIRI and COROICO rivers at Puerto Ballivián; flows N to BENI RIVER at Puerto Pando; 15°10′S 67°44′W. Sometimes called MAPIRI RIVER.

Kakashura (kah-kah-shoo-RAH), village (2005 population 5,710), central DAGESTAN REPUBLIC, NE CAUCASUS, SE European Russia, 15 mi/24 km S of MAKHACHKALA; 42°39′N 47°23′E. Elevation 2,611 ft/795 m. Dairy livestock.

Kakata (KAH-kah-tah), town, ⊙ MARGIBI county, W LIBERIA, 35 mi/56 km NE of MONROVIA, on road; 06°31′N 10°21′W. Palm oil. Site of Booker T. Washington Institute (agricultural and technical school).

Kakaur (kuh-KOUR), town, BULANDSHAHR district, W UTTAR PRADESH state, INDIA, 12 mi/19 km SW of BULANDSHAHR; 28°19′N 77°42′E. Wheat, oilseeds, cotton, barley, corn, jowar. Also spelled Kakor and Kakore.

Kake, town (2007 population 5,182), OROMIYA state, W ETHIOPIA, 23 mi/37 km NE of DAMBI DOLO; 08°49′N 34°56′E.

Kake (KAH-ke), town, Yamagata county, HIROSHIMA prefecture, SW HONSHU, W JAPAN, 16 mi/25 km N of HIROSHIMA; 34°36′N 132°19′E.

Kake, village (2000 population 710), SE ALASKA, on N shore of KUPREANOF ISLAND, 40 mi/64 km WNW of PETERSBURG; 56°58′N 133°55′W. Fishing, fish processing, lumbering.

Kakegawa (kah-KE-gah-wah), city, SHIZUOKA prefecture, central HONSHU, E central JAPAN, 25 mi/40 km SW of SHIZUOKA; 34°36′N 138°01′E. Agriculture center (tea); *kappu* textiles.

Kakenge (kuh-KEN-ge), town, KASAI-OCCIDENTAL province, SW CONGO, on road, 100 mi/160 km ESE of ILEBO; 04°51′S 21°55′E. Elev. 1,532 ft/466 m. Also known as MAKENGE.

Kakeya (kah-KAI-yah), town, Ishi county, SHIMANE prefecture, SW HONSHU, W JAPAN, 24 mi/39 km S of MATSUE; 35°11′N 132°49′E.

Kakhetia (kah-khe-TEE-ah), historic region, GEORGIA; TELAVI was the chief town. Kakhetia was an independent kingdom from the 8th century until 1010, when it became part of Georgia. Again independent between 1468 and 1762, it then became part of the East Georgian kingdom that was joined with RUSSIA in 1801.

Kakhetian Range (kah-khe-TEE-yuhn), S spur (6,530 ft/1,990 m) of the E Greater Caucasus, in E GEORGIA; extends in arc from Mount BARBALO c.75 mi/121 km S and SE to area of GURDZHAANI; forms watershed between IORI and ALAZAN rivers. Vineyards on Alazan River slopes.

Kakhi (kahk-HEE), town and administrative center of KAKHI region, N AZERBAIJAN, on S slope of the Greater CAUCASUS, on road and 22 mi/35 km NW of Nukha. Rice, tobacco, sericulture; lumbering. Fruit cannery; creamery.

Kakhib (kah-HEEP), village (2004 population 2,325), central DAGESTAN REPUBLIC, N CAUCASUS, SE European Russia, 8 mi/13 km SW of KHUNZAKH; 42°25′N 46°36′E. Elevation 5,629 ft/1,715 m. Population largely Avar.

Kakhonak, village, SW ALASKA, on S shore of ILIAMNA Lake, at base of ALASKA PENINSULA; 59°26′N 154°54′W.

Kakhovka (kah-KOV-kah) (Ukrainian *Kakhivka*) (Polish *Kachowka*), city (2004 population 31,000), central KHERSON oblast, UKRAINE, on SE shore of KAKHOVKA RESERVOIR (landing), opposite BERYSLAV, 40 mi/64 km ENE of KHERSON; 46°48′N 33°28′E. Elevation 111 ft/33 m. Raion center; manufacturing (electric welding and sprinkler irrigation equipment, agricultural implements), food flavoring. Local historical museum. Known since 1791, city since 1919. Jewish community since the 19th century, numbering 2,441 in 1939; decimated by the pogroms during the Russian civil war of 1918–1921, the survivors destroyed by the Nazis in 1942; fewer than 100 Jews remaining in 2005.

Kakhovka Irrigation System (kah-KOV-kah) (Ukrainian *Kakhovs'ka zroshuval'na systema*), an irrigation system in SE KHERSON oblast, UKRAINE, on the BLACK SEA LOWLAND, extending from KAKHOVKA for 30 mi/ 48 km to N coast of SYVASH SEA (S), and for 90 mi/145 km to W shore of MOLOCHNA LAGOON of the Sea of AZOV (E). The system distributes DNIEPER water, drawn from the KAKHOVKA RESERVOIR at an intake 6 mi/10 km E of Kakhovka city center, by the Kakhovka Mainline Canal (Ukrainian *Kakhovs'ky mahistral'nyy kanal*), extending for 81 mi/130 km SE and E to Malyy Utlyuk River, and then by six branch canals. Begun in 1967, the system irrigated 647,000 acres/262,000 ha in 1991; future projected irrigated area may reach 1,850,000 acres/750,000 ha. Dark chestnut soils have demonstrated susceptibility to salinization; drainage canals (100 mi/160 km) have been built to ameliorate the condition on 77,000 acres/31,200 ha. Sprinkler irrigation is applied to grow grain, feed crops, and vegetables.

Kakhovka Mainline Canal, UKRAINE: see KAKHOVKA IRRIGATION SYSTEM.

Kakhovka Reservoir (kah-KOV-kah) (Ukrainian *Kakhovs'ke vodoskhovyshche*) (Russian *Kakhovskoye vodokhranilishche*), on the DNIEPER River, S central UKRAINE, in borad Dnieper valley from 4 mi/6.4 km S of ZAPORIZHZHYA (NE) to NOVA KAKHOVKA (SW). Formed in 1955–1958 with the construction of the Kakhovka hydroelectric power station. Length, 143 mi/230 km; average width, 5.8 mi/9.4 km; maximum width, 16 mi/25 km; surface area, 832 sq mi/2,155 sq km; average depth, 28 ft/9 m; maximum depth 79 ft/24 m. Banks are generally steep, incised by ravines; few segments of shore with gentle, sandy slopes; islands abound. Water level fluctuates by about 3 ft/1 m, with maximum during spring floods. Reservoir serves hydroelectric power generation, navigation, fisheries, irrigation of 2,510 sq mi/6,501 sq km, and recreation. On its shores are the cities of Nikopol', Enerhodar, Kamyanka-Dniprovs'ka, Kakhovka, Beryslav, and Dniprorudne.

Kakhovs'ka zroshuval'na systema, UKRAINE: see KAKHOVKA IRRIGATION SYSTEM.

Kakhovs'ke vodoskhovyshche, UKRAINE: see KAKHOVKA RESERVOIR.

Kakhovskoye vodokhranilishche, UKRAINE: see KAKHOVKA RESERVOIR.

Kakhovs'kyy mahistral'nyy kanal, UKRAINE: see KAKHOVKA IRRIGATION SYSTEM.

Kakhun (kah-HOON), village (2005 population 7,570), central KABARDINO-BALKAR REPUBLIC, N CAUCASUS, S European Russia, on road and near railroad, 11 mi/18 km ENE of NAL'CHIK; 43°32′N 43°52′E. Elevation 974 ft/296 m. Agricultural products.

Kaki Bukit (KAH-ki BOO-kit), village, PERLIS, NW MALAYSIA, near THAILAND border, 13 mi/21 km N of KANGAR, on slopes of the GUNONG Cina. Tin mining.

Kakinada, city (2001 population 296,329), ⊙ EAST GODAVARI district, ANDHRA PRADESH state, S INDIA, port on BAY OF BENGAL, in E GODAVARI RIVER delta, 300 mi/483 km NNE of CHENNAI (Madras); 16°56′N 82°13′E. Served by railroad spur from SAMALKOT and by delta canal system; exports cotton, peanuts, castor seeds, sugar, tobacco; cotton, rice, and flour milling; iron foundry, saltworks. Seat of college and engineering institute (affiliated with Andhra University). S suburb of Jagannathapuram was site of early Dutch trading station; ceded to English in 1825. Former port of Coringa (important in 18th century; now destroyed by advance of Godavari River silt) is 9 mi/14.5 km S. Boatbuilding 11 mi/18 km S, at village of Tallarevu. Railroad spur continues to Kotipalli village, 22 mi/35 km S, on Gautami Godavari River. Formerly spelled Cocanada.

Kakinoki (kah-KEE-no-kee), village, Kanoashi county, SHIMANE prefecture, SW HONSHU, W JAPAN, 97 mi/157 km S of MATSUE; 34°26′N 131°52′E.

Kakinoura-shima (kah-KEE-no-oo-RAH–shee-mah), island (□ 2 sq mi/5.2 sq km), NAGASAKI prefecture,

SW JAPAN, in EAST CHINA SEA, just off NW coast of SONOGI PENINSULA, NW KYUSHU; 2.5 mi/4 km long, 1 mi/1.6 km wide; sandstone.

Kakira (kah-KEE-ruh), village, JINJA district, EASTERN region, SE UGANDA, 8 mi/12.9 km NE of Jinja. Was part of former BUSOGA province.

Kakizaki (kah-KEE-zah-kee), town, Nakakubiki county, NIIGATA prefecture, central HONSHU, N central JAPAN, 56 mi/90 km S of NIIGATA; 37°16′N 138°23′E.

Kakkapalliya (KAHK-kuh-puhl-lee-uh), village, NORTH WESTERN province, SRI LANKA, 3.5 mi/5.6 km SSE of CHILAW; 07°32′N 79°49′E. Rice, coconut palms.

Kakmozh (kahk-MOZH), village, central UDMURT REPUBLIC, E central European Russia, near highway, 12 mi/19 km NW of UVA; 57°05′N 51°55′E. Elevation 561 ft/170 m. Woodworking.

Kakogawa (kah-KO-gah-wah), city (2005 population 267,100), HYOGO prefecture, S HONSHU, W central JAPAN, on the Kako River, 22 mi/35 km W of KOBE; 34°45′N 134°50′E. Industrial center; manufacturing (steel and apparel).

Kakonko (kah-KON-ko), village, KIGOMA region, NW TANZANIA, 50 mi/80 km SSW of BIHARAMULO, near BURUNDI border; 03°15′S 31°03′E. Timber; tobacco; goats.

Kakontwe (kah-KONT-wai), village, Katenga province, SE CONGO, 8 mi/12.9 km W of LIKASI; 10°59′S 26°40′E. Elev. 4,114 ft/1,253 m. Copper mining, flour milling (manioc, wheat, corn), palm-oil milling, limestone quarrying; explosives manufacturing. Also spelled Kakontwa.

Kakopetria (kah-ko-pet-ree-YAH), town, LEFKOSIA district, W central CYPRUS, on Klarios River, in EVRYKHOU VALLEY, 28 mi/45 km WSW of NICOSIA; 34°59′N 33°54′E. Mount OLYMPUS 4 mi/6.4 km to SW. Fruit, nuts, grapes, vegetables; goats, sheep. Church of Ayios Nikolaos tis Steyis (10th century). Chromium mine (closed 1983) 3 mi/4.8 km to SW.

Kakor, INDIA: see KAKAUR.

Kakori (KAH-ko-ree), town, LUCKNOW district, central UTTAR PRADESH state, N INDIA, 8 mi/12.9 km W of LUCKNOW; 26°53′N 80°48′E. Wheat, rice, gram, millet, oilseeds. Has 16th–17th-century *dargahs* (tombs of important people). Famous for the "Kakori Conspiracy" case against British rule.

Kakoulima (kah-koo-LEE-mah), village, Kindia administrative region, W GUINEA, in Guinée-Maritime geographic region, on railroad, 25 mi/40 km NE of CONAKRY; 09°45′N 13°26′W. Bananas, palm kernels. The Kakoulima mountain (3,688 ft/1,124 m) and resort area adjoins IA.

Kakpin (KAHK-pin), town, Zanzan region, NE CÔTE D'IVOIRE, 21 mi/34 km NW of NASSIAN; 08°39′N 03°48′W. Agriculture (sorghum, corn, beans).

Kakrala (kuhk-RAH-lah), town, BUDAUN district, central UTTAR PRADESH state, N central INDIA, 11 mi/18 km SSE of BUDAUN. Wheat, pearl millet, mustard, barley, gram. Serai.

Kaktovik, village (2000 population 293), on BARTER ISLAND, ALASKA, E BEAUFORT SEA; 70°16′N 143°38′W. School; weather station. Only native settlement E of COLVILLE RIVER; Inuits concentrated here in the 1950s when the now-abandoned DEW LINE was built.

Kakuda (kah-KOO-dah), city, MIYAGI prefecture, N HONSHU, NE JAPAN, on ABUKUMA RIVER, 22 mi/35 km S of SENDAI; 37°38′N 140°47′E. Video equipment.

Kakul (kah-kuhl), village, ABBOTTABAD district, NE NORTH-WEST FRONTIER province, N PAKISTAN, 3 mi/4.8 km NE of ABBOTTABAD; 34°11′N 73°16′E. Site of Pakistan Kakul Military Academy and Ayub Medical College. Now merged with ABBOTTABAD town.

Kakuma, town, RIFT VALLEY province, NW KENYA, 75 mi/121 km NW of LODWAR, near E border of UGANDA; 04°17′N 34°50′E. Refugee camp. Trade center.

Kakunodate (kah-koo-NO-dah-te), town, Senhoku county, AKITA prefecture, N HONSHU, NE JAPAN, 25 mi/40 km S of AKITA; 39°35′N 140°33′E. Woodcrafts.

Kál (KAHL), Hungarian *Kál*, village, HEVES county, N HUNGARY, on TARNA RIVER, 9 mi/14 km N of HEVES; 47°44′N 20°16′E. Wheat, tomatoes, green peppers, grapes; cattle; manufacturing (electrical appliances, aluminum items).

Kala (KAH-lah), village, RUKWA region, W TANZANIA, 43 mi/69 km WSW of SUMBAWANGA, on Lake TANGANYIKA; 08°06′S 30°57′E. Lake port. Fish; goats, sheep; corn, wheat.

Kala (KAH-lah), urban settlement, AZIZBEKOV region, AZERBAIJAN; 40°26′N 50°09′E. Manufacturing (oil industry equipment); machine overhaul plant.

Kala, for Afghan names beginning thus: see under QALA.

Kalaa, for Arabic names beginning thus and not found here: see under QAL'A.

Kalaat, for Arabic names beginning thus and not found here: see under QAL'AT.

Kalaat M'gouna (ka-LAHT m-GOO-nuh), town, Ouarzazate province, Souss-Massa-Draâ administrative region, MOROCCO, at the mouth of the M'goun River, 50 mi/80 km ENE of OUARZAZATE, in the Oued DADES valley.

Kalabagh (kah-lah-bah-gah), town, MIANWALI district, NW PUNJAB province, central PAKISTAN, on INDUS River, 26 mi/42 km N of MIANWALI; 32°58′N 71°34′E. MARI INDUS is 1 mi/1.6 km S, across river. Headworks of THAL canal irrigation project 4 mi/6.4 km S. Proposed dam to be constructed nearby.

Kalabahi, INDONESIA: see ALOR.

Kalabak, Bulgaria and Greece: see RADOMIR.

Kalabak, Greece: see KALAMBAKA.

Kalabaki, Greece: see KALAMBAKI.

Kalabo, township, WESTERN province, W ZAMBIA, 40 mi/64 km NW of MONGU on Luanginga River, ANGOLA border 45 mi/72 km to W. Airstrip; road terminus. Agriculture (corn, cashews); cattle; timber.

Kalabsha (kah-LAHB-shuh), former village, ASWAN province, S EGYPT, on W bank of the NILE, 40 mi/64 km S of ASWAN; 23°33′N 32°52′E. Site of Roman temple built during reign of Augustus on site of earlier sanctuary built by Thutmose III (c.1450 B.C.E.). The area was flooded with the inauguration of the ASWAN HIGH DAM; village was evacuated and is now under Lake Nasser.

Kalach (kah-LAHCH), city (2006 population 20,420), E VORONEZH oblast, S central European Russia, on road and railroad, 182 mi/293 km SE of VORONEZH, and 33 mi/53 km SE of BUTURLINOVKA; 50°25′N 41°01′E. Elevation 278 ft/84 m. Railroad terminus; agricultural center; meat packing, flour milling, sugar refining. Became city in 1945.

Kalachevs'ke, UKRAINE: see LENINA, IMENI.

Kalachevskiy Rudnik, UKRAINE: see LENINA, IMENI.

Kalachevskoye, UKRAINE: see LENINA, IMENI.

Kalachevs'kyy Rudnyk, UKRAINE: see LENINA, IMENI.

Kalachinsk (kah-LAH-cheensk), city (2006 population 24,000), SE OMSK oblast, SW SIBERIA, RUSSIA, on the left bank of the OM' RIVER, on highway junction and the TRANS-SIBERIAN RAILROAD (Kalachinskaya station), 55 mi/89 km E of OMSK; 55°03′N 74°34′E. Elevation 357 ft/108 m. Mechanical shops, ceramic wall materials, footwear, special cloth weaving (carpets, rugs, and towels); food industries (meat-packing plant, dairy, bakery, confectionery). Known since 1795; in 1830, expanded with settlers from European provinces; railroad station since 1896; made city in 1952.

Kalach-na-Donu (kah-LACH–nah–duh-NOO), city (2006 population 27,755), N central VOLGOGRAD oblast, SE European Russia, on the left bank of the DON River, on highway and railroad, 53 mi/85 km W of VOLGOGRAD, at the W terminus of the VOLGA-DON CANAL; 48°41′N 43°32′E. Elevation 127 ft/38 m. Port on the TSIMLYANSK Reservoir. Ship construction and repair, bridge-building materials, woodworking; food industries. Made city in 1951.

Kalach Upland (kah-LAHCH), hills in VORONEZH, VOLGOGRAD, and ROSTOV oblasts, SE European Russia, S Eastern European Plain, on the left bank of the DON RIVER; maximum elevation 787 ft/240 m. Has numerous valleys, ravines, and gulches. Main rivers include Osered', Tolucheevka, and Peskovatka (all left tributaries of the Don). Steppes produce wheat, rye, millet, and sunflowers.

Kaladan River (kah-lah-DAHN), main stream of N Arakan coast, 200 mi/322 km long, Myanmar, formed in Chin Hills on India border at 22°47′N by junction of BOINU and TYAO rivers. Flows through India's Lushai Hills and Myanmar's Chin State, past Paletwa (head of navigation) and Kyauktaw, to Bay of Bengal at Sittwe, where it is linked by tidal creeks with Mayu and LEMRO rivers. Navigable for 96 mi/154 km.

Kaladgi (kuhl-AHD-gee), village, BIJAPUR district, KARNATAKA state, S INDIA, on GHATPRABHA RIVER, 13 mi/21 km W of BAGALKOT. Peanuts, cotton. Known for sandstone, slate, and flagstone quarried nearby.

Ka Lae (kah LEI) [=south point], S extremity of HAWAII island, HAWAII county, HAWAII, 65 mi/105 km SSW of HILO, 220 mi/354 km SE of HONOLULU; 18°54′N 155°41′W. Southernmost point in the UNITED STATES. Ka Lae Park at W side; Green Sand Beach on E side.

Kalagarh (KAH-lah-guhr), village, GARHWAL district, N UTTAR PRADESH state, N central INDIA, on RAMGANGA RIVER, 22 mi/35 km SE of KOTDWARA, at foot of SHIWALIK RANGE. Irrigation and power dam (340 ft/104 m high) is just N.

Kalagi (kah-LAH-jee), town, WESTERN DIVISION, THE GAMBIA, near the SENEGAL border, 30 mi/48 km E of BANJUL, on main S road near bridge over Bintang River; 13°15′N 15°50′W.

Kalahandi (kah-lah-HUHN-dee), district (□ 3,165 sq mi/8,229 sq km), SW ORISSA state, E central INDIA; ⊙ KALAHANDI. Bordered W by MADHYA PRADESH state; crossed (N-S) by several ranges of Eastern GHATS. Rice is chief crop; forests yield sal and bamboo. Created 1949 by merger of former princely state of Kalahandi of CHHATTISGARH STATES and S subdivision of original SAMBALPUR district, which is now Balengir district. Kalahandi district is repeatedly afflicted hunger and destitution caused by scarcities of basic goods.

Kalahandi (kah-lah-HUHN-dee), city, ⊙ KALAHANDI district, ORISSA state, E central INDIA; 19°45′N 83°00′E.

Kalahari Desert (kah-lah-HAH-ree), desert, arid plateau region (□ 100,000 sq mi/260,000 sq km), in BOTSWANA, NAMIBIA, and the Republic of SOUTH AFRICA. The Kalahari, covered largely by reddish sand, lies between the ORANGE and ZAMBEZI rivers and is studded with dry lake beds. Yearly rainfall varies from 5 in/12.7 cm in the SW, where there are active sand dunes, to 20 in/51 cm in the NE. Grass grows throughout the Kalahari in the rainy season, and some parts also support low thorn scrub and forest. Grazing and a little agriculture are possible in certain areas. Many game animals live in the Kalahari. Its human inhabitants are mainly San, who are nomadic hunters, and Khoikhoi, who are hunters and farmers. Tswana and Herero herders have moved into the area, and the Kalahari has become a popular tourist destination.

Kalaheo (kah-LAH-HAI-o), town (2000 population 3,913), S KAUAI, KAUAI county, HAWAII, 10 mi/16 km WSW of LIHUE, 2.5 mi/4 km N of S coast; 21°55′N 159°31′W. Sugarcane plantations to SE and S, referred to as the "Cane Coast"; Numila Cane Mill to SW, on Kaumualii Highway; Pohakea Mountain (1,429 ft/436 m) to NW; Ipuolono Reservoir to W; Elua Reservoir to N; Lihue-Koloa Forest Reserve to N.

Kalah Shergat, IRAQ: see SHARQAT.

Kalaikhumb (kah-lei-KOOMB), village, W BADAKHSHAN AUTONOMOUS VILOYAT, TAJIKISTAN, in the PAMIR, on PANJ RIVER (AFGHANISTAN border), 45 mi/72 km SE of GARM; 38°28′N 70°46′E. Gold placers; sheep, goats; tertiary level administrative center.

Kalai-Mirzabai, TAJIKISTAN: see KALININABAD, village.

Kala-i-Mor (ku-lu-ee-MOR), town S MARY weloyat, SE TURKMENISTAN, on railroad to MARY, on Gushgy River (tributary of the MURGAB RIVER), 30 mi/48 km NE of KUSHKA; 35°39′N 62°33′E. Karakul state farm.

Kalai-Vamar, TAJIKISTAN: see RUSHAN.

Kalajoki (KAH-lah-YO-kee), village, OULUN province, W FINLAND, 40 mi/64 km NE of KOKKOLA; 64°15′N 23°57′E. On GULF OF BOTHNIA at mouth of KALAJOKI. Seaside resort.

Kalajoki (KAH-lah-YO-kee), river, 80 mi/129 km long, W FINLAND; rises in lake region E of KOKKOLA; flows NW, past HAAPAJÄRVI and YLIVIESKA, to GULF OF BOTHNIA at KALAJOKI.

Kalakan (KAH-lah-kahn), village, NW CHITA oblast, E SIBERIA, RUSSIA, on the VITIM River, 250 mi/402 km NE of CHITA; 55°08′N 116°45′E. Elevation 2,011 ft/612 m. Lumbering. Was capital of former Vitim-Olekma National Okrug.

Kalakh, IRAQ: see CALAH.

Kalakhiri Range (KAH-LAH-KI-REE), on THAILAND-Malaysia border, on MALAY PENINSULA, forming S continuation of SITHAMMARAT RANGE; rises to 5,035 ft/1,535 m. Crossed by Malaysia-THAILAND highway between KROH and BETONG.

Kalakh, Tell, SYRIA: see TELL KALAKH.

Kalalè, town, BORGOU department, NE BENIN, 100 mi/161 km NE of PARAKOU, 15 mi/24 km from NIGERIA border; 11°10′N 02°36′E. Cotton; livestock; shea-nut butter.

Kalam (kah-LUHM), village and tehsil, SWAT district, NORTH-WEST FRONTIER PROVINCE, N PAKISTAN; 35°30′N 72°40′E. Popular summer tourism and recreation destination, with more than fifty hotels. High mountains, extensive deodar forests, and numerous trout streams attract visitors. Major source of seed potatoes for NORTH-WEST FRONTIER PROVINCE and PUNJAB. Some nuts and fruit also exported. Extensive commercial logging is now developed. Old pre-Islamic religious structure. Local language is a branch of Kohestani languages with residents now converted to Sunni Islam. Heavy snow in winter isolates Kalam from S SWAT district.

Kalama (kuh-LAM-uh), town (2006 population 2,039), COWLITZ county, SW WASHINGTON, 10 mi/16 km S of KELSO, on COLUMBIA River near mouth of KALAMA RIVER; 46°01′N 122°51′W. River and railroad shipping point, wheat export center; strawberries, vegetables; fish; manufacturing (lumber, chemicals, beverages); logging. To NE are two state fish hatcheries on Kalama River. Founded 1870, was capital of Cowlitz county until 1932.

Kalámai, Greece: see KALAMATA.

Kalamaki (kah-lah-MAH-kee), village, S coast of ZÁKINTHOS island, ZÁKINTHOS prefecture, IONIAN ISLANDS department, off W shore of GREECE, 3 mi/4.8 km from ZÁKINTHOS and adjacent to LAGANA (W); 37°44′N 20°55′E. Popular beach resort. Breeding ground for the loggerhead turtle (*Caretta caretta*).

Kalama River (kuh-LAM-uh), c.45 mi/72 km long, COWLITZ county, SW WASHINGTON; rises SW of MOUNT ST. HELENS, flows SW to COLUMBIA River 3 mi/4.8 km NW of KALAMA. Lower river has two salmon hatcheries.

Kalamas River, Greece: see THYAMIS RIVER.

Kalamata (kah-lah-MAH-tah), city (2001 population 54,184), ⊙ MESSENIA prefecture, SW PELOPONNESE department, extreme SW GREECE, port on the Gulf of MESSENIA; 37°02′N 22°07′E. Agricultural trade center; ships olives, olive oil, and fruits. Silk, flour, and liquor are manufactured. Airport. Developed after c.1205, when it became a fief of the Villehardouin family. Later came under the rule of VENICE and (1459–1821) the Ottoman Turks. Destroyed (1825) by Ibrahim Pasha during the Greek War of Independence. Also called Kalámai.

Kalamata, Gulf of, Greece: see MESSENIA, GULF OF.

Kalamazoo (KAL-ah-muh-zoo), county (□ 580 sq mi/1,508 sq km; 2006 population 240,720), SW MICHIGAN; 42°14′N 85°31′W; ⊙ KALAMAZOO. Cattle, hogs, poultry, dairy products; apples, grapes, cherries, strawberries, corn, wheat, oats, soybeans; manufacturing at Kalamazoo, PORTAGE. Drained by KALAMAZOO RIVER and short PORTAGE RIVER. GULL LAKE in NE; numerous small lakes in SW. Organized 1830.

Kalamazoo (KAL-ah-muh-zoo), city (2000 population 77,145), ⊙ KALAMAZOO county, SW MICHIGAN, on the KALAMAZOO RIVER at its confluence with Portage Creek; 42°16′N 85°35′W. Industrial and commercial center in a fertile farm area. Agriculture (celery, peppermint, fruit); important paper industry; manufacturing (paper products, hydraulic equipment, handling devices, meat products, furniture, concrete, motor vehicle parts, printing plates, sheet metal products, pharmeceuticals); winery. Railroad junction. KALAMAZOO is the seat of Western Michigan University, Kalamazoo College, Nazareth College, Kalamazoo Valley Community College, and a state mental hospital. The city has a natural history museum, an art institute, Kalamazoo Public Museum and Planetarium, Aviation History Museum, and a symphony orchestra. Timber Ridge State Area to NW; numerous lakes to S and SW; municipal airport SE. County fairgrounds. Incorporated in 1883.

Kalamazoo River (KAL-ah-muh-zoo), c.138 mi/222 km long, S and SW MICHIGAN; formed by junction of North and South branches at ALBION. North Branch rises in S JACKSON county and flows c.25 mi/40 km NW; South Branch rises in NE HILLSDALE county and flows c.35 mi/56 km in a separate course, generally NNW. Main river continues NW, past MARSHALL, BATTLE CREEK, KALAMAZOO, OTSEGO, and ALLEGAN, to LAKE MICHIGAN at SAUGATUCK; 42°14′N 84°45′W. Power development at Winkler Lake, below Otsego, and Allegan Dam Pond below Allegan.

Kalamb (KUHL-uhmb), town, PUNE district, MAHARASHTRA state, W central INDIA, on NIRA RIVER, 16 mi/26 km SE of BARAMATI; 19°03′N 73°57′E. Agriculture market (sugarcane, peanuts, millet); sugar and oilseed milling, dairy farming, soap manufacturing.

Kalamba (kuh-LAHM-buh), village, Équateur province, NW CONGO, on road 30 mi/48 km S of MBANDAKA; 00°26′S 18°17′E. Elev. 1,499 ft/456 m.

Kalambaka (kah-lahm-BAH-kah), town, TRIKKALA prefecture, THESSALY department, N GREECE, 14 mi/23 km NW of TRIKKALA, on PENEIOS River where it enters Thessalian plain; 39°42′N 21°38′E. Narrow-gauge railroad terminus; trade in corn, wheat, vegetables, cheese. The ancient city of Aeginium, it was known as Stagous in Byzantine times and received present name under Turkish rule. Meteora monasteries are just N. Also spelled Kalampaka and Kalabaka.

Kalambaki (kah-lahm-BAH-kee), town, DRÁMA prefecture, EAST MACEDONIA AND THRACE department, NE GREECE, 8 mi/12.9 km S of DRÁMA; 41°03′N 24°11′E. Tobacco, barley; wine; olive oil. Also called Kalabaki.

Kalambo Falls (kah-LAHM-bo), falls on KALAMBO RIVER, on ZAMBIA-TANZANIA border, 2 mi/3.2 km E of river's entrance to LAKE TANGANYIKA, 20 mi/32 km NW of MBALA (Abercorn), Zambia; 08°35′S 31°12′E. Falls descend 3,000 ft/914 m in 6 mi/9.7 km course through gorge of volcanic rock; largest single drop is 704 ft/215 m. Road access.

Kalambo River (kah-LAHM-bo), c. 85 mi/137 km long, S TANZANIA and NE ZAMBIA; rises c. 15 mi/24 km NW of SHINYANGA, Tanzania, flows S through SHINYANGA region and KALAMBO FALLS, and enters Lake TANGANYIKA 25 mi/40 km NW of MBALA, Zambia.

Kalambur, INDIA: see KOLAMBUR.

Kalamit Bay (kah-lah-MEET) (Ukrainian *Kalamits'ka zatoka*) (Russian *Kalamitskiy zaliv*), N inlet of BLACK SEA 8 mi/13 km into SW side of CRIMEA, extending 25 mi/40 km from Cape YEVPATORIYA in NNW to Cape

---

Cross-references are shown in SMALL CAPITALS. The pronunciation guide is shown on page xix. The sources of population figures are shown on page xvii.

LUKULL in S. Depth increases from 15 ft/5 m offshore to 100 ft/30 m; receives two small rivers, ALMA and BULHANAK ZAKHIDNYY (Russian *Zapadnyy Bulganak*); coastal area used for recreation; in N port and resort YEVPATORIYA.

Kalamits'ka zatoka, UKRAINE: see KALAMIT BAY.

Kalamitskiy zaliv, UKRAINE: see KALAMIT BAY.

Kalamnuri (kah-lahm-NOO-ree), town, PARBHANI district, MAHARASHTRA state, W central INDIA, 36 mi/58 km N of NANDED; 19°40′N 77°20′E. Cotton, millet, wheat.

Kalamos (KAH-lah-mos), Italian *Calamo*, island (□ 7 sq mi/18.2 sq km), AKARNANIA prefecture, WESTERN GREECE department, GREECE, in IONIAN SEA 1.5 mi/2.4 km off mainland (E); 38°37′N 20°55′E. Rises to 2,445 ft/745 m; 7 mi/11.3 km long, 1 mi/1.6 km–3 mi/4.8 km wide. Chief town, Kalamos, is on NE shore. Fisheries.

Kalampaka, Greece: see KALAMBAKA.

Kalamunda (KA-luh-MUN-duh), residential town, SW WESTERN AUSTRALIA state, W AUSTRALIA, 11 mi/18 km ESE of PERTH and on W slopes of DARLING RANGE; 31°57′S 116°03′E. Kalamunda National Park on N border. Orchards.

Kalamunda National Park (KA-luh-MUN-duh), national park (□ 1 sq mi/2.6 sq km), SW WESTERN AUSTRALIA state, W AUSTRALIA, 25 km from PERTH, and near KALAMUNDA town. Diverse native plants. DARLING RANGE forest country traversed by Piesse Brook; jarrah, marri, wandoo, and butter gum (eucalyptus) woods. Rarely seen Grey Kangaroo, Echidna, Short-nosed Bandicoot; also birdlife. Walking tracks, including the initial stage of the ancient Aboriginal walking track, the Bibbulmun Trail. Established 1964.

Kalan, TURKEY: see TUNCELI.

Kalana, town, THIRD REGION/SIKASSO, MALI, 126 mi/210 km S of BAMAKO, near border with Republic of GUINEA; 10°47′N 08°12′W. Nearby gold, silver, and platinum mines.

Kalanaur (kuh-LAH-nour), village, GURDASPUR district, NW PUNJAB state, N INDIA, 14 mi/23 km W of GURDASPUR. Akbar proclaimed emperor here, in 1556.

Kalanchak (kah-lahn-CHAHK), town, S KHERSON oblast, UKRAINE, near PEREKOP Isthmus, 40 mi/64 km SE of KHERSON, alongside the NORTH CRIMEAN CANAL; 46°15′N 33°18′E. Raion center; grain, butter, food-flavoring factory. Established in 1794, town since 1967.

Kalangadoo (kuh-lang-guh-DOO), village, SE SOUTH AUSTRALIA state, S central AUSTRALIA, 220 mi/354 km SSE of ADELAIDE; 37°34′S 140°42′E. On NARACOORTE–MOUNT GAMBIER railroad. Dairy products; livestock. In red-gum (eucalyptus) timber area.

Kalangala (kah-lahn-GAH-luh), administrative district (□ 3,501 sq mi/9,067 sq km [of which only 167 sq mi/432 sq km is land]; 2002 population 34,766; 2005 estimated population 44,300), CENTRAL region, S UGANDA, comprising the SESE ISLANDS in LAKE VICTORIA; ⊙ KALANGALA; 00°25′S 32°20′E. Includes forests. Experiences high humidity due to high evaporation of Lake Victoria. Fishing is a major industry, also lumbering; some fruits are grown.

Kalangala (kah-lahn-GAH-luh), town (2002 population 2,943), ⊙ KALANGALA district, CENTRAL region, S UGANDA, on main island of SESE archipelago, 50 mi/80 km SSW of KAMPALA; 00°19′S 32°18′E. Fishing; sugarcane. Was part of former SOUTH BUGANDA province.

Kalang River (kah-LAHNG), arm of the BRAHMAPUTRA RIVER, 100 mi/161 km long, central ASSAM state, NE INDIA; leaves river 8 mi/12.9 km E of SILGHAT, flows SW, through rice, jute, tea, rape, and mustard area, past SAMAGURI, NAGAON, and RAHA, returning to the Brahmaputra 15 mi/24 km NE of GUWAHATI. Left tributaries drain NAGAON district.

Kalanguy (kah-lahn-GOO-yee), town (2005 population 2,640), SE CHITA oblast, E SIBERIA, RUSSIA, in the NERCHINSK RANGE, on the TUNGA RIVER (AMUR

basin), 40 mi/64 km E of OLOVYANNAYA; 51°01′N 116°31′E. Elevation 2,555 ft/778 m. Fluorite mine; ore refining.

Kalaniya, ISRAEL: see MOTZA.

Kalaniya, SRI LANKA: see KELANIYA.

Kalansawe (kah-lahn-SAH-we) or **Qalansawe**, Arab township, SE of NETANYA and 3.1 mi/5 km SW of TULKARM, on SHARON PLAIN, central ISRAEL; 32°17′N 34°58′E. Elevation 147 ft/44 m. The township includes a small *tell* that shows remains of Crusader fortress, church, and mausoleum. Founded in 17th century (on site of ancient settlement). In 1936–1939 Arab revolt, served as a base for Arab fighters, and in 1948 War as Iraqi army base. Included in Israel's territory as part of the 1949 armistice agreement.

Kalao (KAH-lou), coral island, INDONESIA, in FLORES SEA, 70 mi/113 km N of FLORES, near BONERATE; 07°18′S 120°58′E. Island is 18 mi/29 km long, 4 mi/6.4 km wide. Fishing.

Kalaoa (KAH-lah-O-ah), town, W HAWAII island, HAWAII county, HAWAII, 6 mi/9.7 km N of KAILUA-KONA, 55 mi/89 km W of HILO, 5 mi/8 km inland from the KONA (W) Coast; 19°43′N 156°01′W. Cattle-ranching area. Coffee. Keahole Airport to W. Kaupulehu Forest Reserve to E; HONOKOHAU NATIONAL HISTORIC PARK to SW.

Kalaotoa (kah-loo-TO-wah), coral island (6 mi/9.7 km long, 6 mi/9.7 km wide), INDONESIA, in FLORES SEA, 75 mi/121 km N of FLORES; 07°22′S 121°47′E. Fishing. Also spelled Kalaotowa.

Kala Oya (KUH-lah O-yuh), river, 97 mi/156 km long, NORTH WESTERN PROVINCE, NW SRI LANKA; rises in NW extension of SRI LANKA HILL COUNTRY, SW of DAMBULLA; flows N and WNW, along border of NORTH CENTRAL PROVINCE, to DUTCH BAY NNE of KALPITIYA, entering it in two streams.

Kalaque (kah-LAH-kai), canton, OMASUYOS province, LA PAZ department, W BOLIVIA, NW of LA PAZ. Elevation 12,543 ft/3,823 m. Clay, limestone, and gypsum deposits. Agriculture (potatoes, yucca, bananas, rye); cattle.

Kälarne (KE-lahr-ne), village, JÄMTLAND county, N central SWEDEN, on small Ansjön, 40 mi/64 km WSW of SOLLEFTEÅ; 62°59′N 16°05′E.

Kalashnikovo (kah-LAHSH-nee-kuh-vuh), town (2006 population 4,980), central TVER oblast, W European Russia, on highway branch and railroad, 38 mi/61 km NW of TVER; 57°17′N 35°14′E. Elevation 695 ft/211 m. Manufacturing (electric lighting fixtures).

Kalasin (GAH-LAH-SIN), province (□ 2,028 sq mi/5,272.8 sq km), NE THAILAND; ⊙ KALASIN; 16°30′N 103°30′E. Rice, tobacco, tapioca, sugarcane, peanuts, cotton, buffalo, and horses.

Kalasin (GAH-LAH-SIN), town, ⊙ KALASIN province, E THAILAND, on KORAT PLATEAU, on PHAO RIVER, 20 mi/32 km N of MAHA SARAKHAM; 16°26′N 103°30′E. Rice, tobacco; buffalo and horse raising. Sometimes spelled Kalasindhu. Formed in 1940s out of MAHA SARAKHAM province

Kalat (kuh-LUHT), former princely state, now part of BALUCHISTAN province, SW PAKISTAN. The capital was MASTUNG; residence of khan was at Kalat. Was bounded E by SIND, W by KHARAN and MAKRAN. Comprised three divisions: Sarawan, JHALAWAN, and KACHHI. Established 17th century; made treaty (1876) with British by which the khan was recognized as leader of confederacy of princely states (Kalat, LAS BELA, KHARAN, MAKRAN). Acceded 1948 to PAKISTAN. Sometimes spelled KHELAT.

Kalat (kuh-LUHT), town, KALAT district, KALAT division, BALUCHISTAN province, SW PAKISTAN, 85 mi/137 km SSW of QUETTA; 29°02′N 66°35′E. Sometimes spelled KHELAT.

Kalat, AFGHANISTAN: see KALAT-I-GHILZAI.

Kalata, RUSSIA: see KIROVGRAD.

Kalat-e-Naderi (kah-laht-ei-nah-de-REE), village, Khorāsān province, NE IRAN, 50 mi/80 km NNE of

MASHHAD, near TURKMENISTAN border. Wheat; sheep raising, sheepskins. Nearby is natural fortress called Kalat-i-Nadiri used as treasure cache by Nadir Shah. Also called Kabūd Gonbad or Kalat.

Kalat-i-Ghilzai (kah-lat-ee-geel-ZAI) or **Qalat**, town, ⊙ ZABUL province, SE AFGHANISTAN, on TARNAK R., 80 mi/129 km NE of Kandahar; 32°07′N 66°54′E. Elev. 5,543 ft/1,690 m. On highway to KABUL; road junction in oasis. Irrigated agriculture. Fort was occupied in 1842 and 1879–1880 by British garrison.

Kalatinskiy Zavod, RUSSIA: see KIROVGRAD.

Kalaupapa (kah-LOU-PAH-pah), peninsula, extending out from N MOLOKAI island, terminates at Kahiu Point, former leper colony, which is part of MAUI county, HAWAII. Isolated by rock cliff (elevation 1,600 ft/488 m) to S; leper colony, open 1886–1969, now part of KALAUPAPA NATIONAL HISTORIC PARK. Formerly known as Makanalua Peninsula.

Kalaupapa National Historic Park (□ 17 sq mi/44 sq km), N MOLOKAI island, MAUI county, HAWAII, authorized 1980. Site of Kalaupapa colony for people suffering from leprosy (Hansen's disease), located at W base of KALAUPAPA peninsula; separated from the island by 2,000-ft/610-m cliff. Park includes village of Kalawao, on E side of peninsula with monument to Belgian priest Father Damien (1886–1969), who organized the colony and gave spiritual comfort. Treatment has eliminated the need for isolation, but a few sufferers chose to remain. There are also ruins of 300 structures relating to early Hawaiian settlement. Extension 6 mi/9.7 km SE into island interior includes Waikolu Valley and Puu Kaeo (3,702 ft/1,128 m).

Kalaus River (kuh-luh-OOS), 245 mi/394 km long, STAVROPOL TERRITORY, SE European Russia; rises 10 mi/16 km NE of KURSAVKA, flows N past PETROVSKOYE (railroad bridge) and IPATOVO, and E, disappearing N of ARZGIR into marshes of the Manych Depression. Floods in spring; dries up during summer in upper and lower portions.

Kalavad (KAH-lah-vahd), town, JAMNAGAR district, GUJARAT state, W central INDIA, 26 mi/42 km SE of JAMNAGAR; 22°13′N 70°23′E. Millet, oilseeds, cotton. Formerly part of SAURASHTRA. Sometimes spelled Kalawad.

Kalavasos (kah-lah-VAH-sos), village, LARNACA district, S CYPRUS, on Vasilikos River, 22 mi/35 km WSW of LARNACA; 34°46′N 33°18′E. On mining railroad, ships pyrite and gypsum to MEDITERRANEAN SEA 3 mi/4.8 km to S. Grain, olives, fruit, nuts, grapes, vegetables; goats, sheep, hogs. Governor's Beach resort area (developed in 1990s) to S; ancient ruins to NW.

Kalávrita (kah-LAH-vree-tah), ancient *Cynaetha*, town, AKHAIA prefecture, WESTERN GREECE department, NW PELOPONNESUS, GREECE; 38°02′N 22°07′E. Chiefly a summer resort. At the nearby monastery of Ayia Laura (founded 961) the Greeks first rallied (1821) in the War of Independence. The monastery of Megaspelaion, said to date from the 4th century, is in a vaulted cave just NE of the town. Ski area located on main peak of nearby Helmos.

Kalaw (ka-LAW), township, SHAN STATE, MYANMAR, 35 mi/56 km WSW of TAUNGGYI, on road and railroad to THAZI. Hill resort; gold deposits nearby.

Kalawao (KAH-lah-WOU), county (□ 14 sq mi/36.4 sq km; 2006 population 120), officially district of MAUI county, on KALAUPAPA Peninsula, N MOLOKAI island, HAWAII. Kalaupapa leper settlement was on W side of peninsula; village of Kalawao in E side. Leper colony ended in 1969. KALAUPAPA NATIONAL HISTORIC PARK authorized in 1980; county continues to exist in name only, with Kalaupapa as its seat; Kalaupapa Airport and Molokai Lighthouse at Kahiu Point, N tip of peninsula.

Kalawewa (KUH-lah-WAH-wuh), village, NORTH CENTRAL PROVINCE, SRI LANKA, 25 mi/40 km SSE of ANURADHAPURA; 08°01′N 80°31′E. Rice, coconut, cowpeas, chilies, onions, vegetables. Kalawewa irrigation reservoir is just E; 6 mi/9.7 km long, 3 mi/4.8 km wide. Additional water from the Bowatenna Reservoir has enabled expansion of cultivated area by 24,000 acres/9,717 ha in the 1980s.

Kalba (KIL-beh), sheikdom (□ 25 sq mi/65 sq km), UNITED ARAB EMIRATES, on GULF OF OMAN, forming an enclave in FUJAIRAH sheikdom; agriculture (dates, tobacco, wheat). Consists of town of Kalba (airfield) and environs. Long a dependency of SHARJAH; joined trucial league in 1937.

Kalbacar, urban settlement, SW AZERBAIJAN, in the Lesser CAUCASUS, in the KURDISTAN, on TERTER RIVER, 70 mi/113 km SW of YEVLAKH; 40°06′N 46°02′E. Wheat, barley, livestock; lumbering. Administrative center of Kel'badzhar region. Formerly called Kelbadzhar.

Kalba Range (kahl-BAH), W branch of ALTAI Mountains, in E. KAZAKHSTAN; extends c.150 mi/241 km NW from IRTYSH River (in East Kazakhstan region) to TURK-SIB RAILROAD; rises to 5,250 ft/1,600 m. Extensive gold, tin, and tungsten deposits. Also called Kolba Range.

Kalbarri (kal-BE-ree), town, W central WESTERN AUSTRALIA state, W AUSTRALIA, 80 mi/129 km NW of GERALDTON, 44 mi/71 km NNW of North West Highway, on MURCHISON RIVER estuary of INDIAN OCEAN; 27°40′S 114°10′E. Area called Batavia Coast. The first landing of Europeans in Australia occurred here in 1629 when 2 Dutchmen were put ashore for their part in the Batavia mutiny. Headquarters and service center for KALBARRI NATIONAL PARK. Crayfish, herring. Tourism.

Kalbarri National Park (kal-BE-ree) (□ 718 sq mi/ 1,866.8 sq km), W central WESTERN AUSTRALIA state, W AUSTRALIA, 390 mi/628 km NNW of PERTH; 27°45′S 114°25′E. Park is 40 mi/64 km long, 35 mi/56 km wide. Layered sandstone cliffs rise 100 ft/30 m above pounding surf of INDIAN OCEAN. Sea rock formations, arches, stacks; Nature's Window formation overlooks Murchison Gorge. MURCHISON RIVER runs length of park before entering ocean at KALBARRI townsite. Banksias, grevillias, melaleuca, black boys (grass trees), kangaroo paw flowers. Red and gray kangaroos, euros; emus; 170 other bird species. Picnicking, hiking, swimming. Established 1953.

Kalbe an der Milde (KAHL-be ahn der MIL-de), town, SAXONY-ANHALT, central GERMANY, on small Milde River, 9 mi/14.5 km N of GARDELEGEN; 52°40′N 11°22′E. Agricultural center with food-processing industry. Has remains of old castle. Also spelled CALBE.

Kalburgi, INDIA: see GULBARGA, city.

Kal'chyk River (KAHL-chik) (Russian *Kal'chik*), river in S UKRAINE. Rises around DONETS'K and flows into the Sea of AZOV around MARIUPOL'. Part of the AZOV IRRIGATION SYSTEM. Formerly known as Kalka River. In 1223, the site of the first major military engagement between the Mongols, led by Genghis Khan, and the Slavic tribes of KIEVAN RUS'.

Kaldi Dag, (Turkish=*Kaldi Dağ*) highest peak (12,251 ft/ 3,734 m) of the ALA DAG range, S TURKEY, 30 mi/48 km SE of NIGDE.

Kaldzhir River, KAZAKHSTAN: see MARKAKOL.

Kalé (KAH-lai), village, FIRST REGION/KAYES, SW MALI, on the BAKOY RIVER, on DAKAR-NIGER railroad, 70 mi/113 km SE of KAYES. Peanuts; livestock.

Kale, TURKEY: see KARLIOVA.

Kalecik, village, central TURKEY, 32 mi/51 km ENE of ANKARA; 40°06′N 33°22′E. Wheat, barley; mohair goats.

Kaledupa (kah-LAI-doo-pah), island (□ 30 sq mi/78 sq km), TUKANGBESI ISLANDS, INDONESIA, between FLORES SEA and MALUKU SEA, just SE of WANGIWANGI Islands; 05°32′S 123°47′E. Island is 12 mi/19 km long, 3 mi/4.8 km wide; generally low. Fishing; agriculture (coconuts, sago). Also spelled Kaledoepa.

Kalefeld (KAH-le-felt), village, LOWER SAXONY, central GERMANY, 19 mi/31 km N of GÖTTINGEN; 51°48′N 10°03′E.

Kalehe (kah-LAI-hai), village, SUD-KIVU province, E CONGO, on W shore of LAKE KIVU, 30 mi/48 km N of Bakavu; 02°06′S 28°55′E. Elev. 4,940 ft/1,505 m. Used to be a center of European agricultural settlement; boat landing. Coffee and palm plantations; palm-oil milling.

Kaleida (kuh-LEI-duh), community, S MANITOBA, W central CANADA, 8 mi/13 km S of MANITOU, in PEMBINA rural municipality; 49°07′N 98°28′W.

Kalekovets (kah-le-KO-vets), village, PLOVDIV oblast, MARITSA obshtina, BULGARIA; 42°14′N 24°49′E.

Kalemba, CONGO: see KAMPENE.

Kalemera (kah-lai-MER-ah), village, MWANZA region, NW TANZANIA, 55 mi/89 km ENE of MWANZA, on SPEKE GULF, Lake VICTORIA; 02°19′S 33°44′E. SERENGETI NATIONAL PARK to NE. Fish; cattle, goats, sheep; corn, wheat, millet, cotton.

Kalemie (kuh-LEM-mee), city, KATANGA province, SE CONGO, on Lake TANGANYIKA at the mouth of the LUKUGA RIVER; 05°56′S 29°12′E. Elevation 2,506 ft/763 m. A commercial center and a railroad-steamer transfer point, handling goods moving between the CONGO and TANZANIA. Manufacturing includes textiles and cement. The city was founded in 1892 by Belgians as a military post in their campaign against Arab traders. Previously called ALBERTVILLE.

Kalemyo (kah-LEM-yo), town, SAGAING division, MYANMAR, on MYITTHA RIVER, 15 mi/24 km W of KALEWA, on road to CHIN HILLS, in KABAW VALLEY.

Kalenga (kah-LAIN-gah), village, IRINGA region, central TANZANIA, 5 mi/8 km WSW of IRINGA, on LITTLE RUAHA RIVER; 06°48′S 35°36′E. Cattle, sheep, goats; tobacco, pyrethrum, corn, wheat.

Kalesija (kah-LE-shyah), village, NE BOSNIA, BOSNIA AND HERZEGOVINA.

Kaleva (kal-AI-vuh), village (2000 population 509), MANISTEE county, NW MICHIGAN, 18 mi/29 km NE of MANISTEE; 44°22′N 86°00′W. Apples, apple juice, and cider.

Kalevala (KAH-lee-vah-lah), Finnish *Uhtua*, town (2005 population 5,385), W central Republic of KARELIA, NW European Russia, on Middle KUITO Lake, 105 mi/169 km WNW of KEM' (linked by road); 65°12′N 31°10′E. Elevation 347 ft/105 m. Sawmilling, woodworking. Ceded by FINLAND to the Soviet Union in 1940 following the Winter War; renamed in 1948.

Kalewa (kah-LE-wah), town, SAGAING division, MYANMAR, on W bank of CHINDWIN RIVER (landing) at mouth of MYITTHA RIVER, 140 mi/225 km NW of MANDALAY. Coal deposits nearby. Head of roads to MANIPUR (INDIA) via TIDDIM and TAMU.

Kaleykino (kah-LYAI-kee-nuh), village, SE TATARSTAN Republic, E European Russia, on the ZAY River (left tributary of the KAMA River), on road and near railroad, 5 mi/8 km W of AL'MET'YEVSK, to which it is administratively subordinate; 54°55′N 52°12′E. Elevation 416 ft/126 m. Logging, lumbering, woodworking.

Kalga (kahl-GAH), village, SE CHITA oblast, E SIBERIA, RUSSIA, 110 mi/177 km ENE of BORZYA; 50°55′N 118°54′E. Elevation 2,273 ft/692 m. In agricultural area.

Kalgan, China: see ZHANGJIAKOU.

Kalghatgi (KUHL-guht-gee), town, tahsil headquarters, DHARWAD district, KARNATAKA state, SW INDIA, 19 mi/31 km S of DHARWAD; 15°11′N 74°58′E. Rice milling; millet, cotton, rice.

Kalgin Island, S ALASKA, in COOK INLET, 90 mi/145 km SW of ANCHORAGE; 60°26′N 151°57′W. Island is 13 mi/ 21 km long, 2 mi/3.2 km–4 mi/6.4 km wide.

Kalgo (KAHL-go), town, KEBBI state, NW NIGERIA, on ZAMFARA River, 85 mi/137 km SW of SOKOTO. Market town. Fish; millet, guinea corn.

Kalgoorlie (kal-GOOR-lee), town, WESTERN AUSTRALIA state, S AUSTRALIA, 371 mi/597 km E of PERTH; 30°45′S 121°28′E. It is the chief mining town of the state and the center of the EAST COOLGARDIE GOLDFIELD. Gold was found at nearby Coolgardie in 1892; nickel is also mined. The Western Australia

School of Mines (1902) was transferred (1903) from Coolgardie to Kalgoorlie. Adjacent towns of Kalgoorlie and Boulder (sometimes called Boulder City), combined in 1947.

Kalhat (kuh-HAT), township, E OMAN, on GULF OF OMAN, 80 mi/129 km SE of MUSCAT, at foot of EASTERN HAJAR hill country; 22°42′N 59°22′E. Was major port in 15th and 16th centuries prior to rise of Muscat under Portuguese. Sometimes spelled Qalhat.

Kalia, town (2001 population 18,430), JESSORE district, SW EAST BENGAL, BANGLADESH, 30 mi/48 km ESE of JESSORE; 23°00′N 89°38′E. Trades in rice, jute, linseed.

Kaliakra, Cape (kah-lee-AHK-rah), Romanian *Caliacra*, on the BLACK SEA, NE BULGARIA, 30 mi/48 km NE of VARNA; 43°22′N 28°29′E. Medieval ruins and lighthouse are on 200-ft/61-m cliffs.

Kalibia (kai-lee-BYAH), ancient *Clupea*, village (2004 population 43,209), NABUL province, NE TUNISIA, on the MEDITERRANEAN SEA, near E tip of CAPE BON peninsula, 63 mi/101 km E of TUNIS. Fishing port; tourism; muscat grapes. Founded by Phoenicians in 3rd century B.C.E.

Kalibo (kah-LEE-bo), town, ⊙ AKLAN province, N PANAY island, PHILIPPINES, on small inlet of SIBUYAN SEA, 28 mi/45 km WNW of ROXAS; 11°41′N 122°22′E. Agricultural center (tobacco, rice). Known for handicrafts made from abaca. Ati-Atihan, a major festival, is held here every January. Airport.

Kalida (kuh-LEI-duh), village (□ 1 sq mi/2.6 sq km; 2006 population 1,134), PUTNAM county, NW OHIO, 17 mi/27 km NNW of LIMA; 40°59′N 84°11′W. Livestock; grain.

Kali Gandaki River (KAH-lee guhn-DUH-kee), c. 200 mi/322 km long, central NEPAL; rises in the N NEPAL HIMALAYA c. 30 mi/48 km NNE of MUKTINATH, near Nepal-TIBET (CHINA) border; flows SSW past BAGLUNG and KUSMA, and E past Riri Bazar, joining TRISULI RIVER 55 mi/89 km W of KATHMANDU to form NARAYANI (GANDAK) River at 27°46′N 84°25′E. Also called KRISHNA GANDAKI.

Kaliganj (KAH-lee-gawnj), village (2001 population 36,733), DHAKA district, E EAST BENGAL, BANGLADESH, on tributary of DHALESWARI RIVER, 16 mi/26 km NE of DHAKA; 21°44′N 91°49′E. Agriculture (rice, jute, oilseeds); sugar milling.

Kaliganj, village, KHULNA district, SW EAST BENGAL, BANGLADESH, in the SUNDARBANS, on JAMUNA RIVER, 40 mi/64 km SW of KHULNA; 21°44′N 91°49′E. Trades in rice, jute, oilseeds; manufacturing of pottery, cutlery, horn implements.

Kalihiwai (kah-LEE-hee-WEI), village (2000 population 717), KAUAI island, KAUAI county, HAWAII, on Kalihiwai Bay, N coast, 16 mi/26 km N of LIHUE; 22°13′N 159°25′W. Sugarcane; fish. Princeville Airport to W; Kilauea Point National Wildlife Refuge to NE; Anini Beach Park to W.

Kalikino (kah-LEE-kee-nuh), village (2006 population 3,345), E LIPETSK oblast, S central European Russia, on the W bank of the VORONEZH RIVER, on local highway branch, 6 mi/10 km N of DOBROYE; 52°57′N 39°49′E. Elevation 377 ft/114 m. Fisheries.

Kalikot (KAH-lee-kot), district, W NEPAL in KARNALI zone; ⊙ MANMA.

Kali Limines (kah-LEE lee-MEE-nes) [Greek=fair havens], ancient *Lasea*, settlement, IRÁKLION prefecture, on SE coast of central CRETE department, S of IRÁKLION, GREECE; 34°56′N 24°48′E. Terminal for oil tankers offloading their cargo to storage tanks on shore. Formerly a sleepy fishing port. Said to be the spot where St. Paul first set foot on CRETE, as described in the New Testament. Memorialized by the name of one of the three offshore islets. Some ruins remain. Also known as Kaloi Limines, Kaloi Limenes.

Kalima-Kingombe (kah-LEE-mah–king-GOM-bai), village, SUD-KIVU province, E CONGO, near ULINDI RIVER, 125 mi/201 km N of KASONGO; 02°34′S 26°37′E. Elev. 2,798 ft/852 m. Center of tin-mining area employing c.11,000 workers; tin concentrating, also tantalite mining. Has hospitals, airports; two hydroelectric power plants nearby. KAMISUKU tin mines are 7 mi/11.3 km NE.

Kalimantan, INDONESIA; see BORNEO.

Kálimnos (KAH-leem-nos), mountainous island (□ 41 sq mi/106.6 sq km), DODECANESE islands, SOUTH AEGEAN department, GREECE, 11 mi/18 km off the coast of SW TURKEY. Formerly an important spongefishing center now in decline. Also produces figs, olives, citrus fruits, and almonds. The main town is Kálimnos, on the SE shore.

Kalimok (kah-lee-MOK), island in the DANUBE, NE BULGARIA; 44°02′N 26°33′E.

Kalimpang (kah-LIM-puhng), town, DARJILING district, N WEST BENGAL state, E INDIA, on TISTA RIVER, 13 mi/21 km E of DARJILING, in extreme SW ASSAM HIMALAYAN foothills; 27°04′N 88°29′E. Terminus of main trade route to TIBET; trade center for Tibetan goods; handicraft cotton-weaving center; manufacturing (beverages). Residence of Bhutanese political officer. Cinchona plantation 6 mi/9.7 km NE. Railroad station 4 mi/6.4 km SW, at Gielkhola. Health resort.

Kalin, village, N central TURKEY, 15 mi/24 km WSW of SIVAS; 39°41′N 36°47′E. Railroad junction.

Kali Nadi (KAH-lee-nuh-DEE), river, c.250 mi/402 km long, W UTTAR PRADESH state, N central INDIA; rises c.20 mi/32 km N of MEERUT, flows S and SE through GANGA-YAMUNA DOAB to the GANGA River SSE of FARRUKHABAD.

Kalinga-Apayao, province (□ 2,721 sq mi/7,074.6 sq km), CORDILERA ADMINISTRATIVE REGION, N central LUZON, PHILIPPINES; ⊙ TABUK; 17°45′N 121°15′E. Population 14.6% urban, 85.4% rural; in 1991, 44% of urban and 37% of rural settlements had electricity. Mountainous province, much unexplored. Predominantly tribal peoples living on subsistence agriculture in river valleys. Major river is the CHICO. Province was created 1966.

Kalingapatam, INDIA: see SRIKAKULAM.

Kalinikha (kah-LEE-nee-hah), village, E NIZHEGOROD oblast, central European Russia, near the VETLUGA RIVER, on road and local railroad spur, 2 mi/3.2 km S, and under administrative jurisdiction, of VOSKRE-SENSKOYE; 56°48′N 45°26′E. Elevation 334 ft/101 m. Logging and lumbering.

Kalinin (ka-LEEN-een), city, NE DASHHOWUZ weloyat, N TURKMENISTAN, on Khiva oasis, 25 mi/40 km NW of DASHHOWUZ; 42°08′N 59°40′E. Tertiary-level administrative center. Cotton. Until 1936, called Porsy. Formerly also called Kalininsk.

Kalinin (kah-LEE-nin), town, TOSHKENT wiloyat, NE UZBEKISTAN; 41°15′N 69°10′E. Tertiary-level administrative center.

Kalinin (kah-LEE-nin), village, KHATLON viloyat, TA-JIKISTAN, 23 mi/37 km S of KHATLON viloyat (linked by narrow-gauge railroad); long-staple cotton. Developed in 1930s. Formerly called Voroshilovabad.

Kalinin, RUSSIA: see TVER, city.

Kalinin, UKRAINE: see ORDZHONIKIDZE, city.

Kalininabad, town, KHATLON viloyat, TAJIKISTAN, Vaksh valley, 9 mi/15 km ENE of KURGAN-TYUBE; 39°45′N 69°08′E.

Kalininabad, village, central LENINOBOD viloyat, TAJI-KISTAN, on N slope of TURKESTAN RANGE, 15 mi/24 km SE of URA-TYUBE; 39°52′N 68°55′E. Wheat; horses. Until c.1935 called Kalai-Mirzabai.

Kalinina, Imeni M. I. (kah-LEE-nee-nah, EE-mye-nee), town (2006 population 2,735), NW NIZHEGOROD oblast, central European Russia, on road and railroad spur, 25 mi/40 km WNW of VETLUGA; 57°59′N 45°07′E. Elevation 410 ft/124 m. Paper-milling center (cardboard products). Before 1938, called Kartonnaya Fabrika (Russian=cardboard factory).

Kalininaul (kah-lee-neen-ah-OOL), village (2005 population 4,865), NW DAGESTAN REPUBLIC, NE CAU-CASUS, SE European Russia, on road, 50 mi/80 km W

of MAKHACHKALA, and 12 mi/19 km S of KHASA-VYURT; 43°04′N 46°34′E. Elevation 2,103 ft/640 m. Part of the territory claimed by Chechen Republic following the withdrawal of Russian troops in 1996; escalating tensions resulted in the renewal of the Russian-Chechen conflict in 1999. Until 1944, known as Yurt-Aukh.

Kalinin Bay, bay, SE ALASKA, on N shore of KRUZOF ISLAND, on Salisbury Sound, 25 mi/40 km NW of SITKA.

Kalinindorf, UKRAINE: see KALININS'KE.

Kaliningrad (kah-lee-neen-GRAHT), city (2005 population 436,760), ⊙ KALININGRAD oblast, W European Russia, 800 mi/1,287 km W of MOSCOW, on the Pregolya River near its mouth on the Vislinskiy Bay, which empties into the Gulf of Kaliningrad on the BALTIC SEA; 54°42′N 20°30′E. A major ice-free Baltic seaport and naval base, and an important industrial, fishing, and commercial center, Kaliningrad has industries that produce ships, railroad cars, machinery, food products (especially fish), wood pulp, and paper. It is the home of the Russian Baltic fleet. There are amber deposits nearby that are considered to be the richest in the world. The city of Königsberg was founded (1255) as a fortress of the Teutonic Knights. It joined the Hanseatic League (1340) and became the seat of the grand master of the Teutonic Order (1457). It was the residence of the dukes of Prussia from 1525 until 1618. As part of the N section of East Prussia, the city was transferred to the USSR in 1945, and the remaining German population was largely expelled. A new Soviet city named Kaliningrad was laid out in the former residential suburbs of Königsberg; its population is almost entirely Russian. The region is separated from the rest of Russia by LITHUANIA and BELARUS.

Kaliningrad (kah-lee-neen-GRAHD), oblast (□ 5,830 sq mi/15,158 sq km; 2004 population 932,000), in W European Russia, westernmost part of the RUSSIAN FEDERATION and Russia's smallest oblast, on the S coast of the BALTIC SEA; ⊙ KALININGRAD. A Russian enclave between LITHUANIA (N, E) and POLAND (S). Its Baltic shore consists of Courland and Vistula lagoons, separated by the Samland peninsula. Largely a humid lowland with moderate continental climate; drained by PREGEL and NEMAN rivers. Agriculture (potatoes, sugar beets, rye, barley, dairy cattle, hogs). Major industries include engineering, electronics, agriculture and forestry processing, and the world's richest amber deposits, processed at YANTARNYY; also, lumbering (paper and pulp milling), fishing. Noted seaside resorts are SVETLOGORSK and ZELENOGRADSK; main industrial centers are Kaliningrad (Königsberg), CHERNYAKHOVSK (Insterburg), SOVETSK (Tilsit), and GUSEV (Gumbinnen). Dense railroad and road network. Formed after World War II out of the N portion of former East Prussia. Assigned (1945) to the USSR by the Potsdam Conference. Was first constituted by the Russians as a special Königsberg okrug; became an oblast in April 1946; renamed Kaliningrad in July 1946. Population overwhelmingly Russian and 77% urban. Focus on trade with Poland and Germany.

Kaliningrad, RUSSIA: see KOROLEV.

Kalinino (kah-LEE-nee-nuh), town, SE KIROV oblast, E central European Russia, on the right bank of the SHOSHMA River near its confluence with the VYATKA River, on road, 2 mi/3.2 km SW, and under administrative jurisdiction, of MALMYZH; 56°30′N 50°38′E. Elevation 498 ft/151 m. In agricultural region; grain and produce processing, seed testing, agricultural machinery and supplies, creamery, distillery.

Kalinino (kah-LEE-nee-nuh), town (2005 population 34,760), central KRASNODAR TERRITORY, S European Russia, 6 mi/10 km W, and under administrative jurisdiction, of KRASNODAR (connected by road and railroad); 45°05′N 39°03′E. Elevation 154 ft/46 m. Most residents work in Krasnodar and neighboring towns.

Kalinino (kah-LEE-nee-nuh), village, SE AMUR oblast, SE SIBERIA, RUSSIAN FAR EAST, on the AMUR River (landing), on road, approximately 4 mi/6.4 km NW of the Chinese border, and 130 mi/209 km ESE of BLA-GOVESHCHENSK; 49°24′N 129°18′E. Elevation 377 ft/114 m. In agricultural area (grain, soybeans). Formerly known as Nikol'skoye.

Kalinino (kah-LEE-nee-nuh), village (2004 population 3,710), central CHUVASH REPUBLIC, central European Russia, on the GREATER TSIVIL RIVER, on road, 18 mi/29 km ENE of SHUMERLYA; 55°36′N 46°51′E. Elevation 370 ft/112 m. Grain. Until 1939, called Norusovo.

Kalinino, ARMENIA: see TASHIR.

Kalinino, UKRAINE: see KALININS'K.

Kalininsk (kah-LEE-neensk), city (2006 population 18,635), SW SARATOV oblast, SE European Russia, on the Balanda River (right affluent of the MEDVEDITSA River), 34 mi/55 km SW of ATKARSK (connected by a railroad spur); 51°30′N 44°28′E. Elevation 498 ft/151 m. Sewing factory; food processing (dairy products, poultry). Known as Balanda until 1962, when it was made a city.

Kalinins'k (kah-LI-neensk) (Russian *Kalinino*), city, central DONETS'K oblast, UKRAINE, in the DONBAS, 3 mi/4.8 km NE of HORLIVKA; 48°21′N 38°08′E. Elevation 846 ft/257 m. Coal-mining center. Until approximately 1935, called Bayrak.

Kalininskaya (kah-LEE-neen-skah-yah), village (2005 population 13,425), W central KRASNODAR TERRI-TORY, S European Russia, on road and railroad, 34 mi/56 km NNW of KRASNODAR; 45°29′N 38°40′E. Flour mill; in agricultural area, producing wheat, sunflowers, hemp, and dairy products. Formerly called Po-povicheskaya.

Kalinins'ke (kah-LYEE-nyin-ske), town, W KHERSON oblast, UKRAINE, on INHULETS' River (head of navigation), 36 mi/58 km NNE of KHERSON; 47°07′N 32°59′E. Flour mill, weaving. Village, established in 1807, was a Jewish settlement called Velyka Seyde-mynukha (Russian *Bol'shaya Seidemenukha*); renamed Kalinindorf in 1927, and became Kalinins'ke (Russian *Kalininskoye*) in 1944. Jewish population majority until the beginning of World War II (close to 3,900 in 1939), wiped out by the Nazis in 1941–1942; fewer than 100 Jews remaining in 2005.

Kalininskiy, RUSSIA: see KOROLEV.

Kalininskiy, UKRAINE: see KALININS'KYY.

Kalininskoye, UKRAINE: see KALININS'KE.

Kalinins'kyy (kah-LYEE-nyin-skyiy) (Russian *Kali-ninskiy*), town, S LUHANS'K oblast, UKRAINE, near source of Nahol'na River, tributary of MIUS RIVER, on road and railroad spur 4 mi/7 km SW and subordinated to SVERDLOVS'K; 48°01′N 39°35′E. Elevation 928 ft/282 m. Coal enrichment plant. Established in 1932; town since 1957.

Kalinjar (kah-LIN-juhr), village, BANDA district, S UTTAR PRADESH state, N central INDIA, 33 mi/53 km SSE of BANDA. Shivaite pilgrimage site. Famous ancient fort and city that figured in the *Mahabharata* epic. Was civil capital of Chandel Rajputs after its transfer (c.1182) from MAHOBA. Noted Afghan, Sher Shah, killed here in 1545. Stormed by Akbar's troops in 1569; occupied by British in 1812. Has extensive cave inscriptions, 12th-century ruins. Also called Tarahti.

Kalinjara (kah-LIN-jah-rah), village, BANSWARA district, RAJASTHAN state, NW INDIA, 16 mi/26 km SSW of BANSWARA. Hand-loom weaving; corn, rice. Ruined Jain temple.

Kalinkovichi (kuh-LEEN-ko-vi-chi), city, GOMEL oblast, BELARUS, 6 mi/10 km NNE of MOZYR, 52°10′N 29°13′E. Railroad junction; manufacturing (furniture, roofing materials, concrete articles, chemical products); meat packing, flour milling, fruit canning, machine overhauling plant.

Kalino (KAH-lee-nuh), town (2006 population 2,455), E PERM oblast, central URALS, E European Russia, on highway, 8 mi/13 km WSW, and under jurisdiction, of Chusovoy; 58°15′N 57°36′E. Elevation 456 ft/138 m. Railroad junction; peat works.

Kalinov, RUSSIA: see BYKOVO, VOLGOGRAD oblast.

Kalinovaya Balka, RUSSIA: see BYKOVO, VOLGOGRAD oblast.

Kalinovik (kah-LI-no-vik), town, upper HERZEGOVINA, BOSNIA AND HERZEGOVINA, 23 mi/37 km S of SARAJEVO; 43°30′N 18°26′E. Nearby TRESKAVICA Mountain rises NNW, LELIJA Mountain S.

Kalinovka (kah-LEE-nuhf-kah), village, N central KALININGRAD oblast, W European Russia, on road and narrow-gauge railroad, 11 mi/18 km N of INSTERBURG; 54°48′N 21°46′E. Elevation 118 ft/35 m. Until 1945, in East Prussia and called Aulowönen or Grossaulowönen and, later (1938–1945), Aulenbach.

Kalinovka, RUSSIA: see CHECHEN-AUL.

Kalinovka, UKRAINE: see KALYNIVKA, Vinnytsya oblast.

Kalinovka, UKRAINE: see KALYNIVKA, KIEV OBLAST.

Kalinovo (kah-LEE-nuh-vuh), town (2006 population 2,885), W central SVERDLOVSK oblast, E central URALS, W Siberian Russia, near Tavatuy Lake, on road and near railroad, 23 mi/37 km S of NEVYANSK, to which it is administratively subordinate; 57°08′N 60°09′E. Elevation 803 ft/244 m. Precious stone mining; fish farming.

Kalinovo, UKRAINE: see KALYNOVE.

Kalinovskaya (kah-LEE-nuhf-skah-yah), village (2005 population 8,315), NW CHECHEN REPUBLIC, S European Russia, on the TEREK River, on road and near railroad, 19 mi/31 km NW of GROZNYY; 43°34′N 45°31′E. Elevation 209 ft/63 m. Agriculture (grain, livestock). Established as a Cossack stronghold in the 19th century.

Kaliparhi, INDIA: see RANIGANJ.

Kalipetrovo (kah-lee-PE-tro-vo), village, RUSE oblast, SILISTRA obshtina, BULGARIA; 44°04′N 27°14′E.

Kali River, NEPAL: see SARDA RIVER.

Kaliro, administrative district, EASTERN region, SE central UGANDA. As of Uganda's division into eighty districts, borders PALLISA (NNE), NAMUTUMBA (ESE), IGANGA (S), and KAMULI (W) districts. Some marsh area. Rural and agricultural area (including bananas and coffee). Secondary railroad runs through S tip of district, connecting it to main railroad between KASESE town (W Uganda) and MOMBASA (SE KENYA). Formed in 2005 from NE portion of former KAMULI district (current Kamuli district formed from all but NE portion).

Kaliro (kah-LEE-ro), town, KALIRO district, EASTERN region, SE UGANDA, on railroad, 35 mi/56 km NE of Jinja. Tobacco, coffee, bananas, corn. Was part of former BUSOGA province.

Kalis (KAH-lees), village, NE central SOMALIA, 40 mi/64 km E of GAROE, in NOGAL (Nuga) Valley; 08°26′N 49°05′E. Market (sheep, camels, dates). Sometimes spelled CALLIS.

Kalisch, POLAND: see KALISZ.

Kali Sindh River (KAH-lee SIND), c.220 mi/354 km long, MAHARASHTRA and RAJASTHAN states, W central INDIA, rises in VINDHYA RANGE in DEWAS district, MAHARASHTRA state, c.30 mi/48 km ESE of INDORE, flows N past SONKACH and SARANGPUR, to CHAMBAL RIVER 12 mi/19 km SE of Kakheri (RAJASTHAN state). Sometimes spelled Kali Sind.

Kalisizo (kah-li-SEE-zo), town, CENTRAL region, S UGANDA, 15 mi/24 km SSW of MASAKA. Cotton, coffee, bananas, corn, millet. Was part of former SOUTH BUGANDA province.

Kalispell (KA-lis-pel), city (2000 population 14,223), ⊙ FLATHEAD county, NW MONTANA, 95 mi/153 km N of MISSOULA, on FLATHEAD River and Ashley Creek, at mouths of STILLWATER and Whitefish rivers; 48°12′N 114°19′W. Tourism; forestry (including Christmas-tree production); Agriculture (fruit, wheat, barley, corn (maize); peas, garbanzo beans, lentils, rapeseed, mustard, mint, potatoes; cattle, horses, hogs,

llamas, poultry); trade; manufacturing (lumber, furniture; food processing; meat and dairy products; semiconductor equipment, glass items; service industries. Flathead Valley Community College, Conrad Mansion (1895), and Hockaday Center for the Arts here. The headquarters of the Flathead National Forest are in KALISPELL, parts of forest to S, W, and E. FLATHEAD LAKE, 9 mi/14.5 km SSE, GLACIER NATIONAL PARK to NW. Creston National Fish Hatchery to E; Lone Pine State Park SW of city. HUNGRY HORSE Dam and Reservoir to E. Incorporated 1892.

Kalisz (KAH-leesh), German *Kalisch*, city (2002 population 109,498), central POLAND. An industrial center, it produces textiles, clothing, chemicals, and paper. One of the oldest Polish towns, it has been identified as the Slavic settlement of Calissia mentioned in the 2nd century by Ptolemy. It flourished as a trade center from the 13th century. At Kalisz, Casimir III signed (1343) the treaty with the Teutonic Knights by which he gave up his rule over East POMERANIA. The city passed to PRUSSIA in 1793, was transferred to RUSSIA in 1815, and was restored to Poland in 1919. In a treaty signed (1813) at Kalisz, Prussia and Russia formed an alliance against Napoleon I.

Kalisz Pomorski (KAH-leesh po-MOR-skee), German *Kallies*, town, Koszalin province, NW POLAND, in lake region, 40 mi/64 km ESE of STARGARD SZCZECINSKI. Paper manufacturing. Until 1945, was in POMERANIA.

Kalita, UKRAINE: see KALYTA.

Kalitva River (kah-leet-VAH), approximately 130 mi/209 km long, ROSTOV oblast, S European Russia; rises E of CHERTKOVO, flows S through steppe, past KRIVOROZHYE and Litvinovka, to the Northern Donets River at BELAYA KALITVA.

Kaliua (ka-lee-OO-ah), village, TABORA region, W central TANZANIA, 70 mi/113 km W of TABORA; 05°05′S 31°48′E. Junction of railroad spur to MPANDA mining district. Timber; subsistence crops (corn, wheat); livestock (sheep, goats). Also spelled Kaliuwa.

Kaliub, EGYPT: see QALYUB.

Kaliurang (kah-LEE-yoo-rahng), resort area, Yogyakarta province, S central JAVA, INDONESIA, 16 mi/25 km N of YOGYAKARTA; 07°36′S 110°25′E. Situated on S slope of Mount MERAPI, the resort is a popular destination for climbers.

Kalivia, Greece: see ASPROPIRGOS.

Kaliwungu (kah-lee-WOONG-goo), town, Java Tengah province, INDONESIA, near JAVA SEA, 10 mi/16 km W of SEMARANG; 06°57′S 110°14′E. Trade center in agricultural area (sugar, rice, peanuts, tobacco, coffee, kapok). Formerly spelled Kaliwoengoe.

Kalix (KAH-leeks), village, NORRBOTTEN county, N SWEDEN, on small inlet of GULF OF BOTHNIA at mouth of KALIX ÄLVEN RIVER, 30 mi/48 km W of HAPARANDA; 65°52′N 23°10′E. Manufacturing (paper machine industry). Church (fifteenth century).

Kalix älven (KAH-leeks-EL-ven), river, 270 mi/435 km long, Lappland, N SWEDEN; rises near NORWEGIAN border WNW of KIRUNA, flows in wide arc SE and S, past ÖVERKALIX, to GULF OF BOTHNIA at KALIX.

Kaliyala, RUSSIA: see KAMENKA, LENINGRAD oblast.

Kalka (kahl-KAH), town, AMBALA district, HARYANA state, N INDIA, on railroad (workshops), 33 mi/53 km NNE of AMBALA, at the foot of the Shiwalik HIMALAYAS; 30°50′N 76°56′E. Trades in spices, wheat, maize, bamboo; manufacturing of aerated water, snuff, millstones. Limestone quarried just E. Acquired 1843 by British from Patiala state as depot for Shimla. Light railroad terminus where passengers transfer from broad-gauge track to continue journey to hill resort at Shimla.

Kalkali Ghat, INDIA: see PATHAPKANDI.

Kalkandelen, MACEDONIA: see TETOVO.

Kalkar (kahl-KAHR), town, NORTH RHINE–WESTPHALIA, W GERMANY, 7 mi/11.3 km SE of KLEVE; 51°44′N 06°18′E. Center of an agricultural region with sugar

refining, dairying, and butchering. It was the site of a center for research on fast-breeding nuclear reactors (abandoned in 1991 due to political pressure). Has 15th-century church and town hall. Gen. Seydlitz born here. Until 1936, spelled CALCAR.

Kalkara (kahl-KAH-rah), town, E MALTA, on inlet of GRAND HARBOUR, c.1 mi/1.6 km E of VALLETTA; 35°53′N 14°31′E. Fishing, boat repairing. Heavily damaged during World War II. Its 19th-century parish church and Villa Bighi, which housed naval hospital, were completely destroyed. Severely damaged also was the adjoining baroque Fort Risacoli (1670) at E entrance of Grand Harbour opposite Fort St. Elmo. Sometimes spelled Calcara.

Kalkaska (kal-KAS-kuh), county (□ 570 sq mi/1,482 sq km; 2006 population 17,330), NW central MICHIGAN; 44°41′N 85°04′W; ⊙ KALKASKA. Drained by North Branch of MANISTEE and BOARDMAN rivers. Agriculture (potatoes, corn, wheat; cattle, hogs); logging, wood products, especially in NE; resorts. Several lakes. Includes a state forest and a state game refuge. Part of Camp Grayling Michigan National Guard Reserve in SE; part of LAKE SKEGEMOG in NW corner (S extension of ELK LAKE). Organized 1871.

Kalkaska (kal-KAS-kuh), town (2000 population 2,226), ⊙ KALKASKA county, NW central MICHIGAN, 22 mi/35 km ESE of TRAVERSE CITY, on BOARDMAN RIVER; 44°43′N 85°10′W. Supply center for farm and lake area; manufacturing (steel parts, crafts, wire products). Incorporated 1887.

Kalken (KAHL-kuhn), village, Dendermonde district, EAST FLANDERS province, NW BELGIUM, on SCHELDT RIVER, 9 mi/14.5 km E of GHENT; 51°02′N 03°55′E. Agricultural market. Gothic church. Formerly spelled Calcken.

Kalkfontein Dam, dam, FREE STATE, SOUTH AFRICA, on RIET RIVER, 65 mi/104 km SW of BLOEMFONTEIN (MANGAUNG). Supplies water to W area of Free State for diamond mining and irrigation farming.

Kalkuda (KAHL-ku-dah), village, EASTERN PROVINCE, SRI LANKA, on E coast, 18 mi/29 km NNW of BATTICALOA; 08°07′N 79°43′E. Coconuts. Excellent beaches and diving. Holiday resort at nearby Passekudah.

Kall (KAHL), village, NORTH RHINE–WESTPHALIA, W GERMANY, 28 mi/45 km SW of BONN; 50°33′N 06°33′E. Limestone quarries; has 12th-century castle and church.

Kallaa of Béni Hammad (kahl-LAH BAI-nee hah-MAHD), historic site, M'SILA wilaya, interior central ALGERIA. An important Roman site; remains exist. In the early 11th century, became capital of the Hammadite dynasty, founded by Hammad, son of Bologhine Ibn Ziri, the founder of ALGIERS. It was abandoned 80 years later by BEJAIA following the Béni Hillal invasion. Large-scale excavations both during and since French occupation have unearthed remains of richly decorated palaces.

Kalladi (KAHL-lah-dee), village, EASTERN PROVINCE, SRI LANKA, on long lagoon near E coast, 1 mi/1.6 km E of BATTICALOA; 07°43′N 81°43′E. Coconuts. Also spelled Kallady.

Kalla, El (kahl-LAH, el), MEDITERRANEAN coastal port town, El TARAF wilaya, NE ALGERIA, E of ANNABA, 11 mi/18 km W of TUNISIA border; 36°54′N 08°27′E. In a heavily wooded region with many lakes. Cork oak trees cover an area of 270 sq mi/699 sq km. A national park is in the area. Renowned for its coral fishing. In the mid-16th century, a trading and fishing company from MARSEILLE won the right from the Dey (Turkish ruler of Algeria) to set up a base here. It was the first French foothold in N AFRICA. Formerly called La Calle.

Kallafo, ETHIOPIA: see K'ELAFO.

Kallakkurichchi (KUHL-luh-KUHR-i-chee), town, VILLUPURAM RAMASAMY PADAYATCHIAR district, TAMIL NADU state, S INDIA, 55 mi/89 km W of CUDDALORE; 11°44′N 78°58′E. Trades in timber and gum, arabic from KALRAYAN HILLS (W; magnetite deposits). Cotton and woolen weaving; rice, millet. Formerly also spelled Kallakurchi.

Kallam (kuhl-LUHM), town, OSMANABAD district, MAHARASHTRA state, INDIA, on MANJRA RIVER, 28 mi/45 km N of OSMANABAD. Cotton ginning; millet, wheat, rice.

Kallands, locality, central ALASKA, on YUKON River, 30 mi/48 km W of TANANA.

Kallang (KAHL-lahng), NE section of SINGAPORE city, S Singapore island, SINGAPORE, 3 mi/4.8 km NE of downtown, at mouth of small Kallang River; 01°18′N 103°52′E. On Mass Rapid Transit (MRT) line. Manufacturing (apparel, semiconductors, computer peripherals, electronic components, household audio and video equipment, shipbuilding). Paya Lebar Airport to NE. National Stadium here.

Kallaste (KAH-lah-stai), city, E ESTONIA, on Lake Peipu, 25 mi/40 km NE of TARTU; 58°39′N 27°09′E. Agricultural market (flax, oats); fishing. Population largely Russian.

Kallavesi (KAHL-lah-ve-see), lake, S central FINLAND. On W shore is KUOPIO. Dotted with numerous islands, it forms part of SAIMAA lake system, through which it drains into LAKE LADOGA (RUSSIA). Lake is 30 mi/48 km long, 1 mi/1.6 km–10 mi/16 km wide.

Kallayi, INDIA: see KOZHIKODE.

Kalletal (KAHL-le-tahl), commune, NORTH RHINE–WESTPHALIA, N GERMANY, 17 mi/27 km NE of BIELEFELD; 52°07′N 08°55′E. Commune consists of sixteen villages. Woodworking; tourism; manufacturing of machinery. There is a 16th-century Renaissance palace in nearby suburb.

Källfallet (SHEL-fahl-let), village, VÄSTMANLAND county, central SWEDEN, in BERGSLAGEN region, 15 mi/24 km SW of FAGERSTA; 59°40′N 15°32′E.

Kalli, INDIA: see KOZHIKODE.

Kallia, settlement in WEST BANK, at N end of DEAD SEA, 16 mi/26 km E of JERUSALEM; 1,286 ft/392 m below sea level. In West Bank since 1948; was entirely destroyed (1948) when region was invaded by JORDAN forces. It had been site of important potash-extracting and -refining plant. Just SW was site of popular health resort. Reestablished 1968 as a kibbutz near original site on hill 1 mi/1.6 km from the shores of the Dead Sea. Growing dates and tropical fruit.

Kallidaikurichi (KUHL-li-DEI-kuhr-i-chee), town, Tirunelveli Kattabomman district, TAMIL NADU state, S INDIA, on TAMBRAPARNI River opposite AMBASAMUDRAM, 15 mi/24 km WSW of TIRUNELVELI; 08°40′N 77°29′E. Cotton-weaving center; sesame oil extraction. Also spelled Kallidaikurichchi, Kulladakurichi.

Kallidromon (kah-LEE-[th]ro-mon), mountain massif in FTHIOTIDA prefecture, CENTRAL GREECE department, GREECE; rises to 4,589 ft/1,399 m 12 mi/19 km SSE of LAMÍA; 38°44′N 22°34′E. Also called Koukos; also spelled Kallidhromon.

Kallies, POLAND: see KALISZ POMORSKI.

Kallifiton, Greece: see KALLIFYTOS.

Kallifytos (kah-LEE-fee-tos), town, DRÁMA prefecture, EAST MACEDONIA AND THRACE department, NE GREECE, 4 mi/6.4 km ENE of DRÁMA; 41°10′N 24°13′E. Tobacco, barley; olive oil. Formerly called Ravika; sometimes Kallifiton or Kalliphyton.

Kallinge (KAHL-leeng-e), village, BLEKINGE county, S SWEDEN, 12 mi/19 km WNW of KARLSKRONA; 56°15′N 15°17′E.

Kallipefki (kah-lee-PEF-kee), village, LÁRISSA prefecture, NE THESSALY department, N GREECE, on slope of the Lower OLYMPOS, 16 mi/26 km ENE of ELASSON; 39°59′N 22°28′E. Chromite mining. Formerly called Nezeros. Also spelled Kallipeuke, Kallipevki.

Kallista (kuh-LI-stuh), town, VICTORIA, SE AUSTRALIA, 27 mi/43 km E of MELBOURNE, on SE edge of Dandenong Ranges National Park; 37°53′S 145°22′E. Potatoes, berry fruits; flowers, bulbs.

Kallithea (kah-lee-THE-ah), SW suburb of ATHENS, ATTICA prefecture, ATTICA department, E central GREECE, 2.5 mi/4 km from city center; 38°19′N 23°27′E. Cotton textile manufacturing.

Källö-Knippla (CHEL-UH-KNIP-lah), fishing village, GÖTEBORG OCH BOHUS county, SW SWEDEN, on islet (□ 123 acres/50 ha) of same name in SKAGERRAK, 12 mi/19 km WNW of GÖTEBORG.

Kalloni, Gulf of (kah-lo-NEE), inlet of AEGEAN SEA in S LESBOS island, LESBOS prefecture, NORTH AEGEAN department, GREECE; 13 mi/21 km long, up to 5 mi/8 km wide. Kalloni near N shore. Also spelled Gulf of Kallone.

Kallurkot (kahl-luhr-KOT), town, MIANWALI district, W PUNJAB province, central PAKISTAN, 33 mi/53 km SSW of MIANWALI; 32°09′N 71°16′E. Wheat, millet, dates. Also spelled KALUR KOT.

Kallurkotte, INDIA: see HOSANAGARA.

Kalmak Kure, China: see ZHAOSU, Zhaosu county.

Kalmalo (KAHL-MAHL-o), town, SOKOTO state, extreme NW NIGERIA, on NIGER border opposite BIRNI NKONNI (Niger), 45 mi/72 km N of SOKOTO; 13°43′N 05°15′E. Tobacco, millet, rice; cattle, skins. Sometimes spelled Kalmaloo.

Kalmaloo, NIGERIA: see KALMALO.

Kalmanka (kahl-MAHN-kah), village, central ALTAI TERRITORY, S central SIBERIA, RUSSIA, on the OB′ RIVER, on road, 40 mi/64 km S of BARNAUL; 52°54′N 83°32′E. Elevation 442 ft/134 m. In agricultural area (rye, oats, flax; livestock raising).

Kalmar (KAHL-MAHR), county (□ 4,484 sq mi/11,658.4 sq km), SE SWEDEN, on BALTIC SEA; ⊙ KALMAR; 57°20′N 16°00′E. Includes ÖLAND island; mainland section of county includes E part of SMÅLAND province. Surface is undulating, generally wooded. Marshy in W; coastal region has fertile soil with large meadow areas. Agriculture (sheep; dairying); manufacturing (woodworking, furniture, glass, metal products). Important towns are Kalmar, VÄSTERVIK, OSKARSHAMN (nuclear power station), NYBRO, VIMMERBY, and BORGHOLM.

Kalmar (KAHL-MAHR), town, ⊙ Kalmar county, SE SWEDEN, on Kalmarsund (arm of BALTIC SEA) opposite ÖLAND Island (with EUROPE's longest bridge); 56°41′N 16°21′E. Commercial and tourist center; airport; manufacturing (machinery, glass, processed food). Important trade center since eighth century Kalmar Union negotiated here (1397). Kalmar Slott (thirteenth-century castle) withstood numerous sieges during Danish-Swedish wars (sixteenth–seventeenth centuries).

Kalmar Sound (KAHL-MAHR), strait of BALTIC SEA, between SE SWEDEN and ÖLAND island, 85 mi/137 km long, 4 mi/6.4 km–14 mi/23 km wide. Chief ports are KALMAR and OSKARSHAMN (mainland); BORGHOLM and FÄRJESTADEN (Öland island).

Kalmeshwar (kuhl-MAISH-wuhr), town, NAGPUR district, central MAHARASHTRA state, central INDIA, 12 mi/19 km WNW of NAGPUR; 21°14′N 78°54′E. Cotton ginning; millet, wheat, oilseeds.

Kalmit, GERMANY: see HARDT MOUNTAINS.

Kal′mius River (KAHL-mee-oos), 130 mi/209 km long, DONETS′K oblast, UKRAINE; rises near YASYNUVATA 6 mi/9.7 km N of DONETS′K, flows S, past Donets′k, to Sea of AZOV at MARIUPOL′.

Kalmthout (KAHLMT-hout), commune (2006 population 17,508), Antwerp district, ANTWERPEN province, N BELGIUM, near NETHERLANDS border, 12 mi/19 km NNE of ANTWERP; 51°23′N 04°28′E. Market gardening, in reclaimed polders.

Kalmuck, in Russian names: see KALMYK.

Kalmunai, town, EASTERN PROVINCE, SRI LANKA, on E coast, 24 mi/39 km SSE of BATTICALOA; 07°25′N 81°49′E. Resort.

Kalmykia, RUSSIA: see KALMYKIA-KHALMG-TANGEH.

Kalmykia-Khalmg-Tangeh, Republic of (kulh-MI-kee-yah–HAHLMG–TAHN-ge) or **Kalmykia**, constituent republic (□ 29,400 sq mi/76,440 sq km; 2004 population 319,000), SE European Russia, on the CASPIAN SEA; ⊙ ELISTA. Lying mostly in the vast depression of the N Caspian Lowland, the republic is largely a steppe and desert area. There are salt lakes but no permanent waterways. Livestock raising (horses, cattle, sheep, goats, and some pigs and camels) is by far the leading economic activity, and fishing is important. Irrigation has made limited agriculture possible; winter wheat, maize, and fodder crops are grown. Industry revolves primarily around the processing of agricultural products, fish, and minerals. The population, most of which lives in rural areas, is primarily Kalmyk (45%) and Russian (37%), with small Gard, Chechen, Kazakh, German, Ukrainian, and Avar minorities. Largest population centers are ELISTA, LAGAN, and GORODOVIKOVSK. A seminomadic branch of the Oirat Mongols, the Kalmyks migrated from Chinese Turkistan to the steppe W of the VOLGA RIVER mouth in the mid-17th century. They became allies of the Russians and were charged by Peter I with guarding the E frontier of the Russian Empire. Under Catherine II, however, the Kalmyks became vassals. In 1771, about 300,000 Kalmyks E of the Volga set out to return to China but were decimated en route by Russian, Kazak, and Kyrgyz attacks. The Kalmyks W of the Volga remained in Russia, where they retained their Lamaist Buddhist religion and their seminomadic life. The word *Kalmyk* in Turkish means "remnant," referring to those who stayed behind. The Kalmyk Autonomous Region was established in 1920; it became an autonomous republic in 1936. During World War II, Kalmyk units fought the Russians in collaboration with the Germans. As a result, the Kalmyks were deported to SIBERIA in 1943, and their republic was dissolved. In 1956, Nikita Khrushchev denounced the deportation as a Stalinist crime, and the following year about 6,000 Kalmyks were returned. The Kalmyk Autonomous SSR was officially reestablished in 1958. The republic was a signatory, under the name Republic of Kalmykia-Khalmg-Tangeh, to the March 31, 1992, treaty that created the Russian Federation. Separatist movement continued to be a strong presence in the republic until local elections in 1994. Formerly called the Kalmyk Republic.

Kalmykovo (kahl-MI-ko-vuh), town, S WEST KAZAKHSTAN region, KAZAKHSTAN, on URAL River, 150 mi/241 km S of ORAL (Uralsk); 49°02′N 51°55′E. Tertiary-rural (raion) administrative center. Millet; cattle.

Kalmytskiy Bazar, RUSSIA: see PRIVOLZHSKIY, ASTRAKHAN oblast.

Kalna (KAHL-nah), town, BARDDHAMAN district, central WEST BENGAL state, E INDIA, on the HUGLI River, 32 mi/51 km E of BARDDHAMAN; 23°13′N 88°22′E. Trades in rice, potatoes, wheat, sugarcane, gram. Rice milling, cotton weaving. Site of 109 Shiva lingam temples (built 1809), Muslim fort ruins.

Kalnibolotskaya (kahl-nee-buh-LOTS-kah-yah), village (2005 population 6,610), NE KRASNODAR TERRITORY, S European Russia, on the YEYA RIVER, on road, 22 mi/35 km NE of TIKHORETSK; 46°04′N 40°27′E. Elevation 357 ft/108 m. In agricultural area (wheat, flax, sunflowers, castor oil; livestock). Has a Holocaust memorial honoring the Kuban Jews exterminated by the Nazis during World War II.

Kalnik (KAHL-neek), mountain (2,109 ft/643 m), central CROATIA, 13 mi/21 km SSE of Varaždin. In lignite area.

Kalnik (KAHL-neek), river, central BULGARIA, tributary of the VIT RIVER; 43°02′N 24°20′E.

Kalni River, BANGLADESH: see SURMA VALLEY.

Kalnitsa (KAHL-nee-tsah), river, SE BULGARIA, tributary of the TUNDZHA RIVER; 42°20′N 26°25′E.

Kal′nivka, UKRAINE: see TROYITS′KE, town.

Kal′novka, UKRAINE: see TROYITS′KE, town.

Kalø Bay, DENMARK: see MOLS.

Kalocsa (KAH-lo-chah), city, BÁCS-KISKUN county, S central HUNGARY, near the DANUBE River, 23 mi/37 km N of BAJA; 46°32′N 19°00′E. Railroad terminal, market center. Agricultural experiment station; grain, red pepper; cattle; manufacturing (flour milling and farinaceous foods, dairy products, industrial equipment, transport vehicles). Beautiful embroidery produced in vicinity; vineyards. Created bishopric by Saint Stephen; became seat of an archbishop in 12th century. Has archepiscopal palace (18th century), library, Roman Catholic academy, observatory.

Kalofer (kah-LO-fer), city, PLOVDIV oblast, KARLOVO obshtina, central Bulgaria, on the S spur of the KALOFER MOUNTAINS, at the Krastets Pass, on the TUNDZHA RIVER, 8 mi/13 km E of Karlovo; 42°36′N 24°59′E. Summer resort; horticulture, wine making, tailoring, manufacturing (electromagnets). Once a Turkish cloth-weaving town; declined following a Bulgarian uprising (1876), when it was burned by Turks. Hristo Botev was born here.

Kalofer Mountains (kah-LO-fer), highest section of the STARA PLANINA Mountains, central BULGARIA, extends c.20 mi/32 km between the Troyan Mountains (W) and the SHIPKA MOUNTAINS (E); 42°45′N 25°00′E. Rise to 7,793 ft/2,375 m at the BOTEV PEAK. Its S spur is crossed by the Krastets Pass.

Kalogrea, Cape (kah-LO-yre-ah), Latin *Canastraeum*, SE extremity of KASSANDRA prong of KHALKIDHIKÍ Peninsula, CENTRAL MACEDONIA department, NE GREECE, on AEGEAN SEA; 39°54′N 23°44′E. Also called Kalogria, Kanastraion, and Paliouri, or Paliuri.

Kalogrea, Cape, Greece: see ARAXOS, CAPE.

Kalohi (kah-LO-hee), channel in PACIFIC OCEAN between LANAI and MOLOKAI islands, MAUI county, HAWAII, 8 mi/12.9 km wide. Splits into PAILOLO CHANNEL (to NE) and AUAU CHANNEL (to SE).

Kaloi Limines, Greece: see KALI LIMINES.

Kaloko-Honokohau National Historic Park (kah-LO-ko HO-no-KO-HOU) (1.81 sq mi/4.69 sq km), W HAWAII island, HAWAII county, HAWAII, on KONA Coast, 3.5 mi/5.6 km NNW of KAILUA-KONA on Honokohau Bay; authorized 1978. Site of important early pre-European settlements; archaeological remains; petroglyphs; three fish ponds.

Kalol (KAH-lol), town, MAHESANA district, GUJARAT state, W central INDIA, 25 mi/40 km SSE of MAHESANA; 22°36′N 73°27′E. Railroad junction; market center; trades in millet, pulses, wheat, oilseeds; cotton milling, handicraft cloth weaving, thread and bobbin manufacturing, bone crushing.

Kalol, town, PANCH MAHALS district, GUJARAT state, W central INDIA, 15 mi/24 km SW of GODHRA; 23°15′N 72°29′E. Local market for timber, millet, wheat.

Kaloleni (kah-lo-LAI-nee), village, COAST province, SE KENYA, 27 mi/43 km N of MOMBASA; 03°49′S 39°38′E. Sugarcane, fruits. Mission.

Kalomo (kah-LO-mo), township, SOUTHERN province, S ZAMBIA, 70 mi/113 km NE of MARUMBA. On railroad. Farming and ranching center; agriculture (tobacco, corn, vegetables); cattle. Manufacturing (paprika processing). Coal mining is at MAAMBA; amethyst. Was capital of Barotseland–North-Western Rhodesia until formation (1911) of Northern Rhodesia. KAFUE NATIONAL PARK to NW.

Kalona, town (2000 population 2,293), WASHINGTON county, SE IOWA, near ENGLISH RIVER, 14 mi/23 km N of WASHINGTON; 41°29′N 91°42′W. Feed, processed turkeys, dairy and metal products. Large Amish settlement.

Kalones, Cape, Greece: see SOUNION, CAPE.

Kalookan, PHILIPPINES: see CALOOCAN.

Kalopanayiotis (kah-lo-pahn-ah-YO-tis), village, LEFKOSIA district, W central CYPRUS, on Setrakhos River, 33 mi/53 km WSW of NICOSIA, 2 mi/3.2 km N of PEDHOULAS; 34°59′N 32°50′E. Elev. 2,350 ft/716 m. Fruit-growing center (grapes, peaches, pears, lychees, olives); vegetables, almonds; goats, sheep. Spa and

resort area known for its sulphur springs. Monastery of Ayios Ioannis Lambadistis.

Kalorama, township, VICTORIA, SE AUSTRALIA, 25 mi/40 km E of MELBOURNE, in Dandenong Ranges, and just N of Olinda State Forest; 37°49′S 145°22′E. Potatoes, fruit; flowers.

Kaloshino (kah-LO-shi-nuh), settlement, NE MOSCOW oblast, central European Russia, near highway, 20 mi/32 km N of (and administratively subordinate to) SERGIYEV POSAD; 56°40′N 38°04′E. Elevation 472 ft/143 m. In agricultural area; produce processing.

Kalotina (kah-LO-teen-ah), village, SOFIA oblast, DRAGOMAN obshtina, W BULGARIA, near the Yugoslav (Serbian) border, on the NISHAVA RIVER, 30 mi/48 km NW of SOFIA; 42°59′N 22°52′E. Rye, potatoes. Has a 16th-century church. Once a Roman settlement, Calotina, on road from Pannonia to Byzantium. Berende, across the Nishava, has a 14th-century church.

Kalovo, Bulgaria: see DYANKOVO.

Kaloyanovo (kah-lo-YAHN-o-vo), village, PLOVDIV oblast, ⊙ Kaloyanovo obshtina (1993 population14,735), W central BULGARIA, 14 mi/23 km N of PLOVDIV; 42°04′N 24°44′E. Rye, potatoes; livestock. Formerly called Seldzhikovo.

Kalpaki (kahl-PAH-kee), village, IOÁNNINA prefecture, EPIRUS department, NW GREECE, 22 mi/35 km N of IOÁNNINA; 39°53′N 20°38′E. At a strategic road junction. The Greeks stopped the invading Italian army here in November 1940.

Kalpentyn, SRI LANKA: see KALPITIYA.

Kalpetta, town, ⊙ WAYANAD district, KERALA state, S INDIA; 11°36′N 76°06′E.

Kalpi (KAHL-pee), town, JALAUN district, S UTTAR PRADESH state, N central INDIA, on the YAMUNA RIVER (bridged), 21 mi/34 km ENE of ORAI; 26°07′N 79°44′E. Trades in gram, wheat, oilseeds, jowar, cotton, ghee. Has 15th-century Afghan ruins. Captured by Afghans in 1196; became an important fortress of Muslim power; taken by Mogul emperor, Humayan, in 1527; stormed by Afghans under Sher Shah in 1540. Headquarters of Marathas in BUNDELKHAND during 18th century. Scene (1477) of great battle between Jaunpur kingdom and Delhi sultan. Formerly an important cotton center of East India Company. Nearby is site of important British victory (1858) over native leaders, including Rani of Jhansi, during Sepoy Rebellion.

Kalpin (KUH-PING), town and oasis, ⊙ Kalpin county, W XINJIANG UYGUR AUTONOMOUS REGION, NW CHINA, 75 mi/121 km SW of AKSU, on road to KASHI; 40°31′N 79°03′E. Livestock; food processing. Also spelled Keping.

Kalpitiya (KUHL-pi-TI-yuh), fishing village, NORTH WESTERN province, SRI LANKA, on Kalpitiya Peninsula, on DUTCH BAY, 15 mi/24 km NNW of PUTTALAM; 08°14′N 79°46′E. Coconuts. Ruins of Dutch fort (built 1646). Sri Lankan Christian pilgrimage center at Saint Anna's Church, 10 mi/16 km SSW. Kalpitiya Peninsula, in SE Gulf of MANNAR, extends 27 mi/43 km N-S, parallel to mainland; forms W shore of Dutch Bay and PUTTALAM LAGOON. Formerly called KALPENTYN.

Kalrayan Hills (kuhl-RAH-yuhn), outlier of S Eastern GHATS, TAMIL NADU state, S INDIA, NE of SALEM; c.25 mi/40 km long, c.20 mi/32 km wide. Rise to over 4,000 ft/1,219 m in SW plateau (magnetite deposits). Timber, gum arabic, tanbark. GADILAM RIVER rises in E foothills.

Kals am Grossglockner (KAHLS ahm gros-GLAWKner), village, EAST TYROL, SW AUSTRIA, at S foot of Mount GROSSGLOCKNER, 12 mi/19 km NNW of LIENZ; 47°00′N 12°39′E. Elevation 4,039 ft/1,231 m. Summer tourism; base for mountain tours to Grossglockner (including the first tour to the mountain in 1853). Hohe Tauern (national park) nearby.

Kalsching, CZECH REPUBLIC: see CHVALSINY.

Kalsdorf (KAHLS-dorf), township, STYRIA, SE AUSTRIA, near MUR River, 7 mi/11.3 km S of GRAZ; 46°58′N 15°29′E. Airport of Graz (Thalerhof) is nearby.

Kalsi (KAHL-see), village, DEHRADUN district, NW Uttarakhand state, N central INDIA, near the YAMUNA RIVER, 19 mi/31 km NNW of DEHRADUN. Large quartz boulder nearby bears edicts of Asoka.

Kalsia (KUHLS-yuh), former princely state of PUNJAB HILL STATES, INDIA. Since 1948, merged with PATIALA AND EAST PUNJAB STATES UNION; part of HARYANA state since it was carved out of PUNJAB state in 1966.

Kalsia, INDIA: see CHHACHHRAULI.

Kalskag, Inuit village, W ALASKA, on lower KUSKOKWIM River, 70 mi/113 km NE of BETHEL; 61°32′N 160°22′W.

Kalsoy, Danish *Kalsø*, island (□ 12 sq mi/31.2 sq km) of the NE FAEROE ISLANDS; c.10 mi/16 km long, 1.5 mi/2.4 km wide. Mountainous terrain; highest point is Nestindur (2,591 ft/790 m), near N tip. Fishing, sheep raising.

Kalsubai Peak (kuhl-SOO-bei) (5,400 ft/1,646 m), in WESTERN GHATS, NASHIK district, MAHARASHTRA state, W central INDIA, 27 mi/43 km SSW of NASHIK; 19°36′N 73°43′E.

Kaltag, village (2000 population 230), W ALASKA, on YUKON River, 60 mi/97 km SW of GALENA; 64°19′N 158°44′W.

Kaltan (kahl-TAHN), city (2005 population 25,895), S KEMEROVO oblast, S central SIBERIA, RUSSIA, on the KONDOMA RIVER (tributary of the TOM' River), on railroad, 210 mi/338 km S of KEMEROVO, 30 mi/48 km S of NOVOKUZNETSK, and 7 mi/11 km S, and under administrative jurisdiction, of OSINNIKI; 53°30′N 87°17′E. Elevation 1,023 ft/311 m. Coal mining in the KUZNETSK coal basin; production of boiler equipment and pipelines. Created in 1946 with the construction of the S Kuzbas regional electric power station. Made city in 1959.

Kaltasy (kahl-tah-SI), village (2005 population 4,460), NW BASHKORTOSTAN Republic, E European Russia, on road, 25 mi/40 km S of YANAUL; 55°58′N 54°48′E. Elevation 452 ft/137 m. Brewery. Oil pipeline in the vicinity. Close to one-quarter of the population are Mari.

Kaltbrunn, commune, ST. GALLEN canton, NE SWITZERLAND, 4 mi/6.4 km E of LAKE ZÜRICH. Elevation 1,447 ft/441 m. Textiles.

Kaltenkirchen (kahl-ten-KIR-khuhn), town, SCHLESWIG-HOLSTEIN, NW GERMANY, 19 mi/31 km N of HAMBURG; 53°50′N 09°57′E. Manufacturing includes electronics, machinery, motor vehicles; food processing. Chartered 1973.

Kaltenleutgeben (kahl-ten-LOIT-gai-ben), township, E LOWER AUSTRIA, in the WIENERWALD of VIENNA at its urban fringe; 48°07′N 16°12′E. Limestone quarries.

Kaltimo (KAHL-ti-mo), village, POHJOIS-KARJALAN province, SE FINLAND, on the Pielisjoki (river), 18 mi/29 km NE of JOENSUU; 62°48′N 30°08′E. Elevation 330 ft/100 m. Pulp and board mills, quartz quarries.

Kaluaggala, village, WESTERN PROVINCE, SRI LANKA; 06°55′N 80°06′E. Trades in rubber. Nearby reservoir supplies water to COLOMBO.

Kaluga (kah-LOO-gah), city (2005 population 340,475), ⊙ KALUGA oblast, central European Russia, on the OKA River, 117 mi/188 km SW of MOSCOW; 54°32′N 36°16′E. Elevation 682 ft/207 m. It is a river port, railroad junction, and an industrial center producing railroad and electrical equipment, turbines, machinery, plastics, chemical pharmaceuticals, clothing, shoes, perfumes, and food products. The founder of astronautics, Konstantin E. Tsiolkovskiy, worked in Kaluga at the end of the 19th century; the city has a museum of the history of astronautics. Known since 1371 as part of Moscow principality; center of the peasant uprising under the leadership of Ivan I. Bolotnikov in 1606–1607; experienced industrial growth in the early 18th century with the establishment of a sail-making plant and a paper mill; became an administrative center of Kaluga gubernia in 1777.

Kaluga (kah-LOO-gah), oblast (□ 11,544 sq mi/30,014.4 sq km; 2004 population 1,097,000) in W central European Russia, approximately 124 mi/200 km SW of MOSCOW; ⊙ KALUGA. In Ugra-Oka river basin, between SMOLENSK-MOSCOW (N) and CENTRAL RUSSIAN uplands (S); in mixed-forest zone; sandy soils. Basic crops, coarse grain, wheat and potatoes (NE), hemp (S); fodder crops, fruit, vegetables. Dairy cattle, hogs raised. Main mineral resources include peat, quartz, phosphorites. Rural industries based on lumber (paper milling at KONDROVO, woodworking), agriculture (fruit canning, dairying, flour milling, distilling), and minerals (peat, glass, porcelain, and phosphate-fertilizer industries). Textile milling at BOROVSK (woolens) and BALABANOVO (cotton); machinery and electronics are made at Kaluga and LYUDINOVO. Population almost 74% urban. Declared the Kaluga Federal Republic for a short time in 1918. Under German occupation for two months in 1941. Formed in 1944 out of TULA and other oblasts. In 1993, the oblast council (soviet) opposed the presidential decree disbanding the Supreme Soviet and was itself disbanded by the governor.

Kalu Ganga (KUH-lu guhn-nguh), river, 70 mi/113 km long, SW SRI LANKA; rises in SABARAGAMUWA PROVINCE, in S SRI LANKA HILL COUNTRY, in several headstreams joining near RATNAPURA at 06°30′N 79°57′E; flows W past Ratnapura and into WESTERN PROVINCE, then SW to INDIAN OCEAN at KALUTARA.

Kalugerovo (kah-LOO-ger-o-vo), village, PLOVDIV oblast, LESICHEVO obshtina, W central BULGARIA, on the TOPOLNITSA RIVER, 11 mi/18 km NW of PAZARDZHIK; 42°20′N 24°11′E. Vineyards, rice, produce. Formerly called Gelvere.

Kalugumalai (KUHL-uh-guh-muh-LEI), town, CHIDAMBARANAR district, TAMIL NADU state, S INDIA, 30 mi/48 km N of TIRUNELVELI; 09°09′N 77°43′E. Seasonal livestock-trade center during temple pilgrimage festivals. Has rock-carved Jain figures.

Kalukhali (kah-luh-KAH-lee), village, FARIDPUR district, S central EAST BENGAL, BANGLADESH, near PADMA RIVER, 24 mi/39 km NW of FARIDPUR. Railroad junction, with spurs to FARIDPUR and to BHATIAPARA GHAT, 39 mi/63 km SSE.

Kalule-Sud (kah-loo-LAI–sud), village, KATANGA province, SE CONGO, on railroad, 75 mi/121 km NW of Likari; 10°08′S 25°57′E. In cattle-raising region.

Kalum (kah-LOOM), town, Plateau state, central NIGERIA, on river, 100 mi/161 km SE of JOS; 13°11′N 04°09′E. Market center. Cotton, guinea corn, cassava.

Kalumburu, community (2006 population 450), N WESTERN AUSTRALIA state, W AUSTRALIA, 395 mi/636 km from DERBY, on King Edward River; 14°14′S 126°38′E. Indigenous settlement in remote area; permits required to visit. Catholic mission. Aboriginal art sites (paintings, engravings, campsites); most are closed to the public.

Kalundborg (KAH-loon-bor), city (2000 population 15,405), VESTSJÆLLAND county, central DENMARK, port on the KALUNDBORG FJORD (an arm of the STORE BÆLT); 55°43′N 11°06′E. It is a commercial, industrial, and communications center. Manufacturing includes chemicals, machinery, and ships. Founded c.1170, the city has a twelfth-century church laid out in the form of a Greek cross. The novelist Sigrid Undset (1882–1949) was born here.

Kalundborg Fjord (KAH-loon-bor), W SJÆLLAND, DENMARK, between ROSNÆS (N) and ASNÆS (S) peninsulas; 8 mi/12.9 km long. Fisheries.

Kalundu, CONGO: see UVIRA.

Kalur Kot, PAKISTAN: see KALLURKOT.

Kalush (KAH-loosh) (Polish *Kalusz*), city (2001 population 67,902), NW IVANO-FRANKIVS′K oblast, UK-RAINE, on Syvka River, left tributary of Lomnytsya River, 17 mi/27 km WNW of IVANO-FRANKIVS′K; 49°01′N 24°22′E. Elevation 931 ft/283 m. Mining center (salt and potash deposits); chemical manufacturing, and machine building for petroleum extraction; cement manufacturing, iron casting, agricultural processing (cereals, fruit, hops), brewery, food flavoring; sawmilling. Known since 1241 as part of Halych-Volyn′ Principality; passed to Poland (1387), to Austria (1772); part of West Ukrainian National Republic (1918); reverted to Poland (1919); ceded to USSR in 1945; part of independent Ukraine since 1991. Jewish community since the 18th century, receiving Austria-Hungary civil rights in 1867; numbering 3,277 in 1939; eliminated during World War II—fewer than 100 Jews remaining in 2005.

Kalusz, UKRAINE: see KALUSH.

Kaluszyn (kah-LOO-sheen), Polish *Kałuszyn*, Russian *Kalushin*, town, Siedlce province, E central POLAND, 34 mi/55 km E of WARSAW. Manufacturing of construction materials, tanning, brewing, flour milling.

Kalutara, district (□ 608 sq mi/1,580.8 sq km; 2001 population 1,066,239), WESTERN PROVINCE, SW SRI LANKA; ⊙ KALUTARA; 06°35′N 80°10′E.

Kalutara (KUH-lu-tuh-ruh) [Sinhalese=black port], town (2001 population 37,451), ⊙ KALUTARA district, WESTERN PROVINCE, SRI LANKA, on SW coast, at mouth of the KALU GANGA River (1,200-ft/366-m bridge), 24 mi/39 km SSE of COLOMBO; 06°35′N 79°57′E. Fishing and trade (tea, rubber, coconuts, rice, mangosteens) center; basket and mat weaving; distillery. Modern Buddhist stupa.

Kalvakurti, INDIA: see KALWAKURTI.

Kalvan, INDIA: see KALWAN.

Kalvarija (kahl-vuh-REE-juh), Polish *Kalwaria*, city, SW LITHUANIA, on upper SHESHUPE RIVER, 10 mi/16 km SSE of MARIJAMPOLE, near Polish border; 54°24′N 23°14′E. Agricultural market; manufacturing (shoes, furniture); sawmilling, oilseed pressing, flour milling. Dates from 13th century (castle ruins); passed 1795 to PRUSSIA, 1815 to Russian POLAND; in Suvalki government until 1920. Also spelled Kalvariya.

Kalvø Bay, DENMARK: see MOLS.

Kalvsund (KAHLV-SUND), fishing village, GÖTEBORG OCH BOHUS county, SW SWEDEN, on Björkön islet in SKAGERRAK, 10 mi/16 km WNW of GÖTEBORG.

Kalwakurti (kuhl-wuh-KUHR-tee), town, MAHBUB-NAGAR district, ANDHRA PRADESH state, INDIA, 33 mi/53 km E of MAHBUBNAGAR; 16°41′N 78°30′E. Millet, rice, castor beans, peanuts. Also spelled Kalvakurthi, Kalvakurti.

Kalwan (kuhl-WAHN), town, tahsil headquarters, NASHIK district, MAHARASHTRA state, W central INDIA, on GIRNA RIVER and 38 mi/61 km NNE of NASHIK. Gur, millet, rice.

Kalwaria, LITHUANIA: see KALVARIJA.

Kalwaria Zebrzydowska (kahl-VAH-ryah zeeb-zee-DOV-skah), town, Bielsko-Biała province, S POLAND, 18 mi/29 km SW of CRACOW. Railroad junction; furniture manufacturing. Pilgrimage center; monastery.

Kal′ya (kahl-YAH), town (2006 population 6,640), N SVERDLOVSK oblast, E URALS, W Siberian Russia, on road, 36 mi/58 km N of KARPINSK; 60°14′N 59°59′E. Elevation 702 ft/213 m. Bauxite-mining center, supplying Krasnoturinsk aluminum works; gold placers. Also known as Kal′yinskiy.

Kalyan (kuhl-YAHN), city, THANE district, MAHAR-ASHTRA state, W central INDIA, on ULHAS RIVER, 26 mi/42 km NE of MUMBAI, in the KONKAN; 19°15′N 73°09′E. Railroad junction (workshops); road center; trade in rice, cloth fabrics, tobacco, fish; rice, cotton, and oilseed milling, bone crushing; manufacturing of ice, leather goods, bricks, tiles, pharmaceuticals. Rayon mill. Substation for power lines of ANDHRA RIVER valley hydroelectric plant; power plant is 1 mi/1.6 km SW, at CHOLE.

Kalyandurg (kuhl-YAHN-druhg), town, ANANTAPUR district, ANDHRA PRADESH state, SE INDIA, 34 mi/55 km WSW of ANANTAPUR; 14°33′N 77°06′E. Oilseed milling, hand-loom woolen weaving. Corundum mining nearby.

Kalyani (KUHL-yuh-nee), town, GULBARGA district, KARNATAKA state, SW INDIA, in enclave within BIDAR district, 38 mi/61 km N of GULBARGA. Millet, rice.

Kalyani, township, NADIA district, WEST BENGAL state, E INDIA, c.37 mi/60 km NE of CALCUTTA, off National Highway 34, in a Notified Area; 22°59′N 88°29′E. The infrastructure of a World War II American base called Rooseveltnagar was expanded into a township, Kalyani, to hold the Indian National Congress Session of 1954 and later developed into a planned satellite community for Calcutta Metropolis. With a rectilinear town plan (numbered, rather than named, streets), abundant open space, and the GANGA River flowing by, Kalyani attracted settlers from congested Calcutta, which is within commuting distance by railroad. Several good hospitals, a residential university, and a school for agricultural sciences. Newer housing complexes, such as Salt Lake City and Lake Town, nearer Calcutta, have attracted new residents away from Kalyani, which had shown signs of stagnation. New industries on the periphery have regenerated the economy.

Kalyazin (kah-LYAH-zeen), city (2006 population 14,600), SE TVER oblast, W central European Russia, on the VOLGA RIVER, 108 mi/174 km NE of TVER, and 40 mi/64 km NNE of KIMRY; 57°15′N 37°52′E. Elevation 380 ft/115 m. Railroad junction, highway terminus; local transshipment point; footwear and clothing manufacturing; flax processing; handicrafts; sawmills. Arose in the 12th century. Site of the strongly fortified Makaryev monastery (15th century), now a museum. Chartered in 1775. Part of the old city was submerged during construction of the Uglich hydroelectric power station and reservoir.

Kalydon (kah-lee-[TH]ON), ancient city of AETOLIA, in what is now WESTERN GREECE department, on Efinos River, 6 mi/9.7 km E of modern MESOLONGI, near Gulf of Calydon (modern Gulf of PATRAS). Here, in Greek legend, Meleager killed the Kalydonian boar.

Kal′yinskiy, RUSSIA: see KAL′YA.

Kalymnos (KAH-leem-nos), Italian *Calino*, Latin *Calymana*, island (□ 41 sq mi/106.6 sq km), DODECANESE prefecture, SOUTH AEGEAN department, GREECE, off Bodrum Peninsula (SW Turkey) and N of KÓS island; 37°00′N 27°00′E. Rises to 2,228 ft/679 m; 12 mi/19 km long, 6 mi/9.7 km wide. Produces olive oil, beans, figs, almonds, citrus fruits. Major sponge-fishing center. Main town, Kalymnos, is on SE shore. Also spelled Kalimnos, Calymnos.

Kalynivka (kah-LI-nif-kah) (Russian *Kalinovka*), city, N VINNYTSYA oblast, UKRAINE, 15 mi/24 km N VIN-NYTSYA; 49°28′N 28°32′E. Elevation 889 ft/270 m. Raion center; railroad junction; machinery, dairy, cannery. Technical school. Established in the first half of the 19th century; city since 1979. Jewish community since the city's formation, numbering 1,097 in 1939; wiped out during the Russian civil war pogroms of 1918–1920 and the Nazi invasion during World War II — fewer than 100 Jews remaining in 2005.

Kalynivka (kah-LI-nif-kah) (Russian *Kalinovka*), town, central KIEV oblast, UKRAINE, on railroad junction 5 mi/8 km NW of VASYL′KIV; 50°25′N 29°48′E. Elevation 515 ft/156 m. Forestry, sawmilling. Established at the beginning of the 19th century as Khutor (Ukrainian=village) Kalynivka; renamed and town status since 1957.

Kalynove (kah-LI-no-ve) (Russian *Kalinovo*), town (2004 population 4,085), W LUHANS′K oblast, UK-RAINE, in the DONBAS, 7 mi/11.3 km SE of POPASNA; 47°45′N 38°41′E. Coal mines, exhausted; most residents work in establishments of nearby cities of

STAKHANOV, TEPLOHIRS'K, and PERVOMAYS'K. Established in 1753, town since 1938.

Kalyta (kah-li-TAH) (Russian *Kalita*), town (2004 population 5,840), E central KIEV oblast, UKRAINE, near the Kiev-Chernihiv highway, 20 mi/33 km NNE of BROVARY; 50°45′N 31°01′E. Elevation 331 ft/100 m. Experimental feed mill, state farm. Town since 1973.

Kalyubiya, EGYPT: see QALYUBIYA.

Kalyvia (kah-LEE-vyah), town, ATTICA prefecture, ATTICA department, E central GREECE, on railroad, 15 mi/24 km SE of ATHENS; 37°55′N 22°18′E. Wheat, vegetables, straw; wine.

Kam, new town, extreme NW ALBANIA, NW of KUKËS. Chrome mining.

Kama (KAH-mah), township, MAGWE division, MYANMAR, on AYEYARWADY RIVER, 20 mi/32 km SSE of THAYETMYO.

Kama (KAH-mah), town (2006 population 3,430), SE UDMURT REPUBLIC, E European Russia, port on the left bank of the KAMA River and lower KAMA Reservoir, on road and near railroad, 3 mi/5 km NNW of KAMBARKA, to which it is administratively subordinate; 56°18′N 54°14′E. Elevation 367 ft/111 m. Formerly known as Butysh; town status and current name since 1966.

Kama (KAH-mah), village, SUD-KIVU province, E CONGO, on ELILA RIVER and 70 mi/113 km NNE of KASONGO; 03°32′S 27°07′E. Elev. 1,971 ft/600 m. Gold-mining center; rice processing.

Kama (KAH-mah), river, approximately 1,260 mi/2,028 km long, E European Russia, the chief left tributary of the VOLGA RIVER. It rises in the foothills of the central URAL Mountains just N of KULIGA (far N UDMURT REPUBLIC), flows N past LOINO (KIROV oblast), then E, and SW past PERM (PERM oblast), SARAPUL' (SE Udmurt Republic), and CHISTOPOL' (TATARSTAN) to join the Volga below KAZAN. The VYATKA River is its principal tributary. The Kama is an important transportation artery. There are hydroelectric power stations at Perm, VOTKINSK (Udmurt Republic), and NIZHNEKAMSK (Tatarstan).

Kamabai (kah-mah-BEI), village, NORTHERN province, N central SIERRA LEONE, 20 mi/32 km NNE of MAKENI; 09°09′N 11°58′W. Palm oil, kernels, peanuts.

Kamachumu (kah-mah-CHOO-moo), village, KAGERA region, NW TANZANIA, 22 mi/35 km SSW of BUKOBA, near SW shore of Lake VICTORIA; Lake IKIMBA to NW; 01°37′S 31°37′E. Coffee; corn; sheep and goats.

Kamae (kah-MAH-e), town, South Amabe county, OITA prefecture, E KYUSHU, SW JAPAN, 34 mi/55 km S of OITA, on HYUGA SEA; 32°47′N 131°55′E. Yellowtail.

Kamagari (kah-MAH-gah-ree), town, Aki county, HIROSHIMA prefecture, SW HONSHU, W JAPAN, 22 mi/35 km S of HIROSHIMA; 34°11′N 132°44′E.

Kamagaya (kah-MAH-gah-yah), city, CHIBA prefecture, E central HONSHU, central JAPAN, 9 mi/15 km N of CHIBA; 35°46′N 140°00′E. Pottery, traditional dolls.

Kamaing (kah-MEING), township, KACHIN STATE, MYANMAR, on Mogaung River, 40 mi/64 km W of MYITKYINA. Scene of fighting in World War II.

Kama-ishi (kah-MAH–ee-shee), city, IWATE prefecture, N HONSHU, NE JAPAN, port on the PACIFIC OCEAN, 50 mi/80 km S of MORIOKA; 39°16′N 141°53′E. Wire and wire material. Marine products (*wakame* sea vegetable, abalone, sea urchin, salmon); shiitake mushrooms. Magnetite ore. Sometimes spelled Kamaisi.

Kamakou, Mount (KAH-mah-KO-oo) (4,970 ft/1,515 m), E MOLOKAI island, MAUI county, HAWAII, 9 mi/14.5 km E of KAUNAKAKAI, in Molokai Forest Reserve. Highest peak on island.

Kamakura (kah-MAH-koo-rah), city (2005 population 171,158), KANAGAWA prefecture, E central HONSHU, E central JAPAN, on SAGAMI BAY, at the base of the MIURA PENINSULA, 31 mi/50 km S of YOKOHAMA; 35°18′N 139°33′E. Computers. Resort and residential area; chiefly noted as a religious center, the site of over

eighty shrines and temples. Especially famous for its *daibutsu* [Japanese=great Buddha], a 42-ft/13-m-high bronze figure of Buddha, cast in 1252, and for a 30-ft/9-m-high gilt-and-camphor statue of Kannon, the goddess of mercy. Seat of Shogun Ashikaga Yoritomo and his descendants (ruled Japan 1192–1333); thereafter, served as Ashikaga Shogunate's E Japan headquarters until 1573. An earthquake in 1923 severely damaged the city.

Kamakwi (kah-MAH-kwee), town, NORTHERN province, NW SIERRA LEONE, on road, 65 mi/105 km NNE of PORT LOKO; 09°30′N 12°14′W. Center of trade with GUINEA. Has hospital. Also called Makwi.

Kamalampa (kah-mah-LAHM-pah), village, RUKWA region, W TANZANIA, 75 mi/121 km SE of KIGOMA; 05°32′S 30°31′E. Timber; corn, wheat; goats, sheep.

Kamalapuram (KAH-muh-lah-puh-ruhm), town, CUDDAPAH district, ANDHRA PRADESH state, SW INDIA, near PENNER RIVER, 13 mi/21 km NW of CUDDAPAH; 14°35′N 78°39′E. Cotton, rice, melons. Limestone quarrying nearby.

Kamalapuram (KAH-muh-lah-puh-ruhm), village, ANANTAPUR district, ANDHRA PRADESH state, SE INDIA, just S of HAMPI. Tourists' bungalows here for visitors of historic sites in Hampi area.

Kamalganj (KUHM-uhl-guhnj), town, FARRUKHABAD district, central UTTAR PRADESH state, N central INDIA, on tributary of the GANGA River, 9 mi/14.5 km SSE of FARRUKHABAD; 27°16′N 79°39′E. Wheat, gram, jowar, corn, potatoes, tobacco.

Kamalia (kah-mah-LYAH), town, FAISALABAD district, E central PUNJAB province, central PAKISTAN, 55 mi/89 km SSW of FAISALABAD; 30°44′N 72°39′E.

Kamamaung (kah-mah-MOUNG), village, PAPUN township, KAYIN STATE, MYANMAR, 60 mi/97 km N of MAWLAMYINE, on THANLWIN RIVER (head of navigation) at mouth of YUNZALIN RIVER.

Kaman (KAH-muhn), town, BHARATPUR district, RAJASTHAN state, NW INDIA, 33 mi/53 km NNW of BHARATPUR. Hand-loom weaving; millet, gram, wheat. A place of Hindu pilgrimage.

Kaman, village (2000 population 27,962), central TURKEY, 28 mi/45 km NW of KIRSEHIR; 39°22′N 33°43′E. Grain, linseed; mohair goats.

Kamaniola (kah-mahn-YO-lah), village, SUD-KIVU province, E CONGO, on RUZIZI RIVER, 20 mi/32 km SSE of Bakavu; 02°46′S 29°00′E. Elev. 2,965 ft/903 m. Terminus of railroad from UVIRA (KALUNDU); cotton growing, cotton ginning.

Kamarajar, district (□ 1,612 sq mi/4,191.2 sq km), TAMIL NADU state, S INDIA; ⊙ Virudhunagar.

Kamaran Islands (kuh-mer-AN) (□ 22 sq mi/57.2 sq km), in RED SEA, off YEMEN coast, 200 mi/322 km N of the strait BAB EL MANDEB; includes main Kamaran Island (10 mi/16 km long, 5 mi/8 km wide) off SALIF and nearby smaller islands; 15°21′N 42°35′E. Kamaran Island has airfield. Formerly a Turkish possession, the Kamaran group was occupied in 1915 by the British. Its status having been left indeterminate by the Treaty of Lausanne (1923), the group was ceded to SOUTH YEMEN with the withdrawal of Britain from Aden (1967). Most of the islands are uninhabited. The small population of the main islands are engaged mostly in fishing.

Kamard, AFGHANISTAN: see KAHMARD.

Kamareddi (kah-mah-RAI-dee), town, NIZAMABAD district, ANDHRA PRADESH state, SE INDIA, 30 mi/48 km SE of NIZAMABAD; 18°19′N 78°21′E. Distillery; rice, oilseeds. Also spelled Kamaredi; formerly called Kamareddipet.

Kama Reservoir (KAH-mah), Russian *Kamskoye Vodokhranilishche*, artificial lake (surface □ 737 sq mi/1,909 sq km) created by expanding the natural course of the KAMA River in the area where the In'va River, Obva River, Iren' River, and a number of smaller tributaries feed into it, W PERM oblast, E European Russia; 58°40′N 56°00′E. Elevation 358 ft/109 m. Mean

depth 20 ft/6 m (maximum depth 95 ft/29 m). Freezes November through April. Wetlands on its E shore. Used primarily for irrigation, industrial production, and electricity generation; pollution levels are a cause for environmental concerns.

Kamarhati (kah-mahr-HAH-tee), city and suburb of CALCUTTA, CALCUTTA Metropolitan District, WEST BENGAL state, E INDIA; 22°40′N 88°22′E. Manufacturing of textiles, rubber goods, cement, paint, pottery, and jute products.

Kamarkhali Ghat, BANGLADESH: see MADHUKHALI.

Kamarlu, ARMENIA: see ARTASHAT.

Kamarlyu, ARMENIA: see ARTASHAT.

Kamaryut (kah-MAHR-YOOT), section of YANGON, MYANMAR, W of INYA LAKE. Location of Yangon University, Institute of Economics, and International School.

Kamas (KA-muhs), town (2006 population 1,493), SUMMIT county, N central UTAH, on diversion canal, 12 mi/19 km E of PARK CITY; 40°38′N 111°16′W. Sheep; logging, lumber. Part of Wasatch National Forest (UINTA Range) to E. Kamas State Fish Hatchery to E; Jordanelle Reservoir and State Park to W. Elevation 6,473 ft/1,973 m. Settled 1857.

Kamashi (kah-mah-SHI), city, central KASHKADARYO wiloyat, UZBEKISTAN, on railroad, 35 mi/56 km E of KARSHI; 38°50′N 66°28′E. Wheat, cotton.

Kamatanda (kah-mah-TAHN-dah), village, KATANGA province, SE CONGO, 10 mi/16 km E of LIKASI; 10°57′S 26°46′E. Railroad junction, copper-mining center.

Kamatapur (kuh-MAH-tah-puhr), ruined city, KOCH BIHAR district, NE WEST BENGAL state, E INDIA, near the JALDHAKA RIVER, 12 mi/19 km SSW of KOCH BIHAR. Founded early 15th century as capital of Khen kingdom; abandoned late 15th century after overthrow of Khens by Afghans. Also called Rajpat. Figured prominently on early maps of India as Comotay.

Kamativi, town, MATABELELAND NORTH province, W central ZIMBABWE, 37 mi/60 km E of HWANGE, near GWAYI RIVER; 18°19′S 27°03′E. Kamativi Tin Mine N of town. Mzola Forest Land to E, Msuna Safari Area to N. Cattle, sheep, goats; corn, wheat, soybeans, tobacco, cotton.

Kamay, unincorporated village, WICHITA county, N TEXAS, 18 mi/29 km WSW of WICHITA FALLS, near WICHITA RIVER; 33°51′N 98°48′W. Oil-field center. Cattle; cotton; light manufacturing. Formerly called Kemp City.

Kamayut (kah-MAH-YOOT), NW suburb of YANGON, MYANMAR, on railroad, 6 mi/9.7 km from central YANGON. Manufacturing of tobacco products, rubber shoes.

Kamba (KAHM-bah), town, KEBBI state, NW NIGERIA, on NIGER border, 60 mi/97 km SW of BIRNIN KEBBI. Market town. Maize, millet.

Kamba, China: see GAMBA.

Kambadianga (kuh-mbuh-dee-AHN-guh), town, KASAI-OCCIDENTAL province, SW CONGO, 50 mi/80 km WSW of LUEBO; 05°34′S 20°28′E. Elev. 2,250 ft/685 m.

Kambaengbejr, THAILAND: see KAMPHAENG PHET, city.

Kambalda (kam-BAL-duh), town, S central WESTERN AUSTRALIA state, W AUSTRALIA, 383 mi/616 km E of PERTH, 24 mi/39 km S of KALGOORLIE; 31°12′S 121°40′E. Former gold-mining center, 1897–1906. Area rebounded with discovery of nickel in 1966. Planned community of Western Mining Nickel Plant. Divided into Kambalda East and Kambalda West, 2 mi/4 km from each other. "Land yachting" on Lake Lefroy salt lake to E. Wheat; cattle.

Kambam (KUHM-buhm), town, MADURAI district, TAMIL NADU state, S INDIA, 60 mi/97 km WSW of MADURAI, in KAMBAM VALLEY; 09°44′N 77°18′E. Rice along SURULI RIVER (E); cardamom, teak, turmeric in CARDAMOM HILLS (W). Formerly spelled Cumbum.

Kambam Valley (KUHM-buhm), along E slopes of S Western GHATS, TAMIL NADU state, S INDIA, between

Palni Hills (N) and VARUSHANAD and Andipatti hills (SE); drained by upper VAIGAI RIVER and its left affluent, the SURULI RIVER; irrigated by PERIYAR LAKE project. Mainly agriculture (rice, grain, tobacco); date and coconut palms. Chief products of nearby hills include tea, coffee, cardamom, timber. Main towns are BODINAYAKKANUR, KAMBAM. Formerly spelled Cumbum.

Kambar (kuhm-BAHR), town, LARKANA district, NW SIND province, SE PAKISTAN, 12 mi/19 km WNW of LARKANA; 27°36′N 68°00′E. Market center for local produce.

Kambara, FIJI: see KABARA.

Kambarka (kahm-BAHR-kah), city (2006 population 12,410), SE UDMURT REPUBLIC, E European Russia, near the KAMA River, on railroad, 72 mi/116 km SE of IZHEVSK, and 20 mi/32 km SE of SARAPUL; 56°15′N 54°11′E. Elevation 275 ft/83 m. Locomotives, motors, metal goods, gas appliances, gardening equipment; flax processing, food processing (bakery). Founded in 1761 around an ironworks factory; became a city in 1945. Formerly called Kambarskiy Zavod.

Kambarskiy Zavod, RUSSIA: see KAMBARKA.

Kambaye, CONGO: see KANDA-KANDA.

Kambia (kahm-BEE-ah), town, ⊙ Kambia district, NORTHERN province, NW SIERRA LEONE, near GUINEA border, on GREAT SCARCIES RIVER (head of navigation), 25 mi/40 km NNW of PORT LOKO; 09°03′N 12°53′W. Palm oil and kernels, peanuts, rice.

Kambileyevskoye (kahm-bee-LYE-eef-skuh-ye), village (2006 population 7,820), E NORTH OSSETIAN REPUBLIC, RUSSIA, in the N central CAUCASUS Mountains, on highway, 4 mi/6 km E of VLADIKAVKAZ; 43°04′N 44°45′E. Elevation 2,086 ft/635 m. Produce processing, livestock raising. Formerly known as Nizhniy (Russian=lower) Dzherakh.

Kambolé (KAHM-bo-LAI), town, CENTRALE region, E central TOGO, 50 mi/80 km SE of SOKODÉ, on BENIN border, near KOUSSOUNTOU; 08°45′N 01°36′E. Largely subsistence farming of manioc and haricot.

Kambos (KAHM-bos), town, LEFKOSIA district, central CYPRUS, 38 mi/61 km WSW of NICOSIA, in NW part of TROODOS MOUNTAINS; 35°02′N 32°44′E.

Kambove (kahm-BO-vai), village, KATANGA province, SE CONGO, 10 mi/16 km NW of LIKASI; 10°52′S 26°32′E. Elev. 4,783 ft/1,457 m. Copper-mining center; railroad terminus; cobalt and platinum-palladium are also mined. Has Protestant mission. Copper deposits of KAMBOVE were exploited by Swahilis of E African coast before the arrival of white explorers.

Kamchatka (kahm-CHAHT-kah), oblast (□ 298,843 sq mi/776,991.8 sq km; 2004 population 444,000), extreme E SIBERIA, RUSSIAN FAR EAST, between the Sea of OKHOTSK and BERING SEA; ⊙ PETROPAVLOVSK-KAMCHATSKIY, the oblast's only seaport. Includes KAMCHATKA Peninsula, Komandorskiy Islands, KORYAK Autonomous Okrug, Penzhina Basin, and the BERING Sea coast as far as Cape Rubikon. There are two parallel mountain ranges on the KAMCHATKA PENINSULA covered by tundra, turning into birch and larch forests near the coast. The region is geologically active, with 300 volcanoes, of which twenty-nine are live. Marginal agriculture in the S central valley. Fishing and fish processing (chief industry, based in Petropavlovsk-Kamchatskiy), fur trapping, lumbering; some agriculture along the KAMCHATKA RIVER; fish canning along the coast. Largest population centers are PETROPAVLOVSK-Kamchatskiy, YELIZOVO, and KLYUCHI (also PALANA in Koryak Autonomous Okrug). Local indigenous peoples, including Itelmen (S), Koryak (N; in KORYAK AUTONOMOUS OKRUG), Yukagir, Even, and Aleut (KOMANDORSKIY ISLANDS), once lived on fishing, hunting, and reindeer breeding. Though often engaged in fur farming, they have suffered much from current economic instability and are in danger of dying out. Formed in 1932 within former Far Eastern Territory. Currently divided into seven districts.

Kamchatka Gulf (kahm-CHAHT-kah), inlet of the PACIFIC OCEAN, on the E coast of KAMCHATKA PENINSULA, KAMCHATKA oblast, extreme E SIBERIA, RUSSIAN FAR EAST, between the Kronotskiy Cape (S) and UST-KAMCHATSK (N), where it receives the KAMCHATKA RIVER, 85 mi/137 km wide; 55°35′N 162°21′E.

Kamchatka Peninsula (kahm-CHAHT-kah) (□ 104,200 sq mi/270,920 sq km), extreme E SIBERIA, RUSSIAN FAR EAST, separating the Sea of OKHOTSK in the W from the BERING SEA and the Pacific Ocean in the E; 750 mi/1,207 km long; 51°00′N–61°00′N. Terminates in the S in Cape LOPATKA, beyond which lie the KURIL ISLANDS. PETROPAVLOVSK-KAMCHATSKIY is the chief city. There are many rivers and lakes, and the E shore is deeply indented by gulfs and bays. The peninsula's central valley, drained by the KAMCHATKA RIVER, is enclosed by 2 parallel volcanic ranges that extend N-S; there are about 300 volcanoes (29 of which are active), and hot springs also indicate tectonic instability. The highest point is Klyuchevskaya Sopka (15,600 ft/4,755 m), itself an active volcano. Kamchatka is covered with mountain vegetation, except in the central valley and on the W coast, which has peat marshes and tundralike moss. The climate is cold and humid. There are numerous forests, mineral springs, and geysers. Kamchatka's mineral resources include coal, gold, mica, pyrites, sulfur, and tufa. Fishing, sealing, hunting, and lumbering are the main occupations. As a Russian fishing area (notably for crabs, which are exported worldwide), Kamchatka is surpassed only by the CASPIAN SEA region. Fur trapping on the peninsula yields most of the furs of the Russian Far East. Some cattle breeding in the S, and farming (rye, oats, potatoes, vegetables) in the Kamchatka valley and around Petropavlovsk-Kamchatskiy. Reindeer are also raised on the peninsula. Industries include fish processing, shipbuilding, and woodworking. Russia's only geothermal power station is here. The majority of the population is Russian, with large minorities of Koryak peoples. The N part of the peninsula is administered as the KORYAK AUTONOMOUS OKRUG (□ 301,500 sq mi/780,885 sq km; 1992 population 39,000); ⊙ PALANA. The Russian explorer Atlasov discovered Kamchatka in 1697. Its exploration and development continued in the early 18th century under Czar Peter I. Russian conquest was complete by 1732. Heavy Russian colonization occurred in the early 19th century. From 1926 to 1938, Kamchatka formed part of the Far Eastern Territory.

Kamchatka River (kahm-CHAHT-kah), 335 mi/539 km long, KAMCHATKA PENINSULA, KAMCHATKA oblast, extreme E SIBERIA, RUSSIAN FAR EAST; rises in S central mountain range of the peninsula, flows N through a wide, populated agricultural valley, past MILKOVO (head of shallow-draught navigation) and Sredne-Kamchatsk, and E, past NIZHNE-KAMCHATSK, to the KAMCHATKA GULF of the Pacific Ocean at UST'-KAMCHATSK.

Kamchiiska (kahm-CHEE-skah), mountain, E STARA PLANINA, BULGARIA; 42°55′N 27°30′E.

Kamchiya Dam (KAHM-chee-yah), dam, E central BULGARIA, on the LUDA KAMCHIYA RIVER; 42°52′N 26°52′E.

Kamchiya River (KAHM-chee-yah), 23 mi/37 km long, E central BULGARIA; formed at TSONEVO by a confluence of the GOLYAMA KAMCHIYA and LUDA KAMCHIYA rivers; flows E through Longoza alluvial flood plain (□ c.65 sq mi/168 sq km), past RAKOVETS, to the BLACK SEA 13 mi/21 km S of VARNA; 43°01′N 27°43′E. Navigable 4 mi/6 km–5 mi/8 km in the lower course. Also called Ticha River.

Kamdoli (kuhm-DO-lee), town, DHARWAD district, KARNATAKA state, SW INDIA, 24 mi/39 km SE of DHARWAD. Local cotton market.

Kameda (KAH-me-dah), town, Nakakanbara county, NIIGATA prefecture, central HONSHU, N central JAPAN, 5 mi/8 km S of NIIGATA; 37°51′N 139°06′E.

Kamehameha, Fort (kah-MAI-hah-MAI-hah), military reservation, on E side of entrance to PEARL HARBOR, S OAHU, HONOLULU county, HAWAII, 6 mi/9.7 km W of HONOLULU. Honolulu International Airport to E; HICKAM Air Force Base to N. Established 1909.

Kameiros (KAH-mee-ros), ancient city of RHODES island, in what is now DODECANESE prefecture, SOUTH AEGEAN department, GREECE, 18 mi/29 km SW of modern RHODES city. One of the members of the Dorian Hexapolis. Also spelled Camirus.

Kamembe (kah-MEM-bai), village, W RWANDA, on S shore of LAKE KIVU, near CONGO border, 85 mi/137 km SW of KIGALI; 02°26′S 28°55′E. Airport.

Kamen (KAH-men), town, NORTH RHINE–WESTPHALIA, W GERMANY, in the RUHR industrial district, 4 mi/6.4 km N of UNNA; 51°35′N 07°40′E. Manufacturing of machinery, plastics, and electrotechnical goods. Coal mining from 13th century until 1983. Has Gothic church from 13th century. Chartered in 1265.

Kamenets (KUH-me-nets), Polish *Kamieniec* or *Kamieniec Litewski*, town, W BREST oblast, BELARUS, on LESNA RIVER, 20 mi/32 km NNE of BREST; 52°23′N 23°46′E. Manufacturing (vegetable drying, butter and cheese production). Has ruins of old tower. Founded in 13th-century.

Kamenetskiy (kah-mee-NYETS-keeyee), settlement (2006 population 2,760), E TULA oblast, W central European Russia, on road, 5 mi/8 km S of NOVOMOSKOVSK, and 5 mi/8 km NW of DONSKOY; 54°00′N 38°13′E. Elevation 761 ft/231 m. In lignite-mining and lumbering region.

Kamenets-Podol'skiy, UKRAINE: see KHMEL'NYTS'KYY, oblast.

Kamenets-Podol'skiy, UKRAINE: see KAM'YANETS'-PODIL'S'KYY, city.

Kamenica (KAH-me-nee-tsah), village, W SERBIA, 9 mi/14.5 km WNW of VALJEVO. Also called Kamenitsa, Kamenica Valjevska, or Kamenitsa Valyevska.

Kamenice nad Lipou (KAH-me-NYI-tse NAHD li-POU), German *kamenitz an der linde*, town, JIHOCESKYprovince, S BOHEMIA, CZECH REPUBLIC, on railroad, 11 mi/18 km NNE of JINDŘICHŮV; 49°18′N 15°05′E. Agriculture (barley, oats); manufacturing (lace, glass, textiles). Has a 13th-century church, 15th-century castle.

Kamenicky, SLOVAKIA: see LIPANY.

Kamenicky Senov (KAH-me-NYITS-kee SHE-nof), Czech *Kamenický Šenov*, German STEINSCHÖNAU, town, SEVEROCESKY province, N BOHEMIA, CZECH REPUBLIC, on railroad, 22 mi/35 km NE of ÚSTÍ NAD LABEM; 50°46′N 14°29′E. Glass manufacturing and refining; a center of Bohemian cut-glass industry, dating from 16th century; glass museum.

Kamenishchi (kah-mee-NEE-shchee), settlement, SE NIZHEGOROD oblast, central European Russia, on the P'YANA RIVER, on highway, 5 mi/8 km NE of (and administratively subordinate to) BUTURLINO; 55°36′N 45°02′E. Elevation 439 ft/133 m. Gravel quarries.

Kamenishki Vruh (kah-men-EESH-kee VRUHK), peak (8,310 ft/2,533 m) in the PIRIN MOUNTAINS, SW BULGARIA; 41°42′N 23°35′E.

Kameniti Chal (kah-men-EE-tee CHAHL), peak (8,501 ft/2,591 m) in the RILA MOUNTAINS, SW BULGARIA.

Kamenititsa (kah-me-NEE-tee-tsah), peak (8,943 ft/2,726 m) in the PIRIN MOUNTAINS, SW BULGARIA; 41°40′N 23°30′E.

Kamenitsa (KAH-mahn-eet-sah), peak (9,239 ft/2,816 m) in the PIRIN MOUNTAINS, SW BULGARIA; 41°41′N 23°28′E.

Kamenitsa, Bulgaria: see VELINGRAD.

Kamenitz an der Linde, CZECH REPUBLIC: see KAMENICE NAD LIPOU.

Kamenka (KAH-myen-kah), city (2005 population 41,100), central PENZA oblast, E European Russia, on railroad (Belinskaya station), 46 mi/74 km W of PENZA; 53°11′N 44°03′E. Elevation 561 ft/170 m. Flour milling, sugar refining, food processing (dairy, bakery, brewery); manufacturing of farm implements; grain elevators. Also known as Kamenka Belinskaya or Belinskaya Kamenka. Made city in 1951.

Kamenka (KAH-myin-kuh), town, NW WEST KAZAKHSTAN region, KAZAKHSTAN, on railroad (Shipovo station), 45 mi/72 km W of ORAL (Uralsk); 51°09′N 50°17′E. Tertiary-level (raion) administrative center. Cattle, horses.

Kamenka (KAH-meen-kuh), town, N ARCHANGEL oblast, N European Russia, on road, 12 mi/19 km NNW of MEZEN′; 65°53′N 44°07′E. River port on the MEZEN′ River. Sawmilling.

Kamenka (KAH-meen-kah), town (2005 population 4,075), NE IVANOVO oblast, central European Russia, on the VOLGA RIVER, on road, 14 mi/23 km NNW of VICHUGA; 57°23′N 41°47′E. Elevation 439 ft/133 m. Cotton mill.

Kamenka (KAH-meen-kah), town (2006 population 9,410), S central VORONEZH oblast, S central European Russia, on road (Yevdakovo station), 18 mi/29 km SE of OSTROGOZHSK; 50°42′N 39°29′E. Elevation 748 ft/227 m. Dairy plant.

Kamenka (KAH-meen-kah), village, SE central TAMBOV oblast, S central European Russia, on road, 45 mi/72 km SSE of TAMBOV; 52°04′N 41°48′E. Elevation 449 ft/136 m. In agricultural area (wheat, flax, oats); produce processing.

Kamenka (KAH-meen-kah), village (2006 population 1,680), E MARITIME TERRITORY, SE RUSSIAN FAR EAST, on the NW coast of the Sea of JAPAN, on highway, 20 mi/32 km E of DAL′NEGORSK; 44°27′N 136°00′E. Fisheries; forestry enterprises.

Kamenka (KAH-meen-kah), village (2005 population 8,060), NW LENINGRAD oblast, NW European Russia, on the NE shore of the Gulf of FINLAND, 10 mi/16 km NNW of PRIMORSK; 60°30′N 28°31′E. Fisheries. Summer homes, sanatoria. Until 1948, called Kaliyala.

Kamenka (KAH-meen-kah), village (2005 population 3,560), central KABARDINO-BALKAR REPUBLIC, S European Russia, in the foothills of the N CAUCASUS Mountains, 5 mi/8 km NW of NAL′CHIK; 43°32′N 43°31′E. Elevation 1,811 ft/551 m. Agriculture (grain, vegetables; sheep raising). Penal colony and Russian military base in the vicinity.

Kamenka (KAH-meen-kah), village (2004 population 1,900), SE BELGOROD oblast, SW European Russia; near highway; 53 mi/85 km S of VORONEZH and 5 mi/8 km WSW of ROSSOSH (VORONEZH oblast); 51°05′N 38°23′E. Elevation 439 ft/133 m. Fish farm.

Kamenka, MOLDOVA: see CAMENCA.

Kamenka, UKRAINE: see KAM′YANKA.

Kamenka, UKRAINE: see KUYBYSHEVE, Zaporizhzhya oblast.

Kamenka Belinskaya, RUSSIA: see KAMENKA, PENZA oblast.

Kamenka-Bugskaya, UKRAINE: see KAM′YANKA-BUZ′KA.

Kamenka-Dneprovskaya, UKRAINE: see KAM′YANKA-DNIPROVS′KA.

Kamenka-na-Dnepre, UKRAINE: see KAM′YANKA-DNIPROVS′KA.

Kamen′-Kashirskiy, UKRAINE: see KAMIN′-KASHYRS′KYY.

Kamenka-Strumilovskaya, UKRAINE: see KAM′YANKA-BUZ′KA.

Kamenna, CZECH REPUBLIC: see LIBINA.

Kamen′-na-Obi (KAH-myen–nah–uh-BEE), city (2005 population 44,600), NW ALTAI TERRITORY, S central SIBERIA, RUSSIA, port on the OB′ RIVER, at the beginning of the Kulundinskiy Canal and near the NOVOSIBIRSK reservoir, 130 mi/209 km WNW of BARNAUL; 53°47′N 81°20′E. Elevation 400 ft/121 m. In agricultural area; railroad, power plant, iron foundry, hardware, food products, grain elevator. Made city in 1915.

Kamenniki (KAH-meen-nee-kee), settlement, central YAROSLAVL oblast, central European Russia, on the S shore of the RYBINSK RESERVOIR, on local railroad spur and highway branch, 7 mi/11 km NW of RYBINSK; 58°09′N 38°40′E. Elevation 383 ft/116 m. Ferro-concrete manufacturing; fisheries.

Kamennogorsk (kah-mee-nuh-GORSK), city (2005 population 6,250), NW LENINGRAD oblast, NW European Russia, on the KARELIAN ISTHMUS, on the VUOKSIJOKI River, on highway and railroad, 97 mi/156 km W of SAINT PETERSBURG, and 20 mi/32 km NE of VYBORG; 60°58′N 29°07′E. Paper mill. Called Antrea while in FINLAND (until 1940) and, until 1948, in USSR.

Kamenno-Millerovo, UKRAINE: see KAM′YANE, Luhans′k oblast.

Kamennomostskiy (kah-myen-nuh-MOST-skeeyee), town, ADYGEY REPUBLIC, NW CAUCASUS, RUSSIA, on the BELAYA RIVER, on road and railroad spur, 20 mi/32 km S of MAYKOP, and 52 mi/84 km N of SOCHI; 44°18′N 40°11′E. Elevation 1,505 ft/458 m. Wood industries, gypsum souvenirs; gypsum quarries. Tourist center for the W Caucasus.

Kamennomostskoye (kah-meen-nuh-MOST-skuh-ye), village (2005 population 6,380), W central KABARDINO-BALKAR REPUBLIC, N CAUCASUS, S European Russia, on highway, 28 mi/45 km WNW of NAL′CHIK, and 10 mi/16 km N of GUNDELEN; 43°44′N 43°03′E. Elevation 2,664 ft/811 m. Agricultural products. Formerly known as Karmovo.

Kamennoye, UKRAINE: see KAM′YANE, Luhans′k oblast; or KAM′YANE, Zaporizhzhya oblast.

Kamennyy Brod, UKRAINE: see KAM′YANYY BRID.

Kameno (KAH-men-o), city (1993 population 5,497), BURGAS oblast, ⊙ Kameno obshtina, E BULGARIA, on railroad, 10 mi/16 km WNW of BURGAS; 42°34′N 27°18′E. In sugar-beet area; sugar refinery. Formerly called Kayalii. City since 1974.

Kamenolomni (kah-mye-nuh-LOM-nee), town (2006 population 12,260), SW ROSTOV oblast, S European Russia, on railroad, 28 mi/45 km N of ROSTOV-NA-DONU, and 4 mi/6 km S of SHAKHTY; 47°39′N 40°11′E. Elevation 242 ft/73 m. Railroad establishments. Established as a settlement of Kamenolomnya, around the building stone quarries in the area, now mostly exhausted; although officially renamed in the 1970s, the old name is still in use locally.

Kamenolomnya, RUSSIA: see KAMENOLOMNI.

Kameno Pole (KAH-men-o po-LAI), village, MONTANA oblast, ROMAN obshtina, BULGARIA; 43°14′N 23°54′E.

Kamen′-Rybolov (KAH-myen–ri-buh-LOF), village (2006 population 11,400), SW MARITIME TERRITORY, SE RUSSIAN FAR EAST, on the SW shore of Lake KHANKA, on crossroads and branch of the TRANS-SIBERIAN RAILROAD, 65 mi/105 km N of USSURIYSK; 44°44′N 132°03′E. Elevation 209 ft/63 m. In rice-growing and cattle-farming area; fisheries. Founded in 1865 by the Cossacks and runaway peasant serfs. Former river-road transfer point on the Khabarovsk-Vladivostok route.

Kamensk (KAH-meensk), village (2005 population 7,920), central BURYAT REPUBLIC, S SIBERIA, RUSSIA, on the NW slopes of the KHAMAR-DABAN RANGE, near the TRANS-SIBERIAN RAILROAD, 42 mi/68 km W of ULAN-UDE, and 5 mi/8 km S, and under administrative jurisdiction, of KABANSK; 51°58′N 106°36′E. Elevation 2,267 ft/690 m. Cement mill; asbestos-cement products.

Kamenskiy (KAH-myen-skeeyee), town (2006 population 3,610), S SARATOV oblast, SE European Russia, on road junction and near railroad, 13 mi/21 km SW of KRASNOARMEYSK; 50°53′N 45°29′E. Elevation 685 ft/208 m. Manufacturing of agricultural implements, flour milling. Established as Grimm (in German Volga Autonomous SSR); in 1941, renamed Lesnoy Karamysh, and the German population evacuated; renamed Kamenskiy in 1954.

Kamenskoye (KAH-meen-skuh-ye), village (2005 population 595), NW KORYAK AUTONOMOUS OKRUG, RUSSIAN FAR EAST, on the right bank of the BELAYA River, on local road, 290 mi/467 km N of PALANA; 62°30′N 166°12′E. Elevation 370 ft/112 m. In lignite area. Trading post in fishing and reindeer-raising area. Formerly known as Koryakskaya Kul′tbaza (Russian=Koryak cultural base).

Kamenskoye, UKRAINE: see DNIPRODZERZHYNS′K.

Kamensk-Shakhtinskiy (KAH-meensk–SHAHKH-teen-skeeyee), city (2006 population 75,905), W central ROSTOV oblast, S European Russia, on the N DONETS River, on road and railroad, 90 mi/145 km NNE of ROSTOV; 48°19′N 40°15′E. Elevation 236 ft/71 m. A mining center of the E Donets coal basin, the city is also an important producer of artificial fibers, mining machinery, chemical fertilizers, and distilled spirits; thermoelectric power plant. Founded in 1671 by Cossack settlers. Made city in 1927.

Kamensk-Ural′skiy (KAH-meensk–oo-RAHL-skeeyee), city (2006 population 180,475), S SVERDLOVSK oblast, E URALS, W Siberian Russia, at the junction of the Kamenka and Iset′ rivers (OB′ RIVER basin), 60 mi/97 km SE of YEKATERINBURG; 56°24′N 61°55′E. Elevation 501 ft/152 m. Railroad and highway junction. A city specializing in metallurgy, alumina, aluminum products, airplane equipment, non-ferrous metal products, tubing, electrical goods, light industries; thermoelectric power station. Has seven art schools, four museums, a conference pavilion, and a park. It was the site of the first iron production of the Urals in 1701, ordered by Peter the Great. Called Kamensk until 1940. Made city in 1935. Some of the heavy manufacturing enterprises have gone out of business or downsized drastically following the introduction of the free-market economy in 1991, and lack of jobs is largely responsible for the recent decrease in population.

Kamenz (KAH-ments), town, SAXONY, E central GERMANY, in UPPER LUSATIA, on the BLACK ELSTER, 25 mi/40 km NE of DRESDEN; 51°17′N 14°06′E. Textile center; manufacturing of glass and machinery. Granite stone quarrying. Gotthold Lessing born here. Town destroyed by fire in 1842. Marienstern Cistercian convent, founded 1264, is 6 mi/9.7 km ESE.

Kamenz, POLAND: see KAMIENIEC ZĄBKOWICKI.

Kameoka (kah-me-O-kah), city, KYOTO prefecture, S HONSHU, W central JAPAN, 9 mi/15 km W of KYOTO; 35°00′N 135°34′E. Beef cattle; matsutake mushrooms, chestnuts, adzuki beans. Bamboo work; confectionery. Hot springs.

Kameruka, village, NEW SOUTH WALES, SE AUSTRALIA, 279 mi/449 km S of SYDNEY, 13 mi/21 km SW of BEGA; 36°45′S 149°42′E. Cheese production.

Kamerun: see CAMEROON, REPUBLIC OF.

Kameshkir Russki, RUSSIA: see RUSSKI KAMESHKIR.

Kameshkovo (KAH-meesh-kuh-vuh), town (2006 population 13,860), N VLADIMIR oblast, central European Russia, on highway and railroad (Derbenevo station), 25 mi/40 km NE of VLADIMIR, and 13 mi/21 km W of KOVROV; 56°21′N 41°00′E. Elevation 331 ft/100 m. Cotton spinning and weaving; electrical goods. In the early 1920s, also called Imeni Sverdlova. Made ciy in 1951.

Kameshli, SYRIA: see QAMISHLIYE, EL.

Kamešnica (KAH-mesh-nee-tsah), mountain (5,935 ft/1,809 m), S CROATIA, in DALMATIA, on border of BOSNIA and HERZEGOVINA, 10 mi/16 km E of SINJ. Country's 2nd-highest mountain.

Kamet (KUH-mait), mountain (25,447 ft/7,756 m), SE ZASKAR RANGE, GARHWAL district, N UTTAR PRADESH state, N central INDIA, near TIBET (border undefined), 14 mi/23 km NE of BADRINATH; 30°55′N 79°35′E. Was first scaled (1931) by English group including noted Himalayan climbers Frank S. Smythe and Eric Shipton; for five years highest mountain climbed by man. A Sherpa porter was first to set foot on summit. Sometimes called Ibi Gamin.

Kameyama (kah-ME-yah-mah), city, MIE prefecture, S HONSHU, central JAPAN, 9 mi/15 km N of TSU; 34°51′N 136°27′E. Candles.

Kami (KAH-mee), town, Taka district, HYOGO prefecture, S HONSHU, W central JAPAN, 32 mi/51 km N of KOBE; 35°05'N 134°54'E. Paper.

Kami (KAH-mee), village, Shimoina county, NAGANO prefecture, central HONSHU, central JAPAN, 87 mi/140 km S of NAGANO; 35°23'N 137°58'E. Minami Alps National Park nearby.

Kami, mining area, AYOPAYA province, COCHABAMBA department, W central BOLIVIA, in Cordillera de CO-CHABAMBA, just SE of ICOYA, on a mountain with elevation 14,206 ft/4,330 m; 17°22'S 66°48'W. Tungsten and tin mining.

Kamiagata (kah-MEE-AH-gah-tah), town, Kamiagata county, NAGASAKI prefecture, NW KYUSHU, SW JAPAN, 133 mi/215 km N of NAGASAKI; 34°38'N 129°23'E.

Kamiah (KAM-e-ei), town (2000 population 1,160), LEWIS county, W IDAHO, 10 mi/16 km E of NEZ PERCE, on South Fork of CLEARWATER river, at mouth of Lawyer's Creek; 46°14'N 116°02'W. Manufacturing (wood products). In E part of Nez Perce Indian Reservation; NEZ PERCE NATIONAL HISTORICAL PARK (East Kamiah Site) to E; Nez Perce National Forest to SE; Clearwater National Forest to NE.

Kamibayashi (kah-MEE-bah-YAH-shee), village, Iwafune county, NIIGATA prefecture, central HONSHU, N central JAPAN, 31 mi/50 km N of NIIGATA; 38°10'N 139°27'E.

Kamien (KAH-myen), Polish *Kamień Pomorski*, German *Cammin*, town, Szczecin province, NW POLAND, on KAMIEN LAGOON (E arm of ODER River estuary mouth), opposite WOLIN island, 40 mi/64 km NNE of SZCZECIN. Seaport; fish and cattle market; health resort with mineral springs; woodworking. Bishopric, moved here (1175) from Wolin, again moved in 1574 to KÖSLIN (Koszalin). Town passed 1679 to BRANDENBURG. In World War II, c.50% destroyed. Seat (1933–1945) of Protestant bishopric. In POMERANIA until 1945.

Kamieniec Ząbkowicki (kah-MYE-nyets zomb-kov-VEETS-kee), German *Kamenz*, town in LOWER SILESIA, Wałbrzych province, SW POLAND, on the GLATZER NEISSE, 6 mi/9.7 km SE of ZABKOWICE. In nickel-mining, grain-growing region; railroad junction. Has 14th-century church of former monastery (1210). Chartered after 1945 and briefly called Kamieniec.

Kamien Koszyrski, UKRAINE: see KAMIN'-KA-SHYRS'KYY.

Kamien Krajeński (KAH-myen krah-YEN-skee), Polish *Kamień Pomorski*, German *Kamin*, town, Bydgoszcz province, NW POLAND, on railroad, 35 mi/56 km NW of BYDGOSZCZ. Flour milling. Tourist resort; 16th-century church.

Kamien Lagoon (KAH-myen), Polish *Zalew Kamieński*, German *Camminer Boden*, expansion (□ 4 sq mi/10.4 sq km) of DIEVENOW RIVER (E arm of ODER River estuary mouth), NW POLAND, between WOLIN island (W) and mainland (E); drains N into BALTIC SEA. Kamien Pomorski (CAMMIN) is on E shore. In POMERANIA until 1945. Also called Plycizna Kamienska, Polish *Płycizna Kamieńska*.

Kamienna Gora (kah-MYE-nah GO-rah), Polish *Kamienna Góra*, German *Landeshut*, town (2002 population 21,984), LOWER SILESIA, Jelenia Góra province, SW POLAND, at E foot of the RIESENGEBIRGE, on BOBRAWA RIVER, 11 mi/18 km W of WAŁBRZYCH (Waldenburg). Coal mining; cotton and linen milling, metalworking, manufacturing of machinery, shoes, leather goods. Has one of six Churches of Grace (1709–1730) allowed Silesian Protestants under Treaty of Altranstädt (1707). In 1745, Prussians defeated Austrians here; in Seven Years War, scene (1760) of Austrian victory over Prussians.

Kamienna River (kah-MYE-nah), c.80 mi/129 km long, E central POLAND; rises 12 mi/19 km SSW of SZYDŁOWIEC, flows generally E, past SKARŻYSKO-KAMIENNA, WIERZBNIK, and OSTROWIEC, to VISTULA River 24 mi/39 km SSW of PUŁAWY.

Kamienna-Skarzysko, POLAND: see SKARZYSKO-KAMIENNA.

Kamienski, Zalew, POLAND: see KAMIEN LAGOON.

Kamifukuoka (kah-MEEF-koo-O-kah), city, SAITAMA prefecture, E central HONSHU, E central JAPAN, 6 mi/10 km W of URAWA; 35°52'N 139°31'E. Computer parts.

Kamifurano (KAH-mee-foo-RAH-no), town, Kamikawa district, central Hokkaido prefecture, N JAPAN, 68 mi/110 km E of SAPPORO; 43°27'N 142°28'E. Potatoes. Hot-springs resort.

Kamigori (kah-mee-GO-ree), town, Ako county, HYOGO prefecture, S HONSHU, W central JAPAN, 47 mi/75 km W of KOBE; 34°52'N 134°21'E.

Kamigoto (kah-mee-GO-to), town, South Matsuura county, NAGASAKI prefecture, NW KYUSHU, SW JAPAN, 50 mi/80 km N of NAGASAKI; 32°58'N 129°04'E.

Kamiichi (kah-MEE-ee-chee), town, Nakanikawa county, TOYAMA prefecture, central HONSHU, central JAPAN, 9 mi/15 km E of TOYAMA; 36°41'N 137°21'E. Chubu-Sangaku National Park, Anan River (water source) nearby.

Kamiishizu (kah-MEE-ee-sheez), town, Yoro county, GIFU prefecture, central HONSHU, central JAPAN, 16 mi/25 km S of GIFU; 35°17'N 136°28'E. Wasabi, shiitake mushrooms; tea; white orchids. Charcoal; pottery; pickles.

Kamiiso (kah-MEE-ee-so), town, Oshima county, SW HOKKAIDO prefecture, N JAPAN, on inlet of TSUGARU STRAIT, 93 mi/150 km S of SAPPORO; 41°49'N 140°39'E. Cement. Trappist abbey in the vicinity.

Kamiita (kah-MEE-ee-tah), town, Itano county, TOKUSHIMA prefecture, SE SHIKOKU, W JAPAN, 9 mi/15 mi/24 km N of TOKUSHIMA; 34°07'N 134°24'E.

Kamiizumi (kah-MEE-ee-ZOO-mee), village, Kodama county, SAITAMA prefecture, E central HONSHU, E central JAPAN, 37 mi/60 km N of URAWA; 36°08'N 139°03'E. Stones for gardens. Sanba Seki Gorge nearby.

Kami-kamakari-jima (kah-MEE–kah-mah-kah-REE-jee-mah), island (□ 9 sq mi/23.4 sq km), HIROSHIMA prefecture, W JAPAN, in HIUCHI SEA (central section of INLAND SEA) just SE of KURE (on SW HONSHU); 5 mi/8 km long, 2 mi/3.2 km wide. Mountainous, fertile (rice, sweet potatoes). Fishery. Also spelled Kami-kamagari-jima.

Kamikatsu (kah-MEE-kahts), town, Katsuura county, TOKUSHIMA prefecture, SE SHIKOKU, W JAPAN, 16 mi/25 km S of TOKUSHIMA; 33°53'N 134°24'E. Onions.

Kamikawa (kah-MEE-kah-wah), town, Kamikawa district, Hokkaido prefecture, N JAPAN, 93 mi/150 km N of SAPPORO; 43°50'N 142°46'E. Mount Daisetsu National Park (Japan's largest), Soun Gorge nearby.

Kamikawa, town, Kodama county, SAITAMA prefecture, E central HONSHU, E central JAPAN, 37 mi/60 km N of URAWA; 36°12'N 139°06'E.

Kamikawa (kah-MEE-kah-wah), village, East Kanbara county, NIIGATA prefecture, central HONSHU, N central JAPAN, 31 mi/50 km N of NIIGATA; 37°37'N 139°27'E.

Kamikawachi (kah-MEE-KAH-wah-chee), town, Kawachi county, TOCHIGI prefecture, central HONSHU, N central JAPAN, 6 mi/10 km N of UTSUNOMIYA; 36°40'N 139°54'E.

Kamikita (kah-MEE-kee-tah), town, Kamikita county, Aomori prefecture, N HONSHU, N JAPAN, near Ogawara Lake, 28 mi/45 km E of AOMORI; 40°43'N 141°15'E. Seafood (clams, eel, fish). Honey.

Kamikitayama (kah-MEE-kee-tah-YAH-mah), village, Yoshino district, NARA prefecture, S HONSHU, W central JAPAN, 40 mi/64 km S of NARA; 34°07'N 136°00'E. Cryptomeria. Sushi, *mochi* rice cakes. Mount Odaigahara (YOSHINO-KUMANO NATIONAL PARK) is nearby.

Kamikoani (kah-MEE-ko-AH-nee), village, North Akita county, AKITA prefecture, N HONSHU, NE JAPAN, 25 mi/40 km N of AKITA; 40°03'N 140°17'E. Cryptomeria, beech; woodworking.

Kamikoshiki (kah-mee-KO-shee-kee), village, KAMI-KOSHIKI-SHIMA island, Satsuma county, KAGOSHIMA prefecture, SW JAPAN, 43 mi/70 km N of KAGOSHIMA; 31°49'N 129°52'E. *Kanoko* lilies. Kai Pond here is known for its fossils.

Kami-koshiki-shima (KAH-mee-ko-shee-KEE–shee-mah), island (□ 21 sq mi/54.6 sq km) of KOSHIKI-RETTO island group, KAGOSHIMA prefecture, SW JAPAN, in EAST CHINA SEA 14 mi/23 km off SW coast of KYUSHU; 9 mi/14.5 km long, 4 mi/6.4 km wide; irregular coastline; mountainous. Produces rice, wheat. Fishing.

Kamikuishiki (KAH-mee-koo-EESH-kee), village, West Yatsushiro county, YAMANASHI prefecture, central HONSHU, central JAPAN, near Motosu Lake, 9 mi/15 km S of KOFU; 35°31'N 138°36'E.

Kamiluk Lake (KA-mi-luhk) or **Kamilukuak Lake** (ka-mi-LUH-kwak), NORTHWEST TERRITORIES, CANADA; drains into DUBAWNT LAKE (S); 62°25'N 101°40'W. Lake is 30 mi/48 km long, 20 mi/32 km wide; on S shore is trading post.

Kamimine (kah-MEE-mee-ne), town, Miyaki county, SAGA prefecture, N KYUSHU, SW JAPAN, 9 mi/15 km N of Saga; 33°18'N 130°25'E.

Kamin', UKRAINE: see TEOFIPOL'.

Kamina (kah-MEE-nah), city, KATANGA province, S CONGO; 08°44'S 24°59'E. Elev. 3,595 ft/1,095 m. It is an administrative and transportation center. A major military airfield is located here. Kamina was used by the Belgians as a center for interventionist actions in the early months of CONGO's independence (1960). It later served as a center for UN operations during the crisis caused by the secession (1960–1963) of Katanga.

Kaminaka (kah-MEE-nah-kah), town, Onyu county, FUKUI prefecture, central HONSHU, W central JAPAN, 50 mi/80 km S of FUKUI; 35°27'N 135°51'E.

Kaminaka, town, Naka county, TOKUSHIMA prefecture, SE SHIKOKU, W JAPAN, 22 mi/35 km S of TOKUSHIMA; 33°48'N 134°22'E. Citrons. Nagayasuguchi Dam nearby.

Kaminak Lake (KA-min-ak), NUNAVUT territory, CANADA, 60 mi/97 km W of TAVANI; 62°12'N 95°00'W. Lake is 40 mi/64 km long, 1 mi/2 km–22 mi/35 km wide; drains E into HUDSON BAY.

Kaminal Juyú (kah-me-NAHL hoo-YOO), archaeological site, GUATEMALA deparment, GUATEMALA, W part of GUATEMALA city; 14°38'N 90°33'W. Partly destroyed by urbanization, an important preclassic site; a trade center. Also spelled Kaminaljuyú.

Kaministikwia (kuh-min-is-TIK-wee-uh), river, c.60 mi/100 km long, CANADA; rising in DOG LAKE, W ONTARIO, and flowing S, then E into Lake SUPERIOR at THUNDER BAY. In fur-trading days, it was the chief alternate to the GRAND PORTAGE–PIGEON RIVER route into the NW. After 1783, when the Pigeon River formed part of the U.S. boundary, it became the main route used by the North West Company to FORT WILLIAM, their W headquarters at the mouth of the river. Kakabeka Falls (130 ft/40 m high), W of THUNDER BAY, is used to generate hydroelectricity.

Kamin'-Kashyrs'kyy (KAH-meen—kah-SHIR-skee) (Polish *Kamien Koszyrski*) (Russian *Kamen'-Kashirskiy*), city, N VOLYN' oblast, UKRAINE, 30 mi/48 km NNE of KOVEL'; 51°38'N 24°58'E. Elevation 538 ft/163 m. Raion center. Railroad terminus; woodworking, butter making, flour milling, canning, flax processing, feed milling; professional-technical school, local museum, wooden church (1723). Known since 1196 in the Volyn' principality as Kosher, later Kamin', Kamin'-Koshyr'skyy; passed to Lithuania in the 14th century; to Poland (1569); to Russia (1795); reverted to Poland (1521); annexed to Ukrainian SSR (1939); renamed Kamin'-Kashyrs'kyy; city since 1939.

Kamin'-Koshyrs'kyy, UKRAINE: see KAMIN'-KA-SHYRS'KYY.

Kaminoho (kah-MEE-no-ho), village, Mugi county, GIFU prefecture, central HONSHU, central JAPAN, 25 mi/40 km N of GIFU; 35°36'N 137°02'E. Tea, miso.

Kaminokawa (kah-MEE-no-KAH-wah), town, Ka-wachi county, TOCHIGI prefecture, central HONSHU, N central JAPAN, 9 mi/15 km S of UTSUNOMIYA; 38°08′N 140°16′E. Motor vehicles. Dried gourd shavings.

Kami no Kuni (KAH-mee-no-KOO-nee), town, Hiyama district, Hokkaido prefecture, N JAPAN, near Daisengendake mountain, 109 mi/175 km S of SAP-PORO; 41°47′N 140°07′E. Pigs. Manganese, silica. Habitat of Kami no Kuni breed of dog. Ancient tombs in the vicinity.

Kaminoseki (kah-MEE-no-SE-kee), town, Kumage county, YAMAGUCHI prefecture, SW HONSHU, W Japan, 43 mi/70 km S of YAMAGUCHI; 33°49′N 132°06′E.

Kaminoyama (kah-MEE-no-YAH-mah), city, YAMA-GATA prefecture, N HONSHU, NE Japan, 9 mi/15 km S of YAMAGATA, near Zao highlands; 38°08′N 140°16′E. Agricultural center (fruits, hops); wine, *konnyaku* (paste from devil's tongue). Traditional papermaking. Known for hot springs since 15th century; skiing area. Site of Kaminoyama Castle.

Kaminskiy (kah-MEEN-skeeyee), town (2005 population 1,625), N IVANOVO oblast, central European Russia, on the TEZA RIVER, near railroad, 11 mi/18 km WNW of RODNIKI; 57°09′N 41°28′E. Elevation 351 ft/106 m. Cotton cloth. Until 1947, called Gorki-Pavlovy.

Kaminuriak Lake (ka-min-YOO-ree-ak) (□ 360 sq mi/936 sq km), NUNAVUT territory, CANADA, near 63°00′N 95°45′W.

Kamioka (kah-mee-O-kah), town, Senhoku county, Akita prefecture, N HONSHU, NE JAPAN, 22 mi/35 km S of AKITA city; 39°29′N 140°25′E.

Kamioka (kah-MEE-o-kah), town, Yoshiki county, GIFU prefecture, central HONSHU, central JAPAN, 71 mi/115 km N of Gifu; 36°19′N 137°18′E. Lacquerware. Nonferrous metals (zinc, lead, silver).

Kamionka Strumilowa, UKRAINE: see KAM'YANKA-BUZ'KA.

Kamisaibara (kah-MEE-SAH-ee-bah-rah), village, Tomata county, OKAYAMA prefecture, SW HONSHU, W JAPAN, 43 mi/70 km N of OKAYAMA; 35°16′N 133°55′E. Mushroom canning; uranium mining. Skiing area.

Kamisato (kah-MEE-sah-to), town, Kodama county, SAITAMA prefecture, E central HONSHU, E central JAPAN, 37 mi/60 km N of URAWA; 36°14′N 139°08′E.

Kamishak Bay, S ALASKA, on W side of mouth of COOK INLET, W of SELDOVIA, 45 mi/72 km wide at mouth, 20 mi/32 km long; 59°18′N 153°30′W. Contains AUGUS-TINE ISLAND and volcano.

Kamishihi (kah-MEE-shee-hee), village, Yoshida county, FUKUI prefecture, central HONSHU, W central JAPAN, 9 mi/15 km E of FUKUI; 36°04′N 136°24′E.

Kamishihoro (kah-MEE-shee-HO-ro), town, Tokachi district, HOKKAIDO prefecture, N JAPAN, 99 mi/160 km E of SAPPORO; 43°13′N 143°17′E. Dairying; lumber.

Kami-shikuka, RUSSIA: see LEONIDOVO.

Kamisuku, CONGO: see KALIMA-KINGOMBE.

Kamisuma (kah-MEE-soo-mah), town, Kashima county, IBARAKI prefecture, central HONSHU, E central JAPAN, 34 mi/55 km S of MITO; 35°53′N 140°40′E. Oil manufacturing. Peppers. Well-known fountain.

Kamisunagawa (kah-mee-soo-NAH-gah-wah), town, Sorachi district, Hokkaido prefecture, N JAPAN, 43 mi/70 km N of SAPPORO; 43°28′N 141°59′E.

Kamitaira (kah-mee-TAH-ee-rah), village, East To-nami county, TOYAMA prefecture, central HONSHU, central JAPAN, 28 mi/45 km S of TOYAMA; 36°23′N 136°54′E.

Kamitakara (kah-mee-TAH-kah-rah), village, Yoshiki county, GIFU prefecture, central HONSHU, central JAPAN, 71 mi/115 km N of GIFU; 36°16′N 137°22′E. Hot springs.

Kamitonda (kah-mee-TON-dah), town, West Muro county, WAKAYAMA prefecture, S HONSHU, W central JAPAN, 40 mi/65 km S of WAKAYAMA; 33°41′N 135°25′E. Traditional *daruma* dolls.

Kamitsue (kah-MEETS-e), village, Hita county, OITA prefecture, E KYUSHU, SW JAPAN, 37 mi/60 km S of OITA; 33°05′N 130°58′E.

Kamitsushima (kah-MEE-TSOO-shee-mah), town, Kamiagata county, NAGASAKI prefecture, NW KYUSHU, SW JAPAN, 136 mi/220 km N of NAGASAKI; 34°38′N 129°27′E.

Kamituga (kah-mee-TOO-gah), village, SUD-KIVU province, E CONGO, 65 mi/105 km SW of BUKAVU; 03°03′S 28°11′E. Elev. 3,963 ft/1,207 m. Center of gold-mining operations employing c.7,000, workers; gold processing, also diamond mining. Hydroelectric power plant.

Kamiura (kah-MEE-oo-rah), town, Ochi county, EHIME prefecture, NW SHIKOKU, W JAPAN, 31 mi/50 km N of MATSUYAMA; 34°15′N 133°02′E.

Kamiura, town, South Amabe county, OITA prefecture, E KYUSHU, SW JAPAN, on HOYO STRAIT, 22 mi/35 km S of OITA; 33°02′N 131°55′E.

Kamiyahagi (kah-MEE-yah-HAH-gee), town, Ena county, GIFU prefecture, central HONSHU, central JAPAN, 43 mi/70 km E of GIFU; 35°17′N 137°30′E.

Kamiyaku (kah-MEE-yah-koo), town, Kumage county, KAGOSHIMA prefecture, SW KYUSHU, SW JAPAN, 81 mi/130 km S of KAGOSHIMA; 30°25′N 130°34′E.

Kamiyama (kah-MEE-yah-mah), town, Myozai county, TOKUSHIMA prefecture, SE SHIKOKU, W JAPAN, 20 mi/32 km S of TOKUSHIMA; 33°57′N 134°21′E. Japanese plums, *sudachi* citrons, flowers (irises, orchids).

Kamiyamada (kah-MEE-yah-MAH-dah), town, Sarashina county, NAGANO prefecture, central HONSHU, central JAPAN, 12 mi/20 km S of NAGANO; 36°28′N 138°09′E. Flowers (carnations, balloon flowers). Skiing area.

Kamiyubetsu (kah-mee-YOO-bets) town, Abashiri district, Hokkaido prefecture, N JAPAN, 136 mi/220 km N of SAPPORO; 44°08′N 143°34′E. Asparagus. Whisky.

Kamloops (KAM-loops), city (□ 53 sq mi/137.8 sq km; 2001 population 77,281), S BRITISH COLUMBIA, W CANADA, at the junction of the North Thompson and South THOMPSON rivers, and in Thompson-NICOLA regional district; 50°40′N 120°20′W. A trading post was first established on the site in 1812. A village grew up at the time of the Cariboo gold rush (1860), and in 1885 the main line of the Canadian Pacific reached Kamloops. A transportation, finance, education, and administrative center, Kamloops supplies the surrounding mining, lumbering, and farming districts. It is also the center of British Columbia's cattle industry and a growing tourist site.

Kamloops Lake (KAM-loops), S BRITISH COLUMBIA, W CANADA, expansion of THOMPSON River, 6 mi/10 km W of KAMLOOPS; 20 mi/32 km long, 20 mi/32 km wide; 50°44′N 120°38′W.

Kammenoye, UKRAINE: see KAM'YANE.

Kammersee, AUSTRIA: see SALZKAMMERGUT.

Kammertal (KAHM-mer-tahl), valley of the rivers LIESING and Palten, central STYRIA, central AUSTRIA; 47°20′N–47°33′N, 14°19′E–15°01′E. Vertex is Schober Pass (2,588 ft/789 m); traffic corridor for Pyhrn Motorway and Selzthal–Sankt Michael railroad in Upper Styria. Major settlements are TRIEBEN, ROTTENMANN, Wald am Schoberpass, and MAUTERN IN STEIERMARK.

Kamnik (KAHM-neek), town (2002 population 12,013), N SLOVENIA, 13 mi/21 km NNE of LJUBLJANA, at S foot of KAMNIK-SAVINJA ALPS; 46°14′N 14°36′E. Terminus of railroad to Ljubljana. Light metallurgy; textiles, furniture, chemical, and electronics industries. Tourist center. Castle ruins and church with fine frescoes nearby.

Kamnik Alps, SLOVENIA: see KAMNIK-SAVINJA ALPS.

Kamnik-Savinja Alps, Slovenian *Kamniŝko-Savinjske Alpe*, spur of main KARAWANKEN mountain range in N SLOVENIA, extending c.15 mi/24 km SE from Austrian border at LJUBELJ (Loibl) Pass. Highest peak is GRINTAVEC, German *Grintouz* (8,390 ft/2,557 m). Tourist area. Sometimes called Kamnik, or Savinja, Alps.

Kamo (KAH-mo), city, NIIGATA prefecture, central HONSHU, N central JAPAN, 16 mi/25 km S of NIIGATA; 37°39′N 139°03′E. Paulownia furniture.

Kamo (KAH-mo), town, Soraku county, KYOTO prefecture, S HONSHU, W central JAPAN, 19 mi/30 km S of KYOTO; 34°44′N 135°52′E.

Kamo, town, Tomata county, OKAYAMA prefecture, SW HONSHU, W JAPAN, 37 mi/60 km N of OKAYAMA; 35°10′N 134°03′E.

Kamo, town, Ohara co., Shimane prefecture, SW Honshu, W Japan, 12 mi/20 km S of Matsue; 35°20′N 132°54′E.

Kamo (KAH-mo), village, Kamo county, SHIZUOKA prefecture, central HONSHU, E central JAPAN, 25 mi/40 km S of SHIZUOKA; 34°50′N 138°46′E. Shiitake mushrooms, carnations. Fish (bonito, saury, mackerel).

Kamo (KAH-mo), N suburb of WHANGAREI, Whangarei district, NORTHLAND region, NEW ZEALAND, 85 mi/137 km NNW of AUCKLAND. Mineral springs.

Kamo (kah-MO) or **Katalanga**, river, 210 mi/339 km long, in EVENKI AUTONOMOUS OKRUG, KRASNOYARSK TERRITORY, Siberian Russia, left tributary of the PODKAMENNAYA TUNGUSKA RIVER. Basin area is 5,598 sq mi/14,500 sq km. The Tokhomo River is the main tributary (left). Also called Levaya (Russian=left) Kamo River.

Kamoenai (kah-MO-e-nah-ee), village, Shiribeshi district, Hokkaido prefecture, N JAPAN, 47 mi/75 km W of SAPPORO; 43°08′N 140°26′E.

Kamogata (kah-MO-gah-tah), town, Asaguchi county, OKAYAMA prefecture, SW HONSHU, W JAPAN, 22 mi/35 km S of OKAYAMA; 34°31′N 133°35′E.

Kamogawa (kah-MO-gah-wah), city, CHIBA prefecture, E central HONSHU, E central JAPAN, on SE coast of BOSO PENINSULA, 25 mi/40 km S of CHIBA; 35°06′N 140°06′E. Dairying.

Kamogawa (kah-MO-gah-wah), town, Mitsu county, OKAYAMA prefecture, SW HONSHU, W JAPAN, 16 mi/25 km N of OKAYAMA; 34°51′N 133°48′E.

Kamojima (kah-MO-jee-mah), town, Oe county, TO-KUSHIMA prefecture, SE SHIKOKU, W JAPAN, 12 mi/20 km W of TOKUSHIMA; 34°04′N 134°21′E.

Kamoke (kah-mo-kuh), town, GUJRANWALA district, E PUNJAB province, central PAKISTAN, 12 mi/19 km S of GUJRANWALA; 31°58′N 74°13′E. Market center.

Kamola (kah-MO-lah), village, KATANGA province, SE CONGO, c.35 mi/56 km SW of MANONO; 07°47′S 27°18′E. Elev. 2,276 ft/693 m. Tin mining; power plant.

Kamoto (KAH-mo-to), town, Kamoto county, KUMA-MOTO prefecture, SW KYUSHU, SW JAPAN, 16 mi/25 km S of KUMAMOTO; 32°59′N 130°44′E.

Kamou (kah-MO-oo), town, Aira county, KAGOSHIMA prefecture, SW KYUSHU, SW JAPAN, 12 mi/20 km N of KAGOSHIMA; 31°45′N 130°34′E.

Kamouraska (kam-oo-RA-skuh), county (□ 871 sq mi/2,264.6 sq km; 2006 population 22,514), Bas-Saint-Laurent region, SE QUEBEC, E Canada, on U.S. (MAINE) border, on the SAINT LAWRENCE RIVER; ⊙ SAINT-PASCAL; 47°30′N 69°30′W. Composed of nineteen municipalities. Formed in 1982.

Kamouraska (kam-oo-RA-skuh), village (□ 16 sq mi/41.6 sq km), Bas-Saint-Laurent region, SE QUEBEC, E CANADA, on the SAINT LAWRENCE RIVER, 15 mi/24 km ESE of LA MALBAIE; 47°34′N 69°52′W. Dairying; resort.

Kampala (kahm-PAH-luh), administrative district (□ 92 sq mi/239.2 sq km; 2005 population 1,337,900), CENTRAL region, S central UGANDA, coextensive with city of KAMPALA; ⊙ Kampala; 00°20′N 32°35′E. As of Uganda's division into eighty administrative districts, Kampala is surrounded by WAKISO district on all sides except SE where it borders LAKE VICTORIA and MU-KONO district. Railroad between KASESE town (W Uganda) and MOMBASA (KENYA) travels W-E through district. Main roads branch out of Kampala to all areas of Uganda.

Kampala (kahm-PAH-luh), city (2002 population 1,189,142), ⊙ KAMPALA district and UGANDA, principle city of CENTRAL region, on LAKE VICTORIA (Victoria Nyanza). It is Uganda's largest city and its administrative, communications, economic, and transportation center. Manufacturing includes processed foods, beverages, furniture, and machine parts. Agricultural exports include coffee, cotton, tea, and sugar. It is on the Trans-African Highway; and linked by railroad with KASESE town, a mining center in SW Uganda, and with MOMBASA (KENYA), on the INDIAN OCEAN coast. Steamers on Lake Victoria link the city to ports in Kenya and TANZANIA. An international airport is nearby, at ENTEBBE. An epicenter of the AIDS virus epidemic. Kampala grew up around a fort constructed (1890) by Captain Frederick Lugard for the British East Africa Company. In 1962, Kampala replaced Entebbe as the capital of Uganda. Despite its proximity (20 mi/32 km) to the equator, the city has a moderate climate, largely because of its elevation (c.4,000 ft/1,220 m). The city is built on and around six hills and has modern government and commercial quarters as well as wide avenues that fan out toward the surrounding suburbs. Much of the city was destroyed after the overthrow (1979) of Idi Amin's dictatorship and subsequent civil war. Under the Museveni regime Kampala has enjoyed relative stability since 1986. Foreign investment has provided funding for the rehabilitation of the city's infrastructure and the restoration of services. Kampala is the seat of the East Africa Development Bank and Makerere University. Was captial of former CENTRAL province.

Kampar (KAHM-pahr), town (2000 population 19,795), S central PERAK, MALAYSIA, at foot of central Malayan range, on railroad, 20 mi/32 km SSE of IPOH. A tin-mining center of Kinta Valley.

Kampar River (KAHM-pahr), c.200 mi/322 km long, central SUMATRA, INDONESIA; rises in PADANG HIGHLANDS in BARISAN MOUNTAINS E of Mount OPHIR; flows generally E through marshy area across RIAU province to Strait of MALACCA near MENDOL Island; 00°32′N 103°08′E.

Kampen (KAHM-puhn), city (2001 population 31,892), OVERIJSSEL province, central NETHERLANDS, on the IJSSEL RIVER, opposite the entrance to Ganzendiep canal (from N), 45 mi/72 km of AMSTERDAM, 8 mi/12.9 km WNW of ZWOLLE; 52°33′N 05°55′E. Mouth of IJssel River on KETELMEER (arm of IJSSELMEER) 4 mi/6.4 km to WNW. Dairying; cattle, sheep; grain, vegetables; manufacturing (trailers, timber, dairy products, liqueurs). Kampen was first mentioned in the thirteenth century, and in the fifteenth century it was a member of the Hanseatic League. Notable structures include the fourteenth-century town hall and several churches and buildings from the fourteenth and fifteenth centuries, including the church of St. Nicholas (c.1400). Seat of Theological University of Reformed Churches. Museum.

Kampene (kahm-PE-nai), village, MANIEMA province, E CONGO, 55 mi/89 km N of KASONGO; 03°36′S 26°40′E. Elev. 2,998 ft/913 m. Center of gold-mining area employing c.3,000 workers; rice processing. Tin and wolfram are also mined in vicinity (E and S). Gold mines of KALEMBA are 25 mi/40 km N, those of KISULU 15 mi/24 km NNW.

Kampengpet, THAILAND: see KAMPHAENG PHET, city.

Kampenhout (KAHM-puhn-hout), commune (2006 population 10,997), Halle-Vilvoorde district, BRABANT province, central BELGIUM, 11 mi/18 km E of BRUSSELS; 50°57′N 04°33′E.

Kampeska, village, CODINGTON county, NE SOUTH DAKOTA, 10 mi/16 km W of WATERTOWN; 44°52′N 97°16′W. Just N is Lake Kampeska (4 mi/6.4 km long, 3 mi/4.8 km wide), used for recreation. Shady Shore State Recreational Area on S side.

Kämpfelbach (KAM-pfel bahkh), suburb of PFORZHEIM, BADEN-WÜRTTEMBERG, SW GERMANY, just NW of city; 48°56′N 08°39′E.

Kamphaeng Phet (GUHM-PANG PET), province (□ 3,436 sq mi/8,933.6 sq km), N THAILAND; ⊙ KAMPHAENG PHET; 16°15′N 99°30′E. Once part of the SUKHOTHAI KINGDOM, there are still ruins of the old walled city in what is now the Kamphaeng Phet Historical Park. In addition to growing rice, corn, and tapioca, the province also produces fine bananas called kluay khai [Thai=egg bananas].

Kamphaeng Phet (GUHM-PANG PET), town, ⊙ KAMPHAENG PHET province, central THAILAND, on PING RIVER, 200 mi/322 km NNW of BANGKOK; 16°28′N 99°30′E. Ancillary walled town built by King Li Thai (1347–1368), in rice-growing region. Provincial museum with fine bronze statue of the Hindu god Siva. Sometimes spelled KAMPENGPET and KAMBAENGBEJR.

Kampil (KUHM-pil), village, FARRUKHABAD district, central UTTAR PRADESH state, N central INDIA, near KAIMGANJ. Was capital of ancient Panchala kingdom (figured in the *Mahabharata*). Jain temple, ruins of fort built by 13th-century Afghan leader. Extensive Hindu ruins nearby.

Kampinoski National Park (kam-pee-no-skee) (□ 130 sq mi/338 sq km), POLAND, NW of WARSAW. Established 1959, it is the largest national park in Poland; 70% of the land is covered by forest (pine, oak). Dunes up to 98 ft/30 m high, swamps and bogs, a variety of animal life, attractive recreational areas for Warsaw's inhabitants, and trails for hiking and cycling. E part accessible by public transportation from Warsaw. One trail in W part ends in Żelazowa WOLA where composer Chopin was born. Also known as Puszcza Kampinowska.

Kampi-ya-Moto (kahm-bee–yah–MO-to), village, RIFT VALLEY province, W KENYA, on railroad, 15 mi/24 km N of NAKURU; 00°09′N 35°55′E. Coffee, wheat, corn.

Kampli (KUHM-plee), town, BELLARY district, KARNATAKA state, SW INDIA, on TUNGABHADRA RIVER, 28 mi/45 km NW of BELLARY; 15°24′N 76°37′E. Rice milling; bamboo coracles; coconut palms, plantain.

Kamp-Lintfort (KAHMP–LINT-fort), city, NORTH RHINE-WESTPHALIA, W GERMANY, in the RUHR, 6 mi/9.7 km NW of MOERS; 51°30′N 06°33′E. Coal mining; manufacturing of machinery, steel products. Has former Cistercian abbey Kamp (founded 1123; destroyed 16th century; rebuilt 1700–1714). Town formed 1934 through incorporated of Kamp, LINTFORT, and four neighboring villages.

Kamponde (kahm-PON-dai), village, KASAI-OCCIDENTAL province, S CONGO, on railroad, 100 mi/161 km WSW of KABINDA; 06°42′S 22°50′E. Elev. 2,723 ft/829 m. Educational center, including noted agricultural school and junior college Roman Catholic missions.

Kampong (kahm-pawng) [Malay=village], for Malaysian names beginning thus and not found here; see under following part of the name.

Kampong Ayer (kahm-PONG AH-yuhr), locality, NE BRUNEI DARUSSALAM, residential area of BANDAR SERI BEGAWAN. Composed entirely of houses built on stilts in the waters of Brunei River, on both sides of river, at mouth of Kianggah River, connected by wooden walkways and water taxis. The district is centuries old, referred to as "Water Village" by British colonialists. Large Omar Ali Saifuddien Mosque immediately adjacent to village. Kampong Buang Sakar, housing development, is upriver from Kampong Ayer.

Kâmpóng Saôm, municipality, city, and port, CAMBODIA: see KOMPONG SOM.

Kampot (KUHM-PUHT), province (□ 1,881 sq mi/4,890.6 sq km; 2007 population 619,088), SW CAMBODIA, borders KOH KONG and KOMPONG SPEU provinces (N), TAKEO province (E), VIETNAM (SE), and GULF OF THAILAND (S and W); ⊙ KAMPOT; 10°45′N 104°15′E. Ecologically diverse, rice and shifting cultivation. Recent deforestation. Khmer population, with Cham, Saoch, and Chinese minorities. KOMPONG SOM and KEP municipalities are located in Kampot province, though they are politically and administratively independent of it.

Kampot (KUHM-PUHT), city, ⊙ KAMPOT province, S CAMBODIA, on the GULF OF THAILAND; 10°37′N 104°11′E. It is a seaport on the PHNOM PENH–KOMPONG SOM railroad and the center of the Cambodian pepper culture and durian fruit production. Food processing and light manufacturing Several agricultural projects were set up in Kampot after 1975. Many buildings and the urban infrastructure were destroyed by the Khmer Rouge, but Kampot remains a charming coastal community that likely will prosper as a center of tourism in the future. Population largely Khmer, with small Cham and Chinese communities.

Kamp River (KAHMP), 68 mi/109 km long, N LOWER AUSTRIA; rises near NW border of Lower Austria in the WALDVIERTEL, flows E and S, past Zwettl, to the Danube,10 mi/16 km E of Krems. Hydroelectric power stations on the middle course; picturesque narrow valley with several castles on the lower run between Kamp and LANGENLOIS.

Kampsville, village (2000 population 302), CALHOUN county, SW ILLINOIS, 9 mi/14.5 km N of HARDIN, on ILLINOIS RIVER; 39°17′N 90°36′W. Riprap Landing Conservation Area to N. Ferry across river.

Kamptee (KAHM-tee), city, NAGPUR district, central MAHARASHTRA state, central INDIA, on KANHAN RIVER, 8 mi/12.9 km NE of NAGPUR. Trade center (mainly cotton); cotton ginning, tea processing; cement works, and tanneries. Explosives factory (mining operations). Sandstone and marble quarries nearby. Has college Municipal waterworks. A part of the city is a military cantonment. Believed to have derived name from "Camp T" of army usage. Formerly spelled Kamthi.

Kampti (kahmp-TEE), village, PONI province, SUDOUEST region, SW BURKINA FASO, 80 mi/129 km SE of BOBO-DIOULASSO; 10°07′N 03°22′W. Near CÔTE D'IVOIRE border. Agriculture (peanuts, shea nuts, sesame) and livestock (cattle, sheep, goats).

Kampung Koh (KAHM-pong KO), village, W PERAK, MALAYSIA, 5 mi/8 km ESE of LUMUT; 04°10′N 100°43′E. Rubber, coconuts; fishing; tourism.

Kampung Kuala Kemaman (KAHM-pong KWAH-lah CUH-mah-mahn), town, S TERENGGANU, MALAYSIA, SOUTH CHINA SEA port at mouth of KEMAMAN RIVER, 3 mi/5 km NE of CHUKAI, 75 mi/121 km SSE of KUALA TERENGGANU; 04°16′N 103°27′E. Fisheries.

Kamrar, town (2000 population 229), HAMILTON county, central IOWA, 7 mi/11.3 km SE of WEBSTER CITY; 42°23′N 93°43′W. In agricultural area.

Kamrej (KAHM-raij), town, SURAT district, GUJARAT state, W central INDIA, on TAPI RIVER, 10 mi/16 km NE of SURAT; 21°17′N 72°59′E. Cotton, millet.

Kamrup (KAHM-roop), district (□ 1,678 sq mi/4,362.8 sq km), W ASSAM state, NE INDIA; ⊙ GUWAHATI. Mainly in BRAHMAPUTRA RIVER valley; spurs of JAINTIA HILLS (S); swamp lakes (center SE); bounded N by BHUTAN, W by MANAS RIVER; traversed by Brahmaputra River and drained by its tributaries. Cotton ginning and baling, flour, rice, and oilseed milling, manufacturing of silk cloth, soap. Mainly alluvial soil; rice, mustard, jute, cotton, tea; lac growing, sal timbering, silk growing in dispersed forest area. University and several well-known colleges at Guwahati. Vishnuite college (founded fifteenth century) at Barpeta. Pilgrimage centers at Kamakhya (Sivaite temple) and HAJO (Vishnuite temple). Guwahati was capital of ancient Hindu kingdom of Kamarupa (from which present district derives its name) and of late eighteenth-century Ahom (Shan) kingdom.

Kamsack (KAM-sak), town (2006 population 1,713), SE SASKATCHEWAN, CANADA, on ASSINIBOINE RIVER, at mouth of WHITESAND RIVER, 35 mi/56 km NE of YORKTON; 51°34'N 101°54'W. Distributing center; grain elevators; lumbering, dairying, livestock raising; natural gas production.

Kamsar (KAM-sahr), town, Boké prefecture, Boké administrative region, W GUINEA, in Guinée-Maritime geographic region, port, 25 mi/40 km SW of BOKÉ; 10°40'N 14°36'W. On railroad; swamp rice, fishing.

Kamsar, IRAN: see QAMSAR.

Kamskiye Polyany (KAHM-skee-ye puh-LYAH-ni), town (2006 population 15,090), N TATARSTAN Republic, E European Russia, on the KAMA River, near highway, 23 mi/37 km W of NIZHNEKAMSK; 55°26'N 51°25'E. Elevation 426 ft/129 m. Materials and equipment for hydroelectric power stations; garment factory.

Kamskoye Ust'ye (KAHM-skuh-ye OOST-ye), town (2006 population 4,465), SW TATARSTAN Republic, E European Russia, port on the KUYBYSHEV Reservoir of the VOLGA RIVER, opposite the former mouth (Russian *ust'ye*) of the KAMA River, near highway, 40 mi/64 km S of KAZAN; 55°12'N 49°16'E. Elevation 442 ft/134 m. Freight-transfer point; oil processing, anhydrite mining; creamery. Until about 1928, known as Bogorodsk. Made city in 1939.

Kamta Rajaula (KAHM-tuh ruh-JOU-luh), former petty state of CENTRAL INDIA Agency, one of the CHAUBE JAGIRS. In 1948, merged with VINDHYA PRADESH, now within BANDA district, UTTAR PRADESH state.

Kamthi, INDIA: see KAMPTEE.

Kamudi (KUHM-uh-dee), town, RAMANATHAPURAM district, TAMIL NADU state, S INDIA, 39 mi/63 km SSE of MADURAI; 09°25'N 78°22'E. In cotton area; cotton weaving, bell metal wares. Also spelled Kamuti.

Kamuela, HAWAII: see WAIMEA.

Kamula, Mount, UKRAINE: see HOLOHORY.

Kamuli (kah-MOO-lee), former administrative district (□ 1,679 sq mi/4,365.4 sq km; 2005 population 776,800), EASTERN region, SE central UGANDA, along S shore of LAKE KYOGA and E shore of VICTORIA NILE RIVER; capital was KAMULI; 01°05'N 33°15'E. Average elevation 3,553 ft/1,083 m. As of Uganda's division into fifty-six districts, was bordered by LIRA, KABERAMAIDO, and SOROTI (N, on opposite shore of Lake Kyoga), PALLISA (E), IGANGA (SE), JINJA (S), and KAYUNGA (W, border formed by Victoria Nile River) districts. Vegetation formed of forest and savannah woodland; marsh area in NW. Was inhabited primarily by Basoga people. Rural and agricultural area with N and S forming distinct agricultural zones (S included bananas, cocoa, coffee, maize, rice, and vanilla; N included primarily cattle, also cotton, groundnuts, maize, and citrus fruit). Fishing in Victoria Nile River and Lake Kyoga. In 2005 NE portion of district was carved out to form KALIRO district and remainder of district was formed into current KAMULI district.

Kamuli (kah-MOO-lee), town (2002 population 11,344), ⊙ KAMULI district, EASTERN region, SE central UGANDA, 35 mi/56 km NNW of Jinja. Near railroad; cotton, tobacco, coffee, bananas, corn. Was part of former BUSOGA province.

Kamuli, administrative district, EASTERN region, SE central UGANDA, on S shore of Lake KYOGA and E shore of VICTORIA NILE RIVER; ⊙ KAMULI. As of Uganda's division into eighty districts, borders AMOLATAR, KABERAMAIDO, and SOROTI (N, on opposite shore of Lake Kyoga), KALIRO (NE), IGANGA (E), JINJA (S), and KAYUNGA (W, formed by Victoria Nile River) districts. Marsh area in N, near Victoria Nile River and Lake Kyoga. Rural and agricultural area (including bananas, coffee, and cotton; livestock). Fishing in Victoria Nile River and Lake Kyoga. Secondary railroad connects district to main railroad between KASESE town (W Uganda) and MOMBASA (SE KENYA). Road runs S from Kamuli town to Jinja town.

Formed in 2005 from all but NE portion of former KAMULI district (Kaliro district formed from NE portion).

Kamuti, INDIA: see KAMUDI.

Kamvali, CONGO: see LUISHIA.

Kamvunia (kahm-VOO-nyah), Latin *Cambunia*, mountain massif on border of WEST MACEDONIA and THESSALY departments, N central GREECE, forming divide between ALIÁKMON and upper Titarisios rivers. Rises to 4,849 ft/1,478 m in the Amarvik, 20 mi/32 km SE of KOZANI; 40°00'N 21°52'E.

Kamwenge, administrative district (2005 population 322,000), WESTERN region, SW UGANDA; ⊙ Kamwenge; 00°10'N 30°25'E. As of Uganda's division into eighty districts, borders KYENJOJO (N and E), KIRUHURA (E), IBANDA (SES), BUSHENYI (SW tip), KASESE (W), and KABAROLE (NW) districts. Primary inhabitants are the Bafumbira, Bakiga, Batagwenda, and Batooro peoples. District includes tropical forests and escarpments. Part of Lake GEORGE in S. Primarily agricultural (including maize, bananas, cassava, groundnuts, millet, potatoes, and sweet potatoes, cash crops include coffee and cotton). Livestock (cattle is most important, also goats, pigs, sheep, and poultry). Railroad between KASESE town (c.20 mi/32 km W of Kamwenge town) and MOMBASA (SE KENYA) runs W-E through Kamwenge town. Created in 2000 from S portion of former KABAROLE district (current Kabarole district was created from NW portion and Kyenjojo district from N and E portions).

Kam'yane (kah-myah-NE) (Russian *Kamennoye*), town (2004 population 15,900), S LUHANS'K oblast, UKRAINE, in the DONBAS, 10 mi/16 km WNW of ROVEN'KY, 5 mi/8 km NE (and under jurisdiction of ANTRATSYT); 48°10'N 39°10'E. Elevation 938 ft/285 m. Coal mine; state farm. Established in the 1770s, town since 1938. Formerly called Kam'yanno-Millerove (Russian *Kamenno-Millerovo*).

Kam'yane (KAHM-yah-ne) (Russian *Kamennoye*), town (2004 population 2,734), N ZAPORIZHZHYA oblast, UKRAINE, 10 mi/16 km ENE of ZAPORIZHZHYA city center; 48°11'N 34°58'E. Elevation 462 ft/140 m. Granite quarry, grading plant; local museum. Established in 1886, town since 1986. Known until 1945 as Yantsivs'kyy Karyer (Russian *Yantsovskiy Karyer*) and until 1957 as Khutir Kam'yanyy (Russian *Khutor Kamennyy*).

Kam'yanets'-Podil's'kyy (kah-myah-NETS-po-DEEL-skee) (Russian *Kamenets-Podol'skiy*), city (2001 population 99,610), KHMELNYTS'KYY oblast (formerly Kam'yanets'-Podil's'kyy oblast), UKRAINE; 48°40'N 26°34'E. Elevation 590 ft/179 m. Raion center; former capital of Kam'yanets'-Podil's'kyy oblast; railroad terminus; food processing, manufacturing (farm and electrical machinery, machine tools, cables, auto parts, building materials, furniture, textiles, clothing). Kam'yanets-Podil's'kyy is known since the 11th century, when it was part of the Halych principality. In the 14th century, it passed to Poland, and to Russia in 1793. During Ukrainian struggle for independence (1917–1920), it was temporarily the seat of Ukrainian government and housed the Kam'yanets-Podil's'kyy Ukrainian State University (established in 1918). When restrictions against Jewish settlements were lifted in mid-19th century, Jewish community grew to nearly half the population on eve of World War I. The Jewish population declined during Soviet period and was eliminated in World War II. A popular tourist site, the city features historic landmarks such as the fortress (14th–17th century), which is now a museum, some cathedrals and monasteries dating from the 14th century, and the oldest (1280) Armenian church in Ukraine.

Kam'yanets'-Podil's'kyy, UKRAINE: see KHMEL'NYTS'KYY, oblast.

Kam'yani Mohyly, UKRAINE: see UKRAINIAN STEPPE NATURE PRESERVE.

Kam'yanka (KAH-myahn-kah) (Russian *Kamenka*), city, S CHERKASY oblast, UKRAINE, 16 mi/26 km SE of SMILA; 49°02'N 32°06'E. Elevation 419 ft/127 m. Raion center; machine building for textile and chemical industries; sugar refining, distilling, lumber milling. Established at the beginning of the 17th century, city since 1956. Called (1930–1944) Kam'yanka-Shevchenkivs'ka (Russian *Kamenka-Shevchenkovskaya*).

Kam'yanka (Russian *Kamenka*), city, SE CHERKASY oblast, UKRAINE, on the TYASMYN River, 27 mi/43 km S of CHERKASY. Raion center; on the Kiev-Dnipropetrovs'k highway and railroad line; machine construction for textile and chemical industry, distillery, sugar refinery. Established at the beginning of 17th century; city since 1956. Literary museum for Pushkin and Tchaikovsky.

Kamyanka, UKRAINE: see KUYBYSHEVE, Zaporizhzhya oblast.

Kam'yanka-Buz'ka (KAH-myahn-kah–BOOZ-kah) (Russian *Kamenka-Bugskaya*), city, central L'VIV oblast, UKRAINE, on BUH RIVER, 20 mi/32 km NE of L'VIV; 50°06'N 24°21'E. Elevation 705 ft/214 m. Raion center; parquetry, sawmilling, clothing, linen, bricks, agricultural processing (grain, potatoes, hops). Has a 17th-century wooden church; ruins of an old monastery. Known since 1406 as village Dymoshyn; renamed Kam'yanka in the 1440s; town in 1471; under Austrian rule (1772–1918); part of West Ukrainian National Republic (1918); reverted to Poland (1919); annexed to Ukranian SSR in 1939; part of independent Ukraine since 1991. From 1485 until 1940, called Kam'yanka-Strumylivs'ka (Russian *Kamenka-Strumilovskaya*, Polish *Kamionka Strumilowa*).

Kam'yanka-Dniprovs'ka (KAH-myahn-kah–dnee-PROV-skah) (Russian *Kamenka-Dneprovskaya*), town, W ZAPORIZHZHYA oblast, UKRAINE, on S shore of KAKHOVKA reservoir, opposite NIKOPOL; 47°29'N 34°24'E. Elevation 104 ft/31 m. Raion center; food processing (fruit packing and canning, dairy, grain). Irrigation research station; museum. Founded in 1793, until 1920 called Mala Znam'yanka (Russian *Malaya Znamenka*); later also Kam'yanka-na-Dnipri (Russian *Kamenka-na-Dnepre*). Remains of Scythian (5th–late 3rd centuries B.C.E.) fortified settlement nearby (now mostly submerged).

Kam'yanka-na-Dnipri, UKRAINE: see KAM'YANKA-DNIPROVS'KA.

Kam'yanka-Strumlivs'ka, UKRAINE: see KAM'YANKA-BUZ'KA.

Kam'yanno-Millerove, UKRAINE: see KAM'YANE, Luhans'k oblast.

Kamyanske, UKRAINE: see DNIPRODZERZHYNS'K.

Kam'yanyy Brid (kah-myah-NEE BREED) (Russian *Kamennyy Brod*), town (2004 population 3,627), W ZHYTOMYR oblast, UKRAINE, 18 mi/29 km SE of NOVOHRAD-VOLYNS'KYY; 50°25'N 27°43'E. Elevation 636 ft/193 m. Ceramic industry, forestry. Established in 1862 as Rogachev, renamed and town status since 1938.

Kamyaran (KUHM-yah-RAHN), town, Kordestân province, W IRAN, 35 mi/55 km SSW of SANANDAJ, on road between KERMANSHAH and SANANDAJ; 34°47'N 46°56'E.

Kamyshany, UKRAINE: see KOMYSHANY.

Kamyshevakah, UKRAINE: see KOMYSHUVAKHA, Luhans'k oblast.

Kamyshevakha, UKRAINE: see KOMYSHUVAKHA, Luhans'k oblast; or KOMYSHUVAKHA, Zaporizhzhya oblast.

Kamyshevatskaya (kah-mi-shi-VAHTS-kah-yah), village (2005 population 5,225), NW KRASNODAR TERRITORY, S European Russia, on the E shore of the Sea of AZOV, on road, 25 mi/40 km SW of YEYSK; 46°24'N 37°57'E. Terminus of three local highways. Fisheries; vegetable growing.

Kamyshin (kah-MI-shin), city (2006 population 128,910), NE VOLGOGRAD oblast, SE European Russia, port on the Volgograd Reservoir of the VOLGA RIVER (opposite Nikolayevskiy), at the mouth of the small Kamyshinka River, on branch railroad, 120 mi/193 km NNE of VOLGOGRAD; 50°06′N 45°24′E. Elevation 255 ft/77 m. Major warehousing and transshipment point for grain, salt, and petroleum; port facilities. Cotton textiles, machinery, tools, furniture, food products; timber, metalworks (ferro-alloys), glassworks. Founded in 1668 in the Moscow principality as a fortress to protect the Volga trade route. Chartered in 1780; called Dmitriyevsk until 1784.

Kamyshla (kah-mi-SHLAH), village (2006 population 4,780), NE SAMARA oblast, E European Russia, on the SOK RIVER, near highway junction, 45 mi/72 km ENE of SERGIYEVSK; 54°07′N 52°10′E. Elevation 721 ft/219 m. Transshipment point for agricultural products; milk processing.

Kamyshlov (kuh-mi-SHLOF), city (2006 population 27,710), S SVERDLOVSK oblast, W Siberian Russia, on the PYSHMA RIVER, on crossroads and railroad, 84 mi/135 km E of YEKATERINBURG; 56°50′N 62°42′E. Elevation 341 ft/103 m. Railroad repair shops; insulators and other electrical goods, clothing, leather goods, glue; metalworking, woodworking, food processing (dairy, bakery). Founded in 1666 as a fortress of Kamyshevskaya Sloboda; chartered in 1781.

Kamyshnya, UKRAINE: see KOMYSHNYA.

Kamysh-Zarya, UKRAINE: see KOMYSH-ZORYA.

Kamyzyak (kah-mi-ZYAHK), city (2005 population 16,240), E ASTRAKHAN oblast, SE European Russia, on the Kamyzyak arm of the VOLGA RIVER delta mouth, on road, 20 mi/32 km S of ASTRAKHAN; 46°07′N 48°04′E. Below sea level. In agricultural area; food industries, fisheries; ceramics; forestry. Made city in 1973.

Kamzë (KAHMS), new town (2001 population 44,553), central ALBANIA, 4 mi/6.4 km NW of TIRANË; 41°22′N 19°45′E. Built for workers at new coal-mining center of VALIAS. In wheat-producing area; agricultural college.

Kan (KAHN), village, Tehrān province, N IRAN, 10 mi/16 km NW of TEHRAN; 35°45′N 51°16′E. Grain, fruit (pomegranates, figs). Formerly called KAND.

Kana, village, MATABELELAND NORTH province, W central ZIMBABWE, 36 mi/ 58 km SW of GOKWE, on KANA RIVER; 18°54′S 28°31′E. Cattle, sheep, goats; corn, subsistence crops.

Kanab (KUH-nab), town (2006 population 3,754), ⊙ KANE county, S UTAH, on KANAB CREEK, and 60 mi/97 km S of SAINT GEORGE, near ARIZONA state line; 37°01′N 112°31′W. Trade center for ranching and agricultural area (alfalfa; cattle); lumbering; manufacturing (rubber stamps); terra-cotta mining. Elevation 4,925 ft/1,501 m. Coral Pink Sands State Park to W; ZION NATIONAL PARK to NW. Kaibab National Forest and GRAND CANYON National Park to SE, in Arizona Kaibab Indian Reservation and PIPE SPRING NATIONAL MONUMENT to SW, both in Arizona Settled 1864. Name means "willow" in Paiute.

Kanab Creek (KUH-nab), UTAH and ARIZONA; rises in NW KANE county, S Utah, flows 90 mi/145 km S, past KANAB, Utah, and into Arizona, through Kaibab Indian Reservation, Kaibab Creek Wilderness Area and canyon in KANAB PLATEAU, to COLORADO RIVER in N part of GRAND CANYON National Park. Artifacts of cliff dwellers have been found on its banks.

Kanabec (kuh-NAI-bek), county (□ 533 sq mi/1,385.8 sq km; 2006 population 16,276), E MINNESOTA; ⊙ MORA; 45°57′N 93°17′W. Drained by SNAKE, Knife, Ann, and Groundhouse rivers. Agricultural area (alfalfa, hay, oats, barley, potatoes; hogs, cattle, sheep, poultry; dairying). Ann Lake Wildlife Area in W center; parts of Rum River State Forest and Mille Lacs Wildlife Area on W boundary; Snake River State Forest in NE; Knife Lake in N; Fish Lake in S. Formed 1858.

Kanab Plateau, tableland (c.6,000 ft/1,829 m) in NE MOHAVE county, NW ARIZONA, extends N from COLORADO RIVER. Bounded E by KANAB CREEK, KAI-

BAB PLATEAU is E of creek. Grand Canyon National Park in S.

Kanacea (kahn-ah-DHAI-ah), volcanic island (□ 5 sq mi/13 sq km), Lau group, FIJI, SW PACIFIC OCEAN; 2 mi/3.2 km long. Copra. Sometimes spelled Kanathea.

Kanada (KAH-nah-dah), town, Tagawa county, FUKUOKA prefecture, N KYUSHU, SW JAPAN, 22 mi/35 km N of FUKUOKA; 33°40′N 130°46′E.

Kanadey (kah-nah-DYAI), village (2006 population 2,070), S ULYANOVSK oblast, E central European Russia, on the Syzranka River (VOLGA RIVER basin), on highway junction and railroad, 9 mi/14 km W of NOVOSPASSKOYE; 53°09′N 47°32′E. Elevation 278 ft/84 m. Agricultural products. Summer homes.

Kanadukattan (KAH-nah-duh-KAH-tuhn), village, PASUMPAN MUTHURAMALINGA THEVAR district, TAMIL NADU state, S INDIA, 7 mi/11.3 km N of KARAIKKUDI. Residential and financial center of Chetty merchant community. Trade in foreign luxury goods. Also spelled Kanadukathan.

Kanaga Island (ka-NAW-guh), ANDREANOF ISLANDS, ALEUTIAN ISLANDS, SW ALASKA, 10 mi/16 km W of ADAK ISLAND; 30 mi/48 km long, 4 mi/6.4 km–8 mi/12.9 km wide; 51°47′N 177°15′W. Kanaga Volcano (N).

Kanagawa (kah-nah-GAH-wah), prefecture (□ 932 sq mi/2,423.2 sq km), E central HONSHU, E central JAPAN, on TOKYO (E) and SAGAMI (S) bays; ⊙ YOKOHAMA. Bordered N by Tokyo prefecture, W by SHIZUOKA and YAMANASHI prefectures. The urban belt of the E part (especially YOKOHAMA and KAWASAKI) of the prefecture merges with TOKYO to the N and is a major industrial and service center. Flowers and dairy products are produced and fishing is an important industry. Important cities include Yokohama, Kawasaki, YOKOSUKA, and KAMAKURA (a religious center).

Kanagi (KAH-nah-gee), town, North Tsugaru county, Aomori prefecture, N HONSHU, N JAPAN, 16 mi/25 km N of AOMORI. Rice, tobacco, vegetables, apples; pigs. Lumber-related industries. Writer Dazai Osamu (pen name of Tsushima Shuji) born here, 1909. Also Kanaki.

Kanagi, town, Naka county, SHIMANE prefecture, SW HONSHU, W JAPAN, 66 mi/107 km S of MATSUE; 34°51′N 132°09′E.

Kanai (kah-NAH-ee), town, Sado county, NIIGATA prefecture, N central JAPAN, 37 mi/60 km N of NIIGATA; 38°00′N 138°22′E. Pottery.

Kanakanak (kah-NUH-kah-nuhk), village, S ALASKA, on W shore of NUSHAGAK BAY, 6 mi/9.7 km SW of DILLINGHAM.

Kanal (kah-NAHL), village, W SLOVENIA, on SOCA River, on railroad, and 10 mi/16 km N of NOVA GORICA; 46°05′N 13°38′E. Part of Italy (1919–1947); then passed to the former YUGOSLAVIA.

Kanal Dnieper-Bug, BELARUS: see DNIEPER-BUH CANAL.

Kanan (KAH-nahn), town, Mono county, MIYAGI prefecture, N HONSHU, NE JAPAN, 25 mi/40 km N of SENDAI; 38°30′N 141°11′E. Rice.

Kanan, town, S Kawachi county, OSAKA prefecture, S HONSHU, W central JAPAN, 10 mi/16 km S of OSAKA; 34°29′N 135°37′E.

Kananaskis Country or **Kananaskis Range**, front range (□ 2,008 sq mi/5,220.8 sq km) of ROCKY MOUNTAINS W of CALGARY, SW ALBERTA, W CANADA; 50°50′N 115°15′W. From rolling sandstone foothills in E, the region grades to montane forest, alpine meadows, and rock and glacier peaks. Ranching, logging, coal mining, and hydroelectricity have made way to conservation and recreation pursuits. Site of 1988 Winter Olympic ski events.

Kananaskis Improvement District (□ 1,626 sq mi/4,227.6 sq km; 2001 population 462), SW ALBERTA, W CANADA; 50°42′N 114°52′W. Established 1945.

Kananga (kuh-NAHN-guh), city, KASAI-OCCIDENTAL province, S central CONGO, on the LULUA RIVER. It is the commercial and transportation center of a region

where cotton is grown, and diamonds are mined; 05°54′S 22°25′E. Elev. 1,998 ft/608 m. The city was founded in 1884 by the German explorer Hermann von Wissmann. In 1895, Batetela troops stationed there revolted after their chief was executed by authorities of the Belgian-run Independent State of the CONGO. Although successful at first, the mutineers were finally defeated in 1901. Kananga grew rapidly in the early 20th century with the coming of the railroad. Many Luba people settled there and became economically dominant over the indigenous Lulua people. After ZAÏRE achieved independence (1960), there were violent clashes between the Luba and Lulua, and many Luba fled to the short-lived (1960–1961) Mining State of South KASAI. In 1961–1962, the city was held by rebel troops from EQUATEUR region. Formerly called LULUABOURG.

Kanangra-Boyd National Park (kuh-NANG-gruh–BOID) (□ 264 sq mi/686.4 sq km), E NEW SOUTH WALES, SE AUSTRALIA, 120 mi/193 km E of SYDNEY; 42 mi/68 km long, 40 mi/64 km wide; 34°02′S 150°03′E. Rugged mountain wilderness in GREAT DIVIDING RANGE. Waterfalls, narrow canyons, limestone caves. Wet and dry sclerophyll forests. Camping, picnicking, trout fishing, hiking. Established 1969.

Kanara (KUH-nuh-ruh), Kanarese *Karnatak*, region (□ 60,000 sq mi/156,000 sq km) of S DECCAN PLATEAU, mainly in UTTAR KANNAD and BELLARY districts (KARNATAKA state) and extending E into ANDHRA PRADESH state, SW INDIA. Named for its Kanarese-speaking people. Was center of sixteenth-century Vijayanagar kings (⊙ was HAMPI), who were forced onto plains SE of the plateau after their defeat (1565) by Deccan sultans at battle of Talikota. Became an important province of the Marathas under Shivaji in 1670s. In eighteenth century, the English restricted the term *Carnatic* to the plains area.

Kanarak, INDIA: see KONARAK.

Kana River, c. 90 mi/145 km long, MATABELELAND NORTH province, W central ZIMBABWE; rises at Samakohwa Escarpment, SW part of MAFUNGABUSI PLATEAU, c.105 mi/169 km N of BULAWAYO; flows W, entering GWAYI RIVER, 63 mi/101 km ESE of HWANGE.

Kanarraville (kuh-NAHR-uh-vil), village (2006 population 305), IRON county, SW UTAH, just N of ZION NATIONAL PARK, 10 mi/16 km SW of CEDAR CITY; 37°32′N 113°10′W. Alfalfa, barley; dairying; cattle ranching. W of Cedar Mountains. Elevation 5,541 ft/1,689 m. Dixie National Forest to W; ZION NATIONAL PARK (KOLOB Canyons Area) to S. Settled 1861.

Kanasago (kah-nah-SAH-go), town, Kuji county, IBARAKI prefecture, central HONSHU, E central JAPAN, 12 mi/20 km N of MITO; 36°33′N 140°28′E. Buckwheat.

Kanash (kah-NAHSH), city (2005 population 49,650), E CHUVASH REPUBLIC, central European Russia, on crossroads, 35 mi/56 km SSE of CHEBOKSARY; 55°30′N 47°29′E. Elevation 633 ft/192 m. Railroad junction; machinery and metal goods (automobile parts, railroad equipment, loading devices), chemicals, food industries. Emerged in 1891 as a railroad center; made city in 1925. Called Shikhrany until 1920.

Kanasín (kah-nah-SEEN), town, ⊙ Kanasín municipio, YUCATÁN, SE MEXICO, on railroad, and 5 mi/8 km SE of MÉRIDA; 20°56′N 89°34′W. Henequen; corn, beans, tropical fruit.

Kanastraion, Cape, Greece, see KALOGREA, CAPE.

Kanata (kuh-NAH-tuh), former city (□ 51 sq mi/132.6 sq km; 2001 population 58,636), SE ONTARIO, E central CANADA, and included in city of OTTAWA; formerly part of dissolved Ottawa-Carleton Regional Municipality (1991 population 678,147); 45°20′N 75°54′W. Borders OTTAWA River on NE. Only E entremity is urbanized, remainder is rural. Manufacturing (chemicals, electronic components, engineering instruments); mixed farming, dairying.

Kanathea, FIJI: see KANACEA.

Kanaud, India: see MAHENDRAGARH, HARYANA state.

Area is shown by the symbol □, and capital city or county seat by ⊙.

Kanauj (kuh-NOUJ), town, FARRUKHABAD district, central UTTAR PRADESH state, N central INDIA, near the GANGA River, 31 mi/50 km SE of FARRUKHABAD; 24°46′N 74°32′E. A market center for tobacco, rose water; noted for its perfume. A very ancient city and a center of Brahmanic culture, its name was Kanya-kubja. Called Kanogiza by Ptolemy, it was important during Gupta empire (320 B.C.E.–C.E. c.480) and again as the capital and brilliant cultural center of Harshah's empire (early 7th century), when it was visited by the Chinese traveller Hsüan-tsang. Underwent a period of decline under successive Indian principalities after Harshah's death. In the 9th century it became the capital of the Pratihara empire and was famous for its poets. Sacked and conquered by Mahmud of Ghazni 1018–1019; was the capital of Rajput (Rathor) kingdom 1019–1194, when it was defeated by Afghans; town declined rapidly thereafter. During 18th century, under nawabs of Oudh and Farrukhabad and, at times, Marathas. Humayan was defeated (1540) nearby, by Sher Shah. Also spelled Kannauj.

Kanaung (KAH-NOUN), village, MYANAUNG township, AYEYARWADY division, MYANMAR, port on right bank of AYEYARWADY RIVER, and 37 mi/60 km NNW of HENZADA. On road to MYANAUNG.

Kanavino (kah-NAH-vee-nuh), former city, European Russia, formed in 1925, incorporated into NIZHNIY NOVGOROD in 1928.

Kanawat (kah-nah-WAHT), township, SUWEIDA district, S SYRIA, just NE of Es SUWEIDA in the mountains; 32°45′N 36°36′E. Elev. 4,000 ft/1,219 m. Extensive remains of ancient Roman city of Kanatha, a member of the DECAPOLIS. Also spelled El QANAWAT.

Kanawha (kuh-NAW-uh), county (□ 911 sq mi/2,368.6 sq km; 2006 population 192,419), W central WEST VIRGINIA; ⊙ CHARLESTON; 38°20′N 81°31′W. On AL-LEGHENY PLATEAU; bounded W, in part, by BIG COAL RIVER; drained by KANAWHA, ELK, and POCATALICO rivers. Bituminous-coal region; natural gas and oil wells. Manufacturing at CHARLESTON, SOUTH CHARLESTON, DUNBAR, BELLE, and NITRO. Agriculture (corn, tobacco, alfalfa, hay); cattle; poultry. Construction. Kanawha State Forest in SW; part of Wallback Wildlife Management Area in NE corner. Formed 1788.

Kanawha (kah-NAW-wah), town (2000 population 739), HANCOCK county, N IOWA, 15 mi/24 km SW of GARNER; 42°55′N 93°47′W. In agricultural area.

Kanawha (kah-NAW-wah), river, 97 mi/156 km long, WEST VIRGINIA, formed by the confluence of the NEW and GAULEY rivers, NW FAYETTE county, S central WEST VIRGINIA, 35 mi/56 km ESE of CHARLESTON. Flows NW past SMITHERS, Charleston, ST. ALBANS, and WINFIELD, to the OHIO RIVER at POINT PLEASANT. There are rich coal, natural gas, and salt-brine deposits in the region, and numerous chemical plants along the river. There are navigation locks and power dams on the river, navigable to Charleston.

Kanaya (KAH-nah-yah), town, Haibara county, SHI-ZUOKA prefecture, central HONSHU, E central JAPAN, 19 mi/30 km S of SHIZUOKA; 34°49′N 138°07′E. Tea-making machines. Tea garden in Makinohara plateau.

Kanaya, town, Arida county, WAKAYAMA prefecture, S HONSHU, W central JAPAN, 14 mi/22 km S of WA-KAYAMA; 34°03′N 135°15′E.

Kanayama (kah-NAH-yah-mah), town, Mashita county, GIFU prefecture, central HONSHU, central JAPAN, 31 mi/50 km N of GIFU; 35°39′N 137°09′E.

Kanayevka (kah-NAH-eef-kah), village, E central PENZA oblast, E central European Russia, on the SURA River, on highway branch and railroad, 9 mi/14 km ENE of ZOLOTAREVKA; 53°06′N 45°34′E. Elevation 508 ft/154 m. Bakery.

Kanazawa (kah-nah-ZAH-wah), city (2005 population 454,607), ⊙ ISHIKAWA prefecture, central HONSHU, central JAPAN, on the SEA OF JAPAN at the mouth of the SAI RIVER; 36°33′N 136°39′E. Agriculture includes fruits, lotus root, bamboo shoots, cinnamon vines. Fish. Produces textiles, machinery, lacquerware, rolling stock, and iron. Traditional crafts include *kaga yuzen* textile dyeing, sword making, inlay work, and the production of gold lacquer and Kanazawa foil (design). The city, built on the site of the old village of Yamazaki, was the seat of the Maeda clan (16th–19th century) and gradually became an industrial center. Well-known tourist destination, especially for Ken-rokuen Park (rebuilt 17th century), with its splendid landscape gardens, a famous No theatre and school, and the Hyakumangoku Festival.

Kanazi (kah-nah-ZEE), village, KAGERA region, NW TANZANIA, 10 mi/16 km SSW of BUKOBA, on Lake VICTORIA; 01°29′S 31°44′E. Coffee, corn, wheat; goats, sheep; fish.

Kanazu (kah-NAHZ), town, Sakai county, FUKUI prefecture, central HONSHU, W central JAPAN, 9 mi/15 km N of FUKUI; 36°12′N 136°13′E.

Kanbalu (KAHN-bah-loo), township, SAGAING division, MYANMAR, on railroad, and 45 mi/72 km NNW of SHWEBO, and 12 mi/19 km E of AYEYARWADY RIVER. Teak forest.

Kanbara (KAHN-bah-rah), town, Ihara county, SHI-ZUOKA prefecture, central HONSHU, E central JAPAN, on N shore of SURUGA BAY, at mouth of FUJI RIVER, 16 mi/25 km N of SHIZUOKA; 35°06′N 138°36′E. Citrus-fruit center.

Kanbauk (kahn-BOUK), village, Yebyu township, TA-NINTHARYI division, MYANMAR, on inlet of ANDAMAN SEA, on road, and 40 mi/64 km NNW of TAVOY. Minor port.

Kanbe (kahm-BAI), town, TWANTE township, YANGON division, MYANMAR, 10 mi/16 km WSW of YANGON.

Kanchanaburi (GAHN-juh-nuh-BU-REE), province, (□ 7,518 sq mi/19,546.8 sq km), W THAILAND; ⊙ KANCHANABURI; 14°38′N 99°06′E. Shares border with MYANMAR. Paper manufacturing; rice, sugarcane, tapioca, corn, cotton; tobacco and kapok plantations; ruby and sapphire mining at BO PHLOI; lead mining. It is most famous as the site of the "Death Railway" to MYANMAR that the Japanese occupational army forced Allied POWs to build at the cost of about 16,000 Allied lives and 90–100,000 Southeast Asian coolies. The film *The Bridge on the river kwai* popularized the history of this railroad. Kanchanaburi also has a number of impressive waterfalls, a large national park, an old border crossing into MYANMAR, and the Ban Kao Neolithic Museum displaying artifacts dating back 3500–4,000 years, found in the village of BAN KAO during the construction of the "Death Railway." The elite 9th Infantry Division is based in this province Sinakarin Dam is here as well.

Kanchanaburi (GAHN-juh-nuh-BU-REE), town, ⊙ KANCHANABURI province, W THAILAND, on MAE KLONG RIVER at mouth of KHWAE NOI RIVER, on railroad (made famous by the film *The Bridge on the river kwai*,) and 79 mi/127 km WNW of BANGKOK; 14°01′N 99°32′E. Paper manufacturing; tobacco and kapok plantations. Ruby and sapphire mining at BO PHLOI (N); lead mining nearby. A walled town, it was built (18th century) as military base against Burmese armies. Allied war cemetary from WWII. Neolithic finds in caves to W of town.

Kanchanpur (kuhn-CHUHN-poor), district, SW NEPAL, in MAHAKALI zone; ⊙ MAHENDRANAGAR.

Kanchenjunga (kuhn-CHUHN-juhng-GAH), Tibetan *Gangchhendzönga*, Nepali *Kumbhkaran Lungur*; 3rd-highest mountain in the world, in SINGALILA RANGE of E NEPAL HIMALAYA, on border between NEPAL and SIKKIM state, NE INDIA, 105 mi/169 km NW of DAR-JILING; 27°42′N 88°09′E. Consists of five peaks; highest is 28,169 ft/8,586 m. Climbed in 1955 by Charles Evans, but the climbers stopped short of the highest point, in deference to the religious sentiments of the Sikkimese, who consider the peak sacred, not to be desecrated by human feet. Usual route is NNE ridge via the NE spur. ZEMU GLACIER, on E slope, descends into Sikkim state. Formerly spelled Kinch-injunga.

Kanchi, India: see KANCHIPURAM.

Kanchipuram (kahn-CHEE-poo-ruhm), city, Chen-galpattu-M.G.R. district, TAMIL NADU state, S INDIA, on PALAR RIVER, and 40 mi/64 km WSW of CHENNAI (Madras); 12°50′N 79°43′E. Famous archaeological site and pilgrimage center, it is one of the most sacred Hindu towns in India. Known as the "golden city" and the "Varanasi of the South." Several temples survive from the period when it was the capital of the Pallava empire (3rd–8th century) of S India and SRI LANKA and a center of Brahmanical and Buddhist culture. Provides a remarkably complete survey of Dravidian architecture over several centuries. Has over 100 Hindu temples, with shrines sacred to vo-taries of both Shiva and Vishnu. Captured (8th century) by the Chalukya dynasty and subsequently passed to the Chola (11th–13th century), to the Vi-jayanagar (early 15th century), and to the Orissa (late 15th century) kingdoms. After 1481 it fell to several different Muslim sultanates. A base of French power in India, it was captured by Robert Clive in 1758. An educational center since Buddhist times; Shankar-acharya (8th-century C.E. Vedantic philosopher) founded one of his chief schools here. Still a center of Brahmanic culture; Sanskrit schools. Local hand-woven silk saris are known throughout India. Ancient name was Kanchi; formerly Conjeeveram.

Kanchow, China: see GANZHOU.

Kanchrapara (kahnch-rah-PAH-ruh), city, NORTH 24-PARGANAS district, SE WEST BENGAL state, E INDIA, near the HUGLI River, 26 mi/42 km NNE of CALCUTTA city center. Large railroad workshops; jute milling. A thriving commercial center, with schools and a college.

Kanchriech (KUHN-CHRICH), village, PREY VENG province, S CAMBODIA, 45 mi/72 km ENE of PHNOM PENH; 11°43′N 105°33′E. Corn- and rice-cultivating area. Khmer population.

Kanczuga (kahn-CHOO-gah), Polish *Kańczuga*, town, Przemyśl province, SE POLAND, 19 mi/31 km ESE of RZESZÓW. Brewing, cement manufacturing.

Kand, IRAN: see KAN.

Kand, PAKISTAN: see TOBA-KAKAR RANGE.

Kanda (KAHN-DAH), town, Miyako county, FUKUOKA prefecture, N KYUSHU, SW JAPAN, on SUO SEA, 37 mi/60 km N of FUKUOKA; 33°46′N 130°58′E. Manu-facturing (motor vehicles). Has feudal castle.

Kandagach, KAZAKHSTAN: see OKTYABRSK.

Kandaghat (kuh-duh-GAHT), town, ⊙ SOLAN district, HIMACHAL PRADESH state, N INDIA, in LESSER HI-MALAYAS, on railroad, and 10 mi/16 km SSW of SHIMLA; 30°59′N 77°07′E. Famous for its breweries. Became the capital of former East Punjab province (1947–1956) for some government departments.

Kandahar (kan-dah-HAHR) or **Qandahar**, province (□ 50,000 sq mi/130,000 sq km; 2005 population 971,400), SE AFGHANISTAN; ⊙ KANDAHAR; 31°00′N 65°45′E. Bounded by HELMAND province (W), OR-UZGAN province (N), ZABUL province (E), and BA-LUCHISTAN province, PAKISTAN (S and SE). Outliers of the Hindu Kush mountains in N and REGISTAN desert on Baluchistan border in S. Population and agricultural economy are concentrated at foot of mountains, in oases watered by ARGHANDAB, TAR-NAK, and ARGHASTAN rivers. These oases, Kandahar and KALAT-I-GHILZAI, are linked by Kabul-Kandahar-Herat highway; Airport (Kandahar city). Agriculture (wheat, rice, cotton, fruit). Semiprecious stone and gold mining; handicrafts. Trade with Pakistan passes through Spinbaldak. Population is nearly all Durani in W, Ghilzai in NE; Baluch nomads in S desert. Scene of fighting in December 2001 between last remnants of the Taliban government and U.S. troops aided by anti-Taliban tribes.

Cross-references are shown in SMALL CAPITALS. The pronunciation guide is shown on page xix. The sources of population figures are shown on page xvii.

Kandahar (kan-dah-HAHR) or **Qandahar**, city, ⊙ KANDAHAR province, S AFGHANISTAN, 285 mi/459 km SW of KABUL; 31°35′N 65°45′E. On level plain between TARNAK and ARGHANDAB rivers, it is Afghanistan's second largest city and a chief trade center. Market for sheep, wool, cotton, food grains, fresh and dried fruit, and tobacco. International airport; linked by road with KABUL, HERAT, PAKISTAN, and TURKMENISTAN. Manufacturing (woolen cloth, felt, silk; dried, canned, and preserved fruit); irrigated agriculture (fruit, especially grapes) in surrounding region. The old city is oblong, and surrounded by a partially preserved 27-ft/8-m-high wall and a wide ditch separating it from the new garden suburbs. Kandahar was founded by Alexander the Great (4th century B.C.E.). INDIA and PERSIA long fought over the city, which was strategically located on the trade routes of central ASIA. Conquered by Arabs (7th century), Turkic Ghaznivids (10th century), Jenghiz Khan (12th century), Timur (14th century). Babur, founder of the Mogul empire of India, took Kandahar in the 16th century was the first capital of the independent kingdom of Afghanistan. British forces occupied the city during the First Afghan War (1839–1842) and from 1879 to 1881. Long a bastion of traditional Islam, the old city was laid out by Ahmad Shah (1724–1773) and is dominated by his octangular, domed mausoleum. There are also numerous mosques (one said to contain the Prophet Muhammad's cloak) and bazaars. Modern Kandahar adjoins the old city; technical college Together with Peshawar, Pakistan, Kandahar is the principal city of the Pashtu people. During the Soviet military occupation of 1979–1989, Kandahar was the site of a Soviet command. It changed hands several times until the fall of the Najibullah government in 1992. In late 1994 it was captured by the Taliban forces, who est. headquarters here. The city was the scene of fighting between the Taliban and U.S. armed forces and anti-Taliban tribes; it was the Taliban's last urban stronghold and fell on December 7, 2001.

Kandahar (kuhn-dah-HAHR), town, NANDED district, MAHARASHTRA state, W central INDIA, 21 mi/34 km SSW of NANDED; 18°52′N 77°12′E. Millet, cotton, wheat. Formerly spelled Khandhar and Qandhar.

Kanda-Kanda (KAHN-duh—KAHN-duh), village, KASAI-ORIENTAL province, S CONGO, on left bank of Luilu River, a headstream of the SANKURU, near railroad, and 80 mi/129 km SW of KABINDA; 06°13′S 22°13′E. Elev. 2,017 ft/614 m. Trading center in area that grows cotton and food staples; cottonseed-oil milling. Cattle farms at KAMBAYE, 15 mi/24 km ENE. THIELEN-SAINT-JACQUES Roman Catholic mission is 7 mi/11.3 km S.

Kandal, province (□ 1,377 sq mi/3,580.2 sq km; 2007 population 1,280,781), S CAMBODIA, borders TAKEO (SWW), KOMPONG SPEU (W), KOMPONG CHHNANG and KOMPONG CHAM (N), and PREY VENG (E) provinces, and VIETNAM (S); ⊙ TAKHMAU; 11°30′N 104°50′E. BASSAC and MEKONG rivers run through the province. Khmer population, with Cham, Vietnamese, and Chinese minorities. PHNOM PENH municipality and capital city is located in Kandal province, though it is politically and administratively independent of it.

Kandalaksha (kahn-dah-LAHK-shah), Finnish *Kannanlahti*, city (2006 population 37,730), SW MURMANSK oblast, NW European Russia, seaport at the mouth of the NIVA RIVER on the KANDALAKSHA BAY of the WHITE SEA, 170 mi/274 km S of MURMANSK; 67°09′N 32°24′E. Railroad; aluminum works; hydroelectric power stations. Settlement since the first half of the 11th century known to the Vikings; since 1517, within Novgorod principality, specializing in fisheries, salt, and lumber. Made city in 1938.

Kandalaksha Bay (kuhn-dah-LAHK-shuh), Russian *Kandalakshskaya Guba*, northwesternmost and dee-

pest section of the WHITE SEA, N European Russia, off MURMANSK oblast (KOLA PENINSULA: NE) and Republic of KARELIA (SW); 130 mi/209 km long, 60 mi/97 km wide at mouth, 1,115 ft/340 m deep; 66°55′N 32°45′E. SW coast is deeply indented and strewn with islands; Port of Kandalaksha lies at the head of the inlet.

Kandalaksha Preserve (kuhn-dah-LAHK-shuh), wildlife refuge, on the islands and small areas of the continental coast of KANDALAKSHA BAY, as well as the Murmansk coast of the BARENTS SEA, MURMANSK oblast, extreme NW European Russia. Covers 143,563 acres/58,100 ha; includes the Sem' Ostrovov [Russian= seven islands] Archipelago, with adjacent areas of the mainland and the Ainov Islands. Established in 1932 to protect the resting areas of nesting birds; local fauna include elk, fox, marten, lynx, brown bear, and wolverine.

Kandangan (KAHN-dahng-gahn), town, ⊙ Hulu Sungai Selatan district, Kalimantan Selatan province, SE BORNEO, INDONESIA, 60 mi/97 km NE of BANJARMASIN; 02°47′S 115°16′E. Trade center of rubber-growing region. Also spelled Kendangan.

Kandanur (kuhn-dah-NOOR), town, PASUMPAN MUTHURAMALINGA THEVAR district, TAMIL NADU state, S INDIA, 5 mi/8 km NE of Karaikudi. A merchant community belonging to the Chettia caste.

Kandau, LATVIA: see KANDAVA.

Kandava (KAHN-dah-vuh), German *kandau*, city, NW LATVIA, in KURZEME, 16 mi/26 km SSE of TALSI; 57°02′N 22°46′E. Grain, fodder, potatoes. Castle ruins.

Kandavu, FIJI: see KADAVU.

Kandé (KAHN-dai), town, ⊙ KÉRAN prefecture, KARA region, N TOGO, 30 mi/48 km N of LAMA–KARA on road; 10°00′N 01°00′E. On S edge of where cattle, sheep, and goats are raised.

Kande Kiniati, CONGO: see KINIATI.

Kandel (KAHN-del), town, RHINELAND-PALATINATE, W GERMANY, 9 mi/14.5 km SSE of LANDAU; 49°05′N 08°12′E. Largely residential; manufacturing of electrical goods.

Kandel (KAHN-del), mountain (4,071 ft/1,241 m) in the BLACK FOREST, S GERMANY, 7 mi/11.3 km NE of FREIBURG; 48°03′N 08°02′E. Tourist center of WALDKIRCH is at NW foot.

Kandelion, Greece: see KANDILION.

Kandergrund, commune, BERN canton, SW central SWITZERLAND, in BERNESE ALPS, in the Kandertal, and 14 mi/23 km S of THUN. Hydroelectric power plant; matches.

Kandern (KAHN-dern), town, BADEN-WÜRTTEMBERG, SW GERMANY, on SW slope of BLACK FOREST, 7 mi/11.3 km N of LÖRRACH; 47°43′N 07°40′E. Elevation 1,155 ft/352 m. Manufacturing of pottery. Tourism; summer resort.

Kander River, drains (□ 435 sq mi/1,126 sq km), S central SWITZERLAND; rises near the BLÜMLISALP in BERNESE ALPS; flows 27 mi/43 km N, past FRUTIGEN, to LAKE THUN. Kandergrund and Spiez hydroelectric power plants are in it. The Kandertal valley leads S to the LÖTSCHBERG tunnel and hence to the SIMPLON pass and ITALY.

Kandersteg, commune, BERN canton, SW central SWITZERLAND, in BERNESE ALPS, at N end of LÖTSCHBERG tunnel, in the Kandertal, and 19 mi/31 km S of THUN. Elevation 3,858 ft/1,176 m. Resort. Has 16th-century church.

Kandhilion, Greece: see KANDILION.

Kandhkot (kahn-dah-KOT), town, JACOBABAD district, N SIND province, SE PAKISTAN, 45 mi/72 km E of JACOBABAD; 28°14′N 69°11′E. Local market.

Kandhla (KAHN-dluh), town, MUZAFFARNAGAR district, N UTTAR PRADESH state, N central INDIA, on EASTERN YAMUNA CANAL, and 28 mi/45 km WSW of MUZAFFARNAGAR; 29°19′N 77°16′E. Trades in wheat, gram, sugarcane, oilseeds. Cotton cloth manufacturing.

Kandi (KAHN-dee), town, ⊙ ALIBORI department, N BENIN, 325 mi/523 km N of PORTO-NOVO, and 70 mi/113 km from NIGER border; 11°08′N 02°56′E. Major road junction N-S and E-W routes. Agricultural center (cotton, kapok, shea nuts, peanuts, corn, millet, palm oil); manufacturing (cotton, kapok ginning, textiles). Regional airfield; customhouse; garrison.

Kandi (KAHN-dee), town, MURSHIDABAD district, central WEST BENGAL state, E INDIA, 17 mi/27 km WSW of BAHARAMPUR; 32°59′N 74°44′E. Metalware manufacturing; rice, gram, oilseeds, jute.

Kandili (kahn-DEE-lee), village, ARKADIA prefecture, E PELOPONNESE department, S GREECE, 16 mi/26 km N of TRÍPOLIS; 37°46′N 22°23′E. Wheat, tobacco; livestock (sheep, goats). Sometimes spelled Kandila or Kandela.

Kandilion (kahn-[TH]EE-lee-on), mountain (3,919 ft/1,195 m), NW ÉVVIA island, ÉVVIA prefecture, CENTRAL GREECE department, extends 10 mi/16 km along W coast SE of LIMNI; highest point is 17 mi/27 km NW of KHALKÍS; 38°41′N 23°27′E. Famous 12th century nunnery Aghin Galataki at base of mountain. Also called Kandili, Kandhilion, or Kandelion.

Kandira, township, NE TURKEY, 23 mi/37 km NE of IZMIT, 5 mi/8 km from BLACK SEA; 41°05′N 30°08′E. Oats, wheat, hemp.

Kandiyohi (kan-dee-YO-hei), county (□ 862 sq mi/2,241.2 sq km; 2006 population 41,088), SW central MINNESOTA; ⊙ WILLMAR; 45°08′N 95°00′W. Watered by several natural lakes, notably GREEN LAKE (NE), and BIG KANDIYOHI LAKE (S); drained by CROW RIVER. Agricultural region (oats, wheat, corn, alfalfa, hay, soybeans, sugar beets, beans, peas; hogs, sheep, cattle, poultry; dairying). Swedish settlement. Sibley State Park in N. Formed 1858.

Kandiyohi (kan-dee-YO-hei), village (2000 population 555), KANDIYOHI county, S central MINNESOTA, 6 mi/9.7 km E of WILLMAR; 45°07′N 94°55′W. Livestock, poultry; grain, sugar beets, alfalfa; dairying. Little Kandiyohi Lake to S, small natural lakes in area.

Kandla (KUHND-luh), town, KACHCHH district, GUJARAT state, W central INDIA, 12 mi/19 km SE of ANJAR; 23°02′N 70°13′E. Port on GULF of KACHCHH (opened early 1930s). Railroad terminus; trades in grain, textiles, timber; saltworks. It has been developed as a major port since 1955 to ease overcrowding at port of MUMBAI (Bombay); centrally administered. A free-trade zone was created at the port—the first of its kind—in 1965. The township of Gandhidham has grown up nearby.

Kandos (KAN-dahs), town, E NEW SOUTH WALES, SE AUSTRALIA, 105 mi/169 km W of NEWCASTLE, in Capertree Valley; 32°51′S 149°58′E. Agriculture and sheep center; cement works.

Kandrian (KAHN-dree-ahn), village, WEST NEW BRITAIN province, SW NEW BRITAIN island, E PAPUA NEW GUINEA, 70 mi/113 km SW of HOSKINS. Located on SOLOMON SEA coast. Airstrip. Cocoa, copra; prawns, tuna.

Kandry (kahn-DRI), town (2005 population 12,110), W BASHKORTOSTAN Republic, E European Russia, on road and railroad, 75 mi/121 km WSW of UFA; 54°34′N 54°07′E. Elevation 462 ft/140 m. Oil exploration center. Formerly known as Kandry-Amineva.

Kandry-Amineva, RUSSIA: see KANDRY.

Kandy, Sinhalese *Mahnuwara* [=great city], city (2001 population 109,343), ⊙ KANDY district and CENTRAL PROVINCE, SRI LANKA (Ceylon), on the KANDY PLATEAU, 72 mi/116 km ESE of COLOMBO; 07°17′N 80°38′E. Hill resort (elevation 1,600 ft/488 m), tourist, and market (tea, rice, vegetables; crafts) center. Mask making, wood carving, silver and brassware. Population a mixture of Sinhalese, Tamils, and Moors. Former capital of the highland kingdom from the 15th century until its conquest by the British in 1815. Parts of the 18th-century royal palace currently used as the Archaeological Museum; the smaller queen's palace is now used as the National Museum. The wooden Audience Hall (1784) is well preserved. The city centers around an artificial lake built by the last king of Kandy in 1806. Near the lake is the Temple of the Tooth (Dalada Maligawa), built to house a tooth relic of the Buddha, said to have been brought to Sri Lanka in the 4th century. A colorful two-week *Esala* procession honoring the relic draws vast crowds in July-August. Several Hindu temples; St. Paul's Anglican Church (1846), with a 120-year-old organ. Formerly spelled CANDY.

Kandy, district (□ 740 sq mi/1,924 sq km; 2001 population 1,279,028), CENTRAL PROVINCE, central SRI LANKA; ☉ KANDY; 07°15′N 80°45′E.

Kandy Plateau (elev. 1,500 ft/457 m–3,000 ft/914 m), in W SRI LANKA HILL COUNTRY, CENTRAL PROVINCE, S central SRI LANKA; c.25 mi/40 km long N-S, up to 10 mi/16 km wide; drained by upper MAHAWELI GANGA River; 07°17′N 80°38′E. Chief centers are KANDY, GAMPOLA.

Kane (KAIN), community, S MANITOBA, W central CANADA, 16 mi/25 km from town of MORRIS, in MORRIS rural municipality; 49°21′N 97°43′W.

Kane, county (□ 524 sq mi/1,362.4 sq km; 2006 population 493,735), NE ILLINOIS; ☉ GENEVA; 41°56′N 88°25′W. Agricultural area (dairy products; corn, soybeans), with industrial centers along FOX RIVER producing varied manufactures. Limestone quarries. E margin of county has experienced urban growth from CHICAGO. Major satellite communities include CARPENTERSVILLE, BATAVIA, ELGIN, Geneva, AURORA (manufacturing centers). Drained by FOX RIVER and small Mill Creek. Formed 1836.

Kane, county (□ 4,108 sq mi/10,680.8 sq km; 2006 population 6,532), S UTAH; ☉ KANAB; 37°17′N 111°53′W. Grazing and agricultural area bordering on ARIZONA, bounded on E and SE by COLORADO RIVER (Lake Powell, formed by GLEN CANYON DAM, just S of Arizona border), and drained, in W half, by PARIA and VIRGIN rivers. Agriculture (apples; gold); timber; terra-cotta; tourism. Part of Paria Canyon–Vermilion Cliffs Wilderness Area in S. KAIPAROWITS PLATEAU is in E, part of PAUNSAUGUNT PLATEAU in NW. Large part of Glen Canyon National Recreation Area in E; parts of Dixie National Forest and part of BRYCE CANYON NATIONAL PARK in NW. Part of ZION NATIONAL PARK on W state line. Coral Pink Sand Dunes State Park in SW; Kodachrome Basin State Park in N. Formed 1864.

Kane, village (2000 population 459), GREENE county, W ILLINOIS, 8 mi/12.9 km SSE of CARROLLTON; 39°11′N 90°20′W. In agricultural area (corn, soybeans).

Kane, village, BIG HORN county, N WYOMING, on BIG-HORN RIVER (BIGHORN LAKE), at mouth of SHOSHONE RIVER, near MONTANA state line, and 9 mi/14.5 km E of LOVELL. BIGHORN CANYON NATIONAL RECREATION AREA immediately E.

Kane, borough (2006 population 3,862), MCKEAN county, N PENNSYLVANIA, 22 mi/35 km SSW of BRADFORD. Oil wells; manufacturing (lumber, fabricated metal products, consumer goods); agriculture (dairying). Natural gas nearby. Health resort. Allegheny National Forest to N, W, and S. Laid out 1860.

Kane Basin, waterway, 110 mi/177 km long, part of the channel between NW GREENLAND and E ELLESMERE ISLAND. The HUMBOLDT GLACIER flows into the basin. It is named for the U.S. explorer Elisha K. Kane.

Kane Fracture Zone, transverse fracture dissecting the MID-ATLANTIC Ridge at 24°N.

Kanegasaki (kah-ne-gah-SAH-kee), town, Isawa county, IWATE prefecture, N HONSHU, NE JAPAN, on KITAKAMI RIVER, 34 mi/55 km S of MORIOKA; 39°11′N 141°07′E. Computer parts.

Kanelovskaya (kah-NYE-luhf-skah-yah), village (2005 population 4,355), N KRASNODAR TERRITORY, S European Russia, on the YEYA RIVER, on road and railroad, 47 mi/76 km SW of ROSTOV-NA-DONU (ROSTOV

oblast), and 10 mi/16 km ENE of STAROMINSKAYA, to which it is administratively subordinate; 46°35′N 39°11′E. In agricultural area (grain, fruits; livestock). Formerly known as Krasnopolyanka.

Kanem (kah-NEM), administrative region, W CHAD; ☉ MAO. Borders BORKOU-ENNEDI-TIBESTI (NNE), BATHA (E), HADJER-LAMIS (S), and LAC (SW) administrative regions and NIGER (WNW). Drained by BAHR EL GHAZAL. KANEM was formerly a prefecture prior to Chad's administrative division reorganization from fourteen prefectures to twenty-eight departments. It was recreated as a region with approximately the same boundaries following a decree in October 2002 that reorganized Chad's administrative divisions into eighteen regions.

Kanem (kah-NEM), former prefecture (2000 population 340,163), W CHAD; capital was MAO; 15°00′N 16°00′E. Was bordered N by BORKOU-ENNEDI-TIBESTI prefecture, E by BATHA prefecture, S by CHARI-BA-GUIRMI and LAC prefectures, and W by NIGER. Drained by BAHR EL GHAZAL. This was a prefecture prior to Chad's administrative division reorganization from fourteen prefectures to twenty-eight departments. It was recreated as a region named KANEM with approximately the same boundaries following a decree in October 2002 that reorganized Chad's administrative divisions into eighteen regions. Also name of former empire in AFRICA in the areas near Lake CHAD that are now part of Chad and N NIGERIA. The empire began in the 9th century, when the Sefawa migrated to the area from the SAHARA desert. The rulers eventually embraced Islam and extended their control to neighboring BORNU. After attacks by the Bulalas forced the rulers of Kanem to shift their capital to Bornu (c.1380), Bornu gradually emerged as the center of a revitalized empire of which Kanem became a protectorate.

Kaneohe (KAH-nei-O-hei), unincorporated city (2000 population 34,970), HONOLULU county, HAWAII, on the E coast of OAHU, 8 mi/12.9 km NE of HONOLULU, across KOOLAU RANGE, on KANEOHE BAY; 21°24′N 157°47′W. Once the site of a pineapple plantation and cannery, it is a residential seaside community. Manufacturing (printing and publishing, jewelry). Windward Community College; Hawaii Loa campus of Hawaii Pacific University. The U.S. Kaneohe Marine Corps Base is nearby; it was attacked by the Japanese on December 7, 1941. Many ancient fishponds built by Hawaiian chiefs are in the area. Kaneohe Bay Marine Corps Base to NE on Mokapu Peninsula; Haiku Naval Reserve to SW; Heeia State Park to N, on coast; Kaneohe Beach County Park to E; Valley of Temples to NW.

Kaneohe Bay (KAH-nai-O-hai), HAWAII, on the E coast of OAHU island, HONOLULU county, 10 mi/16 km NNE of HONOLULU. Protected by coral reefs and dotted with islands. City of KANEOHE on SW end of bay. The shores of the bay are rimmed with ancient fishponds built by the Hawaiian chiefs. University of Hawaii Marine Laboratory on COCONUT ISLAND; state park; Kaneohe U.S. Marine Corps Base is to E, at N end of Mokapu Peninsula.

Kanev, UKRAINE: see KANIV.

Kanevskaya (KAH-nyef-skah-yah), city (2005 population 45,550), NW KRASNODAR TERRITORY, S European Russia, on the CHELBAS River, on road and railroad, 75 mi/121 km N of KRASNODAR; 46°06′N 38°54′E. Elevation 108 ft/32 m. Railroad station, highway hub; local transshipment point. Gas equipment plant. Established initially as a Cossack *stanitsa* (village).

Kanevskiy Zapovednik, UKRAINE: see KANIV NATURE PRESERVE.

Kanevskoye vodokhranilishche, UKRAINE: see KANIV RESERVOIR.

Kaneyama (kah-NE-yah-mah), town, Onuma county, FUKUSHIMA prefecture, N central HONSHU, NE JAPAN, 56 mi/90 km S of FUKUSHIMA city; 37°27′N 139°31′E. Agricultural (paulownia).

Kaneyama, town, Kani county, GIFU prefecture, central HONSHU, central JAPAN, 22 mi/35 km E of GIFU; 35°27′N 137°05′E.

Kaneyama, town, Mogami county, YAMAGATA prefecture, N HONSHU, NE JAPAN, 43 mi/70 km N of YAMAGATA city; 38°52′N 140°20′E.

Kanga (KAHNG-gah), village, ORIENTALE province, NE CONGO, 15 mi/24 km NW of KILO-MINES; gold mining.

Kanga (KAN-gah), village, MOROGORO region, E central TANZANIA, 55 mi/89 km N of MOROGORO; Nguru Mountain to W; 06°03′S 37°48′E. Corn, wheat; goats, sheep.

Kangaamiut (kahn-GAH-myoot), fishing settlement, Maniitsoq (Sukkertoppen) commune, SW Greenland, at S tip of small KANGAAMIUT Island in DAVIS STRAIT, 30 mi/48 km NNW of SUKKERTOPPEN (Maniitsoq); 65°49′N 53°20′W. Original site of Sukkertoppen settlement, founded here 1755.

Kangaatsiaq (kahn-GAHT-syahk), fishing settlement, Kangaatsiag commune, W GREENLAND, on small island in DAVIS STRAIT, 30 mi/48 km SW of EGEDESMINDE (Aasiaat); 68°18′N 53°27′W.

Kangaba, town, SECOND REGION/KOULIKORO, MALI, 62 mi/100 km SW of BAMAKO; 11°56′N 08°25′W. Tourist center.

Kangahun (kahng-gah-HOON), village, SOUTHERN province, SW SIERRA LEONE, 12 mi/19 km ESE of MOYAMBA; 08°04′N 12°23′W.

Kangal, village, central TURKEY, 40 mi/64 km SSE of SIVAS; 39°15′N 37°23′E. Wheat, barley.

Kangalassy (kahn-gah-LAH-si), village (2006 population 1,700), central SAKHA REPUBLIC, central RUSSIAN FAR EAST, on the left bank of the LENA RIVER, on road junction, 18 mi/29 km N of (and administratively subordinate to) YAKUTSK; 62°21′N 129°58′E. Elevation 249 ft/75 m. In agricultural and logging area.

Kangan (kahn-GAHN), town, Būshehr province, S IRAN, near PERSIAN GULF coast, 110 mi/177 km SE of BUSHEHR; 27°50′N 52°05′E. Grain, dates. Airfield. Also called Bandar-e-Kangan.

Kangar (KAHNG-ahr), town (2000 population 54,390), ☉ PERLIS state, MALAYSIA, central MALAY PENINSULA, on the PERLIS RIVER. Center of rice-growing region.

Kangaroo Ground (KAN-guh-roo GROUND), rural locality, VICTORIA, SE AUSTRALIA, 16 mi/26 km NE of MELBOURNE; 37°41′S 145°13′E. Settled 1842.

Kangaroo Island (kan-guh-ROO), small island, SOUTH AUSTRALIA state, S central AUSTRALIA, at the entrance to Gulf SAINT VINCENT; 90 mi/145 km long and 34 mi/55 km wide; 35°50′S 137°15′E. The chief products are barley, sheep, salt, gypsum, eucalyptus oil; Monterey pine plantations. At its W end is Flinders Chase, a large reservation for native flora and fauna. There are many summer resorts. KINGSCOTE (1986 population 3,488) is the principal settlement. Variant name: Karta.

Kangaroo Lake, Wisconsin: see BAILEYS HARBOR.

Kangaroo Valley (KAN-guh-roo), valley, NEW SOUTH WALES, SE AUSTRALIA, 99 mi/159 km S of SYDNEY, between the South Coast and Southern Highlands; 34°44′S 150°32′E.

Kangaruma (kahn-gah-ROO-mah), village, POTARO-SIPARUNI district, GUYANA, 05°20′N 59°12′W. Located on POTARO RIVER, 20 mi/32 km N of MAHDIA.

Kangas (KAHNG-ahs), village, OULUN province, W FINLAND, 40 mi/64 km SSE of RAAHE; 64°09′N 24°43′E. Elevation 248 ft/75 m. Lumber and paper mills.

Kangavar (kahn-gah-VAHR), town (2006 population 48,978), Kermānshāhān, W IRAN, 50 mi/80 km ENE of KERMANSHAH, on road to HAMADAN; 34°30′N 47°58′E. Grain, tobacco cotton; sheep raising.

Kangaz, MOLDOVA: see CONGAZ.

Kangbachen (KAHNG-bah-chen), mountain peak (25,930 ft/7,903 m), NE NEPAL; 27°43′N 88°07′E. First climbed in 1974 by a Polish expedition.

Cross-references are shown in SMALL CAPITALS. The pronunciation guide is shown on page xix. The sources of population figures are shown on page xvii.

Kangbao (KANG-BOU), town, ⊙ Kangbao county, NW HEBEI province, NE CHINA, 70 mi/113 km N of ZHANGJIAKOU, at INNER MONGOLIA border; 41°52′N 114°51′E. Oilseeds, sugar beets, cattle; food processing, textiles and clothing; coal mining.

Kangchungtse (KAHNG-choong-tse), mountain peak (25,190 ft/7,678 m), E NEPAL; 27°55′N 87°05′E. Climbed by a French party in 1954. Also known as Makalu II (muh-KAH-loo).

Kangding (KANG-DING), city, W SICHUAN province, SW CHINA, in the Garze Tibetan Autonomous District, 125 mi/201 km SW of CHENGDU; 30°03′N 102°02′E. It is a transportation center on the main road from Chengdu to LHASA (Tibet). The development of a hydroelectric power plant has allowed Kangding to establish some light industries. Sometimes spelled Kangting; also called Dardo.

Kangdong (GAHNG-DONG), county, SOUTH PYONGAN province, NORTH KOREA, 20 mi/32 km NE of PYONGYANG. Major coal-mining area.

Kangean Islands (KAHNG-ge-ahn), small island group (□ 258 sq mi/670.8 sq km), Java Timur province, INDONESIA, in JAVA SEA, 75 mi/121 km N of BALI; 06°57′S 115°30′E. Comprises three islands surrounded by c.60 islets. Largest island is Kangean (188 sq mi/487 sq km), 25 mi/40 km long, 12 mi/19 km wide, generally low, rising in NE to 1,192 ft/363 m. Chief products are coconuts, teak, salt, fish, cattle. Chief town is Arjasa. The other large islands (SE of Kangean) are Paliat (9 mi/14.5 km long, 3 mi/4.8 km wide) and Sepanjang (10 mi/16 km long, 5 mi/8 km wide).

Kangek, Kangeq (both: KAH-ngek), former fishing settlement, NUUK (Godthåb) commune, SW GREENLAND, on islet in the ATLANTIC, at S end of DAVIS STRAIT, at mouth of GODTHÅBSFJORD, 10 mi/16 km WSW of Nuuk (Godthåb); 64°06′N 52°02′W. Sometimes spelled Kangek.

Kangerluarsuk (kahn-ger-LWAHKH-shook), former settlement, Maniitsoq (SUKKERTOPPEN) commune, SW GREENLAND, on DAVIS STRAIT, at mouth of Sondre Isortoq Fjord, 13 mi/21 km ESE of Sukketoppen (Maniitsoq).

Kangerluarsuk (kahn-ger-LAWAHKH-shook), inlet, Qaqortoq (Greenlandic *julianehåb*) commune, S GREENLAND; 60°52′N 45°15′W. Eudialyte deposits.

Kangerlussuaq (kahng-er-LOOKH-shwahk), inlet of DENMARK STRAIT, SE GREENLAND; 45 mi/72 km long, 3 mi/4.8 km–9 mi/14.5 km wide; 68°20′N 32°15′W. Extends NW between sheer mountains (elevation 8,500 ft/2,591 m) to edge of inland ice, where it receives several large glaciers, including Kangerlussuaq Glacier. Meteorological and radio station at mouth, near 68°10′N 31°20′W. Low grade palladium and gold deposits located nearby.

Kanggye (GAHNG-GYE), city, ⊙ CHAGANG province, NORTH KOREA, 95 mi/153 km W of HUNGNAM; 41°00′N 126°35′E. Collection center for agriculture products, textiles. Graphite mined nearby. Connects to primary railroads.

Kanghwa (GAHNG-HWAH), administrative district of INCHON city (□ 163 sq mi/423.8 sq km), NW KYONGGI province, SOUTH KOREA, on the YELLOW SEA; 37°40′N 126°25′E. Industries include fishing and rice farming; ginseng, barley, beans, potatoes; *Hwamunsok*, reed mats. Located where the HAN, Imjin, and Yesong rivers meet, it consists of about twenty islands, Kanghwa being the largest, and faces KIMPO over Yomha on the E, Kyonggi Bay on the SW, and NORTH KOREA over the cease-fire line N. Kanghwa was briefly the site of the Korean capital in the 13th century. It was early fortified as an outer defense for SEOUL and was stormed by the French in 1866 and by the Americans in 1871.

Kangiqsualujjuaq (KAN-je-SOO-yah-yuk), village (□ 14 sq mi/36.4 sq km), NORD-du-Québec region, NE QUEBEC, E CANADA, on SE shore of UNGAVA BAY, at mouth of GEORGE River; 58°41′N 65°57′W. Replaced

Port-Nouveau-Quebec village. Inuit community. Hunting, fishing, trapping. Scheduled air service.

Kangiqsujuaq (KAN-je-SOO-yak) or **Wakeham Bay** (WAIK-uhm BAI), village (□ 5 sq mi/13 sq km), NORD-du-Québec region, N QUEBEC, E CANADA, on S shore of HUDSON STRAIT, NE end of Ungava Peninsula; 61°35′N 71°57′W. Inuit community. Hunting, fishing, trapping. Scheduled air service.

Kangirsuk or **Payne Bay** (PAIN BAI), village (□ 23 sq mi/59.8 sq km), NORD-du-Québec region, N QUEBEC, E CANADA, W shore of UNGAVA BAY, E side of UNGAVA Peninsula, at mouth of Arnaud River; 60°01′N 70°01′W. Inuit community. Hunting, fishing, trapping. Scheduled air service.

Kangirtuagaapik (KAN-jer-TOO-yah-pik), hamlet, BAFFIN region, NUNAVUT territory, CANADA, W shore of Patricia Bay, E coast of BAFFIN ISLAND, c.1,338 mi/2,153 km NE of YELLOWKNIFE. Hudson's Bay Company post 1922–1987. U.S. Coast Guard weather station.

Kangjin (GAHNG-JIN), county, SOUTH CHOLLA province, SOUTH KOREA, 50 mi/80 km SE of KWANGJU; 34°38′N 126°46′E. Agriculture (rice, barley, soybeans, cotton); textiles, metalwork.

Kangle (KANG-LUH), town, ⊙ Kangle county, SE GANSU province, NW CHINA, 55 mi/89 km S of LANZHOU; 35°16′N 103°39′E. Grain, oilseeds; food and forage processing, engineering.

Kangley, village (2000 population 287), LA SALLE county, N ILLINOIS, on VERMILION RIVER, 2 mi/3.2 km NW of STREATOR, and 16 mi/26 km S of OTTAWA; 41°08′N 88°52′W. Agriculture (dairying; corn, wheat, soybeans).

Kangmar (KANG-MAH), town, S TIBET, SW CHINA, on the NYANG QU river, on main India-Lhasa trade route, and 26 mi/42 km SSE of GYANGZE; 28°30′N 89°45′E. Elevation 13,900 ft/4,237 m. Livestock, grain; crafts. Sometimes spelled Khangmar.

Kangnung (GAHNG-NUHNG), city, KANGWON province, NE SOUTH KOREA, a port on the E coast (SEA OF JAPAN); 37°45′N 128°53′E. An agricultural and fishing center, Kangnung is famed for its beautiful scenery. The largest city in Kangwon province; transportation hub between SEOUL-Kangnung highway and Donghae highway.

Kango (KAHN-go), city, ESTUAIRE province, NW GABON, on N shore of Gabon estuary, 50 mi/80 km ESE of LIBREVILLE; 00°13′N 10°06′E. Agricultural center with coffee, cacao, manioc, groundnuts, corn. Also produces rubber.

Kangpa, China: see GAMBA.

Kangping (KANG-PING), town, ⊙ Kangping county, N LIAONING province, NE CHINA, 60 mi/97 km SW of SIPING; 42°45′N 123°20′E. Grain, oilseeds, tobacco, sugar beets; textiles, food industry, coal mining.

Kangra (KAHN-gruh), district (□ 9,979 sq mi/25,945.4 sq km), HIMACHAL PRADESH state, N INDIA, ⊙ DHARMSHALA. Bordered N by CHAMBA district, S by MANDI, HAMIRPUR, and KULLU districts, E by LAHUL AND SPITI district; crossed by PUNJAB HIMALAYAS (in NE section peaks rise to over 20,000 ft/6,096 m). CHENAB, RAVI, and BEAS rivers rise in district. Hand-loom woolen and cotton weaving, basket making. Main crops are wheat, corn, rice, barley, tea, gram, and spices; fruit grown in Kullu valley. Forests of deodar, pine, oak, bamboo. Chief villages: PALAMPUR, Kullu, Nagar, NURPUR. For century area was dominated by Rajput princes, who long resisted Muslim penetrations. Largely annexed by Ranjit Singh in 1828, after he had overcome the Gorkhas.

Kangra (KAHN-gruh), town, KANGRA district, HIMACHAL PRADESH state, N INDIA, on railroad, and 9 mi/14.5 km SSW of DHARMSHALA; 32°06′N 76°16′E. Local trade center for wheat, rice, tea, corn, wool; handicrafts (enamel work, printed cloths). Annual Hindu festival. For many century a Rajput stronghold, known as Nagarkot; plundered by Mahmud of Ghazni

in 1009; later garrisoned by Moguls. Destroyed by earthquake in 1905.

Kangri, TURKEY: see CANKIRI.

Kangshan, town, S TAIWAN, 10 mi/16 km N of KAOHSIUNG, on railroad; 22°48′N 120°17′E. Agricultural center for sugarcane, rice, and bananas; large airfield, airplane repair and assembly plant.

Kangtega (KAHNG-te-gah), mountain peak (22,241 ft/6,779 m), E NEPAL; 27°48′N 86°49′E.

Kangting, China: see KANGDING.

Kangto (KANG-TO), peak (23,260 ft/7,090 m) in main range of central ASSAM HIMALAYAS, on undefined TIBET-ASSAM (INDIA) border; 27°50′N 92°30′E.

Kangu (KAHNG-goo), village, BAS-CONGO province, W CONGO, on railroad, 40 mi/64 km NNW of BOMA; 04°56′S 13°26′E. Elev. 1,784 ft/543 m. Palm-oil milling, coffee and rubber plantations. Also a mission center, with several trade schools, and seat of vicar apostolic of BOMA.

Kangurt (KAHN-uhrt), village, W KHATLON viloyat, TAJIKISTAN, 28 mi/45 km NW of KULOB, and just W of Nurek reservoir; 38°15′N 69°31′W. Wheat.

Kangwha Island, SOUTH KOREA: see KANGHWA.

Kangwon (GAHNG-WON), province, NORTH KOREA, bounded on E by SEA OF JAPAN and S by SOUTH KOREA; ⊙ WONSAN. Populated areas include Wonsan city, and Chonnae, Muchon, and Anbyon counties. Manufacturing (machinery).

Kangwon (GAHNG-WON), province, N SOUTH KOREA; ⊙ CHUNCHON; 37°45′N 128°15′E. The 38th parallel that divided Korea after World War II ran through Kangwon, but after the Korean War truce of 1953 much of the province returned from NORTH to South Korean rule. Mining (iron, coal, tungsten, fluorite), farming (potatoes, rye, hop), and fishing are chief economic activities in the province. Surak, ODAE, Chiak National Parks.

Kang Xian (KANG-SI-AN), town, ⊙ Kang Xian county, SE GANSU province, NW CHINA, 50 mi/80 km E of WUDU; 33°26′N 105°37′E. Grain, livestock; food processing.

Kanhangad, INDIA: see KASARAGOD, city.

Kanhan River (kahn-HAHN), 120 mi/193 km long, in MADHYA PRADESH and MAHARASHTRA states, central INDIA; rises in central SATPURA RANGE E of BETUL (Madhya Pradesh); flows E and SE, past KHAPA and KAMPTEE (Maharashtra state), to WAINGANGA RIVER, 7 mi/11.3 km SW of BHANDARA; receives PENCH RIVER. Waterworks on Kanhan River at Kamptee supply municipal water to NAGPUR city.

Kanheri, INDIA: see BORIVLI.

Kanhhoa, VIETNAM: see DIEN KHANH.

Kani (KAH-nee), city, GIFU prefecture, central HONSHU, central JAPAN, 19 mi/30 km E of GIFU; 35°25′N 137°03′E. Automotive parts.

Kani (kah-NEE), township, SAGAING division, MYANMAR, on right bank of CHINDWIN RIVER and 30 mi/48 km NW of MONYWA.

Kani (KAH-nee), village, Worodougou region, central CÔTE D'IVOIRE, 37 mi/60 km N of SÉGUÉLA; 08°29′N 06°36′W. Agriculture (coffee, palm kernels, cotton, rice, corn, beans, yams, tobacco, vegetables); livestock.

Kaniama (kahn-YAH-mah), village, KATANGA province, SE CONGO, on railroad, 95 mi/153 km NW of KAMINA; 07°31′S 24°11′E. Elev. 2,913 ft/887 m. Cotton ginning.

Kaniapiskau, CANADA: see CANIAPISCAU.

Kaniapiskau, Lake (kan-yuh-PI-sko) (□ 375 sq mi/975 sq km), N central QUEBEC, E CANADA, on HUDSON BAY–SAINT LAWRENCE watershed; 54°10′N 69°50′W; 20 mi/32 km long, 12 mi/19 km wide. Elevation 1,850 ft/564 m; drained by Kaniapiskau River.

Kanibadam (kah-nee-bah-DAHM), city, E LENINOBOD viloyat, TAJIKISTAN, on Great Fergana Canal, on railroad, and 40 mi/64 km E of KHUJAND; 40°17′N 70°25′E. Fruit- and vegetable-canning center; cotton gins, cottonseed oil milling. Also spelled Konibodom.

Kanie (kah-NEE-e), town, Ama county, AICHI prefecture, S central HONSHU, central JAPAN, 6 mi/10 km S of NAGOYA; 35°07′N 136°47′E. Flowers; hot springs.

Kaniere (KAN-ee-REE), township, W SOUTH ISLAND, NEW ZEALAND, 3.1 mi/5 km SE of HOKITIKA. Sheep, cattle; tourism. Lake Kaniere is nearby. Former gold-mining center.

Kanifing (KAH-nee-feeng), local government area (□ 29 sq mi/75.4 sq km; 2003 population 322,410), THE GAMBIA; 13°25′N 16°39′W. N of WESTERN division and S of GAMBIA RIVER estuary, adjoining SAINT MARY'S ISLAND. Includes BAKAU, Latin Kunda, and SEREKUNDA, rapidly growing suburbs of BANJUL city. From 1973 to 1993, the division population grew from 39,404 to 228,945. Was acquired in 1840 by the British for settlement of liberated slaved. KANIFING and BANJUL local government areas comprise the GREATER BANJUL AREA division. Formerly Kombo Saint Mary Division; sometimes called Kanifing Municipal Council.

Kanifing Municipal Council, THE GAMBIA: see KANIFING, local government area.

Kanigiri (kah-NEE-gee-ree), town, PRAKASAM district, ANDHRA PRADESH state, SE INDIA, 75 mi/121 km NNW of NELLORE; 15°24′N 79°31′E. Manufacturing of spinning instruments, scissors, razors; millet, cotton. Granite quarries nearby.

Kaniguram (kahn-ee-GOO-ruhm), village, in centrally administered tribal area of SOUTH WAZIRISTAN, excluded from NORTH-WEST FRONTIER PROVINCE, W central PAKISTAN, 20 mi/32 km NE of WANA, in N SULAIMAN Range; 37°31′N 69°47′E. Trades in hides and grain. Noted for its ancient Ormuri language, spoken in PAKISTAN only in this area.

Kanima, China: see GYANYIMA.

Kanimekh (kah-nee-MYEK), town, SE BUKHARA wiloyat, UZBEKISTAN, 30 mi/48 km NE of GIJDUVAN; 40°16′N 65°07′E. Cotton; livestock; food processing. Site of 11th century mausoleum of Mir Said Bakhrom and Khwaka Kas.

Kanin (KAH-neen), former town, E central KOMI REPUBLIC, NE European Russia on the right bank of the PECHORA River, on the North Pechora railroad, now a suburb of PECHORA, 4 mi/6 km E of the city center; 65°27′N 57°10′E. Elevation 141 ft/42 m. In a natural gas district. Also known as Kanina Nos.

Kanin (KUH-neen), peninsula, NENETS AUTONOMOUS OKRUG, ARCHANGEL oblast, N European Russia, projecting into the BARENTS SEA between the WHITE SEA (W) and CHESHA (Cheshskaya) Bay (E); 68°00′N 45°00′E. Its northernmost cape is called KANIN NOS. The native Nenets (Samoyed) people engage in fishing, hunting, and reindeer raising.

Kanin, ITALY and SLOVENIA: see CANIN, MOUNT.

Kanina (kuh-NEE-nuh), township, S YEMEN, in the Wadi HAJR; 14°30′N 48°21′E.

Kanin Nos, RUSSIA: see KANIN, suburb of PECHORA.

Kanin Nos (KUH-neen–NOS), NW cape of the KANIN Peninsula, in the BARENTS SEA, in NENETS AUTONOMOUS OKRUG, ARCHANGEL oblast, N European Russia; 68°27′N 43°10′E. Lighthouse.

Kanisah, Jabal al-, LEBANON: see KENISA, JEBEL EL.

Kanita (KAH-nee-tah), town, East Tsugaru county, Aomori prefecture, N HONSHU, N JAPAN, on W AOMORI BAY, 16 mi/25 km N of AOMORI; 41°02′N 140°38′E.

Kaniv (KAH-nif) (Russian *Kanev*), city (2004 population 22,300), N CHERKASY oblast, UKRAINE, port on the DNIEPER River, 65 mi/105 km S of KIEV, and 33 mi/53 km NW of CHERKASY; 49°45′N 31°28′E. Elevation 334 ft/101 m. Raion center; electric power station; manufacturing (electromechanical, medical equipment); food processing (flour, dairy). Educational, professional technical school; St. George's Cathedral. Grave of a renowned Ukrainian writer and poet Taras Shevchenko and a museum dedicated to his memory is 4 mi/6.4 km away, on the Tarasova (formerly Chernecha) hora.

Kaniva (kuh-NEI-vuh), village, W VICTORIA, SE AUSTRALIA, 230 mi/370 km WNW of MELBOURNE; 36°23′S 141°15′E. Wool; wheat, oats, barley; cement works.

Kaniv Nature Preserve (KAH-neef) (Ukrainian *Kanivs'kyy zapovidnyk*) (Russian *Kanevskiy zapovednik*), in N CHERKASY oblast, on high right bank of DNIEPER (Ukrainian *Dnipro*) River, in the wooded steppe zone of UKRAINE, 1 mi/1.6 km–2 mi/3.2 km SE of KANIV. Covers 4,900 acres/1,983 ha. Consists mostly of high hills, deep ravines, clad in broadleaf forests; also includes Kruhlyk Island in the Dnieper River. Contains 333 animal species and 9,125 insect species, of which 31 and 32 species, respectively, are listed in the Red Data Book (former Soviet, now Ukrainian, endangered species list). Purpose is to study and preserve flora and fauna of the middle Dnieper region. Established in 1923; since 1939, used as a field station for students of the Natural Sciences faculties of Taras Shevchenko University in KIEV.

Kaniv Reservoir (KAH-neef) (Ukrainian *Kanivs'ke vodoskhovyshche*) (Russian *Kanevskoye vodokhranilishche*) (□ 261 sq mi/675 sq km), on the DNIEPER (Ukrainian *Dnipro*) River, N central UKRAINE, in broad Dnieper valley between KIEV (NW) and KANIV (SE); length, 76 mi/123 km; width, 5 mi/8 km; average depth, 13 ft/4 m; maximum depth, 69 ft/21 m. Formed in 1972–1978 with the construction of Kaniv hydroelectric power station. Right banks are steep, left banks are gently sloping, sandy. Water level fluctuation does not exceed 2 ft/0.6 m (maximum during spring floods). Reservoir serves hydroelectric power generation, navigation, fisheries, irrigation of 8,648 acres/3,500 ha, water supply, and recreation.

Kanivs'ke vodoskhovyshche, UKRAINE: see KANIV RESERVOIR.

Kanivs'kyy zapovidnyk, UKRAINE: see KANIV NATURE PRESERVE.

Kanjamalai (KUHN-juh-muh-LEI), hill (3,236 ft/986 m), SW outlier of SHEVAROY HILLS, SALEM district, TAMIL NADU state, S INDIA, WSW of SALEM; c.5 mi/8 km long, c.2 mi/3.2 km wide. Extensive magnetite, magnesite, and chromite mining. CHALK HILLS (NNE) are a major source of India's magnesite.

Kanjiroba (kahn-JEE-ro-BAH), mountain peak (22,580 ft/6,882 m), W NEPAL; 29°23′N 82°38′E.

Kanjiža (KAH-nyee-zhah), Hungarian *Magyarkanizsa*, village (2002 population 27,510), VOJVODINA, N SERBIA, on TISZA RIVER, on railroad, and 18 mi/29 km E of SUBOTICA, near Hungarian border; 46°03′N 20°04′E. Mineral baths. NOVI KNEZEVAC (formerly called NOVA KANJIZA), village is E, across the Tisa. Also spelled Kanyiža, Stara Kanjiza, or Stara Kanyizha.

Kankakee (KAN-ka-kee), county (□ 681 sq mi/1,770.6 sq km; 2006 population 109,090), NE ILLINOIS, on INDIANA state line (E); ⊙ KANKAKEE; 41°08′N 87°51′W. Suburban communities of BRADLEY and BOURBONNAIS adjoin city of Kankakee on N. Agriculture (corn, soybeans; livestock; dairying); manufacturing (industrial machinery, wood products, food products, paper products, chemicals, drugs, plastics products, tile, industrial machinery, metal prods). Limestone deposits. Drained by KANKAKEE and IROQUOIS rivers. Kankakee River State Park in NW. Formed 1853.

Kankakee, city (2000 population 27,491), ⊙ KANKAKEE county, E ILLINOIS, on the KANKAKEE river; 41°07′N 87°51′W. It is a trade, processing, and shipping center for a rich agricultural region (corn, soybeans; dairying). Manufacturing (wire, vinyl tiles, consumer goods, printing, chemicals, pet food, polymers, corn milling, industrial metals). Limestone quarries are nearby. A state mental hospital, a state park, and Olivet Nazarene College are nearby. Kankakee Community College is in the city. Incorporated 1855.

Kankakee River, c.135 mi/217 km long, in INDIANA and ILLINOIS; rises near SOUTH BEND, N Indiana; flows SW to a point near KANKAKEE, thence NW, joining

DES PLAINES River SW of JOLIET to form ILLINOIS RIVER.

Kankan, administrative region, ENE GUINEA; ⊙ KANKAN. Bordered N and E by MALI, S by N'ZÉRÉKORÉ administrative region (part of E portion of border formed by Dion River), and SWW by Faranah administrative region (S portion of border formed by Mafou River). BAKOY RIVER originates in N of region, flowing N into Mali (country); MILO RIVER runs through center of the region, joining NIGER RIVER at Niandankoro in N central of region; TINKISSO RIVER runs E across N part of the region, joining Niger River just S of SIGUIRI town in NE of region; Niger River flows NE through the region, entering the central part of the region from W, passing near KOUROUSSA, Balato, Niandankoro, Siguiri, and Bankon towns, before entering Mali (country) in NE; Mafou River forms S part of the border with Faranah administrative region, before joining Niger River as it enters Kankan administrative region; Kokoro River in extreme NE; Fié River in E; Dion, Kouraï, and Sankarani rivers in SE; Kouya and Niandan rivers in SW central. Part of Fon Going ridge in S. Mount Niagouelé (elevation 2,497 ft/761 m) in NE. Kankan Nature Reserve in extreme SE. Towns include DOUAKO, KÉROUANÉ, KINTINIAN, Kouroussa, MANDIANA, NIAGASSOLA and Siguiri. The railroad from CONAKRY to Kankan town enters the central part of the region from the W, passing through Cisséla and Kouroussa towns before terminating at Kankan. Several major roads come out of Kankan town, including one that runs W roughly parallel with the railroad to Conakry; one that heads N through FODÉKARIA, Niandankoro, Siguiri, and Doko towns, before leaving the region in the NE corner and heading into Mali (country); one that heads E through Mandiana town and into Mali (country); one that heads S through Komodou and Kérouané towns before entering N'Zérékoré administrative region; and one that heads SW through Moribaya and Tokounou towns before entering Faranah and then N'Zérékoré adminstrative regions. Airport at Kankan. Includes the prefectures of Siguiri in N, Mandiana in E, Kankan in central, Kérouané in S, and Kouroussa in W. All of Kankan administrative region is in Haute-Guinée geographic region.

Kankan, prefecture, Kankan administrative region, E GUINEA, in Haute-Guinée geographic region; ⊙ KANKAN. Bordered NWN by Siguiri prefecture, N and E by Mandiana prefecture, SE by CÔTE D'IVOIRE, S by Kérouané prefecture, SW by Kissidougou prefecture, and W by Kouroussa prefecture. MILO RIVER flows N through center of prefecture (joining Niger River near Niandankoro) and Dion and Kouraï rivers in SE. Kankan Nature Reserve in SE. Towns include FODÉKARIA, Kankan, Moribaya, Sabadou Baranama, and Tokounou. Railroad originating in CONAKRY and running across the country terminates in Kankan town. Roads branch out of Kankan town to all parts of the prefecture, as well as W Guinea, MALI, and Côte d'Ivoire. Airport at Kankan town.

Kankan, city (2004 population 113,900), ⊙ KANKAN prefecture and KANKAN administrative district and HAUTE-GUINÉE geographic region, E GUINEA, a port on the MILO RIVER, a tributary of the NIGER River; 10°23′N 09°18′W. It is the commercial and administrative center for a farm area where rice, sesame, maize, tomatoes, oranges, mangoes, and pineapples are grown. Diamonds are mined, and the national diamond exchange is here. Bricks and fruit juices are made in Kankan, which also has a tomato-canning factory and a sawmill. The city is connected by railroad with CONAKRY. Kankan was probably founded in the 18th century as a trade center that linked the SUDAN region with the forest belt and the ATLANTIC OCEAN coast. Malinke leader, Samori Touré began (c.1866) his career as a military leader and empire builder in the Kankan district and in 1873 took

Kankan itself. The French occupied the city in 1891. Kankan has a polytechnic institute, a center for research on rice cultivation, a teacher-training school, and the national police school.

Kankanhalli (KAHN-kahn-huh-lee), town, BANGALORE district, KARNATAKA state, SE INDIA, on tributary of KAVERI RIVER, and 30 mi/48 km SSW of BANGALORE. Trades in bamboo, sandalwood from nearby hills; sawmilling, livestock grazing.

Kankari, TURKEY: see CANKIRI.

Kanker (kuhn-KER), town (2001 population 24,485), ⊙ Kanker district, S central Chhattisgarh state, E central INDIA; on small affluent of MAHANADI River, and 30 mi/48 km S of DHAMTARI; 20°16′N 81°29′E. Rice, millet, oilseeds. Sal, bamboo, myrobalan in dense forest area (S). Was the capital of former princely state of Kanker, one of the CHHATTISGARH STATES; incorporated in 1948 into BASTAR district in MADHYA PRADESH; in 1999, northern part of Bastar became Kanker district, with Kanker as capital; became part of Chhattisgarh in 2000.

Kankesanturai (KAHN-kai-sahn-THU-rai), town, NORTHERN PROVINCE, at N tip of SRI LANKA, on N JAFFNA Peninsula, on PALK STRAIT, 10 mi/16 km NNE of JAFFNA; 09°48′N 80°02′E. Railroad terminus; seaport (trades in grain, sugar); cement manufacturing. Lighthouse. Extensive tobacco cultivation; 5 mi/8 km S, near CHUNNAKAM, a betel-leaf market; also spelled CHUNAKAM. Palaly airport nearby (3 mi/4.8 km SSE) serves Jaffna peninsula.

Kan'ko, UKRAINE: see VYNOHRADIV.

Kankossa (kahn-KO-sah), village (2000 population 11,083), Assaba administrative region, S MAURITANIA, 99 mi/160 km N of KAYES (MALI); 15°56′N 11°31′W. Livestock, dates, millet.

Kankroli (kuhn-kuh-RO-lee), village, Rajsamand district, RAJASTHAN state, NW INDIA, on S shore of RAJ SAMAND, 35 mi/56 km NNE of UDAIPUR. Place of Hindu pilgrimage; Vishnuite temple.

Kanlaon, PHILIPPINES: see CANLAON.

Kanmaki (KAHN-mah-kee), town, North Katsuragi district, NARA prefecture, S HONSHU, W central JAPAN, 11 mi/18 km S of NARA; 34°33′N 135°43′E. Footwear.

Kanmon Bridge, suspension bridge (3,504 ft/1,068 m) spanning SHIMONOSEKI STRAIT between Moji (Fukuoka prefecture) and SHIMONOSEKI (Yamaguchi prefecture). Completed 1973. Also Kanmonkyo or Kanmon-kyo.

Kanmon-kaikyo, strait, JAPAN: see SHIMONOSEKI STRAIT.

Kannabe (KAHN-nah-be), town, Fukayasu county, HIROSHIMA prefecture, SW HONSHU, W JAPAN, 53 mi/85 km E of HIROSHIMA; 34°32′N 133°23′E.

Kannad (KUHN-nuhd), town, AURANGABAD district, NW MAHARASHTRA state, W INDIA, 29 mi/47 km NNW of AURANGABAD; 20°16′N 75°08′E. Cotton, millet, oilseeds. Sheep raising in nearby hills.

Kannami (KAHN-nah-mee), town, Tagata county, SHIZUOKA prefecture, central HONSHU, E central JAPAN, 34 mi/55 km N of SHIZUOKA; 35°05′N 138°56′E. Dairy products; watermelons. Hot springs.

Kannanlahti, RUSSIA: see KANDALAKSHA.

Kannapolis (kuh-NAP-uh-lis), city (□ 30 sq mi/78 sq km; 2006 population 40,223), CABARRUS and ROWAN counties, W central NORTH CAROLINA, 25 mi/40 km NNE of CHARLOTTE; 35°29′N 80°37′W. Twin city to CONCORD 5 mi/8 km SSE. It is a planned residential company town once owned by Cannon Mills; known for its production of household linens and textiles in neighboring Concord. As former headquarters of the Pillowtex Corporation, which owned Fieldcrest Cannon, Kannapolis became the center of the largest layoff in North Carolina history when Pillowtex closed facilities in 2003; several business parks and partnerships have since helped attract new business and jobs to the city. Manufacturing (textiles, health and beauty products, light manufacturing); service industries. Rowan-Cabarrus Community College to SW. Cannon Village includes Fieldcrest Cannon Textile Museum and Exhibition Hall, and Dale Earnhardt Tribute Center. Founded c.1905, incorporated 1984.

Kannari (KAHN-nah-ree), town, Kurihara county, MIYAGI prefecture, N HONSHU, NE JAPAN, 37 mi/60 km N of SENDAI; 38°47′N 141°04′E.

Kannauj, INDIA: see KANAUJ.

Kanniyai, SRI LANKA: see TRINCOMALEE.

Kanniyakumari, district, S TAMIL NADU state, southernmost district in INDIA, on GULF OF MANNAR and ARABIAN SEA; ⊙ NAGERCOIL. Bordered N by Tirunelveli Kattabomman district, W by KERALA state.

Kanniyakumari, town, KANNIYAKUMARI district, TAMIL NADU state, extreme S tip of INDIA, where BAY OF BENGAL, INDIAN OCEAN, and ARABIAN SEA meet; 08°05′N 77°34′E. Well-known religious center and tourist spot (known for its beautiful sunrises and sunsets).

Kannod (kuh-NOD), town (2001 population 15,165), DEWAS district, W central MADHYA PRADESH state, central INDIA, 55 mi/89 km ESE of INDORE; 22°40′N 76°44′E. Market center for wheat, cotton, corn, timber; cotton ginning. Kheoni Wildlife Sanctuary (blue bull, barking deer, four-horned antelope).

Kanntorp, SWEDEN: see SKÖLDINGE.

Kannur (kuh-NOOR), city (1991 urban agglomeration population 463,962), ⊙ CANNANORE district, KERALA state, S INDIA; 11°51′N 75°22′E. Formerly the capital of Kolattiri Raja, it traded with Arabia and Persia in the twelfth and thirteenth century Vasco da Gama visited in 1498, and it became a Portuguese settlement. Control passed to the Dutch in the mid-seventeenth century, and to the British in 1783. A military station, it trades in coconut products, rice, pepper, timber products, dried fish, cotton fabrics, and tobacco, and has large textile mills. Also called Cannanore.

Kano (KAH-no), state (□ 7,772 sq mi/20,207.2 sq km; 2006 population 9,383,682), N NIGERIA; ⊙ KANO; 11°30′N 08°30′E. Bounded N and E by JIGAWA state, W by KATSINA state, SW by KADUNA state, and SE by BAUCHI state. Savanna (S), scrub vegetation (N). Mainly agricultural (cotton, peanuts, millet); cattle raising. Wheat production projects. Cotton weaving, metal-, brass-, and leatherworking. Tin mining at TUDUN WADA and RURUWEI (S) and near GWARAM (SE). Population largely Hausa and Fulah. Includes Kano emirate, the successor of an original Hausa state, conquered (c. 1800) by the Fulah; came (1903) under British rule. Established 1968; area reduced with the creation of Jigawa state in 1991.

Kano (KAH-no), city (2004 population 3,410,000), ⊙ KANO state, N NIGERIA. It is the trade and shipping center for an agricultural region where cotton, cattle, and about half of Nigeria's groundnuts are raised. Kano is the major industrial center of N Nigeria, manufacturing peanut flour and oil, cotton textiles, steel furniture, processed meat, concrete blocks, shoes, and soap. The city has long been known for its leatherwork; its tanned goatskins were sent (from about 15th century) to N AFRICA and were known in EUROPE as Morocco leather. One of the 7 Hausa city-states, Kano's written history dates back to C.E. 999, when the city already was several centuries old. It was a cultural, handicraft, and commercial center, with wide trade contacts in W and N Africa. In the early 16th century, Kano accepted Islam. Kano reached the height of its power in the 17th and 18th centuries. In 1809 it was conquered by the Fulani, but it soon regained its leading commercial position. In 1903 a British force captured the city. Local institutions include Abdullahi Bayero College (1960; part of Ahmadu Bello University, ZARIA), Gidan Makama Museum (with examples of local art), and the palace of the emir (the former ruler of the Kano city-state). International airport to N.

Kano (KAH-no), town, Tsuno county, YAMAGUCHI prefecture, SW HONSHU, W JAPAN, 22 mi/35 km E of YAMAGUCHI; 34°13′N 131°49′E.

Kanonji (KAH-non-jee), city, KAGAWA prefecture, NE SHIKOKU, W JAPAN, on the HIUCHI SEA, 25 mi/40 km S of TAKAMATSU; 34°07′N 133°39′E. Lettuce. Medicines; fish processing. Religious center and agricultural market noted for its Kanonji (Buddhist) Temple.

Kanopolis (kan-NAHP-uh-lis), village (□ 22 sq mi/57.2 sq km; 2000 population 543), ELLSWORTH county, central KANSAS, on SMOKY HILL RIVER, and 4 mi/6.4 km ESE of ELLSWORTH; 38°42′N 98°09′W. Railroad junction. In wheat and livestock area. Kanopolis Reservoir (□ 21.7 sq mi/56.2 sq km; 26 mi/42 km long) is formed by Kanopolis Dam, 12 mi/19 km SE on SMOKY HILL RIVER. Old Fort Harker here. Kanopolis State Park on lake. Mushroom Park to E, rock formations.

Kanopolis Lake, reservoir (□ 22 sq mi/57.2 sq km), ELLSWORTH county, central KANSAS, on SMOKY HILL RIVER, 25 mi/40 km SW of SALINA; 26 mi/42 km long; 38°37′N 97°58′W. Maximum capacity 433,000 acre-ft. Formed by Kanopolis Dam (131 ft/40 m high), built (1948) by Army Corps of Engineers for flood control; also used for irrigation, recreation, and water supply. Kanopolis State Park is near the dam; Smoky Hill Air National Guard Range nearby.

Kanorado (kan-or-AID-o), village (2000 population 248), SHERMAN county, NW KANSAS, 17 mi/27 km W of GOODLAND, at COLORADO state line; 39°20′N 102°02′W. Shipping point in agricultural area.

Kanoroba (kan-o-RO-bah), town, NW CÔTE D'IVOIRE, 37 mi/60 km SE of BOUNDIALI; 09°07′N 06°08′W. Agriculture (corn, beans, tobacco, peanuts, cotton).

Kanose (KAH-no-se), town, East Kanbara county, NIIGATA prefecture, central HONSHU, N central JAPAN, 28 mi/45 km S of NIIGATA; 37°41′N 139°29′E.

Kanosh (ka-NAWSH), village (2006 population 481), MILLARD county, W UTAH, 15 mi/24 km SSW of FILLMORE; 38°47′N 112°26′W. Alfalfa, barley, wheat; cattle; manufacturing (apparel). SEVIER LAKE (intermittent) 30 mi/48 km W. Paiute Indian Chief Kanosh buried at Kanosh Cemetery. Fishlake National Forest to E and S, Kanosh Indian Reservation to NE. Elevation 5,125 ft/1,562 m.

Kanowna (kuh-NOU-nuh), ghost town, S central WESTERN AUSTRALIA, AUSTRALIA, 11 mi/18 km NNE of KALGOORLIE; 30°36′S 121°36′E. Former gold-mining town. The town's last hotel closed 1952. Originally called White Feather.

Kanoya (KAH-no-yah), city, KAGOSHIMA prefecture, S KYUSHU, SW JAPAN, on the OSUMI PENINSULA, 22 mi/35 km S of KAGOSHIMA; 31°23′N 130°51′E. Sweet potatoes; pigs, cattle.

Kanpetlet (kahn-PET-let), southernmost township in CHIN STATE, MYANMAR, 70 mi/113 km WSW of PAKOKKU.

Kanpur (kahn-puhr), city (2001 population 2,715,555), ⊙ KANPUR NAGAR (City) district, S UTTAR PRADESH state, N central INDIA, on the GANGA River, 115 mi/185 km NW of ALLAHABAD; 26°28′N 80°21′E. Transportation hub (railroad, road, airport); headquarters of several big industrial houses, a major Indian industrial center for trade (grains, oilseeds, sugarcane, cotton, salt, saltpeter), air and space technology; woolen, cotton, and jute milling, tanneries (chief tanning material is babul bark), boot and shoe manufacturing. Also includes light manufacturing, Seat of agriculture and law colleges, nurses' training center, Central Textile Institute, government experimental farm and a university. Recently added a high-level Institute of Technology; an older one existed long ago, for practical training. Named for a hero from the *Mahabharata* epic. When ceded to the British in 1801 by nawab of OUDH, Kanpur was a village; grew rapidly to its present commercial importance and to be the foremost industrial city of N India.

Suffered a period of decline during the Sepoy Mutiny of 1857. Formerly spelled Cawnpore.

Kanpur Branch, INDIA: see UPPER GANGA CANAL.

Kanpur Dehat, rural district (□ 1,973 sq mi/5,129.8 sq km), UTTAR PRADESH state, central INDIA; ⊙ KANPUR. On GANGA PLAIN and GANGA-YAMUNA DOAB; irrigated by Etawah branch of the LOWER GANGA CANAL and its distributaries. Agriculture (gram, wheat, jowar, barley, mustard, corn, sesame, rice, pearl millet, sugarcane, cotton); dhak jungle, mango and mahua groves. JUHI and GHATAMPUR are important centers. Also called just Kanpur district.

Kanpur Nagar, urban district (□ 411 sq mi/1,068.6 sq km), S UTTAR PRADESH state, central India; ⊙ KANPUR. On GANGA PLAIN and GANGA-YAMUNA DOAB. Distritct is coterminous with Kanpur city and its suburbs and is a major industrial center of the state and country. Also known as just Kanpur district.

Kanra (KAHN-rah), town, Kanra county, GUMMA prefecture, central HONSHU, N central JAPAN, 16 mi/25 km S of MAEBASHI; 36°14′N 138°55′E. Silk cocoons. Rock carving of Buddha image.

Kan River (KAHN), 318 mi/512 km long, S KRASNOYARSK TERRITORY, central SIBERIA, RUSSIA; rises in the EASTERN SAYAN MOUNTAINS; flows N, past IRBEYSKOYE and KANSK, and W to the YENISEY RIVER, 50 mi/80 km NE of KRASNOYARSK.

Kansai (kahn-SEI), town, N LENINOBOD region, TAJIKISTAN, on S slope of KURAMA RANGE, 17 mi/27 km NNE of KHUDJAND; 40°30′N 69°41′E. Lead-zinc mine and works. Also spelled Kansay.

Kansas, state (□ 82,282 sq mi/213,110 sq km; 2000 population 2,688,418; 2005 estimated population 2,565,328), central UNITED STATES, admitted to the Union in 1861 as the 34th state; ⊙ TOPEKA; 38°21′N 98°12′W. Major cities include Topeka, WICHITA (the largest city in the state) and KANSAS CITY. Kansas is known as the "Sunflower State" due to the wild flowers growing on the plains and the officially recognized state flower.

Geography
Almost rectangular in shape, Kansas is bounded on the N by NEBRASKA, on the E by MISSOURI (the MISSOURI River forms the NE boundary for a short distance), on the S by OKLAHOMA, and on the W by COLORADO. The geographical center of the U.S. (exclusive of ALASKA and HAWAII) is located in Kansas between SMITH CENTER and LEBANON. Mostly part of the GREAT PLAINS, Kansas is known for its massive wheat fields. The land rises gradually more than 3,000 ft/914 m from the E alluvial prairies of Kansas to its W semiarid high plains, which stretch toward the foothills of the ROCKY MOUNTAINS. The state is drained by the KANSAS and ARKANSAS rivers, both of which generally run from W to E.

Climate
The average annual rainfall of 27 in/69 cm is not evenly distributed: the E prairies receive up to 40 in/102 cm of rain, while the W plains average 17 in/43 cm. Occasional dust storms plague farmers and ranchers in the W. The climate is continental, with wide extremes—cold winters with blizzards and hot summers with tornadoes. Floods also wreak havoc in the state; hence, flood-control projects, such as dams, reservoirs, and levees, are a major undertaking.

Economy
Kansas was once primarily an agricultural state, but manufacturing and services have surpassed agriculture in economic importance. However, farming is still important to the state's economy, and Kansas is the nation's leading producer of wheat and one of the top producers of sorghum for grain. Corn and hay are also major crops. Cattle and calves are raised on the state's abundant grazing lands and constitute the single most valuable agricultural item. Meatpacking and dairy industries are major economic activities, and the Kansas City stockyards are among the nation's largest. The two leading industries are the

manufacture of transportation equipment and industrial machines. Wichita is a leader in the aircraft industry, especially in the production of private planes. Other important manufactured items are petroleum and coal products and nonelectrical machinery. The state is a major producer of crude petroleum and has large reserves of natural gas and helium. Kansas was once part of a great shallow sea, and salt deposits in commercially profitable quantities still remain.

History to 1803
When the Spanish explorer Francisco Vásquez de Coronado visited (1541) the Kansas area in his search for *quivira*, a fabled kingdom of riches, the area was occupied by various Native American groups of the Plains descent, notably the Kansa, the Wichita, and the Pawnee. In 1601, another Spanish explorer, Juan de Oñate, penetrated the region, resulting in the introduction of the horse, which revolutionized the life of the Native Americans. While not actually exploring the Kansas area, Robert Cavelier, sieur de La Salle, claimed (c.1682) for France all territory drained by the MISSISSIPPI RIVER, including Kansas. By the Treaty of PARIS (1763), ending the French and Indian Wars, France ceded the territory of W LOUISIANA (including Kansas) to Spain. In 1800, Spain secretly retroceded the territory to France, from whom the U.S. acquired it in the Louisiana Purchase in 1803.

History - 1803 to 1853
The region was little known, however, and subsequent explorations include the Lewis and Clark expedition (1803–1806), the Arkansas River journey of Zebulon M. Pike in 1806, and the scientific expedition of Stephen H. Long in 1819. Most of the territory that eventually became Kansas was in an area known as the "Great American Desert," considered unsuitable for U.S. settlement because of its apparent barrenness. In the 1830s the region was designated permanent Native American country, and N and E tribes were relocated there. Forts were constructed for frontier defense and for the protection of the growing trade along the SANTA FE TRAIL, which crossed Kansas. FORT LEAVENWORTH was established in 1827, FORT SCOTT in 1842, and FORT RILEY in 1853.

History - 1853 to 1856
Kansas, at this time mainly a region to be crossed on the way to CALIFORNIA and OREGON, was organized as a territory in 1854. Its settlement, however, was spurred not so much by natural westward expansion as by the determination of both proslavery and antislavery factions to achieve a majority population in the territory. The struggle between the factions was further complicated by conflict over the location of a transcontinental railroad, with proponents of a central route (rather than a S route) eager to resolve the slavery issue in the area and promote settlement. The Kansas-Nebraska Act (1854), an attempted compromise on the extension of slavery, repealed the Missouri Compromise and reopened the issue of extending slavery N of latitude 36°30′ by providing for squatter sovereignty in Kansas and Nebraska, allowing settlers of territories to decide the matter themselves. Meanwhile, the Emigrant Aid Company was organized in MASSACHUSETTS to foster antislavery immigration to Kansas, and proslavery interests in Missouri and throughout the South took counteraction. Towns were established by each faction: LAWRENCE and Topeka by the free-staters, and LEAVENWORTH and ATCHISON by the proslavery settlers. Soon all the problems attendant upon organizing a territory for statehood became subsidiary to the single issue of slavery. The first elections in 1854 and 1855 were won by the proslavery group; armed Missourians intimidated voters and election officials and stuffed the ballot boxes. Andrew H. Reeder was appointed the first territorial governor in 1854. The first territorial legislature ousted (1855) all free-state

members, secured the removal of Governor Reeder, moved the capital to LECOMPTON, and adopted proslavery statutes. In retaliation the abolitionists set up a rival government at Topeka in October, 1855. Violence soon came to the territory. The murder of a free-state man in November 1855, led to the so-called Wakarusa War, a bloodless series of encounters along the WAKARUSA RIVER. The intervention of the new governor, Wilson Shannon, kept proslavery men from attacking Lawrence. However, civil war ultimately turned the territory into "bleeding Kansas."

History - 1856 to 1860
On May 21, 1856, proslavery groups and armed Missourians known as "Border Ruffians" raided Lawrence. A few days later a band led by the abolitionist crusader John Brown murdered five proslavery men in the Pottawatomie massacre. Guerrilla warfare between free-state men called Jayhawkers and proslavery bands—both sides abetted by desperadoes and opportunists—terrorized the land. After a new governor, John W. Geary, persuaded a large group of "Border Ruffians" to return to Missouri, the violence subsided. The Lecompton legislature met in 1857 to make preparations for convening a constitutional convention. Governor Geary resigned after it became clear that free elections would not be held to approve a new constitution. Robert J. Walker was appointed governor, and a convention held at Lecompton drafted a constitution. Only that part of the resulting proslavery constitution dealing with slavery was submitted to the electorate, and the question was drafted to favor the proslavery group. Free-state men refused to participate in the election with the result that the constitution was overwhelmingly approved. Despite the dubious validity of the Lecompton constitution, President James Buchanan recommended (1858) that Congress accept it and approve statehood for the territory. Instead, Congress returned it for another territorial vote. The proslavery group boycotted the election, and the constitution was rejected. Lawrence became de facto capital of the troubled territory until after the Wyandotte constitution (framed in 1859 and totally forbidding slavery) was accepted by Congress.

History - 1860 to 1890
The Kansas conflict and the issue of statehood for the territory became a national issue and figured in the 1860 Republican party platform. Kansas became a state in 1861, with the capital at Topeka. Charles Robinson was the first governor and James H. Lane, an active free-stater during the 1850s, one of the U.S. senators. In the Civil War, Kansas fought with the North and suffered the highest rate of fatal casualties of any state in the Union. The Confederate William C. Quantrill and his guerrilla band burned Lawrence in 1863. With peace came the development of the prairie lands. The construction of railroads made cowtowns such as ABILENE and DODGE CITY, with their cowboys, saloons, and frontier marshals, the shipping point for large herds of cattle driven overland from Texas. The buffalo herds disappeared (some buffalo still roam in state parks and game preserves), and cattle took their place. Pioneer homesteaders, adjusting to life on the timberless prairie and living in sod houses, suffered privation. In 1874, Mennonite emigrants from Russia brought the Turkey Red variety of winter wheat to Kansas. This wheat was instrumental in making Kansas the Wheat State as winter wheat soon came to replace spring wheat. Corn, too, soon became a major cash crop. Agricultural production was periodically disrupted by national depressions and natural disasters. Repeated and prolonged droughts accompanied by dust storms, occasional grasshopper invasions, and floods caused severe economic dislocation. Mortgages weighed heavily on farmers, and discontent was expressed in farmer support of radical farm organizations and third-party movements, such as the Granger

Cross-references are shown in SMALL CAPITALS. The pronunciation guide is shown on page xix. The sources of population figures are shown on page xvii.

movement, Greenback party, and Populist party. Tax relief, better regulation of interest rates, and curbs on the power of the railroads were sought by these organizations.

History - 1890 to 1932

Twice in the 1890s, Populist-Democrats were elected to the governorship. As conditions improved, Kansas returned largely to its allegiance to the Republican party and gained a reputation as a conservative stronghold with a bent for moral reform, indicated in the state's strong support of prohibition; laws against the sale of liquor remained on the books in Kansas from 1880 to 1949. Over the years improved agricultural methods and machines increased crop yield. Irrigation proved practicable in some areas, and winter wheat and alfalfa were cultivated in dry regions. Wheat production greatly expanded during World War I, but the end of the war brought financial difficulties. During the 1920s and 1930s, Kansas was faced with labor unrest and the economic hardships of the depression. As part of the Dust Bowl, Kansas sustained serious land erosion during the long drought of the 1930s. Erosion led to the implementation of conservation and reclamation projects, particularly in the N and W parts of the state. In 1924 an effort of the Ku Klux Klan to gain political control was fought by William Allen White, editor of the Emporia *Gazette*, who supported many liberal causes.

History - 1932 to

Alfred M. Landon, elected governor in 1932, was one of the few Republican candidates in the country to win election in the midst of the sweeping Democratic victory that year. He was nominated as the Republican presidential candidate in 1936, but lost in a landslide to Franklin D. Roosevelt. During World War II agriculture thrived and industry expanded rapidly. The food-processing industry grew substantially, the cement industry enjoyed a major revival, and the aircraft industry boomed. After the war agricultural prosperity once again declined when the state was hit by a severe drought and grasshopper invasion in 1948. Prosperity returned briefly during the Korean War, but afterward farm surpluses and insufficient world markets combined to make the state's tremendous agricultural ability part of the national "farm problem." Kansas has become increasingly industrialized and urbanized, however, and industrial production has surpassed farm production in economic importance. Flood damage in the state, especially after a major flood in 1951, spurred the construction of dams (such as the Tuttle Creek, Milford, and Wilson dams) on major Kansas rivers, and their reservoirs have vastly increased water recreational facilities for Kansans. Since the 1970s, Kansas has become increasingly less rural. Accordingly, the economy has shifted its emphasis to finance and service industries located in and around major urban centers such as Wichita, Topeka, Lawrence, and Kansas City. Robert Dole, unsuccessful 1996 Republican presidential candidate, was a longtime senator for the state. Institutions of higher learning include the University of Kansas (Lawrence), Kansas State University (MANHATTAN), Wichita State University (Wichita), and Washburn University of Topeka (Topeka).

Historical Sites

Points of historical interest in Kansas include the boyhood home of President Dwight D. Eisenhower and the Eisenhower Library in Abilene; the home turned museum of Carry Nation (in MEDICINE LODGE), who became convinced of her divine appointment to destroy the saloons; and Fort Leavenworth (a large Federal penitentiary).

Government

Government in Kansas is based on the constitution of 1859, adopted just before Kansas attained statehood. An elected governor heads the executive branch and serves a term of four years. The current governor is Kathleen Sebelius. The legislature has a house of representatives and a senate, with the 125 members of the house elected for two-year terms and the forty members of the senate elected for four-year terms. Kansas is represented in the U.S. Congress by four representatives and two senators and has six electoral votes in presidential elections. Kansas has long been a Republican stronghold.

Kansas has 105 counties: ALLEN, ANDERSON, ATCHISON, BARBER, BARTON, BOURBON, BROWN, BUTLER, CHASE, CHAUTAUQUA, CHEROKEE, CHEYENNE, CLARK, CLAY, CLOUD, COFFEY, COMANCHE, COWLEY, CRAWFORD, DECATUR, DICKINSON, DONIPHAN, DOUGLAS, EDWARDS, ELK, ELLIS, ELLSWORTH, FINNEY, FORD, FRANKLIN, GEARY, GOVE, GRAHAM, GRANT, GRAY, GREELEY, GREENWOOD, HAMILTON, HARPER, HARVEY, HASKELL, HODGEMAN, JACKSON, JEFFERSON, JEWELL, JOHNSON, KEARNY, KINGMAN, KIOWA, LABETTE, LANE, LEAVENWORTH, LINCOLN, LINN, LOGAN, LYON, MCPHERSON, MARION, MARSHALL, MEADE, MIAMI, MITCHELL, MONTGOMERY, MORRIS, MORTON, NEMAHA, NEOSHO, NESS, NORTON, OSAGE, OSBORNE, OTTAWA, PAWNEE, PHILLIPS, POTTAWATOMIE, PRATT, RAWLINS, RENO, REPUBLIC, RICE, RILEY, ROOKS, RUSH, RUSSELL, SALINE, SCOTT, SEDGWICK, SEWARD, SHAWNEE, SHERIDAN, SHERMAN, SMITH, STAFFORD, STANTON, STEVENS, SUMNER, THOMAS, TREGO, WABAUNSEE, WALLACE, WASHINGTON, WICHITA, WILSON, WOODSON and WYANDOTTE.

Kansas, village (2000 population 842), EDGAR county, E ILLINOIS, 13 mi/21 km WSW of PARIS; 39°32′N 87°56′W. Livestock, grain.

Kansas, village (2006 population 722), DELAWARE county, NE OKLAHOMA, 20 mi/32 km NNE of Tahelquah; 36°12′N 94°47′W. Highway transportation center. Rocky Ford State Park to SW.

Kansas, river, 170 mi/274 km long, KANSAS; formed by the junction of the SMOKY HILL and REPUBLICAN rivers at JUNCTION CITY, in E central KANSAS; flows E past MANHATTAN, TOPEKA, LAWRENCE, and SHAWNEE to the MISSOURI River at KANSAS CITY, KANSAS at MISSOURI border. The system drains parts of KANSAS, NEBRASKA, and COLORADO. Heavy floods (especially in 1951, 1977, and 1993) on the KANSAS and its tributaries caused great damage to the surrounding farms and KANSAS CITY area. Although it's free-flowing, numerous dams and reservoirs have been built on its tributaries to prevent flooding. Locally referred to as Kaw River.

Kansas City, adjacent cities of the same name, one (2000 population 146,866), ⊙ WYANDOTTE county, NE KANSAS (incorporated 1859); the other (1990 population 435,146) in CLAY, JACKSON, and PLATTE counties, NW MISSOURI (incorporated 1850); 39°07′N 94°43′W. They are at the junction of the MISSOURI and KANSAS (or Kaw) rivers and together form a large commercial, industrial, and cultural center. They are a port of entry, the focus of many transportation lines, with markets for wheat, hay, poultry, and seed. Both cities have meat, dairy, and agriculture processing and packaging plants. Among the chief manufactures of the metropolitan area are auto bodies, chemicals, petroleum and paper products, machinery, and transportation equipment; also printing and publishing. During the 1970s and 1980s the outlying towns and cities that comprise Kansas City's suburban area developed their own industries, businesses, and corporate bases for various companies. As a result, the population of the two adjacent cities declined, and nearby suburban communities and housing developments grew. The area was the starting point of many Western expeditions including the SANTA FE and OREGON trails passed through here. Several historic settlements of the early 19th century (including Westport) have become full-fledged cities. Kansas City, Kansas, is the seat of two junior colleges, two theological seminaries, the University of Kansas Medical Center, and a state school for the blind (established 1868). It has an agricultural hall of fame, a Shawnee mission (1839), and several museums. A 19th-century Native American cemetery is being incorporated into a unique center city mall. Kansas City, Missouri, is the site of the noted Nelson Art Gallery, the Atkins Museum of Fine Arts, and the Country Club Plaza (finished in 1922; one of the first U.S. shopping malls). Among its educational institutions are the University of Missouri–Kansas City, Avila College, Park College, Rockhurst College, Kansas City Art Institution, a college of osteopathy and surgery, a conservatory of music, two community colleges, and a number of theological schools. The city has a philharmonic orchestra and several theaters. *The Kansas City Star* has a distinguished history; it was founded (1880) by William Rockhill Nelson and headed by him until 1915. The Kansas City Royals (baseball) and the Kansas City Chiefs (football) are the major sports teams. The city has long been noted for its music history, particularly for jazz and swing, popular here since the 1930s, when black musicians were attracted to the city and made their music nationally famous. Kansas City holds various jazz and blues festivals throughout the year, and a Jazz Museum opened in 1997. Richards-Gebaur Air Force Base lies to the S.

Kansenia (kahn-SEN-yah), village, KATANGA province, SE CONGO, on railroad, and 65 mi/105 km NW of LIKASI; 10°19′S 26°05′E. Elev. 4,783 ft/1,457 m. Agriculture and cattle-raising center. Also a resort. Benedictine mission.

Kansk (KAHNSK), city (2005 population 101,020), SE KRASNOYARSK TERRITORY, SE SIBERIA, RUSSIA, on the KAN′ RIVER (tributary of the YENISEY RIVER), on crossroads and the TRANS-SIBERIAN RAILROAD, 153 mi/246 km E of KRASNOYARSK; 56°12′N 95°42′E. Elevation 675 ft/205 m. Industrial center in lignite-mining area; cotton textile, woodworking, paper-making equipment manufacturing, leather tanning, construction materials, food processing (meat packing, flour mill, brewery, distillery); biochemical plant. Has a drama theater, heritage museum. Founded in 1628 as a small wooden fortress; in 1640, moved 27 mi/43 km downstream to its current location; made a city in 1782; railroad station added in 1899. Former tsarist exile settlement, Kansk developed rapidly during the late-19th century gold boom.

Kansongwe, CONGO: see LUISHIA.

Kansu, China: see GANSU.

Kant (KAHNT) [=sugar], city (1999 population 22,075), CHÜY region, KYRGYZSTAN, in CHU valley, on railroad, and 13 mi/21 km E of BISHKEK; 42°53′N 74°53′E. Beet-sugar refining center; meat- and butter-processing plants, cement works; candy manufacturing. Tertiary-level administrative center. Founded 1932.

Kantagi (kahn-TAH-gee), town, central SOUTH KAZAKHSTAN region, KAZAKHSTAN, in the KARATAU Mountains, 25 mi/40 km NE of TURKESTAN. Lead and zinc mines. Also spelled Khantagy, Qantagi.

Kantalai (KUHN-thuh-lai), town, EASTERN PROVINCE, SRI LANKA, 20 mi/32 km SW of TRINCOMALEE; 08°22′N 81°01′E. Rice and coconut-palm plantations, vegetable gardens; sugarcane; sugar factory. Just W is Kantalai Tank (4.5 mi/7.2 km long, 2 mi/3.2 km wide; built c.275 C.E. by King Maha Sena), a large irrigation lake.

Kantanagar (kahn-tah-naw-gor), village, DINAJPUR district, NW EAST BENGAL, BANGLADESH, on tributary of the MAHANANDA RIVER, and 12 mi/19 km N of DINAJPUR; 25°51′N 88°35′E. Agriculture (rice, jute, sugarcane). Annual fair. Noted 18th-century Vishnuite temple.

Kantang (GUHN-TUHNG), village and district center, TRANG province, S THAILAND, on W coast of MALAY PENINSULA, at mouth of Trang River, 15 mi/24 km

SSW of TRANG; 07°25′N 99°31′E. Small port on STRAIT OF MALACCA; railroad terminus of line from CHA MAI. Rice, rubber, pepper.

Kantara (KAHN-tah-rah) [Turkish=*sinandag*] , village, FAMAGUSTA district, NE CYPRUS, in KYRENIA MOUNTAINS, near base of KARPAS PENINSULA, 20 mi/32 km N of FAMAGUSTA; 35°23′N 33°54′E. MEDITERRANEAN SEA 3 mi/4.8 km to NW. Grapes, grain, fruit, tobacco, cotton; goats, sheep. A 13th century castle lies 2 mi/3.2 km to NE.

Kantara, El (kawng-tah-RAH, el), village, BISKRA wilaya, NE ALGERIA, guarding a beautiful gorge through the AURÈS massif, 25 mi/40 km N of BISKRA; 35°13′N 05°40′E. The El KANTARA defile was used by Romans (remains of bridge) and by caravans to TOUGGOURT, and is now used by BATNA-Biskra road and railroad.

Kantara, El, EGYPT: see Qantara, El.

Kantarawadi (kahn-tah-rah-wah-DEE) or **Kantarawaddy**, former KARENNI STATE (□ 3,161 sq mi/ 8,218.6 sq km) of MYANMAR; capital was LOIKAW. Hilly jungle (teak). Wet paddy. Population includes Karens, Burmese, Shans. Included former states of NAMMEKON and NAUNGPALE. Now part of Demoso township, KAYAH STATE.

Kantchari (kahn-CHAH-ree), town, TAPOA province, EST region, SE BURKINA FASO, on road, and 70 mi/113 km SSW of NIAMEY; 12°37′N 01°37′W. Agriculture (peanuts, shea nuts, beeswax); livestock (cattle, sheep, goats).

Kantemirovka (kahn-tye-MEE-ruhf-kah), town (2006 population 12,715), S VORONEZH oblast, S central European Russia, approximately 12 mi/19 km NE of the Ukrainian border, 38 mi/61 km SSE of ROSSOSH; 49°42′N 39°51′E. Elevation 518 ft/157 m. Railroad station and highway hub; local transshipment point. In agricultural area (wheat, rye, oats, flax).

Kanth, town, MORADABAD district, N central UTTAR PRADESH state, N central INDIA, near RAMGANGA RIVER, 17 mi/27 km NNW of MORADABAD; 29°04′N 78°38′E. Sugar refining, hand-loom cotton weaving; wheat, rice, pearl millet, sugarcane.

Kanth, POLAND: see KATY.

Kanthal, INDIA: see PARTABGARH.

Kanthi, INDIA: see CONTAI.

Kantilo (KUHN-ti-lo), village, NAYAGARH district, E ORISSA state, E central INDIA, on MAHANADI River, and 55 mi/89 km NW of PURI. Local market for sal timber, rice; handicraft brass ware. Historical site.

Kantishna, village, S central ALASKA, N of MOUNT MCKINLEY NATIONAL PARK, 100 mi/161 km SW of NENANA. Tourism; gold mining. Rich mineral region, with lead, zinc, copper, gold, silver, antimony deposits.

Kantishna River, S tributary of TANANA River, 200 mi/ 322 km long, central ALASKA; rises on N slope of Mount MCKINLEY; flows NE, past TOKLAT, to Tanana River at 64°45′N 149°58′W. Upper course called McKinley River. Receives Toklat River.

Kanto (KAHN-to), largest and most densely populated plain (□ 5,000 sq mi/13,000 sq km) of E central JAPAN, in E central HONSHU; contains most of Greater Tokyo and the Tokyo-Yokohama industrial belt. Includes all or parts of Tokyo, KANAGAWA, CHIBA, IBARAKI, SAITAMA, TOCHIGI, and GUMMA prefectures. Rural area produces rice, tea, dry grains, and tobacco. Formerly spelled Kwanto.

Kanton, coral atoll (□ 4 sq mi/10.4 sq km), central PACIFIC, largest of the PHOENIX ISLANDS, which comprise part of KIRIBATI, c.2,000 mi/3,000 km SE of HONOLULU, HAWAII; 02°50′S 171°41′W. Annexed by the British at the end of the 19th century, the island was also claimed by American guano companies. In 1937 the British built a radio station on Kanton, but in 1938 the U.S. formally claimed the island and placed it under the Department of the Interior. British and American colonists were brought to Kanton in 1938 but were evacuated during World War II. In 1939 both

GREAT BRITAIN and the U.S. agreed on joint control of Kanton and nearby ENDERBURY ISLAND. In 1979, Kanton (then Canton), along with most of the PHOENIX ISLANDS, became part of the independent Republic of Kiribati. Kanton is also known as Abariringa.

Kan-Too, KYRGYZSTAN: see KHAN-TENGRY PEAK.

Kantunil (kahn-TOO-neel), town, YUCATÁN, SE MEXICO, 10 mi/16 km S of IZAMAL; 20°48′N 89°02′W. Henequen, sugarcane, corn.

Kantunilkin, city, ⊙ Lázaro Cárdenas municipio, QUINTANA ROO, SE MEXICO, on E YUCATÁN Peninsula, near YUCATÁN border, 42 mi/67 km W of CANCÚN. Chicle, henequen, fruit. Archaeological remains nearby. Also called Kantunil-Kin.

Kanturk, Gaelic *Ceann Toirc*, town (2006 population 1,915), NW CORK county, SW IRELAND, 12 mi/19 km WNW of MALLOW; 52°10′N 08°54′W. Agricultural market. Has uncompleted Elizabethan castle of the MacCarthys, now national monument.

Kantyshevo (KAHN-ti-shi-vuh), village (2005 population 17,990), W INGUSH REPUBLIC, N central CAUCASUS, S European Russia, on highway, on the administrative border with NORTH OSSETIAN REPUBLIC, 14 mi/23 km N of VLADIKAVKAZ, and 4 mi/6 km NE of BESLAN; 43°14′N 44°38′E. Elevation 1,742 ft/530 m. Agriculture (grain, fruits, livestock); food processing. Chechen refugee camp in the vicinity. Formerly called Nartovo.

Kanukov, RUSSIA: see PRIVOLZHSKIY, ASTRAKHAN oblast.

Kanuku Mountains (kah-NOO-koo), mountain range, UPPER TAKUTU–UPPER ESSEQUIBO district, S GUYANA, spur of the GUIANA HIGHLANDS, extending c.80 mi/ 129 km E from BRAZIL border just N of 03°N. A sandstone plateau (rising to c.2,500 ft/762 m) within the RUPUNUNI SAVANNA, mostly covered by grassland, where cattle of the RUPUNUNI valley are raised. Deposits of uranium ore.

Kanuma (KAH-noo-mah), city, TOCHIGI prefecture, central HONSHU, N central JAPAN, 6 mi/10 km W of UTSUNOMIYA; 36°33′N 139°44′E. Industrial center manufacturing wooden housewares, furniture, and *konnyaku* (paste made from devil's tongue); strawberries. Dried gourd shavings; clay. Local attractions include hot springs, Sano Shrine, and the Chausuyama ancient tomb.

Kanungu, administrative district (□ 474 sq mi/1,232.4 sq km; 2005 population 216,400), WESTERN region, SW UGANDA; ⊙ Kanungu; 00°45′S 29°42′E. As of Uganda's division into eighty districts, borders RUKUNGIRI (E) and KABALE and KISORO (S) districts, as well as DEMOCRATIC REPUBLIC OF THE CONGO (W) and Lake EDWARD (N). Vegatation ranges from tropical forest to grasslands; some hilly areas, with Burimbi peak (8,212 ft/2,503 m) the highest; also some wetlands, especially near Lake Edward. Primarily agricultural area (staples include bananas, beans, millet, potatoes, rice, sorghum, and sweet potatoes, cash crops include tobacco and tea). Tea production for national and international markets. Timber-producing area. Formed in 2001 from W portion of former RUKUNGIRI district (current Rukungiri district formed from E portion).

Kanunka (kah-NUNG-kah), village, KATANGA province, SE CONGO, on railroad, and 15 mi/24 km NNW of LIKASI; 10°49′S 26°43′E. Elevation 4,606 ft/1,403 m. Iron mining. Nearby Kalabi (5 mi/8 km NE) has copper mines.

Kanuparti (kuh-noo-PUHR-tee), village, PRAKASAM district, ANDHRA PRADESH state, SE INDIA, on BUCKINGHAM CANAL, and 12 mi/19 km NE of ONGOLE. Salt refining.

Kanwa, INDIA: see KHANUA.

Kanyaboli, KENYA: see YALA.

Kanyankaw (KAN-yuhn-kor), town, WESTERN REGION, GHANA, 22 mi/35 km NW of TAKORADI; 05°03′N

02°02′W. Gold-mining center; cacao, palm oil and kernels. Also known as KAYIANKOR.

Kanyasso (kahn-YAH-so), town, Denguélé region, NW CÔTE D'IVOIRE, 21 mi/34 km N of ODIENNÉ; 09°49′N 07°31′W. Agriculture (sorghum, rice, millet, maize, beans, peanuts, tobacco, cotton).

Kanye (KAH-nyai), town (2001 population 40,628), ⊙ SOUTHERN DISTRICT, SE BOTSWANA, 17.5 mi/28.2 km SW of GABORONE; 25°00′S 25°18′E. Elevation 4,480 ft/1,366 m. A commercial and administrative center. Asbestos and diamonds mined nearby.

Kanyemba (kah-en-YAI-em-bah), village, MASHONALAND WEST province, N ZIMBABWE, 105 mi/169 km ENE of CHIRUNDU, on ZAMBEZI River (ZAMBIA border), opposite mouth of LUANGWA RIVER, near MOZAMBIQUE border (E); 15°39′S 30°24′E. Chewore Safari Area to SW, Mupata Gorge, on ZAMBEZI, 25 mi/40 km to W. FEIRA, ZAMBIA, 3 mi/4.8 km to N; ZUMBO, Mozambique, 4 mi/6.5 km to NE. KAPSUKU HILLS (3,522 ft/1,074 m) to W. Livestock; subsistence crops.

Kanyizha, SERBIA: see KANJIŽA.

Kanyutkwin (kahn-YOOT-KWIN), village, PYU township, BAGO division, MYANMAR, on Yangon-Mandalay railroad, and 40 mi/64 km S of TOUNGOO, in SITTANG VALLEY. Important rice area.

Kanzaki (KAHN-zah-kee), town, Kanzaki district, HYOGO prefecture, S HONSHU, W central JAPAN, 33 mi N of KOBE; 35°04′N 134°46′E. Joss sticks.

Kanzaki, town, Kanzaki county, SAGA prefecture, NW KYUSHU, SW JAPAN, 6 mi/10 km N of SAGA; 33°18′N 130°22′E.

Kao (kah-O), town, Kamoto county, KUMAMOTO prefecture, W KYUSHU, SW JAPAN, 12 mi/20 km N of KUMAMOTO; 32°58′N 130°40′E.

Kao (KAH-aw), now uninhabited island, W HA'APAI group, central TONGA, S PACIFIC OCEAN; 19°40′N 175°01′E. Elev. 3,380 ft/1030 m (highest point in the Tongan group). Extinct volcanic cone.

Kaoge (KOU-gai), town, KEBBI state, WNW NIGERIA, near NIGER River, and 95 mi/153 km S of BIRNIN KEBBI. Market town. Maize, millet.

Kaohsiung (GOU-SIUNG), city, S TAIWAN; second-largest city of Taiwan and its leading port, 246 mi/394 km S of TAIPEI; 22°40′N 120°17′E. Its designation as an export processing zone in the late 1960s has stimulated foreign investment. Leading manufacturing includes petroleum products, steel, cement, aluminum, wood and paper products, fertilizers, metals, machinery, and shipbuilding. International airport to SE. The city grew up from a small fishing village and was developed as a manufacturing center and port by the Japanese, who occupied Taiwan in 1895. Site of Sun Yatsen University An important naval base nearby. Formally called Takou by the Ami aborigines, then Takau by the Japanese. The Dutch called it Tarcoia. In 1920, the city received its present name. First settled by Chinese from the mainland provinces of Fukien (Fujian) and Kuangtung (Guangdong) during the reign of the Ming Dynasty's emperor Yung Lo (1403–1424). Also spelled Kaohsiung.

Kaokab Abu al-Hija (KAW-kahb ah-BUHL hi-JAH), Arab village, N of NAZARETH in LOWER GALILEE, Israel; 32°49′N 35°15′ E. Elevation 1,053 ft/320 m. Mixed farming. Jewish township in biblical and later times. Nearby is the burial site of Sheikh Abu al-Hija, a Muslim saint from the 17th century. Generally known as Kaokab.

Kaokab Abu al-Hija (KAW-kahb ah-BUHL hi-JAH), Arab village, N of NAZARETH in LOWER GALILEE, ISRAEL; 32°49′N 35°15′ E. Elevation 1,053 ft/320 m. Mixed farming. Jewish township in biblical and later times. Nearby is the burial site of Sheikh Abu al-Hija, a Muslim saint from the 17th century. Generally known as Kaokab.

Kaokoveld Mountains (kah-o-KO-velt), range in NW NAMIBIA, extends c.450 mi/724 km SE from ANGOLA border along E edge of NAMIB DESERT, parallel to the

Kaolack (KOU-lak), administrative region (□ 6,181 sq mi/16,070.6 sq km; 2004 population 1,114,292), central SENEGAL; ⊙ KAOLACK; 14°00′N 15°30′E. Bounded on N by FATICK, LOUGA, and MATAM administrative regions, E by TAMBACOUNDA administrative region, S by THE GAMBIA, W by Fatick administrative region. Major peanut-growing region. Formerly part of SINE-SALOUM province.

Kaolack (KOU-lak), city (2007 population 255,334), ⊙ KAOLACK administrative region, W SENEGAL, a port on the SALOUM RIVER; 14°09′N 16°04′W. Lying in a farming area, Kaolack is a major peanut marketing and exporting center; there is a large peanut oil factory in nearby Lyndiane. Leather tanning, cotton ginning, and fish processing are also important industries. Salt is produced from nearby salines. The city has one of the largest marketplaces in Senegal; there is also a significant Lebanese population. A major transportation crossroads. The city is the terminus of a branch line of the Dakar-Niger railroad. Center of the Niassene branch of the Sufi Muslim Tijaniyya brotherhood, whose mosque is on the city's outskirts. Airfield, hospital, high school.

Kaolan, China: see LANZHOU.

Kaolinovo (kah-o-LEEN-o-vo), city, VARNA oblast, obshtina center, NE BULGARIA, in the SW DO-BRUDZHANSKO plateau, 19 mi/31 km N of NOVI PAZAR; 43°37′N 27°06′E. Railroad terminus; porcelain manufacturing; kaolin factory; grain, tobacco; livestock. Kaolin quarried nearby (E). Formerly called Shumnubokhchalav (until 1950) and Bozhidar (1934–1950).

Kaoma, township, WESTERN province, W ZAMBIA, 120 mi/193 km ENE of MONGU, on LUENA RIVER. Junction on Lusaka-Mongu road. Cattle; corn. KAFUE NATIONAL PARK to E; airfield. Formerly called Mankoya.

Kaonde-Lunda, ZAMBIA: see NORTH-WESTERN.

Kaongeshi River (kah-on-GESH-shi), tributary of the KASAI RIVER, CONGO; 07°53′S 22°20′E. Also spelled Kaongweshi.

Kaoping River (GOU-PING), Chinese *Kaoping Hsi* (GOU-PING SHEE), 97 mi/156 km long, second-longest stream on TAIWAN; rises at E foot of YU SHAN (Jade Mountain); flows SSW to TAIWAN STRAIT at TUNGKANG. Lower course (navigable for 30 mi/48 km) is spanned by 5,000-ft/1,524-m bridge carrying PING-TUNG-KAOHSIUNG railroad. Extensive reclamation works in lower reaches. Formerly called Tashui River.

Kaosiung, TAIWAN: see Kaohsiung.

Kapaa (ka-PAH-ah), town (2000 population 9,472), E KAUAI, KAUAI county, HAWAII, 9 mi/14.5 km NNE of LIHUE, on E coast (Coconut Coast), on Kuhio Highway, S of mouth of Kapaa Stream; 22°06′N 159°20′W. Manufacturing (printing and publishing). Kapaa Beach Park is here; Nonou Forest Reserve to SW; Hoopii Falls to W (on Kapaa Stream).

Kapaau (KAH-pah-OU), village (2000 population 1,159), N HAWAII island, HAWAII county, HAWAII, at N end of KOHALA PENINSULA, 2 mi/3.2 km E of HAWI, 58 mi/93 km NW of HILO, 1 mi/1.6 km inland from N coast; 20°13′N 155°48′W. Sugarcane; cattle. King Kamehameha Statue.

Kapaau Halaula, town, N HAWAII island, HAWAII county, HAWAII, on KOHALA PENINSULA, 1 mi/1.6 km inland from N. coast. Kamehameha I born nearby; his statue stands in village. Also called Kohala.

Kapadvanj (KUHP-uhd-vuhnj), town, KHEDA district, GUJARAT state, W central INDIA, 31 mi/50 km E of AHMADABAD; 23°01′N 73°04′E. Railroad terminus; trade center (rice, millet, corn). Industries include cotton ginning, handicraft cloth weaving, oilseed milling; manufacturing (glass bangles, soap, and leather jars for ghee). Agate, onyx found in bed of small stream (N). Formerly spelled Kapadwanj.

Kapal (kah-PAHL), town, E TALDYKORGAN region, KAZAKHSTAN, in the DZUNGARIAN ALATAU Moun-

tains, 35 mi/56 km E of TALDYKORGAN; 45°08′N 79°03′E. Irrigated agriculture (wheat); sheep. Founded (1847) as Russian military post. Nearby (NE) are warm springs and resort of Arasan-Kapal or Arasan-Kopal. Also spelled Qapal.

Kapalaoa (kah-PAH-lah-O-ah), peak (3,310 ft/1,009 m), central KAUAI, HAWAII.

Kapalua (KAH-pah-LOO-ah), village (2000 population 467), MAUI island, MAUI county, HAWAII, 8 mi/12.9 km N of LAHAINA, on Oneloa Bay, NW coast, West Maui Peninsula; 21°00′N 156°39′W. Pineapples.

Kapan, city (2001 population 45,711), SE ARMENIA, on SE slopes of the ZANGEZUR RANGE, in valley of Vokhcha River, a tributary of the ARAS RIVER, 55 mi/89 km NE of NAKHICHEVAN; 39°12′N 46°24′E. Rises to 2,625 ft/800 m. Terminal station on a branch of the Baku-Yerevan railroad line. Center of copper-mining industry; furniture manufacturing; meat processing; cannery; milk processing.

Kapanga (kah-PHANG-gah), village, KATANGA province, S CONGO, on LULUA RIVER, and 165 mi/266 km WNW of KAMINA; 08°21′S 22°35′E. Elev. 2,893 ft/881 m. Cotton ginning. Roman Catholic and Protestant missions.

Kapapa (kah-PAH-pah), island, off E coast of OAHU island, HONOLULU county, HAWAII, 5 mi/8 km NW of Mokapu Peninsula, in KANEOHE BAY. Sea bird sanctuary.

Kapasan (kuh-PAH-suhn), town, CHITTAURGARH district, RAJASTHAN state, NW INDIA, 20 mi/32 km W of CHITTAURGARH. Cotton ginning; agriculture (millet, cotton, wheat).

Kapasi, INDIA: see KAPSI.

Kapasiwin, summer village (2001 population 15), central ALBERTA, W CANADA, 41 mi/65 km from ED-MONTON, in PARKLAND County; 53°33′N 114°27′W. Formed as a village in 1913, and as a summer village in 1993.

Kapchorwa, administrative district, EASTERN region, E UGANDA; ⊙ KAPCHORWA. As of Uganda's division into eighty districts, borders NAKAPIRIPIRIT (N), BUKWO (E), and SIRONKO (S and W) districts. Part of MT. ELGON (14,176 ft/4,321 m) in S. District affected by Mt. Elgon's rainfall zone. Agriculture (including cotton). Road connects district to Sironko and MBALE districts. Formed in 2005 from all but ESE portion of former KAPCHORWA district (Bukwo district formed from ESE portion).

Kapchorwa (kahp-CHOR-wuh), former administrative district (□ 671 sq mi/1,744.6 sq km; 2005 population 217,400), EASTERN region, E UGANDA, along KENYA border (to E); capital was KAPCHORWA; 01°20′N 34°35′E. As of Uganda's division into fifty-six districts, was bordered by MBALE (S tip), SIRONKO (W), and NAKAPIRIPIRIT (N) districts. MT. ELGON (14,176 ft/4,321 m) dominated S section. District affected by Mt. Elgon's rainfall zone. Agricultural region (beans, groundnuts, maize, millet, peas, potatoes, sunflowers, and yams; cash crops included coffee, cotton, and wheat; goats, cattle, and sheep), though land shortage and mountainous terrain caused difficulties. Some areas of district were not accessible in rainy season. In 2005 ESE portion of district was carved out to form BUKWO district and remainder of district was formed into current KAPCHORWA district.

Kapchorwa (kahp-CHOR-wuh), town (2002 population 8,747), ⊙ KAPCHORWA district, EASTERN region, E UGANDA, on road, and 35 mi/56 km NE of MBALE; 01°25′N 34°28′E. Cotton, corn. Was part of former EASTERN province.

Kapela, Velika, CROATIA: see VELIKA KAPELA.

Kapelle (kah-PE-luh), town, ZEELAND province, SW NETHERLANDS, in SOUTH BEVELAND region, and 3 mi/4.8 km ESE of GOES; 51°29′N 03°58′E. Between EASTERN SCHELDT (S) and WESTERN SCHELDT (S) estuaries; SOUTH BEVELAND CANAL passes to E. Dairying; cattle; agriculture (seeds, vegetables, fruit, sugar

beets); manufacturing (farm products; food processing).

Kapelle, NETHERLANDS: see CAPELLE.

Kapelle aan den IJssel, NETHERLANDS: see CAPELLE AAN DEN IJSSEL.

Kapellen (kah-PEL-luhn), French *Capellen-lez-Anvers* (awn-VER), commune (2006 population 26,072), Antwerp district, ANTWERPEN province, N BELGIUM, 7 mi/11.3 km N of ANTWERP; 51°19′N 04°26′E.

Kapelle-op-den-Bos (KAH-pel–op–duhn–BAHS), French *Capelle-au-Bois* (ka-pelle-o-BWA), commune (2006 population 8,847), Hal-Vilvoorde district, BRABANT province, central BELGIUM, near WILLEBROEK CANAL, 12 mi/19 km N of BRUSSELS; 51°01′N 04°22′E. Manufacturing.

Kapemba (kuh-PEM-buh), village, KATANGA province, SE CONGO, near Konde Lungu Mt.; 08°16′S 30°34′E.

Kapenguria (kah-pain-goo-REE-ah), town (1999 population 12,984), RIFT VALLEY province, W KENYA, near UGANDA boundary, on road, and 23 mi/37 km NNE of KITALE; 01°14′N 35°08′E. Sisal, coffee, tea, wheat, corn. Asbestos deposits.

Kapfenberg (KAHP-fen-berg), town, STYRIA, SE central AUSTRIA, on Mürz River, and 3 mi/4.8 km NNE of BRUCK AN DER MUR; 47°26′N 15°17′E. Refined steelworks; manufacturing of metals and tools.

Kaphereus, Cape, Greece: see KAFIREOS, CAPE.

Kaphereus Channel, Greece: see KAFIREOS CHANNEL.

Kapidagi Peninsula, (Turkish=*Kapidağı*) ancient *Cyzicus*, NW TURKEY, extends into Sea of MARMARA, 70 mi/113 km SW of ISTANBUL, between gulfs of Erdek (W) and Bandirma (E); 19 mi/31 km wide, 8 mi/12.9 km long. Rises to 2,640 ft/805 m. Sometimes spelled Kapu Dagh. Township of ERDEK on SW shore. In ancient times the notable city of CYZICUS was at its base.

Kapilavastu (kuh-PEE-luh-vuhs-too), district, S NEPAL; ⊙ TAULIHAWA. Also, the name of an ancient kingdom once ruled by King Suddhodana, father of the Buddha. According to legend, the Buddha was born in nearby LUMBINI c. 563 B.C.E. and passed his early years here.

Kapilmuni (ko-peel-moo-nee), village, KHULNA district, SW EAST BENGAL, BANGLADESH, on river arm of GANGA DELTA, and 18 mi/29 km WSW of KHULNA; 22°09′N 89°06′E. Agriculture (rice, jute, oilseeds). Large annual fair.

Kapingamarangi (kah-PING-ah-mahr-ang-ee), atoll, southernmost of CAROLINE ISLANDS, State of POHNPEI, Federated States of MICRONESIA, W PACIFIC; 6.5 mi/10.5 km long, 4.5 mi/7.2 km wide; 01°4′N 154°46′E. Coral islets (thirty-one) on reef; rises to 12 ft/4 m. Polynesian inhabitants.

Kapiri Mposhi, township, CENTRAL province, central ZAMBIA, 40 mi/64 km NNE of KABWE. Junction of railroad and road to NE Zambia and TANZANIA (the Great North Road, or "Hell Road"). Agriculture (tobacco, cotton, soybeans, corn); cattle, ostriches. Manufacturing (beer mugs, jars, bottles); clay. Lusemfwa (Mita Hills) Dam and Reservoir to SE, hydroelectric station.

Kapisa (kah-PEE-sah), province (□ 5,358 sq mi/13,930.8 sq km; 2005 population 367,400), AFGHANISTAN; ⊙ Mahmud Raqi; 34°58′N 69°17′E. Bounded by KABUL (S), LAGHMAN (E), PANJSHIR (N), and PARWAN (W) provinces. Major rivers are the Nirjab, PANJSHIR, and Tagab. Known for agriculture (mulberries and pomegranates). Major manufacturing at GULBAHAR (textile mills) and JABAL-US-SIRAJ (cement production). Created in 1964 formerly part of PARWAN province. It is named after the ancient town of Kapisa, located at the present BAGRAM, which is said to have been founded by Alexander the Great (4th century B.C.E.).

Kapisillit, fishing settlement, Nuuk (Godthåb) commune, SW GREENLAND, in GODTHÅBSFJORD, near its head, 50 mi/80 km ENE of NUUK (Godthåb); near 64°26′N 50°15′W.

Kapitan-Andreevo (ka-pee-THAN–ahn-DRAI-vo), village, HASKOVO oblast, SVILENGRAD obshtina, BULGARIA; 41°43′N 26°19′E. Borders TURKEY.

Kapitanivka (kah-pee-TAH-neef-kah) (Russian *Kapitanovka*), town (2004 population 4,130), NW KIROVOHRAD oblast, UKRAINE, 34 mi/55 km NW of KIROVOHRAD; 48°55′N 31°43′E. Elevation 639 ft/194 m. Sugar refinery. Vocational technical school. Established in 1774 as Zlatopil′ (Russian *Zlatopol′*); town since 1961, renamed Kapitanivka.

Kapitanivka (Russian *Kapitanovka*), town, N central KIROVOHRAD oblast, UKRAINE, on railroad, 10 mi/16 km NNE of NOVOMYRHOROD. Sugar refinery; vocational school. Established 1774, town since 1961.

Kapitanovka, UKRAINE: see KAPITANIVKA.

Kapitanovtsi (ka-pee-THAN-ov-tsee), village, MONTANA oblast, VIDIN obshtina, BULGARIA; 44°03′N 12°53′E.

Kapiti Island (KA-pi-tee), Kapiti Coast district, facing COOK STRAIT, NEW ZEALAND, 3.7 mi/6 km offshore, N of WELLINGTON; 10 mi/16 km long, 1.9 mi/3 km wide. Forest and bird reserve (1897); now also a Marine Reserve. Formerly called ENTRY ISLAND.

Kaplan (KA-pluhn), city (2000 population 5,177), VERMILION parish, S LOUISIANA, 8 mi/13 km W of ABBEVILLE; 30°00′N 92°17′W. Railroad terminus (from ABBEVILLE). Manufacturing (shirts, milled rice); oil field nearby. Holds Bastille Day Celebration. Incorporated 1902.

Kaplice (KAP-li-TSE), German KAPLITZ, town, JIHOCESKY province, S BOHEMIA, CZECH REPUBLIC, on MALSE RIVER, on railroad, and 9 mi/14.5 km SE of CESKY KRUMLOV; 48°44′N 14°29′E. Manufacturing (pottery, machinery); food processing. Has a 16th-century Gothic church.

Kaplitz, CZECH REPUBLIC: see KAPLICE.

Kapna (KAHP-nah), canton, BAUTISTA SAAVEDRA province, LA PAZ department, W BOLIVIA, N of Moco Moco; 15°12′S 68°57′W. Elevation 11,417 ft/3,480 m. Copper-bearing lode; clay, limestone, gypsum deposits. Agriculture (potatoes, yucca, bananas, rye).

Kapnikbánya, ROMANIA: see CAVNIC.

Kapoeas Mountains, Malaysia: see KAPUAS MOUNTAINS.

Kapoeas River, Malaysia: see KAPUAS.

Kapolei (KAH-po-LAI), town, S OAHU island, HONOLULU county, HAWAII, 11 mi/18 km WNW of HONOLULU, 3 mi/4.8 km inland from S coast, N of entrance to Barbers Point Naval Air Station. Industrial area (concrete products, chemicals, lumber, machinery, fabricated metal products; wood preserving, petroleum refining, beverage processing, printing). BARBERS POINT and Barbers Point Beach Park to SW.

Kápolna (KAH-pol-nah), Hungarian *Kápolna*, village, HEVES county, N HUNGARY, on TARNA river, 12 mi/19 km SW of EGER; 47°46′N 20°15′E. Orchards, vineyards. Hungarians defeated here (1849) by Austrians.

Kapong (GUH-PONG), village and district center, PHANGNGA province, S THAILAND, in MALAY PENINSULA, 16 mi/26 km NNW of PHANGNGA; 08°41′N 98°27′E. Tin mining.

Kapos River (KAH-posh), around 65 mi/105 km long, SW and central HUNGARY; rises 10 mi/16 km W of KAPOSVÁR; flows E, past Kaposvár and DOMBÓVÁR and N to SIÓ RIVER 6 mi/10 km W of SIMONTORNYA. Flow regulated (as Kapos Canal) between Dombóvár and Sió River.

Kaposvár (KAH-posh-vahr), city (2001 population 68,697) with county rank, SW HUNGARY, on the KAPOS RIVER; 46°22′N 17°48′E. It is a road and railroad junction. Manufacturing (textiles, apparel, leather, wood products; sugar refining, meat processing). Landmarks include an 18th-century church, a 19th-century town hall, and the ruins of an old castle. Nearby Tazar village (7 mi/11.3 km W) is site of first U.S. military base in Hungary (established 1996).

Kapotaksha, BANGLADESH: see KOBADAK RIVER.

Kapoutzedes, Greece: see PYLAIA.

Kappa, village (2000 population 170), WOODFORD county, central ILLINOIS, on MACKINAW RIVER, and 12 mi/19 km N of BLOOMINGTON; 40°40′N 89°00′W. Agricultural area.

Kappel am Albis, commune, ZÜRICH canton, N SWITZERLAND, 10 mi/16 km S of ZÜRICH. Zwingli fell in battle here (1531). Also known as Cappel.

Kappeln (KAHP-peln), town, SCHLESWIG-HOLSTEIN, N GERMANY, on the SCHLEI near its mouth on the BALTIC, 18 mi/29 km NE of SCHLESWIG, in the ANGELN; 54°40′N 09°55′E. Fishing and excursion port; food processing (smoked fish). Has baroque church. Was first mentioned 1357.

Kappelrodeck (kahp-pel-RO-dek), village, BADEN-WÜRTTEMBERG, SW GERMANY, on W slope of BLACK FOREST, 11 mi/18 km NE of OFFENBURG; 48°35′N 08°07′E. Noted for its red wine. Has 11th-century castle, renovated in 19th century.

Kappelshamn (KAHP-pels-HAHMN), village, GOTLAND county, SE SWEDEN, on N coast of Gotland island, 25 mi/40 km NE of VISBY; 57°51′N 18°46′E. Remains of prehistoric fortifications.

Kappl (KAHPL), village, TYROL, W AUSTRIA, in RHAETIAN ALPS, on TRISANNA RIVER, and 10 mi/16 km WSW of LANDECK; 47°04′N 10°22′E. Main village of the PAZNAUN valley; winter sports, cable cars to the Diasalpe (elevation 5,678 ft/1,731 m).

Kaprijke (KAHP-rei-kuh), agricultural commune (2006 population 6,147), Eeklo district, EAST FLANDERS province, NW BELGIUM, 3 mi/4.8 km NE of EEKLO; 51°13′N 03°36′E. Formerly spelled Caprycke.

Kapronca, CROATIA: see KOPRIVNICA.

Kaprun (kah-pruhn), village SALZBURG, W central AUSTRIA, on right affluent of the SALZACH river, in the PINZGAU, 6 mi/9.7 km SW of ZELL AM SEE; 47°16′N 12°45′E. Large hydroelectric works generated by the KESSEL FALLS (c.200 ft/61 m high); main installations are at LIMBERG DAM, other reservoirs are upstream at MOSERBODEN. A water tunnel through the GROSSGLOCKNER massif feeds the works with water from CARINTHIA (Margaritzen Reservoir). Winter sports center with summer skiing on the Kitzsteinhorn glacier (cable cars up to an elevation of 9,232 ft/2,814 m).

Kapsabet (kahp-SAH-bait), town (1999 population 17,918), RIFT VALLEY province, W KENYA, 30 mi/48 km NE of KISUMU; 00°12′N 05°35′E. Coffee, tea, wheat, corn; goldfields just W. Headquarters of district inhabited by Nandi people.

Kapsali, Greece: see KÍTHIRA.

Kapshagai (kahp-shah-GEI), city, S ALMATY region, KAZAKHSTAN, on TURKISH-SIBERIAN RAILROAD, 44 mi/70 km N of ALMATY, at W end of KAPSHAGAI RESERVOIR; 43°51′N 77°14′E. Shipyards; manufacturing (construction materials, lumber milling, food processing). Hydroelectric power. Formerly called Ili, Ilisk; also spelled Kapchagai, Kapchagay, Qapshaghai, Qapshaghay.

Kapshagai Reservoir (kahp-shah-GEI), SE ALMATY and S TALDYKORGAN regions, KAZAKHSTAN, on ILI RIVER. Formed (1970) by dam for irrigation and power generation. Also a recreation area. Kapshagai (city) on W shore.

Kapshticë (kahp-SHTEETS), village, SE ALBANIA, on Greek border, 11 mi/18 km E of KORÇË. Customs station on Korçë-KASTORIÁ (Greece) road; 40°36′N 21°01′E. Also spelled Kapeshtica.

Kapsi (KAHP-see), town, KOLHAPUR district, MAHARASHTRA state, central INDIA, 25 mi/40 km S of KOLHAPUR. Tobacco, chili, sugarcane. Also spelled Kapshi or Kapasi.

Kapsowar, town, ⊙ Marakwet district, RIFT VALLEY province, W central KENYA. Agriculture (coffee, maize, beans); livestock.

Kapsukas, LITHUANIA: see MARIJAMPOLĖ.

Kapsuku Hills, ZIMBABWE: see KANYEMBA.

Kapuas (KAH-poo-ahs), river, c.710 mi/1,143 km long, Kalimantan Barat province, INDONESIA; rises in the

mountains of central BORNEO; flows SW through W KALIMANTAN to the SOUTH CHINA SEA near PONTIANAK; 00°25′S 109°40′E. Its valley is intensively cultivated; rice is the chief crop. The river is navigable for c.560 mi/901 km of its length.

Kapuas Mountains (KAH-poo-ahs), central BORNEO Island, on boundary between SE SARAWAK and INDONESIAN BORNEO, c.50 mi/80 km SE of SIBU, Sarawak, E MALAYSIA; 01°15′N 113°30′E. Range extends c.120 mi/193 km generally E-W; highest peak is 5,797 ft/1,767 m. Also spelled KAPOEAS Mountains.

Kapu Dagh, TURKEY: see KAPIDAGI PENINSULA.

Kapunda (kuh-PUN-duh), town, SE SOUTH AUSTRALIA, 45 mi/72 km NNE of ADELAIDE; 34°21′S 138°54′E. Livestock; marble, flagstones. Formerly important copper-mining center.

Kapuni (kah-POO-nei), between MOUNT EGMONT and HAWERA, NORTH ISLAND, NEW ZEALAND. Natural gas, and some oil condensate. In 1994 was producing c.10% of New Zealand's local supply; some used at Kapuni urea plant.

Kapurthala (kuh-POOR-tuh-luh), district (□ 631 sq mi/1,640.6 sq km), PUNJAB state, N INDIA; ⊙ KAPURTHALA.

Kapurthala (kuh-POOR-tuh-luh), town, ⊙ KAPURTHALA district, PUNJAB state, N INDIA, 12 mi/19 km WNW of JALANDHAR; 31°23′N 75°23′E. Trades in wheat, gram, sugarcane, tobacco; manufacturing (pharmaceutical products, chemicals); fruit growing nearby. Has college. Was the capital of former princely state of Kapurthala of PUNJAB STATES, mainly along E bank of BEAS RIVER. In 1948, state merged with PATIALA AND EAST PUNJAB STATES UNION, later becoming part of Punjab state (after 1956).

Kapuskasing (ka-puhs-KAI-sing), town (□ 32 sq mi/83.2 sq km; 2001 population 9,238), central ONTARIO, E central CANADA, on the KAPUSKASING RIVER, N of TIMMINS; 49°25′N 82°26′W. Lumbering and pulp and paper mills; large tourism industry. Federal experimental farm nearby. The name is from the Aboriginal for "place where the river bends."

Kapuskasing River (ka-puhs-KAI-sing), 160 mi/257 km long, N ONTARIO, E central CANADA; rises near 48°N 83°05′W; flows NE, past KAPUSKASING, to MATTAGAMI River at 49°50′N 82°00′W.

Kapustin Yar (kah-POO-steen-YAHR), village (2005 population 6,470), N central ASTRAKHAN oblast, S European Russia, on the E bank of the VOLGA RIVER, near highway, 11 mi/18 km S of Znamensk; 48°24′N 45°42′E. Below sea level. Fisheries. Also known as Korochin.

Kapustin Yar-1, RUSSIA: see ZNAMENSK, ASTRAKHAN oblast.

Kapuvár (KAH-pu-vahr), Hungarian *Kapuvár*, city, SOPRON county, W HUNGARY, 21 mi/34 km ESE of SOPRON; 47°36′N 17°02′E. Agriculture (wheat, corn, alfalfa; cattle); manufacturing (transportation equipment).

Kapyong (GAH-PYOUNG), county (□ 326 sq mi/847.6 sq km), KYONGGI province, SOUTH KOREA, around middle reaches of Pukhan River, bordering KANGWON province on the E, and PYONGCH'ANG, Yangju, and YANGPYONG on the SW; 37°50′N 127°28′E. KWANGJU Mountains in N, Pukhan River in S. Small rivers flowing off the mountains form narrow basins all over the county. Over half of the land under cultivation is used for rice farming; other agriculture includes barley, beans, ginseng, honey, pine nuts, and jujubes. Important crossroad connecting SEOUL to CHUNCHON. Chongpyong and Daesongri resorts and Nam-i Island are famous for sightseeing.

Kaquena (kah-KAI-nah), canton, DANIEL CAMPOS province, POTOSÍ department, W central BOLIVIA. Elevation 12,054 ft/3,674 m. Lead-bearing lode, salt extraction; agriculture (potatoes, yucca, bananas; cattle).

Kara, region (2005 population 689,210), N central TOGO ⊙ LAMA-KARA; 09°40′N 00°55′E. Bordered by SAvanes region (N), benin (E), centrale region (S), and ghana (W). Part of Kéran National Park in NE, part of Fazao Malfakassa National Park in S. Regional airport at niamtougou. Kara region is composed of seven prefectures: assoli, bassar, binah, dankpen, doufelgou, kéran, and kozah. Also called Région de la Kara.

Kara, Togo: see lama–kara.

Kara, town, allahabad district, SE uttar pradesh state, N central india, on the ganga River, and 34 mi/ 55 km WNW of allahabad; 25°42′N 81°21′E. Gram, wheat, barley, pearl millet, oilseeds.

Kara (KAH-rah) or **Ust′-Kara** (OOST–KAH-rah), trading settlement, E nenets autonomous okrug, archangel oblast, NE European Russia, port on the Kara Bay (W inlet of the baydarata Bay), near the mouth of the kara River, 120 mi/193 km N of vorkuta; 69°14′N 64°55′E. Terminus of a local highway. Government observation station.

Kara (KAH-rah), river, approximately 140 mi/225 km long, NE European and NW Siberian Russia; flows N from the N ural Mountains into the baidarata bay (arm of the kara sea), forming part of the traditional border between European and Asian Russia. It is navigable in its lower course. Receives the Silova River (left). It forms the NW border of tyumen oblast (yamalo-nenets autonomous okrug) with komi republic and archangel oblast (nenets autonomous okrug).

Karaach, Bulgaria: see brestovene.

Kara Agach, Bulgaria: see levski.

Kara-alan, Bulgaria: see bryagovo.

Karababa Dagi, (Turkish=*Karababa Daği*) mountain (7,693 ft/2,345 m), central turkey, part of the Akdaglar range, 15 mi/24 km SE of akdagmadeni, overlooking the kizil irmak. Rises to 7,693 ft/2,345 m.

Karabakh Range (kahr-ah-BAHKH), branch of the Lesser caucasus, in SW azerbaijan; extends from armenia, c.70 mi/113 km SE to aras river; rises to 11,000 ft/3,353 m. Forms W border of nagorno-karabakh autonomous oblast. Karabakh Steppe lies E; Karabakh Upland is a plateau between karabakh and zangezur ranges.

Kara Balkan Mountains, Bulgaria: see chernatitsa mountains.

Kara-Balta (kuh-ruh-bahl-TAH), city (1999 population 47,159), chüy region, kyrgyzstan, 50 mi/80 km W of bishkek; 42°41′N 73°55′E. Vodka, wine, sugar, wool carpets and textiles; wheat. Uranium oxide and molybdenum processing. Also spelled Karabalty.

Karabanovo (kah-rah-BAH-nuh-vuh), city (2006 population 15,490), NW vladimir oblast, central European Russia, on railroad, 80 mi/129 km NW of vladimir, and 5 mi/8 km S, and under administrative jurisdiction, of aleksandrov; 56°19′N 38°42′E. Elevation 508 ft/154 m. Cotton textiles. Arose with the construction of a dye factory in 1846; became city in 1938.

Karabas (kah-rah-BAHS), town, karaganda region, kazakhstan, near abai, 21 mi/35 km SSW of karaganda; 49°38′N 73°01′E. Agricultural products; coal. Also spelled Qarabas.

Karabash (kah-rah-BAHSH), city (2005 population 15,530), N central chelyabinsk oblast, russia, on the NE slopes of the S urals, on road and railroad spur, 60 mi/97 km NW of chelyabinsk, and 20 mi/32 km SW of kyshtym; 55°29′N 60°14′E. Elevation 1,453 ft/ 442 m. Major copper center (mining and smelting) based on extensive copper deposits exploited since 1830; first copper works established in 1834. Zinc concentrating; pyrite mining; bakery. Largely developed in the 1920s; became city in 1933. Formerly called Karabashskiy Zavod.

Karabashskiy Zavod, russia: see karabash.

Karabekaul (kah-rah-bye-kuh-OOL), town, central lebap weloyat, E turkmenistan, on the amu darya (river), and 54 mi/90 km SE of charjew; 38°30′N 64°08′E. Tertiary-level adminstrative center. Cotton; vineyards. Also spelled Qorabekovul.

Kara-Bogaz-Gol, turkmenistan: see garabogazkol.

Karabudakhkent (kah-rah-boo-dahkh-KYENT), village (2005 population 13,525), E dagestan republic, NE caucasus, SE European Russia, russia, on the caspian sea coastal plain, on road, 18 mi/29 km SSE of makhachkala; 42°42′N 47°33′E. Elevation 856 ft/ 260 m. Vegetable growing. Experienced significant population growth since 1992 with the influx of Chechen refugees.

Karabük (kah-rah-BUK), city (2000 population 100,749), N turkey, 75 mi/121 km SE of zonguldak, 35 mi/56 km from the black sea; 41°12′N 32°36′E. Established 1939 as the first of Turkey's iron and steel plants. Zonguldak coal fields nearby; also iron ore from the divrigi mines (625 mi/1,006 km SE).

Karabulak (kah-rah-boo-LAHK), city, central taldy-korgan region, kazakhstan, on the karatal River, 8 mi/12.9 km SE of taldykorgan; 44°53′N 78°29′E. Tertiary-level (raion) administrative center. Beet-sugar refinery; metallurgy. Also spelled Qarabulaq.

Karabulak (kah-rah-boo-LAHK), town (2005 population 35,755), central ingush republic, N caucasus, S European Russia, in the sunzha river basin, on road and railroad, 7 mi/11 km NE of nazran′, to which it is administratively subordinate; 43°18′N 44°54′E. Elevation 1,312 ft/399 m. Agriculture (grain, fruits, livestock); food processing (canning). Chechen refugee camp in the vicinity.

Karabunar (kah-RAH-boo-nahr), village, plovdiv oblast, septemvri obshtina, W central bulgaria, 9 mi/15 km NW of pazardzhik; 42°16′N 24°10′E. Vineyards; agriculture (rice, vegetables).

Karabunar, Bulgaria: see sredets.

Karaburun (kah-rah-boo-ROON), village, W turkey, port on gulf of izmir, 40 mi/64 km of izmir, near the NE tip of the karaburun peninsula (60 mi/97 km long, 15 mi/24 km wide) extending into aegean sea between Gulf of Izmir and Chíos island; 38°37′N 26°31′E. Rises to 3,976 ft/1,212 m; cesme (town) on W shore, urla near its base. Mercury; tobacco, olives, raisins. Formerly called Ahirli.

Karaburun (kah-rah-boo-ROON), peninsula of SW albania, at Strait of otranto, on W side of Bay of vlorë; 10 mi/16 km long, 3 mi/4.8 km wide. Forms NW extremity of rrzeza e kanalit (for which it is sometimes named) and terminates in Cape (Kepi) Gjuhëzës; rises to 2,753 ft/839 m. Also spelled Karaburuni.

Karaburun Steppe (kuh-RAH-buh-ROON), mountain, W armenia, SW slope of mount aragats; traversed by Giumri-Yerevan railroad. Site of irrigation project (50 mi/80 km-long canal system from akhuryan S to oktemberyan area).

Karabutak (kah-rah-boo-TAHK), town, NE aktöbe region, kazakhstan, on irgiz river, 125 mi/201 km E of aktöbe (Aktyubinsk); 49°55′N 60°05′E. Tertiary-level (raion) administrative center. Millet; sawmilling. Also spelled Qarabutaq.

Karaca, turkey: see siran.

Karacabey, town (2000 population 40,624), NW turkey, on simav river, and 22 mi/35 km ESE of bandirma; 40°14′N 28°22′E. Agricultural center (wheat, barley, oats, rye, corn, beans, onions); iron nearby. Formerly called Mihalic or Mikhalitch.

Karacadag, (Turkish=*Karacadağ*) (kah-rah-jah-DAH), village, European turkey, at Bulgarian border, 28 mi/ 45 km NE of kirklareli; 41°57′N 27°40′E.

Karaca Dag, (Turkish=*Karaca Dağ*) mountain (5,655 ft/ 1,724 m), central turkey, 27 mi/43 km SE of hay-mana.

Karaca Dag, peak (6,430 ft/1,960 m), S central turkey, 12 mi/19 km ENE of karaman.

Karacali Dag, (Turkish=*Karacali Dağ*) peak (6,296 ft/ 1,919 m), SE turkey, 26 mi/42 km SW of diyarbakir, between the tigris and the euphrates. Part of the Karacali, a range extending N-S for over 30 mi/48 km.

Karacasu, village, W turkey, 26 mi/42 km W of denizli; 37°43′N 28°37′E. Figs, barley, cotton.

Karachala (kah-rah-chah-LAH), town, E azerbaijan, on SE shirvan steppe, on lower kura river, and 7 mi/11 km S of ali-bairamly. Cotton-ginning center for cotton district developed in late 1930s.

Karachayevsk (kah-rah-CHAH-eefsk), city (2005 population 22,380), karachevo-cherkess republic, N caucasus, S European Russia, in the kuban′ River valley, on the Sukhumi Military Road, 37 mi/60 km SW of cherkessk; 43°46′N 41°55′E. Elevation 3,559 ft/ 1,084 m. Manufacturing (instruments; food industries). Formed in 1926–1927. Called Mikoyan-Shakhar until 1944 and Klukhori, 1944–1957. Made city in 1929.

Karachev (kah-RAH-chyef), city (2005 population 19,620), NE bryansk oblast, central European Russia, 27 mi/43 km ESE of bryansk; 53°07′N 34°59′E. Elevation 672 ft/204 m. On railroad; major hemp-milling center. Manufacturing (transportation equipment; electrical equipment); food industries (dairy processing). One of the principal hemp-shipping points in Russia. Dates from 1146.

Karacheva Pustynya, ukraine: see sharhorod.

Karachevka, ukraine: see pokotylivka.

Karachevo-Cherkess Republic (kuh-RAH-chee-vuh-cheer-KYESS), constituent republic (□ 5,444 sq mi/ 14,154.4 sq km), 2004 population 436,000), SE European Russia, in the Greater caucasus Mountains, along the upper kuban′ River; ⊙ cherkessk. The oblast consists of lowland steppe in the N and the Caucasian foothills in the S. Agriculture (grains, fruits, vegetables; livestock); lead, zinc, copper, and gold mines. Industrial products include building materials, foodstuffs, and machinery. The population consists of Russians (40%), Karachay (31%), and Circassians (or Cherkess, as they are pronounced in Russian) and Abazins (16%), with the remainder made up of smaller minorities. The Karachay are Turkic-speaking Muslims who arrived in the region in the 14th century. In the 16th century they became vassals of Kabardinian princes, then passed (1733) to Turkish suzerainty, and in 1828 were conquered by the Russians. The region was included (1921) in the Mountain People's Republic, but in 1922 it became the Karachay-Cherkess autonomous region. In 1924, it was divided into the Karachay autonomous region and the Cherkess National Area; the latter became an autonomous region in 1928 (see circassia). In 1943, the Karachay, accused of collaborating with the Germans in World War II, were deported to Siberia and their autonomous region was abolished. However, the Karachay-Cherkess autonomous region was reestablished in 1957, when the "rehabilitation" of deported peoples was decreed. In 1990, the Supreme Soviet of Karachevo-Cherkess declared the region a full Soviet Socialist Republic. It was a signatory, under the name Republic of Karachayevo-Cherkisiya, to the March 31, 1992, treaty that created the russian federation. In 2004, consisted of 2 cities, 13 towns, and 59 villages.

Karachi, administrative area (□ 812 sq mi/2,111.2 sq km), SE pakistan, on arabian sea, almost entirely surrounded by sind province; ⊙ karachi. Created in 1948 out of British-era karachi district.

Karachi (kuh-rah-chee), city, ⊙ Sind province, largest city and former capital of Pakistan, SE Pakistan, on the Arabian Sea just NW of the Indus River delta, and 550 mi/885 km NW of Mumbai (Bombay), India; 24°52′N 67°03′E. Pakistan's chief seaport and shipping point for agr. produce of extensive irrigated areas in Sind and Punjab; industrial center, a transportation, commercial, and financial hub, and a military headquarters. Has a large motor vehicle assembly plant, an oil refinery, a steel mill, shipbuilding and railroad yards, jute and textile factories, printing and publishing plants, media and entertainment industries, food processing plants, and chemical and engineering works. International airport to N. An old settlement, Karachi was developed as a port and trading center by Hindu merchants in the early 18th century. In 1843 it passed to the British, who made it the seat of the Sind government Steady improvements in harbor facilities made Karachi a leading Indian port by the late 19th century, while agr. development of the hinterland gave it a large export trade. Karachi served as the capital Pakistan from 1947, when the country gained independence (1959), when Rawalpindi became the interim capital pending completion of Islamabad. Seat of a university and several other educational institutions, including the Aga Khan University Hosp.; the national museum, with a fine archaeological collection; and the tomb of Muhammad Ali Jinnah, founder of the Bhutto family, Karachi is troubled by political violence among local Sindhis, the descendants of *muhajirs* (Muslim immigrants who fled India after Partition), and Pashtun tribesmen from the frontier. Rapid population growth in periurban areas has far outpaced the supply of basic infrastructure and services; one remedy involves the self-help development scheme at the Orangi Pilot Project.

Karachi (kah-rah-CHEE), mud bath health resort, W NOVOSIBIRSK oblast, SW SIBERIA, RUSSIA, near the TRANS-SIBERIAN RAILROAD, 40 mi/64 km E of TATARSK, on Lake Karachi.

Karachi, PAKISTAN: see TATTA.

Karachivka, UKRAINE: see POKOTYLIVKA.

Karacurun, TURKEY: see HILVAN.

Karad, town, SATARA district, MAHARASHTRA state, W central INDIA, on KRISHNA RIVER, and 31 mi/50 km SSE of SATARA. Road junction; trade center (millet, wheat, peanuts). Cotton and silk milling, handicraft cloth weaving and dyeing; manufacturing (agriculture implements, cots, soap). Seat of College of Engineering.

Kara Dag, (Turkish=*Kara Dağ*) peak (7,451 ft/2,271 m), S central TURKEY, 16 mi/26 km NNW of KARAMAN.

Kara Dag, (Turkish=*Kara Dağ*) peak (11,910 ft/3,630 m), SE TURKEY, 7 mi/11.3 km N of HAKKARI.

Kara Dag, (Turkish=*Kara Dağ*) peak (9,940 ft/3,030 m), N central TURKEY, 15 mi/24 km W of ERZINCAN.

Kara Dagh, IRAN: see QARA DAGH.

Karadag Nature Preserve (kah-rah-DAHG) (Ukrainian *Karadaz'kyy zapovidnyk*) (Russian *Karadagskiy zapovednik*), on the SE coast of CRIMEA, UKRAINE, 15 mi/24 km SW of FEODOSIYA. Local natural monument (established in 1947), raised (1963) to republican status, and later became a preserve (1979). Covers an area of 7,200 acres/2,914 ha, of which 2,000 acres/809 ha consists of BLACK SEA nearshore marine zone. Studies of structure and dynamics of coastal ecosystems and conservation of mineral, floral, and faunal terrestrial and marine resources.

Karadagskiy zapovednik, UKRAINE: see KARADAG NATURE PRESERVE.

Kara Darya, river, c.170 mi/274 km, headstream of the SYR DARYA (river), in KYRGYZSTAN and UZBEKISTAN; rises in several branches in FERGANA RANGE (Kyrgyzstan), on CHINA border; flows NW, past UZGEN, into FERGANA VALLEY and E Uzbekistan, joining NARYN RIVER near BALYKCHI to form the Syr Darya. Lower course used for irrigation. Also spelled Kara Dar'ya.

Kara Darya, TAJIKISTAN: see ZERAVSHAN RIVER.

Kara Darya, UZBEKISTAN: see PAISHAMBA.

Karadaz'kyy zapovidnyk, UKRAINE: see KARADAG NATURE PRESERVE.

Karadeniz Bogazi, TURKEY: see BOSPORUS.

Karadzhitar, Bulgaria: see ISKRA.

Karadzica (kah-vah-DAHR-tsee), Serbo-Croatian *Karadžica*, mountain, N MACEDONIA; highest peak (8,111 ft/2,472 m) is 17 mi/27 km SSW of SKOPJE. Also spelled Karadzhitsa.

Kara, El, EGYPT: see QARA, EL.

Karafoca, TURKEY: see FOCA.

Karafuto, RUSSIA: see SAKHALIN, island.

Karaga (kuh-RAH-guh), town, NORTHERN region, GHANA, 20 mi/32 km W of GUSHIEGU; 09°55'N 00°27'W. Livestock; groundnuts, shea-nut butter.

Karaga (kah-rah-GAH), village, central KORYAK AUTONOMOUS OKRUG, RUSSIAN FAR EAST, on the NE coast of the KAMCHATKA PENINSULA, 100 mi/161 km E of PALANA; 59°06'N 163°07'E. Reindeer-raising area; fisheries.

Karagach (kah-rah-GAHCH), village (2005 population 5,930), N central KABARDINO-BALKAR REPUBLIC, N CAUCASUS, S European Russia, on road, 18 mi/29 km NNE of NAL'CHIK; 43°48'N 43°46'E. Elevation 898 ft/273 m. Highway junction. In agricultural area; produce processing.

Karagaily (kah-rah-gah-EE-lee), town, KARAGANDA region, KAZAKHSTAN, near KARKARALY, 112 mi/180 km ESE of KARAGANDA; 49°20'N 75°41'E. Light industry; textile manufacturing, wood processing. Agricultural products. Also spelled Qaragaily.

Karaganda (kah-rah-gahn-DAH), region (□ 43,200 sq mi/112,320 sq km), E central KAZAKHSTAN; ⊙ KARAGANDA. Bounded by the regions of AKMOLA (NW), PAVLODAR (N), SEMEY (E), TORGAI (W), and ZHEZKAZGAN (S). In KAZAK HILLS and partially on dry steppe; sharply continental climate. NURA RIVER flows roughly E-W. Karaganda coal basin is most extensively mined bituminous area in the country. Manufacturing includes metalworks; machinery, consumer goods, and building materials. Bisected N-S by Trans-Kazakhstan railroad. Some grains harvested near Karaganda. Pop. consists of Kazaks and Russians. Formed 1932 as Karaganda oblast of Kazakh ASSR; in 1973 most of its territory lost to the newly formed Dzhezkazgan oblast. Also spelled Qaraghandy.

Karaganda (kah-rah-GAHN-dah), city, ⊙ KARAGANDA region, in E central KAZAKHSTAN, on the Trans-Kazakhstan railroad; 49°55'N 73°20'E. The country's 2nd-largest city after ALMATY, it is a leading industrial center, with iron and steel plants, flour mills, food and beverage plants, and factories that produce mining equipment, building materials, machinery, and footwear. Cultural center of Kazakhstan, it has theaters, a university, and professional training institutes. Founded (1857) as a copper-mining settlement. The Karaganda coal basin, developed in the late 1920s, is a major producer of bituminous coal. The Novo-Karaganda power station is nearby. The Irtysh-Karaganda Canal supplies the city with water. Also spelled Qaraghandy.

Karagash (kah-rah-GAHSH), village, SE MOLDOVA, on left bank of DNIESTER (Nistru) River, and 5 mi/8 km SSE of TIRASPOL. Hq. of Karagash irrigation system (vegetable growing).

Karagay (kah-rah-GEI), village (2006 population 6,530), W PERM oblast, W URALS, E European Russia, on a right tributary of the KAMA River, on crossroads, 30 mi/48 km WNW of KRASNOKAMSK; 58°16'N 54°55'E. Elevation 524 ft/159 m. Food processing.

Karagaylinskiy (kah-rah-GEI-leen-skeeye), town (2005 population 4,290), W central KEMEROVO oblast, S central Siberian Russia, on highway and cargo spur of the TRANS-SIBERIAN RAILROAD (Karagaylinskaya station), 19 mi/31 km N of PROKOP'YEVSK; 54°10'N 86°33'E. Elevation 820 ft/249 m. Coal mining.

Karagel (kah-rah-GEL), town, BALKAN weloyat, W TURKMENISTAN, on Cheleken Peninsula of CASPIAN SEA, 3.1 mi/5 km E of CHELEKEN; 39°23'N 53°11'E.

Karaginskiy Island (kah-rah-GEEN-skeeye) (□ 775 sq mi/2,015 sq km), in the SW BERING SEA, in KORYAK AUTONOMOUS OKRUG, kamchatka oblast, extreme E SIBERIA, RUSSIAN FAR EAST; 69 mi/111 km long, up to 25 mi/40 km wide; 58°57'N 164°19'E. Separated by 29-mi/47-km-wide Litke Strait from the NE KAMCHATKA PENINSULA; rises to 3,140 ft/957 m. Fish-canning center on the W coast; fur farms.

Karagiye Sink, KAZAKHSTAN: see BATYR DEPRESSION.

Karagöl, peak (10,154 ft/3,095 m), NE TURKEY, 21 mi/34 km ENE of MESUDIYE.

Karagwe (kah-RAH-gwai), former tribal kingdom, East AFRICA, just W of Lake VICTORIA, mainly in KAGERA region, NW TANZANIA. Included lake port of BUKOBA and KARAGWE, tin mines at MURONGO, 80 mi/129 km W of lake along KAGERA RIVER (UGANDA-Tanzania and RWANDA-BURUNDI borders). Formed by alliance of Bachwezi and Basita tribes in 15th century; came to end (1879) with the Pax Germanica.

Karahisar, TURKEY: see AFYONKARAHISAR.

Karahisar, TURKEY: see SEBINKARAHISAR.

Karahissar, TURKEY: see SEBINKARAHISAR.

Karaidel' (kah-rah-ee-DYEL), village (2005 population 5,190), N BASHKORTOSTAN Republic, in the NW foothills of the S URALS, E European Russia, on the UFA RIVER (landing), on road, 65 mi/105 km SSW of KRASNOUFIMSK; 55°50'N 56°54'E. Elevation 377 ft/114 m. Distilling, lumbering.

Karaidel'skiy (kah-rah-ee-DYEL-skeeye), rural settlement (2005 population 202), N BASHKORTOSTAN Republic, E European Russia, in the NW foothills of the S URALS, on road, 6 mi/10 km E of KARAIDEL'; 55°49'N 57°05'E. Elevation 741 ft/225 m. In agricultural area.

Karaikal (kah-REI-kuhl), district (□ 62 sq mi/161.2 sq km), PONDICHERRY Union Territory, S INDIA; ⊙ KARAIKAL. Also spelled Karikal.

Karaikal (kah-REI-kuhl), town, ⊙ KARAIKAL district, PONDICHERRY Union Territory, S INDIA; 10°55'N 79°50'E. Also spelled Karikal.

Karaikkudi (kah-REI-kuh-dee), city PASUMPON MUTHURAMALINGA THEVAR district, TAMIL NADU state, S INDIA, 45 mi/72 km ENE of MADURAI; 10°04'N 78°47'E. Residential and financial center of Chettia merchant community; trades in foreign luxury goods. Electrochemical research institute.

Karaisali, village, S TURKEY, 24 mi/39 km NW of ADANA; 37°16'N 35°05'E. Wheat, barley, vetch, sesame, cotton. Formerly called Ceceli.

Karaisen (kah-REI-sen), village, LOVECH oblast, PAVLIKENI obshtina, N BULGARIA, 16 mi/26 km S of SVISHTOV; 43°22'N 25°22'E. Wheat, corn; livestock.

Karaitivu (KAH-RAI-thi-woo), island, in NORTHERN PROVINCE, N tip of SRI LANKA, between JAFFNA LAGOON (E) and PALK STRAIT (W); joined by causeway (N) to JAFFNA Peninsula, by ferry to VELANAI island (S); 09°44'N 79°52'E. Extensive rice, coconut, and palmyra-palm plantations. Lighthouse on NW shore. Hospital for infectious diseases at fort off SW coast.

Karaitivu, SRI LANKA: see KARATIVU.

Karaj (kah-RAHJ), city (2006 population 1,386,030), Tehrān province, N IRAN, on the Karaj River, c.25 mi/40 km WNW of TEHRAN; 35°49'N 51°00'E. It is an agricultural market and a transportation center. Chemicals are manufactured here.

Karajgaon, INDIA: see KARASGAON.

Karak, Al (KUH-rek, el), town (2004 population 20,280), W central JORDAN; 31°10'N 35°45'E. It is a road junction and an agricultural trade center. The ancient KIR MOAB (also mentioned in the Bible as Kir Hareseth, Kir Haresh, and Kir Heres), it was the walled citadel of the Moabites. Al Karak played an important role in the Crusades. The town was an archiepiscopal see from the early Christian era until the Christians were massacred or expelled in 1910. A 12th-century castle built by the Crusaders at Al Karak is well preserved. Also known as Krak.

Kara-Kala (kah-rah-kah-LAH), town, SE BALKAN weloyat, SW TURKMENISTAN, near IRAN border, on SUMBAR RIVER (right affluent of the ATREK), and 35 mi/56 km S of GYZYLARBAT; 38°34'N 56°44'E. Subtropical agriculture (guayule, date palms, figs). Also spelled Qoraqala.

Karakalli, TURKEY: see OZALP.

Cross-references are shown in SMALL CAPITALS. The pronunciation guide is shown on page xix. The sources of population figures are shown on page xvii.

Karakalpak Republic, republic (□ 61,000 sq mi/158,600 sq km), W UZBEKISTAN, on the AMU DARYA River, on the KAZAKHSTAN (N and W) and TURKMENISTAN (S) borders; ☉ NUKUS. Comprises parts of the USTYURT plateau, the KYZYL KUM desert, and the AMU DARYA delta on the ARAL SEA. It is the major central Asian producer of alfalfa; other crops are cotton, rice, corn, and jute. Livestock raising (notably cattle and karakul sheep) and silkworm breeding are widespread; numerous light industries. The population, concentrated in the delta, consists of Turkic-speaking Karakalpaks (31%), Uzbeks (31%), Kazaks (26%), Turkmens, Russians, and Tatars. The Karakalpak, known since the 16th century, when they lived along the lower and middle courses of the SYR DARYA River, were partly subjugated by the Kazaks. In the 18th century they migrated to their present homeland and in the 19th century came under the rule of the khanate of KHIVA. The khanate passed under Russian control at the end of the 19th century and under Bolshevik control by 1920. The Karakalpak autonomous region was formed in 1925 within the Kazakh autonomous republic. It became an autonomous republic itself in 1932 and was transferred to the Uzbek SSR in 1936. Karakalpak has a 185-member parliament. The economy and the environment in Karakalpak are deteriorating due to the evaporation of the Aral Sea and misuse of agricultural chemicals on the soil, as well as the effects of the virtual extinction of fishing economy. Also called Karakalpakstan, Qoraqalpoghiston.

Karakelong (kah-RAH-kuh-long), largest island (□ c.380 sq mi/984 sq km) of TALAUD ISLANDS, INDONESIA, 190 mi/306 km NE of MANADO, in North Sulawesi province; 48 mi/77 km long, 12 mi/19 km wide; 04°15'N 126°49'E. Rises to 2,264 ft/690 m. Timber, copra, sago, nutmeg; fishing. On E coast is Beo, chief town and port.

Kara-Kermen, UKRAINE: see OCHAKIV.

Karakilisse, TURKEY: see AĞRI.

Kara-Kirghiz Autonomous Oblast: see KYRGYZSTAN.

Karakishlak, UZBEKISTAN: see SARAI.

Karakocan (Turkish=*Karakoçan*), village (2000 population 23,994), E central TURKEY, 47 mi/76 km ENE of ELAZIG; 38°57'N 40°02'E. Grain, legumes, fruit. Formerly called Tepe or Tepekoy.

Karakol (kuh-ruh-KOL), city (1999 population 64,322), ☉ ISSYK-KOL region, KYRGYZSTAN, near SE shore of ISSYK-KOL lake, 190 mi/306 km E of BISHKEK, on highway from Balakchy; 42°30'N 78°23'E. In wheat area; wines and fruit juices; food processing (flour, sunflower oil), furniture; machine shops. PRISTAN PRZHEVALSK (port), 7 mi/11.3 km NW. Seat of university. Originally called KARA-KOL; renamed PRZHEVALSK (przhe-VAHLSK) for Russian explorer N. M. Przhevalski (died here, 1888); after 1920 again became Kara-Kol and, in 1939, Przhevalsk. After Kyrgyz became independent in 1991, named Karakol again. Founded 1869.

Kara-Köl (kuh-ruh-KOL), city (1999 population 17,977), JALAL-ABAD region, KYRGYZSTAN, located at confluence of NARYN and KARA-SUU rivers, at SW edge of TOKTOGUL reservoir; 41°27'N 72°43'E. Alcohol plant. Also spelled Kara-Kul.

Karakoram (KAH-ruh-KO-ruhm), mountain range, extending c.300 mi/483 km, between the INDUS and YARKANT rivers, N KASHMIR region, S central ASIA; SE extension of the HINDUKUSH mountains. Average elevation 19,685 ft/6,000 m. It covers disputed territory, held by CHINA (N), INDIA (E), and PAKISTAN (W). Karakoram's main range has some of the world's highest peaks, with sixty-one peaks over 22,966 ft/7,000 m, including K2 (GODWIN-AUSTEN; 28,250 ft/8,611 m), the second-tallest peak in the world. Also has several of the world's largest glaciers. Its S slopes are the watershed for many tributaries of the Indus. The valley bottoms receive little precipitation while high

snowfall is characteristic of mountains. Sparse vegetation is found on lower slopes but high elevation grasslands are used for summer grazing. The mountains, the greatest barrier between India and central Asia, are crossed above the perpetual snow line by two natural routes. KARAKORAM PASS (18,290 ft/5,575 m), the chief pass, is on the main LADAKH-China route. Another important pass, KHUNJERAB (15,420 ft/4,700 m), is on the Pakistan-China route and is crossed by the Karakoram Highway. Also spelled Karakorum.

Karakoram Highway, 800 mi/1,300 km all-weather motorway over the KARAKORAM Range from RAWALPINDI, PAKISTAN, to KASHI (KASHGAR), XINJIANG, CHINA. Follows valleys of the INDUS, GILGIT, and HUNZA rivers to 15,420 ft/4,700 m high KHUNJERAB PASS, then down into the TARIM Basin. Built as a joint project by the two countries and known in CHINA as the Great Pakistan–China Friendship Road. Also known to tourists as "the Silk Route" because it retraces part of a medieval trade route.

Karakoram Pass (18,290 ft/5,575 m), Aghil-Karakoram Range, NE JAMMU AND KASHMIR state, extreme N INDIA, 95 mi/153 km N of LEH, on India-CHINA border; important pass on main Ladakh (India)-Yarkand and Kashgar (China) trade route. Also known as Shaksgam Louisiana.

Karakorie, town (2007 population 7,842), AMHARA state, central ETHIOPIA; 10°25'N 39°56'E. It is 60 mi/97 km SE of DESSIE.

Karakoro River (kah-rah-KO-ro), c.180 mi/290 km long; rises in S MAURITANIA, near KIFFA; flows S, most of its course along Mauritania-MALI border, to SÉNÉGAL RIVER 45 mi/72 km WNW of KAYES (Mali). Supports forms of sedentary agriculture, fishing, and livestock.

Karakorum (KA-ruh-KO-ruhm), ruined city, central MONGOLIA, near the ORHON River, SW of ULAANBAATAR, near settlement of ERDENE DZUU; 47°14'N 102°50'E. The area around Karakorum had been inhabited by nomadic Turkic tribes from the 1st century C.E., but the city itself was not laid out until c.1220, when Jenghiz Khan, founder of the Mongol empire, established his residence here. As capital of the Mongols, Karakorum was visited (c.1247) by a papal mission under Giovanni Carpini. A Buddhist monastery was also built in that year. The city was abandoned (and later destroyed) after Kublai Khan, grandson of Jenghiz, transferred (1267) the Mongol capital to Khanbaliq (modern BEIJING). The noted Lamaist monastery of Erdene Dzuu was built near Karakorum in 1586. The ruins of the ancient Mongol city were discovered in 1889 by N. M. Yadrinstev, a Russian explorer, who also uncovered the Orhon Inscriptions (see ORHON GOL). Karakorum is also the name of a nearby site, which in the 8th and 9th century was the capital of the Uigurs.

Karakove, UKRAINE: see NOVOEKONOMICHNE.

Karakovo, UKRAINE: see NOVOEKONOMICHNE.

Karaksha (kah-rahk-SHAH), village, SW KIROV oblast, E central European Russia, on highway, 17 mi/27 km S of YARANSK, to which it is administratively subordinate; 57°02'N 47°56'E. Elevation 465 ft/141 m. In agricultural area (wheat, rye, flax, potatoes); logging and lumbering.

Karakubbud, UKRAINE: see KOMSOMOL'S'KE, Donets'k oblast.

Karakubstroy, UKRAINE: see KOMSOMOL'S'KE, Donets'k oblast.

Karakul (kahr-a-KOOL), city, SW BUKHARA wiloyat, UZBEKISTAN, on TRANS-CASPIAN RAILROAD, on ZERAVSHAN RIVER, and 35 mi/56 km SW of BUKHARA. Cotton-ginning center.

Kara-Kul, mountain lake (□ 140 sq mi/364 sq km), BADAKHSHAN AUTONOMOUS VILOYAT, E TAJIKISTAN, in the PAMIR, near the Chinese border; 39°02'N 73°33'E. It is c.12,840 ft/3,914 m above sea level, and its greatest depth is 780 ft/238 m.

Karakulino (kah-RAH-koo-lee-nuh), village (2002 population 5,109), SE UDMURT REPUBLIC, E European Russia, on the right bank of the KAMA River (landing), on road, 60 mi/97 km SSE of IZHEVSK; 56°00'N 53°42'E. Elevation 183 ft/55 m. Sheep raising; food processing (dairy).

Kara-Kulja (kuh-ruh-kool-JAH), village, E OSH region, KYRGYZSTAN, 17 mi/27 km SE of UZGEN; 40°26'N 73°38'E. Cotton. Tertiary-level administrative center. Also spelled Kara-Kuldzha.

Karakul'skiy, RUSSIA: see KARAKUL'SKOYE.

Karakul'skoye (kah-RAH-kool-skuh-yi), village, E CHELYABINSK oblast, SE URALS, RUSSIA, on the UY RIVER, on road, approximately 2 mi/3.2 km N of the KAZAKHSTAN border, and 32 mi/51 km E of TROITSK; 54°03'N 62°25'E. Elevation 396 ft/120 m. In agricultural area (oats, flax, feed corn). Formerly spelled Karakul'skiy.

Kara Kum (kah-rah-KOOM), two deserts, one in TURKMENISTAN and one in KAZAKHSTAN. The Caspian Kara Kum, the larger desert (□ 115,000 sq mi/299,000 sq km), is W of the AMU DARYA (river) and includes most of Turkmenistan. The MURGAB and Tejen rivers flow out of the HINDUKUSH MOUNTAINS to the S and empty into the desert, providing water for irrigation. The oases of MARY (Merv) and TEJEN are noted for cotton growing. The KARA KUM CANAL carries water from the Amu Darya at KELIF W across the desert to Mary and ultimately to ASHGABAT, a distance of c.500 mi/800 km. The canal water permits irrigated agriculture (especially cotton) and industry along the S margin of the desert. The TRANS-CASPIAN RAILROAD, a leading transportation artery of Central Asia, crosses the desert from TURKMENBASHI, on the CASPIAN SEA, to Ashgabat, Mary, BUKHARA (UZBEKISTAN), and TOSHKENT. Natural gas deposits have been discovered at DARVAZA and Mary. The Aral Kara Kum desert (W Kazakhstan; □ c.15,440 sq mi/40,000 sq km) lies NE of the ARAL SEA. Also spelled Qoraqum.

Kara Kum, China: see YULI.

Kara Kum Canal (kah-rah-KOOM), SE TURKMENISTAN, waterway project diverting water from the AMU DARYA (river) just N of KERKI W across the KARA KUM desert to the MURGAB and TEJEN oases. Also supplies water to ASHGABAT and other cities. Construction begun 1948.

Karakuni-dake, JAPAN: see KIRISHIMA-YAMA.

Karakuwa (kah-rah-KOO-wah), town, Motoyoshi county, MIYAGI prefecture, N HONSHU, NE JAPAN, 59 mi/95 km N of SENDAI; 38°54'N 141°39'E. Seaweed (*wakame*).

Karakuyu, village, W central TURKEY, 2 mi/3.2 km NE of DINAR; 37°20'N 27°41'E. Railroad junction.

Karal (kah-RAHL), village, HADJER-LAMIS administrative region, W CHAD, 60 mi/97 km NW of N'DJAMENA; 12°50'N 14°46'E.

Karalis, TURKEY: see BEYSEHIR, LAKE.

Karama (kahr-a-MAH), city, NAWOIY wiloyat, central UZBEKISTAN, on left bank of ZERAVSHAN RIVER; 40°24'N 67°34'E.

Karamai, China: see KARAMAY.

Karamakhi (kah-rah-MAH-hee), village (2005 population 4,000), central DAGESTAN REPUBLIC, NE CAUCASUS, SE European Russia, near highway, 21 mi/34 km SSW of MAKHACHKALA; 42°37'N 47°15'E. Elevation 3,188 ft/971 m. Agriculture (wheat, barley; meat livestock). Under unofficial control of Chechen separatists between 1997 and 1999, during which time it was turned into an armed enclave; seized and occupied by Russian troops shortly after the beginning of the second Russian-Chechen conflict.

Karaman, town (2000 population 105,384), S central TURKEY, at the N foot of the TAURUS MOUNTAINS, 60 mi/97 km SE of KONYA; 37°11'N 33°13'E. Important road junction. The ancient Laranda, Karaman was renamed after the chieftain of a Turkic tribe who conquered the city c.1250 and set up the independent Muslim state of Karamania (at one time comprising most of ASIA MINOR). A successor state of the Seljuk empire, Karamania existed until its final subjugation by the Ottoman Turks in the late 15th century. Karaman has retained ruins of the Karamanid castle and of two fine mosques.

Karamanovo (kah-rah-MAHN-o-vo), village, RUSE oblast, TSENOVO obshtina, N BULGARIA, 11 mi/18 km ESE of SVISHTOV; 43°33'N 25°32'E. Grain; vineyards; livestock.

Karamay (KUH-LAH-MAH-YEE), city (□ 3,668 sq mi/ 9,500 sq km; 1994 estimated urban population 212,600; 1994 estimated total population 226,900), N XINJIANG UYGUR, NW CHINA, in the Junggarian basin; 45°30'N 84°55'E. Since the discovery here of one of the largest oil fields in China (in 1955, at the foot of the Heiyou Mountains), the city has grown into a petrochemical and oil-producing and -refining center. In the early 1960s, it was the largest oil field in China, but it was later surpassed by the DAQING oil field in HEILONGJIANG province. Since the 1980s, many new oil fields have been discovered in an area surrounding the original Heiyou oil field. Also spelled Kelamayi or Karamai.

Kara-Mazar, mountain range, NE LENINOBOD viloyat, TAJIKISTAN. Part of KURAMA RANGE.

Kara-Mazar, TAJIKISTAN: see ADRASMAN.

Karambakkudi (kuh-RUHM-buh-kuhd-ee), town, PUDUKKOTTAI district, TAMIL NADU state, S INDIA, 23 mi/37 km ENE of PUDUKKOTTAI; 10°28'N 79°08'E. Rice, peanuts, millet. Also spelled Karambakudy.

Karamea Bight (KAR-uh-mee-ah), embayment of TASMAN SEA, BULLER district, SOUTH ISLAND, NEW ZEALAND, extending from CAPE FOULWIND in SW, c.62 mi/100 km to Heaphy Bluff to N. Receives BULLER RIVER mouth, with WESTPORT.

Karameh (ke-RAH-me) or **Karama**, township, NE JORDAN, SW of SALT. Commercial center of S GHOR region, E of the JORDAN River. Market gardening; fruit. Was a refugee camp until 1956.

Karamet-Niyaz (kah-rah-MET-nee-YAHZ), town, LEBAP weloyat, SE TURKMENISTAN, alongside KARA KUM CANAL, in the KELIF UZBOI depression, c.109 mi/ 175 km SE of CHARJEW; 37°45'N 64°34'E. Also spelled Qoramatniyoz.

Karamken (kah-rahm-KYEN), settlement (2006 population 675), S MAGADAN oblast, E RUSSIAN FAR EAST, in the SW KOLYMA RANGE, in the Khasyn River basin, on highway, 39 mi/63 km N of MAGADAN; 60°12'N 151°10'E. Elevation 2,837 ft/864 m. Fisheries. One of the largest gold mining and processing complexes in the region since its opening in 1978 and until the switch to the market-based economy following the breakup of the Soviet Union in 1991; now mostly abandoned.

Karamoja (kah-ruh-MO-juh), former province (□ 10,546 sq mi/27,419.6 sq km), NE UGANDA; capital was MOROTO. Was bordered E by KENYA, W by NORTHERN province, S by EASTERN province, N by SUDAN. Drained by OKOK and OKERE, KOTIPE, Lochemun, Dopeth, and Kapeta rivers. Consisted of MT. NAPAK, MT. KADAM, Mt. Morungoto, and KIDEPO NATIONAL PARK. Cash crops included coffee, cotton. Staple food crops included groundnuts, cassava, maize, sorghum. Industries included local crafts, pottery, weaving, woodworking. Main centers were Moroto, KOTIDO, KAABONG. Game reserves were MATHENIKO, BAKORA CORRIDOR, and PIAN-UPE.

Karamursel (Turkish=Karamürsel), town (2000 population 29,353), NW TURKEY, on Gulf of IZMIT, 17 mi/ 27 km WSW of IZMIT; 40°42'N 29°37'E. Grain, fruit, vegetables.

Kara Musal, Bulgaria: see VINOGRADETS.

Karamysh (kah-rah-MISH), river, 91 mi/147 km long, VOLGOGRAD and SARATOV oblasts, RUSSIA, in the VOLGA Upland, left tributary of the MEDVEDITSA River

(in the DON River basin). Basin area 1,305 sq mi/3,380 sq km. Used for irrigation.

Karamyshevo (kah-rah-MI-shi-vuh), village, central PSKOV oblast, W European Russia, near railroad, 17 mi/27 km ESE of PSKOV; 57°45'N 28°42'E. Elevation 180 ft/54 m. In agricultural area (flax, rye, potatoes, feed grass).

Karan', UKRAINE: see ANDRIYIVKA.

Karand, IRAN: see KEREND.

Karanda, village, MASHONALAND CENTRAL province, NE ZIMBABWE, 19 mi/31 km NE of MOUNT DARWIN; on RUYA RIVER; 16°40'S 31°51'E. Cattle, sheep, goats; corn.

Karandeniya, town, SOUTHERN PROVINCE, SRI LANKA; 06°16'N 80°04'E. Trades in tea, coconuts, rubber, cinnamon, and rice.

Karanfil Dag (Turkish=Karanfil Dağ), mountain peak (10,154 ft/3,095 m), S TURKEY, in the ALA DAG, part of the larger TAURUS range, 32 mi/51 km E of ULUKISLA.

Karangbolong Beach (kah-rahng-BO-long), resort area, Java Barat province, INDONESIA, 19 mi/30 km N of LABUAN; 06°38'S 105°38'E. Large natural archway and beach. Also spelled Karangbolong.

Karangboto, Cape, cape in INDIAN OCEAN, just off SW coast of Java Tengah province, INDONESIA; 07°47'S 109°24'E. Protects CILACAP harbor from open Indian Ocean.

Karanja (KAH-ruhn-juh), town, AKOLA district, MAHARASHTRA state, central INDIA, 36 mi/58 km ESE of AKOLA. Cotton ginning, oilseed milling.

Karanji (KUHR-uhn-jee), village, AHMADNAGAR district, S central MAHARASHTRA state, W central INDIA, 15 mi/24 km E of AHMADNAGAR. Cotton ginning, handicraft cloth weaving; millet, cotton.

Karanpur (kuh-ruhn-POOR), town, GANGANAGAR district, N RAJASTHAN state, NW INDIA, 125 mi/201 km N of BIKANER. Local market for millet, wheat, gram.

Karánsebes, ROMANIA: see CARANSEBEŞ.

Karanskiy Kamennyy Kar'er, UKRAINE: see MYRNE, Donets'k oblast.

Karans'kyy Kam'yanyy Kar'yer, UKRAINE: see MYRNE, Donets'k oblast.

Karantaba (kah-RAHN-tah-bah), village, CENTRAL RIVER division, E central THE GAMBIA, on N bank of GAMBIA RIVER (wharf), and 14 mi/23 km ENE of GEORGETOWN; 13°34'N 14°35'W. Peanuts, beeswax, hides, and skins. A BRITISH trading post established c.1780. Formerly called Pisania; also known as Karantaba Tabokoto.

Karany (KAH-rah-NEE), Czech Kárany, village, STREDOCESKY province, N central BOHEMIA, CZECH, at confluence of JIZERA and ELBE rivers, and 16 mi/26 km ENE of PRAGUE; 50°10'N 14°44'E. Supplies Prague with water (150 wells).

Kara Orman (KAH-rah or-MAHN), mountain (5,084 ft/1,550 m) in the PINDUS system, W MACEDONIA, E of the BLACK DRIN River, 7 mi/11.3 km NNE of STRUGA. Also spelled Karaorman.

Karapcha, Bulgaria: see MALOMIR.

Karapelit (kah-RAHP-el-eet), village, VARNA oblast, DOBRICH obshtina, BULGARIA; 43°38'N 27°35'E.

Karapinar, town (2000 population 35,285), S central TURKEY, 60 mi/97 km ESE of KONYA; 37°44'N 33°34'E. Agriculture (salt, figs, cereals). Formerly called Sultaniye.

Karapiro (KAR-ah-PI-ro), village, N central NORTH ISLAND, NEW ZEALAND, c. 30 mi/48 km upstream from HAMILTON, on the WAIKATO river. Site of a major hydroelectricity-generating dam, with an artificial lake used for recreational rowing events.

Kararim, Al (kuh-ruh-REEM, el), village, TRIPOLITANIA region, NW LIBYA, near Gulf of SURT, on highway, 17 mi/27 km S of MISRATAH; 32°24'N 15°04'E. Agricultural settlement (grain, olives, fodder, vegetables) in fertile, irrigated region. Founded by Italians (1938), who left after World War II, replaced by Libyan population Formerly called Gioda.

Karasburg (kah-ruhs-buhrg), town, S NAMIBIA, 110 mi/ 177 km SSE of KEETMANSHOOP, 160 mi/257 km WNW of UPINGTON. Distributing center for S part of country; on railroad and main road route to Upington, SOUTH AFRICA. In karakul-sheep-raising region. Education center.

Kara Sea, Russian Karskoye More, shallow section of the ARCTIC OCEAN, off N RUSSIA, between SEVERNAYA ZEMLYA and NOVAYA ZEMLYA. It has an average depth of 420 ft/128 m. It receives the OB RIVER, the YENISEI, the PYASINA, and the TAYMYRA rivers, and is important as a fishing ground. Its main ports are NOVY PORT and DIKSON, but the ice-locked sea is navigable only during August and September.

Karasgaon (KUHR-uhs-goun), town, AMRAVATI district, MAHARASHTRA state, central INDIA, 8 mi/12.9 km NE of ACHALPUR; 21°19'N 77°36'E. Cotton ginning; millet, wheat. Sometimes spelled Karajgaon.

Kara Shahr, China: see YANQI.

Karashina (kahr-a-shee-AH), village, SE KASHKADARYO wiloyat, S UZBEKISTAN, on main Karshi-Termez road. Tertiary-level administrative center.

Karash River, ROMANIA and SERBIA: see CARAŞ RIVER.

Karasi (kah-rah-SEE), village, SE TULA oblast, W central European Russia, on local railroad spur, 3 mi/5 km SE of (and administratively subordinate to) VOLOVO; 53°29'N 37°59'E. Elevation 731 ft/222 m. In agricultural area; groats plant.

Karasinskiy, RUSSIA: see BAGAN.

Karasjok (KAH-rah-shawk), Sami village, FINNMARK county, N NORWAY, near Finnish border, on Karasjokka River (tributary of TANA River), and 96 mi/154 km SE of HAMMERFEST. Livestock (reindeer). Sami museum here. First settled by Finns; church established 1807.

Kara-Soo, KYRGYZSTAN: see KARA-SUU.

Karasouli, Greece: see POLIKASTRO.

Karas Region (kuh-RAHS), administrative division, entire S region of NAMIBIA from S of GIBEON to ORANGE RIVER; 27°00'S 17°50'E. Largest political region with most sparse population; largely semi-desert with settlements near water. Major diamond area.

Karaş River, ROMANIA and SERBIA: see CARAŞ RIVER.

Kara Strait, RUSSIA: see KARSKIYE VOROTA.

Kara Su, name sometimes given for the W headstream of the EUPHRATES which combines with the MURAT RIVER in E central TURKEY. Several minor streams of Turkey are also called Kara Su. Also, Turkish names for STRUMA (Bulgaria and Greece) and MESTA (Bulgaria and Greece) rivers.

Karasu (KAH-rahs), town, Ichishi county, MIE prefecture, S HONSHU, central Japan, on W shore of Izu Bay, 5.6 mi/9 km S of TSU; 34°38'N 136°32'E.

Karasu, township, NW TURKEY, port on BLACK SEA at mouth of SAKARYA RIVER, 26 mi/42 km NNE of ADAPAZARI; 41°07'N 30°37'E. Grain, potatoes. Formerly called Incirli.

Kara-Su, Bulgaria and Greece: see MESTA RIVER.

Karasu, UZBEKISTAN: see KURGANTEPA.

Kara-Su, RUSSIA: see KARASUK RIVER.

Karasubazar, UKRAINE: see BILOHIRS'K.

Karasu-Bazar, UKRAINE: see BILOHIRS'K.

Karasufla, TURKEY: see HIZAN.

Karasuk (kah-rah-SOOK), city (2006 population 28,545), SW NOVOSIBIRSK oblast, SW SIBERIA, RUSSIA, at railroad junction and on crossroads, on the KARASUK RIVER, 420 mi/676 km W of NOVOSIBIRSK, and 55 mi/89 km NNW of SLAVGOROD; 53°44'N 78°02'E. Elevation 370 ft/112 m. Food industries (meatpacking, flour milling, canned milk). Made city in 1954.

Karasuk Lakes (kuh-ruh-SOOK), group of salt lakes on NW KULUNDA STEPPE, on the border of RUSSIA (ALTAI TERRITORY) and KAZAKHSTAN (PAVLODAR region); 53°34'N 77°24'E. KARASUK Lake or Kara-Su (□ 2 sq mi/5.2 sq km), in PAVLODAR region (Kazakhstan), is largest of the group and is 60 mi/97 km NW of SLAVGOROD (Russia).

Karasuk River (kuh-ruh-SOOK), 220 mi/354 km long, NOVOSIBIRSK oblast, RUSSIA; rises on the E BARABA STEPPE; flows SW, past KOCHKI, KRASNOZËRSKOYE, and KARASUK, to one of the KARASUK LAKES. Also known in the lower course as Kara-Su.

Karasu Mountains, E central TURKEY, extends 70 mi/113 km E from TERCAN and SW of ERZURUM, E and S of Upper (KARASU) EUPHRATES RIVER, crossed by TUZLA RIVER. Rises to 10,335 ft/3,150 m in SAKSAK DAGI.

Karasusufla, TURKEY: see HIZAN.

Kara-Suu (kuh-ruh-SOO), city (1999 population 19,143), NE OSH region, KYRGYZSTAN, in E FERGANA VALLEY, 12 mi/19 km NNE of OSH; 40°33′N 73°00′E. At junction (Kara-Suu–Uzbekski station) of railroads to Osh, KÖK-JANGGAK, and ANDIJAN (UZBEKISTAN). Tertiary-level administrative center. Cotton-ginning center; butter-processing plant; coal mining nearby. Also spelled Kara-Su, Karasu.

Karasuyama (kah-RAH-soo-YAH-mah), town, Nasu county, TOCHIGI prefecture, central HONSHU, N central JAPAN, 16 mi/25 km N of UTSUNOMIYA; 36°39′N 140°09′E.

Karata (kah-rah-TAH), village (2005 population 4,955), W DAGESTAN REPUBLIC, N CAUCASUS, SE European Russia, on a right branch of the ANDI KOISU, on road, 8 mi/13 km SE of BOTLIKH; 42°35′N 46°20′E. Elevation 3,907 ft/1,190 m. Woolen milling; sheep raising. District inhabited by the Akhvakh, one of the mountain tribes of Dagestan. The village's population increased significantly with the influx of refugees from neighboring Chechnya.

Karata, ETHIOPIA: see WERETA.

Karatal (kah-rah-TAHL), river, 220 mi/354 km long, in TALDYKORGAN region, KAZAKHSTAN; rises in the DZUNGARIAN ALATAU Mountains; flows NNW past TALDYKORGAN, through SARYESIK-ATYRAU (desert), to S Lake BALKASH. Upper course irrigates sugar-beet and rice area; salt lakes and fisheries near mouth. Also spelled Qaratal.

Karatas, Cape (Turkish=*Karataş*), on SE MEDITERRANEAN coast of ASIA MINOR, TURKEY, at W entrance to GULF OF ISKENDERUN; 36°33′N 35°19′E.

Karatau (kah-rah-TOU), city, ZHAMBYL region, KAZAKHSTAN, in foothills of KARATAU range, and on reservoir of Tamdy River, 42 mi/70 km NW of ZHAMBYL; 43°09′N 70°28′E. Phosphorite mining. Also spelled Qaratau.

Karatau (kah-rah-TOU), mountain range, SOUTH KAZAKHSTAN region, KAZAKHSTAN, W branch of TIANSHAN mountain system; extends 210 mi/338 km NW from TURK-SIB RAILROAD, parallel to the SYR DARYA river. Rises to 6,035 ft/1,839 m. Metamorphic schists. Cotton grown on irrigated slopes. Coal and manganese deposits. Large lead-zinc mines at KANTAGI and ASHISAI; extensive phosphorite deposits at CHULAKTAU. Also spelled Qaratau.

Karategin Range (kah-rah-te-GEEN), S Central TAJIKISTAN, oriented SW to NE beginning near DUSHANBE and stretching 60 mi/97 km NE to NAVABAD; rises to c.12,000 ft/3,658 m. The Karategin principality was a division of former Bukhara Khanate.

Kara Tepe, AFGHANISTAN: see TORGHONDI.

Karatina (kah-rah-TEE-nah), town (1999 population 6,852), CENTRAL province, S central KENYA, on railroad, and 12 mi/19 km ESE of NYERI; 00°29′S 08°37′E. Dried vegetables; livestock.

Karativu (KAH-RAI-thi-woo), two low-lying islands, NORTH WESTERN PROVINCE, off W coast of SRI LANKA, separated from mainland by PORTUGAL BAY; 09°35′N 80°11′E. Larger island (N; 8 mi/12.9 km long, 1 mi/1.6 km wide) has fishing camps; smaller island (S; 2 mi/3.2 km long). Also spelled KARAITIVU and Kara Tivu.

Karatöbe (kah-rah-too-BE), village, E WEST KAZAKHSTAN region, KAZAKHSTAN, 135 mi/217 km SE of oral (Uralsk); 49°44′N 53°30′E. Tertiary-level (raion) administrative center. Cattle. Also Kara-Tyube, Qaratöbe.

Karatog, village, W TAJIKISTAN, on S slope of GISSAR RANGE, 22 mi/35 km W of DUSHANBE, in wheat area; 38°37′N 68°20′E. Coal and phosphorite deposits. Also spelled Kara-tag, Qarotog.

Karaton (kah-rah-TON), oil town, SE ATYRAU region, KAZAKHSTAN, near CASPIAN SEA, 31 mi/50 km SSW of KOSCHAGYL, in EMBA oil field; 46°26′N 53°32′E. Also spelled Qaraton.

Karatorgai River, KAZAKHSTAN: see TORGAI RIVER.

Karatoya River, c.150 mi/241 km long, BANGLADESH; formed 43 mi/69 km NW of RANGPUR by junction of headstreams; flows SSE, past BOGRA, SHERPUR, and ULAPARA, to ATRAI RIVER 3 mi/4.8 km NNW of BERA. Another Karatoya River is the upper course of the ATRAI RIVER.

Karatsu (KAH-rahts), city, SAGA prefecture, NW KYUSHU, SW JAPAN, on Karatsu Bay, 25 mi/40 km N of SAGA; 33°26′N 129°58′E. Summer resort and fishing port important historically as Japan's ancient communications point with KOREA. Pottery.

Karatuzskoye (kah-rah-TOOS-skuh-ye), village (2005 population 7,860), S KRASNOYARSK TERRITORY, SE SIBERIA, RUSSIA, on the Amyl River (tributary of YENISEY River), on crossroads, 45 mi/72 km E of MINUSINSK; 53°36′N 92°52′E. Elevation 1,010 ft/307 m. Local transshipment point for agricultural products.

Kara-Tyube, KAZAKHSTAN: see KARATÖBE.

Karau (kah-RAH-yoo), village, EAST SEPIK province, NE NEW GUINEA island, NW PAPUA NEW GUINEA, 50 mi/80 km ESE of WEWAK. Located on PACIFIC OCEAN, 15 mi/24 km W of mouth of SEPIK RIVER. Boat access. Bananas, coconuts, palm oil; prawns, trepang, tuna.

Karaul (kah-rah-OOL), village (2006 population 805), W TAYMYR (DOLGAN-NENETS) AUTONOMOUS OKRUG, KRASNOYARSK TERRITORY, NE SIBERIA, RUSSIA, N of the ARCTIC CIRCLE, on the YENISEY RIVER, 80 mi/129 km NW of DUDINKA; 70°04′N 83°11′E. Fisheries.

Karaul (kahr-ah-OOL), village, KHORAZM wiloyat, NW UZBEKISTAN. Tertiary-level administrative center.

Karauli (kuh-ROU-lee), city, SAWAI MADHOPUR district, RAJASTHAN state, NW India, 80 mi/129 km ESE of JAIPUR; 26°30′N 77°01′E. Agricultural market (millet, gram, wheat, cotton); hand-loom weaving (carpets), biri manufacturing. Founded mid-14th century. Was capital of former princely state of Karauli in RAJPUTANA STATES; state joined union of Rajasthan (1949).

Kara-Unkurt, KYRGYZSTAN: see KAZAN-KUYGAN.

Karauzyak (kah-rah-ooz-YAHK), town, N KARAKALPAK REPUBLIC, W UZBEKISTAN, in the AMU DARYA delta, 60 mi/97 km NNW of NUKUS, near CHIMBAI; 43°02′N 60°01′E. Cotton, rice.

Karavan (kuh-ruh-VUHN), village, OSH region, KYRGYZSTAN, 5 mi/8 km E of KYZYL-KYYA; 40°08′N 72°13′E. Wheat.

Karavan, KYRGYZSTAN: see KERBEN.

Karavansarai, ARMENIA: see IJEVAN.

Karavas (kah-rah-VAHS) [Turkish=Alsançak], town, KYRENIA district, N CYPRUS, 1 mi/1.6 km S of MEDITERRANEAN coast, 15 mi/24 km NW of NICOSIA; 35°20′N 33°12′E. KYRENIA MOUNTAINS to S. Citrus, almonds, carobs, olives; goats, sheep. Akhiropiitos Monastery to N.

Karavassara, Greece: see AMFILOKHIA.

Karavasta Lagoon (kah-rah-VAH-stah), Albanian *Lag e Karavastase*, coastal lagoon in W central ALBANIA, on the ADRIATIC SEA, between DURRËS and VLORË; 9 mi/14.5 km long, 3 mi/4.8 km wide. Adjacent to delta of SEMAN RIVER (S).

Karavayevo (kah-rah-VAH-ee-vuh), town (2005 population 7,410), SW KOSTROMA oblast, central European Russia, on the VOLGA RIVER, on road, 5 mi/8 km ESE of (and administratively subordinate to) KOSTROMA; 57°43′N 41°04′E. Elevation 374 ft/113 m. Oil pipeline service station. Has an agricultural academy.

Karavostasi (kah-rah-vo-STAH-see) [Turkish= Gemikonaği], town, LEFKOSIA district, NW CYPRUS,

on MORPHOU BAY, 30 mi/48 km W of NICOSIA; 35°08′N 32°49′E. Mouth of Setrakhos River to E; Attila Line passes to S. Citrus, almonds, olives, vegetables; goats, sheep; fish. Ruins of ancient city of Soli 1 mi/1.6 km to W.

Karawa (kah-RAH-wah), village, Équateur province, NW CONGO, 100 mi/161 km NW of LISALA; 03°20′N 20°18′E. Elev. 1,489 ft/453 m. Cotton ginning. Protestant mission.

Karawanken (kAH-rah-VAHNG-ken), Slovenian *Karavanke*, Italian *Caravanche*, mountain range, Southern Limestone ALPS between GAILITZ RIVER (W) and BLEIBURG (E) along the Austro (Carinthian)–SLOVENIAN border, extending c.50 mi/80 km, forming continuation of CARNIC Alps; 46°25′–46°35′N 13°39′–14°50′E. A limestone formation, except for schist outcroppings in W, becoming steeper in E. Rises up (6,821 ft/2,079 m) to HOCHSTUHL, (Slovenian *Stol*), 15 mi/24 km SW of KLAGENFURT. The valleys of its Austrian part are populated by the Carinthian Slovene minority. The Karawanken Motorway, the Karawanken railroad, and the road across LOIBL PASS tunnel the range; roads across WURZEN PASS (elevation 3,270 ft/997 m) and SEEBERG (elevation 3,712 ft/1,131 m).

Karayashnyk, UKRAINE: see PETRIVKA.

Karayazi, village, NE TURKEY, 33 mi/53 km ESE of ERZURUM; 39°41′N 42°09′E. Grain. Formerly known as Cullu, Bayrakdar.

Karayazy Steppe (kah-rah-yah-ZEE), in E GEORGIA and W AZERBAIJAN, between IORI and KURA rivers, SE of TBILISI. Dry salt steppe, partially irrigated for wheat and cotton growing; average elevation, 1,000 ft/305 m. Winter pastures.

Karazhal (kah-rah-ZHAHL), city, ZHEZKAZGAN region, KAZAKHSTAN, by SARYSU River, 144 mi/240 km E of ZHEZKAZGAN; 48°00′N 70°55′E. Iron and manganese extraction. Also spelled Qarazhal.

Karb, POLAND: see BOBREK.

Karbala (kahr-BUH-luh), city, ⊙ KERBELA province, central IRAQ, 55 mi/89 km SW of BAGHDAD, at the edge of the SYRIAN DESERT on Bahr al Milh; 32°35′N 44°01′E. Trades in religious objects, hides, wool, and dates. Karbala is the site of the tomb of the Shiite leader Husein, who was killed in the city in 680. It is second only to MAKKA (Mecca) in being a holy place visited by Shiite pilgrims. The tomb, with a gilded dome and three minarets, is the most notable building; it was destroyed by the Wahabis in 1801 but was quickly restored by contributions from Persians and other Shiite Muslims. Iranian pilgrims to Mecca usually begin their journey at Karbala, and many pious Muslims take the bones of their dead for burial there.

Karben (KAHR-ben), town, HESSE, central GERMANY, 9 mi/14.5 km NE of FRANKFURT; 50°15′N 08°46′E. Furniture, metalworking. Has 18th-century palace, 17th-century castle, and 13th-century church. Chartered 1970 after incorporation of several villages.

Karbi-Anglong, district (□ 4,029 sq mi/10,475.4 sq km), ASSAM state, NE INDIA; ⊙ DIPHU.

Karbitz, CZECH REPUBLIC: see CHABAROVICE.

Karcag (KAHR-tsag), city (2001 population 22,574), E HUNGARY; 47°19′N 20°56′E. Railroad junction; wheat, corn; hogs, cattle; manufacturing (tiles, milling, food processing).

Karcal Dagi (Turkish=*Karçal Daği*), mountain range, NE TURKEY, rising to 10,990 ft/3,350 m, running SW-NE from the CORUH RIVER valley (W) to the Georgian border (NE).

Karchana (KUHR-chuh-nah), town, ALLAHABAD district, SE UTTAR PRADESH state, N central INDIA, 13 mi/21 km SSE of ALLAHABAD; 25°17′N 81°56′E. Gram, rice, barley, wheat, sugarcane.

Kardailovo (kahr-dah-EE-luh-vuh), village (2006 population 3,030), S ORENBURG oblast, SE European Russia, on the URAL River, on road, 21 mi/34 km E of (and administratively subordinate to) Ilek, and 9 mi/ 14 km W of KRASNOKHOLM; 51°32′N 53°54′E. Elevation 236 ft/71 m. Agricultural products.

Kardam (kahr-DAHM), village, RUSE oblast, POPOVO obshtina, NE BULGARIA, on the CHERNI LOM RIVER, 2 mi/3.2 km NE of Popovo; 43°24′N 26°18′E. Poultry; oil-bearing plants, vegetables. Formerly called Khaidar.

Kardam (kahr-DAHM), village, VARNA oblast, GENERAL TOSHEVO obshtina, BULGARIA; 43°46′N 28°07′E. Grain, sugar beets; livestock.

Kardamila (kahr-[TH]AH-mee-lah), town, N KHÍOS island, KHÍOS prefecture, NORTH AEGEAN department, GREECE, 11 mi/18 km N of KHÍOS; 38°32′N 26°06′E. Olive oil, wine, wheat, barley. Also Kardhamila.

Kardasch Retschitz, CZECH REPUBLIC: see KARDASOVA RECICE.

Kardasova Recice (KAHR-dah-SHO-vah RZHE-chi-TSE), Czech *Kardašova Řečice*, German KARDASCH RETSCHITZ, town (1991 population 2,105), JIHOCESKY province, SE BOHEMIA, CZECH REPUBLIC, on railroad, and 7 mi/11.3 km WNW of JINDŘICHŮV HRADEC; 49°12′N 14°53′E. Pencils, sealing wax; fish ponds. Has a 14th-century baroque church. A 16th-century Renaissance castle of CERVENA LHOTA, Czech *Červená Lhota*, is 4 mi/6.4 km NNE.

Kardeljevo, CROATIA: see PLOČE.

Kardhamila, Greece: see KARDAMILA.

Karditsa (kahr-[TH]EE-tsah), prefecture (□ 936 sq mi/ 2,424 sq km; 1991 population 126,854), SW THESSALY department, N GREECE; ⊙ Karditsa; 39°20′N 21°45′E. Bordered N by TRIKKALA prefecture, E by LÁRISSA prefecture, SE by CENTRAL GREECE department, SW by WESTERN GREECE department, and W by EPIRUS department (AKHELÓOS River, border). PINDOS mountain system in W; includes S portion of Trikkala lowland (PENEIOS River valley). Corn, wheat, vegetables; livestock. Industry centered at Karditsa, served by narrow-gauge railroad from Vólos. Also Kardhitsa.

Karditsa (kahr-[TH]EE-tsah), city (2001 population 32,031), ⊙ KARDITSA prefecture, SW THESSALY department, N GREECE, on narrow-gauge railroad, and 15 mi/24 km SE of TRIKKALA; 39°22′N 21°55′E. Trading center for corn, vegetables, tobacco, cotton, dairy products; livestock. Developed during Turkish rule (after 15th century). Also Karditza, Kardhitsa, and Carditsa.

Kärdla (KARD-lah) or **Kyardla**, German *Kertel*, city, Baltic port on N shore of HIIUMAA island, ESTONIA, 30 mi/48 km W of HAAPSALU; 58°59′N 22°44′E. Agricultural market (barley, potatoes); manufacturing (woolens, cellulose).

Kardo, SOMALIA: see GARDO.

Kardonikskaya (kahr-DO-neek-skah-yah), village (2005 population 7,845), E central KARACHEVO-CHERKESS REPUBLIC, S European Russia, in the NW CAUCASUS Mountains, on the GREAT ZELENCHUK RIVER, on road, 6 mi/10 km E of ZELENCHUKSKAYA; 43°51′N 41°43′E. Elevation 2,972 ft/905 m. In agricultural area. Originally a Cossack *stanitsa* (village).

Kardymovo (kahr-DI-muh-vuh), town (2006 population 4,910), central SMOLENSK oblast, W European Russia, on road and railroad, 15 mi/24 km ENE of SMOLENSK; 54°53′N 32°26′E. Elevation 570 ft/173 m. In agricultural area (wheat, oats, rye, flax; livestock raising).

Kardzhali, Bulgaria: see KURDZHALI, city.

Kardzhali, Bulgaria: see KURDZHALI, dam.

Kardzhin (kahr-JIN), village (2006 population 3,175), N NORTH OSSETIAN REPUBLIC, RUSSIA, on the slopes of the N central CAUCASUS Mountains, on the TEREK RIVER, on the highway and railroad, 10 mi/16 km WNW of BESLAN, and 6 mi/10 km SSE of EL′KHOTOVO; 43°16′N 44°17′E. Elevation 1,161 ft/353 m. Agriculture (grain, livestock).

Kareedouw, town, EASTERN CAPE province, SOUTH AFRICA, in Langkloof Valley 27 mi/43 km W of HUMANSDORP, near Krom River. Center of lumber industry; famous for high quality furniture made of stinkwood, yellowwood and blackwood. Commercial center for the area.

Kareima (kuh-RAI-muh), town, Northern state, SUDAN, SW of 4th cataract of the NILE River, 5 mi/8 km N of MARAWI; 18°33′N 31°51′E. Terminus of railroad spur from ABU HAMED, and transfer point for steamers to KARMA (above 3d cataract). Tourist base for nearby NAPATA ruins, AL-KURU and NURI pyramid fields. Airfield. Also spelled Karima.

Kareli (kah-RE-lee), city and administrative center of Kareli region, central GEORGIA, on railroad, and 12 mi/ 19 km W of GORI; 42°01′N 43°53′E. Agriculture (orchards); winery.

Kareli (kuh-RAI-lee) or **Kareliganj**, town (2001 population 25,035), NARSINGHPUR district, NW MADHYA PRADESH state, central INDIA, 8 mi/12.9 km W of NARSINGHPUR; 22°55′N 79°04′E. Dal milling, saw-milling; wheat, oilseeds.

Karelian Isthmus, land bridge, LENINGRAD oblast, NW European Russia, connecting RUSSIA and FINLAND. Situated between the Gulf of FINLAND (W) and Lake LADOGA (E), it is 25 mi/40 km to 70 mi/113 km wide, 90 mi/145 km long. ST. PETERSBURG and VYBORG (Viipuri) are its chief cities. Originally part of the Grand Duchy of Sweden, the isthmus passed to Russia in 1721, and—except for its southernmost section—became part of Finland in 1917. The MANNERHEIM Line, which crossed the isthmus, was breached in 1940 by the Russians, who occupied the area. It was briefly held (1941–1944) by Finnish and German units during World War II. The isthmus was formally ceded to the USSR in 1944, and more than 400,000 of its Finnish residents moved to Finland.

Karelia, Republic of, constituent republic (□ 66,564 sq mi/173,066.4 sq km; 2004 population 785,000), NW European Russia, extending from the Finnish border (W) to the WHITE SEA (E), and from the KOLA PENINSULA (N) to Lakes LADOGA and ONEGA (Europe's largest freshwater lakes); ⊙ PETROZAVODSK. A glaciated plateau, Karelia is covered by more than 60,000 lakes and by coniferous forests (85% of the land is state forest); fishing and lumbering are major industries. Other industries include ship building, food-processing; manufacturing (furniture, aluminum, building materials, and textiles). Agriculture, generally hampered by the cold climate and poor soil, is possible only in the S, where some grains, potatoes, fodder grasses, and vegetables are grown; also some dairy farming and livestock raising. Karelia has valuable deposits of iron ore, magnetite, lead, zinc, copper, titanium, marble, pyrite, chrome, vanadium, molybdenum, gold, quartzites, among other minerals. Power for industry is supplied partially by the republic's many short, rapid rivers. The republic is crossed by the Murmansk railroad and by the BALTIC-WHITE SEA CANAL, which is both commercially and strategically important. Russians and Ukrainians constitute a majority of the population, the rest of which consists mainly of Karelians, Finns, and Lapps, who are very closely related and have an identical written language. The Karelians, a major division of the Finns, were first mentioned in the 9th century and formed a strong medieval state. Karelia, the region N and E of Lake Onega, was conquered in the 12th and 13th centuries by the Swedes, who took the W, and by Novgorod, which took the E. In 1617, the E part was taken from Russia by SWEDEN, but later restored (1721) by the Treaty of Nystad. The W part shared the history of FINLAND until 1940. It was from oral traditions among the Karelians that the Finnish national epic, the *Kalevala*, was compiled in the 19th century by Elias Lönnrot. The Karelian area of the Russian Empire was economically backward and was often a place of exile for political prisoners. In 1920, an autonomous oblast, known as the Karelian Workers' Commune, was set up in E Karelia; in 1923, it was made into the Karelian ASSR, which, after the Soviet-Finnish War of 1939–1940, incorporated most of the territory ceded by Finland to the USSR. In March 1940, the region's status was raised to that of a constituent republic, called the Karelo-Finnish SSR. During World War II, the Finns occupied most of Karelia; but it was returned to the USSR in 1944. Karelia reverted to the status of an autonomous republic in 1956. It was a signatory to the March 31, 1992, treaty that created the RUSSIAN FEDERATION. The republic has a 150-member parliament.

Karema (kah-RAI-mah), town, RUKWA region, W TANZANIA, 50 mi/80 km SW of MPANDA, on Lake TANGANYIKA; 06°49′S 30°25′E. Lake port, road terminus. KATAVI PLAINS to E. Fish; agriculture (livestock; subsistence crops). Chromite and gold deposits.

Karenni State, MYANMAR: see KAYAH STATE.

Karera (kuh-RAI-ruh), town (2001 population 23,491), SHIVPURI district, N central MADHYA PRADESH state, central INDIA, 31 mi/50 km E of SHIVPURI; 25°31′N 74°13′E. Agriculture (millet, gram, wheat, corn). Formerly part of Madhya Barat state.

Karera, INDIA: see BHUPALSAGAR.

Karf, POLAND: see BOBREK.

Kargala (kahr-gah-LAH), settlement (2006 population 1,970), central ORENBURG oblast, SE European Russia, on the URAL River, on road and near railroad, 11 mi/18 km N, and under administrative jurisdiction, of ORENBURG; 51°57′N 55°10′E. Elevation 318 ft/96 m. Natural gas processing.

Kargali (kahr-gah-LEE), village, central TATARSTAN Republic, E European Russia, on road, 15 mi/24 km SE of CHISTOPOL′; 55°13′N 50°54′E. Elevation 479 ft/145 m. In agricultural area (oats, wheat, rye, flax).

Kargalinskaya (kahr-gah-LEEN-skah-yah), village (2005 population 5,375), NE CHECHEN REPUBLIC, S European Russia, on road and railroad, on the TEREK RIVER, 13 mi/21 km SW of KIZLYAR (Dagestan Republic); 43°44′N 46°28′E. Originally a Cossack *stanitsa* (fortified village).

Kargapol′ye (kahr-gah-PO-leeye), town (2005 population 8,485), N KURGAN oblast, SW SIBERIA, RUSSIA, on the MIASS River, on road and railroad, 30 mi/48 km ESE of SHADRINSK; 55°57′N 64°26′E. Elevation 282 ft/ 85 m. In agricultural area; food processing.

Kargasok (kahr-gah-SOK), town (2006 population 8,585), central TOMSK oblast, W SIBERIA, RUSSIA, on the OB′ RIVER, near the mouth of the VASYUGAN RIVER, on road, 265 mi/426 km NW of TOMSK, and 25 mi/40 km WNW of NARYM; 59°03′N 80°51′E. Elevation 157 ft/47 m. River port; lumbering industry, fish processing. In oil- and gas-producing area.

Kargat (kahr-GAHT), city (2006 population 10,785), central NOVOSIBIRSK oblast, SW SIBERIA, RUSSIA, on the KARGAT RIVER, on crossroads and the TRANS-SIBERIAN RAILROAD, 110 mi/177 km W of NOVOSIBIRSK; 55°11′N 80°17′E. Elevation 446 ft/135 m. Meat and dairy products. Sand, clay, and peat deposits in the vicinity. Known since the 18th century as Kargatinskiy Forpost; railroad station since 1896; made city in 1965.

Kargatinskiy Forpost, RUSSIA: see KARGAT.

Kargat River (kuhr-GUHT), 180 mi/290 km long, NOVOSIBIRSK oblast, RUSSIA; rises on the NE BARABA STEPPE; flows SW, past KARGAT and ZDVINSK, to SE Lake CHANY. Mostly navigable in the summer season (June–October).

Kargi (Turkish=*Kargi*), village, N TURKEY, near the KIZIL IRMAK, and 40 mi/64 km SE of KASTAMONU; 41°09′N 34°32′E. Cereals.

Kargil (KUHR-gil), district (□ 5,419 sq mi/14,089.4 sq km), JAMMU AND KASHMIR state, extreme N INDIA; ⊙ KARGIL.

Kargil (KUHR-gil), town, ⊙ KARGIL district, JAMMU AND KASHMIR state, extreme N INDIA, in DEOSAI Mountains, on SURU River, and 80 mi/129 km NE of SRINAGAR; 34°34′N 76°06′E. Near northernmost limit of India-administered KASHMIR region. Pulses, wheat, corn. Fort; scene of Indo-Pakistani fighting during 1948. Buddhist cave monastery is 15 mi/24 km SE at Shergol (sometimes spelled Shargol).

Kargilik, China: see YECHENG.

Kargopol' (KAHR-guh-puhl), city (2005 population 10,900), SW ARCHANGEL oblast, N European Russia, on the ONEGA River, at its outlet from Lake LACHA, on road, 265 mi/426 km S of ARCHANGEL, and 60 mi/97 km NW of KONOSHA; 61°30′N 38°58′E. Elevation 419 ft/127 m. Beer, dairy products. Architectural monuments. Chartered in 1380; formerly a salt-trading center on the Vologda-Onega route.

Kargowa (kahr-GO-vah), German *Unruhstadt*, town, Zielona Góra province, W POLAND, 18 mi/29 km NE of ZIELONA GÓRA (Grünberg). Candy manufacturing, flour milling. Chartered 1661; briefly occupied, 1918–1919, by Poles. c.45% destroyed.

Karguéri (kahr-GAI-ree), village, ZINDER province, S NIGER, on NIGERIA border, and 100 mi/161 km SE of ZINDER; 13°27′N 10°25′E. Also spelled Karguiri.

Karhal (KUHR-huhl), town, MAINPURI district, W UTTAR PRADESH state, central INDIA, 16 mi/26 km SSW of MAINPURI; 27°01′N 78°57′E. Wheat, gram, pearl millet, corn, barley.

Karhuisa (kar-HWEE-sah), canton, LOS ANDES province, LA PAZ department, W BOLIVIA, E of LA PAZ. Elevation 12,625 ft/3,848 m. Some gas resources in area; copper-bearing lode; clay, limestone, and gypsum deposits. Agriculture (potatoes, yucca, bananas, rye); cattle.

Karhula (KAHR-hoo-lah), village, KYMEN province, SE FINLAND, near GULF OF FINLAND, 2 mi/3.2 km NNE of KOTKA; 60°31′N 26°57′E. Elevation 66 ft/20 m. Railroad terminus. Glassworks; steel, pulp, paper, and board mills; machine shops.

Karia ba Mohammed (kah-ree-yuh bah-mo-HAM-med), village, Taounate province, Taza-Al Hoceima-Taounate administrative region, N MOROCCO, near the Oued SEBOU river, 25 mi/40 km NW of FES; 34°22′N 05°12′W. Agriculture (wheat, barley, vegetables); horse raising.

Kariai (kah-ree-AI), town, ⊙ and commune of Mount Athos (Ayion Oros) autonomous district, NE GREECE, midway down AKTI Peninsula of the KHALKIDHIKÍ, near its E coast; 40°15′N 24°15′E. Inhabited solely by males, most of whom are monks. The Holy Council, composed of a deputy from each of the twenty monasteries, meets periodically to govern the community in the Council House here. The oldest (10th century) church on Mount ATHOS, the restored Protaton [Greek.=foremost] with fine frescoes. Also spelled Karyes.

Kariba (kah-REE-bah), town, MASHONALAND WEST province, N ZIMBABWE, 175 mi/282 km NW of HARARE, on ZAMBEZI River (ZAMBIA border), at KARIBA DAM, in Charara Safari Area; 16°32′S 28°47′E. Lake KARIBA reservoir to SW; Kariba Gorge to NE; Hurungwe Safari Area to NE. Border crossing at dam; airport to E. Kariba South Hydro Station (666MW) is here. Church of Saint Barbara, built to honor those who lost their lives in constructing dam. Casino. Tourism.

Kariba Dam, hydroelectric project, in Kariba gorge of the ZAMBEZI RIVER, on the ZAMBIA-ZIMBABWE border, S central AFRICA; 16°31′S 28°46′E. Built 1955–1959, it is one of the world's largest dams (420 ft/128 m high, 1,900 ft/579 m long). The creation by the dam of KARIBA LAKE forced resettlement of about 50,000 people living along the Zambezi River. In 1960–1961, Operation Noah captured and removed the animals threatened by the lake's rising waters. The Kariba project supplies electricity to various parts of Zambia and Zimbabwe, including the COPPERBELT region in N central Zambia.

Kariba Dam, ZIMBABWE and ZAMBIA: see KARIBA, LAKE.

Kariba, Lake (□ 2,000 sq mi/5,180 sq km), reservoir, ZIMBABWE and ZAMBIA, formed by KARIBA DAM (completed 1959) on ZAMBEZI River, 175 mi/282 km NW of HARARE, Zimbabwe; 170 mi/274 km long, 5 mi/8 km-20 mi/32 km wide; 16°32′S 28°46′E. Zambezi River enters lake from SW in Devil's Gorge; SANYATI and SENGWA rivers enter from S. Commercial and sport fishing; recreation.

Karibib (kah-ri-beeb), town, W NAMIBIA, 90 mi/145 km NW of WINDHOEK. Tin, gold, and lithium mining; stone cutting works. On main route to coast from Windhoek by road and railroad.

Kariega, river, SOUTH AFRICA: see KENTON-ON-SEA.

Karihaara (KAH-ri-HAH-rah), NW suburb of KEMI, LAPIN province, NW FINLAND; 65°45′N 24°33′E. Elevation 33 ft/10 m. On islet in GULF OF BOTHNIA, at mouth of the KEMIJOKI. Lumber and pulp mills.

Karijini National Park (□ 2,385 sq mi/6,201 sq km), NW WESTERN AUSTRALIA state, W AUSTRALIA, 190 mi/306 km S of PORT HEDLAND, in PILBARA district. Situated at center of HAMERSLEY RANGE (60 mi/97 km long, 50 mi/80 km wide). Rugged mountain range of iron-rich sandstone. Almost 20 gorges; waterfalls; pools support plant and animal communities. Western Australia's highest peak, Mount Meharry (4,085 ft/1,245 m), on SE park boundary. Summer temperatures often exceed 40°C. River red gums, ferns; spinifex grass on plateau. Red kangaroos, rock wallabies, dingoes, echidnas, bats, pebble-mound mice; emus, spinifex pigeon; geckoes, dragons, legless lizards, pythons. Large termite mounds. Camping, picnicking, swimming, boating, hiking. Formerly Hamersley Range National Park.

Karikal, INDIA: see KARAIKAL.

Kari Kari (KAH-ree KAH-ree), canton, SAUCARI province, ORURO department, W central BOLIVIA, SE of TOLEDO; 19°39′S 65°40′W. Elevation 12,126 ft/3,696 m. Kari Kari mountains here. Lead-bearing lodes; clay, limestone deposits. Agriculture (potatoes, yucca, bananas); cattle.

Karima, SUDAN: see KAREIMA.

Karimama, town, ALIBORI department, NE BENIN, 20 mi/32 km NW of MALANVILLE near NIGER RIVER, on Benin-NIGER border; 10°56′N 03°43′E. Cotton; livestock; shea-nut butter.

Karimata Islands (kah-REE-mah-tah), INDONESIA, in KARIMATA STRAIT, off SW coast of BORNEO; 01°36′S 108° 53′E. Comprises two large islands: Karimata (10 mi/16 km long, 7 mi/11.3 km wide) and Serutu (8 mi/12.9 km long, 2 mi/3.2 km wide) surrounded by c.60 islets. Fishing. Padang (chief port) on SE coast of Karimata.

Karimata Strait (kah-ree-MAH-tah), channel, INDONESIA, connects JAVA SEA (S) with SOUTH CHINA SEA (N), between BELITUNG island (W), in Sumatra Selatan province, and SW Kalimantan Barat province (E); c.125 mi/201 km wide; 02°05′S 108°40′E. KARIMATA ISLANDS are here.

Karimganj (kuh-REEM-guhnj), district (□ 698 sq mi/1,814.8 sq km), ASSAM state, NE INDIA; ⊙ KARIMGANJ.

Karimganj (kuh-REEM-guhnj), town, ⊙ KARIMGANJ district, ASSAM state, NE INDIA, in SURMA RIVER valley, on KUSIYARA river, and 28 mi/45 km W of SILCHAR; 24°52′N 92°21′E. Trades in rice, tea, mustard; tea processing. Rice research station. Until 1947, in SYLHET district, which is now in BANGLADESH.

Karimnagar (kuh-REEM-nuh-guhr), district (□ 4,565 sq mi/11,869 sq km), ANDHRA PRADESH state, S INDIA, on DECCAN PLATEAU; ⊙ KARIMNAGAR. Bounded N by GODAVARI RIVER; mainly lowland, drained by tributaries of the Godavari. Rice and oilseed milling, cotton ginning. Largely sandy red soil; millet, rice, oilseeds (especially peanuts, sesame), cotton; poultry farming. Bamboo (used in paper manufacturing), teak, ebony in forests (E, W). Main towns include Karimnagar (noted silver filigree), JAGTIAL. Became part of HYDERABAD during state's formation in eighteenth century and then part of Andhra Pradesh state after Independence.

Karimnagar (kuh-REEM-nuh-guhr), town (2001 population 218,302), ⊙ KARIMNAGAR district, ANDHRA PRADESH state, S INDIA, 85 mi/137 km NNE of HYDERABAD; 18°26′N 79°09′E. Road center in agricultural area (millet, rice, oilseeds, cotton); noted silver filigree.

Karimun Islands (kah-REE-moon), island group, RIAU ARCHIPELAGO, RIAU province, INDONESIA, in SOUTH CHINA SEA at entrance to Strait of MALACCA, between SINGAPORE and SUMATRA; 01°03′N 103°22′E. Largest of several islands is Kundur (18 mi/29 km long, 10 mi/16 km wide); site of Ungar. N of Kundur is Karimun Besar or Great Karimun (12 mi/19 km long, 5 mi/8 km wide), chief port of group is TANJUNGBALAI. Chief products include sago, coconuts, pepper, gambier, firewood; some tin is mined; fishing industry is important. Also spelled Karimoen Islands.

Karimunjawa Islands (kah-ree-moon-JAH-wah), island group (□ 19 sq mi/49.4 sq km), INDONESIA, in JAVA SEA, 80 mi/129 km N of SEMARANG, JAVA; 05°48′S 110°20′E. Comprise c.25 islands, largest being Karimunjawa (4 mi/6.4 km in diameter, rising to 1,660 ft/506 m) and Kemujan or Kemoedjan (4 mi/6.4 km long). Coconut growing and fishing on larger islands; many of smaller islands are barren and uninhabited. Also spelled Karimoendjawa.

Karin (KAH-reen), village, N SOMALIA, minor port on Gulf of ADEN, in lowland, 60 mi/97 km ENE of Berbera, near Cape Kanzir. Gums, resins; fisheries.

Karind, IRAN: see KEREND.

Käringön (SHER-eeng-UHN), fishing village, GÖTEBORG OCH BOHUS county, SW SWEDEN, on island (□ 575 acres/233 ha) of same name in SKAGERRAK, 12 mi/19 km SSW of LYSEKIL; 58°07′N 11°22′E. Seaside resort.

Karino (KAH-ree-nuh), village, central KIROV oblast, E central European Russia, in the CHEPTSA RIVER basin, near highway and railroad, 6 mi/10 km SE of (and administratively subordinate to) SLOBODSKOY; 58°37′N 50°20′E. Elevation 610 ft/185 m. Agricultural products; logging and lumbering.

Karintorf (kah-reen-TORF), town, central KIROV oblast, E European Russia, on the VYATKA River, on railroad spur, 27 mi/43 km E of KIROV, and 4 mi/6 km E of KIROVO-CHEPETSK, to which it is administratively subordinate; 58°33′N 50°11′E. Elevation 354 ft/107 m. Peat works.

Karis (KAH-ris), Finnish *Karjaa*, town, UUDENMAAN province, S FINLAND, near GULF OF FINLAND, 40 mi/64 km W of HELSINKI; 60°05′N 23°45′E. Elevation 132 ft/40 m. Railroad center in lumber and dairy region.

Karisimbi, Mount (kar-ee-SIM-bee), mountain peak and extinct volcano (14,787 ft/4,507 m) highest peak in the VIRUNGA mountain range, CONGO; 01°30′S 29°27′E. Also called Volcan Karisimbi.

Karistos (KAH-rees-tos), Latin *Carystus*, town, SE ÉVVIA island, ÉVVIA prefecture, CENTRAL GREECE department, port on Bay of KARISTOS, 55 mi/89 km SE of KHALKÍS; 38°01′N 24°25′E. Important marble quarries; sheep raising; fisheries. Site of ancient Carystus on slope of Mount OKHI (NE) is occupied by medieval ruins. Present town dates from Greek War of Independence. Also spelled Karystos.

Karistos, Bay of (KAH-rees-tos), inlet of AEGEAN SEA, off SE ÉVVIA island, ÉVVIA prefecture, CENTRAL GREECE department, between capes MANDILI (E) and PAXIMADI (W); 6 mi/9.7 km long. Also called Bay of Karystos.

Karitaina (kah-REE-tai-nah), village, ARKADIA prefecture, central PELOPONNESE department, S mainland GREECE, on road, and 8 mi/12.9 km NW of MEGALOPOLIS; 37°29′N 22°02′E. Remains of noted 13th-century French castle. Also spelled Karytaina.

Kariwa (kah-REE-wah), village, Kariwa county, NIIGATA prefecture, central HONSHU, central JAPAN, 40 mi/65 km S of NIIGATA; 37°25′N 138°37′E.

Kariya (KAH-ree-yah), city, Aichi prefecture, S central Honshu, central Japan, 12 mi/20 km S of Nagoya;

34°59′N 137°00′E. Industrial center; manufacturing (automotive parts and transportation equipment).

Karjaa, FINLAND: see KARIS.

Karjat (KUHR-juht), town, AHMADNAGAR district, MAHARASHTRA state, central INDIA, 40 mi/64 km SE of AHMADNAGAR. Agriculture market (millet, gur).

Karjat, town, RAIGAD district, W MAHARASHTRA state, W central INDIA, 32 mi/51 km E of MUMBAI (Bombay), on Mumbai-PUNE railroad. Timber depot (teak, blackwood); rice.

Karkabat, Eritrea: see KERKEBET.

Karkal (KUHR-kah-luh), town, Dakshin Kannad district, KARNATAKA state, SW INDIA, 26 mi/42 km NNE of MANGALORE; 13°12′N 74°59′E. Rice milling, manufacturing of potstone culinary vessels (steatite quarries nearby). Has monolithic Jain statue. Other noted Jain archaeological remains 10 mi/16 km S at village of Mudbidri. Wildlife sanctuary here.

Karkamb (KUHR-kuhmb), town, SOLAPUR district, MAHARASHTRA state, central INDIA, 42 mi/68 km WNW of SOLAPUR; 17°51′N 75°18′E. Market center; cotton dyeing, handicraft cloth weaving, betel farming.

Karkamis, TURKEY: see CARCHEMISH.

Karkar, volcanic island (□ 140 sq mi/364 sq km), MADANG province, N central PAPUA NEW GUINEA, BISMARCK SEA, SW PACIFIC OCEAN, 9 mi/14.5 km NE of NEW GUINEA; 15 mi/24 km long; rises to 6,007 ft/1,831 m. Active volcano here. Coconuts, copra. Formerly called DAMPIER ISLAND.

Karkaralinsk, Kazakhstan: see KARKARALY.

Karkaraly (kahr-kah-rah-LEE), city, E KARAGANDA region, KAZAKHSTAN, 105 mi/169 km ESE of KARAGANDA; 49°21′N 75°27′E. Tertiary-level (raion) administrative center. Road junction in KAZAK HILLS. Cattle; meatpacking. Lead-silver deposits. Founded (1824) in Russian conquest of Kazakhstan. Also called Karkaralinsk, Qarqaraly.

Karkel'n, RUSSIA: see MYSOVKA.

Karkheh (kahr-KE), ancient *Choaspes*, river, c.350 mi/560 km long, W IRAN; rises in the ZAGROS mountains; flows S into the Khuzistan lowland, where it forms swamps bordering the TIGRIS RIVER. An ancient storage dam on the river at SHUSH made Khuzistan one of the most prosperous agr. regions of SW ASIA until the system fell into disrepair and the irrigated area reverted to desert. The area has been reclaimed as part of the Khuzistan project.

Karkinit Gulf (kahr-kee-NEET), (Ukrainian *Karkinits'ka zatoka*), (Russian *Karkinitskiy zaliv*), gulf, shallow N inlet of BLACK SEA in UKRAINE, between Ukrainian mainland (N) and CRIMEA (S); 15 mi/24 km–50 mi/80 km wide, 65 mi/105 km long. Separated by PEREKOP ISTHMUS (E) from SYVASH lagoon. Main ports: SKADOVS'K, KHORLY (N), and CHORNOMORS'KE (S). Site of offshore exploration for oil and gas.

Karkinits'ka zatoka, UKRAINE: see KARKINIT GULF.

Karkinitskiy zaliv, UKRAINE: see KARKINIT GULF.

Karkkila (KAHRK-kee-lah), Swedish *Högfors*, town, UUDENMAAN province, S FINLAND, 35 mi/56 km NW of HELSINKI; 60°32′N 24°11′E. Elevation 330 ft/100 m. Railroad terminus. Industrial center; metalworking.

Karkloof Falls, on Karkloof River, a tributary of UMGENI RIVER, 6 mi/10 km NE of Howicks, central KWAZULU-NATAL, SOUTH AFRICA; 29°23′S 30°16′E. Drops 350 ft/107 m in a series of cascades in a wooded area SE of Karkloof Nature Reserve.

Karkonoski National Park (kar-kon-os-kee) (□ 22 sq mi/57.2 sq km), SW POLAND, on the border with CZECH REPUBLIC. The park forms a narrow belt approximately 9 mi/14.5 km long, up to the elevation of 4,101 ft/1,250 m. Characteristic of the park are cirques carved by glaciers, and dense spruce forests; popular hiking area. KARPACZ and SZKLARSKA PORĘBA are the most popular tourist centers.

Karkonosze, CZECH REPUBLIC: see GIANT MOUNTAINS.

Karksi, ESTONIA: see KARKSI-NUIA.

Karksi-Nuia (KAHRK-see NOO-iyuh), town, S ESTONIA, 18 mi/29 km S of VILJANDI; 58°06′N 25°33′E. Road junction. Agriculture (flax) and livestock; manufacturing (toys, auto and truck reconstruction). Karksi is just E.

Karkuanah (kahr-KWAH-nah), archaeological site, NABUL province, NE Tunisia, 7 mi/11.3 km N of KALIBIA, on CAPE BON. Former city of Tamazrat, discovered 1957. Founded in 6th century B.C.E. by Phoenicians; principal activities included fishing, dye-making, crafts. City was razed in 2nd century B.C.; Punic ruins have been partially excavated. Also called Kerkouane.

Karla, Lake (KAHR-lah), ancient *Boebeis*, (□ 44 sq mi/114.4 sq km), lake in E THESSALY department, N GREECE, 9 mi/14.5 km NW of VÓLOS; 18 mi/29 km long, 2.5 mi/4 km wide. Formerly called Lake Kavla, Lake Voiveis, or Voiviis.

Karl Alexander Island, in N FRANZ JOSEF LAND, ARCHANGEL oblast, extreme N European Russia, in the ARCTIC OCEAN; 22 mi/35 km long, 10 mi/16 km wide; 81°30′N 57°30′E.

Karla Libknekhta, UKRAINE: see SHYROKOLANIVKA.

Karla Libknekhta, Imeni, UKRAINE: see POKROVS'KE, Dnipropetrovs'k oblast.

Karla Marksa, Imeni, UKRAINE: see KARLO-MARKSOVE.

Karli (KAHR-lee), village, AKOLA district, MAHARASHTRA state, W central INDIA. Nearby are Buddhist caves that may have been excavated as early as the second century B.C.E. The most famous of them measures 124 ft/38 m by 45 ft/14 m and is India's largest cave temple. Its ancient shrine, columns, and ornamentation survive in part.

Karlikova, Greece: see MIKROPOLIS.

Karlin (KAHR-leen), Czech *Karlín*, German *karolinenthal*, industrial district, NE of PRAGUE, PRAGUE-CITY province, central BOHEMIA, CZECH REPUBLIC, on right bank of VLTAVA RIVER. Manufacturing (machinery, electronics).

Karlingen, FRANCE: see CARLING.

Karlino (kahr-LEE-no) or **Korlino**, German *Körlin* or *Körlin an der Persante*, town, in Koszalin province, NW POLAND, on PROSNICA (Persante) River, and 16 mi/26 km SW of KOSZALIN (Köslin). Manufacturing of machinery, bricks; sawmilling, flour milling. Has late-Gothic church.

Karliova, village, E central TURKEY, 40 mi/64 km NE of BINGOL, borders on Georgia, Armenia, and Azerbaijan (Nahichevan autonomous region); 39°17′N 41°02′E. Grain. Formerly known as Kale.

Karlivka (KAHR-lif-kah), (Russian *Karlovka*), city, E POLTAVA oblast, UKRAINE, 27 mi/43 km ESE of POLTAVA; 49°27′N 35°08′E. Elevation 301 ft/91 m. Raion center; railroad station. Sugar refinery, food processing, and distillery, equipment; manufacturing (grain elevators, flour mills, food flavoring, furniture). Two vocational schools. Established in the 1740s, city since 1957.

Karl Libknekht (KAHRL LEEP-nyekht) or **Imeni Karla Libknekhta**, town (2005 population 7,970), W central KURSK oblast, SW European Russia, near the SEYM RIVER, on road and railroad, 30 mi/48 km W of KURSK; 51°39′N 35°34′E. Elevation 564 ft/171 m. Sugar refinery; machinery. Until 1930, called Penskiy Sakharnyy Zavod.

Karl-Marx-Stadt, GERMANY: see CHEMNITZ.

Karlö, FINLAND: see HAILUOTO.

Karlobag (KAHR-lo-bahg), Italian *Carlopago*, village, W CROATIA, on Velebit Channel of ADRIATIC SEA, at W foot of VELEBIT Mountains, 29 mi/47 km N of ZADAR. Small lumber-exporting seaport; resort.

Karlo Libknekhta, Imeni, UKRAINE: see SOLEDAR.

Karlo-Libknekhtivs'k, UKRAINE: see SOLEDAR.

Karlo-Libknekhtovsk, UKRAINE: see SOLEDAR.

Karlo-Marksove (KAHR-lo-MAHRK-so-ve), (Russian *Karlo-Marksovo*), town, E central DONETS'K oblast, UKRAINE, in the DONBAS, between HORLIVKA (Russian *Gorlovka*) and YENAKIYEVE (Russian *Yenakiyevo*), 4

mi/6.4 km NW of Yenakiyeve, and subordinated to the Yenakiyeve city council; 48°16′N 38°08′E. Elevation 721 ft/219 m. Bituminous coal mine. Established in 1783 as Sofiyivka (Russian *Sofiyevka*); known as Imeni Karla Marksa (1924–1965).

Karlo-Marksovo, UKRAINE: see KARLO-MARKSOVE.

Karlovac (KAHR-lo-vahts), German *Karlstadt*, Hungarian *Károlyváros*, city, central CROATIA, near confluence of KUPA and KORANA rivers, 30 mi/48 km SW of ZAGREB, in the POKUPLJE; 45°30′N 15°32′E. Industrial and trade center; manufacturing (textiles, leather goods, chemicals, machinery; lumber, iron); brewery. Former fortress on VOJNA KRAJINA (Austrian-Turk. frontier). Hydroelectric plant and summer resort nearby, on Korana River Founded 1579.

Karlovac, SERBIA: see BANATSKI KARLOVAC.

Karlova Studanka, CZECH REPUBLIC: see VRBNO POD PRADEDEM.

Karlovci (KAHR-lov-tsee), German *Karlowitz*, town (2002 population 8,839), VOJVODINA, N SERBIA, on the DANUBE River, on railroad, and 6 mi/9.7 km SE of NOVI SAD in the SREM region, at NE foot of FRUSKA GORA; 45°11′N 19°56′E. Seat of Serbian patriarch (1690; 1713–1918), when it was a political and cultural center of Serbia. By treaty signed here (1699), TURKEY ceded to AUSTRIA all of HUNGARY (except the BANAT region), TRANSYLVANIA, and CROATIA-SLAVONIA, to VENICE the PELOPONNESUS and DALMATIA, and returned PODOLIA to POLAND. Also spelled Karlovtsi, Sremski Karlovci, Sremski Karlovtsi.

Karlovka, UKRAINE: see KARLIVKA.

Karlovo (KAHR-lo-vo), city (1993 population26,895), PLOVDIV oblast, ⊙ Karlovo obshtina, central BULGARIA, at the S foot of the Troyan Mountains, 34 mi/55 km N of PLOVDIV; 42°38′N 24°50′E. Railroad junction; summer resort. Major rose oil-extracting center in the KARLOVO BASIN; manufacturing (agricultural equipment, textiles; wine making, canning). Technical school. Has an old clock tower and mosque. Under Turkish rule, 15th–19th century. Health resort, Banya, 5 mi/8 km to the S, has thermal springs.

Karlovo Basin (KAHR-lo-vo), valley (□ 110 sq mi/286 sq km), central BULGARIA, between the Troyan Mountains (N) and SREDNA GORA (S); 42°38′N 24°50′E. Average elevation 1,150 ft/351 m; drained by the STRYAMA RIVER. Noted rose-growing region; vineyards, cotton, tobacco, nuts, mint, medicinal herbs. Main center, KARLOVO.

Karlovy Vary (KAHR-lo-VI VAH-ri), German *Karlsbad*, city (2001 population 53,358), ZAPADOCESKY province, CZECH REPUBLIC, W BOHEMIA, at the confluence of the TEPLA and OHRE rivers; 50°13′N 12°54′E. A famous health resort, it is one of the best-known spas of EUROPE; its hot mineral water is taken especially for digestive diseases. The medicinal springs, known for centuries, attracted European aristocrats until World War I. Also noted for its china, glass, and porcelain industries as well as for its distillery (Becherovka). The city was chartered in the 14th century by Emperor Charles IV, who is said to have discovered its springs. Military base.

Karlowitz, SERBIA: see KARLOVCI.

Karlsbad (KAHRLS-bahd), commune, BADEN-WÜRTTEMBERG, SW GERMANY, on N slope of BLACK FOREST, 10 mi/16 km SE of KARLSRUHE; 48°52′N 08°30′E. Electronic industry. Created in 1971 through unification of several villages.

Karlsbad, CZECH REPUBLIC: see KARLOVY VARY.

Karlsbad, LATVIA: see JŪRMALA.

Karlsborg (KAHRLS-BORY), town, SKARABORG county, S central SWEDEN, on NW shore of LAKE VÄTTERN, 18 mi/29 km W of MOTALA; 58°32′N 14°30′E. Terminus (E) of central section of GÖTA CANAL. Garrison town; old fortress.

Karlsborg (KAHRLS-BORY), village, NORRBOTTEN county, N SWEDEN, on small inlet of GULF OF BOTH-

NIA, near mouth of KALIX ÄLVEN RIVER, 25 mi/40 km W of HAPARANDA; 65°48′N 23°18′E. Manufacturing (paper and pulp milling).

Karlsbrunn, CZECH REPUBLIC: see VRBNO POD PRADEDEM.

Karlsburg, ROMANIA: see ALBA IULIA.

Karlsfeld (KAHRLS-felt), suburb of MUNICH, BAVARIA, S GERMANY, in UPPER BAVARIA, 7 mi/11.3 km NW of city center; 48°12′N 11°27′E.

Karlshafen, GERMANY: see BAD KARLSHAFEN.

Karlshafen, Bad, GERMANY: see BAD KARLSHAFEN.

Karlshamn (KAHRLS-HAHMN), town BLEKINGE county, SE SWEDEN, on BALTIC SEA; 56°11′N 14°51′E. Manufacturing (concentrated food); Seat of fishing fleet. Chartered 1664.

Karlshorst (KAHRLS-horst), residential section of LICHTENBERG district, BERLIN, GERMANY, 7 mi/11.3 km E of city center; 52°30′N 13°30′E. Scene (May 8, 1945) of ratification of German surrender at end of World War II.

Karlskoga (KAHRLS-KOO-gah), town, ÖREBRO county, S SWEDEN; 59°15′N 14°15′E. Manufacturing center (seat of BOFORS iron and armaments works; steel, machines, explosives, chemicals; saw mills; distribution of cable and satellite TV. Chartered 1940.

Karlskrona (KAHRLS-KROO-nah), town, ⊙ BLEKINGE county, SE SWEDEN, on BALTIC SEA; 56°12′N 15°38′E. Seaport and fishing center with large modern port; many industries connected to Swedish navy (headquarters here since 1680). Naval museum. An old spelling is Carlscrona.

Karlsruhe (KAHRLS-roo-e), city, BADEN-WÜRTTEMBERG, SW GERMANY, on the N fringes of the BLACK FOREST, connected by canal with a port on the nearby RHINE RIVER; 49°00′N 08°23′E. A transportation, industrial, and cultural center, it is also the seat of the federal constitutional court and the federal court of justice. Manufacturing includes electrical products, building equipment, transportation equipment, perfumes, jewelry, pharmaceuticals, machinery, and refined oil. Pipelines have been constructed leading to MARSEILLE and STRASBOURG, FRANCE, and to INGOLSTADT on the DANUBE in Germany. Karlsruhe was founded in 1715 by Karl Wilhelm, margrave of Baden-Durlach, to replace nearby DURLACH (incorporated into Karlsruhe in 1938) as the margravial residence. After 1771 it was the capital of the duchy (later grand duchy and, after 1919, state) of BADEN. The old part of Karlsruhe, badly damaged in World War II, was laid out as a vast semicircle with the streets converging radially upon the ducal palace (1752–1785; restored after 1945). The city has a university (founded as a technical academy in 1825), a school of fine arts, a school of music, a center for atomic research; theaters, art galleries, and the Schwarzwaldhalle (a large conference center; 1953–1954). Sometimes spelled Carlsruhe.

Karlsruhe (KAHRLZ-roo), village (2006 population 110), MCHENRY CO., central NORTH DAKOTA, 15 mi/24 km ENE of VELVA, on Wintering River; 48°05′N 100°37′W. Founded in 1912 and incorporated in 1927. Named for KARLSRUHE, GERMANY.

Karlstad (KAHRL-stahd), town, ⊙ VÄRMLAND county, S SWEDEN, on LAKE VÄNERN; 59°24′N 13°31′E. Manufacturing (forest products, heavy machinery). Known as Thingvalla (or Tingvalla) in Middle Ages, chartered by Charles IX as Karlstad (1584). Much of city destroyed by fire (1865). Treaty severing union of NORWAY and Sweden negotiated and signed here (1905). Airport.

Karlstad, town (2000 population 794), KITTSON county, NW MINNESOTA, 37 mi/60 km NNW of THIEF RIVER FALLS; 48°34′N 96°31′W. Agriculture (grain, potatoes, sugar beets, flax, sunflowers, beans; sheep); manufacturing (meat processing; pastas). Twin Lakes Wildlife Area to NE.

Karlstadt (KAHRL-shtaht), town, BAVARIA, central GERMANY, in LOWER FRANCONIA, on the Main, 13 mi/21 km NW of WÜRZBURG; 49°57′N 09°46′E. Manufacturing of building materials, musical instruments; metalworking, foundries. Surrounded by walls; has Gothic church and 15th-century town hall. Andreas Bodenstein, Protestant reformer, born here. On opposite side of Main are ruins of castle Karlsburg, founded by Charles Martel and destroyed in Peasants War. Chartered c.1200.

Karlstadt, CROATIA: see KARLOVAC.

Karlstein, CZECH REPUBLIC: see KARLSTEJN.

Karlstein am Main (KAHRL-shtein ahm MEIN), village, BAVARIA, central GERMANY, in LOWER FRANCONIA, on Main River, 6 mi/9.7 km NW of ASCHAFFENBURG; 50°03′N 09°03′E. Wine; tourism.

Karlstejn (KAHRL-shtain), Czech *Karlštejn*, German *karlstein*, village, STREDOCESKY province, W central BOHEMIA, CZECH REPUBLIC, on BEROUNKA RIVER, on railroad, and 15 mi/24 km SW of PRAGUE. Famous for 14 century castle (now a museum), former seat of Charles IV and depository, until 1414, of crown jewels of Holy Roman Empire.

Karlu (KAHR-loo), village, 12 mi/19 km S of PUJEHUN, SOUTHERN province, S SIERRA LEONE; 07°10′N 11°42′W. Chiefdom headquarters. Location of periodic market; cassava, fishing, coconuts.

Karluck, Alaska: see KARLUK.

Karluk, unincorporated townsite (2000 population 27), at mouth of Karluk River, S ALASKA, W KODIAK ISLAND, on SHELIKOF STRAIT, 80 mi/129 km WSW of KODIAK; 57°35′N 154°24′W. Salmon fishing (cannery at UYAK BAY); nearby is salmon hatchery. Largest settlement in the Karluk Indian Reservation. Sometimes spelled Karluck.

Karlyuk (kahrl-YOOK), town, SE LEBAP weloyat, extreme SE TURKMENISTAN, on S spur of the BAISUN-TAU mountains, 65 mi/105 km ESE of KERKI; 37°35′N 66°20′E. Cotton, wheat.

Karma (KAR-mah), canton, ANTONIO QUIJARRO province, POTOSÍ department, W central BOLIVIA. Elevation 12,024 ft/3,665 m. Lead-bearing lode; antimony (CARACOTA) and manganese (Mina Virgén de Copacabana) mining; clay, limestone deposits. Agriculture (potatoes, yucca, bananas); cattle.

Karma (KAHR-mah), town, ALLAHABAD district, SE UTTAR PRADESH state, N central INDIA, 10 mi/16 km S of ALLAHABAD; 24°14′N 85°14′E. Gram, rice, barley, wheat, oilseeds, sugarcane.

Karma (KER-muh), town, Northern state, SUDAN, on right bank of the NILE (head of navigation above 3rd cataract), and 30 mi/48 km N of DONGOLA. Cotton, wheat, barley, corn, fruits, and durra; livestock. Harvard-Boston Expeditions carried out extensive excavations near here 1913–1915. It is an important historical and archaeological site. Sudanese and others are continuously excavating in the area around the city. Also spelled Kerma.

Karma, El (kahr-MAH, el), village, ORAN wilaya, NW ALGERIA, at E edge of the ORAN SEBKHA. Saltworks. Formerly known as Valmy.

Karmakaly, RUSSIA: see KARMASKALY.

Karmala (kuhr-MAH-lah), town, SOLAPUR district, MAHARASHTRA state, W central INDIA, 70 mi/113 km NW of SOLAPUR; 18°25′N 75°12′E. Market center for millet, oilseeds, livestock; handicraft cloth weaving, dal milling, cotton ginning.

Karmalyukova Hora zakaznyk, UKRAINE: see TOVTRY.

Karmana, town, SE NAWOIY wiloyat, UZBEKISTAN, near TRANS-CASPIAN RAILROAD, 35 mi/56 km E of GIJDUVAN, and just N of NAWOIY; 40°05′N 65°24′E. Cotton-ginning center. Town of STANTSIYA KERMINE (ker-MEE-ne) S at railroad station. Mausoleum of Mir Said Bakhrom (11th century) nearby.

Karmanovo (kahr-MAH-nuh-vuh), village, NE SMOLENSK oblast, W European Russia, on road and railroad spur, 20 mi/32 km NNW of GZHATSK; 55°50′N

34°51′E. Elevation 666 ft/202 m. In agricultural area (wheat, rye, oats, flax, vegetables).

Karma, Ras, YEMEN: see QADHUB.

Karmaskaly (kahr-mah-SKAH-li), village (2005 population 7,865), central BASHKORTOSTAN Republic, E European Russia, on road, 27 mi/43 km SSE of UFA; 54°22′N 56°10′E. Elevation 492 ft/149 m. Distilling. Also known as Karmakaly.

Karmé (kahr-MAI), town, W CHAD, 50 mi/80 km NE of N'DJAMENA; 12°33′N 15°54′E. Was part of former CHARI-BAGUIRMI prefecture.

Karme Yosef (kahr-ME yo-SEF) or **Karmei Yosef**, urban suburb, 3.7 mi/6 km SE of RAMLA on coastal plain, central ISRAEL. Principally a commuter suburb of TEL AVIV; formerly an agricultural village. Founded in 1983 and named for Joseph Sappir, a former leading member of the Israeli Liberal Party.

Karmiel (kahr-mee-EL), city (2006 population 44,000), E of AKKO (Acre) on the N edge of the LOWER GALILEE, N ISRAEL; 32° 55′ N 35° 18′ E. Elevation 521 ft/158 m. Founded in 1964 as one of the planned development towns, it has becomes a prosperous city, owing to its large industrial zone (metal, plastics, electronics, and textile factories). Provides surrounding area with commercial and other services.

Karmovo, RUSSIA: see KAMENNOMOSTSKOYE.

Karmøy (KAHRM-uh-u), island (□ 68 sq mi/176.8 sq km) in NORTH SEA at mouth of BOKNAFJORDEN, ROGALAND county, SW NORWAY. Separated from mainland by narrow sound; 18 mi/29 km long (N-S), 6 mi/9.7 km wide. Harbors at KOPERVIK, SKUDENESHAVN, ÅKRAHAMN, AVALDSNES, Torvastad. Fishing (herring, mackerel, lobster); copper mining at Visnes. One of Europe's largest aluminum works near Kopervik. Bronze Age burial sites found at Utrik.

Karnack (KAHR-nak), unincorporated village, HARRISON county, NE TEXAS, 14 mi/23 km NE of MARSHALL. Manufacturing (military ammunition and explosives); timber; cattle. Pre–Civil War structures SW of town; Claudia Taylor (Lady Bird Johnson) born here; Caddo Lake State Park (hunting, fishing, boating) is N, natural lakes formed by Big Cypress Creek, rumored to have been formed by New Madrid earthquake 1811–1812.

Karnak (KAHR-nahk), township, central EGYPT, on the E bank of the NILE River. It is 1 mi/1.6 km NE of LUXOR and occupies part of the site of ancient THEBES. Remains of the pharaohs abound at Karnak. Most notable is the Great Temple of Amon. Although there was an older foundation, the temple was largely conceived and accomplished in the XVIII dynasty, and it is often considered the finest example of New Empire religious architecture. The temple grounds extend about 1,000 ft/300 m. The W half comprises a vast court and the great hypostyle hall (388 ft/118 m by 170 ft/118 m by 52 ft/16 m), with 134 columns arranged in sixteen rows. The E half is a complex of halls and shrines, many of the Middle Kingdom. There are smaller temples at Karnak dedicated to Mut and to Khensu, wife and son respectively of Amon. Karnak is one of Egypt's most important tourist centers.

Karnak (KAHR-nack), village (2000 population 619), PULASKI county, extreme S ILLINOIS, 18 mi/29 km NE of MOUND CITY; 37°17′N 88°58′W. Adjacent to Cache River State Natural Area.

Karnal (kuhr-NAHL), district (□ 759 sq mi/1,973.4 sq km), HARYANA state, N INDIA, ⊙ KARNAL. Bounded E by YAMUNA RIVER; irrigated by WESTERN YAMUNA CANAL system. Hand-loom weaving; agriculture (wheat, gram, millet, maize, rice, cotton). Chief towns are PANIPAT (famous battleground), KARNAL, KAITHAL, SHAHABAD. Area has many ancient sites connected with legends of *Mahabharata*.

Karnal (kuhr-NAHL), city (2001 population 221,236), HARYANA state, N INDIA, on the DELHI-AMBALA railroad; 29°41′N 76°59′E. The town's name believed to be derived from Karna, the rival of Arjuna in the Sanskrit

Mahabharata epic. Market for cotton, salt, wheat, and metal, with manufacturing of leather, rubber, plastics, chemicals, pharmaceuticals, vegetable oil, perfume, and liquor. Livestock-breeding center and the site of the National Dairy Research Institute. The British occupied the town in 1805.

Karnali (kuhr-NAH-lee), administrative zone (2001 population 309,084), W NEPAL. Includes the districts of DOLPA, HUMLA, JUMLA, KALIKOT, and MUGU.

Karnali River (kuhr-NAH-lee), major river of W NEPAL; rises on slopes of MOUNT KAILAS in TIBET (CHINA); flows S through Nepal, receiving the SETI RIVER and later the BHERI RIVER at 27°30′N 81°15′E, continuing to the INDIA border, where it becomes the GHAGHARA RIVER.

Karnaphuli River (KUHR-nuh-puh-lee), 146 mi/235 km long, in NE INDIA (MIZORAM state) and BANGLADESH; rises in W Lushai Hills; flows S, past BARKAL (Bangladesh), and SW, past RANGAMATI and CHITTAGONG, to BAY OF BENGAL 8 mi/12.9 km SSW of Chittagong. Navigable by steamer to Rangamati. Hydroelectric project.

Karnaprayag, INDIA: see CHAMOLI.

Karnataka (kuhr-nah-TUH-kah) (Kannada=region of black soils), state (□ 74,051 sq mi/192,532.6 sq km), S INDIA, on the ARABIAN SEA (W); ⊙ BANGALORE. Known until 1973 as Mysore. Bordered N by the states of GOA and MAHARASHTRA, E by ANDHRA PRADESH, and S by TAMIL NADU and KERALA. The present state was formed from parts of former states of Mysore, BOMBAY, and HYDERABAD. Most of the area is a plateau (elevation 1,000 ft/305 m–3,000 ft/914 m), part of DECCAN PLATEAU and WESTERN GHATS, traversed by the upper KAVERI, TUNGA, and BHADRA Rivers, flowing E. Its many rivers are used for irrigation and hydroelectric power generation (abundant in the state). The coast is part of the KONKAN coastline. Has an excellent road and railroad system. Manufacturing includes steel and steel products, motor vehicles, airplanes, machine tools, watches, computer software, and electronic goods. Coffee is the major crop, but cotton, millet, sugarcane, rice, and fodder are also grown. In the 1990s and 2000s, Bangalore became the center of India's technology boom, with multinational corporations investing heavily in the city as well as many companies starting up there. Most of India's valuable sandalwood forests are here. Produces nearly all of India's gold and chromite and has considerable deposits of iron ore and manganese. The population is largely Hindu with sizable minorities of Muslims and Christians and all speak Kannada. The linguistic uniformity of the state and its excellent education system contribute to one of India's highest literacy rates. Engineering, technological, and medical institutions abound here and provide skilled manpower for a strong industrial base. Several universities and research institutes add to this advantage. The region was part of the empire of the Mauryas (c.325–185 B.C.E.). From the third to the eleventh centuries it was ruled by the Gangas and Chalukyas. In 1313 it was conquered by the DELHI SULTANATE, but it was soon lost to the VIJAYANAGAR kingdom. The region was the site of the earliest European settlements in India. During the eighteenth century the Carnatic (or Karnatic) plains became the arena for the struggle between GREAT BRITAIN and FRANCE for supremacy in India. The early European settlers sometimes applied the term Carnatic to all of S India. In the late eighteenth century the Muslim leaders Haider Ali and his son, Tippoo Sahib, conquered the Hindu rulers of Karnataka, but were defeated in 1799 by the British, who restored the Old Hindu dynasty and thereafter provided protection. In 1947 the state of Mysore acceded to the Indian Union. For centuries Kannada-speaking peoples had been fragmented by division into different regions; in the 1950s Mysore was granted additional territories, doubling its area and largely

consolidating the linguistic group. Governed by a chief minister and cabinet responsible to a bicameral legislature (with one elected house) and by a governor appointed by the president of India.

Karnatic, INDIA: see KARNATAKA.

Karnaticgarh, INDIA: see JAVADI HILLS.

Karnaukhivka (kahr-nah-OO-khif-kah), (Russian *Karnaukhovka*), town, central DNIPROPETROVS'K oblast, UKRAINE, on the right bank of the DNIEPER River, 5 mi/8 km E of DNIPRODZERZHYNS'K and subordinated to its city council; 48°29′N 34°44′E. Elevation 173 ft/52 m. Sand and gravel excavation and grading, cement construction materials; many residents work in neighboring Dniprodzerzhyns'k and Dnipropetrovs'k. Established in 1737, town since 1938.

Karnaukhovka, UKRAINE: see KARNAUKHIVKA.

Karnei Shomron or **Qarnei Shomron**, Jewish settlement, E of KFAR SABA in Samarian highlands, WEST BANK. Most residents work in surrounding industrial areas and in service sector in nearby Israeli cities. The Nahal Kana (the valley of the Kana seasonal stream) nature reserve is nearby. Founded in 1977 by Orthodox ideological Jewish settlers, the name means the "horns of Samaria."

Karnes, county (□ 753 sq mi/1,957.8 sq km; 2006 population 15,270), S TEXAS; ⊙ KARNES CITY; 28°54′N 97°51′W. Drained by SAN ANTONIO RIVER and CIBOLO CREEK, source of Blanco and Medio creeks. Agriculture (corn, wheat, grain sorghum); livestock (cattle); dairying. Oil, natural gas wells, uranium. Formed 1854.

Karnes City, town (2006 population 3,406), ⊙ KARNES county, S TEXAS, c.50 mi/80 km SE of SAN ANTONIO, near SAN ANTONIO RIVER; 28°53′N 97°54′W. Elevation 404 ft/123 m. In sorghum, corn, wheat; cattle area; dairying. Oil and gas, uranium, stone; manufacturing (fiberglass tanks and pipes, steel trailers). Settled 1885, incorporated 1914.

Karni, (KAR-nee) crossing point between ISRAEL and N GAZA STRIP, 6.25 mi/10 km NE of GAZA city. Used mainly as a checkpoint for goods and merchandise. Closed frequently during the second *intifada* following Palestinian terrorist attempts to smuggle explosives through it, and attacks on the crossing itself.

Karnische Alpen, AUSTRIA, ITALY: see CARNIC ALPS.

Karnobat (kahr-no-BAHT), city (1993 population 21,883), BURGAS oblast, ⊙ Karnobat obshtina, E BULGARIA, 26 mi/42 km WNW of BURGAS; 42°39′N 26°59′E. Railroad junction; sheep-raising center, cattle market; grain, cotton, vegetables; vegetable-oil processing; agricultural machinery repair factory; cotton processing, manufacturing metal construction parts for bridges. Has ruins of a Turkish fortress.

Karnobatska (kahr-no-BAHT-ska), mountain, E STARA PLANINA, BULGARIA; 42°55′N 27°00′E

Karnobatski (kahr-no-BAHT-skee), pass in the E STARA PLANINA, BULGARIA; 42°49′N 26°53′E.

Karns City, borough (2006 population 232), BUTLER county, W PENNSYLVANIA, 12 mi/19 km NE of BUTLER. Manufacturing (petroleum refining); oil; agriculture (dairying).

Kärnten, AUSTRIA: see CARINTHIA.

Karo Highlands (kah-ROW), volcanic upland, N. Sumatra province, SSW of MEDAN, SUMATRA, INDONESIA. Includes Karo district, BRASTAGI town, Mount SIBAYAK, and Mount SINABUNG. Karo Batak people in segment of BARISAN MOUNTAINS. Resorts; agricultural area (vegetables, fruits, flowers, and rice) serves lowland N Sumatra, W MALAYSIA, and SINGAPORE.

Karoi, town, MASHONALAND WEST province, N ZIMBABWE, 50 mi/80 km NW of CHINHOYI, near source of MWAMI River; 16°44′S 29°41′E. Road junction; airstrip to SE. Mica mining to NE; agriculture (cattle, sheep, goats; corn, wheat, tobacco, cotton).

Karo La (KAH-LUH-LAH), pass (c.15,000 ft/4,572 m) in E HIMALAYAS, S TIBET, SW CHINA, on main India-

Lhasa trade route, and 15 mi/24 km WSW of NAGARZE. In 1904, scene of battle between Tibetans and British expedition under Younghusband.

Karolinenthal, CZECH REPUBLIC: see KARLIN.

Károlyváros, CROATIA: see KARLOVAC.

Karonga (kah-rah-ngah), administrative center and district (2007 population 250,775), Northern region, MALAWI, port near NW tip of Lake Malawi (LAKE NYASA), 140 mi/225 km NNE of MZIMBA; 09°56′S 33°56′E. Fishing; rice, cotton. Airport. Arab slave-trading center in 1880s, 1890s. Attacked by Germans, 1914.

Karoo, the semiarid plateaus of WESTERN and EASTERN CAPE provinces, SOUTH AFRICA. The SOUTHERN KAROO is located N of the Cape Ranges and extends circa 200 mi/320 km from E to W (elevation 1,000 ft/305 m–2,000 ft/610 m). It is separated from the GREAT KAROO (*circa* 300 mi/480 km long; elevation 2,000 ft/610 m–3,000 ft/914 m) by the SWARTBERG MOUNTAINS. Other mountain ranges dot this vast region, including SNEEUBERG (Snow Mountains) in the SE, whose highest peak is the Kompassberg (6,300 ft/1,920 m). The Karoo, where irrigated, is very fertile. Livestock grazing, particularly sheep and goats, is important here, and farms in the valleys and lowlands produce fruits, vegetables, and alfalfa. The name is also applied to the low scrub vegetation found in semiarid regions and also to an extensive system of sedimentary rocks laid down over central and S Africa during the late Paleozoic and early Mesozoic eras. Called the Karoo system which is extremely rich in fossils, particularly dinosaur remains. *Karoo* is a Khoi word meaning "place of little water." Also spelled Karroo.

Karoola (kuh-ROO-luh), village, N TASMANIA, SE AUSTRALIA, 12 mi/19 km N of LAUNCESTON; 41°15′S 147°09′E. Oil shale.

Karoo National Park (□ 177.5 sq mi/459.7 sq km) EASTERN CAPE province, SOUTH AFRICA, just W of FORT BEAUFORT. Covers 113,620 acres/46,000 ha. This park houses splendid examples of the various levels of vegetation found in the Karoo. It is geologically extremely rich in fossil-bearing sediments and contains the widest range of small mammals and reptiles common to this environment. Hiking and game-viewing. Established 1973.

Karoonda (ka-ROON-duh), village, SE SOUTH AUSTRALIA, AUSTRALIA, 75 mi/121 km ESE of ADELAIDE; 35°06′S 139°54′E. Railroad junction; agriculture center (sheep; wheat, barley). On display is a meteorite that fell near Karoonda in 1930.

Karor (kah-ROR), town, MUZAFFARGARH district, SW PUNJAB province, central PAKISTAN, 80 mi/129 km NNW of MUZAFFARGARH; 31°13′N 70°57′E. Sometimes called Karor Lal Isa.

Karora (kah-ROR-uh), Italian *Carora*, village, SEMENAWI KAYIH BAHRI region, N ERITREA, near Eritrea-Sudan border; 70 mi/113 km SE of TOKAR (SUDAN), and 20 mi/32 km from the RED SEA; 17°42′N 38°21′E. On ASMARA-PORT SUDAN road; frontier and police posts; livestock market.

Karos (KAH-ros), island (□ 6 sq mi/15.6 sq km), CYCLADES department, SOUTH AEGEAN department, GREECE, in Aegean Sea, SE of NAXOS island; 4 mi/6.4 km long, 1.5 mi/2.4 km wide; 36°52′N 25°40′E. Fisheries. Also spelled Karos.

Karoyi, village, MASHONALAND CENTRAL province, NE ZIMBABWE, 15 mi/ 24 km N of MOUNT DARWIN; S of MAVURADONHA MOUNTAINS; 16°38′S 31°38′E. Cattle, goats, sheep; corn.

Karpacz (KAHR-pahch), German *Krummhübel*, town in LOWER SILESIA, Jelenia Góra province, SW POLAND, near CZECH border, in the RIESENGEBIRGE, at N foot of the SCHNEEKOPPE, on small river (irrigation dam), and 9 mi/14.5 km S of JELENIA GÓRA (Hirschberg). Popular health and winter-sports resort. Large orphanage here. Until early 19th century, noted for its trade in medicial herbs.

Karpasia, CYPRUS: see KARPAS PENINSULA.

Karpas Peninsula (kahr-PAHS), Turkish *kirpas peninsula*, NE CYPRUS, juts c. 50 mi/80 km ENE into the MEDITERRANEAN, narrowing from a c. 10 mi/16 km-wide base to Cape APOSTOLOS ANDREAS. Famagusta Bay to S; KYRENIA MOUNTAINS extend into peninsula from N CYPRUS.

K'arpa'talja, UKRAINE: see TRANSCARPATHIAN OBLAST.

Kárpathos (KAHR-pah-thos), Italian *Scarpanto*, Latin *Carpathus*, island (□ 111 sq mi/288.6 sq km), DODECANESE prefecture, SOUTH AEGEAN department, GREECE, in the AEGEAN SEA, 30 mi/48 km SW of RHODES across Kárpathos Strait; 35°37′N 27°10′E. Rises to 3,986 ft/1,215 m; 30 mi/48 km long, 2 mi/3.2 km–6 mi/9.7 km wide (third-largest island in the Dodecanese). Mountainous. Barley, wheat, olive oil, grapes, chickpeas, potatoes; livestock raising (sheep, goats); fishing; gypsum deposits. Some tourism. Village of OLIMBOS famed for colorful, traditional costumes. Airport. Main town, Pegadia or Pigadhia, is on SE shore.

Karpats'ka Ukrayina, UKRAINE: CARPATHIAN UKRAINE.

Karpatskaya Ukraina, UKRAINE: see CARPATHIAN UKRAINE.

Karpatskiy Prirodnyy Natsional'nyy Park, UKRAINE: see CARPATHIAN NATURAL NATIONAL PARK.

Karpatskiy zapovednik, UKRAINE: see CARPATHIAN NATURE PRESERVE.

Karpats'kyy Pryrodnyy Natsional'nyy Park, UKRAINE: see CARPATHIAN NATURAL NATIONAL PARK.

Karpats'kyy zapovidnyk, UKRAINE: see CARPATHIAN NATURE PRESERVE.

Karpenesion (kahr-pe-NEE-see-on), town, ⊙ EVRITANIA prefecture, CENTRAL GREECE department, at S end of PINDOS mountains, NE of AGRINION, and 35 mi/56 km W of LAMÍA (linked by road); 38°55′N 21°47′E. Elevation 3,280 ft/1,000 m (highest prefectural capital in Greece). Livestock center; dairy products (cheese, milk). Seat of bishop. Marco Bozzaris fell here (1823) in battle against the Turks. Also spelled Karpenission, Karpenision.

Kärpf, peak (9,167 ft/2,794 m) in GLARUS ALPS, E central SWITZERLAND, 8 mi/12.9 km S of GLARUS.

Karpinsk (kahr-PEENSK), city (2006 population 29,695), N SVERDLOVSK oblast, extreme W Siberian Russia, in the NE foothills of the central URALS, on railroad (Bogoslovsk station), on the Turya River (OB′ RIVER basin), 267 mi/430 km N of YEKATERINBURG, and 23 mi/37 km WNW of SEROV; 59°45′N 60°00′E. Elevation 583 ft/177 m. Bituminous-coal-mining center; machine-building, specialized clothing, timbering, food industries (nonalcoholic beverages plant, bakery). Founded in 1769. Known as Bogoslovskiy or Bogoslovskiy Zavod prior to World War I, when it produced pig iron and copper; later called Ugol'nyy (1935–1941); became city and renamed in 1941.

Karpogory (kahr-puh-GO-ri), village (2005 population 4,380), S ARCHANGEL oblast, N European Russia, on the PINEGA RIVER, 120 mi/193 km ESE of ARCHANGEL; 64°05′N 44°27′E. Elevation 252 ft/76 m. Lumbering.

Karpuninskiy (kahr-POO-neen-skeeye), town (2006 population 585), central SVERDLOVSK oblast, W SIBERIA, RUSSIA, on the TURA RIVER, on highway and railroad, 28 mi/45 km S of SOS'VA; 58°43′N 61°48′E. Elevation 295 ft/89 m. In copper-, iron-, and gold-mining region; logging.

Karpushikha (kahr-POO-shi-hah), town (2006 population 1,490), SW central SVERDLOVSK oblast, central URALS, W Siberian Russia, on road, 54 mi/87 km N of YEKATERINBURG, and 8 mi/13 km NW (under jurisdiction) of KIROVGRAD; 57°30′N 59°53′E. Elevation 869 ft/264 m. On a railroad spur; copper-mining center, supplying the Kirovgrad refinery.

Karradi (kahr-RAH-dee), village, THI-GAR province, SE IRAQ, on the SHATT AL GHARRAF, 50 mi/80 km N of

AN NASIRIYAH, and 12 mi/20 km N of ancient LAGASH. Grain, dates.

Karrat Fjord (KAH-raht), inlet of BAFFIN BAY, W GREENLAND; 70 mi/113 km long, 3 mi/4.8 km–17 mi/27 km wide; 71° 30′N 53°W. Separated from UUMMANNAQ FJORD (S) by UBEKENDT and UPERNIVIK islands. Receives large glacier at its head. Mountains rise steeply to over 7,000 ft/2,134 m on both shores. Extends 40-mi/64-km arm (N), called UKKUSISSAT FJORD.

Karratha (ka-RATH-uh), town (2001 population 10,057), NW WESTERN AUSTRALIA state, W AUSTRALIA, in PILBARA region, on INDIAN OCEAN, 147 mi/237 km WSW of PORT HEDLAND; 20°44′S 116°51′E. Established 1968 as residential and service center for employees of Hamersley Iron Company, Dampier Salt Works, and North West Shelf oil and gas production.

Kärrgruvan (SHER-GROO-van), village, VÄSTMANLAND county, central SWEDEN, in BERGSLAGEN region, 7 mi/11.3 km NE of FAGERSTA. In former iron-mining region.

Karroo, plateau, SOUTH AFRICA: see KAROO.

Kars, city (2000 population 78,473), E TURKEY, in former Turkish Armenia, on the KARS RIVER; 40°35′N 43°05′E. Manufacturing (textiles, carpets, food products). An old fortified city, well situated in the mountains, Kars was the capital of an Armenian state in the 9th–10th centuries. It was destroyed by Tamerlane in 1386 and then captured and rebuilt by the Ottoman Turks in the 16th century. In 1828, 1855, and 1877 the city was occupied by RUSSIA; and together with the surrounding region was ceded to Russia by the Congress of Berlin in 1878. By a peace treaty (1921) between the nationalist Turkish government of Kemal Atatürk and the USSR, Kars and ARDAHAN were returned to Turkey. Kars has an 11th century Armenian church.

Karsa, ETHIOPIA: see KERSA.

Karsakpai (kahr-sahk-PEI), town, W ZHEZKAZGAN region, KAZAKHSTAN, on railroad, 320 mi/515 km SW of KARAGANDA; 47°47′N 66°43′E. Copper smelter (fed by Zhezkazgan mines), power plant. Also spelled Qarsaqbai.

Karsau, LATVIA: see KĀRSAVA.

Kārsava (KAHR-sah-vuh), German *karsau*, city, E LATVIA, in LATGALE, 23 mi/37 km NE of REZEKNE, near Russian border; 56°47′N 27°40′E. In lumbering area. Russian name (until 1917) was korsovka; was in Russian VITEBSK government until 1920.

Karseong, INDIA: see KARSIYANG.

Karshi, city, KASHKADARYO wiloyat, S UZBEKISTAN, on the KASHKA DARYA; 38°53′N 65°48′E. Center of a fertile oasis that produces wheat, cotton, and silk. Founded (9th century) as a stop on the caravan route between SAMARKAND and AFGHANISTAN. Has a 16th century mosque and mausoleum. Also spelled Qarshi.

Karsiang, INDIA: see KARSIYANG.

Karsiyang, town, DARJILING district, N WEST BENGAL state, NE INDIA, 11 mi/18 km S of DARJILING, in SE Nepal Himalayan foothills; 26°53′N 88°17′E. Health resort; trades in tea, rice, corn, cardamom, oranges, jute; tea processing; general engineering products. Also spelled Karseong; formerly Karsiang, Kurseong.

Karskiye Vorota (KAHR-skeeye-vuh-RO-tuh) [Russian=Kara gates], strait joining BARENTS (W) and KARA (E) seas of the ARCTIC OCEAN, off N ARCHANGEL oblast, NE European RUSSIA, between NOVAYA ZEMLYA (N) and VAYGACH Island (S); 30 mi/48 km wide, 20 mi/32 km long; 70°30′N 58°00′E. Sometimes called Kara Strait.

Karsovay (kahr-suh-VEI), village, N UDMURT REPUBLIC, E European Russia, on road, 22 mi/35 km ENE of GLAZOV; 58°13′N 53°11′E. Elevation 748 ft/227 m. In agricultural area (wheat, oats, rye, flax).

Kars River, 55 mi/89 km long, NE TURKEY; rises 18 mi/29 km NW of SARIKAMIS; flows NE, past KARS, to the ARPA 31 mi/50 km E of Kars.

Karst, SLOVENIA: see KRAS.

Kårstø (KAWRST-uh), gas pipeline terminal on N shore of BOKNAFJORDEN, ROGALAND county, NORWAY, SE of HAUGESUND and N of STAVANGER. Receives natural gas and natural gas liquids from fields in the NORTH SEA. Deep-water port facilities for liquid gas export.

Karsun (kahr-SOON), town (2006 population 8,105), NW ULYANOVSK oblast, E central European Russia, on the BARYSH River (right tributary of the SURA River), on crossroads, 55 mi/89 km WSW of ULYANOVSK; 54°11′N 46°59′E. Elevation 482 ft/146 m. Highway hub; local transshipment point; textile and woodworking industries. Made city in 1780; reduced to status of a village in 1925; made town in 1943.

Karszulkadrive, TURKEY: see KADIRLI.

Karszulkadriye, TURKEY: see KADIRLI.

Karta, AUSTRALIA: see KANGAROO ISLAND.

Kartaba, LEBANON: see QARTABA.

Kartal, town, NW TURKEY in ASIA MINOR, on Sea of MARMARA, 14 mi/23 km SE of ISTANBUL; 40°54′N 29°12′E. Grain market.

Kartalinia, GEORGIA: see KARTLI.

Kartaly (kahr-tah-LI), city (2005 population 28,050), SE CHELYABINSK oblast, SE URALS, RUSSIA, on a left tributary of the TOBOL River, railroad junction on the SOUTH SIBERIAN RAILROAD, approximately 12 mi/19 km NW of the KAZAKHSTAN border, 161 mi/259 km SSW of CHELYABINSK, and 70 mi/113 km ESE of MAGNITOGORSK; 53°03′N 60°40′E. Elevation 984 ft/299 m. Locomotive and railroad car repair shops; spare parts for agricultural machinery; carpeting. Became city in 1944, absorbing anthracite mines of POLTAVKA (2 mi/3.2 km N).

Kartarpur (kuhr-TAHR-puhr), town, JALANDHAR district, PUNJAB state, N INDIA, 8 mi/12.9 km NW of JALANDHAR; 31°27′N 75°30′E. Agricultural market center (wheat, gram, cotton, maize); hand-loom weaving; known for carpentry (chairs, tables, boxes). Visited annually by Sikh pilgrims.

Karthala (kahr-tah-LAH), volcanic mountain (7,746 ft/2,361 m), NJAZIDJA island and district, NW COMOROS, INDIAN OCEAN, 10 mi/16 km SE of MORONI, near S end of island; 11°48′S 43°21′E. Hiking trails. Erupted April 5, 1977, burying village of SINGANI to SW.

Karthalin Range (KAHR-thuh-lin), S spur of the E Greater Caucasus, in NE GEORGIA; extends in arc from Mount BARBALO c.50 mi/80 km SW and S to area NE of TBILISI; forms watershed between Iora and Aragva rivers.

Karthaus, POLAND: see KARTUZY.

Kartli (KAHRT-lee), Russian *Kartliya* or *Kartaliniya*, hilly central region, GEORGIA, between TRIALET RANGE (S) and outliers of the central Greater Caucasus; drained by middle KURA River. Grain, orchards. The heart of Georgia, it contains cities of GORI and TBILISI. Sometimes spelled Karthlia or Karthalinia.

Kartli Mountains (KAHRT-lee), range on the S slopes of the Greater CAUCASUS, between Pshavskaia Araqvi and IORI rivers; 328 mi/100 km long. Elevation 9,843 ft/3,000 m.

Kartsa, RUSSIA: see OKTYABR'SKOYE, NORTH OSSETIAN REPUBLIC.

Kartuzy (kahr-TOO-zee), German *Karthaus*, town, Gdańsk province, N POLAND, 18 mi/29 km W of GDANSK. Manufacturing of bricks, mineral water; sawmilling, brewing, tanning. Resort center.

Karuah (kuh-ROO-uh), town, NEW SOUTH WALES, SE AUSTRALIA, 127 mi/205 km N of SYDNEY, 32 mi/51 km N of NEWCASTLE, at mouth of Karuah River, and at W end of Port STEPHENS; 32°40′S 151°58′E. Forest reserve NW. Resort; fishing. Annual aquatic and oyster festival. The Worimi Aboriginal people inhabited the area before European settlement. Formerly called Sawyers Point.

Karubwe, township, CENTRAL province, central ZAMBIA, 20 mi/32 km NNE of LUSAKA. On railroad.

Agriculture (tobacco, wheat, corn, soybeans, cotton, flowers); cattle.

Karumai (kah-ROO-mah-ee), town, Kunohe county, IWATE prefecture, N HONSHU, NE JAPAN, 47 mi/75 km N of MORIOKA; 40°19′N 141°27′E. Hops, tobacco; lumber; charcoal.

Karumba, port, town, QUEENSLAND, NE AUSTRALIA, 275 mi/443 km N of CLONCURRY, 43 mi/69 km from NORMANTON, on Gulf of CARPENTARIA, and on NORMAN RIVER banks; 17°25′S 140°50′E. Commercial fishing.

Karumo (kah-ROO-moh), village, MWANZA region, NW TANZANIA, 5 mi/8 km W of MWANZA, on Mwanza Gulf, inlet of Lake VICTORIA, opposite Mwanza; 02°41′S 32°38′E. Fish; cotton, corn, wheat; cattle, goats, sheep.

Karumwa (kah-ROO-mwah), village, MWANZA region, NW TANZANIA, 50 mi/80 km SSW of MWANZA; 02°53′S 32°38′E. Timber; cattle, goats, sheep; corn, wheat, millet.

Karun, Lake, EGYPT: see BIRKET QARUN.

Karun River (kah-ROON), c.450 mi/720 km long, W IRAN; rises in the ZAGROS mountains; flows S to the SHATT AL ARAB on the IRAQ border. The Karun is navigable for shallow draft vessels; rapids prevent further upstream passage except during high water in April and May. The river was opened to foreign trade in 1888; but since the construction of a railroad line during World War II between the river port of KHORRAMSHAHR, AHVAZ, and the main Iranian system, this route has lost importance. At SHUSHTAR there is a dam designed to irrigate an area of 500 sq mi/1,295 sq km; it is surmounted by a magnificent bridge (no longer in use), probably built in the 3rd century for Shapur I of PERSIA by captured Roman soldiers.

Karur (kuhr-ROOR), city, Tiruchchirapalli district, TAMIL NADU state, S INDIA; 10°57′N 78°05′E. Milled rice, cotton fabrics, and brass ware are the city's chief products. According to Hindu legend, Brahma began the work of creation in Karur, which is referred to as the "place of the sacred cow." Upon the dissolution of the Hindu Vijayanagar empire in 1565, Karur fell to the Naik kings of Madurai. The British occupied the city in 1760.

Karur, INDIA: see KERUR.

Karuri, INDIA: see KARWI.

Karuzi (kah-ROO-zhee), province (□ 563 sq mi/1,463.8 sq km; 1999 population 384,187), NE central BURUNDI; ⊙ KARUZI; 03°07′S 30°08′E. Borders NGOZI NWN, formed partly by Murarangaro River), MUYINGA (N and E), CANKUZO (ESE), RUYIGI (SES), and GITEGA (S and W) provinces. Ndurumu River flows through province. Includes Buhiga and Karuzi towns.

Karuzi (KAH-roo-zhee), village, ⊙ KARUZI province, NE central BURUNDI, 28 mi/45 km NE of GITEGA; 03°06′S 30°09′E. Cinchona and aleurite plantations; tropical plant nurseries.

Karvasaras, Greece: see AMFILOKHIA.

Karviná (KAHR-vi-NAH), German *Karwin*, city (2001 population 65,141), SEVEROMORAVSKY province, E CZECH SILESIA, CZECH REPUBLIC, near the Polish border; 49°52′N 18°33′E. Industrial center of the Ostrava-Karviná coal-mining region; metallurgy, machinery manufacturing, food processing. Formerly in AUSTRIA, the city became (after 1918) an object of dispute between POLAND and CZECHOSLOVAKIA; after World War I a conference of Allied ambassadors awarded it to Czechoslovakia (1920) despite Polish claims. Seized by Poland in October 1938, but restored to Czechoslovakia in 1945.

Karwar (KAHR-wahr), town, ⊙ UTTAR KANNAD district, KARNATAKA state, SW INDIA, port on ARABIAN SEA, on Karwar Bay (estuary), 305 mi/491 km SSE of MUMBAI (Bombay); 14°48′N 74°08′E. Market center for rice, fish, coconuts, timber; manufacturing (furniture; fish curing, fruit canning). English established factory nearby (trade in pepper); used 1638 to 1752, when captured by Portuguese.

Karwendel (kahr-VEN-del), range of TYROLEAN-BAVARIAN LIMESTONE ALPS, in TYROL, W AUSTRIA, extending c. 20 mi/32 km WNW from the Inn at SCHWAZ to Seefeld Pass. Highest peak, Birkkarspitze (8,379 ft/2,554 m); ISAR RIVER rises here. Subsidiary ranges are NORDKETTE (N of INNSBRUCK), Bettelwurfkette, Mintere Karwendelkette, and Vordere Karwendelkette; 47°27′N 11°20′E. Famous for mountaineering.

Karwi (KUHR-wee), town, BANDA district, S UTTAR PRADESH state, N central INDIA, on tributary of the YAMUNA RIVER, and 40 mi/64 km ESE of BANDA; 25°12′N 80°54′E. Trade center (gram, jowar, wheat, cotton, oilseeds, rice). In 19th century, residence of a Maratha leader who built several temples. Sometimes called Karuri.

Karwin, CZECH REPUBLIC: see KARVINÁ.

Karya (kah-ree-AH), town, near center of LEFKAS island, LEFKAS prefecture, IONIAN ISLANDS department, off W shore of GREECE, 9 mi/14.5 km SW of LEFKAS; 38°22′N 20°35′E. Largest inland village on the island. Famed for embroidery work.

Karyes, Greece: see KARIAI.

Karymskoye (kah-RIM-skuh-ye), town (2005 population 11,730), S central CHITA oblast, E SIBERIA, RUSSIA, on highway and the TRANS-SIBERIAN RAILROAD, on the INGODA RIVER, 62 mi/100 km SE of CHITA; 51°37′N 114°21′E. Elevation 1,870 ft/569 m. Railroad establishments; cement plant; forestry services. Tarskiy junction (RR branch to Manchuria) is just E.

Karystos, Greece: see KARISTOS.

Karytaina, Greece: see KARITAINA.

Kas, (Turkish=*Kaş*) village, SW TURKEY, port on MEDITERRANEAN SEA, 75 mi/121 km SW of ANTALYA; 36°12′N 29°38′E. Surrounded by mountains on three sides. Wheat, barley, sesame, vetch. In ancient times, was an important part of S LYCIA. Remaining ruins include a theater, on the long Kas Peninsula and within walking distance of the town, as well as rock tombs and sarcophagi. Founded in the 4th century B.C.E. as Antiphellos; formerly also known as Andifili.

Kasaan (kuh-SAHN), village (2000 population 39), SE ALASKA, on KASAAN BAY, on E shore of PRINCE OF WALES ISLAND, 35 mi/56 km WNW of KETCHIKAN; 55°32′N 132°24′W. Fishing; cannery. Haida Indian settlement.

Kasaan Bay (kuh-SAHN), SE ALASKA, inlet of CLARENCE STRAIT, E PRINCE OF WALES ISLAND, 25 mi/40 km WNW of KETCHIKAN; 35 mi/56 km long, 2 mi/3.2 km–8 mi/12.9 km wide; near 55°31′N 132°25′W. Fishing. Sparsely settled by Haida Indians. KASAAN village on N shore.

Kasaba, TURKEY: see TURGUTLU.

Kasagi (kah-SAH-gee), town, Soraku county, KYOTO prefecture, S HONSHU, W central JAPAN, 22 mi/35 km S of KYOTO; 34°45′N 135°56′E.

Kasahara (kah-SAH-hah-rah), town, Toki county, GIFU prefecture, central HONSHU, central JAPAN, 25 mi/40 km S of GIFU; 35°17′N 137°09′E. Tiles, dishes.

Kasai (kah-SEI), former province (□ 124,000 sq mi/322,400 sq km), S central CONGO. LULUABOURG (now KANANGA) was the capital of the province and is now capital of KASAI-OCCIDENTAL. Between the KASAI and the SANKURU rivers, the Kuba kingdom of the Shongo people existed from the early 17th century. In the S of the province were the constantly warring Luba and Bena Lulua peoples. This ethnic conflict was partly responsible for the secession (Aug. 1960) of the Baluba-dominated Mining State of South Kasai, headed by Albert Kalonji, who proclaimed himself king of South Kasai. The central government reestablished control over the whole of Kasai in December 1961. Now encompasses the KASAI-ORIENTAL (⊙ MBUJI MAYI) and KASAI-OCCIDENTAL administrative regions. Still has far-reaching autonomy, including its own currency. Center of country's diamond mining.

Kasai (KAH-sah-ee), city, HYOGO prefecture, S HONSHU, W central JAPAN, 24 mi/39 km N of KOBE; 34°55′N 134°50′E. Appliances.

Kasaï-Occidental, province (2004 population 4,342,000), SW central DEMOCRATIC REPUBLIC OF THE CONGO; ⊙ KANANGA. Bordered N by ÉQUATEUR province (W portion of border formed by Lokolo River and central portion formed by LUILAKA RIVER), NEE by Kasaï-Oriental province (very small piece of N portion of border formed by LUKENIE RIVER and small piece of central portion of border formed by Lukibu River), SE by KATANGA province, SW by ANGOLA (small piece of central portion of border formed by TSHIKAPA RIVER), and WNW by Bandundu province (S portion of border formed by KASAI RIVER and small piece of central portion of border formed by Lukibu River). SE of province forms part of Plateau du Kasaï. Lukenie and Lukibu rivers flow across N part of province; TSHIKAPA, Luachimo, and Tshiumbe rivers located in S central portion of province. Towns include KANANGA and LUEBO. Railroad runs from SE (from Kasai-Oriental) to W central part of the province, terminating at ILEBO. Regional airports at KANANGA and TSHIKAPA. Kasaï-Occidental and Kasaï-Oriental were formerly combined as KASAI province.

Kasaï-Occidental, CONGO: see KASAI.

Kasaï-Oriental, province (2004 population 5,237,000), central DEMOCRATIC REPUBLIC OF THE CONGO; ⊙ MBUJI-MAYI. Bordered N by ÉQUATEUR province (part of W portion of border formed by Loto River and part of central portion of border formed by LOMELA RIVER), NE tip by ORIENTALE province, E by MANIEMA province (central portion of border formed by LOMAMI RIVER), SES by KATANGA province, and W by Kasaï-Occidental province (very small piece of N portion of border formed by LUKENIE RIVER and small piece of central portion of border formed by Lukibu River). Marsh area in NE around LOMAMI RIVER, on border with MANIEMA. Réserve de Faune des Hippopotemes also located here. Loto, Lonkonia, LOMELA, TSHUAPA, and LUKENIE rivers in N part of province, and SANKURU RIVER in central part of province. Towns include MBUJI-MAYI, KABINDA, GANDAJIKA, and MWENE-DITU. Regional airports at LODJA and TSHILENGE. Kasaï-Oriental and Kasaï-Occidental were formerly combined as KASAI province.

Kasaï-Oriental, CONGO: see KASAI.

Kasai River (ka-SEI) or **Cassai** (KAH-sei), river, c.1,100 mi/1,770 km long, ANGOLA-CONGO border, S central AFRICA; rising in central Angola; flowing E, N, and NW through W Congo to the CONGO River. The Kasai, navigable for c.475 mi/764 km above its mouth, is an important trade artery. Its tributaries include many navigable streams (some of which are rich in alluvial diamonds). Part of river below MUSHIE (Congo) called the Kwa.

Kasai River (kah-SEI), c.215 mi/346 km long, WEST BENGAL state, E INDIA; rises on E CHOTA NAGPUR PLATEAU, W of PURULIA; flows generally SE, past RAIPUR and MEDINIPUR, to HUGLI River, 15 mi/24 km SW of DIAMOND HARBOR. In lower course, called Haldi. Under the irrigation plan of Kangsabati Project, the river has been dammed about 1 mi/1.6 km upstream of its confluence with Kumari River.

Kasaji (kah-SAH-jee), village, KATANGA province, SE CONGO, on railroad, and 150 mi/241 km SW of KAMINA; 10°22′S 23°27′E. Elev. 3,569 ft/1,087 m. Trading center; rice processing. Protestant mission.

Kasakake (kah-SAH-kah-ke), town, Nitta county, GUMMA prefecture, central HONSHU, N central JAPAN, 16 mi/25 km E of MAEBASHI; 36°23′N 139°17′E.

Kasala (kah-SAH-lah), state (2005 population 1,666,000), Eastern region, E SUDAN; ⊙ KASALA; 16°00′N 36°00′E. Bordered by Gadarif (S and W), Khartoum (W), Nile (NWN), and Red Sea (N) states and ERITREA (E). Created in 1996 following a reorganization of administrative divisions. ATBARA RIVER flows

N through state. Railraod between AL-GADARIF and PORT SUDAN travels through KASALA city. Multiple roads branch out of KASALA city, connecting the state with the rest of SUDAN as well as ERITREA. Regional airport at KASALA city. The area that makes up Kasala state was part of former Kasala province. Also spelled Kassala.

Kasala (kah-SAH-lah), former province (□ 134,450 sq mi/349,570 sq km), NE SUDAN, covering approximately the same area as current Gadarif, Kasala, and Red Sea states: see Eastern, region.

Kasala (kah-SAH-lah), city, ☉ KASALA state, NE SUDAN; 15°28′N 36°24′E. Market center and railroad transport hub; extensive fruit gardens. Founded in 1840 as a military camp for the troops of Muhammad Ali during his conquest of SUDAN, Kasala was captured by the Mahdists in 1885 and by the Italians in 1894. Restored to Egyptian sovereignty in 1897, it became part of the Anglo-Egyptian Sudan. Airfield. Kasala city was capital of former KASALA province. Sometimes spelled Kassala.

Kasala, region, SUDAN: see Eastern, region.

Kasama (kah-SAH-mah), city, IBARAKI prefecture, central HONSHU, central JAPAN, 6 mi/10 km W of MITO; 36°22′N 140°14′E.

Kasama (kah-SAH-mah), township, ☉ NORTHERN province, NE ZAMBIA, 400 mi/644 km NE of LUSAKA; 10°12′S 31°11′E. Road junction; on Tazara railroad; airport. Agriculture (corn, vegetables, fruit; cattle); manufacturing (coffee processing). Chishimba Falls to NW on Lukulu River, hydroelectric station; Isangano National Park to SW. Was capital of former Awemba province.

Kasamatsu (kah-SAH-mahts), town, Hashima county, GIFU prefecture, central HONSHU, central JAPAN, on KISO RIVER, and 3.7 mi/6 km S of GIFU; 35°21′N 136°45′E.

Kasamonte (kah-sah-MON-tai), canton, VALLEGRANDE province, SANTA CRUZ department, E central BOLIVIA, SE of VALLEGRANDE, and W of Highway 9 (YACUIBA–SANTA CRUZ); 18°41′S 64°03′W. Elevation 6,660 ft/2,030 m. Clay, limestone, and gypsum deposits. Agriculture (potatoes, yucca bananas, corn, sweet potatoes, peanuts, tobacco); cattle raising for meat.

Kasan, city, W KASHKADARYO wiloyat, UZBEKISTAN, on railroad, on the KASHKA DARYA River, and 17 mi/27 km NW of KARSHI; 39°02′N 65°35′E. Silk milling, food processing.

Kasan, RUSSIA: see KAZAN′.

Kasana (kah-SAH-nah), village, KIGOMA region, W TANZANIA, 17 mi/27 km WSW of KIBONDO, near BURUNDI border; 03°39′S 30°30′E. Timber; tobacco; goats, sheep.

Kasane (KAH-sah-nai), town (2001 population 7,638), NORTH-WEST DISTRICT, BOTSWANA, 262 mi/422 km NE of FRANCISTOWN; 17°46′S 25°09′E. Former capital of CHOBE District. Close to borders of NAMIBIA, ZAMBIA, and ZIMBABWE. Major customs and tourist center.

Kasanga (kah-SANG-gah), town, RUKWA region, SW TANZANIA, 45 mi/72 km SW of SUMBAWANGA small port on E shore of Lake TANGANYIKA, near its S end; 08°26′S 31°09′E. KALAMBO FALLS on ZAMBIA border to S. Fish; corn; sheep, goats. Known as BISMARCKBURG under German rule.

Kasangulu (kah-sahng-GOO-loo), village, BAS-CONGO province, W CONGO, on railroad, and 18 mi/29 km SSW of KINSHASA; 04°35′S 15°10′E. Elev. 1,115 ft/339 m. Trade center; hardwood lumbering. Roman Catholic and Protestant missions.

Kasanka National Park (□ 183 sq mi/475.8 sq km), NE CENTRAL province, E central ZAMBIA, 100 mi/161 km ENE of NDOLA. CONGO border to W. Drained by Luwomba River; has eight lakes, papyrus swamp. Home of sitatunga antelope and rare shoebill stork (similar to extinct dodo); over one million fruit bats roost here in

the rainy season (Dec.–April); other wildlife include: baboons, vervet monkeys, civets, jackals, hyenas, leopards, and unusual bird species. Privately managed since 1990.

Kasansai (kah-sahn-SEI), city, N NAMANGAN wiloyat, UZBEKISTAN, on S slope of CHATKAL RANGE, 18 mi/29 km NNW of NAMANGAN; 41°15′N 71°31′E. In cotton- and silk-growing area; orchards. KASANSAI reservoir, on the short KASAN-SAY River, is N of here (KYRGYZSTAN). Until 1930s called KASSAN.

Kasan-Say, river, c.62 mi/100 km long, in KYRGYZSTAN and UZBEKISTAN; rises in CHATKAL RANGE, in W JALAL-ABAD region of Kyrgyzstan; flows S past KASANSAI (Uzbekistan) to the NARYN River.

Kasaoka (kah-SAH-o-kah), city, OKAYAMA prefecture, SW HONSHU, W JAPAN, on HIUCHI SEA, 28 mi/45 km S of OKAYAMA; 34°30′N 133°30′E. Nori. Protected king crab habitat.

Kasaragod (KAH-suhr-uh-god), district (□ 769 sq mi/1,999.4 sq km), KERALA state, S INDIA; ☉ KASARAGOD.

Kasaragod (KAH-suhr-uh-god), city, KASARAGOD district, KERALA state, S INDIA, on MALABAR COAST of ARABIAN SEA, 27 mi/43 km SSE of MANGALORE; 12°30′N 75°00′E. Fish curing (sardine, mackerel); coconuts, mangoes, rice. Handicraft spinning and weaving 20 mi/32 km SSE, at village of Nileswaram; pottery-clay pits. Sometimes spelled Kasargod, Kanhangad.

Kasar, Cape (kah-SAHR), headland of NE Afr. on the RED SEA, near border between SUDAN and ERITREA; 18°02′N 38°34′E.

Kasari (KAH-sah-ree), town, at N end of AMAMI-O-SHIMA island, Oshima county, KAGOSHIMA prefecture, SW JAPAN, 223 mi/360 km S of KAGOSHIMA; 28°26′N 129°40′E. Pongee, tea.

Kasasa (kah-SAH-sah), town, Kawanabe county, KAGOSHIMA prefecture, S KYUSHU, SW JAPAN, on SW SATSUMA PENINSULA, on EAST CHINA SEA, 25 mi/40 km S of KAGOSHIMA; 31°24′N 130°11′E. Also spelled Kasasuna.

Kasatka (kah-SAHT-kah) [Russian=orca], porton the E coast of ITURUP ISLAND, S Kuriles, RUSSIAN FAR EAST, port on the Kasatka Bay (best-protected bay of island; 6 mi/9.7 km long, 5 mi/8 km wide), 17 mi/27 km SW of KURIL′SK; 45°00′N 147°44′E. Under Japanese rule (until 1945) called Toshimoe. From here, Japanese fleet sailed (1941) to attack PEARL HARBOR.

Kasauli (kuh-SOU-lee), town, SOLAN district, HIMACHAL PRADESH state, N INDIA, 19 mi/31 km SW of SHIMLA; 30°55′N 76°57′E. Hill resort (elevation c.6,300 ft/1,920 m) in SHIWALIK RANGE. Has Pasteur Institute, medical research institute, food laboratory. Distillery nearby. Formerly in exclave of Ambala district, Indian Punjab.

Kasayevo, RUSSIA: see BAKSANENOK.

Kasba Lake, NORTHWEST TERRITORIES, CANADA, near MANITOBA border; 43 mi/69 km long, 6 mi/10 km–23 mi/37 km wide; 60°20′N 102°15′W. Drains N into KAZAN RIVER through ENNADAI LAKE.

Kasba Tadla (KAHS-buh TAD-luh), town, Béni Mellal province, Tadla-Azilal administrative region, W central MOROCCO, on the upper Oued OUM ER RBIA river, and 105 mi/169 km SE of CASABLANCA, at foot of the Middle ATLAS mountains; 32°35′N 06°17′W. Road junction; agricultural trade center (grain; livestock; lumber). Town has one of Morocco's most imposing kasbahs. Founded end of 17th century by Sultan Moulay Ismail to control the TADLA plain. Occupied by French in 1913, and served as headquarters for pacification (1914–1933) of the Middle Atlas region. Nearby Kasba Tadla dam and hydroelectric project on Oued Oum er Rbia river. Also spelled Kasbah Tadla.

Kascas (kahs-kahs), village, SAINT-LOUIS administrative region, N SENEGAL, on SENEGAL River (MAURITANIA border), and 60 mi/97 km ESE of PODOR. Trad-

ing center; fishing; agriculture. Sometimes spelled Cascas.

Kaschau, SLOVAKIA: see KOŠICE.

Kaseda (kah-SE-dah), city, KAGOSHIMA prefecture, SW JAPAN, on W SATSUMA PENINSULA, 19 mi/30 km S of KAGOSHIMA; 31°24′N 130°19′E. Scallions.

Kasempa (kah-SEI-em-pah), township, NORTH-WESTERN province, NW ZAMBIA, 220 mi/354 km NW of LUSAKA on Lufupa River, near its source. Diesel power station. Agriculture (corn, beeswax); cattle. Transferred 1946 from KAONDE-LUNDA province. Was the capital of former Kasempa province. Busanga Swamp and Plain, and KAFUE NATIONAL PARK to S.

Kasenga (kah-SENG-gah), village, KATANGA province, SE CONGO, on LUAPULA RIVER, on ANGOLA border, and 120 mi/193 km NE of LUBUMBASHI. Customs station and head of navigation on the LUAPULA. Manufacturing of ice; fisheries. Roman Catholic mission.

Kasenyi (kah-SEN-yee), village, ORIENTALE province, E CONGO, on LAKE ALBERT (UGANDA border), and 30 mi/48 km E of IRUMU; 01°24′N 30°26′E. Elev. 2,204 ft/671 m. Terminus of lake navigation from UGANDA side; customs station and fishing port. Has a research center for study of trypanosomiasis and is also known as tourist center and base for excursions to the alligator grounds of the SEMLIKI. Bogoro agricultural center with Protestant mission is 10 mi/16 km W.

Kasese (kah-SES-ai), administrative district (□ 1,237 sq mi/3,216.2 sq km; 2005 population 589,200), WESTERN region, SW UGANDA, along Equator, N of Lake EDWARD along DEMOCRATIC REPUBLIC OF THE CONGO border (to W); ☉ KASESE; 00°10′N 30°00′E. As of Uganda's division into eighty districts, borders BUNDIBUGYO and KABAROLE (N), KAMWENGE (E), BUSHENYI (SE), and RUKUNGIRI and KANUNGU (S, on opposite shore of Lake Edward) districts. Main inhabitants are the Bakonjo (speaking the Lukonja language) and Batooro (speaking the Lutooro language) people. Vegetation primarily of savannah type with mountainous area in N and plains in S, also some wetlands in district. Lake GEORGE in SE and RUWENZORI Mountains in N have copper and tin deposits; sulphur and cobalt also mined; primarily agricultural area (bananas, passion fruit, cassava, finger millet, groundnuts, maize, matooke, potatoes, sorghum, and sweet potatoes; cattle, goats, pigs, sheep); timber and fishing also important. Railroad originating in Kasese town travels E across S Uganda (including Kamwenge, KAMPALA, JINJA, and IGANGA cities), turns SE after entering KENYA, travels through S Kenya (including ELDORET, NAKURU, and NAIROBI cities), and terminates in MOMBASA (SE Kenya on INDIAN OCEAN). Roads also connect district to surrounding Uganda and W into Democratic Republic of the Congo.

Kasese (kah-SES-ai), town (2002 population 53,907), ☉ KASESE district, WESTERN region, W UGANDA, on railroad, and 40 mi/64 km SSW of Kabarole, N of LAKE GEORGE; 00°10′N 30°06′E. Petroleum deposits, electricity generation, cotton, corn. Airfield. Was part of former WESTERN province.

Kasese (kah-SE-sai), village, NORD-KIVU province, E CONGO, 295 mi/475 km SE of KISANGANI; 01°38′S 27°07′E. Elev. 2,158 ft/657 m. Tin-mining center; also wolfram and tantalite mining.

Kasevo, RUSSIA: see NEFTEKAMSK.

Kaseyevo, RUSSIA: see NEFTEKAMSK.

Kasganj (KAHS-guhnj), town, ETAH district, W UTTAR PRADESH state, N central INDIA, 55 mi/89 km NE of AGRA; 27°49′N 78°39′E. Railroad junction; trade center (wheat, pearl millet, barley, corn, jowar, mustard, cotton, sugarcane); sugar refining; coir rope manufacturing.

Kasha, Al, YEMEN: see QASHA, AL.

Kashaf River, 150 mi/241 km long, Khorāsān province, NE IRAN; rises in the BINALUD RANGE 50 mi/80 km

NW of MASHHAD; flows SE, through Mashhad valley, to the HARI RUD (Tedzhen River) on TURKMENISTAN-Iranian border. Used for irrigation (wheat, barley).

Kashagawigamog Lake, S ONTARIO, E central CANADA, extends SW from HALIBURTON; 10 mi/16 km long, 1 mi/2 km wide; 44°59′N 78°35′W.

Kāshān (KASH-ahn), city (2006 population 253,509), Esfahān province, W central IRAN; 33°59′N 51°26′E. The city has long been noted for its silk textiles, carpets, ceramics, copperware, and rose water. The Ardebil carpet and celebrated porcelain tiles were made here in the Safavid period. The present city is also a transportation center. The skyline is dominated by a 13th-century minaret that is 150 ft/45 m high. Nearby are Sialk, a prehistoric site, and the well-known rose fields of QAMSAR, or KAMSAR.

Kashar (kah-SHAWR), village, W central ALBANIA, on Tiranë-Durrës railroad, and 5 mi/8 km W of TIRANË; 41°21′N 19°43′E. Linked (since 1949) by 4-mi/6.4-km railroad spur with W industrial outskirts of Tiranë. Also called Kashari.

Kashary (kah-shah-RI), village (2006 population 6,290), N ROSTOV oblast, S European Russia, on road, 30 mi/48 km ENE of MILLEROVO; 49°02′N 41°01′E. Elevation 272 ft/82 m. Flour milling.

Kashega (kuh-SHEE-guh), village, SW UNALASKA Island, ALEUTIAN ISLANDS, SW ALASKA, 40 mi/64 km SW of DUTCH HARBOR.

Kashegelok (kah-SHEE-ge-lok), Inuit village, SW ALASKA, on lower Holitna River, and 110 mi/177 km SE of HOLY CROSS. Formerly spelled Kashegaluk.

Kashgar, China: see KASHI.

Kashgar River, China: see KAXGAR RIVER.

Kashi (KAH-SHI), city (□ 37 sq mi/97 sq km; 1994 estimated urban population 190,000; 1994 estimated total population 243,000; 2000 total population 214,624), SW XINJIANG UYGUR, NW CHINA, on the KAXGAR RIVER (a tributary of the TARIM); 39°29′N 76°02′E. The city is the hub of an important commercial district, the W terminus of the main road of the province, and a center for caravan trade with INDIA, AFGHANISTAN, TAJIKISTAN, and KYRGYZSTAN. Kashi is the largest commercial market in central Asia. Wheat, corn, cotton, barley, rice, beans, and fruit are grown here. Cotton and wool cloth, rugs, leatherware, and jewelry are manufactured here. A mountain pass from Kashi provides a route to SAMARKAND and then to the MIDDLE EAST. The city, predominantly Uygur in ethnic composition, first came under Chinese rule in the period of the Han dynasty (206 B.C.–A.D. 221). Romans traded here in the 6th century. When Kashi was the capital of the Uygur Turks (750–840), it was also a center of Manichaeism. Visited by Marco Polo in 1275, Kashi was conquered soon after by Jenghiz Khan. It was ruled by hereditary Khojar (Muslim) kings from the 15th to the 17th century. The city passed definitively to China in 1760, but there have been uprisings and periods of contested control since then. Hit by a devastating earthquake in March 1996. Also called Kashgar.

Kashi, INDIA: see VARANASI, city.

Kashihara (kah-SHEE-hah-rah), city, NARA prefecture, S HONSHU, W central JAPAN, near Yamato mountains, 12 mi/20 km S of NARA; 34°30′N 135°47′E. Automotive parts. Mausoleum of first emperor Jimmu; ruins of Fujiwara castle.

Kashima (KAH-shee-mah), city, SAGA prefecture, W KYUSHU, SW JAPAN, on E HIZEN peninsula, 16 mi/25 km S of SAGA; 33°06′N 130°06′E. Agricultural (nori, mandarin oranges); manufacturing (textiles, handicrafts). Site of Yutoku Inari Shrine.

Kashima (KAH-shee-mah), town, Soma county, FUKUSHIMA prefecture, N central HONSHU, NE JAPAN, 31 mi/50 km N of FUKUSHIMA city; 37°41′N 140°58′E. Pears, shiitake mushrooms, cucumbers; pigs.

Kashima (kah-SHEE-mah), town, Kashima county, IBARAKI prefecture, central HONSHU, E central JAPAN,

on E shore of Kitaura lagoon, 28 mi/45 km S of MITO; 35°57′N 140°38′E. Iron, petrochemicals; coal liquefaction. Bonsai pines. Well-known Shinto Kashima shrine said to have been founded in 7th century B.C.E. Swordsman Tsukahara Bokuden born here.

Kashima (KAH-shee-mah), town, Kamimashiki county, KUMAMOTO prefecture, W KYUSHU, SW JAPAN, 3.1 mi/5 km S of KUMAMOTO; 32°44′N 130°45′E.

Kashima (KAH-shee-mah), town, Yatsuka co., Shimane prefecture, SW Honshu, W Japan, 4.3 mi/7 km N of Matsue; 35°31′N 132°59′E. Seaweed (wakame).

Kashima (KAH-shee-mah), village, Satsuma county, KAGOSHIMA prefecture, SW KYUSHU, SW JAPAN, 47 mi/75 km W of KAGOSHIMA; 31°46′N 129°47′E.

Kashimadai (KAH-shee-mah-DEI), town, Shida county, MIYAGI prefecture, NE JAPAN, 19 mi/30 km N of SENDAI; 38°28′N 141°06′E.

Kashimo (kah-SHEE-mo), village, Ena county, GIFU prefecture, central HONSHU, central JAPAN, 43 mi/70 km N of GIFU; 35°42′N 137°22′E. Tomatoes; hinoki.

Kashin (KAH-shin), city (2006 population 16,370), SE TVER oblast, W central European Russia, on the Kashinka River (tributary of the VOLGA RIVER), on crossroads and railroad, 93 mi/150 km ENE of TVER; 57°21′N 37°36′E. Elevation 459 ft/139 m. Flax processing, food processing (dairy, distillery, meat packing, mineral water); manufacturing (electrical equipment; felt boots, textiles). Health resort (ferruginous springs, mud baths) since 1884; has a dental clinic. Dates from 1238.

Kashing, China: see JIAXING.

Kashipur (KAH-shee-puhr), town (2001 population 92,978), UDHAM SINGH NAGAR district, N Uttarakhand state, N central INDIA, on tributary of the RAMGANGA RIVER, and 33 mi/53 km SW of NAINITAL; 29°13′N 78°58′E. Railroad and road junction; trades in rice, wheat, oilseeds, gram, corn, cloth; sugar milling. Every March, hundreds of thousands of Hindu pilgrims go to the Chaiti Devi temple for a fair called Navratras nearby. Also nearby are the ruins of a large settlement and fort of the Chand Dynasty rulers.

Kashira (kah-SHI-rah), city (2006 population 39,820), S MOSCOW oblast, central European Russia, on the right bank of the OKA River, on road and railroad, 71 mi/114 km SSE of MOSCOW; 54°50′N 38°10′E. Elevation 606 ft/184 m. Ship construction and repair, steel structures, machinery, furniture, ceramics. Kashira power plant, 3 mi/5 km to the E, is based on lignite from the Moscow coal basin.

Kashi Range, China: see MUZTAGATA RANGE.

Kashirinskoye, RUSSIA: see OKTYABR'SKOYE, ORENBURG oblast.

Kashirskoye (kah-SHIR-skuh-ye), village (2006 population 4,545), central VORONEZH oblast, S central European Russia, on road, 23 mi/37 km SE of VORONEZH; 51°24′N 39°35′E. Elevation 442 ft/134 m. In agricultural area (wheat, rye, flax, oats, vegetables); food processing.

Kashitu (kah-SHEE-too), township, COPPERBELT province, N central ZAMBIA, 50 mi/80 km S of NDOLA. On railroad. Agriculture (tobacco, wheat, corn, cotton, soybeans, flowers).

Kashiwa (kah-SHEE-wah), city (2005 population 380,963), CHIBA prefecture, E central HONSHU, E central JAPAN, 12 mi/20 km N of CHIBA; 35°51′N 139°58′E. Spring onions.

Kashiwa (KAH-shee-wah), village, West Tsugaru county, Aomori prefecture, N HONSHU, N JAPAN, 19 mi/30 km E of AOMORI; 40°46′N 140°24′E.

Kashiwabara, RUSSIA: see SEVERO-KURIL'SK.

Kashiwara (kah-SHEE-wah-rah), city, OSAKA prefecture, S HONSHU, W central JAPAN, 9 mi/15 km S of OSAKA; 34°34′N 135°37′E.

Kashiwazaki (KAH-shee-wah-ZAH-kee), city, NIIGATA prefecture, central HONSHU, N central JAPAN, on the SEA OF JAPAN, 47 mi/75 km S of NIIGATA; 37°22′N 138°33′E. Resort town known for its hot springs. Ka-

shiwazaki produces automotive parts and natural gas. Has a nuclear power plant. Fukuura Shojo Cave is nearby.

Kashka Darya, river, c.200 mi/322 km long, UZBEKISTAN. It is the basis of a wide network of irrigation canals near the towns of KITAB and KARSHI.

Kashkadarya, town, KASHKADARYO wiloyat, S UZBEKISTAN; 39°30′N 64°34′E. Also called Kashka Darya, Qashqadaryo.

Kashkadaryo, wiloyat (□ 11,300 sq mi/29,380 sq km), SE UZBEKISTAN; ⊙ KARSHI. Bounded E by the BAISUN-TAU mountains; drained by the KASHKA DARYA. Cotton growing, orchards in SHAKHRISABZ-KITAB area; wheat, cotton, sericulture in irrigated Karshi oasis; elsewhere, dry farming (wheat) and karakulsheep breeding. KAGAN-DUSHANBE (TAJIKISTAN) railroad passes NW-SE, with branch to KITAB. Population consists of Uzbeks, Tajiks. Formed 1943. An earlier oblast with this name existed 1924–1926. Also called Qashqadaryo, KASHKA-DARYA.

Kashkhatau (kahsh-hah-TAH-oo), town (2005 population 5,425), SE KABARDINO-BALKAR REPUBLIC, N CAUCASUS, S European Russia, on the CHEREK RIVER, on highway, 26 mi/42 km S of NAL'CHIK; 43°19′N 43°36′E. Elevation 2,585 ft/787 m. Lumbering, woodworking. Called Sovetskoye between 1944 and 1990.

Kashla-koi, Bulgaria: see ZIMNITSA.

Kashmar (kahsh-MAHR), town (2006 population 83,667), Khorāsān province, NE IRAN, on edge of salt desert, 110 mi/177 km SW of MASHHAD; 35°14′N 58°28′E. Agriculture (cotton, grain, raisins, fruit, opium); produces silk. Originally called SULTANABAD; later TURSHIZ or TORSHIZ. Present village of SULTANABAD is 23 mi/37 km E, on road to TORBAT-e-Hey-darigeh.

Kashmir (KASH-mir), region (□ 85,000 sq mi/221,000 sq km) and former princely state, extreme N INDIA and NE PAKISTAN. The region is administered in two sections: the Indian state of JAMMU and KASHMIR and the Pakistani-controlled AZAD KASHMIR. Kashmir is bordered on the W and NW by Pakistan, on the S by India, and on the N and NE by CHINA. A beautiful region of S ASIA, Kashmir is covered with lofty, rugged mountains, including sections of the Himalayan and Karakoram ranges. Rivers, including the Indus, run through relatively narrow but heavily populated valleys. Three of the main rivers, the INDUS, CHENAB, and JHELUM, rise in India and flow into Pakistan. The valley of the Jhelum River, the celebrated Vale of Kashmir, is the most populous area (mostly Muslim population) and the economic heart of the region; it produces abundant crops of wheat and rice. The noted handicraft industry, particularly the making of woolen cloth and shawls (cashmeres) and carpet-making, have, along with tourism, suffered setbacks due to the troubled situation in the region. In the late fourteenth century, after years of Buddhist and Hindu rule, Kashmir was conquered by Muslims who converted most of the population. It became part of the Mogul empire in 1586, but by 1751 the local ruler was independent. After a century of disorder the British pacified Kashmir in 1846 and installed a Hindu prince as ruler of the predominantly Muslim region. When British control ended, the population was 77% Muslim and 20% Hindu (mostly in Jammu and the S), with the remainder being Sikhs and Buddhists. When India was partitioned in 1947, a Muslim revolt flared up against the state government of Kashmir. The Hindu ruler fled to Delhi and there agreed to place Kashmir under the dominion of India. Indian troops were flown to Srinagar to engage the rebels, who were backed by Pakistan. The fighting was ended by a UN cease-fire (1949), but the region was divided between India and Pakistan along the cease-fire line. A constituent assembly in Indian Kashmir voted in 1953 for incorporated into India, but this was delayed by continued Indo-Pakistani disagreement and UN disapproval of annexation without a plebiscite. In 1955 India and Pakistan agreed to keep their respective forces in Kashmir 6 mi/9.7 km apart. A new vote by the assembly in Indian Kashmir in 1956 led to the integration of Kashmir as an Indian state; Azad Kashmir (to the W) remained, however, under the control of Pakistan. India refused to consider subsequent Pakistani protests and UN resolutions calling for a plebiscite. The situation was complicated in 1959, when Chinese troops occupied the Aksai Chin section of the district of Ladakh. Indo-Pakistani relations became more inflamed in 1963 when a Sino-Pakistani agreement defined the Chinese border with Pakistani Kashmir. Serious fighting between India and Pakistan broke out in August 1965. A UN cease-fire took effect in September. In January 1966, President Ayub Khan of Pakistan and Prime Minister Lal Bahadur Shastri of India met at Toshkent (now in Uzbekistan) at the invitation of the Soviet government and agreed to the mutual withdrawal of troops to the positions held before the latest outbreak. In the December 1971 war between India and Pakistan, however, India made some gains in further fighting in Kashmir. In December 1972, a new cease-fire line along the positions held at the end of the 1971 war was agreed to by India and Pakistan. During the 1990s, Indian and Pakistani troops continued to exchange fire over control of the 2,000-ft/610-m Siachen Glacier, an area that, because of its altitude, cannot support human life. Formerly spelled Cashmere.

Kashmir North, INDIA: see BARAMULA.

Kashmir South, INDIA: see ANANTNAG.

Kashmir, Vale of (KASH-mir), valley in the HIMALAYAS, JAMMU AND KASHMIR state, N INDIA, W KASHMIR region, drained by upper JHELUM (Veth) River; 85 mi/137 km long (NW-SE), c.20 mi/32 km wide. Average elevation 5,300 ft/1,615 m. Surrounded by lofty peaks (12,000 ft/3,658 m–16,000 ft/4,877 m) of PIR PANJAL and main Himalayan ranges and containing a wealth of scenic grandeur, the valley is one of the garden spots of the East and a famous tourist attraction. Places of interest include SRINAGAR, largest city and the capital of JAMMU AND KASHMIR state, DAL LAKE (houseboating), WULAR LAKE, ANANTNAG (noted springs), GULMARG (hill resort), and VERNAG (source of JHELUM at nearby spring). Area is the most populous section and the economic heart of the KASHMIR region, as well as the core area for JAMMU AND KASHMIR state. Noted for its wool and silk weaving, and wood carving; fertile alluvial soil yields rice, corn, fruit (apples, pears), and vegetables. Main approaches to valley are by road over BANIHAL Pass (SE) from JAMMU area and through JHELUM gorge near BARAMULA (NW) from PAKISTAN; airport at SRINAGAR. Long known for its beauty, the "happy valley" was visited by MOGUL emperors, especially Jahangir, who built fine gardens and buildings here. Also called Kashmir Valley.

Kashmor (kah-shah-MOR), town, JACOBABAD district, N SIND province, SE PAKISTAN, 70 mi/113 km ENE of JACOBABAD; 28°26'N 69°35'E.

Kashofu (kuh-SHO-foo), village, SUD-KIVU province, E CONGO, on IDJWI ISLAND (L. KIVU); 02°13'S 29°02'E. Mission.

Kashtak (kahsh-TAHK), village, NE CHELYABINSK oblast, SW SIBERIA, RUSSIA, near highway, 10 mi/16 km N of CHELYABINSK; 55°18'N 61°21'E. Elevation 754 ft/229 m. In a wooded area; forestry.

Kashubia, POLAND: see KASSUBIA.

Kasia (KUHS-yuh), town, DEORIA district, E UTTAR PRADESH state, N central INDIA, 34 mi/55 km E of GORAKHPUR; 26°45'N 83°55'E. Reputed site of ancient Kusangara or KUSINAGARA is 1 mi/1.6 km W; scene of Buddha's death and cremation; one of eight ancient great places of Buddhist pilgrimage. Buddhist remains include stupa ruins and a large red sandstone image of the dying Buddha carved at MATHURA during Gupta period.

Kasibugga, INDIA: see PALASA.

Kasigluk (KAH-si-gluhk), village (2000 population 543), SW ALASKA, 32 mi/51 km NW of BETHEL; 60°53'N 162°33'W. Numerous small lakes in area. Fishing, hunting. Also called Akolmiut.

Kasillunca (kah-see-YOON-kah), canton, PACAJES province, LA PAZ department, W BOLIVIA. Elevation 12,989 ft/3,959 m. Gas resources in area; copper (unmined), clay, limestone, and gypsum deposits. Agriculture (potatoes, yucca, bananas, rye); cattle.

Kasilof (KA-si-lof), fishing village, S ALASKA, W KENAI PENINSULA, on COOK INLET, 65 mi/105 km WNW of SEWARD. Site of Russian settlement, established 1786, called Saint George.

Kasimbila (kah-seem-BEE-lah), town, TARABA state, SE NIGERIA, on road, and 175 mi/282 km SW of JALINGO. Market town. Groundnuts, millet, maize.

Kasimov (kah-SEE-muhf), city (2006 population 35,205), NE RYAZAN oblast, central European Russia, on the OKA River, on crossroads and railroad spur, 97 mi/156 km ENE of RYAZAN; 54°56'N 41°23'E. Elevation 465 ft/141 m. Manufacturing (refrigeration machinery; net weaving, sheepskin and fur processing, ship repair). Agriculture (wheat, oats, vegetables; livestock raising). Has several relics of Tatar domination (minaret, mausoleum). Founded in 1152 (named Gorodets-Meschcherskiy until 1471); passed to MOSCOW in 1393; chartered and named Kasimov (1467); was the capital of Tatar principality (15th–17th centuries).

Kasindi (kah-SEN-dee), village, NORD-KIVU province, E CONGO, on N shore of LAKE EDWARD, 90 mi/145 km NNE of BUKAVU; 00°03'S 29°41'E. Elev. 3,687 ft/1,123 m. Customs station on UGANDA border.

Kasioei, INDONESIA: see WATUBELA ISLANDS.

Kasiui, INDONESIA: see WATUBELA ISLANDS.

Kaskaskia, small village (2000 population 9), RANDOLPH county, SE ILLINOIS, on Kaskaskia Island in the MISSISSIPPI RIVER where it is joined by the KASKASKIA RIVER, 5 mi/8 km W of CHESTER; 37°55'N 89°55'W. The settlement was established 1703 by Jesuit missionaries and named for a local Native American group of the Illinois. The French built a fort here in 1721, which was destroyed when Kaskaskia was taken over (1763) by the British. During the American Revolution, George Rogers Clark took (1778) U.S. possession of the village. It thrived as the capital of Illinois Territory (1809–1818) and state capital (1818–1820); the first Illinois newspaper started here in 1814. The community declined after the capital was shifted (1820) to VANDALIA. Flooding from the Mississippi in the late nineteenth century further discouraged growth. Fort Kaskaskia State Park was established in 1927 across the Mississippi River near Chester, Illinois. Frequent floods make permanent settlement difficult.

Kaskaskia River, c.320 mi/515 km long, central and SW ILLINOIS; rises near URBANA; flows generally SW across state to the MISSISSIPPI just NW of KASKASKIA; 39°59'N 88°21'W. The river is dammed in CLINTON county, forming large CARLYLE LAKE, and SHELBY county, forming large Lake SHELBYVILLE.

Kaskelen (kahs-ke-LEN), city, SW ALMATY region, KAZAKHSTAN, on Kaskelen River (left tributary of ILI RIVER), 15 mi/24 km W of ALMATY; 43°10'N 76°35'E. Tertiary-level (raion) administrative center. Sawmilling. Also spelled Qaskelen.

Kaski (kahs-KEE), district, central NEPAL, in GANDAKI zone; ⊙ POKHARA.

Kaskö (KAHS-kuh), Finnish Kaskinen, town, VAASAN province, W FINLAND, 50 mi/80 km S of VAASA; 62°23'N 21°13'E. Elevation 33 ft/10 m. On island in GULF OF BOTHNIA, reached by causeway. Timbershipping port; large pulp mill. Founded 1785.

Kasli (kah-SLEE), city (2005 population 17,520), N CHELYABINSK oblast, in the extreme NE foothills of the S URALS, RUSSIA, between two small lakes, on road, 85 mi/137 km NW of CHELYABINSK; 55°53'N 60°46'E. Elevation 751 ft/228 m. Ironware, machinery, radios, and apparel; art moulding plant; fish processing; bakery. Founded in 1746 as a copper-smelting plant (pig iron produced from 1770); became city in 1942. Formerly called Kaslinskiy Zavod.

Kaslinskiy Zavod, RUSSIA: see KASLI.

Kaslo (KAZ-lo), village (□ 1 sq mi/2.6 sq km; 2001 population 1,032), SE BRITISH COLUMBIA, W CANADA, in SELKIRK MOUNTAINS, on KOOTENAY Lake, 35 mi/56 km NNE of NELSON, in Central KOOTENAY regional district; 49°55'N 116°55'W. Elevation 1820 ft/546 m. Silver, lead, zinc mining and smelting, logging (cedar), shipbuilding, fruit (cherries). Resort. Historical town. Oldest mineral claim in British Columbia. Airfield. First incorporated 1893.

Kasoda (KAH-so-duh), town, JALGAON district, MAHARASHTRA state, W central INDIA, 22 mi/35 km SW of JALGAON; 20°49'N 75°18'E. Local market for cotton, peanuts.

Kasongo (kah-SAWNG-go), town, ⊙ MANIEMA province, E CONGO, on LUALABA RIVER, and 200 mi/322 km SW of BUKAVU; 04°27'S 26°40'E. Elev. 2,053 ft/625 m. Trading center for cotton, palm kernels, ground nuts, and food staples (manioc, yams, plantain); cotton ginning. LUALABA RIVER is navigable between here and KIBOMBO; airport. Mission (R.C.) of Kasongo-Saint-Charles, 7 mi/11.3 km NE, built on site of Arab slave-traders' fort, still preserves remains of fortifications and cemetery of the Arab campaign (1890–1894). Alternative spelling: Kazongo. Also known as TONGONI, TONGONI KAPAYA.

Kasongo Lunda (kah-SAWNG-go LOON-dah), village, BANDUNDU province, SW CONGO, on right bank of KWANGO RIVER (ANGOLA border), and 165 mi/266 km SW of KIKWIT; 06°28'S 16°49'E. Elev. 1,794 ft/546 m. Customs station; trading center; coffee, fibers.

Kasos (KAH-sos), Italian Caso, Latin Casus, island (□ 27 sq mi/70.2 sq km), DODECANESE prefecture, SOUTH AEGEAN department, GREECE, in the AEGEAN SEA, SW of KÁRPATHOS, and 16 mi/26 km NE of CRETE across Kasos Strait; 8 mi/12.9 km long, 4 mi/6.4 km wide; 35°23'N 26°55'E. Rises to 1,972 ft/601 m. Sponge-fishing center; agriculture (olives, fruit, barley, wine); gypsum deposits. Main town, Fry (Fri) or Kastro, is on NW shore. Airport. Participated in Greek War of Independence; conquered (1824) by the Egyptian navy.

Kasota (kuh-SO-duh), village (2000 population 680), LE SUEUR county, S MINNESOTA, 9 mi/14.5 km N of MANKATO, and 2 mi/3.2 km S of ST. PETER, on Shanaska Creek, S of its confluence with MINNESOTA RIVER; 44°17'N 93°58'W. Agricultural area (corn, oats, soybeans; livestock, poultry); manufacturing (building materials, asphalt, injection molding, cut marble); limestone and marble quarries.

Kasperske Hory (KAHSH-per-SKE HO-ree), Czech Kašperské Hory, German bergreichenstein, town, ZAPADOCESKY province, SW BOHEMIA, CZECH REPUBLIC, 6 mi/9.7 km SSE of Susice, in foothills of BOHEMIAN FOREST; 49°09'N 13°34'E. Manufacturing (optics, woodworking); lumbering. Former gold-mining settlement. Summer resort. Has old town hall, churches, museum.

Kaspi (KAHS-pee), city (2002 population 15,233) and center of Kaspi region, central GEORGIA, on railroad, on KURA River, and 22 mi/35 km WNW of TBILISI; 41°55'N 44°25'E. Manufacturing (building materials, electrical equipment; cannery); winemaking.

Kaspichan (KAHS-pee-chahn), city (1993 population 3,907), VARNA oblast, ⊙ Kaspichan obshtina, E BULGARIA, on the PROVADIISKA RIVER, 3 mi/5 km S of NOVI PAZAR; 43°18'N 27°10'E. Railroad junction; grain, vegetables; manufacturing (construction ceramics).

Kaspiysk (kahs-PEE-yeesk), city (2005 population 83,205), E central DAGESTAN REPUBLIC, NE CAUCASUS,

SE European Russia, on the CASPIAN SEA, 9 mi/14 km SE of MAKHACHKALA; 42°53′N 47°38′E. Below sea level. Industrial center (engines and precision machines); thermo-electric power station. Founded in 1932; called Dvigatel'stroy (1936–1947); made city in 1947.

Kaspiyskiy, RUSSIA: see LAGAN'.

Kasplia River, BELARUS and RUSSIA: see KASPLYA RIVER.

Kasplya (KAHS-plyah), village, W SMOLENSK oblast, W European Russia, on the KASPLYA RIVER, on road, 20 mi/32 km NW of SMOLENSK; 54°59′N 31°37′E. Elevation 583 ft/177 m. In agricultural area (wheat, rye, oats, vegetables, potatoes).

Kasplya River (KAHS-plyah), 100 mi/161 km long, in W European Russia and BELARUS; rises in SMOLENSK oblast ENE of RUDNYA; flows SE, N, past KASPLYA, and generally W, past DEMIDOV (head of navigation), to the WESTERN DVINA RIVER at SURAZH (VITEBSK oblast, Belarus).

Kasr, for Arabic and Persian names beginning thus: see QASR.

Kasrik, TURKEY: see GURPINAR.

Kas River (KUHS), Russian *Bol'shoy Kas* [=Greater Kas], 165 mi/266 km long, W KRASNOYARSK TERRITORY, Russia; flows NE to the YENISEY RIVER above YARTSEVO. Receives the **Lesser kas River**, Russian *Malyy Kas* (55 mi/89 km long), which is connected with the KET River (tributary of the OB' RIVER) by the OB'-YENISEI CANAL SYSTEM.

Kassa, village, GUINEA: see LOS ISLANDS.

Kassa, SLOVAKIA: see KOŠICE.

Kassaba, TURKEY: see TURGUTLU.

Kassala, SUDAN: see Eastern (region and former province), Kasala (state), or KASALA (city).

Kassan, UZBEKISTAN: see KASANSAI.

Kassandra (kah-SAHN-[th]rah), ancient *Pallene*, westernmost of the three arms of the KHALKIDHIKÍ Peninsula, CENTRAL MACEDONIA department, NE GREECE, on AEGEAN SEA, between Gulf of Thessaloníki and Gulf of KASSANDRA; 30 mi/48 km long, 5 mi/8 km wide; 40°03′N 23°25′E. Rises to 1,090 ft/332 m. Terminates in Capes KASSANDRA (SW) and KALOGREA (SE). On narrow isthmus at its base stood ancient POTIDAIA. Gold deposits were discovered here in the 1970s; lead and zinc can also be found. Mining area for more than 2,000 years; a plant was modernized in 1997 and adapted to handle gold processing.

Kassandra, Cape (kah-SAHN-[th]rah), Latin *Posidium*, SW cape of KASSANDRA prong of KHALKIDHIKÍ Peninsula in CENTRAL MACEDONIA department, NE GREECE, on AEGEAN SEA; 39°57′N 23°20′E. Also called Cape Poseidion, or Posidhion.

Kassandra, Gulf of (kah-SAHN-[th]rah), ancient *Toronaicus Sinus*, arm of AEGEAN SEA, between KASSANDRA and SITHONIA prongs of KHALKIDHIKÍ Peninsula, off KHALKIDHIKÍ prefecture, CENTRAL MACEDONIA department, NE GREECE; over 30 mi/48 km long, up to 15 mi/24 km wide. Named for ancient Torone, on SE shore. Formerly called Toronaic Gulf.

Kassaro, town, FIRST REGION/KAYES, MALI, on DAKAR-NIGER railroad, and 60 mi/100 km NW of BAMAKO.

Kassel (KAHS-sel), city, HESSE, central GERMANY, on the FULDA RIVER; 51°21′N 09°30′E. Industrial, railroad, and cultural center; manufacturing includes textiles, optical and precision instruments, locomotives, and motor vehicles. Kassel was mentioned in 913 and was chartered in 1198. It became the capital (1567) of the landgraviate of HESSE-KASSEL (raised to an electorate in 1803); also the capital the kingdom of WESTPHALIA (1807–1813). After Electoral Hesse and NASSAU passed (1866) to PRUSSIA and were united as the province of HESSE-NASSAU, Kassel was made the captial. As a center of German airplane and tank production in World War II, Kassel was severely damaged by Allied air raids. Many historic buildings were destroyed, though some were restored after 1945. Kassel has several important museums; international exhibits of modern art are held every five years in the city and are considered the world's largest art exhibitions. A former spelling is CASSEL.

Kasserine, TUNISIA: see QASRAYN, AL.

Kassidiares, Greece: see NARTHAKAN.

Kasson (KA-suhn), town (2000 population 4,398), DODGE county, SE MINNESOTA, 15 mi/24 km W of ROCHESTER, near Middle Branch South Fork ZUMBRO RIVER; 44°01′N 92°45′W. Grain, soybeans, peas; livestock, poultry; dairying; manufacturing (consumer goods, injection molds, concrete). Settled 1865, incorporated 1870.

Kassopi (kah-so-PEE), archaeological site, PREVEZA prefecture, S EPIRUS department, NW GREECE, 19 mi/31 km NW of PREVEZA, on a plateau looking S over the IONIAN SEA and AMBRACIAN GULF. At its height, in the 4th century B.C.E., was the capital the Kassopaian tribe, with a population of c.9,000; some remains of the era still visible. In 1803, on a nearby mountain some sixty women leapt off a cliff with their children to escape capture by the Ottoman troops of Ali Pasha; monumental sculpture dedicated to the victims in nearby village/monastery of Zalonga.

Kassubia (kah-SOO-bee-ah) or **Kashubia**, Polish *Kaszuby* (kah-schoo-bee), region in NW part of Gdańsk province, N POLAND, extends inland from the BALTIC SEA and HEL PENINSULA; centered on GDYNIA, WEJHEROWO, and KARTUZY. Of ethnological importance, it is inhabited by Kashubes, a Slavonic people speaking Kassubian, a W Slavic language.

Kastamonu (kahs-tah-mo-NOO) or **Kastamuni** (kahs-ta-mu-NEE), city (2000 population 64,606), N TURKEY; 41°22′N 33°47′E. It is a manufacturing center, noted for its textiles and copper utensils, and is the chief city of a region with a variety of mineral deposits. Kastamonu was captured by the Ottoman Turks in 1393, was taken by Tamerlane in 1403, and was regained by the Ottomans in 1460. Archeological protected sites include Zimbilli Hill (Pompeipolis) in Taşköprü district, Abeş Castle and Geriş Hill in INEBOLU district, Ginolu Bay in Çatalzeytin district, and Gideros Bay in Cide District.

Kastav (KAH-stahv), Italian *Castua*, village, W CROATIA, 6 mi/9.7 km WNW of RIJEKA (Fiume), on KVARNER gulf. Occupies site of ancient Illyrian settlement. Important medieval fortress; church from 15th century, city lodge from 16th century.

Kasteelbrakel, BELGIUM: see BRAINE-LE-CHÂTEAU.

Kastel (KAHS-tel), former suburb of MAINZ (connected by tramway), after 1945 incorporated into WIESBADEN, HESSE, W GERMANY, on right bank of the RHINE (bridge), and 5 mi/8 km SSE of WIESBADEN city center; 50°00′N 08°17′E. Built on site of Roman *Castellum Mattiacorum*.

Kaštelanska Rivijera (KAH-shte-lahn-skah REE-vee-ye-rah), Dalmatian resort area, S CROATIA, extending c.6 mi/9.7 km along ADRIATIC SEA, between SPLIT and TROGIR. Consists of seven small villages or hamlets (including Kaštel Sućurac), each named after a Venetian castle. Also called Kaštela [Ital. *Castelli*] or Sedam Kaštela [Croatian=seven castles].

Kastellaun (kahs-tel-LOUN), town, RHINELAND-PALATINATE, W GERMANY, 21 mi/34 km S of KOBLENZ; 50°04′N 07°26′E. Summer resort. Woodworking. Chartered 1305; in 1689 destroyed by French; chartered again in 1969.

Kastelli (kahs-TE-lee), village, KHANIÁ prefecture, W CRETE department, GREECE, port on Gulf of KISSAMOS, 14 mi/23 km W of KHANIÁ. Citrus fruits, wheat, carobs; olive oil, dairy products (milk, cheese). Sometimes called Kastellion; formerly called Kisamos.

Kastellorizo (kahs-te-LO-ree-zo), Italian *Castelrosso*, easternmost island (□ 4 sq mi/10.4 sq km), DODECANESE prefecture, SOUTH AEGEAN department, Greece, in MEDITERRANEAN SEA, off small Turkish port of KAS and 80 mi/129 km E of RHODES; 3.5 mi/5.6 km long, 1.5 mi/2.4 km wide; 36°08′N 29°35′E. Olive oil, wine; sponge fishing. Occupied by France during World War I; passed 1921 to Italy and developed as a naval base. Nearby islets ceded to Italy by Turkey in 1932. Awarded to GREECE 1947–1948. Claimed by TURKEY. Sometimes spelled Castellorizo, Kastellorizon.

Kaštel Sućurac (KAH-shtel SUH-chuhr-rahts), town, S CROATIA, in DALMATIA, on ADRIATIC SEA, on railroad, 3 mi/4.8 km N of SPLIT, in the Kaštelanska Riviera. Manufacturing (cement, plastics); seaside resort.

Kasterlee (KAHS-tuhr-lai), commune (2006 population 17,962), Turnhout district, ANTWERPEN province, N BELGIUM, 6 mi/10 km S of TURNHOUT; 51°15′N 04°57′E.

Kastoriá (kahs-to-RYAH), prefecture (□ 667 sq mi/1,734.2 sq km), WEST MACEDONIA department, N central GREECE; ⊙ KASTORÍA; 40°30′N 21°10′E. Bounded NW by ALBANIA, N by FLÓRINA prefecture, E by KOZANI prefecture, SE by GREVENA prefecture, and SW by EPIRUS department. GRAMOS mountains in SW, Vernon mountains in NE. Drained by upper ALIÁKMON RIVER; Lake KASTORÍA located here. Lumbering, charcoal burning; wheat, bean, and tobacco growing.

Kastoriá (kahs-to-RYAH), city (2001 population 14,813), ⊙ KASTORÍA prefecture, WEST MACEDONIA department, N central GREECE, on a peninsula extending into Lake KASTORÍA; 40°31′N 21°16′E. Market for farm produce; fisheries. Airport. Major fur-trade center in the 17th and 18th centuries. Many small Byzantine churches and palatial homes.

Kastoriá, Lake (kahs-to-RYAH) (□ 12 sq mi/31.2 sq km), in WEST MACEDONIA department, N central GREECE, on peninsula jutting out from W shore, and 16 mi/26 km SSW of FLÓRINA; 4 mi/6.4 km long 3 mi/4.8 km wide; 40°31′N 21°18′E. Fisheries. Also called Lake Orestias.

Kastornoye (kah-STOR-nuh-ye), town (2005 population 4,355), NE KURSK oblast, SW European Russia, on the OLYM River (tributary of the SOSNA RIVER, DON River basin), on railroad, 94 mi/151 km WNW of VORONEZH; 51°50′N 38°07′E. Elevation 636 ft/193 m. Railroad junction; site of battles in civil war (1919) and World War II (1943).

Kastos (KAHS-tos), island (□ 3 sq mi/7.8 sq km) in IONIAN SEA, in AKARNANIA prefecture, WESTERN GREECE department, 4 mi/6.4 km off central Greek mainland (E); 4.5 mi/7.2 km long, 0.5 mi/0.8 km wide; 38°35′N 20°55′E. Rises to 520 ft/158 m. Oak trees; fisheries.

Kastraki (kahs-TRAH-kee), town, TRIKKALA prefecture, THESSALY department, N GREECE, 5 mi/0.8 km N of KALAMBAKA, just below the METEORA, the monastery complex perched on rocks rising above it; 39°43′N 21°37′E. Base for visitors to the Meteora, with hotels, campgrounds, and restaurants.

Kastri, Greece: see ERMIONI.

Kástron (KAHS-tron), town, on W coast of LÍMNOS island, LESBOS prefecture, NORTH AEGEAN department, GREECE, in the N AEGEAN SEA; 39°53′N 25°04′E. Seaport trading in local produce. Also spelled Kastro or Castro. In ancient times known as Myrina.

Kastrup (KAH-stroop), town, Copenhagen county, DENMARK, on E coast of AMAGER island, and 4 mi/6.4 km SSE of COPENHAGEN, on the ØRESUND. Shipbuilding; Copenhagen international airport.

Kasuga (KAHS-gah), city, FUKUOKA prefecture, N KYUSHU, SW JAPAN, 5 mi/8 km S of FUKUOKA; 33°31′N 130°27′E.

Kasuga (KAHS-gah), town, Hikami district, HYOGO prefecture, S HONSHU, W central JAPAN, 33 mi/53 km N of KOBE; 35°09′N 135°06′E. Eggplants.

Kasuga (KAHS-gah), village, Ibi county, GIFU prefecture, central HONSHU, central JAPAN, 16 mi/25 km N of GIFU; 35°27′N 136°29′E.

Kasugai (kah-SOO-gah-ee), city (2005 population 295,802), AICHI prefecture, S central HONSHU, central JAPAN, 6 mi/10 km N of NAGOYA; 35°14′N 136°58′E. Suburb of Nagoya. Motors, paper; cacti.

Cross-references are shown in SMALL CAPITALS. The pronunciation guide is shown on page xix. The sources of population figures are shown on page xvii.

Kasugai (kah-SOO-gah-ee), town, E. Yamanashi county, YAMANASHI prefecture, central HONSHU, central JAPAN, 5 mi/8 km E of KOFU; 35°39′N 138°39′E. Peaches.

Kasuga, Mount, JAPAN: see NARA.

Kasukabe (kahs-KAH-BE), city (2005 population 238,506), SAITAMA prefecture, E central HONSHU, E central JAPAN, 9 mi/15 km N of URAWA; 35°58′N 139°45′E. Paulownia furniture, battledores, hats.

Kasukawa (kahs-KAH-wah), village, Seta county, GUMMA prefecture, central HONSHU, N central JAPAN, 9 mi/15 km E of MAEBASHI; 36°24′N 139°12′E.

Kasulu (kah-SOO-loo), village, KIGOMA region, W TANZANIA, 40 mi/64 km NE of KIGOMA, near Ruchugi River; 04°37′S 30°07′E. Livestock; subsistence crops.

Kasumbalesa (kah-soom-bah-LE-sah), village, KATANGA province, SE CONGO, on railroad, and 40 mi/64 km SSE of LUBUMBASHI; 12°12′S 27°48′E. Elev. 4,465 ft/ 1,360 m. Near ANGOLA border. Iron mining.

Kasumi (kah-SOO-mee), town, Kinosaki county, HYOGO prefecture, S HONSHU, W central JAPAN, on SEA OF JAPAN, 71 mi/114 km N of KOBE; 35°38′N 134°37′E. Crabs; wood products. Sake brewing, fish processing. Rias Seacoast nearby.

Kasumi-ga-ura (kah-SOO-mee-gah-OO-rah), lagoon (□ 73 sq mi/189.8 sq km), IBARAKI prefecture, central HONSHU, E central JAPAN, 16 mi/26 km S of MITO; 10 mi/16 km long, 5 mi/8 km wide. Has three arms, the largest of which is 8 mi/12.9 km long, 3 mi/4.8 km wide. Connected by TONE RIVER. TSUCHIURA on W arm. Sometimes called Nishiura.

Kasum-Ismailovo, AZERBAIJAN: see GORANBOY.

Kasumkent (kah-soom-KYENT), village (2005 population 12,520), SE DAGESTAN REPUBLIC, NE CAUCASUS, SE European Russia, on mountain highway, 33 mi/53 km SSW of DERBENT; 41°40′N 48°08′E. Elevation 1,545 ft/470 m. Fruit processing. Population largely Lezghian. Experienced significant population spurt since 1992 with the influx of refugees fleeing armed conflict in CHECHNYA.

Kasumpti (kah-soomp-TEI), town, SHIMLA district, HIMACHAL PRADESH state, N INDIA, an E suburb of SHIMLA town. Carpet making. Tibetan monastery.

Kasungu (kah-suh-nguh), administrative center and district (2007 population 650,103), Central region, MALAWI, 55 mi/89 km NW of DOWA; 13°02′S 33°29′E. Tobacco, cotton, corn.

Kasungu National Park (□ c.770 sq mi/2,000 sq km), KASUNGU district, Central region, MALAWI, 102 mi/164 km NW of LILONGWE. Elephant habitat; hot, rolling woodland areas broken up by occasional rocky crops.

Kasur (kah-SOOR), city, LAHORE district, E PUNJAB province, central PAKISTAN, 32 mi/51 km SSE of LAHORE; 31°07′N 74°27′E. Railroad junction; trade center. A Pathan settlement in 17th and 18th centuries.

Kasuya (KAH-soo-yah), town, Kasuya county, FUKUOKA prefecture, N KYUSHU, SW JAPAN, 4.3 mi/7 km E of FUKUOKA; 33°36′N 130°28′E. Roses; *shochu* (distilled alcoholic beverage).

Kaszuby, POLAND: see KASSUBIA.

Kataghan, Qataghan, or **Qatghan,** former province, NE AFGHANISTAN, now divided into KUNDUZ, TAKHAR, and BAGHLAN provinces.

Katagum (kah-TAH-goom), town, JIGAWA state, N NIGERIA, on Katagum River (branch of the KOMADUGU YOBE River), 25 mi/40 km SE of HADEJIA; 12°17′N 10°21′E. Peanuts, cotton, millet; cattle, skins.

Katahdin (kah-TAHD-uhn), mountain (5,268 ft/1,606 m), N central MAINE, between branches of the PENOBSCOT RIVER. Highest point in Maine. The peak and the beautifully wooded, lake-dotted territory surrounding it constitute Baxter State Park, the gift of Governor Percival P. Baxter in 1931. Mount Katahdin is the N terminus of the Appalachian Trail.

Katahdin Iron Works (kuh-TAHD-uhn), PISCATAQUIS county, central MAINE, on Silver Lake, and 18 mi/29 km NNE of DOVER-FOXCROFT. In lumbering, recreational area. Site of abandoned iron mines.

Katahigashi (kah-TAH-hee-GAH-shee), village, West Kanbara county, NIIGATA prefecture, central HONSHU, N central JAPAN, 9 mi/15 km S of NIIGATA; 37°45′N 138°58′E.

Kataketa (kah-tah-KAI-tah), village, MOROGORO region, S central TANZANIA, 140 mi/225 km SE of IRINGA; 09°09′S 36°44′E. SELOUS GAME RESERVE to S and E. Livestock; grain, sweet potatoes, beans. Also spelled Kiteketa.

Katako Kombe (kah-TAH-ko KOM-bai), village, KASAI-ORIENTAL province, central CONGO, on headstream of LUKENIE RIVER, and 120 mi/193 km NNE of LUSAMBO; 03°24′S 24°25′E. Elev. 1,998 ft/608 m. Trading center in cotton area; cotton ginning.

Katakol, town, BELGAUM district, KARNATAKA state, SW INDIA, 44 mi/72 km ENE of BELGAUM. Agriculture market (cotton, wheat, peanuts). Sometimes spelled Katkola or Katkol.

Katakolo (kah-TAH-ko-lo), village, ILIA prefecture, WESTERN GREECE department, NW PELOPONNESUS, port on Cape Katakolo at N end of Gulf of KIPARISSIA, 7 mi/11.3 km W of PÍRGOS (connected by railroad), for which it is the port. Exports Zante currants. Founded 1857.

Katakwi, former administrative district (□ 1,975 sq mi/ 5,135 sq km; 2005 population 365,900), EASTERN region, E UGANDA; capital was Katakwi; 02°00′N 33°45′E. Elevation c.3,609 ft/1,100 m. As of Uganda's division into fifty-six districts, was bordered by MOROTO (N and E), KUMI (S), SOROTI (SW), KABERAMAIDO (W), and LIRA (NW) districts. Main inhabitants were the Etesots and Langis. District was primarily savannah grassland; also had swamp areas; alluvium deposits. Lake BISINA in S. Agriculture (major crops included cassava, groundnuts, and sorghum). Livestock raising was important (included cattle, goats, sheep, pigs, and poultry). Some fishing. Created in 1997 from NE portion of former SOROTI district (Kaberamaido district was created from SW portion in 2001 and current Soroti district from S portion). In 2005 W portion of district was carved out to form AMURIA district and E portion was formed into current KATAKWI district.

Katakwi, administrative district, EASTERN region, E UGANDA, on N shore of LAKE BISINA; ☉ Katakwi. As of Uganda's division into eighty districts, Katakwi's border districts include MOROTO (N), NAKAPIRIPIRIT (E), and AMURIA (W). Some marsh area in W. Agriculture; fishing. Road connects Katakwi town to MOROTO town to NE and SOROTI town to SW. Formed in 2005 from E portion of former KATAKWI district (Amuria district formed from W portion).

Katalanga, RUSSIA: see KAMO, river.

Katalla (KA-duh-lah), village, S ALASKA, 50 mi/80 km ESE of CORDOVA; 60°12′N 144°32′W. Center of abandoned Katalla oil field; had population of 5,000 in 1908.

Katamatite, township, VICTORIA, SE AUSTRALIA, 142 mi/229 km N of MELBOURNE, 31 mi/50 km NE of SHEPPARTON, on Boosey Creek, and in MURRAY valley irrigation area; 36°05′S 145°42′E. Museum.

Katana (kah-TAH-nah), village, SUD-KIVU province, E CONGO, on W shore of LAKE KIVU, 20 mi/32 km N of BUKAVU; 02°13′S 28°50′E. Elev. 5,236 ft/1,595 m. Noted medical research center with hospitals. Has Roman Catholic mission, various schools (including ones for medical assistants and teachers) and small seminary. Seat of vicar apostolic of Kivu. Katana was previously known as Liége-Saint-Lambert. Kakondo (port) is an agricultural center with coffee and cinchona plantations.

Katandra, township, N VICTORIA, SE AUSTRALIA, GOULBURN valley; 36°14′S 145°37′E. In orchard, dairy district.

Katanga, province (□ 200,000 sq mi/520,000 sq km; 2004 population 8,612,000), SE CONGO; ☉ LUBUMBASHI (formerly ELISABETHVILLE); 11°40′S 27°28′E. It borders ANGOLA on the W and SW, ZAMBIA on the S and SE, and LAKE TANGANYIKA on the E. Drained by the Lualuba (CONGO), LUFIRA, LUVUA, and LUKUGA rivers. The region encompasses the fertile Katanga Plateau (elev. 3,000 ft/914 m-6,000 ft/1,829 m), a rich farming and ranching region of open or park-like savanna grasslands with some scrub-forest. In the E is an enormously rich mining area, which supplies much of the world's cobalt, as well as extensive quantities of copper, tin, radium, uranium, and diamonds. The region's considerable industrial base is largely concerned with the processing of minerals. Katanga is well connected by railroad with the rest of CONGO and with ANGOLA and ZAMBIA. There is also steamer service on LAKE TANGANYIKA between KALEMIE and KIGOMA (TANZANIA). Copper has been mined and exported for centuries. From the 17th to the 19th century, much of the region was controlled by the Luba and Lunda kingdoms. In the late 19th century, M'Siri, a Nyamwezi trader from what is now central TANZANIA, founded a kingdom in the area that lasted until he was killed by the Belgians in 1891. Under Belgian rule (1884–1960), mineral resources were exploited and the region was developed much more rapidly than the rest of the country. The province was called ELISABETHVILLE, 1935–1947. In July 1960, after the country (then also called the DEMOCRATIC REPUBLIC OF THE CONGO) became independent, Katanga proclaimed itself a republic and seceded from the central government. Disorder was widespread, and the central government requested and received help from the UN. In 1960, a small UN force entered Katanga. Later, a considerable number of UN troops, committed to a policy of nonintervention, were stationed in Katanga to oversee the withdrawal of foreign troops. Belgian troops were slowly withdrawn, but white mercenary officers continued to command in the army of Katanga. There was recurrent trouble between the UN force and the Katangese, and attempts at reconciliation with the central government proved fruitless. The situation grew steadily more volatile until early 1961, when the former premier Patrice Lumumba was murdered in Katanga. Under a new, stronger UN mandate, the international force took control (1961) of ELISABETHVILLE and other strongpoints. An agreement (December 1961) for reintegrating Katanga into the country proved abortive. In January 1963, UN troops ended the Katanga secession. In 1966, the central government nationalized Union Minière du Haut Katanga, the Belgian firm that had controlled most of Katanga's mining interests. It was renamed Gécamines. In 1972, Katanga was renamed SHABA. Throughout the 1970s, further insurrections were put down by the government, with help from Moroccan troops in 1977 and the French Foreign Legion in 1978. Gécamines has had difficulty in maintaining its operations and exporting its copper. The province was renamed Katanga c.1998.

Katangi (kuh-TAHN-gee), village (2001 population 14,760), BALAGHAT district, central MADHYA PRADESH state, INDIA, 25 mi/40 km W of BALAGHAT. Railroad spur terminus; shipping point for manganese brought from mines near RAMRAMA TOLA, 10 mi/16 km NE, to BALAGHAT.

Katangli (kah-TAHN-glee), settlement (2006 population 835), N SAKHALIN oblast, on the E coast of central SAKHALIN Island, RUSSIAN FAR EAST, on highway branch and railroad spur, 41 mi/66 km S of CHAIVO; 51°42′N 143°14′E. Oil and natural gas extraction and processing.

Katanino (kah-tah-NEE-no), township, WESTERN province, N central ZAMBIA, 40 mi/64 km SSE of NDOLA. On railroad. Agriculture (tobacco, cotton, soybeans, corn, flowers); cattle.

Katanning (kuh-TA-ning), town, SW WESTERN AUSTRALIA, 155 mi/249 km SE of PERTH; 33°42′S 117°33′E. Railroad junction; agriculture (wheat, oats); wool; stockyards. Flour mill museum.

Katano (kah-TAH-no), city, OSAKA prefecture, S HONSHU, W central JAPAN, 12 mi/20 km N of OSAKA; 34°47′N 135°40′E.

Kataoka, RUSSIA: see BAYKOVO, KURIL ISLANDS.

Kataragama (KUH-thah-rah-GUH-mah), town, UVA PROVINCE, SE SRI LANKA, 45 mi/72 km SSE of BADULLA, and 13 mi/21 km NNE of TISSAMAHARAMA; 06°25′N 81°20′E. Major pilgrim center, with main shrine devoted to the Hindu god Skanda, also known as the god of Kataragama; July–August festival. Buddhists visit the main shrine and Kirivehera, a stupa less than 1 mi/1.6 km away, said to commemorate Buddha's third visit to the island. Muslims pray at the Masjad-ul-Khizr mosque, burial place of an important Muslim saint. Isolated peak of Kataragama (1,390 ft/424 m) is 2 mi/3.2 km S.

Katarnian Ghat (kuh-TUHR-nyahn GAHT), village, BAHRAICH district, N UTTAR PRADESH state, N central INDIA, on Girwa arm of KARNALI (GHAGHARA) River, and 60 mi/97 km NNW of BAHRAICH. Railroad terminus; connected by road with RAJAPUR (Nepal).

Katas (kah-TUHS), village, JHELUM district, N PUNJAB province, central PAKISTAN, in SALT RANGE, 45 mi/72 km WSW of JHELUM; 32°40′N 72°55′E. Sacred pool (elevation c.2,200 ft/671 m), connected with Shivaite legend, visited annually by pilgrims.

Katashina (kah-TAH-shee-nah), village, Tone county, GUMMA prefecture, central HONSHU, N central JAPAN, 34 mi/55 km N of MAEBASHI; 36°46′N 139°13′E. Near Oze area (in NIKKO NATIONAL PARK).

Katavi Plains National Park (kah-TAH-vi), RUKWA region, W TANZANIA, 75 mi/121 km NNW of SUMBAWANGA; drained by Mfulsi River in S, source of Katuma River in N. Marshy area in Katavi Plains known for its wildlife. Formerly known as Katavi Plains Game Reserve.

Katav-Ivanovsk (kah-TAHF ee-VAH-nuhfsk), city (2005 population 19,070), W CHELYABINSK oblast, RUSSIA, in the NW foothills of the S URALS, on the Katav River (left tributary of the YURYUZAN RIVER), on road and railroad, 200 mi/322 km SW of CHELYABINSK, and 50 mi/80 km NNW of BELORETSK; 54°45′N 58°11′E. Elevation 1,404 ft/427 m. Metallurgical center, based on Bakal iron ores; manufacturing (cement, construction equipment); forestry. Founded in 1775, became city in 1939. Initially called Katav-Ivanovskiy Zavod; later, until city status granted, Katav-Ivanovskiy.

Katav-Ivanovskiy, RUSSIA: see KATAV-IVANOVSK.

Kataysk (kah-TEISK), city (2005 population 15,550), NW KURGAN oblast, SW SIBERIA, RUSSIA, on the ISET' RIVER (in the OB' RIVER basin), on crossroads and railroad, 133 mi/214 km NW of KURGAN, and 40 mi/64 km WNW of SHADRINSK; 56°17′N 62°34′E. Elevation 370 ft/112 m. Agricultural center. Manufacturing (water pumps). In the 18th and 19th centuries, was situated on the main access route to Siberia. Founded in 1655 as a village of Katayskoye; city status and current name since 1944.

Kataysk, RUSSIA: see KATAISK.

Katayskoye, RUSSIA: see KATAYSK.

Katcha (KAH-CHAH), town, NIGER state, W central NIGERIA, port on NIGER RIVER opposite EGGAN, on railroad, and 20 mi/32 km S of AGAIE; 08°46′N 06°19′E. Road center. Shea-nut processing, rope- and sackmaking; roofing timber; agriculture: cotton, cassava, yams, and durra. Sometimes spelled Kacha.

Katchall Island (KAH-chuhl), one of NICOBAR ISLANDS, in ANADAMAN AND NICOBAR ISLANDS Union Territory, INDIA, in BAY OF BENGAL, 50 mi/80 km NNW of GREAT NICOBAR ISLAND; 12 mi/19 km long NW-SE, 5 mi/8 km wide; 07°57′N 93°22′E.

Katentania (kah-ten-TAHN-yah), village, Katanga province, SE CONGO, near railroad, 65 mi/105 km NW of Likasi; 09°43′S 25°23′E. Elev. 2,919 ft/889 m. Livestock raising.

Katepwa Beach (kuh-TEHP-wuh), village (2006 population 285), SE SASKATCHEWAN, CANADA, on the FISHING LAKES, 12 mi/19 km N of INDIAN HEAD; 50°43′N 103°37′W. Resort.

Kater, Cape, NW extremity of BAFFIN ISLAND, BAFFIN region, NUNAVUT territory, CANADA, on PRINCE REGENT INLET; 73°53′N 90°11′W.

Katerini (kah-te-REE-nee), city (2001 population 54,941), ☉ PIERÍA prefecture, CENTRAL MACEDONIA department, NE GREECE; 40°16′N 22°30′E. Commercial center for a productive tobacco-growing region. Salt and flour are also produced.

Katerinky (KAH-te-RZHIN-ki), Czech *Kateřinky*, German *katharein*, NE suburb of OPAVA, SEVEROMORAVSKY province, central CZECH SILESIA, CZECH REPUBLIC, on left bank of OPAVA RIVER. Has a 14th century chapel.

Katerinopol', UKRAINE: see KATERYNOPIL'.

Katerynivka, UKRAINE: see ROVEN'KY.

Katerynopil' (kah-te-ri-NO-peel) (Russian *Katerinopol'*), town, S central CHERKASY oblast, UKRAINE, 37 mi/60 km ENE of UMAN'; 48°56′N 30°59′E. Elevation 459 ft/139 m. Raion center; brickworks, feed mill. Known since the mid-16th century as the town of Kanyboloto; from 1648 to 1712, it was a company town in the Korsun' Cossack regiment; annexed by Russia in 1793, renamed in 1795.

Kates Needle, peak (10,002 ft/3,049 m), ALASKA-BRITISH COLUMBIA (CANADA) border, in COAST RANGE, 40 mi/64 km NE of PETERSBURG; 57°03′N 132°03′W.

Katghora (kuht-GO-ruh), village (2001 population 18,534), KORBA district, N CHHATTISGARH STATES, central INDIA, 40 mi/64 km NE of BILASPUR, 22°30′N 82°33′E. Rice, wheat, oilseeds. Coal mining 17 mi/27 km NNW at village of MATIN.

Katha (kah-THAH), district (☐ 7,593 sq mi/19,741.8 sq km), MYANMAR; ☉ KATHA. Astride upper AYEYARWADY RIVER. Yearly rainfall is 58 in/147 cm. Agriculture (rice, tea, cotton, sesame); teak and bamboo forests in hills. Gem mine at MOGOK, coal at PINLEBU, gold near WUNTHO. Served by Myitkyina-Mandalay railroad and Ayeyarwady steamers.

Katha (kah-THAH), village, ☉ KATHA township, SAGAING division, MYANMAR, on W bank of AYEYARWADY RIVER, and 105 mi/169 km SW of MYITKYINA. Head of branch line to Mandalay-Myitkyina railroad.

Katharein, CZECH REPUBLIC: see KATERINKY.

Katharinaberg, CZECH REPUBLIC: see HORA SVATE KATERINY.

Katharinenstadt, RUSSIA: see MARKS.

Katherina, Gebel, highest peak (8,651 ft/2,637 m) of SINAI Peninsula and of EGYPT, 57 mi/92 km NNW of Ras MUHAMMED. The famous convent of St. Catherine is found on the slope of the nearby peak of Gebel MUSA (2 mi/3.2 km N); major tourist attraction. Small airport.

Katherine (KATH-rin), settlement, NW central NORTHERN TERRITORY, AUSTRALIA, 165 mi/266 km SE of DARWIN; 14°28′S 132°16′E. On Darwin-Birdum railroad. Peanuts. NITMILUK NATIONAL PARK.

Katherine Gorge National Park, AUSTRALIA: see NITMILUK NATIONAL PARK.

Kathgodam, INDIA: see HALDWANI.

Kathiawar (KAH-tyah-wahr), peninsula (☐ 25,000 sq mi/65,000 sq km), W central INDIA, between the gulfs of KACHCHH and KHAMBAT. Almost all of Kathiawar is included in GUJARAT state; a small area, DIU, is part of DAMAN AND DIU Union Territory. The region, mostly level, has chemical industries. Agriculture is also important, the major crops being cotton, wheat, and millet. Stone quarries, cement, salt from brine, and bauxite. BHAVNAGAR is the chief port. Under BRITISH rule the region contained numerous princely states.

Kathib, Ras (RAHS KETH-EEB), cape, YEMEN coast of RED SEA, 10 mi/16 km NNW of HODEIDA, sheltering N facing deep bay which forms natural harbor just N of the city. Also spelled Ras Kethib.

Kathima, INDIA: see BANBASA.

Kathiri (kuh-THEE-ree), former sultanate, S YEMEN, and one of the former HADHRAMAUT states; SEIYUN was its capital. Extends in a 30-mi/48-km-wide belt from central main Hadhramaut valley (where population is concentrated) N toward the Rub' AL KHALI; farming, livestock raising; handicraft industries (gold- and silverwork; lime kilns). In power after c.1500, the Kathiri tribe had supremacy in the Hadhramaut until challenged by the rising Quaiti tribe in 19th century In the developing feud, the British sided with the Quaiti and the Kathiri lost control of the seacoast. The Quaiti-Kathiri agreement of 1918, concluded under British pressure, defined territorial control of each tribe and extended the British protectorate to the Kathiri state. Since 1945, the 2 Kathiri centers of Seiyun and TARIM, previously autonomous, have been under a single administration. The sultanate was abolished when SOUTH YEMEN became independent (1967).

Kathiwara (kuh-ti-WAH-ruh), former princely state of CENTRAL INDIA agency, lying in W VINDHYA RANGE, now in JHABUA district, W MADHYA PRADESH state, central India. Merged with MADHYA BHARAT in 1948; part of Madhya Pradesh state since 1956.

Kathleen (KATH-leen), town (☐ 3 sq mi/7.8 sq km), 2000 population 3,280), POLK county, central FLORIDA, 8 mi/12.9 km NW of LAKELAND; 28°07′N 82°01′W. Manufacturing includes cabinets.

Kathleen, town, HOUSTON county, GEORGIA, 8 mi/12.9 km NE of PERRY; 32°29′N 83°36′W. Manufacturing includes food products; glass processing.

Kathmandu (KAHT-mahn-doo), district, central NEPAL, in BAGMATI zone; ☉ KATHMANDU.

Kathmandu (KAHT-mahn-doo), city (2001 population 671,846), ☉ NEPAL, central Nepal, in a fertile valley of the E HIMALAYA; 27°41′N 85°21′E. Elevation 4,390 ft/1,338 m. Administrative, business, and commercial center of Nepal, lying astride an ancient trade and pilgrim route from INDIA to TIBET, CHINA, and MONGOLIA. Originally ruled by the Newars, Kathmandu became independent in the 15th century and was captured in 1768 by the Shah Dynasty, who made it their capital. In the late 18th century, the city became the seat of a British resident. Following the 1951 downfall of the Rana prime ministership, Kathmandu experienced an influx of Western tourists. Tourism and trade with India led to a rapid increase in Kathmandu's population. With PATAN (LALITPUR), Kathmandu forms a twin cities urban system with one single built-up urban area. The inner core of Kathmandu contains a high population density. Surrounding this core lies an outer zone developed between 1847 and 1950. Rana rulers built numerous palaces outside the inner city. Between 1950 and 1970 residential development filled most of the vacant land. Beyond the outer zone lies the urban fringe developed since the 1970s. Landmarks include the elaborate Durbar Square, palaces of the politically dominant Rana family, numerous pagoda-shaped temples, and many small temples. Kathmandu also has a number of colleges and a university. Airport. Also spelled KATMANDU.

Kathmandu Valley (KAHT-mahn-doo), undulating plain (☐ 342 sq mi/889.2 sq km), in central NEPAL HIMALAYA, central NEPAL; formed by bifurcation of ridge running S from Gosainthan peak; surrounded on all sides by mountains; 27°40′N 85°21′E. Elevation c. 4,700 ft/1,433 m. Bounded S by MAHABHARAT LEKH range; c. 20 mi/32 km long, 12 mi/19 km–14 mi/23 km wide. Drained by BAGMATI RIVER. Very fertile; rice, wheat, oilseeds, barley, vegetables, fruit. Contains the three cities of Nepal: KATHMANDU, PATAN, and BHADGAON. Main access route is TRIBHUVAN RAJPATH, the major road from INDIA. Historically the most important area of Nepal and center of Nepalese civilization under Newars; overcome in 18th century by Gorkhas. Pop. mainly Newar. Formerly called Nepal Valley.

Katholisch-Hennersdorf, POLAND: see HENNERSDORF.

Kathryn, village (2006 population 56), BARNES CO., SE NORTH DAKOTA, 17 mi/27 km S of VALLEY CITY; 46°40′N 97°58′W. Clausen Springs to Fort Ransom Park to S. Founded in 1906 and incorporated in 1917. Named for Kathryn Mellon, daughter of Charles S. Mellon, a president of Northern Pacific Railroad.

Kathua (KUHT-wah), district (□ 1,024 sq mi/2,662.4 sq km), JAMMU AND KASHMIR state, N INDIA; ⊙ KATHUA. Bounded SW by PAKISTAN, S and E by RAVI River and PUNJAB state; SHIWALIK Range crosses center (NW-SE). Agriculture (wheat, rice, corn, bajra, barley, oilseeds, sugarcane, cotton, spices, tobacco). Main towns: KATHUA, PAROL, BASOLI. Dogri is the prevailing mother tongue.

Kathua (KUHT-wah), town, ⊙ KATHUA district, JAMMU AND KASHMIR state, N INDIA, near the RAVI River, 45 mi/72 km SE of JAMMU; 32°22′N 75°31′E. Rice, wheat, corn, sugarcane, oilseeds.

Kati (KAH-tee), town, SECOND REGION/KOULIKORO, MALI, on DAKAR-NIGER railroad, and 6 mi/9.7 km NW of BAMAKO; 12°44′N 08°04′W. Peanuts, shea nuts, kapok, vegetables, fruit. Military school, Roman Catholic mission.

Katibougou, village, SECOND REGION/KOULIKORO, MALI, 99 mi/159 km N of BAMAKO. Rural polytechnic institute.

Katif, SAUDI ARABIA: see QATIF.

Katigara, town, CACHAR district, S ASSAM state, NE INDIA, on BARAK (Surma) River, and 15 mi/24 km WNW of SILCHAR. Tea processing; rice, tea, cotton, rape and mustard. Also spelled Katigora, Katigorha.

Katihar (KUHT-i-hahr), district (□ 1,180 sq mi/3,068 sq km; 2001 population 2,389,533), E BIHAR state, E INDIA; ⊙ KATIHAR. Rice, jute and paper milling.

Katihar (KUHT-i-hahr), city (2001 population 175,169), ⊙ KATIHAR district, E BIHAR state, NE INDIA, on GANGA plain, 17 mi/27 km SSE of PURNIA; 25°32′N 87°35′E. Railroad and road junction; trade center (corn, tobacco, wheat, sugarcane); railroad workshops. Agriculture: rice, jute, oilseed and flour milling; match manufacturing; sheep breeding. Formerly called SAIFGANJ.

Katima Mulilo (kah-TEE-muh moo-LEE-lo), town (2001 population 22,694), administrative center of CAPRIVI REGION, NAMIBIA, on ZAMBEZI RIVER, and 110 mi/177 km W of MARAMBA (Livingstone), ZAMBIA. Cattle. Regional center of education and tourism.

Katiola (kah-TYO-lah), town, Vallée du Bandama region, central CÔTE D'IVOIRE, on railroad, and 205 mi/330 km NNW of ABIDJAN; 08°08′N 05°06′W. Agriculture (sisal, cotton, rubber, tobacco, pepper, beans, peanuts, yams). Roman Catholic mission.

Katipunan (kah-tee-POO-nahn), town, ZAMBOANGA DEL NORTE province, N MINDANAO, PHILIPPINES, 45 mi/72 km NNW of PAGADIAN; 08°26′N 123°17′E. Coconuts, corn, rice.

Kativik, Administration régionale, administrative region, (□ 193,114 sq mi/502,096.4 sq km; 2006 population 10,240), NORD-DU-QUEBEC region, QUEBEC, E CANADA; 58°29′N 71°29′W. Composed of twenty-nine municipalities; main center is KUUJJUAQ. Formed in 1978.

Katlabug Lagoon, UKRAINE: see KATLABUH LAGOON.

Katlabuh Lagoon (kaht-lah-BOO) (Romanian *Catlabug* or *Catlapug*), SW ODESSA oblast, UKRAINE, on Kiliya arm of the DANUBE, NE of IZMAYIL; 13 mi/21 km long, up to 4 mi/6.4 km wide (S). Receives minor Katlabuh River.

Katlacherra, INDIA: see KATLICHARA.

Katlanovo (kaht-lah-NO-vo), village, N MACEDONIA, on PCINJA RIVER, and 15 mi/24 km SE of SKOPJE. Health resort with sulphur springs (Katlanovska Banja) nearby.

Katlehong, township, GAUTENG province, SOUTH AFRICA, c.15 mi/24 km S of JOHANNESBURG; 26°20′S 28°09′E. Along with neighboring ALBERTON, has seen tremendous growth since 1990.

Katlichara (kaht-LEE-chah-rah), town, CACHAR district, S ASSAM state, NE INDIA, on tributary of BARAK (Surma) River, and 30 mi/48 km SSW of SILCHAR; 24°27′N 92°34′E. Rice, tea, cotton, rape and mustard, sugarcane. Also spelled Katlicherra, Katlacherra.

Katmai National Park and Preserve (KAT-mei), park, preserve, wilderness, at the N end of the ALASKA PENINSULA on SHELIKOF STRAIT, S ALASKA. Has one of the largest areas in the U.S. National Park System; established 1918 as a national monument; its borders were since expanded, and it was designated a park (3,716,000 acres/1,503,865 ha) and preserve (37,400 acres/15,136 ha) area in 1980; wilderness covers 3,473,000 acres/1,405,523 ha. Mount KATMAI and Novarupta volcanoes and the VALLEY OF THE TEN THOUSAND SMOKES are located in this region, which is the site of one of the greatest volcanic eruptions in recorded history: Novarupta (1912). All plant and animal life in the area was destroyed by the ash and lava, although no people were reported killed. KODIAK ISLAND (100 mi/160 km to the SE) was covered with c.1 ft/0.3 m of ash. As lava beneath Mount Katmai drained W to Novarupta, its top collapsed, forming a crater, 8 mi/12.8 km in circumference and 3,700 ft/1,128 m deep, in which a lake has formed. The Valley of the Ten Thousand Smokes (72 sq mi/186 sq km) has countless holes and cracks through which hot gases passed to the surface; all but a few have become extinct. The park also includes glacier-covered peaks, crater lakes, a coastline with dramatic fjords and waterfalls, dense marshlands, and heavy forests with a variety of wildlife, notably moose and grizzly bears.

Katmai Volcano (6,715 ft/2,047 m), S ALASKA, in KATMAI National Monument, 100 mi/161 km WNW of KODIAK; 58°16′N 154°59′W. Active volcano with crater 8 mi/12.9 km in circumference and 3,700 ft/1,128 m deep. Within crater are lake, small island, and glacier-covered walls. Erupted June 5, 1912, desolated KODIAK ISLAND, covered 53 sq mi/137 sq km with sand and lava, created VALLEY OF THE TEN THOUSAND SMOKES, a fumarole field. Investigated in 1915 by National Geographic Society expedition.

Katmandu, NEPAL: see KATHMANDU.

Katni (KUHT-nee), or **murwara** or **Mudwara**, town (2001 population 186,738), KATNI district, E central MADHYA PRADESH state, central INDIA, 55 mi/89 km NE of JABALPUR; 23°47′N 80°29′E. Railroad junction; industrial center; manufacturing (cement and ordnance, pottery, paints, varnishes); fuller's earth processing, rice milling. Industrial school. Important bauxite, limestone, marble, and other workings nearby.

Kato Akhaia (KAH-to ah-khah-EE-ah), town, AKHAIA prefecture, WESTERN GREECE department, GREECE, N PELOPONNESUS, on GULF OF PATRAS, on railroad, and 13 mi/21 km WSW of PATRAS; 38°09′N 21°33′E. Zante currants, wine, wheat; livestock raising (cattle, sheep). Also called Kato Achaia.

Kato Amiandos (kah-TO ahm-YAHN-dos), village, LIMASSOL district, W central CYPRUS, in TROODOS MOUNTAINS, 31 mi/50 km SW of NICOSIA; 34°55′N 32°56′E. PANO AMIANDOS 2 mi/3.2 km to NW, summit of Mount OLYMPUS, 4 mi/6.4 km to WNW. Livestock; grapes, olives, almonds, peaches, vegetables. Also spelled Kato Amiantos.

Kato Dhikomo (KAH-to [TH]EE-ko-mo) [Turkish= *Asagi Dikmen*], town, KYRENIA district, NE central CYPRUS, 8 mi/12.9 km NNW of LEFKOSIA; 35°16′N 33°19′E. KYRENIA MOUNTAINS to N. Citrus, olives, grapes, grain; livestock.

Kato Klitoria (KAH-to klee-TO-ree-ah), town, AKHAIA prefecture, WESTERN GREECE department, N PELOPONNESUS, GREECE, in LADON RIVER valley, 32 mi/51 km SE of PATRAS; 37°54′N 22°08′E. Livestock raising (sheep, goats); tobacco. Also Kato Kleitoria.

Kato Koufinission (KAH-to koo-fee-NEE-see-on), island (□ 2 sq mi/5.2 sq km), CYCLADES prefecture, SOUTH AEGEAN department, GREECE, 3 mi/4.8 km off the SE end of NAXOS island; 34°55′N 26°10′E. Also Kato Kouphonesos, Kato Koufonisos, or Koufounisio.

Katol (kuh-TO-yah), town, NAGPUR district, MAHARASHTRA state, central INDIA, on tributary of WARDHA RIVER, and 33 mi/53 km WNW of NAGPUR. Cotton ginning; millet, wheat, oilseeds; mango and orange groves.

Kato Lakatamia (KAH-to lah-kah-TAHM-yah), town, LEFKOSIA district, central CYPRUS, 5 mi/8 km SW of NICOSIA; 35°07′N 33°18′E. Citrus, olives, grain, vegetables; livestock.

Katombora (kah-to-em-BO-lah), village, SOUTHERN province, SW ZAMBIA, on ZAMBEZI RIVER above (W of) VICTORIA FALLS, 25 mi/40 km W of MARAMBA. Loading point for teakwood floated down Zambezi River.

Katonah (kuh-TO-nah), unincorporated residential village, WESTCHESTER county, SE NEW YORK, 16 mi/26 km NNE of WHITE PLAINS, on a reservoir of CROTON RIVER system; 41°15′N 73°42′W. Botique shops. John Jay Homestead State Historic Site. Caramoor Music Festival each summer at Caramoor Center for Music and Arts.

Kato Nevrokopion (KAH-to nev-ro-KO-pee-on), town, DRÁMA prefecture, EAST MACEDONIA AND THRACE department, NE GREECE, 20 mi/32 km NW of DRÁMA, on road to GOTSE DELCHEV (BULGARIA); 41°21′N 23°52′E. Tobacco center. Formerly called Zyrnovon or Zirnovon. Sometimes spelled Kato Neurokopi (Nevrokopi) or Kato Neurokopion.

Katonga (kah-TON-guh), river, CENTRAL and WESTERN regions, SW UGANDA; flows c.135 mi/217 km W from LAKE VICTORIA to NE end of LAKE GEORGE. Forms N borders of MASAKA, SEMBABULE, and KIRUHURA districts.

Katonkaragai (kah-ton-kah-rah-GEI), town, E EAST KAZAKHSTAN region, KAZAKHSTAN, near BUKTYRMA RIVER, in N foothills of NARYM RANGE, 150 mi/241 km SE of UST-KAMENOGORSK; 49°15′N 85°33′E. Tertiary-level (raion) administrative center. Tourism. Also spelled Katon-Karagai, Qatonqaragai.

Katoomba (kuh-TOOM-buh), municipality (2001 population 17,700), E NEW SOUTH WALES, AUSTRALIA, 55 mi/89 km WNW of SYDNEY, in BLUE MOUNTAINS; 33°42′S 150°18′E. Tourist resort; orchards.

Kato Pyrgos (KAH-to pree-GOS), town, LEFKOSIA district, NW CYPRUS, 40 mi/64 km W of NICOSIA, on MEDITERRANEAN SEA; 35°11′N 32°41′E. MORPHOU BAY to E; Attila Line meets coast to E, small coastal occupied enclave of KOKKINA to W. Fish; livestock; citrus, olives, vegetables.

Kato Vathia, Greece: see AMARINTHOS.

Katowice (kah-to-VEE-tse), German *Kattowitz*, city (2002 population 327,222), S POLAND; 50°16′N 19°01′E. One of the chief mining and industrial centers of Poland, it has industries producing heavy machinery, steel, food textiles, and chemicals; mines in the region yield coal, iron, zinc, and lead. The city was chartered in 1865 and passed from GERMANY to Poland in 1921. Katowice is also an important educational and cultural center.

Katoya (kuh-TO-yah), town, BARDDHAMAN district, central WEST BENGAL state, E INDIA, on BHAGIRATHI RIVER, at AJAY RIVER mouth, and 33 mi/53 km NNE of Barddhaman; 23°39′N 88°08′E. Railroad workshops; rice milling; rice, jute, gram, sugarcane, and vegetables. Vaishnava saint, Chaitanya, adopted life of an ascetic here.

Kato Zakros (KAH-to ZAHK-ros), archaeological site, LASITHI prefecture, midway down SE coast of CRETE department, GREECE; 35°06′N 26°16′E. Minor remains were found in 1901; excavations here begining in 1962 revealed the fourth of the great Minoan palace-centers. Some 5 acres/2 ha have been found, featuring palatial structures and facilities that were part of a major port.

Kato Zodhia, CYPRUS: see ZODHIA, KATO.

Katra (KUH-trah), town, RAJAURI district, JAMMU AND KASHMIR state, N INDIA, in SHIWALIK Range, 18 mi/29 km NNE of JAMMU; 32°59′N 74°57′E. Corn, wheat, rice, oilseeds.

Katra, town, Partapgarh district, SE UTTAR state, N central INDIA, 5 mi/8 km SSW of BELA. Rice, barley, wheat, gram. Also called Katra Mehduniganj.

Katra, INDIA: see BIRPUR KATRA.

Katra, INDIA: see MIRANPUR KATRA.

Katrancik Dag, (Turkish=*Katrancik Dağ*) peak (7,657 ft/2,334 m), SW TURKEY, 15 mi/24 km W of BUCAK.

Katrime, community, S MANITOBA, W central CANADA, 23 mi/38 km from PORTAGE LA PRAIRIE, in Westbourne rural municipality; 50°04′N 98°47′W.

Katrinah, Jabal, EGYPT: see KATHERINA, GEBEL.

Katrineholm (KAHT-ree-ne-HOLM), town, SÖDERMANLAND county, S SWEDEN; 58°59′N 16°12′E. Commercial, industrial, and transportation center; manufacturing (ball bearings, motor vehicle bodies, telecommunications systems, printing). Chartered 1917.

Katrine, Loch (KAH-trin, LAWK), lake (8 mi/12.9 km long, 1 mi/1.6 km wide; 495 ft/151 m deep), STIRLING, central Scotland, at foot of BEN VENUE, 5 mi/8 km E of LOCH LOMOND, and 10 mi/16 km W of CALLANDER, surrounded by hills and woods; 56°15′N 04°31′W. It drains into TEITH RIVER through LOCH ACHRAY (from which it is separated by the TROSSACHS) and LOCH VENACHAR. When it became source of GLASGOW water supply in 1859, its water level was raised, submerging the Silver Strand of Scott's *The Lady of the Lake* and reducing size of ELLEN'S ISLE, island at E end of the loch.

Katschberg Pass (KAHCH-berg PAHS) (elevation 5,002 ft/1,525 m), at E end of the HOHE TAUERN, S AUSTRIA, on CARINTHIA-SALZBURG border, connecting SPITTAL AN DER DRAU (S) with MUR River valley (N); 47°04′N 13°32′E. Winter sports facilities. The pass is tunneled (3.4 mi/5.5 km) by the Tauern Motorway, one of the most important transit roads between GERMANY, ITALY, and S EUROPE.

Katscher, POLAND: see KIETRZ.

Katse Dam (KAH-tshee), LERIBE and THABA-TSEKA districts, N central LESOTHO, forms a reservoir in Malibamatso River (29°20′S 28°30′E), 65 mi/105 km E of MASERU; 25 mi/40 km long. Lies between CENTRAL RANGE and DRAKENSBERG Mountains. Completed 1996, part of LESOTHO HIGHLANDS WATER SCHEME. Town of Katse developed near dam wall. Mashai Reservoir downstream (to SE). Bokong River enters from NW, 8-mi/12.9-km. Bokong arm joins reservoir near dam, extending eventual size of dam to more than 40 mi/64 km.

Katsena Ala, NIGERIA: see KATSINA ALA.

Katsepy (kah-TSAIP), village, MAHAJANGA province, NW MADAGASCAR, across Bombetaka Bay from MAHAJANGA; 15°44′S 46°14′E.

Katsina (kah-CHEE-nah), former province, one of former Northern Provinces, N NIGERIA, in upper KEBBI RIVER basin; capital was KATSINA. One of Hausa native states. Formed in mid-1930s from N ZARIA and Kano provinces.

Katsina (kah-CHEE-nah), state (□ 9,340 sq mi/24,284 sq km; 2006 population 5,792,578), N NIGERIA; ⊙ KATSINA; 12°15′N 07°30′E. Bordered N by NIGER, NE by JIGAWA state, E by KANO state, S by KADUNA state, and W by ZAMFARA state. In savanna vegetation zone. Drained by SOKOTO River. Agriculture includes groundnuts, cotton, maize, millet, guinea corn. Dairying, metal processing, steel rolling; flour mills; textiles. Main centers are Katsina, DAURA, and Dutsinma.

Katsina (kah-CHEE-nah), city, ⊙ Katsina state, N Nigeria, near the Niger frontier. The city, surrounded by a wall 13 mi/21 km long, is the trade center for an agr. region where guinea corn and millet are grown for home consumption, and groundnuts, cotton, and hides are produced commercially. The city has steel-rolling and vegetable oil mills. Leather handicrafts are made here. In the 17th and 18th centuries it was the largest of the 7 Hausa city-states and the cultural and commercial center of Hausaland. In 1807, Katsina was conquered by the Fulani and lost to Kano its preeminent position among Hausa cities. The city is the site of Katsina Training College and Gobaru Tower mosque.

Katsina Ala (kah-CHEE-nah AH-lah), town, BENUE state, E central NIGERIA, on KATSINA ALA River, and 55 mi/89 km SW of WUKARI; 07°10′N 09°17′E. Shea nuts, sesame, cassava, durra, yams. Also spelled Katsena Ala.

Katsina Ala River, W CAMEROON and E NIGERIA, c.200 mi/322 km NW long; rises in BAMENDAhighlands in Cameroon; flows, into TARABA and BENUE states, Nigeria, past KATSINA ALA town, to Benue River just NE of Abinsi. Navigable for c.90 mi/145 km below Katsina Alabama.

Katsumoto (kah-TSOO-mo-to), town, on N coast of IKI-SHIMA island, Iki county, NAGASAKI prefecture, SW JAPAN, 78 mi/125 km N of NAGASAKI; 33°50′N 129°41′E.

Katsunuma (kahts-NOO-mah), town, East Yamanashi county, YAMANASHI prefecture, central HONSHU, central JAPAN, 9 mi/15 km E of KOFU; 35°39′N 138°43′E. Grapes.

Katsura (KAHTS-rah), town, East Ibaraki county, IBARAKI prefecture, central HONSHU, E central JAPAN, 12 mi/20 km N of MITO; 36°30′N 140°22′E. Lacquering.

Katsuragi (kah-TSOO-rah-gee), town, Ito county, WAKAYAMA prefecture, S HONSHU, W central JAPAN, 21 mi/34 km E of WAKAYAMA; 34°17′N 135°30′E. Fruit (persimmons, mandarin oranges).

Katsurao (kah-TSOO-rou), village, Futaba county, FUKUSHIMA prefecture, N central HONSHU, NE JAPAN, 22 mi/35 km S of FUKUSHIMA city; 37°30′N 140°46′E.

Katsuren (kah-TSOO-ren), town, Nakagami county, Okinawa prefecture, SW JAPAN, on peninsula extending E from SE OKINAWA island, 16 mi/25 km N of NAHA; 26°18′N 127°54′E. Carrots, seaweed (*mozoku*). Travertine. Includes Tsuken-jima and Hamahikajima islands.

Katsushika (kahd-ZOO-shee-kah), ward, in NE corner of TOKYO city, Tokyo prefecture, E central HONSHU, E central JAPAN. Bordered N by SAITAMA prefecture, E by CHIBA prefecture, SE by EDOGAWA ward, SW by SUMIDA and ARAKAWA wards, and NW by ADACHI ward.

Katsuta (KAHTS-tah), city (2005 population 153,639), IBARAKI prefecture, central HONSHU, E central JAPAN, on the Naka River, 3.7 mi/6 km N of MITO; 36°23′N 140°32′E. Commercial center; manufacturing (elevators, appliances); dried potatoes.

Katsuta (KAHTS-tah), town, Katsuta county, OKAYAMA prefecture, SW HONSHU, W JAPAN, 31 mi/50 km N of OKAYAMA; 35°04′N 134°11′E.

Katsuura (kahts-TSOO-rah), city, CHIBA prefecture, E central HONSHU, E central JAPAN, on E (PACIFIC OCEAN) coast of Chiba Peninsula, 25 mi/40 km S of CHIBA; 35°08′N 140°19′E.

Katsuura (kah-TSOO-oo-rah), town, Katsuura county, TOKUSHIMA prefecture, SE SHIKOKU, W JAPAN, 9 mi/15 km S of TOKUSHIMA; 33°55′N 134°30′E. Mandarin oranges.

Katsuyama (kah-TSOO-yah-mah), city, FUKUI prefecture, central HONSHU, W central JAPAN, 16 mi/25 km E of FUKUI; 36°03′N 136°30′E.

Katsuyama (kah-TSOO-yah-mah), town, Miyako county, FUKUOKA prefecture, N KYUSHU, SW JAPAN, 31 mi/50 km E of FUKUOKA; 33°41′N 130°55′E.

Katsuyama, town, Maniwa county, OKAYAMA prefecture, SW HONSHU, W JAPAN, 34 mi/55 km N of OKAYAMA; 35°05′N 133°41′E. Lumber trade, bamboo work. Nearby is Kanba Waterfall (300 ft/91 m high) on small Asashi River; hot springs. Sometimes called Chugoku-Katsuyama.

Katsuyama (kah-TSOO-yah-mah), village, South Tsuru county, YAMANASHI prefecture, central HONSHU, central JAPAN, 16 mi/25 km S of KOFU; 35°30′N 138°44′E.

Kattabomman, Nellai, INDIA: see NELLAI KATTABOMMAN.

Kattakurgan (kah-tah-koor-GAHN), city, W SAMARKAND wiloyat, UZBEKISTAN, on TRANS-CASPIAN RAILROAD, near the KARA DARYA River (S arm of ZERAVSHAN RIVER), and 40 mi/64 km WNW of SAMARKAND; 39°55′N 66°15′E. In cotton area. Cotton ginning, cottonseed-oil extraction, food processing. Just S is KATTAKURGAN storage reservoir (also known as Uzbek Sea; completed 1948), fed by canal from Zeravshan River arm. Maximum capacity 500,000 acre-ft. Population made up of Uzbeks (70.3%), Russians (12.4%), Tatars (6.6%), Crimean Tatars (4.2%). Also spelled Kattaqurghon.

Kattakurganskoye Vodokhranilishche, [Russian= Kattakurgan reservoir], town, W SAMARKAND wiloyat, UZBEKISTAN, just S of KATTAKURGAN, on Kattakurgan reservoir; 39°50′N 66°15′E.

Kattefoss (KAHT-tuh-faws), waterfall (1,125 ft/343 m) in VEST-AGDER county, S NORWAY, 39 mi/63 km N of MANDAL, on the Skjerka at its influx into Mandal River. Hydroelectric plant.

Kattegatt (KAHT-te-GAHT), strait, between SWEDEN and DENMARK; c.140 mi/225 km long, 40 mi/64 km– 100 mi/161 km wide. Connected with NORTH SEA through SKAGERRAK, which begins at N tip of JUTLAND, and with BALTIC SEA by way of ÖRESUND, STORE BAELT, and LILLE BAELT. Chief ports are GÖTEBORG (Sweden) and ÅRHUS (Denmark).

Kattowitz, POLAND: see KATOWICE.

Kattumannarkoil, INDIA: see MANNARGUDI, town.

Katugastota (KUH-tu-gahs-THO-tuh), town and suburb, CENTRAL PROVINCE, SRI LANKA, on KANDY PLATEAU, on the MAHAWELI GANGA River, and 3 mi/4.8 km N of KANDY; 07°19′N 80°38′E. Rice, vegetables. Nearby the 350-ft/107-m long Polgolla dam across the Mahaweli allows diversion of water for power at UKUWELA.

Katul, IRAN: see ALIABAD.

Katun' (kuh-TOON) river, approximately 415 mi/668 km long, ALTAI TERRITORY, S Siberian Russia; rises in the KATUN' ALPS; flows generally N to join the BIYA RIVER, with which it forms the OB' RIVER. Partly navigable.

Katuna (kah-TOO-nah), town, AKARNANIA prefecture, WESTERN GREECE department, 6 mi/9.7 km SSW of AMFILOKHIA. Olive oil; wine; livestock.

Katun' Alps (kuh-TOON), Russian *Katun'skiye Belki*, highest range of the ALTAI Mountains, ALTAI TERRITORY, SW Siberian Russia, in a bend formed by the upper KATUN' River; 85 mi/137 km long, 35 mi/56 km wide; rise to 15,155 ft/4,619 m at Belukha peak; 49°46′N 86°31′E. Includes 15 glaciers.

Katunayaka (KUH-tu-NAH-yah-kuh), town, WESTERN PROVINCE, SRI LANKA, on Negombo Lagoon, 19 mi/31 km N of COLOMBO; 07°10′N 79°52′E. Coconuts and rice. Over 30,000 people (mostly women) work in over fifty factories in the 160-sq-mi/414-sq-km investment promotion zone opened in 1978. Sri Lanka's international airport found here.

Katunga, village, N VICTORIA, SE AUSTRALIA, 6 mi/10 km N of NUMURKAH, in MURRAY valley irrigation area; 36°01′S 145°28′E. Dairying; wheat, grains.

Katunga, NIGERIA: see OLD OYO.

Katunitsa (kah-TOO-nee-tsah), village, PLOVDIV oblast, SADOVO obshtina, S central BULGARIA, on the CHEPELARSKA RIVER, 7 mi/11 km ESE of PLOVDIV; 42°06′N 24°52′E. Fruit (strawberries) and vegetable gardening, wine growing; rice milling, liquor distilling.

Katunki (kah-toon-KEE), town (2006 population 810), W NIZHEGOROD oblast, central European Russia, on the Gorkiy Reservoir of the VOLGA RIVER, near highway, 40 mi/64 km NW of NIZHNIY NOVGOROD, and 4 mi/6 km NNW of CHKALOVSK; 56°50′N 43°13′E.

Elevation 278 ft/84 m. In agricultural area (wheat, rye, oats, vegetables).

Katuntsi (kah-TOON-tsee), village, SOFIA oblast, SANDANSKI obshtina, BULGARIA; 41°27′N 23°26′E.

Katwa (KAHT-wuh), village, NORD-KIVU province, E CONGO, on the outskirts of VIRUNGA NATIONAL PARK; 00°06′N 29°19′E. Elev. 4,999 ft/1,523 m.

Katwe (kaht-wai), village, KASESE district, WESTERN region, SW UGANDA, on NE shore of LAKE EDWARD, 60 mi/97 km SW of Kabarole. Salt-mining center; fishing; coffee, tea, bananas, corn. Was part of former WESTERN province.

Katwijk (KAHT-veik), city, SOUTH HOLLAND province, W NETHERLANDS, on NORTH SEA, at mouth of OLD RHINE RIVER, 5 mi/8 km NW of LEIDEN; 52°12′N 04°26′E. Dairying; cattle; flowers, nursery; livestock, fruit, vegetables; manufacturing (food processing, building materials). Resort area; recreational center to S. Lighthouse on Old Rhine River.

Katwijk aan den Rijn (KAHT-veik ahn duhn REIN), village, SOUTH HOLLAND province, W NETHERLANDS, 5 mi/8 km WNW of LEIDEN, S of Katwijk AAN ZEE; 52°11′N 04°25′E. NORTH SEA 1 mi/1.6 km to NW. Dairying; livestock; vegetables, flowers.

Katy (KAH-tee), Polish *Kąty Wrocławskie*, German *Kanth*, town in LOWER SILESIA, in Wrocław province, SW POLAND, 12 mi/19 km WSW of WROCŁAW (Breslau). Agricultural market (grain, sugar beets, potatoes; livestock).

Katy, town (2006 population 13,561), on FORT BEND–HARRIS–WALLER county line, S TEXAS, suburb 25 mi/40 km W of downtown HOUSTON, on W fringe of large urbanized area; 29°48′N 95°49′W. Manufacturing (signs, concrete roof tile, plastic products); agriculture (cattle, rice, vegetables; nurseries).

Katyk, UKRAINE: see SHAKHTARS′K.

Katyn′ (kah-TIN), village, W SMOLENSK oblast, W central European Russia, near a railroad junction, 12 mi/19 km W of SMOLENSK; 54°47′N 31°44′E. Elevation 551 ft/167 m. In agricultural area; building materials (lumber). Site of a massacre of 4,421 Polish army officers by Soviet security forces on Stalin's orders in 1939.

Katyr-Yurt (kah-TIR–YOORT), village (2005 population 9,210), W central CHECHEN REPUBLIC, S European Russia, on road, 13 mi/21 km SW of GROZNYY; 43°10′N 45°22′E. Elevation 774 ft/235 m. Agriculture (grain, livestock); lumbering. Population increased significantly since 1994 due to influx of refugees from GROZNYY, fleeing military action. Formerly known as Tutovo (1944–1958).

Katyshka, RUSSIA: see GOLYSHMANOVO.

Katzenbuckel (KAHT-tsen-buk-kel), highest peak (2,054 ft/626 m) of the ODENWALD, W GERMANY; 49°27′N 09°04′E. Basalt quarried at EBERBACH (W base).

Katzenelnbogen, village, W GERMANY, in RHINELAND–PALATINATE, 9 mi/14.5 km SW of LIMBURG. Has ancestral castle of counts of Katzenellenbogen.

Katzimo, NEW MEXICO: see ENCHANTED MESA.

Katzrin (kahts-REEN) or **Qazrin**, Jewish township, 12 mi/19 km ENE of ZEFAT (Safed), GOLAN HEIGHTS, ISRAEL. Built on basalt bedrock, Katzrin is the largest Jewish town and administrative center of the Golan Heights. Industry (Israeli aircrafts); winery; mineral water plant. Founded in 1977 on the site of ancient town (probably Qisrin). Nearby ruins include remains of 3rd-century Jewish village with a synagogue (built of basalt) with a mosaic pavement. In use for several hundred years, the synagogue was partly destroyed, possibly by earthquake (747).

Káua (KAH-wah), town, YUCATÁN, MEXICO, 13 mi/21 km WSW of VALLADOLID on the Mexico Highway 180; 20°37′N 88°25′W. Henequen, corn.

Kauai (kou-AH-ee), county (□ 1,266 sq mi/3,279 sq km; 1990 population 51,177; 2000 population 58,463), NW HAWAII, mainly KAUAI Island, also NIIHAU Island,

KAULA Island; ⊙ LIHUE, Kauai; 22°00′N 159°41′W. Locally administered by ten districts, nine on Kauai and one on Niihau; no incorporated cities.

Kauai (kou-AH-ee), island (□ 549 sq mi/1,427.4 sq km), 549 sq mi/1,422 sq km, 32 mi/51 km in diameter, NW HAWAII, separated from OAHU island to the SE by KAUAI CHANNEL, from NIIHAU island to SW by KUMUKAHI CHANNEL. LIHUE (1990 population 5,536) is the largest town and ⊙ KAUAI county, and NAWILIWILI HARBOR the chief port. Administered by nine districts. Geologically, Kauai is the oldest of the larger Hawaiian Islands. It was formed by volcanoes now extinct and deeply worn by heavy rainfall and by assault of surrounding ocean; KAWAIKINI (5,243 ft/1,598 m) and WAIALEALE (5,148 ft/1,569 m) are the tallest peaks. High annual rainfall has eroded deep valleys in Kauai's central mountain mass. WAIMEA CANYON (2,000 ft/610 m–3,000 ft/914 m deep; c.10 mi/16 km long), "Hawaii's Grand Canyon," in W, often compared with Arizona's GRAND CANYON. The northeastern slopes of Waialeale, one of the wettest spots on earth, receive an annual average rainfall of 450 in/1,143 cm. Alakai Swamp occupies the island's NW center, unusual in that it is located in upper elevations and is source of island's W streams, including headstreams of Waimea River. An independent kingdom when visited by English Captain James Cook in 1778, Kauai became part of the Kingdom of Hawaii in 1810. The first major attempt at agricultural development in Hawaii occurred there with the establishment of a sugar plantation in 1835. Most of the island's people live along the coast. Agriculture is a major industry with sugarcane, fruit, vegetables, and coffee the chief crops; tourism is central to the economy. Barking Sands Pacific Missile Range at W end of island; Forest reserves: Moloaa and KEALIA (NE); Nonou and Kalepa (E); Lihue-Koloa, Haleea and NaPali-Kona (center); Pun Ka Pele (W). Hanalei National Wildlife Refuge in N; Huleia National Wildlife Refuge in SE; Kilauea National Wildlife Refuge in NE; National Tropical Botanical Gardens in S; Na Pali Coast State Park (NW); Waimea Canyon and Kokee State Park in W center; Polihale State Park on W coast; Russian Fort Elizabeth State Historic Park in SW; Wailua River and Lydgate State Parks in E; Ahukini State Recraeetional Park in E.

Kauai Channel, PACIFIC OCEAN, between KAUAI and OAHU islands, HAWAII, 63 nautical mi/116.6 km wide. Formerly called KAIEIEWAHO CHANNEL.

Kaub (KOUB), town, RHINELAND-PALATINATE, W GERMANY, on right bank of the RHINE, 16 mi/26 km SSE of Oberlahnstein; 50°06′N 07°46′E. Wine; tourism. Ancient fortress Gutenfels (restored 1886) towers above town. On a reef in the Rhine is the large castle of Pfalzgrafenstein, built in 14th century to protect the Rhine toll. Sometimes spelled CAUB.

Kaubakan, YEMEN: see SHIBAM.

Kaubar, Arab village, Ramallah dist., 6.3 mi/10 km of RAMALLAH, in the Samarian Highlands, WEST BANK; 31°59′N 35°10′E. Agriculture (olives, fruits, cereals, vegetables).

Kaudom Game Reserve (KAW-duhm), national reserve, NE NAMIBIA, along BOTSWANA border, just S of CAPRIVI STRIP, 80 mi/129 km SE of RUNDU. Sandy game reserve with many large herbivores and predators typical of the region. 1,500 sq mi/3,885 sq km. Only accessible by four-wheel-drive vehicles.

Kaufbeuren (kouf-BOI-ruhn), city, BAVARIA, S GERMANY, in SWABIA, on the WERTACH, 18 mi/29 km NE of KEMPTEN; 47°53′N 10°38′E. Railroad junction; manufacturing of glass, jewelry, machinery, electronic equipment Jewelry-making school here. Has two 15th-century churches. Chartered c.1220.

Kaufering (KOU-fe-ring), village, BAVARIA, S GERMANY, in UPPER BAVARIA, on LECH RIVER, 18 mi/29 km S of AUGSBURG; 48°06′N 10°52′E. Baroque church.

Kaufman, county (□ 806 sq mi/2,095.6 sq km; 2006 population 93,241), NE TEXAS; ⊙ KAUFMAN; 32°36′N

96°18′W. Mainly rich blackland prairies, bounded W by TRINITY RIVER and drained by its East Fork (forms LAKE RAY HUBBARD in NW corner) and Cedar Creek. Agriculture (cotton; corn, sorghum, wheat; hay; peaches; nursery products; cattle, horses). Manufacturing at TERRELL, Kaufman. Part of CEDAR CREEK RESERVOIR in SE. Formed 1848.

Kaufman, town (2006 population 8,058), NE TEXAS, 30 mi/48 km ESE of DALLAS; ⊙ KAUFMAN county; 32°34′N 96°18′W. Elevation 438 ft/134 m. Market, shipping center in rich blackland agricultural area (cotton; peaches, corn, wheat; cattle, horses); manufacturing (textiles, fabricated metal products). Founded 1848, incorporated 1873.

Kaufungen (KOU-fung-uhn), town, HESSE, central GERMANY, just E of KASSEL; 51°17′N 09°35′E. Manufacturing of machinery, plastics.

Kauhajoki (KOU-hah-YO-kee), village, VAASAN province, W FINLAND, 50 mi/80 km SSE of VAASA; 62°26′N 22°11′E. Elevation 363 ft/110 m. Road center in lumbering, grain-growing region.

Kauhava (KOU-hah-vah), village, VAASAN province, W FINLAND, 45 mi/72 km E of VAASA; 63°06′N 23°05′E. Elevation 165 ft/50 m. In lumbering region; light manufacturing.

Kauiki Head (KAH-oo-EE-kee), point, E end of MAUI island, MAUI county, HAWAII, 1 mi/1.6 km SE of HANA; 20°45′N 155°59′W. Extinct crater.

Kaukab (kaw-KAB), village, HAMA district, W SYRIA, on railroad, 11 mi/18 km NNE of HAMA; 35°18′N 36°48′E. Cotton, cereals. Also spelled KAWKAB.

Kaukaban, YEMEN: see SHIBAM.

Kaukauna (ka-ka-HOO-nuh), city (2006 population 15,095), OUTAGAMIE county, E WISCONSIN, on the FOX RIVER, a suburb 8 mi/12.9 km E of APPLETON; 44°16′N 88°16′W. Manufacturing (food, fabricated metal products, paper); stone quarries nearby. A fur-trading post was established on the site by Pierre Grignon in 1760. The Grignon mansion, built 1836–1839 on the first land deeded in Wisconsin, has been restored. Outagamie County Teachers College is in Kaukauna. Settled 1793, incorporated 1885.

Kaukehmen, RUSSIA: see YASNOYE.

Kaukopää (KOU-ko-PAH), village, KYMEN province, SE FINLAND, 25 mi/40 km ENE of LAPPEENRANTA; 61°15′N 28°52′E. Elevation 330 ft/100 m. Near Russian border, on E shore of LAKE SAIMAA. Lumber, pulp, cellulose mills.

Kaula (ka-OO-lah), island, westernmost point of main (or Eastern) Hawaiian Islands, KAUAI county, HAWAII, c.20 mi/32 km WSW of NIIHAU. Small, barren rock; light station (550 ft/168 m) for U.S. lighthouse service.

Kaulakahi Channel (kah-OO-lah-KAH-hee), 194 mi/312 km wide, PACIFIC OCEAN, between KAUAI and NIIHAU islands, HAWAII. Formerly Kumukahi Channel.

Kauliranta (KOU-lee-RAHN-tah), village, LAPIN province, NW FINLAND, 45 mi/72 km NNW of TORNIO; 66°27′N 23°41′E. Elevation 165 ft/50 m. On TORNIOJOKI (river at SWEDISH border). Railroad head, starting point of roads to N Lapland.

Kaumajet Mountains, LABRADOR, CANADA: see COD ISLAND.

Kaumakani (KOU-mah-KAH-nee), town (2000 population 607), KAUAI county, HAWAII, 16 mi/26 km WSW of LIHUE, 1 mi/1.6 km inland from S coast; 21°55′N 159°37′W. Sugarcane, fruit; cane sugar milling. Hoary Head Stadium to E. Salt Pond Beach Park to SE.

Kaumalapau (KOU-MAH-la-PAH-oo), port, W LANAI, MAUI county, HAWAII, on Kaumalapau Harbor, 6 mi/9.7 km WSW of LANAI CITY, on W coast. Lanai airport to E; W terminus of Kaumalapau Highway.

Kaunakakai (KOU-nah-kah-KEI), town (2000 population 2,726), S MOLOKAI island, MAUI county, HAWAII; 21°05′N 157°00′W. Port for Molokai island.

Area is shown by the symbol □, and capital city or county seat by ⊙.

Kaunas (KOU-nuhs), Polish *Kowno*, Russian *Kovno*, city (2001 population 378,943), in Lithuania, on the Nemunas River; 54°54′N 23°54′E. It is a river port, major highway and railroad junction, and an industrial center; manufacturing (machinery, chemicals, plastics, textiles). Over 85% of the population is Lithuanian Probably founded as a fortress at the end of the 10th century, Kaunas was a medieval trading center and a Lithuanian stronghold against the Teutonic Knights. It passed to a united Lithuanian-Pol. state in 1569 and to Russia in the 3rd partition of Poland (1795). It was captured (1915) by the Germans in World War I and became the provisional capital of Lithuania-Vilnius (1918–1940) being held by Poland until 1939. During this period, Kaunas was the center for Jewish economic and cultural/religious life in independent Lithuania. During the German occupation (1941–1944) the Jews of Kaunas (about 30% of the prewar population) were virtually exterminated. 50,000 Jews of the Kaunas ghetto were killed at the Ninth Fort Nazi base, located outside the city. Before evacuating at the approach of Soviet troops the Germans destroyed much of the city. Historic buildings and ruins (16th-cent. town hall, castle ruins, Vytautus church, 17th-cent. monastery). The city has a university (founded 1922), a polytechnical institute (founded 1950), a medical institute (founded 1951), and several museum

Kaunchi, UZBEKISTAN: see YANGIYUL.

Kauneonga Lake, NEW YORK: see WHITE LAKE.

Kaunia (kah-oo-nee-ah), village (2001 population 64,516), RANGPUR district, N EAST BENGAL, BANGLADESH, on TISTA RIVER (railroad bridge) opposite TISTA village and 10 mi/16 km ENE of RANGPUR; 25°47′N 89°26′E. Railroad junction. Rice, jute, tobacco.

Kauniainen (KOU -nee-EI-nen), Swedish *Grankulla*, town, UUDENMAAN province, S FINLAND, 8 mi/12.9 km WNW of HELSINKI; 60°13′N 24°45′E. Elevation 100 ft/30 m. Part of Helsinki urban region.

Kauno Marios (KO-nuh MAHR-yos), reservoir (□ 24.5 sq mi/63.5 sq km, max. depth 82 ft/25 m), LITHUANIA, on NEMANUS RIVER, on E edge of Kaunus. Largest inland body of water in Lithuania.

Kaup (KAH-oop), village, EAST SEPIK province, NE NEW GUINEA island, NW PAPUA NEW GUINEA, 30 mi/48 km SE of WEWAK. Located on PACIFIC OCEAN. Boat access. Bananas, coconuts, palm oil; tuna.

Kaur (kah-OOR), town, CENTRAL RIVER division, W central THE GAMBIA, on N bank of GAMBIA RIVER (wharf and ferry) and 85 mi/137 km ENE of Banjul; 13°42′N 15°21′W. Peanut-shipping point; palm oil and kernels, rice. Also Kau-ur.

Kaur (KOR), granitic ranges in ADEN hinterland, extending WSW-ENE in the former frontier zone between N and S YEMEN; consists of the Kaur al AUDHILLA (elev. 8,000 ft/2,438 m–9,000 ft/2,743 m) and the Kaur al AWALIQ (E). Also spelled Kawr.

Kaura Namoda (kou-RAH nah-MO-dah), town, ZAMFARA state, NW NIGERIA, 100 mi/161 km ESE of SOKOTO; 12°36′N 06°35′E. Railroad terminus; cotton-shipping point; millet, peanuts; cattle raising. Gold and diamonds mined nearby.

Kauriaganj (KOUR-yah-guhnj), town, ALIGARH district, W UTTAR state, N central INDIA, 15 mi/24 km E of ALIGARH. Wheat, barley, pearl millet, gram, cotton.

Kauriala Ghat (KOUR-yah-luh GAHT), village, KHERI district, N UTTAR PRADESH state, N central INDIA, on KAURIALA arm of KARNALI (GHAGHARA) River and 27 mi/43 km E of PALIA. Railroad terminus; connected by road with RAJAPUR (Nepal).

Kaurzim, CZECH REPUBLIC: see KOURIM.

Kausala (KOU-sah-lah), village, KYMEN province, SE FINLAND, 25 mi/40 km ESE of LAHTI; 60°54′N 26°22′E. Elevation 248 ft/75 m. Center of commerce in lake region. Part of UUDENMAAN province until 1949.

Kausambi, INDIA: see KOSAM.

Kaushany, MOLDOVA: see CAUSANI.

Kaustinen (KOUS-tee-nen), village, VAASAN province, W central FINLAND; 63°32′N 23°42′E. Located 31 mi/50 km SE of KOKKOLA. A gateway to GULF OF BOTHNIA on the Perhonjoki, known for its folk music; annual music festival. Tourism, horse racing; historic church dates from 1777.

Kautenbach (KOU-tuhn-bahk), village, Kautenbach commune, N central LUXEMBOURG, on WILTZ RIVER, at mouth of Clervé River, and 8 mi/12.9 km NNW of ETTELBRUCK; 49°57′N 06°01′E. Railroad junction; light manufacturing.

Kautokeino (KOU-taw-KAI-naw), Sami village, FINNMARK county, N NORWAY, near Finnish border, on Kautokeinoelva River (upper course of ALTAELVA River) and 70 mi/113 km S of ALTA, 110 mi/177 km ESE of TROMSØ. Airfield. Nordic Sami Institute for Research here. First settled by Finns; church established 1701.

Kautokeinoelva, river, NORWAY: see ALTAELVA.

Kau-ur, GAMBIA: see KAUR.

Kavadarci (kah-vah-DAHR-tsee), village, MACEDONIA, in the TIKVES valley, 50 mi/80 km SSE of SKOPJE. Trade center for wine-growing and opium-producing region. Also spelled Kavadartsi; formerly called Kavadar.

Kavajë (kah-VAW-ee), town (□ 160 sq mi/416 sq km; 2001 population 24,817), W central ALBANIA, near the ADRIATIC SEA, on railroad, and 10 mi/16 km SSE of DURRËS; ⊙ KAVAJË district; 41°11′N 19°33′E. Agricultural center (tobacco, cotton, oilseeds, corn). Agricultural school. Also spelled Kavaja.

Kavak, village, N TURKEY, on SAMSUN-AMASYA railroad and 21 mi/34 km SW of SAMSUN; 41°04′N 36°03′E. Cereals.

Kavakli, Greece: see AYIOS ATHANASIOS.

Kavakli, TURKEY: see MERIC.

Kavaklii, Bulgaria: see TOPOLOVGRAD.

Kavala, Greece: see KAVÁLLA.

Kavalerovo (kah-vah-LYE-ruh-vuh), town (2006 population 16,810), SE MARITIME TERRITORY, SE RUSSIAN FAR EAST, in the SE SIKHOTE-ALIN Mountains, near a pass through the mountains, on the short Takushi River, which drains into the Sea of JAPAN, on road, 4 mi/6 km E of RUDNYY; 44°16′N 135°03′E. Elevation 741 ft/225 m. Tin mining; wood processing.

Kavali (KAH-vuh-lee), town, NELLORE district, ANDHRA PRADESH state, SE INDIA, 34 mi/55 km N of NELLORE, near COROMANDEL COAST of BAY OF BENGAL; 14°55′N 79°59′E. Cashew and casuarina plantations; palmyra sugar. Laterite quarries nearby.

Kaválla (kah-VAH-lah), prefecture (□ 838 sq mi/2,178.8 sq km), EAST MACEDONIA AND THRACE department, NE GREECE; ⊙ KAVÁLLA; 41°00′N 24°30′E. On N AEGEAN SEA between Mount PANGAION (W) and MESTA (Nestos) River mouth (E), it includes offshore island of THÁSOS. Bordered N by RODOPI prefecture, E by XÁNTHI prefecture, and W by DRÁMA prefecture. Agriculture (mainly in KHRISOUPOLIS lowland) specializes in well-known Macedonian tobacco, exported here. Corn and citrus fruit also grown. Served by Dráma-Xánthi highway.

Kaválla (kah-VAH-lah), port city (2001 population 58,663), EAST MACEDONIA AND THRACE department, NE GREECE, at the head of Gulf of Kaválla, an inlet of the N AEGEAN SEA; ⊙ KAVÁLLA prefecture; 40°56′N 24°25′E. Surrounded by a rich tobacco-growing hinterland, it is a leading Greek city for processing and exporting tobacco. Fish and manganese are also shipped, and flour is manufactured. International airport, Megas Alexandros. Known as Neapolis in ancient times, the city was the landing place of St. Paul on his way to Philippi, the ancient site which is nearby. Held by the Ottoman Turks from 1387 to 1913, when it passed to Greece. Mehemet (Muhammad) Ali born here. Also Cavalla, Kavala.

Kavanagh (KA-vuh-nah), hamlet, central ALBERTA, W CANADA, 6 mi/10 km from LEDUC, in LEDUC County; 53°11′N 113°31′W.

Kavango River (ka-VAHN-go), NE NAMIBIA at neck of CAPRIVI STRIP, flows past Andara and BAGANI. Southward-flowing perennial river feeds into inland Kavango Swamps, including Popa Falls.

Kavaratti Island (kah-vah-RAH-tee), coral island (2001 population 10,113) and ⊙ LAKSHADWEEP (LACCADIVE) Union Territory, INDIA, in ARABIAN SEA; 10°35′N 72°35′E. Coconuts.

Kavarna (kah-VAR-na), Romanian *Cavarna*, city, VARNA oblast ⊙, Kavarna obshtina (1993 population17,984), NE BULGARIA, on the BLACK SEA, in S DOBRUDZHANSKO, 25 mi/40 km NE of VARNA; 43°25′N 28°20′E. Port (grain exports). Flour milling; manufacturing of machinery. In Romania (1913–1940). Coal deposits in the area but too deep for extraction. Land oil nearby in SHABLA; current exploration offshore.

Kavathe Piran, INDIA: see KAVTHA PIRAN.

Kaveri (kah-VAI-ree), river, c.475 mi/764 km long, S INDIA; rising in the WESTERN GHATS, KARNATAKA state, and flowing SE across a plateau, through TAMIL NADU state, to the BAY OF BENGAL. The BHAVANI and NOYIL rivers are its main tributaries. In Karnataka the river is divided by Sivasamudram island and drops 320 ft/98 m, forming KAVERI FALLS. On the left falls is India's first hydroelectric plant (built 1902), which supplies most of S India with power. At its mouth is a great, fertile delta that is irrigated by an extensive canal system, one of the oldest in India; the GRAND ANICUT dam and canal were built in the eleventh century by the Chola kings. The KAVERI, India's second-most sacred river, is sometimes called the GANGA of the S. A Ganga-Kaveri link has been considered lately. METTUR Dam was built on the river in 1925 (2,150 acres/870 ha long and 308 ft/94 m high), the largest dam in the world at that time. According to Hindu legend, Vishnumaya, daughter of the god Brahma, was born on earth as the child of a mortal, Kavera Muni. In order to bring beatitude for Kavera Muni, she became a river whose water would purify all sins.

Kaveri Falls (KAH-vai-ree), two series of scenic rapids and cataracts, each on an arm of KAVERI River, MYSORE district, on Karnataka-TAMIL NADU state border, SW INDIA, 38 mi/61 km E of Mysore, 3.5 mi/5.6 km NW of SHIMSA-KAVERI River confluence, surrounding Sivasamudram island; each arm descends c.320 ft/98 m; falls on left arm supply pivotal hydroelectric works (since 1940), augmented by plant at nearby village of Shimshapura. Sacred stream to Hindus; often called GANGA of the S; main pilgrimage center is SRIRANGAM ISLAND. Noted archaeological sites at Seringapatam. Chief tributaries are LAKSHMANTIRTHA, AMARAVATI, BHAVANI, and NOYIL (right) rivers. Formerly spelled CAUVERY.

Kaverino (kah-VYE-ree-nuh), village, E RYAZAN oblast, central European Russia, on a highway branch, 12 mi/19 km SSW of SASOVO; 54°09′N 41°46′E. Elevation 498 ft/151 m. In agricultural area (wheat, rye, oats, flax, vegetables).

Kaveripak (KAH-vai-ree-pahk), town, NORTH ARCOT AMBEDKAR district, TAMIL NADU state, S INDIA, on PALAR RIVER and 22 mi/35 km N of VELLORE; 12°54′N 79°28′E. In canal-irrigated agricultural area growing cotton, rice, peanuts. Also Kaveripakkam.

Kaveripatnam (KAH-vai-ree-PUHT-nuhm), town, DHARMAPURI district, TAMIL NADU state, S INDIA, on PONNAIYAR RIVER and 8 mi/12.9 km S of KRISHNAGIRI. Sesame and castor oil extraction.

Kavieng (KA-vee-ENG), port town (2000 population 11,560), N NEW IRELAND, BISMARCK ARCHIPELAGO, PAPUA NEW GUINEA, SW PACIFIC OCEAN. Chief town of island; airfield nearby.

Kavir Buzurg, IRAN: see DASHT-E KAVIR.

Kavirondo (kah-vee-RON-doh), region in W KENYA along NE shore of Lake VICTORIA; chief town, KISUMU. Inhabited by Nilotic Kavirondo people. Administratively divided into three districts

Cross-references are shown in SMALL CAPITALS. The pronunciation guide is shown on page xix. The sources of population figures are shown on page xvii.

(North, Central, and South Kavirondo) of NYANZA province.

Kavirondo Gulf (kah-vee-RON-doh), shallow NE inlet of Lake VICTORIA, in SW KENYA; 35 mi/56 km long, 15 mi/24 km wide. Connected with lake via 3 mi/4.8 km-wide strait. Its port is KISUMU. Navigable for 8 ft/2.4 m draught.

Kavkazskaya (kahf-KAS-kah-yah), village (2005 population 11,735), E KRASNODAR TERRITORY, S European Russia, on the KUBAN' River, on road and railroad, 4 mi/6 km E of KROPOTKIN (at Kavkazskaya railroad junction); 45°26′N 40°40′E. Elevation 275 ft/83 m. In agricultural area; food processing.

Kavkazskiy (kahf-KAHZ-keeyee), settlement (2005 population 3,060), N KARACHEVO-CHERKESS REPUBLIC, S European Russia, on the slopes of the NW CAUCASUS Mountains, just NW of the KUBAN Reservoir, on highway, 8 mi/13 km E of CHERKESSK; 44°16′N 42°14′E. Elevation 2,119 ft/645 m. In agricultural area (livestock).

Kävlinge (SHEV-leeng-e), town, SKÅNE county, S SWEDEN, on KÄVLINGEÅN RIVER, 7 mi/11.3 km NW of LUND; 55°48′N 13°07′E. Railroad junction.

Kävlingeån (SHEV-leeng-e-ON), river, 60 mi/97 km long, S SWEDEN; rises SE of HÖRBY; flows generally W, past Örtofta and KÄVLINGE, to ÖRESUND 8 mi/12.9 km N of MALMÖ.

Kavrepalanchok (kahv-RAI-puh-lahn-CHOK), district, central NEPAL, in BAGMATI zone; ⊙ DHULIKHEL.

Kavtha Piran (KOU-tah pi-RAHN), town, SANGLI district, MAHARASHTRA state, W central INDIA, 6 mi/9.7 km WNW of SANGLI. Local market for millet, sugarcane, cotton. Sometimes spelled Kavathe Piran.

Kavungo (ka-VOON-go), town, MOXICO province, ANGOLA, on road junction and 30 mi/48 km S of Democratic Republic of the CONGO; 11°31′S 23°03′E. Market center.

Kaw (KO), town, N FRENCH GUIANA, on small Kaw River, near the ATLANTIC OCEAN, on road, and 40 mi/64 km SE of CAYENNE; 04°29′N 52°02′W. Also spelled CAUX.

Kawaba (kah-WAH-bah), village, Tone county, GUMMA prefecture, central HONSHU, N central JAPAN, 25 mi/40 km N of MAEBASHI; 36°41′N 139°06′E.

Kawabe (KAH-wah-be), town, Kawabe county, Akita prefecture, N HONSHU, NE JAPAN, 6 mi/10 km S of AKITA city; 39°38′N 140°13′E. Fish (salmon, trout); mushrooms; manufacturing of food products

Kawabe, town, Kamo county, GIFU prefecture, central HONSHU, central JAPAN, 19 mi/30 km N of GIFU; 35°29′N 137°04′E.

Kawabe (kah-WAH-be), town, Hidaka county, WAKAYAMA prefecture, S HONSHU, W central JAPAN, 22 mi/35 km S of WAKAYAMA; 33°54′N 135°11′E. Mandarin oranges.

Kawabe (KAH-wah-be), village, Kita county, EHIME prefecture, NW SHIKOKU, W JAPAN, 25 mi/40 km S of MATSUYAMA; 33°28′N 132°44′E.

Kawachi, former province in S HONSHU, JAPAN; now part of OSAKA prefecture.

Kawachi (KAH-wah-chee), town, Kawachi county, TOCHIGI prefecture, central HONSHU, N central JAPAN, 5 mi/8 km N of UTSUNOMIYA; 36°37′N 139°56′E.

Kawachi (KAH-wah-chee), village, Inashiki county, IBARAKI prefecture, central HONSHU, E central JAPAN, 37 mi/60 km S of MITO; 35°52′N 140°14′E.

Kawachi, village, Ishikawa county, ISHIKAWA prefecture, central HONSHU, central JAPAN, 12 mi/20 km S of KANAZAWA; 36°24′N 136°38′E.

Kawachinagano (kah-WAH-chee-nah-GAH-no), city, OSAKA prefecture, S HONSHU, W central JAPAN, 16 mi/25 km S of OSAKA; 34°27′N 135°34′E.

Kawage (KAH-wah-ge), town, Age county, MIE prefecture, S HONSHU, central JAPAN, 4.3 mi/7 km N of TSU; 34°46′N 136°32′E.

Kawagoe (kah-WAH-go-e), city (2005 population 333,795), SAITAMA prefecture, E central HONSHU, E

central JAPAN, 9 mi/15 km W of URAWA; 35°55′N 139°29′E. Electronics. Site of Kitain Temple (built 830), famed for its images of the five hundred disciples of Buddha.

Kawagoe (kah-WAH-go-e), town, Mie county, MIE prefecture, S honshu, central JAPAN, 22 mi/35 km N of TSU; 35°01′N 136°40′E.

Kawaguchi (kah-WAH-goo-chee), city (2005 population 480,079) and, SAITAMA prefecture, E central HONSHU, E central JAPAN, on the Ajikawa and Kizagawa rivers, 5 mi/8 km S of URAWA; 35°48′N 139°43′E. TOKYO suburb.

Kawaguchi (kah-WAH-goo-chee), town, N. Uonuma county, NIIGATA prefecture, central HONSHU, N central JAPAN, 43 mi/70 km S of NIIGATA; 37°16′N 138°51′E. Mushrooms, watermelons; beef cattle; freshwater fish. Manufacturing of machinery.

Kawaguchi, town, S. Tsuru county, YAMANASHI prefecture, central HONSHU, central JAPAN, near Kawaguchi Lake, 16 mi/25 km S of KOFU; 35°29′N 138°46′E.

Kawahara (kah-WAH-hah-rah), town, Yazu county, TOTTORI prefecture, S HONSHU, W JAPAN, 3.1 mi/5 km S of TOTTORI; 35°23′N 134°12′E. Also Kawabara.

Kawahigashi (kah-wah-hee-GAH-shee), town, Kawanuma county, FUKUSHIMA prefecture, N central HONSHU, NE JAPAN, 31 mi/50 km S of FUKUSHIMA city; 37°32′N 139°55′E.

Kawai (kah-WAH-ee), town, N KATSURAGI district, NARA prefecture, S HONSHU, W central JAPAN, 9 mi/15 km S of NARA; 34°34′N 135°44′E.

Kawai (kah-WAH-ee), village, Yoshiki county, GIFU prefecture, central HONSHU, central JAPAN, 71 mi/115 km N of GIFU; 36°18′N 137°06′E. Freshwater fish farming.

Kawai, village, Shimohei county, IWATE prefecture, N HONSHU, NE JAPAN, 31 mi/50 km S of MORIOKA; 39°35′N 141°41′E. Lumber; beef cattle; lettuce, daikon.

Kawaihae Bay (kah-WEI-HAH-ei), NW HAWAII, HAWAII, on S Kohala Coast. KOHALA PENINSULA to N. Harbor at Kawaihae is principal port for W Hawaii. Landing for cattle shipment; bathing beach.

Kawaikini (kah-WEI-KEE-nee), peak (5,243 ft/1,598 m), central KAUAI, KAUAI county, HAWAII, 9 mi/14.5 km NW of LIHUE. Highest peak on island.

Kawajima (kah-WAH-jee-mah), town, Hiki county, SAITAMA prefecture, E central HONSHU, E central JAPAN, 12 mi/20 km N of URAWA; 35°58′N 139°29′E. Strawberries.

Kawajiri (kah-WAH-jee-ree), town, Toyota county, HIROSHIMA prefecture, SW HONSHU, W JAPAN, on INLAND SEA just E of KURE and 19 mi/30 km S of HIROSHIMA; 34°13′N 132°41′E.

Kawakami (kah-wah-KAH-mee), town, Kawakami county, OKAYAMA prefecture, SW HONSHU, W JAPAN, 40 mi/64 km N of OKAYAMA; 34°43′N 133°29′E.

Kawakami (kah-WAH-kah-mee), village, Yoshino district, NARA prefecture, S HONSHU, W central JAPAN, 25 mi/41 km S of NARA; 34°19′N 135°57′E. Cryptomeria; woodwork.

Kawakami, village, Maniwa county, OKAYAMA prefecture, SW HONSHU, W JAPAN, 29 mi/47 km N of OKAYAMA; 35°16′N 133°38′E. Dairy farming; daikon. Hiruzen Highlands, in nearby Daisen Oki National Park.

Kawakami, village, Abu county, YAMAGUCHI prefecture, SW HONSHU, W JAPAN, 7 mi/12 km N of YAMAGUCHI; 34°21′N 131°32′E. Abu River Dam nearby.

Kawakami, RUSSIA: see SINEGORSK.

Kawakami-Tandzan, RUSSIA: see SINEGORSK.

Kawakawa (KAH-wah-KAH-wah), township, BAY OF ISLANDS area, NE NORTH ISLAND, NEW ZEALAND, 110 mi/177 km NNW of AUCKLAND. Agricultural and orcharding center.

Kawakita (kah-WAH-kee-tah), town, Nomi county, ISHIKAWA prefecture, central HONSHU, central JAPAN, 9 mi/15 km S of KANAZAWA; 36°27′N 136°32′E. Paper.

Kawamata (kah-WAH-mah-tah), town, Date county, FUKUSHIMA prefecture, N central HONSHU, NE JAPAN, 9 mi/15 km S of FUKUSHIMA city; 37°39′N 140°36′E. Food products; silkworms, leaf tobacco; silk.

Kawambwa (kah-WAI-em-bwah), township, LUAPULA province, N ZAMBIA, 170 mi/274 km WNW of KASAMA, CONGO border 30 mi/48 km to W. Agriculture (corn, coffee, tea, bananas); cattle. Manufacturing (food). Lake LAKE MWERU to NW; Lusenga Plain National Park to N; Lumangwe Falls (Kalungwishi River) to NW. Transferred 1947 from NORTHERN province.

Kawaminami (kah-WAH-mee-NAH-mee), town, Koyu county, MIYAZAKI prefecture, SE KYUSHU, SW JAPAN, 22 mi/35 km N of MIYAZAKI; 32°11′N 131°31′E. Pigs, chicken; pumpkins; sea urchins.

Kawamoto (kah-WAH-mo-to), town, Osato county, SAITAMA prefecture, E central HONSHU, E central JAPAN, 28 mi/45 km N of URAWA; 36°08′N 139°17′E.

Kawamoto, town, Ochi county, SHIMANE prefecture, SW HONSHU, W JAPAN, 46 mi/74 km S of MATSUE; 34°59′N 132°29′E. Corn, melon; wild ducks.

Kawanabe (kah-WAH-NAH-be), town, Kawanabe county, KAGOSHIMA prefecture, SW KYUSHU, SW JAPAN, on central SATSUMA PENINSULA, 16 mi/25 km S of KAGOSHIMA; 31°23′N 130°23′E. Pigs. Buddhist altars. Site of Senkoku Dai Gorinto, a five-ringed pagoda carved into a wall of rock.

Kawane (kah-WAH-ne), town, Haibara county, SHIZUOKA prefecture, central HONSHU, E central JAPAN, 19 mi/30 km W of SHIZUOKA; 34°56′N 138°04′E. Tea; lumber.

Kawanishi (kah-WAH-nee-shee), city (2005 population 157,668), Hyogo prefecture, S HONSHU, central JAPAN, on the INA RIVER, 17 mi/27 km N of KOBE; 34°49′N 135°25′E.

Kawanishi, town, Nakauonuma county, NIIGATA prefecture, central HONSHU, N central JAPAN, 53 mi/85 km S of NIIGATA; 37°00′N 138°42′E. Rice.

Kawanishi, town, E. Okitama county, YAMAGATA prefecture, N HONSHU, NE JAPAN, 25 mi/40 km S of YAMAGATA city; 38°00′N 140°02′E.

Kawanoe (kah-WAH-no-e), city and port, EHIME prefecture, N SHIKOKU, W JAPAN, on HIUCHI SEA, 47 mi/75 km E of MATSUYAMA; 33°59′N 133°34′E. Paper-milling center; *mizuhiki* thread production.

Kawanokami (KAH-wah-no-KAH-mee), village, S. Saku county, NAGANO prefecture, central HONSHU, central JAPAN, 50 mi/80 km S of NAGANO; 35°58′N 138°34′E. Lettuce, Chinese cabbage.

Kawara (KAH-wah-rah), town, Tagawa county, FUKUOKA prefecture, N KYUSHU, SW JAPAN, 28 mi/45 km E of FUKUOKA; 33°39′N 130°50′E.

Kawardha (kuh-WUHR-duh), town (2001 population 31,788), KABIRDHAM district, W central CHHATTISGARH STATES, E central INDIA, ⊙ Kabirdham district, 55 mi/89 km N of DURG; 22°01′N 81°15′E. Cotton, wheat, oilseeds. Forested hills (W, N). Was capital of former princely state of KAWARDHA, one of the Chhattisgarh States; in 1948 incorporated into DURG district; then into RAJNANDGAON district in MADHYA PRADESH in 1973; into Kabirdham district in 1998; joined Chhattisgarh States in 2000.

Kawartha Lakes (kuh-WOR-thuh), city (□ 1,181 sq mi/3,070.6 sq km; 2001 population 69,179), E ONTARIO, E central CANADA; 44°35′N 78°49′W. Created in 2001, replacing VICTORIA county; composed of six communities or service centers: BETHANY, BOBCAYGEON, COBOCONK, KIRKFIELD, LINDSAY (former Victoria county seat), and OMEMEE.

Kawartha Lakes (kuh-WOR-thuh), group of fourteen lakes, in a region c.50 mi/80 km long and c.25 mi/40 km wide, S ONTARIO, E central CANADA, near the towns of LINDSAY and PETERBOROUGH; 44°29′N 78°27′W. BALSAM is the largest lake. They are popular as summer resorts. Many of the lakes form part of the TRENT-SEVERN WATERWAY system.

Area is shown by the symbol □, and capital city or county seat by ⊙.

Kawasaki (kah-wah-SAH-kee), city (2005 population 1,327,011), KANAGAWA prefecture, E central HONSHU, E central JAPAN, on TOKYO BAY, 28 mi/45 km W of YOKOHAMA; 35°31′N 139°42′E. Located in the Tokyo-Yokohama industrial area, it has oil refineries and factories that produce chemicals, consumer goods, electronic equipment, motor vehicles, computer equipment, fabricated metal products. Site of Kawasaki Daishi Temple. Ninth-largest city in the country.

Kawasaki (kah-WAH-sah-kee), town, Tagawa county, FUKUOKA prefecture, N KYUSHU, SW JAPAN, 25 mi/40 km E of FUKUOKA; 33°35′N 130°49′E.

Kawasaki, town, Shibata county, MIYAGI prefecture, N HONSHU, NE JAPAN, 12 mi/20 km S of SENDAI, near Zao mountains; 38°10′N 140°38′E. Lumber. Ruins of Tokugawa-era tollgate at Uyamuya.

Kawasaki (kah-WAH-sah-kee), village, E. Iwai county, IWATE prefecture, N HONSHU, NE JAPAN, 56 mi/90 km S of MORIOKA; 38°53′N 141°16′E. Shiitake mushrooms, apples. Handicrafts; food products. Pine trees.

Kawasato (kah-WAH-sah-to), village, N. Saitama county, SAITAMA prefecture, E central HONSHU, E central JAPAN, 19 mi/30 km N of URAWA; 36°06′N 139°30′E.

Kawashima (kah-WAH-shee-mah), town, Hashima county, GIFU prefecture, central HONSHU, central JAPAN, 6 mi/10 km S of GIFU; 35°21′N 136°49′E.

Kawashima, town, Oe county, TOKUSHIMA prefecture, SE SHIKOKU, W JAPAN, on YOSHINO RIVER, 12 mi/20 km W of TOKUSHIMA; 34°03′N 134°19′E. Garlic.

Kawasoe (kah-WAH-so-e), town, Saga county, SAGA prefecture, N KYUSHU, SW JAPAN, 3.7 mi/6 km S of SAGA; 33°11′N 130°19′E. Nori.

Kawatana (kah-WAH-tah-nah), town, East Sonogi county, NAGASAKI prefecture, W KYUSHU, SW JAPAN, on NW coast of HIZEN PENINSULA, on OMURA BAY, 22 mi/35 km N of NAGASAKI; 33°04′N 129°51′E.

Kawauchi (kah-WAH-oo-chee), town, Shimokita county, Aomori prefecture, N HONSHU, N JAPAN, on MUTSU BAY, 45 mi/72 km N of AOMORI; 41°11′N 141°00′E. Scallops.

Kawauchi, town, Onsen county, EHIME prefecture, NW SHIKOKU, W JAPAN, 9 mi/15 km S of MATSUYAMA; 33°47′N 132°55′E.

Kawauchi (kah-WAH-oo-chee), village, Futaba county, FUKUSHIMA prefecture, N central HONSHU, NE JAPAN, 34 mi/55 km S of FUKUSHIMA city; 37°20′N 140°48′E.

Kawaue (kah-WAH-oo-e), village, Ena county, GIFU prefecture, central HONSHU, central JAPAN, 47 mi/75 km N of GIFU; 35°36′N 137°30′E.

Kawau Island (KAH-wou), in HAURAKI GULF, N NORTH ISLAND, NEW ZEALAND, 30 mi/48 km NE of AUCKLAND; 5 mi/8 km long, 5 mi/8 km wide. Summer resort. Traces of copper and manganese formerly mined. Now much is government-owned, administered by HAURAKI GULF MARITIME PARK board.

Kawaura (kah-WAH-oo-rah), town, on S part of AMAKUSA SHIMO-JIMA island, Amakusa county, KUMAMOTO prefecture, SW JAPAN, 50 mi/80 km S of KUMAMOTO; 32°19′N 130°04′E.

Kawazu (KAH-wahz), town, Kamo county, SHIZUOKA prefecture, central HONSHU, E central JAPAN, 37 mi/60 km S of SHIZUOKA; 34°45′N 138°59′E. Oranges, wasabi; flowers (carnations, chrysanthemums). Amagi mountain pass nearby.

Kaw City, village (2006 population 359), KAY county, N OKLAHOMA, within bend of ARKANSAS RIVER (KAW LAKE reservoir), 11 mi/18 km E of PONCA CITY; 36°45′N 96°51′W. In agricultural area; oil and gas wells; recreation.

Kaweah, Lake, reservoir (□ 30 sq mi/78 sq km), TULARE county, central CALIFORNIA, on KAWEAH RIVER, just W of SEQUOIA NATIONAL PARK in Sequoia National Forest, 15 mi/24 km ENE of VISALIA; 36°25′N 119°00′W. Maximum capacity 143,000 acre-ft. Formed by Terminus Dam (255 ft/78 m high), built (1962) by

the Army Corps of Engineers for flood control, irrigation, recreation, and power generation.

Kaweah River (kah-WEE-ah), c.40 mi/64 km long, central CALIFORNIA; formed in TULARE county at confluence of North and South forks in LAKE KAWEAH reservoir, 20 mi/32 km ENE of VISALIA; flows SW of SEQUOIA NATIONAL PARK, past Farmersville (SE of Visalia) and S of Tulare to floor of San Joaquin Valley, where it divides into many channels, some of which reach TULARE LAKE; St. Johns River (23 mi/37 km long) is a distributary. Used for irrigation. North Fork (c.25 mi/40 km long) flows SW, receives Kaweah Scout Fork (c.15 mi/24 km long, from E) and East Fork c.25 mi/40 km long, also from E, South Fork, c.25 mi/40 km long, flows W. All four branches rise in center of Sequoia National Park, from N to S.

Kawela (kah-WAI-lah), village, N OAHU island, HONOLULU county, HAWAII, in NE corner of island, 28 mi/45 km NNW of HONOLULU, 1 mi/1.6 km inland from Turtle Bay Resort, N coast; 21°41′N 158°00′W. Kawela Bay to NW; Kahuka Point to NE.

Kawempe, UGANDA: see JINJA-KAWEMPE.

Kawerau (KAH-weh-rou), town (2001 population 6,975) and district (□ 8 sq mi/20.8 sq km; 2001 population 6,975), BAY OF PLENTY region, NE NORTH ISLAND, NEW ZEALAND, 37 mi/60 km NE of ROTORUA. A town (and now district) planned by pulp and paper company in farming area, with geothermal power in view, near extinct volcanic Mount Edgecumbe. Dominated by timber milling utilizing plantations of Monterey pine. Viewed as New Zealand's most industrialized town, on railroad to Mount Maunganui, timber products exporter.

Kawhia (KAH-fi-ah), town, OTOROHANGA district (□ 796 sq mi/2,069.6 sq km), WAIKATO region, W NORTH ISLAND, NEW ZEALAND, 80 mi/129 km S of AUCKLAND and on N shore of KAWHIA Harbour (8 mi/12.9 km long, 6 mi/9.7 km wide). Dairy and prime lamb products, limestone. Hot springs on beach.

Kawishiwi River (kuh-WI-shi-wee), 60 mi/97 km long (through North Fork), NE MINNESOTA; rises in NE Lake county, in Lake Polly; flows N through Koma and Malberg lakes, then W through Alice, Insula, and Hudson lakes and four reservoirs (lakes Three, Two, and One, and Carefree Lake), below dam of Carefree Lake river splits into two widely separate channels: North Fork flows c.10 mi/16 km W to Farm Lake; South Fork flows c.20 mi/32 km first SW to NE end of BIRCH LAKE, then from same end, through Birch Lake Dam and WHITE IRON LAKE, joins North Fork at Farm Lake (Birch Lake extends 15 mi/24 km to SW of South Fork); 47°54′N 91°43′W. Kawishiwi River continues N from Farm Lake through FALL LAKE and Nenton Lake, and through Newton Falls (cataracts) to PIPESTONE BAY of BASSWOOD LAKE (on U.S.-CANADA border). Entire river is in Superior National Forest; upper course is in BOUNDARY WATERS Canoe Area.

Kawkab, SYRIA: see KAUKAB.

Kawkareik (KAW-kuh-RAIK), township, KAYIN STATE, MYANMAR, 40 mi/64 km E of MAWLAMYINE, at foot of DAWNA RANGE, on road to MYAWADDY (THAILAND border).

Kaw Lake, reservoir (□ 27 sq mi/70.2 sq km), KAY county, NE OKLAHOMA, on ARKANSAS RIVER, 9 mi/14 km E of PONCA CITY; 36°42′N 96°45′W. Maximum capacity 1,348,00 acre-ft. Extends N to KANSAS state line. Formed by Kaw Dam (125 ft/38 m high), built (1976) by Army Corps of Engineers for flood control; also used for water supply and recreation. Borders on Osage Indian Reservation.

Kawlin (KAW-lin), township, SAGAING division, MYANMAR, on railroad, 50 mi/80 km SW of KATHA. In rice-growing area.

Kawoela or **Kawula**, INDONESIA: see LOMBLEN.

Kawr, YEMEN: see KAUR.

Kawthaung, southernmost village of MYANMAR, in MERGUI district of TENASSERIM. Minor port on AN-

DAMAN SEA at mouth of PAKCHAN RIVER (THAILAND border), opposite RANONG, and 170 mi/274 km S of MERGUI, at ISTHMUS OF KRA. Tin mining (25 mi/40 km N). Formerly VICTORIA POINT and KOW SONG.

Kaxgar River (JAH-SHI-GAH-UHR), flows 300 mi/483 km E, in W XINJIANG UYGUR AUTONOMOUS REGION, NW CHINA; rises in several headstreams just across the KAZAKHSTAN and TAJIKISTAN borders in the TRANS-ALAI (Chong-Alay) range; flows past KASHI, toward the YARKANT RIVER, but disappears into the desert. Intermittent flow in lower course; irrigation. Also known as Kashgar River.

Kaxipet, INDIA: see KAZIPET.

Kay, county (□ 945 sq mi/2,457 sq km; 2006 population 45,889), N OKLAHOMA; ⊙ NEWKIRK; 36°48′N 97°08′W. Bounded N by KANSAS line, bounded in SE by ARKANSAS RIVER (KAW LAKE reservoir in E); intersected by Arkansas and CHIKASKIA rivers and by the SALT FORK OF ARKANSAS RIVER. Includes Lake Ponca in E. Agriculture (wheat, barley, sorghum, alfalfa, soybeans); dairying; cattle, sheep. Manufacturing at PONCA CITY, BLACKWELL, and TONKAWA. Oil and natural gas wells; extensive petroleum refining. Formed 1893.

Kaya (KAH-yah), town (2005 population 41,077), ⊙ SANMATENGA province and CENTRE-OUEST region, central BURKINA FASO, 55 mi/89 km NNE of OUAGADOUGOU; 13°05′N 01°05′W. Located on new railroad line. Growing and processing of shea nuts; livestock raising (cattle, sheep, goats). Market and hospital.

Kaya (KAH-yah), town, Yosa county, KYOTO prefecture, S HONSHU, W central JAPAN, near Mount Oe, 50 mi/80 km N of KYOTO; 35°30′N 135°05′E. Dyeing center; *tango* crepe manufacturing.

Kayah State (KAH-yah), state (□ 4,506 sq mi/11,715.6 sq km), E MYANMAR, on the THAILAND border; ⊙ LOIKAW. In the S are the MAWCHI MINES, an important source of tungsten. Rice, maize, and vegetables are grown, and the forests yield teak. The terrain is mountainous and is traversed by the THANLWIN RIVER, the principal river. The inhabitants of the state are Karens (Padaung, Zayein, Yinbaw, and Bre). Under the 1947 Burmese constitution, the KARENNI STATE was constituted from the three states that had treaty relationships with the British crown. The name was changed to Kayah State in 1952.

Kayakent (kah-ya-KYENT), village (2005 population 12,090), SE DAGESTAN REPUBLIC, E CAUCASUS, extreme SE European Russia, near the Caspian coastal railroad and on mountain highway, 33 mi/53 km NNW of DERBENT; 42°23′N 47°54′E. Elevation 383 ft/116 m. Fisheries. Hot sulphur springs, natural gas deposits. Recent sharp population increase largely because of influx of emigrants from the war-torn CHECHEN REPUBLIC.

Kayak Island (20 mi/32 km long, 1 mi/1.6 km–2 mi/3.2 km wide), S ALASKA, in Gulf of ALASKA, 65 mi/105 km SE of CORDOVA in CHUGACH NATIONAL FOREST; 59°55′N 144°26′W. Cape SAINT ELIAS is S extremity. Bering anchored off shore, 1741. Island has shape of a kayak.

Kayalii, Bulgaria: see KAMENO.

Kayalpattinam (kah-YAHL-puht-TEE-nuhm), town, V.O. CHIDAMBARANAR district, TAMIL NADU state, S INDIA, on GULF OF MANNAR, 50 mi/80 km ESE of TIRUNELVELI; 08°34′N 78°07′E. Population descended from pre-Portuguese Arab settlers. Trades in rice, coconuts, timber, betel, and local palmyra products. Numerous mosques. Salt factory at ARUMUGANERI, 2 mi/3.2 km WNW.

Kaya Mountain National Park, Korean *Kaya-san Kungnip Kongwon*, SOUTH KOREA, in NW KYONGSANG province, 38 mi/62 km W of TAEGU. Highest peak, Sangwangbong (4,692 ft/1,430 m). Scenic area of Hongryu-dong [Korean=red stream village] Valley, extending from foot of mountain to Haein-sa Temple, the location of the Tripitaka Koreana (national treasure No. 32), a set of 81,258 wooden printing blocks engraved on both sides (one of the most comprehensive collection of Buddhist scriptures). Changgyong Pango (national treasure No. 52), the library preserving the Tripitaka Koreana, is on the World Heritage List. Lodges, campgrounds, picnic area, hiking trails. Established in 1972.

Kayan (KAH-yahn), township, YANGON division, MYANMAR, 30 mi/48 km E of YANGON, on small KAYAN RIVER (steamer service from PEGU RIVER) and on Thongwa-Pegu Road.

Kayanga River, GUINEA-BISSAU: see GEBA RIVER.

Kayankulam (kuh-YAHN-kuh-luhm), city, ALAPPUZHA district, KERALA state, S INDIA, 20 mi/32 km NNW of KOLLAM; 09°11′N 76°30′E. Trades in coir rope and mats, rice, cassava; cashew nut processing. Also spelled Kayangulam.

Kayan River (KAH-yahn), c.250 mi/402 km long, E BORNEO Island, INDONESIA; rises in IRAN MOUNTAINS on boundary between SARAWAK and KALIMANTAN (Indonesian Borneo); flows E through mountainous area, turns generally ENE past TANJUNGSELOR to CELEBES SEA, 25 mi/40 km E of Tanjungselor; 02°55′N 117°35′E. Sometimes spelled Kajan.

Kayanza, province (□ 476 sq mi/1,237.6 sq km; 1999 population 458,815), N BURUNDI, on RWANDA border (to N); ⊙ KAYANZA; 03°04′S 29°40′E. Borders NGOZI (E), GITEGA (SE), MURAMVYA (S), BUBANZA (W), and CIBITOKE (NW) provinces.

Kayanza (kah-yahn-ZHA), village, ⊙ KAYANZA province, N BURUNDI, 60 mi/97 km NNE of BUJUMBURA; 02°55′S 29°37′E. Trade center, hospital, secondary schools.

Kayattar (kuh-yah-TUHR), town, TIRUNELVELI KATTABOMMAN district, TAMIL NADU state, S INDIA, 15 mi/24 km NNE of Tirunelveli; 08°57′N 77°48′E. Weaving of grass mats; cotton, palmyra. Also spelled Kayathar, Kayatar.

Kayayak, Alaska: see KAGUYAK.

Kaycee (KAI-see), village (2006 population 282), JOHNSON county, N central WYOMING, on Red Fork of POWDER RIVER, 60 mi/97 km NNW of CASPER; 43°42′N 106°38′W. Elevation 4,660 ft/1,420 m. BIGHORN MOUNTAINS to W.

West of Kaycee is the infamous "Hole-in-the-Wall" country and outlaw cave where the legendary outlaws Butch Cassidy and the Sundance Kid and the rest of the Hole-in-the-Wall gang hid out. The scenery varies from rolling rangeland to red rock buttes to steep canyons and the view is always spectacular. Kaycee was also an important site of the Johnson County War, a significant event in Old West history. The BOZEMAN TRAIL, which linked the OREGON TRAIL to the MONTANA gold mines, can still be viewed east of Kaycee. A military post, Old Fort Reno, was established on the Bozeman Trail to protect travelers.

Kaye, Cape, NW BAFFIN ISLAND, BAFFIN region, NUNAVUT territory, CANADA, on PRINCE REGENT INLET; 72°16′N 89°58′W.

Kayenta, unincorporated town (2000 population 4,922), NAVAJO county, NE ARIZONA, 130 mi/209 km NE of FLAGSTAFF, 18 mi/29 km S of UTAH state line, in Navajo Indian Reservation, on Laguna Creek. Elevation 5,798 ft/1,767 m. Sheep, cattle, hogs; corn, alfalfa.

Kayenzi (kah-YAIN-zee), village, MWANZA region, NW TANZANIA, 55 mi/89 km SSW of MWANZA; 03°18′S 32°33′E. Cattle, goats, sheep; cotton, corn, wheat. One of two villages so named in Mwanza region.

Kayenzi, village, MWANZA region, NW TANZANIA, 15 mi/24 km ENE of MWANZA, on SPEKE GULF, Lake VICTORIA; 02°18′S 33°05′E. Fish; cattle; cotton, corn, wheat, millet, sugarcane. One of two villages so named in Mwanza region.

Kayes, administrative region, MALI: see FIRST REGION.

Kayes, city, ⊙ FIRST REGION/KAYES, W MALI, a port on the SÉNÉGAL RIVER; 14°27′N 11°26′W. It is the administrative and commercial center for a region where

peanuts and gum arabic are produced. The town has tanneries and livestock is raised. Kayes lies at the upper limit of navigation on the Sénégal River and is an important railroad terminus. Radio station; airport; light industry, cement factory.

Kayford, unincorporated village, KANAWHA county, W central WEST VIRGINIA, 25 mi/40 km SSE of CHARLESTON; 38°00′N 81°27′W.

Kayiankor, GHANA: see KANYANKAW.

Kayin State (KAH-yin), state (□ 11,730 sq mi/30,498 sq km), MYANMAR, ⊙ Hpan-an, E MYANMAR, bordering THAILAND for roughly 300 mi/483 km. One of seven states within MYANMAR; contains seven townships; formerly known as Karen State. Home to a majority portion of Myanmar's Karen speaking population. Heavily forested rugged hills comprise much of the area. Cut diagonally by the THANLWIN RIVER and, in its S portion, crossed by the World War II railroad route to the bridge on the RIVER KWAI. This same route is today being developed for a natural gas pipeline leading to BANGKOK.

Kayjay (KAI-jai), village, KNOX county, SE KENTUCKY, in the CUMBERLAND MOUNTAINS, 16 mi/26 km NNW of MIDDLESBORO. In bituminous coal and timber area. Kentucky Ridge State Forest to SE.

Kayl (KAIL), town, Kayl commune, S LUXEMBOURG, 3 mi/4.8 km E of ESCH-SUR-ALZETTE; 49°29′N 06°02′E. Iron mining; manufacturing (transportation equipment).

Kaymakchalan, MACEDONIA and GREECE: see KAJMAKCALAN.

Kaynar, KAZAKHSTAN: see KAINAR.

Kaynardzha, Bulgaria: see KAINARDZHA.

Kaynary, MOLDOVA: see CAINARI.

Kayombo (kuh-YOM-bo), village, KATANGA province, SE CONGO, near MANIKA PLATEAU (UPEMBA NATIONAL PARK); 09°36′S 25°37′E. Elev. 2,477 ft/754 m.

Kayseri (KEI-se-ree) or **Kaisaria** (kei-SAH-ree-ah), city (2000 population 536,392)), central TURKEY, at the N foot of ERCIYAS DAGI; 38°42′N 35°28′E. It is an important commercial center and has textile mills, sugar refineries, and cement factories. Carpets are made there. The ancient CAESAREA MAZACA, it was taken by the Seljuk Turks in the mid-11th century, briefly held (1097) by the Crusaders, and captured (1243) by the Mongols. The city was occupied by the Mamluks of EGYPT in 1419. Sultan Selim I incorporated Kayseri into the Ottoman Empire in 1515. The city has numerous historical remains. Nearby is Kanesh, an archaeological site that dates back to the 3rd millennium B.C.

Kaysersberg (kai-zer-BER), German (KEI-zurs-berk), commune (□ 10 sq mi/26 sq km), HAUT-RHIN department, E FRANCE, at E foot of the VOSGES, in ALSACE and 6 mi/9.7 km NW of COLMAR; 48°08′N 07°15′E. Cotton milling, cheese and pottery manufacturing. Its 12th–15th-century church, 16th-century Renaissance town hall, and fortified bridge across the Weiss (tributary of the FECHT RIVER) give the place a medieval appearance. Albert Schweitzer born here in 1875.

Kayshyadoris, LITHUANIA: see KAIŠIADORYS.

Kaysville, city (2006 population 23,563), DAVIS county, N UTAH, near GREAT SALT LAKE, suburb 13 mi/21 km S of OGDEN, 17 mi/27 km N of SALT LAKE CITY; 41°01′N 111°56′W. Elevation 4,349 ft/1,326 m. Irrigated fruit and vegetable-raising area; dairynig; cattle, sheep; manufacturing (food); coal mining. WASATCH RANGE and National Forest to E. Originally called Kay's Fort. Settled 1849, incorporated 1868.

Kayts (KAH-ITS), town, small seaport on N VELANAI island, NORTHERN PROVINCE, SRI LANKA, on PALK STRAIT shipping route and connected to JAFFNA peninsula by causeway; 09°41′N 79°52′E. Fishing. Ruins of 17th-century Portuguese and Dutch fort.

Kayunga (kuh-YOON-gah), administrative district (□ 696 sq mi/1,809.6 sq km; 2005 population 312,300), CENTRAL region, S UGANDA, on S shore of LAKE

KYOGA and W shore of VICTORIA NILE RIVER; ⊙ KAYUNGA; 01°00′N 32°52′E. As of Uganda's division into eighty districts, borders AMOLATAR (N, on opposite shore of Lake Kyoga), KAMULI and JINJA (E, formed by Victoria Nile River), MUKONO (S), and LUWERO and NAKASONGOLA (W) districts. Area's original inhabitants are the Baganda people. Characterized primarily by savannah, also some wetlands. Population primarily rural and agricultural (including bananas, pineapples, and vanilla, also coffee though growth has been affected by disease; poultry, cattle, goats, pigs, and sheep). Fishing is also economically important. Formed in 2000 from N portion of former MUKONO district (current Mukono district formed from S portion).

Kayunga (kuh-YOON-gah), town (2002 population 19,797), ⊙ KAYUNGA district, CENTRAL region, S central UGANDA, 30 mi/50 km NE of KAMPALA. Road junction in coffee, cocoa producing region.

Kayuyak, Alaska: see KAGUYAK.

Kayyerkan (kei-eer-KAHN), city (2005 population 27,350), W central TAYMYR (DOLGAN-NENETS) AUTONOMOUS OKRUG, administratively in KRASNOYARSK TERRITORY, N SIBERIA, RUSSIA, on road and local railroad, 970 mi/1,561 km N of KRASNOYARSK, and 13 mi/21 km W of NORIL'SK; 69°22′N 87°44′E. Elevation 278 ft/84 m. Metallurgy of nonferrous metals based on local ores and on local coal mined both in open mines and shafts. City distinguished by the construction of 9-story apartment buildings in circle around city center, protecting it from the purga (steady frigid Arctic winter air currents). Founded in 1943 as a mining town; made city in 1982, but administratively subordinate to Noril'sk.

Kayynda, KYRGYZSTAN: see KAYYNGDY.

Kayyngdy (keing-DI), town, CHÜY region, KYRGYZSTAN, in CHU valley, on railroad (Kayyngdy station), just N of PANFILOVKA and 45 mi/72 km W of BISHKEK; 42°51′N 73°30′E. Beet-sugar refinery; meatpacking plant, woodworking plant; bricks, cable. Formerly called Molotovsk. Also spelled KAINDA, KAYYNDA.

Kaz (KAHS), village (2005 population 4,815), S KEMEROVO oblast, S central SIBERIA, RUSSIA, on highway and TASHTAGOL-bound extension of the TRANS-SIBERIAN RAILROAD, 20 mi/32 km S of MUNDYBASH; 52°55′N 87°19′E. Elevation 1,253 ft/381 m. Mineral processing. Has a sanatorium.

Kazabazua, village (2006 population 779), OUTAOUAIS region, SW QUEBEC, E CANADA; 45°57′N 76°01′W.

Kazachinskoye (kah-ZAH-cheen-skuh-ye), village (2005 population 2,560), E IRKUTSK oblast, S central SIBERIA, RUSSIA, on the KIRENGA RIVER, on local road, 105 mi/169 km S of KIRENSK; 56°16′N 107°35′E. Elevation 1,368 ft/416 m. Lumbering.

Kazachinskoye (kah-ZAH-cheen-skuh-ye) village (2005 population 3,965), S central KRASNOYARSK TERRITORY, SE SIBERIA, RUSSIA, on the YENISEY RIVER, on road, 120 mi/193 km N of KRASNOYARSK; 57°41′N 93°16′E. Elevation 337 ft/102 m. In agricultural area. Founded in 1650 as a Cossack fort, called Kazachiy Lug.

Kazachiy Lug, RUSSIA: see KAZACHINSKOYE, KRASNOYARSK Territory.

Kazachka (kah-ZACH-kah), village, SW SARATOV oblast, SE European RUSSIA, on crossroads, 30 mi/48 km E of BALASHOV; 51°28′N 43°56′E. Elevation 574 ft/174 m. In agricultural area (wheat, flax, rye, oats, vegetables, potatoes).

Kazachya Lopan', UKRAINE: see KOZACHA LOPAN'.

Kazachye (kah-ZAHCH-ye), village, N SAKHA REPUBLIC, N Russian Far East, N of the ARCTIC CIRCLE, on the YANA RIVER, on road, 468 mi/753 km NNE of YAKUTSK; 70°44′N 136°13′E. Elevation 127 ft/38 m. Reindeer raising.

Kazakh (kah-ZUKH), city, E AZERBAIJAN, near KURA RIVER, 50 mi/80 km SW of BAKY. Railroad junction. Wheat, cotton. Until 1939, Adzhikabul. Railroad transport enterprises.

Kazak Hills (KAH-zahk) or Kazak Uplands, eroded mountain area, E central KAZAKHSTAN, extending from CHINGIZTAU Mountains (NE of Lake BALKASH) NW to ISHIM River; dry steppes; rises to 4,800 ft/1,463 m at peak of Kyzylrai, 135 mi/217 km SE of KARAGANDA. Extensive coal deposits (Karaganda), copper, lead, zinc, tungsten, molybdenum.

Kazakhstan (kah-zahk-STAHN), republic (□ 1,049,155 sq mi/2,717,311 sq km; 2004 estimated population 15,143,704; 2007 estimated population 15,284,929), central ASIA, ☉ ASTANA. Also spelled Kazakstan, Qazaqstan. In late 1997, the government announced its decision to return to the old spelling of Kazakhstan.

Geography

Bordered N by Russian SIBERIA, E by CHINA, S by KYRGYZSTAN, UZBEKISTAN, and TURKMENISTAN, and W by the CASPIAN SEA, Kazakhstan's major cities include ALMATY, SHYMKENT, SEMEY (Semipalatinsk), AKTÖBE (Aktyubinsk), Astana (Akmolinsk, Tselinograd), and UST-KAMENOGORSK. Former Kazakh SSR, part of the former USSR, it is a vast flatland, bounded by a high mountain belt in the SE. Extends nearly 2,000 mi/3,200 km from the lower VOLGA and the Caspian Sea (W) to the ALTAI Mountains (E). Largely lowland in N and W (W. Siberian, Caspian, and TURAN lowlands), hilly in the center (KAZAK HILLS), and mountainous in S and E (TIANSHAN and Altai ranges). Kazakhstan is a region of inland drainage; the SYR DARYA, ILI, CHU, and other rivers drain into the ARAL SEA and Lake BALKASH. It has a dry continental climate, and most of the region is desert or has limited and irregular rainfall; yearly precipitation averages 8 in/20.3 cm–12 in/30 cm, with extremes of 20 in/51 cm on mountain slopes and less than 4 in/10.2 cm near the Aral Sea. Mean temperature in January varies from 28°F/-2.2°C (S) to 0°F/-17.8°C (N) and in July from 86°F/30°C (S) to 72°F/22.2°C (N).

Population

The population consists of Kazaks (53%), Russians (30%, though falling with emigration), Germans, Ukrainians, Uzbeks, and Tartars. The Russian population is concentrated in the industrialized N and W. The Kazaks speak a Turkic language and are nominally Sunni Muslims, originally converted in the late 18th century at the behest of Catherine the Great in order to foster stability among border nomads. The culture of the Kazak nomads featured in the Central Asian epics, ritual songs, and legends. The 19th century saw the growth of the Kazak intelligentsia. A written literature strongly influenced by Russian culture was then developed. Almaty is the seat of the Kazak Al-Farabi State National University (founded 1934) and the Kazak Academy of Sciences (founded 1946).

Economy: Agriculture

The agricultural economy is determined by latitudinal soil and vegetation differentiation. Black-earth wooded steppe (extreme N) is chief grain-growing area (hard-grained wheat, millet) and also has dairy farming; in chestnut-soil dry steppes (51°–40°N), millet and fat-tailed sheep are important; semi-desert and desert are principal grazing areas (fat-tailed sheep, camels); on piedmont loess plains (S) irrigated agriculture (sugar beets, tobacco, opium poppy, rubber-bearing plants, cotton, rice, orchards) predominates. Dry farming of cereal grains along the N borders was attempted from 1954 through the mid-1960s. The Virgin Lands Program under Krushchev brought hundreds of thousands of Russian, Ukrainian, and German settlers to the area. Kazakhstan produced much of the USSR's wool and cattle and a very great part of its wheat. Soviet agricultural practices employing diversion of rivers for irrigation and extensive over-fertilization have compromised soil quality and left much of the water supply polluted; the poisoning and relatively rapid evaporation of the ARAL SEA is a large and very visible example of this legacy.

Economy: Natural Resources

The Kazak Hills in the core of the region have important mineral resources. Coal is mined at KARAGANDA and EKIBASTUZ, and there are major oil fields in the EMBA Basin (which includes the newly developed TENGIZ fields), at N tip of the Caspian Sea and in the MANGYSHLAK PENINSULA. In 1997 Russia and Kazakhstan signed an agreement to export oil and gas via two pipelines (one to be built) linking the Tengiz field to Russian ports. Well over half the copper, lead, zinc, nickel, chromium, and silver mined in the former Soviet Union comes from this area; in NW and central Kazakhstan there are huge iron-ore deposits and gold is found in various parts of the country. The IRTYSH River hydroelectric stations at UST-KAMENOGORSK and the BUKTYRMA reservoir are a major source of power.

Economy: Industry and Aerospace

The country's industries are located along the margins of the region: TEMIRTAU is the iron and steel center. Manufacturing includes machinery, fertilizers, steel, phosphoric acids, artificial fibers, synthetic rubber, textiles, and medicines. Semey (Semipalatinsk) was the Soviet center of space-related industries, and was also the site of Soviet nuclear testing; radiation pollution is widespread here and in adjoining regions. The BAIKONUR COSMODROME in central Kazakhstan was the Soviet space-operations center, and continues to serve Russian space exploration through an agreement between the two nations.

History

The original Turkic tribes were conquered by the Mongols in the 13th century and ruled by various khanates until the Russian conquest (1730–1840). In 1916, the Kazaks rebelled against Russian domination and were in the process of establishing a Western-style state at the time of the 1917 Bolshevik Revolution. Organized as the Kirghiz ASSR in 1920, it was renamed the Kazakh ASSR in 1925 and became a constituent republic in 1936. Kazakhstan declared its independence from the Soviet Union on December 16, 1991, and Nursultan Nazarbayev became the country's first president. In 1994, Kazakhstan reached a security agreement with the U.S. that called for the latter to help Kazakhstan dispose of nuclear, biological, and chemical weapons that it had inherited from the Soviet Union. Conflicts with parliament and a court decision declaring the 1994 elections illegal led Nazarbayev to dissolve the body in 1995. Subsequently, in two referendums tainted by accusations of fraud, Nazarbayev's term was extended and a constitution strengthening his powers was approved. Voting for a new parliament (1995) and subsequent legislative elections have been criticized as flawed. In 1997 the capital was moved from Almaty to Astana (formerly Aqmola). Nazarbayev was reelected in 1999, after disqualifying the significant opposition candidates, and in 2005, after a campaign and vote observers called undemocratic. Kazakhstan is a member of the Commonwealth of Independent States.

Government

Kazakhstan is governed under the constitution of 1995 as amended. The president, who is head of state, is elected by popular vote to a five-year term (prior to 2007, a seven-year term); government power is disproportionately concentrated in the presidency. There is a two-term limit on the president, except for Nursultan Nazarbayev, as the first president of the republic. The government is headed by the prime minister, who is appointed by the president. There is a bicameral Parliament. Of the thirty-nine members of the Senate, seven are appointed by the president and the rest are elected by local governments; all serve six-year terms. The 107 members of the Mazhilis serve five-year terms; 98 are popular elected, and nine are chosen by the Assembly of the People of Kazakhstan, which represents Kazakhstan's ethnic minorities. A party must receive 7% of the vote to be represented in the Mazhilis.

President Nursultan Nazarbayev has been head of state since 1991. The current head of government is Karim Masimov (January 2007). Administratively, Kazakhstan is divided into fourteen provinces (AKMOLA, AKTÖBE, ALMATY, ATYRAU, EAST KAZAKHSTAN (Shygys Kazakhstan), KARAGANDA, KOSTANAI, KYZYLORDA, MANGYSTAU, NORTH KAZAKHSTAN, PAVLODAR, SOUTH KAZAKHSTAN, WEST KAZAKHSTAN, and ZHAMBYL) and three independent cities (ALMATY, ASTANA, and BAIKONUR).

Kazakhstan (kah-zahk-STAHN), town, N WEST KAZAKHSTAN region, KAZAKHSTAN, on railroad, 65 mi/105 km E of ORAL (Uralsk); 51°11′N 52°52′E. In cattle-raising area. Also Qazakstan, Kazakhstan.

Kazaki (kah-zah-KEE), village (2006 population 3,220), W LIPETSK oblast, S central European Russia, on a left tributary of the SOSNA RIVER, on highway and near railroad, 9 mi/14 km W of YELETS; 52°38′N 38°16′E. Elevation 544 ft/165 m. In agricultural area (wheat, rye, flax).

Kazakovo (kah-zah-KO-vuh), village, S NIZHEGOROD oblast, central European Russia, on road, 15 mi/24 km S of (and administratively subordinate to) ARZAMAS; 55°12′N 43°48′E. Elevation 633 ft/192 m. In agricultural area (wheat, rye, oats, vegetables): collective farm.

Kazakovskiy Promysel (kah-ZAH-kuhf-skeeyee PRO-mi-seel), settlement (2004 population 400), S central CHITA oblast, S SIBERIA, RUSSIA, on the Unda River (tributary of the AGA RIVER), near highway, 13 mi/21 km NE, and under administrative jurisdiction, of BALEY; 51°47′N 117°01′E. Elevation 2,349 ft/715 m. Mining and geological combine.

Kazaly (kah-zah-LEE), Russian *Kazalinsk*, city, NW KYZYLORDA region, KAZAKHSTAN, on the SYR DARYA (river) and 5 mi/8 km SW of ZHANAKAZALY (Novo-Kazalinsk); 45°45′N 62°01′E. Rice; fabricated metal products, food. Former caravan center. Founded (1859) as Russian military post.

Kazamaura (kah-ZAH-mah-OO-rah), village, Shimokita county, Aomori prefecture, N HONSHU, N JAPAN, 47 mi/75 km N of AOMORI; 41°29′N 140°59′E. Sea urchin, abalone farming.

Kazan' (kah-ZAHN), city (2006 population 1,104,800), E European Russia, on the VOLGA RIVER; ☉ TATARSTAN Republic, 494 mi/795 km E of MOSCOW; 55°45′N 49°08′E. Elevation 226 ft/68 m. Railroad junction. It is a major historic, cultural, industrial, and commercial center. Manufacturing include machinery, chemicals, explosives, electronic equipment, building materials, consumer goods, food products. Kazan's port and shipyards on the Volga make it an important water transport center. The city's university is one of the oldest and most prestigious in the nation, with such notable graduates as the great Russian author Leo Tolstoy, and the founder of the Soviet state Vladimir I. Lenin. Birthplace of a renowned Russian opera singer Fyodor Shalyapin. Founded in 1401, Kazan' became the capital of a powerful, independent Tatar khanate (1445), which emerged from the empire of the Golden Horde. The khanate was conquered and the city sacked in 1552 by Ivan IV. It became the capital of the Volga region in 1708 and was an outpost (18th century) of Russian colonization in the E. More than 400 of the city's mosques were destroyed during the period of forced conversion of the Tatars into Christianity in the mid-18th century, under the edict of Empress Elizabeth; some of them were rebuilt later in the century, under the rule of Catherine the Great. The name of the city is sometimes spelled Kasan.

Kazan, Bulgaria: see KOTEL.

Kazán, ROMANIA: see CAZANE DEFILE.

Kazangula, village, MATABELELAND NORTH province, W ZIMBABWE, 40 mi/64 km W of VICTORIA FALLS town, on BOTSWANA border, opposite KAZANGULA; 17°53′S 25°15′E. Livestock.

Kazanka (kah-ZAHN-kah), town, NE MYKOLAYIV oblast, UKRAINE, on VYSUN' RIVER, right tributary of the

INHULETS' River, and on highway and near railroad, 17 mi/28 km ENE of NOVYY BUH; 47°50′N 32°50′E. Elevation 321 ft/97 m. Raion center; cheese factory, feed mill, poultry; vocational school, heritage museum. Established in the beginning of the 19th century; town since 1967.

Kazanka River (kah-ZAHN-kah), 60 mi/97 km long, in TATARSTAN, E European Russia; rises S of BALTASI and flows generally SW, past ARSK and KAZAN', to the VOLGA RIVER 3 mi/5 km SW of Kazan'.

Kazan-Kuygan (kuh-zuhn-koo-ee-GAHN), town, central NARYN region, KYRGYZSTAN, on NARYN-Balakchy highway, and 25 mi/40 km NNW of Naryn; 41°26′N 75°57′E. Tungsten mining; smelter. Until 1945 called KARA-UNKURT.

Kazanlik, Bulgaria: see KAZANLUK.

Kazanluk (kah-zahn-LUHK), city, HASKOVO oblast, central BULGARIA, in the KAZANLUK BASIN; ⊙ Kazanluk obshtina (1993 population 88,434); 42°38′N 25°24′E. Region famous for its rose fields. Kazanluk developed in the 17th century as a manufacturing center for attar of roses, rose research. Other manufacturing includes textiles, machinery, food. Agriculture (livestock; fruit). Hydroelectricity.

Kazanluk Basin (kah-zahn-LUHK) (□ 216 sq mi/561.6 sq km), BULGARIA, part of the TUNDZHA RIVER valley between the central STARA PLANINA Mountains (N) and SREDNA GORA (S); 42°36′N 25°21′E. Has rose, lavender, mint fields, chestnut groves, oak and elm forests.

Kazanovka (kah-ZAH-nuhf-kah), town (2006 population 775), E TULA oblast, central European Russia, on the DON River, in the MOSCOW BASIN, near highway, 5 mi/8 km S of YEPIFAN'; 54°20′N 37°54′E. Elevaiton 875 ft/266 m. Lignite mining.

Kazan-retto, JAPAN: see VOLCANO ISLANDS.

Kazan River (kuh-ZAHN, -ZAN), 455 mi/732 km long, NUNAVUT territory, CANADA; flows through ENNADAI and YATHKYED lakes, to BAKER LAKE (which is drained by CHESTERFIELD INLET).

Kazanshunkur (kah-zahn-shun-KOOR), town, central SEMEY region, KAZAKHSTAN, 45 mi/72 km NNE of ZHANGYZTÖBE; 49°34′N 81°16′E. Gold mining. Also Qazanshunkur.

Kazanskaya (kah-ZAHN-skah-yah), village (2005 population 10,845), E KRASNODAR TERRITORY, S European Russia, on the KUBAN' River, on road and railroad, 6 mi/10 km W of KROPOTKIN; 45°25′N 40°26′E. Elevation 423 ft/128 m. In agricultural area (wheat, sunflowers, castor oil); produce processing. Archaeological artifacts dating back to 2nd and 3rd centuries B.C.E. have been found in the vicinity.

Kazanskaya (kah-ZAHN-skah-yah), village (2006 population 4,945), N ROSTOV oblast, S European Russia, on the DON River, on road, 60 mi/97 km NE of MILLEROVO; 49°47′N 41°09′E. Elevation 272 ft/82 m. Flour mill. Established as a Don Cossack *stanitsa* (village); population remains largely Cossack.

Kazanskoye (kah-ZAHN-skuh-ye), town (2006 population 5,840), SE TYUMEN oblast, W SIBERIA, RUSSIA, on the ISHIM River, on crossroads, 30 mi/48 km S of ISHIM; 55°38′N 69°14′E. Elevation 249 ft/75 m. In agricultural area; fish processing, produce processing, creamery. Local transshipment point for processed foodstuffs.

Kazanskoye (kah-ZAHN-skuh-ye), village, NE MARI EL REPUBLIC, E central European Russia, near the administrative border with KIROV oblast, 60 mi/97 km NE of YOSHKAR-OLA; 57°08′N 49°08′E. Elevation 505 ft/153 m. In agricultural area (vegetables, rye, oats, flax, potatoes).

Kazantip Bay (kah-zahn-TEEP) (Ukrainian *Kazantips'ka zatoka*) (Russian *Kazantipskiy zaliv*), shallow S bay of the Sea of AZOV in KERCH PENINSULA, E CRIMEA, UKRAINE; intrudes 5.6 mi/9 km into N central part of Kerch Peninsula, about 10 mi/16 km wide at the entrance. Up to 28 ft/9 m deep, bottom mostly muddy;

salinity 13–14%; fishing villages on E side; Kazantip Cape (a rocky peninsula) on the W end.

Kazantip Cape, UKRAINE: see KAZANTIP BAY.

Kazantips'ka zatoka, UKRAINE: see KAZANTIP BAY.

Kazantipskiy zaliv, UKRAINE: see KAZANTIP BAY.

Kazarman (kuh-zuhr-MUHN), village, JALAL-ABAD region, KYRGYZSTAN, on E slope of FERGANA RANGE, on the NARYN River, 50 mi/80 km ENE of JALAL-ABAD; 41°16′N 74°03′E. Grain; livestock. Tertiary-level administrative center.

Kazatin, UKRAINE: see KOZYATYN.

Kazatskoye, UKRAINE: see KOZATS'KE.

Kazaure (kah-zou-RAI), town, JIGAWA state, N NIGERIA, 45 mi/72 km NNW of KANO; 12°39′N 08°25′E. Agricultural trade center; cotton, peanuts, millet; cattle, skins. Until 19th century, seat of important native emirate.

Kazbegi (kahz-BE-gee), urban settlement and center of Kazbegi region, N GEORGIA, in central Greater Caucasus, on GEORGIAN MILITARY ROAD, on TEREK RIVER and 65 mi/105 km N of TBILISI, at E foot of Mount KAZBEK. Bottling plant; tourist center. Until c.1940, spelled Kazbek.

Kazbek, Mount (KAHZ-bek), peak (16,541 ft/5,042 m), N GEORGIA, in the Greater Caucasus. An extinct volcano, it rises above the DARYAL GORGE and the GEORGIAN MILITARY ROAD. Its glaciers give rise to the TEREK RIVER; total area covered with ice measures 52 sq mi/135 sq km. High-mountain meteorology station on the Gerget Glacier. Mount Kazbek was first scaled in 1868. Tourist resort. An alternate spelling is Kasbek.

Kaz Daği, ancient Ida Mountains, range, NW TURKEY, SE of the location of ancient TROY. Mount Gargarus (5,797 ft/1,767 m) is the highest point. The mountain was dedicated in ancient times to the worship of Cybele who was therefore sometimes called Idae Mater.

Kazerun (kah-ZE-roon), city (2006 population 87,326), Fārs province, SW IRAN, c.60 mi/97 km NE of BUSHEHR; 29°37′N 51°40′E. It is an agricultural trade center.

Kazgorodok (kahz-go-ruh-DOK), village, S KÖKSHETAU region, KAZAKHSTAN, 5 mi/8 km NNW of STEPNYAK; 52°55′N 70°40′E. Cattle.

Kazhim (KAH-zhim), town, SW KOMI REPUBLIC, NE European Russia, on the SYSOLA RIVER (head of navigation), tributary of the VYCHEGDA River, on road and near railroad spur, 152 mi/245 km SSE of SYKTYVKAR; 60°20′N 51°32′E. Elevation 518 ft/157 m. Wood industries. Arose at an iron mine and foundry (dating from 1756), which closed in 1928.

Kazichene (ka-ZEECH-en-e), village, SOFIA oblast, GRAD SOFIA oblast, PANCHAREVO obshtina, BULGARIA; 42°40′N 23°28′E. Railroad station and junction. Chemicals, textiles, wood products.

Kazikazi (kah-zee-KAH-zee), village, SINGIDA region, central TANZANIA, 60 mi/97 km SW of SINGIDA, on Kapatu River; 05°35′S 34°04′E. Goats, sheep; corn, wheat.

Kazi-Magomed, AZERBAIJAN: see QAZIMMAMMAD.

Kazimayn, Al-, IRAQ: see KADHIMAIN, AL.

Kazimierz, POLAND: see CRACOW.

Kazimierz Dolny (kah-ZEE-myesh DOL-nee), Russian *Kazimerzh*, town, Lublin province, E POLAND, port on the VISTULA and 6 mi/9.7 km S of PUŁAWY. Tanning, flour milling; stone quarrying. Has 14th century Gothic church. Castle ruins nearby. Tourist resort.

Kazimoto (kah-zee-MO-to), village, LINDI region, SE central TANZANIA, 115 mi/185 km W of KILWA MASOKO, near MATANDU RIVER, in SELOUS GAME RESERVE; 09°05′S 36°49′E. Livestock; grain.

Kazim Pasa, TURKEY: see OZALP.

Kazincbarcika (KAH-zents-bahr-tse-kah), city (2001 population 32,356), BORSOD-ABAÚJ-ZEMPLÉN county, NE HUNGARY, on SAJÓ river and 13 mi/21 km NW of MISKOLC; 48°15′N 20°38′E. One of the new

"socialist towns" built during the 1950s, now in deep crisis. Coal chemicals and fertilizers, switched to natural gas in late 1960s; PCV plastics. Lignite production and large electric power plant. All these industries contracted sharply in first half of 1990s. Construction of new industrial town with large lignite-fed power plant was begun nearby in 1951. Formerly called Barcika.

Kazinga Channel (kah-ZEEN-guh), WESTERN region, SW UGANDA, flows 25 mi/40 km from LAKE GEORGE WSW to LAKE EDWARD.

Kazinka (KAH-zeen-kah), village, E LIPETSK oblast, S central European Russia, on the MATYRSKOYE Reservoir (VORONEZH RIVER basin), on road and railroad, 7 mi/11 km SE, and under administrative jurisdiction, of LIPETSK; 52°33′N 39°44′E. Elevation 387 ft/117 m. In agricultural area; meat- and produce-processing complex. Formerly known as Novaya Zhizn' (Russian=new life).

Kazipet (kah-ZEE-pait), town, WARANGAL district, ANDHRA PRADESH state, SE INDIA, 3 mi/4.8 km SW of WARANGAL; 17°58′N 79°30′E. Major railroad junction; branch to CHANDRAPUR (140 mi/225 km N) provides shortest railroad route between MADRAS and DELHI via the well-known Grand Trunk Express. Formerly also spelled Kaxipet.

Kazlozhki Suhodol (kahz-LOZH-kee soo-ho-DOL), peak in the PIRIN MOUNTAINS (8,661 ft/ 2,640 m), SW BULGARIA; 41°50′N 23°22′E.

Kazlų Rūda (kahz-LOO ROOD-ah), city, LITHUANIA, 20 mi/32 km WSW of KAUNAS; 54°46′N 23°30′E. In pine forest area; railroad junction. Sawmilling, metalworks.

Kaznau, CZECH REPUBLIC: see KAZNEJOV.

Kaznejov (KAHZ-nye-YOF), Czech *Kaznějov*, German *kaznau*, village, ZAPADOCESKY province, W BOHEMIA, CZECH REPUBLIC, on railroad and 10 mi/16 km N of PLZEŇ; 49°53′N 13°24′E. Manufacturing (ceramics, chemicals, construction materials); kaolin mining in vicinity.

Kazo (KAH-zo), city, SAITAMA prefecture, E central HONSHU, E central JAPAN, 19 mi/30 km N of URAWA; 36°07′N 139°36′E. Cucumbers. Traditional streamers. Also Kaso.

Kazreti (kahz-RE-tee), urban settlement, GEORGIA; 41°26′N 44°12′E. Ore-dressing combine for processing of the Madneuli deposits of polymetallic ores.

Kaztalovka (kahz-tah-LOF-kuh), village, NW W. KAZAKHSTAN region, KAZAKHSTAN, on Lesser Uzen River, 25 mi/40 km S of ALEKSANDROV-GAI (Russia); 49°45′N 48°42′E. Wheat, millet; cattle, camels.

Kazuma Pan National Park (□ 12 sq mi/31 sq km; 7,731 acres/3,130 ha), MATABELELAND NORTH province, W ZIMBABWE, 55 mi/89 km W of HWANGE, bounded W by BOTSWANA. Kazuma Pan Depression, on border, provides water source for herds of wildlife. Limited access.

Kazumba (kah-ZOOM-bah), village, KASAI-OCCIDENTAL province, S CONGO, 70 mi/113 km SSE of LUEBO; 06°25′S 22°02′E. Elev. 2,253 ft/686 m. Trading center in cotton-growing region.

Kazungula (kah-zoo-eng-GOO-lah), village, SOUTHERN province, SW ZAMBIA, 40 mi/64 km W of MARAMBA on ZAMBEZI RIVER; 17°47′S 25°16′E. On Zambia's 10-mi/16-km-long border with BOTSWANA to S. NAMIBIA's Caprivi Strip to W; ZIMBABWE to SE. Agriculture (corn, millet); fishing; timber. Here David Livingstone first reached the Zambezi River and later turned down the river to discover VICTORIA FALLS in 1855.

Kazusa, former province in central HONSHU, JAPAN; now part of CHIBA prefecture.

Kazusa (KAH-zoo-sah), town, S Takaki county, NAGASAKI prefecture, W KYUSHU, SW JAPAN, on S SHIMABARA PENINSULA, 19 mi/30 km ESE of NAGASAKI across TACHIBANA BAY 32°37′N 130°10′E. Potatoes.

Kazym River (kah-ZIM), approximately 350 mi/563 km W, in N KHANTY-MANSI AUTONOMOUS OKRUG, TYU-

MEN oblast, RUSSIA; rises in a swampy zone, through reindeer-raising area, to an arm of the OB' RIVER, 30 mi/48 km E of BERËZOVO. A 16th-century Ostyak principality was on its shores.

Kbely (KBE-li), NE district of PRAGUE, PRAGUE-CITY province, central BOHEMIA, CZECH REPUBLIC, 6 mi/9.7 km from city center. Manufacturing (machinery). Site of military airport.

Kcynia (KTSEE-nyah), German *Exin*, town, BYDGOSZCZ province, W central POLAND, 23 mi/37 km WSW of BYDGOSZCZ. Railroad junction; brick manufacturing, flour milling.

Kdyne (KDI-nye), Czech *Kdyně*, German *neugedein*, town, ZAPADOCESKY province, SW BOHEMIA, CZECH REPUBLIC, 28 mi/45 km SSW of PLZEŇ; 49°24'N 13°02'E. Manufacturing (machinery, textiles).

Kea (KEE), village (2001 population 1,516), W CORNWALL, SW ENGLAND, 2 mi/3.2 km SW of TRURO; 50°14'N 05°01'W. Agricultural market.

Kéa (KE-ah), Latin *Ceos*, island (□ 61 sq mi/158.6 sq km), NW CYCLADES prefecture, SOUTH AEGEAN department, off E central GREECE, in the AEGEAN SEA; 37°37'N 24°20'E. Fruits, barley, and silk are produced. Kéa, the main town, is on the site of ancient Iulis. The poets Bacchylides and Simonides born here. Pirate haven under Ottoman rule. Also spelled Cea, Keos, Zea or Zia.

Keaau (KAI-ah-OU), town (2000 population 2,010), E HAWAII, HAWAII county, HAWAII, 7 mi/11.3 km SSE of HILO; in interior, 5 mi/8 km WSW of Haena Bay; 19°37'N 155°02'W. Mauna Loa Macadamia Nut Factory and Visitors Center to NE; Nani Mau Gardens to N; Panaewa Forest Reserve to N; Waiakea Forest Reserve to W. Also called Olaa.

Keady (KEE-dee), Gaelic *An Céide*, village (2001 population 1,968), W ARMAGH, Northern Ireland, 7 mi/11.3 km SSW of Armagh; 54°15'N 06°42'W. Center for angling on River Callan.

Kealaikahiki Channel (kai-AH-lah-EE-kah-HEE-kee), 16 nautical mi/29.6 km wide, in PACIFIC OCEAN, between LANAI and KAHOOLAWE islands, MAUI county, HAWAII.

Kealakekua (kai-AH-lah-kai-KOO-ah), town (2000 population 1,645), W HAWAII island, HAWAII county, HAWAII, 52 mi/84 km WSW of HILO, 2 mi/3.2 km N of Captain Cook, 2 mi/3.2 km inland from KONA (W) Coast; 19°32'N 155°52'W. Coffee, fruit. University of Hawaii Agricultural Experimental Station, Kona Historical Society Museum are here. KEALAKEKUA BAY to SW.

Kealakekua Bay (kai-AH-lah-kai-KOO-ah), W HAWAII, HAWAII county, HAWAII. Captain James Cook, first European explorer of the islands, was killed here 1779; Captain Cook Memorial at Cook Point, N end of bay; lighthouse at point. Kealakekua State Underwater Park along NE shore.

Kealia (ke-AH-LEE-ah), village, KAUAI, KAUAI county, HAWAII, on E coast, at mouth of Kapaa Stream on Kuhio Highway.

Keams Canyon, unincorporated village, NAVAJO county, NE ARIZONA, in valley on high tableland (c.6,000 ft/1,829 m), in Hopi Indian Reservation, c.65 mi/105 km N of HOLBROOK. Agriculture (sheep, cattle, hogs; corn, alfalfa). Reservation headquarters here. Awatori ruins to SW.

Keansburg, resort borough (2006 population 10,573), MONMOUTH county, E NEW JERSEY, on RARITAN BAY and 9 mi/14.5 km ESE of PERTH AMBOY, on Point Comfort; 40°28'N 74°09'W. Fishing. Largely residential. Incorporated 1917.

Kearney (KAHR-nee), county (□ 516 sq mi/1,341.6 sq km; 2006 population 6,701), S NEBRASKA; ⊙ MINDEN; 40°30'N 98°57'W. Agricultural region bounded N by PLATTE RIVER. Cattle, hogs; corn, soybean, wheat, sorghum, alfalfa, sunflower seed; dairy products. Fort Kearny State Historical Park in NW; Pioneer Village at Minden. Formed 1860.

Kearney (KAHR-nee), city (2000 population 5,472), CLAY county, W MISSOURI, 10 mi/16 km NE of Liberty, 23 mi/37 km NNE of KANSAS CITY (satellite community); 39°21'N 94°21'W. Ships livestock, grain; coal mines; light manufacturing. A major center of mule breeding and raising in early twentieth century. Outlaw Jesse James' birthplace and farm on E side of town. Platted 1867.

Kearney, city (2006 population 29,385), ⊙ BUFFALO county, S central NEBRASKA, on the PLATTE RIVER, and 125 mi/200 km W of LINCOLN; 40°42'N 99°04'W. Commercial, industrial, and transportation center in an agricultural area. University of Nebraska at Kearney and Kearney State College located here. Manufacturing (plastic products, machinery, fabricated metal products, printing and publishing, transportation equipment, apparel). Fort Kearny State Historical Park to S (named for General Stephen W. Kearny), established 1848 to protect the OREGON TRAIL, was abandoned 1871. Kearney County State Recreation Area to SE. County Fairgrounds, Municipal Airport, Union Pacific State Wayside Area to W. Incorporated 1873.

Kearney (KEER-nee), town (□ 205 sq mi/533 sq km; 2001 population 773), S ONTARIO, E central CANADA, 40 mi/64 km ENE of PARRY SOUND; 45°37'N 79°07'W. Lumbering. First established 1908; restructured 1979.

Kearny, town (2000 population 2,249), PINAL county, S central ARIZONA, 55 mi/89 km N of TUCSON, on GILA River between Tortilla Mountains (SW) and Dripping Springs Mountains (NE). Cattle, sheep. Ray Copper Mine (open pit) to N.

Kearneysville (KAHRN-eez-vil), unincorporated village, JEFFERSON county, NE WEST VIRGINIA, 6 mi/9.7 km SE of MARTINSBURG; 39°23'N 77°53'W. Agriculture (grain, apples); livestock; poultry. Light manufacturing. Leetown U.S. Fish Hatchery to SW. West Virginia University horticultural farm.

Kearns (KUHRNZ), unincorporated town, SALT LAKE county, N UTAH, a suburb 9 mi/14.5 km SW of downtown SALT LAKE CITY; 40°38'N 112°00'W. Alfalfa, barley; cattle, sheep. Kennecott Bingham Copper Mine to SW. Salt Lake City Municipal Airport No. 2 to S. In spite of large population growth, Salt Lake county remains an important agricultural county.

Kearny (KAHR-nee), county (□ 871 sq mi/2,264.6 sq km; 2006 population 4,469), SW KANSAS; ⊙ LAKIN; 38°00'N 101°18'W. Gently rolling plain, drained by ARKANSAS RIVER. Wheat; cattle; sorghum, sugar beets, alfalfa. Mountain/Central time zone boundary, follows W county boundary. KEARNY county was on Mountain time zone, is now on Central. Lake Mckinney Reservoir in E. Formed 1888.

Kearny (KAHR-nee), town (2006 population 38,008), HUDSON county, NE NEW JERSEY, 2 mi/3.2 km NNE of NEWARK; 40°45'N 74°07'W. The town was the site of shipyards (greatly enlarged in 1941) and dry docks, which have been replaced by an industrial complex, warehousing facilities, and distribution center. Kearny contains many tidal wetlands between the PASSAIC and the HACKENSACK rivers that were filled for industrial and recreational purposes. One development project was the construction of the Meadowlands racetrack and arena, located near the town of Kearny. Incorporated 1899.

Kearsarge (KEER-sarj), village, CARROLL county, E NEW HAMPSHIRE, in town of CONWAY, 6 mi/9.7 km N of town center, at S edge of White Mountain National Forest. Resort area. Mount Cranmore Ski area is to E.

Kearsarge, Mount (KEER-sarj), peak (2,931 ft/893 m) in MERRIMACK county, S central NEW HAMPSHIRE, 11 mi/18 km WSW of FRANKLIN, in Winslow State Park.

Kearsarge North Mountain (KEER-sarj), peak (3,268 ft/996 m) of WHITE MOUNTAINS, N CARROLL county, NEW HAMPSHIRE, 9 mi/14.5 km NNE of CONWAY; in White Mountain National Forest.

Kearsarge Pass (11,823 ft/3,604 m), INYO-FRESNO county line, E CALIFORNIA, in the SIERRA NEVADA, c.10 mi/16 km W of INDEPENDENCE. Forms E boundary of KINGS CANYON NATIONAL PARK.

Kearsley (KIRZ-lee), town (2001 population 13,248), GREATER MANCHESTER, central ENGLAND, on IRWELL RIVER and 3 mi/4.8 km SE of BOLTON; 53°32'N 02°23'W. Textiles, chemicals, paper.

Keasbey, NEW JERSEY: see WOODBRIDGE.

Keatchie (KEE-chee), village, DE SOTO parish, NW LOUISIANA, 22 mi/35 km SSW of SHREVEPORT, near Bayou Pierre; 32°11'N 93°54'W. Timber; cattle; oil and natural gas.

Keats Island (KEETS) (□ 3 sq mi/7.8 sq km), SW BRITISH COLUMBIA, W CANADA, in HOWE SOUND, 18 mi/29 km WNW of VANCOUVER, opposite GIBSONS; 49°24'N 123°28'W. Mixed farming.

Keauhou (kai-OU-HO-oo, kai-O-oo-HO-oo), town, W HAWAII island, HAWAII county, HAWAII, on KONA (W) Coast, 55 mi/89 km W of HILO, 6 mi/9.7 km S of KAILUA-KONA. Tourism. Keauhou Bay to S. King Kamehameha III Birthplace State Monument to S. Kona Gardens here. Kahaluu Forest Reserve to E.

Kêb, municipality and town, CAMBODIA: see KEP.

Keban, township, E central TURKEY, on EUPHRATES RIVER near mouth of MURAT RIVER, and 27 mi/43 km WNW of ELAZIG; 38°48'N 38°45'E. Mines (lead ore with silver and gold). The first dam on the Euphrates in Turkey (built in the 1960s) with a hydroelectric power station. The dam forms a large artificial lake known as the Keban Lake.

Kebao Island, VIETNAM: see CAI BAU.

Kebayoran (kuh-bah-YO-rahn), neighborhood, JAKARTA, INDONESIA; 06°14'S 106°47'E. Post-World War II planned residential area in S Jakarta.

Kebbi (KE-bee), state (□ 14,208 sq mi/36,940.8 sq km), NW NIGERIA; ⊙ BIRNIN KEBBI. Bordered NNE by SOKOTO state, E by ZAMFARA state, S by NIGER state, SWW by BENIN, and WNW by NIGER. In guinea (tall grass/wooded) savanna belt and fadama (wet season swamp) lowland. Drained by NIGER and SOKOTO rivers. Agriculture includes groundnuts, cotton, sugarcane, wheat, maize, rice, soybeans, millet, guinea corn, beans, ginger, fruits, and vegetables. Livestock (sheep, cattle, goats) rearing is important. Sugar refining, flour milling, textile manufacturing. Gypsum, limestone, clay, and salt deposits. Main centers are Birnin Kebbi, ARGUNGU (fishing festival), and Yelwa-Yauri. Has many historic places and festivals. Carved out of Sokoto state in 1991. Included 4 emirates: Gwandu, Argungu, Yauri, and Zuru. Named for 14th century Kebbi kingdom, ruled as part of the Songhai empire by Askia the Great.

Kebbi (KE-bee), town, KEBBI state, NW NIGERIA, on ZAMFARA River and 70 mi/113 km SSW of SOKOTO; 12°02'N 04°38'E. Market town. Beans, millet, maize; fish.

Kebbi River (KE-bee), flows around 450 mi/724 km NW, NW NIGERIA and S NIGER; rises in KATSINA state, Nigeria, SE of KATSINA, past JIBIYA and into MARADI province, NIGER, past MARADI, then W, back into Nigeria (SOKOTO state), and SW, past WURNO, Sokoto, and BIRNIN KEBBI (Kebbi state), to the NIGER River at GOMBA. Receives SOKOTO and ZAMFARA rivers (left). Also called Rima River. Sometimes known as Sokoto River below the Sokoto River mouth. Kebbi state named for this river.

Kébèmer (KAI-beh-mer), town (2002 population 14,438), LOUGA administrative region, W SENEGAL, on Dakar-Saint-Louis railroad, 22 mi/35 km SW of LOUGA; 15°22'N 16°27'W. In peanut-growing region.

Kébi (KAI-bee), town, Savanes region, NW CÔTE D'IVOIRE, 16 mi/26 km SW of BOUNDIALI; 09°18'N 06°07'W. Agriculture (manioc, rice, corn, beans, peanuts, tobacco, cotton).

Kebili or **Kébili**, province, TUNISIA: see QABILI, province.

Cross-references are shown in SMALL CAPITALS. The pronunciation guide is shown on page xix. The sources of population figures are shown on page xvii.

Kebin, KYRGYZSTAN: see KEMIN.

Kebnekaise (KEB-ne-ka-i-se) [Lappish=kettle top], mountain peak (6,965 ft/2,123 m), NORRBOTTEN county, N SWEDEN; 67°56′N 18°20′E. Highest peak in Sweden. On slopes are sixteen small glaciers.

Keboemen, INDONESIA: see KEBUMEN.

Kebrabassa Rapids, MOZAMBIQUE: see CABORA BASSA.

K'ebri Dehar (ke-BREE dai-HAR), village (2007 population 37,815), SOMALI state, SE ETHIOPIA; 06°44′N 44°17′E. In the OGADEN, on FAFEN RIVER, 217 mi/350 km SE of HARAR. Occupied in Italo-Ethiopian War by Italians (1935), and in World War II by British, who made it their headquarters in the OGADEN. Has airfield.

Kebumen (kuh-BOO-men), town, S Java Tengah province, INDONESIA, 50 mi/80 km W of YOGYA-KARTA, near S coast; ⊙ Kebumen district; 07°40′S 109°39′E. Trade center in agricultural area (sugar, rice, corn); tile works. Also spelled Keboemen.

Kecel (KE-tsel), village, PEST county, S central HUN-GARY, 33 mi/53 km SW of KECSKEMET; 46°32′N 19°16′E. Market center; flour mills; wheat, barley, to-bacco; horses, cattle.

Kechi (KEECH-ei), village (2000 population 1,038), SEDGWICK county, S central KANSAS, suburb 7 mi/11.3 km NNE of downtown WICHITA; 37°47′N 97°16′W. Residential. Agriculture to N and E (wheat; cattle).

Kech River, PAKISTAN: see DASHT RIVER.

Keciborlu (Turkish=Keçiborlu), village, W central TURKEY, on Isparta-Afyonkarahisar railroad and 19 mi/31 km NW of ISPARTA; 37°57′N 30°18′E. Attar of roses distillery; sulphur mines.

Kecoughtan, Virginia: see HAMPTON.

Kecskemét (KECH-ke-met), city (2001 population 107,749) with county rank, central HUNGARY, in a fruit-growing region; 46°55′N 19°43′E. It is a county administrative center, a road and railroad hub, and a manufacturing center (apparel, electronic equipment, transportation equipment, food, textiles). Known since the 4th century, the city has several churches, a museum, and a law school with a large library. It has a major theater and is a selective higher education center. The city has benefited from its proximity to BUDAPEST and has grown rapidly since 1970. It is among the few cities of the Great Alföld that are relatively successful in restructuring their economic base and attracting foreign investment. The Hungarian dramatist Joseph Katona was born in Kecskemét.

Keda (KE-dah), urban settlement, central Adzhar Autonomous Republic, GEORGIA, on ADZHARIS-TSKHALI RIVER and 15 mi/24 km E of BATUM. Tea factory; dairying; sawmill.

Kedabek, AZERBAIJAN: see GADABAY.

Kedah (KEH-dah), state (□ 3,660 sq mi/9,479 sq km; 2000 population 1,649,756), central MALAY PENINSULA, MALAYSIA, on the Strait of MALACCA. It is bordered on the N and NE by THAILAND; ⊙ and chief city is ALOR SETAR. Sungai Patani is an important town. Along the coast are wide alluvial plains where rice is grown. Kedah is one of the traditional "rice bowls" of the peninsula. S Kedah has rubber plantations, and tin is mined in the hills of the interior. Generally level, Kedah has on its E border a mountain range that rises to 6,600 ft/2,012 m. Several islands are also included in the state; LANGKAWI off the NW coast is the largest. The majority of the inhabitants of Kedah are Malays; there are also many Chinese, Indians working on the rubber plantations, and small groups of aborigines. Kedah was the center of the early Hinduized kingdom of Lang-kasuka, according to Arab and Chinese reports of the 6th–8th century. During the Sri Vijaya domination of the Malay Peninsula (8th–13th century) it was an important naval and trade center. During the 15th century it fell under the domination of Malacca, but maintained substantial independence and a profitable trade with INDIA and INDONESIA. At this time most of the inhabitants were converted to Islam. After the fall of Malacca (1511), Kedah was fought over by the Portu-guese, Dutch, Bugis, Minangkabau, and Siamese. By ceding PINANG (1786) and PROVINCE WELLESLEY (1800) to the British, the sultan of Kedah embittered his relations with the Siamese court, which was not appeased by his subsequent conquest of PERAK for SIAM. A bloody Siamese invasion (1821) drove him into exile until 1842; upon his return PERLIS was created as a separate state. In 1909, Siam transferred sovereignty over Kedah to GREAT BRITAIN. Before the establishment of the FEDERATION OF MALAYA (1948), Kedah was classed as one of the UNFEDERATED MALAY STATES. See MALAYSIA, FEDERATION OF.

Kedah Peak (KEH-dah), isolated mountain (3,978 ft/1,212 m) in W KEDAH, MALAYSIA, near Strait of MA-LACCA, 25 mi/40 km NNE of George Town. Sanatorium; tin and iron ore deposits. Also called Gunong Jerai.

Kedah River (KEH-dah), flows 10 mi/16 km W, in KEDAH, NW MALAYSIA; formed at Alor Star by union of two longer headstreams, flows to Strait of MALACCA at KUALA KEDAH.

Kedainiai (ke-DEI-nai), Polish Kiejdany, city (2001 population 32,048), central LITHUANIA, on NEVEZIS RIVER and 27 mi/43 km N of KAUNAS on the VIA BALTICA HIGHWAY from Tallenn to WARSAW; 55°17′N 23°58′E. Railroad and road junction; leather-working center; tanning; manufacturing (apparel, furniture, chemicals); metalworking, flour-milling; thermal power plant. Dates from 15th century. In Russian KOVNO government until 1920. Also spelled as Ke-daynyay or Kedainyai.

Kedarnath (KAI-dahr-naht), village (2001 population 479), RUDRAPRAYAG district, NE Uttarakhand state, INDIA, in central KUMAON HIMALAYAS, 45 mi/72 km NNE of PAURI. Noted Hindu pilgrimage center, with Shivaite temple. KEDARNATH peak (22,770 ft/6,940 m), 5 mi/8 km N, is a destination for pilgrims who come via BADRINATH and climb up to the temple on top (open May–October), along steep and rugged slopes. One of the four sites of the Char Dham, the most important pilgrimage sites in Hinduism; the other three are BADRINATH, GANGOTRI and Yamu-notri in UTTARKASHI.

Kédékou, town, ATAKORA department, NW BENIN, 35 mi/56 km NE of NATITINGOU; 10°40′N 01°52′E. Cotton; livestock; shea-nut butter.

Kedgwick, village (2001 population 1,184), NEW BRUNSWICK, CANADA, 50 mi/80 km SW of Campellton; 47°38′N 67°20′W. Site of many 19th-century lumber camps. Sawmills, pulp shipment.

Kediet Ijill (ke-DYET ee-JEEL), mountain (rises to c.3,000 ft/914 m and is the highest point in MAUR-ITANIA), Tiris Zemmour administrative region, N central Mauritania, near border with WESTERN SA-HARA, 16 mi/25 km SE of the town of FDERIK; 22°34′N 10°26′W. This mountain top and others in the range yield high-grade iron ore (63–64% iron content). Mining began in the early 1960s and first shipments were made in 1963 when the railroad to NOUADHIBOU was completed.

Kediri (ke-DEE-ree), city (2000 population 244,705), Java Timur province, INDONESIA, on BRANTAS RIVER, and 65 mi/105 km SW of SURABAYA, at foot of WILIS MOUNTAINS; ⊙ Kediri district; 07°49′S 112°01′E. Trade center for agricultural area (sugar, coffee, tobacco, rubber, rice, cassava). Textile and lumber mills, machine shops. Kediri was capital of Hindu-Javanese kingdom of Kediri which flourished eleventh-thirteenth century. Sometimes spelled Kadiri.

Kedoengwoeni, INDONESIA: see KEDUNGWUNI.

Kedoin (ke-DO-een), town, Satsuma county, KA-GOSHIMA prefecture, SW KYUSHU, SW JAPAN, 19 mi/30 km N of KAGOSHIMA; 31°54′N 130°29′E. Nearby Imuta Pond (crater lake) has many peat-forming plants.

Kedong (KUH-DUNG), town, ⊙ Kedong county, W HEILONGJIANG province, NE CHINA, on road, and 20 mi/32 km SW of BEI'AN; 48°04′N 126°17′E. Grain, sugar beets, jute, oilseeds.

Kédougou (KAI-doo-goo), village (2002 population 16,672), TAMBACOUNDA administrative region, SE SENEGAL, near GUINEA border, on GAMBIA RIVER, 130 mi/209 km SE of TAMBACOUNDA; 12°33′N 12°11′W. Region produces shea-nut butter, kapok, beeswax, corn, peanuts; rice; sheep and goats. Market center. Airfield; marble quarry.

Kedrovaya Pad' (kee-DRO-vah-yah–PAHD), wildlife refuge, MARITIME TERRITORY, RUSSIAN FAR EAST, extreme SE Siberian Russia, on the W bank of the AMUR River, primarily in the Kedrovian Basin; 44,230 acres/17,900 ha. Mostly forest. Fauna include deer, boar, leopard, bear, marten, wildcat, tiger, lynx, wolf; salmon, trout. Established as a protected area in 1916; given legal status of a preserve in 1924.

Kedrovka (kee-DROF-kah), town (2005 population 17,605), N KEMEROVO oblast, S central SIBERIA, RUSSIA, near highway and railroad spur, 16 mi/25 km N of KEMEROVO, to which it is administratively subordinate; 55°32′N 86°03′E. Elevation 551 ft/167 m. Coal mining in the Kuznetsk coal basin; parts and equipment for maintenance and repair of mining machinery.

Kedrovka (KYE-druhf-kah), village (2006 population 2,390), W SVERDLOVSK oblast, extreme W Siberian Russia, in the central URALS, on road and railroad, 14 mi/23 km WSW of KUSHVA; 58°10′N 59°22′E. Elevation 1,072 ft/326 m. In ore-mining and -processing region (copper and iron).

Kedrovoye (KYE-druh-vuh-ye), town (2006 population 2,150), SW SVERDLOVSK oblast, extreme W Siberian Russia, E URALS, on short highway branch, 11 mi/18 km NNW, and under administrative jurisdiction, of VERKHNYAYA PYSHMA; 57°09′N 60°34′E. Elevation 885 ft/269 m. Peat works; garment factory.

Kedrovyy (kee-DRO-viyee), city (2006 population 3,065), central TOMSK oblast, W SIBERIA, RUSSIA, in the OB' RIVER basin, near highway, 214 mi/344 km NW of TOMSK; 57°31′N 79°31′E. Elevation 410 ft/124 m. Arose in 1982 as a dormitory settlement for workers in development of the Pudinsk oil and gas district. City status since 1987.

Kedrovyy (KYED-ruh-viyee), industrial settlement, central KHANTY-MANSI AUTONOMOUS OKRUG, W central SIBERIA, RUSSIA, on the right bank of the OB' RIVER, 37 mi/60 km NNW of KHANTY-MANSIYSK; 61°30′N 68°15′E. Elevation 173 ft/52 m. Manufacturing of equipment and machinery for the oil, gas, and hydroelectric energy industries.

Kedrovyy (KYED-ruh-viyee), urban settlement (2005 population 5,115), S KRASNOYARSK TERRITORY, SE Siberian Russia, 26 mi/42 km W of KRASNOYARSK; 56°10′N 91°49′E. Elevation 1,325 ft/403 m. A closed residential settlement for military personnel, and their families, of a Russian strategic missile base. Also known by its secret designation, KRASNOYARSK-66.

Kedumim or **Qedumim,** Jewish settlement, 6 mi/10 km WSW of NABLUS in Samarian highlands, next to the Arab village of Kadun, WEST BANK. Ruins include ritual baths (for purification) and burial caves. Also, several 5th–6th century oil presses, which are displayed at the local Kedem museum. Founded in 1977 by orthodox Jewish settlers on site of an ancient large Samarian settlement.

Kedungwuni (kuh-DUNG-woo-nee), town, N Java Tengah province, INDONESIA, 50 mi/80 km W of SE-MARANG; 06°58′S 109°39′E. Trade center for agricultural area (sugar, rice, tobacco). Also spelled Kedoengwoeni.

Kedzierzyn-Koźle (ken-DEE-zeen KO-zee-le), Polish Kędzierzyn, German Heydebreck, town (2002 population 67,097) in UPPER SILESIA, after 1945 in OPOLE province, S POLAND, near Klodnitz River (Gliwice Canal), 18 mi/29 km N of Ratibor (RACIBORZ). Important railroad junction; fertilizer manufacturing; thermal power plant. Until 1934, called Kędzierzyn.

Keedysville (KEE-DEEZ-vill), town (2000 population 482), WASHINGTON county, W MARYLAND, near Antietam Creek, 11 mi/18 km S of HAGERSTOWN, near Crystal Grottoes and ANTIETAM NATIONAL BATTLEFIELD; 39°29′N 77°42′W. When residents asked the original name of Centerville be changed to avoid confusion with several other towns of the same name in the state, the petition had so many Keedy signatures on it the name Keedysville was decided on.

Keegan, MAINE: see VAN BUREN.

Keego Harbor (KEE-go), village (2000 population 2,769), OAKLAND county, SE MICHIGAN, suburb 4 mi/6.4 km SW of PONTIAC on E end of CASS LAKE and S end of SYLVAN LAKE; 42°36′N 83°20′W.

Keel, IRELAND: see ACHILL ISLAND.

Keele Peak (c.8,500 ft/2,591 m), YUKON TERRITORY, CANADA, in MACKENZIE MOUNTAINS; 63°27′N 130°20′W.

Keeler, unincorporated village, INYO county, E CALIFORNIA, 12 mi/19 km S of Lone Pine on California 136, on E side of Owens Lake. Formerly soda-ash mining; served Cerro Gordo mine in INYO MOUNTAINS (closed early 20th century).

Keeline (KEE-lein), village, NIOBRARA county, E WYOMING, 15 mi/24 km W of LUSK, near Muddy Creek. Elevation 5,289 ft/1,612 m.

Keeling Islands, AUSTRALIA: see COCOS ISLANDS.

Keelung, TAIWAN: see CHI-LUNG.

Keelung River (JEE-LUNG) or **Chi-lung Ho** (JEE-LUNG HUH), 40 mi/64 km long, N TAIWAN; rises near N coast, SE of CHI-LUNG; flows WSW, past CHITU, SHIHCHIH, and SUNGSHAN, to TANSHUI RIVER in N outskirts of TAIPEI; coal mining in valley.

Keene (KEEN), city (2006 population 22,672), SW NEW HAMPSHIRE, 40 mi/64 km W of MANCHESTER; ⊙ CHESHIRE county; 42°57′N 72°17′W. Drained by the ASHUELOT RIVER. It is a trade and manufacturing center in a farming and resort area. Manufacturing (printing and publishing, electronic equipment, pet containment systems, fabricated metal products, glass products); insurance. Seat of Keene State College. Keene Muncipal Airport in S. MONADNOCK MOUNTAIN (3,165 ft/965 m), 10 mi/16 km to SE (state park), is a popular hiking site. Settled 1736, incorporated as a city 1873.

Keene, town (2006 population 6,196), JOHNSON county, N central TEXAS, 23 mi/37 km S of FORT WORTH; 32°23′N 97°19′W. In agricultural area (cotton; grain; dairying); light manufacturing. Grew up around Seventh Day Adventist Academy, now Southwestern Adventist College.

Keene, hamlet, ESSEX county, NE NEW YORK, in ADIRONDACK MOUNTAINS, on E. Branch of AUSABLE RIVER, 35 mi/56 km SSW of PLATTSBURGH; 44°11′N 73°48′W. In resort area.

Keeneland, KENTUCKY: see LEXINGTON.

Keenesburg, village (2000 population 855), WELD county, NE COLORADO, 35 mi/56 km NE of DENVER; 40°06′N 104°31′W. Elevation 4,958 ft/1,511 m. Sugar beets, beans, wheat, cattle. HORSE CREEK RESERVOIR to SW.

Keene Valley, resort village, ESSEX county, NE NEW YORK, in scenic Keene Valley of ADIRONDACK MOUNTAINS, on E. Branch of AUSABLE RIVER, 39 mi/63 km SSW of PLATTSBURGH.

Keeney Knob, mountain (3,927 ft/1,197 m), SUMMERS county, S WEST VIRGINIA, 11 mi/18 km NE of HINTON. Highest point of Keeney Mountain, a ridge of the ALLEGHENY MOUNTAINS.

Keensburg, village (2000 population 252), WABASH county, SE ILLINOIS, 7 mi/11.3 km SW of MOUNT CARMEL; 38°21′N 87°52′W. In agricultural area.

Keephills (KEEP-hilz), hamlet, central ALBERTA, W CANADA, 37 mi/60 km from EDMONTON, in PARKLAND County; 53°26′N 114°21′W.

Keep River National Park (KEEP RI-vuhr) (□ 219 sq mi/569.4 sq km), NW NORTHERN TERRITORY, N central AUSTRALIA, 280 mi/451 km W of KATHERINE on Northern Territory-WESTERN AUSTRALIA border; 22 mi/35 km long, 18 mi/29 km wide; 15°48′S 129°03′E. Administered by Northern Territory Conservation Commission. Sandstone hills, gorges, rock formations. Massive marine deposits and evidence of glaciation. Rock wallabies; emus, rock pigeons, sandstone shrikes. Boab trees, kapok trees, river red gums, bloodwood. Aboriginal art sites of Mitiwung people. Camping, picnicking hiking. Established 1980.

Keeranur (KEE-rah-noor), town, DINDIGUL-ANNA district, TAMIL NADU state, S INDIA, 10 mi/16 km N of PALANI. Rice, grain, tobacco. Also spelled KIRANUR.

Keerbergen (ker-BUHR-gen), commune (2006 population 12,515), BRABANT province, central BELGIUM, 16 mi/25 km NE of BRUSSELS; 51°00′N 04°38′E. Brussels suburb known as a golf/resort center.

Keeseville (KEEZ-vil), village (□ 1 sq mi/2.6 sq km; 2006 population 1,780), on CLINTON-ESSEX county line, NE NEW YORK, on AUSABLE RIVER (bridged), 13 mi/21 km S of PLATTSBURGH; 44°30′N 73°28′W. Some manufacturing; in farm area. Resort. AUSABLE CHASM is nearby. Settled 1806, incorporated 1878.

Keesler Air Force Base, MISSISSIPPI: see BILOXI.

Keeten (KAI-tuhn), channel (4 mi/6.4 km long, 1.5 mi/2.4 km wide), ZEELAND province, SW NETHERLANDS, SW extension of MASTGAT channel to EASTERN SCHELDT estuary. Krabbe Creek estuary extends 4 mi/6.4 km inland to E. DUIVELAND region to NW, THOLEN region to SE.

Keetmanshoop (KAIT-mahns-hop), town (2001 population 15,543), S central NAMIBIA, 300 mi/483 km S of WINDHOEK. It is the trade and tourist center for the S (KARAS REGION) where karakul sheep are raised. Keetmanshoop was founded in 1866 as a German missionary station.

Keewatin (kee-WAH-tuhn), unincorporated town (□ 5 sq mi/13 sq km; 2001 population 2,064), W ONTARIO, E central CANADA, on LAKE OF THE WOODS, and included in KENORA; 49°46′N 94°33′W. Hunting and fishing center, noted for its scenic beauty.

Keewatin (kee-WAW-duhn), town (2000 population 1,164), ITASCA county, NE MINNESOTA, 7 mi/11.3 km WSW of HIBBING in MESABI IRON RANGE; 47°23′N 93°05′W. Light manufacturing. Grew with exploitation of ore deposits after 1909.

Keewatin, CANADA: see KIVALLIQ.

Kef, province, TUNISIA: see KAF, AL, province.

Kefa, former province (□ 22,120 sq mi/57,512 sq km), SW ETHIOPIA; ⊙ JIMMA; 07°00′N 36°20′E. Extended W of OMO RIVER and bordered on SUDAN. Formerly a kingdom, it was Christianized in the 16th century, rivaled ETHIOPIA (18th–early 19th centuries), and was then incorporated into it in 1897. Has Hamite tribes population, considerably reduced by slave trade, which persisted until well into the 20th century. Now part of OROMIYA and SOUTHERN NATIONS states. Also spelled Keffa, Kafa, and Kaffa.

Kefallinía (ke-fah-lee-NEE-ah), prefecture, IONIAN ISLANDS department, off W coast of GREECE, between IONIAN SEA (W) and Gulf of PATRAS (E); 38°15′N 20°30′E. Includes islands of KEFALLINÍA and ITHÁKI, as well as some small islands just off the coast of WESTERN GREECE. LEFKAS prefecture to N, ZÁKINTHOS prefecture to S.

Kefallinía (ke-fah-lee-NEE-ah), island (□ 300 sq mi/780 sq km), KEFALLINÍA prefecture, IONIAN ISLANDS department, off W GREECE, the largest of the Ionian Islands. It has an irregular coastline and is largely mountainous, rising to c.5,340 ft/1,628 m at Mount AINOS; 38°15′N 20°35′E. ARGOSTÓLI, a port, is the island's main town and ships local products such as fruit and wine. Sheep raising and fishing are important occupations. In ancient times, Kefallinía was an ally of ATHENS in the Peloponnesian War and later was a member of the Aetolian League. Taken by ROME in 189 B.C.E. After the division of the Roman Empire (C.E. 395), it was held by the BYZANTINE EMPIRE until its occupation (1126) by VENICE. Subsequently ruled by several Italian families; seized by the Ottoman Turks (1479) and ceded (1499) to Venice, which held it until the Treaty of Campo Formio (1797). Its subsequent history is that of the IONIAN ISLANDS. In 1953 the island was devastated by earthquakes. Tourism is increasingly important. Also Cephalonia, Kephallenia.

Kefalos (KE-fah-los), Italian *Cefalo*, town, KÓS island, DODECANESE prefecture, SOUTH AEGEAN department, GREECE, 20 mi/32 km WSW of KÓS; 36°49′N 26°58′E. Also Kephalos.

Kefar, (Hebrew=village), for names in ISRAEL beginning thus: see KFAR.

Kefar Nahum, ISRAEL: see CAPERNAUM.

Kefermarkt (KAI-fer-mahrkt), township, NE UPPER AUSTRIA, in the Mühlviertel, 16 mi/26 km NE of LINZ; 48°26′N 14°32′E. Brewery; gothic wood-carved altar; Renaissance castle Weinberg.

Keffa, ETHIOPIA: see KEFA.

Keffi (KE-fee), town, BENUE state, central NIGERIA, 25 mi/40 km NNE of NASARAWA. Tin-mining center.

Keffin-Hausa (KE-feen-HOU-sah), town, JIGAWA state, N NIGERIA, on road, 20 mi/32 km SSW of HADEIJA; 12°14′N 09°55′E. Market center. Groundnuts, maize, millet, guinea corn.

Kefissos River, Greece: see KIFISSOS RIVER.

Keflavík (KEP-lah-veek), town (2000 population 7,862), SW ICELAND, on the FAXAFLOI, W of REYKJAVIK. It is a major fishing port, best known for its large international airport, which was built by the U.S. during World War II; in 1951 the U.S. was granted the right to use it as a NATO military base.

Kef Mahmel (KEF mah-MEL), mountain peak (elev. 7,615 ft/2,321 m), third highest in ALGERIA and second highest of the AURÈS massif, BATNA wilaya, NE Algeria, 15 mi/24 km SSE of BATNA.

Kefri, IRAQ: see KIFRI.

Kef Sidi Amar, ALGERIA: see OUARSENIS MASSIF.

Kegalla, district (□ 651 sq mi/1,692.6 sq km; 2001 population 785,524), SABARAGAMUWA PROVINCE, W central SRI LANKA; ⊙ KEGALLA; 07°07′N 80°20′E.

Kegalla (KAH-guhl-luh), town (2001 population 17,139), SABARAGAMUWA PROVINCE, W central SRI LANKA, 22 mi/35 km W of KANDY; ⊙ KEGALLA district; 07°15′N 80°20′E. Trades in rubber, vegetables, rice, coconuts. Elephant orphanage nearby. Also spelled Kegalle.

Keg Creek, flows c.60 mi/97 km SW, SW IOWA; rises near WESTPHALIA in SHELBY county, past GLENWOOD, to MISSOURI RIVER 20 mi/32 km S of COUNCIL BLUFFS.

Kegeili (ke-GAI-lee), town, S KARAKALPAK REPUBLIC, W UZBEKISTAN, in the AMU DARYA delta, 15 mi/24 km N of NUKUS; 42°42′N 59°36′E. Cotton, cattle, camels. Also KEGEYLI.

Kegel, ESTONIA: see KEILA.

Kegen (ke-GYEN), town, SE ALMATY region, KAZAKHSTAN, on KEGEN RIVER, 115 mi/185 km E of ALMATY; 43°01′N 79°13′E. Tertiary-level (raion) administrative center. Sheep.

Kegen River (ke-GYEN), c.180 mi/290 km long, ALMATY region, KAZAKHSTAN; rises near CHINA border; flows W past KEGEN, and NE to ILI RIVER. Used for irrigation.

Kegeyli, UZBEKISTAN: see KEGEILI.

Keggum, LATVIA: see ĶEGUMS.

Kegichevka, UKRAINE: see KEHYCHIVKA.

Kegonsa, Lake, c.3 mi/4.8 km long, 2 mi/3.2 km wide, southernmost of the FOUR LAKES, DANE county, S WISCONSIN, 9 mi/14.5 km SE of MADISON. Connected to LAKE WAUBESA to NW by YAHARA RIVER, drained by Yahara River on E. Lake Kegonsa State Park on E shore.

Kegon Waterfall, JAPAN: see CHUZENJI.

Kegul'ta, RUSSIA: see SADOVOYE, Republic of KALMYKIA-Khalmg-Tangeh.

Ķegums (TCHE-gums), German *keggum*, village, central LATVIA, in VIDZEME, on right bank of the DVINA (DAUGAVA) River, and 6 mi/10 km SE of OGRE; 56°44′N 24°43′E. Large hydroelectric station.

Kegworth (KEG-wuhth), village (2001 population 3,894), NW LEICESTERSHIRE, central ENGLAND, on SOAR RIVER and 5 mi/8 km NW of LOUGHBOROUGH; 52°49′N 01°16′W. Light industry. Has 15th-century church.

Kehancha, town, NYANZA province, W KENYA; ⊙ Kuria district; 01°11′N 34°42′W. Agriculture (tobacco, maize, finger millet; livestock); market and trade center. Formerly part of S Nyamnza district.

Kehl (KEHL), town, BADEN-WÜRTTEMBERG, SW GERMANY, port on right bank of the RHINE, at mouth of KINZIG RIVER, near French border, opposite STRASBOURG; 48°36′N 07°49′E. Railroad junction; customs station. Manufacturing of cellulose, steel; woodworking, paper milling. Active trade in large harbor (built 1842–1900). Founded as French fortress in late-17th century; bridgehead occupied by FRANCE 1919–1930; town under French administration 1945–1956.

Kehlen (KAI-luhn), village, Kehlen commune, SW LUXEMBOURG, 6 mi/9.7 km NW of LUXEMBOURG city; 49°40′N 06°02′E.

Kehra (KE-ruh) or **kehkra**, town, N ESTONIA, 22 mi/35 km ESE of TALLINN; 59°20′N 25°19′E. Sawmilling, light manufacturing; oil-shale-based power plant.

Kehsi-Mansam (kai-SHEE–mahn-SAHM), township (□ 551 sq mi/1,432.6 sq km), S SHAN STATE, MYANMAR, on the NAM PANG, 70 mi/113 km S of LASHIO; ⊙ Kehsi Mansam. Forested hills; trading; cattle breeding.

Kehychivka (ke-hi-CHIF-kah) (Russian *Kegichevka*), town, W KHARKIV oblast, UKRAINE, 15 mi/24 km ESE of KRASNOHRAD; 49°17′N 35°47′E. Elevation 561 ft/170 m. Raion center; on railroad; flour mill; dairying; construction materials. Vocational technical school. Established in the late 18th century, town since 1957.

Keighley (KEETH-lee), town (2001 population 51,429), WEST YORKSHIRE, N central ENGLAND, at the junction of the AIRE and Worth rivers; 53°53′N 01°55′W. The Leeds and Liverpool Canal connects Keighley with Liverpool and Hull. Keighley's products include wool, machinery, textiles, and consumer goods.

Keiheuvel (KEI-huh-vuhl), recreational park near BALEN, ANTWERPEN province, N BELGIUM, 14 mi/23 km SE of TURNHOUT.

Keihin Industrial Zone, JAPAN: see YOKOHAMA

Keihoku (KAI-ho-koo), town, N Kuwata county, KYOTO prefecture, S HONSHU, W central JAPAN, 12 mi/20 km NW of KYOTO; 35°09′N 135°38′E. Ayu; *matsutake* mushrooms.

Kei Islands, INDONESIA: see KAI ISLANDS.

Keila (KAI-luh), German *Kegel*, city, NW ESTONIA, 15 mi/24 km SW of TALLINN; 59°18′N 24°24′E. Agricultural market center (potato, barley); manufacturing (construction materials, fabricated metal products, concrete). Railroad junction.

Keilberg, CZECH REPUBLIC: see KLINOVEC.

Keilor (KEE-luhr), suburb and statistical area of MELBOURNE, S central VICTORIA, SE AUSTRALIA, on Calder Freeway; 37°43′S 144°50′E. Rapid urban growth in 1980s and 1990s, Melbourne Airport to N, ESSENDON Airport to E. Organ Pipes National Park (210 acres/85 ha) to NW, hexagonal columns of basalt rise 65 ft/20 m above Jackson Creek.

Keimoes, town, NORTHERN CAPE province, SOUTH AFRICA, on N bank of ORANGE RIVER and 25 mi/40 km SW of UPINGTON opposite Skaap and Rooikop islands; 28°44′S 20°58′E. Elevation 2,788 ft/850 m. Wheat, peas, sultanas, lucerne, fruit, viticulture. Railroad terminus on N14 highway.

Kei Mouth, town, SOUTH AFRICA: see GREAT KEI RIVER.

Keisen (KAI-sen), town, Kaho county, FUKUOKA prefecture, N KYUSHU, SW JAPAN, 16 mi/25 km E

of FUKUOKA; 33°34′N 130°40′E. Ancient tombs at Otsuka.

Keiser (KEI-suhr), town, MISSISSIPPI county, NE ARKANSAS, 22 mi/35 km SSW of BLYTHEVILLE; 35°40′N 90°05′W. Cotton, rice. Incorporated 1933.

Keiskammahoek or **Keiskamma Hoek**, town, EASTERN CAPE province, SOUTH AFRICA, on KEISKAMMA RIVER, and 20 mi/32 km NW of KING WILLIAM'S TOWN, at foot of short AMATOLA RANGE; 32°36′S 27°06′E. Elevation 3,247 ft/990 m. Livestock; fruit, grain.

Keiskamma River, 160 mi/257 km long, EASTERN CAPE, SOUTH AFRICA; rises in AMATOLA mountain range NE of KEISKAMMAHOEK near Dontsa Pass; flows in winding course first SW past KEISKAMMAHOEK to ALICE, then SE across coastal lowlands past Mount Vale to the INDIAN OCEAN at Hamburg, 30 mi/48 km SW of EAST LONDON.

Keïta (KAI-ee-tuh), town, TAHOUA province, NIGER, 37 mi/60 km ESE of TAHOUA; 14°46′N 05°46′E. Livestock market. Administrative center.

Keitele, Lake (KAI-te-lai) (46 mi/74 km long, 1 mi/1.6 km–8 mi/12.9 km wide), S central FINLAND, 50 mi/80 km W of KUOPIO. Drains S into LAKE PÄIJÄNNE. Chief villages on lake are ÄÄNEKOSKI, SUOLAHTI, and VIITASAARI.

Keith, county (□ 1,109 sq mi/2,883.4 sq km; 2006 population 8,250), SW central NEBRASKA, ⊙ OGALLALA; 41°12′N 101°39′W. Agricultural region drained by NORTH PLATTE and SOUTH PLATTE rivers. LAKE C. W. MCCONAUGHY reservoir, on the North Platte, was created by Kingsley Dam. Cattle, hogs; wheat, corn, alfalfa, beans, sunflower seeds. Central/Mountain time zone boundary follows E boundary and E part of N boundary. Lake Ogallala State Recreation Area N of North Platte River, at center of county (S of Kingsley Dam). Lake McConaughy State Recreation Area at county center and in NW; seven units (three on S shore, four on N shore). Formed 1873.

Keith (KEETH), town (2001 population 4,491), MORAY, NE Scotland, on the ISLA RIVER, and 18 mi/29 km SW of BANFF; 57°32′N 02°56′W. Whiskey distilling, manufacturing of machinery. Previously woolen milling. Remains of ancient Milton Tower, stronghold of the Oliphants. Just W, on opposite bank of Isla River, is Fife Keith, formerly separate town. Just N of Keith is agricultural village of Newmill; James Gordon Bennett born here.

Keith (KEETH), village, SE SOUTH AUSTRALIA, S central AUSTRALIA, 130 mi/209 km SE of ADELAIDE; 36°06′S 140°21′E. On railroad; wheat. Conservation parks to N and S of town.

Keithsburg (KEETHS-burgh), city (2000 population 714), MERCER county, NW ILLINOIS, on the MISSISSIPPI at mouth of POPE CREEK, and 13 mi/21 km SW of ALEDO; 41°06′N 90°56′W. In agricultural area. Mark Twain National Wildlife Refuge nearby. Incorporated 1857.

Keithville (KEETH-vil), unincorporated village, CADDO parish, NW LOUISIANA, 12 mi/19 km SSW of SHREVEPORT; 32°19′N 93°50′W. In agricultural and timber area; oil and natural gas; manufacturing (construction equipment, fabricated metal products).

Keitum (KEI-tum), part of WESTERLAND, SCHLESWIG-HOLSTEIN, NW GERMANY, on SYLT island, 3 mi/4.8 km ESE of city center; 54°53′N 08°23′E. Historic main village of Sylt; now a seaside resort.

Keizer (KEI-zuhr), city (2006 population 35,027), MARION county, NW OREGON, a residential suburb 3 mi/4.8 km N of SALEM, on WILLAMETTE RIVER; 45°00′N 123°01′W. Agriculture (fruit, nuts, vegetables; poultry; dairying). Willamette Mission and Maud Williamson State Parks to N.

Kejimkujik National Park (□ 140 sq mi/363 sq km), S central NOVA SCOTIA, CANADA, near Maitland Bridge. With numerous lakes and rivers, Kejimkujik was central to traditional canoe routes between the Bay of

Fundy and the ATLANTIC coast. Earliest inhabitants, from 4,500 years ago, were the Maritime Archaic Indians. Features petroglyphs made by the Mi'kmaq. European settlement dates from early 19th century. Park name comes from Kejimkujik Lake, the largest lake, in park. Logging, gold mining, and recreation. Established 1968.

Kej River, PAKISTAN: see DASHT RIVER.

Kejser Franz Joseph Fjord, GREENLAND: see FRANZ JOSEF FJORD.

Kekaha (kai-KAH-hah), town (2000 population 3,175), SW KAUAI, KAUAI county, HAWAII, 25 mi/40 km W of LIHUE, 5 mi/8 km W of WAIMEA Bay, on S coast, on Kaumualii Highway; 21°58′N 159°43′W. Barking Sands Pacific Missile Range to W; Puu Ka Pele Forest Reserve to N; Waimea Canyon State Park to NE; Kokole Lighthouse and viewpoint of NIIHAU to W; Kekaha Beach Park.

Kekem (KAI-kaim), town (2001 population 14,700), West province, CAMEROON, 33 mi/53 km SW of BAFOUSSAM; 05°09′N 10°03′E. Coffee growing area.

Kekertak, GREENLAND: see QEQERTAQ.

Kékes, Mount (KAI-kesh), Hungarian *Kékes* (3,330 ft/1,015 m), in MÁTRA MOUNTAINS, HEVES county, N HUNGARY; 47°52′N 20°01′E. Highest point in country. On SW slope (elevation 2,329 ft/710 m) is sanatorium.

Kekhra, ESTONIA: see KEHRA.

Kékkö, SLOVAKIA: see MODRY KAMEN.

Kekri (KAI-kree), town, AJMER district, RAJASTHAN state, NW INDIA, in detached area, 45 mi/72 km SE of Ajmer; 25°58′N 75°09′E. Markets millet, cotton, wheat; cotton ginning, hand-loom weaving.

Keku Strait (KAI-koo), 40 mi/64 km long, SE ALASKA, in ALEXANDER ARCHIPELAGO; extends N to S between KUPREANOF ISLAND (E) and KUIU ISLAND (W), W of KAKE.

Kelaat Sraghna (E-luht se-RAGH-nuh), town, El Kelaâ des Sraghna province, Marrakech-Tensift-Al Haouz administrative region, W MOROCCO, 45 mi/72 km NE of MARRAKECH. In irrigated farm region; olive groves, orchards; trade in wool and olive oil. Also spelled El Kelaâ des Srarhna.

K'elafo, town (2007 population 14,882), SOMALI state, SE ETHIOPIA; 05°37′N 44°08′E. Near the E bank of the WABĒ SHEBELĒ River, on road, c.20 mi/32 km from SOMALIAN border. Agricultural trade center; airfield. Also spelled Callafo, Kallafo.

Kélakam (KAI-luh-kahm), town, DIFFA province, NIGER, 45 mi/72 km NW of MAÏNÉ-SOROA; 13°35′N 11°44′E.

Kelala, town (2007 population 4,689), AMHARA state, central ETHIOPIA; 10°28′N 38°43′E. 80 mi/129 km SW of DESSIE.

Kelamayi, China: see KARAMAY.

Kelamet, Eritrea: see KELHAMET.

Kelan (KUH-LAN), town, ⊙ Kelan county, NW SHANXI province, NE CHINA, 45 mi/72 km SW of NINGWU; 38°42′N 111°34′E. Grain, oilseeds; textiles, food.

Kelang (KUH-lahng), island (10 mi/16 km long, up to 8 mi/12.9 km wide), S MALUKU, INDONESIA, in SERAM SEA, just W of SERAM; 03°13′S 127°44′E. Rises to 2,641 ft/805 m; wooded, hilly.

Kelani Ganga, river, SW Sri Lanka; rises in SABARAGAMUWA province, in S SRI LANKA HILL COUNTRY in two headstreams joining 11 mi/18 km ESE of YATIYANTOTA at 06°57′N 79°55′E; flows WNW past Yatiyantota, and generally W through extensive tea, rubber, and coconut-palm plantations, past RUWANWELLA and into WESTERN PROVINCE, past KELANIYA, to INDIAN OCEAN at COLOMBO.

Kelaniya (KA-lah-ni-yuh), town, WESTERN PROVINCE, SRI LANKA, on the KELANI GANGA River and 4 mi/6.4 km ENE of COLOMBO city center; 06°57′N 79°53′E. Pottery and fertilizer manufacturing; coconuts, rice, vegetables. Ancient Buddhist pilgrimage center; stupa (built 13th century C.E., restored late 18th century). Temple festival in January. Historically, site of city built 3rd century B.C.E. by Sinhalese king. Also spelled KALANIYA.

Kelantan (KEH-lahn-tahn), state (□ 5,780 sq mi/14,970 sq km; 1982 est. population 904,000), central Malay Peninsula, Malaysia, on the South China Sea. It is bordered on the N by Thailand; ⊙ is Kota Baharu. It is drained by the Kelantan River (c.150 mi/240 km long), which flows into the South China Sea. Being one of the traditional "rice bowls" of the peninsula, rice is the most important commercial crop and is grown on the wide coastal plains; other products are rubber and copra. Tin, gold, manganese, and iron are mined on a small scale in the hills of the interior. The people are mainly Malay, but there is a small Chinese minority. Kelantan was ruled by Sri Vijaya until the 13th century; it fell under the sway of Malacca in the 15th century. After the fall of Malacca (1511), conflict among many powers resulted eventually in the establishment by Siam of sovereignty over the area (early 19th century). Kelantan became a protectorate of Great Britain in 1909. Before the establishment of the Federation of Malaya (1948), Kelantan was classed as one of the Unfederated Malay States. Kelantan is the poorest, most rural, and conservative state on the peninsula. See MALAYSIA, FEDERATION OF.

Kelantan River (KEH-lahn-tahn), flows over 150 mi/241 km N, a chief river of KELANTAN, N MALAYSIA; rises in several branches (Galas, Nenggiri, Lebir) in mountain borders of Kelantan, past Kuala Krai, PASIR MAS, and Kota Bharu, to SOUTH CHINA SEA at TUMPAT. Navigable for launches for 80 mi/129 km. Its valley is used by E coast railroad.

Kelayres (kuh-LER-uhs), unincorporated town, Kline township, SCHUYLKILL county, E central PENNSYLVANIA, 6 mi/9.7 km SSW of HAZLETON, and 1 mi/1.6 km W of MCADOO; 40°54′N 76°00′W. Anthracite coal. Tuscarora State Park to S.

Kelbadzhar, AZERBAIJAN: see KALBACAR.

Kelbra (KEL-brah), town, SAXONY-ANHALT, central GERMANY, at N foot of the KYFFHÄUSER, in the GOLDENE AUE, on the HELME, 12 mi/19 km ESE of NORDHAUSEN; 51°27′N 11°02′E. Light manufacturing. Helme dam, a reservoir and excursion resort, is just W.

Kelchterhoef Domein (KEL-tuhr-huf do-MAIN), recreational park near HOUTHALEN-HELCHTEREN, LIMBURG province, NE BELGIUM, 7 mi/11.3 km N of HASSELT.

Këlcyrë (kel-TSYOOR), town, S ALBANIA, on Vijosë River, 30 mi/48 km SSE of BERAT; 40°19′N 20°11′E. Road junction. Also spelled Këlcyra.

Keles, city, N TOSHKENT wiloyat, UZBEKISTAN, on TRANS-CASPIAN railroad, just NW of TOSHKENT, on KAZAKHSTAN border; 41°24′N 69°12′E. Tertiary-level administrative center.

Kelford (KEL-fuhrd), village (2006 population 232), BERTIE county, NE NORTH CAROLINA, 20 mi/32 km NW of WINDSOR, 2 mi/3.2 km S of ROXOBEL, near source of CASHIE RIVER; 36°10′N 77°13′W. Manufacturing; service industries. Named for a fjord in Scotland. Incorporated 1893.

Kelhamet (kah-LAH-met), Italian *Chelamet*, village, N ERITREA, 25 mi/40 km NE of KEREN; 16°04′N 38°38′E. In agricultural region (cereals, fruit, vegetables; livestock); road junction. Also spelled Kelamet.

Kelheim (KEHL-heim), town, BAVARIA, S GERMANY, in LOWER BAVARIA, at confluence of ALTMÜHL and DANUBE rivers, 12 mi/19 km SW of REGENSBURG; 49°55′N 11°52′E. Manufacturing of chemicals, construction materials, engines; brewing. Has 15th-century church, Benedictine abbey and remains of town wall from 13th century. Developed at site of Celtic, then Roman, settlement. On nearby hill (W) is the Befreiungshalle, a rotunda built (1842–1863) by Louis I of Bavaria to commemorate the heroes of the War of Liberation from Napoleonic rule. Chartered 1181.

Kelif (ke-LEEF), village, SE LEBAP weloyat, extreme SE TURKMENISTAN, at the point where the AMU DARYA

(river) enters Turkmenistan, on railroad and 70 mi/113 km SE of KERKI; 37°21′N 66°18′E. On the border with both AFGHANISTAN and UZBEKISTAN. Cotton. Limestone quarries.

Kelkheim (KELK-heim), town, HESSE, central GERMANY, on S slope of the TAUNUS, 10 mi/16 km W of FRANKFURT; 50°08′N 08°27′E. Largely residential; furniture making.

Kelkit, village, NE TURKEY, on KELKIT RIVER and 22 mi/35 km S of GÜMUSHANE; 40°07′N 39°28′E. Wheat, barley. Also called Ciftlik.

Kelkit River, ancient *Lycus*, 220 mi/354 km long, N TURKEY; rises in Gümushane Mountains, 20 mi/32 km SE of GÜMUSHANE; flows WNW, past KELKIT, SUSEHRI, KOYULHISAR, RESADIYE, NIKSAR, and ERBAA, to the YESIL IRMAK 6 mi/9.7 km NW of Erbaa.

Kell, village (2000 population 231), MARION county, S ILLINOIS, 9 mi/14.5 km S of SALEM; 38°29′N 88°54′W. In agricultural and oil area.

Kellé (kel-ai), town, ZINDER province, NIGER, 30 mi/48 km NW of GOURÉ; 14°18′N 10°04′E.

Kellé (kel-LAI), village, CUVETTE-OUEST region, central Congo Republic, c.100 mi/160 km WNW of OWANDO; 00°03′S 14°29′E. Airfield.

Kellé (kel-AI), village, THIÈS administrative region, W SENEGAL, on Dakar-Niger railroad and 65 mi/105 km NE of DAKAR; 15°10′N 16°35′W. Peanut growing.

Keller, city (2006 population 36,925), TARRANT county, N TEXAS, suburb 14 mi/23 km NNE of downtown FORT WORTH; 32°55′N 97°13′W. Remnant agricultural areas being replaced by urban growth. Manufacturing (fabricated metal products, concrete, electronic equipment).

Keller (KE-luhr), town, ACCOMACK county, E VIRGINIA, 8 mi/13 km SSW of ACCOMAC; 37°37′N 75°45′W. Agriculture (grain vegetables; livestock; poultry; dairying).

Kellerberrin (kel-uh-BE-ruhn), town, WESTERN AUSTRALIA state, W AUSTRALIA, c.124 mi/200 km E of PERTH; 31°40′S 117°41′E. Wheat, oats, barley. Folk museum. Passing through the town is a pipeline transporting water 350 mi/563 km from Perth to KALGOORLIE.

Kellerovka (ke-ler-ROF-kuh), town, N KÖKSHETAU region, KAZAKHSTAN, 40 mi/64 km N of KÖKSHETAU; 53°51′N 69°15′E. Tertiary-level (raion) administrative center. Wheat; cattle.

Kellerovo, RUSSIA: see KOL'CHUGINO, VLADIMIR oblast.

Kellerton, town (2000 population 372), RINGGOLD county, S IOWA, 10 mi/16 km E of MOUNT AYR; 40°42′N 94°02′W. In livestock and grain area.

Kellerup, DENMARK: see KJELLERUP.

Kellett Strait, NORTHWEST TERRITORIES, CANADA: see FITZWILLIAM STRAIT.

Kelley, town (2000 population 300), STORY county, central IOWA, 11 mi/18 km WSW of NEVADA; 41°57′N 93°39′W. Livestock; grain.

Kelleys Island (KE-leez), island village (□ 4.5 sq mi/11.7 sq km; 1990 population 172; 2000 population 367), in LAKE ERIE, ERIE county, N OHIO; 41°35′N 82°42′W. Summer resort, fishing, wine growing; Glacial Grooves State Memorial is here. The name WINE ISLANDS is sometimes applied to Kelleys Island and neighboring islands.

Kelliher (KE-li-huhr), village, SE SASKATCHEWAN, CANADA, in the Beaver Hills, 55 mi/89 km W of YORKTON; 51°16′N 103°44′W. Mixed farming.

Kelliher (KE-li-yuhr), village (2000 population 294), BELTRAMI county, MINNESOTA, 38 mi/61 km NE of BEMIDJI, near South Branch Battle River; 47°56′N 94°27′W. Grain, sunflowers; livestock; dairying; manufacturing (hardwood lumber). Red Lake State Forest to N; Pine Island State Forest to E; Upper RED LAKE to NW; Lower Red Lake to W.

Kellinghusen (kel-ling-HOO-suhn), town, SCHLESWIG-HOLSTEIN, N GERMANY, on the STÖR (head of navigation), and 8 mi/12.9 km ENE of ITZEHOE;

53°57′N 09°43′E. Manufacturing of pottery, china; clay quarrying. Site (1765–1840) of noted faïence factory.

Kellington (KEL-ling-tuhn), village (2001 population 991), NORTH YORKSHIRE, N ENGLAND, 6 mi/9.7 km SW of SELBY; 53°42′N 01°09′W. AIRE RIVER 1 mi/1.6 km to N.

Kellits (KEL-its), town, CLARENDON parish, central JAMAICA, 15 mi/24 km N of MAY PEN; 18°10′N 77°14′W. Road junction town.

Kellnersville, village (2006 population 349), MANITOWOC county, E WISCONSIN, 12 mi/19 km NW of MANITOWOC; 44°13′N 87°47′W. Dairying; grain, vegetables, fruit. Hidden Valley Ski Area to N.

Kello (KEL-lo), village, OULUN province, W FINLAND, on GULF OF BOTHNIA, 7 mi/11.3 km NNW of OULU; 65°08′N 25°20′E. Elevation 66 ft/20 m. Timber-shipping port; sawmills.

Kelloe (KE-lo), community, SW MANITOBA, W central CANADA, 8 mi/13 km WNW of town of SHOAL LAKE, in SHOAL LAKE rural municipality; 50°28′N 100°45′W.

Kellogg, town (2000 population 2,395), SHOSHONE county, N IDAHO, on fork of COEUR D'ALENE RIVER and 10 mi/16 km WSW of WALLACE; 47°32′N 116°08′W. In mining district of COEUR D'ALENE MOUNTAINS; former site of lead and zinc smelters and refineries, cadmium plant. Manufacturing (fabricated metal products, apparel). Grew with development of Bunker Hill and Sullivan lead mines (discovered 1885, now combined as one of world's leading producers). Mining Museum, Coeur d'Alene National Forest to N; St. Joe National Forest to S; Sunshine Miners Memorial to E; Silver Mountain Ski Area to S. Entire town has been transformed into Bavarian village. Founded as Milo 1893, renamed 1894, incorporated 1913.

Kellogg, town (2000 population 606), JASPER county, central IOWA, on NORTH SKUNK RIVER and 7 mi/11.3 km E of NEWTON; 41°43′N 92°54′W. Light manufacturing.

Kellogg, village (2000 population 439), WABASHA county, SE MINNESOTA, near MISSISSIPPI RIVER, 5 mi/8 km SSE of WABASHA; 44°18′N 92°00′W. Livestock; poultry; dairying; light manufacturing; timber. Whitewater Wildlife Area to S; Richard J. Dorer Memorial Hardwood State Forest to SW; Lock and Dam Number 4 to NE.

Kellokoski (KEL-lo-KOS-kee), village, UUDENMAAN province, S FINLAND, 25 mi/40 km N of HELSINKI; 60°33′N 25°06′E. Elevation 116 ft/35 m. Metalworking.

Kellomäki, RUSSIA: see KOMAROVO, LENINGRAD oblast.

Kelloselkä (KEL-lo-SEL-kah), village, LAPIN province, NE FINLAND; 66°56′N 28°50′E. Elevation 990 ft/300 m. Border station 8 mi/12.9 km WSW of Kuolayarvi (Kuolajärvi), Russia.

Kells, Gaelic *Ceanannus Mór*, market town (2006 population 2,257), MEATH county, NE IRELAND, on BLACKWATER RIVER, and 9 mi/14.4 km NW of NAVAN; 53°44′N 06°53′W. The relic of an ancient monastery (founded by St. Columba c.550), the round tower, and several ancient crosses are noteworthy. The Book of Kells, now one of the treasures of the Trinity College library in DUBLIN, is generally regarded as the finest example of Celtic illumination.

Kelly Lake (40 mi/64 km long, 2 mi/3 km–4 mi/6 km wide), NORTHWEST TERRITORIES, CANADA, 15 mi/24 km NE of NORMAN WELLS; 65°25′N 126°15′W.

Kelly's Ford (KE-leez FORD), on RAPPAHANNOCK RIVER, FAUQUIER county, N VIRGINIA, 20 mi/32 km NW of FREDERICKSBURG; 38°28′N 77°46′W. Scene of an indecisive Civil War cavalry engagement (March 17, 1863).

Kellyville, town (2006 population 921), CREEK county, central OKLAHOMA, 20 mi/32 km SW of TULSA, 5 mi/8 km SW of SAPULPA; 35°56′N 96°13′W. In agricultural and oil- and natural gas-producing area; light manufacturing.

Kelmė (KEL-mai), Russian *Kelmy* or *Kel'my*, city (2001 population 10,900), W central LITHUANIA, 25 mi/40 km SW of SIAULIAI; 55°38'N 22°56'E. Manufacturing (apparel), flour milling, flax-processing, distilling. In Russian KOVNO government until 1920. Has museum.

Kel'mentsi (KEL-men-tsee) (Russian *Kel'mentsy*) (Romanian *Chelmenti*), town (2004 population 7,600), E CHERNIVTSI oblast, UKRAINE, in BESSARABIA, 14 mi/23 km ESE of KHOTYN, near DNIESTER (Ukrainian *Dnister*) River; 48°28'N 26°50'E. Elevation 744 ft/226 m. Raion center; food processing (butter, cheese); mineral water. Professional technical school. Border customs station. First mentioned in 1559, town since 1960.

Kel'mentsy, UKRAINE: see KEL'MENTSI.

Kelmis (kel-MIS), French *La Calamine*, commune (2006 population 10,424), Verviers district, LIÈGE province, E BELGIUM, near German border, 6 mi/9.7 km SW of AACHEN.

Kelmy, LITHUANIA: see KELMĖ.

Kélo (kai-LO), town, SW CHAD, TANDJILÉ administrative region, 65 mi/105 km NNW of MOUNDOU. Cotton ginning; millet; livestock.

Kelod (KAI-lod), town, NAGPUR district, MAHARASHTRA state, central INDIA, 25 mi/40 km NW of NAGPUR; 21°27'N 78°51'E. Cotton ginning; millet, wheat, oilseeds; mango groves. Manganese deposits nearby.

Keloed, INDONESIA: see KELUD.

Kelowna (kil-O-nuh), city (□ 82 sq mi/213.2 sq km; 2001 population 96,288), S BRITISH COLUMBIA, W CANADA, on OKANAGAN LAKE, and in central OKANAGAN regional district; 49°54'N 119°29'W. Kelowna is a tourist resort and serves as a trade center for a fruit-growing and lumbering area. Manufacturing of wine, machinery.

Kelsall (KEL-suhl), village (2001 population 3,439), W CHESHIRE, W ENGLAND, 7 mi/11.3 km E of CHESTER; 53°13'N 02°43'W.

Kelsey (KEL-see), rural municipality, W MANITOBA, W central CANADA; 54°09'N 101°22'W. Agriculture; forestry; tourism. Consists of the areas of CARROT River Valley, Young's Point, BIG EDDY and Umperville settlements, RALLS Island, WANLESS, and CRANBERRY Portage urban district. Incorporated 1944.

Kelseyville, unincorporated town (2000 population 2,928), Lake county, NW CALIFORNIA, near Clear Lake, 8 mi/12.9 km SE of LAKEPORT; 38°58'N 122°49'W. Pears, grapes, walnuts, oats; cattle; winery. Large Clear Lake State Park to N and E.

Kelso, city (2006 population 12,120), SW WASHINGTON, 45 mi/72 km NNW of PORTLAND, OREGON, and 60 mi/97 km S of OLYMPIA, on the COWLITZ RIVER near the COLUMBIA River; ☉ COWLITZ county; 46°08'N 122°54'W. Twin city with LONGVIEW, adjoins KELSO to SW. In a fertile farm area. Boatbuilding, fishing, and dairy farming are the major industries; manufacturing (machinery, paper products, wood products). Seaquest State Park, on Silver Lake reservoir, to NE. Kelso Longview Airport to S. County Historical Museum here. Settled in 1847, Kelso was an important stopping place for early steamboat travel along the Cowlitz River. Incorporated 1889.

Kelso (KEL-so), town (2001 population 5,116), Scottish Borders, SE Scotland, on the TWEED RIVER at mouth of the TEVIOT RIVER, and 17 mi/27 km ENE of SELKIRK; 55°35'N 02°26'W. Agricultural market. Georgian square and bridge built by John Rennie. The Ballantyne brothers established printing press here, publishing Scott's *The Minstrelsy of the Scottish Border* in 1802.

Kelso, town (2000 population 527), SCOTT county, SE MISSOURI, near MISSISSIPPI RIVER, 3 mi/4.8 km SE of SCOTT CITY; 37°11'N 89°32'W. Feeds.

Kelsterbach (KEL-stuhr-bahkh), town, HESSE, central GERMANY, on left bank of the Main, and 7 mi/11.3 km WSW of FRANKFURT; 50°03'N 08°31'E. Manufacturing of synthetic fiber and plastics. Site of faïence factory (1756–1838). Chartered 1952.

Keltemashat (kel-te-mah-SHAHT), town, SE SOUTH KAZAKHSTAN region, KAZAKHSTAN, in the Talass

ALATAU Mountains, 7 mi/11.3 km SE of SASTÖBE; 42°32'N 70°00'E. Coal mining.

Keltern (KEL-tern), village, BADEN-WÜRTTEMBERG, SW GERMANY, 5 mi/8 km W of PFORZHEIM; 48°54'N 08°36'E.

Kel'tma (keelt-MAH), name of two rivers in E European Russia. Northern Keltma River (approximately 60 mi/97 km long), S KOMI REPUBLIC, is a left affluent of the VYCHEGDA River; in its upper course, 10-mi/16-km-long Yekaterina [Russian=Catherine] Canal connects it with Southern Kel'tma River (50 mi/80 km long), a left affluent of the upper KAMA River in N PERM oblast. The waterway (built 1822) is in disuse.

Keltsy, POLAND: see KIELCE, city.

Kelty (KEL-tee), town (2001 population 5,628), FIFE, E Scotland, 6 mi/9.7 km NNE of DUNFERMLINE; 56°08'N 03°23'W. On N edge of former coalfield.

Keltys, city, ANGELINA county, E TEXAS, just NW of LUFKIN. Lumber milling.

Kelud (KUH-lut), active volcano (elev. 5,643 ft/1,720 m) of central Java Timur province, INDONESIA, SE of KEDIRI, and NE of BLITAR; 07°56'S 112°18'E. Has erupted fifteen times in last 200 years, most recently in 1990. Also spelled Keloed.

Kelve, INDIA: see MAHIM.

Kelvedon (KELV-duhn), village (2001 population 5,019), NE central ESSEX, SE ENGLAND, on BLACKWATER River, and 9 mi/14.5 km WSW of COLCHESTER; 51°51'N 00°42'E. Has 13th–15th-century church.

Kelvin Grove (KEL-vin GROV), NW residential suburb of BRISBANE, SE QUEENSLAND, AUSTRALIA; 27°27'S 153°01'E. One of the campuses of Queensland University of Technology is here.

Kelvington (KEL-ving-tuhn), town (2006 population 866), E SASKATCHEWAN, CANADA, 80 mi/129 km NW of YORKTON; 52°10'N 103°31'W. Dairying, lumbering; grain, livestock.

Kelvin River (KEL-vin), 21 mi/34 km long, GLASGOW, central Scotland; rises in Kilsyth Hills 3 mi/4.8 km NE of KILSYTH, flows SW to CLYDE RIVER in NW Glasgow.

Kelvinside, Scotland: see GLASGOW.

Kelwood (KEL-wud), community, S MANITOBA, W central CANADA, 28 mi/44 km N of NEEPAWA, in ROSEDALE rural municipality; 50°37'N 99°27'W.

Kem' (KYEM), Finnish *Kemi*, city (2005 population 13,575), NE Republic of KARELIA, NW European Russia, on the WHITE SEA, at the mouth of the KEM' River, on the Murmansk railroad, 270 mi/435 km N of PETROZAVODSK; 64°57'N 34°34'E. Manufacturing of construction materials; fishing. Hydroelectric station. Railroad shops. Its port is RABOCHEOSTROVSK, 5 mi/8 km ENE; linked by a railroad spur. Kem' is the administrative center of Pomorye (White Sea coast), a district with a large Russian population. One of oldest settlements of Karelia and a 15th century trading center.

Kem' (KYEM), river, approximately 240 mi/386 km long, NW European Russia (Republic of KARELIA) and FINLAND. It rises SE of KUUSAMO, NE Finland, near the Russian-Finnish border, and flows S and then E through Koito Lakes and into the WHITE SEA at KEM'.

Ké-Macina (KAI–MAH-see-nah), town, FOURTH REGION/SÉGOU, S MALI, landing in fertile MACINA depression of the NIGER RIVER (irrigation), 66 mi/110 km NE of SÉGOU; 13°58'N 05°22'W. Wool market; manufacturing of woolen blankets. Airfield. Also spelled Massina.

Kemah (KEE-mah), town (2006 population 2,475), GALVESTON county, SE TEXAS, residential suburb 12 mi/19 km NNW of TEXAS CITY and 22 mi/35 km SE of HOUSTON, on GALVESTON BAY, at mouth of Clear Creek, in area referred to as CLEAR LAKE CITY; 29°32'N 95°01'W. Manufacturing (transportation equipment, food, fabricated metal products).

Kemah, village, E central TURKEY, on EUPHRATES RIVER, on Erzincan-Sivas railroad, and 25 mi/40 km WSW of ERZINCAN; 39°35'N 39°02'E. Grain.

Kemajoran, INDONESIA: see KEMAYORAN AIRPORT.

Kemaliye, village, E central TURKEY, on W bank of the EUPHRATES RIVER, and 60 mi/97 km N of MALATYA; 39°16'N 38°29'E. Grain. Formerly Egin.

Kemalla (ke-MAH-yah), canton, CERCADO province, ORURO department, W central BOLIVIA; 17°39'S 67°10'W. Elevation 12,372 ft/3,771 m. Antimony-bearing lode; tin mining to E; clay, limestone, and gypsum deposits. Agriculture (potatoes, yucca, bananas, barley, oats; cattle).

Kemalpasa (Turkish=*Kemalpaşa*), village (2000 population 25,448), W TURKEY, 14 mi/23 km E of IZMIR; 41°30'N 41°30'E. Raisins, tobacco. Formerly Nif (Nymphio).

Kemalpasa River, ancient *Rhyndacus*, 115 mi/185 km long, NW TURKEY; rises 6 mi/9.7 km NW of SIMAV, flows N past MUSTAFA KEMALPASA, to Lake ULUBAT, then 5 mi/8 km W to SIMAV RIVER at KARACABEY. Also called Mustafa Kemalpasa; formerly also Adranos or Edrenos.

Kemaman, Malaysia: see CHUKAI.

Kemaman River (KE-mah-mahn), 60 mi/97 km long, S TERENGGANU, MALAYSIA; rises on PAHANG border, flows ESE to SOUTH CHINA SEA at Kuala Kemaman; 04°15'N 103°20'E. Iron mining along course.

Kemanlar, Bulgaria: see ISPERIH.

Kemano Dam, CANADA: see NECHAKO, river.

Kemayoran Airport, in greater JAKARTA metropolitan area (DKI), NW JAVA, INDONESIA; 06°10'S 106°51'E. Former main airport of Jakarta, was superseded by new Soekarno Hatta international airport in 1985. Now primarily handles local air traffic and smaller aircraft. Formerly spelled Kemajoran. Airport Code JKT.

Kembé (kem-BAI), village, BASSE-KOTTO prefecture, S CENTRAL AFRICAN REPUBLIC, on KOTTO RIVER, and 120 mi/193 km SE of BAMBARI; 04°36'N 21°54'E. Cotton ginning.

Kemberg (KEM-berg), town, SAXONY-ANHALT, central GERMANY, 7 mi/11.3 km S of WITTENBERG; 51°48'N 12°38'E. Manufacturing (cement products, wood products).

Kembs (KAHNS), German (KEMPS), commune (□ 6 sq mi/15.6 sq km), HAUT-RHIN department, ALSACE, E FRANCE, on the Grand; 47°41'N 07°30'E. Canal near the RHINE, and 9 mi/14.5 km SE of MULHOUSE. Dam and large hydroelectric plant on the Rhine, 3 mi/4.8 km S. It was an important Roman settlement with remains of a cement bridge across the Rhine.

Kembul (KEM-bool), village, EAST NEW BRITAIN province, central NEW BRITAIN island, E PAPUA NEW GUINEA, 160 mi/257 km SW of RABAUL. Located 15 mi/24 km inland from SOLOMON SEA on S flank of NAKANAI MOUNTAINS. Cocoa, coffee, copra.

Kemchik River, RUSSIA: see KHEMCHIK RIVER.

Kemer, TURKEY: see BURHANIYE.

Kemer Dag (Turkish=*Kemer Dag*), peak (10,660 ft/3,249 m), NE TURKEY, in TRABZON MOUNTAINS, 38 mi/61 km SSE of TRABZON.

Kemeri (TCHE-me-ree), German *kemmern*, city, W central LATVIA, in VIDZEME, 23 mi/37 km W of RĪGA, near S shore of Gulf of RIGA; 56°56'N 23°29'E. Major health resort; sulphur springs, mud baths.

Kemerovo (KYE-mee-ruh-vuh), city (2005 population 474,595), S central SIBERIA, RUSSIA, on the TOM' River, on a railroad junction on a branch of the TRANS-SIBERIAN RAILROAD, 2,164 mi/3,483 km ESE of MOSCOW; 55°20'N 86°05'E. Elevation 446 ft/135 m. ☉ KEMEROVO oblast. It is the center of the KUZNETSK Basin, Russia's largest coal-producing district; produces chemicals (coke by-products, chemical fertilizers, plastics, synthetic fibers), machinery, electrical goods; textile industries (cotton, silk). Made city in 1925. Named Shcheglovsk from 1925 to 1932.

Kemerovo (KYE-mye-ruh-vuh), oblast (□ 36,873 sq mi/95,869.8 sq km; 2004 population 3,063,000) in S central Siberian Russia; ☉ KEMEROVO. Between the SALAIR RIDGE (W) and KUZNETSK ALATAU (E); drained by

TOM and KIYA rivers; continental climate, with cold and long winters and short and hot summers. Most highly developed industrial area of SIBERIA, based mainly on the KUZNETSK BASIN coal and GORNAYA SHORIYA iron mines; includes ferrous metallurgy (NOVOKUZNETSK, GURYEVSK), chemical industry (fertilizers, artificial fiber, plastics), zinc metallurgy (BELOVO). The population (chiefly Russians and, in S, Tatars), concentrated along the TRANS-SIBERIAN RAILROAD (N) and its branches are also engaged in agriculture. In mountainous areas (E, S), there is gold mining (KIYA RIVER), lumbering, fur trapping, cattle raising. Formed in 1943 out of NOVOSIBIRSK oblast.

Kemi (KE-mee), town, LAPIN province, W central FINLAND; 65°44′N 24°34′E. Elevation 17 ft/5 m. A port with large sawmills, pulp and paper mills, and a power station on the GULF OF BOTHNIA at the mouth of the KEMIJOKI. An old trading post chartered in 1869. Airport 4 mi/6.4 km SSE at LAUTIOSAARI.

Kemi, RUSSIA: see KEM', city.

Kemijärvi (KE-mee-YAR-vee), Swedish *Kemiträsk*, village, LAPIN province, N FINLAND, 50 mi/80 km ENE of ROVANIEMI; 66°40′N 27°25′E. Elevation 594 ft/180 m. On Kemijärvi (lake), expansion (20 mi/32 km long, 1 mi/1.6 km–5 mi/8 km wide) of the KEMIJOKI (river). Commercial center of important lumbering region; large pulp mill; railroad terminus.

Kemijoki (KE-mee-YO-kee), river, c.345 mi/560 km long, longest river of FINLAND; rises near Sokosti peak, NE Finland; flows generally SW to Kemijärvi (lake), then W into the GULF OF BOTHNIA at KEMI. With its many tributaries, the Kemijoki drains most of N Finland. Major hydroelectric plant.

Kemin, village (1999 population 11,401), E CHÜY region, KYRGYZSTAN, on KEMIN RIVER, in CHU valley, 20 mi/32 km E of TOKMOK, and 55 mi/89 km E of BISHKEK; 42°46′N 74°41′E. Tertiary-level administrative center. Agricultural orchards, vegetables. Gold refinery, carbide plant; electronics, construction materials, food. Railroad terminus until extension (late 1940s) of line to BALYKCHY. Formerly BYSTROVKA.

Kemin (ke-MIN), name of two glacier-fed rivers, KYRGYZSTAN. GREATER KEMIN RIVER, Kazak *Chong-Kemin*, c.70 mi/113 km long, rises on KAZAKHSTAN border, flows WSW, between the Ilenin Ala-Too TRANS-ILI ALATAU and KÜNGEY ALA-TOO (KUNGEI ALATAU) mountains, to the CHU RIVER N of BOOM GORGE. The LESSER KEMIN RIVER, (Kazak *Kichi-Kemin*) c.35 mi/56 km long, rises in the Ilenin Ala-Too, flows parallel to and N of the GREATER KEMIN, past AK-TÜZ and BOROLDOY, to the Chu River 20 mi/32 km E of Tokmok. Also called KEBIN RIVER.

Kemise, town (2007 population 19,792), AMHARA state, central ETHIOPIA; 10°43′N 39°52′E. Business and trade center 60 mi/97 km NNE of DEBRE BERHAN.

Kemiträsk, FINLAND: see KEMIJÄRVI.

Kemlya (KYEM-lyah), town (2006 population 4,725), NE MORDVA REPUBLIC, central European Russia, near the ALATYR' RIVER, on road and railroad, 38 mi/61 km SE of LUKOYANOV; 54°42′N 45°15′E. Elevation 347 ft/105 m. In agricultural area; vegetable growing, food processing.

Kemmanugundi, INDIA: see BABA BUDAN RANGE.

Kemmarath, THAILAND: see KHEMMARAT.

Kemmerer (KEM-uhr-uhr), town (2006 population 2,525), SW WYOMING, on HAMS FORK, near UTAH state line, and 70 mi/113 km WNW of ROCK SPRINGS; ⊙ LINCOLN county; 41°46′N 110°32′W. Elevation c. 6,927 ft/2,111 m. Shipping point for coal, cattle, sheep; some manufacturing. FOSSIL BUTTE NATIONAL MONUMENT to W. The J.C. Penney Mother Store, first in nation-wide chain, 1902, is located here and is still open. Incorporated 1899.

Kemmern, LATVIA: see KEMERI.

Kemnath (KEM-naht), town, BAVARIA, S GERMANY, in UPPER PALATINATE, at SW foot of the FICHTELGE-BIRGE, 15 mi/24 km ESE of Bayreyth; 49°52′N 11°53′E. Manufacturing of textiles; apparel. Chartered in 14th century.

Kemnay, community, SW MANITOBA, W central CANADA, 8 mi/13 km W of BRANDON, and in WHITEHEAD rural municipality; 49°50′N 100°08′W.

Kemnay (KEM-nai), village (2001 population 3,697), Aberdeenshire, NE Scotland, on DON RIVER, and 4 mi/6.4 km W of KINTORE; 57°14′N 02°27′W. Granite quarrying. Castle Fraser is 2 mi/3.2 km S.

Kémo (KAI-mo), prefecture (□ 6,641 sq mi/17,266.6 sq km; 2003 population 118,420), central CENTRAL AFRICAN REPUBLIC; ⊙ SIBUT. Bordered N by NANA-GRÉBIZI prefecture, E by OUAKA prefecture, S by Democratic Republic of the CONGO, SW by OMBELLA-M'POKO prefecture, and NW by OUHAM prefecture. Drained by GRIBINGUI, KÉMO, Tomi, and UBANGI rivers. Agriculture (cotton, rubber, coffee); textile mills. Main centers are Sibut and DEKOA. Also called Kémo-Gribingui.

Kemoedjan, INDONESIA: see KARIMUNJAWA ISLANDS.

Kémo-Gribingui (KAI-mo–gree-bang-GEE), prefecture, CENTRAL AFRICAN REPUBLIC: see KÉMO, prefecture.

Kemondo Bay (KAI-mon-do), small inlet, W side of Lake VICTORIA, KAGERA region, NW TANZANIA; 03°20′S 31°49′E. Port for bukoba town.

Kémo River (kai-MO), 120 mi/193 km long, KÉMO prefecture, S CENTRAL AFRICAN REPUBLIC; rises 35 mi/56 km E of DEKOA; flows S to UBANGI RIVER just E of KAGA BANDORO.

Kemp, town (2006 population 1,289), KAUFMAN county, NE TEXAS, c.40 mi/64 km SE of DALLAS, near Cedar Creek and headwaters of CEDAR CREEK RESERVOIR (S); 32°26′N 96°13′W. In cotton, corn, cattle area.

Kemp, village, BRYAN county, S OKLAHOMA, near RED RIVER, 16 mi/26 km S of DURANT. In agricultural area.

Kemp City, village, Wichita county, Texas: see KAMAY.

Kemp Coast, ANTARCTICA, E of ENDERBY LAND, on INDIAN OCEAN, between head of EDWARD VIII BAY and William Scoresby Bay; 67°10′S 56°30′E–59°40′E. Discovered 1833 by British captain, Peter Kemp.

Kempen (KEM-pen), city, NORTH RHINE–WESTPHALIA, W GERMANY, 6 mi/9.7 km WNW of KREFELD; 51°22′N 06°25′E. Manufacturing of textiles, fabricated metal products, electronics, furniture, food, printing and publishing. Has Gothic church, 14th-century castle. Was first mentioned 890. Chartered 1294. St. Thomas à Kempis born here.

Kempen, POLAND: see KEPNO.

Kempendyay (kyem-pyen-DYEI), village, SW SAKHA REPUBLIC, W RUSSIAN FAR EAST, on the short Kempendyay River (right affluent of the VILYUY River), terminus of highway branch, 150 mi/241 km SSW of VILYUYSK; 62°02′N 118°39′E. Elevation 567 ft/172 m. Salt mining.

Kempenland (KEM-puhn-lahnd), French *Campine* (kahm-peen), low plateau in N and E BELGIUM, occupying much of ANTWERPEN province, northern LIMBURG province, and extending into S NETHERLANDS. The sandy soils of the region are reflected in the piney woods and meadows with their thin grass cover. The Kempenland was settled early but with moderate density until the nineteenth century. Much of the region has since been drained and cultivated. Coal was discovered near GENK, in E LIMBURG, in 1901, and exploitation only began after World War I. The Kempenland became Belgium's leading coal producer, superseding the older fields of the Sambre-Meuse region. There has been a severe decline in coal production since the 1960s, though not as catastrophic as in the southern coalfields, and the Kempenland has not profited from national attention and subsidies as other coal regions. While agriculture continues to be important, most employment is in various forms of industry ranging from electronics to processing of non-ferrous minerals. HASSELT and TURNHOUT are the region's main cities.

Kemper, county (□ 767 sq mi/1,994.2 sq km; 2006 population 10,108), E MISSISSIPPI, bordering E on ALABAMA; ⊙ DE KALB; 32°45′N 88°39′W. Drained by Sucarnoochee and okatibbee creeks. Agriculture (cotton, corn), cattle raising; catfish; timber. Kemper County Lake (state lake) in center. Formed 1833.

Kemper, FRANCE: see QUIMPER.

Kemperle, FRANCE: see QUIMPERLÉ.

Kempimpi (kem-PEEM-pee), village, BANDUNDU province, W CONGO, NE of BANDUNDU; 03°01′S 18°17′E. Elev. 1,138 ft/346 m.

Kempish Kanaal, BELGIUM: see SCHELDT-MEUSE JUNCTION CANAL.

Kemp, Lake, reservoir (□ 24 sq mi/62.4 sq km), BAYLOR county, N central TEXAS, on Big Wichita Creek, 40 mi/64 km W of WICHITA FALLS; 33°46′N 99°09′W. Maximum capacity 502,900 acre-ft. Formed by Lake Kemp Dam (150 ft/35 m high), built (1923) by Army Corps of Engineers for water supply; also used for irrigation and recreation.

Kemp's Bay, town, W BAHAMAS, on SE shore of ANDROS Island, 75 mi/121 km S of NASSAU; 24°01′N 77°33′W. Fishing, lumbering.

Kempsey (KEMP-see), municipality, E NEW SOUTH WALES, SE AUSTRALIA, on MACLEAY RIVER, and 145 mi/233 km NNE of NEWCASTLE; 31°05′S 152°50′E. About halfway between SYDNEY and QUEENSLAND border, on Pacific Highway. Dairying and agricultural center; hat factory; tourism.

Kempston (KEMP-stuhn), town (2001 population 19,440), W central BEDFORDSHIRE, central ENGLAND, on OUSE RIVER, and 2 mi/3.2 km SW of BEDFORD; 52°07′N 00°29′W. Leather and shoe industry. Has Norman church.

Kempten (KEMP-ten), city, BAVARIA, S central GERMANY, in SWABIA, on the ILLER RIVER, in the ALLGÄU; 47°43′N 10°19′E. It is the center of a dairying region and is widely known for its cheeses. Manufacturing includes machinery, electronics, pharmaceuticals, and beer. Of Celtic origin, Kempten became a flourishing Roman colony called CAMBODUNUM. A free imperial city from the late 13th century, it was sacked (1632) by the Swedes in the Thirty Years War. Passed to Bavaria in 1803. The city is rich in historic architecture, including the abbey church of St. Lorenz (1652), the town hall (1474), and the Church of St. Mang (1426, restored).

Kempt Lake (22 mi/35 km long, 15 mi/24 km wide), SW central QUEBEC, E CANADA, in the LAURENTIANS, 120 mi/193 km NW of MONTREAL; 47°26′N 74°16′W. Elevation 1,372 ft/418 m. Drained (W) by LIÈVRE River. Contains numerous islands.

Kempton, town (2000 population 380), TIPTON county, central INDIANA, near CICERO CREEK, 14 mi/23 km SSW of KOKOMO; 40°17′N 86°14′W. In agricultural area.

Kempton (KEMP-tuhn), village, SE central TASMANIA, SE AUSTRALIA, 24 mi/39 km NNW of HOBART; 42°31′S 147°12′E. Wheat. Also known as Green Ponds.

Kempton, village (2000 population 235), FORD county, E ILLINOIS, 22 mi/35 km WSW of KANKAKEE; 40°56′N 88°14′W. In rich agricultural area (corn, soybeans, livestock).

Kempton Park, town, GAUTENG province, SOUTH AFRICA, 15 mi/24 km NE of JOHANNESBURG, at N edge of WITWATERSRAND and of gold-mining region; 26°06′S 28°15′E. Elevation 5,456 ft/1,663 m. Iron, brick, cement works. Johannesburg International Airport nearby to SE.

Kempton Park, ENGLAND: see HOUNSLOW.

Kemptville (KEMT-vil), unincorporated town (□ 1 sq mi/2.6 sq km; 2001 population 3,667), SE ONTARIO, E central CANADA, near Rideau River and RIDEAU CANAL, 28 mi/45 km S of OTTAWA, and included in North GRENVILLE township; 45°01′N 75°38′W. Dairying center, lumbering. Recreation.

Kemujan, INDONESIA: see KARIMUNJAWA ISLANDS.

Kena, EGYPT: see QENA.

Kenadsa (kai-nah-DZAH), mining town, BÉCHAR wilaya, SW ALGERIA, near MOROCCO border, 13 mi/21 km WSW of BÉCHAR; 31°30′N 02°30′W. Originally the site for an important religious brotherhood (or Zaouia), which grew into a town. It has grown in population recently, with a large residential section. Coal mining, formerly a major industry, has declined considerably.

Kenai (KEE-nai), locality (2000 population 6,942), S ALASKA, W KENAI PENINSULA, on COOK INLET, at mouth of Kenai River, 65 mi/105 km SW of ANCHORAGE; 60°32′N 151°12′W. Oil refineries, urea plant, offshore oil platforms; tourism; fishing, fish processing. University of Alaska has Kenai community college here. Has Russian Orthodox church. U.S. garrison here, 1869. Established 1791 by Russians who built Fort Saint Nicholas here.

Kenai Fjords National Park (KEE-nai) (□ 1,046 sq mi/ 2,709 sq km), S ALASKA; authorized 1980. Wilderness preserve, vast icefields, fjords, and outflowing glaciers.

Kenai Lake (KEE-nai) (25 mi/40 km long, 1 mi/1.6 km wide), S ALASKA, 20 mi/32 km N of SEWARD, on ANCHORAGE-Seward Highway. Tourism; fishing, hiking.

Kenai Mountains (KEE-nai), rise to c.7,000 ft/2,134 m, 35 mi/56 km WSW of SEWARD, part of the COAST RANGES, S ALASKA, extend 150 mi/241 km along SE side of KENAI PENINSULA; slope steeply to Gulf of ALASKA. Range has several glacier fields. Continued NE by CHUGACH MOUNTAINS.

Kenai Peninsula, borough (□ 16,079 sq mi/41,805.4 sq km; 2006 population 52,304), S ALASKA, includes the KENAI PENINSULA except for E section and part of W side of COOK INLET. Bounded on the S by the Gulf of ALASKA and the KENAI MOUNTAINS to S and E. KENAI FJORDS NATIONAL PARK to S; part of KATMAI National Park to S; LAKE CLARK National Park and Preserve to W; part of CHUGACH NATIONAL FOREST to E; Kenai National Wildlife Reserve to N and center; Kachemak Bay State Wilderness Park to S. Main towns are SEWARD, SOLDOTNA, KENAI, and HOMER. Tourism; fishing; timber.

Kenai Peninsula (KEE-nai), S ALASKA, jutting c.150 mi/ 240 km into the Gulf of ALASKA, between PRINCE WILLIAM SOUND and COOK INLET. The KENAI MOUNTAINS, c.7,000 ft/2,130 m high, occupy most of the peninsula. The coastal climate is mild, with abundant rainfall and a growing season adequate for many crops. There are forest, mineral, and fishing resources in the E and, in the W section, petroleum, natural gas, coal, refineries; fishing; tourism; farmland. The Alaska railroad crosses the peninsula from SEWARD. Highways connect ANCHORAGE with Seward, SOLDOTNA, KENAI, HOMER, HOPE. On the S coast of the peninsula is KENAI FJORDS NATIONAL PARK, which has striking glacial formations and contains the breeding areas for a variety of birds and sea mammals.

Kenansville (KEE-nuhnz-vil), town (□ 2 sq mi/5.2 sq km; 2006 population 899), E NORTH CAROLINA, 30 mi/ 48 km SW of KINSTON, 55 mi/88.5 km N of WILMINGTON, 80 mi/128.7 km SE of RALEIGH; ☉ DUPLIN county; 34°57′N 77°58′W. Elevation 127 ft/38.7 m. Service industries; light manufacturing; agriculture (tobacco, cotton, grain; poultry, livestock). James Sprunt Community College. Cowan Museum. Laid out in 1818.

Kenaston (KE-nuhs-tuhn), village (2006 population 259), S central SASKATCHEWAN, CANADA, 36 mi/58 km WSW of WATROUS; 51°30′N 106°17′W. Wheat, livestock.

Kenberma, Massachusetts: see HULL.

Kenbridge (KEN-brij), town (2006 population 1,318), LUNENBURG county, S VIRGINIA, 27 mi/43 km SE of FARMVILLE; 36°57′N 78°07′W. Light manufacturing; agriculture (tobacco, grain, soybeans; cattle); timber. Lunenburg County Airport to W.

Kenbuchi (KEN-boo-chee), town, Kamikawa district, HOKKAIDO prefecture, N JAPAN, 87 mi/140 km N of SAPPORO; 44°05′N 142°21′E. Soybeans.

Kendal (KEN-duhl), town (2001 population 28,030), CUMBRIA, NW ENGLAND, on KENT RIVER, and 20 mi/ 32 km N of LANCASTER; 54°19′N 02°45′W. Manufacturing of apparel, textiles, carpets, paper, and machinery. Famed for the "Kendal Mint Cake," a high-energy food for mountain climbers. Has church dating from 13th–15th centuries, ruins of 14th-century castle, and some 16th-century houses. Known in Middle Ages for manufacturing of wool cloth known as "Kendal Green." Woolen industry introduced by Flemish immigrants in early 14th century. Remains of a Roman station found here.

Kendal (KUHN-dahl), town, N Java Tengah province, INDONESIA, near JAVA SEA, 15 mi/24 km W of SEMARANG; ☉ Kendal district; 06°55′S 110°12′E. Trade center for agricultural area (sugar, rice, peanuts, tobacco, coffee, kapok).

Kendale Lakes (KEN-duhl), unincorporated suburb (□ 8 sq mi/20.8 sq km; 2000 population 56,901) of Miami, SW MIAMI-DADE county, SE FLORIDA, 16 mi/ 26 km SW of MIAMI; 25°42′N 80°24′W.

Kendall (KEN-duhl), county (□ 322 sq mi/837.2 sq km; 2006 population 88,158), NE ILLINOIS; ☉ YORKVILLE; 41°35′N 88°25′W. Rich farming area (corn, soybeans, wheat). Manufacturing: dairy products, fabricated metal products, electronic equipment. Drained by FOX RIVER. Includes Silver Springs State Park. Formed 1841. Initial stages of urban growth, influenced by CHICAGO, have started in extreme NE corner.

Kendall (KEN-duhl), county (□ 663 sq mi/1,723.8 sq km; 2006 population 30,213), S central TEXAS; ☉ BOERNE; 29°56′N 98°42′W. Elevation c.1,000 ft/305 m–2,000 ft/610 m. Generally broken area, on S edge of EDWARDS PLATEAU; drained by GUADALUPE and BLANCO rivers; bounded by CIBOLO CREEK on S. Chiefly sheep, Angora-goat ranching region (wool, mohair marketed); also beef cattle, grain, wheat, oats. Natural gas deposits. Hunting, fishing, caves (CASCADE CAVERNS, others) attract tourists. Formed 1862.

Kendall (KEN-duhl), village, NEW SOUTH WALES, SE AUSTRALIA, 231 mi/371 km NE of SYDNEY, 22 mi/36 km SW of PORT MACQUARIE, and on Camden Haven River; 31°38′S 152°42′E. Middle Brother State Forest to S. Named for Australian poet Henry Kendall. Formerly called Camden Heads.

Kendall, village, MONROE county, W central WISCONSIN, 45 mi/72 km E of LA CROSSE, in dairying and livestock region. Sportswear. On La Crosse State Trail.

Kendall (KEN-duhl), unincorporated suburb (□ 23 sq mi/59.8 sq km; 2000 population 75,226) of MIAMI, MIAMI-DADE county, SE FLORIDA, 11 mi/18 km SW of Miami; 25°40′N 80°20′W. Centered on North Kendall Drive, an E-W highway running from BISCAYNE BAY on E to EVERGLADES about 15 mi/24 km to W. Residential communities predominate, but major retail and business complexes abound, particularly around DADELAND where North Kendall Drive intersects the Palmetto Expressway and U.S. 1.

Kendall Green, Massachusetts: see WESTON.

Kendall Mountain (13,068 ft/3,983 m), SAN JUAN county, SW COLORADO, peak in SAN JUAN MOUNTAINS, 2 mi/3.2 km SE of SILVERTON.

Kendall Peak (13,338 ft/4,065 m), SAN JUAN county, SW COLORADO, in SAN JUAN MOUNTAINS, 3 mi/4.8 km ESE of SILVERTON.

Kendallville, city (2000 population 9,616), NOBLE county, NE INDIANA, on ELKHART RIVER, and 8 mi/ 12.9 km NE of ALBION; 41°26′N 85°16′W. Shipping center in agricultural area (especially onions); also dairy products, livestock, soybeans, grain. Manufacturing (transportation equipment, paper products, printing and publishing, machinery, fabricated metal products, food). Mulholland Museum has Indian and pioneer relics. Settled 1833.

Kendangan, INDONESIA: see KANDANGAN.

Kendangomuwa (KAN-thah-GUH-mu-wuh), town, SABARAGAMUWA PROVINCE, SW central SRI LANKA, 15 mi/24 km NNW of RATNAPURA; 06°51′N 80°15′E. Vegetables, rice, rubber. Graphite mines nearby. Also spelled Kendangamuwa.

Kendari (KUHN-dah-ree), town, SE SULAWESI, INDONESIA, port on BANDA SEA, 230 mi/370 km ENE of UJUNG PANDANG; ☉ Kendari district and Sulawesi Tenggara province; 03°58′S 122°35′E. Trade center shipping timber, resin, rattan. There is an important gold and silver filigree ornament industry here, carried on by Chinese. Monginsidi Airport.

Kendelton (KEND-uhl-tin), village, FORT BEND county, SE TEXAS, 44 mi/71 km SW of HOUSTON, near SAN BERNARD RIVER; 29°26′N 96°00′W. Agricultural area (rice, cotton, vegetables, corn, nurseries). Oil and natural gas.

Kendenup, settlement, WESTERN AUSTRALIA state, W AUSTRALIA, 214 mi/345 km SE of PERTH, between Cranbrook and MOUNT BARKER; 34°28′S 117°35′E. W entrance to STIRLING RANGE NATIONAL PARK.

Kenderes (KEN-de-vesh), village, SZOLNOK county, E central HUNGARY, 13 mi/21 km WSW of KARCAG; 47°15′N 20°41′E. Wheat, corn, tobacco; hogs, sheep.

Kenderlyk River (ken-der-LIK), 105 mi/169 km long, E. KAZAKHSTAN region, KAZAKHSTAN; rises in SAUR RANGE, flows NW to Lake ZAISAN. Coal and oil-shale deposits.

Kendraparha (kain-DRAH-pah-ruh), district (□ 983 sq mi/2,555.8 sq km), ORISSA state, E central INDIA; ☉ KENDRAPARHA.

Kendraparha (kain-DRAH-pah-ruh), town, ☉ KENDRAPARHA district, E ORISSA STATES, E central INDIA, 34 mi/55 km E of CUTTACK; 20°30′N 86°25′E. Biri manufacturing; palm-mat making; hand-loom weaving; center of large rice-growing area. Visited by Hindu pilgrims; Vishnuite shrine.

Kendrick, village (2000 population 369), LATAH county, W IDAHO, 18 mi/29 km SE of MOSCOW, on POTLATCH RIVER; 46°37′N 116°39′W. Agricultural center (cattle, sheep; alfalfa, barley, oats). Nez Perce Indian Reservation to S.

Kendrick, village (2006 population 144), LINCOLN county, central OKLAHOMA, 14 mi/23 km S of CUSHING; 35°47′N 96°46′W. In agricultural area.

Kendrick Peak (10,418 ft/3,175 m), rises from high plateau in COCONINO county, N central ARIZONA, 18 mi/29 km NW of FLAGSTAFF. On boundary of Kaibab (W) and Coconino (E) national forests.

Kendu Bay (KAIN-doo), town, Homa Bay district, NYANZA province, KENYA, on coast of Winam Gulf 19 mi/30 km SSW of KISUMU; 01°03′S 34°28′E. Market and trading center.

Kendujhargarh, district (□ 3,219 sq mi/8,369.4 sq km), N ORISSA STATES, E central INDIA; ☉ KENDUJHARGARH. Consists of hills (2,000 ft/610 m–3,000 ft/914 m) and lowland. Valuable deposits of manganese and iron ore (worked) in N area; sal, bamboo, and lac from forests. Formerly a princely state in ORISSA STATES of EASTERN STATES agency; merged 1948 with Orissa and made a district.

Kendujhargarh, town, N ORISSA state, E central INDIA, 80 mi/129 km NNW of CUTTACK; ☉ KENDUJHARGARH district. Market center for rice, timber, lac, hides; hand-loom weaving. Formerly called Keonjhar, Keonjhargarh.

Kenduskeag (ken-DUHS-keg), town, PENOBSCOT county, S MAINE, lake area, 10 mi/16 km NW of BANGOR; 44°55′N 68°55′W.

Kenduskeag Stream (ken-DUHS-keg), c.30 mi/48 km long, S MAINE; rises in S PENOBSCOT county, flows SE to the PENOBSCOT at BANGOR.

Keneba (KEN-ai-bah), town, LOWER RIVER division, THE GAMBIA, 35 mi/56 km ESE of BANJUL; 13°20′N 15°59′W. On road 5 mi/8 km N of Bitand Bolong River. Site of medical-research center (established 1948).

Kénédougou, province (□ 3,143 sq mi/8,171.8 sq km; 2005 population 261,014), HAUTS-BASSINS region, SW BURKINA FASO; ⊙ ORODARO; 11°25′N 05°00′W. Borders HOUET (E) and COMOÉ and LÉRABA (S) provinces and MALI (W and N). Drained by MOUHOUN RIVER. Agriculture (sorghum, vegetables, cotton). Main center in Orodaro. Sparsely populated.

Kenedy, county (□ 1,945 sq mi/5,057 sq km; 2006 population 402), extreme S TEXAS; ⊙ SARITA; 26°55′N 97°37′W. Flat coastal plain, bordering E on GULF OF MEXICO; PADRE ISLAND (National Seashore), a sand barrier island, separated from mainland by LAGUNA MADRE and Intracoastal Waterway. Large-scale ranching area (cattle, horses), most of county is in huge KING RANCH; some agriculture (watermelons). Oil and gas. County has numerous small lakes. Formed 1921 from parts of WILLACY, HIDALGO, and CAMERON counties.

Kenedy, town (2006 population 3,378), KARNES county, S TEXAS, c.60 mi/97 km SE of SAN ANTONIO; 28°49′N 97°50′W. Oil and natural gas; cattle; dairying; wheat; corn, sorghum; manufacturing (chemicals). Founded 1882, incorporated 1910.

Kenefick (KEN-uh-fik), village, BRYAN county, S OKLAHOMA, 11 mi/18 km N of DURANT; 34°08′N 96°21′W. In agricultural area.

Keneh, EGYPT: see QENA.

Kenema (ke-NE-mah), town (1999 estimated population 99,100; 2004 population 128,402; ⊙ Kenema district (□ 6,005 sq mi/15,613 sq km; 2004 population 128,402) and EASTERN province, E SIERRA LEONE, 40 mi/64 km ESE of BO; 07°52′N 11°12′W. Trade and road center; palm oil and kernels, cacao, coffee. Sawmilling. Chromite mining at HANGHA (7 mi/11.3 km NNE).

Kenenkou, village, SECOND REGION/KOULIKORO, MALI, 28 mi/45 km NE of KOULIKORO. Also spelled Kenenkoun.

Kenesaw (KE-nuh-saw), village (2006 population 943), ADAMS county, S NEBRASKA, 14 mi/23 km WNW of HASTINGS; 40°37′N 98°39′W. Light manufacturing.

Kenet el-Jalil, ISRAEL: see CANA.

Kenge (KENG-gai), village, BANDUNDU province, SW CONGO, on WAMBA RIVER, and 120 mi/193 km WNW of KIKWIT; 04°52′S 16°59′E. Elev. 1,072 ft/326 m. Steamboat landing and center of native trade, palm-oil milling, fiber-growing.

Kengeja (kai-NGAI-jah), village, PEMBA SOUTH region, NE TANZANIA, at S end of PEMBA island, on INDIAN OCEAN, 15 mi/24 km S of CHAKE CHAKE; 05°26′S 39°45′E. Cloves, bananas, copra.

Kengere (keng-GAI-rai), village, Katanga province, SE Congo, near Angola border, 85 mi/137 km WSW of Likasi; 11°10′S 25°28′E. Elev. 4,596 ft/1,400 m. Lead mining.

Kengkok (kang-KAWK), village, SAVANNAKHET province, S LAOS, 30 mi/48 km ESE of SAVANNAKHET; 16°51′N 106°05′E. Rice growing. Sometimes spelled Kengkak.

Kengtung (KENG-TOONG), formerly the easternmost state (sawbwaship) (□ 12,405 sq mi/32,253 sq km), S SHAN STATE, MYANMAR; ⊙ KENGTUNG. Elevation 8,000 ft/2,438 m. Rice and sugarcane in valleys, tea and opium in hills; teak forests; sericulture, weaving, pottery; exports are opium and cotton. The largest and most populous SHAN STATE, bounded NE by Chinese YUNNAN province, E by LAOS along MEKONG RIVER, S by THAILAND, and W by THANLWIN RIVER. Drained by the Nam Hkok (right affluent of the MEKONG). Now divided between six townships in the easternmost part of the SHAN STATE. Sometimes spelled Kentung.

Kengtung, town, SHAN STATE, MYANMAR, 165 mi/266 km E of TAUNGGYI, and 60 mi/97 km E of the THANLWIN RIVER, on road to LAMPANG (THAILAND); ⊙ KENGTUNG township. Trading center. A walled and moated town, it has a palace, pagodas, and monasteries. Founded 1819.

Kenhardt, town, NORTHERN CAPE province, SOUTH AFRICA, on HARTEBEES RIVER, and 60 mi/97 km S of UPINGTON; 29°10′S 21°05′E. Agricultural center (wheat, livestock, fruit); on Sishen-Saldanha iron-ore railroad line.

Kenhorst, borough (2006 population 2,672), BERKS county, SE central PENNSYLVANIA, residential suburb 2 mi/3.2 km S of READING, on Angelica Creek. Manufacturing (fabricated metal products). Incorporated 1931.

Kéniéba (kai-nee-AI-bah), town, FIRST REGION/KAYES, MALI, 108 mi/180 km SSE of KAYES; 12°50′N 11°14′W. Peanuts, rice, millet; livestock. Gold mine; airport.

Kenilworth (KE-nuhl-wuhth), town (2001 population 22,582), WARWICKSHIRE, central ENGLAND, 6 mi/9.7 km N of WARWICK; 52°21′N 01°35′W. A market and residential town, it is famous for the ruins of Kenilworth Castle. In the 13th century, the castle became the property of Simon de Montfort. In the castle's Great Hall, Edward II was forced to relinquish his crown in 1327. The castle then passed by marriage to John of Gaunt. It became royal property through John's son, Henry IV, until Queen Elizabeth I presented it to Robert Dudley, Earl of Leicester. Also in Kenilworth are ruins of an Augustinian priory founded c.1122.

Kenilworth (KE-nuhl-woorth), village (2000 population 2,494), COOK county, NE ILLINOIS, N suburb of CHICAGO, on Lake MICHIGAN, just N of WILMETTE; 42°05′N 87°42′W. Eugene Field is buried here. Incorporated 1896.

Kenilworth, borough (2006 population 7,741), UNION county, NE NEW JERSEY, 4 mi/6.4 km WNW of ELIZABETH; 40°40′N 74°17′W. Manufacturing. Incorporated 1907.

Kenilworth (KE-nuhl-worth), settlement, QUEENSLAND, NE AUSTRALIA, 96 mi/154 km N of BRISBANE; 26°35′S 152°44′E. Dairy products. Walking tracks, gem fossicking. Scarecrow-and-limerick festival, movie museum, Kenilworth State Forest.

Kenisa, Jebel el (ken-EES-eh, JE-bel el), French Kenisse, mountain (elevation 6,660 ft/2,030 m), LEBANON range, central LEBANON, 16 mi/26 km ESE of BEIRUT. Also spelled Jabal al-Kanisah.

Kenitra (ke-NEE-truh), city, Kénitra province, ⊙ GHARB-CHRARDA-BENI HSSEN administrative region, NW MOROCCO, on the OUED SEBOU RIVER; 34°16′N 06°36′W. Its port, which is declining in importance, exports agricultural products and fertilizers. Kenitra is known for its traditional carpets. It has a small public university. The city was developed by the French and called Port Lyautey. American troops landed there in November, 1942, during World War II.

Kenjakura (KAIN-juh-kuhr-uh), village, BANKURA district, W WEST BENGAL state, E INDIA, 11 mi/18 km W of Bankura. Cotton weaving; rice, corn, wheat. Also spelled Kenjiakura.

Ken, Loch, Scotland: see NEW GALLOWAY.

Kenly (KEN-lee), town (□ 1 sq mi/2.6 sq km; 2006 population 1,868), JOHNSTON and WILSON counties, on NE edge of Johnston county, central NORTH CAROLINA, 15 mi/24 km NE of SMITHFIELD, 40 mi/64.4 km E of RALEIGH; 35°35′N 78°07′W. Light manufacturing; service industries; agriculture (tobacco, cotton, grain, sweet potatoes; poultry, livestock). Tobacco Farm Life Museum of North Carolina.

Kenmar (KEN-mahr), unincorporated town, Loyalsock township, LYCOMING county, N central PENNSYLVANIA, residential suburb 2 mi/3.2 km E of WILLIAMSPORT, on West Branch of SUSQUEHANNA RIVER; 41°15′N 76°57′W.

Kenmare (ken-MER), Gaelic Neidín, town (2006 population 1,701), S KERRY county, SW IRELAND, at head of KENMARE RIVER, 13 mi/21 km SSW of KILLARNEY; 51°53′N 09°35′W. Small port; woolen milling, lacemaking. Founded 1670 by Sir William Petty.

Kenmare (KEN-mer), town (2006 population 1,080), WARD CO., N NORTH DAKOTA, 47 mi/76 km NW of MINOT, and on DES LACS RIVER; 48°40′N 102°04′W. Grain farms; dairy products, livestock, potatoes.

Lower Des Lacs Lake to S, Upper Des Lacs Lake to N, Des Lacs National Wildlife Refuge on Des Lacs River and lakes. Founded in 1897 and named for KENMARE, IRELAND.

Kenmare River, Gaelic An Ribhéar, deep inlet (2 mi/3.2 km–6 mi/9.7 km wide) of the ATLANTIC, between SW CORK county and S KERRY county, SW IRELAND, extends 28 mi/45 km inland between LAMB'S HEAD (N) and Dursey Head (S). KENMARE at head of inlet.

Kenmaur, village, MATABELELAND NORTH province, W central ZIMBABWE, 80 mi/129 km NNW of BULAWAYO, near BUBI RIVER; 19°05′S 27°55′E. Livestock; grain.

Kenmore, unincorporated town (2006 population 19,980), KING county, W WASHINGTON, residential suburb 11 mi/18 km NNE of downtown SEATTLE, at N end of LAKE WASHINGTON, at mouth of Sammamish River; 47°46′N 122°14′W. St. Edward State Park to S.

Kenmore (KEN-mor), village, PERTH AND KINROSS, central Scotland, on TAY RIVER at E end of LOCH TAY, 6 mi/9.7 km WSW of ABERFELDY; 56°35′N 04°00′W. Resort. The 19th-century Taymouth Castle (now a hotel) is on site of 16th-century Castle of Balloch (demolished in 1805).

Kenmore, residential village (□ 1 sq mi/2.6 sq km; 2006 population 15,318), ERIE county, NW NEW YORK, contiguous with NW BUFFALO; 42°57′N 78°52′W. Incorporated 1899.

Kennan, village (2006 population 151), PRICE county, N WISCONSIN, 25 mi/40 km ENE of LADYSMITH; 45°31′N 90°35′W. Dairying. A unit of Chequamegon National Forest to S; FLAMBEAU RIVER State Forest to NW.

Kennard, town (2000 population 455), HENRY county, E central INDIANA, 8 mi/12.9 km W of NEW CASTLE; 39°54′N 85°31′W. In agricultural area.

Kennard, village (2006 population 387), WASHINGTON county, E NEBRASKA, 6 mi/9.7 km SW of BLAIR, near MISSOURI RIVER; 41°28′N 96°12′W.

Kennard, village (2006 population 317), HOUSTON county, E TEXAS, 15 mi/24 km ENE of CROCKETT, in Davy Crockett National Forest; 31°21′N 95°10W. Timber area.

Kennebago Lake (ken-uh-BAI-go), FRANKLIN county, W MAINE, lake (5 mi/8 km long, 1 mi/1.6 km wide) 10 mi/16 km N of Rangeley Lake. Hunting, fishing.

Kennebec (KEN-uh-bek), county (□ 951 sq mi/2,472.6 sq km; 2006 population 121,068), S MAINE; ⊙ AUGUSTA; 44°24′N 69°46′W. Manufacturing (apparel, textiles, paper, wood and pulp products) at HALLOWELL, WATERVILLE, and GARDINER on the KENNEBEC RIVER; dairying; canning and shipping of farm, orchard produce. Water power from SEBASTICOOK and Kennebec rivers. Many resorts, notably in BELGRADE and CHINA lakes regions. Formed 1799.

Kennebec, town (2006 population 286), ⊙ LYMAN county, S central SOUTH DAKOTA, 40 mi/64 km SE of PIERRE, and on Medicine Creek; 43°53′N 99°51′W. In farming region (dairy products, livestock, poultry, grain). Dam and game preserve nearby. Lower Brule Indian Reservation to N.

Kennebec, river, 164 mi/264 km long, NW MAINE; rising in MOOSEHEAD LAKE and flowing S to the ATLANTIC. The ANDROSCOGGIN RIVER is its chief tributary. Samuel De Champlain explored the area in 1604–1605; in 1607, George Popham established a short-lived colony, Fort Saint George, at its mouth. Trading posts were established shortly after 1625. In 1775, American general Benedict Arnold's expedition went up the Kennebec en route to QUEBEC. Lumber and, in the nineteenth century, ice, were shipped down the river to the coast, and shipbuilding flourished along its banks. Villages such as AUGUSTA and WATERVILLE, established near power sites, became industrial centers.

Kennebecasis River (ke-ne-buh-KA-sis), 60 mi/97 km long, S NEW BRUNSWICK, CANADA; rises NE of SUSSEX, flows SW, past Sussex and HAMPTON, through 20-mi/

32-km-long estuary called Kennebecasis Bay (4 mi/6 km wide) to St. John River just above SAINT JOHN.

Kennebunk (KEN-uh-buhnk), town, YORK COUNTY, SW MAINE, adjacent to KENNEBUNKPORT, and 10 mi/16 km SE of ALFRED; 43°23′N 70°34′W. The first settlement (c.1650) grew as a trading and, later, a shipbuilding and shipping center with light manufacturing. The Wedding Cake House at Kennebunk is known for its scroll-saw architecture. Incorporated 1820.

Kennebunkport (KEN-uh-buhnk-port), town, YORK county, SW MAINE, on the ATLANTIC coast, and 15 mi/24 km SE of ALFRED; 43°23′N 70°27′W. The early town, called Arundel, appears in Kenneth Roberts's books; the name was changed in 1821. The town is a summer resort, especially for authors, artists, and actors. It is also the vacation home of former President George Bush, dubbed the "summer White House" because of the frequent meetings there between Bush and his political advisers and other world leaders. Settled 1629, incorporated 1653.

Kennebunk River (KEN-uh-buhnk), 15 mi/24 km long, SW MAINE; rises in central YORK county, flows SE to the ATLANTIC at KENNEBUNKPORT.

Kennecott, Alaska: see KENNICOTT.

Kennedale, town (2006 population 6,736), TARRANT county, N TEXAS, suburb 10 mi/16 km SE of downtown FORT WORTH; 32°38′N 97°13′W. Drained by Village Creek (forms Lake Arlington to N). Manufacturing (rubber products, explosives, concrete, ceramics). Incorporated after 1940.

Kennedy, town (2000 population 541), Lamar co., W ALABAMA, c.14 mi/23 km SSE of Vernon. Apparel, lumber.

Kennedy, township, ALLEGHENY county, W PENNSYLVANIA, residential suburb 7 mi/11.3 km NW of PITTSBURGH on OHIO river; 40°28′N 80°05′W.

Kennedy, town, MATABELELAND NORTH province, W central ZIMBABWE, 55 mi/89 km SE of HWANGE, on railroad, on NE boundary of HWANGE NATIONAL PARK; 18°52′S 27°10′E. Cattle, sheep, goats; grain.

Kennedy (KEN-e-dee), village (2006 population 187), SE SASKATCHEWAN, CANADA, 30 mi/48 km WSW of MOOSOMIN; 50°01′N 102°21′W. Mixed farming.

Kennedy, village (2000 population 255), KITTSON county, NW MINNESOTA, 9 mi/14.5 km S of HALLOCK, in RED RIVER valley; 48°38′N 96°54′W. Sunflowers, flax, sugar beets; beans; manufacturing (food).

Kennedy, hamlet, CHAUTAUQUA county, extreme W NEW YORK, on CONEWANGO CREEK, 8 mi/12.9 km NE of JAMESTOWN; 42°09′N 79°06′W. In agricultural area.

Kennedy Channel, sea passage (80 mi/129 km long, up to 24 mi/39 km wide), in the ARCTIC OCEAN, between NE ELLESMERE ISLAND (Canada) and NW GREENLAND; 81°00′N 66°00′W. Opens N to the Arctic via HALL BASIN, S to KANE BASIN. Limited navigation in late summer and fall.

Kennedy Lake (KE-ne-dee) (20 mi/32 km long, 1 mi/2 km–3 mi/5 km wide), SW BRITISH COLUMBIA, W CANADA, on W VANCOUVER ISLAND, 30 mi/48 km WSW of PORT ALBERNI; 49°04′N 125°34′W. Drains W into Tofino Inlet of CLAYOQUOT SOUND.

Kennedy, Mount (13,894 ft/4,235m), mountain peak, SW YUKON TERRITORY, CANADA, in the ST. ELIAS MOUNTAINS near the ALASKAN border; 60°20′N 138°58′S. It was named in honor of U.S. President John F. Kennedy in 1965. Although visited in 1935, the peak was climbed for the first time in 1965 by a team that included Robert F. Kennedy, the President's brother.

Kennedy Peak (7,456 ft/2,273 m), SW GRAHAM county, SE ARIZONA, in GALIURO MOUNTAINS, c.50 mi/80 km NE of TUCSON, in section of Coronado National Forest.

Kennedy Space Center (KEN-uh-dee), NASA's main U.S. launch facility, ORANGE county, E central FLORIDA, 15 mi/24 km E of TITUSVILLE, on the peninsula just N of CAPE CANAVERAL; 28°40′N 80°40′W.

Kennedyville, village, KENT county, E MARYLAND, on the EASTERN SHORE, 7 mi/11.3 km NE of CHESTERTOWN. Originally settled by the Amish. Nearby is Shrewsbury Church, built 1832 on site of original church of the 1600s named for the Earl of Shewsbury. The grave of General John Cadwalader bears an inscription written by Thomas Paine, the pamphleteer of both the American and French Revolutions.

Kenner (KEN-uhr), city (2000 population 70,517), JEFFERSON parish, SE LOUISIANA, a suburb 10 mi/16 km W of downtown NEW ORLEANS; 29°59′N 90°14′W. Bounds Lake PONTCHARTRAIN on N, MISSISSIPPI RIVER on S. Kenner has grown rapidly since the 1970s into an area of moderate- to upper-income family housing developments in the NEW ORLEANS metropolitan region. Commercial activities; retail businesses; manufacturing (electronics, food, chemicals, machinery, lumber, printing, and publishing). New Orleans International Airport within the city limits and Jefferson Downs racetrack nearby. Incorporated 1952.

Kennesaw (KEN-uh-saw), town (2000 population 21,675), COBB county, NW central GEORGIA, 22 mi/35 km NW of ATLANTA; 34°01′N 84°37′W. Suburb of Atlanta; major retail and service center along U.S. Highway 41 and I-75. Manufacturing includes transportation equipment, chemicals, concrete, printing and publishing, consumer goods. KENNESAW MOUNTAIN National Battlefield Park (Civil War) nearby. Kennesaw State University, a unit of the University System of Georgia. Big Shanty Museum includes the General, a famous steam engine in Civil War lore.

Kennesaw Mountain (KEN-uh-saw), peak in COBB county, NW GEORGIA, 15 mi/24 km NW of ATLANTA. Rises in two summits of 1,550 ft/472 m and 1,809 ft/551 m; 33°58′N 84°34′W. Site of major Civil War battle.

Kenneth, village (2000 population 61), ROCK county, SW MINNESOTA, 10 mi/16 km NE of LUVERNE; 43°45′N 96°04′W. Grain; livestock; dairying.

Kenneth City (KEN-uhth), town (2000 population 4,400), PINELLAS county, W central FLORIDA, 8 mi/12.9 km NW of ST. PETERSBURG; 27°49′N 82°43′W.

Kennet River (KEN-et), 44 mi/71 km long, SE ENGLAND; rises on Marlborough Downs in E WILTSHIRE, flows E into West Berkshire, past Hungerford, to the THAMES at READING.

Kennett, city (2000 population 11,260), in the boot heel of extreme SE MISSOURI, near SAINT FRANCIS River, 42 mi/68 km SE of POPLAR BLUFF; ☉ DUNKLIN county; 36°14′N 90°02′W. Rice, cotton, soybeans; manufacturing (apparel, electronic equipment, printing and publishing). Regional commercial and service center. Founded c.1845.

Kennett Square (KE-nuht), borough (2006 population 5,292), CHESTER county, SE PENNSYLVANIA, 12 mi/19 km NW of WILMINGTON, DELAWARE, near Delaware state line. Agricultural area (corn, wheat, mushrooms, soybeans; poultry, livestock, dairying). Manufacturing (apparel, food, concrete, electronic equipment). Major tourist center. Chaddsford Winery and Longwood Gardens to NE; White Clay Creek State Park to SW (Pennsylvania-Delaware border); Brandywine River Museum nearby. Settled c.1750, incorporated 1855.

Kennewick (KEN-uh-wik), city (2006 population 62,276), BENTON county, SE WASHINGTON, 160 mi/257 km SE of SEATTLE, on the COLUMBIA River (Lake Wallala reservoir) near the influx of the Snake River (5 mi/8 km downstream, E); 46°12′N 119°10′W. Railroad junction. In an irrigated farm and vineyard area. One of the Tri-Cities (a fast-growing metropolis), along with RICHLAND (9 mi/14.5 km WNW) and PASCO (2 mi/3.2 km NE); surpassed Richland c.1979 as largest of the three. Among the crops processed and packaged in the city are vegetables, potatoes, sugar beets, cherries, corn, and grapes; wheat; cattle; manufacturing (machinery, fabricated metal products,

textiles, printing and publishing, fertilizers, food, electronics). The Department of Energy's nearby HANFORD WORKS (established during World War II) and various hydroelectric dams and power plants on the Columbia River are also important to the economy. Juniper Dunes Wilderness Area to NE; McNary National Wildlife Refuge to E. Sacajawea State Park to E. Nuclear and other waste storage to N (large tract); has visitors' centers at Test Facility and Plant 2. Incorporated 1904.

Kenney, village (2000 population 374), DE WITT county, central ILLINOIS, 6 mi/9.7 km SW of CLINTON; 40°06′N 89°05′W. In agricultural area. On SALT CREEK.

Kennicott (KE-ni-kaht), village, S ALASKA, 4 mi/6.4 km NE of MCCARTHY, and 120 mi/193 km NE of CORDOVA, at foot of WRANGELL MOUNTAINS; 61°28′N 142°54′W. Site of formerly important Kennecott copper mine, discovered 1898, closed down 1938. Village now practically deserted. Summer tourism; lodge located here.

Kennington, ENGLAND: see LAMBETH.

Kennoway (KEN-o-WAI), village (2001 population 4,628), FIFE, E Scotland, 3 mi/4.8 km NW of LEVEN; 56°12′N 03°03′W.

Kénogami, Lake (kai-NAH-guh-mee) (17 mi/27 km long, 1 mi/3 km–5 km/8 mile wide), central QUEBEC, E CANADA, on CHICOUTIMI River, and 10 mi/16 km SW of CHICOUTIMI; 48°19′N 71°22′W. Water reservoir for Chicoutimi-Jonquière region.

Kenogami River (kuh-NAH-guh-mee), c.200 mi/322 km long, NW ONTARIO, E central CANADA; issues from N end of LONG LAKE, flows E and N to ALBANY RIVER at 51°06′N 84°30′W.

Keno Hill or **Keno City** (KEE-no), village (□ 21 sq mi/54.6 sq km; 2006 population 15), central YUKON, CANADA, near STEWART RIVER, 30 mi/48 km NE of Mayo Landing; 63°55′N 135°19′W. Formerly a booming silver and lead mining center; small mine behind village still operating. Named after the gambling game of chance popular with miners.

Kenong Recreation Park (KE-nong) (□ 49 sq mi/127.4 sq km) or **Kenong Rimba**, N PAHANG, NW MALAYSIA, 15 mi/24 km E of KUALA LIPIS, NE of Jelai River, and S of TAMAN NEGARA NATIONAL PARK; 04°10′N 102°15′E. Mountain streams, waterfalls, lush jungle vegetation; several limestone caves. Traversed by Kenong River access (sampan). Large variety of flora and fauna.

Kenora (kuh-NO-ruh), district (□ 157,208 sq mi/408,740.8 sq km; 2001 population 61,802), W ONTARIO, E central CANADA, on LAKE OF THE WOODS and on MANITOBA border; ☉ KENORA; 51°00′N 90°00′W.

Kenora (kuh-NO-ruh), city (□ 82 sq mi/213.2 sq km; 2001 population 15,838), ☉ KENORA district, W ONTARIO, E central CANADA; 49°48′N 94°26′W.

Kenora (kuh-NO-ruh), unincorporated town (□ 6 sq mi/15.6 sq km; 2001 population 9,742), W ONTARIO, E central CANADA, at the N end of the LAKE OF THE WOODS; 49°46′N 94°28′W. Fish-processing plants and lumber, flour, pulp, and paper mills. Kenora has an airport and serves as a base for fishing, hunting, and canoe trips.

Kenosha (ke-NO-shuh), county (□ 754 sq mi/1,960.4 sq km; 2006 population 162,001), extreme SE WISCONSIN; ☉ KENOSHA; 42°34′N 87°48′W. Bordered E by LAKE MICHIGAN, S by ILLINOIS line. Dairying and farming area in W. Agriculture (wheat, corn, soybeans, hogs, sheep); manufacturing at Kenosha. Small lakes in SW. Bong State Recreation Area in NW. Drained by DES PLAINES and FOX rivers. Formed 1850.

Kenosha (ke-NO-shuh), city (2006 population 96,240), seat of KENOSHA county, SE WISCONSIN, a port of entry on LAKE MICHIGAN, suburb for N ILLINOIS; 42°34′N 87°50′W. Manufacturing (transportation equipment, apparel, printing and publishing, herbicides and fertilizers, electronics, food, machinery, fabricated metal products). The first public school in the state was begun here in 1849. A historical and art museum and the county courthouse (containing the county historical museum) are part of the civic center. Also in the city are Carthage College, Gateway Technical College, and a library designed by Daniel Burnham. Within the vicinity is the University of Wisconsin at Parkside (to N). With RACINE to N, forms urban link between CHICAGO, Illinois, to S, and MILWAUKEE, Wisconsin, to N. Incorporated 1850.

Kenosha Mountains, Park county, central COLORADO, in FRONT RANGE, just N of TARRYALL MOUNTAINS, W of SOUTH PLATTE RIVER. KENOSHA PASS crosses hills in NW tip. North Twin Cone (12,319 ft/3,755 m), South Twin Cone (12,323 ft/3,756 m), and Mount Blaine (12,306 ft/3,751 m) are three peaks of one mountain, 4 mi/6.4 km E of Kenosha Pass.

Kenosha Pass (10,001 ft/3,048 m), Park county, central COLORADO, in Kenosha Hills, Park county, c.45 mi/72 km SW of DENVER, and 18 mi/29 km NE of FAIRPLAY. Crossed by U.S. Highway 285.

Kenova (kuh-NO-vuh), town (2006 population 3,349), WAYNE county, W WEST VIRGINIA, suburb 8 mi/12.9 km W of HUNTINGTON, West Virginia, and 6 mi/9.7 km SSE of ASHLAND, KENTUCKY, on the OHIO RIVER (OHIO state line) at mouth of BIG SANDY RIVER (Kentucky state line); 38°23′N 82°34′W. Railroad junction and trade center in bituminous-coal-mining area. Manufacturing (food, chemicals, lumber); sand and gravel pits. Tri-State Airport (Walker Long Field) to SE. Incorporated 1894.

Kenoza Lake, hamlet, SULLIVAN county, SE NEW YORK, on small Kenoza Lake, 12 mi/19 km WSW of LIBERTY; 41°44′N 74°57′W.

Ken River, c.235 mi/378 km long, in central INDIA; rises in central VINDHYA RANGE 15 mi/24 km WNW of KATNI (MURWARA) in MADHYA PRADESH state, flows generally N, through BANDA district, UTTAR PRADESH state, to YAMUNA RIVER 21 mi/34 km SW of FATEHPUR. Receives SONAR RIVER (left).

Kensal (KEN-sil), village (2006 population 151), STUTSMAN CO., central NORTH DAKOTA, 28 mi/45 km N of Jamestown; 47°17′N 98°43′W. Arrowwood National Wildlife Refuge to W. Founded in 1893 and incorporated in 1907. Named for Kensal, IRELAND.

Kensal Green, ENGLAND: see BRENT.

Kensal Rise, ENGLAND: see BRENT.

Kenscoff (KENS-kahf), town, OUEST department, S HAITI, mountain resort in the MASSIF DE LA SELLE, 6 mi/9.7 km SSE of PORT-AU-PRINCE; 18°27′N 72°17′W. Elevation c.4,400 ft/1,341 m. Coffee growing; bauxite deposits nearby.

Kensett (KEN-set), town (2000 population 1,791), WHITE county, central ARKANSAS, 4 mi/6.4 km ESE of SEARCY; 35°14′N 91°40′W. In agricultural area; manufacturing (hardwood lumber).

Kensett, town (2000 population 280), WORTH county, N IOWA, 6 mi/9.7 km S of NORTHWOOD; 43°21′N 93°12′W. Livestock, grain.

Kensico Reservoir (KEN-si-ko) (□ 3 sq mi/7.8 sq km), WESTCHESTER county, S NEW YORK, on BRONX RIVER, 15 mi/24 km N of YONKERS, and just W of New York-Connecticut border; 41°04′N 73°46′W. Max. capacity 116,560 acre-ft. Formed by Kensico Dam (168 ft/51 m high), built (1916) for water supply to NEW YORK CITY.

Kensington (KEN-zing-tuhn), town (2001 population 1,775), W central PRINCE EDWARD ISLAND, CANADA, near MALPEQUE BAY, 10 mi/16 km ENE of SUMMERSIDE; 46°26′N 63°39′W. Agricultural market in dairying, cattle-raising, and potato-growing region.

Kensington, unincorporated town (2000 population 4,936), CONTRA COSTA county, W CALIFORNIA; residential suburb 6 mi/9.7 km N of downtown OAKLAND, and 2 mi/3.2 km N of BERKELEY, 2 mi/3.2 km E of SAN FRANCISCO BAY; 37°55′N 122°17′W. Charles Lee Tilden Regional Park and San Pablo Ridge to E.

Kensington, town (2000 population 1,873), MONTGOMERY county, central MARYLAND, NNW of WASHINGTON, D.C., on ROCK CREEK; 39°02′N 77°04′W. Originally known as Knowles Station, it was renamed in the 1890s by Brainard H. Warner, a landowner who admired the Kensington district of LONDON. The town has spread out in recent years and the population of South and North Kensington are much larger than the center area. Site of very large Mormon temple.

Kensington, town, ROCKINGHAM county, SE NEW HAMPSHIRE, and 3 mi/4.8 km S of EXETER; 42°55′N 70°57′W. Drained in NW by EXETER RIVER. Light manufacturing; agriculture (nursery crops, vegetables; poultry, cattle; dairying).

Kensington, village, HARTFORD county, CONNECTICUT; postal section of BERLIN.

Kensington, village (2000 population 529), SMITH county, N KANSAS, 13 mi/21 km W of SMITH CENTER; 39°46′N 99°01′W. In corn belt; grain storage.

Kensington, village (2000 population 286), DOUGLAS county, W MINNESOTA, 18 mi/29 km WSW of ALEXANDRIA; 45°46′N 95°42′W. Dairying; poultry; alfalfa, soybeans, grain; manufacturing (feeds). Kensington Rune Stone, with ancient inscription describing journey of Swedish and Norwegian explorers, was found nearby in 1989. Small natural lakes in area, especially to NE.

Kensington, upper-income residential village (2006 population 1,185), NASSAU county, SE NEW YORK, near N shore of W LONG ISLAND, just SSE of GREAT NECK; 40°47′N 73°43′W.

Kensington, AUSTRALIA: see LEOPOLD.

Kensington and Chelsea (KEN-sing-tuhn and CHEHL-see), inner borough (□ 5 sq mi/13 sq km; 2001 population 158,919) of GREATER LONDON, SE ENGLAND; 51°27′N 00°11′W. Kensington is largely residential with fashionable shopping streets and several luxurious hotels. Portobello Road is a well-known street market. The area has undergone extensive urban renewal and contains blocks of large, tall flats. In the borough are three bridges: Battersea, Albert, and Chelsea. A large park, Kensington Gardens, adjoins HYDE PARK. The gardens originally were the grounds of Kensington Palace (Nottingham House), partially designed by Christopher Wren, which was the home of William and Mary, Queen Anne, and George I and George II. Holland House was the residence of the Fox family and, for a time, of William Penn. South Kensington is a center of colleges and museums; it is the site of the natural history section of the British Museum, the Victoria and Albert Museum, the Science Museum, the Royal College of Art, and the Imperial College of Science, among others. Albert Hall, a concert hall, is also here. Chelsea is a literary and artistic quarter. Sir Thomas More, D. G. Rossetti, James Whistler, Charles Dickens, and many others were associated with it. Thomas Carlyle's house is here. Chelsea Old Church, part of which dates from the 13th century, includes the Chapel of Sir Thomas More (1528). The church, as well as the Royal Hospital for Soldiers, also designed (1682–1692) by Wren, was badly damaged in World War II.

Kensington and Norwood (KEN-zeeng-tuhn, NOR-wud), E residential suburb of ADELAIDE, SE SOUTH AUSTRALIA; consists of two adjacent towns; 34°55′S 138°39′E.

Kensworth (KENZ-wuhth), agricultural village (2001 population 1,504), S BEDFORDSHIRE, central ENGLAND, 2 mi/3.2 km SE of DUNSTABLE; 51°51′N 00°30′W. Cement works. Has Norman church.

Kent, former county, ONTARIO, CANADA: see CHATHAM-KENT.

Kent (KENT), district municipality (□ 65 sq mi/169 sq km; 2001 population 4,926), SW BRITISH COLUMBIA, W CANADA, 79 mi/127 km E of VANCOUVER, in upper FRASER RIVER valley; 49°17′N 121°45′W. Dairy and mixed farming.

Kent, county (□ 1,734 sq mi/4,508.4 sq km; 2001 population 31,383), E NEW BRUNSWICK, CANADA, on NORTHUMBERLAND Strait and Gulf of St. Lawrence; ⊙ RICHIBUCTO.

Kent (KENT), county (□ 1,525 sq mi/3,965 sq km; 2001 population 1,621,000), SE ENGLAND; ⊙ MAIDSTONE; 51°13′N 00°39′E. It lies between the THAMES estuary and the STRAIT OF DOVER. The Isle of SHEPPEY is separated from the N coast by the narrow Swale channel. The chalky NORTH DOWNS cross the county E-W, and to the S lies the fertile Weald and Romney Marsh. The chief rivers of the county are the MEDWAY, STOUR, and DARENT. The region, largely agricultural, is a market-gardening center, "the Garden of England." Crops include fruit, grain, and hops. Sheep and cattle grazing, fishing, and dairying are also prevalent. One of London's "Home Counties," Kent is increasingly important industrially because of the encroachment of the London urban area into its W portion. Since Great Britain's entry into the European Economic Community in 1973, warehousing has emerged as a growing enterprise. Paper, pottery, brick, cement, chemicals, and beer are produced, and there is shipbuilding and oil refining.

Because of its strategic location on the path to the Continent through DOVER, Kent has been important throughout English history. Julius Caesar landed at Kent in 55 B.C.E., and Roman roads crossed the county. In 597, St. Augustine founded a Christian mission near the CANTERBURY cathedral. Kent was one of the seven Anglo-Saxon kingdoms. In the Middle Ages, many religious houses were established in the old kingdom of Kent, and Canterbury became the goal of numerous pilgrims such as Chaucer described in the *Canterbury Tales*. The region was intimately associated with the rebellions of Wat Tyler, Jack Cade, and Sir Thomas Wyatt. The other significant towns of the county are Ashford, Folkestone, Margate, Ramsgate, Tonbridge, Tunbridge Wells, and the Medway towns (Chatham, Gillingham, and Rochester).

Kent, county (□ 800 sq mi/2,080 sq km; 2006 population 147,601), central DELAWARE; 39°06′N 75°33′W. Level coastal plain, with some marshland; bounded N in part by SMYRNA RIVER, W by MARYLAND state line, S in part by MISPILLION RIVER, E by DELAWARE RIVER and DELAWARE BAY (both NEW JERSEY state line); drained by Leipsic, CHOPTANK, MURDERKILL, and ST. JONES rivers and MARSHYHOPE CREEK. Agriculture (corn, vegetables, wheat, fruit; poultry; dairying); fishing; oysters; fruit and vegetable canning; processing of dairy products; manufacturing at DOVER. Dover Air Force Base is here. Killen's Pond State Park in S center. Woodland Beach Wildlife Area, Bombay Hook National Wildlife Area, Little Creek Wildlife Area, Harvey Conservation Area, Milford Neck Wildlife Area are all in E (N to S); Blackiston Wildlife Area in NW; Norman G. Wilder Wildlife Area in SW. Formed 1683.

Kent, county (□ 414 sq mi/1,076.4 sq km; 2006 population 19,983), E MARYLAND; ⊙ CHESTERTOWN; 39°14′N 76°06′W. Peninsula on EASTERN SHORE, bounded E by DELAWARE state line, W by CHESAPEAKE BAY. Coastal plain agricultural area (vegetables, fruit, corn, wheat, livestock, dairy products); large seafood industry (esp. oysters); summer resorts; fishing, hunting. Once Kent county took in all of the Eastern Shore; now it is the smallest county in the area. Includes many old churches and houses as well as Washington College, chartered in 1783 and named for George Washington with his express permission. Formed in 1642.

Kent, county (□ 872 sq mi/2,267.2 sq km; 2006 population 599,524), SW MICHIGAN; 43°01′N 85°32′W; ⊙ GRAND RAPIDS. Intersected by GRAND RIVER and drained by FLAT, ROGUE, and THORNAPPLE rivers. Fruit growing (apples, peaches); also agriculture (wheat, oats, barley, soybeans, cucumbers, onions, corn, potatoes, beans; cattle, hogs, sheep; dairy products); manufacturing at Grand Rapids. Gypsum quarries, gravel pits. Numerous lakes in NE ¼. Has a state fish hatchery. Pando and Cannonsburg ski areas at county center; SW quadrant of county is highly urbanized by city of Grand Rapids and its suburbs. Organized 1836.

Kent, county (□ 190 sq mi/494 sq km; 2006 population 170,053), W and central RHODE ISLAND, bounded E by NARRAGANSETT BAY, W by CONNECTICUT state line; ⊙ EAST GREENWICH; 41°40′N 71°35′W. Industrial, resort, and agricultural area, producing chiefly textiles and textile machinery; also fabricated metal products, chemicals; agriculture (dairy products, poultry, corn, potatoes, fruit, mushrooms); fisheries; lumbering; many printing and publishing establishments. Many coast recreation sites. Includes state forests and parks. Drained by PAWTUXET, MOOSUP, FLAT, and Wood rivers. Incorporated 1750.

Kent, county (□ 902 sq mi/2,345.2 sq km; 2006 population 734), NW TEXAS; ⊙ CLAIREMONT; 33°11′N 100°46′W. Elevation 2,000 ft/610 m–2,800 ft/853 m. Rolling plains, with some broken areas; drained by Salt and Double Mount forks of BRAZOS RIVER, and WHITE RIVER. Cattle-ranching region; agriculture (cotton, wheat, sorghum); beekeeping. Some oil and gas; sand and gravel. Formed 1876.

Kent, industrial city (□ 9 sq mi/23.4 sq km; 2006 population 27,946), PORTAGE county, NE OHIO, 15 mi/24 km NE of AKRON; 41°08′N 81°22′W. Settled in 1805 as Franklin Mills, combined with Carthage and renamed as Kent 1863, incorporated as a city 1920. Manufacturing includes machinery and food. Main campus of Kent State University, where four young people were killed by Ohio National Guardsmen during a 1970 protest of the Vietnam War. Liquid-crystal research center sponsored by National Science Foundation and state of Ohio.

Kent, city (2006 population 83,501), KING county, W central WASHINGTON, suburb 15 mi/24 km SSE of downtown SEATTLE, and 13 mi/21 km NE of downtown TACOMA, near PUGET SOUND; 47°23′N 122°14′W. Located in a fertile agricultural area, the city has numerous food- and dairy-processing plants. Manufacturing (chemicals, fabricated metal products, electrical equipment, fixtures, paper products, transportation equipment, apparel, aircraft, plastic products, machinery, printing and publishing, furniture, food). Kent also has a large aerospace industry. The city and its population grew in the 1980s and early 1990s along with the developing Seattle metropolitan area. Puget Sound area 4 mi/6.4 km W. Incorporated 1890.

Kent, resort town, LITCHFIELD county, W CONNECTICUT, on HOUSATONIC River and NEW YORK state line, 19 mi/31 km WSW of TORRINGTON; 41°43′N 73°27′W. In hilly region; summer camps. Kent school (1906) here. Includes South Kent village, seat of South Kent school (1923); site of Bull's Bridge-one of two covered bridges in Connecticut open to vehicular traffic. Industry includes agriculture and manufacturing of machinery and electronic equipment; 3 state parks. Lake WARAMAUG is SE. Settled 1738, incorporated 1739.

Kent, town (2000 population 52), UNION county, S IOWA, near LITTLE PLATTE RIVER, 8 mi/12.9 km SW of CRESTON.

Kent, village, SIERRA LEONE, on ATLANTIC OCEAN, on Cape SHILLING of SIERRA LEONE PENINSULA, and 22 mi/35 km S of FREETOWN; 08°10′N 13°10′W. Fishing. Popular recreation site (beaches).

Kent, village (2000 population 120), WILKIN county, W MINNESOTA, 13 mi/21 km NNW of BRECKENRIDGE; 46°26′N 96°40′W. Grain, sunflowers; livestock.

Kent Acres, unincorporated town (2000 population 1,637), KENT county, central DELAWARE, 1 mi/1.6 km S

of DOVER; 39°08′N 75°31′W. Elevation 19 ft/5 m. Suburb of DOVER.

Kentau (ken-TOU), city (1999 population 55,500), S KAZAKHSTAN region, KAZAKHSTAN, in foothills of KARATAU range, 93 mi/155 km NW of SHYMKENT; 43°28′N 68°36′E. Metallurgy; manufacturing of heavy machinery, food processing; thermoelectric power.

Kent City, village (2000 population 1,061), KENT county, SW MICHIGAN, 18 mi/29 km NNW of GRAND RAPIDS; 43°13′N 85°45′W. In farm area. Fruit growing (apples, cherries, peaches); manufacturing (frozen cherries, applesauce, juice).

Kentei, MONGOLIA: see HENTIY.

Kentei, MONGOLIA: see HENTIYN NURUU.

Kentfield, unincorporated town (2000 population 6,351), MARIN county, W CALIFORNIA; residential suburb 13 mi/21 km NNW of downtown SAN FRANCISCO, 1 mi/1.6 km SW of SAN RAFAEL, on Corte Madera Creek; 37°57′N 122°33′W. College of Marin (two-year).

Kenting National Park (KEN-DING), natural area, SW TAIWAN, situated on a bay near southernmost tip of Taiwan; 21°55′N 120°51′E. White sandy beaches, tropical forests, and warm winter weather. Designated (1984) as Taiwan's first national park, consisting of the Kenting Forest Recreation Area and Sheting Natural Park. Adjacent to the beach is a livestock research station, open to the public.

Kentish Town, ENGLAND: see CAMDEN.

Kent Island, E MARYLAND, largest (c.15 mi/24 km long, 1 mi/1.6 km–6 mi/9.7 km wide) and most historic of CHESAPEAKE BAY islands, in QUEEN ANNES county, at S side of mouth of CHESTER RIVER, E of ANNAPOLIS. Separated from EASTERN SHORE mainland by narrow channel (bridged). Chester (seafood packing) and Stevensville are chief villages; others are Love Point (at N tip; lighthouse); Romancoke (S); Matapeake; Dominion; and Normans. Kent Point (S tip) has lighthouse. Island is E anchor of Chesapeake Bay Bridge. Large fishing and oystering fleet; seafood processing, tourist resort, with fishing (bluefish, rock and black bass), bathing, and duck hunting. William Claiborne established here (1631) first permanent English settlement in Maryland, claiming island for Virginia; Lord Baltimore's grant of 1632 transferred it to Maryland, and his settlers landed at St. Mary's in 1634, but conflicting claims went unsettled until 1657. Near Stevensville is Kent Fort Manor (built c. 1638–1640).

Kent, Kingdom of (KENT), one of the kingdoms of Anglo-Saxon ENGLAND. It was settled in the mid-5th century by aggressive bands of people called Jutes. Historians are in dispute over the authenticity of the traditional belief that Hengist and Horsa landed in 449 to defend the Britons against the Picts and whether Hengist and his son Aesc subsequently turned against their employer, Vortigern. The Jutes, at any rate, soon overcame the British inhabitants and established a kingdom that comprised essentially the same area as the modern county of Kent.

Æthelbert of Kent established his hegemony over England S of the HUMBER RIVER, received St. Augustine of Canterbury's first mission to England in 597, and became a Christian. During the following century, Kent was periodically subjugated and divided by WESSEX and MERCIA and finally became a Mercian province under Offa. A Kentish revolt after Offa's death in 796 was put down. Conquered by Egbert of Wessex in 825, Kent was forced to acknowledge the overlordship of Wessex and became part of that kingdom. Although it suffered heavily from Danish raids, it remained one of the most advanced areas in pre-Norman England because of the archbishopric of Canterbury and because of its steady intercourse with the European continent. The metalwork and jewelry of Kent were distinctive and beautiful.

Kentland, town (2000 population 1,822), NW INDIANA, near ILLINOIS state line, 38 mi/61 km NW of LAFAY-

ETTE; ⊙ NEWTON county; 40°46′N 87°27′W. In grain and farming area. Corn, oats, soybeans, corn; light manufacturing. Railroad junction. Settled 1860. George Ade was born here.

Kenton (KEN-tuhn), community, SW MANITOBA, W central CANADA, 31 mi/50 km WNW of BRANDON, and in WOODWORTH rural municipality; 49°59′N 100°36′W.

Kenton, county (□ 164 sq mi/426.4 sq km; 2006 population 154,911), extreme N KENTUCKY; ⊙ INDEPENDENCE (formerly shared with COVINGTON); 38°55′N 84°32′W. Bounded N by OHIO RIVER (OHIO state line), E by LICKING RIVER. Gently rolling upland area, in N part of BLUEGRASS REGION. Urbanized in N, part of CINCINNATI Metropolitan area, remainder is agricultural (cattle; poultry; corn, burley tobacco, hay, alfalfa; dairying). Manufacturing in Covington and ERLANGER. Formed 1840.

Kenton (KEN-tuhn), city (□ 5 sq mi/13 sq km; 2006 population 8,149), ⊙ HARDIN county, W central OHIO, 26 mi/42 km ESE of LIMA, on SCIOTO RIVER; 40°38′N 83°37′W. Trade center for agricultural area; manufacturing of food, machinery, and transportation equipment. Limestone quarries. Aircraft industries in area. Plotted 1833.

Kenton, town (2006 population 1,306), GIBSON and OBION counties, NW TENNESSEE, on Rutherford Fork of the OBION RIVER, and 15 mi/24 km S of UNION CITY; 36°12′N 89°01′W. In farm area. Near Gooch Wildlife Management Area.

Kenton (KENT-uhn), village (2001 population 1,131), S DEVON, SW ENGLAND, near EXE RIVER estuary, 6 mi/9.7 km SSE of EXETER; 50°38′N 03°28′W. Agricultural market. Has 15th-century church.

Kenton, village (2000 population 237), KENT county, W central DELAWARE, 9 mi/14.5 km NW of DOVER on Leipsic River. MARYLAND state line is 5 mi/8 km to W; 39°13′N 75°40′W. Elevation 49 ft/14 m.

Kenton, village, CIMARRON county, OKLAHOMA Panhandle, on high plains, the westernmost village in Oklahoma, near NEW MEXICO state line, 25 mi/40 km NW of BOISE CITY, on CIMARRON RIVER. Just NW is BLACK MESA (4,973 ft/1,516 m), highest point in Oklahoma. Black Mesa State Park and Lake Etling to SE.

Kenton-on-Sea, town, EASTERN CAPE province, SOUTH AFRICA, 27 mi/43 km SSE of GRAHAMSTOWN on INDIAN OCEAN at twin mouths of Boesmans and Kariega rivers. Serves as retirement and holiday resort. Fixed population quite small. Also acts as local farming community's service center.

Kenton Vale, village (2000 population 156), KENTON county, N KENTUCKY, residential suburb 4 mi/6.4 km S of CINCINNATI, OHIO, 2 mi/3.2 km S of COVINGTON, Kentucky; 39°02′N 84°31′W.

Kent Peninsula, NUNAVUT territory, CANADA, on DEASE STRAIT; 68°30′N 107°00′W. Extends 105 mi/169 km W from narrow isthmus into CORONATION GULF; 7 mi/11 km–29 mi/47 km wide.

Kent Point, MARYLAND: see KENT ISLAND.

Kent River (KENT), 20 mi/32 km long, CUMBRIA, NW ENGLAND; rises 9 mi/14.5 km WSW of Shap, flows S past KENDAL, to its estuary on MORECAMBE BAY.

Kents Hill, MAINE: see READFIELD.

Kentucky, state (□ 40,411 sq mi/105,068.6 sq km; 2006 population 4,206,074), SE central UNITED STATES; ⊙ FRANKFORT; 37°32′N 85°16′W. The commonwealth was admitted as the fifteenth state of the Union in 1792. LOUISVILLE and LEXINGTON are the largest and second-largest cities, respectively. Kentucky is known as the "Bluegrass State" due to the grass that grows in the state's rich soil.

Geography

The N boundary is formed by the OHIO RIVER, separating Kentucky from OHIO (NE), INDIANA (N), and ILLINOIS (NW); the W boundary is formed by the MISSISSIPPI RIVER, forming the MISSOURI state line. At the extreme SW tip of the state, Madrid Bend (□ c.5 mi/13 sq km of Kentucky territory; created by a loop in the course of the Mississippi River), protrudes N from TENNESSEE into Missouri and is entirely separate from remainder of Kentucky. Tennessee borders Kentucky on the S. In the E, the boundary with WEST VIRGINIA is formed by the BIG SANDY RIVER and its tributary, the TUG FORK; the VIRGINIA border runs NE-SW through the Cumberland Mountains, part of the AP-PALACHIAN MOUNTAIN chain. Many rapid creeks in the mountains feed the KENTUCKY, the CUMBERLAND, and the LICKING rivers, which, together with the TENNESSEE and the Ohio, are the chief rivers of the state. The Kentucky Dam on the Tennessee River and Barkley Dam on the Cumberland River (the two dams are only 3 mi/4.8 km apart) near PADUCAH, are a major part of the Tennessee Valley Authority (TVA) hydroelectric system. Land between the Lakes Recreation Area, managed by the Tennessee Valley Authority, lies in a neck of land 7 mi/11.3 km wide between KENTUCKY LAKE and LAKE BARKLEY, and extends S into Tennessee. Rough River Lake, Nolin River Lake, and Bannen River Lake reservoirs, all in the center of state, and LAKE CUMBERLAND reservoir, in S center, are important. The Central/Eastern time zone passes through the center of state.

From elevations of about 2,000 ft/610 m on the CUMBERLAND PLATEAU in the SE, where BLACK MOUNTAIN (4,139 ft/1,262 m) marks the state's highest point, Kentucky slopes to elevations of less than 800 ft/244 m along the W rim. The narrow valleys and sharp ridges of the mountain region are noted for forests of giant hardwoods and scented pine and for springtime blooms of laurel, magnolia, rhododendron, and dogwood. Unfortunately, these forests have suffered from the effects of acid rain. To the W, the plateau breaks in a series of escarpments, bordering a narrow plains region interrupted by many single conical peaks called knobs. Surrounded by the knobs region on the S, W, and E and extending as far W as Louisville is the BLUEGRASS country, the heart and trademark of the state. To the S and W lie the rolling plains and rocky hillsides of the PENNYROYAL PLATEAU, a region that takes its name from a species of mint that grows abundantly in the area. There, underground streams have washed through limestone to form miles of caverns, some of the notable ones being in MAMMOTH CAVE NATIONAL PARK, in SW central Kentucky. NW Kentucky is generally rough, rolling terrain, with scattered but important coal deposits. The isolated far-western region, bounded by the Mississippi, Ohio, and Tennessee rivers, is referred to as the Purchase, or Jackson Purchase (for Andrew Jackson, who was a prominent member of the commission that bought it from the Chickasaw in 1818). Consisting of flood plains, some less than 250 ft/76 m above sea level, and rolling uplands, it is among the largest migratory-bird routes in the U.S. Little remains of Kentucky's great forests that once spread over three-quarters of the state and were renowned for their size and density. Kentucky's climate is generally mild, with few extremes of heat and cold.

Economy

The state is noted for the distilling of Bourbon whiskey, named after BOURBON county, and for the breeding of thoroughbred racehorses. In 1990, the state, especially in the W, N, and center, was also the second-largest U.S. grower of tobacco (both dark and burley varieties), after NORTH CAROLINA. Tobacco has long been the state's chief crop, and it is also the chief farm product, followed by cattle, horses, and dairy products. Hay, corn, and soybeans are other major crops raised in the state. Kentucky's economy derives by far the greatest share of its income from industry. Lexington is one of the world's largest loose-leaf tobacco markets. Louisville, Lexington, BOWLING GREEN, COVINGTON (in CINCINNATI metro area), OWENSBORO, and HOPKINSVILLE are major industrial centers. Manufacturing includes electrical equipment, food, automobiles,

machinery, chemicals, and fabricated and primary metals. Printing and publishing, as well as tourism, have also become important industries. Kentucky is one of the major U.S. producers of bituminous coal, the state's most valuable mineral, especially in the Appalachian counties of E Kentucky. Other mineral products include stone, petroleum, and natural gas.

Tourism

Tourist attractions include the famous Kentucky Derby at Churchill Downs in Louisville and the celebrated horse farms surrounding Lexington in the heart of the Bluegrass region. The ABRAHAM LINCOLN BIRTHPLACE NATIONAL HISTORIC SITE, 45 mi/72 km S of Louisville, and CUMBERLAND GAP HISTORIC PARK (Kentucky, Tennessee, Virginia), in the SE corner, are historic landmarks. There are two national forests: Daniel Boone National Forest, composed of two large sections in the SE center, and part of Jefferson National Forest, on the boundary of Virginia, in the SE. Part of BIG SOUTH FORK NATIONAL RIVER AND RECREATION AREA, on the S boundary, extends into Kentucky from Tennessee. The U.S. Gold Depository is at FORT KNOX. Kentucky was renowned for its former family feuds, such as the notorious Hatfield-McCoy affair in the early 19th century.

History: to 1774

When the Eastern seaboard of NORTH AMERICA was being colonized in the 1600s, Kentucky was part of the inaccessible country beyond the mountains. British interest in the area quickened after Robert Cavelier, Sieur de La Salle, claimed all regions drained by the Mississippi and its tributaries for France. Dr. Thomas Walker, who explored the E mountain region in 1750 for the Loyal Land Company, led the first major expedition to the Tennessee region. Hunters and scouts, including Christopher Gist, soon followed Walker. The last conflict (1754–1763) of the French and Indian Wars, between the French and British for control of North America, and Pontiac's Rebellion, a Native American uprising (1763–1766), interrupted further exploration. With the British victorious in both conflicts, settlers soon entered Kentucky. They came in defiance of a royal proclamation of 1763, which forbade settlement W of the Appalachians. Daniel Boone, the famous American frontiersman, first came to Kentucky in 1767; he returned in 1769 and spent two years in the area.

History: 1774 to 1792

A surveying party under James Harrod established the first permanent settlement at HARRODSBURG in 1774. Boone, as agent for Richard Henderson and the Transylvania Company (a colonizing group of which Henderson was a member), blazed the WILDERNESS ROAD from Tennessee into the Kentucky region and founded BOONESBORO in 1775. Title to this land was challenged by Virginia, whose legislature voided (1778) the Transylvania Company's claims, although individual settlers were confirmed in their grants. Meanwhile, Kentucky was made (1776) a county of Virginia, and new settlers came through the CUMBERLAND GAP and over the Wilderness Road or down the Ohio River. These early pioneers of Kentucky and Tennessee were constantly in conflict with the Native Americans. The growing population of Kentuckians, feeling that Virginia had failed to give them adequate protection, worked for statehood in a series of conventions held at DANVILLE (1784–1791). Others, observing the weaknesses of the U.S. government, considered forming an independent nation. Since trade down the Mississippi and out of Spanish-held NEW ORLEANS was indispensable to Kentucky's economic development, an alliance with Spain was contemplated, and U.S. General James Wilkinson, who lived in Kentucky at the time, worked toward that end.

History: 1792 to 1800

A constitution was finally framed and accepted in 1792, and later that year the Commonwealth of Kentucky (its official designation) was admitted to the

Union, the first state W of the Appalachians. Isaac Shelby was elected the first governor, and Frankfort was chosen capital. U.S. General Anthony Wayne's victory at the battle of FALLEN TIMBERS in 1794 effectively ended Native American resistance in Kentucky. In 1795, Pinckney's Treaty between the U.S. and Spain granted Americans the right to navigate the Mississippi, a right soon completely assured by the Louisiana Purchase of 1803. Enactment by the Federal government of the Alien and Sedition Acts (1798) promptly provoked a sharp protest in Kentucky. The state grew fast as trade and shipping centers developed and river traffic down the Ohio and Mississippi increased. The War of 1812 spurred economic prosperity in Kentucky, but financial difficulties after the war threatened many with ruin. The state responded to the situation by chartering in 1818 a number of new banks that were allowed to issue their own currency. These banks soon collapsed, and the state legislature passed measures for the relief of the banks' creditors. However, the relief measures were subsequently declared unconstitutional by a state court. The legislature then repealed legislation that had established the offending court and set up a new one. The state became divided between pro-relief and anti-relief factions, and the issue also figured in the division of the state politically between followers of Tennessee's Andrew Jackson, then rising to national political prominence, and supporters of the Whig Party of Henry Clay, who was a leader in Kentucky politics for almost half a century.

History: 1800 to 1861

In the first half of the 19th century, Kentucky was primarily a state of small farms rather than large plantations and was not adaptable to extensive use of slave labor. Slavery thus declined after 1830, and, beginning in 1833, the importation of slaves into the state was forbidden for seventeen years. The legislature repealed this restriction in 1850, however, and Kentucky, where slave trading had begun to develop quietly in the 1840s, was converted into a huge slave market for the lower South. Anti-slavery agitation had begun in the state in the late 18th century within the churches, and abolitionists such as James G. Birney and Cassius M. Clay labored vigorously in Kentucky for emancipation before the Civil War. Soon Kentucky, like other border states, was torn by conflict over the slavery issue. In addition to the radical anti-slavery element and the aggressive pro-slavery faction, there was also a conciliatory group in the state. Kentucky attempted to remain neutral at the outbreak of the Civil War. Governor Beriah Magoffin refused to sanction President Lincoln's call for volunteers, but Magoffin's warnings to the Union and the Confederacy not to invade were ignored. Confederate forces invaded and occupied part of S Kentucky, including COLUMBUS and Bowling Green.

History: 1861 to 1900

The state legislature voted to oust the Confederates in September 1861. Ulysses S. Grant crossed the Ohio and took Paducah, and the state was secured for the Union. After battles in MILL SPRINGS, RICHMOND, and PERRYVILLE in 1862, there was no major fighting in the state, although the Confederate cavalryman John Hunt Morgan occasionally led raids into the state. Guerrilla warfare was constant. For Kentucky, it was truly a civil war as neighbors, friends, and even families became bitterly divided in their loyalties. Over 30,000 Kentuckians fought for the Confederacy, while about 64,000 served in the Union ranks. Many in the state opposed Federal Reconstruction policies after the war, and Kentucky refused to ratify the Thirteenth and Fourteenth amendments to the U.S. Constitution. As in the South, an overwhelming majority of Kentuckians supported the Democratic party in the period of readjustment after the war, in many ways as bitter as the war itself. Increased railroad construction aided industrial and commercial recovery after the Civil

War, but farmers were plagued by the liabilities of the one-crop (tobacco) system. After the turn of the century, the depressed price of tobacco gave rise to a feud between buyers and growers, resulting in the Black Patch War. Night riders terrorized buyers and growers in an effort to stage an effective boycott against monopolistic practices of buyers.

History: 1900 to Present

For more than a year general lawlessness prevailed until the state militia forced an agreement in 1908. Coal mining, which began on a large scale in the 1870s, was well established in mountainous E Kentucky by the early 20th century. The mines boomed during World War I, but after the war, when demand for coal lessened and production fell off, intense labor troubles developed. The attempt of the United Mine Workers of America (UMW) to organize the coal industry in HARLAN county in the 1930s resulted in outbreaks of violence, drawing national attention to "bloody" Harlan, and in 1937 a U.S. Senate subcommittee began an investigation into allegations that workers' civil rights were being violated. Further violence ensued, and it was not until 1939 that the UMW was finally recognized as a bargaining agent for most of the state's miners. Labor disputes and strikes have persisted in the state; some are still accompanied by violence. Improvements of the state's highways were made after World War I, and a much-needed reorganization of the state government was carried out in the 1920s and 1930s. Since World War II, construction of turnpikes (toll highways), extensive development of state parks (especially state resort parks with complete tourist facilities), and a marked rise in tourism have all contributed to the development of the state. Kentucky benefited from the energy crisis of the 1970s, when the state's large coal supply was in great demand. A mixture of strip, open-pit, and subsurface mining continues. Coal continues to be the primary (95%) type of fuel used in power plants in Kentucky and surrounding states despite hydroelectricity projects and tighter environmental controls. However, a steady decline in manufacturing has not been offset by a shift to service industries, as is the case in other states. In consequence, Kentucky has grown very little in the past few decades. Institutions of higher learning include the University of Kentucky and Transylvania University, at Lexington; the University of Louisville, at Louisville; Eastern Kentucky University, at Richmond; Murray State University, at MURRAY; Western Kentucky University, at Bowling Green; Kentucky Wesleyan College, at Owensboro; Union College, at BARBOURVILLE, and Kentucky State University, at Frankfort.

Government

Kentucky's state constitution was adopted in 1891. The governor of the state is elected to a four-year term. Ernie Fletcher is the current governor. The general assembly, or legislature, is bicameral with a senate of thirty-eight members and a house of representatives of 100 members. State senators are elected to single four-year terms, and representatives for two-year terms. Kentucky is represented in the U.S. Congress by six Representatives and two Senators, and thus has eight electoral votes in presidential elections.

Kentucky has 120 counties: ADAIR, ALLEN, ANDERSON, BALLARD, BARREN, BATH, BELL, BOONE, BOURBON, BOYD, BOYLE, BRACKEN, BREATHITT, BRECKINRIDGE, BULLITT, BUTLER, CALDWELL, CALLOWAY, CAMPBELL, CARLISLE, CARROLL, CARTER, CASEY, CHRISTIAN, CLARK, CLAY, CLINTON, CRITTENDEN, CUMBERLAND, DAVIESS, EDMONSON, ELLIOTT, ESTILL, FAYETTE, FLEMING, FLOYD, FRANKLIN, FULTON, GALLATIN, GARRARD, GRANT, GRAVES, GRAYSON, GREEN, GREENUP, HANCOCK, HARDIN, HARLAN, HARRISON, HART, HENDERSON, HENRY, HICKMAN, HOPKINS, JACKSON, JEFFERSON, JESSAMINE, JOHNSON, KENTON, KNOTT, KNOX, LARUE, LAUREL, LAWRENCE, LEE, LESLIE, LETCHER, LEWIS, LINCOLN, LIVINGSTON, LOGAN, LYON, MADISON, MAGOFFIN, MARION, MARSHALL, MARTIN, MASON, MCCRACKEN, MCCREARY, MCLEAN, MEADE, MENIFEE, MERCER, METCALFE, MONROE, MONTGOMERY, MORGAN, MUHLENBERG, NELSON, NICHOLAS, OHIO, OLDHAM, OWEN, OWSLEY, PENDLETON, PERRY, PIKE, POWELL, PULASKI, ROBERTSON, ROCKCASTLE, ROWAN, RUSSELL, SCOTT, SHELBY, SIMPSON, SPENCER, TAYLOR, TODD, TRIGG, TRIMBLE, UNION, WARREN, WASHINGTON, WAYNE, WEBSTER, WHITLEY, WOLFE, and WOODFORD.

Kentucky, river, 259 mi/417 km long, formed by the junction of the North and South forks, at BEATTYVILLE, LEE county, central KENTUCKY, and flows NW past IRVINE, BOONESBORO, FRANKFORT to OHIO RIVER at CARROLLTON. The river is navigable by small craft for its entire length by means of locks. The Kentucky's upper course flows through a coal-mining region of the CUMBERLAND MOUNTAINS. The lower course flows through the heart of Kentucky's BLUEGRASS REGION. North Fork Kentucky River, c.125 mi/201 km long, rises in E LETCHER county, SE Kentucky, near VIRGINIA state line. Flows generally NW past HAZARD and JACKSON; receives Middle Fork 1 mi/1.6 km E of confluence with South Fork. South Fork Kentucky River, c.30 mi/48 km long, formed by joining of Red Bird and Goose creeks in N CLAY county, SE Kentucky, flows N to join North Fork. Middle Fork Kentucky River, c.85 mi/137 km rises in NE HARLAN county, SE Kentucky Flows generally NNW past Hayden and through Buckhorn Lake reservoir to North Fork.

Kentucky Lake, reservoir, SW KENTUCKY and NW TENNESSEE, on TENNESSEE RIVER, 8 mi/12.9 km W and parallel to LAKE BARKLEY (linked by channel), and 19 mi/31 km ESE of PADUCAH; c.185 mi/298 km long, maximum 3 mi/4.8 km wide; 37°00′N 88°16′W. Maximum capacity 6,002,600 acre-ft. BIG SANDY RIVER enters from SSW, forms 14-mi/23-km arm; DUCK RIVER enters from E 14 mi/23 km SW of WAVERLY, Tennessee. Formed by Kentucky Lock and Dam (206 ft/63 m high, 8,422 ft/2,567 m long), built (1944) as part of Tennessee Valley Authority system for flood control, navigation, and power generation. Pickwick Dam (Tennessee) at S tip. Surrounded by state parks and recreation areas, including Nathan B. Forrest State Historic Park (Tennessee). Partly in units of Tennessee National Wildlife Refuge.

Kentucky Natural Bridge, KENTUCKY: see NATURAL BRIDGE STATE RESORT PARK.

Kentung, MYANMAR: see KENGTUNG.

Kentville (KENT-vil), town (2001 population 5,610), cap. Kings county, W NOVA SCOTIA, CANADA, on the CORNWALLIS RIVER, NW of HALIFAX; 45°04′N 64°30′W. Elevation 173 ft/52 m. Horton Corner was the original English name. In 1826, it was renamed Kentville in honor of Prince Edward, Duke of Kent, who visited in 1794. It is a tourist and trade center in the Annapolis valley, a fruit-growing region.

Kentwood, city (2000 population 45,255), KENT county, SW MICHIGAN, suburb 6 mi/9.7 km SE of GRAND RAPIDS; 42°52′N 85°35′W. Manufacturing (plastics, transportation equipment, furniture). Drained by Plaster River. Kent County International Airport to SE.

Kentwood, town (2000 population 2,205), TANGIPAHOA parish, SE LOUISIANA, 42 mi/68 km N of PONCHATOULA, on TANGIPAHOA RIVER, near MISSISSIPPI line; 30°56′N 90°31′W. Strawberries, vegetables; catfish; lumber milling; manufacturing (feeds, apparel, food, bricks). Settled in mid-nineteenth century; incorporated 1893.

Kenty, POLAND: see KETY.

Kenvil, village, MORRIS county, N NEW JERSEY, 9 mi/14.5 km NW of MORRISTOWN; 40°52′N 74°37′W. Manufacturing. Large powder plant here (founded 1871) nearly destroyed in 1940 by explosions that took 48 lives.

Kenville (KEN-vil), community, SW MANITOBA, W central CANADA, 8 mi/12 km SSW of town of SWAN RIVER, in SWAN RIVER rural municipality; 52°00′N 101°19′W.

Kenvir (KEN-vuhr), unincorporated town, HARLAN county, SE KENTUCKY, in the CUMBERLAND MOUNTAINS, 9 mi/14.5 km E of HARLAN. Bituminous coal; timber.

Kenwood, unincorporated town, SONOMA county, W CALIFORNIA, 10 mi/16 km NW of SONOMA, on Sonoma Creek. Los Guilicos Warm Springs to S; Annadel State Park to W; Sugarloaf Ridge State Park to NE. Grapes, apples, grain; dairying. Wineries.

Kenwood (KEN-wuhd), unincorporated village (□ 2 sq mi/5.2 sq km; 2000 population 7,423), HAMILTON county, extreme SW OHIO; NE suburb of CINCINNATI, off Interstate Highway 71; 39°12′N 84°22′W.

Kenwood, suburban village, BALTIMORE county, central MARYLAND, 5 mi/8 km WSW of downtown BALTIMORE, near CATONSVILLE. Considered a part of BETHESDA, it is noted for the flowering cherry trees on its curving streets in springtime.

Kenwyn (KEN-win), village (2001 population 4,944), W CORNWALL, SW ENGLAND, just NW of TRURO; 50°13′N 05°16′W. Site of bishop of Truro's palace. Has 15th-century church.

Kenya, republic (□ 224,960 sq mi/582,646 sq km; 2004 estimated population 32,021,856; 2007 estimated population 36,913,721), E AFRICA, ⊙ NAIROBI; other cities include MOMBASA (the chief port), NAKURU, KISUMU, and ELDORET.

Geography

Kenya is divided into seven provinces—CENTRAL, COAST, EASTERN, NORTH EASTERN, NYANZA, RIFT VALLEY, and WESTERN—and the Nairobi Area. The country is bordered by SOMALIA on the E, the INDIAN OCEAN on the SE, TANZANIA on the S, Lake VICTORIA (Victoria Nyanza) on the SW, UGANDA on the W, the SUDAN on the NW, and ETHIOPIA on the N. The country, which lies astride the equator, is made up of several geographical regions. The first is a narrow, coastal strip that is low-lying except for the TAITA HILLS in the S. The second, an inland region of bush-covered plains, constitutes most of the country's land area. In the NW, straddling Lake TURKANA (Rudolf) and the KULAL Mountains, are high-lying scrublands. In the SW are the fertile grasslands, and forests of the Kenya highlands. In the W is the GREAT RIFT VALLEY, an irregular depression that cuts through W Kenya from N to S in two branches. It is also the location of some of the country's highest mountains, including Mount KENYA (17,058 ft/5,199 m). Kenya's main rivers are the TANA and the ATHI.

Population

People of African descent make up about 99% of the population; they are divided into forty-two ethnic groups, of which the Bantu-speaking Kikuyu, Kamba, Gusii, and Luhya and the Nilotic-speaking Luo are predominant. Small numbers of persons of Indian, Pakistani, Goan, and European descent live in the interior, and there are some Arabs along the coast. The national language of Kenya is Swahili, but English is the official language which is spoken widely in commercial affairs. Most of the population is Christian, about one-fifth follows traditional religious beliefs, and the remainder are Muslim or Hindu.

Economy

The great majority of Kenyans are engaged in farming, largely of the subsistence type, but industry is growing. Coffee, tea, sisal, pyrethrum, maize, and wheat are grown in the highlands, mainly on small African-owned farms formed by dividing some of the large, formerly European-owned estates. Coconuts, cashew nuts, cotton, sugarcane, sisal, and maize are grown in the lower-lying areas. Much of the country remains savanna, where large numbers of cattle are pastured. Kenya's manufacture includes chemicals, processed food, cement, textiles, leather goods, and fabricated metal products. Industrial development has been hampered by poor infrastructure, corruption, limits on imports, and shortages in hydroelectric power. The chief minerals produced are limestone, soda ash, gold, and salt. Kenya attracts many tourists, largely lured by its coastal beaches and varied wildlife, which are protected in a number of national parks. Kenya's chief exports are coffee and tea; fluctuations in their world prices have tremendous economic impact. The leading imports are petroleum and petroleum products, chemicals, and machinery; trade is mainly with GREAT BRITAIN, Uganda, and the U.S. The nation's population growth continually exceeds the rate of its economic growth, resulting in large budget deficits and high unemployment. A large portion of the industrial sector is foreign owned. Kenya has a well-developed transportation system and in the 1980s the railroad lines were further improved. Kenya is a member of the Commonwealth and is an ACP (African, Caribbean, and Pacific) member of the EEC. In 1967, Kenya formed the East African Development Bank with Tanzania and Uganda. There are a number of universities including the University of Nairobi, Kenyatta University, Egerton University, and Moi University.

History: 1887
Some of the earliest remains of hominids who lived c. two million years ago have been discovered in Kenya. The existence of farming and domestic herds can be dated to c.1000 B.C.E. Trade between the Kenya coast and ARABIA was brisk by C.E. 100. Arabs settled on the coast during medieval times, and they soon established several autonomous city-states (including Mombasa, MALINDI, and Pate). Farmers and herders traveled S from Ethiopia and settled in Kenya in c.2000 B.C.; there is also evidence that Bantu-speaking people and Nilotic speakers from the S Sudan settled in Kenya between 500 B.C.E. and 500 C.E. The Portuguese first visited the Kenya coast in 1498, and by the end of the 16th century they controlled much of it. However, in 1729, the Portuguese were permanently expelled from Mombasa and were replaced as the leading power on the coast by Arab dynasties. From the early 19th century there was long-distance caravan trading between Mombasa and Lake Victoria. Beginning in the mid-19th century, European explorers (especially John Ludwig Krapf and Joseph Thomson) mapped parts of the interior. The British and German governments agreed upon spheres of influence in East Africa in 1886, with most of present-day Kenya passing to the British.

History: 1887 to 1920
In 1887, a British association received concessionary rights to the Kenya coast from the sultan of ZANZIBAR. The association in 1888 was given a royal charter as the Imperial British East African Company, but severe financial difficulties soon led to its takeover by the British government, which established the East African Protectorate (or British East Africa) in 1895. A railroad was built (1895–1901) from Mombasa to Kisumu on Lake Victoria in order to facilitate trade with the interior and with Uganda. In 1903, the first settlers of European descent established themselves as large-scale farmers in the highlands by taking land from the Kikuyu, Maasai, and others. At the same time, Indian merchants moved inland from the coast.

History: 1920 to 1963
In 1920, the territory was renamed and its administration changed; the interior became Kenya Colony and a coastal strip (10 mi/16 km wide) was constituted the Protectorate of Kenya. From the 1920s to the 1940s, European settlers controlled the government and owned extensive farmlands; Indians maintained small trade establishments and were lower-level government employees; and Africans grew cash crops such as coffee and cotton on a small scale, were subsistence farmers, or were laborers in the towns (especially Nairobi). In the 1920s, Africans began to protest their inferior status. Protest reached a peak

between 1952 and 1956 with the so-called Mau Mau Emergency, a complex armed revolt led by the Kikuyu, which was in part a rebellion against British rule and in part an attempt to reestablish traditional land rights and ways of governance. The British declared a state of emergency and imprisoned many of the colony's nationalist leaders, including Jomo Kenyatta. After the revolt, Britain increased African representation in the colony's legislative council until, in 1961, there was an African majority.

History: 1963 to 1980
On December 12, 1963, Kenya (including both the colony and the protectorate) became independent. In 1964 the country became a republic, with Kenyatta as president. The first decade of independence was characterized by disputes among ethnic groups (especially between the Kikuyu and the Luo), by economic growth and diversification, and by the end of European predominance. Boundary disputes with the Somali Democratic Republic resulted in sporadic fighting (1963–1968). In 1969, Tom Mboya, a leading government official and a possible successor to Kenyatta, was assassinated. More than 70% of the country was affected by the sub-Saharan drought of the early 1970s. Kenyatta's silencing of opponents led to further unrest domestically. Throughout the 1970s relations with neighboring countries deteriorated as well; there was a territorial dispute with Uganda, and Tanzania closed its border with Kenya after Kenya harbored several of Idi Amin's supporters after the fall of his regime. After Kenyatta's death in 1978, Vice President Daniel arap Moi succeeded him as president. Moi promoted the Africanization of industry by placing limits on foreign ownership and by extending credit to African investors. Domestically, he rejected demands for democratization and suppressed opposition. With economic conditions worsening, rumors of a coup led to rioting by students. Moi dismantled the air force and ordered the imprisonment of those suspected of involvement.

History: 1980 to Present
Throughout the 1980s, Moi consolidated power in the presidency and continued to conduct periodic purges of his administration. The worst rioting in the nation's independent history erupted in 1988 after several outspoken proponents of a multiparty democracy were arrested. Moi was reelected in the 1992 election. In 1992, Kenya became a multiparty democracy. Since 1993 liberalization of the economy has been continuing. Moi stepped down in December 2002 following a fair and peaceful election; National Rainbow Coalition (NARC) candidate Mwai Kibaki won the presidency and control of the National Assembly. In July 2005 parliament approved a draft constitution that largely preserved presidential powers, but voters rejected it in a November 2005 referendum.

Government
Kenya is governed under the constitution of 1963 as amended. The president, who is both head of state and head of government, is popularly elected for a five-year term and is eligible for a second term. The unicameral legislature consists of the 224-seat National Assembly or Bunge. There are 210 members who are popularly elected to serve five-year terms, 12 who are appointed by the president, and two ex-officio members. The current head of state is President Mwai Kibaki.

Kenya, Mount, dormant volcano, central KENYA, just S of the equator. Its highest peak, Batian, reaches 17,058 ft/5,199 m, making Mount Kenya the highest mountain in AFRICA after Mount KILIMANJARO. In the heart of Kikuyu country, Mount Kenya was a focal point during the Mau Mau disturbances (1952–1956). The Kikuyu, Meru, and Embu people cultivate Mount Kenya's fertile lower slopes. From 5,000 ft/1,524 m to 15,000 ft/4,572 m are dense woodlands inhabited by elephants, buffalo, and leopards. Snow-capped Mount Kenya has several

glaciers in its uppermost regions. The MOUNT KENYA NATIONAL PARK is located here and Kirin Yaga attracts many mountain climbers from around the world. Occasionally called Kirinyaga or Kirin Yaga.

Kenyasi Mohu (KEN-yah-see MO-hoo), town, local council headquarters, GHANA, 23 mi/37 km S of SUNYANI; ☉ BRONG-AHAFO region; 07°04′N 02°20′W. Located in cocoa- and timber-producing area.

Kenyon (KEN-yuhn), town (2000 population 1,661), GOODHUE county, SE MINNESOTA, on North Fork ZUMBRO RIVER, and 15 mi/24 km E of FARIBAULT; 44°16′N 92°59′W. Grain, soybeans; livestock, poultry; dairying; light manufacturing. Nestrand Big Woods State Park to NW. Settled 1856, incorporated 1885.

Kenyon, village, WASHINGTON county, RHODE ISLAND: see CHARLESTOWN.

Kenzhe (kyen-ZHE), village (2005 population 10,750), central KABARDINO-BALKAR REPUBLIC, S European Russia, in the foothills of the N central CAUCASUS Mountains, 3 mi/5 km W, and under administrative jurisdiction, of NAL'CHIK; 43°30′N 43°33′E. Elevation 1,689 ft/514 m. Agricultural products. Has a mosque.

Kenzingen (KEN-tsing-uhn), town, BADEN-WÜRTTEMBERG, SW GERMANY, at W foot of BLACK FOREST, on the ELZ, 11 mi/18 km SW of LAHR; 48°12′N 07°46′E. Wine. Has 16th-century town hall (formerly Capuchin monastery).

Keo (KEE-o), village (2000 population 235), LONOKE county, central ARKANSAS, 18 mi/29 km ESE of LITTLE ROCK; 34°36′N 92°00′W. Toltoc Mounds State Park to NW.

Keokuk, county (□ 579 sq mi/1,505.4 sq km; 2006 population 11,081), SE IOWA; ☉ SIGOURNEY; 41°20′N 92°10′W. Prairie agricultural area (hogs, cattle, sheep, poultry; corn, oats) drained by SKUNK, NORTH and South Skunk, and South Fork ENGLISH rivers. Limestone quarries, clay pits. Widespread river flooding occurred here in 1993. Formed 1837.

Keokuk, city (2000 population 11,427), seat of LEE county, extreme SE IOWA, on the MISSISSIPPI RIVER at the foot of the DES MOINES RIVER rapids and in a farm area; 40°24′N 91°24′W. Its industries focus on food processing and packaging (turkeys, dairy items, wheat and corn products); manufacturing (fabricated metal products). The city was named for Keokuk, a Sac tribal chief who ceded lands to settlers and who is buried beneath an impressive statue in Rand Park. Because of its location at the foot of the treacherous Des Moines River rapids, Keokuk was a natural rest stop for boats ascending the Mississippi. During the Civil War 5 army hospitals there served the wounded; those who did not survive were buried in the city's national cemetery, where the Unknown Soldier Monument was erected. In 1877 a ship canal (9 mi/14.5 km long) was completed around the rapids; in 1910–1913 the river was dammed, Lock and Dam No. 19, creating Lake Keokuk, and the dam along with a Mississippi River plant furnish hydroelectric power. Mark Twain worked as a printer in Keokuk; mementos of his stay are preserved. Also located here are Southeastern Community College and a Riverboat Museum. Serious flood damage occurred here and in adjacent parts of MISSOURI and ILLINOIS in 1993. Incorporated 1847.

Keomah Village, village (2000 population 97), MAHASKA county, S central IOWA, 5 mi/8 km ESE of OSKALOOSA, at E end of Lake Keomah; 41°17′N 92°32′W. Lake Keomah State Park nearby. Corn; cattle, hogs.

Keon Park, suburb 9 mi/14 km N of MELBOURNE, VICTORIA, SE AUSTRALIA, between Reservoir and Thomastown; 37°43′S 144°50′E. Railway station.

Keonthal, former princely state of PUNJAB HILL STATES, INDIA. Since 1948, merged with HIMACHAL PRADESH state.

Keos, Greece: see KÉA.

Keosauqua, town (2000 population 1,066), SE IOWA, on DES MOINES RIVER, and 30 mi/48 km SE of OT-

TUMWA; ⊙ VAN BUREN county; 40°43′N 91°57′W. Light manufacturing. Lacey Keosauqua State Park to S; W unit of Shimek State Forest to SW. The DES MOINES RIVER flooded this area in 1993. Settled 1836, incorporated 1851.

Keota, town (2000 population 1,025), KEOKUK county, SE IOWA, 14 mi/23 km WNW of WASHINGTON; 41°22′N 91°57′W. Shipping point for livestock. Manufacturing (machinery, food). Settled 1871, incorporated 1873.

Keota (kee-O-tuh), village, WELD county, N COLORADO, 38 mi/61 km NE of GREELEY; 40°42′N 104°04′W. Elevation 4,961 ft/1,512 m. Surrounded by units of Pawnee National Grassland. Near Wild Horse Creek.

Keota, village (2006 population 529), HASKELL county, E OKLAHOMA, 21 mi/34 km NW of POTEAU, on Sansbois Arm (Sansbois Creek) of ROBERT S. KERR reservoir; 35°15′N 94°55′W. In agricultural area; light manufacturing.

Keowee, Lake (KEE-uh-wee), reservoir, OCONEE county and on PICKENS county border, NW SOUTH CAROLINA, on SENECA (Keowee) and LITTLE rivers, 28 mi/45 km W of GREENVILLE. Formed by two dams, the KEOWEE Dam (70 ft/21 m high; built 1970; 34°48′N 82°53′W) on the Seneca (Keowee) River, and the LITTLE RIVER DAM (135 ft high; built 1969; 34°44′N 82°54′W), on the LITTLE RIVER; channel just W of Keowee Dam joins reservoirs. Max. capacity 500,000 acre-ft. Little River section 10 mi/16 km long, extends NW, with 4-mi/6.4-km SW arm; Keowee section 12 mi/19 km long, extends N to JOCASSEE Dam. Both built by the Duke Power Company for power generation.

Keowee River, NW SOUTH CAROLINA; rises in BLUE RIDGE MOUNTAINS of SW NORTH CAROLINA; flows S where it joins SAVANNAH River WSW of ANDERSON, South Carolina; two hydroelectric facilities created when KEOWEE Dam was completed forming Lake KEOWEE in 1971 and JOCASSEE Dam completed in 1973 creating Lake JOCASSEE.

Keowee River, SOUTH CAROLINA: see SENECA RIVER.

Kep (KEP), municipality and town (□ 129 sq mi/335.4 sq km; 2007 population 40,280), SW CAMBODIA, small deep-water port on GULF OF THAILAND, 12 mi/19 km SE of KAMPOT, near VIETNAM border, located in—but politically independent of—KAMPOT province; 10°29′N 104°19′E. Fisheries. Founded 1908 and once a popular seaside resort patronized by the French and Cambodia's Westernized elite, Kep is today almost a ghost town. Its scores of villas were systematically destroyed by the Khmer Rouge and the underground petroleum-storage tank of one gas station became a mass grave for slaughtered townspeople. Although plans have been drawn to rebuild the hotels and recreational facilities that once made Kep a prosperous center of tourism, the development process presently lacks both capital and a potential clientele of wealthy foreigners or Cambodians. Also spelled Kêb; also called Krong Kep and Krong Kaeb.

Kepa, AUSTRIA and SLOVENIA: see MITTAGSKOGEL.

Kephallenia, Greece: see KEFALLINÍA.

Kephalos, Greece: see KEFALOS.

Kephisia, Greece: see KIFISSIA.

Keping, China: see KALPIN.

Kepno (KENP-no), Polish Kępno, German Kempen, town, Kalisz province, SW central POLAND, 40 mi/64 km ENE of WROCŁAW (Breslau). Railroad junction; manufacturing of cement, bricks, machinery, chemicals; sawmilling.

Keppel (KE-puhl), former township (□ 151 sq mi/392.6 sq km; 2001 population 4,111), S ONTARIO, E central CANADA, 9 mi/15 km from OWEN SOUND; 44°40′N 81°02′W. Amalgamated into GEORGIAN BLUFFS township in 2001.

Keppel Bay (KE-puhl), inlet of PACIFIC OCEAN, E QUEENSLAND, AUSTRALIA; sheltered E by CURTIS ISLAND; 30 mi/48 km long, 12 mi/19 km wide; 23°21′S

150°55′E. PORT ALMA (part of port of ROCKHAMPTON) is on SW shore, at mouth of FITZROY RIVER.

Keppel Harbour, channel, 3 mi/4.8 km long, 1 mi/1.6 km wide, SINGAPORE, arm of Straits of SINGAPORE, between Singapore island (N) and SENTOSA (Blakang Mati) island (S); 01°15′N 103°49′E. Divided into 2 channels in E with main channel N, Selat Sengkir channel S, Pulau Brani (see BRANI, PULAU) island between. Port facilities for Singapore, 1 mi/1.6 km SW of city center. Causeway and cable-car line cross channel in W to Sentosa resort; World Trade Centre on N; ferries to outer islands depart from the Centre.

Keppel Island, TONGA: see NIUATOPUTAPU.

Keqiao (KUH-CHI-OU), town, N ZHEJIANG province, SE CHINA, 25 mi/40 km SE of HANGZHOU, and on railroad to NINGBO. In rice-growing region.

Kerakat, INDIA: see KIRAKAT.

Kerala (KAI-ruh-luh), state (□ 15,005 sq mi/39,013 sq km; 2001 population 31,841,374), S INDIA, on the ARABIAN SEA; ⊙ THIRUVANANTHAPURAM. Bordered N and NE by KARNATAKA state, E and SE by TAMIL NADU state. The capital area has many lagoons and channels. A wet tropical climate and coastal lowlands support cultivation of rice, coconuts, tapioca, and spices; the interior hills produce rubber, coffee, and tea and have deposits of lignite, tin, and mica. Densely populated; created in 1956 from the Malayalam-speaking former princely states of COCHIN and TRAVANCORE and Malayalam-speaking areas formerly in MADRAS STATES (now TAMIL NADU state). About 60% of the population is Hindu; Christians and Muslims each make up about 20% of the remaining inhabitants. Although infant mortality is low and the literacy rate is highest (among women, also) in India, Kerala suffers from economic underdevelopment and unemployment. It has recently built up a vigorous trade connection with Western Asia. In 1957, India's first Communist state administration was elected here, and a Communist coalition was again elected in 1967 and 1970. Maoist Naxalite groups were active here. Takes its name from the ancient Tamil kingdom of Kerala (Chera), which traded with the Phoenicians, Greeks, and Romans. The state is governed by a chief minister responsible to an elected unicameral legislature and by a governor appointed by the president of India. Depending on point of view, is considered to be either a socialist or a progressive state; is often cited as a model among Indian states. Communist Party very active in national government here.

Kerama-retto (ke-RAH-mah–RET-to), island group (□ 18 sq mi/46.8 sq km) of OKINAWA ISLANDS, in RYUKYU ISLANDS in EAST CHINA SEA, Shimajiri county, Okinawa prefecture, SW JAPAN, 15 mi/24 km W of OKINAWA. Comprises volcanic islands of TOKASHIKI-SHIMA (largest), ZAMAMI-SHIMA, and several small coral islets. Mountainous, forested. Agriculture (sugarcane, sweet potatoes). Fishing.

Keramoti (ke-rah-mo-TEE), town, KAVÁLLA prefecture, EAST MACEDONIA and THRACE department, NE GREECE, on N AEGEAN SEA, 22 mi/35 km E of KAVÁLLA; 40°52′N 24°42′E. Center of a cultured-oyster industry. Terminal for ferry to THÁSOS, 3 mi/4.8 km offshore.

Kéran, prefecture (2005 population 73,229), KARA region, N TOGO ⊙ KANDÉ; 10°05′N 01°00′E.

Kerang (kuh-RANG), town, N VICTORIA, SE AUSTRALIA, on LODDON RIVER, and 150 mi/241 km NNW of MELBOURNE; 35°44′S 143°55′E. Railroad and commercial center for livestock, agricultural area; irrigation; dairy plant; flour mill. Wetlands, tourism. Ibis rookery.

Kerani (ke-RAH-nee), canton, LOS ANDES province, LA PAZ department, W BOLIVIA, E of LA PAZ; 16°27′S 68°29′W. Elevation 12,625 ft/3,848 m. Copper-bearing lode; clay, limestone, gypsum deposits. Agriculture (potatoes, yucca, bananas, rye); cattle.

Kerasun, TURKEY: see GIRESUN.

Keratea (ke-rah-TAI-ah), town, ATTICA prefecture, ATTICA department, E central GREECE, on railroad, and 19 mi/31 km SE of ATHENS; 37°48′N 23°59′E. Olive oil, wine; summer resort.

Kerava (KE-rah-vah), Swedish Kervo, town, UUDENMAAN province, S FINLAND; 60°24′N 25°07′E. Elevation 99 ft/30 m. Commuter suburb located 16 mi/26 km NNE of HELSINKI. Railroad junction; light industry.

Keraya (ke-RAH-yah), canton, LOAYZA province, LA PAZ department, W BOLIVIA; 17°14′S 67°44′W. Elevation 8,333 ft/2,540 m. Gas resources in area. Tin mining at Mina Viloco, 40 mi/70 km SE of LA PAZ; clay, limestone, and gypsum deposits. Agriculture (potatoes, yucca, bananas, rye, oats); cattle.

Kerbela (kahr-BUH-luh), province (□ 2,275 sq mi/5,915 sq km), W central IRAQ, extending from the EUPHRATES valley WSW into the SYRIAN DESERT (the desert area, which extends to the SAUDI ARABIA border, is not included in the province area); ⊙ KARBALA; 32°15′N 43°07′E. Served by a branch of the BASRA-BAGHDAD railroad. In the E there is date cultivation. In the province are Muslim shrines, at Karbala.

Kerben, village, S central JALAL-ABAD region, KYRGYZSTAN, on UZBEKISTAN border, in SE foothills of CHATKAL RANGE, 28 mi/45 km W of TASH-KÖMÜR, and 70 mi/113 km NW of JALAL-ABAD; 41°19′N 71°36′E. Tertiary-level administrative center. Wheat and dairy products. Also KARAVAN.

Kerbi, RUSSIA: see POLINY OSIPENKO, IMENI.

Kerbi River, RUSSIA: see AMGUN′ RIVER.

Kerch (KERCH), city (2001 population 157,007), the Republic of CRIMEA, UKRAINE, on the KERCH STRAIT of the BLACK SEA, and at the E end of the KERCH PENINSULA, a strip of land between the Sea of AZOV and the Black Sea; 45°21′N 36°28′E. A seaport and major industrial center, it has iron and steel mills, shipyards, fisheries, and canneries. Other industries include manufacturing (glass, building materials, apparel), food processing. Iron ore and vanadium are extracted nearby. The city was founded as Panticapaeum (6th century B.C.E.) by Greek colonists from Miletus and was the forerunner of all Milesian cities in the area. It was a large trade center and a terraced mountanus city with self-government. It became (5th century B.C.–4th century A.D.) the capital of the European part of the Kingdom of Bosporus (see CRIMEA). It was conquered (around 110 B.C.E.) by Mithridates VI of Pontus, then passed under Roman and Byzantine rule, and was devastated by the Huns in 375 C.E. In the 10th century, it became a Slavic settlement, called Korchev, which belonged to the Tmutorokan principality and a center of trade between Kievan Rus′, Crimea, Caucasia, and the Orient. Later (13th century), it became a Genoese trade center called Cherkio and was conquered (1475) by the Crimean Tatars, who called it Cherzeti. It was captured (1771) by the Russians in the first Russo-Turkish War (1768–1774), and the Treaty of Kuchuk Kainarji (1774) formally gave it to Russia. Under Russia, Kerch was a military port and after 1820 became a commercial port. There are ruins of the ancient acropolis on top of the steep hill of Mithridates. Archaeological remains, discovered in catacombs and burial mounds near the city, are in the archaeological museum (founded in 1826), which is famous for its Greco-Scythian antiquities. The Church of St. John the Baptist dates from the 8th century. The city has a marine fishery and oceanographic research institute.

Kerchens′ka protoka, between UKRAINE and RUSSIA: see KERCH STRAIT.

Kerchenskiy proliv, between UKRAINE and RUSSIA: see KERCH STRAIT.

Kerchevskiy (KYER-chyef-skeeye), town (2006 population 2,965), N central PERM oblast, W URALS, E European Russia, on the right bank of the KAMA River, on road, 30 mi/48 km SSW of CHERDYN′; 59°56′N 56°17′E. Elevation 413 ft/125 m. Lumbering.

Kerch Peninsula (KERCH), arid E section of CRIMEA, UKRAINE, between the Sea of AZOV (N) and BLACK SEA (S); separated by KERCH STRAIT (E) from TAMAN PENINSULA; 60 mi/97 km long, 20 mi/32 km–30 mi/48 km wide. Connected with main part of Crimea by 10-mi/16-km-wide isthmus. Irrigated agriculture (wheat, cotton). Limestone, sulphur, and gypsum quarries. KERCH, with iron mines of Kamysh-Burun, is at the E end.

Kerch Strait (KERCH) (Ukrainian *Kerchens'ka protoka*) (Russian *Kerchenskiy proliv*), shallow channel, approximately 25 mi/40 km long, between UKRAINE and RUSSIA, connecting the Sea of AZOV with the BLACK SEA and separating the CRIMEA in the W from the TAMAN peninsula in the E. Its N end, opening into the Sea of Azov, is narrowed to a width of from 2 mi/3.2 km to 3 mi/4.8 km by the narrow Chuska landspit; the S end, opening into the Black Sea, is approximately 9 mi/14 km wide. Its arm, the Taman Gulf, penetrates E into the Taman peninsula. The city of KERCH lies near the middle of the strait, on the Crimean side, in Ukraine. Kerch Strait was the Cimmerian Bosporus (*Bosporus Cimmerius*) of the ancients; it is also known by its Tatar name, Yenikale.

Kerdous (ker-DOOS), pass (3,608 ft/1,100 m), SW MOROCCO, in SW spur of the Anti-ATLAS mountains, 27 mi/43 km SE of TIZNIT, on desert track and caravan route to sub-Saharan AFRICA; 29°33′N 09°19′W.

Kerdylion (ker-[TH]EE-lee-on), mountain of CENTRAL MACEDONIA department, NE GREECE, near ORFANI Gulf, W of STRUMA (Strymon) River mouth; rises to 3,583 ft/1,092 m, 10 mi/16 km SE of NIGRITA; 40°47′N 23°39′E. Also Kerdhillion.

Kerege (kai-RAI-gai), village, PWANI region, E TANZANIA, 20 mi/32 km NW DAR ES SALAAM, near INDIAN OCEAN; 06°35′S 39°01′E. Cashews, corn, sweet potatoes, manioc, sisal; goats, sheep; timber.

Kerekegyháza (KE-rek-ed-yuh-hah-zah), village, PEST county, central HUNGARY, 10 mi/16 km WNW of KECSKEMÉT; 46°56′N 19°29′E. Grain, apricots, apples; cattle; vineyards.

Kerekere, CONGO: see ZANI.

Kereli, TURKEY: see KIRELI.

Kerema (kuh-REE-muh), town (2000 population 5,116), ⊙ GULF province, S PAPUA NEW GUINEA, SE NEW GUINEA island, on Gulf of GULF OF PAPUA, 140 mi/225 km NW of PORT MORESBY. Timber; coconuts, palm oil, sago, yams; crafts (masks and skull racks). Gulf province was especially noted for cannibalism and headhunting until arrival of Christianity (20th century). Selected as province center because of its drier location, E of the lowland delta region. Airstrip, head of short coastal road SE to KUKIPI. Main modes of transportation in this gulf region are coastal freighter and canoe. McAdam National Park circa 55 mi/89 km to N.

Keremeos, village (□ 1 sq mi/2.6 sq km; 2001 population 1,197), S BRITISH COLUMBIA, W CANADA, on fertile bench of SIMILKAMEEN River, 26 mi/42 km S of PENTICTON, 32 mi/52 km from the U.S. (WASHINGTON) border, in Okanagan-SIMILKAMEEN regional district; 49°12′N 119°49′W. Former Hudson's Bay Company post. Mining; wine production, fruit growing, hay; cattle, horse ranching; logging. Incorporated 1956.

Kerempe, Cape, N TURKEY, on BLACK SEA, 22 mi/35 km W of INEBOLU; 42°02′N 33°18′E.

Keren, Italian *Cheren*, former administrative division, N ERITREA, bordered by SUDAN (N) and RED SEA (E); ⊙ KEREN. Difficult upland terrain and aridity render much of area useless. Sparsely settled.

Keren (KER-en), Italian *Cheren*, town (2003 population 57,000), ANSEBA region, N ERITREA, on central plateau, near ANSEBA RIVER, and 45 mi/72 km NW of ASMARA (linked by railroad); 15°45′N 38°27′E. Elevation 4,500 ft/1,372 m. Road junction. Agricultural trade center (agave, tobacco, coffee, citrus fruit, veg-

etables [vegetable ivory]), oilseed pressing, flour milling; incense industry. Has electric power station, fort, churches, mosque. Occupied (1889) and administered by Italians until World War II, when it was taken by British (1941).

Kerend (ke-RUHND), town, Kermānshāhān province, W IRAN, 50 mi/80 km W of KERMANSHAH, and on road to IRAQ border at QASR-E-SHIRIN; 34°15′N 46°14′E. Grain, fruit, dairy products; sheep raising. Metalwork (handmade rifles and knives). Also called KARIND or KARAND.

Kerens, town (2006 population 1,824), NAVARRO county, E central TEXAS, 14 mi/23 km E of CORSICANA; 32°07′N 96°13′W. RICHLAND CHAMBERS RESERVOIR to S. Settled 1881.

Kerensk, RUSSIA: see VADINSK.

Keret' (kee-RYET), Finnish *Kieretti*, village, NE Republic of KARELIA, NW European Russia, a lumber port on the KANDALAKSHA BAY of the WHITE SEA, at the mouth of the Keret' River, 65 mi/104 km SSE of KANDALAKSHA; 66°16′N 33°33′E. Terminus of a local highway. Sawmilling; fishing.

Keret, Lake (kye-RYET), Finnish *Kierettijärvi*, in NE Republic of KARELIA, S of LOUKHI; (surface area □ 77 sq mi/200 sq km; 20 mi/32 km long); contains a large island (□ 15 sq mi/39 sq km) in the S; 65°50′N 33°00′E. Drains N into the Keret River, which flows 35 mi/56 km E to the KANDALAKSHA BAY arm of the WHITE SEA at KERET.

Kerewan (ker-AI-wahn), town, ⊙ NORTH BANK DIVISION, W THE GAMBIA, near GAMBIA RIVER, on short N affluent (wharf and ferry), and 32 mi/51 km E of BANJUL; 13°29′N 16°06′W. Fishing; peanuts, palm oil and kernels, rice. Has prison.

Kergez (kyir-GYES), town in Karadag district of Greater BAKY, AZERBAIJAN, 12 mi/19 km SW of Baky; 40°18′N 49°37′E. Oil fields (developed 1933); stone quarries.

Kerguelen (KUHR-guh-len), subantarctic island of volcanic origin (1,320 sq mi/3,419 sq km), in the S INDIAN OCEAN, c.3,300 mi/5,311 km SE of the S tip of AFRICA; 49°30′S 69°30′E. Largest of the 300 Kerguelen Islands (total area c.2,700 sq mi/7,000 sq km), part of the French Southern and Antarctic Territories. Kerguelen rises in the S to Mount Rose (6,120 ft/1,865 m), and Cook Glacier covers its W third. Glacial lakes, peat marshes, lignite, and guano deposits are found on the island; it also has seals, rabbits, wild hogs, and wild cats and dogs. Cold, windy climate; usually rough seas. The island, famous for the native Kerguelen cabbage, was once a seal-hunting and whaling base, but is now used mainly as a scientific research station. Facilities at Port-Aux-Francais; small permanent population. Kerguelen was discovered in 1772 by the French navigator Yves Joseph de Kerguélen-Trémarec, who named it Desolation Island. It has belonged to France since 1893.

Kerguelen Plateau (kuhr-GUH-len), ANTARCTICA, plateau extending SE from Kerguelen Island toward LEOPOLD AND ASTRID COAST; separates ATLANTIC-INDIAN BASIN from SOUTH INDIAN BASIN; 60°00′S 83°00′E.

Kerhonkson, village (□ 5 sq mi/13 sq km; 2000 population 1,732), ULSTER county, SE NEW YORK, on RONDOUT CREEK, 19 mi/31 km SW of KINGSTON, just W of SHAWANGUNK range; 41°46′N 74°17′W. Residential area, popular summer recreational spot.

Keri (ke-REE), village, SW ZÁKINTHOS island, ZÁKINTHOS prefecture, IONIAN ISLANDS department, off W coast of GREECE, 10 mi/16 km SSW of Zákinthos town; 37°40′N 20°49′E. Mineral pitch deposits.

Kericho (kai-REE-cho), town, RIFT VALLEY province, W KENYA, on road from Lumbwa station, and 40 mi/64 km SE of KISUMU; 00°22′S 35°19′E. Country's leading tea-growing center; other agriculture (wattle bark, coffee, flax; corn); agricultural trade and market center. Manufacturing (dairying). Textile spinning and weaving school.

Kerinchi, Mount, INDONESIA: see KERINCI, MOUNT.

Kerinci, Lake (kuh-REEN-chee) (□ 296 sq mi/769.6 sq km), W Jambi province, INDONESIA, 124 mi/200 km SE of PADANG; 02°09′S 101°30′E. Also known as Danau Kerinci.

Kerinci, Mount (kuh-REEN-chee) (12,467 ft/3,800 m), in the BARISAN MOUNTAINS, Sumatra Barat province, W central SUMATRA, INDONESIA; 01°42′S 101°16′E. It is Sumatra's highest point. KERINCI-SEBLAT Reserve on SE face. Formerly spelled Kerintji, Kerinchi, Kurinchi, Piek van Indrapura (Indrapura Peak).

Kerinci-Seblat National Park (kuh-REEN-chee-suh-BLAHT) (□ 5,732 sq mi/14,846 sq km), SUMATRA, INDONESIA, 80 mi/130 km SE of Padang; 02°00′N 101°20′E. Large national park encompassing parts of four provinces (W Sumatra, S Sumatra, Jambi, Bengkulu). Mount KERINCI, Indonesia's second-highest peak, is located in NE portion of park. Park is habitat for the endangered Sumatran rhinoceros and Malay tapir as well as elephants, tigers, sun bears, and clouded leopards.

Kerintji, INDONESIA: see KERINCI, MOUNT.

Kerio Valley National Reserve (□ 25 sq mi/65 sq km), RIFT VALLEY province, W KENYA, N of ELDORET town; 00°42′N 35°36′E. Designated a national reserve in 1983.

Keriya, China: see YUTIAN, Xinjiang Uygur Autonomous Region.

Keriya River (KUH-REE-YAH), 300 mi/483 km long, SW XINJIANG UYGUR, NW CHINA; rises in the KUNLUN MOUNTAINS on TIBET border; flows N, past Yutian, Misalay, and Tongguzhasti, disappearing into central TAKLIMAKAN DESERT; 36°00′N 82°00′E.

Kerkdriel (kerk-DREEL), town, GELDERLAND province, central NETHERLANDS, on MAAS RIVER, and 6 mi/9.7 km NNE of 's-HERTOGENBOSCH; 51°46′N 05°21′E. Agriculture (dairying; cattle, hogs, poultry; grain, vegetables, sugar beets). Manufacturing (farm equipment, games). Sometimes called Maasdriel.

Kerkebet (KUHR-ke-bet), Italian *Carcabat*, village, ANSEBA region, W ERITREA, on BARKA river, and 60 mi/97 km NW of AKURDET; 16°13′N 37°28′E. In irrigated cotton-growing region. Also spelled KARKABAT.

Kerken (KER-ken), town, NORTH RHINE–WESTPHALIA, W GERMANY, 9 mi/14.5 km NW of KREFELD; 51°26′N 06°25′E. Manufacturing (transportation equipment, electric and electronic equipment). Has remains of abbey with 15th-century church.

Kerkera (ker-ke-RAH), town, SKIKDA wilaya, E ALGERIA, 9 mi/15 km SE of COLLO; 36°55′N 06°35′E. This former 1950s French military regroupment center has evolved into a local administrative center.

Kerkheh River, IRAN: see KARKHEH.

Kerkhoven (kuhrk-HO-ven), village (2000 population 759), SWIFT county, SW central MINNESOTA, 14 mi/23 km WNW of WILLMAR, near Shakopee Creek; 45°11′N 95°19′W. Agriculture (grain; livestock, poultry; dairying); light manufacturing. Monson Lake State Park to N.

Kerki (kyir-KEE), city, SE LEBAP weloyat, SE TURKMENISTAN, port on the W bank of AMU DARYA, opposite KERKICHI, and 125 mi/201 km SE of CHARJEW; 37°50′N 65°12′E. Metalworking, cotton ginning, food processing (meat and flour products). Was capital of former Kerki oblast (c.1943–1947). Founded (late 19th century) as Russian fort on Bukhara-Afghanistan trade route. Also Karki.

Kerkichi (kyer-ki-CHEE), town, SE LEBAP weloyat, SE TURKMENISTAN, on railroad, on the E bank of AMU DARYA, opposite KERKI, and 125 mi/201 km SE of CHARJEW; 37°51′N 65°14′E. Cotton-ginning center.

Kerkini (kee-KEE-nee), village, SÉRRAI prefecture, CENTRAL MACEDONIA department, NE GREECE, 16 mi/26 km W of SIDIROKASTRON, just W of Lake KERKINI; 41°12′N 23°09′E. Formerly called Boutkovon or Butkovon; also spelled Kerkine.

Kerkini, Bulgaria, Greece, and Macedonia: see RADOMIR, mountain.

Kerkini, Lake (ke-RKEE-nee) (□ 16 sq mi/41.6 sq km), SÉRRAI prefecture, CENTRAL MACEDONIA department, NE GREECE, 20 mi/32 km NNW of SÉRRAI, on the Strymon (or STRUMA) River, which enters at the N and exits at the S; 41°12′N 23°09′E. Eighth-largest lake in Greece; natural flood reservoir for the regulated Strymon.

Kerkinitidis, UKRAINE: see YEVPATORIYA.

Kérkira (KER-kee-rah), prefecture, IONIAN ISLANDS department, in extreme NW GREECE, off coasts of NW Greece and SW ALBANIA, in IONIAN SEA; ☉ KÉRKIRA; 39°40′N 19°45′E. Farthest NW of all prefectures of Greece. Includes island of KÉRKIRA, as well as smaller islands N (OTHONI, ERIKUSA) and S (PAXÍ, ANTIPAXOS).

Kérkira (KER-kee-rah) or **Corfu**, ancient *Corcyra*, city (2001 population 39,048), ☉ KÉRKIRA prefecture, chief town on KÉRKIRA island, IONIAN ISLANDS department, extreme NW GREECE, across Channel of CORFU from Greece and SW ALBANIA; 39°37′N 19°55′E. Major port exporting chiefly olive oil; other trade in grain, wine, citrus fruit; manufacturing (soap, paraffin); printing and engraving. Fisheries. Tourist trade; stifled in mid-1990s by unrest in Balkan states, especially Albania. Airport, Ioannis Kapodistrias. Seat of Greek Orthodox and Roman Catholic archbishops. Once surrounded by a wall; largely kept a labyrinthine street layout. A broad esplanade (adjoined N by government palace) separates the town from the old Venetian island fortress (E). Site of ancient town (founded 734 B.C.E.) was on peninsula 2 mi/3.2 km S. Was capital of British protectorate of IONIAN ISLANDS, 1797–1864. Also spelled Kerkyra.

Kérkira (KER-kee-rah) or **Corfu**, Latin *Corcyra*, island (□ 229 sq mi/595.4 sq km), KÉRKIRA prefecture, IONIAN ISLANDS department, extreme NW GREECE, in the IONIAN SEA, the second-largest of the IONIAN ISLANDS, separated by narrow Channel of CORFU from the SW Albanian and NW Greek coasts; ☉ KÉRKIRA; 39°40′N 19°45′E. Though rising 2,980 ft/908 m at Mount Pantokrator in the NE, Kérkira is largely a fertile lowland producing olive oil, figs, wine, and citrus fruit. Livestock raising (poultry, hogs, and sheep) and fishing are important sources of livelihood. Tourism, centered in Kérkira city, has increased dramatically in recent years (but suffered in the mid-1990s because of unrest in ALBANIA); the island is known internationally. International airport. Identified with Scheria, the island of the Phaeacians in Homer's *Odyssey*. Settled c.730 B.C.E. by Corinthian colonists and shared with Corinth in the founding of Epidamnus (DURRËS) on the mainland but became the competitor of CORINTH in the ADRIATIC SEA. The two rivals fought the first recorded (by Thucydides) naval battle in 665 B.C.E. In 435 B.C.E., Kérkira (then Corcyra) made war on Corinth over the control of Epidamnus, and in 433 it concluded an alliance (often renewed) with ATHENS; this alliance helped to precipitate (431) the Peloponnesian War. Passed under Roman rule in 229 B.C.E.; in C.E. 336 became part of the BYZANTINE EMPIRE. Seized from the Byzantines by the Normans of SICILY in the 1080s and 1150s, by VENICE (1206), and later by Epirus (1214–1259), and the Angevins of Naples. In 1386 the Venetians obtained a hold that ended only with the fall of the Venetian republic in 1797. Under Venetian rule, the island had successfully resisted two celebrated Turkish sieges (1537, 1716). Under British protection, 1815; ceded to Greece, 1864. Occupied (1916) by the French in World War I; union of Serbia, Croatia, and Slovenia concluded here (1917). In 1923, after Italian officers trying to establish the Greek-Albanian border were slain in Greece, Kérkira was bombarded and temporarily occupied in retaliation by Italian forces. A major earthquake in 1953 did little damage.

Kerkrade (KERK-rah-duh), city (2001 population 33,957), LIMBURG province, SE NETHERLANDS, 17 mi/27 km E of MAASTRICHT, and 7 mi/11.3 km NNW of AACHEN (GERMANY); 50°50′N 06°04′E. German border to E and SE. Agriculture (dairying; cattle; grain, fruit, vegetables). Manufacturing (plastic products, boilers). Former coal-mining center (from twelfth century through 1974). Castle.

Kerma, SUDAN: see KARMA.

Kermadec Deep (kuhr-MA-dek), ocean depth (32,963 ft/10,047 m) of S PACIFIC OCEAN, E of KERMADEC ISLANDS, and continuing TONGA DEEP to S. Discovered 1895.

Kermadec Islands, volcanic group inhabited only for scientific observation (□ c.13 sq mi/34 sq km), South PACIFIC, 621 mi/1,000 km NE of AUCKLAND, NEW ZEALAND, to which they belong. The largest, Sunday (or Raoul), is crater-shaped. Annexed to New Zealand in 1887, the group adjacent to the KERMADEC-TONGA OCEANIC TRENCH is prone to volcanic action and earthquakes. Includes CURTIS and MACAULEY islands. A nature reserve and now a marine reserve.

Kermān (ker-MAHN), province (□ 69,466 sq mi/180,611.6 sq km), SE IRAN; ☉ KERMAN; 29°00′N 57°00′E. Bounded N by Khorāsān, NW by YAZD, W by Fārs, E by Sīstan va Balūchistān; contains part of the SE ZAGROS ranges (S), and extends onto the central Iranian plateau (N), where it includes the salt desert DASHT-i-Lut. Agriculture (mainly irrigated): wheat, barley, cotton, and millet in higher, cooler regions; rice, corn, henna, dates, and fruit (pomegranates, oranges) on warmer plains. Copper, iron, zinc, and chromite deposits. Sheep and camel raising. Fine wool, which is the main export, also is the basis of the Kermān rug and shawl industry. Access to the PERSIAN GULF is through BANDAR ABBAS. Main inland centers are Kermān, BAM, SIRJAN, and RAFSANJAN. Main highways lead from Kermān to YAZD, SHIRAZ, and ZAHEDAN. The ancient *Carmania*, it was traversed (325 B.C.E.) by Alexander the Great and flourished under the Sassanids. It resisted Arab control, but was ravaged by the Afghans (17th century) and by Aga Mohammed Khan (18th century).

Kermān (ker-MAHN), city (2006 population 515,114), ☉ Kermān province, E central IRAN; 30°17′N 57°04′E. Elev. 6,065 ft/1,830 m. It is noted for making and exporting carpets. Cotton textiles and goats-wool shawls are also manufacturing. Airport. Kerman was under the Seljuk Turks in the 11th and 12th centuries, but remained virtually independent, conquering OMAN and Fārs. Marco Polo visited (late 13th century) and described the city. Kerman changed hands many times in ensuing years, prospering under the Safavid dynasty (16th century) and suffering under the Afghans (17th century). In 1794 its greatest disaster occurred: Aga Muhammad Khan, shah of Persia, ravaged the city, selling 20,000 of its inhabitants into slavery and blinding another 20,000. Reminders of historic Kerman include medieval mosques, the beautiful faïence found among the extensive ruins outside the city walls, and 16th-cent. mosaics with Chinese motifs. Nearby is the shrine of Shah Vali Namatullah, a 15th-cent. Sufi holy man.

Kerman, city (2000 population 8,551), FRESNO county, central CALIFORNIA, in San Joaquin Valley, 15 mi/24 km W of FRESNO; 36°43′N 120°04′W. Agriculture (cotton; grapes, nectarines, figs, raisins, almonds, vegetables; sugar beets, grain, alfalfa); winery. Manufacturing (fertilizers, food products, sanitary food containers). SAN JOAQUIN RIVER to N. Incorporated 1946.

Kermanshah (ker-MAHN-shah), city (2006 population 794,863), Kermānshahān province, W IRAN; 34°17′N 47°04′E. Elev. 1,420 ft/433 m. Trade center for a rich agricultural region that produces grain, rice, vegetables, fruits, and oilseed. Manufacturing (carpets, canvas shoes, textiles, refined petroleum, refined sugar, food products). Airport. Kermanshah has numerous caravansaries that are crowded semiannually with Shiite pilgrims to KARBALA, IRAQ. Kurds form the majority of the population Kermanshah was founded by the Sassanids in the 4th century C.E. and became a secondary royal residence. It was captured by the Arabs in the 7th century. Later it was a frontier fortress against the Ottoman Turks, who occupied it a number of times, including the period from 1915 to 1917. Nearby are the famed Behistun Inscriptions and notable Sassanian rock reliefs. Also called Bakhtaran.

Kermānshshān, province (□ 9,138 sq mi/23,758.8 sq km), W IRAN, along the border with IRAQ, and among the N flanks of the ZAGROS Mountains, borders the provinces of Kordestān to N, Hamadān to E, Lorestān to SE, and Ilām to S; ☉ KERMANSHAH; 34°30′N 47°00′E. Largely an agricultural region (wheat and barley). Pastureland and woodlands. Some oil is produced in fields near the Iraqi border. Nearly 40% of the population lives in the city of KERMANSHAH, a major refining center.

Kermen (KER-men), city, BURGAS oblast, SLIVEN obshtina, E central BULGARIA, 13 mi/21 km W of YAMBOL; 42°30′N 26°15′E. Grain, tobacco.

Kermit (KER-men), town (2006 population 5,204), ☉ WINKLER county, W TEXAS, 38 mi/61 km NE of PECOS, near base of CAPROCK ESCARPMENT; 31°51′N 103°05′W. Elevation 2,890 ft/881 m. Supply center for oil, natural-gas area (Permian Basin), with cattle ranches. Manufacturing (natural-gas processing, well hookups). Incorporated 1938.

Kermit, village (2006 population 226), MINGO county, SW WEST VIRGINIA, on TUG FORK RIVER (bridged), opposite WARFIELD, KENTUCKY; 37°50′N 82°24′W. Natural gas, bituminous coal. Manufacturing (lumber). Cabwaylingo State Forest to NE, Laurel Creek Wildlife Management Area to E.

Kern (KUHRN), county (□ 8,142 sq mi/21,169.2 sq km; 2006 population 780,117), S central CALIFORNIA; ☉ BAKERSFIELD; 35°20′N 118°43′W. Includes S end of San Joaquin Valley, walled in by TEHACHAPI MOUNTAINS (S), S part of the SIERRA NEVADA (E), and COAST RANGES (W). Irrigated agriculture (since 1880s); cotton; potatoes, grains, nuts, grapes, citrus, apples, plums, carrots, tomatoes, peppers; cattle; dairying. Limestone quarrying; gypsum, gold, clay mining; oil and natural gas. State's leading petroleum-producing county, with oil and natural-gas fields along W side of San Joaquin Valley and in Bakersfield district. Part of MOJAVE DESERT (borax, tungsten, silver, gold mines) in E and SE. Drained by KERN RIVER, used for hydroelectric power; irrigation water supplied by FRIANT-KERN CANAL of Central Valley Project. LOS ANGELES AQUEDUCT crosses county N-S in E; California Aqueduct crosses county NW-SE in W. Part of Sequoia National Forest in N; part of Los Padres National Forest in S; Pacific Crest National Scenic Trail crosses county N-S in E center; Fort Tejon State Historical Park in S; Kern National Wildlife Refuge in NW, Tule Elk State Reserve in W center; Red Rock Canyon State Recreation Area in E; Isabella Lake reservoir in NE. Most of EDWARDS AIR FORCE BASE in SE corner, part of China Lake Naval Air Weapons Station in NE corner. Elk Hills Naval Petroleum Reserve in W. Formed 1893.

Kernen im Remstal (KER-nen im REMS-tahl), town, BADEN-WÜRTTEMBERG, SW GERMANY, 12 mi/19 km E of STUTTGART; 48°47′N 09°29′E. Several small industries; wine growing. Created in 1975 through unification of Stetten and Rommelshausen.

Kernersville (KUHR-nuhrs-vil), city (□ 12 sq mi/31.2 sq km; 2006 population 21,862), FORSYTH county, N central NORTH CAROLINA, suburb 9 mi/14.5 km E of WINSTON-SALEM and 15 mi/24 km W of GREENSBORO; 36°07′N 80°04′W. Manufacturing (apparel, printing and publishing, electronic equipment, paper and plastic products, construction materials, textiles, transportation equipment, metal and stone processing, machinery); service industries. High Point Reservoir to SE. Korners Falls, Seven-level house built 1897. Settled before 1770 by Germans, incorporated 1871.

Kern River (KUHRN), 155 mi/249 km long, E CALIFORNIA; rises in the S SIERRA NEVADA, NE TULARE county, in NE corner of SEQUOIA NATIONAL PARK, flows S through Inyo and Sequoia national forests, then SW through Isabella Lake reservoir, and into S part of the San Joaquin Valley, where it passes BAKERSFIELD to N and enters BUENA VISTA LAKE irrigation reservoir. Town of ALTA SIERRA 5 mi/8 km W of Isabella Lake. The river has Isabella Dam as its chief facility. Kern River is the S terminus of the FRIANT-KERN CANAL, constructed between 1945 and 1951 to bring the waters of the SAN JOAQUIN RIVER to the region (see CENTRAL VALLEY); irrigated agriculture (alfalfa, fruit; cotton) and cattle grazing are practiced. U.S. explorer John C. Frémont named the river in honor of Edward M. Kern, the topographer of his third expedition. Gold was discovered along the river in 1853. Water from river's lower course and Buena Vista Lake is used for irrigating the Bakersfield area, also channeled into California Aqueduct and other canals, leaving river with no natural outlet.

Kerns, commune, Obwalden half-canton, central SWITZERLAND, just E of SARNEN. Health resort.

Kerns (KUHRNZ), township (□ 35 sq mi/91 sq km; 2001 population 360), NE central ONTARIO, E central CANADA, 17 mi/28 km from COBALT; 47°37′N 79°49′W. Agriculture (livestock, cash crops). Established 1904.

Kernville, unincorporated town (2000 population 1,736), KERN county, S central CALIFORNIA, 40 mi/64 km NE of BAKERSFIELD, at N end of Isabella Lake reservoir (KERN RIVER), in S part of SIERRA NEVADA; 35°46′N 118°26′W. Parts of Sequoia National Forest surround area. Shirley Meadows Ski Area to S. Timber; cattle. Tourism.

Keroman (kai-ro-mahn), S part of city of LORIENT, MORBIHAN department, W FRANCE, in BRITTANY, on an inlet of the Bay of BISCAY. Formerly a small old fishing village, Keroman was developed as a trawler port after 1920 and as a German submarine base in World War II. The sub base is now an active French naval installation, and Keroman has become a leading home base for the French high-seas fishing fleet, with a large ice-making plant and cold-storage facilities.

Kérou, town, ATAKORA department, N BENIN, 50 mi/80 km NE of NATITINGOU; 10°50′N 02°06′E. Cotton; livestock; shea-nut butter.

Kérouané, prefecture, Kankan administrative region, SE GUINEA, in Haute-Guinée geographic region; ⊙ KÉROUANÉ. Bordered N by Kankan prefecture, ESE by Beyla prefecture, SSW by Macenta prefecture, and W by Kissidougou prefecture. MILO RIVER flows N through center of the prefecture and Dion River in E (forming much of the border with Beyla prefecture). Part of Fon Going ridge in S. Towns include Banankoro, Férédou, Kérouané, Komodou, and Sibiribaro. Main road runs through Kérouané town, connecting it to KANKAN town to N and BEYLA and N'ZÉRÉKORÉ towns to S; several secondary roads also run through the prefecture.

Kérouané (ker-WAH-nai), town, ⊙ Kérouané prefecture, Kankan administrative region, SE GUINEA, in Haute-Guinée geographic region, on road to CÔTE D'IVOIRE, 80 mi/129 km S of KANKAN; 12°14′N 12°20′W. Tobacco, coffee, rice; cattle. Diamond deposits nearby. Airfield.

Kerpen (KER-pen), city, NORTH RHINE–WESTPHALIA, W GERMANY, 10 mi/16 km NE of DÜREN; 50°52′N 06°42′E. In former lignite-mining region, now an industrial city. Manufacturing (motor vehicles; metalworking; food processing; electronics). Was first mentioned in 871, grew intensively after World War II. It was chartered in 1975 after the incorporated of several villages. Several castles nearby from the 16th–18th centuries.

Kerr (KUR), county (□ 1,107 sq mi/2,878.2 sq km; 2006 population 47,254), SW TEXAS; ⊙ KERRVILLE; 30°03′N 99°20′W. Elevation c.1,100 ft/335 m–2,400 ft/732 m. In scenic hill country of EDWARDS PLATEAU; drained by GUADALUPE RIVER, rising in springs here. Ranching (sheep, goats, cattle); agriculture (hay, oats, wheat; pecans, apples) and vacation area, with camps, guest ranches, health resorts; hunting. Kerrville-Schreiner State Park in E center at Kerrville. Formed 1856.

Kerrabë (ko-RAHB), new town, central ALBANIA between SHKUMBI and ERZEN rivers; crossed by Elbasan-Tiranë highway; 41°12′N 19°59′E. Coal mining. Also spelled Krraba.

Kerr Dam, MONTANA: see POLSON.

Kerrera (kuh-RER-ah), island (5 mi/8 km long, 2 mi/3.2 km wide; rises to 617 ft/188 m; 2006 estimated population 30), Argyll and Bute, W Scotland, in the Firth of Lorn, opposite OBAN, and across narrow Sound of Kerrera; 56°23′N 05°34′W. Site of Gylen Castle, built 1587 by MacDougalls. Alexander II died in 1249.

Kerrick, village (2000 population 71), PINE county, E MINNESOTA, 38 mi/61 km SW of DULUTH; 46°20′N 92°35′W. Dairying. Nemadji State Forest to E.

Kerr, Lake (KUHR), reservoir, MARION county, N central FLORIDA, in Ocala National Forest, 23 mi/37 km ENE of OCALA; 3 mi/4.8 km long, 2 mi/3.2 km wide; 29°21′N 81°22′W.

Kerrobert (kuh-RAH-buhrt), town (2006 population 1,001), W SASKATCHEWAN, CANADA, 30 mi/48 km N of KINDERSLEY; 51°55′N 109°08′W. Grain elevators, lumbering, dairying.

Kerrortussok, GREENLAND: see QERRORTUSOQ.

Kerrville (KUR-vil), city (2006 population 22,361), ⊙ KERR county, S central TEXAS, 55 mi/89 km NW of SAN ANTONIO, on the GUADALUPE RIVER; 30°02′N 99°08′W. Elevation 1,645 ft/501 m. Agriculture (cattle, sheep, goats). Manufacturing (transportation equipment, computers, solar products, food processing). Youth camps, dude ranches. It is also a vacation and health resort in the hill country on the edge of the Edwards Plateau. Schreiner University and a number of art galleries are in the city. Kerrville-Schreiner State Park on S edge of city. Settled 1846, incorporated 1942.

Kerry, county (□ 1,828 sq mi/4,752.8 sq km; 2006 population 139,835), SW IRELAND; ⊙ TRALEE. Kerry consists of a series of mountainous peninsulas that extend into the ATLANTIC. The shoreline is deeply indented by DINGLE BAY, Tralee Bay, and the KENMARE RIVER. Bounded by LIMERICK and CORK counties to E. CARRAUNTUOHIL (3,414 ft/1,041 m), in the MACGILLYCUDDY'S REEKS range, is the highest point in Ireland. The streams are short and precipitous, and many bogs exist. The LAKES OF KILLARNEY are a popular tourist attraction; Farranfore Airport is 50 mi/80 km N of KILLARNEY. Farming (hay, oats, and potatoes), fishing, sheep and cattle raising, dairying. Footwear is made in Tralee and Killarney. Many well-preserved dolmens, stone forts, round towers, castles, and abbeys still stand. Gaelic is spoken by inhabitants of the W parts of the Dingle and Iveragh peninsulas. Other significant towns include LISTOWEL, DINGLE, and CAHIRCIVEEN.

Kerry (KER-ee), village (2001 population 1,922), POWYS, E Wales, 3 mi/4.8 km ESE of NEWTOWN; 52°30′N 03°16′W. Previously sawmills.

Kerry Head, Gaelic Ceann Chiarraí, cape, NW KERRY county, SW IRELAND, at mouth of the SHANNON RIVER, 14 mi/23 km NW of TRALEE; 52°25′N 09°57′W.

Kersa (KUHR-suh), town (2007 population 9,469), OROMIYA state, central ETHIOPIA, 30 mi/48 km SSW of ASELA; 07°31′N 39°00′E. In livestock raising area, E of LAKE LANGANO.

Kersa (KUHR-suh), village (2007 population 3,394), OROMIYA state, E central ETHIOPIA, 21 mi/34 km WNW of HARAR; 09°27′N 41°52′E. In highlands, on the road to ADDIS ABABA. Coffee growing. Also Karsa and Carsa.

Kersey, town (2000 population 1,389), WELD county, N COLORADO, on SOUTH PLATTE RIVER, and 8 mi/12.9 km ESE of GREELEY; 40°23′N 104°33′W. Elevation c.4,617 ft/1,407 m. In agricultural region. Manufacturing (agricultural products). Lower Latham Reservoir to SW.

Kersey, unincorporated town, Fox township, ELK county, N central PENNSYLVANIA, 8 mi/12.9 km SE of RIDGWAY; 41°21′N 78°35′W. Manufacturing (machinery; metal finishing). Agriculture (grain; livestock, dairying). Moshannon State Forest to S.

Kersey (KUH-zee), agricultural village (2001 population 350), S SUFFOLK, E ENGLAND, 2 mi/3.2 km NW of HADLEIGH; 52°03′N 00°55′E. Has 15th-century church. The name kersey, a woolen cloth made in England in the 13th century, probably originates here. Just WNW is village of LINDSEY, where linsey-woolsey, another woolen cloth, originated.

Kershaw (kuhr-SHAW), county (□ 740 sq mi/1,924 sq km; 2006 population 57,490), N central SOUTH CAROLINA; ⊙ CAMDEN; 34°20′N 80°35′W. Contains Lake WATEREE and WATEREE River; bounded E by LYNCHES RIVER. Tourist area in Sand Hills; mining of mica, granite, sand, kaolin. Some agriculture (peaches, pecans, sweet potatoes; poultry, cattle; corn, rye, hay, cotton), timber. Formed 1791.

Kershaw (kuhr-SHAW), town (2006 population 1,637), LANCASTER county, N SOUTH CAROLINA, 20 mi/32 km N of CAMDEN, near Little LYNCHES RIVER; 34°32′N 80°35′W. Lumber mills. Manufacturing (machinery, grain products, mica processing, textiles, apparel).

Kershaw Ice Rumples, a large area of disturbed ice between FLETCHER ICE RISE and KORFF ICE RISE, in the SW part of the RONNE ICE SHELF, WEST ANTARCTICA; 78°45′S 75°40′W.

Kerteh (kuhr-TE), town, S TRENGGANU, MALAYSIA, on SOUTH CHINA SEA, 60 mi/97 km SSE of Kuala Trengganu. Fishing; coconut and rubber plantations. Sometimes spelled Kretai.

Kertel, ESTONIA: see KÄRDLA.

Kerteminde (KER-tuh-mi-nuh), city (2000 population 5,542) and port, FYN county, DENMARK, on S FYN island, and 11 mi/18 km ENE of ODENSE, on KERTEMINDE Bugt; 55°30′N 10°40′E. Fishing; fish canning; manufacturing (agricultural machinery).

Kertosono (kuhr-TAW-saw-naw), town, Java Timur province, INDONESIA, on E bank BRANTAS RIVER, 50 mi/80 km SW of SURABAYA; 07°35′S 112°06′E. Trade center for agricultural area (rice, cassava, corn, peanuts); also railroad and highway hub.

Kerugoya (kai-roo-GO-yah), town, Kirinyaga district, CENTRAL province, KENYA, 25 mi/40 km S of Mount KENYA; 00°31′S 37°18′E. Market and trading center.

Kerulen (KER-oo-len), river (Mongolian Herlen Gol, Chinese Herlen He), 784 mi/1,264 km long, E MONGOLIA and N CHINA; rises in the HENTIYN NURUU (Kentei Mountains), NE of ULAANBAATAR; flows S, then E to HULUN NUR (Kulun Lake), HEILONGJIANG province, NE China at 48°48′N 117°00′E. Its drainage basin covers 44,940 sq mi/116,400 sq km. A road from Ulaanbaatar to CHOYBALSAN, its principal city, follows the river. Also spelled Herelen River.

Kerulen, MONGOLIA: see CHOYBALSAN.

Kerumutan Baru Reserve (□ 463 sq mi/1200 sq km), RIAU province, INDONESIA, 121 mi/195 km SE of PEKANBARU; 00°10′N 102°25′E. Lowland forest and swamp reserve.

Kerur (KAI-roor), town, BIJAPUR district, KARNATAKA state, SW INDIA, 60 mi/97 km SSW of BIJAPUR; 16°01′N 75°34′E. Handicraft cloth weaving. Agriculture (cotton, millet, peanuts, wheat). Sometimes spelled Karur.

Kerva (KYER-vah), town (2006 population 2,775), E MOSCOW oblast, central European Russia, on small Svyatoye Lake, on railroad spur, 4 mi/6 km NNE of, and under administrative jurisdiction of, SHATURA; 55°37′N 39°35′E. Elevation 406 ft/123 m. Industrial electronics (lasers); peat production for the Shatura power station.

Kervignac (ker-vee-nyahk), commune (□ 15 sq mi/39 sq km), MORBIHAN department, in BRITTANY, NE FRANCE; 47°46′N 03°14′W. Agricultural village 7 mi/11 km E of LORIENT.

Kervo, FINLAND: see KERAVA.

Kerzers (KER-tzuhrs), commune, FRIBOURG canton, W SWITZERLAND, 12 mi/19 km N of FRIBOURG. Elevation 1,483 ft/452 m.

Kerzhenets (KYER-zhi-nyets), settlement (2006 population 665), central NIZHEGOROD oblast, central European Russia, on highway, 16 mi/26 km NE of NIZHNIY NOVGOROD; 56°28′N 44°25′E. Elevation 364 ft/110 m. Peat works.

Ke Sach (KAI SAHK), town, SOC TRANG province, S VIETNAM, 10 mi/16 km N of SOC TRANG, near Hau River in highly irrigated area; 09°46′N 105°59′E. Rice center. Formerly Kesach.

Kesan, town (2000 population 42,755), European TURKEY, 55 mi/89 km S of EDIRNE; 40°52′N 26°37′E. Agriculture (wheat, rice, barley). Some lignite nearby. Sometimes spelled Keshan.

Kesap, village, N Turkey, on BLACK SEA, 6 mi/9.7 km E of GIRESUN; 40°55′N 38°31′E. Hazelnuts, corn.

Kesarevo (ke-SAHR-e-vo), village, LOVECH oblast, STRAZHITSA obshtina, N BULGARIA, on a headstream of the Bregovitsa River, 14 mi/23 km E of GORNA ORYAHOVITSA; 43°08′N 25°58′E. Horticulture, vegetables, sugar beets.

Ke Sat (KAI SAHT), village, HAI HUNG province, N VIETNAM, 25 mi/40 km ESE of HANOI; 20°54′N 106°09′E. Rice center. Formerly Kesat.

Kesch, Piz, SWITZERLAND: see PIZ KESCH.

Kese (KE-sai), village, BANDUNDU province, W CONGO, on left bank of KASAI RIVER, 20 mi/32 km E of BANDUNDU; 03°24′S 18°05′E. Elev. 843 ft/256 m.

Kesennuma (ke-sen-NOO-mah), city, MIYAGI prefecture, N HONSHU, NE JAPAN, on Kesennuma Bay, 59 mi/95 km N of SENDAI; 38°54′N 141°34′E. A fishing port (spearfish, swordfish, bonito, sharks; squid processing). Rikuchu Kaigan [=seacoast] National Park is nearby.

Kesgrave (KES-graiv), village (2001 population 9,026), SUFFOLK, E ENGLAND, 2 mi/3.2 km E of IPSWICH, near DEBEN RIVER; 52°03′N 01°14′E. Farming.

Keshan (KUH-SHAN), town, ⊙ Keshan county, W HEILONGJIANG province, NE CHINA, 100 mi/161 km NE of QIQIHAR, and on railroad; 48°04′N 125°54′E. Agriculture (grain, sugar beets, jute, oilseeds). Manufacturing (food processing, machinery).

Keshan, TURKEY: see KESAN.

Keshena (ke-SHEE-nuh), village (2000 population 1,394), ⊙ MENOMINEE county, NE WISCONSIN, on WOLF RIVER, and 9 mi/14.5 km WNW of SHAWANO, in the Menominee Indian Reservation; 44°52′N 88°38′W. The Indian Agency buildings and an Roman Catholic mission are here. Oshkosh clan burial plot is nearby. Legend of Spirit lakes to E; Keshena Falls to N.

Keshi (KE-shee), W suburb of ASHGABAT, S central TURKMENISTAN; 37°59′N 58°20′E. Residential section; parks.

Keshiketeng Qi, China: see HEXINGTEN QI.

Keshm, AFGHANISTAN: see KISHM.

Keshod (KAI-shod), town, JUNAGADH district, formerly in SW SAURASHTRA, now in GUJARAT state, W central INDIA, 20 mi/32 km SW of JUNAGADH; 21°18′N 70°15′E. Local market for cotton, millet, oilseeds; hand-loom weaving. Airport.

Keshorai Patan (KAI-sho-rei PAH-tuhn), town, BUNDI district, RAJASTHAN state, NW INDIA, on CHAMBAL RIVER, and 22 mi/35 km SE of Bundi; 25°18′N 75°52′E. Local market for wheat, millet, oilseeds, gram.

Kesi, INDIA: see JHUSI.

Kesigi, INDIA: see KOSIGI.

Kesimpuco (ke-seem-POO-ko), canton, CHAYANTA province, POTOSÍ department, W central BOLIVIA; 18°42′S 66°00′W. Elevation 13,642 ft/4,158 m. Antimony-bearing lode; four small iron deposits here; clay,

limestone. Agriculture (potatoes, yucca, bananas, barley); cattle raising for meat and dairy products.

Kesis Dag (Turkish=*Keşiş Dağ*) peak (11,604 ft/3,537 m), E central TURKEY, in Esene (or Kesis) Mountains, 15 mi/24 km ENE of ERZINCAN.

Keskin, township, central TURKEY, 45 mi/72 km ESE of ANKARA; 39°41′N 33°36′E. Wheat, barley; mohair goats. Formerly Maden.

Keski-Suomen (KES-kee–SOO-o-men), province (□ 7,755 sq mi/20,163 sq km), central FINLAND; ⊙ JYVÄSKYLÄ. In lake region, bordered in N by OULUN province, NE by KUOPION province, SE by MIKKELIN province, SW by HÄMEEN, and NW by VAASAN provinces. PÄIJÄNNE, a large lake in the center of province, is connected to KEITELE-Päijänne Canal, and is a popular vacation spot and tourist area.

Kes'ma (kees-MAH), village, NE TVER oblast, W central European Russia, on road junction, 45 mi/72 km NNE of BEZHETSK; 58°24′N 37°03′E. Elevation 606 ft/184 m. In agricultural area; flax processing.

Kesmark, SLOVAKIA: see KEŽMAROK.

Kesova Gora (KYE-suh-vah guh-RAH), town (2006 population 3,905), E TVER oblast, W central European Russia, on road and railroad, 18 mi/29 km NW of KASHIN; 57°35′N 37°17′E. Elevation 534 ft/162 m. In agricultural area; flax processing.

Kessaria, ISRAEL: see CAESAREA.

Kessariani, Greece: see KAISARIANI.

Kessel (KER-suhl), village in commune of LIER, Mechelen district, ANTWERPEN province, N BELGIUM, 3 mi/4.8 km ENE of LIER; 51°08′N 04°37′E. Agricultural market. Has fourteenth–fifteenth-century church, object of pilgrimage.

Kessel (KE-suhl), village, LIMBURG province, SE NETHERLANDS, on MAAS RIVER (car ferry), and 8 mi/12.9 km NNE of ROERMOND; 51°37′N 06°03′E. German border 3 mi/4.8 km to SE. Castle Kessel to NE; recreational center to S. Agriculture (dairying; cattle; mustard, vegetables, fruit).

Kessel Falls, AUSTRIA: see KAPRUN.

Kessel-Lo (KE-suhl–LO), town in commune of LEUVEN, Leuven district, BRABANT province, central BELGIUM, 2 mi/3.2 km ENE of Leuven; 50°54′N 04°45′E. Metal industry; manufacturing. Site of former Benedictine abbey founded in twelfth century. Kessel-Lo Provincial Recreational Center.

Kessel-Lo (KE-suhl–LO), provincial recreation center near KESSEL-LO, BRABANT province, central BELGIUM, 2 mi/3.2 km ENE of LEUVEN.

Kesselsdorf (KE-suhls-dorf), village, SAXONY, E central GERMANY, 7 mi/11.3 km W of DRESDEN; 51°03′N 13°36′E. Scene (December 1745) of Prussian victory over Saxons in second Silesian War.

Kessingland (KES-sing-LAND), fishing village (2001 population 4,760), NE SUFFOLK, E ENGLAND, near NORTH SEA, 5 mi/8 km SSW of LOWESTOFT; 52°25′N 01°42′E. Resort.

Kestenets (kes-ten-ETS), village, SOFIA oblast, KOSTENETS obshtina, BULGARIA; 42°15′N 23°50′E. Mountain resort.

Kesten'ga (KYES-teen-gah), Finnish *Kiestinki*, town, N Republic of KARELIA, NW European Russia, on TOPOZERO Lake, on crossroads and railroad spur, 35 mi/56 km WSW of LOUKHI; 65°53′N 31°50′E. Elevation 364 ft/110 m. Woodworking.

Kesteren (KES-tuh-ruhn), town, GELDERLAND province, central NETHERLANDS, 7 mi/11.3 km ENE of TIEL; 51°56′N 05°34′E. LOWER RHINE RIVER (bridge) 1 mi/1.6 km to N, WAAL RIVER 3 mi/4.8 km to S. Dairying; cattle, hogs; grain, vegetables, fruit, sugar beets. Manufacturing (food processing).

Keston, ENGLAND: see BROMLEY.

Keswick (KEZ-ik), town (2001 population 5,391), CUMBRIA, NW ENGLAND, in the LAKE DISTRICT, on Greta River near N shore of DERWENT WATER, and 22 mi/35 km SSW of CARLISLE; 54°37′N 03°08′W. Market town and tourist center, with some manufacturing,

especially lead pencils. Has church showing traces of Norman origin, Beatrix Potter Lake District exhibition, and Cumberland Pencil Museum. Keswick was scene of Shelley's honeymoon and attracted such poets and writers as Coleridge, Wordsworth, Southey, Lamb, and Ruskin. Robert Southey is buried here; Greta Hall, his home, is preserved. Nearby is the Druids' Circle.

Keswick, town (2000 population 295), KEOKUK county, SE IOWA, 9 mi/14.5 km N of SIGOURNEY; 41°27′N 92°14′W. Livestock, grain.

Keszthely (KEST-he-i), resort city (2001 population 22,388), ZALA county, W HUNGARY, on Lake BALATON, and 25 mi/40 km NNE of NAGYKANIZSA; 46°46′N 17°15′E. Meat industry, small brewery, brickworks. School of agriculture and economics (the *Georgicon*) founded 1797. Winegrowing, horse breeding nearby. Karl Goldmark born here. Ruins of major Roman fortress nearby.

Ket (KYET), river, approximately 845 mi/1,360 km long, KRASNOYARSK TERRITORY, W central SIBERIA, RUSSIA; rises in central Siberia, just N of KRASNOYARSK, and flows NW and W into the OB' RIVER. Navigable for about 410 mi/660 km. It is connected with the KAS RIVER (a tributary of the YENISEY RIVER) by the OB'-YENISEY CANAL SYSTEM.

Keta (KAI-tuh), town, district headquarters, VOLTA region, GHANA, 85 mi/137 km ENE of ACCRA; 05°55′N 00°59′E. Port on narrow strip of land between KETA LAGOON and Gulf of GUINEA. Fishing; saltworking, copra processing, cotton weaving. Agricultural (shallots, corn, cassava). Danish trading post, founded 17th century; passed to British in 1850. Land reclamation project. Site of Fort Prinsenstein, Danish, established 1714.

Ketaka (ke-TAH-kah), town, Ketaka county, TOTTORI prefecture, S HONSHU, W JAPAN, 11 mi/18 km W of TOTTORI; 35°30′N 134°03′E.

Keta Lagoon (KAI-tuh), SE GHANA; 32 mi/51 km long, 2 mi/3.2 km–7 mi/11.3 km wide; 05°54′N 00°56′E. Closed off from Gulf of GUINEA by sandbar.

Ketama (ke-TAH-muh), rural district, Al Hoceïma province, Taza-Al Hoceima-Taounate administrative region, N MOROCCO, 43 mi/69 km SW of AL HOCEIMA, in highest section of RIF mountains with peaks (snow-covered in winter) rising over 7,000 ft/2,134 m. Located in pine and cedar forest; market for illegal "kif" (cannabis) trade. Claims to be the point of origin for Franco's rebellion in 1936, which led to Spanish civil war. The town of MELILLA makes the same claim.

Ketapang (KUH-tah-pahng), town, ⊙ Ketapang district, Kalimantan Barat province, W BORNEO, INDONESIA, port on KARIMATA STRAIT, 130 mi/209 km SSE of PONTIANAK; 01°52′S 109°59′E. Trade center, shipping rubber and rattan.

Ketchenery (kye-chee-NYE-ri), village (2005 population 3,800), NW Republic of KALMYKIA-KHALMG-TANGEH, S European Russia, on crossroads, 62 mi/100 km N of ELISTA; 47°18′N 44°31′E. Elevation 131 ft/39 m. In agricultural area; produce processing. For a short time (1944–1958), called Sovetskoye.

Ketchikan (KE-chi-kan), city (2000 population 7,922), GATEWAY borough, SE ALASKA, a port of entry on REVILLAGIGEDO ISLAND in the ALEXANDER ARCHIPELAGO; 55°21′N 131°35′W. A supply point for miners in the gold rush of the 1890s, it has become a center of Alaska's fishing (especially salmon, halibut, and abalone), and the now-declining logging and pulp industries. Tourism and fish processing adds to the economy. Major molybdenum deposit nearby under sporadic development. Ferry connections to other panhandle communities and BRITISH COLUMBIA, CANADA. Its excellent ice-free harbor on Tongass Narrows makes it an important port on the INSIDE PASSAGE and a distribution point for a large area. Continuance of pulp mill N of town problematic. Headquarter for TONGASS NATIONAL FOREST; Dolly's

House Museum, preserves some of twenty brothels that operated until 1954, in Creek Street red-light district.

Ketchikan Gateway, borough (□ 13,828 sq mi/35,952.8 sq km; 2006 population 13,384), SE ALASKA, includes REVILLAGIGEDO and GRAVINA Islands, as well as part of the Alaska Panhandle mainland. Most of the borough is located in the TONGASS NATIONAL FOREST. Misty Fjords National Monument to E; Totem Bight State Historical Park to S. Main town is KETCHIKAN. Tourism; fishing; timber.

Ketchum, town (2000 population 3,003), BLAINE county, S central IDAHO, 13 mi/21 km NNW of HAILEY, on BIG WOOD RIVER; 43°42′N 114°23′W. Railroad terminus. Manufacturing (printing and publishing, sporting equipment). Summer and winter resort town; Sun Valley Ski Area to N; Sawtooth National Recreation Area to NW; Sawtooth National Forest to E, N, and W.

Ketchum, village (2006 population 293), CRAIG county, NE OKLAHOMA, 10 mi/16 km SE of VINITA; LAKE OF THE CHEROKEES to SE; 36°31′N 95°01′W. Concrete. Incorporated 1938.

Kete-Krachi (ke-TAI–krah-CHEE), town, KRACHI local council headquarters, VOLTA region, GHANA, on N fork of Lake VOLTA on peninsula, 100 mi/161 km NNW of Ho; 07°46′N 00°03′W. Fishing and trade center. Agricultural (cacao, yams, palm oil, kernels). Emergency airfield linked by road.

Ketelmeer (KAI-tuhl-mair), inlet, FLEVOLAND province, E IJSSELMEER, central NETHERLANDS, between NORTH-EAST POLDER and EASTERN FLEVOLAND Polder.

Kethib, Ras, YEMEN: see KATHIB, RAS.

Keti Bandar (ke-TEE bahn-DAHR), town, TATTA district, SW SIND province, SE PAKISTAN, small port in INDUS River delta mouth, near ARABIAN SEA, 55 mi/89 km SSE of KARACHI; 24°08′N 67°27′E.

Ketmen-Töbö (get-min–doo-WUH), town, JALAL-ABAD region, KYRGYZSTAN, on S shore of TOKTOGUL reservoir; 41°33′N 73°07′E. Also KETMEN-TYUBE.

Ketmen-Tyube, KYRGYZSTAN: see KETMEN-TÖBÖ.

Ketoi-kaikyo, RUSSIA: see RIKORD STRAIT.

Keton, RUSSIA: see SMIRNYKH.

Ketongai, RUSSIA: see SMIRNYKH.

Kétou (KAI-too), town, PLATEAU department, SE BENIN, 60 mi/97 km N of PORTO-NOVO, on road from Porto-Novo to NIGER; 07°40′N 02°40′E. Elevation 450 ft/137 m. Cotton, palm kernels, palm oil; customshouse.

Ketovo (kye-TO-vuh), village (2005 population 6,915), central KURGAN oblast, SW SIBERIA, RUSSIA, on the TOBOL River, near highway, 6 mi/10 km SE of KURGAN; 55°21′N 65°19′E. Elevation 301 ft/91 m. In agricultural area (rye, oats, flax); food processing.

Ketoy Island, island, Asia: see KETOI ISLAND.

Ketoy Island (kye-TO-yee) (□ 35 sq mi/91 sq km), one of central main KURIL ISLANDS chain, SAKHALIN oblast, extreme E SIBERIA, RUSSIAN FAR EAST; separated from USHISHIR ISLANDS (N) by RIKORD STRAIT, from SIMUSHIR ISLAND (S) by DIANA STRAIT; 47°20′N 152°29′E. Nearly circular (6 mi/9.7 km in diameter); rises to 3,945 ft/1,202 m.

Ketrzyn (KENT-zeen), Polish *Kętrzyn*, German *Rastenburg*, town (2002 population 28,743) in EAST PRUSSIA, after 1945 in OLSZTYN province, NE POLAND, in Masurian Lakes region, 40 mi/64 km ENE of OLSZTYN (Allenstein). Agricultural market (grain, sugar beets; livestock); sawmilling, brewing. Teutonic Knights founded castle here in early 14th century. Formerly site of German state stud farm. After World War II, when it was c.50% destroyed, German population left the town.

Ketsch (KETSH), town, BADEN-WÜRTTEMBERG, SW GERMANY, on an arm of the RHINE, 2.5 mi/4 km SW of SCHWETZINGEN; 49°22′N 08°32′E. Many residents commute to MANNHEIM. Agriculture (tobacco, asparagus).

Kétté (KAI-TAI), town, East province, CAMEROON, 65 mi/105 km ENE of BERTOUA; 04°53′N 14°34′E.

Kettering (KE-tuhr-ing), city (□ 19 sq mi/49.4 sq km; 2006 population 54,666), MONTGOMERY county, SW OHIO; suburb of DAYTON; 39°42′N 84°09′W. Manufacturing (electric motors, transportation equipment, machinery). Numerous testing laboratories for auto and electrical products. The city is the seat of the Kettering College of Medical Arts and two major hospitals and research centers. Settled c.1812, incorporated 1952. Kettering's population and economic production has gradually declined since 1970.

Kettering (KE-tuh-reeng), coastal town, TASMANIA, SE AUSTRALIA, 23 mi/37 km S of HOBART, on D'ENTRECASTEAUX CHANNEL; 43°07′S 147°16′E. Opposite BRUNY island; ferry access. Orchards.

Kettering (KET-tuhr-ing), town (2001 population 81,844), NORTHAMPTONSHIRE, central ENGLAND, 12 mi/19.3 km NE of NORTHAMPTON; 52°24′N 00°43′W. Road and railroad center. Manufacturing (footwear, leather products, textiles, food products, computer software, apparel, paper products, and machinery).

Kettle Falls, town (2006 population 1,610), STEVENS county, NE WASHINGTON, 8 mi/12.9 km. NE of COLVILLE, between COLVILLE RIVER (S) and a bay of FRANKLIN D. ROOSEVELT LAKE reservoir (COLUMBIA River); 48°37′N 118°04′W. Junction of railroad spur to MARCUS (N), main line terminates at West Kettle Falls, to W. Agriculture (alfalfa, wheat, barley, oats; hogs). Manufacturing (wood products). Part of Colville National Forest to W; Colville Indian Reservation to SW. COULEE DAM National Recreation Area surrounds lake. Town moved (1938–1939) from site c.4 mi/6.4 km S, now covered by lake.

Kettle Island, village, BELL county, SE KENTUCKY, in the CUMBERLAND MOUNTAINS near PINE MOUNTAIN, 13 mi/21 km NNE of MIDDLESBORO. Bituminous coal; timber.

Kettleman City, unincorporated town (2000 population 1,499), KINGS county, S central CALIFORNIA, 26 mi/42 km SW of HANFORD, on California Aqueduct; 36°01′N 119°58′W. TULARE LAKE irrigation reservoir to NE. Agriculture (pistachios, almonds, peaches, plums, nectarines, cantaloupes; grain; cattle, dairying; poultry). Manufacturing (nut processing).

Kettleman Hills, low range (c.1,300 ft/396 m), mainly KINGS and FRESNO counties, S central CALIFORNIA, along W side of San Joaquin Valley, 50 mi/80 km SW of FRESNO. Important oil and natural-gas field here. KETTLEMAN CITY to E.

Kettle Rapids Dam, CANADA: see NELSON, river.

Kettle River, village (2000 population 168), CARLTON county, E MINNESOTA, on KETTLE RIVER, and 40 mi/64 km SW of DULUTH; 46°29′N 92°52′W. Agriculture (poultry; oats, alfalfa; dairying). Manufacturing (Indian pottery, feeds). Sandstone quarries in area.

Kettle River (KE-tuhl), 175 mi/282 km long, S BRITISH COLUMBIA (Canada) and N WASHINGTON (U.S.); rises in Canada in MONASHEE MOUNTAINS W of Upper Arrow Lake, flows S to vicinity of GRAND FORKS, where it crosses Washington state line twice, finally entering COLUMBIA River 13 mi/21 km NW of COLVILLE; 49°00′N 118°12′W. Fertile agricultural valley (fruits, vegetables).

Kettle River, 80 mi/129 km long, E MINNESOTA; rises in marshy area N of CARLTON county, c.15 mi/24 km W of CLOQUET, flows S, through Kettle Lake, and past villages of KETTLE RIVER and WILLOW RIVER, and town of SANDSTONE to ST. CROIX River, in PINE county, 10 mi/16 km ENE of PINE CITY, on W boundary of St. Croix State Park; 46°41′N 92°47′W.

Kettlersville (KET-luhrz-vil), village (□ 1 sq mi/2.6 sq km; 2006 population 170), SHELBY county, W OHIO, 12 mi/19 km NNW of SIDNEY; 40°26′N 84°16′W. Also spelled Kettlerville.

Ketton (KET-tuhn), village (2001 population 2,469), LEICESTERSHIRE, central ENGLAND, 4 mi/6.4 km WSW

of STAMFORD; 52°37′N 00°33′W. Cement, concrete works; stone quarrying. Church dates from 13th century.

Kettwig (KET-vik), suburb of ESSEN, NORTH RHINE-WESTPHALIA, W GERMANY, on RUHR RIVER, and 7 mi/11.3 km SW of city center; 51°22′N 06°56′E. Incorporated into Essen 1975.

Kety (KEN-tee), Polish *Kęty*, German *Kenty*, town, Bielsko-Biała province, S POLAND, 34 mi/55 km WSW of CRACOW, near SOLA RIVER. Manufacturing (furniture, cement products, woolen textiles), tanning, flour milling, sawmilling; iron foundry, brickworks. Monastery. Until 1951 in Crakow province.

Ketzin (ket-TSEEN), town, BRANDENBURG, E GERMANY, on the HAVEL, and 11 mi/18 km NW of POTSDAM; 52°29′N 12°50′E. Manufacturing (animal feeds; seed nurseries).

Keuka (KYOO-kuh), resort village, STEUBEN county, W central NEW YORK, on E shore of KEUKA LAKE, 27 mi/43 km SSW of GENEVA; 42°29′N 77°07′W.

Keuka Lake (KYOO-kuh), one of FINGER LAKES, W central NEW YORK; 42°28′N 77°10′W. Lake is 18 mi/29 km long and 0.5 mi/0.8 km–2 mi/3.2 km wide; drains NE into SENECA LAKE. Is the only Finger Lake that still retains its distinctive, post-Pleistocene Y-shape. PENN YAN at N end and HAMMONDSPORT at S end are trade centers for surrounding resort, grape-growing, and wine-making region. The vineyard-cloaked slopes (esp. on W side) make this perhaps the loveliest of the Finger Lakes.

Keuka Park (KYOO-kuh), village, YATES county, W central NEW YORK, on KEUKA LAKE, 19 mi/31 km SSW of GENEVA; 42°37′N 77°05′W. Seat of Keuka College.

Keuruu (KAI-oo-roo), Swedish *Keuru*, village, KESKI-SUOMEN province, S central FINLAND, 13 mi/21 km W of JYVÄSKYLÄ; 62°16′N 24°42′E. Elevation 396 ft/120 m. On small lake, in lumbering region. Has eighteenth-century church.

Keutschach am See (KOI-chahkh ahm SAI), village, CARINTHIA, S AUSTRIA, in the Sattnitz 6 mi/9.7 km SW of KLAGENFURT; 46°36′N, 14°11′E. Elevation 1,631 ft/497 m. Summer resort in the center of a series of small lakes (Keutschacher See, Hafnersee, Rauschelesee). Castle.

Keve (KAI-VAI), town, ⊙ AVÉ prefecture, MARITIME region, S TOGO, 35 mi/56 km NW of LOMÉ, on road and railroad to KPALIMÉ; 06°26′N 00°56′E. Cacao, cassava, and coffee.

Kevelaer (kai-ve-LAHR), town, NORTH RHINE–WEST-PHALIA, W GERMANY, in the RUHR industrial district, near the NIERS, 7 mi/11.3 km SSE of GOCH; 51°35′N 06°15′E. Manufacturing (jewelry, musical instruments, crafts; silverworks, printing); flower breeding. Noted pilgrimage place (since 1642) with image of the Virgin.

Kevil (KEV-uhl), village (2000 population 574), BALLARD county, W KENTUCKY, 16 mi/26 km W of PADUCAH; 37°04′N 88°53′W. In agricultural area (tobacco, grain, soybeans; livestock; dairying). Manufacturing (loudspeakers, machinery).

Kevin, village, TOOLE county, N MONTANA, 19 mi/31 km NNW of SHELBY; 48°45′N 111°58′W. Located in Kevin-Sunburst oil and natural gas field. Cattle, sheep, hogs; wheat, barley, hay.

Kew, town, TURKS AND CAICOS ISLANDS, NORTH CAICOS Island; 21°54′N 72°02′W.

Kew (KYOO), residential suburb, S VICTORIA, AUSTRALIA, 4 mi/6 km ENE of MELBOURNE, bounded on N and W by the YARRA RIVER, on S by HAWTHORN, and on E by Balwyn; 37°49′S 145°02′E. In metropolitan area.

Kewagama Lake, W QUEBEC, E CANADA, 22 mi/35 km E of ROUYN-NORANDA; 10 mi/16 km long, 5 mi/8 km wide. Elevation 958 ft/292 m. Drains W into Lake TIMISKAMING.

Kewanee (kee-WAH-nee), industrial city (2000 population 12,944), HENRY county, NW ILLINOIS, 15 mi/24 km ESE of CAMBRIDGE; 41°14′N 89°55′W. A regional livestock, processing, trade, and shipping center. Manufacturing (boilers, metal products, farm machinery). The city holds an annual "Hog Capital Festival." The E campus of Black Hawk College is located 6 mi/9.7 km S of Kewanee. Incorporated 1855.

Kewanna, town (2000 population 614), FULTON county, N INDIANA, 11 mi/18 km SW of ROCHESTER; 41°01′N 86°25′W. Agricultural area. Manufacturing (metal products).

Kewaskum (kee-WAWS-kuhm), town (2006 population 3,970), WASHINGTON county, E WISCONSIN, on branch of MILWAUKEE RIVER, and 7 mi/11.3 km NNW of West Branch; 43°31′N 88°13′W. In dairy and farm area. Manufacturing (food-processing, machinery, cookware). Sunburst Ski Area to S; S end of Kettle Moraine State Forest (N unit) to E.

Kewaunee (kee-WAW-nee), county (□ 1,084 sq mi/ 2,818.4 sq km; 2006 population 20,832), E WISCONSIN; ⊙ KEWAUNEE; situated near base of DOOR PENINSULA; 44°35′N 87°26′W. Farming, dairying, lumbering, woodworking; agriculture (barley, oats, wheat, corn, peas, beans; hay; cattle, hogs). Partly wooded, hilly terrain, drained by Kewaunee River and several small streams. Bounded E by LAKE MICHIGAN, NW by GREEN BAY. Kewaunee nuclear power plant, initial criticality March 7, 1974, is 27 mi/43 km of GREEN BAY, uses cooling water from Lake Michigan, and has a maximum dependable capacity of 511 MWe. Formed 1852.

Kewaunee, city (2006 population 2,868), ⊙ KEWAUNEE county, E WISCONSIN, on LAKE MICHIGAN, on DOOR PENINSULA, at mouth of small Kewaunee River, 25 mi/ 40 km E of GREEN BAY; 44°27′N 87°30′W. In dairying and stock-raising area. Manufacturing (furniture, wood and aluminum products, construction equipment, green-pea combines); dairy plants; breweries. Railroad terminus. A coast guard station is here. The North West Company set up a fur-trading post on site in 1795. Incorporated 1883.

Kewaunee Nuclear Power Plant, WISCONSIN: see KEWAUNEE county.

Keweenaw (KEE-WEE-naw), county (□ 5,959 sq mi/ 15,493.4 sq km; 2006 population 2,183), NW UPPER PENINSULA, MICHIGAN; ⊙ EAGLE RIVER; 47°28′N 88°09′W. On NE Keweenaw Peninsula in LAKE SUPERIOR. Resorts. About 2/3 of residences in county are seasonal. Includes ISLE ROYALE NATIONAL PARK (NW) and is traversed by the COPPER RANGE. Several lakes. One of heaviest snowfall areas in U.S. MANITOU ISLAND off E end of peninsula (3 mi/4.8 km long, 1 mi/ 1.6 km wide). Fort Wilkins State Park, E of COPPER HARBOR. Formed and organized 1861.

Keweenaw (KEE-WEE-naw), peninsula, 60 mi/97 km long, projecting NE from the W UPPER PENINSULA, NW MICHIGAN, into LAKE SUPERIOR; 47°15′N 88°20′W. PORTAGE LAKE and a connecting ship canal cut across the middle of the peninsula, converting its upper portion into an island and creating an important waterway. Consists mainly of parts of HOUGHTON and KEWEENAW counties. The canal is crossed by a highway bridge with one of the world's heaviest lift spans. Main towns are HOUGHTON and HANCOCK. The area is popular with vacationers.

Keweenaw Bay (KEE-WEE-naw), NW UPPER PENINSULA, MICHIGAN, inlet of Lake SUPERIOR (c.30 mi/ 48 km long, 15 mi/24 km wide at entrance, narrows to a point at S end) lying to E of curving KEWEENAW Peninsula, 55 mi/89 km NW of MARQUETTE; 46°55′N 88°19′W. It is E terminus of KEWEENAW WATERWAY. BARAGA, Keweenaw Bay Village, and L'ANSE, all resorts, are at its head.

Keweenaw National Historical Park (KEE-WEE-naw), HOUGHTON, KEWEENAW, and ONTONAGON counties, UPPER PENINSULA, NW MICHIGAN. Commemorates copper mining heritage of Upper Michigan's KEWEENAW Peninsula. Interim boundaries include village of CALUMET, which has remained virtually unchanged from mining days (c.1900), and the old Quincy Mine N of HANCOCK. There are also eight cooperative sites, including KEWEENAW and HOUGHTON county museum, FORT WILKINS and PORCUPINE MOUNTAINS state parks. Established 1992; under development as of 1995.

Keweenaw Range, MICHIGAN: see COPPER RANGE, WISCONSIN.

Keweenaw Waterway (KEE-WEE-naw), navigation channel (c.25 mi/40 km long), NW UPPER PENINSULA, MICHIGAN, intersecting KEWEENAW Peninsula between KEWEENAW BAY (SE) and Lake SUPERIOR (NW). Consists of PORTAGE LAKE (c.20 mi/32 km long, 2 mi/3.2 km wide), with a natural connection with Keweenaw Bay, and a land-cut ship canal (c.2 mi/3.2 km long) across former portage between PORTAGE LAKE and LAKE SUPERIOR. U.S. lighthouse at Lake Superior entrance. HANCOCK, HOUGHTON are ports.

Keweigek (ke-WEI-gek), village, CUYUNI-MAZARUNI district, GUYANA; 05°56′N 60°38′W. Airfield.

Kew Gardens, residential section of central QUEENS borough of NEW YORK CITY, SE NEW YORK; 40°43′N 73°50′W. Upper-middle-class, formerly mostly white but now relatively diverse neighborhood. Originally laid out as an English-style planned community in early 20th century, Contains Queens Borough Hall.

Kew Gardens (KYOO), SURREY, S ENGLAND, on the THAMES just W of London; 51°28′N 00°18′W. Royal Botanic Gardens is the official name. The gardens were founded by the dowager princess of Wales in 1761 and consisted of about 0.01 sq mi/0.04 sq km. They were presented to the nation as a royal gift in 1841. They now cover 0.4 sq mi/1.2 sq km and contain thousands of species of plants, four museums, and laboratories and hothouses. The Chinese Pagoda, c.165 ft/50 m high, was designed by William Chambers in 1761; it is still a famous landmark.

Kexholm, RUSSIA: see PRIOZËRSK.

Keyaluvik (kai-YAH-luh-vik), village, SW ALASKA, on Bethel Bay.

Keya Paha (KEE-yah PAH-hah), county (□ 774 sq mi/ 2,012.4 sq km; 2006 population 892), N NEBRASKA; ⊙ SPRINGVIEW (located N of Sand Hills region); 42°52′N 99°43′W. Grazing and mixed farming area bounded N by SOUTH DAKOTA state line, S by NIOBRARA RIVER; drained by KEYA PAHA RIVER. Cattle, dairying, hogs; corn, alfalfa, wild hay. Formed 1884.

Keya Paha River (KEE-yah PAH-hah), 101 mi/163 km long, S SOUTH DAKOTA and N NEBRASKA; rises in TODD county, South Dakota, flows ESE to NIOBRARA RIVER near BUTTE, Nebraska; 43°13′N 100°23′W.

Key Biscayne (bis-KAIN), town, E MIAMI-DADE county, SE FLORIDA; 25°41′N 80°10′W. Became independent from city of MIAMI in 1990. Site of annual Lipton tennis tournament.

Key Biscayne (bis-KAIN), island (□ 1 sq mi/2.6 sq km), MIAMI-DADE county, SE FLORIDA, 5 mi/8 km SSE of MIAMI; 25°41′N 80°10′W. Island is 5 mi/8 km long. Partly shelters BISCAYNE BAY (W). At S end is CAPE FLORIDA, location of Bill Baggs State Park and Beach.

Key Colony Beach (KEE), island (2000 population 788), MONROE county, central FLORIDA KEYS, 54 mi/ 87 km NE of KEY WEST; 24°43′N 81°01′W.

Keyes, unincorporated town (2000 population 4,575), STANISLAUS county, central CALIFORNIA, 7 mi/11.3 km SSE of MODESTO, in San Joaquin Valley; 37°34′N 120°55′W. Fruit, nuts, vegetables, grapes, melons; grain; dairying, poultry.

Keyes (KEEZ), village (2006 population 362), CIMARRON county, W OKLAHOMA Panhandle, 15 mi/ 24 km ENE of BOISE CITY; 36°48′N 102°15′W. Livestock; grain.

Keyesport, village (2000 population 481), in BOND and CLINTON counties, S central ILLINOIS, 12 mi/19 km NNE of CARLYLE; 38°44′N 89°16′W. In agricultural and oil area. On CARLYLE LAKE.

Keyhole Reservoir, CROOK county, NE WYOMING, on BELLE FOURCHE RIVER, 18 mi/29 km WSW of SUN-DANCE; 10 mi/16 km long; 44°22′N 104°42′W. Maximum capacity 634,000 acre-ft. Has 5 mi/8 km S arm. Formed by Keyhole Dam (113 ft/34 m high), built (1952) by the Bureau of Reclamation for irrigation and flood control. Keyhole State Park at E end.

Key Largo (LAHR-go), town (2000 population 11,886), MONROE county, upper FLORIDA KEYS, 60 mi/97 km S of MIAMI, northernmost island of Florida Keys; 25°06′N 80°26′W. Diversified light manufacturing, especially marine equipment.

Key Largo (LAHR-go), narrow island (□ 22 sq mi/57.2 sq km), c.30 mi/48 km long, off SE FLORIDA, northernmost and largest of the FLORIDA KEYS; 25°06′N 80°26′W. Along with other Florida Key islands, especially KEY WEST, it has become an increasingly popular tourist destination, noted for its scuba diving and beachside resorts. Housing developments and shopping complexes have been recently constructed. A major attraction is John Pennekamp Coral Reef State Park, the first underwater park in the U.S., containing c.78 sq mi/202 sq km of living coral and hundreds of varieties of marine life.

Keyling Inlet, AUSTRALIA: see FITZMAURICE RIVER.

Key, Lough, Gaelic *Loch Cé*, lake (3 mi/4.8 km long, 3 mi/4.8 km wide), N ROSCOMMON county, N central IRELAND, on Boyle River, 2 mi/3.2 km NE of BOYLE; 54°00′N 08°15′W. Noted for its scenic beauty. Remains of Abbey of the Trinity are on an island in lake; the *Annals of Lough Cé*, written here, are in Trinity College library, DUBLIN.

Keymer (KEE-muh), town (2001 population 5,136), West SUSSEX, SE ENGLAND, 7 mi/11.3 km N of BRIGHTON; 50°55′N 00°08′W. Church dates from 13th century.

Keynsham (KAIN-shuhm), town (2001 population 17,972), North Somerset, W ENGLAND, on the Avon River, and 5 mi/8 km SE of BRISTOL; 51°25′N 02°30′W. Manufacturing (chocolate, paper, chemicals, and soap). Has 13th–15th-century church, remains of three Roman villas, and traces of ancient monastery.

Keyport, unincorporated village, KITSAP county, NW WASHINGTON, 8 mi/12.9 km N of BREMERTON, on PORT ORCHARD, arm of PUGET SOUND, at entrance to Liberty Bay. Manufacturing (electronics; computers). U.S. Naval Reservation is here.

Keyport, borough (2006 population 7,471), MONMOUTH county, E NEW JERSEY, 13 mi/21 km NNE of FREE-HOLD; 40°25′N 74°12′W. Resort and fishing center with harbor on RARITAN BAY. Manufacturing; largely residential. Settled before 1700, incorporated 1908.

Keysborough (KEEZ-bruh), residential and industrial suburb 17 mi/27 km SE of MELBOURNE, VICTORIA, SE AUSTRALIA, just N of DANDENONG; 38°01′S 145°10′E. Partly agricultural. Several parks and reserves, including a flora reserve.

Keyser (KEIZ-uhr), town (2006 population 5,334), ⊙ MINERAL county, NE WEST VIRGINIA, 20 mi/32 km SW of CUMBERLAND, MARYLAND, in E PANHANDLE, on North Branch of the POTOMAC RIVER; 39°26′N 78°58′W. Manufacturing (lumber, crushed limestone, glass, apparel, stainless-steel products, air-conditioning products, charcoal, coal processing). Agriculture (grain, tobacco); livestock; poultry. Potomac State College of West Virginia University (2 year). An important supply base in Civil War. Settled 1802.

Keyser, NORTH CAROLINA: see ADDOR.

Keystone, town (2000 population 687), BENTON county, E central IOWA, 14 mi/23 km SW of VINTON; 42°00′N 92°12′W. In agricultural area.

Keystone, village (2006 population 315), PENNINGTON county, SW SOUTH DAKOTA, 10 mi/16 km SW of RAPID CITY, in BLACK HILLS and Black Hills National Forest; 43°53′N 103°25′W. Elevation 4,342 ft/1,323 m. Feldspar mill and pottery market. MOUNT RUSHMORE NATIONAL MEMORIAL nearby. CUSTER STATE PARK to S.

Keystone (KEE-ston), village (2006 population 389), MCDOWELL county, S WEST VIRGINIA, on TUG FORK

RIVER, 7 mi/11.3 km E of WELCH; 37°25′N 81°27′W. Semi-bituminous-coal region. Incorporated 1909.

Keystone Heights (KEE-ston), town (2000 population 1,349), CLAY county, NE FLORIDA, c.45 mi/72 km W of ST. AUGUSTINE; 29°46′N 82°01′W. Resort in lake region.

Keystone Lake, reservoir, TULSA and CREEK counties and on OSAGE-PAWNEE county border, NE central OKLAHOMA, on ARKANSAS RIVER, 17 mi/27 km W of TULSA; c.30 mi/48 km long; 36°08′N 96°14′W. Maximum capacity 1,879,000 acre-ft. CIMARRON RIVER forms c.25-mi/40-km W arm. Formed by Keystone Dam (104 ft/32 m high), built by the Army Corps of Engineers for flood control, water supply, and power generation. Keystone (S shore) and Walnut Creek (N shore) state parks are here.

Keystone Lake, reservoir, ARMSTRONG county, W PENNSYLVANIA, on North Branch of Plum Creek, 10 mi/16 km NW of INDIANA; 5 mi/8 km long; 40°13′N 79°18′W. Maximum capacity 168,400 acre-ft. Formed by Keystone Dam (100 ft/30 m high), built (1965) for water supply. Also called Atwood Keystone Lake.

Keystone State; see PENNSYLVANIA.

Keysville, town, BURKE county, E GEORGIA, 21 mi SW of AUGUSTA, and on BRIER CREEK; 33°14′N 82°14′W.

Keysville (KEEZ-vil), town (2006 population 784), CHARLOTTE county, S VIRGINIA, 45 mi/72 km SE of LYNCHBURG; 37°02′N 78°28′W. Manufacturing (lumber, apparel, furniture). Agriculture (tobacco, grain; dairying; cattle); timber. Southside Virginia Community College (John H. Daniel campus).

Keytesville (KEETS-vil), city (2000 population 533), ⊙ CHARITON county, N central MISSOURI, near MISSOURI River, 28 mi/45 km W of MOBERLY; 39°25′N 92°56′W. Corn, wheat, soybeans; hogs, cattle; lumber products. Site of French Fort Orléans (established 1723) nearby. Plotted 1830.

Key West (KEE), city (□ 6 sq mi/15.6 sq km; 2000 population 25,478), MONROE county, on an island at the SW extremity of the FLORIDA KEYS, c.150 mi/240 km SW of MIAMI (but only 90 mi/145 km from CUBA). The southernmost city of the continental U.S., it is a port of entry (cruise ships), a popular resort with a tropical climate, a shrimping and fishing center, and an artists' colony.

Kez (KYES), town (2006 population 10,755), NE UDMURT REPUBLIC, E European Russia, on road and the TRANS-SIBERIAN RAILROAD, 50 mi/80 km ESE of GLAZOV; 57°53′N 53°43′E. Elevation 554 ft/168 m. Wood industries, flax processing, food processing. Established in 1899 with the construction of the railroad; town status since 1942.

Kezar Falls, MAINE: see PARSONSFIELD.

Kezar Lake (KEI-zuhr), in OXFORD county, W MAINE, in summer resort area near NEW HAMPSHIRE state line; c.7.5 mi/12.1 km long. Drains S, through Kezar Pond, to SACO RIVER.

Kezar Lake, NEW HAMPSHIRE: see SUTTON.

Kézdivásárhely, ROMANIA: see TÎRGU SECUIESC.

Kezhma (kyezh-MAH), village, SE KRASNOYARSK TERRITORY, central SIBERIA, RUSSIA, on the ANGARA RIVER, on road, 235 mi/378 km NE of TAYSHET; 58°58′N 101°07′E. Elevation 534 ft/162 m. Lumbering. Founded in 1665.

Kezi (KE-zee), town, MATABELELAND SOUTH province, SW ZIMBABWE, 55 mi/89 km SSW of BULAWAYO; 20°55′S 28°27′E. MATOBO NATIONAL PARK to N. Cattle, sheep, goats; peanuts, corn, soybeans; cotton.

Kežmarok (kezh-MAH-rok), German *Kaisersmarkt*, Hungarian *Kesmark*, city, VYCHODOSLOVENSKY province, N SLOVAKIA, in SE part of the HIGH TATRAS, on POPRAD RIVER, on railroad; 49°08′N 20°26′E. Long-established trade center; has woodworking; machinery and textile manufacturing. Has picturesque castle, formerly of Tököly family, with five towers and 15th-century chapel; 15th-century Gothic church; 18th-century wooden church; Renaissance buildings;

modern church with remains of Imre Tököly, brought from TURKEY. Settled continuously since Neolithic times. Military base.

Kezmarske Zlaby (kezh-MAHR-skai zhle-BI), Slovak *Kežmarské Žľaby*, village, VYCHODOSLOVENSKY province, N SLOVAKIA, at E foot of LOMNICA PEAK in the HIGH TATRAS, 7 mi/11.3 km NW of KEŽMAROK; 49°12′N 20°18′E. Elevation of 3,215 ft/980 m. Summer resort with sanatorium. Belianska Cave just N.

Kfar Adumim, Jewish settlement, SW of JERICHO and NE of JERUSALEM in the Judean Wilderness, WEST BANK. An outlying suburb of Jerusalem. Some mixed farming (hothouse flowers and orchards). Founded in 1979.

Kfar Azar (ki-FAHR ah-ZAHR), moshav, W ISRAEL, in SHARON plain, 4 mi/6.4 km E of TEL AVIV; 32°03′N 34°50′E. Elevation 147 ft/44 m. Dairying, poultry; vegetables. Rapidly urbanizing. Founded 1932.

Kfar Azza (ki-FAHR AH-zah), kibbutz, ISRAEL, 4 mi/6.4 km ESE of GAZA in NW NEGEV; 31°29′N 34°32′E. Elevation 370 ft/112 m. International plastics manufacturer Kafrit. Mixed farming. Manufacturing (chemicals, paints). Founded in 1951.

Kfar Baruch (ki-FAHR bah-ROOKH) or **Kfar Barukh**, cooperative settlement, NW ISRAEL, in JEZREEL plain, near KISHON RIVER, 6 mi/9.7 km WNW of AFULA; 32°38′N 35°11′E. Elevation 206 ft/62 m. Grain, sunflowers. Founded 1926.

Kfar Bialik (ki-FAHR bee-AH-leek), village on NE outskirts of HAIFA, ISRAEL; 32°49′N 35°05′E. Elevation 39 ft/11 m. Vegetable farming, especially for export. Founded in 1934.

Kfar Bilu (ki-FAHR BEE-loo), moshav on SE outskirts of REHOVOT on coastal plain, central ISRAEL; 31°52′N 34°49′E. Elevation 200 ft/60 m. Mixed farming, citriculture. Ruins of ancient Samarian settlement. Founded in 1932.

Kfar Blum (ki-FAHR BLOOM), kibbutz, N ISRAEL, 3.7 mi/6 km SE of KIRYAT SHMONA in HULA VALLEY; 33°10′N 35°36′E. Elevation 239 ft/72 m. Manufacturing (metal products). Agriculture (cotton; dairy; fruit). Regional school and sports arena. Tourism. Guest house and water sports on the JORDAN RIVER (kayaking and rowing). Summer music festival. Founded in 1943 and named for Leon Blum, the French Jewish socialist and premier.

Kfar Chabad (ki-FAHR khah-BAHD) or **Kfar Habad**, Jewish township, ISRAEL, NW of RAMLA, and W of Ben-Gurion Airport on coastal plain; 31°59′N 34°51′E. Elevation 213 ft/64 m. Founded in 1949 by Russian immigrants who were members of the ultra-orthodox, hassidic Chabad sect. Serves as the Israeli center of Habad and has a replica of the brick-faced Chabad headquarters in BROOKLYN, NEW YORK.

Kfar Ezion or **Kfar Etzion**, kibbutz in Ezion Bloc, WEST BANK, 8 mi/12.9 km N of HEBRON, and 6 mi/10 km SW of BETHLEHEM. Mixed farming, light industry (camping equipment, cookware, kitchen accessories). The first modern Jewish settlement was established here in the late 1920s, but was abandoned twice before a kibbutz took over in 1943 and was destroyed in the 1948 Arab revolt. In 1967 the present kibbutz was established on the site of the earlier settlement by Orthodox Jewish settlers and survivors from the 1943 kibbutz.

Kfar Felix Warburg, ISRAEL: see KFAR WARBURG.

Kfar Gil'adi (ki-FAHR gee-lah-DEE), kibbutz, N ISRAEL, 2 mi/3.2 km N of KIRYAT SHMONA in UPPER GALILEE; 33°14′N 35°34′E. Elevation 826 ft/251 m. Mixed farming. Manufacturing (machinery, fish breeding); quarry; guest house. Nearby, the remains of a mausoleum. Founded in 1916 and temporarily abandoned during World War I and again in 1920 following the battle at Tel Hai. Merged with the survivors of Tel Hai in 1926. A Palmah base camp and transit stop for illegal immigrants from Syria and Lebanon before 1948.

Kfar Hahoresh (ki-FAHR hah-kho-RESH), kibbutz, LOWER GALILEE, N ISRAEL, 2 mi/3.2 km W of NAZARETH; 32°42′N 35°16′E. Elevation 1,548 ft/471 m. Livestock, fruit; afforestation. Founded 1933.

Kfar Hamaccabi (ki-FAHR hah-mah-kah-BEE), kibbutz, NW ISRAEL, in ZEBULUN VALLEY, 7 mi/11.3 km E of HAIFA; 32°47′N 35°06′E. Elevation 239 ft/72 m. Manufacturing (rubber products). Mixed farming, banana growing; carp ponds. Founded 1936.

Kfar Hanasi (ki-FAHR hah-nah-SEE), kibbutz, N ISRAEL, 6 mi/9.7 km E of ZEFAT (Safed) and N of SEA OF GALILEE; 32°58′N 35°36′E. Mixed farming, some industry, and a guest house. Site of the only hydroelectric power plant in Israel. Founded in 1948 and named for Chaim Weizmann, Israel's first president (Hebrew, *nasi*=president).

Kfar Haroeh (ki-FAHR hah-ro-EH), moshav, W ISRAEL, in SHARON plain, 4 mi/6.4 km S of HADERA; 32°23′N 34°55′E. Elevation 32 ft/9 m. Mixed farming, citriculture. Has Talmudic and agriculture college. Founded 1934.

Kfar Hasidim (ki-FAHR khah-see-DEEM), religious kibbutz, N ISRAEL, 6 mi/9.7 km SE of HAIFA; 32°45′N 35°05′E. Elevation 269 ft/81 m. Market gardening, various field crops; dairying. Founded in 1924 by hassidic Jews from Poland.

Kfar Hess (ki-FAHR HES), cooperative settlement, ISRAEL, 6 mi/10 km SE of NETANYA on SHARON plain; 32°15′N 34°56′E. Elevation 288 ft/87 m. Roman cemetery. Founded in 1933, was on front lines in 1948 war and the target of incursions from nearby Arab villages and Iraqi-held areas.

Kfar Hittim (ki-FAHR khee-TEEM), cooperative settlement, LOWER GALILEE, NE ISRAEL, 2 mi/3.2 km WNW of TIBERIAS; 32°48′N 35°30′E. Elevation 45 ft/13 m below sea level. Mixed farming; textile milling. Near the HORNS OF HITTIN, site of the famous Battle of the Horns of Hittin, in which Saladin had a major victory over the Crusaders. Founded 1936.

Kfariye (ke-fah-RI-ye), township, LATTAKIA district, W SYRIA, 20 mi/32 km NE of LATTAKIA; 35°41′N 36°04′E. Solid bitumen deposits exploited here.

Kfar Malal (ki-FAHR mah-LAHL), moshav, W ISRAEL, on SHARON plain, 3 mi/4.8 km E of HERZLIYA; 32°10′N 34°53′E. Elevation 236 ft/71 m. Dairying, mixed farming. Founded 1911; destroyed in World War I and again 1921.

Kfar Menahem (ki-FAHR mi-nah-KHEM), kibbutz, ISRAEL, 12 mi/19 km SE of ASHDOD on coastal plain; 31°44′N 34°50′E. Elevation 429 ft/130 m. Mixed farming. Manufacturing (metal products, ceramics). Founded in 1935, it was only in 1939 that it was firmly established. Nearby, signs of prehistoric and Roman settlements.

Kfar Pines (ki-FAHR PEE-nes), cooperative settlement, ISRAEL, 4.5 mi/7.2 km NE of HADERA; 32°28′N 35°00′E. Elevation 154 ft/46 m. Mixed farming, citriculture. Founded in 1933.

Kfar Rosh Hanikra (ki-FAHR ROSH hah-neek-RAH) or **Kfar Rosh Haniqra**, kibbutz, N ISRAEL, 6 mi/10 km N of NAHARIYA on MEDITERRANEAN coast just S of Lebanese border; 33°05′N 35°06′E. Elevation 0 ft/0 m. Farming. Manufacturing (medical appliances); holiday village. Founded in 1949, name derives from adjacent ROSH HANIKRA promontory.

Kfar Ruppin (ki-FAHR roo-PEEN), kibbutz, NE ISRAEL, on the JORDAN RIVER (border of KINGDOM OF JORDAN), and 4 mi/6.4 km SE of BEIT SHE'AN; 32°27′N 35°33′E. Below sea level 757 ft/230 m. Manufacturing (plastic products); fish breeding, fruit growing. Founded 1938.

Kfar Sava (ki-FAHR SAH-bah) or **Kfar Saba**, city (2006 population 81,100), W ISRAEL, in SHARON plain, 11 mi/18 km NE of TEL AVIV metropolitan area; 32°10′N 34°54′E. Elevation 167 ft/ 50 m. Manufacturing (metal products, pharmaceuticals, food processing). Agriculture (citriculture, mixed farming). Has large hospital Just N is site of ancient locality of *Capharsaba*, founded in era of Second Temple, fortified by Alexander Jannaeus. Received status of local council in 1937 and officially became a city in 1962. Modern settlement founded 1903, destroyed during World War I, and again in subsequent Arab riots; resettled 1924.

Kfar Shmaryahu (ki-FAHR shi-mahr-YAH-hoo), urban settlement, ISRAEL, between TEL AVIV and HERZLIYA on SHARON plain; 32°11′N 34°49′E. Elevation 95 ft/28 m. Founded in 1936 as an agricultural settlement, it is now one of Israel's most exclusive neighborhoods; home also to many diplomats serving in Israel. Site of American School. Caves and some graves from 4th century.

Kfar Sirkin (ki-FAHR SIR-keen), cooperative on eastern outskirts of PETAH TIKVA, central ISRAEL; 32°04′N 34°55′E. Elevation 180 ft/54 m. Agriculture (citrus, deciduous, and subtropical fruit cultivation, market gardening; poultry). Ancient grave found by the village. Founded in 1930s.

Kfar Szold (ki-FAHR SOLD), kibbutz, N ISRAEL, 5 mi/8 km ESE of KIRYAT SHMONA, HULA VALLEY; 33°11′N 35°39′E. Elevation 1,007 ft/306 m. Mixed farming; fish ponds. Manufacturing (electronic products); solar energy factory. Founded in 1940s and named for the U.S.-born Zionist and philanthropist Henrietta Szold, who played a major role in the Youth Aliya (immigration) program that brought thousands of Jewish youths to the area from wartime Europe. In 1948 war, was the first Jewish settlement in the N to be attacked by Syrians.

Kfar Tavor (ki-FAHR tah-VOR) or **Kfar Tabor**, moshav, N ISRAEL, 7 mi/11.3 km ESE of NAZARETH (E of Mt. Tavor) in LOWER GALILEE; 32°41′N 35°25′E. Elevation 452 ft/137 m. Mixed farming, viticulture, olives; grain; dairying. Founded initially in 1901 but officially in 1909.

Kfar Uriya (ki-FAHR oo-ree-YAH) or **Kfar Uria**, village, central ISRAEL, in the JUDEAN FOOTHILLS, 11 mi/18 km E of REHOVOT; 31°47′N 34°57′E. Elevation 1,082 ft/329 m. Grain, mixed farming. Founded 1912. Destroyed during 1929 riots, resettled 1944.

Kfar Vitkin (ki-FAHR VEET-keen), moshav, ISRAEL, 3 mi/4.8 km NNE of NETANYA on SHARON plain; 32°23′N 34°53′E. Elevation 88 ft/26 m. Possibly site of Byzantine settlement. Founded in 1930, the first years were plagued with malaria; later, it developed into the largest agricultural village in the country and retained this status for several decades. Named for Zionist pioneer Joseph Vitkin.

Kfar Vradim (ki-FAHR vi-rah-DEEM) or **Kfar Veradim** (Hebrew=Village of Roses) Jewish town, in WESTERN GALILEE, ISRAEL, 1.2 mi/2 km S of MA'ALOT-TARSHIHA. Established in 1984 as a private initiative by industrialist Steff Wertheimer as an environmentally friendly suburb most of whose residents are professionals and work in nearby cities. TEFEN Industrial Park, Wertheimer's industrial center 1.2 mi/2 km to the S, makes blades for aircraft industry and other products.

Kfar Warburg (ki-FAHR VAHR-boorg), moshav, central ISRAEL, in coastal plain, 13 mi/21 km SSW of REHOVOT; 31°43′N 34°44′E. Elevation 157 ft/47 m. Mixed farming, citriculture. Sometimes called Kfar Felix Warburg. Founded 1939.

Kfar Yehezkel (ki-FAHR yi-KHEZ-kel) or **Kfar Yehezqel**, cooperative settlement, N ISRAEL, 5 mi/8 km SE of AFULA in JEZREEL VALLEY; 32°34′N 35°21′E. Elevation 82 ft/24 m. Mixed farming, citriculture, market gardening. Signs of settlement found here from Chalcolithic era, early and middle Bronze Age, early Iron Age, and Roman and Byzantine times. Founded in 1921.

Kfar Yehoshua (ki-FAHR yi-ho-SHOO-ah), moshav, N ISRAEL, 8 mi/12.9 km WSW of NAZARETH in JEZREEL VALLEY; 32°40′N 35°09′E. Elevation 236 ft/71 m. Mixed farming (dairy, poultry; field crops, orchards; honey, flowers). Local archaeological and natural history museum. Founded 1927.

Kfar Yona (ki-FAHR yo-NAH) or **Kfar Yonah**, township, W ISRAEL, in SHARON plain, 5 mi/8 km E of NETANYA; 32°19′N 34°56′E. Elevation 160 ft/48 m. Former cooperative village, now suburb. Citriculture. Founded 1932.

Kgalagadi District (gah-lah-GAH-dee), administrative division (□ 3,750 sq mi/9,750 sq km; 2001 population 42,049), SW BOTSWANA; ⊙ TSHABONG. Bounded by NAMIBIA W, SOUTH AFRICA SW and S, SOUTHERN and KWENENG districts E, and GHANZI DISTRICT N. Includes GEMSBOK National Park (□ 3,750 sq mi/9,713 sq km) and bird sanctuary. NOSSOB (Nossop) River forms SW border with South Africa.

Kgatleng District (gah-TLAING), administrative division (2001 population 73,507), SE BOTSWANA; ⊙ MOCHUDI. Bordered by SOUTH AFRICA E, CENTRAL DISTRICT N, KWENENG DISTRICT W, and SOUTH-EAST DISTRICT SW. MARICO River forms most of border with South Africa. Crossed (NE-SW) by road and railroad. Formerly called Bakgatla.

Khabarovo (khah-BAH-ruh-vuh), village, NE NENETS AUTONOMOUS OKRUG, ARCHANGEL oblast, NE European Russia, on strait YUGORSKIY Shar, opposite the VAYGACH Island, between the WHITE SEA and the KARA SEA in the ARCTIC OCEAN, 170 mi/274 km NW of VORKUTA; 69°39′N 60°24′E. Trading post; reindeer raising.

Khabarovsk (hah-BUH-ruhfsk), city (2005 population 590,100), ⊙ KHABAROVSK TERRITORY, RUSSIAN FAR EAST, port on the AMUR River near its junction with the USSURI, 5,302 mi/8,533 km E of MOSCOW; 48°30′N 135°06′E. Elevation 324 ft/98 m. An industrial center and a major transportation point on the TRANS-SIBERIAN RAILROAD. Manufacturing (wood processing, farm machinery, transportation equipment, machinery, oil-refinery products); oil refineries, shipyards. Center for import of goods and venture-capital firms from Pacific countries. International airport; connected by regular air service to ALASKA and JAPAN. Has 4 museums and 4 professional theaters. Population mostly Russian (80%), with Korean, Ukrainian, and Armenian minorities. Established in 1858 as a fortified trading post, Khabarovsk (at the time named Khabarovka), prospered greatly due to its strategic location at the confluence of the two greatest rivers in the region. Telegraph connection with VLADIVOSTOK established in 1868. Became a city and capital of the newly formed Priamur'ye gubernia in 1884. Received its current name in 1893. Railroad connection with Vladivostok established in 1897; connected to the European part of the country by the Trans-Siberian Railroad in 1916. Changed hands a number of times between the Red and White armies during the Russian civil war. The city was the capital of the Soviet Far East from 1926 to 1938, and grew into an important industrial center during World War II. Partitioned between 1947 and 1956 to create SAKHALIN, MAGADAN, AMUR, and KAMCHATKA oblasts, and JEWISH AUTONOMOUS OBLAST.

Khabarovsk Territory (khah-BAH-ruhfsk), administrative division (□ 304,480 sq mi/791,648 sq km; 2004 population 1,571,000), in the E and NE extremity of SIBERIA, RUSSIAN FAR EAST; ⊙ KHABAROVSK. Bounded E by the Sea of OKHOTSK, S by the MARITIME TERRITORY and China (HEILONGJIANG province; border, AMUR River), and N by MAGADAN oblast and SAKHA REPUBLIC (border, SUNTAR-Khayata range). Includes the JEWISH AUTONOMOUS OBLAST. The mountainous territory is crossed by the DZHUGDZHUR and BUREYA ranges, where gold, tin, and coal are extracted. Grain and potatoes are grown in the Amur valley, and in the N there are reindeer herds and fur trappers. Herring, flounder, and salmon are caught along the coast. Major cities are Khabarovsk, the industrial center KOMSOMOL'SK-NA-AMURE, and the ports SOVETSKAYA GAVAN' and NIKOLAYEVSK-NA-AMURE; 75% of the total population (Russians,

Ukrainians, Byelorussians, Jews, Tatars, and Yakuts) is concentrated in the cities. Founded in 1938 and reorganized in 1953 and 1957.

Khabary (khah-BAH-ri), village, W ALTAI TERRITORY, S central SIBERIA, RUSSIA, on the BURLA RIVER, on road, 55 mi/89 km NE of SLAVGOROD; 53°38′N 79°31′E. Elevation 465 ft/141 m. In agricultural area (feed corn, flax, rye; livestock raising).

Khabez (hah-BYES), village (2005 population 5,815), central KARACHEVO-CHERKESS REPUBLIC, NW CAUCASUS, S European Russia, on the LITTLE ZELENCHUK RIVER, on road, 20 mi/32 km SW of CHERKESSK; 44°03′N 41°46′E. Elevation 2,257 ft/687 m. Vegetable growing; food processing; woodworking.

Khabis, IRAN: see SHAHDAD.

Khabne, UKRAINE: see POLIS'KE.

Khabnoye, UKRAINE: see POLIS'KE.

Khabour River, SYRIA: see KHABUR RIVER.

Kh. Abu Falah, large Arab village, RAMALLAH district, WEST BANK, 9.3 mi/15 km NE of RAMALLAH, in the E slopes of the SAMARIAN Highlands; 32°01′N 35°18′E. Agriculture (olives, cereals).

Khabur, river, c.200 mi/320 km long, SE TURKEY and NE SYRIA; rises in SE Turkey; flows generally S through NE Syria to enter the EUPHRATES RIVER, 8 mi/12.9 km N of MEYADIN, S of DEIR EZ ZOR. The Khabur River project, begun in the 1960s, involved the construction of a series of dams and canals. The Khabur valley, which now has about 4 million acres/1.6 million ha of farmland, is one of Syria's main wheat-producing areas. In ancient times the Khabur was known as the Habor; it is believed that along its banks in Gozan the Israelite captives from SAMARIA were settled in the 8th century B.C.E. (2 Kings 17.6; 18.11).

Khabura (kuh-BOO-ruh), town, N OMAN, port on GULF OF OMAN, 35 mi/56 km SE of SOHAR. Sometimes called Al Khabura or Al Khaburah.

Khabur River (KUH-ber), c.100 mi/161 km long, largely in NE IRAQ; rises in mountains of Turkish KURDISTAN, flows S and W into Iraq to the TIGRIS at the meeting area of the TURKISH, SYRIAN, and IRAQI boundaries, 65 mi/105 km NW of MOSUL.

Khachmas, AZERBAIJAN: see XACMAZ.

Khachrod (KAHCH-rod), town (2001 population 29,897), UJJAIN district, W central MADHYA PRADESH state, central INDIA, 36 mi/58 km NW of UJJAIN; 23°25′N 75°17′E. Market center for grain, cotton, tobacco; cotton ginning, hand-loom weaving; handicraft woodwork.

Khadara, Al (ku-DAH-ruh, el), village, TRIPOLITANIA region, NW LIBYA, on JABAL NAFUSAH plateau, on main road, 33 mi/53 km WSW of Al KHUMS. Elevation c. 1,300 ft/396 m. Agricultural settlement (grain, olives, nuts; livestock. Founded as Breviglieri by Italians 1938–1939; Italian population left after World War II, replaced by local Arabs and Berbers. Small Roman ruins.

Kh. Adas, Arab village, 2 mi/3 km NE of RAFAH, in the GAZA STRIP. Agriculture (almonds, vegetables, cereals). Most of the inhabitants are refugees and their descendents.

Khader, El, large Arab village, Bethlehem district, WEST BANK, 2.5 mi/4 km SW of BETHLEHEM, in the JUDEAN HIGHLANDS. Elevation 2,887 ft/880 m. Vineyards. Large Greek Orthodox monastery, which was famous until the 1940s for its mental asylum.

Khadirabad, INDIA: see JALNA, city.

Khadkhal, MONGOLIA: see HATGAL.

Khadyzhensk (hah-DI-zhinsk), city (2005 population 21,830), S KRASNODAR TERRITORY, S European Russia, at the NW foot of the Greater CAUCASUS, on crossroads and near railroad, 70 mi/113 km SE of KRASNODAR, and 30 mi/48 km WSW of MAYKOP; 44°25′N 39°32′E. Elevation 429 ft/130 m. Petroleum center linked by pipeline with Krasnodar; natural gas wells. Manufacturing (machinery, wood products). Health resort with baths. Also known as Khadyzhenskiy. Made city in 1949.

Khadyzhenskiy, RUSSIA: see KHADYZHENSK.

Khadzhalmakhi (hah-jahl-MAH-hee), village (2005 population 6,235), central DAGESTAN REPUBLIC, NE CAUCASUS, SE European Russia, in the extreme SE part of the SULAK RIVER basin, on road, 31 mi/50 km SSW of MAKHACHKALA; 42°25′N 47°11′E. Elevation 4,166 ft/1,269 m. Fruit cultivation center.

Khadzhibeyskiy Liman, UKRAINE: see KHADZHYBEY LAGOON.

Khadzhi Dimitrovo, Bulgaria: see HADZHIDIMITROVO.

Khadzhy-Bey, UKRAINE: see ODESSA, city.

Khadzhybey Lagoon (hahd-zhi-BAI) (Ukrainian *Khadzhybeys'kyy Lyman*) (Russian *Khadzhibeyskiy Liman*), BLACK SEA coastal lagoon, E central ODESSA oblast, 3 mi/4.8 km NNW of ODESSA city center; 25 mi/40 km long, 1 mi/1.6 km–2 mi/3.2 km wide, surface area 27 sq mi/70 sq km; up to 44 ft/13 m deep. Steep slopes; fed from N by intermittent Malyy Kuyal'nyk River; separated from sea by sand and shell bar 2.8 mi/4.5 km long. The improved separation by a dam and increased inflow of effluents has lowered the salinity of the lagoon water, resulting in a shift of species to freshwater fishes. Black bottom muds are not used for medicinal purposes because of pollution.

Khadzhybeys'kyy Lyman, UKRAINE: see KHADZHYBEY LAGOON.

Khadzhy-Dere, UKRAINE: see OVIDIOPOL'.

Khaf (KHAHF), town, Khorāsān province, NE IRAN, 70 mi/113 km SE of TORBAT-E-HAYDARIYEH, and on road to Afghan border. Fruit-growing region (figs, grapes, melons). Also spelled Khvāf.

Khafs Dhagara, well of KHARJ oasis, E central Nejd, SAUDI, 13 mi/21 km SSE of DILAM. Center of small model farm (developed in 1940s), irrigated from limestone pit. Agricultural (wheat, millet, vegetables, dates).

Khaga (KAH-guh), town, FATEHPUR district, S UTTAR PRADESH state, N central INDIA, 21 mi/34 km ESE of Fatehpur; 25°47′N 81°07′E. Gram, barley, rice, jowar.

Khagaria (KAHG-ahr-yuh), district (□ 574 sq mi/1,492.4 sq km), BIHAR state, E INDIA; ⊙ KHAGARIA. Rice, wheat, maize.

Khagaria (KAHG-ahr-ee-yuh), town (2001 population 45,126), ⊙ KHAGARIA district, BIHAR state, on BURHI GANDAK RIVER, and 11 mi/18 km NNE of MUNGER; 25°30′N 86°29′E. Road and trade (rice, wheat, corn, gram, barley, oilseeds) center.

Khagaul (kuh-GOUL), town (2001 population 48,330), PATNA district, W central BIHAR state, E INDIA, on PATNA CANAL, and 7 mi/11.3 km WSW of PATNA; 25°35′N 85°03′E. Rice, gram, wheat, oilseeds, barley, corn. Large sugar-milling plant 10 mi/16 km W, at Bihta.

Khagrachhari, town (2001 population 38,879), CHITTAGONG HILLS district, SE EAST BENGAL, BANGLADESH, on Chingi River, and 40 mi/64 km NW of RANGAMATI; 23°00′N 91°46′E. Trades in rice, cotton, oilseeds, tobacco.

Khaibar (KAI-ber), outlying township and oasis, MEDINA province, N Naj'd, SAUDI ARABIA, 80 mi/129 km NNW of Medina. Dates, grain. Formerly considered part of HEJAZ and spelled Kheibar or Khaybar.

Khaibar, PAKISTAN: see KHYBER.

Khaidar, Bulgaria: see KARDAM, RUSE oblast.

Khaidarkan, KYRGYZSTAN: see KHAYDARKAN.

Khair (KEIR), town, ALIGARH district, W UTTAR PRADESH state, N central INDIA, 15 mi/24 km WNW of Aligarh; 27°57′N 77°50′E. Road junction. Agriculture (wheat, barley, pearl millet).

Khairabad (KEI-rah-bahd), town, SITAPUR district, central UTTAR state, N central INDIA, 6 mi/9.7 km SE of Sitapur. Manufacturing (durrie carpet). Agriculture (wheat, rice, barley, oilseeds). Has sixteenth century mosques. Annual fair. Founded eleventh century.

Khairagarh (KEI-rah-guhr), town (2001 population 15,149), RAJ NANDGAON district, W central CHHATTISGARH STATES, central INDIA, 25 mi/40 km NW of DURG; 21°25′N 80°58′E. Manufacturing (biri). Agriculture (rice, wheat). Seat of Indira Kala-Sangeet Vishwvidyalaya (established 1956), reputed music and fine arts university. Sometimes called Khairagarh Raj. Was capital of former princely state of KHAIRAGARH, one of the CHHATTISGARH STATES; in 1948 incorporated into Durg district and finally into RAJ NANDGAON district.

Khairagarh, town, AGRA district, W UTTAR PRADESH state, N central INDIA, on BANGANGA RIVER, and 20 mi/32 km SSW of Agra; 26°57′N 77°49′E. Pearl millet, gram, wheat, barley. Also spelled KHERABGARH.

Khairan (kai-RAHN), inlet of GULF OF OMAN, in OMAN, 13 mi/21 km SE of MUSCAT; 23°30′N 58°44′E. Sheltered natural harbor. Also called Bandar Khairan or Bandar Khayran.

Khairkhana Pass (kah-IR-kah-nah), in S outlier of the Hindu Kush, E AFGHANISTAN, 7 mi/11.3 km NW of KABUL, on highway to MAZAR-I-SHARIF; 34°34′N 69°06′E.

Khairpur (khei-ahr-puhr), city, ⊙ Khairpur district, SIND province, SE PAKISTAN, 11 mi/18 km SSW of SUKKUR, in irrigated tract; 27°32′N 68°46′E. Trades in wheat, cotton, tobacco, oilseeds, dates, and cloth fabrics. Linked by road and railroad to KARACHI. Manufacturing (textiles, carpets, refined sugar, pottery, lacquerware, leather goods). Was capital of the former princely state of Khairpur, which was founded in 1783 and merged into PAKISTAN in 1955. Cultural center with fine historic buildings, notably the Faiz Palace, and with several colleges affiliated with Sind University and an industrial school.

Khairpur (khei-ahr-puhr), town, BAHAWALPUR district, BAHAWALPUR division, PUNJAB province, central PAKISTAN, 35 mi/56 km NE of BAHAWALPUR; 29°35′N 72°14′E.

Khairwara, town, UDAIPUR district, RAJASTHAN state, NW INDIA, 40 mi/64 km S of Udaipur, in SE ARAVALLI RANGE.

Khajuha (KUHJ-uh-huh), village, FATEHPUR district, S UTTAR PRADESH state, N central INDIA, 3 mi/4.8 km WNW of BINDKI. Gram, barley, rice, jowar. Formerly a town, founded 1659 by Aurangzeb.

Khajuraho (khah-joo-RAH-ho), village (2001 population 19,282), CHHATARPUR district, NE MADHYA PRADESH state, central INDIA, 21 mi/34 km ESE of CHHATARPUR; 24°50′N 79°55′E. A religious and archaeological site famous for its beautiful, richly carved Brahmin and Jain temples (only twenty-two out of original eighty-five remain), built during Chandella dynasty (C.E. 950–1050) of sandstone, and containing fine erotic sculptures and valuable inscriptions; Kandariya Temple is decorated with sculptures considered among the greatest masterpieces of Indian art; museum (built 1919). Tourist center; protected as a UNESCO World Heritage Site; site of Khajuraho festival of classical music and dance.

Khakass Republic (hah-KAHZ) or **Khakassia**, constituent republic (□ 23,900 sq mi/62,140 sq km; 2004 population 586,000), S central SIBERIA, RUSSIA, in KRASNOYARSK TERRITORY; ⊙ ABAKAN. Largely consisting of black-earth steppe [Russian *chernozëm*], it is bounded E by the upper YENISEY RIVER and W and S by the wooded KUZNETSK ALATAU and SAYAN ranges, respectively. Approximately 40% of the republic's territory is forested. The ABAKAN (a tributary of the Yenisey River) and CHULYM rivers drain the area. Railroads are the chief mode of transportation. The region's swift-flowing rivers provide hydroelectric power, and many of the numerous lakes are sources of therapeutic mineral waters. Mining, forestry, and food processing are the main industries. Gold, coal, iron ore, barite, copper, lead, and molybdenum are mined, and gypsum, limestone, marble, and other

building stones are quarried. The forests of the taiga zone yield lumber and wood products. Logs are floated down the Abakan River to sawmills in Abakan. CHERNOGORSK (a coal-mining center) is another major city. Although the republic's population is primarily Russian (with some Ukrainians), there is a large Khakass minority. The Khakass are a Turkic-Mongol nationality that inhabited the S Yenisey valley for many centuries. Formerly nomadic herdsmen, they are now mostly settled in farming, hunting, or livestock-breeding collectives. They speak a Turkic language and are Orthodox Christians. The region, known for mining and trade from the 8th to the 11th century, came under Russian control in the 17th century. Numerous Russian settlers were attracted by copper mining in the 18th century. The Khakass sided with counterrevolutionary forces during the Russian civil war. The autonomous region was formed in 1930. The region was given republic status in 1991. It was a signatory, under the name Republic of Khakasiya, to the March 31, 1992, treaty that created the RUSSIAN FEDERATION. Khakassia is divided into eight districts and includes five cities, seventeen towns, and sixty-eight villages.

Khakkulabad (hahk-KOOL-ah-baht), city, SE NAMANGAN wiloyat, UZBEKISTAN, on railroad (Khakkulabad station), and 22 mi/35 km ESE of NAMANGAN; 40°55′N 72°07′E. Cotton ginning. Formerly called NARYN.

Khakurate, RUSSIA: see TAKHTAMUKAY.

Khakurinokhabl', RUSSIA: see SHOVGENOVSKIY.

Khalach (hah-LAHCH), town, SE LEBAP weloyat, SE TURKMENISTAN, on the AMU DARYA (river), 25 mi/40 km NW of KERKI, and c.78 mi/125 km SE of CHARJEW; 38°04′N 64°52′E. Tertiary-level administrative center. Cotton; metalworks.

Khaladiyah (kah-lah-DEE-yah), village, TUNIS province, N TUNISIA, on railroad, and 10 mi/16 km S of TUNIS. Agriculture (fruit and wine growing).

Khalandhri (kah-LAHN-[th]ree), NE suburb of ATHENS, ATTICA prefecture, ATTICA department, E central GREECE, 5 mi/8 km from city center, in Athens metropolitan district; 38°02′N 23°48′W. Summer resort. Also Chalandri, Halandri.

Khalasa (kah-LAH-sah), ancient city, S ISRAEL, in the NEGEV, 12 mi/19 km SW of BEERSHEBA. In Roman times called *Elusa*. Much of city was excavated; Aramaic and Byzantine ruins extant. The Nabotean Khalutza was for some time in 4th century capital of Palestinina III. Sometimes spelled Khalutza.

Khalastra (khah-LAH-strah), town, THESSALONÍKI prefecture, CENTRAL MACEDONIA department, NE GREECE, 10 mi/16 km W of THESSALONÍKI, in Axios (VARDAR) River delta. Agriculture (wheat, silk, cotton). Also Chalastra. Formerly called Kouloukia.

Khalatse (KUH-luht-sai), village, LEH district, JAMMU AND KASHMIR state, in central KASHMIR region, extreme N INDIA, at S foot of LADAKH Range, on the INDUS River (suspension bridge), and 40 mi/64 km WNW of LEH. Agriculture (pulses, wheat, oilseeds). Brahmi and Karoshti inscriptions discovered here; has Dard and Tibetan castle ruins. Fighting here in 1948, during India-PAKISTAN struggle for control. Gold deposits NW. Noted eleventh-century Buddhist monastery with frescoes, carvings, and sculptures is 18 mi/29 km ESE, at ALCHI or Alchi Gompa, village. Also spelled KHALSI.

Khalifat Mountain (khah-lee-fah-TAH), N spur and one of highest points (11,434 ft/3,485 m) of CENTRAL BRAHUI RANGE, in SIBI district, NE central BALUCHISTAN, SW PAKISTAN, S of ZIARAT; 30°15′N 67°40′E. Limestone deposits.

Khalilabad (kuh-LEE-lah-bahd), town, BASTI district, NE UTTAR state, N central INDIA, 17 mi/27 km W of GORAKHPUR, and 25 mi/40 km E of Basti; 26°47′N 83°05′E. Hand-loom cotton-weaving center. Agriculture (rice, wheat, barley).

Khalil, Al, WEST BANK: see HEBRON.

Khalilovo (hah-LEE-luh-vuh), town, E central ORENBURG oblast, SE European Russia, in the foothills of the S URALS, on road and railroad, 22 mi/35 km NW of ORSK; 51°24′N 58°07′E. Elevation 1,118 ft/340 m. In the Orsk-Khalilovo industrial district. Mining center, based on limonite deposits associated with titanium, vanadium, chromite, and nickel ores; supplies NOVOTROITSK steel industry. Manufacturing (chemicals based on local magnesite). Developed in the 1930s.

Khalimbekaul (hah-leem-byek-ah-OOL), village (2005 population 4,395), N central DAGESTAN REPUBLIC, SE European Russia, in the NE foothills of the Greater CAUCASUS Mountains, on road and railroad, 22 mi/35 km SW of MAKHACHKALA, and less than 2 mi/3.2 km NE of BUYNAKSK, to which it is administratively subordinate; 42°50′N 47°08′E. Elevation 1,227 ft/373 m. Livestock raising.

Khalkabad (HAHLK-ah-baht), town, NAMANGAN wiloyat, NE UZBEKISTAN; 40°53′N 71°05′E.

Khalkhal (KHAHL-khahl), town, Ardabīl province, NW IRAN, 45 mi/72 km SSE of ARDEBIL; 37°36′N 48°32′E. Elev. 5,741 ft/1,750 m. Agriculture (grain, fruit); sheep raising. Manufacturing (rug making); copper and lead deposits. Also called Harauabad, Harowabad, or HEROWABAD; formerly Harau or HERAU.

Khalkhim Gol, MONGOLIA and NORTHEAST region (CHINA): see HALHIN GOL.

Khalki (khahl-KEE), Italian *Calchi*, island (□ 11 sq mi/ 28.6 sq km), DODECANESE prefecture, SOUTH AEGEAN department, GREECE, in AEGEAN SEA, off W coast of RHODES; 6 mi/9.7 km long, 2 mi/3.2 km wide; 36°14′N 27°35′E. Rises to 1,954 ft/596 m (W). Produces figs, barley, wheat, olives and olive oil, grapes; sponge fishery. Main village, Chalke, is on E shore. Also Chalke.

Khalkidhikí (khahl-kee-[th]ee-KEE), prefecture (1991 population 92,117), CENTRAL MACEDONIA department, NE GREECE, coterminous with KHALKIDHIKÍ Peninsula, with AKTI, KASSANDRA, and SITHONIA prongs, on N AEGEAN SEA, ORFANI Gulf (E), and Gulf of SALONICA (W); ⊙ Khalkidhikí; 40°25′N 23°30′E. Bordered N by SÉRRAI, KILKIS, and THESSALONÍKI prefectures.

Khalkidhikí (khahl-kee-[th]ee-KEE), peninsula (1981 population 79,036), KHALKIDHIKÍ prefecture, CENTRAL MACEDONIA department, NE GREECE, projecting into the N AEGEAN SEA. Its southern extremity terminates in three peninsulas: KASSANDRA (ancient Greek *Pallene*) in the W, SITHONIA in the center, and AKTI in the E. The region is largely mountainous, dry, and agricultural. Olive oil, wine, wheat, and tobacco are produced; magnesite is mined. In antiquity the peninsula was famous for its timber and wine; Olynthus and Potidaia were the chief towns. Today, POLYGYROS is the leading town and an administrative center. The peninsula was named for Khalkís, which established colonies here in the 8th and 7th centuries B.C.E. In the 4th century B.C.E. the peninsula was conquered by Philip II of Macedon (MACEDONIA), and in the 2nd century B.C.E. by ROME. The subsequent history of Khalkidhikí is essentially that of THESSALONÍKI. Also Chalkidike. Formerly Chalcidice.

Khalki Island, TURKEY: see HEYBELI ISLAND.

Khalkís (khahl-KEES), city (2001 population 67,091), on ÉVVIA (Euboea) island, ⊙ ÉVVIA prefecture, CENTRAL GREECE department, on the ÉVRIPOS Strait, connected to the mainland by a bridge; 38°28′N 23°36′E. Trade center for local products, including wine, cotton, and citrus fruits. Popular resort. Cement and other manufacturing. The chief city of ancient Euboea, Khalkís was settled by the Ionians and early became a commercial and colonizing center. It established (8th–7th centuries B.C.E.) colonies on KHALKIDHIKÍ and in SICILY. Subdued by ATHENS (c.506 B.C.E.); led the revolt of Euboea against Athens in 446

B.C.E. Again defeated, it came under Athenian rule until 411 B.C.E.; passed to MACEDONIA in 338 B.C.E. Aristotle died here (322 B.C.E.). In succeeding centuries, used as a base for invading Greece. In the Middle Ages named Negroponte by the Venetians, who occupied it in 1209. Passed to the Ottoman Turks in 1470; became part of GREECE in 1830. Diamond-shaped Venetian citadel. Also Chalcis, Chalkis, Halkis.

Khal'mer-Sede, RUSSIA: see TAZOVSKIY.

Khal'mer-Yu (HAL-myer-YOO), town, NE KOMI REPUBLIC, NE European Russia, on highway, 40 mi/64 km NE of VORKUTA, to which it is administratively subordinate; 67°58′N 64°50′E. Elevation 738 ft/224 m. Railroad terminus; coal mines (PECHORA basin). Developed in the 1940s.

Khalsi, INDIA: see KHALATSE.

Khalturin, RUSSIA: see ORLOV.

Khalutza, ISRAEL: see KHALASA.

Kham (KAHM), historical E province of TIBET, SW CHINA, between c.93° and c.99°E (Chang Jiang line); main town, QAMDO (Chamdo). Part of XIKANG province after 1914; part of Tibet today. Sometimes spelled Kam.

Khamamatyurt (hah-mah-maht-YOORT), village (2005 population 5,055), NW DAGESTAN REPUBLIC, SE European Russia, on the E bank of the TEREK RIVER, on road, 29 mi/47 km NNW of KHASAVYURT; 43°36′N 46°30′E. In oil- and gas-producing region.

Khamar-Daban Range (hah-MAHR–dah-BAHN), in S BURYAT REPUBLIC, S SIBERIA, RUSSIA, S of Lake BAYKAL; extension of the EASTERN SAYAN MOUNTAINS E to the SELENGA River; 51°15′N 105°00′E. Wooded slopes, rich in minerals, rise to 7,540 ft/2,298 m.

Khamashi, city, KASHKADARYO wiloyat, S UZBEKISTAN, 39 mi/63 km E of KARSHI. Tertiary-level administrative center. Some light industry near Khamashi Chimkurgan reservoir.

Khamasin (kah-mah-SEEN) or **Al Khamasin**, town, S NAJD, SAUDI ARABIA, in the WADI DWASIR, 320 mi/515 km SSW of RIYADH; 20°29′N 44°50′E. Chief town of the Wadi Dwasir. Agriculture (dates, vegetables, fruit), livestock raising; handicrafts. Formerly known as Dam.

Khamba, China: see GAMBA.

Khambat, town, KHEDA district, GUJARAT state, W INDIA, on the MAHI RIVER estuary; 22°18′N 72°37′E. KHAMBAT is a trading center whose industries include textile weaving, carpet making, petroleum refining, and manufacturing of salt and stone ornaments. Once a great port under the Muslim rulers of Gujarat (fourteenth–fifteenth centuries), Khambat lost its importance when the harbor silted up. Until 1948 it was capital of the former princely state of Khambat. The GULF OF KHAMBAT, a shallow arm of the ARABIAN SEA, lies between KATHIAWAR peninsula and GUJARAT.

Khambat, Gulf of, inlet of ARABIAN SEA, W INDIA, between N GUJARAT state (E) and KATHIAWAR peninsula (W); 15 mi/24 km–120 mi/193 km wide between DAMAN AND DIU; 130 mi/209 km long. Receives SABARMATI, MAHI, NARMADA, and TAPI rivers. Ports with coastal trade include DIU, BHAVNAGAR, KHAMBAT, SURAT, and DAMAN. Fishing (pomfrets, jewfish, Bombay duck). Rivers have largely silted up at the N end.

Khambha (KUHM-buh), town, AMRELI district, GUJARAT state, W central INDIA, 31 mi/50 km S of AMRELI; 21°09′N 71°15′E. Millet; livestock breeding.

Khambhaliya (kuhm-BAHL-yuh), town, JAMNAGAR district, in former SAURASHTRA, now in GUJARAT state, W central INDIA, 32 mi/51 km SW of JAMNAGAR; 22°12′N 69°39′E. Railroad junction; trade center (millet; cotton, ghee, salt, oilseeds; cloth fabrics); cotton ginning, flour milling, peanut shelling; metalwork. Also spelled Khambhalia.

Khambi-Irze (hahm-BEE–eer-ZE), village (2005 population 3,135), W central CHECHEN REPUBLIC, S European Russia, on road, 10 mi/16 km WSW of

GROZNYY; 43°14′N 45°27′E. Elevation 574 ft/174 m. Agriculture (barley, buckwheat). Also known as Lermontov-Yurt.

Khamgaon (KAHM-goun), town, BULDANA district, MAHARASHTRA state, W central INDIA, 28 mi/45 km W of AKOLA; 20°41′N 76°34′E. Railroad spur terminus; road and cotton-trade center in major cotton tract; cotton ginning, oilseed milling; chemical works. Industrial school and colleges.

Kham-i-Ab (kahm-EE–ahb), village, JOZJAN province, N AFGHANISTAN, on the Amu Darya River, on TURKMENISTAN border, adjoining Bossaga, 55 mi/89 km NE of ANDKHUI; 37°32′N 65°42′E.

Khami National Monument, MATABELELAND NORTH province, SW ZIMBABWE, 10 mi/16 km W of BULAWAYO, on KHAMI RIVER; 20°09′S 28°25′E. Stone walls date between 15th and 17th centuries. Ruins are remains of the capital of the Torwa state. Declared World Heritage Site, 1986. Also spelled Kame.

Khamir (kuh-MIR), township, N central YEMEN, on central plateau, 50 mi/80 km NNW of SANA; 15°38′N 42°58′E. Was stronghold of Imam during 2nd Turkish occupation (1872–1918) of Sana.

Khami River, c. 65 mi/105 km long, SW ZIMBABWE; rises c.15 mi/24 km SW of BULAWAYO; flows NW, through Khami River Dam, past Khami village and KHAMI RUINS; enters GWAYI RIVER 50 mi/80 km WNW of Bulawayo.

Khami Ruins, ZIMBABWE: see KHAMI NATIONAL MONUMENT.

Khamis Mushait (KAH-mis muh-SHAH-it), town (2004 population 372,695), ASIR province, SAUDI ARABIA, on plateau, 20 mi/32 km ENE of ABHA, in upper reaches of the WADI BISHA; 18°18′N 42°44′E. Agricultural center (dates, millet, sesame, olives). Also called Khmis Mushit.

Kham Keut (KAHM KOO-aht), town, Bolikhamsi province, central LAOS, in TRUONG SON RANGE, 65 mi/105 km WSE of PAKSANE town and 65 mi/105 km WSW of VINH (VIETNAM; linked by road); 18°15′N 104°43′E. District administrative center. Shifting cultivation, agroforestry, forest products. Lao, Thai, Bo, and other peoples. Also spelled Kamkeut (kahm-KOOT).

Khammam (KUHM-muhm), district (□ 6,189 sq mi/ 16,091.4 sq km), ANDHRA PRADESH state, S INDIA; ⊙ KHAMMAM.

Khammam (KUHM-muhm), city, ⊙ KHAMMAM district, ANDHRA PRADESH state, S INDIA, 65 mi/105 km SSE of WARANGAL; 17°15′N 80°09′E. Agriculture trade center (rice); oilseed milling. Sometimes spelled Khammameth, Khammamett.

Khammouone, LAOS: see THA KHAEK.

Khammuan or **cammon**, province (2005 population 336,935), S LAOS, N border with Bolikhamsi province, E border with VIETNAM, S border with SAVANNAKHET province, and W border with THAILAND; ⊙ THA KHAEK; 18°00′N 105°00′E. Rice, shifting cultivation; agroforestry; cattle; tin mining. Lao, Tai, So, and other minorities.

Khammuan Plateau (KAHM-MOO-ahn), KHAMMUAN province, LAOS; 17°49′N 105°09′E. Elevation 3,281 ft/ 1,000 m–4,921 ft/1,500 m.

Khamney (hahm-NYAI), rural settlement (2004 population 450), SW BURYAT REPUBLIC, S Siberian Russia, in the KHAMAR-DABAN RANGE, on the DZHIDA RIVER, approximately 15 mi/24 km N of the Mongolian border, on road, 84 mi/135 km SW of TANKHOY; 50°24′N 103°52′E. Elevation 3,034 ft/924 m. Agricultural cooperative. Also known as Khamneyskiy.

Khamneyskiy, RUSSIA: see KHAMNEY.

Khamseh (kahm-SE), former province of NW IRAN; main town, ZANJAN. Bounded N by AZERBAIJAN, E by GILAN and QAZVIN, S by HAMADAN, and E by GARUS. Drained by Zanjan River (tributary of the QEZEL-UZAN). Agriculture (grapes; fruit). Crossed by TEHRANTABRIZ highway and railroad. Now largely a part of Zanjān province.

Khamza (HAHM-zah), city, E FERGANA wiloyat, NE UZBEKISTAN, on railroad, and 12 mi/19 km WNW of FERGANA; 40°25′N 71°30′E. Cotton ginning and fabric factory. Petroleum-refining center, serving CHIMION and ANDIJAN oil fields. Called Vannovsk or Vannovskii until 1963.

Khamzoren, Bulgaria: see BEZMER.

Khanabad (KAH-nah-bahd), city, in KUNDUZ province, NE AFGHANISTAN, near the TAJIKISTAN border, 110 mi/177 km E of MAZAR-I-SHARIF; 36°41′N 69°07′E. The center of an oasis irrigated by right tributary of KUNDUZ RIVER. Agriculture (cotton, rice). Linked by direct highway to Kabul city.

Khanabad (HAHN-ah-baht), village, E ANDIJAN wiloyat, UZBEKISTAN, on the KARA DARYA River, on railroad, and 30 mi/48 km E of ANDIJAN; 40°48′N 73°02′E. Tanning center. Junction of railroad spur to JALAL-ABAD (KYRGYZSTAN). Also Khonobod.

Khanapur (KAH-nah-puhr), town, BELGAUM district, KARNATAKA state, S INDIA, 14 mi/23 km S of BELGAUM. Rice market. Manufacturing (tiles, bricks, pottery). Kaolin and laterite worked nearby.

Khanaqin (kuh-nuh-KEEN), town, DIYALA province, NE IRAQ, 5 mi/8 km from the IRANIAN border, on a tributary of the DIYALA; 34°21′N 45°22′E. It is located in an oil-producing region and has an oil refinery. Khanaquin was severely damaged in the Iran-Iraq War in the 1980s. Sometimes spelled Khanikin or Khaniqin.

Khan Bela (khahn BAI-lah), town, RAHIMYARKHAN district, BAHAWALPUR division, PUNJAB province, central PAKISTAN, 65 mi/105 km SW of BAHAWALPUR; 28°58′N 70°44′E. Wheat, rice.

Khancoban, village, NEW SOUTH WALES, SE AUSTRALIA, on edge of SNOWY MOUNTAINS; 36°09′S 148°06′E. Power stations nearby; tours available. Formerly called Swampy Plains.

Khandagayty (hahn-dah-GEI-ti), village (2006 population 3,140), SW TUVA REPUBLIC, S Siberian Russia, on the border with MONGOLIA, near road junction, 122 mi/196 km SW of KYZYL; 50°44′N 92°03′E. Elevation 3,920 ft/1,194 m. Logging and lumbering; hardy grain growing, sheep herding.

Khandaparha, town, NAYAGARH district, ORISSA state, E central INDIA, 50 mi/80 km NW of PURI. Timber trade. Formerly Khandpara. Was capital of former princely state of Khandpara in ORISSA STATES, along right bank of MAHANADI RIVER; state incorporated 1949 into Puri district and later into NAYAGARH district.

Khandbari (KAHND-bah-ree), town (2001 population 21,789), ⊙ SANKHUWASABHA district, E NEPAL; 27°22′N 87°13′E. Elevation 3,410 ft/1,040 m.

Khandela (kuhn-DAI-luh), town, SIKAR district, RAJASTHAN state, NW INDIA, 50 mi/80 km NNW of JAIPUR. Lacquered woodwork. Agriculture (millet, gram, wheat).

Khandesh, INDIA: see JALGAON, district.

Khandgiri, village, KHORDHA district, E ORISSA state, E central INDIA, 15 mi/24 km SSW of CUTTACK, 4 mi/6.4 km WNW of BHUBANESHWAR. Old Buddhist and Jain caves in nearby hills of Khandgiri and UDAYAGIRI; oldest from first–second centuries British Columbia.

Khandhar, INDIA: see KANDAHAR.

Khandwa (KUHND-wah), city (2001 population 171,976), ⊙ EAST NIMAR district, SW MADHYA PRADESH state, central INDIA, on MUMBAI (Bombay)-Calcutta railroad; 21°49′N 76°23′E. Market for cotton, timber, grain. There are cotton gins, sawmills, and oilseed mills. Believed by some authorities to be the city of Kognabanda mentioned by the ancient Greek geographer Ptolemy. During the twelfth century it was a center of Jainism. Seat of several colleges.

Khandwa, INDIA: see EAST NIMAR.

Khandyga (hahn-di-GAH), village (2006 population 6,720), E central SAKHA REPUBLIC, central RUSSIAN FAR EAST, on the ALDAN RIVER, on highway, 187 mi/301 km ENE of YAKUTSK; 62°40′N 135°36′E. Elevation

429 ft/130 m. Logging, lumbering. Ecotourism base. Local airfield.

Khanewal (kah-nai-wah-lah), town, MULTAN district, S PUNJAB province, central PAKISTAN, 27 mi/43 km ENE of MULTAN; 30°18′N 71°56′E.

Khan ez Zabib (KAHN ez zuh-BEEB), village, central JORDAN, on HEJAZ RAILROAD, and 35 mi/56 km SSE of AMMAN. Railroad station and bedouin trade center; camel raising. Marble deposits (E). Sometimes spelled Khan ez Zebib.

Khan ez Zebib, JORDAN: see KHAN EZ ZABIB.

Khanfar (KAHN-fahr), village, S YEMEN, near the Wadi BANA, and 28 mi/45 km WSW of SHUQRA. Agricultural oasis. Saltpeter deposits nearby.

Khang, MONGOLIA: see TURTA.

Khangah Dogran (khah-nah-GAH do-GAH-run), village, SHEIKHUPURA district, E central PUNJAB province, central PAKISTAN, 22 mi/35 km WNW of SHEIKHUPURA; 31°50′N 73°37′E. Sometimes spelled Khangah Dogra.

Khangai, MONGOLIA: see HANGAYN NURUU.

Khangarh (kah-nah-GUHR), town, MUZAFFARGARH district, SW PUNJAB province, central PAKISTAN, near CHENAB River, 10 mi/16 km S of MUZAFFARGARH; 29°55′N 71°10′E. Market center.

Khangayin, MONGOLIA: see ARVAYHEER.

Khangmar, China: see KANGMAR.

Khanh Hoa, province (□ 2,029 sq mi/5,275.4 sq km), S central VIETNAM, N border with PHU YEN province, E border on SOUTH CHINA SEA, S border with NINH THUAN province, W border with DAC LAC and LAM DONG provinces; ⊙ NHA TRANG; 12°20′N 109°00′E. Including narrow coastal lowlands, hilly midlands, and a mountainous interior, province exhibits a number of environmental zones and localized economies. Its once heavily forested mountains and hills have suffered widespread deforestation in recent decades. Parts of province also suffered much defoliation and devastating bombing during the Vietnam War and still has damage from the conflict in many areas. Diverse soils and mineral resources (gold, kaolin, construction sands), wet rice farming, shifting cultivation, forest products (resins, medicinals, bamboo, foraged foods), fishing and aquaculture, agroforestry, livestock raising, and commercial agriculture (coconuts, cassava, tobacco, sugarcane, tea, vegetables, fruits). Sawmilling, handicrafts, various industries (textiles, brick and tile, sugar refining, pharmaceuticals, glass, beer and soft drinks, chemicals, alcohol, salt extraction, leather goods, building materials, marine equipment, fish curing, food processing). Light manufacturing, shipbuilding, international and domestic tourism focused on Nha Trang and ancient Cham ruins. Predominantly Kinh population with Raglai, E-de, Chinese, and other minorities.

Khani (HAH-nee), settlement (2006 population 895), SW SAKHA REPUBLIC, SW RUSSIAN FAR EAST, in the W STANOVOY RANGE, near the confluence of the Khani and OLEKMA rivers, on the administrative border with AMUR oblast, on the BAYKAL-AMUR MAINLINE, 138 mi/222 km W of NERYUNGRI; 57°02′N 120°58′E. Elevation 1,482 ft/451 m. Logging, lumbering.

Khaniá (khah-NYAH), Greek *Chaniá* or *Haniá*, prefecture (□ 926 sq mi/2,407.6 sq km), W CRETE department, GREECE, W of Mt. Lefka; ⊙ KHANIÁ; 35°20′N 24°00′E. Agriculture (citrus fruits, olives, wheat, vegetables; wine); livestock products (meat, milk, cheese). Copper, lignite, and iron deposits. Fisheries. Main port, Khaniá, is on N shore; KHORA SFAKION and PALAIOKHORA on S shore. Heavily damaged during German invasion in 1941. Also Hania, Chania.

Khaniá (khah-NYAH), ancient Greek *Cydonia*, city (2001 population 80,909), ⊙ KHANIÁ prefecture, on N coast of CRETE department, Greece, a port on the Gulf of Khaniá, an arm of the Sea of CRETE; 35°31′N 24°02′E. Ships olives, citrus fruits, and wine. Tourism. Airport. One of the oldest Cretan cities. Conquered 69

B.C.E. by the Romans; fell to Arabs in C.E. 826. Reconquered (961) by the BYZANTINE EMPIRE; became (13th century) a Venetian colony. The OTTOMAN EMPIRE took the city in 1645. Was capital of CRETE from 1841 to the mid-20th century. Severely damaged during German invasion in World War II. Has synagogue, mosque, and several churches. Among its historic sites are medieval fortifications and an old Venetian arsenal. Formerly Canea.

Khaniadhana (kuhn-YAH-dah-nuh) or **Khaniyadhana,** town (2001 population 12,595), SE SHIVPURI district, N MADHYA PRADESH state, central INDIA, 40 mi/64 km SE of SHIVPURI; 25°01′N 78°08′E. Was capital of former princely state of Khaniadhana of GWALIOR Residency; in 1948, state was merged with VINDHYA PRADESH and in 1950 was transferred to MADHYA BHARAT, eventually becoming part of Madhya Pradesh state in 1956.

Khanikin, IRAQ: see KHANAQIN.

Khanino (HAH-nee-nuh), town (2006 population 1,245), W TULA oblast, central European Russia, on road and near railroad, 40 mi/64 km W of TULA; 54°13′N 36°37′E. Elevation 616 ft/187 m. Woodworking.

Khaniqin, IRAQ: see KHANAQIN.

Khaniyadhana, INDIA: see KHANIADHANA.

Khanka, village, E KHORAZM wiloyat, UZBEKISTAN, in KHIVA oasis, on railroad, and 10 mi/16 km SSE of URGANCH; 41°28′N 60°47′E. Silk milling; cotton.

Khanka, El (kahn-kuh, el), village, QALYUBIYA province, Lower EGYPT, 13 mi/21 km NE of CAIRO; 30°13′N 31°21′E. Agriculture (cotton, flax; cereals, fruits). Has psychiatric clinic.

Khankala (hahn-kah-LAH), village (2005 population 7,935), central CHECHEN REPUBLIC, S European Russia, on the Groznyy-Gudermes railroad line, less than 3 mi/5 km E of GROZNYY, to which it is administratively subordinate; 43°18′N 45°45′E. Elevation 436 ft/132 m. Site of a Russian military base and command headquarters.

Khanka, Lake (HAHN-kah), Chinese *Xingkai Hu* (SING-KEI-HOO) (□ 1,700 sq mi/4,420 sq km), on CHINA (HEILONGJIANG province)-RUSSIA (MARITIME TERRITORY) border, 100 mi/161 km N of VLADIVOSTOK; 60 mi/97 km long, 45 mi/72 km wide, 33 ft/10 m deep; 45°00′N 132°25′E. Fisheries. Inlets include the Mo and Lefu rivers. The Songacha River, a tributary of the USSURI River, is the outlet. The Khanka lowland (S), in Maritime Territory, is one of the leading agricultural districts of SIBERIA. The lake itself is registered as protected wetland by Russia because of its importance for migratory birds.

Khanka Plain (HAHN-kah), Russian *Prikhankayskaya Nizmennost'*, low, fertile black-earth agricultural area in SW MARITIME TERRITORY, extreme SE SIBERIA, RUSSIAN FAR EAST, E and S of Lake KHANKA. Agriculture (soybeans, kaoliang, millet, rice, sugar beets, fruit; vegetable gardening); beekeeping. Food processing at USSURIYSK. Large Ukrainian population.

Khanki (kah-nah-kee), village, GUJRANWALA district, E PUNJAB province, central PAKISTAN, on CHENAB River, and 9 mi/14.5 km SW of WAZIRABAD; 32°24′N 73°59′E. Headworks of Lower CHENAB CANAL irrigation system here.

Khan Krum, Bulgaria: see HAN KRUM.

Khanlar (khahn-LAHR), city and administrative center of Khanlar region, W AZERBAIJAN, 6 mi/10 km S of GYANDZHA; 40°18′N 49°50′E. Railroad spur terminus in orchard and wine-growing area; distilleries, wineries. Settled by Germans; until 1938, called Yelenendorf.

Khan-Mamed-Kala, RUSSIA: see MAMEDKALA.

Khanna (KUHN-nuh), town, LUDHIANA district, central PUNJAB state, N INDIA, 26 mi/42 km SE of Ludhiana. Agricultural market (wheat, gram; cotton, oilseeds); cotton ginning. Manufacturing (hosiery); flour milling; steel rolling mills. Several schools and colleges. Sometimes called Khanna Kalan.

Cross-references are shown in SMALL CAPITALS. The pronunciation guide is shown on page xix. The sources of population figures are shown on page xvii.

Khanozai (KAH-no-zei), village, QUETTA district, QUETTA division, BALUCHISTAN province, SW PAKISTAN, 32 mi/51 km NNE of QUETTA; 30°37′N 67°19′E.

Khan Piyesak, BOSNIA AND HERZEGOVINA: see HAN PIJESAK.

Khanpur (KHAHN-puhr), town, BULANDSHAHR district, W UTTAR PRADESH state, N central INDIA, 15 mi/24 km NE of Bulandshahr. Wheat, oilseeds, barley, corn, jowar; cotton. Also called Khanpur Gantu.

Khanpur (KHAHN-puhr), town, RAHIMYARKHAN district, BAHAWALPUR division, PUNJAB province, central PAKISTAN, 80 mi/129 km SW of BAHAWALPUR; 28°39′N 70°39′E. Railroad junction. Agricultural trade center (wheat, rice, millet, dates; cotton). Manufacturing (rice husking, cotton ginning, hand-loom weaving; metalware, sugar industry).

Khanqah Sharif (khahn-KA shah-REE-fuh), town, BAHAWALPUR district, BAHAWALPUR division, PUNJAB province, central PAKISTAN, 10 mi/16 km SW of BAHAWALPUR; 29°19′N 71°33′E.

Khan Sheikhun (KAHN SHAI-koon), village, IDLIB district, NW SYRIA, 60 mi/97 km SSW of ALEPPO, on the main highway between DAMASCUS and Aleppo; 35°26′N 36°38′E. Cotton, cereals. Here have been found remains dating back to 20th century B.C.E.

Khanskoye (HAHN-skuh-ye), lake (surface □ 39 sq mi/100 sq km), NW KRASNODAR TERRITORY, S European Russia, near the AZOV SEA coast; 46°08′N 38°23′E. Mostly shallow, with many marshy islands, and fringed with vegetation. Abounds in fish and serves as breeding grounds for dozens of species of marine birds. Silt mud from the lake bottom has curative properties and is used by itself, or mixed into creams, to treat certain locomotor, skin, and nervous system disorders.

Khantaak Island (KAN-tuhk), SE ALASKA, 3 mi/4.8 km N of YAKUTAT; 6 mi/9.7 km long; 59°36′N 139°46′W.

Khantagy, KAZAKHSTAN: see KANTAGI.

Khan Taishirin Khure, MONGOLIA: see ALTAY.

Khantau (kahn-TOU), town, ZHAMBYL region, KAZAKHSTAN, 135 mi/225 km NE of ZHAMBYL; 44°14′N 73°47′E. Wool; agricultural products.

Khan-Tengry Peak [=lord of heaven] (22,944 ft/6,993 m), in SARY-JAZ mountain range of TIANSHAN mountain system, KYRGYZSTAN; the country's third-highest mountain peak. ENGILCHEK glacier is located here. Also called KAN-TOO.

Khanty-Mansi Autonomous Okrug (hahn-TI–mahn-SEE), administrative division (□ 201,969 sq mi/525,119.4 sq km; 2004 population 1,331,000), central TYUMEN oblast, W SIBERIA, RUSSIA; ⊙ KHANTY-MANSIYSK. The region, mostly forest and swamp with numerous lakes and peat bogs, is drained by the lower IRTYSH and Ob′ rivers, which are also important transportation arteries. It has seen spectacular population growth since 1970, mostly because of expansion of petroleum and some natural-gas exploitation; the largest concentrations of people are in the Ob′ and Irtysh valleys. Lumbering, fishing, fur farming and trading, and reindeer breeding are the area's chief native occupations. Some grain and vegetables are grown in the S, and fish processing is carried on. Since 1965, the area has become the largest oil producer in Russia, notably at BERËZOVO, SURGUT, and NIZHNEVARTOVSK. Russians make up two-thirds of the area's population, and there are small minorities of Khanty (Ostyaks) and Mansi (Voguls), both of whom belong to the Finno-Ugric linguistic family. Some Komi and Nenets also inhabit the region. The Khanty, who were under the control of the Siberian Tatars, opposed Russian conquest and rule from the 16th through the 18th centuries. The Mansi have been in the area since the 11th century; they, too, resisted Muscovite domination. The two groups were originally nomads living by breeding reindeer, hunting, and fishing. Pollution and the loss of their traditional subsistence economy have caused social problems and reduced their life expectancy. The area, formed in 1930 for the indigenous peoples, was known until 1940 as the Ostyak-Vogul National Area. In 1977 it was made an autonomous area.

Khanty-Mansiysk (HAHN-ti–mahn-SEEYSK), city (2005 population 60,470), ⊙ central KHANTY-MANSI AUTONOMOUS OKRUG, W central SIBERIA, RUSSIA, port on the IRTYSH River, 9 mi/15 km S of its confluence with the OB′ RIVER, on local road, 1,714 mi/2,758 km E of MOSCOW, and 687 mi/1,106 km N of TYUMEN; 61°02′N 69°01′E. Fish canning, lumbering, furniture; agriculture (poultry farm). Has a university, medical institute, school of continuing education, scientific centers for exploration of natural resources, environmental protection, and epidemiological control. No railroad and very limited highway connection to major population centers; dependent mainly on river and air access. Until 1940, called Ostyako-Vogul′sk. Founded in 1931 as a village. Made city in 1950 with inclusion of the village of Samarovo, known since the 16th century.

Khanua (KAHN-wuh), village, BHARATPUR district, RAJASTHAN state, NE INDIA, 13 mi/21 km SSE of Bharatpur, near BANGANGA RIVER. Site of decisive battle of 1527, when MOGULS under Babur completely routed. Rajput confederacy. Formerly also spelled KANWA.

Khan Yunis, Arab town, S GAZA STRIP, 5 mi/8 km N of RAFAH, 2.5 mi/4 km from the MEDITERRANEAN shore; 31°21′N 34°29′E. Has 13th-century mosque. The large influx of refugees in 1948 increased its population more than 5-fold. The Khan Yunis refugee camp (1995 population 53,800) is located on the outskirts of the town, with which it forms a continuous urban area.

Khao Yai National Park (KOU YEI) [Thai=big mountain], THAILAND, known for its herds of wild elephants and flocks of tropical birds such as hornbills. Poaching has reduced the number and variety of wildlife species residing within its borders.

Khapa (KAH-puh), town, NAGPUR district, MAHARASHTRA state, central INDIA, on KANHAN RIVER, and 20 mi/32 km NNW of NAGPUR. Railroad spur terminus, serving manganese mines between here and SOANER, 5 mi/8 km SW. Power plant.

Khapalu (khahp-LOO), village, Gangche district, NORTHERN AREAS, extreme NE PAKISTAN, in N central KASHMIR region, in LADAKH RANGE, near SHYOK RIVER, 40 mi/64 km ESE of Sakardu; 35°10′N 76°20′E. Agriculture (pulses, wheat). Small gold mines nearby.

Khapcheranga (hahp-chee-RAHN-gah), town, SW CHITA oblast, S SIBERIA, RUSSIA, on the Tyrin River (tributary of the ONON RIVER), approximately 16 mi/26 km N of the Mongolian border, on the E slope of the Onon Range, 175 mi/282 km SSW of CHITA; 49°42′N 112°24′E. Elevation 3,270 ft/996 m. Tin mining. Developed in the 1930s.

Khaptad National Park (KUHP-tuhd) (□ 87 sq mi/226.2 sq km), W NEPAL; 29°16′N 81°12′E. Elevation c. 10,000 ft/3,048 m. Grassland and forest. Established 1984.

Kharab, village, N YEMEN, c.90 mi/145 km SE of Saada; 16°24′N 44°39′E. Desert and frontier station.

Kharabali (hah-rah-bah-LEE), city (2005 population 18,200), E ASTRAKHAN oblast, SE European Russia, on the left bank of the AKHTUBA RIVER, a distributary of the lower VOLGA RIVER, on road and railroad, 88 mi/142 km NNW of ASTRAKHAN; 47°24′N 47°15′E. Below sea level. Vegetable canning, dairy products. Made city in 1974.

Kharaghoda (KAH-ruh-go-duh), village, AHMADABAD district, GUJARAT state, W central INDIA, on SE edge of Little RANN of KACHCHH, 55 mi/89 km WNW of AHMADABAD. Railroad terminus; market center. Salt manufacturing (extensive brine deposits in the Rann); large magnesia works. Saltworks 6 mi/9.7 km N, at OORU.

Khara Gol, MONGOLIA: see HARAA GOL.

Kharagpur (kuh-RUHG-puhr), city (2001 population 272,865), MEDINIPUR district, WEST BENGAL state, E INDIA; 22°20′N 87°20′E. It is a major railroad junction claiming to have the longest platform in the country. Growing into an industrial center with a projected steel mill nearby. Manufacturing (shoes, textiles, chemicals); railroad workshops. The first of the seven prestigious Indian Institute of Technology (IIT) campuses established here after Independence.

Kharakhalpoghistan, town, NW KARAKALPAK RE-PUBLIC, W UZBEKISTAN; 44°45′N 56°05′E. Formerly Karakalpakiia. Also Karakalpovistan, Qoraqpol-ghiston.

Kharan (KHAH-rahn), district (□ 18,508 sq mi/48,120.8 sq km), KALAT division, BALUCHISTAN province, SW PAKISTAN; ⊙ KHARAN KALAT. Bordered by IRAN (W), RAS KOH hills (N), and SIAHAN RANGE (S). Consists mainly of sandy desert wastes with low sand dunes and scant vegetation. Formerly under suzerainty of KALAT; acceded independently in 1948 to PAKISTAN.

Kharan Kalat (khah-ruh-NAH kah-lah-TAH), village, ⊙ KHARAN district, KALAT division, BALUCHISTAN province, SW PAKISTAN, on NE edge of sandy desert, 145 mi/233 km SW of QUETTA; 28°35′N 65°25′E. Sometimes called Kharan.

Khara Nor, MONGOLIA: see HAR NUUR.

Khara Nur, MONGOLIA: see HAR NUUR.

Kharar (kuh-RAHR), town, RUPNAGAR district, PUNJAB state, N INDIA, 26 mi/42 km NNW of AMBALA; 30°45′N 76°39′E. Local market for wheat, corn, gram. Woolen milling, cotton ginning.

Kharar, town, MEDINIPUR district, SW WEST BENGAL state, E INDIA, 31 mi/50 km NE of Medinipur; 22°42′N 87°41′E. Cotton weaving, metalware manufacturing. Agriculture: rice, pulses, potatoes; jute.

Kharari, INDIA: see ABU ROAD.

Kharas, Arab village, Hebron district, 7 mi/11 km NW of HEBRON, in the JUDEAN HIGHLANDS, WEST BANK. Vineyards; cereals; vegetables.

Khara Usu, MONGOLIA: see HAR US NUUR.

Kharbata [Arab.=ruin place], Arab village, Ramallah district, WEST BANK, 8.75 mi/14 km NW of RAMALLAH, in the SAMARIAN Highlands. Agriculture (olives, cereals). Tomb of Sheikh Abu-Jushef nearby.

Kharbata El Mesbah, Arab village, Ramallah district, WEST BANK, S of the road from RAMALLAH to the coastal plain, in the SAMARIAN Highlands. Agriculture (olives, cereal, fruit).

Khardah (KUHR-duh), town, NORTH 24–PARGANAS district, SE WEST BENGAL state, E INDIA, on the HUGLI River, and 11 mi/18 km N of CALCUTTA city center; 22°44′N 88°22′E. Jute-milling center; woodworking. Vishnuite temple (pilgrimage center). Well served by bus and suburban railroad. Home to many immigrant families from former EAST PAKISTAN (now BANGLADESH).

Kharfa (KAHR-fuh), village and oasis, S NAJ′D, RIYADH province, SAUDI ARABIA, in AFLAJ subregion, 6 mi/9.7 km SSW of LAILA. Grain, vegetables, fruit; weaving.

Kharfat, OMAN: see KHARIFUT.

Kharg, IRAN: see KHARK.

Kharga (KAHR-gah), large oasis, S central EGYPT, in the LIBYAN (Western) Desert. Many Arab Bedouins and Arab-speaking Berbers. The irrigated oasis produces cereals, vegetables, olives, dates, citrus fruits, and alfalfa. Cattle and poultry are also raised. Al Kharga, the chief settlement, is a railroad terminus. Airport. The oasis was prosperous in ancient times, and there are ruins of temples built by the Achaemenids of ancient Persia and by the Romans. Part of the "New Valley" scheme of the 1960s that was to irrigate vast portions of the Libyan Desert by tapping into artesian wells. The project failed to meet its goals.

Khargone (kahr-GON), town (2001 population 86,443), ⊙ KHARGONE (WEST NIMAR) district, formerly in MADHYA BHARAT, now in MADHYA PRADESH state, central INDIA, 65 mi/105 km SSW of INDORE; 21°49′N 75°39′E. Market center for cotton, millet, oilseeds, timber; cotton ginning, rice and oilseed milling.

Khargone, INDIA: see WEST NIMAR.

Khargu, IRAN: see KHARKU.

Khargupur (KUHR-guh-puhr), town, GONDA district, NE UTTAR PRADESH state, N central INDIA, 17 mi/27 km N of Gonda; 27°23′N 81°59′E. Rice, wheat, corn, gram, oilseeds.

Kharhial Road, village, NAWAPARA district, W ORISSA state, E central INDIA, on railroad, and 80 mi/129 km NW of BHAWANIPATNA. Soap manufacturing, rice milling.

Kharian (khah-yuhn), village, GUJRAT district, NE PUNJAB province, central PAKISTAN, 20 mi/32 km NW of GUJRAT; 32°49′N 73°52′E.

Kharifut (kah-ree-FOOT), village, DHOFAR governate, OMAN, 55 mi/89 km WSW of SALALA, on Wadi Kharfat, near E limit of YEMEN; 16°43′N 53°18′E. Sometimes Kharfat.

Kharik (HAH-reek), settlement (2004 population 710), SW IRKUTSK oblast, S SIBERIA, RUSSIA, near the TRANS-SIBERIAN RAILROAD (Kharik station, 5 mi/8 km away), 50 mi/80 km SE of TULUN; 54°14′N 101°39′E. Elevation 1,742 ft/530 m. In agricultural area (soybean, rye, oats).

Kharimkotan Island (kuh-reem-kuh-TUHN), Japanese *Harumukotan-to* (□ 16 sq mi/41 sq km), one of N main KURIL ISLANDS group, SAKHALIN oblast, extreme E SIBERIA, RUSSIAN FAR EAST, separated from ONEKOTAN ISLAND (N) by Sixth Kuril Strait and from SHIASHKOTAN ISLAND (S) by SEVERGIN STRAIT; 8 mi/12.9 km long, 5 mi/8 km wide; 49°07′N 154°32′E. Rises to 3,976 ft/1,212 m in volcanic SEVERGIN (syi-vyir-GEEN) Peak, Japanese *Harumukotan-dake.*

Kharino, RUSSIA: see BEREGOVOY.

Kharitonovo (hah-ree-TO-nuh-vuh), village, SW KOSTROMA oblast, central European Russia, on the E bank of the VOLGA RIVER, near highway, 12 mi/19 km SSE of KOSTROMA; 57°37′N 41°14′E. Elevation 456 ft/138 m. Popular vacation and weekend retreat spot for residents of neighboring cities and towns; has a river resort and spa.

Kharj (KAH-rizh), small district and group of oases (2004 population 200,958), E central NAJ'D, RIYADH province, SAUDI ARABIA, on railroad, and 50 mi/80 km SE of Riyadh; 24°00′N 47°00′E. One of chief agricultural areas of Riyadh province, producing dates, wheat, alfalfa, and millet. Modern irrigation from limestone well pits was mechanized and a model farm set up in 1940s. District was formerly called Yamamah. Large Saudi and U.S. air base in area.

Khark (KHAHRK), island, c.4 mi/6 km long and c.2 mi/3 km wide, SW IRAN, in the PERSIAN GULF; 29°14′N 50°18′E. Site of one of the world's largest deep-water oil ports, it is linked to the mainland by pipelines. The oil terminal was a target of frequent attacks during the IRAN-IRAQ War in the 1980s and was partly destroyed. Airport. Also spelled KHARG.

Kharkhira Uul, MONGOLIA: see HARHIRA UUL.

Kharkho, China: see GYANYIMA.

Kharkiv (KHAHR-kif) (Ukrainian *Kharkivs'ka*) (Russian *Khar'kovskaya*), oblast (□ 12,129 sq mi/31,535.4 sq km; 2001 population 2,914,212), NE UKRAINE (□). In SW outspurs of CENTRAL RUSSIAN UPLAND; drained by DONETS, upper ORIL', and lower Oskil rivers. Population has a Ukrainian (62.8%) majority, with a large Russian (33.2%) and small Jewish (1.5%), Belorussian (0.7%), Armenian (0.3%), and other minorities. Rich agricultural area; wheat, corn for grain and silage, sunflowers (S), sugar beets (W), small farm produce and orchards in Kharkiv metropolitan area; cattle, pigs. Rural industries are chiefly food processing (sugar refining in W, flour milling and sunflower-oil extraction in S, distilling, dairying). Machine construction, chemical, food, and clothing industries in Kharkiv and suburbs. Other major cities with industries: VOVCHANS'K, KUP'YANS'K, IZYUM, BARVINKOVE, LOZOVA, KRASNOHRAD, BOHODUKHIV. Good railroad network. Formed 1932. In 1993 had 17 cities, 62 towns, and 26 rural raions.

Kharkiv (HAHR-keef) (Russian *Khar'kov*), city (2005 population 1,479,800), ⊙ KHARKIV oblast E UKRAINE, at the confluence of the Kharkiv, Lopan, and Udy rivers in the upper DONETS valley; 50°00′N 36°15′E. Elevation 374 ft/113 m. Ukraine's second-largest city, Kharkiv is also one of the country's main railroad junctions and economic and cultural centers. Proximity to the iron mines of KRYVYY RIH and the coal of the DONETS BASIN has provided the basis for the largest construction of engineering industries in Ukraine that produce a wide variety of machines, including heavy-metal items such as diesel motors, turbines, tractors, and locomotives. Kharkiv's industries also include food and tobacco processing, printing, and chemical manufacturing. Founded in 1654 by the Ukrainian Cossacks led by I. Kharkach (after whom the city was named) as a military strongpoint to defend Moscow's S frontier, it became the headquarters of the Kharkiv regiment (1659–1765) in the Sloboda Ukraine and, following the Russian government's abolition of the Ukrainian cossack regimental system, the capital of Sloboda Ukraine (1765–1835). It remained loyal to the czar during the Cossack uprisings of the late 17th century, and, as a result, Kharkiv received more autonomy than most other Ukrainian cities. Developing as an intellectual and commercial center, Kharkiv became the site of the Kharkiv college (1734) and of large annual trade fairs, which were held from the second half of the 18th century until the Russian Revolution. Russia's annexation of the Crimea in 1783, the colonization of the Ukrainian steppes, and the development of coal and metallurgical industries further stimulated Kharkiv's economic growth. Kharkiv also became an important center of the 19th-century Ukrainian national and literary movements. The city became the capital of Soviet Ukraine in 1919 (in opposition to non-Communist Ukrainian government in Kiev) but was superseded by Kiev in 1934. During the World War I–World War II period, the city became a center for rapid growth of Jewish community and Jewish intellectual life. Kharkiv's landmarks include the cathedral of the Protectress (1686), the cathedral of the Assumption (1771), and a bell tower that was built to celebrate Napoleon's defeat in 1812. The university dates from 1805, and there are numerous scientific research institutes and some 20 institutions of post-secondary education. Heavy fighting raged in Kharkiv during World War II. Following the war, the city was rebuilt with large avenues and many apartment blocks.

Kharkivs'ka oblast, UKRAINE: see KHARKIV OBLAST.

Khar'kov, UKRAINE: see KHARKIV, city.

Khar'kovskaya oblast, UKRAINE: see KHARKIV, oblast.

Kharku (KHAHR-koo), island in PERSIAN GULF, SW IRAN, just N of KHARG island and 40 mi/64 km NW of BUSHEHR; 29°20′N 50°21′E. Quarantine station. Also spelled KHARGU.

Kharlu (HAHR-loo), Finnish *Harlu*, town, SW Republic of KARELIA, NW European Russia, on road and railroad, 10 mi/16 km NE of SORTAVALA; 61°48′N 30°57′E. Elevation 226 ft/68 m. Paper milling. In FINLAND until 1940.

Kharmanli, Bulgaria: see HARMANLI.

Kharovsk (hah-ROFSK), city (2006 population 11,000), central VOLOGDA oblast, N central European Russia, on the KUBENA RIVER, on road and railroad, 55 mi/89 km N of VOLOGDA; 59°58′N 40°11′E. Elevation 452 ft/137 m. Sawmilling, flax processing; glassworking. Established as a village and formerly called Kharovskaya.

Kharovskaya, RUSSIA: see KHAROVSK.

Kharp (HAHRP), settlement (2006 population 7,310), W YAMALO-NENETS AUTONOMOUS OKRUG, TYUMEN oblast, NW SIBERIA, RUSSIA, on the E slope of the central Polar URALS, on road and railroad spur, 17 mi/27 km NW, and under administrative jurisdiction of LABYTNANGI; 66°48′N 65°48′E. Elevation 285 ft/86 m. Manufacturing of ferro-concrete support structures for oil and gas pipelines.

Kharsawan (kar-SAH-wuhn), town (2001 population 6,790), WEST SINGHBHUM district, S Jharkhand state, E INDIA, 23 mi/37 km W of JAMSHEDPUR; 22°48′N 85°50′E. Rice, oilseed, corn, jowar, sugarcane; lac and silk growing. Sal, kusum, bamboo in forest area (W, N); a major cyanite and copper-mining center. Was the capital of former princely state of KHARSAWAN in ORISSA STATES; state merged 1948 with SINGHBHUM district.

Kharsia (KUHRS-yuh) or **Khursia**, town (2001 population 17,387), RAIGARH district, NE CHHATTISGARH STATES, central INDIA, 20 mi/32 km WNW of RAIGARH; 21°58′N 83°07′E. Rice, oilseeds. Sal forests, coal and iron ore deposits nearby.

Khartoum (kahr-TOOM), region, state, and former province (□ 8,190 sq mi/21,294 sq km; 2005 population 5,761,000), central SUDAN; ⊙ KHARTOUM; 16°00′N 32°30′E. When states were created in 1996 the area of the province remained the same and became Kharthoum state. The confluence of BLUE and WHITE NILE rivers is chief physiographic feature. A strip 0.5 mi/0.8 km–3 mi/4.8 km wide on either river bank is cultivable. Agriculture includes cotton, wheat, barley, corn, fruits, and durra. Livestock includes cattle, sheep, and goats. Main centers are KHARTOUM, OMDURMAN, KHARTOUM NORTH.

Khartoum (kahr-TOOM), city, ⊙ Khartoum state and SUDAN, a port at the confluence of the BLUE NILE and WHITE NILE rivers; 15°36′N 32°32′E. Khartoum is the Sudan's second-largest city and its administrative center. Food, beverages, cotton, gum, and oil-seed processing. Manufacturing includes cotton textiles, knitwear, glass, tiles; has strong military industry. Construction of an oil pipeline between Khartoum and PORT SUDAN was completed in 1977. Khartoum is a railroad hub and is connected by roads to the heart of the adjacent cotton-growing region, Port Sudan in NE, and AL UBAYYID in W. International airport to E. Founded in 1821 as an Egyptian army camp, Khartoum developed as a trade center and slave market. In the war between GREAT BRITAIN and the forces of the Mahdi, Gen. Charles Gordon was killed here (1885) after resisting a long siege and during which the city was severely damaged. Khartoum was retaken by H. H. Kitchener in 1898 and rebuilt. During the 1970s, 1980s, and 1990s thousands of refugees from the W and S and from other African nations (especially CHAD, ETHIOPIA, and UGANDA) settled in Khartoum, and they are continuing to do so. It is estimated that over two million refugees from the S alone have entered the city, most fleeing war. Foreign-aid packages to feed and shelter the refugees are inadequate, resulting in the growth of slums in the city. An educational center, Khartoum is the site of the University of Khartoum (founded 1903 as Gordon Memorial College), a branch of the University of Cairo, and Khartoum Polytechnic. The city's Sudan National Museum has important archaeological holdings. Bridges link Khartoum with KHARTOUM NORTH and OMDURMAN. Sometimes spelled Khartum.

Khartoum North (kahr-TOOM), Arabic *Khartoum Bahri*, town, KHARTOUM state, central SUDAN, on right bank of the BLUE NILE River just above its junction with the WHITE NILE, on railroad, and opposite (N of) KHARTOUM (linked by bridge); 15°38′N 32°38′E. Airfield to N Commercial center; numerous industries, engineering, metal products, dockyard, and marine workshops, with schools of marine and electrical engineering; vegetables. Has markets, commercial stores, and military barracks. The main industrial city of SUDAN.

Khartsyz'k (khahrt-SIZK) (Russian *Khartsyzsk*), city (2004 population 99,400), E central DONETS'K oblast, UKRAINE, in the DONBAS, 9 mi/14.5 km E of MAKIYIVKA; 48°02′N 38°09′E. Elevation 820 ft/249 m. Steel-pipe rolling, wire and cable, machinery works, food concentrates. Established in 1869 as a railroad station and coal mine; city since 1938. Since 1954, incorporated some adjoining settlements, including Voykove (Russian *Voykovo*) in 1989.

Khartsyzsk, UKRAINE: see KHARTSYZ'K.

Khartum, SUDAN: see KHARTOUM.

Khasab (kuh-SAHB), chief town of MUSANDAM region (2003 population 17,730) N OMAN, on the W side of the N tip of OMAN PROMONTORY, 80 mi/129 km NE of SHARJAH; 26°11'N 56°14'E. Date groves; fisheries.

Khasan (hah-SAHN), settlement (2006 population 770), SW MARITIME TERRITORY, extreme SE RUSSIAN FAR EAST, near KHASAN LAKE, 2 mi/3.2 km E of the border with NORTH KOREA, on road and the Vladivostok-Najin (CHINA) railroad, 89 mi/143 km SW of VLADIVOSTOK; 42°25'N 130°39'E. Customs station. In a multinational natural preserve area, particularly aimed at increasing the depleted populations of Amur leopard and Amur tiger.

Khasani, Greece: see ELLINIKON.

Khasan, Lake (hah-SAHN), SW MARITIME TERRITORY, SE RUSSIAN FAR EAST, near the point where RUSSIA, NORTH KOREA, and CHINA meet, S of the POSYET BAY; 42°27'N 130°36'E. Zaozernaya Hill, Chinese *Changkufeng* (W), was the scene of Soviet-Japanese border fighting (1938).

Khasan'ya (hah-sah-nee-YAH), village (2005 population 10,605), central KABARDINO-BALKAR REPUBLIC, S European Russia, in the foothills of the N CAUCASUS Mountains, 5 mi/8 km SW of NAL'CHIK; 43°26'N 43°34'E. Elevation 2,296 ft/699 m. Agricultural products. Formerly known as Prigorodnoye (Russian= suburban).

Khasavyurt (hah-sahf-YOORT), city (2005 population 139,310), NW DAGESTAN REPUBLIC, on the slopes of the NE CAUCASUS Mountains, SE European Russia, on railroad, 50 mi/80 km WNW of MAKHACHKALA; 43°15'N 46°35'E. Elevation 429 ft/130 m. Highway hub; local transshipment point. Center of agricultural area; fruit and vegetable canning, meat packing, wine making, dairy; metalworking, making of furniture and clothing. Developed mainly after 1931, when it was made a city. Sharp population increase is largely because of the influx of refugees fleeing the unrest in neighboring CHECHEN REPUBLIC.

Khāsh (KHASH), town (2006 population 57,811), Sīstan and Balūchestān province, SE IRAN, on road, and 90 mi/145 km SSE of ZAHEDAN; 28°13'N 61°12'W. Elev. 3,684 ft/1,123 m. Wheat, barley; cotton. Military post. Sometimes spelled KWASH or Khwash. Formerly called Vasht.

Khash, village, NIMRUZ province, SW AFGHANISTAN, in S Afghan desert, on the KHASH RUD, 75 mi/121 km SE of FARAH; 31°31'N 62°52'E.

Khashdala, UZBEKISTAN: see DJAMBAI.

Khashm Al-Girba (KAH-shuhm el-GIR-buh), township, KASALA state, NE SUDAN, on left bank of ATBARA RIVER, on railroad, and 45 mi/72 km SW of KASALA; 14°58'N 35°55'E. Cotton center, large sugar factory. Many villagers, evacuated from areas in the Nile valley flooded by Lake NASSER (Lake NUBIA), were settled in its neighborhood. Nearby large dam on the ATBARA RIVER, built in 1964, provides water for irrigation scheme.

Khash Rud, river, 250 mi/402 km long, SW AFGHANISTAN; rises in outlier of the Hindu Kush, 130 mi/209 km NW of KANDAHAR; flows SW, past DILARAM, KHASH, and CHAKHANSUR, into the HELMAND River; 31°11'N 62°05'E.

Khashuri (khah-SHOO-ree), city (2002 population 28,560) and administrative center of Khashuri region, central GEORGIA, on railroad, on KURA River, and 60 mi/97 km WNW of TBILISI; 41°59'N 43°36'E. Industry connected with railroad transport; manufacturing of glass containers; machine shop, cannery, meatpacking; dairy combine; oil-pumping plant and storage facilities. Known as Mikhailovo before c.1920; known since as Khashuri, except in early 1930s, when it was called Stalinissi.

Khasia (KHAH-syah), mountain range on border of WEST MACEDONIA and THESSALY departments, N central GREECE, forming divide between middle ALIÁKMON and upper PENEIOS rivers. Rises to 5,131 ft/ 1,564 m in the Kratsovon, 13 mi/21 km NW of KALAMBAKA; 39°52'N 21°30'E. Also Chasia.

Khasi and Jaintia Hills (KAH-see JEINT-yuh), former autonomous district, in what is now MEGHALAYA state, NE INDIA. Geographically within Khasi Hills and Jaintia Hills (SHILLONG PLATEAU in central section). Original district enlarged 1948 by addition of former KHASI. In 1970, with the formation of Meghalaya state, Khasi Hills district was partitioned into WEST KHASI HILLS and EAST KHASI HILLS districts, with Jaintia Hills remaining a separate district. Also United Khasi and Jaintia Hills.

Khasi Hills (KAH-see), range, central MEGHALAYA state, NE INDIA, part of Meghalaya plateau, rising sharply from SURMA VALLEY (S) to SHILLONG PLATEAU (N), which separates them from JAINTIA HILLS; c.140 mi/225 km long E-W. Rise to c.6,370 ft/1,942 m. Average annual rainfall at CHERRAPUNJI (S slope) is 450 in/1,143 cm.

Khasi States (KAH-see), former federation of 25 petty ASSAM STATES, NE INDIA. Situated largely in KHASI HILLS in what is now MEGHALAYA state. Acceded to India 1947 as a nonviable unit. In 1948, joined original Assam state district of KHASI AND JAINTIA HILLS to form united Khasi and Jaintia Hills district. Now divided among EAST KHASI HILLS, WEST KHASI HILLS, and JAINTIA HILLS districts, Meghalaya state.

Khaskovo, Bulgaria: see HASKOVO.

Khatanga (hah-TAHN-gah), village (2006 population 3,125), E TAYMYR (DOLGAN-NENETS) AUTONOMOUS OKRUG, KRASNOYARSK TERRITORY, NE SIBERIA, RUSSIA, N of the ARCTIC CIRCLE, on the KHATANGA River, near highway, 425 mi/684 km ENE of DUDINKA; 71°58'N 102°30'E. Elevation 180 ft/54 m. Government observation post; airfield; in hunting, fishing, and reindeer-raising area.

Khatanga (hah-TAHN-gah), river, approximately 715 mi/ 1,151 km long, KRASNOYARSK TERRITORY, N central SIBERIA, RUSSIA; formed by the union of the KOTUY and Kheta rivers; flows N through the CENTRAL SIBERIAN PLATEAU past KHATANGA and NE into the KHATANGA GULF of the LAPTEV SEA, forming the SE border of the TAYMYR PENINSULA. The river is navigable.

Khatanga Gulf (hah-TAHN-gah), inlet of the LAPTEV SEA of the ARCTIC OCEAN, in NE KRASNOYARSK TERRITORY, N SIBERIA, RUSSIA, at the mouth of the KHATANGA River, extends along SE side of the TAYMYR PENINSULA to Begichev Island; 175 mi/282 km long; 73°45'N 109°00'E. Salt and petroleum deposits along coasts.

Khatassy (hah-tahs-SI), village (2006 population 3,635), central SAKHA REPUBLIC, central RUSSIAN FAR EAST, on the LENA RIVER, on highway, 8 mi/13 km S of YAKUTSK, to which it is administratively subordinate; 61°54'N 129°38'E. Elevation 337 ft/102 m. In agricultural area; livestock breeding.

Khatauli (kuh-TOU-lee), town, MUZAFFARNAGAR district, N UTTAR PRADESH state, N central INDIA, on UPPER GANGA CANAL, and 13 mi/21 km S of Muzaffarnagar; 29°17'N 77°43'E. Trades in wheat, gram, sugarcane, oilseeds. Large Jain temples; serai built by Shah Jehan. Hydroelectric station (3,000 kw) 5 mi/ 8 km SSW, at village of SALAWA.

Khatayevicha, Imeni Tovarishcha, UKRAINE: see SYNEL'NYKOVE.

Khategaon (KAH-tai-goun), town (2001 population 21,018), DEWAS district, W central MADHYA PRADESH state, central INDIA, 12 mi/19 km ESE of KANNOD; 22°36'N 76°55'E. Cotton, wheat.

Khatkhyl, MONGOLIA: see HATGAL.

Khatlon, viloyat, SW TAJIKISTAN ⊙ KULOB; 37°45'N 69°45'E. Bounded by PANJ RIVER (S, E); drained by the KYZYL-SU and YAKH-SU rivers; DARVAZA RANGE in

NE. Agriculture (wheat, fruit, and vegetables); cotton in lower valleys. Livestock (sheep, goats, and cattle) on mountain slopes; extensive walnut forests (N). Industry (cotton ginning, cottonseed-oil extraction) at Kulyab and PARKHAR. Gold placers along rivers. Formed in early 1990s by joining Kurgan-Tyube (Qurghonteppa) oblast and Kulyab (Kulob) oblast. Population numbers are uncertain due to ongoing civil war; the region has been the center of fighting that has been ongoing since the early 1990s.

Khatmia, EGYPT: see SINAI.

Khatra (KAH-truh), town, tahsil headquarters, BANKURA district, W WEST BENGAL state, E INDIA, 22 mi/35 km SW of Bankura; 22°59'N 86°51'E. Lac trade center; rice, corn, wheat. Extensive lac-growing nearby.

Khatsapetivka, UKRAINE: see VUHLEHIRS'K.

Khatsapetovka, UKRAINE: see VUHLEHIRS'K.

Khattara, El (kaht-TAH-ruh), village, QENA province, Upper EGYPT, on the NILE opposite QUS. Cotton weaving, pottery making; cereals, sugarcane, dates.

Khaur (khaw-RAH), village, ATTOCK district, NW PUNJAB province, central PAKISTAN, on POTWAR PLATEAU, 34 mi/55 km SSE of CAMPBELLPUR; 33°16'N 72°28'E.

Khauz-Khan (HOUZ–HAHN), town, E MARY weloyat, TURKMENISTAN, 43 mi/70 km SW of MARY. Tertiary-level administrative center. Associated with Khauz-Khan Reservoir. Also Khovuzkhon.

Khavast, city, S TOSHKENT wiloyat, UZBEKISTAN, on TRANS-CASPIAN RAILROAD, at junction of branch to KOKAND and ANDIJAN, and 75 mi/121 km S of TOSHKENT; 40°10'N 68°49'E. Railroad workshops; metalworks. Formerly Ursatyevskaya.

Khawak Pass (KAH-wak) (11,640 ft/3,548 m), in Hindu Kush mountains, NE AFGHANISTAN, at head of PANJSHIR valley, 90 mi/145 km NNE of KABUL; 35°42'N 69°31'E. Alexander the Great crossed here (327 B.C.E.) on his way to INDIA.

Khawar, Al, QATAR: see KHOR, AL.

Khawr [Arabic=inlet], for Arabic names beginning thus: see under following part of the name.

Khawr 'Umayrah, YEMEN: see KHOR 'UMEIRA.

Khaybar, SAUDI ARABIA: see KHAIBAR.

Khaybulino, RUSSIA: see AKYAR.

Khaydarkan (hei-duhr-KUHN), town, W central OSH region, KYRGYZSTAN, on N slope of ALAY RANGE, 45 mi/72 km SSE of KOKAND (UZBEKISTAN); 39°50'N 71°25'E. Mercury- and antimony-mining center; calcium chloride. Also spelled KHAIDARKAN, HAIDARKAN.

Khayetinav, town, KHATLON viloyat, SW TAJIKISTAN, in N Vaksh valley near Yavansa River, and 16 mi/25 km NE of QURGHONTEPPA (Kurgan-Tyube); 38°08'N 68°55'E. Also spelled Hayotinav.

Khayredin, Bulgaria: see HAIREDIN.

Khazar (hah-ZAHR), village (2005 population 3,825), SE DAGESTAN REPUBLIC, E CAUCASUS, SE European Russia, on the CASPIAN SEA, on coastal railroad, 57 mi/ 92 km SE of MAKHACHKALA, and 4 mi/6 km SSE of DERBENT; 41°59'N 48°20'E. Below sea level. Fisheries; fruit growing.

Khazarasp (hahz-ahr-AHSP), city, S KHOREZM wiloyat, UZBEKISTAN, in KHIVA oasis, on railroad, and 30 mi/ 48 km SSE of URGANCH; 41°19'N 61°05'E. Tertiary-level administrative center; cotton ginning. Also Hazorasp.

Khazora, village, W TAJIKISTAN, in GISSAR RANGE, on Khudjand-Dushanbe highway, on border with LENINOBOD viloyat near Anzob Pass, and 35 mi/56 km N of DUSHANBE (linked by narrow-gauge railroad); 39°03'N 68°50'E. Coal mining. Formerly called Ziddy.

Kh. Beit Ta'amir, bedouin village, Bethlehem district, WEST BANK, 3.1 mi/5 km SE of BETHLEHEM, on the W fringes of the Judean Wilderness; 31°28'N 35°07'E. Elevation 2,297 ft/700 m. Sheep raising.

Khebda (hyeb-DAH), village (2005 population 2,815), W central DAGESTAN REPUBLIC, NE CAUCASUS, SE

European Russia, on the W bank of the AVAR KOYSU, on road, 44 mi/71 km SW of MAKHACHKALA; 42°26′N 46°33′E. Elevation 4,931 ft/1,502 m. Agriculture (grain, livestock).

Khebibchevo, Bulgaria: see LYUBIMETS.

Khed (KAID), town, PUNE district, MAHARASHTRA state, W central INDIA, 23 mi/37 km N of PUNE. Agricultural market (millet, wheat).

Khed, town, RATNAGIRI district, MAHARASHTRA state, W central INDIA, 50 mi/80 km NNE of RATNAGIRI. Local rice market.

Kheda (KAID-ah), district (□ 2,778 sq mi/7,222.8 sq km), GUJARAT state, W central INDIA; ⊙ KHEDA. Bounded S and E by MAHI RIVER, SW by SABARMATI RIVER, N by SABAR KANTHA district. Hand-loom weaving, cloth printing; metal products. Cotton mills at NADIAD, PETLAD, KHAMBAT. Agriculture (millet, rice, tobacco, cotton, oilseeds). Other important towns: KAPADVANJ, ANAND, BORSAD. Under Rajput dynasties (eighth–thirteenth centuries). Original district enlarged by incorporated (1949) of several former Gujarat states, including Cambay and Balasinor, and parts of former BARODA state.

Kheda (KAID-ah), town, ⊙ KHEDA district, GUJARAT state, W central INDIA, 19 mi/31 km SSE of AHMADABAD; 22°45′N 72°41′E. Market center for rice, millet, cloth fabrics; cotton and rice milling, cotton cloth printing, match manufacturing. Very old site (c.fifth century B.C.E.).

Kheibar, SAUDI ARABIA: see KHAIBAR.

Khekhtsir Preserve (he-HTSIR), Russian *Bol'she-Khekhtsir*, wildlife refuge (□ 17,259 sq mi/44,873.4 sq km), KHABAROVSK TERRITORY, extreme SE SIBERIA, RUSSIAN FAR EAST, in the SW part of the Khekhtsir Range, S of KHABAROVSK, at the point where the USSURI River flows into the AMUR River. Fauna includes red deer, wild boar, sable, marten, otter, fox, raccoon dog, lynx, bear, and badger. Established in 1964.

Khekra (KAI-kruh), town, MEERUT district, NW UTTAR PRADESH state, N central INDIA, 27 mi/43 km WSW of Meerut; 28°52′N 77°17′E. Trade center (wheat, millet, sugarcane, oilseeds).

Khelat, PAKISTAN: see KALAT.

Khelidonia, TURKEY: see KIRLANGIÇ, CAPE.

Khelmos, Greece: see AROANIA.

Kh. El Tayiba, Arab village, Jenin district, WEST BANK, 4.3 mi/7 km NW of JENIN, in the foothills of the SAMARIAN Highlands; 31°44′N 35°13′E. Agriculture (wheat and olives). Believed to be the biblical city Beit Offra. Archaeological excavations found remains of settlements from Canaanite Israelite, Talmudic, and medieval periods.

Khelyulya (HYE-lyoo-lyah), urban settlement (2005 population 3,055), W Republic of KARELIA, NW European Russia, on the N shore of Lake LADOGA, on road, 2 mi/3.2 km N of SORTAVALA, to which it is administratively subordinate; 61°44′N 30°41′E. Manufacturing (furniture, skis). Weather station in the vicinity.

Khemarath, THAILAND: see KHEMMARAT.

Khem-Belder, RUSSIA: see KYZYL.

Khemchik River (KHYEM-cheek), 310 mi/499 km long, in W TUVA REPUBLIC, RUSSIA; rises in the ALTAI Mountains system; flows NE past KYZYL-MAZHALYK, to the YENISEY RIVER, 40 mi/64 km WNW of SHAGONAR. Copper, asbestos along its course. Irrigates an agricultural area. Also known as Kemchik River.

Khemis El Khechna (ke-MEES el kesh-NAH), town, BOUMERDES wilaya, on coastal central ALGERIA, in E part of MITIDJA plain. Elev. 328 ft/100 m. In a very rich agricultural region, where the first agricultural village of the Mitidja was built in 1973. Also, a major marble-producing area. Formerly Fondouk.

Khemis Miliana (ke-MEES mee-lyah-NAH), town, AÏN DEFLA wilaya, N central ALGERIA, in middle CHÉLIFF River valley, on railroad, and 60 mi/97 km SW of ALGIERS; 36°15′N 02°18′E. Agricultural trade center

(cereals, cattle) and commercial outlet for MILIANA (4 mi/6.4 km N) and TÉNIET EL HAAD (30 mi/48 km SSW); some manufacturing. Formerly Affreville.

Khemissa (ke-mee-SAH), village, SOUK AHRAS wilaya, NE ALGERIA, 18 mi/29 km SW of SOUK AHRAS. Important remains of ancient *Thubursicum Numidarum* have been excavated here.

Khemisset, town, Khémisset province, Rabat-Salé-Zemmour-Zaër administrative region, NW central MOROCCO, 30 mi/48 km W of MEKNES; 33°40′N 06°00′W. Agricultural trade center (cereals, livestock, wool); weekly carpet market.

Khem Karan (KAIM KUHR-uhn), town, AMRITSAR district, W PUNJAB state, N INDIA, 37 mi/60 km SSW of Amritsar; 31°09′N 74°34′E. Agricultural market (wheat, gram, rice, oilseeds); hand-loom weaving. Sometimes KHEM KARN.

Khemmarat (KEM-muh-RAHD), village and district center, Ubon province, E THAILAND, on right bank of MEKONG RIVER (LAOS border; rapids), and 60 mi/97 km NNE of Ubon; 16°03′N 105°13′E. Sometimes spelled KHEMARATH and KEMMARATH.

Khenchella (ken-shel-LAH), wilaya, E ALGERIA; ⊙ KHENCHELLA; 35°00′N 07°00′E. Gateway to the SAHARA via El OUED wilaya (S). Renowned for its carpets, wool, and other handicrafts. Grain farming and sheep rearing predominate; trades in esparto grass, cereals, wool, and handicrafts. Main towns are Chechar (S of Kenchela) and Kais, in NW. Established 1984 with territory carved out of BATNA, TEBESSA, and OUM EL BOUAGHI wilaya.

Khenchella (ken-shel-LAH), town, NE ALGERIA, at NE foot of the AURÉS Mountains, 70 mi/113 km SSE of CONSTANTINE; ⊙ KHENCHELLA wilaya; 35°22′N 07°09′E. Occupies a strategic position at the entrance of the gap between the Aurès and the NEMENCHA MOUNTAINS. Oued Barhaa is to W. Its administrative and economic roles have grown since it became a wilaya capital. Its carpets are renowned for their quality. Grain farming; market for sheep, wool, esparto, and cereals.

Khenifra (kai-nee-FRAH), town, Khénifra province, Meknès-Tafilalet administrative region, central MOROCCO, at N foot of the Middle ATLAS mountains, on the upper Oued OUM ER RBIA river, and 65 mi/105 km S of MEKNES; 32°56′N 05°40′W. Berber commercial center; woodworking. Lead mine (16 mi/26 km N) near Aguelmous; antimony mine (N) at Tourtit; zinc and gold also found in area. Founded 17th century, expanded by Sultan Mouley Ismail to control the region's Berber tribes.

Khentsiny, POLAND: see CHECINY.

Kherabgarh, INDIA: see KHAIRAGARH.

Kheralu (kai-RAH-loo), town, MAHESANA district, GUJARAT state, W central INDIA, 23 mi/37 km NNE of MAHESANA; 23°53′N 72°37′E. Agriculture market (millet, pulses, wheat); oilseed milling.

Kheri (KAI-ree), district (□ 2,965 sq mi/7,709 sq km), UTTAR PRADESH state, N central INDIA; ⊙ LAKHIMPUR. Bounded E by GHAGHARA River, N by Nepal; drained by SARDA and GOMATI rivers. A major Indian sugar-milling district. Agriculture (rice, wheat, gram, corn, barley, oilseeds, sugarcane, millet, jute); extensive sal forest (N) containing *asaina* (*Terminalia tormentosa*). Livestock breeding. Main towns: Lakhimpur, Kheri, MUHAMDI, GOLA.

Kheri (KAI-ree), town, KHERI district, N UTTAR PRADESH state, N central INDIA, 3 mi/4.8 km S of LAKHIMPUR; 27°54′N 80°48′E. Rice, wheat, oilseeds, corn, millet. Has sixteenth century Muslim tomb.

Kherla, INDIA: see BETUL.

Kherrata (ke-rah-TAH), village, BEJAÏA wilaya, NE ALGERIA, in BABOR range of Little KABYLIA, 22 mi/35 km SE of BEJAÏA; 36°30′N 05°18′E. Iron mines.

Kherson (kher-SON), city (2005 population 328,900), ⊙ KHERSON oblast, S UKRAINE, on the DNIEPER (Ukrainian *Dnipro*) River near its mouth on the BLACK

SEA; 46°38′N 32°36′E. It is a railroad junction and a sea and river port, exporting grain, timber, and manganese ore and importing oil from the Caucasus. Kherson has one of Ukraine's largest cotton textile mills; the city's other industries include shipbuilding, grain-combine assembly, oil refining, and food processing. It is also a scientific and cultural center, with four institutes, 15 post-secondary vocational technical schools, philharmonic, theater, and history and art museum. Kherson was founded in 1778 by Count Potemkin as a naval station, fortress, and shipbuilding center. Its name derives from its location on the probable site of the Greek colony Chersonesus Heracleotica. The city became the administrative and defense center for Russia's newly acquired holdings along the Black Sea. By the late 19th century, it was an important export center. Cultural and business center for Jewish community from mid-19th century until World War II, mainly engaged in grain trade. The dredging of a deepwater canal along an arm of the Dnieper to the sea in 1901 further stimulated Kherson's growth as a port. The city's importance was enhanced still more with the building of the DNIPROHES power station in 1932 and the development of navigation on the Dnieper. Kherson's landmarks include the fortress with earthen ramparts and stone gates and the 18th-century cathedral that contains Potemkin's tomb.

Kherson oblast (kher-SON) (Ukrainian *Khersons'ka*), (Russian *Khersonskaya*), oblast (□ 10,989 sq mi/ 28,571.4 sq km), S UKRAINE; ⊙ KHERSON. Population chiefly Ukrainian (76%), with a large Russian (20%) minority. Extends N of CRIMEA from Sea of AZOV to BLACK SEA; drained by the DNIEPER (Ukrainian *Dnipro*) River (NW). Flat, dry steppe with wheat, corn, barley, and sunflowers as chief field crops, and orchards in area of Kherson, vineyards on the Dnieper below KAKHOVKA, dairy farming. Industry centered at Kherson, Kakhovka, HENICHES'K. Salt deposits along coasts. Chief ports: Kherson, Heniches'k, and SKADOVS'K. SW area served by Dzhankoy-Tsyurupinsk railroad (built during World War II), subsequently connected via bridge to Kherson. Central part traversed by post-war railroad from SNIHURIVKA through Kakhovka E to Tokmak and Donets'k. Formed 1944; has (1993) nine cities, thirty towns, and eighteen rural raions.

Khersons'ka oblast, UKRAINE: see KHERSON OBLAST.

Khersonskaya oblast, UKRAINE: see KHERSON oblast.

Khertseg Novi, MONTENEGRO: see HERCEG NOVI.

Khe Sanh (KAI SAHN), historic site, QUANG TRI province, N central VIETNAM; 16°37′N 106°44′E. Site of one of the most famous and costly battles of the Vietnam War. Vietnamese, Bru, and other minorities inhabit the area and nearby Khe Sanh town. Economic development is expected when the VIETNAM-LAOS border opens.

Khetia (KE-tee-yah), village (2001 population 14,265), BARWANI district, SW MADHYA PRADESH state, central INDIA, 32 mi/51 km SW of BARWANI, on MAHARASHTRA state border. Cotton ginning.

Khetri (KAI-tree), town, JHUNJHUNUN district, RAJASTHAN state, NW INDIA, 75 mi/121 km N of JAIPUR, in N ARAVALLI Range; 27°59′N 75°48′E. Markets grain, cotton, oilseeds; hosiery and soap manufacturing. Copper, nickel, and cobalt deposits nearby.

Khewra (khaiv-rah), village, JHELUM district, N PUNJAB province, central PAKISTAN, in SALT RANGE, 5 mi/8 km NW of PIND DADAN KHAN, on railroad spur; 32°39′N 73°01′E. Dispatching station for rock salt and coal mined nearby; soda-ash factory. Cement factory; gypsum deposits, limestone quarries 10 mi/16 km NE. Mayo salt mine known to have been worked by Moguls.

Khiav, IRAN: see MESHKINSHAHR.

Khibilii, Bulgaria: see STRAHILOVO.

Khibinogorsk, RUSSIA: see KIROVSK, MURMANSK oblast.

Khibiny Mountains (hee-BEE-ni), central MURMANSK oblast, NW European Russia, on the KOLA PENINSULA between Lake IMANDRA (W) and Lake UMBOZERO (E); 25 mi/40 km in diameter; rises to 3,930 ft/1,198 m at Lyavochorr Mountain; 67°45′N 33°45′E. Rich deposits of apatite and nephelite (mines at KIROVSK and APATITY). More than 60 elements found here, including titanium, vanadium, molybdenum, and rare earths.

Khiching (ki-CHING), village, MAYURBHANJ district, N ORISSA state, E central INDIA, 55 mi/89 km W of BARIPADA. Archaeological museum religious and historical site. Was ancient capital of Mayurbhanj; has remains of tenth–eleventh-century temples with finre sculptures.

Khidirpur, section of CALCUTTA Municipal Corporation, SE WEST BENGAL state, E INDIA, 2.5 mi/4 km SW of city center. Netaji Subhar Bose (formerly King George) Docks, main dockyards of Calcutta; railroad workshop; general engineering works; hosiery manufacturing, sawmilling.

Khidr, Al (KID-er, ahl), township, MUTHANNA province, SE central IRAQ, on the E bank of the EUPHRATES, on railroad, and 20 mi/32 km SE of SAMAWA; rice, dates. Sometimes spelled Al-Khudur.

Khiitola (hee-EE-tuh-lah), Finnish *Hiitola*, SW Republic of KARELIA, near Lake LADOGA, 18 mi/29 km NW of PRIOZËRSK; 61°14′N 29°42′E. Elevation 108 ft/32 m. Railroad junction. In FINLAND until 1940.

Kh. Ikhza'a, Arab village, 3.75 mi/6 km SE of KHAN YUNIS, in the GAZA STRIP. Agriculture (citrus, almonds, vegetables). Inhabitants mainly of bedouin origin.

Khilchipur (KIL-chi-puhr), town (2001 population 15,321), RAJGARH district, W central MADHYA PRADESH state, central INDIA, 26 mi/42 km NW of NARSINGHGARH, 10 mi/16 km W of RAJGARH; 24°03′N 76°23′E. Markets millet, cotton, wheat; cotton ginning; exports ochre and building stone. Was the capital of former princely state of KHILCHIPUR of Central India agency; in 1948, territory merged with MADHYA BHARAT; then incorporated within Madhya Pradesh state 1956, after Reorganisation of States.

Khilgana (heel-gah-NAH), village, central BURYAT REPUBLIC, S SIBERIA, RUSSIA, just E of the BARGUZIN PRESERVE, 110 mi/177 km SSE of KHORINSK; 53°53′N 110°06′E. Elevation 1,578 ft/480 m. Agricultural cooperative.

Khiliodromia, Greece: see ALONNISSOS.

Khilly (KHEE-lee), urban settlement, SE AZERBAIJAN, on KURA RIVER, 15 mi/24 km from its mouth, and 45 mi/72 km NNE of LÄNKÄRÄN; 39°25′N 49°06′E. Cotton; metalworks.

Khilok (hee-LOK), city (2005 population 10,460), SW CHITA oblast, E SIBERIA, RUSSIA, on road and the TRANS-SIBERIAN RAILROAD, on the KHILOK RIVER, 160 mi/257 km WSW of CHITA; 51°21′N 110°27′E. Elevation 2,647 ft/806 m. Railroad establishments; wood industry. Made city in 1951.

Khilok River (hee-LOK), 380 mi/612 km long, RUSSIA; rises in SW CHITA oblast, in the YABLONOVYY RANGE NW of CHITA; flows SW, past MOGZON and KHILOK, along the TRANS-SIBERIAN RAILROAD, into S BURYAT REPUBLIC and N to the SELENGA River below NOVOSELENGINSK. Abounds in fish.

Khimki (HEEM-kee), city (2006 population 143,195), central MOSCOW oblast, central European Russia, administratively a suburb of MOSCOW, 19 mi/30 km NW of the city center; port on the MOSCOW CANAL; 55°54′N 37°26′E. Elevation 620 ft/188 m. Aluminum works, glassworks, steel structures, agricultural and earth-moving machinery, plastic products, gaming equipment, sawmilling and lumbering, furniture; mechanical and electronic equipment repair; cannery. Has a number of research institutes devoted to industrial technologies, as well as Moscow State University of Culture and Arts. Developed after 1937; made city in 1939.

Khingan, Great, China: see HINGGAN, GREATER.

Khingansk (heen-GAHNSK), town (2005 population 2,070), NW JEWISH AUTONOMOUS OBLAST, S RUSSIAN FAR EAST, on highway, 8 mi/13 km N of OBLUCHYE;

49°07′N 131°11′E. Elevation 1,778 ft/541 m. Tin mining and processing. Formerly known as Mikoyanovsk (mee-kuh-YAH-nuhfsk).

Khinganskiy Preserve (heen-GAHN-skeeyee), wildlife refuge (□ 378 sq mi/982.8 sq km), AMUR oblast, extreme SE SIBERIA, RUSSIAN FAR EAST, on the left bank of the AMUR River, between the Urilo and Mutnaya rivers, just S of the main line of the TRANS-SIBERIAN RAILROAD. Fauna includes deer, elk, boar, bear, lynx, sable, marten, raccoon dog. Established in 1963. Also called Khingan Preserve.

Khíos (KHEE-os), prefecture, NORTH AEGEAN department, GREECE, in AEGEAN SEA, off the coast of W central TURKEY, including the islands of KHÍOS, ANTIPSARA, PSARA, and Oinussa; ⊙ KHÍOS (on KHÍOS island); 38°25′N 26°00′E. Also spelled Hios, Chios.

Khíos (KHEE-os), island (□ 355 sq mi/923 sq km), KHÍOS prefecture, NORTH AEGEAN department, GREECE, in the AEGEAN SEA, just W of TURKEY; 8 mi/12.9 km–15 mi/24 km wide; 38°22′N 26°00′E. Mountainous and famous for its scenic beauty and good climate. Highest point is Mount Elias (c.4,160 ft/1,270 m). Produces figs; olives, wine, mastic; marble quarries, lignite and nickel deposits; sulfur springs. Sheep and goats are raised. Some tourism. Airport. Colonized by Ionians; held (494–479 B.C.E.) by the Persians. Independent in 479 B.C.E. and joined the Delian League. Rebelled several times against Athenian ascendancy in the league. On good terms with Rome, maintaining its independence until the reign of Vespasian (1st century C.E.). Became part of the Byzantine Empire; later passed (1204) to the Latin emperors of Constantinople and then (1261) to the Genoese. Conquered by Ottoman Turks (1566); held until the First Balkan War (1912), when it was taken by Greece. A rebellion against Turkish rule resulted (1822) in a ruthless massacre of the population. Claims to be the birthplace of Homer.

Khíos, seaport (2001 population 31,361), chief town of KHÍOS island and capital of KHÍOS prefecture, GREECE. Also Chios.

Khiov, IRAN: see MESHKINSHAHR.

Khipro (kheep-ro), town, THAR PARKAR district, E central SIND province, SE PAKISTAN, 39 mi/63 km NW of UMARKOT; 25°50′N 69°22′E. Local market (millet, rice, dates). Natural salt deposits SE, on W edge of THAR DESERT; saltworks.

Khíraddinah (kee-rah-DEE-nah), village, TUNIS province, N TUNISIA, on land tongue between GULF OF TUNIS and LAKE OF TUNIS, just N of HALQ AL WADI. Airport.

Khirasra (ki-KUHS-ruh), town, RAJKOT district, in former SAURASHTRA, now in GUJARAT state, W central INDIA, 10 mi/16 km WSW of RAJKOT; 21°44′N 69°52′E. Was the capital of former WEST KATHIAWAR state of Khirasra of WEST INDIA STATES agency; state merged 1948 with Saurashtra and finally with Gujarat state when it was created in 1960. Sometimes spelled Khirasara.

Khirbet Anab, WEST BANK: see ANAB.

Khirbet-et-Tell, WEST BANK: see AI.

Khirbet Ruheiba, Israel: see RUHEIBA.

Khirgis Nur, MONGOLIA: see HYARGAS NUUR.

Khirokitia, village, LARNACA district, S CYPRUS, 31 mi/50 km S of Nicosia, in SE foothills of TROODOS MOUNTAINS; 34°48′N 32°25′E. MEDITERRANEAN SEA 5 mi/8 km to S. Grain, fruit, grapes, vegetables; livestock. KHIROKITIA archaeological site is here, oldest settlement excavated on CYPRUS, dating to 5800–5250 B.C.E. Also spelled Choirokoitia.

Khirpai (kir-PEI), town, MEDINIPUR district, SW WEST BENGAL state, E INDIA, 27 mi/43 km NE of Medinipur; 22°42′N 87°37′E. Rice, wheat, pulses.

Khislavichi (hee-SLAH-vee-chee), town (2006 population 4,455), SW SMOLENSK oblast, W European Russia, on the SOZH RIVER (tributary of the DNIEPER River), on crossroads, 40 mi/64 km S of SMOLENSK;

54°11′N 32°09′E. Elevation 518 ft/157 m. Distilling, hemp milling.

Khiv (HEEF), village (2005 population 2,460), S DAGESTAN REPUBLIC, E CAUCASUS, SE European Russia, on mountain highway, approximately 23 mi/37 km NW of the AZERBAIJAN border, 13 mi/21 km WNW of KASUMKENT; 41°45′N 47°55′E. Elevation 3,018 ft/919 m. Carpet weaving. Heritage theater. Population largely Lezghian.

Khiva (hee-VAH), city, KHORAZM wiloyat, S UZBEKISTAN, in the KHIVA oasis and on the AMU DARYA River; 41°24′N 60°22′E. Industries include metalworking, cotton and silk spinning, wood carving, and carpetmaking. The city, in existence by the 6th century, was the capital of the Khorezm (Chorasima) kingdom in the 7th and 8th centuries. From the late 16th until the early 20th centuries, Khiva was the capital of the KHANATE of the same name. Significant trade and handicraft center in the late 18th and early 19th centuries. Passed to RUSSIA in 1873. Served as the capital of the Khorezm Soviet People's Republic from 1920 to 1923 and of the Khorezm SSR in 1923 and 1924. The ancient ¼ of the city has been set aside to preserve such landmarks as an 18th century fort, the khan's palace (now a museum), and a 19th century mausoleum and minaret.

Khiva, khanate of, former state of central ASIA, based on the KHIVA (KHOREZM) oasis along the AMU DARYA River. The khanate lay S of the ARAL SEA and included large areas of the KYZYL KUM and KARA KUM deserts. Founded c.1511 as part of the Khorezm state, Khiva rose in the late 16th century as a Muslim Uzbek state. It flourished in the early 19th century but was conquered by RUSSIA in 1873; the khans subsequently continued to rule under Russian protection. Khiva's economy was based on agriculture, livestock breeding, brigandage, and handicrafts. The territory comprised the Khorezm Soviet People's Republic from 1920 to 1924, when the area was divided between the Uzbek SSR and the TURKMEN SSR. For earlier history, see KHWARAZM.

Khizar, Bulgaria: see HISARYA.

Khizroyevka, RUSSIA: see SULAK, DAGESTAN Republic.

Khizroyevskiy, RUSSIA: see SULAK, DAGESTAN Republic.

Khizy (khee-ZEE), village and center of Khizy region, E AZERBAIJAN, at SE end of the Greater CAUCASUS, 50 mi/80 km NW of BAKY. Rug manufacturing; livestock.

Khlebnikovo (HLYEB-nee-kuh-vuh), town, central MOSCOW oblast, central European Russia, on the MOSCOW CANAL, on road and railroad, 3 mi/5 km N of (and administratively subordinate to) DOLGOPRUDNYY; 55°58′N 37°30′E. Elevation 508 ft/154 m. Industrial machinery; brick works.

Khlebnikovo (HLYEB-nee-kuh-vuh), village, N TAMBOV oblast, S central European Russia, on the TSNA RIVER, on highway, 8 mi/13 km E, and under administrative jurisdiction, of SOSNOVKA; 53°14′N 41°35′E. Elevation 485 ft/147 m. Fisheries.

Khlebnikovo (HLYEB-nee-kuh-vuh), village, E MARI EL REPUBLIC, E central European Russia, on road, 65 mi/105 km NNE of KAZAN'; 56°39′N 49°56′E. Elevation 488 ft/148 m. In agricultural area (grains, vegetables, potatoes).

Khlebnoye, RUSSIA: see KHLEVNOYE.

Khlebodarskoye, UKRAINE: see KHLIBODARS'KE.

Khlevnoye (HLYEV-nuh-ye), village (2006 population 6,135), S LIPETSK oblast, S central European Russia, on the DON River, on crossroads, 27 mi/43 km SW of LIPETSK; 52°12′N 39°05′E. Elevation 528 ft/160 m. In agricultural area; grain storage, bakery. Also known as Khlebnoye.

Khlibodars'ke (khlee-bo-DAHR-ske) (Russian *Khlebodarskoye*), town, central ODESSA oblast, UKRAINE, at highway intersection 7 mi/12 km W of ODESSA city center. The cite of Odessa Agricultural Research Station. Established in 1987.

Khlong Toei (KLAWNG TUH-ee), SE suburb of BANGKOK, Phra Nakhon province, S THAILAND, on CHAO PHRAYA RIVER, on railroad spur, and 5 mi/8 km SE of city center. Railroad-river transfer wharf, chief port for ships drawing less than 12 ft/4 m. Seaplane base; oil refinery. Also spelled KLONG TOI.

Khlorakas (khlo-rah-KAHS), suburb, of PAPHOS, PAPHOS district, W CYPRUS, 2 mi/3.2 km NNW of city; 34°48′N 32°25′E. MEDITERRANEAN SEA to W. Cotton, grain, fruit, vegetables; livestock. Also spelled Chlorakas.

Khlynov, RUSSIA: see KIROV, KIROV oblast.

Khlyupino (HLYOO-pee-nuh), settlement, central MOSCOW oblast, central European Russia, near the MOSKVA River, on road and railroad, 12 mi/19 km W of ODINTSOVO, to which it is administratively subordinate; 55°40′N 36°57′E. Elevation 646 ft/196 m. Agricultural chemicals, polymer-based construction materials.

Khmelevitsy (hmye-lye-VEE-tsi), village, NE NIZHEGOROD oblast, central European Russia, on highway, 23 mi/37 km ESE of VETLUGA; 57°44′N 46°22′E. Elevation 374 ft/113 m. In agricultural area (wheat, flax, rye, oats, vegetables); grain processing.

Khmelevoye, UKRAINE: see KHMEL′OVE.

Khmelinets (hmye-lee-NYETS), village, central LIPETSK oblast, S central European Russia, on the W bank of the DON River, on road, 5 mi/8 km N of (and administratively subordinate to) ZADONSK; 52°28′N 38°53′E. Elevation 383 ft/116 m. Agricultural products.

Khmelnik, POLAND: see CHMIELNIK.

Khmel′nik, UKRAINE: see KHMIL′NYK.

Khmel′nitskaya oblast, UKRAINE: see KHMEL′NYTS′KYY oblast.

Khmel′nyts′ka oblast, UKRAINE: see KHMEL′NYTS′KYY OBLAST.

Khmel′nyts′kyy (khmel-NITS-kee) (Ukrainian *Khmel′nyts′ka*) (Russian *Khmel′nitskaya*), oblast (□ 7,971 sq mi/20,724.6 sq km; 2001 population 1,430,775), W UKRAINE; ⊙ KHMEL′NYTS′KYY. In PODOLIAN UPLAND; bounded S by the DNIESTER (Ukrainian *Dnister*) River, W by ZBRUCH RIVER; wooded steppe region. Population principally Ukrainian (90%), with Russian (5.8%), Polish (2.4%), and Jewish (0.7%) minorities. Chiefly agricultural (except extreme N), with sugar beets, wheat, fruit orchards, tobacco, corn; dairy farming. Phosphorite and peat deposits. Industry includes sugar refining, fruit canning, dairying (S), lumbering, paper milling, ceramics (N). Chief centers: Khmel′nyts′kyy, KAMYANETS′-PODIL′S′KYY, SHEPETIVKA. Formed 1937. Until 1954 called Kam′yanets′-Podil′s′kyy (Russian *Kamenets-Podol′skiy*); has (1993) thirteen cities, twenty four towns, and twenty rural raions.

Khmel′nyts′kyy (khmel-NITS-kee) (Polish *Chmielnicki*) (Russian *Khmel′nitskiy*), city (2003 population 269,100), ⊙ KHMEL′NYTS′KYY oblast, UKRAINE, on the upper reaches of the Southern BUH RIVER; 49°25′N 27°00′E. Elevation 935 ft/284 m. Railroad and highway node, industrial center; machine-building, metalworking, food processing (dairy, cheese; sugar, meat), light manufacturing (leather and shoe, sewing, knitting), chemical, construction materials (cement, brick, reinforced-concrete fabrications), technological institute, technical schools of various specializations; oblast musical-drama theater, oblast doll theater, philharmonic orchestra. Known since the end of the 15th century as defense town Ploskyriv (Polish *Ploskirów*); 1780 renamed Proskuriv (Polish *Proskurów*); in 1793, became part of the Russian Empire; in 1918–1920, mostly in Ukrainian National Republic; subsequently part of Ukrainian SSR; in 1954, re-named Khmel′nyts′kyy. City since 1795.

Khmel′ove (KHME-lo-ve) (Russian *Khmelevoye*), village, W KIROVOHRAD oblast, UKRAINE, 40 mi/64 km W of KIROVOHRAD; 48°34′N 31°24′E. Elevation 646 ft/196 m. Dairy; wheat, sugar beets.

Khmer Empire, ancient kingdom of SE ASIA. In the 6th century the Cambodians, or Khmers, established an empire roughly corresponding to modern CAMBODIA and LAOS. Divided during the 8th century, it was reunited under the rule of Jayavarman II in the early 9th century; the capital was established in the area of ANGKOR by the king Yasovarman I (882–900). The Angkor period (889–1434), the golden age of Khmer civilization, saw the empire at its greatest extent; it held sway over the valleys of the lower Menam (in present-day THAILAND) and the lower Mekong (present-day Cambodia and SOUTH VIETNAM), as well as N into Laos. The Khmer civilization was largely formed by Indian cultural influences. Buddhism flourished side by side with the worship of Shiva and of other Hindu gods, while both religions coalesced with the cult of the deva-raja, or deified king. In the Angkor period many Indian scholars, artists, and religious teachers were attracted to the Khmer court, and Sanskrit literature flourished with royal patronage. The great achievement of the Khmers was in architecture and sculpture. The earliest known Khmer monuments, isolated towers of brick, probably date from the 7th century Small temples set on stepped pyramids next appeared. The development of covered galleries led gradually to a great elaboration of plan. Brick was largely abandoned in favor of stone. Khmer architecture reached its height with the construction of Angkor Wat by Suryavarman II (1113–1150) and Angkor Thom by the great warrior and king Jayavarman VII (1181–c.1218), who is also remembered for founding hospitals, constructing resthouses for religious pilgrims, and guaranteeing food for the poor. Sculpture, which also prospered at Angkor, showed a steady development from relative naturalism to a more conventionalized technique. Bas-reliefs, lacking in the earliest monuments, came to overshadow in importance statues in the round; in the later stages of Khmer art, hardly a wall was left bare of bas-reliefs, which conveyed in the richness of their detail and vitality a vivid picture of Khmer life. The Khmers fought repeated wars against the Annamese and the Chams; in the early 12th century they invaded CHAMPA, but, in 1177, Angkor was sacked by the Chams. After the founding of AYUTHIA (c.1350), Cambodia was subjected to repeated invasions from Thailand, and the Khmer power declined. In 1434, after the Thai captured ANGKOR, the capital was transferred to PHNOM PENH; this event marks the end of the brilliance of the Khmer civilization.

Khmer Republic: see CAMBODIA.

Khmil′nyk (KHMEEL-nik) (Russian *Khmel′nik*), city (2004 population 26,700), NW VINNYTSYA oblast, UKRAINE, on the Southern BUH, and 32 mi/51 km NW of VINNYTSYA; 49°33′N 27°58′E. Elevation 862 ft/262 m. Raion center; dairying, food processing; auto repair, furniture, sewing. Sanatoriums and health resorts with radioactive mineral waters and peat baths, used to treat physical disorders. First mentioned in 1362, when it came under Lithuanian rule; granted rights of Magdeburg law in 1448; raion center since 1923. Population largely Jewish (58%) in 1926; dated back to the early 16th century. Suffered the Cossack Chmielnicki massacres (1648), and partially emigrated, but mostly persisted until destroyed by the Nazis—fewer than 1,000 Jews remaining in 2005.

Khobdo, MONGOLIA: see HOVD.

Khobi (KHO-bee), urban settlement and center of Khobi region, W GEORGIA, in RIONI River lowland, on railroad, and 15 mi/24 km NE of POTI. Tea factory, cannery.

Khobotovo (HO-buh-tuh-vuh), village, W TAMBOV oblast, S central European Russia, on road and near railroad, 9 mi/14 km NNW of MICHURINSK; 53°02′N 40°24′E. Elevation 544 ft/165 m. Logging, lumbering, woodworking.

Khodech, POLAND: see CHODECZ.

Khodja (hod-JAH), village, SE SAMARKAND wiloyat, Uzbekistan, c.5 mi/8 km from SAMARKAND; 39°48′N

66°25′E. Formerly KHODZHA-AKHRAR. Also KHODZHA, Khodzhakishlak.

Khodjaabad (hod-JAH-ah-baht), village, central ANDIJAN wiloyat, UZBEKISTAN, 15 mi/24 km SE of ANDIJAN; 40°53′N 72°01′E. Cotton, sericulture; metalworks. Also Khadzhaabad.

Khodjakul, village, NAWOIY wiloyat, central UZBEKISTAN; 40°06′N 65°48′E. Also Khodzhakul.

Khodjent, TAJIKISTAN: see KHUDJAND.

Khodoriv (KHO-do-reef) (Russian *Khodorov*) (Polish *Chodorów*), city, L′VIV oblast, UKRAINE, on tributary of the DNIESTER (Ukrainian *Dnister*) River, and 23 mi/37 km NE of STRYY. Railroad junction; agricultural market; metalworking, food processing (sugar, meat, cereals, vegetable oils), sawmilling, tile and brick manufacturing. Has an old monastery and palace. First mentioned in 1394 as the village of Khodoriv-Stav, when acquired by Poland. Passed from Poland to Austria (1772); part of West Ukrainian National Republic (1918); reverted to Poland (1919); annexed to Ukrainian SSR in 1939; part of independent Ukraine since 1991.

Khodorov, UKRAINE: see KHODORIV.

Khodzha-Akhrar, UZBEKISTAN: see KHODJA.

Khodzhamar, Bulgaria: see RUDNIK.

Khodzhambas (hod-ZHAHM-bahs), town, SE LEBAP weloyat, SE TURKMENISTAN, on the AMU DARYA (river), 22 mi/35 km NW of KERKI, and 99 mi/159 km SE of CHARJEW; 38°07′N 65°00′E. Tertiary-level administrative center. Cotton. Formerly spelled Khodzhambass; also Hojambas.

Khodzheili (hod-ZHAI-lee), city, W KARAKALPAK REPUBLIC, W UZBEKISTAN, on railroad, near the AMU DARYA River, and 10 mi/16 km WSW of NUKUS; 42°25′N 59°28′E. Cotton ginning, metalworking. Also Khodzheylie, KHUJAYLI.

Khodzhent, TAJIKISTAN: see KHUDJAND.

Khoi, IRAN: see KHOY.

Khoiniki (HOI-ni-ki), town, GOMEL oblast, BELARUS, ⊙ KHOINIKI region, on railroad spur, and 33 mi/53 km ESE of Mozvr. Manufacturing (wood products, peat processing, reinforced concrete products); automotive repair plant.

Khojak Pass, PAKISTAN: see TOBA-KAKAR RANGE.

Khokand, UZBEKISTAN: see KOKAND.

Khokhlovo (huh-HLO-vuh), village (2006 population 2,465), SW VOLOGDA oblast, N central European Russia, near the SUDA RIVER where it flows into the RYBINSK RESERVOIR, on railroad spur, 5 mi/8 km ESE of KADUY; 59°09′N 37°24′E. Elevation 357 ft/108 m. Agricultural products.

Khokhol (huh-HOL), village (2006 population 4,855), NW VORONEZH oblast, S central European Russia, in the DON River basin, on highway junction and local railroad spur, 9 mi/14 km SW of LATNAYA, and 1 mi/1.6 km W of KHOKHOL′SKIY, to which it is administratively subordinate; 51°34′N 38°45′E. Elevation 364 ft/110 m. In agricultural area (grain, potatoes; livestock).

Khokhol, RUSSIA: see KHOKHOL′SKIY.

Khokhol′skiy (huh-HOL-skeeyee), town (2006 population 7,740), W VORONEZH oblast, S central European Russia, on road and railroad spur, 20 mi/32 km WSW of VORONEZH; 51°34′N 38°46′E. Elevation 390 ft/118 m. In agricultural area; food processing. Formerly known as Khokhol.

Khokhryaki (hukh-ryah-KEE), village (2006 population 3,530), central UDMURT REPUBLIC, E central European Russia, on road and railroad, 5 mi/8 km NE of IZHEVSK; 56°55′N 53°19′E. Elevation 472 ft/143 m. Forestry, woodworking, flax milling, food processing.

Kholbon (huhl-BON), town (2005 population 2,500), central CHITA oblast, E SIBERIA, RUSSIA, on the TRANS-SIBERIAN RAILROAD, 115 mi/185 km E of CHITA; 51°53′N 116°15′E. Elevation 1,620 ft/493 m. Power plant. Coal mines here and at nearby ARBAGAR (3 mi/5 km to the S).

Cross-references are shown in SMALL CAPITALS. The pronunciation guide is shown on page xix. The sources of population figures are shown on page xvii.

Kholm (HOLM), city (2006 population 4,195), S NOV-GOROD oblast, NW European Russia, on the LOVAT' RIVER, on road, 125 mi/201 km S of NOVGOROD, and 45 mi/72 km NNW of TOROPETS; 57°08′N 31°10′E. Elevation 209 ft/63 m. Sawmilling. Known since 1144. Chartered in 1777.

Kholmogory (hul-muh-GO-ri), village (2005 population 4,360), ARCHANGEL oblast, NW European Russia, at the mouth of the NORTHERN DVINA River, on road, 47 mi/76 km SE of ARCHANGEL; 64°58′N 46°18′E. Elevation 124 ft/37 m. Known since 1355, Kholmogory was a major trade center for NOVGOROD merchants in the 15th and 16th centuries and became a shipping and cattle-raising center in the 18th century. Its significance declined with the rise of Archangel. Made city in 1784, but reduced to village status in 1925.

Kholmsk (HOLMSK), city (2006 population 31,880), SW SAKHALIN oblast, RUSSIAN FAR EAST, ice-free port on the TATAR STRAIT, Sea of JAPAN, on the W coastal highway and railroad, 50 mi/81 km W of YUZHNO-SAKHALINSK; 47°03′N 142°03′E. Elevation 436 ft/132 m. Railroad junction; ocean fishing; pulp and paper milling, fish canning; ship repairing; cold-storage plant. Port developed in the early 1920s. Under Japanese rule (1905–1945), called Maoka.

Kholmskiy (HOLM-skeeyee), town (2005 population 17,580), S KRASNODAR TERRITORY, S European Russia, on highway from KRASNODAR to NOVOROSSIYSK, 4 mi/6 km from railroad (Akhtyrskaya station), 34 mi/55 km SW of KRASNODAR; 44°51′N 38°23′E. Elevation 173 ft/52 m. Lumbering. Population mostly Meskhetian Turks.

Kholmy (khol-MEE), town (2004 population 3,150), NE CHERNIHIV oblast, UKRAINE, 28 mi/45 km WSW of NOVHOROD-SIVERS'KYY; 51°22′N 32°14′E. Elevation 380 ft/115 m. Distilling; bricks, sawmilling; butter. Heritage museum.

Kholm-Zhirkovskiy (KHOLM–zhir-KOF-skeeyee), town (2006 population 3,695), N SMOLENSK oblast, W European Russia, near the DNIEPER River, on road, 40 mi/64 km NW of VYAZ'MA; 55°31′N 33°28′E. Elevation 816 ft/248 m. Dairying, flax retting, logging. Formerly known as Kholm.

Kholodna Balka (kho-LOD-nah BAHL-kah) (Russian *Kholodnaya Balka*), former town, central DONETS'K oblast, UKRAINE, in the DONBAS, 3 mi/4.8 km SE of MAKIYIVKA and subordinated to Makiyivka city council. Coal mines. Formerly Novovoykove (Russian *Novo-Voykovo*). Emerged as site of mines in 1909; town since 1938. Annexed into Makiyivka by 1987.

Kholodnaya Balka, UKRAINE: see KHOLODNA BALKA.

Kholodnyy (huh-LOD-niyee) [Russian=cold], town (2006 population 1,215), W MAGADAN oblast, E RUSSIAN FAR EAST, in the E central CHERSKIY RANGE, in the KOLYMA River basin, on road, 8 mi/13 km SW of SUSUMAN; 62°43′N 147°57′E. Elevation 2,037 ft/620 m. In gold-mining region.

Kholomon (kho-lo-MON), mountain (3,819 ft/1,164 m), on KHALKIDHIKÍ Peninsula, in CENTRAL MACE-DONIA department, NE GREECE, 7 mi/11.3 km NE of POLYGYROS. Also Cholomon.

Kholopenichi (hah-lah-PE-nee-chee), urban settlement, NE MINSK oblast, BELARUS, 27 mi/43 km NE of BORISOV; 54°80′N 28°58′E. Manufacturing (clothing, shoes); creamery.

Kholshcheviki (huhl-shchye-vee-KEE), village, W central MOSCOW oblast, central European Russia, near the ISTRA RIVER, on road and near railroad, 7 mi/11 km W, and under administrative jurisdiction, of ISTRA; 55°55′N 36°41′E. Elevation 675 ft/205 m. In agricultural area; bakery.

Kholst (HOLST), town, S central NORTH OSSETIAN REPUBLIC, Russia, in the N CAUCASUS Mountains, on local road, 16 mi/26 km S of ALAGIR; 42°50′N 44°09′E. Elevation 4,963 ft/1,512 m. Formerly a major zinc-mining town, now mostly abandoned. The majority of remaining residents work for the North Ossetian

Preserve, within the boundaries of which the town is located.

Kholtoson (HOL-tuh-suhn), village (2005 population 1,035), SW BURYAT REPUBLIC, S SIBERIA, RUSSIA, on the border with MONGOLIA, on highway, 5 mi/8 km S of ZAKAMENSK; 50°17′N 103°18′E. Elevation 4,383 ft/1,335 m. Tungsten mining.

Kholuy (HO-looyee), town (2004 population 985), S IVANOVO oblast, central European Russia, on crossroads, 7 mi/11 km W of YUZHA; 56°34′N 41°52′E. Elevation 255 ft/77 m. Handicraft center (miniature painting on papier-mâché, wood carving).

Khomain, IRAN: see KHUMEIN.

Khomas Hochland, range, NAMIBIA, extending into NAMIB DESERT, outrider of escarpment. Approximately 8,000 ft/2,220 m high, comprising part of the central highlands. Highland savanna.

Khomas Region (ko-mahs), administrative division (2001 population 250,262), central NAMIBIA. WIND-HOEK is the highest city of plateau; 22°30′S 17°00′E. Most developed political region with high urban population density. Hub of Namibian political, administrative, and social life.

Khombole (khom-BO-lai), town (2004 population 12,136), THIÈS administrative region, W SENEGAL, on Dakar-Niger railroad, and 50 mi/80 km E of DAKAR; 14°46′N 16°42′W. Peanut growing.

Khomein (KHO-main), town, Markazī province, W central IRAN, 20 mi/32 km NW of GOLPAYEGAN, and on highway to ARAK; 33°37′N 50°05′E. Wheat, barley, cotton; wool weaving, rug making. Also spelled Khomeyn.

Khomeir, Bandar, IRAN: see BANDAR KHUMAIR.

Khomeynishahr (KHO-mai-nee-shahr), city (2006 population 223,071), Esfahān province, central IRAN, 13 mi/21 km W of ESFAHAN; 32°40′N 51°32′E. Barley cultivation.

Khomutivs'kyy step, UKRAINE: see UKRAINIAN STEPPE NATURE PRESERVE.

Khomutiv Steppe, UKRAINE: see UKRAINIAN STEPPE NATURE PRESERVE.

Khomutovka (huh-moo-TOF-kah), town (2005 population 4,900), NW KURSK oblast, SW European Russia, near the administrative border with BRYANSK oblast, on crossroads, 25 mi/40 km SW of DMITRIYEV-L'GOVSKIY; 51°55′N 34°33′E. Elevation 639 ft/194 m. Hemp processing. Formerly called Khomutovo.

Khomutovo (huh-moo-TO-vuh), town (2006 population 4,630), central ORËL oblast, SW European Russia, on road and railroad, 55 mi/89 km E of ORËL; 52°51′N 37°26′E. Elevation 826 ft/251 m. Vegetable canning. Formerly known as Khomutovskiy.

Khomutovo (huh-MOO-tuh-vuh), village (2005 population 4,875), SE IRKUTSK oblast, S central Siberian Russia, on a short left tributary of the ANGARA RIVER, on road, 11 mi/18 km N, and under administrative jurisdiction, of IRKUTSK; 52°28′N 104°24′E. Elevation 1,502 ft/457 m. In agricultural area (soybeans, rye, oats, feed corn); agricultural machinery.

Khomutovo, RUSSIA: see KHOMUTOVKA.

Khomutovskiy, RUSSIA: see KHOMUTOVO, OREL oblast.

Khone (KON), village, CHAMPASAK province, S LAOS, 8 mi/12.9 km SE of KHONG, and 180 mi/290 km NNE of PHNOM PENH, at Cambodia border, on island in ME-KONG River above KHONE Falls. The island, an extension of DANGREK Range, extends 9 mi/14.5 km across river. A small railroad (c.5 mi/8 km long) spans Khone island for transshipment of goods above falls.

Khone (KON), rapids and waterfall, CHAMPASAK province, S LAOS, 8 mi/12.9 km stretch of MEKONG beginning near Ban Thakho town.

Khong (KAWNG), town, CHAMPASAK province, S LAOS, port on KHONG Island (20 mi/32 km long, 5 mi/8 km wide) in MEKONG River, and 65 mi/105 km S of PAKSE, at CAMBODIA border. Trading center; rice growing; imports cloth, exports rice, tobacco, lac, kapok, timber, cattle. Sometimes spelled Kong.

Khonghirat, UZBEKISTAN: see QUNGHIROT.

Khong Sedone (KAWNG sai-DAWN), town, SALAVAN province, S LAOS, 35 mi/56 km N of PAKSE, on the Se Done (left tributary of the MEKONG). Market center; rice growing. Sometimes spelled Kong Sedone.

Khoni (KHO-nee), city (2002 population 11,315), W GEORGIA, 15 mi/24 km WNW of KUTAISI; 42°19′N 42°25′E. Silk-spinning center; sawmilling, tea processing, dairying. Formerly Tsulukidze.

Khon Kaen (KAWN GAN), province (□ 5,332 sq mi/13,863.2 sq km), NE THAILAND; ⊙ KHON KAEN; 16°30′N 102°05′E. Rice, corn, tapioca, tobacco, cotton, peanuts, handicrafts, horse raising. Chonnabot district is a silk-production center.

Khon Kaen (KAWN GAN), town, ⊙ KHON KAEN province, E THAILAND, in KORAT PLATEAU, 278 mi/447 km NE of BANGKOK and on railroad, near CHI RIVER; 16°26′N 102°50′E. Rice, corn; tobacco, horse raising. Sometimes spelled KONKEN. University town famous for silk.

Khonko V. San Salvador de Machaca (KON-ko sahn sahl-vah-DOR dai mah-CHAH-kah), canton, INGAVI province, LA PAZ department, W BOLIVIA; 16°46′S 68°40′W. Elevation 12,641 ft/3,853 m. Some gas resources in area; copper-bearing lode, clay, limestone, gypsum deposits. Agriculture (potatoes, yucca, bananas, rye, barley); cattle.

Khonoma (ko-NO-muh), town, KOHIMA district, NA-GALAND state, NE INDIA, in Naga Hills, 5 mi/8 km WSW of Kohima; 25°39′N 94°02′E. Rice, cotton, oranges. Former stronghold of Naga tribes.

Khonsa, town, ⊙ TIRAP district, ARUNACHAL PRADESH state, in extreme NE INDIA, near MYANMAR border; 27°01′N 95°34′E.

Khonsar (KHUHN-sahr), town, Esfahān Prov., W central IRAN, 15 mi/24 km S of GOLPAYEGAN, and on highway to ESFAHAN. Grain, tobacco, fruit, tamarisk.

Khonuu (huh-NOO-oo), village (2006 population 2,385), E SAKHA REPUBLIC, central RUSSIAN FAR EAST, on the INDIGIRKA River (head of navigation), 265 mi/426 km ESE of VERKHOYANSK; 66°27′N 143°06′E. Elevation 659 ft/200 m. Trading post; reindeer farms. Formerly known as Moma, or Mota.

Khoper (HO-pyer), river, approximately 625 mi/1,006 km long, S European Russia; rises SW of PENZA (PENZA oblast); flows SW, through SARATOV and VORONEZH oblasts, then S across NW VOLGOGRAD oblast into the DON River. It is partly navigable.

Khoper Preserve (HO-pyer), wildlife refuge (□ 62 sq mi/161.2 sq km), VORONEZH oblast, European Russia, on the KHOPER River. Established in 1935.

Khopoli (KO-po-lee), town, RAIGARH district, MAHAR-ASHTRA state, W central INDIA, 36 mi/58 km ESE of MUMBAI (Bombay); 18°47′N 73°20′E. Important transportation node; railroad terminus. Hydroelectric plant (in operation since 1915) supplied by dams on two reservoirs just NE; powers (via PANVEL) industries and railroad in Mumbai and suburban areas.

Khor (HOR), town (2005 population 11,390), SW KHABAROVSK TERRITORY, RUSSIAN FAR EAST, on Khor River (right tributary of the USSURI River), near its mouth, on road and the TRANS-SIBERIAN RAILROAD, 45 mi/72 km S of KHABAROVSK; 47°53′N 134°57′E. Elevation 196 ft/59 m. Hydrolysis plant; building materials. Established as a railroad station in 1899; town status since 1938. Formerly known as Khorskiy.

Khor (KOR), village, Esfahān province, central IRAN, 135 mi/217 km NNE of YAZD, in the DASHT-e-KAVIR. Barley, gums; camel breeding.

Khor [Arabic=inlet], for Arabic names beginning thus and not found here: see under following part of the name.

Khora (KHO-rah), town, MESSENIA prefecture, SW PELOPONNESE department, extreme SW GREECE, 23 mi/37 km W of KALAMATA; 37°41′N 21°50′E. Livestock raising (goats, hogs). Also Chora. Formerly Lygou-dista or Lygudista.

Khor Abu Habil, seasonal stream, central SUDAN; rises in NUBA MOUNTAINS; flows E alongside the railroad between Umm Rowaba amd KOSTI. It marks the boundary between the Goz land (sandy soil), and the clay soil in N KURDOFAN. Main crops are cotton, watermelon, and durra.

Khoraiba, YEMEN: see KHOREIBA.

Khor, Al, town, QATAR, SE ARABIA, 75 mi/121 km N of DOHA; 25°41′N 51°25′E. Fishing; historical site; beaches. Sometimes called Al Khawr.

Khor Al-Gash, Eritrea and Sudan: see GASH RIVER.

Khorāsān (KHUH-rah-SAHN), province (□ 120,980 sq mi/314,548 sq km), NE IRAN; ⊙ MASHHAD; 35°00′N 59°00′E. Largest province in IRAN, Mashhad is chief city; other cities include SABZEVAR, BOJNURD, and NEYSHABUR. Bordered in N by TURKMENISTAN, W by Mazandoran, Semnān, and Esfahān provinces, SW by YAZD province, S by Kermān province Sīstān-o-Balūshestān and E by AFGHANISTAN. Mountainous and arid with some salt deserts. Wheat and barley are grown in drier NE area, some grazing in E area and, in the fertile valley, the province produces large quantities of fruit, nuts, sugar beets, and cotton. Products include agricultural goods, refined sugar, textiles, carpets, turquoise, and wool. Khorāsān was occupied by the Arabs in the mid-7th century, and it was here that Abu Muslim began (8th century) his campaign against the Umayyads. The province contributed to the military power of the early Abbasid caliphs. Khorāsān was devastated by the Orghuz (or Ghuz), Turks in 1153 and 1157 and by the Mongols from 1220 to 1222. In 1383 the province was invaded by Tamerlane. It is also known as KHURASAN.

Khora Sfakion (KHO-rah sfah-KEE-on), officially *Chora Sphakíon*, village, KHANIÁ prefecture, W CRETE department, GREECE, port on S coast, 26 mi/42 km SSE of KHANIÁ; 35°12′N 24°09′E. Allied troops were evacuated from here following battle of Crete (1941) in World War II.

Khorasgan (khuh-rahs-GAHN), town (2006 population 87,282), Esfahān prov., central IRAN, 5 mi/8 km E of ESFAHAN city; 32°39′N 51°45′E. E suburb; also spelled Khvorāsgān.

Khora Sphakion, Greece: see KHORA SFAKION.

Khorat Plateau, THAILAND: see KORAT PLATEAU.

Khorazm (khor-AH-zuhm), wiloyat (□ 1,900 sq mi/4,940 sq km), NW UZBEKISTAN; ⊙ URGANCH. In KHIVA oasis, on left bank of the AMU DARYA River, between KARAKALPAK REPUBLIC (NE) and TURKMENISTAN (SW). Irrigated cotton region; sericulture; rice; cattle. Cotton ginning, cotton and silk milling. Chief cities are Urganch and Khiva. Population is mainly Uzbek. Formed 1938. An earlier Russian oblast of the same name existed 1924–1926.

Khor Baraka, Eritrea and Sudan: see BARKA RIVER.

Khordha, INDIA: see KHURDA.

Khorefto (kho-re-FTO), beach, MAGNESIA prefecture, central THESSALY department, N GREECE, on E coast of Mount PELION peninsula, 18 mi/29 km E of VÓLOS; 39°27′N 23°07′E. Former fishing village; now, with its sandy beach and clear waters, a tourist attraction. Also spelled Horefto.

Khoreiba (kor-EE-buh), township, S YEMEN, in the former QUAITI state, in the upper Wadi Duan, 65 mi/105 km NW of MUKALLA. Site of airfield, sometimes known as Qa' Ba Nua. Also spelled Khoraiba, Khuraiba, and Khuraybah.

Khor Fakan, town, S FUJAIRAH emirate, UNITED ARAB EMIRATES, near city of FUJAIRAH on PERSIAN GULF coast; 25°21′N 56°22′E. Growing beach-resort area.

Khorgosh, SERBIA: see HORGOŠ.

Khorinsk (HO-reensk), settlement (2005 population 7,940), S central BURYAT REPUBLIC, S SIBERIA, RUSSIA, on the UDA RIVER (tributary of the SELENGA River), on road, 103 mi/166 km ENE of ULAN-UDE; 52°10′N 109°46′E. Elevation 2,148 ft/654 m. Sawmilling; printing house. Town status revoked in 1990.

Khoristi (kho-ree-STEE), town, DRÁMA prefecture, EAST MACEDONIA AND THRACE department, NE GREECE, 3 mi/4.8 km ESE of DRÁMA; 41°08′N 24°13′E. Tobacco, barley; olive oil. Also spelled Choriste. Under Turkish rule, Chatalja.

Khorixas, town, NW NAMIBIA, 120 mi/193 km W of OTJIWARONGO, in KUNENE REGION. Central town of DAMARALAND, administrative center for the region. Cattle, sheep, and goats. Tourism.

Khor Kalba (KOR KIL-be), village, UNITED ARAB EMIRATES, in FUJAIRAH sheikdom, on GULF OF OMAN, 6 mi/9.7 km S of Fujairah town; 25°N 56°20′E. Marks boundary between United Arab Emirates and OMAN sultanate.

Khorlovo (HOR-luh-vuh), town (2006 population 3,845), E central MOSCOW oblast, central European Russia, on road and railroad, 9 mi/14 km WSW of YEGORYEVSK; 55°20′N 38°49′E. Elevation 498 ft/151 m. Cotton milling, chemical plant.

Khorly (khor-LEE), village, S KHERSON oblast, UKRAINE, port on KARKINIT GULF of BLACK SEA, near PEREKOP ISTHMUS, 50 mi/80 km SE of KHERSON; 46°05′N 33°18′E. Ships, grain.

Khormaksar (KOR-muh-KES-ser), low isthmus linking rocky Aden peninsula to mainland; 0.75 mi/1.21 km wide at narrowest point; bounded N by small inlet KHORMAKSAR. S YEMEN's main international airport was built here.

Khormal, IRAQ: see KHURMAL.

Khormuj (khuhr-MOOJ), town, Būshehr providence, S IRAN, 40 mi/64 km ESE of BUSHEHR; 29°30′N 56°53′E. Grain, cotton, tobacco.

Khorol (kho-ROL), city, central POLTAVA oblast, UKRAINE, near KHOROL RIVER (right tributary of the PSEL), 19 mi/31 km SE of LUBNY; 49°47′N 33°17′E. Elevation 265 ft/80 m. Raion center; flour milling, food processing, milk canning; light manufacturing, construction materials (bricks, asphalt). Agricultural technical school; historic museum. Peat bogs (S). First mentioned in 1083, Cossack town since 1648, raion center since 1923. Sizeable Jewish community since the 18th century, numbering close to 2,100 in 1926; wiped out by the Nazis during World War II—fewer than 100 Jews remaining in 2005.

Khorol' (huh-ROL), village (2006 population 11,155), SW MARITIME TERRITORY, SE RUSSIAN FAR EAST, on crossroads and a branch of the TRANS-SIBERIAN RAILROAD, 45 mi/72 km N of USSURIYSK; 44°25′N 132°06′E. Elevation 347 ft/105 m. In agricultural area (rice); brewery. Military airbase in the vicinity.

Khorol River (kho-ROL), river, 190 mi/306 km long, N UKRAINE; rises 35 mi/56 km WSW of SUMY; flows S, past MYRHOROD and KHOROL, and E to PSEL RIVER 30 mi/48 km NNE of KREMENCHUK. Banks near Myrhorod are flat and sandy, creating a natural beach that, along with mineral springs, is part of Myrhorod's recreational and health tourism attractions.

Khoroshch, POLAND: see CHOROSZCZ.

Khorosheve (KHO-ro-she-ve) (Russian *Khoroshevo*), town (2004 population 14,300), N central KHARKIV oblast, UKRAINE, on Udy River, 9 mi/14.5 km S of KHARKIV city center; 49°52′N 36°10′E. Elevation 593 ft/180 m. Quarrying, state farm; many residents work in Kharkiv. Has an 18th-century church. Established in the mid-17th century; town since 1938.

Khoroshevo, UKRAINE: see KHOROSHEVE.

Khorostkiv (kho-ROST-kif) (Russian *Khorostkov*) (Polish *Chorostków*), city (2004 population 6,900), E TERNOPIL' oblast, UKRAINE, on right tributary of ZBRUCH RIVER, and 12 mi/19 km SE of TEREBOVLYA; 49°14′N 25°55′E. Elevation 1,013 ft/308 m. Grain milling, sugar refining, distilling; stone quarry. Agricultural research station. Has an old palace, Gothic church. First mentioned in 1564. Passed from Poland to Austria (1772); part of West Ukrainian National Republic (1918); reverted to Poland (1919); annexed to Ukrainian SSR in 1939; city since 1977; part of independent Ukraine since 1991. Jewish community since 1765, mostly small artisans, reaching its peak in 1900 (2,075), gradually declining due to pogroms of 1905 and 1918–1920, but still numbering 1,980 in 1939; completely eliminated during World War II.

Khorostkov, UKRAINE: see KHOROSTKIV.

Khorramābād (khuhr-rahm-AH-BAHD), city (2006 population 333,945), ⊙ Lorestān province, W central IRAN. It is the trade center of a mountainous region where fruit, grain, and wool are produced. Airport.

Khorramdarreh (khuhr-rahm-dahr-RE), town, Zanjān province, NW IRAN, 55 mi/90 km SE of Zanjān city, on road and railroad, midway between Zanjān and QAZVIN; 36°12′N 49°11′E.

Khorramshahr (kuhr-rahm-SHAHR), city (2006 population 125,859), KHUZESTĀN province, SW IRAN, at confluence of KARUN RIVER and the SHATT AL ARAB, near PERSIAN GULF; 30°26′N 48°10′E. Elevation 9.8 ft/3 m. Airport. Its development dates to the late 19th century, when steam navigation on the Karun River was started. The city was known as Muhammerah until the mid-1920s, when Reza Shah took it out of the hands of a semi-independent local sheikh and placed it under the control of the central government as Khorramshahr. It was a busy port at the S terminus of a railroad until severely attacked and partially destroyed during the Iran-IRAQ war in the 1980s; in 1976 the city had 140,490 residents, most of whom relocated during this period. The port has since been refurbished and the local economy revived. The city has lost some of its importance due to the shallowness of the Shatt al Arab, which is inaccessible to large ocean-going merchant vessels.

Khorsabad (kor-SAH-bad), village, NINEVEH province, NE IRAQ, 10 mi/16 km from E bank of the TIGRIS RIVER, and 12 mi/19 km NE of MOSUL. Airport. It is built on the site of Dur Sharrukin (KHORSABAD) an Assyrian city (founded 8th century B.C.E. by Sargon), which covered 1 sq mi/3 sq km. Its mounds were excavated by P. E. Botta in 1842 and 1851; statues of Sargon and of huge, winged bulls that guarded the gates of the royal palace were taken to the Louvre. In 1932 hundreds of cuneiform tablets in the Elamite language and a list of kings ruling from c.2200 B.C.E. to 730 B.C.E. were discovered here.

Khorskiy, RUSSIA: see KHOR.

Khortiatis (khor-tee-AH-tees), village, THESSALONÍKI prefecture, CENTRAL MACEDONIA department, NE GREECE, at base of KHALKIDHIKÍ Peninsula, c.10 mi/16 km NE of THESSALONÍKI, on N slope of KHORTIATIS mountain; 40°36′N 23°06′E. Also Chortiatis, Hortiatis.

Khortiatis (khor-tee-AH-tees), mountain peak, THESSALONÍKI prefecture, CENTRAL MACEDONIA department, NE GREECE, at base of KHALKIDHIKÍ Peninsula, c.10 mi/16 km NE of THESSALONÍKI; rises to 3,940 ft/1,201 m (closed to public above 2,500 ft/762 m). Wooded slopes are popular for tourists. Also called Chortiatis, Hortiatis.

Khortitsa, UKRAINE: see VERKHNYA KHORTYTSYA.

Khortitsa Island, UKRAINE: see KHORTYTSYA ISLAND.

Khortytsya, UKRAINE: see VERKHNYA KHORTYTSYA.

Khortytsya Island (KHOR-ti-tsyah) (Russian *Khortitsa*), UKRAINE, part of ZAPORIZHZHYA city, between arms of the DNIEPER (Ukrainian *Dnipro*) River, just below the DNIPROHES dam; 6 mi/9.7 km long, 2 mi/3.2 km wide. Has an important agricultural research station. In the 17th century, headquarters (*sich*) of the Zaporozhzhian Cossacks. In 1965 became a state historical-cultural preserve, with a museum, with archaeological and historical objects on display; park and recreation area.

Khorugh (khor-OOG), city, ⊙ BADAKHSHAN AUTONOMOUS VILOYAT, E TAJIKISTAN, in the PAMIR, at junction of GUNT RIVER and PANJ RIVER on border with AFGHANISTAN; 37°30′N 71°36′E. Manufacturing of shoes, metal goods, processed food, and construction materials. Major opium-production center from local poppies. Also spelled Khorog.

Khor 'Umeira (KOR oo-MEI-ruh), village, in SUBEIHI tribal area, S YEMEN, 60 mi/97 km W of ADEN, on the Khor 'Umeira (inlet of Gulf of ADEN). Airfield. Was center of the former Barhimi sheikdom. Also spelled Khawr 'Umayrah.

Khorzhele, POLAND: see CHORZELE.

Khosedakhard (huh-sye-dah-HART), settlement, S NENETS AUTONOMOUS OKRUG, ARCHANGEL oblast, NE European Russia, in the BOLSHEZEMEL'SKAYA TUNDRA, on local road, 175 mi/282 km ESE of NARYAN-MAR; 67°02′N 59°24′E. Elevation 328 ft/99 m. Cultural center for the Nentsy people; trading post; reindeer raising.

Khosrovi (kuhs-ro-VEE), village, Kermānshāhān province, W IRAN, 90 mi/145 km W of KERMANSHAH; 34°23′N 45°29′W. Customs station on IRAQ border, on main TEHRAN-BAGHDAD route.

Khost, province, AFGHANISTAN: see KHOWST.

Khost, town, ⊙ KHOWST province, E AFGHANISTAN, 50 mi/80 km ESE of GARDEZ, near PAKISTAN border; 33°22′N 69°57′E. Trade center in lumbering area (pine); livestock raising. Also called Matun.

Khost, village, Sibi dist., NE central Baluchistan prov., SW Pakistan, in Central Brahui Range, 50 mi/80 km NNW of Sibi; 30°13′N 67°35′E.

Khosta (HO-stah), resort settlement, S KRASNODAR TERRITORY, NW CAUCASUS, RUSSIA, on the BLACK SEA coast of the Caucasus, on coastal road and railroad, 8 mi/13 km SE, and under administrative jurisdiction, of SOCHI; 43°31′N 39°52′E. Elevation 495 ft/150 m. Subtropical seaside and health resort.

Khotang (KO-tahng), district, E NEPAL, in SA-GARMATHA zone; ⊙ DIKTEL.

Khotchino, RUSSIA: see GATCHINA.

Khoten, UKRAINE: see KHOTIN'.

Khothor, town, SURAT district, GUJARAT state, W central INDIA, on TAPI RIVER, and 8 mi/12.9 km NE of SURAT. Local market center for millet, cotton, rice; handicraft cloth weaving, calico printing. Sometimes spelled Kathore or Kathor.

Khotimsk (hah-TEEMSK), urban-type settlement, SE MOGILEV oblast, BELARUS, 40 mi/64 km SSW of RO-SLAVL; 53°24′N 32°36′E. Manufacturing (construction materials); creamery.

Khotin' (kho-TEEN) (Russian *Khoten*), town (2004 population 4,500), E SUMY oblast, UKRAINE, 11 mi/18 km N of SUMY; 51°04′N 34°46′E. Elevation 534 ft/162 m. Flour mill, brickworks, food-flavoring plant; vocational school, museum. Known since the end of the 14th century; town since 1968.

Khotin, UKRAINE: see KHOTYN.

Khot'kovo (huht-KO-vuh), city (2006 population 20,300), NE MOSCOW oblast, central European Russia, on the TRANS-SIBERIAN RAILROAD, 37 mi/60 km NE of MOSCOW, and 6 mi/10 km SW of (and administratively subordinate to) SERGIYEV POSAD; 56°15′N 38°00′E. Elevation 620 ft/188 m. Cotton milling, machine building, manufacturing (chemicals, electric and heat insulators, art objects). Formed around the Veil Khot'kovo monastery in 1308; a workers settlement since 1939; city since 1949. Nearby is Abramt-sevo, a 19th century country residence (now a museum) of many literary figures and painters.

Khotyn (kho-TIN) (Russian *Khotin*), city (2004 population 9,700), NE CHERNIVTSI oblast, UKRAINE, on the DNIESTER River, 31 mi/50 km ENE of CHERNIVTSI; 48°29′N 26°30′E. Elevation 629 ft/191 m. A raion center, it lies in Bessarabia in an agricultural district and has agricultural and food-processing industries and enterprises that produce wood products, kilims, textiles, clothing, and handicrafts. Located on the site of an ancient fortified Slavic settlement, the city is named for Kotizon, a 3rd-century Dacian chief. It was included in Kievan Rus' in the 10th century and later

became part of the Halych and Halych-Volyn duchies. Khotyn developed into an important trade and craft center and in the 13th century was the site of a Genoese trading colony. The city was included in the Hungarian and Moldavian states in the 14th and 15th centuries. Its strategic location at an important Dniester River crossing caused the city to change hands frequently among the Poles, Ukrainian Cossacks, Moldovans, and Turks from the 16th to 18th centuries. Seized by Russia in 1739, Khotyn was incorporated into the Russian Empire in 1812 as part of Bessarabia. The city was under Romanian rule from 1918 to 1940 and under German occupation from 1941 to 1944. Khotyn has remains of an imposing fortified castle that was built (13th century) by the Genoese, enlarged (14th–15th centuries) by the Moldovans, and restored (18th century) by the Ottoman Turks.

Khotynets (huh-ti-NYETS), town (2006 population 4,165), W ORËL oblast, SW European Russia, on road and railroad, 30 mi/48 km WNW of ORËL; 53°07′N 35°24′E. Elevation 869 ft/264 m. In agricultural area (grains, potatoes); hemp processing.

Khouribga, town, CHAOUIA-OUARDIGHA administrative region, W central MOROCCO, on railroad and 65 mi/105 km SE of CASABLANCA; 32°53′N 06°54′W. Morocco's chief phosphate-mining center. Phosphate drying, superphosphate manufacturing.

Khovaling (hov-ah-LEENG), village, NE KHATLON viloyat, TAJIKISTAN, 30 mi/48 km NNE of KULOB; 38°21′N 69°58′E. Wheat; horses. Tertiary level administrative center.

Khovd, MONGOLIA: see HOVD.

Khovrino, RUSSIA: see KRASNOOKTYABR'SKIY.

Khovu-Aksy (HO-voo–ahk-SI), village (2006 population 3,665), S central TUVA REPUBLIC, S SIBERIA, RUSSIA, on highway, 36 mi/58 km S of KYZYL; 51°08′N 93°36′E. Elevation 4,186 ft/1,275 m. Cobalt mining; woodworking.

Khovuzkhon, TURKMENISTAN: see KHAUZ-KHAN.

Khowai, town, tahsil headquarters, WEST TRIPURA district, TRIPURA state, NE INDIA, 20 mi/32 km NNE of AGARTALA; 24°06′N 91°38′E. Trades in rice, cotton, tea, mustard, jute. Railroad-spur terminus at BALLA station, just W; rice milling.

Khowst, province (2005 population 478,100), E AFGHANISTAN; ⊙ KHOST; 33°23′N 69°53′E. Borders PA-KISTAN (N, E, and S), PAKTIKA province (SW), and PAKTIA province (WNW). Regional airport near Khost. Formed from Paktia province. Also speleld Khost.

Khoy or **Khoi**, city (2006 population 181,465), Āzer-bāyjān-e Gharbi province, NW IRAN. Trade center for a fertile, irrigated farm region that produces grain, fruit, and timber. Because of its strategic location near TURKEY and the former SOVIET UNION, control of the city has frequently been in dispute. Khvoy was attacked by RUSSIA in 1827, occupied by Turkey in 1911, and held by the Soviet Union during World War II. The 1514 battle of Chaldiran, when Selim I, an Ottoman sultan, defeated Shah Ismail of PERSIA, took place nearby. Also called Khvoy.

Khoyto-Gol (HO-yee-tuh–GOL), rural settlement (2004 population 420), SW BURYAT REPUBLIC, S Siberian Russia, in the S foothills of the Tunkinskiye Gol'tsy range, near the IRKUT RIVER, 24 mi/39 km W of KYREN; 51°43′N 101°30′E. Elevation 3,087 ft/940 m. Agricultural cooperative.

Khozdar, PAKISTAN: see KHUZDAR.

Khram River (KHRAHM), Georgian *Khrami*, 122 mi/196 km long, S GEORGIA; rises NE of AKHALKALAKI in TRIALET RANGE at 7,950 ft/2,423 m; flows ESE, past TSALKA (reservoir) and MOLOTOVO (site of Khramges hydroelectric station), to KURA River on Georgia-Azerbaijan line, S of RUSTAVI. Its basin covers 3,220 sq mi/8,340 sq km. There are three hydroelectric power plants along the river.

Khrapunovo, RUSSIA: see VOROVSKOGO, IMENI, MOS-cow oblast.

Kh. Raqa'a, Arab village, Hebron district, WEST BANK, 5 mi/8 km S of HEBRON, in the JUDEAN HIGHLANDS; 31°28′N 35°07′E. Inhabited by bedouin tribes that have become sedentary in recent decades.

Khrebtovaya (hryep-TO-vah-yah), village (2005 population 2,000), W IRKUTSK oblast, E central Siberian Russia, on road and branch line of the BAYKAL-AMUR MAINLINE, 9 mi/14 km N of ZHELEZNOGORSK-ILIMS-KIY; 56°42′N 104°15′E. Elevation 1,896 ft/577 m. Formerly known as Izbushechnaya (ees-BOO-shich-nah-yah).

Khrenovo, RUSSIA: see KHRENOVOYE.

Khrenovoye (HRYE-nuh-vuh-ye), village (2006 population 5,450), central VORONEZH oblast, S central European Russia, on crossroads and railroad, 12 mi/19 km E of BOBROV; 51°07′N 40°17′E. Elevation 479 ft/145 m. In virgin steppe (Russian *tselina*); woodworking. Large horse-breeding station (founded in 1775). Formerly also called Khrenovo.

Khrestivka, UKRAINE: see KIROVS'KE, city.

Khrisos, Greece: see CHRYSOS.

Khrisoupolis (khree-SOO-po-lees), town, KAVÁLLA prefecture, EAST MACEDONIA AND THRACE department, NE GREECE, on river, and 15 mi/24 km E of KAVÁLLA; 40°59′N 24°42′E. Agricultural center of lowland W of NESTOS (Mesta) River mouth; tobacco, cotton, wheat. Also spelled Chrysoupolis and Chrys-soupolis. Formerly called Sapai (Sapaioi) and, under Turkish rule, Sari Shaban.

Khristinovka, UKRAINE: see KHRYSTYNIVKA.

Khristoforovo (three-stuh-FO-ruh-vuh), town, NW KIROV oblast, E European Russia, on railroad, 13 mi/21 km NW of LAL'SK; 60°54′N 47°15′E. Elevation 597 ft/181 m. Highway terminus; lumbering.

Khrompik, RUSSIA: see DVURECHENSK.

Khromtau (krom-TOU), city, N AKTÖBE region, KA-ZAKHSTAN, on railroad, 37 mi/60 km E of AKTÖBE (Aktyubinsk); 50°15′N 58°22′E. Tertiary-level (raion) administrative center. Chrome mining. Developed in early 1940s. Also Khrom-Tau.

Khromtsovo (HROM-tsuh-vuh), industrial settlement (2005 population 1,165), NW IVANOVO oblast, central European Russia, near railroad, 4 mi/6 km WSW, and under administrative jurisdiction, of FURMANOV; 57°13′N 40°58′E. Elevation 501 ft/152 m. Sand and gravel quarrying. Also spelled Khromtsevo (HROM-tsi-vuh).

Khroub, El (KROOB, el), town, CONSTANTINE wilaya, NE ALGERIA, 8 mi/12.9 km SE of CONSTANTINE; 36°15′N 06°40′E. Railroad junction; agricultural market in wheat-growing region. Flour milling. Has regional (wilaya) airport.

Khrumiriyah (kroo-mee-REE-yah), mountainous region of NW TUNISIA, JANDUBA province, in the E High Tell, between the MEDITERRANEAN SEA (N) and MAJARDAH RIVER (S); extends from ALGERIAN border (W) to JABAL ABYAD area (E). Has abundant rainfall and extensive cork oak forests. Chief towns are AYN AD-DARAHAM and TABARQAH (port). Berber population. Also spelled Kroumirie.

Khrushchev, UKRAINE: see SVITLOVODS'K.

Khrushchëvo (hroo-SHCHO-vuh), village, central RYAZAN oblast, central European Russia, on highway and railroad, 30 mi/48 km SE of RYAZAN, and 8 mi/13 km ESE of STAROZHILOVO, to which it is administratively subordinate; 54°11′N 39°59′E. Elevation 472 ft/143 m. Dairying.

Khrushchov, UKRAINE: see SVITLOVODS'K.

Khrustal'ne (khroos-TAHL-ne) (Russian *Khrustal'-noye*), town, S LUHANS'K oblast, UKRAINE, in the DONBAS, 3 mi/4.8 km W of KRASNYY LUCH and subordinated to its city council; 48°09′N 38°50′E. Elevation 554 ft/168 m. Coal mines; most residents work in Krasnyy Luch.

Khrustal'noye, UKRAINE: see KHRUSTAL'NE.

Khrustal'nyy (hroos-TAHL-niyee), settlement (2006 population 3,550), E central MARITIME TERRITORY, SE RUSSIAN FAR EAST, in the SE SIKHOTE-ALIN RANGE, near highway, 23 mi/37 km WSW of DAL'NEGORSK; 44°21′N 135°06′E. Elevation 1,505 ft/458 m. Tin mining, quartz quarrying.

Khryashchevka (HRYAH-shchyef-kah), village (2006 population 3,205), W SAMARA oblast, E European Russia, on the left bank of the VOLGA RIVER, terminus of local highway, 25 mi/40 km NW of TOL'YATTI; 53°49′N 49°05′E. Elevation 246 ft/74 m. In agricultural area (wheat, rye, sunflowers); produce processing.

Khryplin, UKRAINE: see KHRYPLYN.

Khryplyn (KHRIP-lin) (Russian *Khryplin*) (Polish *Chryplin*), railroad junction in central IVANO-FRAN-KIVS'K oblast, UKRAINE, 2 mi/3.2 km SSE of IVANO-FRANKIVS'K; 48°53′N 24°44′E. Elevation 846 ft/257 m.

Khrysokhou Bay (kree-SO-khoo), NW CYPRUS, arm of MEDITERRANEAN SEA, bounded W by Akamas Peninsula and Cape AKAMAS; c. 15 mi/24 km long, 6 mi/9.7 km wide. Town of POLIS near its S shore; Baths of Aphrodite on SW shore; Khrysokhou River enters from S.

Khrystynivka (khris-TI-nif-kah) (Russian *Khristinovka*), city, SW CHERKASY oblast, UKRAINE, 13 mi/21 km NW of UMAN; 48°50′N 29°58′E. Elevation 784 ft/238 m. Raion center; railroad junction; railway servicing; metalworking, food processing; feed milling; asphalt. First mentioned in 1574 as the town of Khrystyhorod; declined and re-named; revived in the 19th century; city since 1956.

Khubar (koo-BAHR), town (2004 population 165,799), in AL AHSA region, E SAUDI ARABIA, port on PERSIAN GULF, 5 mi/8 km ESE of DHAHRAN, opposite BAHRAIN; 26°17′N 50°12′E. Fishing, pearling. Connected by causeway with Bahrain. Serves traffic between Bahrain and HASA region of Saudi Arabia.

Khubsugul, MONGOLIA: see HÖVSGÖL.

Khuchni (hooch-NEE), village (2005 population 3,725), SE DAGESTAN REPUBLIC, E CAUCASUS, SE European Russia, on road, 53 mi/85 km SSE of MAKHACHKALA, and 20 mi/32 km WSW of DERBENT; 41°57′N 47°57′E. Elevation 2,870 ft/874 m. Woolen milling. Population predominantly Tabasaran, with a sizeable Chechen minority, most of them arriving after 1992.

Khudat (khoo-DAHT), town, NE AZERBAIJAN, on railroad, and 105 mi/169 km NNW of BAKY, near CASPIAN SEA coast. Metalworks, vegetable and fish canning.

Khudian (khood-YUHN), town, LAHORE district, E PUNJAB province, central PAKISTAN, 39 mi/63 km S of LAHORE; 30°59′N 74°17′E.

Khudjand (hoo-JAHND), city, NW TAJIKISTAN, on the SYR DARYA River at its exit from the FERGANA VALLEY; ⊙ LENINOBOD region; 40°17′N 69°37′E. Airport. A major center for silk production and one of the oldest centers of Tajik decorative and applied arts; manufacturing includes clothing, footwear, and food products. Located on an ancient caravan route from CHINA to the Mediterranean, Khudjand was a famous town marking the farthest expansion of Alexander the Great. There he founded a new fortress called Alexandria Eskhat [=outermost Alexandria]. It was plundered (711) by Arab forces and later (1220) was razed by Jenghiz Khan. As part of the Kokand khanate (early 19th century), it was annexed (1866) by Russia. The city and surrounding area belonged to UZBEKISTAN from 1924 to 1929. Formerly called Leninabad (1936–1992). Also spelled Khudzhand, Khodzhent, and Khodjent.

Khudur, Al, IRAQ: see KHIDR, AL.

Khudzhand, TAJIKISTAN: see KHUDJAND.

Khujayli, UZBEKISTAN: see KHODZHEILI.

Khukhut Waterfowl Park, THAILAND. Bird sanctuary about 30 mi/48 km N of SONGKHLA town on E shore of the SONGKHLA inner sea.

Khuldabad (kuhl-dah-BAHD), town, AURANGABAD district, MAHARASHTRA state, W central INDIA, 13 mi/21 km NW of AURANGABAD; 20°03′N 75°11′E. Muslim place of pilgrimage; site of Aurangzeb's tomb. Famous Ellora caves are just NW. Formerly also called RAUZA.

Khulm, town and district, SAMANGAN province, N AFGHANISTAN, about 33 mi/53 km E of MAZAR-I-SHARIF on the Kabul–Mazar-i-Sharif highway; 36°42′N 67°41′E. Elev. 1,495 ft/456 m. Population consists primarily of Tajiks and Uzbeks. Irrigated by the Khulm (Samangan) River. Market for wool and sheep. Agriculture (fruit orchards; vegetables; rice); flour milling. Handicrafts include leatherwork, iron implements, copperware, and wooden utensils. The notable covered bazaar is comprised of 1,200 shops; damaged in the 1976 earthquake. Commonly identified with ancient "Aornos" on line of march (330 B.C.–329 B.C.E.) of Alexander the Great. Later became known as Khulm, destroyed in mid-16th century. An autonomous khanate, the area came under Kabul rule in 1850. In recent years the name Khulm replaced Tashkurgan (Turkish=stone or brick fort).

Khulm River, 130 mi/209 km long, SAMANGAN province, N AFGHANISTAN; rises in the Hindu Kush; flows N, past HAIBAK and KHULM, and disappears into desert N of Khulm; 36°16′N 68°01′E. Used for irrigation in lower course. Also called Samangan River.

Khulna (kool-nah), district, W EAST BENGAL, BANGLADESH, in GANGA DELTA; ⊙ KHULNA. Bounded E by MADHUMATI RIVER, S by BAY OF BENGAL; drained by river arms of GANGA DELTA; central portion of the SUNDARBANS in S (sundari timber; habitat of Bengal tigers, leopards, wild buffalo). Alluvial soil; extensive groves of date, palms (betal and coconut); agriculture (rice, jute, oilseeds, sugarcane, tobacco); milling (rice, flour, oilseed); cotton-cloth manufacturing at KHULNA (trade center for the SUNDARBANS) and BAGHERHAT; shipbuilding at KHULNA; fishing. Ruins (15th-century) of Gaur kingdom at BAGHERHAT. IS-WARIPUR was 16th-century independent Muslim kingdom until defeated in 1576 by Akbar's Hindu general. Part of former British Bengal province in INDIA until 1947, when it was incorporated into the Pakistani province of East Bengal.

Khulna (kool-nah), city (□ 267 sq mi/694.2 sq km; 2001 population 1,172,831), SW EAST BENGAL, BANGLADESH, on the PUSUR RIVER; 22°53′N 89°29′E. River port; trade and processing center for the products of the SUNDARBANS; a swampy, forested coastal region. Processing of agricultural products (rice, oilseed, and cotton); weaving; wood processing and shipbuilding industries. A power station is the center of an irrigation project. Khulna has several colleges affiliated with Rajshahi University.

Khulo (KHOO-lo), urban settlement and center of Khulo region, E Adzhar Autonomous Republic, GEORGIA, 34 mi/55 km E of BATUMI. Manufacturing of garments.

Khumair, Bandar, IRAN: see BANDAR KHUMAIR.

Khumalag (hoo-mah-LAHK), village (2006 population 3,930), central NORTH OSSETIAN REPUBLIC, in the N CAUCASUS Mountains, RUSSIA, on mountain highway, 6 mi/10 km WNW of BESLAN; 43°14′N 44°28′E. Elevation 1,476 ft/449 m. In agricultural district (grain, livestock).

Khumbu (KOOM-boo), mountainous region of NE NEPAL, near CHINA (TIBET) border. Includes many of the country's highest mountains, such as EVEREST, LHOTSE, NUPTSE, and CHO OYU; also NAMCHE BAZAAR village and TENGPOCHE monastery.

Khums, Al (KUMS, el), town, TRIPOLITANIA region, NW LIBYA, port on MEDITERRANEAN SEA, 65 mi/105 km ESE of TRIPOLI; 32°39′N 14°16′E. Highway junction; commercial and tourist center; manufacturing (esparto-grass processing; olive oil). Important Phoenician port, later Roman colony known as Leptis Magna; Septimius Severus was born here. Bombed (1942) in World War II. Extensive Roman ruins (harbor, forum, temples, baths, theater); archaeological museum. Formerly Lebda, Homs. Was capital of former Al Khums province.

Khungari (hoon-gah-REE), town, S KHABAROVSK TERRITORY, RUSSIA, on the Khungari River (right tributary of the AMUR), near a branch of the TRANS-SIBERIAN RAILROAD, 60 mi/97 km ESE of KOMSOMOL'SK-NA-AMURE, and 20 mi/32 km S of OKTYABR'SKIY (AMUR oblast); 50°24′N 138°10′E. Elevation 1,381 ft/420 m. Sawmilling.

Khunjerab National Park (khun-jai-rub), major protected mountainous area, NORTHERN AREAS, extreme NE PAKISTAN, on border with CHINA; 36°40′N 75°00′E. Diverse flora and fauna unique to this part of KARAKORAM mountains, including black and brown bear, and several caprids, e.g., Marco Polo sheep, urial, and ibex. Affords largest range of mountain wildlife in PAKISTAN. Local herders preserve grazing rights in some parts of the park. Major foreign tourist destination. KARAKORAM HIGHWAY bisects the park. Established 1975.

Khunjerab Pass (khun-jai-rub), NORTHERN AREAS, extreme NE point of PAKISTAN, on Chinese (XINJIANG) border; 36°50′N 75°20′E. Elevation 15,420 ft/4,700 m. Transit to XINJIANG negotiated in 1964 boundary treaty with CHINA. Former transit routes, Mintaka and KILIK passes, abandoned with completion (1979) of KARAKORAM HIGHWAY through here in 1979. Major import and export routes between CHINA and PAKISTAN. Daily passenger-bus service during snow-free season between SOST and Pir Ali over the pass.

Khunsar, IRAN: see KHONSAR.

Khun Tan Range (KUH TAHN), N THAILAND, between PING and WANG rivers (headstreams of the CHAO PHRAYA). Rises to 6,601 ft/2,012 m, 30 mi/48 km NE of CHIANG MAI. Crossed by LAMPANG-CHIANG MAI railroad in 1 mi/1.6 km long tunnel.

Khunzakh (hoon-ZAHKH), village (2005 population 4,050), central DAGESTAN REPUBLIC, NE CAUCASUS, SE European Russia, on the E bank of the AVAR KOISU, on road, 30 mi/48 km SW of BUYNAKSK; 42°32′N 46°42′E. Elevation 5,095 ft/1,552 m. Clothing handicrafts. Population largely Avar. A former capital of the Avar khanate; has ruins of an ancient palace and fortress. Experienced significant population increase since 1992 largely due to influx of Chechen refugees fleeing military hostilities.

Khur, IRAN: see KHOR.

Khurai (kuh-REI), town (2001 population 31,887), SAGAR district, central MADHYA PRADESH state, central INDIA, 30 mi/48 km NW of SAGAR; 24°03′N 78°19′E. Trades in wheat, oilseeds, millet; livestock market.

Khuraiba, YEMEN: see KHOREIBA.

Khurasan, IRAN: see KHORĀSĀN.

Khurba (hoor-BAH), settlement (2005 population 5,485), S central KHABAROVSK TERRITORY, RUSSIAN FAR EAST, in the AMUR River basin, on road and branch connecting the BAYKAL-AMUR MAINLINE at KOMSOMOL'SK-NA-AMURE with the TRANS-SIBERIAN RAILROAD at VOLOCHAYEVKA (JEWISH AUTONOMOUS OBLAST), 11 mi/18 km SSW of Komsomol'sk-na-AMURE; 50°24′N 136°53′E. Elevation 134 ft/40 m. Russian air force base in the vicinity.

Khurda (KUHR-duh), district (□ 1,115 sq mi/2,899 sq km), ORISSA state, E central INDIA; ⊙ KHURDA. Sometimes spelled Khordha.

Khurda (KUHR-duh), town, ⊙ KHURDA district, E ORISSA state, E central INDIA, 28 mi/45 km NNW of PURI. Road junction; market center (rice, forest produce); handicraft cloth weaving. Khurda Road, 5 mi/8 km ESE, is railroad junction (workshops) for journey to PURI on coast; rice mill, biri factory. Sometimes spelled Khordha.

Khurdalan (khoor-dah-LAHN), town, in KIROV district of Greater BAKY, AZERBAIJAN, 7 mi/11 km NW of BAKY, on W APSHERON Peninsula. Oil fields.

Khurd Kabul (KHOORD KAH-bool), village, KABUL province, E AFGHANISTAN, 15 mi/24 km SE of KABUL; 34°23′N 69°23′E. Coal mining. The Khurd Kabul Pass (N), scene of massacre (1842) of British troops in first Afghan War, was formerly used by Peshawar-Kabul road.

Khurja, town, BULANDSHAHR district, W UTTAR PRA-
DESH state, N central INDIA, 45 mi/72 km SE of DELHI;
28°15′N 77°51′E. Road junction; trade center (wheat,
oilseeds, barley, jowar, cotton, sugarcane, ghee);
pottery manufacturing. Large Jain temple.

Khurma, SAUDI ARABIA: see TURABA, WADI.

Khurmal (KUR-mel), township, SULAIMANIYA province, NE IRAQ, in KURDISTAN, near IRAN border, 35
mi/56 km SE of SULAIMANIYA; 35°18′N 46°02′E. Tobacco, fruit, livestock. Sometimes spelled Khormal.

Khurmuj, IRAN: see KHORMUJ.

Khursia, INDIA: see KHARSIA.

Khusf (KHOOSF), village, Khorāsān province, NE
IRAN, 20 mi/32 km WSW of BIRJAND; 32°47′N 58°54′E.

Khushab (khoo-shub), town, SARGODHA district, PUN-
JAB province, central PAKISTAN, on JHELUM RIVER, and
23 mi/37 km NW of SARGODHA; 32°18′N 72°21′E.

Khust (KHOOST) (Czech *Chust* or *Husté*) (Hungarian
Huszt), city, S TRANSCARPATHIAN oblast, UKRAINE,
near TYSA River, on railroad, and 32 mi/51 km SE of
MUKACHEVE; 48°11′N 23°18′E. Elevation 544 ft/165 m.
Raion center; trading center; manufacturing (headwear, clothing, footwear, ceramics, food products).
Airfield. Ruins of an 11th-century Hungarian castle
nearby. Part of Halych-Volyn duchy (1281–1321), then
reverted to Hungary; fought over by Hapsburg and
Transylvanian princes and beseiged by the Tatars (17th
century); in Austria-Hungary until 1918, Ruthenian
congress in Khust (1919) voted to unite Transcarpathia
with Ukraine; passed 1920 to Czechoslovakia, 1939 to
Hungary, and 1945 to USSR. In 1938–1939, it was the
capital of the autonomous Czech province called
Carpatho-Ukraine, which, upon invasion by Hungarian troops (March 1939), fought back and declared
an independent Carpatho-Ukraine. Large Jewish
community since the mid-19th century, numbering
over 11,000 in 1939; destroyed by the Nazis in World
War II—fewer than 100 Jews remaining in 2005.

Khutir Dubrova, UKRAINE: see DIBROVA.

Khutir Zabolochans'kyy, UKRAINE: see NOVOSELIVKA.

Khutor Dubrova, UKRAINE: see DIBROVA.

Khutorok, RUSSIA: see NOVOKUBANSK.

Khutse Game Reserve (oo-TSAI), park (□ 1,000 sq mi/
2,600 sq km), central BOTSWANA, S of CENTRAL KA-
LAHARI GAME RESERVE, and N of KWENENG DISTRICT.
Unique ecosystem of non-perennial pans. Gemsboks,
giraffes, and lions, among other game, live here.

Khuzdar (khooz-dahr), village, KHUZDAR district,
KALAT division, BALUCHISTAN province, SW PAKI-
STAN, 85 mi/137 km S of KALAT; 27°48′N 66°37′E.
Market center. Sometimes spelled KHOZDAR.

Khuzestān (KHOO-zhest-ahn), province (□ 25,978 sq
mi/67,542.8 sq km), SW IRAN, bordering on IRAQ in
the W and the PERSIAN GULF in the S; ⊙ AHVAZ;
31°00′N 49°00′E. Its major cities include AHVAZ,
KHORRAMSHAHR, DEZFUL, and ABADAN. Khuzestān
has large petroleum deposits. Its major oil refineries
were bombed in the early 1980s during the Iran-Iraq
War. Mountainous in the E, it has a hot, dry climate;
agricultural products include dates, citrus fruit, rice,
and vegetables. Dams on the Dez River in the N part
of the province provide water for irrigation and hydroelectricity. The W part includes the SE portion of
the Mesopotamian lowlands. There is a railroad line
and an extensive road network. The area was conquered (7th century) by the Arabs and invaded (13th
century) by the Mongols; it passed to Tamerlane in
the 14th century. Development of the oil industry in
the 20th century has led to growth of Khuzestān's
population and economy. The province was formerly
called ARABISTAN, and is still refered to as such by
Arab countries and Arab League publications.

Khuzhir (hoo-ZHIR), town (2005 population 1,110), SE
IRKUTSK oblast, S SIBERIA, RUSSIA, on the W shore of
the OLKHON ISLAND in Lake BAYKAL; 53°11′N 107°20′E.
Elevation 1,637 ft/498 m. Fish processing. Founded in
1942. Formerly known Khuzhirtuy.

Khuzhirtuy, RUSSIA: see KHUZHIR.

Khvalynsk (hvah-LINSK), city (2006 population
13,335), N SARATOV oblast, SE European Russia, grain
port on the W bank of the SARATOV RESERVOIR on the
VOLGA RIVER, on road, 145 mi/233 km NE of SARATOV;
52°29′N 48°06′E. In picturesque chalk hills; aparatus
for hydroelectric projects; fruit and vegetable processing; flour and oilseed mills. In the 18th and 19th
centuries, religious and cultural center of the Old
Believers; chartered in 1780.

Khvastiv, UKRAINE: see FASTIV.

Khvastovichi (hvah-STO-vee-chee), village (2005
population 4,530), S KALUGA oblast, central European
Russia, 23 mi/37 km N of KARACHEV; 53°28′N 35°06′E.
Elevation 659 ft/200 m. Highway junction; local
transshipment point. Hemp processing, rubber footwear manufacturing; dairy plant.

Khvatovka (HVAH-tuhf-kah), town (2006 population
2,295), N SARATOV oblast, SE European Russia, on
crossroads and railroad, 8 mi/13 km NE of BAZARNYY
KARABULAK; 52°21′N 46°34′E. Elevation 816 ft/248 m.
Glassworks.

Khvorostyanka (hvuh-ruh-STYAHN-kah), village
(2006 population 4,890), SW SAMARA oblast, E European Russia, on road junction, 40 mi/64 km SW of
CHAPAYEVSK; 52°37′N 48°58′E. Elevation 183 ft/55 m.
In agricultural area; bakery, creamery.

Khvorostyanka (hvuh-ruhs-TYAN-kah), village, SE
LIPETSK oblast, S central European Russia, on road
and railroad, 16 mi/26 km SE of GRYAZI; 52°19′N
40°14′E. Elevation 534 ft/162 m. In agricultural area
(grains, hemp, potatoes).

Khvoynaya (HVO-yee-nah-yah), town (2006 population 6,585), NE NOVGOROD oblast, NW European
Russia, on railroad and crossroads, 40 mi/64 km NNE
of BOROVICHI; 58°54′N 34°32′E. Elevation 511 ft/155 m.
Transformer-coil factory, railroad establishments,
food processing (bakery, brewery, dairy). Agricultural
cooperatives. Has a music school for children.

Khwae Noi River (KWA NOI) [Thai=lesser Khwae],
150 mi/241 km long, W central THAILAND; rises in
TENASSERIM RANGE on MYANMAR border near THREE
PAGODAS PASS; flows SE through densely forested
valley to MAE KLONG RIVER at KANCHANABURI. Followed by BAN PONG-Thambyuzayat railroad and road,
built during World War II. Made famous by the
novel, The BRIDGE ON THE RIVER KWAI, written in
1954 by Pierre Boulle and made into a Hollywood
movie. The bridge in question was actually built and
spanned the KHWAE NOI. The railroad line was never
used and was soon torn up by local residents. In 1997,
the line is being reconstructed by a joint Thai-Burmese effort.

Khwae Yai River, THAILAND: see MAE KLONG RIVER.

Khwarazm, former state of central ASIA, situated in and
around the basin of the lower AMU DARYA River in
what is now NW UZBEKISTAN. Khwarazm is one of the
oldest centers of civilization in Central ASIA. Part of
the Achaemenid empire of Cyrus the Great in the 6th
century B.C.E.; became independent in the 4th century B.C.E. Later inhabited by Indians who adhered to
Zoroastrianism and used Aramaic script. Conquered
by the Arabs in the 7th century and converted to Islam.
In 995 the country was united under the emirs of N
Khwarazm. Their capital city of URGANCH became a
major seat of Arab learning, a center of agriculture and
trade, and the residence of the ruling shahs. In the late
12th century, Khwarazm gained independence from
the Seljuk Turks. It then expanded; at the height of its
power (early 13th century) the state extended from the
CASPIAN SEA to BUKHARA and SAMARKAND. It was
conquered in 1221 by Jenghiz Khan and included in the
GOLDEN HORDE. The development of caravan trade by
the Mongols was profitable to Khwarazm. In the late
14th century, Khwarazm, along with its vast irrigation
system, was destroyed by Timur. A century of struggle
over Khwarazm between the Timurids, the descen-

dants of Timur, and the Golden Horde was followed
by the Uzbek conquest in the early 16th century.
Khwarazm became an independent Central Asian
state and was known as the khanate of KHIVA, with its
capital at KHIVA. There are ruins of ancient forts, one
of which dates back to the 6th century B.C.E. Also
called Khorezm.

Khyargas Nuur, MONGOLIA: see HYARGAS NUUR.

Khyber (KHEI-buhr), tribal region, centrally administered tribal area, W NORTH-WEST FRONTIER PRO-
VINCE, NW PAKISTAN. Comprises mountainous
territory along Afghan border, crossed S by E SAFED
KOH Range, N by S offshoots of the HINDU KUSH
mountains. Contains most of KHYBER PASS. Wheat
and barley grown in valleys; principal commercial
activity is smuggling. Chief tribes are Mohmands and
Afridis. Sometimes spelled KHAIBAR.

Khyber Pass (KHEI-buhr), 28 mi/45 km long, winding
through the SAFED KOH Mountains, in the centrally
administered KHYBER region of NORTH-WEST FRON-
TIER PROVINCE, W central PAKISTAN, on the AFGHA-
NISTAN border, from JAMRUD (11 mi/18 km W of
PESHAWAR) to LANDI KOTAL; highest point is rela-
tively low at 3,500 ft/1,067 m. The routes through it
link the cities of PESHAWAR (PAKISTAN) and KABUL
(AFGHANISTAN). The major land link between central
ASIA and the INDIAN SUBCONTINENT. For centuries a
trade and invasion route, the Khyber Pass was one of
the principal approaches of the armies of central
Asians in their invasions of what are now INDIA and
PAKISTAN from at least 1000 B.C.E. Terrain and local
banditry made it less important as trade route until
British built road during First Afghan War (1839–
1842). The Khyber Pass is now often closed although
traversed by an asphalt road and an old caravan route.
An alternative route to AFGHANISTAN is over the
Nawa Pass, 50 mi/80 km to the N in BAJAUR, which is
less susceptible to closure due to recalcitrant tribesmen in frontier tribal territory. A railroad (built 1920–
1925), which passes through thirty-four tunnels and
over ninety-two bridges and culverts, runs from
JAMRUD to the Afghan border at TORKHAM (AFGHA-
NISTAN). PAKISTAN controls the entire pass. LANDI
KOTAL, the principal town in the pass, is noted for
smuggling of illicit drugs, armaments, and consumer
goods.

Khyrim (KEI-rim), village, East Khasi district, ME-
GHALAYA state, NE INDIA, on SHILLONG PLATEAU, 5
mi/8 km S of Shillong. Rice, sesame, cotton.

Khyriv (KHI-rif) (Polish *Chyrów*) (Russian *Khyrov*),
city (2004 population 7,000), SW L'VIV oblast, UK-
RAINE, 56 mi/90 km WSW of L'VIV and 15 mi/24 km
W of SAMBIR, on the Stryvihor River, tributary of the
DNIESTER River; 49°32′N 22°51′E. Elevation 1,125 ft/342
m. Minor railroad junction, sawmilling, furniture and
brick manufacturing. Has a 16th-century church, old
monastery; Jesuit school until 1939. It was first mentioned in 1374; gained town rights in 1528; city since
1940. Battles between the Polish and Ukrainian armed
forces in winter 1918–1919. Jewish community since
the 17th century, numbering 1,000 in 1939; wiped out
by the Nazis in 1941.

Khyrov, UKRAINE: see KHYRIV.

Khzar-abad (kha-zahr-ah-BAHD), town, Māzandarān
province, N IRAN, on CASPIAN SEA, 15 mi/24 km N of
SARI.

Kia, town, JAVA, INDONESIA, 40 mi/64 km E of JA-
KARTA. Motor-vehicle assembly.

Kiafu (KYAH-foo), village, BAS-CONGO province, W
CONGO, on railroad, and 60 mi/97 km N of BOMA. Agricultural center; palm-oil milling, soap manufacturing.

Kialineq Bay (KYAH-li-nek), Greenlandic *Kialiip
imaa*, Danish *Skrækkensbugt*, small inlet of DENMARK
STRAIT, SE GREENLAND; 66°55′N 33°05′W. Gives its
name to surrounding region, known for its good
hunting.

Kialing, China: see JIALING.

Area is shown by the symbol □, and capital city or county seat by ⊙.

Kialla, village, N VICTORIA, SE AUSTRALIA, c.6 mi/10 km S of SHEPPARTON, and in GOULBURN valley irrigation area. Mining of sand.

Kiama (kei-A-muh), municipality and port (2001 population 11,711), E NEW SOUTH WALES, SE AUSTRALIA, on coast, 55 mi/89 km S of SYDNEY; 34°41′S 150°52′E. Coastal resort. Fishing. Area attractions include a lighthouse and nearby blowhole created by a fault in the cliffs, and native gardens S of town. Exports dairy products. Coal mines nearby.

Kiama or **Kiamma**, AUSTRALIA: see CROOKWELL.

Kiambi (KYAHM-bee), village, KATANGA province, SE CONGO, on LUVUA RIVER (head of navigation) and 130 mi/209 km SW of KALEMIE; 07°20′S 28°01′E. Elev. 1,994 ft/607 m. Trading center in tin-mining region. Protestant mission.

Kiambu (kee-AHM-boo), town (1999 population 13,814), CENTRAL province, S central KENYA, 5 mi/8 km N of NAIROBI; 01°10′S 51°36′E. Major coffee center; wheat, corn. Market center.

Kiamichi River, c.165 mi/266 km long, SE OKLAHOMA; rises in E LE FLORE county. near ARKANSAS state line in the OUACHITA MOUNTAINS; flows SW and S, past Muse and CLAYTON, to ANTLERS, then SE through HUGO LAKE reservoir to RED RIVER S of FORT TOWSON. Hugo Lake (for flood control in Red River basin) is 7 mi/11.3 km E of HUGO.

Kiamika, village (□ 134 sq mi/348.4 sq km; 2006 population 761), LAURENTIDES region, SW QUEBEC, E CANADA; 46°25′N 75°20′W.

Kiamusze, China: see JIAMUSI.

Kian, China: see JI'AN, Jiangxi province.

Kiana (kee-YAH-nuh), village (2000 population 388), NW ALASKA, on lower KOBUK RIVER, and 60 mi/97 km E of KOTZEBUE, N of SELAWIK LAKE; 66°58′N 160°27′W. Scene of gold rush, 1910.

Kiandra (kei-AN-druh), ghost town, SE NEW SOUTH WALES, AUSTRALIA, 50 mi/80 km SSW of CANBERRA, in GREAT DIVIDING RANGE, near Muniong Range; 35°53′S 148°30′E. Elevation 4,600 ft/1,402 m. Snow falls May–September. Was briefly a gold-mining town. Originally called Giandara or Giandara Plain.

Kiangsi, China: see JIANGXI.

Kiangsu, China: see JIANGSU.

Kiang West National Park (KEE-ahng), The Gambia's largest park, in LOWER RIVER DIVISION, THE GAMBIA, 45 mi/72 km ESE of BANJUL; 13°23′N 15°55′W. Mangroves, forests; over 300 species of birds. Established 1987.

Kianly (kee-AHN-lee), town, W AHAL weloyat, W TURKMENISTAN, fishing port on CASPIAN SEA, 12 mi/19 km W of TURKMENBASHI; 40°12′N 52°46′E. Canneries. Until c.1945, called Tarta.

Kiantone, hamlet, CHAUTAUQUA county, extreme W NEW YORK, 6 mi/9.7 km SSE of JAMESTOWN; 42°01′N 79°12′W.

Kiaton, Greece: see SIKYON.

Kiawah Island (KEE-uh-wah) (2006 population 1,108), CHARLESTON county, SE SOUTH CAROLINA, one of SEA ISLANDS, 12 mi/19 km SW of CHARLESTON; c.10 mi/16 km long; 32°37′N 80°03′W. Resort center separated by narrow channel from JOHNS ISLAND (N); STONO RIVER is E, ATLANTIC OCEAN S.

Kibaale, administrative district (□ 1,699 sq mi/4,417.4 sq km; 2005 population 478,300), WESTERN region, W UGANDA, SE of LAKE ALBERT; ⊙ KIBALE; 00°55′N 31°10′E. As of Uganda's division into eighty districts, borders MASINDI (N), KIBOGA (NEE), MUBENDE (ESE), KYENJOJO (S), and KABAROLE and BUNDIBUGYO (W) districts and Lake Albert (NW). Rural region where sweet potato and cassava farming is important. Formed from S part of Hoima district in 1991. Also spelled Kibale.

Kibala (kee-BAH-luh), town, ZAIRE province, ANGOLA, on road, and 55 mi/89 km NE of AMBRIZ; 07°25′S 13°45′E. Market center.

Kibale (kee-BAH-lai), town (2002 population 4,762), ⊙ KIBAALE district, WESTERN region, W UGANDA, on

road, and 55 mi/89 km ENE of Kabarole, S of Nkusi river; 00°50′N 31°06′E. Corn, millet. Was part of former WESTERN province. Also spelled Kibaale.

Kibali (kee-BAH-lee), river, CONGO. One of the tributaries of the CONGO RIVER, discharges into Lake Victoria in UGANDA. Dense tropical forest and abundant wildlife alongside both banks. DUNGU, a village at the confluence of DUNGU and KIBALI rivers, was the transportation hub for the area due to the presence of private Canadian airline until 2000, when military hostilities forced the company to relocate.

Kibanga (kee-BAHN-guh) or **Kibanga Port**, town, MUKONO district, CENTRAL region, S UGANDA, small port on LAKE VICTORIA, 22 mi/35 km SE of KAMPALA; 00°10′N 32°52′E. Cotton, coffee, sugarcane; fisheries. Was part of former NORTH BUGANDA province.

Kibangou (kee-bahng-GOO), village, NIARI region, SW Congo Republic, on NIARI RIVER, and 50 mi/80 km NNW of LOUBOMO.

Kibara (kee-BAH-rah), village, MARA region, NW TANZANIA, 45 mi/72 km NE of MWANZA, on SPEKE GULF, Lake Victoria; 02°10′S 33°26′E. Fish; cattle, goats, sheep; cotton, corn, wheat.

Kibarty, LITHUANIA: see KYBARTAI.

Kibata (kee-BAH-tah), village, LINDI region, E TANZANIA, 45 mi/72 km NNW of KILWA MASOKO, near INDIAN OCEAN; 08°24′S 39°00′E. Cashews, corn, bananas, manioc, sisal; goats, sheep; timber.

Kibati (kee-BAH-tee), village, TANGA region, E central TANZANIA, 60 mi/97 km NNW of MOROGORO, on Boruma River, in NGURU Mountains; 05°54′S 37°25′E. Cattle, goats, sheep; wheat, corn.

Kibati, CONGO: see GOMA.

Kibau (kee-BAH-oo), village, IRINGA region, S central TANZANIA, 60 mi/97 km SSW of IRINGA; 08°35′S 35°16′E. Tobacco, pyrethrum, corn, wheat; cattle, sheep, goats.

Kibaya (kee-BAH-yah), village, Manyara region, N central TANZANIA, 83 mi/134 km NE of DODOMA, at S edge of MASAI STEPPE; 05°18′S 36°34′E. Road junction. Cattle, goats, sheep; wheat, corn.

Kibbi, GHANA: see KIBI.

Kibbutz Mahar, ISRAEL: see GEVARAM.

Kibbutz Mash'abbim, ISRAEL: see MASH'ABBE SADE.

Kibby Mountain (KIB-ee) (3,638 ft/1,109 m), W MAINE, peak in extreme N FRANKLIN county, 20 mi/32 km SW of JACKMAN.

Kiberege (kee-bai-RAI-gai), village, MOROGORO region, E central TANZANIA, 90 mi/145 km SSW of MOROGORO, on railroad; 07°56′S 36°52′E. SELOUS GAME RESERVE to E. Cattle, sheep, goats; corn, wheat, tobacco, pyrethrum; timber.

Kiberen, FRANCE: see QUIBERON.

Kiberi (kee-BAI-ree), village, GULF province, S central NEW GUINEA island, SW PAPUA NEW GUINEA, 30 mi/48 km W of KIKORI. Located 5 mi/8 km E of Turama River, about 30 mi/48 km NW of its mouth and the Gulf of PAPUA. Access by boat and walking track. Bananas, coconuts, sago; timber.

Kibet, town (2007 population 6,487), SOUTHERN NATIONS state, central ETHIOPIA, 10 mi/16 km S of BUTAJIRA in livestock raising area; 07°55′N 38°19′E. Historic carved stelae nearby.

Kibi (KEE-BEE), town, district headquarters, EASTERN region, GHANA; 06°10′N 00°33′W. Located 25 mi/40 km WNW of KOFORIDUA; gold- and diamond-mining center (in BIRIM RIVER valley). Sometimes spelled Kibbi.

Kibi (KEE-bee), town, Arida county, WAKAYAMA prefecture, S HONSHU, W central JAPAN, 12 mi/20 km S of WAKAYAMA; 34°03′N 135°13′E. Mandarin oranges.

Kibigori (kee-bee-GO-ree), village, NYANZA province, W KENYA, on railroad, and 20 mi/32 km E of KISUMU; 00°04′S 35°01′E. Agriculture (sugarcane, cotton, peanuts, sesame, corn). Also spelled Kibigore.

Kibish River (ki-BEESH), c.135 mi/217 km long, SOUTHERN NATIONS state, ETHIOPIA; rises in high-

lands of SW ETHIOPIA near MAJI; flows S, forming part of boundary with SUDAN, then drains into marsh area N of LAKE TURKANA. Generally dry.

Kibiti (kee-BEE-tee), village, PWANI region, E TANZANIA, 60 mi/97 km SSW of DAR ES SALAAM; 07°42′S 38°55′E. Corn, bananas, manioc, sisal, sweet potatoes; goats, sheep; timber.

Kibo, TANZANIA: see KILIMANJARO, mountain.

Kiboga (kee-BO-guh), administrative district (□ 1,562 sq mi/4,061.2 sq km; 2005 population 259,400), CENTRAL region, central UGANDA, midway between KAMPALA city and LAKE ALBERT; ⊙ KIBOGA; 01°00′N 31°45′E. Elevation averages 4,593 ft/1,400 m–5,905 ft/1,800 m. As of Uganda's division into eighty districts, borders MASINDI (N), NAKASEKE (E), MITYANA and MUBENDE (S), KIBAALE (W), and HOIMA (NW) districts. Area originally inhabited by Baganda people. Vegetation is primarily savannah grasslands, including hills, valleys, and seasonal wetlands. Rural agricultural area (including bananas, cassava, maize, and sweet potatoes; cattle). Highway between HOIMA town and Kampala city travels NW-SE through Kiboga town. Formed in 1991.

Kiboga (kee-BO-guh), town (2002 population 11,956), ⊙ KIBOGA district, CENTRAL region, central UGANDA, on road, and 75 mi/121 km NW of KAMPALA; 00°56′N 31°48′E. Millet, cotton. Was part of former NORTH BUGANDA province.

Kibola (kee-BO-luh), village, KATANGA province, SE CONGO, near MANIKA PLATEAU (UPEMBA NATIONAL PARK); 09°42′S 27°05′E. Elev. 2,775 ft/845 m.

Kibombo (kee-BOM-bo), village, MANIEMA province, central CONGO, on the LUALABA, on railroad, and 55 mi/89 km NW of KASONGO; 03°54′S 25°55′E. Elev. 1,712 ft/521 m. Agricultural center (rubber, cotton, rice); lumbering, rice processing, cotton ginning. Has Roman Catholic mission. LUALABA RIVER is navigable between here and KASONGO.

Kibondo (kee-BON-do), village, KIGOMA region, W TANZANIA, 120 mi/193 km NNE of KIGOMA, near BURUNDI border, crossing 15 mi/24 km to WNW; 3°37′S 30°42′E. Airstrip here. Tobacco, grain; livestock; timber.

Kibongoto (kee-bon-GO-to), village, KILIMANJARO region, N TANZANIA, 18 mi/29 km NW of MOSHI; KILIMANJARO Mountain and MOUNT KILIMANJARO NATIONAL PARK to NE; 03°12′S 37°05′E. Coffee, tea, grain; cattle, sheep, goats.

Kibos (KEE-bos), village, NYANZA province, W KENYA, on railroad, and 8 mi/12.9 km ENE of KISUMU. Sugarcane center; cotton, peanuts, sesame, corn. Area inhabited by Kavirondo people. Cotton and sugar research institutes.

Kibouendé, town, POOL region, S Congo Republic, 30 mi/48 km W of BRAZZAVILLE.

Kibrai (kee-BREI), town, TOSHKENT wiloyat, NE UZBEKISTAN, just NE of TOSHKENT; 41°23′N 69°28′E. In orchard area; metalworks. Tertiary-level administrative center. Until c.1935, LUNACHARSKOYE. Also Kibray.

Kibre Mengist, township (2007 population 37,772), OROMIYA state, S ETHIOPIA; 05°53′N 38°59′E. On road from ADDIS ABABA to NW SOMALIA, and 95 mi/153 km E of ARBA MINCH. The town is in a forested highland area. Gold mining in nearby Awata Valley. Formerly called Adola.

Kibungo, former province (2002 population 702,248), SE RWANDA. Was bordered by KIGALI RURALE (W) and UMUTARA (N) provinces, as well as TANZANIA (E and S) and BURUNDI (SW). KIBUNGO town was the capital. The province was further divided into ten districts: Cyarubare, Kayonza, Kibungo, Kigarama, Mirenge, Muhazi, Nyarubuye, Rukira, Rusumo, and Rwamagana.

Kibungo (ki-BOON-go), town (pop. 5,000), E RWANDA, 35 mi/56 km ESE of KIGALI; 02°10′S 30°32′E. Established 1933; administrative center near TANZANIA border. Seat of Bishop of Roman Catholic Church. Was the capital of former KIBUNGO province.

Kibunzi (kee-BOON-zee), village, BAS-CONGO province, W CONGO, 100 mi/160 km S of KINSHASA; 05°10′S 13°53′E. Elev. 1,269 ft/386 m.

Kiburg, SWITZERLAND: see KYBURG.

Kibuye, former province (2002 population 469,016), W RWANDA. Was bordered by GISENYI (N), GITARAMA (E), GIKONGORO (SES), and CYANGUGU (SSW) provinces and LAKE KIVU (W). KIBUYE town was the capital. The province was further divided into six districts: Budah, Gisunzu, Itabire, Kibuye, Rusenyi, and Rutsiro.

Kibuye (ki-boo-YE), town, RWANDA, on shore of LAKE KIVU, 85 mi/137 km W of KIGALI; 02°03′S 29°20′E. Established 1928; administrative center, hospital, market center, tourist site. Scene of 1994 massacre of Tutsi and moderate Hutus. Was the capital of former Kibuye province.

Kibwesa (kee-BWE-sah), village, KIGOMA region, W TANZANIA, 110 mi/177 km SSE of KIGOMA, on Lake TANGANYIKA; Makari Mountains (MAHALE MOUNTAINS NATIONAL PARK) to N; 06°29′S 29°57′E. Lake port. Fish; goats, sheep; corn, wheat.

Kibwezi (kee-BWAI-zee), town, Makueni district, EASTERN province, S central KENYA, on railroad, and 110 mi/177 km SE of NAIROBI; 02°25′S 37°57′E.

Kicevo (kee-CHAI-vo), Serbo-Croatian *Kičevo*, village, W MACEDONIA, on TRESKA RIVER, on narrow-gauge railroad, and 40 mi/64 km SW of SKOPJE. Market center for cattle region. Iron. Also spelled Kichevo.

Kichera (KEE-chee-rah), village (2005 population 1,585), NW BURYAT REPUBLIC, S central SIBERIA, RUSSIA, in the UPPER ANGARA RIVER basin, on the BAYKAL-AMUR MAINLINE, 104 mi/167 km N of KURUMKAN; 55°56′N 110°06′E. Elevation 1,522 ft/463 m. Agricultural cooperative; logging and lumbering.

Kichha (KICH-uh), town, tahsil headquarters, NAINI TAL district, N UTTAR PRADESH state, N central INDIA, 32 mi/51 km S of Naini Tal; 28°55′N 79°30′E. Sugar processing; rice, wheat, mustard, gram.

Kichma (keech-MAH), village, S KIROV oblast, E European Russia, in the Nemda River (tributary of the VYATKA River) basin, on road, 55 mi/89 km NE of YOSHKAR-OLA (MARI EL REPUBLIC), and 25 mi/40 km S of SOVETSK; 57°11′N 48°55′E. Elevation 380 ft/115 m. In agricultural area (wheat, flax, vegetables, potatoes); agricultural machinery servicing station.

Kichmengskiy Gorodok (KEECH-myen-skeeyee guh-ruh-DOK), village (2006 population 6,580), E VOLOGDA oblast, N central European Russia, on the YUG RIVER, on the Sharya-Kotlas road, 55 mi/89 km SSW of VELIKIY USTYUG; 59°59′N 45°48′E. Elevation 334 ft/101 m. Flax processing.

Kickamuit River (KIK-ah-meyoo-it), c.4 mi/6.4 km long, E RHODE ISLAND; rises in SWANSEA, MASSACHUSETTS town, near Massachusetts-R.I. state line; flows generally SE, between BRISTOL and WARREN, to MOUNT HOPE BAY.

Kickapoo Creek, c.60 mi/97 km long, in central ILLINOIS; rises in MCLEAN county; flows SW to SALT CREEK SE of LINCOLN; 40°27′N 88°46′W.

Kickapoo Creek, c.25 mi/40 km long, in N central ILLINOIS; rises NW of PEORIA; flows SE to ILLINOIS RIVER just below Peoria; 40°57′N 89°38′W.

Kickapoo, Lake, Texas: see LITTLE WICHITA RIVER.

Kickapoo River (KI-kuh-poo), c.90 mi/145 km long, SW WISCONSIN; rises S of TOMAH; flows SSW though Wildcat Mountain State Park to WISCONSIN RIVER 13 mi/21 km E of PRAIRIE DU CHIEN.

Kickelhahn, GERMANY: see ILMENAU.

Kick 'em Jenny, volcano, GRENADA, WEST INDIES. Underwater volcano, 663 ft/202 m below surface located N of ISLE LA RONDE.

Kicking Horse (KIK-eeng HORS), river, SE BRITISH COLUMBIA, W CANADA; rising in the ROCKY MOUNTAINS; flowing SW and NW to GOLDEN, where it enters the COLUMBIA River; 51°18′N 116°59′W. Its course is rapid, with several high falls. Kicking Horse Pass, 5,339 ft/1,627 m high, NW of Lake LOUISE, in BANFF NATIONAL PARK, connects the BOW RIVER with the Kicking Horse and is one of the principal railroad and highway passes over the CONTINENTAL DIVIDE.

Kicking Horse Pass, CANADA: see KICKING HORSE.

Kidachi (kee-DAH-chee), village, DODOMA region, central TANZANIA, 40 mi/64 km SSW of DODOMA; 06°48′S 35°34′E. Cattle, goats, sheep; wheat, corn, sweet potatoes, beans.

Kidal (KEE-dahl), town, ⊙ EIGHTH REGION/KIDAL, E MALI, at S foot of ADRAR DES IFORAS, 170 mi/274 km NE of GAO; 18°26′N 01°24′W. Livestock. Military camp; airport.

Kidal, administrative region, MALI: see EIGHTH REGION.

Kidapawan, town (2000 population 101,205), ⊙ NORTH COTABATO province, central MINDANAO, PHILIPPINES, 34 mi/55 km WSW of DAVAO; 07°02′N 125°05′E. Agricultural center of region; tourism (departure point for treks on MOUNT APO, the country's highest mountain). Manobu tribal area; a 10-day multi-tribal culture fair is held here each year.

Kidatu (kee-DAH-too), village, MOROGORO region, E central TANZANIA, 70 mi/113 km SSW of MOROGORO, on GREAT RUAHA RIVER (bridged), near NW corner of SELOUS GAME RESERVE; 07°42′S 36°59′E. Udzungwa Mountains National Park to SW. Cattle, goats, sheep; corn, wheat; timber.

Kidder, county (□ 1,357 sq mi/3,528.2 sq km; 2006 population 2,453), central NORTH DAKOTA; ⊙ STEELE; 46°58′N 99°46′W. Agricultural area watered by several small lakes, largest being E part of Long Lake and Horsehead Lake. Diversified farming (cattle, wheat, barley, oats, hay, flax; dairying). Long Lake National Wildlife refuge in SW. Slade National Wildlife Refuge in S. Lake George National Wildlife Refuge and state park in SE. Formed 1873 and organized in 1881. Named for early setler Jefferson Parrish Kidder (1816–1883).

Kidder, town (2000 population 271), CALDWELL county, NW MISSOURI, 8 mi/12.9 km N of CAMERON; 39°46′N 94°05′W. Livestock area (cattle, hogs).

Kidderminster (KID-uhr-min-stuhr), town (2001 population 55,182), Worcestershire, W central ENGLAND, about 20 mi/51.8 km SW of BIRMINGHAM; 52°23′N 02°15′W. The city had a prosperous cloth trade from the 14th to 18th centuries. Kidderminster carpets have been produced since 1735.

Kidepo National Park (kee-DAI-po) (□ 519 sq mi/1,349.4 sq km), KAABONG district, NORTHERN region, NE UGANDA; 03°52′N 33°52′E. One of the largest national parks in Uganda. Wild animals include many types of birds, buffaloes, bush bucks, cheetahs, elands, elephants, giraffes, jackals, Jackson's hartebeest, leopards, lions, oribi, ostriches, roan antelopes, warthogs, water bucks, and zebras. Was part of former KARAMOJA province. Established 1958.

Kidero (KEE-dee-ruh), village (2005 population 670), SW DAGESTAN REPUBLIC, central CAUCASUS, SE European Russia, approximately 7 mi/11 km N of the RUSSIA-GEORGIA border, on the short Kidero river (ANDI KOYSU basin), near mountain highway, 72 mi/116 km SW of MAKHACHKALA; 42°11′N 45°57′E. Elevation 6,519 ft/1,986 m. Sometimes spelled Kidiro (same pronounciation).

Kides, FINLAND: see KITEE.

Kidete (kee-DAI-tai), town, DODOMA region, central TANZANIA, 60 mi/97 km SE of DODOMA, on railroad; 06°41′S 36°33′E. Corn, wheat, millet; cattle, sheep, goats.

Kidira (ki-DEE-rah), village, TAMBACOUNDA administrative region, E SENEGAL, on FALÉMÉ RIVER (MALI border), on Dakar-Niger railroad, 50 mi/80 km W of KAYES (Mali); 14°28′N 12°13′W. Hardwood, peanuts. Customs house.

Kidlington (KID-ling-tuhn), suburb (2001 population 13,719), central OXFORDSHIRE, central ENGLAND, on CHERWELL RIVER, and 5 mi/8 km N of OXFORD; 51°50′N 01°17′W. Agricultural market. Has 13th-century church.

Kidnappers, Cape, S cliffed extremity of HAWKE BAY, E NORTH ISLAND, NEW ZEALAND, 15 mi/24 km SE of NAPIER; 39°38′S 177°06′E. Gannet sanctuary.

Kidodi (kee-DO-dee), village, MOROGORO region, E central TANZANIA, 55 mi/89 km SSW of MOROGORO, on MKATA River, on railroad; 07°37′S 37°05′E. MIKUMI NATIONAL PARK to N. Tobacco, pyrethrum, corn, wheat; cattle, goats, sheep; timber.

Kidričevo (KEED-ree-che-vo), village, NE SLOVENIA, 4 mi/6.4 km W of PTUJ; 46°24′N 15°47′E. In lignite area; aluminium industry. Formerly called Strnišče.

Kidron Stream or **Nahal Kidron** (Hebrew) or **Wadi en-Nar** (Arabic=stream of fire), ISRAEL, in the JUDAEAN DESERT; rises in JERUSALEM; flows SE to Mar Saba, where it turns E and flows into the NW DEAD SEA. It is a sporadic stream which flows only on a few days annually; deep canyon. Mentioned in the Bible and in other ancient documents. Ancient tombs and caves used by refugees were found along the walls of the canyon.

Kidsgrove (KIDZ-grov), town (2001 population 7,150), NW STAFFORDSHIRE, central ENGLAND, 6 mi/9.7 km NNW of STOKE-ON-TRENT; 53°05′N 02°14′W. Manufacturing chemicals, textiles. Former coal-mining site.

Kidston (KID-stuhn), mining zone, NE QUEENSLAND, AUSTRALIA, 175 mi/282 km SW of CAIRNS; 18°52′S 144°10′E. Gold-mining center. Gold first discovered in area in 1880s. Sometimes called The Oaks Goldfield.

Kidugallo (kee-doo-GA-lo), village, MOROGORO region, E TANZANIA, 37 mi/60 km E of MOROGORO, on railroad. Sisal, corn millet.

Kidwelly (kid-WE-lee), Welsh *Cydweli*, town (2001 population 3,289), CARMARTHENSHIRE, SW Wales, on CARMARTHEN BAY of Bristol Channel, and 8 mi/12.9 km S of CARMARTHEN; 51°44′N 04°18′W. Previously manufacturing of optical glass and silica bricks. Has ruins of 12th-century castle. Formerly in DYFED, abolished 1996.

Kiedrich (KEE-drikh), village, HESSE, W GERMANY, in the RHEINGAU, 7 mi/11.3 km W of WIESBADEN; 50°02′N 08°05′E. Wine growing. Has one of Germany's most beautiful late-Gothic churches, with Germany's oldest working organ.

Kief (KEEF), village (2006 population 12), MCHENRY co., central NORTH DAKOTA, 20 mi/32 km SE of VELVA; 47°51′N 100°30′W. Kruger Lake and other small lakes to S. Founded in 1907 and named for KIEV by settlers from South RUSSIA.

Kiefer (KEEF-uhr), town (2006 population 1,403), CREEK and TULSA counties, central OKLAHOMA, residential suburb 15 mi/24 km SSW of TULSA, 5 mi/8 km SE of SAPULPA; 35°56′N 96°02′W. In agricultural and oil area. Bethesda Airport to S.

Kiefersfelden (KEE-fers-fel-den), village, BAVARIA, S GERMANY, in UPPER BAVARIA, on E slope of the Bavarian Alps, on the INN, and 2 mi/3.2 km N of KUFSTEIN, on Austrian border; 47°37′N 12°11′E. Manufacturing includes cementworks; marble quarries.

Kiejdany, LITHUANIA: see KEDAINIAI.

Kiel (KEEL), city (2005 population 234,433), ⊙ SCHLESWIG-HOLSTEIN, N central GERMANY, on Kiel Bay, an arm of the BALTIC SEA; 54°19′N 10°07′E. Situated at the head of the KIEL CANAL, the city was Germany's chief naval base from 1871 to 1945, when the naval installations were dismantled. Kiel is now a shipping and industrial center; the major industries are shipbuilding and engineering. Service, administrative, and educational center of state. There are large shipyards and factories that make textiles, metal products, and printed materials. It is the largest and economically the most important city in Schleswig-Holstein. Chartered in 1242; joined the HANSEATIC LEAGUE in 1284. It became the residence of the dukes of HOLSTEIN. Passed to DENMARK in 1773; with Holstein it was annexed by PRUSSIA in 1866. The sailors' mutiny that began at Kiel at the end of World War I touched off a socialist revolution in Germany. In World War II the city suffered severe damage from Allied air attacks. The city is the seat of a university (founded 1665) and several museums, including the oldest art gallery and botanical gardens in Germany. The sailing and yachting events of the 1972 Olympic summer games were held there. The city holds a yearly regatta that draws visitors from around the world.

Kiel (KEEL), town (2006 population 3,512) on MANI-TOWOC-CALUMET county line, E WISCONSIN, on SHEBOYGAN RIVER, and 20 mi/32 km NW of SHE-BOYGAN; 43°55′N 88°01′W. Trade center for dairying area; woodworking; manufacturing (cheese processing, cheesemaking equipment, metal fabricating, wood tables, food-industry packaging equipment, machinery, concrete pipe). Incorporated 1920.

Kiel Canal (KEEL), artificial waterway, SCHLESWIG-HOLSTEIN, N central GERMANY, connecting the NORTH SEA with the BALTIC SEA; 61 mi/98 km long; 53°54′N 09°08′E–54°22′N 10°09′E. At sea level, the canal extends from KIEL on the Baltic to BRUNSBÜTTEL at the mouth of the ELBE RIVER. Locks at each end of the canal minimize tidal variation. Built (1887–1895) to facilitate movement of the German fleet, the Kiel Canal was widened and deepened from 1905 to 1914. Large oceangoing ships can pass through the canal. Because of its great military and commercial importance the canal was internationalized by the Treaty of Versailles (1919), though its direct administration was left with the Germans. Hitler repudiated its international status in 1936, but free navigation in the canal was returned after World War II. The canal is also known as the KAISER WILHELM CANAL (KEI-ser VIL-helm), for William II of Germany, and as the NORTH SEA–BALTIC CANAL (German *Nord-Ostsee-Kanal*). Today the canal is a major passage for shipping in the Baltic region.

Kielce (KEEL-tse), province (□ 7,545 sq mi/19,617 sq km), SE central POLAND; ⊙ Kielce. Upland region with rolling hills rising to 2,004 ft/611 m in Gory SWIETO-KRZYSKIE range; drained by VISTULA (S and E boundary), PILICA (N boundary), NIDA, and KAMIENNA rivers. Textile milling, metalworking, manufacturing of leather goods, oats, wheat, barley, flax; livestock. Largest cities: RADOM, KIELCE, OSTROWIEC, WIERZBNIK, Skarzysko. Boundaries of pre–World War II province (□ 9,880 sq mi/25,589 sq km) were changed by transfer of territory to Katowice and KRAKOW provinces. Includes greater part of former KELTSY (Kielce) and Radom governments of Russian Poland.

Kielce (KEEL-tse), city (2002 population 212,429), S central POLAND; 50°50′N 20°40′E. It is a railroad junction and manufacturing center where metals, machinery, and foodstuffs are produced. It also has marble quarries. Founded in 1173, Kielce obtained municipal rights in the 14th century. It belonged to the bishops of KRAKÓW until 1789. The city passed to AUSTRIA in 1795 and to RUSSIA in 1815 and reverted to Poland in 1919. By the late 1930s, most of the city's Jewish population had been deported to German-run concentration camps. Four such camps were located in Kielce during World War II. In 1946, Jews returning to Poland after the war were massacred here. Its most notable buildings are a 12th century cathedral and a 17th century palace.

Kieldrecht (KEEL-drekht), agricultural village in commune of SINT-GILLIS-WAAS, Sint-Niklaas district, EAST FLANDERS province, N BELGIUM, 9 mi/14.5 km N of SINT-NIKLAAS, near NETHERLANDS border; 51°17′N 04°10′E.

Kiel Firth (KEEL FIRT), German *Kieler Förde* (KEEL-er fur-der), estuarine inlet and best natural harbor of the BALTIC, NW GERMANY; 10 mi/16 km long, 4 mi/6.4 km wide at mouth; 54°19′N 10°07′E–54°28′N 10°16′E. KIEL city extends on both sides of inner firth (5 mi/8 km

long, 1.5 mi/2.4 km wide, 26 ft/8 m–52 ft/16 m deep); industrial districts with wharves are on E bank; free port (built 1924) and mouth of KIEL CANAL are on W bank.

Kiembe Samaki (kee-aim-bai sah-MAH-kee), village, ZANZIBAR SOUTH region, E TANZANIA, near W coast of ZANZIBAR island, 4 mi/6.4 km SSE of ZANZIBAR city; 06°12′S 39°13′E. Fish; copra, cloves, bananas. Airport nearby.

Kien An (kee-EN AHN), city, HAI PHONG urban region, N VIETNAM, in THAI BINH RIVER delta, 55 mi/89 km ESE of HANOI; 20°48′N 106°38′E. Nearby tobacco plantations; silk spinning, salt extraction. Commercial center, administrative activities, light manufacturing. Becoming suburbanized as well. Meteorological observatory. Formerly Kienan or PHULIEN.

Kienberg, CZECH REPUBLIC: see LOUCOVICE.

Kien Giang, province (□ 2,410 sq mi/6,266 sq km), S VIETNAM, in MEKONG Delta, N border with CAMBODIA, NE border with AN GIANG province, E border with CAN THO province, S border with MINH HAI province, W border on Gulf of THAILAND; ⊙ RACH GIA; 10°00′N 105°10′E. Crisscrossed with drainage canals and irrigation systems, province is ecologically diverse region of marshes, mangroves, and forested landscapes formed by multiple distributaries of the Mekong River. Rich alluvial soils and bounteous agriculture (wet rice cultivation, vegetables, coconuts, cattle raising). Food processing, riverine and maritime fisheries, animal-feed production, distilling, light manufacturing. Kinh population with significant Khmer minority.

Kie-Ntem (KEE-ntem), province (□ 1,522 sq mi/3,957.2 sq km; 2001 population 167,279), NE Equatorial Guinea; ⊙ EBEBIYIN; 02°02′N 11°10′E. Bounded on S by WELE-NZAS PROVINCE, E by GABON, N by CAMER-OON, W by CENTRO-SUR PROVINCE. A heavily forested section of the country.

Kieretti, RUSSIA: see KERET'.

Kierettijärvi, RUSSIA: see KERET LAKE.

Kierspe (KEER-spe), town, NORTH RHINE–WESTPHA-LIA, W GERMANY, 6 mi/9.7 km SSW of LÜDENSCHEID; 51°07′N 07°36′E. Manufacturing includes plastics; chemical, electronics, and technology industry. Tourism.

Kiester (KEES-tuhr), village (2000 population 540), FARIBAULT county, S MINNESOTA, 18 mi/29 km SW of ALBERT LEA, on Brush Creek, near IOWA state line; 43°32′N 93°42′W. Agricultural area (grain, soybeans; livestock); manufacturing (feeds).

Kiestinki, RUSSIA: see KESTEN'GA.

Kieta (KYAI-tah), town on SE coast of BOUGAINVILLE island, ⊙ of NORTH SOLOMONS province, E PAPUA NEW GUINEA, on PACIFIC OCEAN. It was headquarters as capital of Kieta district by SOHANO on Sohano Island after World War II; reestablished 1980s. Fish, coconuts, yams, rice, bananas. PANGUNA copper district 15 mi/24 km to SW.

Kietrz (keetz), German *Katscher*, town in UPPER SILE-SIA, after 1945 in OPOLE province, S POLAND, near CZECH border, 10 mi/16 km W of RACIBORZ (*Ratibor*). Woolen milling, gypsum quarrying.

Kiev (KEE-yev) (Ukrainian *Kyyivs'ka oblast'*) (Russian *Kievskaya oblast'*), oblast (□ 11,185 sq mi/29,081 sq km; 2001 population 4,439,221), N central UKRAINE; ⊙ KIEV (new official spelling Kyiv). In DNIEPER LOWLAND (N) and DNIEPER UPLAND (S); drained by DNIEPER River and its affluents, TETERIV, DESNA, and Ros' rivers. Polissya (forested) in N, wooded steppe in middle, clearing toward S. Population majority is Ukrainian (79.7%), with Russian (15.6%), Jewish (2.4%), Belo-russian (0.8%), Polish (0.3%), and other minorities. Flax, potatoes, buckwheat, forests in N part; truck produce around Kiev; wheat and sugar beets in S; dairy cattle. Peat cutting NW of Kiev. Sugar refining, flour milling, distilling. Chief manufacturing centers: Kiev, BILA TSERKVA, FASTIV. Formed in 1932. In 1993 had twenty five cities, thirty one towns, and twenty five rural raions.

Kiev (KEE-yev), now officially spelled Kyiv (Ukrainian *Kyyiv*) (Russian *Kiyev*), city (2001 population 2,611,327), ⊙ UKRAINE and of KIEV oblast, a port on the DNIEPER (Ukrainian *Dnipro*) River; 50°26′N 30°31′E. Elevation 554 ft/168 m. The largest city in Ukraine, Kiev is the leading industrial, commercial, political, religious, scientific, and cultural center. Compex and precision machine construction and metalworking (notably instrument and machine tool making, electronic, optical and transport equipment), chemicals, building materials, food processing, and textiles are the major industries. Kiev is one of the oldest towns in E EUROPE. It probably existed as a commercial center as early as the 5th century. A Slavic settlement on the great trade route between SCANDINAVIA and CON-STANTINOPLE, Kiev was tributary to the Khazars when the Varangians under Oleh (Russian *Oleg*, Norse *Helgi*) established themselves there in 882. Under Oleh's successors it became the capital of medieval Kievan Rus (the first glorious imperial state claimed by Ukranians, Belarusians and Russians) and was a leading European cultural and commercial center. It was also an early seat of Christianity in Eastern Europe. The city reached its apogee in the 11th century, but by the late 12th century it had begun to decline. From 1240, when it was devastated by the Mongols, until the 14th century, the city paid tribute to the Golden Horde. Kiev then passed under the control of LITHUANIA, which in 1569 was united with POLAND. With the establishment of the Kiev-Mohyla Academy in 1632, the city became a center of Ukrainian learning and scholarship. In 1648, when the Ukrainian Cos-sacks under Bohdan Khmelnytskyy (Polish *Chmiel-nicki*) rose against Poland, Kiev served as the religious, cultural, and commercial center of a Ukrainian state and headquarters of the Kiev regiment. Following the Treaty of PEREYASLAV in 1654, however, the city was acquired (1686) by MOSCOW. Much of its cultural and architectural development (1687–1709) owes to support of I. Mazepa, hetman of Left Bank Ukraine, protectorate of Russia. Despite subsequent Russifi-cation and oppression, a Ukrainian national movement arose in Kiev in the 19th century. In January 1918 Kiev became the capital of the newly proclaimed Ukrainian republic; but in the ensuing civil war (1918–1920), it was occupied in succession by German, White Russian, Polish, and Soviet troops. While Jewish merchants were attracted to Kiev from its earliest days, Jews only were permitted to settle in the city in the 19th century. By 1923, one-third of the city's population was Jewish, and Kiev became the major center in Ukraine for Jewish culture. In 1934 the capital of the Ukrainian SSR was transferred from KHARKIV to Kiev. In an effort to Sovietize the city and clear the area for the new government center, over two dozen churches and other landmarks were destroyed, including the priceless Tithes Church (989), Three Saints Church (1183), and St. Michael's Golden Domed Monastery (1108–1113). German forces held the city during World War II and massacred thousands of its inhabitants, including 50,000 Jews. The mass murder of Jews by the Nazis in the nearby ravine of Babi Yar (Russian *Babiy Yar*) was memorialized in Yevtushenko's poem, and by a monument at the site. Post-war reconstruction of the heavily damaged city was not completed until about 1960. Lying amid hills along the Dnieper and filled with gardens and parks, Kiev is one of Europe's most beautiful cities, as well as a treasury of medieval art and architecture. Its most outstanding buildings include the restored ruins of the Golden Gate (11th century), and the 11th-century Cathedral of St. Sophia (now a museum), which was modeled on Hagia Sophia in Constantinople and contains splendid mosaics, frescoes, and icons. The Dormition Cathedral, dynamited by retreating So-viets during World War II, remains in ruins within the celebrated Lavra cave monastery (11th century), which is partly preserved as a museum and a sacred place of pilgrimage and partly an active monastery. The St. Vladimir Cathedral (1862–1896) is famed for its murals. Among the city's educational and cultural institutions are the University of Kiev (1833) and the Ukrainian Academy of Sciences (1918), and the renewed (1991) National University of the "Kiev-Mohyla Academy."

Kievan Rus' (ROOS), medieval state in Eastern Europe, encompassing mostly the E Slavs. Flourishing from the 10th to the 13th century, it included nearly all of present-day UKRAINE and BELARUS and, at its maximum extent, part of NW European Russia, reaching as far N as Novgorod and Vladimir. According to the Rus' *Primary Chronicle*, a medieval history, the Varangian Rurik established himself at Novgorod around 862 and founded a dynasty. His successor, Oleh (Russian *Oleg*; died around 912), shifted his attention to the S, seized Kiev (around 879), and established the new Kievan state. According to the Normanists, the Varangians were also known as *Rus* or *Rhos*; it is possible that this name was early extended to the Slavs of the Kievan state, which became known as Kievan Russia. Other theories trace the name *Rus* or *Ros* to a Slavic origin. Oleh united the E Slavs and freed them from the suzerainty of the Khazars. His successors were Ihor (Russian *Igor*; reigned 912–945) and Ihor's widow, St. Olha (Russian *Olga*), who was regent until about 962. Under Olha's son, Svyatoslav (died in 972), the Khazars were crushed, and Kievan power was extended to the lower Volga and N Caucasus. Christianity was introduced by Volodymyr (Russian *Vladimir I*; reigned 980–1015), who adopted (around 988) Greek Orthodoxy from the Byzantines. The reign (1019–1054) of Volodymyr's son, Yaroslav the Wise, represented the political and cultural apex of Kievan Russia. After his death, the state was divided into principalities ruled by his sons; this soon led to civil strife. The last effort for unity was made by Volodymyr II (reigned 1113–1125), but the perpetual princely strife and the devastating raids of the nomadic Cumans soon ended the supremacy of Kiev. In the middle of the 12th century, a number of local centers of power developed: Halych in the W, Novgorod in the N, Vladimir-Suzdal in the NW, and Kiev in the S. In 1169, Kiev was sacked and pillaged by the armies of Andrei Bogolubsky of Suzdal, and the final blow to the Kievan state came with the Mongol invasion (1237–1240). The economy of the Kievan state was based on agriculture and extensive trade with Byzantium, Asia, and Scandinavia. Culture, as well as religion, was drawn from Byzantium; Church Slavonic was the literary and liturgical language of the state. According to Soviet and some Western scholars, the history of the Kievan state is the common heritage of modern Russians, Ukrainians, and Belarussians, although their existence as separate peoples has been traced as far back as the 12th century. Ukrainian scholars consider Kievan Rus' to be central to the history of Ukraine.

Kievan Russia, UKRAINE: see KIEVAN RUS'.

Kiev Polissya, UKRAINE: see POLISSYA, UKRAINIAN.

Kiev Reservoir (KEE-yev) (Ukrainian *Kyyivs'ke vodoskhovyshche*) (Russian *Kiyevskoye vodokhranilishche*), on the DNIEPER (Dnipro) River, N central UKRAINE, N of KIEV city center. Formed in 1964–1966 with the construction of Kiev hydroelectric station; 68 mi/109 km to 7.5 mi/12.1 km wide; surface area, 356 sq mi/922 sq km; average depth 13 ft/4 m; max. depth, 47 ft/14 m. Right banks of reservoir are somewhat elevated, left banks are low, abutted by shallow water. Water level rises during spring flood, exceeding low level by 5 ft/1.5 m. Reservoir serves hydroelectric power generation, navigation, and water supply. Since the Chernobyl accident (1986), contamination occurred in the reservoir. To reduce increasing contamination, bottom catchment basins were built on the UZH and TETERIV (2 main right bank affluents) rivers, barrages

with filtration points on inflowing streams and drainage canals, and an underwater catchment upstream of Kiev hydroelectric station.

Kievskaya oblast, UKRAINE: see KIEV oblast.

Kifaya (KEE-fahr-yah), town, Gaoual prefecture, Boké administrative region, NW GUINEA, in Moyenne-Guinée geographic region, NE of GAOUAL; 12°10′N 13°04′W. Peanuts.

Kiffa (KEE-fah), village (2000 population 32,033), ☉ Assaba administrative region, S MAURITANIA, on NOUAKCHOTT-NÉMA HIGHWAY, 150 mi/241 km N of KAYES (MALI); 16°37′N 11°24′W. Millet, gum arabic; sheep, goats, cattle; big game. Airport.

Kifissia (kee-fee-SYAH), town, ATTICA prefecture, ATTICA department, E central GREECE, on spur of the PENTELIKÓN, 9 mi/14.5 km NE of ATHENS, in Athens metropolitan district; 38°04′N 23°49′E. Summer tourist and residential center; archaeological museum. Cherry orchards. Also Kephisia.

Kifissos River (kee-fee-SOS), 17 mi/27 km long, ATTICA prefecture, ATTICA department, one of several streams in E central GREECE; rises in the PATERA massif; flows E and S to Bay of ELEUSIS of SARONIC GULF at ELEUSIS. Also called Sarandapotamos or Sarantapotamos; Kephisos, Kifisos, or Kephissos River. Formerly spelled Cephisus and Cephissus River.

Kifissos River (kee-fee-SOS), 60 mi/97 km long, one of several streams in CENTRAL GREECE department; rises in PHOCIS prefecture, on N slopes of Mount PARNASSÓS; flows N, then ESE, through FTHIOTIDA and BOEOTIA prefectures, to N arm of Gulf of ÉVVIA 10 mi/16 km WNW of KHALKÍS. Forms lakes ILIKI and PARALIMNI in lower course. Formerly the Cephisus ended in Lake COPAIS (drained since 1880s). Also Kephisos, Kifisos, or Kephissos River. Formerly called Mavroneri; formerly spelled Cephisus, Cephissus.

Kifri (KIF-ree), township, DIYALA province, NE IRAQ, 60 mi/97 km SE of KIRKUK; 34°42′N 44°58′E. Barley, wheat, sheep raising. Sometimes spelled Kefri or Kufri; also called Salahiya and Zeng Abad.

Kift, EGYPT: see QIFT.

Kifuankese (kee-foo-ahn-KES-sai), town, KASAI-ORIENTAL province, SW CONGO, 55 mi/89 km SE of LUSAMBO; 05°24′S 24°13′E. Elev. 2,001 ft/609 m.

Kigali, province, central RWANDA. Bordered by NORTH (N), EAST (E and S), and SOUTH (W) provinces. Contains KIGALI city. Created in 2006 following a reorganization that replaced Rwanda's twelve provinces with five provinces. Sometimes called Ville de Kigali.

Kigali (ki-GAH-lee), city (2002 population 603,049), KIGALI province, ☉ RWANDA; 01°57′S 30°03′E. Located in the center of the country, it is Rwanda's main administrative and economic hub. The city has an international airport to S and road access to all of the country's borders. Iron ore (cassiterite) is mined nearby and the city built a smelting plant in the 1980s. Founded in 1907 under German colonial rule, Kigali was made the capital when Rwanda gained independence in 1962.

Kigali Rurale, former province (2002 population 789,330), central RWANDA. Was bordered by BUTARE (SW tip), GITARAMA and KIGALI-VILLE (W), RUHENGERI (NW), BYUMBA (N), UMUTARA (NE), and KIBUNGO (E) provinces and by BURUNDI (S). The province was further divided into ten districts: Bicumbi, Buriza, Gasabo, Gashora, Kabuga, Ngenda, Nyamata, Rulindo, Rushashi, and Shyorongi. Kigali-ville was part of Kigali Rurale until c.1998, when Kigali-ville became a separate province. Also called Kigali Ngali.

Kigali-ville, former province (2002 population 603,049), central RWANDA. Was bordered by KIGALI RURALE (N, E, and S) and GITARAMA provinces. The province was further divided into eight districts: Butamwa, Gikondo, Gisozi, Kacyiru, Kanombe, Kicukiro, Nyamirambo, and Nyarugenge. Coterminous with KIGALI city. Kigali-ville was part of Kigali Rurale

province until c.1998, when Kigali-ville became a separate province. Also called City of Kigali.

Kigandu (kee-GAHN-doo), village, BANDUNDU province, W CONGO, 35 mi/56 km E of KIKWIT; 05°30′S 18°25′E. Elev. 2,099 ft/639 m. Also spelled KINGANDU.

Kigi (Turkish=*Kiği*) village, E central TURKEY, 30 mi/48 km NNW of BINGOL, on the PERI RIVER; 39°19′N 40°20′E. Grain.

Kigoma (kee-GO-mah), region (2006 population 1,971,000), W TANZANIA; ☉ KIGOMA; 4°52′S 29°37′E. Bounded W by Lake TANGANYIKA (CONGO border), NW by BURUNDI. MALAGARASI RIVER crosses region in S center. GOMBE STREAM NATIONAL PARK in W, MAHALE MOUNTAINS (Makari Hills) in far SW, Moyowosi Game Reserve in E. Fish; rice, subsistence crops; sheep, goats.

Kigoma (kee-GO-mah), town, ☉ KIGOMA region, W TANZANIA, 625 mi/1,006 km WNW of DAR ES SALAAM, port on Kigoma Bay, Lake TANGANYIKA; 04°52′S 29°37′E. Terminus of the railroad from Dar es Salaam (completed 1914) and is connected by ship with CONGO and BURUNDI. GOMBE STREAM NATIONAL PARK to N. Manufacturing (food processing); agriculture (fish; corn, wheat, rice; sheep, goats). Kigoma was an important settlement of Arab and Swahili ivory and slave traders, c. 1850–1890. The region was occupied by the Germans in the 1890s. Formerly Kigoma-Ujiji; UJIJI now a separate town to SE. Town was a major focus for refugees in the RWANDA crisis 1994–1996.

Kigombe (kee-GOM-bai), village, TANGA region, NE TANZANIA, on PEMBA CHANNEL of INDIAN OCEAN, 15 mi/24 km S of TANGA; 04°18′S 39°02′E. Cashews, sisal; livestock; fish.

Kigozi River (kee-GO-zee), c. 125 mi/201 km, NW TANZANIA; rises in NW SHINYANGA region c. 75 mi/121 km SW of MWANZA; flows SSW, receives Moyowosi River in marshy area 3 mi/4.8 km N of its joining the GOMBE RIVER to form MALAGARASI RIVER.

Kigwa (KEE-gwah), village, TABORA region, NW central TANZANIA, 23 mi/37 km ESE of TABORA; 04°11′S 33°07′E. Timber; corn, wheat; goats, sheep.

Kigwe (KI-gwe), village, DODOMA region, central TANZANIA, 18 mi/29 km WNW of DODOMA, on railroad; 06°07′S 35°28′E. Peanuts, grain; cattle, sheep, goats.

Kihei (KEE-HAI), city (2000 population 16,749), MAUI island, MAUI county, HAWAII, 10 mi/16 km S of KAHULUI, on Maui's SW coast; 20°45′N 156°26′W. Tourism. Cattle, fish, prawns. Manufacturing (communications equipment, printing). Captain Vancouver Monument.

Kihnu (KEE-noo), island (□ 8 sq mi/20.8 sq km), ESTONIA, in GULF OF RIGA, 25 mi/40 km SW of Pärnu; 4.5 mi/7.2 km long, 2 mi/3.2 km wide; 58°07′N 23°57′E.

Kiho (KEE-HO), town, S. Muro county, MIE prefecture, S HONSHU, central JAPAN, 74 mi/120 km S of TSU; 33°43′N 135°59′E.

Kihoku (KEE-ho-koo), town, Soo county, KAGOSHIMA prefecture, SW KYUSHU, SW JAPAN, 19 mi/30 km E of KAGOSHIMA; 31°32′N 130°51′E.

Kihurio (kee-hoo-REE-o), village, KILIMANJARO region, NE TANZANIA, 85 mi/137 km SE of MOSHI, near KENYA border; 04°28′S 38°05′E. PARE MOUNTAINS to W, MKOMAZI Game Reserve to NE. Cattle, sheep, goats. Corn; timber.

Kii, former province in S HONSHU, JAPAN; now WAKAYAMA prefecture and part of MIE prefecture.

Kii Channel (KEE), Japanese *Kii-suiro* (KEE–swee-RO), W JAPAN, strait connecting OSAKA BAY (NE) and HARIMA SEA of INLAND SEA (E) with PHILIPPINE SEA (S); between E coast of SHIKOKU and S coast of HONSHU; c.30 mi/48 km N–S, c.35 mi/56 km E–W. Contains oyster beds.

Kiima, KAZAKHSTAN: see KIMA.

Kiinagashima (KEE-nah-GAH-shee-mah), town, N Muro county, MIE prefecture, S HONSHU, central JAPAN, 37 mi/60 km S of TSU; 34°11′N 136°19′E. Lobster, abalone; dried fish. Miso. Mandarin oranges.

Area is shown by the symbol □, and capital city or county seat by ☉.

Kii Peninsula (KEE), Japanese *Kii-hanto* (KEE–HAHN-to), S HONSHU, W central JAPAN, between KII CHANNEL (W) and KUMANO SEA (E); terminates S at SHIO POINT; comprises WAKAYAMA prefecture and parts of NARA and MIE prefectures; 80 mi/129 km E-W, 60 mi/97 km N-S. Forested interior, drained by KUMANO RIVER SE section is a resort area; contains YOSHINO-KUMANO NATIONAL PARK (□ 214 sq mi/554 sq km; established 1936), with scenic rapids, waterfalls, and ancient temples and shrines. Hot springs in SW.

Kiire (kee-ee-RE), town, Ibusuki county, KAGOSHIMA prefecture, SW KYUSHU, SW JAPAN, 16 mi/25 km S of KAGOSHIMA; 31°22′N 130°32′E. Rape seeds. Oil reserve base. N limit of *himerugi* plant.

Kijabe (kee-JAH-beh), town, CENTRAL province, S central KENYA, on railroad and 25 mi/40 km NNW of NAIROBI, on E slope of GREAT RIFT VALLEY; 00°56′S 36°34′E. Elevation 6,787 ft/2,069 m. Agriculture (coffee, wheat, corn, fruits); dairy farming. Africa Inland Mission. Junction for road to NAROK and Mara wildlife reserve. Base for ascent of Longonot volcano. Main hospital.

Kijimadaira (kee-JEE-mah-DAH-ee-rah), village, Shimotakai county, NAGANO prefecture, central HONSHU, central JAPAN, 19 mi/30 km N of NAGANO; 36°51′N 138°24′E.

Kijo (KEE-jo), town, Koyu county, MIYAZAKI prefecture, SE KYUSHU, SW JAPAN, 19 mi/30 km N of MIYAZAKI; 32°09′N 131°28′E.

Kijungu (kee-JOON-goo) or **Kijungu Well**, locality, Manyara region, NE central TANZANIA, 100 mi/161 km NNW of MOROGORO, in Kitwei Plain; 05°25′S 37°11′E. Sheep, goats; grain.

Kikagati (kee-kuh-GAH-tee), village, ISINGIRO district, WESTERN region, SW UGANDA, on KAGERA RIVER. Opposite MURONGO (TANZANIA) and 30 mi/48 km S of MBARARA. Tin-mining center. Was part of former SOUTHERN province.

Kikai (kee-KAH-ee), largest town on SW coast of KIKAI-SHIMA of AMAMI-GUNTO island group, in RYUKYU ISLANDS, Oshima county, KAGOSHIMA prefecture, SW JAPAN, 229 mi/370 km S of KAGOSHIMA; 28°19′N 129°56′E. Sugarcane, sweet potatoes.

Kikai-shima (kee-KAH-ee-shee-mah), island, in AMAMI-GUNTO island group, in RYUKYU ISLANDS, in the PHILIPPINE SEA, Oshima county, KAGOSHIMA prefecture, S JAPAN, 15 mi/24 km E of AMAMI-O-SHIMA; 9 mi/14.5 km long, 3.5 mi/5.6 km wide; hilly; fertile (sugarcane, sweet potatoes). Chief town, KIKAI (SW). Also Kikaigashima.

Kikale (kee-KAH-lai), village, PWANI region, E TANZANIA, 65 mi/105 km S of DAR ES SALAAM, on N edge of RUFIJI RIVER delta, near MAFIA CHANNEL, INDIAN OCEAN; 07°50′S 39°12′E. Fish; cashews, bananas, manioc, copra; goats, sheep.

Kikerino (kee-KYE-ree-nuh), town (2005 population 1,990), W LENINGRAD oblast, NW European Russia, on railroad, 27 mi/43 km W of SAINT PETERSBURG, and 6 mi/10 km ENE of volosovo; 59°28′N 29°38′E. Elevation 426 ft/129 m. Porcelain center (electrical appliances).

Kikhchik (KEEKH-cheek), town, SW KAMCHATKA oblast, RUSSIAN FAR EAST, on the W coast of the KAMCHATKA PENINSULA, on the Sea of OKHOTSK, near highway, 40 mi/65 km N of UST'-BOL'SHERETSK; 53°25′N 156°12′E. Elevation 114 ft/34 m. Fish cannery.

Kikinda (KEE-keen-dah), Hungarian *Nagykikinda*, city, VOJVODINA, N SERBIA, 50 mi/80 km NE of NOVI SAD, in the BANAT region; 45°49′N 20°27′E. Railroad junction (BELGRADE, BUDAPEST, and BUCHAREST lines); major wheat-trading center. Oil and natural gas found nearby. Manufacturing (ceramic products); flour milling. Until c.1947, called Velika Kikinda.

Kikladhes, Greece: see CYCLADES.

Kiknur (KEEK-noor), town (2005 population 5,195), SW KIROV oblast, E European Russia, on the KOKSHAGA RIVER (tributary of the VOLGA RIVER), on

crossroads, 27 mi/43 km W of YARANSK; 57°18′N 47°12′E. Elevation 364 ft/110 m. Agricultural equipment, chemicals, and supplies; flax processing, bakery, creamery; livestock (veterinary station, pedigree farm); woodworking.

Kikole (kee-KO-lai), village, KATANGA province, SE CONGO, near MANIKA PLATEAU (UPEMBA NATIONAL PARK); 09°27′S 28°21′E. Elev. 3,018 ft/919 m.

Kikombo (kee-KOM-bo), town, DODOMA region, central TANZANIA, 15 mi/24 km ESE of DODOMA, on railroad; 06°16′S 36°00′E. Peanuts, grain; livestock.

Kikonai (koo-KO-nah-ee), town, Oshima county, SW Hokkaido prefecture, N JAPAN, on TSUGARU STRAIT, 105 mi/170 km S of SAPPORO; 41°40′N 140°26′E.

Kikondja (kee-KON-jah), village, KATANGA province, SE CONGO, on W shore of LAKE KISALE, 100 mi/161 km ENE of KAMINA; 08°11′S 26°26′E. Elev. 1,994 ft/607 m. Manioc, palm oil, fish. Roman Catholic and Protestant missions. Also spelled KIKONDJI.

Kikondji, CONGO: see KIKONDJA.

Kikonzi (kee-KON-zee), village, BAS-CONGO province, W CONGO, 60 mi/97 km N of BOMA; 05°01′S 13°02′E. Elev. 1,095 ft/333 m. Also spelled KINKONZI.

Kikori (kee-KAW-ree), town, GULF province, S PAPUA, PAPUA NEW GUINEA, SE NEW GUINEA island, 245 mi/394 km NW of PORT MORESBY on Kikori River, near its multi-channeled delta on Gulf of PAPUA. Coconuts, palm oil, yams, taro; crocodile skins; crafts. Airstrip; accessed also by freighter and canoe.

Kikuchi (kee-KOO-chee), city, KUMAMOTO prefecture, W KYUSHU, SW JAPAN, 12 mi/20 km N of KUMAMOTO; 32°58′N 130°48′E. Tobacco, rice; *kasumi so* plants. Kikuchi Gorge is nearby.

Kikugawa (kee-KOO-gah-wah), town, Ogasa county, SHIZUOKA prefecture, central HONSHU, E central JAPAN, 22 mi/35 km S of SHIZUOKA; 34°45′N 138°05′E. Tea.

Kikugawa, town, Toyoura county, YAMAGUCHI prefecture, SW HONSHU, W JAPAN, 25 mi/40 km W of YAMAGUCHI; 34°06′N 131°02′E.

Kikuka (KEE-koo-kah), town, Kamoto county, KUMAMOTO prefecture, W KYUSHU, SW JAPAN, 16 mi/25 km N of KUMAMOTO; 33°01′N 130°45′E.

Kikuma (KEE-koo-mah), town, Ochi county, EHIME prefecture, NW SHIKOKU, W JAPAN, on IYO SEA, 12 mi/20 km N of MATSUYAMA; 34°01′N 132°50′E. Tile.

Kikungiri, UGANDA: see KABALE, district.

Kikusui (kee-KOO-swee), town, Tamana county, KUMAMOTO prefecture, W KYUSHU, SW JAPAN, 16 mi/25 km N of KUMAMOTO; 32°58′N 130°36′E.

Kikuyo (kee-KOO-yo), town, Kikuchi county, KUMAMOTO prefecture, W KYUSHU, SW JAPAN, 6 mi/10 km N of KUMAMOTO; 32°51′N 130°49′E. Manufacturing of traditional dolls.

Kikuyu (kee-KOO-yooh), town (1999 population 4,104), CENTRAL province, S central KENYA, on railroad and 10 mi/16 km W of NAIROBI; 01°15′S 36°40′E. Agriculture (coffee, wheat, corn). Church of Scotland Mission.

Kikuyu Escarpment (kee-KOO-yooh), section of E rim of GREAT RIFT VALLEY, in S central KENYA, just W of NAIROBI; extends 60 mi/97 km NS between KIJABE and KAJIADO. Elevation c.8,000 ft/2,440 m. Kikuyu Plateau (E) is densely settled and descends to 5,000 ft/1,525 m in Nairobi area.

Kikvidze (keek-VEE-dze), town, NW VOLGOGRAD oblast, SE European Russia, on the BUZULUK RIVER, on road, 22 mi/35 km NE of NOVOANNINSKIY; 50°44′N 43°03′E. Elevation 308 ft/93 m. In agricultural area (grains, vegetables); produce processing. Until 1936, called Preobrazhenskaya.

Kikwetu (kee-KWAI-too), village, LINDI region, SE TANZANIA, 7 mi/11.3 km N of LINDI, near INDIAN OCEAN; 09°54′S 39°47′E. Lindi Airport is here. Peanuts, cashews, corn, bananas, copra; goats, sheep.

Kikwit (KEE-kweet), town, BANDUNDU province, SW CONGO, on left bank of KWILU RIVER and 250 mi/402 km ESE of KINSHASA; 05°04′S 18°52′E. Elev. 1,440 ft/

438 m. Commercial center, terminus of steam navigation; palm-oil milling (in the 1970s, had world's largest oil-palm plantation), fiber growing. Has Roman Catholic missions, hospital, trade and teachers' schools. Airport. Ebola virus reappeared here in mid-1990s.

Kil (SHEEL), town, VÄRMLAND county, W SWEDEN, 10 mi/16 km NW of KARLSTAD; 59°30′N 13°19′E. Railroad center. First Swedish railroad built (1849) between here and Fryksta (FrIksta), a village at S end of LAKE FRYKEN, 2 mi/3.2 km NE.

Kila, for Afghan names beginning thus: see QALA.

Kila Didar Singh (ki-lah dee-dahr SING-uh), town, GUJRANWALA district, E PUNJAB province, central PAKISTAN, 10 mi/16 km W of GUJRANWALA; 32°08′N 74°01′E. Wheat, rice. Kila also spelled as Qila and Killa.

Kilakkarai (ki-luh-kuh-REI), town, RAMANATHAPURAM district, TAMIL NADU state, S INDIA, port on GULF OF MANNAR, 10 mi/16 km SSW of Ramanathapuram; 09°14′N 78°47′E. Fishing center; exports coconuts, coir products, coral.

Kilambé (kee-lahm-BAI), peak (5,580 ft/1750 m) in N spur of CORDILLERA ISABELIA, N NICARAGUA, near COCO RIVER, 25 mi/40 km ENE of QUILALÍ.

Kila Saifulla, PAKISTAN: see KILLA SAIFULLA.

Kilasevalpatti (ki-luh-SAI-vuhl-puht-tee), town, Ramnathapuram district, TAMIL NADU state, S INDIA, 7 mi/11.3 km NE of TIRUPPATTUR. In cotton area. Also spelled Kilasevalpatti.

Kilauea (KEE-LOU-AI-ah), town (2000 population 2,092), NE KAUAI, KAUAI county, HAWAII, 1 mi/1.6 km inland from N coast, on Kilauea Stream, on Kuhio Highway, 16 mi/26 km NNW of LIHUE; Mokolea Point to NE; Kalihiwai Bay to NW. Princeville Airport to W; Kapinao Heiau (temple) to SE, on Kilauea Bay; Kilauea Point National Wildlife Refuge to N; Mokuaeae Island Seabird Sanctuary N of Point: Moloaa Forest Reserve to S; Anini Beach Park to NW.

Kilauea (KEE-LOU-AI-ah), caldera, 3,412 ft/1,040 m deep, SE HAWAII island, HAWAII county, HAWAII, 22 mi/35 km SW of HILO, 20 mi/32 km ESE of summit of MAUNA LOA; in HAWAII VOLCANOES NATIONAL PARK; 22°12′N 159°24′W. Kilauea is at the volcanically active SE end of the Hawaiian chain, sitting atop the stationary Hawaiian vent. HALEMAUMAU crater in W part of caldera. One of the largest active craters in the world, Kilauea has a circumference of c.8 mi/12.9 km and is surrounded by a wall of volcanic rock 200 ft/61 m to 500 ft/152 m high. Elevation at Hawaiian Volcano Observatory, 4,078 ft/1,243 m. Elevation of crater floor 3,412 ft/1,040 m; maximum depth, 666 ft/203 m. Last erupted in August 1997. Lava from the eruption destroyed a 700-year-old temple formerly used in human sacrifices.

Kilbarchan (kil-BAHR-kuhn), town (2001 population 3,622), Renfrewshire, W Scotland, on BLACK CART WATER, and 2 mi/3.2 km W of JOHNSTONE; 55°50′N 04°33′W. Previously manufacturing of sewing cotton, leather, and handloomed woolens.

Kilbirnie (kil-BUHR-nee), town (2001 population 7,280), North Ayrshire, SW Scotland, on Garnock River, and 10 mi/16 km N of IRVINE; 55°45′N 04°42′W. Former steel-milling town. Previously lace and rubber production. Just E is Kilbirnie Loch, a lake 1 mi/1.6 km long and 0.5 mi/0.8 km wide. Former steel-milling town of Glengarnock is at S end of lake.

Kilbourn, WISCONSIN: see WISCONSIN DELLS.

Kilbourne (KILL-born), village (2000 population 375), MASON county, central ILLINOIS, 14 mi/23 km NNW of PETERSBURG; 40°08′N 90°00′W. In agricultural area. Near Sand Prairie-Scrub Oak State Natural Area.

Kilbourne (KIL-buhrn), village, WEST CARROLL parish, NW LOUISIANA, 12 mi/19 km NNW of LAKE PROVIDENCE town, on ARKANSAS boundary; 33°00′N 91°19′W. In agricultural area (cotton, rice, vegetables; cattle, hogs); timber.

Kilbrannan Sound (kil-BRA-nuhn SOUND) or **Kilbrennan Sound**, arm of the Firth of Clyde, Argyll and Bute, W Scotland, extending c.25 mi/40 km N-S, separating KINTYRE peninsula (W) and ARRAN island (E); 3 mi/4.8 km–8 mi/12.9 km wide; 55°32'N 05°25'W.

Kilbride, East (KIL-breid) town (2001 population 73,796), South Lanarkshire, S central Scotland, 11 mi/18 km S of GLASGOW; 55°46'N 04°10'W. Established in 1946 under the New Towns Act to absorb the overspill population of Glasgow, East Kilbride has engineering works and produces motor vehicle and aircraft engines, as well as electronic equipment. Center for engineering research is here. The town also bottles milk.

Kilbride, West (KIL-breid) town (2001 population 4,393), North Ayrshire, SW Scotland, 4 mi/6.4 km NNW of ARDROSSAN; 55°41'N 04°51'W. Just W, on Firth of Clyde, is resort of SEAMILL, near FARLAND HEAD. Law Castle is nearby.

Kilburn, ENGLAND: see BRENT.

Kilchberg, commune, ZÜRICH canton, N SWITZERLAND, on LAKE ZÜRICH, and 4 mi/6.4 km S of ZÜRICH. Textiles.

Kilchoan, Scotland: see ARDNAMURCHAN.

Kilchu (GEEL-JU), county, NORTH HAMGYONG province, NORTH KOREA, 60 mi/97 km SSW of CHONGJIN. Commercial center for livestock, lumbering, and agricultural area; makes celluloid products. Pulp milling, magnesite mining. Connects primary railroads.

Kilcolman Castle, IRELAND: see DONERAILE.

Kilcoy (kil-KOI), town, QUEENSLAND, NE AUSTRALIA, 47 mi/75 km NW of Caboolture; 26°58'S 152°30'E.

Kilcreggan, Scotland: see COVE.

Kilcullen (kil-KUH-luhn), Gaelic *Cill Chuillinn an Droichid*, town (2006 population 2,985), central KILDARE county, E central IRELAND, on the LIFFEY RIVER (bridge dates from 1319), and 7 mi/11.3 km E of KILDARE, at E end of THE CURRAGH; 53°07'N 06°44'W. Agricultural market. Remains of ancient round tower and abbey nearby.

Kilcunda (kil-KUN-duh), village, S VICTORIA, AUSTRALIA, on BASS STRAIT and 55 mi/89 km SSE of MELBOURNE, near WONTHAGGI; 38°33'S 145°29'E. Old coal mine. Agriculture; tourism. Trestle bridge.

Kildare, county (□ 654 sq mi/1,700.4 sq km; 2006 population 186,335), E central IRELAND; ⊙ NAAS. Borders MEATH county to N, DUBLIN and WICKLOW counties to E, CARLOW county to S, and LAOIGHIS and OFFALY counties to W. The region is a flat plain, containing the greater portion of the BOG OF ALLEN, as well as THE CURRAGH. The principal rivers are the LIFFEY, Greese, and BARROW. Agriculture is the chief occupation; the breeding of racehorses is also significant. The county is named for the oak (*Cill Dara*) under which St. Bridget constructed her cell. Pre-Christian and early-Christian relics remain, including a 13th-century castle and monastery.

Kildare (kil-DER), Gaelic *Cill Dara*, town, KILDARE county, E central IRELAND, 30 mi/48 km WSW of DUBLIN; 53°10'N 06°55'W. Agricultural market; paper mills. The Irish National Stud (horse-breeding facility) is just E. First church was established here 490 by St. Bridget; it was succeeded by many other religious establishments, which were attacked by Danes and, later, in the Elizabethan wars. Cathedral, built 1229, was destroyed 1641 by Cromwell and rebuilt 1683. An old round tower adjoins it. There are remains of a 13th-century castle and of an ancient Carmelite monastery.

Kildare (kil-DER), village (2006 population 90), KAY county, N OKLAHOMA, 7 mi/11.3 km N of PONCA CITY; 36°48'N 97°02'W. In agricultural area.

Kildare, Cape, (KILD-air), NW PRINCE EDWARD Island, Canada, on the Gulf of St. Lawrence, 6 mi/9.7 km SE of TIGNISH; 46°53'N 63°58'W. Harvesting of Irish moss. Reputed landfall (1534) of Jacques Cartier. Named by Samuel Holland in 1765 after James, 20th earl of Kildare.

Kildeer, village (2000 population 3,460), Lake county, NE ILLINOIS, residential suburb 28 mi/45 km NW of downtown CHICAGO, 3 mi/4.8 km SE of LAKE ZURICH; 42°10'N 88°02'W.

Kil'din Island (KEEL-deen), in the BARENTS SEA, in MURMANSK oblast, NW European Russia, off the N KOLA PENINSULA, 20 mi/32 km NE of MURMANSK; 13 mi/21 km long, 5 mi/8 km wide. Kil'din village at the E end; fisheries.

Kil'dinstroy (keel-deen-STRO-yee), town (2006 population 2,645), NW MURMANSK oblast, NW European Russia, on the KOLA PENINSULA, on the KOLA River, on road and the Murmansk railroad, 12 mi/19 km S of MURMANSK; 68°48'N 33°06'E. Waste-water treatment plant.

Kildonan, town, MASHONALAND WEST province, N ZIMBABWE, 45 mi/72 km NW of HARARE, in MVURWI RANGE; 17°20'S 30°36'E. Terminus of railroad spur from MARYLAND JUNCTION. Chromite mining. Livestock; grain, cotton, tobacco.

Kildonan (kil-DO-nuhn), unincorporated village, SW BRITISH COLUMBIA, W CANADA, on S central VANCOUVER ISLAND, on an inlet of BARKLEY SOUND, 18 mi/29 km SSW of PORT ALBERNI, in ALBERNI-Clayoquot regional district; 49°00'N 125°00'W. Fishing (salmon, herring).

Kilemary (kee-lye-MAH-ri), village (2006 population 3,890), NW MARI EL REPUBLIC, E central European Russia, on the GREATER KUNDYSH RIVER, on road, 44 mi/70 km N of CHEBOKSARY, and 40 mi/64 km WNW of YOSHKAR-OLA; 56°47'N 46°52'E. Elevation 413 ft/125 m. Lumbering.

Kilembe (kee-LEM-bai), village, KASESE district, WESTERN region, UGANDA, on E slopes of the RUWENZORI MOUNTAINS, 35 mi/56 km SSW of Kabarole; 00°14'N 30°01'E. Copper deposits nearby. Was part of former WESTERN province.

Kilengwe (kee-LAIN-gwai), village, MOROGORO region, E central TANZANIA, 50 mi/80 km S of MOROGORO, near MGETA River; SELOUS GAME RESERVE to S; 07°34'S 37°32'E. Corn, wheat; goats, sheep; timber.

Kilfinane (kil-FI-nain) or **Kilfinnane**, Gaelic *Cill Fhionáin*, town (2006 population 727), SE LIMERICK county, SW IRELAND, 15 mi/24 km SE of TIPPERARY; 52°22'N 08°28'W. Agricultural market. Has notable ancient rath.

Kilfinnane, IRELAND: see KILFINANE.

Kilgore (kil-GOR), city (2006 population 12,040), on GREGG-RUSK county line, E TEXAS, 24 mi/39 km E of TYLER; 32°23'N 94°52'W. Elevation 371 ft/113 m. Oil-producing and oil field supply center, in E Texas field. Manufacturing (oil field equipment, plastic products, fiberglass boats, satellite antennas). Seat of Kilgore College (two year). Lake Cherokee reservoir to E. Settled 1872, incorporated 1931. Rapid growth followed oil discovery, 1930.

Kilgore, village (2006 population 96), CHERRY county, N NEBRASKA, 20 mi/32 km W of VALENTINE, near SOUTH DAKOTA state line; 42°56'N 100°57'W. Manufacturing (feed-handling equipment, flatbeds for pickups).

Kili (KEE-lee), coral island (1999 population 774), RALIK CHAIN, MAJURO district, MARSHALL ISLANDS, W central PACIFIC, c.500 mi/800 km SE of BIKINI; 05°38'N 169°07'E; 1 mi/1.6 km long. Coconuts. Bikini inhabitants moved here in 1949 from UJELANG. Some later returned to Bikini. Sometimes called Hunter Island.

Kilifarevo (kee-tee-FAHR-o-vo), city, LOVECH oblast, VELIKO TURNOVO obshtina, N BULGARIA, on a branch of the YANTRA RIVER, 5 mi/8 km SSW of VELIKO TURNOVO; 43°00'N 25°36'E. Flour milling, vegetable canning; agricultural machinery workshops for foodstuffs. Has a 14th-century church and a 19th-century monastery.

Kilifi (kee-LEE-fee), town (1999 population 30,394), COAST province, SE KENYA, small land-locked port on INDIAN OCEAN, 30 mi/48 km NNE of MOMBASA; 03°37'S 39°50'E. Sisal center; cotton, copra. Fisheries.

Kilik Pass (KI-LEEK), in N extension of KARAKORAM mountain system, on KASHMIR-CHINA border (undefined), near AFGHANISTAN panhandle, 21 mi/34 km NNW of MISGAR (Kashmir). Elevation 15,600 ft/4,755 m. On important trade route from GILGIT to KASHI.

Kilimafeza (kee-lee-mah-FAI-zah), village, MARA region, N TANZANIA, 95 mi/153 km SE of MUSOMA, near Oranji River; 02°18'S 34°56'E. Former gold-mining town near center of SERENGETI NATIONAL PARK. Also spelled Kilmafeza.

Kilimane, MOZAMBIQUE: see QUELIMANE.

Kilimanjaro (kee-lee-mahn-JAH-ro), region (2006 population 1,503,000), NE TANZANIA; ⊙ MOSHI, bounded N and NE by KENYA, PANGANI RIVER forms part of W boundary. KILIMANJARO Mountain, in KILIMANJARO NATIONAL PARK, in N; PARE MOUNTAINS in S; MKOMAZI Game Reserve in SE. Coffee, tea, sisal, rice, sugarcane; cattle, sheep, goats; timber. Tourism. Part of former TANGA province.

Kilimanjaro (ki-li-mahn-JAH-ro), mountain, KILIMANJARO region, N TANZANIA, near KENYA border, 18 mi/29 km of MOSHI; 03°05'S 37°21'E. The highest point in AFRICA, it rises to two peaks, UHURU (KIBO) (19,340 ft/5,895 m), and MAWENZI (17,564 ft/5,354 m), which are joined by a broad saddle (elevation c. 15,000 ft/4,600 m). It is a conical, snow-capped volcano, recently extinct with remnant activity. MOUNT KILIMANJARO NATIONAL PARK protects the mountain generally above 6,562 ft/2,000 m level. It and the SERENGETI plain to W are two of Africa's most popular tourist attractions.

Kilimatinde (ki-li-mah-TIN-de), village, SINGIDA region, central TANZANIA, 8 mi/12.9 km SSE of MANYONI; BAHI Swamp to E; 05°52'S 34°54'E. Cattle, goats, sheep; corn, wheat.

Kilimli, town, N TURKEY, on BLACK SEA, on ZONGULDAK-ANKARA railroad and 3 mi/4.8 km NE of Zonguldak; 41°30'N 31°51'E. Coal mines.

Kilinailau, atoll, Papua New Guinea: see TULUN.

Kilindi (kee-LEEN-dee), village, PWANI region, E TANZANIA, 80 mi/129 km N of DAR ES SALAAM, on RUFIJI RIVER (ferry), near its entrance to INDIAN OCEAN; 08°08'S 39°11'E. Cashews, bananas, sisal, copra, rice. Goats; timber.

Kilindoni (kee-leen-DO-nee), village, PWANI region, E TANZANIA, 75 mi/121 km SW of DAR ES SALAAM, S end of MAFIA ISLAND, INDIAN OCEAN; 07°56'S 39°11'E. Road junction. Airport; tourist lodge. Tourism; sport fishing. Fish; livestock. Copra, bananas.

Kilingi-Nõmme (KEE-lin-gee–NUH-mai), German *Kurkund*, city, SW ESTONIA, on railroad and 25 mi/40 km SE of Pärnu; 58°09'N 24°57'E. Agricultural market; fodder crops, flax, and livestock.

Kilinochchi, district (□ 465 sq mi/1,209 sq km), NORTHERN PROVINCE, SRI LANKA; ⊙ KILINOCHCHI; 09°30'N 80°15'E.

Kilinochchi, town, ⊙ KILINOCHCHI district, NORTHERN PROVINCE, SRI LANKA, 32 mi/51 km SSE of JAFFNA; 09°24'N 80°24'E. Rice and vegetables. Held by Tamil Tiger rebels until recapture by government forces in 1996. Site of major refugee camp in 1996.

Kilis, city (2000 population 70,670), S TURKEY, near Syrian border, 29 mi/47 km SSW of GAZIANTEP; 38°42'N 35°28'E. Olives, grain, sesame, vetch, cotton.

Kiliya (kee-lee-YAH) (Romanian *Chilia-Nouă*), [=new Kiliya], city, SW ODESSA oblast, UKRAINE, in BESSARABIA, port on Kiliya (N) arm (Romanian border) of the DANUBE River delta, 22 mi/35 km ENE of IZMAYIL; 45°27'N 29°16'E. Raion center. Fishing center and ship repairs; flour milling, food processing (dairy, juice, meat). Manufacturing (bricks, asphalt). Across the Danube arm is Chilia-Veche [=old Kiliya], Romanian fishing village with ruins of a 14th-century Moldavian fortress. First mentioned in the late 7th century B.C.E. as the Greek polis Licostomo; part of Kievan Rus' (10th century C.E.) and Halych-Volyn duchy (13th century); changed hands between Hungary and Moldova (14th-15th century), and passed to Turkey (1484–1812) and Russia. Most of the population in 1897 was Ukrainian. Ceded to Romania (1918). Kiliya was (1941–1944) the capital of Romanian Chilia department. Its cession to USSR in 1940 was confirmed in 1947. Since 1991, part of independent Ukraine.

Kilkee (kil-KEE), Gaelic *Cill Chaoi*, town (2006 population 1,325), SW CLARE county, W IRELAND, on Moore Bay (small inlet of the Atlantic), 8 mi/12.9 km WNW of KILRUSH; 52°41′N 09°38′W. Seaside resort.

Kilkeel (kil-KEEL), Gaelic *Cill Chaoil*, town (2001 population 6,338), S DOWN, SE Northern Ireland, on IRISH SEA at mouth of Kilkeel River, 15 mi/24 km ESE of NEWRY; 54°04′N 06°01′W. Fishing center, seaside resort.

Kilkenny (kil-KE-nee), Gaelic *Cill Chainnigh*, county (□ 800 sq mi/2,080 sq km; 2006 population 87,558), SE IRELAND; ⊙ KILKENNY. Borders LAOIGHIS county to N, CARLOW and WEXFORD counties to E, WATERFORD county to S, and TIPPERARY county to W. The region is mainly a rolling plain, part of the central plain of Ireland, with low hills to the S. The principal rivers are the SUIR, NORE, and BARROW. Grains, vegetables; livestock. Industries include food processing, brewing, agricultural engineering, clothing, and handicrafts. The county has maintained reforestation programs for the past several years. Kilkenny is roughly coextensive with the ancient kingdom of OSSORY; the county is rich in antiquities. Its chief towns are CASTLECOMER, CALLAN, THOMASTOWN, and GRAIGUENAMANAGH-TINNAHINCH.

Kilkenny (kil-KE-nee), town (2006 population 8,661), ⊙ KILKENNY county, S IRELAND, on the NORE RIVER, and 18 mi/29 km SW of CARLOW; 52°39′N 07°15′W. The districts of Irishtown and Englishtown, separated by a stream, were legally united in 1843. Strife between the inhabitants of the two districts, to the near destruction of both, may have given rise to the stories of the Kilkenny cats, who ate each other up. A third district is High Town. Industries include brewing, printing, and crafts. Kilkenny was the seat of the kings of OSSORY. The first earl of Pembroke founded a castle there in the 12th century (restored c.1835) overlooking the Nore. Parliaments and assemblies were held in the 14th, 16th, and 17th centuries. Among noted pupils at the Protestant school of Kilkenny were Jonathan Swift, Bishop Berkeley, and William Congreve. In Irishtown is the great Cathedral of St. Canice (13th century), the seat of the Protestant dioceses of the United Dioceses of Ossory, Ferns, and Leighlin. The Roman Catholic Cathedral of St. Mary (seat of the diocese of Ossory), a round tower, and remains of Dominican and Franciscan monasteries (mostly 13th century) are noteworthy.

Kilkenny (kil-KE-nee), town, COOS county, N central NEW HAMPSHIRE, 8 mi/12.9 km E of LANCASTER. Wilderness area in White Mountain National Forest; MOUNT CABOT (4,160 ft/1,268 m) at center. Timber.

Kilkenny (kil-KE-nee), village (2000 population 148), LE SUEUR county, S MINNESOTA, 15 mi/24 km W of FARIBAULT; 44°18′N 93°34′W. Dairying. Numerous small lakes in area; Diamond Lake to NW.

Kilkhampton (kilk-HAM-tuhn), village (2001 population 1,193), NE CORNWALL, SW ENGLAND, 3 mi/4.8 km NNE of BUDE; 50°53′N 04°29′W. Agricultural market. Has 15th-century church.

Kilkich, Greece: see KILKIS.

Kilkieran Bay (kil-KEE-ruhn), inlet (10 mi/16 km long) of GALWAY BAY, SW GALWAY county, W IRELAND; 53°19′N 09°43′W. Contains numerous islands, including GORUMNA, LETTERMORE, and LETTERMULLEN. MWEENISH Island at entrance.

Kilkis (keel-KEES), prefecture (□ 968 sq mi/2,516.8 sq km), CENTRAL MACEDONIA department, NE GREECE; ⊙ KILKIS; 41°00′N 22°40′E. Bordered N by the Republic of MACEDONIA, E by SÉRRAI prefecture, S by KHALKIDHIKÍ and THESSALONÍKI prefectures, and W by EMATHEIA and PELLA prefectures. PAIKON massif in W; drained by VARDAR (Axios) River. Agriculture includes silkgrowing; cotton, tobacco, rice, wine, red peppers. Main centers are Kilkis and GOUMENISSA; served by railroads from THESSALONÍKI to BELGRADE (SERBIA) and to EDIRNE (TURKEY). Formed in 1930s.

Kilkis (keel-KEES), Macedonian *Kukush*, city (2001 population 17,430), ⊙ KILKIS prefecture, CENTRAL MACEDONIA department, NE GREECE, on highway and railroad, 25 mi/40 km N of THESSALONÍKI; 41°00′N 22°52′E. Road junction; trading center for cotton, tobacco, silk, wine. Called Kilkich under Turkish rule, from 15th century to Balkan Wars (1912–1913). Greek victory here (1913) over Bulgarians.

Kilkivan (kil-KEE-vuhn), village, SE QUEENSLAND, NE AUSTRALIA, 110 mi/177 km NNW of BRISBANE, 32 mi/52 km NW from GYMPIE, in E hills of GREAT DIVIDING RANGE; 26°05′S 152°14′E. Beef, dairy cattle service center. Copper smelter ruins. Folk museum.

Killa Didar Singh, PAKISTAN: see KILA DIDAR SINGH.

Killala (ki-LA-luh), Gaelic *Cill Ala*, town (2006 population 569), NW MAYO county, NW IRELAND, at head of KILLALA BAY, 25 mi/40 km N of CASTLEBAR; 54°12′N 09°13′W. Fishing port. It was formerly seat of bishopric, reputedly founded by St. Patrick. Has cathedral dating from c.1670 and ancient round tower.

Killala Bay (ki-LA-luh), Gaelic *Cuan Cill Ala*, inlet (7 mi/11.3 km long, 6 mi/9.7 km wide) of the ATLANTIC, between N MAYO county and NW SLIGO county, NW IRELAND. Receives MOY RIVER at head of bay; KILLALA is on SW shore.

Killaloe (ki-la-LOO), Gaelic *Cill Dalúa*, town (2006 population 1,035), SE CLARE county, W IRELAND, on the SHANNON (bridged), and at S end of LOUGH DERG, 13 mi/21 km NE of LIMERICK; 52°48′N 08°27′W. Agricultural market (potatoes, grain; dairying). Cathedral, probably built 1182 by King Donal O'Brien on site of church founded in 7th century. Killaloe or its vicinity was site of Kincora, 10th-century palace of Brian Boru.

Killaloe (kil-uh-LOO), former village (□ 3 sq mi/7.8 sq km; 2001 population 661), SE ONTARIO, E central CANADA, near Golden Lake, 23 mi/37 km SW of PEMBROKE; 45°33′N 77°24′W. Dairying, lumbering. Formerly Killaloe Station. Amalgamated into KILLALOE, Hagarty and Richards township in 2000.

Killaloe, Hagarty and Richards (kil-uh-LOO, HA-guhr-tee, RI-chuhrdz), township (□ 153 sq mi/397.8 sq km; 2001 population 2,492), SE ONTARIO, E central CANADA, 24 mi/39 km from PEMBROKE; 45°36′N 77°30′W. Formed in 2000 from KILLALOE village and the township of HAGARTY and Richards. Also written Killaloe, Hagarty & Richards.

Killam (KI-luhm), town (□ 2 sq mi/5.2 sq km; 2001 population 1,004), E ALBERTA, W CANADA, 40 mi/64 km ESE of CAMROSE, in FLAGSTAFF County; 52°47′N 111°51′W. Grain elevators; oil and gas. Established as a village in 1906; became a town in 1965.

Killamarsh (KIL-uh-mahsh), village (2001 population 9,627), NE DERBYSHIRE, central ENGLAND, 8 mi/12.9 km SE of SHEFFIELD; 53°19′N 01°19′W. Former coal-mining site. Church has 15th-century tower.

Killarney (ki-LAHR-nee), town, QUEENSLAND, NE AUSTRALIA, 122 mi/196 km SW of BRISBANE, 21 mi/34 km E of WARWICK, on Condamine (BALONNE) River banks, and near NEW SOUTH WALES border; 28°18′S 152°15′E. Queen Mary Falls National Park 6 mi/10 km from town. Mostly rebuilt since a 1968 cyclone.

Killarney (kil-AHR-nee), town (□ 2 sq mi/5.2 sq km; 2001 population 2,221), SW MANITOBA, W central CANADA, on Killarney Lake (4 mi/6 km long), 50 mi/80 km SSE of BRANDON, in TURTLE MOUNTAIN rural municipality; 49°11′N 99°40′W. Manufacturing of agricultural implements, cement; grain elevators; dairying, lumbering, livestock. Resort. Site of experimental fruit farm. Incorporated 1906.

Killarney (ki-LAHR-nee), town (□ 584 sq mi/1,518.4 sq km; 2001 population 428), SE central ONTARIO, E central CANADA, 25 mi/40 km from FRENCH RIVER; 46°03′N 80°58′W. Contains Killarney, French River provincial parks. Formed in 1999 from George Island and Rutherford township; the unorganized townships of Hansen, Goshen, Sale, Attlee, Kilpatrick, Struthers, Allen, and Travers; and portions of Bigwood, Humboldt, and Carlyle. Several N Georgian Bay islands were added in 2001.

Killarney (ki-LAHR-nee), Gaelic *Cill Airne*, town (2006 population 13,497), KERRY county, SW IRELAND; 52°03′N 09°31′W. Mineral-water bottling, footwear, lace, hosiery, woolens, and ornamental-ironwork industries. Tourist center for the LAKES OF KILLARNEY. The Roman Catholic cathedral (Gothic Revival) of the diocese of Kerry, designed by A. W. Pugin, is here.

Killarney, Lakes of (ki-LAHR-nee), three lakes in central KERRY county, SW IRELAND, surrounded by mountains and famed for their beauty. Lough Leane or Lower Lake (□ 8 sq mi/20 sq km; c.5 mi/8 km long) contains Ross island, with castle of the O'Donoghues, and INNISFALLEN island, scene of Moore's poem and site of ruins of 6th-century abbey founded by St. Finian, where the 11th-century *Annals of Innisfallen* were composed. On N shore of Muckross Lake, Middle Lake or Lough Torc (□ 1.1 sq mi/2.8 sq km) are remains of 15th-century Muckross Abbey. Upper Lake (□ 0.7 sq mi/1.7 sq km) is S of Lough Leane. MACGILLYCUDDY'S REEKS tower over the lakes.

Killary Harbour (KI-luh-ree), Gaelic *An Caoláire Rua*, narrow inlet (12 mi/19 km long) of the ATLANTIC, between MAYO and GALWAY counties, W IRELAND; 53°37′N 09°52′W.

Killa Saifulla (ki-LAH sei-fuh-LAH), village, ZHOB district, NE BALUCHISTAN province, SW PAKISTAN, near ZHOB RIVER, 80 mi/129 km SW of Fort Sandeman; 30°43′N 68°21′E. Also spelled KILA SAIFULLA, Killa Saif-Ullah, and QILA SAIFULLAH.

Killashandra, IRELAND: see KILLESHANDRA.

Killbuck (KIL-buk), village (2006 population 893), HOLMES county, central OHIO, on KILLBUCK CREEK, 33 mi/53 km ESE of MANSFIELD; 40°29′N 81°59′W.

Killbuck Creek (KIL-buk), N OHIO; rises in region W of AKRON; flows c.75 mi/121 km S, past WOOSTER and MILLERSBURG, to WALHONDING RIVER 5 mi/8 km NW of COSHOCTON, and into the Muskingum River; 40°19′N 81°56′W.

Killburg, GERMANY: see KYLLBURG.

Killdeer, village (2006 population 683), DUNN co., W central NORTH DAKOTA, 35 mi/56 km N of DICKINSON and on Spring Creek; 47°22′N 102°45′W. Oil fields; livestock; dairy products; grain. Little Missouri State Primitive Park to N. Killdeer Battlefield Site to NW. Lake Ilo to SE. Founded and incorporated in 1915 and named for nearby Killdeer Mountains.

Killdeer Mountains, series of lofty buttes in DUNN co. W NORTH DAKOTA, NW of KILLDEER; they extend 10 mi/16 km E-SW and rise 600 ft/183 m above surrounding countryside. S of LITTLE MISSOURI RIVER, Killdeer, and Battlefield Historic Site; 47°26′N 102°55′W. Elev. 3,000 ft/914 m. Name most likely comes from the Sioux Indian phrase "tak-kah-p-kuty" meaning "the place where they kill deer."

Kill Devil Hills (KIL DEV-uhl HILZ), town (□ 5 sq mi/13 sq km; 2006 population 6,614), DARE county, NE NORTH CAROLINA, 39 mi/63 km SE of ELIZABETH CITY, on ATLANTIC OCEAN, on OUTER BANKS; 36°01′N 75°40′W. ALBEMARLE SOUND to W. Town of KITTY HAWK 4 mi/6.4 km to NW. Beach resort area. Service industries; retail trade; light manufacturing. Jockey's Ridge State Park to S. Site of Wright Brothers' flight experiments (1900–1903). WRIGHT BROTHERS NATIONAL MEMORIAL to W.

Killeen (kil-EEN), city (2006 population 102,003), BELL county, central TEXAS, 21 mi/34 km W of TEMPLE, half of Temple-Killeen metropolitan statistical area;

31°06′N 97°43′W. Elevation 833 ft/254 m. In a ranching and cotton region. The city has some manufacturing (concrete, printing, wooden cabinets), but adjacent FORT HOOD is the major source of employment. Founded in 1882 and named for a Santa Fe Railroad official, Killeen remained a small farming and ranching village until the establishment (1942) of Camp Hood, mainly to N, also to W of city. The camp's redesignation (1950) as a fort with a permanent status spurred a great population growth in the city. Fort Hood Army Airfield to N, Robert Gray Army Airfield to W, both in Fort Hood Military Reservation. Central Texas College (two year) is in Killeen and University of Central Texas and American Educational Complex (Fort Hood) to W. Nearby BELTON (NE) and STILLHOUSE HOLLOW (SE) lakes provide recreational facilities. Site of mass murder in Luby's Restaurant (1991). Incorporated 1893.

Killegray (KIL-ah-grai), uninhabited island (less than 1 mi/1.6 km long, 0.5 mi/0.8 km wide), OUTER HEBRIDES, Eilean Siar, NW Scotland, in the Sound of Harris, between Harris, LEWIS AND HARRIS, and NORTH UIST; 57°43′N 07°04′W.

Killeshandra (ki-luh-SHAN-druh) or **Killashandra**, Gaelic *Cill na Seanrátha*, town (2006 population 411), W CAVAN county, N central IRELAND, on W shore of LOUGH OUGHTER, 7 mi/11.3 km W of CAVAN; 54°01′N 07°32′W. Agricultural market, angling center. Ancestors of Edgar Allan Poe came from nearby Kildallan.

Killian (KIL-ee-an), town (2000 population 1,053), LIVINGSTON parish, LOUISIANA, 10 mi/16 km SW of PONCHATOULA; 30°21′N 90°35′W.

Killiecrankie, Pass of (KIL-ee-KRAN-kee), wooded pass, N PERTH AND KINROSS, central Scotland, through which the GARRY RIVER flows, and 3 mi/4.8 km NW of PITLOCHRY; 56°45′N 03°47′W. Site of battle of 1689 in which Jacobites under "Bonnie Dundee" defeated the army of William III.

Killik River, NW ALASKA; rises in BROOKS RANGE near 67°47′N 154°35′W; flows c.125 mi/201 km N to COLVILLE RIVER at 69°N 153°55′W.

Killin (KIL-in), town (2001 population 666), STIRLING, central Scotland, at W end of LOCH TAY, at mouth of DOCHART RIVER, and 16 mi/26 km N of CALLANDER; 56°28′N 04°19′W. Previously woolen (tartan plaid) milling.

Killinek Island (20 mi/32 km long, 2 mi/3 km–9 mi/14 km wide), NUNAVUT territory, CANADA, at SE entrance of HUDSON STRAIT, off N extremity of LABRADOR. NE extremity is Cape CHIDLEY (60°23′N 64°26′W), usually considered to be N tip of LABRADOR. On W coast is KILLINIQ trading post.

Killiney (ki-LEI-nee), Gaelic *Cill Inghean Léinín*, residential town and resort (2006 population 9,588), SE DUBLIN county, E IRELAND, on small inlet of the IRISH SEA, 3 mi/4.8 km SSE of DÚN LAOGHAIRE; 53°15′N 06°07′W. Large public park and fine beach.

Killingly, town, WINDHAM county, NE CONNECTICUT, on the QUINEBAUG RIVER and near the RHODE ISLAND border, in a farm area; settled 1693, incorporated 1708; 41°49′N 71°50′W. Dairy and livestock-trading center; fruits and vegetables; manufacturing (textiles, machinery, and synthetic materials). Winter skiing.

Killington Peak (4,241 ft/1,293 m), S central VERMONT, E of RUTLAND, in recreational area with major downhill ski resort; one of highest summits of GREEN MOUNTAINS.

Killingworth, town, MIDDLESEX county, S CONNECTICUT, 20 mi/32 km ENE of NEW HAVEN; 41°22′N 72°34′W. Industry includes agriculture and steel fabrication; summer camps. Mainly residential. Has 18th-century houses, fine church (1817). State forest here.

Killini (kee-LEE-nee), Latin *Cyllene*, village and port, ILIA prefecture, WESTERN GREECE department, on westernmost PELOPONNESE, on IONIAN SEA, 23 mi/37 km NW of PÍRGOS. On Cape Killini (37°56′N 21°09′E) at S end of Gulf of Killini, a broad but slight inlet of

Ionian Sea, 20 mi/32 km wide. Railroad terminus. Port of departure for KEFALLÍNIA. Remains of 13th-century French castle. Promontory here known as Glarentza or Klarentza (Italian *Chiarenza*). Sometimes spelled Kyllini or Kyllene.

Killini Mountains (kee-LEE-nee), Latin *Cyllene*, in KORINTHIA prefecture, N PELOPONNESE department, S mainland GREECE; 15 mi/24 km long; rise to 7,792 ft/2,375 m; 27 mi/43 km NW of ARGOS; 37°55′N 22°26′E. Formerly called Zeria or Ziria. Formerly spelled Kyllene.

Killiniq, former trading post, W Killiniq Island, NUNAVUT territory, CANADA, on UNGAVA BAY; 60°25′N 64°49′W; Royal Canadian Mounted Police post. Formerly PORT BURWELL.

Killinkoski (KIL-lin-KOS-kee), village, HÄMEEN province, W FINLAND, 60 mi/97 km N of TAMPERE; 62°24′N 23°52′E. Elevation 462 ft/140 m. In lake region; lumber and pulp mills.

Killisnoo, Alaska: see HOOD BAY.

Killough (kil-LAHK), fishing village and resort (2001 population 845), SE DOWN, SE Northern Ireland, on Killough Bay, small inlet of the IRISH SEA, 6 mi/9.7 km SSE of DOWNPATRICK; 54°15′N 05°38′W.

Kill Van Kull (KIL-van-KUHL), channel, 4 mi/6.4 km long and 0.5 mi/0.8 km wide, connecting Upper NEW YORK BAY with NEWARK BAY, between BAYONNE, NEW JERSEY, and STATEN ISLAND, NEW YORK. It is the main route for ships docking at the busy Port Newark-Elizabeth Marine Terminal in New Jersey. Bayonne Bridge (1931; 1,652 ft/504 m long), the second-longest steel-arch bridge in the United States, spans the channel.

Killybegs (ki-lee-BEGZ), Gaelic *Na Cealla Beaga*, town (2006 population 1,280), S DONEGAL county, N IRELAND, on inlet of DONEGAL BAY, 15 mi/24 km W of DONEGAL; 54°38′N 08°27′W. Fishing port; carpet making. Seat of the MacSweeneys.

Killyleagh (ki-lee-LAI), Gaelic *Cill Ó Laoch*, town (2001 population 2,483), E DOWN, SE Northern Ireland, on W shore of STRANGFORD LOUGH, 6 mi/9.7 km NNE of DOWNPATRICK; 54°24′N 05°43′W. Fishing port. Modern castle of the Hamilton Rowans includes remains of ancient structure.

Kilmacolm (KIL-mah-kolm), town (2001 population 4,000), Inverclyde, W Scotland, 10 mi/16 km W of RENFREW; 55°53′N 04°37′W. Commuter town to Glasgow.

Kilmainham, IRELAND: see DUBLIN, city.

Kilmallock (kil-MAH-luhk), Gaelic *Cill Mocheallóg*, town (2006 population 1,443), S LIMERICK county, SW IRELAND, 18 mi/29 km S of LIMERICK; 52°24′N 08°35′W. Agricultural market. Two gates remain of ancient fortifications razed by Cromwell. Remains of old church incorporate a round tower. Town was once of great importance as the seat of the Desmond Fitzgeralds. Remains of 13th-century Dominican abbey nearby.

Kilmanstahl, UKRAINE: see PYS'MENNE.

Kil'manstal', UKRAINE: see PYS'MENNE.

Kilmarnock (kil-MAHR-nahk), town (2001 population 43,588), ⊙ East Ayrshire, S Scotland, 7 m/11.2 km E of IRVINE; 55°38′N 04°30′W. Light manufacturing; whiskey. Previously a major producer of carpets and shoes. Robert Burns's first poems were published here; the Burns Monument has a museum. To S is Riccarton; to ESE across IRVINE RIVER is HURLFORD.

Kilmarnock (kil-MAHR-nuhk), town (2006 population 1,201), LANCASTER and NORTHUMBERLAND counties, E VIRGINIA, 18 mi/29 km NNE of GLOUCESTER; Fleets Bay, arm of CHESAPEAKE BAY, to SE; 37°42′N 76°22′W. Manufacturing (printing and publishing, crabmeat processing); fish, oysters, crabs. Christ Church (1732) to SW.

Kilmaurs (kil-MORS), village (2001 population 2,601), East Ayrshire, S Scotland, 2 mi/3.2 km NW of KILMARNOCK; 55°39′N 04°32′W. Nearby is ancient Rowallan Castle, newest parts of which date from c.1560.

Kilmer, Camp, NEW JERSEY: see NEW BRUNSWICK.

Kil'mez' (keel-MYES), town (2005 population 5,820), SE KIROV oblast, E central European Russia, on the KIL'MEZ' RIVER (tributary of the VYATKA River), on road, 50 mi/80 km N of VYATSKIYE POLYANY; 56°56′N 51°04′E. Oil and gas processing; seed inspection, livestock veterinary station; logging; lumbering, woodworking; food processing (dairy). Has a hotel.

Kil'mez' River (keel-MYEZ), E central European Russia; rises 15 mi/24 km W of SERGIYEVSKIY (UDMURT REPUBLIC); flows 118 mi/190 km SW and W, past KIL'MEZ' (KIROV oblast), to the VYATKA River opposite SHURMA. Navigable for 75 mi/121 km; timber floating. Receives Loban' (right) and Vala (left) rivers.

Kilmichael (KIL-mei-kuhl), town (2000 population 830), MONTGOMERY county, central MISSISSIPPI, 10 mi/16 km ESE of WINONA; 33°26′N 89°34′W. Agriculture (cotton, corn, soybeans; cattle). Manufacturing (men's and women's slacks).

Kilmorack (kil-MOR-ak), village, central Highland, N Scotland, on BEAULY RIVER (Falls of Kilmorack) and 10 mi/16 km W of INVERNESS; 57°28′N 04°24′W. Nearby Beaufort Castle is seat of earls of Lovat, chiefs of Clan Fraser.

Kilmore (kil-MOR), town, S central VICTORIA, SE AUSTRALIA, 36 mi/58 km N of MELBOURNE, of which it is a satellite; 37°18′S 144°57′E. Railroad junction in livestock area; dairy plants. Horse races.

Kilmuir, Scotland: see SKYE.

Kiln, unincorporated town (2000 population 2,040), HANCOCK county, SE MISSISSIPPI, 9 mi/14.5 km NW of BAY SAINT LOUIS, near Jordan River; 30°25′N 89°25′W. Cotton, corn. Manufacturing (signs, wooden reels). Stennis Space Center (NASA) to W.

Kilo-État (kee-LO–ai-TAH), village, ORIENTALE province, NE CONGO, 28 mi/45 km NE of IRUMU; 01°50′N 30°08′E. Elev. 4,904 ft/1,494 m. Gold mining and trading center, with Roman Catholic mission, school for teachers. Lost its former importance when the seat of gold operations transferred to KILO-MINES (1919). Sometimes called Vieux Kilo.

Kilofoss, NORWAY: see NÆRØYFJORDEN.

Kilómetro 100, ARGENTINA: see CASTELLI, Chaco province.

Kilómetro 101, ARGENTINA: see CONSCRIPTO BERNARDI.

Kilómetro 1082, ARGENTINA: see JOAQUÍN V. GONZÁLEZ.

Kilómetro 1172, ARGENTINA: see CAMPO QUIJANO.

Kilómetro 1308, ARGENTINA: see MUÑANO.

Kilómetro 511, ARGENTINA: see LLAJTA MAUCA.

Kilómetro 642 (kee-LO-me-tro), village, ⊙ Patiño department, central FORMOSA province, ARGENTINA, on BERMEJO RIVER and 190 mi/306 km NW of FORMOSA; 24°57′S 60°57′W. Livestock.

Kilómetro 924, ARGENTINA: see ANGACO NORTE, town.

Kilo-Mines (kee-LO), village, ORIENTALE province, NE CONGO, near right bank of SHARI RIVER and 35 mi/56 km NE of IRUMU; 01°48′N 30°14′E. Elev. 4,812 ft/1,466 m. Center of major gold-mining and processing area employing c.13,000 native workers, with main gold fields at MONGBWALU, NIZI, TSI, ISURU, and KANGA. Power for smelting is supplied by three large hydroelectric plants on SHARI RIVER, notably at SOLENIAMA, 8 mi/12.9 km SSE. Kilo-Mines is also a commercial center and has several missions with educational institutions and hospitals. Gold was first discovered in area in 1895; mining started in 1905. Seat of gold operations was transferred to Kilo-Mines from KILO-ÉTAT in 1919. Local activities are coordinated with those at WATSA into a single concern called KILO-MOTO. Sometimes called BAMBU.

Kilo-Moto, CONGO: see KILO-MINES.

Kilosa (kee-LO-sah), town, MOROGORO region, E central TANZANIA, 43 mi/69 km W of MOROGORO, on Mkondoa River, at W edge of Makata Plain; 06°50′S 37°01′E. Railroad junction. MIKUMI NATIONAL PARK to S;

Area is shown by the symbol □, and capital city or county seat by ⊙.

Tendigo Swamp to E. Cotton, sisal, grain; cattle, sheep, goats. An Arab settlement in mid-19th century.

Kilrenny (kil-RE-nee), village, FIFE, E Scotland, on the Firth of Forth, near its mouth on the NORTH SEA, 9 mi/14.5 km SE of ST. ANDREWS; 56°15′N 02°41′W. With adjacent Anstruther Easter and Anstruther Wester (see ANSTRUTHER), it forms royal burgh. Fishing port and seaside resort.

Kilronan (kil-RO-nuhn), Gaelic *Cill Rónáin*, town (2006 population 164), SW GALWAY county, W IRELAND, on NE coast of INISHMORE (Aran Islands), and 28 mi/45 km WSW of GALWAY; 53°07′N 09°40′W. Fishing port and chief settlement. Numerous remains of ancient churches dating from early Christian era and a large prehistoric fort are nearby.

Kilrush (kil-RUHSH), Gaelic *Cill Rois*, town (2006 population 2,657), SW CLARE county, W IRELAND, on the SHANNON estuary, 26 mi/42 km SW of ENNIS; 52°38′N 09°29′W. Port and seaside resort, with dock installations; agricultural market. Slate and flagstone are quarried nearby. Hog Island and SCATTERY ISLAND offshore.

Kilsyth (KIL-seith), town (2001 population 9,816), North Lanarkshire, central Scotland, 7 mi/11.3 km SW of DENNY and on Forth and Clyde Canal; 55°58′N 04°03′W. Previously hosiery knitting. The ANTONINE WALL (named after Antoninus Pius) passes through town. Montrose and his Royalist forces defeated the Covenanters under Baillie in 1645.

Kilsyth (KIL-seith), suburb 20 mi/32 km E of MELBOURNE, VICTORIA, SE AUSTRALIA, between Mooroolbark and Boronia, and in Dandenong Ranges foothills; 37°48′S 145°19′E.

Kilsyth Hills, Scotland: see LENNOX HILLS.

Kiltamagh, IRELAND: see KILTIMAGH.

Kiltan Island (kil-TAHN), coral island of Amin Divi group of LAKSHADWEEP (LACCADIVE) Islands, LAKSHADWEEP Union Territory, INDIA, in ARABIAN SEA; 11°30′N 73°00′E. Coconuts; manufacturing of coir and copra. Lighthouse.

Kiltimagh (kil-chi-MAH) or **Kiltamagh**, Gaelic *Coillte Amach*, town (2006 population 1,096), E central MAYO county, NW IRELAND, 13 mi/21 km E of CASTLEBAR; 53°51′N 09°00′W. Agricultural market. It is after this place that men from Mayo are called "Culshies."

Kiltu Kara, town (2007 population 4,967), OROMIYA state, W ETHIOPIA; 09°43′N 35°13′E. 10 mi/16 km SE of MENDI. Iron deposits nearby.

Kilwa (KEEL-wah), village, KATANGA province, SE CONGO, on SW bank of LAKE MWERU, at ANGOLA border, 175 mi/282 km NNE of LUBUMBASHI; 09°18′S 28°25′E. Elev. 3,015 ft/918 m. Customs station, steamboat landing; fisheries. Franciscan mission.

Kilwa Kisiwani (KEEL-wah kee-see-WAH-nee), historic site, SE TANZANIA, on small Kilwa Island, in Sanguru Haven, inlet of INDIAN OCEAN, immediately S of KILWA MASOKO; 08°59′S 39°31′E. Founded in 10th century by Persians, it is one of the oldest settlements in Tanzania; was a prosperous slave and ivory trading center. Also called Kilwa.

Kilwa Kivinje (KEEL-wah kee-VEEN-jai), town, LINDI region, SE TANZANIA, 15 mi/24 km NNW of KILWA MASOKO, on INDIAN OCEAN; 08°45′S 39°25′E. Cashews, peanuts, bananas; sisal, copra, timber. Goats; fish.

Kilwa Masoko (KEEL-wah mah-SO-ko), town, LINDI region, SE TANZANIA, 70 mi/113 km N of LINDI, on Sangurungu Haven (bay), INDIAN OCEAN; 08°58′S 39°32′E. Agricultural (cashews, bananas, copra, manioc, sweet potatoes); Fish; goats, sheep; timber. KILWA KISIWANI, historical ruins, on island to S.

Kilwinning (kil-WI-neeng), town (2001 population 15,908), North Ayrshire, SW Scotland, on Garnock River and 3 mi/4.8 km N of IRVINE; 55°39′N 04°43′W. Previously machinery mills, ironworks, and fireclay works. Included in area of Irvine New Town. Traditional birthplace of Scottish freemasonry. There are remains of abbey founded 1140, dedicated to St. Winnin, who is reputed to have lived here. Kilwinning, long celebrated for archery meets, was formerly scene of annual shooting match, described in Scott's *Old Mortality*.

Kim (keem), town, NE LENINOBOD viloyat, N TAJIKISTAN, 6 mi/9.7 km SSE of KANIBADAM; 40°12′N 70°28′E. Oil field (producing since 1913). Until 1929 called Santo (an acronym for the Central Asian Oil Trading Organization).

Kim, village (2000 population 65), LAS ANIMAS county, SE COLORADO, 65 mi/105 km E of TRINIDAD, near NEW MEXICO line; 37°15′N 103°20′W. Elevation 5,686 ft/1,733 m. Livestock, wheat, sorghum, hay. Parts of Comanche National Grassland to E and W.

Kima (KEE-mah), village, ORIENTALE province, E CONGO, 170 mi/274 km SSE of KISANGANI; 01°26′S 26°43′E. Elev. 2,290 ft/697 m. In tin-mining area; gold mining; also wolfram mining.

Kima (kee-MAH), village, NW AKMOLA region, KAZAKHSTAN, on ISHIM River, 35 mi/56 km SW of ATBASAR; 51°37′N 67°31′E. Cattle. Also Kiima, Kiyma.

Kima (KEE-mah), village, CENTRAL province, S central KENYA, on railroad and 60 mi/97 km SE of NAIROBI; 01°58′S 16°37′E. Sisal center; rubber, wheat, corn.

Kimaam (KEE-mah-ahm), town, IRIAN JAYA, INDONESIA, 132 mi/212 km NE of MERAUKE; 07°58′S 138°53′E.

Kimamba (kee-MAHM-bah), village, MOROGORO region, E central TANZANIA, 33 mi/53 km W of MOROGORO, in Makata Plain, on railroad; 06°48′S 37°06′E. Corn, wheat, sisal; sheep, goats; timber.

Kimba (KIM-buh), village, S SOUTH AUSTRALIA, on E central EYRE PENINSULA, 115 mi/185 km NNE of PORT LINCOLN; 33°09′S 136°26′E. On Port Lincoln-BUCKLEBOO railroad; wheat, wool. Kimba & Gawler Ranges Historical Society Museum. A range of flora and fauna at GAWLER RANGES, N of town.

Kimball, county (□ 952 sq mi/2,475.2 sq km; 2006 population 3,710), W NEBRASKA, in High Plains; ⊙ KIMBALL; 41°12′N 103°42′W. Borders COLORADO on S and WYOMING on W; drained by LODGEPOLE CREEK. Agriculture (cattle, hogs, wheat, potatoes, sunflower seed). Highest point (unnamed) in Nebraska, in SW corner 5,426 ft/1,654 m. Oliver Reservoir State Recreation Area near county center, W of Kimball. Formed 1888.

Kimball or **Kimball Prairie**, city (2006 population 2,299), ⊙ KIMBALL county, W NEBRASKA, 42 mi/68 km S of SCOTTSBLUFF, and on LODGEPOLE CREEK; 41°13′N 103°39′W. Trade and grain-shipping center in GREAT PLAINS region; livestock, grain, beans, sunflower seed, potatoes, dried fruit and nuts. Manufacturing (oil-field machinery and buildings, pipe threading, polyethylene pipe, burglar switches, printing, hay equipment, computer keyboards). Oliver Reservoir and Lodgepark State Wayside Area to W; Pawnee National Grassland to S (in Colorado). Incorporated 1885.

Kimball (KIM-buhl), village, STEARNS county, S central MINNESOTA, 19 mi/31 km SSW of ST. CLOUD, near CLEARWATER RIVER. Grain; livestock, poultry; dairying; manufacturing (gravel-truck bodies and hoists, concrete products, sausages); 45°18′N 94°18′W. Powder Ridge Ski Area to NW. Also called Kimball Prairie.

Kimball, village (2006 population 686), BRULE county, S central SOUTH DAKOTA, 20 mi/32 km ESE of CHAMBERLAIN; 43°45′N 98°57′W.

Kimball, village (2006 population 356), MCDOWELL county, S WEST VIRGINIA, on TUG FORK RIVER, 4 mi/6.4 km E of WELCH; 37°25′N 81°30′W. Semibituminous-coal mining. Agriculture (fruit, tobacco); livestock. Manufacturing (concrete). State mine rescue station here. Incorporated 1911.

Kimball, Mount (c.9,500 ft/2,896 m), E ALASKA, in ALASKA RANGE, 40 mi/64 km W of TANACROSS; 63°15′N 144°41′W.

Kimballton, town (2000 population 342), AUDUBON county, W central IOWA, 10 mi/16 km SW of AUDUBON; 41°37′N 95°04′W.

Kimbambili (keem-bahm-BEE-lee), village, BANDUNDU province, W CONGO, 10 mi/16 km NNW of BANDUNDU; 03°10′S 17°15′E. Subsistence farming; fishing.

Kimbe, town (2000 population 14,656), WEST NEW BRITAIN province, N central NEW BRITAIN island, E PAPUA NEW GUINEA, 25 mi/40 km W of HOSKINS. Located on ASTROLABE (Kimbe) BAY, BISMARCK SEA. Road access; airstrip at Hoskins. Commercial center for region. Cocoa, copra, coconuts, palm oil; tuna, mackerel, prawns. Timber; tourism.

Kimberley, city, ⊙ NORTHERN CAPE province, SOUTH AFRICA; 28°44′S 24°44′E. Elevation 4,002 ft/1,220 m. The city is primarily a diamond-mining center, although textiles, construction materials, and machinery are manufactured. The city is also an important railroad junction and has an airport. Kimberley was founded in 1871 when diamonds were discovered on a nearby farm. The De Beers Consolidated Mines, organized by Cecil Rhodes, assumed control of the diamond fields in 1888. In 1899–1900, during the South African War, the city was besieged by Boer (Afrikaner) forces. Became a city in 1913. N Cape Technical College and the Alexander McGregor Memorial Museum are in Kimberley. This city grew up around a Kimberlite volcanic pipe which was mined out to depth of 2,624 ft/800 m between 1871 and 1914 to form the Big Hole.

Kimberley (KIM-buhr-lee), town (□ 22 sq mi/57.2 sq km; 2001 population 6,484), SE BRITISH COLUMBIA, W CANADA, in EAST KOOTENAY regional district; 49°41′N 115°59′W. One of Canada's highest cities (3,660 ft/1,115 m), it is the site of the now-closed Sullivan mine, where large quantities of silver, lead, and zinc were mined, 1909–2001. Incorporated in 1968.

Kimberley (KIM-buhr-lee), geographical area (□ 162,723 sq mi/421,451 sq km), WESTERN AUSTRALIA, NW AUSTRALIA, and on S margins of SOUTHEAST ASIA; 16°00′S 126°00′E. Bounded by TIMOR SEA (N), INDIAN OCEAN (W), GREAT SANDY DESERT (S), Tanami Desert (SE), and the NORTHERN TERRITORY (E). The KIMBERLEY GOLDFIELD was the site (1882) of the first major Western Australian gold strike. The Kimberley Diamond Mine is one of the largest producers in the world. Cattle and sheep raising, pearling, fishing, retail trade, and tourism are also important industries. The region contains 8 national parks and 1 marine park.

Kimberley Goldfield (KIM-buhr-lee GOLD-feeld) (□ 47,000 sq mi/122,200 sq km), NE WESTERN AUSTRALIA. Mining center is HALLS CREEK. Gold discovered here 1882; area placed (1886) under government control and leased to mining interests.

Kimberly, town (2000 population 2,614), TWIN FALLS county, S IDAHO, 5 mi/8 km ESE of TWIN FALLS; 42°32′N 114°22′W. Livestock; wheat, beans, potatoes; manufacturing (fertilizers).

Kimberly, town (2006 population 6,398), OUTAGAMIE county, E WISCONSIN, on FOX RIVER, suburb 4 mi/6 km E of APPLETON; 44°16′N 88°20′W. Manufacturing (paper milling, coated paper, paper machine components, gumball vending machines). Incorporated 1910.

Kimberly, unincorporated village, FAYETTE county, S central WEST VIRGINIA, near KANAWHA RIVER, 24 mi/39 km SE of CHARLESTON; 38°08′N 81°18′W. Coal-mining region. Agriculture (grain); livestock. Manufacturing (machining).

Kimbiji (keem-BEE-jee), village, PWANI region, E TANZANIA, 20 mi/32 km ESE of DAR ES SALAAM, on INDIAN OCEAN, at Ras (Cape) Kimbiji; 07°00′S 39°30′E. Lighthouse. Cashews, corn, wheat, bananas, copra, manioc, sweet potatoes. Goats, sheep. Fish; timber.

Kimble (KIM-buhl), county (□ 1,251 sq mi/3,252.6 sq km; 2006 population 4,570), W central TEXAS; ⊙ JUNCTION; 30°29′N 99°45′W. In scenic EDWARDS PLATEAU. Elevation 1,800 ft/549 m–2,400 ft/732 m. Drained by N

Llano and S Llano rivers, which join to form LLANO RIVER at Junction, in center of county. Ranching (especially goats; also sheep, cattle); a leading U.S. county in wool, mohair production. Sorghum, pecan growing. Sand and gravel; oil and gas. Scenery, hunting, fishing attract tourist trade. Formed 1858.

Kimbolton, township, MANAWATU district (□ 1,013 sq mi/2,633.8 sq km), NORTH ISLAND, NEW ZEALAND, c.16 mi/25 km N of PALMERSTON NORTH. Agricultural center, sheep, dairy.

Kimbolton (kim-BOL-tuhn), village (2001 population 1,432), CAMBRIDGESHIRE, E central ENGLAND, 9 mi/14.5 km NW of ST. NEOTS; 52°18′N 00°24′W. Reservoir at Grafham Water 2 mi/3.2 km to W. Kimbolton Castle (16th century), formerly home of Catherine of Aragon, now a school.

Kimbolton (KIM-buhl-tuhn), unincorporated village (2000 population 190), GUERNSEY county, E OHIO, 9 mi/14 km N of CAMBRIDGE; 40°09′N 81°34′W.

Kimchaek (GEEM-CHAIK), city, NORTH HAMGYONG province, NORTH KOREA, port at mouth of small Susong River, on SEA OF JAPAN, 80 mi/129 km SSW of CHONGJIN; 40°45′N 129°10′E. Industrial center (pig iron, steel, and magnesium). Paper milling. Port was opened in 1899 to foreign trade; exports fish, livestock, graphite, and soybeans. Previously called Songjin.

Kimchon (GEEM-CHUHN), city, N. Kyongsangprovince, S. Korea, 40 mi/64 km NW of Taegu; 36°07′N 128°06′E. Railroad junction; on the Kyongbu Expressway; commercial center for agr. products (rice, soy beans, barley, cotton); paper, raw silk.

Kimhae, city, SE SOUTH KYONGSANG province, SOUTH KOREA, on W bank of Nakgong River N of KOREA STRAIT; 35°14′N 128°52′E. Mountainous in N and W; wide field along NAKDONG RIVER in S. Heavy agriculture (rice, fruits, vegetables, husbandry) for urban markets in PUSAN; manufacturing (food processing, textiles, chemistry; machinery); fishery (sole, trout, seaweed). Kimhae International Airport in suburbs; NAMHAE Expressway. Prehistoric remains (Kimhae Shell Mound); Silla Dynasty tombs.

Kimi (KEE-mee), Latin *Cyme*, town, on E coast of ÉVVIA island, ÉVVIA prefecture, CENTRAL GREECE department, port on AEGEAN SEA, 29 mi/47 km ENE of KHALKÍS. Port of departure for Sporadic Islands (Sprades). Figs, walnuts; wines. Lignite deposits (S) and near Andronianoi, 3 mi/4.8 km WNW. Cape Kyme is 3 mi/4.8 km ENE at 38°38′N 24°10′E. Also spelled Kymi and Kyme; formerly Koumi or Kumi.

Kimilili (kee-mee-LEE-lee), town (1999 population 10,261), Bungoma district, WESTERN province, KENYA, 31 mi/50 km SSE of Mount ELGON; 00°47′N 34°42′E. Market and trading center.

Kimina (KEE-mee-nah), town, THESSALONÍKI prefecture, CENTRAL MACEDONIA department, NE Greece, W of THESSALONÍKI, in the Axios (VARDAR) River delta; 40°37′N 22°42′E. Oil refinery. Formerly Divala or Diavata.

Kimisi (kee-MEE-see), village, KAGERA region, NW TANZANIA, 65 mi/105 km SW of BUKOBA; Lake BURIGI to E, Burigi Game Reserve to S; 02°08′S 31°07′E. Timber; tobacco, corn, wheat; goats, sheep.

Kimita (KEE-mee-tah), village, Futami county, HIROSHIMA prefecture, SW HONSHU, W JAPAN, 40 mi/65 km N of HIROSHIMA; 34°52′N 132°51′E.

Kimitsu (KEE-meets), city, CHIBA prefecture, E central HONSHU, E central JAPAN, on W BOSO PENINSULA, on TOKYO BAY, 19 mi/30 km S of CHIBA; 35°19′N 139°54′E. Iron, steel products, toothpicks. Chicken.

Kimje (KEEM-JEI), city, W NORTH CHOLLA province, SOUTH KOREA, in center of Kimje county, and surrounded by its neighborhoods; 35°48′N 126°53′E. Rice farming with highly developed irrigation system main economic activity. Other agricultural products include rye, mushrooms, vegetables. Sericulture. Orchards. Kimje (1404), a local school with Confucian shrine; Honam railroad.

Kimmswick, town (2000 population 94), JEFFERSON county, E MISSOURI, on MISSISSIPPI RIVER, S of mouth of the MERAMEC, residential suburb 18 mi/29 km S of ST. LOUIS; 38°22′N 90°22′W. Nineteenth-century summer resort. Historical town; tourism. Mastodon State Park nearby.

Kimobetsu (kee-MO-bets), town, Shiribeshi district, HOKKAIDO prefecture, N JAPAN, 28 mi/45 km S of SAPPORO; 42°47′N 140°56′E. Asparagus canning.

Kimolos (KEE-mo-los), Latin *Cimolus*, island (□ 16 sq mi/41.6 sq km), CYCLADES prefecture, SOUTH AEGEAN department, GREECE, just NE of MÍLOS island; 5 mi/8 km long, 4 mi/6.4 km wide; 36°48′N 24°34′E. Produces wine, olive oil, wheat; chalk and iron ore mining; fisheries. Main town, Kimolos, is on SE shore. Formerly called Argentiera.

Kimovsk (KEE-muhfsk), city (2006 population 31,555), E TULA oblast, central European Russia, in the coal-producing MOSCOW basin, on the DON River, on highway junction and railroad, 48 mi/77 km SE of TULA, and 12 mi/19 km SE of NOVOMOSKOVSK; 53°58′N 38°32′E. Elevation 715 ft/217 m. Manufacturing (radio equipment, garments), food processing (bakery). Lignite mining, once the chief industry, lost its importance with the shutdown of many mines following the post-Soviet switch to market economy. Until 1948, called Mikhaylovka or Bol'shaya Mikhaylovka. Made city in 1952.

Kimpako (keem-PAH-ko), village, BAS-CONGO province, W CONGO, 40 mi/64 km SE of KINSHASA; 05°01′S 15°16′E. Elev. 1,961 ft/597 m.

Kimpangu (keem-PAHNG-goo), village, BAS-CONGO province, W CONGO, on Angola boundary, 125 mi/201 km E of BOMA; 05°51′S 15°01′E. Elev. 2,132 ft/649 m. Customs station and trading center. Roman Catholic mission.

Kimparana, town, FOURTH REGION/SÉGOU, MALI, 89 mi/149 km SE of SÉGOU; 12°50′N 04°56′W. Cotton; crossroads. Also spelled Kemparana.

Kimpersaiski, KAZAKHSTAN: see BADAMSHA.

Kimpese (keem-PE-sai), village, BAS-CONGO province, W CONGO, on railroad and 90 mi/145 km N of BOMA; 05°33′S 14°26′E. Elev. 990 ft/301 m. Agricultural center for food staples. Has Roman Catholic and Baptist missions, mission schools, hospital.

Kimpo (KEEM-PO), county (□ 123 sq mi/319.8 sq km), northwesternmost KYONGGI province, SOUTH KOREA, S and W of PAJU and KOYANG city across the HAN RIVER, contiguous to SEOUL, PUCHON, INCHON (SE), and KANGHWA (W); 37°38′N 126°42′E. Located in delta on lower Hang River (Kimpo Peninsula). Relatively flat despite three mountains within its boundaries. Agriculture (high-quality rice, ginseng, staples).

Kimry (KEE-mri), city (2006 population 51,555), E TVER oblast, W central European Russia, port on the VOLGA RIVER, on road and railroad, 82 mi/132 km E of TVER; 56°52′N 37°21′E. Elevation 321 ft/97 m. Important center of the shoe industry since the 17th century; produces shoe-making machinery, thermal equipment, building materials, furniture.

Kimvula (keem-VOO-luh), village, BAS-CONGO province, W CONGO, 120 mi/193 km SSE of KINSHASA; 05°44′S 15°58′E. Elev. 2,257 ft/687 m.

Kin (KEEN), town, central OKINAWA island, Kunigami county, Okinawa prefecture, SW JAPAN, 22 mi/35 km N of NAHA; 26°27′N 127°55′E.

Kinabalu (kin-ah-BAH-loo) or **Kinibalu. Mount**, peak (13,455 ft/4,101 m), N SABAH state, MALAYSIA, NE of KOTA KINABALU; 06°05′N 116°36′E. Highest peak on BORNEO and in SE ASIA.

Kinabalu National Park (KIN-ah-BAH-loo) (□ 296 sq mi/769.6 sq km), N SABAH, E MALAYSIA, 40 mi/64 km ENE of KOTA KINABALU; 06°05′N 116°33′E. Includes slopes and summit of Mount Kinabalu (13,455 ft/4101 m), highest mountain in SE ASIA. Large variety of bird life, unique flora. Hot spring. Two-day mountain climb. Park lodge. Road access.

Kinabatangan River (KIN-ah-BAH-tahng-ahn), largest river of BORNEO, in SABAH, MALAYSIA; rises in central highlands of EAST COAST residency; flows c.350 mi/563 km NE to SULU 30 mi/48 km ESE of SANDAKAN. Navigable for 75 mi/121 km by small craft. Has wide delta.

Kinaion, Cape (kee-NE-on), NW extremity of ÉVVIA island, ÉVVIA prefecture, CENTRAL GREECE department, between N Gulf of ÉVVIA (E) and MALIAKOS GULF (W). Formerly called Cape Lithada. Also Cape Kynaion.

Kinak (KEE-nak), Inuit village, W ALASKA, on KUSKOKWIM River, near its mouth on KUSKOKWIM BAY, 40 mi/64 km SW of BETHEL.

Kinango (kee-NAHN-goh), town, COAST province, SE KENYA, on road and 26 mi/42 km WSW of MOMBASA; 04°07′S 39°18′E. Sugarcane, fruits.

Kinapusan Islands (kee-nah-POO-sahn), island group (□ 0.7 sq mi/1.8 sq km), in TAWITAWI group, TAWITAWI province, PHILIPPINES, in SULU ARCHIPELAGO, 23 mi/37 km E of Tawitawi Island; 05°13′N 120°38′E. Included in South Ubian municipality.

Kinards (KEI-nuhrds), unincorporated town, NEWBERRY and LAURENS counties, NW central SOUTH CAROLINA, 11 mi/18 km NW of NEWBERRY. Manufacturing of wood chips, timber. Agriculture includes livestock, poultry, grain.

Kinaros (kee-NAH-ros), Italian *Chinaro*, island (□ 2 sq mi/5.2 sq km), DODECANESE prefecture, SOUTH AEGEAN department, GREECE, NE of AMORGOS; 36°59′N 26°17′E. Considered part of Greece during Italian rule of the Dodecanese. Also Kynaros.

Kinasa (kee-NAH-sah), village, Kamiminochi county, NAGANO prefecture, central HONSHU, central JAPAN, 9 mi/15 km W of NAGANO; 36°40′N 138°00′E.

Kinbrae (KIN-buh-ree), village (2000 population 21), NOBLES county, SW MINNESOTA, 16 mi/26 km NNE of WORTHINGTON; 43°49′N 95°28′W. Grain; livestock; dairying. Small natural lakes in area.

Kinburn Kosa (keen-BOORN), W sandspit (Ukrainian and Russian *kosa*), extremity of peninsula in S UKRAINE, jutting W into the BLACK SEA and forming S shore of the DNIEPER LIMAN; 6.2 mi/10 km long, 300 ft/100 m—5,000 ft/1,500 m wide. Westernmost tip (opposite OCHAKIV) was the site of an 18th-century Kinburn fortress.

Kincaid (kin-KAID), village (2006 population 135), SW SASKATCHEWAN, CANADA, 45 mi/72 km W of ASSINIBOIA; 49°40′N 107°00′W. Wheat.

Kincaid, village (2000 population 1,441), CHRISTIAN county, central ILLINOIS, on South Fork of SANGAMON RIVER (bridged here) and 19 mi/31 km SE of SPRINGFIELD; 39°35′N 89°24′W. In agricultural and bituminous-coal area. Incorporated 1915. Near Sangchris Lake State Park.

Kincaid (kin-KAID), village (2000 population 178), ANDERSON county, E KANSAS, 14 mi/23 km S of GARNETT; 38°04′N 95°09′W. In livestock, grain, and dairy region.

Kincaid, unincorporated village, FAYETTE county, S central WEST VIRGINIA, 9 mi/14.5 km W of FAYETTEVILLE; 38°02′N 81°16′W. Agriculture (grain); livestock.

Kincardine (kin-KAHR-din), township (□ 223 sq mi/579.8 sq km; 2001 population 11,029), SW ONTARIO, E central CANADA, on Lake HURON, W of WALKERTON; 44°10′N 81°38′W. Resort that depends largely on jobs provided by the Bruce Nuclear Power Development to the N. Includes TIVERTON, UNDERWOOD, INVERHURON, ARMOW, GLAMMIS, BERVIE, and MILLARTON.

Kincardine and Deeside (kin-KAHR-deen and DEE-seid), former administrative district, Aberdeenshire, NE Scotland. Agriculture, fishing, food processing, paper, whiskey, and oil.

Kincardine-on-Forth (kin-KAHR-deen–ahn–FORTH), town (2001 population 3,035), FIFE, E Scotland, on the Firth of Forth, 5 mi/8 km SE of ALLOA; 56°04′N

03°44′W. Small port. Power station (ceased operation in 2001) was downstream. Former coal-mining area. Nearby are remains of 15th-century Tulliallan Castle.

Kinchafoonee River (kinch-uh-FOO-nee), SW GEORGIA; rises near BUENA VISTA; flows S and SE c.75 mi/ 121 km, past PRESTON, to FLINT RIVER at ALBANY.

Kinchega National Park (KIN-chuh-guh), (□ 171 sq mi/444.6 sq km), W NEW SOUTH WALES, SE AUSTRALIA, 610 mi/982 km W of SYDNEY, 67 mi/108 km N of BROKEN HILL; 20 mi/32 km long, 18 mi/29 km wide; 32°30′S 142°20′E. Red sand dunes and black soil flats; complex system of saucer-shaped overflow lakes of DARLING RIVER. Prehistoric Aboriginal campsites. Red and gray kangaroos, euros; snakes, blue-tongue lizards; extensive waterfowl. Camping, picnicking, hiking, boating, swimming. Established 1967.

Kinchil (KEEN-cheel), town, YUCATÁN, SE MEXICO, 22 mi/35 km W of MÉRIDA; 20°55′N 89°57′W. Tropical fruit, corn, henequen.

Kinchow, China: see JINZHOU.

Kincumber (kin-KUHM-buhr), settlement, NEW SOUTH WALES, SE AUSTRALIA, 57 mi/91 km N of SYDNEY, and at S end of Central Coast; 33°29′S 151°25′E. Resort region.

Kindaba, town, POOL region, S Congo Republic, 60 mi/ 97 km NW of BRAZZAVILLE. Tobacco. Airfield.

Kinda Dam (kin-DAH), YENGAN township, SHAN STATE, MYANMAR. Hydroelectric station and focus of watershed management area.

Kindat (kin-DAHT), village, MAWLAIK township, SAGAING division, MYANMAR, on E bank of CHINDWIN RIVER, and 6 mi/9.7 km N of MAWLAIK. Low-water head of navigation.

Kindberg (KINDT-berg), town STYRIA, E central AUSTRIA, on Mürz River and 11 mi/18 km NE of BRUCK AN DER MUR. Steelworks; 47°30′N 15°27′E.

Kinde (KEIND), village (2000 population 534), HURON county, E MICHIGAN, 10 mi/16 km N of BAD AXE; 43°56′N 82°59′W. In farm area.

Kinder (KIN-duhr), town (2000 population 2,148), ALLEN parish, SW LOUISIANA, near CALCASIEU RIVER, 28 mi/45 km NE of LAKE CHARLES city; 30°30′N 92°51′W. Railroad junction; in rice, dairying, and cattle-raising area. Agriculture (peaches, soybeans, pecans). Manufacturing (life ring buoys, lumber); oilfield nearby. Large casino complex (Grand Casino Coushatta).

Kinderdijk (KIN-duhr-DEIK), town, SOUTH HOLLAND province, W NETHERLANDS, on NOORD RIVER and 5 mi/8 km NNW of DORDRECHT; 51°53′N 04°39′E. At mouth of LEK RIVER, which joins with Noord River to form NEW MAAS RIVER (ROTTERDAM harbor). Dairying; cattle, hogs, poultry. Vegetables, fruit, sugar beets. Formerly spelled Kinderdyk.

Kinderhook (KIN-der-huk), village (2000 population 249), PIKE county, W ILLINOIS, 20 mi/32 km WNW of PITTSFIELD; 39°42′N 91°09′W. In agricultural area.

Kinderhook, village, BRANCH county, S MICHIGAN, 10 mi/16 km. S of COLDWATER, near INDIANA line, in lake area; 41°47′N 85°00′W. Large Coldwater Lake to NE.

Kinderhook, village (□ 1 sq mi/2.6 sq km; 2006 population 1,330), COLUMBIA county, SE NEW YORK; 42°23′N 73°42′W. Some manufacturing. Settled before the American Revolution. Richard Upjohn designed St. Paul's Church (1851) here. President Martin Van Buren was born and is buried in Kinderhook; the Van Buren homestead, "Lindenwald," is S of the village. The House of History, maintained by the county historical society, occupies an early-19th-cent. mansion. Incorporated 1838.

Kinderhook Creek, c.45 mi/72 km long, SE NEW YORK; rises in S RENSSELAER county in the TACONIC MOUNTAINS; flows generally SW, past VALATIE and KINDERHOOK, to the HUDSON RIVER 4 mi/6.4 km N of HUDSON.

Kinder Scout, ENGLAND: see PEAK DISTRICT.

Kindersley (KIN-duhrz-lee), town (2006 population 4,412), W SASKATCHEWAN, CANADA, 100 mi/161 km NW of SWIFT CURRENT; 51°28′N 109°08′W. Grain elevators, lumbering, livestock raising.

Kindia, administrative region, WSW GUINEA; ⊙ KINDIA. Bordered W and N by Boké administrative region (part of S portion of border formed by KONKOURÉ RIVER), NE tip by Labé administrative region, E by Mamou administrative region (N portion of border formed by Kakrima River), SES by SIERRA LEONE (central part of border formed by KOLENTÉ RIVER [called Great Scarcies River in Sierra Leone]), and SW by CONAKRY and the ATLANTIC OCEAN. Fatala River in N central, Konkouré River runs E-W to the Atlantic Ocean through central part of the region, Badi River in central, and Kolenté River in S. Sangaréa Bay on N coast. Marsh area along much of the coast. Towns include BENTY, COYAH, DUBRÉKA, FORÉCARIAH, FRIA, and TÉLIMÉLÉ. Main railroad runs roughly W-E through region, originating at Conakry and going through Kindia, KOLENTÉ, Souguéta, and LINSAN towns before entering Mamou administrative region (terminates at KANKAN city in E Guinea); secondary railroad branches off E of Conakry and S of Dubréka town, going N through Dubréka town and terminating at Fria town. Main road runs roughly parallel to the main railroad, passing through Coyah, Kindia, Kolenté, and Souguéta towns before entering Mamou administrative region (the road runs parallel to the railroad to Kankan city, with part branching off near KOUROUSSA town and going into MALI). Two roads branch off this main road—one runs NW through Dubréka and WASSOU towns into Boké administrative region (terminates at BOKÉ town) and another runs S from Coyah town into Sierra Leone. Airport at Fria town. Includes the prefectures of Télimélé in N, Kindia in E, Forécariah in S, Coyah in SW, Dubréka in W, and Fria in NW. All of Kindia administrative region is in Guinée-Maritime geographic region.

Kindia, prefecture, Kindia administrative region, SW GUINEA, in Guinée-Maritime geographic region; ⊙ KINDIA. Bordered NW by Télimélé prefecture, N by Pita prefecture, NE tip by Dalaba prefecture, E by Mamou prefecture, SE by SIERRA LEONE, S by Forécariah prefecture, SW by Coyah prefecture, and W by Dubréka prefecture. Kolenté River (called GREAT SCARCIES RIVER in Sierra Leone) originates here and flows SW. Summit (elevation 3,688 ft/1,124 m) in S of prefecture. Towns include Kindia, KOLENTÉ, LINSAN, Madina Woula, and Souguéta. The major railroad between CONAKRY (SW Guinea) and KANKAN town (E Guinea) travels SW-NE through center of the prefecture, running through Kindia, Kolenté, Souguéta, and Linsan towns. Multiple roads branch out of Kindia town connecting it to the rest of the prefecture and administrative region, as well as W Guinea (including Conakry), NW Guinea, and E Guinea (including MAMOU, DABOLA, and Kankan towns), and Sierra Leone.

Kindia (KIN-dyah), town, ⊙ Kindia prefecture and administrative region and Guinée-Maritime geographic region, SW GUINEA; 10°00′N 52°12′W. A railroad and road hub, Kindia is the administrative and trade center for an area where bananas, manioc, rice, fruits, vegetables are grown, and bauxite is mined. Wood is processed for use in furniture factories outside CONAKRY. Kindia has a fruit research center, a medical center, and a school of agriculture.

Kindongo (keen-DON-go), village, BANDUNDU province, W CONGO, 20 mi/32 km S of BANDUNDU; 03°54′S 17°32′E. Elev. 1,204 ft/366 m. Subsistence farming.

Kindred, village (2006 population 579), CASS CO., SE NORTH DAKOTA, 19 mi/31 km SSW of FARGO, near SHEYENNE RIVER; 46°38′N 97°01′W. Grain, livestock, dairy. Founded in 1880 and incorporated in 1920. Named for William S. Kindred, a Fargo realtor.

Kindu (KEEN-doo), town, ⊙ MANIEMA province, central CONGO, on both banks of the LUALABA and 115 mi/185 km NNW of KASONGO; 02°57′S 25°57′E. Elev. 1,656 ft/ 504 m. River port and commercial center. Important trans-shipment point, terminus of railroad from KALEMIE and head of steam navigation to Ubundu. Shipyards, railroad workshops; rice and tin processing, pharmaceuticals manufacturing. Airport, catholic mission and mission schools, including business school; hospital Formerly known as Kindu-Port-Empain.

Kinel′ (kee-NYEL), city (2006 population 34,335), central SAMARA oblast, E European Russia, near the confluence of GREATER KINEL′ and SAMARA rivers, 25 mi/40 km E of SAMARA; 53°14′N 50°39′E. Elevation 141 ft/42 m. Center of agricultural area; railroad junction (repair shops); petroleum-extracting center; gypsum quarrying, flour milling, food industries. Founded in 1837, developed during World War II. Became town in 1930, city in 1944.

Kinel′-Cherkassy (kee-NYEL–cheer-KAH-si), village (2006 population 18,350), E SAMARA oblast, E European Russia, on the GREATER KINEL′ RIVER, on road and railroad, 39 mi/63 km ENE of KINEL′; 53°28′N 51°30′E. Elevation 236 ft/71 m. Agricultural center (grains, vegetables, potatoes); flour milling, food processing.

Kineo, Mount (kin-EE-o), peak (1,789 ft/545 m), PISCATAQUIS county, central MAINE, on peninsula extending W into MOOSEHEAD LAKE and 17 mi/27 km NNW of GREENVILLE. Summer resort.

Kineo, Mount (KI-nee-yo), peak (3,320 ft/1,012 m) of WHITE MOUNTAINS, GRAFTON county, central NEW HAMPSHIRE, 11 mi/18 km N of PLYMOUTH, in White Mountain National Forest.

Kineshma (KEE-neesh-mah), city (2005 population 92,255), NE IVANOVO oblast, central European Russia, port on the VOLGA RIVER, 65 mi/105 km NE of IVANOVO; 57°26′N 42°07′E. Elevation 370 ft/ 112 m. Old cotton textile center with machinery, wood, chemical, and auto parts plants. Meat processing; grain milling. Has a drama theater, federal museum of arts history. First mentioned in 1429; city since 1777. Historical sites include Voznesenskiy (built in 1745) and Troitskiy (1836) cathedrals and Voznesenskaya (built in 1760) church.

Kinesi (kee-NAI-see), village, MARA region, NW TANZANIA, 6 mi/9.7 km NE of MUSOMA, on Mara Bay, Lake VICTORIA; 01°26′S 33°52′E. Fish; cattle, goats, sheep; cotton, corn, wheat, millet.

Kineton (KEIN-tuhn), village (2001 population 2,276), S WARWICKSHIRE, central ENGLAND, 9 mi/14.5 km SSE of WARWICK; 52°10′N 01°30′W. Former agricultural market. Has church with 14th-century tower. Here Charles I assembled his army (1642) before the battle of EDGEHILL.

King, county (□ 913 sq mi/2,373.8 sq km; 2006 population 287), NW TEXAS, ⊙ GUTHRIE; 33°36′N 100°21′W. Rolling plains area. Elevation 2,000 ft/ 610 m–2,500 ft/762 m. Drained by tributaries of S and Middle WICHITA rivers, and BRAZOS River in extreme SE. Large-scale cattle ranching county; horses, also some agriculture (vegetables). Oil and natural-gas wells; lime, copper deposits. Formed 1876.

King, county (□ 2,306 sq mi/5,995.6 sq km; 2006 population 1,826,732), W central WASHINGTON; ⊙ SEATTLE; 47°28′N 121°51′W. SNOQUALMIE RIVER rises in CASCADE RANGE in E. Bounded on W by PUGET SOUND and Colvos Passage, on E by crest of Cascade Range, on S in part by WHITE RIVER. Strawberries, lettuce, hay; dairying; cattle, poultry; lumber, clays, coal (at BLACK DIAMOND). Muckleshoot Indian Reservation in S, SE of AUBURN; rapidly growing urban area in W, between mountains and sound, around Seattle and S toward TACOMA (just SW of county line). Includes part of Mount Baker–Snoqualmie National Forest and Alpine Lake Wilderness. Federation Forest, Olallie, Nolte and Kanaskat-Palmer state parks in S; LAKE SAMMAMISH, St. Edward, and Bridle Trails state parks in W; Salt Water and Dash Point state parks in SW. County includes VASHON and Maury islands, near Kitsap Peninsula, in W in Puget Sound. Formed 1852.

Cross-references are shown in SMALL CAPITALS. The pronunciation guide is shown on page xix. The sources of population figures are shown on page xvii.

King (KEENG), township (□ 129 sq mi/335.4 sq km; 2001 population 18,533), York region, S central ONTARIO, E central CANADA, suburb 22 mi/35 km NNW of downtown TORONTO, E of Highway 400 Freeway. Urban growth area, primarily rural. Mixed farming, dairying, cattle, equestrian farms. Includes the communities of KING CITY, NOBLETON, SCHOMBERG, LASKAY, and POTTAGEVILLE. Incorporated 1850.

King (KEENG), town (□ 5 sq mi/13 sq km; 2000 population 5,952), FORSYTH and STOKES counties, N NORTH CAROLINA, 15 mi/24 km NNW of WINSTON-SALEM; 36°16′N 80°21′W. Manufacturing (printing and publishing, blood flow meters, automotive parts, burial vaults, wood products); service industries; agriculture (tobacco, grain, soybeans; livestock). Pilot Mountain State Park to NW; Hanging Rock State Park to NE. Founded as a train depot in 1888, incorporated 1983.

Kingait, Northwest Territories, CANADA: see META INCOGNITA.

King Alexander Canal, SERBIA: see NOVI SAD-MALI STAPAR CANAL.

King and Queen (KEENG, KWEEN), county (□ 326 sq mi/847.6 sq km; 2006 population 6,903), E VIRGINIA; ⊙ KING AND QUEEN COURT HOUSE; 37°43′N 76°54′W. In Tidewater region; bounded SW by MATTAPONI and YORK rivers. Agriculture (especially tomatoes; also barley, hay, soybeans, corn, wheat, legumes); cattle; oysters. Formed 1691.

King and Queen Court House (KEENG, KWEEN), unincorporated village, ⊙ KING AND QUEEN county, E VIRGINIA, near MATTAPONI RIVER, 34 mi/55 km ENE of RICHMOND; 37°40′N 76°52′W. Agriculture (grain, soybeans, tomatoes; cattle). Mattaponi Indian Reservation to S. Sometimes called King and Queen.

Kingandu, CONGO: see KIGANDU.

Kinganga (keen-GAHN-guh), town, BAS-CONGO province, W CONGO, on the left bank of CONGO RIVER; 05°01′S 13°48′E. Elev. 1,833 ft/558 m.

Kinganga (keen-GAHN-guh), village, BAS-CONGO province, W CONGO, on left bank of CONGO RIVER, 100 mi/160 km SW of KINSHASA; 04°29′S 14°06′E. Elev. 1,804 ft/549 m.

Kingani River, river, E TANZANIA: see RUVU RIVER.

Kingaroy (king-uh-ROI), town, SE QUEENSLAND, NE AUSTRALIA, 100 mi/161 km NW of BRISBANE; 26°32′S 151°50′E. Agriculture center (peanuts, wheat, corn). Bunya Mountains National Park 37 mi/60 km SW.

King Cays, Spanish *Cayos King*, small islands and reefs, NICARAGUA, 5 mi/8 km E of PEARL LAGOON.

King Channel (KEENG), arm of the PACIFIC, in CHONOS ARCHIPELAGO, S CHILE, at c.44°30′S; extends c.50 mi/80 km E-W between the ocean and MORALEDA CHANNEL. Forms CHILOÉ-AISÉN province line.

King Charles Islands, SVALBARD: see KONG KARLS LAND.

King Charles Land, island, NORWAY: see KONG KARLS LAND.

King Charles South Land, CHILE-ARGENTINA: see TIERRA DEL FUEGO.

King Chiang Saen, THAILAND: see CHIANG SAEN.

King Christian Island (17 mi/27 km long, 9 mi/14 km wide), NUNAVUT territory, CANADA, in Maclean Strait, PRINCE GUSTAV ADOLPH SEA, off ELLEF RINGNES ISLAND; 77°45′N 102°10′W.

King Christian IX Land, Danish *Kong Christian IX Land*, coastal region of SE GREENLAND; 65°–70°N.

King Christian X Land, Danish *Kong Christian X Land*, coastal region of E GREENLAND; 70°–75°N.

King City, city (2000 population 11,094), MONTEREY county, W CALIFORNIA, on SALINAS RIVER, and 45 mi/72 km SE of SALINAS; 36°13′N 121°08′W. Trade and shipping center near S end of rich irrigated Salinas valley. Agriculture (beans, grain, some sugar beets, vegetables); dairying; cattle. Manufacturing (dehydrated vegetables); asbestos mining and milling; printing and publishing. Hunter Liggett Military Reservation to SW. Los Padres National Forest to W. Founded 1868, incorporated 1911.

King City, city (2000 population 1,012), GENTRY county, NW MISSOURI, 26 mi/42 km NE of SAINT JOSEPH; 40°02′N 94°31′W. Corn, soybeans, wheat; cattle, hogs; native bluestem prairie sod and seed. Plotted 1869.

King City, town (2006 population 2,225), WASHINGTON county, NW OREGON, suburb 10 mi/16 km SW of downtown PORTLAND, on TUALATIN RIVER; 45°24′N 122°47′W. Dairying, poultry, fruit, vegetables, grapes, berries. Retirement center. Bald Peak State Park to W.

King City (KEENG), unincorporated village, York region, S ONTARIO, E central CANADA, 15 mi/24 km N of TORONTO, and included in the township of KING; 43°55′N 79°31′W.

King Country, The, occupies the Western Uplands of the NORTH ISLAND, NEW ZEALAND, an indefinite term commonly accepted as extending along the TASMAN coast from KAWHIA Harbour to the Urenui River in N TARANAKI, and extending inland to the WAIPA RIVER, W LAKE TAUPO, and the E margin of the Tengariro volcanic cluster. Its Maori name "Rohe Potae" traditionally refers to an area over which the Maori King, Tawhiao, threw his hat. Following accommodation in the 1880s, it lost political connotation, but remains in common usage. Once the geographical manifestation of a Maori movement to establish a Maori King, it nevertheless retains some regional distinction. It is composed essentially of soft rock deposited in the Tertiary period, with older limestones in the N replaced by softer sandstones and mudstones to the S. Rivers, including branches of the WHANGANUI and WAIKATO, which branch dendritically: a maze of ridges and gullies is characteristic. Farm settlement is sporadic along the Main Trunk railroad, with smaller timber milling and agricultural service centers along the line. At times, especially at WHANGAMOMONA, farming based on extensive sheep, beef cattle grazing, and dairying, has been established, but reversion to forest often signals retreat; patches of virgin bush remain. Small quantities of coal are mined, and the dark ironsands of Lake Taharoa S of Kawhia are exported to JAPAN. Small towns, usually Maori and European, as in Taumaranui and TE KUITI, hold most of a limited population.

King Cove, village (2000 population 792), SW ALASKA, near SW extremity of ALASKA PENINSULA, 10 mi/16 km WSW of BELKOFSKI; 55°03′N 162°18′W. Fish canneries.

Kingdom City, village (2000 population 121), CALLAWAY county, central MISSOURI, 7 mi/11.3 km N of FULTON; 38°57′N 91°56′W. Important highway service and trucking center.

King Edward, Mount (keeng ED-wuhrd) (11,400 ft/3,475 m), on ALBERTA–BRITISH COLUMBIA border, W CANADA, in ROCKY MOUNTAINS, on S edge of JASPER NATIONAL PARK, 55 mi/89 km SSE of JASPER; 52°09′N 117°31′W.

King Edward VIII Falls, central GUYANA; 05°35′N 59°40′W. Located on King Edward River, an affluent of the MAZARUNI, and 30 mi/48 km NW of KAIETEUR FALLS.

King Edward VIII Gulf, ANTARCTICA: see EDWARD VIII BAY.

King Edward VII Land, ANTARCTICA: see EDWARD VII PENINSULA.

King Edward VII Point, SE extremity of ELLESMERE ISLAND, BAFFIN region, NUNAVUT territory, CANADA, on JONES SOUND; 76°08′N 81°09′W.

Kingersheim (KING-uhrs-heim), town (□ 3 sq mi/7.8 sq km), HAUT-RHIN department, E FRANCE, a N suburb of MULHOUSE in the ALSACE lowland; 47°48′N 07°20′E.

King Fahad Causeway, 16 mi/25 km long, connects SAUDI ARABIA to BAHRAIN in the PERSIAN GULF. It begins in Aziziah area, S Khobar and passes over Umm Nassan Island in Bahrain (customs point) and ends W of Manamma. Opened in 1986. Largest and longest in MIDDLE EAST; second largest in the world. Dual three-lane road. A causeway authority building, two mosques, and two restaurants are located on a man-made island in the center. The causeway has had positive economic effects for both countries.

Kingfield, town, FRANKLIN county, W central MAINE, on the CARRABASSETT, and 20 mi/32 km N of FARMINGTON; 45°00′N 70°10′W. Wood products. Sugar Loaf Ski resort is NW. Settled 1805, incorporated 1816.

Kingfisher, county (□ 906 sq mi/2,355.6 sq km; 2006 population 14,316), central OKLAHOMA; ⊙ KINGFISHER; 35°56′N 97°56′W. Intersected by CIMARRON RIVER and TURKEY CREEK. Diversified agriculture (wheat, alfalfa, barley, oats; cattle, sheep; dairying). Manufacturing of industrial machinery is chief industry. Oil and natural gas deposits. Oil field. Formed 1890.

Kingfisher, town (2006 population 4,497), ⊙ KINGFISHER county, central OKLAHOMA, 35 mi/56 km NW of OKLAHOMA CITY; 35°50′N 97°56′W. Elevation 1,056 ft/322 m. In rich agricultural area (mainly wheat; also alfalfa, livestock); oil and natural gas; dairying; light manufacturing CHISHOLM TRAIL Museum; Seay Mansion here. Founded 1889.

King Frederik VI Coast, Danish *Kong Frederik VI Kyst*, coastal region of SE GREENLAND; 60°–65°N.

King Frederik VIII Land, Danish *Kong Frederik den VIII Land*, coastal region of NE GREENLAND; 75°–81°N.

King George (KEENG JORJ), county (□ 187 sq mi/486.2 sq km; 2006 population 21,780), E VIRGINIA; ⊙ KING GEORGE; 38°15′N 77°09′W. At base of NORTHERN NECK peninsula; bounded by POTOMAC RIVER (shore forms Virginia-MARYLAND state line), RAPPAHANNOCK RIVER on S. Rolling dairying and agricultural region (corn, barley, wheat, hay, soybeans, alfalfa; cattle, hogs); hunting; commercial fishing. Includes DAHLGREN Naval Surface Weapons Center in NE. Caledon Natural Area in N. Formed 1720.

King George (KEENG JORJ), unincorporated village, ⊙ KING GEORGE county, E VIRGINIA, 16 mi/26 km E of FREDERICKSBURG; 38°16′N 77°11′W. Manufacturing (concrete, bacon processing); in agricultural area (grain, soybeans; cattle, hogs; dairying).

King George Bay (c.20 mi/32 km long, 2 mi/3.2 km–7 mi/11.3 km wide), inlet, WEST FALKLAND ISLAND; 51°40′S 60°30′W.

King George Island (50 mi/80 km long, c.19 mi/31 km wide), SOUTH SHETLAND ISLANDS, off GRAHAM LAND, ANTARCTICA; 62°S 58°15′W. Several scientific stations. Also known as Waterloo Island.

King George Islands, NUNAVUT territory, CANADA, group of fifteen small islands and islets in HUDSON BAY off W Ungava Peninsula; 57°20′N 78°30′W. Covers area c.30 mi/48 km long, 20 mi/32 km wide.

King George Islands, TUAMOTU ARCHIPELAGO: see TAKAROA.

King George IV Lake (6 mi/10 km long, 5 mi/8 km wide), SW NEWFOUNDLAND and LABRADOR, CANADA, on Lloyds River and 55 mi/89 km S of CORNER BROOK, at SW end of the ANNIEOPSQUOTCH MOUNTAINS. Outdoor recreation area.

King George, Mount (KEENG JORJ) (11,226 ft/3,422 m), SE BRITISH COLUMBIA, W CANADA, near ALBERTA border, 40 mi/64 km S of BANFF (Alberta); 50°36′N 115°24′W.

King George Sound (KEENG JORJ), inlet of INDIAN OCEAN, SW WESTERN AUSTRALIA, between BALD HEAD (SW) and Cape VANCOUVER (NE); 5 mi/8 km wide (at mouth, N-S), 10 mi/16 km long (E-W); 35°02′S 117°58′E. PRINCESS ROYAL HARBOUR (W inlet) is site of ALBANY.

King George VI Falls, falls, CUYUNI-MAZARUNI district, W GUYANA; 05°42′N 61°05′W. Located near BRAZIL border, on affluent of the MAZARUNI, and 45 mi/72 km NW of Mount RORAIMA.

King George VI Sound, ANTARCTICA: see GEORGE VI SOUND.

King George V Land, ANTARCTICA: see GEORGE V COAST.

Kinghorn (KEENG-horn), village (2001 population 2,835), FIFE, E Scotland, on Firth of Forth, 4 mi/6.4 km S of KIRKCALDY; 56°04′N 03°11′W. Coastal roadside monument (1883) commemorates spot where Alexander III (1241–1286) was killed in riding accident.

Kingisepp (KEEN-gee-syep), city (2005 population 50,660), W LENINGRAD oblast, NW European Russia, 85 mi/137 km SW of SAINT PETERSBURG, approximately 14 mi/22 km E of the Estonian border; 59°22′N 28°36′E. Elevation 108 ft/32 m. Port on the LUGA RIVER, highway junction, and railroad station; mining of phosphorite; wood, furniture, leather, and shoe industries. The site was settled in the 9th century, and the fortress of Yam was founded there in 1384 as a frontier post of Novgorod. The fortress was taken by SWEDEN in 1585 and passed to RUSSIA in 1703, when it was renamed Yamburg; current namse since 1922.

King Island (2.5 mi/4 km long, 1.5 mi/2.4 km wide), NW ALASKA, in BERING SEA, SW of SEWARD PENINSULA, 90 mi/145 km WNW of NOME; 64°59′N 168°03′W. Rises to 700 ft/213 m. UKIVOK, an Inuit village on S shore, was abandoned; people moved to Nome.

King Island, largest island (□ 170 sq mi/442 sq km) of MERGUI ARCHIPELAGO, peninsular MYANMAR, in ANDAMAN SEA, 10 mi/16 km W of MERGUI town; 25 mi/40 km long, 4 mi/6.4 km–10 mi/16 km wide; 12°36′N 98°19′E. Mountainous (highest point, 2,125 ft/648 m); mangrove swamps on coast; rubber plantations on coastal plain (SE); manganese, galena, glass sands.

King Island (KEENG), island (□ 425 sq mi/1,105 sq km), in BASS STRAIT, 55 mi/89 km off NW coast of TASMANIA, SE AUSTRALIA; 39 mi/63 km long, 16 mi/26 km wide; 39°55′S 144°00′E. Many shipwrecks off coast due to volatile Bass Strait weather conditions; several lighthouses built to try to prevent accidents. Significant annual rainfall. Dairy products; beef cattle; peas; kelp processing, crayfish, abalone. Tungsten (scheelite) mines at GRASSY. CURRIE is its port and largest town. Birds, wildlife.

Kingittoq, island (3 mi/4.8 km long, 1 mi/1.6 km wide), W GREENLAND, in BAFFIN BAY, 14 mi/23 km NNW of UPERNAVIK; 72°58′N 56°25′W. Stone with runic inscription (last half of thirteenth century) found on the neighboring island of Kingittorsuaq.

King Karl Islands, SVALBARD: see KONG KARL'S LAND.

King Kirkland (KEENG KUHRK-luhnd), unincorporated village, NE ONTARIO, E central CANADA, 4 mi/6 km E of town of KIRKLAND LAKE, and included in unorganized W part of TIMISKAMING; 48°09′N 79°52′W. Gold mining.

Kinglake (KING-laik), town, VICTORIA, SE AUSTRALIA, 29 mi/46 km N of MELBOURNE, at top of GREAT DIVIDING RANGE; 37°31′S 145°21′E. Annual raspberry fair. The surrounding Kinglake National Park is the largest national park near Melbourne.

King, Lake (KEENG), lagoon (□ 35 sq mi/91 sq km), SE VICTORIA, AUSTRALIA, 145 mi/233 km E of MELBOURNE; separated from TASMAN SEA by sandspit, with opening at town of LAKES ENTRANCE; merges SW with Lake REEVE and Lake VICTORIA; 9 mi/14 km long, 6 mi/10 km wide. Irregularly shaped. Contains small island. Receives MITCHELL RIVER. Tourism.

Kinglassie (KEENG-lah-see), town (2001 population 1,320), FIFE, E Scotland, 5 mi/8 km NW of KIRKCALDY; 56°10′N 03°14′W.

King Lear, NEVADA: see JACKSON MOUNTAINS.

King Leopold Ranges (KEENG LEE-uh-pold), N WESTERN AUSTRALIA, extend 180 mi/290 km SE from shores of COLLIER BAY; 17°30′S 125°45′E. Mount Broome (3,040 ft/927 m), highest peak.

Kingman, county (□ 866 sq mi/2,251.6 sq km; 2006 population 7,975), S KANSAS; ☉ KINGMAN; 37°33′N 98°08′W. Plains region, watered by CHIKASKIA RIVER and South Fork of NINNESCAH RIVER. Wheat, sheep,

cattle, hogs, rye, sorghum, strawberries. Kingman State Fishing Lake in NW; CHENEY Reservoir and Dam on NE corner, also Cherry State Park. Formed 1874.

Kingman, city (2000 population 20,069), ☉ MOHAVE county, W ARIZONA, E of BLACK MOUNTAINS, 88 mi/142 km SE of LAS VEGAS, NEVADA, and 163 mi/262 km NW of PHOENIX; 35°12′N 114°01′W. Elevation 3,325 ft/1,013 m. Road and railroad hub. Agriculture (cattle; grain, alfalfa; melons); manufacturing (fishing lures, metal products, boat construction, aquarium supplies). Silver and copper mines nearby. HUALAPAI MOUNTAINS extend SE from Kingman, Corbat Mountains extend NW; Wabayuma Peak Wilderness Area to S; Mount Nuttf and Warm Springs wilderness areas to SW; Hualapai Peak (8,417 ft/2,566 m), in Hualapai Mountains Park to SE; large main section of Hualapai Indian Reservation to NE. LAKE MEAD NATIONAL RECREATION AREA to W and N; DAVIS DAM, on COLORADO RIVER (forms Lake Mohave), 30 mi/48 km W; HOOVER DAM, also on Colorado River, 67 mi/108 km NW (forms Lake MEAD). Founded 1882.

Kingman, town (2000 population 538), FOUNTAIN county, W INDIANA, 14 mi/23 km SSE of COVINGTON; 39°58′N 87°17′W. Agriculture; manufacturing (fiberglass gloves, steel products, and safety clothing).

Kingman, town (2000 population 3,387), ☉ KINGMAN county, S KANSAS, on South Fork of NINNESCAH RIVER and 40 mi/64 km W of WICHITA; 37°38′N 98°06′W. Railroad junction. Market center for wheat region. Manufacturing (metal stampings, meat packing, auto accessories). Kingman State Fishing Lake to W. Founded c.1872, incorporated 1883.

Kingman Reef, uninhabited reef (□ 1 sq mi/2.6 sq km), central PACIFIC, 1,075 mi/1,730 km SW of HONOLULU. It was discovered by Americans in 1798 and annexed by the U.S. in 1922. Formerly an airport on the route from Honolulu to PAGO PAGO, Kingman Reef is now under the jurisdiction of the U.S. Navy.

King, Mount (17,130 ft/5,221 m), SW YUKON, CANADA, near ALASKA border, in ST. ELIAS MOUNTAINS, 190 mi/306 km W of WHITEHORSE; 60°35′N 140°40′W.

King Mountain (3,141 ft/957 m), W TEXAS, near PECOS RIVER, 5 mi/8 km N of MCCAMEY. A landmark for pioneers.

King, Mount Clarence, peak (12,909 ft/3,935 m) of the SIERRA NEVADA, E FRESNO county, E CALIFORNIA, in KINGS CANYON NATIONAL PARK, 14 mi/23 km W of INDEPENDENCE. MOUNT STARR KING is in YOSEMITE NATIONAL PARK.

King of Prussia, unincorporated city (2000 population 18,511), Upper Merion township, MONTGOMERY county, SE PENNSYLVANIA; 40°06′N 75°22′W. Manufacturing (glass and steel fabricating, food, linen textiles, printing and publishing, liquified petroleum gas, water treatment equipment, electrical equipment, motors and gears, roller bearings, security systems, concrete reinforcing). Site of King of Prussia Plaza, one of the largest shopping malls in U.S. Villanova University is nearby. Valley Forge National Park, to NW, contains the Freedom Foundation national shrine that honors medal recipients from U.S. wars. Revolutionary War Museum to SE.

Kingolwira (keen-go-LWEE-rah), village, MOROGORO region, E central TANZANIA, 8 mi/12.9 km E of MOROGORO, on railroad; 06°48′S 37°45′E. Road junction. Cotton, sisal, grain; livestock. Mica mining.

Kingombe, CONGO: see KALIMA-KINGOMBE.

Kingoonya, village, S central SOUTH AUSTRALIA, 220 mi/354 km NW of PORT PIRIE; 30°54′S 135°18′E. On Trans-Australian railroad; wool, salt.

King Oscar Fjord, Danish Kong Oscar Fjord, inlet (90 mi/145 km long, 8 mi/12.9 km–15 mi/24 km wide) of GREENLAND SEA, E GREENLAND; 72°–73°N, 22°15′–24°50′W. Extends NW between mainland (SW) and TRAILL Island (NE); from its head several arms radiate to edge of inland ice and to FRANZ JOSEF FJORD. Mouth on Greenland Sea called DAVY SOUND.

King Oscar Land, BAFFIN region, NUNAVUT territory, CANADA, SW part of ELLESMERE ISLAND.

King Peninsula, ice-covered peninsula jutting out into the AMUNDSEN SEA between the ABBOTT and the COSGROVE Ice Shelves, WALGREEN COAST, WEST ANTARCTICA, 100 mi/160 km long, and 20 mi/30 km wide; 73°12′S 101°00′W.

King Peter Canal, SERBIA: see DANUBE-TISZA CANAL.

King Ranch (c.825,000 acres/333,878 ha), S TEXAS, between CORPUS CHRISTI and HARLINGEN, headquarted and centered around KINGSVILLE, Texas; 27°31′N 97°55′W. Located mainly in KLEBERG and KENEDY counties, occupying most of their land area, ranch extends N into NUECES county and S into WILLACY county. One of the largest ranches in the world, largest ranch in continental UNITED STATES (ranch on island of HAWAII is larger). It has several divisions, of which the best known is Santa Gertrudis, the "home" ranch. The Santa Gertrudis, the only true cattle breed developed in NORTH AMERICA, was developed here. Thoroughbred racehorses are also raised. The ranch was founded in 1853 by Richard King, a steamboat captain. After King's death, the giant holdings were managed by his son-in-law, Robert Kleberg; later, Kleberg's son succeeded to the management. The property was divided in 1935, but the central ranches are still large enough to resemble a semifeudal domain. Profits from oil and natural gas rights and farming have been added to income gained from the great beef herds. King Ranch Museum at Kingsville.

Kings, county (□ 1,374 sq mi/3,572.4 sq km; 2001 population 64,208), S NEW BRUNSWICK, CANADA extending N from SAINT JOHN, drained by St. John River; ☉ HAMPTON.

Kings, county (□ 842 sq mi/2,189.2 sq km; 2001 population 58,866), NW NOVA SCOTIA, CANADA, on the BAY OF FUNDY; ☉ KENTVILLE. Split from Halifax county in 1759.

Kings, county (□ 641 sq mi/1,666.6 sq km; 2001 population 19,180), in E PRINCE EDWARD ISLAND, CANADA; ☉ GEORGETOWN. Forestry and fishing.

Kings, county (□ 1,391 sq mi/3,616.6 sq km; 2006 population 146,153), S central CALIFORNIA; ☉ HANFORD; 36°04′N 119°49′W. Level irrigated farm land of San Joaquin Valley, drained by KINGS and TULE rivers. TULARE LAKE irrigation reservoir in center. KETTLEMAN HILLS (oil and natural-gas field here) in SW. Cotton, oats, wheat, barley; fruits and vegetables; dairying; turkeys. Processing of farm products, oil refining. Gypsum quarrying. Part of Lemoore Naval Air Station in NW. Part of DIABLO RANGE in far SW. Formed 1893.

Kings, NEW YORK: see BROOKLYN.

King Salmon, village (2000 population 442), SW ALASKA, 320 mi/515 km SW of ANCHORAGE, 15 mi/24 km E (by road) from Naknek River; 58°44′N 156°32′W. Gateway to KATMAI NATIONAL PARK AND PRESERVE, to E. Outfitting center. Salmon fishing.

King Salmon River, S ALASKA, on ALASKA PENINSULA; rises in SW part of KATMAI National Monument near 58°09′N 155°25′W; flows 90 mi/145 km W to BRISTOL BAY at EGEGIK.

Kings Bay, SPITSBERGEN, NORWAY: see KONGSFJORDEN.

Kings Beach, unincorporated town (2000 population 4,037), PLACER county, E CALIFORNIA, 20 mi/32 km SSW of RENO (Nevada), at end of LAKE TAHOE, on Nevada state boundry, opposite Crystal Bay, Nevada; 39°15′N 120°01′W. Tahoe National Forest to W; Toiyabe National Forest (Nevada) to E. Tourism.

Kingsbridge (KINGZ-brij), urban district (2001 population 5,845), S DEVON, SW ENGLAND, on inlet of the CHANNEL, 10 mi/16 km WSW of DARTMOUTH; 50°17′N 03°46′W. Agricultural market known for breed of large cattle (South Devons), and scene of annual cattle fair. Has 17th-century grammar school, 15th-century church, 16th-century market house, and museum.

Kingsbridge, section of W BRONX borough of NEW YORK city, SE NEW YORK, along HARLEM RIVER opposite N

MANHATTAN; 40°52′N 73°55′W. The name comes from an old bridge that, spanning the Spuyten Duyvil (see HARLEM RIVER), connected the "mainland" (then YONKERS) to Manhattan's MARBLE HILL as part of the Boston Post Road. By the late 19th century the Kingsbridge area was predominantly Irish, but in recent decades it has become much more ethnically diverse.

Kingsburg, city (2000 population 9,199), FRESNO county, central CALIFORNIA, in San Joaquin Valley, near KINGS RIVER, 20 mi/32 km SE of FRESNO; 36°31′N 119°33′W. Cantaloupes, citrus, nectarines, grain, sugar beets, almonds, vegetables; cotton; manufacturing (dehydrated as well as canned fruits and vegetables; animal feeds, glass containers). Settled in 1870s, incorporated 1908.

Kingsbury, county (□ 863 sq mi/2,243.8 sq km; 2006 population 5,464), E central SOUTH DAKOTA, ⊙ DE SMET; 44°22′N 97°29′W. Agricultural area watered by several lakes in E half of county, including LAKE WHITEWOOD, Lake Henry, LAKE PRESTON, and LAKE THOMPSON. Corn, wheat, flax, soybeans; dairy products; cattle; abundance of ducks and pheasants. Formed 1873.

Kingsbury (KEENGZ-buh-ree), town (□ 2 sq mi/5.2 sq km; 2006 population 151), ESTRIE region, S QUEBEC, E CANADA, 6 mi/9 km from RICHMOND; 45°35′N 72°09′W. Incorporated 1896.

Kingsbury, town (2000 population 229), LA PORTE county, NW INDIANA, 5 mi/8 km S of LA PORTE; 41°32′N 86°42′W. Manufacturing (chemicals, desulfurization products, fireworks, plastic moldings, metal coatings).

Kingsbury (KINGZ-buh-ree), village (2001 population 7,421), N WARWICKSHIRE, central ENGLAND, 11 mi/18 km NE of BIRMINGHAM; 52°35′N 01°40′W. Church, with 14th-century tower, dates from Norman times.

Kingsbury (kings-BUH-ree), unincorporated village, GUADALUPE county, S central TEXAS, c.45 mi/72 km ENE of SAN ANTONIO. Manufacturing (zinc electroplating); agriculturally diverse area; cattle.

Kingsbury, plantation, PISCATAQUIS county, central MAINE, 20 mi/32 km W of DOVER-FOXCROFT; 45°08′N 69°35′W. Lumbering, recreational area.

Kingsbury (KINGZ-buh-ree), suburb 7 mi/12 km NNE of MELBOURNE, VICTORIA, SE AUSTRALIA. A campus of LaTrobe University here.

Kingsbury, ENGLAND: see BRENT.

Kings Canyon National Park (□ 722 sq mi/1,869 sq km), FRESNO and TULARE counties, E central CALIFORNIA, c.60 mi/97 km E of FRESNO. General Grant Grove section detached from park, adjoins SEQUOIA NATIONAL PARK. Two large canyons of KINGS RIVER, numerous peaks of SIERRA NEVADA, and giant sequoia trees. Sequoia National Park adjoins to S. Authorized 1890. Originally called General Grant National Park, name changed 1940.

Kings Canyon National Park, AUSTRALIA: see WATARRKA NATIONAL PARK.

Kingsclere (KINGZ-clir), town (2001 population 4,944), N HAMPSHIRE, S ENGLAND, 8 mi/12.9 km NW of BASINGSTOKE; 51°19′N 01°15′W. Agricultural market. Has Norman church.

Kingscliff (KINGZ-klif), resort village, NEW SOUTH WALES, SE AUSTRALIA, 522 mi/840 km N of SYDNEY, 21 mi/34 km S of Surfers Paradise, and just S of GOLD COAST; 28°16′S 153°34′E. Beaches. Tropical fruit plantations and tourist centers. Sometimes called Tweed Coast.

Kingscote (KINGZ-kot), village, port, NE KANGAROO ISLAND, SOUTH AUSTRALIA, on NW shore of NEPEAN BAY; 35°40′S 137°38′E. Tourist resort; sheep, barley, eucalyptus oil. Fishing.

King's County, IRELAND: see OFFALY.

Kingscourt, Gaelic *Dún an Rí*, town (2006 population 1,748), E CAVAN county, N central IRELAND, 18 mi/29 km NNW of NAVAN; 53°54′N 06°48′W. Agricultural market; china-clay mining; manufacturing of gypsum products.

Kings Creek, unincorporated village, CHEROKEE county, N SOUTH CAROLINA, 12 mi/19 km E of GAFFNEY. KINGS MOUNTAIN NATIONAL MILITARY PARK is NE. Manufacturing of industrial minerals.

King Sejong Station (sai-JUHNG), ANTARCTICA, Korean station on KING GEORGE ISLAND; 62°13′S 58°47′W.

Kingsey Falls (KEENG-zee FAHLZ), village (□ 27 sq mi/70.2 sq km), CENTRE-DU-QUÉBEC region, S QUEBEC, E CANADA, on NICOLET RIVER and 18 mi/29 km E of DRUMMONDVILLE; 45°51′N 72°04′W. Paper milling, dairying; cattle, pigs.

Kingsford, town (2000 population 5,549), DICKINSON county, SW UPPER PENINSULA, MICHIGAN, suburb 2 mi/3.2 km SW of IRON MOUNTAIN city, and on the MENOMINEE RIVER; 45°48′N 88°05′W. Manufacturing (wood products, iron castings, cutting tools, truck mounted loaders, concrete, furniture). Ford Airport nearby. Incorporated as city 1947.

Kingsford Heights, town (2000 population 1,453), LA PORTE county, NW INDIANA, 10 mi/16 km S of LA PORTE; 41°29′N 86°41′W.

King's Heath, ENGLAND: see MOSELEY AND KING'S HEATH.

King's Island, Maldive Islands: see MALÉ ISLAND.

Kingsland, town (2000 population 10,506), CAMDEN county, extreme SE GEORGIA, 27 mi/43 km SSW of BRUNSWICK, near FLORIDA line; 30°47′N 81°40′W. Originally a railroad town and the site of a rice plantation. Manufacturing includes trusses, missile parts, chemical plant, sawmilling. Crooked River State Park nearby. Growth of the town has been affected by the construction of Kings Bay submarine base in St. Mary's, Georgia, just to the E of KINGSLAND. Founded 1894.

Kingsland, town (2000 population 4,584), LLANO county, S central TEXAS, 50 mi/80 km WNW of AUSTIN, on COLORADO RIVER, at head of Lake Lyndon B. Johnson reservoir and at mouth of LLANO RIVER; 30°39′N 98°27′W. Elevation 856 ft/261 m. Agriculture: cattle, peaches, grain. Manufacturing (chalk and tack boards). Recreation. LONGHORN CAVERNS STATE PARK to E; Inks Dam National Fish Hatchery to NE.

Kingsland, village (2000 population 449), CLEVELAND county, S central ARKANSAS, 29 mi/47 km SW of PINE BLUFF; 33°51′N 92°17′W. In agricultural area; manufacturing (hardwood lumber). Marks Mills Battleground State Historical Monument to SE.

King's Landing, historic settlement, SW NEW BRUNSWICK, CANADA, 17 mi/27 km WSW of FREDERICTON, on SW bank of St. John River. Construction of dam at Mactaquac, 1967, flooded pioneer farms in valley. Historical settlement built in mid-1970s has sixty buildings depicting period 1783–1900. Mixed farming, potatoes, dairying, apples, timber. Sport fishing, tourism. Formerly Lower Prince William.

Kings Langley (KINGZ LAING-lee), village (2001 population 5,072), SW HERTFORDSHIRE, E ENGLAND, on GADE RIVER and 4 mi/6.4 km NNW of WATFORD; 51°43′N 00°28′W. Manufacturing of paper, pharmaceuticals; previously flour mills. Has 15th-century church.

Kingsley (KEENGZ-lee), community, S MANITOBA, W central CANADA, 10 mi/16 km from MANITOU, and in PEMBINA rural municipality; 49°20′N 98°42′W.

Kingsley, town (2000 population 1,245), PLYMOUTH county, NW IOWA, near West Fork Little Sioux River, 17 mi/27 km SW of LE MARS; 42°35′N 95°58′W. Manufacturing (livestock remedies, troughs, millwork products). Incorporated 1884.

Kingsley, village (2000 population 428), JEFFERSON county, N KENTUCKY, a suburb 4 mi/6.4 km E of downtown LOUISVILLE; 38°13′N 85°40′W. Bownan Field airport to NE. Farmington Historical Home (1810) to S.

Kingsley, village (2000 population 1,469), GRAND TRAVERSE county, NW MICHIGAN, 13 mi/21 km SSE of TRAVERSE CITY; 44°34′N 85°32′W.

Kingsley Dam, Nebraska: see C. W. MCCONAUGHY, LAKE.

King's Lynn or **Lynn Regis** (KINGZ LIN), town (2001 population 40,921), NORFOLK, E ENGLAND, on the Great OUSE River near its influx into The WASH, an inlet of the NORTH SEA; 52°45′N 00°23′E. Its large harbor serves foreign as well as coastal trade and is the base for a fishing fleet. A farm market, King's Lynn is a center for fertilizer production, canning, beet-sugar refining, shipbuilding, metalworking, and light engineering. The town dates from Saxon times. Red Mount Chapel was visited by pilgrims in the 15th and 16th centuries. A Norman church also remains, as do relics of a moat that surrounded the town in the 15th century. King's Lynn was the birthplace of the novelist Fanny Burney. Town includes West Lynn.

Kings Manor, unincorporated village, Upper Merion township, MONTGOMERY county, SE PENNSYLVANIA, suburb 14 mi/23 km NW of downtown PHILADELPHIA, and 2 mi/3.2 km S of NORRISTOWN, near SCHUYLKILL RIVER; 40°05′N 75°20′W. Light manufacturing.

Kingsmill Group: see GILBERT Islands.

Kings Mills, unincorporated village, WARREN county, SW OHIO, 22 mi/35 km NE of CINCINNATI, on Little Miami River; 39°21′N 84°15′W. Produces communications equipment, chemicals, and electrical equipment. Major theme park.

Kings Mountain (KEENGZ MOUN-tuhn), city (□ 8 sq mi/20.8 sq km; 2006 population 10,919), CLEVELAND and GASTON counties, SW NORTH CAROLINA, 10 mi/16 km W of GASTONIA near SOUTH CAROLINA state line; 35°14′N 81°20′W. Manufacturing (textiles and apparel, heat-recovery modules, roof and floor trusses, fire extinguishers, tire cords, welding materials, filters, mica processing, bricks, truck cabs, textile machinery, aircraft parts, compact discs); service industries; agriculture (cotton, grain, hay; poultry, livestock). KINGS MOUNTAIN NATIONAL MILITARY PARK (South Carolina) 8 mi/12.9 km to SSW, Crowders Mountain State Park to SE. Incorporated 1874.

Kings Mountain, isolated ridge (1,040 ft/317 m) in YORK county, N SOUTH CAROLINA, near NORTH CAROLINA state line, 30 mi/48 km WSW of CHARLOTTE, North Carolina in KINGS MOUNTAIN NATIONAL MILITARY PARK and state park.

Kings Mountain National Military Park, YORK and CHEROKEE counties, N SOUTH CAROLINA, authorized 1931, 6 sq mi/15.5 sq km. Site of a crucial American victory over the British during the American Revolution (Oct. 7, 1780). Marked battlefield trail.

King's Norton, ENGLAND: see BIRMINGHAM.

King Sound (KEENG), inlet of INDIAN OCEAN, N WESTERN AUSTRALIA; Cape LEVEQUE at E end of entrance; 40 mi/64 km E-W, 80 mi/129 km N-S; 16°50′S 123°25′E. Broken into small bays. BUCCANEER ARCHIPELAGO at NE entrance; DERBY to SE.

Kings Park (KEENGZ PAHRK), unincorporated town, FAIRFAX county, NE VIRGINIA, residential suburb 12 mi/19 km WSW of WASHINGTON, D.C.; 38°48′N 77°14′W. Lake Accotink Park to S.

Kings Park, village (□ 6 sq mi/15.6 sq km), SUFFOLK county, SE NEW YORK, shore of W LONG ISLAND, 9 mi/14.5 km E of HUNTINGTON; 40°53′N 73°15′W. Developed as utopian community in 1872; became farm for insane in 1885. State psychiatric center opened 1892 (closed 1996). Nearby is Sunken Meadow State Park (□ 520 acres/210 ha; bathing, hiking, picnicking). N terminus of the Long Island Greenbelt trail; S terminus is at HECKSCHER STATE PARK. Trail follows Connetquot and NISSEQUOGUE River valleys for 34 mi/55 km.

Kings Park West (KEENGZ PAHRK WEST), unincorporated town, FAIRFAX county, NE VIRGINIA, residential suburb 14 mi/23 km WSW of WASHINGTON, D.C., 3 mi/5 km S of FAIRFAX, near Pohick Creek.

Kings Peak, DUCHESNE county, NE UTAH, 45 mi/72 km N of DUCHESNE: see UINTA MOUNTAINS. In Uinta Mountains, High Uinta Wilderness Area of Ashley National Forest. 13,528 ft/4,123 m, highest point in Utah.

Kings Point, upper-income residential village (2006 population 5,178), NASSAU county, SE NEW YORK, on NW LONG ISLAND, on GREAT NECK peninsula, 8 mi/12.9 km N of JAMAICA; 40°49′N 73°44′W. Seat of U.S. Merchant Marine Academy (established 1942). Incorporated 1924.

Kingsport, city (2006 population 44,191), HAWKINS and SULLIVAN counties, NE TENNESSEE, on the HOLSTON River near the VIRGINIA line, 21 mi/34 km NNW of JOHNSON CITY; 36°33′N 82°34′W. Manufacturing and telecommunications. The city, encircled by mountains, stands on the site of forts Robinson (1761) and Patrick Henry (1775) on the old WILDERNESS ROAD. Warriors Path State Park is nearby. Incorporated 1917.

Kings River, in NW ARKANSAS and SW MISSOURI; rises in the OZARKS in SE MADISON county (Arkansas); flows c.115 mi/185 km N to WHITE RIVER, lower 10 mi/16 km is arm of TABLE ROCK LAKE reservoir; 15 mi/24 km SE of CASSVILLE, Missouri.

Kings River, 125 mi/201 km long, FRESNO county, central CALIFORNIA; begins with joining of Middle and South forks, 50 mi/80 km E of FRESNO, W of KINGS CANYON NATIONAL PARK; flows W, receives North Fork from NE, 35 mi/56 km E of Fresno; flows through PINE FLAT LAKE reservoir, then SW, crossed by FRIANT-KERN CANAL as it enters San Joaquin Valley, where it passes through a network of irrigation canals before entering TULARE LAKE irrigation reservoir.

Kings River, in NW NEVADA; rises in N HUMBOLDT county near OREGON line; flows c.40 mi/64 km S to QUINN RIVER NE of JACKSON MOUNTAINS.

King's Seat, Scotland: see OCHIL HILLS.

King's Somborne (KINGZ SAHM-bawn), village and parish (2001 population 1,573), W HAMPSHIRE, S ENGLAND, 8 mi/12.9 km W of WINCHESTER; 51°05′N 01°29′W. Church dates from 12th century.

King's Stanley (KINGZ STAN-lee), village (2001 population 3,757), GLOUCESTERSHIRE, central ENGLAND, near FROME RIVER, 3 mi/4.8 km WSW of STROUD; 51°44′N 02°16′W. Previously woolen milling. Church dates from Norman times. Roman remains found here; it was the residence of a Mercian king.

Kingsteignton (KINGZ-stain-tuhn), town (2001 population 11,192), S DEVON, SW ENGLAND, on TEIGN RIVER, and 2 mi/3.2 km NE of NEWTON ABBOT; 50°33′N 03°35′W. Clay quarrying. Has 15th-century church.

Kingston (KEENG-stuhn), city (□ 174 sq mi/452.4 sq km; 2001 population 114,195), SE ONTARIO, E central CANADA, on Lake ONTARIO, near the head of the SAINT LAWRENCE RIVER and at the end of RIDEAU CANAL from OTTAWA; 44°14′N 76°30′W. It is halfway between MONTREAL and TORONTO. Kingston has probably the best harbor on the lake. Industries include the manufacturing of locomotives, ships, vehicle parts, mining equipment, textiles, aluminum products, synthetic yarn, and ceramics. On the site stood Fort Frontenac, which was of great importance in the French and Indian Wars. The present city was founded by United Empire Loyalists in 1783 and prospered during the War of 1812 as the Canadian naval base for operations against the Americans. From 1841 to 1844 it served as the capital of Canada. Fort Henry, built during the War of 1812 and rebuilt from 1832 to 1836, is now a museum. Kingston is the seat of Queen's University (1841), of the Royal Military College, of Canadian National Defense College, and of Anglican and Roman Catholic bishoprics and cathedrals. Restructured in 1998 to include the former city of Kingston (1991 population 56,597) and the townships of Kingston and Pittsburgh.

Kingston, city (2001 population 96,052) and parish, ⊙ JAMAICA and its largest city, SE Jamaica; 17°57′N 76°44′W. Bounded on S by Kingston Harbor Waterfront, it spreads N from the harbor over LIGUANEA PLAIN, bounded W and NW by ST. ANDREW parish, N by Lower St. Andrew, NE by St. Andrew along the Long Mountain range, continuing from the Fort Nugent Tower S through Harbor View toward the CARIBBEAN SEA. This chief port was founded in 1693 on a land-locked harbor as the finest harbor in the WEST INDIES and the seventh-largest natural harbor in the world. In 1692 the former capital, PORT ROYAL, at the tip of the long narrow peninsular forming the harbor, was inundated by an earthquake. The capital moved first to SPANISH TOWN, 13 mi/21 km to the W and then in 1872 to Kingston, now the island's leading commercial city. During this century, the city was ravaged by hurricanes and a 1907 earthquake, as well as periods of severe urban unrest. By 1968 the Urban Development Corporation modernized the entire waterfront and commercial areas. Historic sites include the Cenotaph within the National Heroes Park—burial place of Prime Ministers and National Heroes of the island including Marcus Garvey, Norman Manley, and Sir Alexander Bustamante—the Institute of Jamaica, and the Jamaica Conference Center. Railroad terminus Kingston to MONTEGO BAY Station. Served by Norman Manley International Airport at PALISADOES. Lighthouse at Great Plumb Point entrance to Kingston Harbor at 17°44′N 77°10′W. Exports include sugar, rum, molasses, bananas. Major industries include tourism, clothing manufacturing, tobacco processing, oil refining, flour milling, cement manufacturing plants at Rockfort.

Kingston, city (2000 population 287), NW MISSOURI, ⊙ CALDWELL county, 48 mi/77 km NE of KANSAS CITY; 39°38′N 94°02′W. Corn; cattle, hogs. Plotted 1843.

Kingston, city (□ 8 sq mi/20.8 sq mi; 2006 population 22,828), ⊙ ULSTER county, SE NEW YORK, on the HUDSON RIVER at the mouth of RONDOUT CREEK; 41°55′N 74°00′W. A tourist hub for the CATSKILL-SHAWANGUNK recreational area; it makes a small variety of manufactured goods and trades in agricultural products from the surrounding area. The first permanent settlement (called Wiltwyck) was established in 1652. Kingston served as the first capital of New York state until it was burned by the British in October 1777. Its growth in the early 19th century was stimulated by the construction of the DELAWARE AND HUDSON CANAL. It has undergone major economic changes: once a major producer of sandstone, then cigars, then shirts, and then computer mainframes (IBM closed in 1985). Among notable landmarks are many old Dutch stone houses; the Senate house (1676), meeting place of the first New York state legislature; the old Dutch church (1659) and cemetery (1661); the burial place of James Clinton; and "Slabsides," former cottage of John Burroughs, located 9 mi/14.5 km S at WEST PARK. Incorporated as a village 1805, and as a city through the union (1872) of Kingston and Rondout. The conductor and music writer Robert Craft was born here.

Kingston, city (2006 population 5,553), ⊙ ROANE county, E TENNESSEE, on arm of WATTS BAR Reservoir of CLINCH River and 34 mi/55 km WSW of KNOXVILLE; 35°52′N 84°31′W. Large coal-fired steam plant built by Tennessee Valley Authority (1955) generates electricity. Manufacturing; waterfront recreation.

Kingston (KEENG-stuhn), former township (□ 81 sq mi/210.6 sq km; 2001 population 45,946), SE ONTARIO, E central CANADA; 44°18′N 76°33′W. Amalgamated into city of Kingston in 1998.

Kingston, township, Queenstown-Lakes district, SOUTH ISLAND, NEW ZEALAND, on S shore of LAKE WAKATIPU and 23 mi/37 km S of QUEENSTOWN. Lake port serving tourism; extensive livestock industry.

Kingston (KING-stuhn), town, suburb 7 mi/11 km S of HOBART, SE TASMANIA, on DERWENT RIVER estuary; 42°58′S 147°19′E. Orchards.

Kingston, town, including Kingston village, PLYMOUTH county, SE MASSACHUSETTS, on PLYMOUTH BAY and 5 mi/8 km NW of PLYMOUTH; 41°59′N 70°45′W. Publishing. Settled 1620, incorporated 1726.

Kingston, town, ROCKINGHAM county, SE NEW HAMPSHIRE, 6 mi/9.7 km SW of EXETER; 42°54′N 71°04′W. Manufacturing (kitchen cabinets, lumber); timber; agriculture (nursery crops, vegetables; cattle, poultry; dairying). Josiah Bartlett's 18th-century home (remodeled). Kingston State Park, between Great Pond (W) and Powwow Pond (SE) in S. Country Pond in S. Incorporated 1694.

Kingston, town (2006 population 1,538), MARSHALL county, S OKLAHOMA, 8 mi/12.9 km SSE of MADILL. Lake TEXOMA to S and E (RED RIVER is S; WASHITA River is E); 34°00′N 96°43′W. Lake Texoma State Park to E.

Kingston, unincorporated town (2000 population 1,611), KITSAP county, W WASHINGTON, 5 mi/8 km W of EDMONDS, on Appletree Cove of PUGET SOUND, at NE end of Kitsap Peninsula; 47°48′N 122°30′W. W terminus of ferry from Edmonds. Fishing; fruit, dairying. City of SEATTLE to SE. Port Madison Indian Reservation to S.

Kingston (KING-stuhn) or **Kingston South East**, village and port, SE SOUTH AUSTRALIA, on E shore of LACEPEDE BAY and 150 mi/241 km SSE of ADELAIDE; 34°14′S 140°21′E. Railroad terminus; dairy products, livestock; fishing (lobster). Sundial.

Kingston (KING-stuhn), ⊙, principal village of NORFOLK ISLAND, AUSTRALIA, S PACIFIC, on Sydney Bay, S coast; 29°04′S 167°57′E. Commercial and tourist center; produces citrus and passion fruit. Convict ruins, swimming at Emily Bay.

Kingston (KEENGS-tuhn), fishing village, MORAY, NE Scotland, on Spey Bay of MORAY FIRTH, at mouth of the SPEY river, and 8 mi/12.9 km ENE of ELGIN; 57°42′N 03°07′W. Established 1784 by natives of KINGSTON UPON HULL.

Kingston, village (2000 population 659), BARTOW county, NW GEORGIA, 13 mi/21 km E of ROME, near ETOWAH River; 34°14′N 84°57′W. Manufacturing of bird baths. Confederate cemetery.

Kingston, village, DE KALB county, N ILLINOIS, on South Branch of KISHWAUKEE RIVER (bridged here) and 9 mi/14.5 km NNW of SYCAMORE; 42°05′N 88°45′W. In rich agricultural area.

Kingston, village, SOMERSET county, SE MARYLAND, 22 mi/35 km SSW of SALISBURY. In vegetable farm area; lumber. Notable for Kingstown Hall, built in the early 1800s and the birthplace in 1815 of Anna Ella Carroll, daughter of Thomas King Carroll, governor of Maryland (1830–1831).

Kingston, village (2000 population 450), TUSCOLA county, E MICHIGAN, 12 mi/19 km SE of CARO; 43°24′N 83°11′W. In agricultural area. Shay Lake to S; Evergreen and Cat lakes to SW.

Kingston, village (2000 population 120), MEEKER county, S central MINNESOTA, 25 mi/40 km S of ST. CLOUD, on North Fork CROW RIVER; 45°11′N 94°18′W. Fish processing; agriculture (grain, livestock, dairying). Lake Francis to NE.

Kingston (KINGZ-stuhn), village (2006 population 1,040), ROSS county, S OHIO, 10 mi/16 km NNE of CHILLICOTHE; 39°28′N 82°55′W. Grain products; gas wells.

Kingston, village (2006 population 131), PIUTE county, S UTAH, 2 mi/3.2 km S of JUNCTION, just E of confluence with SEVIER River, on E Fork; 38°12′N 112°10′W. Dairying. Pointe Reservoir to N; parts of Fishlake National Forest to N and NW, Dixie National Forest to S.

Kingston, residential village (2000 population 5,446) in S. KINGSTON town, WASHINGTON county, S Rhode Island, 24 mi/39 km S of PROVIDENCE; 41°28′N 71°31′W. In agricultural area. University of Rhode Island here. Has many historic 18th- and 19th-century houses.

Cross-references are shown in SMALL CAPITALS. The pronunciation guide is shown on page xix. The sources of population figures are shown on page xvii.

Kingston (KEENGZ-tuhn), unincorporated village, FAYETTE county, S central WEST VIRGINIA, 31 mi/50 km SE of CHARLESTON; 37°58′N 81°18′W. In coal-mining region.

Kingston, village (2006 population 292), GREEN LAKE county, central WISCONSIN, on GRAND RIVER and 35 mi/56 km W of FOND DU LAC; 43°41′N 89°07′W. In lake, wildlife, dairy and farm area.

Kingston, borough (2006 population 13,131), LUZERNE county, NE PENNSYLVANIA, suburb 1 mi/1.6 km NW of WILKES-BARRE, on the SUSQUEHANNA RIVER; 41°15′N 75°53′W. Although chiefly residential, it has varied manufactures (food products, textiles, machinery, upholstered furniture, apparel, conveyors). Settled 1769, incorporated 1857.

Kingston, AUSTRALIA: see HEATHERTON.

Kingston-by-Sea, ENGLAND: see SHOREHAM-BY-SEA.

Kingston Mines, village, PEORIA county, central ILLINOIS, on ILLINOIS RIVER and 12 mi/19 km SW of PEORIA; 42°06′N 88°45′W. In agricultural and bituminous-coal-mining area; gravel pits.

Kingston-on-Murray (KING-stuhn–awn–MU-ree), township, SOUTH AUSTRALIA state, S central AUSTRALIA, 133 mi/214 km NE of ADELAIDE, on MURRAY RIVER. In agricultural area; orchards, vineyards. Formerly named Thurk.

Kingston upon Hull, ENGLAND: see HULL.

Kingston upon Thames (KING-stuhn UP-ahn TEMZ), outer borough (☐ 14 sq mi/36.4 sq km; 2001 population 147,243) of GREATER LONDON, SE ENGLAND; 51°25′N 00°18′W. Mainly residential, the borough has light engineering works and manufacturing (electronic equipment). It also contains one of the largest shopping centers in outer London. In the 10th century, several Anglo-Saxon kings were crowned at Kingston upon Thames; the stone believed to have been used during the coronations is preserved in the market place. Modernization of the area began in the early 19th century. Kingston College of Further Education and Kingston University are here. Kingston Grammar School (where historian Edward Gibbon attended briefly) was founded in 1561. Districts include Coombe, The Maldens, and Surbiton.

Kingstown, town, ⊙ St. Vincent and the Grenadines, West Indies; 13°08′N 61°13′W. The chief port of St. Vincent, Kingstown is an export center for the island's agr. industry as well as a port of entry for tourists. Airport.

Kingstown, IRELAND: see DÚN LAOGHAIRE.

Kingstown, Rhode Island: see NORTH KINGSTOWN and SOUTH KINGSTOWN.

Kingstree, town (2006 population 3,352), ⊙ WILLIAMSBURG county, E central SOUTH CAROLINA, on BLACK RIVER and 38 mi/61 km S of FLORENCE; 33°40′N 79°49′W. Manufacturing includes ethyl alcohol, sportswear and other fabric clothing, hardwood veneers, enzymes, processed meats, rubber and plastic products. Trade center for agricultural products such as tobacco, soybeans, grains, livestock, timber, cotton. Hunting and fishing resort. Settled 1732.

Kingsville, city (2006 population 24,394), ⊙ KLEBERG county, S TEXAS, 35 mi/56 km SW of CORPUS CHRISTI; 27°30′N 97°51′W. Elevation 66 ft/20 m. It is headquarters of the gigantic KING RANCH; city is located in middle of ranch. The city is a processing center for cotton, cattle, hogs, horses, poultry, sorghum, vegetables and dairy products in a farm, oil, uranium, stone, and gas area; manufacturing (tortillas, horseback riding equipment, printing). Large petrochemical and gas plants are in the vicinity. Seat of Texas A&M University–Kingsville. Head of Cayo del Grullo, arm of BAFFIN BAY to SE. Incorporated 1911.

Kingsville (KEENGZ-vil), town (☐ 95 sq mi/247 sq km; 2001 population 19,619), S ONTARIO, E central CANADA, on Lake ERIE, 25 mi/40 km SE of WINDSOR; 42°02′N 82°45′W. Food processing. Nearby is large bird sanctuary.

Kingsville, town, JOHNSON county, W central MISSOURI, 18 mi/29 km W of WARRENSBURG; 38°44′N 94°04′W. Manufacturing (aluminum mold castings, casting machines).

Kingsville, hamlet (2000 population 4,214), BALTIMORE county, N MARYLAND, 15 mi/24 km NE of downtown BALTIMORE; 39°27′N 76°25′W. Nearby Jerusalem Mill (begun 1772), where guns were made during the Revolutionary War, is a part of Gunpowder State Park. The Kingsville Inn, encompassing a small house believed to have been built before 1740, is now incorporated into a funeral home.

Kingswood (KINGZ-wuhd), town (2001 population 62,679), South Gloucestershire, SW ENGLAND, 4 mi/6.4 km E of BRISTOL; 51°28′N 02°30′W. Residential suburb of Bristol; printing and light engineering industries. The area is noted for its open-air chapel, which marks the site of Methodist open-air sermons on HANHAM MOUNT by John Wesley and George Whitefield in the 18th century.

Kingswood (KINGZ-wuhd), village (2001 population 2,025), S GLOUCESTERSHIRE, central ENGLAND, 15 mi/24 km NE of BRISTOL; 51°37′N 02°21′W. Previously silk milling and manufacturing of elastic.

Kington (KING-tuhn), town (2001 population 2,597), NW Herefordshire, W ENGLAND, on ARROW RIVER and 17 mi/27 km NW of HEREFORD; 52°12′N 03°01′W. Agricultural market. Limestone quarries nearby. Has 14th-century church.

Kingussie (KENG-uh-see), small town (2001 population 1,410), HIGHLAND, N Scotland, on SPEY RIVER, and 29 mi/47 km S of INVERNESS; 57°05′N 04°03′W. Health resort and tourist center. Previously woolen mills. Houses Highland Folk Museum. Poet James Macpherson born here. Kingussie was the chief town of BADENOCH AND STRATHSPEY, a former district of Highland (abolished 1996). Just S of Kingussie, on the Spey, is Ruthven, with ruins of fortifications built against the Highlanders in 1718, destroyed by them in 1746. After battle of CULLODEN MOOR (1746), the Highlanders assembled here for the last time.

King Wilhelm Land (VIL-helm), Danish *Kong Wilhelm Land*, region, NE GREENLAND, on GREENLAND SEA, extends c.150 mi/241 km N-S; 74°30′–76°30′N 22°00′W.

King William (KEENG WIL-yuhm), county (☐ 285 sq mi/741 sq km; 2006 population 15,381), E VIRGINIA; ⊙ KING WILLIAM; 37°42′N 77°05′W. In Tidewater region; bounded SW by PAMUNKEY RIVER, NE by MATTAPONI RIVER, which join at SE tip of county to form YORK RIVER estuary, arm of CHESAPEAKE BAY. Manufacturing (pulp and paper) at WEST POINT; agriculture (corn, barley, wheat, soybeans, hay, legumes; cattle; dairying); timber; fish and shellfish industries. Pamunkey Indian Reservation in S; Mattaponi Indian Reservation in SE. Zoar State Forest in NW. Formed 1702.

King William (KEENG WIL-yuhm), unincorporated village, ⊙ KING WILLIAM county, E VIRGINIA, 27 mi/43 km ENE of RICHMOND; 37°41′N 77°00′W. Agriculture (grain, soybeans; cattle; dairying). Zoar State Forest to NW. Mattaponi Indian Reservation to SE; Pamunkey Indian Reservation to S.

King William Island, part of the Arctic Archipelago, in the Arctic Ocean, central Northwest Territories, Canada, between Boothia Peninsula and Victoria Island. The N coast of the island was explored (1831) by Sir James C. Ross. In 1837, Thomas Simpson of the Hudson's Bay Co. traced the S coast. The ships of the expedition of Sir John Franklin were wrecked off the W coast, and the island was further explored by searchers for Franklin, notably John Rae and Sir Francis Lake McClintock. Roald Amundsen wintered here in 1903–1904 while on his way through the NW Passage. See P. F. Cooper, *Island of the Lost* (1961).

King William Island (☐ 5,062 sq mi/13,111 sq km), KITIKMEOT region, NUNAVUT territory, CANADA, in ARCTIC ARCHIPELAGO, between VICTORIA ISLAND and BOOTHIA PENINSULA. Lakes abound on the low (elevation 449 ft/137 m), rolling plain. Caribou summer range. The N coast of the island was explored (1831) by Sir James C. Ross. In 1837, Thomas Simpson of the Hudson's Bay Company traced the S coast. The ships of the expedition of Sir John Franklin were wrecked off the W coast, and the island was further explored by searchers for Franklin, notably John Rae and Sir Francis L. McClintock. Remains of Franklin expedition found here. Roald Amundsen wintered there in 1903–1904 while on his way through the NORTHWEST PASSAGE.

King William's Town, town, BUFFALO CITY municipality, EASTERN CAPE province, SOUTH AFRICA, on BUFFALO RIVER, and 30 mi/48 km WNW of EAST LONDON on the N2 highway; 32°52′S 27°22′E. Elevation 1,820 ft/555 m. Cotton milling, tanning of hides, soap manufacturing; center of livestock-raising, grain-growing region. Has Kaffrarian Museum (1889), botanical gardens. Education center for the district. Founded 1825 as mission station, it was capital of what was then called BRITISH KAFFRARIA until 1865. Popularly called simply King. Close to two large black residential towns Bisho (3 mi/5 km NE) and zwelitsha (2 mi/4 km SE) where there is a textile plant.

Kingwood, town (2006 population 2,954), ⊙ PRESTON county, N WEST VIRGINIA, 19 mi/31 km SE of MORGANTOWN; 39°28′N 79°40′W. Coal-mining, timber, and agricultural area (grain, apples, grapes; livestock; poultry; dairying). Manufacturing (lumber, bronze products, coal processing). West Virginia Northern railroad excursion trains to TUNNELTON. Briery Mountain Wildlife Management Area to SE; Alpine Lake Ski Resort to E. Founded 1811.

Kingwood (king-WUHD), unincorporated village, HARRIS county, SE TEXAS, industrial suburb 25 mi/40 km NNE of downtown HOUSTON, on N shore of W arm of LAKE HOUSTON (West Fork of SAN JACINTO RIVER). Manufacturing (calibration equipment, atmospheric sensors, chart recorders).

Kinhwa, China: see JINHUA.

Kinia, town, THIRD REGION/SIKASSO, MALI, on border with BURKINA FASO, 99 mi/159 km NE of SIKASSO.

Kiniati (keen-YAH-tee), village, BAS-CONGO province, W CONGO, on railroad and 40 mi/64 km NNW of BOMA; 05°20′S 12°56′E. Elev. 734 ft/223 m. Coffee, rubber, and cacao plantations; palm-oil milling, coffee processing. Also known as KANDE KINIATI.

Kinistino (ki-NI-sti-no), town (2006 population 643), central SASKATCHEWAN, CANADA, 35 mi/56 km ESE of PRINCE ALBERT; 52°57′N 105°02′W. Grain elevators, mixed farming.

Kinkaid Lake, reservoir, JACKSON county, S ILLINOIS, on Kincaid Creek, 7 mi/11.3 km W of MURPHYSBORO; 10 mi/16 km long; 37°47′N 89°27′W. Maximum capacity 79,000 acre-ft. Formed by CRISENBERRY DAM (87 ft/27 m high), built (1970) for water supply. Partly in Shawnee National Forest.

Kinkala (keen-kah-LAH), town, ⊙ POOL region, S Congo Republic, near railroad, 30 mi/48 km WSW of BRAZZAVILLE. Market center; food processing.

Kinki, JAPAN: see HONSHU.

Kinkony, Lake (keen-KOON) (☐ 55 sq mi/143 sq km), MAHAJANGA province, NW MADAGASCAR, near the coast S of MITSINJO; c.18 mi/29 km wide, 10 mi/16 km–15 mi/24 km long; 16°10′S 45°50′E. Rice region. Drains into MAHAVAVY RIVER.

Kinkonzi, CONGO: see KIKONZI.

Kinkosi (keen-KO-see), town, BAS-CONGO province, W CONGO, 100 mi/160 km S of KINSHASA; 05°38′S 15°40′E. Elev. 2,545 ft/775 m.

Kinloch (KIN-lahk), town (2000 population 449), residential suburb, E MISSOURI, 12 mi/19 km NW of downtown ST. LOUIS; 38°44′N 90°19′W. An African-American suburb that was a resettlement area for victims of the race riots in EAST SAINT LOUIS in the 1920s.

Area is shown by the symbol ☐, and capital city or county seat by ⊙.

Kinlochleven (KIN-lawk-LE-vin), town (2001 population 897), S HIGHLAND, N Scotland, at head of LOCH LEVEN, at mouth of Leven River, and 9 mi/14.5 km SE of FORT WILLIAM; 56°42'N 04°58'W. Former site of important hydroelectric station and aluminum works.

Kinloss (kin-LAHS), former township, hamlet (□ 75 sq mi/195 sq km; 2001 population 1,471), SW ONTARIO, E central CANADA; 44°05'N 81°25'W. Amalgamated into Huron-Kinloss in 1999.

Kinloss (KIN-lahs), fishing village (2001 population 1,931), MORAY, NE Scotland, on Findhorn Bay of MORAY FIRTH, and 3 mi/4.8 km NE of FORRES; 57°37'N 03°34'W. Has remains of Cistercian abbey founded 1150 by David I.

Kinmen, TAIWAN: see JINMEN.

Kinmount (KIN-mount), unincorporated village, S ONTARIO, E central CANADA, 40 mi/64 km NNW of PETERBOROUGH, and included in city of Kawartha Lakes; 44°47'N 78°39'W. Dairying, farming.

Kinmundy (KIN-mun-dee), city (2000 population 892), MARION county, S central ILLINOIS, 13 mi/21 km NNE of SALEM; 38°46'N 88°50'W. In oil-producing area; food processing. Agriculture (corn, wheat, livestock).

Kinna (SHIN-nah), town, Älvsborg county, SW SWEDEN, on Viskan River, 17 mi/27 km SW of BORÅS; 57°30'N 12°41'E. Ruins of medieval castle.

Kinnairds Head, Scotland: see FRASERBURGH.

Kinnaur, district (□ 2,471 sq mi/6,424.6 sq km), NW HIMACHAL PRADESH state, N INDIA; ⊙ REKONG.

Kinnear's Mills (ki-NEERZ MILZ), village (□ 36 sq mi/93.6 sq km; 2006 population 354), Chaudière-Appalaches region, S QUEBEC, E CANADA, 9 mi/14 km from THETFORD MINES; 46°13'N 71°23'W.

Kinnelon (KE-nuh-luhn), borough (2006 population 9,681), MORRIS county, N NEW JERSEY, 12 mi/19 km NW of PATERSON and 14 mi/23 km NNE of MORRISTOWN; 40°58'N 74°23'W. In suburban area.

Kinneret (kee-NE-ret) or **Kinnereth**, two adjoining agricultural settlements, LOWER GALILEE, NE ISRAEL, on SW shore of SEA OF GALILEE, 5 mi/8 km SSE of TIBERIAS; 32°44'N 35°34'E. Kinneret Jewish village (Moshava), village, small private farms; growing fruit and vegetables, dairying. Established 1908. Kinneret (Kibbutz), mixed farming, tropical fruit; agriculture; manufacturing building materials. Founded 1913.

Kinneret, Lake, ISRAEL: see GALILEE, SEA OF.

Kinney (ki-NEE), county (□ 1,365 sq mi/3,549 sq km; 2006 population 3,342), SW TEXAS; ⊙ BRACKETTVILLE; 29°21'N 100°25'W. On S edge of EDWARDS PLATEAU and crossed E-W by BALCONES ESCARPMENT; RIO GRANDE (Mexican border) is SW boundary. Drained by NUECES RIVER and Sycamore (forms part of W boundary). Ranching area (cattle, sheep, goats; cotton; oats, sorghum); wool, mohair marketed; hunting. Formed 1850.

Kinney, village (2000 population 199), ST. LOUIS county, NE MINNESOTA, in MESABI IRON RANGE, 9 mi/14.5 km W of VIRGINIA; 47°31'N 92°43'W. Large open-pit iron mine nearby. Superior National Forest to N.

Kinoe (kee-NO-e), largest town on E coast of OSAKI-KAMI-SHIMA, Toyota county, HIROSHIMA prefecture, W JAPAN, 28 mi/45 km E of HIROSHIMA; 34°13'N 132°55'E. Mandarin oranges, prunes. Fishery (sea bream, flattfish, shrimp).

Kinomoto (kee-NO-mo-to), town, Ika county, SHIGA prefecture, S HONSHU, central JAPAN, on N shore of LAKE BIWA, 40 mi/65 km N of OTSU; 35°29'N 136°13'E. Kinomoto Jizo [=Buddha statue] (19.7 ft/6 m high) is in the Joshin Temple here.

Kinosaki (kee-NO-sah-kee), town, Kinosaki county, HYOGO prefecture, S HONSHU, W central JAPAN, 47 mi/76 km N of KOBE; 35°37'N 134°48'E. Hot-springs resort; winter sports. Fruits, chestnuts. Straw products; mulberry tree work. Well-known historic streetscape. Sanin Kaigan [=seacoast] National Park nearby.

Kinosota, community, S MANITOBA, W central CANADA, 35 mi/57 km ESE of MAKINAK, and in ALONSA rural municipality; 50°54'N 98°51'W.

Kinpo (KEEN-po), town, Hioki county, KAGOSHIMA prefecture, SW KYUSHU, SW JAPAN, 16 mi/25 km S of KAGOSHIMA; 31°27'N 130°20'E. Scallions.

Kinrooi (KEEN-roi), commune (2006 population 12,054), Maaseik district, LIMBURG province, NE BELGIUM, 5 mi/8 km NNW of MAASEIK, in NE corner of BELGIUM; 51°09'N 05°45'E.

Kinross (KIN-rahs), town (2001 population 4,681), PERTH AND KINROSS, E Scotland, on LOCH LEVEN, and 9 mi/14.5 km N of DUNFERMLINE; 56°12'N 03°25'W. Previously woolen and linen manufacturing. Mary, Queen of Scots was imprisoned in the nearby Loch Leven Castle for eleven months (1567–1568). Kinross House, in the style of an Italian Renaissance mansion, was built for James II of England in 1685.

Kinross, town (2000 population 80), KEOKUK county, SE IOWA, 14 mi/23 km NE of SIGOURNEY; 41°27'N 91°59'W. Livestock; grain.

Kinsale (kin-SAIL), Gaelic *Cionn tSáile*, town (2006 population 2,298), CORK county, SW IRELAND, on BANDON RIVER estuary, and 13 mi/21 km S of CORK; 51°42'N 08°31'W. Fishing port and seaside resort. Manufacturing (sheet steel, electrical components, yacht equipment). Kinsale was an Anglo-Norman settlement. Hugh O'Neill was defeated here 1601 trying to relieve his Spanish allies. James II landed at Kinsale in 1689. The town surrendered to John Churchill in 1690. It was a British naval base in the 17th and 18th centuries. St. Multose Church dates from the 12th century.

Kinsale (KIN-sail), unincorporated village, WESTMORELAND county, E virginia, 50 mi/80 km ESE of FREDERICKSBURG, on Yeocomico River, inlet of POTOMAC RIVER estuary; 38°01'N 76°34'W. Manufacturing (oyster processing, lumber, fertilizer); agriculture (grain, soybeans; cattle).

Kinsale (KIN-sail), village, W MONTSERRAT, WEST INDIES, just SE of PLYMOUTH. Sea-island cotton, fruit. Abandoned after 1997 eruption of SOUFRIÉRE Hills volcano; now lies in exclusion zone.

Kinsarvik (KIN-sahr-veek), village, HORDALAND county, SW NORWAY, on S shore of HARDANGERFJORDEN, at mouth of SØRFJORD, 22 mi/35 km NNE of ODDA. Terminus of ferry across Hardangerfjord to Kvanndel. Has stone church (c.1290).

Kinsele (keen-SE-lai), village, BAS-CONGO province, W CONGO, 80 mi/129 km SE of KINSHASA; 04°08'S 16°26'E. Elev. 1,581 ft/481 m.

Kinsella (kin-SE-luh), hamlet, E ALBERTA, W CANADA, 17 mi/28 km N of SEDGEWICK, in BEAVER County; 53°00'N 111°32'W.

Kinsey, town (2000 population 1,796), Houston co., SE Alabama, 6 mi/9.7 km NNE of Dothan; 31°17'N 85°20'W. In peanut-growing region. Named for Eliza Kinsey, the first postmaster, appointed in 1866. Inc. in 1957.

Kinshan, TAIWAN: see CHINSHAN.

Kinshasa (keen-SHAH-suh), city (2004 population 7,017,000), ⊙ CONGO, Kinshasa federal district (ville), W CONGO, a port on the CONGO RIVER; 04°18'S 15°18'E. Elev. 583 ft/177 m. It is the country's largest city and its administrative, communications, and commercial center. Major industries are food and beverage processing, tanning, construction, ship repairing, and the manufacturing of chemicals, mineral oils, textiles, and cement. A transportation hub, it is the terminus of the railroad from MATADI and of navigation on the CONGO RIVER from KISANGANI; the international airport is a major link for African air traffic with EUROPE and the Americas. There is motorboat service to BRAZZAVILLE, CONGO REPUBLIC, on the opposite bank of MALEBO POOL. In 1881 Henry M. Stanley, the Anglo-American explorer, renamed Kinshasha LEOPOLDVILLE after his patron, Leopold II, king of the Belgians. In 1898 the railroad link with MATADI was completed, and in 1926 the city succeeded BOMA as the capital of the BELGIAN

CONGO. Its main growth occurred after 1945. A major anti-Belgian rebellion that took place here in January 1959 started the country on the road to independence (June 1960). In 1966 the city's name was changed from LEOPOLDVILLE to Kinshasa, the name of one of the villages that occupied the site in 1881. Modern Kinshasa is an educational and cultural center and is the seat of Lovanium University of Kinshasa (1954), which has an archaeological museum, the National School of Law and Administration, a telecommunications school, a research center for tropical medicine, and a museum of Africana. Historical buildings include the chapel of the American Baptist Missionary Society (1891) and a Roman Catholic cathedral (1914). There is a large stadium (seating capacity about 70,000). The city is famous as a center for modern African music.

Kinshui, village, NW TAIWAN, 13 mi/21 km SSW of S HSINCHU; 24°37'N 120°53'E. Natural-gas producing center; benzine and carbon black.

Kins'ka River (KEEN-skah) (Russian *Konka*), approximately 93 mi/150 km long, left tributary of the DNIEPER River, in ZAPORIZHZHYA oblast, Ukraine; rises in AZOV UPLAND 17 mi/27 km SE of POLOHY, flows generally W, past Polohy, ORIKHIV, and KOMYSHUVAKHA, into KAKHOVKA RESERVOIR of the Dnieper River, 12 mi/19 km SSE of ZAPORIZHZHYA. Formed (around 1770) a Russian defense line against the Crimean Tatars.

Kinsley, town (2000 population 1,658), ⊙ EDWARDS county, SW central KANSAS, on ARKANSAS RIVER, and 35 mi/56 km ENE of DODGE CITY; 37°55'N 99°24'W. Elevation 2,160 ft/658 m. Railroad junction. Trade center for wheat and cattle region. Manufacturing (power cylinders; concrete). SANTA FE TRAIL passed through here. Incorporated 1878.

Kinsman, village (2000 population 109), GRUNDY county, NE ILLINOIS, 15 mi/24 km SSW of MORRIS; 41°11'N 88°34'W. In agricultural area.

Kinsman (KINZ-muhn), village, TRUMBULL county, NE OHIO, 18 mi/29 km NE of WARREN, near PENNSYLVANIA line; 41°26'N 80°35'W. Brass products, lumber, animal feed.

Kinsman Mountain, peak (4,363 ft/1,330 m) of WHITE MOUNTAINS, W New Hampshire, just N of KINSMAN NOTCH and 8 mi/12.9 km NW of North Woodstock village, in White Mountain National Forest. APPALACHIAN TRAIL crosses summit.

Kinsman Notch, pass in WHITE MOUNTAINS, Grafton county, W New Hampshire, 5 mi/8 km W of North Woodstock, in White Mountain National Forest. Lost River (Tributary of Moosilauke River), running through caverns, is scenic feature. Source of Wild Ammonoosuc River flows NW. APPALACHIAN TRAIL crosses notch. State Highway 112 passes through.

Kinston (KIN-stuhn), city (□ 17 sq mi/44.2 sq km; 2006 population 22,729), ⊙ LENOIR county, E NORTH CAROLINA, 70 mi/113 km SE of RALEIGH on the NEUSE RIVER; 35°16'N 77°35'W. It is a market for bright leaf tobacco and other agricultural products. Service industries; manufacturing (concrete, boats, dairy products, lumber, textiles and apparel, industrial chemicals, appliances, food processing). Lenoir Community College Harmony Hall (1772). Richard Caswell Memorial. CSS Neuse State Historical Site, Confederate gunboat. Settled c.1740, incorporated 1849.

Kinston, town (2000 population 602), Coffee co., S Alabama, 16 mi/26 km SSW of Elba, near Pea River. Originally known as 'Cross Trains' because of the intersection of two roads, it was changed to 'Boone' in 1900, the maiden name of Pink Hickman, wife of the donor of land for the town's depot. It was renamed 'Pink' for Mr. Hickman, and then changed to 'Kinston' for Kinston, NC, former home of another resident of the town. Inc. in 1920.

Kinta, village (2006 population 249), HASKELL county, E OKLAHOMA, 12 mi/19 km SSW of STIGLER, on Sansbois Creek; 35°07'N 95°14'W. In farm area. Sansbois Mountains to S.

Kintail Forest (KIN-tal FOR-est), deer forest (□ 40 sq mi/104 sq km), HIGHLAND, N Scotland, at head of LOCH DUICH, at foot of Beinn Fhada or Ben Attow (3,383 ft/1,031 m).

Kintampo (kin-TAHM-po), town, local council headquarters, BRONG-AHAFO region, GHANA, on road, and 35 mi/56 km NE of WENCHI; 08°03′N 01°43′W. Hardwood, rubber, cacao.

Kinta River (KIN-tah), right affluent of PERAK RIVER, 60 mi/97 km long, S central PERAK, MALAYSIA; rises in central Malayan range N of CAMERON HIGHLANDS; flows W and S, past Tanjung Rambutan, IPOH, and BATU GAJAH, to Perak River, 10 mi/16 km N of Teluk Intan; 04°30′N 101°03′E. The Kinta Valley, along middle course, between KLEDANG RANGE(W) and central Malayan range, is leading tin-mining area of MALAY.

Kintbury (KINT-buh-ree), agricultural village (2001 population 4,898), West Berkshire, S ENGLAND, on KENNET RIVER, and 6 mi/9.7 km W of NEWBURY; 51°24′N 01°27′W. Tile manufacturing. Has Norman church. Saxon burial ground found here.

Kintinian, town, Siguiri prefecture, Kankan administrative region, NE GUINEA, near MALI border, 20 mi/32 km NW of SIGUIRI; 11°36′N 09°23′W.

Kintinku (keen-TEEN-koo), town, SINGIDA region, central TANZANIA, 40 mi/64 km WNW of DODOMA, on railroad; 05°56′S 35°13′E. Cattle, sheep, goats; corn, wheat.

Kintla Peak, MONTANA: see LEWIS RANGE.

Kintore (KIN-tor), village (2001 population 1,696), Aberdeenshire, NE Scotland, on DON RIVER, and 11 mi/18 km NW of ABERDEEN; 57°14′N 02°21′W. Nearby is Hallforest Castle of great antiquity.

Kintus, RUSSIA: see SALYM.

Kintyre (KIN-tei-uhr), peninsula (42 mi/68 km long, 10 mi/16 km wide), Argyll and Bute, W Scotland, joined to the mainland at the isthmus of TARBERT between EAST LOCH TARBERT and WEST LOCH TARBERT; 55°30′N 05°35′W. The MULL OF KINTYRE, at the SW tip, is 13 mi/21 km from IRELAND. The majority of the terrain is hilly and uncultivated. CAMPBELTOWN is the main town. Airfield at Machrihanish.

Kinu (KEE-noo) or **Kin-u**, township, SAGAING division, MYANMAR, on railroad, and 15 mi/24 km WNW of SHWEBO. Important irrigated rice area.

Kinuso (ki-NOO-so), village (2001 population 231), central ALBERTA, W CANADA, 26 mi/42 km from town of SLAVE LAKE, in BIG LAKES municipal district; 55°20′N 115°25′W. Incorporated 1949.

Kinver (KIN-vuh), town (2001 population 6,805), S STAFFORDSHIRE, W ENGLAND, 5 mi/8 km WSW of BRIERLY HILL; 52°27′N 02°14′W. Manufacturing of radio parts, metalworking. Has church dating from 14th century. Just W is the hill of Kinver Edge (543 ft/166 m) with remains of a Saxon camp.

Kinwarton (KIN-wuh-tuhn), village (2001 population 1,164), WARWICKSHIRE, central ENGLAND, 1 mi/1.6 km E of ALCESTER; 52°13′N 01°51′W. Kinwarton Dovecote (14th century) has a fine ogee doorway.

Kinwat (kin-waht), village, ADILABAD district, ANDHRA PRADESH state, central INDIA, on PENGANGA RIVER, and 22 mi/35 km W of ADILABAD. Millet, rice; cotton ginning. Sometimes spelled Kinvat.

Kinyeti, Mount (kin-YE-tee) (10,456 ft/3,187 m), SE SUDAN, near boundary with UGANDA. Highest mountain in SUDAN.

Kinzambi (keen-ZAHM-bee), village, BANDUNDU province, SW CONGO, on KWILU RIVER, and 25 mi/40 km NNW of KIKWIT; 04°59′S 18°47′E. Elev. 1,026 ft/312 m. Seat of River. Catholic mission, seminary. Business school.

Kinzao (keen-ZAH-o), village, BAS-CONGO province, CONGO, 30 mi/48 km NW of MATADI; 05°48′S 15°57′E. Elev. 2,198 ft/669 m. Also spelled KINZAW.

Kinzaw, CONGO: see KINZAO.

Kinzenga (keen-ZEN-guh), village, BAS-CONGO province, W CONGO, 100 mi/160 km S of KINSHASA; 05°50′S 15°17′E. Elev. 2,322 ft/707 m.

Kinzhal (keen-ZHAHL) [Russian=dagger], peak (9,281 ft/2,829 m) in N rocky front range of the central Greater CAUCASUS Mountains, KABARDINO-BALKAR REPUBLIC, SE European Russia, 40 mi/64 km W of NAL'CHIK, to the right of the MALKA RIVER.

Kinzia (keen-ZEE-yah), village, BANDUNDU province, W CONGO, on left bank of KASAI RIVER; 03°36′S 18°26′E. Elev. 748 ft/227 m. Palm-oil production.

Kinzig River (KIN-tsik), 60 mi/97 km long, in BADEN-WÜRTTEMBERG, S GERMANY, rises in the BLACK FOREST 3 mi/4.8 km SSE of FREUDENSTADT; 48°25′N 08°25′E. Flows SW, W, and NW, past OFFENBURG, to the RHINE at KEHL.

Kinzig River, 45 mi/72 km long, in HESSE, W GERMANY, rises in the RHÖN MOUNTAINS, flows WSW to the Main at HANAU; source at 50°17′N 09°38′E.

Kinzua Dam, Pennsylvania: see ALLEGHENY RESERVOIR.

Kioa (kee-O-ah), volcanic island (□ 9 sq mi/23.4 sq km), FIJI, SW PACIFIC OCEAN, c.2 mi/3.2 km E of VANUA LEVU; 16°39′S 179°55′E. Inhabited by Tuvaluans. Formerly Tate Island

Kioga, UGANDA: see KYOGA.

Kionga, MOZAMBIQUE: see QUIONGA.

Kion-Khokh (KEE-uhn–HOKH), highest peak (11,230 ft/3,423 m) in the N rocky front range of the central Greater CAUCASUS Mountains, NORTH OSSETIAN REPUBLIC, SE European Russia, 35 mi/56 km WSW of VLADIKAVKAZ, W of the OSSETIAN MILITARY ROAD.

Kiowa, county (□ 1,785 sq mi/4,641 sq km; 2006 population 1,413), E COLORADO; ⊙ EADS; 38°25′N 102°44′W. Borders on KANSAS; watered by BIG SANDY and ADOBE creeks and reservoirs in S. Cattle; wheat, sunflowers, sorghum. SAND CREEK Massacre site in N; Adobe Creek Reservoir (BLUE LAKE) on S boundary, in SW. Group of reservoirs in S center; Nee So Pah, Nee Nashe, Nee Grande, Nee Shah (the Nee Reservoirs); others extend into PROWERS county to S, formed from Kicking Bend, Santanta, and other irrigation canals of ARKANSAS RIVER. Formed 1889.

Kiowa (KEE-uh-wah), county (□ 722 sq mi/1,877.2 sq km; 2006 population 2,969), S KANSAS; ⊙ GREENSBURG; 37°33′N 99°17′W. Rolling plain, located in Red Hills region, watered by RATTLESNAKE and Mule creeks and MEDICINE LODGE RIVER. Wheat, cattle, sorghum, corn, hay, soybeans. Formed 1886.

Kiowa (KEE-uh-wah), county (□ 1,030 sq mi/2,678 sq km; 2006 population 9,778), SW OKLAHOMA; ⊙ HOBART; 34°55′N 98°58′W. Bounded N by NORTH FORK OF RED RIVER; drained by ELK CREEK; WASHITA River forms far E part of N boundary. Part of low WICHITA MOUNTAINS in E. Agriculture (wheat, cotton, oats, sorghum; cattle, sheep). Manufacturing (rubber and plastic products) at Hobart. Granite and marble quarrying; oil wells. Great Plains State Park and Tom Steed Lake reservoir in S; ALTUS LAKE reservoir on W boundary. Formed 1901.

Kiowa (KEE-uh-wah), town (2000 population 1,055), BARBER county, S KANSAS, near OKLAHOMA border, 18 mi/29 km S of MEDICINE LODGE; 37°01′N 98°28′W. In cattle region. In 1900, Carry Nation damaged saloon here. Incorporated 1885.

Kiowa, village (2000 population 581), ⊙ ELBERT county, central COLORADO, on KIOWA CREEK, 40 mi/64 km SE of DENVER; 39°21′N 104°27′W. Elevation 6,363 ft/1,939 m. Dairy products; cattle; wheat, oats, sunflowers.

Kiowa (KEE-uh-wah), village (2006 population 704), PITTSBURG county, SE OKLAHOMA, 17 mi/27 km SSW of MCALESTER; 34°43′N 95°54′W. In agricultural area (corn, oats); manufacturing of metal products McAlester U.S. Army Ammunition Plant to N.

Kiowa Creek, central COLORADO, rises in N EL PASO county, flows intermittently 111 mi/179 km N past KIOWA and BENNETT, then NNE to SOUTH PLATTE RIVER 4 mi/6.4 km WNW of FORT MORGAN.

Kipampwa (kee-PAHM-pwah), village, SINGIDA region, W central TANZANIA, 100 mi/161 km SW of MANYONI, in RUNGWA GAME RESERVE; 07°05′S 34°03′E. Livestock; grain.

Kiparissia (kee-pah-ree-SEE-ah), Latin *Cyparissiae*, town, MESSENIA prefecture, SW PELOPONNESE department, extreme SW GREECE, port and railroad head on Gulf of KIPARISSIA, 28 mi/45 km WNW of KALAMATA; 37°27′N 22°04′E. Fisheries; wheat, olive oil. Lignite mine just NE. Castle on cliff above town. Known as Arcadia from Middle Ages until destruction (1825) by Turks. Also Kyparissia.

Kiparissia, Gulf of (kee-pah-ree-SEE-ah), inlet of IONIAN SEA, off SW PELOPONNESE department, SW GREECE, SE of Cape KATAKOLO; 35 mi/56 km wide, 10 mi/16 km long. Kyparissia village is on SE shore. Formerly called Gulf of Arcadia; formerly spelled Kyparissia.

Kipawa (KI-puh-wah), village (□ 18 sq mi/46.8 sq km; 2006 population 577), Abitibi-Témiscamingue region, SW QUEBEC, E CANADA, 7 mi/11 km from TÉMISCAMING; 46°47′N 78°59′W.

Kipchak, village, SE KARAKALPAK REPUBLIC, UZBEKISTAN, on the left bank of the AMU DARYA River and 30 mi/48 km SE of NUKUS; 42°12′N 60°05′E. Metalworks.

Kipemba (kee-PEM-buh), town, BAS-CONGO province, W CONGO, on road 80 mi/129 km S of KINSHASA; 05°33′S 15°30′E.

Kipembawe (kee-paim-BAH-wai), village, MBEYA region, SW central TANZANIA, 75 mi/121 km N of MBEYA, near LUPA RIVER (source to N); 07°54′S 33°21′E. Cattle, goats, sheep; corn, wheat.

Kipengere Range (kee-pain-GAI-rai), IRINGA region, S TANZANIA, SE of MBEYA and N of Lake NYASA; 50 mi/80 km long; rises to 9,708 ft/2,959 m.

Kipercheny, MOLDOVA: see CHIPERCENI.

Kiperere Chini (kee-pai-RAI-rai CHEE-nee), village, LINDI region, SE TANZANIA, 65 mi/105 km W of LINDI, on Mbeweburu River; 09°57′S 38°41′E. Timber; corn, wheat; sheep, goats.

Kipfenberg (KIP-fen-berg), village, BAVARIA, S GERMANY, in MIDDLE FRANCONIA, on the Atmühl, and 10 mi/16 km ENE of EICHSTÄTT; 48°57′N 11°24′E. Has early-17th-century baroque church. Towered over by ruins of medieval castle with 13th-century watchtower.

Kipili (kee-PEE-lee), town, RUKWA region, W TANZANIA, 70 mi/113 km WNW of SUMBAWANGA, on Lake TANGANYIKA; 07°27′S 30°36′E. Lake port. Fish; goats, sheep; corn, wheat.

Kipini (kee-PEE-neeh), town, COAST province, SE KENYA, small port at mouth of TANA River on INDIAN OCEAN, 120 mi/193 km NNE of MOMBASA; 02°31′S 40°31′E. Sisal center; cotton copra, sugarcane; fisheries.

Kipkabus (KEEP-kah-bahs), village, RIFT VALLEY province, W KENYA, on UASIN GISHU plateau, on railroad and 22 mi/35 km SSE of ELDORET; 00°18′N 35°31′E. Coffee, wheat, corn, tea, wattle.

Kipkarren River (keep-KAH-rain), village, RIFT VALLEY province, W KENYA, on railroad and 22 mi/35 km WNW of ELDORET; 00°37′N 34°58′E. Coffee, tea, sisal, corn.

Kiplelion, town, RIFT VALLEY province, W KENYA, on railroad and 20 mi/32 km WSW of LONDIANI; elevation 6,339 ft/1,932 m. Dairy plant; corn, coffee, wheat. Junction for road to KERICHO and SOTIK tea-growing areas.

Kipling (KIP-ling), town (2006 population 973), SE SASKATCHEWAN, CANADA, 65 mi/105 km NE of WEYBURN; lumbering, mixed farming.

Kipnuk, village (2000 population 644), W ALASKA, near BERING SEA and ETOLIN Strait, 100 mi/161 km SW of BETHEL; 59°55′N 164°05′W.

Kippax (KIP-aks), village (2006 population 10,200), WEST YORKSHIRE, N ENGLAND, 7 mi/11.3 km E of LEEDS; 53°46′N 01°22′W. Has Norman church.

Kippen (ki-PEN), agricultural village (2001 population 934), STIRLING, central Scotland, near the FORTH RIVER, and 9 mi/14.5 km W of Stirling; 56°07′N 04°10′W. Vineyards.

Area is shown by the symbol □, and capital city or county seat by ⊙.

Kippenheim (KIP-pen-heim), village, BADEN-WÜRTTEMBERG, SW GERMANY, 21 mi/34 km N of FREIBURG; 48°18′N 07°49′E.

Kippford (KIP-fuhrd) or **Scaur**, village, DUMFRIES AND GALLOWAY, S Scotland, on small inlet of SOLWAY FIRTH, and 4 mi/6.4 km S of DALBEATTIE; 54°52′N 03°49′W. Seaside resort. Previously granite quarrying.

Kippure (ki-PYOOR), mountain (2,473 ft/754 m), N WICKLOW county, E IRELAND, on DUBLIN county border, 12 mi/19 km S of DUBLIN; 53°10′N 06°20′W.

Kip's Bay, district of MANHATTAN borough of NEW YORK city, SE NEW YORK, along EAST RIVER S of 42nd Street.

Kiptsy (keep-TSI), village, NW SARATOV oblast, SE European Russia, near highway and railroad, 2 mi/3.2 km NNW of YEKATERINOVKA, to which it is administratively subordinate; 52°05′N 44°20′E. Elevation 777 ft/236 m. Forestry services.

Kipushi (kee-POO-shee), town, KATANGA province, SE CONGO, 15 mi/24 km WSW of LUBUMBASHI; 11°46′S 27°14′E. Elev. 4,363 ft/1,329 m. Major copper- and zinc-mining center, railroad terminus, and customs station on ANGOLA border. Has copper and zinc concentrating plants as well as chemical works producing sulphuric acid, glycerin, fatty acids, sodium chlorate, caustic soda, and lubricants. Its noted Prince Leopold Mine is the only deep-shaft mine in SHABA. Benedictine mission.

Kira (KEE-rah), town, Hazu county, AICHI prefecture, S central HONSHU, central JAPAN, 28 mi/45 km S of NAGOYA; 34°48′N 137°04′E. Strawberries.

Kirakat (ki-rah-kaht), town, JAUNPUR district, SE UTTAR PRADESH state, N central INDIA, on the GOMATI RIVER, and 16 mi/26 km SE of Jaunpur; 25°38′N 82°55′E. Barley, rice, corn, wheat. Also spelled Kerakat.

Királyháza, UKRAINE: see KOROLEVE.

Király Hegy, SLOVAKIA: see KRALOVA HOLA.

Királyhelmec, SLOVAKIA: see KRALOVSKY CHLMEC.

Kirama, village, SOUTHERN PROVINCE, SRI LANKA; 06°13′N 80°40′E. Fishing center. Produces coir rope; trades in coconuts.

Kirando (kee-RAHN-do), village, RUKWA region, W TANZANIA, 75 mi/121 km NW of SUMBAWANGA, on Lake TANGANYIKA, 4 mi/6.4 km N of KIPILI; 06°25′S 30°35′E. Fish; grain, subsistence crops; livestock.

Kiranur (KEE-ruh-noor), town, headquarters of Kulattur Tahsil, in PUDUKKOTAI district, TAMIL NADU state, S India, 13 mi/21 km N of Pudukkottai. Millet, peanuts, rice. Also spelled Keeranur.

Kiranur, INDIA: see ULUNDURPETTAI.

Kiraoli (ki-ROU-lee), town, tahsil headquarters, AGRA district, W UTTAR Pradesh state, N central INDIA, 14 mi/23 km WSW of Agra; 27°09′N 77°47′E. Pearl millet, gram, wheat, barley.

Kira Panayia, Greece: see PELAGONESI.

Kiratpur (KI-ruht-puhr), town, BIJNOR district, N UTTAR PRADESH state, N central INDIA, 10 mi/16 km NNE of Bijnor. Sugar refining; rice, wheat, gram, sugarcane. Ruins of eighteenth century Pathan fort.

Kirby (kir-BEE), town (2006 population 8,574), BEXAR county, S central TEXAS, residential suburb 6 mi/9.7 km ENE of SAN ANTONIO; 29°27′N 98°23′W. Fort Sam Houston Military Reservation to W.

Kirby, town, CALEDONIA CO., NE VERMONT, just NE of St. Johnsbury. Population peaked in 1840 at 540 residents. In agricultural area.

Kirby, village, PIKE county, SW ARKANSAS, 38 mi/61 km WSW of HOT SPRINGS. Cinnabar mine. Lake Greerson reservoir to SW; Daily State Park on N shores to W.

Kirby (KUHR-bee), village (2006 population 126), WYANDOT county, N central OHIO, 21 mi/34 km NW of MARION; 40°49′N 83°25′W.

Kirby, village (2006 population 55), HOT SPRINGS county, N central WYOMING, on BIGHORN RIVER, and 12 mi/19 km N of THERMOPOLIS; 43°47′N 108°10′W. Elevation 4,270 ft/1,301 m. Coal-shipping point.

Kirby, Lake (kir-BEE), in S part of ABILENE, TAYLOR county, W central TEXAS, impounded by dam in small CEDAR CREEK (a S tributary of CLEAR FORK OF BRAZOS RIVER); c.2.5 mi/4 km long; capacity 8,500 acre-ft.

Kirby Muxloe (KUH-bee MUHKS-lo), village (2001 population 4,523), LEICESTERSHIRE, central ENGLAND, 4 mi/6.4 km NW of LEICESTER; 52°37′N 01°13′W. Light manufacturing. Has remains of a moated, brick-built castle begun in 1480 but never completed.

Kirbyville (KUR-bee-vil), town (2006 population 2,028), JASPER county, E TEXAS, near NECHES RIVER, c.40 mi/ 64 km NNE of BEAUMONT; 30°39′N 93°54′W. Railroad junction; lumbering; cattle, horses; vegetables; light manufacturing. Founded 1895, incorporated 1926.

Kirby, West (KUH-bee), town (2001 population 12,869), MERSEYSIDE, NW ENGLAND, on WIRRAL Peninsula at seaward end of DEE RIVER estuary; 53°22′N 03°09′W. Resort. To N is HOYLAKE.

Kirchberg, commune, BERN canton, NW central SWITZERLAND, on EMME RIVER, and 11 mi/18 km NE of BERN. Metal- and woodworking, textiles.

Kirchberg, commune, ST. GALLEN canton, NE SWITZERLAND, 16 mi/26 km W of ST. GALLEN. Embroideries, cotton textiles.

Kirchberg (KIRKH-berg), town, RHINELAND-PALATINATE, W GERMANY, in the HUNSRÜCK, 22 mi/35 km W of BINGEN; 49°57′N 07°25′E. Wine; manufacturing (plastics, glass). Chartered 1259.

Kirchberg, town, SAXONY, E central GERMANY, at N foot of the ERZGEBIRGE, 7 mi/11.3 km S of ZWICKAU; 50°38′N 12°32′E. Textiles.

Kirchberg an der Jagst (KIRKH-berg ahn der YAHGST), town, BADEN-WÜRTTEMBERG, S GERMANY, on the JAGST RIVER, 6 mi/9.7 km NW of CRAILSHEIM; 49°12′N 09°59′E. Has 16th–18th-century castle; now a retirement home.

Kirchberg an der Pielach (KIRKH-berg ahn der PEE-lahkh), township, central LOWER AUSTRIA, 15 mi/24 km SSW of Sankt Pölten; 48°02′N 15°26′E. Dairy farming, summer resort.

Kirchberg in Tirol (KIRKH-berg in ti-ROL), village, TYROL, W AUSTRIA, 3 mi/4.8 km W of Kitzbühel; 47°27′N 12°19′E. Tourism center. Pilgrimage church.

Kirchbichl (KIRKH-bikhl), village, TYROL, W AUSTRIA, on the INN River, and 6 mi/9.7 km SSW of KUFSTEIN; 47°31′N 12°05′E. Cement; hydroelectric station on Inn River.

Kirchdorf, GERMANY: see POEL.

Kirchdorf am Inn (KIRKH-dorf ahm IN), village, BAVARIA, S GERMANY, in UPPER BAVARIA, on the INN River, 5 mi/8 km S of ROSENHEIM; 47°47′N 12°07′E. Tourism.

Kirchdorf an der Krems (KIRKH-dorf ahn der KREMS), town, SE central UPPER AUSTRIA, 15 mi/24 km E of GMUNDEN; 47°54′N 14°07′E. Market center; tools, automobile parts, cement; brewery.

Kirchen (KIR-khuhn), village, RHINELAND-PALATINATE, W GERMANY, on the SIEG RIVER, and 7 mi/11.3 km SW of SIEGEN; 50°48′N 07°53′E. Incorporated WEHBACH in 1942.

Kirchenlamitz (kir-khuhn-LAH-mits), town, BAVARIA, E central GERMANY, in UPPER FRANCONIA, in the FICHTELGEBIRGE, on small Lamitz River, and 8 mi/12.9 km W of SELB; 50°09′N 11°57′E. Manufacturing of porcelain, sporting equipment, plastics; granite quarrying.

Kirchentellinsfurt (kir-khuhn-TEL-lings-furt), village, BADEN-WÜRTTEMBERG, SW GERMANY, on NECKAR River, 4 mi/6.4 km NW of REUTLINGEN; 48°32′N 09°09′E.

Kirchenthumbach (kir-khuhn-TUM-bahkh), village, BAVARIA, S GERMANY, in UPPER PALATINATE, 16 mi/26 km SE of BAYREUTH; 49°45′N 11°44′E.

Kirchhain im Bezirk Kassel (KIRKH-hein im be-TSIRK KAHS-sel), town, HESSE, central GERMANY, near the OHM RIVER, 6 mi/9.7 km E of MARBURG; 50°50′N 08°55′E. Manufacturing of carpets; metalworking. Has many half-timbered houses, a half-

timbered town hall (1562), and churches from the 13th, the 14th, and the 15th centuries.

Kirchheim (KIRKH-heim), village, HESSE, central GERMANY, 20 mi/32 km N of FULDA; 50°50′N 09°36′E.

Kirchheim am Neckar (KIRKH-heim ahm NEK-kahr), village, BADEN-WÜRTTEMBERG, SW GERMANY, 8 mi/ 12.9 km S of HEILBRONN; 49°03′N 09°09′E.

Kirchheim bei München (KIRKH-heim bei MOOIN-khen), suburb of MUNICH, BAVARIA, S GERMANY, in UPPER BAVARIA, 9 mi/14.5 km E of city center; 48°10′N 11°46′E.

Kirchheim-Bolanden (KIRKH-heim–bo-LAHND-en), town, RHINELAND-PALATINATE, W GERMANY, 16 mi/ 26 km W of WORMS; 49°40′N 08°00′E. Manufacturing of motor vehicles, containers, toys; meat processing; hardstone quarrying; wine. Has 18th-century castle; remains of town wall and two town gates from 14th century.

Kirchheim unter Teck (KIRKH-heim un-tuhr TEK), city, BADEN-WÜRTTEMBERG, S GERMANY, at N foot of the TECK, 9 mi/14.5 km SE of ESSLINGEN; 48°38′N 09°27′E. Largely industrial; machinery, electronics, textiles and paper; foundries. Tourism. Has 14th-century church and former ducal palace, now local museum. Chartered 1270.

Kirchhein, GERMANY: see DOBERLUG-KIRCHHEIN.

Kirchhellen (KIRKH-hel-len), district 5 mi/8 km N of BOTTROP, NORTH RHINE–WESTPHALIA, W GERMANY, in the RUHR industrial district; 51°38′N 06°55′E. Site of a large amusement park which opened in 1996.

Kirchhundem (KIRKH-hund-em), town, NORTH RHINE–WESTPHALIA, W GERMANY, 15 mi/24 km NNE of SIEGEN; 51°06′N 08°06′E. Manufacturing of paper; wood- and metalworking. Tourism.

Kirchlengern (kirkh-LENG-gern), town, NORTH RHINE–WESTPHALIA, W GERMANY, 6 mi/9.7 km N of HERFORD; 52°12′N 08°38′E. Metal- and woodworking; manufacturing of furniture and machinery; power plant.

Kirchlinteln (kirkh-LING-teln), village, LOWER SAXONY, N GERMANY, 25 mi/40 km SE of BREMEN; 52°56′N 09°18′E.

Kirchseeon, village, BAVARIA, S GERMANY, in UPPER BAVARIA, 17 mi/27 km SE of MUNICH; 48°04′N 11°55′E.

Kirchwärder, GERMANY: see VIERLANDE.

Kirchzarten (kirkh-TSAHR-tuhn), village, BADEN-WÜRTTEMBERG, SW GERMANY, in BLACK FOREST, 5 mi/8 km ESE of FREIBURG; 47°58′N 07°58′E. Paper milling, woodworking. Summer resort.

Kircubbin (kuhr-KUH-bin), fishing village (2001 population 1,214), NE DOWN, SE Northern Ireland, on E shore of STRANGFORD LOUGH, 10 mi/16 km SE of NEWTOWNARDS; 54°29′N 05°32′W. Sailing center.

Kirda, UZBEKISTAN: see GULBAKHOR.

Kirdimi (keer-dee-MEE), town, BORKOU-ENNEDI-TI-BESTI administrative region, N CHAD, 45 mi/72 km NW of FAYA.

Kireka (ki-RAI-kah), town, WAKISO district, CENTRAL region, S UGANDA, suburb 5 mi/8km NE of KAMPALA; 00°20′N 32°38′E.

Kireli, village, W central TURKEY, near NE shore of Lake BEYSEHIR, 50 mi/80 km W of KONYA. Formerly also Kereli.

Kirenga River (kee-RYEN-gah), 340 mi/547 km long, in E central IRKUTSK oblast, RUSSIA; rises in the BAYKAL RANGE, flows N, past KAZACHINSKOYE, to the LENA RIVER at KIRENSK. Navigable for 90 mi/145 km above its mouth.

Kirensk (KEE-reensk), city (2005 population 13,180), N central IRKUTSK oblast, E central SIBERIA, RUSSIA, on the LENA RIVER, at the mouth of the KIRENGA RIVER, on road, 440 mi/708 km NNE of IRKUTSK, 150 mi/241 km NNW of Lake BAYKAL; 57°47′N 108°06′E. Elevation 869 ft/264 m. Ship repairing; port facilities; lumbering. Founded in 1630, made city in 1775.

Kireyevsk (kee-RYE-eefsk), city (2006 population 25,260), E TULA oblast, central European Russia, on crossroads and railroad, 25 mi/40 km SE of TULA; 53°56′N 37°55′E. Elevation 734 ft/223 m. Iron ore and lignite mining (MOSCOW BASIN); mica processing; metal goods, furniture, artificial furs, stockings; dairy plant. Made city in 1956.

Kirghiz Range, Russian *Aleksandrovskiy Khrebet*, W branch of TIANSHAN mountain system, on KAZAKH-STAN-KYRGYZSTAN border; extends from BOOM GORGE on CHU RIVER (Kyrgyzstan) 225 mi/362 km W to area of ZHAMBYL (Kazakhstan); rises to 14,800 ft/4,511 m. Most important passes are Merke, Kara-Balty, and Shamsi. Gives rise to many affluents of Chu and Talass rivers. KUNGEI ALATAU (Küngey Ala-Too) forms E extension. Also Kyrgyz Range; formerly called Alexander Range.

Kirghiz Soviet Socialist Republic: see KYRGYZSTAN.

Kirgis Nor, MONGOLIA: see HYARGAS NUUR.

Kirgiz-Mayaki, RUSSIA: see KIRGIZ-MIYAKI.

Kirgiz-Miyaki (keer-GEES–mee-yah-KEE), village (2005 population 7,680), W BASHKORTOSTAN Republic, E European Russia, on road, 50 mi/80 km W of STERLITAMAK; 53°38′N 54°47′E. Elevation 715 ft/217 m. In agricultural area (wheat, rye, flax, hemp). Also called Kirgiz-Mayaki.

Kir Hareseth, JORDAN: see KARAK, AL.

Kir Haresh, JORDAN: see KARAK, AL.

Kir Heres, JORDAN: see KARAK, AL.

Kiriat-arba or **Kiryat-Arba**, settlement, Jewish suburb NE of HEBRON, WEST BANK. Tourism; some industry; wine-making. Founded in 1970 by Jewish religious settlers. Also spelled Kirjath-arba or Qiryat Arba.

Kiribathgoda, town, WESTERN PROVINCE, SRI LANKA; 06°57′N 79°53′E. Residential suburb of COLOMBO. Tire factory. Trades in rice and vegetables.

Kiribati (KEE-RAH-BAHS), officially the Republic of Kiribati (□ 342 sq mi/889.2 sq km; 2007 population 107,817), consists of thirty-three islands scattered across 2,400 sq mi/3,800 km of the PACIFIC OCEAN astride the equator; ⊙ TARAWA; 05°00′S 170°00′W.

Geography
It includes eight of the eleven LINE ISLANDS, including KIRITIMATI (formerly Christmas Island), as well as the GILBERT and PHOENIX groups and BANABA Island (one of the three great phosphate rock islands in the Pacific).

Population
The population is nearly all Micronesian, with about 30% concentrated on Tarawa. Overcrowding has been a problem, and in 1988 it was announced that 4,700 residents of the main island group would be resettled in less populated islands. Languages spoken are English (official) and Gilbertese, a Micronesian dialect.

Economy
Fishing and the growing of taro and bananas form the basis of the mainly subsistence economy. Fish and copra became the chief exports after the mining of Banaba's once thick phosphate deposits ended in 1979.

History
The islands were administered (1892–1916) with the Gilbert Islands as a British protectorate that became (1916) the British GILBERT AND ELLICE ISLANDS Colony. They gained self-rule in 1971 and, after the Ellice Islands gained (1978) independence as TUVALU, the remaining islands were granted independence (1979) as Kiribati. UNITED STATES claims to several islands, including KANTON (formerly Canton) and ENDERBURY, were abandoned in 1979. In 1994 Teburovo Tito was elected president. In 1995 Kiribati moved the international date line to the eastern border of the sprawling island nation so that it would no longer be divided by it. Tito was reelected in 1998 and 2003, but in March 2003, he was removed from office by a no-confidence vote, and replaced by a Council of State. Anote Tong was elected president in July 2003.

Government
Kiribati is governed under the constitution of 1979. The president, who is both head of state and head of government, is elected by popular vote for a four-year term and is eligible for two more terms. The unicameral House of Parliament has forty-two members, most elected by popular vote, who serve four-year terms. The current head of state is President Anote Tong (since July 2003). Administratively the country is divided into three units (the Gilbert, Line, and Phoenix islands), and subdivided into six districts. There are also twenty-one island councils, one for each of the inhabited islands.

Kirigalpotta (KI-RI-guhl-poth-thah), peak (7,857 ft/2,395 m) in S CENTRAL PROVINCE, S central SRI LANKA, on HORTON PLAINS, 11 mi/18 km S of NUWARA ELIYA; second highest peak in the country; 06°48′N 80°46′E.

Kirikhan, township, S TURKEY, 23 mi/37 km NNE of ANTAKYA; 39°32′N 41°20′E. Grain.

Kirikkale, town, TURKEY, near the E bank of the KIZIL IRMAK, on ANKARA-KAYSERI railroad, and 35 mi/56 km E of Ankara; 39°51′N 33°32′E. Agricultural and manufacturing center; grain, fruit; mohair.

Kirikovka, UKRAINE: see KYRYKIVKA.

Kirillov (kee-REE-luhf), city (2006 population 8,240), W central VOLOGDA oblast, N central European Russia, on a canal joining the VOLGA basin with the NORTHERN DVINA basin, on road, 80 mi/129 km NW of VOLOGDA; 59°52′N 38°23′E. Elevation 374 ft/113 m. Center of an agricultural region; wood and food industries. Tourist center. Has a 14th century monastery, now museum of old Russian art, and other historical and architectural monuments. Chartered in 1776.

Kirillovka, UKRAINE: see KYRYLIVKA, town.

Kirillovo (kee-REE-luh-vuh), village, W LIPETSK oblast, S central European Russia, on a short left tributary of the SOSNA RIVER, on road, 11 mi/18 km NW of YELETS; 52°44′N 38°14′E. Elevation 600 ft/182 m. In agricultural area; distillery.

Kirilly (kee-REE-li), village, SW SMOLENSK oblast, W European Russia, on the OSTER River (tributary of the SOZH RIVER, DNIEPER River basin), near highway, 3 mi/5 km NE, and under administrativer jurisdiction, of ROSLAVL'; 53°59′N 32°55′E. Elevation 623 ft/189 m. Logging, woodworking.

Kirin, village, BENISHANGUL-GUMUZ state, W central ETHIOPIA, 12 mi/19 km WSW of ASOSA, and 8 mi/12.9 km from SUDANESE border; 09°57′N 34°22′E.

Kirin, China: see JILIN.

Kirinda, village, SOUTHERN PROVINCE, SRI LANKA, 11 mi/18 km SE of TISSAMAHARAMA; 06°12′N 81°30′E. Fishing. Buddhist pilgrimage site.

Kirindi Oya, river, S SRI LANKA, 73 mi/117 km long, rises in UVA PROVINCE, S UVA BASIN, just ESE of BANDARAWELA, at 06°13′N 81°17′E; flows N and S into SOUTHERN PROVINCE, to INDIAN OCEAN 6 mi/9.7 km S of TISSAMAHARAMA. Forms part of border between Uva and Southern provinces. Land reclamation project (□ 37 sq mi/96 sq km) on lower course near Tissamaharama.

Kiri River (KEE-ree), c.30 mi/48 km long, in N ALBANIA, rises in North Albanian Alps, flows SSW to the DRIN River just S of SHKODËR. Site of village of Mes, with its famous 18th-century Turkish bridge.

Kirirom (KI-REE-RUHM), town, central CAMBODIA, small hill station; 11°20′N 104°05′E. Future tourism plans.

Kirishi (KEE-ree-shi), city (2005 population 56,390), S LENINGRAD oblast, NW European Russia, on the VOLKHOV RIVER (landing), on road, 70 mi/113 km SE of SAINT PETERSBURG, and 32 mi/51 km SSW of VOLKHOV; 59°27′N 32°01′E. Railroad junction; petroleum products (oil pipeline from Almatyevsk); electric power station fueled by mazut (fuel oil); also sawmilling. Made city in 1965.

Kirishima (kee-REE-shee-mah), town, Aira county, KAGOSHIMA prefecture, SW KYUSHU, SW JAPAN, 22 mi/35 km N of KAGOSHIMA; 31°48′N 130°50′E. Nearby is Kirishima Yaku National Park (Japan's oldest).

Kirishima-yama (KEE-ree-SHEE-mah–YAH-mah), collective name for two volcanic peaks on MIYAZAKI-KAGOSHIMA prefecture border, S KYUSHU, SW JAPAN, near TAKAHARU, 33 mi/53 km W of MIYAZAKI, in Kirishima National Park (□ 83 sq mi/215 sq km). Higher peak is Karakuni-dake (5,610 ft/1,710 m), lower is Takachiho-dake (5,194 ft/1,583 m). Hot springs on slopes of both peaks.

Kirit (KEE-rith), village, NW SOMALIA, on road, and 55 mi/89 km SE of BURAO (Buro). Camels, sheep, goats.

Kiritimati (ki-REE-ti-mah-tee), largest atoll in the PACIFIC (□ 222 sq mi/577.2 sq km), in the LINE ISLANDS of the Republic of KIRIBATI; 01°52′N 157°25′W. Known as Christmas Island, it was associated with the GILBERT AND ELLICE ISLANDS. The island's chief agricultural product is copra; coconut, timber, fish farming, solar salt; tourism. The atoll was explored by Captain James Cook in 1777, annexed by Great Britain in 1888, and included in the Gilbert and Ellice Islands Colony in 1919. British nuclear tests were conducted on the atoll in 1957 and 1958 and United States' tests in 1962. Recent rapid population growth, as the government resettles people from crowded TARAWA and other islands

Kiriwina (ki-ri-WEE-nuh), a raised atoll, largest of TROBRIAND ISLANDS, MILNE BAY province, SE PAPUA NEW GUINEA, SOLOMON SEA, SW PACIFIC OCEAN, 95 mi/153 km SE of NEW GUINEA; 30 mi/48 km long, 6 mi/9.7 km wide. Chief town and port is LOSUIA, on W coast. Yams, coconuts, palm oil; fish. Annual Yam Festival in Trobriands, July and August.

Kirjali, Bulgaria: see KURDZHALI, city.

Kirjath-arba, WEST BANK: see KIRIAT-ARBA.

Kirjath-jearim, ISRAEL: see ABU GHOSH.

Kirkagac, (Turkish=*Kirkağaç*) town, W TURKEY, on railroad, and 36 mi/58 km NNE of MANISA; 39°06′N 27°40′E. Raisins, tobacco, wheat, barley, vetch, chickpeas, cotton.

Kirkburton (kuhk-BUH-tuhn), town (2001 population 16,773), WEST YORKSHIRE, N ENGLAND, 4 mi/6.4 km SE of HUDDERSFIELD; 53°36′N 01°42′W. Previously woolen mills. Chemical works.

Kirkby (KUHK-bee), village (2001 population 40,006), MERSEYSIDE, NW ENGLAND, 6 mi/9.7 km NE of LIVERPOOL; 53°29′N 02°54′W. Some light industry; electrical engineering. Has modern church with 12th-century font.

Kirkby-in-Ashfield (KUHK-bee–in–ASH-feeld), suburb (2001 population 19,585), NOTTINGHAMSHIRE, central ENGLAND, 4 mi/6.4 km SW of MANSFIELD; 53°06′N 01°14′W. Former coal-mining site. Hosiery produced here. NEWSTEAD ABBEY (home of Lord Byron) 2 mi/3.2 km to S.

Kirkby Lonsdale (KUHK-bee LUNZ-dail), town (2001 population 2,534), CUMBRIA, NW ENGLAND, near LANCASHIRE border, on LUNE RIVER, and 10 mi/16 km SE of KENDAL; 54°12′N 02°35′W. Market town in cattle-raising and agricultural area. Has remains of Stone Age graves and excavations of Roman camp. Bridge dating back to 14th century.

Kirkbymoorside (KUHK-bee-MAW-seid), town (2001 population 3,283), NORTH YORKSHIRE, N ENGLAND, on DOVE RIVER, and 11 mi/18 km NNW of MALTON; 54°16′N 00°56′W. Has 12th–13th-century church, re-built by Sir Gilbert Scott. George Villiers, second duke of Buckingham, died here (1687). Nearby are several stone quarries. Located 3 mi/4.8 km WSW is Kirkdale, site of cave discovered in 1821 that yielded remains of ancient species of animals, as well as Stone Age implements and weapons.

Kirkby Stephen (KUHK-bee STEF-en), town (2001 population 2,495), E CUMBRIA, NW ENGLAND, on EDEN RIVER, and 8 mi/12.9 km SSE of APPLEBY; 54°28′N 02°21′W. Cattle and sheep raising. Has many 17th-century houses, and church of Saxon and Norman origin with 16th-century tower.

Kirkcaldy (kuhr-KAHL-dee), town (2001 population 46,912), FIFE, E Scotland, on the Firth of Forth opposite EDINBURGH; 56°07′N 03°10′W. Furniture manufacturing, light electrical engineering. Its port engages in coastal trade. Pathhead is the docks district. Several Flemish-style structures from the later Middle Ages are found near Kirkcaldy Harbor. There are several villages, including DYSART, strung along the shore, giving rise to the town's nickname "The Lang Toun." Adam Smith born here.

Kirkconnel (kuhr-KAH-nul), town (2001 population 2,074), DUMFRIES AND GALLOWAY, S Scotland, on NITH RIVER, and 3 mi/4.8 km WNW of SANQUHAR; 55°23′N 04°00′W. Medicinal springs nearby.

Kirkcudbright (kuhr-KOO-bree), small town (2001 population 3,447), DUMFRIES AND GALLOWAY, SW Scotland, at head of DEE RIVER estuary, and 11 m/17.6 km SW of DALBEATTIE; 54°50′N 04°02′W. It has granaries and creameries. Kirkcudbright is a market town and artists' colony.

Kirkdale, ENGLAND: see LIVERPOOL.

Kirkee, town (1991 cantonment population 61,432), PUNE district, MAHARASHTRA state, W central INDIA, 2 mi/3.2 km N of PUNE. Military station; large ordnance factories and workshops, chemical works; copper, brass, and iron manufacturing; dairy farm. Brewery 2 mi/3.2 km NW. School of Military Engineering. Scene of British victory (1817) over Marathas. Also spelled Khadki.

Kirkel (KIR-kel), village, SAARLAND, SW GERMANY, 12 mi/19 km E of SAARBRÜCKEN; 49°17′N 07°15′E.

Kirkenes (KIR-kuh-nais), village, FINNMARK county, NE NORWAY, near Russian border, on Bøkfjorden (S arm of VARANGERFJORDEN), at mouth of PASVIKELVA River, 160 mi/257 km ESE of HAMMERFEST, 90 mi/145 km NW of MURMANSK; 69°43′N 30°03′E. Port and commercial center for Norway's main iron-mining region, until the mines closed in 1997. A shipbuilding industry and medical services remain. Seaplane base and airport; terminus for the coastal steamer from BERGEN. Village grew since beginning (1910) of mining operations in region. German base in World War II, it was largely destroyed by air raids. Captured by USSR (Oct., 1944) and occupied by the Russians until shortly after the end of hostilities. Has central library and hospital.

Kirkersville (KUHRK-uhrz-vil), village (□ 2 sq mi/5.2 sq km; 2006 population 541), LICKING county, central OHIO, 12 mi/19 km SW of NEWARK; 39°57′N 82°35′W.

Kirkfield (KUHRK-feeld), former village, S ONTARIO, E central CANADA, 20 mi/32 km NW of LINDSAY; 44°33′N 78°58′W. Dairying; mixed farming. Amalgamated into KAWARTHA LAKES when that city was created in 2001.

Kirkham (KUHK-uhm), town (2001 population 7,127), W LANCASHIRE, NW ENGLAND, 7 mi/11.3 km W of PRESTON. Previously cotton milling. A Roman road is town's main street.

Kirkintilloch (kuhr-KIN-ti-lawk), town (2001 population 20,281), ⊙ East Dunbartonshire, W Scotland, on the Forth and Clyde Canal, and 6 m/9.6 km NNE of GLASGOW; 55°55′N 04°10′W. An engineering center, the town has factories that produce mining machinery, valves, and concrete. The electrical grid of S Scotland is controlled from Kirkintilloch. On the line of the ANTONINE WALL.

Kirk-Kilise, TURKEY: see KIRKLARELI.

Kirkkonummi, FINLAND: see KYRKSLÄTT.

Kirkland (KUHRK-luhnd), city (□ 4 sq mi/10.4 sq km; 2006 population 21,735), Montréal administrative region, S QUEBEC, E CANADA, on W MONTREAL ISLAND; 45°27′N 73°49′W. Incorporated 1961; was part of the city of MONTREAL 2002–2005, but voted to regain its independence in 2006. Part of the Metropolitan Community of Montreal (*Communauté Metropolitaine de Montréal*).

Kirkland, city (2006 population 46,476), KING county, W WASHINGTON, a suburb 6 mi/9.7 km ENE of SEATTLE, on E shore of LAKE WASHINGTON; 47°41′N 122°11′W. Manufacturing (semiconductors, transformers, prefabricated metal buildings, heating equipment, computer peripherals, motor vehicles, apparel, navigation

equipment, aircraft parts, medical products, machinery parts, molded plastics, petroleum refining, plywood). In the 1980s and early 1990s, Kirkland grew rapidly along with the Seattle metropolitan area. It is the seat of Northwest College (Assemblies of God). Bridle Trails State Park on S side of city. Incorporated 1905.

Kirkland, village (2000 population 1,166), DE KALB county, N ILLINOIS, on South Branch of KISHWAUKEE RIVER (bridged here), and 12 mi/19 km NW of SYCAMORE; 42°05′N 88°50′W. In rich agricultural area.

Kirkland (kuhrk-LAND), unincorporated village, CHILDRESS county, extreme N TEXAS, 10 mi/16 km ESE of CHILDRESS; 34°22′N 100°03′W. In cotton, wheat, livestock region.

Kirkland, Scotland: see MAXWELTON HOUSE.

Kirkland Lake (KUHRK-luhnd), mining town (□ 101 sq mi/262.6 sq km; 2001 population 8,616), NE ONTARIO, E central CANADA; 48°09′N 80°02′W. An important gold-mining center. Gold was discovered here in 1911 and again in the 1980s at Harker. Iron ore mining; tourism.

Kirklar, CYPRUS: see TYMBOU.

Kirklar Dag, (Turkish=*Kirklar Dağ*) peak (11,348 ft/3,459 m), NE TURKEY, in RIZE MOUNTAINS, 35 mi/56 km S of RIZE.

Kirklareli, city (2001 population 53,221), NW TURKEY; 41°45′N 27°12′E. Transportation hub, trade center for butter and cheese. During the First Balkan War the Bulgarians defeated (1912) the Turks here. The city has numerous mosques and Greek churches. Formerly Kirk-Kilise.

Kirklees (kuhk-LEEZ), locality (2001 population 388,567), WEST YORKSHIRE, N ENGLAND, 4 mi/6.4 km NNE of HUDDERSFIELD; 53°35′N 01°50′W. Has remains of a convent built 1155. Legend states that Robin Hood died here.

Kirklin, town (2000 population 766), CLINTON county, central INDIANA, near SUGAR CREEK, 10 mi/16 km SE of FRANKFORT; 40°11′N 86°22′W. In agricultural area.

Kirkliston (kuhrk-LIS-tuhn), town (2001 population 3,043), EDINBURGH, E Scotland, 8 mi/12.9 km W of Edinburgh; 55°57′N 03°24′W.

Kirkman, town (2000 population 76), SHELBY county, W IOWA, on WEST NISHNABOTNA RIVER, and 6 mi/9.7 km NE of HARLAN; 41°43′N 95°16′W.

Kirkmansville, village, TODD county, S KENTUCKY, on POND RIVER, and 17 mi/27 km NE of HOPKINSVILLE.

Kirk Mountains, on MOZAMBIQUE-MALAWI border, NW of BLANTYRE, Malawi; extend c.40 mi/64 km N-S between Ncheu and NENO; rise to 5,000 ft/1,524 m.

Kirkmuirhill (kuhrk-MEER-hil), village (2001 population 3,717), South Lanarkshire, S SCOTLAND, 6 mi/9.7 km W of LANARK; 55°40′N 03°55′W. Farming; sheep.

Kirknewton (kuhrk-NOO-tuhn), agricultural village (2001 population 1,648), WEST LOTHIAN, central Scotland, 9 mi/14.5 km WSW of EDINBURGH; 55°32′N 02°07′W.

Kirk of Mochrum, Scotland: see MOCHRUM.

Kirkoswald (kuhr-KAWS-wahld), agricultural village, South Ayrshire, SW Scotland, 4 mi/6.4 km WSW of MAYBOLE; 55°20′N 04°46′W. Here are graves of Burns's Tam o' Shanter and Souter Johnnie, and the latter's cottage, housing Burns relics. On Firth of Clyde, 2 mi/3.2 km W, is small seaside resort of Maidens, with the Shanter farm just S.

Kirkovo (KEER-ko-vo), village (1993 population 553), HASKOVO oblast, ⊙ Kirkovo obshtina, BULGARIA; 41°20′N 25°23′E.

Kirkøy, NORWAY: see HVALER.

Kirkpatrick (kuhrk-PA-trik), locality, S ALBERTA, W CANADA, on RED DEER River, and 6 mi/10 km WNW of DRUMHELLER, in KNEEHILL county; 51°30′N 112°50′W. Coal mining.

Kirkpatrick, Mount (14,856 ft/4,528 m), highest peak of QUEEN ALEXANDRA RANGE, ANTARCTICA; 84°20′S 166°25′E. Discovered 1908 by Sir Ernest Shackleton. Also known as Mount Kilpatrick.

Kirk Sandall (KUHK SAN-dul), suburb (2006 population 4,954), WEST YORKSHIRE, N ENGLAND, on DON RIVER and 4 mi/6.4 km NE of DONCASTER; 53°33′N 01°04′W. Glass-manufacturing center. Just NE is agricultural village of BARNBY DUN.

Kirkstall, ENGLAND: see LEEDS, city.

Kirkstead, ENGLAND: see WOODHALL SPA.

Kirksville, city (2000 population 16,988), ⊙ ADAIR county, N MISSOURI; 40°12′N 92°34′W. A processing, trade, and shipping center for a large agricultural area. Corn, soybeans; sheep, cattle, hogs; light manufacturing. Andrew Taylor Still founded the first school of osteopathy here in 1892; it is now the Kirksville College of Osteopathic Medicine. Truman State University. Thousand Hills State Park is nearby. Incorporated 1857.

Kirkton of Largo, Scotland: see LARGO.

Kirktown, CANADA: see SOUTH LANCASTER.

Kirkuk (kir-KOOK), city, ⊙ TA'MEEM province, NE IRAQ; 35°28′N 44°23′E. It is the center of IRAQ's oil industry and is connected by pipeline to ports on the MEDITERRANEAN SEA in LEBANON, SYRIA, and TURKEY, closed as a result of the PERSIAN GULF War. Oil production throughout the 1980s was reduced due to the IRAN-Iraq War. The city is a market for the region's produce, including cereals, olives, fruits, and cotton. There is a small textile industry. Airport. Kirkuk is built on a mound containing the remains of a settlement dating back to 3000 B.C.E. The majority of the population are Kurds. There is a large Turkmen minority.

Kirkville, town (2000 population 214), WAPELLO county, SE IOWA, 10 mi/16 km NNW of OTTUMWA; 41°08′N 92°30′W. Livestock; grain.

Kirkwall (KUHRK-wal), town (2001 population 6,206), ⊙ ORKNEY Islands, N Scotland, on the E coast of MAINLAND ISLAND; 58°58′N 02°57′W. It is the trading center and administrative seat of the Orkney Islands, and exports eggs, fish, whiskey, cattle, and sheep. Food processing and packing. Town was founded sometime prior to 1046 (when it was mentioned in a saga) and became important as a port on the N trade route to Scandinavia and the Baltic states. St. Magnus Cathedral dates from 1137. Airport 3 mi/4.8 km SE.

Kirkwood, city (2000 population 27,324), SAINT LOUIS county, E MISSOURI, a suburb 15 mi/24 km of ST. LOUIS; on MERAMEC RIVER; 38°34′N 90°25′W. Primarily residential, it has some light manufacturing. Meramec Community College. Incorporated 1865.

Kirkwood, town, EASTERN CAPE, SOUTH AFRICA, on SUNDAYS RIVER, and 40 mi/64 km NNW of PORT ELIZABETH (NELSON MANDELA METROPOLE), at foot of Zuurberg mountain range; 33°25′S 25°24′E. Elevation 236 ft/ 71 m. Railroad terminus; irrigated farming, citrus production.

Kirkwood, village (2000 population 794), WARREN county, W ILLINOIS, 6 mi/9.7 km SW of MONMOUTH; 40°52′N 90°45′W. In agricultural area.

Kirlampudi, INDIA: see SAMALKOT.

Kirlangiç, Cape, TURKEY, on SW coast of ASIA MINOR, on the MEDITERRANEAN, at entrance of Gulf of ANTALYA (Adalia); 36°12′N 30°24′E. Formerly also Khelidonia. Just S are tiny Besadalar islands, just E small SULU or Granbusa Island.

Kirloskarvadi (kir-los-kuhr-VAH-dee), village, SANGLI district, MAHARASHTRA state, W central INDIA, 22 mi/35 km NW of MIRAJ, and on railroad. Manufacturing of agriculture machinery. Named after the Kirloskar family which opened a factory making agricultural implements here in the 1940s.

Kirmasti, TURKEY: see MUSTAFA KEMALPASA.

Kirmir River, 75 mi/121 km long, in N central TURKEY, rises in KOROGLU MOUNTAINS, flows S and SW, past KIZILCAHAMAM, into an artificial lake formed by the Sariyar Dam on the SAKARYA RIVER 15 mi/24 km SW of BEYPAZARI.

Cross-references are shown in SMALL CAPITALS. The pronunciation guide is shown on page xix. The sources of population figures are shown on page xvii.

Kir Moab, JORDAN: see KARAK, AL.

Kirn, town, Rhineland-Palatinate, on the NaheRiver, 18 mi/29 km WSW of Bad Kreuznach; 49°48′N 07°27′E. Leatherworking; plastics; stone quarrying. Evidence of prehistoric settlement in the area.

Kirn (KUHRN), resort, Argyll and Bute, W Scotland, on Firth of Clyde, adjacent to DUNOON; 55°57′N 04°56′W.

Kirnasovka, UKRAINE: see KYRNASIVKA.

Kiron, town (2000 population 273), CRAWFORD county, W IOWA, 13 mi/21 km N of DENISON; 42°11′N 95°19′W. In agricultural area.

Kirongwe (kee-RON-gwai), village, PWANI region, E TANZANIA, 75 mi/121 km SE of DAR ES SALAAM, on MAFIA ISLAND, on MAFIA CHANNEL, INDIAN OCEAN; 07°48′S 39°50′E. Fish; copra; bananas; livestock.

Kirov (KEE-ruhf), city (2005 population 457,355), ⊙ KIROV oblast, E central European Russia, on the VYATKA River, 557 mi/896 km ENE of MOSCOW; 58°36′N 49°39′E. Elevation 547 ft/166 m. It is a river port, a major railroad station on the TRANS-SIBERIAN RAILROAD, and an industrial center; machinery and metalworking, chemicals (including agrochemicals and synthetic building materials), woodworking (furniture, matches), armaments, consumer electronics, and musical instruments. Regional educational and cultural center; pedagogical, agricultural, and polytechnic institutes, as well as branches of All-Russian institute of finance and economics and Perm medical institute; drama theater, museums of arts, heritage, and aviation and cosmonautics. The repose cathedral of the Trifonov monastery (1689), the Trinity Church (1775), and a library (1837) founded by Alexander Hertzen, who was an exile in the city, are notable. Founded in 1374 as Khlynov by Novgorod colonists, it was fortified against Votyak (Udmurt) and Cheremiss (Mari) attacks. It soon became the capital of an independent republic which was annexed to Moscow by Ivan III in 1489. Its location made for favorable trade conditions with Ustyug, the Volga region, and Archangelsk. In the 17th century, it grew in importance due to its location on the road from Moscow to Siberia. The city was renamed Vyatka in 1781, and Kirov in 1934. In the 19th century, it was used as a place of political exile. A railroad station opened in 1899. Administrative capital of Kirov oblast since 1936.

Kirov (KEE-ruhf), city (2005 population 39,480), SW KALUGA oblast, central European Russia, on the Bolva River, on railroad (Fayansovyy station), 100 mi/161 km SW of KALUGA; 54°04′N 34°18′E. Elevation 695 ft/211 m. Iron foundry; bathtubs and sinks, porcelain, bricks, reinforced concrete components; garment factory; dairy plant. Arose in 1745 as an iron-working center. Until 1936, known as Pesochnya.

Kirov (KEE-ruhf), oblast (□ 46,641 sq mi/121,266.6 sq km), N central European Russia; ⊙ KIROV. Rolling lowland in the VYATKA River basin; bordered N by the NORTHERN UVALS; crossed N-S by the VYATKA UVAL. Continental climate; podzolic soils; forests and peat marshes (NE) - 60% of the oblast's territory is forested. Main mineral resources (NE) are iron and phosphorites (mining at RUDNICHNY). Metallurgy in OMUTNINSK area, machine manufacturing (Kirov), fur processing, tanning, shoe manufacturing, food processing (Kirov-Slobodskoy area). Sawmills at river-railroad junctions (Kirov, KOTEL'NICH, VYATSKIYE POLYANY), match manufacturing, paper milling, wood distillation. Widespread handicraft industries (woodworking; toys, lace, felt boots). Flax grown on the right bank, rye and oats on the left bank, of the Vyatka River; dairy farming (Kirov-Kotelnich area). Trade oriented toward the Ural Mountains and Volga region. Population 70% urban. Originally constituted as Vyatka government, incorporated (1929) in Nizhegorod Territory and made (1936) a separate oblast. Divided into 39 districts and contains 18 cities, 58 towns, and over 5,000 villages and rural settlements.

Kirovabad, AZERBAIJAN: see GYANDZHA.

Kirova, Imeni (KEE-ruh-vah, EE-min-yee), town, in Surakhansky district of greater BAKY, E AZERBAIJAN, on central APSHERON Peninsula, c.10 mi/16 km NE of Baky; 40°25′N 50°01′E. Vineyards; produce.

Kirova, Imeni, UKRAINE: see KIROVE.

Kirova, Imeni, UZBEKISTAN: see BESHARYK.

Kirovakan, ARMENIA: see VANADZOR.

Kirove (KEE-ro-ve) (Russian *Kirovo*), town, central DONETS'K oblast, UKRAINE, 3 mi/5 km E of and subordinated to DZERZHYNS'K; 48°24′N 37°55′E. Elevation 869 ft/264 m. Railroad station Mahdalynivka (Russian *Magdalinovka*). Coal mine. Established in the 18th century as Severnyy Rudnik; town since 1938, called Imeni Kirova (1938–1965).

Kirovgrad (kee-ruhf-GRAT), city (2006 population 22,520), SW SVERDLOVSK oblast, extreme W Siberian Russia, in the E foothills of the central URALS, on highway, 60 mi/97 km NNW of YEKATERINBURG; 57°26′N 60°03′E. Elevation 885 ft/269 m. Railroad junction; a major copper-smelting center, based on nearby mines of BELORECHKA, KARPUSHIKHA, and LEVIKHA; chemical processing, bronze production; pyrite mining. Founded in the 18th century. Made city in 1932. Originally called Kalatinskiy Zavod and, later (1928–1935), Kalata.

Kirov Island, RUSSIA: see SERGEY KIROV ISLANDS.

Kirovo (KEE-ruh-vuh), village, central KURGAN oblast, SW SIBERIA, RUSSIA, on the MIASS River, on road, 15 mi/24 km NNW of MISHKINO; 55°33′N 63°45′E. Elevation 285 ft/86 m. In agricultural area. Until 1939, called Voskresenskoye.

Kirovo, UKRAINE: see KIROVE, Donets'k oblast.

Kirovo, UKRAINE: see KIROVOHRAD.

Kirovo, UZBEKISTAN: see BESHARYK.

Kirovo-Chepetsk (KEE-ruh-vuh–chee-PYETSK), city (2005 population 90,235), central KIROV oblast, E central European Russia, on the VYATKA River, at the mouth of the CHEPTSA River, on highway branch and railroad, 25 mi/40 km E of KIROV; 58°33′N 50°02′E. Elevation 416 ft/126 m. Electrical machinery, metalworking; agricultural chemicals and supplies; food and textile industries. Thermo-electric station. Has a theater, sanatorium. Made city in 1955. Formerly known as Cheptsa.

Kirovograd, UKRAINE: see KIROVOHRAD.

Kirovograd oblast, UKRAINE: see KIROVOHRAD oblast.

Kirovogradskaya oblast, UKRAINE: see KIROVOHRAD oblast.

Kirovohrad (kee-ro-vo-HRAHD) (Russian *Kirovograd*), city (2001 population 254,103), ⊙ of KIROVOHRAD oblast, UKRAINE, on the INHUL River; 48°30′N 32°16′E. Elevation 410 ft/124 m. It is an agricultural trade center, with large farm machinery plants (seeders, tractor parts); other manufacturing (radio components, typewriters, food processing, apparel, building materials, furniture). It has numerous post-secondary institutions, including institutes of teaching, agricultural machine building, and civil aviation; theaters, an orchestra, and a museum. Jewish population increased rapidly in the 19th century and Jews owned majority of flour mills, distilleries, and tobacco factories until they were nationalized. Founded as a fortress in 1754, it became a major garrison city named Yelyzavethrad (Russian *Yelisavetgrad*) in honor of Empress Elizabeth. In the late 1860s it became a center of agricultural machinery building, flour milling and distilling, and seasonal farm labor. Between 1881 and 1919, it was the scene of several pogroms. It was renamed Zinovi'yevs'k (Russian *Zinovyesk*) in 1924, Kirove (Russian *Kirovo*) in 1934, and Kirovohrad in 1939, when it became capital of a new oblast.

Kirovohrad (kee-ro-vo-HRAHD) (Ukrainian *Kirovohrads'ka*) (Russian *Kirovogradskaya*), oblast (□ 9,493 sq mi/24,681.8 sq km; 2001 population 1,133,052), central UKRAINE; ⊙ KIROVOHRAD. In DNIEPER UPLAND; bounded NE by DNIEPER River; drained by INHUL and INHULETS rivers. Chiefly agri-

cultural region, with wheat, sugar beets (NW, E), sunflowers (S), and potatoes in moist wooded region (N); livestock raising. Manufacturing at Kirovohrad, OLEKSANDRIYA (lignite mining), and ZNAM'YANKA. Sugar refining, flour milling, dairying. Deposits of kaolin and refractory clays; granite quarries. Formed in 1939; had (1993) twelve cities, twenty six towns, twenty one rural raions.

Kirovohrads'ka oblast, UKRAINE: see KIROVOHRAD oblast.

Kirov Peak, TAJIKISTAN: see PETER THE FIRST RANGE.

Kirovsk (KEE-ruhfsk), city (2006 population 28,960), central MURMANSK oblast, NW European Russia, on the KOLA peninsula, in the S KHIBINY MOUNTAINS, on highway branch and railroad spur, 125 mi/201 km S of MURMANSK, and 15 mi/24 km from APATITY; 67°37′N 33°40′E. Elevation 1,207 ft/367 m. The city is the center of a mining complex that produces and concentrates apatite and nepheline, raw materials for the superphosphate and aluminum industries. Its botanical garden, founded in 1931, is considered to be the northernmost in the world. It was founded in 1929, made a city in 1931, and called Khibinogorsk until 1934. Held by the Finns between 1941 and 1944, during which time it was called Hiipina.

Kirovsk (KEE-ruhfsk), city (2005 population 24,785), central LENINGRAD oblast, NW European Russia, on the NEVA RIVER, near Lake LADOGA, just W of MGA, and 35 mi/56 km E of SAINT PETERSBURG; 59°52′N 31°00′E. Railroad (Nevdubstroy station); electric power station. Also called Imeni Kirova. Until 1953, when made a city, called Nevdubstroy. Arose in 1929 with the construction of the power plant.

Kirovs'k (KEE-rovsk) (Russian *Kirovsk*), city (2004 population 64,400), SW LUHANS'K oblast, UKRAINE, in the DONBAS, 6 mi/10 km N of STAKHANOV; 48°38′N 38°39′E. Elevation 646 ft/196 m. Coal mines, sewing factory. Mining transport technical school, other professional schools. Founded as Holubivs'kyy Rudnyk (Russian *Golubovskiy Rudnik*) in 1764, shortened to Holubivka (Russian *Golubovka*) at the beginning of the 19th century; renamed Kirovs'k in 1944; city since 1962.

Kirovsk (KEE-ruvsk), town, SE AZERBAIJAN, near CASPIAN coast, 5 mi/8 km NW of LÄNKÄRÄN; 38°47′N 48°43′E. Subtropical agriculture (tea); tea processing factory.

Kirovs'k (KEE-rovsk) (Russian *Kirovsk*), town (2004 population 2,500), N DONETS'K oblast, UKRAINE, on the Zherebets' River, left tributary of DONETS River, 6 mi/10 km ENE of KRASNYY LYMAN; 49°01′N 37°56′E. Elevation 252 ft/76 m. Oil seed procesing. Established in the 1730s, town since 1938. Named originally Popivka (Russian *Popovka*); renamed (1941) Imeni Kirova; renamed (1965) Kirovs'k.

Kirovsk (KEE-rahfsk), urban settlement, MOGILEV oblast, BELARUS, ⊙ Kirovsk region, 14 mi/23 km NE of Bobruisk. Manufacturing (building materials, flax processing, vegetable-drying); dairy. Until c.1940, STARITSY.

Kirovsk, TURKMENISTAN: see BABADAYHAN.

Kirovsk, UKRAINE: see KIROVS'K, Lunans'k oblast; or KIROVS'K, Donets'k oblast.

Kirovskaya (KEE-ruhf-skah-yah), village (2006 population 5,915), SW ROSTOV oblast, S European Russia, on road and railroad, 25 mi/40 km SE of ROSTOV-NA-DONU; 46°59′N 40°04′E. Elevation 196 ft/59 m. Agricultural products.

Kirovs'ke (KEE-rov-ske) (Russian *Kirovskoye*), city (2004 population 33,700), E central DONETS'K oblast, UKRAINE, in the DONBAS, 9 mi/14 km SE of YENAKIYEVE and 5 mi/8 km NNW of SHAKHTARS'K; subordinated to the Shakhtars'k city council; 48°09′N 38°21′E. Elevation 744 ft/226 m. The city has five coal-mining enterprises, footwear manufacturing, and a sewing and knitting factory. Established as a worker settlement in 1953 and called Nova Khrestivka (Russian *Novaya Krestovka*) until 1956, when it was renamed Kirovs'ke; city since 1958.

Kirovs'ke (KEE-rov-ske) (Russian *Kirovskoye*), town, DNIPROPETROVS'K oblast, UKRAINE, suburb 10 mi/16 km WNW of DNIPROPETROVS'K, on left bank of the DNIEPER River; 48°32′N 34°51′E. Elevation 193 ft/58 m. River port; broiler factory, woodworking. Until 1938, known as Obukhivka (Russian *Obukhovka*).

Kirovs'ke (KEE-rov-ske) (Russian *Kirovskoye*), town (2004 population 6,200), E Republic of CRIMEA, UKRAINE, on railroad, and 15 mi/24 km NW of FEODOSIYA; 45°14′N 35°13′E. Flour mill. Until 1944, known as Islam-Terek.

Kirovski (KEE-ruvs-kee), town, in Binagadinsky district of Greater BAKY, AZERBAIJAN, on APSHERON Peninsula, 5 mi/8 km N of Baky.

Kirovski (kee-ROF-skee), town, central TALDYKORGAN region, KAZAKHSTAN, 10 mi/16 km SE of TALDY-KORGAN; 44°54′N 78°13′E. Tertiary level (raion) administrative center. Irrigated agriculture (sugar beets); sugar mill. Also called Imeni Kirova.

Kirovskii, town, KHATLON viloyat, TAJIKISTAN, in Vakhsh valley, 5 mi/8 km ESE of QURGHONTEPPA (Kurgan-Tyube); 37°49′N 68°51′E. In long-staple cotton area; cotton ginning. Formerly called Imeni Kirova. Also spelled Kirovskiy.

Kirovskiy (KEE-ruhf-skeeyee), town (2005 population 2,250), E ASTRAKHAN oblast, SE European Russia, on the Kamyzyak arm of the VOLGA RIVER delta mouth, on highway branch, 35 mi/56 km S of ASTRAKHAN; 45°51′N 48°07′E. Below sea level. Fish-processing center. Until 1934, known as a village of Bol'shaya Pavlova.

Kirovskiy (KEE-ruhf-skeeyee), town (2005 population 3,020), S central KURSK oblast, SW European Russia, in the SEYM RIVER basin, near highway, 25 mi/40 km SE of KURSK, and 4 mi/6 km W of SOLNTSEVO; 51°25′N 36°40′E. Elevation 649 ft/197 m. In agricultural area; food processing.

Kirovskiy (KEE-ruhf-skeeyee) town (2006 population 9,395), SW MARITIME TERRITORY, SE RUSSIAN FAR EAST, on the USSURI River, on road and near the TRANS-SIBERIAN RAILROAD (9 mi/14 km to the W), 120 mi/193 km NE of USSURIYSK; 45°05′N 133°31′E. Elevation 249 ft/75 m. Lumbering. In agricultural area. Tourist base. Prior to 1939, called Uspenka or Uspenovka.

Kirovskiy (KEE-ruhf-skeeyee), settlement, NW AMUR oblast, RUSSIAN FAR EAST, 40 mi/64 km N of SKO-VORODINO; 54°21′N 124°24′E. Elevation 3,405 ft/1,037 m. Gold mining.

Kirovskiy (KEE-ruhf-skeeyee), settlement, W central KAMCHATKA oblast, RUSSIAN FAR EAST, on the W coast of the KAMCHATKA PENINSULA, on the Sea of OKHOTSK, 95 mi/153 km N of UST'-BOL'SHERETSK; 54°13′N 155°49′E. Fish-processing plant. Town status revoked in 1987.

Kirovskoye (KEE-ruhf-skuh-ye), village, N central SA-KHALIN, RUSSIAN FAR EAST, on road and near railroad, 27 mi/43 km ESE of ALEKSANDROVSK-SAKHALINSKIY; 50°42′N 142°43′E. Elevation 452 ft/137 m. In agricultural area. Until 1937, known as Rykovskoye.

Kirovskoye, KYRGYZSTAN: see KYZYL-ADYR.

Kirovskoye, UKRAINE: see KIROVS'KE, Donets'k oblast; KIROVS'KE, Dnipropetrovs'k oblast; or KIROVS'KE, Republic of Crimea.

Kirovskoye, Crimean Republic: see KIROVS'KE.

Kirpas Peninsula, CYPRUS: see KARPAS PENINSULA.

Kirpili River (keer-pee-LEE), approximately 94 mi/151 km long, in KRASNODAR TERRITORY, S European Russia; rises in central Krasnodar Territory, less than 8 mi/13 km WNW of LADOZHSKAYA, and flows generally W, past KIRPIL'SKAYA and PLATNIROVSKAYA; turns more to the N past MEDVEDOVSKAYA and flows in a long NW-to-W arch, with many sharp twists and turns, past TIMASHEVSK and ROGOVSKAYA, to the Kirpil'skiy Liman, a lagoon through which it connects to the Sea of AZOV. Its lower course is shallow and marshy and often dries out in the summer.

Kirpil'skaya (keer-PEEL-skah-yah), village (2005 population 5,755), E central KRASNODAR TERRITORY, S European Russia, on the KIRPILI River, on highway junction, 30 mi/48 km NE of KRASNODAR; 45°22′N 39°42′E. Elevation 183 ft/55 m. In agricultural area; produce processing.

Kirriemuir (KIR-ree-mir), small town (2001 population 5,963), ANGUS, E Scotland, 5 mi/8 km WNW of FOR-FAR; 56°40′N 03°00′W. Former weaving center. Inverquharity Castle is 3 mi/4.8 km NNE.

Kirs (KEERS), city (2005 population 11,255), NE KIROV oblast, E European Russia, on railroad, on the VYATKA River, 175 mi/282 km NE of KIROV, and 45 mi/72 km N of OMUTNINSK; 59°20′N 52°16′E. Elevation 547 ft/166 m. Cable plant based on a metallurgical works in operation since 1728; logging, lumbering, timbering; produce processing, livestock, food processing (confectionary). Made city in 1965.

Kirsanov (keer-SAH-nuhf), city (2006 population 17,805), E TAMBOV oblast, S central European Russia, on the VORONA RIVER, on highway junction and railroad, 60 mi/97 km E of TAMBOV; 52°39′N 42°43′E. Elevation 426 ft/129 m. Center of an agricultural region; textile machinery, garment factories, woodworking; food industries (flour milling, meat packing, vegetable drying, sugar refining, dried skim milk). Chartered in 1779.

Kirşehir, Turkish *Kırşehir*, ancient *Andrapa*, town (2000 population 88,105), TURKEY, 85 mi/137 km SE of ANKARA; 39°09′N 34°08′E. Noted for its carpet manufacturing; grain, linseed, vetch, potatoes; mohair goats. Sometimes spelled Kir-Shehr.

Kirstenbosch National Botanical Gardens (□ 3.2 sq mi/8.3 sq km), WESTERN CAPE province, SOUTH AFRICA, 5 mi/8 km S of CAPE TOWN. One of the world's most famous reserves of indigenous flora, lying on E slopes of TABLE mountain, watered by the Liesbeeck River; this relatively small park contains 4,700 species of flora, constituting nearly 25% of all of the republic's indigenous plant species. Established 1913.

Kirthar Range (KIR-thar), mountain system along SIND-BALUCHISTAN province border, S PAKISTAN; extends c.220 mi/354 km S from MULA RIVER in E KALAT to point NE of KARACHI; c.20 mi/32 km wide; rises to 7,430 ft/2,265 m in N peak. Drained W by HAB RIVER, E by seasonal streams. SE offshoots, including LAKHI HILLS, form hilly region (KOHISTAN) in SW SIND. Range consists of limestone ridges with several hot springs, some having medicinal properties. Little cultivation, but ibex and mountain sheep are found.

Kirtipur (KIR-tee-poor), town (2001 population 40,835), central NEPAL, in KATHMANDU VALLEY, on tributary of the BAGMATI RIVER, and 4 mi/6.4 km SW of KATHMANDU; 27°40′N 85°17′E. Elevation 4,429 ft/1,350 m. Originally under suzerainty of PATAN kingdom; fell (1768) to Gurkha leader Prithvi Narayan Shah after two severe defeats of the Gurkhas.

Kirtland (KUHRT-luhnd), city (□ 17 sq mi/44.2 sq km; 2006 population 7,309), LAKE county, NE OHIO, 20 mi/32 km ENE of CLEVELAND; 41°36′N 81°20′W. Seat of Lakeland Community College. Village of KIRTLAND HILLS just N. One of the headquarters for the Latter Day Saint movement in 19th century.

Kirtland, unincorporated town (2000 population 6,190), SAN JUAN county, NW NEW MEXICO, 8 mi/12.9 km W of FARMINGTON, on SAN JUAN RIVER; 36°44′N 108°20′W. Cattle, sheep; dairying; grain, blue corn, alfalfa, potatoes, pumpkins. Navajo Indian Reservation to S and N; Ute Mountain Indian Reservation (New Mexico/COLORADO) to N.

Kirtland Air Force Base, NEW MEXICO: see ALBU-QUERQUE.

Kirtland Hills (KUHRT-luhnd HILZ), village (□ 6 sq mi/15.6 sq km; 2006 population 765), Lake county, NE OHIO, 20 mi/32 km ENE of CLEVELAND; 41°38′N 81°20′W. The first Mormon temple was built here

(1833–1836) by Joseph Smith and his followers. Settled 1808, incorporated 1926.

Kirton-in-Lindsey (KUH-tuhn), town (2001 population 2,694), North Lincolnshire, NE ENGLAND, 9 mi/14.5 km NE of GAINSBOROUGH; 53°28′N 00°36′W. Former agricultural market, with limestone quarries and light industry. Has church dating from 13th century. Airfield to SE.

Kirtorf (KIR-torf), town, HESSE, central GERMANY, 15 mi/24 km E of MARBURG; 50°46′N 09°08′E. Completely destroyed by fire in 18th century; has half-timbered town hall from 1781.

Kiruhura, administrative district, WESTERN region, SW UGANDA. As of Uganda's division into eighty districts, Kiruhura's border districts include KYENJOJO (N), SEMBABULE (E), ISINGIRO (S), MBARARA and IBANDA (W), and KAMWENGE (NW). Agricultural area. Formed in 2005 from N, E, and central portions of former MBARARA district (current Mbarara district formed from W portion, Ibanda district from NW portion, and Isingiro district from S portion).

Kiruna (KI-roo-nah), town (□ 7,500 sq mi/19,500 sq km), NORRBOTTEN county, N SWEDEN; 67°51′N 20°14′E. Northernmost city in Sweden, center of Lappland iron-mining region. Ore shipped on Lappland railroad (completed 1902) either to ice-free port of NARVIK, NORWAY on Atlantic or to LULEÅ, Sweden, on GULF OF BOTHNIA. Kiruna became world's largest municipality with the incorporation of several distant mining villages (1948). Geophysical institute; ES-RANGE space center. Winter sports center. Airport.

Kirundo, province (□ 658 sq mi/1,710.8 sq km; 1999 population 502,171), NE BURUNDI, on RWANDA border (to W and N); ☉ KIRUNDO; 02°35′S 30°10′E. Borders MUYINGA (E and S) and NGOZI (SSW) provinces. Lakes Cohoha and RWERU in N on Rwanda border. Regional airport at Kirundo town.

Kirundo (ki-roon-DO), town, ☉ KIRUNDO province, NE BURUNDI, 125 mi/201 km NE of BUJUMBURA; 03°42′S 29°02′E. Trading center.

Kirundu (kee-ROON-doo), village, ORIENTALE province, E CONGO, on LUALABA RIVER, and 90 mi/145 km SSE of KISANGANI; 00°44′S 25°32′E. Elev. 1,492 ft/454 m. Steamboat landing and trading post; palm-oil milling. Has Protestant mission. Former headquarters of Arab slave-traders. CONGO FREE STATE troops defeated the Arabs here (1893).

Kirurumo (kee-roo-ROO-mo), village, SINGIDA region, central TANZANIA, 75 mi/121 km SSW of SINGIDA, near Kapata River; 05°51′S 34°09′E. Sheep, goats; corn, wheat.

Kirvin (KER-vin), village (2006 population 127), FREE-STONE county, E central TEXAS, 23 mi/37 km SSE of CORSICANA; 31°46′N 96°19′W. Vegetables, peaches, melons; cattle.

Kirwan Escarpment (KUHR-wuhn), EAST ANTARC-TICA, extends c.90 mi/145 km from 73°25′S 01°00′E to c.74°10′S 07°00′E. Constitutes NW side of polar plateau in QUEEN MAUD LAND, S of PENCK TROUGH; highest elevation over 8,200 ft/2,499 m. Discovered 1939 by German expedition.

Kirwin, village (2000 population 229), PHILLIPS county, N KANSAS, on North Fork SOLOMON RIVER and 12 mi/19 km ESE of PHILLIPSBURG; 39°40′N 99°07′W. Railroad junction. Corn, livestock. Kirwin Reservoir and National Wildlife Reserve to SW; dam just S of town.

Kirwin Reservoir (□ 17 sq mi/44 sq km), PHILLIPS county, N central KANSAS, on North Fork of SOLOMON and BOW rivers, in Kirwin National Wildlife Refuge, 54 mi/87 km N of HAYS; 39°39′N 99°07′W. Maximum capacity 513,020 acre-ft. Formed by Kirwin Dam (169 ft/52 m high), built (1955) by the Bureau of Reclamation for irrigation; also used for flood control.

Kirya (keer-YAH), town (2005 population 2,395), S central CHUVASH REPUBLIC, central European Russia, on road and railroad, 20 mi/32 km NE of ALATYR';

55°05′N 46°52′E. Elevation 639 ft/194 m. Sawmilling, woodworking.

Kirya or **Kiryah**, ISRAEL: see HAKIRYA.

Kiryas Joel, incorporated village (□ 1 sq mi/2.6 sq km; 2006 population 20,071), ORANGE county, SE NEW YORK, 50 mi/80 km NW of NEW YORK CITY; 41°20′N 74°10′W. Founded in 1974 by the Satmar sect of Hasidic Jews to accommodate, in part, their burgeoning population in the WILLIAMSBURG section of BROOKLYN. A 1994 Supreme Court ruling declared the school in the community unconstitutional, because it violated the establishment clause of the Constitution. The state subsequently set up a school district in the area, but tensions with adjoining communities remain as Kiryas Joel continues to grow.

Kiryat Amal, ISRAEL: see KIRYAT TIVON.

Kiryat Anavim (kir-YAHT ah-nah-VEEM) or **Qiryat Anavim**, kibbutz, in JUDEAN HIGHLANDS, 6 mi/9.7 km WNW of JERUSALEM, E ISRAEL; 31°48′N 35°07′E. Elevation 1,922 ft/585 m. Insulation products; silicone technologies; water and building projects; dairying, fruit growing. Summer resort guest house. Modern village founded 1920; important Israeli defense and transportation base (1948) during siege of Jerusalem. Nearby is site of biblical locality of same name.

Kiryat Ata (kir-YAHT AH-tah) or **Qiryat Ata**, city, 9 mi/14 km E of HAIFA in ZEVULUN VALLEY, N ISRAEL; 32°48′N 35°06′E. Elevation 203 ft/61 m. Diversified industries; many residents commute to jobs in Haifa. Founded in 1925, it was destroyed by Arabs in the riots of 1929. Reestablished, it became an important industrial center and the site of Israel's largest textile plant, established 1935 and closed in the 1980s. Scene of serious fighting in 1948. Adjacent Kiryat Binyamin was integrated into its borders in 1965.

Kiryat Bialik (keer-YAHT bi-YAH-leek) or **Qiryat Bialik**, city, NW ISRAEL, in ZEBULUN VALLEY, near BAY OF ACRE, satellite town 5 mi/8 km E of HAIFA; 32°49′N 35°05′E. Elevation 36 ft/10 m. Population almost entirely Jewish. Part of "Garden City Plan" for Haifa. Light manufacturing. Just E is KFAR BIALIK. Founded 1934. Named for the Hebrew poet Haim Nahman Bialik.

Kiryat Binyamin, ISRAEL: see KIRYAT ATA.

Kiryat Ekron (kir-YAHT ek-RON) or **Qiryat Eqron**, town, ISRAEL, 1.2 mi/2 km S of REHOVOT, on coastal plain; 31°51′N 34°49′E. Elevation 252 ft/76 m. Name derives from ancient city of Ekron that may have been nearby.

Kiryat Gat (kir-YAHT GAHT) or **Qiryat Gat**, city, ISRAEL, 12 mi/20 km SE of ASHKELON, on coastal plain; 31°36′N 34°46′E. Elevation 413 ft/125 m. Population almost entirely Jewish. Named for ancient Philistine town of Gat presumed to have been nearby. Scene of heavy fighting between Israel and EGYPT in 1948. Founded in 1950s, it is a growing industrial center with textile plants and brass, metal, and electronics works.

Kiryat Haroshet (keer-YAHT khah-RO-shet), village, NW ISRAEL, at NW end of PLAIN OF JEZREEL, at SE foot of MOUNT CARMEL, on KISHON RIVER, 10 mi/16 km SE of HAIFA; 32°41′N 35°06′E. Elevation 242 ft/73 m. Became part of KIRYAT TIVON in 1979. Founded 1935 as workers settlement.

Kiryat Hayim (keer-YAHT khah-YEEM) or **Qiryat Haim**, suburb 5 mi/8 km NE of HAIFA, NW ISRAEL, in ZEBULUN VALLEY, near BAY OF ACRE, and on railroad; 32°49′N 35°03′E. Elevation 65 ft/19 m. Food processing. Incorporated into the Haifa area in 1950. Founded 1933. Part of Abercrombie's Garden City Plan. Named for Labor Zionist leader Haim Lazaroff.

Kiryat Malachi (kir-YAHT mah-LAH-khee) or **Qiryat Malakhi**, town, ISRAEL, 6 mi/10 km SE of ASHDOD, on the coastal plain; 31°44′N 34°44′E. Elevation 246 ft/74 m. As of 2005, some 20% of the population were first- or second-generation Ethiopian immigrants. Named

for the Jewish community in LOS ANGELES [Hebrew, angels=malakhim], which lent its support to this town, which began as an immigrant camp in 1951. Various light industries, center for surrounding rural settlement.

Kiryat Motzkin (keer-YAHT MOTS-keen), city, NW ISRAEL, in ZEBULUN VALLEY, near BAY OF ACRE; satellite town 5 mi/8 km E of HAIFA; 32°50′N 35°04′E. Elevation 75 ft/22 m. Diverse light manufacturing. Part of Abercrombie's Garden City Plan. Founded 1934. Named for early Zionist leader Leo Motzkin.

Kiryat Ono (kir-YAHT O-no), city, ISRAEL; part of TEL AVIV metropolitan area; 32°03′N 34°51′E. Elevation 118 ft/35 m. It has grown by absorbing surrounding neighborhoods over the years. Founded 1933.

Kiryat Shmona (kir-YAHT shmo-NAH) or **Qiryat Shemona**, city, ISRAEL, NE of ZEFAT (Safed), in UPPER GALILEE, in NW fringes of the HULA VALLEY, and near border with LEBANON; 33°12′N 35°34′E. Elevation 298 ft/90 m. Named for the militant Jewish leader Josef Trumpeldor and his seven comrades [Hebrew, eight=shmona] who fought to the death to defend nearby TEL HAI in 1920. Began as a town for immigrants in 1949 on the site of the abandoned Arab village of Halsa. It is now the main commercial and service center for NE Upper Galilee with hotels, shopping center, and a high-tech industrial park. Kiryat Shmona has also been a prime target for Katyusha rocket attacks, first by the PLO and, more recently, by Hizbullah in Lebanon. Also attacked in 1974 by PLO terrorists who infiltrated from Lebanon, entered an apartment building, and gunned down its eighteen residents, including nine children. Tel Hai College adjoins city.

Kiryat Tivon (keer-YAHT teev-ON), township, NW ISRAEL, between ZEBULUN VALLEY and PLAIN OF JEZREEL, SE foothills of LOWER GALILEE, near KISHON RIVER, 9 mi/14.5 km SE of HAIFA; 32°43′N 35°07′E. Elevation 656 ft/199 m. Founded as garden suburb 1947. Has a youth village. Merged with two other settlements in 1958 (ELRO-I and Kiryat Amal) to become a city (adding KIRYAT HAROSHET) in growing suburb of Haifa. Formerly Tivon.

Kiryat Yam (kir-YAHT YAHM), city, N ISRAEL, NE of HAIFA in ZEVULUN VALLEY, near MEDITERRANEAN; 32°51′N 35°04′E. At sea level. Founded after World War II as a settlement for demobilized soldiers, it developed rapidly after Israel's independence.

Kiryat Ye'arim (kir-YAHT yi-ah-REEM), Jewish town and youth village, ISRAEL, 8 mi/12.9 km WNW of JERUSALEM, and 0.4 mi/1 km W of ABU GHOSH; 31°49′N 35°06′E. Elevation 2,591 ft/789 m. The youth village (1994 population 323) was founded in 1952 and the nearby (religious) settlement (1994 population 1,950) in 1975. Both are named for biblical Kirjath-jearim, located in this area.

Kiryu (KEE-ryoo), city, GUMMA prefecture, central HONSHU, N central JAPAN, 19 mi/30 km E of MAEBASHI; 36°24′N 139°20′E. A major center of silk production since the 8th century.

Kirzhach (keer-ZHAHCH), city (2006 population 21,945), W VLADIMIR oblast, central European Russia, on the Kirzhach River (left affluent of the KLYAZ'MA River), on railroad, 78 mi/126 km W of VLADIMIR, and 18 mi/29 km SSE of ALEKSANDROV; 56°09′N 38°52′E. Elevation 475 ft/144 m. Center of an agricultural region; silk milling, clothing, furniture, instruments, food industries. Founded in 1328.

Kisa (KEE-sah), town, Futami county, HIROSHIMA prefecture, SW HONSHU, W JAPAN, 37 mi/60 km N of HIROSHIMA; 34°42′N 132°59′E. Some four hundred ancient tombs in the vicinity.

Kisa (SHEE-sah), village, ÖSTERGÖTLAND county, SE SWEDEN, 25 mi/40 km E of TRANÅS; 57°59′N 15°38′E. Papermilling. Has thirteenth-century church.

Kisač (KEE-sahch), Hungarian *Kiszács*, village, VOJVODINA, N SERBIA, 8 mi/12.9 km NW of NOVI SAD, in the BACKA region. Also spelled Kisach.

Kisai (KEE-sah-ee), town, North Saitama county, SAITAMA prefecture, E central HONSHU, E central JAPAN, 16 mi/25 km N of URAWA; 36°05′N 139°34′E.

Kisakata (kee-SAH-kah-tah), town, Yuri county, Akita prefecture, N HONSHU, NE JAPAN, on SEA OF JAPAN, near Mount Chokai, 37 mi/60 km S of AKITA city; 39°12′N 139°54′E. Electronics. Also Kisagata.

Kisaki (kee-SAH-kee), village, MOROGORO region, E central TANZANIA, 43 mi/69 km S of MOROGORO, on MGETA River; 07°29′S 38°41′E. Road terminus. SELOUS GAME RESERVE to S. Livestock; corn, wheat; timber. Mica mining in area.

Kisale, Lake (kee-SAH-lai), expansion of LUALABA RIVER in SE CONGO, N of LAKE UPEMBA; 10 mi/16 km long, c.12 mi/19 km wide, surface area 77 sq mi; 08°15′S 26°27′E. Swampy and overgrown with papyrus. Receives LUFIRA RIVER. Its S and SE shores are part of UPEMBA NATIONAL PARK.

Kis Alföld, HUNGARY and SLOVAKIA: see LITTLE ALFÖLD.

Kisamos, Greece: see KASTELLI.

Kisamos, Gulf of, Greece: see KISSAMOS, GULF OF.

Kisanga (kee-SAHNG-gah), village, KATANGA province, SE CONGO, 10 mi/16 km NW of LIKASI. Iron mining.

Kisangani (ki-sahng-GAH-nee), city, ⊙ ORIENTALE province, N central CONGO, a once wealthy trading port on the CONGO RIVER; 00°31′N 25°12′E. Elev. 1,469 ft/447 m. The city is the terminus of steamer navigation on the CONGO RIVER from KINSHASA and is a transportation center for NE CONGO. Major air base. It is on a short railroad line (to Ubundi) that skirts the BOYOMA FALLS. Manufacturing includes metal goods, furniture, and beer; cotton and rice are shipped from here. The diamond trade became important in 1993. Kisangani has an international airport, and hydroelectricity is produced on a nearby tributary of the CONGO. Founded (originally on a nearby island in the river) in 1883 by the explorer Henry M. Stanley, it is a city of boom-and-bust cycles. Late in the colonial period, Belgian planters prospered from coffee, cotton, and rubber. Later, uranium, tropical wood, and gold became its economic base. As STANLEYVILLE, the city where he had served as a postal clerk, it became the stronghold of Patrice Lumumba in the late 1950s. After the assassination of Lumumba in 1961, Antoine Gizenga set up a government there that rivaled the central government in LEOPOLDVILLE (now KINSHASA). Gizenga's regime was quashed in 1962, but in 1964, 1966, and 1967 the city was the site of temporarily successful revolts against the central government. When Mobutu Sese Seku seized power, he nationalized foreign-owned plantations. City was pillaged in 1991, 1993, and 1997 by rampaging government troops. Has a campus of the National University and a museum.

Kisangire (kee-sahn-GEE-rai), village, PWANI region, E TANZANIA, 55 mi/89 km SW of DAR ES SALAAM, near Luhule River; 07°29′S 38°41′E. Corn, wheat, cashews, bananas, manioc, sweet potatoes; poultry, goats, sheep.

Kisangiro (kee-sahn-GEE-ro), village, KILIMANJARO region, N TANZANIA, 25 mi/40 km SE of MOSHI, near KENYA border, on railroad; 03°38′S 37°31′E. Airstrip. Sugarcane, sisal; cattle, sheep, goats.

Kisantu (kee-SAHN-too), village, BAS-CONGO province, W CONGO, on right bank of INKISI RIVER, on railroad, and 140 mi/225 km ENE of BOMA; 05°07′S 15°05′E. Elev. 1,384 ft/421 m. Major Roman Catholic missionary center. Experimental farms and gardens, livestock raising, various schools, including junior college Seat of vicar apostolic. In 1940s, Kisantu became, under sponsorship of several African welfare agencies, one of the main centers in CONGO for training agricultural and medical assistants. Hospitals. Also called INKISI-KISANTU.

Kisar (KEE-sahr), island, S MALUKU, INDONESIA, SE of WETAR Island, 20 mi/32 km NE of TIMOR Island; 08°05′S 127°10′E. Hilly. Coconuts; fishing. Main town is Wonreli.

Kisaran (KEE-sah-rahn), plantation and tranportation center, SE Sumatra Utara province, INDONESIA, 48 mi/77 km SE of MEDAN; 02°59′N 99°37′E. Designated in 1980s as an administrative center and capital of Asahan district.

Kisarawe (kee-sah-RAH-wai), village, PWANI region, E TANZANIA, 10 mi/16 km SW of DAR ES SALAAM, on railroad; 06°56′S 39°10′E. Cashews, copra, bananas, corn; poultry, goats, sheep. Dar es Salaam International Airport to E.

Kisarazu (kee-sah-RAHZ), city, CHIBA prefecture, E central HONSHU, E central JAPAN, on TOKYO BAY, 16 mi/25 km S of CHIBA; 35°22′N 139°55′E. Residential and industrial suburb of TOKYO noted for its Shojoji (Buddhist) temple. Electronics.

Kisawa (kee-SAH-wah), village, Naka county, TOKUSHIMA prefecture, SE SHIKOKU, W JAPAN, 22 mi/35 km S of TOKUSHIMA; 33°49′N 134°18′E.

Kisbér (KISH-bair), city, KOMÁROM county, N HUNGARY, 22 mi/35 km SE of Györ; 47°30′N 18°02′E. Railroad center; cattle; manufacturing (bricks, tiles, flour milling). Stud farm established 1853.

Kisbey (KIZ-bee), village (2006 population 185), SE SASKATCHEWAN, CANADA, near Moose Mountain Creek, 40 mi/64 km NNE of ESTEVAN; 49°38′N 102°40′W. Mixed farming.

Kis Duna River, SLOVAKIA: see LITTLE DANUBE RIVER.

Kisei (kee-SAI), town, Watarai county, MIE prefecture, S HONSHU, central JAPAN, 31 mi/50 km S of TSU; 34°17′N 136°23′E.

Kiselëvsk (kee-sye-LYOFSK), city (2005 population 100,915), W central KEMEROVO oblast, S central SIBERIA, RUSSIA, in the foothills of the SALAIR range, on the Aba River, on railroad, 150 mi/241 km S of KEMEROVO; 54°00′N 86°39′E. Elevation 1,095 ft/333 m. Coal-mining center in the KUZNETSK BASIN; manufacturing mining machinery, building materials; footwear factory; food processing (bakery, brewery, confectionaries). Founded in 1932; made city in 1936.

Kiseljak (kee-sel-YAHK), town, central BOSNIA, BOSNIA AND HERZEGOVINA, 5 mi/8 km WSW of VISOKO. Mineral waters and baths. Also spelled Kiselyak.

Kis Fátra, SLOVAKIA: see LESSER FATRA.

Kish (KISH), ancient city of MESOPOTAMIA, in the EUPHRATES valley, 8 mi/13 km E of BABYLON, and 12 mi/19 km E of the modern city of HILLAH, IRAQ. It was occupied from very ancient times, and its remains go back as far as the protoliterate period in Mesopotamia. In the early third millennium B.C.E., Kish was a Semitic city. Although it was one of the provincial outposts of SUMERIAN civilization, it had a cultural style of its own. There is an excavated palace of Sargon I of Agade, a native of Kish, and a great temple built by Nebuchadnezzar and Nabonidus in the later Babylonian period. The site also yielded a complete sequence of pottery from the Sumerian period to that of Nebuchadnezzar (6th century B.C.E.).

Kish (KEESH), island in PERSIAN GULF, off S IRAN, 50 mi/80 km W of BANDAR-e Lengeh; 9 mi/14.5 km long, 4 mi/6.4 km wide; 26°33′N 54°00′E. Fishing; pearl trade; agriculture. Airport; national tourist resort; small aquarium. Was great trade center (11th–14th century) in Middle Ages until rise of HORMOZ island; 11th–12th-cent. palace complex on N coast of the old city. Iran's first and only free port.

Kishan, TAIWAN: see Chishan.

Kishanda (kee-SHAN-dah), village, KAGERA region, NW TANZANIA, 30 mi/48 km SSW of BUKOBA; 01°44′S 31°37′E. Coffee, corn, wheat; goats, sheep.

Kishanganga River (ki-SHUHN-guhn-gah), c.150 mi/241 km long, in JAMMU AND KASHMIR state, extreme N INDIA and AZAD KASHMIR, NE PAKISTAN; rises in several headstreams in main range of GREATER HIMALAYAS, flows WNW, SSW past Tithwal (PAKISTAN), and W to JHELUM River at MUZAFFARABAD. Also spelled Kishenganga.

Kishanganj (ki-SHUHN-guhnj), district (□ 728 sq mi/1,892.8 sq km), NE BIHAR state, E INDIA; ⊙ KISHANGANJ. Became district in 1990 from NE part of PURNIA district.

Kishanganj (ki-SHUHN-guhnj), town, ⊙ KISHANGANJ district, NE BIHAR state, NE INDIA, near MAHANANDA RIVER, 37 mi/60 km NE of PURNIA; 25°42′N 86°57′E. Trade center (rice, jute, corn, tobacco, rape, mustard, wheat); jute pressing. Annual livestock fair.

Kishangarh (ki-SHUHN-guhr), former princely state in RAJPUTANA STATES, INDIA. Established in early 17th century; chiefs (Rathor Rajputs) entered Mogul service; treaty made with BRITISH in 1818. In 1948, merged with union of RAJASTHAN. Sometimes spelled Kishengarh.

Kishangarh (ki-SHUHN-guhr), city, AJMER district, central RAJASTHAN state, NW INDIA, 65 mi/105 km WSW of JAIPUR; 26°34′N 74°52′E. Trades in cotton, millet, barley, oilseeds, textiles; manufacturing of soap, woolen carpets, shawls; hand-loom weaving, dyeing, precious-stone cutting. Cotton mill at suburb of MADANGANJ, 1 mi/1.6 km N. Mica, marble, and sandstone deposits worked in vicinity. Was capital of former RAJPUTANA state of Kishangarh. Founded 1611.

Kishanpur (KI-shuhn-puhr), town, FATEHPUR district, S UTTAR PRADESH state, N central INDIA, on the YAMUNA RIVER, and 10 mi/16 km SSW of KHAGA; 30°42′N 78°30′E. Seed crops, barley, rice, jowar.

Kishcha (keesh-CHAH), village (2005 population 3,915), central DAGESTAN republic, NE CAUCASUS, SE European Russia, near highway, 38 mi/61 km S of MAKHACHKALA; 42°10′N 47°35′E. Elevation 3,943 ft/1,201 m. Also known as Verknyaya Kishcha (Russia=upper Kishcha).

Kishen'-Aukh, RUSSIA: see NOVOLAKSKOYE.

Kishenganga River, INDIA and PAKISTAN: see KISHANGANGA RIVER.

Kishengarh, INDIA: see KISHANGARH.

Kishen'ki, UKRAINE: see KYSHEN'KY.

Kishert', RUSSIA: see POSAD.

Kishi (KEE-shee), town, OYO state, SW NIGERIA, 60 mi/97 km NW of ILORIN. Cotton weaving, shea nut processing; cattle.

Kishigawa (kee-shee-GAH-wah), town, Naga county, WAKAYAMA prefecture, S HONSHU, W central JAPAN, 9 mi/15 km E of WAKAYAMA; 34°12′N 135°19′E.

Kishiku (kee-SHEE-koo), town, on N coast of Fukaeshima island of GOTO-RETTO island group, South Matsuura county, NAGASAKI prefecture, SW JAPAN, 65 mi/105 km W of NAGASAKI; 32°44′N 128°45′E.

Kishimoto (kee-SHEE-mo-to), town, Saihaku county, TOTTORI prefecture, S HONSHU, W JAPAN, 48 mi/78 km W of TOTTORI; 35°23′N 133°24′E.

Kishinev, MOLDOVA: see CHIŞINĂU.

Kishiwada (kee-SHEE-wah-dah), city (2005 population 201,000), OSAKA prefecture, S HONSHU, W central JAPAN, on OSAKA BAY, 19 mi/30 km S of OSAKA; 34°27′N 135°22′E. Industrial and residential suburb of Osaka.

Kishkareny, MOLDOVA: see CHISCARENI.

Kishm (KEE-shem), town, BADAKHSHAN province, NE AFGHANISTAN, 40 mi/64 km SW of FAIZABAD, near road to KHANABAD; 36°48′N 70°06′E. Also spelled Keshm.

Kishon (kee-SHON) [Hebrew Qishon=tortuous], river, c.45 mi/72 km long, N ISRAEL, rises below MOUNT GILBOA; flows NW through the PLAINS OF JEZREEL, to the MEDITERRANEAN SEA at HAIFA; only the lower 7-mi/11.3-km section is a permanent stream. In the Hebrew Bible, the defeat of Sisera and the slaying by Elijah of the prophets of Baal occurred on the river bank. Also spelled Kison.

Kishpek (keesh-PYEK), rural settlement (2005 population 4,370), central KABARDINO-BALKAR REPUBLIC, N CAUCASUS, S European Russia, on road, 8 mi/13 km N of NAL'CHIK; 43°39′N 43°38′E. Elevation 1,171 ft/356 m. Sheep breeding. The area was the site of inter-

mittent fighting between German and Soviet army units in 1942, during World War II.

Kishtwar (kisht-WAHR), town, DODA district, JAMMU AND KASHMIR state, N INDIA, in SW KASHMIR region, in PIR PANJAL RANGE, near CHENAB River, 45 mi/72 km NNE of UDHAMPUR; 33°19′N 75°46′E. Corn, wheat, barley, rice, fruit. Was capital of former independent Rajput raja. Shah Shuja stayed here, 1814–1816, in his exile from AFGHANISTAN. Important deodar forest nearby.

Kishwaukee River (kish-WAH-kee), N ILLINOIS, c.60 mi/97 km long, rises in MCHENRY county, flows generally W and SW, past BELVIDERE, to ROCK RIVER 6 mi/9.7 km below ROCKFORD; 42°17′N 88°26′W. Its South Branch rises near SHABBONA in DE KALB county, flows c.56 mi/90 km generally N and NW to Kishwaukee River, SE of Rockford.

Kisielice (kee-see-LEE-tse), German *Freystadt*, town, in EAST PRUSSIA, Elbląg province, NE POLAND, 17 mi/27 km ESE of KWIDZYN (Marienwerder). Grain and cattle market. Has 14th-century church. Until 1919, in WEST PRUSSIA province.

Kisigo River (kee-SEE-go), c. 160 mi/257 km long, in central TANZANIA, rises in central SINGIDA region, 85 mi/137 km WNW of DODOMA; flows generally S, through E end of Kisigo Game Reserve; receives NJOMBE RIVER 65 mi/105 km SW of Dodoma; then continues E, entering GREAT RUAHA RIVER in MTERA RESERVOIR 55 mi/89 km S of Dodoma.

Kisii (kee-SEEH), town (1999 population 25,634), NYANZA province, W KENYA, 40 mi/64 km S of KISUMU; 00°40′S 34°47′E. Agricultural and trade center; tea, coffee, cotton, peanuts, sesame, corn. Soapstone deposits. Headquarters of Seventh Day Adventists' mission.

Kisiju (kee-SEE-joo), village, PWANI region, E TANZANIA, 40 mi/64 km S of DAR ES SALAAM, on MAFIA CHANNEL, INDIAN OCEAN, at mouth of Luhule River; 07°25′S 39°19′E. Fish; sisal, cashews, bananas, copra, manioc, corn; goats, sheep.

Kisiklik, Greece: see NEOS SKOPOS.

Kisilwa (kee-SEEL-wah), village, IRINGA region, central TANZANIA, 35 mi/56 km NW of IRINGA, on GREAT RUAHA RIVER, at mouth of LITTLE RUAHA RIVER; 07°19′S 34°27′E. Timber; tobacco, corn, wheat, sweet potatoes; cattle, sheep, goats. RUAHA NATIONAL PARK to W.

Kisjenö, ROMANIA: see CHIŞINEU-CRIŞ.

Kiska, island: see ALEUTIAN ISLANDS.

Kiskalota, ROMANIA: see CĂLĂŢELE.

Kiskapus, ROMANIA: see COPŞA MICĂ.

Kis Karpatok, SLOVAKIA: see LITTLE CARPATHIAN MOUNTAINS.

Kiskatom, hamlet, GREENE county, SE NEW YORK, in the CATSKILL foothills, 5 mi/8 km WSW of CATSKILL; 42°12′N 73°58′W. In summer resort area.

Kiskiminetas River (KIS-kuh-MEH-nuh-tus), 27 mi/43 km long, in SW PENNSYLVANIA, formed at SALTSBURG by confluence of CONEMAUGH RIVER and LOYALHANNA CREEK; flows NW past AVONMORE, VANDERGRIFT, and LEECHBURG to ALLEGHENY RIVER 1 mi/1.6 km E of FREEPORT; 40°29′N 79°27′W.

Kiskissink, unorganized territory, S QUEBEC, E CANADA; 47°55′N 72°09′W. Forms part of La TUQUE agglomeration.

Kiskittogisu Lake (□ 99 sq mi/256 sq km), central MANITOBA, W central CANADA, 18 mi/29 km N of Lake WINNIPEG; 30 mi/48 km long, 8 mi/13 km wide; 54°15′N 98°13′W. Drains N into NELSON River.

Kiskitto Lake (□ 65 sq mi/168 sq km), central MANITOBA, W central CANADA, 26 mi/42 km N of Lake WINNIPEG; 20 mi/32 km long, 6 mi/10 km wide; 54°16′N 98°30′W. Drains N into NELSON River.

Kisko (KIS-ko), village, TURUN JA PORIN province, SW FINLAND, 40 mi/64 m ESE of TURKU; 60°14′N 23°29′E. In lake region; mining (copper, zinc, lead, silver).

Kiskörös (KISH-kuh-vuhsh), city, PEST county, central HUNGARY, 27 mi/43 km SW of KECSKEMÉT; 46°37′N

19°18′E. Railroad and road junction; wheat, rye, apples, cherries, apricots; manufacturing (agricultural machinery, flour milling). Birthplace of poet Sándor Petöfi.

Kisköszeg, CROATIA: see BATINA.

Kiskundorozsma (KISH-kun-do-rozh-mah), village, CSONGRÁD county, S HUNGARY, suburb 5 mi/8 km WNW of SZEGED. Wheat, grapes, peaches; hogs, sheep.

Kiskunfélegyháza (KISH-kun-fai-ledy-hah-zah), city (2001 population 32,632), PEST county, S central HUNGARY; 46°43′N 19°51′E. It is a railroad and road junction; apples, apricots, grapes, hops; manufacturing (poultry and egg processing, agricultural machinery, power station equipment, plastic foam, baked goods, farinaceous foods).

Kiskunhalas (KISH-kun-hah-lahsh), city (2001 population 29,954), PEST county, S central HUNGARY, 34 mi/55 km SSW of KECSKEMÉT; 46°26′N 19°30′E. Railroad and road center; wheat, grapes, peaches; hogs, cattle; manufacturing (furniture, textiles, pumps, flour milling, knitted wear, farinaceous foods). Exports Halas lace. Formerly called Halas. Small Lake Halas nearby.

Kiskunlacháza (KISH-kun-lats-hah-zah), village, PEST county, central HUNGARY, 21 mi/34 km S of BUDAPEST; 47°12′N 19°01′E. Grain, apples, apricots, grapes; cattle, hogs.

Kiskunmajsa (KISH-kun-mahy-sah), city, PEST county, S central HUNGARY, 29 mi/47 km S of KECSKEMÉT; 46°29′N 19°45′E. Corn, wheat, paprika, tobacco; cattle, hogs; farinaceous foods.

Kiskunság (KISH-kun-shahg) [=little Cumania], section of the Alföld plain, S central HUNGARY, between the DANUBE and TISZA rivers. Sandy soils, largely anchored by vegetation. Fruit orchards, especially apricots; vineyards. The region took its name from the Cuman tribes, settled here by King Béla IV in the early 13th century.

Kislotnyy, RUSSIA: see PERM, city.

Kislovodsk (kee-sluh-VOTSK) [Russian=sour water], city (2006 population 133,835), S central STAVROPOL TERRITORY, N CAUCASUS, S European Russia, in the N Caucasus Mountains, on road, 145 mi/233 km SE of STAVROPOL, and 24 mi/39 km S of MINERAL'NYYE VODY; 43°55′N 42°43′E. Elevation 2,755 ft/839 m. Famous health resort with mineral springs, sanatoriums, and a physical therapy institute; bottling plant for "Narzan" mineral water; porcelain factory. Founded in 1803, made city in 1830.

Kislyakovskaya (kees-LYAH-kuhf-skah-yah), village (2005 population 5,525), N KRASNODAR TERRITORY, S European Russia, on the YEYA RIVER, on road and near railroad, 49 mi/79 km S of ROSTOV-NA-DONU (ROSTOV oblast), and 8 mi/13 km S of KUSHCHEVSKAYA, to which it is administratively subordinate; 46°26′N 39°40′E. In agricultural area. Founded as a Cossack *stanitsa* (village).

Kismayo (kees-MAH-yo), town, ⊙ JUBBADA HOOSE region, SW SOMALIA, port on the INDIAN Ocean; 00°22′N 42°32′E. Small food-processing, clothing, and leather industries. Airport. Founded 1872 by the sultan of Zanzibar. Several mosques and a palace located here.

Kismet (KIZ-met), village (2000 population 484), SEWARD county, SW KANSAS, 15 mi/24 km NE of LIBERAL; 37°12′N 100°42′W. In grain region.

Kiso (kee-SO), village, Kiso county, NAGANO prefecture, central HONSHU, central JAPAN, 56 mi/90 km S of NAGANO; 35°55′N 137°47′E. Canvas frames, combs.

Kisofukushima (KEE-so-foo-KOO-shee-mah), town, Kiso county, NAGANO prefecture, central HONSHU, central JAPAN, 62 mi/100 km S of NAGANO; 35°50′N 137°42′E.

Kisogawa (kee-SO-gah-wah), town, Haguri county, AICHI prefecture, S central HONSHU, central JAPAN, on KISO RIVER just NW of ICHINOMIYA and 12 mi/20 km N of NAGOYA; 35°20′N 136°46′E.

Kiso-koma-ga-take (kee-SO–ko-MAH–GAH–tah-ke), mountain (9,240 ft/2,816 m), NAGANO prefecture, cen-

tral HONSHU, central JAPAN, 20 mi/32 km NNW of IDA, Thickly wooded. Sometimes called Koma-ga-take.

Kison, ISRAEL: see KISHON.

Kiso River (KEE-so), Japanese *Kiso-gawa* (kee-SO-gah-wah), 135 mi/217 km long, central HONSHU, central JAPAN; rises near ON-TAKE peak in NAGANO prefecture, flows E, then generally SW, past AGEMATSU, INUYAMA, KASAMATSU, and OKU, to ISE BAY 10 mi/16 km WSW of NAGOYA. Many hydroelectric stations provide power to industrial areas. Scenic rapids. Chief tributaries include the Nagara (scene of protests over dams) and Ibi rivers.

Kisoro (kee-SOR-o), administrative district (□ 282 sq mi/733.2 sq km; 2005 population 226,900), WESTERN region, extreme SW corner of UGANDA, at border with RWANDA (S) and DEMOCRATIC REPUBLIC OF THE CONGO (W); ⊙ KISORO; 01°10′S 29°40′E. Elevation 6,499 ft/1,981 m. As of Uganda's division into eighty districts, borders KANUNGU (N) and KABALE (E) districts. Primary inhabitants are Bafumbira, Bakiga, and Batwa people. Contains rain forests, grasslands, and volcanic soils, with mountainous area in N and lowlands in S. Population primarily rural and agricultural (beans, sweet potatoes); some tin and tungsten deposits. Roads connect district to surrounding Uganda, as well as Rwanda and Democratic Republic of the Congo.

Kisoro (kee-SOR-o), town, ⊙ KISORO district, WESTERN region, SW UGANDA, on road, and 60 mi/97 km S of LAKE EDWARD, 5 mi/8 km from RWANDA border; 01°17′S 29°42′E. Was part of former SOUTHERN province.

Kisosaki (kee-SO-sah-kee), town, Kuwana county, MIE prefecture, S HONSHU, central JAPAN, 28 mi/45 km N of TSU; 35°04′N 136°44′E. Tomatoes.

Kisra-Sumei (KIS-rah–SOO-mei), Druze township, N ISRAEL, E of AKKO (Acre) in UPPER GALILEE. Mixed farming, olive growing.

Kissamos, Gulf of (KEE-sah-mos), AEGEAN inlet, off KHANIÁ prefecture, extreme NW CRETE department, GREECE, between capes VOUXA and SPATHA; 6 mi/9.7 km wide, 10 mi/16 km long. KASTELLI is on S shore. Also Gulf of Kisamos.

Kissármás, ROMANIA: see SĂRMAŞ.

Kisseraing Island (KEE-se-REING), in central MERGUI ARCHIPELAGO, peninsular MYANMAR, in ANDAMAN SEA, 45 mi/72 km S of MERGUI; 20 mi/32 km long, 10 mi/16 km wide. Irregular coast with mangrove swamps; forested hills.

Kissidougou, prefecture, Faranah administrative region, SE GUINEA, in Guinée-Forestière geographic region; ⊙ KISSIDOUGOU. Bordered N by Kouroussa prefecture, NE by Kankan prefecture, E by Kérouané prefecture, SE by Macenta prefecture, SSW by Guékédou prefecture, and W by Faranah prefecture. Kouya River in N and Niandan River in center and NE of prefecture. Towns include Bandama, Kissidougou, and Yèndè Milimou. Several main roads branch out of Kissidougou town connecting it to FARANAH town to NW, KANKAN town to NE, and GUÉKÉDOU town to S; multiple secondary roads also run through the prefecture.

Kissidougou (kee-see-DOO-goo), town, ⊙ Kissidougou prefecture, Faranah administrative region, SE GUINEA, in Guinée-Forestière geographic region, in interior highland, near SIERRA LEONE, 100 mi/161 km SW of KANKAN; 09°05′N 00°10′W. Livestock raising (cattle, sheep, goats) and agricultural center. Exports rice, palm oil, palm kernels, coffee, kola nuts. Churches.

Kissimmee (kuh-SIM-ee), city (□ 13 sq mi/33.8 sq km; 2000 population 47,814), OSCEOLA county, central FLORIDA, on LAKE TOHOPEKALIGA; 28°18′N 81°24′W. Located in an important agricultural area, it is a major processing, packaging, and shipping center for the surrounding KISSIMMEE RIVER basin, where citrus products (especially oranges) and beef are raised. The city also lies in a popular vacation area. Among the

largest contributors to Kissimmee's economic development is nearby WALT DISNEY WORLD (12 mi/19 km NW), the theme park and entertainment complex that draws millions of visitors each year. Kissimmee has motels and restaurants that serve many of the park's visitors.

Kissimmee, Lake (kuh-SIM-ee), OSCEOLA county, central FLORIDA, c.40 mi/64 km S of ORLANDO; c.15 mi/24 km long, 5 mi/8 km wide; 27°54′N 81°16′W. Entered (NW) and drained (S) by KISSIMMEE RIVER. Contains several small islands.

Kissimmee River (kuh-SIM-ee), c.140 mi/225 km long, central FLORIDA, rises in LAKE TOHOPEKALIGA in OSCEOLA county, flows SSE to N end of LAKE OKEECHOBEE 7 mi/11.3 km SW of OKEECHOBEE. In upper course, connects chain of lakes (Tohopekaliga, HATCHINEHA, KISSIMMEE). River basin has large cattle range (Kissimmee Prairies), some citrus-fruit growing, and large wilderness tracts. River was channelized by the U.S. Army Corps of Engineers in 1970s and is now being restored.

Kissing (KIS-sing), village, BAVARIA, S GERMANY, in SWABIA, 6 mi/9.7 km SE of AUGSBURG; 48°17′N 11°00′E. Ironworking, manufacturing of steel and machinery. Has three churches from 17th–18th centuries.

Kissingen, Bad, GERMANY: see BAD KISSINGEN.

Kississing Lake (□ 141 sq mi/365 sq km), W MANITOBA, W central CANADA, near SASKATCHEWAN border, 24 mi/39 km NE of FLIN FLON; 17 mi/27 km long, 16 mi/26 km wide; 55°10′N 101°20′W. Drains into CHURCHILL River.

Kisslegg (KIS-leg), village, BADEN-WÜRTTEMBERG, S GERMANY, in the ALLGÄU, 12 mi/19 km E of RAVENSBURG; 47°48′N 09°53′E. Tourism; health resort. Has two ancient castles, 18th-century church.

Kissufim Pass, (kee-soo-FEEM) or **Kisufim**, crossing point between ISRAEL and the GAZA STRIP. Situated on the SE part of the Gaza Strip, about 6.8 mi/14 km SE of GAZA city. Closed by Israel in September 2005 as part of its unilateral disengagement plan.

Kissy (KEE-see), town, SIERRA LEONE PENINSULA, 5 mi/8 km ESE of FREETOWN; 08°28′N 13°11′W. Fishing; cassava, corn, ginger. Psychiatric hospital. Also called Kissy Mess.

Kisszeben, SLOVAKIA: see SABINOV.

Kisszombor (KISH-som-bor), village, Csanád county, S HUNGARY, 4 mi/6 km SW of MAKÓ. Market center; grain; cattle.

Kistanje (KEE-stahn-ye), Italian *Chistagne*, town, S CROATIA, near KRKA RIVER, 17 mi/27 km N of SIBENIK, in DALMATIA, on Knin-Zadar railroad. Part of Srpska (Serb) Krajina 1991–1995. Heavily damaged during the fighting.

Kistelek (KISH-te-lek), city, CSONGRÁD county, S HUNGARY, 17 mi/27 km NNW of SZEGED; 46°28′N 19°59′E. Market center. Apples, peaches, paprika, wheat, corn; hogs; vineyards.

Kistendey (kees-tyen-DYAI), village, W SARATOV oblast, SE European Russia, on railroad, 11 mi/18 km SSW of RTISHCHEVO; 52°06′N 43°40′E. Elevation 652 ft/198 m. In agricultural area (grains, potatoes); flour milling.

Kistler (KIS-luhr), village, LOGAN county, SW WEST VIRGINIA, 9 mi/14.5 km SE of LOGAN; 37°45′N 81°51′W. Railroad junction. Coal mining area.

Kistler (KIST-luhr), borough (2006 population 331), MIFFLIN county, central PENNSYLVANIA, 1 mi/1.6 km E of MOUNT UNION, on JUNIATA RIVER opposite ALLENPORT (to SW). There is an unincorporated village of KISTLER 24 mi/39 km to E, in Northeast Madison township, PERRY county, PENNSYLVANIA.

Kistna, INDIA: see KRISHNA river.

Kisufim, (kee-soo-FEEM) kibbutz, ISRAEL, 2.5 mi/4 km N of NIRIM, in NW NEGEV region; 31°22′N 34°23′E. Elevation 305 ft/92 m. Wheat, fruit, vegetables. A local museum contains mosaics and remains from the Hellenistic, Roman, and Byzantine periods.

Area is shown by the symbol □, and capital city or county seat by ⊙.

Kisújszállás (KISH-uy-sahl-lash), city, SZOLNOK county, E central HUNGARY, 11 mi/18 km SW of KARCAG; 47°13′N 20°46′E. Railroad center; wheat, corn, hemp; cattle, sheep; dairy; manufacturing (earthmoving equipment, apparel).

Kisuki (kees-KEE), town, Ohara county, SHIMANE prefecture, SW HONSHU, W JAPAN, 17 mi/27 km S of MATSUE; 35°17′N 132°54′E. Also Kitsugi.

Kisulu, CONGO: see KAMPENE.

Kisumu (kee-SOO-moo), city (2004 population 227,100), ⊙ NYANZA province, SW KENYA, on KAVIRONDO GULF (an arm of Lake VICTORIA), 00°08′S 34°47′E. It is the principal lake port of Kenya, its third-largest city, and the commercial center of a prosperous farm region. Manufacturing (refined sugar, frozen fish, textiles, beer, and processed sisal). An ethanol plant was built in the 1980s. The city is developing a tourist industry with the attractions of Lake Victoria and nearby wildlife. The railroad from MOMBASA reached Kisumu in 1901. The city was formerly called Port Florence. Airport; hospital.

Kisundi (kee-SOON-dee), village, BAS-CONGO province, W CONGO, on road, 30 mi/48 km N of BOMA; 05°34′S 13°03′E. Elev. 872 ft/265 m.

Kisvárda (KISH-vahr-dah), city, SZABOLCS-SZATMÁR county, NE HUNGARY, 25 mi/40 km NE of NYIREGYHÁZA; 48°13′N 22°05′E. Railroad junction, market center. Apples, alfalfa, tobacco, corn, wheat; cattle, sheep; manufacturing (aluminum articles, apparel, milling, farinaceous foods). Teachers college here.

Kiswe (KIS-we), village, DAMASCUS district, SW SYRIA, 11 mi/18 km SSW of Damascus; 33°21′N 36°15′E. Cereals, fruits.

Kiswere (kee-SWAI-rai), village, LINDI region, SE TANZANIA, 40 mi/64 km N of LINDI, on INDIAN OCEAN, NW of Grant Point; 09°27′S 39°33′E. Cashews, peanuts, corn, bananas; goats; fish.

Kiszucaújhely, SLOVAKIA: see KYSUCKE NOVE MESTO.

Kiszuczahegyeshely, SLOVAKIA: see VYSOKA NAD KYSUCOU.

Kiszuczakarásznó, SLOVAKIA: see KRASNO NAD KYSUCOU.

Kita (KEE-tah), town, FIRST REGION/KAYES, SW MALI, on DAKAR-NIGER railroad, and 100 mi/161 km WNW of BAMAKO; 13°03′N 09°29′W. Peanut-growing center; also shea-nut butter, rice, corn, potatoes, manioc, mango; goats, sheep, cattle. Oil factory. Radio station.

Kita (kee-TAH), ward, N TOKYO, Tokyo prefecture, E central HONSHU, E central JAPAN. Bordered N by SAITAMAprefecture, E by ADACHI and ARAKAWA wards, S by TAITO, BUNKYO, and TOSHIMA wards, and W by ITABASHI ward.

Kita-aidzu (kee-TAH–AH-eedz), village, North Aidzu county, FUKUSHIMA prefecture, N central HONSHU, NE JAPAN, 37 mi/60 km S of FUKUSHIMA; 37°29′N 139°52′E.

Kitaaiki (kee-TAH-ah-ee-kee), village, South Saku county, NAGANO prefecture, central HONSHU, central JAPAN, 47 mi/75 km S of NAGANO; 36°03′N 138°33′E.

Kitaarima (kee-TAH-AH-ree-mah), town, South Takaki county, NAGASAKI prefecture, NW KYUSHU, SW JAPAN, 22 mi/35 km E of NAGASAKI; 32°39′N 130°15′E.

Kitab, city, NE KASHKADARYO wiloyat, UZBEKISTAN, 60 mi/97 km ENE of KARSHI, 3 mi/4.8 km NE of terminus of railroad spur from Karshi; 39°08′N 66°52′E. Cotton-ginning center in fertile oasis (cotton, rice, fruit). Has International Latitude Observatory.

Kitadaito (kee-TAH-DAH-ee-TO), village, on KITA-DAITO-JIMA island, Shimajiri county, Okinawa prefecture, SW JAPAN, 226 mi/365 km E of NAHA; 25°56′N 131°17′E.

Kita-daito-jima, JAPAN: see DAITO-JIMA.

Kitagata (kee-TAH-gah-tah), town, Motosu county, GIFU prefecture, central HONSHU, central JAPAN, 3.7 mi/6 km N of GIFU; 35°26′N 136°41′E.

Kitagata, town, Kishima county, SAGA prefecture, N KYUSHU, SW JAPAN, 16 mi/25 km S of SAGA; 33°12′N 130°04′E.

Kitagawa (kee-TAH-gah-wah), town, East Usuki county, MIYAZAKI prefecture, SE KYUSHU, SW JAPAN, near Mount Okue, 56 mi/90 km N of MIYAZAKI; 32°41′N 131°41′E.

Kitagawa (kee-TAH-gah-wah), village, Aki county, KOCHI prefecture, S SHIKOKU, W JAPAN, 31 mi/50 km E of KOCHI; 33°26′N 134°02′E. Lumber; citron, and citron processing. Hot springs in the area. Nakaoka Shintaro born here.

Kitagi-shima (kee-TAH-gee–SHEE-mah), island (□ 3 sq mi/7.8 sq km), OKAYAMA prefecture, W JAPAN, in HIUCHI SEA, just off S coast of HONSHU, 8 mi/12.9 km E of Tomo; 2.5 mi/4 km long, 2 mi/3.2 km wide. Raw silk, wheat, sweet potatoes. Building-stone quarries.

Kitago (kee-TAH-go), town, South Naka county, MIYAZAKI prefecture, SE KYUSHU, SW JAPAN, 16 mi/25 km N of MIYAZAKI; 31°40′N 131°22′E.

Kitago (kee-TAH-go), village, East Usuki county, MIYAZAKI prefecture, SE KYUSHU, SW JAPAN, 40 mi/65 km N of MIYAZAKI; 32°29′N 131°26′E.

Kitahata (kee-TAH-hah-tah), village, East Matsuura county, SAGA prefecture, N KYUSHU, SW JAPAN, 22 mi/35 km N of SAGA; 33°22′N 129°57′E.

Kitahiyama (kee-TAH-hee-YAH-mah), town, Hiyama district, Hokkaido prefecture, N JAPAN, 87 mi/140 km S of SAPPORO; 42°24′N 139°52′E.

Kitai, China: see QITAI.

Kita-ibaraki (kee-TAH–ee-BAH-rah-kee), city, IBARAKI prefecture, central HONSHU, E central JAPAN, 31 mi/50 km N of MITO; 36°47′N 140°45′E. Fish processing, tile manufacturing; pumpkins. Izura coast nearby.

Kitai Lagoon, UKRAINE: see KYTAY LAGOON.

Kitajima (kee-TAH-jee-mah), town, Itano co., TOKUSHIMA prefecture, SE SHIKOKU, W JAPAN, 3.7 mi/6 km N of TOKUSHIMA; 34°07′N 134°32′E. Synthetic fibers.

Kitakami (kee-tah-KAH-mee), city, IWATE prefecture, N HONSHU, NE JAPAN, 28 mi/45 km S of MORIOKA; 39°17′N 141°07′E. Computer components.

Kitakami (kee-tah-KAH-mee), town, Mono county, MIYAGI prefecture, N HONSHU, NE JAPAN, 80 mi/60 km N of SENDAI; 34°38′N 141°27′E. Seaweed (*wakame*, kombu), scallops, abalone. Habitat of the Inuwashi eagle.

Kitakami River (kee-TAH-kah-mee), Japanese *Kitakami-gawa* (kee-TAH-kah-mee–GAH-wah), 152 mi/245 km long, N HONSHU, NE JAPAN, in IWATE and MIYAGI prefectures; rises in mountains 25 mi/40 km N of MORIOKA; flows generally S, past Morioka, HANAMAKI, Kurosawajiri, MIZUSAWA, and Tome, to ISHINOMAKI BAY at ISHINOMAKI. Irrigates extensive rice-growing area.

Kitakata (kee-TAH-kah-tah), city, FUKUSHIMA prefecture, N central HONSHU, NE JAPAN, 34 mi/55 km W of FUKUSHIMA; 37°38′N 139°52′E. Hops. Lacquer and lacquerware, paulownia products (furniture, clogs, folk art). Warehouse center.

Kitakata (kee-TAH-kah-tah), town, East Usuki county, MIYAZAKI prefecture, SE KYUSHU, SW JAPAN, 43 mi/70 km N of MIYAZAKI; 32°33′N 131°31′E.

Kitakawabe (kee-TAH-KAH-wah-be), town, North Saitama county, SAITAMA prefecture, E central HONSHU, E central JAPAN, 22 mi/35 km N of URAWA; 36°11′N 139°39′E. Crucian carp; persimmons.

Kita-kozawa, RUSSIA: see TEL'NOVSKIY.

Kitakyushu (kee-TAH-KYOO-shoo), city (2005 population 993,525), FUKUOKA prefecture, N KYUSHU, SW JAPAN, on the Kanmon Strait between the INLAND SEA and the KOREA STRAIT, 31 mi/50 km N of FUKUOKA; 33°52′N 130°52′E. Japan's 10th-largest city, it is one of the country's most important manufacturing regions and one of its chief ports and railroad centers. Has a great variety of industries, the chief of which produce iron and steel (especially in Yawata ward), machinery, computer components, metal fittings, industrial robots, cement, plywood, industrial medical and hygiene equipment, and industrial furnaces. Bamboo shoots. Its ports (especially in Moji and Wakamatsu wards) receive raw materials and export manufactured goods. Kokura ward is the city's commercial and financial center. Tobata ward has a major coal-handling facility. A deep-sea fishing fleet (sea urchins, globefish) is based here. There are several institutions of higher learning here. Connected by Kanmon Tunnel and bridge to SHIMONOSEKI on HONSHU. Hirao Dai karst highlands are nearby. Formed in 1963 by the union of the cities of Kokura, Moji, Tobata, Wakamatsu, and Yawata (or Yahata), which are now wards.

Kitale (kee-TAH-lai), town (1999 population 63,245), administrative center of Trans-Nzoia district (□ 1,155 sq mi/3,003 sq km; 1999 population 63,245), RIFT VALLEY province, W KENYA, at E foot of Mount ELGON, near UGANDA border, 146 mi/235 km NW of NAKURU; 00°58′N 34°57′E. Elevation 6,220 ft/1,896 m. Railroad spur terminus. Agricultural trade center (coffee, pyrethrum, sisal, wheat, corn); cattle raising. Site of Stoneham Museum and Research Center, European hospital. Airport. Starting point for Mount Elgon ascents.

Kitami (kee-TAH-mee), city, Hokkaido prefecture, NE HOKKAIDO, N JAPAN, on the Tokoro River, 136 mi/220 km E of SAPPORO; 43°48′N 143°53′E. Onions.

Kitamimaki (kee-TAH-mee-MAH-kee), village, South Saku county, NAGANO prefecture, central HONSHU, central JAPAN, 25 mi/40 km S of NAGANO; 36°19′N 138°20′E.

Kitamoto (kee-TAH-mo-to), city, SAITAMA prefecture, E central HONSHU, E central JAPAN, 9 mi/15 km N of URAWA; 36°01′N 139°32′E.

Kitamura (kee-tah-MOO-rah), village, Sorachi district, Hokkaido prefecture, N JAPAN, 22 mi/35 km N of SAPPORO; 43°15′N 141°42′E.

Kitanakagusuku (kee-TAH-nah-kah-GOOS-koo), village, S OKINAWA island, Nakagami county, Okinawa prefecture, SW JAPAN, 9 mi/15 km N of NAHA; 26°17′N 127°47′E.

Kita-nayoshi, RUSSIA: see LESOGORSK.

Kitangari (kee-tahn-GAH-ree), village, MTWARA region, SE TANZANIA, on MAKONDE PLATEAU, on Mambi River, 65 mi/105 km WSW of MTWARA; 10°41′S 39°19′E. Timber; corn, cashews, sweet potatoes, manioc; goats, sheep.

Kitano (kee-TAH-no), town, Mii county, FUKUOKA prefecture, N central KYUSHU, SW JAPAN, 19 mi/30 km S of FUKUOKA; 33°20′N 130°35′E.

Kitaryu (kee-tah-RYOO), town, Sorachi district, Hokkaido prefecture, N JAPAN, 53 mi/85 km N of SAPPORO; 43°43′N 141°52′E. Melons, sweet corn, sunflower seeds.

Kitashigeyasu (kee-TAH-shee-GE-yahs), town, Miyaki county, SAGA prefecture, N KYUSHU, SW JAPAN, 9 mi/15 km N of SAGA; 33°19′N 130°27′E.

Kitashiobara (kee-TAH-SHYO-bah-rah), village, Yama county, FUKUSHIMA prefecture, N central HONSHU, NE Japan, near Mount Bandai and Bandai Highlands, 28 mi/45 km W of FUKUSHIMA; 37°39′N 139°56′E.

Kita-shiretoko-misaki, RUSSIA: see TERPENIYE, CAPE.

Kitatachibana (kee-TAH-tah-CHEE-bah-nah), village, Seta county, GUMMA prefecture, central HONSHU, N central JAPAN, 6 mi/10 km N of MAEBASHI; 36°28′N 139°02′E.

Kitaura (kee-TAH-oo-rah), town, East Usuki county, MIYAZAKI prefecture, SE KYUSHU, SW JAPAN, 56 mi/90 km N of MIYAZAKI; 32°42′N 131°49′E. N limit of *biro ju* palm.

Kitaura (kee-TAH-oo-rah), village, Namegata county, IBARAKI prefecture, central HONSHU, E central JAPAN, 22 mi/35 km S of MITO; 36°04′N 140°32′E. Tobacco.

Kita-uruppu-suido, RUSSIA: see BOUSSOLE STRAIT.

Kitaya (kee-TAH-yah), village, MTWARA region, SE TANZANIA, 21 mi/34 km S of MTWARA, near RUVUMA River (MOZAMBIQUE border); 10°38′S 40°08′E. Timber; cashews, corn, sweet potatoes, manioc, bananas; goats, sheep.

Cross-references are shown in SMALL CAPITALS. The pronunciation guide is shown on page xix. The sources of population figures are shown on page xvii.

Kitayama (kee-TAH-yah-mah), village, East Muro county, WAKAYAMA prefecture, S HONSHU, W central JAPAN, 51 mi/83 km S of WAKAYAMA; 33°55′N 135°58′E. Bellows.

Kitay Lagoon, UKRAINE: see KYTAY LAGOON.

Kit Carson, county (□ 2,161 sq mi/5,618.6 sq km; 2006 population 7,590), E COLORADO; ⊙ BURLINGTON; 39°18′N 102°35′W. Grain and livestock area bordering on KANSAS. Drained by SOUTH FORK OF REPUBLICAN and NORTH FORK OF SMOKY HILL rivers and Spring, SAND, Landsman, and BEAVER creeks. Cattle; wheat, hay, sunflowers, beans, sorghum, oats, corn. Flagler Reservoir and Flagler State Wildlife Area in W. Formed 1889.

Kit Carson, village (2000 population 253), CHEYENNE county, E COLORADO, on BIG SANDY CREEK, and 24 mi/39 km W of CHEYENNE WELLS; 38°45′N 102°47′W. Elevation c.4,285 ft/1,306 m. Cattle; wheat, sunflowers, sorghum, corn. Sand Creek Massacre Site to SE. Incorporated 1931.

Kit Carson Pass, California: see CARSON PASS.

Kit Carson Peak (14,165 ft/4,317 m), in SANGRE DE CRISTO MOUNTAINS, E SAGUACHE county, S COLORADO; summit is just W of CUSTER county line, in Rio Grande National Forest.

Kitchener (KI-chuh-nuhr), city (□ 53 sq mi/137.8 sq km; 2001 population 190,399), WATERLOO region, SW ONTARIO, E central CANADA, in the GRAND RIVER valley; 43°27′N 80°30′W. Packaged meats, metal and leather goods, spirits, appliances, furniture, and rubber products. Due to close ties between Kitchener and the city of WATERLOO, the area is commonly known as Kitchener-Waterloo. Woodside National Historic Park commemorates the birthplace of former Canada prime minister W.L. MacKenzie King. Settled largely by Mennonites from PENNSYLVANIA in 1806, the city was known as Berlin until 1916, when it was renamed in memory of Lord Kitchener.

Kitchener, Mount (KI-che-nuhr), (11,500 ft/3,505 m), SW ALBERTA, W CANADA, near BRITISH COLUMBIA border, in ROCKY MOUNTAINS, on S edge of JASPER NATIONAL PARK, 55 mi/89 km SE of JASPER; 52°13′N 117°20′W.

Kitchener-Waterloo, CANADA: see KITCHENER, WATERLOO.

Kitchi, Mount (9,352 ft/2,850 m), E BRITISH COLUMBIA, W CANADA, near ROCKY MOUNTAINS, 100 mi/161 km E of PRINCE GEORGE; 53°58′N 120°24′W.

Kite, village, JOHNSON county, E central GEORGIA, 12 mi/19 km WNW of SWAINSBORO, and on OHOOPEE RIVER; 32°41′N 82°31′W. Light manufacturing.

Kite, unincorporated village, KNOTT county, E KENTUCKY, 9 mi/14.5 km E of HINDMAN, on right fork of BEAVER CREEK, in CUMBERLAND MOUNTAINS. Bituminous coal. Manufacturing (machinery).

Kitee (KI-tai), Swedish *Kides*, village, POHJOIS-KARJALAN province, SE FINLAND, 40 mi/64 km SSE of JOENSUU; 62°06′N 30°09′E. Elevation 330 ft/100 m. Near Russian border, in SAIMAA lake region. Lumbering, sawmilling; tourism.

Kitgum, district (□ 3,774 sq mi/9,812.4 sq km; 2005 population 320,000), NORTHERN region, N UGANDA; ⊙ KITGUM; 03°30′N 33°00′E. As of Uganda's division into eighty districts, borders KAABONG (E), KOTIDO (SE), PADER, GULU (SW), and AMURU (W) districts and SUDAN (N). ACHWA River forms much of W border, PAGER River flows W through center of district to Achwa River. District primarily characterized by savannah woodlands. Population primarily agricultural (including beans, cassava, groundnuts, maize, millet, simsim, and sorghum, cotton production increasing). Towns include Kitgum and Padibe. Formed in 2001 from N portion of former KITGUM district (Pader district was formed from S portion).

Kitgum (kit-goom), former administrative district (□ 6,230 sq mi/16,198 sq km), N UGANDA, along SUDAN border (to N); capital was KITGUM; 03°30′N 33°00′E.

Was largest district of Uganda. As of Uganda's division into thirty-nine districts, was bordered by KOTIDO (E), LIRA (S), APAC (SW), and GULU (W) districts. Agricultural area (millet, cassava). Kitgum was only large town. In 2001 N portion of district was formed into current KITGUM district and S portion formed into PADER district.

Kitgum (kit-goom), town (2002 population 41,821), ⊙ KITGUM district, NORTHERN region, N UGANDA, 50 mi/80 km NE of GULU. Road center; cotton, peanuts, sesame, sweet potatoes, millet. Gold deposits. Was part of former NORTHERN province.

Kithairon (kee-the-RON), Greek *Kithairón*, mountain range, c.10 mi/16 km long, CENTRAL GREECE department, in the SE corner of BOEOTIA prefecture, at the border with ATTICA department (S); rises to 4,623 ft/1,409 m; 38°12′N 23°15′E. Scene of many events in Greek mythology; especially sacred to Dionysius. Also spelled Cithaeron.

Kit Hill, ENGLAND: see CALLINGTON.

Kíthira (KEE-thee-rah), island (□ 109 sq mi/283.4 sq km), ATTICA prefecture, ATTICA department, southernmost of IONIAN ISLANDS, in the MEDITERRANEAN SEA off S PELOPONNESUS, SE mainland GREECE; 36°15′N 23°00′E. Mostly rocky with many streams, it produces wine, goat cheese, olives, corn, and flax. Airport. Some tourism. Kapsali (1971 population 349), village, formerly called Kíthira, on S shore. Ancient Kíthira was a center of the cult of Aphrodite. Passed to Greece in 1864. Also spelled Cythera, Kythera.

Kithnos (KEETH-nos), Latin *Cythnus*, island (□ 38 sq mi/98.8 sq km), CYCLADES prefecture, SOUTH AEGEAN department, GREECE, between KÉA and SERIFOS, W of SÍROS island; 11 mi/18 km long, 4 mi/6.4 km wide; rises to 965 ft/294 m; 37°25′N 24°28′E. Produces barley, wine, almonds, figs, olive oil. Hematite deposits. Main town, Kithnos (pop. 1,243), is near E shore. Colonized by Ionians. Ruled (1537–1832) by TURKEY. Also called Thermia for hot springs on E coast. Also spelled Kythnos.

Kiti (KEE-tee), town, LARNACA district, S CYPRUS, near the MEDITERRANEAN, 6 mi/9.7 km SW of LARNACA; 34°50′N 33°34′E. Citrus, olives, grapes, vegetables, potatoes; goats, sheep, hogs. Church of Panayia Angelektistos (7th century; rebuilt 12th century). Cape KITI is 3 mi/4.8 km SE.

Kiti, Cape, promontory, S central CYPRUS, on MEDITERRANEAN SEA, 3 mi/4.8 km SE of KITI town, and 7 mi/11.3 km SW of LARNACA; 34°49′N 33°36′E. Lighthouse. Formerly called Cape DADES.

Kitikmeot (ki-TIC-mee-aht), region (□ 176,530 sq mi/458,978 sq km; 2001 population 4,816), W NUNAVUT territory, CANADA. CAMBRIDGE BAY on VICTORIA ISLAND is the regional headquarters. The region extends E from the GREAT BEAR LAKE to the BOOTHIA PENINSULA, and N from the FORT SMITH and KIVALLIQ regions to include the S and E parts of Victoria Island and the lower half of PRINCE OF WALES ISLAND, as well as KING WILLIAM ISLAND. Part of the NORTHWEST TERRITORIES until 1999; created in 1981 by the territorial government, the region is inhabited mostly by Inuit people who rely on traditional activities such as fishing, sealing, and crafts.

Kitimat (KIT-i-mat), town, port (□ 94 sq mi/244.4 sq km; 2001 population 10,285), W BRITISH COLUMBIA, W CANADA, at the head of Douglas Channel, and in Kitimat-Stikine regional district; 54°00′N 128°42′W. Huge aluminum smelter (opened 1954), pulp and paper mills, and a petrochemical plant. Deep-water anchorage.

Kitimat-Stikine (KI-tuh-mat–sti-KEEN), regional district (□ 35,487 sq mi/92,266.2 sq km; 2001 population 40,876), NW BRITISH COLUMBIA, W CANADA; 55°20′N 129°00′W. Includes the municipalities of KITIMAT, TERRACE, STEWART, HAZELTON, and New Hazelton.

Kition, CYPRUS: see CITIUM.

Kito (kee-TO), village, Naka county, TOKUSHIMA prefecture, SE SHIKOKU, W JAPAN, 28 mi/45 km S of

TOKUSHIMA; 33°46′N 134°12′E. Lumber; chopticks; citron and citron processing.

Kitobola (kee-to-BO-lah), village, BAS-CONGO province, W CONGO, near railroad, 105 mi/169 km ENE of BOMA; 05°22′S 14°31′E. Elev. 1,715 ft/522 m. Agricultural center (palms, sugarcane, cattle).

Kitombe (kee-TOM-bai), village, BANDUNDU province, W CONGO, 17 mi/27 km S of KIKWIT; 05°24′S 18°58′E. Elev. 1,679 ft/511 m. Oil palms.

Kitona (kee-TO-nuh), village, BAS-CONGO province, W CONGO, on road, 15 mi/24 km E of MUANDA; 05°58′S 12°28′E.

Kitovo (kee-TO-vuh), village (2005 population 2,900), central IVANOVO oblast, central European Russia, on road and railroad, less than 3 mi/5 km W of SHUYA, to which it is administratively subordinate; 56°52′N 41°18′E. Elevation 344 ft/104 m. In agricultural area (grain, sunflowers, hemp, sugarbeets).

Kitoy (kee-TO-yee), town (2005 population 3,920), SE IRKUTSK oblast, E central SIBERIA, RUSSIA, on road and the TRANS-SIBERIAN RAILROAD, 30 mi/48 km NW of IRKUTSK; 52°36′N 103°54′E. Elevation 1,391 ft/423 m. Sawmilling.

Kitoy, RUSSIA: see KITOI.

Kitsap (KIT-suhp), county (□ 565 sq mi/1,469 sq km; 2006 population 240,604), W WASHINGTON; ⊙ PORT ORCHARD; 47°38′N 122°39′W. Occupies N part of Kitsap peninsula, bounded W by HOOD CANAL and E by PUGET SOUND, deeply indented on E by Sinclair and Dyes inlets and Liberty Bay; BAINBRIDGE ISLAND separated from mainland by Port Orchard passage. BREMERTON is important seaport and site of Puget Sound Navy Yard. County directly opposite city of SEATTLE, ferry service to Seattle and EDMONDS. Illahee State Park in center; Scenic Beach State Beach in W; Blake Island, Fay Bainbridge, and Fort Ward state parks in E, on Bainbridge Island U.S. Naval Station at Bangor, in W on Hood Canal, has Trident ballistic missile submarine base with half of all U.S. long-defense missiles. Includes Port Gamble and Port Madison Indian reservations in N. Fruits, nuts, poultry; fish, crabs, oysters, clams. Formed 1857.

Kitscoty, village (□ 1 sq mi/2.6 sq km; 2001 population 671), E ALBERTA, W CANADA, near SASKATCHEWAN border, 21 mi/34 km E of VERMILION, in VERMILION RIVER COUNTY NO. 24; 53°20′N 110°20′W. Dairying; grain. Incorporated 1911.

Kitsman' (KEETS-man) (German *Kotzman*) (Romanian *Cozmeni*), town (2004 population 12,200), N CHERNIVTSI oblast, UKRAINE, in N BUKOVYNA, 13 mi/21 km NNW of CHERNIVTSI; 48°26′N 25°45′E. Elevation 705 ft/214 m. Raion center; flour, oilseeds. Agricultural school. Known since 1413; part of Moldova to the mid-16th century; then, until 1774, belonged to Turkey; under Austria (1774–1918) developed into a Ukrainian cultural and educational center; November 1918, briefly capital Ukrainian-Bukovynian government; then annexed by Romania and in 1940 by the USSR; since 1991, part of independent Ukraine.

Kitsuki (KEETS-kee), city, OITA prefecture, NE KYUSHU, SW JAPAN, on S KUNISAKI PENINSULA, 12 mi/20 km N of OITA across BEPPU BAY; 33°24′N 131°37′E. Prawns; mandarin oranges.

Kitsuregawa (kee-TSOO-re-GAH-wah), town (pop.11,434), Shioya county, TOCHIGI prefecture, central HONSHU, N central JAPAN, 12 mi/20 km N of UTSUNOMIYA; 36°42′N 140°01′E.

Kittam River (kee-TAHM), tidal inlet, c.30 mi/48 km long; SW SIERRA LEONE; formed 30 mi/48 km ESE of BONTHE by union of WAANJE and SEWA rivers; flows WNW, past GBAP, to SHERBRO River 7 mi/11.3 km S of Bonthe. Its course forms N side of Turner's Peninsula. Sometimes called BUM RIVER or BUM KITTAM RIVER.

Kittanning (KI-tah-ning), borough (2006 population 4,411), ⊙ ARMSTRONG county, W central PENNSYLVANIA, 35 mi/56 km NE of PITTSBURGH, on ALLEGHENY RIVER (bridged). Manufacturing (agricultural chemicals, fabricated metal products, construction materials, apparel, sand and gravel processing). Agriculture (corn, hay; livestock; dairying); bituminous coal; limestone; sand and gravel. Original site was an Indian village. Lock and Dam Number 7 here. Settled 1796; laid out 1804; incorporated 1821.

Kittatinny Mountain (KI-tuh-TI-nee), a ridge of the APPALACHIAN system, extending across NW NEW JERSEY from SHAWANGUNK MOUNTAIN, SE NEW YORK, to BLUE MOUNTAIN, E central PENNSYLVANIA; rises to HIGH POINT (1,803 ft/550 m), the highest peak in New Jersey. Kittatinny Mountain is a major resort and recreation area; the APPALACHIAN Trail lies atop the ridge. The DELAWARE RIVER cuts through the E part of the ridge forming the DELAWARE WATER GAP.

Kittery (KIT-uhr-ee), town, YORK county, extreme SW MAINE, at the mouth of the PISCATAQUA RIVER, 27 mi/43 km S of ALFRED; 43°06'N 70°42'W. Its economy centers around tourism and the Portsmouth Naval Shipyard, which services nuclear-powered submarines and is located on two islands owned by the Federal government and connected with Kittery by two bridges. NEW HAMPSHIRE also lays claim to these two islands Shipyard was first public shipyard in U.S., established in 1800 as colonial shipyard, built first warship in America, HMS *Falkland*. In 1917, first U.S. submarine built. Yard built submarines (first public shipyard to build nuclear submarine) until 1971. The oldest town in Maine (settled c.1623), it grew as a trading, fishing, lumber-shipping, and shipbuilding center. John Paul Jones's ship *Ranger* (1777) and the *Kearsarge* of Civil War fame were both built there. Several eighteenth-century houses remain. William Whipple, a signer of the Declaration of Independence, was born in Kittery. Incorporated 1647.

Kittigazuit, locality, NORTHWEST TERRITORIES, CANADA, on Refuge Cove, bay of the BEAUFORT SEA, at mouth of E channel of MACKENZIE RIVER delta, 90 mi/145 km NE of AKLAVIK; 69°21'N 133°43'W. Site of Hudson's Bay Company post, 1915–1929.

Kittilä (KIT-ti-lah), village, LAPIN province, NW FINLAND, on the OUNASJOKI (river), and 80 mi/129 km NNW of ROVANIEMI; 67°40'N 24°54'E. Elevation 660 ft/200 m. Lapp trading center. Large iron deposits centered in Porkonen and Pahtavaara localities.

Kittim, CYPRUS: see CITIUM.

Kittitas (KIT-i-tuhs), county (□ 2,332 sq mi/6,063.2 sq km; 2006 population 37,189), central WASHINGTON; ⊙ ELLENSBURG; 47°07'N 120°41'W. NACHES and Little Naches rivers form SW boundary; bounded on E by COLUMBIA RIVER (WANAPUM LAKE reservoir). Mount area drained by Yakima River. Potatoes, alfalfa, hay, wheat, barley, oats, peas; cattle, sheep; lumber; coal, gold, silica; dairying. Parts of Wenatchee National Forest in NW, including part of Alpine Lakes Wilderness Area; part of Gifford Pinchot National Forest in SW; Olmstead Place State Park in center; Gingko Petrified Forest and Wanapum state parks in E; Lake Easton and Iron Horse state parks in W. Kencholus, Haches, and Clo Elum reservoirs in W; PRIEST RAPIDS LAKE reservoir, SE. Part of U.S. Military Reservation Yakima Training Center, in SE. Formed 1883.

Kittitas (KIT-i-tuhs), town (2006 population 1,183), KITTITAS county, central WASHINGTON, 6 mi/9.7 km E of ELLENSBURG; 47°00'N 120°25'W. Wheat, barley, oats, potatoes, peas, hay. Olmstead Place State Park to W; Gingko Petrified Forest and Wanapum state parks to E. U.S. Military Reservation Yakima Training Center to SE.

Kitt Peak (6,875 ft/2,096 m), in E edge of Tohono O'odham (Papago) Indian Reservation in the N end of BABOQUIVARI MOUNTAINS, central PIMA County, S ARIZONA, 40 mi/64 km WSW of TUCSON. It is the site of KITT PEAK NATIONAL OBSERVATORY. Mexican border 29 mi/47 km SSW.

Kitt Peak National Observatory, central PIMA county, S ARIZONA, astronomical observatory located 40 mi/64 km WSW of TUCSON, ARIZONA, at summit of KITT PEAK (6,875 ft/2,096 m). It was founded in 1958 under contract with the National Science Foundation and is administered by the Association of Universities for Research in Astronomy. Its principal instrument is the Mayall 158 in/4 m reflector. The observatory's equipment also includes 84 in/2.1 m, 50 in/1.3 m, 36 in/0.9 m, and 16 in/0.4 m reflecting telescopes as well as a planned 138-in/3.5-m telescope. Used for wide angle photographs and electronic images of the sky, the Burrell Schmidt telescope is operated jointly with Case Western Reserve University. The 60-in/1.5-m Robert McMath Solar Telescope is the largest instrument of its kind in the world. Stellar research, now part of the National Optical Astronomy Observatories, includes basic research on galaxies, stars, nebulae, and the solar system. The solar division, now part of the National Solar Observatory, using the solar telescope in coordination with a vacuum spectrograph, analyzes the composition, magnetic field strength, motion, and physical nature of the sun. Other telescopes are located on Kitt Peak, notably those of the National Radio Astronomy Observatory and Steward Observatory. Any astronomer can apply for time on the telescopes. A telescope allocation committee of astronomers selects the best proposals and time is assigned every six months.

Kittrell (ki-TRELL), village (2006 population 148), VANCE county, N NORTH CAROLINA, 8 mi/12.9 km SSW of HENDERSON; 36°13'N 78°26'W. Manufacturing (corn feed); agriculture (grain, tobacco; livestock).

Kitts, unincorporated village, HARLAN county, SE KENTUCKY, 3 mi/4.8 km E of HARLAN, in the CUMBERLAND MOUNTAINS, on CLOVER FORK of CUMBERLAND RIVER. In coal mining area.

Kittsee (KIT-sai), Croatian *Gieca*, village N BURGENLAND, E AUSTRIA, 4 mi/6.4 km S of BRATISLAVA, Slovakia; 48°05'N 17°04'E. Vineyards. Has Croatian minority. Baroque castle with an ethnographic museum.

Kittson, county (□ 1,103 sq mi/2,867.8 sq km; 2006 population 4,691), extreme NW MINNESOTA; ⊙ HALLOCK; 48°46'N 96°46'W. Drained by TWO RIVERS and its North, Middle, and South branches; bounded W by RED RIVER (NORTH DAKOTA state line), N by CANADA (MANITOBA). Agricultural area (wheat, alfalfa, hay, flax, oats, barley, sugar beets, beans, sunflowers; sheep). Lake Bronson State Park in E center, on Lake Bronson reservoir; Twin Lakes Wildlife Area in SE; Skull Lake Wildlife Area in NE. Formed 1878.

Kitty, town, DEMERARA-MAHAICA district, N GUYANA; 06°49'N 58°08'W. A NE suburb of GEORGETOWN, on the ATLANTIC OCEAN.

Kittybrewster, Scotland: see ABERDEEN.

Kitty Hawk or **Kittyhawk** (KIT-ee HAWK), town (□ 8 sq mi/20.8 sq km; 2006 population 3,332), DARE county, NE NORTH CAROLINA, 35 mi/56 km SE of ELIZABETH CITY, on ALBEMARLE SOUND, at entrance to CURRITUCK SOUND (N); 36°04'N 75°43'W. ATLANTIC OCEAN 3 mi/4.8 km to E. Bridge to mainland to NW. Construction; service industries; light manufacturing. Residential developments in area. To SE is KILL DEVIL HILLS and WRIGHT BROTHERS NATIONAL MEMORIAL, where the Wright brothers experimented successfully (1900–1903) with gliders and airplanes. First successful flight December 17, 1903. Established 1981.

Kitui (kee-TOO-eeh), town (1999 population 13,244), district administrative center, EASTERN province, S central KENYA, on road, and 80 mi/129 km E of NAIROBI; 01°22'S 38°01'E. Agriculture (sisal, corn).

Kitumbeine (kee-toom-bai-EE-nai), village, ARUSHA region, N TANZANIA, 40 mi/64 km NNW of ARUSHA, in Engaruka Basin; 02°49'S 36°19'E. Kitumbeine Mountain (9,653 ft/2,942 m) to SW. Corn, wheat, millet; cattle, sheep, goats. Also spelled Ketumbaine.

Kitunda (kee-TOON-dah), town, TABORA region, W central TANZANIA, 120 mi/193 km SSE of TABORA; 06°49'S 33°11'E. Road junction. Timber; corn, wheat; goats, sheep.

Kitwe, city (2000 population 363,734), COPPERBELT province, N central ZAMBIA, 170 mi/274 km N of LUSAKA, near CONGO border; 12°49'S 28°12'E. On railroad. Founded 1936; main copper-mining center; some commercial activity. Manufacturing (ceramics, clothing and textiles, automotive equipment, hardware, food products, agricultural products processing, wood products, consumer goods); agriculture (cotton, tobacco, peanuts, coffee, soybeans, flowers); cattle. Mining (copper, emeralds, clay). Airport. Hospital and secondary school for miners' children. The Zambia Institute of Technology is here.

Kitzbühel (KITS-byooi-el), town, in TYROL province, W AUSTRIA, in the Kitzbühel Alps; 47°27'N 12°24'E. Funiculars to the Hahnenhamm at an elevation of 5,044 ft/1,537 m and Kitzbüheler Horn at an elevation of 6,084 ft/1,854 m with ski slopes. It is a famous winter sports and resort center and a summer spa. Market center; manufacturing (machinery, apparel). An old silver and copper-mining town. Remarkable churches.

Kitzbühel Alps (KITS-byooi-el), German *Kitzbüheler Alpen*, range of Eastern Alps in TYROL and W SALZBURG, W AUSTRIA, extending c. 40 mi/64 km E from ZILLER RIVER to SAALACH River, bounded S by the Salzach River. Divided into two sections by THURN PASS. W range rises to 7,797 ft/2,377 m in the KREUZJOCH; E range to 7,202 ft/2,195 m in the GAISSTEIN. HAHNENKAMM (5,044 ft/1,537 m) and Kitzbüheler Horn (6,084 ft/1,854 m) in N, Schmittenhöhe (5,989 ft/1,825 m) in E are tourist and winter sports centers.

Kitzingen (KIT-tsing-uhn), city, BAVARIA, central GERMANY, in LOWER FRANCONIA, on the MAIN, and 11 mi/18 km ESE of WÜRZBURG; 49°44'N 10°10'E. Manufacturing of machinery, glass, chemicals; metalworking. Wine-trading center. Site of research center for viticulture. Has late 15th-century tower; mid-16th-century city hall; former Benedictine convent (founded c.750) with 17th-century church. Chartered mid-13th century; captured several times by Swedes in Thirty Years War.

Kitzmiller, town, GARRETT county, W MARYLAND, in the ALLEGHENIES, on North Branch of the POTOMAC (bridged to WEST VIRGINIA), and 30 mi/48 km SW of CUMBERLAND; 39°23'N 79°11'W. Bituminous-coal mining. Named for Ebenezer Kitzmiller, an early settler in what was orginally a logging and later a mining area. Bloomington Dam and reservoir have recently been constructed here. Nearby are Potomac State Park and BACKBONE MOUNTAIN. Sometimes called Kitzmillersville.

Kitzscher (KIT-tshuhr), town, SAXONY, E central GERMANY, 4 mi/6.4 km NE of BORNA; 51°10'N 12°34'E. In lignite-mining region.

Kiu (KEE-ooh), village, EASTERN province, S central KENYA, on railroad and 50 mi/80 km SSE of NAIROBI; 01°54'S 37°10'E. Sisal, rubber, wheat, corn.

Kiunga (kee-OONG-gah), village (2000 population 8,265), WESTERN province, central NEW GUINEA island, W central PAPUA NEW GUINEA, 215 mi/346 km NW of KIKORI, 22 mi/35 km E of Indonesian (IRIAN JAYA) border. Situated on NW side of Fly River Airstrip. S terminus of 20 mi/32 km road from OK TEDI mining district; road construction cost $90 million. Sago; crocodile skins; timber.

Kiunga Marine National Park (□ 97 sq mi/251 sq km), COAST province, SE KENYA, N of MOMBASA near MALINDI; 01°59'N 41°23'E. Designated a national park in 1979.

Kiunglang, China: see NITI PASS.

Kiuzeli-Gyr (kyoo-ZEL-ee-goor), archaeological site, KARAKALPAK REPUBLIC, W UZBEKISTAN. Ruins of a fortress (6th–5th century B.C.E.). Earliest walled settlement of ancient KHWARAZM. Also spelled Kiuzelikyr.

Kivach Falls, RUSSIA: see SUNA RIVER.

Kivach Preserve (kee-VAHCH), environmental reserve (□ 40 sq mi/104 sq km), in Republic of KARELIA, NW European Russia, in the basin of the SUNA RIVER

(Kivach Falls on this river). Established in 1931 to protect the tayga ecosystem. Elk, brown bear.

Kivalina (ki-vuh-LEE-nuh), Inuit village (2000 population 377), NW ALASKA, on CHUKCHI SEA, 80 mi/129 km NW of KOTZEBUE; 67°43′N 164°31′W. Reindeer herding.

Kivalliq, region (□ 171,692 sq mi/446,399.2 sq km; 2001 population 7,557), NUNAVUT territory, N of MANITOBA and W of HUDSON BAY; 65°00′N 95°00′W. Its boundaries are almost entirely within the CANADIAN SHIELD. Includes SOUTHAMPTON and COATS islands. RANKIN INLET is the regional headquarters. Before 1999, when Nunavut was established, Kivalliq existed under slightly different boundaries as Keewatin (kee-WAH-tin), which was created as a region of the NORTHWEST TERRITORIES in the early 1970s with the combining of the former Keewatin and E MACKENZIE districts. Fur trapping, sealing, and craft-making are important economic activities. Also called Keewatin.

Kivdinskiy (keev-DEEN-skeeyee), formerly a town, SE AMUR oblast, RUSSIAN FAR EAST, now a suburb of RAYCHIKHINSK, less than 6 mi/10 km N of the city center; 49°51′N 129°41′E. Elevation 446 ft/135 m. On a spur of the TRANS-SIBERIAN RAILROAD. Lignite mines. Originally formed as a residential settlement for the miners and called Kivdinskiye Kopi [Russian=Kivdin Mines].

Kivdinskiye Kopi, RUSSIA: see KIVDINSKIY.

Kivennapa, RUSSIA: see PERVOMAYSKOYE, LENINGRAD oblast.

Kiverichi (kee-VYE-ree-chee), village, E central TVER oblast, W central European Russia, on crossroads, 42 mi/67 km N of TVER, and 30 mi/48 km S of BEZHETSK; 57°22′N 36°35′E. Elevation 544 ft/165 m. In agricultural area (grain, hemp, vegetables).

Kivertsi (KEE-ver-tsee) (Russian *Kivertsy*) (Polish *Kiwerce*), city, SE VOLYN' oblast, UKRAINE, 8 mi/13 km NE of LUTS'K; 50°50′N 25°28′E. Elevation 606 ft/184 m. Raion center; railroad junction; sawmilling; railroad shops, food canning, building materials. Kivertsi village is 4 mi/6 km SW. Village known since 1583; town arose at a railroad junction, built in 1873; city since 1951.

Kivertsy, UKRAINE: see KIVERTSI.

Kivik (SHEE-veek), fishing village, SKÅNE county, S SWEDEN, on BALTIC SEA, 10 mi/16 km NNW of SIMRISHAMN; 55°41′N 14°41′E. Agriculture (fruits). Seaside resort. Site of Bronze Age grave nearby.

Kiviõli (KEE-vee-uh-lee) [English=shale oil], city, NE ESTONIA, on railroad and 22 mi/35 km E of RAKVERE; 59°21′N 26°58′E. Oil-shale mining center; chemicals.

Kivsharivka (kif-SHAH-rif-kah) (Russian *Kovsharovka*), town, E KHARKIV oblast, UKRAINE, 10 mi/16 km SSE of KUP'YANS'K, and 1.2 mi/2 km S of KUP'YANS'K-VUZLOVYY, on the left bank of the Chervonooskol'ske Reservoir of the Oskil River (Russian *Krasnooskol'skoe* Reservoir of the *Oskol* River); 49°37′N 37°42′E. Elevation 360 ft/109 m. Subordinated to the Kup'yans'k city council. Kup'yans'k foundry, reinforced concrete building materials factory. Professional-technical school. Established in 1970.

Kivu (KEE-voo), region (□ 89,000 sq mi/231,400 sq km), E CONGO; ⊙ BUKAVU. It borders on UGANDA, RWANDA, BURUNDI, and LAKE TANGANYIKA on the E. Kivu is divided into three administrative regions, NORD-KIVU, SUD-KIVU, and MANIEMA. Coffee, cotton, rice, and palm-oil are produced, and tin and some gold are mined. The RUWENZORI mountains and VIRUNGA NATIONAL PARK, a vast game preserve, are in the region. Most of Kivu was controlled (1961–1962) by the breakaway regime of Antoine Gizenga, which was centered at KISANGANI (then STANLEYVILLE). The central government re-established control over Kivu in 1962 and maintained it after rebel activity subsided in the later 1960s. The E part of the region is still the gateway to RWANDA, and during the 1994 conflict in that country, large numbers of Rwandan refugees sought refuge in Kivu.

Kivu, Lake (KEE-voo), (□ 1,042 sq mi/2,709.2 sq km), 55 mi/89 km long, on the CONGO-RWANDA border, E central Africa; highest lake in Africa (elev. 4,788 ft/1,459 m); 2°13′S 29°10′E. It is drained by the RUZIZI RIVER, which flows S into LAKE TANGANYIKA. Beneath the lake lie vast reserves of methane gas, which have not been exploited. Tourist center.

Kiwa (KEE-wah), town, South Muro county, MIE prefecture, S HONSHU, central JAPAN, 68 mi/110 km S of TSU; 33°52′N 135°55′E.

Kiwalik (ki-WAH-lik), village, NW ALASKA, on SE shore of KOTZEBUE SOUND, N SEWARD PENINSULA, at mouth of KIWALIK RIVER and 60 mi/97 km SSE of KOTZEBUE. Port and transfer point for CANDLE and Kiwalik River valley. Sometimes spelled Keewalik.

Kiwalik River (ki-WAH-lik), NW ALASKA, on N side of SEWARD PENINSULA; rises near 65°20′N 161°30′W; flows c.60 mi/97 km N, past CANDLE, to KOTZEBUE SOUND at KIWALIK.

Kiwerce, UKRAINE: see KIVERTSI.

Kiwindi (kee-WEEN-dee), village, RUVUMA region, S TANZANIA, 70 mi/113 km SW of SONGEA, on Lake NYASA, at MOZAMBIQUE border; LUKOMA BAY to NW; 10°34′S 34°57′E. Timber; corn, sweet potatoes; goats.

Kiyaly-Kurgancha, UZBEKISTAN: see DANGARA.

Kiyama (KEE-yah-mah), town, Miyaki county, SAGA prefecture, NW KYUSHU, SW JAPAN, 19 mi/30 km N of SAGA; 33°24′N 130°31′E.

Kiya River (kee-YAH), Russia; rises in KUZNETSK ALATAU E of Pezas (KEMEROVO oblast); flows N, past MARYINSK (on the TRANS-SIBERIAN RAILROAD), and NW into TOMSK oblast to the CHULYM RIVER at ZYRYANSKOYE; 315 mi/507 km long. Gold deposits along the upper course, in Kemerovo oblast (BERIKUL'SKIY, TSENTRAL'NYY mines).

Kiya-Shaltyr', RUSSIA: see BELOGORSK, KEMEROVO oblast.

Kiyasovo (kee-YAH-suh-vuh), village (2006 population 3,255), S UDMURT REPUBLIC, E European Russia, on road, 35 mi/56 km S of IZHEVSK; 56°20′N 53°07′E. Elevation 521 ft/158 m. In agricultural area (grains, vegetables, potatoes; livestock).

Kiye, TURKEY: see GEMLIK.

Kiyev, UKRAINE: see KIEV, city.

Kiyevka (kee-YEF-kuh), town, N KARAGANDA region, KAZAKHSTAN, near NURA RIVER, 70 mi/113 km NW of KARAGANDA; 50°15′N 71°33′E. In cattle area. Tertiary-level (raion) administrative center.

Kiyevskoye (KEE-eef-skuh-ye), settlement (2005 population 4,750), SW KRASNODAR TERRITORY, S European Russia, in the KUBAN' River basin, on road and railroad, 23 mi/37 km NE of ANAPA; 45°02′N 37°53′E. Population largely Meskhetian Turks.

Kiyevskoye vodokhranilishche, UKRAINE: see KIEV RESERVOIR.

Kiyma, KAZAKHSTAN: see KIMA.

Kiyokawa (kee-YO-kah-wah), village, Aiko county, KANAGAWA prefecture, E central HONSHU, E central JAPAN, 22 mi/35 km W of YOKOHAMA; 35°28′N 139°16′E.

Kiyokawa (kee-yo-KAH-wah), village, Ono county, OITA prefecture, E KYUSHU, SW JAPAN, 19 mi/30 km S of OITA; 32°57′N 131°30′E.

Kiyomi (KEE-yo-mee), village, Ono county, GIFU prefecture, central HONSHU, central JAPAN, 56 mi/90 km N of GIFU; 36°07′N 137°11′E.

Kiyone (kee-YO-ne), village, Tsukubo county, OKAYAMA prefecture, SW HONSHU, W JAPAN, 12 mi/20 km W of OKAYAMA; 34°38′N 133°44′E.

Kiyosato (kee-YO-sah-to), village, Nakakubiki county, NIIGATA prefecture, central HONSHU, N central JAPAN, 71 mi/115 km S of NIIGATA; 37°04′N 138°20′E.

Kiyose (KEE-yo-se), city, Tokyo prefecture, E central HONSHU, E central JAPAN, 12 mi/20 km N of SHINJUKU; 35°46′N 139°31′E.

Kiyosu (KEE-yos), town, West Kasugai county, AICHI prefecture, S central HONSHU, central JAPAN, 5 mi/8 km N of NAGOYA; 35°12′N 136°50′E. Manufacturing (electronic components).

Kiyotake (kee-YO-tah-ke), town, Miyazaki county, MIYAZAKI prefecture, SE KYUSHU, SW JAPAN, 3.7 mi/6 km N of MIYAZAKI; 31°51′N 131°23′E. Computer components. Oranges.

Kiyskoye, RUSSIA: see MARIINSK.

Kizel (KEE-zyel), city (2006 population 21,160), central PERM oblast, W URALS, E European Russia, on the Kizel River, on road and railroad, 100 mi/161 km NE of PERM; 59°03′N 57°38′E. Elevation 843 ft/256 m. Coal mining (Kizel bituminous-coal basin); repair of coal mining equipment; manufacturing (machinery, clothing, wood, and food products). Founded in the late 18th century as a coal mining town, but the industry has been losing its importance as a source of revenue for city residents since the post-Soviet switch to market economy. Made city in 1926.

Kizelstroy, RUSSIA: see GUBAKHA.

Kizema (KEE-zee-mah), village (2005 population 3,600), S ARCHANGEL oblast, central European Russia, on road and railroad, 65 mi/105 km W of KOTLAS; 61°07′N 44°49′E. Elevation 400 ft/121 m. Also spelled Kizima.

Kizhaba (kee-ZHA-bah), urban settlement, ASTARA region, AZERBAIJAN, 7 mi/12 km NW of Astara railroad station, terminus of line from BAKY; 38°32′N 48°49′E. Tea factory; subtropical crops.

Kizhinga (kee-zhin-GAH), village (2005 population 6,420), SE BURYAT REPUBLIC, S central SIBERIA, RUSSIA, on road and railroad spur, 95 mi/153 km E of ULAN-UDE; 51°50′N 109°55′E. Elevation 2,290 ft/697 m. In agricultural area (hemp, flax, feed corn; livestock).

Kizi (KEE-ZEE), river, W UGANDA; rises from N WESTERN region; flows c.35 mi/56 km SSW to KAFU RIVER.

Kizil, RUSSIA: see KYZYL.

Kizil Adalar, group of nine small islands (□ 4 sq mi/10.4 sq km), NW TURKEY, in the Sea of MARMARA, SE of ISTANBUL. The islands are a popular resort area. They were used as places of exile in Byzantine times. There are several old monasteries and churches. BÜYÜK ISLAND is the largest of the group. Formerly known as Princes Islands.

Kizilbás, CYPRUS: see TRAKHONAS.

Kizilcahamam, village, central TURKEY, on KIRMIR RIVER and 38 mi/61 km NNW of ANKARA. Hot springs; grain, fruits, chickpeas, vetch, mohair goats; 40°28′N 32°37′E. Formerly called Yabanabat and Chorba.

Kizil Dag, (Turkish = *Kizil Dağ*) peak (9,892 ft/3,015 m), N central TURKEY, 22 mi/35 km W of REFAHIYE, near source of the KIZIL IRMAK.

Kizildere River, N TURKEY; rises in GIRESUN MOUNTAINS 20 mi/32 km E of MESUDIYE; flows 75 mi/121 km W and N, past Mesudiye, to BLACK SEA near ORDU.

Kizilhisar Dag, (Turkish=*Kizilhisar Dağ*) peak (7,352 ft/2,241 m), SW TURKEY, in MENTESE MOUNTAINS, 10 mi/16 km ESE of TAVAS.

Kizil Irmak River, ancient *Halys*, longest river of TURKEY, c.715 mi/1,150 km long; rises in the KIZIL DAĞ, N central TURKEY; flows in a wide arc SW, then N, and then NE into the BLACK SEA. It has an irregular volume and is not used for navigation. The river is an important source of water and hydroelectric power. The Hirfanli Dam (SW of Kaman) and Lake (reservoir) are the main sources of electricity and water, which the river provides.

Kizil Kum, KAZAKHSTAN: see KYZYL KUM.

Kizil'skoye (kee-ZEEL-skuh-ye), village, SW CHELYABINSK oblast, S URALS, RUSSIA, on the URAL River, on crossroads and railroad spur, 50 mi/80 km SSW of MAGNITOGORSK; 52°44′N 58°54′E. Elevation 941 ft/286 m. Anthracite deposits.

Kiziltash Liman (kee-zeel-TAHSH), lagoon of the BLACK SEA, KRASNODAR TERRITORY, S European Russia, SE of the TAMAN' PENINSULA; 12 mi/19 km long, 6 mi/10 km wide. Receives the marshy Old Kuban' arm of the KUBAN' River.

Kiziltepe, township, SE TURKEY, 11 mi/18 km SW of MARDIN; 37°12′N 40°36′E. Grain, legumes; mohair goats. Also called Kochisar.

Kizilyurt (kee-zeel-YOORT), city (2005 population 30,045), NW DAGESTAN REPUBLIC, SE European Russia, on the E bank of the SULAK RIVER, on road and railroad, 41 mi/66 km WNW of MAKHACHKALA, and 14 mi/23 km E of KHASAVYURT; 43°12′N 46°52′E. Elevation 160 ft/48 m. Railroad depots; agricultural machinery manufacturing and repair; food processing. Formerly known as Chir-Yurt.

Kizilyurt, RUSSIA: see KUMTORKALA.

Kizima, RUSSIA: see KIZEMA.

Kizimbani (kee-zeem-BAH-nee), village, LINDI region, SE TANZANIA, 14 mi/23 km SW of KILWA MASOKO, near Mavuji River; 09°05′S 39°09′E. Cashews, corn, bananas, sisal, copra; goats, sheep; timber.

Kizimkazi (kee-zeem-KAH-zee), village, ZANZIBAR SOUTH region, E TANZANIA, near S end of ZANZIBAR island, 27 mi/43 km SE of ZANZIBAR city; 06°27′S 39°27′E. Fishing center. Former ⊙ Zanzibar and site of 12th-century Persian mosque (Shirazi Mosque). Also spelled Kizinkasai.

Kizlyar (kee-ZLYAHR), city (2005 population 51,300), NE DAGESTAN REPUBLIC, NE CAUCASUS, SE European Russia, in the Caspian depression on the left bank of the TEREK RIVER, 105 mi/169 km NW of MAKHACHKALA, and 60 mi/97 km NE of GROZNYY (CHECHEN REPUBLIC); 43°51′N 46°43′E. Highway and railroad junction; local transshipment point. Wine making (cognac, liquors); electrical equipment; food industries. Founded in 1735 as a fortress; became an important center for trade with Dagestan mountain tribes. Acquired large Armenian population in the early 19th century during the separatist movements within the Ottoman Empire; another sizeable population increase, begun in the early 1990s, is due to the influx of refugees from war-torn Chechnya.

Kizlyar (keez-LYAHR), village (2006 population 8,600), N NORTH OSSETIAN REPUBLIC, SE European Russia, on the TEREK RIVER, on highway and near railroad, 3 mi/5 km SW of MOZDOK; 43°42′N 44°35′E. Elevation 508 ft/154 m. Since 1992, experienced almost continuous civil unrest due to ethnic clashes and incursions of Chechen rebels across the border to the E.

Kizner (keez-NYER), town (2006 population 9,815), SW UDMURT REPUBLIC, E European Russia, on the Lyaga River (tributary of the VYATKA River), on railroad, 75 mi/121 km SW of IZHEVSK, and 47 mi/76 km N of NIZHNEKAMSK (TATARSTAN); 56°16′N 51°31′E. Elevation 291 ft/88 m. Lumbering, flax processing, food processing.

Kizu (KEEZ), town, Soraku county, KYOTO prefecture, S HONSHU, W central JAPAN, 19 mi/30 km S of KYOTO; 34°44′N 135°49′E. Also spelled Kitsu.

Kizukuri (kee-ZOO-koo-ree), town, W Tsugaru county, Aomori prefecture, N HONSHU, N JAPAN, 19 mi/30 km E of AOMORI; 40°48′N 140°23′E. Watermelons.

Kizumbi (kee-ZOOM-bee), village, RUKWA region, W TANZANIA, 55 mi/89 km WNW of SUMBAWANGA, on Lake TANGANYIKA; 07°47′S 30°47′E. Fish; corn, wheat; goats, sheep.

Kizu River, JAPAN: see YODO RIVER.

Kizyl, in Russian names: see also Kizil, KYZYL, Kzyl.

Kizyl-Arvat, TURKMENISTAN: see GYZYLARBAT.

Kizyl-Asker, KYRGYZSTAN: see KYZYL-ASKER.

Kizyl-Atrek (KI-zuhl-ah-TREK), town, SW BALKAN weloyat, SW TURKMENISTAN, on SUMBAR RIVER (tributary of the ATREK; IRAN border), and 195 mi/314 km WSW of ASHGABAT; 38°20′N 55°00′E. Tertiary-level administrative center. Subtropical agriculture (olives; wheat); cattle, camels. Also spelled Kizilatrek.

Kizyl-Ayak (KI-zuhl-ah-YAHK), town, SE LEBAP weloyat, SE TURKMENISTAN, on the AMU DARYA (river), and 15 mi/24 km SE of KERKI; 37°40′N 65°23′E. Cotton.

Kizyl-Kaya (KI-zuhl-kah-YAH), town, E BALKAN weloyat, TURKMENISTAN, on E spur of TRANS-CASPIAN

RAILROAD, 68 mi/110 km E of GARABOGAZKOL gulf of CASPIAN SEA and 145 mi/242 km E of TURKMENBASHI. Also spelled Qizilqiya.

Kizyl-Kup (KI-zuhl–KOOP), town, W BALKAN weloyat, NW TURKMENISTAN, on S shore of GARABOGAZKOL Gulf, 65 mi/105 km NE of TURKMENBASHI. Glauber's salt extraction; fish canneries. Endangered by rising level of CASPIAN SEA.

Kizyl-Su (KI-zuhl–SOO), town, W BALKAN weloyat, W TURKMENISTAN, on CASPIAN SEA sandspit, on Krasnovodsk Gulf, 15 mi/24 km S of TURKMENBASHI; 39°48′N 53°01′E. Fisheries; shipbuilding (fishing vessels).

Kjeller (KEL-luhr), village, AKERSHUS county, SE NORWAY, 10 mi/16 km ENE of OSLO. Site of military airfield (Norway's oldest airfield). Has nuclear reactor and military research institute. Part of continued built-up area, stretching eastward from Oslo.

Kjellerup or **Kellerup** (KE-luh-roop), town, VIBORG county, central JUTLAND, DENMARK, 12 mi/19 km S of VIBORG; 56°20′N 09°30′E. Cattle; bricks, furniture, timber.

Kjoge, DENMARK: see KØGE.

Kjøpsvik (KUHPS-veek), village, NORDLAND county, N NORWAY, on Tyssfjorden (inlet of VESTFJORDEN), 40 mi/64 km SW of NARVIK. Cementworks.

Kjosarsýsla (KYAW-sahr-sees-lah), county, SW ICELAND ⊙ Dagafell, immediately NE of REYKJAVIK. Extends around Kollarfjörður and along SE shore of Hvalfjörður, branches of Flaxafloi Bay. Mountains in N, dominated by ESJA (2,982 ft/909 m), lava plains in S. Fishing; sheep, cattle; dairying; vegetables.

Kjosfoss (KOS-faws), waterfall (722 ft/220 m) in SOGN OG FJORDANE county, W NORWAY, in the FLÅMSDAL just N of Myrdal. Site of hydroelectric station.

Kladanj (klah-DAHN-yuh), town, central BOSNIA, BOSNIA AND HERZEGOVINA, on Drinjaca River and 28 mi/45 km NNE of SARAJEVO; 44°13′N 18°41′E. Local trade in cereals; poultry.

Klädesholmen (KLE-des-HOLM-en), fishing village (□ 1 sq mi/2.6 sq km), GÖTEBORG OCH BOHUS county, SW SWEDEN, on island (410 acres/166 ha) of same name in SKAGERRAK, 4 mi/6.4 km N of MARSTRAND; 57°57′N 11°33′E.

Kladno (KLAHD-no), city (2001 population 71,132), STREDOCESKY province, W central BOHEMIA, CZECH REPUBLIC; 50°09′N 14°06′E. An industrial center of the Kladno coal-mining region, it has large iron and steel plants; manufacturing includes machinery and steel alloys. Founded in 1315, Kladno grew rapidly with the opening of its first coal mine in 1846. Has mining museum and Baroque castle.

Kladovo (KLAH-do-vo), village (2002 population 23,613), E SERBIA, on the DANUBE, opposite (7 mi/11.3 km W of) TURNU SEVERIN (ROMANIA); 44°36′N 22°33′E. Sturgeon fishing.

Klaeng (GLANG), village, RAYONG province, S THAILAND, 26 mi/42 km ENE of RAYONG; 12°47′N 101°39′E. Fisheries (S) on GULF OF THAILAND. Birthplace of the poet Sunthorn Phu.

Klagenfurt (KLAH-gen-furt), city, CARINTHIA province, S AUSTRIA, on the GLAN RIVER and near Wörther See; 46°37′N 14°18′E. Situated in a mountain-lake region, it is the market center of Lower Carinthia and of a region with intensive summer tourism. Manufactures include electrical apparati, electronics, machines, foodstuffs, chemicals, pharmaceuticals, and leather goods. Of note among the city's annual fairs is the timber fair. The city boasts a university, railroad and road junction, airport, and a center of higher education for the Slovene minority in Carinthia. Klagenfurt was chartered about the mid-13th century, became the capital of Carinthia in 1514, and became an Episcopal see in the late 18th century. It was a center of Protestantism before the counter-reformation. The city has a 16th century cathedral that originally was built as a Protestant church, a theological seminary, a palace of the provincial government (1574–1594), and several museums.

Klagshamn (KLAHGS-hahmn), village, SKÅNE county, S SWEDEN, on ÖRESUND, 6 mi/9.7 km SSW of MALMÖ. Portuguese.

Klaipe da (KLEI-pe-duh), city (2001 population 192,954), third largest city in LITHUANIA, on the BALTIC SEA, at the entrance to the Courland Lagoon (Kuršių Marios); 55°43′N 21°07′E. An ice-free seaport and an industrial center; manufacturing (textiles, fertilizers, wood products); center for fishing and fish processing. One of the oldest cities of Lithuania, Klaipeda was the site of a settlement as early as the 7th century. It was conquered and burned in 1252 by the Teutonic Knights, who built a fortress and named it Memelburg. The city was ceded (1629) by PRUSSIA to SWEDEN but reverted to Prussia in 1635. In the Napoleonic Wars, the city was the refuge and residence of Frederick William III of Prussia (1807). From 1919 it shared the history of the Memel Territory. Also spelled Klaypeda.

Klakksvig (KLAHGS-veeg), Danish *Klaksvig*, town, and port in FAEROE ISLANDS, on SW Bordoy island, 16 mi/26 km NNE of Tórshavn.

Klamath (KLAM-uhth), county (□ 6,136 sq mi/15,953.6 sq km; 2006 population 66,438), S OREGON, in CASCADE foothills; ⊙ KLAMATH FALLS; 42°40′N 121°39′W. Borders on CALIFORNIA to S. Drained by KLAMATH, WILLIAMSON, SPRAGUE, and LOST rivers. Livestock raised in Klamath irrigation project, which extends S from Lost River into SISKIYOU county, N California Agriculture (wheat, oats, barley, alfalfa, potatoes; poultry, hogs, sheep, cattle); dairy products. Millwork. Recreation. Bear Valley National Wildlife Refuge in S; parts of Winema National Forest in center and W; part of Deschutes National Forest in E; parts of Fremont National Forest in E; part of Rogue River National Forest in SW; Klamath Forest National Wildlife Refuge in N center, including Klamath Marsh; Upper Klamath National Wildlife Refuge, on UPPER KLAMATH LAKE, in SW center; part of Lower Klamath National Wildlife Refuge on S boundary; Collier Memorial and Kimball state parks in W center; CRATER LAKE NATIONAL PARK in W. Formed 1882.

Klamath, unincorporated town (2000 population 651), DEL NORTE county, NW CALIFORNIA, 18 mi/29 km SSE of CRESCENT CITY, 2 mi/3.2 km E of PACIFIC OCEAN, and on KLAMATH RIVER (here crossed by Douglas Memorial Bridge) near its mouth on the Pacific; 41°31′N 124°02′W. Game fishing; timber; cattle; dairying. Hoopa Valley Indian Reservation to SE. PRAIRIE CREEK REDWOODS STATE PARK to S; Six Rivers National Forest to E; REDWOOD NATIONAL PARK to W; Rattlesnake Mountain (3,568 ft/1,088 m) to N.

Klamath Agency (KLAM-uhth), unincorporated village, KLAMATH county, S OREGON, 30 mi/48 km NNW of KLAMATH FALLS; elevation 4,170 ft/1,271 m. Collier Memorial State Park to E, located in small unit of Winema National Forest. Fish hatchery to NW.

Klamath Falls (KLAM-uhth), city (2006 population 19,785), ⊙ KLAMATH county, SW OREGON, 55 mi/89 km ESE of MEDFORD and 17 mi/27 km N of CALIFORNIA boundary, at the southern tip of UPPER KLAMATH LAKE, at exit of KLAMATH RIVER; 42°13′N 121°46′W. Elevation 4,105 ft/1,251 m. A processing and distributing center of a lumber, livestock, and farm area. Timber, dairy products, and tourism are central to the city's economy. Agriculture (grain, potatoes; poultry; dairy products; nurseries; manufacturing (wood chips, millwork, plywood, furniture, printing, publishing, concrete, horsedrawn carriages, ice cream carts); geothermal energy for heating. Klamath Falls was settled in 1867 as Linkville. The Klamath Irrigation Project (1900) and the coming of the railroad (1909) stimulated its growth from a hamlet to a thriving city. Site of Oregon Institute of Technology and Favell Museum of Western Art and Indian Artifacts. Kingsley Field airport to S. CRATER LAKE NATIONAL PARK is 55 mi/89 km to NNW; LAVA BEDS

NATIONAL MONUMENT is 32 mi/51 km SSE (California); parts of Winema National Forest to N and NW, including Mountain Lakes Wilderness Area to NW; Lower Klamath National Wildlife Refuge to S; Upper Klamath National Wildlife Refuge to NW; Bear Valley National Wildlife Refuge to SW. Incorporated 1905.

Klamath Lake, California: see LOWER KLAMATH LAKE.

Klamath Lake, Oregon: see UPPER KLAMATH LAKE.

Klamath Mountains (KLAM-uhth), part of Pacific COAST RANGE extending c.240 mi/386 km from SW OREGON, above the TRINITY RIVER, to NW CALIFORNIA. The range is covered by four national forests: SISKIYOU in Oregon, Six Rivers, Klamath, and SHASTA-TRINITY in California, and contains wildlife preserves and scenic portions of the KLAMATH RIVER, rising in UPPER KLAMATH LAKE, and the SACRAMENTO RIVER. Tourism and timber industries are chief economic activity of the region. Hiking, camping.

Klamath River (KLAM-uhth), 263 mi/423 km long, in S OREGON and N CALIFORNIA; rises in small LAKE EWAUNA connected to UPPER KLAMATH LAKE by LINK RIVER, 2 mi/3.2 km long, at KLAMATH FALLS; flows generally SW through John Boyle Reservoir, Oregon, and Capco Lake Reservoir, California, receives Shartin River from E, then flows through KLAMATH MOUNTAINS. Receives Trinity River from SE, then turns NW through Klamath Mountains, California, turning NW in HUMBOLDT county, and emptying into the PACIFIC NW of KLAMATH, California, 50 mi/80 km N of EUREKA, at REDWOOD NATIONAL PARK. The river is connected by Klamath Strait, 7 mi/11.3 km long, (largely in Oregon) with LOWER KLAMATH LAKE, in California. Supplies water for Klamath Irrigation Project, serving agricultural area in KLAMATH county, Oregon, and SISKIYOU county, California. Capco No. 1 Dam, completed in 1922 (227 ft/69 m high, 415 ft/126 m long), in California section of stream, just S of Oregon line, is used for power. TRINITY RIVER, in California, is chief tributary.

Klang (klahng), city, W SELANGOR, MALAYSIA, on railroad and KLANG RIVER, 20 mi/32 km WSW of KUALA; 03°02′N 101°28′E. Former sultan's residence; a traditional center of rubber- and fruit-growing (bananas, pineapples, durian, mangos) district; manufacturing (rubber footwear, pineapple canning, palm oil, flour milling, aluminum and steel products). Malaysia's busiest international port, Port Klang 03°00′N 101°25′E. CONNAUGHT BRIDGE (2 mi/3 km E) is site of large thermal power station.

Klang, THAILAND: see SONG.

Klang Island (klahng), flat mangrove island in Strait of MALACCA, SELANGOR, MALAYSIA, off Port Klang at KLANG RIVER mouth; 03°00′N 101°18′E; 8 mi/13 km long and 4 mi/6 km wide. North Klang Strait, between Klang Island and mainland, is chief access route to Port Klang.

Klang River (klahng) or **Kelang**, in SELANGOR, W MALAYSIA; rises on PAHANG line in central Malayan range; flows 60 mi/97 km S and W, past KUALA LUMPUR and Klang, to Strait of MALACCA at Port Klang; 03°02′N 101°30′E. Mangrove islands (KLANG ISLAND) off mouth.

Klanjec (KLAH-nyets), village, central CROATIA, in HRVATSKO ZAGORJE, on SUTLA RIVER and 20 mi/32 km NNW of ZAGREB, on Slovenia border; 46°04′N 15°46′E. Local trade center. Center of a municipality with characteristics of a town, despite its small population. Several castles and monasteries nearby.

Kläppa (KLEP-pah), residential village, GÄVLEBORG county, E SWEDEN, on LJUSNAN RIVER, just SE of LJUSDAL.

Klarentza, Greece: see KILLINI.

Klaster, CZECH REPUBLIC: see TEPLA.

Klasterec nad Ohri (KLAHSH-te-RETS NAHT O-rzhee), Czech *Klášterec nad Ohří*, German *klösterle an der eger*, town, SEVEROCESKY province, W BOHE-

MIA, CZECH REPUBLIC, ON OHRE RIVER, on railroad and 17 mi/27 km NE of KARLOVY VARY; 50°23′N 13°09′E. Manufacturing (porcelain, machinery, bearings). Mineral springs. Has a 16th century castle with museum of porcelain.

Klaster-Hradiste, CZECH REPUBLIC: see MNICHOVO HRADISTE.

Klastor pod Znievom (klahst-TOR POT-znye-VOM), Slovak *Kláštor pod Znievom*, HUNGARIAN *Znióváralja*, village, STREDOSLOVENSKY province, W central SLOVAKIA, on railroad and 17 mi/27 km SSE of ŽILINA; 48°58′N 18°48′E. Known for its trade in medicinal herbs; has a 13th-century castle, church, and monastery.

Klaten (KLAH-ten), town, E Java Tengah province, Indonesia, 17 mi/27 km ENE of YOGYAKARTA, at SE foot of Mount MERAPI; 07°42′S 110°35′E. Trade center for agricultural area (tobacco, rice, peanuts); textile mills. Agricultural experiment station and a branch of JAKARTA meterological station are here. Dutch fort built in 1807.

Klatovy, German *Klattau*, city (2001 population 23,033), ZAPADOCESKY province, SW BOHEMIA, CZECH REPUBLIC; 49°24′N 13°18′E. Railroad junction. Noted tanning industry; manufacturing (machinery, linen goods, leather goods, furniture, cheese). Located in foothills of BOHEMIAN FOREST, it is a center of floriculture (annual exhibitions of carnations and roses) and point of departure for excursions into SUMAVA MOUNTAINS. Has a 13th century cathedral with a 16th century tower, 17th century Jesuit church with catacombs, and a 16th century city hall.

Klattau, CZECH REPUBLIC: see KLATOVY.

Klausberg, POLAND: see MIKULCZYCE.

Klausdorf (KLOUS-dorf), suburb of KIEL, SCHLESWIG-HOLSTEIN, N GERMANY, just E of city; 54°18′N 10°14′E.

Klausenburg, ROMANIA: see CLUJ-NAPOCA.

Klausen Pass (6,391 ft/1,948 m), in GLARUS ALPS, Uri canton, central SWITZERLAND. Klausen Road, over the pass, joins towns of GLARUS and ALTDORF.

Klaustal-Zellerfeld, GERMANY: see CLAUSTHAL-ZELLERFELD.

Klavdiyeve-Tarasove (KLAHF-dee-ye-ve-tah-RAH-so-ve) (Russian *Klavdiyevo-Tarasovo*), town (2004 population 7,300), N central KIEV oblast, UKRAINE, on railroad, 23 mi/37 km WNW of KIEV; 50°35′N 30°01′E. Elevation 498 ft/151 m. Experimental woodworking, cotton textiles, knitting, Christmas tree decorations. Established in 1900; town since 1938.

Klavdiyevo-Tarasovo, UKRAINE: see KLAVDIYEVE-TARASOVE.

Klavreström (KLAHV-re-STRUHM), village, KRONOBERG county, S SWEDEN, 20 mi/32 km NE of VÄXJÖ; 57°08′N 15°08′E.

Klawak or **Klawock** (KLAH-wahk), village, SE ALASKA, on W coast of PRINCE OF WALES ISLAND, 5 mi/8 km N of CRAIG. Fishing, fish processing. First salmon cannery in Alaska was located here.

Klawak Island (KLAH-wahk) (4 mi/6.4 km long, 1 mi/1.6 km wide), SE ALASKA, in ALEXANDER ARCHIPELAGO, W of KLAWAK village on PRINCE OF WALES ISLAND; 55°33′N 133°06′W.

Klawer, town, NW WESTERN CAPE province, SOUTH AFRICA, on OLIFANTS RIVER, 31 mi/50 km NW of CLANWILLIAM, on road and railroad, near N end of CEDARBERG RANGE. 31°46′S 18°37′E. Elevation 216 ft/65 m. Center of irrigation farming area, producing fruits, grapes, and wheat under irrigation.

Klawock, village, Alaska: see KLAWAK.

Klazienaveen (klah-ZEE-nah-vain), town, DRENTHE province, NE NETHERLANDS, 6 mi/9.7 km SE of EMMEN; 52°43′N 06°59′E. On Hoogeveensevaart canal; German border 3.5 mi/5.6 km to E and 6 mi/9.7 km to S. Dairying; cattle, hogs, poultry; vegetables, grain; manufacturing (cast iron).

Kleberg, county (□ 1,090 sq mi/2,834 sq km; 2006 population 30,353), S TEXAS; ⊙ KINGSVILLE; 27°25′N

97°39′W. Bounded on E by GULF OF MEXICO, S by BAFFIN BAY and Los Olmos Creek. N part of PADRE ISLAND (Natl. Seashore) separated from mainland by LAGUNA MADRE, park headquarters and road access from N. Ranching area, including part of huge KING RANCH, headquarters at Kingsville; beef cattle, horses, also hogs, poultry, some agriculture (cotton, sorghum; vegetables); oil, natural gas; uranium; stone. Coast resorts, fishing, wildlife sanctuary. Kingsville Naval Air Station Arms of Baffin Bay, Alazan Bay (NE), and Cayo del Grullo (NW) almost divide county into three sections. Inland lake, Laguna Largo, is in NE. Formed 1913.

Klechev, POLAND: see KLECZEW.

Klecko (KLE-tsko), Polish *Klecko*, German *Kletzko*, town, Poznań province, W central POLAND, on railroad and 26 mi/42 km NE of POZNAŃ. Brick manufacturing, flour milling.

Kleczew (KLE-tchev), Russian *Klechev*, town, Konin province, W central POLAND, 50 mi/80 km E of POZNAŃ.

Kledang Range (KLEH-dahng), in central PERAK, MALAYSIA, forming divide between PERAK and KINTA rivers; 04°40′N 101°00′E; 25 mi/40 km long; rises to 3,469 ft/1,057 m. At E foot are the large Kinta Valley tin-mining centers of Teronoh, BATU GAJAH, and IPOH. Rice, tea, fruits, tobacco.

Kleidi, Greece: see RUPEL PASS.

Klein, village, MUSSELSHELL county, central MONTANA, on Halfbreed Creek, 3 mi/4.8 km S of ROUNDUP. Coal; sheep, cattle.

Kleinbettingen (KLEIN-BE-ting-uhn), hamlet, STEINFORT commune, SW LUXEMBOURG, 10 mi/16 km WNW of LUXEMBOURG city; 49°39′N 05°55′E. Frontier station on Belgian border. Manufacturing (paints, varnishes, farinaceous food products).

Kleinblittersdorf (klein-BLIT-ters-dorf), town, SAARLAND, SW GERMANY, on SAAR RIVER (French border) and 3 mi/4.8 km NNW of SARREGUEMINES; 49°09′N 07°02′E. Part of SAARBRÜCKEN conurbation. Manufacturing includes plastics; sparkling wine. GROSSBLIEDERSTROFF (FRANCE) is on the opposite bank of the SAAR.

Klein Bonaire, DUTCH WEST INDIES: see LITTLE BONAIRE.

Kleinburg (KLEIN-buhrg), unincorporated village, S ONTARIO, E central CANADA, 14 mi/23 km from TORONTO, and included in city of VAUGHAN; 43°50′N 79°37′W. First settled c.1848 around saw, grist mill operations. Variant spelling: Klineburg.

Klein Curaçao, NETHERLANDS ANTILLES: see LITTLE CURAÇAO.

Kleine Dommel River (KLEI-nuh DAW-muhl), S NETHERLANDS, rises 9 mi/14.5 km SE of EINDHOVEN, flows c.13 mi/21 km NNW past Geltrop, joins DOMMEL RIVER 3 mi/4.8 km NE of Eindhoven.

Kleine Emme River, SWITZERLAND: see EMME RIVER, KLEINE.

Kleineislingen, GERMANY: see EISLINGEN.

Kleine Mühl River (KLEI-ne MYOOIL), N UPPER AUSTRIA; rises near German border in the Mühlviertel; flows 15 mi/24 km S to the Danube, 3 mi/4.8 km W of Partenstein.

Kleine Nete River (KLEI-nuh NAI-tuh), French *Petite Nèthe* (Net), N BELGIUM; rises 7 mi/11.3 km N of MOL; flows 32 mi/51 km W, joining GROTE NETE RIVER at LIER to form NETE RIVER.

Kleine Scheidegg, SWITZERLAND: see GROSSE SCHEIDEGG.

Kleines Walsertal (klei-nes VAHL-ser-tahl), high alpine valley of Breitach River, a tributary of ILLER RIVER, VORARLBERG, W AUSTRIA; 47°19′N–49°23′N, 10°05′E–10°14′E. Isolated by a mountain range from the rest of Austria, German currency and customs area; MITTELBERG and RIEZLERN are the main settlements; center of winter sports. Settled by Walsers in the 14th century.

Kleinheubach (klein-HOI-bahkh), village, BAVARIA, central GERMANY, in LOWER FRANCONIA, on the MAIN, and 2 mi/3.2 km NW of MILTENBERG; 49°43′N 09°12′E. Hydroelectric station; distilling. Has early 18th-century Baroque castle.

Kleinkirchheim, AUSTRIA: see BAD KLEINKIRCHHEIM.

Kleinmacher-Bech (KLEIN-mahk-uhr), village, WELLENSTEIN commune, SE LUXEMBOURG, on MOSELLE RIVER, and 11 mi/18 km SE of LUXEMBOURG city, just SSW of REMICH, on German border; 49°32′N 06°22′E. Vineyards. Museum of folklore and viticulture.

Kleinmachnow (klein-MAHKH-nou), residential village, BRANDENBURG, E GERMANY, on TELTOW CANAL, 10 mi/16 km SW of BERLIN city center; 52°25′N 13°13′E. Metalworking.

Klein Matterhorn, peak (12,566 ft/3,830 m) in PENNINE ALPS, VALAIS canton, SW SWITZERLAND, almost on crest at Italian border, and 7 mi/11 km S of ZERMATT; reached by overhead tramway from Zermatt; panoramic views.

Kleinostheim (klein-OST-heim), village, BAVARIA, central GERMANY, in LOWER FRANCONIA, on MAIN RIVER, 5 mi/8 km NW of ASCHAFFENBURG; 50°00′N 09°04′E.

Klein-Schlatten, ROMANIA: see ZLATNA.

Kleinsee (KLAYN-SEAR), town, NORTHERN CAPE province, SOUTH AFRICA, on NAMAQUALAND coast of Atlantic, 48 mi/78 km W of SPRINGBOK, on Buffels River mouth; 29° 40′ S 17°04′ E; elevation 232 ft/70 m. Serves as holiday resort. Popular surf fishing spot. Point from which viewing of famous Namaqualand spring flowers occurs.

Kleinseite, CZECH REPUBLIC: see LITTLE QUARTER.

Klein Swartberg, SOUTH AFRICA: see SWARTBERG.

Klein Vet, river, SOUTH AFRICA: see ALLANRIDGE.

Kleinwallstadt (klein-VAHL-shtaht), village, BAVARIA, central GERMANY, in LOWER FRANCONIA, on right bank of MAIN RIVER, 14 mi/23 km N of MILTENBERG; 49°53′N 09°10′E. 14th-century church and graveyard.

Klein Zwartberg, SOUTH AFRICA: see SWARTBERG.

Klekovača (kle-ko-VAH-chah), mountain in DINARIC ALPS, W BOSNIA, BOSNIA AND HERZEGOVINA; highest point, Velika Klekovča (6,432 ft/1,960 m), is 7 mi/11.3 km NE of TITOV DRVAR. Also spelled Klekovacha.

Klemënovo (klye-MYO-nuh-vuh), town, E MOSCOW oblast, central European Russia, on road and railroad, 4 mi/6 km E, and administratively a suburb of YEGOR′YEVSK; 55°23′N 39°09′E. Elevation 495 ft/150 m. Food processing (confectionery). Also spelled Klemenovo.

Klemme, town (2000 population 593), HANCOCK county, N IOWA, 6 mi/9.7 km S of GARNER; 43°00′N 93°35′W. Animal feed.

Klemtu, unincorporated village, W BRITISH COLUMBIA, W CANADA, in CENTRAL COAST regional district, on E coast of SWINDLE ISLAND, in Kitasoo/Xai′xais (hei heis) First Nations territory; 52°35′N 128°31′W. Situated on a sheltered harbor. Commercial fishing, fish processing; forestry; tourism.

Klenócz, SLOVAKIA: see KLENOVEC.

Klenovec (kle-NO-vets), Hungarian Klenócz, village, STREDOSLOVENSKY province, S central SLOVAKIA, in SLOVAK ORE MOUNTAINS and 15 mi/24 km NW of RIMAVSKÁ SOBOTA; 48°36′N 19°54′E. Manufacturing of clothing; summer resort with dam (drinking-water supply). Folk architecture and customs.

Klenovnik (KLE-nov-neek), hamlet, E SERBIA, 5 mi/8 km N of POZAREVAC. Lignite mine.

Klenovyy (kle-NO-vee), town, S LUHANS′K oblast, UKRAINE, on road and on railroad spur 4 mi/7 km NE of and subordinated to ROVEN′KY; 48°07′N 39°27′E. Elevation 961 ft/292 m. Bituminous coal mining. Established in 1955; town since 1956.

Klepp (KLEP), village, ROGALAND county, SW NORWAY, on railroad and 14 mi/23 km S of STAVANGER. Quarries; agriculture, livestock, and related industries. Several hundred findings from the Stone Age and many from the Bronze Age and later, including burial sites, can be found here.

Klerksdorp, city, NORTH-WEST province, N central SOUTH AFRICA, on the Schoonspruit River, a tributary of the VAAL River; 26°52′S 26°40′E. Elevation 4,461 ft/1,360 m. The town, which has grain elevators, lumberyards, food-processing and beverage-making industries, is the mining and processing center for major gold and uranium deposits and is also the distribution center for neighboring farms. There are railroad and road connections with CAPE TOWN (N2) and JOHANNESBURG. Founded in 1837 by Afrikaner farmers (Boers), and with POTCHEFSTROOM, was one of the first European towns founded in the TRANSVAAL. Gold mining began in 1886 but declined in the late 1890s. Heavy fighting occurred in the area during the South African War (1899–1902). Gold mining revived in 1932, and the town underwent an economic revival, which accelerated after World War II. Klerksdorp has a training school for nurses. One of the world's largest grain cooperatives is located here.

Kleshcheli, POLAND: see KLESZCZELE.

Klesiv (KLE-sif) (Russian Klesov) (Polish Klesów), town (2004 population 4,300), NE RIVNE oblast, UKRAINE, on railroad, 12 mi/19 km E of SARNY; 51°20′N 26°56′E. Elevation 524 ft/159 m. Sawmilling; stone quarrying, peat bricketing; kaolin deposits nearby. Established around 1900; town since 1940.

Klesov, UKRAINE: see KLESIV.

Klesow, UKRAINE: see KLESIV.

Kleszczele (kle-sh-CHE-le), Russian Kleshcheli, town, Białystok province, E POLAND, on railroad and 40 mi/64 km S of BIAŁYSTOK, near BELARUS border. Pottery making, flour milling.

Kletino (KLYE-tee-nuh), settlement, N RYAZAN oblast, central European Russia, at a confluence of the Kolp′ and OKA Rivers, on road, 12 mi/19 km NW of (and administratively subordinate to) KASIMOV, and 3 mi/5 km S of GUS′-ZHELEZNYY; 55°00′N 41°12′E. Elevation 364 ft/110 m. Machine-building plant.

Kletnya (klyet-NYAH), town (2005 population 13,570), N BRYANSK oblast, central European Russia, on the Navda River (DNIEPER River basin), on railroad spur, 25 mi/40 km WSW of ZHUKOVKA; 53°24′N 33°16′E. Elevation 590 ft/179 m. Sawmills; woodworking (prefabricated houses, furniture).

Kletsk (KLETSK), Polish Kleck, city, MINSK oblast, BELARUS, ☉ KLETSK region, 26 mi/42 km ESE of BARANOVICHI. Manufacturing (fruit canning, tanning, flax carding, woolen weaving, rope, bricks, flour milling, sawmilling). Assaulted by Tatars (14th–16th century); passed (1793) from POLAND to RUSSIA; reverted (1921) to Poland; ceded to USSR in 1945.

Kletskaya (KLYETS-kah-yah), village (2006 population 5,370), W VOLGOGRAD oblast, SE European Russia, on the right bank of the DON River, on crossroads, 75 mi/121 km NW of VOLGOGRAD; 49°19′N 43°03′E. Elevation 193 ft/58 m. Agricultural products. Population largely Cossack.

Klettgau (KLET-gou), village, BADEN-WÜRTTEMBERG, SW GERMANY, near Swiss border, 10 mi/16 km SW of SCHAFFHAUSEN (SWITZERLAND); 47°39′N 08°25′E.

Klettgau (KLET-gou), region, GERMANY and SWITZERLAND, along WUTACH RIVER, W of SCHAFFHAUSEN; 47°45′N 08°20′E–47°45′N 08°30′E.

Kletzko, POLAND: see KLECKO.

Klevan′ (KLE-vahn) (Polish Klewań), town (2004 population 6,900), W RIVNE oblast, UKRAINE, 13 mi/21 km NW of RIVNE; 50°45′N 25°59′E. Elevation 685 ft/208 m. Flour milling, sawmilling, lumbering. Has old churches, ruins of a 15th-century castle, park; sanatorium; vocational technical school. Known since 1458; chartered in 1654; town since 1940. Close to half of the population Jewish in 1939; community eliminated by the Nazis in 1941–1943, with fewer than 100 Jews remaining in 2005.

Kleve (KLEH-ve), city, North Rhine-Westphalia, W GERMANY, near the Dutch border. Tourism is important in the city, and its manufacturing includes foodstuffs, clothing, tobacco, and chemical products. It is a railroad junction and popular resort. Among its noteworthy buildings are the collegiate church (14th–15th century), which contains the tombs of the dukes of Kleve, and the 11th-century Schwanenburg [Ger.=swans' castle], which is associated with the legend of Lohengrin. Also spelled Cleve or Cleves.

Klewan, UKRAINE: see KLEVAN′.

Klichev (KLEE-chev), urban settlement, MOGILEV oblast, BELARUS, ☉ KLICHEV region, 25 mi/40 km NNE of Babruysk. Lumbering.

Klichka (kleech-KAH), village (2005 population 2,480), S central CHITA oblast, S Siberian Russia, 13 mi/21 km S of KOKUY; 50°26′N 118°00′E. Elevation 2,342 ft/713 m. Calcite mine in the vicinity. Also known as Klichki.

Klichki, RUSSIA: see KLICHKA.

Klickitat (KLIK-uh-tat), county (□ 1,904 sq mi/4,950.4 sq km; 2006 population 20,335), S WASHINGTON, ☉ GOLDENDALE; 45°52′N 120°48′W. Agricultural, lumbering area, rising toward CASCADE RANGE in N. Carrots, potatoes, alfalfa, hay, wheat, oats, grapes, pears, apples; cattle. Bounded by COLUMBIA River (S boundary and OREGON state boundary), drained by KLICKITAT and WHITE SALMON (part of W boundary) rivers. Includes part of Yakima Indian Reservation along N boundary; Conboy Lake National Wildlife Refuge in NW; Brooks Memorial and Goldendale Observatory state parks, at center; Maryhill and Horse Thief Lake state parks in S. Formed 1860.

Klickitat (KLIK-uh-tat), unincorporated town, KLICKITAT county, S WASHINGTON, 16 mi/26 km W of GOLDENDALE, on KLICKITAT RIVER. Cattle; wheat, vegetables. Klickitat Springs to E. Klickitat Wildlife Area to N.

Klickitat River (KLIK-uh-tat), S WASHINGTON, rises in CASCADE RANGE in Yakima Indian Reservation in W YAKIMA county, flows c.85 mi/137 km generally S to COLUMBIA River, 10 mi/16 km NW of the DALLES, OREGON. Receives West Fork (15 mi/24 km long) from slopes of Mount Adams; receives Little Klickitat River (c.30 mi/48 km long) from E, rises in N central KLICKITAT county, flows SW and W, past GOLDENDALE, joins KLICKITAT RIVER 11 mi W of Goldendale.

Klidhi, Greece: see RUPEL PASS.

Klimentina, Bulgaria: see TRUD.

Klimentinovo, Bulgaria: see TRUD.

Klimkovice (KLIM-ko-VI-tse), German königsberg, town, SEVEROMORAVSKY province, central SILESIA, CZECH REPUBLIC, 7 mi/11.3 km WSW of OSTRAVA, on railroad; elevation 827 ft/252 m; 49°48′N 18°08′E. Agriculture (sugar beets, barley); manufacturing (furniture). Health resort. Has a 16th century Renaissance castle.

Klimovichi (KLEE-mo-vee-chee), city, SE MOGILEV oblast, BELARUS, ☉ KLIMOVICHI region, 45 mi/72 km SW of ROSLAVL; 53°36′N 31°58′E. Railroad station; manufacturing (construction materials, metal goods, vodka and liqueurs, butter, dry milk).

Klimovo (KLEE-muh-vuh), town (2005 population 14,390), SW BRYANSK oblast, central European Russia, on railroad, approximately 13 mi/21 km NW of the Ukrainian border, and 14 mi/23 km SE of NOVOZYBKOV; 52°22′N 32°11′E. Elevation 482 ft/146 m. In agricultural area (starch, dried vegetables, dairy products); clothing. Formerly called Klimov.

Klimovsk (KLEE-muhfsk), city (2006 population 55,075), S central MOSCOW oblast, central European Russia, on road and railroad (Grivno station), 34 mi/55 km SW of MOSCOW, and 5 mi/8 km S of PODOL′SK; 55°21′N 37°32′E. Elevation 593 ft/180 m. Textile machinery (since 1883), small hardware, agricultural machinery, toys. Called Klimovskiy after 1928, until acquiring city status in 1940.

Klimovskiy, RUSSIA: see KLIMOVSK.

Klin (KLEEN), city (2006 population 79,995), N MOSCOW oblast, central European Russia, on the Sestra River (VOLGA RIVER basin), 55 mi/89 km NW of

MOSCOW; 56°20′N 36°44′E. Elevation 587 ft/178 m. Railroad and highway junction; refrigerators, synthetic fibers, heat pipes, laboratory equipment, machine tools, glass, woodworking, textiles and haberdashery; food processing (bakery, brewery). Chartered in 1318. Site of the Pyotr I. Tchaikovskiy museum (home of the world-renowned composer).

Klin, UKRAINE: see YAMPIL', Sumy oblast.

Klinaklini River, SW BRITISH COLUMBIA, W CANADA; rises in COAST MOUNTAINS near 51°50′N 125°W; flows 120 mi/193 km in a wide arc SW and S to head of KNIGHT INLET.

Klin-Dmitrov Ridge (KLEEN–DMEE-truhf), E extension of the SMOLENSK-MOSCOW UPLAND, in N MOSCOW oblast, central European Russia; forms a watershed between the VOLGA and KLYAZ'MA rivers; rises to 960 ft/293 m. Crossed N-S by the MOSCOW CANAL.

Kline, village (2006 population 235), BARNWELL county, SW SOUTH CAROLINA, 8 mi/12.9 km S of BARNWELL; 33°07′N 81°20′W. Industry includes poultry, livestock; grain, cotton, peanuts.

Klinge, NETHERLANDS: see CLINGE.

Klingenberg am Main (KLING-uhn-berg ahm MEIN), town, BAVARIA, central GERMANY, in LOWER FRANCONIA, on the MAIN, and 13 mi/21 km S of ASCHAFFENBURG; 49°47′N 09°11′E. Known for its red wine. Manufacturing includes measuring instruments, artificial dyes. Nearby ruined castle was built on Roman foundations. Chartered 1276.

Klingenthal (KLING-uhn-tahl), town, SAXONY, E central GERMANY, in the ERZGEBIRGE, 18 mi/29 km ESE of PLAUEN; frontier station on Czech border, opposite KRASLICE; 50°23′N 12°28′E. Manufacturing center for musical instruments, and site of research institute for acoustics. Winter-sports center. Has 17th-century palace.

Klingnau, commune, AARGAU canton, N SWITZERLAND, on AARE RIVER, and 7 mi/11.3 km NNW of BADEN, near German border. Hydroelectric plant; plywood.

Klingnau, SWITZERLAND: see ZURZACH.

Klinovec (KLEE-no-VETS), Czech *Klínovec*, German *keilberg*, highest peak (4,080 ft/1,244 m) of the ORE MOUNTAINS, ZAPADOCESKY province, in W BOHEMIA, CZECH REPUBLIC, near German border, 3 mi/4.8 km NE of JACHYMOV; 50°24′N 12°58′E. Winter-sports center.

Klintehamn (KLIN-te-HAHMN), village, GOTLAND county, SE SWEDEN, on W coast of Gotland island, on bay of BALTIC SEA, 17 mi/27 km S of VISBY; 57°24′N 18°12′W. Seaside resort. Nearby is thirteenth-century church.

Klintsovka (kleen-TSOF-kah), village, E SARATOV oblast, SE European Russia, on road, 29 mi/47 km SE of PUGACHËV; 51°40′N 49°11′E. Elevation 229 ft/69 m. In agricultural area (grains, rice, vegetables).

Klintsy (kleen-TSI), city, (2005 population 66,020), W BRYANSK oblast, central European Russia, on the Turosha River, on railroad, 105 mi/169 km SW of BRYANSK; 52°46′N 32°14′E. Elevation 577 ft/175 m. Highway junction; local transshipment point. Fine woolen cloth, leather goods (tanning), clothing, textile machinery; telephones. Dates from the 18th century; became city in 1925.

Klipheuwel, village, WESTERN CAPE province, SOUTH AFRICA, 22 mi/35 km NE of CAPE TOWN; 33°42′S 18°42′E. Elevation 354 ft/108 m. Radio transmitter station, on railroad from NORTHERN CAPE.

Klippan (KLIP-pahn), town, SKÅNE county, SW SWEDEN, near Rönneå River, 18 mi/29 km ENE of HELSINGBORG; 56°08′N 13°09′E. Railroad junction; manufacturing (paper milling, tanning; machinery). Site of first Swedish paper mill (1637).

Klipplaat, village, EASTERN CAPE, SOUTH AFRICA, on Henningkrip River and 55 mi/89 km S of GRAAFF REINET; 33°01′S 24°18′E. Elevation 2,099 ft/640 m. Railroad junction; livestock; animal feed; fruits.

Klip River, W KWAZULU-NATAL, SOUTH AFRICA; formed 4 mi/6.4 km WNW of LADYSMITH by 2 short tributaries rising in DRAKENSBERG range near the E TRANSVAAL border; flows 30 mi/48 km SE, past Ladysmith, to TUGELA RIVER 13 mi/21 km SE of Ladysmith. Near Ladysmith, it supplies water to a major coal-burning power station.

Klipspruit, river, SOUTH AFRICA: see JAMESTOWN.

Klirou (KLEE-ro), town, LEFKOSIA district, central CYPRUS, 15 mi/24 km SW of NICOSIA, in NE foothills of TROODOS MOUNTAINS, near Serachis River; 35°01′N 33°10′E. Grain, vegetables, olives; livestock.

Klis (KLEES), Italian *Clissa*, village, S CROATIA, on railroad, and 5 mi/8 km NE of SPLIT, in DALMATIA, at pass (elev. 1,200 ft/366 m) on W slope of Mosor Mountain. Has ruins of castle dating from Roman period, once a Turkish strongpoint menacing Venetian-held Split. First mentioned in Croatian history in C.E. 850.

Klisura (klee-SOO-rah), city, PLOVDIV oblast, KARLOVO obshtina, central BULGARIA, at the E end of the KOZNITSA PASS, on the STRYAMA RIVER, 19 mi/31 km W of Karlovo; 42°41′N 24°27′E. On railroad; rose growing; livestock; wood processing. Once a Turkish commercial town; declined after Bulgarian uprising in 1876.

Klisura Pass, Bulgaria: see SAPAREVA BANYA.

Kljajičevo (KLEi-yee-che-vo), village, VOJVODINA, N SERBIA, 7 mi/11.3 km E of SOMBOR, in the BACKA region. Until 1948, called Krnjaja or Krnyaya (Hungarian *Kerény*). Also spelled Klyayichevo.

Ključ (KLYOOCH), town, central BOSNIA, BOSNIA AND HERZEGOVINA, on SANA RIVER and 26 mi/42 km SW of Banja Luka. Local trade center. First mentioned in 1325. Also spelled Klyuch.

Klobouk, CZECH REPUBLIC: see KLOBOUKY.

Klobouky (KLO-bou-KI), German *klobouk*, town, JIHOMORAVSKY province, S Moravia, Czech Republic, on railroad and 18 mi/29 km SE of Brno; 48°59′N 16°52′E. Agr. center (wheat, barley, sugar beets); wine production.

Klobuck (KLO-bootsk), Polish *Kłobuck*, Russian *Klobutsk*, town, Częstochowa province, S POLAND, 10 mi/16 km NW of częstochowa. Tanning, sawmilling.

Klodawa (klo-DAH-vah), Polish *Kłodawa*, Russian *Klodava*, town, Konin province, central POLAND, 28 mi/45 km ENE of KONIN. Brick manufacturing, flour milling. Thirteenth century churches.

Kłodnica River (klod-NEE-tsah), German *Klodnitz*, 47 mi/76 km long, S POLAND; rises SSW of KATOWICE, flows WNW past GLIWICE (Gleiwitz) and Łabędy, and W past UJAZD and Blachownia, to ODER River just SE of KOŹLE. In lower course, beyond Gliwice, parallels Gliwice Canal.

Kłodzkie, Gory, POLAND: see KRALICKY SNEZNIK.

Kłodzko (KLOTS-ko), German *Glatz*, town (2002 population 29,173), Wałbrzych province, SW POLAND. It is a commercial center with lumber and textile mills, metalworks, and sugar refineries. Founded in the 10th century, it was capital of a country created in 1462. It was seized by Frederick II of PRUSSIA in the War of the Austrian Succession and was formally ceded to Prussia in 1745. It was returned to Poland in 1945.

Kloengkoeng, INDONESIA: see KLUNGKUNG.

Kloetinge (KLOO-ting-uh), village, ZEELAND province, SW NETHERLANDS, in SOUTH BEVELAND region, and 2 mi/3.2 km E of GOES; 51°30′N 03°55′E. Dairying; cattle; fruits, vegetables, sugar beets.

Klokocov (klo-KO-chou), Slovak *Klokočov*, Hungarian *Hajagos*, village, STREDOSLOVENSKY province, NW SLOVAKIA, by CZECH border, and 17 mi/27 km NW of ŽILINA; 49°27′N 18°34′E. Lumbering; sheep breeding.

Klokotnitsa (klo-KOT-nee-tsah), village, HASKOVO oblast and obshtina, S central BULGARIA, 5 mi/8 km NW of HASKOVO; 41°59′N 25°30′E. Cotton, tobacco,

vineyards. Theodore of Epirus was defeated here (1230) by Bulgarian King Ivan Asen.

Klondike (KLAHN-deik), region of YUKON TERRITORY, CANADA, just E of the ALASKA border. It lies around KLONDIKE RIVER, a small stream that enters the YUKON RIVER from the E at DAWSON. The discovery in 1896 of rich placer gold deposits in BONANZA (Rabbit) Creek, a tributary of the Klondike, caused the Klondike stampede of 1897–1898. News of the discovery reached the U.S. in July 1897, and within a month thousands of people were rushing N. Most landed at SKAGWAY at the head of LYNN CANAL and crossed by Chilkoot or WHITE PASS to the Upper Yukon, which they descended to Dawson. Others went in by the Copper River Trail or over the Teslin Trail by STIKINE RIVER and Teslin Lake, and some by the all-Canadian Ashcroft and Edmonton trails. The rush continued by these passes all the following winter. The other main access route was up the Yukon River, c.1,600 mi/2,575 km, by steamer. Many of those using this route late in 1897 were caught by winter ice below Fort Yukon and had to be rescued. With unexpected thousands coming in, the region was threatened by a food famine, and supplies were commandeered and rationed. The number in the Klondike in 1898 was c.25,000. Thousands of others who did not find claims drifted down the Yukon and found placer gold in Alaskan streams, notably at NOME, to which there was a new rush. Others went back to the U.S. Gold is still mined in the area.

Klondike Gold Rush Historical Park (KLAWN-deik) (□ 20 sq mi/52 sq km), SW ALASKA and N WASHINGTON; 59°28′N 135°19′W (SKAGWAY portion). Authorized 1976. Site of the historic goldfields including Visitors' Center in Seattle's Pioneer Square, which served as the miners' point of departure. Historic buildings in Skagway; goldfields along CHILKOOT and WHITE PASS Trail.

Klondike River (KLAHN-deik), W YUKON, CANADA; rises E of DAWSON; flows 100 mi/161 km W to YUKON RIVER at Dawson. Receives BONANZA CREEK just SE of Dawson. Noted for its salmon. It gives its name to surrounding region.

Klong Toi, THAILAND: see KHLONG TOEI.

Klöntalersee (KLUHN-tal-ehr-say), lake (□ 1 sq mi/2.6 sq km), E central SWITZERLAND, W of GLARUS. Elevation 2,779 ft/847 m. Inlet: Klön River; outlet: Löntsch River. The GLÄRNISCH range rises along S shore. Regulated by dam, lake feeds Löntsch hydroelectric plant.

Klopeiner See (klaw-PEI-ner SAI), lake, CARINTHIA, S AUSTRIA, in the JAUNTAL, 13 mi/21 km E of KLAGENFURT; 46°36′N 14°35′E. Elevation 1,359 ft/414 m. Warmest lake in Austria. Summer tourism.

Klos (KLAWS), village, central ALBANIA, near upper MAT RIVER, 22 mi/35 km SW of PESHKOPI; 41°30′N 20°06′E. Also spelled Klosi.

Kloster, GERMANY: see HIDDENSEE.

Klosterlausnitz, Bad, GERMANY: see BAD KLOSTERLAUSNITZ.

Klösterle an der Eger, CZECH REPUBLIC: see KLASTEREC NAD OHRI.

Klosterneuburg (klaw-ster-NOI-burg), town, LOWER AUSTRIA province, NE AUSTRIA, on the right bank of the DANUBE River and the N slope of the WIENERWALD, near VIENNA; 48°18′N 16°19′E. It is the site of a wealthy Augustinian monastery (consecrated in 1136), the oldest in Austria and called the "Austrian Escorial." The monastery has an extensive library, enormous wine cellars, and the famous Verduner Altar (1181) by Nicholaus of Verdun. Klosterneuburg is also home to the Federal College of Viticulture and Pomology (1860), one of the only schools in the world with a focus on wine-making. Manufacturing of metals and electronic chips; vineyards; cultural tourism.

Klosters, commune, GRISONS canton, E SWITZERLAND, on LANDQUART RIVER, and 16 mi/26 km E of CHUR.

Elevation 3,957 ft/1,206 m. Valley resort among mountains; hydroelectric plant.

Klostertal (KLAW-ster-tahl), valley of Alfenz River, a right tributary of ILL RIVER, VORARLBERG, W AUSTRIA; 47°07′N–47°09′N 09°50′E–10°10′E. Thinly populated, high alpine valley, mainly a traffic corridor between BLUDENZ and Arlberg Pass (railroad, highway). Major settlements are Innerbraz, DALAAS, and Klösterle.

Klosterwappen, AUSTRIA: see SCHNEEBERG.

Kloster Zinna (KLOAW-ster TSIN-nah), village, BRANDENBURG, E GERMANY, 6 mi/9.7 km SSW of LUCKENWALDE; 52°02′N 13°06′E. Site of former Cistercian monastery (founded 1170 by monks from COLOGNE; dissolved 1547); extant buildings include 12th–13th-century church and 15th-century abbots' house.

Kloten (KLOH-tuhn), town (2000 population 17,190), ZÜRICH canton, N SWITZERLAND, 6 mi/9.7 km NNE of ZÜRICH. Elevation 1,463 ft/446 m. Metal products. Site of Zürich International Airport.

Kloto, prefecture (2005 population 192,763), PLATEAUX region, S TOGO; ⊙ KPALIMÉ; 07°05′N 00°45′E. Also spelled Klouto.

Kloto, town, TOGO: see KLOUTO, town.

Klötze (KLOE-tse), town, SAXONY-ANHALT, central GERMANY, 15 mi/24 km S of SALZWEDEL; 52°37′N 11°09′E. Agricultural market; dairying; woodworking.

Klouto, prefecture, TOGO: see KLOTO, prefecture.

Klouto (KLOO-to), town, PLATEAUX region, S TOGO; 06°57′N 00°34′E. Near GHANA border, 70 mi/113 km NW of LOMÉ, 6 mi/10 km SSW of KPALIMÉ. Agricultural center (cacao, coffee, palm kernels and oil, and cotton). Customhouse; iron smelter. Also spelled Kloto.

Kluane (kloo-AIN) native village, SW YUKON, CANADA, near ALASKA border, on KLUANE LAKE, at foot of ST. ELIAS MOUNTAINS, 120 mi/193 km W of WHITEHORSE, and on Alaska Highway; 61°02′N 138°24′W. Center of Kluane Game Sanctuary (established 1943).

Kluane Lake (kloo-AIN) (□ 184 sq mi/478.4 sq km), SW YUKON, CANADA, near ALASKA border, at foot of ST. ELIAS MOUNTAINS, 120 mi/193 km W of WHITEHORSE; 61°15′N 138°44′W. It is 60 mi/97 km long, 1 mi/2 km–6 mi/10 km wide; drains N into YUKON RIVER.

Kluane National Park (kloo-AIN) (□ c.8,500 sq mi/22,000 sq km), SW YUKON TERRITORY, CANADA, between KLUANE LAKE and the BRITISH COLUMBIA and Alaska borders; established 1972. Located in the ST. ELIAS MOUNTAINS, the park contains some of Canada's highest mountains (including MOUNT LOGAN, the nation's highest peak) and one of the world's largest nonpolar systems of ice fields. Great variety of wildlife.

Kluang (CLUE-ahng), town, central Johore, MALAYSIA, on railroad and 50 mi/80 km NW of JOHOR Baharu. Rubber and oil-palm plantations; airport.

Kluczbork (KLOOCH-bork), German *Kreuzburg*, town (2002 population 26,595) in UPPER SILESIA, after 1945 in OPOLE province, S POLAND, 25 mi/40 km NNE of OPOLE (Oppeln). Railroad junction; textiles, machinery, leather, bricks. Novelist Freytag born here.

Kluisbergen (KLOIS-ber-kuhn), commune (2006 population 6,198), Oudenaarde district, EAST FLANDERS province, BELGIUM; 50°47′N 03°31′E.

Kluisbos (KLOIS-bahs), recreational park near KLUISBERGEN, EAST FLANDERS province, NW BELGIUM, 12 mi/19 km SSE of KORTRIJK.

Klukhori (kloo-KHO-ree), city, NW GEORGIA, on N slope of the Greater Caucasus, on KUBAN River, at mouth of TEBERDA RIVER, and 70 mi/113 km NE of SUKHUMI, on SUKHUMI MILITARY ROAD. Brickworks; dairy products. Founded c.1920; until 1943, called Mikoyan–Shakhar as capital. Former Karachai Autonomous Oblast. During World War II, held (1942–1943) by Germans.

Klukhori, RUSSIA: see KARACHAYEVSK.

Klukhori Pass (kloo-KHO-ree), in main range of the W Greater Caucasus, NW GEORGIA. SUKHUMI MILITARY ROAD runs across the pass 45 mi/72 km ENE of SUKHUMI; elevation 9,239 ft/2,816 m.

Klukwan (KLOOK-wahn), village, SE ALASKA, on CHILKAT RIVER and 20 mi/32 km NW of HAINES, on Haines Highway; 59°24′N 135°54′W. Area contains iron ore.

Klundert (KLUHN-dert), town, NORTH BRABANT province, SW NETHERLANDS, 10 mi/16 km NNE of ROOSENDAAL; 51°39′N 04°32′E. HOLLANDS DIEP river channel 2 mi/3.2 km to N, Tonnekreek Pumping Station to NW; industrial and port facility to E. Dairying; cattle, hogs; grain, vegetables, sugar beets.

Klungkung (KLOONG-koong), town, ⊙ Klungkung District, SE BALI, INDONESIA, 15 mi/24 km NE of DENPASAR; 08°32′S 115°24′E. Wood carving, metalworking. Nearby are several Hindu temples. Was capital of Bali from seventeenth century to 1908. Also spelled Kloengkoeng.

Kluterhöhle, GERMANY: see ENNEPETAL.

Klutina River (KLOO-tuh-nuh), S ALASKA; rises in CHUGACH MOUNTAINS near 61°44′N 145°45′W; flows 60 mi/97 km NE, through Klutina Lake (17 mi/27 km long), to COPPER RIVER at Cooper Center.

Klütz (KLOOITS), town, Mecklenburg-Western Pomerania, N GERMANY, near WISMAR BAY of the BALTIC, 13 mi/21 km WNW of WISMAR; 53°58′N 11°10′E. Seaside resort. Has 13th-century church. Nearby is 18th-century Bothmer castle.

Klyavlino (KLYAH-vlee-nuh), village (2006 population 6,305), NE SAMARA oblast, E European Russia, on railroad, 35 mi/56 km WSW of BUGUL'MA; 54°15′N 52°01′E. Elevation 777 ft/236 m. Highway junction; local transshipment point. Construction materials; grain storage, flour milling.

Klyayichevo, SERBIA: see KLJAJIČEVO.

Klyaz'ma (KLYAHZ-mah), town, central MOSCOW oblast, central European Russia, on the MOSCOW CANAL, on highway, 4 mi/6 km N of which it is administratively subordinate; 55°58′N 37°27′E. Elevation 570 ft/173 m. Service enterprises. Sheremetyevo international airport is 1 mi/1.6 km to the W.

Klyaz'ma (KLYAHZ-mah), urban settlement (2006 population 3,430), central MOSCOW oblast, central European Russia, on road and railroad, 3 mi/5 km S of (and administratively subordinate to) PUSHKINO; 55°58′N 37°51′E. Elevation 518 ft/157 m. Mostly residential.

Klyaz'ma River (KLYAHZ-mah), in central European Russia; rises in MOSCOW oblast, on the KLIN-DMITROV RIDGE, E of SOLNECHNOGORSK; flows 390 mi/628 km E, through important textile-manufacturing region, past SCHCHELKOVO, NOGINSK, OREKHOVO-ZUYEVO, VLADIMIR (Vladimir oblast), KOVROV, and VYAZNIKI, to the OKA River WSW of Volodary (NIZHEGOROD oblast border). Navigable in the lower course below SOBINKA (VLADIMIR oblast). Upper course connected with the MOSCOW CANAL, which passes through the Klyaz'ma Reservoir (hydroelectric station at PIROGOVSKIY, Moscow oblast). Receives UCHA, Kirzhach, NERL, Uvod', Teza (left) and Sudogda (right) rivers.

Klyn, UKRAINE: see YAMPIL', Sumy oblast.

Klyuch (KLOOCH), village, SOFIA oblast, PETRICH obshtina, BULGARIA; 41°22′N 23°01′E.

Klyuch (KLYOOCH), village, S central RYAZAN oblast, central European Russia, on the Ramova River (OKA River basin), 6 mi/10 km ENE of, and under administrative jurisdiction of; 53°59′N 40°12′E. Elevation 374 ft/113 m. In agricultural area; winery.

Klyuchevsk (KLYOO-chyefsk), town (2006 population 1,945), S SVERDLOVSK oblast, E central URALS, W Siberian Russia, on railroad and near highway, 15 mi/24 km NNE (under jurisdiction) of BERËZOVSKIY; 57°07′N 60°56′E. Elevation 761 ft/231 m. Lumber mills. Developed in the 1930s; called Tëplyy Klyuch until 1933.

Klyuchevskaya Sopka (klyoo-chyef-SKAH-yah–SOP-kuh), highest active volcano (15,912 ft/4,850 m), in the E mountain range of the KAMCHATKA PENINSULA, KAMCHATKA oblast, extreme E SIBERIA, RUSSIAN FAR EAST, on the right bank of KAMCHATKA River, 220 mi/354 km NNE of PETROPAVLOVSK-KAMCHATSKIY. Has a perfect conic shape; the crater is 650 ft/198 m in diameter. Frequent eruptions.

Klyuchevskiy (KLYOO-cheef-skeeyee), town (2005 population 1,600), NE CHITA oblast, S SIBERIA, RUSSIA, on highway, 16 mi/26 km SW of, and administratively subordinate to, MOGOCHA; 53°33′N 119°26′E. Elevation 2,650 ft/807 m. In protected old-growth forest area. Mineral mine in the vicinity.

Klyuchi (klyoo-CHEE), city (2005 population 6,310), E central KAMCHATKA oblast, RUSSIAN FAR EAST, on the E central KAMCHATKA PENINSULA, on the KAMCHATKA RIVER, at the NE foot of the KLYUCHEVSKAYA SOPKA, on local road, 280 mi/451 km N of PETROPAVLOVSK-KAMCHATSKIY, and 60 mi/97 km W of UST'-KAMCHATSK; 56°18′N 160°51′E. Elevation 298 ft/90 m. Sawmilling. Volcanic research station. Founded in 1731; made city in 1979.

Klyuchi (klyoo-CHEE), village (2005 population 9,200), W ALTAI TERRITORY, S central SIBERIA, RUSSIA, on crossroads and railroad, approximately 14 mi/22 km E of the KAZAKHSTAN border, 55 mi/89 km SSE of SLAVGOROD; 52°16′N 79°10′E. Elevation 462 ft/140 m. Flour mill.

Klyuchi (klyoo-CHEE), village (2006 population 3,950), NE OMSK oblast, SW Siberian Russia, in the IRTYSH River basin, on road, 24 mi/39 km NNE of TARA; 57°23′N 74°50′E. Elevation 32ft/97 m. In agricultural area; grain processing.

Klyuchinskiy, RUSSIA: see KRASNOMAYSKIY.

Klyuchishchi, RUSSIA: see BOL'SHIYE KLYUCHISHCHI.

Klyukvennaya, RUSSIA: see UYAR.

Knabstrup (KNAHB-stroop), town, VESTSJÆLLAND county, SJÆLLAND, DENMARK, 7 mi/11.3 km SW of HOLBÆK; 55°40′N 11°33′E.

Knapp, village (2006 population 405), DUNN county, W WISCONSIN, 12 mi/19 km WNW of MENOMONIE; 44°57′N 92°04′W. In dairying area. Creamery; cheese.

Knapp Creek, hamlet, CATTARAUGUS county, W NEW YORK, 7 mi/11.3 km SW of OLEAN. In oil-producing area on PENNSYLVANIA line.

Knäred (KNE-RED), village, HALLAND county, SW SWEDEN, on LAGAN RIVER, 20 mi/32 km ESE of HALMSTAD; 56°31′N 13°19′E.

Knaresborough (NERZ-buh-ruh), town (2001 population 22,651) NORTH YORKSHIRE, N ENGLAND, on NIDD RIVER and 3 mi/4.8 km ENE of HARROGATE; 54°00′N 01°28′W. Agricultural market; resort. Has remains of castle built by Henry I, including 14th-century keep; church dating from 13th century; and "Dropping Well" with petrifying waters.

Knaresborough Spa, ENGLAND: see HARROGATE.

Knebworth (NEB-wuhth), town (2001 population 5,034), central HERTFORDSHIRE, SE ENGLAND, 8 mi/12.9 km NW of HERTFORD; 51°52′N 00°12′W. Agricultural market. Knebworth House was home of Edward Bulwer-Lytton. Has 13th-century church.

Kneehill County (NEE-hil), municipal district (□ 1,305 sq mi/3,393 sq km; 2001 population 5,319), S ALBERTA, W CANADA; 51°39′N 113°15′W. Agriculture; oil and gas. Includes THREE HILLS, CARBON, TROCHU, Linden, ACME, and the hamlets of BIRCHAM, SUNNYSLOPE, WIMBORNE, HESKETH, SWALWELL, HUXLEY, and TORRINGTON. Formed in 1912.

Knee Lake (NEE), NE MANITOBA, W central CANADA, on HAYES River, 50 mi/80 km long, 5 mi/8 km wide; 55°03′N 94°44′W.

Knešpolje (knes-POL-ye), village, W HERZEGOVINA, BOSNIA AND HERZEGOVINA, on road, and 9 mi/14.5 km W of MOSTAR, on NW border of MOSTAR LAKE. Bauxite mine. Also spelled Knezhpolye.

Cross-references are shown in SMALL CAPITALS. The pronunciation guide is shown on page xix. The sources of population figures are shown on page xvii.

Knesselare (KNE-suh-lah-ruh), commune (2006 population 7,898), Ghent district, EAST FLANDERS province, W central BELGIUM, 10 mi/16 km SE of BRUGES; 51°08′N 03°25′E.

Knetzgau (KNETS-gou), village, BAVARIA, central GERMANY, in LOWER FRANCONIA, on MAIN RIVER, 17 mi/27 km NW of BAMBERG; 50°00′N 10°34′E. Wine.

Kneuttingen, FRANCE: see KNUTANGE.

Knevichi (KNYE-vee-chee), town (2006 population 4,570), S MARITIME TERRITORY, SE RUSSIAN FAR EAST, on road and near railroad, 18 mi/29 km NNE of VLADIVOSTOK; 43°24′N 132°11′E. The region's main international airport is located here.

Knezha (kne-ZHAH), city, MONTANA oblast, ⊙ KNEZHA obshtina (1993 population 18,502), N BULGARIA, on a small tributary of the ISKUR RIVER, 9 mi/15 km ENE of BYALA SLATINA; 43°30′N 24°05′E. Agricultural and livestock trading center; manufacturing: vegetable oil, construction materials. Has an agricultural school and a corn-breeding experimental station. Formerly spelled Knizha.

Knićanin (NEECH-ah-neen), Hungarian *Rezsdháza*, village, VOJVODINA, N SERBIA, on left bank of TISZA RIVER, at BEGEJ RIVER mouth opposite TITEL, and 24 mi/39 km E of NOVI SAD, in the BANAT region. Also spelled Knichanin.

Knierim (NEAR-um), town (2000 population 70), CALHOUN county, central IOWA, 10 mi/16 km ENE of ROCKWELL CITY; 42°27′N 94°27′W.

Knife Lake, Lake county, NE MINNESOTA and RAINY RIVER district, W ONTARIO, 35 mi/56 km NE of ELY, in chain of lakes on CANADA-U.S. border; U.S. part is in Superior National Forest (BOUNDARY WATERS Canoe Area); Canadian part is in Quetico Provincial Park; 10 mi/16 km long, average width 1 mi/1.6 km; 48°06′N 91°13′W. Fed from Ottertrack Lake to NE through short stream; drains into Seed Lake to SW through short stream.

Knife River, NORTH DAKOTA; rises in KILLDEER MOUNTAINS, DUNN co., W central NORTH DAKOTA; flows E 120 mi/193 km to MISSOURI River near STANTON; 47°20′N 103°10′W.

Knife River Indian Villages National Historic Site, at STANTON, MERCER CO., central NORTH DAKOTA, on MISSOURI River. Authorized in 1974, 2 sq mi/5.2 sq km. Contains the ruins of villages of Hidasta and Mandan Native Americans that were last occupied in 1845.

Knightdale (NEIT-dail), town (□ 2 sq mi/5.2 sq km; 2006 population 6,479), WAKE county, central NORTH CAROLINA, suburb 8 mi/12.9 km E of RALEIGH, near NEUSE RIVER; 35°47′N 78°29′W. Service industries; manufacturing (sheet metal fabricating, tractor equipment, motor controls, transformers, crushed granite); agriculture (tobacco, grain; livestock). Established 1927.

Knight Inlet (NEIT), SW BRITISH COLUMBIA, W CANADA arm (75 mi/121 km long, 1 mi/2km–4 mi/6 km wide) of QUEEN CHARLOTTE STRAIT, opposite VANCOUVER ISLAND; receives KLINAKLINI RIVER at head; 50°47′N 125°38′W.

Knight Island, S ALASKA, in PRINCE WILLIAM SOUND, E of KENAI PENINSULA, 40 mi/64 km SE of WHITTIER (25 mi/40 km long, 2 mi/3.2 km–9 mi/14.5 km wide); 60°20′N 147°44′W.

Knight Island (4 mi/6.4 km long), SE ALASKA, in YAKUTAT BAY, 10 mi/16 km NW of YAKUTAT; 59°43′N 139°34′W.

Knight Island Passage, S ALASKA, SW entrance to PRINCE WILLIAM SOUND, between KNIGHT ISLAND (E) and KENAI PENINSULA and CHENEGA ISLAND (W); 20 mi/32 km long, 3 mi/4.8 km–5 mi/8 km wide; 60°18′N 148°00′W.

Knighton (NEI-tuhn), Welsh *Trefyclo*, town (2001 population 3,043), POWYS, E Wales, on TEME RIVER at English border, and 14 mi/23 km NW of LUDLOW; 52°21′N 03°03′W. Previously woolen milling, agricultural engineering; forestry, sawmilling. Just S is

well-preserved section of OFFA'S DYKE, 8th-century boundary between Anglo-Saxon and Welsh territory.

Knightsbridge, road junction, CYRENAICA region, NE Libya, 28 mi/45 km SW of TOBRUK, near Bir Hakim. Scene of major World War II tank battles (1942) between Axis and British. Name used during and shortly after World War II.

Knights Landing, unincorporated village, YOLO county, central CALIFORNIA, 18 mi/29 km NW of SACRAMENTO, and on SACRAMENTO RIVER (bridged here). Fruits, nuts, grain, sugar beets.

Knightstown, town (2006 population 2,007), HENRY county, E central INDIANA, on BIG BLUE RIVER and 34 mi/55 km E of INDIANAPOLIS; 39°48′N 85°32′W. In livestock and grain area; manufacturing. Laid out 1827.

Knightstown, IRELAND: see VALENCIA.

Knightsville, town (2000 population 624), CLAY county, W INDIANA, just E of BRAZIL; 39°32′N 87°05′W. In agricultural and bituminous-coal area. Residential community.

Knightville Reservoir, HAMPSHIRE county, W central MASSACHUSETTS, on WESTFIELD RIVER, 11 mi/18 km W of NORTHAMPTON; 43°18′N 72°52′W. Maximum capacity 64,000 acre-ft. Formed by Knightville Dam (145 ft/44 m high), built (1941) by Army Corps of Engineers for flood control. Also known as Dry Reservoir.

Knik (kuh-NIK), village, S ALASKA, on W side of KNIK ARM, 18 mi/29 km N of ANCHORAGE; 61°27′N 149°44′W. Tourism.

Knik Arm (kuh-NIK), N arm (30 mi/48 km long, 2 mi/3.2 km–6 mi/9.7 km wide) of COOK INLET, S ALASKA; extends NE from Cook Inlet, just N of ANCHORAGE.

Knin (KNEEN), town, S CROATIA, on KRKA RIVER, and 26 mi/42 km NNE of SIBENIK, in DALMATIA, near BOSNIA-HERZEGOVINA border, in ZAGORA region. Railroad junction (Split-Zadar-Zagreb line); road hub; local trade center; gypsum deposits. Roman Catholic bishopric. Has museum of Croatian art, old castle, old churches (some ruined), palace ruins. Former capital of medieval Croatia. Passed (1699) from Turkish rule to Venetian Dalmatia. Was capital of Serb (Srpska) Krajina, 1991–1995.

Knippa (KAH-nip-pah), unincorporated village, UVALDE county, SW TEXAS, 10 mi/16 km NE of UVALDE, on FRIO RIVER. Cattle ranching, also sheep, goats; cotton, vegetables. Crushed stone, sand and gravel.

Knislinge (KNIS-leeng-e), town, SKÅNE county, S SWEDEN, near HELGE Å RIVER, 11 mi/18 km NNW of KRISTIANSTAD; 56°11′N 14°5′E. Has thirteenth-century church.

Knittelfeld (KNIT-tel-feld), town central STYRIA, S central AUSTRIA, on MUR River and 30 mi/48 km NW of GRAZ; market center; workshop of the Austrian Federal Railways; manufacturing of metal tableware, machines; dairying; 47°13′N 14°50′E. Heavily bombed in World War II.

Knittlingen (KNIT-ling-uhn), town, BADEN-WÜRTTEMBERG, SW GERMANY, 10 mi/16 km SE of BRUCHSAL; 49°02′N 08°26′E. Traditional birthplace of Dr. Faust.

Knivskjellodden (kuh-NEEV-shel-LAWD-duhn) or *Knivskjerodden*, low cape on BARENTS SEA of ARCTIC OCEAN, on island of MAGERØY, FINNMARK CO., N NORWAY, 60 mi/97 km NE of HAMMERFEST; 71°11′N 25°43′E. The northernmost point of EUROPE, though NORTH CAPE, 4 mi/6.4 km ESE, is popularly celebrated as such. Also spelled Knivskjelodden or Knivskjaerodden.

Knivsta (KNEEV-stah), village, STOCKHOLM county, E SWEDEN, 10 mi/16 km SE of UPPSALA; 59°43′N 17°48′E.

Knizha, Bulgaria: see KNEZHA.

Knjaževac (KNYAHZH-e-vahts), town (2002 population 37,172), E SERBIA, near BELI TIMOK RIVER, on railroad, and 24 mi/39 km NE of NIŠ, near Bulgarian

border; 43°35′N 22°12′E. Coal mines in vicinity. Also spelled Knyazhevats.

Knobel (NO-buhl), village (2000 population 358), CLAY county, extreme NE ARKANSAS, 19 mi/31 km NNW of PARAGOULD; 36°19′N 90°35′W. Manufacturing (hardwood, lumber). Dave Donaldson-Black River Wildlife management area to NW.

Knob Hill, village, EL PASO county, E central COLORADO, 2 mi/3.2 km E of downtown COLORADO SPRINGS. Part of Colorado Springs.

Knobly Mountain (NAHB-lee), c.50 mi/80 km, MINERAL and GRANT counties, NE WEST VIRGINIA, a ridge of the APPALACHIAN MOUNTAINS, in E PANHANDLE; from North Branch of the POTOMAC RIVER, opposite CUMBERLAND, MARYLAND, extends SW to a point W of PETERSBURG, VIRGINIA, rising to 2,000 ft/610 m–3,000 ft/914 m, with knoblike summits to over 3,000 ft/914 m.

Knob Noster (nahb NAHS-tuhr), city (2000 population 2,462), JOHNSON county, W central MISSOURI, 10 mi/16 km E of WARRENSBURG; 38°46′N 93°34′W. Grain; livestock, egg processing. WHITEMAN AIR FORCE BASE (former Minuteman missile site, now bomber base) 3 mi/4.8 km S. Knob Noster State Park to SW.

Knock, Gaelic *Cnoc Mhuire*, town (2006 population 745), MAYO county, NW IRELAND, 7 mi/11.3 km NE of CLAREMORRIS; 52°37′N 09°20′W. Apparitions of the Virgin Mary in 1879–1890 in Roman Catholic church have made this a place of pilgrimage. Airport. Modern basilica with spire.

Knockalongy (nahk-uh-LAHNG-gee), Gaelic *Sliabh Gamh*, mountain (1,778 ft/542 m), W SLIGO county, NW IRELAND, 13 mi/21 km WSW of SLIGO; 54°12′N 08°45′W. Highest peak in the SLIEVE GAMPH range.

Knockanaffrin (nahk-uh-NA-fruhn), mountain (2,504 ft/763 m), N WATERFORD county, S IRELAND, 9 mi/14.5 km SE of CLONMEL; 52°17′N 07°36′W. Highest point of COMERAGH MOUNTAINS.

Knockando (NAHK-kan-do), town, MORAY, NE Scotland, on the SPEY RIVER, and 5 mi/8 km W of Aberlour; 57°27′N 03°21′W. Whiskey distilling. Previously woolen milling. Village of Carron is 2 mi/3.2 km E.

Knocklayd (nahk-LAID), mountain (elevation 1,695 ft/517 m), N ANTRIM, NE Northern Ireland, 3 mi/4.8 km S of BALLYCASTLE; 55°09′N 06°13′W.

Knockmealdown Mountains (nahk-meel-DOUN), Gaelic *Cnoc Mhaoldonn*, range in MUNSTER province, S IRELAND, extending 15 mi/24 km E-W along border between NW WATERFORD county and SW TIPPERARY county; 52°14′N 07°55′W. Rises to 2,609 ft/795 m, 6 mi/9.7 km N of LISMORE.

Knokke-Heist (KNAH-kuh-HEIST), commune (2006 population 33,953), Bruges district, WEST FLANDERS province, NW BELGIUM, near NORTH SEA, 10 mi/16 km NNE of BRUGES; 51°21′N 03°16′E. Nearby, on North Sea, are popular seaside resorts: Heist-aan-Zee; Duinbergen; Het Zoute, French *Le Zoute*; and Albertstrand, French *Albert-Plage*. Butterfly park and Het Zwijn Nature Preserve nearby.

Knollwood Park, village, JACKSON county, S MICHIGAN, unincorporated suburb 4 mi/6.4 km E of JACKSON, near GRAND RAPIDS; 42°16′N 37°85′W.

Knossos (kno-SOS), ancient city, near N coast of CRETE, near modern IRÁKLION, in what is now IRÁKLION prefecture, CRETE department, GREECE; 35°18′N 25°10′E. Site occupied long before 3000 B.C.E.; center of an important Bronze Age culture. It is from a study of the great palace here, as well as other sites in Crete, that knowledge of the Minoan civilization has been drawn. Destroyed before 1500 B.C.E. (possibly by earthquake); splendidly rebuilt only to be destroyed again c.1400 B.C.E., probably at the hands of invaders from the Greek mainland. This event marked the end of Minoan culture. Later became an ordinary but flourishing Greek city; continued to exist through the Roman period until the 4th century C.E.

In Greek legend, capital of King Minos and the site of the labyrinth. The name also appears as Cnosus, Cnossus, and Knossus.

Knott, county (□ 353 sq mi/917.8 sq km; 2006 population 17,536), E KENTUCKY; ⊙ HINDMAN; 37°21′N 82°57′W. Drained by Carr Fork of KENTUCKY RIVER and Troublesome and Buckhorn creeks. Bituminous coal mining and agricultural (livestock; some tobacco) area, in CUMBERLAND foothills. Part of Robinson Forest Preserve (University of Kentucky) in NW. Carr Fork Lake reservoir in S. Formed 1884.

Knott (NOT), unincorporated village, HOWARD county, W central TEXAS, 14 mi/23 km NW of BIG SPRING. Oil and natural gas. Agricultural area (cattle; cotton, vegetables, black-eyed peas).

Knottingley (NOT-ting-lee), town (2001 population 13,503), WEST YORKSHIRE, N ENGLAND, on AIRE RIVER and 14 mi/23 km SE of LEEDS; 53°42′N 01°14′W. Glass and chemicals.

Knowles (NOLZ), village (2006 population 30), BEAVER county, E OKLAHOMA Panhandle, 19 mi/31 km ENE of BEAVER, midway between CIMARRON RIVER (N) and CANADIAN River (S); 36°52′N 100°11′W. In wheat and livestock area; manufacturing (steel fabrication).

Knowlton (NOL-tuhn), unincorporated village, S QUEBEC, E CANADA, at S end of Brome Lake, 16 mi/26 km SE of GRANBY, and included in Lac-Brome; 45°13′N 72°31′W. Dairying center; resort. Was seat of historic BROME county. Formerly known as Coldbrook.

Knowsley (NOZ-lee), village (2001 population 11,343), MERSEYSIDE, NW ENGLAND, 3 mi/4.8 km NW of PRESCOT; 53°27′N 02°51′W. Site of LIVERPOOL reservoir. Industrial estate to N.

Knox, county (□ 719 sq mi/1,869.4 sq km; 2006 population 52,906), N central ILLINOIS; ⊙ GALESBURG; 40°56′N 90°12′W. Agriculture (corn, wheat, soybeans; livestock; dairy products). Clay, gravel. Diversified manufacturing, chiefly at Galesburg. Drained by SPOON RIVER and POPE and HENDERSON creeks. Formed 1825.

Knox, county (□ 524 sq mi/1,362.4 sq km; 2006 population 38,241), SW INDIANA; ⊙ VINCENNES; 38°41′N 87°25′W. Bounded W by WABASH RIVER (here forming ILLINOIS boundary), E by West Fork of WHITE RIVER, and S by White River. Drained by Maria Creek, Deschee River, and Pond Creek. Agricultural area (fruits, vegetables, watermelons, grain); also produces oil, natural gas, bituminous coal, sand and gravel. Manufacturing at Vincennes; fruit-packing plants, nurseries, creameries. Indiana Territory State Memorial and GEORGE ROGERS CLARK NATIONAL MEMORIAL (25.5 acres/10.3 ha) at Vincennes. Southwest-Purdue Agricultural Center near Emison; White Oak State Fishing Area near center of county. Cypress Pond, oxbow lake of Wabash River, in SW corner. Indiana's first county, formed 1790.

Knox, county (□ 387 sq mi/1,006.2 sq km; 2006 population 32,527), SE KENTUCKY; ⊙ BARBOURVILLE; 36°53′N 83°51′W. In the CUMBERLAND MOUNTAINS; drained by CUMBERLAND RIVER and Stinking Creek. Includes Doctor Thomas Walker State Historic Site in SW. Bituminous coal mining; agriculture (corn, hay, burley tobacco; cattle); timber, oil wells. Manufacturing at CORBIN and Barbourville. Formed 1799.

Knox, county (□ 1,142 sq mi/2,969.2 sq km; 2006 population 41,096), S MAINE, on PENOBSCOT BAY; ⊙ ROCKLAND; 44°06′N 69°07′W. Drained by SAINT GEORGE RIVER. Coastal and island area has resorts, fishing, shipping of seafood, cement, and lime. Inland agricultural and lake area produces poultry; apples. Resorts. Lake Saint George State Park. Formed 1860.

Knox, county (□ 512 sq mi/1,331.2 sq km; 2006 population 4,093), NE MISSOURI; ⊙ EDINA; 40°08′N 92°09′W. Drained by North Fork of SALT RIVER and MIDDLE and SOUTH Fabius rivers. Corn, wheat, hay, soybeans; hogs; lumber; limestone quarry. Formed 1845.

Knox, county (□ 1,139 sq mi/2,961.4 sq km; 2006 population 8,812) NE NEBRASKA; ⊙ CENTER; 42°38′N 97°53′W. Agricultural area bounded N by MISSOURI RIVER (dammed [Gavins Point Dam] in NE corner of county to form LEWIS AND CLARK LAKE reservoir) and SOUTH DAKOTA; drained by NIOBRARA RIVER. Cattle, hogs; corn, wild hay, alfalfa, soybeans; dairying. Niobrara State Park in NW; Lewis and Clark Lake State Recreation Area in NE; Santer Indian Reservation in N central part of county.

Knox (NAHKS), county (□ 532 sq mi/1,383.2 sq km; 2006 population 58,561), central OHIO; ⊙ MOUNT VERNON; 40°24′N 82°22′W. Drained by KOKOSING and MOHICAN rivers and North Fork of LICKING RIVER. Mostly in the Glaciated Plain physiographic region; NE portion in the Unglaciated Plain region. Agricultural area (sheep, hogs, poultry; corn); manufacturing at Mount Vernon (engines, air and gas compressors); motor vehicle equipment and parts. Formed 1808.

Knox, county (□ 517 sq mi/1,344.2 sq km; 2006 population 411,967), E TENNESSEE; ⊙ KNOXVILLE; 36°00′N 83°57′W. In GREAT APPALACHIAN VALLEY; drained by TENNESSEE RIVER (here formed by junction of HOLSTON and FRENCH BROAD rivers); bounded SW by CLINCH River. Includes part of FORT LOUDOUN Reservoir. Manufacturing. Formed 1792.

Knox (NAHKS), county (□ 855 sq mi/2,223 sq km; 2006 population 3,702), N TEXAS; ⊙ BENJAMIN; 33°36′N 99°44′W. Plains area, bounded N in part by N WICHITA RIVER, drained by BRAZOS RIVER, North and South forks of the Wichita. Irrigated agriculture and livestock-raising (cotton, grain, sorghum, wheat, vegetables; cattle). Oil and gas. Formed 1841.

Knox, city (2000 population 3,721), ⊙ STARKE county, NW INDIANA, on YELLOW RIVER and 33 mi/53 km SW of SOUTH BEND; 41°17′N 86°37′W. In agricultural area producing chiefly mint and onions; manufacturing (automotive stampings, paper containers, molded plastics, timber products, wood and aluminum windows). Bass Lake State Fish Hatchery nearby to SE. Laid out 1851.

Knox, village (2006 population 59), BENSON co., N central NORTH DAKOTA, 15 mi/24 km E of RUGBY; 48°20′N 99°41′W. Dairy; livestock; wheat, oats. Railroad junction of spur to Wolford to E (at York). Founded in 1887 and incorporated in 1906.

Knox, borough (2006 population 1,118), CLARION county, W central PENNSYLVANIA, 8 mi/12.9 km WNW of CLARION. Agriculture (corn, hay, potatoes; livestock; dairying); manufacturing (corrugated boxes); bituminous coal; timber. Name changed from Edenburg to its post office name in 1933.

Knox Atoll, uninhabited atoll, RATAK CHAIN, MARSHALL ISLANDS, W central PACIFIC, c.5 mi/8 km SE of MILI; 10 islets; 11°07′N 166°32′E. Also called Narik.

Knox, Cape (NAHKS), W BRITISH COLUMBIA, W CANADA, NW extremity of GRAHAM ISLAND, on the PACIFIC, at SW end of DIXON ENTRANCE; 54°11′N 132°54′W.

Knox City, town (2000 population 223), KNOX county, NE MISSOURI, between MIDDLE and SOUTH Fabius rivers, 9 mi/14.5 km E of EDINA; 40°08′N 92°00′W. Grain; livestock; lumber.

Knox City (NAHKS), town (2006 population 1,063), KNOX county, N TEXAS, c.80 mi/129 km WSW of WICHITA FALLS, near BRAZOS River; 33°25′N 99°49′W. In cotton, grain; cattle area. Incorporated 1916.

Knox Coast, ANTARCTICA, part of WILKES LAND, on INDIAN OCEAN, between 100°31′ and 109°16′E. Discovered 1840 by Charles Wilkes, U.S. explorer. Also called Knox Land.

Knoxfield (NAHKS-feeld), residential and commercial suburb 17 mi/27 km ESE of MELBOURNE, VICTORIA, SE AUSTRALIA, between Ferntree Gully and Scoresby. Light industry (food; graphic design and printing houses).

Knox, Fort, KENTUCKY: see FORT KNOX.

Knoxville, city (2000 population 3,183), KNOX county, NW central ILLINOIS, 4 mi/6.4 km SE of GALESBURG; 40°54′N 90°17′W. In agricultural area; dairy products. Incorporated 1832. Formerly the county seat.

Knoxville, city (2000 population 7,731), ⊙ MARION county, S central IOWA, 33 mi/53 km ESE of DES MOINES, near WHITEBREAST CREEK; 41°19′N 93°05′W. Ships coal and livestock; manufacturing (clothing, dairy and meat products, printing, industrial tapes, feed, concrete blocks). Limestone quarries nearby. Has large U.S. veterans hospital. Red Rock Reservoir (DES MOINES RIVER) to N and NE. Settled 1845; incorporated 1854.

Knoxville, city (2006 population 182,337), ⊙ KNOX county, E TENNESSEE, on the TENNESSEE RIVER, 100 mi/161 km NE of CHATTANOOGA, in the GREAT SMOKY MOUNTAINS; 36°00′N 83°57′W. A port of entry, it is a trade and shipping center; diversified industries. Tourism adds to the economy. The city is surrounded by mountains and lakes, and the Great Smoky Mountains National Park and several state parks are nearby. A house was built on this site c.1785, followed by a fort and then a town, named for General Henry Knox. Knoxville was the capital of the Territory of the United States South of the River Ohio from 1792 to 1796 and twice (1796–1812, 1817–1818) served as the state capital. During the Civil War the area was torn by divided loyalties; Federals occupied the city in September 1863, and successfully withstood a Confederate siege (November–December, 1863). The city is the seat of the University of Tennessee, Knoxville College, and the Tennessee School for the Deaf. It was also the site of the 1982 World's Fair, which introduced permanent new structures, such as the Sunsphere and the Tennessee Amphitheatre. Knoxville is headquarters of the Tennessee Valley Authority. Norris Dam, from which the city procures its power, is nearby. Confederate Memorial Hall, the William Blount Mansion (1792), a replica of the old fort, Chisholm's Tavern (1792), and many other historic buildings. Incorporated 1876.

Knoxville, agricultural village, ⊙ CRAWFORD county, central GEORGIA, 20 mi/32 km WSW of MACON; 32°43′N 83°59′W.

Knoxville, village, FREDERICK county, W MARYLAND, on the POTOMAC, 15 mi/24 km WSW of FREDERICK. Boasting of a select school for young ladies from 1864 to 1879, Knoxville's greatest claim to fame being the only place in the local polling district where liquor was allowed in the election of 1904.

Knoxville, borough (2006 population 593), TIOGA county, N PENNSYLVANIA, 15 mi/24 km NNW of WELLSBORO, on Cowanesque River, at mouth of Troups Creek, near NEW YORK state line. Agriculture (corn, hay; dairying); manufacturing (feeds, ornamental irons, magnet assembly).

Knuckles Group, extension of NE SRI LANKA HILL COUNTRY, CENTRAL PROVINCE, S central SRI LANKA; irregular in shape; 25 mi/40 km long (N-S), up to 15 mi/24 km wide; rises to 6,112 ft/1,863 m in Knuckles Peak, 9.5 mi/15.3 km NE of WATTEGAMA; 07°24′N 80°48′E. Agriculture (tea, rubber, rice, vegetables, cardamom); average rainfall, 100 in/254 cm to c.200 in/508 cm.

Knüllwald (KNUL-vahlt), village, HESSE, central GERMANY, 22 mi/35 km S of KASSEL; 51°00′N 09°30′E.

Knutange (nu-tahnzh), German *Kneuttingen*, commune, MOSELLE department, in LORRAINE, NE FRANCE, 7 mi/11.3 km WSW of THIONVILLE; 49°20′N 06°02′E. Iron mines, near an active iron- and steel-making complex.

Knutsford (NUHTS-fuhd), town (2001 population 12,656), N central CHESHIRE, W ENGLAND, 11 mi/18 km SW of STOCKPORT; 53°18′N 02°22′W. Some manufacturing. Has many 16th- and 17th-century houses and a 17th-century Unitarian chapel.

Cross-references are shown in SMALL CAPITALS. The pronunciation guide is shown on page xix. The sources of population figures are shown on page xvii.

Knyaginin, RUSSIA: see KNYAGININO.

Knyaginino (knye-GEE-nee-nuh), town (2006 population 6,955), E central NIZHEGOROD oblast, central European Russia, on the Imza River (SURA River basin), on road, 65 mi/105 km SE of NIZHNIY NOVGOROD, and 15 mi/24 km S of LYSKOVO; 55°49'N 45°02'E. Elevation 501 ft/152 m. Headdress factory, machine shop, dry milk plant; woodworking, leather products. Made city in 1779; reduced to status of a village in 1925; made town in 1968. As a city, called Knyaginin.

Knyazepetrovskiy, RUSSIA: see NYAZEPETROVSK.

Knyaze-Volkonka, RUSSIA: see KNYAZE-VOLKONSKOYE.

Knyaze-Volkonskoye (KNYAH-zye-vuhl-KON-skuh-ye), village (2005 population 8,850), S central KHABAROVSK TERRITORY, RUSSIAN FAR EAST, at the edge of the AMUR River basin, on road, 16 mi/26 km E of KHABAROVSK; 48°28'N 135°28'E. Elevation 137 ft/41 m. In agricultural area; produce processing. Also known as Knyaze-Volkonka. Named after Count Sergey Volkonsky, a decorated Russian military officer who participated in the Napoleonic Wars of the early 19th century and later was exiled E for sedition.

Knyazha Luka, UKRAINE: see SHARHOROD.

Knyazheva Polyana, Bulgaria: see SMIRNENSKI, RUSE oblast.

Knyazhevats, SERBIA: see KNJAŽEVAC.

Knyazhevo (KNYAH-zhe-vo), SW suburb of SOFIA, SOFIA oblast and Greater Sofia, W BULGARIA; 42°43'N 23°20'E. Paper manufacturing, brewing; trucks; dairying; poultry. Thermal springs. Formerly called Bali-yefendi.

Knyazhnoye, RUSSIA: see NABEREZHNOYE.

Knyazhpogost, RUSSIA: see YEMVA.

Knyaz Simeonovo, Bulgaria: see OBEDINENIE.

Knysna, town, WESTERN CAPE province, SOUTH AFRICA, on Knysna Lagoon, on the INDIAN OCEAN, 150 mi/241 km W of PORT ELIZABETH (NELSON MANDELA METROPOLE), 35 mi/56 km E of GEORGE on the famous Garden Route on the N2 highway between Port Elizabeth and CAPE TOWN; 34°01'S 23°02'E. Railroad terminus; furniture manufacturing and lumbering center, boat building. Popular resort. Founded 1806.

Knysna National Lake Area (□ 37,050 acres/15,000 ha) WESTERN CAPE province, SOUTH AFRICA. Lagoon area on the coast SE of Knysna and part of the lakeland area stretching from Wilderness National Park in the W through SEDGEFIELD to the Knysna Heads. This area constitutes the second most popular lakeland area in the republic.

Knyszyn (KNEE-shin), Russian *Knyshin*, town, Białystok province, NE POLAND, 16 mi/26 km NW of BIAŁYSTOK. Near railroad; cement, bricks; tanning; flour milling.

Ko [Thai=island], for names in THAILAND beginning thus and not found here: see under following part of the name.

Koalla, town, GNAGNA province, EST region, E BURKINA FASO, 120 mi/193 km NE of OUAGADOUGOU; 13°24'N 00°08'W. Also spelled Koala.

Koani (ko-AH-nee), village, ⊙ ZANZIBAR SOUTH region, E TANZANIA, 4 mi/6.4 km E ZANZIBAR city; 06°08'S 39°17'E. Copra; cloves; livestock.

Kobadak River, a branch of Bhairab River of KUSHTIA, JESSORE, and KHULNA, W of EAST BENGAL, BANGLADESH; flows S to the forest outpost of Kobadak at the edge of the SUNDARBANS; joins the Dholpetua to form the Arpangasia River. Also called Kapotaksha River.

Kobarid (ko-BAH-reed), Italian *Caporetto*, village, NW SLOVENIA, on Soča River, 45 mi/72 km WNW of LJUBLJANA, near Italian border; 46°14'N 13°34'E. Part of Italy (1919–1947).

Kobayashi (ko-BAH-yah-shee), city, MIYAZAKI prefecture, S central KYUSHU, SW JAPAN, near Mount Kirishima, 28 mi/45 km W of MIYAZAKI; 31°59'N 130°58'E. Beef cattle.

Kobbelbude, RUSSIA: see SVETLOYE.

Kobbermine Bay (KO-buhr-MEE-nuh), Danish *Kobberminebugten*, inlet (20 mi/32 km long, 1mi/1.6 km–10 mi/16 km wide) of the ATLANTIC OCEAN, SW GREENLAND, 70 mi/113 km WNW of Qaqortoq (JULIANEHÅB); 60°55'N 48°10'W. Copper formerly mined here.

Kobbo, ETHIOPIA: see KOBO.

Kobdo, MONGOLIA: see HOVD.

Kobdo Gol, MONGOLIA: see HOVD GOL.

Kobdo Lakes, MONGOLIA: see HOVD LAKES.

Kobe (KO-BE), city (2005 population 1,525,393), ⊙ HYOGO prefecture, S HONSHU, W central JAPAN, on OSAKA BAY, near Mount Rokko; 34°41'N 135°11'E. Second largest port of Japan, major transfer point for containers to SOUTHEAST ASIA and PACIFIC RIM. A trading center to CHINA and ASIA since 8th century. Sixth busiest port in the world and fourth in tonnage handled. Also a major industrial center and railroad hub. Part of a transportation network, which includes express trains and highways, that links it to OSAKA, KYOTO, and NAGOYA. It has shipbuilding yards, vehicle factories; manufacturing of iron, tires, furniture, footwear, holiday decorations; food processing (confections). Beef cattle; dairying. Wine, sake. A cultural center, Kobe has several colleges and universities and many temples and shrines. Since 1878, the city has included HYOGO, an ancient port that was prominent during the Ashikaga period (14th–16th century) and regained importance after it was reopened to foreign trade in 1868. Heavily bombed during World War II; it has since been rebuilt and enlarged. In 1995, Kobe suffered extensive damage during the Great Hanshin earthquake. Area attractions include nearby hot springs and the Sumanoura coast. Akashi Bridge (E) crosses Akashi Strait to Awaji-jima island

Kobelyaki, UKRAINE: see KOBELYAKY.

Kobelyaky (ko-be-LYAH-kee) (Russian *Kobelyaki*), city, SE POLTAVA oblast, Ukraine, on VORSKLA RIVER and 35 mi/56 km SSW of POLTAVA; 49°09'N 34°12'E. Elevation 239 ft/72 m. Raion center; woolen milling, clothing manufacturing, food processing, grain milling; bricks. Professional technical school. Known since the early 17th century; city since 1803. Jewish community since the 19th century, numbering 1,400 in 1939; destroyed during World War II.

København (kuh-ben-HOUN) or **Copenhagen**, county (□ 201 sq mi/522.6 sq km), NE SJÆLLAND, E DENMARK, includes SALTHOLM and part of the island of AMAGER; excludes independent cities of KØBENHAVN and FREDERIKSBERG; ⊙ København; 55°40'N 12°35'E. Transitional area (24 mi/39 km long, 20 mi/32 km wide), highly urbanized in E to gently rolling farmland in W. Main cities are BALLERUP, GLOSTRUP, GENTOFTE, Lyngby. Bounded E by ØRESUND, S by Koge Bugt (BALTIC SEA), SW by ROSKILDE county, NW by FREDERIKSBORG county. Manufacturing (food products, pharmaceuticals, fabrics, paperboard, furniture); fruit (apples, strawberries); dairying; wheat, barley, fishing; tourism.

København, DENMARK: see COPENHAGEN.

Kobenni (ko-BE-nee), village (2000 population 6,291), Hodh El Gharbi administrative region, SE MAURITANIA, 171 mi/275 km NE KAYES (MALI), and 62 mi/100 km S of AIOÛN EL ATROÛS; 15°55'N 09°27'W. Livestock; gum arabic, millet.

Koberice (KO-be-RZHI-tse), Czech *Kobeřice*, German *köberwitz*, village, SEVEROMORAVSKY province, central SILESIA, CZECH REPUBLIC, 7 mi/11.3 km ENE of OPAVA. Manufacturing (building materials); basalt and gypsum quarries. Folk architecture. Under German rule until 1920.

Köberwitz, CZECH REPUBLIC: see KOBERICE.

Koblenz (KOB-lents), city, RHINELAND-PALATINATE, W GERMANY, at the confluence of the RHINE and the MOSELLE (Ger. *Mosel*) rivers; 50°22'N 07°36'E. Manufacturing includes furniture, pianos, clothing, and chemicals; important trade center for Rhine and Moselle wines. The merging rivers at Koblenz also make it a center for river traffic; the outlying countryside, with its abundance of forests and lakes, attracts many tourists. The city was founded (9 B.C.E.) as Castrum ad Confluentes by Drusus. It was prominent in Carolingian times as a residence of the Frankish kings and as a meeting place for churchmen. Held by the archbishops of TRIER from 1018 to the late 18th century. In 1794 it was occupied by French troops and in 1798 was annexed by FRANCE and made the capital of the Rhine and Moselle department. Passed to PRUSSIA in 1815. After World War I it was occupied by Allied troops from 1919 to 1929. Noteworthy buildings in Koblenz include the Church of St. Castor (founded 836; rebuilt c.1200), the fortress of EHRENBREITSTEIN, and an 18th-century castle. Part of the state archives of the former WEST GERMANY are located in the city. Sourh of the city is Schren, site of a U.S. Air Force base that was closed in 1994 and then used to house German refugees from the former USSR. Sometimes spelled Coblenz.

Kobo, village (2007 population 37,859), AMHARA state, NE ETHIOPIA; 12°09'N 39°38'E. Near source of GOLIMA RIVER, on road and 17 mi/27 km N of WELDIYA; cereals, corn. Also spelled Kobbo.

Koboko, administrative district, NORTHERN region, extreme NW UGANDA, on DEMOCRATIC REPUBLIC OF THE CONGO (W) and SUDAN (N) borders. As of Uganda's division into eighty districts, borders YUMBE (E) and MARACHA-TEREGO (S) districts. Primarily agricultural. Formed in 2005 from extreme N portion of former ARUA district created in 2000 (current ARUA district formed from central and S portions and Maracha-Terego district from N central portion in 2006).

Kobra (KO-brah), village, NW KIROV oblast, E central European Russia, near the MOLOMA River, on crossroads, 12 mi/19 km ENE and, under administrative jurisdiction, of DAROVSKOY; 58°53'N 48°18'E. Elevation 377 ft/114 m. In agricultural region; produce processing; sawmilling, lumbering. Has a hospital.

Kobra (KO-brah), village, NE KIROV oblast, E European Russia, in the VYATKA River basin, on road and cargo railroad spur, 30 mi/48 km N of NAGORSK, to which it is administratively subordinate; 59°46'N 50°51'E. Elevation 485 ft/147 m. Sawmilling, timbering, logging.

Kobrin (KO-brin), Polish *Kobryn*, city, central BREST oblast, BELARUS, on MUKHAVETS RIVER and 30 mi/48 km ENE of BREST; 52°16'N 24°22'E. Highway junction and railroad station. Manufacturing (construction materials, tools, reinforced concrete structural elements, clothing, furniture, weaving); motor vehicle repair, flax mill, creamery, cannery; poulty. Founded in thirteenth-century; passed (1795) from POLAND to RUSSIA; reverted (1921) to Poland; ceded to USSR in 1945.

Kobrino, RUSSIA: see KOBRINSKOYE.

Kobrinskoye (KO-breen-skuh-ye), town (2005 population 1,155), W central LENINGRAD oblast, NW European Russia, on road and railroad, 23 mi/37 km S of SAINT PETERSBURG, and 11 mi/18 km S of GATCHINA; 59°25'N 30°07'E. Elevation 262 ft/79 m. Peat works. Formerly known as Kobrino.

Kobroor, INDONESIA: see ARU ISLANDS.

Kobu (ko-BOO), urban settlement, APSHERON region, AZERBAIJAN; 40°24'N 49°42'E. Sheep breeding; carpet weaving.

Kobuchizawa (ko-BOO-chee-ZAH-wah), town, N Koma county, YAMANASHI prefecture, central HONSHU, central JAPAN, 19 mi/30 N of KOFU; 35°51'N 138°19'E. Pottery; trout farming (famous for boiled trout). Otaki spring nearby.

Kobuk (KO-buhk), village (2000 population 109), NW ALASKA, on KOBUK RIVER and 6 mi/9.7 km E of SHUNGNAK; 66°54'N 156°54'W.

Kobuk River (KO-buhk), NW ALASKA; rises in BROOKS RANGE near 67°05'N 154°15'W; flows c.300 mi/483 km

W, past KOBUK, SHUNGNAK, KIANA, and NOORVIK, to HOTHAM INLET 30 mi/48 km ESE of KOTZEBUE.

Kobuk Valley National Park (KO-buhk) (□ 2,735 sq mi/7,084 sq km), NW ALASKA; authorized 1980. A wildlife preserve N of the ARCTIC CIRCLE; archaeological remnants of 10,000 years of human habitation.

Kobuleti (ko-boo-LE-tee), city (2002 population 18,556) and center of Kobuleti region, N Adzhar Autonomous Republic, Georgia, on BLACK SEA, on railroad and 15 mi/24 km NNE of BATUMI; 41°48′N 41°53′E. Cannery; knitwear manufacturing.

Koburg, GERMANY: see COBURG.

Kobuta Station, unincorporated village, BEAVER county, W PENNSYLVANIA, 4 mi/6.4 km WSW of MONACA, on OHIO RIVER, at mouth of Raccoon Creek; 40°39′N 80°21′W. Synthetic rubber. Nuclear power plant to W.

Kobyakovo (kuh-byah-KO-vuh), settlement, central MOSCOW oblast, central European Russia, near highway, 6 mi/10 km WSW, and under administrative jurisdiction, of ODINTSOVO; 55°35′N 37°02′E. Elevation 666 ft/202 m. Willow wicker products.

Kobyay (kuh-BYEI), village, central SAKHA REPUBLIC, central RUSSIAN FAR EAST, in the Central Yakutsk Lowland, on road, 130 mi/209 km NW of YAKUTSK; 63°35′N 126°30′E. Elevation 334 ft/101 m. In agricultural area.

Kobylets'ka Polyana (ko-bi-LETS-kah po-LYAH-nah) (Russian *Kobyletskaya Polyana*), town (2004 population 4,470), SE TRANSCARPATHIAN oblast, UKRAINE, in a deep stream valley of the SVYDOVETS MOUNTAINS, 6 mi/10 km W of RAKHIV; 48°04′N 24°04′E. Elevation 2,073 ft/631 m. Armature-making; forestry, sawmilling; woodworking; tourist base. First mentioned in the 15th century; town since 1971.

Kobyletskaya Polyana, UKRAINE: see KOBYLETS'KA POLYANA.

Kobylin (ko-BEE-leen), town, Leszno province, W POLAND, 9 mi/14.5 km W of KROTOSZYN. Railroad junction; flour milling, sawmilling. Monastery.

Kocane (ko-CHAH-ne), Macedonian *Kocani* or *Kochani*, Serbo-Croatian *Kočane* or *Kočani*, village, Republic of MACEDONIA, near BREGALNICA RIVER, 50 mi/80 km E of SKOPJE. In fertile rice-growing valley. Terminus of railroad to TITOV VELES; commercial center for region producing poppies for opium and roses for attar. Also spelled Kochane.

Koca River, 105 mi/169 km long, NW TURKEY; rises 13 mi/21 km SE of BURHANIYE; flows NW to Lake MANYAS, then NW to SIMAV RIVER at KARACABEY.

Koca River, 55 mi/89 km long, in N TURKEY; rises in ISFENDIYAR (KURE) MOUNTAINS, a section of the Pontus mountain system, 12 mi/19 km SE of KURE; flows W and NW to BLACK SEA near CIDE.

Koca River, 75 mi/121 km long, in SW TURKEY; rises S of Lake SOGUT in the TAURUS MOUNTAINS; flows S to the MEDITERRANEAN. Near its mouth was ancient Xanthus, a name that was also given to the river.

Kocarli, Turkish, *Koçarli*, village, W TURKEY, near BÜYÜK MENDERES and 10 mi/16 km SW of AYDIN; 37°45′N 27°42′E. Figs, olives.

Kočevje (ko-CHEV-ye), town (2002 population 8,868), S SLOVENIA, 34 mi/55 km SSE of LJUBLJANA; 45°38′N 14°51′E. Terminus of railroad spur to Ljubljana. Manufacturing of woolen textiles, wood products, transportation equipment; chemical industry. Mining of lignite until 1970; lake covers old open pit mine.

Kočevski Rog (ko-CHEV-skee ROG), mountain, S SLOVENIA, 15.5 mi/24.9 km SW of Novo Mesto and 7 mi/11.3 km ENE of Kočevje, part of DINARIC ALPS; 21.7 mi/34.9 km long, 9.3 mi/15 km wide. Karst topography. Highest peak is Veliki Rog (3,606 ft/1,099 m). Timber exploitation since 19th century; several virgin forest reserves. Military stronghold during World War I.

Kocgiri, TURKEY: see ZARA.

Kochan (ko-CHAN), village, SOFIA oblast, SATOVCHA obshtina, BULGARIA; 41°35′N 24°02′E.

Kochane, MACEDONIA: see KOCANE.

Kochang (GUH-CHANG), county, S. KYONGSANG province, S. KOREA, 40 mi/64 km WSW of TAEGU; 35°45′N 127°55′E. Agricultural center (rice, soybeans, hemp, cotton, ramie); makes grass linen. It is on the route of Olympic Expressway since 1984.

Koch Bihar (koch bee-HAHR), district (□ 1,308 sq mi/3,400.8 sq km), WEST BENGAL state, E INDIA; ⊙ KOCH BIHAR.

Koch Bihar (koch bee-HAHR), town (2001 population 76,812), ⊙ KOCH BIHAR district, NE WEST BENGAL state, E INDIA, on central plain, on TORSA River and 88 mi/142 km SE of DARJILING; 26°19′N 89°26′E. Trade center (rice, jute, tobacco, oilseeds, sugarcane); manufacturing of leather goods. Has college and an airport. Formerly spelled COOCH BEHAR.

Kochek, Tell, SYRIA: see KOJAK, TELL.

Kochel (KO-khel) or **Kochel am See**, village, BAVARIA, S GERMANY, in UPPER BAVARIA, on N slope of the Bavarian Alps, at NE tip of the KOCHELSEE, 11 mi/18 km SW of BAD TÖLZ; 47°40′N 10°23′E. Summer resort. Manufacturing of machinery. Nearby is hydroelectric plant WALCHENSEE.

Kochelsee (KO-khel-seh), lake (□ 2 sq mi/5.2 sq km), BAVARIA, S GERMANY, in UPPER BAVARIA, on N slope of the Bavarian Alps, 11 mi/18 km SW of BAD TÖLZ; 2.5 mi/4 km long, 1.5 mi/2.4 km wide, 213 ft/65 m deep; elevation 1,968 ft/600 m; 47°39′N 11°20′E. Fed and drained by the LOISACH. Large hydroelectric plant on S shore; resort of KOCHEL at N tip.

Kochendorf, GERMANY: see BAD FRIEDRICHSHALL.

Kochenëvo (kuh-chye-NYO-vuh), town (2006 population 16,230), E central NOVOSIBIRSK oblast, SW SIBERIA, RUSSIA, on the BARABA STEPPE, on the TRANS-SIBERIAN RAILROAD, 25 mi/40 km W of NOVOSIBIRSK; 55°01′N 82°11′E. Elevation 492 ft/149 m. Dairying; poultry. Railroad station since 1896; town status since 1960.

Kocherezhky, UKRAINE: see NOVOMYKOLAYIVKA, Zaporizhzhya oblast.

Kocherinovo (ko-che-REEN-o-vo), city, SOFIA oblast, ⊙ Kocherinovo obshtina (1993 population 7,260), W BULGARIA, in the STRUMA River valley, 10 mi/16 km S of DUPNITSA; 42°05′N 23°03′E. Railroad junction; tobacco, fruits, vineyards, poppies; livestock. Paper manufacturing, sawmilling (2 mi/3.2 km S) in Barakovo.

Kocher River (KO-khuhr), BADEN-WÜRTTEMBERG, SW GERMANY; rises 3 mi/4.8 km S of AALEN; meanders generally NW, past SCHWÄBISCH HALL, to the NECKAR at Bad Friedrichshall; length, over 90 mi/145 km; source at 48°47′N 10°07′E.

Kochetok (ko-che-TOK), town, N central KHARKIV oblast, UKRAINE, on DONETS River and 4 mi/6 km NE of CHUHUYIV, in Kharkiv metropolitan area; 49°53′N 36°44′E. Elevation 528 ft/160 m. Water purification plant for Kharkiv; sawmilling. Forestry school. Known since 1641; town since 1938.

Kochetovka (kuh-chye-TOF-kah), town, W TAMBOV oblast, S central European Russia, on road and railroad, 6 mi/10 km N of MICHURINSK; 52°59′N 40°29′E. Elevation 501 ft/152 m. Railroad shops; fruit and vegetable juices and juice concentrates.

Kochevo (KO-chee-vuh), village (2005 population 3,220), central KOMI-PERMYAK Autonomous Okrug, PERM oblast, E European Russia, on road, 42 mi/68 km NNW of KUDYMKAR; 59°36′N 54°19′E. Elevation 515 ft/156 m. In agricultural area (flax, rye, oats, feed corn).

Kochi (KO-chee), Japanese *ken*, prefecture (□ 2,743 sq mi/7,104 sq km; 1990 population 825,063), S SHIKOKU, W JAPAN, on TOSA BAY (S), HOYO STRAIT (SW), and PHILIPPINE SEA (SE); chief port and ⊙ KOCHI. Bordered NW by EHIME prefecture and NE by TOKUSHIMA prefecture; includes offshore island of OKINO-SHIMA. Mountainous, with fertile coastal plains; shores warmed by Japan Current. NIYODO and Shimando rivers drain

large forested and rice-growing areas. Extensive forests (pine, Japanese cedar); orange trees, mulberry groves. Widespread cultivation of rice, tea, soybeans, sweet potatoes. Cattle raising; lumbering. Produces agricultural tools, machinery, paper, silk textiles, sake, soy sauce, charcoal. Numerous fishing ports mainly produce dried bonito and ornamental coral. Kochi is only large center.

Kochi (KO-chee), city (2005 population 333,484), ⊙ KOCHI prefecture, S SHIKOKU, W JAPAN; 33°33′N 133°32′E. Coral processing; traditional papermaking. Katsura Beach is nearby.

Kochi (KO-chee), town, Kamo county, HIROSHIMA prefecture, SW HONSHU, W JAPAN, 25 mi/40 km E of HIROSHIMA; 34°28′N 132°53′E. Seat of Chikurin temple (Buddhist Shingon Shu sect).

Kochi, INDIA: see COCHIN, region.

Kochinda (ko-CHEEN-dah), town, S OKINAWA island, Shimajiri county, Okinawa prefecture, SW JAPAN, 5 mi/8 km S of NAHA; 26°08′N 127°43′E.

Kochisar, TURKEY: see HAFIK.

Kochisar, TURKEY: see KIZILTEPE.

Kochisar, TURKEY: see SEREFLI KOCHISAR.

Kochisarbala, TURKEY: see ILGAZ.

Kochishevo (KO-chee-shi-vuh), rural settlement, N UDMURT REPUBLIC, E central European Russia, 10 mi/16 km NW of BALEZINO, and 8 mi/13 km SE of GLAZOV; 58°03′N 52°45′E. Elevation 485 ft/147 m. In agricultural area; food processing.

Ko-chiu, China: see GEJIU.

Kochkar' (kuch-KAHR), village, central CHELYABINSK oblast, SE URALS, RUSSIA, on crossroads, 56 mi/90 km S of CHELYABINSK, and 5 mi/8 km N of PLAST, to which it is administratively subordinate; 54°27′N 60°48′E. Elevation 889 ft/270 m. Gold mines and placers nearby. Formerly called Kochkarskiy.

Kochkarskiy, RUSSIA: see KOCHKAR'.

Kochki (KOCH-kee), village (2006 population 4,085), S NOVOSIBIRSK oblast, SW SIBERIA, RUSSIA, on the KARASUK River, on crossroads, 60 mi/97 km S of KARGAT; 54°20′N 80°29′E. Elevation 482 ft/146 m. In dairy-farming area; bakery. Formerly known as Kochkovskoye.

Kochkor (kuhch-KAWR), village, NARYN region, KYRGYZSTAN, on CHU RIVER, and 28 mi/45 km SW of ISSYK-KOL lake; 42°13′N 75°45′E. Wheat; salt and coal mining. Tertiary-level administrative center.

Kochkor-Ata (kuhch-kawr–ah-TAH), town (1999 population 16,104), JALAL-ABAD region, KYRGYZSTAN, in E FERGANA VALLEY; 40°53′N 72°25′E. Agriculture; oil.

Kochkovskoye, RUSSIA: see KOCHKI.

Kochkurovo (kuhch-KOO-ruh-vuh), village (2006 population 3,160), SE MORDVA REPUBLIC, central European Russia, on road, 14 mi/23 km SE of SARANSK; 54°26′N 46°07′E. Elevation 688 ft/209 m. In hemp-growing area; bast fiber processing.

Koch Peak, MONTANA: see MADISON RANGE.

Kochubey (kuh-choo-BYAI), settlement (2005 population 6,980), N DAGESTAN REPUBLIC, SE European Russia, near the CASPIAN SEA coast, on road and railroad, 47 mi/76 km N of KIZLYAR; 44°23′N 46°33′E. Below sea level. Highway hub and railroad station; local transshipment point. In oil-producing region. Formerly known as Chërnyy Rynok [Russian=black market].

Kochura (kuh-choo-RAH), town, S KEMEROVO oblast, S central SIBERIA, RUSSIA, in the GORNAYA SHORIYA, on road and railroad spur, 7 mi/11 km SW, and under administrative jurisdiction, of TASHTAGOL; 52°45′N 87°52′E. Elevation 1,541 ft/469 m. Iron mining.

Kock (kotsk), Russian *Kotsk*, town, Lublin province, E POLAND, on TYSMIENICA RIVER, near its mouth, and 28 mi/45 km N of LUBLIN. Starch manufacturing, flour milling; brickworks. Before World War II, population 75% Jewish.

Kocs (KOCH), village, KOMÁROM county, N HUNGARY, 22 mi/35 km ESE of Győr; 47°36′N 18°13′E. Wheat, tobacco; cattle, hogs. Nearby springs contain sulphate of magnesium.

Koda (KO-DAH), town, Takata county, HIROSHIMA prefecture, SW HONSHU, W JAPAN, 28 mi/45 km N of HIROSHIMA; 34°41′N 132°45′E.

Kodagu (ko-dah-goo), district (□ 1,584 sq mi/4,118.4 sq km), KARNATAKA state, SW INDIA; ⊙ MADIKERI.

Kodaikanal (ko-dei-KAH-nuhl), city, NAGAPATTINAM QUAID-E-MILLETH district, TAMIL NADU state, S INDIA, 50 mi/80 km WNW of MADURAI; 10°14′N 77°29′E. Elevation c.7,000 ft/2,134 m. Climatic health resort (sanatorium) on scenic wooded plateau in Palni Hills. Fruit (oranges, lemons, limes) and eucalyptus plantations. Site of solar photographic observatory (Elevation 7,700 ft/2,347 m), established 1899. Christian missionary colony.

Kodaikanal, INDIA: see AMMAYANAKKANUR.

Kodaira (ko-DAH-ee-rah), city (2005 population 183,796), Tokyo prefecture, E central HONSHU, E central JAPAN, 93 mi/150 km W of SHINJUKU; 35°43′N 139°28′E. Suburb of TOKYO. Tires.

Kodali, INDIA: see KODOLI.

Kodama (ko-DAH-mah), town, Kodama county, SAITAMA prefecture, E central HONSHU, E central JAPAN, 37 mi/60 km N of URAWA; 36°11′N 139°07′E. Eggplants, onions. Tokugawa-era scholar Hanawaho Kiichi born here.

Kodari (ko-DAH-ree), village, N central NEPAL, near CHINA (TIBET) border; 27°57′N 85°58′E. Border crossing point from Nepal to Khasa (Zhangmu), Tibet.

Koddiyar Bay (KOD-di-yahr), EASTERN PROVINCE, NE SRI LANKA, semicircular inlet of INDIAN OCEAN, between FOUL POINT (S) and Trincomalee promontory (N); 8 mi/12.9 km long, 7 mi/11.3 km wide. Trincomalee Inner Harbour (3 mi/4.8 km long, 3 mi/4.8 km wide) is N inlet; TAMBALAGAM BAY (5 mi/8 km long, 3 mi/4.8 km wide; pearl banks) is W inlet. Lighthouse on island in middle of bay. Was much used by sailing ships before the modern era. Receives the MAHAWELI GANGA River (S).

Kodera (KO-de-rah), town, Kanzaki county, HYOGO prefecture, S HONSHU, W central JAPAN, 29 mi/46 km N of KOBE; 34°54′N 134°44′E.

Koderma (koh-DHER-mah), district (□ 931 sq mi/2,420.6 sq km; 2001 population 540,901), N Jharkhand state, E India; ⊙ KODERMA. Was part of BIHAR state until it joined other districts in 2000 to form Jharkhand. Mica, quartz, granite deposits. Heavily forested area.

Koderma (koh-DHER-mah), town (2001 population 17,160), ⊙ KODERMA district, N Jharkhand state, E INDIA, 50 mi/80 km WNW of GIRIDIH, 33 mi/53 km NNE of HAZARIBAG. Mica-mining center; railroad station.

Kodiak, city (2000 population 6,334), NE KODIAK ISLAND, S ALASKA, on CHINIAK BAY of Gulf of ALASKA; 57°47′N 152°24′W. Major fishing port; beef cattle; hunting; numerous canneries. Connected to SEWARD and HOMER by Alaska Ferry system. Saint Herman's Orthodox seminary. Russians under Shelekhov here moved (1792) their main settlement, originally established at THREE SAINTS BAY, named it Saint Paul's Harbor and later Kadiak or Kodiak. Headquarter of Russian-American Company before it was moved to SITKA. In World War II Fort Greeley, major army supply base, was built nearby. WOMENS BAY coast guard base is 10 mi/16 km SW.

Kodiak Island, borough (□ 6,463 sq mi/16,803.8 sq km; 2006 population 13,072), S ALASKA, includes the E part of the ALASKA PENINSULA, KODIAK ISLAND, and adjacent islands (notably AFOGNAK and Shuyak islands to the NE and the TRINITY ISLANDS to the S). Main town is KODIAK. Part of KATMAI National Park and parts of Becharof and Alaska Peninsula National Wildlife Reserves to W; Kodiak National Wildlife Reserve in center of Kodiak Island, part of CHUGACH NATIONAL FOREST on Afognak Island; Fort Aber-

crombie State Historical Park at Kodiak. Kodiak Indian Reservation in W part of Kodiak. Fishing; timber; furs.

Kodiak Island (□ 11,845 sq mi/30,679 sq km; 1990 population 13,309), off S ALASKA, separated from the ALASKA PENINSULA by SHELIKOF STRAIT; 57°42′N 153°46′W. Alaska's largest island (c.100 mi/160 km long and 10 mi/16 km–60 mi/97 km wide), Kodiak is mountainous and heavily forested in the N and E; the native grasses in the S offer good pasturage for beef cattle and sheep. The island has many ice-free, deeply penetrating bays that provide sheltered anchorages and transportation routes. The Kodiak bear and the Kodiak king crab are native to the island. Most of the island is a national wildlife refuge. In 1912 the eruption of Mount KATMAI on the mainland blanketed the island with volcanic ash, causing widespread destruction and loss of life. Explored in 1763 by Russian fur trader Stepan Glotov, the island was the scene of the first permanent Russian settlement in Alaska, founded by Grigori Shelekhov, a fur trader, on THREE SAINTS BAY in 1784. The settlement was moved to Kodiak village in 1792 and became the center of Russian fur trading. The largest town on the island is KODIAK (1990 population 6,365). Salmon fishing is a major occupation. Livestock farms, numerous canneries, and hunting. Many Aleut settlements and Orthodox churches.

Kodikkarai, village, THANJAVUR district, TAMIL NADU state, S India, projected port on PALK STRAIT just W of POINT CALIMERE, 33 mi/53 km S of Nagappattinam. Railroad terminus (extension from VEDARANNIYAM, 7 mi/11.3 km NNE; built 1935). Comprises the two original villages of KODIYAKADU (or Kodiyakkadu) and Kodiyakkarai (or Kodikkarai). Formerly called Point Calimere.

Kodinar (ko-DEE-nahr), town, AMRELI district, GUJARAT state, W central INDIA, on S KATHIAWAR peninsula, 65 mi/105 km SSW of AMRELI; 20°47′N 70°42′E. Railroad terminus; trades in cotton, millet, ghee, oilseeds; handicraft cloth weaving and printing, cotton ginning. Pomfrets and Bombay duck caught in ARABIAN SEA (S).

Kodino (KO-dee-nuh), town, NW ARCHANGEL oblast, N European Russia, on the Kodina River (tributary of the ONEGA), on railroad and highway branch, 45 mi/72 km ESE of ONEGA; 63°43′N 39°41′E. Elevation 121 ft/36 m. Sawmilling, wood pulp. Developed after 1941.

Kodinsk (kuh-DEENSK), city (2005 population 15,995), E KRASNOYARSK TERRITORY, SE SIBERIA, RUSSIA, on the ANGARA RIVER (tributary of the YENISEY RIVER), on road, 465 mi/748 km NNW of IRKUTSK, and 455 mi/732 km NE of KRASNOYARSK; 58°41′N 99°11′E. Elevation 577 ft/175 m. Arose in 1977 in connection with the construction of the Boguchansk dam to back up the reservoir (233 mi/375 km long) and build a hydroelectric power station on the middle Angara River. Wood industries and building materials. Known as Kodinskiy until 1989, when granted city status and renamed.

Kodiyakadu, INDIA: see KODIKKARAI.

Kodok (KO-dok), township, SE SUDAN, on the left bank of WHITE NILE River; 09°53′N 32°07′E. In 1898 it was the scene of the Fashoda Incident, which brought Britain and France to the brink of war and resulted, in 1899, in an Anglo-French agreement establishing the frontier between the SUDAN and the FRENCH CONGO along the watershed between the CONGO and NILE basins. The formation of an Anglo-French entente in 1904 prompted the British to change the town's name in hopes of obliterating the memory of the incident. Also known as Fashoda.

Kodoli (ko-DO-lee), town, KOLHAPUR district, MAHARASHTRA state, INDIA, 12 mi/19 km NNW of KOLHAPUR; 16°53′N 74°12′E. Market center for millet and rice. Sometimes spelled KODALI.

Kodomari (ko-DO-mah-ree), village, N Tsugaru county, Aomori prefecture, N HONSHU, N JAPAN, on Cape Kodomari, 31 mi/50 km N of AOMORI; 41°07′N 140°18′E. Processing of marine products.

Kodori Pass (kuh-DOR-ee) (7,760 ft/2,365 m), through VODORAZDEL'NYI RANGE of the Greater Caucasus, connecting the valley of the upper course of the Andiiskoe Koisu River in DAGESTAN with ALAZAN' valley in GEORGIA.

Kodori Range (kuh-DOR-ee) or **Panavi Range**, S spur of the W Greater Caucasus, in NW GEORGIA; extends c.50 mi/80 km SW to BLACK SEA; rises to 10,856 ft/3,309 m. Forms left watershed of KODOR River. Coal deposits at TKVARCHELI.

Kodori River (kuh-DOR-ee) or **Kodor River**, 45 mi/72 km long, ABKHAZ Autonomous Republic, GEORGIA; rises in the W Greater Caucasus, flows SW to BLACK SEA SSE of SUKHUMI. Formed by the confluence of the Sakeni and Gvandra rivers, which rise on the SW slopes of the Greater Caucasus. Used for floating timber. SUKHUMI MILITARY ROAD passes through its mountain valley.

Kodra (KOD-rah), town (2004 population 4,400), W KIEV oblast, UKRAINE, on right tributary of TETERIV RIVER, 40 mi/64 km WNW of KIEV; 50°36′N 29°34′E. Elevation 495 ft/150 m. Glass; forestry. Evangelical church, hospital. Small Jewish community since 1880, decimated by the 1919 pogroms, the survivors emigrated.

Kodry, MOLDOVA: see CODRI.

Kodugallur, town, Trichur district, KERALA state, S INDIA, near PERIYAR RIVER mouth, 15 mi/24 km N of ERNAKULAM; 10°13′N 76°13′E. Coir products (rope, mats), copra; fishing (in coastal lagoons). Trade with PERSIA in 15th century. A historical and religious site. Also spelled Cranganore.

Kodumuru (ko-doo-MOO-roo), town, KURNOOL district, SW ANDHRA PRADESH state, SE INDIA, 21 mi/34 km SW of KURNOOL; 15°41′N 77°47′E. Rice and oilseed (peanut) milling; millet, cotton, mangoes.

Kodyma (KO-dee-mah), city, NW ODESSA oblast, UKRAINE, on Zhmerynka-Odessa railroad trunkline, 25 mi/40 km WNW of BALTA; 48°06′N 29°07′E. Elevation 912 ft/277 m. Raion center; fruit canneries, dairy, grain mill; limestone works, fabrication of reinforced concrete building materials. Professional technical school; heritage museum. Established in 1754, city since 1979. Sizeable Jewish community since the city's founding, numbering 1,970 in 1939; eliminated by the Nazis in 1941.

Koealakapoeas, INDONESIA: see KUALAKAPUAS.

Koealalangsa, INDONESIA: see KUALALANGSA.

Koeandang, INDONESIA: see KUANDANG.

Koeantan River, INDONESIA: see INDRAGIRI RIVER.

Koeberg Nuclear Power Station, SOUTH AFRICA: see BLOUBERGSTRAND.

Koedoes, INDONESIA: see KUDUS.

Koedoespoort, village, GAUTENG, SOUTH AFRICA, 5 mi/8 km E of TSHWANE; 25°43′S 28°17′E. Elevation 4,494 ft/1,370 m. Important railroad workshops, close to N1 highway.

Koegas, SOUTH AFRICA: see MARYDALE.

Koekelare (KOO-kuh-lah-ruh), commune (2006 population 8,303), Diksmuide district, WEST FLANDERS province, W BELGIUM, 6 mi/9.7 km ENE of DIKSMUIDE; 51°05′N 02°58′E. Agriculture. Formerly spelled Couckelaere.

Koekelberg (KOO-kuhl-berk), commune (2006 population 18,287), Brussels district, BRABANT province, central BELGIUM, residential and industrial NW suburb of BRUSSELS. Textile and flour mills.

Koemamba Islands, INDONESIA: see KUMAMBA ISLANDS.

Koepang, INDONESIA: see KUPANG.

Koerich (KUHR-rik), village, Koerich commune, SW LUXEMBOURG, 9 mi/14.5 km WNW of LUXEMBOURG city, near Belgian border; 49°41'N 05°57'E. Church, castle ruins.

Koersel (KOOR-suhl), village, BERINGEN commune, Hasselt district, LIMBURG province, NE BELGIUM, 10 mi/16 km NNW of HASSELT; 51°04'N 05°16'E. Formerly spelled Coursel.

Koesan (KO-SAHN), county (□ 372 sq mi/967.2 sq km), NORTH CHUNGCHONG province, SOUTH KOREA. Mountainous in E, small rivers form agricultural basin (barley, wheat, corn, tobacco, ginseng) in center. Hwayangdong Valley and Hanchon mineral water springs. Hakdongjosuji reservoir.

Koesfeld, GERMANY: see COESFELD.

Koetai River, Malaysia: see MAHAKAM RIVER.

Koetoardjo, INDONESIA: see KUTOARJO.

Koettlitz Glacier, in EAST ANTARCTICA, flowing from the ROYAL SOCIETY RANGE, S VICTORIA LAND, into the MCMURDO ICE SHELF; 78°20'S 164°30'E.

Koevorden, NETHERLANDS: see COEVORDEN.

Kofa Mountains (KO-fah) (4,828 ft/1,472 m), YUMA and LA PAZ counties, SW ARIZONA, N of GILA River; extend c.20 mi/32 km NE from CASTLE DOME MOUNTAINS; rise to their highest point in W tip. In Kofa National Wildlife Refuge.

Kofelē, town (2007 population 13,753), OROMIYA state, SW central ETHIOPIA, 22 mi/35 km E of LAKE AWASA; 07°04'N 38°47'E.

Koffiefontein, town, W FREE STATE province, SOUTH AFRICA, on RIET RIVER, 47 mi/75 km SSE of KIMBERLEY; 29°26'S 25°00'E. Elevation 4,067 ft/1,240 m. Stock, grain, feed. Former diamond-mining center; mines closed 1932.

Köflach (KUHF-lahkh), town, STYRIA, S AUSTRIA, 16 mi/26 km W of GRAZ; 46°04'N 15°05'E. Railroad terminus; manufacturing of metals (tools, machines). Lignite mines nearby. In the surroundings are the baroque pilgrim church Maria Lankowitz with an abandoned monastery and the stud farm Piber, where the Lipizan horses for the Spanish Riding School in Vienna are bred.

Koforidua (ko-fo-REE-dwuh), town, ⊙ EASTERN REGION, GHANA; 06°05'N 00°15'W. Commercial center for region producing cocoa, palm oil, cassava, and corn; also serves as a road and railroad center. Koforidua was founded (c.1875) by refugees from ASHANTI. Also called NEW JUABEN. West African Cocoa Research Institute nearby at TAFO.

Kofu (KO-FOO), city (2005 population 200,100), ⊙ YAMANASHI prefecture, central HONSHU, central JAPAN; 35°39'N 138°34'E. Industrial center manufacturing of crystal and leather goods. Grapes; wine production. Was (16th century) the castle town of the Takeda family. Shosenkyo Gorge is nearby.

Kofu (KO-foo), town, Hino county, TOTTORI prefecture, S HONSHU, W JAPAN, 45 mi/73 km S of TOTTORI; 35°16'N 133°29'E. Daikon, wasabi, shiitake mushrooms; miso. Skiing area.

Koga (ko-GAH), city, IBARAKI prefecture, central HONSHU, E central JAPAN, 43 mi/70 km SW of MITO; 36°10'N 139°42'E. Umbrellas.

Koga (KO-gah), town, Kasuya county, FUKUOKA prefecture, N KYUSHU, SW JAPAN, on GENKAI SEA, 9 mi/15 km N of FUKUOKA; 33°43'N 130°28'E. Oranges.

Kogal'nik River, UKRAINE: see KOHYLNYK RIVER.

Kogaly (ko-gah-LEE), town, SE TALDYKORGAN region, KAZAKHSTAN, in the DZUNGARIAN Alatau Mountains, 37 mi/60 km SSE of TALDYKORGAN; 44°30'N 78°41'E. Irrigated agriculture (wheat, poppies). Also spelled Kugaly, Qogaly.

Kogalym (kuh-gah-LIM), city (2005 population 58,695), N central KHANTY-MANSI AUTONOMOUS OKRUG, central SIBERIA, in the middle of the OB' RIVER lowland, on the Inguyagun River; on the railroad line connecting SURGUT to NOVYY URENGOY, 536 mi/863 km NNE of TYUMEN, 200 mi/322 km NE of KHANTY-MANSIYSK, and 125 mi/201 km NW of NIZHNEVARTOVSK; 61°15'N 73°30'E. Oil and gas ex-

traction and production; clothing plant; food processing. Has a heritage museum and museum of fine arts. Arose around 1975, with the development of nearby oil and gas fields. Made city in 1985.

Koganei (ko-gah-NAI), city, Tokyo prefecture, E central HONSHU, E central JAPAN, 68 mi/110 km S of SHINJUKU; 35°41'N 139°30'E. Suburb of TOKYO. Seat of Tokyo University of Liberal Arts.

Kogarah (KAH-guh-ruh), residential suburb 9 mi/15 km from SYDNEY, E NEW SOUTH WALES, AUSTRALIA, 10 mi/16 km SW of city center, on BOTANY BAY; 33°59'S 151°07'E. Commercial, retail, and light industries. In metropolitan area. Sydney international airport is nearby.

Køge (KU-yuh), city (2000 population 32,996), ROSKILDE county, on SJÆLLAND, DENMARK, port on KØGE BUGT of the ØRESUND, 22 mi/35 km SSW of COPENHAGEN; 55°27'N 12°11'E. Manufacturing (woodworking, wood products, rubber products, glue, meat canning); agriculture (dairying, grain). Formerly spelled Kjoge.

Koge (KO-GE), town, Yazu county, TOTTORI prefecture, S HONSHU, W JAPAN, 2.5 mi/4 km S of TOTTORI; 35°24'N 134°15'E. Fruit (pears, persimmons).

Koge Bugt (KU-yuh), bay, on E coast of SJÆLLAND island, DENMARK, on the ØRESUND. AMAGER island in N.

Kogi (KO-gee), state (□ 11,518 sq mi/29,946.8 sq km; 2006 population 3,278,487), central NIGERIA; ⊙ LOKOJA. Bordered N by NIGER state (NIGER river), NNE by ABUJA FEDERAL CAPITAL TERRITORY, NE by NASARAWA state (BENUE River), E by BENUE state, SE by ENUGU state, S by ANAMBRA state, SW by EDO state, W by ONDO and EKITI states, and NW by KWARA state. Agriculture includes soybeans, groundnuts, cocoa, oil palms, coffee, yams, cotton, rice, cassava, maize, millet, tobacco, kola nuts; livestock, fishing. Steel complex and palm oil mill; marble and wood processing. Mineral resources include coal, limestone, iron ore, feldspar, clay. Main centers are LOKOJA, AJAOKUTA, Itakpe, IDAH, Okura-Olasa.

Kogilnik River, UKRAINE: see KOHYLNYK RIVER.

Kogluktok, CANADA: see KOGLUKTOK.

Kogota (ko-GO-tah), town, Tooda county, MIYAGI prefecture, NE HONSHU, NE JAPAN, 22 mi/35 km N of SENDAI; 38°32'N 141°03'E. Rice.

Kogyei (KO-jai), town, BRONG-AHAFO region, GHANA, on Tombe River, 20 mi/32 km SW of BUI DAM, on Côte d'Ivoire border; 08°09'N 02°31'W.

Koh (Persian, *mountain*), for names in AFGHANISTAN, IRAN, and PAKISTAN beginning thus and not found here: see under KUH, or under following part of the name.

Koh [Thai=island], for names in THAILAND beginning thus and not found here: see under following part of the name.

Kohala Peninsula (ko-HAH-lah), N HAWAII island, HAWAII county, HAWAII. Flanked by PACIFIC OCEAN to NE and SW; ALENUIHAHA CHANNEL, separating Hawaii and MAUI, off NW tip. Rich in relics of ancient Hawaii. Kohala Mountains rise to 5,489 ft/1,673 m; large Kohala Forest Reserve occupies lower E quarter of peninsula; includes small Paoakalani Mokupuku and Paalaea islands, seabird sanctuaries; main town is HAWI, near N end.

Köhalom, ROMANIA: see RUPEA.

Koháryháza, SLOVAKIA: see POHORELA.

Kohat (ko-HUT), district (□ 2,707 sq mi/7,038.2 sq km), NORTH-WEST FRONTIER PROVINCE, N PAKISTAN; ⊙ KOHAT. Bordered NW by SE outliers of SAFED KOH Range, NE by INDUS River; crossed E-W by broken hill ranges. Subsistence agriculture (wheat, millet, corn, gram, barley); handicrafts (cloth, baskets, palm mats, leather, and metal goods). Hills (S center) have large deposits of rock salt (quarries at BAHADUR KHEL and JATTA ISMAIL KHEL) and gypsum. Raided 1505 by Babur; under Sikh rule in early 19th century.

District exercises political control over adjoining tribal area.

Kohat (ko-HUT), town, ⊙ KOHAT district, NORTHWEST FRONTIER PROVINCE, N PAKISTAN, 28 mi/45 km S of PESHAWAR, on the Kohat Toi River; 25°07'N 67°55'E. The town, enclosed by a wall with fourteen gates, is noted for its cotton fabrics and lungis; market center (wheat, millet, corn, rock salt); fodder milling, handicrafts, cloth weaving, cement, and sugar industry. Has a 19th century British fort built on the site of an old Sikh fortress. A military station and air force base, ordnance factory, and engineering workshop.

Koh Daman (ko DAH-mahn), district, KABUL province, E AFGHANISTAN, N of KABUL, beyond KHAIRKHANA Pass, occupying S portion of the Samt-i-Shimali plain; 35°00'N 69°15'E. Name is sometimes extended to entire plain.

Koh-i-Baba (ko-hee–BAH-bah), mountain range, W outlier of the Hindu Kush, in central AFGHANISTAN, extends 125 mi/201 km E-W along 34°40'N between 66°30'E and 68°30'E; rises to 16,872 m/5,143 m in the Shah Fuladi, 17 mi/27 km SW of BAMIAN. Continued W by the SAFED KOH and SIAH KOH. Meaning the Grand Father Mt., Koh-i-Baba is the main ridge of Afghanistan that continues W. Koh-i-Fuladi serves as a watershed for 4 great rivers that flow through Afghanistan in four different directions—KABUL RIVER, the HARI RUD, the Helmand River, and KUNDUZ RIVER.

Koh-i-Dalil, PAKISTAN: see DAMODIM.

Kohila (KO-ee-luh), township, NW ESTONIA, 19 mi/31 km S of TALLINN; 59°10'N 24°45'E. Manufacturing (paper).

Kohima (ko-HEE-mah), district (□ 1,560 sq mi/4,056 sq km; 2001 population 314,366), NAGALAND state, NE INDIA; ⊙ KOHIMA. Nearby Dzükou Valley, one of the most popular hiking spots in NE India due to its wide variety of flora, immortalized by writer Vikram Seth in children's poem "The Elephant and the Tragopan" as Bingle Valley.

Kohima (ko-HEE-mah), hill town (2001 population 78,584), ⊙ KOHIMA district and NAGALAND state, NE INDIA, in Naga Hills, 139 mi/224 km E of SHILLONG; 25°40'N 94°07'E. Elevation 4,921 ft/1,500 m. Trades in rice, cotton, oranges, potatoes, lac. Northernmost point of Japanese penetration (1944) into India.

Koh-i-Malik Siah (ko-hee-mah-leek see-yah), peak (5,390 ft/1,643 m), NW BALUCHISTAN province, at junction of SW PAKISTAN, SE IRAN, and S AFGHANISTAN borders; 29°51'N 60°52'E. Fixed as a frontier point in 1905. Also spelled KUH-I-MALIK SIAH.

Kohinggo (ko-HING-go), island, SOLOMON ISLANDS, SW Pacific; 08°13'S 157°09'E. Flat coral-formed island separated from W of NEW GEORGIA by Hathorn Sound and Diamond Narrows. Also called Arundel.

Kohir, town, BIDAR district, KARNATAKA state, SW INDIA, 25 mi/40 km SSE of Bidar; 17°36'N 77°43'E. Cotton, rice, sugarcane, mangoes.

Koh-i-Sabz, PAKISTAN: see SIAHAN RANGE.

Kohistan (ko-hee-STAHN), mountainous district, NE AFGHANISTAN, on S slopes of the Hindu Kush and W of NURISTAN; 35°07'N 69°18'E.

Kohistan (KO-his-tahn), former district, in former E PATIALA AND EAST PUNJAB STATES UNION, NW INDIA, now within HIMACHAL PRADESH state.

Kohistan (ko-HEE-stahn), hilly region, SW SIND province, SE PAKISTAN, comprising S offshoots of KIRTHAR RANGE; lies between KARACHI (SSW) and SEHWAN in DADU district (NE); drained by HAB (W) and BARAN (E) rivers. LAKHI HILLS are E offshoot. Little cultivation; some sheep, goat, and camel breeding. Hot sulphur springs, small coal deposits.

Koh-i-Sultan (ko-hee-suhl-TAHN), extinct, oval-shaped volcano, CHAGAI district, NW BALUCHISTAN province, SW PAKISTAN, W of CHAGAI HILLS; c.20 mi/32 km long, 12 mi/19 km wide; 29°07'N 62°50'E. Consists of three cones; rises to 7,650 ft/2,332 m in

---

Cross-references are shown in SMALL CAPITALS. The pronunciation guide is shown on page xix. The sources of population figures are shown on page xvii.

MIRI peak. Deposits of aluminum and sulphur ore on S slope; barite is found in a dike.

Koh Ker (KUH KAI), ruins, N CAMBODIA. Jayavarman IV made this his capital in the 10th century. Potential for tourism.

Kohkīlūyeh va Būyer Ahmadī, province (□ 5,506 sq mi/14,315.6 sq km), IRAN, in a mountainous area dominated by the ZAGROS Mountains in N and bordering on the provinces of Fārs (E), Khuzestān (SW, W), and Chahār Mahāll va Bākhtīarī (N); ⊙ Yasūj; 30°50′N 50°40′E. A rural province of pastures and woodlands. Livestock raising. Oil production in S; DOGONBADAN and Yasūj are the only large towns.

Koh Kong (KUH KONG), province (□ 4,308 sq mi/11,200.8 sq km; 2007 population 207,474), SW CAMBODIA, borders PURSAT (N), KOMPONG SPEU (E), and KAMPOT (SES) provinces, and GULF OF THAILAND (S and W) and small tip of THAILAND (NW); ⊙ KOH KONG; 11°30′N 103°30′E. Ecologically diverse; the CARDAMOM and ELEPHANT Ranges are still forested (though deforestation is a growing problem). Shifting cultivation, fishing.

Koh Kong (KUH KONG), town, ⊙ KOH KONG province, SW CAMBODIA, minor port on GULF OF THAILAND, 85 mi/137 km NW of KAMPOT, near boundary with THAILAND. Coastal trade and fishing center. Khmer population, with tiny Chinese minority.

Koh Kong (KUH KONG), island, SW CAMBODIA, KOH KONG province, 50 mi/80 km SE of Ko Chang Island (THAILAND); 11°20′N 103°05′E. Key point of entry for smuggled Thai and Singaporean goods into Cambodian territory.

Kohler, town (2006 population 1,984), SHEBOYGAN county, E WISCONSIN, suburb 2 mi/3.2 km W of SHEBOYGAN, on the SHEBOYGAN RIVER; 43°44′N 87°46′W. Manufacturing (plumbing fixtures). The Kohler plumbing-fixtures plant here, which still produces its famous stainless steel and porcelain products, has been the scene of some of the longest and most bitter labor disputes in U.S. history. The last strike began in 1954 and ended in 1962. Incorporated 1912.

Kohler (KO-luhr), unincorporated village, S ONTARIO, E central CANADA, 23 mi/37 km from SIMCOE, and included in HALDIMAND; 42°54′N 79°51′W.

Kohler Range (KUHR-luh), mountain range, ANTARCTICA, on WALGREEN COAST, 75°S between 113° and 115°W. Discovered 1940 by Richard E. Byrd.

Kohlfurt, POLAND: see WEGLINIEC.

Kohlgrub (KOL-groob) or **bad kohlgrub**, village, BAVARIA, S GERMANY, in UPPER BAVARIA, on N slope of the Bavarian Alps, 12 mi/19 km N of GARMISCH-PARTENKIRCHEN; 47°40′N 11°03′E. Summer and winter resort (elevation 2,953 ft/900 m), with mud baths.

Kohlgrub, Bad, GERMANY: see KOHLGRUB.

Kohl Janowitz, CZECH REPUBLIC: see UHLIRSKE JANOVICE.

Kohlsaat, Cape (kuhl-SAHT), easternmost point of FRANZ JOSEF LAND, ARCHANGEL oblast, extreme N European Russia, in the ARCTIC OCEAN, on the GRAHAM BELL ISLAND; 81°25′N 65°00′E.

Kohoku (KO-HO-koo), town, Kishima county, SAGA prefecture, N KYUSHU, SW JAPAN, 9 mi/15 km W of SAGA; 33°13′N 130°09′E.

Kohoku (KO-ho-koo), town, East Asai county, SHIGA prefecture, S HONSHU, central JAPAN, 37 mi/60 km N of OTSU; 35°26′N 136°14′E. Hot springs.

Koh Rong (KUH RONG), island, KOH KONG province, E CAMBODIA, at the mouth of the KOMPONG SOM BAY; 10°46′N 103°14′E.

Kohsan, village and district, HERAT province, NW AFGHANISTAN, 60 mi/97 km WNW of HERAT, on right bank of the HARI RUD, near IRAN border; 34°39′N 61°12′E. Tirpul oil field SE. Also spelled Kuhsan.

Koh Sichang, THAILAND: see SICHANG, KO.

Kohtla-Järve (KOT-lah-YAHR-vai) or **kohtla-yarve**, city, NE ESTONIA, 32 mi/51 km E of RAKVERE; 59°23′N 27°16′E. Several railroad lines run through the area; oil-shale distilling center; produces shale oil, gas; power plant. Linked (1947) by gas pipeline to LENINGRAD. Developed largely after World War II. Nearby are oil-shale mines of Kava (S) and Kukruse (E), and large mining residue mounds. Polluted ground water.

Kohung, county (□ 228 sq mi/592.8 sq km), SE SOUTH CHOLLA province, SOUTH KOREA, consisting of the peninsula of KOHUNG between SUNCH'ON and POSONG bays and 160 islands. A wide tideland around Haech'ang Bay and Oma Island was reclaimed and is used for agricultural production (rice, barley, garlic, citron, tangerines). Fishery (scallops, shrimp, ark shells, seaweed). Well-developed beaches.

Kohylnyk River (ko-HIL-nik), Russian *Kogilnik*, *Kogal'nik*, Romanian *Cogâlnic*, 150 mi/241 km long, MOLDOVA and SW UKRAINE; rises in the CODRI Hills of central Moldova, 11 mi/18 km WSW of Straşeni, flows SSE, past Hânceşti, Basarabaesca, and into SW ODESSA oblast, Ukraine, past ARTSYZ, to SASYK LAGOON at TATARBUNARY. Intermittent in lower reaches. Also known as Kunduk.

Koide (ko-ee-DE), town, North Uonuma county, NIIGATA prefecture, central HONSHU, N central JAPAN, near Uonuma River, 47 mi/75 km S of NIIGATA; 37°13′N 138°57′E. Rice; ayu.

Koil, INDIA: see ALIGARH, city.

Koilkuntla (ko-il-KUHNT-luh), town, KURNOOL district, ANDHRA PRADESH state, SE INDIA, 45 mi/72 km SSE of KURNOOL; 15°14′N 78°19′E. Road center; cotton ginning; lacquerware; millet, rice.

Koilovtsi (KOI-lov-tsee), village, LOVECH oblast, PLEVEN obshtina, N BULGARIA, 9 mi/15 km NE of Pleven; 43°28′N 24°46′E. Agriculture. Also spelled Koylovtsi.

Koilpatti, INDIA: see KOVILPATTI.

Koimbani (kwahm-BAH-nah), village, Njazidja island and district, NW Comoros Republic, 10 mi/16 km NE of Moroni, 2 mi/3.2 km from E coast of island; 11°39′S 43°22′E. Livestock; ylang-ylang, vanilla. Lake Hantsangoma, lake on N flank of Mt. Karthala volcano, to S.

Koina (KOI-nah), easternmost village of THE GAMBIA, near SENEGAL border, in UPPER RIVER division, head of navigation on GAMBIA RIVER (wharf), 25 mi/40 km ENE of BASSE SANTA SU; 13°28′N 13°52′E. Peanuts; beeswax, hides and skins.

Koinadugu, SIERRA LEONE: see KABALA.

Koinare (koi-NAH-re), city, Pleven oblast, CHERVEN BRYAG obshtina, N BULGARIA, on the ISKUR River, 13 mi/21 km SE of BYALA SLATINA; 43°12′N 24°09′E. Flour milling; livestock; grain. Also spelled Koynare.

Köi Sanjaq (KO-ee SUHN-juhk), town, ERBIL province, N IRAQ, in KURDISTAN, 35 mi/56 km ESE of ERBIL. Sesame, millet, corn, fruit. Sometimes spelled Kuway Sandjaq.

Koishiwara (ko-ee-shee-WAH-rah), village, Asakura county, FUKUOKA prefecture, N KYUSHU, SW JAPAN, 25 mi/40 km S of FUKUOKA; 33°27′N 130°49′E. Shiitake mushrooms, pears, apples, yams; pottery.

Koisu (koee-SOO), name applied to several headstreams of the SULAK RIVER, in DAGESTAN REPUBLIC, SE European Russia. The ANDI Koisu, 100 mi/161 km long, rises on Mount BARBALO in NE GEORGIA, flows E and NE past AGVALI (Dagestan) and BOTLIKH, joining the AVAR Koisu (right) near GIMRY to form the SULAK River. The Avar Koisu, 100 mi/161 km long, rises on the DYULTY-DAG (S Dagestan), flows NW past TLYARATA, and generally N to confluence with the Andi Koisu. Also spelled Koysu.

Koitash, town, W JIZZAKH wiloyat, UZBEKISTAN, 17 mi/27 km NW of GALLYAARAL, at E end of the NURATAU mountains; 40°11′N 69°19′E. Molybdenum and tungsten mining. Also spelled KOYTASH.

Koivisto, RUSSIA: see PRIMORSK, LENINGRAD oblast.

Köja (SHUH-yah), village, VÄSTERNORRLAND county, NE SWEDEN, on ÅNGERMANÄLVEN RIVER estuary, 20 mi/32 km NNW of HÄRNÖSAND; 62°59′N 17°49′E.

Kojak, Tell (ko-JAHK, tel), township, HASEKE district, NE SYRIA, on IRAQ border, on railroad, 75 mi/121 km ENE of El HASEKE; 36°48′N 42°03′E. Cereals. The inhabitants of this township and its region are mainly Kurds. Also spelled Tell KOCHEK or Tall Kujik. Official name Al Ya'rubiyeh or Yarubiya. Archaeological site Tell HAMOUKAR 5 mi/8 km W.

Koje (GUH-JE), city (□ 142 sq mi/369.2 sq km), SOUTH KYONGSANG province, SOUTH KOREA, on South Sea, adjacent to TONGYONG over the Koje Channel; 34°51′N 128°35′E. Consists of KOJE ISLAND (second-largest in Korea), and sixty smaller islands. Mountains mark the end of Sobaek and TAEBAEK ranges. Agriculture (rice, citron, tangerine, pineapple); fishery (oysters). Okpo and Chukpo shipbuilding yards. Koje bridge (1971) connects island to mainland. Tourism.

Koje Island (GUH-JE), Korean *Koje-do* (□ 148 sq mi/384.8 sq km), SOUTH KYONGSANG province, SOUTH KOREA, just off S coast, sheltering CHINHAE BAY; deeply indented W coast; 23 mi/37 km long, 2 mi/3.2 km–15 mi/24 km wide. The major island of KOJE city. Hilly, fertile. Fishing and agriculture (rice, barley, wheat, soybeans).

Kojetein, CZECH REPUBLIC: see KOJETIN.

Kojetin (KO-ye-TYEEN), Czech *Kojetín*, German *kojetein*, town, SEVEROMORAVSKY province, central MORAVIA, CZECH REPUBLIC, on MORAVA RIVER, 17 mi/27 km S of OLOMOUC; 49°21′N 17°19′E. Railroad junction. Agricultural center (wheat, barley, sugar beets, grapes); food processing; sugar refining.

Ko-jima (KO–jee-mah), volcanic island (□ 1 sq mi/2.6 sq km) of IZU-SHICHITO island group, Tokyo prefecture, SE JAPAN, in PHILIPPINE SEA, just W of Hachijojima; 2 mi/3.2 km long, 1 mi/1.6 km wide. Fishing, farming; livestock; silk-worm culture.

Kojonup (KO-juh-nup), town, SW WESTERN AUSTRALIA state, W AUSTRALIA, 150 mi/241 km SSE of PERTH; 33°50′S 117°09′E. Wool, cattle, and grain center. Wildflowers, especially orchids, and jarrah trees in area.

K'ok'a, town (2007 population 7,055), OROMIYA state, central ETHIOPIA, 20 mi/32 km SW of NAZRĒT, along shore of Koko Reservoir; 08°24′N 39°00′E.

Koka (KO-kah), town, Koka county, SHIGA prefecture, S HONSHU, central JAPAN, 22 mi/35 km S of OTSU; 34°53′N 136°13′E. Medicine. Hinoki. Site of eleven-headed Koga Ninjutsu, the biggest wooden seated Kannon (Buddhist goddess of mercy) statue in the country.

Kokadjo (kah-KAHJ-o), resort village, PISCATAQUIS county, central MAINE, on FIRST ROACH POND (formerly Kokadjo Lake), 37 mi/60 km NNW of DOVER-FOXCROFT. Hunting, fishing.

Kokalyane, Bulgaria: see ISKUR DAM.

Kokand, city, FERGANA wiloyat, E UZBEKISTAN, in the FERGANA VALLEY; 40°30′N 70°57′E. Manufacturing center (fertilizers, chemicals, machinery, and cotton and food products). Sulphur deposits to W. Important since the 10th century, Kokand was capital of an Uzbek khanate which became independent of the emirate of BUKHARA in the middle of the 18th century and flowered in the 1820s and 1830s. Taken by the Russians in 1876 and became part of RUSSIAN TURKISTAN. Was (1917–1918) capital of the anti-Bolshevik autonomous government of Turkistan. Has a ruined palace of the last khan, working mosques, and royal mausoleums. Also spelled Khokand, Quqon.

Kokanee Peak (ko-KA-nee) (9,400 ft/2,865 m), SE BRITISH COLUMBIA, W CANADA, in SELKIRK MOUNTAINS, 18 mi/29 km NNE of NELSON; 49°45′N 117°08′W.

Kokan-Kishlak, UZBEKISTAN: see PAKHTAABAD, ANDIJAN wiloyat.

Area is shown by the symbol □, and capital city or county seat by ⊙.

Kokava nad Rimavicou (ko-KAH-vah NAHD ri-MAH-vi-TSOU), Hungarian *Rimakokova*, village, STREDOSLOVENSKY province, S central SLOVAKIA, in S foothills of SLOVAK ORE MOUNTAINS, on railroad, 33 mi/53 km ESE of BANSKÁ BYSTRICA; 48°34′N 19°50′E. Glassworks; woodworking, textile manufacturing. Has 16th-century church. Folk architecture.

Kokawa (KO-kah-wah), town, Naga county, WAKAYAMA prefecture, S HONSHU, W central JAPAN, on W KII PENINSULA, 15 mi/24 km E of WAKAYAMA; 34°16′N 135°24′E.

Kokayty, UZBEKISTAN: see KAKAIDY.

Kok-Bulak, village, E KASHKADARYO wiloyat, UZBEKISTAN, c.25 mi/40 km S of SHAKHRISABZ; 38°36′N 66°51′E. Grain; livestock.

Kokcha River or **Kokchah River**, 200 mi/322 km long, BADAKHSHAN province, NE AFGHANISTAN; rises in the Hindu Kush on PAKISTAN line; flows N and W, past ZEBAK and FAIZABAD, to PANJ RIVER (TAJIKISTAN border) 65 mi/105 km W of Faizabad; 37°10′N 69°23′E. Bulk of region's population and irrigated agriculture system is in its valley, which also has salt and sulphur deposits and, near its source, the noted lapis lazuli of Badkhshan.

Kokemäenjoki (KO-ke-mah-en-YO-kee), river, 90 mi/145 km long, SW FINLAND; issues from Lake Pyhä 8 mi/12.9 km WSW of TAMPERE; flows in a wide arc generally W, through several small lakes, past NOKIA, HARJAVALTA (major power station), NAKKILA, ULVILA, and PORI, to GULF OF BOTHNIA 5 mi/8 km NW of Pori.

Kokemäki (KO-ke-MAH-kee), Swedish *Kumo*, village, TURUN JA PORIN province, SW FINLAND, on KOKEMÄENJOKI (river), 25 mi/40 km SE of PORI; 61°15′N 22°21′E. In lumbering region.

Kokenhusen, LATVIA: see KOKNESE.

Kokes, TURKEY: see SERIK.

Ko Kha (GO KAH), village and district center, LAMPANG province, N THAILAND, on WANG RIVER and 10 mi/16 km SW of LAMPANG, in sugarcane region; 18°11′N 99°24′E. Sugar mill.

Kokhanovo (ko-HUH-no-vo), urban-type settlement, GOMEL oblast, BELARUS, 17 mi/27 km WSW of ORSHA, 54°29′N 29°29′E. Manufacturing (flax processing, reinforced concrete).

Ko Khao Tapu (GO KOU TUH-POO) [Thai=Nail Mountain Island], island, THAILAND, with its striking shape and its prominent role in *The Man with the Golden Gun*, is a popular tourist destination in AO PHANG NGA NATIONAL PARK.

Kokhav Ya'ir (ko-KHAHV yah-IR) or **Kochav Yair**, village, ISRAEL, NE of KFAR SAVA, and on E fringe of the SHARON PLAIN. Founded in 1981 and named after Yair Stern, the leader of the militant anti-British Stern Gang of the 1940s (prior to independence).

Kokhma (KOKH-mah), city (2005 population 29,070), central IVANOVO oblast, central European Russia, on the UVOD' RIVER (a tributary of the KLYAZ'MA RIVER), on road and railroad, 4 mi/6 km SE of IVANOVO; 56°56′N 41°05′E. Elevation 380 ft/115 m. Railroad station. Cotton and linen textiles, construction machinery, enamelware. Established in 1619 as a village of Rozhdestvenskoye-Kokhma; city status and renamed in 1925.

Kokhtla-Yarve, ESTONIA: see KOHTLA-JÄRVE.

Koki (KO-KEE), town, S CAMBODIA, on MEKONG River and 9 mi/14.5 km SE of PHNOM PENH. Corn, rice. Beach resort, tourism. Khmer population, with tiny Chinese minority.

Kokilamukh, INDIA: see JORHAT, town.

Kokiu, China: see GEJIU.

Kök-Janggak (guhk–yuhng-GUHK), city (1999 population 10,727), S JALAL-ABAD region, KYRGYZSTAN, in SW foothills of FERGANA RANGE, 14 mi/23 km NE of JALAL-ABAD; 40°54′N 73°10′E. Coal-mining center; power plant; terminus (since 1932) of railroad from KARA-SUU. Developed after 1930. Also spelled Kok-Yangak.

Kokkina (KO-ki-nah) [Turkish=*Erenkoy*], village, LEFKOSIA district, NW CYPRUS, 44 mi/71 km W of NICOSIA, on MEDITERRANEAN SEA, in small enclave of Turkish occupied zone; 35°11′N 32°37′E. Attila Line to W, S, and E. Olives, vegetables, citrus; fish.

Kokkini Trimithia (KO-ki-nee tri-mee-THYAH), town, LEFKOSIA district, N central CYPRUS, 10 mi/16 km W of NICOSIA; 35°09′N 33°12′E. Citrus, grain, vegetables; livestock. Nicosia International Airport to E; Attila Line passes to N.

Kokkinovrachos (ko-kee-NO-vrah-khos), village, MAGNESIA prefecture, SE THESSALY department, N GREECE, 14 mi/23 km E of PHARSALA; 38°11′N 21°55′E. Chromite mining. Also spelled Kokkinovrakhos; formerly called Ardouan or Arduan.

Kokkola (KOK-ko-lah), Swedish *Gamlakarleby*, town, VAASAN province, W FINLAND, on the GULF OF BOTHNIA; 63°50′N 23°07′E. It is a port with lumber, machine, chemical, and leather industries. Chartered in 1620. Airport.

Kokkulam, INDIA: see TIRUMANGALAM.

Köklaks (CHUHRK-lahks), Finnish *Kauklahti*, village, UUDENMAAN province, S FINLAND, 13 mi/21 km W of HELSINKI; 60°11′N 24°37′E. Manufacturing includes glass, zinc oxide.

Koknese (KOK-nai-sai), German *Kokenhusen*, village, S central LATVIA, in VIDZEME, on right bank of the DVINA (DAUGAVA) River (scenic course), 12 mi/19 km W of PLAVINAS; 56°39′N 25°26′E. Tourist resort; castle ruins.

Koknur, INDIA: see KUKNUR.

Koko (KO-ko), town, DELTA state, S NIGERIA, port on BENIN River, 23 mi/37 km SSW of BENIN. Palm oil processing; hardwood, rubber, cacao, kola nuts.

Kokoda (ko-KO-dah), town, NORTHERN province, SE PAPUA NEW GUINEA, SE NEW GUINEA island, on YODDA River, 55 mi/89 km NE of PORT MORESBY, near mt. pass through OWEN STANLEY RANGE. NE terminus of KOKODA TRAIL from near Port Moresby; W terminus of road from POPONDETTA and coast of SOLOMON SEA. In World War II, taken 1942 by the Japanese in their unsuccessful drive on Port Moresby.

Kokoda Trail or **Kokoda Track** (both: ko-KO-dah), c. 56-mi/90-km jungle trail, CENTRAL and NORTHERN provinces, SE NEW GUINEA island, SE PAPUA NEW GUINEA. Originates in S at SOGERI, 29 mi/47 km by road E of PORT MORESBY, extends NNE across OWEN STANLEY RANGE to KOKODA, Northern province. Originally built during 1890s gold rush to reach gold fields of YODDA and KOKODA. During World War II, Japanese made their assault toward Port Moresby along the trail only to be defeated in bitter fighting by Australian forces backed by American air support. It is now a popular walking track requiring 5 to 10 days with moderate to hazardous difficulty.

Kokofata, village, FIRST REGION/KAYES, MALI, 34 mi/55 km WSW of KITA.

Koko Head, promontory and peak (642 ft/196 m), SE OAHU island, HONOLULU county, HAWAII, 9 mi/14.5 km ESE of HONOLULU, E of MAUNALUA BAY; Nonoula and Ihehelauakea craters to NE and E; Koko Crater is 2.5 mi/4 km to NE (botanic garden to N of crater) and Kohelepelepe peak (1,208 ft/368 m) is 2 mi/3.2 km to NE. Both overlook Halona Blowhole, water spout formed by incoming ocean waves. All of these features are within Koko Head Regional Park.

Kokolik River (ko-KO-lik), c.150 mi/241 km long, NW ALASKA; rises in NW BROOKS RANGE near 68°35′N 162°W, flows N to CHUKCHI SEA at 69°52′N 162°43′W.

Kokomo, city (2000 population 46,113), ⊙ HOWARD county, N central INDIANA, on WILDCAT CREEK; 40°29′N 86°08′W. Manufacturing (glass, motor vehicle parts, metal products, plastics, food and beverages, plumbing fixtures). The first commercially built automobile was invented and tested in Kokomo in 1894 by Elwood Haynes. Of interest is the Elwood Haynes Museum. Indiana University has a campus at Kokomo; branch of Indiana Vocational and Technical College (Ivy Tech). GRISSOM AIR BASE is nearby to N. Railroad junction. Incorporated 1865.

Kokonoe (ko-KO-NO-E), town, Kusu county, OITA prefecture, E KYUSHU, SW JAPAN, 25 mi/40 km W of OITA; 33°14′N 131°10′E.

Koko Nor, China: see QINGHAI HU and QINGHAI.

Kokopo, town (2000 population 20,262), NE NEW BRITAIN, BISMARCK ARCHIPELAGO, ⊙ EAST NEW BRITAIN province, PAPUA NEW GUINEA, SW PACIFIC OCEAN, on BLANCHE BAY, 15 mi/24 km SE of RABAUL. Coconut plantations. Was German capital of the archipelago. Provincial capital relocated here from Rabaul when that town suffered damage from volcanic activity in 1994. Formerly called HERBERTSHÖHE.

Kokorevka (KO-kuh-reef-kah), town, SE BRYANSK oblast, central European Russia, on road and railroad, 20 mi/32 km E of TRUBCHEVSK, and 37 mi/60 km S of BRYANSK; 52°35′N 34°16′E. Elevation 633 ft/192 m. Sawmilling; furniture.

Kokorinsko (KO-ko-RZHEEN-sko), Czech *Kokořínsko*, protected landscape region (□ 104 sq mi/270.4 sq km), SEVEROCESKY and STREDOCESKY provinces, N central BOHEMIA, CZECH REPUBLIC, 25 mi/40 km N of PRAGUE. It has sandstone rocks and cliffs in which water has hollowed out a number of canyon-type valleys. The rock formations have bizarre shapes. Another feature are so-called lids, an example of uneven disintegration of rocks. Noted 14th century Gothic castle Kokorin, Czech *Kokořín*, is located here.

Kokoro, town, COLLINES department, S central BENIN, 25 mi/40 km NE of SAVÉ near NIGERIA border, on road and railroad; 08°24′N 02°37′E. Cotton, groundnuts.

Kokoshkino (kuh-KOSH-kee-nuh), town (2006 population 9,795), central MOSCOW oblast, central European Russia, on railroad, 20 mi/32 km SW of MOSCOW; 55°35′N 37°10′E. Elevation 547 ft/166 m. Mostly residential.

Kokosing River (ko-KO-sing), c.50 mi/80 km long, central OHIO; rises in MORROW county, flows generally SE, past CHESTERVILLE, MOUNT VERNON, and GAMBIER, joining MOHICAN RIVER to form WALHONDING RIVER 16 mi/26 km NW of COSHOCTON; 40°21′N 82°09′W.

Kokoumbo (ko-KOOM-bo), town, Lacs region, S CÔTE D'IVOIRE, 120 mi/193 km NW of ABIDJAN; 06°33′N 05°15′W. Agriculture (beans, yams, vegetables); Gold mining; livestock (sheep, goats); manufacturing (palm oil, coffee, cacao).

Kökpekty (kuk-pek-TEE), village, E SEMEY region, KAZAKHSTAN, on KÖKPEKTY RIVER, 150 mi/241 km SE of SEMEY (Semipalatinsk); 48°47′N 82°28′E. Tertiary-level (raion) administrative center. Livestock.

Kökpekty River (kuk-pek-TEE), c.100 mi/161 km long, SE SEMEY region, KAZAKHSTAN; rises in KALBA RANGE, flows SE, past KÖKPEKTY, to Lake ZAISAN. Gold mining along upper course.

Kokrajhar (KO-kruh-juhr), district (□ 2,881 sq mi/7,490.6 sq km), ASSAM state, NE INDIA; ⊙ KOKRAJHAR.

Kokrajhar (KO-kruh-juhr), town, ⊙ KOKRAJHAR district, W ASSAM state, NE INDIA, on ROAD 11 mi/37 km NE of DHUBRI; 26°24′N 90°16′E. Rice, mustard.

Kokrek (kuh-KRYEK), village (2005 population 4,280), NW DAGESTAN REPUBLIC, SE European Russia, on railroad and near highway, 47 mi/76 km NW of MAKHACHKALA, and 6 mi/10 km E of KHASAVYURT; 43°14′N 46°43′E. Elevation 275 ft/83 m. In oil- and gas-producing area.

Kokrines (KAHK-rins), village, W ALASKA, on YUKON River, 80 mi/129 km WSW of TANANA, in Kokrines Hills.

Kok River, Thai *Mae Nam Kok*, Burmese *Nam Hkok*, c.150 mi/241 km long, BURMA and THAILAND, right affluent of MEKONG River Rises S of KENGTUNG (SHAN

State). Flows S, E, and NE, past CHIANG RAI, to the MEKONG SE of CHIANG SAEN.

Koksan (GOK-SAHN), county, NORTH HWANGHAE province, S NORTH KOREA, 50 mi/80 km ESE of PYONGYANG. Tobacco area.

Kokshaal-Tau (KUK-SHAH-AHL-TOU), Kyrgyz *Kaksha Ala-Too*, branch of TIANSHAN mountain system, on CHINA-KYRGYZSTAN border; extends from POBEDA PEAK 300 mi/483 km SW to CHATYR-KÖL (lake) area; rises to 17,380 ft/5,297 m. Main crossings are on TURUGART PASS and BEDEL PASS, both of which are China-Kyrgyzstan trade routes.

Kokshaga River, RUSSIA: see GREATER KOKSHAGA RIVER or LESSER KOKSHAGA RIVER.

Kökshetau (kuk-she-TOU), region (□ 30,150 sq mi/78,390 sq km), N KAZAKHSTAN; ⊙ KÖKSHETAU. On wooded steppe merging (S) into picturesque hilly lake region. Dry, sharply continental climate. Chiefly agriculture (wheat, oats, millet), with emphasis, in E, on livestock raising (cattle, sheep). Manufacturing includes building materials; metalworking. Trans-Kazakhstan railroad crosses region N-S. Pop. consists of Russians, Ukrainians, and Kazaks. Formed (1944) as Kokchetav oblast, Kazakh SSR. Also spelled Kokchetav.

Kökshetau (kuk-she-TOU), city (1999 population 123,389), ⊙ KÖKSHETAU region, KAZAKHSTAN, at S edge of wooded steppe, on railroad, 770 mi/1,239 km NNW of ALMATY; 53°18′N 69°25′E. Cattle- and sheep-raising area; cattle market. Tanning, food processing, distilling, consumer goods manufacturing; mechanical-repair plants. Founded (1824) during Russian conquest of Kazakhstan. Also spelled Kokchetav.

Koksijde (KAHK-sei-duh), commune (2006 population 21,117), Veurne district, WEST FLANDERS province, W BELGIUM, near NORTH SEA, 5 mi/8 km SW of NIEUWPOORT; 51°06′N 02°39′E. Seaside resort of Koksijde-Bad is 2 mi/3.2 km NNW, on NORTH SEA. Formerly spelled Coxyde.

Koksoak River (KAHK-so-ahk), 90 mi/145 km long, N QUEBEC, E CANADA; formed by CANIAPISCAU and LARCH rivers, 50 mi/80 km SW of KUUJJUAQ (FORT-CHIMO), flows NE and N, past KUUJJUAQ, to UNGAVA BAY 30 mi/48 km N of Kuujjuaq; 3 mi/5 km wide at mouth; 58°32′N 68°10′W.

Koksong (GOK-SUHNG), county (□ 210 sq mi/546 sq km), NE SOUTH CHOLLA province, SOUTH KOREA, adjacent to KURYE on E, TAMYANG and HWASUN to W, SUNGJU on S, and NORTH CHOLLA province on N. Mountainous (maximum elevation 2,297 ft/700 m) in SW; fields form in N along Sobaek and Okka river basins. Mainstream Sobaek starts here. Some agriculture (rice, barley, beans, cotton, sweet potatoes, hemp, apples, medical herbs, honey; mushrooms, chestnuts); fishery (sweet-smelt, mandarin fish, carp) in rivers. Aprok Resort at junction of SOMJIN and POSONG rivers.

Koksovyy (KOK-suh-viyee), town (2006 population 8,455), W central ROSTOV oblast, S European Russia, on the NORTHERN DONETS RIVER, on road and railroad spur, 5 mi/8 km NW of BELAYA KALITVA; 48°11′N 40°36′E. Elevation 400 ft/121 m. Coal mining; railroad terminus. Developed in 1932 to work coal deposits in SE part of the DONBAS.

Kokstad, town, KWAZULU-NATAL province, SOUTH AFRICA, on UMZIMHLAVA RIVER, 88 mi/140 km SW of PIETERMARITZBURG (MSUNDUZI), in foothills of DRAKENSBERG range; 30°30′S 29°27′E. Elevation 4,625 ft/1,420 m. Railroad terminus; just off N2 highway near KwaZulu-Natal border. Center of dairying, stock-raising region. Founded by Adam Kok III (d. 1875), last Griqua chief. Was capital of former GRIQUALAND EAST district. Boy Scout war memorial on Mount Currie (7,295 ft/2,224 m) 5 mi/8 km N of town.

Koktal (kok-TAHL), village, SE TALDYKORGAN region, KAZAKHSTAN, near Chinese border, on highway from

saryozek, 10 mi/16 km W of ZHARKENT. Irrigated agriculture (cotton, poppies); sheep.

Kök-Tash (guhk-TAHSH), town, JALAL-ABAD region, KYRGYZSTAN, 7 mi/12 km S of MAYLUU-SUU; 41°02′N 72°19′E.

Kokubu (ko-KOO-boo), city, KAGOSHIMA prefecture, S KYUSHU, SW JAPAN, 16 mi/25 km N of KAGOSHIMA; 31°44′N 130°46′E. Electronic equipment. Tobacco.

Kokubunji (ko-koo-BUN-jee), city, Tokyo prefecture, E central HONSHU, E central JAPAN, 121 mi/195 km W of SHINJUKU; 35°42′N 139°27′E. Suburb of Tokyo noted for its Kokubunji (Buddhist) temple founded in 1588.

Kokubunji (ko-koo-BUN-jee), town, Ayauta county, KAGAWA prefecture, NE SHIKOKU, W JAPAN, 5.6 mi/9 km S of TAKAMATSU; 34°17′N 133°57′E.

Kokubunji, town, Shimotsuga co., Tochigiprefecture, central Honshu, N central Japan, 12 mi/20 km S of Utsunomiya; 36°23′N 139°50′E.

Kokufu (KOK-foo), town, Yoshiki county, GIFU prefecture, central HONSHU, central JAPAN, 62 mi/100 km N of GIFU; 36°12′N 137°12′E.

Kokufu (KO-koo-foo), town, Iwami county, TOTTORI prefecture, S HONSHU, W JAPAN, 2.5 mi/4 km S of TOTTORI; 35°28′N 134°16′E. Fruits, daikon.

Kokuy (kuh-KOO-yee), town (2005 population 7,650), central CHITA oblast, SE SIBERIA, RUSSIA, on the SHILKA River, near highway, 7 mi/11 km W of SRETENSK; 50°38′N 117°53′E. Elevation 2,329 ft/709 m. Shipbuilding.

Kok-Yangak, KYRGYZSTAN: see KÖK-JANGGAK.

Kola (KO-lah), city (2006 population 10,025), NW MURMANSK oblast, NW European Russia, on the KOLA PENINSULA, at mouths of KOLA and TULOMA rivers, on the KOLA GULF of the BARENTS SEA, 7 mi/11 km S of MURMANSK; 68°53′N 33°01′E. Murmansk railroad junction (spur to Polyarnyy); in reindeer-raising area; furniture, food industries (beer, macaroni). Has wooden 17th-century churches. One of the oldest settlements on the Kola Peninsula; founded in 1264 in Novgorod principality.

Kola (KO-lah), town, GRAND GEDEH county, LIBERIA, 15 mi/24 km NE of ZWEDRU, close to GUINEA border; 06°18′N 08°00′W. Bauxite.

Kola (KO-lah), village, PWANI region, E TANZANIA, 20 mi/32 km WSW of DAR ES SALAAM; 06°59′S 38°58′E. Cashews, corn, sweet potatoes, bananas, vegetables; sisal; poultry, goats, sheep.

Kola, INDONESIA: see ARU ISLANDS.

Kolab River, INDIA: see SABARI RIVER.

Kolachel, INDIA: see COLACHEL.

Kola Deba, ETHIOPIA: see KOLADIBA.

Koladiba, town (2007 population 16,035), AMHARA state, NW ETHIOPIA, 15 mi/24 km SW of GONDAR; 12°25′N 37°19′E. Formerly spelled Kola Deba.

Kola Gulf (KO-luh), fjord-like ice-free inlet of the BARENTS SEA, NW KOLA PENINSULA, MURMANSK oblast, NW European Russia; 50 mi/80 km long, 0.5 mi/0.8 km wide; receives TULOMA and KOLA rivers. KOLA and MURMANSK on the E shore. Inlet N of Polyarny (W shore) was formerly known as Yekaterina [Russian=Catherine] Harbor.

Kolahun (KO-lah-hoon), town, LOFA county, LIBERIA, 150 mi/241 km NNE of MONROVIA; 08°15′N 10°04′W. Palm oil and kernels, cotton, pineapples; cattle raising. Radio station.

Kolaka (ko-LAH-kah), town, ⊙ Kolaka district, Sulawesi Tenggara province, 90 mi/145 km E of KENDARI. Naval base.

Kolambugan (koo-lahm-BOO-gahn), town, LANAO DEL NORTE province, W central MINDANAO, PHILIPPINES, on inlet of ILIGAN BAY, opposite Ozamiz (MISAMIS OCCIDENTAL); 08°05′N 123°54′E. Port; has sawmill. Rice, corn.

Kolambur (kol-ahm-buhr), town, TIRUVANNAMALAI district, TAMIL NADU state, S INDIA, on railroad (ARANI Road station), 5 mi/8 km SW of Arani. Trades in products (sandalwood, tanbark, tamarind) of JA-

VADI HILLS (W). Formerly called Aliabad (also spelled ALIYABAD or Alliyabad). Also spelled KALAMBUR.

Kola Peninsula (KO-lah) (□ 50,000 sq mi/130,000 sq km), MURMANSK oblast, extreme NW European Russia. Forming an E extension of the Scandinavian peninsula, it lies between the BARENTS SEA (N) and the WHITE SEA (S). In the NE, part is tundras; the SW area is forested. The peninsula has rich mineral deposits in the Khibiny mountains, which rise to approximately 4,000 ft/1,219 m. Hydroelectric plants have been built along the Tuloma, Voronya, and Niva rivers. The port of MURMANSK and the mining center of KIROVSK (iron, bauxite, phosphate, nickel) are the major cities of the peninsula. Along the coasts and in the mining centers, including the nickel and copper centers along the Norwegian and Swedish borders, the population is primarily Russian; in the interior are Lapps, who, until the Chernobyl disaster, subsisted largely on reindeer raising. Near Murmansk is the ancient town of Kola, founded in 1264 by the Slavs from Novgorod.

Kolapur, INDIA: see KOLHAPUR.

Kolar (KO-lahr), district (□ 3,175 sq mi/8,255 sq km), KARNATAKA state, S INDIA; ⊙ KOLAR. On DECCAN PLATEAU; undulating tableland, rising in W to c.4,800 ft/1,463 m at NANDI (health resort). Tobacco curing, tanning, tile and match manufacturing, goldsmithing; handicrafts (silk, woolen, and cotton weaving, pottery, glass bangles, wickerwork). Agriculture (silk, tobacco, sugarcane, cotton); dispersed kaolin and graphite workings. Chief towns are Kolar Gold Fields, Kolar, CHIK BALLAPUR, and CHINTAMANI.

Kolar (KO-lahr), city, ⊙ KOLAR district, KARNATAKA state, S INDIA, 115 mi/185 km NE of MYSORE, 15 mi/24 km NW of KOLAR GOLD FIELDS; 13°08′N 78°08′E. Road center; tobacco curing, tanning, hand-loom silk and cotton weaving, goldsmithing. Founded in the late 19th century, it is the center of the Indian gold-mining industry. The first hydroelectric project in S India was built in 1902 to provide electricity for the goldfields. Silk research station and sheep farm nearby.

Kolaras (ko-LAH-ruhs), town (2001 population 15,674), SHIVPURI district, MADHYA PRADESH state, N central INDIA, 14 mi/23 km SSW of SHIVPURI; 25°14′N 77°36′E. Millet, gram, wheat.

Kolar Gold Fields (KO-lahr), city (□ 30 sq mi/78 sq km), KOLAR district, KARNATAKA state, S INDIA, 45 mi/72 km E of BANGALORE. Railroad spur terminus; goldsmithing, tanning, tobacco curing; slaughterhouse. Center of INDIA's gold-mining industry; produces over 95% of India's gold (average annual output, 300,000 oz/8,505 kg) and some silver. Mines opened c.1885; powered since 1902 by hydroelectric plant near Sivasamudram Island (80 mi/129 km SW). City limits include mining areas of CHAMPION REEF (electrical transmission works), NANDIDRUG or Nandydroog (another Nandidrug, health resort, is near CHIK BALLAPUR), and OORGAUM or Urigam, industrial area of MARIKUPPAM (tile and brick manufacturing), and residential area of ROBERTSONPET (commercial college).

Kola River (KO-lah), approximately 44 mi/71 km long, in W central and NW MURMANSK oblast, NW European Russia. Rises at the N tip of Kolozero Lake and flows N, past Pulozero, Taybola, Kitsa, Magnetity, Shonguy, KIL′DINSTROY, and Molochnyy, to feed into the KOLA GULF of the BARENTS SEA at KOLA. Main source of drinking water for over 400,000 people. Abounds in salmon and trout.

Kolarovgrad, Bulgaria: see SHUMEN.

Kolarovo (ko-LAH-ro-VO), Slovak *Kolárovo*, town, ZAPADOSLOVENSKY province, SW SLOVAKIA, on GREAT RYE ISLAND, at junction of LITTLE DANUBE and VÁH rivers; 47°55′N 18°00′E. Railroad terminus; major agricultural center (barley, wheat, corn, sugar beets). Manufacturing of textiles, machinery, furniture; food processing. Formerly called Guta (gu-TAH), HUNGARIAN *Gúta*.

Kolarovo (ko-LAH-ro-vo), village, SOFIA oblast, PETRICH obshtina, BULGARIA; 41°22′N 23°06′E.

Kolarov Vruh (ko-LAH-rov VRUHK), peak (8,619 ft/ 2,627 m) in the E RILA MOUNTAINS, W BULGARIA, 14 mi/23 km SE of SAMOKOV; 42°11′N 23°47′E. Called Belmeken until 1949.

Kolasin (ko-LAHS-een), Serbian *Kolašin*, town (2003 population 2,989), E MONTENEGRO, on TARA RIVER, 30 mi/48 km NNE of PODGORICA. Local trade center; health and sport resort. BJELASICA mountain to NE. Under Turkish rule until 1878. Also spelled Kolashin.

Kolay, UKRAINE: see AZOVS′KE.

Kolayat (ko-LAH-yaht), town, tahsil headquarters, BIKANER district, N RAJASTHAN state, NW India, 26 mi/42 km SW of Bikaner; 27°50′N 72°57′E. Railroad spur terminus. Also spelled SRIKOLAYATJI.

Kolbäck (KOOL-BEK), town, VÄSTMANLAND county, central SWEDEN, on KOLBÄCKSÅN RIVER, on STRÖMSHOLM CANAL, 8 mi/12.9 km ENE of KÖPING; 59°34′N 17°48′E. Railroad junction.

Kolbäcksån (KOOL-BEKS-ON), river, 100 mi/161 km long, central SWEDEN; rises NW of LUDVIKA, flows generally SE, past Ludvika, SMEDJEBACKEN, FAGERSTA, and HALLSTAHAMMAR, to LAKE MÄLAREN 9 mi/ 14.5 km E of KÖPING. Canalized below Smedjebacken, it forms STRÖMSHOLM CANAL (68 mi/109 km long; built 1777–1795, enlarged 1842–1860); serves BERGSLAGEN region.

Kolba Range, KAZAKHSTAN: see KALBA RANGE.

Kolberg, POLAND: see KOŁOBRZEG.

Kolbermoor (kuhl-ber-MAWR), town, BAVARIA, S GERMANY, in UPPER BAVARIA, on Mangfall River; 47°51′N 12°05′E. Manufacturing (electrical machinery and equipment, tools, distilling). Chartered in 1963.

Kolbuszowa (kol-boo-SHO-vah), town, Rzeszów province, SE POLAND, 17 mi/27 km NNW of RZESZÓW. Flour milling, lumbering, distilling.

Kolchedan (kuhl-chee-DAHN), settlement, S SVERDLOVSK oblast, W Siberian Russia, in the ISET′ RIVER basin, on highway and railroad spur, 11 mi/18 km E, and under administrative jurisdiction of KAMENSKURAL′SKIY; 56°22′N 62°11′E. Elevation 377 ft/114 m. Manufacturing (concrete structures).

Kol′chino, UKRAINE: see KOL′CHYNE.

Kol′chugino (kuhl-CHOO-gee-nuh), city (2006 population 47,105), W VLADIMIR oblast, central European Russia, on the Peksha River (tributary of the KLYAZ′MA River), on crossroads and railroad (Peksha station), 47 mi/76 km WNW of VLADIMIR; 56°18′N 39°23′E. Elevation 495 ft/150 m. Nonferrous metallurgical center, based on the Urals copper ore; copper alloys, electrical cables, garments. Known as Kellerovo until 1931, when became a city and was renamed.

Kol′chugino, RUSSIA: see LENINSK-KUZNETSKIY.

Kol′chyne (KOL-chi-ne) (Russian *Kol′chino*), town, W central TRANSCARPATHIAN oblast, UKRAINE, on road and railroad spur, 2.5 mi/4 km NNE of MUKACHEVE; 48°28′N 22°46′E. Elevation 400 ft/121 m. Manufacturing (machine tools, asphalt, cement, stone crushing, food). Known since 1430; town since 1979.

Kolda, administrative region (□ 8,112 sq mi/21,091.2 sq km; 2004 population 893,867), S central SENEGAL; ⊙ KOLDA; 13°00′N 15°00′W. Bounded on N by THE GAMBIA, E by TAMBACOUNDA administrative region, S by GUINEA and GUINEA-BISSAU, W by ZIGUINCHOR administrative region. Agriculture (peanuts, cotton); forest products; livestock. Formerly part of CASAMANCE province.

Kolda, town, ⊙ KOLDA administrative region (2004 population 57,573), S SENEGAL, on upper CASAMANCE River, 90 mi/145 km ENE of ZIGUINCHOR; 12°53′N 14°57′W. Peanut, rice, cotton production; livestock raising. Airfield, zoological research station.

Kolding (KOL-ding), city (2000 population 53,447), VEJLE county, S central DENMARK, port on KOLDING FJORD, arm of the LILLE BÆLT; 55°32′N 09°28′E. It is a commercial, industrial, and fishing center that pro-duces ships, machinery, and textiles. Of note in the city are Koldinghus, a royal castle built in 1248 that now houses a historical museum, and the oldest stone church (built in the thirteenth century) in Denmark.

Kolding Fjord (KOL-ding-fyor), inlet of the LILLE BÆLT, E JUTLAND, DENMARK, c.8 mi/13 km long. FÆNØ island at mouth. KOLDING city at head.

Koldyban′, RUSSIA: see KRASNOARMEYSKOYE, SAMARA oblast.

Kole (KO-lai), village, KASAI-ORIENTAL province, central CONGO, on LUKENIE RIVER, 125 ft/38 m NNW of LUSAMBO; 03°47′S 22°38′E. Elev. 2,001 ft/609 m. Steamboat landing; cotton ginning.

Koléa (ko-lai-AH), town, TIPAZA wilaya, N central ALGERIA, in the *sahel* (coastal area), 20 mi/32 km SW of ALGIERS; 36°42′N 02°46′E. Agricultural trade center (fruits, vegetables); embroidering. Founded 1550, town is a Muslim pilgrimage center. Destroyed by earthquake in 1825.

Kolendo (kuh-lyen-DO), settlement, NW SAKHALIN oblast, in the N central part of SAKHALIN Island, RUSSIAN FAR EAST, on local road, 13 mi/21 km NNE of OKHA; 53°46′N 142°47′E. On-shore oil drilling. Has a museum of native culture.

Koleno, RUSSIA: see YELAN′-KOLENOVSKIY.

Kolenté (KO-len-tai), town, Kindia prefecture, Kindia administrative region, SW GUINEA, in Guinée-Maritime geographic region, 20 mi/32 km E of KINDIA, on road and railroad; 10°06′N 12°37′W. Bananas.

Kolenté River, Afica: see GREAT SCARCIES RIVER.

Kolguyev (kul-GOO-yef), island (□ 1,350 sq mi/3,510 sq km), off NENETS AUTONOMOUS OKRUG, ARCHANGEL oblast, NE European Russia, in the BARENTS SEA, E of the KANIN peninsula, 50 mi/80 km from the mainland. A tundra region inhabited mainly by the Nenets (Samoyeds), who engage in fishing, seal hunting, reindeer raising, and trapping. Burgino is the major settlement.

Kolhapur (KOL-hah-puhr), former princely state in what is now MAHARASHTRA state, W central INDIA. A center of the Marathas, Kolhapur was an important state of the DECCAN. Transferred to Maharashtra state in 1960.

Kolhapur (KOL-hah-puhr), district (□ 2,967 sq mi/ 7,714.2 sq km), MAHARASHTRA state, W central INDIA; ⊙ KOLHAPUR. On W edge of DECCAN PLATEAU; crossed N-S by WESTERN GHATS; bounded N by SATARA district, NE by KRISHNA River, S by KARNATAKA state. A hilly region with extensive forests. Hand-loom weaving, pottery and hardware manufacturing. Agriculture (rice, millet, sugarcane, tobacco, oilseeds). Teak, sandalwood, blackwood, myrobalan; iron ore and bauxite deposits (N). Kolhapur is a large trade center; other market towns are ICHALKARANJI, Gadhinglaj, and SHIROL. Formed in 1949 out of most of former Kolhapur state.

Kolhapur (KOL-hah-puhr), city (2001 population 505,541), ⊙ KOLHAPUR district, MAHARASHTRA state, W central INDIA, 180 mi/290 km SSE of MUMBAI; 16°42′N 74°13′E. Railroad terminus and road junction; commercial center and agriculture market (millet, rice, sugarcane, tobacco); trades in grain, cloth fabrics, timber; cotton, sugar, and oilseed milling, hand-loom weaving, manufacturing of pottery, matches; tobacco factory, motion picture studios. Religious and cultural center; seat of college, university, and law and technical schools. Has many Buddhist remains, notably a stupa, or shrine (third century B.C.E.), with inscriptions in characters of the Asoka period. Well-known Ambabai (Durga) temple here. Was capital of former princely state of Kolhapur belonging to descendants of Maratha leader Shivaji.

Kolhapur (KOL-hah-puhr), town, MAHBUBNAGAR district, ANDHRA PRADESH state, SE INDIA, near KRISHNA RIVER, 50 mi/80 km SE of MAHBUBNAGAR; 16°06′N 78°18′E. Rice and oilseed milling. Also spelled Kolapur and Kollapur.

Kolhapur and Deccan States Agency, INDIA: see DECCAN STATES.

Kolhumadulu Atoll, S central group of MALDIVES, in INDIAN OCEAN, between 02°10′N 72°18′E and 02°33′N 73°54′E. Coconuts, fishing.

Koliganek (ko-li-GAH-nek), village (2000 population 182), SW ALASKA, on NUSHAGAK RIVER, 65 mi/105 km NE of DILLINGHAM; 59°49′N 157°26′W.

Kolijnsplaat, NETHERLANDS: see COLIJNSPLAAT.

Kolín (KO-leen), city (2001 population 30,258), STREDOCESKY province, N central CZECH REPUBLIC, in BOHEMIA, on the ELBE RIVER; 50°02′N 15°12′E. It is a river port with metal and chemical industries. Has a petroleum refinery, a hydroelectric station, and brewery (established in 1547). Founded in the 13th century, Kolín grew rapidly after the construction (19th century) of the Vienna-Prague railroad. The 13th century Church of St. Bartholomew is noted for its Gothic choir. Site of annual music festival.

Kolindros (ko-leen-[TH]ROS), town, THESSALONÍKI prefecture, CENTRAL MACEDONIA department, NE GREECE, 25 mi/40 km WSW of THESSALONÍKI; 40°29′N 22°29′E. Wheat, cotton, tobacco. Winery.

Kolka, Cape (KOL-kuh), Latvian *kolkasrags*, Swedish *Domesnäs*, headland of W LATVIA, on Gulf of RIGA, at E end of IRBE STRAIT; 57°45′N 22°36′E.

Kolkasrags, LATVIA: see KOLKA, CAPE.

Kolkata, INDIA: see CALCUTTA.

Kolkhozabad (kuhl-HOZ-ah-bahd), town, KHATLON viloyat, TAJIKISTAN, in Vakhsh valley, 18 mi/29 km SSW of QURGHONTEPPA (linked by narrow-gauge railroad); long-staple cotton, vegetables; metalworks. Developed in 1930s. Until 1950s, called Kaganovichabad.

Kolkhozabad, TAJIKISTAN, see VOSE.

Kolkhoznoye, RUSSIA: see ARGUN.

Kolki, UKRAINE: see KOLKY.

Kolkwitz (KOLK-vits), village, SAXONY, E GERMANY, just W of COTTBUS; 51°45′N 14°16′E.

Kolky (kol-KEE) (Russian *Kolki*) (Polish *Kolki*) town (2004 population 4,500), E VOLYN′ oblast, UKRAINE, on STYR RIVER (head of navigation), 30 mi/48 km NNE of LUTS′K; 50°00′N 28°04′E. Elevation 797 ft/242 m. Lumber-trading center; woodworking, feed milling, flour milling, canning. Established in the 16th century, town since 1940.

Kollaimalai Hills (kol-LEI-muh-lei), outlier of S EASTERN GHATS, TAMIL NADU state, S INDIA, SE of SALEM, separated from PACHAIMALAI HILLS (E) by river valley; 18 mi/29 km long, 12 mi/19 km wide. Rise to over 4,600 ft/1,402 m. Magnetite mined here and in S outlying hill of TALAMALAI. Agriculture products (rice, cotton, plantains, oranges).

Kollam (ko-LUHM), district (□ 962 sq mi/2,501.2 sq km), KERALA state, S INDIA; ⊙ KOLLAM. Formerly spelled Quilon.

Kollam (ko-LUHM), city (2001 population 380,091), ⊙ KOLLAM district, KERALA state, S INDIA, on the ARABIAN SEA; 08°53′N 76°36′E. Market for coconut products, spices, tea, coffee, and rice. Mineral processing and other manufacturing are important. The oldest city on the MALABAR COAST. In the 7th century it was noted by a Nestorian patriarch as the southernmost point of Christian influence in India. By the time the Dutch occupied Kollam in 1662, the PORTUGUESE had already established a factory here. Soon after the Dutch came, the British East India Company took control. Formerly spelled QUILON.

Kollapur, INDIA: see KOLHAPUR.

Kölleda (kuh-LAI-dah), town, SAXONY-ANHALT, central GERMANY, 15 mi/24 km NNW of WEIMAR; 51°12′N 11°44′E. Electronics and technology industry. Has Gothic church. Formerly spelled Cölleda.

Kollegal (ko-LE-gahl), town, MYSORE district, KARNATAKA state, SW INDIA, on KAVERI RIVER, 33 mi/53 km ESE of MYSORE; 12°09′N 77°07′E. Major silk-growing and manufacturing center; supplies raw silk to S Indian filatures; produces silk of parachute quality; gold cloth manufacturing, cotton milling. Noted Cauvery Falls and hydroelectric works 9 mi/14.5 km NE, at Sivasamudram.

Kolleru Lake, INDIA: see COLAIR LAKE.

Kölliken (KUH-li-kuhn), commune, AARGAU canton, N SWITZERLAND, 4 mi/6.4 km S of AARAU. Textiles, tiles.

Kollipara (ko-LEE-pah-rah), town, GUNTUR district, ANDHRA PRADESH state, SE INDIA, in KRISHNA RIVER delta, 20 mi/32 km E of GUNTUR. Rice milling; tobacco.

Kollum (KAW-lum), town, FRIESLAND province, N NETHERLANDS, 7 mi/11.3 km ESE of DOKKUM; 53°17′N 06°09′E. Dairying; cattle, sheep; vegetables, grain.

Kollur, INDIA: see SATTENAPALLE.

Kolluru (ko-LOO-roo), town, GUNTUR district, ANDHRA PRADESH state, SE INDIA, in KRISHNA RIVER delta, on navigable irrigation canal, 25 mi/40 km ESE of GUNTUR. Rice milling; tobacco. KOLLURU Road railroad station is 3 mi/4.8 km W. Also spelled Kollur.

Kolmar, FRANCE: see COLMAR.

Kolmar, POLAND: see CHODZIEZ.

Kolmården (KOOL-MORD-en), mountainous region, on border of SÖDERMANLAND and ÖSTERGÖTLAND counties, E central SWEDEN; 58°42′N 16°29′E. Safari park (natural habitat and amusement park).

Kolmården (KOOL-MORD-en), village, ÖSTERGÖTLAND county, SE SWEDEN, on N coast of Bråviken, 30-mi/48-km-long inlet of BALTIC SEA, 8 mi/12.9 km NE of NORRKÖPING; 58°43′N 16°11′E. Wildlife park and zoo nearby.

Kolmogorovo (kuhl-muh-GO-ruh-vuh), village, NW KEMEROVO oblast, S central Siberian Russia, on the TOM′ River, on highway, 29 mi/47 km NW of KEMEROVO; 55°38′N 85°40′E. Elevation 360 ft/109 m. Poultry processing.

Köln, Germany: see COLOGNE.

Kolno (KOL-no), Russian *Kol′no*, town, ŁOMŻA province, NE POLAND, 18 mi/29 km N of ŁOMŻA. Railroad spur terminus; flour milling, cement manufacturing.

Kolo, town, TILLABÉRY province, NIGER, 22 mi/35 km E of NIAMEY; 13°14′N 02°20′E. Administrative center.

Kolo (KO-lo), Polish *Koło*, town (2002 population 23,552), Konin province, central POLAND, on WARTA River, 45 mi/72 km NW of ŁÓDŹ. Railroad junction; manufacturing (cement, pottery, agricultural machinery, roofing materials, hosiery, flour). In World War II, under German control, called Warthbrücken.

Kolo (KO-lo), village, BAS-CONGO province, W CONGO, on railroad, 100 mi/161 km ENE of BOMA; 05°27′S 14°52′E. Elev. 1,755 ft/534 m. Agricultural center with cattle farms, coffee and elaeis-palm plantations. Roman Catholic mission. Airfield. Cattier railroad workshops are 10 mi/16 km ENE.

Kolo (KO-lo), village, DODOMA region, N central TANZANIA, 100 mi/161 km N of DODOMA, near Bubu River; 04°44′S 35°48′E. Cattle, sheep, goats; corn, wheat.

Koloa (KO-LO-ah), town (2000 population 1,942), SE KAUAI, KAUAI county, HAWAII, 7 mi/11.3 km WSW of LIHUE, 1.5 mi/2.4 km N S coast at point where Waihohonu and Omao streams join to form Waikomo Stream; 21°54′N 159°27′W. Koloa Landing at coast at mouth of Waikomo; Koloa Mill, Hawaii's oldest (1837) sugar plantation, 1 mi/1.6 km to E; Lihue-Koloa Forest Reserve to NW; National Tropical Botanical Garden to W at Lawai; Waita Reservoir to NE; Mauka Reservoir to N; Poipa Beach Park at POIPU and SE; Spouting Horn, a coastal water spout, is at Spouting Horn Beach Park to SW. Oldest Catholic church in Hawaii.

Kolob Canyon, UTAH: see KOLOB TERRACE.

Kolobovo (KO-luh-buh-vuh), town (2005 population 2,875), S IVANOVO oblast, central European Russia, on a highway branch and near railroad (Ladyginskaya station), 10 mi/16 km SSW of SHUYA; 56°42′N 41°20′E. Elevation 341 ft/103 m. Cotton textile factory.

Kołobrzeg (ko-WOB-zeg) or **Kolberg**, town and resort (2002 population 44,947), NW POLAND, on the BALTIC SEA at the mouth of the PROŚNICA River. It is a seaport, seaside resort, and railroad junction. A salt-trading center in the Middle Ages, it was chartered in 1255. It was besieged three times by the Russians in the Seven Years War before it fell in 1761. Kołobrzeg was virtually obliterated during World War II.

Kolob Terrace (KO-lahb), deeply dissected plateau, primarily in NE WASHINGTON county, SW UTAH, extends N into IRON county, extending W from MARKAGUNT PLATEAU and S from CEDAR CITY; bounded W by jagged 3,000-ft/914-m escarpment. Rises to maximum elevation of 9,000 ft/2,743 m in N; includes colorful Kolob Canyon (1,500 ft/457 m–2,500 ft/762 m) and Horse Pasture Plateau in S, both part of ZION NATIONAL PARK. Also called Kolob Plateau.

Koločep Island (KO-lo-chep), Italian *Calamotta*, ancient *Calaphodia*, Dalmatian island in ADRIATIC SEA, S CROATIA, 5 mi/8 km WNW of DUBROVNIK; c.2 mi/3.2 km long. A portion (□ 110 sq mi/310 sq km) of its forest has been designated a park.

Kolodeznyy (kuh-LO-deez-niyee), settlement (2006 population 5,830), W central VORONEZH oblast, S central European Russia, on the VORONEZH RIVER, near railroad spur, 2 mi/3.2 km W of NOVOVORONEZH; 51°20′N 39°11′E. Elevation 344 ft/104 m. Agricultural processing, fishing. Summer homes.

Kolodnya (kuh-LO-dnyah), former town, W central SMOLENSK oblast, W European Russia, now a suburb of SMOLENSK, 4 mi/6 km E of the city center; 55°48′N 34°04′E. Elevation 748 ft/227 m.

Kolofata (ko-lo-FAH-tah), town, Far-North province, CAMEROON, 42 mi/68 km NNW of MAROUA; 11°09′N 14°03′E.

Kologriv (kuh-luh-GREEF), city (2005 population 3,575), N KOSTROMA oblast, central European Russia, on the UNZHA RIVER (landing), a tributary of the VOLGA RIVER, on road, 210 mi/338 km NE of KOSTROMA, and 75 mi/121 km ENE of GALICH; 58°49′N 44°18′E. Elevation 456 ft/138 m. Sawmilling, dairy. Chartered in 1609.

Kolokani, town, SECOND REGION/KOULIKORO, S MALI, 77 mi/124 km N of BAMAKO; 13°35′N 08°02′W. Peanuts, shea nuts; livestock. Manufacturing of native pottery.

Kolomak (ko-lo-MAHK), town (2004 population 3,830), W KHARKIV oblast, UKRAINE, on Kolomak River, left tributary of VORSKLA RIVER, 40 mi/64 km WSW of KHARKIV; 49°50′N 35°18′E. Elevation 436 ft/132 m. Raion center. Sugar refining, food processing; bricks. Known since 1571, town since 1959.

Kolombangara (kaw-lawm-BAHNG-ah-rah), extinct volcanic island (□ 263 sq mi/683.8 sq km), NEW GEORGIA, SOLOMON ISLANDS, SW Pacific, 5 mi/8 km W of main island of New Georgia, across KULA GULF; 08°00′S 157°05′E. Island is 20 mi/32 km long, 15 mi/24 km wide; rises to 5,449 ft/1,661 m. Timber logging. Sometimes spelled Kulambangra. Also called Nduke.

Kolomea, UKRAINE: see KOLOMYYA.

Kolomna (kuh-LOM-nah), city (2006 population 146,895), SE MOSCOW oblast, central European Russia, at the confluence of the MOSKVA and OKA rivers, on crossroads and railroad, 70 mi/113 km SE of MOSCOW; 55°05′N 38°46′E. Elevation 472 ft/143 m. Locomotives, heavy machine tools, textile machinery, furniture, asphalt and concrete, industrial chemicals (paints, varnishes), synthetic rubber, food industries (meat packing, bakery, confectionery). Known since 1177 as a border fortress in Ryazan′ principality; became a Muscovite outpost in 1301; rally point for the Russian armies in 1301 and 1472 during campaigns against the Golden Horde, and in 1547 during the campaign against the Kazan′ khanate. A stone kremlin (fortifications) was built in 1525–1531, the remains of which are still standing. An industrial center since 1863, with the construction of the Moscow-Saratov railroad that

passed through the city. Lost more than half of its population during the 1918–1921 civil war and rebuilt in the 1930s through amalgamation with six neighboring villages. An important troops training center and rally point for the Soviet army during World War II. Considered by many to be a museum city due to the presence of 420 federally and regionally recognized historical landmarks and monuments within its limits.

Kolomyya (ko-lo-MEE-yah) (German *Kolomea*) (Polish *Kolomyja*), city (2001 population 254,103), SW IVANO-FRANKIVS′K oblast, W UKRAINE, on the PRUT RIVER, in the CARPATHIAN foothills; 48°32′N 25°02′E. Elevation 915 ft/278 m. It is a raion center, railroad junction, and agricultural trade center. Industries include food processing, woodworking, and light manufacturing. It is especially known for Hutsul crafts (wood carving, weaving, kilims, embroidery, ceramics). Museum of Hutsul folk art; four vocational-technical schools; wooden church (1587). First mentioned in 1240, Kolomyya was a Ukrainian settlement in the Halych-Volyn principality. It passed in the 14th century to the Poles, who fortified it; received the rights of the Magdeburg Law in 1405. Kolomyya was taken by Austria during the Polish partition of 1772 and became part of the newly independent West Ukrainian National Republic in 1918 but reverted to Poland in 1920. It was incorporated into the Ukrainian Soviet Socialist Republic (SSR) in 1939, and is now part of independent Ukraine (since 1991). Jewish community since the 16th century, received full citizenship rights under Austria-Hungary rule in 1867. The 1890–1900 wave of immigration and pogroms in 1905 and 1918–1920 decimated the community, but it still numbered more than 18,000 in 1939; destroyed by the Nazis in 1941—fewer than 1,000 Jews remaining in 2005.

Kolondiéba (ko-lon-JAI-ba), town, THIRD REGION/SIKASSO, MALI, 78 mi/130 km WSW of SIKASSO; 11°05′N 06°54′W.

Kolonedale, INDONESIA: see KOLONODALE.

Kolonelsdiep (kaw-law-NELZ-deep), canal, FRIESLAND province, N NETHERLANDS; extends 7 mi/11.3 km E-W from E shore of BERGUMERMEER (W) to the HOENDIEP canal at Stroobos, 10 mi/16 km ESE of DOKKUM. Now part of PRINSES MARGRIET CANAL.

Kolones, Cape, Greece: see SOUNION, CAPE.

Kolonje, ALBANIA: see ERSEKË.

Kolonnawa (KO-LON-nah-wuh), E suburb of COLOMBO, WESTERN PROVINCE, SRI LANKA, on tributary of the KELANI GANGA River, 3 mi/4.8 km E of city center; 06°55′N 79°53′E. Oil refinery.

Kolonodale (ko-LO-no-dah-lai), town, E Sulawesi Tengah province, INDONESIA, port at head of Towarr Bay (small inlet of Gulf of TOLO), 250 mi/402 km NE of UJUNG PANDANG; 02°00′S 121°19′E. Trade center, shipping resin and rattan. Sometimes spelled Kolonedale.

Kolonyama (ko-LON-yah-mah), village, LERIBE district, NW LESOTHO, 23 mi/37 km NNE of MASERU, 6 mi/9.7 km N of TEYATEYANENG, near Phuthiatsana River; 29°04′S 27°45′E. Kolonyama Handicraft Centre is here; manufacturing (pottery, handicrafts). Cattle, goats, sheep; corn, sorghum, vegetables.

Kolosevka, UKRAINE: see KUDRYAVTSIVKA.

Kolosia (ko-LO-see-ah), village, RIFT VALLEY province, W KENYA, 70 mi/113 km NE of KITALE; 01°40′S 35°47′E. Livestock raising; cotton, peanuts, sesame, corn.

Kolosivka, UKRAINE: see BEREZIVKA.

Kolosivka, UKRAINE: see KUDRYAVTSIVKA.

Kolosjoki, RUSSIA: see NIKEL′.

Kolosovka (kuh-luh-SOF-kah), village (2006 population 5,735), central OMSK oblast, SW SIBERIA, RUSSIA, on the OSHA RIVER, on crossroads, 40 mi/64 km SW of TARA; 56°28′N 73°36′E. Elevation 285 ft/86 m. In dairy-farming area; grain storage. Until about 1940, called Nizhne-Kolosovskoye.

Area is shown by the symbol □, and capital city or county seat by ☉.

Kolozs, ROMANIA: see COJOCNA.

Kolozsborsa, ROMANIA: see BORŞA.

Kolozsvár, ROMANIA: see CLUJ-NAPOCA.

Kolp' (KOLP) river, 157 mi/254 km long, NW European Russia, right tributary of SUDA RIVER (VOLGA RIVER basin); rises in LENINGRAD oblast in the Vepsovskaya Elevation and flows through the Mologa-Sheksna Lowland in VOLOGDA oblast, passing BABAYEVO. Basin area, 1,440 sq mi/3,730 sq km. Timber floating.

Kolpashevo (kuhl-PAH-shi-vuh), city (2006 population 27,690), central TOMSK oblast, W SIBERIA, RUSSIA, port on the OB' RIVER, on road, 200 mi/322 km NW of TOMSK; 58°19′N 82°54′E. Elevation 209 ft/63 m. Fish canning; metalworks, lumbering, ship building and repair. Arose in the 17th century. Was capital of former Narym okrug. Made city in 1938.

Kölpin Lake (KUHL-pin), German *Kölpinsee* (KUHL-pin-ZAI), lake (□ 8 sq mi/20.8 sq km), Mecklenburg-Western Pomerania, N GERMANY, just NW of MÜRITZ LAKE, 3 mi/4.8 km W of WAREN; 53°30′N 12°35′E. Traversed by ELDE RIVER; 5 mi/8 km long, 1 mi/1.6 km–3 mi/4.8 km wide; greatest depth 100 ft/30 m, average depth 13 ft/4 m.

Kolpino (KOL-pee-nuh), city (2005 population 135,895), central LENINGRAD oblast, NW European Russia, on the IZHORA RIVER, 16 mi/26 km SE of SAINT PETERSBURG, of which it is a suburb (connected by railroad); 59°45′N 30°36′E. The largest industrial suburb of St. Petersburg; machine manufacturing; site of Izhora naval construction works (founded in 1722) producing machine tools, steam engines (since the early 19th century), ships (since 1863), equipment for atomic power plants; steel-rolling mills; mica factory; sewing mill.

Kolpna, RUSSIA: see KOLPNY.

Kolpny (KOLP-ni), town (2006 population 7,050), S ORËL oblast, SW European Russia, on the SOSNA RIVER (tributary of the DON), on road, 39 mi/63 km N of KURSK (KURSK oblast), and 26 mi/42 km SW of LIVNY; 52°13′N 37°02′E. Elevation 495 ft/150 m. Railroad terminus; cardboard and printing combine, sugar refining, dairy products, canning, hemp milling. Also known as Kolpna.

Kolpur (kol-puhr), village, KALAT district, BALUCHISTAN province, SW PAKISTAN, at W end of BOLAN PASS, 21 mi/34 km SSE of QUETTA; 29°54′N 67°08′E. Coal mined nearby (E).

Kolsva (KOOLS-vah), town, VÄSTMANLAND county, central SWEDEN, 7 mi/11.3 km NW of KÖPING; 59°36′N 15°51′E. Manufacturing (steelworks since sixteenth century).

Koltakenges, RUSSIA: see BORISOGLEBSK, MURMANSK oblast.

Kol'tsovo (kuhl-TSO-vuh), town (2006 population 14,415), S SVERDLOVSK oblast, E URALS, W Siberian Russia, on railroad and near highway, 12 mi/20 km SW of YEKATERINBURG; 56°46′N 60°49′E. Elevation 813 ft/247 m. Railroad station; airport. Asphalt and reinforced concrete forms.

Kol'tsovo (kuhl-TSO-vuh), urban settlement (2006 population 9,410), E NOVOSIBIRSK oblast, SW Siberian Russia, on the OB' RIVER, on railroad spur, 12 mi/19 km E of NOVOSIBIRSK, to which it is administratively subordinate; 54°55′N 83°15′E. Elevation 659 ft/200 m. Advanced virology and biotechnology research enterprises.

Koltubanovskiy (kuhl-too-BAH-nuhf-skeeye), town (2006 population 3,825), W ORENBURG oblast, SE European Russia, on the Borovka River (tributary of the SAMARA River), on the administrative border with SAMARA oblast, on railroad (Koltubanka station) and local road, 13 mi/21 km NE of BUZULUK; 52°57′N 52°02′E. Elevation 203 ft/61 m. Sawmilling; furniture making.

Koltur (KUL-tuhr), Danish *Kolter*, island (□ 1 sq mi/2.6 sq km) of the W FAEROE Islands, separated from SW STREYMOY by Hestfjørður. Rises to 1,568 ft/478 m.

Kolubara, Serbian *Kolubarski Okrug*, district (□ 955 sq mi/2,483 sq km; 2002 population 192,204), ⊙ VALJEVO, W central SERBIA; 44°15′N 19°53′E. Includes municipalities (*opštinas*) of LAJKOVAC, Ljig, MIONICA, Osecina, UB, and Valjevo. Agriculture, food processing.

Kolubara River (KOL-u-ba-rah), c.50 mi/80 km long, W SERBIA; rises SW of VALJEVO; flows NNE, past OBRENOVAC, to SAVA River 2 mi/3.2 km E of OBRENOVAC.

Koluk, TURKEY: see KAHTA.

Kolushkino, RUSSIA: see YEFREMOVO-STEPANOVKA.

Koluszki (ko-LOOSH-kee), Russian *Kolyushki*, town, Piotrków Trybunalski province, central POLAND, 15 mi/24 km E of Łódź railroad junction.

Koluton (ko-loo-TON), town, central AKMOLA region, KAZAKHSTAN, on Koluton River (tributary of the ISHIM RIVER), on South Siberian railroad, 40 mi/64 km E of ATBASAR; 51°45′N 69°45′E. Bauxite deposits nearby.

Kolva River (kuhl-VAH), approximately 290 mi/467 km long, PERM oblast, RUSSIA; rises in the N URAL Mountains, on NW slope of the ISHERIM Mountain; flows S and generally W, through coniferous taiga, and S, past CHERDYN', to the Vishera River just S of CHERDYN'. Timber floating; seasonal tug navigation for approximately 160 mi/257 km in the lower course. Receives Visherka (right) and Berezovaya (left) rivers. Once part of a trade route connecting Pechora and Kama rivers.

Kolva River (kuhl-VAH), 339 mi/546 km long, passing through NENETS AUTONOMOUS OKRUG, ARCHANGEL oblast, and KOMI REPUBLIC, N European Russia; right tributary of the USA RIVER (PECHORA River basin); originates in the Yaneymusyur Upland in the NE corner of Archangel oblast and takes a meandering course S through the BOL'SHEZEMEL'SKAYA TUNDRA into N KOMI Republic. Basin area 6,988 sq mi/18,100 sq km.

Kolvitsa (KOL-vee-tsah), lake (□ 47 sq mi/122.2 sq km), MURMANSK oblast, in SW KOLA PENINSULA, extreme NW European Russia. Elevation 200 ft/61 m; average depth 39 ft/12 m, maximum depth 66 ft/20 m. The Kolvitsa River originates here, its flow regulated by a dam for timber floating.

Kolwezi (kol-WE-zee), city, KATANGA province, SE CONGO; 10°43′S 25°28′E. Elev. 4,753 ft/1,448 m. It is a center for copper and cobalt mining. There are copper-ore concentration plants, a zinc refinery, and a brewery. SHABA rebels attacked the city in 1978, flooding the mines. Sharp declines in mineral production.

Kolyanur (kuh-lyah-NOOR), town, S KIROV oblast, E central European Russia, in the VYATKA River basin, on road, 9 mi/14 km SSW of SOVETSK, to which it is administratively subordinate; 57°25′N 48°48′E. Elevation 452 ft/137 m. Agriculture (wheat, oats, flax, potatoes, vegetables). Sand and gravel quarries in the vicinity.

Kolybel'skoye (kuh-li-BYEL-skuh-ye), village, E LIPETSK oblast, S central European Russia, in the VORONEZH RIVER basin, on road, 11 mi/18 km SSW of CHAPLYGIN; 53°06′N 39°56′E. Elevation 383 ft/116 m. In agricultural area (grains, vegetables, potatoes, sunflowers); food processing.

Kolyberovo (kuh-li-BYE-ruh-vuh), former town, E central MOSCOW oblast, Russia, on the MOSKVA River, on railroad (Moskvoretskaya station), 29 mi/46 km SE of MOSCOW, and 12 mi/19 km N of KOLOMNA; 55°16′N 38°44′E. Elevation 406 ft/123 m. Cement-making center. Now incorporated into VOSKRESENSK.

Kolyma (kuh-li-MAH), river, approximately 1,500 mi/2,414 km long, NE SIBERIA, RUSSIAN FAR EAST; rising in several headstreams in the KOLYMA and Cherskogo ranges. Rising in MAGADAN oblast, it flows generally N, across the E edge of the SAKHA REPUBLIC, entering the ARCTIC OCEAN at NIZHNIYE KRESTY. Navigable (June–October) for approximately 1,000 mi/1,609

km. Its upper course crosses the rich KOLYMA Gold Fields, which supplied much of the gold for Soviet foreign trade. Gold mining was begun in the 1930s, and both the fields and the surrounding area were developed using the labor of prisoners from Stalin's Gulag camp system.

Kolyma Gold Fields, RUSSIA: see KOLYMA, river.

Kolyma Highlands or **Kolyma Range** (kuh-li-MAH), system of ranges, plateaus, and ridges dividing the KOLYMA River basin (W) from the headwaters of the ANADYR' and Penzhina river watersheds (E), MAGADAN oblast, NE SIBERIA, RUSSIAN FAR EAST, extending NNE from just NE of MAGADAN to the headwaters of the Anyui River at the ARCTIC CIRCLE. The S part follows the W shore of SHELEKHOV GULF. Plateaus prevail in the SW (elevation 4,921 ft/1,500 m–5,906 ft/1,800 m) and the Omzukchan Range reaches 6,234 ft/1900 m. The Maimandzhinskiy Range in the NE reaches 5,906 ft/1,800 m. The Oloy Plateau (N) is crossed by the Kolyma Highway. In the W are gold, tin, rare metallic ores; hard coal and lignite; thermal springs. Continental climate, with dry summers. About two-thirds of the area is tundra without forest.

Kolyma Lowland (kuh-li-MAH), low-lying plain, NE SAKHA REPUBLIC, N RUSSIAN FAR EAST, in the basin of the Alazeya and Bol'shaya Chukoch'ya rivers and on the left bank of the lower reaches of the KOLYMA River, stretching 465 mi/750 km along the Kolyma, from the EAST SIBERIAN SEA to the Cherskiy Mountains, between Alazeya and Yukagirskoye uplands. Elevation 328 ft/100 m. Subarctic climate, with permafrost; tundra. Reindeer.

Kolyma Range, RUSSIA: see KOLYMA HIGHLANDS.

Kolyshley (kuh-lish-LYAI), town (2005 population 8,340), S PENZA oblast, E European Russia, on road and railroad, 40 mi/64 km SSW of PENZA; 52°42′N 44°32′E. Elevation 688 ft/209 m. Metal goods, dried milk.

Kolyubakino (kuh-lyoo-BAH-kee-nuh), town (2006 population 2,720), W central MOSCOW oblast, central European Russia, on crossroads, 34 mi/55 km W of MOSCOW, and 5 mi/8 km NNE of TUCHKOVO; 55°40′N 36°31′E. Elevation 587 ft/178 m. Light manufacturing (needle factory).

Kolyuchin Bay (kuh-LYOO-cheen), inlet of the CHUKCHI SEA in N CHUKCHI PENINSULA, CHUKCHI AUTONOMOUS OKRUG, extreme NE SIBERIA, RUSSIAN FAR EAST; 37 mi/60 km long. Kolyuchin Island, site of a government arctic station, is 25 mi/40 km N, off the bay entrance.

Kolyushki, POLAND: see KOLUSZKI.

Kolyvan' (kuh-li-VAHN), town, S ALTAI TERRITORY, S central SIBERIA, RUSSIA, 60 mi/97 km ESE of RUBTSOVSK; 51°19′N 82°34′E. Elevation 1,630 ft/496 m. Silver, lead, copper, tungsten mining; polishing of precious stones mined here. One of oldest mines (founded in 1726) in Siberia. Made city in 1783, but reduced to village status in 1917; made town in 1939. Formerly called Gornaya Kolyvan'.

Kolyvan' (kuh-li-VAHN), town (2006 population 10,765), NE NOVOSIBIRSK oblast, SW SIBERIA, RUSSIA, on the OB' RIVER, on crossroads, 32 mi/51 km N of NOVOSIBIRSK; 55°18′N 82°44′E. Elevation 406 ft/123 m. River port. Founded in 1713; a former commercial center on the old Siberian Road at the Ob' River crossing; declined after construction (1890s) of the TRANS-SIBERIAN RAILROAD (S) and the rise of Novosibirsk. Made city in 1822; reduced to status of a village in 1925; made town in 1964.

Kolyvan' (kuh-li-VAHN), range (3,937 ft/1,200 m) in the NW ALTAI Mountains, ALTAI TERRITORY, RUSSIA; approximately 62 mi/100 km long.

Kolyvan' Lake (kuh-li-VAHN), S ALTAI TERRITORY, S central Siberian Russia, 15 mi/24 km W of KOLYVAN'; 2.5 mi/4 km long, 1.5 mi/2.4 km wide. Known for its picturesque granite cliffs.

Kom (KOM), peak (6,613 ft/2,016 m) in the BERKOVITSA MOUNTAINS, NW BULGARIA, 5 mi/8 km SSW of

BERKOVITSA; 43°11′N 23°03′E. NISHAVA RIVER rises at the S foot.

Komadougou Yobé River, NIGERIA: see KOMADUGU YOBE.

Komadugu Yobe (ko-mah-DOO-goo YO-bah), French *Komadougou Yobé* (KO-muh-doo-goo YO-bai), river, c. 200 mi/322 km long, in NE NIGERIA and SE NIGER; formed NE of GORGORAM, YOBE state, Nigeria, by union of Hadejia and Katagum rivers, flows ENE, past GEIDAM, to Lake CHAD at BOSSO. Forms border with Niger (DIFFA province) in its lower course. Seasonal river.

Komae (ko-MAH-e), city, Tokyo prefecture, E central HONSHU, E central JAPAN, immediately SW of TOKYO city; 35°37′N 139°34′E.

Komagane (ko-MAH-gah-ne), city, NAGANO prefecture, central HONSHU, central JAPAN, near MOUNT KISO-KOMA-GA-TAKE, 65 mi/105 km S of NAGANO; 35°43′N 137°56′E.

Koma-ga-take (ko-MAH-gah-TAH-ke), volcanic peak (3,740 ft/1,140 m), SW Hokkaido prefecture, N JAPAN, near SW entrance to UCHIURA BAY, 22 mi/35 km SW of MURORAN. Long thought to be extinct, it erupted violently in 1929.

Koma-ga-take, JAPAN: see KISO-KOMA-GA-TAKE.

Komai, China: see GOGEN.

Komaki (KO-mah-kee), city, AICHI prefecture, S HONSHU, S central JAPAN, on the Nobi Plain, 6 mi/10 km N of NAGOYA; 35°17′N 136°54′E. Suburb of Nagoya. Manufacturing helicopters, airplane parts, and rubber. Peaches.

Koman (ko-MAHN), new town, N ALBANIA, on DRIN River, E of SHKODËR; 42°05′N 19°49′E. Dam, reservoir, and hydroelectric plant on river to N.

Komandjari, province, BURKINA FASO: see KOMONDJARI.

Komandjoari, province, BURKINA FASO: see KOMONDJARI.

Komandorski Islands (kuh-mahn-DOR-skee), Russian *Komandorskiye Ostrova* [=Commander Islands], group of tree-less islands, off E KAMCHATKA PENINSULA, KAMCHATKA oblast, extreme E SIBERIA, RUSSIAN FAR EAST, in SW BERING SEA. They consist of BERING Island, MEDNYY Island, and two islets. These hilly, foggy islands often have earthquakes. Their inhabitants, Russians and Aleuts, are engaged in fishing, hunting, and whaling. The largest village is Nikol'skoye on Bering Island.

Komandorskiye Ostrova, RUSSIA: see KOMANDORSKI ISLANDS.

Komao, China: see GOGEN.

Komarapalaiyam (kuh-MAH-ruh-PAH-lei-yuhm), town, SALEM district, TAMIL NADU state, S INDIA, on KAVERI River (bridged) opposite BHAVANI, 36 mi/58 km WSW of Salem. Rice, peanuts, castor beans. Corundum deposits nearby. Also spelled Komarapalayam, Kumarapalaiyam.

Komarevo (ko-ma-RE-vo), village, LOVECH oblast, DOLNA MITROPOLIYA obshtina, BULGARIA; 43°38′N 24°38′E.

Komari, village, EASTERN PROVINCE, SRI LANKA, on E coast, 50 mi/80 km SSE of BATTICALOA; 06°59′N 81°52′E. Rice and coconut-palm plantations nearby. Sangamankanda promontory (formerly spelled Sankamankandimunai), 1 mi/1.6 km NNE, is easternmost point of Sri Lanka. Lighthouse. Also spelled Komariya.

Komarichi (kuh-MAH-ree-chee), town (2005 population 7,040), SE BRYANSK oblast, central European Russia, on crossroads and railroad, 60 mi/97 km SSE of BRYANSK; 52°25′N 34°47′E. Elevation 702 ft/213 m. Railroad; food industries (sugar refining, meat packing, dairying, fruit processing).

Komarin (ko-MUH-rin), urban settlement, GOMEL oblast, BELARUS, on DNIEPER River, 30 mi/48 km W of CHERNIGOV; 51°25′N 30°25′E. Agricultural products; cloth milling.

Komarivka (ko-mah-RIF-kah) (Russian *Komarovka*), village, central CHERNIHIV oblast, UKRAINE, 17 mi/27 km NE of NIZHYN. Grain.

Komarivka, UKRAINE: see PIVDENNE.

Komarne (ko-MAHR-ne) (Polish *Komarno*) (Russian *Komarno*), city (2004 population 10,560), central L'VIV oblast, UKRAINE, 20 mi/32 km SW from L'VIV; 49°38′N 23°42′E. Elevation 846 ft/257 m. Natural gas distribution and compressor stations, wood working, grain-milling, instrument-making, dairy, fish-processing. Architectural monuments include a Roman Catholic church (1656) and belfry and a monument to victory over the Turks and Tatars (1663). Established in the 12th–13th century, granted town rights in 1473, city since 1940. Jewish community since the 17th century, numbering 2,550 in 1931; destroyed by the Nazis in 1941–1943—fewer than 100 Jews remaining in 2005.

Komarno (kuh-MAHR-no), community, S MANITOBA, W central CANADA, 43 mi/70 km N of WINNIPEG, and in ROCKWOOD rural municipality; 50°30′N 97°15′W.

Komarno (KO-mahr-no), Slovak *Komárno*, German *Komorn*, Hungarian *Komárom*, city on the Danube (Slovak-Hungarian border), 55 mi/89 km SE of BRATISLAVA, 42 mi/68 km NW of BUDAPEST, with twin city on the Slovak side. Slovak *Komarno*, railroad, shipping, and trade center (lumber, coal), is on left bank of the Danube, at VAH River mouth, at E extremity of Great Schütt island; manufacturing of machinery and textiles (notably silk), shipbuilding, fishing, flour milling; still retains part of 15th–17th century fortifications. Hungarian *Komarom* (1991 estimated population 19,552), railroad center and river port on right bank, has lumberyards, sawmills, textile mills, foundry; manufacturing of earth-moving equipment. Developed as suburb of Komarno. The twin cities were originally one autonomous Hungarian city; noted in 18th century for its grain trade, birthplace of Mór Jókai, famed 19th-century Hungarian writer; divided by Treaty of Trianon (1920), reunited in 1939, divided again in 1945, after Hitler's dismemberment of CZECHOSLOVAKIA.

Komárno (ko-MAHR-no), German *Komorn*, city, ZAPADOSLOVENSKY province, SLOVAKIA, on left bank of the Danube River, at its confluence with the NITRA and VÁH rivers, opposite KOMÁROM (HUNGARY); 47°46′N 18°08′E. Transport center; machinery, and textile plants. Has port installations and docks. For most of its history, it was joined with Komárom. The site was fortified by the Romans. It became a free city in 1331. Still retains part of 15th–17th century fortifications. Later a part of the Austro-Hungary monarchy, and noted for its grain trade, it was partitioned in 1920 between Hungary and CZECHOSLOVAKIA. Under Hungarian rule between 1938–1945, it was redivided after World War II.

Komarno, UKRAINE: see KOMARNE.

Komárom (KO-mah-rom), county (□ 765 sq mi/1,989 sq km), N HUNGARY; ⊙ TATABÁNYA; 47°35′N 18°20′E. Mountainous in E (GERECSE MOUNTAINS), level in W (part of the Little Alföld); bounded N by the DANUBE River. Industrial and mining region; brown coal at Tatabánya and DOROG; small oil refinery at Szöny; production of cement, concrete, and lime at LÁBATLAN and Dorog; iron foundry at Komárom; large electric power plants at OROSZLÁNY and Tatabánya. The Suzuki automobile plant at ESZTERGOM was one of the largest foreign investment projects in Hungary during the first half of the 1990s. In the Communist period, Komárom county was the most industrialized in all of Hungary, but its mineral- and energy-intensive industries in the 1990s were contracting and suffering the pains of restructuring. The county's central location near BUDAPEST and relative proximity to Western EUROPE, however, eased the difficulty of structural change, and made the county among the most attractive for foreign investment. Grain, pota-

toes; dairy farming; hogs. Stud farms and research stations for animal husbandry at BÁBOLNA and KISBÉR. Area once part of the Roman "limes," guarding the province of PANNONIA. Formed in 1920 from parts of Komárom and Ezstergom counties.

Komárom, city, KOMÁROM county, HUNGARY, on right bank of DANUBE River, opposite KOMÁRNO (SLOVAKIA), 55 mi/89 km SE of BRATISLAVA, 42 mi/68 km NW of BUDAPEST; 47°44′N 18°07′E. Has lumber yards, sawmills, textile plants, and port installations. Separated from Slovak city by Treaty of Trianon in 1920, reunited in 1939, divided again in 1945. Mór Jókai, famed 19th century Hungarian writer, born here. For earlier history, see KOMÁRNO.

Komarov (KO-mah-ROF), Czech *Komárov*, German *komorau*, village, STREDOCESKY province, W central BOHEMIA, CZECH REPUBLIC, 32 mi/51 km SW of PRAGUE, on railroad. Manufacturing (machinery); museum of steel- and ironworks.

Komarovka, UKRAINE: see PIVDENNE.

Komarovka, UKRAINE: see KOMARIVKA.

Komarovo (kuh-mah-RO-vuh), town (2005 population 1,030), N LENINGRAD oblast, N European Russia, on the Gulf of FINLAND, on road and railroad, 4 mi/6 km SE of ZELENOGORSK, and 26 mi/42 km NW (and administratively a suburb) of SAINT PETERSBURG; 60°11′N 29°49′E. Elevation 134 ft/40 m. Railroad station; seaside resort. Called Kellomäki until 1948.

Komarovo (kuh-mah-RO-vuh), town (2005 population 830), E central NOVGOROD oblast, NW European Russia, near the MSTA River, 22 mi/35 km NW of BOROVICHI; 58°24′N 30°10′E. Elevation 131 ft/39 m. Lignite mining.

Komarovo (kuh-mah-RO-vuh), settlement (2004 population 600), central BASHKORTOSTAN Republic, SW URALS, RUSSIA, near mountain highway and railroad, 3 mi/5 km SW of TUKAN; 53°48′N 57°23′E. Elevation 1,847 ft/562 m. In Komarovo-Zigazinskiy iron ore district

Komarovskiy (kuh-mah-ROF-skeeyee), urban settlement (2006 population 8,285), E ORENBURG oblast, extreme SE European Russia, in the SE URALS, on the Kumak River (tributary of the URAL River) basin, 18 mi/29 km ESE of NOVOORSK; 51°17′N 59°38′E. Elevation 816 ft/248 m. Formerly part of the Soviet nuclear weapons research complex and known under its secret designation of Dombarovskiy-3.

Komatipoort, town, MPUMALANGA, SOUTH AFRICA, on KOMATI RIVER near the MOZAMBIQUE border, opposite RESSANO GARCIA, at influx of CROCODILE RIVER, 55 mi/89 km NW of MAPUTO, near SE edge of KRUGER NATIONAL PARK on a gorge through the LEBOMBO MOUNTAINS; 25°27′S 31°58′E. Elevation 689 ft/210 m. Airfield. Customs station on JOHANNESBURG-MAPUTO railroad and N4 highway. In South African War, occupation of KOMATIPOORT (October 1900) by British marked end of organized Boer resistance.

Komati River, Portuguese *Incomati*, circa 500 mi/805 km long, in SOUTH AFRICA, SWAZILAND, and MOZAMBIQUE; rises in N DRAKENSBERG range near Breyton circa 10 mi/16 km N of ERMELO; flows in a winding course NE through NOOITGEDACHT Nature Reserve then E past Tjakastad, to enter N Swaziland 5 mi/8 km SSW of BULEMBU. It continues NE, reenters South Africa, and flows past KOMATIPOORT, where it turns SE and enters Mozambique; continues NE past MAGUDE, then turns sharply S; passes to E of MANHICA, enters INDIAN OCEAN 15 mi/24 km NE of MAPUTO at Ponta de Macaneta. Receives CROCODILE (Krokodil) River from NW at Komatispoort. Called Rio Incomati River in Mozambique.

Komatsu (ko-MAHTS), city, ISHIKAWA prefecture, central HONSHU, central JAPAN, 16 mi/25 km S of KANAZAWA; 36°24′N 136°26′E. Bulldozers, textiles. *Kutani yaki*-style pottery. Hot springs nearby. Tokugawa-era tollgate at Ataka no Seki.

Komatsu (ko-MAHTS), town, Syuso county, EHIME prefecture, N SHIKOKU, W JAPAN, 19 mi/30 km E of MATSUYAMA; 33°53′N 133°06′E.

Komatsushima (ko-MAHTS-shee-mah), city, TOKUSHIMA prefecture, E SHIKOKU, W JAPAN, port on KII CHANNEL, 5 mi/8 km SSE of TOKUSHIMA (its outer port); 34°00′N 134°35′E.

Komavangard, RUSSIA: see SOBINKA.

Komba (KOM-buh), village, ORIENTALE province, N CONGO, 25 mi/40 km WNW of BUTA; 02°53′N 23°59′E. Elev. 1,377 ft/419 m. Railroad junction in cotton area.

Kombai (KOM-bei), town, MADURAI district, TAMIL NADU state, S INDIA, 7 mi/11.3 km N of KAMBAM, in KAMBAM VALLEY; 09°50′N 77°19′E. Cardamom, bamboo, turmeric in CARDAMOM HILLS (W); grain, livestock raising. Livestock market at VIRAPANDI village, 12 mi/19 km NE. Sometimes spelled COMBAI.

Kombe (KOM-bai), village, TABORA region, W TANZANIA, 85 mi/137 km W of TABORA, on railroad; 05°06′S 31°18′E. Timber; tobacco, corn; goats, sheep.

Kombissiri, town, ⊙ BAZÉGA province, CENTRE-SUD region, central BURKINA FASO, on road 25 mi/40 km S of OUAGADOUGOU; 12°01′N 01°27′W. Rice, fruits, cotton.

Kombolcha, town (2007 population 72,024), AMHARA state, E central ETHIOPIA, 8 mi/12.9 km SE of DESSIE; 11°05′N 39°44′E.

Kombolcha, town (2007 population 4,774), OROMIYA state, W central ETHIOPIA; 100 mi/161 km NW of ADDIS ABABA; 09°29′N 37°28′E. Near Fincha'a Dam and power station.

Kombo Saint Mary, THE GAMBIA: see KANIFING, local government area.

Kome Island (KOM-ai), MWANZA region, NW TANZANIA, in Lake VICTORIA, 30 mi/48 km WNW of MWANZA, 1 mi/1.6 km N of mainland; 12 mi/19 km long, 7 mi/11.3 km wide; 02°10′S 32°27′E.

Kom el Ahmar, El (KOOM el AH-mahr), village, QENA province, Upper EGYPT, 9 mi/14.5 km NW of IDFU. Site of the extensive ruins and tombs of ancient Hieraconpolis.

Komenda (KO-men-duh), town (2000 population 12,278), CENTRAL REGION, GHANA, on Gulf of GUINEA, 15 mi/24 km W of CAPE COAST; 05°03′N 01°29′W. Fishing. British trading station from 1663. Fort Vrendeaburg established by Dutch in 1689. Formerly spelled KOMMENDA or COMMENDA.

Kom en Nur (KOOM en NOOR), village, DAQAHLIYA province, Lower EGYPT, 2 mi/3.2 km NE of MIT GHAMR; 29°01′N 30°53′E. Cotton, cereals.

Komga, town, EASTERN CAPE province, SOUTH AFRICA, near the source of a tributary of the GREAT KEI RIVER, 30 mi/48 km N of EAST LONDON (BUFFALO CITY); 32°05′S 29°55′E. Elevation 2,296 ft/700 m. Livestock, dairying, grain. On railroad route from KING WILLIAM'S TOWN to UMTATA.

Kom Hamada (KOOM hah-MAH-duh), village, BEHEIRA province, Lower EGYPT, on RASHID branch of the NILE, 23 mi/37 km SE of DAMANHUR; 30°46′N 30°42′E. Cotton, rice, cereals.

Komilla, BANGLADESH: see COMILLA.

Kominio (ko-MEEN-o), town, EDO state, SW NIGERIA, on road, 85 mi/137 km NE of BENIN; 07°17′N 06°22′E. Market town. Cassava, plantains, yams.

Komintern, RUSSIA: see NOVOSHAKHTINSK.

Komintern, UKRAINE: see MARHANETS'.

Kominternivs'ke (ko-meen-TER-neef-ske) (Russian *Kominternovskoye*), town (2004 population 5,000), SE ODESSA oblast, UKRAINE, 25 mi/40 km NNE of ODESSA; 46°49′N 30°56′E. Elevation 219 ft/66 m. Raion center. Brickworks; grain and feed mill. Until 1933, Antonove-Kodyntseve (Russian *Antono-Kodintsevo*). Established in 1802; town since 1965.

Kominternovskiy (kuh-meen-TER-nuhf-skeeye), former town, N central KIROV oblast, E European Russia, on road and railroad, across the VYATKA River from KIROV, into which it is now incorporated;

58°38′N 49°44′E. Elevation 400 ft/121 m. Shoe manufacturing.

Kominternovskoye, UKRAINE: see KOMINTERNIVS'KE.

Komi-Permyak (KO-mee–pyer-MYAHK) autonomous okrug (□ 12,703 sq mi/33,027.8 sq km; 2004 population 157,000), Russian *Komi-Permyatskiy Avtonomnyy Okrug*, NW PERM oblast, E central European Russia, in the basin of the upper KAMA River; ⊙ KUDYMKAR. Borders KOMI REPUBLIC to the N. The terrain is slightly hilly and heavily forested and is drained by the Kama River and its tributaries, the Veslyana, Kosa, and Inva rivers. The navigable Kama is also the area's chief transportation artery. Lumbering is the major (and only important) industry, as tayga forests cover most of the okrug. Agriculture (poorly developed) includes rye, oats, spring wheat, and flax. The Komi and the Permyaks, both Finno-Ugric peoples, are an ethnic subgroup of the Komi people, to whom they are closely related. They make up around 60% of the population; the rest are mostly Russians. Originally they lived by breeding livestock and growing grain, along with hunting furs to trade. Their literature has a venerable tradition, dating back to medieval times, when the Komi came into contact with Eastern Orthodox missionaries. The area was organized in 1925, and given autonomous status in 1977. As a result of a regional referendum in October 2004, the okrug was officially incorporated into Perm oblast.

Komi Republic (KO-mee), constituent republic (□ 160,580 sq mi/417,508 sq km; 2004 population 1,185,000), NE European Russia; ⊙ SYKTYVKAR. The region is a wooded lowland, stretching across the PECHORA and VYCHEGDA river basins and the upper reaches of the MEZEN' River. The N part is permanently frozen, wooded tundra. Mining is the most important economic activity. There are major coalfields in the Pechora basin, yielding heating and coking coal and providing most of the coal for SAINT PETERSBURG. Syktyvkar is a major lumber center; VORKUTA is a coal-mining center; and there is extensive lumbering, livestock raising, fishing, and hunting. Russians (58%), Komi (23%), and Ukrainians constitute the majority of the population. The Komi, formerly called Zyrians, speak a Finno-Ugric language and adhere to the Russian Orthodox religion. The area underwent a spectacular economic advance after the opening (1942) of the Kotlas-Vorkuta railroad to transport the area's coal and oil; current oil stock is estimated at 213 billion tons. It also supplies the rest of the Russian Federation, as well as Europe, with bauxite and manganese ores. The area belonged to the Novgorod principality from the 13th century. The Zyrian autonomous region was constituted in 1921; it became an autonomous republic in 1936. In 1990, Komi declared its sovereignty, and the word "autonomous" was dropped from its name. It was a signatory to the March 31, 1992, treaty that created the RUSSIAN FEDERATION. The Komi Republic has a 180-member parliament.

Komisarivka (ko-mee-SAH-rif-kah) (Russian *Komissarovka*), town, W central LUHANS'K oblast, UKRAINE, near highway, and on railroad, 14 mi/22 km WSW of ALCHEVS'K, and 7 mi/11 km NE of DEBAL'TSEVE; 48°28′N 39°30′E. Elevation 213 ft/64 m. Machine building. Poultry; vegetable experimental farm. Established in 1765 as Holubivka (Russian *Golubovka*), renamed in 1917, town since 1963.

Kom Ishqaw (KOOM ISH-kaw), village, SOHAG province, central Upper EGYPT, 7 mi/11.3 km NW of TAHTA. Cotton, cereals, dates, sugarcane. Site of ancient Aphroditopolis. Also spelled Kum Ishqaw.

Komissarovka, UKRAINE: see KOMISARIVKA.

Komiža, CROATIA: see VIS.

Komját, SLOVAKIA: see KOMJATICE.

Komjatice (kom-YAH-tyi-TSE), Hungarian *Komját*, village, ZAPADOSLOVENSKY province, S SLOVAKIA, on railroad, and 11 mi/18 km N of NOVÉ ZÁMKY; 48°09′N

18°11′E. Wheat, corn, sugar beets; food processing; vineyards. Paleolithic-era archaeological site. Large Hungarian minority. Under Hungarian rule between 1938–1945. Summer resort of POLNY KESOV (pol-yuh-NEE ke-SOU), Slovak *Pol'ný Kesov*, is 4 mi/6.4 km W.

Komló (KO-lo), Hungarian *Komló*, city (2001 population 27,081), BARANYA county, S HUNGARY, in MECSEK MOUNTAINS, 8 mi/13 km N of PÉCS; 46°12′N 18°16′E. Coal mines, now bankrupt; government wants to close them down, but these mines are almost the sole local employer.

Kommenda, GHANA: see KOMENDA.

Kommuna Imeni Stalina, RUSSIA: see PERTOMINSK.

Kommunar (kuh-moo-NAHR), town (2005 population 2,715), W KHAKASS REPUBLIC, S SIBERIA, RUSSIA, in the KUZNETSK ALATAU mountains, on road, 90 mi/145 km NW of ABAKAN; 54°21′N 89°17′E. Elevation 2,670 ft/813 m. Gold mining; boomed in the early 1930s. Until 1932, known as Bogomdarovannyy (Russian=God's gift).

Kommunar (kuhm-moo-NAHR), town (2005 population 17,430), central LENINGRAD oblast, NW European Russia, on the IZHORA RIVER (tributary of the NEVA RIVER), on railroad, 18 mi/29 km S of SAINT PETERSBURG; 59°37′N 30°24′E. Elevation 190 ft/57 m. Manufacturing (paper and cardboard).

Kommunarka (kuh-moo-NAHR-kah), settlement (2006 population 4,675), central MOSCOW oblast, central European Russia, on local railroad spur, 12 mi/19 km SSW of MOSCOW; 55°34′N 37°29′E. Elevation 606 ft/184 m. Research center for ecologically clean technologies for natural gas industry.

Kommunarsk, UKRAINE: see ALCHEVS'K.

Kommunisticheskiy (kuh-moo-nees-TEE-cheesk-keeyee) [Russian=communist (adj.)], settlement (2005 population 2,710), N KHANTY-MANSI AUTONOMOUS OKRUG, W central Siberian Russia, on road and railroad, 162 mi/261 km WNW of KHANTY-MANSIYSK; 61°40′N 64°29′E. Elevation 439 ft/133 m. In oil- and gas-producing and transporting region. Formerly called Samza.

Kommunizm (kom-moo-NEEZM), town, SE LEBAP weloyat, SE TURKMENISTAN, near the AMU DARYA (river), and 60 mi/97 km SE of CHARJEW; 38°28′N 64°23′E. Cotton. Formerly called Burdalyk.

Komo, China: see GOGEN.

Komochi (ko-MO-chee), village, North Gumma county, GUMMA prefecture, central HONSHU, N central JAPAN, 12 mi/20 km N of MAEBASHI; 36°30′N 139°00′E. Devil's tongue paste (*konnyaku*).

Komodo (kuh-MO-do), island (□ c.184 sq mi/477 sq km), Nusa Tenggara Timur province, INDONESIA, between FLORES SEA (N) and SUMBA STRAIT (S), between FLORES (E) and SUMBAWA (W) islands; 20 mi/32 km long, 12 mi/19 km wide; 08°36′S 119°30′E. Hilly. Copra; fish. One of few habitats of the endangered Komodo dragon, a unique meat-eating lizard.

Komoé River, BURKINA FASO and CÔTE D'IVOIRE: see COMOÉ RIVER.

Kom Ombo (KOOM UHM-boo), town, ASWAN province, S EGYPT, on railroad, 27 mi/43 km NNE of ASWAN, near the E bank of the NILE River; 24°28′N 32°57′E. Sugar milling, cotton ginning. Site of ruins of several temples. Also spelled Kum Umbu.

Komondjari, province (□ 1,947 sq mi/5,062.2 sq km; 2005 population 60,300), EST region, E BURKINA FASO; ⊙ Gayéri; 12°40′N 00°40′E. Bordered N by YAGHA province, NE by NIGER, E tip by TAPOA, SES by GOURMA, and W by GNAGNA. Established in 1997 with fourteen other new provinces. Also spelled Komandjoari and Komandjari.

Komono, town, LÉKOUMOU region, SW Congo Republic, 180 mi/290 km WNW of BRAZZAVILLE; 03°16′S 13°13′E. Food processing.

Komono (ko-MO-no), town, Mie county, MIE prefecture, S HONSHU, central JAPAN, 19 mi/30 km N of TSU; 35°00′N 136°30′E. Hot springs.

Cross-references are shown in SMALL CAPITALS. The pronunciation guide is shown on page xix. The sources of population figures are shown on page xvii.

Komono, rubber plantations, Congo Republic: see SIBITI.

Komorany (KO-mo-RZHAH-ni-yuh), Czech *Komořany*, village, SEVEROCESKY province, NW BOHEMIA, CZECH REPUBLIC, 3 mi/4.8 km W of MOST. Coal mines; power plants. Mining village of TREBUSICE, Czech *Třebušice*, is just SE on railroad.

Komorau, CZECH REPUBLIC: see KOMAROV.

Komoro (ko-MO-ro), city, NAGANO prefecture, central HONSHU, central JAPAN, near MOUNT ASAMA, 25 mi/40 km S of NAGANO; 36°19′N 138°25′E. Lettuce.

Komoshtitsa (ko-MOSH-tee-tsah), village, MONTANA oblast, YAKIMOVO obshtina, NW BULGARIA, 9 mi/15 km SE of LOM; 43°44′N 23°19′E. Grain, legumes.

Komotau, CZECH REPUBLIC: see CHOMUTOV.

Komotiní (ko-mo-tee-NEE), city (2001 population 43,326), ⊙ RODOPI prefecture, EAST MACEDONIA AND THRACE department, NE GREECE, E of XÁNTHI; 41°07′N 25°24′E. Commercial center for a region that produces grains, silk, and tobacco. Sizable Muslim minority.

Komovi (KO-mo-vee), mountain in DINARIC ALPS, E MONTENEGRO, near ALBANIA border, in the BRDA. Highest point (8,148 ft/2,484 m) is 8 mi/12.9 km WSW of ANDRIJEVICA.

Kompaneyevka, UKRAINE: see KOMPANIYIVKA.

Kompaniyivka (kom-pah-NEE-yif-kah) (Russian *Kompaneyevka*), town, S KIROVOHRAD oblast, UKRAINE, on road, and 18 mi/29 km S of KIROVOHRAD; 48°14′N 32°12′E. Elevation 524 ft/159 m. Raion center. Mixed feed mill; poultry hatchery. Forest reclamation station; veterinary technical school. Founded in the mid-18th century, town since 1965.

Kompian (KAHM-pee-ahn), village, ENGA province, E central NEW GUINEA island, N central PAPUA NEW GUINEA, 25 mi/40 km NW of WABAG, in Great Plateau. Accessible by walking track. Bananas, coffee, copra, sweet potatoes.

Kompienga, province (□ 2,702 sq mi/7,025.2 sq km; 2005 population 63,278), EST region, SE BURKINA FASO; ⊙ PAMA; 11°25′N 00°55′E. Bordered N by GOURMA province, ENE by TAPOA province, ESE by BENIN, S by TOGO, and W by KOULPÉLOGO province. Established in 1997 with fourteen other new provinces.

Kompong Cham (KAWM-PAWNG CHAHM), province (□ 3,783 sq mi/9,835.8 sq km; 2007 population 1,914,152), E CAMBODIA, on VIETNAM (E) border; ⊙ KOMPONG CHAM; 12°00′N 105°30′E. Borders PREY VENG (S), KANDAL (SW), KOMPONG CHHNANG (W), KOMPONG THOM (N), and KRATIE (NEE) provinces. Agriculture (rubber plantations; rice, tree crops); fishing. Deforestation is a growing problem. Khmer population, with Cham, Vietnamese, and Chinese minorities.

Kompong Cham (KAWM-PAWNG CHAHM), city, ⊙ KOMPONG CHAM province, SE CAMBODIA, a port on the MEKONG River; 12°00′N 105°27′E. The third-largest city in Cambodia, it has a large textile factory, built with Chinese aid; transportation hub (roads and water). Food processing and light manufacturing. Potential for tourism. Major irrigation projects, started after the revolution in 1975, have improved the area's agricultural production. Rice growing; rubber plantations. Khmer population with Cham and Vietnamese minorities. Nearby archaeological sites include Wat Nakor (13th-century monument), PREAH THEAT PREAH SREI, and Preah Nokor, 7th-century KHMER capital. Five mass graves of Khmer Rouge victims c.19 mi/30 km N.

Kompong Chhnang (KAWM-PAWNG CHNUHNG), province (□ 2,131 sq mi/5,540.6 sq km; 2007 population 538,163), central CAMBODIA; ⊙ KOMPONG CHHNANG; 12°00′N 104°30′E. Borders PURSAT (WNW), KOMPONG THOM (NNE), KOMPONG CHAM (E), KANDAL (SE), and KOMPONG SPEU (SSW) provinces. TONLÉ SAP lake in N. Rice; cattle raising; fishing.

Khmer population with Cham and Vietnamese minorities.

Kompong Chhnang (KAWM-PAWNG CHNUHNG), city, ⊙ KOMPONG CHHNANG province, central CAMBODIA, at outlet of Tonlé Sap, 50 mi/80 km NNW of PHNOM PENH, on main highway; 12°15′N 104°40′E. Trade center; manufacturing (bricks, pottery); distillery, food processing; cattle market; rice growing. Light manufacturing head of navigation on Tonlé Sap River at low water. Khmer population with Cham and Vietnamese minorities.

Kompong Chrey (KAWM-PAWNG CHRAI), town, TAKEO province, SW CAMBODIA, 50 mi/80 km S of PHNOM PENH. Rice center. Khmer population with Vietnamese minority.

Kompong Kleang (KAWM-PAWNG KLENG), village, SIEM REAP province, W CAMBODIA, near TONLÉ SAP lake, 25 mi/40 km SE of SIEM REAP; 13°06′N 104°08′E. Fisheries. Khmer population with N Pear and Vietnamese minorities.

Kompong Luong (KAWM-PAWNG LOONG), village, KANDAL province, S CAMBODIA, on TONLÉ SAP River (ferry), and 20 mi/32 km NNW of PHNOM PENH. Agricultural market; distillery (rice alcohol). Rice-growing and gardening area. Khmer population with Cham minority.

Kompong Som (KAWM-PAWNG SAWM) or **Sihanoukville, municipality** (□ 335 sq mi/868 sq km; 2004 estimated population 201,981; 2007 estimated population 223,608), city (1998 population 66,723), and seaport, located in, but politically independent of, KAMPOT province, S CAMBODIA, on the GULF OF THAILAND; 10°38′N 103°30′E. Although a new city (completed 1960), it is the principal deepwater port and commercial outlet of Cambodia. The city and port were built on mud flats, with French aid, and grew with the construction of a highway and railroad to PHNOM PENH. The docks and warehouses have been greatly expanded with U.S. aid. The country's only oil refinery, located here, was destroyed (1971) by insurgent Khmer Rouge troops. Has an international airport. With fine beaches and skin diving nearby, Kompong Som has much potential as a tourist center. Khmer population with Cham and Chinese minorities. Also spelled Kâmpóng Saôm; also called Preah Seihanu and Krong Preah Sihanouk.

Kompong Som Bay (KAWM-PAWNG SAWM), inlet of GULF OF THAILAND, SE CAMBODIA; 15 mi/24 km–20 mi/32 km wide, 35 mi/56 km long; receives small Kompong Som River (N). Excellent beaches. Fishing.

Kompong Speu (KAWM-PAWNG SPUH), province (□ 2,709 sq mi/7,043.4 sq km; 2007 population 762,500), central CAMBODIA; ⊙ KOMPONG SPEU; 11°30′N 104°30′E. Borders PURSAT and KOMPONG CHHNANG (N), KANDAL (E), TAKEO (SE), KAMPOT (S), and KOH KONG (W) provinces. Ecologically diverse. Shifting cultivation, rice growing, gardening; tree crops. Recent problems with deforestation. Khmer population with Chinese minority.

Kompong Speu (KAWM-PAWNG SPUH), city, ⊙ KOMPONG SPEU province, S CAMBODIA, 25 mi/40 km WSW of PHNOM PENH, at foot of CARDAMOM Range; 11°27′N 104°32′E. Transportation hub; silk spinning, timber trading, food processing, light manufacturing. Rice-growing, market-gardening area, tree crops. Recent problems with forest exploitation. Khmer population with Chinese minority.

Kompong Thom (KAWM-PAWNG TAWM), province (□ 5,333 sq mi/13,865.8 sq km; 2007 population 708,398), central CAMBODIA; ⊙ KOMPONG THOM; 13°00′N 105°00′E. Borders SIEM REAP (WNW), PREAH VIHEAR (N), STUNG TRENG (NE), KRATIE (E), KOMPONG CHAM (S), and KOMPONG CHHNANG (SW) provinces and TONLÉ SAP lake (W). Rice growing; forest products; shifting cultivation; fishing. Khmer population with Vietnamese and Kurj minorities.

Kompong Thom (KAWM-PAWNG TAWM), city, ⊙ KOMPONG THOM province, central CAMBODIA, 75 mi/121 km N of PHNOM PENH, on the Stung Sen (affluent of lake TONLÉ SAP; navigable at high water); 12°42′N 104°54′E. In forested (gum, gutta percha, lac) big-game region; iron-ore, jet, and gem deposits. Food processing, light manufacturing. Rice growing; shifting cultivation; fishing. Ruins of SAMBOR PREI KUK (ancient Isanapura), KOH KER, and PREAH KHAN are nearby. Khmer population with Vietnamese and Chinese minorities.

Kompong Trabek (KAWM-PAWNG TRAW-BEK), town, PREY VENG province, S CAMBODIA, on main highway to HO CHI MINH CITY (VIETNAM), and 50 mi/80 km SE of PHNOM PENH; 11°09′N 105°28′E. Rice, corn; fisheries. Khmer population with small Vietnamese minority.

Kompong Trach (KAWM-PAWNG TRAHCH), village, KAMPOT province, S CAMBODIA, 18 mi/29 km ESE of KAMPOT, on small isolated ridge of ELEPHANT RANGE; 10°34′N 104°28′E. Grottoes.

Kompong Tralach (KAWM-PAWNG TRAW-LAHCH), village, KOMPONG CHHNANG province, central CAMBODIA, on TONLÉ SAP River, and 25 mi/40 km NNW of PHNOM PENH; 11°54′N 104°47′E. In forested (precious wood) area; fisheries. Khmer population with Cham and Vietnamese minorities.

Komrat, MOLDOVA: see COMRAT.

Komsberg Escarpment, WESTERN CAPE and NORTHERN CAPE provinces, SOUTH AFRICA, extends *circa* 60 mi/97 km E-W in semicircle along SW edge of the NORTHERN KAROO on the provincial border, between ROGGEVELD ESCARPMENT (W) and NUWEVELD RANGE (E), 38 mi/60 km N of LAINGSBURG; rises 2,000 ft/610 m–3,000 ft/914 m steeply from GREAT KAROO.

Komsomolabad, village, central TAJIKISTAN, on VAKHSH RIVER, and 23 mi/37 km SW of GARM; 38°52′N 69°57′E. Tertiary level administrative center; area under direct republic (not direct viloyat administrative) supervision. Wheat, vegetables; cattle; gold placers. Until c.1935, called Pombachi.

Komsomolets (kom-so-mo-LYETS), town, NW KOSTANAI region, KAZAKHSTAN, on railroad, 22 mi/35 km SE of TROITSK (Russia); 53°47′N 62°01′E. Tertiary-level administrative center.

Komsomolets (kuhm-suh-MO-lyets), village (2006 population 3,640), central TAMBOV oblast, S central European Russia, near highway branch and railroad, 8 mi/13 km W of TAMBOV; 52°46′N 41°14′E. Elevation 518 ft/157 m. Agricultural products.

Komsomolets Island (kuhm-suh-MO-lyets), N island (3,570 sq mi/9,246 sq km) of the SEVERNAYA ZEMLYA archipelago, in the ARCTIC OCEAN, off KRASNOYARSK TERRITORY, N Russia. Glaciers cover 65% of the island.

Komsomol Preserve (kuhm-suh-MOL), wildlife refuge (□ 249 sq mi/647.4 sq km), KHABAROVSK TERRITORY, extreme E SIBERIA, RUSSIAN FAR EAST, on the right bank of the AMUR River, opposite KOMSOMOL'SK-NA-AMURE, extending from Pivan' to Lake Bel'go and including the lake's water area. A branch of the reserve is located in the middle course of the Khungari River. Local fauna include brown and black bear, raccoon dog, elk, roe, and musk deer. Established in 1963.

Komsomol'sk (kuhm-suh-MOLSK), city (2005 population 9,015), W central IVANOVO oblast, central European Russia, on road and railroad spur, 37 mi/60 km W of IVANOVO; 57°01′N 40°22′E. Elevation 459 ft/139 m. Electrical equipment, building machinery equipment, textiles; wood processing; electric power station. Food processing, bakery, dairy plant. Agriculture includes wheat, rye, oats, peas, vegetables, potatoes; livestock raising. Arose in 1931 with the construction of the power station; made city in 1950. Historical landmarks include the Rozhdestvenskaya (Christmas) church, built in 1771.

Area is shown by the symbol □, and capital city or county seat by ⊙.

Komsomol's'k (kom-so-MOLSK) (Russian *Komsomol'sk*), city (2001 population 51,740), SW POLTAVA oblast, UKRAINE, on the left bank of the DNIPRODZERZHYNS'K RESERVOIR of the DNIEPER (Ukrainian *Dnipro*) River, 10 mi/16 km ESE of KREMENCHUK; 49°02'N 33°40'E. Elevation 249 ft/75 m. Fabrication of reinforced concrete building materials, stone crushing, grain milling. Iron-ore-beneficiation plant, knitting factory. Mining-metallurgy technical school, professional-technical schools. N of the city are iron ore open pit mines. Established in 1960; city since 1972.

Komsomol'sk (kuhm-suh-MOLSK), town (2005 population 2,545), NE KEMEROVO oblast, S central SIBERIA, RUSSIA, on road, 7 mi/11 km N of BERIKUL'SKIY; 55°38'N 88°10'E. Elevation 1,233 ft/375 m. Coal and mineral mining in the vicinity.

Komsomolsk, town, NAWOIY wiloyat, central UZBEKISTAN, W of NAWOIY, between Nawoiy and KYZYLTEPA; 40°10'N 65°17'E.

Komsomol'sk (kuhm-suh-MOLSK), village (2005 population 3,600), W KALININGRAD oblast, NW European Russia, near the PREGEL RIVER, on railroad, 11 mi/18 km ESE of KALININGRAD; 54°37'N 20°45'E. Elevation 111 ft/33 m. Nearby is an 18th-century palace. Until 1945, in German-administered East Prussia and called Löwenhagen.

Komsomol's'k (kom-so-MOLSK) (Russian *Komsomol'sk*), NW suburb of HORLIVKA, central DONETS'K oblast, UKRAINE, in the DONBAS, 4 mi/6 km NW of Horlivka city center; 48°21'N 37°58'E. Elevation 744 ft/226 m. Coal mines.

Komsomol'sk, UKRAINE: see KOMSOMOL's'K, Donets'k oblast; or KOMSOMOL's'K, Poltava oblast.

Komsomol'skaya Pravda Islands (kuhm-suh-MOLskah-yah–PRAHV-dah), in the LAPTEV SEA of the ARCTIC OCEAN, 10 mi/16 km–15 mi/24 km off the N TAYMYR PENINSULA, in KRASNOYARSK TERRITORY, N RUSSIA; 77°15'N 107°00'E.

Komsomol's'ke (kom-so-MOL-ske) (Russian *Komsomol'skoye*), city, SE DONETS'K oblast, UKRAINE, in the DONBAS, on KAL'MIUS RIVER, 27 mi/43 km SE of DONETS'K; 47°40'N 38°04'E. Elevation 406 ft/123 m. Limestone-quarrying center; food processing. Technical school. Until 1949, Karakubbud (Russian *Karakubstroy*). Established in 1933, city since 1957.

Komsomol's'ke (kom-so-MOL-ske) (Russian *Komsomol'skoye*), town, central KHARKIV oblast, UKRAINE, on the Kharkiv-Donbas railroad trunkline, 28 mi/45 km SE of KHARKIV, 12 mi/19 km SE of ZMIYIV; 49°35'N 36°31'E. Elevation 341 ft/103 m. Site of the Zmiyiv regional thermal-electric station (1956); factories of construction materials and electrical machinery; dairy; vegetable processing. Energy-construction technical school. Established in 1956, town since 1960.

Komsomol's'ke (kom-so-MOL-ske) (Russian *Komsomol'skoye*), village (2004 population 7,200), N VINNYTSYA oblast, UKRAINE, 34 mi/55 km N of VINNYTSYA; 49°43'N 28°40'E. Elevation 879 ft/267 m. Flour mill. Until about 1935, known as Makhnivka (Russian *Makhnovka*).

Komsomol'skiy (kuhm-suh-MOL-skeeyee), town, W KIROV oblast, E central European Russia, on railroad, 12 mi/19 km WSW of (and administratively subordinate to) KOTEL'NICH; 58°11'N 48°00'E. Elevation 459 ft/139 m. Peat works. Has a hospital.

Komsomol'skiy (kuhm-suh-MOLS-keeyee), town (2006 population 3,190), central SAMARA oblast, E European Russia, on the left bank of the VOLGA RIVER, at lower navigation locks of the KUYBYSHEV DAM, adjoining the village of Kuneyevka (less than 2 mi/3.2 km to the E), opposite ZHIGULEVSK (5 mi/8 km to the S), and 33 mi/53 km NW of SAMARA; 53°29'N 49°28'E. Elevation 213 ft/64 m. Founded in 1950, at which time it was called Komsomol'sk-na-Volge; the name changed in the 1970s.

Komsomol'skiy (kuhm-suh-MOL-skyee), town (2005 population 3,780), NE KOMI REPUBLIC, NE European Russia, on road and railroad, 12 mi/20 km NW of VORKUTA; 67°33'N 63°47'E. Elevation 600 ft/182 m. Coal mining (Pechora coal basin). Sharp population decline due mainly to closing of some mines following the post-Communist introduction of market economy.

Komsomol'skiy (kuhm-suh-MOL-skeeyee), town (2006 population 13,800), E MORDVA REPUBLIC, central European Russia, on road and near railroad (Nuya station), on the Nuya River (basin of the SURA River, tributary of the VOLGA RIVER), 30 mi/48 km NE of SARANSK; 54°27'N 45°49'E. Elevation 652 ft/198 m. Manufacturing (cement, asbestos cement products). Formerly known as Zavodskoy.

Komsomolskiy, town, E FERGANA wiloyat, NE UZBEKISTAN, just S of MARGHILON, on Kokand-Andijan railroad; 40°25'N 71°45'E. Junction of railroad spur to FERGANA and KYZYL-KYYA (KYRGYZSTAN); cotton ginning. Formerly STANTSIYA GORCHAKOVO (gorCHAH-kah-vah). Also spelled Komsomol'skii, Komsomolskii.

Komsomol'skiy (kuhm-suh-MOL-skeeyee), town (2005 population 545), N CHUKCHI AUTONOMOUS OKRUG, N RUSSIAN FAR EAST, N of the ARCTIC CIRCLE, 148 mi/238 km ENE of BILIBINO; 69°10'N 172°41'E. Elevation 964 ft/293 m. Gold mining.

Komsomol'skiy (kuhm-suh-MOL-skeeyee), settlement (2005 population 2,815), N DAGESTAN REPUBLIC, SE European Russia, at the edge of the NE Caspian lowlands, on road and railroad, 7 mi/11 km N of KIZLYAR; 43°59'N 46°42'E. Below sea level. Agriculture (grain, fruits, grapes).

Komsomol'skiy (kuhm-suh-MOL-skeeyee), settlement (2005 population 3,845), S Republic of KALMYKIA-KHALMG-TANGEH, SE European Russia, on crossroads, 62 mi/100 km W of LAGAN; 45°20'N 46°02'E. In agricultural area (wheat, oats, vegetables). Formerly called Krasnyy Kamyshanik.

Komsomol'skiy, UKRAINE: see KOMSOMOL's'KYY.

Komsomol'sk-na-Amure (kuhm-suh-MOLSK–nah-ah-MOO-rye), city (2005 population 273,205), central KHABAROVSK TERRITORY, RUSSIAN FAR EAST, on the AMUR River, on crossroads and the BAYKAL-AMUR MAINLINE, 220 mi/354 km NE of KHABAROVSK; 50°33'N 137°00'E. Elevation 127 ft/38 m. Railroad junction, with the main line of the BAM continuing E to terminate at SOVETSKAYA GAVAN', and a S branch connecting with the TRANS-SIBERIAN RAILROAD at VOLOCHAYEVKA (JEWISH AUTONOMOUS OBLAST). Manufacturing center producing steel, machinery, ships, aircraft, electrical goods, refined oil, and wood products. Tin mines nearby. Has pedagogical and polytechnic institutes, drama theater, heritage and art museums. Founded in 1860 as a village of Permskoye; expanded to accommodate the growing steel industry and incorporated under current name in 1932. While the name suggests that it was built by members of the Komsomol (Communist youth organization), more than 70% of manpower was provided by convicts, a great number of whom died during construction.

Komsomol'sk-na-Volge, RUSSIA: see KOMSOMOL'SK, SAMARA oblast.

Komsomol'skoye (kuhm-suh-MOL-skuh-ye), village (2005 population 7,150), S central CHECHEN REPUBLIC, NE CAUCASUS, S European Russia, on road, 16 mi/26 km S of GROZNYY; 43°03'N 45°36'E. Elevation 1,253 ft/381 m. Meat and dairy livestock raising. Formerly known as Dubayeva.

Komsomol'skoye (kuhm-suh-MOL-skuh-ye), village (2005 population 3,115), E central CHECHEN REPUBLIC, S European Russia, near the SUNZHA RIVER, on road and near local railroad spur, 3 mi/5 km NW of and under administrative jurisdiction of GUDERMES; 43°23'N 46°09'E. Elevation 104 ft/31 m. Agriculture (oats, barley, buckwheat, soybean).

Komsomol'skoye (kuhm-suh-MOL-skuh-ye), village (2005 population 4,780), E CHUVASH REPUBLIC, central European Russia, on a left tributary of the SVIYAGA RIVER, on road, 18 mi/29 km S of KANASH; 55°15'N 47°33'E. Elevation 377 ft/114 m. In agricultural area (grains, vegetables, sunflowers); agricultural products; industrial ceramics. Until 1939, called Bol'shoy Kosheley.

Komsomol'skoye (kuhm-suh-MOL-skuh-ye), village (2005 population 8,920), W central DAGESTAN REPUBLIC, SE European Russia, in the SULAK RIVER basin, on road and railroad, 40 mi/64 km NW of MAKHACHKALA, and 15 mi/24 km E of KHASAVYURT; 43°11'N 46°54'E. Elevation 232 ft/70 m. In oil- and gas-producing region.

Komsomol'skoye (kuhm-suh-MOL-skuh-ye), village, SE SARATOV oblast, SE European Russia, 12 mi/19 km SSE of KRASNYY KUT; 50°46'N 47°03'E. Elevation 193 ft/58 m. Flour milling. Until 1941 (in German VOLGA ASSR), known as Fridenfeld.

Komsomol'skoye, UKRAINE: see KOMSOMOL's'KE, Donets'k oblast; KOMSOMOL's'KE, Kharkiv oblast; or KOMSOMOL's'KE, Vinnytsya oblast.

Komsomolsk-Ustyurt, town, KARAKALPAK REPUBLIC, W UZBEKISTAN, on W shore of ARAL SEA; 44°10'N 58°15'E. Soviet-era research station. Also Komsomolsk-na-Ustyurte.

Komsomol's'kyy (kom-so-MOL-skee) (Russian *Komsomol'skiy*), town, SE LUHANS'K oblast, UKRAINE, on road, 4 mi/6 km N of SVERDLOVS'K and subordinated to its city council; 48°07'N 39°40'E. Elevation 941 ft/286 m. Coal mine, enrichment plant, stone quarry. Established in 1905; until 1943, known as Tsentrosoyuz; town since 1954.

Komunars'k, UKRAINE: see ALCHEVS'K.

Komyshany (ko-mi-SHAH-nee) (Russian *Kamyshany*), town, W KHERSON oblast, UKRAINE, on the DNIEPER River delta, right bank of the Dnieper, 5 mi/8 km W of and subordinated to KHERSON; 46°38'N 32°30'E. Known since 1795; town since 1963.

Komyshnya (ko-MISH-nyah) (Russian *Kamyshnya*), town, N POLTAVA oblast, UKRAINE, near KHOROL RIVER, 32 mi/51 km NE of LUBNY; 50°11'N 33°41'E. Elevation 531 ft/161 m. Flour mill, food-flavoring plant; brickworks; forestry. Established before the 14th century, town since 1957.

Komyshuvakha (ko-mi-shoo-VAH-khah) (Russian *Kamyshevakha*), town (2004 population 7,900), W central LUHANS'K oblast, UKRAINE, in the DONBAS, at road junction and on railroad, 5 mi/8 km N of POPASNA; 48°27'N 39°01'E. Elevation 508 ft/154 m. Railroad station near a junction. Building materials. Established in 1853, town since 1938.

Komyshuvakha (ko-mi-shoo-VAH-hah) (Russian *Kamyshevakha*), town, N ZAPORIZHZHYA oblast, UKRAINE, on KINS'KA River, and 15 mi/24 km SE of ZAPORIZHZHYA; 47°43'N 35°31'E. Elevation 170 ft/51 m. Railroad station. Manufacturing of road-building machinery, transportation equipment, building materials. Historical museum. Established in 1770, town since 1957.

Komysh-Zorya (ko-MISH–zor-YAH) (Russian *Kamysh-Zarya*), town (2004 population 2,300), E ZAPORIZHZHYA oblast, UKRAINE, 38 mi/61 km N of BERDYANS'K; 47°19'N 36°41'E. Elevation 708 ft/215 m. Railroad junction; grain elevator, feed mill, and grading station. Established in 1905, town since 1938.

Kona (KO-nah), district, S and central part of HAWAII island's W coast, HAWAII county, HAWAII; c.50 mi/80 km long. Includes the Kona's coffee belt. The Kona coast, with fine deep-sea fishing offshore, is a favorite tourist spot. Growth of tourism has led to development of formerly agriculture-oriented towns of KAILUA-KONA, CAPTAIN COOK, HOLUALOA, HONAUNAU, among others. On Kealakekua Bay stands a monument to English explorer Captain James Cook, killed here by natives in 1779. KAILUA BAY to the N was the landing site in 1820 of the first U.S. missionaries to Hawaii.

Kona, town, FIFTH REGION/MOPTI, MALI, 47 mi/76 km NE of MOPTI; 14°57'N 03°53'W. Fishing.

Cross-references are shown in SMALL CAPITALS. The pronunciation guide is shown on page xix. The sources of population figures are shown on page xvii.

Konagai (ko-NAH-gah-ee), town, North Takaki county, NAGASAKI prefecture, NW KYUSHU, SW JAPAN, 22 mi/35 km N of NAGASAKI; 32°55′N 130°11′E.

Konagkend, urban settlement, NE AZERBAIJAN, at SE end of the Greater CAUCASUS, 20 mi/32 km S of KUBA; 40°41′N 48°42′E. Wheat; livestock. Rug manufacturing.

Konahuanui (KO-NAH-HOO-ah-NOO-ee), peak (3,105 ft/946 m), E OAHU, HAWAII, of KOOLAU RANGE. Unnamed point nearby reaches 3,150 ft/960 m.

Konakovo (kuh-nah-KO-vuh), city (2006 population 42,215), SE TVER oblast, W central European Russia, on the Ivankovo Reservoir of the VOLGA RIVER (landing), on highway branch and railroad spur, 50 mi/80 km ESE of TVER; 56°42′N 36°46′E. Elevation 449 ft/136 m. Porcelain center since 1809; metal goods, power tools, industrial lubricants; food processing (bakery, fish factory). Electric power station. Until 1930, called Kuznetsovo. Received the population evacuated (1936–1937) from KORCHEVA, flooded by the construction of the reservoir. Made city in 1937.

Konakry, GUINEA: see CONAKRY.

Konan (KO-nahn), city, AICHI prefecture, S central HONSHU, central JAPAN, 12 mi/20 km N of NAGOYA; 35°19′N 136°52′E. Textiles.

Konan (KO-NAHN), town, Kagawa county, KAGAWA prefecture, NE SHIKOKU, W JAPAN, 6 mi/10 km S of TAKAMATSU; 34°14′N 134°00′E.

Konan (KO-nahn), town, Osato county, SAITAMA prefecture, E central HONSHU, E central JAPAN, 25 mi/40 km N of URAWA; 36°06′N 139°20′E. Air conditioners.

Konan (KO-NAHN), town, Koka county, SHIGA prefecture, S HONSHU, central JAPAN, 19 mi/30 km S of OTSU; 34°55′N 136°10′E. Origin of *Koga Ryu Ninja* martial art. Formed in early 1940s by combining former villages of Tatsuike, Terasho, and two smaller villages.

Konanur (KO-nuh-noor), town, HASSAN district, KARNATAKA state, S INDIA, on KAVERI RIVER, and 25 mi/40 km S of HASSAN; 12°38′N 76°03′E. Handicraft wickerwork; grain, rice.

Konar, AFGHANISTAN: see KUNAR.

Konarak (ko-NAH-ruhk), village, PURI district, E ORISSA state, E central INDIA, 18 mi/29 km ENE of Puri, near BAY OF BENGAL. A religious site, with famous Black Pagoda, ruined thirteenth century temple dedicated to sun god, whose chariot it represents; fine carvings of wheels and horses about base. Has a lake. Sometimes spelled Kanarak, Konark.

Kona Shahr, China: see WENSU.

Konawa (KAHN-uh-wah), town (2006 population 1,424), SEMINOLE county, central OKLAHOMA, 15 mi/24 km NNW of ADA; 34°57′N 96°45′W. Trade center and shipping point for agriculture (corn, alfalfa, peanuts; livestock) and oil and natural gas area; manufacturing (orthopedic devices). Lake Konawa reservoir to NE.

Konda (KON-dah), town, Taki county, HYOGO prefecture, S HONSHU, W central JAPAN, 22 mi/35 km N of KOBE; 35°00′N 135°06′E. Pottery.

Konda, INDIA: see KONTA.

Konda, Russia: see ZELENOBORSK.

Kondagaon (KON-dah-goun), also **Kondegaon, Konda**, or **Konta**, town (2001 population 26,772), BASTAR district, S CHHATTISGARH state, E central INDIA, on tributary of the GODAVARI RIVER 40 mi/64 km NNW of JAGDALPUR, 60 mi/97 km NNW of RAJAHMUNDRY; 19°36′N 81°40′E. In forest area (sal, bamboo, myrobalan; lac cultivation). Rice, oilseeds, mangoes. Mica deposits (E).

Kondalwadi (KON-dahl-wah-dee), town, NANDED district, MAHARASHTRA state, W central INDIA, 37 mi/60 km SE of NANDED, 7 mi/11.3 km W of GODAVARI-MANJRA river confluence; 18°48′N 77°46′E. Millet, rice. Also spelled Kundalwadi.

Kondapalle (kon-dah-pah-LE), town, KRISHNA district, ANDHRA PRADESH state, S INDIA, near KRISHNA RIVER, 8 mi/12.9 km NW of VIJAYAWADA; 16°37′N 80°32′E. Noted hand-carved wooden toys. Rice, oilseeds, cotton. Has fortress from 14th century, taken by Aurangzeb in 1687, fell to English in 1766. Chromite mining in nearby forested hills (bamboo, myrobalan).

Konda River (kuhn-DAH), 715 mi/1,151 km long, W KHANTY-MANSI AUTONOMOUS OKRUG, TYUMEN oblast, RUSSIA; rises in swampy area; flows W, SSE, E, past NAKHRACHI, and NE to the IRTYSH River at Repolovo, 30 mi/48 km SSE of KHANTY-MANSIYSK. Large sable and beaver reserve in its upper reaches.

Kondi, PAKISTAN: see NOK KUNDI.

Kondinskoye (KON-deen-skuh-ye), town (2005 population 4,195), SW KHANTY-MANSI AUTONOMOUS OKRUG, W SIBERIA, RUSSIA, near the terminus of a local road, 92 mi/148 km SW of KHANTY-MANSIYSK; 59°39′N 67°24′E. Elevation 108 ft/32 m. In oil- and gas-producing area. Formerly called Nakhrachi.

Kondinskoye, RUSSIA: see OKTYABR'SKOYE, KHANTY-Mansi Autonomous Okrug.

Kondoa (kon-DO-ah), town, DODOMA region, N central TANZANIA, 85 mi/137 km N of DODOMA, near Bubu River; 04°55′S 35°45′E. Road junction. Grain; livestock.

Kondofrei (kahn-do-FREE), village, SOFIA oblast, RADOMIR obshtina, BULGARIA; 42°25′N 24°00′E. Said to be named after a Belgian Crusader.

Kondol' (KON-duhl), village, S central PENZA oblast, E European Russia, on road, 25 mi/40 km S of PENZA; 52°49′N 45°03′E. Elevation 688 ft/209 m. In wheat-growing area; livestock; dairy processing (butter and cheese plant).

Kondoma River (kuhn-duh-MAH), 265 mi/426 km long, SW KEMEROVO oblast, RUSSIA; rises in the ABAKAN RANGE; flows N, through the GORNAYA SHORIYA, past TASHTAGOL and KUZEDEYEVO, to the TOM′ River at NOVOKUZNETSK. Lower course lies in the KUZNETSK coal basin.

Kondopoga (KON-duh-puh-gah), Finnish *Kontupohja*, city (2005 population 34,200), S Republic of KARELIA, NW European Russia, on shore of the Kondopoga Bay of Lake ONEGA, on the Murmansk railroad, 33 mi/53 km N of PETROZAVODSK; 62°12′N 34°17′E. Elevation 203 ft/61 m. Wood pulp and paper (one-third of Russia's total newsprint paper production); hydroelectric plant on the SUNA RIVER. Made city in 1938.

Kondowe, MALAWI: see LIVINGSTONIA.

Kondrashivska, UKRAINE: see STANYCHNO-LUHANS'KE.

Kondrativs'kyy (kon-DRAH-teev-skee) (Russian *Kondratyevskiy*), SE suburb of HORLIVKA, central DONETS'K oblast, UKRAINE, in the DONBAS, 6 mi/9.7 km SE of Horlivka city center; 48°18′N 38°12′E. Elevation 816 ft/248 m. Coal mines.

Kondratyevskiy, UKRAINE: see KONDRATIVS'KYY.

Kondrovo (KON-druh-vuh), city (2005 population 17,145), N central KALUGA oblast, central European Russia, on the Shanya River (OKA River basin), on road and railroad, 25 mi/40 km NW of KALUGA; 54°48′N 35°55′E. Elevation 459 ft/139 m. Paper mill; medical equipment; food packaging; bakery. Has a theater. In 1790, a paper mill was built, which in the 19th century produced high-quality paper; wood pulp and paper (especially writing paper). Became city in 1938; until then, a village of Troitskoye.

Konduga (kahn-DOO-gah), town, BORNO state, extreme NE NIGERIA, 25 mi/40 km SE of MAIDUGURI; 11°39′N 13°25′E. Cassava, millet, durra, gum arabic; cattle, skins.

Kondurcha River (kuhn-door-CHAH), 75 mi/121 km long, SAMARA oblast, SE European Russia; rises approximately 6 mi/10 km E of SHENTALA; flows generally W, past ZUBOVKA, and S, past YELKHOVKA, to the SOK RIVER at KRASNYY YAR.

Konduz, AFGHANISTAN: see KUNDUZ.

Koné (ko-nai), village, W NEW CALEDONIA, 130 mi/209 km NW of NOUMÉA; 21°04′S 164°50′E. Agriculture (coffee; livestock). Provides nickel for DONIAMBO smelter.

Koneprusy Caves, CZECH REPUBLIC: see BOHEMIAN KARST.

Konetspol, POLAND: see KONIECPOL.

Koneurgench, TURKMENISTAN: see KUNYA-URGENCH.

Konevo (kuh-NYO-vuh), village, SW ARCHANGEL oblast, NW European Russia, on the ONEGA River, on road, 50 mi/80 km SW of PLESETSK; 62°08′N 39°20′E. Elevation 344 ft/104 m. Coarse grain.

Konezavodskiy (kuh-nye-zah-VOT-skeeye), settlement, S OMSK oblast, SW Siberian Russia, near highway and the TRANS-SIBERIAN RAILROAD, 18 mi/29 km W of OMSK, and 8 mi/13 km E of MAR′YANOVKA, to which it is administratively subordinate; 54°57′N 72°51′E. Elevation 341 ft/103 m. Horse breeding.

Konfodah, SAUDI ARABIA: see QUNFIDHA.

Konfogsi, GHANA: see KUNFOGSI.

Konfosi, GHANA: see KUNFOGSI.

Konfusi, GHANA: see KUNFOGSI.

Kong (KAWNG), largely abandoned town, Savanes region, N CÔTE D'IVOIRE, in uplands 260 mi/418 km N of ABIDJAN, 70 mi/113 km ESE of KORHOGO; 09°09′N 04°37′W. Was capital of ancient Kong kingdom. Founded in 11th century. Has monument to commemorate Marchand mission (which led to the FASHODA incident, 1898).

Konganevik Point (kon-gah-NAI-vik), NE ALASKA, on CAMDEN BAY of BEAUFORT SEA, 100 mi/161 km ESE of BEECHEY POINT; 70°02′N 145°11′W. Inuit settlement here.

Köngen (KUHNG-uhn), village, BADEN-WÜRTTEMBERG, SW GERMANY, on the NECKAR, 4.5 mi/7.2 km SE of ESSLINGEN; 48°41′N 09°22′E. Has excavated Roman castrum.

Kongju (GONG-JOO), city, SOUTH CHUNGCHONG province, SOUTH KOREA, on KUM RIVER, and 19 mi/31 km NW of TAEJON; 36°26′N 127°07′E. Agricultural center (rice, barley, sweet potatoes, tobacco); sericulture. Old capital of Paekche Kingdom; many historical sites. Part of KYERYONG MOUNTAIN NATIONAL PARK.

Kongka, China: see GONGGAR.

Kong Karls Land or **King Charles Land**, group of 3 islands and several islets (□ 128 sq mi/332.8 sq km), part of the Norwegian possession of SVALBARD, in BARENTS SEA of ARCTIC OCEAN, E of NORTHEAST LAND and W of SPITSBERGEN; 78°36′–79°00′N 26°20′–30°02′E. Largest island is KONGSOYA, Norwegian *Kongsøya*; 30 mi/48 km long, 2 mi/3.2 km–7 mi/11.3 km wide; 78°54′N 28°50′E. Rises to 1,050 ft/320 m. Other islands include Svenskøya and Abeløya. Also called King Charles, or Karl, Islands

Konglu (kawng-LOO), village, Nagmung township, KACHIN STATE, MYANMAR, 33 mi/53 km E of PUTAO, on pack trail to CHINA (YUNNAN province).

Kongo, Kingdom of, former state of W central Africa, founded in the 14th century. In the 15th century, the kingdom stretched from the CONGO (N) to the Loje (S) rivers and from the ATLANTIC OCEAN (W) to beyond the KWANGO RIVER (E). Received tribute from several smaller autonomous states to the S and E. Kongo was ruled by the *manikongo*, or king, who appointed governors over each of the six provinces. In 1482, Diogo Cão, a Portuguese explorer, visited the kingdom, and the reigning *manikongo*, Nzinga Nkuwu, was favorably impressed with Portuguese culture. In 1491, Portuguese missionaries, soldiers, and artisans were welcomed at MBANZA, the capital. The missionaries soon gained converts, including Nzinga Nkuwu, and the soldiers helped suppress an internal rebellion. The next *manikongo*, Afonso I (reigned 1505–1543), was raised as a Christian and attempted to convert the kingdom to Christianity and European ways. However, the Portuguese residents in Kongo were primarily interested in increasing their private fortunes (especially through capturing Africans and selling them into slavery), and their continued rapaciousness played a major part in weakening the kingdom and reducing the hold of the capital (renamed SÃO SAL-VADOR) over the provinces. After the death of Afonso, Kongo declined rapidly and suffered major civil wars. The Portuguese shifted their interest S to the kingdom of Ndongo and helped Ndongo defeat Kongo in 1556. The slave trade, which undermined the social structure of Kongo, continued to weaken the authority of the *manikongo*. In 1641, Manikongo Garcia II allied himself with the Dutch in an attempt to control Portuguese slave traders, but in 1665 a Portuguese force decisively defeated the army of Kongo. From that time onward the *manikongo* was little more than a vassal of PORTUGAL. The kingdom disintegrated into a number of small states, all controlled to varying degrees by the Portuguese. The area of Kongo was incorporated mostly into ANGOLA and partly into the Independent State of the Congo (now the DEMOCRATIC REPUBLIC OF THE CONGO) in the late 19th century.

Kongolo (kawng-GO-lo), town, KATANGA province, E CONGO, on LUALABA RIVER, on railroad, and 160 mi/257 km WNW of KALEMIE; 05°23'S 27°00'E. Elev. 1,755 ft/534 m. Transshipment point (steamer-railroad) and market for produce (manioc, yams, palm oil, plantains); cotton ginning. Airport. Has Roman Catholic mission and schools. Seat of vicar apostolic of N SHABA.

Kongoussi, town (2005 population 22,195), ⊙ BAM province, CENTRE-NORD region, N central BURKINA FASO, on road, 65 mi/105 km N of OUAGADOUGOU; 13°19'N 01°31'W. Agriculture (shea nuts, vegetables) and livestock (cattle). Gold deposits nearby.

Kongque He, CHINA (TIBET), NEPAL, and INDIA: see GHAGHARA River.

Kongsberg (KAWNGS-ber), city (2007 population 23,644), BUSKERUD county, S E NORWAY, on the Numedalslågen River; 59°39'N 09°39'E. Commercial, industrial, and winter sports center and has a hydroelectric power plant. Formerly a silver-mining center, Kongsberg has old mines and a great church (1761) that are tourist attractions.

Kong, Se (KAWNG, SAI), river, 300 mi/483 km long, in LAOS and CAMBODIA; rises in TRUONG SON Range SW of HUE at c.16°00'N; flows SSW, past MUONG MAY and SIEM PANG, and after receiving the SE SAN (or San River) enters the MEKONG River at STUNG TRENG.

Kongsfjorden (KAWNGS-fyawr-uhn) [Norwegian= kings bay], inlet of the ARCTIC OCEAN, 14 mi/23 km long, NW SPITSBERGEN, SVALBARD, NORWAY. NY-ÅLESUND is on the inlet. The scenic fjord is often visited by tourist vessels.

Kongsøya, NORWAY: see KONG KARLS LAND.

Kongsvinger (KAWNGS-vin-guhr), city, HEDMARK county, SE NORWAY, near Swedish border, on GLOMMA River, and 50 mi/80 km ENE of OSLO; 60°12'N 12°00'E. Railroad junction. Agriculture (wheat, potatoes); manufacturing (furniture, skis, gloves); industry and growing service sector. Has fortress, built 1683; and a county hospital.

Kongsvoll (KAWNGS-vawl), village, SØR-TRØNDELAG county, central NORWAY, on DRIVA River, on railroad, and 22 mi/35 km NE of DOMBÅS. Winter-sports center. Called Hullet until visit of Fredrik IV in 1704.

Kongwa (KON-gwah), village, DODOMA region, central TANZANIA, 60 mi/97 km E of DODOMA; 06°12'S 36°25'E. Airstrip here. Peanuts, corn, wheat; cattle, sheep, goats. Center of failed peanut-growing scheme of late 1940s and early 1950s.

Koni, CONGO: see MWADINGUSHA.

Koniakari (ko-NEE-a-kar-ee), town, FIRST REGION/KAYES, MALI, 50 mi/80 km ENE of KAYES; 14°34'N 10°54'W.

Konice (KO-nyi-TSE), German *konitz*, town, JIHO-MORAVSKY province, W central MORAVIA, CZECH RE-PUBLIC, on railroad, and 16 mi/26 km W of OLOMOUC; 49°35'N 16°53'E. Agriculture (barley, oats); manufacturing (clothing); tanning. Has a baroque castle.

Koniecpol (ko-NEETS-pol), Russian *Konetspol* or *Konetspol'*, town, Częstochowa province, S central POLAND, on PILICA RIVER, and 22 mi/35 km SSE of RADOMSKO. Copper rolling.

König, Bad, GERMANY: see BAD KÖNIG.

Königgrätz, CZECH REPUBLIC: see HRADEC KRÁLOVÉ.

Königinhof, CZECH REPUBLIC: see DVUR KRALOVE NAD LABEM.

Königsaal, CZECH REPUBLIC: see ZBRASLAV.

Königsbach-Stein (KO-niks-bahkh–SHTEIN), village, BADEN-WÜRTTEMBERG, SW GERMANY, 6 mi/9.7 km WNW of PFORZHEIM; 48°58'N 08°48'E. Wine. Has fortified Gothic church.

Königsberg, CZECH REPUBLIC: see KLIMKOVICE.

Königsberg, POLAND: see CHOJNA.

Königsberg, RUSSIA: see KALININGRAD, city, KALININGRAD oblast.

Königsberg an der Eger, CZECH REPUBLIC: see KYN-SPERK NAD OHRI.

Königsberg in Bayern (KO-niks-berg in BEI-ern), town, BAVARIA, central GERMANY, in LOWER FRAN-CONIA, 15 mi/24 km ENE of SCHWEINFURT; 50°05'N 10°33'E. Manufacturing of electrical goods, lamps, concrete. Regiomontanus (1436–1476), German astronomer, born here. Has many half-timbered houses.

Königsbronn (ko-niks-BRUHN), village, BADEN-WÜRTTEMBERG, S GERMANY, 4 mi/6.4 km NNW of HEIDENHEIM; 48°45'N 10°07'E. Site of ironworks since 14th century.

Königsbrück (ko-niks-BROOK), town, SAXONY, E GERMANY, in UPPER LUSATIA, 17 mi/27 km NNE of DRESDEN; 51°17'N 13°55'E. Manufacturing of electronics. Spa. Chartered 1351.

Königsbrunn (ko-niks-BRUN), town, BAVARIA, S GERMANY, in SWABIA, 7 mi/11.3 km S of AUGSBURG; 48°16'N 10°23'E. Manufacturing of machinery, tools, and electrical goods. Chartered in 1967.

Königsee (KO-niks-zai), town, THURINGIA, central GERMANY, at foot of THURINGIAN FOREST, 12 mi/19 km W of SAALFELD; 50°40'N 11°06'E. Manufacturing of toys, pharmaceuticals; brewing.

Königsfeld, CZECH REPUBLIC: see KRALOVO POLE.

Königsfeld im Schwartzwald (KO-niks-feld im SHVAHRTS-vahlt), village, BADEN-WÜRTTEMBERG, SW GERMANY, in BLACK FOREST, 6 mi/9.7 km N of VILLINGEN-SCHWENNINGEN; 48°09'N 08°25'E. Climatic health resort. 12th-century church.

Königshof, CZECH REPUBLIC: see KRALUV DVUR.

Königshofen im Grabfeld, Bad, GERMANY: see BAD KÖNIGSHOFEN IM GRABFELD.

Königshütte, POLAND: see CHORZÓW.

Königslutter am Elm (KO-niks-luht-tuhr ahm ELM), town, LOWER SAXONY, N GERMANY, 8 mi/12.9 km WNW of HELMSTEDT; 52°15'N 10°49'E. Manufacturing of tobacco products, machinery; sugar processing. Has Romanesque basilica with tomb of Emperor Lothair III.

Königsee (KO-niks-zai) or **Bartholomäussee** [Ger.= lake of St. Bartholomew], lake (□ 2 sq mi/5.2 sq km), UPPER BAVARIA, GERMANY, in SALZBURG Alps, 10 mi/16 km SSE of BAD REICHENHALL; 47°27'N 12°59'E. One of Germany's most beautiful lakes, it lies amid magnificent Alpine scenery (elevation 1.975 ft/0.602 m). Lake is 5 mi/8 km long, 1 mi/1.6 km–1.5 mi/2.4 km wide, 617 ft/188 m deep. Protected as a national park.

Königstadtl, CZECH REPUBLIC: see MESTEC KRALOVE.

Königstein (KO-nig-shtein), town, SAXONY, E central GERMANY, in SAXONIAN SWITZERLAND, on ELBE RIVER, 7 mi/11.3 km ESE of PIRNA; 50°55'N 14°05'E. Paper milling, woodworking; summer resort. Towered over by fortress built before 1150 by kings of BOHEMIA. Town passed 1406 to margraves of MEIS-SEN. Used in both World Wars as general-officers' prisoner-of-war camp.

Königstein im Taunus (KO-nig-shtein im TOU-nus), town, HESSE, central GERMANY, in the TAUNUS, 5 mi/8 km W of OBERURSEL; 50°11'N 08°28'E. Summer resort

and winter-sports center. On nearby hill (W) are ruins of 13th-century fortress, blown up by French in 1796.

Königstuhl (KO-nig-shtool), massif (1,857 ft/566 m), BADEN-WÜRTTEMBERG, GERMANY, at HEIDELBERG; 49°43'N 08°45'E. Observatory; tower on summit affords excellent view of Neckar valley and Rhine plain.

Königswart, Bad, CZECH REPUBLIC: see LAZNE KYNZ-VART.

Königswiesen (KUH-niks-vee-sen), township, NE UPPER AUSTRIA, in the Mühlviertel, 26 mi/42 km ENE of LINZ, near the Lower Austrian line; 48°24'N 14°50'E. Summer resort; dairy farming. The parish church is an excellent example of late Gothic in Austria.

Königswinter (ko-niks-WIN-ter), city, North Rhine-Westphalia, W GERMANY, in the SIEBENGEBIRGE, at N foot of the DRACHENFELS (rack-and-pinion railroad), on right bank of the RHINE (landing), 6 mi/9.7 km SE of BONN; 50°41'N 07°12'E. Seat of a number of political institutions, including an academy for training civil servants. Manufacturing includes motor vehicle parts, electronic instruments, ceramics, and furniture. Basalt is quarried nearby. It is a wine trading center. Summer resort, tourist center. It was chartered in 1898 and after World War II developed rapidly due to its proximity to BONN, which had just become West Germany's capital. Noteworthy buildings are a Romanesque castle from the 12th century and a 12th-century Benedictine monastery with a 12th–13th-century church. Nearby is "Petersberg," the government's guest house for foreign visitors.

Königs-Wusterhausen (KO-niks–vus-tuhr-HOU-suhn), town, BRANDENBURG, E GERMANY, on DAHME RIVER, 17 mi/27 km SE of BERLIN; 52°19'N 13°38'E. Largely residential, linked to BERLIN by railroad. Had one of Germany's largest radio broadcasting stations during Hitler's reign and during the existence of the German Democratic Republic. Hunting lodge (18th century) was favorite retreat of Frederick William I. Sometimes written as Königswusterhausen.

Konin (KO-neen), town (2002 population 82,353), Konin province, W central POLAND, on WARTA River, and 60 mi/97 km NW of ŁÓDŹ railroad junction. Manufacturing of bricks, furniture; brewing, flour milling, tanning; retailing center. Vital pre-World War II Jewish secular and religious community destroyed by Nazis.

Konispol (ko-nees-POL), town, in southernmost AL-BANIA, on Greek border, 30 mi/48 km S of GJIR-OKASTËR, near Channel of CORFU; 39°39'N 20°10'E. Greek border crossing closed recently. Archaeological site nearby. Also spelled Konispoli.

Konitsa (KO-nee-tsah), town, IOÁNNINA prefecture, EPIRUS department, NW GREECE, near Albanian border, 27 mi/43 km N of IOÁNNINA, on Aoos (VIJOSË) River; 40°03'N 20°45'E. Barley, corn; olive oil; timber; livestock (dairy products, meat, hides). Tourist center. Bishopric with Byzantine churches. Also spelled Konitza.

Konitz, CZECH REPUBLIC: see KONICE.

Konitz, POLAND: see CHOJNICE.

Konitza, Greece: see KONITSA.

Köniz (KUH-nitz), town (2000 population 37,782), BERN canton, W central SWITZERLAND; SW suburb of BERN. Elevation 1,877 ft/572 m. The Romanesque-Gothic church, founded in the 10th century by Rudolph II of Burgundy, has noteworthy 14th-century stained glass and frescoes.

Konjic (kon-YEETS), town, upper HERZEGOVINA, BOS-NIA AND HERZEGOVINA, on the NERETVA RIVER, on railroad, and 28 mi/45 km SW of SARAJEVO; 43°38'N 17°57'E. Local trade center. Also spelled Konyich.

Konjuh (kon-YOO), mountain (4,356 ft/1,328 m) in DINARIC ALPS, central BOSNIA, BOSNIA AND HERZE-GOVINA, 9 mi/14.5 km NW of KLADANJ; 44°18'N 18°31'E. Drinjača River rises on E slope. Also spelled Konyukh.

Cross-references are shown in SMALL CAPITALS. The pronunciation guide is shown on page xix. The sources of population figures are shown on page xvii.

Konkan (kon-KAHN), coastal plain in W MAHARASH-TRA state, W central INDIA, between WESTERN GHATS (E) and ARABIAN SEA (W); from DAMAN, DAMAN AND DIU Union Territory (N) extends c.330 mi/531 km S, through districts of THANE, MUMBAI (BOMBAY) Suburban (port of Mumbai just S), RAIGARH, and RATNAGIRI, SINDHUDURG to GOA state; 30 mi/48 km–60 mi/97 km wide. Has outliers of WESTERN GHATS, including MATHERAN (health resort) and JAWHAR; drained by several mountain streams. Fertile coast (rice, coconuts, mangoes), with fishing centers (mackerel, pomfrets) at Bassein, MURUD, Ratnagiri, and MALVAN; inland rises to rocky and rugged spurs. Annual rainfall, over 100 in/254 cm. Crossed by new Mumbai-Goa railroad line. Portuguese dominated coastal trade in sixteenth and seventeenth century; numerous creeks and harbors (as at VIJAYADURG) were retreats for Maratha pirates in eighteenth century.

Konka River, UKRAINE: see KINS'KA RIVER.

Konken, THAILAND: see KHON KAEN.

Konko (KON-ko), town, on KONO-SHIMA island, Asaguchi county, OKAYAMA prefecture, SW HONSHU, W JAPAN, 19 mi/30 km WSW of OKAYAMA; 34°32′N 133°37′E. Headquarters of Konko-kyo new religious section.

Konkouré (kon-KOO-rai), village, Mamou prefecture, Mamou administrative region, W central GUINEA, in Moyenne-Guinée geographic region, on CONAKRY-KANKAN railroad, and 110 mi/177 km ENE of Conakry. Also called Kouraia Konkouré.

Konkouré River, c.160 mi/257 km long, GUINEA; rises in FOUTA DJALLON mountains W of MAMOU; flows generally W, past WASSOU, to the ATLANTIC OCEAN at Sangaréa Bay, just N of CONAKRY; 09°48′N 13°46′W.

Konkweso (kahn-KWAI-so), town, NIGER state, WNW NIGERIA, near BENIN border, and 160 mi/257 km NNW of Illorin; 10°51′N 04°06′E. Market town. Rice, yams, and cassava.

Konnagar (KON-nuh-guhr), town, HUGLI district, S central WEST BENGAL state, E INDIA, on HUGLI River, and 9 mi/14.5 km N of CALCUTTA city center; 22°42′N 88°21′E. Manufacturing of chemicals, glass; cotton milling, jute pressing, liquor distilling. Oilseed milling at HATIRKUL, in N area.

Könnern (KON-nern), town, SAXONY-ANHALT, central GERMANY, near the SAALE, 9 mi/14.5 km S of BERN-BURG; 51°41′N 11°46′E.

Konnur (ko-NOOR), town, BELGAUM district, KARNA-TAKA state, S INDIA, 28 mi/45 km NE of BELGAUM; 16°12′N 74°45′E. Markets cotton, chili, millet; wool spinning. Noted cell tombs nearby. Irrigation dam 1 mi/1.6 km N, on GHATPRABHA River.

Kono (KO-no), village, Nanjo county, FUKUI prefecture, central HONSHU, W central JAPAN, 19 mi/30 km S of FUKUI; 35°49′N 136°04′E.

Kono, SIERRA LEONE: see SEFADU.

Konobeyevo-Lesnoye, RUSSIA: see LESNOYE KONO-BEYEVO.

Konobougou, village, FOURTH REGION/SÉGOU, MALI, 31 mi/50 km N of DIOÏLA.

Konocti, Mount, peak (3,967 ft/1,209 m), Lake county, NW CALIFORNIA, on W shore of Clear Lake, c.80 mi/129 km N of SAN FRANCISCO.

Konokovo (kuh-NO-kuh-vuh), village (2005 population 7,545), E KRASNODAR TERRITORY, S European Russia, in the NW foothills of the Greater CAUCASUS Mountains, on the KUBAN' River, on road and railroad, 11 mi/18 km SE of ARMAVIR; 44°51′N 41°19′E. Elevation 787 ft/239 m. Thermal hydroelectric power plant in the vicinity. Sometimes spelled Kanokovo (same pronunciation).

Konolfingen, commune, BERN canton, W central SWITZERLAND, 10 mi/16 km ESE of BERN.

Konolfingen, district, central BERN canton, W central SWITZERLAND. Main town is MÜNSINGEN; population is German-speaking and Protestant.

Konongo (ko-NON-go), town, ASHANTI region, GHANA, on railroad, and 7 mi/11.3 km WNW of JUASO; 06°37′N 01°13′W. Gold mining; cacao, cassava, corn.

Konopischt, CZECH REPUBLIC: see BENESOV.

Konopiste, CZECH REPUBLIC: see BENESOV.

Konoplyanka, UKRAINE: see HEORHIYIVKA.

Konosha (KO-nuh-shuh), town (2005 population 12,200), SW ARCHANGEL oblast, N European Russia, on road, 90 mi/145 km N of VOLOGDA; 60°58′N 40°15′E. Elevation 682 ft/207 m. Junction of Moscow-Archangel and North Pechora railroads; railroad establishments. Sawmilling.

Kono-shima (ko-NO–shee-mah), island (□ 7 sq mi/18.2 sq km), Asaguchi county, OKAYAMA prefecture, off SW HONSHU, W JAPAN, in HIUCHI, nearly connected with S coast of Honshu, near KASAOKA; 4 mi/6.4 km long, 2 mi/3.2 km wide. Produces rice, rushes.

Konosu (KO-nos), city, SAITAMA prefecture, E central HONSHU, E central JAPAN, 16 mi/25 km N of Urawa; 36°03′N 139°31′E.

Konotop (ko-no-TOP), city (2001 population 92,657), W SUMY oblast UKRAINE, on the Yezuch River, left tributary of the SEYM RIVER; 51°14′N 33°12′E. Elevation 492 ft/149 m. Raion center, railroad junction, and agricultural center. Railroad repair shops; machine building, electromechanics; food processing, manufacturing (clothing, construction materials). Three vocational technical schools; heritage museum. Founded in 1634 and made into a fortress. Ruled briefly by the Ukrainian hetman Khmel'nyts'kyy (Polish *Chmielnicki*) and served in 1654–1781 as a fortified company town of the Nizhyn regiment of the Ukrainian Cossacks; beseiged by the Russian army, which was defeated by Ukrainian Cossack Hetman I. Vyhovs'kyy (1659); destroyed by the Poles (1664); okruha center (1923–1932), raion center of Chernihiv oblast (1932–1939), subsequently, in Sumy oblast. Sizeable Jewish community since the 19th century; reduced by the 1881 pogroms and subsequent emigration, as well as the 1905 and 1919 pogroms, but still numbering 5,800 in 1939; eliminated by the Nazis in 1941—fewer than 1,000 Jews remaining in 2005.

Konoura (ko-NO-oo-rah), town, Yuri county, Akita prefecture, N HONSHU, NE JAPAN, fishing port on SEA OF JAPAN, 34 mi/55 km S of AKITA city; 39°15′N 139°55′E.

Konovo Mountains, Bulgaria: see KONYAVSKA MOUNTAINS.

Konradshof, POLAND: see SKAWINA.

Konskie (KON-skee), Polish *Końskie*, Russian *Konski*, town (2002 population 21,338), KIELCE province, E central POLAND, on railroad, and 36 mi/58 km WSW of RADOM. Manufacturing of agricultural machinery; tanning, sawmilling. Castle ruins nearby.

Konstadt, POLAND: see WOLCZYN.

Konstantin (kon-stahn-TEEN), village, LOVECH oblast, ELENA obshtina, BULGARIA; 42°56′N 26°14′E.

Konstantinograd, UKRAINE: see KRASNOHRAD.

Konstantinovka (kuhn-stahn-TEE-nuhf-kah), village, SE KIROV oblast, E central European Russia, on the VYATKA River, 12 mi/19 km NE, and under administrative jurisdiction, of MALMYZH; 56°41′N 50°52′E. Elevation 423 ft/128 m. Logging, lumbering, sawmilling.

Konstantinovka (kuhn-stahn-TEE-nuhf-kah), village, SW AMUR oblast, SE SIBERIA, RUSSIAN FAR EAST, less than 8 mi/13 km NE of Yanjiang, across the Chinese border, on road, 53 mi/85 km S of BLAGOVESHCHENSK; 49°37′N 127°59′E. Elevation 334 ft/101 m.

Konstantinovka, UKRAINE: see KOSTYANTYNIVKA, Donets'k oblast; KOSTYANTYNIVKA, MYKOLAYIV oblast; or KOSTYANTYNIVKA, KHARKIV oblast.

Konstantinovka, UKRAINE: see YUZHNOUKRAYINS'K.

Konstantinov Kamen' (kuhn-stahn-TEE-nuhf-KAH-myen), northernmost point (1,480 ft/451 m) of the URAL Mountains, in NW Siberian Russia, rising in isolated site amid the tundra, 25 mi/40 km from the BAYDARATA BAY of the KARA SEA; 68°30′N 66°20′E.

Konstantinovo (kuhn-stahn-TEE-nuh-vuh), village (2006 population 3,555), NE MOSCOW oblast, central European Russia, on the slopes of the N central KLIN-DMITROV RIDGE, on road, 17 mi/27 km NNW of SERGIYEV POSAD; 56°33′N 38°02′E. Elevation 531 ft/161 m. In agricultural area; food processing (dairy).

Konstantinovsk (kuhn-stahn-TEE-nuhfsk), city (2006 population 18,905), central ROSTOV oblast, S European Russia, on the right bank of the DON River, on road, 105 mi/169 km NE of ROSTOV, and 40 mi/64 km E of SHAKHTY; 47°35′N 41°05′E. Elevation 187 ft/56 m. Center of agricultural area; food industries (fish processing, flour milling; dairy products). Arose in the 17th century as a Cossack settlement (*stanitsa*) called Babskiy. In the 1860s, became a village of Konstantinovskaya, later town of Konstantinovskiy. Made city and renamed in 1967.

Konstantinovskaya (kuhn-stahn-TEE-nuhf-skah-yah), village (2005 population 4,170), E KRASNODAR TERRITORY, S European Russia, on the LABA RIVER, on road, 6 mi/10 km SE of KURGANINSK; 44°50′N 40°43′E. Elevation 705 ft/214 m. In agricultural area; has an agricultural (land survey) vocational school.

Konstantinovskaya, RUSSIA: see KONSTANTINOVSK.

Konstantinovskiy (kuhn-stahn-TEE-nuhf-skeeyee), town, S central MOSCOW oblast, central European Russia, on road and railroad, 34 mi/55 km S of MOSCOW, and 7 mi/11 km E of PODOL'SK; 55°25′N 37°42′E. Elevation 465 ft/141 m. Woolen milling.

Konstantinovskiy (kuhn-stahn-TEE-nuhf-skeeyee), town (2006 population 5,680), E YAROSLAVL oblast, central European Russia, on the VOLGA RIVER, on road and railroad spur, 18 mi/29 km NW of YAROSLAVL; 57°49′N 39°36′E. Elevation 419 ft/127 m. Petroleum refinery.

Konstantinovskiy, RUSSIA: see KONSTANTINOVSK.

Konstantinovy Lazne, CZECH REPUBLIC: see BEZDRUZICE.

Konstantynow (kon-stan-TEE-nov), Polish *Konstantynów Łódzki*, Russian *Konstantinov*, town, ŁÓDŹ province, central POLAND, on NER River, and 6 mi/9.7 km W of ŁÓDŹ city center. Cloth weaving and finishing, manufacturing of dyes, shawls, tiles; flour milling.

Konstanz (KON-stahnts), French *Constance*, city, BADEN-WÜRTTEMBERG, SW GERMANY, on the RHINE RIVER, at the W end of LAKE CONSTANCE, and near the Swiss border; 47°40′N 09°10′E. Its industries include the manufacturing of textiles, machinery, chemicals, and electrical equipment. Also a tourist center. Founded as a Roman fort in 260; became an episcopal see at the end of the 6th century. The bishops became powerful and held large territories, including much of what are now BADEN-WÜRTTEMBERG and SWITZERLAND, as princes of the Holy Roman Empire. Located on a trade route between GERMANY and ITALY, Konstanz became a free imperial city in 1192. In 1531 the city, which had accepted the Reformation, joined the Schmalkaldic League. Emperor Charles V, after defeating the League, deprived Konstanz of its free imperial status and gave it to his brother, later Emperor Ferdinand I. Passed to AUSTRIA in 1548; ceded to BADEN in 1805. The bishopric was suppressed in 1821, and the diocese was abolished in 1827. Among the numerous historic buildings are the cathedral (11th century; additions in the 15th and 17th century); the Council building (1388); and a former Dominican convent (now a hotel), the birthplace (1838) of Graf von Zeppelin, the soldier and aviator. Seat of a university and of an ecological research institute.

Konta (kon-TAH), or **Konda**, village, BASTAR district, S CHHATTISGARH state, E central INDIA, on tributary of the GODAVARI RIVER, and 60 mi/97 km NNW of RAJAHMUNDRY. Rice, oilseeds, mangoes. Sal forests nearby.

Kontagora (kahn-TAH-goo-RAH), town, NIGER state, W NIGERIA, 90 mi/145 km NW of MINNA. Gold mining; shea-nut processing; cassava, millet, and durra. Peanut-growing scheme (S).

Kontarne (kon-TAHR-ne) (Russian *Kontarnoye*), town, E central DONETS'K oblast, UKRAINE, in the DONBAS, on road and railroad spur, 4 mi/6 km NE of, and subordinated to, SHAKHTARS'K; 48°05'N 38°31'E. Elevation 734 ft/223 m. Coal mine. Established in 1932, town since 1956.

Kontarnoye, UKRAINE: see KONTARNE.

Kontich (KAHN-tikh), commune (2006 population 20,286), Antwerp district, ANTWERPEN province, N BELGIUM, 5 mi/8 km SSE of ANTWERP; 51°08'N 04°27'E. Agriculture market. Formerly spelled Contich.

Kontiolahti (KON-tee-o-LAH-tee), Swedish *Kontiolaks*, village, POHJOIS-KARJALAN province, SE FINLAND, 10 mi/16 km NNE of JOENSUU; 62°46'N 29°51'E. Elevation 330 ft/100 m. On lake of SAIMAA system, in lumbering region.

Kontiomäki (KON-tee-o-MAH-kee), village, OULUN province, central FINLAND, near E end of OULUJÄRVI (lake), 14 mi/23 km NE of KAJAANI; 64°21'N 28°09'E. Elevation 413 ft/125 m. Railroad junction. In lumbering region.

Kon Tum (KON DUHM), province (□ 3,835 sq mi/9,971 sq km), central VIETNAM, in the CENTRAL HIGHLANDS, on border of LAOS and CAMBODIA (both W); ⊙ KON TUM; 14°40'N 108°00'E. Bordered N by QUANG NAM-DA NANG province, E by QUANG NGAI province, and S by GIA LAI province. Drained by numerous rivers flowing E directly into the SOUTH CHINA SEA and W into the MEKONG River running through Laos and Cambodia. Situated on a high plateau marking the W slope of the TRUONG SON RANGE (average elevation 2,000 ft/610 m) and impressive basaltic formations reaching over 6,562 ft/2,000 m. Grand mountain scenery, misty valleys and ridges, swiftly flowing rivers, lakes and waterfalls, and cooler climate than rest of country. Embraces broad range of regional habitats, many minority communities, and localized economies. Once densely forested, its landscape has been transformed through continuing deforestation by loggers and migrants from the lowlands who have settled in scattered New Economic Zones. Also subjected to defoliation and repeated bombings during Vietnam War; scars from the conflict still remain in some places. As a result of contemporary resettlement schemes, wartime depredations, widespread timber exploitation, and rapidly growing population, the forests here now cover less than half the province and continue to suffer from serious overcutting. Diverse soils and mineral resources (gold, copper, tin, titanium, chromite, bauxite, granite, mineral water); shifting cultivation, animal raising, forest products (gums, resins, medicinals, bamboo, rattan, honey, beeswax, exotic decorative plants, foraged foods), agro-forestry, and commercial agriculture (coffee, tea, sugarcane, rubber, ginger, sesame, vegetables, fruits). Lumbering, sawmilling, sugar milling, paper production, furniture making and woodworking, animal feed blending, brick and tile manufacturing, handicraft production, meat and food processing, light manufacturing. Tourism potential. Minorities include Kinh, Jarai, Bahnar, Rongao, Sedang, and Thai.

Kon Tum (KON DUHM), city, KON TUM province, central VIETNAM, on the highlands on KON TUM PLATEAU, 140 mi/225 km SSE of HUE; 14°20'N 108°02'E. Elevation 1,720 ft/524 m. Area was heavily damaged during the Vietnam War. After 1975, the Vietnamese government resettled many lowland Vietnamese (Kinh) in New Economic Zones to alleviate population pressure throughout the lowlands and develop the uplands. Today, Kon Tum is a major administrative, transportation, educational, and trading center. Produces hides; wax; rubber; light manufacturing. Has sesame, coffee and tea farms. Formerly big-game hunting; tourism. Development plans include sugar-, wood-, cassava-, and pine-resin- processing plants, in addition to furniture and paper factories. Indigenous population is largely Bahhar, Jarai, Renago, and Sedang. Formerly Kontum.

Kon Tum Plateau (KON DUHM), central VIETNAM, one of six plateaus of the CENTRAL HIGHLANDS, W of TRUONG SON RANGE; c. 70 mi/113 km long, c. 60 mi/97 km wide; 13°55'N 108°05'E. Elevation c. 2,000 ft/610 m, with impressive basaltic formations rising up to 2,500 ft/762 m. Formerly extensive forests increasingly destroyed by shifting cultivators and lumbering activities that produce grassy plains. Agriculture (tea, coffee, vegetables, fruit; agro-forestry). Area of European colonization in the past and Kinh resettlement today. Main cities Play Ku and KON TUM. Indigenous peoples include Jarai, Bahmar, Rengao, and other ethnic minorities. The migration of lowland Vietnamese into the area, beginning in 1975, has transformed this area. Many of the indigenous people were encouraged to forgo swidden production and join large state farms, especially coffee and rubber plantations. Also called JARAI, or GIA LAI, Plateau.

Kontupohja, RUSSIA: see KONDOPOGA.

Konu (KON), town, Konu county, HIROSHIMA prefecture, SW HONSHU, W JAPAN, 40 mi/65 km N of HIROSHIMA; 34°42'N 133°04'E.

Konuma, RUSSIA: see NOVOALEKSANDROVSK, SAKHALIN oblast.

Konya, city (2000 population 742,690), S central TURKEY, on ISTANBUL-ADANA railroad; 37°51'N 32°30'E. Important road junction; airport. It is the trade center of a rich agricultural and livestock-raising region. Manufacturing includes cement, carpets, and leather, cotton, and silk goods. As the ancient ICONIUM, the city was important in Roman times, but it reached its peak after the victory (1071) of Alp Arslan over the Byzantines at Manzikert, which resulted in the establishment (1099) of the sultanate of Iconium, or Rum (so called after Rome), a powerful state of the Seljuk Turks. In the late 13th century the Seljuks of Iconium were defeated by the Mongols, and their territories subsequently passed to Karamania. In the 15th century the whole region was annexed to the Ottoman Empire by Sultan Muhammad II, the conqueror of CONSTANTINOPLE. Konya lost its political importance but remained a religious center as the chief seat of the Whirling Dervishes (Mawlawiyya, or Mevlevi, Sufi sect that uses dancing and music as part of its spiritual method), whose order was founded here in the 13th century by the poet and mystic Celaleddin Rumi. The Mawlawiyya order is still centered in Konya, and the tomb of the founder, several medieval mosques, and the old city walls have been preserved. In 1832 an Egyptian army under Ibrahim Pasha completely routed the Turks here. The Armenian population of the town, once very numerous, was largely deported during World War I. Konya province, the largest in Turkey, has important mineral resources.

Konyaereglisi, TURKEY: see EREGLI.

Konyár (KO-nyahr), Hungarian *Konyár*, village, HAJDU-BIHAR county, E HUNGARY, 14 mi/23 km S of DEBRECEN; 47°19'N 21°40'E. Konyar salt lake and baths nearby.

Konyavska Mountains (KON-yahv-skah), W BULGARIA; bordered by the upper STRUMA (N, W, and S), and DZHERMAN (SE) rivers; c.25 mi/40 km long, c.20 mi/32 km wide; rise to 4,877 ft/1,487 m at Mount Viden, 9 mi/15 km ENE of KYUSTENDIL; 42°25'N 22°46'E. Formerly called Konovo Mountains.

Könye (KUH-nye), village, KOMÁROM county, HUNGARY, 4 mi/6.4 km W of TATABÁNYA. Large plant making food formulas and additives.

Konyich, BOSNIA AND HERZEGOVINA: see KONJIC.

Konyovo (KON-yo-vo), village, BURGAS oblast, NOVA ZAGORA obshtina, YAMBOL district, E central BULGARIA, 8 mi/13 km NE of Nova Zagora; 42°32'N 26°10'E. Horse-raising center; grain. Formerly known as Atloolu (until 1906).

Konyrat (ko-nee-RAHT), Russian *Kounradski*, town, SE ZHEZKAZGAN region, KAZAKHSTAN, 10 mi/16 km N of BALKASH (linked by railroad); 46°58'N 74°59'E. Copper- and molybdenum-mining center. Also Qonyrat.

Konyshevka (KO-ni-shif-kah), town (2005 population 4,010), W KURSK oblast, SW European Russia, on road and railroad, 20 mi/32 km SSE of DMITRIYEV-L'GOVSKIY; 51°51'N 35°17'E. Elevation 711 ft/216 m. In agricultural area (grains, sunflowers, potatoes; livestock); food processing.

Konyukh, BOSNIA AND HERZEGOVINA: see KONJUH.

Konyukovo (kon-YOO-ko-vuh), town, NE NORTH KAZAKHSTAN region, KAZAKHSTAN, c.70 mi/113 km E of PETROPAVLOVSK; 55°08'N 70°44'E. In wheat area.

Konz (KOHNTS), town, RHINELAND-PALATINATE, W GERMANY, at confluence of SAAR and MOSEL rivers, 5 mi/8 km SW of TRIER, near LUXEMBOURG border; 49°42'N 06°35'E. Manufacturing (machinery, plastics). Has remains of Roman imperial villa. Formerly spelled Conz. Chartered 1959.

Konza (KON-zah), town, RIFT VALLEY province, S central KENYA, railroad junction for MAGADI, 40 mi/64 km SE of NAIROBI; 01°45'S 37°07'E. Elevation 5,427 ft/1,654 m. Wheat, corn; dairy farming.

Konzhakovskiy Kamen' (kuhn-zhah-KOF-skeeye KAH-myen'), highest peak (5,154 ft/1,571 m) in the central URAL Mountains, SVERDLOVSK oblast, RUSSIA; 59°40'N 59°10'E.

Koocanusa Lake (koo-kuh-NOO-suh), reservoir, KOOTENAI River (*Kootenay* in CANADA), NW MONTANA and SE BRITISH COLUMBIA, largely within Kootenai National Forest in Montana; c.85 mi/137 km long, including 40 mi/64 km N extension into Canada, 1 mi/1.6 km–2 mi/3.2 km wide; 48°23'N 115°19'W. Formed by multipurpose Libby Dam (420 ft/128 m high) in LINCOLN county (Montana), 10 mi/16 km E of LIBBY. TOBACCO RIVER enters from E in Montana, Elk River enters from NE in British Columbia. Bounded by the PURCELL Range to the W, and the Salish Mountains to the E.

Koochiching (KOO-chi-cheeng), county (□ 3,154 sq mi/8,200.4 sq km; 2006 population 13,658), N MINNESOTA; ⊙ INTERNATIONAL FALLS; 48°15'N 93°46'W. Bounded N by CANADA (ONTARIO) border, formed mainly by Rainy River and RAINY LAKE in NE; drained by BIG FORK and LITTLE FORK rivers. Agricultural area (alfalfa; cattle; some dairying); timber; some resorts in NE. Lumber and paper milling at International Falls. Area is noted for its record cold temperature. Includes much of NETT LAKE and Nett Lake Indian Reservation in SE. Large Koochiching State Forest covers much of SE part of county; Pine Island State Forest covers most of W part; Smokey Bear State Forest is in N; part of VOYAGEURS NATIONAL PARK in NE corner. Formed 1906.

Koog aan de Zaan (KAWKH ahn duh ZAHN), district of ZAANSTAD, NORTH HOLLAND province, W NETHERLANDS, on ZAAN RIVER, 1 mi/1.6 km NW of city center, in ZAANSTREEK industrial area; 52°28'N 04°50'E. Dairying; cattle; flowers, fruit, vegetables; manufacturing (food processing; steel drums, wax paper).

Koolan Island (KOO-lan), island of BUCCANEER ARCHIPELAGO, off N coast of WESTERN AUSTRALIA, in Yampi Sound between COLLIER BAY and KING SOUND, 70 mi/113 km N of DERBY; 7 mi/11 km long, 2 mi/3 km wide; rises to 670 ft/204 m; 16°08'S 123°45'E. Valuable iron deposits.

Koolau Range (KO-o-LOU), mountain chain, extending c.30 mi/48 km NW-SE, nearly the length of OAHU island in HONOLULU county, HAWAII; rises to 3,105 ft/946 m in KONAHUANUI. It is cut by 2 scenic passes, NUUANU PALI (State Highway 61, Pali Highway to KAILUA) and another at the head of Kaliki Valley (State Highway 63, Likelike Highway to KANEOHE), which shorten the route between E and W Oahu; a third route across range is Interstate H3, under construction to NW of HONOLULU; protected by Kahuka and Kawailoa forest reserves (N), Ewa Forest Reserve (center), Honolulu Watershed Forest Reserve (S), others; Kahana Valley and Sacred Falls State Parks to NE. Also known as Koolau Mountains.

Koonap River, SOUTH AFRICA: see ADELAIDE.

Koondrook (KOON-druk), town, VICTORIA, SE AUSTRALIA, 180 mi/289 km NW of MELBOURNE, and on MURRAY RIVER; 35°39′S 144°11′E. Dairy products; timber; citrus fruits.

Koontz Lake, village (2000 population 1,554), STARKE county, N central INDIANA, 10 mi/16 km NW of PLYMOUTH, on Koontz Lake; 41°25′N 86°29′W. Recreation.

Koonya (KOON-yuh), settlement, TASMANIA, SE AUSTRALIA, 59 mi/95 km from HOBART; 43°03′S 147°51′E. Established as a convict outstation; now a restored penitentiary; holiday accommodation. Formerly called Cascades.

Koorawatha (KOOR-uh-WAH-thuh), village, SE central NEW SOUTH WALES, SE AUSTRALIA, 155 mi/249 km WSW of SYDNEY; 34°02′S 148°33′E.

Koorda (KOR-duh), town, WESTERN AUSTRALIA state, W AUSTRALIA, 146 mi/235 km NE of PERTH; 30°50′S 117°51′E. Wheat, sheep. Local tradition of making dolls from corn husks and stalks. Museum housed in an old hospital. Wildflower area.

Kooringa, AUSTRALIA: see BURRA.

Koosharem, village (2006 population 290), SEVIER county, central UTAH, 20 mi/32 km SE of RICHFIELD, on OTTER CREEK; 38°30′N 111°52′W. Agriculture (cattle). Elevation 6,914 ft/2,107 m. Site of treaty conference between Native Americans and settlers, 1873. Fish Lake and Johnson reservoirs to NE; parts of Fishlake National Forest to W and NE. Koosharem Reservoir (Cutter Creek) to N. SEVIER PLATEAU SSW.

Kooskia (KOOS-kee-uh), village (□ 1,315 sq km; 2000 population 675), IDAHO county, W central IDAHO, 12 mi/19 km SE of NEZPERCE, at junction of Middle Fork and South Fork of CLEARWATER RIVER, forming Clearwater River, in SE part of Nez Perce Indian Reservation; 46°08′N 115°58′W. Agriculture; lumber; cheese; manufacturing (lumber products). Loscha Historic Ranger Station 47 mi/76 km ENE. Nearby are Clearwater (NE) and Nez Perce (SE) national forests. CLEARWATER MOUNTAINS to SE; NEZ PERCE NATIONAL HISTORICAL PARK (East Kamiah Site) to N.

Kootenai, county (□ 1,315 sq mi/3,419 sq km; 2006 population 131,507), N IDAHO, in Panhandle Region; ⊙ COEUR D'ALENE; 47°41′N 116°31′W. Rolling, wooded area bordering on WASHINGTON (W), watered by COEUR D'ALENE LAKE and SPOKANE and COEUR D'ALENE rivers. SPOKANE (Washington), 15 mi/24 km W of state line, has partially influenced county's growth. Lumbering; agriculture (wheat, oats, alfalfa; cattle); manufacturing (furniture); retail trade. Old Mission at Gataldo (1853) Historical State Park in SE. Includes part of COEUR D'ALENE MOUNTAINS, in Coeur d'Alene National Forest in E, including Fourth of July Canyon (COEUR D'ALENE RIVER). Part of Coeur d'Alene Indian Reservation in S. Farragut State Park, at S end of LAKE PEND OREILLE in NE corner; several small lakes in county. Formed 1864.

Kootenai (KOOT-en-ee), village (2000 population 441), BONNER county, N IDAHO, 3 mi/4.8 km NE of SANDPOINT, 1 mi/1.6 km E of PONDERAY, and on N shore of LAKE PEND OREILLE; 48°19′N 116°31′W. Schweitzer Basin Ski Area to NW; Kaniksu National Forest to E.

Kootenai (KOO-tin-ai), river, 407 mi/655 km long, NW U.S. and SW CANADA; rises in the ROCKY MOUNTAINS, SE BRITISH COLUMBIA; flows S into NW MONTANA, NW through N IDAHO, then N back into Canada, through Kootenay Lake (64 mi/103 km long; □ 191 sq mi/495 sq km), an expansion of the river, before joining the COLUMBIA River at CASTLEGAR;

49°19′N 117°39′W. The river is used to generate hydroelectricity. The Canadian name is spelled Kootenay.

Kootenay Boundary (KOOT-uh-nai BOUN-duh-ree), regional district (□ 3,126 sq mi/8,127.6 sq km; 2001 population 31,843), S BRITISH COLUMBIA, W CANADA; 49°20′N 118°45′W. Consists of five electoral areas and eight municipalities (FRUITVALE, GRAND FORKS, GREENWOOD, MIDWAY, MONTROSE, ROSSLAND, TRAIL, and WARFIELD). Incorporated 1966.

Kootenay Lake, CANADA: see KOOTENAI.

Kootenay National Park (KOOT-uh-nai (□ 543 sq mi/1,411.8 sq km), SE BRITISH COLUMBIA, SW CANADA, in the ROCKY MOUNTAINS near Kootenay Lake; 50°57′N 116°02′W. Contains high peaks, glaciers, deep canyons, and hot springs. The Banff-Windermere Highway crosses the park. Established 1920.

Kootenay River, U.S. and CANADA: see KOOTENAI.

Kootwijk (KAWT-veik), village, GELDERLAND province, central NETHERLANDS, 8 mi/12.9 km WSW of APELDOORN; 51°26′N 04°07′E. National Park De Hoge Veluwe to SE. Dairying; cattle; grain, vegetables. Formerly spelled Kootwyk.

Koo-Wee-Rup (koo–wee–RUHP), town, S VICTORIA, SE AUSTRALIA, 38 mi/61 km SE of MELBOURNE; 38°12′S 145°30′E. Railroad junction in agricultural area; asparagus, potatoes; dairying. Extensive area of reclaimed swampland.

Kooyong (koo-YAHNG), locality, suburb 4 mi/6 km SE of MELBOURNE, VICTORIA, SE AUSTRALIA, on S side of Gardiners Creek valley.

Kop, TURKEY: see BULANIK.

Kopačevski Rit Nature Park (□ 68 sq mi/176 sq km), wetlands, E CROATIA, in BARANJA region, NE of OSIJEK. Includes Kopačevsko jezero [=lake], swamps, and forest. Winter bird sanctuary; fishing.

Kopačevsko Lake, Croatian *Kopačevsko jezero* (□ 1.4 sq mi/3.6 sq km), E CROATIA, in BARANJA region, in the wetlands of Kopačevski Rit Nature Park; linked to the Danube (DUNAV) River by a canal through which water flows in either direction depending on the Danube water level. Croatia's fourth-largest natural lake.

Kopaganj (KO-pah-guhnj), town, AZAMGARH district, E UTTAR PRADESH state, N central INDIA, 24 mi/39 km E of Azamgarh; 26°01′N 83°34′E. Trades in rice, barley, wheat, sugarcane; cotton weaving, saltpeter processing.

Kopaigorod, UKRAINE: see KOPAYHOROD.

Kopais, Lake, Greece: see COPAIS, LAKE.

Kopanskaya (kuh-PAHN-skah-yah), village (2005 population 4,025), NW KRASNODAR TERRITORY, S European Russia, on the N shore of Lake KHANSKOYE, on road, 29 mi/47 km SSE of YEYSK; 46°16′N 38°29′E. Fisheries; summer homes.

Kopaonik (kop-AH-o-neek), mountain range, S central SERBIA, extends c.40 mi/64 km S along right bank of middle IBAR RIVER, and into KOSOVO; 43°15′N 20°50′E. Highest point (6,616 ft/2,017 m) is 13 mi/21 km SW of BRUS. Large lead, zinc, silver, and pyrite deposits mined at TREPCA, at S foot. A major wintersports region.

Kopargaon (KO-pahr-goun), town, AHMADNAGAR district, MAHARASHTRA state, W central INDIA, on GODAVARI River, and 60 mi/97 km NNW of AHMADNAGAR; 19°53′N 74°29′E. Trade center for sugar, millet, cotton; sugar milling, gur manufacturing.

Kopatkevichi (ko-paht-KE-vee-chee), urban settlement, GOMEL oblast, BELARUS, on PTICH RIVER, and 27 mi/43 km NW of MOZYR. Creamery; woodworking.

Kopaygorod, UKRAINE: see KOPAYHOROD.

Kopayhorod (ko-PEI-ho-rod) (Russian *Kopaigorod*), town (2004 population 6,100), W VINNYTSYA oblast, UKRAINE, on Nemyya River, left tributary of DNIESTER (Ukrainian *Dnister*) River, 27 mi/43 km N of MOHYLIV-PODIL'S'KYY; 48°52′N 27°47′E. Elevation 833 ft/253 m. Flour milling; sugar beets, wheat, fruit. Also spelled Kopaygorod. Jewish community since 1600,

growing progressively numerous despite the Khmel'nyts'kyy pogroms of 1648–1649, the Barsky Confederation pogroms of 1768–1772, and the civil war pogroms of 1918–1921, numbering 5,000 and accounting for the majority of the population in 1939; confined to a ghetto and gradually exterminated by the Nazis between 1941 and 1944.

Köpenick (KO-pe-nik), district of BERLIN, E GERMANY, at the confluence of the SPREE and DAHME rivers; 52°26′N 13°32′E. Industrial center and a tourist spot, with forests and large lakes. Scene of the trial (1730) of Crown Prince Frederick (later Frederick II).

Koper (KO-per), Italian *Capodistria*, town (2002 population 23,285), SW SLOVENIA, on the Istrian peninsula in the Gulf of TRIESTE; 45°32′N 13°43′E. Principal seaport of Slovenia and main port for Austrian and Hungarian exports; has salt mines and a radio factory; fishing. Was capital of Istria (1278–1797) under Venetian rule. The Treaty of Campo Formio, which dissolved the republic of Venice, transferred town to Austria. Passed to ITALY after World War I and became part of the Free Territory of Trieste in 1947. Annexed to the former YUGOSLAVIA (1954). Preserves aspects of an old Venetian town, with a Romanesque cathedral and campanile, a Gothic loggia, and a pinnacled town hall.

Kopervik (KAW-puhr-veek), town, ROGALAND county, SW NORWAY, on E shore of KARMØY, 9 mi/14.5 km S of HAUGESUND. Fishing center for herring and sardines; canneries; one of Europe's largest aluminum works nearby. Many Norse relics found in graves and barrows nearby.

Kopet Dag (ko-PET DAHG), one of the Turkmen-Khurasan mountain ranges, in SW TURKMENISTAN and in NE IRAN (Khorāsān), extending c.200 mi/322 km NW-SE between 56°00′E and 60°00′E and along Turkmenistan-Iran border; rises to c.10,000 ft/3,048 m. Steppe vegetation in higher reaches. Resort of FIRYUZA (Turkmenistan) is on N slopes; range is crossed by ASHGABAT (Turkmenistan)-QUCHAN (Iran) highway at GAUDAN (Turkmenistan) and BAJGIRAN (Iran) frontier posts. TRANS-CASPIAN RAILROAD hugs N foot in Turkmenistan.

Kopeysk (kuh-PYAISK), city (2005 population 69,990), E CHELYABINSK oblast, SW SIBERIA, RUSSIA, on road, 12 mi/19 km SE of CHELYABINSK; 55°06′N 61°39′E. Elevation 702 ft/213 m. On railroad (Dobycha Uglya coal mining station). Lignite-mining center; manufacturing and repair of mining machinery, excavators, tin cookware; food processing (dairy, bakery). Known as Ugol'nyye Kopi (OO-guhl-ni-ye KO-pee); subsequently called Kopi until 1933; became city of Kopeisk in 1933. Absorbed the town of GORNYAK (3 mi/5 km NNE). Coal mining began in 1907, but main development came in the 1930s (in the First Five-Year Plan), and during World War II.

Kopeysk, RUSSIA: see KOPEISK.

Kopi, RUSSIA: see ALEKSANDROVSK, PERM oblast.

Kopi, RUSSIA: see KOPEISK.

Kopilovtsi (KO-pee-lov-tsee), village, MONTANA oblast, GEORGI-DAMYANOVO obshtina, BULGARIA; 43°20′N 22°55′E. Border crossing to MACEDONIA soon to be reopened.

Köping (SHUHP-eeng), town, VÄSTMANLAND county, S central SWEDEN, at W end of LAKE MÄLAREN; 59°31′N 15°60′E. Important lake port and commercial and industrial center; manufacturing (machinery, motor vehicles, fertilizer; cement).

Köpingebro (SHUHP-eeng-e-BROO), village, SKÅNE county, S SWEDEN, near BALTIC SEA, 5 mi/8 km ENE of YSTAD; 55°27′N 13°56′E.

Kopli (KOP-lee), peninsula, extending in the GULF OF TALLINN. Harbor activity of largely industrial traffic; much of the activity is being moved to MUUGA Harbor.

Koplik (ko-PLEEK), town, ⊙ Malesia e Madhe district (□ 410 sq mi/1,066 sq km), N ALBANIA, 10 mi/16 km

NNW of SHKODËR; near Lake SHKODËR; 42°12′N 19°26′E. Also spelled Kopliku.

Köpmanholmen (SHUHP-mahn-HOLM-en), village, VÄSTERNORRLAND county, NE SWEDEN, on GULF OF BOTHNIA, 9 mi/14.5 km SW of ÖRNSKÖLDSVIK; 63°10′N 18°34′E.

Kopomá (ko-po-MAH), town, YUCATÁN, SE MEXICO, 28 mi/45 km SW of MÉRIDA; 20°39′N 89°54′W. Henequen, corn, tropical fruit.

Kopondei Point (ko-PON-dai) or **Cape Tanjung**, promonotory at NE tip of FLORES Island, INDONESIA, on FLORES SEA.; 08°04′S 122°52′E. Formerly called Flores Head.

Kopor'ye Gulf (kuh-POR-ree-ye), in the S shore of the Gulf of FINLAND, in the BALTIC SEA, LENINGRAD oblast, NW European Russia; runs inland for 7 mi/12 km; 10 mi/16 km wide at mouth; 12 mi/20 m deep. Low-lying rocky banks covered with forest. Freezes in winter. Named after an ancient Kopor'ye fortress, to the S.

Koppa (ko-PAH), town, CHIKMAGALUR district, KARNATAKA state, S INDIA, 37 mi/60 km NW of CHIKMAGALUR; 12°42′N 76°57′E. Rice milling. Coffee, cardamom, and pepper estates in nearby hills (bamboo, sandalwood).

Koppal (ko-PAHL), town, RAICHUR district, KARNATAKA state, S INDIA, 55 mi/89 km WNW of BELLARY; 15°21′N 76°09′E. Oilseed milling, cotton ginning. Nearby hill fort remodeled 1786 by French engineers under Tipu Sultan. Sometimes spelled Kuppal.

Koppang (KAWP-pahng), village, HEDMARK county, E NORWAY, in ØSTERDALEN, on GLOMMA River, on railroad, and 56 mi/90 km N of ELVERUM. Lumbering; agriculture.

Kopparberg (KOP-per-BER-yuh), county (□ 11,648 sq mi/30,284.8 sq km), central SWEDEN, on NORWEGIAN border; ⊙ FALUN; 61°00′N 14°30′E. Almost coextensive with DALARNA province. Undulating surface rises toward mountain range along Norwegian border; drained by DALÄLVEN RIVER and its tributaries. Manufacturing (steel; paper and pulp milling, lumbering, sawmilling, woodworking, tanning, textile milling). Large LAKE SILJAN, in central part of county, in agriculture and dairying region. GRÄNGESBERG was center of one of Sweden's largest mining (iron, copper, lead, and zinc) regions. Important towns are Falun, LUDVIKA, BORLÄNGE (with Domnarvet metalworks), AVESTA, HEDEMORA, and SÄTER.

Kopparberg (KOP-per-BER-yuh), town, ÖREBRO county, S central SWEDEN, on Hörksälven River, 19 mi/31 km SSW of LUDVIKA; 59°52′N 15°02′E. Former copper- and zinc-mining and smelting center. Has seventeenth-century church.

Koppel (KAH-puhl), borough (2006 population 796), BEAVER county, W PENNSYLVANIA, 3 mi/4.8 km SW of ELLWOOD CITY, near BEAVER RIVER. Agriculture (corn, hay, alfalfa; dairying); manufacturing (fabricated metal products, wood products).

Koppenbrügge, GERMANY: see COPPENBRÜGGE.

Kopperston, unincorporated village, WYOMING county, S WEST VIRGINIA, 19 mi/31 km W of BECKLEY; 37°44′N 81°34′W. In coal and gas region. Railroad terminus. Manufacturing (coal processing).

Koppies, town, N FREE STATE province, SOUTH AFRICA, on Rhenoster River, near N1 highway, on main N-S railroad, and 35 mi/56 km NE of KROONSTAD; 27°15′S 27°35′E. Elevation 4,661 ft/1,421 m. Center of irrigated-farming region watered from Koppies Dam 6 mi/10 km E of town, producing wheat, corn, and fruit.

Kopreinitz, CROATIA: see KOPRIVNICA.

Koprinka (ko-PREEN-kah), dam, HASKOVO oblast, KAZANLUK obshtina, BULGARIA, on the TUNDZHA RIVER; 42°47′N 25°17′E. Formerly called Georgi Dimitrov; renamed after 1989.

Koprinka Reservoir (ko-PREEN-kah), reservoir created by the KOPRINKA Dam in central BULGARIA, on the TUNDZHA RIVER at Koprinka, near KAZANLUK. Formerly called Georgi Dimitrov; renamed after 1989.

Koprivnica (KO-preev-nee-tsah), Hungarian *Kaproncza*, German *Kopreinitz*, town, central CROATIA, in the PODRAVINA region, 25 mi/40 km ESE of Varaždin, at NW foot of BILOGORA Mountains; 46°10′N 16°50′E. Manufacturing (food processing; chemicals, superphosphates, and sulphuric acid). Railroad junction (Zagreb-Budapest and Varaždin-Osijek lines). Former lignite mine nearby. First mentioned in 13th century; fortress during Turkish invasions (16th–17th century). Hlebine, village, 6 mi/9.7 km SE.

Kopřivnice (KOP-rzhiv-NYI-tse), German *Nesselsdorf*, city (2001 population 23,747), SEVEROMORAVSKY province, NE MORAVIA, CZECH REPUBLIC, on railroad; 49°36′N 18°09′E. Major engineering works produce rolling stock, motor vehicles. Has museum of Tatra and Lassko region.

Koprivshtitsa (ko-PREEV-shtee-tsah), city, SOFIA oblast, ⊙ Koprivshtitsa obshtina (1993 population 3,066), W central BULGARIA, in the SREDNA GORA, on the TOPOLNITSA RIVER, 13 mi/21 km NE of PANAGYURISHTE; 42°38′N 24°22′E. Health resort; livestock center; dairying, meat preserving; manufacturing (carpets, knitting, construction materials). Museum village. Has a museum and historic buildings (Bulgarian Renaissance). Important commercial center (cattle exports to TURKEY and EGYPT) under Turkish rule (15th–19th centuries). Declined following a Bulgarian uprising in 1876.

Koprsko Primorje (ko-PER-sko PREE-mor-ye), SW SLOVENIA, region extending along entire Slovenian coastline, N ADRIATIC SEA, from Italian border just S of TRIESTE to Croatian border; c.12 mi/19 km long. The only Slovenian access to the sea. Koper, the country's major seaport, is located here, as are ANKARAN, PIRAN, Portorož, IZOLA, and LUCIJA. Tourism. Also called Koper Riviera.

Kopru River (Turkish=*Köprü*), ancient *Eurymedon*, 97 mi/156 km long, S Turkey; rises SE of Lake EGRIDIR; flows S to MEDITERRANEAN SEA 25 mi/40 km E of ANTALYA (Adalia). In ancient Pamphylia, Cimon defeated (c.467 B.C.E.) the Persians near its mouth.

Kopryukoi, Bulgaria: see GROZDYOVO.

Kopryu-Koi, Bulgaria: see IVANSKI.

Kopstal (KOPS-tahl), village, Kopstal commune, SW LUXEMBOURG, 4 mi/6.4 km NNW of LUXEMBOURG city; 49°40′N 06°05′E.

Koptelovo (kuhp-TYE-luh-vuh), village, S central SVERDLOVSK oblast, E URAL Mountains, W Siberian Russia, on the REZH RIVER, on crossroads, 12 mi/19 km SSE of ALAPAYEVSK; 57°42′N 61°52′E. Elevation 314 ft/95 m. Lumbering.

Koptsevichi (kop-TSE-vee-chee), village, GOMEL oblast, BELARUS, 45 mi/72 km WNW of MOZYR; 52°12′N 28°14′E. Plywood mill.

Kopychintsy, UKRAINE: see KOPYCHYNTSI.

Kopychyntsi (ko-PI-chin-tsee) (Russian *Kopychintsy*), Polish *Kopyczyńce*, city (2004 population 7,000), E TERNOPIL' oblast, UKRAINE, on left tributary of the DNIESTER (Ukrainian *Dnister*) River, and 8 mi/13 km NE of CHORTKIV; 49°06′N 25°56′E. Elevation 1,082 ft/329 m. In deciduous woodland. Railroad junction; agricultural trading center; rubber toy factory, cannery, fishery. Agricultural technical school. Has an old castle. First mentioned in the early 14th century; passed from Poland to Austria (1772); part of West Ukrainian National Republic (1918), reverted to Poland (1919); to Ukrainian SSR in 1939; part of independent Ukraine since 1991.

Kopyczyńce, UKRAINE: see KOPYCHYNTSI.

Kop'yevo (KOP-yee-vuh), village (2005 population 4,840), N KHAKASS REPUBLIC, S central Siberian Russia, on local road and railroad branch connecting the TRANS-SIBERIAN RAILROAD at ACHINSK (KRASNOYARSK TERRITORY) with the SOUTH SIBERIAN RAILROAD at ABAKAN, 30 mi/48 km NNW of SHIRA; 54°59′N 89°49′E. Elevation 1,279 ft/389 m. Railroad and auto depots.

Kopyl (ko-PIL), town, MINSK oblast, BELARUS, 22 mi/35 km WNW of SLUTSK. Creamery; agricultural-produce processing.

Kopylovo (kuh-PI-luh-vuh), settlement, SE TOMSK oblast, S central Siberian Russia, on railroad, 8 mi/13 km NE (and administratively subordinate to) TOMSK; 56°36′N 85°07′E. Elevation 465 ft/141 m. Silicate building materials, ceramics. Also spelled Kopylova.

Kopys (ko-PIS), urban settlement, S VITEBSK oblast. BELARUS, on DNIEPER River, and 15 mi/24 km SSW of ORSHA, 52°20′N 30°17′E. Manufacturing (tile); workshop of the Orsha combine, which produces silicate articles.

Kora (KO-rah), town, Inukami county, SHIGA prefecture, S HONSHU, central JAPAN, 25 mi/40 km N of OTSU; 35°12′N 136°15′E.

Kora, INDIA: see FATEHPUR.

Kora, RUSSIA: see KYRA.

Korab, MACEDONIA: see KORABIT.

Korabit (ko-RAH-beet) or **Korab** (ko-RAHB), mountain, on Macedonian-Albanian border, E of the BLACK DRIN RIVER. Its peak (9,066 ft/2751 m), highest in both countries and second-highest in former YUGOSLAVIA, is 20 mi/32 km W of GOSTIVAR (MACEDONIA) and 15 mi/24 km NE of PESHKOPI (ALBANIA).

Korablino (kuh-RAH-blee-nuh), city (2006 population 14,400), central RYAZAN oblast, central European Russia, on road and railroad, 55 mi/89 km S of RYAZAN, and 14 mi/23 km N of RYAZHSK; 53°55′N 40°01′E. Elevation 452 ft/137 m. Silk textiles; dairy plant. Made city in 1965.

K'orahē, village, SOMALI state, SE ETHIOPIA; 06°37′N 44°18′E. In the OGADEN, on FAFEN RIVER, on road, 125 mi/201 km SE of DEGEH BUR. Occupied by Italians (1935) after aerial bombardment in the Italo-Ethiopian War and by British (1941) in World War II. Sometimes spelled Gorrahei or Korrahei.

Koraka, Cape, W TURKEY, on AEGEAN SEA, 40 mi/64 km SW of IZMIR; 38°05′N 26°37′E.

Korakkasy (kuh-rahk-KAH-si), village (2005 population 3,035), W central CHUVASH REPUBLIC, central European Russia, near highway, 30 mi/48 km WSW of CHEBOKSARY; 55°45′N 46°40′E. Elevation 666 ft/202 m. Poultry farm.

Koralpe (KOR-ahl-pe), small range of central Alps on STYRIA-CARINTHIA border, Austria, extending 25 mi/40 km N from DRAVA RIVER, E of LAVANT RIVER; rises to 6,523 ft/1,988 m in the Grosser Speikkogel. Continued N by PACKALPE mountains. Lignite mined at NE foot. Winter sports facilities in its central part.

Korana River (KO-rah-nah), c.70 mi/113 km long, W CROATIA; rises in PLITVICE LAKES, in LIKA; flows N, past SLUNJ, to KUPA RIVER just NE of KARLOVAC. Receives Mrežnica River Hydroelectric plant near Karlovac.

Kora National Park (□ 690 sq mi/1,787 sq km), COAST province, E KENYA, N of MOMBASA; 00°17′N 38°47′E. Lion habitat. Designated a national park in 1989.

Korangal, town, MAHBUBNAGAR district, ANDHRA PRADESH state, S INDIA, 35 mi/56 km NW of MAHBUBNAGAR; 17°06′N 77°38′E. Road center in millet and rice area.

Korapun (KO-rah-poon), village, EAST NEW BRITAIN province, E NEW BRITAIN island, E PAPUA NEW GUINEA, on SOLOMON SEA coast, 10 mi/16 km W of Crater Point, 90 mi/145 km SSW of RABAUL. Boat access. Cocoa, coffee; tuna; timber.

Koraput (KO-rah-puht), district (□ 3,295 sq mi/8,567 sq km), SW ORISSA state, E central INDIA; ⊙ KORAPUT. Bordered E by ANDHRA PRADESH state, W by MADHYA PRADESH state; crossed (NE-SW) by EASTERN GHATS. Plateaus and hill ranges (2,000 ft/610 m–3,000 ft/914 m average; several peaks over 4,000 ft/1,219 m); thickly forested (sal, teak, bamboo). Rice is chief crop; some sugarcane and oilseeds grown. Trade centers include Japur, GUNUPUR. Area under Chola kings in tenth–eleventh century. Formerly part of VIZAGAPATAM

Cross-references are shown in SMALL CAPITALS. The pronunciation guide is shown on page xix. The sources of population figures are shown on page xvii.

district, MADRAS; in 1936, transferred to newly-created Orissa province and incorporated as separate district.

Koraput (KO-rah-puht), town, ⊙ KORAPUT district, SW ORISSA state, E central INDIA, in EASTERN GHATS, 11 mi/18 km ESE of JAYPUR; 18°49′N 82°43′E. Rice; timber. Aluminum deposits nearby (N).

Korat, THAILAND: see NAKHON RATCHASIMA.

Korata, ETHIOPIA: see WERETA.

Koratagere (KO-ruh-tuh-gir-ee), town, TUMKUR district, KARNATAKA state, S INDIA, 15 mi/24 km NE of TUMKUR; 13°32′N 77°14′E. Biri manufacturing, goldsmithing; oilseeds, millet, rice. Annual temple-festival market. Granite and corundum quarrying in nearby hills (sandalwood, lac).

Koratla (ko-RAHT-lah), town, KARIMNAGAR district, ANDHRA PRADESH state, S INDIA, 13 mi/21 km W of JAGTIAL; 18°49′N 78°43′E. Handmade paper (bamboo forests).

Korat Plateau (KO-RAHD), saucer-shaped tableland of E THAILAND, bounded by MEKONG RIVER (LAOS border; N and E), Petchabun and DONG PHAYA YEN RANGES (W), SAN KAMPHAENG and DANGREK ranges (S); 250 mi/402 km in diameter; mean elevation 600 ft/183 m, tilting toward SE. Drained by MUN, CHI, and PHAO rivers, it has impermeable soils flooded during rainy season (Apr.–Nov.) and waterless during dry season. Agriculture (rice, corn, cotton, hemp, peanuts); sericulture; cattle, horse, and hog raising. Served by railroad from BANGKOK. The largest centers are NAKHON RATCHASIMA (KORAT; SW), Udon (N), and Ubon (SE).

Korb (KORB), village, BADEN-WÜRTTEMBERG, S GERMANY, 2 mi/3.2 km E of WAIBLINGEN; 48°51′N 09°22′E. Wine.

Korba (KOR-ba), city (2001 population 315,695), ⊙ Korba district, CHHATTISGARH state, central INDIA, 20 mi/32 km S of CHAMPA; 22°21′N 82°41′E. One of India's major coalfields is nearby, supporting 300-MW Korba thermal power station. Transmission lines supply power to various places in BILASPUR and RAIPUR districts. Was part of Bilaspur district, MADHYA PRADESH state, until 1998, when the new district of Korba was formed with Korba city as capital. Joined Chhattisgarh state upon formation in 2000.

Korba, TUNISIA: see QURBAL.

Korbach (KOR-bahkh), town, HESSE, central GERMANY, 26 mi/42 km W of KASSEL; 51°16′N 08°53′E. Railroad junction; manufacturing (tires, metal furniture). Was first mentioned in 980; chartered 1075. Member of the HANSEATIC LEAGUE; devastated in the Thirty Years War. Noteworthy buildings are two churches from 14th and 15th century, the town hall (1377), and many half timbered houses.

K'orbeta (kor-BAI-tah), village, N ETHIOPIA, 7 mi/11.3 km E of MAYCH'EW; 12°48′N 39°39′E. In agricultural region (cereals, coffee, cotton); salt market.

Korbu, Gunong (KOR-boo, GOO-nong) or **Kerban** (KER-bahn), highest peak (7,160 ft/2,182 m) of central Malayan range, MALAYSIA, in E central PERAK, 15 mi/24 km NE of IPOH.

Korçë (KORCH), city (2001population 55,130) and region, ⊙ KORÇË district (□ 670 sq mi/1,742 sq km), SE ALBANIA, near the Greek border; 40°40′N 20°45′E. In an agriculture region, it is a commercial and industrial center producing foodstuffs, rugs, and knitwear. Lignite deposits mined nearby. Early Albanian educational center, with a rich cultural tradition. Seat of a Greek Orthodox metropolitan. Known in 1280, it was destroyed (1440) by the Turks but developed again after the 16th century. Ever since Albania gained independence in the Balkan Wars, Korçë has been claimed by Greece. Greek troops occupied it 1912–1913 during the Balkan Wars and again early in World War I. Center of early Communist Party activity. From 1916 to 1920 it was occupied and administered by the French, and in World War II it was held again (Nov. 1940–April 1941) by the Greeks. Origin of many immigrants to the U.S. Has a large 15th-century mosque and several modern government buildings. Also spelled Korça and Koritsa.

Korcheva (KOR-chee-vah), former city in TVER oblast, W central European Russia, on the VOLGA RIVER, 40 mi/64 km E of TVER. Population evacuated (1936–1937) to KONAKOVO and the city flooded by the filling of the VOLGA RESERVOIR.

Korčula (KOR-chuh-lah), Ital. *Curzola*, island (□ 107 sq mi/278.2 sq km), S CROATIA, in the ADRIATIC SEA, in DALMATIA. Popular tourist resort; covered with pine forests, pastures, and vineyards. Most of the inhabitants are sailors, farmers, or fishermen. Colonized by the Greeks in the 4th cent. B.C.E. and occupied, succesively, by Romans, Byzantines, Venetians, and Austrians. Passed to the former Yugoslavia in 1918. Chief town is KORCULA (1991 population 3,232); 42°56′N 16°50′E. It has retained its fine medieval cathedral and fortifications. According to some sources, Marco Polo was born here.

Kordestān (kor-de-STAHN), province (□ 9,652 sq mi/25,095.2 sq km), NW IRAN, bordering IRAQ (W), and the provinces of Āzyarbāyjān-e Gharbi (N), Zanjān and Hamadān (E), and Kermānshahān (S); ⊙ SANANDAJ; 35°40′N 47°00′E. A largely mountainous region, wheat is the major agricultural crop. Lead and zinc deposits found in N. Sanandaj and Saqqez are the largest cities. Also spelled KORDESTAN.

Kordestan, IRAN and TURKEY: see KORDESTĀN.

Kordkuy (KORD-koo-EE), town, Māzandarān province, NE IRAN, 20 mi/32 km WSW of GORGAN, near CASPIAN SEA; 36°47′N 54°07′E. Road junction. Formerly called KURD MAHALLEH; also spelled KURD KUI.

Kordofan, region and former province, SUDAN: see KURDOFAN.

Kordun (KOR-duhn), region in central CROATIA, between KARLOVAC (NW) and BOSNIA and HERZEGOVINA (SE) border. Hilly, wooded, and sparsely populated area; one of Croatia's poorest areas. Livestock raising (hogs, dairy cattle) and agriculture (rye; plums, apples, potatoes) are main activities. Some manufacturing (lumber and wood processing) in centers of VOJNIC, VRGINMOST, and SLUNJ. Predominantly Serb population, with Croatian enclaves around Slunj and other small towns. Part of Austrian VOJNA KRAJINA frontier (16th–19th centuries). Scene of fierce confrontations between Yugoslav partisans and Germans during World War II. Captured by Croatian army in 1995. Chief center is Vojnić.

Kore, town (2007 population 4,803), OROMIYA state, S central ETHIOPIA; 07°13′N 38°55′E.

Korea (ko-REE-ah), Korean *Choson* in NORTH KOREA, *Taehan Minguk* in SOUTH KOREA, country, E Asia (□ 85,049 sq mi/220,277 sq km), E ASIA; ⊙ traditionally SEOUL. Korean peninsula is 600 mi/966 km long, and separates the YELLOW SEA (and KOREA BAY, a N arm of the Yellow Sea) on W from the Sea of JAPAN (or East Sea) on E. On S it is bounded by KOREA STRAIT (connecting the Yellow Sea and the Sea of Japan) and on the N its land borders with CHINA (c.500 mi/800 km) and with RUSSIA (only c.11 mi/18 km) are marked chiefly by the great YALU RIVER and Tumen River.

Geography

The Korean peninsula is largely mountainous; the principal series of ranges, extending along the E coast, rises (in the NE) to 9,003 ft/2,744 m at PAEKTU, the highest peak in Korea. Most rivers are relatively short and many are unnavigable, filled with rapids and waterfalls; important rivers, in addition to the Yalu and Tumen, are the HAN RIVER, KUM RIVER, TAEDONG RIVER, NAKDONG RIVER, and the Youngsan River. Off the heavily indented coast (c.5,400 mi/8,690 km long) lie some 3,420 islands, most of them rocky and uninhabited (of the inhabited islands, about half have a population of less than 100); the main island group is in the Korean Archipelago in the W KOREA STRAIT. The climate of Korea ranges from dry and extremely cold winters in the N to almost tropical conditions in parts of the S. The country once had large timber resources. Most of the remaining stands are in the N, where, despite excessive cutting during the Japanese occupation (1910–1945), timber remains an important resource. Korea has mineral resources, most of it (80%–90%) concentrated in the N. Of the peninsula's five major minerals—gold, iron ore, coal, tungsten, and graphite—only the tungsten and amorphous graphite are found principally in the S. Because of the mountainous and rocky terrain, only about 20% of Korean land is arable. Rice is the major crop, with wet paddy fields constituting about half of the farmland. Paddies are found along the coasts, in claimed tidal areas, and in river valleys. Barley, wheat, corn, soybeans, and grain sorghums are also extensively cultivated, especially in the uplands; other crops include cotton, tobacco, fruits, potatoes, beans, and sweet potatoes. Before the country was divided (1945), the colder and less fertile N depended heavily upon the S for food.

History to 1231

Chinese influences have been strong throughout Korean history, but the Koreans, descended from Tungusic tribal peoples, are a distinct racial and cultural group. The documented history of Korea begins in the 12th century B.C.E., when a Chinese scholar, Ki-tze (Kija), founded a colony at PYONGYANG. After 100 B.C.E. the Chinese colony of Lolang, established near Pyongyang, exerted a strong cultural influence on the Korean tribes on the peninsula. The kingdom of Koguryo, the first native Korean state, arose in the N near the Yalu River. In the 1st century C.E., and by the 4th century it had conquered Lolang. In the S, two kingdoms emerged, that of Paekche (C.E. c.250) and the powerful kingdom of Silla (C.E. c.350). With Chinese support, the kingdom of Silla conquered Koguryo and Paekche in the 7th century and unified the peninsula. Under Silla rule, Korea prospered and the arts flourished; Buddhism, which had entered Korea in the 4th century, became dominant in this period. In 935 the Silla dynasty was peacefully overthrown by Wang Kon, who established the Koryo dynasty (the name was selected as an abbreviated form of Koguryo). During the Koryo period, literature was cultivated, and although Buddhism remained the state religion, Confucianism—introduced from China during the Silla years—controlled the pattern of government.

History - 1231 to 1876

In 1231, Mongol forces invaded from China, initiating a war that was waged intermittently for some thirty years. Peace came when the Koryo kings accepted Mongol rule, and a long period of Koryo-Mongol alliance followed. In 1392, Yi Songgye overthrew the Koryo dynasty and established Choson ruled by the Yi dynasty. The dynasty, which was to rule until 1910, built a new capital at Seoul and adopted Confucianism as the official religion. During King Seijong's rule (mid-15th century), Hangul, an efficient Korean phonetic alphabet) as well as printing with movable metal type were developed. In 1592 an invasion by the Japanese conqueror Hideyoshi was driven back by the Yi dynasty with Chinese (Ming dynasty) help, but only after six years of great devastation and suffering. Manchu invasions in the first half of the 17th century resulted in Korea being made (1637) a vassal of the Manchu dynasty (later becoming the Ching dynasty, which replaced the Ming). Korea attempted to close its frontiers and became so isolated from other foreign contact as to be called the "Hermit Kingdom."

History - 1876 to 1948

All non-Chinese influences were excluded until 1876, when Japan forced a commercial treaty with Korea. To offset the Japanese influence, trade agreements were also concluded (1880s) with the U.S., and the countries of Europe. Japan's control was tightened after the First Sino-Japanese War (1894–1895) and the Russo-

Japanese War (1904–1905), when Japanese troops moved through Korea to attack Manchuria. These troops were never withdrawn, and in 1905 Japan declared a virtual protectorate over Korea and in 1910 formally annexed the country. The Japanese instituted vast social and economic changes, building modern industries and railroads, but their rule (1910–1945) was harsh and exploitative. Sporadic Korean attempts to overthrow the Japanese were unsuccessful, and after 1919 a provisional Korean government-in-exile, under Syngman Rhee, was established at Shanghai. In World War II, at the Cairo Conference (1943), the U.S., Great Britain, and China promised Korea independence. At the end of the war Korea was arbitrarily divided into two zones as a temporary expedient; Soviet troops were N and Americans S of the line of 38°00′N latitude.

History - 1948 to 1980
The Soviet Union thwarted all UN efforts to hold elections and reunite the country under one government. When relations between the Soviet Union and the U.S. worsened, trade between the two zones ceased, and great economic hardship resulted, since the regions were economically interdependent, industry and trade being concentrated in the N and agriculture in the S. In 1948 two separate regimes were formally established—the Republic of Korea in the S, and the Democratic People's Republic of Korea under communist rule in the N. By mid-1949 all Soviet and American troops were withdrawn, and two rival Korean governments were in operation, each eager to unify the country under its own rule. In June 1950, the North Korean army launched a surprise attack against the South, initiating the Korean War, and with it, severe hardship, loss of life, and enormous devastation. After the war the border was stabilized along a line running from the Han estuary NE across the 38th parallel, with a "no-man's land," 1.24 mi/2 km wide and occupying a total of 487 sq mi/1,261 sq km, on either side of the border. Throughout the 1950s and 1960s an uneasy truce prevailed; thousands of soldiers were poised on each side of the demilitarized zone, and there were occasional shooting incidents. In 1971 negotiations between North and South Korea provided the first hope for peaceful reunification of the peninsula; in November 1972, an agreement was reached for the establishment of joint machinery to work toward unification.

History - 1980 to Present
The countries met several times during the 1980s to discuss reunification, and in 1990 there were three meetings between the two prime ministers. These talks have yielded some results, such as the exchange of family visits organized in 1989. The problems blocking complete reunification, however, continue to be substantial. Two incidents of terrorism against South Korea were widely attributed to North Korea: a 1983 bombing that killed several members of the South Korean government, and the 1987 destruction of a South Korean airliner over the Thai-Burmese border. The South Koreans also have protested a dam built near the border that they perceive to be a flooding threat. The emotional appeal of reunification continues to be a factor. Militant students in the S have staged several protests demanding the withdrawal of U.S. military forces from Korea and the reunification of the peninsula. Despite the problems involved, it is likely that the N-S dialogue will continue. For further information see NORTH KOREA and SOUTH KOREA.

Korea (KO-ree-ah), or **Koriya**, former princely state, one of CHHATTISGARH STATES, INDIA. Incorporated in 1948 into SURGUJA district of MADHYA PRADESH state; territory was made into KOREA district in 1998; joined other districts to form Chhattisgarh in 2000.

Korea Bay, inlet of YELLOW SEA, between Liaodong peninsula of LIAONING province, NE CHINA (W) and NORTH KOREA (E); maximum length 200 mi/320 km. Receives YALU RIVER. Its main ports are DANDONG and ZHUANGHE.

Korea Strait, channel connecting Sea of JAPAN (NE) and EAST CHINA SEA (S), between S coast of South Korea (NW) and SW Japan (SE); c.110 mi/177 km wide. The large island TSUSHIMA divides strait into 2 sections, the E channel being the Tsushima strait. The name Korea Strait sometimes indicates only the W channel (c.60 mi/97 km wide).

Koregaon (ko-RAI-goun), town, SATARA district, S MAHARASHTRA state, W central INDIA, 12 mi/19 km E of SATARA; 17°43′N 74°10′E. Agriculture market; peanut milling.

Koreiz, UKRAINE: see KOREYIZ.

Korelichi (kah-RE-lee-chee), Polish *Korelicze*, urban settlement, GRODNO oblast, BELARUS, ⊙ KORELICHI region, on the Rutka River, 13 mi/21 km E of NOVOGRUDOK. Butter and dry milk plant.

Korem (kor-EM), town (2007 population 30,706), TIGRAY state, N ETHIOPIA, S of LAKE ASHANGE, 70 mi/113 km S of MEK'ELĒ; 12°30′N 39°32′E. In cereal and coffee growing region. Formerly Quoram.

Koré Mayroua (KOR-ai MEI-roo-uh), town, DOSSO province, NIGER, 119 mi/192 km E of NIAMEY, near border with NIGERIA; 13°18′N 03°55′E.

Koren, RUSSIA: see KYREN.

Korenëvo (kuh-rye-NYO-vuh), town (2005 population 6,430), W KURSK oblast, SW European Russia, on the SEYM RIVER, on road and railroad, 15 mi/24 km SE of RYL'SK; 51°23′N 34°55′E. Elevation 538 ft/163 m. Railroad junction; depots. Electrical appliances, food products (groats, canned goods). Also spelled Korenevo.

Korenëvo (kuh-ree-NYO-vuh), town (2006 population 7,300), central MOSCOW oblast, central European Russia, near the MOSKVA River, on road and railroad, 4 mi/6 km ESE of LYUBERTSY, to which it is administratively subordinate; 55°40′N 38°00′E. Elevation 456 ft/138 m. Light manufacturing. Research institute for starch products.

Korenica, CROATIA: see TITOVA KORENICA.

Korenov (KO-rzhe-NOF), Czech *Kořenov*, village, SEVEROCESKY province, N BOHEMIA, CZECH REPUBLIC, in the JIZERA MOUNTAINS (Isergebirge), on railroad, on JIZERA RIVER, 17 mi/27 km E of LIBEREC, on Polish border. Manufacturing (furniture, toys); food processing. Winter resort of POLUBNY (elevation 2,379 ft/725 m) is just NW.

Korenovsk (kuh-rye-NOFSK), city (2005 population 41,965), central KRASNODAR TERRITORY, S European Russia, on the Beysuzhek River (tributary of the BEYSUG RIVER), on road and railroad, 40 mi/64 km NNE of KRASNODAR; 45°28′N 39°27′E. Elevation 108 ft/32 m. Center of agricultural area; sugar refining, beer brewing, sunflower oil extraction, dairy products. Founded in 1794 as a village of Korenovskaya. City status and renamed in 1961.

Korenovskaya, RUSSIA: see KORENOVSK.

Koren Pass, AUSTRIA and SLOVENIA: see WURZEN PASS.

Korensko Sedlo, AUSTRIA and SLOVENIA: see WURZEN PASS.

Koréra Koré (ko-RAI-rah ko-RAI), village, FIRST REGION/KAYES, MALI, 39 mi/65 km E of NIORO.

Korets' (KO-rets) (Polish *Korzec*), city, E RIVNE oblast, UKRAINE, on highway, 38 mi/61 km E of RIVNE; 50°37′N 27°10′E. Elevation 652 ft/198 m. Raion center. Plastic and cement manufacturing, sugar refining, flour milling. Granite outcroppings; radon waters used for healing in sanatorium. Kaolin deposits nearby. Flourished in the 18th century as a textile- and porcelain-manufacturing center. Known since 1150; site of defeat of the Tatar army in 1494; capital of Korets'kyy princes in Grand Duchy of Lithuania; passed to Poland (1569), to Russia (1793); part of Ukraine (1917–1921); reverted to Poland in 1921; ceded to USSR in 1945; part of independent Ukraine since 1991. One of the oldest Jewish communities in Ukraine, settling here in the 13th century and the community was a majority of the population until World War II.

Koreyiz (ko-re-YEEZ) (Russian *Koreiz*), town, S Republic of CRIMEA, UKRAINE. A BLACK SEA beach resort adjoining (W) GASPRA, 5 mi/8 km SW and under jurisdiction of YALTA; 44°26′N 34°05′E. Elevation 301 ft/91 m.

Korf (KORF), town, NE KORYAK AUTONOMOUS OKRUG, RUSSIAN FAR EAST, on the NW shore of the Korf Bay of the BERING SEA, 220 mi/354 km NE of PALANA, and 10 mi/16 km SW of TILICHIKI; 60°19′N 165°50′E. Fish canning. Lignite mines in the vicinity.

Korff Ice Rise, in the SW part of the RONNE ICE SHELF, WEST ANTARCTICA, 50 mi/80 km ENE of SKYTRAIN ICE RISE; 80 mi/130 km long and 20 mi/30 km wide; 79°00′S 69°30′W.

Korfovskiy (KOR-fuhf-skeeye), town (2005 population 5,595), SW KHABAROVSK TERRITORY, RUSSIAN FAR EAST, on road and the TRANS-SIBERIAN RAILROAD, 20 mi/32 km S of KHABAROVSK; 48°13′N 135°03′E. Elevation 613 ft/186 m. Stone and sand quarrying. Weather station in the vicinity.

Korgen (KAWR-guhn), village, NORDLAND county, N central NORWAY, on Røssåga River, and 19 mi/31 km SSW of Missouri. Lead, zinc mining. Site of German prison camp (1942–1944) for Yugoslav prisoners.

Korhogo (kor-HO-go), town (2003 population 115,000), ⊙ Savanes region, N CÔTE D'IVOIRE; 09°27′N 05°38′W. Administrative and processing center for a mountainous region. Agriculture (cotton, kapok, rice, millet, sorghum, beans, tobacco, vegetables, groundnuts, and maize); cotton ginning; livestock (cattle). Diamonds are mined in the area. In pre-colonial times, Korhogo was on an important trade route, which led to the ATLANTIC OCEAN. Airport.

Kori (KO-ree), town, Date county, FUKUSHIMA prefecture, N central HONSHU, NE JAPAN, 4.3 mi/7 km N of FUKUSHIMA city; 37°50′N 140°31′E. Fruit.

Koria, district, INDIA: see KOREA.

Koribundu (ko-ree-BOON-doo), town, SOUTHERN province, S central SIERRA LEONE, 17 mi/27 km SSE of BO; 07°42′N 11°41′W. Road center; palm oil and kernels, cacao, coffee.

Kori Creek, INDIA: see LAKHPAT.

Korientzé (ko-REE-uhn-zai), town, FIFTH REGION/MOPTI, MALI, 66 mi/110 km NNE of MOPTI; 15°24′N 03°47′W. Also spelled Korienzé.

Korinthia (ko-reen-THEE-ah), prefecture, NE corner of PELOPONNESE department, SE mainland GREECE, on Gulf of CORINTH and SARONIC GULF; ⊙ CORINTH; 37°55′N 22°40′E. Include S section of Isthmus of CORINTH, with CORINTH CANAL. Borders ATTICA department on NE, ARGOLIS prefecture on S, and WESTERN GREECE department on W.

Korinthos, Greece: see CORINTH.

Köris, Mount (KUH-rish), Hungarian *Kőrishegy* (2,339 ft/713 m), highest point in BAKONY MOUNTAINS, NW central HUNGARY; 47°12′N 17°49′E. Heavily forested.

Koritnik (kor-EET-neek), peak (7,854 ft/2,394 m) in SAR MOUNTAINS, KOSOVO province, on SERBIA-ALBANIA border, 13 mi/21 km SW of PRIZREN (Serbia).

Koritschan, CZECH REPUBLIC: see KORYCANY.

Koriya, district, INDIA: see KOREA.

Koriya (KO-ree-ah), or **Korea**, city, ⊙ KORIYA district, N central CHHATTISGARH state, central INDIA. Originally part of SURGUJA district in MADHYA PRADESH, became capital of Koriya district in 1998; became part of Chhattisgarh in 2000.

Koriyama (KO-ree-yah-mah), city (2005 population 338,834), FUKUSHIMA prefecture, N central HONSHU, NE JAPAN, near LAKE INAWASHIRO, on the ABUKUMA RIVER, 25 mi/40 km N of FUKUSHIMA city; 37°23′N 140°21′E. Major commercial and communications center with industries producing textiles, computers, and electronic goods. Tomatoes. Traditional dolls, wooden molds.

Koriyama (ko-REE-yah-mah), town, Hioki county, KAGOSHIMA prefecture, SW KYUSHU, SW JAPAN, 9 mi/15 km N of KAGOSHIMA; 31°40′N 130°28′E.

Cross-references are shown in SMALL CAPITALS. The pronunciation guide is shown on page xix. The sources of population figures are shown on page xvii.

Korkeakoski (KOR-kai-a-KOS-kee), village, HÄMEEN province, SW FINLAND, 30 mi/48 km NE of TAMPERE; 61°48′N 24°22′E. Elevation 413 ft/125 m.

Korkino (KOR-kee-nuh), city (2005 population 39,600), E CHELYABINSK oblast, SW SIBERIA, Russia, on road and railroad, 26 mi/42 km S of CHELYABINSK; 54°53′N 61°24′E. Elevation 839 ft/255 m. Railroad spur terminus. Lignite mining (Chelyabinsk basin); cement, glassworks, furniture; auto repair; sewing factory; food processing (dairy, bakery). Became city in 1942.

Korkmaskala (kuhrk-mahs-kah-LAH), village (2005 population 6,555), E central DAGESTAN REPUBLIC, SE European Russia, on road and near railroad, 20 mi/32 km WNW of MAKHACHKALA; 43°01′N 47°17′E. Elevation 183 ft/55 m. In oil- and gas-producing region.

Korkudeli, TURKEY: see KORKUTELI.

Korkuteli, township, SW TURKEY, 30 mi/48 km NNW of ANTALYA; 37°07′N 30°11′E. Wheat, barley. Sometimes spelled Korkudeli. Formerly Istanos.

Korla (KUH-UHR-LUH), oasis and city (□ 2,876 sq mi/ 7,449 sq km; 1994 estimated urban population 177,800; estimated total population 295,500), central XINJIANG UYGUR AUTONOMOUS REGION, NW CHINA, on the highway to the TIANSHAN mountains, and 160 mi/ 257 km SW of URUMQI (linked by railroad); 41°48′N 86°10′E. The S Xinjiang railroad reached here in 1984, linking it with E China. Construction is currently underway to link the city to KASHI. A transportation hub in S Xinjiang, Korla has flights to regional (Urumqi, Qiemo, and Kuche) and national (Beijing and Jinan) centers. The oil pipelines extend to Lunnan and Shanshan. Major petrochemical center in Xinjiang. Other industries include textiles, paper, and utilities. Crop growing, animal husbandry. Main agricultural product is fruit. Designated a city in 1979.

Körlin, POLAND: see KARLINO.

Korlino, POLAND: see KARLINO.

Korlyaki (kuhr-lyah-KEE), village, SW TVER oblast, W European Russia, on road, 15 mi/24 km NW of SANCHURSK; 57°06′N 46°55′E. Elevation 374 ft/113 m. In agricultural area (wheat, rye, sunflowers).

Korma (kor-MUH), urban-type settlement, N GOMEL oblast, BELARUS, ⊙ KORMA region, near SOZH RIVER, 50 mi/80 km N of GOMEL, 53°08′N 30°47′E. Manufacturing (building materials; fruit and vegetable processing, flax milling).

Kormakiti (kor-MAH-kee-tee) [Turkish=Koruçam] town, KYRENIA district, N CYPRUS, 2 mi/3.2 km to S of MEDITERRANEAN SEA, 24 mi/39 km NW of LEFKOSIA; 35°20′N 33°01′E. MORPHOU BAY 5 mi/8 km to W, Cape KORMAKITI 7 mi/11.3 km to NW. Citrus, olives, grapes, grain; livestock.

Kormakiti, Cape (kor-MAH-kee-tee), Turkish Cape Koruçam, ancient Crommyou, N CYPRUS, in MEDITERRANEAN Sea at NE gate to MORPHOU BAY, 30 mi/48 km NW of NICOSIA; 35°22′N 32°55′E. Lighthouse.

Kormantyn, GHANA: see SALTPOND.

Körmend (KUHR-mend), village, VAS county, W HUNGARY, on RÁBA RIVER, 15 mi/24 km S of SZOMBATHELY; 47°01′N 16°36′E. Road center; manufacturing (pharmaceuticals, footwear, bricks, tiles, flour milling, farinaceous food products). Battle against Ottoman forces, 1664.

Kormilovka (kuhr-MEE-luhf-kah), town (2006 population 10,130), SE OMSK oblast, SW SIBERIA, RUSSIA, on the OM′ RIVER, on road and the TRANS-SIBERIAN RAILROAD, 27 mi/43 km E of OMSK; 55°00′N 74°06′E. Elevation 387 ft/117 m. Dairy products, meat processing. Railroad station since 1896; town since 1960.

Körmöcbánya, SLOVAKIA: see KREMNICA.

Kormovishche (kuhr-muh-VEE-shchye), SE PERM oblast, W URALS, E European Russia, on highway junction and railroad, 14 mi/23 km SSE of LYS′VA; 57°52′N 58°01′E. Elevation 918 ft/279 m. Sawmilling, timbering.

Korna, IRAQ: see QURNA, AL.

Kornaka, town, NIGER, NNW of MARADI; 14°05′N 06°54′E.

Kornati National Park (□ 86 sq mi/223 sq km), Kornati Islands, DALMATIA, S CROATIA. Comprises 125 islands (including reefs). No permanent settlement here. Residents from other islands grow olives, figs, and grapes (for wine) on patches of red soil here; also, fishing and tourism. Est. 1980.

Kornat Island (KOR-naht), Italian Incoronata (eenKO-ro-NAH-tah), Dalmatian island in ADRIATIC SEA, S CROATIA; 15 mi/24 km long, up to 1.5 mi/2.4 km wide; largest island in the Kornat archipelago; northernmost point is 12 mi/19 km WSW of BIOGRAD NA MORU. Venetian ruins, swimming beaches. KORNATI NATIONAL PARK is here.

Kornelimünster (kor-ne-le-MYOON-stuhr), suburb of AACHEN, North Rhine-Westphalia, W GERMANY, 5 mi/8 km SE of city center; 50°44′N 06°11′E. Has late-Gothic buildings of former abbey. Formerly also spelled Cornelimünster.

Korneshty, MOLDOVA: see Corneşti or CORNESTI.

Korneuburg (kor-NOI-boorg), town, E LOWER AUSTRIA, on the left bank of the DANUBE River, and 7 mi/ 11.3 km N of VIENNA; 48°21′N 16°20′E. Railroad junction. Developed as the opposite bridgehead of KLOSTERNEUBURG. Large thermoelectric power station; manufacturing of paper, machines.

Kornik (kor-neek), Polish Kórnik, German Kurnik, town, Poznań province, W POLAND, 13 mi/21 km SSE of POZNAŃ. Flour milling, cider making, sawmilling. Castle with art treasures and noted library.

Kornin, UKRAINE: see KORNYN.

Kornisi (kor-NEE-see), village, SW South Ossetian Autonomous Oblast, GEORGIA, 10 mi/16 km WSW of TSKHINVALI. Orchards; grain. Znauri until 1991.

Kornoq, GREENLAND: see QOORNOQ.

Kornoukhovo (kuhr-nuh-OO-huh-vuh), village, W central TATARSTAN Republic, E European Russia, on road, 21 mi/34 km SE of KAZAN; 55°33′N 49°53′E. Elevation 295 ft/89 m. Distilling.

Kornsjø (KAWRN-shuh), village, ØSTFOLD county, SE NORWAY, on railroad, frontier station on Swedish border, opposite Møn (SWEDEN),17 mi/27 km SE of HALDEN.

Korntal-Münchingen (KORN-tahl–MYOON-khinguhn), suburb of STUTTGART, BADEN-WÜRTTEMBERG, SW GERMANY, just E of city; 48°50′N 09°07′E. Manufacturing of tools and chemicals. Chartered in 1975 after unification of KORNTAL and Münchingen.

Kornwestheim (korn-VEST-heim), town, BADEN-WÜRTTEMBERG, SW GERMANY, 2 mi/3.2 km S of LUDWIGSBURG; 48°52′N 09°11′E. Railroad junction; shoe-manufacturing center; foundry; also manufacturing of machinery. Chartered 1931.

Környe (KUHR-nye), village, KOMÁROM co., N HUNGARY, 34 mi/55 km W of BUDAPEST; 47°33′N 18°20′E. Wheat, alfalfa, potatoes; cattle. Ruins of Roman settlement nearby.

Kornyn (KOR-nin) (Russian Kornin), town (2004 population 3,700), SE ZHYTOMYR oblast, UKRAINE, on Irpin′ River, 38 mi/61 km ESE of ZHYTOMYR; 50°05′N 29°22′E. Elevation 718 ft/218 m. Sugar refining, flour milling, cheese making; bricks, granite quarry. Known since 1550, town since 1938.

Koro, town, FIFTH REGION/MOPTI, MALI, on border with BURKINA FASO, 99 mi/159 km SE of MOPTI; 14°04′N 03°05′W. Commercial center.

Koro (KAW-ro), basaltic volcanic island (□ 40 sq mi/ 104 sq km), FIJI, SW PACIFIC OCEAN, c.30 mi/50 km S of VANUA LEVU; 11 mi/18 km long, 5 mi/8 km wide. Densely wooded range rises to 1,850 ft/564 m. Copra, bananas, yagona (kava).

Korobkovo, RUSSIA: see GUBKIN.

Korocha (kuh-RO-chah), city (2005 population 6,100), central BELGOROD oblast, S central European Russia, on road, 35 mi/56 km NE of BELGOROD; 50°49′N 37°11′E. Elevation 538 ft/163 m. Fruit canning, flour

milling, sunflower oil extraction, dairying. Chartered in 1638.

Korochin, RUSSIA: see KAPUSTIN YAR.

Koroglu Mountains, (Turkish=Köroğlu) N central TURKEY, in the W part of the Pontus mountain system, extending 230 mi/370 km W from BILECIK; ZAMANTI and DEVREZ rivers to N and the SAKARYA, KIRMIR, and KIZIL IRMAK rivers to S; rise to 7,802 ft/ 2,378 m. Towns of NALLIHAN, BEYPAZARI, KIZILCAHAMAM, CANKIRI, and ISKILIP on S slope. Copper in central area.

Korogwe (ko-RO-gwai), town, TANGA region, NE TANZANIA, 40 mi/64 km WSW of TANGA, on PANGANI RIVER (bridged), and on railroad; 05°11′S 38°27′E. Highway junction. USAMBARA Mountains to NE, Mount Korogwe (3,704 ft/1,129 m) to E. Sisal, rice, corn; livestock; timber. A busy caravan center until mid-19th century Seat of College of National Education.

Koroit (kuh-ROIT), municipality, SW VICTORIA, SE AUSTRALIA, 145 mi/233 km WSW of MELBOURNE, 11 mi/18 km NW of WARRNAMBOOL, near coast, and on N slopes of Tower Hill, an extinct volcano; 38°18′S 142°32′E. Railroad junction; commercial center for livestock area; dairying; potatoes and other vegetables. Botanic gardens.

Korolëv (kuh-ruh-LYOF), city (2006 population 138,870), central MOSCOW oblast, central European Russia, on railroad (Podlipki station), 12 mi/19 km NE of (and administratively subordinate to) MOSCOW, and 1 mi/1.6 km NE of MYTISHCHI; 55°55′N 37°49′E. Elevation 505 ft/153 m. Center of Russian space research by satellites; textiles, woodworking; machinery; amber trade. Ice-free river port; shipyard. Has a naval college. Oil deposits in the vicinity. Formed in 1938 on the basis of Kalinskiy settlement and further expanded by amalgamation with the city of KOSTINO in 1960 and PERVOMAYSKIY, Tekstilshchiki, and Bol′shevo in 1963. Historical landmarks include the Königsberg cathedral. The resting place of Immanuel Kant. Originally called Podlipki, then Kalinskiy, 1928–1938, and Kaliningrad until 1996, when renamed in honor of Sergey Korolev, the late leader of the Soviet space program.

Koroleve (KO-ro-le-ve) (Russian Korolevo) (Czech Královo nad Tisou) (Hungarian Királyháza), town (2004 population 15,600), S TRANSCARPATHIAN oblast, UKRAINE, on TISZA RIVER, and 9 mi/15 km SW of KHUST; 48°09′N 23°08′E. Elevation 449 ft/136 m. Railroad junction and depot near the Romanian border. Quarry and stone crushing; sewing; handicrafts; food processing. Known since 1262 (Hungarian castle), town since 1947.

Korolevo, UKRAINE: see KOROLEVE.

Koromogawa (ko-RO-mo-GAH-wah), town, Isawa county, IWATE prefecture, N HONSHU, NE JAPAN, 47 mi/75 km S of MORIOKA; 39°02′N 141°04′E.

Koronadal (ko-ro-nah-DAHL), town (2000 population 133,786), ⊙ SOUTH COTABATO province, S MINDANAO, PHILIPPINES, 16 mi/26 km WNW of Buayan, and NW of head of SARANGANI BAY; 06°28′N 124°53′E. Rice, coconuts. Majority of population is tribal: Maguindanao and B′laan. Called Marbel by the B′laan.

Koroneia (ko-ro-NEE-ah), ancient town of BOEOTIA, in what is now CENTRAL GREECE department, GREECE, 7 mi/11.3 km SE of LEVADIA and 6 mi/9.7 km SW of Lake COPAIS. Here the Boeotians defeated (447 B.C.E.) the Athenians. Later, the Spartans under Agesilaus II won a victory here (394 B.C.E.) over the Thebans and their allies. On site is modern village of Koroneia or Koronia, formerly called Koutoumoula. Also Coronea.

Korongo (ko-RON-go), village, SINGIDA region, central TANZANIA, 60 mi/97 km SW of DODOMA, on KISIGO RIVER; 06°28′S 34°51′E. Kisigo Game Reserve to S. Timber; cattle, goats, sheep; corn, sweet potatoes.

Koroni (ko-RO-nee), ancient Asine, town, MESSENIA prefecture, SW PELOPONNESE department, extreme

SW GREECE, port on Gulf of MESSENIA, 17 mi/27 km SSW of KALAMATA; 36°48′N 21°57′E. Developed (c.1300) as Venetian supply port of Coron, captured (c.1500) by Turks. Has ruins of Venetian castle. Ancient Corone was 10 mi/16 km N, at site of modern village of Petalidhion or Petalidi. Formerly spelled Korone.

Koronia, Lake (ko-ro-NEE-ah), Latin *Coronea*, lake (□ 22 sq mi/57.2 sq km), in CENTRAL MACEDONIA department, NE GREECE, at base of KHALKIDHIKÍ Peninsula, 7 mi/11.3 km ENE of THESSALONÍKI; 8 mi/12.9 km long, 3 mi/4.8 km wide; 40°41′N 23°09′E. Also called Lankada or Langadha. Formerly spelled Koroneia.

Koroni, Gulf of, Greece: see MESSENIA, GULF OF.

Koronin-kaikyo, RUSSIA: see GOLOVNIN STRAIT.

Koronis, Lake (kuh-RO-nis), STEARNS county, SE central MINNESOTA, extends S into MEEKER county, 2 mi/3.2 km S of PAYNESVILLE, 21 mi/34 km NE of WILLMAR; 4 mi/6.4 km long, 3 mi/4.8 km wide. Boating and fishing resorts. Fed (from NE and RICE LAKE) and drained (SE, through dam) by North Fork CROW RIVER; both inflow and outflow are at SE end. Paynesville Municipal Park on N shore. Sometimes known as Cedar Lake.

Koronowo (ko-ro-NO-vo), German *Krone* or *Krone an der Brahe*, town, Bydgoszcz province, N central POLAND, on BRDA RIVER, 13 mi/21 km NNW of BYDGOSZCZ. Manufacturing of bricks, ceramic products, furniture, mattresses; distilling, flour milling, sawmilling. German name also spelled Crone or Crone an der Brahe.

Korop (KO-rop), town, E CHERNIHIV oblast, UKRAINE, near DESNA River, 27 mi/43 km NNE of BAKHMACH; 51°34′N 32°58′E. Elevation 439 ft/133 m. Raion center. Flax and food processing. Known since 1153, town since 1924.

Koropets' (ko-ro-PETS) (Polish *Koropiec*) town (2004 population 7,560), SW TERNOPIL' oblast, UKRAINE, on the DNIESTER River, and 12 mi/19 km SW of BUCHACH; 48°56′N 25°11′E. Elevation 721 ft/219 m. Household goods manufacturing; sewing. Has a 19th-century park. Established in 1421, town since 1984.

Koropi (ko-ro-PEE), town (2001 population 15,860), ATTICA prefecture, ATTICA department, E central GREECE, on railroad, and 9 mi/14.5 km SE of ATHENS; 37°54′N 23°53′E. Winery center; vegetables, wheat, olive oil.

Koropiec, UKRAINE: see KOROPETS'.

Koror, Republic of Palau: see OREOR.

Körös (KUH-ruhsh) [Hungarian=triple Körös], Romanian *Criş*, river, around 345 mi/560 km long, formed in E HUNGARY by the junction of three headstreams that rise in TRANSYLVANIA's BIHOR MOUNTAINS, NW ROMANIA. It meanders W through farmland to the TISZA RIVER at CSONGRÁD. The Körös is used for irrigation. Also called Hármas Körös River.

Körös, CROATIA: see KRIŽEVCI.

Körösbánya, ROMANIA: see BAIA DE CRIŞ.

Koro Sea (KAW-ro), section of the PACIFIC OCEAN, in FIJI islands; bounded S by Viti Levu, N by VANUA LEVU, E by Lau group.

Koroška (ko-ROSH-kah), German *Kärnten*, historical region, N SLOVENIA (ancient Carniola), part of medieval Hapsburg province of CARINTHIA, part of Austria-Hungary (1848–1918). Encompasses W part of Karawanken Alps; principal mountains are POHORJE (5,062 ft/1,543 m), Kozjak (4,537 ft/1,383 m), PECA (6,972 ft/2,125 m), Uršlja gora (5,574 ft/1,699 m). Drained by DRAVA, Mislinja, and Meža rivers. Mining of zinc, lead, and iron ores since Roman times; timber and tourism. Principal towns are DRAVOGRAD, Ravne na Koroškem, Leše, Mežica, and SLOVENJ.

Korosko (KO-ros-ko), former village, ASWAN province, S EGYPT. Former trade center and starting point for caravans crossing the NUBIAN DESERT. The village and its lands were flooded under Lake NASSER after the inauguration of the ASWAN HIGH DAM.

Körösladány (KUH-ruhsh-lah-dah-nyuh), Hungarian *Köröladány*, village, BÉKÉS county, SE HUNGARY, on the Rapid Körös, 22 mi/35 km E of Mezőtúr; 46°58′N 21°05′E. Grain, sunflowers; cattle, hogs; manufacturing (bricks, flour milling, farinaceous foods, vehicle components).

Körösmező, UKRAINE: see YASINYA.

Korosten' (KO-ros-ten), city (2001 population 66,669), N central ZHYTOMYR oblast, UKRAINE, 48 mi/77 km N of ZHYTOMYR, on the UZH River; 50°57′N 28°39′E. Elevation 531 ft/161 m. Railroad and road junction. Raion center. Construction machinery, woodworking, porcelain, chemical processing, granite working, textiles, railroad shops, food processing. Technical school; heritage museum. First mentioned in 883 as Iskorosten'; capital of the land of Dervlyane; gained commercial importance in the 19th century as a junction on the Kiev-Warsaw and St. Petersburg–Odessa railroad lines. City since 1926. Until World War II, when city was held (1941–1943) by the Germans, population was 50% Jewish.

Korostyshev, UKRAINE: see KOROSTYSHIV.

Korostyshiv (ko-ros-TI-shif) (Russian *Korostyshev*), city, E ZHYTOMYR oblast, UKRAINE, on TETERIV RIVER, and 17 mi/27 km E of ZHYTOMYR, on main road from KIEV; 50°19′N 29°04′E. Elevation 554 ft/168 m. Raion center. Paper and textile mills, flax, distilling; reinforced concrete, bricks; food processing. Granite and limestone quarries. Has a 19th-century park, health resort; heritage museum. Known since the 13th century; in 1649, became town of Bila Tserkva. Cossack regiment; chartered 1779; city since 1938.

Koro Toro (ko-RO to-RO), town, BORKOU-ENNEDI-TIBESTI administrative region, N CHAD, 170 mi/274 km SW of FAYA; 16°05′N 18°30′E.

Korotovo (KO-ruh-tuh-vuh), village, SW VOLOGDA oblast, N central European Russia, on the NW shore of the RYBINSK RESERVOIR, near highway, 20 mi/32 km SW of CHEREPOVETS; 58°57′N 37°28′E. Elevation 321 ft/97 m. In agricultural area (wheat, barley, rye, sunflowers; livestock raising).

Korotoyak (kuh-ruh-tuh-YAHK), village, W central VORONEZH oblast, S central European Russia, on the DON River, on railroad spur, 49 mi/79 km S of VORONEZH, and 9 mi/14 km NNW of OSTROGOZHSK; 50°59′N 39°11′E. Elevation 262 ft/79 m. Sunflower oil extraction, flour milling; chalk quarrying.

Korotych (ko-RO-tich), town (2004 population 14,600), NW KHARKIV oblast, UKRAINE, 9 mi/15 km WSW of KHARKIV; 49°57′N 36°02′E. Elevation 666 ft/202 m. Experimental farm of the Ukrainian Institute of Soil Science and Agrochemistry. Most residents work in Kharkiv. Founded in the late 17th century; town since 1938.

Korovanitu or **Mount Evans** (3,921 ft/1,195 m), NW Viti Levu, FIJI.

Korovin Volcano (KOR-ro-vin) (5,030 ft/1,533 m), on E ATKA Island, ALEUTIAN ISLANDS, SW ALASKA; 52°23′N 174°11′W. Active.

Korpilombolo (KOR-pee-loom-boo-lo), village, NORRBOTTEN county, N SWEDEN, near FINNISH border, 75 mi/121 km NNW of HAPARANDA; 66°51′N 23°03′E.

Korpo (KOR-po), Finnish *Korppoo*, village, TURUN JA PORIN province, SW FINLAND, 30 mi/48 km WSW of TURKU; 60°10′N 21°34′E. On Korpo island (7 mi/11.3 km long, 5 mi/8 km wide) in strait between BALTIC SEA and GULF OF BOTHNIA. Fishing; limestone quarrying. Has thirteenth-century church.

Korpona, SLOVAKIA: see KRUPINA.

Korrahei, ETHIOPIA: see K'ORAHĒ.

Kor River (KUHR), ancient *Cyrus*, 200 mi/322 km long, in ZAGROS ranges of S central IRAN; rises 120 mi/193 km NW of SHIRAZ; flows SE, past RAMJIRD (irrigation headworks), to Negriz Lake. Receives PULVAR RIVER (left) on MARVDASHT plain. Called BANDAMIR RIVER in lower course. Also spelled KUR RIVER.

Korsakov (KOR-sah-kuhf), city (2006 population 34,580), S SAKHALIN oblast, RUSSIAN FAR EAST, port on the ANIVA GULF, on the coastal highway and railroad, 26 mi/42 km S of YUZHNO-SAKHALINSK; 46°38′N 142°47′E. Southern terminus of the E coastal railroad. Oceanic fishing, fish processing; distilling (whisky, brandy); corrugated containers. Chief port (opened in 1909) of S SAKHALIN; when icebound in winter, replaced by KHOLMSK. First Russian military post (founded in 1853) of SAKHALIN. Under Japanese rule (1905–1945), called Otomari, and served as capital of Karafuto administrative region.

Korsakovo (kuhr-SAH-kuh-vuh), village (2006 population 1,460), N OREL oblast, SW European Russia, 45 mi/72 km N of KURSK (KURSK oblast), and 31 mi/50 km E of MTSENSK; 53°16′N 37°21′E. Elevation 711 ft/216 m. Distilling.

Korschenbroich (KOR-shuhn-broikh), town, North Rhine-Westphalia, W GERMANY, just E of MÖNCHENGLADBACH. Textiles, metalworking, brewing, gravel quarrying, flower-growing. Many residents commute to MÖNCHENGLADBACH. Has a moated castle begun in the 14th century and the suburb of Liedberg has a castle from the 14th–17th centuries. First mentioned in 1127; incorporated adjoining villages in 1975; chartered in 1981.

Korshev, UKRAINE: see KORSHIV.

Korshik (KOR-shik), village, S central KIROV oblast, E central European Russia, on road, 15 mi/24 km SSE, and under administrative jurisdiction, of ORICHI; 58°12′N 49°17′E. Elevation 557 ft/169 m. In agricultural area (wheat, flax, potatoes, vegetables); logging and lumbering. Has a hospital.

Korshiv (KOR-shif) (Russian *Korshev*) (Polish *Korszów*), town, central IVANO-FRANKIVS'K oblast, UKRAINE, on railroad, 8 mi/13 km NNW of KOLOMYYA; 48°39′N 25°01′E. Elevation 1,059 ft/322 m. Wheat, rye, potatoes; lumbering. Town since 1992.

Korsholm (KOSH-HOLM), Finnish *Mustasaari*, village, VAASAN province, W FINLAND, on GULF OF BOTHNIA, 4 mi/6.4 km SE of VAASA; 63°05′N 21°43′E. Formerly site of fourteenth-century castle (no remains). Original city of Vaasa, destroyed by fire in 1852, was just S of here.

Korshunikha-Angarskaya, RUSSIA: see ZHELEZNOGORSK-ILIMSKIY.

Korsnäs (KORSH-NES), village, KOPPARBERG county, central SWEDEN, at N end of Runn Lake (10 mi/16 km long), 3 mi/4.8 km SE of FALUN; 60°35′N 15°41′E.

Korsør (kor-SUHR), city (2000 population 14,714), VESTSJAELLAND county, S central DENMARK, a seaport on the STORE BÆLT; 55°20′N 11°15′E. Fisheries; glass, food processing.

Korsovka, LATVIA: see KĀRSAVA.

Korsten, W residential suburb of PORT ELIZABETH (NELSON MANDELA METROPOLE), EASTERN CAPE province, SOUTH AFRICA, on fringe of industrial section; 33°56′S 25°34′E; elevation 52 ft/15 m.

Korsun' (KOR-soon), town, central DONETS'K oblast, UKRAINE, in the DONBAS, 6 mi/10 km WSW of Yenakiyeve and subordinated to its city council; 48°12′N 38°05′E. Elevation 469 ft/142 m. Farming; vegetables. Established in 1622, town since 1938.

Korsun', UKRAINE: see KORSUN'-SHEVCHENKIVS'KYY.

Korsun'-Shevchenkivs'kyy (KOR-soon–shev-CHEN-keev-skee) (Russian *Korsun-Shevchenkovskiy*), city, central CHERKASY oblast, UKRAINE, on Ros' River, and 35 mi/56 km NW of CHERKASY; 49°26′N 31°15′E. Elevation 347 ft/105 m. Raion center. Automotive and tractor parts, machine tools, apparel, furniture, building, materials, metalworking, fruit canning, feed milling. Established in 1032; destroyed by the Mongols in 1240; Polish fortress (1584); taken by Cossack forces (1630); razed by Polish forces (1637); site of the Battle of Korsun' (1648) in Cossack-Polish War; center of Korsun' regiment (1648–1712); passed to Russia (1793); site of a Red Army victory in bat-

---

Cross-references are shown in SMALL CAPITALS. The pronunciation guide is shown on page xix. The sources of population figures are shown on page xvii.

tle over German forces (1944). Called Korsun' until 1944.

Korsun'-Shevchenkovskiy, UKRAINE: see KORSUN'-SHEVCHENKIVS'KYY.

Korszów, UKRAINE: see KORSHIV.

Kortemark (KAHR-tuh-mahrk), commune (2006 population 11,945), Sint-Niklaas district, WEST FLANDERS province, W BELGIUM, 4 mi/6.4 km SW of TORHOUT; 51°02′N 03°02′E. Agriculture market; food processing. Has thirteenth-century church. Formerly spelled Cortemarck.

Korten (kor-TEN), village, BURGAS oblast, NOVA ZAGORA obshtina, BULGARIA; 42°32′N 25°59′E.

Kortenaken (KAHR-tuh-nah-kuhn), commune (2006 population 7,553), Leuven district, BRABANT province, central BELGIUM, 5 mi/8 km S of DIEST; 50°55′N 05°04′E.

Kortenberg (KAHR-tuhn-berk), commune (2006 population 18,410), Leuven district, BRABANT province, central BELGIUM, 9 mi/14 km ENE of BRUSSELS; 50°53′N 04°32′E.

Kortes Dam (kor-TEZ), CARBON county, S central WYOMING, on North Platte River, 35 mi/56 km NE of RAWLINS; 42°11′N 106°52′W. Minor reservoir extends 4 mi/6.4 km S to Seminoe Dam: headwaters of PATHFINDER RESERVOIR below dam (N). Construction of Kortes Dam was started in 1946 and completed in 1951.

Kortessem (KAHR-tes-suhm), commune (2006 population 8,124), Tongeren district, LIMBURG province, NE BELGIUM, 6 mi/10 km SSE of HASSELT; 50°52′N 05°22′E.

Korti (KOR-tee), town, Northern state, SUDAN, on left bank of the NILE, and 30 mi/48 km SSW of MARAWI; 18°07′N 31°33′E. Wheat; livestock.

Kortkeros (kuhrt-kee-ROS), village (2005 population 4,355), SW KOMI REPUBLIC, NE European Russia, on the VYCHEGDA River, near the confluence of branches of three local highways, 25 mi/40 km ENE of SYKTYVKAR; 61°48′N 51°35′E. Elevation 295 ft/89 m. In agricultural area (flax, potatoes, vegetables); logging and lumbering.

Kortrijk (KAHR-treik), French *Courtrai* (coor-tray), city (□ 156 sq mi/405.6 sq km; 2006 population 73,694), ⊙ Kortrijk district, WEST FLANDERS province, SW BELGIUM, on the Leie River; 50°50′N 03°16′E. It is an important linen, lace, and textile-manufacturing center. One of the earliest (fourteenth century) and most important cloth-manufacturing towns of medieval FLANDERS. In 1302, Flemish burghers defeated French knights here in the first Battle of the Spurs. The Church of Notre Dame (thirteenth century) contains Anthony Van Dyck's *Elevation of the Cross* (1631). The Gothic city hall dates from the sixteenth century.

Koruçam, CYPRUS: see KORMAKITI.

Koruçam, Cape, CYPRUS: see KORMAKITI, CAPE.

Korumburra (kah-ruhm-BUHR-uh), town, S VICTORIA, AUSTRALIA, 65 mi/105 km SE of MELBOURNE; 38°26′S 145°49′E. Livestock center; cheese. Old coal mines nearby; Coal Creek Historical Village.

Koru Mountains, coastal range, TURKEY, extending c.20 mi/32 km along shore of Gulf of SAROS from Greek border; rising to 2,379 ft/725 m in Koru Dag, 9 mi/14.5 km S of MALKARA.

Korup National Park (KOOR-up), South-West province, CAMEROON, near the NIGERIA border, NW of MUNDEMBA. Anc. rainforest with great biological diversity.

Korvey, GERMANY: see HÖXTER.

Koryak Autonomous Okrug (kuh-RYAK), Russian *Koryatskiy Avtonomnyy Okrug*, administrative region (□ 116,410 sq mi/302,666 sq km), extreme NE SIBERIA, RUSSIAN FAR EAST; ⊙ PALANA. Occupies the N half of the KAMCHATKA PENINSULA, the adjoining part of the Asian mainland, and the KARAGINSKIY ISLAND. Borders KAMCHATKA oblast in the S, MAGADAN oblast in

the NW, and CHUKCHI AUTONOMOUS OKRUG in the N. Soils are mostly tundra and peaty-boggy permafrost; vegetation consists lowland moss and lichen, mountain tundra, and cedar-alder forests. Population mostly Russian (62%), with the Koryaks (16%) and Ukrainians (7%) representing the largest minorities. The region's forests abound in sable, fox, weasel, brown bear, and snow sheep, and fur trading is among the most important industries. Fishing is the main occupation of the population in the coastal areas. Mineral resources include brown coal (KORF) and mercury ore (Cape Olyutorsk). Agriculture is poorly developed, with only 0.2% of the region under cultivation, and consists mostly of greenhouse vegetable growing; reindeer herding, on the other hand, is widespread. At the time of the region's discovery by Russian explorers in the mid-17th century, Koryak tribes lived in large patriarchal family communities and consisted of two distinctive groups of nomadic herders and settled communities of hunter-gatherers. As the Russian exploration and settlement of the region progressed, reindeer herding became widespread because it provided the most valuable commodity used to trade for European goods. Divided into four administrative districts, with centers in OSSORA, TILICHIKI, KAMENSKOYE, and TIGIL.

Koryak Range (kuh-RYAHK), NE KAMCHATKA oblast, extreme E SIBERIA, RUSSIAN FAR EAST, extends from neck of the KAMCHATKA PENINSULA NNE to the ANADYR' GULF; rises to approximately 4,500 ft/1,372 m.

Koryakskaya Kul'tbaza, Russia: see KAMENSKOYE.

Koryazhma (kuh-RYAZH-mah), city (2005 population 42,600), central ARCHANGEL oblast, N European Russia, on the VYCHEGDA River, near its confluence with the NORTHERN DVINA River, near railroad station (Nizovka), on road, 515 mi/829 km SE of ARCHANGEL, and 25 mi/40 km E of KOTLAS; 61°19′N 47°07′E. Elevation 272 ft/82 m. Chemical industry (chlorine); paper milling. Arose in 1961 with the construction of the Kotlas wood pulp and paper combine. Became city in 1985.

Korycany (KO-ri-CHAH-ni), Czech *Koryčany*, German *koritschan*, town, JIHOMORAVSKY province, SE MORAVIA, CZECH REPUBLIC, 24 mi/39 km WSW of ZLÍN; 49°07′N 17°11′E. Railroad terminus. Manufacturing of bentwood furniture. Baroque castle. Neolithic-era archaeological site.

Koryo (KO-ryo), town, N Katsuragi district, NARA prefecture, S HONSHU, W central JAPAN, 9 mi/14 km S of NARA; 34°32′N 135°45′E. Socks.

Koryo (ko-RYO), town, Hikawa county, SHIMANE prefecture, SW HONSHU, W JAPAN, 24 mi/39 km S of MATSUE; 35°19′N 132°40′E.

Koryong (GO-RYOUNG), county (□ 148 sq mi/384.8 sq km), SW NORTH KYONGSANG province, SOUTH KOREA. Gaya Mountain, of the Sobaek Mountains, is on border with SOUTH KYONGSANG province; the NAKDONG RIVER separates the county from TALSONG in the E. Agriculture (rice, beans, peanuts, vegetables, watermelon, musk melon) along the W bank of the Nakdong and other small rivers. Pottery; medical herbs. Relics. 88 Expressway connects North and South Kyongsang to NORTH and SOUTH CHOLLA provinces.

Korytnica, SLOVAKIA; see LIPTOVSKA LUZNA.

Koryukivka (kor-YOO-kif-kah) (Russian *Koryukovka*), town, CHERNIHIV oblast, UKRAINE, 45 mi/72 km NE of CHERNIHIV; 51°46′N 32°16′E. Elevation 475 ft/144 m. Raion center. Paper, printing, furniture, building materials; food processing.

Koryukovka, UKRAINE: see KORYUKIVKA.

Korzec, UKRAINE: see KORETS'.

Korzhava (kuhr-ZHAH-vah), settlement, E central NOVGOROD oblast, NW European Russia, on road, 12 mi/19 km NNW of OKULOVKA, to which it is administratively subordinate; 58°32′N 33°13′E. Elevation 498 ft/151 m. Logging, lumbering.

Korzhevskiy (KOR-zhif-skeeyee), settlement (2005 population 3,945), W central KRASNODAR TERRITORY, S European Russia, in the KUBAN' River delta, on road, 19 mi/31 km NE of ANAPA; 45°12′N 37°43′E. In agricultural area (sunflowers, fruit, grapes).

Kós (KOS), Italian *Coo*, Latin *Cos*, main city (2001 population 17,890) of KÓS island, DODECANESE prefecture, SOUTH AEGEAN department, GREECE, at NE end of island, across CHANNEL OF KÓS from Bodrum Peninsula (TURKEY); 38°54′N 27°16′E. Trade center for tobacco, olive oil, wine, figs; manufacturing of tobacco products and brandy; sponge fisheries. Some tourism; medieval castle. Hippocrates born here.

Kós (KOS), Latin *Cos*, island (□ 111 sq mi/288.6 sq km), DODECANESE prefecture, SOUTH AEGEAN department, GREECE, in the AEGEAN SEA; second-largest of the DODECANESE islands, near the Bodrum Peninsula (Turkey); 36°50′N 27°10′E. Although it rises to c.2,870 ft/875 m in the SE, the island is mostly low-lying. Fishing, sponge diving, and tourism are important industries. Port of entry for TURKEY. Grain, tobacco, olive oil, and wine are produced, and cattle, horses, and goats are raised. Mineral deposits; several sulphur springs. Airport. Main town is KÓS, on the NE shore. In ancient times controlled in turn by ATHENS, Macedon (MACEDONIA), SYRIA, and EGYPT. A cultural center; the site of a school of medicine founded in the 5th century B.C.E. by island native Hippocrates. Later enjoyed great prosperity as a result of its alliance with the Ptolemaic dynasty of Egypt, which valued the island as a naval base. Became part of modern Greece in 1947.

Kosa (KO-sah), town, Kamimashiki county, KUMAMOTO prefecture, W KYUSHU, SW JAPAN, 9 mi/15 km SE of KUMAMOTO; 32°38′N 130°48′E.

Kosa (KO-sah), Italian *Cossa*, village, SW ETHIOPIA, 12 mi/19 km NNE of JIMMA; 07°51′N 36°50′E. Coffee growing. Also spelled KOSSA.

Kosa (kuh-SAH), village (2005 population 2,170), E KOMI-PERMYAK Autonomous Okrug, PERM oblast, E European Russia, on road, 65 mi/105 km NNE of KUDYMKAR; 59°56′N 54°59′E. Elevation 564 ft/171 m. Lumbering.

Kosa, RUSSIA: see KOSINO, KIROV oblast.

Kosagi (ko-SAH-gee), town, on Ojika-jima island, of GOTO-RETTO island group, North Matsuura county, NAGASAKI prefecture, SW JAPAN, 37 mi/60 km N of NAGASAKI; 33°12′N 129°34′E. Pearls; fish processing. Saikai National Park nearby.

Kosai (KO-sah-ee), city, SHIZUOKA prefecture, central HONSHU, E central JAPAN, near HAMANA LAKE, 53 mi/85 km S of SHIZUOKA; 34°42′N 137°32′E. Motor vehicles and automotive parts. Plane trees.

Kosai (KO-sah-ee), town, Nakakoma county, YAMANASHI prefecture, central HONSHU, central JAPAN, 6 mi/10 km S of KOFU; 35°35′N 138°28′E.

Kosaka (KO-sah-kah), town, Kadzuno county, Akita prefecture, N HONSHU, NE Japan, near LAKE TOWADA, 53 mi/85 km N of AKITA; 40°19′N 140°45′E. Mining center (gold, silver, copper, lead, zinc, sulphurous minerals); barium salts; acacia products.

Kosala (KO-sah-lah), ancient Indian kingdom, corresponding roughly in area with the region of OUDH; capital was AYODHYA. It was a powerful state in the 6th century B.C.E. but was weakened by a series of wars with the neighboring kingdom of MAGADHA and finally (4th century B.C.E.) absorbed by it. Kosala was the setting of much Sanskrit epic literature including the Ramayana. Gautama Buddha and Mahavira, founder of Jainism, taught here.

Kosam (KO-suhm), village, ALLAHABAD district, SE UTTAR PRADESH state, N central INDIA, on the YAMUNA RIVER, and 28 mi/45 km WSW of Allahabad. Extensive ancient Hindu fort ruins; eleventh century Jain sculptures and coins found here. Was capital of Hindu kingdom (first or second century B.C.E.) founded by Kuru noble from KURUKSHETRA. Identified as famous ancient city of KAUSAMBI. Important cave inscriptions found 5 mi/8 km WNW, at PABHOSA. Name applied to two contiguous villages of Kosam Inam and Kosam Khiraj.

Ko Samet Marine National Park (GO SUH-MED) [Thai=Samet Island], THAILAND, named for the tree that grows in abundance on the island. Close to BANGKOK and sporting numerous white-sand beaches. Well-known tourist spot.

Kosaya Gora (kuh-SAH-yah guh-RAH), town (2006 population 17,455), central TULA oblast, W central European Russia, on the Voronka River (tributary of the OKA River), on road and railroad, 6 mi/10 km S of (and administratively subordinate to) TULA; 54°07′N 37°33′E. Elevation 652 ft/198 m. Metallurgy and ironworking. Connected by tramway with Tula.

Koschagyl (kos-chah-GIL), oil town, SE ATYRAU region, KAZAKHSTAN, on railroad spur, and 90 mi/145 km SE of ATYRAU, in Emba oil fields (pipeline to Orsk); 46°52′N 53°48′E. Also spelled Koshchagyl, Kosshagyl, Qosshagyl.

Kós, Channel of (KOS), arm of Aegean Sea, in the DODECANESE, between KÓS. island (SOUTH AEGEAN department), GREECE, and Bodrum Peninsula, TURKEY; 2.5 mi/4 km wide; 36°55′N 27°15′E. KÓS city is a port on S shore.

Koschmin, POLAND: see KOZMIN.

Koscian (KO-see-tee-ahn), Polish *Kościan*, German *Kosten*, town (2002 population 24,098), Leszno province, W POLAND, on OBRA RIVER, and 25 mi/40 km SSW of POZNAN. Railroad junction; agricultural trade center; sugar-beet and flour milling; manufacturing of liqueur, furniture; sawmilling.

Koscierzyna (kos-tee-ZEE-nah), Polish *Kościerzyna*, German *Berent*, town (2002 population 23,105), Gdansk province, N POLAND, 31 mi/50 km SW of GDANSK. Railroad junction; agricultural center; lumbering, agricultural tools, tannery, brewery. Tourist resort.

Kosciusko (kuh-see-UH-sko), county (□ 554 sq mi/1,440.4 sq km; 2006 population 76,541), N INDIANA, ⊙ WARSAW; 41°14′N 85°52′W. Center of NE Indiana's lake region; includes lakes WAWASEE and WINONA. More than twenty natural lakes, glacial in origin, distributed across county from NE corner to SW corner; largest is Lake Wawasee. Drained by TIPPECANOE and EEL rivers and small Turkey Creek. Agriculture (poultry, hogs, cattle; dairy products; corn, soybeans, vegetables); timber. Manufacturing at Warsaw. Wawasee State Fishing Area and Tri-County State Fish and Wildlife Area in NE. Formed 1835.

Kosciusko, town (2000 population 7,372), ⊙ ATTALA county, central MISSISSIPPI, 58 mi/93 km NE of JACKSON, near YOCKANOOKANY RIVER; 33°03′N 89°35′W. railroad spur terminus. Cotton, corn, soybeans; cattle; dairying; timber; light manufacturing. Kosciusko Museum. NATCHEZ TRACE PARKWAY passes to SE. Settled in early 1830s on old NATCHEZ TRACE; incorporated 1836.

Kosciusko Island (kah-zee-UH-sko), SE ALASKA, in ALEXANDER ARCHIPELAGO, W of PRINCE OF WALES ISLAND, c.40 mi/64 km NW of CRAIG; 25 mi/40 km long, 5 mi/8 km–12 mi/19 km wide; 56°03′N 133°30′W.

Kosciusko, Mount (kah-zee-UH-sko), (7,310 ft/2,228 m), SE NEW SOUTH WALES, highest peak of AUSTRALIA; 36°27′S 148°16′E. Tourism developed significantly in the 1980s.

Kosciusko National Park (kah-zee-UH-sko) (□ 2,498 sq mi/6,494.8 sq km), SE NEW SOUTH WALES, SE AUSTRALIA, 270 mi/435 km SW of SYDNEY, 40 mi/64 km SW of CANBERRA; 36°30′S 148°16′E. Australia's largest alpine region, including 10 peaks over 6,890 ft/2,100 m high. Glacial lakes, treeless uplands, open snowgum woodlands. Winter snow on upper elevs. Skiing and accommodations at Thredbo, Charlotte Pass, Perisher Valley, Smiggin Holes, Guthega, and Selwyn resorts. Yarrangobilly Caves. Camping, cabins, picnicking, swimming, boating, hiking. Established 1944.

Kose (KO-se), town, Koka county, SHIGA prefecture, S HONSHU, central JAPAN, 12 mi/20 km E of OTSU; 35°00′N 136°05′E.

Kose Dag (Turkish=*Köse Dağ*) peak (9,190 ft/2,801 m), N central TURKEY, 13 mi/21 km NE of ZARA, NW of source of the KIZIL IRMAK.

Kösely River (KUH-she-i), 100 mi/161 km long, E HUNGARY; rises 5 mi/8 km N of DEBRECEN, flows S and W, past HAJDUSZOBOSZLO and NÁDUDVAR, to HORTOBÁGY RIVER 5 mi/8 km SW of Nádudvar.

Kösen, Bad, GERMANY: see BAD KÖSEN.

Kosgi (KOS-gee), town, MAHBUBNAGAR district, ANDHRA PRADESH state, SE INDIA, 25 mi/40 km NW of MAHBUBNAGAR; 16°59′N 77°43′E. Hand-woven silks; millet, rice.

Kosh, IRAQ: see ALQOSH.

Koshaba (ko-SHA-bah), island in the DANUBE, NW BULGARIA; 44°04′N 23°03′E.

Kosh-Agach (KOSH-ah-GAHCH), village, SE ALTAI REPUBLIC, S central SIBERIA, RUSSIA, on the CHUYA RIVER, on the Chuya highway (Russian *Chuyskiy Trakt*), 180 mi/290 km SE of GORNO-ALTAYSK; 50°00′N 88°40′E. Elevation 5,734 ft/1,747 m. Ecotourism.

Koshava (ko-SHA-vah), village, MONTANA oblast, VIDIN obshtina, BULGARIA; 40°05′N 22°02′E. Gypsum.

Koshchagyl, KAZAKHSTAN: see KOSCHAGYL.

Koshe, town (2007 population 6,369), SOUTHERN NATIONS state, central ETHIOPIA, 15 mi/24 km NW of LAKE ZIWAY; 08°00′N 38°32′E. In livestock raising area.

Koshedary, LITHUANIA: see KAIŠIADORYS.

Koshekhabl' (kuh-she-HAHBL), mountain village, E ADYGEY REPUBLIC, NW CAUCASUS, RUSSIA, on the LABA RIVER, on road and railroad, 30 mi/48 km NE of MAYKOP; 44°54′N 40°30′E. Elevation 508 ft/154 m. In agricultural area producing wheat, sunflowers, hemp, essential oils (Kazanlik roses, coriander).

Koshestan (ko-shai-stahn), district, NORTHERN AREAS, NE PAKISTAN; ⊙ Dassu. Drained by INDUS River (crossed by KARAKORAM HIGHWAY at Dassu). Subsistence farming; lumbering (deodar, pine). Formerly governed as tribal territory. Formed from E part of former SWAT state and British-administered CHILAS (N).

Koshi (KO-SHEE), town, Kikuchi county, KUMAMOTO prefecture, W KYUSHU, SW JAPAN, 6 mi/10 km N of KUMAMOTO; 32°53′N 130°47′E.

Koshigaya (ko-SHEE-gah-yah), city (2005 population 315,792), SAITAMA prefecture, E central HONSHU, E central JAPAN, on the Motoara River, 9 mi/15 km E of URAWA; 35°53′N 139°47′E. Suburb of TOKYO manufacturing paulownia items, paper tumblers, and traditional dolls.

Koshiji (ko-SHEE-jee), town, Santo county, NIIGATA prefecture, central HONSHU, N central JAPAN, 37 mi/60 km S of NIIGATA; 37°23′N 138°47′E. Sake.

Koshiki-retto (ko-SHEE-kee–RET-to), island group (□ 47 sq mi/122.2 sq km), KAGOSHIMA prefecture, SW JAPAN, in EAST CHINA SEA off SW coast of KYUSHU. Comprises SHIMO-KOSHIKI-SHIMA (largest island), KAMI-KOSHIKI-SHIMA, and scattered islets. Mountainous. Rice, wheat, sweet potatoes; fishing.

Koshimbanda (kaw-sheem-BAHN-duh), town, BANDUNDU province, W CONGO, 50 mi/80 km E of KIKWIT; 05°10′S 19°52′E. Elev. 1,502 ft/457 m. Also known as PATA-MISUMBA.

Koshimizu (ko-SHEE-meez), town, Abashiri district, Hokkaido prefecture, N JAPAN, near Toufutsu Lake, 164 mi/265 km E of SAPPORO; 43°51′N 144°27′E. Sugar beets. Wildflower garden in nearby Abashiri quasi national park.

Koshino (KO-shee-no), village, Nyu county, FUKUI prefecture, central HONSHU, W central JAPAN, 12 mi/20 km W of FUKUI; 36°02′N 136°00′E.

Koshioku (KO-shyo-koo), city, NAGANO prefecture, central HONSHU, central JAPAN, 9 mi/15 km S of NAGANO; 36°31′N 138°07′E. Apricots.

Koshi Tappu Wildlife Reserve (ko-SHEE tah-POO), (□ 68 sq mi/176.8 sq km), wildlife (wild buffalo) and bird sanctuary on banks of the SAPTA KOSI RIVER, NEPAL; 26°42′N 87°07′E.

Koshk, AFGHANISTAN: see KUSHK.

Koshkar (kosh-KAHR), oil town, Atyrau region, 66 mi/110 km ENE of Atyrau, and 55 mi/90 km NE of Caspian Sea; 47°28′N 53°24′E. Oil industry support. Also Qoshkar.

Koshkel'dy (kuhsh-kyel-DI), village (2005 population 5,225), E CHECHEN REPUBLIC, S European Russia, just W of the administrative border with DAGESTAN REPUBLIC, on road, 14 mi/23 km ESE of GUDERMES; 43°15′N 46°21′E. Elevation 462 ft/140 m. Agriculture (grain, livestock).

Koshki (KOSH-kee), village (2006 population 8,070), NW SAMARA oblast, E European Russia, near the KONDURCHA RIVER, on road and near railroad, 43 mi/69 km N of SAMARA, and 33 mi/53 km WNW of SERGIYEVSK; 54°12′N 50°28′E. Elevation 154 ft/46 m. Flour milling; bakery, creamery.

Koshkonong (kosh-kuh-NAWNG), town, OREGON county, S MISSOURI, in the OZARKS near SPRING RIVER, 15 mi/24 km SW of ALTON; 36°35′N 91°39′W. Grain; livestock. Wood pallets, ties, lumber. Grand Gulf State Park to S.

Koshkonong (KAHSH-kah-nahng), village, JEFFERSON county, S WISCONSIN, 12 mi/19 km NNE of JANESVILLE, on S shore of LAKE KOSHKONONG.

Koshkonong, Lake, reservoir, ROCK county, SE WISCONSIN, on ROCK RIVER, 27 mi/43 km SE of MADISON; 42°49′N 89°05′W. Maximum capacity 107,000 acre-ft. Formed by Indian Lake Dam (13 ft/4 m high), built (1932) for recreation.

Koshkupyr (kuhsh-koo-PIR), city, W KHORAZM wiloyat, UZBEKISTAN, in KHIVA oasis, 14 mi/23 km W of URGANCH; 41°32′N 60°21′E. Cotton. Tertiary-level administrative center. Also Qushkupir.

Koshrabad, village, SAMARKAND wiloyat, UZBEKISTAN, in Aktau mountains, c.35 mi/56 km NE of KATTA-KURGAN; 40°18′N 66°32′E.

Koshtan-Tau (kuhsh-tuhn–TAH-oo), peak (approximately 16,880 ft/5,145 m), in the central Greater CAUCASUS Mountains, KABARDINO-BALKAR REPUBLIC, S European Russia.

Koshu-Kavak, Bulgaria: see KRUMOVGRAD.

Koshurnikovo (kuh-SHOOR-nee-kuh-vuh), village (2005 population 3,725), S KRASNOYARSK TERRITORY, SE SIBERIA, RUSSIA, on the Kizir River (tributary of the YENISEY RIVER), near road and railroad, 13 mi/21 km S of ARTËMOVSK; 54°10′N 93°18′E. Elevation 1,932 ft/588 m. Coal and mineral mines in the vicinity.

Kosi (ko-SEE), administrative zone (2001 population 2,110,664), E NEPAL; includes the districts of BHOJPUR, DHANKUTA, MORANG, SANKHUWASABHA, SUNSARI, and TERHATHUM.

Kosi (KO-see), town, MATHURA district, W UTTAR PRADESH state, N central INDIA, on AGRA CANAL, and 25 mi/40 km NW of Mathura; 27°48′N 77°26′E. Large livestock market; trades in gram, jowar, wheat, cotton, oilseeds. Has sixteenth-century serai.

Košice (ko-SHI-tse), German *Kaschau, Hungarian Kassa*, city center of VYCHODOSLOVENSKY province, E SLOVAKIA; 48°43′N 21°15′E. Major industrial center, transportation hub, and a market for the surrounding agricultural area. Manufacturing includes steel- and ironworks, food processing, brewing and distilling; machinery and ceramics. International and military airports (air force base and headquarters). Originally a fortress town, KOŠICE was chartered in 1241 and became an important trade center during the Middle Ages. Frequently occupied by AUSTRIAN, Hungarian, and TURKISH forces. By the Treaty of TRIANON (1920) the city passed from Hungary to CZECHOSLOVAKIA, though Hungary occupied it from 1938–1945. Most notable historic buildings are the Gothic Cathedral of St. Elizabeth (14th–15th century), the 14th-century

Franciscan monastery and church, and an 18th-century town hall. Seat of a university and several cultural institutions.

Ko Sichang, THAILAND: see SICHANG, KO.

Kosigi (KO-si-gee), town, KURNOOL district, ANDHRA PRADESH state, S INDIA, 16 mi/26 km N of ADONI; 15°51′N 77°16′E. Cotton ginning, peanut milling; tannery. Granite quarries nearby. Also spelled Kosgi or Kesigi.

Kosikha (kah-SYEE-khuh), village, NE ALTAI TERRITORY, SE West SIBERIA, RUSSIA, 30 mi/48 km E of BARNAUL. In agricultural area.

Kosino (KO-see-nuh), town (2005 population 2,305), E KIROV oblast, E European Russia, on the Kosa River (tributary of the VYATKA), on railroad and highway branch, 7 mi/11 km ESE of ZUYEVKA; 58°24′N 51°16′E. Elevation 492 ft/149 m. Paper mill, bakery, pedigree livestock breeding farm. Has a hospital. Until 1938, called Kosa.

Kosino (KO-see-nuh), former town, central MOSCOW oblast, central European Russia, on railroad, now a suburb of MOSCOW, 21 mi/34 km SE of the city center, and 3 mi/5 km NW of LYUBERTSY; 55°43′N 37°51′E. Elevation 452 ft/137 m. Knitting mills.

Kosi River (ko-SEE), c. 200 mi/322 km long, N tributary of the GANGA River, in E NEPAL and E INDIA (BIHAR state); formed by confluence of three headstreams (SUN KOSI, ARUN KOSI, and TAMUR KOSI rivers) 13 mi/21 km WSW of DHANKUTA (Nepal) at 26°53′N 87°10′E. As the SAPTA KOSI River, it flows S across Indian border (26°30′N 86°56′E), where name changes to Kosi River, through SAHARSA Plain in N Bihar state—here dividing into two main arms and many shifting channels, joining again E of KHAGARIA—and E to the Ganga River 25 mi/40 km SSW of Purnea Dam (770 ft/235 m high) at Barakahshetra, below confluence of headstreams, just S of the Kosi Thapu Wildlife Sanctuary, and barrage at CHATARA, with two canals extending from both banks, irrigate 4,688 sq mi/12,142 sq km–6,250 sq mi/16,188 sq km in Nepal and Bihar, providing power, navigation, control of destructive floods in Bihar, drainage, malarial control, land reclamation, fish hatcheries, and recreation facilities.

Kosiv (KO-sif) (Russian *Kosov*) (Polish *Kosów*), city (2004 population 9,900), SE IVANO-FRANKIVS'K oblast, UKRAINE, on right tributary of PRUT RIVER, and 15 mi/24 km SSE of KOLOMYYA; 48°19′N 25°06′E. Elevation 1,151 ft/350 m. Raion center. Summer resort. Center of folk (Hutsul) art (ceramics, wood carvings, metal and leather goods, weaving, kilims, embroidery). Formerly important for salt extraction. Known since 1424, city since 1939.

Kosjerić (KOS-yer-eech), village (2002 population 14,001), W SERBIA, 10 mi/16 km NNE of UZICE; 45°11′N 19°56′E. Also spelled Kosyerich.

Koskullskulle (KOOS-kuls-kul-le), village, NORRBOTTEN county, N SWEDEN, just E of MALMBERGET; 67°11′N 20°04′E.

Koslan (KOS-lahn), village (2005 population 2,505), W KOMI REPUBLIC, NE European Russia, on the MEZEN' River, on road, 135 mi/217 km NNW of SYKTYVKAR; 63°27′N 48°54′E. Elevation 390 ft/118 m. Lumbering, sawmilling.

Koslanda, town, UVA PROVINCE, S central SRI LANKA, in SRI LANKA HILL COUNTRY, 18 mi/29 km SSW of BADULLA; 06°44′N 81°01′E. Extensive tea and rubber plantations; rice, vegetables.

Köslin, POLAND: see KOSZALIN.

Kosmach (kos-MAHCH) (Polish *Kosmacz*), village, S IVANO-FRANKIVS'K oblast, UKRAINE, in E BESKYDS, 17 mi/27 km SW of KOLOMYYA; 48°45′N 24°22′E. Elevation 1,476 ft/449 m. Petroleum and natural gas extraction. Population largely Hutsuls. Wood carving, weaving; sheep raising.

Kosmaj (KOS-mei), mountain (2,060 ft/628 m) in the SUMADIJA region, central SERBIA, 18 mi/29 km S of BELGRADE. Several lead mines date from Roman times. Also spelled Kosmai or Kosmay.

Kosmanos, CZECH REPUBLIC: see KOSMONOSY.

Kosmet, SERBIA: see KOSOVO.

Kosmonosy (KOS-mo-NO-si), German *kosmanos*, N suburb of MLADÁ BOLESLAV, STREDOCESKY province, N BOHEMIA, CZECH REPUBLIC; 50°26′N 14°56′E. Has a 16th century castle, and a 17th century monastery.

Kosmynino (kuhs-MI-nee-nuh), town (2005 population 1,655), SW KOSTROMA oblast, central European Russia, on railroad, 15 mi/24 km SSW of KOSTROMA; 57°35′N 40°46′E. Elevation 492 ft/149 m. Peat working and peat briquets.

Kosogol, province, MONGOLIA: see HÖVSGÖL.

Kosogol, lake, MONGOLIA: see HÖVSGÖL NUUR.

Kosolapovo (kuh-suh-LAH-puh-vuh), village, E MARI EL REPUBLIC, E central European Russia, on the Buy River (tributary of the VYATKA River), on road, 65 mi/105 km ENE of YOSHKAR-OLA; 56°55′N 49°36′E. Elevation 456 ft/138 m. In agricultural area (wheat, barley, oats, flax, potatoes).

Kosong (GO-SUHNG), county, KANGWON province, S NORTH KOREA, on SEA OF JAPAN, 60 mi/97 km SE of WONSAN, near Kumkang mountains. Rice, soybeans, and honey; livestock.

Kosong, county (□ 240 sq mi/624 sq km), NE KANGWON province, SOUTH KOREA, just S of cease-fire line on SEA OF JAPAN. TAEBAEK mountains in W. Wildlife (bears, wild boars, roe deer). Some agriculture (vegetables, beans, corn, potatoes); fishery (pollack, squid, pike). Chinburyong ski resort. Beaches.

Kosong (GO-SUHNG), county (□ 198 sq mi/514.8 sq km), S central SOUTH KYONGSANG province, SOUTH KOREA (facing HALLYO MARINE NATIONAL PARK). Haean Mountains run through region. Well irrigated fields in S and NE produce rice, barley, beans, sweet potatoes, cotton, sesame, mushrooms. Fishing (anchovies, sea bream, stingrays, shells). Important copper mine. Many relics.

Kosov, UKRAINE: see KOSIV.

Kosovo (KO-so-vo), Serbian *Kosovo i Metohija* and *Kosmet*, Albanian *Kosova*, province (4,126 sq mi/10,686 sq km), S SERBIA; ☉ PRISTINA. The largely mountainous region includes the fertile valleys of Kosovo and METOHIJA and is drained by the Southern MORAVA RIVER. Kosovo's population is 80% Albanian; Serbs and Muslims are the dominant minorities. Farming, livestock raising, forestry, and the mining of lead and other metals are the major occupations. Settled by the Slavs in the 7th century, the region passed to BULGARIA in the 9th century and to Serbia in the 12th century. At Kosovo Field (Serbian *Kosovo Polje*=field of the blackbirds), in 1389, the Turks under Sultan Murad I defeated Serbia and its Bosnian, Montenegrin, and other allies. The battle of Kosovo Field broke the power of Serbia and BULGARIA, which soon passed under Ottoman rule. The battle figures prominently in Serbian poetry. From the battle until the Balkan War of 1913, Kosovo was under Turkish rule. Partitioned in 1913 between Serbia and MONTENEGRO, it was incorporated into the former YUGOSLAVIA after World War I. Following World War II, Kosovo became an autonomous region within Serbia. In 1990, demands for greater autonomy were rebuffed by Serbia, which imposed direct rule and rescinded Kosovo's status as an autonomous region. The strife between ethnic Albanians and Serbs in the province escalated in the late 1990s into a military occupation of the province by Serb/Yugoslav troops, with the expulsion of some ethnic Albanians and the flight of hundred of thousands of other refugees. NATO forces forced the withdrawal of Serb/Yugoslav troops in 1999 after an intensive bombing campaign in Kosovo and Serbia, after which the province was occupied by NATO and Russian peacekeeping forces. The current premier is Bajram Kosumi, a former environment minister, who was sworn in March 2005. Kosovo declared unilateral independence from Serbia in 2008.

Kosovo (KO-so-vo), village, MONTANA oblast, BREGOVO obshtina, BULGARIA; 44°06′N 22°39′E.

Kosovo Polje (KO-so-vo POL-yai) [Serbian=field of blackbirds], fertile valley in DINARIC ALPS, S SERBIA; since 1946 part of KOSOVO province. Drained by SITNICA RIVER and its tributary and (in S) by upper SOUTHERN MORAVA and LEPENAC rivers. Lignite, magnesite, and asphalt deposits; lead and zinc mining in mountains N and E of valley. Largely agricultural (wheat). Chief towns are PRISTINA and KOSOVSKA MITROVICA. Kosovo Polje, railroad station, 5 mi/8 km WSW of Priština, is junction of Belgrade-Skoplje and Priština-Peć railroads; site of Turkish victories over Serbs (1389) and Hungarians (1448), with tombs of Turkish Sultan Murad I and Serb hero Milosh Obilich. Under Turkish rule (until 1913), valley was part of Kosovo (also spelled Kossovo) vilayet; capital was Prizren. W section of vilayet passed (1913) to Albania, where its main town is Kukës. Also Kosovo.

Kosovska Mitrovica (KOS-ov-skah MEET-ro-veetsah), city, in KOSOVO, S SERBIA, on IBAR RIVER at SITNICA RIVER mouth, on railroad, and 22 mi/35 km NNW of PRISTINA; 42°53′N 20°52′E. Flour milling, stone working. Magnesite deposits. Mining, milling, and smelting of metal ores at nearby TREPCA and ZVECAN. Until 1913, under Turkish rule. Also called Mitrovica Kosovska, or Mitrovitsa.

Kosów, UKRAINE: see KOSIV.

Kospash (KOS-pahsh), former city, E central PERM oblast, W URALS, RUSSIA, now a suburb of KIZEL, 4 mi/6 km E of the city center; 59°03′N 57°46′E. Elevation 1,328 ft/404 m. A major mining center in the Kizel bituminous coal basin. Developed in 1941; became city in 1949, before being incorporated into Kizel. Formerly known as Kospashskiy.

Kospashskiy, RUSSIA: see KOSPASH.

Kosrae (kos-REI), volcanic island and state (□ 42 sq mi/109.2 sq km; 2000 population 7,686), easternmost of CAROLINE ISLANDS, Federated States of MICRONESIA, W PACIFIC, 780 mi/1,255 km ESE of CHUUK ISLANDS; ☉ Lele Harbor; 5°19′N 162°59′E. The island is 8 mi/12.9 km in diameter, enclosed by a narrow fringing reef. The central mass, called UALAN, has 2 peaks, the higher being Mount Crozer (2,079 ft/634 m). Copra. Site of ruins of ancient stone walls, dikes. Guano deposits in caves. Also Kusaie or Kuseie.

Kossa, ETHIOPIA: see KOSA.

Kosse, village (2006 population 516), LIMESTONE county, E central TEXAS, 35 mi/56 km SE of WACO; 31°18′N 96°37′W. In farm area (cotton, corn; cattle); silica sand and kaolin clay. Twin Oak Reservoir to SE.

Kosseir, EGYPT: see QUSAYR, AL.

Kössen (KOS-sen), village, TYROL, W AUSTRIA, on Tiroler Ache River, 12 mi/19 km NE of KUFSTEIN, near German border; 47°40′N 12°24′E. Elevation 1,795 ft/547 m. Border stations; summer and winter tourism.

Kosshagyl, KAZAKHSTAN: see KOSCHAGYL.

Kossi, province (□ 2,829 sq mi/7,355.4 sq km; 2005 population 279,730), BOUCLE DU MOUHOUN region, W BURKINA FASO; ☉ NOUNA; 12°55′N 03°50′W. Borders SOUROU (E, formed by Sounou River), MOUHOUN (SE, formed by MOUHOUN RIVER), and BANWA (S) provinces and MALI (W and N). Drained by MOUHOUN RIVER. Agriculture includes cotton and vegetables. A portion of this province was excised in 1997 when fifteen additional provinces were formed.

Kossovo (KO-sah-vah), city, BREST oblast, BELARUS. Furniture manufacturing.

Kossuth, county (□ 974 sq mi/2,532.4 sq km; 2006 population 16,011), N IOWA, on MINNESOTA line; ☉ ALGONA; 43°12′N 94°12′W. Largest county in land area in Iowa; part of Iowa lakes district. Rolling prairie agricultural area (cattle, hogs, poultry; corn, oats, soybeans) drained by East DES MOINES RIVER, Middle Branch BLUE EARTH RIVER, and Union Slough. Has sand and gravel pits. Ambrose A. Call State Park SW

Area is shown by the symbol □, and capital city or county seat by ☉.

of Algona; Union Slough National Wildlife Refuge at center. Widespread flooding in 1993. Formed 1851.

Kossuth (kah-SOOTH), village (2000 population 170), ALCORN county, NE MISSISSIPPI, 8 mi/12.9 km WSW of CORINTH; 34°52′N 88°38′W. In agricultural area (cotton, corn, soybeans; cattle).

Kosta (KOOS-tah), village, KRONOBERG county, S SWEDEN, 20 mi/32 km E of VÄXJÖ; 56°51′N 15°24′E. Noted for glass manufacturing. Founded 1742.

Kosta-Khetagurovo, RUSSIA: see NAZRAN'.

Kostakoz, TAJIKISTAN: see CHKALOVSK.

Kostanai (kos-tah-NEI), region (□ 44,200 sq mi/114,920 sq km), NW KAZAKHSTAN; ⊙ KOSTANAI. Drained by TOBOL and OBAGAN rivers (N); steppe plateau with black earth (N), dry steppe (S). Sharply continental climate. Wheat, millet, and oats grown N of South SIBERIAN RAILROAD; cattle and sheep raising in S. Iron, bauxite, nickel, titanium deposits. Gold mining at ZHETYKARA. Meat and grain processed at Kostanai. Pop. consists of Kazaks, Russians, Ukrainians. Formed (1936) as Kustanai Oblast (Kazakh SSR). Also spelled Kostany, Kustanai, Qostanay, Qostanai.

Kostanai (kos-tah-NEI), city, ⊙ KOSTANAI region, N KAZAKHSTAN, on the TOBOL River; 53°15′N 63°40′E. Agricultural center and producer of chemical fibers. Manufacturing includes consumer goods, building materials; food processing. Rich iron deposits nearby at RUDNYI. Grew dramatically in the 1960s as a result of the VIRGIN LANDS campaign. Also spelled Kustanay, Kustanai, Qostanay, Qostanai.

Kostandenets (kos-tahn-den-ETS), village, RUSE oblast, TSAR KALOYAN obshtina, NE BULGARIA, 20 mi/32 km SSE of RUSE; 43°34′N 26°12′E. Wheat, rye, sunflowers.

Kostandovo (kos-TAHN-do-vo), village, PLOVDIVoblast, RAKITOVO obshtina, BULGARIA; 42°01′N 24°06′E.

Kostany (KOSH-tyah-NI), Czech Košt'any, German kosten, village, SEVEROCESKY province, NW BOHEMIA, CZECH REPUBLIC, 3 mi/4.8 km WNW of TEPLICE; 49°47′N 15°33′E. Railroad junction. Manufacturing (textiles); glassworks; lignite mining in vicinity.

Kostek (kuhs-TYEK), village (2005 population 4,235), W central DAGESTAN REPUBLIC, SE European Russia, at the S edge of the SW Caspian lowlands, on highway, 47 mi/76 km NW of MAKHACHKALA, and 14 mi/23 km ENE of KHASAVYURT; 43°20′N 46°51′E. Below sea level. Originally a Cossack settlement.

Kostel, CZECH REPUBLIC: see PODIVIN.

Kostelec, CZECH REPUBLIC: see JIHLAVA.

Kostelec nad Cernymi Lesy (KOS-te-LETS NAHT cher-NEE-mi LE-si), Czech Kostelec nad Černými Lesy, German schwarz kosteletz, town (1991 population 3,164), STREDOCESKY province, W central BOHEMIA, CZECH REPUBLIC, 20 mi/32 km ESE of PRAGUE; 49°59′N 14°52′E. Popular excursion center. Tanning; food processing; earthenware manufacturing. Sanatorium for patients with tuberculosis. Has 16th century castle. Fishing in vicinity, notably at JEVANY, 3 mi/4.8 km SW, also a summer and winter resort.

Kostelec nad Labem (KOS-te-LETS NAHD LAH-bem), German kosteletz an der elbe, town, STREDO-CESKY province, N central BOHEMIA, CZECH REPUB-LIC, on railroad, on ELBE RIVER, and 11 mi/18 km NE of PRAGUE; 50°14′N 14°36′E. Agriculture (sugar beets, vegetables); sugar refinery. Has a 15th-century Gothic church.

Kostelec nad Orlici (KOS-te-LETS NAHD OR-li-TSEE), Czech Kostelec nad Orlicí, German ad-lerkosteletz, town, VYCHODOCESKY province, E BOHE-MIA, CZECH REPUBLIC, on Divoka Orlice River, on railroad, and 18 mi/29 km SE of HRADEC KRÁLOVE; 50°08′N 16°14′E. Manufacturing (textile machinery, automotive parts, rubber); woodworking. Has a 17th-century castle with large park.

Kostelec na Hane (KOS-te-LETS NAH HAH-ne), Czech Kostelec na Hané, German kosteletz in der hanna, town, JIHOMORAVSKY province, central MORAVIA,

CZECH REPUBLIC, on railroad, and 4 mi/6.4 km NW of PROSTĚJOV; 49°34′N 17°04′E. Agriculture (wheat, sugar beets, fruit, vegetables). Has an 18th-century church and museum. Neolithic-era archaeological site.

Kosteletz an der Elbe, CZECH REPUBLIC: see KOSTELEC NAD LABEM.

Kosteletz in der Hanna, CZECH REPUBLIC: see KOSTE-LEC NA HANE.

Kosten, CZECH REPUBLIC: see KOSTANY.

Kosten, POLAND: see KOSCIAN.

Kostenets (KOS-ten-ets), city, SOFIA oblast, ⊙ Koste-nets obshtina (1993 population 16,367), W central BULGARIA, on the E slope of the RILA MOUNTAINS, 10 mi/16 km S of IHTIMAN; 42°19′N 23°51′E. Sawmilling, manufacturing (cutlery, chemical dyes), food pro-cessing. Health resort with thermal radioactive baths. Oil shale found nearby. Its railroad station, 3 mi/5 km NW, on the MARITSA RIVER, has match, paper mills. Kostenets village is 4 mi/6 km to the SW. Formed in 1956 from former villages of Momina Banya and Gara Kostenets. Momin Prohod joined Gara Kostenets to form Kostenets in 1964.

Kostenkovo (kuhs-TYEN-kuh-vuh), industrial settle-ment, W Kemerovo oblast, S central Siberian Russia, in the SALAIR RIDGE, near the administrative border with ALTAI TERRITORY, 14 mi/23 km SW of NOVO-KUZNETSK; 53°37′N 86°49′E. Elevation 885 ft/269 m. Coal mining.

Koster, town, NORTH-WEST province, SOUTH AFRICA, on Koster River, on railroad, 25 mi. SW of RUSTEN-BURG, at foot of NW WITWATERSRAND; 25°49′S 26°54′E. Elevation 5,209 ft/1,588 m. Fruit, wheat, oats, cotton; grain elevator.

Kosterevo (KOS-tye-ree-vuh), city (2006 population 9,265), W VLADIMIR oblast, central European Russia, on the Lipnya River (tributary of the KLYAZ'MA RIVER), on railroad, 25 mi/40 km ENE of OREKHOVO-ZUYEVO (moscow oblast), and 5 mi/8 km E of PET-USHKI, to which it is administratively subordinate; 55°56′N 39°37′E. Elevation 387 ft/117 m. Woodwork-ing; plastics. Made city in 1981. Also spelled Koster-ovo.

Kosti (KO-stee), town, White Nile state, E central SUDAN, on W bank of the WHITE NILE River, and 170 mi/274 km S of KHARTOUM; 13°10′N 32°43′E. railroad-steamer transfer point for KHARTOUM (N), JUBA (S), and AL UBAYYID (W); agriculture and trade center; cotton, wheat, rice, fruits, durra, and gum arabic; livestock. Airfield. Kosti Bridge (completed 1910, 2 mi/3.2 km upstream; 470 yd/430 m long) carries railroad and road traffic, and was the first bridge built on the NILE insudan. A 2nd bridge was built in the 1970s to cope with agriculture expansion in the area and to facilitate transportation between W, central, and E SUDAN.

Kostinbrod (KO-steen-brod), city, SOFIA oblast, ⊙ Kostinbrod obshtina (1993 population 17,800), W BULGARIA, on a left tributary of the ISKUR River, 9 mi/15 km NW of SOFIA; 42°49′N 23°13′E. Freight railroad station. Fertilizer, soap, cosmetics; sunflower oil ex-tracting.

Kostino (KOS-tee-nuh), city, N central MOSCOW oblast, central European Russia, near highway and railroad, 38 mi/61 km NE of MOSCOW, and 3 mi/5 km NE of PUSHKINO; 56°02′N 37°55′E. Elevation 570 ft/173 m. Formed in 1940.

Kostino (KOS-tee-nuh), village (2005 population 3,430), NE KIROV oblast, E European Russia, on the E bank of the KAMA River, on highway, 11 mi/18 km S of BISEROVO, and 3 mi/5 km N of Afanasyevo; 58°54′N 53°16′E. Elevation 757 ft/230 m. Poultry factory, ped-igree livestock breeding farm.

Kostino (KOS-tee-nuh), settlement, W IRKUTSK oblast, E central SIBERIA, RUSSIA, on road, 85 mi/137 km SW of BRATSK; 55°37′N 99°31′E. Elevation 961 ft/292 m. Logging, lumbering, woodworking. Also known as Kostina (same pronounciation).

Kostolac (KOS-to-lahts), town, E SERBIA, near MLAVA RIVER and an arm of the DANUBE River, 8 mi/12.9 km N of POZAREVAC. Railroad terminus; lignite mines. Also spelled Kostolats.

Kostomuksha (kuhs-tuh-MOOK-shah), city (2005 population 30,235), W central Republic of KARELIA, NW European Russia, on a highway branch, 285 mi/459 km W of PETROZAVODSK, and 125 mi/201 km W of BELOMORSK; 64°41′N 30°49′E. Elevation 511 ft/155 m. Arose in 1977 with the development of mining and concentrating of the Kostomuksha iron-ore deposit, in association with Finland. Made a city in 1983.

Kostopil' (ko-STO-peel) (Russian Kostopol'), city, central RIVNE oblast, UKRAINE, 20 mi/32 km NNE of RIVNE; 50°53′N 26°27′E. Elevation 541 ft/164 m. Raion center. Sawmilling center, producing plywood and prefabricated houses; glass working, basalt, and heat insulating materials; food processing. Founded in the late 18th century; site of battles against Bolshevik forces (1919, 1921). City since 1939.

Kostopol', UKRAINE: see KOSTOPIL'.

Köstritz, Bad, GERMANY: see BAD KÖSTRITZ.

Kostrizhevka, UKRAINE: see KOSTRYZHIVKA.

Kostroma (kuh-struh-MAH), oblast (□ 23,205 sq mi/60,333 sq km; 2004 population 806,000) in N central European Russia; ⊙ KOSTROMA. Bordered N by the NORTHERN UVALS; drained by left affluents of the VOLGA River (KOSTROMA, UNZHA, and upper VETLUGA rivers); in forested region, partly cleared (W). Tem-perate continental climate. Flax is the main crop throughout the region; wheat near SOLIGALICH (N) and NEREKHTA (SW), potatoes near MAKARYEV (S); dairy farming centered at VOKHMA (NE). Beekeeping and livestock raising also are well developed and widespread. Lumbering is chief industry; sawmilling, veneering at SHARYA, MANTUROVO, NEYA, GALICH, and BUY, along the Vologda-Kirov railroad. Wood-working, distilling, flax processing in rural areas; linen milling at Nerekhta. Kostroma (SW) is the main in-dustrial center. Formed in 1944 out of YAROSLAVL oblast. Divided into 24 administrative districts and consists of 11 cities, 18 towns, and 250 villages.

Kostroma (kuh-struh-MAH), city (2005 population 277,335), ⊙ KOSTROMA oblast, central European Rus-sia, port on the VOLGA RIVER at the mouth of the KOSTROMA RIVER, 231 mi/372 km NE of MOSCOW; 57°46′N 40°56′E. Elevation 347 ft/105 m. A major linen-milling, textile machinery, shipbuilding, and woodworking center. Other industries include jewelry making; agricultural machinery repair, oil and gas pipeline equipment, industrial rubber, gravel plant, electromechanical plant; peat works; agricultural chemicals, livestock feed, poultry farm, flour mill, fish and meat processing. Has a polytechnic college, a military academy for defense against chemical and biological weapons, and a regional institute of animal diseases. Founded in 1152, it was the capital of a principality in the 13th and 14th centuries, and became an important commercial center. Destroyed almost completely in a fire in 1773; rebuilt by 1784 as an ad-ministrative center of Kostroma gubernia. Historical monuments include the Uspenskiy Cathedral (around 1250) and the Ipatyev and Bogoyavlenskiy monasteries (16th century).

Kostroma River (kuh-struh-MAH), approximately 175 mi/282 km long, in KOSTROMA oblast, RUSSIA; rises NW of SUDAY in the N URALS; flows W, past SOLI-GALICH, and S, past Bui (head of navigation), to the VOLGA RIVER at KOSTROMA.

Kostryzhivka (kos-tri-ZHEEV-kah) (Russian Kos-trizhevka), town (2004 population 9,800), N CHER-NIVTSI oblast, UKRAINE, on the right bank of the DNIESTER River, on railroad, 1 mi/2 km W of and across the river form ZALISHCHYKY; 48°39′N 25°42′E. Eleva-tion 485 ft/147 m. Sugar refinery, building materials.

Kostrzyn (KOST-zeen), German Kostschin, town, Poznań province, W central POLAND, 13 mi/21 km E of

POZNAŃ. Flour milling, sawmilling, gingerbread making. Also called Kostrzyń Wielkopolski.

Kostrzyń, POLAND: see KÜSTRIN.

Kostui (kos-TOI), island in the DANUBE River, NE BULGARIA; 44°04′N 26°42′E.

Kostyantyniv, UKRAINE: see STAROKOSTYANTYNIV.

Kostyantynivka (kos-tyahn-TI-nif-kah) (Russian *Konstantinovka*), city (2001 population 95,111), N central DONETS'K oblast, UKRAINE, in the DONBAS, 27 mi/43 km N of DONETS'K; 48°32′N 37°43′E. Elevation 406 ft/123 m. Raion center. It is an iron, zinc, steel, chemical, and glass-making center. Food processing, manufacturing (leather, building materials); railroad servicing. Has auto glass products research institute, as well as medical, industrial, and agricultural technical schools. City was rebuilt after World War II. Established in 1859, city since 1932.

Kostyantynivka (kos-tyahn-TI-neef-kah) (Russian *Konstantinovka*), town (2004 population 3,700), NW KHARKIV oblast, UKRAINE, on road, and on railroad spur, 9 mi/14 km SSW of KRASNOKUTS'K; 50°21′N 36°02′E. Elevation 554 ft/168 m. Sugar refinery. Established in 1780, town since 1938.

Kostyantynivka (kos-tyahn-TI-neef-kah) (Russian *Konstantinovka*), town, NW MYKOLAYIV oblast, UKRAINE, on the left bank of the Southern BUH RIVER, and on road adjacent to and just W of YUZHNOUKRAYINS'K; 47°50′N 31°09′E. Elevation 311 ft/94 m. Dairy. Known since the late 18th century; town since 1976.

Kostyantynivka, UKRAINE: see YUZHNOUKRAYINS'K.

Kostyantynohrad, UKRAINE: see KRASNOHRAD.

Kostyshchentsi, UKRAINE: see STAROKOSTYANTYNIV.

Kostyukovichi (ko-styoo-KO-vi-chi), city, SE MOGILEV oblast, BELARUS, on Zhadunka River (tributary of DNIEPER), and 55 mi/89 km SW of ROSLAVL; 53°20′N 32°01′E. Distillery, flax processing plant, creamery.

Kostyukovka (ko-styoo-KOV-kah), urban settlement, S GOMEL oblast, BELARUS, 7 mi/11 km NNW of GOMEL; 52°32′N 30°57′E. Railroad station; glassworks; peat.

Kosudo (ko-SOO-do), town, Nakakanbara county, NIIGATA prefecture, central HONSHU, N central JAPAN, on SHINANO RIVER, 9 mi/15 km S of NIIGATA; 37°45′N 139°04′E.

Kosuge (ko-SOO-ge), village, North Tsuru county, YAMANASHI prefecture, central HONSHU, central JAPAN, 22 mi/35 km N of KOFU; 35°45′N 138°56′E.

Kosugi (ko-SEE-gee), town, Imizu county, TOYAMA prefecture, central HONSHU, central JAPAN, 6 mi/10 km W of TOYAMA; 36°42′N 137°06′E.

Kosulino, RUSSIA: see VERKHNEYE DUBROVO.

Kosumberk, CZECH REPUBLIC: see LUZE.

Kos'va River (KOS-vah), 190 mi/306 km long, in PERM oblast, RUSSIA, rises in the central URAL Mountains approximately 10 mi/16 km S of KYTLYM, flows generally SW, past GUBAKHA, and W to the KAMA River opposite (NE of) CHERMOZ. Timber floating; seasonal navigation for approximately 60 mi/97 km in the lower course.

Kos'ya (KOS-yah), town (2006 population 500), W SVERDLOVSK oblast, RUSSIA, in the central URALS, extreme W Siberian Russia, on the Is River (left tributary of the TURA RIVER), on road, 13 mi/21 km W of Is; 58°48′N 59°21′E. Elevation 977 ft/297 m. Near a railroad spur (Kryuchkovka station); gold and platinum placers.

Kosyerich, SERBIA: see KOSJERIC.

Kosyorove, UKRAINE: see APOSTOLOVE.

Kosyorove, UKRAINE: see STANYCHNO-LUHANS'KE.

Kosyorovo, UKRAINE: see APOSTOLOVE; also STANYCHNO-LUHANS'K.

Kos'yu (KOS-yoo), river, approximately 150 mi/241 km long, in KOMI REPUBLIC, RUSSIA; rises in the N URAL Mountains at about 65°10′N; flows NNE, through the PECHORA coal basin, to USA River at Kos'ya-Vom, 140 mi/225 km SW of VORKUTA. Receives Kozhym and Inta rivers (right).

Koszalin (ko-SHAH-leen), German *Köslin*, city (2002 population 108,709), NW POLAND, near the BALTIC SEA. In farming area; light manufacturing, timber milling. It was founded in 1188, prospered from 14th to 16th century, but suffered greatly in the Thirty Years War. Has 14th century Gothic cathedral. Transferred from GERMANY to Poland by the POTSDAM Conference (1945).

Köszeg (KUH-seg), city, VAS county, W HUNGARY, 11 mi/18 km NNW of SZOMBATHELY, near Austrian border; 47°23′N 16°33′E. Footwear; distilling, manufacturing wool, cotton, rugs, soap, candles; small brewery. Heroic stand here, in 1532, against Turks.

Kota (KO-tah), district (□ 4,802 sq mi/12,485.2 sq km), RAJASTHAN state, NW INDIA; ⊙ KOTA. Formerly Kotah.

Kota (KO-tah), city (2001 population 703,150), ⊙ KOTA district, RAJASTHAN state, NW INDIA, on the CHAMBAL RIVER; 25°11′N 75°50′E. Market for sugarcane, oilseeds, wheat, cotton, millet, cloth, and building stone. Airport. Nearby Chambal Dam supplies power for its diversified industries (textiles, precision instruments, electric cable, rubber products, paper, processed foods; distilleries). The city is enclosed by a massive wall. The Mathureshi temple is here. Was capital of former RAJPUTANA state of Kotah.

Kota (KO-TAH), town, Nukata county, AICHI prefecture, S central HONSHU, central JAPAN, 28 mi/45 km S of NAGOYA; 34°51′N 137°10′E. Videocassette recorders. Persimmons.

Kotaagung (ko-tah-AH-goong), town, S SUMATRA, INDONESIA, port on SEMANGKA BAY, 45 mi/72 km W of BANDA ACEH; 04°04′S 103°27′E. Trade center shipping copra, pepper, coffee; timber. Also spelled Kotaagoeng.

Kotabaru (ko-tah-BAH-roo), town, ⊙ Kotabaru district, on N coast of Pulu Laut, port on MACASSAR STRAIT, INDONESIA, off SE coast of BORNEO; 03°12′S 116°10′E. Exports coal, rubber, pepper. Former coaling station for steam ships. Also called Kotabarupululaut or Kotabaroe.

Kota Batu (ko-TAH BAH-too), town, BRUNEI MUARA district, NE BRUNEI DARUSSALAM, suburb 4 mi/6.4 km ESE of BANDAR SERI BEGAWAN, on Brunei River Has arts and handicrafts center, Brunei Museum, Malay Technical Museum, House of Twelve Roofs (1906), and ancient tomb of Sultan Bolkiah.

Kotadaik, INDONESIA: see LINGGA.

Kot Adu (kot ud-oo), town, MUZAFFARGARH district, SW PUNJAB province, central PAKISTAN, on railroad, and 30 mi/48 km NW of MUZAFFARGARH; 30°28′N 70°58′E. Also spelled Kot Addu.

Kotagede (KO-tah-GUH-dai), neighborhood of YOGYAKARTA, Yogyakarta province, INDONESIA; 07°49′S 110°23′E. Established 1579 by Penembahau Senopati, founder of the second Mataram dynasty, who is buried near central market. Area is famed for its silver workshops and architecturally eclectic buildings. The royal cemetery is a popular pilgrimage site during Ramadan (Pasua). Sometimes spelled Kota Gede.

Kotagiri (KO-tah-gi-ree), town, NILGIRI district, TAMIL NADU state, S INDIA, in Nilgiri Hills, 11 mi/18 km E of OOTY. Scenic resort; eucalyptus oil extraction; millet, barley, peas, potatoes. Tea and coffee estates nearby.

Kotah (KO-tah), former princely state in RAJPUTANA States, NW INDIA. Formed in early seventeenth century out of original BUNDI state; treaty with BRITISH in 1817. In 1948, merged with RAJASTHAN union; now within Rajasthan state, as KOTA district.

Kotake (KO-tah-ke), town, Kurate county, FUKUOKA prefecture, N KYUSHU, SW JAPAN, 19 mi/30 km N of FUKUOKA; 33°41′N 130°42′E. Also Kodake.

Kota Kinabalu (KO-tah KIN-ah-BAH-loo), city (2000 population 305,382), ⊙ SABAH, Malaysia, in N BORNEO, and on a small inlet of the SOUTH CHINA SEA; 05°59′N 116°04′E. It is the chief port of the state and is connected by road and railroad with the interior.

Rubber is exported. It was founded in 1899 and in 1947 replaced SANDAKAN as the capital of what was then BRITISH NORTH BORNEO. Destroyed during World War II and rebuilt. Called JESSELTON during the colonial period.

Kota Kinabalu, city, WEST COAST residency, N BORNEO, SABAH state, MALAYSIA; chief port on small inlet of SOUTH CHINA SEA, 1,000 mi/1,609 km ENE of SINGAPORE; 05°59′N 116°04′E. Railroad and trade center for agricultural and livestock-raising area; exports rubber. Has rice mills, fisheries. Trade is largely controlled by the Chinese. Founded 1899. Severely damaged in World War II. Became capital of the colony in June 1947, supplanting SANDAKAN. Called JESSELTON until 1968.

Kotamaobagu (ko-TAH-mou-BAH-goo), town, ⊙ Bolaangmongondow district, Sulawesi Utara province, INDONESIA, 100 mi/161 km W of GORONTALO.

Kotaraja, INDONESIA: see BANDA ACEH.

Kota Tinggi (KO-tah TING-gee), town (2000 population 39,006), SE Johore, MALAYSIA, 20 mi/32 km NE of Johore Bharu, on Johore River. Center of rubber and pineapple (canning) district. Tin mining (iron). Became (1511) capital of Johore-Riouw kingdom. Site of royal mausoleum; domestic tourism based on nearby waterfalls.

Kotchandpur (kot-chand-poor), town (2001 population 32,025), JESSORE district, W EAST BENGAL, BANGLADESH, 21 mi/34 km NNW of JESSORE; 23°25′N 89°01′E. Trades in rice, jute, oilseeds, sugarcane.

Kotcho Lake (□ 90 sq mi/233 sq km; 15 mi/24 km long, 1 mi/2 km–7 mi/11 km wide), NE BRITISH COLUMBIA, W CANADA, near ALBERTA border; 59°04′N 121°09′W. Drains S into HAY RIVER.

Kot Chutta (kot chu-TAH), town, DERA GHAZI KHAN district, SW PUNJAB province, central PAKISTAN, 11 mi/18 km S of DERA GHAZI KHAN; 29°53′N 70°39′E. Sometimes spelled Kot Chhutta.

Kotda Sangani (KOT-dah SUHN-gah-nee), town, RAJKOT district, GUJARAT state, W central INDIA, 18 mi/29 km SSE of RAJKOT. Was capital of former WEST KATHIAWAR state of Kotda Sangani of WEST INDIA STATES agency; merged 1948 with SAURASHTRA and ultimately with Gujarat state.

Kotdwara (KOT-dwah-rah), town, tahsil headquarters, GARHWAL district, N UTTAR PRADESH state, N central INDIA, on tributary of the RAMGANGA RIVER, and 11 mi/18 km SW of LANSDOWNE, at foot of SHIWALIK RANGE. railroad spur terminus; trades in borax from TIBET, wheat, barley, sugar, rice, cloth, oilseeds.

Kotel (KO-tel), city (1993 population 7,778), BURGAS oblast, ⊙ Kotel obshtina, E central BULGARIA, in the KOTEL MOUNTAINS, at the S end of the Kotel Pass, 15 mi/24 km NNE of SLIVEN; 42°54′N 26°26′E. Carpet and woodworking shops; upholstery textiles; fruit, livestock. Children's sanatorium, health resort. Developed as a wool-processing and cultural center under Turkish rule (15th–19th century), when it was called Kazan. Declined after Bulgarian independence.

Kotel Mountains (KO-tel), E central BULGARIA, part of the E BALKANS, extend c.12 mi/19 km from the VRATNIK PASS E to the Varbitsa Mountains; 42°51′N 26°24′E. Crossed by the Kotel Pass (2,451 ft/747 m), just N of KOTEL, on a highway to OMURTAG. On the main road between the DANUBE and ISTANBUL during Turkish rule (15th–19th century), when it was called Kazan Pass.

Kotel'nich (kuh-TYEL-neech), city (2005 population 26,375), W central KIROV oblast, E European Russia, on the VYATKA River, on crossroads and the TRANS-SIBERIAN RAILROAD, 77 mi/124 km WSW of KIROV; 58°18′N 48°19′E. Elevation 439 ft/133 m. Railroad junction of lines to KIROV, VOLOGDA, and NIZHNIY Novgorod. Flour milling, dairying, meat packing, canning; woodworking (furniture, sailing masts), timber floating; agricultural chemicals; machine building and repair; knitting factory. Founded in 1143 by Mari

(Cheremiss) population and called Koksharov; current name since 1181. Russian colonization began after the 14th century; annexed to MUSCOVY in 1489, and made city in 1780. Railroad station since 1906.

Kotel'niki (kuh-TYEL-nee-kee), town (2006 population 17,560), central MOSCOW oblast, central European Russia, on the MOSKVA River, on road and railroad, 31 mi/50 km SE of MOSCOW, and 2 mi/3.2 km S of LYUBERTSY, to which it is administratively subordinate; 55°39′N 37°51′E. Elevation 521 ft/158 m. Ore mining and processing; light manufacturing (clothing, rugs, linens); construction and servicing enterprises.

Kotel'nikovo (kuh-TYEL-nee-kuh-vuh), city (2006 population 19,350), S VOLGOGRAD oblast, SE European Russia, on crossroads and railroad, 118 mi/190 km SW of VOLGOGRAD; 47°38′N 43°09′E. Elevation 183 ft/55 m. Center of an agricultural area; agricultural machinery, steel products, food industries (dairy products, cannery). Until 1929, known as Kotel'nikovskaya. Made city in 1955.

Kotel'nikovo (kuh-TYEL-nee-kuh-vuh), village, W central LENINGRAD oblast, NW European Russia, near highway and railroad, 25 mi/40 km S of SAINT PETERSBURG, and 3 mi/5 km SSW of GATCHINA; 59°30′N 30°04′E. Elevation 324 ft/98 m. Called Salyuzi until 1949.

Kotel'nikovskaya, RUSSIA: see KOTEL'NIKOVO, VOLGOGRAD oblast.

Kotel'nya (ko-TEL-nyah), village, SE ZHYTOMYR oblast, UKRAINE, on the Huyva River, right tributary of the TETERIV RIVER, 14 mi/23 km SE of ZHYTOMYR. Sugar beets. First mentioned as Kotel'nych (1143), also known as Kotel'nytsya; until 1930s a small town; since then, a village. Ruins of medieval fortifications.

Kotel'nyy Island (kuh-TYEL-niyee), largest island of the Anjou group of the NEW SIBERIAN ISLANDS, approximately 100 mi/161 km long and approximately 60 mi/97 km wide, off SAKHA REPUBLIC, RUSSIAN FAR EAST. Sighted in 1773 by Ivan Lyakhov, a Russian merchant. Polar foxes and reindeer.

Kotel'va (ko-tel-VAH), town, NE POLTAVA oblast, UKRAINE, near VORSKLA RIVER, on a highway, and 35 mi/56 km NNE of POLTAVA; 50°04′N 34°45′E. Elevation 400 ft/121 m. Raion center. Flour milling, feed milling, dairy; sewing. Has the Holy Trinity Church (1812). Founded in the late 16th century; until the 1930s, one of the largest villages in Ukraine (approximately 20,000), with well developed cottage industry (principally weaving); town since 1971.

Kotgarh (KOT-guhr), town, SHIMLA district, central HIMACHAL PRADESH state, N INDIA, 24 mi/39 km NE of SHIMLA. One of the most productive apple belts of the state; apple trees were first introduced here from the U.S. Also wheat, corn, tea, barley.

Kothagudium, INDIA: see KOTTAGUDEM.

Kothapeta, INDIA: see KOTTAPETA.

Köthen (KO-ten), city, SAXONY-ANHALT, central GERMANY; 51°46′N 11°59′E. Lignite mines, sugar refineries, and heavy engineering industries. The city is also a railroad junction, and has an airport. It is home to a school of chemical engineering and a teachers' college Known in 1115, Köthen was from 1603 to 1847 the residence of the dukes of Anhalt-Köthen, at whose court Johann Sebastian Bach was musical director from 1717 to 1723. Formerly spelled Cöthen.

Kother, INDIA: see ANANTNAG.

Kothi (KO-tee), village (2001 population 7,710), SATNA district, MADHYA PRADESH state, central INDIA, 13 mi/21 km NNW of SATNA. Millet and wheat market; small ochre works. Was capital of former petty state of Kothi of CENTRAL INDIA agency; in 1948, merged with VINDHYA PRADESH, then with Madhya Pradesh state.

Kotido (ko-TEE-do), administrative district, NORTHERN region, NE UGANDA; ⊙ KOTIDO. As of Uganda's division into eighty districts, borders KAABONG (N), MOROTO (E and S), ABIM (W), and PADER and KIT-

GUM (NW) districts. Primary inhabitants are the Jie people. Agricultural area. Formed in 2005-2006 from central portion of former KOTIDO district (Kaabong district formed from N portion in 2005 and Abim district from S portion in 2006).

Kotido (ko-TEE-do), former administrative district (□ 5,100 sq mi/13,260 sq km; 2005 population 774,400), NORTHERN region, NE UGANDA, along borders with KENYA (NEE) and SUDAN (NWN); capital was KOTIDO; 03°20′N 34°00′E. As of Uganda's division into fifty-six distrcts, was bordered by MOROTO (SES), LIRA (SW), and PADER and KITGUM (W) districts. Primarily inhabited by Jie, Dodoth, and Labwor peoples (each with a corresponding county within the district). Mountainous areas, including Mount MORUNGOLE (9,020 ft/2,749 m) in N; district also contained forests. KIDEPO NATIONAL PARK in N. Towns included Kotido and KAABONG. Sparsely populated area with some agriculture (mainly sorghum, also beans, maize, millet, peas, and simsim, cotton was cash crop). Cattle raising was important (also sheep, goats, and poultry). Desertification was a problem. District emblem was an ostrich. In 2005 N portion of district was carved out to form KAABONG district; in 2006 S portion was carved out to form ABIM district and central portion was formed into into current KOTIDO district.

Kotido (ko-TEE-do), town (2002 population 13,694), ⊙ KOTIDO district, NORTHERN region, NE UGANDA, road junction, and 50 mi/80 km NW of MOROTO, between OKOK and OKERE rivers; 03°02′N 34°07′E. Was part of former KARAMOJA province.

Kotipalli, INDIA: see KAKINADA.

Kotipe (ko-TEE-pai), river, 40 mi/64 km long, rises in NE UGANDA, flows W to join OKERE RIVER.

Kotka (KOT-kah), city, KYMEN province, SE FINLAND, on two small islands (railroad and road bridge to mainland) in the GULF OF FINLAND near Russian border, at mouth of the KYMIJOKI River, and 70 mi/113 km ENE of HELSINKI; 60°28′N 26°55′E. A major export center for paper, pulp, and timber, with chemical and woodworking industries. Destroyed by British fleet in 1855, the city was rebuilt and chartered in 1878. Distribution point for VYBORG-FINLAND natural-gas pipeline.

Kot Kapura (KOT kuh-POO-rah), town, FARIDKOT district, PUNJAB state, N INDIA, 8 mi/12.9 km SSE of Faridkot; 30°35′N 74°54′E. Railroad junction; trade center (gram, wheat, cotton, oilseeds, millet); cotton ginning, oilseed pressing, hand-loom weaving. Formerly in princely state of Faridkot.

Kotkhai (KOT-kei), town, central HIMACHAL PRADESH state, N INDIA, 36 mi/58 km E of SHIMLA. Apples and hill fruits are the mainstay of the economy. Also wheat, barley, potatoes; handicraft basket weaving. A carton-making factory at Pragatinagar, 6 mi/10 km away. JUBBAL-Kothkhai is an important political constituency of the state.

Kotlas (KOT-lahs), city (2005 population 58,700), S ARCHANGEL oblast, N European Russia, on NORTHERN DVINA River, at the mouth of the VYCHEGDA River, on the North Pechora railroad, 500 mi/805 km SE of ARCHANGEL, and 210 mi/338 km NW of KIROV; 61°15′N 46°39′E. Elevation 246 ft/74 m. Junction of railroad to Kirov. Shipyards; sawmilling; prefabricated houses, silicate bricks, ceramics, concrete products; electromechanical plant. Arose in 1890 as a railroad terminus. Chartered in 1917. Sometimes spelled Kotlass.

Kotlass, RUSSIA: see KOTLAS.

Kotlenski (KOT-len-skee), pass in the E STARA PLANINA, E BULGARIA; 42°54′N 26°27′E.

Kotli (KOT-lee), town, Kotli district, AZAD KASHMIR, NE PAKISTAN, on tributary of the JHELUM RIVER, and 24 mi/39 km NNE of MIRPUR; 33°31′N 73°55′E. Agriculture (wheat, pearl millet, corn, pulses). Largely destroyed in 1947 during Indo-Pakistani struggle for control.

Kotlik (KOT-lik), village (2000 population 591), W ALASKA, on SW shore of NORTON SOUND, 55 mi/89 km SW of SAINT MICHAEL, at edge of YUKON River delta; 63°02′N 163°33′W.

Kotli Loharan (kot-lee lo-HAHR-ahn), village, SIALKOT district, E PUNJAB province, central PAKISTAN, 7 mi/11.3 km NNW of SIALKOT; 32°35′N 74°29′E. Metal products.

Kotlin Island (KOT-leen), W central LENINGRAD oblast, NW European Russia, in the Gulf of FINLAND, approximately 20 mi/32 km W of SAINT PETERSBURG; □ 56 sq mi/145 sq km. The location of KRONSHTADT, an important port and naval base.

Kotluban' (kuht-loo-BAHN), village, S central VOLGOGRAD oblast, SE European Russia, near highway and railroad, 17 mi/27 km NW of (and administratively subordinate to) GORODISHCHE; 49°01′N 44°14′E. Elevation 206 ft/62 m. Livestock breeding.

Kotlyarevskaya (kuht-lyah-RYEF-skah-yah), rural settlement (2005 population 3,930), E central KABARDINO-BALKAR REPUBLIC, N CAUCASUS, S European Russia, on road and railroad, 18 mi/29 km ENE of NAL'CHIK, and 4 mi/6 km S of MAYSKIY, to which it is administratively subordinate; 43°34′N 44°04′E. Elevation 751 ft/228 m. Railroad station. Agricultural products.

Kotmale, reservoir, CENTRAL PROVINCE, SRI LANKA, 25 mi/40 km S of KANDY; 07°00′N 80°35′E. Formed by dam (280 ft/85 m high) across the Kotmale River. Produces 76.5 MW. Created 1985.

Kot Moman (kot mo-muhn), town, SARGODHA district, PUNJAB province, central PAKISTAN, 22 mi/35 km ENE of SARGODHA; 32°11′N 73°02′E. Also spelled Kot Mumin and Kot Momin.

Kot Najibullah (kot na-jee-BUHL-lah), town, ABBOTTABAD district, NORTH-WEST FRONTIER PROVINCE, N PAKISTAN, 24 mi/39 km SW of ABBOTTABAD; 33°56′N 72°51′E. Sometimes written Kot Najib Ullah.

Koto (KO-to), town, Echi county, SHIGA prefecture, S HONSHU, central JAPAN, 25 mi/40 km N of OTSU; 35°07′N 136°14′E. Rice. Temple-bell production.

Koto (KO-TO), SE ward of TOKYO, Tokyo prefecture, E central HONSHU, E central JAPAN, on TOKYO BAY (S) and Arakawa River. Bordered N by SUMIDA ward, E by EDOGAWA ward, and W by CHUO ward.

Kotobi (ko-TO-bee), town, N'zi-Comoé region, E CÔTE D'IVOIRE, 7 mi/11 km NE of BONGOUANOU; 06°41′N 04°07′W. Agriculture (manioc, bananas, maize, palm oil, coffee, cacao).

Kotohira (ko-TO-hee-rah), town, Nakatado county, KAGAWA prefecture, N SHIKOKU, W JAPAN, 16 mi/25 km S of TAKAMATSU; 34°11′N 133°49′E. Railroad junction. Site of Kotohira-gu, one of Japan's most important Shinto shrines. Its grounds are densely wooded with cryptomeria, pine, and camphor.

Kotonami (ko-TO-nah-mee), town, Nakatado county, KAGAWA prefecture, NE SHIKOKU, W JAPAN, 16 mi/25 km S of TAKAMATSU; 34°09′N 133°55′E.

Kotonkoro (KO-ton-KO-ro), town, NIGER state, NW NIGERIA, on road, and 110 mi/177 km NW of MINNA; 11°02′N 05°57′E. Market town. Kenaf, sorghum, yams, and cassava.

Kotonu, BENIN: see COTONOU.

Koto-oka (ko-TO-o-kah), town, Yamamoto county, Akita prefecture, N HONSHU, NE JAPAN, 22 mi/35 km N of AKITA; 40°02′N 140°05′E.

Kotor (KO-tor), Italian *Cattaro*, town (2003 population 5,341), in MONTENEGRO, on the Bay of KOTOR, an inlet of the ADRIATIC SEA. It is a seaport and a tourist center. Airport. A naval base of former Serbia-Montenegro. The town was colonized by Greeks (3rd century B.C.E.) and later belonged to the Roman and Byzantine empires. In 1797 it passed to Austria and became an important naval base; in 1918 it was transferred to the former YUGOSLAVIA. It has a medieval fort and town walls and a 16th-cent. cathedral. As the oldest town in Montenegro, it is a state-protected historical monument.

Cross-references are shown in SMALL CAPITALS. The pronunciation guide is shown on page xix. The sources of population figures are shown on page xvii.

Kotor, Gulf of (KO-tor), Serbian *Boka Kotorska*, Italian *Bocche di Cattaro*, winding inlet of the ADRIATIC SEA, largely in MONTENEGRO, but touched (SW) by CROATIA (DALMATIA). Well-protected harbor; 25 ft/50 ft–8 m/15 m deep, up to c.20 mi/32 km long from entrance channel; accommodates vessels of any size. Includes bays of TOPLA (NW), TIVAT (S), and RISAN (N), and Kotor (SE); all connected by channels. On its shores are HERCEG NOVI, RISAN, KOTOR, and TIVAT (all in Montenegro). LOVCEN mountain rises SE. Croatia claims historic rights to the gulf.

Kotoriba (ko-to-REE-bah), Hungarian *Kotor*, village, N CROATIA, on MURA RIVER, and 18 mi/29 km E of ČAKOVEC, in the Međimurje region, on Hungarian border. On Čakovec-Nagykanizsa (Hungary) railroad.

Koto River (ko-TO), 395 mi/636 km long, in E and S CENTRAL AFRICAN REPUBLIC, rises in HAUTE-KOTTO region, on SUDAN border; flows S, SW, and again S, past BRIA, forming the border between M'BOMOU and BASSE-KOTTO prefectures, to UBANGI RIVER 60 mi/97 km E of MOBAYE. Several rapids along its course. Sometimes called Kotto River.

Kotorosl' River (KO-tah-ruh-seel), 72 mi/116 km long, in YAROSLAVL oblast, RUSSIA, formed by two headstreams just E of Lake NERO, flows ENE, past GAVRILOV-YAM, and N to the VOLGA RIVER at YAROSLAVL. Logging.

Kotor Varoš (KO-tor VAH-rosh), town, N BOSNIA, BOSNIA AND HERZEGOVINA, on VRBANJA RIVER, and 15 mi/24 km SSE of Banja Luka. Local trade center. Also spelled Kotor Varosh.

Kotosh (ko-TOSH), archaeological site, HUÁNUCO region, central PERU, near city of HUÁNUCO. Known for its earliest temple structures. These earliest buildings are over 4,000 years old. Includes the famous Temple of Crossed Hands.

Kotovka (ko-TOV-kah), village (2004 population 2,600), N DNIPROPETROVS'K oblast, UKRAINE, near ORIL' RIVER, and along the DNIEPER-DONBAS CANAL, 45 mi/72 km N of DNIPROPETROVS'K; 49°08′N 34°57′E. Elevation 291 ft/88 m. Wheat, sunflowers.

Kotovo (KO-tuh-vuh), city (2006 population 27,060), N VOLGOGRAD oblast, SE European Russia, on road, 140 mi/225 km N of VOLGOGRAD, and 30 mi/48 km NW of KAMYSHIN; 50°18′N 44°49′E. Elevation 419 ft/127 m. In Korabkovskoye oil and gas district; gas processing, drilling equipment, food processing (dairy, bakery). Made city in 1966.

Kotovsk (kuh-TOFSK), city (2006 population 32,760), central TAMBOV oblast, S central European Russia, on the right bank of the TSNA RIVER, terminus of local highway branch and on railroad spur (Tambov II station), 8 mi/13 km SSE of TAMBOV; 52°35′N 41°30′E. Elevation 495 ft/150 m. Developed around a chemical plant (gunpowder, explosives, plastics, paint and varnish, synthetic fur). As part of Tambov, called Porokhovatyy Zavod or (after 1930) Krasnyy Boyevik. Became (1940) a separate city, although still an industrial suburb and satellite of Tambov.

Kotovs'k (ko-TOFSK) (Russian *Kotovsk*), city, W ODESSA oblast, UKRAINE, 14 mi/23 km SSW of BALTA; 47°45′N 29°32′E. Elevation 718 ft/218 m. Railroad shops; winery; light manufacturing, woodworking, food processing. Medical, technical schools; heritage museum. First mentioned in 1779; until 1935, called Birzula. In Moldavian Autonomous SSR (1924–1940), where it was temporary capital (1928–1930). City since 1938.

Kotovsk, UKRAINE: see KOTOVS'K.

Kotovskoye, MOLDOVA: see HÂNCEŞTI.

Kot Putli (KOT POOT-lee), town, JAIPUR district, E RAJASTHAN state, NW INDIA, 60 mi/97 km NNE of Jaipur; 27°43′N 76°12′E. Millet, oilseeds; marble sculpting (quarries 8 mi/12.9 km SW). Sometimes written Kotputli.

Kotra (KO-trah), village, UDAIPUR district, S RAJASTHAN state, NW INDIA, 35 mi/56 km WSW of Udaipur, in S ARAVALLI Range.

Kotrang (KO-truhng), town, HUGLI district, S central WEST BENGAL state, E INDIA, on HUGLI RIVER, and 8 mi/12.9 km N of CALCUTTA. Manufacturing of bricks, tiles, rope.

Kotri (KOT-ree), town, DADU district, SW SIND province, SE PAKISTAN, on INDUS River (bridged), and 3 mi/4.8 km W of HYDERABAD; 25°22′N 68°18′E. Railroad junction. Trade center for overland and river traffic; grain market; handicraft cloth weaving, jute industry, fishing, soap manufacturing; distillery. Nearby irrigation barrage (lowest on the INDUS), depleting flow of fresh water to delta. Gypsum deposits (NW).

Kot Sabzal (kot sub-zul), town, RAHIMYARKHAN district, BAHAWALPUR division, PUNJAB province, central PAKISTAN, near SIND province border, 28 mi/45 km SW of RAHIMYARKHAN; 28°13′N 69°54′E. Also SABZAL KOT.

Kot Samaba (kot suhm-AH-bah), town, RAHIMYARKHAN district, BAHAWALPUR division, PUNJAB province, central PAKISTAN, on railroad, and 13 mi/21 km NE of RAHIMYARKHAN; 28°33′N 70°28′E. Also spelled Kot Samba, Kotsamba, Kot Somaba.

Kötschach-Mauthen (KOT-shahkh–MOU-ten), township, CARINTHIA, S AUSTRIA, near GAIL RIVER, 15 mi/24 km SE of LIENZ; 46°40′N 12°59′E. Railroad terminus, lumberyards, summer resort, and base for mountain tours to the Plöcken massif. Has late-Gothic church. Museum of the high Alpine war (1915–1917) in the Carnic Alps.

Kotsk, POLAND: see KOCK.

Kotsyubinskoye, UKRAINE: see KOTSYUBYNS'KE.

Kotsyubyns'ke (ko-tsyoo-BIN-ske) (Russian *Kotsyubinskoye*), NW suburb of KIEV, UKRAINE, 9 mi/15 km WNW of Kiev and 3 mi/4.8 km N of SVYATOSHYNE; 50°30′N 30°22′E. Elevation 508 ft/154 m. Until 1941, Berkovets'. Furniture, insulation, sound-proofing; woodworking, stoneworking.

Kotta, SRI LANKA: see KOTTE.

Kottagoda, SRI LANKA: see KOTTEGODA.

Kottagudem (kot-tah-GOO-daim), city, Khamman district, ANDHRA PRADESH state, S INDIA, 75 mi/121 km ESE of WARANGAL; 18°04′N 80°28′E. Mining center in Singareni coalfield; on railroad (BHADRACHALAM ROAD station) from DORNAKAL, 32 mi/51 km W. Mines opened in 1920s; center of operations moved here from YELLANDU in 1941. Also called Kottaguda, Kothagudium.

Kottai Malai (KOT-tei muh-LEI), peak (6,624 ft/2,019 m), in S WESTERN GHATS, S INDIA, on TAMIL NADU/KERALA state border, 55 mi/89 km WSW of MADURAI and 9 mi/14.5 km E of PERIYAR lake; 09°30′N 77°24′E. VAIGAI RIVER rises on N slope.

Kottaiyur (kot-TEI-yoor), town, PASUMPON MUTHURAMALINGA THEVAR district, TAMIL NADU state, S INDIA, 4 mi/6.4 km NNE of KARAIKKUDI; 10°07′N 78°49′E. Residence of Chetty merchant community.

Kottapatnam, INDIA: see ALLURU KOTTAPATNAM.

Kottapeta (KOT-tah-pet-tah), town, EAST GODAVARI district, ANDHRA PRADESH state, SE INDIA, on GAUTAMI GODAVARI RIVER, in GODAVARI delta, and 27 mi/43 km SW of KAKINADA. Rice milling; sugarcane, tobacco, coconuts. Also called Kothapet, also spelled Kothapeta.

Kottarakara (ko-TAH-rah-kah-rah), town, KOLLAM district, KERALA state, S INDIA, 15 mi/24 km NE of KOLLAM; 09°00′N 76°48′E. Road center; trades in coir rope and mats, rice, cassava; cashew nut processing.

Kottayam (KO-tuh-yuhm), district (□ 851 sq mi/2,212.6 sq km), KERALA state, S INDIA; capital of KOTTAYAM.

Kottayam (KO-tuh-yuhm), city, ⊙ KOTTAYAM district, KERALA state, S INDIA, 50 mi/80 km N of KOLLAM; 09°35′N 76°31′E. Processing and packing of rubber, manufacturing of copra, coir rope and mats, plywood, tiles; engineering workshops (motor vehicle repairing), sawmills; rice milling, cashew nut processing. College (affiliated with University of Travancore).

Kottbus, GERMANY: see COTTBUS.

Kotte (KOT-te), town, WESTERN PROVINCE, SRI LANKA, 5 mi/8 km SE of COLOMBO city center; 06°53′N 79°54′E. New Parliament building; many government ministries. Rice, vegetables, coconuts. Was the capital of Kotte kingdom (15th–16th century). Also spelled COTTA and KOTTA.

Kottegoda, village, SOUTHERN PROVINCE, SRI LANKA, on S coast, 7 mi/11.3 km E of MATARA; 05°56′N 80°38′E. Fishing center; vegetables, rice. Also spelled KOTTAGODA.

Kottingbrunn (kaht-ting-BRUN), township, E LOWER AUSTRIA, 4 mi/6.4 km S of BADEN. Vineyards, orchards; 48°57′N 16°14′E. Manufacturing of plastics and pump stations. Known for its tire-test course.

Kotto River, CENTRAL AFRICAN REPUBLIC: see KOTO RIVER.

Kottuchcheri (kot-TOO-cher-ree), French *Cotchéry*, town, KARAIKAL district, PONDICHERRY Union Territory, S INDIA, 3 mi/4.8 km NW of Karaikal; 10°57′N 79°49′E. Also spelled Kottucheri, Karikal.

Kottur (KO-toor), town, COIMBATORE district, TAMIL NADU state, S INDIA, 9 mi/14.5 km SSW of POLLACHI. Cotton pressing; tea factory. Marichinaickenpalayam adjoins (S).

Kottur, INDIA: see KOTTURU.

Kotturu (ko-TOO-roo), town, BELLARY district, KARNATAKA state, SW INDIA, 32 mi/51 km SSW of HOSPET; 18°37′N 80°22′E. Railroad spur terminus; peanut milling, hand-loom cotton weaving. Bamboo, fiber, gum arabic in nearby forests. Sometimes spelled Kottur.

Kotur, IRAN: see QUTUR.

Koturdepe (ko-toor-de-PE), town, BALKAN weloyat, W TURKMENISTAN, 50 mi/84 km W of JEBEL.

Kotuy River, RUSSIA: see KHATANGA, river.

Kotwa, village, MASHONALAND EAST province, NE ZIMBABWE, 150 mi/241 km NE HARARE; 16°59′S 32°42′E. Livestock; grain.

Kotyuzhany, MOLDOVA: see COTIUGENII-MARI.

Kotzebue (KAHT-se-boo), city (2000 population 3,082), NW ALASKA, on KOTZEBUE SOUND at the tip of BALDWIN PENINSULA; 66°53′N 162°39′W. It is one of the largest settlements of Inuit (Eskimos) in Alaska. A regional trade and supply center with local government offices, Kotzebue has a tourist industry. Fishing is economically important. The city, set on a tundra, began in the eighteenth century as an Eskimo trading post for arctic Alaska and part of Siberia; reindeer station established 1897. Regional center for NOATAK, KOBUK, and SELAWIK river valleys. University of Alaska Extension Center. NANA Museum of the Arctic, owned by native Inuit corporation. Red Dog Zinc Mine to N. Incorporated 1958.

Kotzebue Sound (KAHT-se-boo), NW ALASKA, arm of CHUKCHI SEA, on N side of SEWARD PENINSULA, bounded E by BALDWIN PENINSULA; 100 mi/161 km long, 70 mi/113 km wide; 66°40′N 163° W. KOTZEBUE, KIWALIK, and DEERING villages on shores. ESCHSCHOLTZ BAY is SE arm. Discovered and named 1816 by Count Otto von Kotzebue while searching for NORTHWEST PASSAGE. Knud Rasmussen here completed (1924) overland crossing from REPULSE BAY, NORTHWEST TERRITORIES.

Kotzenau, POLAND: see CHOCIANOW.

Kotzman, UKRAINE: see KITSMAN'.

Kötzschenbroda, GERMANY: see RADEBEUL.

Kötzting (KOTS-ting), town, BAVARIA, SE GERMANY, in LOWER BAVARIA, in BOHEMIAN FOREST, on the WHITE REGEN, 24 mi/39 km NNW of DEGGENDORF; 49°10′N 12°51′E. Manufacturing of metal products, lumber milling; granite quarrying. Chartered before 1344.

Kou'an (KO-AN), town, S JIANGSU province, E CHINA, 15 mi/24 km S of TAIZHOU, on CHANG JIANG (Yangzi River). Commercial port; shipbuilding.

Kouandé (KWAHN-dai), town, ATAKORA department, BENIN, 16 mi/26 km E of NATITINGOU; 10°20′N

01°42′E. Elevation 2,108 ft/643 m. Peanuts, shea nuts, livestock.

Kouango (kwahng-GO), village, OUAKA prefecture, S CENTRAL AFRICAN REPUBLIC, on UBANGI RIVER (ZAÏRE border), at mouth of OUAKA RIVER and 65 mi/ 105 km SSW of BAMBARI; 04°59′N 19°59′E. Center of trade in cotton region.

Kouango River, CENTRAL AFRICAN REPUBLIC: see OUAKA RIVER.

Kouara (koo-AH-rah), town, Savanes region, N CÔTE D'IVOIRE, 9 mi/15 km N of OUANGOLODOUGOU; 10°06′N 05°12′W. Agriculture (sorghum, rice, millet, maize, beans, peanuts, tobacco, cotton).

Kouba (koo-BAH), town, ALGIERS wilaya, N central ALGERIA, now an outer SE suburb of ALGIERS, adjoining (S) HUSSEIN DEY. Boiler making, woodworking, manufacturing of vinegar, soap, ink, agricultural equipment. Vegetable gardens. View of Algiers Bay.

Koubia, prefecture, Labé administrative region, N central GUINEA, in Moyenne-Guinée geographic region; ⊙ Koubia. Bordered NWN by Mali prefecture, NNE by MALI, E and S by Tougué prefecture, and SWW by Labé prefecture. Gambie River in N and W of prefecture. Part of FOUTA DJALLON massif here. Roads connect the prefecture to the rest of the administrative region (including LABÉ town), as well as Sélouma (in Dinguiraye prefecture) to SE.

Koubia, town, ⊙ Koubia prefecture, Labé administrative region, N central GUINEA, in Moynee-Guinée geographic region; 11°35′N 11°53′W. On road between LABÉ town to W and Sélouma to E.

Kouchibouguac (koo-shee-buh-KWAK), village, E NEW BRUNSWICK, CANADA, on KOUCHIBOUGUAC RIVER, near its mouth on NORTHUMBERLAND STRAIT, 10 mi/16 km NW of RICHIBUCTO. Lumbering, national park.

Kouchibouguac National Park (koo-shee-buh-KWAK) (□ 87 sq mi/226.2 sq km), on Kouchibouguac Bay, E NEW BRUNSWICK, CANADA, near RICHIBUCTO; established 1969. The park's scenic features include lagoons, bays, and offshore sandbars.

Kouchibouguac River (koo-shee-buh-KWAK), 45 mi/ 72 km long, in E NEW BRUNSWICK, CANADA; rises 30 mi/48 km S of NEWCASTLE; flows ENE to NORTHUMBERLAND STRAIT 12 mi/19 km NNW of RICHIBUCTO.

Koudougou (koo-DOO-goo), town (2005 population 81,477), ⊙ BOULKIEMDÉ province, CENTRE-OUEST region, central BURKINA FASO, 55 mi/89 km W of OUAGADOUGOU; 12°15′N 02°23′W. On railroad. Agricultural center (shea nuts, peanuts, millet, rice, manioc, onions); processing of shea-nut butter, cotton ginning; stock raising (cattle, sheep, goats). Some chromium and manganese deposits nearby.

Koufalia (koo-FAH-lyah), town, Thessaloníki prefecture, Central MACEDONIA department, NE GREECE, 21 mi/34 km NW of Thessaloníki; 40°47′N 22°35′E. Wheat, cotton, vegetables. Also Kouphalia.

Koufinission, Greece: see ANO KOUFINISSION and KATO KOUFINISSION.

Kougnohou, town, TOGO: see KOUNIOHOU.

Kouibli (kou-EEB-lee), town, Dix-Huit Montagnes region, W CÔTE D'IVOIRE, 25 mi/40 km SE of Man; 07°15′N 07°14′W. Agriculture (manioc, bananas, rice, palm oil).

Kouif, El (KWEEF, el), mining town, TÉBESSA wilaya, NE ALGERIA, near Tunisian border and 13 mi/21 km NE of TÉBESSA; 35°30′N 08°19′E. Once Algeria's richest phosphate-producing region, it has been surpassed by Djebel ONK. Linked to Annaba by railroad.

Kouilou (kwee-LOO), region (2007 population 806,670), SW Congo Republic, bordered by GABON to the N, NIARI region to the NE and E, the Angolan enclave of CABINDA to the S, and the Atlantic Ocean to the S and W; ⊙ POINTE-NOIRE; 04°00′S 12°00′E. The KOUILOU RIVER flows E-W across to the Atlantic. The CRYSTAL Mountains are located on the N

edge of the region. Manufacturing (plywood, aluminum ware, soap, starch, tapioca); shipbuilding and food processing at Pointe-Noire. Other important elements of the economy are oil refining and offshore oil drilling, gold mining, palm mills. Sport fishing.

Kouilou River (kwee-LOO), circa 200 mi/322 km long, Congo Republic; rises in CRYSTAL MOUNTAINS 50 mi/ 80 km SSW of DJAMBALA; flows S, W past LOUDIMA, NW, and SW past KAKAMOEKA, to the Atlantic 32 mi/ 51 km NNW of POINTE-NOIRE. Known as N'DOUO in its upper course, and as NIARI in its middle course. Navigable in lower course for circa 40 mi/64 km below KAKAMOEKA. Many rapids. Niari River basin is noted as mining area (copper, lead, zinc). Also spelled Kwilu.

Kouinine (kwee-NEEN), Saharan village, El OUED wilaya, E ALGERIA, one of the SOUF oases, 4 mi/6.4 km NNW of El OUED. Deglet Nour dates.

Koukdjuak, Great Plain of the, tundra region of SW BAFFIN ISLAND, BAFFIN region, NUNAVUT territory, CANADA; c.120 mi/193 km long, 60 mi/97 km–90 mi/ 145 km wide. N boundary is Koukdjuak River (50 mi/ 80 km long), which drains NETTLING LAKE into FOXE BASIN.

Kouklia (KOOK-li-yah), ancient *Palaepaphos* [Greek=old PAPHOS], town, PAPHOS district, SW CYPRUS, on Dhiarizos River, 10 mi/16 km ESE of PAPHOS; 35°06′N 33°45′E. MEDITERRANEAN SEA 1.5 mi/2.4 km to SSW, Asprokemmos Reservoir to NW. Wine grapes, peaches, bananas, grain, cotton, vegetables, avocados; goats, sheep, hogs. Palaia Paphos, ruins of ancient city 1 mi/1.6 km to W, includes remains of temple to Aphrodite (6th century). After an earthquake, town was restored (15 B.C.E.) by Augustus. Paphos International. Airport 5 mi/8 km to W.

Kouklia (KOOK-li-yah), village, FAMAGUSTA district, E CYPRUS, 12 mi/19 km W of FAMAGUSTA; 34°12′N 32°34′E. Attila Line passes 4 mi/6.4 km to S. Fruit, grain, vegetables; livestock. Kouklia Reservoir to N.

Koukos, Greece: see KALLIDROMON.

Koulamoutou or **Koula-Moutou** (koo-luh-MOO-too), city, ⊙ OGOOUÉ-LOLO province, SE GABON, 256 mi/414 km ESE of PORT-GENTIL; 01°06′S 12°29′E. Coffee- and cocoa-producing region. Gold deposits nearby.

Koulikoro (koo-LEE-ko-ro), city, ⊙ SECOND REGION/ KOULIKORO, S MALI, landing on NIGER RIVER (irrigation area), terminus of DAKAR-NIGER railroad and 37 mi/60 km ENE of BAMAKO. Trading and processing center (peanuts, shea nuts, kapok). Factory oil and soap. Port, military camp, administrative center.

Koulikoro, administrative region, MALI: see SECOND REGION.

Kouloukia, Greece: see KHALASTRA.

Koulpeleogo River, 120 mi/193 km long, S BURKINA FASO; rises in KOURITENGA province; flows SE through GOURMA province to GHANA border near Bittau where it joins OTI RIVER.

Koulpélogo, province (□ 2,065 sq mi/5,369 sq km; 2005 population 232,305), CENTRE-EST region, SE BURKINA FASO; ⊙ Ouargaye; 11°25′N 00°10′E. Bordered N by GOURMA province, E by KOMPIENGA province, S by TOGO, SW by GHANA, and WNW by BOULGOU province. Established in 1997 with fourteen other new provinces.

Koumac (KOO-MAHK), village, NEW CALEDONIA, on NW coast, 180 mi/290 km NW of NOUMÉA; 20°32′S 164°20′E. Agricultural products; livestock; chrome mining.

Koumbia (koom-bee-ah), town, Gaoual prefecture, Boké administrative region, NW GUINEA, in Moyenne-Guinée geographic region, 25 mi/40 km NW of GAOUAL; 11°48′N 13°30′W. Peanuts, millet, sorghum.

Koumi (KO-oo-mee), town, South Saku county, NAGANO prefecture, central HONSHU, central JAPAN, 43 mi/70 km N of NAGANO; 36°05′N 138°29′E. Smelts.

Koumi, Greece: see KIMI.

Koumra (koom-RAH), town, ⊙ MANDOUL administrative region, S CHAD, 60 mi/97 km WSW of SARH;

08°55′N 17°33′E. Cotton ginning. Until 1946 in Ubangi-Chari colony.

Koundara, prefecture, Boké administrative region, extreme NW GUINEA, in Moyenne-Guinée geographic region; ⊙ KOUNDARA. Bordered N by SENEGAL, E by Mali prefecture, S by Gaoual prefecture, and W by GUINEA-BISSAU. Koulountou River runs from SE to N central of prefecture. Part of FOUTA DJALLON massif here. Badiar National Park in N. Towns include Koundara, SAMBAILO, and YOUKOUNKOUN. Several roads run through the prefecture connecting it to W Guinea (including LABÉ and MALI towns), Senegal (road terminates near the ATLANTIC OCEAN in the GAMBIA), and Guinea-Bissau.

Koundara (KOON-dah-rah), town, ⊙ Koundara prefecture, Boké administrative region, NW GUINEA, in Moyenne-Guinée geographic region; 12°29′N 13°18′W. Millet, sorghum, and peanuts. Airport.

Koundian (KOON-jahn), village, FIRST REGION/KAYES, MALI, 99 mi/159 km SE of KAYES. Archaeological site.

Koun-Fao (KOON-FOU), town, Zanzan region, E CÔTE D'IVOIRE, 25 mi/40 km N of AGNIBILÉKROU; 07°28′N 03°05′W. Agriculture (taro, bananas, corn, peanuts, yams, cacao, coffee, kola nuts); lumbering.

Koungheul (KOON-gel), town (2007 population 17,332), KAOLACK administrative region, S central SENEGAL, on Dakar-Niger railroad and 75 mi/121 km WNW of TAMBACOUNDA; 13°59′N 14°48′W. Peanuts; livestock.

Koung-Khi, department (2001 population 121,794), WEST province, CAMEROON.

Koungou (koong-GOO), mountain, HAUTE-M'BOMOU prefecture, SE CENTRAL AFRICAN REPUBLIC.

Kounié, NEW CALEDONIA: see PINES, ISLE OF.

Kouniohou (KOO-nyo-HOO), town, PLATEAUX region, S central TOGO, 10 mi/16 km WNW of ATAKPAMÉ; 07°40′N 00°48′E. Also called Kougnohou.

Kounradski, KAZAKHSTAN: see KONYRAT.

Kountze, town (2006 population 2,165), ⊙ HARDIN county, SE TEXAS, 25 mi/40 km NNW of BEAUMONT; 30°22′N 94°19′W. Elevation 85 ft/26 m. railroad junction; timber area; cattle, hogs; forage crops; egg products; manufacturing (hardwood lumber, paper cores and shorting). Calls itself (referred to as) the Big Lights in the Big Thicket, surrounded by several units of BIG THICKET NATIONAL PRESERVE, visitor information station to N. Incorporated after 1940.

Koupéla, town (2005 population 21,288), ⊙ KOURITENGA province, CENTRE-EST region, E central BURKINA FASO; 12°07′N 00°21′W. On road, 85 mi/137 km ESE of OUAGADOUGOU. Road junction to S and E.

Kouphalia, Greece: see KOUFALIA.

Kouraia Konkouré, GUINEA: see KONKOURÉ.

Kourémalé (KOR-ai-ma-lai), township, SECOND REGION/KOULIKORO, MALI, on GUINEA border, 70 mi/113 km SW of BAMAKO. Gold fields.

Kouri (KOR-ee), town, THIRD REGION/SIKASSO, MALI, on border with BURKINA FASO, 81 mi/135 km NE of SIKASSO; 12°11′N 04°48′W. Also spelled Koury.

Kourim (KOU-rzhim), *Czech Kouřim*, German *kaurzim*, town, STREDOCESKY province, central BOHEMIA, CZECH REPUBLIC, 13 mi/21 km WNW of KUTNÁ HORA; 50°00′N 14°59′E. Manufacturing (veterinary medicines, machinery, bricks). Medieval regional center. Important Neolithic-era archaeological site.

Kourion (kor-YON), ruins, LIMASSOL district, SW CYPRUS, near EPISKOPI BAY, 9 mi/14.5 km W of LIMASSOL. Founded by Argives of Peloponnese in 12th century B.C.E. Constant raids by Saracens finally destroyed town in 7th century. Greco-Roman theater dating to 2nd century is used for performances; also here is the Roman villa House of Eustolios, the foundation of a 5th century Christian basilica. Also spelled Curium.

Kouritenga, province (□ 1,012 sq mi/2,631.2 sq km; 2005 population 302,204), CENTRE-EST region, E central

BURKINA FASO; ⊙ KOUPÉLA; 12°00′N 00°10′W. Borders NAMENTENGA (N), GNAGNA (NE), GOURMA (E), BOULGOU (S), and OUBRITENGA (W) provinces. Drained by KOULPELEOGO RIVER. Major road intersection. Thermal center. Main center is KOUPÉLA.

Kourou (koo-ROO), town, N FRENCH GUIANA; 05°09′N 52°39′W. Located on small Kourou River, on the coast, and 26 mi/42 km NW of CAYENNE. Population is mainly French. European Space Agency rocket-launching base is nearby and has provided much employment.

Kouroussa, prefecture, Kankan administrative region, E GUINEA, in Haute-Guinée geographic region; ⊙ KOUROUSSA. Bordered NWN by Dinguiraye prefecture, NNE by Siguiri prefecture, E by Kankan prefecture, S by Kissidougou prefecture, SWW by Faranah prefecture, and W by Dabola prefecture. NIGER RIVER flows NE through center of prefecture, Niandan River in extreme E and joins Niger River SE of Kouroussa town, Kouya River in SE, and Mafou River in extreme W and joins Niger River SW of Kouroussa town (forms much of border with Faranah prefecture). Towns include Balato, Banfèlè, Cisséla, DOUAKO, Kouroussa, and Saraya. Railroad between CONAKRY and KANKAN town travels NW-SE through the center of the prefecture, going through Cisséla and Kouroussa towns. Main road runs through Kouroussa town, connecting it to Saraya to the W (continuing across country to Conakry), NE Guinea and MALI to NE, and Kankan town to SE; secondary road also connects Kouroussa town to Banfèlè, Douako, and SE Guinea.

Kouroussa (koo-ROO-sah), town, ⊙ Kouroussa prefecture, Kankan administrative region, central GUINEA, in Haute-Guinée geographic region, road junction and landing on upper NIGER (crossed by railroad) and 275 mi/443 km ENE of CONAKRY, 190 mi/306 km SW of BAMAKO, MALI; 10°39′N 09°53′W. Trading and agricultural center (rice, subsistence crops). Cotton gin.

Kourwéogo, province (□ 613 sq mi/1,593.8 sq km; 2005 population 140,597), PLATEAU CENTRAL region, central BURKINA FASO; ⊙ Boussé; 12°35′N 01°48′W. Bordered N by PASSORÉ province, ENE by OUBRITENGA province, ESE by KADIOGO province, and WSW by BOULKIEMDÉ province. Established in 1997 with fourteen other new provinces.

Kousséri (KOO-ser-ee), city (2001 population 76,200), ⊙ LOGONE-ET-CHARI department, FAR NORTH province, CAMEROON, on left bank of LOGONE River at its confluence with the CHARI, opposite N'djamena (CHAD) and 115 mi/185 km NNE of MAROUA; 12°06′N 14°59′E. Pottery making, experimental nurseries, and fishing. Hospital. Final defeat (1900) of Rabah el Zobeir, Sultan of Bornu, opened for the FRENCH the way to native kingdoms of CHAD area. Refugees from Chad's civil conflicts have swelled the population in recent years. Kalamaloué National Game Reserve nearby. Formerly called FORT FOUREAU.

Koussountou (KOO-soon-TOO), town, CENTRALE region, E central TOGO, c.30 mi/48 km SE of SOKODÉ, on secondary road near BENIN border; 08°50′N 01°31′E. Mainly subsistence agriculture of manioc, maize, and millet.

Kouta, RUSSIA: see KOVDA.

Koutajoki, RUSSIA: see KOVDA RIVER.

Koutiala (KOO-tee-a-la), town, THIRD REGION/SIKASSO, MALI, 170 mi/274 km E of BAMAKO; 12°23′N 05°28′W. Millet, peanuts, cotton, kapok, shea-nut butter; cotton gin and research institute. Iron deposits nearby. Important cross road.

Kouto (KOO-to), town, Savanes region, NW CÔTE D'IVOIRE, 25 mi/40 km N of BOUNDIALI; 09°53′N 06°05′W. Agriculture (corn, beans, tobacco, peanuts, cotton).

Koutoumoula, Greece: see KORONEIA.

Kouts, town (2000 population 1,698), PORTER county, NW INDIANA, 11 mi/18 km S of VALPARAISO; 41°19′N 87°02′W. Manufacturing (animal feed, spring wire). Laid out 1864.

Kouvola (KO-vo-lah), town, ⊙ KYMEN province, SE FINLAND, near the KYMIJOKI (river), 30 mi/48 km NNW of KOTKA; 60°52′N 26°42′E. Elevation 165 ft/50 m. Railroad junction. Pulp and paper milling center; power station in NW. Distribution point for VYBORG-FINLAND natural-gas pipeline.

Kovachevo (KO-vah-CHEV-o), village, PLOVDIV oblast, SEPTEMVRI obshtina, BULGARIA; 42°12′N 24°11′E.

Kovachevtsi (ko-vah-CHEV-tsee), village, SOFIA oblast, ⊙ Kovachevtsi obshtina (1993 population 2,450), BULGARIA; 42°34′N 22°48′E. Electronics manufacturing; spare parts.

Kovachitsa (ko-VACH-eet-sah), village, MONTANA oblast, LOM obshtina, BULGARIA; 43°48′N 23°23′E.

Kovačica (ko-VAH-chee-chah), Hungarian *Antalfalva*, village (2002 population 27,890), VOJVODINA, N SERBIA, 21 mi/34 km NNE of BELGRADE, in the BANAT region. Railroad junction. Also spelled Kovachitsa.

Kovacova, SLOVAKIA: see ZVOLEN.

Koval, POLAND: see KOWAL.

Kovalevskoye (kuh-vah-LYEF-skuh-ye), settlement (2005 population 3,155), E KRASNODAR TERRITORY, S European Russia, on the KUBAN' River, on road and near railroad, 11 mi/18 km N of NOVOKUBANSK; 45°11′N 40°58′E. Elevation 482 ft/146 m. In agricultural area (wheat, flax, sunflowers).

Kovalima, EAST TIMOR: see SUAI.

Kovászna, ROMANIA: see COVASNA.

Kovda (KOV-dah), Finnish *Kouta*, village, SW MURMANSK oblast, NW European Russia, lumber port on the KANDALAKSHA BAY of the WHITE SEA, 35 mi/56 km SSE of KANDALAKSHA; 66°41′N 32°52′E. Pegmatite quarries nearby. Sawmilling at LESOZAVODSKIY (N).

Kovda River (kuhv-DAH), Finnish *Koutajoki*, efflux of TOPOZERO lake, 137 mi/220 km long, in Republic of KARELIA, NW European Russia; flows N, through PYAOZERO lake, and E, through KOVDOZERO lake (MURMANSK oblast), to the KANDALAKSHA BAY of the WHITE SEA at KOVDA. Frozen October through May. Rapids along its course.

Kovdor (KOV-duhr), city (2006 population 19,075), W MURMANSK oblast, NW European Russia, terminus of local highway branch and railroad spur, 230 mi/370 km SW of MURMANSK, and 80 mi/129 km W of APATITY; 67°33′N 30°28′E. Elevation 685 ft/208 m. Mining and concentrating combine for iron ore, mica. Made city in 1965.

Kovdozero (kuhf-DO-zee-ruh), Finnish *Koutajärvi*, lake (□ 154 sq mi/400.4 sq km) in SW MURMANSK oblast, NW European Russia, W of KOVDA; 66°40′N 32°00′E; 30 mi/48 km long, 25 mi/40 km wide; maximum depth 207 ft/63 m. Deeply dissected shores and islands. Remains frozen between November and May. Traversed (W-E) by the KOVDA RIVER.

Kovel' (KO-vel) (Polish *Kowel*), city (2001 population 66,401), NW VOLYN' oblast, UKRAINE, on the TURA RIVER; 51°13′N 24°43′E. Elevation 554 ft/168 m. Raion center. Railroad junction and agricultural center. Railroad shops; agricultural machines; flax. Sewing; food processing (cheese, starch, flour, meat); feed milling; woodworking. First mentioned in the 14th century; it belonged to Lithuania and passed to Poland when the two states were united in 1569. Taken by Russia during the third partition of Poland in 1795. Briefly part of the Ukraine (1917–1920), it was again under Polish rule 1921–1939; then annexed by the USSR; since 1991, part of independent Ukraine. Jews, active in trade and light industry, formed a majority of the population until World War II.

Kovernino (kuh-VYER-nee-nuh), village (2006 population 6,720), NW NIZHEGOROD oblast, central European Russia, in the Uzola River (left tributary of the VOLGA RIVER) basin, on road junction, 35 mi/56 km NW of SEMËNOV; 57°07′N 43°49′E. Elevation 436 ft/132 m. Logging, lumbering; silk milling; flax processing.

Kovilj (KO-veel), Hungarian *Kabol*, village, VOJVODINA, N SERBIA, near the DANUBE River, 9 mi/14.5 km ESE of NOVI SAD, in the BACKA region. Also spelled Kovil.

Koviljača (KO-veel-yah-chah), Serbian *Banja Koviljača*, village, W SERBIA, on the DRINA RIVER (Bosnia-Herzegovina border), and 3 mi/4.8 km WSW of LOZNICA. Railroad terminus; health resort with warm springs and hot mud; noted for treatment of rheumatism. Also spelled Kovilyacha and Banya Kovilyacha.

Kovilpatti, town, TIRUNELVELI KATTABOMMAN district, TAMIL NADU state, S INDIA, 32 mi/50 km NNE of Tirunelveli; 09°10′N 77°52′E. Major cotton-milling center. Also Koilpatti.

Kovin (KO-veen), Hungarian *Kevevára*, village (2002 population 36,802), VOJVODINA, E SERBIA, on the DANUBE River, and 24 mi/39 km E of BELGRADE, in the BANAT region; 44°44′N 20°59′E. Railroad terminus.

Kovlar, urban settlement, AZERBAIJAN.

Kovno, LITHUANIA: see KAUNAS.

Kovrov (kuhv-ROF), city (2006 population 153,825), E central VLADIMIR oblast, central European Russia, on the KLYAZ'MA RIVER (tributary of the OKA River), 40 mi/64 km NW of VLADIMIR; 56°21′N 41°19′E. Elevation 469 ft/142 m. Railroad and highway junction; local transshipment point. Excavating machines, motorcycle engines, panel houses, wood products, linen textiles, food industries (bakeries, meat packing). Arose in the 12th century out of the village of Yelifanovka; later called Rozhdestvennoye.

Kovsharovka, UKRAINE: see KIVSHAROVKA.

Kovur (ko-VOOR), town, NELLORE district, ANDHRA PRADESH state, S INDIA, 4 mi/6.4 km N of NELLORE; 14°29′N 79°59′E. Rice and oilseed milling; cashew and casuarina plantations. Also Kovvur or Kovvuru.

Kovur, INDIA: see KOVVUR.

Kovvur (ko-VOOR), town, WEST GODAVARI district, ANDHRA PRADESH state, S INDIA, on right bank of GODAVARI RIVER (railroad bridge), at head of delta, and 4 mi/6.4 km NW of (opposite) RAJAHMUNDRY; 17°01′N 81°44′E. Rice milling; silk growing. Building-stone quarries nearby. Also spelled Kovur.

Kovvur, INDIA: see KOVVUR.

Kovyagi, UKRAINE: see KOV'YAHY.

Kov'yahy (ko-VYAH-hi) (Russian *Kovyagi*), town (2004 population 3,800), NW KHARKIV oblast, UKRAINE, on road and railroad, 30 mi/48 km WSW of KHARKIV, and 6 mi/9 km NW of VALKY; 49°54′N 35°33′E. Elevation 692 ft/210 m. Asphalt manufacturing; dairy, feed mill. Established at the end of the 17th century; town since 1968.

Kovylkino (kuh-VIL-kee-nuh), city (2006 population 21,830), S MORDVA REPUBLIC, central European Russia, on the MOKSHA River (tributary of the OKA River), 72 mi/116 km W of SARANSK, and 40 mi/64 km W of RUZAYEVKA; 54°02′N 43°55′E. Elevation 475 ft/144 m. Road and railroad junction; local transshipment point. Electromechanical plant; motor vehicle parts, silicate bricks; distilling. Grain elevator. Made city in 1960.

Kovzha River (kuhv-ZHAH), 52 mi/84 km long, in VOLOGDA oblast, RUSSIA; issues from S Kovzha Lake, 25 mi/40 km SW of VYTEGRA; flows S, past ANNENSKIY MOST and Kovzhinskiy Zavod (sawmill) to lake BELOYE OZERO at Kovzha village. Canalized lower course forms part of the Maryinsk canal system; joined to the VYTEGRA River (NW) by the MARYINSK Canal. Receives the Shola River (right).

Kowait: see KUWAIT.

Kowal (KO-vahl), Russian *Koval* or *Koval'*, town, Włocławek province, central POLAND, 9 mi/14.5 km SSE of WŁOCŁAWEK. Brick manufacturing, sawmilling, flour milling.

Kowalewo (ko-vah-LEE-vo), German *Schönsee*, town, Toruń province, N central POLAND, on railroad, and 15 mi/24 km NE of TORUŃ. Flour and beet-sugar milling, sawmilling. Ruins of 13th-century castle. Also called Kowalewo Pomorskie.

Kowary (ko-VAH-ree), German *Schmiedeberg* (SHMEE-den-berg), town, in LOWER SILESIA, Jelenia Góra province, SW POLAND, near CZECH border, at N foot of the Giant Mountains (RIESENGEBIRGE), 9 mi/14.5 km SE of JELENIA GÓRA (Hirschberg). Magnetite mining and smelting, woolen milling, metalworking; manufacturing of carpets, china, industrial ceramics; health resort. Has 13th-century church. After 1945, briefly called Krzyzatka, Polish *Krzyż atka*.

Kowel, UKRAINE: see KOVEL'.

Kowloon, China: see HONG KONG.

Kowno, LITHUANIA: see KAUNAS.

Kow Song, MYANMAR: see KAWTHAUNG.

Koya (KO-yah), town, Ito county, WAKAYAMA prefecture, S HONSHU, W central JAPAN, on N central KII PENINSULA, 25 mi/41 km E of WAKAYAMA, at foot of MOUNT KOYA (3,600 ft/1,097 m); 34°12′N 135°35′E. Kongobu temple. Sometimes called Koyasan.

Koya (KO-yah), peak (2,858 ft/871 m), WAKAYAMA prefecture, S HONSHU, W central JAPAN. On its summit is a Buddhist monastery, with many tombs, founded in 816. The monastery has one hundred and twenty temples and is visited by over a million pilgrims annually. The peak is also known as Koyasan.

Koyadaira (ko-yah-DAH-ee-rah), village, Mima county, TOKUSHIMA prefecture, SE SHIKOKU, W JAPAN, 22 mi/35 km S of TOKUSHIMA; 33°55′N 134°12′E. Citrons; dried Japanese plums.

Koyagi (KO-yah-gee), town, West Sonogi county, NAGASAKI prefecture, NW KYUSHU, SW JAPAN, 5 mi/8 km S of NAGASAKI; 32°41′N 129°48′E. Shipbuilding.

Koyaguchi (KO-yah-GOO-chee), town, Ito county, WAKAYAMA prefecture, S HONSHU, W central JAPAN, on N central KII PENINSULA, 24 mi/38 km ENE of WAKAYAMA; 34°18′N 135°33′E. Textiles.

Koyama (KO-yah-mah), town, Kimotsuki county, KAGOSHIMA prefecture, SW KYUSHU, SW JAPAN, 28 mi/45 km S of KAGOSHIMA; 31°20′N 130°56′E. Mandarin oranges. Abacuses; Buddhist religious items.

Koyambattur, INDIA: see COIMBATORE.

Koyang (GO-YAHNG), city (□ 103 sq mi/267.8 sq km), NW KYONGGI province, SOUTH KOREA, NW of SEOUL, near Pukhan Mountain. Area well known for high quality rice. Agriculture (barley, potatoes, sweet potatoes); animal husbandry (cows, hens, pigs); horticulture for metropolitan markets. Served by Kyong-ui railroad between Seoul and UIJU, and circular railroad around Seoul. Tourism (Pukhan Mountain, Yi dynasty ruins).

Koyasan, JAPAN: see KOYA.

Koycegiz, Turkish *Köycegcaroniz*, village, SW TURKEY, on N shore of Lake KOYCEGIZ, 25 mi/40 km SE of MUGLA, 10 mi/16 km from the MEDITERRANEAN coast; 36°57′N 28°40′E. Chromium, manganese, emery, asbestos mines; sesame, millet. Formerly called Yuksekkum.

Koycegiz, Lake (Turkish=*Köycegiz*) (□ 20 sq mi/52 sq km), SW TURKEY, 24 mi/39 km SE of MUGLA; 10 mi/16 km long, 3 mi/4.8 km wide. It drains by a small river to the MEDITERRANEAN SEA, 10 mi/16 km away.

Koyelga (kuh-yel-GAH), village (2004 population 1,940), central CHELYABINSK oblast, on the NE slope of the S URALS, SW Siberian Russia, on highway, 44 mi/71 km SW of CHELYABINSK; 54°39′N 60°55′E. Elevation 856 ft/260 m. Marble works.

Koygorodok (ko-ee-guh-ruh-DOK), village (2005 population 2,875), SW KOMI REPUBLIC, NE European Russia, on the SYSOLA RIVER, on road and railroad spur, 95 mi/153 km SSE of SYKTYVKAR; 60°26′N 51°01′E. Elevation 666 ft/202 m. In agricultural area; bakery. Wood processing. Phosphorite deposits.

Koygorodok, RUSSIA: see KOIGORODOK.

Koylovtsi, Bulgaria: see KOILOVTSI.

Koynare, Bulgaria: see KOINARE.

Koyp (KO-eep), peak (3,707 ft/1,130 m) in the N URAL Mountains, SE KOMI REPUBLIC, RUSSIA; 62°10′N 59°10′E. PECHORA River rises here.

Koysu, RUSSIA: see KOISU, river.

Koytash, UZBEKISTAN: see KOITASH.

Koyuk (KOI-yuhk), Inuit village (2000 population 297), W ALASKA, on SE SEWARD PENINSULA, at head of NORTON BAY, at mouth of KOYUK RIVER, 130 mi/209 km ENE of NOME; 64°55′N 161°09′W. Sometimes called Inglestat.

Koyuk River (KOI-yuhk), 120 mi/193 km long, W ALASKA; rises on SEWARD PENINSULA near 65°25′N 163°00′W; flows SE to NORTON BAY at KOYUK.

Koyukuk (KOI-yuh-kuhk), Native American village (2000 population 101), W ALASKA, on YUKON River at mouth of KOYUKUK RIVER, 110 mi/274 km W of TANANA; 64°53′N 157°43′W. Airfield.

Koyukuk River (KOI-yuh-kuhk), c.500 mi/805 km long, N central ALASKA; rises on S slope of BROOKS RANGE near 67°58′N 151°15′W; flows generally SW, past BETTLES, to YUKON River at KOYUKUK. Principal tributaries are ALATNA and John rivers. Placer gold deposits on upper course. Partially explored (1855) by H. T. Allen.

Koyulhisar, village, central TURKEY, near the N bank of KELKIT RIVER, and 60 mi/97 km NE of SIVAS; 40°17′N 37°51′E. Lead mines yield ore with traces of silver. Formerly Misaz.

Koyva River (KO-ee-vah), 112 mi/180 km long, PERM oblast, Russia; rises on the W slopes of the central URAL Mountains NW of KOSYA, flows S, past PROMYSLA and TËPLAYA GORA, and SW, past BISER, to the CHUSOVAYA River, 15 mi/24 km ESE of CHUSOVOY.

Koyvisto, RUSSIA: see PRIMORSK, LENINGRAD oblast.

Koyyeri, TURKEY: see SARIZ.

Koza, town, Far-North province, CAMEROON, 36 mi/58 km NW of MAROUA; 10°52′N 13°54′E.

Koza (KO-zah), town, East Muro county, WAKAYAMA prefecture, S HONSHU, W central JAPAN, on S KII PENINSULA, on KUMANO SEA, 63 mi/101 km S of WAKAYAMA; 33°30′N 135°49′E. Fishing port (dried fish).

Kozacha Lopan' (ko-ZAH-chah LO-pahn) (Russian *Kazach'ya Lopan'*), town (2004 population 10,600), N KHARKIV oblast, UKRAINE, on railroad, and 25 mi/40 km N of KHARKIV; 50°20′N 36°11′E. Elevation 574 ft/174 m. Broiler factory. Established in the 17th century, town since 1938.

Kozagawa (ko-ZAH-gah-wah), town, East Muro county, WAKAYAMA prefecture, S HONSHU, W central JAPAN, 61 mi/99 km S of WAKAYAMA; 33°31′N 135°49′E. Citrons. Koza Gorge is nearby.

Kozah, prefecture (2005 population 220,763), KARA region, N TOGO ⊙ LAMA-KARA; 09°35′N 01°10′E.

Kozakai (ko-ZAH-kah-ee), town, Hoi county, AICHI prefecture, S central HONSHU, central JAPAN, 37 mi/60 km S of NAGOYA; 34°47′N 137°21′E.

Kozaki (KO-zah-kee), town, Katori county, CHIBA prefecture, E central HONSHU, E central JAPAN, 19 mi/30 km N of CHIBA; 35°53′N 140°24′E.

Kozak Mountains, NW TURKEY, extend 25 mi/40 km NW from BERGAMA; rise to 4,390 ft/1,338 m. Antimony in W; lignite in E.

Kozakov (KO-zah-KOV), Czech *Kozákov*, mountain (2,441 ft/744 m) in LUSATIAN MOUNTAINS, VYCHODOCESKY province, N BOHEMIA, CZECH REPUBLIC, 5 mi/8 km ENE of TURNOV. Stone quarries, deposits of Bohemian garnets, agates, amethysts.

Kozákov, CZECH REPUBLIC: see KOZAKOV.

Kozan (KO-zahn), town, Sera county, HIROSHIMA prefecture, SW HONSHU, W JAPAN, 37 mi/60 km E of HIROSHIMA; 34°35′N 133°03′E.

Kozan, town, S TURKEY, 40 mi/64 km NE of ADANA. Wheat, barley, oats, cotton; 37°27′N 35°47′E. Has medieval church. Formerly called Sis.

Kozani (ko-ZAH-nee), prefecture (□ 2,372 sq mi/6,167.2 sq km), WEST MACEDONIA department, N central GREECE; ⊙ KOZANI; 40°20′N 21°43′E. Bounded N by FLÓRINA prefecture, E by CENTRAL MACEDONIA department, S by THESSALY department, and W by GREVENA and KASTORÍA prefectures. PINDOS mountain system in W; VERMION and PIÉRIA massifs in E; drained by ALIÁKMON RIVER (hydroelectric plant). Mainly agricultural, producing wheat, tobacco, wine. Saffron crocus cultivated. Timber, charcoal. Livestock products (milk, cheese, skins). Lignite mining (N of Kozani). Main centers are Kozani, PTOLEMAÏS, GREVENA, and SIATISTA. Formerly spelled Kozane.

Kozani (ko-ZAH-nee), city (2001 population 35,242), ⊙ KOZANI prefecture, WEST MACEDONIA department, N central GREECE, 65 mi/105 km WSW of THESSALONÍKI; 40°18′N 21°47′E. Elevation 2,360 ft/719 m. Agricultural trading center (wheat, barley, skins; livestock). Lignite mining to N; chromite mining at VOURINOS to S. Hydroelectric plants E on ALIÁKMON RIVER. Bishopric. Philippos Airport. Founded in Middle Ages; flourished under Turkish rule. Held by Germans during World War II. Formerly called Kozhani; formerly spelled Kozane.

Kozara Mountains (ko-ZAH-rah), in DINARIC ALPS, NW BOSNIA, BOSNIA AND HERZEGOVINA; c. 15 mi/24 km long NW-SE; partly bounded by Sava (W), Una (N), and VRBAS (SE) Rivers; 45°00′N 17°00′E. Highest peak (3,208 ft/978 m) is 17 mi/27 km NW of Banja Luka. KOZARA National Park protects 13 sq mi/34 sq km of the range. Scene of intense fighting during World War II.

Kozara National Park (ko-ZAH-rah) (□ 13 sq mi/33.8 sq km), N BOSNIA, BOSNIA AND HERZEGOVINA.

Kozats'ke (ko-SAHTS-ke) (Russian *Kazatskoye*), town (2004 population 15,600), central KHERSON oblast, UKRAINE, on right bank of DNIEPER River, at the Kakhovka Dam, 2.5 mi/4 km N of and across the river from NOVA KAKHOVKA; 46°47′N 33°20′E. Railroad station, river port. Reinforced concrete manufacturing. Established in 1782, town since 1960.

Koz-bunar, Bulgaria: see DULBOK IZVOR.

Kozel, UKRAINE: see MYKHAYLO-KOTSYUBYNS'KE.

Kozelets' (ko-ze-LETS), town, SW CHERNIHIV oblast, UKRAINE, 40 mi/64 km S of CHERNIHIV; 50°55′N 31°07′E. Elevation 347 ft/105 m. Raion center. Road junction. Flax; bricks; butter; automotive repair depot. Veterinary technical school. Has an 18th-century baroque cathedral and town hall. Known since 1098; Cossack company town (17th–19th centuries). Jewish community since the end of the 18th century, accounting for over 20% of the town's population at the beginning of the 20th century; reduced by the pogroms of 1905 and 1907, and the subsequent emigration, to 750 by 1926; wiped out by the Nazis in 1941.

Kozel'shchina, UKRAINE: see KOZEL'SHCHYNA.

Kozel'shchyna (ko-ZEL-shchi-nah) (Russian *Kozel'shchina*), town, SE POLTAVA oblast, UKRAINE, along railroad and highway, 22 mi/35 km NE of KREMENCHUK; 49°13′N 33°51′E. Elevation 406 ft/123 m. Raion center. Crushed stone gravel, building materials; feed mill; lignite deposits. Known since the 18th century, town since 1938.

Kozel'sk (kuh-ZYELSK), city (2005 population 19,785), E KALUGA oblast, central European Russia, on the ZHIZDRA RIVER (tributary of the OKA River), 45 mi/72 km SSW of KALUGA; 54°02′N 35°47′E. Elevation 446 ft/135 m. Railroad and highway junction; mechanical shops and industrial glassworks; men's clothing factory; bakery. Known since 1146 as part of the Chernigov-Severskiy principality; sacked by the Tatar-Mongols in 1246; part of LITHUANIA throughout the 14th century; passed back to Russia in 1408 and retaken by Lithuania in 1445; ceded to Moscow in 1494. Architectural monuments.

Kozenitsy, POLAND: see KOZIENICE.

Kozhani, Greece: see KOZANI.

Kozhanka (KO-zhahn-kah), town (2004 population 4,800), central KIEV oblast, UKRAINE, on road and railroad, 10 mi/16 km SW of FASTIV and 37 mi/60 km S of KIEV; 49°58′N 29°46′E. Elevation 659 ft/200 m. Sugar refinery, brick works. Established in the 14th century; town since 1972.

Kozhevnikovo (kuh-ZHEV-nee-kuh-vuh), village (2006 population 7,980), S TOMSK oblast, W SIBERIA, RUSSIA, on the OB' RIVER, on highway branch, 40 mi/ 64 km WSW of TOMSK; 56°15′N 83°58′E. Elevation 282 ft/85 m. In a flax-growing area.

Kozhikode, city (2001 population 880,247), ⊙ Kozhikode district, KERALA state, S INDIA, on the MALABAR coast of the ARABIAN SEA, 330 mi/531 km WSW of CHENNAI (Madras); 11°15′N 75°46′E. Once the leading port of S India, it declined in the 19th century but remains the center of India's timber trade. Coir, copra, coconuts, spices, tea, coffee, and rubber are exported. Manufacturing includes wood products, tiles, hosiery, soap, perfumes and cosmetics; shark and vegetable oil processing. KALLAYI or Kalli, just S (connected by canal system with rivers from inland teak and rosewood forests), is timber depot and has cotton milling and manufacturing of hosiery, plywood, coir rope and mats, ties, and electrical supplies. Marine fishery research station in N suburban area of WEST HILL. City has polytechnic school (chemical and electrical engineering). Climate is mild but humid. Extensive jack and mango groves disguise city's growing industrial character. Handicraft calico industry, to which city gave its name, has been important in production since the 17th century; now almost extinct. A center of trade with Arabia in 13th century CALICUT (as it was then called) was (1498) Vasco da Gama's first Indian port of call and soon became a center for European traders. Trading posts established 1511 by PORTUGESE (abandoned 1525), 1644 by ENGLISH, 1698 by FRENCH, and 1752 by Danes. Entire area passed to British rule in 1792; small coastal plot (loge) allotted to the French (1819). Has international airport. Many of its citizens employed as skilled laborers in W ASIA.

Kozhimugol', RUSSIA: see KOZHYM.

Kozhuf, mountain, MACEDONIA and GREECE: see KOZUF.

Kozhva (KOZH-vah), town (2005 population 3,315), E central KOMI REPUBLIC, NE European Russia, on the left bank of the PECHORA River opposite KANIN, on the North Pechora railroad, 145 mi/233 km NE of UKHTA; 65°06′N 57°03′E. Elevation 173 ft/52 m. Sawmilling; in reindeer-raising area. Formerly called Ust'-Kozhva.

Kozhym (KO-zhim), settlement, NE KOMI REPUBLIC, NE European RUSSIA, on the Kozhym River (tributary of the KOS'YU River, USA RIVER basin), on local road and the North Pechora railroad, 26 mi/42 km S of INTA; 65°43′N 59°31′E. Elevation 262 ft/79 m. Founded as a coal mining settlement and called Kozhimugol'; with the closing of the mine, the remaining population is mostly involved in lumbering.

Koziakas, mountain, Greece and Macedonia: see KOZYAK.

Kozienice (ko-zee-NEE-tse), Russian Kozenitsy, town (2002 population 19,036), Radom province, E central POLAND, 21 mi/34 km NE of RADOM, near the VISTULA River, railroad spur terminus. Horse-breeding station; flour milling, sawmilling, manufacturing of bricks, hats, wooden cartons. Kozienice Forest, Polish Puszcza Kozienicka, is SW.

Kozi Hradek, CZECH REPUBLIC: see TÁBOR.

Kozin, UKRAINE: see KOZYN, Rivne oblast; or KOZYN, Kiev oblast.

Kozi Vruh (KO-zee VRUHK), peak (8,491 ft/2,588 m) in RILA MOUNTAINS, SW BULGARIA; 42°11′N 23°37′E.

Kozjak (koz-yahk), mountain, on border between Serbia and former Yugoslav Republic of MACEDONIA, along left bank of PCINJA RIVER; rises to 4,212 ft/1,284 m 15 mi/24 km NE of KUMANOVO (Macedonia), 32 mi/51 km NE of SKOPJE.

Kozjak, Greece and Macedonia: see KOZYAK.

Kozlan, CZECH REPUBLIC: see KOZLANY.

Kozlany (KOZH-lah-NI), Czech Kožlany, German kozlan, town, ZAPADOCESKY province, W BOHEMIA,

CZECH REPUBLIC, 25 mi/40 km WSW of PRAGUE; 50°12′N 14°18′E. Manufacturing (earthenware). Eduard Beneš, second president of CZECHOSLOVAKIA, was born here in 1884.

Kozle (KO-zlee), Polish Kędzierzyn-Koźle, German Cosel, town in UPPER SILESIA, OPOLE province, S POLAND, port on the ODER River, just opposite KLODNICA RIVER mouth, near W end of Gliwice Canal, and 25 mi/40 km SSE of OPOLE (Oppeln). Sugar refining, paper and flour milling, metal and woodworking. N terminus of canal (under construction) between the Oder and the Danube River. In Seven Years War, withstood several Austrian sieges. Within Poland since 1945.

Kozlikha, RUSSIA: see SITNIKI.

Kozliv (koz-LEEV) (Russian Kozlov) (Polish Kozlów), town, central TERNOPIL' oblast, UKRAINE, on left tributary of STRYPA RIVER, and 11 mi/18 km W of TERNOPIL'; 49°33′N 25°21′E. Elevation 1,187 ft/361 m. Distillery; brick works; bakery. Known since 1467, town since 1961.

Kozlodui (ko-zlo-DOO-iee), city, MONTANA oblast, ⊙ Kozlodui obshtina (1993 population 23,959), NW BULGARIA, on the DANUBE River, 11 mi/18 km WNW of ORYAHOVO; 43°47′N 23°47′E. Vineyards; vegetables; fisheries. Atomic power station on the Danube is nearby. Monument marks the landing (1876) of Bulgarian revolutionary troops led by Hristo Botev against the Turks. Sometimes spelled Kozloduy.

Kozlodui (ko-zlo-DOI), island in the DANUBE River, N central BULGARIA; 43°47′N 23°44′E.

Kozlov, RUSSIA: see MICHURINSK.

Kozlov, UKRAINE: see YEVPATORIYA.

Kozlov, UKRAINE: see KOZLIV, Ternopil' oblast.

Kozlovets (koz-LOV-ets), village, LOVECH oblast, SVISHTOV obshtina, N BULGARIA, 7 mi/11 km S of Svishtov; 43°30′N 25°22′E. Grain, vegetables; livestock. Formerly known as Tursko Slivo.

Kozlovka (kuhz-LOF-kah), city (2005 population 13,080), NE CHUVASH REPUBLIC, central European Russia, landing on the KUYBYSHEV Reservoir on the VOLGA River, opposite VOLZHSK (4 mi/6 km to the N), near railroad (Tyurlema station), 60 mi/97 km SE of CHEBOKSARY, and 30 mi/48 km W of KAZAN'; 55°50′N 48°15′E. Elevation 341 ft/103 m. Motor vehicle factory; food processing (dairy, bakery). Made city in 1967.

Kozlovka (kuh-ZLOF-kah), village, NE MORDVA REPUBLIC, central European Russia, near highway and railroad, 16 mi/26 km SW of ARDATOV; 54°29′N 45°27′E. Elevation 656 ft/199 m. In hemp area.

Kozlovka (kuhz-LOF-kah), village (2006 population 3,330), E VORONEZH oblast, S central European Russia, on crossroads and railroad, 38 mi/61 km NW of BORISOGLEBSK; 50°51′N 40°27′E. Elevation 623 ft/189 m. In agricultural area (grains, vegetables, potatoes, sunflowers).

Kozlovo, RUSSIA: see NIZHNEANGARSK.

Kozlovshchina (ko-ZLOV-shchi-nah), Polish Kozlowszczyzna, urban settlement, GRODNO oblast, BELARUS, 16 mi/26 km N of SLONIM. Furniture factory; winery, creamery.

Kozlów, UKRAINE: see KOZLIV.

Kozlu, township, N TURKEY, 6 mi/9.7 km SW of ZONGULDAK; 41°23′N 31°44′E. Coal mines; power plant.

Kozludzha, Bulgaria: see SUVOROVO.

Kozluk, village (2000 population 27,109), SE TURKEY, 30 mi/48 km NW of SIIRT; 38°11′N 41°29′E. Grain. Formerly Hazo.

Kozmin (KOZ-meen), Polish Koźmin, German Koschmin, town, Kalisz province, W central POLAND, 45 mi/72 km SSE of POZNAŃ. Railroad junction. Manufact-

uring of agricultural machinery, flour milling, sawmilling; trades in cattle and grain. Sanatorium; monastery.

Koz'modem'yansk (kuhz-muh-deem-YAHNSK), city (2006 population 22,455), SW MARI EL REPUBLIC, E central European Russia, port on the right bank of the VOLGA River, near the mouth of the VETLUGA River (lumber floating), on highway branch, 65 mi/105 km SW of YOSHKAR-OLA; 56°20′N 46°34′E. Elevation 246 ft/74 m. Major lumber-trading center; sawmilling, woodworking, manufacturing (furniture, musical instruments, pipes), food processing (beer, butter). Founded in 1583 as a Muscovite stronghold; chartered in 1781.

Koz'modem'yansk (kuhz-muh-deem-YAHNSK), village, SE YAROSLAVL oblast, central European Russia, on the KOTOROSL' River, on local railroad and near highway, 7 mi/11 km SE, and under administrative jurisdiction of YAROSLAVL; 57°29′N 39°41′E. Elevation 370 ft/112 m. Swine breeding.

Koznitsa Pass (KOZ-neet-sah) (3,542 ft/1,080 m), central BULGARIA, between the central STARA PLANINA Mountains (N) and SREDNA GORA (S), 4 mi/6 km W of KLISURA; 42°43′N 24°23′E. Crossed by a highway to PIRDOP, connecting KARLOVO (E) and ZLATITSA (W) basins.

Kozova (ko-zo-VAH) (Russian Kozovo) (Polish Kozowa), town, W TERNOPIL' oblast, UKRAINE, on Koropets' River, 22 mi/35 km WSW of TERNOPIL', and 10 mi/16 km E of BEREZHANY; 49°26′N 25°09′E. Elevation 1,200 ft/365 m. Raion center. Dairy; sugar refining, feed and flour milling; woodworking, brickworking. Known since 1440, town since 1958.

Kozovo, UKRAINE: see KOZOVA.

Kozowa, UKRAINE: see KOZOVA.

Kozuchow (ko-ZOO-hoov), Polish Kož uchów, German Freystadt, town in LOWER SILESIA, Zielona Góra province, W POLAND, 14 mi/23 km SSE of ZIELONA GÓRA (Grünberg). Textile milling (linen, hemp, jute). Chartered 1291. One of six churches of Grace allowed Silesian Protestants under Treaty of ALTRANSTÄDT (1707). Established here in 18th century. In World War II, c.50% destroyed. Passed from GERMANY to Poland in 1945.

Kozuf (ko-ZUF), Serbo-Croatian Kožuf, mountain massif on Macedonian-Greek border, W of VARDAR (Axios) River; highest peak (7,013 ft/2,138 m) is 18 mi/29 km W of DJEVDJELIJA (MACEDONIA). Also called Kozuh or Kozhukh (Serbo-Croatian Kožuh); also spelled Kozhuf.

Kozuke, former province in central HONSHU, JAPAN; now GUMMA prefecture.

Kozuki (KO-zoo-kee), town, Sayo county, HYOGO prefecture, S HONSHU, W central JAPAN, 52 mi/84 km N of KOBE; 34°58′N 134°19′E.

Kozul'ka (KO-zool-kah), town (2005 population 8,620), SW KRASNOYARSK TERRITORY, SE SIBERIA, RUSSIA, on road and the TRANS-SIBERIAN railroad, 55 mi/89 km W of KRASNOYARSK; 56°09′N 91°23′E. Elevation 1,010 ft/307 m. In agricultural area; food processing. Founded in 1899 with the construction of the railroad; made town in 1962.

Kozushima (kod-ZOO-shee-mah), village, on KOZUSHIMA island, Oshima district, Tokyo prefecture, SE JAPAN, 31 mi/50 km S of SHINJUKU; 34°12′N 139°08′E.

Kozu-shima (KO-ZOO–shee-mah), island (□ 7 sq mi/18.2 sq km) of IZU-SHICHITO island group, Tokyo prefecture, SE JAPAN, in PHILIPPINE SEA, 35 mi/56 km SSW of O-SHIMA island; 3 mi/4.8 km long, 1 mi/1.6 km–3.5 mi/5.6 km wide. Hilly, with extinct volcano rising to 1,971 ft/601 m. Produces raw silk, camellia oil; fishing, farming. Formerly sometimes called Kantsu.

Kozyak, Greek Koziakas, mountain, on border of N central GREECE (WEST MACEDONIA department) and former Yugoslav Republic of MACEDONIA; rises to 5,951 ft/1,814 m; 11 mi/18 km NW of ARIDAIA (Greece),

Area is shown by the symbol □, and capital city or county seat by ⊙.

65 mi/105 km SSE of SKOPJE (Macedonia); 40°23′N 20°54′E. Formerly Kozjak.

Kozyatyn (ko-ZYAH-tin) (Russian *Kazatin*), city, N VINNYTSYA oblast, UKRAINE, 37 mi/60 km NNE of VINNYTSYA; 49°43′N 28°50′E. Elevation 938 ft/285 m. Raion center. Railroad junction; railroad servicing. Food processing (flour, meat, milk); sewing. Vocational technical school. Known since 1734, city since 1939. Site of battles of Ukrainian nationalist forces against the Bolsheviks during the Ukrainian war for independence (1919).

Kozyn (ko-ZIN) (Russian *Kozin*), town (2004 population 5,000), central KIEV oblast, UKRAINE, on right bank of DNIEPER (Ukrainian *Dnipro*) River, and 11 mi/18 km SSE of KIEV city center; 50°14′N 30°39′E. Elevation 291 ft/88 m. Sawmilling, woodworking, plastic parts manufacturing. Recreation grounds and youth camps for residents of Kiev. Known since the 11th century; town since 1958.

Kozyn (ko-ZIN) (Russian *Kozin*), village (2004 population 5,000), SW RIVNE oblast, UKRAINE, 16 mi/26 km SW of DUBNO; 50°16′N 25°28′E. Elevation 705 ft/214 m. Grain processing, tile manufacturing.

Kozyrëvsk (kuh-zi-RYOFSK), village (2005 population 1,420), N central KAMCHATKA oblast, RUSSIAN FAR EAST, on the KAMCHATKA RIVER, in a valley between the SREDINNYY Range (W) and KLYUCHEVSKAYA SOPKA (E), on local road, 33 mi/53 km SW of KLYUCHI; 56°04′N 159°52′E. Elevation 118 ft/35 m. Fisheries. Base camp for ecotourists and mountain climbers.

Kpadafe (BAH-dah-FAI), town, PLATEAUX region, SW TOGO, 10 mi/16 km SW of KPALIMÉ, on GHANA border; 06°51′N 00°36′E. Manioc and maize grown.

Kpalimé (BAH-lee-mai), town (2003 population 75,200) ⊙ KLOTO prefecture, PLATEAUX region, S TOGO, 3 mi/4.8 km SE of KLOUTO, and 65 mi/105 km NW of LOMÉ, at S end of the ATAKORA MOUNTAINS; 07°00′N 00°45′E. railroad terminus. Agricultural center (cacao, palm oil and kernels, and cotton); cotton ginning; agricultural station. Customhouse, hospital. Formerly spelled Palimé.

Kpandae (kuh-PAHN-dai), town, NORTHERN REGION, GHANA, 45 mi/72 km N of KETE-KRACHI, on main N-S road; 08°28′N 00°01′W. Shea nuts, durra, millet, yams.

Kpandu (kuh-PAHN-doo), town, VOLTA region, near VOLTA RIVER, 30 mi/48 km NNW of Ho; 07°00′N 00°18′E. An important ferry port on Lake VOLTA. Road access from Ho. Cacao center; pottery making, ivory and ebony carving. Also spelled Kpando.

Kpendjal, prefecture (2005 population 120,612), SAVANES region, N TOGO ⊙ MANDOURI; 10°52′N 00°37′E.

Kpessi (BE-see), town, PLATEAUX region, central TOGO, 40 mi/64 km N of ATAKPAMÉ, on MONO RIVER; 08°10′N 01°09′E. Small-scale irrigation of cotton.

Kpong (kuh-PAWNG), town, EASTERN REGION, GHANA, 6 mi/9.7 km NW of AKUSE; 06°09′N 00°04′E. Road junction. Cacao, palm oil and kernels, cassava.

Kra, THAILAND: see KRABURI.

Kraaifontein, town, WESTERN CAPE province, SOUTH AFRICA, on N1 highway, 18 mi/29 km ENE of CAPE TOWN; 33°50′S 18°44′E. Elevation 394 ft/120 m. Railroad junction; agriculture market (grain, fruit).

Kraainem, commune (2006 population 13,141), Halle-Vilvoorde district, BRABANT province, central BELGIUM, 6 mi/10 km ENE of BRUSSELS; 50°52′N 04°28′E.

Kraai River, SOUTH AFRICA: see ABERDEEN.

Kraai River, river, SOUTH AFRICA: see ABERDEEN.

Krabbe Creek, estuary, NETHERLANDS: see KEETEN.

Krabbendijke (KRAH-buhn-DEI-kuh), village, ZEELAND province, SW NETHERLANDS, in SOUTH BEVELAND region, 11 mi/18 km ESE of GOES; 51°26′N 04°07′E. EASTERN SCHELDT estuary 1 mi/1.6 km to NE, WESTERN SCHELDT 2 mi/3.2 km to SW. Dairying; cattle; vegetables, sugar beets.

Krabi (GRUH-BEE), province (□ 1,531 sq mi/3,980.6 sq km), S THAILAND, on W coast of the MALAY PENINSULA; ⊙ KRABI; 08°15′N 99°10′E. Rubber, palm oil,

coconuts, rice, fruit. Tin and coal mining. There are several beach resorts and four national parks.

Krabi (GRUH-BEE), town, ⊙ KRABI province, S THAILAND, in MALAY PENINSULA, port on Strait of MALACCA, 130 mi/209 km NW of SONGKHLA; 08°04′N 98°55′E. Coconuts; mining nearby (coal and tin). Linked by road with railroad station of HUAI YOT. Noppharat Beach here.

Krabin, THAILAND: see KABINBURI.

Kraburi (GRUH-BUH-REE), village and district center, RANONG province, S THAILAND, in ISTHMUS OF KRA, 30 mi/48 km WSW of CHUMPHON (linked by road across isthmus), at head of PAKCHAN RIVER estuary (MYANMAR border); 10°24′N 98°47′E. Sometimes called KRA.

Krâchéh, CAMBODIA: see KRATIE, province, or KRATIE, city.

Krachen, CAMBODIA: see KRATIE, province, or KRATIE, city.

Krachi, GHANA: see KETE-KRACHI.

Kragerø (KRAH-guh-ruh), city and port, TELEMARK county, S NORWAY, at mouth of TOKKE River, on small inlet of the SKAGERRAK, 70 mi/113 km NE of KRISTIANSAND; 58°52′N 09°25′E. Lumbering center, with shipyards, sawmills, and mechanical industries, and manufacturing of wood pulp and knit goods. Major summer tourist resort with harbor. Founded in seventeenth century.

Kraggenburg (KRAH-khuhn-burkh), village, FLEVOLAND province, N central NETHERLANDS, 20 mi/32 km ENE of LELYSTAD, in SE part of NORTH-EAST POLDER (formerly part of OVERIJSSEL province); 52°40′N 05°54′E. ZWARTEMEER to SE. Cattle; apples, sugar beets, wheat, corn, vegetables, potatoes.

Kragujevac (KRAH-goo-ye-vahts), city (2002 population 175,802), central SERBIA, on railroad, and 60 mi/97 km SSE of BELGRADE; 44°01′N 20°55′E. Economic and cultural center of the SUMADIJA region. Largest motor vehicle and munitions industry in Serbia; vegetable canning, flour milling; ships plums. Has suffered heavy decline in employment and production since the breakup of YUGOSLAVIA. Was first mentioned in 17th century; seat (1818–1839) of Milosh Obrenovich I and the cultural and political center of Serbia. Also spelled Kraguyevats.

Krai, Malaysia: see KUALA KRAI.

Kraiburg am Inn (KREI-burg ahm IN), village, BAVARIA, S GERMANY, in UPPER BAVARIA, on the INN RIVER, and 6 mi/9.7 km SW of MÜHLDORF; 48°11′N 12°26′E. Tourism.

Kraichbach (KREIKH-bahkh), river, 35 mi/56 km long, in BADEN-WÜRTTEMBERG, SW GERMANY; rises 1.5 mi/2.4 km SE of SULZFELD; flows NW, past HOCKENHEIM, to an arm of the RHINE RIVER, just SW of KETSCH; source at 49°02′N 08°50′E.

Kraichtal (KREIKH-tahl), town, BADEN-WÜRTTEMBERG, SW GERMANY, 15 mi/24 km NE of KARLSRUHE; 49°08′N 08°40′E. Manufacturing of plastics, precision instruments, and textiles. Has 16th-century castle and many half-timbered houses; chartered in 1971 after unification of GOCHSHEIM and Unteröwisheim.

Krain, SLOVENIA: see KRANJSKA.

Krainitsi (krai-NEET-sah), village, SOFIA oblast, DUPNITSA obshtina, BULGARIA; 42°19′N 23°12′E.

Krainka (KRAH-een-kah), settlement, W TULA oblast, central European Russia, on the OKA River, near highway, 8 mi/13 km W, and under administrative jurisdiction, of SUVOROV; 54°06′N 36°17′E. Elevation 482 ft/146 m. Health resort (mineral springs); sanatorium, summer homes. Mineral water plant.

Kraishte (KRAISH-te), Serbo-Croatian *Krajište*, highland, in W BULGARIA and SERBIA, between BESNA KOBILA mountain (W) and upper STRUMA River (E), S of TRUN; includes MILEVSKA Mountains; 42°41′N 22°28′E. Scattered magnetite, galena, zinc, gold, and lignite deposits. Highest area in SERBIA; rises to 5,686 ft/1,733 m in the Milevska Mountains, on

the border. Deforested; little vegetation. Also spelled Krayishte.

Kraissos, Gulf of, inlet of Gulf of CORINTH, off PHOCIS prefecture, CENTRAL GREECE department; 8 mi/12.9 km wide, 8 mi/12.9 km long. ITEA (port of AMPHISSA) on N shore; GALAXIDHI W shore. Also called Gulf of Amphissa, Galaxidhi, Itea, Salona, and Crisa.

Kra, Isthmus of (GRUH), narrow neck of the MALAY PENINSULA, c.40 mi/64 km wide, SW THAILAND, between the Bay of BENGAL (W) and the GULF OF THAILAND (E). It has long been the proposed site of a ship canal that would bypass the congested STRAIT OF MALACCA.

Krajenka (krah-YEN-kah), German *Krojanke*, town in POMERANIA, Piła province, NW POLAND, 15 mi/24 km NE of Piła (Schneidemühl). Grain and cattle market; woodworking. Until 1938, in former Prussian province of Grenzmark Posen–Westpreussen.

Krajište, BULGARIA and SERBIA: see KRAISHTE.

Krak, JORDAN: see KARAK, AL.

Krakatau (KRAH-kah-tou), volcanic island (c.5 sq mi/13 sq km), S Lampung province, W INDONESIA, in SUNDA STRAIT between JAVA and SUMATRA; 06°07′S 105°24′E. Rising to 2,667 ft/813 m. A terrific volcanic explosion in 1883 blew up most of the island and altered the configuration of the strait; the accompanying tsunami caused great destruction and killed c.36,000 people along the nearby coasts of Java and Sumatra. The explosion is classed as one of the largest volcanic eruptions in modern times; so great was the outpouring of ashes and lava that new islands were formed and debris was scattered across the INDIAN OCEAN as far as MADAGASCAR. Eruptions in 1928 resulted in the appearance of ANAK KRAKATAU, Rakata, Rakata Kecil, and Sertung islands. Included in the UJUNG KULON NATIONAL PARK in 1980. Also spelled Krakatoa.

Krakatau, Anak, INDONESIA; see ANAK KRAKATAU.

Krakatoa, INDONESIA: see KRAKATAU.

Krakau, POLAND: see CRACOW.

Krakor (KRAW-KAW), town, PURSAT province, central CAMBODIA, on road, and 22 mi/35 km ESE of PURSAT, near TONLÉ SAP lake; 12°32′N 104°12′E. Rice-growing area; fisheries. Food processing. Khmer population with Cham minority.

Krakovets' (krah-ko-VETS), (Polish *Krakowiec*), town (2004 population 8,300), W L'VIV oblast, UKRAINE, on the Polish border, 10 mi/16 km W of YAVORIV; 49°58′N 23°10′E. Elevation 659 ft/200 m. Sawmilling; brick making. Known since the beginning of the 15th century, town since 1940. Jewish community since 1640, eliminated in 1941–1942.

Kraków, POLAND: see CRACOW.

Krakowiec, UKRAINE: see KRAKOVETS'.

Kraksaan (KRAHK-sah-ahn), town, East Java province, INDONESIA, on MADURA Strait, 60 mi/97 km SE of SURABAYA; 07°46′S 113°25′E. Trade center for agricultural area (sugar, rice, corn). Also spelled Kraksan or Krakshan.

Kralanh (KRAW-LUHN), town, W CAMBODIA, near border of SIEM REAP and BANTEAY MEAN CHEAY provinces, border town at major road junction; 13°35′N 103°15′E.

Kralendijk (KRAH-luhn-DEIK), chief town of BONAIRE, NETHERLANDS ANTILLES, on W inlet of the island, guarded by LITTLE BONAIRE island, 45 mi/72 km E of WILLEMSTAD, CURAÇAO; 12°09′N 68°16′W. Beachfront harbor. Fishing; shipbuilding; salt, aloe. The region has a desalinization plant, petroleum storage, and a wildlife sanctuary established on former plantations.

Kralice na Hane (KRAH-li-TSE nah HAH-ne), Czech *Kralice na Hané*, German KRALITZ, village, JIHOMORAVSKY province, central MORAVIA, CZECH REPUBLIC, just E of PROSTĚJOV, in the HANA region; 49°28′N 17°11′E. The main Bohemian version of the Bible was first printed here in the 16th century.

Kralicky Sneznik (KRAH-lits-KEE SHYEZH-nyeek), Czech *Králický Snéžník*, German *glatzer schneegebirge*, Polish *góry klodzkie*, mountain group of the SUDETES, in NE BOHEMIA and W CZECH SILESIA, on CZECH REPUBLIC-POLAND border, NW of STARE MESTO (Czech Republic); highest peak (4,669 ft/1,423 m) is KRALICKY SNEZNIK, German *Glatzer Schneeberg*, Polish *snieznik*. Important watershed; GLATZER NEISSE and MORAVA rivers rise on S slope.

Kraliky (KRAH-lee-KI), Czech *Králíky*, German *gru-lich*, town, VYCHODOCESKY province, E BOHEMIA, CZECH REPUBLIC, in SE foothills of the Eagle Mountains, 34 mi/55 km ENE of PARDUBICE, near Polish border; 50°05'N 16°46'E. Railroad junction. Woodworking; manufacturing (toys, bulbs); agriculture (oats).

Kralingen (KRAH-ling-uhn), suburb of ROTTERDAM, SOUTH HOLLAND province, W NETHERLANDS, on NEW MAAS RIVER, 2 mi/3.2 km of city center; 51°55'N 04°30'E.

Kralitz, CZECH REPUBLIC: see KRALICE NA HANE.

Kraljevica (KRAHL-ye-vee-tsah), Italian *Porto Re*, village, W CROATIA, on ADRIATIC SEA, on KVARNER gulf, 8 mi/12.9 km ESE of RIJEKA (Fiume), on channel leading to Bay of Bakar. Shipbuilding; fishing. Seaside resort near wooded hills; sanatorium. Has 2 castles.

Kraljevo (KRAHL-ye-vo), city (2002 population 121,707), central SERBIA, on IBAR RIVER, just above its mouth, and 75 mi/121 km S of BELGRADE; 43°44'N 20°43'E. Railroad junction; manufacturing (railroad cars, aircraft); apple growing. Chromium and magnesite mining (magnesite-brick manufacturing), lead smelter nearby. Monastery of Zica SW of town. Until 1949, called Kraljevo or Kralyevo, then Rankovicevo until the mid-1960s, when the earlier name was restored.

Kralova Dam, SLOVAKIA: see SOPORNA.

Kralova hola (krah-LYO-vah ho-LYAH), Slovak *Kráľova hoľa*, HUNGARIAN *Király Hegy*, second-highest peak (6,391 ft/1,948 m) of the LOW TATRAS, STREDOSLOVENSKY province, central SLOVAKIA, 23 mi/37 km NW of ROZNAVA; 48°53'N 20°09'E. HRON RIVER rises on SE, CIERNY VÁH RIVER on N slopes.

Kralovany (krah-LYO-vah-NI), Slovak *Kraľovany*, Hungarian *Kralován*, village, STREDOSLOVENSKY province, N SLOVAKIA, on VÁH RIVER, at ORAVA RIVER mouth, and 19 mi/31 km ESE of ŽILINA; 49°09'N 19°08'E. Railroad junction; excursion center. Summer and winter health resort of LUBOCHNA (lu-BOKH-nyah) (elevation of 1,460 ft/445 m), Slovak *Lubochňa*, is c.3 mi/4.8 km SE, along left bank of Váh River.

Kralovice (KRAH-lo-VI-tse), German *Kralowitz*, town, ZAPADOCESKY province, W BOHEMIA, CZECH REPUBLIC, on railroad, and 17 mi/27 km NNE of PLZEŇ; 49°59'N 13°29'E. In a barley- and oat-growing region; manufacturing (optics); food processing (milk).

Královo nad Tisou, UKRAINE: see KOROLEVE.

Kralovo Pole (KRAH-lo-VO PO-le), Czech *Královo Pole*, German *königsfeld*, N district of BRNO, JIHO-MORAVSKY province, S MORAVIA, CZECH REPUBLIC. Manufacturing (machinery).

Kralovsky Chlmec (krah-LOU-skee khuhl-METS), Slovak *Kráľovský Chlmec*, Hungarian *Királyhelmec*, town, VYCHODOSLOVENSKY province, SE SLOVAKIA, on railroad, and 38 mi/61 km SE of KOŠICE; 48°25'N 21°59'E. Agricultural center (wheat, sugar beets, corn, grapes). Manufacturing of machinery, furniture; food processing. Has 14th-century church, castle ruins, and 8th-century Slavonic tumulus. Under Hungarian rule from 1938–1945.

Kralup an der Moldau, CZECH REPUBLIC: see KRALUPY NAD VLTAVOU.

Kralupy nad Vltavou (KRAH-lu-PI nahd VUHL-tah-VOU), German *Kralup an der Moldau*, town, STREDOCESKY province, N central BOHEMIA, CZECH REPUBLIC, on VLTAVA RIVER, and 12 mi/19 km NNW

of PRAGUE; 50°14'N 14°19'E. Railroad hub. Large oil refinery; manufacturing (dyes, rubber, plastic, drugs); beer brewing; food processing.

Kraluv Dvur (KRAH-loof DVOOR), Czech *Králův Dvůr*, German *Königshof*, suburb of BEROUN, STREDOCESKY province, central BOHEMIA, CZECH REPUBLIC; 49°56'N 14°03'E. Iron foundries, steel mills, large cement works.

Kralyevo, SERBIA: see KRALJEVO.

Kramators'k (krah-mah-TORSK), city (2001 population 181,025), N DONETS'K oblast, UKRAINE, in the DONBAS, on the Kazennyy Torets' River, right tributary of the DONETS RIVER; 48°43'N 37°32'E. Elevation 200 ft/60 m. Major center of heavy machinery manufacturing (ceramics and coal derivatives, building materials, food, apparel; casting and forging). Scientific research institute of machine building, industrial institute, two technical schools; local museum. Established in 1868, city since 1932.

Kramer, village (2006 population 41), BOTTINEAU co., N NORTH DAKOTA, near SOURIS, or Mouse, River, 15 mi/24 km SW of BOTTINEAU; 48°41'N 100°42'W. J. Clark Salyer National Wildlife Refuge to W and S. Founded in 1905 and incorporated in 1908.

Kramfors (KRAHM-FORSH), town, VÄSTERNORR-LAND county, NE SWEDEN, on ÅNGERMANÄLVEN RIVER estuary, 20 mi/32 km NNW of HÄRNÖSAND; 62°56'N 17°48'E. Airport.

Krammer (KRAH-muhr), channel, SW NETHERLANDS, part of MAAS RIVER estuary, formed 12 mi/19 km NW of ROOSENDAAL as continuation of VOLKERAK channel from NE, at joining of Steenbergse Vliet River from E, flows 7 mi/11.3 km W. Divides into Grevelingen estuary (now blocked by Grevelingen Dam, forming GREVELINGENMEER, W) and MASTGAT channel (SSW), blocked by Philips Dam. Scheldt-Rhine canal enters from S. Forms part of boundary between SOUTH HOLLAND province (N) and NORTH BRABANT and ZEELAND provinces (S). Maximum width 3 mi/4.8 km.

Kramolin (kra-mo-LEEN), village, LOVECH oblast, SEVLIEVO obshtina, BULGARIA; 43°08'N 25°05'E. Wine producing.

Kramsach (KRAHM-sakh), village, TYROL, W AUSTRIA, on the INN River and 26 mi/42 km NE of INNSBRUCK; 47°27'N 11°52'E. Glass manufacturing; summer resort.

Kranach, GERMANY: see KRONACH.

Kranenburg (KRAH-nen-burg), village, North Rhine-Westphalia, W GERMANY, 5 mi/8 km W of KLEVE; customs station near Dutch border; 51°48'N 06°01'E. Until 1936, spelled CRANENBURG.

Kranevo (KRAH-ne-vo), village, VARNA oblast, BALCHIK obshtina, BULGARIA, on the BLACK SEA. Resort; 43°21'N 28°04'E.

Krångede (KRONG-ED-e), locality, JÄMTLAND county, N central SWEDEN, on INDALSÄLVEN RIVER, 35 mi/56 km W of SOLLEFTEÅ; 63°08'N 16°03'E. Major hydro-electric station.

Kranichfeld (KRAH-nikh-felt), town, THURINGIA, central GERMANY, on ILM RIVER, 10 mi/16 km SW of WEIMAR; 50°53'N 11°12'E. Woodworking. Has two medieval castles.

Kranidion (krah-NEE-[th]ee-on), town, ARGOLIS prefecture, E PELOPONNESE department, S mainland GREECE, on ARGOLIS PENINSULA, 23 mi/37 km SE of NÁVPLION; 37°23'N 23°09'E. Olive-oil production; fisheries and sponge fishing. Also Kranidhion.

Kranj (KRAHN-ye), ancient *Carnium*, city (2002 population 35,237), N SLOVENIA, on Sava River, and 15 mi/24 km NNW of LJUBLJANA; 46°14'N 14°21'E. railroad junction; manufacturing of woolen textiles, electronic equipment, leather. Ljubljana International Airport to E. Site of former castle of dukes of Carniola.

Kranjska (KRAHN-ye-skah), Latin *Carniola*, historic region, NW and central SLOVENIA. The first known inhabitants, a Celtic tribe called the Carni, were displaced by the Romans, who made Carniola part of their province of Pannonia. Slovenes settled here in

the 6th century Later incorporated into Charlemagne's empire. Under Austrian Hapsburg rule (1282–1918); made (1364) a titular duchy and, eventually, a crown land (1849). LJUBLJANA was its chief city. Divided after World War I between ITALY and the former YUGOSLAVIA, but the Italian part passed to the former Yugoslavia in 1947.

Kranjska Gora (KRAHN-ye-skah GO-rah), village, NW SLOVENIA, on the SAVA DOLINKA RIVER, on railroad, 25 mi/40 km WNW of JESENICE and 2.5 mi/4 km E of Rateče, at N foot of JULIAN ALPS; 46°29'N 13°47'E. Part of a resort area especially known for winter sports.

Krannon (krah-NON), ancient town in what is now central THESSALY department, N GREECE, 13 mi/21 km SW of modern LÁRISSA; 39°31'N 22°19'E. One of leading Thessalian cities, rivaling Lárissa under the Scopadae family. Here Antipater won his final victory (322 B.C.E.) over the League of Cities in Central Greece in Lamian War. On site is modern Greek village of Krannon or Kranon (1940 population 296). Also Crannon.

Kranz, RUSSIA: see ZELENOGRADSK.

Krapanj Island (□ 4/10 sq mi/1 sq km), S CROATIA, in DALMATIA, in ADRIATIC SEA, near SIBENIK. Tourism. Known in the past for its corals and sea sponges.

Krapina (KRAH-pee-nah), town, central CROATIA, on railroad, and 25 mi/40 km N of ZAGREB, near SLOVENIA border; 46°10'N 15°50'E. Local trade center; manufacturing (textiles, apparel, ceramic tile); sulphur deposits. Chief town in HRVATSKO ZAGORJE region. Former lignite mine. First mentioned in 1193. Krapinske Toplice, 5 mi/8 km SSW, is health resort with mineral springs. Nearby is Hušnjakovo Cave Nature Monument, where skeletal hominid fossils from 50,000–200,000 years ago were found (1899).

Krapina River, c.40 mi/64 km long, central CROATIA; rises 11 mi/18 km SSW of Varaždin; flows SW to Sava River 12 mi/19 km SE of Zaprešić.

Krapinske Toplice [Croatian, *toplice*=spa], resort town, central CROATIA, in HRVATSKO ZAGORJE region, 15.5 mi/24.9 km N of ZAGREB. Tourism.

Krapivinskiy (krah-PEE-veen-skeeyee), town (2005 population 7,845), central KEMEROVO oblast, S central SIBERIA, RUSSIA, in the KUZNETSK BASIN, on the TOM' River, on road, 30 mi/48 km NE of LENINSK-KUZNETSKIY, and 27 mi/43 km E of KEMEROVO; 55°03'N 86°48'E. Elevation 416 ft/126 m. Coal mining; logging and lumbering.

Krapivna (krah-PEEV-nah), village, central TULA oblast, W central European Russia, on the UPA RIVER, on crossroads, 25 mi/40 km SW of TULA, and 15 mi/24 km WSW of SHCHEKINO, to which it is administratively subordinate; 53°56'N 37°09'E. Elevation 518 ft/157 m. In agricultural area (wheat, oats, hemp, vegetables; livestock); confectionery, bakery.

Krapkowice (krahp-ko-VEE-tse), German *Krappitz*, town in UPPER SILESIA, after 1945 in OPOLE province, SW POLAND, on the ODER and 13 mi/21 km S of OPOLE (Oppeln). Cellulose manufacturing, sawmilling, limestone quarrying. After 1945, briefly called Chrapkowice.

Krapotkin, RUSSIA: see KROPOTKIN, KRASNODAR Territory, or KROPOTKIN, IRKUTSK oblast.

Krappfeld (KRAHP-feld), CARINTHIA, S AUSTRIA, fertile region between ALTHOFEN and SANKT VEIT AN DER GLAN, drained by GURK RIVER.

Krappitz, POLAND: see KRAPKOWICE.

Kra River, THAILAND and MYANMAR: see PAKCHAN RIVER.

Kras (KRAHS), Italian *Carso*, German *Karst*, limestone plateau, in the DINARIC ALPS, SW Slovenia, N of ISTRIA region, and extending c.50 mi/80 km SE from the lower SOCA River valley. Characterized by deep gullies, caves, sinkholes, and underground drainage, all the result of carbonation-solution. The best-known caves are at POSTOJNA (see POSTOJNA CAVES). The barren nature of the plateau deters human settlement. Rough pasture or forest covers much of the surface, and there is little arable land. The term *karst* is used to describe any area where similar geological formations are found.

Krasavino (krah-SAH-vee-nuh), city (2006 population 7,815), NE VOLOGDA oblast, N central European Russia, on the LESSER NORTHERN DVINA RIVER, on road and railroad, 400 mi/644 km NW of VOLOGDA, and 14 mi/23 km NNE of VELIKIY USTYUG; 60°57′N 46°29′E. Elevation 193 ft/58 m. One of the oldest textile centers of RUSSIA; flax processing. Made city in 1947, but administratively subordinate to VELIKIY USTYUG.

Krašić (KRAH-sheech), village, central CROATIA, in foothills of Žumberačka gora, 10 mi/16 km N of KARLOVAC.

Krasilov, UKRAINE: see KRASYLIV.

Krasilow, UKRAINE: see KRASYLIV.

Krasivka (krah-SEEF-kah), village, E TAMBOV oblast, S central European Russia, on the VORONA RIVER, on road and near railroad, 25 mi/40 km SSW of KIRSANOV; 52°16′N 42°31′E. Elevation 465 ft/141 m. In agricultural area (grains, hemp, feed corn).

Kraskino (KRAHS-kee-nuh), town (2006 population 3,340), SW MARITIME TERRITORY, SE RUSSIAN FAR EAST, on the Expedition Bay (POSYET BAY) in the Peter the Great Gulf of the Sea of JAPAN, on the coastal highway, 175 mi/282 km SW of VLADIVOSTOK; 42°42′N 130°47′E. Terminus of a branch of the TRANS-SIBERIAN RAILROAD; fisheries; food processing. Formerly known as Novokiyevsk.

Kraskino (KRAHS-skee-nuh), camp on the S coast of the S island of NOVAYA ZEMLYA, NENETS AUTONOMOUS OKRUG, ARCHANGEL oblast, NE European RUSSIA; 70°45′N 54°30′E. Trading post.

Kraskovo (KRAHS-kuh-vuh), town (2006 population 11,805), central MOSCOW oblast, central European Russia, near the MOSKVA River, on road and railroad, 24 mi/39 km SE of MOSCOW, and 4 mi/6 km E of LYUBERTSY, to which it is administratively subordinate; 55°39′N 37°59′E. Elevation 465 ft/141 m. Manufacturing (industrial scales and electronic precision instruments, construction materials, uniforms). Town status since 1961.

Krāslava (KRAH-slah-vah), German *krasslau*, city (2000 population 11,412), SE LATVIA, in LATGALE, on right bank of the DVINA (DAUGAVA) River, and 25 mi/40 km E of DAUGAVPILS; 55°54′N 27°10′E. Flour milling, wool processing; rye, flax. In Russian VITEBSK government until 1920. Until 1917, known as KRESLAVKA in Russian.

Kraslice (KRAHS-li-TSE), German *Graslitz*, town, ZAPADOCESKY province, W BOHEMIA, CZECH REPUBLIC, in ORE MOUNTAINS, on railroad and 18 mi/29 km WNW of KARLOVY VARY, near German border opposite KLINGENTHAL (GERMANY); 50°20′N 12°31′E. Manufacturing (light manufacturing, lace, embroidery, woolen textiles).

Krasna, CZECH REPUBLIC: see AS.

Krasna Horka, SLOVAKIA: see ROZNAVA.

Krasna Lipa (KRAHS-nah LEE-pah), Czech *Krásná Lípa*, German *Schönlinde*, town, SEVEROCESKY province, N BOHEMIA, CZECH REPUBLIC, in W SUDETES, 27 mi/43 km NE of ÚSTÍ NAD LABEM. Railroad junction. Manufacturing (machinery, medical products). Founded in 13th century by German colonists.

Krasnaya [Russian=red], in Russian names: see also KRASNO- [Russian combining form], KRASNOYE, KRASNY, KRASNYE.

Krasnaya Armiya Strait (KRAHS-nah-yah–AHR-mee-yah) [Russian=Red Army], 90 mi/145 km long, 5 mi/8 km–10 mi/16 km wide; joins KARA and LAPTEV seas of the ARCTIC OCEAN, at approximately 80°00′N 92°00′E–80°00′N 98°00′E. In KRASNOYARSK TERRITORY, N central Siberian Russia; separates KOMSOMOLETS and PIONER islands from OKTYABRSKAYA

REVOLYUTSIYA ISLAND of the SEVERNAYA ZEMLYA archipelago.

Krasnaya Baltika, RUSSIA: see GOSTILITSY.

Krasnaya Gora (KRAH-snah-yah guh-RAH) [Russian=red mountain], town (2005 population 6,425), W BRYANSK oblast, central European Russia, near highway, approximately 12 mi/19 km SE of the RUSSIA-BELARUS border, 30 mi/48 km NW of KLINTSY; 53°04′N 31°36′E. Elevation 426 ft/129 m. In agricultural area (grains, flax, potatoes, vegetables; livestock); food processing, flour milling.

Krasnaya Gora, RUSSIA: see KRASNOGORSK, MOSCOW oblast.

Krasnaya Gorbatka (KRAHS-nah-yah guhr-BAHT-kah), town (2006 population 9,155), E VLADIMIR oblast, central European Russia, on crossroads and railroad, 23 mi/37 km NNW of MUROM; 55°53′N 41°45′E. Elevation 396 ft/120 m. Machinery production, sawmilling.

Krasnaya Gorka (KRAHS-nah-yah GOR-kah), town (2006 population 1,080), SE NIZHEGOROD oblast, central European Russia, on the SURA River where it makes the natural border with CHUVASH REPUBLIC, on highway junction, 27 mi/43 km ESE of SERGACH; 55°23′N 46°06′E. Elevation 360 ft/109 m. In agricultural area (sunflowers, flax, hemp).

Krasnaya Gorka (KRAHS-nah-yah GOR-kah), village (2005 population 4,085), NE BASHKORTOSTAN Republic, E European Russia, on the UFA RIVER (landing), 45 mi/72 km NE of UFA, in woodland; 56°08′N 54°09′E. Elevation 219 ft/66 m. Terminus of a local highway branch. Lumbering.

Krasnaya Gorka (KRAHS-nah-yah GOR-kah), village, central MOSCOW oblast, central European Russia, on the MOSCOW CANAL, on highway, 9 mi/14 km WNW, and under administrative jurisdiction, of MYTISHCHI; 55°59′N 37°31′E. Elevation 567 ft/172 m. In agricultural area (poultry).

Krasnaya Gorka, RUSSIA: see KRASNOGORSKIY.

Krasnaya Kamenka, UKRAINE: see CHERVONA KAMYANKA.

Krasnaya Luka (KRAHS-nah-yah LOO-kah), village, E central NIZHEGOROD oblast, central European Russia, on the Sundovik River (right tributary of the VOLGA RIVER), on road, 42 mi/68 km ESE of NIZHNIY NOVGOROD, and 3 mi/5 km W of LYSKOVO, to which it is administratively subordinate; 56°01′N 44°57′E. Elevation 393 ft/119 m. In agricultural area; poultry plant.

Krasnaya Pakhra (KRAHS-nah-yah pah-HRAH), village, S central MOSCOW oblast, central European Russia, on the PAKHRA River (MOSKVA River basin), 23 mi/37 km SW of MOSCOW, and 10 mi/16 km W of PODOL'SK, to which it is administratively subordinate; 55°26′N 37°17′E. Elevation 485 ft/147 m. Mechanical repair plant. In agricultural area (produce, poultry, fruit).

Krasnaya Polyana (KRAHS-nah-yah puh-LYAH-nah), town (2005 population 7,750), SE KIROV oblast, E central European Russia, on the VYATKA River, on road and railroad, 6 mi/10 km W of SOSNOVKA, and 3 mi/5 km E of VYATSKIYE POLYANY, to which it is administratively subordinate; 56°14′N 51°08′E. Elevation 223 ft/67 m. Gas pipeline service station, oil storage facilities; building materials.

Krasnaya Polyana (KRAHS-nah-yah puh-LYAH-nah), town (2005 population 4,040), SE KRASNODAR TERRITORY, S European Russia, on the S slope of the NW Greater CAUCASUS, 6 mi/10 km N of the Russia-Georgia border, near the MZYMTA RIVER, on road, 32 mi/51 km NNE of ADLER (connected by road), and 17 mi/27 km N of SOCHI; 43°40′N 40°12′E. Climatic mountain resort (elevation 2,562 ft/780 m). Hydroelectric station, wood combine. Mineral springs nearby. Tourist base.

Krasnaya Polyana (KRAHS-nah-yah puh-LYAH-nah), former town, central MOSCOW oblast, central European Russia, on railroad, 20 mi/32 km NNW of

MOSCOW. Railroad; cotton mill (since 1848); electronic equipment, ceramics. Incorporated into LOBNYA in 1975.

Krasnaya Polyana (KRAHS-nah-yah puh-LYAH-nah), village (2006 population 3,730), S ROSTOV oblast, S European Russia, on the YEGORLYK RIVER, on local road junction, 21 mi/34 km S of SAL'SK; 46°07′N 41°30′E. Elevation 108 ft/32 m. Agricultural products.

Krasnaya Polyana (KRAHS-nah-yah puh-LYAH-nah), settlement (2005 population 3,105), E KRASNODAR TERRITORY, S European Russia, on the KUBAN' River, on road, 5 mi/8 km N, and under administrative jurisdiction, of ARMAVIR; 45°03′N 41°05′E. Elevation 518 ft/157 m. In agricultural area (wheat, flax; livestock).

Krasnaya Poyma (KRAHS-nah-yah PO-ee-mah), village (2006 population 3,105), SE MOSCOW oblast, central European Russia, on the OKA River, near highway, 11 mi/18 km E of KOLOMNA; 55°04′N 39°05′E. Elevation 370 ft/112 m. Agricultural products.

Krasnaya Rechka (KRAHS-nah-yah–RYECH-kah), former town, S KHABAROVSK TERRITORY, RUSSIAN FAR EAST, on the TRANS-SIBERIAN RAILROAD, 8 mi/13 km S of KHABAROVSK, into which it has been incorporated; 48°21′N 135°04′E. Elevation 216 ft/65 m.

Krasnaya Shapochka (KRAHS-nah-yah SHAH-puhch-kah) [Russian=little red hat], hill on the E slopes of the central URAL Mountains, SVERDLOVSK oblast, W Siberian Russia; 60°10′N 59°50′E. Bauxite mining was begun here around 1940, at SEVEROURAL'SK.

Krasnaya Sloboda (KRAHS-nei-ah SLUH-bud-ah), urban settlement, NE AZERBAIJAN; 41°22′N 48°34′E. Orchards; fruit canning.

Krasnaya Sloboda (KRUH-nah-yah slo-bo-DUH), urban, MINSK oblast, BELARUS, on Vyznianka River (Pripiat basin), and 20 mi/32 km SW of SLUTSK; 52°51′N 27°09′E. Fruit canning factory.

Krasnaya Sloboda, RUSSIA: see KRASNOSLOBODSK, VOLGOGRAD oblast.

Krasnaya Sloboda, RUSSIA: see KRASNOSLOBODSK, MORDVA Republic.

Krasnaya Strelka, RUSSIA: see STRELKA, KRASNODAR Territory.

Krasnaya Yaruga (KRAHS-nah-yah–yah-ROO-gah), town (2005 population 7,960), S BELGOROD oblast, S central European Russia, on railroad spur and near highway, approximately 13 mi/21 km E of the Ukrainian border, and 40 mi/64 km WNW of BELGOROD; 50°48′N 35°39′E. Elevation 702 ft/213 m. Sugar refinery.

Krasnaya Zarya (KRAHS-nah-yah zah-RYAH), settlement (2006 population 1,795), central ORËL oblast, SW European Russia, on road and railroad, 39 mi/63 km N of KURSK (KURSK oblast), and 25 mi 40 km NNE of LIVNY; 52°47′N 37°41′E. Elevation 826 ft/251 m. In agricultural area (wheat, vegetables, sunflowers; livestock raising).

Krasnaya Zor'ka, UKRAINE: see CHERVONA ZORYA.

Krasne (KRAHS-ne) (Russian *Krasnoye*) (Polish *Krasne*), town (2004 population 6,000), central L'VIV oblast, UKRAINE, 25 mi/40 km ENE of L'VIV; 49°55′N 24°37′E. Elevation 688 ft/209 m. Railroad junction; sugar refining; sawmilling, feed milling. Has ruins of an 18th-century castle and church. First mentioned in 1476; served as an air base of the Ukrainian Halych army during the Ukrainian-Polish War (1918–1919); a Jewish community flourished here in the 13th century, later forced to resettle in LUTS'K; town since 1953.

Krasne Lake, UKRAINE: see PEREKOP LAKES.

Krasnen'kaya (KRAHS-neen-kah-yah) [Russian=little red one], settlement (2006 population 4,445), central TAMBOV oblast, S central European Russia, on the TSNA RIVER, on highway, less than 2 mi/3.2 km N of TAMBOV, to which it is administratively subordinate; 52°45′N 41°27′E. Elevation 419 ft/127 m. Geological survey enterprise. Many residents work in Tambov.

Cross-references are shown in SMALL CAPITALS. The pronunciation guide is shown on page xix. The sources of population figures are shown on page xvii.

Krasnik (KRAH-sneek), Polish *Kraśnik*, town (2002 population 36,648), Lublin province, E POLAND, 27 mi/43 km SSW of LUBLIN, on medieval route from LITHUANIA to CRACOW. Flour milling, tanning; brickworks. Before World War II, population 40% Jewish.

Krasni Okny (KRAHS-nee OK-nee) (Russian *Krasnyye Okny*), town, W ODESSA oblast, UKRAINE, 28 mi/45 km SSW of BALTA; 47°32′N 29°27′E. Elevation 479 ft/145 m. Raion center; food processing (grain, dairy). Heritage museum. Established in the last quarter of the 18th century, town since 1959. Until 1919, known as Okny. Large Jewish community since the end of the 19th century, numbering 1,972 in 1939; wiped out during World War II.

Krasno- [Rus. combining form=red], in Russian names: see also KRASNAYA, KRASNOYE, KRASNY, KRASNYE.

Krasnoarmeisk (krahs-no-ahr-MAISK), city, N KÖKSHETAU region, KAZAKHSTAN, on railroad, 40 mi/64 km NNE of KÖKSHETAU; 53°52′N 69°51′E. Tertiary-level (raion) administrative center. Coal mining. Formerly called Taincha.

Krasnoarmeysk (krah-snuh-ahr-MYAISK), city (2006 population 25,570), E MOSCOW oblast, central European Russia, on railroad spur and near highway branch, 31 mi/50 km NE of MOSCOW; 56°06′N 38°08′E. Elevation 564 ft/171 m. Cotton milling; construction materials. Established as Voznesenkaya Manufaktura; renamed Krasnoarmeyskiy in 1928; city status and current name since 1947.

Krasnoarmeysk (krah-snuh-ahr-MYAISK), city (2006 population 25,985), S SARATOV oblast, SE European Russia, near the VOLGA RIVER, on road junction, 46 mi/74 km SSW of SARATOV; 51°01′N 45°42′E. Elevation 718 ft/218 m. Industrial center; cotton milling, apparel, food processing. Founded in 1766. Called Golyy Karamysh until 1926, then Baltser or Balzer in 1926–1942, while in former German VOLGA ASSR.

Krasnoarmeysk (krah-snuh-ahr-MYAISK), S suburb of VOLGOGRAD, SE VOLGOGRAD oblast, SE European Russia, on the VOLGA RIVER, on railroad, 14 mi/23 km S of Volgograd city center, at the N end of the SARPA LAKES valley; 48°32′N 44°38′E. Below sea level. Shipbuilding; locomotive- and car-building works. Founded around 1770 as a German colony; called Sarepta until 1920. Incorporated into Volgograd in 1931.

Krasnoarmeysk, RUSSIA: see KRASNOARMIYS′K.

Krasnoarmeyskaya (krahs-nuh-ahr-MYAIS-skah-yah), village (2005 population 29,150), W central KRASNODAR TERRITORY, S European Russia, on crossroads and railroad, 42 mi/67 km NE of ANAPA, and 44 mi/70 km WNW of KRASNODAR; 45°21′N 38°13′E. Flour mill. Established as a Cossack *stanitsa* (village) of Poltavskaya; renamed in the 1930s.

Krasnoarmeyskiy (krahs-nuh-ahr-MYAIS-keeyee), settlement (no permanent population), N CHUKCHI AUTONOMOUS OKRUG, N RUSSIAN FAR EAST, N of the ARCTIC CIRCLE, 165 mi/266 km NE of BILIBINO; 69°35′N 172°00′E. Elevation 889 ft/270 m. Seasonal base for indigenous family groups.

Krasnoarmeyskiy (krahs-nuh-ahr-MYAIS-keeyee), rural settlement (2006 population 3,715), SE ROSTOV oblast, S European Russia, in the SAL RIVER basin, on railroad spur and highway, 12 mi/19 km SW of ZIMOVNIKI; 47°03′N 42°12′E. Elevation 209 ft/63 m. Agricultural products.

Krasnoarmeyskiy, RUSSIA: see KRASNOARMEYSK, MOSCOW oblast.

Krasnoarmeyskiy Rudnik, UKRAINE: see SVYATOHORIVKA.

Krasnoarmeyskoye (krahs-nuh-ahr-MYAIS-kuh-ye), village (2006 population 5,390), S central SAMARA oblast, E European Russia, near highway, 46 mi/74 km S of SAMARA, and 23 mi/37 km SE of CHAPAYEVSK; 52°44′N 50°02′E. Elevation 160 ft/48 m. In agricultural area; dairy plant. Formerly known as Koldyban′.

Krasnoarmeyskoye (krahs-nuh-ahr-MYAI-skuh-ye), village (2005 population 4,285), N central CHUVASH REPUBLIC, central European Russia, on road, 25 mi/40 km S of CHEBOKSARY; 55°46′N 47°10′E. Elevation 446 ft/135 m. Dairy plant. Until 1939, called Peredniye Traki.

Krasnoarmeyskoye, RUSSIA: see URUS-MARTAN.

Krasnoarmeyskoye, UKRAINE: see VIL′NYANS′K.

Krasnoarmeyskoye, UKRAINE: see KRASNOARMIYS′K.

Krasnoarmiys′k (krahs-no-ahr-MEESK) (Russian *Krasnoarmeysk*), city (2001 population 69,154), W DONETS′K oblast, UKRAINE, in the DONBAS, 35 mi/56 km NW of DONETS′K; 48°17′N 37°11′E. Elevation 672 ft/204 m. Railroad junction; coal-mining center; metalworks. Established in the 1880s; until 1935, Hryshyne (Russian *Grishino*); 1934–1938 Postyshene (Russian *Postyshevo*), city since 1938, until 1964 Krasnoarmiys′ke (Russian *Krasnoarmeyskoye*).

Krasnoarmiys′ke, UKRAINE: see KRASNOARMIYS′K.

Krasnoarmiys′kyy Rudnyk, UKRAINE: see SVYATOHORIVKA.

Krasnoborsk (krahs-nuh-BORSK), village (2005 population 4,780), SE ARCHANGEL oblast, N European Russia, on the NORTHERN DVINA River, on road, 530 mi/853 km SE of ARCHANGEL, and 34 mi/55 km NW of KOTLAS; 61°33′N 45°56′E. Elevation 200 ft/60 m. Made city in 1780; reduced to status of a village in 1917.

Krasnobrodskiy (krahs-nuh-BROT-skeeyee), town, W KEMEROVO oblast, S central SIBERIA, RUSSIA, on the upper reaches of the Uskat River (tributary of the TOM′ River), 3 mi/5 km E of railroad (Trudarmeyskaya station), 70 mi/113 km NNW of NOVOKUZNETSK, and 16 mi/26 km S of BELOVO, to which it is administratively subordinate; 54°10′N 86°28′E. Elevation 948 ft/288 m. Coal mining in the KUZNETSK coal basin; industrial explosives manufacturing. Has a profilactic sanatorium. Formerly known as Krasnyy Brod.

Krasnodar (krahs-nuh-DAHR), city (2005 population 651,250) (cap). KRASNODAR TERRITORY, S European Russia, on the KUBAN′ River, 956 mi/1,538 km SSE of MOSCOW; 45°02′N 38°58′E. Elevation 118 ft/35 m. River port and railroad junction; industrial, transport, cultural, and administrative center; petroleum refineries, machinery, metalworking, textiles, glassworks, chemicals, tobacco products, and food processing. Founded in 1793 by the Zaporozhye Cossacks as their administrative and military center protecting Russia's Caucasian frontier. After 1918 it was for a time the capital of the Kuban-Black Sea Soviet Republic. Called Yekaterinodar, or Ekaterinodar, 1793-1920.

Krasnodarskiy, RUSSIA: see KALININO, KRASNODAR Territory.

Krasnodar Territory (krahs-nuh-DAHR), administrative division (□ 29,237 sq mi/76,016.2 sq km; 2005 population 5,125,220), S European Russia, extending E from the Sea of AZOV and the BLACK SEA into the Kuban steppe and occupying the W part of the N CAUCASUS Mountains; ⊙ KRASNODAR. The territory includes the ADYGEY REPUBLIC. The main agricultural section is in the Kuban steppe and along the lower KUBAN River. The N two-thirds of the region is an extensive plain with high-quality black soil and is used for intensive agriculture, which dominates the regional economy. The territory is one of Russia's principal regions for growing winter wheat, sugar beets, maize, rice, and tobacco. The subtropical Black Sea littoral, stretching from the Sea Azov to the GEORGIA (Abkhazia) border, produces fruit, tea, and wine and is dotted with health resorts, of which SOCHI is the best known. There are petroleum and gas (TAMAN PENINSULA), coal, refractory clays, machinery, cement, and lumber industries. Krasnodar, MAYKOP, and ARMAVIR are the chief industrial centers; NOVOROSSIYSK and TUAPSE are important ports. More than 90% of the population is Russian and assimilated Ukrainian; the region's dialect is a mixture of the two languages; Armenians represent the largest ethnic minority (5%). The rest of the population is Adygey or Circassian. The population is split almost evenly between urban (54%) and rural (46%) communities. The area N of the Kuban belonged to the Crimean Khanate and was annexed by RUSSIA in 1783. The Kuban Cossacks, who settled here, gradually displaced the native nomadic Nogay Tatars. The Black Sea littoral was ceded to Russia by TURKEY in the Treaty of Adrianople (1829). The remainder, known as Circassia, was annexed in 1864. Krasnodar Territory was formed in September 1937.

Krasnodon (krahs-no-DON), city (2001 population 50,560), SE LUHANS′K oblast, UKRAINE, in the DONBAS, on Velyka Kamyanka River (right affluent of the DONETS) and 30 mi/48 km SE of LUHANS′K; 48°18′N 39°44′E. Elevation 383 ft/116 m. Raion center; coal-mining center; manufacturing (transportation equipment, construction materials, food). Five vocational technical schools. Founded (1914) as Sorokyns′kyy Rudnyk (Russian *Sorokinskiy Rudnik*) or Sorokyne (Russian *Sorokino*) until 1938; city since 1938. Since 1954 it absorbed some adjacent settlements, like Pervomayka, with its coal mines.

Krasnodon (krahs-no-DON), town, SE LUHANS′K oblast, UKRAINE, in the DONBAS, 9 mi/15 km WNW of KRASNODON; 48°19′N 39°34′E. Elevation 459 ft/139 m. Subordinated to the Krasnodon city council. Coal mines. Established in 1910 as Yekaterynodon (Russian *Yekaterinodon*), since 1922 Krasnodons′kyy Rudnyk (Russian *Krasnodonskiy Rudnik*), town since 1938, known until recently as Krasnodons′kyy (Russian *Krasnodonskiy*).

Krasnodonetskaya (krahs-nuh-duh-NYETS-kah-yah), village, central ROSTOV oblast, S European Russia, on road and near railroad spur, 19 mi/30 km S of MAYSKIY, and 3 mi/5 km E of SINEGORSKIY; 48°01′N 40°55′E. Elevation 180 ft/54 m. Coal mining.

Krasnodonskiy, UKRAINE: see KRASNODON, town.

Krasnodonskiy Rudnik, UKRAINE: see KRASNODON, town.

Krasnodons′kyy, UKRAINE: see KRASNODON, town.

Krasnodons′kyy Rudnyk, UKRAINE: see KRASNODON, town.

Krasnofarfornyy (krahs-nuh-fahr-FOR-niyee), town (2006 population 1,600), N NOVGOROD oblast, NW European Russia, on the VOLKHOV RIVER, near highway, 6 mi/10 km E of CHUDOVO; 59°08′N 31°51′E. Porcelain manufacturing.

Krasnogorodsk, RUSSIA: see KRASNOGORODSKOYE.

Krasnogorodskoye (krahs-nuh-guh-ROT-skuh-ye), town (2006 population 4,520), W PSKOV oblast, W European Russia, on crossroads, 57 mi/92 km S of PSKOV, and 17 mi/27 km NW of OPOCHKA; 56°50′N 28°16′E. Elevation 354 ft/107 m. Flax processing. Also known as Krasnogorodsk.

Krasnogorovka, UKRAINE: see KRASNOHORIVKA.

Krasnogorsk (krah-snuh-GORSK), city (2006 population 93,080), central MOSCOW oblast, central European Russia, on the left bank of the MOSKVA River, 15 mi/24 km NW of MOSCOW (connected by highway and railroad); 55°49′N 37°19′E. Elevation 488 ft/148 m. Metalworking, pharmaceuticals, consumer goods; food industries (bakery, confectionery). Established in the mid-19th century as Ban′ki. With the move here of an optical factory in 1926–1927, the settlement of KRASNAYA GORA arose. In the 1930s, Ban′ki and Krasnaya Gora were united as Optikogorsk, renamed Krasnogorsk in 1932 (railroad station Pavshino). Became city in 1940.

Krasnogorsk (krahs-nuh-GORSK), city (2006 population 3,205), on the W coast of S SAKHALIN oblast, RUSSIAN FAR EAST, at the SW foot of Mount Krasnaya Gora, 180 mi/290 km NW of YUZHNO-SAKHALINSK, and 45 mi/72 km N of TOMARI; 48°24′N 142°06′E. Fisheries; ship repair, lumbering. Founded in 1905. Under Japanese rule (1905–1945), called Chinnai. Made a city in 1947.

Krasnogorskii, town, TOSHKENT wiloyat, NE UZBEKISTAN; 41°08′N 60°45′E. Also Krasnogorskiy.

Krasnogorskiy (krahs-nuh-GOR-skeeyee), town (2006 population 7,040), S MARI EL REPUBLIC, E central European Russia, on the ILET' RIVER (tributary of the VOLGA), on road, 40 mi/64 km SSE of YOSHKAR-OLA; 56°09′N 48°19′E. Elevation 242 ft/73 m. On railroad spur; manufacturing (electric motors, cargo vans), sawmilling. Until 1938, known as Ilet'.

Krasnogorskiy (krahs-nuh-GOR-skeeyee), town (2005 population 13,570), central CHELYABINSK oblast, SW SIBERIA, RUSSIA, on highway branch and the Chelyabinsk-Troitsk railroad (Krasnoselka station), 43 mi/70 km S of CHELYABINSK, and 12 mi/20 km S of YEMANZHELINSK, to which it is administratively subordinate; 54°36′N 61°14′E. Elevation 793 ft/241 m. Coal mining.

Krasnogorskiy (krahs-nuh-GORS-keeyee), town (2005 population 2,970), central KEMEROVO oblast, S central SIBERIA, RUSSIA, in the TOM' River basin, on road and railroad, 3 mi/5 km S of POLYSAYEVO; 54°33′N 86°16′E. Elevation 541 ft/164 m. Coal mining. In protected old-growth forest region. Formerly a settlement of Krasnaya Gorka.

Krasnogorskiy Rudnik, UKRAINE: see KRASNOHORIVKA.

Krasnogorskoye (krahs-nuh-GOR-skuh-ye), village, E ALTAI TERRITORY, S central SIBERIA, RUSSIA, on road, 45 mi/72 km ESE of BIYSK; 52°18′N 86°12′E. Elevation 836 ft/254 m. Dairy farming. Earlier called Staraya Barda, or Staro-Bardinskoye.

Krasnogorskoye (krahs-nuh-GOR-skuh-ye), village (2006 population 4,495), W UDMURT REPUBLIC, W central URALS, E European Russia, on crossroads, 30 mi/48 km SSW of GLAZOV; 57°42′N 52°30′E. Elevation 797 ft/242 m. Flax processing; woodworking. Until 1938, called Baryshnikovo.

Krasnograd, UKRAINE: see KRASNOHRAD.

Krasnogradskaya, RUSSIA: see YEKATERINOGRADSKAYA.

Krasnogrigoryevka, UKRAINE: see CHERVONOHRYHORIVKA.

Krasnogvardeisk, UZBEKISTAN: see BULUNGHUR.

Krasnogvardeyets (krahs-nuh-gvahr-DYE-yets) [Russian=soldier of the Red Guard], village (2006 population 3,830), W ORENBURG oblast, SE European Russia, in the SAMARA River basin, 10 mi/16 km S of BUZULUK; 52°38′N 52°19′E. Elevation 387 ft/117 m. In oil-processing and -transporting region. Formerly called Umnovskiy (oom-NOF-skeeyee).

Krasnogvardeysk, RUSSIA: see GATCHINA.

Krasnogvardeyskiy (krahs-snuh-gvahr-DYAIS-keeyee), town (2006 population 4,720), S central SVERDLOVSK oblast, W SIBERIA, RUSSIA, on the Irbit River (right tributary of the NITSA RIVER), on road and railroad (Talyy Klyuch station), 37 mi/60 km SSE of ALAPAYEVSK; 57°23′N 62°19′E. Elevation 351 ft/106 m. Manufacturing (cranes); peat works. Founded in 1776; until 1938, called Irbitskiy Zavod.

Krasnogvardeyskiy (krahs-nuh-gvahr-DYAIS-keeyee), former town, E TULA oblast, central European Russia, in the MOSCOW BASIN, near BOLOKHOVO. Abandoned after the shutdown of its lignite mine in the mid-1990s.

Krasnogvardeyskoye (kruh-snuh-gvahr-DYE-yee-skuh-ye), town (2005 population 8,200), BELGOROD oblast, S central European Russia, on the TIKHAYA SOSNA RIVER, 32 mi/51 km SW of OSTROGOZHSK; 50°39′N 38°23′E. Elevation 495 ft/150 m. Fruit and vegetable processing, flour milling. Arose as a village of Biryuch in the 17th century; renamed Budennoye from the early 1920s to 1960s; then Krasnogvardeyskoye.

Krasnogvardeyskoye (krahs-no-gvahr-DAIS-ko-ye), town, N central Autonomous Republic of CRIMEA, UKRAINE, 15 mi/24 km SSW of DZHANKOY; 45°30′N 34°18′E. Elevation 127 ft/38 m. Raion center; on railroad; wine and grape juice factories, flour mill, butter creamery. Until 1945, Kurman-Kemelchi.

Krasnogvardeyskoye (krah-snuh-gvahr-DYAI-skuh-ye), village, N ADYGEY REPUBLIC, NW CAUCASUS, S European Russia, on the E shore of the TSHCHIKSKOYE Reservoir, on road, 7 mi/11 km SW of UST'-LABINSK, and 16 mi/ E of KRASNODAR; 45°06′N 39°33′E. Elevation 114 ft/34 m. Livestock raising; fisheries; food processing.

Krasnohirs'kyy Rudnyk, UKRAINE: see KRASNOHORIVKA.

Krasnohorivka (krahs-no-HO-rif-kah) (Russian *Krasnogorovka*), city, central DONETS'K oblast, UKRAINE, in the DONBAS, 14 mi/23 km W of DONETS'K; 48°01′N 37°37′E. Elevation 449 ft/136 m. Quartzite and gypsum quarries; fire bricks; auto repair. Established in the 1870s as Krasnohirs'kyy Rudnyk (Russian *Krasnogorskiy Rudnik*); city since 1938.

Krasnohrad (krahs-no-HRAHD) (Russian *Krasnograd*), city, W KHARKIV oblast, UKRAINE, 50 mi/80 km SW of KHARKIV; 49°22′N 35°27′E. Elevation 291 ft/88 m. Raion center; railroad junction; food processing; manufacturing of textiles, fur, furniture. Post-secondary agricultural mechanization, health care; vocational technical schools; museum. Established in 1731; city since 1784; until 1922, known as Kostyantynohrad (Russian *Konstantinograd*).

Krasnoil'sk, UKRAINE: see KRASNOYIL's'K.

Krasnokamensk (krahs-nuh-KAH-meensk), city (2005 population 53,770), SE CHITA oblast, E SIBERIA, RUSSIA, on road, 380 mi/612 km SE of CHITA; 50°03′N 118°01′E. Elevation 2,083 ft/634 m. Mining and chemical combine; transportation equipment; food processing (bakery). Krasnokamensk iron ore deposits discovered in 1943. Power station. Made a city in 1969. Until then, was called Chindachi.

Krasnokamensk (krahs-nuh-KAH-meensk), settlement (2005 population 4,750), S KRASNOYARSK TERRITORY, SE Siberian Russia, near highway, 6 mi/10 km SW of ARTËMOVSK; 54°20′N 93°15′E. Elevation 1,627 ft/495 m. Gold and non-ferrous ore mining.

Krasnokamsk (krahs-nuh-KAHMSK), city (2006 population 52,355), central PERM oblast, W URALS, E European Russia, port on the right bank of the KAMA River, near road and railroad, 23 mi/37 km NW of PERM; 58°04′N 55°44′E. Elevation 380 ft/115 m. Center of petroleum collection district and refining; wooden toys, wood pulp and paper, machinery, oil industry equipment, chemicals; food industries (meat processing, bakery, dairy, macaroni factory). Developed in 1929 around a paper mill. Exploitation of oil fields began in 1937. Became city in 1938.

Krasnokholm (krahs-nuh-HOLM), village (2006 population 6,030), S ORENBURG oblast, SE European Russia, near the URAL River, on crossroads, 40 mi/64 km WSW of ORENBURG; 51°35′N 54°09′E. Elevation 242 ft/73 m. Agricultural products. Also known as Krasnyy Kholm (Russian=red hill).

Krasnokholmskiy (krahs-nuh-HOLM-skeeyee), village (2005 population 8,125), N BASHKORTOSTAN Republic, E European Russia, on a short tributary of the BYSTRYY TANYP RIVER, on road, 29 mi/47 km NNE of DYURTYULI; 55°59′N 55°03′E. Elevation 337 ft/102 m. Textiles.

Krasnokokshaysk, RUSSIA: see YOSHKAR-OLA.

Krasnokutivka, UKRAINE: see MALOKATERYNIVKA.

Krasnokutovka, UKRAINE: see MALOKATERYNIVKA.

Krasnokuts'k (krahs-no-KOOTSK), town, NW KHARKIV oblast, UKRAINE, 45 mi/72 km W of KHARKIV; 50°03′N 35°10′E. Elevation 426 ft/129 m. Raion center; food processing; manufacturing (furniture, bricks). Experimental fruit growing farm. Established in 1651, town since 1925. Until 1780, known as Krasnyy Kut.

Krasnolesnyy (krahs-nuh-LYES-niyee), town (2006 population 5,280), N central VORONEZH oblast, S central European Russia, 20 mi/32 km NE, and under administrative jurisdiction, of VORONEZH; 51°52′N 39°35′E. Elevation 531 ft/161 m. Railroad junction

(Grafskaya station); lumber mill, railroad establishments. Popular holiday outing spot for residents of neighboring cities.

Krasnomayskiy (krahs-nuh-MEIS-keeyee), town (2006 population 5,550), W central TVER oblast, W European Russia, on railroad (Leontyevo station), at the mouth of the Shlina River in the Vyshniy Volochek reservoir, 6 mi/10 km NW of VYSHNIY VOLOCHEK; 57°37′N 34°25′E. Elevation 472 ft/143 m. Glassworking center since 1859. Until 1940, called Klyuchinskiy.

Krasno-Medvedovskaya, RUSSIA: see RAYEVSKAYA.

Krasno nad Kysucou (krahs-NO NAHT-ki-su-TSOU), Slovak *Krásno nad Kysucou*, Hungarian *Kiszuczakarásznó*, village, STREDOSLOVENSKY province, NW SLOVAKIA, on railroad, and 12 mi/19 km NNE of ŽILINA; 49°23′N 18°50′E. Woodworking; wooden folk architecture.

Krasnoobsk (krahs-nuh-OPSK), town (2006 population 16,825), central NOVOSIBIRSK oblast, SW SIBERIA, RUSSIA, on the W bank of the OB' RIVER, 8 mi/13 km S, and under administrative jurisdiction, of NOVOSIBIRSK; 54°55′N 82°58′E. Elevation 347 ft/105 m. Refrigerating equipment, agricultural machinery. Made town in 1976.

Krasnooktyabr'skiy (krahs-nuh-uhk-TYAHBR-skeeyee), town (2006 population 11,090), SE VOLGOGRAD oblast, SE European Russia, on the E shore of the Volgograd Reservoir of the VOLGA RIVER, terminus of railroad spur and highway branch, 5 mi/8 km N of VOLZHSKIY, to which it is administratively subordinate; 48°52′N 44°45′E. Chemicals.

Krasnooktyabr'skiy (krahs-nuh-uhk-TYAHBR-skeeyee), town (2006 population 4,240), central MARI EL REPUBLIC, E central European Russia, on highway and railroad junction, 8 mi/13 km W of YOSHKAR-OLA; 56°40′N 47°39′E. Elevation 406 ft/123 m. In agricultural area (wheat, flax, hemp, potatoes); produce processing.

Krasnooktyabr'skiy (krah-snuh-uhk-TYABR-skeeyee), former town, central MOSCOW oblast, RUSSIA, adjoining MOSCOW on the NW, 18 mi/29 km from the city center; now part of MOSCOW; 55°52′N 37°30′E. Elevation 518 ft/157 m. Mostly residential. Building materials. Until 1928, called Khovrino.

Krasnooktyabr'skoye, RUSSIA: see ALKHAST.

Krasnooskol'skoye vodokhranilishche, UKRAINE: see CHERVONOOSKIL RESERVOIR.

Krasnoostrovskiy (krahs-nuh-uh-STROF-skeeyee), town, NW LENINGRAD oblast, NW European Russia, on the Bol'shoy Berezovyy Island in the Gulf of FINLAND, 4 mi/6 km S of PRIMORSK; 60°18′N 28°40′E. Wood pulp and paper factories. Until 1940, in Finland and called Saarenpaa; ceded to the USSR after the Winter War and renamed Byerkskiy; current name since 1948.

Krasnopartizansk, RUSSIA: see BELOGORSK.

Krasnopavlivka (krahs-no-PAHV-leef-kah) (Russian *Krasnopavlovka*), town, S KHARKIV oblast, UKRAINE, on highway and railroad 17 mi/27 km N of LOZOVA; 49°08′N 36°19′E. Elevation 515 ft/156 m. Food products. Krasnopavlivka water reservoir on the Dnieper-Donbas canal 2.5 mi/4 km E. Established in 1869, town since 1972.

Krasnopavlovka, UKRAINE: see KRASNOPAVLIVKA.

Krasnoperekops'k (krahs-no-pe-re-KOPSK) (Russian *Krasnoperekopsk*), city (2004 population 55,600), N Republic of CRIMEA, UKRAINE, on PEREKOP ISTHMUS, on railroad and 8 mi/12.9 km SE of ARM'YANS'K; 45°57′N 33°47′E. Raion center; bromine works, food processing, construction materials. Founded in 1932. Made a city in 1966.

Krasnoperekopsk, UKRAINE: see KRASNOPEREKOPS'K.

Krasnopillya (krahs-no-PEEL-lyah) (Russian *Krasnopolye*), town, SE SUMY oblast, UKRAINE, on railroad, 22 mi/35 km ESE of SUMY; 50°46′N 35°15′E. Elevation 564 ft/171 m. Raion center; food processing, woodworking, furniture. Established in 1640; town since 1956.

Krasnopillya (krahs-no-PEEL-lyah) (Russian *Krasnopolye*), central DNIPROPETROVS'K oblast, UKRAINE, S suburb of DNIPROPETROVS'K, 6 mi/9.7 km SW of city center; 48°24′N 34°56′E. Elevation 426 ft/129 m.

Krasnopillya, UKRAINE: see BRYANKA.

Krasnopol', UKRAINE: see SOLOTVYN.

Krasnopolyanka, RUSSIA: see KANELOVSKAYA.

Krasnopolye (kruhs-no-DO-lyie), urban settlement, S MOGILEV oblast, BELARUS, 60 mi/97 km SE of MOGILEV; ⊙ KRASNOPOLYE region; 53°20′N 31°24′E. Manufacturing (construction materials, food).

Krasnopol'ye, RUSSIA: see LIPOVTSY.

Krasnopolye, UKRAINE: see BRYANKA.

Krasnopolye, UKRAINE: see KRASNOPILLYA, Sumy oblast; or KRASNOPILLYA, Dnipropetrovs'k oblast.

Krasnorechenskiy (krahs-nuh-RYE-cheen-skeeyee), settlement (2006 population 3,800), S central MARITIME TERRITORY, SE RUSSIAN FAR EAST, in the SE SIKHOTE-ALIN RANGE, on highway, 11 mi/18 km WNW of DAL'NEGORSK; 44°37′N 135°21′E. Elevation 1,781 ft/542 m. Forestry services. Tin and lead mining in the vicinity.

Krasnorechenskoye, UKRAINE: see KRASNORICHENS'KE.

Krasnorichens'ke (krahs-no-REE-chen-ske) (Russian *Krasnorechenskoye*), town, NW LUHANS'K oblast, UKRAINE, on the left bank of the Krasna River, left tributary of the DONETS River, on road and on railroad 11 mi/18 km N of KREMINNA; 49°13′N 38°13′E. Elevation 347 ft/105 m. Fabricated metal products, construction materials, feeds. Known since 1701; town since 1957.

Krasnosel'kup (krahs-nuh-syel-KOOP), village (2006 population 4,030), S YAMALO-NENETS AUTONOMOUS OKRUG, TYUMEN oblast, NW SIBERIA, RUSSIA, on the TAZ River, 180 mi/290 km SW of IGARKA; 65°42′N 82°28′E. In reindeer-raising area; trading post. Population largely Sel'kups (Ostyak-Samoyeds). Also known as Krasnosel'kupsk.

Krasnosel'kupsk, RUSSIA: see KRASNOSEL'KUP.

Krasno Selo (KRAS-no SAI-lo), village, BULGARIA, now part of SOFIA; 42°42′N 23°17′E.

Krasnoselsk (kruhs-no-SELSK), urban settlement, GRODNO oblast, BELARUS. Cement factory, lime plant, asbestos and cement items plant.

Krasnosel'skiy (krahs-nuh-SYEL-skeeyee), settlement (2005 population 7,595), E KRASNODAR TERRITORY, S European Russia, on the KUBAN' River, on road and railroad, 3 mi/5 km S of KROPOTKIN; 45°23′N 40°36′E. Elevation 265 ft/80 m. Railroad depots. Agriculture (wheat, flax; livestock); produce processing.

Krasnoshchëkovo (krahs-nuh-SHCHO-kuh-vuh), village, S ALTAI TERRITORY, S central SIBERIA, RUSSIA, on the CHARYSH RIVER, on road, 65 mi/105 km E of RUBTSOVSK; 51°40′N 82°43′E. Elevation 777 ft/236 m.

Krasnoslobodsk (krah-snuh-sluh-BOTSK), city (2006 population 14,335), SE VOLGOGRAD oblast, SE European Russia, on the left bank of the VOLGA RIVER, opposite VOLGOGRAD; 48°42′N 44°34′E. Ship construction and repair; timbering and woodworking; fish factory. Made city in 1955. Formerly called Krasnaya Sloboda.

Krasnoslobodsk (krahs-nuh-sluh-BOTSK), city (2006 population 10,720), W central MORDVA REPUBLIC, central European Russia, on the MOKSHA River (OKA River basin), on road, 66 mi/106 km WNW of SARANSK; 54°25′N 43°47′E. Elevation 531 ft/161 m. In agricultural area (hemp); bast fiber processing, spinning and weaving; food industry (creamery); electronics. Founded in 1571. Made city in 1780. Called Krasnaya Sloboda until 1780.

Krasnostav, POLAND: see KRASNYSTAW.

Krasnotorka (krahs-nuh-TOR-kah), town, N DONETS'K oblast, UKRAINE, in the DONBAS, on the Kazennyy Torets' River, 2 mi/3.2 km S of KRAMATORS'K and

subordinated to its city council; 48°41′N 37°31′E. Elevation 216 ft/65 m. State farm. Established in 1861; town since 1938.

Krasnoturansk (krahs-nuh-too-RAHNSK), village (2005 population 5,815), SW KRASNOYARSK TERRITORY, SE SIBERIA, RUSSIA, on the YENISEY RIVER, on road, 35 mi/56 km N of MINUSINSK; 54°19′N 91°34′E. Elevation 1,272 ft/387 m. Dairy farming. Formerly called Abakanskoye.

Krasnotur'insk (krahs-nuh-toor-YEENSK), city (2006 population 63,995), W SVERDLOVSK oblast, extreme W Siberian Russia, in the NE foothills of the central URALS, on the Turya River (OB' RIVER basin), on railroad (Vorontsovka station), 215 mi/346 km N of YEKATERINBURG, 18 mi/29 km NW of SEROV, and 5 mi/8 km E of KARPINSK; 59°46′N 60°11′E. Elevation 685 ft/208 m. Aluminum refining (developed during World War II), based on Severoural'sk, Cheremukhovo, and Kalya bauxite mines; electricity-generating station; food processing (bakery). Arose in the second half of the 18th century, but largely developed after the construction of aluminum works. Until 1944, called Tur'inskiy.

Krasnoufimsk (krahs-nuh-oo-FEEMSK), city (2006 population 42,970), SW SVERDLOVSK oblast, extreme E European Russia, in the SW foothills of the central URALS, on the right bank of the UFA RIVER, 140 mi/225 km WSW of YEKATERINBURG; 56°36′N 57°45′E. Elevation 669 ft/203 m. On railroad (repair shops); manufacturing center (machinery, furniture, food). Founded in 1736 as a stronghold; chartered in 1781.

Krasnoural'sk (krahs-nuh-oo-RALSK), city (2006 population 27,235), W SVERDLOVSK oblast, extreme W Siberian Russia, in the E foothills of the central URALS, on road, 75 mi/120 km N of YEKATERINBURG, and 30 mi/48 km N of NIZHNIY TAGIL; 58°21′N 60°02′E. Elevation 738 ft/224 m. Railroad spur terminus; copper mining, concentrating, and smelting; manufacturing of chemicals and fertilizers; grain-processing and bread-baking complex. Developed in 1925 with mining of copper ore. Originally called Bogomolstroy; renamed Uralmed'stroy in 1929; city status and current name since 1932.

Krasnoural'skiy Rudnik, RUSSIA: see NOVOASBEST.

Krasnousol'skiy (krahs-nuh-oo-SOL-skeeyee), town (2005 population 11,830), central BASHKORTOSTAN Republic, SW URALS, RUSSIA, on the Usol'ka River (KAMA River basin), on road, 44 mi/71 km S of UFA, and 30 mi/48 km NNE of STERLITAMAK; 53°53′N 56°28′E. Elevation 419 ft/127 m. Arose in 1752 with the opening of a copper smelter, which in 1893 was converted to a glass factory (window glass); glass manufacturing. Health resort with mud baths. Quartz sand quarries nearby. Formerly called Krasnousol'skiy Zavod.

Krasnousol'skiy Zavod, RUSSIA: see KRASNOUSOL'SKIY.

Krasnovishersk (krahs-nuh-VEE-shersk), city (2006 population 18,165), N PERM oblast, W central URALS, E European Russia, on the Vishera River (landing), on road, 300 mi/483 km N of PERM, and 55 mi/89 km NNE of SOLIKAMSK; 60°24′N 57°05′E. Elevation 439 ft/133 m. Wood pulp and paper combine, lumber, fiber board; food processing (dairy, bakery). Arose in 1930 with the construction of a paper combine. Became city in 1942.

Krasnovka (krahs-NOF-kah), rural settlement (2006 population 3,965), W ROSTOV oblast, S European Russia, in the Severskiy DONETS River basin, near the border with UKRAINE, on road and railroad, 12 mi/19 km SW of MILLEROVO; 48°49′N 40°06′E. Elevation 357 ft/108 m. In anthracite-mining and stone-quarrying region. Formerly known as Donskoy.

Krasnovo (KRAS-no-vo), village, PLOVDIV oblast, HISARYA obshtina, W central BULGARIA, at the S foot of the central SREDNA GORA, 25 mi/40 km NNW of PLOVDIV; 42°27′N 24°30′E. Health resort; rye, potatoes; livestock. Formerly called Krastovo.

Krasnovodsk, TURKMENISTAN: see TURKMENBASHI.

Krasnoyarka (krahs-nuh-YAHR-kah), village (2006 population 3,990), central OMSK oblast, SW SIBERIA, RUSSIA, on the IRTYSH River, on highway branch, 25 mi/40 km NNW of OMSK; 55°20′N 73°07′E. Elevation 334 ft/101 m. In agricultural area (grain, vegetables; livestock).

Krasnoyarsk (krahs-nuh-YAHRSK), city (2005 population 906,950), SE SIBERIA, RUSSIA, on the YENISEY RIVER, 2,458 mi/3,956 km E of MOSCOW; 56°06′N 92°47′E. Elevation 754 ft/229 m. ⊙ KRASNOYARSK TERRITORY. River port, railroad-river transshipment point on the TRANS-SIBERIAN RAILROAD, industrial and administrative center; heavy machinery and equipment for the Trans-Siberian Railroad. Manufacturing of construction materials, lumber, machinery, chemicals, wood products, ordance. Plutonium production plant from nuclear wastes. Large hydroelectric plant on the Yenisey River. Has an opera house, drama and comedy theaters, concert and organ-music halls, museums and art galleries. Architectural landmarks include the Veil church (built in 1795) and Annunciation church (1812). Founded in 1628 as the Cossack outpost of Krasnyy Yar. Grew in importance with the construction of a trading route from Siberia to Moscow (1735-1741); made city in 1822. It grew rapidly after the discovery of gold and the construction of the Trans-Siberian Railroad (late 19th century).

Krasnoyarsk-26 (Atomgrad) (krahs-nuh-YAHRSK), city (2005 population 93,860), S KRASNOYARSK TERRITORY, SE SIBERIA, on the small Kantat and Baykal rivers, near their mouths in the YENISEY RIVER, on branch railroad N from the TRANS-SIBERIAN RAILROAD, 40 mi/64 km N of KRASNOYARSK; 56°15′N 93°32′E. Elevation 547 ft/166 m. Arose as a result of a decision in the 1950s by the government of the former USSR to build a uranuim-graphite reactor to produce plutonium-239 (material for the atom bomb and nuclear weapons) in Siberia in the area of the middle Yenisey River at a maximum distance from the external borders of the country, under the secret name of K-26, or Zheleznogorsk. Complex of industrial plants. The existence and location of the city was a state secret until 1994.

Krasnoyarsk-45 (krahs-nuh-YAHRSK) or **Zelenogorsk** (zye-lye-nuh-GORSK), city (2005 population 72,080), S central KRASNOYARSK TERRITORY, SE SIBERIA, RUSSIA, on a spur railroad from Zaozernaya station of the TRANS-SIBERIAN RAILROAD, on the KAN RIVER (E tributary of the YENISEY RIVER), 110 mi/180 km E of KRASNOYARSK; 56°07′N 94°35′E. Elevation 613 ft/186 m. Electrochemical plant, regional electric power station. Established in 1955. Existence and location of the city was a state secret until 1994 due to nuclear fuel production here.

Krasnoyarsk-66, RUSSIA: see KEDROVYY.

Krasnoyarskaya (krahs-nuh-YAHR-skah-yah), village (2006 population 5,195), E ROSTOV oblast, S European Russia, near the S tip of the TSIMLYANSK RESERVOIR on the DON River, on road junction and railroad, 2 mi/3.2 km SE of VOLGODONSK; 47°29′N 42°11′E. Elevation 150 ft/45 m. Agricultural products.

Krasnoyarskiy (krahs-nuh-YAHR-skeeyee), settlement (2006 population 3,215), NE ORENBURG oblast, extreme SE European Russia, in the SE URALS, 4 mi/6 km W of the border with KAZAKHSTAN, on highway and railroad junction, 27 mi/43 km N of ADAMOVKA; 51°58′N 59°53′E. Elevation 1,112 ft/338 m. In mining region producing ferrous and precious metals.

Krasnoyarsk Territory (krahs-nuh-YAHRSK), administrative division (□ 903,363 sq mi/2,348,743.8 sq km; 2005 population 2,831,720), central Siberian Russia, extending from the SAYAN MOUNTAINS and the Minusinsk basin in the S across the Siberian wooded steppe, taiga, and tundra to the ARCTIC OCEAN; ⊙ KRASNOYARSK. The territory stretches along the entire course of the YENISEY RIVER, comprising parts of the West Siberian lowland on the left bank and the central Siberian Plateau on the right bank. The Yenisey and its tributaries are important transportation routes and electric power sources. Surplus amounts of hydroelectric power are generated. The TRANS-SIBERIAN RAILROAD crosses the S section of the territory. There are deposits of brown coal, graphite, iron ore, manganese, gold, copper, nickel, bauxite, chromite, cobalt, platinum, antimony, uranium, and mica. In the N is an extensive lumber industry. Agriculture (grain; cattle and reindeer); fur trapping. Chief cities include Krasnoyarsk, KANSK, ACHINSK, NORIL'SK, MINUSINSK, and IGARKA. The territory includes Krasnoyarsk proper (S and E of the Yenisey; □ 277,698 sq mi/719,235 sq km), KHAKASS REPUBLIC (SW), the EVENKI AUTONOMOUS OKRUG (in the E central section), and the TAYMYR (DOLGAN-NENETS) AUTONOMOUS OKRUG and the TAYMYR PENINSULA (N of the ARCTIC CIRCLE). The S part of the territory contains 90% of the population, which includes Russians, Ukrainians, Byelorussians, Khakass, Tatars, Evenki, Yakuts, and Nenets. The territory was organized in 1934. During Stalin's rule and after, the area was the site of labor camps.

Krasnoye (KRUHS-no-ye), Polish *Krasne*, town, MINSK oblast, BELARUS, on USHA RIVER (left tributary of VILIYA RIVER) and 12 mi/19 km ESE of MOLODECHNO. Tanning, flour milling, sawmilling; tile and brick manufacturing.

Krasnoye (KRAHS-nuh-ye), village (2006 population 4,010), central LIPETSK oblast, S central European Russia, on road and railroad, 19 mi/31 km NE of YELETS; 52°51'N 38°48'E. Elevation 679 ft/206 m. Agricultural chemicals; grain storage; food processing (dairy, bakery).

Krasnoye (KRAHS-nuh-ye), village (2005 population 3,400), W KEMEROVO oblast, S central Siberian Russia, on highway, 26 mi/42 km SW of PROMYSHLENNAYA; 54°36'N 85°22'E. Elevation 656 ft/199 m. In agricultural area; produce processing.

Krasnoye (KRAHS-nuh-ye), settlement (2005 population 3,520), N KRASNODAR TERRITORY, S European Russia, on road and railroad, 33 mi/53 km S of ROSTOV-NA-DONU (ROSTOV oblast), and 12 mi/19 km N of KUSHCHEVSKAYA; 46°44'N 39°34'E. Elevation 285 ft/86 m. In agricultural area (wheat, sunflowers, castor beans).

Krasnoye [Russian=red], in Russian names: see also KRASNAYA, KRASNO- [Russian combining form], KRASNY, KRASNYE.

Krasnoye, RUSSIA: see KRASNYY, SMOLENSK oblast.

Krasnoye, RUSSIA: see ULAN ERGE.

Krasnoye, UKRAINE: see CHERVONE, Zhytomyr oblast.

Krasnoye, UKRAINE: see KRASNE.

Krasnoye Ekho (KRAHS-nuh-ye E-huh) [Russian=red echo], town (2006 population 2,225), S central VLADIMIR oblast, central European Russia, on road junction, 11 mi/18 km N of GUS'-KHRUSTAL'NYY; 55°48'N 40°42'E. Elevation 436 ft/132 m. Glassworks. Until 1925, called Novogordino.

Krasnoye-na-Volge (KRAHS-nuh-ye–nah–VOL-gye), village (2005 population 7,885), SW KOSTROMA oblast, central European Russia, on the VOLGA RIVER, on road, 20 mi/32 km SSE of KOSTROMA; 57°30'N 41°14'E. Elevation 265 ft/80 m. Flax processing; handicrafts (jewelry).

Krasnoye Selo (KRAHS-nuh-ye sye-LO), town (2005 population 42,830), central LENINGRAD oblast, NW European Russia, a suburb of SAINT PETERSBURG, 15 mi/24 km S of the city center (connected by road and railroad); 59°44'N 30°05'E. Elevation 206 ft/62 m. A favorite summer resort before 1917, with palaces at nearby NAGORNOYE and ROPSHA, and park complexes from the 19th century. Called Krasnyy for a few years following the Communist takeover of power. Paper and plastics factories; fur and feather down factories. Incorporated into St. Petersburg in 1973.

Krasnoye Vereshchagino, RUSSIA: see VERESHCHAGINO.

Krasnoye Znamya (KRAHS-nuh-yuh ZNAHM-yah), town, S MARY weloyat, TURKMENISTAN, on Kushka railroad and MURGAB RIVER, 60 mi/97 km SE of MARY; 36°58'N 62°30'E. Irrigated agriculture.

Krasnoyil's'k (krahs-no-YEELSK) (Russian *Krasnoil'sk*), town (2004 population 7,400), SW CHERNIVTSI oblast, UKRAINE, in the foothills of CARPATHIAN Mountains on road and railroad spur 25 mi/40 km SW of CHERNIVTSI and 5 mi/8 km N of the border with Romania; 48°01'N 25°35'E. Elevation 1,453 ft/442 m. Resort with springs and park; woodworking. Known since 1431.

Krasnozatonskiy (krahs-nuh-zah-TON-skeeye), settlement (2005 population 7,940), W central KOMI REPUBLIC, NE European Russia, on road and railroad, across the SYSOLA RIVER (4 mi/6 km to the E) from SYKTYVKAR; 61°40'N 50°59'E. Elevation 416 ft/126 m. Timbering.

Krasnozavodsk (krahs-nuh-zah-VOTSK), city (2006 population 11,120), NE MOSCOW oblast, central European Russia, on the Kunya River (VOLGA RIVER basin), 55 mi/88 km NE of Moscow, and 10 mi/16 km NNE of SERGIYEV POSAD; 56°27'N 38°13'E. Elevation 748 ft/227 m. Terminus of a local railroad spur. Chemical works, bakery. Developed in the late 1930s; until 1940, called Zagorskiy, when granted city status and renamed.

Krasnozavodsk, city, NE Moscow oblast, central European Russia, on Kunya River (VOLGA River basin), 15 mi/24 km N of SERGIYEV POSAD (Zagorsk). Bakery. Made city in 1940.

Krasnozerskoye (krahs-nuh-ZYER-skuh-ye), town (2006 population 10,295), S NOVOSIBIRSK oblast, SW SIBERIA, RUSSIA, on the KARASUK RIVER, on road, 50 mi/80 km ENE of KARASUK; 54°01'N 79°14'E. Elevation 498 ft/151 m. Dairy products, flour milling.

Krasnoznamensk (krah-snuh-ZNAH-myensk), city (2005 population 3,715), NE KALININGRAD oblast, W European Russia, on the SHESHUPE River (tributary of the NEMAN RIVER), on road, 100 mi/161 km NE of KALININGRAD, and 22 mi/35 km N of NESTEROV; 54°56'N 22°29'E. Elevation 134 ft/40 m. On a narrow-gauge railroad spur; agricultural market. Founded in 1734. Until 1945, in EAST PRUSSIA where it was called Lasdehnen and, between 1938 and 1945, Haselberg.

Krasnoznamenskoe (krahs-noz-nah-MYEN-sko-ye), town, AKMOLA region, KAZAKHSTAN, 80 mi/135 km W of Akmola; 53°03'N 69°27'E. Tertiary-level (raion) administrative center. Agricultural products.

Krasnoznam'yanka Irrigation System (krahs-no-ZNAHM-yahn-kah) (Ukrainian *Krasnoznam'yans'ka zroshuval'na systema*), an irrigation system in SW KHERSON oblast, UKRAINE, on the BLACK SEA LOWLAND, extending from the westernmost bend of the NORTH CRIMEAN CANAL for 50 mi/80 km W to village Krasnoznam'yanka, and from 10 mi/16 km–14 mi/23 km inland (N) to the BLACK SEA coast (S). Water is obtained from the DNIEPER River (KAKHOVKA RESERVOIR) via the North Crimean Canal and then conveyed W for 63 mi/101 km by the Krasnoznam'yanka Mainline Canal (Ukrainian *Krasnoznam'yans'kyy mahistral'nyy kanal*; built between 1956 and 1966); a parallel distributary canal (begun 1976) continues N for 6 mi/10 km and is planned to exend to the E coast of YAHORLYT BAY. The two canals provide irrigation for 239,000 acres/96,800 ha in 1991. Chestnut soils are prone to salinization, especially in S, where solonchaks abound; areas with high water level are ameliorated with drainage. To enhance soil fertility gypsum and fertilizer are applied. Crops grown are mostly grain and feed, and rice in paddies in the S (near the Black Sea coast).

Krasnoznam'yanka Mainline Canal, UKRAINE: see KRASNOZNAM'YANKA IRRIGATION SYSTEM.

Krasnoznam'yans'ka zroshuval'na systema, UKRAINE: see KRASNOZNAM'YANKA IRRIGATION SYSTEM.

Krasnoznam'yans'kyy mahistral'nyy kanal, UKRAINE: see KRASNOZNAM'YANKA IRRIGATION SYSTEM.

Krasny [Russian=red], in Russian names: see also KRASNAYA, KRASNO- [Russian combining form], KRASNOYE, KRASNYE.

Krasnye [Russian=red], in Russian names: see also KRASNAYA, KRASNO- [Russian combining form], KRASNOYE, KRASNY.

Krasny Gorodok, UKRAINE: see RAYHORODOK.

Krasnyi Yar (KRAHS-nee YAHR), town, KÖKSHETAU region, KAZAKHSTAN, 9 mi/15 km W of KÖKSHETAU; 53°22'N 69°16'E. Tertiary-level (raion) administrative center. Agricultural products.

Krasnystaw (krah-SNEE-stahv), Russian *Krasnostav*, town (2002 population 19,748), Chełm province, E POLAND, on WIEPRZ RIVER and 30 mi/48 km SE of LUBLIN. Tanning, flour milling, sawmilling; apparel, construction materials.

Krasnyy (KRAHS-niyee), town (2006 population 4,550), W SMOLENSK oblast, W European Russia, near highway, 28 mi/45 km WSW of SMOLENSK; 55°34'N 32°32'E. Elevation 629 ft/191 m. Formerly a village of KRASNOYE.

Krasnyy (KRAHS-niyee), rural settlement (2006 population 1,670), SW ROSTOV oblast, S European Russia, on the Western MANYCH River approximately 5 mi/8 km E of its confluence with the DON River, near highway, 7 mi/11 km S, and under administrative jurisdiction, of BAGAYEVSKAYA; 47°12'N 40°21'E. In agricultural area producing wheat, sunflowers, grapes, and livestock.

Krasnyy, RUSSIA: see MOZHGA, UDMURT Republic.

Krasnyy Bazar (KRAH-snee bah-ZAHR), town, S NAGORNO-KARABAKH AUTONOMOUS Oblast, AZERBAIJAN, 14 mi/23 km SE of STEPANAKERT. Winery; creamery. Also called Karmir Bazar.

Krasnyy Bogatyr' (KRAHS-niyee buh-gah-TIR), town (2006 population 1,010), central VLADIMIR oblast, central European Russia, on road and railroad spur, 11 mi/18 km ENE of SUDOGDA; 56°01'N 41°08'E. Elevation 577 ft/175 m. Glassworks.

Krasnyy Bor (KRAHS-niyee BOR), town (2005 population 4,845), central LENINGRAD oblast, NW European Russia, on railroad, 18 mi/29 km SE of SAINT PETERSBURG; 59°42'N 30°40'E. Glassworks; chemicals.

Krasnyy Bor (KRAHS-niyee BOR), village, NE TATARSTAN Republic, E European Russia, on the KAMA River, on road, 38 mi/61 km E of NIZHNEKAMSK, and 11 mi/18 km N of MENZELINSK; 55°53'N 53°05'E. Elevation 164 ft/49 m. Bronze relics (5th–7th centuries) excavated nearby. Formerly called Pyanyy Bor.

Krasnyy Boyevik, RUSSIA: see KOTOVSK.

Krasnyy Brod, RUSSIA: see KRASNOBRODSKIY.

Krasnyy Chikoy (KRAHS-niyee chee-KO-yee), village (2005 population 6,725), SW CHITA oblast, E SIBERIA, RUSSIA, on the CHIKOY RIVER, on road, 60 mi/97 km S of PETROVSK-ZABAIKAL'SKIY; 50°22'N 108°45'E. Elevation 2,585 ft/787 m. Gold mining; tanning; forestry.

Krasnyye Baki (KRAHS-ni-ye BAH-kee), town (2006 population 7,765), N central NIZHEGOROD oblast, central European Russia, on the VETLUGA RIVER, on road junction and near railroad, 70 mi/113 km NE of NIZHNIY NOVGOROD; 57°07'N 45°09'E. Elevation 265 ft/80 m. Wood products; dairying.

Krasnyye Barrikady (KRAHS-ni-ye bah-ree-KAH-di) (Russian=red barricades), village (2005 population 6,300), S ASTRAKHAN oblast, S European Russia, on the VOLGA RIVER, near highway, 6 mi/10 km SSE of ASTRAKHAN; 46°12'N 47°54'E. Below sea level. Shipbuilding yard. Formerly known as Bertyul'.

Krasnyye Chetai (KRAHS-ni-ye chee-TAH-ee), village (2005 population 2,945), W CHUVASH REPUBLIC, central European Russia, on highway, 16 mi/26 km NW of SHUMERLYA; 55°41'N 46°09'E. Elevation 564 ft/

171 m. In agricultural area (grains, vegetables, hemp; livestock raising).

Krasnyye Okny, UKRAINE: see KRASNI OKNY.

Krasnyye Tkachi (KRAHS-ni-ye tkah-CHEE), town (2006 population 4,025), N central YAROSLAVL oblast, central European Russia, on the KOTOROSL' RIVER (tributary of the VOLGA RIVER), on road and near railroad, 13 mi/21 km SSW of YAROSLAVL; 57°29′N 39°45′E. Elevation 351 ft/106 m. Linen weaving. Has a sanatorium.

Krasnyy Gorodok, UKRAINE: see RAYHORODOK.

Krasnyy Gulyay (KRAHS-niyee goo-LYEI), settlement (2006 population 3,070), central ULYANOVSK oblast, E central European Russia, on railroad, 22 mi/35 km S of ULYANOVSK, and 21 mi/34 km WNW of SENGILEY, to which it is administratively subordinate; 54°01′N 48°20′E. Elevation 833 ft/253 m. Glassworks.

Krasnyy Horodok, UKRAINE: see RAYHORODOK.

Krasnyy Kamyshanik, RUSSIA: see KOMSOMOL'SKIY, Republic of KALMYKIA-Khalmg-Tangeh.

Krasnyy Kholm (KRAHS-niyee HOLM), city (2006 population 6,030), E TVER oblast, W central European Russia, on the Neledina River (VOLGA RIVER basin), on crossroads and railroad, 110 mi/177 km NE of TVER, and 25 mi/40 km NE of BEZHETSK; 58°03′N 37°07′E. Elevation 538 ft/163 m. Center of agricultural area (dairying; livestock; flax); meat packing. Made city in 1776.

Krasnyy Kholm, RUSSIA: see KRASNOKHOLM.

Krasnyy Klyuch (KRAHS-niyee KLYOOCH), town (2005 population 2,350), NE BASHKORTOSTAN Republic, on the W slopes of the S URALS, RUSSIA, on the UFA RIVER (KAMA River basin) (landing), on road, 13 mi/21 km N of KRASNAYA GORKA, in woodland; 55°23′N 56°39′E. Elevation 469 ft/142 m. Paper products; lumber industry. Formerly known as Belyy Klyuch.

Krasnyy Kommunar (KRAHS-niyee kuh-moo-NAHR), village (2006 population 4,160), N central ORENBURG oblast, SE European Russia, on the URAL River, on road and railroad, 16 mi/26 km NE of ORENBURG, and 2 mi/3.2 km S of SAKMARA; 51°57′N 55°22′E. Elevation 419 ft/127 m. Railroad enterprises. One of the region's two major airports is 10 mi/16 km to the S.

Krasnyy Kurgan (KRAHS-niyee koor-GAHN), village (2005 population 3,250), NE KARACHEVO-CHERKESS REPUBLIC, S European Russia, in the N CAUCASUS Mountains, on the administrative border with STAVROPOL TERRITORY, on road, 13 mi/21 km SW of YESSENTUKI; 43°56′N 42°36′E. Elevation 2,933 ft/893 m. In agricultural area (vegetables, livestock).

Krasnyy Kut (KRAHS-niyee KOOT), city (2006 population 14,835), S SARATOV oblast, SE European Russia, on the YERUSLAN RIVER (tributary of the VOLGA RIVER), 73 mi/117 km SE of SARATOV; 50°57′N 46°58′E. Elevation 144 ft/43 m. Railroad and highway junction; center of agricultural area; railroad establishments, food industries. Until 1941, in German VOLGA Autonomous SSR. Made city in 1966.

Krasnyy Kut (KRAHS-nee KOOT), town, S LUHANS'K oblast, UKRAINE, in the DONBAS, 7 mi/11.3 km NW of KRASNYY LUCH; 48°12′N 38°47′E. Elevation 538 ft/163 m. Coal mine; forestry.

Krasnyy Kut, UKRAINE: see KRASNOKUTS'K.

Krasnyy Liman (KRAHS-niyee lee-MAHN), village, central VORONEZH oblast, S central European Russia, on road, 27 mi/43 km ESE of VORONEZH; 51°32′N 39°50′E. Elevation 498 ft/151 m. Agricultural products. Formerly known as Staroye Khrenovoye.

Krasnyy Liman, UKRAINE: see KRASNYY LYMAN.

Krasnyy Luch (KRAHS-nee LOOCH), city (2001 population 94,875), S LUHANS'K oblast, UKRAINE, in the DONBAS, 34 mi/55 km SSW of LUHANS'K; 48°08′N 38°56′E. Elevation 820 ft/249 m. Coal mining center; manufacturing of machinery, construction materials, furniture, apparel, food. Two technical colleges, nine

vocational-technical schools. Incorporated a number of adjacent coal mining settlements since 1954, including Shakhta No. 5, Shakhta No. 7/8, and Shterivskyy Zavod Imeny Petrovs'koho (Russian *Shterovskiy Zavod Imeni Petrovskogo*). Established as a mining settlement in the 1880s, called Kryndachivka (Russian *Kryndachovka*), until 1920; city since 1926.

Krasnyy Luch (KRAHS-niyee LOOCH) [Russian=red lightbeam], settlement (2006 population 1,440), E PSKOV oblast, W European Russia, in a marshy area near the upper POLIST' RIVER, on highway, 8 mi/13 km NE of (and administratively subordinate to) BEZHANITSY; 57°04′N 30°05′E. Elevation 305 ft/92 m. Glassworks, including light fixtures. Formerly known as Novyy Svet (Russian=new light).

Krasnyy Lyman (KRAHS-nee li-MAHN) (Russian *Krasnyy Liman*), city (2004 population 32,600), N DONETS'K oblast, UKRAINE, in the DONBAS, 65 mi/105 km N of DONETS'K and 12 mi/19 km NE of SLOV'YANSK; 48°59′N 37°49′E. Elevation 354 ft/107 m. Raion center; railroad junction; metalworks; manufacturing (construction materials, food). Technical school, medical school. Established in 1667, called Lyman (Russian *Liman*, after a small lake on the shore of which it is located); renamed and city status since 1938.

Krasnyy Mayak (KRAHS-niyee mah-YAHK) [Russian=red lighthouse], town (2006 population 850), E central VLADIMIR oblast, central European Russia, 22 mi/35 km S of KOVROV; 56°03′N 41°23′E. Elevation 406 ft/123 m. Glassworks.

Krasnyy Oktyabr' (KRAHS-niyee uhk-TYAHBR) [Russian=red October], town (2005 population 4,055), central KURGAN oblast, SW SIBERIA, RUSSIA, on crossroads and the TRANS-SIBERIAN RAILROAD (Kosobrodsk station), 22 mi/35 km NW of KURGAN; 55°39′N 64°49′E. Elevation 367 ft/111 m. Sawmilling and woodworking.

Krasnyy Oktyabr' (KRAHS-niyee uhk-TYAHBR) [Russian=red October], town (2006 population 9,925), W VLADIMIR oblast, central European Russia, on road and railroad, 3 mi/5 km SE of KIRZHACH; 56°07′N 38°53′E. Elevation 567 ft/172 m. Glassworks. Until 1919, called Voznesenskiy.

Krasnyy Oktyabr' (KRAHS-nee ok-TYAH-buhr), town (2004 population 3,700), E DONETS'K oblast, UKRAINE, in the DONBAS, 2 mi/3 km NW of YENAKIYEVE city center; 48°05′N 38°08′E. Elevation 557 ft/169 m. Coal mines. Formerly Narniyevskiy Rudnik, now merged with and part of Yenakiyeve.

Krasnyy Oktyabr' (KRAHS-niyee uhk-TYAH-buhr) [Russian=Red October], village (2006 population 3,295), central SARATOV oblast, SE European Russia, on highway, 13 mi/21 km W of SARATOV; 51°32′N 45°42′E. Elevation 935 ft/284 m. In agricultural area (wheat, rye, flax, sunflowers).

Krasnyy Oktyabr' (KRAHS-niyee uhk-TYAHBR) [Russian=red October], village, NE VOLGOGRAD oblast, SE European Russia, on road junction, 14 mi/23 km W, and under administrative jurisdiction, of PALLASOVKA; 49°59′N 46°33′E. Livestock breeding.

Krasnyy Oktyabr' (KRAHS-niyee uhk-TYAHBR), urban settlement (2006 population 1,180), N VLADIMIR oblast, central European Russia, near railroad spur, 19 mi/31 km SSE of KOVROV, to which it is administratively subordinate; 56°00′N 41°27′E. Elevation 429 ft/130 m. Glassworks. Until 1925, called Yakunchikov.

Krasnyy Oktyabr', RUSSIA: see KYZYL-OKTYABR'SKIY.

Krasnyy Profintern (KRAHS-nee prof-een-TERN), town, central DONETS'K oblast, UKRAINE, in the DONBAS, 2 mi/3 km NW of YENAKIYEVE city center; 48°16′N 38°11′E. Elevation 695 ft/211 m. Coal mines. Formerly called Verovskiy Rudnik, now merged with and part of Yenakiyeve.

Krasnyy Profintern (KRAHS-niyee pruhf-een-TERN), town (2006 population 1,460), E YAROSLAVL oblast, central European Russia, on the VOLGA RIVER, on road and near railroad spur, 21 mi/34 km ENE of

YAROSLAVL; 57°45′N 40°26′E. Elevation 295 ft/89 m. Food processing (starch and syrup). Until 1926, called Ponizovkino, then, until 1945, Guzitsino.

Krasnyy Steklovar (KRAHS-niyee stye-kluh-VAHR), town, SE MARI EL REPUBLIC, E central European Russia, near the ILET' RIVER, on road, 45 mi/72 km SE of YOSHKAR-OLA; 56°13′N 48°47′E. Elevation 308 ft/93 m. Glassworks. Until 1939, called Il'yinskoye.

Krasnyy Stroitel' (KRAHS-niyee–struh-EE-tyel), former town, central MOSCOW oblast, central European Russia, now a suburb of MOSCOW, 24 mi/39 km S of the city center (connected by local railroad); 55°35′N 37°37′N. Elevation 574 ft/174 m. Wool; building materials.

Krasnyy Sulin (KRAHS-niyee soo-LEEN), city (2006 population 44,130), W central ROSTOV oblast, S European Russia, in the E DONETS BASIN, on road, approximately 12 mi/19 mi E of the Ukrainian border, and 60 mi/97 km NNE of ROSTOV; 47°53′N 40°04′E. Elevation 459 ft/139 m. On railroad (Sulin station); metallurgical center; iron and steelworks, firebrick manufacturing; bakery. Coal-fed electric power plant. Founded in 1797. Called Sulin or Sulinovskoye until 1926, when granted city status and renamed.

Krasnyy Tekstil'shchik (KRAHS-niyee tyek-STEEL-shcheek), town (2006 population 3,270), central SARATOV oblast, SE European Russia, on the right bank of the VOLGA RIVER (landing), 22 mi/36 km SW, and under administrative jurisdiction, of SARATOV; 51°21′N 45°50′E. Terminus of a highway branch. Cotton spinning. Called Saratovskaya Manufaktura until 1929, when given town status and renamed.

Krasnyy Tkach (KRAHS-niyee TKACH), town (2006 population 3,160), SE MOSCOW oblast, central European Russia, on road, 6 mi/10 km NNE of YEGOR'YEVSK; 55°28′N 39°05′E. Elevation 449 ft/136 m. Cotton weaving.

Krasnyy Ural, RUSSIA: see URALETS.

Krasnyy Uzel, RUSSIA: see ROMODANOVO.

Krasnyy Vatras (KRAHS-niyee VAHT-rahs), settlement, E NIZHEGOROD oblast, central European Russia, in the Urga River (tributary of the SURA River) basin, on road, 5 mi/8 km NNE of (and administratively subordinate to) SPASSKOYE; 55°56′N 45°44′E. Elevation 410 ft/124 m. Garment factory.

Krasnyy Yar (KRAHS-niyee YAHR), town (2006 population 7,735), N VOLGOGRAD oblast, SE European Russia, on road and railroad, on the MEDVEDITSA River, 60 mi/97 km NW of KAMYSHIN; 50°42′N 44°43′E. Elevation 367 ft/111 m. Fur factory; lumbering and woodworking; meat-packing plant.

Krasnyy Yar (KRAHS-niyee–YAHR), village, E ASTRAKHAN oblast, SE European Russia, on the BUZAN' arm of the VOLGA RIVER delta mouth, on road, approximately 11 mi/18 km W of the KAZAKHSTAN border, and 22 mi/35 km NE of ASTRAKHAN; 46°32′N 48°20′E. Below sea level. Fisheries; dairying; forestry. Made a city in 1785; reduced to village status in 1925.

Krasnyy Yar (KRAHS-niyee YAHR), village (2006 population 7,645), central SAMARA oblast, SE European Russia, on the SOK RIVER (landing), opposite the mouth of the KONDURCHA, on road, 7 mi/11 km E of SAMARA; 53°30′N 50°22′E. Elevation 187 ft/56 m. In agricultural (hemp, flax, sunflowers) and logging area; bakery, dairy.

Krasnyy Yar (KRAHS-niyee YAHR), village (2006 population 3,090), central SARATOV oblast, SE European Russia, near the VOLGA RIVER, on highway branch, 14 mi/22 km NE of SARATOV; 51°38′N 46°25′E. Elevation 147 ft/44 m. Flour mill.

Krasnyy Yar (KRAHS-niyee YAHR), village, NE VOLGOGRAD oblast, central European Russia, on the W bank of the VOLGA RIVER, near highway, 57 mi/92 km S of SARATOV; 50°37′N 45°49′E. Elevation 190 ft/57 m. In agricultural area (grains, sugar beets, sunflowers).

Krasnyy Yar (KRAHS-niyee YAHR), village (2006 population 5,300), S central OMSK oblast, SW Siberian

Russia, on the IRTYSH River, on road and local railroad spur, 23 mi/37 km NNW of OMSK, and 9 mi/14 km NE of LYUBINSKIY, to which it is administratively subordinate; 55°14′N 72°56′E. Elevation 213 ft/64 m. Condensed milk plant.

Krasnyy Yar (KRAHS-niyee YAHR), village (2006 population 3,130), S TOMSK oblast, S central Siberian Russia, on the OB' RIVER, on road junction, 21 mi/34 km SE of KRIVOSHEINO, to which it is administratively subordinate; 57°07′N 84°32′E. Elevation 226 ft/68 m. Lumbering, woodworking.

Krasnyy Yar (KRAHS-nee YAHR), E central LUHANS'K oblast, UKRAINE, NE suburb of LUHANS'K, on the DONETS, and 8 mi/13 km NE of city center; 48°39′N 39°24′E. Elevation 121 ft/36 m. Vegetables.

Krasnyy Yasyl (KRAHS-niyee YAH-sil), settlement, S PERM oblast, W central URALS, E European Russia, on the Iren' River (KAMA River basin), on road, 22 mi/35 km SSW of KUNGUR, and 13 mi/21 km SW of ORDA, to which it is administratively subordinate; 57°05′N 56°39′E. Elevation 459 ft/139 m. Stone crushing and forming.

Krasslau, LATVIA: see KRĀSLAVA.

Kraste, new town, central ALBANIA, SE of BURREL. Chrome mining.

Krastovo, Bulgaria: see KRASNOVO.

Krasyliv (krah-SI-leev), city, (Russian *Krasilov*) (Polish *Krasilow*), central KHMEL'NYTS'KYY oblast, UKRAINE, on SLUCH RIVER, right tributary of HORYN' River, 16 mi/26 km N of KHMEL'NYTS'KYY; 49°39′N 26°58′E. Elevation 964 ft/293 m. Raion center; manufacturing of food, fabricated metal products, machinery, feeds, construction materials. Heritage museum. Known since 1444; owned by Prince Ostroz'kyy and his family (16th–early 17th centuries); city since 1964. Sizeable Jewish community since the 17th century, reduced by the Civil War pogroms of 1918–1921, but still numbering 1,550 in 1926; eliminated by the Nazis in 1941—fewer than 100 Jews remaining in 2005.

Krasyukovskaya (krah-SYOO-kuhf-skah-yah), village (2006 population 4,315), SW ROSTOV oblast, S European Russia, in the DON River basin, on road and railroad, 8 mi/11 km N of NOVOCHERKASSK; 47°33′N 40°06′E. Elevation 118 ft/35 m. In agricultural area (wheat, sunflowers; livestock).

Kraszna, ROMANIA: see CRASNA.

Kraszna River, ROMANIA: see CRASNA RIVER.

Krathis River (KRAH-thees), Latin *Crathis*, flows c.15 mi/24 km N, in AKHAIA prefecture, WESTERN GREECE department, GREECE; rises in AROANIA mountains, flows past N PELOPONNESUS, parallel to border with PELOPONNESE department, to Gulf of CORINTH below AKRATA; site of hydroelectric plants.

Kratie (KRAW-TEE), province (☐ 4,283 sq mi/11,135.8 sq km; 2007 population 351,549), E CAMBODIA, N border with STUNG TRENG province, E border with MONDOLKIRI province, WSW borders with KOMPONG THOM and KOMPONG CHAM provinces, S border with VIETNAM; ☉ KRATIE; 12°30′N 106°00′E. Agriculture (rice, rubber, tree, and garden crops); forest products. Recent deforestation problems. Khmer population with Cham, Biet, Kurj, Lao, and other minorities. Also spelled Krachen and Krâchéh.

Kratie (KRAW-TEE), city, ☉ KRATIE province, central CAMBODIA, on left bank of MEKONG River (head of regular navigation) below the Prek Patang Rapids, and 95 mi/153 km NE of PHNOM PENH; 12°29′N 106°01′E. Transportation hub and market center. Food processing; light manufacturing. In forested (hardwoods, big game) region; agricultural center; corn, tobacco, rubber, cotton, castor beans, rice, kapok; trade in timber, horns, and hides. Slate quarries nearby. Sambor ruins (6th and 7th century Sambhupura) nearby covering 0.4 sq mi/1 sq km. Khmer population with Cham minority. Also spelled Krachen and Krâchéh.

Kratke Range, rises to c. 10,000 ft/3,048 m. in NE PAPUA NEW GUINEA.

Kratovo (krah-TO-vo), village, FORMER YUGOSLAV REPUBLIC OF MACEDONIA, 38 mi/61 km E of SKOPJE, at W foot of OSOGOV MOUNTAINS. Light manufacturing. Copper and lead mines; also traces of gold, silver, iron, and zinc. Mining center in Roman times.

Kratovo (KRAH-tuh-vuh), settlement (2006 population 6,780), central MOSCOW oblast, central European Russia, near the MOSKVA River, on road and railroad, 8 mi/13 km ESE of KRASKOVO, and 2 mi/3.2 km N of RAMENSKOYE; 55°36′N 38°11′E. Elevation 469 ft/142 m. Holiday retreat for residents of neighboring cities; spas, summer homes.

Kratske, RUSSIA: see PODCHINNYY.

Kratsovon, Greece: see KHASIA.

Krattske, RUSSIA: see PODCHINNYY.

Kratzau, CZECH REPUBLIC: see CHRASTAVA.

Kratzke, RUSSIA: see PODCHINNY, VOLGOGRAD oblast.

Kraubath an der Mur (KROU-baht ahn der MUR), village, STYRIA, SE central AUSTRIA, near MUR River, 9 mi/14.5 km SW of LEOBEN; 47°18′N 14°56′E.

Krauchenwies (KROU-khuhn-vees), village, BADEN-WÜRTTEMBERG, SW GERMANY, 5 mi/8 km S of SIGMARINGEN; 48°02′N 09°15′E.

Krauchmar (KROCH-MAH), town, KOMPONG CHAM province, central CAMBODIA, small port on left bank of MEKONG River and 20 mi/32 km NE of KOMPONG CHAM. Transportation hub and market center. In forested (hardwoods) and agricultural (rice, cotton, tobacco, corn, peanuts) area; alcohol distillery. Khmer with Cham and Vietnamese minorities.

Kraul Mountains, a chain of mountains and nunataks in NEW SCHWABENLAND, EAST ANTARCTICA, that extend N from VESTSTRAUMEN GLACIER, 70 mi/113 km long; 73°30′S 14°10′W.

Kraulshavn (KROULS-houn), Greenlandic *Nutaarmiut* (1995 population 63), hunting settlement, Upernavik commune, W GREENLAND, on peninsula in BAFFIN BAY, 93 mi/151 km NNW of UPERNAVIK; 74°08′N 57°05′W.

Krautheim (KROUT-heim), town, BADEN-WÜRTTEMBERG, S GERMANY, on the JAGST, 9 mi/14.5 km SW of MERGENTHEIM; 49°23′N 09°38′E. Textile manufacturing. Has medieval fortifications and a ruined castle from 13th century.

Krauthem, LUXEMBOURG: see CRAUTHEM.

Kravare (KRAH-vah-RZHE), Czech *Kravaře*, German *Krawarn*, town, SEVEROMORAVSKY province, central SILESIA, CZECH REPUBLIC, on OPAVA RIVER and 5 mi/8 km E of OPAVA; 49°56′N 18°00′E. Railroad junction. Manufacturing (textiles, machinery, food); poultry. Has a noted 18th-century baroque castle with park. Part of GERMANY until 1920.

Kravoder (KRAH-vo-der), village, MONTANA oblast, KRIVODOL obshtina, BULGARIA; 43°18′N 23°25′E.

Krawang (kah-RAH-wang), town, Java Barat province, INDONESIA, 32 mi/51 km ESE of JAKARTA; ☉ Krawang district; 06°19′S 107°17′E. Trade center for rice-growing region; machine shops. Nearby are major irrigation works of CITARUM RIVER. Also spelled Karawang.

Krawarn, CZECH REPUBLIC: see KRAVARE.

Krayishte, highland, BULGARIA and SERBIA: see KRAISHTE.

Kraynovka (krah-yee-NOF-kah), village (2004 population 1,300), N central DAGESTAN REPUBLIC, NE CAUCASUS, SE European Russia, at the mouth of the N (Old Terek) arm of the TEREK RIVER delta, on the CASPIAN SEA coastal road, 35 mi/56 km ENE of KIZLYAR; 43°58′N 47°22′E. Below sea level. In agricultural area. Fisheries.

Kraynovka, RUSSIA: see KRAINOVKA, DAGESTAN Republic.

Krdzhali, Bulgaria: see KURDZHALI.

Kreamer (KREE-muhr), unincorporated village, SNYDER county, central PENNSYLVANIA, 5 mi/8 km W of SELINSGROVE on Middle Creek; 40°48′N 76°57′W. Manufacturing of lumber, wood products; agriculture includes dairying; timber.

Krebs, town (2006 population 2,131), PITTSBURG county, SE OKLAHOMA, 3 mi/4.8 km E of MCALESTER; 34°55′N 95°43′W. In agricultural area. Once a coal mine boom town, coal mining has now declined. Lake Arrowhead reservoir to E. Settled c.1880, incorporated 1903.

Krechevitsy (krye-chye-VEE-tsi), town (2006 population 3,260), NW NOVGOROD oblast, NW European Russia, on the VOLKHOV RIVER, near highway, 7 mi/11 km NNE of NOVGOROD; 58°37′N 31°24′E. Military airbase in the vicinity. Residents work mainly in Novgorod.

Krefeld (KREH-feld), city, North Rhine-Westphalia, W GERMANY, a port on the RHINE RIVER; 51°20′N 06°34′E. Center of the German silk and velvet industry and a major railroad hub. Other manufacturing includes fabricated metal products, machinery, apparel, chemicals. Site of a zoological research institute, a zoo, and botanical garden. Chartered in 1373 and was an important linen-weaving center until it passed (1702) to PRUSSIA. The silk industry, encouraged by a monopoly given to the city by Frederick II of PRUSSIA, soon replaced linen weaving; and in the 20th century the manufacturing of artificial silk became important. In 1929 the neighboring town of Ürdingen was incorporated into Krefeld. Formerly sometimes spelled CREFELD.

Kreider, El (krai-DER, el), village, El BAYADH wilaya, NW ALGERIA, in the High Plateaus, at N edge of the Chott ech CHERGUI, on railroad to BÉCHAR and 50 mi/80 km S of SAÏDA.

Kreiensen (KREI-en-sen), village, LOWER SAXONY, NW GERMANY, on the LEINE, 3 mi/4.8 km WSW of BAD GANDERSHEIM; 51°52′N 09°58′E. Railroad junction.

Kreka (KRAI-kah), town, N BOSNIA, BOSNIA AND HERZEGOVINA, just W of TUZLA, on railroad; 44°31′N 18°38′E. Lignite and salt mines; saltworks (evaporating plant, built 1892); stone quarry.

Kremasti (kre-mah-STEE), Italian *Cremasto*, town, N RHODES island, DODECANESE prefecture, SOUTH AEGEAN department, GREECE, on NW shore, 6 mi/9.7 km SW of Rhodes. Formerly Kremaste.

Kremaston (kre-mah-STON), lake and dam, AKARNANIA prefecture, WESTERN GREECE department, 32 mi/51 km N of AGRINION. Largest artificial lake in the country, formed behind dam at the confluence of the Tavropos and Trikeriotis rivers, where they form the AKHELÓOS River.

Kremenchug, UKRAINE: see KREMENCHUK.

Kremenchugskoye Vodokhranilishche, UKRAINE: see KREMENCHUK RESERVOIR.

Kremenchuk (kre-men-CHOOK), city (2001 population 234,073), S POLTAVA oblast, UKRAINE, on the DNIEPER (Ukrainian *Dnipro*); 49°04′N 33°25′E. Elevation 229 ft/69 m. Raion center and node of an industrial complex based on a hydroelectric plant; construction of the dam upstream at SVITLOVODS'K created the large KREMENCHUK RESERVOIR nearby. Kremenchuk has a heavy machine industry (truck assembly and freight car plants, roadbuilding machines); petrochemicals, small oil refinery, manufacturing (construction materials, apparel, furniture, leather goods, food products). It has the freight car construction institute, a branch of the Kharkiv Polytechnical Institute, a civil aviation pilots' school and a museum. It was founded in 1571, a Polish fortress added in 1596; in 17th–18th centuries, it was a Cossack company town and briefly (1661–1663) capital of the Kremenchuk regiment. Under Russian rule it became the capital of Novorossiya guberniya (1765–1783) and administrative center of Katerynoslav vicegerency (1784–1789).

Kremenchuk Reservoir (kre-men-CHOOK) (Ukrainian *Kremenchuts'ke vodoskhovyshche*) (Russian *Kremenchugskoye vodokhranilishche*) on the DNIEPER (Ukrainian *Dnipro*) River, central UKRAINE, in broad Dnieper valley, from KANIV (NW) to KREMENCHUK

(SE). Formed in 1959–1961 with the construction of Kremenchuk hydroelectric station (HES). Length, 93 mi/150 km; width, up to 17 mi/27 km; surface area, 869 sq mi/2,250 sq km; average depth, 20 ft/6 m; maximum depth, 69 ft/21 m; volume, 3.2 cubic mi/13.5 cubic km. Banks are high and steep, strong erosion processes. Level rises with spring flood and falls with winter depletion. Shallows (to 6 ft/2 m) occupy 18% of area, principally along the left bank, provides for aquatic plants, wildlife. Reservoir serves hydroelectric power generation, navigation (48 million tons/52 million metric tons/year), fisheries (6258 tons/6900 metric tons fish/year), irrigation (29,700 acres/12,000 ha), water supply (1.8 cubic km water/year), and recreation. On its shores are the cities of CHERKASY and SVITLOVODS'K.

Kremenchuts'ke vodoskhovyshche, UKRAINE: see KREMENCHUK RESERVOIR.

Kremenets' (KRE-me-nets) (Polish *Krzemieniec*), city, N TERNOPIL' oblast, UKRAINE, 37 mi/60 km N of TERNOPIL'; 50°06'N 25°43'E. Elevation 1,076 ft/327 m. Raion center; on railroad spur end; food processing (sugar, flour), tobacco; powder metallurgy, furniture manufacturing. Medical, teachers and vocational technical schools. Founded in the 11th century, Kremenets' was part of the Kievan duchy and in the 13th century became a fortified city of Halych-Volyn. In 1536–1566, it belonged to the Polish queen Bona. The city passed to Russia during the third partition of Poland in 1795. Briefly as part of Ukraine (1918–1919), it was again under Polish rule from 1919 to 1939, when it was annexed by the USSR. Since 1991, part of independent Ukraine. Also known as Kremyanets'. An early Jewish community, which was greatly reduced by massacres of 1648–1649, revived in the 19th century until World War II. It constituted over half of the population by 1941.

Kremenets Hills (KRE-me-nets) (Ukrainian *Kremenets'ki Hory*) (Polish *Góry Krzemienieckie*), in N rim of PODOLIAN UPLAND, W UKRAINE, extends approximately 45 mi/72 km SW-NE, from IKVA RIVER near KREMENETS' to HORYN' River near OSTROH. Average elevation 1,000 ft/305 m. Chalk and peat deposits.

Kremenets'ki Hory, UKRAINE: see KREMENETS HILLS.

Kremenki (KRYE-meen-kee), city (2005 population 11,955), central KALUGA oblast, central European Russia, in the OKA River basin, on highway, 13 mi/21 km W of KALUGA; 54°53'N 37°07'E. Elevation 436 ft/132 m. Printing house. Site of intense fighting between German and Soviet armies in autumn and winter 1941.

Kremennaya, UKRAINE: see KREMINNA, Luhans'k oblast.

Kremennaya, UKRAINE: see OLEKSANDRIVKA.

Kremges, UKRAINE: see SVITLOVODS'K.

Kremhes, UKRAINE: see SVITLOVODS'K.

Kremikovtsi (kre-MEE-kov-tsee), metallurgical plant, GRAD SOFIA oblast, SOFIA obshtina, W BULGARIA, at the S foot of the MURGASH MOUNTAINS, 10 mi/16 km NE of Sofia; 42°47'N 23°30'E. Large metallurgical works depend highly on import through ports of LOM on the DANUBE (N) and BURGAS, a BLACK SEA port (E). Machine building. Iron, lead, and zinc deposits nearby. Has a 15th-century convent, church with 11th-century paintings.

Kreminna (kre-meen-NAH) (Russian *Kremennaya*), city, W LUHANS'K oblast, UKRAINE, in the DONBAS, near the DONETS River, 6 mi/10 km NW of RUBIZHNE; 49°03'N 38°13'E. Elevation 219 ft/66 m. Raion center; coal mining, woodworking; construction materials, furniture. Company town of Izyum regiment since 1688; coal mining since the late 19th century. Formerly also called Novohlukhiv (Russian *Novo-Glukhov*). Established in 1680.

Kreminna, UKRAINE: see OLEKSANDRIVKA, Donets'k oblast.

Kremlëv, RUSSIA: see SAROV.

Kremlin, village (2006 population 230), GARFIELD county, N OKLAHOMA, 10 mi/16 km N of ENID; 36°32'N 97°49'W. In agricultural area; oil and gas.

Kremlin-Bicêtre, Le (krem-lan–bee-se-truh, luh), town, VAL-DE-MARNE department, ÎLE-DE-FRANCE region, N central FRANCE, an immediate S suburb of PARIS, 3 mi/4.8 km from Notre Dame Cathedral, between GENTILLY (W) and IVRY-SUR-SEINE (E); 48°49'N 02°22'E. Manufacturing (chemicals, electronic equipment). Its hospice for the elderly, founded 1632, is now the Bicêtre psychiatric hospital.

Kremmling, town (2000 population 1,578), GRAND county, N central COLORADO, on COLORADO RIVER, just N of GORE RANGE, and 80 mi/129 km. WNW of DENVER; 40°03'N 106°22'W. Elevation 7,362 ft/2,244 m. Market center for lumber and livestock region; sawmill; manufacturing (lumber, food). Routt National Forest nearby; parts of Arapaho National Forest to W, NE, and SE; WILLIAMS FORK RESERVOIR to E.

Kremnica (krem-NYI-tsah), German *Kremnitz*, Hungarian *Körmöcbánya*, town, W STREDOSLOVENSKY province, central SLOVAKIA, in S spur of the GREAT FATRA, on railroad and 11 mi/18 km WSW of BANSKÁ BYSTRICA; 48°42'N 18°55'E. Former mining center (gold, silver, lead, zinc); manufacturing (machinery). Noted for picturesque surroundings; still retains much of 14th-century fortifications; has Gothic church, 15th-century castle chapel, and square marketplace. Underground hydroelectric power plant. Old gold mint, hot radioactive spring, and museum also here. Skiing facilities in vicinity.

Krems an der Donau (KREMS ahn der DO-nou), city, LOWER AUSTRIA, on the Danube, at mouth of KREMS RIVER and E exit of the WACHAU, a scenic, narrow valley of the Danube, 35 mi/56 km WNW of VIENNA; 48°25'N 15°36'E. Railroad and road junction, river port; furniture, textiles, plastic products, fabricated metal products; thermoelectric power station Theib nearby; market and school center, university for postgraduate studies; excellent wine, wine fair and summer tourism. Stein an der Donau (with 14th century church) and a famous penal institution is part of the city. Krems was an important commercial city from the Middle Ages to the 19th century. Remnants of medieval fortifications (Steinertor). Several remarkable ecclesiastical and secular buildings from Gothic to Renaissance styles. The baroque painter Martin Johann Schmidt (Kremser Schmidt) 1718–1801 lived here.

Kremsier, CZECH REPUBLIC: see KROMĚŘÍŽ.

Kremsmünster (krems-MYOOIN-ster), township, E central UPPER AUSTRIA, on KREMS RIVER and 13 mi/21 km W of STEYR; 48°03'N 14°08'E. Manufacturing includes glass, plastics; oil fields nearby. Notable Benedictine abbey (founded 777) has large library and observatory. The treasury keeps the oldest and most remarkable chalice of the early Middle Ages in Bavaria and Austria (Tassilokelch).

Krems River (KREMS), flows c. 35 mi/56 km N, in E central UPPER AUSTRIA; rises 13 mi/21 km E of Lake TRAUN, near KIRCHDORF AN DER KREMS; flows past Kremsmünster, to Traun River (2 mi/3.2 km E of TRAUN).

Kremyanets', UKRAINE: see KREMENETS'.

Krenitsyn Peak, RUSSIA: see ONEKOTAN ISLAND.

Krenitsyn Strait, RUSSIA: see KURILE STRAIT.

Krenitzin Islands (kre-NIT-sin), group of five islands of the Fox Islands, E ALEUTIAN ISLANDS, SW ALASKA, between UNALASKA (SW) and UNIMAK (ENE); 54°01'N 165°23'W. Main islands are AKUTAN and AKUN.

Krepenskiy, UKRAINE: see KRIPENS'KYY.

Kreslavka, Latvia: see KRĀSLAVA.

Kresna (KRES-nah), city, SOFIA oblast, ⊙ KRESNA obshtina (1993 population 6,956), BULGARIA. Railroad station.

Kresnensko Defile (KRES-nen-sko dee-FEE-le), gorge, on the STRUMA River, SOFIA oblast, BULGARIA; 41°44'N 23°10'E.

Kress, village (2006 population 785), SWISHER COUNTY, NW TEXAS, in LLANO ESTACADO, 12 mi/19 km N of PLAINVIEW; 34°22'N 101°45'W. Agriculture (wheat, cotton, corn, sorghum; cattle).

Kressbronn am Bodensee (kres-BRUHN uhm BO-den-seh), village, BADEN-WÜRTTEMBERG, S GERMANY, on N shore of LAKE CONSTANCE, 7 mi/11.3 km SE of FRIEDRICHSHAFEN; 47°36'N 08°36'E. Tourism; agriculture (fruit, hops).

Krestena (kre-STE-nah), town, ILIA prefecture, WESTERN GREECE department, W PELOPONNESUS, 10 mi/16 km ESE of PÍRGOS; 37°37'N 21°38'E. Road center; livestock, Zante currants, wine. Formerly called Selinous. Also spelled Krestaina.

Krestovaya Guba (kree-STO-vah-yah–goo-BAH), settlement on the W coast of the N island of NOVAYA ZEMLYA on the BARENTS SEA, and in NENETS AUTONOMOUS OKRUG, ARCHANGEL oblast, NE European RUSSIA; 74°05'N 55°40'E. Trading post.

Krestovskiy, RUSSIA: see UST'-DONETSKIY.

Krestovyi Pass (kres-TOV-yee), **Gudaurskii Pass**, Russian *Krestovy Pereval* or *Krestovyy Pereval*, pass (elevation 7,815 ft/2,382 m) through the VODORAZDEL'NYI RANGE in main range of the central Greater Caucasus, N GEORGIA, on GEORGIAN MILITARY ROAD and 55 mi/89 km NNW of TBILISI. Connects TEREK (N) and Aragva (S) river valleys. Formerly Pass of the Cross.

Krestovy Pereval, GEORGIA: see KRESTOVYI PASS.

Kresttsy (kryes-TSI), town (2006 population 9,665), central NOVGOROD oblast, NW European Russia, on the Kholova River (Lake IL'MEN basin), on road, 53 mi/85 km SE of NOVGOROD; 58°15'N 32°31'E. Elevation 147 ft/44 m. Railroad spur terminus and highway; sawmilling; agricultural chemicals, furniture, textiles, apparel; dairying, bread making. Made a city in 1776; reduced to status of a village in 1917; made town in 1938.

Kretai (KRUH-tie), Malaysia: see KERTEH.

Krete, Greece: see CRETE.

Kretinga (KRE-ting-uh), German *Krottingen*, city (2001 population 21,423), W LITHUANIA, 13 mi/21 km NNE of Klaipėda; 55°53'N 21°15'E. Railroad and road junction; manufacturing (linen goods, woolens, candles, furniture), oilseed pressing, flour milling, sawmilling. Dates from 13th century; became Russo-Prussian frontier town after 1795; in Russian Kovno government until 1920.

Kreuth (KROIT), village, UPPER BAVARIA, BAVARIA, S GERMANY, in Bavarian Alps, 12 mi/19 km SE of BAD TÖLZ; 47°39'N 11°45'E. Elevation 2,533 ft/772 m. WILDBAD KREUTH (elevation 2,717 ft/828 m), 1.5 mi/2.4 km S, has sulphur springs known since 16th century.

Kreuz, CROATIA: see KRIŽEVCI.

Kreuz, POLAND: see KRZYZ.

Kreuzau (KROITS-ou), town, North Rhine-Westphalia, W GERMANY, on RUHR RIVER, and N slope of EIFEL MOUNTAINS, 4 mi/6.4 km S of DÜREN; 50°45'N 06°28'E. Paper production and processing; has 13th-century church.

Kreuzberg (KROITS-berg), residential district, central BERLIN, GERMANY; 52°32'N 13°25'E. Densely populated area. Home to many immigrants; tradition of self-help development efforts.

Kreuzburg, LATVIA: see KRUSTPILS.

Kreuzburg, POLAND: see KLUCZBORK.

Kreuzburg, RUSSIA: see SLAVSKOYE.

Kreuzeckgruppe (KROITS-ek-grup-pe), mountain massif of the Central Alps, CARINTHIA, S AUSTRIA, between Möll River (N) and DRAU RIVER (S); 46°44'-56'N, 12°51'–13°22'E. Rises up to 8,486 ft/2,587 m in Polinik. Part of the hydropower system Reisseck-Kreuzeck; cable car from Kolbnitz to the storage station Rosswiese (elevation 3,645 ft/1,111 m).

Kreuzingen, RUSSIA: see BOL'SHAKOVO.

Kreuzjoch, AUSTRIA: see KITZBÜHEL ALPS.

Kreuzlingen (KROYTZ-ling-uhn), district, N THURGAU canton, NE SWITZERLAND. Main town is KREUZLIN-GEN; population is German-speaking and Protestant, but many Roman Catholics in Kreuzlingen.

Kreuzlingen (KROYZ-ling-uhn), town (2000 population 17,118), THURGAU canton, NE SWITZERLAND, on LAKE CONSTANCE, between the main lake and the UNTERSEE (Lower Lake). Elevation 1,322 ft/403 m. The town is contiguous with the German city of KON-STANZ. It is an industrial center with the oldest aluminum rolling mill in Switzerland. Food, chemicals, and motor vehicles are also manufactured. The Augustinian monastery, founded in the 13th century and now a school, has a noted baroque church (1650).

Kreuznach, Bad, GERMANY: see BAD KREUZNACH.

Kreuztal (KROITS-tahl), town, North Rhine-Westphalia, W GERMANY, on N edge of SIEGEN; 50°58'N 08°00'E. Manufacturing of machinery, steel; ironworking, brewing. Chartered in 1969 after incorporated of several villages.

Kreuzwald, FRANCE: see CREUTZWALD.

Kreuzweg, Col du (kruts-veg, kol dyoo), pass (2,520 ft/768 m), in the VOSGES MOUNTAINS of E FRANCE, 10 mi/16 km SW of OBERNAI. Splendid views of RHINE valley (E) and of Germany's BLACK FOREST beyond. The resort of Le HOHWALD is 2 mi/3.2 km N.

Kreuzwertheim (kroits-vert-heim), district of WER-THEIM, BAVARIA, central GERMANY, in LOWER FRAN-CONIA, on the MAIN, 18 mi/29 km W of WÜRZBURG; 49°46'N 09°31'E.

Krian (CREE-ahn), coastal district (□ 331 sq mi/860.6 sq km) of NW PERAK, MALAYSIA, on PINANG and KEDAH (KRIAN RIVER) borders. One of Malaya's leading rice-producing districts; watered by Kuran River with irrigation headworks at Bukit Merah. Coconuts along Strait of MALACCA coast. Main centers are PARIT BUNTAR and BAGAN SERAI.

Krian River (CREE-ahn), MALAYSIA; flows 60 mi/97 km SW on KEDAH-PERAK line, along boundary, past Selama, PARIT BUNTAR, Bandar Bharu, and Nibong Tebal, to Strait of MALACCA in S PROVINCE WELL-ESLEY. Forms N border of rich KRIAN rice district.

Kria Vrissi (KREE-ah VREE-see), village, PELLA prefecture, CENTRAL MACEDONIA department, NE GREECE, 16 mi/26 km SE of EDHESSA, near drained Lake YIANNITSA; 40°41'N 22°18'E. Cotton, wheat; wine. Also spelled Kria Vrisi, Krya Vrissi and Krya Vryse.

Kribi (KREE-bee), town (2001 population 48,800), ☉ OCÉAN department, South province, CAMEROON, on Gulf of Guinea, 80 mi/129 km S of DOUALA; 02°59'N 09°56'E. Cocoa exporting center. Lumber-shipping and fishing port, trading, and tourist center. Iron and bauxite deposits nearby. Beach resort area. Hydroelectric plant; hospital.

Krichev (KRI-chev), city, E MOGILEV oblast, BELARUS, on SOZH RIVER (tributary of DNIEPER) and 55 mi/89 km E of MOGILEV; ☉ KRICHEV region; 53°40'N 31°44'E. Railroad junction; manufacturing (rubber products, food); railroad-transport enterprises.

Krichim (KREE-cheem), city, PLOVDIV oblast, Rodopska obshtina, S central BULGARIA, at the N foot of the W RODOPI Mountains, on the VUCHA RIVER (power plant), 16 mi/26 km WSW of PLOVDIV; 42°02'N 24°29'E. Fruit and vegetable canning, vegetable oil extracting, vineyards, winery, construction supplies for nearby hydroelectric complex (Dospat-Devin-Krichim).

Krichim (KREE-cheem), dam on the VUCHA RIVER, BULGARIA; 41°58'N 24°28'E. Formerly known as Antonivanovtsi Dam.

Krichim, Bulgaria: see STAMBOLIISKI, city.

Krichim Gara, Bulgaria: see STAMBOLIISKI, city.

Krichim River, Bulgaria: see VUCHA RIVER.

Krieglach (KREEG-lahkh), township, STYRIA, E central AUSTRIA, on Mürz River, and 16 mi/26 km NE of BRUCK AN DER MUR; 47°33'N 15°34'E. Summer resort. Home and burial place of the poet Peter Rosegger.

Kriegstetten (KREEG-shte-tuhn), district, S SO-LOTHURN canton, N SWITZERLAND. Main town is ZUCHWIL; population is German-speaking and mixed Roman Catholic and Protestant.

Kriekouki, Greece: see ERITHRAI.

Kriens (KREE-ehns), town (2000 population 24,742), LUCERNE canton, central SWITZERLAND; SW suburb of LUCERNE, at foot of PILATUS. Elevation 1,575 ft/480 m.

Kriewen, POLAND: see KRZYWIN.

Krilon, RUSSIA: see CRILLON, CAPE.

Krim (KREEM), peak (3,632 ft/1,107 m) in DINARIC ALPS, W SLOVENIA, 10 mi/16 km SSW of LJUBLJANA.

Krimmitschau, GERMANY: see CRIMMITSCHAU.

Krimml, village, SW SALZBURG, W central AUSTRIA, near TYROL border, 30 mi/48 km WSW of ZELL AM SEE, on the Krimmler Ache (short tributary of the upper SALZACH River); 47°13'N 12°10'E. Railroad terminus. Just S are the noted Krimml Falls, dropping 1,250 ft/381 m in 3 sects. Summer and winter tourism. National Park Hohe Tauern is nearby.

Krimpen aan den IJssel (KRIM-puhn ahn duhn EI-suhl), suburb, SOUTH HOLLAND province, W NETH-ERLANDS, 5 mi/8 km E of ROTTERDAM, on HOLLANDSE IJSSEL RIVER; 51°55'N 04°35'E. Mouth of NEW MAAS RIVER 2 mi/3.2 km to SW; recreational center to NE. Fishing; dairying; cattle, hogs; sugar beets, vegetables, fruit; manufacturing (doors, shipbuilding, food processing).

Krimpen aan den Lek (KRIM-puhn ahn duhn LEK), town, SOUTH HOLLAND province, W NETHERLANDS, on LEK RIVER (vehicle ferry), at mouth of NOORD RIVER, and 7 mi/11.3 km ESE of ROTTERDAM; 51°54'N 04°38'E. Dairying; cattle, hogs; grain, vegetables, fruit. Light manufacturing.

Kringsjå (KRING-shaw), waterfall (43 ft/13 m) on OTRA RIVER, AUST-AGDER county, S NORWAY, 11 mi/18 km N of KRISTIANSAND. Hydroelectric plant.

Krinichanskiy, UKRAINE: see CHERVONOGVARDIYS'KE.

Krinichki, UKRAINE: see KRYNYCHKY.

Krinichnaya, UKRAINE: see KRYNYCHNA.

Krinichnoye, UKRAINE: see BIRYUKOVE.

Krioneri (kree-o-NE-ree), village, AKARNANIA prefecture, WESTERN GREECE department, deepwater port of MESOLONGI, on Gulf of PATRRAS, 10 mi/16 km ESE of MESOLONGI (linked by railroad); 38°21'N 21°36'E. Fisheries. Also spelled Kryoneri.

Kripens'kyy (KREE-pen-skee) (Russian *Krepenskiy*), town, S LUHANS'K oblast, UKRAINE, in the DONBAS, 3 mi/5 km SW (and under jurisdiction) of ANTRATSYT; 48°04'N 39°03'E. Elevation 669 ft/203 m. Anthracite mining, enrichment plant. Established in 1777; town since 1938.

Krishna, district (□ 3,369 sq mi/8,759.4 sq km), ANDHRA PRADESH state, S INDIA; ☉ MACHILIPATNAM. Lies between EASTERN GHATS (NW) and BAY OF BENGAL (SE); SW portion in KRISHNA River delta. Agriculture includes rice, millet, oilseeds, tobacco, sugarcane, sunn hemp (jute substitute), cotton. Main towns are Bezwada, MACHILIPATNAM, GUDIVADA. E portion of original district separated (1925) to form WEST GOD-AVARI district.

Krishna, river, c.800 mi/1,287 km long, INDIA; rising in MAHARASHTRA state, W India, in the WESTERN GHATS; flows SE through KARNATAKA and ANDHRA PRADESH states to the BAY OF BENGAL. Supplies water for both hydroelectric power and irrigation of extensive areas in all three states; its flow fluctuates according to seasonal monsoon rains. Its source is sacred to Hindus; the river is named for the god Krishna. Formerly spelled Kistna.

Krishna Gandaki River, NEPAL: see KALI GANDAKI RIVER.

Krishnagar (KRISH-nuh-guhr), city, ☉ NADIA district, WEST BENGAL state, E INDIA, on the JALANGI RIVER; 23°24'N 88°30'E. Road and railroad junction. Sugar milling is the largest industry. The main products of the area are rice, jute, sugar, ceramics, and plywood. KRISHNA NAGAR (former spelling) was the residence of the rajas of the former princely state of Nadia. A center of art, sculpture, and music, the city has some well-known educational institutions.

Krishnagiri (KRISH-nuh-gee-ree), town, DHARMAPURI district, TAMIL NADU state, S INDIA, 50 mi/80 km SE of BANGALORE (Karnataka state). Road and trade center in agricultural area; castor, peanut, and sesame oil extraction, tanning; grapes, mangoes.

Krishna Nagar, INDIA: see KRISHNAGAR.

Krishnaraja Nagara (KRISH-nuh-RAH-jah nuh-gah-rah), town, tahsil headquarters, MYSORE district, KARNATAKA state, S INDIA, near KAVERI RIVER, 10 mi/16 km NW of MYSORE city center. KRISHNAR-AJASAGARA, reservoir, lies just NW. Handicrafts (biris, pottery); rice, tobacco, millet. Also spelled Krishnarajnagar. Formerly YEDATORE.

Krishnarajasagara (KRISH-nuh-RAH-juh-SAH-guh-rah), reservoir (□ 50 sq mi/130 sq km), MYSORE district, KARNATAKA state, S INDIA, on KAVERI River, immediately SE of Krishnarajanagara and 20 mi/32 km NW of MYSORE; 14 mi/23 km long, up to 7 mi/11.3 km wide. Impounded by masonry dam (140 ft/43 m high, 8,600 ft/2,621 m long) across the KAVERI River; dam, crossed by motor road, was begun 1911 and completed in present form in 1931. Reservoir supplies IRWIN CANAL (important irrigation system) and furnishes auxiliary water supply to major hydroelectric works near Sivasamudram island. BRIN-DAVAN GARDENS, below the dam, are famous for their landscaped terraces with diversiform fountains (floodlit at night), which have made the place a popular tourist resort.

Krishnarajpet (krish-nuh-RAHJ-pet), town, MANDYA district, KARNATAKA state, S INDIA, 28 mi/45 km WNW of MANDYA; 12°40'N 76°30'E. Trades in millet, sugarcane, rice; hand-loom weaving. Also spelled Krishnarajpete. Formerly called ATTIKUPPA.

Kristdala (KRIST-DAHL-ah), village, KALMAR county, SE SWEDEN, 13 mi/21 km NW of OSKARSHAMN; 57°24'N 16°12'E.

Kristiania, NORWAY: see OSLO.

Kristianiafjord, Norway: see: OSLOFJORDEN.

Kristianopel (kris-TYAHN-OOP-el), village, BLEKINGE county, S SWEDEN, on BALTIC SEA, at S end of KALMAR SOUND, 18 mi/29 km ENE of KARLSKRONA; 56°15'N 16°02'E. Founded in seventeenth century as Danish fortress opposite KALMAR. Old walls; seventeenth-century church.

Kristiansand (krist-yahn-SAHN), city (2007 population 77,840), ☉ VEST-AGDER county, S NORWAY, commercial and passenger port on the SKAGERRAK strait; 58°10'N 08°00'E. Manufacturing includes ships, textiles, metal and wood products, canned fish, and beer. Founded (1641) by Christian IV of DENMARK and Norway and became an episcopal see in 1682; rebuilt after destructive fire (1892). Its Christiansholm Fortress (1662–1672) now houses a restaurant. The Var-odden Bridge (1956), one of the largest suspension bridges in N Europe, spans the nearby Randesund. There is a naval station, and an airport at Kjevik (6 mi/10 km NE). Vehicle ferry to HIRTSHALS (Denmark). Has nineteenth-century cathedral and a county museum. Formerly spelled Christiansand.

Kristianstad, county, SWEDEN: see SKÅNE.

Kristianstad (kris-TYAHN-STAHD), town, SKÅNE county, was capital of former Kristianstad county, SE SWEDEN, on HELGE Å RIVER; 56°02'N 14°10'E. Its nearby seaport, ÅHUS, is on BALTIC SEA. Commercial and industrial center; manufacturing (machinery, processed food); in fertile agricultural region. Founded by Christian IV of DENMARK (1614), Kristianstad changed hands frequently before passing definitively to Sweden (1678). Earliest example of Renaissance town planning in N EUROPE. Birthplace of Swedish film industry; museum. Church built by Archbishop Absalon (twelfth century) nearby. International airport to S.

Kristiansund (krist-yahn-SOON), city, MØRE OG ROMSDAL county, W NORWAY, a port on the ATLANTIC OCEAN; 63°07′N 07°45′E. Site of a large trawler fleet; shipbuilding, fish processing, forest production, manufacturing. Chartered in 1742, Kristiansund was destroyed (1940) by bombardment in World War II and has since been rebuilt on three islands enclosing the harbor.

Kristiinankaupunki, FINLAND: see KRISTINESTAD.

Kristina, FINLAND: see RISTIINA.

Kristineberg (kris-TEE-ne-BER-yuh), village, VÄSTERBOTTEN county, N SWEDEN, 70 mi/113 km WNW of SKELLEFTEÅ; 65°04′N 18°35′E. Mining center (copper, zinc, silver, gold, sulphur). Ore shipped by cable railroad (60 mi/97 km long) to BOLIDEN, thence to smelters at RÖNNSKÄR.

Kristinehamn (kris-TEE-ne-HAHMN), town, VÄRMLAND county, S central SWEDEN, port on LAKE VÄNERN; 59°19′N 14°07′E. Manufacturing (machinery, chemicals for paper mills). First chartered as Bro (1582); rechartered and renamed by Queen Christina (1642).

Kristinestad (kris-TEE-nuh-STAHD), Finnish *Kristiinankaupunki*, town, VAASAN province, W FINLAND, on GULF OF BOTHNIA, 55 mi/89 km S of VAASA; 62°17′N 21°23′E. Elevation 33 ft/10 m. Formerly a major port, now has light industry. Population is largely Swedish-speaking. Founded 1649.

Kritsa (KREET-sah), town, LASITHI prefecture, E CRETE department, 4 mi/6.4 km SW of AYIOS NIKOLAOS; 35°10′N 25°39′E. Carob, raisins, olive oil. Formerly Kretsa.

Kriva (KREE-vah), river, NE BULGARIA; 43°27′N 27°00′E.

Kriva (kree-VAH), river, PLOVDIV oblast, S BULGARIA, tributary of the MARITSA RIVER; 42°54′N 23°12′E.

Krivaja (KREE-vei-yah), village (2002 population 13,494), VOJVODINA, N SERBIA, 10 mi/16 km N of VRBAS, in the BACKA region. Until 1947, called Mali Idjos or Mali Idyosh (Serbo-Croatian *Mali Idoš,* Hungarian *Kishegyes*). Also spelled Krivaya.

Krivaja River (kree-vah-YAH), c. 50 mi/80 km long, central BOSNIA, BOSNIA AND HERZEGOVINA; rises in two headstreams joining 9 mi/14.5 km SW of KLADANJ; flows NW to BOSNA RIVER at Zavidovići. Also spelled Krivaya River.

Krivandino (kree-VAHN-dee-nuh), settlement, E MOSCOW oblast, central European Russia, on road, 7 mi/11 km E of SHATURA; 55°33′N 39°41′E. Elevation 446 ft/135 m. Railroad junction. Glassworks.

Krivan Peak (kri-VAHN-yuh), SLOVAK *Kriváň,* HUNGARIAN *Kriván,* POLISH *Krywan* (8,182 ft/2,494 m) of the HIGH TATRAS, VYCHODOSLOVENSKY province, N Slovakia, 15 mi/24 km NW of POPRAD, near Polish border; 49°29′N 19°34′E. STRBA LAKE is on SE slope.

Kriva Palanka (KREE-vah pah-LAHN-kah), Turkish *Eğridere,* village, FORMER YUGOSLAV REPUBLIC OF MACEDONIA, on the KRIVA REKA (river), and 50 mi/80 km ENE of SKOPJE; at N foot of OSOGOV MOUNTAINS. Local trade center, linked via VELBAZH PASS with KYUSTENDIL (BULGARIA). First mentioned in 1633.

Kriva Reka (KREE-vah RAI-kah) [Serbo-Croatian= crooked river], river, c.45 mi/72 km long, FORMER YUGOSLAV REPUBLIC OF MACEDONIA; rises on N slope of OSOGOV MOUNTAINS, 6 mi/9.7 km SE of KRIVA PALANKA; flows W, past Kriva Palanka, to PCINJA RIVER 6 mi/9.7 km E of KUMANOVO.

Krivaya, SERBIA: see KRIVAJA.

Krivets (kree-VETS), peak (8,888 ft/2,709 m) in the PIRIN MOUNTAINS, SW BULGARIA.

Krivichi (KRI-vi-chi), Polish *Krzywicze,* urban settlement, MINSK oblast, BELARUS, on Servech River (tributary of Viliia River), and 22 mi/35 km NE of VILEIKA. Agricultural processing.

Krivko (kreef-KO), village, N central LENINGRAD oblast, NW European Russia, on the KARELIAN ISTHMUS, 42 mi/68 km N of SAINT PETERSBURG, and 2 mi/3.2 km N of SOSNOVO; 60°35′N 30°15′E. Elevation 134 ft/40 m. Livestock breeding.

Krivodanovka (kree-vuh-DAH-nuhf-kah), village (2006 population 8,970), E NOVOSIBIRSK oblast, SW SIBERIA, RUSSIA, in the OB′ RIVER basin, on local railroad spur, 13 mi/21 km WNW, and under administrative jurisdiction of NOVOSIBIRSK; 55°05′N 82°39′E. Elevation 291 ft/88 m. Plastics factory. Pig farming.

Krivodol (kree-vo-DOL), city, MONTANA oblast, ☉ Krivodol obshtina (1993 population 13,706), BULGARIA, at the mouth of the OGOSTA RIVER; 43°23′N 23°29′E. Railroad station. Ceramics; wood processing, foodstuff manufacturing; machine-building workshops.

Krivoklat (KRZHI-vo-KLAHT), Czech *Křivoklát,* German *Pürglitz,* village, STREDOCESKY province, W central BOHEMIA, CZECH REPUBLIC, on BEROUNKA RIVER, on railroad, and 24 mi/39 km W of PRAGUE; 50°02′N 13°53′E. Has a noted 12th-century castle, which burned down in 1422 and was later rebuilt (today it houses a museum); royal hunting residence in medieval times.

Krivopolyan′ye (kree-vuh-puh-LYAHN-ye), village (2006 population 5,065), NE LIPETSK oblast, S central European Russia, on the VORONEZH RIVER, on road, 2 mi/3.2 km SW of CHAPLYGIN, to which it is administratively subordinate; 53°13′N 39°55′E. Elevation 403 ft/122 m. Food processing, woodworking.

Krivorozhye (kree-vuh-ROZH-ye), village, W central ROSTOV oblast, S European Russia, on the KALITVA RIVER, on road, 17 mi/27 km ESE of MILLEROVO; 48°51′N 40°45′E. Elevation 236 ft/71 m. Flour mill.

Krivorozhye, UKRAINE: see BRYANKA.

Krivoshchekovo (kree-vuh-SHCHYO-kuh-vuh), industrial section of NOVOSIBIRSK, E NOVOSIBIRSK oblast, SW SIBERIA, RUSSIA, on the left bank of the OB′ RIVER, 9 mi/14 km E of the city center; 54°59′N 82°53′E. Elevation 456 ft/138 m. Machine building, sawmilling, food processing.

Krivosheino (kree-vuh-SHE-ee-nuh), village (2006 population 6,020), SE TOMSK oblast, W SIBERIA, RUSSIA, on the OB′ RIVER, on highway junction, 70 mi/113 km NNW of TOMSK; 57°20′N 83°55′E. Elevation 213 ft/64 m. In flax-growing area.

Krivoye Ozero, UKRAINE: see KRYVE OZERO.

Krivoy Rog, UKRAINE: see KRYVYY RIH.

Krivyanskaya (kree-VYAHN-skah-yah), village (2006 population 10,200), SW ROSTOV oblast, S European Russia, on the right bank of the DON River, on highway and near railroad, 20 mi/32 km ENE of ROSTOV-NA-DONU; 47°24′N 40°10′E. Fish-breeding and -processing combine. Population mostly Cossack.

Križevci (KREE-zhev-tsee), Hungarian *Körös,* German *Kreuz,* town, central CROATIA, 16 mi/26 km WNW of BJELOVAR; 46°02′N 16°34′E. Trade and railroad center for wine-growing region; meat industry; manufacturing of steel and construction materials. Agriculture and forestry school. Known since 1253.

Krk (KUHRK), Italian *Veglia* (VAI-lyah), town, W CROATIA, on W coast of KRK Island, on ADRIATIC SEA, 22 mi/35 km S of RIJEKA (Fiume), on KVARNER Gulf; 45°05′N 14°35′E. Seaport; sea resort; fishing. Wine growing in vicinity. Has medieval walls and castle, 13th-century cathedral, and many Venetian houses. First mentioned (3rd century B.C.E.) as Greek *Kurikta*; later known as Roman *Curicum*; became Roman Catholic bishopric in 9th century.

Krk (KUHRK), Italian *Veglia* (VAI-lyah), island (☐ 158 sq mi/410.8 sq km), W CROATIA, in KVARNER Gulf of the ADRIATIC SEA. The largest of Croatia's islands in the Adriatic, it has several small seaside resorts; tourism. The chief town, Krk, has retained its medieval walls and castle and has a 13th-century Roman

Catholic cathedral. The main oil pipeline connects Omišalj with SISAK.

Krka National Park (☐ 55 sq mi/142 sq km), along lower course of KRKA RIVER and its tributaries, DALMATIA, S CROATIA, NE of SIBENIK. Includes several scenic canyons, two connected lakes, and many waterfalls; winter bird sanctuary. Visovac monastery (1445) built by Bosnian Franciscans on small island.

Krka River (KUHR-kah), 46 mi/74 km long, in W CROATIA, in DALMATIA; rises N of KNIN; flows SSW, past Knin, SKRADIN, and SIBENIK, to ADRIATIC SEA just below Šibenik. Navigable for 31 mi/50 km; links Šibenik with sea. Forms series of small lakes in lower course; on island in one of its lakes is a Franciscan convent. Forms fifty waterfalls (up to 280 ft/85 m high, up to 300 ft/91 m wide) utilized by five hydroelectric plants, including one at Skradin. Receives CIKOLA River. KRKA NATIONAL PARK on lower course.

Krka River (KER-kah), c.60 mi/97 km long, S SLOVENIA; rises 7 mi/11.3 km SE of GROSUPLJE; flows E, past Novo Mesto, to Sava River opposite Brežice.

Krkonoše, CZECH REPUBLIC and POLAND: see GIANT MOUNTAINS.

Krkonose, CZECH REPUBLIC: see GIANT MOUNTAINS.

Krn (KERN), peak (7,365 ft/2,245 m) in JULIAN ALPS, NW SLOVENIA, 4 mi/6.4 km ENE of KOBARID. Part of Italy (1919–1947); then passed to the former YUGOSLAVIA.

Krndija (KUHRN-dee-yah), mountain, E CROATIA, 11 mi/18 km SW of Našice, in SLAVONIA; rises to 2,296 ft/700 m, sloping c.10 mi/16 km ESE from peak.

Krnjaja (KUHRN-yei-yah), village, central SERBIA, 5 mi/8 km NE of Palanka. Also spelled Krnyevo.

Krnjaja, SERBIA: see KLJAJICEVO.

Krnov (KUHR-nof), German *Jägerndorf,* city (2001 population 25,764), SEVEROMORAVSKY province, in SILESIA, CZECH REPUBLIC, on the OPAVA RIVER, near the POLAND border; 50°06′N 17°43′E. Railroad junction. An industrial center; manufacturing (textiles, especially woolens; machinery; musical instruments, notably organs); food processing. Founded in 1221, it was as an independent duchy from 1377 to 1523. The city has an 18th-century castle and several fine churches and abbeys.

Krnyaya, SERBIA: see KLJAJICEVO.

Krobia (KRO-bee-ah), German *Kröben,* town, Leszno province, W POLAND, 8 mi/12.9 km S of GOSTYN. Machinery manufacturing, sawmilling, flour milling. Cannery nearby.

Krocehlavy (KRO-che-HLAH-vi), Czech *Kročehlavy,* German *Krocehlaw,* suburb of KLADNO, STREDOCESKY province, W central BOHEMIA, CZECH REPUBLIC.

Krocehlaw, CZECH REPUBLIC: see KROCEHLAVY.

Krøderen (KRUH-druhn), lake expansion (☐ 16 sq mi/41.6 sq km) of DRAMMENSELVA River, SE NORWAY, at S end of the HALLINGDAL, 35 mi/56 km NW of OSLO; 27 mi/43 km long, 1 mi/1.6 km–2 mi/3.2 km wide. Fisheries.

Kroh (crow), town, northernmost PERAK, MALAYSIA, 35 mi/56 km ENE of SUNGEI PATANI, in KALAKHIRI Mountains (THAILAND line) and near KEDAH border. Tin mining. Malay border post on highway to BETONG (Thailand), strategic during Communist insurgency of 1950s and 1960s.

Krojanke, POLAND: see KRAJENKA.

Krokeai (kro-ke-AI), Latin *Croceae,* LAKONIA prefecture, S PELOPONNESE department, extreme SE mainland GREECE, 15 mi/24 km SSE of SPARTA; 36°53′N 22°33′E. Citrus fruits, olives, wheat. Formerly called Levetsova.

Krokodil River or **Krokodilrivier,** REPUBLIC OF SOUTH AFRICA: see CROCODILE RIVER.

Krokom (KROO-kum), town, JÄMTLAND county, N central SWEDEN, on NE arm of STORSJÖN LAKE, at outflow of INDALSÄLVEN RIVER, 11 mi/18 km NW of ÖSTERSUND; 63°20′N 14°27′E. Includes Hissmofors village.

Krolevets' (kro-le-VETS) (Russian *Krolevets*), city, W SUMY oblast, UKRAINE, on road near railroad, 22 mi/35 km NNE of KONOTOP; 51°33′N 33°23′E. Elevation 501 ft/152 m. Raion center; metal parts and moldings, food processing (including dairy, cannery, flour milling), hemp processing, weaving factory, peat cutting, sawmilling. Phosphorite deposits nearby. Established in 1601, company center of Nizhen regiment (1650–1781). Since the 16th century, known for decorative weaving and embroidery. Sizeable Jewish community since the early 19th century, largely involved in dairy farming and tailoring; decimated during the 1905 and 1919 pogroms, with 1,300 remaining in 1926; community eliminated by the Nazis in 1941—fewer than 100 Jews remaining in 2005.

Królewska Huta, POLAND: see CHORZÓW.

Kroměříž (KRO-mnye-RZHEEZH), German *Kremsier*, city (2001 population 29,225), JIHOMORAVSKY province, CZECH REPUBLIC, central MORAVIA, on the MORAVA RIVER; 49°18′N 17°24′E. Agricultural center; manufacturing (generators and gasoline engines); food processing. Chartered in 1290, it served as the residence of the bishops of OLOMOUC. Site of a meeting (November 1848–March 1849) of the first Austrian constituent parliament. Among the city's present-day landmarks is an 18th century palace with a large library and a ceremonial hall. Military base.

Krommenie (KRAW-muh-nee), suburb of WORMER, NORTH HOLLAND province, W NETHERLANDS, 1 mi/1.6 km W of city center and 5 mi/8 km NW of ZAAN-STAD, in ZAANSTREEK industrial area; 52°30′N 04°45′E. Dairying; cattle, sheep; flowers, vegetables, fruit; manufacturing (linoleum).

Kromme Rijn, NETHERLANDS: see CROOKED RHINE RIVER.

Krompachy (krom-PAH-khi), Hungarian *Korompa*, town, VYCHODOSLOVENSKY province, E central SLO-VAKIA, on HERNAD RIVER, on railroad and 22 mi/35 km NW of KOŠICE; 48°55′N 20°52′E. Copper metallurgy, machinery; former iron foundries. Has 18th-century baroque church and museum of metallurgy. Ironworks at MARGECANY (mahr-GE-tsah-NI), 6 mi/9.7 km ESE, and at SLOVINKY (slo-VIN-ki), 4 mi/6.4 km S.

Krom River, SOUTH AFRICA: see KAREEDOUW.

Kromy (KRO-mi), town (2006 population 7,070), W OREL oblast, SW European Russia, on the Kroma River (tributary of the OKA River), on road and near railroad, 25 mi/40 km SW of OREL; 52°41′N 35°46′E. Elevation 561 ft/170 m. Hemp milling. Known since 1147. Made city in 1778; reduced to status of a village in 1917; town since 1959.

Kronach (KRON-ahkh), town, BAVARIA, E central GERMANY, in UPPER FRANCONIA, on small Hasslach River, 24 mi/39 km NW of BAYREUTH; 50°14′N 11°20′E. Noted for Rosenthal porcelain; manufacturing of chemicals, electrical goods; woodworking. Has 14th–16th-century walls; Gothic church. Cranach, the elder, was born here. On hill (N) is 15th–18th-century castle Rosenberg, with 13th-century watchtower. Formerly also spelled KRANACH.

Kronau (KRON-ou), village, BADEN-WÜRTTEMBERG, W GERMANY, on the KRAICHBACH, 7 mi/11.3 km N of BRUCHSAL; 49°13′N 08°38′E. Asparagus.

Kronau, UKRAINE: see VYSOKOPILLYA.

Kronberg im Taunus (KRON-berg im TOU-nus), town, HESSE, central GERMANY, on S slope of the TAUNUS, 9 mi/14.5 km NW of FRANKFURT; 50°11′N 08°30′E. Electronics industry; mineral water; agriculture (fruit). Has ancient castle, rebuilt 1897–1900. Sometimes spelled CRONBERG.

Kronborg castle, DENMARK: see HELSINGØR.

Krone, POLAND: see KORONOWO.

Krongjawng Pass, INDIA and MYANMAR: see KUM-JAWNG PASS.

Krong Kaeb, municipality and town, CAMBODIA: see KEP.

Krong Kep, municipality and town, CAMBODIA: see KEP.

Krong Pailin, municipality and town, CAMBODIA: see PAILIN.

Krong Preah Sihanouk, municipality, city, and port, CAMBODIA: see KOMPONG SOM.

Kronoberg (KROO-no-BERY), county (□ 3,826 sq mi/9,947.6 sq km), S SWEDEN; ⊙ VÄXJÖ; 56°40′N 14°40′E. Forms S part of SMÅLAND province. Rolling area of woods, marshland, and numerous lakes (Åsnen and MÖCKELN are largest); drained by MÖRRUMSÅN, HELGE Å, LAGAN, and many smaller rivers. Manufacturing (glass at ORREFORS and KOSTA, machinery; lumbering, sawmilling, woodworking, paper milling). Chief towns are Växjö and LJUNGBY.

Kronotskaya River (kruh-nuht-SKAH-yah), 30 mi/48 km long, E central KAMCHATKA PENINSULA, KAMCHATKA oblast, extreme E SIBERIA, RUSSIAN FAR EAST; rises in KRONOTSKOYE Lake; flows SE to KRONOTSKIY Gulf of the PACIFIC Ocean. Receives the BOGACHEVKA RIVER.

Kronotskaya Sopka (kruh-nuht-SKAH-yah–SOP-kah), active volcano (11,909 ft/3,630 m), on the E KAMCHATKA PENINSULA, KAMCHATKA oblast, extreme E SIBERIA, RUSSIAN FAR EAST; terminates (S) on the KRONOTSKIY GULF. KRONOTSKOE LAKE to the W.

Kronotskiy Gulf (kruh-NOT-skeeyee), off the E (Pacific) coast of KAMCHATKA PENINSULA, KAMCHATKA oblast, extreme E SIBERIA, RUSSIAN FAR EAST, between Shipunskiy and Kronotskiy peninsulas; 42.5 mi/68.5 km long, 143 mi/231 km wide at its mouth, 4,921 ft/1,500 m deep. Kronotskiy Peninsula separates Kronotskiy and Kamchatka gulfs. Mostly low-lying banks, with some steep points. Freezes in winter.

Kronotskiy Preserve (kruh-NOT-skeeyee), wildlife refuge (□ 4,219 sq mi/10,969.4 sq km), in central region of the E KAMCHATKA PENINSULA, KAMCHATKA oblast, extreme E SIBERIA, RUSSIAN FAR EAST, in the system of mountain ranges descending toward the KAMCHATKA and KRONOTSKIY gulfs. Includes both forest and seashore. Fauna include sable, lynx, snow ram, reindeer, seal, and walrus. Established in 1934 to restore Kamchatka sable. Grizzly bears, poached into extinction in the reserve during the Soviet era, were reintroduced in 1992.

Kronotskoe Lake (kruh-NOT-skuh-ye) (□ 116 sq mi/301.6 sq km), on the E (Pacific) coast of the KAMCHATKA PENINSULA, KAMCHATKA oblast, extreme E SIBERIA, RUSSIAN FAR EAST, W of the KRONOTSKAYA SOPKA and within the KRONOTSKIY PRESERVE; 54°48′N 160°13′E. Depth, 420 ft/128 m; elevation 1,220 ft/372 m. In a volcanic caldera. Source of the KRONOTSKAYA RIVER.

Kronprins Christian Land, GREENLAND: see CROWN PRINCE CHRISTIAN LAND.

Kronprinsen Island (KRON-PRINS-uhn), Danish *Kronprinsens Ejland*, main island of group of seven islets, W GREENLAND, in DAVIS STRAIT, at mouth of DISKO BAY. Abandoned settlement is Imerissoq, 17 mi/28 km SSE of GODHAVN (Qeqertarssuaq); 69°01′N 53°18′W.

Kronshagen (krons-HAH-gen), town, SCHLESWIG-HOLSTEIN, NW GERMANY, 2 mi/3.2 km W of KIEL city center; 54°20′N 10°41′E. Largely residential.

Kronshlot, RUSSIA: see KRONSHTADT.

Kronshtadt (kruhn-SHTAHT), city (2005 population 42,860), NW LENINGRAD oblast, NW European Russia, on the small island of KOTLIN in the Gulf of FINLAND, 30 mi/48 km W of SAINT PETERSBURG; 59°59′N 29°46′E. It is the chief naval base for the Russian Baltic fleet. The harbor is icebound for several months each year. Industries include food processing (meat packing, bakery). Has a naval hospital, Baltic Fleet drama theater, and art school for children. It was founded (1703) by Peter I as a port and fortress to protect the site of Saint Petersburg, and it was the commercial harbor of Saint Petersburg until the 1880s. The port lost its commercial value after the development of Saint Petersburg. Administratively subordinate to Saint Petersburg. A revolt of the sailors in March 1921 was instrumental in establishing Lenin's new economic policy. General unrest among peasants and workers touched off this mutiny of the naval garrison that had been loyal to the Bolsheviks during the revolution. It is also spelled Cronstadt. Until 1723, called Kronshlot.

Krönten, peak (10,197 ft/3,108 m) in ALPS of the FOUR FOREST CANTONS, central SWITZERLAND, 7 mi/11.3 km ESE of ENGELBERG.

Kroonstad, city, NE FREE STATE province, SOUTH AFRICA, on the Vals River; 27°38′S 27°12′E. Elevation 4,723 ft/1,440 m. It is an agricultural and industrial center. There are grain elevators, and grain is shipped from the city. Kroonstad is also an important railroad junction and has large marshaling yards. The town's chief industries are clothing manufacturing and mineral processing, and the production of machine parts. Kroonstad was founded in 1855. Its growth was stimulated by the discovery of gold in the region in the late 19th century. After the fall of Bloemfontein (now also called MANGAUNG) during the South African War, it was (March 13–May 11, 1900) the capital of the ORANGE FREE STATE. Kroonstad Technical College is in the city. Cattle, sheep, wheat, and fruit are produced in the area. Has a large black residential suburb called Maokeng.

Kropachëvo (kruh-pah-CHO-vuh), town, W CHELYA-BINSK oblast, SW URALS, E European Russia, on road and railroad, 27 mi/43 km E of ASHA; 55°01′N 57°59′E. Elevation 1,302 ft/396 m. Railroad establishments.

Kröpelin (KRUH-pe-lin), town, Mecklenburg-Western Pomerania, NE GERMANY, near MECKLENBURG BAY of the BALTIC SEA, 14 mi/23 km W of ROSTOCK; 54°04′N 11°47′E. Chartered 1250.

Kropotkin (kruh-POT-keen), city (2005 population 79,760), E KRASNODAR TERRITORY, S European Russia, on the right bank of the KUBAN' River, 84 mi/135 km ENE of KRASNODAR; 45°26′N 40°34′E. Elevation 229 ft/69 m. On the Rostov-Baku railroad (Kavkazskaya station); railroad center with branches to STAV-ROPOL, KRASNODAR, and NOVOROSSIYSK; center of an agricultural area; railroad workshops; machinery and chemical industries; agricultural processing (flour, vegetable oils, meat, milk, beer, canned goods). Arose after 1771 as a fortress. Developed in 1880 as a railroad station and transport center. Called Romanovskiy Khutor until 1921; became city in 1926. Formerly spelled Krapotkin.

Kropotkin (kruh-POT-keen), town (2005 population 1,750), NE IRKUTSK oblast, E central SIBERIA, RUSSIA, on the PATOM plateau, near road and railroad, 84 mi/135 km NE of BODAYBO; 58°30′N 115°19′E. Elevation 2,742 ft/835 m. Gold mining. Formerly spelled Krapotkin.

Kropp (KRUHP), village, SCHLESWIG-HOLSTEIN, NW GERMANY, 7 mi/11.3 km S of SCHLESWIG; 54°25′N 09°30′E.

Krosniewice (kros-nee-VEE-tse), Polish *Krośniewice*, Russian *Krosnevitse*, town, Płock province, central POLAND, 8 mi/12.9 km W of KUTNO. Brick manufacturing, flour milling.

Krosno (KRO-sno), town (2002 population 48,372), Krosno province, SE POLAND, on WISLOK RIVER, on railroad, and 26 mi/42 km SSW of RZESZOW. Center of region producing petroleum and natural gas (gas pipeline to OSTROWIEC, RADOM, and SANDOMIERZ); petroleum refinery; manufacturing of machinery, glass, rubber footwear; flour mills, sawmills, brickworks; airport.

Krosno Odrzańskie (KRO-sno od-ZAHN-skee), Polish *Krosno Odrzańskie*, German *Crossen*, town in BRAN-DENBURG, after 1945 in Zielona Gora province, W POLAND, port on the ODER, at BOBRAWA RIVER mouth, and 18 mi/29 km ENE of GUBIN, in LOWER LUSATIA.

Woolen milling, metal- and wood-working; grape growing. First mentioned c.1000; chartered 1201; passed to Brandenburg 1482. After World War II, during which it was c.50% destroyed, its German population left.

Krotoszyn (kro-TO-sheen), German *Krotoschin*, town (2002 population 29,351), Kalisz province, W central POLAND, 12 mi/19 km WNW of OSTROW. Railroad junction; manufacturing of bricks, furniture, ceramics, roofing materials, agricultural machinery; brewing, distilling, flour milling, fruit canning.

Krotovka (KRO-tuhf-kah), village (2006 population 5,385), E central SAMARA oblast, E European Russia, near the GREATER KINEL RIVER, on railroad and near terminus of local highway, 22 mi/35 km ENE of KINEL; 53°18′N 51°10′E. Elevation 190 ft/57 m. In agricultural and petroleum-producing area.

Krottingen, LITHUANIA: see KRETINGA.

Krotz Springs, town (2000 population 1,219), SAINT LANDRY parish, S central LOUISIANA, 20 mi/32 km E of OPELOUSAS, and on ATCHAFALAYA RIVER; 30°32′N 91°45′W. Shallow-draft port; in oil and natural-gas area; gasoline processing. Sherburne State Wildlife Area and Atchafalaya National Wildlife Refuge to SE.

Krousson (kroo-SON), town, Iráklion prefecture, central CRETE department, 11 mi/18 km SSW of Iráklion; 35°14′N 24°59′E. Carob, raisins, olive oil.

Krozingen, Bad, GERMANY: see BAD KROZINGEN.

Krško (KERSH-ko), town (2002 population 6,866), S SLOVENIA, on Sava River, on railroad, and 45 mi/72 km E of LJUBLJANA; 45°57′N 15°29′E. Center of wine-growing region; health resort. First mentioned in 895. Nuclear power plant built (1992) to SE.

Krstača (KUHRST-ah-chah), mountain in DINARIC ALPS, on SERBIA-MONTENEGRO border; highest point (5,756 ft/1,754 m) is 10 mi/16 km W of TUTIN (Serbia). Also spelled Krstacha.

Kruft (KRUFT), village, RHINELAND-PALATINATE, W GERMANY, 11 mi/18 km W of KOBLENZ; 50°23′N 07°20′E. Wine.

Kruger National Park, game reserve (□ 8,000 sq mi/20,800 sq km), LIMPOPO and MPUMALANGA provinces (former E Transvaal), NE SOUTH AFRICA, borders onto E MOZAMBIQUE. One of the world's largest and oldest wildlife sanctuaries, it has almost every species of game in South Africa. Vegetation in the park ranges from savanna to dense riverine bush and woodland; has an extensive (*circa* 5,000 mi/13,000 km long) road system. Rich in animal and bird life, with an estimated 250,000 mammals. Originally founded as the Sabi Game Reserve (1898) by S. J. P. Kruger; it was enlarged and made a national park in 1926. Includes Timbavati Game Preserve. Skukuza is the park's largest tourist camp, located close to its S reaches, 50 mi/180 km NE of NELSPRUIT; acts as main administrative center and shopping area for the park, plus staff housing.

Krugersdorp [=kruger's town], city, GAUTENG province, NE SOUTH AFRICA. The chief industrial city of the W WITWATERSRAND, 15 mi/24 km W of JOHANNESBURG; 26°06′S 27°45′E. Elevation 5,674 ft/1,730 m. Krugersdorp is the center for a region where gold, manganese, asbestos, lime, and uranium are mined. The city has uranium extraction plants. It also serves as the trade center for the surrounding farming area. Founded in 1887, it was named for Paul Kruger, president of the Transvaal republic. The Paardekraal monument marks the spot where in 1880 Boers (Afrikaners) pledged themselves to end British rule in the Transvaal. Nearby are the STERKFONTEIN Caves (an important archaeological site), Kromdraai Paleontological Reserve, and Krugersdorp Game Reserve. The city has a technical college and a nature preserve (E of the city). Krugersdorp was incorporated into the MOGALE CITY municipality, created in 2001 as part of a movement to revive tribal names. The city is known colloquially as Mogale.

Krugloye (KROOG-lo-ye), urban settlement, NW MOGILEV oblast, BELARUS, ⊙ KRUGLOYE region, on DRUT RIVER (tributary of DNIEPER) and 33 mi/53 km NW of MOGILEV, 54°15′N 29°50′E. Manufacturing (construction materials); creamery.

Kruglyakov (kroo-glyah-KOF), settlement, S VOLGOGRAD oblast, SE European Russia, on road and railroad (near Zhutovo station), 65 mi/105 km SW of VOLGOGRAD; 47°58′N 43°38′E. In agricultural area (sunflowers, grains, hemp).

Kruglyzhi (kroog-li-ZHI), village, W KIROV oblast, E central European Russia, on road, 18 mi/29 km NNE of (and administratively subordinate to) SVECHA; 58°30′N 47°40′E. Elevation 521 ft/158 m. In agricultural area (wheat, flax, vegetables, potatoes); agricultural chemicals manufacturing, agricultural machinery storage and repair. Logging and lumbering enterprise. Also known as Kruglyzhskoye (kroog-LIZH-skuh-ye).

Kruglyzhskoye, RUSSIA: see KRUGLYZHI.

Kruhlyy Island, UKRAINE: see YAHORLYT BAY.

Kruibeke (KROI-bai-kuh), commune (2006 population 15,288), Sint-Niklaas district, EAST FLANDERS province, N BELGIUM, on SCHELDT RIVER and 5 mi/8 km SW of ANTWERP; 51°10′N 04°19′E. Agricultural market (dairying, vegetables). Has eighteenth-century church. Sometimes spelled Cruybeke.

Kruiningen (KROI-ning-uhn), town, ZEELAND province, SW NETHERLANDS, in SOUTH BEVELAND region, 7 mi/11.3 km ESE of GOES; 51°27′N 04°02′E. EASTERN SCHELDT estuary 2 mi/3.2 km to NE, WESTERN SCHELDT estuary 1 mi/1.6 km to S (car ferry to Perkpolder, on S shore). Dairying; cattle, sheep; flowers, vegetables, fruit; manufacturing (food processing).

Kruisfontein, town, EASTERN CAPE province, SOUTH AFRICA, 50 mi/80 km W of NELSON MANDELA METROPOLE (PORT ELIZABETH), 3 mi/5 km NW of HUMANSDORP on N2 highway; 34°00′S 24°45′E. Elevation 984 ft/300 m. Stock, wheat, fruit, vegetables. On Longkloof railroad route.

Kruishoutem (KROIS-hou-tem), agricultural commune (2006 population 8,135), Oudenaarde district, EAST FLANDERS province, W central BELGIUM, 5 mi/8 km NNW of OUDENAARDE; 50°54′N 03°31′E. Formerly spelled Cruyshautem.

Krujë (KROO-ee), town, ⊙ Krujë district (□ 120 sq mi/312 sq km), N central ALBANIA, 12 mi/19 km N of TIRANË, situated on 2,000-ft/610-m mt. spur above ISHM RIVER valley; 41°33′N 19°45′E. Handicraft industry (metals, pottery, wood, national costumes); olive groves. Bauxite deposits nearby. Population is largely Muslim. The 14th–15th-century fortress was (1443–1468) the center of Scanderbeg's resistance against Turks. Was bishopric (1246–1694). Also spelled Kruja.

Krukenichi, UKRAINE: see KRUKENYCHI.

Krukenitsa, UKRAINE: see KRUKENYCHI.

Krukenychi (KROO-ke-ni-chee) (Russian *Krukenichi*), (Polish *Krkienice*), village, NW L'VIV oblast, UKRAINE, 13 mi/21 km NNW of SAMBIR; 49°41′N 23°10′E. Elevation 885 ft/269 m. Potatoes, oats, rye; lumbering. Also called Krukenytsya (Russian *Krukenitsa*).

Krukenytsya, UKRAINE: see KRUKENYCHI.

Krum, town (2006 population 3,700), DENTON county, N TEXAS, 5 mi/8 km WNW of DENTON; 33°16′N 97°13′W. Agricultural area (cotton, wheat, sorghum, peanuts; cattle, horses). Light manufacturing.

Krumbach (KRUM-bahkh), town, BAVARIA, S GERMANY, in SWABIA, on small Kamlach River, and 20 mi/32 km SE of ULM; 48°15′N 10°21′E. Manufacturing of textiles, machinery, rubber goods, glass. Has rococo church. Mineral springs nearby. Chartered 1380.

Krumë, town, ⊙ Krumë district (□ 150 sq mi/390 sq km), extreme NE ALBANIA, on border of KOSOVO province, SERBIA.

Krummau, CZECH REPUBLIC: see CESKY KRUMLOV.

Krummhörn (KRUM-hoern), commune, LOWER SAXONY, NW GERMANY, in EAST FRIESLAND, 17 mi/27 km W of AURICH; 53°26′N 07°06′E. Fishery; tourism.

Krummhübel, POLAND, see KARPACZ.

Krumovgrad (KROO-mov-grahd), city, HASKOVO oblast, ⊙ Krumovgrad obshtina (1993 population 31,068), S BULGARIA, in the E RODOPI Mountains, on a tributary of the ARDA RIVER, 18 mi/29 km SE of KURDZHALI; 41°29′N 25°37′E. Market center (tobacco, woolen cloth). Tobacco experimental station. Chrome, lead, zinc, and perlite deposits to the N. Until 1934, known as Koshu-Kavak.

Krumovitsa (KROO-mo-veet-sah), river, S central BULGARIA. Tributary of the ARDA RIVER; 41°28′N 25°39′E.

Krumovo (KROO-mo-vo), village, PLOVDIV oblast, Rodopska obshtina, S central BULGARIA, 6 mi/10 km SE of PLOVDIV; 42°05′N 24°49′E. Railroad junction; tobacco, rice, sugar beets. Iron ore mines; iron benification. Formerly called Pashamahla.

Krumpendorf (KRUM-pen-dorf), village, CARINTHIA, S AUSTRIA, on N shore of the Wörthersee and 4 mi/6.4 km W of KLAGENFURT; 46°38′N 14°13′E. Summer resort.

Krun (KROON), village, HASKOVO oblast, KAZANLUK obshtina, BULGARIA; 42°40′N 25°23′E.

Krung Kao, THAILAND: see AYUTTHAYA.

Krupanj (KROO-pahn), village (2002 population 20,192), W SERBIA, 26 mi/42 km W of VALJEVO. Antimony smelter. There are two antimony mines nearby. Center of lead-mining district, with mines dating from Middle Ages. Also spelled Krupan.

Krupets (kroo-PYETS), village, W KURSK oblast, SW European Russia, on road and railroad, 49 mi/79 km S of KURSK, and 14 mi/23 km WNW of RYL'SK; 51°05′N 35°29′E. Elevation 538 ft/163 m. In agricultural area; food processing.

Krupina (kru-PI-nah), German *Karpfen*, Hungarian *Korpona*, town, STREDOSLOVENSKY province, S SLOVAKIA, on railroad and 27 mi/43 km S of BANSKÁ BYSTRICA; 48°21′N 19°04′E. Manufacturing (machinery; brick kilns; food processing, cheese making). Has 13th-century church; museum. Mining of gold and silver ore until 15th century. Military training base at Lest (LESHT-yuh), Slovak *Lešt'*, 9 mi/14 km NE.

Krupka (KRUP-kah), German *Graupen*, town, SEVEROCESKY province, NW BOHEMIA, CZECH REPUBLIC, in the ORE Mountains, on railroad, 3 mi/4.8 km NNE of TEPLICE; 50°41′N 13°52′E. Manufacturing (electrical and ceramics). Pilgrimage church and Jesuit monastery in suburb of Bohosudov (BO-ho-SU-dof). VRCHOSLAV metalworks are just SSW, on railroad.

Krupki (KROOP-ki), town, NE MINSK oblast, BELARUS, ⊙ Krupki region, on Bobr River (tributary of BEREZINA) and 25 mi/40 km ENE of BORISOV, 59°19′N 29°05′E. Fruit and vegetable processing, woodworking.

Krupnik (KROOP-neek), village, SOFIA oblast, SIMITLI obshtina, BULGARIA; 41°51′N 23°06′E.

Krupp, town (2006 population 64), GRANT county, E central WASHINGTON, 25 mi/40 km ENE of EPHRATA, on Crab Creek; 47°25′N 119°00′W. In COLUMBIA basin agricultural region.

Krusadi Island (kroo-SAH-dee), Tamil *Krusadi Tivu*, in GULF OF MANNAR, off SW coast of RAMESWARAM ISLAND, TAMIL NADU state, S INDIA, 29 mi/47 km E of Kilakarai; 1.5 mi/2.4 km long, 0.75 mi/1.21 km wide. Marine research station; pearl farm.

Kruschwitz, POLAND: see KRUSZWICA.

Krusenstern, Cape (KROO-zuhn-stuhrn), NUNAVUT territory, CANADA, on CORONATION GULF, at E end of DOLPHIN AND UNION STRAIT; 68°23′N 113°55′W. Trading post.

Krusenstern Island, Alaska: see LITTLE DIOMEDE ISLAND.

Krusenstern Island, FRENCH POLYNESIA: see TIKEHAU.

Krusenstern Strait, RUSSIA: see KRUZENSHTERN STRAIT.

Kruševac (KROO-she-vahts), city (2002 population 131,368), central SERBIA; 43°34′N 21°20′E. A commercial center, it has a hydroelectric plant and an important chemical industry. The seat of the kings of Serbia until 1389, it has retained the ruins of a medieval castle.

Krusevo (kroo-SHE-vo) Serbo-Croatian *Kruševo*, village, W MACEDONIA, 45 mi/72 km S of SKOPJE. Marketing center; homemade carpets. Pop. of Roman provincial origin; uses a Romance dialect. The center of Macedonian uprising against Turkish rule in 1903 (Krusevo Republic). Also spelled Krushevo.

Krushare (kroo-SHAH-re), village, BURGAS oblast, SLIVEN obshtina, BULGARIA; 42°34′N 26°23′E.

Krushari (kroo-SHAH-re), city (1993 population 1,690), VARNA oblast, Krushari obshtina, BULGARIA. Agriculture (grain, sugar beets; livestock); 43°48′N 27°45′E.

Krushovene (kroo-sho-VE-ne), village, LOVECH oblast, DOLNA MITROPOLIYA obshtina, N BULGARIA, on the ISKUR River, 23 mi/37 km ESE of ORYAHOVO; 43°38′N 24°25′E. Wheat, corn, livestock.

Krušné hory, CZECH REPUBLIC: see ORE MOUNTAINS.

Krusovice, CZECH REPUBLIC: see NOVE STRASECI.

Krustets (kroos-TETS), village, LOVECH oblast, TRYAVNA obshtina, BULGARIA; 42°46′N 25°24′E.

Krustpils (KROOST-peels), German *kreuzburg*, city, S central LATVIA, in LATGALE, on right bank of the DVINA (DAUGAVA) River, opposite JEKABPILS, and 75 mi/121 km ESE of Rīga; 56°32′N 25°53′E. Railroad junction; manufacturing (leather, knitwear), sugar refining. In Russian VITEBSK government until 1920.

Kruszwica (kroosh-VEE-tsah), German *Kruschwitz*, town, Bydgoszcz province, central POLAND, on W shore of Lake GOPLO, 9 mi/14.5 km S of INOWROCLAW. Beet-sugar and flour milling.

Krutaya Gorka (kroo-TAH-yah GOR-kah), town (2006 population 6,755), S central OMSK oblast, SW SIBERIA, RUSSIA, on the IRTYSH River, on highway branch and railroad spur, 22 mi/35 km N of OMSK; 55°21′N 73°15′E. Elevation 416 ft/126 m. Manufacturing (machinery).

Krut, Ban, THAILAND: see BAN KRUT.

Krutikha (kroo-TEE-hah), village, NW ALTAI TERRITORY, S central SIBERIA, RUSSIA, near the OB′ RIVER, on road, 13 mi/21 km N of KAMEN-NA-OBI; 53°58′N 81°12′E. Elevation 403 ft/122 m. In dairy farming area.

Krutinka (kroo-TEEN-kah), town (2006 population 7,725), W OMSK oblast, SW SIBERIA, RUSSIA, on Lake Ik, on crossroads, 30 mi/48 km N of NAZYVAYEVSK; 56°04′N 71°30′E. Elevation 305 ft/92 m. In dairy farming area; fish processing.

Kruzenshtern Island, Alaska: see LITTLE DIOMEDE ISLAND.

Kruzenshtern Strait (kroo-zen-SHTERN), Japanese *Mushiru-Kaikyo*, in the N main KURIL ISLANDS group, RUSSIA, SAKHALIN oblast, extreme E SIBERIA, RUSSIAN FAR EAST, between LOVUSHKI (N) and RAIKOKE ISLANDS (S); 31 mi/50 km wide. Name sometimes extended to the entire strait between SHIASHKOTAN and RAIKOKE islands, including FORTUNA STRAIT. Named after Ivan F. Kruzenshtern, a Russian explorer who first navigated it in the early 19th century.

Kruzof Island (KROO-zawf), SE ALASKA, in ALEXANDER ARCHIPELAGO, 10 mi/16 km W of SITKA; 23 mi/37 km long, 8 mi/12.9 km wide; 57°10′N 135°42′W. Mount EDGECUMBE (S); fishing.

Krya Vryse, Greece: see KRIA VRISSI.

Krylbo (KREEL-BOO), town, KOPPARBERG county, central SWEDEN, on DALÄLVEN RIVER, agglomerated with AVESTA; 60°8′N 16°13′E. Railroad junction.

Krylos (KRI-los), village (2004 population 10,300), N IVANO-FRANKIVS′K oblast, UKRAINE, 11 mi/18 km N of IVANO-FRANKIVS′K, on right bank of the Lukva River, tributary of the DNIESTER; 49°05′N 24°42′E. Elevation 1,003 ft/305 m. Site of princely Halych, a 9th-century Ukranian settlement that became the capital of the duchy of Halych in 1144 and then the capital of Halych-Volyn principality in 1199. Destroyed by the Tatars in 1241. Archaeological excavations in the 20th century revealed remains of the Dormition Cathedral

(built in 1157), city walls, castle moats, stone buildings of the lower town, and the sarcophagus and skeleton of Prince Yaroslav Osmomysl.

Krylovskaya (kri-LOFS-kah-yah), village (2005 population 14,675), N central KRASNODAR TERRITORY, S European Russia, on the YEYA RIVER, on road, 56 mi/90 km S of ROSTOV-NA-DONU (ROSTOV oblast), and 23 mi/37 km SE of KUSHCHEVSKAYA; 46°19′N 39°58′E. Flour mill; in agricultural area (wheat, sunflowers, essential oils). Formerly known as Yekaterinovskaya.

Krylovskaya (kri-LOFS-kah-yah), village (2005 population 6,865), central KRASNODAR TERRITORY, S European Russia, on the CHELBAS RIVER, on road, 17 mi/27 km E of KANEVSKAYA; 46°06′N 39°18′E. In agricultural area (wheat, fruits, vegetables, grapes; livestock); produce processing.

Krym (KRIM), rural settlement (2006 population 4,475), SW ROSTOV oblast, S European Russia, on road, 8 mi/13 km NW of ROSTOV-NA-DONU, and 2 mi/3.2 km NE of CHALTYR′; 47°18′N 39°31′E. Elevation 219 ft/66 m. Agricultural products.

Krym, UKRAINE: see CRIMEA.

Krym, UKRAINE: see STARYY KRYM.

Krymne Lake, UKRAINE: see SHATS′K LAKES.

Krymsk (KRIMSK), city (2005 population 57,895), S KRASNODAR TERRITORY, S European Russia, at the extreme NW foot of the Greater CAUCASUS, on the Adagum River (tributary of the KUBAN′ River), on road and railroad, 54 mi/87 km W of KRASNODAR, and 17 mi/27 km NE of NOVOROSSIYSK; 44°55′N 37°59′E. Elevation 111 ft/33 m. Railroad junction and highway hub; local transshipment point. Food industries (canned goods, dairy products, feed, wine, beer). Founded in 1862. Made city in 1958. Earlier, as a village, called Krymskaya.

Kryms′ka Nyzovyna, UKRAINE: see CRIMEAN LOWLAND.

Krymskaya, RUSSIA: see KRYMSK.

Kryms′ke Zapovidno-Myslyvs′ke Hospodarstvo, UKRAINE: see CRIMEAN HUNTING PRESERVE.

Krymskoye gosudarstvennoye zapovedno-okhotnich′ye khozyaystvo, UKRAINE: see CRIMEAN HUNTING PRESERVE.

Kryndachivka, UKRAINE: see KRASNYY LUCH.

Kryndachovka, UKRAINE: see KRASNYY LUCH.

Krynica (kree-NEE-tsah), town, Nowy Sącz province, S POLAND, in the CARPATHIANS, 18 mi/29 km SSE of NOWY SACZ, near Slovakian border. Railroad spur terminus; health and winter sports center (elevation 1,922 ft/586 m) with mineral springs and hot baths; dry-ice processing. Sometimes called Krynica Zdroj, Polish *Krynica Zdrój* or *Krynica Górska*. During World War II, under German rule, called Bad Krynica.

Krynka (KRIN-kah), river, right tributary of the MIUS RIVER, in E DONETS′K oblast, UKRAINE; rises 6 mi/10 km E of HORLIVKA in the Donets Upland; flows SSE past ZUHRES and then SE for 112 mi/180 km to join Mius River in ROSTOV oblast, Russia, 5 mi/8 km beyond the border of Ukraine and 3 mi/5 km NW of MATVEYEV KURGAN. In Ukraine it has five water reservoirs and many ponds; water is used for domestic and industrial needs; fishing.

Krynychky (kri-NICH-kee) (Russian *Krinichki*), town, central DNIPROPETROVS′K oblast, UKRAINE, near highway, 12 mi/19 km SW of DNIPRODZERZHYNS′K; 48°22′N 34°27′E. Elevation 223 ft/67 m. Raion center; auto repair; fish farming. Established in the 18th century, town since 1958. Formerly called Krynychevate (Russian *Krinichevatoye*).

Krynychna (kri-NICH-nah) (Russian *Krinichnaya*), town, central DONETS′K oblast, UKRAINE, in the DONBAS, 8 mi/12.9 km NNE of MAKIYIVKA; 48°08′N 38°01′E. Elevation 862 ft/262 m. Cement works, mechanical repair factory. Established in 1879, town since 1933.

Krynychne, UKRAINE: see BIRYUKOVE.

Kryoneri Greece: see KRIONERI.

Krypton (KRIP-tahn), village, PERRY county, SE KENTUCKY, 8 mi/12.9 km NW of HAZARD, in CUMBERLAND foothills on North Fork KENTUCKY RIVER. Bituminous coal-mining. Daniel Boone National Forest to SW.

Krystynopil′, UKRAINE: see CHERVONOHRAD.

Krystynopol, UKRAINE: see CHERVONOHRAD.

Kryuchkovka, RUSSIA: see KOS′YA.

Kryukovo (KRYOO-kuh-vuh), former town, central MOSCOW oblast, central European Russia, now a suburb of MOSCOW, on road and railroad, 23 mi/37 km NW of the city center; 55°58′N 37°09′E. Elevation 682 ft/207 m. Reinforced concrete forms, metal goods, glass, clothing. In the late fall of 1941, Marshall Zhukov's troops stopped the Nazi advance on MOSCOW here. Incorporated into Moscow in 1987.

Kryukovo (KRYOO-kuh-vuh), settlement, S MOSCOW oblast, central European Russia, near the Lopasnya River (tributary of the OKA River), near railroad, 4 mi/6 km SE, and under administrative jurisdiction, of CHEKHOV; 55°05′N 37°33′E. Elevation 469 ft/142 m. Manufacturing (fans).

Kryukovskiy, RUSSIA: see SOVETSK, TULA oblast.

Kryulyany, MOLDOVA: see CRIULENI.

Kryve Ozero (kri-VE O-ze-ro) (Russian *Krivoye Ozero*), town, NW MYKOLAYIV oblast, UKRAINE, at crossroads, 24 mi/39 km WSW of PERVOMAYS′K; 47°56′N 30°21′E. Elevation 360 ft/109 m. Food processing (dairy, broiler, food flavoring), asphalt making. Technical-vocational school; heritage museum. Established in the 1750s, town since 1970.

Kryvorizhzha, UKRAINE: see BRYANKA.

Kryvoriz′kyy zalizorudnyy baseyn, UKRAINE: see KRYVVY RIH IRON ORE BASIN.

Kryvyy Rih (kree-VEE REE) (Russian *Krivoy Rog*), city (2001 population 668,980), SW DNIPROPETROVS′K oblast, UKRAINE, at the confluence of the INHULETS′ and SAKSAHAN′ rivers; 47°55′N 33°21′E. Elevation 285 ft/86 m. It is a raion center, a railroad junction, an industrial center, and a metallurgical and coking center of one of the world's richest iron-mining regions. The city extends 20 mi/32 km in a narrow belt paralleling the iron ore deposits along the Saksahan River Burial mounds in the area indicate that Scythians inhabited it and used the iron deposits. Founded in the 17th century by the Zaporizhzhya Cossacks, the city received its name (Ukrainian=Crooked Horn) because of the shape of the iron-mining area. Kryvyy Rih's industrial growth dates from 1881, when French, Belgian, and other foreign interests founded a mining syndicate. The city has mining and pedagogical institutes, eighteen vocational technical schools, music, drama and puppet theaters, a circus, and a heritage museum. An active Jewish community since the 19th century, affected by the 1918–1920 civil war pogroms but still numbering 6,430 in 1939; destroyed during World War II—fewer than 1,000 Jews remaining in 2005.

Kryvyy Rih Iron Ore Basin (kree-VEE REE) (Ukrainian *Kryvoriz′kyy zalizorudnyy baseyn*), mostly in W DNIPROPETROVS′K oblast but also reaching into E KIROVOHRAD oblast, UKRAINE. The largest iron ore-bearing basin in Ukraine and one of the largest in the world; extends in a 1 mi/2 km–4 mi/7 km wide belt for about 62 mi/100 km from S of INHULETS′ NNE along the INHULETS′ River, through KRYVYY RIH along the W bank of Saksahan′ River, and then at TERNY veers N, along the W bank of Zhovta River, though ZHOVTI VODY, to Zhovte; covers an area of about 115 sq mi/300 sq km; lies in the center of the Ukrainian Shield, in the Kryvyy Rih-Kremenchuk tectonic zone. Most of the commercial iron ore deposits are found in five horizons of the Saksahan suite. Known since their use by the Scythians (6th to 2nd century B.C.E.), their industrial development began in the 1870s, systematically explored in the 1920s, and expanded in the post-World War II period. Exploration in the 1970s revealed proven reserves of 16 billion tons/18 billion metric tons. In 1988, there were seventeen mines, five beneficiation plants. The ore mined (about 181 million tons/200 million metric tons/year) serves ferrous metallurgy of Ukraine, Russia, and countries of E central Europe.

Krywan, SLOVAKIA: see KRIVAN PEAK.

Kryzhopil' (kri-ZHO-peel) (Russian *Kryzhopol'*), town, S VINNYTSYA oblast, UKRAINE, on railroad, 50 mi/80 km E of MOHYLIV-PODIL'S'KYY; 48°23′N 28°53′E. Elevation 951 ft/289 m. Food processing, woodworking, furniture manufacturing, sewing. Vocational technical school. Established in 1866; site of battles during the struggle for Ukrainian independence (1918–1920) against the Red Army and General Denikin's volunteer army; town since 1938. Sizeable Jewish community since the 19th century, predominantly merchants and craftsmen; reduced during the civil war pogroms and numbering 1,540 in 1939; further devastated by the establishment of a ghetto and gradual extermination by the Nazis between 1941 and 1944. Despite these hardships, the population of Kryzhopil' was almost 10% Jewish in February 2005.

Kryzhopol', UKRAINE: see KRYZHOPIL'.

Krzemieniec, UKRAINE: see KREMENETS'.

Krzepice (kshe-PEE-tse), Russian *Krzhepitse*, town, Częstochowa province, S POLAND, 20 mi/32 km NW of CZĘSTOCHOWA. Flour milling.

Krzeszowice (kshe-sho-VEE-tse), town, Kraków province, S POLAND, on railroad and 15 mi/24 km WNW of KRAKÓW. Health resort with sulphur springs and hot baths; stone and clay quarrying; manufacturing of chemicals, distilling. Has castle with park and art works.

Krzhepitse, POLAND: see KRZEPICE.

Krzhizhanovsk, RUSSIA: see GUBAKHA.

Krzna River (KSHNAH), c.40 mi/64 km long, Lublin province, E POLAND; rises just W of Łuków; flows generally E, past ŁUKÓW, MIĘDZYRZEC, and Biała Podlaska, to BUG RIVER 8 mi/12.9 km WNW of BREST.

Krzywin (KZEE-veen), Polish *Krzywiń* or *Krzywin*, German *Kriewen* (kree-ven), town, Leszno province, W POLAND, on OBRA RIVER and 11 mi/18 km SE of KOŚCIAN. Flour milling, cement.

Krzyż (kzeesh), Polish *Krzyż*, German *Kreuz*, town in POMERANIA, Piła province, W central POLAND, near NOTEĆ River, 35 mi/56 km WSW of Piła (Schneidemühl). Railroad junction. Founded 1850 during construction of BERLIN-EAST PRUSSIA railroad; chartered 1936. Until 1939, German frontier station on Polish border, 50 mi/80 km NW of POZNAŃ. Until 1938, in former Prussian province of Grenzmark Posen-Westpreussen. Also called Krzyż Wielkopolski.

Krzyzatka, POLAND: see KOWARY.

Ksamil, resort, S ALBANIA, S of BUTRINT and on Greek border. Citrus trees.

K'Sar Chellala (KSAHR shel-lah-LAH), town, TIARET wilaya, W ALGERIA, 35°12′N 02°20′E. Important market for wool products; sheep rearing; major carpet-producing region. The government (together with Australian experts) has begun an experimental rehabilitation of 193 sq mi/500 sq km for sheep grazing in the town.

Ksar El Boukhari (KSAHR el boo-kah-REE), town, MÉDÉA wilaya, N central ALGERIA, on railroad to DJELFA and 40 mi/64 km S of BLIDA at N edge of the High Plateaus where the CHÉLIFF River cuts its valley across the TELL ATLAS; 35°55′N 02°47′E. Trade center (sheep, wool, skins, cereals, esparto grass) for nomads of the interior. Formerly Boghari.

Ksar el Kebir (KSAR el kuh-BEER), city, Larache province, Tanger-Tétouan administrative region, MOROCCO, on road to LARACHE and TANGER; 35°01′N 05°54′W. Market town in center of fertile agricultural area. Near the city on August 4, 1578, the Moroccans soundly defeated the Portuguese, who had invaded Morocco in support of a pretender to the Moroccan throne. Abd al-Malik, ruler of Morocco, the Portu-

guese King, Sebastian, and the Moroccan pretender, Muhammad, all died in the fighting. As a result of the battle, a weakened PORTUGAL soon passed (1580) to Philip II of SPAIN, and the new Moroccan ruler, Ahmad al-Mansur, began his reign with tremendous prestige. Also called Alcazarquivir.

Ksar-es-Souk, MOROCCO: see ERRACHIDIA.

Ksel, Djebel (KSEL, JE-bel), highest peak (elev. 6,588 ft/2,008 m) of the Djebel AMOUR (range of the Saharan ATLAS) in El BAYADH wilaya, N central ALGERIA, just NE of El BAYADH; 33°44′N 01°05′E.

Ksenievskiy, RUSSIA: see ASINO.

Kseur, El (KSUHR, el), town, BEJAÏA wilaya, N central ALGERIA, in Oued SOUMMAM valley, 14 mi/23 km SW of BEJAÏA; 36°42′N 04°50′E. Craft shops and market.

Kshen', RUSSIA: see KSHENSKIY.

Kshenskiy (KSHEN-skeeyee), town (2005 population 6,455), E central KURSK oblast, SW European Russia, on the Kshen' River (DON River basin), on road and railroad, 75 mi/121 km E of KURSK; 51°50′N 37°43′E. Elevation 656 ft/199 m. Sugar factory. Established as a village of Kshen', later renamed Lipovchik; since 1924, called Sovetskiy; current name since 1991.

Kshtu, TAJIKISTAN: see ANZOB.

Ksiaz (ksee-ahsh), Polish *Książ Wielkopolski*, town, Poznań province, W POLAND, 27 mi/43 km SSE of POZNAŃ.

Ksour-Essaf, TUNISIA: see QUSUR ASSAF.

Ksour Mountains (KSOOR), range of the Saharan ATLAS, in NAAMA and El BAYADH wilaya, W ALGERIA, extending c.100 mi/161 km NE from the Moroccan border at FIGUIG (MOROCCO), between the High Plateaus (N) and SAHARA Desert (S). Highest peaks are Djebel AÏSSA (elev. 7,336 ft/2,236 m) just N of AÏN SEFRA, and Djebel MZI (elev. 6,988 ft/2,130 m). Traversed at Aïn Sefra by narrow-gauge railroad to BÉCHAR.

Kstinino (KSTEE-nee-nuh), village, central KIROV oblast, E central European Russia, on road, 6 mi/10 km SW of KIROVO-CHEPETSK, to which it is administratively subordinate; 58°26′N 49°47′E. Elevation 623 ft/189 m. Gas pipeline service station; furniture making. In agricultural area (wheat, oats, rye, vegetables); bakery.

Kstovo (KSTO-vuh), city (2006 population 67,365), central NIZHEGOROD oblast, central European Russia, port on the VOLGA RIVER, on road and railroad spur, 18 mi/29 km SE of NIZHNIY NOVGOROD; 56°08′N 44°11′E. Elevation 410 ft/124 m. Petroleum refining; pre-cast concrete; furniture, cotton products, food processing (dairy, bakery, vitamins and dietary supplements); thermoelectric power station. Arose in the 14th century. Made city in 1957.

Ktypas (KTEE-pas), mountain (3,350 ft/1,0210 m), in BOEOTIA prefecture, E CENTRAL GREECE department, near Gulf of Évvia, 6 mi/9.7 km W of KHALKÍS; 38°27′N 23°27′E. Sometimes called Messapion; also spelled Ktipas.

Kuah (cooah), town (2000 population 20,862) and port on S shore of LANGKAWI ISLAND, KEDAH, MALAYSIA; 06°10′N 99°50′E. Fisheries.

Kuala (KWAH-lah) [Malay=river mouth], place name (generic) meaning river mouth or confluence of 2 rivers.

Kuala Belait (koo-WAH-la BUH-leit), town, ⊙ BELAIT district, W BRUNEI DARUSSALAM, N BORNEO, near SOUTH CHINA SEA, on Belait River (20 mi/32 km long) and 55 mi/89 km WSW of BANDAR SERI BEGAWAN; 04°35′N 114°11′E. Malaysia (SARAWAK) border 1 mi/1.6 km to SW, BARAM RIVER 2 mi/3.2 km SW. Supply center for SERIA oil fields (10 mi/16 km E). Agriculture (cassava, sago, rice); livestock, poultry; fishing. Town has modern facilities; one of Brunei's two major ports. Tamu Kuala Belait, open-air market; Silver Jubilee Park.

Kuala Besut (KWAH-lah) or **Besut**, town, N TRENGGANU, MALAYSIA, SOUTH CHINA SEA port at mouth

of the small Besut River, 50 mi/80 km NW of Kuala Trengganu. Coconuts, rice; fisheries.

Kuala Dungun (KWAH-lah) or **Dungun**, town (2000 population 50,166), E central TERENGGANU, MALAYSIA, SOUTH CHINA SEA port at mouth of small Dungun River, 40 mi/64 km SSE of KUALA TERENGGANU; 04°47′N 103°26′E. Sawmilling, boatbuilding; BUKIT BESI (BOO-kit BEH-see) iron mine (15 mi/24 km W) closed 1970. Tourism based upon sea turtle nesting area; scuba area.

Kuala Kangsar (KWAH-lah), town (2000 population 34,690), N central PERAK, MALAYSIA, on railroad and 14 mi/23 km ESE of TAIPING, on PERAK RIVER. Sultan's residence; center of rice and rubber district; Malay college. Former residence of high commissioner of FEDERATED MALAY STATES.

Kualakapuas (KWAH-lah-kah-POO-wahs), town, ⊙ Kapuas district, Kalimantan Tengah province, SE BORNEO, INDONESIA, on Barito River delta, 28 mi/45 km NW of BANJARMASIN; 02°59′S 114°17′E. Trade center for rubber-growing region. Also spelled Koealakapoeas.

Kuala Kedah (KWAH-lah), town, N KEDAH, MALAYSIA, Strait of MALACCA small port at mouth of KEDAH RIVER, 5 mi/8 km W of Alor Star. Coconuts, rice; fisheries.

Kuala Kerai (KWAH-lah KREI) or **Krai** (KREI), town, N central KELANTAN, MALAYSIA, on KELANTAN RIVER, on interior railroad to E coast (TUMPAT), and 40 mi/64 km S of Kota Bharu; 05°31′N 102°13′E. Some rice growing; tin mining.

Kuala Kubu Bharu (KWAH-lah KOO-boo BAH-roo) or **Kuala Kubu Bahru**, town, NE SELANGOR, MALAYSIA, on SELANGOR River, 29 mi/47 km N of KUALA LUMPUR and on railroad (Kuala Kubu Road station). Road-RR transfer point for Fraser's Hill and PAHANG via SEMANGKO GAP. Original town of KUALA KUBU, located 2 mi/3 km SSW on lower ground, was abandoned (early 1930s) for present site because of floods.

Kualalangsa (kwa-lah-LAHNG-sah), port for LANGSA (city) and East Aceh district, ACEH province, NE SUMATRA, INDONESIA; 04°32′N 98°01′E. Exports oil from East Aceh onshore and offshore oil fields; wood; palm oil products. Also spelled Koealalangsa.

Kuala Lipis (KWAH-lah LEE-piece), town, ⊙ PAHANG, MALAYSIA, on interior railroad to E coast and on road, 75 mi/121 km NNE of KUALA LUMPUR, on Jelai River (a headstream of PAHANG RIVER); 04°09′N 102°02′E. Rubber, rice, oil-palm, gutta percha (used in manufacturing of golf balls); handicrafts, especially songket (brocades) and mats. Important agicultural service center. Kening Rimba Recreational Park to NE; TAMAN NEGARA NATIONAL PARK to NE.

Kuala Lumpur (KWAH-lah LOOM-poor) [=muddy confluence], city (2000 population 1,297,526), ⊙ FEDERATION OF MALAYSIA, S MALAY PENINSULA, at the confluence of the KLANG and Gombak rivers; 04°09′N 101°43′E. The chief inland city of Malaysia, Kuala Lumpur was the commercial center of a tin-mining and rubber-growing district and is a modern transportation hub and major industrial center. It was founded in 1857 by Chinese tin miners and superseded KLANG. In 1880 the British government transferred their headquarters from Klang to Kuala Lumpur, and in 1896 it became the capital of the FEDERATED MALAY STATES. Under the leadership of Sir Frank Swettenham, streets were enlarged, modern materials were used to build offices and new structures, and construction began on the Klang-Kuala Lumpur railroad. In 1957 British rule ended, and Kuala Lumpur became the capital of the independent FEDERATION OF MALAYA. The city became the capital of Malaysia in 1963. The Federal Territory of Kuala Lumpur (Wilayah Persekutuan) was created in 1974. Manufacturing (RR equipment and cars, rubber products, metal products, tin smelting, textiles, handicrafts). Kuala Lumpur is the home of several colleges and universities, including the University of Malaya and the Federal Technical College, as well as many hospitals, museums, and the National Zoo. Among the notable structures is the modern parliament building in Moorish style and a newly constructed government administrative center. Once 2/3 Chinese, the city has become "Malayanized" since the 1980s. Institute of Medical Research; international banking center. National Museum and National Museum of Art; National Library; international Crafts Museum. Sabang Airport 18 mi/29 km W. A second airport is currently under construction S of the city.

Kuala Muda (KWAH-lah MOO-dah) or **Kota Kuala Muda**, town, SW KEDAH, MALAYSIA, on Strait of MALACCA, at mouth of MUDA RIVER (PENANG line), 10 mi/16 km SW of SUNGEI PATANI; 05°34′N 100°22′E. Coconuts, rice, rubber; fisheries. Formerly Kota. Industrial spillover from Pulau Pinang (Pinang Island).

Kuala Perlis (KWAH-lah PER-liss), village, PERLIS, NW MALAYSIA, minor port on Strait of MALACCA at mouth of PERLIS RIVER, 5 mi/8 km SW of KANGAR, near THAILAND line. Fisheries.

Kuala Pilah (KWAH-lah PEE-lah), second-largest town of NEGERI SEMBILAN, SW MALAYSIA, 21 mi/34 km E of SEREMBAN, near MUAR RIVER; 02°45N 102°17′E. In rice and rubber-growing area.

Kualapuu (koo-AH-lah-POO-oo), town (2000 population 1,936), N MOLOKAI, MAUI county, HAWAII, in interior, 2 mi/3.2 km from N coast and 4 mi/6.4 km NNW of KAUNAKAKAI; 21°09′N 157°03′W. KALAUPAPA NATIONAL HISTORIC PARK; pineapples formerly grown here; coffee. Kualapuu Reservoir to W; Molokai Forest Reserve to E; Palaau State Park to NE.

Kuala Selangor (KWAH-lah seh-LAHNG-or), village (2000 population 33,816), W SELANGOR, MALAYSIA, fishing port on Strait of MALACCA, 35 mi/56 km NW of KUALA LUMPUR, at mouth of SELANGOR RIVER; 03°20′N 101°16′E. In rubber and coconut district; cocoa, pineapples, tea.

Kuala Sepetang (KWAH-lah seh-PEH-tahng), town, NW PERAK, MALAYSIA, port on inlet of Strait of MALACCA, 8 mi/13 km W of TAIPING (linked by railroad). Fisheries. Formerly known as PORT WELD.

Kuala Terengganu (KWAH-lah TER-eng-GAH-noo) or **Kuala Trengganu**, city (2000 population 255,109), ⊙ TERENGGANU state, MALAYSIA, central MALAY PENINSULA, on the SOUTH CHINA SEA at the mouth of the TERENGGANU RIVER; 05°19′N 103°09′E. It is a small port and has a weaving industry. The residence of the sultan of Terengganu is in the city. Manufacturing (handicrafts, batik), boatbuilding on island in river; fishing. Airstrip; state museum. Tourism (scuba diving, sea turtle watching). Offshore oil and gas exploration begun in 1981.

Kuandang (KWAN-dahng), town, N SULAWESI, INDONESIA, port on inlet of CELEBES SEA, 25 mi/40 km NNW of GORONTALO; 00°52′N 122°55′E. Ships timber, resin, rattan; hides; copra; kapok. Also spelled Koeandang.

Kuandian (KUAN-DIAN), town, ⊙ Kuandian county, SE LIAONING province, NE CHINA, 45 mi/72 km NNE of DANDONG, on railroad; 40°47′N 124°43′E. Tobacco, tussah; boron-ore mining.

Kuang-hsi, China: see GUANGXI ZHUANG AUTONOMOUS REGION.

Kuanhsi (GUAN-SHEE), town, NW TAIWAN, 13 mi/21 km E of HSINCHU; 24°48′N 121°10′E. Rice, vegetables, tea, oranges; livestock; wood products. Sometimes spelled Kwansi or Kuansi.

Kuansi, TAIWAN: see Kuanhsi.

Kuantan (KWAHN-tahn), largest city (2000 population 289,395) of E PAHANG, MALAYSIA, port on SOUTH CHINA SEA, at mouth of minor Kuantan River and 90 mi/145 km ESE of KUALA LIPIS (linked by highway); 03°50′N 103°20′E. Chief port of PAHANG and E coast; ships tin (from SUNGEI LEMBING, 20 mi/32 km NW), rubber, copra; rubber processing, rubber products,

handicrafts, batik. Tourism. Airport 8 mi/12.9 km SW at Padang Geroda. Offshore oil and gas exploration.

Kuantan River, INDONESIA: see INDRAGIRI RIVER.

Kuantzuling (GUAN-ZU-LING), village, W central TAIWAN, 10 mi/16 km SSE of CHIAI, 9 mi/14.5 km E of HOUPI; 23°20′N 120°30′E. Elevation 891 ft/272 m. Hot-springs resort with famous spa, hotels. Agriculture (tangerines, ginger, honey, bamboo); handicrafts (wood and bamboo items). Attractions also include a well-known grotto (Shuihuotung) and two temples: Biyunsi (1701) and Sanbaotien.

Kuan Yin Shan (GUAN-YIN SHAN) [Chinese is *Goddess of Mercy Mountain*], mountain, TAIWAN, 7.2 mi/11.6 km W of TAIPEI. From a distance, the shape of the mountain is said to resemble the head of Kuan Yin, Goddess of Mercy. Tangerines, bamboo shoots, and tea are produced here. Buddhist temples.

Kuanza River, ANGOLA: see CUANZA.

Kuba (KOO-bah), city, NE AZERBAIJAN, on N slope of the Greater CAUCASUS, 95 mi/153 km NW of BAKY. In orchard district; fruit and vegetable canning, sawmilling; manufacturing (microelectric motors, rugs). Hydroelectric power plant. Teachers' college; Scientific Research Institute of Horticulture, Viticulture, and Subtropical Crops.

Kubachi (koo-bah-CHEE), village (2005 population 3,050), S central DAGESTAN REPUBLIC, NE CAUCASUS, SE European Russia, near mountain road, 43 mi/69 km S of MAKHACHKALA; 42°05′N 47°36′E. Elevation 4,402 ft/1,341 m. Indigenous crafts (goldsmithing, dagger- and sword-making).

Kuban' (koo-BAHN), river, approximately 570 mi/917 km long, S European Russia; rising in the Greater CAUCASUS Mountains on the W slopes of Mount ELBRUS, KARACHEVO-CHERKESS REPUBLIC; flowing N in a wide arc past KARACHAYEVSK, CHERKESSK, and ARMAVIR (KRASNODAR TERRITORY) then W past Krasnodar, entering the Sea of AZOV through two arms. Its upper course is precipitous and leads through several gorges; it then meanders slowly through the Kuban Steppe, a rich black-earth area and one of the major grain and sugar-beet districts of the RUSSIAN FEDERATION. The last 150 mi/241 km are navigable. RUSSIA annexed the khanate of CRIMEA, of which the Kuban area was a part, in 1783. Now mainly within the Krasnodar Territory, the Kuban region was the territory of the Kuban Cossacks from about the mid-18th century to 1920. After Catherine II defeated (1775) the Zaporozhye Cossacks in the UKRAINE, some of them emigrated to TURKEY, but in 1787 they were allowed to return and settle along the Black Sea between the DNIEPER and Bug rivers. Then known as the Black Sea Cossacks, they were in 1792 resettled in the Kuban region. Though they lost much of their freedom and their rights were restricted, they were granted local self-government in return for military service. In 1860, they were renamed the Kuban Cossacks, while defending the Kuban region from hostile Circassian mountaineers to the S. After the Bolshevik Revolution of 1917, the Kuban Cossacks proclaimed an independent republic and fought against the Bolsheviks. After the civil war of 1918–1920, the Soviet regime abolished their government, and their traditional privileges were abrogated.

Kuban Reservoir (koo-BAHN), Russian *Kubanskoye Vodokhranilishche*, an artificial lake (☐ 89 sq mi/231.4 sq km), in NE KARACHEVO-CHERKESS REPUBLIC, S European Russia. Residents of 20 Circassian villages were forcefully evacuated in the 1970s to create the reservoir. Source of continuing dispute since the formation of the RUSSIAN FEDERATION in 1992 between local government, which wants the reservoir drained, and federal authorities, which plan to reroute Great ZELENCHUK and Little ZELENCHUK rivers into the reservoir.

Kubanskiy (koo-BAHNS-keeyee), settlement (2005 population 3,215), NE KRASNODAR TERRITORY, S Eu-

ropean Russia, in the YEYA RIVER basin, on road and railroad, 54 mi/87 km NE of KRASNODAR, and 5 mi/8 km SW of NOVOPOKROVSKAYA, to which it is administratively subordinate; 45°55′N 40°35′E. Elevation 311 ft/94 m. Agricultural products.

Kuba-Taba (KOO-bah–TAH-bah), village (2005 population 3,210), N central KABARDINO-BALKAR REPUBLIC, S European Russia, in the foothills of the N CAUCASUS Mountains, on road, 17 mi/27 km NNW of NAL'CHIK; 43°46′N 43°26′E. Elevation 1,541 ft/469 m. Agricultural products. Seismological station in the vicinity.

Kubatly (koo-BAHT-lee), town and administrative center of Kubatly region, S AZERBAIJAN, in KURDISTAN, 20 mi/32 km S of LACHIN; 39°20′N 46°40′E. Wheat, rice; livestock.

Kubena River (koo-BYE-nah), 215 mi/346 km long, in NW European Russia; rises near KONOSHA, near the border of ARCHANGEL and VOLOGDA oblasts; flows SE and SW, past KHAROVSK, to KUBENO Lake, forming a delta mouth below UST'YE. Formerly spelled Kubina.

Kubeno Lake (KOO-bee-nuh), Russian *Kubenskoye Ozero*, (☐ 89 sq mi/231.4 sq km), in VOLOGDA oblast, NW European Russia, NNW of VOLOGDA; 59°45′N 39°30′E; 37 mi/60 km long, 2 mi/3.2 km wide, 43 ft/13 m deep. Receives the POROZOVITSA River (N; part of the NORTHERN DVINA CANAL system) and the KUBENA RIVER (E). The SUKHONA River is the outlet (SE). Formerly spelled Kubino, Russian *Kubinskoye Ozero*.

Kubenskoye (KOO-been-skuh-ye), village, S VOLOGDA oblast, N central European Russia, on the SW shore of KUBENO LAKE, on road, 16 mi/26 km NW of VOLOGDA; 59°26′N 39°40′E. Elevation 416 ft/126 m. In dairying area; bakery. Also known as Ust'-Kubenskoye.

Kubikenborg (KOO-BEEK-en-BORY), ESE suburb of SUNDSVALL, VÄSTERNORRLAND county, NE SWEDEN, on inlet of GULF OF BOTHNIA, opposite Alnön island. Manufacturing (aluminum-smelting works).

Kubiki (koo-BEE-kee), village, Nakakubiki county, NIIGATA prefecture, central HONSHU, N central JAPAN, 62 mi/100 km S of NIIGATA; 37°11′N 138°19′E.

Kubina River, RUSSIA: see KUBENA RIVER.

Kubinka (KOO-been-kah), town (2006 population 25,880), W central MOSCOW oblast, central European Russia, on road and railroad, 21 mi/34 km W of ODINTSOVO, to which it is administratively subordinate; 55°35′N 36°41′E. Elevation 639 ft/194 m. Telecommunication equipment.

Kubino Lake, RUSSIA: see KUBENO LAKE.

Kubkub, Italian *Cub Cub*, village, SEMENAWI KAYIH BAHRI region, N ERITREA, 25 mi/40 km SE of NAKFA; 16°20′N 38°38′E. In stock-raising region; road junction.

Küblis, commune, GRISONS canton, E SWITZERLAND, on LANDQUART RIVER, 12 mi/19 km ENE of CHUR. Elevation c.2,690 ft/820 m. Hydroelectric plant. Has 15th-century church.

Kubokawa (koo-BO-kah-wah), town, Takaoka county, KOCHI prefecture, S SHIKOKU, W JAPAN, 31 mi/50 km S of KOCHI; 33°12′N 133°08′E.

Kubota (KOO-bo-tah), town, Saga county, SAGA prefecture, N KYUSHU, SW JAPAN, 3.7 mi/6 km W of SAGA; 33°13′N 130°14′E.

Kubrat (koo-BRAHT), city (1993 population 9,965), RUSE oblast, Kubrat obshtina, NE BULGARIA, 25 mi/40 km E of RUSE; 43°48′N 26°30′E. Market center; grain, cattle, vegetables, sunflowers; agricultural machine building, canning, wine making. Until 1934, known as Balbunar. City since 1949.

Kučevo (KOOCH-e-vo), village (2002 population 18,808), E SERBIA, 26 mi/42 km ESE of POZAREVAC; 45°11′N 19°56′E. Railroad terminus. Gold and silver mines nearby. Also spelled Kuchevo.

Kuchan (koo-CHYAHN), town, Shiribeshi district, W Hokkaido prefecture, N JAPAN, 31 mi/50 km W of SAPPORO; 42°53′N 140°45′E. Potatoes; confections, local sake. Skiing area.

Kuchan, IRAN: see QUCHAN.

Kuchawan (koo-CHAH-wuhn), town, NAGAUR district, N central RAJASTHAN state, NW INDIA, 130 mi/209 km ENE of JODHPUR. Trades in salt, metal, millet; handicraft sword making. Brine salt deposits in small lake, just S. KUCHAWAN Road, railroad station, is 12 mi/19 km SE, on SAMBHAR LAKE Also Kuchaman.

Kuche, CHINA: see KUQA.

Kuchen (KOO-khen), village, BADEN-WÜRTTEMBERG, S GERMANY, on the FILS, 2 mi/3.2 km NW of GEISLIN-GEN; 48°38′N 09°48′E.

Kuchesfahan (kuch-es-fah-HAHN), town, Gīlān province, N IRAN, 10 mi/16 km E of RASHT, and on W bank of the SEFID RIVER, 10 mi/16 km from CASPIAN SEA; 37°16′N 49°46′E. Agricultural center; rice, berries, tobacco; silk.

Kuchevo, SERBIA: see KUCEVO.

Kuch-i-Isfahan, IRAN: see KUCHESFAHAN.

Kuching (KOO-ching), city (2000 population 423,873), ⊙ SARAWAK, MALAYSIA, in W BORNEO and on the SARAWAK RIVER; 01°32′N 110°19′E. It is the largest city in the state and a seaport. Sago flour, rice, pepper, and birds' nests are exported; petroleum, gas, timber, copra. Manufacturing (textiles, metal products, boats, cement). Surawak Museum and Aquarium (Borneon ethnology and archaeology); State Library; BAKO NATIONAL PARK to NE, on coast. Old Fort Margherita (1879) served as defense for town. It was founded in 1839 by James Brooke. In the city are mosques, Buddhist temples, Anglican and Roman Catholic cathedrals, and a museum of Borneo folklore. The population is about ⅔ Chinese.

Kuchino (KOO-chee-nuh), former residential town, E central MOSCOW oblast, central European Russia, now part of ZHELEZNODOROZHNYY (3 mi/5 km W of the city center), and 13 mi/21 km E of MOSCOW; 55°45′N 37°58′E. Elevation 498 ft/151 m. Geophysical observatory.

Kuchinotsu (koo-CHEE-nots), town, S. Takaki county, NAGASAKI prefecture, NW KYUSHU, SW JAPAN, on S tip of SHIMABARA PENINSULA, 19 mi/30 km ESE of NAGASAKI across TACHIBANA BAY; 32°36′N 130°11′E.

Kuchiwa (koo-CHEE-wah), town, Hiba county, HIR-OSHIMA prefecture, SW HONSHU, W JAPAN, 43 mi/70 km N of HIROSHIMA; 34°54′N 132°54′E. Livestock (cattle).

Kuchki (KOOCH-kee), village, central PENZA oblast, S central European Russia, on the KHOPER River, on road, 32 mi/51 km SW of PENZA; 53°01′N 44°29′E. Elevation 800 ft/243 m. Bakery. Formerly known as Kuchki Mikhaylovskiye.

Kuchki Mikhaylovskiye, RUSSIA: see KUCHKI.

Kuchl (KOOKHL), township, SALZBURG, W AUSTRIA, on the SALZACH river and 13 mi/21 km SSE of SALZ-BURG; 47°38′N 13°09′E. Gypsum quarry and processing; timber, school of timber trade and timber processing.

Kuchuk Kainarji, Bulgaria: see KAINARDZHA.

Kuchuk Lake, RUSSIA: see KULUNDA STEPPE.

Kuckerneese, RUSSIA: see YASNOYE.

Kuçovë (koo-CHO-vuh), new town, ⊙ Kuçovë district (☐ 40 sq mi/104 sq km), S central ALBANIA, on DE-VOLL river, and 8 mi/12.9 km NNW of BERAT; 40°48′N 19°54′E. Petroleum center, linked by pipeline with VLORË; refinery; power plant; airport. A small village until after World War II, when it was developed as an industrial center. Named Stalin or Qyteti Stalin [Russian=city of Stalin] in 1950; renamed after 1990 revolution. Also Kuçova.

Kucukcekmece, (Turkish=Küçükçekmece) lake (☐ 5 sq mi/13 sq km), European TURKEY, 9 mi/14.5 km W of ISTANBUL; 5 mi/8 km long, 3 mi/4.8 km wide. Connected with Sea of MARMARA.

Küçük Menderes (koo-CHOOK–MEN-de-res), ancient *Scamander*, c.60 mi/95 km long, NW TURKEY; flows W and NW from the KAZ DAĞI through the vicinity of TROY (Troas) into the AEGEAN SEA. Scamander is frequently mentioned in the *Iliad*.

Küçük Stambul, UKRAINE: see SHARHOROD.

Kucukyozgat, TURKEY: see ELMADAGI.

Kuda (KOO-dah), village, KACHCHH district, GUJARAT state, W INDIA, on S edge of Little RANN of Kachchh, 35 mi/56 km NNW of WADHWAN. Railroad spur terminus; saltworks.

Kudal (koo-DAHL), town, ⊙ SINDHUDURG district, MAHARASHTRA state, W central INDIA, 70 mi/113 km SSE of RATNAGIRI; 16°02′N 73°41′E. Rice, coconuts, cashew nuts. Sometimes spelled Kudol.

Kudali (KOO-dah-lee), village, SHIMOGA district, KARNATAKA state, SE INDIA, 7 mi/11.3 km NE of SHIMOGA, at confluence of TUNGA and BHADRA rivers to form the TUNGABHADRA. Annual livestock fair.

Kudalur, INDIA: see CUDDALORE.

Kudamatsu (koo-DAH-mahts), city, YAMAGUCHI prefecture, SW HONSHU, W JAPAN, port on SUO SEA, just SE of TOKUYAMAand 25 mi/40 km E of YAMAGUCHI; 34°00′N 131°52′E. Tinware.

Kudara (koo-DAH-rah), village (2004 population 1,440), W central BURYAT REPUBLIC, S SIBERIA, RUS-SIA, near Lake BAYKAL, on road, 38 mi/61 km NW of ULAN-UDE; 52°13′N 106°39′E. Elevation 1,515 ft/461 m. Wheat, livestock; fish-processing. Petroleum and natural-gas deposits nearby. Until 1954, part of Stalin's Gulag prison-camp system.

Kudara-Somon (koo-DAH-rah-suh-MON), village (2004 population 1,820), SE BURYAT REPUBLIC, S SI-BERIA, RUSSIA, in the CHIKOY RIVER basin, less than 9 mi/14 km N of the Russia-MONGOLIA border, on highway branch, 24 mi/39 km SSW of BICHURA; 50°09′N 107°24′E. Elevation 2,431 ft/740 m. Dairy processing.

Kudasa, Gulf of, TURKEY: see KUSADASI, GULF OF.

Kudat (KOO-daht), town (2000 population 26,746), SABAH, E MALAYSIA, port on NW shore of MARUDU BAY, 70 mi/113 km NE of KOTA KINABALU; 06°53′N 116°46′E. Trade center for area; produces rubber, copra, coconuts; fishing; prawns.

Kudeikha (koo-DYE-ee-hah), village (2005 population 1,310), central CHUVASH REPUBLIC, central European Russia, on road, 60 mi/97 km S of CHEBOKSARY; 55°13′N 46°25′E. Elevation 328 ft/99 m. In agricultural area (grain, hemp, livestock); starch factory.

Kudelka, RUSSIA: see ASBEST.

Kudever' (KOO-dee-vyer), village, S PSKOV oblast, W Eu-ropean Russia, on the BEZHANITSY Upland, near highway, 18 mi/29 km S of NOVORZHEV; 56°47′N 29°26′E. Elevation 636 ft/193 m. In flax-growing area.

Kudeyevskiy (koo-DYE-eef-skeeyee), town (2005 population 2,875), E central BASHKORTOSTAN Re-public, E European Russia, on road and railroad, 36 mi/58 km ENE of UFA, and 12 mi/19 km E of IGLINO; 54°52′N 56°45′E. Elevation 580 ft/176 m. In agricultural area; food processing.

Kudi Boma (KOO-dee—BO-muh), village, BAS-CONGO province, W CONGO, on road 30 mi/48 km W of BOMA; 05°44′S 12°47′E. Elev. 160 ft/48 m.

Kudinovo (koo-DEE-nuh-vuh), village (2005 population 3,370), N KALUGA oblast, central European Russia, near highway, 7 mi/11 km WSW of MALOYAROSLAVETS, to which it is administratively subordinate; 55°01′N 36°15′E. Elevation 603 ft/183 m. In agricultural area (wheat, hemp, vegetables).

Kudinovo, RUSSIA: see ELEKTROUGLI.

Kudligi (KOOD-lee-gee), town, tahsil headquarters, BELLARY district, KARNATAKA state, SW INDIA, 25 mi/40 km S of HOSPET; 14°54′N 76°23′E. Road center; peanuts, tamarind; silk growing.

Kudol, INDIA: see KUDAL.

Kudowa Zdroj (koo-DO-vah zdrooy), Polish *Kudowa Zdrój*, town in LOWER SILESIA, Wałbrzych province, SW POLAND, on CZECH border, at W foot of HEU-SCHEUER MOUNTAINS, 20 mi/32 km W of KŁODZKO (Glatz). Health resort; cotton milling. Mineral springs known since 16th century. Chartered after 1945.

Kudoyama (koo-DO-yah-mah), town, Ito county, WAKAYAMA prefecture, S HONSHU, W central JAPAN, on W KII PENINSULA, 24 mi/39 km E of WAKAYAMA; 34°17′N 135°33′E.

Kudryashovskiy (koo-dree-SHOF-skeeyee), settlement (2006 population 3,920), E NOVOSIBIRSK oblast, SW Siberian Russia, on the OB' RIVER, on highway, 9 mi/14 km NW of NOVOSIBIRSK; 55°06′N 82°46′E. Elevation 272 ft/82 m. Sawmilling, woodworking.

Kudryavtsevka, UKRAINE: see KUDRYAVTSIVKA.

Kudryavtsivka (kood-RYAHV-tsif-kah), (Russian *Ku-dryavtsevka*), town (2004 population 2,400), W MY-KOLAYIV oblast, UKRAINE, at railroad junction 23 mi/37 km SW of VOZNESENS'K; 47°18′N 31°01′E. Elevation 272 ft/82 m. Grain elevator. Established in 1910 as Kolosivka (Russian *Kolosevka*); renamed in 1920; town since 1976.

Kudus (KOO-doos), town, ⊙ Kudus district, E Java Tengah province, INDONESIA, 30 mi/48 km ENE of SEMARANG, at foot of Mount Muria; 06°48′S 110°50′E. Trade center for agricultural area (sugar, rice, cas-sava); textile mills. Also spelled Koedoes.

Kudymkar (koo-dim-KAHR), city (2005 population 31,490), ⊙ KOMI-PERMYAK autonomous okrug, PERM oblast, E European Russia, on the Inva River (short right tributary of the KAMA River), 866 mi/1,394 km ENE of MOSCOW, and 125 mi/201 km NW of PERM; 59°01′N 54°39′E. Elevation 577 ft/175 m. Highway hub; local transshipment point. Agricultural, administra-tive, and cultural center; lumber, metal, and food industries (meat packing, dairy). Has a drama theater; vocational schools in agriculture, forestry, and edu-cation. Arose in the 16th century. Developed mainly after World War I. Became city in 1938.

Kudzsir, ROMANIA: see CUGIR.

Kuei-lin, CHINA: see GUILIN.

Kuei-yang, CHINA: see GUIYANG.

Kuelap, archaeological site, LUYA province, AMAZONAS region, N PERU; 06°25′S 77°48′W. Immense building complex constructed by the CHACHAPOYAS culture. Conservation project developed to protect advanced deterioration.

Kues, GERMANY: see BERNKASTEL-KUES.

Kufa (KOO-fuh), former MESOPOTAMIAN city, NAJAF province, on the W bank of the EUPHRATES RIVER, c.90 mi/145 km S of BAGHDAD. Founded in 638, Kufa soon rivaled BASRA in size. The Arab governor of IRAQ resided here until 702. For a time, Kufa was the seat of the Abbasid caliphate, and Ali, the fourth caliph, was murdered here. Celebrated as a major seat of Arab learning, the city was also a continual source of po-litical and religious unrest. It was repeatedly plun-dered by the Karmathians in the 10th century and lost its importance.

Kufara, LIBYA: see KUFRA.

Kufra (KU-fruh), Arabic *Wahat al Kufarah*, Italian *Cufra*, group of oases in basin (☐ 457 sq mi/1,188.2 sq km), near center of LIBYAN DESERT, CYRENAICA re-gion, SE LIBYA, 575 mi/925 km SE of BANGHAZI; 30 mi/48 km long (NE–SW), 12 mi/19 km wide; 24°00′N 23°00′E. Chief oasis is Al JAWF. Manufacturing (olive oil; textiles, baskets; leather goods, silver goods); agriculture (fruit, grain, olives; camels). Airport. Former center for caravan trade between coast and points S and Tibesti Mountains. Former Senusiyah strong-hold. Occupied and held by Italy 1931–1943. Some-times Kufara.

Kufre, TURKEY: see SIRVAN.

Kufri, IRAQ: see KIFRI.

Kufrinja (ku-FRIN-zhuh) or **Kufranjeh**, township, N JORDAN, 5 mi/8 km W of 'AJLUN; 32°17′N 35°42′E. Vineyards; olives, fruit.

Kufstein (KOOF-shtein), town, in TYROL province, W AUSTRIA, on the INN River, near the German border; 47°35′N 12°10′E. Market center. Manufacturing (skis, rackets, glass, armatures, machines, weapons, metalware). Summer tourism. The fortress of Geroldseck, rebuilt by Emperor Maximilian I in the 16th century on 12th century foundations, contains a modern organ famous for its great size and power.

Kuga (KOO-gah), town, Kuga county, YAMAGUCHI prefecture, SW HONSHU, W JAPAN, 37 mi/60 km E of YAMAGUCHI; 34°05′N 132°04′E.

Kugaaruk, hamlet and Roman Catholic mission station, KITIKMEOT region, NUNAVUT territory, CANADA, at head of Pelly Bay, inlet (75 mi/121 km long, 10 mi/16 km–40 mi/64 km wide) of the Gulf of BOOTHIA; 68°24′N 89°12′W. Crafts; seal hunting, fishing. Radio, TV. Scheduled air service. Remains of Roman Catholic church built of hand-quarried stone. Mission station established 1935. Formerly called Pelly Bay.

Kugaly, KAZAKHSTAN: see KOGALY.

Kugesi (koo-GYE-see), village (2005 population 11,515), N CHUVASH REPUBLIC, central European Russia, on road, 8 mi/13 km SSE of CHEBOKSARY; 56°01′N 47°18′E. Elevation 547 ft/166 m. Agricultural chemicals and equipment; agricultural machinery repair; gas pipeline equipment.

Kugino (koo-GEE-no), village, Aso county, KUMAMOTO prefecture, W KYUSHU, SW JAPAN, 16 mi/25 km E of KUMAMOTO; 32°49′N 131°02′E.

Kugitang, TURKMENISTAN: see SVINTSOVII RUDNIK.

Kugluktuk, town (□ 212 sq mi/551.2 sq km; 2001 population 1,212), KITIKMEOT region, NUNAVUT territory, CANADA, on CORONATION GULF of the BEAUFORT SEA, at mouth of COPPERMINE RIVER; 67°40′N 115°05′W. Rises to 74 ft/22.6 m. Fishing, hunting, trapping. Trading post; government radio and meteorological station, Royal Canadian Mounted Police post; site of Anglican and Roman Catholic missions. Extensive copper deposits in region. Hudson's Bay Company store opened 1927; Anglican mission built 1928. Scheduled air service. Name means "place of rapids." Formerly called Coppermine; a variant spelling is Kogluktok; a variant name is Coronation.

Kugul'ta (koo-gool-TAH), village (2006 population 5,600), central STAVROPOL TERRITORY, N CAUCASUS, S European Russia, on road junction, 19 mi/30 km NE of STAVROPOL; 45°22′N 42°23′E. Elevation 741 ft/225 m. In agricultural area (wheat, grapes; livestock). Also known as Kugul'tinskoye.

Kugul'tinskoye, RUSSIA: see KUGUL'TA.

Kuguno (KOO-goo-no), town, Ono county, GIFU prefecture, central HONSHU, central JAPAN, 56 mi/90 km N of GIFU; 36°02′N 137°16′E.

Kuh, (Persian, *mountain*), for names in AFGHANISTAN, IRAN, and PAKISTAN beginning thus and not found here: see under KOH, or under following part of the name.

Kuha, town (2007 population 17,942), TIGRAY state, N ETHIOPIA; 13°29′N 39°33′E. On Addis Ababa–Asmara (ERITREA) road and 5 mi/8 km E of MEK'ELĒ. Cereals, livestock. Also called Quiha or Kwiha.

Kuhak (KOO-ahk), town, Sīstān va Balūchestān province, SE IRAN, on PAKISTAN border; 27°08′N 63°15′E. Some grazing.

Kühbach (KOO-bahkh), village, BAVARIA, S GERMANY, in UPPER BAVARIA, 20 mi/32 km NW of Dachau; 48°30′N 11°12′E. Has 17th-century church.

Kuhbanan (KOO-bahn-NAHN), town, Kermān province, SE IRAN, 90 mi/145 km NNW of KERMAN, at foot of KUH Banan mountains, one of the S Iranian ranges. Road junction. Also spelled Kūhbonān or Kuh Banan.

Kuhdasht (KOO-dahst), town (2006 population 85,997), Lorestān province, SW IRAN, 45 mi/72 km W of KHORRAMABAD, in mountainous tribal region; 33°31′N 47°36′E. Also spelled KUH-E-DASHT or Kuh-i-Dasht.

Kuh-e-Dasht, IRAN: see KUHDASHT.

Kuh e Khaje (KOO ei khah-JE), island, Sīstān va Balūchestān province, SE IRAN. Located in Daryache ye Hamun Lake (seasonal). Largely uninhabited with two ancient forts overlooking the lake at S and SE points.

Kuh-e-Rang (KOO-ei–RAHNG), mountain (14,000 ft/ 4,267 m) in ZAGROS Mountain ranges, SW IRAN, in BAKHTIARI country, 100 mi/161 km W of ESFAHAN and S of KHONSAR. Also called KUH-I-RANG.

Kuh-i-Aleh (KOO-ee-ah-LAI), Turkish *Ala Dagh*, one of the TURKMEN-KHURASAN ranges, in NE IRAN; rises to 10,000 ft/3,050 m in the Shah Jehan, 110 mi/177 km NW of MASHAD. Continued SE by BINALUD RANGE.

Kuh-i-Malik Siah, PAKISTAN: see KOH-I-MALIK SIAH.

Kuh-i-Rang, IRAN: see KUH-E-RANG.

Kühlungsborn (KOO-lungs-born) or OSTSEEBAD KUHLUNGSBORN, town, Mecklenburg-Western Pomerania, N GERMANY, on MECKLENBURG BAY of the BALTIC, 16 mi/26 km WNW of ROSTOCK; 54°09′N 11°45′E. Seaside resort. Emerged through the joining of the villages of ARENDSEE, BRUNSHAUPTEN, and Fulgen in 1937.

Kuhmo (KOO-mo), village, OULUN province, E FINLAND, 55 mi/89 km E of KAJAANI; 64°08′N 29°31′E. Elevation 578 ft/175 m. Near Russian border, in lake region and wilderness area; lumbering. Chamber music festival in July.

Kühpāyeh (KOO-pah-ye), town, Esfahān province, W central IRAN, 45 mi/72 km E of ESFAHAN, on Esfahan-YAZD road. Wheat, cotton, madder root, almonds.

Kuhsan, AFGHANISTAN: see KOHSAN.

Kui (koo-EE), town, Mitsugi county, HIROSHIMA prefecture, SW HONSHU, W JAPAN, 34 mi/55 km E of HIROSHIMA; 34°30′N 133°01′E.

Kuibishevsk (koo-ee-bee-SHEVSK), town, KHATLON viloyat, SW TAJIKISTAN, in Vakhsh valley, 9 mi/15 km N of QURGHONTEPPA (Kurgan-Tyube); 37°57′N 68°49′E. Also spelled Kuibyshevsk.

Kuibyshevabad, TAJIKISTAN: see KUIBYSHEVSKII.

Kuibyshevka-Vostochnaya, RUSSIA: see BELOGORSK.

Kuibyshevo (koo-ee-bi-shi-vuh), village, central MARY weloyat, S TURKMENISTAN, on railroad (Talkhatan-Baba station), and 20 mi/32 km SE of MARY. Cotton. Sometimes spelled Kuybyshevo; also Imeni Kuibysheva. Until c.1940, called Talkhatan-Baba.

Kuibyshevo, KAZAKHSTAN: see ZHYNGYLDY.

Kuibyshevo, UZBEKISTAN: see RISHTAN.

Kuibyshevskii, town, KHATLON viloyat, TAJIKISTAN, in Vakhsh valley, 8 mi/12.9 km N of QURGHONTEPPA (Kurgan-Tyube); 37°52′N 68°44′E. Cotton; metalworks. Formerly called Aral and (c.1935–1940) Kuibyshevabad. Also spelled Kuibyshevskiy.

Kuiganiar (kwee-gahn-YAHR), town, ANDIJAN wiloyat, UZBEKISTAN, 40°52′N 72°19′E. Also Kuyganyar.

Kuik, NETHERLANDS: see CUIJK.

Kuilenburg, NETHERLANDS: see CULEMBORG.

Kuils River, Afrikaans *Kuilsrivier*, suburb, WESTERN CAPE, SOUTH AFRICA, 15 mi/24 km E of CAPE TOWN, and 5 mi/8 km SE of BELLVILLE; 33°25′S 18°42′E. Fruit, tobacco, viticulture; tin deposits.

Kuilyuk (kweel-YOOK), district of TOSHKENT, N TOSHKENT wiloyat, UZBEKISTAN, on CHIRCHIK RIVER, just SE of Toshkent; 41°14′N 69°17′E. Food processing (fruits, vegetables). Also KUYLYUK.

Kuinjipad, INDIA: see KURINJIPPADI.

Kuiseb River (KWEE-seb), W central NAMIBIA; rises in KHOMAS HOCHLAND; flows in wide SW arc through NAMIB DESERT to Walyis Bay; episodic, deep canyon in mid-course marks divide between Sand Desert to N and Rock Desert to S.

Kuittijärvi, RUSSIA: see KUYTO LAKES.

Kuiu Island (KOO-wee-yoo), SE ALASKA, in ALEXANDER ARCHIPELAGO, between KUPREANOF ISLAND (E) of BARANOF ISLAND (W), 30 mi/48 km W of PETERSBURG; 65 mi/105 km long, 6 mi/9.7 km–23 mi/37 km wide; 56°28′N 134°01′W. Rises to c.3,000 ft/914 m. Fishing.

Kuivastu (KOO-ee-vuh-stoo), village, on E coast of MUHU island, ESTONIA, 6 mi/10 km SE of Muhu village; 58°34′N 23°23′E. Connected by ferry across MUHU SOUND with VIRTSUON mainland.

Kujawy (koo-YAH-vee), region, N central POLAND, in SE Bydgoszcz and NW Warszawa provinces, along left bank of the VISTULA, centered on INOWROCŁAW and WŁOCŁAWEK. Dotted with lakes, it is noted for its fertility. Principal crops are rye, potatoes, oats, wheat, sugar beets; stock raising. Sometimes called Cuyavia.

Kuji (koo-JEE), city, IWATE prefecture, N HONSHU, NE JAPAN, 47 mi/75 km N of MORIOKA; 40°11′N 141°46′E. Amber. Northern limit of female *ama* pearl divers.

Kuju (KOO-joo), town, Naoiri county, OITA prefecture, central KYUSHU, SW JAPAN, 25 mi/40 km SW of OITA, near Mount Taisen and MOUNT KUJU (alpine plants); 33°01′N 131°17′E.

Kujukuri (koo-JOO-koo-ree), town, Sanbu county, CHIBA prefecture, E central HONSHU, E central JAPAN, 12 mi/20 km E of CHIBA; 35°31′N 140°26′E. Sardines. Kujukuri Beach is here.

Kuju, Mount (KOO-joo), Japanese *Kuju-san* (koo-JOO-sahn), peak, in OITA prefecture, E KYUSHU, SW JAPAN, 24 mi/39 km WSW of OITA, in ASO KUJU NATIONAL. Winter sports. Hot-spring resorts in NE foothills.

Kuka (KOO-kah), town, on N coast of OSHIMA island, Oshima county, YAMAGUCHI prefecture, W JAPAN, 50 mi/80 km E of YAMAGUCHI; 33°56′N 132°15′E. Mandarin oranges. Also Kuga.

Kukalaya River (koo-kah-LAH-yah), c.100 mi/161 km long, E NICARAGUA; rises in E outlier of CORDILLERA ISABELIA 8 mi/12.9 km NE of BONANZA; flows SE, through hardwood area, to CARIBBEAN SEA at WOUNTA, here forming WOUNTA LAGOON. Also called Wounta River.

Kukarka, RUSSIA: see SOVETSK, KIROV oblast.

Kukawa (KOO-kah-WAH), town, headquarters of Local Government Area, BORNO state, extreme NE NIGERIA. It is in a farming and salt-mining region. Fish market. Kukawa was founded in 1814 by Muhammad al-Kanemi of the state of BORNU. The capital and chief commercial center of Bornu, Kukawa also was the S terminus of a trans-Saharan caravan route to TRIPOLI, LIBYA. In 1893, Kukawa was conquered and destroyed by forces under Rabih, a Sudanese slave trader. It was rebuilt by the British in 1902 as a garrison town. Also Kuka.

Kukenaam (koo-ken-AHN), mountain and waterfall (drop c.2,000 ft/610 m), on GUYANA-VENEZUELA border, just NW of Mount RORAIMA. Springs from a table top mountain of the same name. Also spelled Cuquenán.

Kukës (KOOKS), new town (2001 population 16,686), ⊙ Kukës district (□ 360 sq mi/936 sq km; 2001 population 16,686), N ALBANIA, near SERBIA (KOSOVO province) border, 45 mi/72 km E of Shkodër, and on Drin River where it is formed by union of WHITE DRIN and BLACK DRIN rivers; 42°05′N 20°20′E. Commercial center of Albanian Kosovo; airport; tourist hotel; hospital. Copper mining; iron and chrome deposits nearby. Built on site of old Kukës, drowned by damming (1970) of DRIN River. Also spelled Kukēsi.

Kukes, TURKEY: see SERIK.

Kuki (KOO-kee), city, SAITAMA prefecture, E central HONSHU, E central JAPAN, 12 mi/20 km N of URAWA; 36°03′N 139°40′E.

Kukipi (kyoo-KEE-pee), village, GULF province, SE NEW GUINEA island, S central PAPUA NEW GUINEA, 110 mi/177 km NW of PORT MORESBY. On GULF OF PAPUA at the mouths of TAURI and LAKEKAMU rivers. Road access from KEREMA. Fish; cattle; timber.

Kukisvumchorr (KOO-kyis-voom-chuhr), former town, S central MURMANSK oblast, NW European Russia, on the KOLA PENINSULA, in the central KHIBINY MOUNTAINS, 3 mi/5 km N of KIROVSK, into which it has been incorporated as a suburb; 67°39′N 33°43′E. Elevation 1,765 ft/537 m. Apatite mines in the vicinity.

Kukizaki (koo-KEE-zah-kee), town, Inashiki county, IBARAKI prefecture, central HONSHU, E central JAPAN, 34 mi/55 km S of MITO; 35°59′N 140°07′E. Peanuts.

Kukkus, RUSSIA: see PRIVOLZHSKOYE.

Kuklen (KOO-klen), village, PLOVDIV oblast, Rodopska obshtina, S central BULGARIA, at the N foot of the W RODOPI Mountains, 5 mi/8 km WNW of ASENOVGRAD; 42°02′N 24°47′E. Nonferrous metallurgy, wine making; tobacco, sericulture. Site of an old monastery.

Kuklite (KOO-klee-te), peak in the PIRIN MOUNTAINS, SW BULGARIA; 41°38′N 23°27′E. Elev. 8,812 ft/2,686 m.

Kukmor (KOOK-muhr), town (2006 population 16,810), N TATARSTAN Republic, E European Russia, near the VYATKA River, on the border with KIROV oblast, on road and railroad, 14 mi/22 km WSW of VYATSKIYE POLYANY; 56°11′N 50°53′E. Elevation 321 ft/97 m. Metalwear, building ceramics, felt boot, fur, clothing, wool-weaving; mixed fodder plant, dairy, bakery.

Kuknur (kook-NOOR), town, RAICHUR district, ANDHRA PRADESH state, SE INDIA, 15 mi/24 km NW of KOPPAL; 15°30′N 76°00′E. Oilseed milling; millet, cotton. Also spelled Koknur.

Kukoboy (koo-kuh-BO-yee), village, N YAROSLAVL oblast, central European Russia, 31 mi/50 km NE of POSHEKHONYE; 58°42′N 39°54′E. Elevation 485 ft/147 m. In flax-growing area; forestry services.

Kukrahill (kook-RAH-heel), town, SOUTH ATLANTIC COAST AUTONOMOUS REGION, ZELAYA department, NICARAGUA, 20 mi/32 km N of BLUEFIELDS, near PEARL LAGOON. Administrative center for a largely Creole area. English widely spoken here. Fishing, rice, bananas. Also known as Kukra Hill.

Kuks (KUKS), German *Kukus*, village, VYCHODOCESKY province, NE BOHEMIA, CZECH REPUBLIC, on ELBE RIVER, 4 mi/6.4 km SE of DVUR KRALOVE; 50°24′N 15°54′E. Has church with notable sculptures and paintings. Rock sculptures depicting biblical scenes nearby. Summer resort.

Kukshi (KOOK-shee) or **Kuksi**, town (2001 population 24,317), DHAR district, SW MADHYA PRADESH state, central INDIA, 45 mi/72 km SW of DHAR; 22°12′N 74°45′E. Markets maize, wheat, cotton, millet; cotton ginning.

Kukshik, RUSSIA: see PERVOMAYSKIY, BASHKORTOSTAN Republic.

Kukuihaele (koo-KOO-ee-hah-AI-lai), village (2000 population 317), N HAWAII island, HAWAII county, HAWAII, on Waipio Bay, 45 mi/72 km NW of HILO, on Hamakua (NE) Coast; 20°07′N 155°34′W. Macadamia nuts; fish. Hamakua Forest Reserve to S.

Kuku Nor, CHINA: see QINGHAI HU.

Kukuom (koo-koo-ahm), town, BRONG-AHAFO region, GHANA, 10 mi/16 km SE of GOASO; 06°41′N 02°27′W. Cacao, timber, kola nuts.

Kukup (KOO-koop), village, S Johore, MALAYSIA, on Strait of MALACCA, 22 mi/35 km WSW of Johore Bharu. Fisheries; coconuts; tourism center based on traditional Chinese fishing settlement.

Kukus, CZECH REPUBLIC: see KUKS.

Kukushtan (koo-koo-SHTAHN), village (2006 population 4,830), S central PERM oblast, E European Russia, in W central URALS, in the KAMA River basin, on highway and railroad, 30 mi/48 km SSE of (and administratively subordinate to) PERM; 57°39′N 56°29′E. Elevation 446 ft/135 m. In agricultural area; yeast factory.

Kula (KOO-lah), city (1993 population 4,519), MONTANA oblast, Kula obshtina, NW BULGARIA, 19 mi/31 km SW of VIDIN, near the Serbian border; 43°53′N 22°32′E. Agricultural center (grain, livestock, truck); manufacturing synthetic rubber goods, hemp textiles, honey. Linked with Zaichar (SERBIA) by the Vrashka-chuka Pass.

Kula (KOO-lah), town (2002 population 48,353), VOJVODINA, NW SERBIA, on DANUBE-TISZA CANAL, and 27 mi/43 km NNW of NOVI SAD, in the BACKA region; 45°37′N 19°32′E. Railroad junction. Woolen and felt-hat making.

Kula (KOO-lah), village (2000 population 23,217), W TURKEY, 65 mi/105 km E of MANISA; 38°33′N 28°38′E. Carpets, wool; wheat, barley, tobacco.

Kulachi (kuh-lah-CHEE), town, DERA ISMAIL KHAN district, S NORTH-WEST FRONTIER PROVINCE, N PAKISTAN, 26 mi/42 km WNW of DERA ISMAIL KHAN; 31°56′N 70°27′E.

Kulageron, urban settlement, at confluence of the Pambak and Dzoraget rivers (KURA Basin), ARMENIA. Railroad station (Tumanian) on the TBILISI-Leninakan border. It is the site of the Dzorajet Hydroelectric Power Plant.

Kulagino (koo-lah-GEE-no), village, N ATYRAU region, KAZAKHSTAN, on URAL RIVER, 85 mi/137 km N of ATYRAU; 48°22′N 51°35′E. Cattle.

Kula Gulf (KOO-lah), SOLOMON ISLANDS, SW Pacific, between KOLOMBANGARA (W) and main island of NEW GEORGIA (E); 5 mi/8 km wide. In World War II, scene of two successful U.S. naval battles against JAPAN.

Kulal, Mount (koo-LAL), volcano (7,520 ft/2,292 m), EASTERN province, N KENYA, at E rim of GREAT RIFT VALLEY, near S tip of Lake TURKANA; 02°44′N 36°54′E. Cone is split into two sections by a cleft 3,000 ft/914 m deep.

Kulambangra, SOLOMON ISLANDS: see KOLOMBANGARA.

Kulanak (koo-luh-NUHK), village, central NARYN region, KYRGYZSTAN, on NARYN RIVER, and 23 mi/37 km W of NARYN; 41°12′N 75°37′E. Wheat; cattle breeding.

Kulangar (koo-LAHN-gahr), village and district, LOGAR province, E AFGHANISTAN, on LOGAR RIVER, 35 mi/56 km S of KABUL, on road to GARDEZ; 34°02′N 69°01′E.

Kulanly (koo-lahn-LEE), village, TURKMENISTAN. Tertiary-level administrative center.

Kulary (koo-LAH-ri), village (2005 population 4,680), central CHECHEN REPUBLIC, S European Russia, near highway and railroad, 8 mi/13 km WSW, and under administrative jurisdiction, of GROZNYY; 43°14′N 45°30′E. Elevation 524 ft/159 m. Has a hospital, mosque. For a short time in the mid-20th century, called Naberezhnoye.

Kulasekharapatnam (koo-lah-SAI-kah-rah-PUHT-nuhm), town, NELLAI KATTABOMMAN district, TAMIL NADU state, S INDIA, port on GULF OF MANNAR, 35 mi/56 km SE of TIRUNELVELI; 08°24′N 78°03′E. Produces alcohol, sugar, salt; exports palmyra fiber, oils and oil cake, jaggery, tobacco. Roman Catholic village of MANAPAD (formerly Manappadu), reputed residence (1540s) of St. Francis Xavier, and lighthouse are on headland 2 mi/3.2 km S. Formerly spelled Kulasekarapatnam.

Kulata (KOO-lah-tah), village, SOFIA oblast, PETRICH obshtina, SW BULGARIA, railroad station and highway crossing on the Greek border, 7 mi/11 km E of Petrich; 41°23′N 23°22′E. Tobacco, oil-bearing plants.

Kulaura (koo-lah-oo-rah), village (2001 population 20,934), SYLHET district, E EAST BENGAL, BANGLADESH, 27 mi/43 km SSE of SYLHET; 24°35′N 91°58′E. Railroad junction (spur to SYLHET). Agriculture (rice, tea, oilseeds); tea processing nearby.

Kulautuva (koo-lou-too-VAH), town, S central LITHUANIA, on right bank of the Nemunas and 12 mi/19 km WNW of KAUNAS; 54°56′N 23°36′E. Summer resort; sanatorium.

Kuldīga (KOOL-dee-guh), German *goldingen*, town (2000 population 13,678), W LATVIA; 56°58′N 21°59′E. Forest products, furniture manufacturing. Founded in 1244, Kuldiga was a residence of the dukes of COURLAND and since it escaped destruction during the war, it retains a medieval character. The city has two 17th-century churches.

Kuldiha (KUL-dee-hah), village, MAYURBHANJ district, NE ORISSA state, E central INDIA, on railroad and 38 mi/61 km WNW of BARIPADA. Pottery manufacturing. Iron ore mined in nearby hills (SE).

Kul'dur (kool-DOOR), health resort (2005 population 1,875), NW JEWISH AUTONOMOUS OBLAST, N central KHABAROVSK TERRITORY, RUSSIAN FAR EAST, on railroad, 18 mi/29 km W of LONDOKO, and 12 mi/19 km N of BIRAKAN; 49°13′N 131°38′E. Elevation 1,207 ft/367 m. Hot alkaline-sulphur springs.

Kulebaki (koo-lee-BAH-kee), city (2006 population 37,155), SW NIZHEGOROD oblast, central European Russia, on road and railroad, 117 mi/188 km SW of NIZHNIY NOVGOROD, and 22 mi/35 km ESE of MUROM; 55°25′N 42°32′E. Elevation 357 ft/108 m. Ferrous metallurgy, metal goods, radios, woodworking, food processing (dairy, bakery). Became city in 1932.

Kulebivka (koo-le-BEEF-kah) (Russian *Kulebovka*), SW suburb of NOVOMOSKOVS'K, N DNIPROPETROVS'K oblast, UKRAINE; 48°38′N 35°12′E. Elevation 187 ft/56 m.

Kulebovka, UKRAINE: see KULEBIVKA.

Kulemborg, NETHERLANDS: see CULEMBORG.

Kuleshovka (koo-lye-SHOF-kah), town (2006 population 14,665), SW ROSTOV oblast, S European Russia, in the DON River delta, on road and railroad, 9 mi/14 km WSW of BATAYSK, and 4 mi/6 km E of AZOV, to which it is administratively subordinate; 47°04′N 39°33′E. Mechanical repair plant; infant food factory.

Kulgam (kul-GAHM), town, ANANTNAG, or KASHMIR SOUTH, district, JAMMU AND KASHMIR state, N INDIA, in SW central KASHMIR region, in VALE OF KASHMIR, near the JHELUM River, 10 mi/16 km SW of ANANTNAG; 33°39′N 75°01′E. Rice, corn, oilseeds, wheat.

Kuli (koo-LEE), village (2005 population 3,985), S DAGESTAN REPUBLIC, Russia, in the E Greater CAUCASUS, at the SE edge of the SULAK RIVER basin, 36 mi/58 km S of MAKHACHKALA, and 12 mi/19 km SSE of KUMUKH; 42°18′N 47°13′E. Elevation 5,344 ft/1,628 m. Hardy grain; sheep. Power station in the vicinity. Population largely Lak, with a sizeable Chechen minority from the influx of refugees from neighboring CHECHNYA's war zones.

Kuliga (koo-LEE-gah), village, NE UDMURT REPUBLIC, E central European Russia, on road, 40 mi/64 km ENE of GLAZOV; 58°11′N 53°46′E. Elevation 869 ft/264 m. Lumbering. Also known as Kuligi.

Kuligi, RUSSIA: see KULIGA.

Kulikov, UKRAINE: see KULYKIV.

Kulikovka, UKRAINE: see KULYKIVKA.

Kulików, UKRAINE: see KULYKIV.

Kulim (KOO-lim), town (2000 population 117,454), S KEDAH, MALAYSIA, 15 mi/24 km E of George Town, near PENANG line; 05°22′N 100°34′E. Rubber-growing center, rice. Tin and tungsten mines to E.

Kulin (KOO-lin), town, WESTERN AUSTRALIA state, W AUSTRALIA, 176 mi/283 km SE of PERTH; 32°42′S 118°08′E. Wheat. Area of wildflowers, especially orchids and the *Macrocarpa* flowering gum (eucalyptus). East of town are rock formations, including a large granite boulder and the pink-and-white Buckley's Breakaways.

Kulinichi, UKRAINE: see KULYNYCHI.

Kulithura (kuh-LI-tuh-ruh), city, TRIVANDRUM district, KERALA state, S INDIA, 20 mi/32 km SE of Trivandrum. Coir rope and mats, palmyra jaggery. Monazite workings nearby. Also Kuzhittura.

Kulittalai (kuh-LIT-uh-lei) or **Kulitalai**, town, TIRUCHCHIRAPPALLI district, TAMIL NADU state, S INDIA, on KAVERI RIVER opposite MUSIRI (ferry) and 18 mi/29 km WNW of Tiruchchirappalli; 10°55′N 78°25′E.

Cotton textiles, wicker coracles; rice, plantains, coconut palms.

Kuliyapitiya, town, NORTH WESTERN PROVINCE, SRI LANKA, 22 mi/35 km W of KURUNEGALA; 07°28′N 80°03′E. Vegetables, rice, coconut palms. DANDAGAMUWA village c.2 mi/3.2 km E.

Kulkuduk, village, NAWOIY wiloyat, central UZBEKISTAN; 42°30′N 63°15′E.

Kulladakurichi, INDIA: see KALLIDAIKURICHI.

Kullen (KUL-len), peninsula, SW Sweden, SKÅNE county, extends 15 mi/24 km into KATTEGATT at mouth of ÖRESUND, 15 mi/24 km NNW of HELSINGBORG; 1mi/1.6 km–10 mi/16 km wide. Forms S shore of Skälderviken Bay; hilly surface rises to 617 ft/188 m (N). Kullen lighthouse at extremity (56°18′N 12°27′E). MÖLLE and ARILD are seaside resorts.

Kullu (KUHL-loo), district (□ 2,125 sq mi/5,525 sq km), HIMACHAL PRADESH state, N INDIA, in PUNJAB HIMALAYAS; ⊙ KULLU. Forests and mountain wastes, with fertile, picturesque valley of upper BEAS RIVER (called Kullu valley). Hand-loom weaving; wheat, corn, barley, potatoes; beekeeping. Kullu valley noted for tea and fruit growing; willow plantation. Strong Rajput state in medieval India.

Kullu (KUHL-loo), town, ⊙ KULLU district, HIMACHAL PRADESH state, N INDIA, in valley on BEAS RIVER, 50 mi/80 km ESE of DHARMSHALA; 31°58′N 77°06′E. Trades in grain, wool, fruit, tea, honey, timber; hand-loom weaving, basket making. Has an airport. Important tourist site. Formerly called SULTANPUR.

Küllük, TURKEY: see GULLUK.

Kulm, district, S AARGAU canton, N SWITZERLAND. Main town is REINACH; population is German-speaking and Protestant.

Kulm (KUHLM), village (2006 population 380), LA MOURE CO., S NORTH DAKOTA, 32 mi/51 km W of LA MOURE; 46°17′N 98°57′W. Livestock, grain, dairy products. Whitestone Battlefield Historic Site to SE. Founded in 1893 and incorporated in 1906.

Kulm, CZECH REPUBLIC: see CHLUMEC.

Kulm, SWITZERLAND: see RIGI.

Kulmbach (KULM-bahkh), town, BAVARIA, central GERMANY, on the WHITE MAIN RIVER; 50°06′N 11°27′E. Noted for its brewing since 1349. Today there are four breweries in the town. Other manufacturing includes machinery, steel, and pharmaceuticals. Known in 1035, Kulmbach became (1340) the residence of the margraves of Kulmbach (later known as the margraves of Bayreuth) of the house of Hohenzollern. Passed to PRUSSIA in 1791; annexed by FRANCE in 1807, by BAVARIA in 1810 and made part of UPPER FRANCONIA. On a nearby hill is the fortress (now a museum) of Plassenburg (12th century; rebuilt in Renaissance style 1560–1570), which served as a prison from 1808 to the early 20th century.

Kulob (koo-LOB), city (2000 population 78,000) ⊙ KHATLON viloyat, TAJIKISTAN, on the YAKH-SU, and 70 mi/113 km SE of DUSHANBE; 37°55′N 69°46′E. In cotton area; cotton ginning, cottonseed-oil extraction; metalworking. Salt deposits. Formerly spelled Kulyab.

Kulotino (koo-LO-tee-nuh), town (2006 population 3,325), E central NOVGOROD oblast, NW European Russia, on the Peretna River (Lake IL′MEN′ basin), on railroad and highway branch, 20 mi/32 km NW of BOROVICHI; 58°27′N 33°23′E. Elevation 301 ft/91 m. Linen spinning and weaving.

Kuloy (koo-LO-yee), town (2005 population 6,500), S ARCHANGEL oblast, N European Russia, on railroad and highway branch, 5 mi/8 km S of VEL′SK; 61°02′N 42°29′E. Elevation 351 ft/106 m. Railroad establishments. Developed in the early 1940s.

Kuloy, in Russian names: see KULOI.

Kuloy River (koo-LO-yee), approximately 150 mi/241 km long, in ARCHANGEL oblast, NW European Russia; rises near PINEGA; flows N to the MEZEN′ BAY of the WHITE SEA at DOLGOSHCHELYE. Called Sotka River in

its upper course. Connected by canal with the PINEGA RIVER.

Kulp, village, E TURKEY, 60 mi/97 km NE of DIYARBAKIR; 38°32′N 41°01′E. Cereals. Formerly Pasur.

Külpenberg (KOOL-puhn-berg), highest point (1,565 ft/477 m) of the KYFFHÄUSER, central GERMANY, 11 mi/18 km WSW of Sangershausen; 51°24′N 11°05′E.

Kulpmont, borough (2006 population 2,796), NORTHUMBERLAND county, E central PENNSYLVANIA, 4 mi/6.4 km E of SHAMOKIN. Manufacturing (textiles, apparel); anthracite coal. Settled 1905, incorporated 1916.

Kulpsville, unincorporated town (2000 population 8,005), MONTGOMERY county, SE PENNSYLVANIA, 20 mi/32 km NW of PHILADELPHIA on Towamencin Creek; 40°14′N 75°20′W. Manufacturing includes fabricated metal products, sealing devices, bearings, machinery; agriculture includes dairying, livestock; grain.

Kulsary (kul-sah-REE), oil town, SE ATYRAU region, KAZAKHSTAN, 10 mi/16 km NE of KOSCHAGYL, in Emba oil fields; 46°59′N 54°02′E. Tertiary-level (raion) administrative center.

Külsheim (KOOLS-heim), town, BADEN-WÜRTTEMBERG, central GERMANY, 6 mi/9.7 km S of WERTHEIM; 49°40′N 09°31′E. Summer resort.

Kul′tbaza Alygdzher, RUSSIA: see ALYGDZHER.

Kulti (kul-TEE), town (1991 Notified Area population 108,930), BARDDHAMAN district, W WEST BENGAL state, E INDIA, in DAMODAR valley and RANIGANJ coalfield, 9 mi/14.5 km WNW of ASANSOL; 23°44′N 86°51′E. A major iron- and steelworks center; brick and tile manufacturing. Coal-mining centers at DISHERGARH (2.5 mi/4 km S) and SALANPUR (3 mi/4.8 km NE) villages. Also Kulti Barakar.

Kultuk (kool-TOOK), town (2005 population 4,115), SE IRKUTSK oblast, E central SIBERIA, RUSSIA, on the TRANS-SIBERIAN RAILROAD, at the SW end of Lake BAYKAL, 41 mi/66 km SW of IRKUTSK, and 4 mi/6.4 km N of SLYUDYANKA, on highway to KYREN and TURTU; 51°43′N 103°41′E. Elevation 1,561 ft/475 m. Meat combine; timber storage. River port. Established as a settlement of Zabaykal′sk in 1647.

K′ulubi, village (2007 population 4,690), OROMIYA state, E central ETHIOPIA, in highlands, 30 mi/48 km W of HARAR; 09°26′N 41°41′E. Barley, corn, durra. A church here has become a religious pilgrimage site. Also spelled Kulubi and Culubi.

Kulubi, ETHIOPIA: see K′ULUBI.

Kulukak (KOO-loo-kak), Inuit village, SW ALASKA, on Kulukak Bay, N arm (7 mi/11.3 km long, 4 mi/6.4 km–6 mi/9.7 km wide) of BRISTOL BAY, 50 mi/80 km W of DILLINGHAM. Fishing.

Kulun, MONGOLIA: see ULAANBAATAR.

Kulunda (koo-loon-DAH), town (2005 population 15,350), W ALTAI TERRITORY, S central SIBERIA, RUSSIA, on crossroads and the SOUTH SIBERIAN RAILROAD, approximately 18 mi/29 km E of the Russia-KAZAKHSTAN border, 33 mi/53 km SSE of SLAVGOROD; 52°35′N 78°57′E. Elevation 426 ft/129 m. Railroad junction; railroad establishments; dairy products, prefabricated concrete forms. Glauber's salt deposits nearby.

Kulunda Lake (koo-loon-DAH), Russian *Kulundinskoye Ozero*, largest (□ 282 sq mi/733.2 sq km) of salt lakes in the KULUNDA STEPPE, NW ALTAI TERRITORY, SW Siberian Russia, 40 mi/64 km E of SLAVGOROD; 53°00′N 79°33′E; up to 13 ft/4 m deep. Salt deposits in narrow inlets. Frozen November through May. Smaller KUCHUK Lake is just S. Kulunda River, which rises near the OB′ RIVER, flows 125 mi/201 km generally SW, past SHARCHINO and BAYEVO, to KULUNDA Lake

Kulunda Steppe (koo-loon-DAH), S extension of the BARABA STEPPE, in ALTAI TERRITORY, SW Siberian Russia. A glacial lake bed drained by the OB′ RIVER. Flat topography, rich soils, and tolerable climate make this the best farming area in SIBERIA (spring wheat,

sunflowers, sugar beets; dairying). Has several small freshwater (e.g., Topolnoye Lake) and many saltwater (e.g., KULUNDA, Kuchuk) lakes. Glauber's salts and soda are extracted. Chief cities include BARNAUL, BIYSK, RUBTSOVSK.

Kuluyevo (koo-LOO-ee-vuh), village, central CHELYABINSK oblast, SE URAL Mountains, W Siberian Russia, on the MIASS River, near highway, 31 mi/50 km W of CHELYABINSK, and 25 mi/40 km NE of MIASS; 55°12′N 60°36′E. Elevation 853 ft/259 m. In agricultural area. Gold placers nearby.

Kulykiv (koo-li-KEEF) (Russian *Kulikov*) (Polish *Kulików*), town (2004 population 8,200), central L′VIV oblast, UKRAINE, on Dumnytsya River, and on road 9 mi/15 km N of L′VIV; 49°59′N 24°05′E. Elevation 793 ft/241 m. Manufacturing (leather, footwear). Heritage museum. Known since the end of the 14th century, town since 1940.

Kulykivka (koo-li-KEEF-kah) (Russian *Kulikovka*), town, central CHERNIHIV oblast, UKRAINE, 16 mi/26 km ESE of CHERNIHIV; 51°22′N 31°39′E. Elevation 387 ft/117 m. Food processing; flax; dry goods. Vocational technical school; heritage museum. Known since the second half of the 17th century, town since 1960.

Kulynychi (koo-LI-ni-chee) (Russian *Kulinichi*), town, N KHARKIV oblast, UKRAINE, within Kharkiv's circular highway 6 mi/10 km E of KHARKIV city center; 49°59′N 36°22′E. Elevation 574 ft/174 m. Agricultural research institute for livestock of the Ukrainian Forest-Steppe and Polissya; feed mill, non-alcoholic beverage bottling plant. Established in 1679, town since 1957.

Kul′zeb (kool-ZYEP), settlement (2005 population 3,350), central DAGESTAN REPUBLIC, SE European Russia, on road and railroad, 35 mi/56 km NW of MAKHACHKALA; 43°10′N 47°00′E. Elevation 170 ft/51 m. In oil- and gas-producing area.

Kum (KUM), peak (3,998 ft/1,219 m) in DINARIC ALPS, SLOVENIA, 27 mi/43 km E of LJUBLJANA.

Kum, IRAN: see QOM.

Kuma (KOO-mah), town, Kamiukena county, EHIME prefecture, NW SHIKOKU, W JAPAN, 16 mi/25 km S of MATSUYAMA; 33°39′N 132°54′E.

Kuma (KOO-mah), town, TARABA state, SE NIGERIA, on road, 175 mi/282 km SW of YOLA. Market town. Millet, maize; livestock.

Kuma (KOO-mah), village, Kuma county, KUMAMOTO prefecture, W KYUSHU, SW JAPAN, 37 mi/60 km S of KUMAMOTO; 32°14′N 130°39′E. Chestnuts, pears; pottery, *shochu* (distilled alcoholic beverage). Kusen-do, Japan's second-longest limestone cave, is nearby.

Kumagaya (koo-MAH-gah-yah), city (2005 population 191,107), SAITAMA prefecture, E central HONSHU, E central JAPAN, 25 mi/40 km N of URAWA; 36°08′N 139°23′E. Wheat. Magnetic tape, computer chips.

Kumage (koo-MAH-ge), town, Kumage county, YAMAGUCHI prefecture, SW HONSHU, W JAPAN, 31 mi/50 km E of YAMAGUCHI; 34°02′N 131°58′E.

Kumai (kuh-MEI), village, DARJILING district, N WEST BENGAL state, E India, 35 mi/56 km E of Darjiling. Tea, rice, corn, millet, cardamom. Limestone deposits nearby. Coal and copper pyrite deposits 7 mi/11.3 km WNW, near DALING.

Kumaishi (koo-MAH-ee-shee), town, Hiyama district, Hokkaido prefecture, N JAPAN, 93 mi/150 km S of SAPPORO; 42°07′N 139°59′E.

Kumait, Al (koo-MAIT, ahl), township, MAYSAN province, E IRAQ, on TIGRIS RIVER, and 22 mi/35 km NW of ′AMARA. Dates, rice, corn, millet, sesame, cotton. Also spelled Al-Kumayt.

Kumak (koo-MAHK), town, E ORENBURG oblast, SE URAL Mountains, extreme SE European Russia, on the Kumak River (left tributary of the URAL River), on road junction, 63 mi/101 km E of ORSK; 51°10′N 60°08′E. Elevation 951 ft/289 m. In ORSK-KHALILOVO industrial district; metalworking, fireproof clay quarrying.

Cross-references are shown in SMALL CAPITALS. The pronunciation guide is shown on page xix. The sources of population figures are shown on page xvii.

Kumamba Islands (KOO-mahm-bah), IRIAN JAYA province, INDONESIA, W NEW GUINEA island, in the PACIFIC OCEAN, just off N coast, 150 mi/241 km WNW of JAYAPURA; 01°40′S 138°47′E. Consists of Liki (4 mi/6.4 km long) and Nirumoar (or Niroemoar; 3 mi/4.8 km long). Also spelled Koemamba Islands.

Kumamoto (koo-MAH-mo-to), prefecture [Jap. *ken*] (□ 2,872 sq mi/7,438 sq km; 1990 population 1,840,383), W central KYUSHU, SW JAPAN, on ARIAKE, SHIMABARA, and YATSUSHIRO bays (all W); ⊙ KUMA-MOTO. Bordered NE by FUKUOKA prefecture, E by OITA prefecture, SE by MIYAZAKI prefecture, and SW by KAGOSHIMA prefecture. Includes SHIMO-JIMA, KAMI-SHIMA, and OYANO-SHIMA islands, and scattered islets of AMAKUSA ISLANDS. Mountainous terrain rises to 5,240 ft/1,597 m in volcanic Aso-san in Aso National Park. Hot springs in NE area. Drained by KUMA and CHIKUGO rivers. Primarily agricultural, producing rice, sweet potatoes, fruit (oranges, pears, plums); silk, tobacco. Extensive lumbering in interior; extensive coastal fishing. Products include cotton textiles, porcelain ware, paper, dolls, camellia oil, cement, chemicals, and processed foods. Principal centers (all on W coast) include Kumamoto, YAT-SUSHIRO (chief port), HITOYOSHI, and ARAO.

Kumamoto (koo-MAH-mo-to), city (2005 population 669,603), ⊙ KUMAMOTO prefecture, W KYUSHU, SW JAPAN; 32°48′N 130°42′E. Computer components. Lotus root. Traditional *zogan* metal carving. Known for *basashimi* (raw horsemeat). Important castle town in the 17th century; one of its castles (built 1651) still stands. There are also several shrines here.

Kumane (KOO-mah-nai), Hungarian *Kumán*, village, VOJVODINA, N SERBIA, 14 mi/23 km NW of ZRENJA-NIN, in the BANAT region.

Kumano (koo-MAH-no), city, MIE prefecture, S HON-SHU, central JAPAN, 62 mi/100 km S of TSU; 33°53′N 136°06′E. Mandarin oranges; pickles, dried fish. Go stones. Nearby scenic coasts at Onigajo and Shichiri Mihama.

Kumano (koo-MAH-no), town, Aki co., HIROSHIMA prefecture, SW HONSHU, W JAPAN, 9 mi/15 km S of HIROSHIMA; 34°20′N 132°35′E. Writing brushes.

Kumanogawa (koo-MAH-no-GAH-wah), town, East Muro county, WAKAYAMA prefecture, S HONSHU, W central JAPAN, 51 mi/83 km S of WAKAYAMA; 33°48′N 135°52′E. Citron vinegar, miso. Kumano Kodo here is known for its historic streetscape.

Kumano River (KOO-mah-no), Japanese *Kumano-gawa* (KOO-mah-no-GAH-wah), 100 mi/161 km long, on KII PENINSULA, NARA prefecture, S HONSHU, central JAPAN; rises in mountains near YOSHINO; flows generally S, past HONGU, and SE to KUMANO SEA at SHINGU. Log drives; navigable for c.30 mi/48 km by excursion motorboats to Hongu. Called Totsu (TOTS) River in upper course.

Kumano Sea (koo-mah-no), Japanese *Kumano-nada* (KOO-mah-no-NAH-dah), N arm of PHILIPPINE SEA; forms wide bight along E coast of KII PENINSULA, NARA prefecture, S HONSHU, central JAPAN; c.80 mi/129 km long.

Kumanovo (koo-mah-NO-vo), city, in MACEDONIA; 42°08′N 21°43′E. In the center of a tobacco-growing region; its major industries are tobacco processing and canning. The Serbs won a decisive victory over the Turks here in 1912.

Kumaon (kuh-MOUN), division (□ 18,273 sq mi/47,509.8 sq km), N UTTAR PRADESH state, N central INDIA; ⊙ NAINI TAL. Comprises Naini Tal, AL-MORA, CHAMOLI, GARHWAL, PITHORAGARH, TEHRI GARHWAL, and UTTARKASHI districts. Former division was enlarged 1949 by incorporation of former princely state of TEHRI. Largely within KUMAON Himalaya (N) and SHIWALIK RANGE (S). Also spelled Kumaun.

Kumara (KOO-muh-rah), township, WESTLAND district, SOUTH ISLAND, NEW ZEALAND, 11 mi/18 km SSE of GREYMOUTH. Timber; sawmilling. Occasional gold dredging.

Kumara (koo-MAH-rah), village, SE AMUR oblast, RUSSIAN FAR EAST, across the AMUR River from the Chinese border, on highway, 95 mi/153 km NNW of BLAGOVESHCHENSK; 51°35′N 126°43′E. Elevation 652 ft/198 m. In agricultural area. Cossack village, founded in 1858.

Kumarapalaiyam. INDIA: see KOMARAPALAIYAM.

Kumardhubi, INDIA: see KUMHARDHUBI.

Kumarhata, INDIA: see HALISAHAR.

Kuma River (KOO-mah), Japanese *Kuma-gawa* (koo-MAH-GAH-wah), 75 mi/121 km long, KUMAMOTO prefecture, W KYUSHU, SW JAPAN; rises in mountains 25 mi/40 km E of YATSUSHIRO; flows SSW, past HI-TOYOSHI, W, and N to YATSUSHIRO BAY at YAT-SUSHIRO. Scenic rapids.

Kuma River (KOO-mah), 360 mi/579 km long, in the N CAUCASUS Mountains, SE European Russia; rises on the N slope of the central Greater Caucasus, ENE of KARACHAYEVSK (NE KARACHEVO-Cherkess Republic); flows generally ENE, across SE STAVROPOL TER-RITORY, past Suvorovskaya, MINERAL'NYYE VODY, SOLDATO-ALEKSANDROVSKOYE, BUDËNNOVSK, and LEVOKUMSKOYE, losing itself on the ASTRAKHAN ob-last-DAGESTAN border, in swamps and reed lakes, approximately 50 mi/80 km from the CASPIAN SEA, which it reaches only during spring floods and in years of above-average precipitation. Low water level, in the lower course, caused by dry climate, evaporation, and loss through irrigation in the Budënnovsk area. Receives Zolka and PODKUMOK rivers (right).

Kumarkhali (koo-mahr-klah-lee), town (2001 population 19,707), KUSHTIA district, WEST BENGAL, BAN-GLADESH, on upper MADHUMATI (Garai) River, and 8 mi/12.9 km ESE of KUSHTIA; 22°25′N 89°48′E. Trade center (rice, jute, linseed, sugarcane). Until 1947 Ku-markhali was in the NADIA district of British Bengal.

Kumasi (koo-MAH-see), city (2005 population 1,517,000), ⊙ ASHANTI region, GHANA; 06°41′N 01°37′W. Ghana's second-largest city; commercial and transportation center in a cocoa-producing region. Large central market. Founded c. 1700 as Ashanti confederacy capital. Although the British destroyed the Ashanti palace in 1874, the city remains the seat of Ashanti kings. Seat of the University of Science and Technology (UST) and other schools. Domestic airport en route from ACCRA to SUNYANI and TAMALE.

Kumatori (koo-MAH-to-ree), town, Sennan county, OSAKA prefecture, S HONSHU, W central JAPAN, 22 mi/35 km S of OSAKA; 34°23′N 135°21′E.

Kumaun, INDIA: see KUMAON.

Kumawu (koo-MOU-woo), town, ASHANTI region, GHANA, 27 mi/43 km NE of KUMASI; 06°54′N 01°17′W. Cacao, kola nuts, hardwood, rubber.

Kumayama (koo-MAH-yah-mah), town, Akaiwa county, OKAYAMA prefecture, SW HONSHU, W JAPAN, 12 mi/20 km N of OKAYAMA; 34°47′N 134°05′E.

Kumayri, city, ARMENIA, near the TURKISH border. Varied light manufacturing. The old craft of rug making is practiced. Most important Armenian industrial center after YEREVAN. Founded (1837) as Aleksandropol on the site of the Turkish fortress of Gumri. Called Leninakan, 1924–1990. The city was leveled in a December 1988 earthquake. Formerly known as Giumri or Gyumri.

Kumayt, Al-, IRAQ: see KUMAIT, AL.

Kumba (KOOM-buh), city (1998 estimated population 110,860; 2001 estimated population 125,600), ⊙ MEME department, South-West province, CAMEROON, on road, and 38 mi/61 km NNE of Buea; 04°39′N 09°27′E. Road junction. Trade center; coffee, cacao, bananas, palm oil, and kernels; sawmills. Has hospital and many schools.

Kumbakonam (kuhm-bah-KO-nuhm), city, THANJA-VUR district, TAMIL NADU state, S INDIA, on the KA-VERI RIVER; 10°58′N 79°23′E. Its district, in the richest part of the river delta, has one of the highest population densities in India. An agricultural trading center in an area known for its rice and betel leaves. Manufacturing includes brass ware and textiles. Brah-manic cultural center. The many Hindu temples along the river are visited by pilgrims every twelve years.

Kumbashi (koom-BAH-shee), town, NIGER state, NW NIGERIA, on road, and 110 mi/177 km NW of MINNA; 10°55′N 05°42′E. Market town. Sorghum, rice, and cassava.

Kumbe (KOOM-bai), town, IRIAN JAYA, INDONESIA, 28 mi/45 km NE of MERAUKE; 08°21′S 140°13′E.

Kumberg (KUM-berg), township, STYRIA, SE AUSTRIA, 7 mi/11.3 km NE of GRAZ; 47°10′N 15°32′E. Summer resort.

Kumbhakarna (koom-BUH-kuhr-NUH) or **Jannu** (jah-NUH), mountain peak (25,300 ft/7,710 m), E NEPAL; 27°41′N 88°03′E. First climbed by a French expedition in 1962.

Kumbhalgarh (KUHM-bahl-guhr), town, UDAIPUR district, S RAJASTHAN state, NW INDIA, 40 mi/64 km NNW of Udaipur, in Aravalli Range. Elevation c.3,500 ft/1,067 m. Has walled hill fortress built in mid-15th century.

Kumbhkaran Lungur, NEPAL and INDIA: see KAN-CHENJUNGA.

Kumbhraj (kuhm-BRAHJ), village (2001 population 13,999), GUNA district, MADHYA PRADESH state, central INDIA, 25 mi/40 km SW of GUNA. Wheat, millet, corn, gram.

Kumbo (KOOM-bo), town (2001 population 116,500), ⊙ BUI department, North-West province, CAMEROON, in BAMENDA hills, 40 mi/64 km NE of Bamenda; 06°11′N 10°40′E. Cattle raising; durra, millet. Exports hides.

Kumbum (KAHM-BAHM), large lamasery at Huang-chang, NE QINGHAI province, CHINA, c.12 mi/19 km SW of XINING. Long a renowned pilgrimage center, it stands on the spot where Tsong-kha-pa, the great Tibetan reformer of Lamaism, is said to have been born in 1417. Its Living Buddha became (1952) the tenth Panchen Lama of Tibet. The lamasery is sometimes spelled Gumbum.

Kumchon (GUM-CHUHN), county, NORTH HWAN-GHAE province, S NORTH KOREA, 50 mi/80 km NW of SEOUL. In coal-mining area.

Kum-Dag (KOOM–dahg), town, BALKAN weloyat, TURKMENISTAN, 22 mi/35 km SSE of NEBITDAG, and 12 mi/20 km E of smaller BALKAN range. Oil production nearby. Also Gumdaq.

Kume (KOO-me), town, Kume county, OKAYAMA prefecture, SW HONSHU, W JAPAN, 28 mi/45 km N of OKAYAMA; 35°03′N 133°54′E.

Kumenan (koo-ME-nahn), town, Kume county, OKAYAMA prefecture, SW HONSHU, W JAPAN, 19 mi/30 km N of OKAYAMA; 34°55′N 133°57′E. Grapes. Honen, founder of the Buddhist Jodo sect, was born at the Tanjo Temple here in the 12th century.

Kumeny (KOO-mye-ni), village (2005 population 5,040), central KIROV oblast, E central European Russia, on road, 29 mi/47 km SSE of KIROV; 58°06′N 49°54′E. Elevation 597 ft/181 m. Gas pipeline service station. In agricultural area; seed inspection, produce processing, livestock veterinary station, agricultural machinery service and repair; logging and lumbering.

Kumertau (koo-meer-TAH-oo), city (2005 population 65,430), SW BASHKORTOSTAN Republic, E European Russia, on road and railroad, 160 mi/257 km S of UFA, 65 mi/105 km S of STERLITAMAK, and 59 mi/95 km N of ORENBURG; 52°46′N 55°47′E. Elevation 987 ft/300 m. Coal mining (BABAYEVO deposits), briquette plant, natural gas deposits in area, heat and electric power plant, manufacturing (industrial machines, aircraft). Founded in 1948. Made city in 1953.

Kume-shima (koo-ME-shee-MAH), westernmost island (□ 26 sq mi/67.6 sq km) of OKINAWA ISLANDS, in

RYUKYU ISLANDS, Okinawa prefecture, SW JAPAN, in EAST CHINA SEA, 55 mi/89 km W of OKINAWA; 8 mi/12.9 km long, 7 mi/11.3 km wide; 26°21′N 126°47′E. Volcanic, mountainous; rises to 1,102 ft/336 m. Pine and oak forests. Produces sugarcane, charcoal, rice.

Kum-gang, SOUTH KOREA: see KUM RIVER.

Kumgang, Mount, Korean *Kumgamg-san*, SE NORTH KOREA, rising to 6,030 ft/1,838 m. There are scenic ravines and caverns and many ancient Buddhist temples.

Kumhardhubi (kuhm-AHR-doo-bee) or **Kumardhubi**, village, DUMKA district, NE Jharkhand state, E INDIA, 13 mi/21 km WNW of ASANSOL. Firebrick and general engineering works, iron- and steel-rolling mill.

Kumharsain (kuhm-HAHR-sein), town, SHIMLA district, central HIMACHAL PRADESH state, N INDIA, near SUTLEJ River, 22 mi/35 km NE of SHIMLA; 31°19′N 77°27′E. Was capital of former princely state of Kumharsain of PUNJAB HILL STATES; since 1948, merged with Himachal Pradesh state.

Kumher (KOOM-her), town, BHARATPUR district, E RAJASTHAN state, NW INDIA, 9 mi/14.5 km NW of Bharatpur; 27°19′N 77°22′E. Agriculture (millet, oilseeds, gram).

Kumi, administrative district, EASTERN region, E UGANDA; ⊙ KUMI. Laki BISINA to N and Lake KYOGA to W. Agricultural area (including cotton and rice; livestock). Secondary railroad travels through district, NW to LIRA and GULU towns and SE to join main railroad between KASESE town (W Uganda) and MOMBASA (SE KENYA). Kumi town on main road running NW-SE between SOROTI and MBALE towns. Formed in 2006 from part of former KUMI district (BUKEDEA district also formed from part of former Kumi district).

Kumi (KOO-mee), former administrative district (□ 1,089 sq mi/2,831.4 sq km; 2005 population 434,600), EASTERN region, E central UGANDA, E of Lake KYOGA; capital was KUMI; 01°25′N 34°00′E. As of Uganda's division into fifty-six districts, was bordered by KATAKWI (N), NAKAPIRIPIRIT and SIRONKO (E), MBALE (SE), PALLISA (S), and SOROTI (W) districts. Primary inhabitants were Iteso people. District was mostly savannah grassland with few tress; primarily flat, with inselbergs, lakes, and wetlands. Lake BISINA in N, on border with Katakwi district. Agricultural region (beans, cassava, cotton, groundnuts, maize, millet, peas, rice, sorghum, sunflowers, and sweet potatoes; cattle, goats, pigs, poultry, and sheep); also some fishing. In 2006 district divided into BUKEDEA and current KUMI districts.

Kumi (GOO-MEE), city (□ 49 sq mi/127.4 sq km), SW NORTH KYONGSANG province, SOUTH KOREA. 32 mi/51 km N of TAEGU, adjacent to Sonsan (N), CHILGOK (E and S), and KUMNUNG (W); 36°31′N 128°46′E. Home of Kumi Industrial Complex, the largest inland industrial complex in South Korea. Kumi is split into E and W parts by the NAKDONG RIVER. Textile and electronic industries centered in W, semiconductors industry in E. Agriculture includes mushrooms, paddy farming, and horticulture. Kum-oh Park; Kyongbu railroad and expressway.

Kumi (KOO-mee), town (2002 population 8,807), ⊙ KUMI district, EASTERN region, E UGANDA, on railroad, and 27 mi/43 km SE of SOROTI, near LAKE KYOGA. Cotton, peanuts, sesame, rice; livestock. Was part of former EASTERN province.

Kumi, Greece: see KIMI.

Kumihama (koo-MEE-hah-mah), town, Kumano county, KYOTO prefecture, S HONSHU, W central JAPAN, on SEA OF JAPAN, 62 mi/100 km N of KYOTO; 35°36′N 134°53′E.

Kuminskiy (KOO-meen-skeeye), settlement (2005 population 2,880), SW KHANTY-MANSI AUTONOMOUS OKRUG, W central Siberian Russia, on the YEKATERINBURG-bound railroad, 56 mi/90 km NNE of TAVDA (SVERDLOVSK oblast); 58°40′N 66°34′E. Elevation 252 ft/76 m. Woodworking.

Kumisheh, IRAN: see QOMSHEH.

Kum Ishqaw, EGYPT: see KOM ISHQAW.

Kumiyama (koo-MEE-yah-mah), town, Kuse county, KYOTO prefecture, S HONSHU, W central JAPAN, 9 mi/15 km S of KYOTO; 34°52′N 135°44′E. Vegetable seedlings; daikon.

Kumjawng Pass (KUHM-joung) (9,613 ft/2,930 m) or **Krongjawng**, on INDIA-MYANMAR border, between ARUNACHAL PRADESH state (extreme NE INDIA) and KACHIN STATE (extreme N MYANMAR) 35 mi/56 km NNW of PUTAO; 27°50′N 97°10′E. Difficult route, rarely used by local population.

Kumkale (KOOM-kah-LAI), village, NW TURKEY, at AEGEAN entrance to the DARDANELLES, on Asian side. In World War I its fortifications were stormed by the Allies in Gallipoli (GELIBOLU) campaign.

Kumla (KOOM-lah), town, ÖREBRO county, S central SWEDEN, 10 mi/16 km S of ÖREBRO; 59°07′N 15°09′E. Railroad junction; manufacturing (footwear, food). Center of toxic waste disposal. Incorporated 1942 as city.

Kummerow Lake (KUM-muhr-ou), German *Kummerower See* or *Cummerower See*, lake (□ 13 sq mi/33.8 sq km), Mecklenburg–Western Pomerania, N GERMANY, 3 mi/4.8 km NE of MALCHIN; 7 mi/11.3 km long, 2 mi/3.2 km–3 mi/4.8 km wide, greatest depth 98 ft/30 m; 53°48′N 12°52′E. Drained by PEENE RIVER. Connected by canal with MALCHIN LAKE, 7 mi/11.3 km SW. Also spelled CUMMEROW LAKE.

Kümmersbrück (KUM-muhrs-brook), village, BAVARIA, S GERMANY, in UPPER PALATINATE, 29 mi/47 km NW of REGENSBURG; 49°23′N 11°50′E.

Kumnung, county (□ 365 sq mi/949 sq km), W NORTH KYONGSANG province, SOUTH KOREA. Bordered E by Sonsan and CHILGOK, S by SOUTH KYONGSANG, W by CHUNGCHONG and NORTH CHOLLA provinces. Surrounded by mountains (elevation 4,265 ft/1,300 m), with Kam and Chikji rivers in NE. Agriculture (rice, onion, sesame, mushrooms, tobacco, food crops) in river basin. Traditional paper; rush mats. Kyongbu railroad and expressway. Chikji Buddhist temple.

Kumo, FINLAND: see KOKEMÄKI.

Kumon Range (koo-MON), in KACHIN STATE, MYANMAR, extends 100 mi/161 km N of MYITKYINA between HUKAWNG valley (W) and Mali headstream of the AYEYARWADY RIVER; rises to 11,190 ft/3,411 m, 20 mi/32 km WNW of SUMPRABUM.

Kumrabai Mamila (koom-rah-BEI mah-MEE-lah), village, NORTHERN province, central SIERRA LEONE, 15 mi/24 km SW of MAGBURAKA. Kumrabai Matuku village is 1 mi/1.6 km SW; terminus of road from BO; 08°32′N 12°06′W. Sometimes spelled Kumrabai Mamilla.

Kum River, Korean *Kum-gang*, 247 mi/397 km long, SOUTH KOREA; rises in mountains c.25 mi/40 km SE of CHONJU; flows N, turns NW and generally SW past KONGJU, PUYO, and KANGGYONG to YELLOW SEA at KUNSAN. Navigable c.80 mi/129 km by small craft. Drains agricultural area.

Kumsan (KOOM-SAHN), county, SOUTH CHUNGCHONG province, SOUTH KOREA, 45 mi/72 km ENE of KUNSAN. Agricultural center (rice, barley, soybeans, cotton); silk cocoons. Gold mined nearby. Mountainous region. Due to the reform of the local administrative system, Kunsan-gun was transferred to South Chungchong province from NORTH CHOLLA province in 1963. Best known for its ginseng and ginseng market.

Kumsi (KUHM-see), town, SHIMOGA district, KARNATAKA state, S INDIA, on railroad, and 15 mi/24 km NW of SHIMOGA; 14°04′N 75°24′E. Nearby manganese mines and limestone quarries supply steel plant at BHADRAVATI.

Kumta (KUHM-tah), town, UTTAR KANNAD district, KARNATAKA state, S INDIA, port on ARABIAN SEA, 32 mi/51 km SSE of KARWAR; 14°25′N 74°24′E. Trade center for rice, betel nuts, coconuts, gur, cotton, spi-

ces; fish-curing yards (mackerel, sardines, catfish). Lighthouse (NW).

Kumtorkala (koom-tuhr-kah-LAH), city, N central DAGESTAN REPUBLIC, NE CAUCASUS, RUSSIA, on the SULAK RIVER, on railroad, 21 mi/34 km NW of MAKHACHKALA; 43°01′N 47°15′E. Elevation 196 ft/59 m. Electric equipment; phosphates; KIZILYURT hydroelectric station. Population originally largely Kumyk. Made city in 1963. Until 1996, known as Kizilyurt.

Kumukahi, Cape (KOO-moo-KAH-hee), E extremity of HAWAII island, HAWAII county, HAWAII; 19°31′N 154°48′W. Easternmost point in Hawaii and Hawaiian Island chain.

Kumukahi Channel, HAWAII: see KAULAKAHI CHANNEL.

Kumukh (koo-MOOKH), mountain village (2005 population 2,720), S DAGESTAN REPUBLIC, NE CAUCASUS, SE European Russia, on road, 45 mi/72 km S of MAKHACHKALA; 42°10′N 47°07′E. Elevation 5,193 ft/1,582 m. Clothing handicraft. Population largely Lak.

Kum Umbu, EGYPT: see KOM OMBO.

Kumya (GUM-YAH), county, SOUTH HAMGYONG province, NORTH KOREA, 28 mi/45 km SW of HUNGNAM. Lumbering; livestock raising; agriculture: soybeans, grains, and hemp. Graphite is mined nearby. Previously called Yonghung.

Kumylzhenskaya (koo-MIL-zhin-skah-yah), village (2006 population 8,070), W VOLGOGRAD oblast, SE European Russia, on the KHOPER River, on road, 30 mi/48 km SW of MIKHAYLOVKA; 49°53′N 42°35′E. Elevation 232 ft/70 m. In agricultural area (sunflowers, wheat, oats, vegetables); poultry factory, flour mill, mineral water plant, sawmill.

Kumysh (koo-MISH), village (2005 population 4,315), E central KARACHEVO-CHERKESS REPUBLIC, S European Russia, in the NW CAUCASUS Mountains, on road, 15 mi/24 km E of ZELENCHUKSKAYA; 43°52′N 41°54′E. Elevation 2,647 ft/806 m. Stone quarrying and mineral mining. Formerly known as Podgornoye.

Kuna, town (2000 population 5,382), ADA county, SW IDAHO, suburb 15 mi/24 km SW of BOISE; 43°29′N 116°25′W. In grain and livestock area (dairying; fruit, vegetables). Boise Municipal Airport (Gowen Field) to NE. LAKE LOWELL RESERVOIR (Deer Flat National Wildlife Refuge to W).

Kunar (koo-NAHR), province (□ 3,742 sq mi/9,729.2 sq km; 2005 population 374,700), AFGHANISTAN; ⊙ ASADABAD (formerly called Kunar); 34°39′N 70°54′W. Bounded by LAGHMAN province (W), NURISTAN province (N), PAKISTAN (E), and NANGARHAR province (S). Agriculture (corn, rice, wheat). Created as a province in the 1960s, Kunar was briefly part of Nanghar province before becoming a separate unit in 1977. Revolt against the Soviet-dominated government began here in 1978. After the fall of the Marxist regime in 1992, the radical Islamist Wahhabi Republic was established, headed by Maulawi Jamilur Rahman. In 1996 the Taliban captured the area. Also spelled Konar.

Kunar River (kuh-nahr), 250 mi/402 km long, in W PAKISTAN and E AFGHANISTAN; rises in the E HINDU KUSH mountains in CHITRAL district of PAKISTAN'S NORTH-WEST FRONTIER PROVINCE; flows SW, past MASTUJ and CHITRAL, and into AFGHANISTAN, past ASMAR and KUNAR, to KABUL RIVER just below JALALABAD; source at 34°25′N 70°32′E. Used for logging and for irrigation in lower course. Called CHITRAL RIVER in upper reaches. Also spelled Konar River.

Kunashak (koo-nah-SHAHK), village, NE CHELYABINSK oblast, E URAL Mountains, E European Russia, on road and near railroad, 35 mi/56 km NNE of CHELYABINSK; 55°42′N 61°33′E. Elevation 587 ft/178 m. In agricultural area (feed corn, flax, potatoes; dairy livestock).

Kunashiri-kaikyo, RUSSIA: see YEKATERINA STRAIT.

Kunashir Island (koo-nah-SHIR), Japanese *Kunashir-ishima*, southernmost and second-largest (□ 1,548 sq mi/4,024.8 sq km) of the main KURIL ISLANDS chain, SAKHALIN oblast, extreme E SIBERIA, RUSSIAN FAR EAST; separated from ITURUP ISLAND (NE) by YEKATERINA STRAIT, from HOKKAIDO (Japan; SW), by NEMURO STRAIT; 44°20′N 146°00′E. Consists of five volcanic massifs connected by lower ridges; rises to 5,978 ft/1,822 m in the volcano Tyatya, second-highest of the Kuriles. Fishing, sealing, fish processing, lumbering, hunting, and sulphur mining are chief economic activities; some garden farming. Main centers are YUZHNO-KURILSK, Golovnino, TYATINO. Already known to the Japanese when visited (1713) by the Russians. Japanese colonization began in the mid-18th century; formalized in 1855. Occupied by Russia in 1945; claimed by Japan as part of the Northern Territories.

Kunbugu (koom-BOON-goo), town, NORTHERN REGION, GHANA, 13 mi/21 km NW of TAMALE; 09°33′N 00°57′W. Shea nuts, cotton; cattle, skins.

Kunbuth (KUN-buhth), village, CYRENAICA region, NE LIBYA, 33 mi/53 km ESE of TOBRUK. Scene of fighting (1941–1942) between Axis and British in World War II. Formerly Gambut.

Kunch (KOONCH), town, JALAUN district, S UTTAR PRADESH state, N central INDIA, 20 mi/32 km W of ORAI. Railroad spur terminus; trade center (gram, wheat, oilseeds, jowar, rice, ghee).

Kund, village, MAHENDRAGARH district, HARYANA state, N INDIA, on railroad, and 14 mi/23 km WSW of REWARI. Metalworks. Slate quarried nearby.

Kunda (KUN-duh), city, N ESTONIA, port on GULF OF FINLAND, 12 mi/19 km NNE of RAKVERE (linked by railroad spur); 59°29′N 26°31′E. Cement-milling center.

Kunda (KOON-dah), town, PARTABGARH district, SE UTTAR PRADESH state, N central INDIA, on railroad, and 27 mi/43 km WNW of ALLAHABAD. Rice, barley, wheat, gram, mustard. railroad station called HARNAMGANJ.

Kundalwadi, INDIA: see KONDALWADI.

Kundapur, INDIA: see KUNDAPURA.

Kundapura, town, DAKSHINA KANNADA district, KARNATAKA state, S INDIA, on MALABAR COAST of ARABIAN SEA, 55 mi/89 km NNW of MANGALORE, at estuary mouth. Fish curing (sardines, mackerel); coconuts, mangoes. Site of 17th–18th century PORTUGUESE and Dutch trading posts. Commercial suburb of GANGOLI (or Ganguli) is just N, across estuary. Clay pits (kaolin) nearby. Also spelled Kundapur.

Kundarkhi (KUHN-dahr-kee), town, MORADABAD district, central UTTAR PRADESH state, N central INDIA, 11 mi/18 km S of Moradabad; 28°42′N 78°47′E. Wheat, rice, pearl millet, sugar.

Kundgol (KUHND-gol), town, DHARWAD district, KARNATAKA state, S INDIA, 22 mi/35 km SE of DHARWAD; 15°15′N 75°15′E. Local cotton center; handicraft cloth weaving.

Kundi, PAKISTAN: see NOK KUNDI.

Kundiawa (KUN-dee-AH-wah), town (2000 population 8,147), ☉ CHIMBU province, E central NEW GUINEA island, N central PAPUA NEW GUINEA, in the Great Plateau, 140 mi/225 km WNW of LAE. Commercial center. Airstrip; road access. Mount WILHELM (Chimbu), highest point in Papua New Guinea (13,432 ft/4,094 m); 10 mi/16 km N. Coffee, tea, sweet potatoes; timber.

Kundima (kun-DEE-mah), village, EAST SEPIK province, NE NEW GUINEA island, NW PAPUA NEW GUINEA, on E bank of YUAT River, 10 mi/16 km S of confluence with SEPIK RIVER, 50 mi/80 km SSE of WEWAK. Marshy region accessible by boat. Bananas, sago, taro, yams; timber.

Kundla (KUHND-luh), town, BHAVNAGAR district, GUJARAT state, W INDIA, 60 mi/97 km SW of BHAVNAGAR; 21°20′N 71°18′E. Agriculture market (cotton, millet, wheat); cotton ginning, handicraft cloth weaving. Sometimes called Savar Kundla.

Kundravy (KOON-drah-vi), village, W CHELYABINSK oblast, central URAL Mountains, E European Russia, on crossroads, 11 mi/18 km SSE of MIASS; 54°49′N 60°13′E. Elevation 1,230 ft/374 m. Flour milling.

Kundryuchya River (koon-DRYOO-chyah), river, a right-bank tributary of the lower DON River, in W UKRAINE. Navigable only by high-draft barges. Extremely polluted by wastes from cities and machine-building, food, and chemical industries along its shores.

Kunduk Lagoon, UKRAINE: see SASYK LAGOON.

Kunduk River, UKRAINE: see KOHYLNYK RIVER.

Kundur, INDONESIA: see KARIMUN ISLANDS.

Kunduz (koon-DOOZ), province (2005 population 817,400), N AFGHANISTAN, ☉ KUNDUZ; 36°45′N 68°51′E. Bordered N by TAJIKISTAN, E by TAKHAR province, S by BAGHLAN province, and W by SAMANGAN province. Watered by KUNDUZ RIVER (called Surkhab River at its source). Population largely Uzbek with some Pashtuns and Tajiks. Cotton industry started here in 1940s. Manufacturing (Spin Zar industrial complex grows and gins cotton; edible oil, soap; silk, textiles, clothing). Colorful silk fabrics, Uzbek garments (the *japan*), long-sleeved caftans, soft-soled boots, and embroidered caps are major export items. Domestic airport S of Kunduz city. Also spelled Konduz or Qunduz.

Kunduz (koon-DOOZ), town, NE AFGHANISTAN, ☉ KUNDUZ province, 15 mi/24 km WNW of Khanabad; 36°45′N 68°51′E. Center of oasis irrigated by KUNDUZ RIVER. Agriculture (cotton, rice); manufacturing (cottonseed-oil, soap); rice and flour milling, cotton ginning. Held by the Taliban during the civil war, it fell to the Northern Alliance, supported by U.S. armed forces, November 26, 2001. Domestic airport S of the city. Also spelled Konduz or Qunduz.

Kunduz River (koon-DOOZ), 250 mi/402 km long, KUNDUZ province, NE AFGHANISTAN; rises W of BAMIAN in the Hindukush Mountains; flows E, through Bamian valley, then N, through Shikari Pass, past DOAB MEKH-I-ZARIN, DOSHI, PUL-I-KHUMRI, BAGHLAN, and KUNDUZ, to the AMU DARYA (river; TAJIKISTAN border) near mouth of VAKHSH RIVER; 37°00′N 68°16′E. Known as Surkhab River in upper course; important cotton and sugar beet irrigation along lower course. Middle course, where it joins Saighan, Kamard, and Andarab valleys, is used by Kabul-Mazar-i-Sharif highway. Also spelled Qunduz River

Kundysh River, RUSSIA: see GREATER KUNDYSH RIVER or LESSER KUNDYSH RIVER.

Kunene Region (KOO-NAI-nai), administrative division (2001 population 68,735), NW NAMIBIA, extending from Ugab River to Angolan border; 19°30′S 14°30′E. Political region includes DAMARALAND and Kaokoland; 2nd-largest region; sparsely populated.

Kunene River, ANGOLA and NAMIBIA: see CUNENE.

Kunerma (koo-neer-MAH), settlement (2005 population 155), central IRKUTSK oblast, E central Siberian Russia, near the BAYKAL-AMUR MAINLINE, 204 mi/328 km ESE of ZHELEZNOGORSK-ILIMSKIY; 55°45′N 108°29′E. Elevation 2,749 ft/837 m. Railroad station.

Kunersdorf, POLAND: see KUNOWICE.

Künes, CHINA: see XINYUAN.

Kunfogsi (kun-FOG-see), town NORTHERN REGION, GHANA, 40 mi/64 km N of BOLE; 09°31′N 02°34′W. Also spelled KONFOGSI, KONFOSI, KONFUSI.

Kungälv (KUNG-ELV), town, GÖTEBORG OCH BOHUS county, SW SWEDEN, on Nordre älv River, near GÖTA ÄLV RIVER, residential suburb 11 mi/18 km N of GOTEBORG; 57°52′N 11°58′E. Manufacturing (food). Important center since twelfth century (fortified 1120), called Kungahälla in Middle Ages; mentioned in sagas. Plundered by Wends (1135); captured by Hanseatic League (1368); later suffered Danish attacks. Swedish territory since 1658. Remains of Bohus Castle (early fourteenth century) on Nordre älv River island.

Kungei Alatau, KYRGYZSTAN and KAZAKHSTAN: see KÜNGEY ALA-TOO, ALATAU.

Küngey Ala-Too, Kazak *Kungei Alatau*, mountain range of N TIANSHAN mountain system, on KYRGYZSTAN/KAZAKHSTAN border, N of LAKE ISSYK-KOL; has numerous peaks over 13,123 ft/4,000 m, rising to 15,646 ft/4,770 m at Mount Chok-Tal, 24 mi/40 km NW of CHOLPON-ATA. Pastures for livestock. Source of KEMIN River. Also spelled Künggö Ala-Too.

Kunghit Island (KUHN-git), (□ 83 sq mi/215.8 sq km), W BRITISH COLUMBIA, SW CANADA, southernmost of the QUEEN CHARLOTTE ISLANDS, 140 mi/225 km NW of VANCOUVER ISLAND, separated from MORESBY ISLAND (N) by Houston Stewart Channel (1 mi/2km–2 mi/3 km wide); 15 mi/24 km long, 1 mi/2 km–8 mi/13 km wide; 52°06′N 131°04′W. At S extremity is Cape Saint James (51°56′N 131°01′W).

Kung-kuan (GUNG-GWAHN), town, NW TAIWAN, 4 mi/6.4 km S of MIAOLI. Oranges, persimmons. Sometimes spelled Kungkwan.

Kungkwan, TAIWAN: see Kung-kuan.

Kungrad, UZBEKISTAN: see QUNGHIROT.

Kungsbacka (KUNGS-BAHK-kah), town, HALLAND county, SW SWEDEN, near KATTEGATT strait, 15 mi/24 km SSE of GÖTEBORG; 57°29′N 12°05′E. Manufacturing (printing, construction). Chartered 1557.

Kungsgården (KUNGS-GOR-den), village, GÄVLEBORG county, E SWEDEN, on N shore of STORSJÖN LAKE, 5 mi/8 km W of SANDVIKEN; 60°36′N 16°37′E.

Kungsör (KUNGS-UHR), town, VÄSTMANLAND county, central SWEDEN, at W end of LAKE MÄLAREN, at mouth of ARBOGAÅN RIVER, 6 mi/9.7 km SE of KÖPING; 59°25′N 16°06′E. Lake port; manufacturing (concrete, metal products).

Kungu (KOON-goo), village, ÉQUATEUR province, NW CONGO, on road, 40 mi/64 km SW of GEMENA; 02°47′N 19°12′E. Elev. 1,499 ft/456 m.

Kungur (koon-GOOR), city (2006 population 65,555), SE PERM oblast, W URAL Mountains, E European Russia, on the SYLVA RIVER (KAMA River basin), on road and railroad, 63 mi/101 km SSE of PERM; 57°26′N 56°57′E. Elevation 452 ft/137 m. In oil-producing district. Machinery, drilling equipment, telephones, leather shoes, furniture; woodworking, stone-cutting handicrafts, ceramics. Gypsum quarries 9 mi/14 km WNW, at YERGACH station; alabaster-walled caves nearby. Founded in 1648 as a fortress; developed as a trading town in the 18th century.

Kungurtug (koon-goor-TOOK), settlement (2006 population 1,815), SE TUVA REPUBLIC, S Siberian Russia, near a small lake, on highway, 156 mi/251 km SE of KYZYL; 50°36′N 97°31′E. Elevation 4,291 ft/1,307 m. Logging and lumbering; livestock (sheep, yak).

Kungutasi (koon-goo-TAH-see), town, MBEYA region, SW TANZANIA, 30 mi/48 km NNW of MBEYA, near LUPA RIVER, in LUPA GOLDFIELD (active c. 1930–1956); 08°30′S 33°14′E. Grain; livestock.

Kungwe Bay (KOON-gwai), KIGOMA region, W TANZANIA, arm of Lake TANGANYIKA, on E side, c. 65 mi/105 km S of KIGOMA; 02°40′S 31°59′E. Village of LAGOSSA on SE shore.

Kungwe, Mount (KOON-gwai), peak (8,250 ft/2,515 m), KIGOMA region, W TANZANIA, in Makari Hills, MAHALE MOUNTAINS NATIONAL PARK, overlooking Lake TANGANYIKA, opposite Kalemie (ZAIRE); 06°14′S 29°52′E.

Kungyangon (koong-yahn-GON), township, YANGON division, MYANMAR, in AYEYARWADY RIVER delta, 25 mi/40 km SSW of YANGON, near ANDAMAN SEA. Important rice-producing area.

Kunhegyes (KUN-he-dyesh), city, SZOLNOK county, E central HUNGARY, 14 mi/23 km NW of KARCAG; 47°22′N 20°38′E. Wheat, corn; cattle; manufacturing (telecommunications equipment, aluminum articles, machinery; flour milling.

Kuni (KOO-nee), village, Agatsuma county, GUMMA prefecture, central HONSHU, N central JAPAN, 34 mi/55 km N of MAEBASHI; 36°35′N 138°37′E.

Area is shown by the symbol □, and capital city or county seat by ☉.

Kuniamuthem, INDIA: see KUNIYAMUTHUR.

Kunié, NEW CALEDONIA: see PINES, ISLE OF.

Kunigal (KUH-ni-gahl), town, TUMKUR district, KARNATAKA state, S INDIA, 22 mi/35 km SSW of TUMKUR; 13°01′N 77°01′E. Rice milling, handicrafts (woolen blankets, biris, pottery); silk growing. Kaolin worked nearby.

Kunigami (koo-NEE-gah-mee), village, at N tip of OKINAWA island, Kunigami county, Okinawa prefecture, SW JAPAN, 47 mi/75 km N of NAHA; 26°44′N 128°10′E.

Kunihar (kuh-ni-HAHR), former princely state of PUNJAB HILL STATES, N INDIA. Since 1948, merged with HIMACHAL PRADESH state.

Kunimi (koo-NEE-mee), town, Date county, FUKUSHIMA prefecture, N central HONSHU, NE JAPAN, 9 mi/15 km N of FUKUSHIMA city; 37°52′N 140°33′E.

Kunimi (koo-NEE-mee), town, South Takaki county, NAGASAKI prefecture, NW KYUSHU, SW JAPAN, 28 mi/45 km E of NAGASAKI; 32°52′N 130°18′E.

Kunimi (koo-NEE-mee), town, East Kunisaki county, OITA prefecture, E KYUSHU, SW JAPAN, 31 mi/50 km N of Oita; 33°40′N 131°35′E.

Kuninkaanristi, RUSSIA: see ROMASHKI.

Kunino (ko-nee-NO), village, MONTANA oblast, ROMAN obshtina, BULGARIA; 43°11′N 24°00′E.

Kunisaki (koo-NEE-sah-kee), town, East Kunisaki county, OITA prefecture, NE KYUSHU, SW JAPAN, on E KUNISAKI peninsula, on IYO SEA, 25 mi/40 km N of OITA; 33°33′N 131°44′E. Dried shiitake mushrooms; seaweed (*wakame*). Also spelled Kunizaki.

Kunisaki Peninsula (koo-NEE-sah-kee), Japanese *Kunisaki-hanto* (koo-nee-SAH-kee–HAHN-to), NE KYUSHU, SW JAPAN, in OITA prefecture, between Suo Sea (N) and BEPPU BAY (S); 24 mi/39 km N-S, 18 mi/29 km E-W. Mountainous; rises to 2,365 ft/721 m. KITSUKI on S coast. Fertile coastal strip produces grain.

Kunitachi (koo-nee-TAH-chee), city, Tokyo prefecture, E central HONSHU, E central JAPAN, 124 mi/200 km W of SHINJUKU; 35°40′N 139°26′E.

Kunitomi (koo-NEE-TO-mee), town, E Morokata county, MIYAZAKI prefecture, SE KYUSHU, SW JAPAN, 9 mi/15 km N of MIYAZAKI; 31°59′N 131°19′E. Tobacco, sweet pepper, daikon. Site of Hokkegoku Yakushi Temple.

Kuniyamuthur (kuh-ni-YUH-muh-toor), town, COIMBATORE district, TAMIL NADU state, S INDIA, suburb (3 mi/4.8 km S) of Coimbatore. Cotton milling. Sometimes called Kuniamuthem; also spelled Kuniyamuttur.

Kunjabangarh (KUHN-juh-buhng-GUHR), village, PURI district, central ORISSA state, E central INDIA, 75 mi/121 km WNW of Puri. Markets timber, rice, bamboo. Was capital of former princely state of DASPALLA. Formerly called Kunjaban.

Kunjah (koon-JAH), town, GUJRAT district, NE PUNJAB province, central PAKISTAN, 7 mi/11.3 km SW of GUJRAT; 32°32′N 73°59′E.

Kunkels Pass (4,432 ft/1,351 m), in the GLARUS ALPS, E SWITZERLAND, 6 mi/9.7 km W of CHUR. Road over the pass leads along TAMINA RIVER valley to Rhine valley.

Kunkurgan, city, SURKHANDARYO wiloyat, UZBEKISTAN. Tertiary-level administrative center. Established in the 1930s at the time of the construction of the Khumkhurgon canal, which was followed by a reservoir as well in the 1950s. Little industry, some agricultural processing. Also Khumkhurghon.

Kunlong (KOON-LONG), township, SHAN STATE, MYANMAR, on E bank of THANLWIN RIVER (ferry), and 65 mi/105 km NE of LASHIO, on route to CHINA (YUNNAN province). Site of small hydrostation built in 1993.

Kunlun Mountains (KUN-LUN), in XINJIANG, QINGHAI, and TIBET autonomous regions, W CHINA, a major mountain range in ASIA, extending from the PAMIR Plateau (W) to NW SICHUAN province. The range is 1,553 mi/2,499 km long, mostly with elevation above 16,000 ft/4,877 m. Has mountain glaciers. As the NW boundary of the Tibetan Plateau, the W portion of the KUNLUN extends along the KASHMIR border (bet. India/PAKISTAN and China), and the Tibet/Xinjiang border, encircling the S edge of the Tarim Basin, parallel to the KARAKORAM Mountains (SW). Runs NW-SE and then in a SW-NE direction; rises to above 23,966 ft/7,305 m at the peaks ULUGH MUZTAG (25,340 ft/7,724 m) and MUZTAG (23,890 ft/7,282 m). In S QINGHAI province, the E portion of the Kunlun turns ESE in direction, borders the SW of the QAIDAM basin, and is generally lower; E of the Kunlun Shankou (the Kunlun Pass), the Kunlun splits into 2 branches: A'nyemaqen (N) and the BAYAN HAR (S) mountains, between which are the lakes GYARING and Ngoring, and the uppermost headwaters of the HUANG HE (here called the Ma Qu). To the E, both the A'nyemaqen and the BAYAN HAR mountains blend into the mountains W of Sichuan province. Sometimes spelled Kwenlun Mountains.

Kunming (KUN-ming), city (□ 803 sq mi/2081 sq km; 1994 estimated urban population 1,240,000; estimated total population 1,623,900), ⊙ YUNNAN province, S CHINA, on the N shore of DIAN CHI Lake; 25°04′N 102°41′E. Major administrative, commercial, and cultural center of S China and leading transportation hub (air, road, railroad), with railroad connections to North VIETNAM. China's largest producer of copper. Coal is mined, and the city has an iron and steel complex. Other manufacturing includes phosphorus, chemicals, machinery, textiles, paper, and cement. Long noted for its scenic beauty and equable climate. It consists of an old walled city, a modern commercial suburb, and a residential and university section Although it was often the seat of kings in ancient times, Kunming's modern prosperity dates only from 1910, when the railroad from HANOI was built. In World War II, Kunming was important as the Chinese terminus of the BURMA Road. Has an astronomical observatory; seat of Yunnan University and a medical college. On the outskirts is a famed bronze temple, dating from the Ming dynasty (1368–1644). Formerly called Yunnanfu.

Kunnamkulam (kuh-NUHM-kuh-luhm), town, Trichur district, KERALA state, S INDIA, 45 mi/72 km NNW of ERNAKULAM; 10°39′N 76°05′E. Coir products (rope, mats), copra; rice and oilseed milling.

Kunneppu (koo-NEP), town, Abashiri district, Hokkaido prefecture, N JAPAN, 127 mi/205 km E of SAPPORO; 43°43′N 143°44′E. Potatoes.

Kunnersdorf, CZECH REPUBLIC: see LAZNE KUNDRATICE.

Kunohe (koo-no-HE), village, Kunohe county, IWATE prefecture, N HONSHU, NE JAPAN, 37 mi/60 km N of MORIOKA; 40°12′N 141°25′E.

Kunovice (KU-no-VI-tse), German *Kunowitz*, village, JIHOMORAVSKY province, SE MORAVIA, CZECH REPUBLIC, on railroad, and just S of UHERSKÉ HRADIŠTĚ. Manufacturing (aircraft). Summer resort; museum of aviation.

Kunowice, German *Kunersdorf*, village, Gorzów province, W POLAND, 4 mi/6.4 km E of FRANKFURT. Formerly part of GERMANY (BRANDENBURG). In Seven Years War, scene (August 1759) of critical defeat of Prussians under Frederick the Great by Austrians under Loudon and Russians under Soltikov. Under Polish administration since 1945.

Kunowitz, CZECH REPUBLIC: see KUNOVICE.

Kunoy, Danish *Kunø*, island (□ 14 sq mi/36.4 sq km) of the NE FAEROE ISLANDS; c.8 mi/13 km long, 2 mi/3.2 km wide. Mountainous (highest point 2,726 ft/831 m). Fishing; sheep raising.

Kunpo (GOON-PO), city, central KYONGGI province, SOUTH KOREA, S of SEOUL; 37°21′N 126°57′E. Manufacturing (machinery, metals, electronics, textiles, chemicals; food processing); some agriculture. Residential area of Seoul due to its location on major road, railroad, and subway line. Site of traditional folk festival honoring Suri Mountain god.

Kunsan (KOON-SAHN), city, NORTH CHOLLA province, SW SOUTH KOREA, on the YELLOW SEA, at the KUM RIVER estuary; 35°59′N 126°43′E. It was a major port, especially for rice shipments, and was a commercial center for the rice grown in the Kum basin during the period of Japanese rule (1905–1945). Rice processing, fishing, paper manufacturing, and lumbering are major industries. Originally a poor fishing village, Kunsan gained importance with the development of its port, which was opened to foreign trade in 1899. In 1980, the port expanded.

Kunshan (KUN-SHAHN), city (□ 334 sq mi/868.4 sq km; 2000 population 568,994), SE JIANGSU province, CHINA, on TIANJIN-SHANGHAI railroad, 34 mi/55 km WNW of Shanghai; 31°21′N 120°57′E. Light industry is the largest sector of the city's economy. Heavy industry and agriculture are also important. Main industries include food processing, manufacturing (textiles, apparel, leather, chemicals, pharmaceuticals, synthetic fibers, plastics, iron and steel, machinery, and electrical and electronic equipment).

Kunstadt, CZECH REPUBLIC: see KUNSTAT.

Kunstat (KUNSH-taht), Czech *Kunštát*, German *Kunstadt*, village, JIHOMORAVSKY province, W MORAVIA, CZECH REPUBLIC, 22 mi/35 km NNW of BRNO. Agriculture (oats); graphite mining; pottery.

Kunszentmárton (KUN-sant-mahr-ton), city, SZOLNOK county, E central HUNGARY, on the Körös River, and 28 mi/45 km S of SZOLNOK; 46°50′N 20°17′E. Corn, wheat, barley, sunflowers; hogs, cattle; manufacturing (bricks, tile).

Kunszentmiklós (KUN-sant-mik-losh), city, BÁCS-KISKUN county, central HUNGARY, 28 mi/45 km WNW of KECSKEMÉT; 47°02′N 19°08′E. Railroad and road junction; manufacturing (electric appliances; flour milling).

Kuntahasi (koon-TAH-ah-see), town, ASHANTI region, GHANA, 20 mi/32 km SE of KUMASI, 2 mi/3.2 km from Lake BOSUMTWI; 06°32′N 01°29′W. Tourism; cacao, coffee, timber.

Kuntaur (koon-tah-OOR), town, CENTRAL RIVER division, central THE GAMBIA, on N bank of GAMBIA RIVER (wharf and ferry), 12 mi/19 km NW of GEORGETOWN; 13°38′N 14°54′W. Peanut-shipping point; palm oil and kernels, rice. Also Kunta-ur.

Kuntsevo (KOON-tsi-vuh), former city, central MOSCOW oblast, central European Russia, now a suburb on the WSW edge of MOSCOW, approximately 14 mi/23 km from the city center; 55°44′N 37°26′E. Elevation 544 ft/165 m. Woolen milling, metalworking; aluminum products. Incorporated into Moscow in 1960.

Kuntzig, LUXEMBOURG: see CLÉMENCY.

Kununurra (kuhn-uh-NUH-ruh), town, NE WESTERN AUSTRALIA state, W AUSTRALIA, KIMBERLEY region, on Victoria Highway, and on ORD RIVER, 70 mi/113 km S of TIMOR SEA, on road 1,977 mi/3,182 km NE of PERTH, 547 mi/880 km SW of DARWIN (NORTHERN TERRITORY), and 14 mi/23 km W of Northern Territory border; 15°46′S 128°44′E. Established 1960 as service and residential center for Ord River Irrigation Scheme. Kununurra Dam and Lake here; Lake Argyle Dam 20 mi/32 km upstream (S). Diamond mine. Rice, sugarcane; tourism.

Kununurra Dam, AUSTRALIA: see KUNUNURRA.

Kunvald (KUN-vahlt), village, VYCHODOCESKY province, E BOHEMIA, CZECH REPUBLIC, in foothills of the Eagle Mountains, 30 mi/48 km ESE of HRADEC KRÁLOVÉ; 50°08′N 16°30′E. Former headquarters of Bohemian-Moravian church assembly (1457).

Kunwi (GOON-WEE), county (□ 381 sq mi/990.6 sq km), central NORTH KYONGSANG province, SOUTH KOREA. Bordered N by UISONG province, E by YONGCHON province, S by TAEGU city and CHILGOK province, and W by Sonsan province. Palgong Mountain on S border with Taegu, Wi River flows E through county, joining the NAKDONG RIVER. Agriculture (potatoes, sweet potatoes, beans, red peppers, garlic, apples, medical herbs) in river basin. Inkak Buddhist temple, where history book of the Three Kingdoms was completed, located here.

Kunya (KOON-yah), town, KANO state, N NIGERIA, on road, 25 mi/40 km N of KANO; 12°13′N 08°32′E. Market center. Groundnuts, sorghum.

Kun′ya (KOON-yah), town (2006 population 3,395), SE PSKOV oblast, W European Russia, on road and railroad, 17 mi/27 km E of VELIKIYE LUKI; 56°17′N 30°58′E. Elevation 482 ft/146 m. Dairying, flax processing.

Kunyang, CHINA: see JINNING.

Kunya-Urgench (KOON-yah–uhr-GENCH), city, DASHHOWUZ weloyat, N TURKMENISTAN, 25 mi/40 km W of AMU DARYA (river) in the delta region, 85 mi/137 km NW of URGANCH (UZBEKISTAN); 42°19′N 59°10′E. Tertiary-level administrative center. Major trade and craft center 10th–13th century; became capital of khanate of KHWARAZM in the 12th century. Destroyed by the Mongols in the early 13th century, partially rebuilt, and finally abandoned in the 16th century. Ruins of an 11th-century minaret and mosque, mausoleums, shops, and the portal of the Caravanserai Gates (14th century) have been uncovered. Also spelled Koneurgench.

Kunzela, town (2007 population 6,071), AMHARA state, NW ETHIOPIA, in low-lying area SW of LAKE TANA, 25 mi/40 km NW of BAHIR DAR; 11°41′N 37°02′E. Also spelled Kunzila.

Künzell (KUN-tsel), village, HESSE, central GERMANY, 2.5 mi/4 km SE of FULDA; 50°33′N 09°44′E. Agricultural center.

Künzelsau (KOON-tels-ou), town, BADEN-WÜRTTEMBERG, SW GERMANY, on the KOCHER, 15 mi/24 km SSW of MERGENTHEIM; 48°13′N 09°41′E. Manufacturing of machinery, motors; printing. Has 17th-century castle and half-timbered town hall from 1548. Seat of engineering and technical school.

Kunzila, ETHIOPIA: see KUNZELA.

Kunzulu (koon-ZOO-loo), village, BANDUNDU province, W CONGO, along CONGO RIVER; 03°29′S 16°09′E. Elev. 938 ft/285 m. Fishing; subsistence farming.

Kuocang Mountains (KWO-ZAHNG), SE ZHEJIANG province, CHINA, near coast, between LING (N) and WU (S) rivers; rises to 4,655 ft/1,419 m SW of LINHAI; 28°36′N 120°30′E.

Kuokkala, RUSSIA: see REPINO.

Kuolajärvi, RUSSIA: see KUOLOYARVI.

Kuoloyarvi (koo-o-luh-YAHR-vee), Finnish *Kuolajärvi*, village, N MURMANSK oblast, NW European Russia, less than 5 mi/8 km E of the Finnish border, on road and near railroad, 90 mi/145 km W of KANDALAKSHA; 66°58′N 29°12′E. Elevation 626 ft/190 m. Reindeer raising. Called Salla (1937–1940); ceded (1940) by FINLAND to the USSR and Finnish population resettled in Kursu (Finland), which was renamed Salla.

Kuopio (KOO-o-pee-o), city, ⊙ KUOPION province, central FINLAND; 62°54′N 27°41′E. Elevation 330 ft/100 m. Forest region with industries based on timber. At the head of the SAIMAA lake system, it is a tourist and inland-navigation center. Chartered in 1654 and rechartered in 1782. Site of principal Finnish Orthodox monastery, library, and museum which holds a collection of religious art; seat of Finnish Orthodox archibishop. Seat of Kuopio University (established 1972). Airport.

Kuopion (KOO-o-pee-on), Finnish *Lääni*, province (□ 7,704 sq mi/20,030.4 sq km), E FINLAND; ⊙ KUOPIO. Bordered E by RUSSIA; forms part of KARELIA region. Land is generally marshy; partly wooded; with numerous lakes, including PIELINEN and Kallajärvi, and N part of SAIMAA lake system. Agriculture (rye, oats, barley, potatoes), livestock raising, dairy farming; lumbering and woodworking industries (manufacturing of plywood, wallboard, spools, bobbins) are

important; also has molybdenum and asbestos mines, limestone quarries. Major towns are Kuopio, JOENSUU, and IISALMI.

Kuortane (KOO-o-tah-nai), village, VAASAN province, FINLAND, 60 mi/90 km SE of VAASA; 62°48′N 23°30′E. Elevation 330 ft/100 m. Quartz and feldspar quarries. Noted for its folk music. Architect Alvar Aalto born here 1898.

Kupang (KOO-pahng), town, ⊙ Nusatenggara province and Kupang district, Timor, near SW tip of TIMOR Island, port on small Kupang Bay of SAVU SEA; 10°10′S 123°34′E. Refueling port and trading port; ships copra, hides, sandalwood, pearls, trepang. Its airport is on JAVA-AUSTRALIA route and serves as area gateway to E INDONESIA. Also spelled Koepang.

Kupari, resort village, S CROATIA, in DALMATIA, on ADRIATIC SEA, in Dubrovnik Riviera, S of DUBROVNIK.

Kupa River (KOO-pah), 184 mi/296 km long, W and central CROATIA; rises 16 mi/26 km NE of RIJEKA (Fiume); flows generally E, past BROD NA KUPI, KARLOVAC, and PETRINJA, to Sava River at SISAK. Navigable for 84 mi/135 km. In upper course forms part of Croatia-Slovenia border; Pokuplje region extends along middle course. Hydroelectric plant at OZALJ, 7 mi/11.3 km N of Karlovac. Receives DOBRA, Mrežnica, KORANA, and GLINA rivers. W of SKRAD, narrow canyon (c.0.6 mi/1 km long) on one of its tributaries forms Varžji Prolaz Special Reserve. Also called Kolpa River in Slovenia.

Kuparuk River (KOO-pah-ruhk), c.140 mi/225 km long, N ALASKA; rises in N BROOKS RANGE, near 68°40′N 149°00′W, flows N to BEAUFORT SEA of ARCTIC OCEAN at 70°23′N 148°47′W.

Kupavna (koo-PAHV-nah), village (2006 population 6,620), central MOSCOW oblast, central European Russia, on road and railroad, 4 mi/6 km E of (and administratively subordinate to) ZHELEZNODOROZHNYY; 55°45′N 38°08′E. Elevation 488 ft/148 m. Agricultural products.

Kupavna, RUSSIA: see STARAYA KUPAVNA.

Kupé-Manengouba, department (2001 population 123,011), SOUTH-WEST province, CAMEROON.

Kupferzell (KU-pfer-tsel), village, BADEN-WÜRTTEMBERG, SW GERMANY, 22 mi/35 km NE of HEILBRONN; 49°12′N 09°41′E.

Kupino (KOO-pee-nuh), city (2006 population 16,155), SW NOVOSIBIRSK oblast, SW SIBERIA, RUSSIA, on road and railroad, 360 mi/579 km W of NOVOSIBIRSK, and 75 mi/121 km SE of TATARSK; 54°21′N 77°17′E. Elevation 400 ft/121 m. In agricultural area; meat and fish canning, dairying. Founded in 1886. Industrialized after 1935. Made city in 1944.

Kupiškis (KUH-pish-kis), Russian *Kupishki*, city, NE LITHUANIA, 25 mi/40 km ENE of PANEVEZYS; 55°50′N 24°58′E. Flour milling; concrete. In Russian KOVNO government until 1920.

Kuppal, INDIA: see KOPPAL.

Kuppam (KUH-puhm), town, CHITTOOR district, ANDHRA PRADESH state, S INDIA, 60 mi/97 km SW of CHITTOOR; 12°45′N 78°22′E. Processing of essential oils (sandalwood), lemon grass; silk growing. Exports chrysanthemum and jasmine flowers.

Kuppenheim (KUP-pen-heim), village, BADEN-WÜRTTEMBERG, SW GERMANY, on the MURG RIVER, 3 mi/4.8 km SE of RASTATT; 48°50′N 08°15′E. Manufacturing (wire, cleaning products). Has early-18th-century town hall.

Kupreanof Island (KOO-pree-yuh-nawf), SE ALASKA, in ALEXANDER ARCHIPELAGO, W of PETERSBURG; 52 mi/84 km long, 20 mi/32 km–30 mi/48 km wide; 56°48′N 133°25′W. Rises to c.4,000 ft/1,219 m (NE). Lindenberg Peninsula (30 mi/48 km long, 12 mi/19 km wide) is separated from SE side of island by DUNCAN CANAL. Fishing, fish processing. KAKE village is in NW. Island named after Captain Kupreanov, gover-

nor of Russian America who succeeded (1836) Baron Wrangell.

Kupreanof Point (KOO-pree-yuh-nawf), promontory, SW ALASKA, on SW ALASKA PENINSULA, on E side of STEPOVAK BAY; 55°34′N 159°37′W.

Kupreanof Strait (KOO-pree-yuh-nawf), S ALASKA, between KODIAK (S) and AFOGNAK (N) islands, connects Gulf of ALASKA (E) and SHELIKOF STRAIT (W), 25 mi/40 km WNW of KODIAK; 20 mi/32 km long, 2 mi/3.2 km–3 mi/4.8 km wide.

Kupres, town, W BOSNIA, BOSNIA AND HERZEGOVINA, center of Kupreško polje.

Kupreško Polje (koo-PRESH-ko POL-ye) or **Kupres Plain**, plain and historical region, W BOSNIA, BOSNIA AND HERZEGOVINA, in karst. Of the few regions within in the W Dinaric limestone desert that sustains agriculture. Principal town is KUPRES.

Küps (KOOPS), village, BAVARIA, central GERMANY, in UPPER FRANCONIA, on the RODACH, 4 mi/6.4 km SW of KRONACH; 50°11′N 11°16′E.

Kupwara (koop-wah-rah), district (□ 919 sq mi/2,389.4 sq km), JAMMU AND KASHMIR state, N INDIA; ⊙ KUPWARA.

Kupwara (koop-wah-rah), town, ⊙ KUPWARA district, JAMMU AND KASHMIR state, N INDIA.

Kup′yans′k (KOOP-yahnsk), (Russian *Kupyansk*), city (2004 population 47,400), E KHARKIV oblast, UKRAINE, on Oskil River, 60 mi/97 km ESE of KHARKIV; 49°43′N 37°36′E. Elevation 429 ft/130 m. Raion center. Metal casting and working; manufacturing (machines, building materials); food processing (meat, milk, food flavoring, sugar, flour). Railroad junction of Kup′yans′k-Vuzlovyy (Russian *Kupyansk-Uzlovoy*) is 5 mi/8 km SSE. Transport, medical schools, three vocational technical schools. Heritage museum. Established in 1655, city since 1779.

Kupyansk, UKRAINE: see KUP′YANS′K.

Kupyansk-Uzlovoy, UKRAINE: see KUP′YANS′K-VUZLOVYY.

Kup′yans′k-Vuzlovyy (KOOP-yahnsk–vooz-lo-VEE), (Russian *Kupyansk-Uzlovoy*), town, E KHARKIV oblast, UKRAINE, 5 mi/8 km SE of KUP′YANS′K and subordinated to its city council; 49°39′N 37°40′E. Elevation 344 ft/104 m. Railroad junction; railroad servicing, including railroad car depot. Manufacturing of building materials. Town since 1925.

Kuqa (KOO-CHAH), town and oasis, ⊙ Kuqa county, W central XINJIANG UYGUR Autonomous Region, NW CHINA, 140 mi/225 km ENE of AKSU, and on highway S of the TIANSHAN Mountains; 41°43′N 82°54′E. Grain, cotton; livestock; food processing, manufacturing (textiles, chemicals, building materials); coal mining. Also appears as Kuche.

Kur, IRAN: see KOR RIVER.

Kura-Araks Lowland, extensive plain in E TRANSCAUCASIA, in AZERBAIJAN, along the lower reaches of KURA and ARAS (Araks) rivers, between Greater and Lesser CAUCASUS; washed by CASPIAN SEA; 155 mi/250 km long, 93 mi/150 km wide. Central and E parts are below sea level. Rich in oil; cotton plantations, orchards, vineyards, winter pastures.

Kurabuchi (koo-RAH-boo-chee), village, Gumma county, GUMMA prefecture, central HONSHU, N central JAPAN, 19 mi/30 km W of MAEBASHI; 36°25′N 138°47′E. Ginger, *nameko* and shiitake mushrooms.

Kuragino (koo-RAH-gee-nuh), town (2005 population 14,005), S KRASNOYARSK TERRITORY, S central SIBERIA, RUSSIA, in the Minusinsk basin, on the TUBA RIVER (YENISEY RIVER basin), on road and railroad spur, 40 mi/64 km ENE of MINUSINSK; 53°54′N 92°40′E. Elevation 948 ft/288 m. Gravel; dairy products; hemp processing, lumbering. Mining of iron ore and gold in the area.

Kurahashi (koo-RAH-hah-shee), town, on S section of KURAHASHI-JIMA island, Aki county, HIROSHIMA prefecture, off SW HONSHU, W JAPAN, 22 mi/35 km S of HIROSHIMA; 34°06′N 132°30′E. Oysters; mandarin oranges.

Kurahashi-jima (koo-rah-ha-SHEE–jee-mah), island (□ 30 sq mi/78 sq km), HIROSHIMA prefecture, off SW HONSHU, W JAPAN, at E side of entrance to HIROSHIMA BAY, just SE of NOMI-SHIMA island and opposite KURE; 9 mi/14.5 km long. Mountainous, fertile. Produces rice, building stone.

Kuraima, JORDAN: see KUREIMA.

Kuraishi (koo-RAH-ee-shee), village, Sannohe county, Aomori prefecture, N HONSHU, N JAPAN, 34 mi/55 km S of AOMORI; 40°30′N 141°15′E. Garlic.

Kurakh (koo-RAHKH), village (2005 population 3,445), S DAGESTAN REPUBLIC, E CAUCASUS, extreme S European Russia, on road, approximately 31 mi/50 km NW of the RUSSIA-AZERBAIJAN border, 20 mi/32 km WSW of KASUMKENT; 41°35′N 47°47′E. Elevation 4,327 ft/1,318 m. Rug weaving. Population largely Lezghian.

Kurakhivbud, UKRAINE: see KURAKHOVE.

Kurakhivhres, UKRAINE: see KURAKHOVE.

Kurakhivka (koo-rah-KHEEF-kah), (Russian *Kurakhovka*), town (2004 population 30,600), central DONETS′K oblast, UKRAINE, in the DONBAS, on VOVCHA RIVER, and 14 mi/22.5 km WNW of DONETS′K; 48°02′N 37°23′E. Elevation 360 ft/109 m. Coal mine enrichment plant. Established in 1924, town since 1938.

Kurakhove (koo-RAH-kho-ve), (Russian *Kurakhovo*), city, W central DONETS′K oblast, UKRAINE, in the DONBAS, 25 mi/40 km W of DONETS′K city center; 47°59′N 37°16′E. Elevation 492 ft/149 m. Site of the Kurakhove state regional thermal-electric power station; also boiler making, manufacturing of reinforced concrete structural materials and fiberglass wool, grain milling. Energy technical school. Established in 1936 as Kurakhivbud (Russian *Kurakhovstroy*), known 1943–1956 as Kurakhivhres; renamed Kurakhove and city status in 1956.

Kurakhovka, UKRAINE: see KURAKHIVKA.

Kurakhovo, UKRAINE: see KURAKHOVE.

Kurakhovstroy, UKRAINE: see KURAKHOVE.

Kuralovo (koo-RAH-luh-vuh), village, NW TATARSTAN Republic, E European Russia, on a short right tributary of the VOLGA River, on road, 16 mi/26 km SW of KAZAN; 55°38′N 48°44′E. Elevation 337 ft/102 m. Starch and syrup plant.

Kurama, Mount (KOO-rah-mah), Japanese *Kuramayama* (koo-RAH-mah–YAH-mah) (1,800 ft/549 m), KYOTO prefecture, S HONSHU, W central JAPAN, 10 mi/16 km N of KYOTO. Site of Buddhist temple founded in 8th century.

Kurama Range, branch of TIANSHAN Mountain system, on UZBEKISTAN-TAJIKISTAN border; extends c.100 mi/161 km. SW from CHATKAL RANGE to the SYR DARYA River, part of the N perimeter of the FERGANA VALLEY; rises to 7,550 ft/2,301 m. Rich in mineral deposits, mainly lead, zinc, antimony, tungsten, and radioactive ores.

Kuranda (koo-RAN-duh), town, QUEENSLAND, NE AUSTRALIA, 17 mi/27 km NW of CAIRNS; 16°49′S 145°39′E. Tourism; rainforests, scenic railway, wildlife noctarium, butterfly sanctuary.

Kurandvad (koo-ruhnd-wahd), town, KOLHAPUR district, MAHARASHTRA state, central INDIA, near KRISHNA RIVER, 12 mi/19 km S of SANGLI; 16°41′N 74°35′E. Agriculture market (millet, cotton, wheat, oilseeds, sugarcane); handicraft cloth weaving. Also spelled Kurundwad, Kurundvad. Was capital of former DECCAN state of Kurandvad Senior.

Kurandvad Junior (koo-ruhnd-wahd), former princely state in DECCAN STATES, India. Incorporated 1949 into SOLAPUR district (MAHARASHTRA state) and BELGAUM district (KARNATAKA state).

Kurandvad Senior (koo-ruhnd-wahd), former princely state in DECCAN states, INDIA. Incorporated 1949 into former SATARA SOUTH, BELGAUM, and BIJAPUR districts of former BOMBAY state, now within KOLHAPUR district (MAHARASHTRA state) and Belgaum and Bijapur districts (KARNATAKA state).

Kuraoli, INDIA: see KURAULI.

Kura River (KOO-rah), ancient *Cyrus*, Georgian *Mhtvari*, c.950 mi/1,530 km long, chief river of GEORGIA and AZERBAIJAN; rises in NE Turkey, NW of KARS; flows NE into Georgia, then SE, parallel to the CAUCASUS MOUNTAINS, to the CASPIAN SEA. Hydroelectric plants are on the river near TBILISI and MINGECHAUR; the extensive reservoir at Mingechaur is used for irrigation. The lower KURA River, joined by the ARAS RIVER, its chief tributary, flows through an irrigated plain that extends into NW IRAN. Cotton is the chief industrial crop of the region, which lies partly below sea level. The Kura is navigable c.300 mi/480 km upstream. Its basin covers 72,587 sq mi/188,000 sq km.

Kurashiki (koo-RAH-shee-kee), city (2005 population 469,377), OKAYAMA prefecture, SW HONSHU, W JAPAN, 24 mi/15 km WSW of OKAYAMA; 34°34′N 133°46′E. Railroad junction. Oil, iron manufacturing; petrochemicals, motor vehicles, textiles. Nori. Many businesses operate in Mizushima Rinkai industrial zone. Seto-o-hashi bridge here to SAKAIDE on SHIKOKU.

Kuratake (koo-RAH-tah-ke), town, Amakusa county, on S coast of AMAKUSA KAMI-SHIMA island, KUMAMOTO prefecture, W KYUSHU, SW JAPAN, 34 mi/55 km S of KUMAMOTO; 32°24′N 130°20′E.

Kurate (koo-RAH-te), town, Kurate county, FUKUOKA prefecture, N KYUSHU, SW JAPAN, 22 mi/35 km N of FUKUOKA; 33°47′N 130°40′E.

Kurauli (koo-ROU-lee), town, MAINPURI district, W UTTAR PRADESH state, N central INDIA, on distributary of LOWER GANGA CANAL, and 12 mi/19 km NNW of Mainpuri; 27°24′N 78°59′E. Wheat, gram, pearl millet, corn, barley. Also spelled Kuraoli.

Kurau River (KOO-rahu), 50 mi/80 km long, NW PERAK, MALAYSIA; rises E of TAIPING; flows W, through rich KRIAN rice district, to Strait of MALACCA at Kuala Kurau. Irrigation headworks at Bukit Merah.

Kuraymah, JORDAN: see KUREIMA.

Kurayoshi (koo-RAH-yo-shee), city, TOTTORI prefecture, S HONSHU, W JAPAN, on the Tenjin River, 24 mi/38 km W of TOTTORI; 35°25′N 133°49′E. Agriculture (fruit; beef cattle; wine) and communications center manufacturing textiles, pottery, bamboo work, and dumplings. A distinctive cloth called *Kurayoshi-kasuri* ("Kurayoshi Splashed Pattern") is identified with the city. Hot springs nearby.

Kurba (KOOR-bah), village, E central YAROSLAVL oblast, central European Russia, on road, 15 mi/24 km WSW of YAROSLAVL; 57°33′N 39°30′E. Elevation 482 ft/146 m. In agricultural area (hemp, flax, vegetables, potatoes; dairy livestock).

Kurba River (KOOR-buh), approximately 110 mi/177 km long, central BURYAT REPUBLIC, S Siberian Russia; rises in E extension of KHAMAR DABAN Range; flows SW to the UDA RIVER, below UNEGETEY. Extensive iron-ore deposits in Balbagar (bahl-bah-GUHR) mines along upper course.

Kurbnesh, New Town, central ALBANIA, NE of BURREL. Copper mining and processing.

Kurca (KUR-tsah), by-channel of TISZA RIVER, 20 mi/32 km long, S HUNGARY, leaving Tisza near CSONGRÁD; flows S, past SZENTES, rejoining Tisza River above MINDSZENT. Formerly spelled Kurcza.

Kurchaloy (koor-chah-LO-yee), city (2005 population 20,925), E central CHECHEN REPUBLIC, S European Russia, on crossroads, 9 mi/14 km S of GUDERMES; 43°12′N 46°05′E. Elevation 606 ft/184 m. Gas pipeline. Agriculture (grain, livestock). Russian military base and operational headquarters. Formerly known as Chkalovo (1944–1959).

Kurchanskaya (koor-CHAHN-skah-yah), village (2005 population 6,390), W KRASNODAR TERRITORY, S European Russia, in the KUBAN′ River delta, on road, 20 mi/32 km NNE of ANAPA; 45°13′N 37°35′E. Fisheries; agriculture (fruits, grapes). Archaeological digs in the vicinity have produced Sarmatian and Scythian artifacts.

Kurchatov, research city, NE SEMEY region, on IRTYSH RIVER, 72 mi/116 km NW of SEMEY; 50°50′N 78°25′E. Associated with nearby nuclear testing site (□ over 6,930 sq mi/17,949 sq km; in operation 1949–1991) to S and SW. Site now contaminated by radiation, which also affects soil in neighboring regions of EAST KASAKSTAN, PAVLODAR, and KARAGANDA. City in decline since site was closed by Kazak government.

Kurchatov (koor-CHAH-tuhf), city (2005 population 47,910), W KURSK oblast, SW European Russia, on the SEYM RIVER (tributary of the DESNA River), on road and near railroad (Luskashëvka station), 30 mi/48 km W of KURSK; 51°39′N 35°39′E. Elevation 574 ft/174 m. Pipelines and special equipment manufacturing; building materials. Arose with the construction of the Kursk nuclear power station. Made city in 1983.

Kurchum (koor-CHUM), city, EAST KAZAKHSTAN region, KAZAKHSTAN, on E shore of BUKTYRMA RESERVOIR, 51 mi/85 km SE of UST-KAMENOGORSK; 48°35′N 83°39′E. Tertiary-level administrative center. Fishing; meat and dairy processing.

Kurchum River (koor-CHUM), 125 mi/201 km long, EAST KAZAKHSTAN region, KAZAKHSTAN; rises in NARYM RANGE N of MARKAKOL (lake); flows W through narrow, wooded valley, to lower IRTYSH RIVER. Gold mining.

Kurdai (koor-DEI), town, ZHAMBYL region, KAZAKHSTAN, in Kindiktas hills, 30 mi/50 km NNE of BISHKEK (KYRGYZSTAN); 43°22′N 75°04′E. Agricultural products. Also spelled Kurday.

Kurdistan (KUHR-dis-tahn) [Persian=country of the Kurds], Farsi *Kordestan*, indefinite plateau and mountain region, shared by TURKEY, IRAQ, and IRAN. The region lies astride the ZAGROS range (IRAN) and the eastern extension of the TAURUS MOUNTAINS (Turkey) and extends in the south across the Mesopotamian plain and includes the upper reaches of the TIGRIS and EUPHRATES rivers. Kurdistan is inhabited by the Kurds, a strongly Sunni Muslim, seminomadic pastoral people speaking an Iranian language. Known as fierce and predatory, the Kurds have traditionally resisted subjugation by other nations in spite of their tribal political organization. Living primarily in the mountains of SE ANATOLIA and the NW Zagros, the Kurds migrate between summer mountain pastures and winter valley quarters, where they engage in rug weaving and some agriculture. Commonly identified with the ancient *Corduene*, inhabited by the Carduchi (mentioned by Xenophon), Kurdistan was converted to Islam (7th century) by the Arabs, held by the Seljuk Turks (11th century) and the Mongols (13th–15th century); and was disputed by Turkey and PERSIA, during 19th century, with Turkey holding the greater part. Kurdish claims for national autonomy were reaffirmed after World War I at the Paris Peace Conference, and the Treaty of Sèvres (1920) provided for an autonomous Kurdistan; however, the Treaty of Lausanne (1923), which superseded it, failed to mention Kurdistan. Subsequent revolts, particularly in 1925 and 1930 in Turkey, and in 1946 in Iran, were suppressed. Agitation among Iraq's Kurds for a unified and autonomous Kurdistan led in the 1960s to prolonged warfare between Iraqi troops and the Kurds. In 1974 the Iraqi government sought to impose its plan for limited autonomy in Kurdistan, which the Kurds rejected. Attacks on the Kurds continued throughout the Iran-Iraq War (1980–1988), culminating in Iraq using poison gas on Kurdish villages to quash resistance. The situation of the nonsovereign Kurds was brought to worldwide attention as a result of the PERSIAN GULF WAR (1991), when they were actively pursued by Iraqi forces and many were forced to flee to the Turkish border and into Iran. Thousands of Kurds returned to their homes after the war, improved their military strength, and held a general election in their region in May 1992. However, the Kurds were split into two opposed groups, the Kurdistan Democratic party and the Patriotic Union of Kurdistan, which engaged in sporadic warfare. In 1999 the two groups agreed to end hostilities; control of the region is divided between them. Kurdish forces aided the U.S.-led invasion of Iraq in 2003, joining with U.S. and British forces to seize the traditionally Kurdish cities of KIRKUK and MOSUL. As of the late 1990s, there were estimated to be more than 20 million Kurds, about half of them in Turkey, where, making up more than 20 percent of the population, they dwell near the Iranian border around Lake Van, as well as in the vicinity of DIYARBAKIR and ERZURUM. The Kurds in Iran, who constitute some 10 percent of its people, live principally in Azerbaijan and KHORASAN, with some in FARS. The Iraqi Kurds, about 23% of its population, live mostly in the vicinity of DAHUK (Dohuk), Mosul, ERBIL, Kirkuk, and Sulaimaniyah. Name of Kurdistan is also applied to a small mountain district (chief town, LACHIN) of SW AZERBAIJAN in TRANSCAUCASIA; and Kurds are settled on the slopes of Mount ARAGATS in Nahkechevan region of Azerbaijan.

Kurd Kui, IRAN: see KORDKUY.

Kurd Mahalleh, IRAN: see KORDKUY.

Kurdofan (kor-do-FAN), region (2005 population 4,013,000), central SUDAN; 13°00′N 29°30′E. The region, formerly a large province also named Kurdofan (with its capital at AL UBAYYID), was divided into three states in the reorganization of administrative divisions in 1996: N, S, and W Kurdofan (W Kurdofan was incorporated into N and S Kurdofan c.2007). The terrain, generally level in the N, rises in the S to the NUBA MOUNTAINS. Agriculture, with millet as the staple crop. The government has sponsored many irrigation projects. Conquered for EGYPT in 1821, Kurdofan was under Turko-Egyptian rule until 1882, when the Mahdi fomented revolt. With the defeat of Mahdist forces in 1898, Kurdofan became part of Anglo-Egyptian SUDAN. The name is also spelled Kordofan.

Kurdufan, region and former province, SUDAN: see KURDOFAN.

Kurduvadi (kuhr-duh-VAH-dee), town, SOLAPUR district, MAHARASHTRA state, W central INDIA, 45 mi/72 km NW of Solapur; 18°05′N 75°26′E. Railroad junction (workshops); trades in millet, cotton, wheat, peanuts; cotton ginning. Also spelled Kurduwadi.

Kurdzhali (KUHRD-zhah-lee), city, HASKOVO oblast, Kurdzhali obshtina (1993 population 77,164), S BULGARIA, in the E RODOPI Mountains, on the ARDA RIVER, 21 mi/34 km SSW of HASKOVO; 41°39′N 25°22′E. Railroad station. Commercial center; tobacco and food processing, lead and zinc smelting; manufacturing manganese casts, motor vehicle engines, textiles, knitting. Teachers college, research station. Dam nearby. Sometimes spelled Kardzhali, Kirjali, or Krdzhali.

Kurdzhali (KUHRD-zhah-lee), dam, HASKOVO oblast, BULGARIA, on the ARDA RIVER; 41°40′N 25°17′E.

Kurdzhinovo (koor-JI-nuh-vuh), village (2005 population 4,240), W KARACHEVO-CHERKESS REPUBLIC, S European Russia, in NW CAUCASUS Mountains, on the administrative border with KRASNODAR TERRITORY, on the E bank of the LABA RIVER, on road, 11 mi/18 km W of PREGRADNAYA; 43°59′N 40°57′E. Elevation 2,775 ft/845 m. Highway junction. In agricultural area; produce processing.

Kure (KOO-re), city (2005 population 251,003), HIROSHIMA prefecture, SW HONSHU, W JAPAN, on HIROSHIMA BAY, 12 mi/20 km S of HIROSHIMA; 34°14′N 132°34′E. Major naval base and port, and vessels such as merchant ships and oil tankers are built here. In addition to steel and machinery, semiconductor manufacturing devices and micrometers are also manufactured. Also whetstone; dried cuttlefish. Ondo Bridge crosses the Ondo-no-Seto strait.

Kure, Turkish=*Küre*, village, N TURKEY, 30 mi/48 km N of KASTAMONU; 41°48′N 33°44′E. Grain, sugar beets.

Kure Atoll or **Kure Island**, circular coral atoll with 2 small islands, N PACIFIC, in NW part of Hawaiian Islands, HAWAII county, HAWAII, c.60 mi/97 km NW of MIDWAY, International Date Line c.100 mi/161 km to W, c.350 mi/563 km N of TROPIC OF CANCER; northernmost and westernmost point in Hawaii and Hawaiian chain. Elevation 20 ft/6 m. Annexed 1886 by Hawaii and worked for bird guano, placed (1936) under U.S. navy, coast guard base. Sometimes written Cure; formerly Ocean Island.

Kure Beach (KYUHR-ee), village (2006 population 2,311), NEW HANOVER county, SE NORTH CAROLINA, 17 mi/27 km S of WILMINGTON, on PLEASURE ISLAND, W of ATLANTIC OCEAN, and E of CAPE FEAR RIVER estuary; 34°00′N 77°54′W. Swimming resort. Pioneering extraction of bromine from sea water at plant that operated near here, 1934–1945. Retail trade; service industries. FORT FISHER State Recreation Area and State Historical Site and North Carolina State Aquarium to S. Toll ferry across Cape Fear River to SOUTHPORT 3 mi/4.8 km SSW.

Kureima (koo-REE-muh), village, N JORDAN, 18 mi/29 km NNW of SALT. Market gardening; vineyards; citrus fruits. Also spelled Kuraima and Kuraymah.

Kure Mountains, TURKEY: see ISFENDIYAR MOUNTAINS.

Kuressaare (KOO-re-sah-rai), town, administrative center on the S coast of Saaremaa. Port cityspecializing in manufacturing (agr. machinery; food processing; juniper woodworks).

Kureyka (koo-RYAI-kah), village, N KRASNOYARSK TERRITORY, N central SIBERIA, RUSSIA, on the KUREYKA RIVER, N of the ARCTIC CIRCLE, 65 mi/105 km SE of IGARKA; 66°29′N 87°06′E. Graphite mines nearby.

Kureyka, RUSSIA: see KUREIKA.

Kureyka River (koo-RYAI-kuh), 500 mi/805 km long, N KRASNOYARSK TERRITORY, N Russia; rises in PUTORANA Mountains; flows W to the YENISEY RIVER at UST′-KUREYKA. Passes through graphite-mining region 60 mi/97 km above its mouth.

Kurgaldzhin, Lake (koor-gahl-JIN), salt lake (□ 200 sq mi/520 sq km), AKMOLA region, KAZAKHSTAN, just E of TENGIZ (lake), on dry steppe. NURA RIVER flows through.

Kurgaldzhinskoye (koor-gahl-JIN-sko-ye), Kazak *Korgalzhin*, village, SW AKMOLA region, KAZAKHSTAN, on NURA RIVER, 70 mi/113 km SW of Akmola; 50°35′N 70°03′E. In agriculture and cattle area. Tertiary-level (raion) administrative center. Until 1937, Kazgorodok. Also Kurgaldzhino, Korgalzhino.

Kurgan (koor-GAHN), oblast (□ 27,765 sq mi/72,189 sq km; 2005 population 989,075), in SW Siberian Russia, on the KAZAKHSTAN border; ⊙ KURGAN. In SW part of WEST SIBERIAN PLAIN; drained by middle TOBOL, TSET, and MIASS rivers. Most of the region is in a steppe zone, with severe continental climate of very cold winters and warm, dry summers. Population mostly Russian (92%), with Tatar, Bashkir, Kazakh, and Ukrainian minorities. Economy mainly agricultural (replacing former steppe vegetation), with emphasis on dairy farming and spring wheat growing, with some rye, oats, and corn. Chief towns (Kurgan, SHADRINSK, SHCHUCHYE) engage in food processing (flour milling, meatpacking, tanning, dairying) and manufacturing machinery and transportation equipment. Served by the TRANS-SIBERIAN RAILROAD, forming two branches W of Kurgan (links to YEKATERINBURG, OMSK, and Chelyabinsk) and by the TOBOL River, which connects to the shipping routes of the Ob-Irtysh system. Radioactive contamination in the area is because of nuclear accidents in adjacent CHELYABINSK oblast. Formed in February 1943 out of Chelyabinsk oblast.

Kurgan (koor-GAHN), city (2005 population 342,365), ⊙ KURGAN oblast, SW SIBERIA, Russia, on the TOBOL River, on the TRANS-SIBERIAN RAILROAD, 1,226 mi/1,973 km E of MOSCOW; 55°27′N 65°20′E. Elevation 265 ft/80 m. Important railroad junction of the W branches of the Trans-Siberian Railroad. Its factories produce agricultural and chemical machinery, road-building equipment, machine tools, and wood and food products. Founded in 1662, became a town in the 17th century, and became a city in 1782. Until 1782, called Tsarёva Sloboda. There are many ancient burial mounds in the area (called *Kurgan* in Turkic).

Kurganinsk (koor-GAH-neensk), city (2005 population 48,065), E KRASNODAR TERRITORY, S European Russia, on the LABA RIVER, on the administrative border with ADYGEY REPUBLIC, 153 mi/246 km E of KRASNODAR, and 25 mi/40 km W of ARMAVIR; 44°53′N 40°36′E. Elevation 561 ft/170 m. Highway and railroad junction; food industries (meatpacking, sugar refining, canning; dairy products). Established as Kurgannaya Stanitsa by the Kuban′ Cossacks; city status and renamed in 1961.

Kurgannaya Stanitsa, RUSSIA: see KURGANINSK.

Kurgannoye, RUSSIA: see BELGATOY.

Kurganovka (koor-GAH-nuhf-kah), former settlement, NW KEMEROVO oblast, W SIBERIA, RUSSIA, on road and railroad, 19 mi/31 km N of KEMEROVO, now part of BERЁZOVSKY, 1 mi/1.6 km S of the town center; 55°35′N 86°12′E. Elevation 751 ft/228 m. Gold mining. Called Zaboyshchik until 1944.

Kurgantepa, village, SE ANDIJAN wiloyat, UZBEKISTAN, on KYRGYZSTAN border (just N of KARA-SUU River); 40°45′N 72°45′E. Cotton. Formerly called KARASU and, later (1937–c.1940), Imeni VOROSHILOVA, then VOROSHILOVO.

Kurgan-Tyube, TAJIKISTAN: see QURGHONTEPPA.

Kurhessen, GERMANY: see HESSE.

Kuri (GOO-REE), city, central KYONGGI province, SOUTH KOREA, E of SEOUL; 37°35′N 127°08′E. Rapid urbanization since the mid-1980s. Horticulture, livestock breeding; metals, machinery, chemicals, electronics and textile industries.

Kur′i (KOOR-yee), village (2006 population 4,155), S SVERDLOVSK oblast, W Siberian Russia, on highway, 73 mi/117 km ENE of YEKATERINBURG, and 3 mi/5 km E of SUKHOY LOG; 56°55′N 62°07′E. Elevation 354 ft/107 m. Refractory-clay quarrying, paper milling.

Kuria (KOO-ree-ah), island (□ 5 sq mi/13 sq km; 2005 population 1,082), N GILBERT ISLANDS, KIRIBATI, W central PACIFIC OCEAN; 14°00′N 173°24′E. Formerly Woodle Island.

Kuria Muria Bay (KOO-ree-yuh), inlet of ARABIAN SEA, on SW OMAN coast, between capes RAS NAUS (SW) and RAS SHARBATAT (NE); 80 mi/129 km wide. Contains KURIA MURIA ISLANDS.

Kuria Muria Islands, Arabic *Jaza′ir Bin Ghalfan*, group of five islands (□ 28 sq mi/72.8 sq km) in KURIA MURIA BAY of ARABIAN SEA, off SW OMAN coast; 17°30′N 56°00′E. The islands, of granite formation and extending 50 mi/80 km E-W, are the summits of a submarine ridge; group includes (W-E): HASIKIYA, SUDA, HALLANIYA, JIBLIYA, and GHARZAUT. HALLANIYA, the largest, is the only inhabited island. Depopulated (1818) following pirate raids from TRUCIAL OMAN (now UNITED ARAB EMIRATES), the group was seized by the Bin Ghalfan family of the Arab Mahra (Mahrah) tribe on adjoining mainland. The group later passed to the sultan of Oman, who ceded it (1854) to Britain for purposes of a cable station. Considered part of British colony of Aden until 1967 when they were ceded back to Oman. SOUTH YEMEN has claimed the islands (the claim is now maintained by united YEMEN). Guano was worked here 1857–1859 and a telegraph station operated 1859–1860.

Kurichchi (kuh-RI-chee), town, COIMBATORE district, TAMIL NADU state, S INDIA, on NOYIL RIVER, and 3 mi/

4.8 km S of Coimbatore; 11°34′N 77°42′E. Cotton milling. Another Kurichchi, near KAVERI RIVER, is 9 mi/14.5 km N of BHAVANI.

Kurigram (koo-ree-grahm), town (2001 population 66,392), RANGPUR district, N EAST BENGAL, BANGLADESH, on DHARLA (JALDHAKA) River, and 26 mi/42 km ENE of RANGPUR; 26°49′N 89°39′E. Terminus of railroad spur from TISTA village. Trades in rice, jute, tobacco, oilseeds, sugarcane.

Kurihashi (koo-REE-hah-shee), town, North Katsushika county, SAITAMA prefecture, E central HONSHU, E central JAPAN, 19 mi/30 km N of URAWA; 36°07′N 139°41′E.

Kurikoma (koo-REE-ko-mah), town, Kurihara county, MIYAGI prefecture, N HONSHU, NE JAPAN, 40 mi/65 km N of SENDAI; 38°49′N 140°59′E.

Kurile Lake (koo-REEL), Russian *Kuril'skoye Ozero*, crater lake (surface □ 39 sq mi/100 sq km) near the S tip of the KAMCHATKA PENINSULA, KAMCHATKA oblast, extreme E SIBERIA, RUSSIAN FAR EAST, 130 mi/209 km SSW of PETROPAVLOVSK-KAMCHATSKIY; 51°28′N 157°06′E; 7 mi/11 km long, 1,004 ft/306 m deep.

Kurile Strait (koo-REEL), name applied to 6 straits in N section of main KURIL ISLANDS group, SAKHALIN oblast, extreme E SIBERIA, RUSSIAN FAR EAST, connecting the SEA OF OKHOTSK and PACIFIC Ocean. First Kuril Strait, Japanese *Shimushu-kaikyo* (7.7 mi/12.4 km wide), separates SHUMSHU ISLAND from CAPE LOPATKA (S extremity of Kamchatka Peninsula). Second Kurile Strait, Japanese *Paramushiru-kaikyo*, narrowest (little more than 1 mi/1.6 km wide) of entire group, separates Shumshu and PARAMUSHIR islands. Third Kuril Strait, Japanese *Shirinki-kaikyo* (5.5 mi/8.9 km wide), separates Paramushir and SHIRINKI islands. Fourth Kurile Strait, Japanese *Onnekotan-kaikyo* (27 mi/43 km wide), separates Paramushir and Shirinki islands (N) from ONEKOTAN and MAKANRU islands (S). Fifth Kuril Strait, Japanese *Yamato-suido* (17 mi/27 km wide), separates Onekotan and Makanru islands. Sixth Kurile Strait or Krenitsyn Strait, Japanese *Harumukotan-kaikyo* (9 mi/14.5 km wide), separates Onekotan and KHARIMKOTAN islands. Straits first reached by Europeans and named in numerical order in the early 18th century by Russian Cossack explorers from KAMCHATKA.

Kuril Islands (koo-REEL) or **Kuriles**, Japanese *Chishima-retto* (chee-SHEE-mah–RET-to), Russian *Kuril'skiye Ostrova*, island chain, (□ 6,020 sq mi/15,652 sq km), SAKHALIN oblast, extreme E SIBERIA, RUSSIAN FAR EAST; 45°10′N 147°51′E. They stretch approximately 775 mi/1,247 km between S KAMCHATKA PENINSULA and NE HOKKAIDO (Japan), and separate the Sea of OKHOTSK from the PACIFIC Ocean. There are 30 large and numerous small islands; ITURUP is the largest. Atlasova volcano (7,674 ft/2,339 m) on Atlasova Island is the highest point of the chain. The islands are mainly of volcanic origin. Active volcanoes are present and earthquakes are frequent. Mostly subarctic flora and climate (low temperature; also, high humidity, and persistent fog). There are, however, communities engaged in sulfur mining, hunting, and fishing. Claimed by both Russians and Japanese in the 18th century. In 1875, Japan gave up Sakhalin in return for Russian withdrawal from the Kuriles, and the Japanese held the islands until the end of World War II. The Yalta Conference ceded the islands to the USSR, and Soviet forces occupied the chain in September 1945. Japan challenged the Soviet right to the Kuriles and demanded the return of the four southernmost islands, which it calls its "Northern Territories." The failure to resolve the impasse remained a major stumbling block in Russo-Japanese relations in the second half of the 20th century. The disputed islands offer deepwater, ice-free natural harbors and space for military bases, as well as rich fisheries in the surrounding waters. Strictly speaking, of the four

southernmost islands, only Iturup and Kunashiri are actually considered part of the Kuril group; subtropical Shikotan and the Habomai Island group are geographically distinct. Population is a mixture of Russians, Japanese, and indigenous groups. Sometimes spelled Kurile Islands.

Kurilo (koo-RE-lo), former village, SOFIA oblast, W BULGARIA, in the SOFIA BASIN, near the S end of the ISKUR River gorge, 9 mi/15 km N of SOFIA; 42°46′N 23°20′E. Now constituent of NOVI ISKUR (since 1974), with Gnilayne and Alexander Voikov.

Kurilovka (koo-REE-luhf-kah), village, N SARATOV oblast, S central European Russia, on road and near railroad, 46 mi/74 km N of SARATOV; 52°09′N 46°53′E. Elevation 141 ft/42 m. In agricultural area (flax, wheat, sunflowers; livestock raising).

Kurilovka, UKRAINE: see KURYLIVKA.

Kuril'sk (koo-REELSK), city (2006 population 2,170), SE SAKHALIN oblast, RUSSIAN FAR EAST, on the W coast of ITURUP ISLAND, S KURIL ISLANDS, on the OKHOTSK Sea, on the coastal highway; 45°14′N 147°53′E. Fish-processing center; whale oil factory. Settled in the 18th century by Russian explorers. Under Japanese rule (until 1945), called Shana.

Kuril'skoye Ozero, RUSSIA: see KURILE LAKE.

Kurim (KU-rzhim), Czech *Kuřim*, German *Gurein*, town, JIHOMORAVSKY province, S MORAVIA, CZECH REPUBLIC, on railroad, and 8 mi/12.9 km NNW of BRNO; 49°18′N 16°32′E. Manufacturing (machinery, machine tools, building materials). Has a 13th century church, and a 16th century Renaissance castle.

Kurimoto (koo-REE-mo-to), town, Katori county, CHIBA prefecture, E central HONSHU, E central JAPAN, 19 mi/30 km N of CHIBA; 35°48′N 140°30′E. Sweet potatoes.

Kurinchi, Mount, INDONESIA: see KERINCI, MOUNT.

Kuringen (KOO-reen-guhn), village in commune of HASSELT, Hasselt district, LIMBURG province, NE BELGIUM, 1 mi/1.6 km WNW of HASSELT; 50°57′N 05°18′E.

Ku-Ring-Gai (koo-RING-gei), N suburb of SYDNEY, E NEW SOUTH WALES, AUSTRALIA. Has domestic gardens, scenic national park, KU-RING-GAI CHASE NATIONAL PARK (□ 55 sq mi/142 sq km, or 35,000 acres/14,165 ha).

Kuring-gai Chase, AUSTRALIA: see KU-RING-GAI CHASE NATIONAL PARK.

Ku-Ring-Gai Chase National Park (koo-RING-gei CHAIS) (□ 57 sq mi/148.2 sq km), E NEW SOUTH WALES, SE AUSTRALIA, 20 mi/32 km N of SYDNEY, at edge of metropolitan area; 12 mi/19 km long, 9 mi/14 km wide; 33°38′S 151°12′E. S side of HAWKESBURY RIVER estuary near PACIFIC OCEAN. Open eucalypt forests, woodlands, scrub, heath. Wallabies, possums, echidnas; sulphur-crested cockatoos, rainbow lorikeets, kookaburras, lyrebirds, currawongs. Aboriginal rock engravings of Guringai people. Visitor center, kiosks. Camping, picnicking, hiking, swimming, fishing. Ferry service from PALM BEACH across Pitt Water to The Basin Campground. Established 1894. Also spelled Kuring-gai Chase.

Kurinjippadi (kuh-rin-ji-PAH-dee), town, SOUTH ARCOT VALLALUR district, TAMIL NADU state, S INDIA, 16 mi/26 km SW of CUDDALORE; 11°34′N 79°36′E. Rice milling, cotton and silk weaving. Cashew plantations nearby. Also called Kuinjipad.

Kurino (koo-REE-no), town, Aira county, KAGOSHIMA prefecture, S KYUSHU, SW JAPAN, 25 mi/40 km N of KAGOSHIMA; 31°56′N 130°43′E.

Kur'inskoye, RUSSIA: see KUR'YA.

Kurinwás River (koo-reen-WAHS), c.80 mi/129 km long, E NICARAGUA; rises in E outlier of HUAPI MOUNTAINS; flows E to PEARL LAGOON off CARIBBEAN SEA.

Kurisawa (koo-ree-SAH-wah), town, Sorachi district, Hokkaido prefecture, N JAPAN, 22 mi/35 km E of SAPPORO; 43°07′N 141°44′E.

Kurische Aa, LATVIA: see LIELUPE RIVER.

Kuriyama (koo-REE-yah-mah), town, Sorachi district, Hokkaido prefecture, N JAPAN, 22 mi/35 km E of SAPPORO; 43°03′N 141°47′E.

Kuriyama (koo-REE-yah-mah), village, Shioya county, TOCHIGI prefecture, central HONSHU, N central JAPAN, 25 mi/40 km N of UTSUNOMIYA; 36°54′N 139°32′E. Hot springs, Kinu numa marsh (alpine plants) nearby.

Kurk, TURKEY: see SIVRICE.

Kurkent (koor-KYENT), village (2005 population 3,300), SE DAGESTAN REPUBLIC, in the foothills of the E Greater CAUCASUS Mountains, extreme SE European Russia, on road, 69 mi/111 km SSE of MAKHACHKALA, and 3 mi/5 km N of KASUMKENT; 41°43′N 48°06′E. Elevation 1,912 ft/582 m. Indigenous handicrafts.

Kurkino (KOOR-kee-nuh), town (2006 population 5,890), SE TULA oblast, central European Russia, on road and railroad, 35 mi/56 km SE of BOGORODITSK; 53°26′N 38°40′E. Elevation 708 ft/215 m. In agricultural area; bakery, creamery.

Kurkund, ESTONIA: see KILINGI-NÕMME.

Kurla (koor-LAH), town, MUMBAI (BOMBAY) SUBURBAN district, MAHARASHTRA state, W central INDIA, on SALSETTE ISLAND, 10 mi/16 km NNE of MUMBAI city center; 19°05′N 72°53′E. Railroad junction. Cotton milling, dyeing, manufacturing (matches, ink, rope, binding tape); motor vehicle assembly factory, glass- and metalworks. Old KURLA (N) has tanneries. First electric railroad in India built (1925) here from Mumbai (then Bombay).

Kurland, LATVIA: see COURLAND.

Kurlovo, RUSSIA: see KURLOVSKIY.

Kurlovskiy (koor-LOF-skeeye), town (2006 population 7,210), S VLADIMIR oblast, central European Russia, on road and railroad, 12 mi/19 km SSW of GUS'-KHRUSTALNYY; 55°27′N 40°36′E. Elevation 465 ft/141 m. Glassworking; woodworking (furniture). Also known as Kurlovo.

Kurlukovo (koor-LOO-ko-vo), village, LOVECH oblast, LUKOVIT obshtina, BULGARIA; 43°10′N 24°04′E.

Kurmanayevka (koor-mah-NAH-eef-kah), village (2006 population 4,545), W ORENBURG oblast, SE European Russia, on the BUZULUK RIVER, on road, 20 mi/32 km SSW of BUZULUK; 52°31′N 52°04′E. Elevation 252 ft/76 m. In agricultural area. Oil shale and phosphorite deposits nearby. Also known as Kurmanayevo.

Kurmanayevo, RUSSIA: see KURMANAYEVKA.

Kurman-Kemelchi, UKRAINE: see KRASNOGVARDEYSKOYE.

Kurmysh (koor-MISH), village, E NIZHEGOROD oblast, central European Russia, on the SURA River, on road, 42 mi/68 km ESE of LYSKOVO; 55°50′N 46°03′E. Elevation 272 ft/82 m. Agricultural products. Founded at the end of the 14th century as a Muscovite frontier settlement.

Kurna, El, EGYPT: see QURNA.

Kurnalovo (koor-NAHL-o-vo), village, SOFIA oblast, PETRICH obshtina, BULGARIA; 41°28′N 23°15′E.

Kurnik, POLAND: see KORNIK.

Kurnool (kuhr-NOOL), district (□ 6,818 sq mi/17,726.8 sq km), ANDHRA PRADESH state, S INDIA, ⊙ KURNOOL. On DECCAN PLATEAU; crossed (E) by EASTERN GHATS. Agriculture includes oilseeds (extensive peanut growing), millet, cotton, rice, chili. Timber, bamboo, dyewood (red sandalwood), and fibers in dispersed forests. Barite, ocher, and slate mines. Main towns are Kurnool, NANDYAL. Original district enlarged 1948 by incorporation of former MADRAS state of BANGANAPALLE.

Kurnool (kuhr-NOOL), city (2001 population 342,973), ⊙ KURNOOL district, ANDHRA PRADESH state, S INDIA, at the confluence of the TUNGABHADRA and Hindri rivers; 15°50′N 78°03′E. Market for grain, hides, and cotton. Ruins of a fort built by the Hindu kings of VIJAYANAGAR in the 16th century. Overrun by Muslims in 1565 and ceded to the BRITISH by the nizam of HYDERABAD in 1800. Was Andhra Pradesh capital 1953–1956, when it was superseded by Hyderabad. Hindu pilgrimage center surrounded by hill resorts.

Kurobane (koo-RO-bah-ne), town, Nasu county, TOCHIGI prefecture, central HONSHU, N central JAPAN, 25 mi/40 km N of UTSUNOMIYA; 36°51′N 140°07′E.

Kurobe (koo-RO-be), city, TOYAMA prefecture, central HONSHU, central JAPAN, 19 mi/30 km N of TOYAMA; 36°52′N 137°27′E. Zippers; glass and aluminum sliding doors. Kuroyon Dam is nearby.

Kurobe River (KOO-ro-be), 60 mi/97 km long, TOYAMA prefecture, central HONSHU, central JAPAN; rises in mountains NW of peak Yari-ga-take, flows N and NNW to TOYAMA BAY 8 mi/12.9 km WSW of Tomari; forms delta mouth. Several hydroelectric plants. Known for scenic gorge.

Kurodashio (koo-ro-DAH-shyo), town, Taka county, HYOGO prefecture, S HONSHU, W central JAPAN, 25 mi/40 km N of KOBE; 35°01′N 134°59′E.

Kurohone (koo-RO-ho-ne), village, Seta county, GUMMA prefecture, central HONSHU, N central JAPAN, 19 mi/30 km N of MAEBASHI; 36°29′N 139°17′E. Pigs.

Kuroishi (koo-RO-ee-shee), city, Aomori prefecture, N HONSHU, N JAPAN, 16 mi/25 km S of AOMORI; 40°38′N 140°35′E. Apples and apple juice. Origin of traditional *Kuroishi yosare* dance.

Kuroiso (koo-RO-ee-so), city, TOCHIGI prefecture, central HONSHU, N central JAPAN, 28 mi/45 km N of UTSUNOMIYA; 36°57′N 140°02′E. Tires. Dairying.

Kurokawa (koo-RO-kah-wah), village, North Kanbara county, NIIGATA prefecture, central HONSHU, N central JAPAN, 25 mi/40 km N of NIIGATA; 38°04′N 139°26′E. Buckwheat. Skiing area.

Kuroki (KOO-ro-kee), town, Yame county, FUKUOKA prefecture, N central KYUSHU, SW JAPAN, 31 mi/50 km S of FUKUOKA; 33°12′N 130°40′E. Mandarin oranges, grapes; tea. Site of feudal castle. Also Kurogi.

Kuromatsunai (KOO-ro-mahts-NAH-ee), town, Shiribeshi district, Hokkaido prefecture, N JAPAN, 59 mi/95 km W of SAPPORO; 42°39′N 140°18′E. Potatoes. Ayu. N limit forest of natural beech trees.

Kurort Arshan, RUSSIA: see ARSHAN.

Kurort-Darasun (koo-RORT–dah-rah-SOON), resort town (2005 population 3,230), S CHITA oblast, S SIBERIA, RUSSIA, on road, 51 mi/82 km S of CHITA; 51°12′N 113°42′E. Elevation 2,549 ft/776 m. Carbonated mineral water. Hot springs health spas.

Kurortne (koo-ROR-tne), (Russian *Kurortnoye*), town (2004 population 3,800), SE Republic of CRIMEA, UKRAINE, on the BLACK SEA coast, 2 mi/3.2 km S of SHCHEBETOVKA; 44°54′N 35°09′E. A beach resort.

Kurortnoye, UKRAINE: see KURORTNE.

Kurort Oberwiesenthal, GERMANY: see OBERWIESENTHAL.

Kurosaki (koo-RO-sah-kee), town, West Kanbara county, NIIGATA prefecture, central HONSHU, N central JAPAN, 3.7 mi/6 km S of NIIGATA; 37°50′N 139°01′E. Flowers, tea.

Kurose (koo-RO-se), town, Kamo county, HIROSHIMA prefecture, SW HONSHU, W JAPAN, 12 mi/20 km N of HIROSHIMA; 34°19′N 132°40′E.

Kuroshio, JAPAN: see JAPAN CURRENT.

Kurotaki (koo-ro-TAH-kee), village, Yoshino district, NARA prefecture, S HONSHU, W central JAPAN, 27 mi/43 km S of NARA; 34°18′N 135°51′E. Lumber, logs.

Kurovskoy (koo-ruhfs-KO-yee), village (2005 population 3,635), W central KALUGA oblast, central European RUSSIA, in the OKA River basin, on highway and near railroad, 11 mi/18 km W of KALUGA; 54°32′N 36°01′E. Elevation 528 ft/160 m. Marble quarrying; clay processing.

Kurovskoye (koo-ruhf-SKO-ye), city (2006 population 19,065), E MOSCOW oblast, central European RUSSIA, on road, 57 mi/92 km ESE of MOSCOW, and 14 mi/23 km S of OREKHOVO-ZUYEVO, to which it is administratively subordinate; 55°34′N 38°54′E. Elevation 390

ft/118 m. Railroad junction. Agricultural chemicals. Made city in 1952.

Kurow (KOO-ou), township, WAITAKI district, E SOUTH ISLAND, NEW ZEALAND, c.43 mi/70 km NE of OAMARU, and on WAITAKI RIVER. Wool, grain. Site of earliest Waitaki hydroelectric power station.

Kurrajong (KOOR-uh-jahng), town, E NEW SOUTH WALES, SE AUSTRALIA, 40 mi/64 km NW of SYDNEY, in BLUE MOUNTAINS foothills; 33°33′S 150°40′E. Railroad terminus. Apple orchards, citrus groves, vineyards.

Kurram (koo-RUM), centrally administered tribal region (□ 738 sq mi/1,918.8 sq km), NORTH-WEST FRONTIER PROVINCE, NW PAKISTAN, W of KOHAT district and S of KHYBER tribal region; ⊙ PARACHINAR. Bordered W by AFGHANISTAN (separated by Durand Line), N by SAFED KOH range; drained by KURRAM RIVER. Agriculture (wheat, rice, corn, barley); fruit growing (apples, pears, peaches); hand-loom weaving. Administered as political agency with tribal law and total local control.

Kurram River, c.200 mi/322 km long, in E AFGHANISTAN and S North-West Frontier province, W PAKISTAN; rises in W SAFED KOH range, SSE of KABUL (Afghanistan); flows SE, past Thal and Bannu, to Indus River 10 mi/16 km. W of Mianwali. Drains valley of Kurram agency and plain of Bannu district; waters wheat, corn, and rice fields.

Kurri Kurri (KUHR-ee kuh-ree), town (2001 population 12,555), E NEW SOUTH WALES, AUSTRALIA, 16 mi/26 km WNW of NEWCASTLE; 32°49′S 151°29′E. Coal-mining center; aluminum smelter. Residential area for CESSNOCK. Laid out 1902. Increasingly known as Kurri Kurri-Weston.

Kurri Kurri-Weston, AUSTRALIA: see KURRI KURRI.

Kur River (KOOR), approximately 250 mi/402 km long, in KHABAROVSK TERRITORY, extreme E SIBERIA, RUSSIAN FAR EAST; rises in the E outlier of the BUREYA Range; flows S, past NOVOKUROVKA, and E, past Volochayevka II, to the AMUR River just below KHABAROVSK. Timber floating.

Kursavka (koor-SAHF-kah), village (2006 population 11,115), W STAVROPOL TERRITORY, N CAUCASUS, S European Russia, on the STAVROPOL PLATEAU, on road and railroad, 42 mi/68 km S of STAVROPOL, and 28 mi/45 km W of MINERAL'NYYE VODY; 44°27′N 42°30′E. Elevation 1,322 ft/402 m. Highway junction; local transshipment point. Sometimes also pronounced as Kurshavka.

Kuršenai (KOOR-shai-nei), Russian *Kurshany*, city (2001 population 14,197), N central LITHUANIA, on VENTA RIVER, and 15 mi/24 km WNW of SIAULIAI; 56°00′N 22°56′E. Sugar refinery, flour mill. In Russian KOVNO government until 1920. Also spelled Kurshenay or Kurshenai.

Kurseong, INDIA: see KARSIYANG.

Kurshab, river, c.53 mi/85 km long, S OSH region, SW KYRGYZSTAN; rises in ALAY RANGE; flows N to KARA DARYA (river) near UZGEN, at the Anjiyan Reservoir.

Kurshany, LITHUANIA: see KURŠENAI.

Kurshavka, RUSSIA: see KURSAVKA.

Kurshenai, LITHUANIA: see KURŠENAI.

Kuršių Marios, LITHUANIA: see KURSKY ZALIV.

Kuršių Nerija National Park (KOOR-shoo NE-ree-yah), LITHUANIA, on COURLAND SPIT, 0.2 mi/0.4 km–2.5 mi/4 km wide. Resort area with sandy beaches and high sand dunes and pine forests. The spit was formed by SW winds and longshore currents.

Kursk (KOORSK), oblast (□ 11,617 sq mi/30,204.2 sq km; 2005 population 1,198,100) in SW European Russia, on the Ukrainian border, approximately 310 mi/500 km S of MOSCOW; ⊙ KURSK. In CENTRAL RUSSIAN UPLAND; drained by upper Seim, Psel, Vorskla, and Oskol rivers; black-earth steppe; moderate, humid climate. Once covered with forest, its rich soil has made this an important agricultural region, with intensive farming producing sugar beets (W) and

winter wheat (E). Also hemp (NW), potatoes (extreme NE), vegetables, orchard products (near cities), legumes. Raising of hogs (based on sugar-refining by-products) and poultry are important. Industry (flour milling, sugar refining, distilling, tanning, fruit canning, meatpacking) based on agriculture; also machine building and chemical enterprises. Mineral industries. Kursk Magnetic Anomaly, an extensive iron ore region, between SHCHIGRY (NW) and VALUYKI (SE), was developed after World War II with large open-pit mine at Mikhaylovka. Main industrial centers are Kursk, ZHELEZNOGORSK. Large nuclear power plant at Kursk. Site of scientific steppe nature preserve. The largest tank battle in history was fought here in 1943, toward the end of World War II. Consists of 10 cities, 23 towns, and 2,807 villages.

Kursk (KOORSK), city (2005 population 408,475), ⊙ KURSK oblast, central black-earth (Russian *chernozëm*) region, SW European Russia, at the confluence of the Tuskor and SEYM rivers (DESNA River basin), 333 mi/536 km S of MOSCOW; 51°44′N 36°11′E. Elevation 603 ft/183 m. An important railroad junction and highway hub, it has machinery, metalworking, chemical, and synthetic fiber plants, and light industry (including food). A large iron ore deposit, the Kursk Magnetic Anomaly, is S of the city. First noted in 1095, Kursk was destroyed by the Mongols in 1240 and was rebuilt as a Muscovite fortress in 1586. Site of fierce and protracted battles between the Soviet and German forces during World War II, between August and October 1943.

Kurskaya (KOOR-skah-yah), village (2006 population 12,115), SE STAVROPOL TERRITORY, N CAUCASUS, SE European Russia, on road, 51 mi/82 km N of VLADIKAVKAZ (NORTH OSSETIAN REPUBLIC), 47 mi/76 km E of NAL'CHIK (KABARDINO-BALKAR REPUBLIC), and 34 mi/37 km SSE of ARZGIR; 44°03′N 44°27′E. Elevation 639 ft/194 m. In agricultural area (sugar beets); food processing.

Kursky Zaliv or **Courland Lagoon**, Lithuanian *Kuršių Marios*, lagoon, in LITHUANIA and NW European RUSSIA; 56 mi/90 km long and 28 mi/45 km wide. Separated from the BALTIC SEA by COURLAND SPIT, a sandspit c.60 mi/100 km long and 1 mi/1.6 km to 2 mi/3.2 km wide, which leaves only a narrow opening at the Klaipeda Channel in the N. The Nemunas River empties into the lagoon.

Kurslak, GERMANY: see VIERLANDE.

Kursu, FINLAND: see SALLA.

Kuršumlija (koor-SHOOM-lee-yah), village (2002 population 21,608), S SERBIA, on TOPLICA RIVER, on railroad, and 34 mi/55 km WSW of NIS; 43°08′N 21°17′E. Kursumlijska Banja, health resort, is 6 mi/9.7 km S. Also spelled Kurshumliya.

Kursunlu, Turkish=*Kurşunlu*, village, N central TURKEY, on ANKARA-ZONGULDAK railroad, near DEVREZ RIVER, and 24 mi/39 km NW of CANKIRI; 40°51′N 33°16′E. Grain; mohair goats.

Kurtalan, village, SE TURKEY, 17 mi/27 km NW of SIIRT; 37°58′N 41°36′E. Railroad terminus from DIYARBAKIR. Grain. Formerly Ayinkasir.

Kurtamysh (koor-tah-MISH), city (2005 population 17,865), SW KURGAN oblast, SW SIBERIA, RUSSIA, on the Kurtamysh River (TOBOL River basin), on road, 56 mi/90 km SW of KURGAN; 54°54′N 64°26′E. Elevation 410 ft/124 m. Light industries; sawmills. City since 1956.

Kürten (KUR-ten), town, North Rhine-Westphalia, W GERMANY, 11 mi/18 km E of LEVERKUSEN; 51°03′N 07°16′E. Agricultural center (cattle), and diverse small industries.

Kurtik Dag, TURKEY: see HAÇREŞ DAG.

Kurtistown, town (2000 population 1,157), E HAWAII island, HAWAII county, HAWAII, 8 mi/12.9 km S of HILO, 1 mi/1.6 km SE of KEAAU, 5 mi/8 km inland from E coast; 19°35′N 155°04′W. Fruit, macadamia nuts. Walakea Forest Reserve to NW. Kulani Honor Camp to W. Several residential and vacation home developments to S and E.

Kurtlak (koort-LAHK), river, left tributary of the CHIR RIVER (DON River basin), 93 mi/150 km long, basin area, 1,066 sq mi/2,760 sq km, S European Russia; rises in VOLGOGRAD oblast; flows SW through hilly steppe terrain into E ROSTOV oblast. Used for irrigation.

Kürtös, ROMANIA: see CURTICI.

Kurtovo Konare (KOOR-to-vo ko-NAH-re), village, PLOVDIV oblast, Rodopska obshtina, S central BULGARIA, on the VUCHA, and 13 mi/21 km WSW of PLOVDIV; 42°05′N 24°30′E. Tomato- and pepper-growing center; tobacco, rice.

Kuru (KOO-voo), village, HÄMEEN province, SW FINLAND, on NW shore of NÄSIJÄRVI (lake), 25 mi/40 km N of TAMPERE; 61°52′N 23°44′E. Elevation 330 ft/100 m. National park.

Kuru-cheshme, Bulgaria: see GORSKI IZVOR.

Kuruksai (kurh-ook-SEI), town, LENINOBOD region, TAJIKISTAN, W of KURAMA mountain range, near UZBEKISTAN border, 28 mi/45 km NW of KHUDJAND, and 43 mi/70 km S of TOSHKENT; 40°35′N 69°24′E. Also spelled Kuruksay.

Kurukshetra (kuh-ruhk-SHAI-trah), district (□ 470 sq mi/1,222 sq km), HARYANA state, N INDIA; ⊙ KURUKSHETRA.

Kurukshetra (kuh-ruhk-SHAI-trah), town, ⊙ KURUKSHETRA district, HARYANA state, N INDIA, 20 mi/32 km NNW of KARNAL; 30°00′N 76°45′E. Railroad junction. Important Hindu pilgrimage center; large sacred bathing tank just W attracts several thousands of bathers during solar eclipses. Wildfowl sanctuary. Nearby area, extending roughly SW to JIND district, was among earliest Aryan settlements in India (c.2000 B.C.E.–1500 B.C.E.) and is associated with legends of *Mahabharata* epic (here was one site of Kurukshetra War). The discourse of the sacred Hindu *Bhagvad-Gita* text was narrated here.

Kuruman, town, NORTHERN CAPE province, SOUTH AFRICA, in KURUMAN HILLS, on upper KURUMAN RIVER, and 120 mi/193 km NW of KIMBERLEY; 27°28′S 23°27′E. Elevation 4,271 ft/1,302 m. Livestock-raising, dairying center. Resort town. Founded 1801 as mission station. The starting point of the mission efforts in Africa. Livingstone's journeys also began here. Robert Moffat Church (1833). Nearby is noted spring, called "Eye of Kuruman," producing 4,300,000 gallons/20,000,000 liters per day of fresh water.

Kuruman Hills, NORTHERN CAPE province, SOUTH AFRICA, N extension of ASBESTOS MOUNTAINS. Extend 50 mi/80 km N-S from W of KURUMAN to DANIELSKUIL, rising to 6,084 ft/11,855 m on Gakarosa (27°53′S 28°03′E).

Kuruman River, intermittent stream, 250 mi/402 km long, NORTHERN CAPE and NORTH-WEST provinces, SOUTH AFRICA; rises in KURUMAN HILLS SE of KURUMAN; flows in a wide arc NW and W, past KURUMAN, Batlharo and Tsineng, parallel to S border of BOTSWANA, to then MOLOPO RIVER near Witdraai at 26°57′S 20°40′E.

Kurume (KOO-roo-me), city (2005 population 306,434), FUKUOKA prefecture, W KYUSHU, SW JAPAN, on the Chikugo Plain, 19 mi/30 km S of FUKUOKA; 33°18′N 130°30′E. Commercial and agricultural center, manufacturing rubber (tires, shoes) and cotton goods ("kasuri" splashed-pattern textiles) and lacquerware. Azaleas. Former castle town.

Kurumkan (koo-room-KAHN), village (2005 population 5,815), NW BURYAT REPUBLIC, S central SIBERIA, RUSSIA, on the BARGUZIN RIVER (Lake BAYKAL basin), near highway, 200 mi/322 km NNE of ULAN-UDE; 54°18′N 110°18′E. Elevation 1,633 ft/497 m. Dairy products; printing.

Kurumoch (koo-roo-MOCH), village (2006 population 6,430), central SAMARA oblast, E European Russia, near the VOLGA RIVER, on junction of major highway and local roads, and near railroad, 19 mi/31 km N of Samara; 53°30′N 50°02′E. Elevation 183 ft/55 m. In oil- and gas-processing area. Administrative and service facilities for the region's main airport (bearing the same name), 4 mi/6 km to the E.

Kurundwad, INDIA: see KURANDVAD.

Kurunegala, district (□ 1,785 sq mi/4,641 sq km; 2001 population 1,460,215), NORTH WESTERN PROVINCE, SRI LANKA; ⊙ KURUNEGALA; 07°45′N 80°15′E.

Kurunegala (KOO-ROO-nah-GUH-luh), town (2001 population 28,401), ⊙ KURUNEGALA district, NORTH WESTERN PROVINCE, W central SRI LANKA; 07°29′N 80°22′E. Road junction and administrative and commercial center of a coconut-, rice-, and rubber-plantation district. Overlooking the town is Elephant Hill, a stronghold in the 14th century, when Kurunegala was capital of a Sinhalese kingdom, then called HASTHIGIRIPURA (Sinhalese=city of the elephant rock).

Kurunjang, suburb 22 mi/35 km WNW of MELBOURNE, VICTORIA, SE AUSTRALIA, just N of MELTON. Arnold Creek runs through the area. Nondenominational college.

Kurush (koo-ROOSH), village (2005 population 7,995), S DAGESTAN REPUBLIC, E CAUCASUS, extreme SE European Russia, approximately 5 mi/8 km N of the RUSSIA-AZERBAIJAN border, on road, 27 mi/43 km SSW of KASUMKENT; 41°17′N 47°50′E. Elevation 7,168 ft/2,184 m. Rest and convenience station for mountain climbers. Local handicrafts (metal smithing, rug weaving, ceramics).

Kuruwita, town, SABARAGAMUWA PROVINCE, SW central SRI LANKA, 7 mi/11.3 km NNW of RATNAPURA; 06°47′N 80°22′E. Precious-stone mining (notably star sapphires); rice, vegetables, rubber.

Kurvelesh Upland, hilly area, SW ALBANIA, SW of VIJOSË RIVER; 40°11′N 19°56′E.

Kurwai (KUHR-wei) or **Korwai**, town (2001 population 13,737), VIDISHA district, central MADHYA PRADESH state, central INDIA, on BETWA RIVER, and 45 mi/72 km NNE of Bhilsa; 24°08′N 78°03′E. Local agriculture market (wheat, millet). Was capital of former princely state of Korwai of Central India agency; in 1948, state merged with MADHYA BHARAT and later with Madhya Pradesh state.

Kurwai, INDIA: see KORWAI.

Kur'ya (KOOR-eeyah), village, S ALTAI TERRITORY, S central SIBERIA, RUSSIA, on crossroads, 45 mi/72 km E of RUBTSOVSK; 51°36′N 82°19′E. Elevation 777 ft/236 m. Dairy products. Also known as Kur'inskoye.

Kurye (GOO-RYE), county (□ 170 sq mi/442 sq km), NE SOUTH CHOLLA province, SOUTH KOREA. Bordered E by SOUTH KYONGSANG province, S by KWANGYANG and SUNGJU provinces, and W by KOKSONG province. Rugged mountain area (CHIRI Mountain in NE); SOMJIN RIVER forms Kuryebunji basin in center. Agriculture (rice, barley, cotton, medical herbs, mushrooms) in narrow fields and hills. CHIRI MOUNTAIN NATIONAL PARK; Hwaom and Yonkok Buddhist temples; sweet-smelt fishing in Somjin River.

Kurylivka (koo-RI-leef-kah) (Russian *Kurilovka*), town, central DNIPROPETROVS'K oblast, UKRAINE, on left bank of the DNIEPER River, 20 mi/32 km W of DNIPROPETROVS'K, and across the river from DNIPRODZERZHYNS'K; 48°34′N 34°35′E. Elevation 177 ft/53 m. Fish farm; forestry.

Kurzeme (KOOR-ze-mai), W LATVIA. Originally settled by the Livs, a Finno-Ugric tribe in the 10th century B.C.E., it is home to Latvia's largest ports, VENTSPILS and LIEPAJA. The VENTA RIVER drains a traditional wine-growing area and includes one of the widest (361 ft/110 m) waterfalls in Europe.

Kus, EGYPT: see QUS.

Kusa (koo-SAH), city (2005 population 19,660), W CHELYABINSK oblast, extreme E European Russia, in the S URAL Mountains, at the confluence of Kusa and Ai rivers (UFA RIVER basin), on highway and railroad spur, 150 mi/241 km W of CHELYABINSK, and 13 mi/21 km NW of ZLATOUST; 55°20′N 59°26′E. Elevation 1,266 ft/385 m. Metalworking; boilers; household appliances; logging and lumbering. Founded in 1787 as an iron-mining and smelting settlement. Became city in 1943. Formerly called Kusinskiy Zavod.

Kusada (koo-SEI-dah), town, KATSINA state, N NIGERIA, 50 mi/80 km NW of KANO; 12°28′N 07°59′E. Peanuts, cotton; cattle, skins.

Kusadak (KOO-sah-dahk), village, central SERBIA, 8 mi/12.9 km WNW of SMEDEREVSKA PALANKA.

Kusadasi, Turkish=Kuşadasi town (2000 population 47,611), W TURKEY, port on Gulf of KUSADASI, 38 mi/61 km W of IZMIR; 37°50′N 27°16′E. Olives, tobacco, figs. Site of ancient EPHESUS is nearby. Formerly Scalanuova, Scalanova, or Seleuk.

Kusadasi, Gulf of, inlet of AEGEAN SEA in W TURKEY, S of IZMIR, 32 mi/51 km W of AYDIN; 25 mi/40 km wide, 19 mi/31 km long. Town of KUSADASI on E shore. Island of SAMOS on its S side. Formerly Gulf of Scalanuova. Also Gulf of Kudasa.

Kusaie, CAROLINE ISLANDS, Federated States of MICRONESIA: see KOSRAE.

Kusanagara, INDIA: see KASIA.

Kusary (KOO-SAH-ree), city and administrative center of Kusary region, NE AZERBAIJAN, on N slope of the Greater CAUCASUS, 7 mi/11 km NW of Kuba. Orchards; manufacturing (canning; rugs, asphalt, dairy products).

Kusatsu (koo-SAHTS), city, SHIGA prefecture, S HONSHU, central JAPAN, near SE shore of LAKE BIWA, 5.6 mi/9 km E of OTSU; 35°00′N 135°57′E. Railroad junction. Appliances. Gourds. Bamboo work.

Kusatsu (KOO-sahts), town, Agatsuma county, GUMMA prefecture, central HONSHU, N central JAPAN, near SHIRANE volcano, 37 mi/60 km N of MAEBASHI; 36°37′N 138°35′E. As early as the 12th century its hot sulfur springs were known for their medicinal properties. Skiing area.

Kuse (KOO-se), town, Maniwa county, OKAYAMA prefecture, SW HONSHU, W JAPAN, 31 mi/50 km N of OKAYAMA; 35°04′N 133°05′E.

Kuseie, CAROLINE ISLANDS, Federated State of MICRONESIA: see KOSRAE.

Kuseifa (koo-SAI-fah), Beduin township, ISRAEL, 17 mi/27 km E of BEERSHEBA, in N NEGEV desert. Remains of churches and other signs of settlement (including mosaic floors) from Roman and Byzantine periods nearby.

Kusel (KOO-sel), town, RHINELAND-PALATINATE, W GERMANY, 17 mi/27 km NW of KAISERSLAUTERN; 49°33′N 07°24′E. Manufacturing of printing machines; brewing; tourism. Excavations of a Roman villa began in 1988.

Kush, SUDAN: see CUSH.

Kushagi (koo-shah-GEE), village, W NOVOSIBIRSK oblast, SW SIBERIA, RUSSIA, on road, 6 mi/10 km N of UST'-TARKA; 55°40′N 75°45′E. Elevation 367 ft/111 m. Livestock raising (meat and dairy). Also known as Kushagovskoye.

Kushagovskoye, RUSSIA: see KUSHAGI.

Kushalgarh (kuh-SHUHL-guhr), town, BANSWARA district, S RAJASTHAN state, NW INDIA, 23 mi/37 km S of Banswara; 23°10′N 74°27′E. Hand-loom weaving; maize, millet, rice. Was capital of former petty state of Kushalgarh in RAJPUTANA STATES, a dependency of Banswara state. In 1948, state merged with union of Rajasthan.

Kushalino (koo-SHAH-lee-nuh), village, S central TVER oblast, W European Russia, near highway junction, 18 mi/29 km NNE of TVER; 57°07′N 36°05′E. Elevation 485 ft/147 m. Flax. Also called Bol'shoye Kushalino.

Kushchevskaya (koo-SHCHYEF-skah-yah), city (2005 population 30,060), N KRASNODAR TERRITORY, S European Russia, on the YEYA RIVER, on road and railroad, 45 mi/72 km S of ROSTOV-NA-DONU, and 26 mi/42 km E of STAROMINSKAYA; 46°33′N 39°38′E. Railroad junction; railroad depots. City status since 1996.

Kushchi (KOOSH-chee), village, W AZERBAIJAN, 4 mi/ 6 km N of DASHKESAN. Terminus of railroad spur (from Alabashly station on main BAKY-Tiflis railroad) for Dashkesan magnetite mines.

Kushh, El (KOSH, el), village, SOHAG province, central Upper EGYPT, 5 mi/8 km E of El BALYANA; 26°14′N 32°05′E. Cotton, cereals, dates, sugar.

Kushibiki (koo-SHEE-bee-kee), town, East Tagawa county, YAMAGATA prefecture, N HONSHU, NE JAPAN, 40 mi/65 km N of YAMAGATA city; 38°40′N 139°51′E. Persimmons, pickles.

Kushigata (koo-SHEE-gah-tah), town, Nakakoma county, YAMANASHI prefecture, central HONSHU, central JAPAN, 6 mi/10 km S of KOFU; 35°36′N 138°28′E. Alpine plants.

Kushihara (koo-SHEE-hah-rah), village, Ena county, GIFU prefecture, central HONSHU, central JAPAN, 40 mi/65 km S of GIFU; 35°14′N 137°23′E. Pigs; *konnyaku* (paste made from devil's tongue). Traditional drums.

Kushikino (KOO-shee-kee-NO), city, KAGOSHIMA prefecture, SW KYUSHU, SW JAPAN, on NW Satsuma Peninsula, 19 mi/30 km N of KAGOSHIMA, on EAST CHINA SEA; 31°42′N 130°16′E. Commercial center; fish processing (fish paste; sardines, tuna). Gold and silver mining. Fukiage sand dune nearby.

Kushima (KOO-shee-mah), city, MIYAZAKI prefecture, SE KYUSHU, SW JAPAN, 31 mi/50 km S of MIYAZAKI; 31°27′N 131°13′E. Sweet potatoes. Includes Cape TOI (wild horses, cycads) and Ko islet.

Kushimoto (koo-SHEE-mo-to), town, West Muro county, WAKAYAMA prefecture, S HONSHU, W central JAPAN, on KUMANO SEA, on S KII PENINSULA, 64 mi/ 104 km S of WAKAYAMA, near SHIO POINT; 33°28′N 135°47′E. Whaling port. Harbor is sheltered by O-SHIMA island (E).

Kushira (KOO-shee-rah), town, Kimotsuki county, KAGOSHIMA prefecture, S KYUSHU, SW JAPAN, on central OSUMI PENINSULA, 28 mi/45 km S of KA-GOSHIMA; 31°23′N 130°57′E. Pigs.

Kushiro (KOO-shee-ro), city (2005 population 190,478), SE Hokkaido prefecture, N JAPAN, on the PACIFIC OCEAN, on the Kushiro River, 152 mi/245 km E of SAPPORO; 42°58′N 144°23′E. Main port of E HOKKAIDO and the island's only ice-free trading port. Manufacturing includes paper, kombu, salmon, and coal; also marine-product processing. Major base for fishermen. Center of the huge Kushiro coalfield, which extends far out to sea; mining is carried out in the sea (though declining). Kushiro marshland, habitat of the *tancho* crane, is nearby.

Kushiro (KOO-shee-ro), town, Kushiro district, Hokkaido prefecture, N JAPAN, immediately E of KUSHIRO city, and 140 mi/225 km E of SAPPORO; 42°59′N 144°28′E. Daikon (N limit).

Kushk (koo-SHEK), town, HERAT province, NW AF-GHANISTAN, on KUSHK RIVER, on highway to HERAT, 40 mi/64 km NE of Herat, and 30 mi/48 km SE of KUSHKA (TURKMENISTAN); S railroad terminus; 34°57′N 62°15′E. Elev. 3,800 ft/1,158 m. Ships cotton and dried fruit. Population is largely Jamshidi. Also spelled Koshk.

Kushka (KOOSH-kah), southernmost city, in the former USSR, in S MARY weloyat, S TURKMENISTAN, on KUSHKA RIVER (left affluent of MURGAB RIVER), on AFGHANISTAN border, and 160 mi/257 km. S of MARY (connected by railroad); 35°38′N 62°19′E. Wheat; pistachio woods. Founded in late 19th century as Russian fort; reached (1898) by railroad. Also Gushgy.

Kushka River, AFGHANISTAN and TURKMENISTAN: see KUSHK RIVER.

Kushk-i-Nakhud (KOOSHK–EE–nah-KHOOD), town, KANDAHAR province, S AFGHANISTAN, on Kushk-i-Nakhud River (right tributary of the ARGHANDAB RIVER), 40 mi/64 km W of KANDAHAR; 31°37′N 65°05′E. Oasis center on highway to GIRISHK; irrigated agriculture Maiwand (NE) was scene of severe defeat (1880) of British by Afghan forces of Akbar Khan.

Kushk River (koo-SHEK), Russian *Kushka*, 150 mi/241 km long, NW AFGHANISTAN and SE TURKMENISTAN; rises in Paropamisus Mountains 55 mi/89 km ENE of HERAT (Afghanistan); flows NW, past KUSHK, then N, past Kushka (Turkmenistan), and NNE, along Kushka railroad branch, to MURGAB RIVER at Tash-kepristroi (dam); 36°03′N 62°47′E. Irrigation along middle course.

Kushmurun (koosh-muh-ROON), town, E KOSTANAI region, KAZAKHSTAN, 60 mi/97 km SE of KOSTANAI, and on SOUTH SIBERIAN railroad, at S end of LAKE KUSHMURUN; 52°30′N 64°37′E. Bauxite and lignite deposits nearby. Also spelled Kusmuryn, Qusmuryn.

Kushmurun, Lake (kush-muh-RUN) (□ 400 sq mi/ 1,040 sq km), E KOSTANAI region, KAZAKHSTAN, on OBAGAN RIVER, N of KUSHMURUN (town). Formerly Lake Obagan. Also spelled Kusmuryn, Qusmuryn.

Kushnarenkovo (koo-shnah-RYEN-kuh-vuh), village (2005 population 10,660), central BASHKORTOSTAN Republic, E European RUSSIA, on the BELAYA RIVER (landing), near highway, 38 mi/61 km NW of UFA; 55°06′N 55°21′E. Elevation 265 ft/80 m. Originally called Topornino; renamed in the 1930s. Sometimes pronounced Kushnarënkovo (koo-shnah-RYON-kuh-vuh).

Kushnitsa, Greece: see PANGAION.

Kushtagi (KUSH-tah-gee), town, tahsil headquarters, RAICHUR district, KARNATAKA state, S INDIA, 28 mi/45 km N of KOPPAL; 15°46′N 76°12′E. Millet, oilseeds.

Kushtia (koosh-tee-ah), district (□ 1,300 sq mi/3,380 sq km), W EAST BENGAL, BANGLADESH; ⊙ KUSHTIA. In GANGA DELTA; bounded N by PADMA RIVER, W by INDIA (WEST BENGAL state); drained by distributaries of the delta. Alluvial plain (rice, jute, linseed, sugarcane, wheat, tobacco, chili, turmeric); scattered swamps (bamboo, moringa, and areca palm groves). Large sugar-processing factory near CHUADANGA; sugar milling at KUSHTIA; food canning, liquor distilling, cotton-cloth weaving; manufacturing (brick, hosiery). Originally part of Sen kingdom, overcome in 13th century by Afghans. Formed in 1947 from E area of NADIA district, British Bengal province, following creation of PAKISTAN.

Kushtia (koosh-tee-ah), town (2001 population 83,658), ⊙ KUSHTIA district, W EAST BENGAL, BANGLADESH, on upper MADHUMATI (GARAI) River, and 11 mi/18 km WSW of PABNA; 23°54′N 89°05′E. Sugar milling, cotton-cloth weaving; textile manufacturing; trades (rice, jute, linseed, sugarcane, wheat). Until 1947, in NADIA district.

Kushugum, UKRAINE: see KUSHUHUM.

Kushugumovka, UKRAINE: see KUSHUHUM.

Kushuhum (koo-shoo-HOOM) (Russian *Kushugum*), town, NW ZAPORIZHZHYA oblast, UKRAINE, on E shore of the KAKHOVKA RESERVOIR, 10 mi/16 km S of ZAPORIZHZHYA; 47°43′N 35°12′E. Elevation 249 ft/75 m. Lime plant; flour mill; fish farm. Established in 1770, formerly called Kushuhumivka (Russian *Kushugumovka*) and then Velyka Katerynivka (Russian *Veli-kaya Yekaterinovka*); town since 1938.

Kushuhumivka, UKRAINE: see KUSHUHUM.

Kushunnai, RUSSIA: see IL'INSKIY, SAKHALIN oblast.

Kushva (KOOSH-vah), city (2006 population 33,570), W SVERDLOVSK oblast, E URAL Mountains, extreme W Siberian RUSSIA, at the SW foot of BLAGODAT' Mountain, on railroad, 60 mi/97 km N of YEKATERINBURG, and 27 mi/43 km NNW of NIZHNIY TAGIL; 58°17′N 59°45′E. Elevation 925 ft/281 m. Metallurgical center (pig iron, steel bars); machinery and machine tools, repair of locomotives, metalworking, woodworking, food processing (bakery). Founded in 1735, following the discovery of Blagodat' Mountain magnetite deposits; became city in 1926.

Kusieh, El, EGYPT: see QUSIYA, EL.

Kusinagara, INDIA: see KASIA.

Kusinskiy Zavod, RUSSIA: see KUSA.

Kusiyara River (kuh-si-YAH-rah), arm of lower SURMA RIVER, c.90 mi/145 km long, in NE INDIA (E ASSAM state) and E central BANGLADESH; leaves the Surma 15 mi/24 km W of SILCHAR (Assam state); flows WSW, past KARIMGANJ and FENCHUGANJ (Bangladesh), dividing into numerous arms and returning to lower Surma (KALNI) River.

Kuskokwim, river, c.650 mi/1,046 km long; rises on the NW slopes of the ALASKA RANGE, central ALASKA; flows SW to the BERING SEA. The shores of the river are mostly forested or uninhabited. BETHEL is major settlement near mouth of the river.

Kuskokwim Bay, SW ALASKA, NW of BRISTOL BAY; 100 mi/161 km long, 100 mi/161 km wide at mouth; center near 59°30′N 162°30′W. Receives KUSKOKWIM River.

Kuskokwim Mountains, SW ALASKA, W of ALASKA RANGE, SE of YUKON River; extend 250 mi/402 km NE-SW; 61°00′–64°00′N 155°00′–159°00′W. Rise to c.5,000 ft/1,524 m.

Kuskovo (koo-SKO-vuh), former city, central MOSCOW oblast, central European RUSSIA, now a suburb of MOSCOW, 8 mi/13 km ESE of the city center; 55°44′N 37°49′E. Elevation 456 ft/138 m. Known for its museums, of Russian musical history and of Russian ceramics industry, the latter located in the former estate of Count Sheremetyev (18th century). Established in 1925.

Kusluyan, TURKEY: see GOLKOY.

Kusma (KOOS-mah), town, ⊙ PARBAT district, central NEPAL, on the KALI GANDAKI RIVER, 18 mi/29 km W of POKHARA; 28°14′N 83°41′E. Elevation 3,363 ft/1,025 m. Rice, wheat, millet, vegetables.

Kusmuryn, KAZAKHSTAN: see KUSHMURUN.

Küsnacht (KOOS-nahkt), town (2000 population 12,484), ZÜRICH canton, N SWITZERLAND, on NE shore of LAKE ZÜRICH, 4 mi/6.4 km SSE of ZÜRICH. Metal products; printing.

Kusong (KOO-SUHNG), city, NORTH PYONGAN province, NORTH KOREA, 45 mi/72 km ESE of SINUIJU; 39°55′N 125°15′E. Agricultural area (rice, millet, and soybeans). Gold mines nearby.

Kussabat, LIBYA: see QASABAT, AL.

Küssnacht am Rigi, commune and district (2000 population 10,704), SCHWYZ canton, central SWIT-ZERLAND, on LAKE LUCERNE. Elevation 1,447 ft/441 m. A resort known chiefly as the scene of the killing of Gessler by William Tell. A nearby 17th-century chapel commemorates the legendary exploit. The district is German-speaking and Roman Catholic.

Küssnacht, Lake, SWITZERLAND: see LUCERNE, LAKE.

Kustanay, KAZAKHSTAN: see KOSTANAI.

Kustendil, Bulgaria: see KYUSTENDIL.

Küsten Kanal, GERMANY: see EMS-HUNTE CANAL.

Küstenland (KOOS-ten-lahnd) [German=coastland], former province (□ 3,077 sq mi/8,000.2 sq km) of AUSTRIA, consisting of Istria, Görz-Gradisca, and Trieste; capital was Trieste. Created in 1849, it passed to ITALY in 1919 and is today subdivided among Italy, SLOVENIA, and CROATIA.

Kustö, FINLAND: see KUUSISTO.

Küstrin, Polish *Kostrzyń* (kos-tree-nee), town, Gorzów province, W POLAND, on the ODER (German border), at WARTA River mouth, 18 mi/29 km NNE of FRANK-FURT. Paper and pulp milling; frontier station. Was first mentioned in early 13th century; passed 1445 to BRANDENBURG. Fortified in 16th century Frederick the Great, as crown prince, was imprisoned here (1730) by his father. In Seven Years War, heavily shelled (1758) by Russians; subsequently rebuilt by Frederick the Great. Occupied (1806), by French. Admiral Tirpitz born here. After World War II, during which it was virtually obliterated, it was evacuated by its German population. Was in Brandenburg until 1945. Formerly spelled Cüstrin.

Kusu (KOOS), town, Mie county, MIE prefecture, S HONSHU, central JAPAN, on W shore of ISE BAY, 16 mi/ 25 km N of TSU; 34°54′N 136°37′E. Sake.

Kusu, town, Kusu county, OITA prefecture, E KYUSHU, SW JAPAN, 28 mi/45 km W of OITA; 33°16′N 131°09′E. Shiitake mushrooms.

Kusunoki (koo-SOO-no-kee), town, Asa county, YA-MAGUCHI prefecture, SW HONSHU, W JAPAN, 16 mi/25 km S of YAMAGUCHI; 34°02′N 131°13′E. Computer components.

Kusy-Aleksandrovskiy Zavod, RUSSIA: see KUS'YE-ALEKSANDROVSKIY.

Kus'ye-Aleksandrovskiy (KOOS-ye–ah-leek-SAHN-druhf-skeeye), town (2006 population 1,880), E PERM oblast, W URAL Mountains, E European Russia, on the KOYVA River (KAMA River basin), on road, 19 mi/31 km E of CHUSOVOY; 58°17′N 58°22′E. Elevation 659 ft/200 m. Woodworking, knitting. Until 1946, known as Kusy-Aleksandrovskiy Zavod.

Kut (KUT) or **Kut al Imara**, city, ⊙ WASIT province, E IRAQ, on the E bank of the TIGRIS RIVER (here throwing off a branch, the SHATT AL GHARRAF), on railroad, and 100 mi/161 km SE of BAGHDAD. Licorice processing; rice, corn, millet, sesame, dates. A barrage is on the Tigris just upstream, diverting water into the Shatt al Gharraf canal-branch.

Kuta (koo-TAH), town, W central NIGERIA, 20 mi/32 km NNE of Minna, near railroad. Formerly gold-mining center. Shea-nut processing; ginger, cassava, durra.

Kuta Beach (KOO-tah), resort area, S BALI, INDONESIA, 4 mi/7 km SW of DENPASAR, 08°43′S 115°10′E. Popular tourist area, first developed in 1930s, but experienced rapid growth in 1960s and 1970s. Popular surfing and windsurfing area.

Kutacane (KOO-tah-KAH-nai), city, ⊙ S Aceh district, ACEH province, N SUMATRA, INDONESIA, 59 mi/95 km E of MEDAN; 03°30′N 97°48′E. Elevation 3,250 ft/991 m. Center of large rice-producing region. Situated on border of MOUNT LEUSER NATIONAL PARK. Formerly spelled Kutatjane.

Kütahya, city (2000 population 166,665), W central TURKEY. An agricultural market center producing sugar beets, fruit, cotton, and textiles. Iron, lignite, mercury, and chromium are mined nearby. The city has been famous since the 16th century for the manufacturing of ceramics. It has a hydroelectric plant. Known in ancient history as Cotyaeum, it was occupied by the Seljuk Turks soon after the battle of Manzikert (1071). In the 15th century it passed to the Ottomans. A former spelling is Kutaiah.

Kutai National Park (koo-TEI) (□ c.772 sq mi/2,000 sq km), Kalimantan Timur province, INDONESIA, 43 mi/70 km NNE of SAMARINDA; 00°00′ 116°30′E. Indonesia's second-largest reserve. Originally established 1930s (□ 1,160 sq mi/3,000 sq km); park size was reduced by one third in 1971 to allow logging. Established 1982 as a national park. A fertilizer plant and coal mine are within the park borders, and a large liquified natural gas plant is in BONTANG, just outside park borders. Fires in 1982–1983 destroyed over 60% of the park forests. Habitat for orangutan, proboscis monkey, rhinoceros, civet cat, and the peacock-like *tumbau* bird.

Kutai River, Malaysia: see MAHAKAM RIVER.

Kutais (koo-tah-EES), town (2004 population 3,300), S KRASNODAR TERRITORY, extreme NW CAUCASUS, S European Russia, at the NW foot of the Greater Caucasus, 37 mi/60 km W of MAYKOP, and 31 mi/50 km S of KRASNODAR; 44°31′N 39°18′E. Elevation 734 ft/223 m. Oil and gas wells. Established as a settlement for drilling stations workers, named Kutaisskiy; further developed, given town status, and renamed after World War II.

Kutaisi (koo-tah-EE-see), city (2002 population 185,965), W GEORGIA, on the RIONI River; 42°15′N 42°41′E. Georgia's second-largest city. Manufacturing includes machinery, chemicals, motor vehicles and parts, equipment for petroleum and gas industry, and building materials; food processing, barrite processing; cannery, silk combine, leather footwear combine. Industry is aided by the Rioni and Gumati hydroelectric power plants. Pedagogical Institute, other schools for mining, automotive training, timber technology. Was capital of ancient COLCHIS (8th century B.C.E.), and capital of IMERITIA in the C.E. 13th, 15th, and 16th century. Taken by the Russians in 1810. There is some notable medieval architecture, including the ruins of the 11th-century Saint George Cathedral.

Kutaisskiy, RUSSIA: see KUTAIS.

Kut-al-Amara, IRAQ: see AL KUT.

Kutaradja, INDONESIA: see BANDA ACEH.

Kutaraja, INDONESIA: see BANDA ACEH.

Kutcharo, Lake (koot-CHAH-ro), Japanese *Kutcharo-ko* (koot-CHAH-ro–ko) (□ 31 sq mi/80.6 sq km), E Hokkaido prefecture, N JAPAN, in AKAN NATIONAL PARK, 22 mi/35 km S of ABASHIRI; 9 mi/14.5 km long, 5 mi/8 km wide; hot springs on shores. Sometimes spelled Kussharo or Kuttyaro.

Kutchuk Kainarji, Bulgaria: see KAINARDZHA.

Kutdligssat, Greenland: see QULLISSAT.

Kut-el-Amara, IRAQ: see AL KUT.

Kutelo (koo-TE-lo), peak, PIRIN MOUNTAINS, SW BULGARIA; 41°48′N 23°23′E. Elevation 9,534 ft/2,906 m.

Kutenholz (KOO-tuhn-hohlts), village, LOWER SAXONY, N GERMANY, 28 mi/45 km W of HAMBURG; 53°28′N 09°19′E.

Kuteynikovo, UKRAINE: see KUTEYNYKOVE.

Kuteynykove (koo-TAI-ni-ko-ve) (Russian *Kuteyni-kovo*), town (2004 population 4,150), SE DONETS'K oblast, in the DONBAS, 8 mi/12.9 km W of AMVROSIYIVKA; 47°49′N 38°17′E. Elevation 741 ft/225 m. Railroad junction; flour mill. Established in 1878, town since 1938.

Kuthanallur, INDIA: see KUTTANALLUR.

Kuthar (kuh-TAHR), former princely state of PUNJAB HILL STATES, N INDIA. Since 1948, merged with HI-MACHAL PRADESH state.

Kuther, INDIA: see ANANTNAG.

Kuthodaw Pagoda (KOO-tho-DAW), religious site, MANDALAY, MYANMAR. A central pagoda, built in 1857 and measuring 98 ft/30 m in height, is surrounded by 729 smaller pagodas, which were added in 1872 during the Fifth Buddhist Synod. The complete Buddhist scripture is recorded on marble slabs housed here.

Kutina (KOO-tee-nah), town, central CROATIA, on railroad, and 20 mi/32 km E of SISAK, in MOSLAVINA; 45°30′N 16°45′E. Local trade center; manufacturing (carbon black, enamelware, fertilizers). Natural-gas (at Gojlo) and oil fields (at Stružec) nearby, connected by pipeline. LONJSKO POLJE Nature Park to S, near Sava River.

Kutina (KOO-tee-nah), village, SOFIA oblast, Grad SOFIA obshtina, BULGARIA; 42°51′N 23°30′E. Lignite deposits.

Kutiyana (kuh-tee-YAH-nah), town, JUNAGADH district, GUJARAT state, W central INDIA, on BHADAR RIVER, and 32 mi/51 km WNW of JUNAGADH; 21°38′N 69°59′E. Agriculture market center (millet, cotton, wheat, oilseeds, sugarcane); handicrafts (cotton cloth, metalware).

Kutkai (koot-KEI), township, SHAN STATE, MYANMAR, on BURMA ROAD, and 35 mi/56 km NNE of LASHIO. Known for vegetable and fruit production. Airport.

Kut, Ko (GUD, GO), southernmost Thai island, on E coast of GULF OF THAILAND, in TRAT province, SE THAILAND, 15 mi/24 km off coast; 11°40′N 102°35′E. Rises to 1,033 ft/315 m.

Kutlu-Bukash (koot-LOO–boo-KAHSH), village, central TATARSTAN Republic, E European Russia, on road, 18 mi/29 km N of CHISTOPOL'; 55°38′N 50°37′E. Elevation 439 ft/133 m. In agricultural area (sunflowers, sugar beets, grains).

Kutná Hora (KUT-nah HO-rah), German *Kuttenberg*, city (2001 population 21,453), STREDOCESKY province, CZECH REPUBLIC, central E BOHEMIA; 49°57′N 15°16′E. Manufacturing (machinery, cigarettes, textiles); brewery (established 1573). It was an important silver-mining center in the Middle Ages. Its famous mint played a large role in building the power and greatness of the medieval kings of Bohemia. Between 1421–1424, Kutná Hora was captured by the Hussites, recaptured by Emperor Sigismund, and captured again and burned by John Žižka. Until then a stronghold of Catholicism, but after it was burned it became the center of Bohemian Protestantism for two centuries. The city suffered again in the Thirty Years War (1618–1648). The city lost its importance after the silver mines closed in the 17th century. Rich in medieval architecture, including the 14th century Church of St. Barbara (Bohemian Gothic) and the Gothic Cathedral of St. James (14th century), with its 266-ft/81-m tower. The "Italian Court," begun in the 13th century, is a palace once used both as a mint and as a residence for the kings of Bohemia. Part of the city is a UNESCO Cultural and Natural Heritage site. Kacina castle is 5 mi/8 km NE.

Kutno (KOOT-no), city (2002 population 48,741), Płock province, central POLAND, 33 mi/53 km N of ŁÓDŹ; 52°14′N 19°22′E. Railroad junction; trade center; manufacturing of agricultural machinery and tools, cement, bricks, soap; beet-sugar and flour milling, chicory drying, sawmilling, distilling, brewing. Lignite deposits in vicinity. Before World War II, population was 50% Jewish. Nearby, Poles fought (1939) one of longest and bloodiest battles against Germans.

Kutnozero, RUSSIA: see KUYTO LAKES.

Kutoarjo (koo-taw-AHR-jaw), town, Java Tengah province, INDONESIA, near INDIAN OCEAN, 30 mi/48 km W of YOGYAKARTA; 07°43′S 109°54′E. Trade center for agricultural area (rice, sugar, peanuts). Also called Kutoardjo or Koetoardjo.

Kutovo (KOO-to-vo), island in the DANUBE River, NW BULGARIA; 44°01′N 23°00′E.

Kutsa Balka, UKRAINE: see NOVYY BUH.

Kutsivka, UKRAINE: see NOVHORODKA.

Kutsovka, UKRAINE: see NOVHORODKA.

Kutsuki (KOOTS-kee), village, Takashima county, SHIGA prefecture, S HONSHU, central JAPAN, 25 mi/40 km N of OTSU; 35°21′N 135°55′E.

Kuttalam (kuh-TUHL-uhm), town, THANJAVUR district, TAMIL NADU state, S INDIA, on arm of KAVERI RIVER delta, and 4 mi/6.4 km WSW of Mayiladuturai. Rice, peanuts, millet.

Kuttalam (kuh-TUHL-uhm), village, NELLAI KATTA-BOMMAN district, TAMIL NADU state, S INDIA, on CHITTAR RIVER (falls), 3 mi/4.8 km SW of TENKASI. Health resort (sanatorium) at E foot of WESTERN GHATS; cool climate caused by SW monsoon blowing through gap in mountains.

Kuttanallur (kuh-TAHN-ahl-LOOR), town, THANJA-VUR district, TAMIL NADU state, S INDIA, in KAVERI River delta, 9 mi/14.5 km SW of Thiruvarur; 10°42′N 79°32′E. Mat weaving; rice, coconut palms, bamboo. Also spelled Kuthanallur.

Kuttawa (KUHT-uh-wai), village (2000 population 596), LYON county, W KENTUCKY, 25 mi/40 km E of PADUCAH, and 3 mi/4.8 km WSW of EDDYVILLE, on CUMBERLAND RIVER (LAKE BARKLEY reservoir); 37°03′N 88°06′W. In agricultural area; manufacturing (catfish processing; plastic parts, plastic molding); limestone quarry; hardwood timber area. Recreation and tourism. Town was relocated 2 mi/3.2 km to W from original site of c.1960 with construction of Barkley Dam (3 mi/4.8 km to W). Kentucky Dam (TENNESSEE RIVER) 6 mi/9.7 km to W.

Kuttenberg, CZECH REPUBLIC: see KUTNÁ HORA.

Küttigen, commune, AARGAU canton, N SWITZERLAND, 2 mi/3.2 km N of AARAU. Elevation 1,339 ft/408 m. Textiles.

Kutu (KOO-too), village, BANDUNDU province, W CONGO, on right bank of FIMI RIVER, at outlet of LAKE MAI-NDOMBE, 55 mi/89 km SSW of INONGO; 02°44′S 18°09′E. Elev. 1,049 ft/319 m. Trading and agricultural center, steamboat landing; rice, palm products, copal. Roman Catholic mission.

Kutubu, oil and natural gas field at LAKE KUTUBU, SOUTHERN HIGHLANDS province, S central PAPUA NEW

GUINEA, 25 mi/40 km SW of Mendi. Hodinia/Agogo joint venture opened 1992. Pipeline extends to GULF OF PAPUA. Tourist facilities at Lake Kutubu Lodge.

Kutubu, Lake (koo-TOO-boo), SOUTHERN HIGHLANDS province, SE PAPUA NEW GUINEA; 12 mi/19 km long, 2 mi/3.2 km–3 mi/4.8 km wide; 06°24′S 143°20′E. 2,700 ft/823 m. Population in area, 400. Site of major oil and natural gas development.

Kutulik (koo-too-LEEK), town (2006 population 5,380), SW UST-ORDYN-BURYAT AUTONOMOUS OKRUG, in IRKUTSK oblast, E central SIBERIA, RUSSIA, on road and the TRANS-SIBERIAN RAILROAD, 18 mi/29 km NW of CHEREMKHOVO; 53°21′N 102°47′E. Elevation 1,751 ft/ 533 m. In agricultural area; logging. Founded in 1899 with the construction of the railroad; town since 1944.

Kutum (kuh-TUHM), township, N DARFUR state, W SUDAN, on road, and 60 mi/97 km NW of AL-FASHER; 14°12′N 24°40′E. Trade center; gum arabic; corn, durra; livestock.

Kutur, IRAN: see QUTUR.

Kutus (KOO-too), town, Kirinyaga district, CENTRAL province, KENYA, on road, and 62 mi/100 km NE of NAIROBI; 00°32′S 37°21′E. Market center.

Kutuzove, UKRAINE: see VOLODARS′K-VOLYNS′KYY.

Kutuzovo (koo-TOO-zuh-vuh), town (2005 population 2,500), E KALININGRAD oblast, W European Russia, on the SHESHUPE river (Lithuanian border) opposite NAUMIESTIS (LITHUANIA), on road, 15 mi/24 km NE of NESTEROV; 54°47′N 22°50′E. Elevation 160 ft/48 m. Agricultural market. First mentioned in 1563; chartered in 1725. Until 1945, part of GERMANY, in EAST PRUSSIA, and called Schirwindt; then Germany's easternmost town.

Kutuzovo, UKRAINE: see VOLODARS′K-VOLYNS′KYY.

Kuty (koo-TEE), town (2004 population 8,250), SE IVANO-FRANKIVS′K oblast, UKRAINE, on CHEREMOSH RIVER, and 20 mi/32 km SSE of KOLOMYYA, in coniferous woodland; 48°15′N 25°10′E. Elevation 1,072 ft/ 326 m. Sawmilling; extraction of pitch, turpentine, tar, colophony, resin; manufacturing (furniture, ceramics). Vocational technical school. Known since 1448, town since 1940.

Kuty (koo-TI), Slovak *Kúty*, Hungarian *Jókút*, village, ZÁPADOSLOVENSKÝ province, W Slovakia, 16 mi/26 km W of SENICA; 48°40′N 17°01′E. Sugar beets. Important railroad junction. Manufacturing of machinery, textiles. Baroque church, Neolithic-era archaeological site.

Kutztown (KUHTS-toun), borough (2006 population 5,038), BERKS county, E central PENNSYLVANIA, 14 mi/ 23 km NNE of READING. Agriculture (livestock, poultry; grain, apples; dairying); manufacturing (printing and publishing, trusses, apparel, food, bottled water; railroad steel foundry); limestone. Seat of Kutztown University of Pennsylvania. Crystal Cave to W. Two covered bridges to NW. Doe Mountain Ski Area to E. Settled 1733 by Germans, incorporated 1815.

Kuujjuaq (KOO-jwak), village (□ 151 sq mi/392.6 sq km), Nord-du-Quebec region, N QUEBEC, E CANADA, on the KOKSOAK RIVER near its mouth at UNGAVA BAY; 58°06′N 68°24′W. A Hudson's Bay Company post established here in 1830. In region are rich iron deposits. Variant names are Fort-Chimo (shee-MO), Chimo, Chapel Hill.

Kuujjuarapik or **Great Whale River** (GRAIT HWAIL), village (□ 3 sq mi/7.8 sq km), NORD-DU-QUÉBEC region, NW QUEBEC, E CANADA, on SE shore of HUDSON BAY, at mouth of Grande Rivière de la Baleine (Great Whale River); 55°17′N 77°45′W. JAMES BAY Hydro Project to E. Populated by Cree and Inuit peoples. Hunting, fishing, trapping. Scheduled air service.

Kuuli-Mayak (koo-lee–mah-YAHK), town, W BALKAN weloyat, W TURKMENISTAN, port on CASPIAN SEA, 22 mi/35 km NNW of TURKMENBASHI; 40°14′N 52°42′E. Fish canneries; salt extraction.

Kuurne (KOOR-nuh), commune (2006 population 12,581), Kortrijk district, WEST FLANDERS province, NW BELGIUM, NE suburb of KORTRIJK; 50°51′N 03°17′E.

Kuusamo (KOO-sah-mo), town, OULUN province, NE FINLAND, near Russian border, 100 mi/161 km ESE of ROVANIEMI; 65°58′N 29°11′E. Elevation 910 ft/275 m. In lake region; road center in lumbering region. Reindeer herding, fur trading, fish processing; timber; tourist resort. Airport.

Kuusankoski (KOO-sahn-KOS-kee), town, KYMEN province, SE FINLAND, on the KYMIJOKI (river), and 4 mi/6.4 km NW of KOUVOLA; 60°54′N 26°38′E. Elevation 165 ft/50 m. Railroad terminus; pulp, cellulose, and paper mills; hydroelectric station. Part of UUDENMAAN province until 1949.

Kuusisto (KOO-sis-to), Swedish *Kustö*, island, TURUN JA PORIN province, SW FINLAND, in inlet of GULF OF BOTHNIA, 6 mi/9.7 km SE of TURKU; 6 mi/9.7 km long, 1 mi/1.6 km–2 mi/3.2 km wide; 60°23′N 22°25′E. Elevation 33 ft/10 m. Remains of fourteenth-century castle of Roman Catholic bishops of Finland, destroyed in 1528 by Gustavus Vasa.

Kuusjärvi (KOOS-YAHR-vee), village), POHJOIS-KARJALAN province, SE FINLAND, 25 mi/40 km WNW of JOENSUU; 62°42′N 28°56′E. Elevation 396 ft/120 m. In SAIMAA lake region; lumbering. OUTOKUMPU copper mines are nearby.

Kuva, city, E FERGANA wiloyat, NE UZBEKISTAN, 18 mi/ 29 km NE of FERGANA, railroad (FEDCHENKO station); 40°18′N 71°59′E. Tertiary-level administrative center. Textile- (cotton ginning; sericulture) and food-processing industries; printing plant.

Kuvakino (koo-VAH-kee-nuh), village (2004 population 1,300), SW CHUVASH REPUBLIC, central European Russia, near highway, 12 mi/19 km NNW of ALATYR′; 54°58′N 46°25′E. Elevation 492 ft/149 m. Wheat; livestock.

Kuvam River, INDIA: see COOUM RIVER.

Kuvandyk (koo-vahn-DIK), city (2006 population 28,720), N ORENBURG oblast, Russia, in S foothills of the SW URAL Mountains, on the SAKMARA RIVER (URAL River basin), on crossroads and railroad, 120 mi/193 km E of ORENBURG, and 10 mi/16 km NW of MEDNOGORSK; 51°28′N 57°21′E. Elevation 846 ft/257 m. In Orsk-Khalilovo industrial district; cryolite works, mechanical presses; railroad shops. Made a city in 1953.

Kuvango (koo-VAHN-go), town, HUÍLA province, S central ANGOLA, on Cubango River, and 120 mi/193 km SSE of HUAMBO. Cattle rearing; hides and skins. Formerly Vila da Ponte (VEE-lah dah PON-te).

Kuvasai (KOO-vah-sei), city, SE FERGANA wiloyat, UZBEKISTAN, on railroad spur, and 12 mi/19 km SE of FERGANA; 40°18′N 71°59′E. Cement works, power plant. Sometimes spelled Kuvasay.

Kuvshinovo (koof-SHI-nuh-vuh), city (2006 population 10,860), W central TVER oblast, W European Russia, on the VALDAY HILLS and the Osuga River (VOLGA RIVER basin), near highway and on railroad, 82 mi/132 km W of TVER, and 30 mi/48 km W of TORZHOK; 56°59′N 32°08′E. Elevation 623 ft/189 m. Paper and cardboard manufacturing, food processing. Made a city in 1938.

Kuwait (koo-WAIT), nominal constitutional monarchy (□ 6,177 sq mi/17,818 sq km, including islands; 2004 estimated population 2,257,549; 2007 estimated population 2,505,559), NE Arabian peninsula, at the head of the PERSIAN GULF; ⊙ AL-KUWAIT, or Kuwait City. Al-Kuwait is a modern city; its port, MINA AL AHMADI, is a trade center with shipyards and oil refineries. Kuwait is sometimes spelled Kowait, Kuweit, or Kuwayt.

Geography

A low, sandy region, mostly flat in the E and hilly in the W rising to 830 ft/290 m, generally barren and sparsely settled, Kuwait is bounded by SAUDI ARABIA (S) and by IRAQ (N and W). It has a warm climate, dry inland and humid along the coast.

Population

The population is predominantly Arab; however, most are non-Kuwaitis. Since the development of the oil industry, large numbers of foreigners have found employment here. Ethnic groups include South Asians, Iranians, Indians, Pakistanis, Yemenis, and Palestinians. Arabic is the official language, but English is widely spoken. Over 92% of the population are Muslims.

Economy

Kuwait's traditional exports were pearls and hides, but since 1946 it has become a major petroleum producer. Kuwait has only small agricultural industry. The country is mostly desert, with some fertile, irrigated (by local wells) areas near the Persian Gulf coast. Oil dominates the economy, accounting for over 90% of Kuwait's revenue (before the Persian Gulf War). Kuwait has the third-largest oil reserves in the world after Saudi Arabia and Iraq. Other industries include food processing, petrochemical industries, fertilizers, various consumer goods, electronics, cement, desalinization, and textiles. Huge amounts of natural gas, liquidized and exported, complement Kuwait's oil production. Food and clothing are the principal imports. Kuwait's major trading partners are the U.S., Japan, and Germany. The main concession for oil exploitation was held by a joint British-American firm until 1974 when Kuwait took control of most of the operations; it had previously retained a large part of the oil profits. Much of the profits have been devoted to the modernization of living conditions and education in the country. Native Kuwaitis have an extremely high per capita income, pay no taxes, and enjoy numerous social services. To provide against the possible future exhaustion of the oil reserves, the government in the 1960s launched a program of industrial diversification and overseas investment.

History: to 1963

The present ruling dynasty was founded by Sabah abu Abdullah (ruled 1756–1772). In the late 18th and early 19th centuries, Kuwait, nominally an Ottoman province, was frequently threatened by the Wahabis. In 1897, Kuwait was made a British protectorate. By this the British prevented Kuwait from becoming linked to the German-controlled Berlin-Baghdad railroad project. In June 1961, the British ended their protectorate, and Kuwait became an independent nation. However, the British supplied troops in July at the request of the Emir when Iraq claimed sovereignty over Kuwait. A short time later the British troops were replaced by detachments from the Arab League. Kuwait was admitted to the UN in 1963. In October 1963, Iraq officially recognized the nation of Kuwait. Kuwait was a founding member of the Organization of Petroleum Exporting Countries (OPEC), and later, in 1981, of the Gulf Cooperation Council (GCC). The country has been a large donor of financial aid to the other Arab countries and has been a supporter of Palestinian causes. Kuwait also maintains strong ties with both Arab and Western nations.

History: 1963 to Present

Kuwait sided with Iraq in the Iran-Iraq War, which caused the country's oil income to decrease by nearly 50%. An oil refinery was attacked by Iran in 1982. In addition, Kuwaiti oil tankers in the Persian Gulf came under fire by Iran, and Iran instigated terrorist activity in Kuwait through radical Muslim groups advocating support of the Iranian leader Khomeini. By 1987, Kuwait sought U.S. protection over its oil tankers in the Persian Gulf until the end of the war in 1988. Iraq invaded Kuwait on August 2, 1990, and declared its annexation. Native Kuwaitis, along with the royal family, fled as Iraqi forces devastated the country, setting fire to over 500 Kuwaiti oil wells. Over 80% of all wells were destroyed or damaged by Iraq, causing phenomenal environmental hazards. Western and Arab coalition forces, the large part of which were American, drove Iraqi forces from Kuwait in an air and ground war in February 1991, after UN economic sanctions and over a month of aerial bombing of Iraq failed to disengage the Iraqis. Thousands of foreign workers who were based in Kuwait fled to the countryside. An estimated 500,000 Palestinians in Kuwait were expelled, after Kuwait regained its independence, because of the Palestinian Liberation Organization's support of Iraq in the Gulf War. Al Sabah returned to Kuwait from his refuge in Saudi Arabia in March 1991. The country has since been rebuilt and the oil industry restored to its full productive capacity. After a security-invoked martial law was lifted in 1991, Al Sabah promised parliamentary elections and democratic reforms. Elections for a national assembly were held beginning in 1992. An edict (1999) by the Emir giving Kuwaiti women the right to vote and run for office failed to win parliamentary ratification; not until 2005 was the change effected. Emir Jaber Al Ahmed Al Sabah died January 15, 2006 and was automatically replaced by crown prince Saad Al Abdullah Al Sabah, who was in ill health. On January 24, 2006, the cabinet voted to replace the new Emir with the prime minister (and half-brother to the former Emir) Sabah Al Ahmed Al Sabah.

Government

Kuwait is a monarchy governed under the constitution promulgated in 1962. The emir, the hereditary monarch of the Mubarak line of the ruling al-Sabah family, serves as head of state. The government is headed by the prime minister, who is appointed by the monarch; until 2003 the prime minister traditionally was the crown prince. The unicameral legislature consists of the fifty-seat National Assembly, whose members are elected by popular vote to serve four-year terms. There are no official political parties, although several political groups act as de facto parties. The current head of state is Emir Sabah Al Ahmed Al Sabah (since 2006). The current head of government Prime Minister Nasir Muhammad Al Ahmad Al Sabah (since 2007). Kuwait is divided into six administrative divisions (governates): CAPITAL (Al Asimah), HAWALLI, Mubarak Al Kabir, Al FARWANIA, Al JAHRA, and Al AHMADI.

Kuwait, Al- (koo-WAIT, el), city (2005 population 32,403), ⊙ KUWAIT, in CAPITAL governate, SE coast of Kuwait Bay; 29°24′N 47°42′E. The city and suburbs cover nearly half of Kuwait's total area. International airport. Serviced by SHUWAIKH port N of the city. Center of government administration, banking. Hospital; Kuwait University. Also called Kuwait City.

Kuwana (koo-WAH-nah), city, MIE prefecture, S HONSHU, central JAPAN, on ISE BAY, 25 mi/40 km N of TSU; 35°03′N 136°41′E. Important port and industrial center manufacturing bearings.

Kuway Sandjaq, IRAQ: see KÖI SANJAQ.

Kuweira, JORDAN: see QUWEIRA, AL.

Kuweit: see KUWAIT.

Kuyal'nitskiy Liman, UKRAINE: see KUYAL'NYK.

Kuyal'nyk (koo-YAHL-nik), (Ukrainian *Kuyal'nyts'kyy lyman*), (Russian *Kuyal'nitskiy Liman*), (surface ☐ 22 sq mi/56 sq km–23 sq mi/60 sq km), BLACK SEA coast lagoon, E central ODESSA oblast, UKRAINE, 12 mi/19 km N of ODESSA city center; 17 mi/28 km long, up to 1.5 mi/2.5 km wide; depth to 10 ft/3 m. Fed by Velykyy Kuyal'nyk (Russian *Bol'shoy Kuyal'nik*), intermittent stream, from N; separated from sea by a shell and sandbar 1.5 mi/2.4 km long and 5 ft/1.5 m–7 ft/2 m high. Water level generally 17 ft/5 m below sea level; contains salts of calcium, manganese, potassium, iodine and bromine; high salinity has limited organic life in aquatic ecosystem. Black muds from bottom are used in Odessa resorts for medicinal purposes; its mineral water is drawn from adjacent groundwater for drinking. A 328-ft/100-m shoreline zone restricts land uses and activities for sanitary purposes.

Kuyal'nyts'kyy Lyman, UKRAINE: see KUYAL'NYK.

Kuyamazar (koo-yah-mah-ZAHR), large reservoir, just inside NAWOIY wiloyat, UZBEKISTAN, NE of BUKHARA on edge of desert; completed 1947; stores ZERAVSHAN RIVER overflow for irrigation of BUKHARA wiloyat.

Kuyanovo (koo-YAH-nuh-vuh), village (2005 population 3,900), N BASHKORTOSTAN Republic, E European Russia, on highway, 23 mi/37 km NNE of VERKHNEYARKEYEVO; 55°48′N 54°32′E. Elevation 275 ft/83 m. In agricultural area; food processing.

Kuybyshev (KOO-ee-bi-shef), city (2006 population 48,135), central NOVOSIBIRSK oblast, SW SIBERIA, RUSSIA, on the BARABA STEPPE, on the OM' RIVER (IRTYSH River basin), 195 mi/314 km W of NOVOSIBIRSK, and 5 mi/8 km N of BARABINSK (linked by railroad spur); 55°27′N 78°19′E. Elevation 377 ft/114 m. Chemicals, meat and milk combines, distillery; regional electric power station. Founded 1722; became city in 1782. Once a commercial transit center on the old Siberian Road; now supplanted by Barabinsk on the TRANS-SIBERIAN RAILROAD. Until 1935, called Kainsk.

Kuybyshev, RUSSIA: see BULGAR, TATARSTAN Republic.

Kuybyshev, RUSSIA: see SAMARA, city.

Kuybyshev Dam (KOO-ee-bi-shef), on the middle VOLGA RIVER, SAMARA oblast, E European Russia, 30 mi/48 km NW of SAMARA, on the N side of the SAMARA BEND. Construction project (begun 1950; completed 1955) includes right-bank hydroelectric plant (2,000,000 kw capacity) near TOL'YATTI, main earth dam section (85 ft/26 m high) and left-bank concrete spillway with navigation canal and locks. The Volga, whose level at the dam is raised by 80 ft/24 m, is backed up 300 mi/483 km and reaches a width of 25 mi/40 km upstream. A multipurpose project, the dam improves navigation of the middle Volga, furnishes power to MOSCOW, Samara, and SARATOV, and activates irrigation pumps for the dry lands E of the Volga.

Kuybysheve (KOO-ee-bi-she-ve), (Russian *Kuybyshevo*), town, E ZAPORIZHZHYA oblast, UKRAINE, on railroad, 40 mi/64 km NNW of BERDYANS'K; 47°21′N 36°39′E. Elevation 770 ft/234 m. Raion center; manufacturing (electronic instruments, building materials, bricks, asphalt); food products (flour, canned goods); granite quarry. Vocational technical school. Established in 1782 as Kamyanka (Russian *Kamenka*), renamed Tsarekostyantynivka (Russian *Tsarekonstantinovka*) in 1845, then Pershotravneve (Russian *Pervomayskoye*); Kuybysheve since 1935; town since 1957.

Kuybysheve (KOO-ee-bi-she-ve), (Russian *Kuybyshevo*), town (2004 population 5,360), S Republic of CRIMEA, UKRAINE, on road, in the valley of the BEL'BEK River, and 8 mi/13 km S of BAKHCHYSARAY; 44°39′N 33°52′E. Elevation 810 ft/246 m. Local agricultural industries; flour milling. Until 1944, known as Albat. Sometimes called Kuybyshevs'ke (Russian *Kuybyshevskoye*). Town since 1960. Ruins of the cave city of Mangup-Kale (6th–15th century) 3.2 mi/6 km to SW.

Kuybyshevka-Vostochnaya, RUSSIA: see BELOGORSK.

Kuybyshevo (KOO-ee-bi-shi-vuh), village (2006 population 5,790), SW ROSTOV oblast, S European Russia, on the MIUS RIVER, on crossroads, approximately 3 mi/5 km S of the Ukrainian border, and 40 mi/64 km N of TAGANROG; 47°49′N 38°54′E. Elevation 265 ft/80 m. Flour mill. Until about 1936, called Golodayevka or Golodayevskoye.

Kuybyshevo (KOO-ee-bi-shi-vuh), fishing village, SAKHALIN oblast, RUSSIA, on the ITURUP Island, S KURIL ISLANDS, on the W coast, 12 mi/19 km SW of KURIL'SK. Under Japanese rule (until 1945), called Rubetsu.

Kuybyshevo, RUSSIA: see SAMARA.

Kuybyshevo, UKRAINE: see KUYBYSHEVE, Zaporizhzhya oblast; or KUYBYSHEVE, Republic of Crimea.

Kuybyshev Reservoir, RUSSIA: see SAMARA RESERVOIR.

Kuybyshevskiy Zaton (KOO-ee-bi-shif-skeeyee zah-TON), town (2006 population 2,920), W TATARSTAN Republic, E European Russia, on the left bank of the VOLGA RIVER, on road, 27 mi/43 km S of KAZAN', and 4 mi/6 km SW of KAMSKOYE UST'YE, to which it is administratively subordinate; 55°09′N 49°10′E. Elevation 177 ft/53 m. Winter anchorage; freight depot; shipyards. Has a mosque. Until 1935, called Spasskiy Zaton.

Kuyeda (koo-ye-DAH), town (2005 population 9,600), S PERM oblast, W URAL Mountains, E European Russia, on road and railroad, on the Buy River (KAMA River basin), 60 mi/97 km S of OSA; 56°26′N 55°35′E. Elevation 442 ft/134 m. In oil-producing area.

Kuyera (koo-YER-ah), town (2007 population 13,382), OROMIYA state, S central ETHIOPIA, 7 mi/11.3 km NE of SHASHEMENE; 07°17′N 38°39′E.

Kuylyuk, UZBEKISTAN: see KUILYUK.

Kuyten-Uul, MONGOLIA: see MONGOLIAN ALTAY.

Kuyto Lakes (koo-yee-TO), Finnish *Kuittijärvi*, chain of lakes in W Republic of KARELIA, NW European Russia, 75 mi/121 km W of KEM; 70 mi/113 km long, up to 6 mi/10 km wide. Include Upper Kuito Lake (W), Middle Kuito Lake (UKHTA on the N shore), and Lower Kuito Lake (E). Crossed W-E by the KEM River. Frozen November through April. Formerly called Kutnozero.

Kuyto Lakes, RUSSIA: see KUITO LAKES.

Kuytun (kooyee-TOON), town (2005 population 10,475), SW IRKUTSK oblast, E central SIBERIA, RUSSIA, on road and the TRANS-SIBERIAN RAILROAD, 40 mi/64 km ESE of TULUN; 54°20′N 101°30′E. Elevation 1,709 ft/520 m. Lumber industry. Established in 1899 with the construction of the railroad; town since 1957.

Kuytun, RUSSIA: see KUITUN.

Kuzbas, RUSSIA: see KUZNETSK BASIN.

Kuze (KOO-ze), village, Ibi county, GIFU prefecture, central HONSHU, central JAPAN, 16 mi/25 km N of GIFU; 35°33′N 136°30′E.

Kuzedeyevo (koo-zye-DYE-ee-vuh), town (2005 population 3,420), SW KEMEROVO oblast, S central SIBERIA, RUSSIA, on the edge of the KUZNETSK BASIN in the GORNAYA SHORIYA Mountains, on the KONDOMA RIVER (TOM' River basin), on road and near railroad, 30 mi/48 km S of NOVOKUZNETSK; 53°20′N 87°10′E. Elevation 744 ft/226 m. Toy production.

Kuzhendeyevo (koo-zhin-DYE-ee-vuh), village, SW NIZHEGOROD oblast, central European Russia, on local road, 2 mi/3.2 km S, and under administrative jurisdiction, of ARDATOV; 55°12′N 43°05′E. Elevation 492 ft/149 m. In agricultural area; collective farm.

Kuzhener (koo-zhi-NYER), town (2006 population 5,775), E MARI EL republic, central European Russia, 40 mi/64 km ENE of YOSHKAR-OLA; 56°48′N 48°55′E. Elevation 682 ft/207 m. In agricultural area (grains, vegetables, sugar beets; dairy livestock); food processing.

Kuzhenkino (koo-ZHEN-kee-nuh), town (2006 population 3,385), NW TVER oblast, W European Russia, on road, 9 mi/14 km SSW of BOLOGOYE; 57°45′N 33°57′E. Elevation 597 ft/181 m. In agricultural area (oats, rye, flax, potatoes; dairy livestock); food processing.

Kuzhittura, INDIA: see KULITHURA.

Kuzino (KOO-zee-nuh), town (2006 population 3,195), SW SVERDLOVSK oblast, extreme E European Russia, in the central URAL Mountains, 20 mi/32 km NW of PERVOURAL'SK (where some of the town's population is employed); 57°01′N 59°26′E. Elevation 1,056 ft/321 m. Railroad junction; railroad establishments.

Kuzino (KOO-zee-nuh), village (2006 population 1,415), W central VOLOGDA oblast, N central European Russia, on the SHEKSNA River, on highway, 7 mi/11 km

---

SW of (and administratively subordinate to) KIR-ILLOV; 59°47'N 38°17'E. Elevation 413 ft/125 m. Agricultural products.

Kuz'michi (kooz-mee-CHEE), village, S VOLGOGRAD oblast, SE European Russia, near highway and railroad, 8 mi/13 km WNW of (and administratively subordinate to) GORODISHCHE; 48°54'N 44°21'E. Elevation 413 ft/125 m. In agricultural area producing livestock, horses.

Kuz'molovskiy (KOOZ-muh-luhf-skeeyee), town (2005 population 9,665), central LENINGRAD oblast, NW European Russia, on railroad, 16 mi/25 km N of SAINT PETERSBURG; 60°05'N 30°29'E. Home to the Scientific-Research Institute of Hygiene, Occupational Pathology and Human Ecology.

Kuznechikha (kooz-NYE-chee-hah), urban settlement (2006 population 3,205), E central YAROSLAVL oblast, central European Russia, near the VOLGA RIVER, on road and railroad, 6 mi/10 km N of (and administratively subordinate to) YAROSLAVL; 57°42'N 39°54'E. Elevation 419 ft/127 m. Agricultural and industrial equipment, chemicals, ceramics, plastics, petroleum products.

Kuznechikha (kooz-NYE-chee-hah), village, S TATARSTAN Republic, E European Russia, near highway, 54 mi/89 km S of KAZAN, 34 mi/55 km E of ULYANOVSK, and 25 mi/40 km SSE of BULGAR; 54°43'N 49°38'E. Elevation 492 ft/149 m. In agricultural area (wheat, oats, rye, vegetables, sunflowers).

Kuznechnoye (kooz-NYECH-nuh-ye), settlement (2005 population 4,710), N LENINGRAD oblast, NW European Russia, on the KARELIAN ISTHMUS, on the NW shore of Lake LADOGA, just S of the administrative border with Republic of KARELIA, on railroad and near highway, 14 mi/23 km NW of PRIOZËRSK; 61°09'N 29°52'E. Elevation 108 ft/32 m. Granite quarrying and processing. Until 1940, in FINLAND and called Kaarlakhti.

Kuznetsk (kooz-NYETSK), city (2005 population 90,000), E PENZA oblast, E European Russia, on railroad, 75 mi/121 km E of PENZA; 53°07'N 46°36'E. Elevation 813 ft/247 m. Major highway hub and transshipment center. Instruments, condensers, polymer machinery, textile machinery, shoes, leather goods. As Naryshkino, in the 18th century, was habitually used as safe haven by highway robbers; chartered and renamed in 1780.

Kuznetsk, RUSSIA: see NOVOKUZNETSK.

Kuznetsk Alatau (kooz-NYETSK–ah-lah-TAH-oo), mountain range, KEMEROVO oblast and KHAKASS REPUBLIC, S Siberian Russia, E of NOVOKUZNETSK (Kemerovo oblast), rising to about 6,900 ft/2,103 m. Part of the great mountain system of central ASIA, the range is composed mainly of metamorphic rocks and yields such minerals as iron, manganese, and gold.

Kuznetsk Basin (kooz-NYETSK), coal basin (□ 10,000 sq mi/26,000 sq km), KEMEROVO oblast, W Siberian Russia, between the KUZNETSK ALATAU mountains and the SALAIR RIDGE. Its abbreviated name is Kuzbas. With extensive coal deposits, particularly of high-grade coking coal, the KUZNETSK Basin is the largest coal producer in RUSSIA. The main fields are around ANZHERO-SUDZHENSK, KEMEROVO, LENINSK-KUZNETSKIY, KISELEVSK, and PROKOPYEVSK. The first iron-smelting works were founded in 1697. Coal deposits were discovered in 1721 and first mined in 1851. The area's industries grew rapidly in the late 19th century, and new heavy industry was started from 1930 to 1932 when the Ural-Kuznetsk industrial combine was formed. With major plants at NOVOKUZNETSK, the Kuznetsk industrial region (□ 27,000 sq mi/69,930 sq km) produces iron and steel, zinc, aluminum, heavy machinery, and chemicals. Ores were brought from E SIBERIA for processing, and during World War II the basin's industrial importance was surpassed only by that of the Ural Mountains. Strikes by Kuznetsk coal miners since 1989 have often involved anti-government protests, and have accelerated the basin's production decline.

Kuznetsk-Sibirskiy, RUSSIA: see NOVOKUZNETSK.

Kuznetsovo (kooz-nye-TSO-vuh), village, SW MOSCOW oblast, central European Russia, near highway and railroad, 10 mi/16 km NE, and under administrative jurisdiction of NARO-FOMINSK; 55°27'N 36°56'E. Elevation 603 ft/183 m. In agricultural area; livestock feed.

Kuznetsovo (kooz-nye-TSO-vuh), village, central MOSCOW oblast, central European Russia, near the MOSKVA River, on road and railroad, 5 mi/8 km SE of (and administratively subordinate to) RAMENSKOYE; 55°30'N 38°21'E. Elevation 406 ft/123 m. Agricultural machinery; food processing (confectionery).

Kuznetsovo, RUSSIA: see KONAKOVO.

Kuznetsovs'k (kooz-ne-TSOVSK), (Russian *Kuznetsovsk*), city, NW RIVNE oblast, UKRAINE, on the right bank of the STYR River, 50 mi/80 km NNW of RIVNE; 51°41'N 25°52'E. Elevation 459 ft/139 m. Site of the Rivne (Russian *Rovno*) Nuclear Power Station; grain mill, hothouse complex. Established in 1977 on the site of the village of Varash; city since 1984.

Kuznetsovsk, UKRAINE: see KUZNETSOVS'K.

Kuznica (koos-NEE-tsah), Polish *Kuźnica*, village, Białystok province, NE POLAND, frontier station on WARSAW–SAINT PETERSBURG railroad, and 9 mi/14.5 km NE of SOKÓŁKA, 13 mi/21 km SE of GRODNO.

Kuzomen' (KOO-zuh-myen), village, S MURMANSK oblast, NW European Russia, local harbor port on the WHITE SEA, on the KOLA PENINSULA, on local coastal road, 135 mi/217 km ESE of KANDALAKSHA; 66°17'N 36°52'E. Fisheries. Also known as Kuzomen'skiy.

Kuzomen'skiy, RUSSIA: see KUZOMEN'.

Kuzovatovo (koo-zuh-VAH-tuh-vuh), town (2006 population 8,580), central ULYANOVSK oblast, E central European Russia, on road junction and railroad, 22 mi/35 km ESE of BARYSH; 53°33'N 47°37'E. Elevation 662 ft/201 m. Sawmilling; furniture, wood chemicals, mixed fodder factory.

Kuzumaki (koo-ZOO-mah-kee), town, Iwate county, IWATE prefecture, N HONSHU, NE JAPAN, 28 mi/45 km N of MORIOKA; 40°02'N 141°26'E. Dairying; lumber; charcoal.

Kuzuu (KOO-zoo), town, Nasu county, TOCHIGI prefecture, central HONSHU, N central JAPAN, 19 mi/30 km S of UTSUNOMIYA; 36°23'N 139°36'E. Cement, lime, dolomite.

Kvænangen (KVAHN-ahng-uhn), inlet of NORWEGIAN SEA, TROMS county, N NORWAY, 60 mi/97 km ENE of TROMSØ; 45 mi/72 km long, 1 mi/1.6 km–12 mi/19 km wide. At mouth are several islands, including SKJERVØY.

Kvaisi (kvah-EE-see), urban settlement on the Dzhedzhara River, South OSSETIA, GEORGIA; 42°31'N 43°39'E. Lead and zinc ores are mined from the Kvaisi deposit. Ore-dressing plant.

Kvaløya (KVAHL-uh-yah), island (□ 127 sq mi/330.2 sq km) in NORWEGIAN SEA, FINNMARK county, N NORWAY; 17 mi/27 km long, 10 mi/16 km wide. Rises to 2,047 ft/624 m. HAMMERFEST city is on W coast, near FUGLENES cape.

Kvaløya, island (□ 284 sq mi/738.4 sq km) in NORWEGIAN SEA, TROMS county, N NORWAY, 6 mi/9.7 km W of TROMSØ, from which it is separated by arm of Tromsøysundet; 30 mi/48 km long, 2 mi/3.2 km–14 mi/23 km wide. Rises to 3,392 ft/1,034 m in Skittind mountain (W). HILLESØY fishing village is on SW coast, 20 mi/32 km W of Tromsø.

Kvalsund (KVAHL-soon), fishing village, FINNMARK county, N NORWAY, on narrow Kvalsund, which separates KVALØYA from mainland, opposite STAL-LOGARGGU (ferry), 13 mi/21 km SW of HAMMERFEST.

Kvalvik, NORWAY: see FREI.

Kvam (KVAHM), village, OPPLAND county, S central NORWAY, in the GUDBRANDSDAL, on LÅGEN River, on railroad, and 50 mi/80 km NW of LILLEHAMMER. In World War II, scene (April 1940) of fighting between Anglo-Norwegian and German forces.

Kvänum (KVEN-OOM), village, SKARABORG county, SW SWEDEN, 10 mi/16 km SW of SKARA; 58°18'N 13°11'E.

Kvareli (kvah-RE-lee), urban settlement (2002 population 9,045) and center of Kvareli region, NE GEORGIA, in KAKHETIA, on S slope of the Greater Caucasus, near ALAZAN RIVER, 17 mi/27 km E of TELAVI. Winery, cognac distillery; essential-oils manufacturing; brickyard. Also spelled Qvareli.

Kvarkeno (KVAHR-kye-nuh), village (2006 population 4,155), NE ORENBURG oblast, extreme SE European Russia, in the E foothills of the S URAL Mountains, on a right tributary of the URAL River, on road, 75 mi/121 km NE of ORSK; 52°09'N 59°43'E. Elevation 1,030 ft/313 m. Gold deposits nearby.

Kvarner (KVAHR-nuhr), region, W CROATIA, encompassing a narrow strip of coast around Kvarner Gulf (Gulf of RIJEKA), in N ADRIATIC SEA. Includes CRES, LOSINJ, KRK, RAB, and PAG isls. Surrounded by high mts. (UCKA, RISNJAK) and dominated by karst topography, the region has provided limited conditions for agr. (olives, grapes, figs, vegetables; sheep). RIJEKA (Fiume) is main port of Croatia. Oil terminal on Krk isl., oil refinery at Rijeka (pipeline to SISAK); mfg. includes electronics, metals, textiles. OPATIJA Riviera attracts tourists. Traditionally seafaring population (fishing, shipbuilding, sailing, trading); the best sailors are said to come from here. One of best-known and oldest naval acads. at BAKAR. Oldest cities include Osor (Cres isl.), Krk (Krk isl.), and Rab (Rab isl.). SENJ and Rijeka grew in importance as linkages with the hinterland expanded during the Middle Ages. Hung. capital built (1870s) Rijeka-Zagreb-Budapest (Hungary) RR and Rijeka became the principal port for Hung. agr. exports prior to World War I. Region divided bet. Italy and the former Yugoslavia bet. the world wars.

Kvarner Gulf (KVAHR-nuhr), Croatian *Riječki zaljev* [=Gulf of Rijeka], Italian *Quarnero* (kahr-NER-ro), *Quarnaro* or *Carnaro* (both: kahr-NAH-ro), largest gulf of ADRIATIC SEA, W CROATIA, in KVARNER region, between ISTRIA (W) and CRES and KRK islands (E); RIJEKA (Fiume) on NE shore. Major ports include Rijeka, SENJ, BAKAR, and Mali Lošinj. KVARNERIC, Italian *Quarnerolo* (kahr-ner-RO-lo), *Quarnarolo* or *Carnarolo* (both: kahr-nah-RO-lo), a smaller gulf, lies between Cres and Lošinj (W) and Krk, Rab, and Pag (E) islands.

Kvarnerić, CROATIA: see KVARNER.

Kvarnsveden (KVAHRNS-VE-den), village, KOPPARBERG county, central SWEDEN, on DALÄLVEN RIVER, agglomerated with BORLÄNGE; 60°31'N 15°24'E.

Kvasiny, CZECH REPUBLIC: see SOLNICE.

Kvernaland (KVAHR-nah-lahn), village, ROGALAND county, SW NORWAY, 14 mi/23 km S of STAVANGER. Manufacturing of agricultural tools and machinery.

Kvernes (KVAHR-nais), village, MØRE OG ROMSDAL county, W NORWAY, on E tip of AVERØY, 7 mi/11.3 km S of KRISTIANSUND. Fishing; lumbering. Has medieval stone church.

Kvichak Bay (kah-VI-chak), S ALASKA, NE arm of BRISTOL BAY, on E side of base of ALASKA PENINSULA; 50 mi/80 km long, 30 mi/48 km wide; 58°40'N 157°34'W. Base for Bristol Bay salmon fleet. Receives KVICHAK River.

Kvichak River (kah-VI-chak), 65 mi/105 km long, S ALASKA, at base of ALASKA PENINSULA; issues from ILIAMNA Lake; flows SW to head of KVICHAK BAY.

Kvidinge (KVEE-deeng-e), village, SKÅNE county, SW SWEDEN, 15 mi/24 km ENE of HELSINGBORG; 56°08'N 13°05'E.

Kvikne (KVIK-nuh), agricultural village, HEDMARK county, E NORWAY, on small ORKLA River (tributary of GLOMMA River), and 35 mi/56 km W of Røros. Bjørnson was born here; his house is national monument. Has seventeenth-century church.

Kvillsfors (KVILS-FORSH), village, JÖNKÖPING county, S SWEDEN, on EMÅN RIVER, 15 mi/24 km E of VETLANDA; 57°24'N 15°30'E.

Kvina (KVEE-nah), river, 45 mi/72 km long, VEST-AGDER county, S NORWAY; rises in the mountains bordering AUST-AGDER county; flows S to inlet of NORTH SEA 10 mi/16 km E of FLEKKEFJORD. Forms several falls, including TRELANDSFOSS.

Kvinesdal (KVEE-nuhs-dahl), village, VEST-AGDER county, S NORWAY, on the KVINA River, 3 mi/4.8 km from its mouth, and 11 mi/18 km E of FLEKKEFJORD. Lumber milling nearby. Sarons Dal, a Pentecostal center, is close by, and is the site of religious conventions in the summer.

Kvineshei Tunnel (KVEE-nuhs-hai), VEST-AGDER county, S NORWAY, on KRISTIANSAND-STAVANGER railroad, and 30 mi/48 km WNW of Kristiansand; 29,739 ft/9,064 m long. Adjoins HÆGEBOSTAD TUNNEL (27,801 ft/8,474 m long). Both extend E from SNARTEMO.

Kvirila River (kvee-REE-lah), 80 mi/129 km long, W GEORGIA; rises in the central Greater Caucasus W of DZHAVA; flows SW, past SACHKHERE and CHIATURA, and W, past ZESTAFONI, to Rion River S of KUTAISI. Used for rafting timber. Its basin covers 1,402 sq mi/3,630 sq km. Rapids. Manganic ore deposit.

Kvissleby (KVIS-sle-BEE), town, VÄSTERNORRLAND county, NE SWEDEN, on LJUNGAN RIVER, near its mouth on GULF OF BOTHNIA, 7 mi/11.3 km SSE of SUNDSVALL.

Kviteseid (KVEE-tuhs-aid), village, TELEMARK county, S NORWAY, 41 mi/66 km WNW of SKIEN, on KVITESEIDVATN Lake on BANDAK-NORSJØ CANAL. Agriculture; fruit and wild berries; lumbering; large tourist traffic.

Kviteseidvatn (KVEE-tuhs-aid-VAH-tuhn), lake, TELEMARK county, S NORWAY, 35 mi/56 km WNW of SKIEN, part of BANDAK-NORSJØCANAL; 6 mi/9.7 km long, 1.5 mi/2.4 km wide, 692 ft/211 m deep; elevation 236 ft/72 m. Fishing; tourist resort.

Kvitok (KVEE-tuhk), village (2005 population 3,130), W IRKUTSK oblast, E central SIBERIA, RUSSIA, in the YENISEY RIVER basin, on road and short spur of the BAYKAL-AMUR MAINLINE, 18 mi/29 km SSE of SHITKINO; 56°04'N 98°28'E. Elevation 833 ft/253 m. Railroad depot; forestry services. A forced-labor camp for Japanese POW following completion of World War II was formerly in the vicinity.

Kvitoya (KVEE-tuh-yah), Norwegian *Kvitøya*, island (□ 102 sq mi/265.2 sq km) of the Norwegian possession SVALBARD, in BARENTS SEA of ARCTIC OCEAN, between NORTHEAST LAND (W) and FRANZ JOSEF LAND (E); 26 mi/42 km long (ENE-WSW), 3 mi/4.8 km–5 mi/8 km wide; 80°09'N 32°30'E. Rises to 886 ft/270 m. S. A. Andrée (1845–1897), Swedish balloonist, and two companions perished here (autumn 1897) after crashing in attempt to reach North Pole by balloon; bodies recovered in 1930.

Kvitsøy (KVITS-uh-oo), island (□ 1 sq mi/2.6 sq km), in NORTH SEA at entrance to BOKNAFJORDEN, ROGALAND county, SW NORWAY, 14 mi/23 km NW of STAVANGER. Fishing village of Leiasundet in NE. Lighthouse in SW (established 1700). Broadcasting station for medium- and short-wave radio signals.

Kwa (KWAH), town, AKWA IBOM state, SE NIGERIA, near Gulf of GUINEA, 35 mi/56 km S of UYO; 09°28'N 11°38'E. Market center. Black-eyed peas, maize, cassava; livestock.

Kwa, CONGO: see KASAI RIVER.

Kwaadmechelen (KWAHD-me-guh-len), village in commune of TESSENDERLO, Hasselt district, LIMBURG province, NE BELGIUM, near junction of ALBERT and DESSEL-KWAADMECHELEN canals, 9 mi/14.5 km NNE

of DIEST; 51°06'N 05°08'E. Chemicals. Formerly spelled Quaedmechelen.

Kwachon (GWAH-CHUHN), city, central KYONGGI province, SOUTH KOREA, just S of SEOUL; 37°26'N 127°00'E. In ancient times, it was a S gate to Seoul where trading markets were established. In 1978 this region was designated as the site to be developed into a new town where the second government complex and a huge amusement park would be located. It has rapidly grown into a satellite of Seoul, becoming a bedroom community since urbanization began in the 1980s. Horticulture; famous flower market. Connected to Seoul by subway line.

Kwadwokurom (kwah-JO-kur-uhm), town, BRONG-AHAFO region, GHANA, port on Lake VOLTA, 50 mi/80 km E of Atebubu, and 5 mi/8 km W of KETE-KRACHI; 07°46'N 00°10'W. Road terminus.

Kwa Ibo River (kwah EE-bo), 70 mi/113 km long, SE NIGERIA, in forest belt; rises near IKOT EKPENE; flows SSE to Gulf of GUINEA below EKET.

Kwai River, THAILAND: see KHWAE NOI RIVER.

Kwajalein (KWAH-jah-lain), coral atoll (□ 6 sq mi/15.6 sq km; 1999 population 10,902), central PACIFIC, in the RALIK CHAIN of the MARSHALL ISLANDS; 08°43'N 167°44'E. The largest atoll of the Marshalls, Kwajalein, consisted of a group of ninety-seven islets (including Roi and Namur, now connected by reclamation) surrounding a lagoon. A large Japanese naval and air base was located here during World War II, and after the U.S. conquest of the Marshalls (1944) U.S. military bases were established. An antimissile missile installation under control of the U.S. Navy is here.

Kwakoegron (KWAH-koo-grohn), village, BROKO-PONDO district, N SURINAME, on SARAMACCA RIVER, 40 mi/64 km SSW of PARAMARIBO; 05°13'N 55°23'W.

Kwaksan (GWAHK-SAHN), county, NORTH PYONGAN province, NORTH KOREA, 46 mi/74 km SE of SINUIJU. Agriculture, gold mining.

Kwakwani (kwah-KWAH-nee), village, UPPER DEMERARA–BERBICE district, NE GUYANA, on BERBICE RIVER, and 85 mi/137 km SSW of NEW AMSTERDAM; 05°17'N 58°03'W. Bauxite mining.

Kwale (KWAH-lai), town (1999 population 4,196), COAST province, SE KENYA, on road, and 17 mi/27 km WSW of MOMBASA; 04°10'S 39°27'E. Copra, sugarcane, fruits, dairy products.

Kwale (kwah-LE), town, DELTA state, S NIGERIA, on headstream of BENIN River, 35 mi/56 km NE of WARRI. Road center; palm oil and kernels, hardwood, rubber, cacao, kola nuts.

Kwall (KWAWL), town, Plateau state, central NIGERIA, 18 mi/29 km WSW of JOS, E of Kwall Falls (800 ft./244 m high). Tin-mining center.

Kwaluseni (kwah-loo-SEH-ni), village, MANZINI district, central SWAZILAND, 4 mi/6.4 km WNW of MANZINI; 26°29'S 31°17'E. Seat of University of Swaziland and Royal Swaziland Society of Science and Technology.

Kwaman (KWAH-mahn), town, ASHANTI region, GHANA, 10 mi/16 km SE of MAMPONG. Cocoa, coffee, gold. Also spelled KWAMANG.

Kwamang, GHANA: see KWAMAN.

KwaMashu, town, KWAZULU-NATAL, SOUTH AFRICA, 10 mi/16 km N of DURBAN, and 5 mi/8 km inland from INDIAN OCEAN coast. Began as one of a group of areas acting as residences for Black workers commuting into Durban's industrial areas. Home to former deputy president Jacob Zuma.

Kwamouth (kwah-MOOT), village, BANDUNDU province, SW CONGO, on left bank of CONGO RIVER (CONGO REPUBLIC border) at mouth of the KWA RIVER, and 180 mi/290 km SW of INONGO; 03°10'S 16°12'E. Elev. 482 ft/146 m. Trading center, steamboat landing; palm products.

Kwa Mtoro (kwahm TO-ro), village, DODOMA region, central TANZANIA, 62 mi/100 km NNW of DODOMA;

05°14'S 35°26'E. Road junction. Cattle, sheep, goats; corn, wheat.

Kwando River, ANGOLA and ZAMBIA: see CUANDO RIVER.

Kwandruma, CONGO: see NIOKA.

Kwangju (GWAHNG-JOO), county (□ 167 sq mi/434.2 sq km), central KYONGGI province, SOUTH KOREA, E of SEOUL, near Kwangju Mountains; 35°10'N 126°55'E. A narrow basin in W used for rice farming forms along the HAN and Kyongan rivers flowing to the N. Agriculture; manufacturing for export. Notable sites include Namhansansong (old fortress), Punwon (pottery), and Chonjinam (birthplace of the Korean Roman Catholic church).

Kwangju (GWAHNG-JOO), city, ⊙ SOUTH CHOLLA province, SW SOUTH KOREA, in the Yongsan River lowland, c.372 mi/599 km S of SEOUL; 35°08'N 126°55'E. The country's fifth-largest city, it is also its own province. A regional, administrative, educational, and commercial center. Manufacturing (machinery, motor vehicles, textiles, and chemicals). Railroad, road, and air hub. In the hills around Kwangju are ancient tombs and temples. Center of the Honam region; historically part of the Paekche kingdom; Called the "City of Art," it is the home of *Pansori*, a type of Korean traditional music, and the Kwangju Biennale, an international art exhibition. Seat of Chonnam National University (1952), CHOSON University (1946), and several other colleges. The Moodung Mountains arise in the E part of the city and the W part faces the Chonnam plain. Known as Muchinju during Paekche kingdom and renamed MUJU during the Silla (757); the name Kwangju was introduced during the Koryo dynasty (940). The 1980 massacre of student demonstrators in Kwangju by government forces provided a symbol that helped to topple Choi Kyu-hah's military government.

Kwangmyong (GWAHNG-MYUHNG), city, KYONGGI province, SOUTH KOREA, a satellite city of SEOUL, near ANYANG River. Promoted to city status in 1981. Anyang River often causes floods in rainy season. An irrigation system allows for agricultural production. Manufacturing includes metals, chemicals, and electronics. Connected to Seoul by subway line.

Kwango River, ANGOLA and Democratic Republic of the CONGO. See CUANGO RIVER.

Kwangsi Chuang Autonomous Region, CHINA: see GUANGXI ZHUANG AUTONOMOUS REGION.

Kwangtung, CHINA: see GUANGDONG.

Kwangyang (GWANG-YAHNG), city (□ 147 sq mi/382.2 sq km), SE SOUTH CHOLLA province, SOUTH KOREA, on Bay of Kwang-yang (S); 35°01'N 127°32'E. Bordered E by SOUTH KYONGSANG province (border, SOMJIN RIVER), N by KURYE province, and W by SUNGJU province. Hilly area, with agriculture (rice, barley, vegetables, chestnuts) in Tong and So river basins. Bay of KWANGYANG, with thirteen inhabited islands, provides for fishing and cultivation of seaweed, scallops, and ark shells. Paekun and Indok reservoirs.

Kwania, Lake (kwah-NEE-uh), S NORTHERN region, central UGANDA, NW arm of LAKE KYOGA; c.40 mi/64 km long. Partly filled with papyrus-reed swamps.

Kwansi, TAIWAN: see Kuanhsi.

Kwanto, JAPAN: see KANTO.

Kwantung, CHINA: see GUANDONG.

Kwanza River, ANGOLA: see CUANZA.

Kwara (KWAH-rah), state (□ 14,218 sq mi/36,966.8 sq km; 2006 population 2,371,089), W NIGERIA, ⊙ ILORIN; 08°30'N 05°00'E. Bordered N by NIGER state (NIGER River, border), E by KOGI state, S by EKITI, OSUN, and OYO states, and W by BENIN. In savanna zone. Agriculture includes cocoa, cashews, coffee, maize, guinea corn, yams, groundnuts, cassava; livestock. Breweries; sugar processing and manufacturing of furniture, soap, candle, and paper. Mineral resources include iron ore, feldspar, limestone, marble, kaolin, columbite. Main centers are Ilorin, Bacita, Ijagbo, Kajola, and Rorin. Owa Falls is here.

Kwash, IRAN: see KHASH.

Kwatta (KWAH-tah), village, WANICA district, N SURINAME, 6 mi/9.7 km W of PARAMARIBO; 05°49′N 55°15′W. Coffee, rice, sugarcane.

Kwazulu, region (□ 12,140 sq mi/31,564 sq km), current name of former ZULULAND region, NE KWAZULU-NATAL province, SOUTH AFRICA, on INDIAN OCEAN (E); ⊙ ULUNDI. Extends from the TUGELA RIVER in SW to S border of SWAZILAND and MOZAMBIQUE in the NE; bounded W and NW by a line extending N from DUNDEE, parallel with the foothills of the DRAKENSBERG range, to the PONGOLA RIVER, N of PAULPIETERSBURG. Rolling hills and wooded valleys with numerous SE-flowing perennial rivers, such as the Black MFOLOZI, White Mfolozi, and the Mhlatuze, carving deep courses to the ocean. Named KwaZulu [=land of the Zulu] and designated by the government of South Africa. in accordance with the Bantu Self-Government Act of 1959, to be the Zulu Bantustan (black "homeland") it is made up of isolated tracts of land, forming only a part of historical Zululand. It is, therefore, neither geographically unified nor territorially homogeneous. The area N of the Tugela River, where the largest tracts of Zulu territory lie, forms the hub of Kwazulu. Slightly over half of South Africa's Zulu population lives here, along with Xhosa, Sotho, and Swazi minorities. The Inkatha movement, an indigenous association whose membership consists primarily of migrant workers from Kwazulu, has played an important and controversial role in the political life of South Africa since 1975. Longstanding hostilities between Inkatha and the African National Congress (ANC) have led to a great deal of bloodshed in the black townships of Kwazulu.

KwaZulu-Natal, province (□ 33,578 sq mi/86,967 sq km; 2004 estimated population 9,738,305; 2007 estimated population 10,014,500), E SOUTH AFRICA, on INDIAN OCEAN (to E); ⊙ PIETERMARITZBURG and ULUNDI; 29°00′S 30°00′E. DURBAN is the largest city. Bounded N by MPUMALANGA province, SWAZILAND, and MOZAMBIQUE; W by FREE STATE province and LESOTHO, and S by EASTERN CAPE province. Rises from a narrow (except in the N around ST. LUCIA) coastal belt to an inland region fringed in the W by the DRAKENSBERG range, whose highest point in KwaZulu-Natal (Champagne Castle) is c.11,075 ft/3,376 m. The TUGELA RIVER flows W to E across central KwaZulu-Natal. Sugar refining and coal mining are two of the main industries. Sheep, cattle; citrus fruits, maize, sorghum, cotton, bananas, and pineapples are also raised. Industries, located mainly in and around Durban, include, besides sugar refineries, manufacturing (textiles, clothing, rubber, fertilizer, detergent, paper), and food-processing plants, tanneries, and oil refineries. KwaZulu-Natal produces considerable coal (especially coking coal) and timber. Good railroad network; Durban is one of South Africa's major ports. About 75% of the population is black of Zulu origin with rest of European and Asian origin. The original homelands operated on a subsistence economy based on cattle raising and corn growing. The main institutions of higher education are the Universities of Natal (Durban and Pietermaritzburg campuses) and of Durban-Westville. Natal National Park in the Drakensberg Range includes falls (c.2,800 ft/850 m) of the Tugela River. From the 17th century to the early 19th century, Bantu-speaking Zulu people became the primary inhabitants of Natal. In the 1820s and 1830s the British acquired much of Natal from the Zulu chiefs Shaka and Dingane. Afrikaner farmers (Boers) arrived in 1837 and, after battles with the Zulu (notably the Boer victory over Dingane at BLOOD RIVER in 1838), established 1838–1839 a republic. In 1843, Britain annexed Natal to Cape Colony, and a Boer exodus followed. In 1856, Natal became a separate colony. Zululand was

annexed in 1887. Sugar cane cultivation began c.1860, and many Indians (mostly indentured laborers) came to work in the sugar industry. Many Indians remained in Natal as free men after their term of indenture expired; and by 1900 they outnumbered whites and moved into the commercial sector. In 1893, Natal was given internal self-goverment, and in 1910 it became a founding province of the Union of South Africa. It was an area of considerable violence between Inkatha and African National Congress (ANC) supporters in the period prior to the birth of new South Africa (1992–1994). Formerly called Natal.

Kwei, CHINA: see GUI RIVER.

Kweichow, CHINA: see GUIZHOU.

Kwei Kiang, CHINA: see GUI RIVER.

Kweilin, CHINA: see GUILIN.

Kweishan (GWAI-SHAHN), township, N TAIWAN, 9 mi/14.5 km SW of TAIPEI, near TAOYUAN. Rapidly growing township along the main high-tech and suburban residential corridor between Taipei and CHUNGLI.

Kweishan Island (GWAI-SHAHN), volcanic island off NE TAIWAN, 13 mi/21 km NE of ILAN. Sulphur deposits; fisheries. Has caves and warm springs.

Kweiyang, CHINA: see GUIYANG.

Kwekwe (kwai-kwai), city, MIDLANDS province, central ZIMBABWE, 110 mi/177 km S of HARARE, near SEBAKWE RIVER (Lower Zimbabwe Dam to N); 18°56′S 29°49′E. Gold- and iron-mining center, focal point of Zimbabwe's iron and steel industry. Manufacturing (steel wire). Agricultural area (corn, tobacco, wheat, cotton, soybeans; cattle, sheep, goats). GLOBE AND PHOENIX Gold Mine to W. Airstrip to N. Mlezu Institute of Agriculture, Gold Mining Museum, and Midlands Museum are here. Founded 1900. Sometimes spelled Kwe Kwe and Que Que; formerly Que Quebec.

Kwenda, town, MIDLANDS province, E central ZIMBABWE, 38 mi/61 km ENE of CHIVHU; 18°49′S 31°27′E. Agriculture (corn, wheat, soybeans, tobacco, citrus); cattle, sheep, goats; dairying.

Kweneng District (KWE-neng), administrative division (2001 population 230,335), SE BOTSWANA, ⊙ MOLEPOLOLE. Bordered N by GHANZI District, S by SOUTHERN DISTRICT, SE by KGATLENG and SOUTHEAST districts, and NE by CENTRAL DISTRICT. Includes Kutse Game Reserve in N; KALAHARI DESERT in W.

Kwenlun, CHINA: see KUNLUN MOUNTAINS.

Kwenlun Mountains, CHINA: see KUNLUN MOUNTAINS.

Kwethluk, Inuit village (2000 population 713), SW ALASKA, near KUSKOKWIM River, on KWETHLUK River, and 12 mi/19 km E of BETHEL; 60°46′N 161°23′W. Formerly spelled Quithlook.

Kwidjwi Island, CONGO: see IDJWI ISLAND.

Kwidzyń (KVEE-tseen), town (2002 population 37,439), Elblag province, N POLAND, near the VISTULA River, 45 mi/72 km SSE of GDAŃSK, 35 mi/56 km SSW of ELBING. Sawmilling; power station. In World War II, c.50% destroyed; Gothic cathedral and 14th century castle remain. Teutonic Knights founded castle (1233) on nearby island in the Vistula; later rebuilt on present site; town remained one of their centers until 1526. Became seat (1254) of bishops of POMERANIA; bishopric dissolved in 1526 when Reformation was introduced in region. Until 1919 under German rule, and capital of WEST PRUSSIA province. Also called Marienwerder.

Kwigamiut (KWI-ga-mee-yoot), village, W ALASKA, at S end of NUNIVAK Island; 59°48′N 166°05′W. Traditional Native culture maintained; carving, fishing.

Kwigillingok (kwi-GI-leeng-gok), village (2000 population 338), SW ALASKA, on N side of KUSKOKWIM BAY, 65 mi/105 km NW of PLATINUM; 59°43′N 162°52′W. Formerly sometimes spelled Quillingok.

Kwiguk (KWI-guhk), village, W ALASKA, on YUKON River, 30 mi/48 km NE of its mouth, 120 mi/193 km SSE of NOME.

Kwiha, ETHIOPIA: see KUHA.

Kwilu-Ngongo (KWEE-loo–NGAWN-go), village, BAS-CONGO province, W CONGO, on railroad, and 120 mi/193 km ENE of BOMA; 05°30′S 14°41′E. Elev. 1,410 ft/429 m. Sugar plantations, sugar mills. Also known as MOERBEKE.

Kwilu River, ANGOLA and Democratic Republic of the CONGO: see CUILO RIVER.

Kwilu River, REPUBLIC OF THE CONGO: see KOUILOU RIVER.

Kwinana (kwi-NAH-nuh), city (2001 population 17,375), WESTERN AUSTRALIA state, SW AUSTRALIA, a suburb 27 mi/43 km S of PERTH; 32°15′S 115°46′E. A new industrial city, Kwinana has oil refineries, steelworks, a cement factory, and a large wheat storage terminal. "The Spectacles" twin lakes wetlands area, a haven for many bird species; public swimming beaches in COCKBURN SOUND.

Kwinella (kwee-NAL-lah), town, LOWER RIVER division, THE GAMBIA, 50 mi/80 km E of BANJUL, on S main road; 13°24′N 15°47′W. Site of battles between Islamic Marabouts and local population (1863).

Kwinhagak (KWI-nuh-gak), Inuit village, SW ALASKA, on E shore of KUSKOKWIM BAY, 50 mi/80 km N of PLATINUM; 59°45′N 161°52′W. Site of Moravian mission. Also spelled Quinhagak.

Kwinji (KWEEN-jee), village, TANGA region, NE TANZANIA, 95 mi/153 km WSW of TANGA, on MSANGASI RIVER; 05°27′S 37°45′E. Corn, wheat; cattle, goats, sheep.

Kwisa River (KVEE-sah), German Queis, 65 mi/105 km long, SW POLAND; rises in the JIZERA MOUNTAINS WSW of JELENIA GÓRA (Hirschberg); flows generally N, past GRYFOW SLASKI, LESNA (irrigation dam 2 mi/3.2 km E), and LUBAN, to BOBRAWA RIVER 5 mi/8 km ESE of ZAGAN (Sagan). Formerly in LOWER SILESIA, in German territory until 1945.

Kwoka, Mount (KWO-kah), peak (9,840 ft/3,000 m), IRIAN JAYA, INDONESIA, 79 mi/128 km ENE of SORONG; 00°31′S 132°27′E.

Kyabé (kyah-BAI), town, MOYEN-CHARI administrative region, S CHAD, 45 mi/72 km ENE of SARH. Cotton ginning. Until 1946 in French-administered Ubangi-Chari colony.

Kyabra (kei-A-bruh), settlement, SW QUEENSLAND, NE AUSTRALIA, 30 mi/48 km SW of WINDORAH, near Kyabra Creek and Goonaghooheeny Billabong; 26°17′S 143°08′E. Opal mine.

Kyabra Creek (kei-A-bruh), stream, SW QUEENSLAND, NE AUSTRALIA, near KYABRA settlement, Goonaghooheeny Billabong; 25°36′S 142°55′E.

Kyabram (kei-A-bruhm), town, N VICTORIA, SE AUSTRALIA, 105 mi/169 km N of MELBOURNE, between CAMPASPE, MURRAY, and GOULBURN rivers; 36°19′S 145°03′E. Agriculture; fruit-growing center; canned and dried fruit; clothing; production of containers. At S end of town is the community-owned Kyabram Fauna Park (□ 0.1 sq mi/0.3 sq km, or 55 ac/22 ha) featuring reptiles, water birds, an aviary, observation tower, and various free-roaming animals.

Kyaikkami (CHEI-kah-MEE), village, MON STATE, peninsular MYANMAR, in MAWLAMYINE, minor ANDAMAN SEA port on headland at S mouth of THANLWIN RIVER, and 30 mi/48 km SSW of MOULMEIN. Pilot station and seaside resort; small coastal trade. Site of Yele-Paya pagoda containing Buddhist hair relics. Formerly called Amherst.

Kyaiklat (cheik-LAHT), township, AYEYARWADY division, MYANMAR, on Bogale River (arm of AYEYARWADY RIVER delta), and 40 mi/64 km SW of YANGON. Significant deep-water rice production.

Kyaikpi (cheik-PEE), village, MYAUNGMYA township, MYANMAR, in AYEYARWADY delta, 25 mi/40 km SW of MAUBIN. Major rice area.

Kyaikthanlan Pagoda (CHEIK-thahn-LAHN), religious site, MAWLAMYINE, MYANMAR. It occupies a hilltop location that offers a magnificent view of the town and harbor.

Area is shown by the symbol □, and capital city or county seat by ⊙.

Kyaiktiyo Pagoda (CHEIK-tee-YO) [=The Golden Rock], religious site, KYAIKTO, MYANMAR, c. 56 mi/90 km NNW of MAWLAMYINE. This 18-ft/5-m high pagoda is built on a huge gold-plated boulder balanced atop a high cliff 6 mi/9.7 km E of town. Legend has it that the boulder can never fall as its balance is maintained by a hair of Buddha that is enshrined in the pagoda.

Kyaikto (CHEIK-to), township, northwesternmost township of the MON STATE, MYANMAR, on E bank of SITTANG RIVER estuary, and 70 mi/113 km NW of MAWLAMYINE on Bago-Martaban railroad. Rice-growing area.

Kya-in Seikkyi (chah-IN saik-CHEE), southernmost township in the KAYIN STATE, MYANMAR, 45 mi/72 km SE of MAWLAMYINE, on ATARAN RIVER (head of navigation). Also spelled Kya-in Seikgyi.

Kyaka (KYAH-kah), village, KAGERA region, NW TANZANIA, 25 mi/40 km W of BUKOBA, on KAGERA RIVER (bridged), N of Lake IKIMBA; 01°18′S 31°27′E. Highway junction. Coffee, corn; goats, sheep.

Kyakhta (KYAHKH-tah), city (2005 population 18,445), S central BURYAT REPUBLIC, S SIBERIA, RUSSIA, near the Mongolian border (less than 3 mi/5 km to the S), on the highway from ULAN-UDE to ULAANBAATAR (MONGOLIA), 145 mi/233 km S of Ulan-Ude; 50°21′N 106°27′E. Elevation 2,434 ft/741 m. Major transit point for Russian-Mongolian trade. Textile (spinning and knitting); forestry services; food-processing (bakery). Founded in 1728, it was a trading point between Russia and CHINA; later a trading center between Russia and OUTER MONGOLIA. Founded in 1934 by combining the city of Troitskosavsk with the village of Kyakhta.

Kyakhulay (kyah-hoo-LEI), settlement (2005 population 5,870), E central DAGESTAN REPUBLIC, SE European Russia, on road and railroad, less than 2 mi/3.2 km W of MAKHACHKALA, to which it is administratively subordinate; 42°58′N 47°29′E. In oil- and gas-producing region. Agriculture (fruits, vegetables, grapes).

Kyakisalmi, RUSSIA: see PRIOZËRSK.

Kyancutta, town, SOUTH AUSTRALIA state, S central AUSTRALIA, 152 mi/244 km W of PORT AUGUSTA, at N of EYRE PENINSULA; 33°08′S 135°31′E. Wheat.

Kyangin (CHANG-gin), township, AYEYARWADY division, MYANMAR, on W bank of AYEYARWADY RIVER, and 35 mi/56 km S of PROME. Head of railroad to HENZADA. 75% of the land in the area is devoted to rice growing.

Kyaning Co, CHINA: see GYARING CO.

Kyardla, ESTONIA: see KÄRDLA.

Kyaring Tso, CHINA: see GYARING CO.

Kyaukme (CHOUK-mai), township, SHAN STATE, MYANMAR, on Mandalay-Lashio railroad, 55 mi/89 km SW of LASHIO, 12 mi/19 km N of GOKTEIK VIADUCT.

Kyaukmedaung (chouk-mai-DOUNG), village, TAVOY township, TANINTHARYI, MYANMAR, on road, and 15 mi/24 km ENE of DAWEI, on ridge of central TENASSERIM RANGE. Mining center for tin, tungsten, molybdenum.

Kyaukmyaung (chouk-MYOUN), village, SHWEBO township, SAGAING division, MYANMAR, on W bank of AYEYARWADY RIVER, and 15 mi/24 km E of SHWEBO. Handles SHWEBO river trade.

Kyaukpadaung (chouk-mai-DOUNG), township, MANDALAY division, MYANMAR, 35 mi/56 km S of PAKOKKU, and 20 mi/32 km from the oil fields at CHAUK. Head of railroad to PYINMANA, on N plateau of PEGU YOMA, at SW foot of MOUNT POPA. Sesame, palm sugar. Pagoda is site of yearly pilgrimage.

Kyaukpyu (CHOUK-pyoo), coastal township (□ 4,793 sq mi/12,461.8 sq km) ⊙ KYAUKPYU, RAKHINE STATE, MYANMAR. Between BAY OF BENGAL and ARAKAN YOMA; includes RAMREE and CHEDUBA islands. Drained by AN RIVER. Mostly hilly and forested; narrow coastal strip is cut up by many islands in mangrove swamps. Petroleum on main offshore islands. Served by coastal steamers and AN PASS route.

Kyaukpyu, town, ⊙ KYAUKPYU township, RAKHINE STATE, MYANMAR, on N end of RAMREE ISLAND. One of best natural harbors on Arakan coast, linked by steamer with SITTWE, CHEDUBA ISLAND, SANDOWAY, and lower AN RIVER; rice trade.

Kyaukse (CHOUK-SAI), township (□ 1,241 sq mi/3,226.6 sq km), MANDALAY division, MYANMAR; ⊙ KYAUKSE. Irrigated agriculture (rice [2–3 crops yearly], sesame, beans). In dry zone (annual rainfall 30 in/76 cm). It lies S of MYITNGE RIVER in SAMON RIVER plain and Shan hills. Served by Yangon-Mandalay railroad. Population is 90% Burmese.

Kyaukse, town, ⊙ KYAUKSE township, MYANMAR, on Yangon-Mandalay railroad, and 20 mi/32 km S of MANDALAY. Trade with SHAN STATE. Pagoda built 1028.

Kyauktan (chouk-TAHN), township, YANGON division, MYANMAR, near RANGOON RIVER, 14 mi/23 km SE of YANGON.

Kyauktan Pagoda (chouk-TAHN), religious site, SYRIAM, MYANMAR. One of the first pagodas seen by travelers arriving by ship, this pagoda occupies a small island in a tiny tributary of the RANGOON RIVER c. 14 mi/23 km downstream from YANGON. Pilgrims feed the huge catfish that crowd the island. Also known as the Ye Le Paya Pagoda.

Kyauktaw (CHOUK-taw), township, RAKHINE STATE, MYANMAR, on KALADAN RIVER, and 45 mi/72 km N of SITTWE.

Kyaunggon (choung-GON), township, AYEYARWADY division, MYANMAR, in AYEYARWADY RIVER delta, 35 mi/56 km NE of PATHEIN.

Kybartai (kee-BAHR-tei), Polish and Russian *Kibarty*, city, S LITHUANIA, on railroad (*Virbalis* station), and 3 mi/5 km WNW of VIRBALIS city, adjoining Chernyshevskoye (KALININGRAD oblast); 54°39′N 22°45′E. Foundry, tin-metal works; manufacturing (knitwear, shoes, furniture, chemicals); sawmilling, oilseed pressing. Developed in 19th century around railroad customs station on German (EAST PRUSSIA)-RUSSIA border; in Russian Suvalki government until 1920.

Kyburg, tiny commune, ZÜRICH canton, N SWITZERLAND, 3 mi/4.8 km S of WINTERTHUR. Medieval castle (10th–11th century) of counts of Kyburg now historical museum. Sometimes spelled Kiburg.

Kyebogyi (chai-bo-JEE), former KARENNI STATE (□ 790 sq mi/2,046 sq km), MYANMAR; ⊙ Kyebogyi (village 20 mi/32 km S of LOIKAW in Heru-so township). Population is Red Karens. Since 1947, part of the KAYAH STATE.

Kyeintali (CHAIN-tah-lee), village, Gwe township, RAKHINE STATE, MYANMAR, on Arakan coast, 30 mi/48 km S of SANDOWAY.

Kyelang (KYAI-luhng), town, ⊙ LAHUL AND SPITI district, HIMACHAL PRADESH state, N INDIA, in PUNJAB HIMALAYA (c.10,200 ft/3,109 m), on Bhaga River (headstream of the CHENAB River), and 50 mi/80 km NE of DHARMSHALA; 32°35′N 77°02′E.

Kyenjojo, administrative district (□ 1,567 sq mi/4,074.2 sq km; 2005 population 420,300), WESTERN region, W UGANDA; ⊙ Kyenjojo; 00°30′N 30°50′E. As of Uganda's division into eighty districts, borders KIBAALE (N), MUBENDE (E), SEMBABULE (SE tip), KIRUHURA (S), KAMWENGE (SW), and KABAROLE (W) districts. Primarily agricultural (tea is most important, also cassava, coffee, finger millet, maize, potatoes, sorghum, sweet potatoes, and yams). Tea and timber respectively provide main revenue sources. Road between FORT PORTAL and KAMPALA city travels W-E through Kyenjojo town; road also travels NNE out of Kyenjojo town to HOIMA town. Formed in 2000 from N and E portions of former KABAROLE district (current Kabarole district was formed from NW portion and Kamwenge district from S portion).

Kyerong, CHINA: see GYIRONG.

Kyeryong Mountain National Park, Korean, *Kyeryong-san Kungnip Kongwon*, SE SOUTH CHUNGCHONG province, SOUTH KOREA, 15 mi/24 km W of TAEJON. Many peaks, valleys (notably Kapsa and Sangbong), streams, and waterfalls. The mountain (2,772 ft/845 m) was named *Kyeryong* [=cock dragon] because of its shape resembling a dragon with a rooster's comb. Sanctuary of many religious sects which blend Shamanism, Confucianism, Buddhism, and Christianity. Buddhist temples (Kap-sa and Tonghak-sa), hermitages, cultural and religious relics. Has 700 species of plants and 160 species of birds in area. Lodges, campgrounds, hiking trails.

Kyffhäuser (KIF-hoi-suhr), forested mountain (1,565 ft/477 m), SAXONY-ANHALT, central GERMANY; 51°24′N 11°05′E. Crowned by the two ruined castles of ROTHENBURG (7th century) and Kyffhausen (12th century) and by a huge monument to Emperor William I (erected 1896). According to legend, Emperor Frederick I (Frederick Barbarossa) sleeps bewitched in a limestone cave in the mountain, sitting at a stone table around which his beard has grown; there he awaits the time when he will go forth to restore German greatness. The legend, treated in poems by Uhland, Heine, and others, probably originally applied to Emperor Frederick II (reigned 1220–1250).

Kyi Chu, CHINA: see LHASA HE.

Kyimyindine (CHEEM-yin-di-NAI), section of YANGON, MYANMAR, NW of SHWE DAGON PAGODA. Site of several embassies, national radio station, and department of higher education.

Kyi Qu, CHINA: see LHASA HE.

Kyirong, CHINA: see GYIRONG.

Kyiv, UKRAINE: see KIEV, CITY.

Kyjov (KI-yof), German *Gaya*, town, JIHOMORAVSKY province, S MORAVIA, CZECH REPUBLIC, on railroad, and 26 mi/42 km ESE of BRNO; 49°01′N 17°07′E. Manufacturing (glass, wine, flour); lignite mining, natural-gas production in vicinity. Extensive orchards and vineyards in vicinity. Picturesque regional folkways attract many tourists. Has an ethnographic museum.

Kyklades, Greece: see CYCLADES.

Kyk-over-all (kei-KO-vuhr-awl), islet, CUYUNI-MAZARUNI district, N GUYANA, at confluence of CUYUNI and MAZARUNI rivers, 4 mi/6.4 km SW of BARTICA; 06°23′N 58°41′W. One of first Dutch settlements and once a flourishing trading post. Old fort.

Kyle (KEI-uhl), town (2006 population 423), SW SASKATCHEWAN, W CANADA, 40 mi/64 km NNW of SWIFT CURRENT; 50°50′N 108°02′W. Wheat.

Kyle (KEIL), town (2006 population 20,655), HAYS county, S central TEXAS, near BLANCO RIVER, 21 mi/34 km SSE of AUSTIN; 29°59′N 97°52′W. Agriculture (cattle, sheep; cotton; fruit, vegetables); manufacturing (steel foraging, steel fabrication).

Kyle, Lake, ZIMBABWE: see KYLE RECREATIONAL PARK.

Kyle of Lochalsh (KEIL uhv LAWK-ahlsh), fishing village (2001 population 739), W HIGHLAND, N Scotland, at mouth of LOCH ALSH, and 60 mi/97 km WSW of INVERNESS; 57°17′N 05°42′W. It was terminal of car ferry to Kyleakin, ISLE OF SKYE, but now ferry has been replaced with a bridge (built 1995). In narrow entrance to Loch Alsh from the INNER SOUND (N), called Kyle Akin, is small Gillean Island, site of lighthouse.

Kyle of Sutherland, Scotland: see OYKEL RIVER.

Kyle of Tongue, Scotland: see TONGUE.

Kyle Recreational Park (□ 351 sq mi/909 sq km), MASVINGO province, SE central ZIMBABWE, 10 mi/16 km SE of MASVINGO. Surrounds Lake Mtirikwe (formerly Lake KYLE) reservoir, on Mtirikwe River. GREAT ZIMBABWE NATIONAL MONUMENT adjoins park on SW. Wildlife includes antelope, giraffe, buffalo, hippo, and white rhino. Accommodations include a luxury rest camp.

Kyle Rhea, Scotland: see LOCH ALSH.

Cross-references are shown in SMALL CAPITALS. The pronunciation guide is shown on page xix. The sources of population figures are shown on page xvii.

Kyles of Bute (KEILZ uhv BYOOT), arm (10 mi/16 km long, c.1 mi/1.6 km wide) of Firth of Clyde, Argyll and Bute, W Scotland; extending around N part of ISLE OF BUTE, separating it from COWAL peninsula; 55°56′N 05°12′W.

Kyllburg (KIL-boorg), village, RHINELAND-PALATINATE, W GERMANY, 19 mi/31 km N of TRIER; 50°02′N 06°35′E. Tourist center in the EIFEL. Has Gothic church. Formerly also spelled KILLBURG.

Kyllene, Greece: see KILLINI.

Kyme, Greece: see KIMI.

Kymen (KUH-muhn), Swedish *Kymmene*, province (□ 4,951 sq mi/12,872.6 sq km), SE FINLAND; ⊙ KOTKA. On GULF OF FINLAND (S) and RUSSIAN border (E). Includes S part of SAIMAA lake system. Drained by KYMIJOKI, VUOKSIJOKI, and several smaller rivers, and by SAIMAA CANAL. Noted falls and hydroelectric station at IMATRA. Agriculture (rye, oats, barley), cattle raising, dairy farming. Kymijoki valley is an industrial center; lumbering, timber processing, and woodworking are important. Other industries are metalworking, iron, textile, and pulp paper milling; manufacturing (glass, cement, bricks). Granite and limestone quarries. Major towns Kotka, LAPPEENRANTA, and HAMINA. Created 1945, comprising the part of former Viipuri province not ceded to Russia.

Kymi (KUH-mee), Swedish *Kymmene*, village, KYMEN province, SE FINLAND, on KYMIJOKI (river), and 5 mi/8 km N of KOTKA; 60°30′N 26°55′E. Elevation 63 ft/25 m. Railroad junction; pulp and paper mills.

Kymi, Greece: see KIMI.

Kymijoki (KUH-mee-YO-kee), Swedish *Kymmene älv*, river, 90 mi/145 km long, SE FINLAND; issues from SE end of LAKE PÄIJÄNNE; flows through several small lakes in winding course generally SE to KUUSANKOSKI; then flows S, over several rapids, past INKEROINEN, ANJALANKOSKI, and KYMI, to GULF OF FINLAND at KOTKA. Provides hydroelectric power for one of Finland's chief industrial regions.

Kymmene, FINLAND: see KYMI.

Kyn (KIN), town, E PERM oblast, W URAL Mountains, E European Russia, near the CHUSOVAYA River, on road and railroad, 33 mi/53 km SE of LYSVA; 57°47′N 58°30′E. Elevation 944 ft/287 m. Lumber industry. Formerly known as Ust'-Kamenka.

Kynaion, Cape, Greece: see KINAION, CAPE.

Kynance Cove (KEI-nuhnz), inlet of the CHANNEL, W CORNWALL, SW ENGLAND, just NW of LIZARD Head; 49°58′N 05°13′W. Site of interesting caves and serpentine rocks.

Kynaros, Greece: see KINAROS.

Kyneton (KEIN-tuhn), town, S central VICTORIA, SE AUSTRALIA, on CAMPASPE RIVER, and 50 mi/80 km NNW of MELBOURNE; 37°15′S 144°26′E. Livestock; flour, woolen mills; dairy plant. Old gold mines nearby. Annual Daffodil and Arts festival.

Kynos Kephalai (kee-NOS ke-fah-LAI) [Greek=dog's heads], hills in SE THESSALY department, N GREECE, rising to 2,382 ft/726 m, 14 mi/23 km W of VÓLOS. Here the Theban general Pelopidas was killed in battle in defeating (364 B.C.E.) Alexander, tyrant of Pherae. The Roman consul Flamininus overwhelmed (197 B.C.E.) the army of Philip V of Macedon here. Formerly Cynoscephalae. Also spelled Kinos Kefalai.

Kynsperk nad Ohri (KIN-shperk NAHT O-hrzhee), Czech *Kynšperk nad Ohří*, German *Königsberg an der Eger*, town, ZAPADOCESKY province, W BOHEMIA, CZECH REPUBLIC, on OHRE RIVER, on railroad, and 7 mi/11.3 km NE of CHEB; 50°07′N 12°32′E. Woodworking; manufacturing (furniture and textiles); food processing. Ruins of a 13th century castle.

Kynuna, settlement, W central QUEENSLAND, AUSTRALIA, 121 mi/195 km SE of CLONCURRY, 103 mi/166 km NW of WINTON, on Landsborough Highway; 21°35′S 141°54′E. Combo Waterhole to SE, billabong mentioned in ballad "Waltzing Matilda." Opals; cattle, sheep.

Kyoga (kee-O-guh), lake (□ 790 sq mi/2,054 sq km), formed by VICTORIA NILE RIVER, primarily EASTERN region, S central UGANDA; c.100 mi/161 km long. Elevation 3,389 ft/1,033 m. It occupies part of the same depression as LAKE VICTORIA, to which it was once joined. The shallow lake has large areas of papyrus swamp. Lake Kyoga provides transportation for a large cotton-growing region. Also spelled Kioga.

Kyogase (KYO-gah-se), village, North Kanbara county, NIIGATA prefecture, central HONSHU, N central JAPAN, 9 mi/15 km S of NIIGATA; 37°50′N 139°11′E.

Kyogle (kei-O-guhl), town, NE NEW SOUTH WALES, SE AUSTRALIA, 80 mi/129 km S of BRISBANE, 20 mi/32 km N of CASINO, near QUEENSLAND border, and on WILSONS RIVER; 28°37′S 153°00′E. Timber; dairying; beef cattle; mixed farming; bananas, pineapples, mangos. Botanic gardens. Toonumbar State Forest to W; rain forest, pine plantations.

Kyogoku (KYO-go-koo), town, Shiribeshi district, Hokkaido prefecture, N JAPAN, 28 mi/45 km S of SAPPORO; 42°51′N 140°53′E. Potatoes, asparagus, sweet corn. Spring water.

Kyokushi (KYOK-SHEE), village, Kikuchi county, KUMAMOTO prefecture, W KYUSHU, SW JAPAN, 12 mi/20 km N of KUMAMOTO; 32°56′N 130°51′E. Dairying.

Kyomipo, NORTH KOREA: see SONGNIM.

Kyonan (KYO-nahn), town, Awa county, CHIBA prefecture, E central HONSHU, E central JAPAN, 28 mi/45 km S of CHIBA; 35°06′N 139°50′E. Carnations, rape, okra; yellowtail. *Daibutsu* (Buddha image) here is 102 ft/31 m high.

Kyondo (CHON-do), village, Myawadi township, KAYIN STATE, MYANMAR, on GYAING RIVER (head of navigation), and 27 mi/43 km ENE of MAWLAMYINE.

Kyong (CHONG), former W state (*ngegunhmu*) (□ 24 sq mi/62.4 sq km), SHAN STATE, MYANMAR, ⊙ Kyong, village 25 mi/40 km W of TAUNGGYI. Grassy downs.

Kyonggi (GYUHNG-GEE), province (□ 3,923 sq mi/10,199.8 sq km), NW SOUTH KOREA on YELLOW SEA (W); ⊙ SUWON; 37°36′N 127°15′E. Surrounds SEOUL; has become part of its industrial region, producing ships, iron, steel, and plate glass, electronics, machinery, and textiles. NE and E are mountainous. YOJU, ICHON, and KIMPO plains on HAN RIVER; KYONGGI plain on Imjin River; PYONGTAEK and ANSONG plains. Agriculture still plays a big part in the economy; rice, wheat, barley, pulses, fruits, and vegetables are the chief crops. Fishing and dairy farming. Gold, iron ore, graphite are mined. Over 150 islands on the Kyonggi gulf offer abundant salt fields. Inchon, SUWON, Ansong, UIJONGBU are satellites of Seoul, and linked to it by subways and expressways. The N part of the old Kyonggi province became part of NORTH KOREA after World War II. Capital moved from Seoul to Suwon 1967.

Kyongju (GYUHNG-JOO), city, NORTH KYONGSANG province, SOUTH KOREA, 35 mi/56 km E of TAEGU; 35°50′N 129°12′E. Railroad junction, in coal-mining area; tourist center. Museum containing historical relics, 7th-century temple, 6th-century monastery, and ancient observatory. Nearby is a cave with walls sculptured in 8th century. Was capital (57 B.C.E.–C.E. 935) of kingdom of Silla.

Kyongju National Park, Korean, *Kyongju Kungnip Kongwon*, SE NORTH KYONGSANG province, SOUTH KOREA. KYONGJU capital of the Silla Kingdom (57 B.C.E.–935 C.E.), is Korea's tourist mecca. Numerous historic sites, cultural relics, legendary monuments. Showcase of national treasures, valuable antiques and Buddhist culture. Major historic relics include Pulguksa Temple, Sokkuram Grotto, Chomsongdae Observatory, tombs of Chonmachong, Kumgwanchong, General Kim Yu Shin, King Muyol, Onung (five tombs of ancient kings and queens), Posokchong Bower, Kyongju National Museum, Pomun Lake Resort, and Yangdong Folk Village.

Kyongsan (GYUHNG-SAHN), city (□ 15 sq mi/39 sq km), S NORTH KYONGSANG province, in Kyongsan county, SOUTH KOREA, just E of TAEGU city; 35°49′N 128°44′E. Low hills; Nam River flows through city, forming a flood field used for agriculture (vegetables, fruits; dairy). Center of Korean apple production. Industry (food processing; textiles, machinery, chemicals). Served by Kyongbu railroad and Expressway. Seat of Yongnam University. Has old tombs; Koryo Dynasty school with Confucian shrine.

Kyongsang-namdo, SOUTH KOREA: see SOUTH KYONGSANG.

Kyongsang-pukdo, SOUTH KOREA: see NORTH KYONGSANG.

Kyongsong, SOUTH KOREA: see SEOUL.

Kyongsong, NORTH KOREA: see ORANG.

Kyonpyaw (chon-PYOU), township, AYEYARWADY division, MYANMAR, on Daga River (arm of AYEYARWADY RIVER delta), and 45 mi/72 km NE of BASSEIN. Head of navigation for steamers.

Kyosato (KYO-sah-to), town, Abashiri district, Hokkaido prefecture, N JAPAN, 174 mi/280 km E of SAPPORO; 43°49′N 144°35′E. Potatoes, sugar beets; distilled alcoholic beverages (*shochu*).

Kyotera (CHO-ter-ah), town (2002 population 7,590), CENTRAL region, S UGANDA, 22 mi/35 km SW of MASAKA; 00°39′S 31°32′E. Agricultural center (coffee, bananas).

Kyoto (KYO-to), prefecture (Japanese *fu*) (□ 1,784 sq mi/4,621 sq km; 1990 population 2,602,520), S HONSHU, W central JAPAN, on the SEA OF JAPAN (N) and LAKE BIWA (E); ⊙ KYOTO. Bordered E by SHIGA prefecture, S by NARA prefecture, SW by OSAKA prefecture, and W by HYOGO prefecture. Covered predominantly by Tamba Mountains. Principally urban, with major part of population centered in Kyoto area. Scattered rice fields; lumbering in mountainous interior. Manufacturing (textiles, rubber goods), woodworking. Kyoto city is center of artistic crafts. Other centers include FUKUCHIYAMA, MAIZURU, UJI. Had the largest industrial production of any prefecture in Japan until World War II.

Kyoto (KYO-to), city (2005 population 1,474,811), ⊙ KYOTO prefecture, S HONSHU, W central JAPAN, on the Kamo River; 35°00′N 135°46′E. Yodo is its port. Japan's sixth-largest city and an important cultural and spiritual center. Key city in Japan's transportation system, and a major center of tourism. Industries include metal engineering and the manufacturing of computer components, precision machines, condensers, lingerie; lumber. Traditional products include textiles, sake, confections, pottery, Buddhist altar fittings, fans, ceramics; also, *yuzen* silk printing. Famous for its cloisonné, bronzes, damascene work, porcelain, and lacquerware, and its renowned silk industry dates from 794. Founded in the 8th century as Uda and named Heian-kyo when it became capital of Japan in 794, the city was popularly called Miyako or Kyoto (sometimes Kioto). After 1192 it lost its de facto political power to other centers; but since 1868, when TOKYO became the official national capital, Kyoto has often been referred to as Saikyo [=western capital]. For centuries it has been the cultural heart of Japan; it contains magnificent art treasures and is the seat of Kyoto University, Doshisha University (founded in 1873 as an American mission college), and other institutions of higher education. Site of the Kyoto Municipal Museum of Art and the Kyoto National Museum. Rich in historic interest, Kyoto is the site of the tombs of many famous Japanese; the old imperial palace as well as Nijo Castle (former palace of the shoguns), with their fine parks and gardens, are also here. In addition, Kyoto is a religious center, noted especially for the Buddhist Kinkaku, Ginkaku, and Kiyomizu temples, its Heian shrine (a Shinto holy place), and its 59-ft/18-m *daibutsu* (Buddha image). The Gion New Bridge is another important site. The Gion, Aoi, and Jidai festivals here are among the country's most famous. Just outside the city are Kuramayama, Sannen slope, and Daimonji mountains.

Kyowa (KYO-wah), town, Senhoku county, Akita prefecture, N HONSHU, NE JAPAN, 12 mi/20 km S of AKITA city; 39°36′N 140°19′E.

Kyowa, town, Shiribeshi district, Hokkaido prefecture, N JAPAN, 37 mi/60 km W of SAPPORO; 42°58′N 140°36′E. Watermelons.

Kyowa, town, Makabe county, IBARAKI prefecture, central HONSHU, E central JAPAN, 43 mi/70 km S of MITO; 36°19′N 140°02′E. Watermelon.

Kyparissia, Greece: see KIPARISSIA.

Kyperounda (kee-pee-ROON-dah), town, LIMASSOL district, central CYPRUS, in TROODOS MOUNTAINS; 34°56′N 32°56′E. Wine grapes, fruit, nuts, vegetables; livestock. Mountain resort area. Also spelled Kyperounta.

Kypros: see CYPRUS.

Ky Qu, CHINA: see LHASA HE.

Kyra (ki-RAH), village (2005 population 4,390), SW CHITA oblast, S SIBERIA, RUSSIA, less than 10 mi/16 km N of the Mongolian border, on road, 185 mi/298 km SSW of CHITA; 49°35′N 111°58′E. Elevation 2,998 ft/913 m. Gold mines; forestry services. Sometimes also spelled, and pronounced, Kora.

Kyra Panagia, Greece: see PELAGONESI.

Kyren (ki-RYEN), settlement (2005 population 5,300), SW BURYAT REPUBLIC, S SIBERIA, RUSSIA, on the IRKUT RIVER, on highway between IRKUTSK and NW MONGOLIA, 95 mi/153 km WSW of Irkutsk; 51°41′N 102°07′E. Elevation 2,447 ft/745 m. In agricultural and livestock-raising area. Formerly a town; reduced to status of rural settlement in 1990. Sometimes also pronounced Koren.

Kyrenia (kee-REN-yah), Turkish *Girne*, district (□ 247 sq mi/642.2 sq km), N CYPRUS, bounded on N and W by the MEDITERRANEAN SEA; ⊙ KYRENIA. Traversed E-W by the KYRENIA MOUNTAINS; Cape KORMAKITI in NW. Predominantly agricultural, producing carobs, olives, almonds, citrus, grapes, grain; goats, sheep, poultry. Principal agricultural center is LAPITHOS.

Kyrenia (kee-REN-yah), [Turkish=*Girne*], town, ⊙ KYRENIA district, N CYPRUS, 12 mi/19 km N of NICOSIA; 35°20′N 33°19′E. Minor MEDITERRANEAN port; KYRENIA MOUNTAINS to S. Manufacturing (food processing; vegetable oils). Fish; goats, sheep, cattle, poultry; grapes, vegetables, potatoes, cotton. Dominated by 12th century castle; 4 mi/6.4 km SE is the Abbey of Bellapaise; Ayios Hilarion castle (10th century) 3 mi/4.8 km SSW; Chapel of St. George (10th century). Town founded 1205; most buildings date from 14th century Casino; ferry to Tasucu (Turkey).

Kyrenia Mountains (kee-REN-yah), Turkish *Girne*, N CYPRUS, extend c. 100 mi/161 km along MEDITERRANEAN coast from Cape KORMAKITI E to Cape APOSTOLOS ANDREAS on KARPAS PENINSULA. Rises to 3,360 ft/1,024 m at Mt. Kyparissovouni, in W. Also called PENTADAKHTYLOS MOUNTAINS.

Kyrgyz Republic: see KYRGYZSTAN.

Kyrgyzstan (kuhr-guhz-STAHN), republic (□ 77,199 sq mi/199,945 sq km; 2004 estimated population 5,081,429; 2007 estimated population 5,284,149), central ASIA; ⊙ BISHKEK. Also spelled Kyrghyzstan.

Geography
Borders on CHINA (SE), KAZAKHSTAN (N), UZBEKISTAN (SW), and TAJIKISTAN (S). Bishkek and OSH are the chief cities. Mountainous country in the TIANSHAN and PAMIR mountain systems, rising to 24,403 ft/7,438 m at POBEDA PEAK on the Chinese border. Ninety-four percent of the country is over 3,281 ft/1,000 m above sea level, with an average elevation of 9,022 ft/2,750 m. Rich pasturage for goats, sheep, cattle, and horses, and vertical vegetation zones: cold summer pastures of the short-grass alpine meadows (above 10,000 ft/3,048 m; summer temperature 32°F/0°C–50°F/10°C); a high-grass subalpine zone (6,000

ft/1,829 m–10,000 ft/3,048 m; summer temperature 50°F/10°C–60°F/15.5°C), which includes the upper NARYN River valley and constitutes the most important grazing area; a warm agricultural zone (4,000 ft/1,219 m–6,000 ft/1,829 m; summer temperature 60°F/15.5°C–72°F/30.8°C), which has fall and winter pastures, irrigated agriculture, and dry farming; and a hot agricultural zone (below 4,000 ft/1,219 m; summer temperature 72°F/30.8°C–82°F/75°C). The climate is extremely continental with great regional variation. Summer temperatures reach 104°F/40°C in the S, while daytime winter temperatures average from 35.6°F/2°C in Bishkek to −40°F/−40°C in the mountains. Rainfall averages 17.7 in/450 mm per year.

Population
The Kyrgyz, a Muslim, Turkic-speaking pastoral people, constitute 64.9% of the population; the rest are Russians (12.5%) and Uzbeks (13.8%), with smaller minorities of Ukranians, Tatars, Kazaks, Germans, Tajiks, Koreans, and Muslim Chinese (Dungans and Uighurs). The population is two-thirds rural. The republic's cultural life stresses epic poems, tales, and folk songs, and the Kyrgyz are know for such traditional crafts as wood carving, rug weaving (especially distinctive felt carpets known as shyrdaks) and jewelry making. Kyrgyzstan State University (now Kyrgyz State National University) was established in 1951 and the Kyrgyzstan Academy of Sciences in 1954; by 1994 there were twenty-one universities throughout the country.

Economy
Over 80% of the cultivated area is irrigated. Cotton, sugar beets, tobacco, wheat, potatoes, fruit, and grapes are grown; sericulture is carried on; and grain crops are cultivated in the nonirrigated areas; horse, sheep, and cattle stockbreeding. There are antimony, gold, molybdenum, tin, coal, tungsten, mercury, uranium, petroleum, and natural gas deposits. Industries include food processing, sugar refining, nonferrous metallurgy, and the manufacture of agricultural machinery, textiles, and building materials. Hydropower is exported.

History: to 1978
Formerly known by the Russians as Kara [=black] Kyrgyz to distinguish them from the Kazaks (at one time called Kirghiz), the Kyrgyz migrated here from the region of the upper YENISEI River, where they had lived from the 7th to the 17th centuries. The area came under the rule of the Kokand khanate in the 19th century and was gradually annexed by RUSSIA between 1855 and 1876. The nomadic Kyrgyz resisted conscription into the czarist army in 1916. Many Kyrgyz were killed, especially in the area near the W end of ISSYK-KOL lake, and many fled to CHINESE TURKISTAN at this time. Bolshevik control was established 1917–1921. The area was formed into the Kara-Kirghiz Autonomous Region within the RSFSR in 1924, an autonomous republic in 1926, and a constituent republic (Kirghiz SSR) in 1936. Most of the Kyrgyz Soviet leaders died in a political purge (1936).

History: 1978 to Present
In 1978 a new constitution was adopted. Kyrgyzstan became sovereign in 1990 and fully independent in 1991 with the breakup of the former Soviet Union. Askar Akayev became (1991) its first president, and it was a signatory to the treaty establishing the COMMONWEALTH OF INDEPENDENT STATES. A new constitution established the Kyrgyz Republic in 1993. Akayev sought more radical economic and political reforms than the other Central Asian leaders, but like them he used referendums (1996, 2003) to expand his powers. In 1998 Kyrgyzstan became the first former Soviet republic to join the World Trade Organization. Parliamentary elections in 2005 resulted in a lopsided government win, and sparked opposition protests that led to Akayev's fleeing the country. Opposition leader Kurmanbek Bakiyev became prime minister and act-

ing president. Bakiyev won election to the presidency in July 2005, but by the following year a lack of progress on reform and crime had led to widespread unhappiness with his presidency. In late 2006 the opposition and government battled over constitutional revisions, leading to amendments that first reduced and then restored some of the president's powers, but that amendment process was declared invalid by the constitutional court in 2007. A referendum in October 2007 finally approved the amendments, but independent observers said the turnout was inflated and there was evidence of ballot fraud.

Government
Kyrgyzstan is governed under the constitution of 1993 as amended. The president, who is head of state, is elected by popular vote for a five-year term and is eligible for a second term. The government is headed by the prime minister, who is appointed by the president. The unicameral legislature consists of the ninety-member Supreme Council (*Jogorku Kenesh*); members are popularly elected on a proportional basis to five-year terms. The current head of state is President Kurmanbek Bakiyev; the current head of government is Prime Minister Almaz Atambayev. Administratively, the country is divided into seven regions (Batken, CHÜY, ISSYK-KOL, JALAL-ABAD, NARYN, OSH, and TALAS) and one city (BISHKEK).

Kyritz (KYOO-rits), town, BRANDENBURG, E GERMANY, 25 mi/40 km N of RATHENOW; 52°57′N 12°23′E. Has late-Gothic church, many half-timbered houses. Swedes under Torstensson here defeated (1635) Saxons.

Kyrkebyn (SHIR-ke-BEEN), village, VÄRMLAND county, W SWEDEN, near NW shore of LAKE VÄNERN, 17 mi/27 km WSW of KARLSTAD.

Kyrkerud, SWEDEN: see SÄFFLE.

Kyrkesund (SHIR-ke-SUND), fishing village, GÖTEBORG OCH BOHUS county, SW SWEDEN, on Härön island (2 mi/3.2 km long) in SKAGERRAK strait, 18 mi/29 km S of LYSEKIL.

Kyrksæterøra (CHUHRK-sah-tuh-RUH-rah), village, SØR-TRØNDELAG county, central NORWAY, at head of Hemnefjorden (12 mi/19 km inlet of NORTH SEA), 40 mi/64 km WSW of TRONDHEIM. Light manufacturing and municipal administration center.

Kyrkslätt (CHUHRK-slet), Finnish *Kirkkonummi*, village, UUDENMAAN province, S FINLAND, 17 mi/27 km W of HELSINKI; 60°07′N 24°26′E. Elevation 83 ft/25 m. In former Soviet-leased defence region of PORKKALA. Has fourteenth-century church.

Kyrleuts'ke Lake, UKRAINE: see PEREKOP LAKES.

Kyrnasivka (kir-NAH-seef-kah), (Russian *Kirnasovka*), town (2004 population 6,600), S central VINNYTSYA oblast, on road and railroad, 7 mi/12 km SE of TUL'CHYN; 48°35′N 28°58′E. Elevation 889 ft/270 m. Sugar refinery; reinforced concrete fabrication. Fowl husbandry, forestry. Heritage museum. Known since the 18th century; town since 1971.

Kyróskoski (KUH-ruh-KOS-kee), Swedish *Kyrófors*, village, HÄMEEN province, SW FINLAND, 20 mi/32 km NW of TAMPERE; 61°40′N 23°11′E. Elevation 330 ft/100 m. In lake region; manufacturing (pulp and paper mills).

Kyrykivka (ki-RI-keef-kah), (Russian *Kirikovka*), town (2004 population 3,600), SE SUMY oblast, UKRAINE, on left bank of the VORSKLA RIVER, and on railroad, 10 mi/17 km ENE of OKHTYRKA; 50°21′N 35°06′E. Elevation 403 ft/122 m. Sugar refinery. Established at the beginning of the 17th century; town since 1956.

Kyrylivka (ki-RI-leef-kah), (Russian *Kirillovka*), town (2004 population 1,100), SW ZAPORIZHZHYA oblast, UKRAINE, on the coast of the Sea of AZOV, at the SW corner of MOLOCHNA LAGOON, 32 mi/52 km S of MELITOPOL; 46°22′N 35°22′E. Fish cannery; seaside resort; sanatorium, with use of medicinal muds of Liman; youth camps, park, and nature preserves nearby.

Kyrylivka, UKRAINE: see SHEVCHENKOVE.

Kyselka, CZECH REPUBLIC: see OSTROV.

Cross-references are shown in SMALL CAPITALS. The pronunciation guide is shown on page xix. The sources of population figures are shown on page xvii.

Kyserike, hamlet, ULSTER county, SE NEW YORK, on RONDOUT CREEK, 12 mi/19 km SW of KINGSTON; 41°48′N 74°10′W. On border of MOHONK LAKE Preserve. Also called ALLIGERVILLE, which in fact lies just W. Summer recreation.

Kyshen'ky (ki-shen-KEE), (Russian *Kishen'ki*), town, SE POLTAVA oblast, UKRAINE, on NE shore of DNIPRODZERZHYNS'K reservoir, at the mouth of the VORSKLA RIVER, 35 mi/56 km ESE of KREMENCHUK, and 1.4 mi/2.3 km E of SVITLOHIRS'KE; 48°53′N 34°08′E. Elevation 200 ft/60 m. Former raion center; river port; wheat, corn.

Kyshtovka (kish-TOF-kah), village (2006 population 5,700), NW NOVOSIBIRSK oblast, SW SIBERIA, RUSSIA, on the TARA RIVER (head of navigation), in the IRTYSH River basin, on road, 85 mi/137 km N of CHANY; 56°33′N 76°38′E. Elevation 272 ft/82 m. In dairy-farming area.

Kyshtym (kish-TIM), city (2005 population 41,530), N CHELYABINSK oblast, RUSSIA, on the extreme NE slope of the S URAL Mountains, amongst several small lakes, 55 mi/89 km NW of CHELYABINSK; 55°42′N 60°34′E. Elevation 862 ft/262 m. Railroad junction. Major center of copper smelting and refining, based on Karabash deposits; copper-refining machinery, boilers, construction machinery, radios, abrasive materials; mining, refractory, and processing of numerous minerals, especially graphite and kaolin; furniture factory; food industries (bakery, dairy, fish processing). Founded in 1757 as ironworks of Verkhniy Kyshtym. Industry shifted from ferrous to copper metallurgy after World War I. Became city in 1934; absorbed the town of Severnyy (N) in 1948.

Ky Son or **muong xen**, town, NGHE AN province, N central VIETNAM, on VINH–LUANG PRABANG (LAOS) highway, and 105 mi/169 km NW of Vinh; 19°23′N 104°10′E. Market center. Shifting cultivation and agro-forestry. Thai, H'mong, and other minority peoples. Formerly Muangsen (Muong Xen).

Kysperk, CZECH REPUBLIC: see LETOHRAD.

Kysucke Nove Mesto (ki-SUTS-kai no-VAI mes-TO), Slovak *Kysucké Nové Mesto*, Hungarian *Kiszucaújhely*, town, ZAPADOSLOVENSKY province, NW Slovakia, 5 mi/8 km NNE of ŽILINA; 49°18′N 18°47′E. Manufacturing (machinery, footwear, building materials; food processing, woodworking); lumber. Has a 17th-century Renaissance brewery. Founded in the 13th century.

Kytay Lagoon (ki-TEI) (Russian *Kitay*) (Romanian *Chitai*), SW ODESSA oblast, UKRAINE, near Kiliya arm of the DANUBE River delta, NW of KILIYA; 15 mi/24 km long, 1 mi/1.6 km–2 mi/3.5 km wide and narrows in the middle 0.4 mi/0.6 km across. Receives minor Kytay River (N).

Kyte River, c.40 mi/64 km long, N ILLINOIS; rises in E LEE county; flows generally NW to ROCK RIVER 3 mi/4.8 km S of OREGON; 42°00′N 89°04′W.

Kythera, Greece: see KÍTHIRA.

Kythnos, Greece: see KITHNOS.

Kythrea (keth-ree-YAH), [Turkish=*degirmenlik*], town, LEFKOSIA district, N CYPRUS, at S foot of KYRENIA MOUNTAINS, 9 mi/14.5 km NE of NICOSIA, 1 mi/1.6 km S of MEDITERRANEAN coast; 35°14′N 33°38′E. KYRENIA MOUNTAINS to S. Grain, vegetables, citrus, carobs, grapes, olives; goats, sheep.

Kytlym (kit-LIM), town (2006 population 870), W SVERDLOVSK oblast, extreme W Siberian Russia, in the central URAL Mountains, on the upper Lobva River (OB′ RIVER basin), on road, 18 mi/29 km N of PAVDA; 59°30′N 59°12′E. Elevation 1,099 ft/334 m. Platinum mining.

Kytmanovo (KIT-mah-nuh-vuh), village, NE ALTAI TERRITORY, S central SIBERIA, RUSSIA, on the CHUMYSH RIVER (OB′ RIVER basin), on road, 70 mi/113 km E of BARNAUL; 53°28′N 85°28′E. Elevation 574 ft/174 m. In agricultural area (flax, hemp, corn; dairy livestock).

Kyulyunken (kyoo-lyoon-KYEN), village, E SAKHA REPUBLIC, E central RUSSIAN FAR EAST, in the VERKHOYANSK RANGE, on the Tompo River (right affluent of the ALDAN RIVER), on highway, 16 mi/26 km N of TOMPO; 64°09′N 136°03′E. Elevation 2,106 ft/641 m. Trading post; reindeer raising.

Kyunhla (choon-LAH), township, SAGAING division, MYANMAR, on MU RIVER, and 60 mi/97 km NNW of SHWEBO.

Kyupriya, Bulgaria: see PRIMORSKO.

Kyur (KYOO-uhr), urban settlement, Shamkir region, AZERBAIJAN; 40°59′N 46°05′E.

Kyuragi (KYOO-rah-gee), town, East Matsuura county, SAGA prefecture, KYUSHU, SW JAPAN, 16 mi/25 km N of SAGA; 33°19′N 130°04′E.

Kyurdakhany (kyuhr-dah-KHAHN-NEE), urban settlement, Sabuchinsky district of Greater BAKY, AZERBAIJAN; 40°32′N 49°55′E.

Kyurdamir (kyuhr-DAH-mir), city, central AZERBAIJAN, on railroad, and 85 mi/137 km W of BAKY. In cotton and wheat district; wine making; metalworking, carpet weaving.

Kyushu (KYOO-shoo), island (□ 13,760 sq mi/35,776 sq km), SW JAPAN. The third-largest, southernmost, and most densely populated of the major islands of Japan. Separated from SHIKOKU by the BUNGO STRAIT and from HONSHU by the SHIMONOSEKI STRAIT; a railroad tunnel under the strait and a bridge link Kyushu with Honshu. Mainly of volcanic origin, the island has a mountainous interior rising to 5,886 ft/1,794 m in Kuju-san; Aso-san, Japan's largest active volcano, is here, and there are many hot springs. The CHIKUGO RIVER (88 mi/142 km long), the island's longest, waters an extensive rice-growing area in the NW. Has a subtropical climate and receives much precipitation. Rice, tea, tobacco, sweet potatoes, fruits, wheat, and soybeans are major crops. Coal, zinc, and copper are mined and raw silk is extensively produced. Noted for its porcelain (Satsuma and Hizen ware). The famous Imari ware was manufactured at the ancient town of ARITA. Heavy industry is concentrated in N Kyushu, near Japan's oldest coalfield; KITAKYUSHU, FUKUOKA, and OMUTA are major industrial centers. NAGASAKI, the chief port of Kyushu, was the first Japanese port to receive Western trade. There are four national parks on the island. Divided into FUKUOKA, KAGOSHIMA, KUMAMOTO, MIYAZAKI, NAGASAKI, OITA, and SAGA prefectures.

Kyustendil (kyoo-sten-DAIL), city, SOFIA oblast, Kyustendil obshtina (1993 population 77,716), SW BULGARIA, near the Yugoslav (Serbian) border; 42°16′N 22°41′E. Famous for its mineral springs used to heat hothouses, Kyustendil is a market city for fruit and other agricultural produce. There are varied light industries; manufacturing (sheet iron, wire, furniture, knitting, shoes). The city's history dates to Roman times. Pautalia then, it was the capital of an independent Bulgarian principality when the Turks took it in the 14th century. The city remained under Turkish rule until 1878, when it became part of Bulgaria. Also spelled Kustendil.

Kyusyur (kyoo-SYOOR), settlement, N SAKHA REPUBLIC, N RUSSIAN FAR EAST, N of the ARCTIC CIRCLE, on the lower LENA RIVER, opposite BULUN (3 mi/5 km to the N), on road, 260 mi/418 km NW of VERKHOYANSK; 70°41′N 127°22′E. Elevation 213 ft/64 m. Trading post; fishing; reindeer farms.

Kywebwe (chwe-BWE), village, OKTWIN township, BAGO division, MYANMAR, on Yangon-Mandalay railroad, and 17 mi/27 km S of TOUNGOO. Road to ferry across SITTANG RIVER, 2 mi/3.2 km E.

Kyyiv, UKRAINE: see KIEV.

Kyyivs'ka oblast, UKRAINE: see KIEV oblast.

Kyyivs'ke vodoskhovyshche, UKRAINE: see KIEV RESERVOIR.

Kyzburun Pervyy (kiz-boo-ROON PYER-viyee) [Russian=Kyzburun the first], village (2005 population 5,895), N central KABARDINO-BALKAR REPUBLIC, N CAUCASUS, S European Russia, on the BAKSAN RIVER, on highway, 13 mi/21 km NW of NAL'CHIK, and 7 mi/11 km W of BAKSAN, to which it is administratively subordinate; 43°39′N 43°23′E. Elevation 1,922 ft/585 m. In agricultural area; produce processing.

Kyzburun Tretiy, RUSSIA: see DUGULUBGEY.

Kyzburun Vtoroy, RUSSIA: see ISLAMEY.

Kyzl-Adyr, village, NW TALAS region, KYRGYZSTAN, on TALAS RIVER, and 32 mi/51 km W of TALAS, on SW shore of reservoir of same name; 42°28′N 71°33′E. Wheat, tobacco; dairy products. Motor vehicle repair. Tertiary-level administrative center. Until 1937, aleksandrovskoye. Also KIROVSKOYE.

Kyzyl (ki-ZIL), city (2006 population 109,680), ⊙ TUVA REPUBLIC, S SIBERIA, on the YENISEY RIVER, 2,913 mi/4,688 km E of MOSCOW; 51°42′N 94°27′E. Elevation 2,073 ft/631 m. Center of a basin between the SAYAN MOUNTAINS to the N and Mongolian border to the S. Major regional highway hub; services motor transport and has brickyards, sawmills, furniture factories, and food-processing plants. Founded in 1914, the city was called Belotsarsk until 1918, and Khem-Beldyr between 1918 and 1926. Center of Tanu Tuva and Tuva, an independent country, 1921–1944. It has a Tuvinian language, history, and literature research institute (founded in 1953) and a Buddhist temple complex. Coal mining nearby. It is approximately at the geographical center of the continent of Asia, but is in a relatively isolated peripheral position with respect to transport lines and centers of economic development. Also called Kyzyl-Khoto.

Kyzylagach Preserve (KEE-zeel-ahg-ahch) (□ 340 sq mi/884 sq km), LÄNKÄRÄN region, AZERBAIJAN. Includes KIROV Gulf and N part of the Malyi Kyzylagach Gulf. Established 1929 to protect largest migrating grounds of waterfowl in former USSR.

Kyzyl-Art Pass, TAJIKISTAN: see TRANS-ALAI.

Kyzyl-Asker (kuh-zuhl–uhs-KER), village and suburb of BISHKEK, Bishkek region, KYRGYZSTAN, in CHU valley, on railroad. Sugar beets; orchards. Until 1944 called Chalakazaki. Also spelled KIZYL-ASKER.

Kyzyl-Dash (KEE-zeel–DUSH), urban settlement, Karadag district of Greater BAKY, AZERBAIJAN; 40°17′N 49°34′E.

Kyzyl-Dzhar, KYRGYZSTAN: see KYZYL-JAR.

Kyzyl-Jar (kuh-zuhl–JAHR), town, JALAL-ABAD region, KYRGYZSTAN, in FERGANA VALLEY, 14 mi/22 km SW of TASH-KÖMÜR; 41°06′N 72°05′E. Also spelled KYZYL-DZHAR.

Kyzyl-Khoto, RUSSIA: see KYZYL.

Kyzylkoga (kee-zeel-ko-GAH), town, NW ATYRAU region, KAZAKHSTAN, 100 mi/161 km NNW of ATYRAU; 48°28 N 53°05 E. Cattle breeding. Also Kzyl-Kuga, Qyzylqoga.

Kyzylkuduk, village, NAWOIY wiloyat, central UZBEKISTAN, on NAWOIY-UCHKUDUK railroad, just E of ZERAVSHAN; 41°40′N 63°58′E. Also Kyzyl-Kuduk.

Kyzyl Kum (kee-ZEEL KUM), [Turk.=red sand], desert (□ 115,000 sq mi/299,000 sq km), SW KAZAKHSTAN and central UZBEKISTAN. This vast region SE of the ARAL SEA between the AMU DARYA and SYR DARYA rivers consists mainly of rocky areas covered by sparse vegetation and shifting sand dunes. Cotton, rice, and wheat are grown in river valleys and irrigated oases. Seminomadic tribesmen raise karakul sheep and camels. Gold and natural-gas deposits are exploited. Also Kyzyl-Kum, Kizil Kum, Qyzyl Qum, Qizilkum.

Kyzyl-Kum, KAZAKHSTAN: see KYZYL KUM.

Kyzyl-Kyya (kuh-zuhl–kee-YAH), city (1999 population 31,844), N OSH region, KYRGYZSTAN, in FERGANA VALLEY, on railroad spur, 38 mi/61 km SW of OSH; 40°07′N 72°10′E. Important coal-mining center; power plant.

Kyzyl-Mazhalyk (ki-ZIL–mah-zhah-LIK), town (2006 population 4,990), SW TUVA REPUBLIC, S SIBERIA, on the KHEMCHIK RIVER (YENISEY RIVER basin), on road, 160 mi/257 km WSW of KYZYL; 51°08′N 90°36′E. Elevation 2,791 ft/850 m. In agricultural area (feed corn, hardy grain cultures).

Kyzyl-Oktyabr'skiy (ki-ZIL–uhk-TYABR-skeeyee), settlement (2005 population 3,610), central KARACHEVO-CHERKESS REPUBLIC, S European Russia, in the NE CAUCASUS Mountains, on highway branch, 7 mi/11 km NW of KARACHAYEVSK; 43°49′N 41°47′E. Elevation 3,435 ft/1,046 m. In agricultural region. Also known as Kyzyl-Oktyabr'. Formerly known as Krasnyy Oktyabr' (Russian=red October).

Kyzylorda (kee-zee-lor-DAH), region (□ 88,100 sq mi/ 229,060 sq km), SW KAZAKHSTAN, on ARAL SEA; ⊙ KYZYLORDA. In TURAN Lowlands; drained by the SYR DARYA (river); includes sandy deserts (Aral KARA KUM and KYZYL KUM). Dry, sharply continental climate. Irrigated agriculture (chiefly rice) in the Syr Darya valley; sheep and camel raising in desert. Extensive saltpeter, salt, and phosphorite deposits. Fisheries along (receding) Aral Sea now abandoned. BAIKONUR COSMODROME in NW. TRANS-CASPIAN RAILROAD runs along the Syr Darya. Pop. chiefly Kazak. Formed (1938) as Kzyl-Orda oblast of Kazakh SSR. Current spelling adopted after Ka-

zakhstan became independent in 1991. Also Qyzylorda.

Kyzylorda (kee-zee-lor-DAH) [Kazak=red capital], city, ⊙ KYZYLORDA region, KAZAKHSTAN, on the SYR DARYA (river), on TRANS-CASPIAN RAILROAD, and 585 mi/941 km WNW of Almaty; 44°52′N 65°27′E. Rice processing, metalworking, meatpacking. Seat of teachers college nearby is TASBOGET, site of Kyzylorda irrigation reservoir. Originally the KOKAND fortress of Ak-Mechet, later Fort Perovskii; stormed by Russians (1853) and renamed Perovsk; again named Ak-Mechet (c.1920) and Kzyl-Orda (1925); was capital of Kazakh Autonomous SSR (1925–1929). Current spelling adopted after Kazakhstan independence (1991). Also Qyzylorda.

Kyzyl-Ozek (ki-ZIL–uh-ZYEK), village (estimated population 3,200), N ALTAI REPUBLIC, S central SIBERIA, RUSSIA, near highway, 5 mi/8 km SE of GORNO-ALTAYSK; 51°52′N 86°04′E. Elevation 1,899 ft/578 m. Environmental monitoring station.

Kyzylrai, KAZAKHSTAN: see KAZAK HILLS.

Kyzyl-Su, river, c.120 mi/193 km long, in KHATLON viloyat, SE TAJIKISTAN; rises in several branches N of Boldzhuan; flows S, past Boldzhuan and SOVETSKII, to PANJ RIVER below PARKHAR. Receives the YAKH-SU (river).

Kyzyl-Suu (kuh-zuhl–SOO), village, central ISSYK-KOL region, KYRGYZSTAN, near SE shore of the ISSYK-KOL lake, 20 mi/32 km WSW of KARAKOL; 42°20′N 78°00′E. Wheat; fisheries. Tertiary-level administrative center. Seat of Institute of Glaciology. Formerly POKROVKA (puhk-ROV-kuh).

Kyzyl-Suu (kuh-zuhl–SOO), river, c.130 mi/209 km long, in S OSH region, KYRGYZSTAN; rises in E ALAY RANGE; flows W, through fertile Alay valley (wheat, pastures), past DAROOT-KORGON into TAJIKISTAN, joining the Muk-Su River to form SURKHAB RIVER, 28 mi/45 km ENE of Khait.

Kyzyltepa (kiz-il-TYEP-ah), city, S BUKHARA wiloyat, UZBEKISTAN, on TRANS-CASPIAN RAILROAD, near KYZYLTEPA station, and 8 mi/12.9 km SE of GIJDUVAN; 40°02′N 64°51′E. Metalworks. Formerly BUSTON, Bustan-Tyube.

Kyzyltu (kee-zeel-TOO), city, NE KÖKSHETAU region, KAZAKHSTAN, 120 mi/193 km ENE of KÖKSHETAU; 47°43′N 75°41′E. In cattle area. Also Kzyl-Tuu, Qyzyltu.

Kzyl, in Russian names: see also KIZIL, KIZYL, KYZYL.

Kzyl-Kuga, KAZAKHSTAN: see KYZYLKOGA.

Kzyl-Orda, KAZAKHSTAN: see KYZYLORDA.

Kzyl-Tuu, KAZAKHSTAN: see KYZYLTU.

L

Laa an der Thaya (LAH ahn der TEI-yah), town, NE LOWER AUSTRIA, near CZECH border, 17 mi/27 km ESE of ZNOJMO, Czech Republic; 48°43'N 16°23'E. Old border town, market center border station. Manufacturing of chemicals, trailers. Brewery and beer mus; fair; remnants of medieval fortification.

Laaber (LAH-buhr), village, BAVARIA, S GERMANY, in UPPER PALATINATE, on Black Laaber River, 10 mi/16 km NW of REGENSBURG; 49°03'N 11°54'E.

La Adela (lah ah-DAI-lah), village (1991 population 1,217), ⊙ Caleu-Caleu department, SE LA PAMPA province, ARGENTINA, on left bank of RÍO COLORADO, across from RÍO COLORADO town in RÍO NEGRO province, on railroad, and 100 mi/161 km W of BAHÍA BLANCA. In agricultural area. Also known as Río Colorado.

Laage (LAH-guh), town, Mecklenburg-Western Pomerania, N GERMANY, on the RECKNITZ, 12 mi/19 km NE of GÜSTROW; 53°56'N 12°21'E. Food processing. Entirely destroyed (1637–1638) in Thirty Years War.

La Aguada (lah ah-GWAH-dah), canton, VALLEGRANDE province, SANTA CRUZ department, E central BOLIVIA. Elevation 6,660 ft/2,030 m. Gas resources (undrilled) in area. Agriculture (potatoes, yucca, bananas, corn, sweet potatoes, peanuts, tobacco, coffee); cattle for meat and dairy products.

La Aguada (lah ah-GWAH-dah), suburb of CORRAL, VALDIVIA province, LOS LAGOS region, S central CHILE; 34°14'S 71°42'W.

Laaiplek, town, SOUTH AFRICA: see GREAT BERG RIVER.

Laakdal (LAHK-dahl), commune (2006 population 14,928), Turnhout district, ANTWERPEN province, N BELGIUM, 6 mi/10 km S of GEEL.

Laakirchen (lah-KIR-khuhn), town, central UPPER AUSTRIA, on TRAUN RIVER, and 4 mi/6.4 km N of GMUNDEN; 47°59'N 13°49'E. Paper mill; manufacturing of motors.

Laaland, DENMARK: see LOLLAND.

La Altagracia (lah alt-tah-grah-SEE-ah), province (□ 1,191 sq mi/3,096.6 sq km; 2002 population 182,020), SE DOMINICAN REPUBLIC; ⊙ HIGUEY; 18°35'N 68°38'W. At E tip of HISPANIOLA island. Off S coast is SAONA island. Tropical lowlands with some outliers of Cordillera CENTRAL. Sugarcane is grown extensively in SW. Cattle raising. Products include bananas, coffee, rice, cacao, corn, hides, timber. Higuey is known for its shrine. Province was set up 1944; formerly part of SEIBO province.

La Amistad National Park, formed by 2 national parks that adjoin along the international border between Costa Rica and Panama. In Costa Rica it covers 492,557 acres/193,920 ha; in Panama it covers 525,780 acres/207,000 ha. The combined parks comprise one of Central Amer.'s largest wilderness areas and the largest undisturbed cloud forest in Latin Amer. Neither park is well developed, although there is access to the park borders by road in both countries. In Panama Guaymí and Teribe Indians live in areas adjacent to the park, as do the related Bribri Indians of Costa Rica.

La Angostura, ARGENTINA: see VILLA LA ANGOSTURA.

La Antigua, MEXICO: see JOSÉ CARDEL.

La Arena (LAH ah-RAI-nah), town (2005 population 30,886), PIURA province, PIURA region, NW PERU, on coastal plain, near PIURA RIVER, and on highway from PIURA to SECHURA, 11 mi/18 km SSW of Piura, in irrigated area (cotton, corn, and plantains); 04°53'S 80°45'W. Periodically devastated by El Niño floods.

La Argentina (LAH ahr-hen-TEE-nah), town, ⊙ La Argentina municipio, HUILA department, S central COLOMBIA, 70 mi/113 km SW of NEIVA on La Plata River; 02°12'N 75°58'W. Coffee, plantains; livestock. Sometimes called Plata Vieja.

Laarne (LAHR-nuh), commune (2006 population 11,760), Dendermonde district, EAST FLANDERS province, NW BELGIUM, 6 mi/9.7 km ESE of GHENT; 51°02'N 03°51'E. Textile industry; agricultural market. Has sixteenth-century Gothic church.

Laasphe, Bad, GERMANY: see BAD LAASPHE.

La Asunción (LAH ah-soon-see-ON), city, ⊙ NUEVA ESPARTA state, NE VENEZUELA, on MARGARITA ISLAND (E), in fertile interior valley, 210 mi/338 km ENE of CARACAS; 11°02'N 63°52'W. Tourist site. Agricultural region (cotton, sugarcane, cassava, coconuts); cotton ginning, corn and sugarmilling, *aguardiente* (liquor) distilling. First Spanish settlement 1524. PARQUE NACIONAL CERRO EL COPEY nearby.

La Asunta (lah ah-SOON-tah), canton, SUD YUNGAS province, LA PAZ department, W BOLIVIA, 10 mi/16 km NE of CHULUMANI, on the SANTA ANA–LA PAZ road; 16°02'S 67°10'W. Elevation 5,686 ft/1,733 m. Tungsten mining at Mina Reconquisada, 75 mi/120 km E of La Paz, Mina CHOJLLA, Chambilaya, Enramada, and Mina Bolsa Negra. Agriculture (potatoes, yucca, bananas, rye, soy, tobacco, coffee, tea, citrus fruits); cattle.

Laatokka, RUSSIA: see LADOGA, LAKE.

Laatzen (LAH-tsuhn), suburb of HANOVER, LOWER SAXONY, N GERMANY, on LEINE RIVER, just S of city; 52°19'N 09°47'E. Small industries; agriculture.

Laâyoune (lei-YOON), city (1999 population 169,000), Laâyoune province and ⊙ LAÂYOUNE-BOUJDOUR-SAKIA EL HAMRA administrative region of MOROCCO, ⊙ WESTERN SAHARA territory (which is claimed by Morocco), S of OUED DRA, 100 mi/161 km SW of TIZNIT; 28°31'N 10°42'W. Port; on the road to MAURITANIA; connected by road to BOUKRAA phosphate mines; international airport to W. Large Moroccan investment has created a city in the desert. Formerly called Al Aïoun Dra.

Laâyoune-Boujdour-Sakia El Hamra, administrative region (2004 population 256,152), MOROCCO; ⊙ LAAYOUNE. Lies mainly in N WESTERN SAHARA, which is claimed and administered by Morocco (though the territory's status is considered unresolved by much of the world, including the U.S. and UN), with N part of the region in SW Morocco proper. Bordered by Oued Eddahab-Lagouira (S) and Guelmim-Es Smara (N) administrative regions, as well as MAURITANIA (E) and ATLANTIC OCEAN (W). Several main roads branch out of Laayoune, including one S along the coast into Oued Eddahab-Lagouira administrative region, one SE through Western Sahara and into N Mauritania (and SAHARA DESERT), and one N into Morocco proper. International airport at Laayoune. Further divided into two secondary administrative divisions called provinces: Boujdour and Laâyoune.

La Azulita (LAH ah-soo-LEE-tah), town, ⊙ Andrés Bello municipio, MÉRIDA state, W VENEZUELA, in ANDEAN spur, 20 mi/32 km WNW of MÉRIDA; 08°42'N 71°27'W. Elevation 3,724 ft/1,135 m. Sugarcane, coffee, cereals.

Labadieville (la-buh-DEE-vil), unincorporated town (2000 population 1,811), ASSUMPTION parish, SE LOUISIANA, 20 mi/32 km NW of HOUMA, on Bayou LAFOURCHE; 29°49'N 90°58'W. Agriculture (sugarcane, rice; cattle); crawfish, alligators; manufacturing (towboats, electronic instruments).

La Baie (lah BAI), former city, borough (French *arrondissement*), of SAGUENAY, SAGUENAY—LAC-SAINT-JEAN region, S QUEBEC, E CANADA, on HA HA BAY, an arm of the SAGUENAY River; 48°20'N 70°52'W. The city was formed by the amalgamation of Bagotville, PORT ALFRED, and the parishes of GRANDE-BAIE and Bagotville; it has a natural harbor that services its paper and aluminum industries.

La Bajada (lah bah-HAH-dah), town, SW PINAR DEL RÍO province, extreme W CUBA, on Corrientes Bay; 21°55'N 84°28'W. Marks W limit of highway, although minor road continues W.

La Bajada (LAH buh-HAH-duh), unincorporated village, SANTA FE county, N central NEW MEXICO, suburb 13 mi/21 km WSW of SANTA FE, W of LA CIENEGA, on Santa Fe River. Cattle, sheep; grain, corn, alfalfa. Part of Santa Fe National Forest to NW.

La'ban (LUH-BAN), village, S central JORDAN, 16 mi/26 km S of AL KARAK. Poultry and vegetable farm (S); copper deposits (W).

Laband, POLAND: see LABEDY.

La Banda (lah BAHN-dah), town (□ 1,285 sq mi/3,341 sq km), ⊙ Banda department (□ 1,285 sq mi/3,328 sq km; 1991 population 104,664), W central SANTIAGO DEL ESTERO province, ARGENTINA, 5 mi/8 km NNE of Santiago del Estero across RÍO DULCE (irrigation area). Railroad junction.

La Barca (lah BAHR-kah), city and township, JALISCO, central MEXICO, E of Lake CHAPALA, on LERMA River, and 60 mi/97 km SE of GUADALAJARA; 20°20'N 102°33'W. Agricultural center (grain, vegetables, oranges; livestock; dairying); tanning, beverage processing.

Laba River (lah-BAH), 219 mi/352 km long, KRASNODAR TERRITORY, S European Russia; rises in the main range of the W Greater CAUCASUS Mountains in two headstreams, the Great LABA (E) and Little LABA (W), joining SSE of MOSTOVSKOYE; flows NW, past LABINSK and KOSHEKHABL (Adygey Republic), and W, past TEMIRGOYEVSKAYA (Krasnodar Territory), to the KUBAN' River at UST'-LABINSK. Has braided middle and lower course. Forms the NE border of ADYGEY Republic and Krasnodar Territory.

Labarthe-Sur-Lèze (lah-bahrt–syoor–lez), commune (□ 4 sq mi/10.4 sq km; 2004 population 150), HAUTE-GARONNE department, MIDI-PYRÉNÉES region, SW FRANCE, on Lèze River (a tributary of the ARIÈGE), and 10 mi/16 km S of TOULOUSE; 43°27'N 01°24'E. Agriculture market for Toulouse metropolitan area.

Labasa (lahm-BAH-sah), town, N Vanua Levu, FIJI, SW PACIFIC OCEAN, on S bank of Labasa River. Sugar center. Includes Indian settlement called Nasea. Sometimes spelled Lambasa.

Labastide-Clairence, FRANCE: see BASTIDE-CLAIRENCE, LA.

Labastide-Rouairoux (lah-bah-steed–roo-e-roo), commune (□ 8 sq mi/20.8 sq km), TARN department, MIDI-PYRÉNÉES region, S FRANCE, on Thoré River, on N slope of the MONTAGNE NOIRE, and 22 mi/35 km SE of CASTRES; 43°28'N 02°39'E. Manufacturing (woolens). Built as a fortified stronghold known as a bastide in Middle Ages.

Labastide-Saint-Amans, FRANCE: see SAINT-AMANS-SOULT.

Labateca (lah-bah-TAI-kah), town, ⊙ Labateca municipio, NORTE DE SANTANDER department, N COLOMBIA, 38 mi/61 km S of CÚCUTA; 07°18'N 72°29'W. Coffee, corn; livestock.

Lábatlan (LAH-baht-lahn), village, KOMÁROM county, N HUNGARY, on the DANUBE River, and 12 mi/19 km W of ESZTERGOM; 47°45'N 18°30'E. Cement works; quicklime; large paper industry. Limestone quarry nearby.

L'Abattoir (lah-bah-TWAHR), suburb, PAMANDZI island, MAYOTTE territory (France), Comoros islands, MOZAMBIQUE CHANNEL, INDIAN OCEAN, 1 mi/1.6 km SE of DZAOUDZI, on W side of island; 12°47'S 45°17'E. Manufacturing (meat processing). Pamandzi Airport to SE. Former capital of Mayotte.

La Baye, GRENADA: see GRENVILLE.

Labé, administrative region, N central GUINEA, ⊙ LABÉ. Bordered N by SENEGAL, NE by MALI, E by Faranah administrative region (part of S portion of border formed by BAFING RIVER), S by Mamou administrative region, SW tip by Kindia administrative region, and WNW by BOKÉ administrative region. Much of FOUTA DJALLON massif located here. TAMGUÉ MASSIF (elevation 4,970 ft/1515 m) in N and Mount Kokou (elevation 4,232 ft/1,290 m) in S central. Gambie and Liti rivers in N central and Koulountou River in NW. Towns include BALAKI, Koubia, Lélouma, MALI, and TOUGUÉ. Several main roads run through Labé: one runs S into Mamou administrative region and SE Guinea and another runs NW through POPODARA into Boké administrative region and Senegal (terminating near the ATLANTIC OCEAN IN THE GAMBIA). Several smaller roads also run out of Labé town, including one heading N through YAMBÉRING and Mali towns, two heading E through Koubia and Tougué towns (before conjoining to enter and terminate in Faranah administrative region), and one heading W through Lélouma (before joining the main road into Senegal). Includes the prefectures of Mali in N, Koubia in E central, Tougué in ESE, Labé in S central, and Lélouma in SWW. All of Labé administrative region is in Moyenne-Guinée geographic region.

Labé, prefecture, Labé administrative region, W central GUINEA, in Moyenne-Guinée geographic region; ⊙ LABÉ. Bordered N by Mali prefecture, NEE by Koubia prefecture, E tip by Tougué prefecture, SE by Dalaba prefecture, SSW by Pita prefecture, and W by Lélouma prefecture. Part of FOUTA DJALLON massif here. Mount Kokou (elevation 4,232 ft/1,290 m) in center of prefecture, N of Labé town. Towns include Labé and POPODARA. Roads branch out of Labé town to all parts of the prefecture, administrative region, and country.

Labé, town, ⊙ Labé prefecture and administrative region and Moyenne-Guinée geographic region, W central GUINEA, in the FOUTA DJALLON; 11°24′N 12°16′W. It is the market and administrative center for a farm region where subsistence crops are grown and cattle are raised. Labé was incorporated in the Mali empire in the early 13th century. From the 16th to the 18th century, after the decline of Mali, it was of commercial and political importance and served as a center of Islam. The Fulani settled here in the second half of the 18th century, displacing the original inhabitants. Labé is today a leading center of Islam in Guinea. Airport here.

Labedy (lah-BEN-dee), Polish *Łabędy*, German *Laband*, district of GLIWICE, UPPER SILESIA, Katowice province, S POLAND, on KLODNICA RIVER (Gliwice Canal), and 5 mi/8 km NW of GLIWICE (Gleiwitz). Railroad junction. In coal-mining region. Foundries; steel milling, machining.

Labégude, FRANCE: see VALS-LES-BAINS.

Labelle (lah-BEL), former county (□ 2,392 sq mi/6,219.2 sq km), SW QUEBEC, E CANADA, on Lake BASKATONG; the county seat was MONT-LAURIER; 46°30′N 75°10′W.

La Belle, city, LEWIS county, NE MISSOURI, near MIDDLE FABIUS RIVER, 12 mi/19 km W of MONTICELLO; 40°07′N 91°54′W. Corn, soybeans; hogs; lumber.

La Belle (LAH BEL), town (□ 2 sq mi/5.2 sq km; 2000 population 4,210), ⊙ HENDRY county, SW FLORIDA, on CALOOSAHATCHEE RIVER, and 27 mi/43 km ENE of FORT MYERS; 26°45′N 81°26′W. Shipping center for sugarcane, cattle, and watermelons.

Labelle (lah-BEL), village (□ 84 sq mi/218.4 sq km), LAURENTIDES region, S QUEBEC, E CANADA, in the LAURENTIANS, 27 mi/43 km NW of SAINTE-GATHE-DES-MONTS; 46°17′N 74°44′W. In garnet-mining region; dairying.

La Belle Lake, WISCONSIN: see LAC LA BELLE.

Laberge, Lake (□ 87 sq mi/225 sq km), S YUKON, CANADA, expansion of LEWES RIVER, 20 mi/32 km N of WHITEHORSE; 30 mi/48 km long, 1 mi/2 km–4 mi/6 km wide.

Laberinto de las Doce Leguas, CUBA: see DOCE LEGUAS, CAYOS DE LAS.

Labe River, CZECH REPUBLIC and GERMANY: see ELBE.

Labes, POLAND: see LOBEZ.

Labette (luh-BET), county (□ 653 sq mi/1,697.8 sq km; 2006 population 22,203), SE KANSAS, ⊙ OSWEGO; 37°11′N 95°17′W. Level area located in the Prairie region, bordering S on OKLAHOMA; drained in E by NEOSHO River. Agriculture (cattle, hogs; strawberries, wheat, sorghum, soybeans, hay; dairying). Oil and gas fields; coal deposits. Tri-city airport in NW corner, serves region between INDEPENDENCE, PARSONS, and CHANUTE. Big Hill Lake reservoir in NW. Formed 1867.

Labette (luh-BET), village (2000 population 68), LABETTE county, SE KANSAS, 9 mi/14.5 km SSE of PARSONS; 37°13′N 95°10′W. Dairying, agriculture.

Labiau, RUSSIA: see POLESSK.

Labin, Italian *Albona d'Istria*, town, W CROATIA, 23 mi/37 km SSW of RIJEKA (Fiume), in ISTRIA region; 45°05′N 14°08′E. Raša coal mine nearby.

Labinsk (lah-BEENSK), city (2005 population 62,135), E KRASNODAR TERRITORY, S European Russia, on the LABA RIVER (tributary of the KUBAN' River), on road and railroad, 175 mi/282 km E of KRASNODAR, and 30 mi/48 km SW of ARMAVIR; 44°38′N 40°44′E. Elevation 915 ft/278 m. Food processing industries (sugar, sunflower oil extraction, meat, dairy products, canning, flour); lumber milling, light industry, machinery. Founded in 1841 as a fortress, which grew into the village of Labinskaya. City status and renamed in 1947.

Labinskaya, RUSSIA: see LABINSK.

Labis (luh-BIS), town, NW JOHOR, MALAYSIA, on railroad, and 17 mi/27 km SE of SEGAMAT; 02°23′N 103°02′E. Agricultural center.

Labiszyn (lah-BEE-sheen), Polish *Łabiszyn*, German *Labischin*, town, Bydgoszcz province, W central POLAND, on NOTEC River, and 13 mi/21 km SSW of BYDGOSZCZ; 52°57′N 17°56′E. Brick manufacturing, distilling, flour milling.

La Blanca, MEXICO: see GENERAL PÁNFILO NATERA.

Labná (lahb-NAH), historic site, Tekax municipio, YUCATÁN, SE MEXICO, 7 mi/11.3 km SW of OXKUTZ-CAB; 20°13′N 89°34′W. Ancient Mayan city; one of best examples of Puuc style architecture. Most famous structure is an arch. The site is partly restored.

Labo (LAH-bo), town, CAMARINES NORTE province, SE LUZON, PHILIPPINES, 9 mi/14.5 km WNW of DAET; 14°08′N 122°45′E. Agricultural center (coconuts; rice); garnet mining.

La Boca (lah BO-kah), SE section of BUENOS AIRES, ARGENTINA, on the RÍO DE LA PLATA. Oldest section of the city, restored, still preserving the character of old port and fishing village; former slum now tourist attraction with cantinas, artist colony, and red light district.

La Boca (lah BO-kah) or **Boca Bío-Bío**, village, CONCEPCIÓN province, BÍO-BÍO region, S central CHILE, at mouth of BÍO-BÍO RIVER on the PACIFIC, 2 mi/3 km inland, 5 mi/8 km W of CONCEPCIÓN.

La Boca (lah BO-kah), village, SANTIAGO province, METROPOLITANA DE SANTIAGO region, central CHILE, minor port on the PACIFIC at mouth of MAIPO RIVER, 3 mi/5 km S of SAN ANTONIO. Railroad terminus. Sometimes called Boca de Maipo.

Laboe (LAH-bo), village, SCHLESWIG-HOLSTEIN, N GERMANY, at mouth of KIEL FIRTH, 7 mi/11.3 km NNE of KIEL; 54°24′N 10°13′E. Seaside resort. Site of World War I navy memorial (280 ft/85 m high).

Laboeandeli, INDONESIA: see LABUANDELI.

Laboeha, INDONESIA: see LABUHA.

Laboehandeli, INDONESIA: see LABUANDELI.

La Bolt, village (2006 population 79), GRANT county, NE SOUTH DAKOTA, 12 mi/19 km S of MILBANK; 45°02′N 96°40′W.

La Bonanza, NICARAGUA: see BONANZA.

La Boquita (lah bo-KEE-tah), turtle nesting beach in MANAGUA department, NICARAGUA.

Laborde (lah-BOR-dai), town, SE CÓRDOBA province, ARGENTINA, 60 mi/97 km SSE of VILLA MARÍA; 33°09′S

62°51′W. Agriculture (rye, flax, corn, oats; livestock; dairying); flour milling.

Laborec River (lah-BO-rets), HUNGARIAN *Laborc*, 85 mi/137 km long, VYCHODOSLOVENSKY province, E SLOVAKIA; rises on S slope of E BESKIDS, near POLISH border, 24 mi/39 km ENE of BARDEJOV; flows S, past HUMENNÉ and MICHALOVCE, to LATORICA RIVER; 12 mi/19 km E of TREBIŠOV. ZEMPLINSKA SIRAVA DAM is just E of Michalovce.

Labores, Las (lahs lah-VO-res), town, CIUDAD REAL province, S central SPAIN, near TOLEDO province border, 30 mi/48 km NE of CIUDAD REAL; 39°16′N 03°31′W. Olives, grain, grapes; olive-oil extracting.

Laborie (lah-buhr-EE), village (2001 population 2,638), S SAINT LUCIA, 17 mi/27 km S of CASTRIES; 13°45′N 61°00′W. Fishing.

Labor, La, HONDURAS: see LA LABOR.

La Bostonnais (lah bos-to-NAI), village (□ 114 sq mi/296.4 sq km; 2006 population 559), MAURICIE region, S QUEBEC, E CANADA; 47°31′N 72°42′W. An independent municipality, it forms part of La TUQUE agglomeration.

Labouheyre (lah-boo-er), commune (□ 13 sq mi/33.8 sq km), LANDES department, AQUITAINE region, SW FRANCE, 30 mi/48 km NW of MONT-DE-MARSAN; 44°13′N 00°55′W. Local wood utilization industry (railroad ties, wood pulp, and turpentine).

Laboulaye (lah-bo-LAH-ye), town (1991 population 18,854), ⊙ Presidente Roque Sáenz Peña department, S CÓRDOBA province, ARGENTINA, 90 mi/145 km SE of RÍO CUARTO; 34°07′S 63°24′W. Railroad junction on BUENOS AIRES–MENDOZA line. Agricultural center (grain, flax, alfalfa, soybeans, sunflowers; sheep and cattle; dairying); flour milling; manufacturing (cement).

Labourd (lah-boor), small historical region of SW FRANCE, surrounding and generally S of BAYONNE (PYRÉNÉES-ATLANTIQUES department, AQUITAINE region). Once a dependency of GASCONY; now part of the French BASQUE country, extending S to the Pyrenean foothills; 43°25′N 01°32′W. Saint-Pée-sur-Nivelle and CAMBO-LES-BAINS are small resort communities.

Labourdonnais, village, Rivière du REMPART district, MAURITIUS, 0.6 mi/1 km NE of MAPOU. Fruit orchard, sugarcane, vegetables.

Labrador, CANADA: see LABRADOR-UNGAVA.

Labrador City, town (□ 15 sq mi/39 sq km; 2006 population 7,240), W NEWFOUNDLAND AND LABRADOR, E CANADA, 185 mi/298 km N of SEPT-ILES, at S end of Wabash Lake. W terminus of spur of Quebec North Shore and Labrador railroad. Connected to WABUSH by project road from Happy Valley-Goose Bay via CHURCHILL FALLS. Iron mining district developed in early 1960s; tourism.

Labrador Current, cold ocean current formed off LABRADOR COAST (CANADA) by currents descending from BAFFIN BAY and W GREENLAND, between 45°N to 58°N; flows S along Labrador coast and E NEWFOUNDLAND, meets the GULF STREAM off the GRAND BANKS; meeting of cold and warm streams results in the fogs for which this region is noted. The Labrador Current carries ice into main shipping lanes between America and Europe. The famous ocean liner *Titanic*, sailing from LIVERPOOL to New York on her maiden voyage, struck an iceberg at 41°46′N 50°14′W and sank on the night of April 14–15, 1912. Also called Arctic Current or Arctic Stream.

Labrador Sea, part of North ATLANTIC OCEAN, between LABRADOR-UNGAVA and SW GREENLAND, linked by DAVIS STRAIT with BAFFIN BAY. The WEST GREENLAND CURRENT flows N along the Greenland coast, and the LABRADOR CURRENT S along the Labrador-Ungava coast.

Labrador-Ungava, peninsular region of E CANADA (□ c.550,000 sq mi/1,424,500 sq km), bounded on the W by HUDSON BAY, on the N by HUDSON STRAIT and

UNGAVA BAY, on the E by the ATLANTIC OCEAN, and on the S by the SAINT LAWRENCE river. It is almost completely unpopulated. The W 80% of the peninsula belongs to Nouveau Québec (UNGAVA) and SAGUENAY counties of QUEBEC province. The E 20%, called simply Labrador, is part of Newfoundland Inuit communities who have lived along the coastline for centuries. The region S of Ungava Bay, originally a possession of the Hudson's Bay Company, was made a part of the NORTHWEST TERRITORIES in 1869, and later (1895) became a separate district. In 1912 it was added to Quebec province, but in 1927 the E coast was awarded to NEWFOUNDLAND by the British Privy Council. The N part of the region is a cold, barren tundra; the S part is covered by coniferous forests. Geologically part of the CANADIAN SHIELD, the glaciated peninsula has many lakes and streams. There are vast and largely untapped mineral, hydroelectric, and timber resources on the peninsula. Since the mid-1950s the region's development has been aided by the construction of new ports and railheads at SEPT-ÎLES and Port Cartier on the Saint Lawrence, which provide outlets for rich, new iron ore mines in the interior; other mining includes asbestos, titanium, and copper. Variant names are Ungava Peninsula, Labrador Peninsula.

Labrang, China: see XIAHE.

Labranzagrande (lah-brahn-sah-GRAHN-dai), town, BOYACÁ department, central COLOMBIA, in valley of Cordillera ORIENTAL, 51 mi/82 km ENE of TUNJA; 05°33'N 72°33'W. Elevation 4,904 ft/1,494 m. Agriculture includes coffee, corn, potatoes.

Labra, Peña (LAH-vrah, PAI-nyah), massif in the CANTABRIAN MOUNTAINS, CANTABRIA and LEÓN provinces, N SPAIN, 15 mi/24 km WNW of REINOSA. The Pico de las Aguas (7,136 ft/2,175 m) is highest point. The CORDILLERA IBÉRICA extends SE from this mountain knot. It is also the source of EBRO RIVER, flowing SE to the MEDITERRANEAN SEA, and of the PISUERGA RIVER, flowing S to the DUERO.

Lábrea (LAH-brai-ah), city (2007 population 36,705), S AMAZONAS, BRAZIL, on right bank of Rio PURUS (navigable) and 120 mi/193 km NNW of PÔRTO VELHO; 07°25'S 64°45'W. Rubber, mica, iron, cattle ranching. Founded 1871.

La Brea (lah-BRAI-ah), village, VALLE department, S HONDURAS; minor Pacific port on GULF OF FONSECA, 7 mi/11.3 km SW of NACAOME. Saltworks.

La Brea (LAH BRAI-ah), village, TALARA province, PIURA region, NW PERU, in W outliers of CORDILLERA OCCIDENTAL, 10 mi/16 km ESE of TALARA; 04°15'S 80°54'W. Petroleum-production center. Petroleum is shipped via pipeline 6 mi/10 km S to Talara.

La Brea, village, SW TRINIDAD, TRINIDAD AND TOBAGO, on the Gulf of PARIA, 10 mi/16 km WSW of SAN FERNANDO. Adjoins famous PITCH LAKE, where asphalt is worked and then exported through La Brea pier at Pitch Point (10°15'N 61°37'W). Petroleum deposits nearby. Sir Walter Raleigh caulked his ships with asphalt here (1595).

La Brea, locality, LOS ANGELES county, S CALIFORNIA. The La Brea asphalt pits (commonly referred to as La Brea Tar Pits), which yielded prehistoric animal and plant remains, are in SE corner of Hancock Park, on Wilshire Boulevard, 6 mi/9.7 km W of downtown Los Angeles. The first fossils were found in 1875, and since 1906 the pits have been extensively researched. Formerly in Rancho La Brea.

Labrecque (la-BREK), village (□ 57 sq mi/148.2 sq km; 2006 population 1,264), SAGUENAY—LAC-SAINT-JEAN region, S central QUEBEC, E CANADA, 10 mi/16 km from ALMA; 48°40'N 71°32'W.

Labrède, FRANCE: see BRÈDE, LA.

La Brenne Natural Regional Park (lah BRE-nuh), INDRE department, CENTRE administrative region, central FRANCE. Marshy region with many ponds between CREUSE and CLAISE rivers.

La Brévine, SWITZERLAND: see BRÉVINE, LA.

La Broquerie (lah bro-kuh-REE), rural municipality (□ 223 sq mi/579.8 sq km; 2001 population 2,894), S MANITOBA, W central CANADA, 39 mi/62 km SE of WINNIPEG; 49°20'N 96°57'W. Agriculture (livestock); forestry; tourism. Includes communities of LA BROQUERIE, MARCHAND. Settled 1877.

La Broquerie (lah bro-kuh-REE), community (2006 population 505), SE MANITOBA, W central CANADA, 39 mi/62 km SE of WINNIPEG, and in LA BROQUERIE rural municipality; 49°31'N 96°29'W.

Labruguière (lah-bru-gyer), industrial town (□ 23 sq mi/59.8 sq km), TARN department, MIDI-PYRÉNÉES region, S FRANCE, on Thoré River, 5 mi/8 km S of CASTRES; 43°32'N 02°16'E. Woodworking industry. Has old quarter built in concentric circles, surrounded by modern district.

la Bruyère, BELGIUM: see Bruyère.

Labské pískovce, CZECH REPUBLIC: see ELBE SANDSTONE ROCKS.

Labuan (LAH-boo-ahn), island (□ 35 sq mi/91 sq km), off the NW coast of Sabah, Malaysia, in the South China Sea; 05°18'N 115°14'E. Coconuts, rubber, and fishing are the main agr. products Victoria (renamed Labuhan), the chief town, has a fine harbor and an airport and was the traditional shipping center for much of N Borneo when settlements were dissolved. Labuan was ceded to Great Britain by the sultan of Brunei in 1846 and became a crown colony in 1848. It was included in the STRAITS SETTLEMENTS in 1906, and in 1946 was joined to British North Borneo, which became SABAH in 1963. It was at Labuan that the Japanese surrendered to Australia in 1945. Separated from Sabah by the Malaysian government in 1984 and designated as a fledgling international financial center; duty-free zone, small seaport, airport at Victoria. Botanical garden.

Labuandeli (lah-BWAN-dai-lee), town, Sumatra Utara province, NE SUMATRA, INDONESIA, on DELI RIVER near its mouth on Strait of MALACCA, 10 mi/16 km N of MEDAN; 03°45'N 98°41'E. Historically shipped rubber, tobacco, fibers, palm oil, spices, tea, copra, gambir. Formerly important as port for Medan, it was superseded by nearby BELAWAN and is now little used. Railroad connection with Medan and Belawan. Also spelled Laboeandeli or Laboehandeli.

Labugama, village, WESTERN PROVINCE, SRI LANKA; 06°51'N 80°10'E. Reservoir here supplies COLOMBO with water.

Labuha (lah-BOO-hah), chief port of BACAN island, INDONESIA, on W coast, on inlet of MALUKU PASSAGE; 00°49'N 127°18'E. Exports timber, resin, bamboo, copra, spices. In World War II, Japanese naval base was here. Also spelled Laboeha.

Labuk Bay (LAH-book), inlet of SULU SEA, NE BORNEO, MALAYSIA, NW of SANDAKAN; 30 mi/48 km long, 20 mi/32 km wide; 06°07'N 117°46'E.

Labuk River (LAH-book), 200 mi/322 km long, NE BORNEO, MALAYSIA; rises in mountains S of Mount Kinabalu; flows generally ENE to LABUK BAY (inlet of SULU SEA), 40 mi/64 km WNW of SANDAKAN; 05°57'N 117°20'E. Navigable by small craft.

Laburnum, locality, suburb 10 mi/16 km E of MELBOURNE, VICTORIA, SE AUSTRALIA, between Blackburn and BOX HILL. Railway station.

Labutta (lah-BOOT-tah), township, AYEYARWADY division, MYANMAR, in AYEYARWADY delta, 45 mi/72 km S of PATHEIN; 16°09'N 94°46'E.

Labytnangi (lah-bit-NAHN-gee), city (2006 population 26,970), W YAMALO-NENETS AUTONOMOUS OKRUG, TYUMEN oblast, NW SIBERIA, RUSSIA, on the E slope of the Polar URALS, on the W bank of the OB' RIVER (landing), on road and the Seyda-Salekhard railroad, 9 mi/14 km W of SALEKHARD; 66°39'N 66°25'E. Timber shipping base; bakery. Gas fields nearby. Ecological research laboratory for Russian Academy of Sciences. Made city in 1975.

Lac, administrative region, W central CHAD, includes NE Lake CHAD; ⊙ BOL. Borders CAMEROON (S), NIGERIA (W), and NIGER (NW) on opposite shores of Lake Chad, as well as KANEM (N and E) and HADJER-LAMIS (SE) administrative regions. LAC was formerly a prefecture prior to Chad's administrative division reorganization from fourteen prefectures to twenty-eight departments. It was recreated as a region with approximately the same boundaries following a decree in October 2002 that reorganized Chad's administrative divisions into eighteen regions.

Lac (LAHK), former prefecture (2000 population 310,890), W central CHAD, NE of Lake CHAD; capital was BOL; 13°30'N 14°30'E. Was bordered N and E by KANEM prefecture, SE by CHARI-BAGUIRMI prefecture, SW by CAMEROON, and W by NIGER and NIGERIA. Main centers included Bol, N'GOURI, and MONDO. This was a prefecture prior to Chad's administrative division reorganization from fourteen prefectures to twenty-eight departments, at which point it became a department with approximately the same boundaries. The department became a region named LAC following a decree in October 2002 that reorganized Chad's administrative divisions into eighteen regions.

Laç (LAHCH), new town, ⊙ Laç district, central ALBANIA, on railroad, 22 mi/35 km N of TIRANË; 41°38'N 19°43'E. Industrial center (fertilizer, building materials); chemical manufacturing; copper smelting.

Lac (lahk) or **Lac de** (lahk duh) [French=lake, lake of]: see proper noun following this term.

L'Acadie (lah-kah-DEE), former village, S QUEBEC, E CANADA, on RICHELIEU River, and 5 mi/8 km W of Saint Jean; 45°19'N 73°20'W. Dairying; grain, cattle. Amalgamated into SAINT-JEAN-SUR-RICHELIEU in 2001.

Lacahahuira River (lah-kah-hah-WEE-rah), 70 mi/113 km long, in ORURO department, W BOLIVIA; rises in marshes SW of Lake POOPÓ, which it drains; flows intermittently W across the ALTIPLANO, to Lake COIPASA; 19°21'S 67°54'W. Sometimes spelled Lacaja-huira.

Laca Laca Quita Quita (LAH-kah LAH-kah KEE-tah KEE-tah), canton, CARANGAS province, ORURO department, W central BOLIVIA. Elevation 12,448 ft/3,794 m. Gas wells in area. Clay, limestone, and gypsum deposits. Agriculture (potatoes, yucca, bananas); cattle.

Lac-à-la-Croix (LAHK–ah-lah-KWAH), former village, central QUEBEC, E CANADA; 48°24'N 71°47'W. Merged with METABETCHOUAN in 1998 to form MÉTABETCHOUAN–LAC-À-LA-CROIX.

Lacalahorra (lah-kah-lah-O-rah), town, GRANADA province, S SPAIN, 9 mi/14.5 km SSE of GUADIX; 37°11'N 03°04'W. Cereals, potatoes, sugar beets. Its 16th-century castle, with four round corner towers, is decorated in Italian Renaissance style.

La Calama (lah kah-LAH-mah), canton, MÉNDEZ province, TARIJA department, S BOLIVIA; 21°26'S 64°49'W. Elevation 6,555 ft/1,998 m. Undrilled gas resources in area; phosphates mined at ISCAYACHI Mine, 18 mi/30 km W of TARIJA; clay, gypsum deposits. Agriculture (potatoes, yucca, bananas, corn, wheat, barley, sweet potatoes); cattle.

La Calamine, BELGIUM: see KELMIS.

Lac-à-la-Tortue (LAHK–ah-lah-tor-TYOO), former village, S QUEBEC, E CANADA; 46°37'N 72°38'W. Amalgamated into SHAWINIGAN in 2002.

La Caldera, ARGENTINA: see CALDERA.

Lac a l'Eau Claire (LAHK ah lo-KLER), lake (□ 410 sq mi/1,066 sq km), N QUEBEC, E CANADA; 56°00'N 74°30'W; 45 mi/72 km long, 20 mi/32 km wide. Elevation 790 ft/241 m; contains several islands. Drains into HUDSON BAY.

La Calera (lah kah-LAI-rah), city, ⊙ Calera comuna (2002 population 47,836), QUILLOTA province, VALPARAISO region, N central CHILE, on ACONCAGUA RIVER, 7 mi/11 km N of QUILLOTA, and 32 mi/51 km NE of VALPARAISO; 32°47′S 71°16′W. Railroad junction; grapes, fruit, vegetables, cereals. Cement. Furnishes limestone for blast furnaces at CORRAL (VALDIVIA region); phosphate production.

La Calera (lah kah-LAI-rah) or **Calera**, town, NW central CÓRDOBA province, ARGENTINA, 10 mi/16 km WNW of CÓRDOBA. Railroad junction, resort center. Dam on RÍO PRIMERO nearby provides hydroelectric power.

La Calera (lah kah-LAI-rah), town, ⊙ Calera comuna, QUILLOTA province, VALPARAISO region, N central CHILE, 7 mi/11 km N of QUILLOTA. On railroad. Fruit, vegetables, cereals, grapes. Food processing.

La Calera (LAH kah-LAI-rah), town, ⊙ La Calera municipio, CUNDINAMARCA department, central CO-LOMBIA, 12 mi/19 km NNE of BOGOTÁ; 04°43′N 73°58′W. Elevation 8,917 ft/2,718 m. Cement plant nearby.

La Campa (lah KAHM-pah), town, LEMPIRA depart-ment, SW HONDURAS, 8 mi/13 km S of GRACIAS, on paved road; 14°28′N 88°17′W. Pottery center; colonial church.

La Campaña (lah kahm-PAH-nah), national park, in VALPARAISO region, 20 mi/32 km SE of QUILLOTA. Protecting native forest.

La Cañada (kahn-YAH-dah), town, ⊙ El Marqués municipio, QUERÉTARO, central MEXICO, on Queré-taro River (affluent of APASEO River), and 4 mi/6.5 km E of QUERÉTARO; 20°37′N 100°19′W. Produces flowers, fruit. Resort noted for thermal springs. Also known as Villa de Marqués.

La Cañada (lah kahn-YAH-dah), village (1991 popula-tion 1,064), ⊙ Figueroa department (□ 2,540 sq mi/ 6,604 sq km), central SANTIAGO DEL ESTERO province, ARGENTINA, on railroad, and 30 mi/48 km ENE of SANTIAGO DEL ESTERO. Corn, alfalfa, cotton; livestock.

La Cañada Flintridge (lah kahn-YAH-dah FLINT-rij), city (2000 population 20,318), LOS ANGELES county, S CALIFORNIA, in S foothills of SAN GABRIEL MOUN-TAINS; residential suburb 11 mi/18 km NNE of downtown LOS ANGELES; 34°13′N 118°12′W. Printing and publishing. Descanso Gardens; Angeles National Forest to NE.

Lacanau (lah-kah-no), commune, resort in GIRONDE department, AQUITAINE region, SW FRANCE, 24 mi/39 km N of ARCACHON, on Lacanau Lake just inland (behind dunes) from Bay of BISCAY; 44°59′N 01°05′W. Pine forests provide lumber and resins. Lacanau-Océan (just W) is a bathing resort with miles of rec-tilinear beach.

La Candelaria, ARGENTINA: see CANDELARIA, village.

Lacantún River (lah-kahn-TOON), MEXICO; c.100 mi/ 161 km long, rises in several branches in CUCHUMA-TANES MOUNTAINS of GUATEMALA; flows NW to USUMACINTA River, receives JATATÉ River in CHIAPAS.

Lacapelle-Marival (lah-kah-pel–mah-ree-vahl), com-mune (□ 4 sq mi/10.4 sq km), LOT department, MIDI-PYRÉNÉES region, W FRANCE, 10 mi/16 km NW of FIGEAC; 44°43′N 01°56′E. Has a 13th–15th-century castle with a massive square keep; Gothic church and 15th-century market place (enclosed).

La Capilla (lah kah-PEE-yah), town, central ENTRE RÍOS province, ARGENTINA, 12 mi/19 km SE of VIL-LAGUAY. Wheat, corn, flax; livestock.

La Capilla (LAH kah-PEE-yah), town, ⊙ La Capilla municipio, BOYACÁ department, central COLOMBIA, 30 mi/48 km S of TUNJA; 05°42′N 73°28′W. Coffee, corn, sugarcane; livestock.

Lácar, ARGENTINA: see SAN MARTÍN DE LOS ANDES, department.

Lacar, Lake (lah-KAHR) (□ 20 sq mi/52 sq km), in the ANDES, SW NEUQUÉN province, ARGENTINA, in Ar-gentinian lake district, E of ILPELA PASS; extends 16 mi/26 km E from CHILE border to SAN MARTÍN DE LOS ANDES. Elevation 2,103 ft/641 m.

La Carlota (lah kahr-LO-tah), town (1991 population 10,264), ⊙ Juárez Celman department, S central

CÓRDOBA province, ARGENTINA, on the RÍO CUARTO, and 70 mi/113 km S of VILLA MARÍA. Railroad junc-tion. Agriculture center (grain, alfalfa, soybeans, sunflowers; cattle); some manufacturing.

La Carlota, PHILIPPINES: see CARLOTA, LA.

Lacarne, Ohio: see PORT CLINTON.

La Carrasquilla (kah-rahs-KEE-yah), residential town, PANAMA province, central PANAMA, 4 mi/6.4 km NNE of PANAMA city. Highway junction.

La Carrera (lah kahr-RAI-rah), village, SE CATAMARCA province, ARGENTINA, on the RÍO DE VALLE, and 12 mi/19 km NNE of CATAMARCA; 28°22′S 65°43′W. Hydroelectric station; grain farming and livestock raising in irrigated area.

La Carreta (lah kah-RE-tah), canton, NOR CHICHAS province, POTOSÍ department, W central BOLIVIA, 6 mi/10 km S of COTAGAITA, and the COTAGAITA RIVER; 20°57′S 65°34′W. Elevation 8,596 ft/2,620 m. Copper-bearing load; antimony mining at Mina Churquina; clay and minor gypsum deposits. Agriculture (pota-toes, yucca, bananas); cattle.

La Carrière, TRINIDAD AND TOBAGO: see POINTE-À-PIERRE.

La Castèllana, PHILIPPINES: see CASTÈLLANA, LA.

Lacaune (lah-kon), commune (□ 35 sq mi/91 sq km; 2004 population 2,844), TARN department, MIDI-PYRÉNÉES region, S FRANCE, in the LACAUNE MOUN-TAINS, 24 mi/39 km ENE of CASTRES; 43°43′N 02°42′E. A children's health resort. Salting of meat products, cutting of slate roof-tiles. Goat milk made into Roquefort cheese.

Lacaune Mountains (lah-KON), plateau region at S edge of MASSIF CENTRAL, in TARN department, MIDI-PYRÉNÉES region, S FRANCE, extending c.25 mi/40 km E of CASTRES, N of AGOUT RIVER valley; 43°40′N 02°36′E. Rises to 4,155 ft/1,266 m. Sheep and cattle grazing; dairying, cheese making. Plateau lies partly within the regional park of HAUT LANGUEDOC [French=upper Languedoc]. Established 1973.

Lac-au-Saumon (lahk-o-so-MON), village (□ 31 sq mi/80.6 sq km), BAS-SAINT-LAURENT region, E QUE-BEC, E CANADA, on Salmon Lake, 30 mi/48 km SSE of MATANE; 48°25′N 67°20′W. Lumbering; dairying. Lake was formerly salmon spawning ground. Re-structured in 1997.

Lac-aux-Sables (lahk-o-SAH-bluh), village (□ 110 sq mi/286 sq km), MAURICIE region, S central QUEBEC, E CANADA, on BATISCAN RIVER, at S end of Lac-aux-Sables (3 mi/5 km long), 27 mi/43 km from SHAWI-NIGAN; 46°52′N 72°23′W. Agriculture.

Lac-Beauport (lahk–bo-POR), resort village (□ 24 sq mi/62.4 sq km), CAPITALE-NATIONALE region, S QUEBEC, E CANADA, on small Lake Beauport, 10 mi/16 km NNW of QUEBEC city; 46°58′N 71°18′W. Skiing. Part of the Metropolitan Community of Quebec (*Communauté Metropolitaine de Québec*).

Lac-Berlinguet (LAHK–ber-lan-GAI), unorganized territory, S QUEBEC, E CANADA; 48°41′N 74°07′W. Forms part of La TUQUE agglomeration.

Lac-Bouchette (lahk-boo-SHET), village (□ 355 sq mi/ 923 sq km), SAGUENAY—LAC-SAINT-JEAN region, S central QUEBEC, E CANADA, on Bouchette Lake (5 mi/8 km long), on Quiatchouanish River, and 17 mi/27 km SSE of ROBERVAL; 48°16′N 72°10′W. Elevation 1,135 ft/ 346 m. Quartz mining.

Lac-Brome (LAHK–BROM) or **Brome Lake** (BROM LAIK), city (□ 81 sq mi/210.6 sq km; 2006 population 5,597), MONTÉRÉGIE region, S QUEBEC, E CANADA, 12 km N of U.S. (MAINE) border; 45°13′N 72°31′W. Formed by the villages of KNOWLTON, FOSTER, Bondville, Fulford, Iron Hill, East Hill, and West Brome.

Laccadive, Minicoy, and Amindivi Islands, INDIA: see LAKSHADWEEP.

Lacchiarella (lahk-kyah-REL-lah), town, MILANO prov-ince, LOMBARDY, N ITALY, 10 mi/16 km S of MILAN; 45°19′N 09°08′E. Fabricated metals, machinery.

Lacco Ameno (LAHK-ko ah-MAI-no), village, NAPOLI province, CAMPANIA, S ITALY, on NW coast of ISCHIA island; 40°45′N 13°54′E. Resort with warm mineral springs and baths.

Lac-Delage (LAHK-duh-LAHZH), city (□ 1 sq mi/2.6 sq km; 2006 population 496), CAPITALE-NATIONALE region, S QUEBEC, E CANADA, 14 mi/22 km from QUEBEC city; 46°58′N 71°24′W. Part of the Me-tropolitan Community of Quebec (*Communauté Metropolitaine de Québec*).

Lac de la Gruyère, SWITZERLAND: see GRUYÈRE, LA.

Lac-des-Aigles (lahk-daiz-E-gluh), village (□ 33 sq mi/ 85.8 sq km; 2006 population 629), BAS-SAINT-LAUR-ENT region, S QUEBEC, E CANADA, 33 mi/53 km from RIMOUSKI; 47°59′N 68°41′W.

Lac-des-Cinq (LAHK-dai-SANK), unorganized terri-tory, S QUEBEC, E CANADA; 46°51′N 72°58′W. Amal-gamated into SHAWINIGAN in 2002.

Lac-des-Écorces (LAHK-dai–zai-KORS), town (□ 56 sq mi/145.6 sq km; 2006 population 2,770), LAUREN-TIDES region, S QUEBEC, E CANADA, 6 mi/10 km from MONT-LAURIER; 46°34′N 75°22′W. Pink granite mine; fish farm. Formed in 2002 from Beaux-Rivages, Val-Barrette.

Lac-des-Moires (LAHK-dai—MWAHR), unorganized territory, S QUEBEC, E CANADA; 47°42′N 72°02′W. Forms part of La TUQUE agglomeration.

Lac-des-Plages (LAHK–dai–PLAHZH), village (□ 47 sq mi/122.2 sq km; 2006 population 412), OUTAOUAIS region, SW QUEBEC, E CANADA, midway between MONT-TREMBLANT and MONTEBELLO; 46°00′N 74°54′W.

Lac-des-Seize-Îles (LAHK-dai–SEZ–EEL), munici-pality (□ 4 sq mi/10.4 sq km; 2006 population 227), LAURENTIDES region, S QUEBEC, E CANADA, 9 mi/14 km from BARKMERE; 45°54′N 74°28′W.

Lac-Drolet (LAHK–dro-LAI), village (□ 48 sq mi/124.8 sq km; 2006 population 1,163), ESTRIE region, S QUEBEC, E CANADA, 9 mi/15 km from LAC-MÉGANTIC; 45°43′N 70°51′W.

Lac du Bonnet (lahk dyoo bo-NAI), rural municipality (□ 425 sq mi/1,105 sq km; 2001 population 2,405), SE MANITOBA, W central CANADA, 51 mi/81 km N of WINNIPEG; 50°15′N 96°10′W. Agriculture (grain, oil-seed); mining; forestry; tourism. Incorporated 1917.

Lac-du-Cerf (LAHK–dyoo–SERF), village (□ 31 sq mi/ 80.6 sq km; 2006 population 453), LAURENTIDES re-gion, S QUEBEC, E CANADA, 7 mi/10 km from NOTRE-DAME-DE-PONTMAIN; 46°18′N 75°30′W.

Lac du Flambeau (lak de FLAM-bo), village (2000 population 1,646), VILAS county, N WISCONSIN, on small Lac du Flambeau, 33 mi/53 km NW of RHINE-LANDER; 45°58′N 89°54′W. Manufacturing (electrical measuring instruments). Art center of the Lac du Flambeau Indian Reservation. Fish hatchery nearby; surrounded by numerous lakes.

Lacedaemon, Greece: see SPARTA.

Lacedonia (lah-che-DO-nyah), village, AVELLINO province, CAMPANIA, S ITALY, 19 mi/31 km ESE of ARIANO IRPINO; 41°03′N 15°25′E. In cereal-growing region. Bishopric. Heavily damaged by earthquakes (1456, 1930).

Lac-Édouard (LAHK–ai-DWAHRD), village (□ 278 sq mi/722.8 sq km; 2006 population 137), MAURICIE re-gion, S QUEBEC, E CANADA; 47°39′N 72°16′W. An in-dependent municipality, it forms part of the La TUQUE agglomeration.

La Ceiba (lah SAI-bah), city (2001 population 114,585), ⊙ ATLÁNTIDA department, N HONDURAS, on the CARIBBEAN SEA, 70 mi/113 km E of PUERTO CORTÉS; 15°47′N 86°48′W. Primary departure point for the ISLAS DE LA BAHÍA. It is the commercial and proces-sing center of a rich agricultural region dominated by pineapple and banana plantations that have largely recovered from being nearly ruined by disease in the 1930s. Coconuts and citrus fruits are also exported. Manufacturing of footwear, cigars, consumer products; flour- and sawmilling, brewing, tanning. More than 1,000 people were killed when Hurricane Fifi hit the city in 1974. International airport to SW. Hospitals, college, radio stations.

La Ceiba (LAH SAI-bah), town, TRUJILLO state, W VENEZUELA, landing on Lake MARACAIBO, 33 mi/53 km WNW of VALERA; 09°27′N 71°04′W. Terminus of railroad from MOTATÁN; sugarmilling.

La Ceiba, DOMINICAN REPUBLIC: see HOSTOS.

La Ceja (LAH SAI-hah), town, ⊙ La Ceja municipio, ANTIOQUIA department, NW central COLOMBIA, in Cordillera CENTRAL, 18 mi/29 km SSE of MEDELLÍN. Elevation 7,273 ft/2,217 m. Agricultural center (coffee, corn, beans, sugarcane, plantains; livestock); food processing.

La Cejita (LAH sai-HEE-tah), town, TRUJILLO state, W VENEZUELA, 3 mi/5 km NE of VALERA; 09°22′N 70°34′W. Elevation 1,122 ft/341 m.

La Celia (LAH SAI-lee-ah), town, ⊙ La Celia municipio, RISARALDA department, W central COLOMBIA, 24 mi/39 km NW of PEREIRA, in the CORDILLERA OCCIDENTAL; 05°02′N 76°01′W. Elevation 5,810 ft/1,770 m. Coffee, plantains, sugarcane; livestock.

La Center, town (2000 population 1,038), BALLARD county, W KENTUCKY, 21 mi/34 km W of PADUCAH; 37°04′N 88°58′W. In agricultural area (tobacco, grain, soybeans; hogs, cattle, poultry; dairying); light manufacturing.

La Center, village (2006 population 1,907), CLARK county, SW WASHINGTON, on East Fork of LEWIS RIVER, and 15 mi/24 km N of VANCOUVER; 45°52′N 122°40′W. In agricultural region: lettuce, berries, potatoes, oats. Paradise Point State Park to SW; COLUMBIA River 6 mi/9.7 km W.

Lacepede Bay (LA-suh-peed), bight of INDIAN OCEAN, SE SOUTH AUSTRALIA, between Cape JAFFA (SW) and Granite Rocks; 19 mi/31 km long; 36°47′S 39°45′E. KINGSTON on E shore.

Lac-Etchemin (lahk–E-chuh-min), town (□ 62 sq mi/161.2 sq km), ⊙ LES ETCHEMINS county, CHAUDIÈRE-APPALACHES region, S QUEBEC, E CANADA, on Lake Etchemin (3 mi/5 km long), 45 mi/72 km SE of QUEBEC city; 46°24′N 70°30′W. Dairying; lumbering; pig raising. Restructured in 2001.

Lacey, city (2006 population 35,412), THURSTON county, W WASHINGTON, suburb 5 mi/8 km E of OLYMPIA, near PUGET SOUND; 47°02′N 122°49′W. Agricultural area (poultry; dairying; vegetables, fruit). Manufacturing (paper products, mobile homes, structural wood, and sheet metal products). Seat of St. Martins College. Fort Lewis Military Reservation to E and SE; Nisqually Indian Reservation to E; Nisqually National Wildlife Refuge and Tolmie State Park to NE, on Puget Sound.

Lacey, township (□ 80 sq mi/208 sq km), OCEAN county, E central NEW JERSEY, 5 mi/8 km S of LAKEHURST; 39°51′N 74°16′W. Incorporated 1871.

Lacey-Lakeview (LAI-see laik-VYOO), town (2006 population 151), MCLENNAN county, E central TEXAS, residential suburb 4 mi/6.4 km NNE of downtown WACO, near Brazos River Texas State Technical College and T.S.T.C. Airport to NE.

Laceyville, borough (2006 population 376), WYOMING county, NE PENNSYLVANIA, 31 mi/50 km WNW of SCRANTON, on SUSQUEHANNA RIVER (bridged). Agriculture (hay; dairying; timber); manufacturing (wooden products).

Lac-Frontière (LAHK–fron-TYER), village (□ 20 sq mi/52 sq km; 2006 population 167), CHAUDIÈRE-APPALACHES region, S QUEBEC, E CANADA, 44 mi/71 km from SAINTE-LUCIE-DE-BEAUREGARD; 46°42′N 70°00′W.

Lacha, Lake (lah-CHAH) (□ 116 sq mi/301.6 sq km), SW ARCHANGEL oblast, NW European Russia, 50 mi/80 km WNW of KONOSHA; 22 mi/35 km long, 6 mi/10 km wide; 61°25′N 38°45′E. Low, marshy banks; frozen November through May. Receives the SVID River (S). ONEGA River is its outlet (Kargopol'). Fisheries.

La Chamiza, CHILE: see CHAMIZA.

Lachanokepos, Greece: see LAKHANOKIPOS.

Láchar (LAH-chahr), town, GRANADA province, S SPAIN, on GENIL RIVER, and 13 mi/21 km W of GRANADA; 37°12′N 03°50′W. Sugar mill; agriculture (olive oil, cereals; livestock); lumber.

La Charqueada, URUGUAY: see GENERAL ENRIQUE MARTÍNEZ.

La Chaux-de-Fonds, SWITZERLAND: see CHAUX-DE-FONDS, LA.

Lachen, commune, SCHWYZ canton, NE central SWITZERLAND, on LAKE ZÜRICH, and 15 mi/24 km NNE of SCHWYZ. Silk textiles; wood- and metalworking. Rococo church. Hydroelectric plants of Etzel and Siebnen nearby.

Lachenaie (lah-shuh-NAI), former city, S QUEBEC, E CANADA; 45°42′N 73°32′W. Amalgamated into TERREBONNE in 2001.

Lachen Qu (LAH-CHEN CHOO), river, c.35 mi/56 km long, TIBET and INDIA; rises in S Tibet, on NE slope of SINGALILA RANGE; flows SSE, joining Lachung Qu (LACHUNG CHU) at CHUNTHANG (SIKKIM, India) to form TISTA RIVER.

Lachhmangarh (LUHCH-muhn-guhr), town, SIKAR district, E RAJASTHAN state, NW INDIA, 15 mi/24 km NNW of Sikar. Trades locally in millet, livestock, hides. Sometimes spelled Lachmangarh.

Lachin (lah-CHEEN), city and center of LACHIN region, S AZERBAIJAN, in the Lesser CAUCASUS, in the KURDISTAN, 75 mi/121 km SSW of YEVLAKH. Gateway to highway to NAGORNO-KARABAKH. Agriculture (livestock; wheat; dairying); chromite mines, mineral springs. Formerly called Abdalyar.

Lachine (luh-SHEEN), former city, borough (French *arrondissement*) of MONTREAL (□ 8 sq mi/20.8 sq km), S QUEBEC, E CANADA, on S central MONTREAL ISLAND, at the E end of Lake SAINT LOUIS; 45°25′N 73°40′W. Industries include iron and steel foundries; manufacturing (tires, electrical appliances, and electronics). Lachine was settled in 1675 and in 1689 was the scene of a battle between the French and the Iroquois. The borough is the SW terminal of the Lachine Canal, connecting Lake Saint Louis with the SAINT LAWRENCE RIVER. Constructed between 1821 and 1825 (later enlarged) to bypass the Lachine Rapids of the Saint Lawrence, the canal has been superceded by the SAINT LAWRENCE SEAWAY canals.

Lachine Canal (luh-SHEEN kuh-NAL), French, *Canal de Lachine* (kah-NAHL duh luh-SHEEN), on the SAINT LAWRENCE RIVER, S QUEBEC, E CANADA, extends SW-NE between the Saint Lawrence at MONTREAL and Lake SAINT LOUIS at LACHINE, bypassing LACHINE RAPIDS; 9 mi/14 km long; 45°26′N 73°20′W. Opened c.1825, later enlarged; there are 5 locks.

Lachine Canal, CANADA: see LACHINE, city.

Lachine Rapids (luh-SHEEN RA-pidz), on the SAINT LAWRENCE RIVER, S QUEBEC, E CANADA, at S end of MONTREAL ISLAND, between S part of MONTREAL and LASALLE; 3 mi/5 km long, with total drop of 42 ft/13 m. At E end of rapids is Heron islet. Rapids are bypassed by LACHINE CANAL. Hydroelectric power.

Lachinovo (lah-CHEE-nuh-vuh), settlement, NE KURSK oblast, SW European Russia, on road and railroad, 35 mi/56 km N of STARYY OSKOL; 51°48′N 37°55′E. Elevation 780 ft/237 m. In agricultural area producing sugar beets.

Lachish (lah-KHEESH), ancient city, CANAAN (present-day ISRAEL), SW of JERUSALEM, in the coastal plain. It is mentioned in the Tell-el-Amarna letters and was one of the Amorite cities allied against the Gibeonites and destroyed by Joshua. Rehoboam fortified it, and Amaziah was murdered here. It was besieged (701 B.C.E.) by Sennacherib. Later, Micah denounced it. Excavations were begun in 1935; they show that Lachish had been populated since c.3200 B.C.E. and was a thriving community as early as the 17th century B.C.E. The finds include twenty-one

ostraca, or potsherds, written in ink. They were written (c.589 B.C.E.) in Hebrew by local commanders to their officers when Lachish was being threatened by the Babylonians under Nebuchadnezzar. The letters are of great linguistic and historic value. The modern Lachish is a cooperative settlement established in 1955 near the site of the ancient city. It is also the name of a subregion in the S part of the coastal plain.

Lachish (la-KHEESH), subregion, S ISRAEL, between the JUDEAN HIGHLANDS in the E and ASHKELON in the W. Named after the ancient city of LACHISH. The development of this subregion, which was completely depopulated during the 1948 war, was planned in the 1950s as a model for rural settlements; twenty new villages with urban centers were established in the following years (1955–1961).

Lachlan River (LAK-luhn), 922 mi/1,484 km long, S central NEW SOUTH WALES, AUSTRALIA; rises in GREAT DIVIDING RANGE N of GUNNING; flows NW, past WYANGALA (dam), COWRA, FORBES, and CONDOBOLIN; thence SW, past LAKE CARGELLIGO and HILLSTON, to MURRUMBIDGEE RIVER 40 mi/64 km NE of BALRANALD; 34°21′S 143°57′E.

La Chorrera (cho-RAI-rah), city (2000 population 55,871), ⊙ La Chorrera district, PANAMA province, central PANAMA, in PACIFIC lowland, on INTERAMERICAN HIGHWAY, and 16 mi/26 km WSW of PANAMÁ; 08°53′N 79°47′W. Road and agricultural center (coffee; orange groves); livestock raising. Its port on GULF OF PANAMA is PUERTO CAIMITO.

La Chulla (lah CHOO-lah), canton, QUILLACOLLO province, COCHABAMBA department, central BOLIVIA, SW of COCHABAMBA, on the Cochabamba-ORURO highway. Elevation 8,343 ft/2,543 m. Clay, limestone, and gypsum deposits. Agriculture (potatoes, yucca, bananas, corn, rye, coffee); cattle raising for meat and dairy products.

Lachung (LAH-chuhng), village, North Sikkim district, SIKKIM state, NE INDIA, on the LACHUNG CHU, and 27 mi/43 km NNE of GANGTOK, in extreme W ASSAM. Agriculture (corn, pulses, oranges; apple orchards). Weaving school; Buddhist monastery nearby.

Lachung Chu (LAH-chuhng CHOO), river, c.30 mi/48 km long, SIKKIM state, NE INDIA; rises in NW DONGKYA RANGE; flows S, past Lachung, joining Lachen Chu at CHUNGTANG to form TISTA RIVER.

Lachute (luh-SHOOT), town (□ 43 sq mi/111.8 sq km), ⊙ ARGENTEUIL county, LAURENTIDES region, S QUEBEC, E CANADA, on the NORTH RIVER, 35 mi/55 km W of MONTREAL; 45°39′N 74°20′W. It is at the foot of the LAURENTIAN MOUNTAINS. Diverse manufacturing includes textiles, lumber, wood, and paper products. Formed 1966.

La Ciénaga (lah SYAI-nah-gah), canton, SUD CINTI province, CHUQUISACA department, SE BOLIVIA. Elevation 7,575 ft/2,309 m. Clay and limestone deposits. Agriculture (potatoes, yucca, bananas, corn, wheat, oats, rye, peanuts); cattle and pig raising.

La Ciénaga (lah SYAI-nah-gah), town, TOMINA province, CHUQUISACA department, SE BOLIVIA, 10 mi/15 km SE of PADILLA, on the SUCRE–SANTA CRUZ highway; 19°19′S 64°21′W. Elevation 6,824 ft/2,080 m. Gas resources to the E and S. Clay, limestone, gypsum deposits. Agriculture (potatoes, yucca, bananas, corn, rye, sweet potatoes, peanuts); cattle and pig raising.

La Ciénaga (LAH see-AI-nah-gah), town, FALCÓN state, NW VENEZUELA, 3 mi/4.8 km SSE of PUERTO CUMAREBO. Cassava, divi-divi; 11°14′N 69°36′W. Petroleum deposits nearby. Also spelling La Ciénega.

La Ciénaga (lah SYAI-nah-gah), village, S central CATAMARCA province, ARGENTINA, 10 mi/16 km N of BELÉN. Dam near BELÉN RIVER. Grain, alfalfa, wine; livestock.

La Cienega (LAH see-EN-i-guh), unincorporated town (2000 population 3,007), SANTA FE county, N central NEW MEXICO, residential suburb 12 mi/19 km WSW of

SANTA FE, on Santa Fe River; 35°34′N 106°06′W. Cattle, sheep; corn, grain. Santa Fe County Municipal Airport and Santa Fe Downs Racetrack to E. Part of Santa Fe National Forest to NE.

La Ciénega, MEXICO: see CIÉNEGA DE ZIMATLÁN.

La Ciénega, VENEZUELA: see LA CIÉNAGA.

Lac Île-à-la-Crosse (lahk-eel-ah-lah–KROS), NW central SASKATCHEWAN, CANADA, expansion of CHURCH-ILL RIVER, 180 mi/290 km NW of PRINCE ALBERT; 60 mi/97 km long, 5 mi/8 km wide; 55°32′N 107°45′W.

La Cisterna (lah sees-TER-nah), town and comuna, METROPOLITANA DE SANTIAGO region, central CHILE, on railroad, and 6 mi/10 km S of SANTIAGO. Residential suburb of Santiago.

Lac-Jérôme (LAHK–zhai-ROM), unorganized territory, CÔTE-NORD region, E QUEBEC, E CANADA; 50°51′N 63°33′W. Formed 1986.

Lackawanna (LAK-ah-WAH-nah), county (☐ 464 sq mi/1,206.4 sq km; 2006 population 209,728), NE PENNSYLVANIA; ☉ SCRANTON. Bounded SW by SUS-QUEHANNA RIVER, SE by LEHIGH RIVER; drained by LACKAWANNA RIVER. Hilly upland rises in N, part of Pocono plateau in S, Lackawanna valley in center. Agriculture (corn, oats, hay, alfalfa, vegetables; cattle; dairying); former major anthracite-coal-mining area. Manufacturing at Scranton, Olyphant, Danmore, Carbondale, and Clark's Summit. Part of Lackawanna State Forest in SE; Lackawanna State Park in NW; Archbald Pothole State Park in NE center; Merli Sarnoski Park in NE. Formed 1878.

Lackawanna, city (☐ 6 sq mi/15.6 sq km; 2006 population 17,926), ERIE county, W NEW YORK, on Lake ERIE; 42°49′N 78°49′W. Some manufacturing. Formerly a major steel-making center, Lackawanna experienced the rapid and total decline of its foremost industry in the 1970s and 1980s. A distinguished city landmark is the elaborate Basilica of Our Lady of Victory, a Roman Catholic shrine. Home of the "Lackawana Six," convicted in 2003 for providing material support to foreign terrorists. Incorporated 1909.

Lackawanna (LAK-ah-WAH-nah), river, 35 mi/56 km long, NE PENNSYLVANIA; rises on SUSQUEHANNA-WAYNE county line, c.15 mi/24 km N of CARBONDALE in West and East branches (both c.10 mi/16 km long and closely parallel each other, flowing S), which join at Stillwater Lake reservoir; 41°41′N 75°29′W. Flows generally SW past FOREST CITY, Carbondale, OLY-PHANT, SCRANTON, and OLD FORGE to SUSQUEHANNA RIVER 8 mi/12.9 km WSW of SCRANTON. Former major anthracite-coal-mining area, now depleted.

Lackawaxen (LAK-uh-WAK-sen), unincorporated village, Lackawaxen township, PIKE county, NE PENNSYLVANIA, 18 mi/29 km WNW of Post Services, NEW YORK, on the DELAWARE RIVER (New York state line), at mouth of LACKAWAXEN RIVER; 41°28′N 74°59′W. Delaware State Forest to SW.

Lackawaxen River (LAK-uh-WAK-sen), 25 mi/40 km long, NE PENNSYLVANIA; formed by joining of West Branch (c.15 mi/24 km long); flows through Belmont Lake and Johnson Creek, 5 mi/8 km ENE of FOREST CITY; flows SE, through Prompon Lake reservoir, past HONESDALE and HAWLEY, to DELAWARE RIVER at LACKAWAXEN (41°34′N 75°19′W); receives Wallenpaupack Creek, 2 mi/3.2 km E of Hawley.

Lac-Kénogami (LAHK–kai-NAH-guh-mee), former village, S central QUEBEC, E CANADA; 48°21′N 71°19′W. Amalgamated into SAGUENAY in 2002.

Lackey (LAK-ee), unincorporated village, FLOYD county, E KENTUCKY, in CUMBERLAND foothills, 17 mi/27 km W of PIKEVILLE. In bituminous coal area; manufacturing (concrete).

Lackland Air Force Base, UNITED STATES military installation, BEXAR county, S TEXAS, in W part of SAN ANTONIO. Covers c.6,835 acres/2,766 ha; major air force training center. Established 1941.

Lac La Belle, village (2006 population 431), WAUKESHA county, SE WISCONSIN, 10 mi/16 km ESE of WATER-

TOWN on N end of La Belle Lake, near OCONOMOWOC; 43°08′N 88°31′W.

Lac La Biche (lahk lah BEESH), town (☐ 2 sq mi/5.2 sq km; 2001 population 2,776), E ALBERTA, on Lac La BICHE, 50 mi/80 km E of ATHABASCA, in LAKELAND COUNTY; 54°46′N 111°58′W. Tanning, lumber and flour milling; dairying; oil; fishing; tourism. Incorporated as a village in 1919; became a town in 1951.

Lac La Ronge, CANADA: see RONGE, LAC LA.

Laclede (luh-KLEED), county (☐ 770 sq mi/2,002 sq km; 2006 population 35,091), S central MISSOURI; ☉ LEBANON; 37°39′N 92°35′W. In the OZARKS; drained by GASCONADE RIVER. Cattle; wheat, corn, fruit growing, hay; oak timber; manufacturing at Lebanon (pleasure boats). Part of Mark Twain National Forest in E. Formed 1849.

Laclede (luh-KLEED), city (2000 population 415), LINN county, N central MISSOURI, 5 mi/8 km W of BROOKFIELD; 39°47′N 93°10′W. Corn, wheat, soy-beans; cattle, hogs. General John J. Pershing boyhood home state historic site. Pershing State Park in SW. Locust Creek Covered Bridge (1868) state historic site to W.

Laclede, unincorporated village, BONNER county, N IDAHO, 12 mi/19 km SW of SANDPOINT, on PEND OREILLE RIVER, Round Lake State Park across river to E; 48°10′N 116°45′W. Agriculture (cattle); timber; manufacturing (log homes, lumber).

Laclubar (lah-KLOO-bahr), town, EAST TIMOR, central TIMOR ISLAND, 27 mi/43 km SE of DILI; 08°45′S 125°55′E. Agricultural (wheat, fruit); cattle raising. Formerly called Vila de Ourique.

Lac Masson, CANADA: see SAINTE MARGUERITE.

Lac-Mégantic (lahk–mai-gahn-TEEK) or **Mégantic**, town (☐ 8 sq mi/20.8 sq km; 2006 population 5,949), ☉ LE GRANIT county, ESTRIE region, SE QUEBEC, E CANADA, on CHAUDIÈRE River, at N end of Lake MEGANTIC, 50 mi/80 km ENE of SHERBROOKE, near U.S. (MAINE) border; 45°35′N 70°53′W. Railroad center; pulp milling, lumbering, dairying; metal products; resort. Was seat of historic FRONTENAC county.

Lac-Montanier (LAHK–mon-tahn-YAI), unorganized territory, SW QUEBEC, E CANADA; 48°02′N 78°31′W. Amalgamated into ROUYN-NORANDA in 2002.

La Cocha (lah KO-chah), town, SW TUCUMÁN province, ARGENTINA, 70 mi/113 km SSW of TUCUMÁN. Railroad terminus. Industrial, lumbering, and agriculture center. Manufacturing (alcohol, charcoal, and wood products); agriculture (corn, alfalfa, sugarcane, tobacco, cotton, olives, fruit; livestock).

Lacock (LAI-cok), village (2001 population 1,000), NW WILTSHIRE, central ENGLAND, on the AVON River, and 3 mi/4.8 km S of CHIPPENHAM; 51°25′N 02°08′W. Agricultural market in dairying region. Has abbey (dating from 13th century; now private residence), 14th–15th-century church, and many 15th-century houses.

Lacolle (lah-KUHL), village (☐ 19 sq mi/49.4 sq km), MONTÉRÉGIE region, S QUEBEC, E CANADA, S of MONTREAL and near the U.S. (NEW YORK) border; 45°05′N 73°22′W. During the War of 1812, the British defeated an invading American army here on March 30, 1814. Restructured in 2001.

La Colmena, town, Paraguarí department, S PARA-GUAY, 70 mi/113 km SE of ASUNCIÓN, and 15 mi/24 km SSW of Sapucci; 25°53′S 56°50′W. Fruit; livestock; lumbering. Sometimes called Colonia La Colmena.

La Coloma, CUBA: see COLOMA.

La Colorada (lah ko-lo-RAH-dah), town, SONORA, NW MEXICO, 32 mi/51 km SE of HERMOSILLO; 28°41′N 110°15′W. Elevation 1,270 ft/387 m. Copper mining; silver, lead, zinc, graphite.

La Colorada, village and minor civil division of San-tiago District, VERAGUAS province, W central PA-NAMA, in PACIFIC lowland, 5 mi/8 km S of SANTIAGO. Coffee, sugarcane; livestock.

La Columna (LAH ko-LOOM-nah), ANDEAN peak (16,411 ft/5,002 m) in MÉRIDA state, W VENEZUELA, highest peak of Cordillera de MÉRIDA and of Vene-zuela, 8.0 mi/12.9 km ESE of MÉRIDA; 08°30′N 71°02′W. BOLÍVAR is name commonly given to highest of La Columna's twin peaks.

Lacombe (luh-KOM), unincorporated city (2000 population 7,518), SAINT TAMMANY parish, LOUISI-ANA, 8 mi/12.9 km SE of MANDEVILLE; 30°18′N 89°56′W. Bryon Lacombe Crab Festival held here. Originally settled by French slave owners.

Lacombe (luh-KOM), town (☐ 7 sq mi/18.2 sq km; 2001 population 9,384), S central ALBERTA, W CANADA, near GULL LAKE, 14 mi/23 km N of RED DEER, in LA-COMBE COUNTY; 52°28′N 113°44′W. Railroad junction. Grain elevators; lumber, flour, grist mills; dairying. Dominion experimental farm.

Lacombe County (luh-KOM), municipality (☐ 1,074 sq mi/2,792.4 sq km; 2001 population 10,159), S central ALBERTA, W CANADA; 52°26′N 113°43′W. Agriculture. Includes BENTLEY, LACOMBE, BLACKFALDS, ALIX, ECKVILLE, CLIVE; the summer villages of Birchcliff, Sunbreaker Cove, Gull Lake, Half Moon Bay; and the hamlets of HAYNES, MORNINGSIDE, JOFFRE, TEES, and MIRROR. Established as a municipal district in 1944.

La Compañia (lah kom-pahn-YEE-ah), town, in the S central part of the state of OAXACA, MEXICO, 31 mi/50 km S of OAXACA DE JUÁREZ, 6 mi/10 km W of EJUTLA DE CRESPO. Agriculture. Zapotec Indian area.

La Comuna (lah ko-MOO-nah), town and canton, ARQUE province, COCHABAMBA department, central BOLIVIA; 17°48′S 66°23′W. Elevation 9,219 ft/2,810 m. Iron mining; clay, limestone, and gypsum deposits. Agriculture (potatoes, yucca, bananas, corn, barley, oats, rye, soy, coffee); cattle for meat and dairy products.

Lacon (LAI-kahn), city (2000 population 1,979), ☉ MARSHALL county, N central ILLINOIS, on ILLINOIS RIVER (bridged), and 24 mi/39 km NNE of PEORIA; 41°01′N 89°24′W. In agricultural area; grain elevator, textiles. Laid out as Columbia in 1826; incorporated 1839.

Lacona, town (2000 population 360), WARREN county, S central IOWA, 15 mi/24 km SE of INDIANOLA; 41°11′N 93°22′W.

Lacona, village (☐ 1 sq mi/2.6 sq km; 2006 population 572), OSWEGO county, N central NEW YORK, 25 mi/40 km NE of OSWEGO; 43°38′N 76°01′W. Some manufacturing and quarrying. Incorporated 1880.

La Concepción (lah kon-sep-see-ON), town (2005 population 12,234), MASAYA department, SW NICAR-AGUA, 7 mi/11.3 km WSW of MASAYA. Coffee and sugarcane center.

La Concepción (lah kon-sep-see-ON), town, ☉ Bugaba District, CHIRIQUÍ province, W PANAMA, on INTER-AMERICAN HIGHWAY, terminus of railroad, and 13 mi/21 km NW of DAVID. Coffee, oranges, cacao, sugar-cane; livestock. Also known as Concepción.

La Conception (lah kon-sep-SYON), village (☐ 55 sq mi/143 sq km; 2006 population 1,132), LAURENTIDES region, S QUEBEC, E CANADA, 9 mi/15 km from LA-BELLE; 46°09′N 74°42′W.

La Concha (LAH KON-chah), ANDEAN peak (16,148 ft/4,922 m) in MÉRIDA state, W VENEZUELA, in COR-DILLERA DE MÉRIDA, near LA COLUMNA peak, 7 mi/11.3 km ESE of MÉRIDA; 08°33′N 71°01′W.

La Concordia (lah kon-KOR-dee-ah), town, CHIAPAS, S MEXICO, in GRIJALVA River valley S of Presa Belisario Domínguez Reservoir; 16°08′N 92°38′W. Elevation 1,805 ft/550 m. Cereals, sugarcane, fruit, livestock.

La Concordia (lah kon-KOR-dee-ah), town, JINOTEGA department, W NICARAGUA, 13 mi/21 km WNW of JINOTEGA; 13°11′N 86°10′W. Sugarcane, coffee, corn.

Laconi (LAH-ko-nee), village, NUORO province, central SARDINIA, ITALY, 45 mi/72 km N of CAGLIARI; 39°51′N 09°03′E. Clay, lignite deposits. Castle ruins; camp-ground, parks.

Area is shown by the symbol ☐, and capital city or county seat by ☉.

Laconia (luh-KO-nee-yuh), city (2006 population 17,060), ⊙ BELKNAP county, central NEW HAMPSHIRE, 22 mi/35 km N of CONCORD; 43°34′N 71°28′W. Bounded SW by WINNISQUAM LAKE, NE by LAKE WINNIPESAUKEE. City center in S on Winnipesaukee River, between OPECHE BAY of LAKE WINNIPESAUKEE (N) and WINNISQUAM LAKE (SW). It is a popular summer and winter resort, and the industrial and trade center of a lake resort and farming region. Manufacturing (computer equipment, machinery, building materials, fabricated metal products, transportation equipment, paper products, textiles, electronics, fabricated metal products, medical equipment; printing and publishing, machining, aluminum foundry). N part is rural, some agriculture (vegetables; cattle, poultry; dairying). Laconia Technical College, New Hampshire College of OPECHE BAY of Continued Education. Laconia Airport to NE. Large Paugus Bay of LAKE WINNIPESAUKEE in N. Winnipesaukee Scenic railroad follows lakeshore; cruise boat at WEIRS BEACH (N). Settled c.1761; incorporated as a city 1893.

Laconia, town (2000 population 29), HARRISON county, S INDIANA, near OHIO River, 12 mi/19 km S of CORYDON; 38°02′N 86°05′W. In agricultural area. Laid out 1816.

Laconia, Greece: see LAKONIA.

La Conner, village (2006 population 791), SKAGIT county, NW WASHINGTON, 7 mi/11.3 km WSW of MOUNT VERNON, and on Swinomish Channel, slough between mainland and FIDALGO, near Skagit Bay, arm of PUGET SOUND; 48°23′N 122°29′W. Agriculture (vegetables, grain, berries; dairying); fish, oysters; manufacturing (boatbuilding, hardware, wood products). Bridges (two) to FIDALGO ISLAND. Swinomish Indian Reservation to W.

La Conquista (lah kon-KEES-tah), town, CARAZO department, SW NICARAGUA, 6 mi/9.7 km S of JINOTEPE; 11°44′N 86°12′W. In sugarcane zone.

La Consulta (lah kon-SOOL-tah), town, NW central MENDOZA province, ARGENTINA, in TUNUYÁN RIVER valley (irrigation area), on railroad, and 60 mi/97 km SSW of MENDOZA; 33°44′S 69°07′W. Fruit, wine, vegetables, potatoes.

La Convención (LAH kon-ven-see-ON), province (□ 40,588 sq mi/105,528.8 sq km), CUSCO region, S central PERU; ⊙ QUILLABAMBA; 12°00′S 73°00′W. Northernmost and largest province of Cusco region, centered on URUBAMBA RIVER. Tropical environment; sugarcane, rice, tropical fruit.

Lacoochee (lah-KOO-chee), town (□ 2 sq mi/5.2 sq km; 2000 population 1,345), PASCO county, W central FLORIDA, 7 mi/11.3 km N of DADE CITY; 28°28′N 82°10′W.

La Corey (lah KO-ree), hamlet, E central ALBERTA, W CANADA, 13 mi/20 km N of BONNYVILLE, in BONNYVILLE NO. 87 municipal district; 54°27′N 110°46′W.

La Corne (lah KORN), village (□ 128 sq mi/332.8 sq km), ABITIBI-TÉMISCAMINGUE region, W QUEBEC, E CANADA, 18 mi/29 km SE of AMOS; 48°21′N 78°00′W. Molybdenum, lithium mining. Also written Lacorne.

La Corona (LAH ko-RO-nah), ANDEAN peak in MÉRIDA state, W VENEZUELA, in CORDILLERA DE MÉRIDA N of LA COLUMNA peak, 10 mi/16 km ESE of MÉRIDA; 08°33′N 71°01′W. Consists of two peaks: Humboldt (16,214 ft/4,942 m), a name sometimes given to entire massif; and Bompland (16,040 ft/4,889 m).

Lacosta Island (luh-KOS-tuh), narrow barrier island, GULF OF MEXICO, SW FLORIDA, N of Captiva Pass, at entrance to CHARLOTTE HARBOR and Pine Island Sound; c.7 mi/11.3 km long; 26°40′N 82°14′W. Sometimes called Cayo Costa.

La Coste (luh CAHST), town, MEDINA county, SW TEXAS, 20 mi/32 km WSW of SAN ANTONIO, on MEDINA RIVER. Oil and natural gas, sand and gravel. Agriculture (cattle, sheep, goats; cotton, vegetables, peanuts). Manufacturing (animal feeds).

Lacoste (lah-KOST), unincorporated village, SW QUEBEC, E CANADA, near Lake Nominingue, 30 mi/48 km ESE of MONT-LAURIER, and included in RIVIÈRE-ROUGE; 46°26′N 74°55′W. Mica mining.

La Coulée (lah koo-LAI), hamlet, SE MANITOBA, W central CANADA, 30 mi/47 km from WINNIPEG, and in SAINTE ANNE rural municipality; 49°39′N 96°35′W.

Lacovia, town, former capital of SAINT ELIZABETH parish, SW JAMAICA, stretches 2 mi/3.2 km along BLACK RIVER, and 36 mi/58 km WNW of MAY PEN; 18°04′N 77°45′W. Used as an inland port for shipping sugar, logwood, and fustic down the river. In agricultural region (cassava, corn, vegetables, spices; livestock).

Lac-Pellerin (LAHK–pe-luh-RAN), unorganized territory, S QUEBEC, E CANADA; 47°47′N 74°37′W. Forms part of La TUQUE agglomeration.

Lac-Poulin (LAHK–poo-LAN), village (2006 population 94), CHAUDIÈRE-APPALACHES region, S QUEBEC, E CANADA, 7 mi/11 km from BEAUCEVILLE; 46°06′N 70°49′W.

Lacq (LAHK), commune (□ 6 sq mi/15.6 sq km), PYRÉNÉES-ATLANTIQUES department, AQUITAINE region, SW FRANCE, on Gave de Pau River, 17 mi/27 km NW of PAU; 43°25′N 00°38′W. Natural gas, discovered here in 1951, is recovered from depths of more than 11,000 ft/3,350 m under great pressure. There are more than 30 wells that yield almost a quarter of France's requirements. Large quantities of sulfur are extracted from the gas. Chemical plants surround the commune (methanol, ammonia, and plastics).

Lac qui Parle (LA kee PAHRL), county (□ 778 sq mi/2,022.8 sq km; 2006 population 7,464), SW MINNESOTA; ⊙ MADISON; 45°00′N 96°10′W. Bounded by SOUTH DAKOTA on W, bounded N and NE by MINNESOTA RIVER (flows through MARSH and LAC QUI PARLE reservoirs; both in Lac qui Parle Wildlife Area), and drained by LAC QUI PARLE RIVER and its West Fork and by Yellow Bank River. Agricultural area (corn, oats, wheat, soybeans, hay, alfalfa; hogs, cattle, sheep). Formed 1871.

Lac qui Parle Lake (LA kee PAHRL), reservoir, on boundary of LAC QUI PARLE and CHIPPEWA counties, extends N into SWIFT county, SW MINNESOTA, on MINNESOTA RIVER, 10 mi/16 km NW of MONTEVIDEO; 8 mi/12.9 km long, 1 mi/1.6 km wide; 45°02′N 95°53′W. Elevation 934 ft/285 m. Receives LAC QUI PARLE RIVER, from SW at SE end (dam), in Lac qui Parle State Park. Remainder of lake and MARSH LAKE (upstream) are in Lac qui Parle Wildlife Area.

Lac qui Parle River (LA kee PAHRL), c.70 mi/113 km long; rises in small lake in DEUEL county, E SOUTH DAKOTA, enters SW MINNESOTA near CANBY; flows NE, past DAWSON, to MINNESOTA RIVER at SE end of LAC QUI PARLE LAKE; 44°45′N 96°32′W.

La Crescent, town (2000 population 4,923), HOUSTON county, extreme SE MINNESOTA, on MISSISSIPPI RIVER opposite LA CROSSE, WISCONSIN (4 mi/6.4 km WNW of La Crosse; two highway bridges); 43°49′N 91°17′W. Agricultural area (grain, soybeans; livestock, poultry; dairying); manufacturing (burial vaults, septic tanks, concrete, wood trusses). Richard J. Dorer Memorial Hardwood State Forest to S; Lock and Dam Number 7 to N; small Winnebago Indian Reservation to S.

La Crescenta–Montrose, unincorporated city (2000 population 18,532), LOS ANGELES county, S CALIFORNIA, in La Crescenta Valley, in S foothills of SAN GABRIEL MOUNTAINS; residential suburb 12 mi/19 km N of downtown LOS ANGELES, NE of Glendale; 34°14′N 118°14′W. Light manufacturing. Site of Mount Waterman Ski Area; Angeles National Forest to NE.

La Crête (lah KRET), hamlet, NW ALBERTA, W CANADA, W of the PEACE River, and included in the specialized municipality of MACKENZIE NO. 23; 58°11′N 116°24′W. Agriculture. Mennonite heritage; established in the 1930s.

La Crête des Cerfs (lah KRE-tuh DE SER), animal park near BOUILLON, LUXEMBOURG province, SE BELGIUM, on SEMOIS RIVER, near French border.

La Croix, Lac (lak luh-KROI), long lake on U.S.-CANADA border, ST. LOUIS county, NE MINNESOTA and Rainy Lake district, W ONTARIO (Canada), 30 mi/48 km N of ELY; c.30 mi/48 km long, maximum width 4 mi/6.4 km; 48°21′N 92°10′W. Elevation 1,184 ft/361 m. U.S. portion is in Superior National Forest (BOUNDARY WATERS Canoe Area); E part of Canadian portion in Quetico Provincial Park. Lake is fed from Iron Lake to SE, through two widely separate channels, the first flows NW through Bottle Lake (on U.S./Canada border), the second flows N through McAree Lake in Ontario, then W to Lac St. Croix; two channels form large Irving Island (Ontario, Canada); 5 mi/8 km long, 4 mi/6.4 km wide. Lake is also drained by two separate rivers. The Namakan River (c.30 mi/48 km long) flows NW and W through Ontario to NAMAKAN LAKE; it is also drained through Loon River (on U.S.-Canada border), first S through Loon Lake, then W and NW through Little Vermilion and Sand Point lakes, passing from Sand Point to Namakan lakes through Namakan Narrows. Resort area.

Lacroix-Saint-Ouen (lah-kwah–san–too-AHN), commune (□ 8 sq mi/20.8 sq km), OISE department, PICARDIE region, N FRANCE, near OISE RIVER, at W edge of Forest of Compiègne, 4 mi/6.4 km SSW of COMPIÈGNE; 49°21′N 02°47′E. Woodworking (furniture, toys).

La Crosse, county (□ 479 sq mi/1,245.4 sq km; 2006 population 109,404), W WISCONSIN; ⊙ LA CROSSE; 43°54′N 91°06′W. Agriculture (corn, soybeans; cattle, hogs, sheep; dairying); lumber milling; other manufacturing at La Crosse. Drained by LA CROSSE and BLACK rivers. Mount La Crosse Ski Area in SW; La Crosse State Trail and Great River State Trail traverse county. Formed 1851.

La Crosse, city (2006 population 50,266), seat of LA CROSSE county, W WISCONSIN, c.100 mi/161 km NW of MADISON, at the foot of high bluffs on the MISSISSIPPI RIVER, where the LA CROSSE and BLACK rivers meet; 43°49′N 91°13′W. Manufacturing (fabricated metal products, machinery, building materials, apparel, dairy products, transportation equipment, consumer goods; food and beverage processing, printing). A French fur-trading post was there in the late 18th century. Later, the city contained a thriving lumber industry. The University of Wisconsin at La Crosse, Viterbo College, Western Wisconsin Technical Institute, and a U.S. fish hatchery and experimental farm are in La Crosse. The city also has a zoo, an aquarium, a historical museum, and a wildlife project. Terminus (E) of La Crosse State Trail and S terminus of Great River State Trail are here. Part of Upper Mississippi River National Wildlife Refuge is along the river here. Mount La Crosse Ski Area to S; Lock and Dam No. 7 to NW. Incorporated 1856.

La Crosse (luh KRAWS), town (2000 population 561), LA PORTE county, NW INDIANA, 22 mi/35 km SSW of LA PORTE; 41°19′N 86°53′W. Agricultural area.

La Crosse, town (2000 population 1,376), ⊙ RUSH county, W central KANSAS, 30 mi/48 km WNW of GREAT BEND; 38°31′N 99°18′W. Shipping point for wheat and cattle. Barbed Wire Museum here, over 500 examples of barbed wire fencing. Founded 1876, incorporated 1886.

La Crosse (lah KRAHS), town (2006 population 598), MECKLENBURG county, S VIRGINIA, suburb 2 mi/3.2 km SE of SOUTH HILL; 36°42′N 78°05′W. Manufacturing (prefabricated steel buildings, textiles); in agricultural area (tobacco, peanuts, cotton, grain; livestock); timber. Mechlenburg Brunswick Regional Airport to E.

La Crosse (lah KRAWS), village (2006 population 344), WHITMAN county, SE WASHINGTON, 32 mi/51 km

WNW of PULLMAN; 46°49′N 117°53′W. Ships wheat, barley, oats, rye; cattle, sheep. PALOUSE FALLS State Park and Lyons Ferry State Park to SW.

La Crosse River, c.50 mi/80 km long, W WISCONSIN; rises in MONROE county; flows generally SW, past SPARTA, to the MISSISSIPPI RIVER at LA CROSSE.

La Cruz (lah KROOS), city, ⊙ La Cruz comuna, QUILLOTA province, VALPARAISO region, N central CHILE, 10 mi/16 km N of QUILLOTA city. On railroad. Center of Chilean fruit growing.

La Cruz (lah KROOS), town (1991 population 5,048), ⊙ San Martin or General San Martín department, E CORRIENTES province, ARGENTINA, port on URUGUAY RIVER (BRAZIL border), and 90 mi/145 km E of MERCEDES. Livestock and agriculture center (corn, rice, manioc, peanuts, maté, olives, watermelons, tobacco, citrus fruit). Founded by Jesuits (1630).

La Cruz (lah KROOS), town, VALPARAISO province, VALPARAISO region, central CHILE, on railroad, on ACONCAGUA RIVER, and 28 mi/45 km NE of VALPARAISO. Agriculture (grain, fruit, wine, hemp, tobacco; livestock).

La Cruz (LAH KROOS), town, ⊙ La Cruz municipio, NARIÑO department, SW COLOMBIA, in Cordillera CENTRAL, on road, and 33 mi/53 km NE of PASTO; 01°03′N 77°34′W. Elevation 9,252 ft/2,820 m. Coffee, cereals, potatoes, cacao, sugarcane; livestock.

La Cruz (lah KROOS), town, ⊙ La Cruz canton, Guanacaste province, NW Costa Rica, on Inter-American Highway, near Nicaragua border, and 33 mi/53 km NNW of Liberia; 11°00′N 85°35′W. Livestock raising; port (Puerto Soley).

La Cruz (lah KROOS), town, CHIHUAHUA, N MEXICO, on railroad and Mexico Highway 45, and 75 mi/121 km SE of CHIHUAHUA; 27°50′N 105°11′W. Elevation 4,062 ft/1,238 m. Cotton, corn, beans, cattle.

La Cruz, town, ⊙ Elota municipio, SINALOA, NW MEXICO, on ELOTA RIVER, in coastal lowland, on railroad, and 70 mi/113 km SE of CULIACÁN ROSALES; 23°53′N 106°53′W. Lumbering and agricultural center (corn, chick peas, fruit); dye-extract factory. Also called La Cruz de Elota.

La Cruz (lah KROOS), town, FLORIDA department, S central URUGUAY, in the CUCHILLA GRANDE INFERIOR, on railroad, and 13 mi/21 km NNW of FLORIDA; 33°56′S 56°15′W. Viticulture center.

La Cruz, NICARAGUA: see LA CRUZ DEL RÍO GRANDE.

La Cruz, VENEZUELA: see LA CRUZ DE TARATARA.

La Cruz de Elota, MEXICO: see LA CRUZ.

La Cruz de la India (lah KROOS dai lah EEN-dee-ah) or **Mina La India**, village, LEÓN department, NICARAGUA, 7 mi/11.3 km SE of SANTA ROSA; 12°45′N 86°18′W. Gold- and silver-mining center.

La Cruz del Río Grande (lah KROOS del REE-o GRAHN-dai), town and township (1995 municipio population 9,064), SOUTH ATLANTIC COAST AUTONOMOUS REGION, ZELAYA department, E NICARAGUA, port on RÍO GRANDE, and 85 mi/137 km NNW of BLUEFIELDS. English widely spoken. Livestock. Also called La Cruz.

La Cruz de Tarara, VENEZUELA: see LA CRUZ DE TARATARA.

La Cruz de Taratara (LAH KROOS dai tah-rah-TAH-rah), town, FALCÓN state, NW VENEZUELA, on S slopes of Sierra de SAN LUIS, 25 mi/40 km S of CORO; 11°04′N 69°42′W. Coffee, cacao. Also known as La Cruz de Tarara, or La Cruz.

Lacs, prefecture (2005 population 234,762), MARITIME region, S TOGO ⊙ ANÉHO; 06°22′N 01°40′E.

Lacs, region (□ 3,450 sq mi/8,970 sq km; 2002 population 597,500), central CÔTE D'IVOIRE; ⊙ YAMOUSSOUKRO; 06°50′N 05°10′W. Bordered N by Vallée du Bandama region, E by N'zi-Comoé region, S by Lagunes region, SW by Fromager region, and W by Marahoué region and Lake Kossou. Part of Lake

Kossou in W. Towns include TIÉBISSOU, TOUMODI, and the country's capital, Yamoussoukro. Railroad runs N-S through NE part of the region, from N'zi-Comoé into Vallée du Bandama. Regional airport at Yamoussoukro.

Lac-Saguay (LAHK–sah-GAI), town (□ 68 sq mi/176.8 sq km; 2006 population 409), LAURENTIDES region, S QUEBEC, E CANADA, 26 mi/42 km from LABELLE; 46°32′N 75°09′W.

Lac-Saint-Charles (LAHK–san–SHAHRL), former village, QUEBEC, E CANADA; 46°53′N 71°22′W. Amalgamated into QUEBEC city in 2002.

Lac Sainte Anne County (lahk sent AHN), municipal district (□ 1,099 sq mi/2,857.4 sq km; 2001 population 8,948), central ALBERTA, W CANADA; 53°50′N 114°41′W. Includes MAYERTHORPE, SANGUDO, ONOWAY, ALBERTA BEACH; the summer villages of NAKAMUN PARK, SILVER SANDS, SUNSET POINT, YELLOWSTONE, BIRCH COVE, ROSS HAVEN, SOUTH VIEW, VAL QUENTIN, CASTLE ISLAND, SANDY BEACH, SUNRISE BEACH, and WEST COVE; and the hamlets of CHERHILL, GUNN, GLENEVIS, RICH VALLEY, GREEN COURT, and ROCHFORT BRIDGE. Formed in 1944.

Lac-Sainte-Marie (LAHK–sent–mah-REE), town (□ 82 sq mi/213.2 sq km; 2006 population 520), OUTAOUAIS region, SW QUEBEC, E CANADA, 3 mi/5 km from KAZABAZUA; 45°57′N 75°57′W.

Lac-Saint-Jean-Est (lahk–san–zhahn–EST), county (□ 905 sq mi/2,353 sq km), SAGUENAY–LAC-SAINT-JEAN region, S central QUEBEC, E CANADA, on Lake SAINT JOHN; ⊙ ALMA; 48°30′N 71°40′W. Composed of eighteen municipalities. Formed in 1982.

Lac-Saint-Jean-Ouest (LAHK–san–zhahn–WEST), former county (□ 22,818 sq mi/59,326.8 sq km), central QUEBEC, E CANADA, on Lake SAINT JOHN; the county seat was ROBERVAL; 49°00′N 73°00′W.

Lac-Saint-Joseph (LAHK–san–zho-SEF), city (□ 13 sq mi/33.8 sq km; 2006 population 198), CAPITALE-NATIONALE region, S QUEBEC, E CANADA, 22 mi/35 km from QUEBEC city; 46°55′N 71°39′W. Part of the Metropolitan Community of Quebec (*Communauté Metropolitaine de Québec*).

Lac Saint Louis, CANADA: see MELOCHEVILLE.

Lac-Saint-Paul (LAHK–san–POL), village (□ 67 sq mi/174.2 sq km; 2006 population 456), LAURENTIDES region, S QUEBEC, E CANADA, 4 mi/6 km from MONT-SAINT-MICHEL; 46°44′N 75°19′W.

Lacs des Arcs (LAHK daiz AHRK), hamlet, SW ALBERTA, W CANADA, 9 mi/14 km from CANMORE, in BIGHORN NO. 8 municipal district; 51°03′N 115°09′W.

Lac-Sergent (LAHK–ser-ZHAHN), city (□ 2 sq mi/5.2 sq km; 2006 population 270), CAPITALE-NATIONALE region, S QUEBEC, E CANADA, 13 mi/20 km from PORTNEUF; 46°51′N 71°44′W.

Lac Simard (lahk see-MAHR) (□ 59 sq mi/153.4 sq km), SW QUEBEC, E CANADA, 90 mi/145 km NNE of NORTH BAY; 13 mi/21 km long, 10 mi/16 km wide; drained W by OTTAWA River; 47°38′N 78°40′W. Formerly known as Lake Expanse.

Lac-Simon (LAHK–see–MON), village (□ 38 sq mi/98.8 sq km; 2006 population 722), OUTAOUAIS region, SW QUEBEC, E CANADA, 3 mi/4 km from CHÉNÉVILLE; 45°54′N 75°06′W.

Lac-Supérieur (LAHK–syoo-PAI-ree-UHR), village (□ 147 sq mi/382.2 sq km; 2006 population 1,469), LAURENTIDES region, S QUEBEC, E CANADA, 14 mi/23 km from LABELLE; 46°12′N 74°28′W.

Lac-Surimeau (LAHK–syu-ree-MO), unorganized territory, SW QUEBEC, E CANADA. Amalgamated into ROUYN-NORANDA in 2002.

Lac-Tourlay (LAHK–toor-LAI), unorganized territory, S QUEBEC, E CANADA; 47°49′N 71°58′W. Forms part of La TUQUE agglomeration.

Lac-Tremblant-Nord (lahk–trahn-blahnt–NOR), municipality, LAURENTIDES region, S QUEBEC, E CANADA; 46°13′N 74°37′W. An independent municipality, it is part of the MONT-TREMBLANT agglomeration.

La Cuchilla, URUGUAY: see RÍO BRANCO.

La Cuesta (lah KWE-stah), village, Puntarenas province, S Costa Rica, near Panama border, 14 mi/23 km N of Puerto Armuelles.

La Cueva (lah KWE-vah), canton, SUD CINTI province, CHUQUISACA department, SE BOLIVIA, 6 mi/10 km SE of CULPINA; 21°05′S 65°46′W. Elevation 7,575 ft/2,309 m. Bismuth-bearing lode. Agriculture (potatoes, yucca, bananas, corn, wheat, oats, rye, peanuts); cattle and pig raising.

La Cueva, canton, Burnet O'Conner province, TARIJA department, S central BOLIVIA, 6 mi/10 km N of SALINAS, on the Tam River; 21°41′S 64°12′W. Elevation 4,035 ft/1,230 m. Abundant gas resources in area; clay, limestone, and gypsum deposits. Agriculture (potatoes, yucca, bananas, corn, sweet potatoes, tobacco); cattle.

La Cueva (lah KWE-vah), unincorporated village, MORA county, N NEW MEXICO, on MORA RIVER, and 5 mi/8 km SE of MORA, E of SANGRE DE CRISTO MOUNTAINS. Elevation 7,070 ft/2,155 m. In irrigated agricultural region; livestock, fruit, grain, alfalfa.

Lacui Peninsula (lah-KOO-ee), NW headland of CHILOÉ ISLAND, S CHILE, on the PACIFIC, just NW of ANCUD. HUECHUCUICUI POINT is its NW tip. The peninsula was fiercely disputed during war of independence and has ruins of Spanish forts.

La Cumbre (lah KOOM-brai), town, NW CÓRDOBA province, ARGENTINA, on SW slope of SIERRA CHICA, 35 mi/56 km NW of CÓRDOBA; 30°58′S 64°30′W. Tourist resort in N Córdoba hills; elevation 4,000 ft/1,220 m. Livestock raising. Adjoining is health resort of Cruz Chica.

La Cumbre (LAH KOOM-brai), town, ⊙ La Cumbre municipio, VALLE DEL CAUCA department, W COLOMBIA, in CORDILLERA OCCIDENTAL, on railroad, and 14 mi/23 km NNW of CALI; 03°39′N 76°34′. Elevation 5,184 ft/1,580 m. Sugarcane, coffee, sorghum, soybeans, flowers; livestock.

La Cumbre, ARGENTINA-CHILE: see USPALLATA PASS.

Lacu Roșu, ROMANIA: see GHEORGHENI.

Lacu Sărat (LAH-koo suh-RAHT), resort, BRĂILA county, SE ROMANIA, 6 mi/9.7 km SW of BRAILA. Noted saline, iodine, and sulphurous springs.

Lac Vieux Desert, lake (□ 10 sq mi/26 sq km) on WISCONSIN-MICHIGAN state line, largely in VILAS county (Nicolet National Forest), N WISCONSIN, and partly in GOGEBIC county (Ottoawa National Forest), W UPPER PENINSULA, Michigan, in forested area. Muskellunge fishing; drained by WISCONSIN RIVER.

Lac-Wapizagonke, unorganized territory, S QUEBEC, E CANADA; 46°43′N 73°01′W. Amalgamated into SHAWINIGAN in 2002.

La Cygne (luh SEEN), city (2000 population 1,115), LINN county, E KANSAS, on MARAIS DES CYGNES River, and 18 mi/29 km SSE of PAOLA; 38°21′N 94°45′W. Livestock and poultry raising, general agriculture. Limestone quarries. Marais des Cygnes Massacre Park to SE; Marais des Cygnes Waterfowl Refuge to S; La Cygne Lake reservoir to E.

La Cygne Lake (luh SEEN), LINN and MIAMI counties, E KANSAS, on North Sugar Creek, 11 mi/18 km NNE of PLEASANTON, and 1 mi/1.6 km W of MISSOURI state line; 4 mi/6.4 km long; 38°19′N 94°39′W. Maximum storage capacity 60,000 acre-ft. Formed by dam (37 ft/11 m high), built (1971) for flood control and water supply. La Cygne State Park is here.

Lacy Isles, AUSTRALIA: see NUYTS ARCHIPELAGO.

Lada (LAH-dah), village, NE MORDVA REPUBLIC, central European Russia, on INSAR RIVER, on road, 30 mi/48 km NNE of SARANSK; 54°35′N 45°25′E. Elevation 380 ft/115 m. Oil shale deposits nearby. Formerly called Lady (LAH-di).

Ladainha (LAH-dei-een-yah), city (2007 population 16,444), E central MINAS GERAIS, BRAZIL, 26 mi/42 km NNE of TEÓFILO OTONI; 17°37′S 41°46′W.

Ladakh (luh-DAHK), region in extreme N INDIA (JAMMU AND KASHMIR state) and extreme NE PAKI-

STAN (NORTHERN AREAS), on the border of CHINA. Referred to sometimes as LITTLE (OR INDIAN) TIBET, the region is allied ethnologically and geographically with TIBET and has a predominantly Lamaist Buddhist population; nominally a dependency of Tibet. After 1531 it was invaded periodically by Muslims from KASHMIR; came under Mogul sovereignty in 1664. Annexed to Kashmir in the mid-19th century. Divided between India (the S half) and Pakistan (the N half) since 1948. Indian LADAKH became a district, later renamed LEH district. Region became focus of a border dispute between CHINA and India, leading to fighting in 1959 and 1962. Ladakhians have a folk culture of their own, due to their historic geographical inaccessibility and isolation; however, with increased tourism and communication with outside world, traditional practices, such as polyandry, are quickly dying out.

Ladakh Range (luh-DAHK), trans-Shyok lateral range of KARAKORAM mountain system, JAMMU AND KASHMIR state, extreme N INDIA and NORTHERN AREAS, extreme NE PAKISTAN; from SHYOK RIVER mouth (in Pakistan) extends c.230 mi/370 km SE, parallel with right bank of the INDUS River (which separates it from ZASKAR RANGE), to CHINA (TIBET; border undefined); rises to crest line of c.20,000 ft/ 6,100 m. DEOSAI MOUNTAINS, a W section of GREATER HIMALAYAS in Pakistan, are sometimes regarded as a trans-Indus extension of LADAKH RANGE.

Ladan (LAH-dahn), town, SE CHERNIHIV oblast, UKRAINE, on UDAY RIVER, and 10 mi/16 km SE of PRYLUKY; 50°31'N 32°35'E. Elevation 331 ft/100 m. Machine building. Technical school. Known since 1619; town since 1938.

Ladbergen (LAHD-ber-gen), village, North Rhine-Westphalia, NW GERMANY, 13 mi/21 km N of MÜNSTER; 52°08'N 07°43'E.

Ladce (lah-TSE), Hungarian *Lédec*, village, ZAPADOSLOVENSKY province, W SLOVAKIA, near VÁH RIVER, on railroad, and 24 mi/39 km SW of ŽILINA; 49°02'N 18°17'E. Major cement works; large hydroelectric plant; limestone quarry.

Ladd, village (2000 population 1,313), BUREAU county, N ILLINOIS, 14 mi/23 km E of PRINCETON; 41°22'N 89°12'W. In agricultural area. Railroad junction. Incorporated 1890.

Ladder of Tyre, series of capes or headlands (c.6 mi/10 km long) jutting into the MEDITERRANEAN SEA and rising in progression from ROSH HANIKRA (Ras en-Naqura) [Arabic=cape of crevices] on the ISRAEL-LEBANON border (N) to RAS EL ABYADH (Ras al Bayadah) [Arabic=the white cape], which lies at the S end of the Tyre plains and 9 mi/15 km from TYRE. A steep, step-like ancient road was carved into the wave-cut white limestone and chalk cliffs that mark the narrow coast. The road was part of the route from EGYPT to PHOENICIA and was later rebuilt by the Romans, who called it the Via Maris [Latin=the sea road]. It served the Egyptians, Assyrians, Phoenicians, Alexander the Great, the Romans, the Crusaders, the Saracens, the Ottomans, and the British. In World War I, the British used the route to penetrate Lebanon. Starting with World War I and continuing on through World War II, a railroad was built in stages by the British and later the Allies along the route from Kantara in Egypt to HAIFA, BEIRUT, and TRIPOLI, and during World War II tunnels were cut through Ras-en-Naqura and Ras el Abyadh as part of the project. In ancient times, the snails found in the Ladder's beaches and coves yielded a purple dye that became the basis for Tyre's wealth.

Ladd Field, Alaska: see FORT WAINWRIGHT.

Laddonia, city (2000 population 620), AUDRAIN county, NE central MISSOURI, 13 mi/21 km ENE of MEXICO; 39°14'N 91°38'W. Corn, wheat, soybeans; cattle, hogs; lumber.

Lade (LAH-duh), industrial area in SØR-TØNDELAG county, part of TRONDHEIM city, central NORWAY, on TRONDHEIMSFJORDEN, on railroad just E of Trondheim. Manufacturing area. Seat of Norwegian Geological Survey; was tenth-century seat of Viking kings, when it was called Hladir (Old Norse *Hlaoir*); church is preserved.

Ladek Zdroj (LON-dek zdroo-ye), Polish *Lądek Zdrój*, German *Bad Landeck*, town in LOWER SILESIA, Wałbrzych province, SW POLAND, near CZECH border, at W foot of REICHENSTEIN MOUNTAINS, 11 mi/18 km SE of KLODZKO (Glatz); 50°21'N 16°53'E. Health resort.

Ladelle (luh-DEL), village, DREW county, SE ARKANSAS, 11 mi/18 km S of MONTICELLO.

La Democracia (lah de-mo-KRAH-see-ah), town, ESCUINTLA department, S GUATEMALA, in Pacific piedmont, 12 mi/19 km SW of ESCUINTLA; 14°14'N 90°57'W. Elevation 541 ft/165 m. Sugarcane, grain. Citronella-oil press nearby.

La Democracia, town, HUEHUETENANGO department, GUATEMALA, 34 mi/55 km NW of HUEHUETENANGO on INTER-AMERICAN HIGHWAY; 15°34'N 91°52'W. Elevation 3,015 ft/919 m. Highway junction for NENTÓN c.5 mi/8 km W of town. Last town before MEXICO border; a predominantly Ladino township in a largely Mam-speaking area.

La Democracia (LAH dai-mo-KRAH-see-ah), town, MIRANDA state, N VENEZUELA, 37 mi/60 km SSE of CARACAS; 10°01'N 66°45'W. Elevation 2,896 ft/882 m. Coffee, cacao, sugarcane.

Ladenburg (LAH-den-burg), ancient *Lopodunum*, town, BADEN-WÜRTTEMBERG, GERMANY, on the NECKAR, 6 mi/9.7 km E of MANNHEIM; 49°28'N 08°37'E. Chemical and technology industries; nurseries; hydroelectric station. Site of biological research center. Has early-Gothic church. Town is of Celtic-Roman origin; has retained medieval appearance.

Ladendorf (LAH-den-dorf), township, NE LOWER AUSTRIA, 5 mi/8 km SW of MISTELBACH; 48°32'N 16°29'E. Corn, vineyards. Castle.

Ladera Heights, unincorporated town (2000 population 6,568), SAN MATEO county, W CALIFORNIA; residential suburb 30 mi/48 km SSE of downtown SAN FRANCISCO, near San Francisquito Creek. SANTA CRUZ MOUNTAINS (SW); Portola and Castle Rock state parks (S). Sometimes called Ladera.

Ladera Heights, suburb of LOS ANGELES, LOS ANGELES county, S CALIFORNIA, 3 mi/4.8 km N of INGLEWOOD in Baldwin Hills; 34°00'N 118°22'W.

La Descubierta (lah des-koo-bee-ER-tah), town, BAHORUCO province, SW DOMINICAN REPUBLIC, on N shore of Lake ENRIQUILLO, 15 mi/24 km W of NEIBA; 18°37'N 71°41'W. Agricultural (coffee, grapes, fruit); timber.

Lădești (luh-DESHT), village, VÎLCEA county, S central ROMANIA, 11 mi/32 km SW of RÎMNICU VÎLCEA; 44°53'N 24°03'E. Agricultural center; orchards.

Ladhon River, Greece: see LADON RIVER.

La Digue (lah DEEG), district and island (2002 population 2,099), part of the MAHÉ ISLAND group, SEYCHELLES, in INDIAN OCEAN, 35 mi/56 km NE of VICTORIA (on Mahé Island); 4 mi/6.4 km long, 2 mi/ 3.2 km wide; 4°22'S 55°50'E. Covers 2,500 acres/1,012 ha; of granite formation. Tourism; fishing; spices. District formed c.1979.

Ladik, Turkish *Lâdik*, ancient *Laodicea Combusta*, village, N TURKEY, 33 mi/53 km SW of SAMSUN; 40°54'N 35°54'E. Cereals.

Ladislao Cabrera (LAH-dees-lah-o kah-BRE-rah), province, ORURO department, W central BOLIVIA; ⊙ Salinas de Garcia MENDOZA; 19°30'S 67°40'W.

Ladislao Cabrera (LAH-dees-lah-o kah-BRE-rah), canton, PACAJES province, LA PAZ department, W BOLIVIA. Elevation 12,989 ft/3,959 m. Gas wells in area. Copper resources (unmined); limestone, gypsum deposits. Agriculture (potatoes, yucca, bananas, rye); cattle.

Ladismith, town, LITTLE KAROO, WESTERN CAPE province, SOUTH AFRICA, 55 mi/89 km W of OUDTSHOORN, at foot of KLEIN SWARTBERG range, 10 mi/16 km SW of SEVEN WEEKS POORT pass; 33°30'S 21°46'E. Elevation 2,263 ft/690 m. Railroad terminus; agriculture center (dairying; grain, alfalfa, fruit; sheep, ostriches); brandy distilling.

Ladner (LAD-nuhr), unincorporated town, SW BRITISH COLUMBIA, W CANADA, near mouth of S branch of FRASER River delta on Strait of GEORGIA, 11 mi/18 km S of VANCOUVER, and included in city of DELTA; 49°05'N 123°05'W. Dairying; milk, fruit, vegetable, and fish canning; fishing, lumber shipping; flowers, nurseries. Urban enclave of Vancouver.

Ladnun (LAHD-nun), town, NAGAUR district, N central RAJASTHAN state, NW INDIA, 125 mi/201 km NE of JODHPUR; 27°39'N 74°23'E. Local trade in hides, livestock, wool, grain, salt; hand-loom weaving; gold ornaments, leather work. Sandstone quarried 1 mi/1.6 km S.

Lado (LAH-do), village, central EQUATORIA state, S SUDAN, on left bank of the BAHR AL-GABAL (WHITE NILE River), on road, and 10 mi/16 km N of JUBA; 05°02'N 31°41'E. Cotton, sesame, and durra; livestock. Founded 1874 by Charles Gordon (British); was the capital of LADO ENCLAVE (□ c.15,000 sq mi/38,850 sq km, now in S SUDAN and N UGANDA), leased (1894–1910) to BELGIAN CONGO; enclave included village of AL-RAJAF.

Ladoeiro (lah-doo-AI-roo), village, CASTELO BRANCO district, central PORTUGAL, 12 mi/19 km E of CASTELO BRANCO; 39°50'N 07°16'W. Olives, vegetables, and fruits (especially melons).

Lado Enclave, SUDAN and UGANDA: see LADO.

Ladoga (luh-DO-guh), town (2000 population 1,047), MONTGOMERY county, W central INDIANA, on RACCOON CREEK, and 11 mi/18 km SE of CRAWFORDSVILLE; 39°55'N 86°48'W. Agricultural area (corn, soybeans; livestock).

Ladoga, Lake (LAH-duh-gah), Finnish *Laatokka*, Russian *Ladozhskoye Ozero* (□ 6,332 sq mi/16,400 sq km), the largest lake in Europe, Republic of KARELIA, NW European Russia, NE of SAINT PETERSBURG; approximately 130 mi/209 km long and 80 mi/129 km wide, with a maximum depth of 755 ft/230 m; 61°00'N 31°00'E. Located on the heavily glaciated BALTIC SHIELD, the lake has shores that are low and marshy in the S, rocky and indented in the N. It is subject to autumn storms and freezes every year for two months in the N and four months in the S. Chief among the many rivers that feed the lake are the Svir, descending from Lake Onega; the Vuoska, which forms the outlet of the Saimaa lake system (Finland); and the Volkhov, coming from Lake Il'men. The main outlet is the NEVA RIVER, which flows W into the Gulf of FINLAND at St. Petersburg. The fortress at Petrokrepost' commands the Neva's exit from the lake. Among the many islands in the N part of the lake is Valaam (Finnish *Valama* or *Valamo*), with a famous Russian monastery dating from the 12th century or earlier. Until the Russo-Finnish War of 1939–1940, the N part of the lake belonged to Finland; cession of the Finnish shore to the USSR was confirmed by the peace treaty of 1947. During the defense of St. Petersburg (then Leningrad), which was surrounded and beseiged by the Germans in World War II, the frozen Lake Ladoga was the lifeline by which Leningrad was supplied in the winters from 1941 to 1943. Because of the difficulties of navigation, the S shore of Lake Ladoga is paralleled by the Ladoga Canals (approximately 100 mi/161 km long) connecting the Svir and Neva rivers and forming part of the Mariinsk System (see Volga-Baltic Waterway) and the BALTIC-WHITE SEA CANAL system.

Ladol (LAH-dol), town, MAHESANA district, GUJARAT state, W central INDIA, 22 mi/35 km E of MAHESANA; 23°37'N 72°44'E. Local trade in millet, pulses, wheat; tanning.

La Dôle, SWITZERLAND: see DÔLE, LA.

Ladonia (luh-DON-yuh), village (2000 population 3,229), RUSSELL county, E ALABAMA, 3 mi/4.8 km W of PHENIX CITY; 32°28′N 85°05′W.

Ladonia (lah-DO-nee-ah), village (2006 population 701), FANNIN county, NE TEXAS, 27 mi/43 km SW of PARIS; 33°25′N 95°57′W. Near North Sulphur River. Agriculture (cotton, grain, peanuts, soybeans); manufacturing (adhesive products). Historic downtown restored.

Ladon River (LAH–[th]-[th]on), c.40 mi/64 km long, in ARKADIA prefecture, central PELOPONNESE department, S mainland GREECE; rises in AROANIA mountains; flows SW and S to the Alpheus River 10 mi/16 km ESE of OLYMPIA, on border of WESTERN GREECE department. Hydroelectric plant. Formerly known as Roufias, Rouphia, or Ruphia River, a name then also applied to lower course of the Alpheus. Also spelled Ladhon.

Ladora, town (2000 population 287), IOWA county, E central IOWA, 6 mi/9.7 km WSW of MARENGO; 41°45′N 92°11′W. Livestock; grain.

La Dorada (LAH do-RAH-dah), town, ⊙ La Dorada municipio, CALDAS department, central COLOMBIA, river port on the MAGDALENA, on railroad, and 70 mi/113 km NW of BOGOTÁ; 05°27′N 74°40′W. Head of navigation of the lower Magdalena. Trading and communication center.

La Doré (lah do-RAI), parish (□ 109 sq mi/283.4 sq km; 2006 population 1,511), SAGUENAY—LAC-SAINT-JEAN region, S central QUEBEC, E CANADA, 24 mi/38 km from ROBERVAL; 48°43′N 72°39′W. Agriculture.

La Dormida (lah dor-MEE-dah), town, NE MENDOZA province, ARGENTINA, on TUNUYÁN RIVER (irrigation), and 60 mi/97 km SE of MENDOZA. Wine, fruit, grain; wine making, sawmilling. Oil production nearby.

Ladozhskaya (LAH-duhsh-skah-yah), village (2005 population 14,975), E central KRASNODAR TERRITORY, S European Russia, on the KUBAN' River, on road and railroad, 36 mi/58 km NE of KRASNODAR; 45°18′N 39°56′E. Elevation 360 ft/109 m. Railroad depots. In agricultural area; produce processing.

Ladrones Islands: see NORTHERN MARIANA ISLANDS.

Ladson, unincorporated city (2000 population 13,264), CHARLESTON and BERKELEY counties, SE SOUTH CAROLINA, 16 mi/26 km NW of CHARLESTON; 33°00′N 80°06′W. Manufacturing (cabinets, power transformers; printing and publishing, steel fabricating); agriculture (poultry, livestock; grain, cotton, soybeans).

La Due, unincorporated community, HENRY county, W central MISSOURI, 7 mi/11.3 km SW of CLINTON. Strip coal mines.

Ladue, city (2000 population 8,645), SAINT LOUIS county, E MISSOURI, a suburb 10 mi/16 km W of downtown ST. LOUIS; 38°38′N 90°22′W. Exclusive residential area. Manufacturing (vending machines); limestone quarry. Incorporated 1936 as a consolidation of former towns of Ladue, Deer Creek, and McKnight.

La Durantaye (lah dyoo-rahn-TEI), parish (□ 13 sq mi/33.8 sq km; 2006 population 772), CHAUDIÈRE-APPALACHES region, S QUEBEC, E CANADA, 4 mi/6 km from SAINT-MICHEL-DE-BELLECHASSE; 46°50′N 70°51′W.

Ladushkin (LAH-doosh-keen), city (2005 population 3,975), SW KALININGRAD oblast, W European Russia, near the BALTIC SEA coast, on road and railroad, 17 mi/27 km SW of KALININGRAD; 54°34′N 20°10′E. Elevation 111 ft/33 m. Summer resort. Founded in 1314. Until 1945, in East Prussia and called Ludwigsort.

Ladva (LAH-dvah), town, S Republic of KARELIA, NW European Russia, on road and near railroad, 30 mi/48 km SSE of PETROZAVODSK; 61°21′N 34°37′E. Elevation 190 ft/57 m. Lumbering.

Lady, RUSSIA: see LADA.

Ladybank (LAI-dee-BANK) or **Ladybank and Monkston**, small town (2001 population 1,487), FIFE, E Scotland, 5 mi/8 km SW of CUPAR; 56°16′N 03°08′W. Asphalt works.

Ladybrand, town, E central FREE STATE province, SOUTH AFRICA, near LESOTHO border in the CALEDON RIVER valley, 75 mi/121 km E of MANGAUNG (Bloemfontein); 29°13′S 27°28′E. Elevation 5,241 ft/1,597 m. Established in 1867 in area taken from the Sotho called the Conquered Territories, named for Boer president Jan Brand's mother. Agricultural center (dairying, wheat, rye, oats, potatoes); freestone quarrying. Airfield. Nearby are numerous caves with old Bushman paintings, Stone Age implements, and fossils.

Lady Elliot Island (LAI-dee EL-yuht), coral islet at S end of GREAT BARRIER REEF, in CORAL SEA, off coast of QUEENSLAND, NE AUSTRALIA, c.50 mi/80 km NE of BUNDABERG; 24°07′S 152°42′E. Resort. Automated lighthouse. Airstrip.

Lady Evelyn Lake (E-vuh-lin), SE central ONTARIO, E central CANADA, 16 mi/26 km W of COBALT; 20 mi/32 km long, 4 mi/6 km wide; 47°18′N 80°11′W. Elevation 930 ft/283 m. Drains S into Lake TIMAGAMI; fishing resort.

Lady Franklin Bay, NE ELLESMERE ISLAND, BAFFIN region, NUNAVUT territory, CANADA, inlet of ROBESON CHANNEL, at NW end of HALL BASIN; 25 mi/40 km long, 6 mi/9.7 km–10 mi/16 km wide; 81°40′N 65°00′W. On N shore, near entrance of bay, is small inlet of DISCOVERY HARBOUR, site of FORT CONGER, both of importance in late-19th-century exploration of the Arctic.

Lady Franklin, Cape, NE extremity of BATHURST ISLAND, BAFFIN region, NUNAVUT territory, CANADA, at N end of WELLINGTON CHANNEL; 76°40′N 98°42′W.

Lady Franklin Point, SW extremity of VICTORIA ISLAND, NUNAVUT territory, CANADA, on CORONATION GULF, at E entrance of DOLPHIN AND UNION STRAIT; 68°31′N 113°09′W.

Lady Frere, town, EASTERN CAPE province, SOUTH AFRICA, in STORMBERG range, on Cacadu River a tributary of White Kei River (see GREAT KEI RIVER), and 25 mi/40 km NE of QUEENSTOWN; 31°44′S 27°14′E. Elevation 3,969 ft/1,210 m. Livestock; dairying; grain. Landing strip.

Lady Grey, town, EASTERN CAPE province, SOUTH AFRICA, in WITTEBERGE range near Jouberts Pass, 30 mi/48 km E of ALIWAL NORTH; 30°44′S 27°14′E. Elevation 6,084 ft/1,855 m. Livestock, dairying, grain. Named for wife of Sir George Grey, governor of the Cape Colony (now CAPE PROVINCE) from 1854 to 1859.

Lady Isle (LAI-dee EI-uhl), islet in Firth of Clyde, off coast of South Ayrshire, SW Scotland, 3 mi/4.8 km WSW of TROON; 55°32′N 04°45′W. Lighthouse.

Lady Lake (LAI-dee), town (□ 6 sq mi/15.6 sq km; 2000 population 11,828), Lake county, central FLORIDA, c.20 mi/32 km SE of OCALA, in citrus-fruit and watermelon area; 28°55′N 81°55′W. Fruit packing.

Lady's Island, island, BEAUFORT county, S SOUTH CAROLINA, one of SEA ISLANDS, c.35 mi/56 km NE of SAVANNAH, GEORGIA, connected by land to SAINT HELENA ISLAND (S) and PORT ROYAL ISLAND (W); c.10 mi/16 km long. Wilkins village on NE shore. Oyster and vegetable canning.

Ladysmith (LAI-dee-smith), town (□ 3 sq mi/7.8 sq km; 2001 population 6,587), SW BRITISH COLUMBIA, W CANADA, on SE VANCOUVER ISLAND, on inlet of Strait of GEORGIA, 13 mi/21 km SSE of NANAIMO, in COWICHAN VALLEY regional district; 48°58′N 123°49′W. Port and shipping point for nearby fishing, logging area; fruit, vegetable growing; dairying. Founded as a home for nearby coal mine workers. Incorporated 1904.

Ladysmith, town, W KWAZULU-NATAL province, SOUTH AFRICA, on the KLIP RIVER; 28°32′S 29°47′E. Elevation 4,018 ft/1,225 m. The town has railroad yards and food-processing, textile, and tire factories. It is the distribution center for the surrounding agricultural and coal-mining region on railroad and road links between JOHANNESBURG and DURBAN. Airport. Ladysmith was founded in 1851 by Boers who had been persuaded by British governor Sir Harry Smith to remain in Natal rather than join the trek to other areas. The town, named for Smith's wife, grew after a railroad to Durban was opened in 1886 with the discovery of gold on the WITWATERSRAND. During the South African War, Sir George White's British forces at Ladysmith were under siege by Boers (November 1899–February 1900), when British reinforcements arrived. Nearby battlefields associated with the siege include Wagon Hill, Nicholson's Nek, and Spioen Kop.

Ladysmith, town (2006 population 3,636), ⊙ RUSK county, N WISCONSIN, on FLAMBEAU RIVER, and c.50 mi/80 km NE of EAU CLAIRE; 45°27′N 91°05′W. In lake-resort area. Dairy products; manufacturing (furniture, aluminum die parts, paper, wooden wares, canned vegetables). A large cooperative creamery is here; has hydroelectric plant. Railroad junction. Big Falls Flowage to NE. Incorporated 1905.

Ladyzhin, UKRAINE: see LADYZHYN.

Ladyzhinka, UKRAINE: see LADYZHYNKA.

Ladyzhyn (lah-DI-zhin) (Russian *Ladyzhin*), city, SE VINNYTSYA oblast, UKRAINE, 52 mi/84 km SE of VINNYTSYA on the Southern BUH River; 48°40′N 29°15′E. Elevation 662 ft/201 m. Site of the Ladyzhyn Regional Thermal-Electric Station, stone-crushing, fabrication of reinforced concrete construction components, silicate brick making, grain milling, canning. First mentioned as a fort of the Halych principality, which the Tatars failed to take in 1240; known during wars of Khmel'nyts'kyy (1648), and during war for Ukraine's independence (1919, 1920). City since 1973.

Ladyzhynka (lah-DI-zhin-kah) (Russian *Ladyzhinka*), village, SW CHERKASY oblast, UKRAINE, on highway, 13 mi/21 km S of UMAN'; 48°33′N 30°15′E. Elevation 643 ft/195 m. Dairy; sugar beets.

Lae, city, E central PAPUA NEW GUINEA, on NE NEW GUINEA island, at the head of the HUON GULF; ⊙ MOROBE province; 06°44′S 147°00′E. Lae is an important commercial and transportation center of Papua New Guinea. Fish, lobster, prawns; coconut, copra, rice, sugar cane; cattle; timber; manufacturing (food products). Located at head of road systems to SW, W, and to NE coast. Founded in 1927 to serve air transport into the Morobe gold fields at WAU and BULOLO in the mountainous interior to SW, no longer active. Occupied by Japanese during World War II, and became an important strategic base against Allied Forces. In 1937 aviator Amelia Earhart departed from Lae for HOWLAND ISLAND and was never seen again. Nadzab International Airport; Lae War cemetery; Botanic Gardens, Papua New Guinea University of Technology (Unitech), including Matheson Library Gold deposits in the mountains above the Botanic Gardens.

Lae (LEI), atoll (1999 population 322), RALIK CHAIN, Kwajalein district, MARSHALL ISLANDS, W central PACIFIC, 75 mi/121 km SW of KWAJALEIN; c.5 mi/8 km long; 08°55′N 166°16′E. Consists of seventeen islets on lagoon.

Laeken (LAH-kuhn), part of N BRUSSELS, BRABANT province, central BELGIUM; 50°52′N 04°21′E. The palace built here (early nineteenth century) by Napoleon I is used as a Belgian royal residence. Just NW is Brupark theme park.

La Emilia (lah e-MEE-lee-ah), town, San Nicolás department, N BUENOS AIRES province, ARGENTINA, on railroad, 37 mi/60 km SE of city of ROSARIO; 33°21′S 60°19′W. Suburb of SAN NICOLÁS. Manufacturing.

Laem Sing (LAM SING), village and district center, CHANTHABURI province, S THAILAND, port at mouth of CHANTHABURI RIVER, on GULF OF THAILAND, serving as deep-water harbor for CHANTHABURI (9 mi/14.5 km N); 12°29′N 102°04′E. Rice milling. Occupied by French troops for eleven years, early in 20th century. Also spelled LAEM SINGH.

Laem Singh (LAM SING) or Laem Sing [Thai=Cape Sing], THAILAND. Small rocky beach on the W side of PHUKET ISLAND, just S of the larger SURIN Beach.

La Encarnación (lah en-kar-nah-see-ON), town, OCOTEPEQUE department, W HONDURAS, 20 mi/32 km NNE of NUEVA OCOTEPEQUE; 14°40′N 89°05′W. Tobacco, rice.

Laeo Kailiu (LAH-ai-O KAH-ee-LEE-oo), cape, on NW coast, KAUAI island, KAUAI county, HAWAII, S of Kāhala Point, in Haena State Park, on PACIFIC OCEAN.

Laer (LER), village, North Rhine-Westphalia, NW GERMANY, 14 mi/23 km NW of MÜNSTER; 52°04′N 07°10′E. Church (15th-century).

Laer, Bad, GERMANY: see BAD LAER.

Lærdal (LAWR-dahl), valley, SOGN OG FJORDANE county, W NORWAY, drained by Lærdalselvi River, and extending from W part of the HEMSEDALSFJELL c.15 mi/24 km W to head of SOGNEFJORDEN, a 5-mi/8-km-long arm of SOGNEFJORDEN. Lærdalsøyri village, at head of fjord, is tourist center and was road terminus for traffic to OSLO before completion of BERGEN railroad.

La Esmeralda (LAH es-mai-RAHL-dah), village, ⊙ Municipio Autónomo Alto Orinoco, AMAZONAS state, S VENEZUELA, landing on the ORINOCO RIVER S of Cerro DUIDA, and 215 mi/346 km SE of PUERTO AYACUCHO; 03°10′N 65°31′W. In tropical forest region (rubber, balata, vanilla). CANAL CASIQUIARE, linking the Orinoco with the AMAZON, branches off 20 mi/32 km W. Airport.

La Esmeralda, CUBA: see ESMERALDA.

Læsø (LE-suh), island (□ 44 sq mi/114.4 sq km), DENMARK, in the KATTEGAT strait, 12 mi/19 km E of NE JUTLAND; 13 mi/21 km long, 4 mi/6.4 km wide; 57°16′N 11°01′E. Highest point is 79 ft/24 m; cliffs along N coast, flat in S.

La Esperanza (lah es-pai-RAHN-sah), city, ⊙ INTIBUCÁ department, SW HONDURAS, 75 mi/121 km WNW of TEGUCIGALPA; 14°47′N 88°10′W. Elevation 4,951 ft/1,509 m. Commercial center in fruit-growing area (peaches, quinces, figs); peach liquor; fruit processing, lumbering. Founded in early 19th century, next to old Indian town of INTIBUCÁ (to N). Also known as Nueva Esperanza.

La Esperanza (lah es-pai-RAHN-sah), town, SE JUJUY province, ARGENTINA, 30 mi/48 km E of JUJUY. Sugar-refining center; manufacturing.

La Esperanza (lah es-pai-RAHN-sah), town, QUEZALTENANGO department, SW GUATEMALA, 2 mi/3.2 km NW of QUEZALTENANGO; 14°52′N 91°34′W. Elevation 8,087 ft/2,465 m. Corn, wheat, fodder grasses; livestock.

La Esperanza, CUBA: see PUERTO ESPERANZA.

La Estanzuela, MEXICO: see GARCÍA DE LA CADENA.

La Estrella (lah es-TRAI-yah), town and canton, SARAH province, SANTA CRUZ department, central BOLIVIA, 65 mi/105 km NNW of PORTACHUELO; 16°30′S 63°45′W.

La Estrella (LAH es-TRAI-yah), town, ⊙ La Estrella municipio, ANTIOQUIA department, NW central COLOMBIA, 3 mi/5 km S of MEDELLÍN; suburb of Medellín; 06°09′N 75°38′W. Coffee, plantains.

La Estrella (lah es-TRAI-yah) or **Estrella**, village, ⊙ La Estrella comuna, CARDENAL CARO province, LIBERTADOR GENERAL BERNARDO O'HIGGINS region, central CHILE, near the coast, 45 mi/72 km NW of SAN FERNANDO. Cereals, vegetables, fruit, grapes; livestock.

La Estrelleta, DOMINICAN REPUBLIC: see ELÍAS PIÑA, province.

La Fábrica (lah FAH-bree-kah), village, LLANQUIHUE province, LOS LAGOS region, S central CHILE, resort on S shore of LAKE LLANQUIHUE, 10 mi/16 km NNE of PUERTO MONTT.

Lafagu (LAH-fah-goo), town, KEBBI state, NW NIGERIA, near NIGER River, 90 mi/145 km S of BIRNIN KEBBI; 11°10′N 04°18′E. Market town. Sorghum, maize, millet.

Lafaiete Coutinho (LAH-fah-E-tai ko-CHEEN-yo), town (2007 population 3,627), E central BAHIA, BRAZIL, 24 mi/38 km NNW of Jequié; 13°40′S 40°13′W.

La Falda (lah FAHL-dah), town (1991 population including VALLE HERMOSO, HUERTA GRANDE, and Villa Giardino 27,292), NW CÓRDOBA province, ARGENTINA, 30 mi/48 km NW of CÓRDOBA. Popular tourist resort in N Córdoba hills; elevation 3,000 ft/914 m. Quarrying (granite, marble, lime) and livestock-raising center. Picturesque setting. Grottoes and waterfalls nearby.

La Farge, village (2006 population 779), VERNON county, SW WISCONSIN, on KICKAPOO River, and 35 mi/56 km SE of LA CROSSE; 43°34′N 90°38′W. In dairying and tobacco-growing area. Timber; cheese, wood pallets.

La Fargeville, village, JEFFERSON county, N NEW YORK, 16 mi/26 km N of WATERTOWN; 44°12′N 75°58′W. In timber and lumbering area.

Lafayette (LAH-fai-ET), county (□ 545 sq mi/1,417 sq km; 2006 population 7,896), SW ARKANSAS, ⊙ LEWISVILLE; 33°14′N 93°36′W. Bounded S by LOUISIANA state line; W by RED RIVER; E, in part, by DORCHEAT Bayou; drained by BODCAU Creek and Bayou. Oil and agricultural area (wheat, soybeans; cattle, hogs, chickens). Timber; gravel; oil and gas. Lake Erling Forest (Bodcau Bayou) in SE; Lafayette Wildlife Management Area in S, E of Lake ERLING; Conway Cemetery State Park in S.

Lafayette (la-FAI-yet), county (□ 547 sq mi/1,422.2 sq km; 2006 population 8,045), N central FLORIDA, bounded E and NE by SUWANNEE River; ⊙ MAYO; 28°46′N 81°43′W. Swampy flatwoods area, with many small lakes. Farming (corn, peanuts, tobacco), cattle raising, lumbering. Limestone and phosphate deposits. Formed 1856.

Lafayette, county (□ 679 sq mi/1,765.4 sq km; 2006 population 40,865), N MISSISSIPPI; ⊙ OXFORD; 34°21′N 89°29′W. Drained by YOCONA and TALLAHATCHIE (forms part of N boundary) rivers; part of SARDIS RESERVOIR (Tallahatchie River) is in NW. Agriculture (cotton, corn, soybeans; cattle; dairying); pine, hardwood timber. Part of Holly Springs National Forest in NE. Formed 1836.

Lafayette (LAH-fee-ET), county (□ 634 sq mi/1,648.4 sq km; 2006 population 33,186), W central MISSOURI; ⊙ LEXINGTON; 39°03′N 93°46′W. Bounded N by MISSOURI River. Agriculture (wheat, soybeans, corn, apples, peaches, sorghum; cattle, hogs); limestone rock quarries; manufacturing at Lexington, ODESSA, CONCORDIA, and HIGGINSVILLE. Confederate Memorial State Park in center. Formed 1834.

Lafayette, county (2006 population 16,298), S WISCONSIN, bordered S by ILLINOIS; ⊙ DARLINGTON; 42°39′N 90°08′W. Dairying; agriculture (barley, oats, wheat, corn, soybeans; cattle, hogs, sheep); zinc mining. Area formerly had important lead-mining industry. Drained by PECATONICA and GALENA rivers. First Capitol State Park is in NW at Leslie; Yellowstone Lake State Park in NE; E part of Pecatonica State Trail in NW. Formed 1846.

Lafayette (lah-FAI-it), city (2000 population 3,234), ⊙ Chambers co., E Alabama, 18 mi/29 km N of Opelika. Manufacturing (rubber, yarns, lumber milling, meat packing). Settled 1883. Originally named 'Chambersville' for the county, it was incorporated in 1835 as Lafayette in honor of the French general Marquis de Lafayette, who had visited AL ten years earlier.

Lafayette (LAH-fai-ET), city (2000 population 23,908), CONTRA COSTA county, W CALIFORNIA; residential suburb 8 mi/12.9 km NE of downtown OAKLAND, in the San Francisco-Oakland area; 37°54′N 122°07′W.

Diversified light manufacturing, including food processing and transportation equipment. The city is a horse-raising and agricultural-trading center, especially for walnuts. Mokelumne Aqueduct runs E-W through city. Las Trampas Regional Park to SE; Briones Regional Park to N; Lafayette Reservoir in SW. Settled 1848, incorporated 1968.

Lafayette, city (2000 population 23,197), BOULDER county, N central COLORADO, suburb, 20 mi/32 km NNW of downtown DENVER; 39°59′N 105°05′W. Elevation c.5,237 ft/1,596 m. In coal, poultry, dairy, grain, and sugar-beet area. Manufacturing (roof trusses, computer ribbons). Incorporated 1890.

La Fayette (luh FAI-it), city (2000 population 6,702), ⊙ WALKER county, NW GEORGIA, 31 mi/50 km NNW of ROME; 34°43′N 85°17′W. Manufacturing includes textiles, apparel, machinery and equipment, wire harnesses, transportation equipment, plastic products, limestone processing. Old courthouse now restored as museum. Confederate General Braxton Bragg planned Battle of Chickamauga here. CHATTAHOOCHEE National Forest nearby. Founded 1835.

Lafayette, city (2000 population 56,397), ⊙ TIPPECANOE county, W central INDIANA, on the WABASH RIVER; 40°25′N 86°52′W. Railroad junction. A manufacturing city in a grain, livestock, and dairy area. Manufacturing (building materials, electrical equipment, fabricated metal products, consumer goods, wire, food processing, transportation equipment, heating equipment, chemicals, animal health and feed products, paperboard, pharmaceuticals, and rubber products; publishing). Motor vehicles are also assembled here. Tippecanoe Battlefield State Memorial (November 1811) 5 mi/8 km N. Of interest is the rebuilt blockhouse of Fort Ouiatenon (1719), 4 mi/6.4 km W. Purdue University located in adjacent WEST LAFAYETTE. Laid out 1825. Incorporated 1853.

Lafayette, city (2000 population 110,257), ⊙ LAFAYETTE parish, S central LOUISIANA, 48 mi/77 km WSW of BATON ROUGE, on the VERMILION RIVER (which is linked to the INTRACOASTAL WATERWAY to S); 30°13′N 92°02′W. Known as the hub of cajun country. It is a commercial, shipping, and medical center for the area; agriculture (sugarcane, rice, cotton; dairy cattle, livestock); manufacturing (petroleum, apparel, jewelry, building materials, and boxes); industries (oil equipment and service, printing and publishing, food processing). The area's oil and natural gas boom has contributed to a large increase and diversity of population and an influx of new businesses. The Heymann Oil Center is headquarters for several oil companies. Of interest are St. John's Cathedral (site 1821; third structure 1916), a Carmelite monastery, a planetarium, Lafayette Natural History Museum, Children's Museum, and the University Art Center. Cajun Dome stadium is here. Lafayette is the seat of the University of Southwestern Louisiana and the scene of an annual Mardi Gras, along with numerous other cultural festivals such as the Festival Acadiens. Evangeline Downs thoroughbred racetrack to N. LONGFELLOW Evangeline State Commemorative Area to SE. Settled 1760s by Acadians; incorporated 1836.

Lafayette, city (2006 population 4,238), ⊙ MACON county, N TENNESSEE, 27 mi/43 km NNE of LEBANON; 36°31′N 86°02′W. Trade center for farm area; agriculture; manufacturing. Incorporated since 1940.

Lafayette (LAH-fai-ET), town (2006 population 3,208), YAMHILL county, NW OREGON, 5 mi/8 km NE of MCMINNVILLE, on YAMHILL river; 45°15′N 123°06′W. Agriculture (vegetables, berries, grapes; poultry); manufacturing (dairy products; bakery products); wineries. Maud Williamson State Park to S.

La Fayette (LAH-fai-ET), village (2000 population 227), STARK county, N central ILLINOIS, 6 mi/9.7 km W of TOULON; 41°06′N 89°58′W. In agricultural and bituminous-coal area.

La Fayette (lah-FAI-it), village, CHRISTIAN county, SW KENTUCKY, near TENNESSEE state line, 17 mi/27 km SW of HOPKINSVILLE. Agriculture (tobacco; livestock). FORT CAMPBELL MILITARY RESERVATION to S and W. Also spelled Lafayette.

Lafayette (LAH-fee-ET), village (2000 population 529), NICOLLET county, S MINNESOTA, 9 mi/14.5 km NNE of NEW ULM; 44°27′N 94°23′W. Agricultural area (grain, soybeans, peas; livestock, poultry; dairying); manufacturing (feeds, fertilizers, meat processing).

Lafayette (LAH-fai-ET), village (2006 population 295), ALLEN county, W OHIO, 8 mi/13 km E of LIMA, near OTTAWA RIVER; 40°45′N 83°57′W. Limestone quarrying.

Lafayette, village, NORTH KINGSTOWN town, WASHINGTON county, S central Rhode Island, 18 mi/29 km S of PROVIDENCE, 2 mi/3.2 km E of WICKFORD; 41°34′N 71°28′W. Residential and manufacturing center. Cocumscussoc State Park (undeveloped) is nearby.

Lafayette (LAH-fai-ET), parish (□ 283 sq mi/735.8 sq km; 2006 population 203,091), S LOUISIANA, ⊙ LAFAYETTE; 30°13′N 92°02′W. Bounded N by short Bayou Carenero, drained by VERMILION RIVER (navigable). Agriculture (home gardens, corn, sorghum, hay, nursery crops, rice, sugarcane, sweet potatoes, cucumbers, other vegetables; honey; cattle, horses, exotic fowl; dairying); crawfish; lumber; oil, natural gas. Varied manufacturing at LAFAYETTE. Formed 1823.

LaFayette, residential hamlet, ONONDAGA county, central NEW YORK, 11 mi/18 km S of SYRACUSE; 42°54′N 76°06′W. Light manufacturing and agriculture.

Lafayette, Brazil: see CONSELHEIRO LAFAIETE.

Lafayette, Mount, NEW HAMPSHIRE: see FRANCONIA MOUNTAINS.

Lafayette Springs (LAH-fai-ET), village, LAFAYETTE county, N MISSISSIPPI, 15 mi/24 km ESE of OXFORD. In agricultural and timber area (cotton, corn; cattle). Holly Springs National Forest to N.

La Fé (lah FAI), town, ISLA DE LA JUVENTUD, SW CUBA, 10 mi/16 km SSE of NUEVA GERONA; 21°45′N 82°45′W. At major crossroads. Resort with mineral springs. Kaolin quarries to E and W. Small airstrip S.

La Fé (lah FAI), town, PINAR DEL RÍO province, W CUBA, on Palencia Bay; 22°03′N 84°17′W. Citrus production and ranching. Fishing.

La Feria (luh FAI-ree-uh), town (2006 population 6,856), CAMERON county, extreme S TEXAS, 27 mi/43 km NW of BROWNSVILLE, and 9 mi/14.5 km W of HARLINGEN, in lower RIO GRANDE valley; 26°09′N 97°49′W. Near Arroyo Colorado and Willacy Canal. In rich irrigated area (cotton; vegetables, citrus); manufacturing (Mexican herbs and spices, plastic molding, maps).

Lafia (lah-FEE-ah), town, ⊙ Nasawara state, central NIGERIA, on railroad, 50 mi/80 km N of MAKURDI. Lignite-mining center. Sometimes called Lafia Beriberi.

Lafiagi (lah-FYAH-jee), town, KWARA state, W central NIGERIA, near the NIGER River, 60 mi/97 km NE of ILORIN. Agricultural trade center; shea-nut processing, cotton weaving; agriculture (cassava, yams, corn; cattle, skins). Limestone deposits.

Lafitte (lah-FEET), unincorporated village (2000 population 1,576), JEFFERSON parish, extreme SE LOUISIANA, on INTRACOASTAL WATERWAY, and 18 mi/29 km S of NEW ORLEANS; 29°42′N 90°06′W. Shrimp fishing, seafood processing (shrimp, oysters, crab meat). Lafitte Village, on site of old settlement of Lafitte's nineteenth-century pirate band, is 5 mi/8 km SSE. Hosts the Jean Lafitte Seafood Festival. Hunting, fishing.

Laflèche or **La Flèche** (both: lah FLESH), town (2006 population 370), S SASKATCHEWAN, CANADA, 27 mi/43 km W of ASSINIBOIA; 49°44′N 106°32′W. Grain elevators, flour mills.

Laflin, borough (2006 population 1,503), LUZERNE county, NE central PENNSYLVANIA, residential sub-

urb, 5 mi/8 km ENE of WILKES-BARRE, and 11 mi/18 km SW of SCRANTON.

La Florencia (lah flo-RAIN-see-ah), village, ⊙ Matacos department, W FORMOSA province, ARGENTINA, on BERMEJO RIVER, and 110 mi/177 km WNW of LAS LOMITAS; 24°12′S 62°01′W. Corn and rice growing; lumbering. Near Formosa Natural Reserve.

La Floresta (lah flo-RE-stah), town, CANELONES department, S URUGUAY, beach resort at mouth of the Río de la PLATA, on railroad, and 29 mi/47 km ENE of MONTEVIDEO; 34°45′S 55°41′W. Sometimes called Costa Azul for an adjoining resort. La Floresta station is 7 mi/11.3 km NNW, near SOCA.

La Florida (LAH flo-REE-dah), town, ⊙ La Florida municipio, NARIÑO department, SW COLOMBIA, 9.0 mi/14.5 km NW of PASTO; 01°18′N 77°24′W. Wheat, sugarcane, vegetables; livestock.

Lafnitz River (LAHF-nits), c.50 mi/80 km long, SE AUSTRIA and SW HUNGARY; rises in SE FISCHBACH ALPS; flows S to Raab River at SZENTGOTTHÁRD, Hungary, just over Hungarian border, forming a long section of the border between Styria and BURGENLAND.

La Follette (luh FAW-lit), city (2006 population 8,183), CAMPBELL county, NE TENNESSEE, near NORRIS Reservoir, 30 mi/48 km NNW of KNOXVILLE, in E foothills of the CUMBERLAND MOUNTAINS; 36°23′N 84°07′W. Coal-mining center; manufacturing. Incorporated 1897.

La Fonciere, PARAGUAY: see PUERTO FONCIERE.

Lafontaine (lah-fon-TEN), former town, S QUEBEC, E CANADA; 45°49′N 74°01′W. Amalgamated into SAINT-JÉRÔME in 2002.

La Fontaine, town (2000 population 900), WABASH county, NE central INDIANA, near MISSISSINEWA RIVER, 10 mi/16 km SSE of WABASH; 40°40′N 85°43′W. In agricultural area.

Laforce (lah-FORS), village (□ 237 sq mi/616.2 sq km; 2006 population 459), ABITIBI-TÉMISCAMINGUE region, SW QUEBEC, E CANADA, 11 mi/17 km from BELLETERRE; 47°32′N 78°44′W.

Laforce, FRANCE: see FORCE, LA.

Lafourche (lah-FOOSH), parish (□ 1,157 sq mi/3,008.2 sq km; 2006 population 93,554), extreme SE LOUISIANA; ⊙ THIBODAUX; 29°48′N 90°49′W. Bounded S by Gulf of MEXICO, SW by Bayou POINTE AU CHIEN, E partly by BARATARIA BAY, N by Bayou DES ALLEMANDS and Lac DES ALLEMANDS; intersected by Bayou LAFOURCHE (navigable); crossed by INTRACOASTAL WATERWAY and Southwestern Louisiana Canal. Lake Boeuf in W, Lake Fields in center. Agriculture (home gardens, sugarcane, cotton, vegetables; cattle, horses; quail and pheasants); crawfish, alligators, shrimp, crabs, finfish; natural gas, oil; logging; manufacturing, including food processing, paper products, industrial machinery, shipbuilding. Part of Pointe au Chien State Wildlife Area in SW, Wisner State Wildlife Area in S. Formed 1805.

Lafourche, Bayou (lah-FOOSH, BEI-yoo), c. 75 mi/121 km long, NE LOUISIANA; rises in MOREHOUSE parish; flows generally S to BOEUF RIVER c. 13 mi/21 km WSW of WINNSBORO. Flows through Russell Sage State Wildlife Area.

Lafourche, Bayou, 107 mi/172 km long, SE LOUISIANA, formerly a right bank distributary of the MISSISSIPPI RIVER, from which it is now cut off by dam at DONALDSONVILLE, ASCENSION parish; extends SE, through ASSUMPTION and LAFOURCHE parishes, to Gulf of MEXICO between TIMBALIER and CAMINADA bays. Called the longest line village in the world because of the 65 mi/105 km of development on its levee. Navigable to NAPOLEONVILLE. Crossed at LEEVILLE by Southwestern Louisiana Canal, and intersected at LAROSE by INTRACOASTAL WATERWAY.

Lafragua, MEXICO: see SALTILLO, PUEBLA.

Lafrançaise (lah-frahn-SEZ), commune (□ 19 sq mi/49.4 sq km), TARN-ET-GARONNE department, MIDI-PYRÉNÉES region, SW FRANCE, near the TARN, 9 mi/14.5 km NW of MONTAUBAN; 44°08′N 01°15′E. Flour milling, fruit and vegetable growing.

La France, unincorporated town, ANDERSON county, NW SOUTH CAROLINA, 10 mi/16 km NNW of ANDERSON. HARTWELL LAKE reservoir to SW. Light manufacturing; agriculture (poultry, livestock; dairying; grain, soybeans).

La Francia (lah FRAHN-see-ah), town, E CÓRDOBA province, ARGENTINA, 32 mi/51 km W of SAN FRANCISCO; 31°24′S 62°38′W. Wheat, flax, oats, sunflowers; livestock; dairy products.

Lagadia, Greece: see LANGADIA.

La Gaiba, town and canton, ANGEL SANDOVAL province, SANTA CRUZ department, E central BOLIVIA; 17°45′S 57°43′W.

La Gallareta (lah gah-yah-RAI-tah), town, N central SANTA FE province, ARGENTINA, 70 mi/113 km NE of SAN CRISTÓBAL; 29°34′S 60°23′W. Agriculture (corn, flax, wheat; livestock); lumbering center, tannin factory.

Lagamar (LAH-gah-mahr), town (2007 population 7,636), W central MINAS GERAIS, BRAZIL, 62 mi/100 km NW of PATOS DE MINAS; 18°15′S 46°48′W.

Laga, Monti della (LAH-gah, MON-tee del-lah), mountain group, S central ITALY, NW of GRAN SASSO D'ITALIA, between TRONTO and VOMANO rivers. Chief peaks include Monte GORZANO (8,040 ft/2,451 m) and PIZZO DE SEVO (7,946 ft/2,422 m).

Lagan' (lah-GAHN), town, SE Republic of KALMYKIA-KHALMG-TANGEH, SE European Russia, on the CASPIAN SEA, on road, 70 mi/113 km SW of ASTRAKHAN, and 192 mi/309 km SE of ELISTA; 45°23′N 47°22′E. Below sea level. Port facilities. Sturgeon breeding farm. Fish canning, meat packing; machinery. After the Kalmyk population was deported in 1944 for cooperating with the German army, the town was renamed Kaspiyskiy; the old name was returned in the late 1950s, upon repatriation of the Kalmyks. Made city in 1963.

Lagan (LAH-gahn), river, 170 mi/274 km long, SW SWEDEN; rises S of JÖNKÖPING; flows S, past VÄRNAMO and LJUNGBY, to MARKARYD; flows W, past LAHOLM, to KATTEGATT strait 4 mi/6.4 km WNW of Laholm. Hydroelectric plants at KNÄRED and TRARYD. Salmon fishing.

Laganas (lah-gah-NAHS), beach resort, on S coast of ZÁKINTHOS island, ZÁKINTHOS prefecture, IONIAN ISLANDS department, off W coast of GREECE, and 5 mi/8 km SW of Zákinthos town; 37°50′N 21°34′E. Very popular and developed tourist resort; almost 4 mi/6.4 km long. Along with neighboring KALAMAKI (just E), a breeding ground for the loggerhead turtle (*Caretta caretta*) and thus a source of contention between preservationists and developers.

Lagan River (LA-guhn), Gaelic *Abhainn an Lagáin*, c.40 mi/64 km long; rising in Slieve Croob, SE DOWN, SE Northern Ireland; flows NW, then NE past LISBURN to BELFAST LOUGH at BELFAST. The port of Belfast and its shipbuilding yards are located at the mouth of the Lagan; a canal joins the river to Lough Neagh.

Lagares (lah-GAH-resh) or **Lagares da Beira**, village, COIMBRA district, N central PORTUGAL, 34 mi/55 km ENE of COIMBRA. Agriculture (corn, grain, beans, wine; pine forests).

Lagarfljöt (LAH-gahr-fuh-LYAWT), river, c.100 mi/161 km long, E ICELAND; rises at E edge of VATNAJOKULL; flows NE to ATLANTIC, 30 mi/48 km NNW of Seydisfjördur. Lower course is navigable.

Lagarina, Val (lah-gah-REE-nah, vahl), valley of ADIGE RIVER, N ITALY; extends from Calliano S to RIVOLI VERONESE. Agriculture (grapes, raw silk, fruit, tobacco). Chief centers: ROVERETO, MORI, Alabama.

Lagartera (lah-gahr-TAI-rah), town, TOLEDO province, central SPAIN, 20 mi/32 km W of TALAVERA DE LA REINA; 39°54′N 05°12′W. Cereals, vegetables, grapes, olives; cattle, sheep. Known for its old Toledan customs.

Lagarto (LAH-gahr-to), city (2007 population 88,989), central SERGIPE state, NE BRAZIL, 40 mi/64 km W of ARACAJU; 10°54′S 37°41′W. Corn, tobacco, manioc.

Lagarto River, PANAMA: see PALMAS BELLAS.

Lagartos Lagoon (lah-GAHR-tos), inlet of Gulf of MEXICO in YUCATÁN-QUINTANA ROO border, N YUCATÁN Peninsula, 60 mi/97 km NE of VALLADOLID; linked with ocean by narrow, 12-mi/19-km channel. Also called Río Lagartos.

Lagash (lai-GASH), ancient city of SUMER, S MESOPOTAMIA, in what is now THI-GAR province, S IRAQ, 40 mi/65 km N of AN NASIRIYAH. Lagash was flourishing by c.2400 B.C.E., but traces of habitation go back at least to the 4th millennium B.C.E. After the fall of AKKAD (2180 B.C.E.), when the rest of Mesopotamia was in a state of chaos, Lagash was able to maintain peace and prosperity under its ruler Gudea. Excavations begun on the site in 1877 revealed the beautiful sculptures of Gudea, which had been dedicated to the city's patron goddess, Ningirsu. Thousands of inscribed tablets were also found at the site. Also known as Shirpurla and Tellah.

La Gaulette, village (2000 population 1,917), BLACK RIVER district, MAURITIUS, on coast, and 27 mi/44 km SW of PORT LOUIS. Sugarcane; fishing.

Lagawe, town, ⊙ IFUGAO province, N LUZON, PHILIPPINES, 43 mi/70 km NE of BAGUIO; 16°47′N 121°14′E. Agricultural region (rice), home territory of the Atifulo highland people. The Imperial Japanese Army made its final stand in the Philippines in this area (1945). General Yamashita surrendered at Kiangan town, 6 mi/10 km SW; his treasure is supposedly buried on Mount Napalauan, 6 mi/10 km NW.

Lagdo (LAHG-do), reservoir, North province, CAMEROON, 35 mi/56 km SE of GAROUA; 08°48′N 13°55′E. The village of LAGDO, a cotton production center, is at the dam site at the NW corner of the reservoir.

Lage (LAH-guh), town, North Rhine-Westphalia, NW GERMANY, at N foot of TEUTOBURG FOREST, on the WERRE, and 5 mi/8 km NW of DETMOLD; 51°59′N 08°48′E. Railroad junction; manufacturing (furniture); woodworking; sugar refining. Was first mentioned in 1274; chartered 1791. Grew into a city in 1970 when neighboring villages were incorporated. Has late-Gothic church and baroque church with tower from 1100.

Lage (LAH-hai), town, LA CORUÑA province, NW SPAIN, fishing port on the ATLANTIC, 31 mi/50 km WSW of LA CORUÑA; 43°13′N 09°00′W. Lumbering, cattle raising; cereals, vegetables, fruit. Kaolin quarries nearby.

Lageado, Brazil: see LAJEADO.

Lågen (LAW-guhn) or **Numedalslågen**, river, 210 mi/338 km long, in SE NORWAY; rises in the E HARDANGERVIDDA N of RØDBERG; flows generally SSE, through the Numedal, past Rødberg, KONGSBERG, and Sandsvær, to the SKAGERRAK at LARVIK. Below Rødberg it falls over 1,150 ft/351 m into expansion of Norefjorden (10 mi/16 km long); major power station here supplies OSLO and TØNSBERG.

Lågen or **Gudbrandsdalslågen**, river, 125 mi/201 km long, in OPPLAND county, S central NORWAY; rises on N slope of JOTUNHEIMEN Mountains; flows generally SE, through the GUDBRANDSDAL, past DOMBÅS, DOVRE, OTTA, KVAM, VINSTRA, RINGEBU, and TRETTEN, to Lake MJØSA 4 mi/6.4 km NNW of LILLEHAMMER. Receives Otta and Vinstra rivers (right).

Lagens, PORTUGAL: see LAJES.

Lagens do Pico, PORTUGAL: see LAJES DO PICO.

Lageri, UKRAINE: see BALAKLIYA.

Lägern (LAI-guhrn), extreme E spur of the Folded Jura, on border of AARGAU and ZÜRICH cantons, N SWITZERLAND; 47°29′N 08°21′E. Burghorn, its highest peak (2,841 ft/866 m), is 4.5 mi/7.2 km E of BADEN.

Lages, in Brazilian and Portuguese names: see LAJES.

Lage Tatra, SLOVAKIA: see LOW TATRAS.

Lage Vaart (LAH-khuh FAHRT), canal, EASTERN and SOUTHERN Flevoland polders, FLEVOLAND province, central NETHERLANDS. Connects Kettlemeer, N of DRONTEN, with Oostvaardersdijk, c.5 mi/8 km N of ALMERE. Crosses NW parts of polders, paralleling HOGE VAART canal for much of its course.

Lage Zwaluwe (LAH-khuh ZVAH-luh-vuh), town, NORTH BRABANT province, SW NETHERLANDS, 8 mi/12.9 km NNW of BREDA; 51°42′N 04°43′E. AMER and NEW MERWEDE river channels join 1 mi/1.6 km to N to form HOLLANDS DIEP channel; MOERDIJK BRIDGES (railroad, highway) 3 mi/4.8 km to WNW. Dairying; cattle, hogs; grain, vegetables, sugar beets.

Laggan (LA-guhn), unincorporated village, SE ONTARIO, E central CANADA, 25 mi/40 km from CORNWALL, and included in NORTH GLENGARRY township; 45°23′N 74°42′W.

Laggan (LA-guhn), agricultural village, HIGHLAND, N Scotland, on the SPEY RIVER, and 10 mi/16 km WSW of KINGUSSIE; 57°02′N 04°16′W. Loch Laggan (7 mi/11.3 km long, 1 mi/1.6 km wide) is 6 mi/9.7 km SW on SPEAN RIVER. Just W of the loch is a new 4 mi/6.4 km long reservoir connected by tunnel with LOCH TREIG, forming part of the Lochaber hydroelectric power system.

Laggan (LA-guhn), agricultural village, HIGHLAND, N Scotland, on CALEDONIAN CANAL, at head of LOCH LOCHY, and 9 mi/14.5 km SW of FORT AUGUSTUS; 57°02′N 04°49′W. Site of canal locks, and scene (1544) of the Battle of the Shirts, between the Frasers and the MacDonalds.

Laggan, Loch (LAG-ahn), lake (c.12 mi/19.2 km long), in LOCHABER district of HIGHLAND, N Scotland; 56°57′N 04°28′W. Has served as a reservoir since 1930s.

Lagginhorn, peak (13,156 ft/4,010 m) in PENNINE ALPS, S SWITZERLAND, 11 mi/18 km S of BRIG, between FLETSCHHORN and WEISSMIES peak.

Laghman (lah-gah-MAHN), province (2005 population 371,000), E AFGHANISTAN; ⊙ MEHTARLAM; 35°00′N 70°00′E. Bordered by PANJSHIR and NURISTAN (N), KUNAR (E), NANGARHAR (S), and KABUL and KAPISA (W) provinces. Mountainous and inaccessible; peaks rise to 15,000 ft/4,570 m in the Hindu Kush mountains. Irrigated valley formed by ALISHANG and ALINGHAR rivers (both rise in the Hindu Kush and flow S to KABUL RIVER) in S central part of province. Agriculture includes rice and wheat. Population Nuristani, Ghilzais, Kuhistanis, and Tajiks.

Laghouat (lah-GWAHT), wilaya, in N Saharan ALGERIA; ⊙ LAGHOUAT; 33°35′N 02°40′E. Until 1984, one of the largest wilaya in the country. Includes the AFLOU oasis and Algeria's largest gas field at HASSI R'MEL. In 1984, its vast S part became GHARDAÏA wilaya.

Laghouat (lah-GWAHT), oasis town, ⊙ LAGHOUAT wilaya, S ALGERIA, on the N edge of the SAHARA Desert, 68 mi/110 km S of DJELFA and 119 mi/192 km N of GHARDAÏA; 33°49′N 02°55′E. The main oasis on the edge of the Sahara between FIGUIG (MOROCCO) and BISKRA. A craft center selling locally made carpets; vegetables and other crops; new industries. It has become the focus of regional administrative, social, and educational activities in its role as wilaya capital. Renowned for its palm groves irrigated partly by Oued Mzi. Laghouat has expanded in recent years, especially to the S. Founded after the Béni Hillal invasions in the 11th century. A French colonial military base for efforts to control Saharan tribespeople. Has a hotel designed by French architect Fernand Pouillon.

Lagich (lah-GEECH), town, NE AZERBAIJAN, on S slope of the Greater CAUCASUS, 20 mi/32 km NW of SHEMAKHA; 40°50′N 48°22′E. Livestock, wheat. Manufacturing (carpets, copperware).

La Gloria (LAH GLO-ree-ah), town, ⊙ La Gloria municipio, CÉSAR department, N COLOMBIA, port on MAGDALENA River, and 132 mi/212 km SSW of VALLEDUPAR; 08°37′N 73°48′W. Coffee, cacao; livestock.

La Gloria (la-GLO-ree-uh), unincorporated village, STARR county, extreme S TEXAS, 40 mi/64 km NNW of MCALLEN. Cattle; cotton; oil and natural gas.

Lagnieu (lah-nyu), town (□ 10 sq mi/26 sq km), AIN department, RHÔNE-ALPES region, E FRANCE, at foot of the S JURA MOUNTAINS, near Rhône River, 27 mi/43 km NE of LYON; 45°54′N 05°21′E. Glass manufacturing. The Bugey nuclear power plant (capacity 4,200 MW) on the RHÔNE RIVER is 7 mi/11.3 km S. Four cooling towers recirculate the warmed water from the steam condenser system.

Lagny (lah-nyee) or **Lagny-sur-Marne** (lah-nyee-syur–mahrn), town (□ 2 sq mi/5.2 sq km), SEINE-ET-MARNE department, ÎLE-DE-FRANCE region, N central FRANCE, on left bank of the MARNE (canalized), and 16 mi/26 km E of PARIS; 48°53′N 02°43′E. Printing, food processing center, metalworking; manufacturing (photographic equipment, optical instruments). Has unfinished 13th-century abbatial church.

Lago, MOZAMBIQUE: see LICHINGA.

Lagoa (lah-GO-ah), town (2007 population 4,792), NW PARAÍBA, BRAZIL, 18 mi/29 km NW of POMBAL; 06°35′S 37°55′W.

Lagoa (lah-GO-ah), town, PONTA DELGADA district, E AZORES, PORTUGAL, on S shore of SÃO MIGUEL ISLAND, 5 mi/8 km E of PONTA DELGADA. Agricultural center (wine, pineapples, sugar beets); fishing, flour milling, pottery manufacturing.

Lagoa, town, FARO district, S PORTUGAL, 5 mi/8 km E of PORTIMÃO. In fertile agriculture region (almonds, figs, carob, wine, vegetables); cork processing, pottery manufacturing.

Lagoa da Canoa (lah-GO-ah dah kahn-O-ah), city (2007 population 18,103), E central ALAGOAS state, BRAZIL, 8 mi/13 km SW of ARAPIRACA on SALVADOR-RECIFE railroad; 09°50′S 36°43′W.

Lagoa da Prata (lah-GO-ah dah PRAH-tah), city (2007 population 44,032), W central MINAS GERAIS state, BRAZIL, 37 mi/60 km E of BAMBUÍ, on Rio SÃO FRANCISCO and railroad; 20°08′S 45°40′W.

Lagoa de Dentro (lah-GO-ah dee DEN-tro), town (2007 population 7,258), NE PARAÍBA state, BRAZIL, 3.1 mi/5 km NE of DUAS ESTRADAS; 06°41′S 35°22′W.

Lagoa do Ouro (lah-GO-ah dos O-ro), city (2007 population 11,900), S PERNAMBUCO state, BRAZIL, 16 mi/26 km NE of BOM CONSELHO; 09°08′S 36°28′W. Manioc, sugar, corn, fruit.

Lagoa dos Gatos (lah-GO-ah dos GAH-tos), city (2007 population 16,278), E PERNAMBUCO state, BRAZIL, 78 mi/126 km SW of RECIFE; 08°39′S 35°54′W. Corn, sugar, aloe, fruit.

Lagoa Dourada (lah-GO-ah do-RAH-dah), city (2007 population 11,792), S central MINAS GERAIS state, BRAZIL, 19 mi/31 km NE of SÃO JOÃO DEL REI; 20°45′S 44°00′W. Tin deposits.

Lagoa Formosa (lah-GO-ah FOR-mo-sah), city (2007 population 16,520), W central MINAS GERAIS, BRAZIL, 14 mi/23 km SE of PATOS DE MINAS; 18°37′S 46°28′W.

Lago Agrio, ECUADOR: see NUEVA LOJA.

Lagoa Nova (lah-GO-ah NO-vah), city (2007 population 13,095), central RIO GRANDE DO NORTE state, BRAZIL, 91 mi/146 km WSW of NATAL; 06°06′S 36°29′W. Cotton; livestock.

Lagoa Real (lah-GO-ah RAI-ahl), city (2007 population 13,661), S central BAHIA state, BRAZIL, 28 mi/45 km E of CAETITÉ; 14°01′S 42°08′W.

Lago Argentino, ARGENTINA: see EL CALAFATE.

Lagoa Salgada (lah-GO-ah SAHL-gah-dah), town (2007 population 7,177), E RIO GRANDE DO NORTE state, BRAZIL, 28 mi/45 km SW of NATAL; 06°07′S 35°29′W. Cotton, aloe.

Lagoa Santa (lah-GO-ah SAHN-tah), city (2007 population 44,932), S central MINAS GERAIS state, BRAZIL, on lake of same name, 18 mi/29 km N of BELO HORIZONTE; 19°32′S 43°52′W. Belo Horizonte International Airport nearby.

Lagoa Seca (lah-GO-ah SE-kah), city (2007 population 24,937), E central PARAÍBA state, BRAZIL, 7 mi/11.3 km N of CAMPINA GRANDE; 07°09′S 35°51′W.

Lagoa Sêca (lah-GO-ah SE-kah), city, S Paraná state, BRAZIL, 23 mi/37 km W of GUARAPUAVA; 25°27′S 51°52′W.

Lagoa Vermelha (lah-GO-ah VER-mel-yah), city (2007 population 27,434), NE RIO GRANDE DO SUL state, BRAZIL, in the Serra GERAL, 30 mi/48 km E of PASSO FUNDO; 28°13′S 51°32′W. Grain; livestock slaughtering; by-products processing. Pinewoods lumbering (N).

Lago Buenos Aires, ARGENTINA: see PERITO MORENO.

Lago da Pedra (LAH-go dah PE-drah), city (2007 population 42,753), N central MARANHÃO state, BRAZIL, 29 mi/47 km W of BACABAL; 04°16′S 45°16′W.

Lago de Camécuaro (LAH-go dai kah-MAI-kwah-ro), a national park, in S MICHOACÁN, MEXICO, 6 mi/10 km SE of ZAMORA DE HIDALGO. Huge cypress trees in crystal clear water are the distinctive feature of the park. Swimming and boating are allowed.

Lagodei, El, SOMALIA: see GARDO.

Lagodekhi (lah-go-DE-khee), city and administrative center of Lagodekhi region, E GEORGIA, in KAKHETIA, on Lagodekhi River, and on S slope of the Greater Caucasus, 40 mi/64 km E of TELAVI, near AZERBAIJAN border; 41°49′N 46°16′E. Tobacco processing; wine making; cannery; essential-oils manufacturing.

Lagodekhi Preserve (lah-go-DE-khee), ecology sanctuary, near LAGODEKHI city, GEORGIA, established 1912. Covers 32,850 acres/13,300 ha. Animals include Caucasian species, deer, chamois, lynx; museum of flora and fauna; high-altitude meteorological station.

Lago District, MOZAMBIQUE: see NIASSA, province.

Lago do Coco (LAH-go do KO-ko), city, W TOCANTINS state, BRAZIL, 186 mi/299 km SW of PALMAS, on COCO RIVER; 09°30′S 49°59′W.

Lago do Junco (LAH-go do ZHOON-ko), town (2007 population 9,640), N central MARANHÃO state, BRAZIL, 37 mi/60 km NW of PEDREIRAS, near RIO MEARIM; 04°30′S 44°52′W.

Lago Maggiore, ITALY: see MAGGIORE, LAKE.

La Gomera (lah go-MAI-rah), town, ESCUINTLA department, S GUATEMALA, in coastal plain, 23 mi/37 km SW of ESCUINTLA; 14°05′N 91°03′W. Elevation 141 ft/43 m. Sugarcane, sorghum, corn; livestock.

Lagonegro (lah-go-NAI-gro), town, POTENZA province, BASILICATA, S ITALY, 22 mi/35 km SSE of SALA CONSILINA; 40°07′N 15°46′E. Agricultural center; manufacturing (woolen textiles, tower clocks).

Lagonoy (lah-go-NOI), town, CAMARINES SUR province, SE LUZON, PHILIPPINES, near LAGONOY GULF, 24 mi/39 km ENE of NAGA; 13°48′N 123°29′E. Chromeore mining and agricultural center (abaca, rice).

Lagonoy Gulf (lah-go-NOI), large inlet of PHILIPPINE SEA, SE LUZON, between RUNGUS POINT peninsula and CATANDUANES island in N, and SAN MIGUEL, CAGRARAY, BATAN, and RAPU-RAPU islands in S; c.50 mi/80 km E-W, c.20 mi/32 km N-S. Merges N with MAQUEDA CHANNEL.

Lagoon Islands: see TUVALU.

Lago Ranco (LAH-go RAHN-ko), village, ⊙ Lago Ranco comuna, VALDIVIA province, LOS LAGOS region, S central CHILE, on S shore of LAKE RANCO, in Chilean lake district, 55 mi/89 km SE of VALDIVIA. Tourist resort; fishing, lumbering.

Lagord (lah-gord), N town (□ 3 sq mi/7.8 sq km) suburb of LA ROCHELLE, CHARENTE-MARITIME department, POITOU-CHARENTES region, W FRANCE,

along highway leading to bridge-viaduct to the Île de RÉE, just offshore in Bay of BISCAY; 46°11′N 01°08′W.

Lagos (LAH-guhs), state (□ 1,291 sq mi/3,356.6 sq km; 2006 population 9,013,534), SW NIGERIA, on BIGHT OF BENIN; ⊙ Ikeja; 06°35′N 03°45′E. Bordered N and E by OGUN state, W by BENIN. Country's economic core. Main center are LAGOS, IKORODU, Muskin, and EPE.

Lagos (LAH-guhs), city (2004 population 8,680,000), LAGOS state, SW NIGERIA, on the Gulf of GUINEA, comprising the island of LAGOS; 06°27′N 03°23′E. Lagos is Nigeria's largest city, its administrative and economic center, its chief port, and is its former national capital. Industries include railroad repair, motor vehicle assembly, food processing; manufacturing of metal products, textiles, beverages, chemicals, pharmaceuticals, soap, and furniture. The city is a road and railroad terminus and has an international airport to W. Its main docks and warehouses are at APAPA on mainland, to W side of LAGOS LAGOON. An old Yoruba town, Lagos, beginning in the 15th century, grew as a trade center and seaport. From the 1820s until it became a British colony, Lagos was a notorious center of the slave trade. Lagos became a British Crown Colony in 1861, both to tap the trade in palm products and other goods with the interior and to suppress the slave trade. It was Britain's first possession on the Slave Coast. As the British absorbed the adjoining Egba kingdom in nearby Ogun (which had relocated from Ibadan region in 1830s). In 1906, Lagos was joined with the British protectorate of South Nigeria, and, in 1914, when South and North Nigeria were amalgamated, it became part of the small coastal Colony of Nigeria. In 1954 most of the colony was merged with the rest of Nigeria, but Lagos was made a separate federal territory. From the late 19th century to independence in 1960, Lagos was the center of the Nigerian nationalist movement. From independence until the late 1980s, Lagos was the capital of Nigeria, but then the capital was moved to ABUJA, although some governmental departments remain. Most of the people living here now are Egba. The University of Lagos (1962), the College of Technology (1948), the National Museum, and a large sports stadium are here.

Lagos (LAH-goosh), city, FARO district, S PORTUGAL, in ALGARVE province, on the ATLANTIC OCEAN. The excellent harbor shelters much coastwise trade and an important sardine and tuna fishing fleet. Tourism. Sancho I with the help of bands of Crusaders captured (1189) the city from the Moors; in 1191 recaptured by the Moors but soon (c.1250) restored to the Portuguese. Starting port for Portuguese navigators in the time of Prince Henry the Navigator, who was first buried here. The disastrous expedition of King Sebastian set out from here. Severely damaged in the 1755 earthquake. The British under Admiral Boscawen defeated the French off LAGOS in 1759.

Lagos de Moreno (lah-gos dai mo-RE-no), city and township, NE JALISCO, central MEXICO, in SIERRA MADRE OCCIDENTAL, on railroad, and 23 mi/37 km NW of León; 21°21′N 101°55′W. Elevation 6,371 ft/1,942 m. Resort; silver-mining and agricultural center (beans, chili, corn; livestock) dairy industry, shoe manufacturing. Colonial churches.

Lagos Lagoon (LAH-guhs), W part of coastal lagoon, LAGOS state, SW NIGERIA, N of LAGOS; 11 mi/18 km long, 5 mi/8 km wide. Opens into Gulf of GUINEA S of Lagos. Receives OGUN River.

Lagos, Los, department, ARGENTINA: see NAHUEL HUAPÍ, village.

Lagos, Los, CHILE: see LOS LAGOS.

Lagossa (lah-GO-sah), village, KIGOMA region, W TANZANIA, 75 mi/121 km SSE of KIGOMA, on Lake TANGANYIKA; 05°59′S 29°46′E. Lake port. Fish; goats, sheep; corn, wheat.

Lagosta, CROATIA: see LASTOVO ISLAND.

Lagouira (lah-goo-WEE-ruh), village, Oued Eddahab-Lagouira administrative region of MOROCCO, southwesternmost point of WESTERN SAHARA (claimed by Morocco). Last stop on road across Western Sahara, on border with MAURITANIA.

La Goulette, TUNISIA: see HALQ AL WADI.

Lago Verde (LAH-go VER-zhe), city (2007 population 14,574), N central MARANHÃO state, BRAZIL, N of BACABAL in MEARIM delta; 03°55′S 44°50′W.

Lago Vista (LAH-go VIS-tuh), town (2006 population 5,794), TRAVIS county, S TEXAS, suburb 19 mi/31 km NW of downtown AUSTIN, lakeside development on 11 mi/18 km of N shore of LAKE TRAVIS reservoir (COLORADO RIVER); 30°27′N 97°59′W. Elevation 1,230 ft/375 m. Agriculture area; light manufacturing. Recreation. Airpower Museum.

La Grande, city (2006 population 12,318), ⊙ UNION county, NE OREGON, 40 mi/64 km SE of PENDLETON, on GRANDE RONDE RIVER, E of BLUE MOUNTAINS, W of Wallowa Mountains; 45°19′N 118°05′W. Elevation 2,788 ft/850 m. Railroad junction; trade and shipping point for livestock, fruit, timber. Manufacturing (particleboard; printing, publishing); agriculture (potatoes, grain). Site of Eastern Oregon State College, Hilgard Junction, and Red Bridge State Parks to W; parts of Wallowa-Whitman National Forest to NW, SW, and E. Founded 1861; incorporated 1886.

La Grande Dam, WASHINGTON: see NISQUALLY RIVER.

La Grandeza (lah grahn-DAI-sah), town, ⊙ La Grandeza municipio, CHIAPAS, S MEXICO, in SIERRA MADRE, 11 mi/18 km N of MOTOZINTLA DE MENDOZA; 15°32′N 92°14′W. Elevation 6,201 ft/1,890 m. Sugarcane, fruit. Formerly called San Antonio La Grandeza.

Lagrange, county (□ 386 sq mi/1,003.6 sq km; 2006 population 37,291), NE INDIANA; ⊙ LAGRANGE; 41°38′N 85°26′W. Bounded N by MICHIGAN state line; drained by PIGEON and short Little Elkhart rivers. Dairying; soybeans, oats, wheat, corn, grain; cattle, sheep, hogs; poultry hatcheries; processing of dairy products. Pigeon River State Fish and Wildlife Area (including Curtis Creek Trout Station) in NE. Scott Millpond State Fishing Area in NW. About twenty-seven small natural lakes, glacial in origin, mostly concentrated in SE quarter of county. Formed 1832.

La Grange (luh GRAINJ), city (2000 population 25,998), ⊙ TROUP county, W central GEORGIA; 33°02′N 85°02′W. Former cotton mill town, now an industrial center; manufacturing (lumber, plastics, textiles, building materials, transportation equipment; printing and publishing). The city is also a processing and shipping center for a rich agricultural area. Many classic Greek revival houses and restored historic buildings; seat of La Grange College, including the Lamar Dodd Art Center. Named for the French estate of Marquis de Lafayette, whose statue, along with a fountain in Lafayette Square, is the focal point of the town. Important retail hub. The Troup County Archive and Historical Society; Chattahoochee Valley Art Museum; West Georgia Technical College. Incorporated 1828.

La Grange, city (2000 population 1,000), LEWIS county, NE MISSOURI, on the MISSISSIPPI, at mouth of WYACONDA RIVER, and 6 mi/9.7 km S of CANTON; 40°03′N 91°30′W. Grain; hogs; manufacturing (iron castings). Plotted 1830. Wakona State Park on S side of town.

Lagrange (luh-GRAINJ), town (2000 population 2,919), ⊙ LAGRANGE county, NE INDIANA, on tributary of PIGEON RIVER, and 28 mi/45 km E of ELKHART; 41°38′N 85°25′W. Manufacturing (electrical equipment, transportation equipment, livestock feed, food products and processing). Plotted 1836; incorporated 1855.

La Grange (luh GRAINJ), town (2000 population 5,676), ⊙ OLDHAM county, N KENTUCKY, 25 mi/40 km ENE of LOUISVILLE; 38°23′N 85°22′W. In agricultural area (burley tobacco, grain; dairying); manufacturing (magnet wire, crushed limestone, stationery; steel fabricating). Rob Morris Home (1830); County Historical Society.

Lagrange, town, PENOBSCOT COUNTY, central MAINE, 25 mi/40 km N of BANGOR; 45°10′N 68°48′W. Agriculture; lumbering.

La Grange (luh GRAINJ), town (□ 2 sq mi/5.2 sq km; 2006 population 2,789), LENOIR COUNTY, E central NORTH CAROLINA, 12 mi/19 km SE of GOLDSBORO; 35°18′N 77°47′W. Service industries; manufacturing (apparel, electrical equipment, food processing, metal containers); agriculture (tobacco, cotton, grain, soybeans, sweet potatoes; poultry, livestock). Cliffs of the Neuse State Park to SW. Incorporated 1869.

La Grange, town (2006 population 145), FAYETTE COUNTY, SW TENNESSEE, near WOLF RIVER, 45 mi/72 km E of MEMPHIS; 35°03′N 89°15′W. Antebellum plantation homes in area.

La Grange (luh GRAINJ), town (2006 population 4,645), ⊙ FAYETTE COUNTY, S central TEXAS, on COLORADO RIVER and c.55 mi/89 km ESE of AUSTIN; 29°54′N 96°52′W. Elevation 272 ft/83 m. Agriculture (dairying; cattle; cotton, peanuts, pecans, corn, sorghum); manufacturing (food processing, safety signs, wood products); oil and natural gas; sand and gravel. Lake Fayette reservoir to E; Monument Hill State Park to SW and Kreische State Historic Site, both to SW. Established 1831; became county seat 1837.

La Grange, town (2006 population 329), GOSHEN COUNTY, SE WYOMING, on HORSE CREEK, near NEBRASKA state line, and 48 mi/77 km NE of CHEYENNE; 41°38′N 104°09′W. Elevation 4,587 ft/1,398 m. Hawk Springs Reservoir and State Recreation Area to NW.

La Grange (luh GRAINJ), village (2000 population 15,608), COOK COUNTY, NE ILLINOIS, a W suburb of CHICAGO; 41°48′N 87°52′W. Settled 1830s, incorporated 1879. It is primarily residential with some manufacturing. Limestone quarries are nearby.

Lagrange (luh-GRAHNJ), village (□ 2 sq mi/5.2 sq km; 2006 population 1,963), LORAIN COUNTY, N OHIO, 8 mi/13 km S of ELYRIA; 41°14′N 82°07′W. In agricultural area.

La Grange Park, village (2000 population 13,295), COOK COUNTY, NE ILLINOIS, a W suburb of CHICAGO; 41°49′N 87°52′W. Manufacturing (pens). Incorporated 1892.

La Granja, SPAIN: see SAN ILDEFONSO.

La Gran Sabana (LAH GRAHN sah-BAH-nah), plateau, in the SE of BOLÍVAR state, SE VENEZUELA; 05°30′N 61°30′W. Largest plateau of Venezuela. Cut by numerous affluents of the CARONÍ RIVER. Diverse wildlife.

La Grita (LAH GREE-tah), town, ⊙ Jáuregui municipio, TÁCHIRA state, W VENEZUELA, in ANDEAN spur, 31 mi/50 km NE of SAN CRISTÓBAL; 08°08′N 71°59′W. Elevation 4,715 ft/1,437 m. Agricultural center (potatoes, wheat, cacao, coffee, sugarcane; livestock).

Lagro (LAI-gro), town (2000 population 454), WABASH COUNTY, NE central INDIANA, on WABASH RIVER opposite mouth of SALAMONIE RIVER, and 5 mi/8 km ENE of WABASH; 40°50′N 85°44′W. Manufacturing (ceiling tile). Salamonie River State Forest nearby to SE. Laid out 1838.

La Grulla (GROO-lah), town (2006 population 1,826), STARR COUNTY, extreme S TEXAS, 25 mi/40 km W of MCALLEN, and on the RIO GRANDE (Mexican border). In rich irrigated agricultural area (vegetables; cotton; cattle). Also called Grulla.

La Guadalupe (lah gwah-duh-LOOP), village (□ 12 sq mi/31.2 sq km), CHAUDIÈRE-APPALACHES region, S QUEBEC, E CANADA, 27 mi/43 km N of LAC-MÉGANTIC; 45°57′N 70°56′W. Dairying; lumbering; manufacturing (food preparations, apparel).

La Guaira (LAH GWEI-rah), city, ⊙ VARGAS state, N VENEZUELA, on the CARIBBEAN SEA, NW of CARACAS, linked by freeway; 10°36′N 66°56′W. It is the principal international port of Venezuela; chief exports include cacao, coffee, and tobacco; frequented by cruise ships. SIMÓN BOLÍVAR INTERNATIONAL AIRPORT for CARACAS nearby (MAIQUETÍA).

La Guajira (LAH gwah-HEE-rah), department (□ 4,726 sq mi/12,287.6 sq km), N COLOMBIA, on CARIBBEAN SEA, the northernmost part of Colombia, and of the South American continent; ⊙ RÍOHACHA; 11°30′N 72°30′W. Located on a peninsula situated NE of Sierra Nevada de SANTA MARTA, and separated by RANCHERÍA RIVER (SW) from MAGDALENA department, and by the Montes de OCA (S) from VENEZUELA. Apart from low outliers of the Cordillera ORIENTAL, it consists of arid plains, having a desert appearance and high temperatures. On indented coast are rich pearl banks, and marine salt is also worked here. Mineral resources include gold, copper, phosphates, gypsum, petroleum, and coal. Most important crop is sorghum; some livestock grazing (goats, sheep, cattle). At W foot of Montes de Oca, coffee, corn, and beans are grown and cattle is raised. Consists of 14 municipios (2004). Most people live in the main cities, which include MAICAO, BARRANCAS, FONSECA, VILLANUEVA, and the capital.

La Guajira (LAH gwah-HEE-rah), peninsula, c.100 mi/161 km long, N COLOMBIA, extending into the CARIBBEAN SEA; 12°00′N 71°40′W. Punta GALLINAS, at the tip, is the northernmost point of South America. On the sparsely populated peninsula are outliers of the Cordillera ORIENTAL, surrounded by hot, arid plains. Open pit coal mines at El Correjón, and maritime saltworks.

La Guardia (lah GWAHR-dee-ah), canton, Andres de Ibáñez province, Santa Cruz department, 12 mi/20 km SW of Santa Cruz, on the Santa Cruz–Cochabamba highway; 17°41′S 63°05′W. Elevation 1,329 ft/405 m. Abundant gas wells in area. Clay and limestone deposits. Agriculture (potatoes, yucca, bananas, corn, rice, peanuts, soy, coffee); cattle.

Laguardia (lah-GWAHR-dhyah), town, ÁLAVA province, N SPAIN, 9 mi/14.5 km NW of LOGROÑO; 42°33′N 02°35′W. Wine producing center; also flour mills; cereals, olive oil, fruit. Has remains of medieval walls.

LaGuardia Airport, borough of QUEENS, NEW YORK city, SE NEW YORK, along FLUSHING BAY and Bowery Bay; 8 mi/13 km from downtown MANHATTAN; elevation 13 ft/ 4 m; 40°47′N 73°52′W. One of three NEW YORK city metropolitan area airports operated by the PORT AUTHORITY OF NEW YORK AND NEW JERSEY (see also JFK INTERNATIONAL AIRPORT and NEWARK LIBERTY INTERNATIONAL AIRPORT), it has five passenger terminals (including a marine air terminal), handling an average volume of 22 million passengers annually, along with 14,000 tons/ 12,700 tonnes of cargo and 18,000 tons/16,330 tonnes of mail. Two runways, each 7,000 ft/2,134 m in length. Opened in 1939 as New York Municipal Airport, New York city's second airport, 70% of its $22-million cost was paid for by the WPA. It was leased to the Port Authority in June 1947 and renamed Laguardia Airport, after New York Mayor Fiorello LaGuardia. After opening, the airport quickly became the city's main airport, eclipsing for a time Idlewild Airport (now JFK International). The airport also had a marine terminal that was built to handle the international "flying boats" of the 1930s; with its outstanding Art Deco architecture, the terminal continues to be used by land-based private aircraft. Airport Code LGA.

LaGuardia Airport, NEW YORK city, NEW YORK, serving the New York metropolitan region and located on FLUSHING BAY in the borough of QUEENS; elevation 13 ft/4 m; 40°47′N 73°52′W. Operated by the Port Authority of New York and New Jersey; opened in June 1947. Covers 690 acres/275 ha. Has two runways and 5 passenger terminals (over 20 million passengers annually). Airport Code LGA.

La Guardia Airport, New York: see LAGUARDIA AIRPORT.

La Guata (lah GWAH-tah), town, OLANCHO department, N central HONDURAS, 28 mi/44 km NW of JUTICALPA, on unpaved road; 15°05′N 86°25′W. Elevation 3,281 ft/1,000 m. Local airport at MANTO.

Laguiole (lah-gwee-OL), commune (□ 24 sq mi/62.4 sq km), AVEYRON department, MIDI-PYRÉNÉES region, S FRANCE, in MASSIF CENTRAL, on W slope of AUBRAC mountains, 26 mi/42 km SE of AURILLAC; 44°41′N 02°51′E. Specializes in manufacturing of pocket knives; also a cheese-making center. Winter sports.

Laguna (lah-GOO-nah), province (□ 679 sq mi/1,765.4 sq km), S LUZON, PHILIPPINES, bounded N by LAGUNA DE BAY; ⊙ SANTA CRUZ; 14°11′20′E. Population 73.9% urban, 26.1% rural. Fertile terrain, drained by many small streams; rises in S to 3,750 ft/ 1,143 m at MOUNT MAKILING, and in SE to 7,177 ft/2,188 m at MOUNT BANAHAO on border of QUEZON province. Agriculture (rice, coconuts, sugarcane), fishing. SAN PABLO city is in the province.

Laguna (lah-GOO-nah), city (2007 population 50,452), SE SANTA CATARINA state, BRAZIL, on the ATLANTIC near entrance to a shallow inlet (lagoon), 65 mi/105 km SSW of FLORIANÓPOLIS; 28°29′S 48°47′W. Coal-shipping port serving mines in TUBARÃO area (just SW; linked by railroad). Coal goes to RIO DE JANEIRO for VOLTA REDONDA steel mill. Also ships timber, skins, clay, alcohol. Founded c.1720. Headquarters of 1893 insurrection. Resort, fishing. S terminus of TORDESILLAS line from mouth of AMAZON to Laguna, 370 leagues W of CAPE VERDE Islands.

Laguna, unincorporated city (2000 population 34,309), SACRAMENTO COUNTY, central CALIFORNIA; residential suburb 9 mi/14.5 km SSE of downtown SACRAMENTO, on Laguna Creek; 38°25′N 121°25′W. Agriculture (grain, alfalfa, fruit, vegetables; dairying; poultry) to S and E.

Laguna (lah-GOO-nah), village, ALAJUELA province, W central COSTA RICA, on SAN CARLOS road, and 1 mi/1.6 km N of ZARCERO. Potatoes, fruit, flowers.

Laguna (luh-GOO-nuh), pueblo (2000 population 423), CIBOLA COUNTY, central NEW MEXICO, at center of Laguna Indian Reservation (1990 population 3,731), 27 mi/43 km ESE of GRANTS, on RIO SAN JOSE; 35°02′N 107°23′W. Established on its present site 1699. Its inhabitants are Pueblo of the Keresan linguistic stock; many farms in outlying areas. Sheep, cattle. The pueblo is used essentially for ceremonial purposes. SAN MATEO MOUNTAINS to NW; Cañoncito Indian Reservation to E.

Laguna Apoyeque (lah-GOO-nah ah-po-YAI-kai), lake, Chiltepe Peninsula, MANAGUA department, NICARAGUA. Sometimes called Lake Apoyo.

Laguna Beach, city (2000 population 23,727), ORANGE COUNTY, S CALIFORNIA; suburb 40 mi/64 km SE of downtown LOS ANGELES, on PACIFIC OCEAN, and at mouth of Laguna Canyon Wash; 33°33′N 117°46′W. Irrigated agricultural area. It is a residential and resort city with noted art colony and many cultural attractions. Laguna Beach has profited from the rapid growth and prosperity of Orange county. Manufacturing (navigation equipment, pottery products). Laguna Beach Museum of Art. San Joaquin Hills to NE. Founded 1887, incorporated 1927.

Laguna Beach (luh-GOO-nuh), town (□ 2 sq mi/5.2 sq km; 2000 population 2,909), BAY COUNTY, NW FLORIDA, 20 mi/32 km WNW of PANAMA CITY; 30°15′N 85°57′W.

Laguna Blanca (lah-GOO-nah BLAHN-kah), canton, PACAJES province, LA PAZ department, W BOLIVIA. Elevation 12,989 ft/3,959 m. Gas wells in area. Lead-bearing lode, tungsten and copper (unmined) deposits; clay, limestone, and gypsum deposits. Agriculture (potatoes, yucca, bananas, rye); cattle.

Laguna Blanca (lah-GOO-nah BLAHN-kah), town (1991 population 4,927), Pilcomayo department, E FORMOSA province, ARGENTINA, 38 mi/61 km WNW of ASUNCIÓN (PARAGUAY). Livestock-raising center;

cotton. Site of Franciscan mission. Near Río Pilcomayo National Park.

Laguna Blanca (lah-GOO-nah BLAHN-kah), village, E
Chaco province, ARGENTINA, on railroad, and 30 mi/
48 km NW of RESISTENCIA. Cotton; livestock center;
cotton ginning, forest industry.

Laguna de Bay, PHILIPPINES: see BAY, LAGUNA DE.

Laguna de Brus, HONDURAS: see BRUS LAGOON.

Laguna de Catemaco National Park (lah-GOO-nah
dai kah-te-MAH-ko) (□ 50 sq mi/130 sq km), in SE
VERACRUZ, MEXICO, 3.1 mi/5 km SE of SAN ANDRÉS
TUXTLA. This small park is located on the inside of
Laguna CATEMACO (3,959 mi/10,059 km), a volcanic
caldera located in the TUXTLA Volcanoes. A popular
weekend spot for visitors from the nearby GULF
Coastal Lowlands.

Laguna de Duero (lah-GOO-nah dhai DHWAI-ro),
town, VALLADOLID province, N central SPAIN, near
the DUERO RIVER, 5 mi/8 km S of VALLADOLID, and on
edge of saltwater lagoon; 41°35′N 04°43′W. Cattle
raising; lumbering.

Laguna de la Laja (lah-GOO-nah dai lah LAH-hah),
national park, in BÍO-BÍO region, CHILE, 50 mi/80 km
E of LOS ÁNGELES. Includes lake and its surroundings,
with rare trees, lava fields, and many wild animals and
birds.

Laguna de la Restinga, Parque Nacional (lah-GOO-
nah dai lah res-TEEN-gah PAHR-kai nah-see-o-
NAHL), national park (□ 73 sq mi/189.8 sq km), on
MARGARITA Island, NUEVA ESPARTA state, N VENE-
ZUELA; 11°00′N 64°10′W. Centered on a mangrove
lagoon, protecting wide variety of bird life. Created
1974.

Laguna del Perro (luh-GOO-nuh del PE-ro), lake,
TORRANCE county, central NEW MEXICO, E of MAN-
ZANO RANGE, just SE of ESTANCIA; 12 mi/19 km long, 1
mi/1.6 km wide. Semi-dry; surrounded by numerous
small, intermittent lakes.

Laguna del Tigre–Río Escondido Biotope (lah-GOO-
nah del TEE-gre-REE-o es-kon-DEE-do), sanctuary
(□ 175 sq mi/455 sq km), NW PETÉN department,
GUATEMALA, 10 mi/16 km N of Río SAN PEDRO;
17°16′N 90°53′W. One of several specially protected
areas within the MAYA BIOSPHERE RESERVE. Protects a
large area of wetlands, an important wintering ground
for migratory birds.

Laguna de Negrillos (lah-GOO-nah dai nai-GREE-
lyos), town, LEÓN province, NW SPAIN, 26 mi/42
km SSW of LEÓN; 42°14′N 05°40′W. Beans, cereals,
wine.

Laguna de Perlas (lah-GOON-ah dai PER-lahs), town,
SOUTH ATLANTIC COAST AUTONOMOUS REGION, ZE-
LAYA department, NICARAGUA, 20 mi/32 km N of
BLUEFIELDS, located at the S end of PEARL LAGOON;
12°21′N 83°40′W. English is widely spoken here.
Fishing; rice, bananas.

Laguna de Tacarigua, Parque Nacional (lah-GOO-
nah dai tah-kah-REE-gwah PAHR-kai nah-see-o-
NAHL), national park (□ 151 sq mi/392.6 sq km),
MIRANDA state, N VENEZUELA; 10°16′N 65°49′W.
Protects beach and mangrove environment. Site of
Club Miami, which predates law forbidding private
beaches in Venezuela. Created 1974.

Laguna de Temazcal National Park (lah-GOO-nah
dai te-MAHZ-kahl), N OAXACA, MEXICO, 3.1 mi/5 km
NE of the President Miguel Aleman Lake (Lago Presidente Miguel Alemán).

Laguna District [Span.=lake], irrigated area in E
DURANGO and W COAHUILA states, N central MEXICO.
Originally a 900,000 acres/364,230 ha tract, consisting
of large estates, the land was reapportioned (1936)
under President Lázaro Cárdenas and distributed to
Mexican farmers under the Ejido system. It was a
successful experiment in agrarian reform until 1952,
when a severe drought scorched more than half the
district, turning 200,000 acres/80,940 ha of wheat and
cotton fields into a dust bowl and obliging the gov-

ernment to take emergency measures to avert a famine. Settlement has continued there, but on a greatly
reduced scale; water for irrigation comes from wells
and from dams on the NAZAS and AGUANAVAL rivers.
Parts of the irrigated area have gone out of production
because of salinization of soil.

Laguna Diversion, dam, on border of extreme SW
ARIZONA (YUMA county) and SE CALIFORNIA (IMPER-
IAL county), on COLORADO RIVER, 7 mi/11.3 km NNE
of YUMA, and c.14 mi/23 km from Mexican (BAJA
CALIFORNIA) border; 43 ft/13 m high; 32°52′N
114°29′W. Built (1909) by the Bureau of Reclamation
for debris control and water supply. Forms Laguna
Reservoir; extends N Fort Yuma (Quechan) Indian
Reservation at S end of reservoir. Also known as La-
guna Dam.

Laguna Guaimoreto National Park (lah-GOO-nah
GWAI-mo-RE-to) (□ 19 sq mi/49.4 sq km), COLÓN
department, N HONDURAS, ENE of TRUJILLO; 15°58′N
85°55′W. Includes coastal lake, mangrove forests, bird
life.

Laguna Heights (la-GOO-nuh), unincorporated town
(2000 population 1,990), CAMERON county, extreme S
TEXAS, residential and resort community at S end of
LAGUNA MADRE, arm of GULF OF MEXICO, 18 mi/29 km
NE of BROWNSVILLE; 26°04′N 97°15′W.

Laguna Hills, city (2000 population 31,178), ORANGE
county, S CALIFORNIA; residential suburb 40 mi/64
km SE of downtown LOS ANGELES, and 6 mi/9.7 km
NE of LAGUNA BEACH; 33°36′N 117°43′W. El Toro
Marine Corps Air Station, to N, closed in 1999. Rapidly growing urban fringe area. Diverse light
manufacturing.

Laguna, La (lah-GOO-nah, lah) or **San Cristóbal de la
Laguna**, city (2001 population 128,822), on SANTA
CRUZ DE TENERIFE province, TENERIFE island, CAN-
ARY ISLANDS, SPAIN; 28°29′N 16°19′W. On an elevated
plain, 1,805 ft/550 m above sea level. The center of a
fertile farm area producing cereals, grapes, fruits, and
vegetables; tourist resort. The University of La Laguna
is here. Seat of the diocese of Tenerife.

Laguna Lachuá National Park (lah-GOO-nah lah-
choo-AH) (□ 20 sq mi/52 sq km), N ALTA VERAPAZ
department, GUATEMALA; 15°55′N 90°40′W. Protects
lowland forest and a nearly untouched lake (□ c.45 sq
mi/117 sq km). Access by trail only.

Laguna Larga (lah-GOO-nah LAHR-gah), town, central CÓRDOBA province, ARGENTINA, 33 mi/53 km SE
of CÓRDOBA. Peanuts, corn, soybeans, wheat, fruit,
vegetables; livestock.

Laguna Madre (lah-GOO-nah MAH-drai), NE MEX-
ICO and S TEXAS, narrow, shallow lagoon along GULF
coast; from CORPUS CHRISTI BAY, Texas, it extends
c.120 mi/193 km S to mouth of the RIO GRANDE, whose
delta interrupts it for c.40 mi/64 km, then continues
c.100 mi/161 km to point 10 mi/16 km N of mouth of
SOTO LA MARINA River, TAMAULIPAS. Sheltered from
the Gulf by narrow barrier islands (notably PADRE
ISLAND, Texas). UNITED STATES section is traversed by
GULF INTRACOASTAL WATERWAY.

Laguna Mountains, wooded range in central SAN
DIEGO county, S CALIFORNIA, extends c.35 mi/56 km
NW from Mexican border (N end of Sierra de Juárez).
Maximum elevation c.6,300 ft/1,920 m. Recreational
region. Largely in Cleveland National Forest. CARRIZO
GORGE at SE end; ANZA-BORREGO DESERT STATE PARK
to E, Cuyamaca Rancho State Park in center; several
small Indian reservations in range.

Laguna Niguel (lah-GOO-nah nee-GEL), city (2000
population 61,891), ORANGE county, S CALIFORNIA;
residential suburb 43 mi/69 km SE of downtown LOS
ANGELES, and 4 mi/6.4 km E of LAGUNA BEACH, near
Aliso Creek, in San Joaquin Hills; 33°32′N 117°42′W.
Manufacturing (plastic products, irrigation equipment).

Laguna Paiva (lah-GOO-nah PEI-vah), town, E SANTA
FE province, ARGENTINA, 22 mi/35 km N of SANTA FE;

31°19′S 60°39′W. Agricultural center (flax, corn, sunflowers, wheat, fruit; livestock).

Laguna Park (la-GOO-nuh), unincorporated village,
BOSQUE county, central TEXAS, 25 mi/40 km NW of
WACO; 31°51′N 97°22′W. Residential and recreational
community at Whitney Dam, on BRAZOS RIVER.
Manufacturing (sportswear); agricultural area.

Lagunas (lah-GOO-nahs), town, ALTO AMAZONAS
province, LORETO region, N central PERU, landing on
HUALLAGA RIVER, in AMAZON basin, and 50 mi/80 km
NE of YURIMAGUAS; 05°20′S 75°40′W. Bananas, yucca,
plantains; forest products.

Lagunas (lah-GOO-nahs), village, TARAPACÁ region, N
CHILE, on railroad, and 60 mi/97 km SE of IQUIQUE;
20°59′S 69°41′W. Salt deposits nearby.

Laguna San Rafael (lah-GOO-nah sahn rah-fah-EL),
national park, in Aisén del General Carlos Ibañez del
Campo region, 100 mi/161 km SW of Puerto Aisen.
Protecting fjord, glacier, and mountain region.

Lagunas de Chacahua National Park (lah-GOO-nahs
dai chah-KAH-wah), a national park, in S OAXACA,
MEXICO, on the PACIFIC coast, 15 mi/24 km W of
PUERTO ESCONDIDO, and 14 mi/23 km SE of SANTIAGO
PINOTEPA NACIONAL. Covers 25,000 acres/10,118 ha.
This park consists of numerous lagoons, bamboo
groves, mangrove swamps, and a beach. There is
varied plant and animal life. A dirt road off Mexico
Highway 200 accesses the park.

Lagunas de Montebello (lah-GOO-nahs dai mon-te-
BE-yo), a national park (□ 23 sq mi/59.8 sq km), in SE
CHIAPAS, MEXICO, 27 mi/44 km E of Mexican Highway 190 from LA TRINITARIA, on the GUATEMALA-
Mexico border. Elevation 5,000 ft. Mountainous
terrain with rain forest vegetation (ferns, pines, orchids, oaks, vanilla, cacao, and hule). The Chincultik
archaeological zone is nearby with Mayan ruins.

Lagunas de Zempoala National Park (lah-GOO-nahs
dai zem-po-AH-lah), in NW MORELOS, MEXICO, near
the state of MÉXICO. It is on the border S of the Sierra
del Ajusco, 9 mi/15 km W of Tres Cumbres. Elevation
9,514 ft/2,900 m. Covers 11,673 acres/4,724 ha. There
are three ancient volcanic craters filled with water. A
branch of the Mexico-CUERNAVACA road goes here.
From the E to the W, they are the Laguna Zempoala,
Compila, Tonatihua, Seca, Ocoyotongo, Quila, and
Hueyapan. The last provides water to the towns of
HUITZILAC and COAJOMULCO. Cool humid climate
surrounded by pine forests. Mexico Highway 95
provides access to this park.

Laguna Vista (la-GOO-nuh VIS-tuh), town (2006
population 2,843), CAMERON county, extreme S
TEXAS, 18 mi/29 km NE of BROWNSVILLE, on S end of
Laguna Largo (s sic) to S; 26°06′N 97°17′W. Recreation.
Laguna Atascosa National Wildlife Refuge is to N.

Laguna Woods, city (□ 4 sq mi/10.4 sq km; 2000
population 16,507), ORANGE county, S CALIFORNIA, c.5
mi/8 km E of MISSION VIEJO; 33°36′N 117°44′W. Almost entirely comprised of a gated senior citizen
community called Leisure World, where the average
age of residents is seventy-eight years. Development
started in the 1960s. Incorporated 1999.

Laguna Yaxhá (lah-GOO-nah yah-SHAH), lake,
PETÉN department, N GUATEMALA, 34 mi/55 km ENE
of FLORES; 6 mi/9.7 km long, 2 mi/3.2 km wide;
17°04′N 89°24′W. In Yaxhá Archaeological Reserve;
nearby is YAXHÁ site, a large and important Maya
ruin.

Lagundi Bay and Beach (LAH-goon-dee), resort, N
SUMATRA, INDONESIA, on S coast of NIAS island, 7 mi/
12 km W of TELUKDALEM; 00°34′N 97°44′E. Popular
surfing location, with waves often higher than 10 ft/3
m and which travel over 720 ft/220 m. Once the main
port of S Nias, it was destroyed by tidal waves resulting from the 1883 eruption of KRAKATAU.

Lagunes, region (□ 5,480 sq mi/14,248 sq km; 2002
population 4,210,200), S central CÔTE D'IVOIRE;
⊙ ABIDJAN; 05°25′N 04°20′W. Bordered NW tip by

Fromager region, NWN by Lacs region, N by N'zi-Comoé region, NNE by Agnéby region (part of E portion of border formed by Mé River), NE tip by Moyen-Comoé region (border formed by KOMOÉ RIVER), E by Sud-Comoé region (S portion of border formed by Komoé River, which also touches at N part of border), S by the GULF OF GUINEA of the ATLANTIC OCEAN, and W by Sud-Bandama region. Mé and Komoé rivers in E and Taabo Lake in extreme NW. ORUMBO BOKA mountain in NW, on border with Lacs. Banco National Park in E and Asagny National Park in SW. Towns include ABIDJAN, ANYAMA, and TIASSALÉ. Railroad in E of region (terminus at Abidjan) running N into Agnéby and through central Côte d'Ivoire. International airport at Abidjan.

Lagunilla (lah-goo-NEE-lyah), village, SALAMANCA province, W SPAIN, 12 mi/19 km WSW of BÉJAR; 40°19′N 05°58′W. Olive-oil processing; livestock; chestnuts, wine.

Lagunillas (lah-goo-NEE-yahs), canton, BAUTISTA SAAVEDRA province, LA PAZ department, W BOLIVIA, E of Peruvian border, and N of PUERTO ACOSTA. Elevation 11,417 ft/3,480 m. Limestone and gypsum deposits. Agriculture (potatoes, yucca, bananas, rye); cattle.

Lagunillas, canton, SEBASTIÁN PAGADOR province, ORURO department, W central BOLIVIA, 41 mi/65 km SE of CHALLAPATA and UYUNI-ORURO railroad and highway, 3 mi/5 km N of POTOSÍ border; 19°14′S 66°20′W. Elevation 12,139 ft/3,700 m.

Lagunillas, canton, VALLEGRANDE province, SANTA CRUZ department, E central, BOLIVIA, 19 mi/30 km N of VALLEGRANDE; 18°16′S 64°09′W. Elevation 6,660 ft/2,030 m. Abundant gas wells in area. Clay, limestone deposits. Agriculture (corn, potatoes, bananas, yucca, sweet potatoes, peanuts, soy, tobacco, coffee); cattle.

Lagunillas (lah-goo-NEE-yahs), town and canton, ⊙ CORDILLERA province, SANTA CRUZ department, SE BOLIVIA, 135 mi/217 km SSW of SANTA CRUZ, on road; 19°38′S 62°36′W. Elevation 3,949 ft/1,204 m. Subtropical fruit (oranges, lemons, limes, melons, bananas), corn.

Lagunillas (lah-goo-NEE-yahs), town, SAN LUIS POTOSÍ, N central MEXICO, 37 mi/60 km SE of RÍO VERDE. Grain, fruit; livestock.

Lagunillas, town, in E central MICHOACÁN, MEXICO, 16 mi/25 km E of PÁTZCUARO and N of HUIRAMBA. On Grande de Morelia River to SW. Agriculture (wheat and corn). Connects with railroad, on Mexico Highway 120. ACÁMBARO-MORELIA-PÁTZCUARO and URUAPAN. Many small lagoons in the area.

Lagunillas (lah-goo-NEE-yahs), town, ⊙ Sucre municipio, MÉRIDA state, W VENEZUELA, in ANDEAN spur, on transandine highway, and 17 mi/27 km WSW of MÉRIDA; 08°31′N 71°24′W. Elevation 3,540 ft/1,079 m. Sugarcane, coffee, fruit, cereals.

Lagunillas (lah-goo-NEE-yahs), town, ZULIA state, NW VENEZUELA, on E shore of Lake MARACAIBO, 40 mi/64 km SE of MARACAIBO; 10°08′N 71°15′W. Petroleum center in what is considered the largest oil-producing field in SOUTH AMERICA; some oil derricks are in lake. Terminus of oil pipeline from fields nearby.

Lagunitas, unincorporated town (2000 population 1,835), MARIN county, W CALIFORNIA, 10 mi/16 km WNW of SAN RAFAEL; 38°01′N 122°42′W. Also on San Geronimo Creek at its entrance to Lagunitas Creek, which rises on N slope of MOUNT TAMALPAIS and flows generally NW c.30 mi/48 km to TOMALES BAY. Dam impounds Kent Lake (c.2 mi/3.2 km long; for water supply) 5 mi/8 km W of San Rafael. GOLDEN GATE NATIONAL RECREATION AREA to S, POINT REYES NATIONAL SEASHORE to W; both parks are on Pacific Ocean. Unincorporated town of Forest Knolls adjacent to Lagunitas.

La Habana (lah ah-BAH-nah), province (□ 2,197 sq mi/ 5,712.2 sq km; 2002 population 77,066), W CUBA;

⊙ HAVANA. Coastal plains average 31 mi/50 km in width between the CARIBBEAN SEA and ATLANTIC OCEAN; beaches and some low mountains in the N. Drained by the ALMENDARES and the Jaruco rivers; several large aquifers (Vento, Sur, Gato) provide Havana with freshwater even though they suffer from saltwater intrusion along the Caribbean. Principal crops include tobacco, sugarcane, and vegetables; food processing is the leading industry. The majority of the population and industry is concentrated in and around Havana; about one-fourth of the population resides in cities with more than 20,000 people. After Havana, the largest towns include Güines, ARTEMISA, SAN JOSÉ DE LAS LAJAS, GUANAJAY, and GUIRA DE MELENA. International tourism was economically important until the U.S. and Cuba severed relations in 1960; tourism with countries besides the U.S. has been encouraged again since 1990.

La Habra, city (2000 population 58,974), ORANGE county, S CALIFORNIA; suburb 18 mi/29 km SE of downtown LOS ANGELES, 6 mi/9.7 km N of ANAHEIM; 33°56′N 117°57′W. La Habra was settled in the 1860s by Basque sheepherders. Manufacturing (computer equipment, fabricated metal products, plumbing fixtures, rubber products, flavoring extracts; printing). Oil-research center nearby. The city has grown along with the surrounding Los Angeles metropolitan area. Major housing developments have been constructed for the increasing population, and the city's economic base has been strengthened by high-technology and electronic manufacturing in the area. Incorporated 1925.

La Habra Heights, city (2000 population 5,712), LOS ANGELES county, S CALIFORNIA; residential suburb 17 mi/27 km ESE of downtown LOS ANGELES, E of WHITTIER; 33°58′N 117°57′W. Puente Hills to N.

Lahad Datu (LAH-hahd DAH-too), town (2000 population 74,601), SABAH, MALAYSIA, on N shore of DARVEL BAY, 55 mi/89 km S of SANDAKAN; 05°03′N 118°19′E. Hemp, tobacco-growing center; timber, oil palm, manilla, cocoa, copra; prawns, fisheries. Processing of edible oils.

Lahaina (lah-HEI-nah), city (2000 population 9,118), MAUI county, on the W coast of MAUI island, Maui county, HAWAII, on AUAU CHANNEL, 13 mi/21 km W of KAHULUI, 80 mi/129 km ESE of HONOLULU; 20°53′N 156°40′W. In a sugarcane and pineapple region. Manufacturing (sugarcane processing). Terminus (S) of scenic train roads from KAANAPALI, 3.5 mi/5.6 km N. It was the scene of the first European settlement in the islands and served as capital from 1810 until the seat of government was moved (1845) to Honolulu. Hawaii's first newspaper was printed in Lahaina, and the island's first school was established here in 1831. A whaling port in the mid-19th century, Lahaina was also an important anchorage for the U.S. Pacific Fleet in the 20th century Lahaina Cannery Mall; Seaman's Hospital (Historical); Jodo Mission Buddhist Cultural Center; Wahikuli State Wayside Park to N; Puamana Beach Park and Launiupoko State Park to SE; West Maui Forest Reserve to E. Whale watching December–April.

Lahan (luh-HAHN), town (2001 population 27,654), central NEPAL; 26°43′N 86°29′E.

Lahar (luh-HAHR), village (2001 population 28,250), BHIND district, N central MADHYA PRADESH state, central INDIA, 50 mi/80 km E of LASHKAR; 26°12′N 78°56′E. Gram, millet.

La Harpe (luh HARP), city (2000 population 1,385), HANCOCK county, W ILLINOIS, 16 mi/26 km NNE of CARTHAGE; 40°34′N 90°58′W. In agricultural area (corn, soybeans, livestock; dairy products). Railroad junction. Incorporated 1859.

La Harpe (luh HAHRP), village (2000 population 706), ALLEN county, SE KANSAS, 6 mi/9.7 km E of IOLA; 37°55′N 95°17′W. Livestock, grain; limestone quarry.

Laharpur (LAH-nuhr-puhr), town, SITAPUR district, central UTTAR PRADESH state, N central INDIA, 17 mi/ 27 km NE of Sitapur; 27°43′N 80°54′E. Trades in wheat, rice, gram, barley, oilseeds. Founded 1374 by Firoz Shah Tughlak. Raja Todar Mal, Akbar's finance minister and general, b. here.

Lahave River (luh-HAIV), 60 mi/97 km long, W NOVA SCOTIA, CANADA; issues from small Lahave Lake, 30 mi/48 km ENE of ANNAPOLIS ROYAL; flows SE, through several small lakes, past BRIDGEWATER, to the ATLANTIC OCEAN 7 mi/11 km SSW of LUNENBURG. Navigable below Bridgewater. Salmon fisheries.

Lahavoth Habashan (lah-bah-VOT hah-bah-SHAHN), kibbutz, NE ISRAEL, in HULA VALLEY, at W foot of the GOLAN HEIGHTS, 15 mi/24 km NE of ZEFAT; 33°08′N 35°38′E. Elevation 193 ft/58 m. Manufacturing of fire-fighting equipment. Mixed farming. Founded 1945.

La Haye, NETHERLANDS: see HAGUE, THE.

Lähden (LAH-den), village, LOWER SAXONY, NW GERMANY, 12 mi/19 km ENE of MEPPEN; 52°44′N 07°33′E.

Lahdenpohja, RUSSIA: see LAKHDENPOKHYA.

Lahej (LAH-hezh), town, S YEMEN, 20 mi/32 km WNW of ADEN, in fertile agricultural area irrigated by the Wadi TIBAN; 13°04′N 44°53′E. Principal commercial center in Aden hinterland, trading with tribal areas; formerly entrepôt on caravan route to Yemen; native handicrafts (metalwork). Sultan's palace on town's S edge. Was the capital of the former ABDALI sultanate. Held by Turks during World War I, Lahej was later temporarily linked by railroad with Aden. Also spelled Lahj.

Lahemaa National Park (LUH-he-maw) (□ 70 sq mi/ 182 sq km), N coast of ESTONIA, 43.5 mi/70 km E of TALLINN. Covers 25,291 acres/64,915 ha of land area. Forested conservation area with rocky coastline.

Laheri, UKRAINE: see BALAKLIYA.

La Herradura (lah ER-rah-DOO-rah), village, COQUIMBO region, N central CHILE, S suburb of COQUIMBO, minor port on small inlet of PACIFIC; resort; 29°59′S 71°22′W.

La Herradura (lah er-rah-DOO-rah), village, LA PAZ department, S SALVADOR, on JALTEPEQUE Lagoon, 28 mi/45 km SE of SAN SALVADOR; 13°21′N 88°57′W. Resort (hunting, fishing). Salt extraction nearby.

La Herradura, PERU: see CHORRILLOS.

Lahic, isolated village, in the CAUCASUS, AZERBAIJAN; 40°40′N 46°13′E. Dedicated almost exclusively to handicrafts; best known for coppersmithing, but leatherwork, carpets, knives, and firearms are also made by hand. The population uses an ancient Persian-based language called "Lahic," spoken nowhere else. The only connection to the outside world is by a dirt path, but their products have been sold for centuries in Middle Eastern cities and displayed in Europe's museums.

La Higuera (lah ee-GE-rah), canton, VALLEGRANDE province, SANTA CRUZ department, E central BOLIVIA; 18°47′S 64°13′W. Elevation 6,660 ft/2,030 m. Abundant gas wells; clay, limestone, and gypsum deposits. Agriculture (potatoes, yucca, bananas, corn, sweet potatoes, peanuts, tobacco); cattle.

La Higuera (lah ee-GE-rah), village, ⊙ La Higuera comuna, ELQUI province, COQUIMBO region, N central CHILE, near the coast, 28 mi/45 km NNW of LA SERENA. Copper and silver mining.

Lahijan (lah-ee-JAHN), town (2006 population 72,950), Gīlān province, N IRAN, 23 mi/37 km E of RASHT, and E of the SEFID Rud; 37°11′N 49°59′E. Trade center; known for its tea, rice, silk, oranges.

Lahinch, IRELAND: see LEHINCH.

Lahiya, Arab town, 3.7 mi/6 km NE of GAZA, in the GAZA STRIP. Agriculture (citrus, wheat, vegetables). The town has grown considerably with the influx of Palestinian refugees since 1948.

Lahj, YEMEN: see LAHEJ.

Cross-references are shown in SMALL CAPITALS. The pronunciation guide is shown on page xix. The sources of population figures are shown on page xvii.

Lahn River, 160 mi/257 km long, W GERMANY; rises on the EDERKOPF (50°53'N 08°14'E); flows E, S, and W, past MARBURG, GIESSEN (head of navigation), WETZLAR, LIMBURG, and BAD EMS, to the RHINE at Oberlahnstein. Receives the DILL (right).

Lahnstein (LAHN-shtein), town, RHINELAND-PALATINATE, W GERMANY, at confluence of LAHN and RHINE rivers, 3 mi/4.8 km S of KOBLENZ; 50°18'N 07°36'E. Port on RHINE RIVER; climatic health resort. Manufacturing (machinery, plastics); industries (paper and chemical). Has two ancient castles.

Laholm (lah-HAWLM), town, HALLAND county, SW SWEDEN, on LAGAN RIVER (falls), near its mouth on KATTEGATT strait, 13 mi/21 km SE of HALMSTAD; 56°30'N 13°02'E. Manufacturing (light industries); fish (salmon). Hydroelectric station. Medieval trade center; suffered heavily in seventeenth-century Swedish-Danish wars. Remains of seventeenth-century castle on site of earlier fortification.

Lahoma (lah-HO-muh), village (2006 population 564), GARFIELD county, N OKLAHOMA, 12 mi/19 km W of ENID; 36°23'N 98°05'W. In agricultural area.

La Honda, unincorporated village, SAN MATEO county, W CALIFORNIA, 11 mi/18 km SSW of REDWOOD CITY, 21 mi/34 km W of SAN JOSE, and 7 mi/11.3 km E of Pacific Ocean. On W side of Santa Cruz Mountains, on La Honda Creek, and in canyon among redwood groves. San Mateo State Beaches to W; Portola and Castle Rock state parks to SE.

Lahontan Dam, NEVADA: see CARSON RIVER.

Lahontan, Lake, extinct lake, W NEVADA and NE CALIFORNIA. Formed during wet periods coinciding with glacial surges during the Pleistocene epoch and, with Lake BONNEVILLE (UTAH). Occupied a part of the GREAT BASIN region. Lake Lahontan retreated shortly after the Pleistocene epoch, but PYRAMID, WINNEMUCCA, and WALKER lakes and CARSON SINK are its remnants. Marine reptile fossils at Berlin-Ichthyosaur State Park, 70 mi/113 km SE of FALLON, (Nev.; archaeological dig). LAHONTAN RESERVOIR, on CARSON RIVER, in NW.

Lahontan Reservoir (luh-HAHN-tuhn), CHURCHILL and STOREY counties, W NEVADA, on CARSON RIVER, in Lahontan State Recreation Area, 13 mi/21 km W of FALLON; 39°27'N 119°04'W. Max. capacity 426,500 acre-ft.; 18 mi/29 km long. Carson River enters on SW, drains NE to CARSON SINK. Formed by LAHONTAN DAM (earth construction; 115 ft/35 m high), built (1915) by the Bureau of Reclamation for irrigation, power generation, and recreation.

Lahore (lah-HOR), district (□ 2,191 sq mi/5,696.6 sq km), E PUNJAB province, central PAKISTAN; ⊙ LAHORE. In Bari Doab, between RAVI (NW) and SUTLEJ (SE) rivers; bordered E by INDIA (PUNJAB state); irrigated by UPPER BARI DOAB, and (S) by DIPALPUR canal systems. Agriculture (wheat, cotton, gram, oilseeds, rice); cotton textile manufacturing, cotton ginning. A great trade center at LAHORE. Original district (□ 2,595 sq mi/6,721 sq km) was divided in 1947 between East (INDIA) and West (PAKISTAN) PUNJAB.

Lahore (lah-HOR), city, Lahore district, ⊙ Punjab province, E central Pakistan, on the Ravi River, 640 mi/1,030 km NE of KARACHI; 31°35'N 74°18'E. Pakistan's 2nd-largest city, once known as "the Paris of the subcontinent." Railroad and air transport center near the Indian border. Banking and commercial city that markets the products of the surrounding fertile agr. area. Home to c.20% of Pakistan's industrial producers; manufacturing (textiles, rubber, iron, cement, and steel); also handicrafts, metalworks, especially gold and silver work. According to Hindu legend, Lahore was founded by Loh, or Lava, son of Rama, the hero of the Sanskrit epic *Ramayana*. In 1036 it was conquered from a Brahman dynasty by the Muslim Turkish Ghaznivids, who made it the capital of their empire in 1106. It passed in 1186 to the Ghori sultans,

also from Afghanistan. India's 1st Muslim emperor, Qutb-ud-din Aibak, was crowned here in 1206 and is also buried here. An 11th-cent. Sufi mystic, Ali ibn Uthman al-Hujwiri, known as Dataganj Bakhsh, is also buried here (his shrine is a major S Asian pilgrimage center). The city, which suffered Mongol raids in the 13th and 14th century, entered the period of its greatest glory in the 16th century, when it became 1 capital of the Mogul empire. Declined after the reign of Aurangzeb; it was annexed in 1767 by the Sikhs, who, under Ranjit Singh, made it their capital. Passed to the British in 1849. When Pakistan won independence in 1947, Lahore became the capital of West Punjab province; the capital of the entire province of West Pakistan (1955–1970) reverting to the capital of the Punjab province thereafter. The architectural remains of the Mogul period, although imperfectly preserved, are among the most splendid of Mogul art. Esp. notable are Lahore Fort, the Badshahi Masjid, the tomb-garden of Emperor Jahangir, and (just outside the city) the Shalamar garden, built on 3 terraces with hundreds of fountains. Other landmarks include the Pearl and Golden mosques, the tomb of Ranjit Singh, the Walled City gates, and the Wazir Khan mosque, which contains the finest known examples of *kashi* (inlaid tilework). Lahore's museum of Indian antiquities, which figures in Rudyard Kipling's *Kim*, is among the most noted in East Asia. Seat of the University of the Punjab (1882), Pakistan's oldest university; several affiliated colleges; a university of engineering and technology; and an atomic research institute.

Lahorighat (lah-ho-ree-gaht), town, NAGAON district, central ASSAM state, NE INDIA, 22 mi/35 km WNW of NAGAON; 26°26'N 92°21'E. Rice, jute, rape and mustard, tea.

Lahou Lagoon (lah-OO), along coast of CÔTE D'IVOIRE, 70 mi/113 km W of ABIDJAN; 05°10'N 05°00'W. Separated from ATLANTIC OCEAN by narrow spit. At its entrance (BANDAMA RIVER mouth) is GRAND-LAHOU.

Lahr, town, BADEN-WÜRTTEMBERG, SW GERMANY, at W foot of BLACK FOREST, 22 mi/35 km N of FREIBURG; 48°21'N 07°52'E. Industrial city, chief products are steel machinery and motor vehicles; other manufacturing includes cigarettes, chemicals, and electronics; printing. Has 12th- and 13th-century churches and 16th-century town hall (rebuilt 1855). Chartered 1279.

Lahri (LAH-ree), village, SIBI district, SIBI division, BALUCHISTAN province, SW PAKISTAN, 100 mi/161 km SE of QUETTA, on KACHHI plain; 29°11'N 68°13'E. Local market.

Lahstedt (LAH-stet), town, LOWER SAXONY, N GERMANY, 11 mi/18 km W of BRAUNSCHWEIG; 52°13'N 10°16'E.

Lahti (LAH-tee), city, HÄMEEN province, S central FINLAND; 61°00'N 25°40'E. Elevation 363 ft/110 m. Connected with S end of PÄIJÄNNE lake system, it is an important lake port, transportation center, and winter resort area. Has many large factories and is a center of the Finnish wood-production industry; other industries include glassworks, furniture manufacturing, breweries, metalworking, and clothing factories. Many Karelians came to Lahti after the Finnish-SOVIET armistice of 1944. The city hall was designed in 1912 by Eliel Saarinen. Distribution point for VYBORG-FINLAND natural gas pipeline. University Research and Training Center. Founded in 1878; incorporated 1905.

La Huaca (LAH HWAH-kah), town, PAITA province, PIURA region, NW PERU, on coastal plain, on CHIRA RIVER, 16 mi/26 km NE of PAITA; 05°03'S 80°13'W. In irrigated area; cotton, rice, fruit.

La Huaca Arco Iris (LAH HWAH-kah AHR-ko EE-ris), pyramidal temple, LA LIBERTAD region, NW PERU. One of main sites at the archaeological site of CHAN CHAN. Also known as La Huaca El Dragón.

La Huaca El Dragón, PERU: see LA HUACA ARCO IRIS.

La Huaca Esmeralda (LAH HWAH-kah es-mai-RAHL-dah), small pyramidal temple, LA LIBERTAD region, NW PERU. One of main sites at the ruins of CHAN CHAN.

La Huacana (lah wah-KAH-nah), town, MICHOACÁN, W MEXICO, 35 mi/56 km S of URUAPAN; 18°58'N 101°49'W. Cereals, sugarcane, fruit; silver deposits.

La Huerta (lah WER-tah), town, in SW JALISCO, MEXICO on PACIFIC coast of Sierra Cacoma, 36 mi/58 km SW of AUTLÁN DE NAVARRO, on Mexico Highway 80; 19°29'N 104°40'W. Irrigated by the Purificación River. The name reflects great agricultural production (rice, sugarcane, tobacco, beans, peanuts, corn, fruit, plantains, watermelon, coconuts; fine woods and lumber).

La Huerta (lah WER-tah) or **Laprida**, village, SE SAN JUAN province, ARGENTINA, at SE foot of SIERRA DE LA HUERTA, 80 mi/129 km ENE of SAN JUAN. Sulphur deposits.

Lahul and Spiti (LAH-hool, SPIT-ee), district (□ 5,342 sq mi/13,889.2 sq km), HIMACHAL PRADESH state, N INDIA, PUNJAB and GREAT HIMALAYAS; ⊙ Kyelong. Highest peak is 23,050 ft/7,026 m; several peaks rise to over 20,000 ft/6,096 m. Bordered W by CHAMBA district, S by KULLU and KINNAUR districts, E by CHINA (TIBET), and N by JAMMU AND KASHMIR state. Consists of rugged mountain ridges, snowfields, glaciers, rock-strewn valleys; forests, animal wildlife. Some barley grown in lower valleys. Antimony ore deposits near SHIGRI (SE). Several Buddhist monasteries.

La Hulpe, BELGIUM: see HULPE, LA.

La Hune, Cape (luh HOON), S NEWFOUNDLAND AND LABRADOR, CANADA, on E side of La Hune Bay (extending 7 mi/11 km inland), 40 mi/64 km E of BURGEO; 47°32'N 56°32'W. Opposite is fishing village of La Hune.

Lahun, El- (lah-HOON, el), village, FAIYUM province, Upper EGYPT, on the BAHR YUSUF, on railroad, and 11 mi/18 km SE of FAIYUM; 29°13'N 30°59'E. Cotton, cereals, sugarcane, fruits. Site of the pyramid built (c.1900 B.C.E.) by Sesostris II.

Lahuy Island (lah-WEE) (□ 7 sq mi/18.2 sq km), CAMARINES SUR province, PHILIPPINES, in PHILIPPINE SEA; 13°56'N 123°50'E. Just off S coast of LUZON, 15 mi/24 km NW of RUNGUS POINT; 6 mi/9.7 km long, 2 mi/3.2 km wide. Rises to 613 ft/187 m. Gold mining.

Laï, town, ⊙ TANDJILÉ administrative region, SW CHAD, on LOGONE RIVER, and 60 mi/97 km NNE of MOUNDOU; 09°24'N 16°18'E. Center of rice cultivation. Former military outpost. Until 1946, in Ubangi-Chari colony. Was capital of former Tandjilé prefecture.

Laiagam (LEI-ah-gam), village, ENGA province, E central NEW GUINEA island, W central PAPUA NEW GUINEA, 18 mi/29 km W of WABAG; 55°17'S 143°46'E. Located on N side of Lagaip River, near its source. Road access. Coffee, copra, sweet potatoes; timber.

Lai'an (LEI-AN), town, ⊙ Lai'an county, NE ANHUI province, CHINA, near JIANGSU border, 30 mi/48 km NW of NANJING; 32°27'N 118°25'E. Rice, oilseeds, jute, cotton; food processing, engineering, building materials.

Laibach, SLOVENIA, see LJUBLJANA.

Laibin (LEI-BIN), town, ⊙ Laibin county, central GUANGXI ZHUANG AUTONOMOUS Region, CHINA, on HONGSHUI RIVER, and 40 mi/64 km S of LIUZHOU, and on railroad; 23°42'N 109°16'E. Rice, sugarcane, oilseeds; food processing, iron-ore mining, coal mining.

Lai Chau (LAI CHOU), province (□ 6,612 sq mi/17,191.2 sq km), N VIETNAM; N border with CHINA, E border with LAO CAI province, SE border with SON LA province, W border with LAOS; ⊙ DIEN BIEN PHU; 22°00'N 103°00'E. Drained by the Song Da (BLACK RIVER); province exhibits a number of environmental zones, has an ethnic diversity, and is the contemporary scene of considerable forest exploitation. Diverse soils and mineral resources, shifting cultivation, opium growing, forest products (medicinals, gums, resins, honey, beeswax, exotic plants, foraged foods), and commercial agriculture (pine-tree plantations, coffee, tea, vegetables, fruits). Lumbering, handicrafts, food processing, domestic tourism. Ta Bu area is set to be the site of a major hydroelectric project designed to dam the Song Da above the HOA BINH reservoir. Population is Thai, H'mong, Dao, Mang, Kho Mu, Tibeto-Burmese, and other minorities.

Lai Chau (LAI CHOU), city, LAI CHAU province, N VIETNAM, along Da River (Song Da), and 60 mi/97 km SW of LAO CAI; 22°02′N 103°10′E. Former capital of province. Transportation and commercial center; shifting cultivation, with key cash crops, timber, and opium facing government restrictions. Major dam may be constructed in Ta Bu area (N of Da River reservoir). If project is approved, area will be flooded and population transferred. Thai, H'mong, Dao, and other ethnic minorities. Formerly called Laichau.

Laichingen (LEI-king-uhn), village, BADEN-WÜRT-TEMBERG, GERMANY, in SWABIAN JURA, 15 mi/24 km NW of ULM; 48°29′N 09°41′E. Manufacturing (textiles, tools); electronics industry. A linen-weaving center since Middle Ages. Nearby (S) is the deepest cave opened to the public in GERMANY. Chartered in 1364.

Laidlaw (LAID-law), unincorporated village, S BRITISH COLUMBIA, W CANADA, on FRASER River, and 19 mi/31 km NE of CHILLIWACK, and in FRASER VALLEY regional district; 49°18′N 121°36′W. Lumbering; dairying.

Laidley, town, SE QUEENSLAND, AUSTRALIA, 40 mi/64 km W of BRISBANE; 27°38′S 152°23′E. Agriculture (corn, sugarcane, oil seed, vegetables; beef cattle; dairying; cotton).

Laie (LAH-EE-ai), town (2000 population 4,585), OAHU, HONOLULU county, HAWAII, on NNE coast, 24 mi/39 km NNW of HONOLULU; 21°38′N 157°55′W. Laie Point to SE; Makahoa Point to N: KOOLAU RANGE to SW. Laie Temple, a large Mormon temple (1919), Brigham Young University Hawaii Campus; Polynesian Cultural Center; Malaekahana State Recreational Area.

Laifeng (LEI-FENG), town, ☉ Laifeng county, SW HUBEI province, CHINA, on HUNAN border, 50 mi/80 km S of ENSHI; 29°31′N 109°18′E. Tobacco; tobacco processing.

Laigle, FRANCE: see AIGLE, L'.

La Iguala (lah ee-GWAH-lah), town, LEMPIRA department, W central HONDURAS, on unpaved road, 8 mi/12.9 km NE of GRACIAS; 14°37′N 88°29′W. Elevation 768 ft/234 m.

Laihka (LEI-kah), former central state (sawbwaship), now township (☐ 1,559 sq mi/4,053.4 sq km), SHAN STATE, MYANMAR, ☉ Laihka, village on Hsipaw-Loilem road, and 55 mi/89 km NW of TAUNGGYI; 25°03′N 97°37′E. Iron ore, lacquer; irrigated rice on valley floor. Mountain range (W) and plateau (E).

Laijun, YEMEN: see GHEIL BIN YUMEIN.

Laikipia Escarpment (lah-ee-KEE-pee-ah), section of E rim of GREAT RIFT VALLEY, in W central KENYA, between RUMURUTI and Baringo. Laikipia Plateau is E.

Laikovats, SERBIA: see LAJKOVAC.

Laila (LAI-lah), town and oasis (2004 population 27,026), S NAJ'D, Riyadh province, SAUDI ARABIA, in AFLAJ district, 170 mi/274 km S of RIYADH. Grain (sorghum, barley), dates, alfalfa, vegetables, fruit. Also spelled Laylah.

Lainate (lei-NAH-te), town, MILANO province, LOMBARDY, N ITALY, 10 mi/16 km NNW of MILAN; 45°34′N 09°02′E. Mills (silk, cotton, flax); food processing; manufacturing (chemicals, plastics, fabricated metals, machinery).

La Independencia (lah een-de-pen-DEN-see-yah), town, CHIAPAS, S MEXICO, in SIERRA MADRE, 8 mi/12 km E of COMITÁN DE DOMÍNGUEZ. Corn, fruit. Tojolabal-speaking Mayan Indians live in rural areas.

Laingsburg, town, WESTERN CAPE province, SOUTH AFRICA, in WITTEBERGE mountains, on Buffels River, and 80 mi/129 km WNW of OUDTSHOORN, on N1 highway between CAPE TOWN and JOHANNESBURG; 33°10′S 20°51′E. Elevation 3,116 ft/ 950 m. Livestock; fruit. Lower section of town destroyed by floodwaters of Buffels River in January 1981; significant loss of property, historic buildings, and lives.

Laingsburg, town (2000 population 1,223), SHIA-WASSEE county, S central MICHIGAN, 15 mi/24 km NE of LANSING; 42°53′N 84°20′W. In farm area (livestock; grain, soybeans, hay; dairy products). Sleepy Hollow State Park to NW (CLINTON county).

Laing's Nek, locality, NW KWAZULU-NATAL, SOUTH AFRICA, near MPUMALANGA border, 7 mi/11.3 km S of VOLKSRUST, in DRAKENSBERG range, near N11 highway; 27°31′S 29°55′E. Elevation 5,399 ft/1,646 m. Scene of battle in Transvaal revolt (Jan. 28, 1881). Sometimes spelled Laingsnek.

Laingsnek, locality, SOUTH AFRICA: see LAING'S NEK.

Lainsitz River, CZECH REPUBLIC and AUSTRIA: see LUZNICE RIVER.

Laird (LERD), township (☐ 39 sq mi/101.4 sq km; 2001 population 1,021), ALGOMA district, ONTARIO, E central CANADA.

Laird (lerd), village (2006 population 207), central SASKATCHEWAN, CANADA, 40 mi/64 km N of SASKA-TOON; 52°43′N 106°35′W. Mixed farming; dairying.

La Isabela, CUBA: see ISABELA DE SAGUA.

La Isabela, DOMINICAN REPUBLIC: see ISABELA.

La Isabelal, town, MONTE CRISTI province, NW DO-MINICAN REPUBLIC, on highway, and 13 mi/21 km SE of MONTE CRISTI; 19°48′N 71°05′W. Agricultural (rice, onions, potatoes). Until 1938, VILLA VÁSQUEZ. Also called Villa Isabel.

Laish (lah-EESH), village, S SAMARKAND wiloyat, UZ-BEKISTAN, near the Ak Darya River (N arm of ZER-AVSHAN RIVER), 20 mi/32 km NW of SAMARKAND; 39°53′N 66°45′E. Cotton. Formerly spelled AK-DARYA.

Laish, ISRAEL: see DAN.

Laishev, RUSSIA: see LAISHEVO.

Laishevo (lah-EE-shi-vuh), town (2006 population 7,750), W central TATARSTAN Republic, E European Russia, on the KUYBYSHEV Reservoir of the VOLGA RIVER, on road, 38 mi/61 km SSE of KAZAN'; 55°24′N 49°33′E. Elevation 177 ft/53 m. Agricultural chemicals; food processing (lard, fish, poultry, grain products, potatoes). Founded in 1555; one of the oldest Russian settlements in the lower KAMA River valley. Until 1926, a city (Laishev); reduced to status of a village, 1926–1950; made town in 1950.

Laishi, ARGENTINA: see HERRADURA.

Laisholm, ESTONIA: see JÕGEVA.

Laishui (LEI-SHWAI), town, ☉ Laishui co., N HEBEI province, CHINA, 50 mi/80 km SW of BEIJING, and on spur of Beijing-WUHAN railroad; 39°23′N 115°42′E. Cotton, grain, oilseeds, fruits.

Laiskiy Zavod, RUSSIA: see LAYA.

La Isla, MEXICO: see SAN ANTONIO LA ISLA.

Laissac (le-sahk), village (☐ 7 sq mi/18.2 sq km), AVEYRON department, MIDI-PYRÉNÉES region, S FRANCE, near AVEYRON River, 13 mi/21 km E of RODEZ; 44°23′N 02°50′E. Dairying. Noted for its fairs.

Laitec Island (LEI-tek) (☐ 13 sq mi/33.8 sq km), off SE coast of CHILOÉ ISLAND, CHILOÉ province, LOS LAGOS region, S CHILE, 50 mi/80 km SSE of CASTRO; 8 mi/13 km long, c.1 mi/2 km wide; 43°13′S 73°37′W.

Laitlyngkot (LEIT-ling-kot), village, EAST KHASI HILLS district, MEGHALAYA state, NE INDIA, in KHASI hills, 9 mi/14.5 km SSW of SHILLONG. Trades in rice, cotton, betel nuts.

Laituri (lah-ee-TOO-ree), urban settlement, GEORGIA, 3.1 mi/5 km from the Meria railroad station, on a branch of the Samtredia-Batumi line. Tea factory.

Laives (LEI-ves), German *Leifers*, town, BOLZANO province, TRENTINO–ALTO ADIGE, N ITALY, near ADIGE RIVER, 5 mi/8 km S of BOLZANO; 46°26′N

11°20′E. Diversified secondary manufacturing center includes metal and wool products, machinery, paper.

Laiwu, city (☐ 864 sq mi/2,238 sq km; 1994 estimated urban population 370,200; 1994 estimated total population 1,187,500), W central SHANDONG province, CHINA, in the Tai Mountains, 45 mi/72 km SE of JINAN; 36°14′N 117°40′E. Agriculture and heavy industry are the largest sectors of the city's economy. Agriculture (grain, jute, oilseeds); manufacturing (utilities, iron and steel, machinery); mining (coal, ferrous mineral).

Laixi (LEI-SEE), city (☐ 1,522 sq mi/3,942 sq km; 1994 estimated urban population 102,600; 1994 estimated total population 711,800), E SHANDONG province, CHINA, on the SHANDONG Peninsula, 60 mi/97 km N of QINGDAO, on railroad to YANTAI; 36°50′N 120°40′E. Agriculture and heavy industry are the largest sectors of the city's economy; light industry is also important. Main industries include non-metallic mineral mining and food processing. Also called Shuiji.

Laiyang (LEI-YANG), city (☐ 669 sq mi/1,733 sq km; 1994 estimated urban population 161,600; 1994 estimated total population 898,600), E SHANDONG province, China, 70 mi/113 km N of QINGDAO; 36°58′N 120°40′E. Agriculture and light industry are the largest sectors of the city's economy; heavy industry is also important. Agriculture (grain, oilseeds, fruits); manufacturing (textiles, chemicals, machinery, transportation equipment; food processing).

Laiyuan (LEI-YU-AN), town ☉ Laiyuan county, NW HEBEI province, China, in TAIHANG MOUNTAINS, 50 mi/80 km NW of BAODING; 39°19′N 114°41′E. Grain; iron-ore mining and dressing.

Laizhou (LEI-JO), city (☐ 701 sq mi/1,816 sq km; 1994 estimated urban population 246,500; 1994 estimated total population 876,300), SHANDONG province, CHINA, on SHANDONG PENINSULA; 37°10′N 119°55′E. Agriculture and light industry are the largest sectors of the city's economy; heavy industry is also important. Manufacturing includes non-ferrous mineral mining, food processing, beverages, chemicals, plastics, and machinery. Laizhou has maintained the ancient deep canals that surround its high city walls. Well known for raising high-quality Chinese roses. Formerly called Ye Xian.

Laja (LAH-hah), town and canton, LOS ANDES province, LA PAZ department, W BOLIVIA, in the ALTI-PLANO, 17 mi/27 km WSW of LA PAZ; 16°32′S 68°23′W. Potatoes; sheep.

Laja, La, PANAMA: see LA LAJA.

Laja, Lake (LAH-hah) (☐ 40 sq mi/104 sq km), in BÍO-BÍO region, BÍO-BÍO region, S central CHILE, in the ANDES near ARGENTINA border, 50 mi/80 km E of LOS ÁNGELES; 20 mi/32 km long; 37°21′S 71°19′W. ANTUCO VOLCANO on W shore.

La Jalca (LAH HAHL-kah) or **Jalca**, town, AMAZONAS region, CHACHAPOYAS province, N PERU, in E AN-DEAN foothills, 20 mi/32 km SSE of CHACHAPOYAS; 06°29′S 77°43′W. Sugar growing, rice, plantains, yucca.

Lajamina (lah-hah-MEE-nah), village, minor civil division of Pocrí district, LOS SANTOS province, S central PANAMA, in PACIFIC lowland, 4 mi/6.4 km SSW of POCRÍ; 07°35′N 80°08′W. Sugarcane, coffee; livestock.

Lajão, Brazil: see CONSELHEIRO PENA.

La Jara, town (2000 population 877), CONEJOS county, S COLORADO, on La Jara Arroyo, near La Jara Creek, E of SAN JUAN MOUNTAINS, and 14 mi/23 km SSW of ALAMOSA, in SAN LUIS VALLEY; 37°16′N 105°57′W. Elevation 7,602 ft/2,317 m. Cattle, sheep; wheat, oats, barley, hay products. Rio Grand National Forest to W.

Laja River (LAH-hah), c.100 mi/161 km long, S central CHILE; rises in LAKE LAJA in the ANDES near ARGEN-TINA border; flows W, past ANTUCO and TUCAPEL, to BÍO-BÍO river at SAN ROSENDO. Has several rapids, of which the best known is Salto del Laja (45 mi/72 km SE of CONCEPCIÓN), a noted tourist site. It is used for hydroelectric power, supplying coal mines and steel mills of Concepción.

Laja River, c.85 mi/137 km long, GUANAJUATO, central MEXICO; rises near SAN FELIPE in SIERRA MADRE OCCIDENTAL; flows S, and past COMONFORT, to APASEO River (affluent of LERMA River) 3 mi/4.8 km SE of CELAYA.

Lajas (LAH-hahs), town, CIENFUEGOS province, central CUBA, on railroad, and 21 mi/34 km W of SANTA CLARA; 22°25'N 80°18'W. Agricultural center (sugarcane, fruit). The sugar center of Caracas is 4 mi/6 km SE.

Lajas (LAH-hahs), town (2006 population 27,583), SW PUERTO RICO, 11 mi/18 km SSE of Mayagüez, near the coast. Farmland; livestock, pineapples. Beach resort, tourism, fishing nearby (S) in LA PARGUERA.

Lajas, Las, PANAMA: see LAS LAJAS.

Laje (LAH-zhe), city (2007 population 21,175), E central BAHIA state, BRAZIL, 30 mi/48 km SSW of SANTO ANTÔNIO DE JESUS; 13°12'S 39°36'W.

Lajeadão (LAH-zhai-ah-DOUN), town, SE BAHIA state, BRAZIL, on MINAS GERAIS border in SERRA DOS AIMORES; 17°38'S 40°20'W.

Lajeado (LAH-zhai-ah-do), city, E central RIO GRANDE DO SUL state, BRAZIL, head of navigation on TAQUARI RIVER, and 4 mi/6.4 km N of ESTRÊLA; 29°27'S 51°58'W. Agriculture (wheat, potatoes; poultry) and livestock center (hog raising); lard manufacturing Amethysts and agates found nearby. Formerly spelled Lageado.

Lajeado, Brazil: see GUIRATINGA.

Lajeado Bonito (LAH-zhe-ah-do BO-nee-to), city, central PARANÁ state, BRAZIL, 61 mi/98 km SE of LONDRINA; 24°10'S 50°47'W.

Lajeado Novo (LAH-zhai-ah-do NO-vo), town (2007 population 6,729), W central MARANHÃO state, BRAZIL, 68 mi/110 km W of BALSAS, near TOCANTINS border; 07°45'S 46°47'W.

Lajedinho (LAH-zhe-ZHEEN-yo), town (2007 population 4,248), central BAHIA state, BRAZIL, 37 mi/60 km W of RUY BARBOSA; 12°21'S 40°51'W.

Lajedo Alto (LAH-zhe-do AHL-to), village, on SALVADOR–BELO HORIZONTE railroad, 35 mi/56 km W of CASTRO ALVES; 12°37'S 39°08'W.

Lajedo do Tabucal (LAH-zhe-de do tah-BAH-kahl), town, central BAHIA state, BRAZIL, 24 mi/38 km E of MARACAS; 13°32'S 40°14'W.

Laje do Muriaé (LAH-zhe do MOO-ree-ei), town, N RIO DE JANEIRO state, BRAZIL, 13 mi/21 km E of ITAPERUNA; 21°12'S 42°07'W. Sugar.

Laje Island (LAH-zhe), low islet in entrance to GUANABARA BAY, RIO DE JANEIRO state, SE BRAZIL, 4 mi/6.4 km from center of RIO DE JANEIRO; 22°57'S 43°09'W. Ship channel is E of it.

Lajemmerais (lah-zhe-muh-RAI), county (□ 134 sq mi/348.4 sq km; 2006 population 68,191), MONTÉRÉGIE region, S QUEBEC, E CANADA; ⊙ VERCHÈRES; 45°41'N 73°26'W. Composed of six municipalities. Formed in 1982.

Lajeosa (lah-jai-O-sah), village, VISEU district, N PORTUGAL, 10 mi/16 km SSW of VISEU. Wine, oranges, corn, rye.

Lajes (LAH-zhes), city, S PIAUÍ state, BRAZIL, 72 mi/116 km SE of FLORIANO, on PIAUÍ RIVER; 07°55'S 42°45'W.

Lajes, city (2000 population 153,582), central SANTA CATARINA state, BRAZIL, in the Serra do MAR, 110 mi/177 km WSW of FLORIANÓPOLIS, and on CURITIBA-PÔRTO ALEGRE highway; 27°48'S 50°19'W. Cattle-raising center supplying RIO GRANDE DO SUL meat-processing plants. Founded 17th century by Portuguese. Formerly spelled Lages.

Lajes (LAH-zhes), town (2007 population 10,392), central RIO GRANDE DO NORTE state, NE BRAZIL, 70 mi/113 km W of NATAL, on railroad; 05°41'S 36°14'W. Livestock; cheese, carnauba wax; gypsum and marble quarries. Also called Itaretama.

Lajes (LAH-zhish) or **Lagens** (LAH-jensh), village, ANGRA DO HEROÍSMO district, central AZORES, POR-TUGAL, on NE shore of TERCEIRA ISLAND, 10 mi/16 km NE of ANGRA DO HEROÍSMO; 38°45'N 27°06'W. Air base built during World War II serves military and commercial flights.

Lajes das Flores (LAH-zhish dahs FLO-resh), town, HORTA district, W AZORES, PORTUGAL, on S shore of FLORES ISLAND, 6 mi/9.7 km SSW of SANTA CRUZ; 39°22'N 31°11'W. Cattle raising; dairying. Formerly spelled Lages or Lagens das Flores.

Lajes do Pico (LAH-zhish dah PEE-koo), town, HORTA district, central AZORES, PORTUGAL, on S coast of PICO ISLAND, 22 mi/35 km SE of HORTA (on FAIAL ISLAND); 38°24'N 28°16'W. Dairying, grain milling. Formerly spelled Lagens do Pico.

Lajes Pintadas (LAH-zhish PEEN-tah-dahs), town (2007 population 4,253), E RIO GRANDE DO NORTE state, BRAZIL, 66 mi/106 km SW of NATAL; 06°09'S 36°07'W. Cotton, aloe; livestock.

Lajes, Ribeirão das, Brazil: see RIBEIRÃO DAS LAJES.

La Jigua (lah HEE-gwah), town, COPÁN department, NW HONDURAS, on paved road, 17 mi/28 km N of SANTA ROSA DE COPÁN; 15°02'N 88°48'W. Small farming, corn, beans.

Lajina (LAH-zheen-yah), city, E central MINAS GERAIS state, BRAZIL, near border with ESPÍRITO SANTO, 32 mi/51 km ENE of MANHUAÇU; 20°15'S 41°35'W.

Lajkovac (LEI-ko-vahts), village, central SERBIA, on KOLUBARA RIVER (railroad bridge), and 15 mi/24 km ENE of VALJEVO; 44°23'N 20°10'E. Also spelled Laikovats or Laykovats.

Lajma (LAH-mah), canton, CERCADO province, ORURO department, W central BOLIVIA; 17°54'S 67°40'W. Elevation 12,372 ft/3,771 m. Clay and dolomite deposits. Agriculture (potatoes, bananas, yucca, barley, oats); cattle.

La Jolla (lah HOI-yah), suburban section of SAN DIEGO, SAN DIEGO county, S CALIFORNIA, 12 mi/19 km NW of downtown San Diego. Manufacturing (measuring devices, cereals, medical instruments); printing and publishing. Ocean beaches, in particular La Jolla shores and Black's Beach, and sea-washed caves; tourism. The city has become a favorite retirement center as well as a prestigious residential and recreational area for San Diego professionals. The Scripps Institution of Oceanography to NE, the University of California at San Diego to NE, and La Jolla Museum of Contemporary Art; Salk Institute to NE. Torrey Pines State Reservoir to N; Mission Bay to S. Founded 1869.

Lajosmizse (LAH-yosh-mi-zhe), village, BÁCS-KISKUN county, central HUNGARY, 10 mi/16 km NNW of KECSKEMÉT; 47°01'N 19°33'E. Agriculture (grain, apricots, apples; hogs, cattle); manufacturing (metal and plastic packaging materials, baked goods, pasta).

La Joya (lah HOI-ah), town and canton, CERCADO province, ORURO department, W BOLIVIA, in the ALTIPLANO, on DESAGUADERO RIVER, and 30 mi/48 km NW of ORURO; 17°46'S 67°29'W. Elevation 12,542 ft/3,823 m. Potatoes; sheep.

La Joya (lah HOI-yah), town (2006 population 4,625), HIDALGO county, S TEXAS, 15 mi/24 km W of MCALLEN, on RIO GRANDE (Mexican border); 26°15'N 98°28'W. Rich irrigated agriculture area (citrus, vegetables, cotton). Bentsen-Rio Grande Valley State Park to SE.

Lajta River, AUSTRIA, HUNGARY: see LEITHA RIVER.

La Junta (lah HUHN-tah), town (2000 population 7,568), ⊙ OTERO county, SE COLORADO, on ARKANSAS RIVER, and 60 mi/97 km ESE of PUEBLO; 37°58'N 103°32'W. Elevation c.4,066 ft/1,239 m. Trade and railroad center, with repair shops, in grain and sugar-beet region; agriculture (vegetables, wheat, corn; cattle, poultry); manufacturing (consumer goods, food processing; publishing and printing, milling). Fort Bent Museum has fossils and scale model of BENT'S FORT, noted trading post that flourished nearby on the Arkansas River from 1820s to 1850s. Otero Junior College here. BENT'S OLD FORT NATIONAL HISTORIC SITE to NE; Comanche National Grassland to S. City founded 1875, incorporated 1881.

La Junta (lah HOON-tah), village, Limón province, E Costa Rica, former railroad junction on Reventazón River, and 2 mi/3.2 km NW of Siquirres. Bananas, cacao.

La Kafubu, CONGO: see KAFUBU.

Lakagígar, ICELAND: see LAKI.

Lakamané, town, FIRST REGION/KAYES, MALI, 96 mi/160 km E of KAYES; 14°31'N 09°55'W.

Lakamti, ETHIOPIA: see NEKEMTE.

Lakash (LAH-kahsh), village, central RYAZAN oblast, central European Russia, near the OKA River, on highway, 26 mi/42 km NE of SPASSK-RYAZANSKIY, to which it is administratively subordinate; 54°40'N 40°53'E. Elevation 311 ft/94 m. In agricultural area; winery.

Lakatizha Rita (lah-KAH-tee-zhah RE-tah), mountain ridge, near SAMOKOV, SW BULGARIA; 42°18'N 23°25'E.

Lakatnik (lah-KAHT-neek), village, SOFIA oblast, SVOGE obshtina, W BULGARIA, in the ISKUR River gorge, 26 mi/42 km N of SOFIA; 43°04'N 23°25'E. Livestock, sheep. Karstlike land formations and caves attract tourists. Scattered copper and zinc deposits nearby (E).

Lak Dera (lahk DAI-rah), seasonal river, 110 mi/177 km long, NORTH EASTERN province, KENYA; originates from the Lorian Swamp; flows E to the SOMALIA border. Also spelled as Lagh Dera.

Lake, county (□ 1,259 sq mi/3,273.4 sq km; 2006 population 65,933), NW CALIFORNIA; ⊙ LAKEPORT; 39°05'N 122°46'W. Mountain and valley region, in the COAST RANGES; drained by Cache Creek and Eel River. Oats, pears, grapes, walnuts; cattle; timber. Anderson Springs 15 mi/24 km S of CLEAR LAKE. Scenic recreational region; fishing, hunting, camping; mineral and hot springs (resorts). Mineral-water bottling; a leading quick-silver producing county of California; also sand and gravel quarrying. HULL MOUNTAIN (6,873 ft/2,095 m) on N boundary; part of Mendocino National Forest in N; Clear Lake State Park in center; Lake Pillsbury in N. Several small Indian reservations (or rancherias). Formed 1861.

Lake, county (□ 383 sq mi/995.8 sq km; 2006 population 7,814), central COLORADO; ⊙ LEADVILLE; 39°13'N 106°20'W. Mining, dairying, and livestock-grazing area, drained by headwaters of ARKANSAS RIVER. Gold, silver, lead, copper, zinc; molybdenum mines at CLIMAX. Large part of county in San Isabel National Forest, except for Arkansas Valley (NE); extreme NE corner in Arapaho National Forest. SAWATCH MOUNTAINS extend N-S; includes MOUNT MASSIVE (14,421 ft/4,396 m; second highest in U.S. ROCKY MOUNTAINS) and MOUNT ELBERT (14,433 ft/4,399 m; highest point in Colorado and in U.S. Rocky Mountains). TURQUOISE LAKE reservoir in NW. Formed 1861.

Lake (LAIK), county (□ 1,156 sq mi/3,005.6 sq km; 2006 population 290,435), central FLORIDA, bounded NE by ST. JOHNS RIVER; ⊙ TAVARES; 28°46'N 81°43'W. Rolling terrain with hundreds of lakes, including HARRIS, GRIFFIN, DORA, EUSTIS, Yale, and part of APOPKA. Ocala National Forest extends into NE corner. Citrus fruit-growing area, with canneries and many packing houses; also watermelons, corn, peanuts, cotton, and poultry. Formed 1887.

Lake, county (□ 457 sq mi/1,188.2 sq km; 2006 population 713,076), extreme NE ILLINOIS, bounded E by Lake MICHIGAN and N by WISCONSIN state line; ⊙ WAUKEGAN; 42°20'N 88°00'W. Drained by FOX and Des Plaines rivers. Includes many N suburbs of CHICAGO; diversified manufacturing; rapid urban growth area. Sand, gravel, stone deposits. Dairying, remnant agriculture. Two ski resorts, Illinois Beach State Park, and CHAIN O' Lakes State Park; fishing, duck hunting. Amusement park near ZION. Formed

1839. Nuclear power plants: ZION 1 (initial criticality June 19, 1973) and Zion 2 (initial criticality December 24, 1973) are 40 mi/64 km N of Chicago. Use cooling water from Lake Michigan, and each has a maximum dependable capacity of 1040 MWe. Major military presence at Great Lakes Naval Training Station and Philip Sheridan Arms Reserve Center.

Lake, county (□ 626 sq mi/1,627.6 sq km; 2006 population 494,202), extreme NW INDIANA; ☉ CROWN POINT; 41°29′N 87°23′W. Bounded N by Lake MICHIGAN, W by ILLINOIS state line, S by KANKAKEE RIVER; traversed by Grand CALUMET and Little Calumet rivers. Heavily industrialized CALUMET region, part of metropolitan area of CHICAGO, is one of world's most important steel-manufacturing centers. Harbors at GARY and EAST CHICAGO (Indiana Harbor). Agricultural areas of county produce vegetables, dairy products, poultry, soybeans, corn. LaSalle State Fish and Wildlife Area in SW corner. Formed 1836.

Lake, county (□ 574 sq mi/1,492.4 sq km; 2006 population 11,793), W central MICHIGAN; ☉ BALDWIN; 43°59′N 85°48′W. Cattle, dairy products, forage crops; lumbering; resort and recreation area. Drained by PERE MARQUETTE and LITTLE MANISTEE rivers, and small South Branch of MANISTEE RIVER. Part of Manistee National Forest in W and S parts of county and NE corner; numerous lakes, especially NW and SW; Wood Hills Ski Area in W. Organized 1871.

Lake, county (□ 2,991 sq mi/7,776.6 sq km; 2006 population 10,966), NE MINNESOTA; ☉ TWO HARBORS; 47°38′N 91°25′W. Extensively watered area; bounded SE by Lake SUPERIOR; N by chain of lakes (W-E: CROOKED LAKE, Basswood River, BASSWOOD LAKE, KNIFE LAKE, and Ottertrack Lake, and other small lakes); along CANADA (ONTARIO) border; S half of county is drained by small streams, which feed into Lake Superior, N half by many lakes, linked by rivers, notably KAWISHIWI, Island, and CLOQUET (source) rivers. Timber; some dairying. Part of Superior National Forest (including part of BOUNDARY WATERS Canoe Area) in N of county and is part of famous recreational region known as "Arrowhead Country." (Refers to triangular shape of NE Minnesota, area also called "North Shore"). State parks (N-S) on or near Lake Superior shore include Caribou Falls State Park and George H. Crosby Manitou State Park (in E), Tettegouche State Park, Split Rock Lighthouse State Park, and Gooseberry Falls State Park (in SE), Flood Bay State Park, at TWO RIVERS (in S); Finland State Forest in E. Formed 1866.

Lake, county (□ 1,653 sq mi/4,281 sq km; 1990 population 21,041; 2000 population 26,507), NW MONTANA; ☉ POLSON; 47°39′N 114°05′W. Mountain region drained by FLATHEAD (forms part of W boundary) and SWAN rivers. Wheat, barley, corn, oats, potatoes, rape seed, hay; cattle, sheep, hogs, horses, llamas, fruit (cherries); dairying. Five of six units of FLATHEAD LAKE STATE PARK in N, along shore of FLATHEAD LAKE: West Shore Unit, Wild Horse Island Unit, Big Arm Unit, Finley Point Unit, and Yellow Bay Unit (E shore). Lake Mary Ronan State Park in NW; Ninepipe and Pablo National Wildlife refuges in center; Swan River National Wildlife Refuge and Swan Lake in NE; part of National Bison Range in SW; part of Mission Mountains Tribal Wilderness in SE. Flathead Indian Reservation in all but N and NE margin of county. Flathead Lake in N. Formed 1923.

Lake (LAIK), county (□ 232 sq mi/603.2 sq km; 2006 population 232,892), NE OHIO; ☉ PAINESVILLE; 41°42′N 81°15′W. Bounded N by LAKE ERIE; drained by GRAND and CHAGRIN rivers. In the Till Plains and Lake Plains physiographic regions. Smallest county in Ohio. Fruit growing (apples, grapes); poultry. Manufacturing at Painesville, WICKLIFFE, and WILLOUGHBY (paper products, industrial organic chemicals, iron and steel foundries, precision-measuring devices). Lake resorts. Formed 1840. Perry

nuclear power plant (7 mi/11 km NE of Painesville), initial criticality June 6, 1986, uses cooling water from Lake Erie, and has a maximum dependable capacity of 1166 MWe.

Lake, county (□ 8,359 sq mi/21,733.4 sq km; 2006 population 7,473), S OREGON, mountainous area containing SUMMER LAKE (W center), LAKE ABERT in center, Silver Lake and part of GOOSE LAKE (S boundary), borders CALIFORNIA and NEVADA; ☉ LAKEVIEW; 42°47′N 120°23′W. Wheat, oats, barley; sheep, cattle. Clay; lumber. Hart Mountain National Antelope Refuge in SE. Warm Valley, a geyser basin with numerous dry lakes in SE (including Old Perpetual Geyser that spouts every ninety seconds). Parts of Fremont National Forest in W; part of GREAT SANDY DESERT in NE; Fort Rock State Monument in NW; Chandler and Booth State Waysides in S; Goose Lake State Recreation Area on S boundary.

Lake, county (□ 575 sq mi/1,495 sq km; 2006 population 11,170), E South Dakota; ☉ MADISON, 44°02′N 97°09′W. LAKE MADISON in SE, Lake Herman State Park in S. Agriculture (corn, soybeans; dairy products); cattle, hogs, poultry; honey. Drained by East Fork of VERMILLION RIVER. Formed 1873.

Lake, county (□ 164 sq mi/426.4 sq km; 2006 population 7,406), extreme NW TENNESSEE; ☉ TIPTONVILLE; 36°20′N 89°30′W. Bounded N by KENTUCKY; W by the MISSISSIPPI (MISSOURI state line); NE by REELFOOT LAKE (hunting, fishing). Timber; agriculture; some manufacturing. Formed 1870.

Lake, village (2000 population 408), SCOTT and NEWTON counties, central MISSISSIPPI, 35 mi/56 km W of MERIDIAN; 32°20′N 89°19′W. Agriculture (cotton, corn; poultry, cattle; dairying); manufacturing (apparel). Bienville National Forest to W and S.

Lake Alfred (AL-fred), town (□ 6 sq mi/15.6 sq km; 2000 population 3,890), POLK county, central FLORIDA, 6 mi/9.7 km W of HAINES CITY, in lake region; 28°05′N 81°43′W. Citrus fruit-shipping center, with packing houses, canneries, and experiment station; manufacturing (citrus oil, pulp feed, and molasses; fertilizer, insecticides).

Lake Almanor Dam, California: see ALMANOR, LAKE.

Lake Alpine, California: see ALPINE, LAKE.

Lake Aluma (uh-LOOM-uh), village (2006 population 93), OKLAHOMA county, central OKLAHOMA, residential suburb 7 mi/11.3 km NE of downtown OKLAHOMA CITY; 35°32′N 97°27′W.

Lake Andes, city (2006 population 784), ☉ CHARLES MIX county, S SOUTH DAKOTA, 47 mi/76 km SW of MITCHELL, near LAKE ANDES, in Yankton Indian Reservation; 43°09′N 98°32′W. Lake Andes National Wildlife Refuge at E end of lake. Fishing resort; trading point for agricultural area; wheat, alfalfa, livestock.

Lake and Peninsula, borough (□ 23,632 sq mi/61,443.2 sq km; 2006 population 1,548), SW ALASKA, includes the base of the ALASKA PENINSULA and several large lakes (Lake CLARK, ILIAMNA Lake, part of NAKNEK LAKE, BECHAROF LAKE, and Upper and Lower Ugaskik Lakes). Bounded on the NW by BRISTOL BAY, SE by PACIFIC OCEAN, and SHELIKOF STRAIT. Part of LAKE CLARK National Park and Preserve and KATMAI NATIONAL PARK AND PRESERVE to N; Becharof National Wildlife Reserve in center and S; ANIAKCHAK NATIONAL MONUMENT AND PRESERVE to SW. Fishing; crabs. Seals; pelts.

Lake Angelus, village (2000 population 326), OAKLAND county, SE MICHIGAN, residential suburb 3 mi/4.8 km NW of PONTIAC, suburb surrounds small Lake Angelus; 42°41′N 83°19′W. Numerous lakes in area.

Lake Ann, village (2000 population 276), BENZIE county, NW MICHIGAN, on NE end of Ann Lake, 12 mi/19 km SW of TRAVERSE CITY; 44°43′N 85°50′W.

Lake Argyle Village (LAIK AHR-geil), town, VICTORIA, SE AUSTRALIA, just W of NORTHERN TERRITORY border. Adjacent large dam.

Lake Ariel, unincorporated town, WAYNE county, NE PENNSYLVANIA, 16 mi/26 km ENE of SCRANTON; 41°27′N 75°22′W. Manufacturing (sand and gravel processing, poultry processing, asphalt, industrial power brushes, wood products, printing, lumber, machinery); agriculture (dairying, livestock; grain); timber.

Lake Arissa, ETHIOPIA: see AFAMBO, LAKE.

Lake Arrowhead, unincorporated town (2000 population 8,934), SAN BERNARDINO county, S CALIFORNIA; residential suburb 11 mi/18 km NE of SAN BERNARDINO; 34°16′N 117°11′W. Surrounds LAKE ARROWHEAD reservoir, in San Bernardino Mountains and San Bernardino National Forest. Rim of the World Drive overlooks Los Angeles area. Manufacturing (printing and publishing).

Lake Arrowhead, California: see ARROWHEAD, LAKE.

Lake Arthur, town (2000 population 3,007), JEFFERSON DAVIS parish, SW LOUISIANA, on LAKE ARTHUR (widening of navigable MERMENTAU RIVER), 34 mi/55 km ESE of LAKE CHARLES city; 30°05′N 92°41′W. In rice and cattle area; also catfish, crawfish; oil and natural-gas fields. Recreation in vicinity. Incorporated 1909.

Lake Arthur, village (2006 population 437), CHAVES county, SE NEW MEXICO, on PECOS RIVER, and 32 mi/51 km SSE of ROSWELL; 33°00′N 104°21′W. Cattle, sheep, dairying; agriculture (triticale, alfalfa, pecans, cotton, melons, hay).

Lake Atescatempa (ah-tes-kah-TEM-pah), lake (□ 2 sq mi/5.2 sq km), JUTIAPA department, GUATEMALA, near ATESCATEMPA and SAN CRISTÓBAL FRONTERA; 14°12′N 89°42′W. On INTER-AMERICAN HIGHWAY. No outlet.

Lakeba (lah-KEM-bah), composite volcanic and limestone island (□ 21 sq mi/54.6 sq km), Lau group, FIJI, SW PACIFIC OCEAN; 5 mi/8 km long. Central mountain range rises to 720 ft/219 m; fertile coast. Site (1835) of 1st missionary settlement in Fiji. Sometimes spelled Lakemba.

Lake Babadag, ROMANIA: see BABADAG, LAKE.

Lake Barcroft (LAIK BAHR-krahft), unincorporated town, FAIRFAX county, NE VIRGINIA, residential suburb 7 mi/11 km WSW of WASHINGTON, D.C., 6 mi/10 km NW of ALEXANDRIA, centered on Lake Barcroft reservoir (Holmes Run creek); 38°51′N 77°09′W.

Lake Barrington, village (2000 population 4,757), LAKE county, NE ILLINOIS, residential suburb 33 mi/53 km NW of downtown CHICAGO, 7 mi/11.3 km ESE of CRYSTAL LAKE (town); 42°12′N 88°10′W. Manufacturing (printing equipment).

Lake Bathurst (LAIK BA-thurst), village, NEW SOUTH WALES, SE AUSTRALIA, 138 mi/222 km SW of SYDNEY, 20 mi/32 km S of GOULBURN; 35°01′N 149°39′E. Lake of same name near village. Lavender gardens.

Lake Bato, PHILIPPINES: see BATO, LAKE.

Lake Bellaire, MICHIGAN: see BELLAIRE.

Lake Benton (village (2000 population 703), LINCOLN county, SW MINNESOTA, 15 mi/24 km S of IVANHOE, near SOUTH DAKOTA state line, at SW end of Lake BENTON; 44°15′N 96°17′W. Agriculture (grain, soybeans; poultry; dairying); manufacturing (printing, animal feeds). Resort area.

Lake Beulah (BYOO-lah), village, WALWORTH county, SE WISCONSIN, E of small Lake Beulah (c.4 mi/6.4 km long), 10 mi/16 km NNW of BURLINGTON.

Lake Bluff, village (2000 population 6,056), Lake county, extreme NE ILLINOIS, N residential suburb of CHICAGO, on Lake MICHIGAN, 17 mi/27 km NNW of EVANSTON; 42°16′N 87°50′W. Manufacturing (steel and nonferrous wool, industrial chemicals, metallurgical equipment, furniture, paper products, freight car components). GREAT LAKES NAVAL TRAINING STATION is just N. Incorporated 1895.

Lake Bolac, settlement, VICTORIA, SE AUSTRALIA, 62 mi/100 km W of BALLARAT, and on shores of freshwater Lake Bolac (□ 6 sq mi/15 sq km); 37°42′S

142°51′E. Agriculture. Recreational fishing; annual yachting regatta. GRAMPIANS NATIONAL PARK 37 mi/60 km N.

Lake Bombon, PHILIPPINES: see TAAL, LAKE.

Lake Bonaparte, NEW YORK: see BONAPARTE, LAKE.

Lake Borgne Canal (born), SE LOUISIANA, joins Lake BORGNE with the MISSISSIPPI RIVER at VIOLET (lock here), 10 mi/16 km ESE of NEW ORLEANS; c. 7 mi/11 km long. Eliminated by MISSISSIPPI RIVER GULF OUTLET, which parallels the lake's SW shore, linking INTRACOASTAL WATERWAY (NW) with Breton Sound, Gulf of MEXICO (SE). Also known as Ship Island Canal.

Lake Bronson, village (2000 population 246), KITTSON county, NW MINNESOTA, 13 mi/21 km ESE of HALLOCK, on South Branch TWO RIVERS; 48°43′N 96°39′W. Agriculture (grain, potatoes, sugar beets, beans, flax, sunflowers; livestock); manufacturing (fertilizer). Twin Lakes Wildlife Area to SE; Lake Bronson (State Park) to E.

Lake Buena Vista (BWAI-nuh VIS-tuh), town (□ 5 sq mi/13 sq km), ORANGE county, central FLORIDA, 15 mi/24 km SW of ORLANDO; 28°23′N 81°31′W. Tourist center located adjacent to WALT DISNEY WORLD.

Lake Buhi, PHILIPPINES: see BUHI, LAKE.

Lake Butler (BUHT-luhr), town (□ 1 sq mi/2.6 sq km; 2000 population 1,927), ⊙ UNION county, N central FLORIDA, 25 mi/40 km N of GAINESVILLE; 30°01′N 82°20′W.

Lake Cargelligo (kahr-JE-li-go), town, S central NEW SOUTH WALES, SE AUSTRALIA, on LACHLAN RIVER, and 210 mi/338 km NW of CANBERRA, near Lake Cargelligo; 3 mi/5 km long, 1.5 mi/2.4 km wide; 33°18′S 146°23′E. Railroad terminus; sheep and agriculture center. Lake has abundant birdlife.

Lake Carmel, village (□ 5 sq mi/13 sq km; 2000 population 8,663), PUTNAM county, SE NEW YORK; 41°27′N 73°40′W. On small artificial Lake CARMEL.

Lake Catherine, unincorporated village (2000 population 1,490), Lake county, NE ILLINOIS, suburb 49 mi/79 km NW of downtown CHICAGO, 1 mi/1.6 km N of ANTIOCH, 1 mi/1.6 km S of WISCONSIN state line; 42°29′N 88°07′W. On narrow isthmus between Lake Catherine (N) and Bluff Lake (S).

Lake Charles, city (2000 population 71,757), ⊙ CALCASIEU parish, SW LOUISIANA, 60 mi/97 km W of LAFAYETTE; 30°13′N 93°12′W. Railroad junction. Located on Lake CHARLES, a widening of the CALCASIEU RIVER, in a rice, timber, oil, and natural-gas region. Fishing (crawfish, shrimp, crabs); manufacturing (machinery, petroleum products, chemicals, concrete, transportation equipment, food products, fiberglass pipe lining, oil field equipment; barge and tugboat construction); printing and publishing; casinos. Lake Charles is an important ship and barge port and port of entry, connected with Gulf of MEXICO and INTRACOASTAL WATERWAY by Calcasieu Ship Channel. A 30-mi/48-km-long channel connects it with the Gulf of MEXICO. Petroleum products, chemicals, rice, and cotton shipped from the port. Cajun Music and Food Festival and Marshland Festival held here. Historical district has Victorian and Queen Anne houses. McNeese State University; Creole Nature Trail. Sam Houston Jones State Park to N. Incorporated 1867.

Lake Charles Canal, c. 25 mi/40 km long, SW LOUISIANA, deepwater E-W land cut connecting SABINE River just below ORANGE (TEXAS) with CALCASIEU RIVER c. 15 mi/24 km below LAKE CHARLES city; section of INTRACOASTAL WATERWAY.

Lake Chelan Dam, WASHINGTON: see CHELAN, LAKE.

Lake Chelan National Recreation Area (shuh-LAN), N WASHINGTON, Stehekin Valley, and in N part of fjord-like LAKE CHELAN; covers 61,883 acres/25,044 ha. Bounded by Wenatchee National Forest on W and S, by Okanogan National Forest on E (Lake Chelan–Sawtooth Wilderness Area, E). NORTH CASCADES NATIONAL PARK (S section) adjoins recreation area to NW. Road access from end of lake; passenger ferry from town of CHELAN at SE end of lake (55 mi/89 km). Pacific Coast Trail passes through park's W corner; Rainbow Falls on Stenekin River above its entrance to Lake Chelan. Authorized 1968.

Lake Chivero Recreational Park, ZIMBABWE: see CHIVERO, LAKE.

Lake City, city (2000 population 1,787), CALHOUN county, central IOWA, 10 mi/16 km SSW of ROCKWELL CITY; 42°16′N 94°43′W. Manufacturing (pipe organs, fiberglass products). Sand and gravel pits nearby. Incorporated 1887.

Lake City, town (2000 population 1,956), shares capital functions with JONESBORO, CRAIGHEAD county, NE ARKANSAS, 12 mi/19 km E of Jonesboro, on SAINT FRANCIS River; 35°49′N 90°27′W. In area of sunken lands caused by 1811–1812 earthquakes. Saint Francis Sunken Lands Wildlife Management Area on river.

Lake City (LAIK), town (□ 11 sq mi/28.6 sq km; 2000 population 9,980), ⊙ COLUMBIA county, N central FLORIDA; 30°11′N 82°38′W. It was founded in the 1830s as a military post. LAKE CITY is located in a farm and cattle area and produces tobacco, lumber products. The city also has airplane repair centers and is the headquarters to Ocala National Forest. Key junction point in N Florida where I-75 and I-10 intersect. Incorporated 1921.

Lake City, town (2000 population 4,950), WABASHA county, SE MINNESOTA, 15 mi/24 km SE of RED WING on Lake PEPIN (natural lake on MISSISSIPPI RIVER between mouths of Miller and Gilbert creeks); 44°27′N 92°16′W. Milling point; agriculture (grain; poultry; livestock; dairying); manufacturing (chemical processing, wind tunnels, food processing, circuit boards, hand trucks, consumer goods; printing). Hydraulics Research Facility. Part of Richard J. Dorer Memorial Hardwood State Forest to NW; Frontenac State Park to NW; Upper Mississippi River National Wildlife Refuge along river margins. Incorporated 1872.

Lake City, town (2006 population 6,666), FLORENCE county, E central SOUTH CAROLINA, 25 mi/40 km S of FLORENCE; 33°52′N 79°45′W. Manufacturing (consumer goods, charcoal, textiles, apparel, building materials); agriculture (cotton, grain, vegetables; poultry, hogs).

Lake City, town (2006 population 1,859), ANDERSON county, E TENNESSEE, 21 mi/34 km NW of KNOXVILLE, near NORRIS Reservoir; 36°13′N 84°09′W. In coal-mining region; manufacturing; small business enterprises. Until 1939, called COAL CREEK.

Lake City, village (2000 population 375), ⊙ HINSDALE county, SW central COLORADO, on Lake Fork of GUNNISON RIVER, at mouth of Henson Creek, in SAN JUAN MOUNTAINS, and 45 mi/72 km SE of MONTROSE; 38°01′N 107°18′W. Elevation 8,671 ft/2,643 m. UNCOMPAHGRE PEAK 8 mi/12.9 km WNW. Lake San Cristobal reservoir to S (on Lake Fork); Uncompahgre National Forest to NW; Gunnison National Forest to E and S; Rio Grand National Forest to S, beyond CONTINENTAL DIVIDE.

Lake City, village (2000 population 923), ⊙ MISSAUKEE county, N central MICHIGAN, 11 mi/18 km NE of CADILLAC, on E side of LAKE MISSAUKEE (c.3 mi/4.8 km long, 2 mi/3.2 km wide); 44°19′N 85°12′W. In resort and farm area. Manufacturing (plastic products, steel die forgings, lumber). Native American earthworks and a state park are nearby. Incorporated in 1932.

Lake City, village (2006 population 44), MARSHALL county, NE SOUTH DAKOTA, 17 mi/27 km ESE of BRITTON, just W of Lake Traverse (Sisseton Wahpeton) Indian Reservation; 45°43′N 97°24′W. Small lakes in area; supply point for hunters and fishermen. Roy Lake State Park to SW; Clear Lake State Lakeside Use Area to SE.

Lake City, borough (2006 population 2,963), ERIE county, NW PENNSYLVANIA, 16 mi/26 km WSW of ERIE, LAKE ERIE 1 mi/1.6 km to NW; 42°01′N 80°20′W. Manufacturing (transportation equipment, machinery, steel molds, fabricated metal products, food products, wood products); agriculture (dairying; livestock). Borough formally called North Girard.

Lake City, suburb (2000 population 2,886) of ATLANTA, CLAYTON county, GEORGIA, 1 mi/1.6 km SW of FOREST PARK; 33°37′N 84°20′W. Manufacturing includes signs, tool and die, commercial printing, feeds.

Lake Clark Park and Preserve, park (□ 2,735 sq mi/7,111 sq km) and preserve (1,407,293 acres/569,531 ha), S ALASKA. Covers 2,636,839 acres/1,067,129 ha. Authorized 1980, proclaimed Lake Clark National Monument 1978. Waterfalls, tundra, and active volcanoes; grizzly and black bears, caribou, wolves, mink. Chigmit Range, red salmon spawning area. REDOUBT Volcano, 10,197 ft/3,108 m.

Lake Clear, resort village, FRANKLIN county, NE NEW YORK, on Lake Clear in the ADIRONDACK MOUNTAINS, 5 mi/8 km NW of SARANAC LAKE village; c.2 mi/3.2 km long; 44°22′N 74°14′W.

Lake Constance, German *Bodensee* (BO-duhn-zai), region (□ 122 sq mi/317.2 sq km), BADEN-WÜRTTEMBERG, S GERMANY. Bordered by SWITZERLAND on south. Chief town is KONSTANZ. RHINE flows through it. Specialized agriculture along the shore owing to mild climate.

Lake Constance, German *Bodensee*, French *Lac de Constance*, lake (□ 208 sq mi/540.8 sq km), bordering on SWITZERLAND, GERMANY, and AUSTRIA; 42 mi/68 km long; maximum depth of 827 ft/252 m. Elevation 1,299 ft/396 m. The lake is fed and drained by the RHINE RIVER and divides into two arms, UNTERSEE and Überlinger See, near the city of Constance. The Rhine flows through the main body of the lake, called the Obersee, and the Untersee, leaving the lake at STEIN AM RHEIN. Fruit is grown on the lakeshore, and wine making and fishing are major industries. The chief towns and cities of the lake are Constance (German *Konstanz*), FRIEDRICHSHAFEN, and LINDAU, all in Germany; BREGENZ in Austria; and RORSCHACH, ROMANSHORN, ARBON, and Kreutzlingen in Switzerland. Remains of lake dwellings have been found. Marbach Castle, on the shore, is used as a conference center.

Lake Constance–Upper Swabia, German *Bodensee-Oberschwaben* (BO-duhn-zai–O-buhr-shwah-buhn), region (□ 1,352 sq mi/3,515.2 sq km), BADEN-WÜRTTEMBERG, S GERMANY. Chief town is RAVENSBURG.

Lake Country (LAIK kuhn-tree), district municipality (□ 47 sq mi/122.2 sq km; 2001 population 9,267), S BRITISH COLUMBIA, W CANADA, in CENTRAL OKANAGAN regional district; 50°06′N 119°22′W. Fruit growing and packing (Macintosh apples, blueberries). Includes the communities of Oyama, Winfield, Carr's Landing, and Okanagan Centre. Incorporated in 1995. Once inhabited by First Nations people.

Lake Cowichan (KOU-ich-uhn), village (□ 3 sq mi/7.8 sq km; 2001 population 2,827), SW BRITISH COLUMBIA, W CANADA, on S VANCOUVER ISLAND, on COWICHAN LAKE at outlet of Cowichan River, 40 mi/64 km NW of VICTORIA, and in COWICHAN VALLEY regional district; 48°49′N 124°02′W. In lumbering region. Incorporated 1944.

Lake Crystal, town (2000 population 2,420), BLUE EARTH county, S MINNESOTA, 12 mi/19 km SW of MANKATO; 44°06′N 94°13′W. Resort; agricultural trading point (corn, oats, soybeans, peas, hogs; sheep, cattle); manufacturing (fertilizers, dump bodies, metal fabricating). Minneopa State Park to NE; Lake Crystal and Loon Lake to E, Lily Lake to N. Plotted 1857; incorporated as village 1870; as city 1930.

Lake Dallas (DAL-uhs), town (2006 population 7,261), DENTON county, N TEXAS, suburb 27 mi/43 km NW of downtown DALLAS, and 30 mi/48 km NE of downtown FORT WORTH, located on W shore of LEWISVILLE LAKE reservoir (Elm Fork, TRINITY RIVER); 33°07′N

97°01′W. Agricultural area (cotton, peanuts; cattle); manufacturing (electronic assembly).

Lake Débo (DAI-bo), lake, FIFTH REGION (MOPTI), 54 mi/90 km N of MOPTI; 15°18′N 04°09′W. Located in frequently inundated inland delta of NIGER RIVER.

Lake Delton, town (2006 population 2,835), SAUK county, S central WISCONSIN, 9 mi/14.5 km NNW of BARABOO, on LAKE DELTON, near the WISCONSIN DELLS; 43°35′N 89°46′W. Manufacturing (metal fabricating); resort area; tourism. Rocky Arbor State Park to N, Mirror Lake State Park to SW.

Lake District (LAIK DIS-trikt), region of mountains and lakes, c.30 mi/48 km in diameter, CUMBRIA, NW ENGLAND. It includes the CUMBRIAN MOUNTAINS and part of the FURNESS peninsula. The Lake District contains over fifteen lakes, among them Ullswater, Windermere, Derwent Water, and BASSENTHWAITE LAKE; several beautiful falls; and some of England's highest peaks—Scafell Pike (3,210 ft/978 m), Scafell, and HELVELLYN. Many of the region's valleys were deforested following Roman and Norse invasions. Numerous ancient relics remain, such as the stone circle near Keswick and the ruins of old castles and churches. This scenic district is a favorite resort of artists and writers. William Wordsworth, Samuel Taylor Coleridge, and Robert Southey were known as the Lake Poets. Herdwick sheep, a local breed, are raised. Tourism is a major source of income. The chief tourist centers are KESWICK, Kendal, Penrith, Bowness-on-Windermere, and Cockermouth. The Forestry Commission has actively planted pine trees in the Lake District National Park (☐ c.125 sq mi/324 sq km) established in 1951.

Lake Elmo, town (2000 population 6,863), WASHINGTON county, E MINNESOTA, residential suburb 9 mi/14.5 km E of downtown ST. PAUL; 45°00′N 92°54′W. Corn, soybeans; cattle, sheep; manufacturing (wood products, sand and gravel, wood pallets, wood moldings). Lake Elmo Airport to E. Lake Elmo Regional Park at center of city. Lake St. Croix (formed by ST. CROIX River) to E, several small lakes in city, notably Eagle Point Lake at center, Olson Lake and Lake De Montreville in NW.

Lake Elmore, VERMONT: see ELMORE.

Lake Elsinore, city (2000 population 28,928), SW RIVERSIDE county, S CALIFORNIA; suburb 22 mi/35 km S of RIVERSIDE; 33°40′N 117°19′W. In rapidly growing area. Resort on NE shore of Lake Elsinore reservoir (c.6 mi/9.7 km long), near hot mineral springs. Cattle, poultry; grain; manufacturing (nuts and bolts, concrete, bricks); clay quarrying. Part of Cleveland National Forest to SW, Lake Elsinore State Recreation Area to W; Railroad Canyon Reservoir to E. Incorporated 1888.

Lake Eyre National Park (LAIK ER) (☐ 4,740 sq mi/ 12,324 sq km), NE SOUTH AUSTRALIA state, S central AUSTRALIA, 260 mi/418 km N of PORT AUGUSTA; 120 mi/193 km long, 100 mi/161 km wide; 28°30′S 137°30′E. Large desert wilderness surrounding and including Lake EYRE, its S extension, and a large areas E of the lake. The lake fills with water an average of twice every 100 years; when it does, it teems with pelicans, gulls, cormorants, egrets, other water birds. Also, dormant fish and amphibians come to life. Red kangaroos, emus, falcons, lizards, snakes. Primitive camping. No facilities. Established 1986.

Lake Faguibine (fah-GEE-been), lake, SIXTH REGION/ TIMBUKTU, central MALI, 54 mi/90 km W of TIMBUKTU; 16°45′N 03°54′W. Located on N fringe of the inland delta of NIGER RIVER.

Lakefield, town (2000 population 1,721), JACKSON county, SW MINNESOTA, 10 mi/16 km WNW of JACKSON, at SE tip of South HERON LAKE, near IOWA state line; 43°40′N 95°10′W. Agricultural trading point (grain; livestock); manufacturing (gas fireplaces). Kilen Woods State Park to NE. Settled 1879; incorporated 1887.

Lakefield (LAIK-feeld), former village (☐ 1 sq mi/2.6 sq km; 2001 population 2,612), S ONTARIO, E central CANADA, on small Katchiwano Lake, at mouth of Otonabee River, 9 mi/14 km NNE of PETERBOROUGH; 44°26′N 78°16′W. Dairying; woodworking; resort. Amalgamated into SMITH-ENNISMORE-LAKEFIELD township in 2001.

Lakefield National Park (LAIK-feeld) (☐ 2,073 sq mi/ 5,389.8 sq km), N QUEENSLAND, NE AUSTRALIA, 240 mi/386 km NNW of CAIRNS, on CAPE YORK PENINSULA, E side; 83 mi/134 km long, 60 mi/97 km wide; 15°00′S 144°05′E. Touches PACIFIC OCEAN on N at Queen Charlotte Bay. Large area of open woodland interspersed with rainforest pockets. Large rivers and billabongs provide haven for plants and wildlife during long dry periods. Cyrus cranes, brolga cranes, galahs, magpie geese, jabiurs; freshwater and saltwater crocodiles. Tall, fan-shaped termite mounds. Cattle-range areas are being reverted back to their natural state. Camping, picnicking, fishing. Established 1979.

Lake Forest, city (2000 population 58,707), ORANGE county, S CALIFORNIA; residential suburb 40 mi/64 km SE of downtown LOS ANGELES, near Aliso Creek; 33°39′N 117°41′W. Irrigated agriculture. U.S. Marine Corps Air Station (to NW) closed 1999. SANTA ANA MOUNTAINS and Cleveland National Forest to NE. Previously called El Toro; the name changed when the city incorporated in 1991.

Lake Forest, city (2000 population 20,059), Lake county, NE ILLINOIS, residential suburb of CHICAGO, on Lake MICHIGAN; 42°14′N 87°51′W. The city is known for its scenic lakefront and impressive estates. It is the seat of Lake Forest College, Barat College, and two preparatory schools. Settled 1835, incorporated 1861.

Lake Forest, unincorporated town, GREENVILLE county, NW SOUTH CAROLINA, residential suburb 3 mi/4.8 km ENE of downtown GREENVILLE. Greenville Downtown Airport to S. Bob Jones University to NW.

Lake Forest, unincorporated village, PLACER county, E CALIFORNIA, 2 mi/3.2 km NE of TAHOE CITY, on NW shore of LAKE TAHOE. Tourism; manufacturing (industrial instruments, computer peripherals, sand and gravel processing).

Lake Forest Park, town (2006 population 12,548), KING county, W WASHINGTON, residential suburb 11 mi/18 km NNE of downtown SEATTLE, at N end of LAKE WASHINGTON, at mouth of McAlee Creek; 47°46′N 122°17′W.

Lake Fork, stream, c.50 mi/80 km long; rising in UINTA Mountains near TOKEWANNA PEAK, N DUCHESNE county, NE UTAH, in High Uintas Wilderness Area of Ashley National Forest; flows SE through Mona Lake reservoir past ALTAMONT to DUCHESNE RIVER, 8 mi/ 12.9 km SW of ROOSEVELT. MOON LAKE DAM (110 ft/34 m high, 1,108 ft/338 m long; completed 1938) is chief unit in Moon Lake irrigation project, which supplies water to c.70,000 acres/28,330 ha in NE Utah.

Lake Fork Reservoir (LAIK FORK), Rains, Wood, and Hopkins counties, NE TEXAS, on Lake Fork Creek, 15 mi/24 km S of SULPHUR SPRINGS; 32°48′N 95°32′W. Maximum capacity of 1,048,480 acre-ft. Formed by Lake Fork Dam (82 ft/25 m high), built (1980) for water supply; also used for recreation.

Lake Fremont, MINNESOTA: see ZIMMERMAN.

Lake Gamarri, ETHIOPIA: see GEMERI, LAKE.

Lake Geneva (juh-NEE-vah), town (2006 population 8,155), WALWORTH county, SE WISCONSIN, on NE shore of LAKE GENEVA at its outlet (WHITE RIVER), 38 mi/61 km SW of MILWAUKEE, and 55 mi/89 km NW of CHICAGO; 42°35′N 88°25′W. Recreation and resort area for both cities; year-round lake sports; manufacturing (consumer goods, dies and stampings, metal spinnings, transportation equipment, machinery, custom seals and moldings; screen printing). Has many estates of residents of Chicago, and a hotel designed by Frank Lloyd Wright. Bird sanctuary of University of Chicago and Big Foot Beach State Park is to S. Settled before 1845, incorporated 1883.

Lake George, village (2006 population 990), ☉ WARREN county, E NEW YORK, on S tip of Lake GEORGE in the foothills of the ADIRONDACK MOUNTAINS; 43°25′N 73°43′W. It is a year-round tourist and sports center. Vestiges of FORT WILLIAM HENRY, built by Sir William Johnson, and Fort George are in the village. Incorporated 1903.

Lake Grace (LAIK GRAIS), township, WESTERN AUSTRALIA state, W AUSTRALIA, 219 mi/353 km SE of PERTH; 33°06′S 118°23′E. Wheat.

Lake Grove, village (☐ 2 sq mi/5.2 sq km; 2006 population 10,621), SUFFOLK county, SE NEW YORK, 7 mi/ 11.3 km NE of central ISLIP, 25 mi/40 km WSW of RIVERHEAD; 40°51′N 73°07′W.

Lake Hamilton (HAM-uhl-tuhn), town (2000 population 1,304), POLK county, central FLORIDA, 4 mi/ 6.4 km S of HAINES CITY, on LAKE HAMILTON; 2 mi/ 3.2 km long; 28°03′N 81°37′W. Citrus-fruit-packing houses.

Lake Harbour or **Kimmirut**, trading post, S BAFFIN ISLAND, BAFFIN region, E NUNAVUT territory, N CANADA, at head of North Bay, inlet of HUDSON STRAIT; 62°15′N 69°53′W. Manufacturing (crafts; greenstone sculpting); fishing, sea mammal hunting. Scheduled air service. Royal Canadian Mounted Police post. Site of Anglican mission. Fur trading post was established 1911.

Lake Hauroko, SE corner of FIORDLAND NATIONAL PARK, a lake of glacial origin, NEW ZEALAND; 1,516 ft/ 462 m deep. The deepest lake in New Zealand.

Lake Havasu City, city (2000 population 41,938), MOHAVE county, W ARIZONA, 145 mi/233 km WNW of PHOENIX, and 51 mi/82 km SSW of KINGMAN, on COLORADO RIVER (Lake Havasu, formed by PARKER DAM 18 mi/29 km SE); 34°30′N 114°18′W. Elevation 482 ft/147 m. Area referred to as Arizona's "West Coast." Manufacturing (stationery, wood millwork, cotton finishing, electrical apparatus, concrete blocks, machine tools; boat building, printing and publishing). Warm Springs Wilderness Area to N. Separate sections of Havasu National Wildlife Refuge to N and SE; MOHAVE MOUNTAINS to NE. Chemehuevi Indian Reservation across lake (W) in CALIFORNIA. Original name Site Six. Planned retirement and recreational community developed in late 1960s. Population increased six times by 1990. London Bridge purchased and moved here 1969–1972, spans artificial lake (channel of Lake Havasu called Little Thames River) at English village theme area. Lake Havasu State Park to S.

Lake Heights, unincorporated town, KING county, W WASHINGTON, residential suburb 8 mi/12.9 km SE of downtown SEATTLE, on E shore of LAKE WASHINGTON, on East Channel, separating MERCER ISLAND from mainland.

Lake Helen (HEL-uhn), town (☐ 4 sq mi/10.4 sq km; 2000 population 2,743), VOLUSIA county, E central FLORIDA, 20 mi/32 km SW of DAYTONA BEACH, in citrus-fruit-growing area; 28°58′N 81°13′W.

Lake Henry, village (2000 population 90), STEARNS county, central MINNESOTA, 32 mi/51 km WSW of ST. CLOUD; 45°27′N 94°47′W. Grain; livestock, poultry; dairying.

Lake Hicpochee, Florida: see CALOOSAHATCHEE RIVER.

Lakehills (LAIK-hilz), town (2000 population 4,668), BANDERA county, SW TEXAS, residential and recreational community, 28 mi/45 km WNW of SAN ANTONIO, in bend on N shore of Medina Lake (MEDINA RIVER); 29°36′N 98°56′W.

Lake Hopatcong, NEW JERSEY: see HOPATCONG, LAKE.

Lake Huntington, resort village, SULLIVAN county, SE NEW YORK, near PENNSYLVANIA state line, on small Lake Huntington, 15 mi/24 km W of MONTICELLO; 41°41′N 74°59′W.

Lakehurst, borough (2006 population 2,674), OCEAN county, E central NEW JERSEY, 6 mi/9.7 km NW of

TOMS RIVER; 40°00′N 74°19′W. Early 20th-century resort area. It is important as the site of the Lakehurst Naval Air Station (established 1919) and Air Warfare Center. The *Shenandoah* (1923) was the first airship to use the station, and transatlantic airships made it their U.S. terminal from 1924. The crash and burning of the *Hindenburg*, which took 36 lives, occurred here (May 6, 1937) as the hydrogen-filled German zeppelin was being moored. Center for U.S. Navy blimps until 1940s. Incorporated 1921.

Lake in the Hills, village (2000 population 23,152), MCHENRY county, NE ILLINOIS, suburb 3 mi/4.8 km S of CRYSTAL LAKE; 42°11′N 88°19′W. Lake in the Hills Airport. Manufacturing (protective clothing, warning signs).

Lake Isabella, unincorporated town (2000 population 3,315), KERN county, S central CALIFORNIA, 33 mi/53 km ENE of BAKERSFIELD, at Isabella Dam (forms Isabella Lake in KERN RIVER), in S part of SIERRA NEVADA; 35°37′N 118°28′W. Area surrounded by parts of Sequoia National Forest. Shirley Meadows Ski Area to N; Miracle Hot Springs to SW. Timber, cattle, grain.

Lake Ishkal, TUNISIA: see ISHKAL, LAKE.

Lake Jackson (JAK-suhn), city (2006 population 27,614), Brazoria county, SE TEXAS, 9 mi/14.5 km NW of FREEPORT on the BRAZOS RIVER; 29°02′N 95°27′W. Near the GULF OF MEXICO; drained by OYSTER CREEK in N. It is a trading and shipping center for the many dairy and fruit farms in the area. Manufacturing (loudspeakers). Lake Jackson is the seat of Brazosport College (two-year). Brazoria County Airport to N; in Brazosport Area twin city to Freeport. Founded 1941.

Lake Junaluska (joon-uh-LUHS-kuh), unincorporated town (□ 5 sq mi/13 sq km; 2000 population 2,675), HAYWOOD county, W NORTH CAROLINA, 3 mi/4.8 km NNE of WAYNESVILLE, on PIGEON RIVER; 35°31′N 82°58′W. GREAT SMOKY MOUNTAINS National Park to NW; parts of Pisgah National Forest to N and SE. Service industries; manufacturing.

Lake Koka, lake, OROMIYA state, central ETHIOPIA, along AWASH RIVER, 50 mi/80 km SE of ADDIS ABABA; 08°28′N 39°09′E. Formed by Koka Dam.

Lakeland (LAIK-luhnd), community, S MANITOBA, W central CANADA, 8 mi/13 km S of LANGRUTH, in LAKEVIEW rural municipality; 50°16′N 98°40′W.

Lakeland (LAIK-luhnd), city (2000 population 78,452), POLK county, central FLORIDA, 30 mi/48 km E of TAMPA; 28°02′N 81°57′W. It is an important processing and shipping center for a citrus-fruit-growing and phosphate-mining region and is a retail trade center that serves a large area. Manufacturing (motor vehicle parts). The Florida Citrus Commission, other state citrus organizations, and the Florida Phosphate Council have their headquarters in Lakeland, which is also a winter resort and recreation center. Points of interest include a major sports and convention complex and the Frank Lloyd Wright buildings at Florida Southern College. The Detroit Tigers baseball team comes here for spring training. Incorporated 1885.

Lakeland, town (2000 population 2,730), ☉ LANIER county, S GEORGIA, 19 mi/31 km NE of VALDOSTA, between ALAPAHA RIVER and Banks Lake; 31°02′N 83°04′W. Agricultural center; manufacturing of lumber and wood products. Nearby Banks Lake National Wildlife Refuge contains one of the best stands of pond cypress in the E UNITED STATES.

Lakeland, town (2000 population 1,917), WASHINGTON county, E MINNESOTA, suburb 14 mi/23 km E of downtown ST. PAUL, on ST. CROIX River (Lake St. Croix; bridged), opposite HUDSON, WISCONSIN; 44°57′N 93°46′W. Manufacturing (plastic products, water agitators, sand and gravel processing, paving machines); agriculture (cattle, sheep; corn, soybeans). LOWER ST. CROIX NATIONAL SCENIC RIVERWAY on St. Croix River.

Lakeland, village, JEFFERSON county, N KENTUCKY, residential suburb 11 mi/18 km E of LOUISVILLE. Nearby is Ormsby Village.

Lakeland (1990 population 1,204; 2000 population 6,862), suburb NE of MEMPHIS, SHELBY county, SW TENNESSEE; 35°13′N 89°44′W. Incorporated 1977.

Lakeland County (LAIK-luhnd), district municipality (□ 6,291 sq mi/16,356.6 sq km; 2001 population 4,959), NE ALBERTA, W CANADA; 55°08′N 110°44′W. Includes LAC LA BICHE and the hamlets of BEAVER LAKE, VENICE, HYLO, and PLAMONDON. Formed in 1998.

Lake Landing (LAIK LAN-deeng), unincorporated village (2000 population 1,852), HYDE county, E NORTH CAROLINA, 29 mi/47 km E of BELHAVEN on SE end of Lake MATTAMUSKEET; 35°28′N 76°04′W.

Lakeland Shores, village (2000 population 355), WASHINGTON county, E MINNESOTA, residential suburb 14 mi/23 km E of downtown ST. PAUL, on shore of Lake St. Croix; 44°57′N 92°45′W. LOWER ST. CROIX NATIONAL SCENIC RIVERWAY on ST. CROIX River.

Lakeland Village, unincorporated town (2000 population 5,626), RIVERSIDE county, S California, 23 mi/37 km S of RIVERSIDE, on SW shore of Lake Elsinore reservoir; 33°39′N 117°20′W. Part of Cleveland National Forest and Elsinore Mountains to SW; Lake Elsinore State Recreation Area to NW.

Lake Lansing, MICHIGAN: see HASLETT.

Lake Lenore (luh-NOR), village (2006 population 306), central SASKATCHEWAN, near LENORE LAKE, 15 mi/24 km NE of HUMBOLDT; 17 mi/27 km long, 4 mi/6 km wide; 52°24′N 104°59′W. Resort.

Lake Lillian, village (2000 population 257), KANDIYOHI county, S central MINNESOTA, 15 mi/24 km SE of WILLMAR, on South Fork CROW RIVER; 44°57′N 94°52′W. Livestock; grain; dairying; manufacturing (agricultural equipment). Lake Lillian to N, BIG KANDIYOHI LAKE to NW.

Lake Linden, town (2000 population 1,081), HOUGHTON county, NW UPPER PENINSULA, MICHIGAN, 9 mi/14.5 km. NE of HOUGHTON, on TORCH LAKE; 47°12′N 88°24′W. Former copper region. Incorporated in 1885.

Lakeline (LAIK-lein), village (2006 population 162), Lake county, NE OHIO, on LAKE ERIE, 17 mi/27 km NE of CLEVELAND; 41°39′N 81°27′W.

Lake Los Angeles, unincorporated town (2000 population 11,523), LOS ANGELES county, S CALIFORNIA, residential suburb 11 mi/18 km ESE of LANCASTER, in MOJAVE DESERT; 34°37′N 117°50′W. Small reservoir in S. Saddleback Butte State Park to NE. Antelope Valley Indian Museum is here. Fruit, vegetables, nuts, cotton, grain, cattle.

Lake Lotawana (lah-tuh-WAH-nuh), village (2000 population 1,872), JACKSON county, E MISSOURI, residential suburb 17 mi/27 km SE of downtown KANSAS CITY; 38°55′N 94°15′W. Village surrounds LAKE LOTAWANA.

Lake Louise (LAIK loo-EEZ), unincorporated village, SW ALBERTA, W CANADA, near BRITISH COLUMBIA border, in ROCKY MOUNTAINS, in BANFF NATIONAL PARK, on BOW RIVER, 30 mi/48 km NW of BANFF, and included in IMPROVEMENT DISTRICT NO. 9; 51°26′N 116°11′W. Elevation 5,050 ft/1,539 m. Famous tourist resort, with several large hotels, heated swimming pools, and extensive entertainment and sports facilities. Near foot of Mount Victoria, 2 mi/3 km SW, is small Lake Louise.

Lake Loveland Dam, California: see SWEETWATER RIVER.

Lake Lugano, SWITZERLAND: see LUGANO.

Lake Lure (LOOR), village (□ 14 sq mi/36.4 sq km; 2006 population 1,023), RUTHERFORD county, SW NORTH CAROLINA, 23 mi/37 km SE of ASHEVILLE and on BROAD RIVER, forms Lake Lure reservoir to N; hydroelectric plant; 35°26′N 82°12′W. Service industries; agriculture (grain, soybeans; poultry, livestock).

Lake Luzerne, year-round resort village, WARREN county, E NEW YORK, in the ADIRONDACK MOUNTAINS, between the HUDSON RIVER, and small Lake Luzerne; 43°19′N 73°48′W. Ski trails nearby. Also called Luzerne.

Lake Mahopac, NEW YORK: see MAHOPAC.

Lake Malawi National Park, MANGOCHE district, Southern region, S central MALAWI, at the S part of Lake Malawi (LAKE NYASA), encompassing the Nankumba peninsula, 90 mi/145 km N of ZOMBA. First park in the world to give protection to the marine life of a tropical, deep water, rift valley lake. The lake contains an amazing diversity of mbuna, the dazzling tropical freshwater fish.

Lake Manawa, resort village, POTTAWATTAMIE county, SW IOWA, on LAKE MANAWA (c.2 mi/3.2 km long), just S of COUNCIL BLUFFS. Lake Manawa State Park is here.

Lake Manjirenji Recreational Park, ZIMBABWE: see MANJIRENJI, LAKE.

Lake Manyame Recreational Park, ZIMBABWE: see MANYAME, LAKE.

Lake Manyara National Park (mah-NYAH-rah), ARUSHA region, N central TANZANIA, 60 mi/97 km W of ARUSHA, on NW shore of Lake MANYARA; 03°30′S 35°45′E. Includes N half of the brackish Lake Manyara. Lions, leopards, other wildlife. Tourism.

Lake Mary (MER-ee), town (□ 9 sq mi/23.4 sq km; 2000 population 11,458), SEMINOLE county, central FLORIDA, 18 mi/29 km N of ORLANDO; 28°45′N 81°19′W. Manufacturing includes electrical equipment, computer peripherals, medical equipment, plastic molding, telephone switching equipment. Rapid growth since 1990.

Lake Mead National Recreation Area (□ 2,336 sq mi/ 6,050 sq km), MOHAVE county, NW ARIZONA, CLARK county, SE NEVADA, includes all margins of Lake MEAD (HOOVER DAM) and Lake Mohave (DAVIS DAM) reservoirs on COLORADO RIVER. The first national recreation area of U.S. National Park System. Adjoin W end of Grand Canyon National Park on NE; separates NE, sections protect S part of SHIVWITS PLATEAU, N of Grand Canyon National Park. Originally named Hoover Dam National Recreation Area; name changed 1947. Authorized 1936.

Lake Meredith National Recreation Area (ME-redith), HUTCHINSON, MOORE and POTTER counties, NW TEXAS; covers 73 sq mi/188 sq km; 35°42′N 101°33′W. Includes Lake Meredith reservoir, on the CANADIAN RIVER, a popular water sports area in the Southwest. Was named Sanford National Recreation Area (1965–1972); Alibates Flint Quarries National Monument to SE. Authorized 1965.

Lake Mills, town (2000 population 2,140), WINNEBAGO county, N IOWA, near MINNESOTA state line, 11 mi/18 km NNW of FOREST CITY; 43°25′N 93°31′W. Vegetable cannery, feed mill. Incorporated 1880. Rice Lake State Park is SE.

Lake Mills, town (2006 population 5,401), JEFFERSON county, S WISCONSIN, on ROCK LAKE, 24 mi/39 km E of MADISON; 43°04′N 88°54′W. In dairying, farming, and resort region. Manufacturing (dairy equipment, food processing and refrigation equipment, magnetic switches, liquid crystal displays, metal components; food processing). Lake Mills State Fish Hatchery to S. Aztalan State Park to E contains remains of 12th century Indian village, rare platform mounds, and a museum with ancient Indian relics. Glacial Drumlin State Trail passes E-W to S. Settled c. 1836, incorporated 1905.

Lake Milton (MIL-tuhn), unincorporated village, MAHONING county, NE OHIO, 16 mi/26 km W of YOUNGSTOWN, on E shore of Milton Reservoir; 41°06′N 80°58′W. Incorporated 1930, disincorporated 1947.

Lake Minchumina (min-CHOO-mi-nuh), village (2000 population 32), E central ALASKA, on Lake Minchumina, 150 mi/241 km SW of FAIRBANKS; 63°52′N 152°24′W. Airfield.

Area is shown by the symbol □, and capital city or county seat by ☉.

Lake Minnewaska, former private resort, ULSTER county, SE NEW YORK, on small Lake Minnewaska, in the SHAWANGUNK range, 17 mi/27 km SW of KINGSTON; 41°44′N 74°13′W. Two large hotels (Cliff House and Wildmere) existed here for nearly a century. After their demise in the 1970s, the state eventually bought the land (1993) and established Minnewaska State Park. Today it encompasses 14,500 acres/5,873 ha.

Lake Mohawk, NEW JERSEY: see SPARTA.

Lake Monroe (MUHN-ro), town, SEMINOLE county, central FLORIDA, 22 mi/35 km N of ORLANDO; 28°49′N 81°19′W. Light manufacturing.

Lakemont, town, RABUN county, extreme NE GEORGIA, 6 mi/9.7 km S of CLAYTON, near Lake Rabun, in the BLUE RIDGE; 34°46′N 83°24′W. Recreation area; manufacturing (apparel).

Lake Montezuma, unincorporated town (2000 population 3,344), YAVAPAI county, central ARIZONA, 37 mi/60 km SSW of FLAGSTAFF, on Wet Beaver Creek, in Prescott National Forest. Timber; cattle, sheep. Units of MONTEZUMA CASTLE NATIONAL MONUMENT to NE and SW.

Lakemoor, village (2000 population 2,788), MCHENRY and Lake counties, NE ILLINOIS, residential suburb 42 mi/68 km NW of downtown CHICAGO, 4 mi/6.4 km ESE of MCHENRY, at E end of LILY LAKE; 42°20′N 88°12′W. Volo Bog State Natural Area to N.

Lakemore (LAIK-mor), village (□ 2 sq mi/5.2 sq km; 2006 population 2,749), SUMMIT county, NE OHIO, 6 mi/10 km SE of downtown AKRON; 41°01′N 81°25′W. Incorporated 1921.

Lake Mutirikwi Recreational Park (moo-tee-REE-kwee), reservoir, MASVINGO province, S central ZIMBABWE, formed on MUTIRIKWI RIVER by Mutirikwi Dam, 18 mi/29 km SE of MASVINGO; 20°15′S 31°02′E. Surrounded by KYLE RECREATIONAL PARK. Mutirikwi River forms E arm, c.10 mi/16 km long; Mukurumidzi River forms W arm, c.12 mi/19 km long; arms meet at dam. Also spelled Mtirikwe; formerly known as Lake Kyle.

Lake Mykee Town (MEI-kee), village (2000 population 326), CALLAWAY county, central MISSOURI, 8 mi/12.9 km NE of JEFFERSON CITY; 38°40′N 92°05′W. Residential community surrounds Lake Mykee.

Lake Nacimiento, unincorporated town (2000 population 2,176), SAN LUIS OBISPO county, SW CALIFORNIA; residential community 12 mi/19 km NW of PASO ROBLES, on SE arm of NACIMIENTO RESERVOIR, 2 mi/3.2 km S of Nacimiento Dam; 35°44′N 120°53′W. San Antonio Reservoir and Hunter Liggett Military Reservation to NW.

Lake Nakuru National Park (□ 73 sq mi/189.8 sq km), W central KENYA, in RIFT VALLEY province N of Lake NAIVASHA, within NAKURU municipality. Features flamingoes, water buck, warthogs, buffaloes, and gazelles.

Lake Naujan, PHILIPPINES: see NAUJAN, LAKE.

Lake Nebagamon (ne-BAG-uh-mawn), resort village (2006 population 1,037), DOUGLAS county, NW WISCONSIN, on small Lake Nebagamon, 22 mi/35 km SE of SUPERIOR; 46°30′N 91°41′W. Was lumbering center in late 1800s.

Lake Neusiedl (NOI-seedl), German *Neusiedler See*, Hungarian *Fertő*, in the AUSTRIA-HUNGARY border SE of VIENNA; 47°50′N 16°46′E. The lake lies 85 mi/137 km in Austria and 25 sq mi/65 sq km in Hungary. Westernmost steppe lake in Europe. Due to high evaporation, the lake has a high salt content. The water level is set by yearly precipitation. The lake's area and depth (average 5.5 ft/1.7 m) vary considerably with the seasons and throughout history; it's depth has fluctuated greatly, ranging from its present level to that of a cultivated riverbed between the years 1864 and 1870. Recorded history shows the lake as being completely dried out during other times. The heavy growth of lake reeds (40 mi/64 km) supplies the Austrian construction industry. Carp and other fish

are in the lake. Its lonely and desolate salt marshes attract a variety of birds and other wildlife and have been protected since 1935. Since 1994, a common Austrian-Hungarian NATIONAL PARK LAKE NEUSIEDLSEEWINKEL comprises the SE parts of the lake and reed belt as well as the salt marshes in the Seewinkel. The Neusiedler region has noted resorts and is by far the main tourist region of BURGENLAND. It has domestic as well as international tourism; bathing, water sports, and bicycling prevail. Main wine region centers are Rust and ILLMITZ. There are remains of prehistoric lake dwellers in the vicinity.

Lakenham, ENGLAND: see NORWICH.

Lakenheath (LAIK-uhn-heeth), agricultural village (2001 population 4,490), NW SUFFOLK, E ENGLAND, 5 mi/8 km SW of BRANDON; 52°25′N 00°12′E. Has medieval church. Site of largest U.S. air force base in England.

Lake Norden, village (2006 population 424), HAMLIN county, E SOUTH DAKOTA, 23 mi/37 km S of WATERTOWN, near Lake Norden; 44°34′N 97°12′W. Dairy products. Lake Poinsett State Recreational Area to SE.

Lake Odessa, town (2000 population 2,272), IONIA county, S central MICHIGAN, 30 mi/48 km W of LANSING; 42°46′N 85°08′W. In resort and agricultural area; poultry. Manufacturing (food processing, wire products). Settled c.1870, incorporated 1889.

Lake of Bays (LAIK, BAIZ), township (□ 259 sq mi/673.4 sq km; 2001 population 2,900), S ONTARIO, E central CANADA; 45°18′N 79°00′W. Tourism, service industries, forestry, aggregate extraction. Additional seasonal population. Formed in 1971 from FRANKLIN, RIDOUT, MCLEAN and SINCLAIR/Finlayson.

Lake of the Arbuckles, OKLAHOMA: see ARBUCKLE RESERVOIR.

Lake of the Cherokees, reservoir; MAYES, DELAWARE, and OTTAWA counties; NE OKLAHOMA; on NEOSHO (Grand) River; 60 mi/97 km NE of TULSA; c.45 mi/72 km long; 36°27′N 95°01′W. ELK RIVER enters from MISSOURI to form 8-mi/12.9-km E arm. Formed by Pensacola Dam (145 ft/44 m high, 6,500 ft/1,981 m long), built 1938–1941 by the state of Oklahoma for power generation, flood control, and water supply. Four state parks on its shores.

Lake of the Ozarks (□ 93 sq mi/241 sq km); MILLER, CAMDEN, MORGAN, and BENTON counties; central MISSOURI; on OSAGE RIVER; 33 mi/53 km SW of JEFFERSON CITY; c.85 mi/137 km long, with shoreline of 1,375 mi/2,213 km; 38°12′N 92°36′W. Maximum capacity 1,428,000 acre-ft. Extends W through OZARKS in serpentine course to Harry S. Truman Dam, near WARSAW. Three arms formed by NIANGUA RIVER (S), Grand Auglaize Creek (SE), and Gravois Creek (N). Formed by Bagnell Dam (2,543 ft/775 m long; 148 ft/45 m high), built (1929–1931) for power generation and recreation by Union Electric Company. Lake of the Ozarks and Ha Ha Tonka state parks are here. Resort and residential communities of OSAGE BEACH, LAKE OZARK, Hurricane Deck, Sunrise Beach, Gravois Mills, Lakeland, and Lakeview Heights on the lake's shores.

Lake of the Pines, unincorporated town, NEVADA county, E central CALIFORNIA, 12 mi/19 km S of GRASS VALLEY, and 9 mi/14.5 km N of AUBURN; 39°02′N 121°04′W. Surrounds Lake of the Pines reservoir (Magnolia Creek), near Bear River.

Lake of the Woods, county (□ 1,774 sq mi/4,612.4 sq km; 2006 population 4,327), NW MINNESOTA; ⊙ BAUDETTE; 48°46′N 94°54′W. Bounded NE by RAINY river, and N and NW by LAKE of the WOODS, both form CANADA-U.S. border (province of MANITOBA to NW, province of ONTARIO to N and NE). Agricultural region (oats, wheat, barley, alfalfa, potatoes, flax, sunflowers); timber. Northwest Angle State Forest in far N; part of Beltrami Island State Forest in S and SW; including part of Red Lake Wildlife Management Area in S and SW. Widely

scattered parts of Red Lake Indian Reservation in far N and in S; Zippel Bay State Park in N center, on S shore of Lake of the Woods. Formed 1922.

Lake of the Woods (LAIK, WUDZ), township (□ 291 sq mi/756.6 sq km; 2001 population 330), NW ONTARIO, E central CANADA, 63 mi/100 km from FORT FRANCES; 49°09′N 94°29′W. Tourism; also logging, farming, commercial fishing, trapping. Formed in 1998 from Morson, McCrosson-Tovell.

Lake of the Woods, unincorporated town (2000 population 3,026), CHAMPAIGN county, E central ILLINOIS, 10 mi/16 km NW of CHAMPAIGN, on SANGAMON RIVER; 40°12′N 88°22′W. Residential area; covered bridge.

Lake of the Woods (LAIK, WUDZ) (□ 1,485 sq mi/3,861 sq km), on the U.S.-CANADA border in the pine forest region of N MINNESOTA (U.S.), SE MANITOBA (Canada), and SW ONTARIO (Canada); c.70 mi/110 km long; 49°05′N 95°14′W. More than 66% of the lake is in Canada. A remnant of former glacial Lake AGASSIZ, it is fed by the Rainy River, and drained to the NW by the WINNIPEG River. It has a very irregular shoreline and approximately 14,000 islands Lake of the Woods separates the NW Angle, the northernmost land of the coterminous U.S., from the rest of Minnesota. Abundant in fish and game, the region is a resort area.

Lake Orion, town (2000 population 2,715), OAKLAND county, SE MICHIGAN, suburb 11 mi/18 km NNE of PONTIAC, on E end of small LAKE ORION; 42°46′N 83°14′W. Agriculture (grain; cattle); manufacturing (tool and die, hydraulic and pneumatic components, rubber products). Satellite community of Detroit. Bald Mountain State Recreation Area to S; ski area to E. Incorporated in 1859.

Lake Orion Heights, village, OAKLAND county, SE MICHIGAN, residential suburb 8 mi/12.9 km N of PONTIAC, near LAKE ORION; 42°46′N 83°15′W.

Lake Oswego (ah-SWEE-go), city (2006 population 36,713), CLACKAMAS county, NW OREGON, suburb 6 mi/9.7 km S of downtown PORTLAND on the WILLAMETTE RIVER, 4 mi/6.4 km NW of OREGON CITY; 45°24′N 122°42′W. Oswego Lake (2.5 mi/4 km long) runs E and W through the center of the city. Railroad junction. Manufacturing (consumer goods, bakery products, textiles, apparel, lumber, transportation equipment; printing and publishing). Marylhurst College to SE. Tryon Creek State Park to N; Mary S. Young State Park to S. Founded c. 1850; incorporated 1909.

Lake o' the Cherokees, OKLAHOMA: see LAKE OF THE CHEROKEES.

Lake o' the Pines (PEINZ), reservoir (□ 29 sq mi/75.4 sq km), CAMP, UPSHUR, and MARION counties, NE TEXAS, on Cypress Creek, 22 mi/51 km NE of LONGVIEW; 32°46′N 94°30′W. Maximum capacity 1,998,740 acre-ft. Formed by Feprells Bridge Dam (97 ft/30 m high), built (1958) by Army Corps of Engineers for water supply; also used for recreation and power generation.

Lake Ozark, village (2000 population 1,489), MILLER county, central MISSOURI, 4 mi/6.4 km NNW of OSAGE BEACH, on LAKE OF THE OZARKS (OSAGE RIVER); 38°11′N 92°37′W. Residential and recreation area. Bagnell Dam 3 mi/4.8 km NE.

Lake Panasoffkee (pan-nah-SOF-kee), town (□ 4 sq mi/10.4 sq km; 2000 population 3,413), SUMTER county, central FLORIDA, 30 mi/48 km S of OCALA; 28°47′N 82°07′W. Coleman Correctional Facility located here.

Lake Park (PAHRK), town (□ 1 sq mi/2.6 sq km; 2000 population 8,721), PALM BEACH county, SE FLORIDA, 5 mi/8 km N of WEST PALM BEACH, on LAKE WORTH lagoon; 26°48′N 80°04′W.

Lake Park, town (2000 population 1,023), DICKINSON county, NW IOWA, 11 mi/18 km WNW of SPIRIT LAKE; 43°27′N 95°19′W. In resort area; feed mill. State park, sand and gravel pits nearby.

Cross-references are shown in SMALL CAPITALS. The pronunciation guide is shown on page xix. The sources of population figures are shown on page xvii.

Lake Park, village (2000 population 549), LOWNDES county, S GEORGIA, 11 mi/18 km SE of VALDOSTA, near FLORIDA, state line; 30°41′N 83°11′W. Manufacturing of paper bags; industrial sandblasting. Location of one of the first discount name-brands outlet malls located on Interstate 75; attracts tourists traveling in and out of Florida.

Lake Park, village (2000 population 782), BECKER county, W MINNESOTA, 13 mi/21 km WNW of DE-TROIT LAKES; 46°52′N 96°05′W. Agriculture (grain, sugar beets, sunflowers; livestock, poultry; dairying); manufacturing (fishing tackle). Numerous small natural lakes in area. White Earth Indian Reservation to NE.

Lake Peekskill, hamlet, PUTNAM county, SE NEW YORK, on Peekskill Lake (c.0.5 mi/0.8 km long), 5 mi/8 km NE of Peekskill; 41°21′N 73°53′W. Lakeside recreation area.

Lake Placid (PLA-sid), town (□ 1 sq mi/2.6 sq km; 2000 population 1,668), HIGHLANDS county, central FLOR-IDA, 15 mi/24 km S of SEBRING, in lake region; 27°18′N 81°22′W. Citrus fruit.

Lake Placid, village (□ 1 sq mi/2.6 sq km; 2006 population 2,814), ESSEX county, NE NEW YORK, in the ADIRONDACKS, surrounding Mirror Lake, a small body of water adjacent to the larger Lake Placid; 44°16′N 73°59′W. Elevation 1,800 ft/549 m. Famous resort and sports center; Winter Olympics (1932, 1980) and the World Bobsled Championships (1969) were held here. Lake Placid has a summer theater and music festival, a figure-skating school, and annual winter sports competitions. The farm and burial place of the abolitionist John Brown are nearby. Terminus (N), at Old Military Road, of 133-mi/214-km Northville–Lake Placid trail connecting Adirondack foothills and High Peaks region. Settled 1850, incorporated 1900.

Lake Pleasant, village, ⊙ HAMILTON county, E central NEW YORK, on Lake Pleasant (□ c.2 sq mi/5.2 sq km; 3.5 mi/5.6 km long), 50 mi/80 km NE of UTICA, in the ADIRONDACKS; 43°28′N 74°25′W. Summer and winter recreation; summer residences.

Lake Pleasant, ARIZONA: see PLEASANT LAKE.

Lake Pleasant, MASSACHUSETTS: see MONTAGUE.

Lakeport, city (2000 population 4,820), ⊙ Lake county, NW CALIFORNIA, 17 mi/27 km ESE of UKIAH, at W end of CLEAR LAKE; 39°02′N 122°55′W. Resort, farm-trade center. Agriculture (pears, grapes, oats, walnuts; cattle); manufacturing (glass products). Mendocino National Forest to N; Clear Lake State Park to SE. Founded 1861, incorporated 1888.

Lakeport, village, BELKNAP county, central NEW HAMPSHIRE, industrial section of LACONIA (city), 1 mi/1.6 km NE of city center, on Lake Paugus, E of channel connecting PAUGUS and OPECHE bays of LAKE WINNIPESAUKEE. Manufacturing (outerwear, commercial printing, metal doors).

Lakeport (LAIK-port), village (2006 population 944), GREGG county, E TEXAS, residential suburb 7 mi/11.3 km SSE of downtown LONGVIEW, on SABINE RIVER; 32°23′N 94°42′W. Cattle; oil and natural gas. Lake Cherokee reservoir to SE. Airport to S.

Lakeport Dam, LACONIA, BELKNAP county, SE NEW HAMPSHIRE, on the Winnipesaukee River; 44°32′N 71°28′W. Maximum capacity 2,400,000 acre-ft; 14 ft/4 m high. Gravity dam built in 1851 for textile mills. Forms LAKE WINNIPESAUKEE (recreation area).

Lake Preston, town (2006 population 666), KINGSBURY county, E central SOUTH DAKOTA, 9 mi/14.5 km E of DE SMET, near LAKE PRESTON; 44°21′N 97°22′W. LAKE WHITEWOOD to SE. In farming region (livestock, poultry; dairy products; grain); manufacturing (cutting tools).

Lake Providence, town (2000 population 5,104), ⊙ EAST CARROLL parish, extreme NE LOUISIANA, on MISSISSIPPI RIVER, and 36 mi/58 km NNW of VICKSBURG (MISSISSIPPI); 32°49′N 91°11′W. In cotton-growing area; agriculture (cattle; rice, pecans, soy-

beans, sorghum); manufacturing (wood products, pepper sauces). Lake Providence (c. 5 mi/8 km long; fishing), an oxbow lake, formed by the MISSISSIPPI, is just N. Bayou Macon State Wildlife Area to W. One of oldest towns in LOUISIANA; settled c. 1812, incorporated 1876.

Lake Purdy, village (2000 population 5,799), JEFFERSON and SHELBY counties, N central ALABAMA, 7 mi/11.3 km E of BIRMINGHAM; 33°25′N 86°41′W.

Lake Quivira (kwuh-VER-uh), village (2000 population 932), JOHNSON county, E KANSAS, residential suburb, 7 mi/11.3 km WSW of KANSAS CITY, near KANSAS RIVER; 39°02′N 94°46′W. Quivira Lake reservoir at center of community.

Lake Range, WASHOE county, W NEVADA, largely in Lake Indian Reservation, between PYRAMID LAKE (W), and WINNEMUCCA and MUD lakes (E). Rises to 8,182 ft/2,494 m in TOHAKUM Peak and to 7,608 ft/2,319 m in PAH-RUM PEAK.

Lake Ridge (LAIK RIJ), unincorporated city, PRINCE WILLIAM county, NE VIRGINIA, residential suburb 21 mi/34 km SW of WASHINGTON, D.C., 12 mi/19 km SE of MANASSAS, near OCCOQUAN RIVER, Occoquan Dam to NE; 38°41′N 77°17′W.

Lakeridge, unincorporated town, KING county, W WASHINGTON, residential suburb of SEATTLE, at S end of LAKE WASHINGTON.

Lake Robertson Recreational Park, ZIMBABWE: see MANYAME, LAKE.

Lake Ronkonkoma, unincorporated residential village (□ 4 sq mi/10.4 sq km; 2000 population 19,701), SUF-FOLK county, SE NEW YORK, on central LONG ISLAND, 7 mi/11.3 km NW of PATCHOGUE, on Lake Ronkonkoma (c.0.75 mi/1.21 km in diameter); 40°49′N 73°06′W. The lake is the largest lake on Long Island.

Lake Roosevelt National Recreation Area (KOO-lee) (□ 157 sq mi/408.2 sq km), NE WASHINGTON, in FERRY, GRANT, LINCOLN, OKANOGAN, and STEVENS counties. FRANKLIN D. ROOSEVELT LAKE, formed by the GRAND COULEE DAM in the COLUMBIA River Basalt rock formations, includes both shores of 130-mi/209-km reservoir. Authorized 1946.

Lake Rushford, NEW YORK: see CANEADEA.

Lakes, state, SUDAN: see Bohairat.

Lake Saint Croix Beach (KROI), town, WASHINGTON county, E MINNESOTA, residential suburb 14 mi/23 km E of downtown ST. PAUL, on shore of Lake St. Croix, natural lake on ST. CROIX River (WISCONSIN state line); 44°55′N 92°46′W. LOWER ST. CROIX NATIONAL SCENIC RIVERWAY on St. Croix River.

Lake Saint Louis, city, SAINT CHARLES county, E MIS-SOURI, residential suburb 14 mi/23 km W of SAINT CHARLES; 38°47′N 90°46′W. Residential area surrounds Lake Saint Louis. Light manufacturing.

Lake San Marcos, unincorporated town, SAN DIEGO county, S CALIFORNIA; residential suburb 30 mi/48 km N of downtown SAN DIEGO, 7 mi/11.3 km W of ES-CONDIDO, 6 mi/9.7 km E of Pacific Ocean, on Lake San Marcos reservoir; 33°08′N 117°13′W.

Lakes District, suburban area, PIERCE county, W WA-SHINGTON, residential suburb 7 mi/11.3 km SW of TACOMA, near PUGET SOUND, drained by Chambers and Clover creeks. General name for area of small lakes, including Steilacoom, Gravelly, and American lakes, and including communities of Lakewood Center, Clover Park, Lakeview, Lake City, and Interlaken. Three fish hatcheries on creeks. Fort Lewis Military Reservation to S; MCCHORD AIR FORCE BASE to E.

Lakes Entrance (LAIKS EN-truhns), town, VICTORIA, SE AUSTRALIA, on SE coast, 165 mi/266 km E of MELBOURNE, at entrance to Lake KING; 37°53′S 147°59′E. Seaside resort; tourism; commercial fishing; service center for offshore oil rigs. Access to the Gippsland Lakes. Formerly called Cunninghame.

Lake Sevan, lake (□ 540 sq mi/1,404 sq km), NE AR-MENIA; 40°21′N 45°20′E. Average depth 95 ft/29 m, maximum depth, 292 ft/83 m. Rises to 6,280 ft/1,914

m. The largest lake of the CAUCASUS, it is fed by some thirty streams, but the HRAZDAN RIVER is its only outlet. Lake Sevan is free of ice in winter. The SEVAN HYDROELECTRIC SYSTEM has drained part of the lake. A tunnel to divert water from the ARPA RIVER to the lake was built in the 1970s. The steep gradient of the river and the great volume of Lake Sevan make possible an extensive hydroelectric system. On Sevan Island, in the NW, stands an ancient Armenian monastery, now a rest home. Also known as Gokcha.

Lakeshire, town (2000 population 1,375), SAINT LOUIS county, E MISSOURI, residential suburb 10 mi/16 km SSW of downtown ST. LOUIS; 38°32′N 90°20′W.

Lakeshore (LAIK-shor), town (□ 205 sq mi/533 sq km; 2001 population 28,746), S ONTARIO, E central CA-NADA; 42°14′N 82°14′W. Formed in 1999 with the amalgamation of the former town of BELLE RIVER and the townships of MAIDSTONE, ROCHESTER, TILBURY NORTH, and TILBURY WEST.

Lakeshore, unincorporated village, FRESNO county, E central CALIFORNIA, 45 mi/72 km NE of FRESNO, in SIERRA NEVADA and Sierra National Forest, at E end of HUNTINGTON LAKE reservoir (Big Creek). Resort area; China Peak Ski Area; McKinley Grove Big Trees to SE.

Lake Shore, village (2000 population 966), CASS county, central MINNESOTA, 15 mi/24 km NW of BRAINERD, and 3 mi/4.8 km WSW of NISSWA at NE end of GULL LAKE RESERVOIR; 46°30′N 94°21′W. Resorts. Dairying; poultry; oats, alfalfa. Pillsbury State Forest to SW.

Lakeshore, unincorporated village, HANCOCK county, SE MISSISSIPPI, 7 mi/11.3 km SW of BAY SAINT LOUIS, near MISSISSIPPI SOUND, Gulf of MEXICO. Manufacturing (boats and barges, boat docks).

Lakeside, unincorporated city (2000 population 19,560), SAN DIEGO county, S CALIFORNIA; residential suburb 17 mi/27 km NE of downtown SAN DIEGO, on SAN DIEGO RIVER; 32°51′N 116°55′W. Agriculture (fruit, avocados; dairying; poultry; flowers, ornamental plants); manufacturing (building materials, fabricated metal products). Part of Cleveland National Forest (E); Capitan Grande Indian Reservation (E); Barona Ranch Indian Reservation (NE); El Capitan Reservoir (E); San Vicente Reservoir (N). Lakeside is S terminus of SAN DIEGO AQUEDUCT.

Lakeside (LAIK-seid), city (□ 17 sq mi/44.2 sq km; 2000 population 30,927), LEON county, NW FLORIDA, 3 mi/4.8 km S of TALLAHASSEE; 30°07′N 81°46′W.

Lakeside (LAIK-seid), unincorporated city, HENRICO county, E central VIRGINIA, residential suburb 5 mi/8 km N of downtown RICHMOND; 37°37′N 77°28′W. J. Sargeant Reynolds Community College (Parham Campus) is here.

Lakeside, town, JEFFERSON county, N central COLOR-ADO. Residential suburb 5 mi/8 km NW of downtown DENVER, near CLEAR CREEK, bounded by Denver (and DENVER county) on N and E, WHEAT RIDGE on W, MOUNTAIN VIEW on S. Elevation 5,360 ft/1,634 m.

Lakeside, resort town (2000 population 484), BUENA VISTA county, NW IOWA, just SE of STORM LAKE; 42°37′N 95°10′W. Incorporated 1933.

Lakeside, town (2006 population 1,486), COOS county, SW OREGON, 12 mi/19 km N of North Bend, 3 mi/4.8 km E of PACIFIC OCEAN, at joining of North Tenmile and South Tenmile lakes; 43°34′N 124°10′W. Lumber. Tourism. Eel Lake and William M. Tugman State Park (N); Elliott State Forest (E); Oregon Dunes National Recreation Area (W).

Lakeside (LAIK-seid), town (2000 population 333), TARRANT county, N TEXAS, residential suburb 10 mi/16 km NW of downtown FORT WORTH, near LAKE WORTH (E) and West Fork of TRINITY river; 28°06′N 97°51′W. Fort Worth Nature Center and Refuge to N.

Lakeside, village, LITCHFIELD county, CONNECTICUT, 10 mi/16 km SW of TORRINGTON, resort town on S side of BANTAM LAKE. Postal section of MORRIS.

Area is shown by the symbol □, and capital city or county seat by ⊙.

Lakeside, village, BERRIEN county, extreme SW MICHIGAN, 14 mi/23 km NE of MICHIGAN CITY, INDIANA, on LAKE MICHIGAN; 41°50′N 86°40′W. Resort area. Warren Dunes State Park to NE.

Lakeside, village, FLATHEAD county, NW MONTANA, 13 mi/21 km SSE of KALISPELL, on W shore of FLATHEAD LAKE, near N end of lake. Residential and recreational community. Flathead National Forest to W.

Lakeside (LAIK-seid), resort village, OTTAWA county, N OHIO, on Marblehead Peninsula in LAKE ERIE, 7 mi/11 km NNW of SANDUSKY; 41°32′N 82°44′W.

Lakeside, village, CHELAN county, central WASHINGTON, 2 mi/3.2 km W of CHELAN, and on LAKE CHELAN.

Lakeside, ARIZONA: see PINETOP-LAKESIDE.

Lakeside City (LAIK-seid), town (2006 population 1,052), ARCHER county, N TEXAS, residential suburb 6 mi/9.7 km SW of downtown WICHITA FALLS, on S shore of Lake Wichita (Holiday Creek); 33°49′N 98°32′W. Cattle; oil and natural gas.

Lakeside Park, town, KENTON county, N KENTUCKY, residential suburb 6 mi/9.7 km SW of CINCINNATI, OHIO, and 4 mi/6.4 km SW of COVINGTON, Kentucky; 39°01′N 84°34′W. Thomas More College to S. Incorporated 1930.

Lakes National Park (LAIKS), national park (□ 9 sq mi/23.4 sq km), VICTORIA, SE AUSTRALIA, in Gippsland Lakes area; 38°05′S 147°40′E. Consists of Sperm Whale Head peninsula and the islands of Rotamah and Little Rotamah. Home to more than 190 recorded bird species, including the White-Bellied Sea Eagle and endangered Little Tern; mammals include kangaroos, wallabies, possums, and bats. The park's sandy soils support such vegetation as eucalypt, salt marsh, seasonal wildflowers, and native orchids. Fishing. Established 1956.

Lake Solai (so-LAH-ee), village, RIFT VALLEY province, W central KENYA, 22 mi/35 km N of NAKURU. Railroad spur terminus in GREAT RIFT VALLEY. Settled after 1909 by cattle breeders and corn farmers. Also called Solai.

Lake Spaulding Dam, California: see YUBA RIVER.

Lake Station, city, Lake county, extreme NW INDIANA, just SE of GARY. In the CALUMET industrial region; manufacturing (surgical instruments, cement blocks).

Lake Stevens, town (2006 population 8,007), SNOHOMISH county, NW WASHINGTON, on LAKE STEVENS, residential suburb 5 mi/8 km E of EVERETT, and 30 mi/48 km NNE of SEATTLE; 48°01′N 122°04′W. Timber; manufacturing (millwork, cedar products, wood fencing); recreation. Mount Baker-Snoqualmie National Forest to E.

Lake Success, prosperous residential village (□ 1 sq mi/2.6 sq km; 2006 population 2,829), NASSAU county, SE NEW YORK, on NW LONG ISLAND; 40°46′N 73°42′W. Some industry (electronics). Lake Success was the temporary home of the UN from 1946 to 1950. Settled c.1730, incorporated 1926.

Lake Taal, PHILIPPINES: see TAAL, LAKE.

Lake Tapawingo, village (2000 population 843), JACKSON county, W MISSOURI, suburb 13 mi/21 km ESE of downtown KANSAS CITY, SE of INDEPENDENCE; 39°01′N 94°18′W. Residential.

Lake Tara, town, CLAYTON county, NW central GEORGIA, N of JONESBORO; 33°32′N 84°21′W.

Laketon, village, WABASH county, N central INDIANA, 12 mi/19 km N of WABASH, on EEL RIVER. Corn, soybeans; cattle. Manufacturing (asphalt, fuel oil, petroleum products, gasoline). Laid out 1836.

Laketown, village (2000 population 188), RICH county, N UTAH, at S end of BEAR LAKE, 28 mi/45 km ENE of LOGAN; 41°49′N 111°19′W. Rendezvous Beach State Park is NW; Circo Beach State Park to NNE (E shore); Wasatch National Forest (W).

Lake Umzingwani Recreational Park, ZIMBABWE: see UMZINGWANI, LAKE.

Lakeview (LAIK-vyoo), rural municipality (□ 219 sq mi/569.4 sq km; 2001 population 384), S MANITOBA, W central CANADA, on W shore of Lake MANITOBA; 50°19′N 98°45′W. Commercial fishing. Main center is Langruth; also includes communities of Lakeland, Embury. Incorporated 1920.

Lakeview, unincorporated town, RIVERSIDE county, S CALIFORNIA, 17 mi/27 km SE of RIVERSIDE, on San Jacinto River; 33°50′N 117°07′W. COLORADO RIVER AQUEDUCT passes to S. Manufacturing (pharmaceuticals).

Lake View, resort town (2000 population 1,278), SAC county, W IOWA, 8 mi/12.9 km SSW of SAC CITY, near BLACK HAWK LAKE (state park here); 42°18′N 95°02′W. Manufacturing (popcorn; concrete). Gravel pits in vicinity. Settled 1875; incorporated 1887.

Lakeview, town, MONTCALM county, central MICHIGAN, 38 mi/61 km NNE of GRAND RAPIDS, on W end of small Tamarack Lake (resort); 43°26′N 85°16′W. Agricultural area (livestock; potatoes, grain, fruit, beans, apples, dairy products); manufacturing (brass fittings). Wintersköl Ski Area to SW.

Lakeview, town (2006 population 2,419), ⊙ LAKE county, S OREGON, 70 mi/113 km E of KLAMATH FALLS, N of GOOSE LAKE; 42°11′N 120°20′W. Elevation 4,798 ft/1,462 m; railroad terminus. Sawmills, woodworking plants. Ranching; tourism. Meat processing. Wheat, oats, barley; sheep, cattle. Booth State Wayside to W; Chandler State Wayside and LAKE ABERT to N; Drews Reservoir to W; parts of Fremont National Forest to W and E; Goose Lake State Recreation Area to S; Warner Valley (geyser basin) to E. Founded 1876, incorporated 1884.

Lake View, town (2006 population 789), DILLON county, NE SOUTH CAROLINA, 35 mi/56 km ENE of FLORENCE, near NORTH CAROLINA state line; 34°20′N 79°10′W. Manufacturing of upholstery material; agriculture (livestock; cotton, grain, soybeans). Formerly called Pages Mill.

Lakeview (LAIK-vyoo), summer village (2001 population 15), central ALBERTA, W CANADA, 41 mi/65 km from EDMONTON, in PARKLAND county; 53°34′N 114°27′W. Formed 1913.

Lakeview, village (2000 population 1,112), CALHOUN county, S MICHIGAN, just SSW of BATTLE CREEK; 43°27′N 85°16′W. Part of city of BATTLE CREEK.

Lakeview, residential village (2000 population 5,607), NASSAU county, SE NEW YORK, on W LONG ISLAND, just SW of HEMPSTEAD; 40°41′N 73°39′W. State park nearby.

Lakeview (LAIK-vyoo), resort village (2006 population 1,079), LOGAN county, W central OHIO, 12 mi/19 km NW of BELLEFONTAINE, on INDIAN LAKE reservoir; 40°29′N 83°55′W.

Lakeview (LAIK-vyoo), village (2000 population 152), HALL county, NW TEXAS, 32 mi/51 km NW of CHILDRESS; 34°40′N 100°42′W. Peanuts; cattle.

Lakeview (LAIK-vyoo), village, MCLENNAN county, E central TEXAS, 6 mi/9.7 km N of WACO; 31°37′N 97°06′W.

Lake View, plantation, PISCATAQUIS county, central MAINE, on SCHOODIC LAKE, and 19 mi/31 km ENE of DOVER-FOXCROFT; 45°21′N 68°53′W. In recreational area.

Lake Villa, village (2000 population 5,864), Lake county, extreme NE ILLINOIS, 13 mi/21 km WNW of WAUKEGAN; 42°25′N 88°04′W. In lake-resort area.

Lake Village, town (2000 population 2,823), ⊙ CHICOT county, extreme SE ARKANSAS, 14 mi/23 km WSW of GREENVILLE (MISSISSIPPI), on Lake CHICOT; 33°19′N 91°16′W. Commercial fishing and fish farming; manufacturing (processed catfish and tilapia [African freshwater fish], hospital and industrial apparel). Founded in 1850s; incorporated 1901.

Lake Village, village, NEWTON county, NW INDIANA, 11 mi/18 km SSW of LOWELL. La Salle State Fish and Wildlife Area to NW. Agricultural area.

Lakeville, city (2000 population 43,128), DAKOTA county, SE MINNESOTA, suburb 18 mi/29 km S of MINNEAPOLIS, and 22 mi/35 km SW of ST. PAUL; 44°40′N 93°14′W. Agricultural area (grain, soybeans; livestock, poultry; dairying); manufacturing (patterns and prototypes, machinery, paper products, building materials, plastic products, furniture; millwork, food and beverage processing, printing and publishing). Airlake Park Airport to S; Crystal Lake on N boundary, Marion Lake in W center.

Lakeville, resort town in SALISBURY town, LITCHFIELD county, NW CONNECTICUT, on Lake Wononscopomuc. Hotchkiss preparatory school (1892) here. Ethan Allen made Revolutionary War munitions here.

Lakeville, town (2000 population 567), SAINT JOSEPH county, N INDIANA, 11 mi/18 km S of SOUTH BEND; 41°32′N 86°16′W. Manufacturing (transportation equipment); agriculture. Potato Creek State Park nearby (W).

Lakeville, agricultural town, PLYMOUTH county, SE MASSACHUSETTS, on ASSAWOMPSETT POND, and 15 mi/24 km N of NEW BEDFORD; 41°50′N 70°58′W. Manufacturing (paper products). Nearby ponds include LONG, GREAT QUITTACAS, and SNIPATUIT. Headquarters of Ocean Spray Cranberries, Inc. Settled 1717, incorporated 1853.

Lakeville (LAIK-vil), village, ASHTABULA county, extreme NE OHIO, surrounding city of CONNEAUT; 41°57′N 80°34′W. Incorporated 1944.

Lakeville (LAIK-vil), unincorporated village, HOLMES county, central OHIO, near MOHICAN RIVER, 27 mi/43 km ESE of MANSFIELD; 40°39′N 82°07′W.

Lakeville, plantation, PENOBSCOT county, E central MAINE, 50 mi/80 km NNE of BANGOR, in wilderness area; 45°18′N 68°06′W. Hunting, fishing.

Lake Waccamaw (WAK-uh-maw), town (□ 3 sq mi/7.8 sq km; 2006 population 1,458), COLUMBUS county, SE NORTH CAROLINA, 10 mi/16 km E of WHITEVILLE, on N side of LAKE WACCAMAW (c.5 mi/8 km long, 3 mi/4.8 km wide); 34°19′N 78°30′W. Hunting, fishing. Service industries; agriculture (tobacco, peanuts, grain; livestock); manufacturing (tobacco processing). Lake Waccanaw State Park to SE.

Lake Wales (WAILZ), town (□ 6 sq mi/15.6 sq km; 2000 population 10,194), POLK county, central FLORIDA, 12 mi/19 km SE of WINTER HAVEN; 27°53′N 81°35′W. Citrus fruit-shipping center; packing houses, large cannery, tannery; sand and gravel pits. Retirement community in the area. A once-noted resort that includes the Singing Tower (with a carillon of seventy-one bells) and bird sanctuary/park established 1929 by Edward W. Bok. Plotted 1911.

Lake Wallula (wah-LAH-luh), reservoir (□ 6 sq mi/15.6 sq km), in N central OREGON (UMATILLA county), and S central WASHINGTON (Benson county), on Columbia River, at E end of the Columbia Gorge, just N of UMATILLA; 45°56′N 119°18′W. Maximum capacity 1,350,000 acre-ft. Formed by MCNARY DAM (220 ft/67 m high), built (1954) by Army Corps of Engineers for navigation; also used for power generation, irrigation, flood control, and recreation.

Lake Washington Ship Canal, WASHINGTON: see WASHINGTON LAKE.

Lake Waukomis (waw-KO-mis), village (2000 population 917), PLATTE county, W MISSOURI, residential suburb, 12 mi/19 km NNW of downtown KANSAS CITY; 39°13′N 94°38′W.

Lakeway, Village of (LAIK-wai), town (2006 population 9,545), TRAVIS county, S TEXAS, residential suburb 15 mi/24 km WNW of downtown AUSTIN; 30°22′N 97°59′W. Agricultural area (cotton, grain, pecans; cattle; dairying); recreation. LAKE TRAVIS (COLORADO RIVER) to N.

Lake Wilson, village (2000 population 270), MURRAY county, SW MINNESOTA, 10 mi/16 km W of SLAYTON, near Beaver Creek; 43°59′N 95°57′W. Agriculture (grain; livestock, poultry; dairying); manufacturing (feeds).

Lake Windermere, CANADA: see ATHALMER.

Lake Winnebago (wi-nuh-BAI-go), village (2000 population 902), CASS county, W MISSOURI, residential suburb 20 mi/32 km SSE of downtown KANSAS CITY; 38°49′N 94°21′W.

Lake Wobegon, fictional lake and incorporated town, Stolid county, central MINNESOTA. Based on the hometown of philosopher-raconteur Garrison Keillor and featured on his radio show *A Prairie Home Companion*. Elevation 1,418 ft/432 m. The 678-acre/274-ha spring-fed lake (1 mi/1.6 km in diameter, and 100 ft/30 m deep) flows into the Wobegon River, which in turn flows into the SAUK RIVER and then the MISSISSIPPI RIVER. Lake fills meteorite crater with high magnetic iron content, precluding the fixing of true geographical coordinates, and therefore appearing on no maps. Extreme continental climate, with icy-cold winters and cool summers. Settled in 1866 by immigrants from Lagebehynde county in N central NORWAY, who were attracted to the "New Scandinavia" as homesteaders. Dairying supports patented cholesterol-rich ice cream. Ice-fishing, hunting, hiking, and contemplation are major activities and the economy is supplemented by a century of cash savings. Lutheran church and annual ice-fishing tournament are visitor attractions. Celebrated for its demographic uniqueness in that "all the women are strong, all the men are good-looking, and all the children are above average."

Lakewood, city (2000 population 79,345), LOS ANGELES county, S CALIFORNIA; suburb 15 mi/24 km SSE of LOS ANGELES, and 5 mi/8 km NE of LONG BEACH; 33°51′N 118°07′W. Bounded on E by SAN GABRIEL RIVER. Extensive aerospace, high-technology, and electronic industries in area; light manufacturing. Archtypical American suburb; first of 17,500 houses built on assembly-line principles between 1950 and 1952. Long Beach Airport (Daugherty Field) to SW. Site of Lakewood Center Mall, one of the largest shopping centers in U.S. Incorporated 1954.

Lakewood, city, JEFFERSON county, N central COLORADO, a growing suburb 7 mi/11.3 km SW of downtown DENVER; 39°42′N 105°06′W. Elevation 5,440 ft/1,658 m. Drained by Bear Creek in S. The city has become a major suburban business center with the development of high-technology industries and corporate offices, including the huge Denver Federal Center. Manufacturing (medical equipment, laboratory equipment, fabricated metal products, consumer goods, soda ash; printing and publishing). Belmar Museum here; Jefferson county stadium at center of city. Bear Creek Reservoir in SW; FRONT RANGE of ROCKY MOUNTAINS to W. County fairgrounds to W; Red Rock Natural Amphitheater to SW. Incorporated 1969.

Lakewood (LAIK-wuhd), city (□ 7 sq mi/18.2 sq km; 2006 population 52,194), CUYAHOGA county, NE OHIO; suburb of CLEVELAND, on LAKE ERIE; 41°29′N 81°48′W. Many varied industries. City was settled as East Rockport and, renamed in 1889. Incorporated 1911. Population has declined gradually since late 20th century.

Lakewood, unincorporated city (2006 population 57,575), PIERCE county, W WASHINGTON, suburb 6 mi/9.7 km SW of downtown TACOMA, near PUGET SOUND, in LAKES DISTRICT; 47°10′N 122°32′W. Service and residential area for MCCHORD AIR FORCE BASE and Fort Lewis Military Reservation, both to S. Steilacoom and Gravelly lakes are in city. Site of Lakewood Mall, one of the largest shopping centers in U.S.

Lakewood, town (2000 population 36,065), OCEAN county, E central NEW JERSEY, 9 mi/14.5 km N of TOMS RIVER, on the METEDECONK RIVER; 40°05′N 74°13′W. A resort in a scenic region near the ATLANTIC coast and center for Hassidic schools. It has plants making a variety of products. Lakewood was the site of early

ironworks and of the Rockefeller estate, which has become a state arboretum. Georgian Court University is here. Settled 1800, incorporated 1892.

Lakewood, village (2000 population 2,337), MCHENRY county, NE ILLINOIS, residential satellite community of CHICAGO, W of village of CRYSTAL LAKE (town), near Crystal Lake, 8 mi/12.9 km SE of WOODSTOCK; 42°13′N 88°22′W. In agricultural area (dairying; corn).

Lakewood, unincorporated village, EDDY county, SE NEW MEXICO, on W shore of LAKE MCMILLAN reservoir (PECOS RIVER), and 17 mi/27 km NNW of CARLSBAD. In irrigated region; grain, cotton, pecans. BRANTLEY Lake State Park to SE.

Lakewood, resort village (□ 1 sq mi/2.6 sq km; 2006 population 3,091), CHAUTAUQUA county, extreme W NEW YORK, on CHAUTAUQUA LAKE, 5 mi/8 km W of JAMESTOWN; 42°06′N 79°19′W. Some manufacturing. Settled 1809, incorporated 1893.

Lakewood, MAINE: see SKOWHEGAN.

Lakewood Bluff, WISCONSIN: see MAPLE BLUFF.

Lakewood Park (LAIK-wood), town (□ 6 sq mi/15.6 sq km; 2000 population 10,458), ST. LUCIE county, FLORIDA, 6 mi/9.7 km S of VERO BEACH; 27°32′N 80°23′W.

Lake Worth (WUHRTH), city (□ 6 sq mi/15.6 sq km; 2000 population 35,133), PALM BEACH county, SE FLORIDA, 7 mi/11.3 km S of WEST PALM BEACH, on LAKE WORTH (a lagoon); 26°37′N 80°03′W. It is a residential suburb and resort center popular for its bathing and fishing facilities. Manufacturing includes sports equipment, tents and awnings, apparel, and food products Palm Beach Community College is here. Incorporated 1913.

Lake Worth (WUHRTH), town (2006 population 4,718), TARRANT county, N TEXAS, residential suburb 7 mi/11.3 km NW of downtown FORT WORTH, near LAKE WORTH (W Fork Trinity River); 32°48′N 97°25′W. Monroe Creek Lake to SE; Fort Worth Nature Center and Refuge to NW. Incorporated after 1940.

Lake Wylie, unincorporated town (2000 population 3,061), YORK county, N SOUTH CAROLINA, 13 mi/21 km SW of CHARLOTTE, NORTH CAROLINA, and 13 mi/21 km N of ROCK HILL, South Carolina; 35°06′N 81°02′W. Residential community on W shore of L WYLIE reservoir. North Carolina state line to W and N. CATAWBA 1 AND 2 NUCLEAR POWER PLANTS adjoin.

Lake Wyonah, unincorporated town, SCHUYLKILL county, E central PENNSYLVANIA, residential community 6 mi/9.7 km SSE of POTTSVILLE on Plum Creek; 40°35′N 76°10′W.

Lake Zurich (ZUHR-ick), village (2000 population 18,104), Lake county, NE ILLINOIS, suburb 31 mi/50 km NW of downtown CHICAGO, and 17 mi/27 km SW of WAUKEGAN; 42°11′N 88°05′W. Rapid urban growth area; remnant agriculture; manufacturing (fabricated metal products, appliances, textiles, food processing and equipment, aluminum die, water treatment products).

Lakhanokipos (lah-kah-NO-kee-pos), town, THESSALONÍKI prefecture, CENTRAL MACEDONIA department, NE GREECE, on railroad (junction), and 6 mi/9.7 km NW of THESSALONÍKI; 40°27′N 21°13′E. Cotton, wheat, silk. Also called Lachanokepos; formerly Araple or Arapli.

Lakhaoti, INDIA: see AURANGABAD SAIYID.

Lakhapani, INDIA: see LIKHAPANI.

Lakhdaria (lahk-dah-RYAH), town, BOUÏRA wilaya, N central ALGERIA, on the Oued ISSER; 36°34′N 03°36′E. Named after Commandant Si Lakhdar, a War of Independence hero. Has the largest paint production unit in Algeria, which produces most of the country's supply; craft production, notably pottery. Formerly called Palestro.

Lakhdenpokhya (lahkh-dyen-POKH-eeyah), Finnish *Lahdenpohja*, city (2005 population 8,350), SW Re-

public of KARELIA, NW European Russia, port on the NW shore of Lake LADOGA, on road and railroad, 205 mi/330 km W of PETROZAVODSK, and 20 mi/32 km SW of SORTAVALA; 61°31′N 30°11′E. Center of agricultural area; plywood mill. Steamer connection to Valaam Island. Until 1940, in Finland.

Lakheri (luh-KAI-ree), town, BUNDI district, E RAJASTHAN state, NW INDIA, 37 mi/60 km NE of Bundi; 25°40′N 76°10′E. Markets wheat, millet, barley, gram; large cement works.

Lakhi Hills (LAH-khee), most easterly offshoot of KIRTHAR RANGE, in DADU district, W SIND province, SE PAKISTAN; narrow ridge extending c.70 mi/113 km S from INDUS River, 5 mi/8 km S of SEHWAN; 1,500 ft/457 m–2,000 ft/610 m high. S end drained by BARAN RIVER. Little vegetation; hot sulphur springs (N) noted for medicinal properties.

Lakhimpur (luh-KIM-puhr), district (□ 879 sq mi/2,285.4 sq km), NE ASSAM state, NE INDIA; ⊙ NORTH LAKHIMPUR. Mainly in BRAHMAPUTRA RIVER valley; bounded N by DAFLA and ABOR hills, W by Dhansari River, E by undefined India-MYANMAR border (PATKAI RANGE), and S and E by Brahmaputra River. Mainly alluvial soil; tea (major Assam tea garden district), rice, jute, sugarcane, rape and mustard; silk growing; cotton trees in forest area, E.

Lakhimpur (luh-KIM-puhr), town, ⊙ KHERI district, N UTTAR state, N central INDIA, on tributary of the SARDA RIVER, and 75 mi/121 km N of LUCKNOW; 27°57′N 80°46′E. Road junction; trade center (rice, wheat, gram, corn, barley, oilseeds, sugarcane); sugar milling. District headquarters since 1859.

Lakhipur (LUHK-i-puhr), town, GOALPARA district, W ASSAM state, NE INDIA, on left tributary of the BRAHMAPUTRA RIVER, and 22 mi/35 km E of Dhuburi. Rice, mustard, jute.

Lakhisarai, INDIA: see LUCKEESARAI.

Lakhna (LAHK-nah), town, ETAWAH district, W UTTAR PRADESH state, N central INDIA, on LOWER GANGA CANAL, and 12 mi/19 km SSE of Etawah. Pearl millet, wheat, barley, corn, gram, cotton.

Lakhnadon (lahk-NAH-don), town (2001 population 14,343), SEONI district, central MADHYA PRADESH state, central INDIA, 36 mi/58 km N of Seoni; 22°36′N 79°36′E. Sunn hemp retting, essential oil extraction (*rosha* or Andropogon); wheat, rice, oilseeds. Livestock raising, lac cultivation in nearby forested hills.

Lakhnau, INDIA: see HATHRAS.

Lakhnau, INDIA: see LUCKNOW.

Lakhpat (LUHK-puht), town, KACHCHH district, GUJARAT state, W central INDIA, at head of KORI CREEK (inlet of ARABIAN SEA; c.35 mi/56 km long), on SW edge of RANN of Kachchh, 70 mi/113 km NW of BHUJ; 23°49′N 68°47′E. Trades in salt, wheat.

Lakhsas, village, Tiznit province, Souss-Massa-Draâ administrative region, MOROCCO. Center of mountainous region of the same name in the former Spanish enclave of IFNI.

Lakhsetipet, INDIA: see LAKSHETTIPET.

Lakhtar (LUHK-tahr), town, SURENDRANAGAR district, GUJARAT state, W central INDIA, 13 mi/21 km NNE of WADHWAN; 22°51′N 71°47′E. Agriculture market (millet, cotton); cotton ginning, oilseed milling. Was the capital of former EAST KATHIAWAR state of Lakhtar of WEST INDIA STATES agency; state merged 1948 with SAURASHTRA, becoming part of Gujarat state in 1960 when the state was formed.

Lakhtinskiy (LAHKH-teen-skeeyee), former town, NW LENINGRAD oblast, NW European Russia, on Gulf of FINLAND, now a suburb of SAINT PETERSBURG, 7 mi/11 km NW of the city center; 59°59′N 30°09′E. Also called Lakhta.

Laki (LAH-kee) or **Skafta**, volcano (2,684 ft/818 m), S ICELAND, at SW edge of VATNAJOKULL; 64°03′N 18°17′W. Major eruption (1783) was one of most devastating volcanic eruptions on record. Surrounding crater are the Lakagígar series of c.100 volcanic rifts. Surrounding lava field (220 sq mi/570 sq km) is called Skaftáreldahraun.

Laki Marwat (LAH-khee muhr-WAHT), town, BANNU district, SE NORTH-WEST FRONTIER PROVINCE, N PAKISTAN, 32 mi/51 km SE of BANNU; 32°36′N 70°55′E. Also spelled Lakki.

Lakin (LAIK-uhn), town (2000 population 2,316), ☉ KEARNY county, SW KANSAS, on ARKANSAS RIVER, and 21 mi/34 km W of GARDEN CITY; 37°56′N 101°15′W. Wheat, cattle. Lake McKinney, irrigation reservoir to NE.

Lakinsk (LAH-keensk), city (2006 population 16,580), W VLADIMIR oblast, central European Russia, on road and railroad (Undol station), 20 mi/32 km WSW of VLADIMIR; 56°01′N 39°57′E. Elevation 374 ft/113 m. Cotton spinning and weaving (since 1889); brewery. Became city in 1969. Formerly called Lakinskiy.

Lakinskiy, RUSSIA: see LAKINSK.

Lakmos, massif in central PINDOS mountains, S EPIRUS department, NW GREECE, 15 mi/24 km E of IOÁNNINA; 39°40′N 21°07′E. Rises to 7,529 ft/2,295 m in the Peristeri.

Laknauti, INDIA: see GAUR.

Lakonia (lah-ko-NEE-ah), prefecture, SE PELOPONNESE department, extreme SE mainland GREECE, on Mirtoan Sea (E), Gulf of LAKONIA (S), and Gulf of MESSENIA (W), opposite island of KÍTHIRA (ATTICA prefecture); ☉ SPARTA; 37°00′N 22°35′E. Bordered N by ARKADIA prefecture and W by MESSENIA prefecture. Drained by EVROTÁS River. Includes ELAFONISOS island. Also, name of ancient region, S PELOPONNESUS, bounded by Messenia (on the W) and Arcadia and ARGOLIS (on the N). Sparta, the capital, was on the Eurotas (now Evrotás), the principal river. Sparta dominated the region, despite the existence of many other towns, until the rise of the Second Achaean League in the 3rd and 2nd century B.C.E. Also known as Laconia and Lacedaemon.

Lakonia, Gulf of (lah-ko-NEE-ah), off LAKONIA prefecture, SE PELOPONNESE department, extreme SE mainland GREECE, inlet of IONIAN SEA, between capes MATAPAN and Malea (ARGILIOS); 35 mi/56 km wide, 30 mi/48 km long. GYTHION (formerly called Marathonisi) is on NW shore. Formerly called Gulf of Marathonisi; formerly spelled Laconia.

Lakor (LAH-kawr), island, LETI ISLANDS, S MALUKU, Indonesia, in BANDA SEA, just E of MOA, 55 mi/89 km ENE of TIMOR. 10 mi/16 km long, 5 mi/8 km wide; 08°14′S 128°10′E. Coconuts; fishing.

Lakota (lah-KO-tah), town, Sud-Bandama region, S CÔTE D'IVOIRE, 65 mi/105 km NW of GRAND-LAHOU; 05°51′N 05°41′W. Agriculture (coffee, taro, bananas, rice, corn, cacao, palm kernels); livestock (sheep, goats).

Lakota, town (2000 population 255), KOSSUTH county, N IOWA, 22 mi/35 km NNE of ALGONA; 43°22′N 94°05′W.

Lakota, town (2006 population 726), ☉ NELSON CO., NE central NORTH DAKOTA, 24 mi/39 km ESE of DEVILS LAKE; 48°02′N 98°20′W. Livestock; wheat. Railroad junction; STUMP LAKE to S. Founded in 1883 and the name is from the Teton Sioux Indian word for allies.

Laksam, BANGLADESH: see LAKSHAM.

Laksapana Falls, SRI LANKA: see LAXAPANA FALLS.

Laksevåg (LAHK-suh-vawg), village, HORDALAND county, part of BERGEN city, SW NORWAY, on Puddefjorden, just WSW of Bergen center; 60°23′N 05°17′E. Mechanical and metal manufacturing; coastal fortifications.

Lakshadweep (LUHK-sha-dweep), LAKSHADWEEP Union Territory, island group (□ 12 sq mi/31.2 sq km; 2001 population 60,595), SW INDIA, in the ARABIAN SEA off the coast of KERALA state; ☉ KAVARATTI. Of this group of thirty six islands, ten are inhabited and MINICOY is the largest. The islands consist of coral atolls enclosing lagoons. The population, mainly Muslim, engages in fishing and copra production.

Malayalam is the main language except on Minicoy, where Mahl is spoken. The union territory is administered by the home minister in the central government of India with an appointed local advisory council. Also known as CANNANORE ISLANDS. Formerly known as LACCADIVE, MINICOY, AND AMINDIVI ISLANDS.

Laksham (loks-shahm), town (2001 population 54,118), COMILLA district, SE EAST BENGAL, BANGLADESH, on tributary of MEGHNA RIVER, and 16 mi/26 km SSW of COMILLA; 23°17′N 91°02′E. Railroad junction (lines to CHITTAGONG, CHANDPUR, and NOAKHALI); rice, jute, oilseeds. Also spelled Laksam.

Lakshettipet (luhk-SHET-tee-pet), town, ADILABAD district, ANDHRA PRADESH state, central INDIA, on GODAVARI RIVER, and 70 mi/113 km SSE of ADILABAD; 18°52′N 79°13′E. Millet, oilseeds, rice; sawmilling nearby. Sometimes spelled Lakshattipet, Luxhattipet; formerly spelled LAKHSETIPET.

Lakshmantirtha River (luhk-shmuhn-tir-TAH), c.70 mi/113 km long, in KODAGU and MYSORE districts, KARNATAKA state, SW INDIA, on COORG plateau; rises in WESTERN GHATS, SE of MERCARA; flows NNE, past HUNSUR, to KRISHNARAJA NAGARA (reservoir on KAVERI River), 16 mi/26 km NW of MYSORE.

Lakshmeshwar (luhk-SHMAI-shwuhr), town, DHARWAD district, KARNATAKA state, SW INDIA, 38 mi/61 km SE of DHARWAD; 15°08′N 75°28′E. Road center; cotton market; cotton ginning, handicraft cloth weaving. Annual fair. Also spelled Lakshmeshvar or Laxmishwar.

Lakshmikantapur, INDIA: see JAYNAGAR.

Lakshmipur (LUKSH-mee-poor), town (2001 population 63,995), NOAKHALI district, SE EAST BENGAL, BANGLADESH, near MEGHNA RIVER, 19 mi/31 km WNW of NOAKHALI; 22°07′N 90°49′E. Trades in rice, jute, oilseeds; manufacturing (textiles).

Laktaši (lahk-TAW-shee), town, N BOSNIA, BOSNIA AND HERZEGOVINA, near VRBAS RIVER, 11 mi/18 km NNE of Banja Luka; 44°54′N 17°18′E. Hot baths. Also spelled Laktashi.

Lal, town, GHOR province, central AFGHANISTAN, on Lal River (tributary of the HARI RUD), 45 mi/72 km W of PANJAO; 34°30′N 66°17′E. On central KABUL-HERAT road, in the HAZARAJAT region.

La Labor (lah lah-BOR), town, OCOTEPEQUE department, W HONDURAS, 12 mi/20 km ENE of NUEVA OCOTEPEQUE; 14°29′N 89°00′W. Corn, beans.

Lalaghat, INDIA: see HAILAKANDI.

La Laguna, municipality and town, CHALATENANGO department, EL SALVADOR, N of CHALATENANGO city near HONDURAN border.

La Laguna, CANARY ISLANDS: see LAGUNA, LA.

La Laja (lah LAH-hah), canton, VALLEGRANDE province, SANTA CRUZ department, E central BOLIVIA; 18°52′S 64°26′W. Elevation 6,660 ft/2,030 m. Gas wells in area; clay, limestone, and gypsum deposits. Agriculture (potatoes, yucca, bananas, corn, sweet potatoes, peanuts, tobacco); cattle.

La Laja (lah LAH-hah), town, ☉ Laja comuna (2002 population 16,288), BÍO-BÍO province, BÍO-BÍO region, S central CHILE, at confluence of BÍO-BÍO RIVER and LAJA RIVER, on railroad, and 22 mi/35 km NW of LOS ÁNGELES. Agricultural center (cereals, grapes, fruit); food processing, lumbering.

La Laja (lah LAH-hah), village, S SAN JUAN province, ARGENTINA, at S foot of SIERRA de VILLICÚN, 15 mi/24 km NNE of SAN JUAN; 31°21′S 68°28′W. Resort, with mineral springs. Marble, travertine quarries in mountains nearby. Sometimes called Baños de la Laja.

La Laja (lah LAH-hah), village and minor civil division of Los Santos province, LOS SANTOS province, S central PANAMA, in pacific lowland, 1 mi/1.6 km S of LAS TABLAS. Sugarcane, coffee; livestock.

Lala Musa (LAH-lah MOO-sah), town, GUJRAT district, NE PUNJAB province, central PAKISTAN, 11 mi/18 km NW of GUJRAT; 32°42′N 73°58′E. Railroad junc-

tion (workshop). Wheat, millet; cotton-cloth handicrafts, sugar factory.

Lalapansi (lah-lah-PAHN-see), town, MIDLANDS province, central ZIMBABWE, 25 mi/40 km ENE of GWERU, on railroad; 19°20′S 30°10′E. Elevation 4,907 ft/1,496 m. Chromite mining; agriculture (livestock; corn, soybeans, tobacco). Also spelled Lalapansi.

Lalapasa, Turkish *Lalapaşa*, village, European TURKEY, 15 mi/24 km NE of EDIRNE; 41°52′N 26°44′E. Wheat, rice, rye, sugar beets. Sometimes spelled Lalapasha.

Lalapasha, TURKEY: see LALAPASA.

Lalara (lah-LAHR-ah), town, WOLEU-NTEM province, central GABON, 60 mi/97 km NE of NDJOLÉ; 00°23′N 11°26′E. Cocoa-producing center.

La Lata, URUGUAY: see CARDONA.

Lalaua District, MOZAMBIQUE: see NAMPULA.

La Laura Malenga, village (2000 population 1,230), MOKA district, MAURITIUS, 13 mi/20.8 km SE of PORT LOUIS. Sugarcane, vegetables; livestock.

La Lava (lah LAH-vah), canton, JOSÉ MARÍA LINARES province, POTOSÍ department, W central BOLIVIA; 19°55′S 65°38′W. Elevation 10,748 ft/3,276 m. Clay, limestone, minor phosphates, and gypsum deposits. Agriculture (potatoes, yucca, bananas, wheat); cattle.

Lalbagh, INDIA: see MURSHIDABAD.

La Leche River (LAH LAI-chai), LAMBAYEQUE region, NW peru; flows past ILLIMO, 06°27′S 79°53′W. Irrigation.

La Leonesa (lah le-o-NAI-sah), town, Bermejo department, Chaco province, ARGENTINA, 35 mi/56 km NE of city of RESISTENCIA; 27°03′S 58°43′W. Tobacco; livestock.

Laleston (LAIL-stuhn), village (2001 population 8,475), BRIDGEND, S Wales, 7 mi/11.3 km S of MAESTEG; 51°30′N 03°40′W. Previously stone quarrying. Formerly in MID GLAMORGAN, abolished 1996.

Lalganj (LAHL-guhnj), town (2001 population 29,847), VAISHALI district, W central BIHAR state, E INDIA, on GANDAK RIVER, and 23 mi/37 km SW of MUZAFFARPUR; 25°52′N 85°11′E. River trade center; rice, wheat, barley, corn, sugarcane, oilseeds, hides. BASARH, ancient VAISALI, village, 7 mi/11.3 km NW, was the capital of a kingdom during MAGADHA period; one of eight great ancient Buddhist pilgrimage centers. Clay seals discovered here.

Lalganj, town, RAE BARELI district, central UTTAR PRADESH state, N central INDIA, 16 mi/26 km WSW of Rae Bareli. Rice, wheat, barley, gram.

Lalgudi (LAHL-goo-dee), town, TIRUCHIRAPPALLI district, TAMIL NADU state, S INDIA, 9 mi/14.5 km ENE of Tiruchchirappalli; 10°52′N 78°50′E. Rice, plantains, coconut palms, millet. Cement factory at village of DALMIAPURAM (10 mi/16 km NE).

Lālī (LAH-LEE), oil township, Khuzestān province, SW IRAN, on KARUN RIVER, and 20 mi/32 km NE of SHUSHTAR, 140 mi/225 km NNE of ABADAN (linked by pipeline); 32°07′N 48°44′E. Oil field opened 1948.

Lalian (LAHL-yuhn), town, JHANG district, central PUNJAB province, central PAKISTAN, 45 mi/72 km NNE of JHANG-MAGHIANA; 31°49′N 72°48′E.

Lalibela (lah-lee-BAI-lah), town (2007 population 15,363), AMHARA state, NE ETHIOPIA, 12°02′N 39°02′E. Near ABUNE YOSEF mountains, 75 mi/121 km NW of DESSIE. Elevation is 8,629 ft/2,630 m. A major religious and pilgrimage center with 10 famous monolithic churches built in 12th century. Located on a steep slope, the town is an important tourist site. Has airfield.

La Libertad, department (□ 843 sq mi/2,191.8 sq km), W central EL SALVADOR, on the PACIFIC; ☉ NUEVA SAN SALVADOR. Bounded N by LEMPA RIVER; crossed E-W by coastal range (center). Agriculture (coffee, sugarcane, grain), livestock raising; balsam of PERU extraction and hardwood lumbering near coast. Served by railroad lines and INTER-AMERICAN HIGHWAY (N section). Main centers are Nueva San Salvador, QUEZALTEPEQUE; major Pacific port (La

Libertad). Contains SAN SALVADOR volcano. Formed 1865.

La Libertad, municipality and town, La Libertad department, S of Nueva San Salvador on the coast.

La Libertad (lah lee-ber-TAHD), town (2001 population 77,646), GUAYAS province, W ECUADOR, port on S shore of SANTA ELENA BAY, largest town on SANTA ELENA PENINSULA, just E of SALINAS and 75 mi/121 km W of GUAYAQUIL; 02°14′S 80°54′W. Petroleum and refining center. Connected by railroad and pipe line with ANCÓN oil wells, 10 mi/16 km SE.

La Libertad (lah lee-ber-TAHD), town, La Libertad department, SW EL SALVADOR, second largest port of El Salvador, on the PACIFIC, 13 mi/21 km S of NUEVA SAN SALVADOR (connected by road); 13°29′N 89°19′W. Fisheries; grain, livestock raising; popular all-year resort. Exports coffee, sugar, balsam (from PERU); hardwood. Developed in 19th century.

La Libertad, town, HUEHUETENANGO department, GUATEMALA, WNW of HUEHUETENANGO; N of CUILCO, off INTER-AMERICAN HIGHWAY; 15°30′N 91°50′W. Elevation 4,790 ft/1,460 m. Predominantly Ladino town in a largely Mam-speaking region; subsistence farming (grain, sugarcane). Formerly known as Trapichillo (until 1922).

La Libertad (lah lee-ber-TAHD), town, PETÉN department, N GUATEMALA, 18 mi/29 km SW of FLORES; 16°47′N 90°07′W. Elevation 591 ft/180 m. Sugarcane, grain; livestock.

La Libertad (lah lee-ber-TAHD), town, COMAYAGUA department, central HONDURAS, on paved road, 24 mi/39 km N of COMAYAGUA; 14°43′N 87°36′W. Elevation 1,969 ft/600 m. Coffee production.

La Libertad, town, FRANCISCO MORAZÁN department, central HONDURAS, 33 mi/53 km SW of TEGUCIGALPA; 13°42′N 87°28′W. Small farming, grain, beans; livestock.

La Libertad (lah lee-ber-TAHD), town, CHIAPAS, S MEXICO, in USUMACINTA River lowland, and 4 mi/6.4 km SSE of EMILIANO ZAPATA. Elevation 177 ft/54 m. Rubber; fruit.

La Libertad (lah lee-ber-TAHD), town, CHONTALES department, S NICARAGUA, 20 mi/32 km NE of JUIGALPA; 12°13′N 85°10′W. Gold- and silver-mining center. Nearby are San Juan and Esmeralda mines.

La Libertad (LAH lee-ber-TAHD), region (□ 9,851 sq mi/25,612.6 sq km), NW PERU; ⊙ TRUJILLO; 08°00′S 78°30′W. Bordered by the PACIFIC (W) and Cordillera CENTRAL of the ANDES (E). E section includes the CORDILLERA OCCIDENTAL, separated from Cordillera CENTRAL by MARAÑÓN RIVER. Its coastal region, drained by JEQUETEPEQUE, CHICAMA, Moche, and Virú rivers, contains a large irrigation system, one of Peru's main sugarcane districts, also producing rice and cotton. Barley and wheat are cultivated in mountains, where cattle and sheep are also raised. Coca grows on E and W slopes of the Andes. Gold, silver, copper mined at SALPO and QUIRUVILCA. Region is crossed by PAN-AMERICAN HIGHWAY. Other centers: CHEPÉN, Trujillo, SAN PEDRO DE LLOC, SANTIAGO DE CHUCO, OTUZCO, ASCOPE, HUAMACHUCO, BOLÍVAR. Site of the RESERVA NACIONAL CALIPUY. Also known as Libertad.

La Ligua (lah LEE-gwah) or **Ligua**, town, ⊙ La Ligua comuna PETORCA province, VALPARAISO region, central CHILE, on LIGUA RIVER and railroad, 50 mi/80 km NNE of VALPARAISO. Railroad and agricultural center (vegetables, cereals).

La Lima (lah LEE-mah), city (2001 population 42,391), CORTÉS department, NW HONDURAS, on CHAMELECÓN RIVER, and 7 mi/11.3 km SE of SAN PEDRO SULA; 13°38′N 87°05′W. In banana zone; consists of old commercial section (left bank) and new residential section (right bank); headquarters of nearby banana plantations.

Lalín (lah-LEEN), city, PONTEVEDRA province, NW SPAIN, in GALICIA, 26 mi/42 km SE of SANTIAGO;

42°39′N 08°07′W. Agricultural trade (cereals, potatoes; livestock). Lignite deposits nearby.

Lalin (LAH-LIN), town, S HEILONGJIANG province, Northeast, CHINA, 40 mi/64 km SSE of HARBIN, and on railroad, near Lalin River (short affluent of SONGHUA River); 45°13′N 126°56′E. Grain, sugar beets, jute.

Lalinde (lah-lan-duh), commune (□ 10 sq mi/26 sq km), DORDOGNE department, AQUITAINE region, SW FRANCE, on the DORDOGNE, and 12 mi/19 km E of BERGERAC; 44°50′N 00°44′E. Wine-growing area; food processing (goose liver pâté, truffles), woodworking. Founded as a medieval stronghold (a *bastide*).

La Línea, SPAIN: see LÍNEA, LA.

Lalitpur (luh-LIT-puhr), district (□ 1,946 sq mi/5,059.6 sq km), UTTAR PRADESH state, N central INDIA; ⊙ Lalitpur.

Lalitpur (luh-LEET-poor), district, central NEPAL, in BAGMATI zone; ⊙ PATAN.

Lalitpur (luh-LIT-puhr), town, ⊙ Lalitpur district, UTTAR PRADESH state, N central INDIA; 24°41′N 78°25′E.

Lallaing (lah-lan), residential town (□ 2 sq mi/5.2 sq km; 2004 population 6,631), NORD department, NORD-PAS-DE-CALAIS region, N FRANCE, on the SCARPE (canalized), and 4 mi/6.4 km ENE of DOUAI; 50°23′N 03°10′E. In metalworking and chemical industry district.

Lalla Khadidja, ALGERIA: see LELLA KHEDIDJA.

Lalla Takerkoust, MOROCCO: see N'FISS, OUED.

Lal-lo (LAHL-lo), town, CAGAYAN province, N LUZON, PHILIPPINES, on CAGAYAN RIVER, and 11 mi/18 km S of APARRI; 18°11′N 121°53′E. Agricultural center (rice, tobacco).

Lalmanir Hat (lahl-mo-nir-haht), town (2001 population 57,236), RANGPUR district, BANGLADESH; 25°58′N 89°02′E. Railroad workshops; trades in rice, jute, tobacco, oilseeds. Rice and oilseed milling, soap manufacturing. Also called Lalmonirhat.

Lalmatie, village (2000 population 9,532), FLACQ district, MAURITIUS, 18 mi/28.8 km E of PORT LOUIS. Agriculture (sugarcane; livestock); light industry.

La Loma (lah LO-mah), canton, SUD CINTI province, CHUQUISACA department, SE BOLIVIA, 12 mi/20 km SE of CULPINA; 20°53′S 65°23′W. Elevation 7,575 ft/2,309 m. Clay, limestone, and gypsum deposits. Agriculture (potatoes, yucca, bananas, corn, wheat, oats, rye, peanuts); cattle and pig raising.

La Loma, DOMINICAN REPUBLIC: see LOMA DE CABRERA.

Lalor (LAI-luhr), suburb 11 mi/17 km N of MELBOURNE, VICTORIA, SE AUSTRALIA; 37°40′S 145°01′E.

Lalouvesc (lah-loo-vesk), commune (□ 4 sq mi/10.4 sq km), ARDÈCHE department, RHÔNE-ALPES region, SE central FRANCE, 20 mi/32 km NW of VALENCE, in the Haut VIVARAIS hills (W of RHÔNE RIVER valley); 45°07′N 04°32′E. Elevation 3,450 ft/1,052 m. It is a climate resort, which also attracts pilgrims to its 19th-century basilica where St. Regis is buried and local saints are honored.

La Louvière, BELGIUM: see LOUVIÈRE, LA.

Lalsanga, INDIA: see CHAMOLI.

Lal'sk (LAHLSK), town, NW TVER oblast, central European Russia, on the Lala River (tributary of the LUZA RIVER), on road and railroad, 327 mi/526 km NW of KIROV, and 45 mi/72 km SE of KOTLAS; 60°44′N 47°35′E. Elevation 291 ft/88 m. Dairy products. Made city in 1780; reduced to status of a village, 1917–1927; made town in 1927.

Lal'sk (LAHLSK), town (2005 population 4,370), NW KIROV oblast, E central European Russia, on the LUZA RIVER, on road and local railroad, 12 mi/19 km ENE, and under administrative jurisdiction, of LUZA; 60°44′N 47°35′E. Elevation 291 ft/88 m. Highway junction; road construction and repair concern. In agricultural area; food processing (bakery, creamery). Logging and lumbering enterprise. Has a hospital.

Lalsot, town, JAIPUR district, E RAJASTHAN state, NW INDIA, 40 mi/64 km SE of Jaipur; 26°34′N 76°20′E. Markets millet, barley, gram. In nearby battle (1787), combined forces of Jaipur and JODHPUR defeated Marathas.

La Luz (lah LOOS), unincorporated town (2000 population 1,615), OTERO county, S central NEW MEXICO, residential suburb 5 mi/8 km N of ALAMOGORDO; 32°58′N 105°56′W. Cattle, sheep; pecans, fruit, alfalfa. SACRAMENTO MOUNTAINS and Lincoln National Forest to E; MESCALERO Apache Indian Reservation to NE.

La Luz (lah LOOS), mining settlement, GUANAJUATO, central MEXICO, in SIERRA MADRE OCCIDENTAL, 7 mi/11.3 km WNW of GUANAJUATO; 23°04′N 102°52′W. Elevation 7,644 ft/2,330 m. Some mercury mining. Also called Mineral de la Luz.

La Luz, NICARAGUA: see SIUNA.

Lama (LAH-mah), lake (□ 174 sq mi/452.4 sq km), in the NW part of the CENTRAL SIBERIAN PLATEAU, KRASNOYARSK TERRITORY, RUSSIA; 62 mi/100 km long, 12 mi/20 km wide, up to 66 ft/20 m deep; 69°30′N 90°00′E.

Lama (LAH-mah), river (basin □ 900 sq mi/2,330 sq km), 86 mi/139 km long, European Russia; rises in MOSCOW oblast; passes VOLOKOLAMSK, flows into the Shosha Bay or Ivan'kovskiy Reservoir (TVER oblast). The ancient waterways between the VOLGA and MOSKVA rivers followed its course.

La Macarena, Serranía de (LAH mah-kah-RAI-nah, ser-rah-NEE-ah dai), E outlier of the Cordillera ORIENTAL, META department, S central COLOMBIA; extends c.80 mi/130 km N-S; rises to over 7,000 ft/2,135 m; 02°45′N 73°55′W. Also called Cordillera Macarena.

La Macaza, village, S central QUEBEC, E CANADA; 46°22′N 74°46′W. An independent municipality, it is part of the Rivière-Rouge agglomeration.

Lamadelaine (lah-mah-DLEN), German *Rollingen*, town, Pétange commune, SW LUXEMBOURG, just ESE of RODANGE, near French border; 49°33′N 05°52′E. Iron-mining center.

La Madrid (lah mah-DREED) or **Lamadrid**, town, S TUCUMÁN province, ARGENTINA, on MARAPA River, and 55 mi/89 km S of TUCUMÁN; 27°38′S 65°15′W. Railroad junction and agricultural center (tobacco, alfalfa, wheat; livestock). Near Río Hondo Reservoir. Sometimes called General Lamadrid.

Lamadrid (lah-mah-DREED), town, COAHUILA, N MEXICO, 26 mi/42 km WNW of MONCLOVA, off Mexico Highway 130, on railroad; 27°05′N 101°49′W. Elevation 2,362 ft/720 m. Cereals, cattle.

Lamadrid, for Argentine names not found here: see under General LAMADRID.

La Magdalena, Colombia: see PUERTO NARE.

La Magdalena Contreras (LAH mahg-dah-LAI-nah), delegación, in SW of Federal Distrito, MEXICO, 5.6 mi/9 km W of TLALPAN. Includes Sierra de las Cruces and Sierra de Ajusco. Site of some of MEXICO City's most exclusive residential areas, along with middle-class housing developments. Conflicts have developed between real estate promoters and conservationists over preservation of forested areas, of which few remain in the Federal Distrito, Site of an important battle (August 19–20, 1847) of the Mexican-American War.

La Magdalena Tlatlauquitepec (lah mahg-dah-LAI-nah tlah-tlah-oo-KEE-te-pek), town, PUEBLA, central MEXICO, 21 mi/34 km SSE of PUEBLA; 18°45′N 98°06′W. Cereals; livestock. Also La Magdalena.

Lama-Kara (lah-MAH–kah-RAH), town (2003 population 49,800), ⊙ Kozah prefecture and Kara region, N Togo, 75 mi/121 km N of Sokodé on Kara River; 09°35′N 01°10′E. Cotton, peanuts, shea nuts; cattle, sheep, and goat raising. International airport. Sometimes known as Kara.

La Malbaie (lah mahl-BAI) or **Murray Bay** (MUH-ree BAI), village (□ 182 sq mi/473.2 sq km), ⊙ CHARLE-

VOIX-EST county, Capitale-Nationale region, S central QUEBEC, E CANADA, at the confluence of the MALBAIE (or Murray) River with the SAINT LAWRENCE RIVER; 47°39′N 70°10′W. Well-known resort in dairy-farming country; lumber. Restructured in 1999.

La Malinche National Park (lah mah-LEEN-che), on the border between the states of PUEBLA and TLAXCALA, Mexico, 9 mi/14 km E of APIZACO then 8 mi/13 km S to Albergue La Malintzin in SE Tlaxcala. The park is located at the foot of the huge MALINCHE Volcano. (Elevation 14,636 ft/4,461 m) within the Mexical Volcanic Axis; covers 114,277 acres/46,248 ha.

Lamalou-les-Bains (lah-mah-loo–lai–ban), commune (□ 2 sq mi/5.2 sq km), HÉRAULT department, LANGUEDOC-ROUSSILLON region, S FRANCE, in the Monts de l'ESPINOUSE, near the ORB RIVER, 15 mi/24 km SW of LODÈVE; 43°36′N 03°04′E. Spa with mineral springs. Old-fashioned railroad circuit in picturesque Orb valley is a tourist attraction.

La Mamacoma, PERU: see PACHACAMAC.

Lamanai National Park, archaeological site, ORANGE WALK district, BELIZE, 50 mi/80 km NW of BELIZE CITY on new River; 17°45′N 88°40′W. May have been continuously occupied for 3,000 years up to mid-17th century. Partly reconstructed.

La Mancha, SPAIN: see MANCHA, LA.

Lamangan, INDONESIA: see LAMONGAN.

Lamaní (lah-mah-NEE), town, COMAYAGUA department, W central HONDURAS, 8 mi/12.9 km S of LA PAZ, on secondary road; 14°08′N 87°35′W. Small farming, grain; livestock.

Lamanon Mountains (luh-muh-NON), ancient volcanic massif, W Sakhalin Mountains, forming the outcrop of the W shore of SAKHALIN Island (Cape Lamanon), SAKHALIN oblast, extreme E SIBERIA, RUSSIAN FAR EAST; 48°42′N 142°06′E. Consists of lava plateaus (elevation 984 ft/300 m–2,625 ft/800 m) and groups of volcanic domes (elevation 1,969 ft/600 m–3,281 ft/1,000 m). Tayga. Named by the French navigator J. La Pérouse.

La Mansión (lah mahn-see-YON), village, GUANACASTE province, NW COSTA RICA, 6 mi/9.7 km ESE of NICOYA. Grain, sugarcane; livestock.

La Manzanilla de la Paz (lah mahn-sahn-EE-yah dai la pahs), town, ⊙ La Manzanilla de la Paz municipio, JALISCO, central MEXICO, in SIERRA MADRE OCCIDENTAL, 30 mi/48 km ENE of SAYULA, and 10 mi/16 km S of Lake CHAPALA; 20°00′N 103°09′W. Grain, beans, fruit; livestock.

La Mar (LAH MAHR), province, AYACUCHO region, S central PERU; ⊙ SAN MIGUEL. Situated between ANDAHUAYLAS, HUANTA, and HUAMANGA provinces and the APURÍMAC RIVER.

Lamar (luh-MAHR), county (□ 605 sq mi/1,573 sq km; 2006 population 14,548), W ALABAMA; ⊙ Vernon. Level region bordering on Mississippi, drained by Buttahatchee River Corn, timber; crude-oil and natural-gas production. Created in 1867 as Jones County after E. P. Jones of Fayette Co., it was recreated the same year as Sanford County in honor of H. C. Sanford of Cherokee County. The name was changed to Lamar on Feb. 8, 1877, for Lucius Quintus Cincinnatus Lamar, a Georgia-born soldier and statesman.

Lamar, county (□ 181 sq mi/470.6 sq km; 2006 population 16,679), W central GEORGIA; ⊙ BARNESVILLE; 33°04′N 84°08′W. Piedmont agriculture featuring cotton, corn, soybeans, wheat; cattle, hogs, poultry; lumber and wood products. Formed 1920.

Lamar, county (□ 500 sq mi/1,300 sq km; 2006 population 46,240), SE MISSISSIPPI; ⊙ PURVIS; 31°12′N 89°30′W. Drained by WOLF River, and by BLACK and RED creeks, source of Wolf River in SW. Agriculture (cotton, corn, pecans; poultry, cattle); pine timber; oil and natural gas. Outskirts (W) of HATTIESBURG in NE. Formed 1904.

Lamar (luh-MAHR), county (□ 932 sq mi/2,423.2 sq km; 2006 population 49,863), NE TEXAS; ⊙ PARIS;

33°40′N 95°34′W. Bounded N by RED RIVER (here the OKLAHOMA state line), S by North Fork of SULPHUR RIVER. Diversified agriculture including cotton, hay, peanuts; cattle; timber; manufacturing at PARIS. Pat Mayse Lake in N; Lake Crook reservoir near center. Formed 1840.

Lamar (luh-MAHR), city (2000 population 4,425), ⊙ BARTON county, SW MISSOURI, on branch of SPRING RIVER, and 32 mi/51 km NNE of JOPLIN; 37°29′N 94°16′W. Agriculture (wheat, corn, hay, sorghum; cattle); manufacturing (furniture, metal products). Founded c.1856. Harry S. Truman born here 1884. TRUMAN State Historic Site.

Lamar (luh-MAHR), town (2000 population 1,415), JOHNSON county, NW ARKANSAS, 14 mi/23 km NW of RUSSELLVILLE, near ARKANSAS RIVER, and Lake DARDANELLE; 35°26′N 93°23′W. Manufacturing (wooden truss parts).

Lamar, town (2000 population 8,869), ⊙ PROWERS county, SE COLORADO, on ARKANSAS River, and 50 mi/80 km ENE of LA JUNTA; 38°04′N 102°37′W. Elevation 3,622 ft/1,104 m. Junction of railroad spurs to WILEY and HARTMAN. Food-processing center for grain and cattle region; manufacturing (transit buses, forage tubs, clay products, alfalfa products). Lamar Community College, Madonna of the Trail Monument, Big Timbers Museum here. Cluster of reservoirs to N; Thurston, King, Nee Shah, Nee Nashe, Nee So Pah, and Nee Grande (the Nee reservoirs). Incorporated 1886.

Lamar, unincorporated town, CLINTON county, N central PENNSYLVANIA, 9 mi/14.5 km SW of LOCK HAVEN, on Fishing Creek; 41°00′N 77°31′W. Agriculture (livestock; dairying). Laman National Fish Hatchery to S. Bald Eagle State Park to NW.

Lamar, town (2006 population 1,003), DARLINGTON county, NE central SOUTH CAROLINA, 17 mi/27 km W of FLORENCE; 34°10′N 80°04′W. Manufacturing (tables, metal fabrication, apparel, machinery); agriculture (poultry, livestock; dairying; grain, soybeans, tobacco).

Lamar (luh-MAHR), village, SPENCER county, SW INDIANA, 3 mi/4.8 km S of SANTA CLAUS. Bituminous-coal mining and processing.

Lamar, village, BENTON county, N MISSISSIPPI, 12 mi/19 km NNE of HOLLY SPRINGS. In agricultural area.

Lamar, village (2006 population 18), CHASE county, SW NEBRASKA, 17 mi/27 km W of IMPERIAL, near COLORADO state line; 40°34′N 101°58′W.

Lamar, village (2006 population 170), HUGHES county, central OKLAHOMA, 16 mi/26 km E of HOLDENVILLE, near CANADIAN River; 35°06′N 96°07′W. In agricultural area; manufacturing (consumer goods).

Lamarão (LAH-mah-ROUN), city (2007 population 10,533), E BAHIA, BRAZIL, on SALVADOR-JUAZEIRO railroad, 16 mi/25 km SE of SERRINHA; 11°47′S 38°52′W.

Lamarche (lah-MAHRSH), village (□ 37 sq mi/96.2 sq km; 2006 population 522), SAGUENAY—LAC-SAINT-JEAN region, S central QUEBEC, E CANADA, 10 mi/17 km from BÉGIN; 48°48′N 71°26′W.

Lamar Heights, village (2000 population 216), BARTON county, SW MISSOURI, suburb 1 mi/1.6 km W of LAMAR; 37°29′N 94°17′W. Residential and commercial area.

La Marina (lah mah-REE-nah), village, ALAJUELA province, N COSTA RICA, 4 mi/6.4 km ENE of CIUDAD QUESADA; 10°23′N 84°23′W.

La Mariscala, URUGUAY: see MARISCALA.

La Marque (luh-MAHRK), city (2006 population 14,033), GALVESTON county, SE TEXAS, suburb 5 mi/8 km SW of TEXAS CITY, on Highland Bayou; 29°22′N 94°59′W. In an agricultural and oil area; manufacturing (machining, machinery, electronic equipment, cable wire). Originally a farm settlement, it later became a railroad shipping point between HOUSTON and GALVESTON. Modern La Marque is primarily a residential suburb for workers in

Texas City and other nearby industrial centers. It is also an oil refining center. Settled c.1860, incorporated 1953.

La Martre (lah MAHR-truh), village (□ 72 sq mi/187.2 sq km; 2006 population 255), GASPÉSIE—ÎLES-DE-LA-MADELEINE region, SE QUEBEC, E CANADA, 6 mi/9 km from MARSOUI; 49°10′N 66°10′W.

Lamas (LAH-mahs), province (□ 770 sq mi/2,002 sq km), SAN MARTÍN region, N central PERU; ⊙ LAMAS; 06°20′S 76°40′W. N central province, bordering on LORETO region.

Lamas (LAH-mahs), city, ⊙ LAMAS province, SAN MARTÍN region, N central PERU, near MAYO RIVER, in E ANDEAN outliers, 45 mi/72 km SE of MOYOBAMBA, 11 mi/18 km WNW of TARAPOTO; 06°25′S 76°32′W. Cotton, sugarcane, tobacco, coffee, coca, bananas, plantains, yucca, cacao, rice.

Lamas, Cerro, ARGENTINA: see PUNTIAGUDO, CERRO.

La Masica (lah mah-SEE-kah), town, ATLÁNTIDA department, N central HONDURAS, on Highway 50, on railroad, and 13 mi/21 km SW of LA CEIBA; 15°37′N 87°07′W. Tropical fruit (bananas).

Lamastre (lah-mah-struh), town (□ 9 sq mi/23.4 sq km), ARDÈCHE department, RHÔNE-ALPES region, S FRANCE, on small Doux, and 13 mi/21 km W of VALENCE; 44°59′N 04°35′E. Small industry (consumer goods, furniture, and regional handicrafts).

La Mata (lah MAH-tah), town, 15 mi/23 km SE of SAN FRANCISCO DE MACORÍS, SÁNCHEZ RAMÍREZ province, central DOMINICAN REPUBLIC; 19°06′N 70°10′W.

La Maurice National Park (lah mo-REES) (□ 210 sq mi/546 sq km), S QUEBEC, E CANADA, near TROIS-RIVIÈRES. In heavily wooded part of the LAURENTIAN MOUNTAINS. Established 1970.

La Maya (lah MEI-yah), town (2002 population 21,278), SANTIAGO DE CUBA province, E CUBA, on railroad, and 17 mi/27 km NE of SANTIAGO DE CUBA; 20°10′N 75°39′W. Fruit; cattle.

Lamayuru (LAH-mah-yoo-roo), town, LEH district, JAMMU AND KASHMIR state, extreme N INDIA, in central KASHMIR region, in NW ZASKAR Range, near left tributary of the INDUS River, 45 mi/72 km WNW of LEH; 34°18′N 76°46′E. Pulses, wheat. A religious and historic site and tourist center. Large Buddhist monastery, founded 9th or 10th century; former center of Bon religion. FOTU LA pass (13,432 ft/4,094 m) in Zaskar Range, is 4 mi/6.4 km W, on Kargil-Leh trade route. Also spelled as Lamayuru Gompa.

Lamb (LAM), county (□ 1,017 sq mi/2,644.2 sq km; 2006 population 14,244), NW TEXAS; ⊙ LITTLEFIELD; 34°04′N 102°20′W. On the LLANO ESTACADO; elevation c.3,500 ft/1,070 m–3,900 ft/1,190 m. Drained by intermittent Blackwater Draw (center) and Running Water Draw (NE). Rich agriculture and livestock region (cattle, sheep) with irrigated areas; leads in TEXAS production of grain sorghum; also grows cotton, corn, vegetables; hay; soybeans. Oil, natural-gas wells; potash deposits; stone. Formed 1876.

Lambach (LAHM-bahkh), township, central UPPER AUSTRIA, on TRAUN river, and 7 mi/11.3 km SW of WELS; 48°06′N 13°53′E. Railroad junction; manufacturing of machines. Benedictine abbey (founded 1056) with large library whose holdings include treasured manuscripts and paintings, as well as the oldest Romanesque frescoes in AUSTRIA.

Lambaesis, ALGERIA: see TAZOULT-LAMBÈZE.

Lamballe (lahn-bahl), town (□ 29 sq mi/75.4 sq km), CÔTES-D'ARMOR department, W FRANCE, in BRITTANY, 12 mi/19 km ESE of SAINT-BRIEUC; 48°28′N 02°31′W. Commercial center; site of national stud farms and equestrian training center. Site of Norman Gothic church of Notre Dame, originally part of a castle, offers a fine view of the surroundings.

Lambaré (lahm-bah-RAI), city (2002 population 119,795), Central department, PARAGUAY, near Paraguay River; 25°21′S 57°39′W. Liquor distilling; saltworks. Now largely a suburb of Asunción.

Lambaréné (lahm-bah-RAI-nai), city (2003 population 9,000), W central GABON, 100 mi/161 km SE of LIBREVILLE; ⊙ MOYEN OGOOUÉ province; 00°42'S 10°12'E. In palm oil and coffee-growing region. Site of the Schweitzer Hospital and Museum.

Lambari (lahm-bah-ree), city (2007 population 18,547), S MINAS GERAIS state, BRAZIL, on railroad, and 32 mi/51 km N of ITAJUBÁ; 21°59'S 45°25'W. Elevation 2,965 ft/904 m. Resort noted for its cool, dry climate and mineral springs. Tourist season: February–April. Old spelling Lambary. Formerly also called Aguas Virtuosas.

Lambari d'Oeste (LAHM-bah-ree DO-es-chee), town, SW MATO GROSSO state, BRAZIL, 62 mi/100 km NW of CACERES; 15°15'S 58°00'W.

Lambasa, FIJI: see LABASA.

Lambate (lahm-BAH-tai), canton, SUD YUNGAS province, LA PAZ department, W BOLIVIA, 19 mi/30 km SE of LA PAZ, on a branch of the La Paz–Tirata road; 16°37'S 67°36'W. Elevation 5,686 ft/1,733 m. Copper-bearing lode; tungsten mining at Mina Reconquistada, Mina Chojlla, and the cooperative Mina Bolsa Negra; clay, limestone, and gypsum deposits. Agriculture (potatoes, yucca, soy, tobacco, coffee, tea, citrus fruits, bananas, rye); cattle.

Lambayeque (lahm-bah-YAI-kai), region (□ 5,488 sq mi/14,268.8 sq km), NW PERU; ⊙ CHICLAYO; 06°10'S 80°00'W. Bordered by the PACIFIC (W) and CORDILLERA OCCIDENTAL (E). Includes deserts of MÓRROPE, Palo Grueso, Salitre, and Reque, and LOBOS ISLANDS. Situated mainly on coastal plain, little rainfall; drained by Reque, LAMBAYEQUE, and SAÑA rivers and Taimi Canal, which feed numerous irrigation channels. Region produces 25% of Peru's rice and 33% of Peru's sugarcane; also yields cotton, corn, vegetables; livestock; apiculture. Sugar milling and distilling (CAYALTÍ, TUMÁN), rice milling (CHICLAYO, PUCALÁ, and FERREÑAFE). Main cities are Lambayeque, Ferreñafe, and Chiclayo.

Lambayeque (lahm-bah-YAI-kai), province (□ 3,615 sq mi/9,399 sq km), LAMBAYEQUE region, NW PERU; ⊙ LAMBAYEQUE; 06°05'S 80°05'W. Largest and northernmost province of Lambayeque region. Crossed by the PAN-AMERICAN HIGHWAY.

Lambayeque (lahm-bah-YAI-kai), city (2005 population 46,351), ⊙ LAMBAYEQUE province, LAMBAYEQUE region, NW PERU, on coastal plain, on LAMBAYEQUE RIVER, and 7 mi/11 km NW of CHICLAYO, on PAN-AMERICAN HIGHWAY; 06°42'S 79°55'W. Trade center for irrigated cotton and rice area; sugarcane, corn; livestock.

Lambayeque River (lahm-bah-YAI-kai), 70 mi/113 km long, LAMBAYEQUE region, NW PERU; formed by confluence of CHANCAY RIVER and Cumbil River 9 mi/15 km ENE of CHONGOYAPE (06°43'S 79°54'W); flows W through major sugarcane region, past LAMBAYEQUE, to the PACIFIC 5 mi/8 km SW of Lambayeque. Feeds numerous irrigation channels in middle and lower course. Called Chancay River in upper course, up to branches of Taymi Canal and Eten River.

Lambay Island (LAM-bai), Gaelic *Reachrainn*, NE DUBLIN county, E IRELAND, in the IRISH SEA, 8 mi/12.9 km ENE of SWORDS; 2 mi/3.2 km long, 1 mi/1.6 km wide; 53°29'N 06°01'W. Rises to 418 ft/127 m. Has remains of 16th-century fortifications. It is now a bird sanctuary and coastguard station. Retreat of Archbishop James Ussher.

Lamberhurst (LAM-buh-huhst), village (2001 population 1,491), SW KENT, SE ENGLAND, 6 mi/9.7 km ESE of TUNBRIDGE WELLS; 51°05'N 00°23'E. Former agricultural market. Was a center of Wealden iron industry. Has 14th-century church and remains of abbey (1260).

Lambersart (lahm-ber-sahr), NW suburb (□ 2 sq mi/5.2 sq km) of LILLE, NORD department, NORD-PAS-DE-CALAIS region, N FRANCE; 50°39'N 03°02'E. Textile dyeing; manufacturing (electrical equipment, tiles). Suburb lies N of Lille's citadel.

Lambert, town (2000 population 1,967), QUITMAN county, NW MISSISSIPPI, 15 mi/24 km E of CLARKSDALE; 34°12'N 90°16'W. Agriculture (cotton, rice, sorghum, soybeans); manufacturing (apparel). O'Keefe Wildlife Management Area to E.

Lambert, village, RICHLAND county, NE MONTANA, on Fox Creek, tributary of YELLOWSTONE RIVER, and 22 mi/35 km W of SIDNEY. Wheat, barley, hay; cattle. Fox Lake Wildlife Management Area to W. Originally called Fox Lake.

Lambert, village (2006 population 8), ALFALFA county, N OKLAHOMA, 15 mi/24 km ESE of ALVA; 36°40'N 98°25'W. In grain-growing area.

Lambert Glacier, glacier, EAST ANTARCTICA; 250 mi/400 km long, up to 40 mi/65 km wide; 71°00'S 70°00'E. One of the largest glaciers in the world. Together with its tributary glaciers it funnels ice from a major section of the E Antarctic plateau into the AMERY ICE SHELF.

Lamberton, town (2000 population 859), REDWOOD county, SW MINNESOTA, near COTTONWOOD RIVER, 23 mi/37 km SSW of REDWOOD FALLS; 44°13'N 95°15'W. Grain, soybeans; livestock; dairying.

Lambert's Bay, Afrikaans *Lambertsbaai*, town, WESTERN CAPE province, SOUTH AFRICA, on Lambert's Bay on the Atlantic, 130 mi/209 km N of CAPE TOWN on Jakkals River; 32°06'S 18°17'E. Elevation 114 ft/34 m. Resort; fishing port; crayfish canning. Bird Island, islet with guano deposits, just offshore. Airfield.

Lambertville, city (2006 population 3,808), HUNTERDON county, W NEW JERSEY, 14 mi/23 km NW of TRENTON, and on DELAWARE RIVER (bridged here) opposite NEW HOPE, PENNSYLVANIA; 40°22'N 74°56'W. Former industrial area replaced by art galleries and retail shops. Delaware and Raritan Canal State Park here. Founded 1732; incorporated 1849.

Lambesc (lahn-besk), town (□ 25 sq mi/65 sq km), BOUCHES-DU-RHÔNE department, SE FRANCE, in PROVENCE, and 13 mi/21 km NW of AIX-EN-PROVENCE; 43°39'N 05°16'E. Market for olive oil, almonds, and aromatic plants.

Lambeth (LAM-buhth), inner borough (□ 10 sq mi/26 sq km; 2001 population 266,169) of GREATER LONDON, SE ENGLAND, on the THAMES RIVER; 51°30'N 00°05'W. Here opposite Westminster is Lambeth Palace, the residence of Archbishop of Canterbury. Also Vauxhall Gardens, the 18th-century place of entertainment. Largely residential but important as an area of governmental and commercial offices. Major transportation hub with several railroad stations, including Waterloo, London's largest. Lambeth is connected to WESTMINSTER borough across the Thames by five bridges. The National Theatre (the Old Vic), the National Film Theatre, and the Royal Festival Hall are in Lambeth, as are the Imperial War Museum, Morley College, and eight hospitals, including two medical schools (St. Thomas's and King's College). Districts in the borough include Brixton, Clapham, Kennington, Streatham, and Herne Hill.

Lambir Hills National Park (LAHM-bir), NE SARAWAK, E MALAYSIA, on BRUNEI border, 25 mi/40 km SE of MIRI; 04°11'N 114°20'E. Sandstone hills with rugged cliffs. Dipterocarp forests; heath on ridges; several species of palm. Borneo gibbons, barking deer, bearded pigs; over 100 bird species. Walking tracks include high suspension bridge; waterfall. Road access.

Lambityeco (lahm-beet-YE-ko), a historic site, in central OAXACA, MEXICO, 19 mi/30 km S of OAXACA DE JUÁREZ, off Highway 190. This Zapotec site is still being excavated and has not yet been restored.

Lambo, town, NORTHERN REGION, GHANA, on MOUHOUN RIVER, 30 mi/48 km NW of KINTAMPO; 08°37'N 01°53'W. Fishing area.

Lambourn (LAM-bawn), town (2001 population 5,967), West Berkshire, S ENGLAND, 6 mi/9.7 km N of HUNGERFORD; 51°31'N 01°32'W. Agricultural market, center of racehorse training. Has 12th–15th-century church and ancient market cross.

Lambrama (lahm-BRAH-mah), town, ABANCAY province, APURÍMAC region, S central PERU, in ANDEAN spur, 18 mi/29 km SSE of ABANCAY; 13°52'S 72°46'W. Sugarcane, corn; livestock.

Lambrecht (LAHM-brekht), village, RHINELAND-PALATINATE, W GERMANY, in HARDT MOUNTAINS, on the SPEYER, 3 mi/4.8 km WNW of NEUSTADT; 49°22'N 08°05'E. Manufacturing of paper, electronics; metalworking. Built around a monastery (977); developed into a textile center in 16th century; chartered 1888. Has 14th-century church.

Lambres-lez-Douai (lahn-bruh-lai-doo-ai), town S suburb of DOUAI, NORD department, NORD-PAS-DE-CALAIS region, N FRANCE; 50°21'N 03°04'E. Auto industry plant.

Lambro River (LAHM-bro), 80 mi/129 km long, LOMBARDY, N ITALY; rises between two arms of LAKE COMO, 5 mi/8 km S of BELLAGIO; flows SSE, past MONZA, MELEGNANO, and Sant'Angelo Lodigiano, to PO RIVER 9 mi/14 km NW of PIACENZA.

Lambs Crove, village, JASPER county, central IOWA, suburb 2 mi/3.2 km W of NEWTON; 41°42'N 93°04'W. Corn, oats; cattle, hogs, sheep.

Lamb's Head, Gaelic *Ceann an Uain*, promontory on the ATLANTIC, SW KERRY county, SW IRELAND, on N shore of entrance to KENMARE RIVER, 26 mi/42 km WSW of KENMARE; 51°44'N 10°08'W. Offshore are Hog Islands.

Lambsheim (LAHMS-heim), village, RHINELAND-PALATINATE, W GERMANY, at NE foot of HARDT MOUNTAINS, 3 mi/4.8 km WSW of FRANKENTHAL; 49°31'N 08°17'E. Wine. Remains from Stone Age excavated nearby.

Lambton (LAM-tuhn), county (□ 1,159 sq mi/3,013.4 sq km; 2001 population 126,971), SW ONTARIO, E central CANADA, on Lake HURON, and on SAINT CLAIR and THAMES rivers, on U.S. (MICHIGAN) border; ⊙ SARNIA; 42°50'N 82°05'W. Composed of SARNIA, LAMBTON SHORES, PETROLIA, BROOKE-ALVINSTON, DAWN-EUPHEMIA, ENNISKILLEN, PLYMPTON-WYOMING, Saint Clair, WARWICK, OIL SPRINGS, and POINT EDWARD.

Lambton (LAM-tuhn), village (□ 41 sq mi/106.6 sq km), Estrie region, S QUEBEC, E CANADA, near Lake SAINT FRANCIS, 20 mi/32 km NNW of LAC-MÉGANTIC; 45°50'N 71°05'W. Lumbering, dairying. Dam across S end of Lake Saint Francis nearby.

Lambton (LAM-tuhn), W residential suburb of NEWCASTLE, NEW SOUTH WALES, AUSTRALIA; 32°55'S 151°42'E.

Lambton, Cape, S extremity of BANKS ISLAND, NUNAVUT territory, CANADA, on AMUNDSEN GULF; 71°04'N 123°09'W.

Lambton Shores (LAM-tuhn SHORZ), city (□ 128 sq mi/332.8 sq km; 2001 population 10,571), SW ONTARIO, E central CANADA, 25 mi/41 km from SARNIA; 43°10'N 81°53'W. Composed of the communities of ARKONA, BOSANQUET, FOREST, GRAND BEND, and THEDFORD.

Lambunao (lahm-BOO-nou), town, ILOILO province, central PANAY island, PHILIPPINES, 25 mi/40 km NNW of ILOILO; 11°07'N 122°24'E. Rice-growing center.

Lam Dong, province (□ 3,927 sq mi/10,210.2 sq km), S central VIETNAM, in S CENTRAL HIGHLANDS; ⊙ DALAT; 11°45'N 108°20'E. It borders DAC LAC province (N), KHANH HOA province (NE), NINH THUAN province (E), BINH THUAN province (S), DONG NAI province (SW), and SONG BE province (W). Drained mainly by the SONG DONG NAI and streams flowing E into the SOUTH CHINA SEA, and W into the MEKONG River (through CAMBODIA). Includes a complex of mountains and plateaus that comprise the S flank of the TRUONG SON RANGE (average elevation 3,280 ft/1,000

m) and basaltic formations reaching higher than 6,560 ft/2,000 m; also varied environmental regions, many minority communities, and localized economies. Subjected to defoliation and bombings during Vietnam War. As a result of contemporary resettlement schemes, an expanding agricultural frontier, the growth of urbanization, widespread timber exploitation, and rapidly growing population, the once dense forests now cover less than half the province and continue to suffer from serious overcutting. Diverse soils and mineral resources (gold, copper, tin, kaolin, granite, mineral water). Industries include shifting cultivation, animal raising, forest products, agro-forestry, and commercial agriculture (coffee, tea, sugarcane, cashews, tobacco, vegetables, fruits). Lumbering, sawmilling, sugar milling, wood chip and pulp production, textiles and garments, furniture making and woodworking, brick and tile, handicrafts, meat and food processing, other light manufacturing, and tourism focused mainly on Dalat. Population includes Kinh, Ma, Koho, Mnong, Lat, Charo, and other minorities.

Lame (LAH-me), town, BAUCHI state, N central NIGERIA, on road, 40 mi/64 km WNW of BAUCHI; 10°27′N 09°15′E. Market center. Millet, cassava; cattle raising.

Lame Deer, town (2000 population 2,018), ROSEBUD county, SE MONTANA, on Lame Deer Creek, and 48 mi/77 km ESE of HARDIN; 45°37′N 106°37′W. Agriculture (cattle, sheep, horses; wheat, barley, oats); manufacturing (lumber products, apparel). Headquarters of Northern Cheyenne Indian Reservation, Dull Knife Memorial College.

Lamego (lah-MAI-goo), city, VISEU district, TRÁS-OS-MONTES E ALTO DOURO province, N PORTUGAL, near left bank of the DOURO RIVER, 45 mi/72 km E of OPORTO; 41°06′N 07°49′W. Commercial center in vineyard region producing fine sparkling wines; figs, almonds, olives. Its cathedral retains part of original Romanesque tower and has Gothic facade (c.1500). Overlooking city are Santuário de Nossa Senhora (famous pilgrimage church), the 7th-century Igreja de São Pedro de Balsemão (Portugal's oldest church), and a ruined Moorish castle. Taken from the Moors by Ferdinand I of Castile in 1057. Scene of Cortes of 1143.

La Mendieta (lah men-dee-AI-tah), town, SE JUJUY province, ARGENTINA, on railroad, on the RÍO GRANDE DE JUJUY, and 22 mi/35 km ESE of JUJUY; 24°19′S 64°58′W. Sugar-refining center; iron mines nearby.

Lamentin (lah-mawng-TANG), town, NE BASSE-TERRE island, GUADELOUPE, French WEST INDIES, 6 mi/9.7 km WNW of POINTE-Á-PITRE; 16°16′N 61°38′W. In sugar-growing region; manufacturing of alcohol (Grosse-Montague Distillery).

Lamentin (lah-mawng-TANG), town, W MARTINIQUE, French WEST INDIES, at head of Fort-de-France Bay, 3 mi/4.8 km E of FORT-DE-FRANCE; 14°37′N 61°01′W. Trade center on fertile plain (bananas, cacao); rum distilling; limekiln; industrial center.

Lamentin, Plain of (lah-mawng-TANG), W MARTINIQUE, French WEST INDIES, along FORT-DE-FRANCE Bay, only extensive level land on the island. Sugar-growing area. Site of Lamentin International Airport.

La Merced (lah mer-SED), canton, ANICETO ARCE province, TARIJA department, S central BOLIVIA, 38 mi/60 km SE of TARIJA on the Tarija–FORTÍN CAMPERO road; 22°01′S 64°44′W. Elevation 6,549 ft/1,996 m. Abundant gas wells in area. Agriculture (potatoes, yucca, bananas, corn); cattle.

La Merced (lah mer-SED), town, central SALTA province, ARGENTINA, on railroad, and 13 mi/21 km SSW of SALTA. Agricultural center (corn, tobacco; livestock); lime works; forest products.

La Merced (LAH mer-SED), town, ⊙ La Merced municipio, CALDAS department, W central COLOMBIA, 23 mi/37 km NNW of MANIZALES; 05°24′N 75°53′W. Coffee, plantains, sugarcane; livestock.

La Merced (LAH mer-SED), town, ⊙ CHANCHAMAYO province, JUNÍN region, central PERU, on CHANCHAMAYO RIVER, 32 mi/51 km NE of TARMA, on road to OXAPAMPA (PASCO region); 11°03′S 75°18′W. Coffee, sugarcane, rice, cotton, fruit; lumbering. Airport.

La Merced (lah mer-SED), village (1991 population 1,274), ⊙ Paclín department (□ 255 sq mi/663 sq km), SE CATAMARCA province, ARGENTINA, 23 mi/37 km NNE of CATAMARCA. In cattle and tobacco area. Forest industry.

Lameroo (lam-uh-ROO), village, SE SOUTH AUSTRALIA state, S central AUSTRALIA 115 mi/185 km ESE of ADELAIDE, and c.25 mi/40 km from the VICTORIA border; 35°20′S 140°31′E. Mixed farming; wheat, wool. An underground basin helps sustain the village.

La Mesa, city (2000 population 54,749), SAN DIEGO county, S CALIFORNIA; residential suburb 8 mi/12.9 km ENE of downtown SAN DIEGO; 32°46′N 117°01′W. It is a retail-trade center in suburban San Diego. Manufacturing (machinery, diversified light manufacturing). La Mesa has become a popular residence for upper- and middle-income professionals in the San Diego area. Points of interest include the McKinney House, the Pacific Southwest Railway Museum, and nearby Mount Helix, with its impressive view of the surrounding region. Murray Reservoir to N. Incorporated 1912.

Lamesa (luh-MEE-suh), city (2006 population 9,259), ⊙ DAWSON county, NW TEXAS, in the LLANO ESTACADO; 32°43′N 101°57′W. Elevation 2,975 ft/907 m. Processing and shipping center for an irrigated area where cattle and hogs are raised; cotton, sorghum, wheat. Agribusiness center. Manufacturing (electric motors, sheet metal fabricating, cottonseed oil and cotton products, apparel). The city has several oil- and natural-gas wells. Incorporated 1917.

La Mesa (LAH MAI-sah), town, ⊙ La Mesa municipio, CUNDINAMARCA department, central COLOMBIA, 19 mi/31 km W of BOGOTÁ; 04°38′N 73°27′W. Coffee, corn, sugarcane; livestock.

La Mesa (lah MAI-sah), town, ⊙ LA MESA district, VERAGUAS province, W central PANAMA, in SAN PABLO RIVER valley, 13 mi/21 km WNW of SANTIAGO. Coffee.

La Mesa (lah MAI-sah), unincorporated town, DONA ANA county, S NEW MEXICO, 13 mi/21 km S of LAS CRUCES, on RIO GRANDE. In irrigated agricultural area (cattle, sheep; vegetables, fruit, nuts).

La Mesa (LAH MAI-sah), town, TRUJILLO state, W VENEZUELA, in ANDEAN spur, 20 mi/32 km SSW of VALERA. Elevation 5,715 ft/1,742 m. Wheat, corn, potatoes.

La Mesilla (LAH muh-SEE-yuh), town (2006 population 2,201), DONA ANA county, S NEW MEXICO, suburb 2 mi/3.2 km SW of downtown LAS CRUCES, on the RIO GRANDE; 32°16′N 106°48′W. Gadsden Purchase was signed here in 1853. The whole Mesilla Valley became part of the U.S. under the agreement. Mesilla was a central station on the overland mail route. From July 1861, to August 1862, it was headquarters for College John River Baylor of the Confederate army, who proclaimed Mesilla the capital of the new Confederate territory. A museum commemorates Billy the Kid, who once stood trial here. Las Cruces Airport to W; La Mesilla State Historical Park is here. Settled c.1850.

Lamesley (LAIMZ-lee), village (2001 population 7,812), TYNE and WEAR, NE ENGLAND, 4 mi/6.4 km S of NEWCASTLE UPON TYNE; 54°55′N 01°31′W.

Lamía (lah-MEE-ah), city (2001 population 46,406), ⊙ FTHIOTIDA prefecture, E CENTRAL GREECE department; 38°54′N 22°26′E. Transportation hub and agricultural center. Founded c.5th century B.C.E., it was the chief city of the small region of MALIS and developed as an ally of ATHENS. Gave its name to the Lamian War (323–322 B.C.E.), waged by the confederate Greeks against Antipater, the Macedonian general, who took refuge here and was besieged for several months. Antipater conquered (322 B.C.E.) the confederates at Krannon, near Larissa. Known as Zituni, or Zeitun, from the 10th to the 19th century.

Lamiaco, SPAIN: see LEJONA.

Lamía, Gulf of, Greece: see MALIAKOS GULF.

L'Amiante (lah-MYAHNT), county (□ 736 sq mi/1,913.6 sq km; 2006 population 43,538), CHAUDIÈRE-APPALACHES region, S QUEBEC, E CANADA; ⊙ THETFORD MINES; 46°05′N 71°18′W. Composed of nineteen municipalities. Formed in 1982.

Lamidanda (lah-mee-DAHN-dah), village, central NEPAL; 27°15′N 86°41′E. Elevation 4,100 ft/1,250 m. Airport.

Lamin (LAH-meen), town, WESTERN DIVISION, THE GAMBIA, 10 mi/16 km SSW of BANJUL; 13°21′N 16°38′W. Near ABUKO NATURE RESERVE and Yundum International Airport.

Lamine River (luh-MEEN), c.70 mi/113 km long, central MISSOURI; rises in several branches in MORGAN and PETTIS counties; flows N to the MISSOURI 6 mi/9.7 km W of BOONVILLE.

La Minerve (lah mee-NERV), village (□ 115 sq mi/299 sq km; 2006 population 1,133), LAURENTIDES region, S QUEBEC, E CANADA, 10 mi/16 km from LABELLE; 46°15′N 74°56′W.

Lamington, Mount, volcano, N central NORTHERN province, SE PAPUA NEW GUINEA, in OWEN STANLEY RANGE (c.5,200 ft/1,585 m), 75 mi/121 km NE of PORT MORESBY; 08°58′S 148°08′E. Erupted violently January 1951, killing many thousands, and destroying town of HIGATURU, the district capital at the time.

La Mirada, city (2000 population 46,783), LOS ANGELES county, S CALIFORNIA; suburb 15 mi/24 km SE of downtown LOS ANGELES, and 12 mi/19 km NE of LONG BEACH; 33°54′N 118°01′W. Manufacturing (metal, plastic, paper, and concrete products). La Mirada derives from the Spanish for "the view," referring to the panoramic view of the surrounding valleys from atop the city's hills. It was the original site of California's olive industry. The city is the seat of Biola University. Incorporated 1960.

La Misión or **Misión** (mee-SYON), town, HIDALGO, central MEXICO, 7 mi/11.3 km NNE of JACALA, and on the QUERÉTARO border. Grain, beans, livestock.

Lamitan (lah-MEE-tahn), town, on BASILAN ISLAND, BASILAN province, PHILIPPINES, off SW tip of MINDANAO; 06°38′N 122°06′E. Coconuts, citrus fruits. Large, multi-ethnic market with Samal, Badjao, Chinese, Chavacanos, Visayan, and especially Yakan vendors and patrons.

Lamjung (LUHM-joong), district, central NEPAL, in GANDAKI zone; ⊙ Sahar.

Lamki, RUSSIA: see VTORYYE LEVYYE LAMKI.

Lamlash (LAM-lash), village (2001 population 1,010) on E coast of ARRAN island, North Ayrshire, SW Scotland, 3 mi/4.8 km S of BRODICK; 55°32′N 05°09′W. Fishing port and resort. HOLY ISLAND shelters the bay.

Lamma, Cantonese *Pokliu Chau*, island, the NEW Territories, HONG KONG, CHINA; 4.5 mi/7.2 km long, 2.5 mi/4 km wide; 22°12′N 114°08′E. Separated from Hong Kong island (NE) by 1-mi/1.6-km-wide East Lamma Channel, and from LAN TAO island by 6-mi/9.7-km-wide West Lamma Channel. Mountainous. Main towns include Tai Wan Tsuer (N) and Tung O (S). Site of a major coal-fired power plant.

Lamme Fjord, DENMARK: see ISE FJORD.

Lammermuir Hills (LAH-muhr-mir), range of hills, East Lothian and Scottish Borders, SE Scotland; 55°55′N 02°44′W. Meikle Says Law (1,749 ft/533 m) is the highest point. Livestock raising (sheep).

Lammhult (LAHM-HULT), village, KRONOBERG county, S SWEDEN, on small Lammen Lake, 20 mi/32 km E of VÄRNAMO; 57°10′N 14°35′E.

Lämmijärv (LAM-mee-yarv), lake, on the Russian-Estonian border. Provides the connection between Lake PEIPSI and Lake Pihkva (Russian *Pskovskoe Ozero*).

Lamogai (LAM-o-gei), village, WEST NEW BRITAIN province, W NEW BRITAIN island, E PAPUA NEW GUINEA, 80 mi/129 km SW of HOSKINS, 15 mi/24 km inland from SOLOMON SEA; 05°51′S 149°22′E. Cocoa, copra, coffee; timber.

Lamoille (lah-MOIL), county (□ 463 sq mi/1,203.8 sq km; 2006 population 24,592), N central VERMONT, with GREEN MOUNTAINS in W; ⊙ HYDE PARK; 44°36′N 72°39′W. Includes Mt. Mansfield (winter sports), highest peak of Green Mountains; resorts. Drained by LAMOILLE RIVER. Dairying; manufacturing (machinery, wood products, textiles); talc, asbestos, lumber; maple sugar. Organized 1835.

Lamoille, unincorporated town, ELKO county, NE NEVADA, lies in the W foothills of the RUBY MOUNTAINS. One of the older communities in N Nevada; a stopping place along the CALIFORNIA Trail in the 1840s (nearby Camp Halleck provided a military presence and security). Originally established as a community of small land-holding ranchers and grain farmers. Lamoille has become increasingly gentrified since the 1980s; ranches sold to movie moguls, lawyers, doctors, and other well-to-do. Heli-skiing in the Ruby Mountains; wildlife inand; Ruby and Franklin marshes; cultural amenities in ELKO. Voted "the best small place to live in America" in 1990. Elko airport (12 mi/ 19 km distant).

La Moille (luh MOIL), village (2000 population 773), BUREAU county, N ILLINOIS, near BUREAU CREEK, and 15 mi/24 km NE of PRINCETON; 41°31′N 89°16′W. In agricultural area.

La Moille, village, MARSHALL county, central IOWA, 7 mi/11.3 km W of MARSHALLTOWN. Feed.

Lamoille River (lah-MOIL), c.70 mi/113 km long, NW VERMONT; rises near Hardwick; flows generally W, through the range, past MORRISVILLE (dam forms Lake Lamoille here), to Lake CHAMPLAIN 10 mi/16 km N of Burlington. Receives North Branch (c.20 mi/32 km long) near Cambridge.

Lamoine (luh-MOIN), fishing and resort town, HANCOCK county, S MAINE, on FRENCHMAN BAY just N of MOUNT DESERT ISLAND, and 8 mi/12.9 km SE of ELLSWORTH; 44°29′N 68°18′W.

La Moine River (luh MOIN), W ILLINOIS; rises in SW WARREN county; flows c.100 mi/161 km SW, S, and SE, to ILLINOIS RIVER below BEARDSTOWN; 40°38′N 90°45′W.

Lamoka Lake, SCHUYLER county, W central NEW YORK, in FINGER LAKES region, 10 mi/16 km W of WATKINS GLEN. 1.5 mi/2.4 km long; 42°24′N 77°05′W. Connected by stream to WANETA LAKE (N). Stocked for fishing.

Lamon Bay (LAH-mon), arm of PHILIPPINE SEA, in PHILIPPINES, bounded W by E coast of LUZON, N by POLILLO ISLANDS, and S by ALABAT ISLAND. Merges NW with POLILLO STRAIT; c.50 mi/80 km long, c.25 mi/40 km wide. Its S arms are LOPEZ and CALAUAG bays.

Lamone River (lah-MO-ne), c.60 mi/97 km long, N central ITALY; rises in ETRUSCAN APENNINES 6 mi/10 km SW of Marradi; flows NNE, past Marradi and FAENZA, and E to the ADRIATIC at MARINA DI RAVENNA. Canalized in lower course; used for irrigation.

Lamongan (lah-MAWNG-gahn), town, East Java province, NE JAVA, INDONESIA, 24 mi/39 km WNW of SURABAYA; 07°07′S 112°25′E. Trade center for agricultural area (tobacco, rice, corn, cassava); textile mills. Oil fields nearby.

Lamoni (lah-MO-nei), town (2000 population 2,444), DECATUR county, S IOWA, near MISSOURI state line 14 mi/23 km SW of LEON; 40°37′N 93°56′W. Manufacturing (concrete blocks, dairy products). Has Graceland Jr. College Plotted 1879, incorporated 1885.

Lamont, unincorporated city (2000 population 13,296), KERN county, S central CALIFORNIA; residential suburb 9 mi/14.5 km SE of BAKERSFIELD; 35°16′N 118°55′W. Agriculture (cotton, potatoes, vegetables,

---

tomatoes, sugar beets, beans, melons, fruit, nuts; dairying; cattle). Sequoia National Forest to NE.

Lamont (luh-MAHNT), town (□ 2 sq mi/5.2 sq km; 2001 population 1,692), central ALBERTA, W CANADA, 30 mi/48 km ENE of EDMONTON, in LAMONT COUNTY; 53°45′N 112°47′W. Grain elevators, lumbering. Gateway to ELK ISLAND NATIONAL PARK (S). Incorporated as a village in 1910; became a town in 1968.

Lamont, town (2000 population 503), BUCHANAN county, E IOWA, near MAQUOKETA RIVER, 15 mi/24 km NE of INDEPENDENCE; 42°36′N 91°38′W. Sand and gravel pits nearby.

Lamont, village (2006 population 425), GRANT county, N OKLAHOMA, 26 mi/42 km NE of ENID, near SALT FORK OF ARKANSAS RIVER; 36°41′N 97°33′W. In agricultural area.

Lamont, village (2006 population 96), WHITMAN county, SE WASHINGTON, 40 mi/64 km SW of SPOKANE; 47°12′N 117°54′W. In agicultural region; wheat, barley; cattle.

La Montañita (LAH mon-tahn-YEE-tah), town, ⊙ La Montañita municipio, CAQUETÁ department, S COLOMBIA, in the Oriente, 15 mi/24 km SE of FLORENCIA; 01°28′N 75°26′W. Elevation 1,614 ft/491 m. Sugarcane, corn, plantains, cassava; livestock.

Lamont County (luh-MAHNT), municipality (□ 927 sq mi/2,410.2 sq km; 2001 population 4,167), SW ALBERTA, W CANADA; 49°16′N 113°09′W. Shares a border with ELK ISLAND NATIONAL PARK. Agriculture. Includes BRUDERHEIM, ANDREW, LAMONT, CHIPMAN, MUNDARE; and the hamlets of HILLIARD, WHITFORD, SAINT MICHAEL, WOSTOK, and STAR. Formed as an improvement district in 1912.

La Monte, city (2000 population 1,064), PETTIS county, central MISSOURI, 11 mi/18 km WNW of SEDALIA; 38°46′N 93°25′W. Soybeans, corn; cattle; wood products.

Lamorandière (lah-mo-rahnd-YER), village (□ 166 sq mi/431.6 sq km; 2006 population 271), ABITIBI-TÉMISCAMINGUE region, W QUEBEC, E CANADA, 22 mi/35 km ENE of AMOS; 48°37′N 77°38′W. Gold mining.

La Morandière (lah mo-rahn-DYER), village (□ 166 sq mi/431.6 sq km; 2006 population 271), ABITIBI-TÉMISCAMINGUE region, SW QUEBEC, E CANADA, 7 mi/10 km from CHAMPNEUF; 48°37′N 77°38′W.

Lamorlaye (lah-mor-lai), town (□ 5 sq km/13 sq km), OISE department, PICARDIE region, N central FRANCE, 16 mi/26 km NE of PONTOISE; 49°09′N 02°26′E.

Lamotrek, atoll, State of YAP, W CAROLINE ISLANDS, Federated States of MICRONESIA, W PACIFIC, c.560 mi/ 901 km ESE of Yap; 8 mi/12.9 km long, 3.5 mi/5.6 km wide; 07°28′N 146°23′E. Wooded islets (three) on triangular reef. Formerly Swede Island.

La Mott, Pennsylvania: see CHELTENHAM.

La Motte, town (2000 population 272), JACKSON county, E IOWA, 15 mi/24 km S of DUBUQUE; 42°17′N 90°37′W. In agricultural area.

La Motte (la MUHT), village (□ 87 sq mi/226.2 sq km), ABITIBI-TÉMISCAMINGUE region, W QUEBEC, E CANADA, on Lake La MOTTE, 16 mi/26 km S of AMOS; 48°23′N 78°07′W. Gold, copper, zinc, lead; logging.

Lamotte-Beuvron (lah-mot–bu-vron), town (□ 9 sq mi/23.4 sq km), LOIR-ET-CHER department, CENTRE administrative region, N central FRANCE, in the SOLOGNE district, on BEUVRON RIVER, and 21 mi/34 km SSE of ORLÉANS; 47°01′N 01°12′E. Agriculture trade center (sheep and cattle market); porcelain factory; sawmilling. Region, once an extensive wetland with many ponds, has been partly drained and reforested and currently attracts huntsmen.

La Motte, Isle, VERMONT: see ISLE LA MOTTE.

La Moure (luh MOR), county (□ 1,150 sq mi/2,990 sq km; 2000 population 4,701), SE central NORTH DAKOTA; ⊙ LA MOURE. Rich agricultural area drained by James and MAPLE rivers and Cottonwood Creek. Wheat, corn, barley, rye; cattle, hogs; dairy products. Lake Lanoar in S. Formed in 1873 and government

---

formed in 1881. Named for Judson LaMoure (1839–1918) who served in state legislature. GRAND RAPIDS was the county seat from 1881–1886.

La Moure (luh MOR), town (2000 population 944), ⊙ LA MOURE CO., SE central NORTH DAKOTA, 45 mi/72 km SSE of Jamestown, and on JAMES RIVER; 46°21′N 98°17′W. Trade center; livestock, dairying. Lake La Moure to S. Settled 1882; incorporated 1905; became the county seat in 1886. Named for Judson La Moure (1839–1918), who served in state legislature.

Lampa (LAHM-pah), province (□ 2,853 sq mi/7,417.8 sq km), PUNO region, SE PERU; ⊙ LAMPA; 15°25′S 70°35′W. Western province of Puno region, bordering on CUZCO and AREQUIPA regions.

Lampa (LAHM-pah), town, ⊙ Lampa comuna, CHACABUCO province, METROPOLITANA DE SANTIAGO region, central CHILE, 15 mi/24 km NW of SANTIAGO; 33°17′S 70°54′W. Agricultural center (alfalfa, cereals, fruit, grapes; livestock); stone quarries. Ancient Inca settlement. Formerly a gold-mining town.

Lampa (LAHM-pah), town, ⊙ LAMPA province, PUNO region, SE PERU, on the ALTIPLANO, and the Lampa River, 41 mi/66 km NW of PUNO; 15°21′S 70°22′W. Elevation 12,648 ft/3,855 m. Silver, zinc, lead, and copper mines; cereals, potatoes; livestock.

Lampang (LUHM-PAHNG), province (□ 4,833 sq mi/12,565.8 sq km), N THAILAND; ⊙ LAMPANG; 18°20′N 99°30′E. Rice, sugarcane, garlic, tobacco, and coffee production, cotton weaving, tanning, and sugar milling. Copper, lead, and iron deposits. Capital has been inhabited since around the 7th century in the Dvaravati period. The province also boasts a walled temple, WAT PHRA THAT LAMPANG LUANG, which dates back to the 10th–11th century in the Hariphunchai period, as well as a young elephant-training center 33 mi/54 km NE of Lampang town.

Lampang (LUHM-PAHNG), town, ⊙ Lampang province, N THAILAND, on WANG RIVER, on railroad, 373 mi/600 km NNW of BANGKOK, and 45 mi/72 km SE of CHIANG MAI, E of KHUN TAN RANGE, in forested area; 18°18′N 99°31′E. Highway N to KENGTUNG (MYANMAR). Cotton weaving, tanning; sugar mill at Ko Kha (SW); rice, sugarcane, tobacco, coffee. Copper, lead, and iron deposits nearby. Airport. Original city, founded 6th–7th century, was located 10 mi/16 km SW of present site. Ruins of Burmese style temples. Thai Elephant Conservation Center here.

Lampa River, PERU: see LAMPA.

Lampasas (LAHM-pa-suhs), county (□ 714 sq mi/1,856.4 sq km; 2006 population 20,758), central TEXAS; ⊙ LAMPASAS; 31°11′N 98°14′W. Bounded W by COLORADO RIVER; drained by LAMPASAS RIVER. Wool, mohair shipped; agriculture (grain sorghum, corn, watermelons, peanuts; cattle). Glass-sand mining; sand and gravel; stone. Mineral springs (health resort) at Lampasas. Formed 1856.

Lampasas (LAHM-pah-suhs), town (2006 population 7,828), ⊙ LAMPASAS county, central TEXAS, on a tributary of LAMPASAS RIVER, and c.60 mi/97 km NNW of AUSTIN; 31°03′N 98°10′W. Elevation 1,025 ft/ 312 m. Cattle ranching; agriculture (peanuts, corn, watermelons); manufacturing (livestock feed, food processing, rubber products, cut stone). Originally named Burleson. Settled 1854; incorporated 1874.

Lampasas River (LAHM-pah-suhs), c.110 mi/177 km long, central TEXAS; rises in HAMILTON county; flows SE and E through STILLHOUSE HOLLOW LAKE reservoir 8 mi/12.9 km SE of KILLEEN, to join LEON RIVER c.9 mi/14.5 km S of TEMPLE, to form LITTLE RIVER.

Lampazos de Naranjo (lahm-PAH-sos dai nah-RAHN-ho), city and township, ⊙ Lampazos de Naranjo municipio, NUEVO LEÓN, N MEXICO, on railroad, and 70 mi/113 km SW of NUEVO LAREDO on Mexico Highway 1; 27°01′N 100°30′W. Elevation 1,115 ft/340 m. Agricultural center (cotton, wheat, livestock); iron ore. Old colonial town; former Spanish mission.

Lampedusa, island (□ 8 sq mi/20.8 sq km), S SICILY, ITALY, in the MEDITERRANEAN SEA between MALTA and TUNISIA; 35°31′N 12°35′E. The largest of the PELAGIE ISLANDS. Il Porto is the only town on the island. Sponge and sardine fishing are the main occupations. Lampedusa was settled in the 18th century. In World War II bombed by the Allies. There is a penal colony on the island.

Lampertheim (LAHM-pert-heim), town, HESSE, central GERMANY, on an arm of the RHINE, 7 mi/11.3 km N of MANNHEIM; 49°35′N 08°28′E. An agricultural center until the middle of the 20th century (especially tobacco). Today it is an industrial town with manufacturing of pharmaceuticals; metalworking; printing. Numerous half-timbered houses. Chartered 1951.

Lampeter (LAM-puh-tuh), Welsh *Llanbedr Pont Steffan*, town (2001 population 2,894), CEREDIGION, W Wales, on TEIFI RIVER, and 25 mi/40 km E of CARDIGAN; 52°07′N 04°04′W. Previously woolen mills and tanneries. The University of Wales, Lampeter (formerly St. David's College), a constituent college of the University of Wales, is here. Village of Cwmann to SSE, on Teifi River. Formerly in DYFED, abolished 1996.

Lamphun (LUHM-POON), province (□ 1,740 sq mi/4,524 sq km), N THAILAND; ⊙ LAMPHUN; 18°15′N 98°55′E. Agriculture (rice, garlic tobacco, lac); known for producing some of the sweetest lam yai (longan) in THAILAND. Cotton weaving is important, especially at Pasang village; silver deposits in the SW part of the province. The capital was founded in the Dvaravati period around the 6th century, and temples and chedi (stupas) in CHANGWAT LAMPHUN date as far back as the 8th or 9th century Lamphun holds an annual Lam Yai Festival in the 2nd week of August.

Lamphun (LUHM-POON), town, ⊙ Lamphun province, N Thailand, near PING RIVER, on railroad, 427 mi/687 km NNW of BANGKOK, and 15 mi/24 km S of CHIANG MAI, in wide valley, E of THANON THONG CHAI RANGE; 18°35′N 99°01′E. Teak and agricultural center (rice, lac, tobacco); cotton weaving. Silver deposits (SW). Famous for handicrafts and orchid gardens. One of oldest cities N THAILAND (founded c.6th century), it flourished until rise of CHIANG MAI in 13th century. Famous for handicrafts and orchid gardens. Sometimes spelled LAMPOON.

Lampman (LAMP-muhn), town (2006 population 634), SE SASKATCHEWAN, CANADA, 20 mi/32 km NE of ESTEVAN; 49°23′N 102°45′W. Railroad junction; grain elevators.

Lampoeng Bay, INDONESIA: see LAMPUNG BAY.

Lampong Bay, INDONESIA: see LAMPUNG BAY.

Lampoon, THAILAND: see LAMPHUN, town.

Lamporecchio (lahm-po-REK-kyo), town, PISTOIA province, TUSCANY, central ITALY, 8 mi/13 km S of PISTOIA; 43°50′N 10°53′E. Diversified secondary manufacturing center.

Lamprechtshausen (lahm-prekhts-HOU-sen), village N SALZBURG, W AUSTRIA, in the FLACHGAU, 14 mi/23 km N of SALZBURG, near German border; 47°59′N 12°57′E. In nearby ARNSDORF, the song Silent Night, Holy Night was composed. Dairy farming; pilgrim church.

Lamprey River (LAM-prai), c.40 mi/64 km long, SE NEW HAMPSHIRE; rises in Pawtuckaway Pond in NOTTINGHAM; flows S, then generally E past RAYMOND, EPPING, and Lee, to GREAT BAY, tidal inlet of ATLANTIC OCEAN at NEWMARKET.

Lampsacus, ancient Greek city of NW ASIA MINOR, on the Hellespont (now DARDANELLES) opposite Callipolis (now GELIBOLU). It was colonized in the 7th century B.C.E. by Greeks from PHOCAEA. Artaxerxes I assigned the city to Themistocles. After the battle of MYCALE (479 B.C.E.) the citizens joined with the Athenians, and the city continued to flourish under the Greeks and the Romans. It was the seat of the cult of Priapus.

Lampung Bay (LAHM-pong), inlet of SUNDA STRAIT, Lampung province, INDONESIA, at S end of SUMATRA, opposite JAVA; 25 mi/40 km N-S, 35 mi/56 km E-W; 05°40′S 105°20′E. BANDAR LAMPUNG (formerly called Telukbetung-Tanjungkarang), is at head of bay; Tanjung Tua is at E side of entrance. Formerly spelled Lampong or Lampoeng Bay.

Lamskoye (lahm-SKO-ye), village, W LIPETSK oblast, S central European Russia, near highway, 12 mi/19 km SSW of YEFREMOV; 52°59′N 38°03′E. Elevation 682 ft/207 m. In agricultural area (wheat, hemp, potatoes). Formerly known as Spasskoye.

Lamtar (lahm-TAHR), village, SIDI BEL ABBÈS wilaya, NW ALGERIA, 12 mi/19 km SW of SIDI BEL ABBÈS. Cereals; vineyards.

Lamu (LAH-moo), town (□ 2,632 sq mi/6,843.2 sq km), COAST province, SE KENYA, on SE shore of Lamu Island in INDIAN OCEAN just off coast, and 150 mi/241 km NNE of MOMBASA; 02°17′S 40°54′E. Minor shipping center; native shipbuilding, shark-liver processing; sisal (rope and bag making); cotton, copra, sugarcane; fisheries. Tourist center. Once a Persian colony and a possession of the sultan of ZANZIBAR, it rivaled Mombasa as an entrepôt of gold, ivory, spice, and slaves until end of 19th century. Major center for Islamic learning with twenty-two mosques. Large Arab population. The Lamu archipelago includes Lamu, MANDA, and Pata islands. Airport.

Lamud (lah-MOOD), city (□ 3,289 sq mi/8,551.4 sq km), ⊙ LUYA province, AMAZONAS region, N PERU, in E ANDEAN foothills, 7 mi/11 km NNW of CHACHAPOYAS; 06°09′S 77°55′W. Agriculture (corn, potatoes, cereals, alfalfa, sugar; livestock).

La Muralla National Park (lah moo-RAH-yah), park (□ 58 sq mi/150.8 sq km), OLANCHO department, central HONDURAS, 5 mi/8 km S of LA UNIÓN; 15°08′N 86°43′W. Cloud forest, quetzal viewing. Maximum elevation 6770 ft/2,064 m.

Lamuria (lah-MOO-ree-ah), village, Laikipia district, RIFT VALLEY province, W KENYA, on road, and 55 mi/89 km ENE of NAKURU. Coffee, tea, wheat, corn.

Lam Vien Plateau (LAHM vee-EN), S central VIETNAM, one of the six plateaus to the W of S TRUONG SON RANGE, c. 30 mi/48 km long; 12°00′N 108°25′E. Average elevation c. 5,400 ft/1,650 m. Once covered with pine forests, today a realm of farms growing mid-latitudinal vegetables, tea, and coffee; healthful climate. DALAT is main city. Population includes Koho, Chil, Lat, Gar, and other minorities. Formerly called LANG BIANG PLATEAU.

Lamy (LAI-mee), village, SANTA FE county, N central NEW MEXICO, in foothills of SANGRE DE CRISTO MOUNTAINS, 16 mi/26 km S of SANTA FE. Elevation c.6,482 ft/1,976 m. Railroad junction, spur N to Santa Fe; cattle, sheep. Santa Fe National Forest to E; Glorieta Mesa to E.

Lana (LAH-nah), town, BOLZANO province, TRENTINO–ALTO ADIGE, N ITALY, 3 mi/5 km S of MERANO, near ADIGE RIVER; 46°37′N 11°09′E. Lumber and paper mills, fruit cannery, chemical works.

Lana, CZECH REPUBLIC: see LANY.

Lanagan (LA-nuh-guhn), town (2000 population 411), MCDONALD county, extreme SW MISSOURI, on branch of ELK RIVER, 5 mi/8 km W of PINEVILLE; 36°36′N 94°27′W. Printing.

Lanagh Lough, IRELAND: see CASTLEBAR.

Lanai (LAH-NAH-ee), island (□ 141 sq mi/366.6 sq km), MAUI county, central HAWAII, W of MAUI island across the AUAU CHANNEL, separated from MOLOKAI to N by KALOHI Channel; from KAHOOLAWE to SE by KEALAIKAHIKI CHANNEL. Mount Lanaihale (3,370 ft/1,027 m), in E, is the island's highest point. Administered as a district of Maui county. For many years the island was used for sugarcane raising and cattle grazing. In 1922 it was purchased by Dole pineapple company and developed as a pineapple-growing center. LANAI CITY (1990 population 2,400) is

near center of island; and KAUMALAPAU port is on W coast; village of MANELE on S coast. In 1961, 98% of the island became the property of the Castle and Cook company, with plans to limit tourism and to designate a large majority of Lanai for the development of forests, meadows, and some agriculture. PALAWAI BASIN in S center; Garden of the Gods rock formation in NW.

Lanai City (LAH-NAH-ee), town (2000 population 3,164), central LANAI, MAUI county, HAWAII, 70 mi/113 km SE of HONOLULU; 20°49′N 156°55′W. Manufacturing (food preparations). Lanai Airport to SW; PALAWAI BASIN to S; Garden of the Gods to NW (by trail); junction of Kaumalapau Highway and Manele Road; Lanaihale (3,370 ft/1,027 m), highest point on Lanai, to SE; Hulopoe Beach Park to S. Founded 1922 by Dole pineapple company.

Lanaja (lah-NAH-hah), town, HUESCA province, NE SPAIN, 25 mi/40 km SSE of HUESCA; 41°46′N 00°20′W. Sawmilling; agriculture trade (wine, cereals; livestock); honey).

Lanaken (LAH-nah-kuhn), commune (2006 population 24,601), Tongeren district, LIMBURG province, NE BELGIUM, near MEUSE RIVER, and the ZUID-WILLEMSVAART CANAL, 3 mi/4.8 km NNW of MAASTRICHT; 50°53′N 05°39′E. Market center.

Lanalhue, Lake (lah-NAHL-wai) (□ 25 sq mi/65 sq km), ARAUCO province, BÍO-BÍO region, S central CHILE, at W foot of CORDILLERA DE NAHUELBUTA, 25 mi/40 km SE of LEBU; c.10 mi/16 km long; 37°55′S 73°18′W. Resorts.

Lanao (LAH-nou), former province, PHILIPPINES, divided into two provinces in 1959: LANAO DEL NORTE and LANAO DEL SUR.

Lanao del Norte (LAH-nou del NOR-te), province (□ 1,194 sq mi/3,104.4 sq km), in CENTRAL MINDANAO region, N central MINDANAO, with coasts on MINDANAO SEA and MORO GULF, PHILIPPINES; ⊙ ILIGAN; 08°00′N 124°00′E. Population 24.5% urban, 75.5% rural; in 1991, 88% of urban and 64% of rural settlements had electricity. Coastal swamps, plains, and foothills rising to 984 ft/300 m. Agriculture for local use (corn, rice, coconuts, bananas). Logging is declining as timber reserves fall. Iligan has been developed as industrial center (steel, chemicals, cement, grain processing), based on power supplied from hydroelectric projects on the AGUS RIVER. It is also the main commercial center, airport, and port on Mindanao Sea.

Lanao del Sur (LAH-nou del SOOR), province (□ 1,495 sq mi/3,887 sq km), in AUTONOMOUS REGION IN MUSLIM MINDANAO, N central MINDANAO, PHILIPPINES, between MORO GULF and MINDANAO SEA; ⊙ MARAWI; 07°55′N 124°20′E. Population 22.4% urban, 77.6% rural; in 1991, 97% of urban and 37% of rural settlements had electricity. Largely mountainous, with coastal strip and settlement around LAKE LANAO. In S, two active volcanoes: MOUNT RAGANG (ten eruptions between 1756 and 1916) and MOUNT MAKATURING. Agriculture (rice, corn) largely dependent on irrigation from Lake Lanao. Handicrafts. Secondary airport at Malabang. Population largely Muslim.

Lanao, Lake (LAH-nou del SOOR) (□ 131 sq mi/340.6 sq km), W central MINDANAO, PHILIPPINES, S of ILIGAN BAY; 20 mi/32 km long, 10 mi/16 km wide. Feeds AGUS RIVER. MARAWI near N shore.

Lanark (LA-nuhrk), county (□ 1,150 sq mi/2,990 sq km; 2001 population 62,495), SE ONTARIO, E central CANADA, on OTTAWA River, and on RIDEAU LAKE; ⊙ PERTH; 45°00′N 76°15′W. Composed of SMITHS FALLS, CARLETON PLACE, MISSISSIPPI MILLS, PERTH, BECKWITH, DRUMMOND-NORTH ELMSLEY, LANARK HIGHLANDS, MONTAGUE, and TAY VALLEY.

Lanark, city (2000 population 1,584), CARROLL county, NW ILLINOIS, near ROCK CREEK, 8 mi/12.9 km E of MOUNT CARROLL; 42°06′N 89°49′W. Trade and shipping center in rich agricultural area (grain; livestock); manufacturing of food products. Incorporated 1867.

Lanark (LAH-nuhrk), former township (□ 99 sq mi/257.4 sq km; 2001 population 1,847), SE ONTARIO, E central CANADA, 16 mi/25 km from PERTH; 45°06′N 76°23′W. Amalgamated into LANARK HIGHLANDS in 1997.

Lanark (LA-nahrk), town (2001 population 8,253), South Lanarkshire, S central Scotland, on the CLYDE RIVER, and 22 mi/35.2 km SE of GLASGOW; 55°40′N 03°47′W. Hydroelectric power stations at the Falls of Clyde, just S of Lanark. Sir William Wallace's first act of rebellion (1297) was the murder of the English sheriff of Lanark and the burning of the town. Robert Owen conducted industrial and social experiments at the nearby New Lanark mills, founded by his father-in-law, David Dale, in 1785.

Lanark (LAH-nuhrk), former village (□ 2 sq mi/5.2 sq km; 2001 population 869), SE ONTARIO, E central CANADA, on CLYDE RIVER, and 10 mi/16 km NW of PERTH; 45°01′N 76°21′W. Knitting and lumber mills; dairying. Amalgamated into LANARK HIGHLANDS in 1997.

Lanark, Dalhousie, and North Sherbrooke (LAH-nuhrk, dal-HOU-zee, SHUHR-bruk), former township (□ 212 sq mi/551.2 sq km; 2001 population 1,622), SE ONTARIO, E central CANADA. Amalgamated into LANARK HIGHLANDS in 1997.

Lanark Highlands (LAH-nuhrk), township (□ 399 sq mi/1,037.4 sq km; 2001 population 4,795), SE ONTARIO, E central CANADA, 19 mi/31 km N of PERTH; 45°06′N 76°31′W. Created in 1997 from LANARK township, LANARK village, Darling, and the township of LANARK, DALHOUSIE, AND NORTH SHERBROOKE.

Lanaudière (lah-no-DYER), region (□ 4,754 sq mi/12,360.4 sq km; 2005 population 424,223), S central QUEBEC, E CANADA; 46°46′N 73°50′W. Composed of seventy two municipalities, centered on TERREBONNE. Agriculture, including horticulture, plant biotechnology; tourism.

Lancang (LAN-ZANG), town, ⊙ Lancang county, SW YUNNAN province, CHINA, near MYANMAR border, 60 mi/97 km WSW of PU'ER, in mountain region, W of the MEKONG; 22°32′N 99°56′E. Grain; lead and zinc ore mining.

Lancang Jiang, CHINA: see MEKONG.

Lancashire (LANG-kuh-shir), county (□ 1,188 sq mi/3,088.8 sq km; 2001 population 1,414,727), NW ENGLAND, on the IRISH SEA; ⊙ LANCASTER; 53°42′N 02°40′W. In the W and S are lowlands (the Lancashire plain) and occasional moors, with deposits of coal, slate, and sandstone. The E part of the county consists of the PENNINES with two westward extensions, the Forests of Bowland and Rossendale, separated by the valley of the Ribble. The principal rivers are the MERSEY (which forms much of the county's S border), the Lune, the Wyre, and the Ribble. The coastline is low and broken by estuaries. MORECAMBE BAY separates the county from CUMBRIA. Lancashire's principal cities are MANCHESTER and LIVERPOOL. Manufacturing (textiles, paper, chemicals, rubber goods, and glass). Vegetables and dairy products; market gardening. Lancaster is the administrative center. Other towns include Blackburn, Blackpool, Burnley, Morecambe, Preston, along with parts of Fleetwood and Heysham. Lancashire in Anglo-Saxon times was part of the kingdom of NORTHUMBRIA. In 1351 it was made a county palatine, and in 1399 the palatine rights were vested in the king. Lancashire's economic growth began in medieval times with the introduction of the woolen industry. The process was accelerated by the Industrial Revolution, coal, and cotton, and the population increased rapidly in the 19th and early 20th centuries. Coal mining, cotton spinning, and weaving have dwindled to insignificance since the mid 20th century.

Lancaster (LANG-KA-stuhr), county (□ 846 sq mi/2,199.6 sq km; 2006 population 267,135), SE NEBRASKA; ⊙ LINCOLN; 40°46′N 96°41′W. Commercial and agricultural region; party; urbanized. Part of the Loess-Drift Hills of SE Nebraska. Drainage is NE to the PLATTE through SALT CREEK, and SE to the NEMAHA. Diversified agriculture, with corn, wheat, sorghum, soybeans, cattle hogs, dairying. The seven state recreational areas are Stagecoach Lake and Wagon Train in SE; Bluestem Lake and Olive Creek Lake in SW; Conestoga Lake and Pawnee Lake in W; and Branched Oak Lake in NW. Formed 1859.

Lancaster (LANG-ki-stuhr), county (□ 983 sq mi/2,555.8 sq km; 2006 population 494,486), SE PENNSYLVANIA; ⊙ LANCASTER; 40°15′N 07°61′W. Bounded S by MARYLAND state line, W and SW by SUSQUEHANNA RIVER (dams form CONOWINGO RESERVOIR, Lake Aldre and Lake Clarke), SE by OCTORARO CREEK and its East Branch, NW by Conewago Creek. Rich agricultural area, gently rolling landscape (corn, wheat, oats, barley, hay, alfalfa, soybeans, potatoes, apples; sheep, hogs, cattle, chickens and eggs; dairying). Manufacturing at Lancaster, LITITZ, DENVER, EPHRATA, LEOLA, NEW HOLLAND. Major center for tourism. First settled 1709 by Swiss and French, later by Germans, Welsh, English, and Scotch-Irish. Conestoga wagon and Kentucky rifle were early products. County has several covered bridges. Muddy Run Reservoir in SW corner. Formed 1729.

Lancaster, county (□ 555 sq mi/1,443 sq km; 2006 population 63,628), N SOUTH CAROLINA; ⊙ LANCASTER; 34°41′N 80°42′W. Bounded W by CATAWBA RIVER, N by NORTH CAROLINA. Manufacturing (mica, clay, sand, shale, gold, silver); agriculture (corn, wheat, oats, hay, soybeans; turkeys, cattle; timber) Formed 1785.

Lancaster (LANG-ki-stuhr), county (□ 231 sq mi/600.6 sq km; 2006 population 11,519), E VIRGINIA; ⊙ LANCASTER; 37°42′N 76°24′W. On NORTHERN NECK peninsula; bounded N and S by RAPPAHANNOCK RIVER estuary, SE by CHESAPEAKE BAY (estuary enters bay at SE tip of county); shores indented by many inlets. Agriculture (especially tomatoes, vegetables, barley, wheat, corn, soybeans; poultry, cattle); timber (pine); fish, oysters, crabs. Resort area. Formed 1652.

Lancaster (LANG-kuh-stuh), city (2001 population 45,952) and district, LANCASHIRE, NW ENGLAND, on LUNE RIVER; 54°04′N 02°49′W. The city's products include furniture, textiles, synthetic fiber, linoleum, and soap. It also has an active livestock market. Lancaster Castle occupies the site of a Roman station. The castle has a Norman keep and tower (built 1170) with a turret called John o' Gaunt's Chair. St. Mary's Church dates from the 15th century. Lancaster has a university and a civic and regimental museum.

Lancaster (LAN-KA-stuhr), city (2000 population 118,718), LOS ANGELES county, S CALIFORNIA; suburb 42 mi/68 km NNE of downtown LOS ANGELES, in Antelope Valley and MOJAVE DESERT; 34°42′N 118°10′W. It developed as a trade center for an irrigated farming area (vegetables, fruit, nuts; cotton), and has since become an important site for electronic, aerospace, aircraft, and defense industries. Manufacturing (business forms, metal stampings); printing and publishing. Local borax mining and EDWARDS AIR FORCE BASE (c.20 mi/32 km NE), major military installation, add to Lancaster's economy. The city is the seat of Antelope Valley College (two-year) and Antelope Valley Indian Museum, with prehistoric artifacts. California Poppy Preserve to W; Saddleback Butte State Park to E. Population has grown steadily since late 20th century. Laid out 1894.

Lancaster (LANG-ki-stuhr), city (2000 population 737), ⊙ SCHUYLER county, N MISSOURI, near CHARITON RIVER, 23 mi/37 km N of KIRKSVILLE; 40°31′N 92°31′W. Corn, soybeans; sheep, cattle, hogs. Incorporated 1856.

Lancaster (LAN-KA-stuhr), city (□ 18 sq mi/46.8 sq km; 2006 population 36,507), ⊙ FAIRFIELD county, S central OHIO, on the HOCKING RIVER; 39°43′N 82°35′W. In a livestock and dairying area; founded 1800 by Ebenezer Zane, incorporated as a village 1831. Manufacturing (glassware, heating equipment, automotive parts). The birthplace of the brothers General William T. Sherman and Senator John Sherman has been preserved. In the area are many covered bridges and a Native American mound in the form of a cross. The city contains a campus of Ohio University.

Lancaster (LANG-ki-stuhr), city (2006 population 54,779), ⊙ LANCASTER county, SE PENNSYLVANIA, 33 mi/53 km ESE of HARRISBURG, on the CONESTOGA RIVER, in the heart of the Pennsylvania Dutch country; 40°02′N 76°17′W. It is the commercial center for a productive agricultural county with a huge farmer's market. Agricultural area (livestock, poultry; grain, potatoes, soybeans, alfalfa, apples; dairying); manufacturing (electrical, concrete, and aluminum products, security products, automotive parts, medical equipment, food products, advertising specialties; commercial printing). Lancaster is the seat of Franklin and Marshall College and a theological seminary, and it is noted for its large Amish and Mennonite communities. The area was settled by German Mennonites c.1709 and was a starting point for W-bound pioneers. The famous Conestoga wagon was developed here. The borough of Lancaster was laid out in 1730 and was one of the first inland cities in the country. A munitions center during the Revolution, it was briefly (1777) a meeting place of the Continental Congress and served as capital of the state for more than 10 years before 1812. Robert Fulton was born nearby. Lancaster Airport to N. Tourist center. Points of interest include Wheatland, the home of President James Buchanan to W (built in 1828; a national historic site since 1962), and the Fulton Opera House (1854), a historic monument; also Heritage Center Museum, North Museum of Natural History and Science, Pennsylvania Victorian Wine Cellars to W; Pennsylvania Dutch Visitors Bureau to E, Pennsylvania Farm Museum to NE; Dutch Wonderland Amish Homestead, Mennonite Information Center, to E; several covered bridges in area. Incorporated as a city 1818.

Lancaster (LANG-KA-stuhr), city (2006 population 33,790), DALLAS county, in NE TEXAS, suburb 11 mi/18 km S of downtown DALLAS; 32°36′N 96°46′W. Elevation 512 ft/156 m. Tenmile Creek drains S section the is a processing and shipping center for a fruit, vegetable, and cotton region; agriculture to S (dairying; cattle; peanuts, cotton, vegetables). Manufacturing (chemicals, transportation equipment, bricks, brass valves, fabricated metal products, packaging). Lancaster has grown significantly as part of the expanding Dallas—FORT WORTH metropolitan area. Lancaster Municipal Airport to SE. Seat of Cedar Valley College (two-year). Town destroyed by 1994 tornado. Settled 1846, incorporated 1886.

Lancaster (LAN-ki-stuhr), town (2000 population 3,734), ⊙ GARRARD county, central KENTUCKY, 31 mi/50 km S of LEXINGTON, in BLUEGRASS REGION; 37°37′N 84°34′W. Agriculture (burley tobacco, corn, wheat, hay; horses, cattle, hogs); manufacturing (wheels, apparel, machinery, crushed stone and lime, lumber, fiberglass boats). Cottage industry (hooked rugs) in vicinity. Carry Nation Birthplace; Historic Garrard County Jail. HERRINGTON LAKE reservoir (DIX RIVER) to NW.

Lancaster (LANG-KA-stuhr), town, WORCESTER county, N central MASSACHUSETTS, on NASHUA RIVER, and 15 mi/24 km NNE of WORCESTER, N of WACHUSETT RESERVOIR; 42°29′N 71°41′W. Luther Burbank born here. Has fine Bulfinch church (1817). Includes South Lancaster village (1990 population 1,772). State prison. Settled 1643, incorporated 1653.

Area is shown by the symbol □, and capital city or county seat by ⊙.

Lancaster, town, ☉ COOS COUNTY, NW NEW HAMPSHIRE, 19 mi/31 km W of BERLIN; 44°28′N 71°32′W. Bounded W by CONNECTICUT RIVER (VERMONT state line), drained by ISRAEL RIVER (flows into CONNECTICUT RIVER). Manufacturing (lumber, inductors, food processing, polyethylene bags; printing and publishing); agriculture (vegetables, nursery crops; poultry, livestock; dairying). Part of White Mountain National Forest in E; Weeks State Park in S; two covered bridges. Includes village of Grange, in center. Incorporated 1764.

Lancaster (LANG-ki-stuhr), town (2006 population 8,374), ☉ LANCASTER COUNTY, N SOUTH CAROLINA, 23 mi/4.8 km SE of ROCK HILL, near the WATEREE; 34°43′N 80°46′W. Manufacturing (metal fabrication, apparel, textiles, fabricated metal products, batteries, asphalt; printing and publishing); agriculture (poultry, livestock; grains, soybeans). Branch campus of University of South Carolina here. Andrew Jackson State Park to N. Courthouse and jail date from 1823. Bordered by several mill villages.

Lancaster (LAN-KA-stuhr), town (2006 population 3,893), ☉ GRANT COUNTY, extreme SW WISCONSIN, 23 mi/37 km N of DUBUQUE, IOWA; 42°51′N 90°42′W. In livestock and dairying area. Dairy products; manufacturing (canned foods, beverages, veterinary remedies, feed, timber; ships, transportation equipment, flexible packaging). Agricultural research station. Settled before 1840, incorporated 1878.

Lancaster, town, MATABELELAND SOUTH province, S central ZIMBABWE, 45 mi/72 km N of BULAWAYO, near Nsiza River; 20°08′S 29°17′E. Cattle, sheep, goats; tobacco, corn, cotton, soybeans.

Lancaster (LANG-kuh-stuhr), unincorporated village (2001 population 797), SE ONTARIO, E central CANADA, on the SAINT LAWRENCE RIVER, 15 mi/24 km NE of CORNWALL, and included in SOUTH GLENGARRY township; 45°08′N 74°30′W. Dairying; mixed farming.

Lancaster (LANG-KA-stuhr), village (2000 population 291), ATCHISON COUNTY, NE KANSAS, 10 mi/16 km W of ATCHISON, near CORN BELT; 39°34′N 95°17′W.

Lancaster (LAN-KA-stuhr), village (2000 population 363), KITTSON COUNTY, NW MINNESOTA, 8 mi/12.9 km NE of HALLOCK, near MANITOBA (CANADA) border, on North Branch TWO RIVERS; 48°51′N 96°47′W. Wheat, potatoes, sugar beets, alfalfa, sunflowers; sheep; manufacturing (agricultural equipment). Skull Lake Wildlife Area to NE; port of entry of U.S.-Canada border, 9 mi/14.5 km N.

Lancaster, village (□ 2 sq mi/5.2 sq km; 2006 population 11,328), ERIE COUNTY, W NEW YORK; 42°53′N 78°40′W. Light manufacturing and services. Historic district features notable buildings. Incorporated 1849.

Lancaster (LANG-ki-stuhr), unincorporated village, ☉ LANCASTER COUNTY, E VIRGINIA, 24 mi/39 km NNE of GLOUCESTER; CHESAPEAKE BAY to E; 37°46′N 76°28′W. Manufacturing (shellfish and seafood processing); agriculture (tomatoes, vegetables, grain, soybeans; cattle; timber; fish, oysters, crabs from nearby coastal towns.

Lancaster Mills (LANG-ki-stuhr), unincorporated village, LANCASTER COUNTY, N SOUTH CAROLINA, residential suburb 3 mi/4.8 km WNW of LANCASTER; near CATAWBA RIVER; 34°42′N 80°47′W.

Lancaster Sound, arm of BAFFIN BAY, c.200 mi/320 km long and 40 mi/60 km wide, BAFFIN region, NUNAVUT territory, CANADA; c.74°13′N 84°00′W. It extends W between DEVON and BAFFIN islands and is part of the shortest water route across N Canada to the BEAUFORT SEA. William Baffin, the English explorer, visited here in 1616.

Lance Amour, CANADA: see ANSE AMOUR, L'.

Lance au Loup, CANADA: see ANSE AU LOUP, L'.

Lance Creek, village, NIOBRARA county, on Lance Creek, E WYOMING, 21 mi/34 km NNW of LUSK. In oil field.

Lancefield (LANS-feeld), township, VICTORIA, SE AUSTRALIA, 45 mi/73 km N of MELBOURNE; 37°17′S 144°45′E. Skeletons of c.10,000 animals were found at Lancefield Swamp S of town. Settled by Europeans c.1837. Also spelled Lance Field.

Lance Field, AUSTRALIA: see LANCEFIELD.

Lancelin (LAN-suh-lin), resort village, WESTERN AUSTRALIA state, W AUSTRALIA, 79 mi/127 km N of PERTH; 31°01′S 115°19′E. White sand beaches, dunes; fishing, windsurfing. Wildflower displays.

Lancetilla Experimental Station (lahn-se-TEE-yah), ATLÁNTIDA department, N HONDURAS, 2.5 mi/4 km S of TELA; 15°42′N 87°28′W. Established by United Fruit Company in 1926; collection of over 1,000 tropical plants from around the world.

Lanchin, UKRAINE: see LANCHYN.

Lanchkhuti (lunch-KHOO-tee), town and center of Lanchkhuti region, W GEORGIA, in COLCHIS LOWLAND, on railroad, and 15 mi/24 km WSW of SAMTREDIA; 42°05′N 42°02′E. Meat and dairy combine; cannery; brick and tile manufacturing. Oil-pumping plant (oil tanks, farms, and storage facilities). Until 1936, spelled Lanchkhuty.

Lan-chou, CHINA: see LANZHOU.

Lanchow, CHINA: see LANZHOU.

Lanchyn (LAHN-chin) (Russian *Lanchin*) (Polish *Lanczyn*), town (2004 population 7,125), S central IVANO-FRANKIVS'K oblast, UKRAINE, on PRUT RIVER, and 14 mi/23 km W of KOLOMYYA; 48°33′N 24°45′E. Elevation 1,164 ft/354 m. Health resort with mineral springs; stone crushing; lumbering. Known since the 10th century then site of salt mining; town since 1940.

Lanciano (lahn-CHAH-no), ancient *Anxanum*, town, CHIETI province, ABRUZZI, S central ITALY, 6 mi/10 km from the ADRIATIC, 15 mi/24 km SE of CHIETI; 42°14′N 14°23′E. On three hills, two of them connected by a Roman bridge. Highly diversified manufacturing center (linen, cotton, and woolen textiles, agricultural machinery, furniture, shoes, pottery, chemicals); food processing. Archbishopric. Has Roman ruins (aqueduct, theater); cathedral (rebuilt 18th century), and 12th-century church. Badly damaged by heavy fighting (1943–1944) in World War II.

L'Ancienne-Lorette (lahn-see-EN–lo-RET), village (□ 3 sq mi/7.8 sq km; 2006 population 16,618), CAPITALE-NATIONALE region, S central QUEBEC, E CANADA, on short Lorette River (waterfalls); 46°47′N 71°23′W. Shoe manufacturing; agriculture, dairying; resort. Founded 1673 by Jesuits. Near by is LORETTEVILLE. An independent municipality, it is centrally located within the QUEBEC city agglomeration. Part of the Metropolitan Community of Quebec (*Communauté Metropolitaine de Québec*).

Lancing (LAHN-sing), town (2001 population 18,700), WEST SUSSEX, SE ENGLAND, on the CHANNEL, and 3 mi/4.8 km ENE of WORTHING; 50°48′N 00°19′W. Seaside resort. Site of Lancing College, a public school founded 1848. Has Norman church and Roman temple.

Lanco (LAHN-ko), town, ☉ Lanco comuna, VALDIVIA province, LOS LAGOS region, S central CHILE, on railroad and PAN-AMERICAN HIGHWAY, 37 mi/60 km NE of VALDIVIA; 39°27′S 72°47′W. Agricultural center (cereals, vegetables; livestock); lumbering, dairying, food processing.

Lancones (lahn-KO-nes), village, SULLANA province, PIURA region, NW PERU, on CHIRA RIVER, and 25 mi/40 km NE of SULLANA; 04°38′S 80°32′W. Sugarcane; vineyards. Near Reservorio de Poechas.

Lançon-Provence (lahn-son–pro-VAHNS), town (□ 26 sq mi/67.6 sq km), BOUCHES-DU-RHONE department, PROVENCE-ALPES-CÔTE D'AZUR region, SW FRANCE, 25 mi/40 km NNW of MARSEILLE, and on road to AVIGNON; 43°35′N 05°08′E.

Lancut (LAHN-tsoot), Polish *Łańcut*, town, Rzeszów province, SE POLAND, on railroad, and 10 mi/16 km E of RZESZÓW, near WISLOK RIVER; 50°04′N 22°14′E.

Distilling, brewing, tanning, flour milling, ether manufacturing. Castle.

Lancy (lahn-SEE), town (2000 population 25,688), GENEVA canton, SW SWITZERLAND; suburb adjacent (SW) to GENEVA; 46°12′N 06°08′E.

Lanczyn, UKRAINE: see LANCHYN.

Landa, village (2006 population 26), BOTTINEAU CO., N NORTH DAKOTA, 22 mi/35 km W of BOTTINEAU; 48°53′N 100°54′W. J. Clark Salyer National Wildlife Refuge to W. Near SOURIS River (Mouse River). Founded in 1904 and incorporated in 1922. Named for cousins, Daniel D. and Theodore T. Landa, early settlers from Norway.

Landaburu, SPAIN: see BARACALDO.

Landa de Matamoros (LAHN-dah dai mah-tah-MO-ros), town, NE QUÉRETARO, MEXICO, 7 mi/12 km E of JALPAN DE SERRA on Mexico Highway 120; 21°13′N 99°20′W. Elevation 3,363 ft/1,025 m. Very steep, rugged terrain crossing the Sierra Gorda. Hot climate. Agriculture (coffee, sugarcane, corn); forestry; mining (magnesium, silver, and lead on a small scale).

Landaff (LAN-daf), town, GRAFTON county, NW NEW HAMPSHIRE, 11 mi/18 km SW of LITTLETON; 44°08′N 71°52′W. Drained by AMMONOOSUC RIVER in NW corner, by Wild Ammonoosuc River in S. In mountain recreational area. Timber. Part of White Mountain National Forest in S.

Landana (lan-DAH-nuh), town, CABINDA province (enclave of Angola), on the ATLANTIC OCEAN, 24 mi/39 km N of CABINDA. Ships palm oil and kernels; sawmilling.

Landau (LAHN-dou), city, RHINELAND-PALATINATE, W GERMANY, on the QUEICH, 19 mi/31 km NW of KARLSRUHE; 49°12′N 08°07′E. A center for wine making and trading; manufacturing of rubber goods and automobile parts. Chartered 1274. World War II damage (c. thirty percent) included 15th-century church. Under French occupation, 1680–1815. Captured by U.S. troops in March 1945.

Landau (LAHN-dou) or **Landau an der Isar** (LAHN-dou ahn der EE-sahr), town, LOWER BAVARIA, BAVARIA, S GERMANY, on the ISAR, and 16 mi/26 km SSE of STRAUBING; 48°40′N 12°42′E. Manufacturing of electrical goods, automobile parts, and vehicles. Has rococo church. Chartered c.1224.

Landau, UKRAINE: see SHYROKOLANIVKA.

Land Between the Lakes, recreation area, STEWART county in NW TENNESSEE and Tipton county in SW KENTUCKY, bordered E by BARKLEY Reservoir (CUMBERLAND RIVER), W by KENTUCKY Reservoir (TENNESSEE RIVER); 36°45′N 82°02′W. Covers 170,000 acres/68,799 ha; managed by Tennessee Valley Authority for fishing, hunting, hiking, camping. Living history farm (The Homeplace, 1850) is here.

Landeck (LAHN-dek), town, TYROL, W AUSTRIA, on the INN river, 40 mi/64 km WSW of INNSBRUCK, at N end of RHAETIAN ALPS; 47°08′N 10°34′E. Important road junction; market center; cotton mills; manufacturing (carbide). Notable Gothic church with striking altar. Old Landeck castle on nearby hill. Cable cars to Krahberg (elevation 6,730 ft/2,051 m).

Landeck, POLAND: see LEDYCZEK.

Landeck, POLAND: see LADEK ZDROJ.

Landen (LAHN-duhn), commune (2006 population 14,749), Leuven district, BRABANT province, E central BELGIUM, 8 mi/12.9 km ESE of TIENEN; 50°45′N 05°05′E. Railroad junction.

Lander (LAN-duhr), county (□ 5,519 sq mi/14,349.4 sq km; 2006 population 5,272), N central NEVADA; ☉ BATTLE MOUNTAIN; 39°57′N 117°01′W. Drained by HUMBOLDT and REESE rivers. Gold, silver, lead, copper deposits; cattle and sheep ranches. Formed 1863. Parts of Toiyabe Natl Forest in S cover N-S running SHOSHONE, TOIYABE, and TOQUIMA mountain ranges. Geographic center of Nevada in SE corner.

Lander, town (2006 population 7,047), ☉ FREMONT county, W central WYOMING, on POPO AGIE RIVER, just E of WIND RIVER RANGE, and 120 mi/193 km W of CASPER; 42°49′N 108°43′W. Elevation c. 5,357 ft/1,633 m. Resort and trade center for Popo Agie Valley; headquarters for Washakie National Forest. Agriculture (cattle, sheep; sugar beets); manufacturing (fabrics, concrete, metal art castings, underground mining equip, lumber; log homes; printing and publishing). Oil wells, gold and coal mines in vicinity. Agricultural experiment station of state university, large Wind River Indian Reservation to N. Fremont County Pioneer Museum, Shoshone National Forest, and Sinks Canyon State Park to SW. To SE is site of first oil well drilled in Wyoming (1884). Settled in 1870s around military post, incorporated 1890.

Landerneau (lahn-duhr-no), town (□ 5 sq mi/13 sq km), FINISTÈRE department, W FRANCE, small part on ELORN RIVER estuary, and 12 mi/19 km ENE of BREST, in BRITTANY; 48°27′N 04°15′W. Active livestock and fish market; agricultural cooperative; wood and leatherworking. Has attractive old mansions built on 16th-century bridge and along banks of the river.

Landeron, Le, commune, NEUCHÂTEL canton, W SWITZERLAND, at SW end of LAKE BIEL, near THIELLE CANAL, and on BERN canton border.

Landero y Coss (lahn-DE-ro ee KOS), town, VERACRUZ, E MEXICO, 15 mi/24 km NE of XALAPA ENRÍQUEZ. Fruit, corn. In Totanac Indian area. Also known as San Juan.

Landes (lahn-duh), region (□ 4,600 sq mi/11,960 sq km), SW FRANCE, vast, nearly triangular tract of sand and former marshland, stretching along the ATLANTIC coast for more than 100 mi/161 km between the ADOUR RIVER and the MÉDOC region near BORDEAUX, and reaching inland as far as 40 mi/64 km; 45°59′N 00°36′W. It thus covers most of LANDES department and part of GIRONDE department. The straight coastline, indented only by the ARCACHON BASIN, is paralleled by a belt of sand dunes (up to 10 mi/16 km wide and 200 ft/61 m high), which separate several marshy lagoons and tidal lakes from the Bay of BISCAY. Formerly, sheep grazing was the only occupation in this insalubrious region, but much of the land has been reclaimed through drainage and the planting of pine and oak forests. Agriculture is relatively minor; manufacturing (wood-pulp resins, cork). The chief towns are MONT-DE-MARSAN, DAX, and ARCACHON (a popular seaside resort). The N central part of the Landes is included in the Landes de Gascony (LANDES DE GASCOGNE) Natural Regional Park, established 1970.

Landes (lahn-duh), department (□ 3,569 sq mi/9,279.4 sq km), in GASCONY, SW FRANCE; ⊙ MONT-DE-MARSAN; 44°00′N 00°50′W. Bounded by BAY OF BISCAY (W) and foothills of W PYRENEES (S), the department is drained by the ADOUR and its tributaries (MIDOUZE, GAVE D'OLORON, GAVE DE PAU) all rising in the Pyrenees. It occupies about ÂF of LANDES region noted for its sand dunes and pine forests. Only the CHALOSSE district (S of Adour River) has intensive agriculture (wheat, corn, and tobacco; poultry; wine). Chief industries are lumber milling and the preparation of resinous products and cork. Experimental missile grounds between Biscarosse and MIMIZAN in dune and lake country face entrenchment. Main towns are Mont-de-Marsan and DAX (commercial center and spa). CAPBRETON, Vieux-Boucau-les-Bains, and Mimizan-Plage are small seaside resorts, backed by high pine-covered dunes. The regional park of the Landes of Gascony occupies the interior N part of the department. Administratively, the department forms part of the AQUITAINE Region. Its SW comprimises in the French part of the BASQUE country.

Landes de Gascogne Natural Regional Park (LAHND duh gahs-KO-nyuh), GIRONDE and LANDES departments, AQUITAINE region, SW FRANCE; corresponds essentially to the basin of the Eyre River, which rises in the low hills (c. 330 ft/100 m) of the *pays*

d'Albret and reaches the ATLANTIC in the ARCACHON BASIN. It includes part of the forest land reclaimed by fixing the coastal sand dunes with plantations of pines in the 19th century. Pulp and paper manufacturing.

Landes de Lanvaux (lahn-duh duh lahn-vo), wooded ridge in MORBIHAN department, BRITTANY, NW FRANCE; extends WNW-ESE across the department c.15 mi/24 km inland from the ATLANTIC OCEAN; c.30 mi/48 km long, 3 mi/4.8 km wide. Rises to 525 ft/160 m. Grazing, hunting. Megalithic monuments discovered here.

Landes du Mené (lahnd dyoo mai-nai), chain of hills, part of the ARMORICAN MASSIF, in BRITTANY, NW FRANCE, c.15 mi/24 km SSE of SAINT-BRIEUC, extending NW-SE. Provides panoramic views of the Channel coast; 48°15′N 02°32′W. Rises to 1,115 ft/340 m. Sometimes spelled Ménez.

Landeshut, POLAND: see KAMIENNA GORA.

Landete (lahn-DAI-tai), town, CUENCA province, E central SPAIN, 40 mi/64 km ESE of CUENCA; 39°54′N 01°22′W. Grain, sheep; apiculture; manufacturing of plaster and tiles.

Landfill, village, WASHINGTON county, E MINNESOTA, residential suburb 6 mi/9.7 km E of downtown ST. PAUL between Tanners Lake (NW) and Battle Creek Lake (SW).

Land Glacier, in MARIE BYRD LAND, WEST ANTARCTICA, flowing into Land Bay on the RUPPERT COAST; 35 mi/60 km long; 75°40′S 141°45′W.

Landgraaf (LAHNT-khrahf), city, LIMBURG province, SE NETHERLANDS, 15 mi/24 km ENE of MAASTRICHT, and 8 mi/12.9 km N of AACHEN (GERMANY); 50°54′N 06°02′E. German border 1 mi/1.6 km to E; railroad junction. Agr. (dairying; cattle, hogs; vegetables, potatoes, grain, fruit); manufacturing (food processing).

Landgrove, town, BENNINGTON co., S central VERMONT, in GREEN MOUNTAINS, 19 mi/31 km W of SPRINGFIELD; 43°15′N 72°50′W. Population peaked at 400 residents in 1830.

Landhi (LUHN-dhee), village, KARACHI administration area, SE PAKISTAN, suburb 13 mi/21 km E of KARACHI; 24°52′N 67°14′E. Saltworks; pharmaceutical works.

Landi Khana (LUHN-dhee KHAH-nah), fort and village, KHYBER PASS, Khyber tribal region, North-West Frontier prov., W central Pakistan, 27 mi/43 km WNW of Peshawar, on Afghan border opposite Torkham (Afghanistan); 34°07′N 71°06′E. RR terminus.

Landi Kotal (LUHN-dhee KO-tahl), fort and town in KHYBER PASS, Khyber tribal region, North-West Frontier prov., W central Pakistan, 25 mi/40 km WNW of Peshawar, near Afghan border; 34°06′N 71°09′E. Major mfg. center for heroin, and smuggling entrepot of consumer goods to Pakistan.

Landi Muhammad Amin Khan (LUHN-dee moo-HUH-muhd uh-MEEN KHAHN), town and oasis, HELMAND province, S AFGHANISTAN, on HELMAND River, 100 mi/161 km SSW of GIRISHK; 30°31′N 63°47′E. Irrigated agriculture. In Garmsel or Garmser district, for which it is sometimes named.

Landing, resort village, MORRIS county, N NEW JERSEY, at S end of LAKE HOPATCONG, 12 mi/19 km NW of MORRISTOWN; 40°54′N 74°39′W. In suburbanizing area. Boating; state park nearby.

Landing, The, AUSTRALIA: see COSSACK.

Landingville, borough (2006 population 171), SCHUYLKILL county, E central PENNSYLVANIA, 7 mi/11.3 km SE of POTTSVILLE, on SCHUYLKILL RIVER. Agriculture (dairying); manufacturing (contract sewing).

Landis (LAN-duhs), town (□ 2 sq mi/5.2 sq km; 2006 population 3,073), ROWAN county, central NORTH CAROLINA, suburb 4 mi/6.4 km N of KANNAPOLIS; 35°32′N 80°36′W. Manufacturing (meat processing, brake linings, textiles); service industries; agriculture (grain, soybeans; livestock; dairying). Incorporated 1901.

Landisburg, borough (2006 population 191), PERRY county, S central PENNSYLVANIA, 24 mi/39 km WNW of HARRISBURG, near Sherman Creek. Agriculture (dairying).

Landisville, unincorporated town, LANCASTER county, SE PENNSYLVANIA, 7 mi/11.3 km NW of LANCASTER near Chickies Creek; 40°05′N 76°24′W. Manufacturing (concrete, rare earth magnets, fabricated metal products; millwork); agriculture (dairying; livestock, poultry and eggs; grain, potatoes, soybeans). Covered bridges in area. SALUNGA is 1 mi/1.6 km to NW.

Landisville, village, ATLANTIC county, S NEW JERSEY, 12 mi/19 km WNW of MAYS LANDING. Clothing, poultry, vegetables.

Landivisiau (lahn-dee-vee-syo), town (□ 7 sq mi/18.2 sq km; 2004 population 8,739), FINISTÈRE department, BRITTANY, W FRANCE, 21 mi/34 km ENE of BREST; 48°31′N 04°04′W. Naval airbase. Horse-breeding center and active cattle market.

Landmark (LAND-mahrk), community, SE MANITOBA, W central CANADA, 21 mi/33 km from WINNIPEG, and in TACHÉ rural municipality; 49°40′N 96°49′W.

Lando, unincorporated town, CHESTER county, N SOUTH CAROLINA, 10 mi/16 km S of ROCK HILL. Manufacturing (blankets); agriculture (livestock; dairying; grain, sorghum).

Land of 10,000 Lakes: see MINNESOTA.

Lándok, SLOVAKIA: see LENDAK.

Land o' Lakes (LAND O LAIKS), town (□ 12 sq mi/31.2 sq km), PASCO county, W central FLORIDA, 18 mi/29 km N of TAMPA; 28°13′N 82°27′W. Industries include lumber and light manufacturing.

Land o' Lakes, village, VILAS county, N WISCONSIN, on Wisconsin-MICHIGAN state line, 37 mi/60 km NNE of RHINELANDER. Manufacturing (fishing tackle, lumber); resort center in wooded lake region. Gateway Ski Area to S; Ottawa National Forest beyond state boundary.

Landon, unincorporated village, HARRISON county, SE MISSISSIPPI, suburb 5 mi/8 km N of GULFPORT, on Bernard Bayou, at intersection of Interstate Highway 10 and U.S. Highway 49; 30°30′N 89°07′W.

Landour (LUHN-dour), town, DEHRADUN district, N UTTAR PRADESH state, N central INDIA, in W KUMAON Himalaya foothills, 10 mi/16 km NNE of Dehra. Residential suburb of MUSSOORIE.

Landover, town, PRINCE GEORGES county, central MARYLAND, suburb 7 mi/11.3 km NE of WASHINGTON, D.C. Incorporated after 1940, it has outgrown the adjacent Landover Hills in population and is a major food distribution point. Jack Kent Cooke Stadium (now known as FedExField) opened in 1997 for the Washington Redskins football team.

Landover Hills, MARYLAND: see LANDOVER.

Landquart or **Landquart Dorf**, village of IGIS commune, GRISONS canton, E SWITZERLAND, on the RHINE RIVER, at mouth of the LANDQUART; 46°57′N 09°34′E. Paper milling, metalworking.

Landquart River, 27 mi/43 km long, E SWITZERLAND; rises in SILVRETTA mountain group (GRISONS canton) near Italian border; flows WNW through the PRÄTIGAU valley to the RHINE at LANDQUART. Drains 239 sq mi/619 sq km.

Landrail Point, village, S BAHAMAS, on NW CROOKED ISLAND, 8 mi/12.9 km NW of COLONEL HILL; 22°50′N 74°20′W.

Landrecies (lahn-dre-see), town (□ 8 sq mi/20.8 sq km), NORD department, NORD-PAS-DE-CALAIS region, N FRANCE, port on the SAMBRE, at N terminus of OISE-SAMBRE CANAL, and 11 mi/18 km W of AVESNES-SUR-HELPE; 50°08′N 03°42′E. A small industrial center in dairying region. A feudal stronghold; rebuilt in 18th century.

Landrienne (lahn-dree-EN), village (□ 107 sq mi/278.2 sq km), ABITIBI-TÉMISCAMINGUE region, W QUEBEC, E CANADA, 8 mi/13 km E of AMOS; 48°33′N 77°57′W. Gold mining.

Landri Sales (LAHN-zhee SAH-les), town (2007 population 5,618), SW PIAUÍ state, BRAZIL, 70 mi/113 km SW of FLORIANO; 07°15′S 43°57′W.

Landrum (LAND-ruhm), town (2006 population 2,544), SPARTANBURG county, NW SOUTH CAROLINA, 22 mi/35 km NW of SPARTANBURG, at NORTH CAROLINA state line; 35°10′N 82°11′W. Manufacturing (textile, fabricated metal products, electronics. paper products; machining); agriculture (dairying; livestock; grain, soybeans, sorghum, peaches, apples).

Landsberg (LAHNDS-berg), town, SAXONY-ANHALT, central GERMANY, 9 mi/14.5 km ENE of HALLE; 51°32′N 12°09′E. Sugar refining, malting; porphyry quarrying. Towered over by remains of 12th-century castle. In 12th century, chief town of a margraviate; passed to MEISSEN in early 13th century, to PRUSSIA in 1815.

Landsberg, POLAND: see GOROWO ILAWECKIE.

Landsberg am Lech (LAHNDS-berg uhm LEKH), city, BAVARIA, SW GERMANY, in UPPER BAVARIA, on the LECH RIVER; 48°03′N 10°52′E. Manufacturing (electronics, metal goods, and paper). Has remains of medieval town walls. Its fortress served as a political prison; Adolf Hitler wrote *Mein Kampf* while imprisoned there in 1923–1924, and numerous convicted Nazi war criminals were held there after 1945.

Landsberg an der Warthe, POLAND: see GORZÓW WIELKOPOLSKI.

Landsborough (LANDZ-buh-ro), town, QUEENSLAND, NE AUSTRALIA, 47 mi/76 km N of BRISBANE, N of GLASS HOUSE MOUNTAINS, and at S end of Blackall Ranges; 26°48′S 152°58′E. Local history museum. Australia Zoo just S.

Landsbro (LAHNTS-BROO), village, JÖNKÖPING county, S SWEDEN, 7 mi/11.3 km WSW of VETLANDA; 57°22′N 14°54′E.

Land's End (LANDZ END), promontory, SW CORNWALL, SW ENGLAND, forming the westernmost extremity of the English mainland; 50°03′N 05°44′W. Wave-carved granite; cliffs c.60 ft/18 m high. Offshore are reefs and rocky islets, on one of which is Longships Lighthouse. Land's End is a major tourist attraction in a privately-owned commercial park.

Landser (lahn-dzer), commune (☐ 1 sq mi/2.6 sq km), HAUT-RHIN department, in ALSACE, E FRANCE, 5 mi/8 km SSE of MULHOUSE; 47°41′N 07°23′E.

Landshut (LAHNDS-hoot), city, BAVARIA, SE GERMANY, in LOWER BAVARIA, on the ISAR RIVER; 48°33′N 12°09′E. Once the capital of LOWER BAVARIA, it is now a transportation and industrial center. Manufacturing (textiles, furniture, beer, chocolate, tobacco, and chemicals); a large influx of more than 12,000 refugees after World War II sparked new industries such as electronics and machine-building. Founded in 1204, Landshut became the residence of the dukes of Bavaria-Landshut in 1255. The city suffered heavily in the Thirty Years War (1618–1648). From 1802 to 1826 it was the seat of the Bavarian university (now at MUNICH). A 13th-century castle, Burg Trausnitz, overlooks the city. St. Martin's Church (1389) has one of the world's highest brick steeples (436 ft/133m).

Landshut, CZECH REPUBLIC: see LANZHOT.

Landskron (LAHNDS-kron), part of VILLACH, CARINTHIA, S AUSTRIA, 3 mi/4.8 km NNE of city center; near Lake Ossiach; 46°39′N 13°54′E. Impressive ruins of a fortress nearby.

Landskron, CZECH REPUBLIC: see LANSKROUN.

Landskrona (lahnts-KROO-nah), town, SKÅNE county, SW SWEDEN, seaport on ÖRESUND; 55°53′N 12°50′E. Commercial and industrial center; manufacturing (metal products, electric fittings). Devastated in sixteenth–seventeenth century Danish-Swedish wars. Swedish naval victory over Danes offshore (1677). Largely rebuilt mid-eighteenth century. Nearby Ven island, residence of sixteenth-century astronomer Tycho Brahe, annexed 1959. Chartered 1413.

Lands Lokk, island, NUNAVUT territory, CANADA, in the ARCTIC OCEAN, at N end of NANSEN SOUND, off NW

ELLESMERE ISLAND; 20 mi long, 8 mi wide; 81°45′N 91°45′W.

Landstrasse (LAHND-shtrahs-se), district (☐ 3 sq mi/ 7.8 sq km) of VIENNA, AUSTRIA, just SE of city center. Belvedere Palace and the diplomatic quarter of Vienna are located here.

Landstuhl (LAHND-shtool), town, RHINELAND-PALATINATE, W GERMANY, 9 mi/14.5 km W of KAISERSLAUTERN; 49°25′N 07°34′E. Manufacturing (china, electronics). Has ruined ancestral castle of Franz von Sickingen, who was killed here during a siege in 1523.

Landusky (lan-DUHS-kee), village, PHILLIPS county, N central MONTANA, in Lewis and Clark National Forest, 48 mi/77 km SW of MALTA. Gold; livestock. Little Rocky Mountains and Fort Belknap Indian Reservation to N.

Landwarów, Lithuania: see LENTVARIS.

Lane, county (☐ 717 sq mi/1,864.2 sq km; 2006 population 1,797), W central KANSAS; ⊙ DIGHTON; 38°28′N 100°28′W. Located in Smokey Hill region, rolling plain, with low hills in E and NE. Drained by WALNUT CREEK. Wheat, sorghum; cattle. Formed 1886.

Lane, county (☐ 4,721 sq mi/12,274.6 sq km; 2006 population 337,870), W OREGON, ⊙ EUGENE; 43°57′N 122°52′W. Mountainous area in COAST and CASCADE ranges, bounded on W by PACIFIC OCEAN, crossed by WILLAMETTE and SIUSLAW rivers. Agriculture (wheat, barley, oats, corn, brans, strawberries, blackberries, blueberries, grapes, apples, cherries, pears, plums, peaches; hogs, sheep, cattle); wineries, nurseries. Salmon fishing. Part of Willamette National Forest in E; part of Umpqua National Forest in SE; part of Siuslaw National Forest in W; part of Oregon Dunes National Recreation Area in extreme SW. FERN RIDGE RESERVOIR in center. County has more than twelve state parks, notably Elijah Bristow and Armitage in center, Morton in NE, and Devils Elbow on coast. Formed 1851.

Lane, village (2000 population 256), FRANKLIN county, E KANSAS, on POTTAWATOMIE CREEK, and 16 mi/ 26 km SE of OTTAWA; 38°26′N 95°04′W. Livestock; grain.

Lane, village (2006 population 539), WILLIAMSBURG county, E central SOUTH CAROLINA, 10 mi/16 km S of KINGSTREE; 33°31′N 79°52′W. Industry includes timber, livestock, grain, soybeans, cotton.

Lane, village (2006 population 53), JERAULD county, SE central SOUTH DAKOTA, 8 mi/12.9 km E of WESSINGTON SPRINGS, near Firesteel Creek; 44°04′N 98°25′W.

La Neblina, VENEZUELA: see SERRANÍA LA NEBLINA, PARQUE NACIONAL.

Lane Cove (LAIN KOV), suburb, E NEW SOUTH WALES, AUSTRALIA, 4 mi/6 km NW of SYDNEY across Port JACKSON; 33°49′S 151°10′E. In metropolitan area; high-tech industries.

Lane Island, KNOX county, S MAINE, in Carvers Harbor; bridged to Vinalhaven island; 0.5 mi/0.8 km long.

Lanesboro, town (2000 population 152), CARROLL county, W central IOWA, near RACCOON RIVER, 12 mi/ 19 km NE of CARROLL; 42°10′N 94°41′W.

Lanesboro or **Lanesborough**, agricultural town, BERKSHIRE county, NW MASSACHUSETTS, 5 mi/8 km N of PITTSFIELD, near PONTOOSUC LAKE Resort; 42°32′N 73°14′W. Settled c.1753; incorporated 1765. Includes state forest and villages of Berkshire and Balance Rock.

Lanesboro, town (2000 population 788), FILLMORE county, SE MINNESOTA, 32 mi/51 km SE of ROCHESTER, on Duschee Creek, S of its confluence with ROOT RIVER; 43°43′N 91°58′W. Grain, soybeans; livestock, poultry; dairying; manufacturing (corrugated dunnage, gopher traps). Richard J. Dorer Memorial Hardwood State Forest to N and E.

Lanesboro, borough (2006 population 570), SUSQUEHANNA county, NE PENNSYLVANIA, 1 mi/1.6 km NE of SUSQUEHANNA depot, near NEW YORK state line.

Agriculture (dairying); manufacturing (electronic coils).

Lanester (lah-ne-ster), NE suburb (☐ 5 sq mi/13 sq km) of LORIENT, MORBIHAN department, in BRITTANY, W FRANCE, on an inlet of the ATLANTIC OCEAN; 47°45′N 03°21′W.

Lanesville, town (2000 population 614), HARRISON county, S INDIANA, 8 mi/12.9 km E of CORYDON; 38°14′N 85°59′W. In agricultural area. Laid out 1817.

Lanesville, hamlet, GREENE county, SE NEW YORK, in the CATSKILL MOUNTAINS, 20 mi/32 km NW of KINGSTON; 42°16′N 74°16′W. In summer-recreation area.

Lanesville, MASSACHUSETTS: see GLOUCESTER.

Lanett, city (2000 population 7,897), Chambers co., E Alabama, on Chattahoochee River, 13 mi/21 km E of Lafayette, and 20 mi/32 km NE of Opelika. Adjoins West Point, Georgia Manufacturing and processing of cotton fabrics. Fort Tyler, site of one of the last Civil War battles E of the Mississippi River (April 1865), nearby. Originally incorporated in 1865 as Bluffton due to its location on a bluff, it was changed to Lanett in 1895. The name is a blend of the two last names of early textile mill developers Lafayette Lanier and Theodore Bennett.

La Neuveville, SWITZERLAND: see NEUVEVILLE, LA.

Laneuveville-devant-Nancy (lah-nu-VEEL–duh-vahn–nahn-SEE), town (☐ 4 sq mi/10.4 sq km) outer SE suburb of NANCY, MEURTHE-ET-MOSELLE department, in LORRAINE, NE FRANCE, at junction of MARNE-RHINE CANAL and Canal de l'EST; 48°39′N 06°14′E. Miscellaneous manufacturing and chemical works. Salt deposits nearby.

Lanexa, unincorporated village, NEW KENT county, E VIRGINIA, 14 mi/23 km NW of WILLIAMSBURG, near CHICKAHOMINY RIVER; 37°25′N 76°54′W. Manufacturing (catfish processing, apparel); agriculture (grain, soybeans; poultry, cattle); catfish.

Lanezi Lake, E BRITISH COLUMBIA, W CANADA, in CARIBOO MOUNTAINS, 65 mi/105 km E of QUESNEL, S of Issac Lake; 12 mi/19 km long, 1 mi/2 km wide; 53°04′N 120°55′W. Drained W by Cariboo River through Sandy Lake (5 mi/8 km long) into QUESNEL RIVER.

Lang (LANG), village (2006 population 172), S SASKATCHEWAN, CANADA, 40 mi/64 km SSE of REGINA; 49°55′N 104°22′W. Wheat.

Langå (LAHNG-o), town, ÅRHUS county, N JUTLAND, DENMARK, 19 mi/31 km ESE of VIBORG; 56°22′N 09°58′E. Agriculture (dairying; hogs); paperboard mills.

Lánga Co (LAH-ANG-ZO), sacred lake (☐ 140 sq mi/ 364 sq km) in W HIMALAYAS, SW TIBET, SW CHINA, 80 mi/129 km SE of GARYARSA; 18 mi/29 km long, 3 mi/4.8 km–13 mi/21 km wide; connected with MAPAM YUMCO (E) by 6-mi/9.7-km-long Ganga Qu river. Elevation 14,900 ft/4,542 m. SUTLEJ RIVER (here called Xiangquan He or Langqen Kanbab) rises in area at the lake; Kongque River (Gogra or KARNALI River) rises to SW. Also called Rakas Lake, Langak Tso, Landak Tso, or Rakshas Tal.

Langada (lahn-GAH–[th]-ah), pass and gorge, LAKONIA and MESSENIA prefectures, SE PELOPONNESE department, SE mainland GREECE, in the TAIYETOS Mountains; 31 mi/50 km long; 37°42′N 22°03′E. An E-W road (SPARTA-KALAMATA) runs through the gorge; regarded as one of the country's most scenic rides, with its hairpin curves and dramatic views. Ancient Spartans said to have left their deformed or weak babies on the slopes here to die.

Langadas (lahn-gah-[TH]AHS), town (1991 population 6,113), THESSALONÍKI prefecture, CENTRAL MACEDONIA department, NE GREECE, 12 mi/19 km NE of THESSALONÍKI; 40°45′N 23°04′E. One of four locales in Macedonia where, on May 21, *anastenaridhes* [Greek=groaners] walk barefoot over hot embers with no harmful effects. *Anastenaria* also held in AYIA

Cross-references are shown in SMALL CAPITALS. The pronunciation guide is shown on page xix. The sources of population figures are shown on page xvii.

ELENI, Ayios Petros (near SÉRRAI) and Meliki (12 mi/ 19 km E of VÉROIA). Also called Langadhas.

Langa de Duero (LAHNG-gah dhai DHWAI-ro), town, SORIA province, N central SPAIN, on DUERO RIVER, on railroad, and 17 mi/27 km ESE of ARANDA DE DUERO; 41°37′N 03°24′W. Cereals, grapes, sugar beets; livestock. Excavation center.

Langadha, Lake, Greece: see KORONIA, LAKE.

Langadhas, Greece: see LANGADAS.

Langadia (lahn-GAH–[th]-yah), town, ARKADIA prefecture, central PELOPONNESE department, S mainland GREECE, 23 mi/37 km NW of TRÍPOLIS; 37°41′N 22°02′E. Livestock (goats, sheep), wine. Summer resort. Also spelled Lagadia, Langadhia, or Lankadia.

Langanes (LOUNG-gah-nes), peninsula, NE ICELAND, extends 30 mi/48 km NE into GREENLAND SEA, between THISTILFJÖRÐUR (NW) and BAKKAFLOI (SE); 66°15′N 15°00′W. Rises to 2,359 ft/719 m. THORSHOFN fishing village (SW).

Langano, Lake (□ c.80 sq mi/207 sq km), OROMIYA state, S central ETHIOPIA; 13 mi/21 km long, 9 mi/14.5 km wide; 07°40′N 38°45′E. In the GREAT RIFT VALLEY, between lakes ZIWAY and SHALA, and parallel to LAKE ABIYATA (2 mi/3.2 km W), 90 mi/145 km S of ADDIS ABABA. Receives many short tributaries (W, S); during flood periods overflows into LAKE ABIYATA. Lake Langano is a weekend resort for fishing and water sports. Its water is saline.

Langao (LAN-GOU), town, ⊙ Langao county, S SHAANXI province, CHINA, in the DABA SHAN, 22 mi/35 km SSW of ANKANG; 32°18′N 108°53′E. Grain, oilseeds, medicinal herbs.

Langar, village, E KASHKADARYO wiloyat, UZBEKISTAN, 25 mi/40 km S of SHAKHRISABZ. Grain; livestock.

Langarud (lahn-gah-ROOD), town (2006 population 65,614), GĪLĀN province, N IRAN, 31 mi/50 km E of RASHT, near CASPIAN SEA; 37°11′N 50°09′E. Subtropical agriculture. Also called Langrud.

Langasian, PHILIPPINES: see PROSPERIDAD.

Langatabiki (lung-gah-TAH-bi-chi), landing in SIPALIWINI district, E SURINAME, on upper MARONI RIVER (FRENCH GUIANA border), and 40 mi/64 km NE of AFOBAKA; 05°00′N 54°28′W. In tropical forests (timber, rubber).

Langat River (LAHNG-aht), 100 mi/161 km long, SELANGOR, W MALAYSIA; rises in central Malayan range at PAHANG-Negri SEMBILAN line; flows SW and W, past KAJANG, to Strait of MALACCA SW of Port Klang. Mangrove islands off mouth.

Lang Biang Plateau, VIETNAM: see LAM VIEN PLATEAU.

Langdai (LANG-DEI), town, SW GUIZHOU province, CHINA, 40 mi/64 km W of ANSHUN. Grain, tobacco.

Langdale (LANG-dail), unincorporated village, SW BRITISH COLUMBIA, W CANADA, on W shore of HOWE SOUND, 20 mi/33 km from VANCOUVER, and in SUNSHINE COAST regional district; 49°26′N 123°28′W.

Langdale Fell (LANG-dail FEL), mountain area of the PENNINES, CUMBRIA, NW ENGLAND, 5 mi/8 km N of SEDBERGH; 54°24′N 02°31′W.

Langdale Pikes (LANG-dail PEIKS), two Cumbrian peaks of the LANGDALE FELL, NW ENGLAND, 7 mi/11.3 km WNW of AMBLESIDE; 54°30′N 03°02′W. Harrison Stickle is 2,401 ft/732 m, Pike o' Stickle 2,323 ft/708 m.

Langdon, town, SULLIVAN county, SW NEW HAMPSHIRE, 13 mi/21 km S of CLAREMONT, near VERMONT state line; 43°10′N 72°22′W. Drained by COLD RIVER. Agriculture (nursery crops, hay; cattle, poultry). Two covered bridges in S.

Langdon, town (2006 population 1,766), ⊙ CAVALIER co., NE NORTH DAKOTA, 85 mi/137 km NW of GRAND FORKS; 48°45′N 98°22′W. Durum wheat, barley, grain-distribution center, dairy products, livestock; manufacturing (farm equipment, concrete, lawn and garden sprayers). Seat of state agriculture experiment station. Founded in 1886 and incoporated in 1888. Named for Robert Bruce Langdon (1826–1895), a railroad official who donated a bell for the school.

Langdon, village (2000 population 72), RENO county, S central KANSAS, 26 mi/42 km SW of HUTCHINSON; 37°51′N 98°19′W. In wheat region.

Langeac (lahn-zhahk), town (□ 13 sq mi/33.8 sq km; 2004 population 4,004), HAUTE-LOIRE department, S central FRANCE, in AUVERGNE region, between MARGERIDE MOUTAINS (W) and Monts du Velay (E), on the ALLIER, and 19 mi/31 km W of Le PUY-EN-VELAY; 45°06′N 03°29′E. Road and fruit and vegetable market center; manufacturing of plastics. Tourism in gorge of upper Allier River.

Langeais (lahn-zhai), commune (□ 23 sq mi/59.8 sq km), INDRE-ET-LOIRE department, CENTRE administrative region, W central FRANCE, in LOIRE RIVER valley, and 14 mi/23 km WSW of TOURS; 47°20′N 00°24′E. Manufacturing (agricultural implements, building materials). Vineyards. Its 15th-century feudal castle built by Louis XI was scene of marriage (1491) of Charles VIII and Anne of BRITTANY. The château was bequeathed to the French Inst. in 1904 and has been beautifully restored. Its interior conveys the lifestyle of noblemen at the beginning of the Renaissance.

Langeberg, mountain range, WESTERN CAPE PROVINCE, SOUTH AFRICA, extends 130 mi/209 km E from HEX RIVER MOUNTAINS near WORCESTER to GOURITZ RIVER valley S of CALITZDORP. E part of range parallels INDIAN OCEAN coast. Rises to 6,809 ft/2,075 m on Keeromsberg at W end of range, 11 mi/18 km NE of Worcester. Continued E by OUTENIQUA MOUNTAINS. Forms part of the S rim of the SOUTHERN KAROO. Two nature reserves toward its E end: Marloth Natural Reserve and Boosmansbos Wilderness Area.

Langedijk (LANG-uh-DEIK), town, NORTH HOLLAND province, NW NETHERLANDS, 5 mi/8 km NNE of ALKMAAR, on branch of NORTH HOLLAND CANAL; 52°42′N 04°49′E. Railroad junction to SE. Cattle, sheep; plants, vegetables.

L'Ange-Gardien (LAHNZH–gahrd-YEN), parish (□ 20 sq mi/52 sq km; 2006 population 2,946), CAPITALE-NATIONALE region, S QUEBEC, E CANADA, on the SAINT LAWRENCE RIVER, 11 mi/18 km NE of QUEBEC city; 46°55′N 71°06′W. Dairying. Part of the Metropolitan Community of Quebec (Communauté Métropolitaine de Québec).

L'Ange-Gardien (LAHNZH–gahrd-YEN), village (□ 87 sq mi/226.2 sq km; 2006 population 3,862), OUTAOUAIS region, SW QUEBEC, E CANADA, 5 mi/8 km from FARNHAM; 45°35′N 75°27′W. Timber, pulp and paper industries. Established 1979.

Langeh, IRAN: see LENGEH.

Langeland (LAHNG-uh-lan), narrow island (□ 110 sq mi/286 sq km), S DENMARK, between FYN and LOLLAND; 55°00′N 10°50′E. RUDKØBING is the main town; other towns include Bagenkop and Lohals. The island is largely agriculture, and grain is the chief product. It is noted for its magnificent beech trees.

Langeland Bælt (LAHNG-uh-lan belt), strait (c.8 mi/ 13 km wide), DENMARK, between LANGELAND (W) and LOLLAND (E) islands; joins the STORE BÆLT (N) and BALTIC SEA (S).

Langeloth (LAN-guh-lawth), unincorporated town, Smith township, WASHINGTON county, SW PENNSYLVANIA, 16 mi/26 km NW of WASHINGTON; 40°21′N 80°24′W. Agriculture (corn, hay; livestock, dairying). Manufacturing (sporting goods); coal. Hillman State Park to N; Cross Creek County Park to S.

Langelsheim (LAHNG-gels-heim), town, LOWER SAXONY, NW GERMANY, at NW foot of the upper HARZ, on the INNERSTE, and 3 mi/4.8 km NW of GOSLAR; 51°56′N 10°20′E. Manufacturing (chemicals); metalworking; tourism. Mining began here as early as the 13th century, but has now ceased. Chartered 1951.

Langemark-Poelkapelle (LAHN-guh-mahrk–pulkah-PEL-luh), commune (2006 population 7,790), Ypres district, WEST FLANDERS province, W BELGIUM, 5 mi/8 km NNE of YPRES. Site of battle (1914) in World War I.

Langen (LAHNG-uhn), town, HESSE, central GERMANY, 8 mi/12.9 km N of DARMSTADT; 49°59′N 08°11′E. Within commuting distance of FRANKFURT and DARMSTADT, the town is largely residential. Electronics and high technology industries; metalworking. Site of a biological research institute (vaccines).

Langen, town, LOWER SAXONY, NW GERMANY, 4 mi/6.4 km N of BREMERHAVEN; 53°36′N 08°35′E. Summer resort; largely residential.

Langen am Arlberg (LAH-en ahm AHRL-berg), village VORARLBERG, W AUSTRIA, 14 mi/23 km E of BLUDENZ; 47°08′N 10°07′E. Winter-sports center (elevation 3,718 ft/1,133 m); W entrance to Arlberg Tunnel.

Langenargen (LAHNG-uhn ahr-guhn), village, BADEN-WÜRTTEMBERG, GERMANY, on N shore of LAKE CONSTANCE, 5 mi/8 km SE of FRIEDRICHSHAFEN; 47°36′N 09°33′E. Agriculture (hops). Has institute for lake research, founded 1920. On nearby small peninsula is Montfort castle, built 1858 in Moorish style, today a hotel.

Langenau (LAHNG-uhn-ou), town, BADEN-WÜRTTEMBERG, GERMANY, 8 mi/12.9 km NE of ULM; 48°30′N 10°07′E. Metalworking. Has Gothic-baroque church.

Langenberg (LAHNG-uhn-berg), district of VELBERT, North Rhine-Westphalia, W GERMANY, 3 mi/4.8 km E of city center; 51°21′N 07°08′E. Manufacturing (machinery). Incorporated into VELBERT in 1975.

Langenberg (LUHNG-uhn-berg), village, North Rhine-Westphalia, GERMANY, 19 mi/31 km SW of BIELEFELD; 51°47′N 08°18′E. Metalworking.

Langenbielau, POLAND: see BIELAWA.

Langenburg (LANG-uhn-buhrg), town (2006 population 1,048), SE SASKATCHEWAN, CANADA, 40 mi/64 km SE of YORKTON, near MANITOBA border; 50°50′N 101°42′W. Mixed farming.

Langenburg (LAHNG-uhn-buhrg), town, BADEN-WÜRTTEMBERG, GERMANY, on the JAGST, 12 mi/19 km NW of CRAILSHEIM. Manufacturing of machinery; tourism. Has Renaissance castle, renovated in 17th century; today a museum.

Langendorf, commune, SOLOTHURN canton, NW SWITZERLAND, just NW of SOLOTHURN. Elevation 1,542 ft/470 m. Watches, metal products. The S end of the Weissenstein railroad tunnel (2 mi/3.2 km; opened 1908) is just NW at Oberdorf.

Langeness, GERMANY: see NORDMARSCH-LANGENESS.

Langenfeld (LAHNG-uhn-felt), city, North Rhine-Westphalia, W GERMANY, 7 mi/11.3 km SW of SOLINGEN; 51°06′N 06°56′E. Manufacturing (steelware, iron pipes and fittings, motor vehicles, medical equipment). Called RICHRATH-REUSRATH until 1936. Chartered 1948.

Langenhagen (lahng-uhn-HAH-guhn), city, LOWER SAXONY, W GERMANY, 5 mi/8 km N of HANOVER city center; 52°27′N 09°44′E. Largely industrial; manufacturing (electronics, cigarettes, foods). Site of airport serving Hanover metropolitan area. Chartered 1959.

Langenlois (LAHNG-uhn-lois), town, central LOWER AUSTRIA, near KAMP RIVER, 5 mi/8 km NE of KREMS AN DER DONAU; 48°29′N 15°40′E. Wine. Remarkable main square; baroque castle Gobelsburg nearby.

Langensalza, Bad, GERMANY: see BAD LANGENSALZA.

Langenschwalbach, GERMANY: see BAD SCHWALBACH.

Langenselbold (lahng-uhn-SEL-bolt), town, HESSE, W GERMANY, on the KINZIG, and 6 mi/9.7 km ENE of HANAU; 50°10′N 09°03′E. Manufacturing (machinery, metalworking). Has 18th-century castle.

Langenthal (LAHNG-uhn-tahl), town (2000 population 14,078), BERN canton, NW central SWITZERLAND, on small Langetedel River, and 12 mi/19 km E of SOLOTHURN; 47°13′N 07°47′E. Textiles (linen, woolen), porcelain, beer; metal- and woodworking, printing.

Langenwang (LAHNG-uhn-vahng), township, STYRIA, E central AUSTRIA, on Mürz River, and 19 mi/31 km ENE of BRUCK AN DER MUR; 47°34′N 15°37′E.

Area is shown by the symbol □, and capital city or county seat by ⊙.

Langenzenn (LAHNG-uhn-tsen), town, MIDDLE FRANCONIA, N central BAVARIA, GERMANY, on small Zenn River, 8 mi/12.9 km WNW of FÜRTH; 49°29′N 10°47′E. Ceramics; wine. Has 15th-century church and former Augustine monastery. Chartered 1442.

Langenzersdorf (lahn-GEN-tsers-dorf), township, NE LOWER AUSTRIA, AUSTRIA, on the Danube at the urban fringe of VIENNA, 7 mi/11.3 km N of city center; 48°19′N 16°22′E. Shopping malls; vineyards, vegetables.

Langeoog (LAHNG-uh-ok), NORTH SEA island (□ 7 sq mi/18.2 sq km) of E Frisian group, GERMANY, 7 mi/11.3 km N of ESENS; 7 mi/11.3 km long (E-W), c.1 mi/1.6 km wide (N-S); 53°45′N 07°27′E. Nordseebad Langeoog (W) is a resort. Has airport, and is connected by ferry to ESENS.

Langepas (lahn-gee-PAHS), city (2005 population 41,075), N central KHANTY-MANSI AUTONOMOUS OKRUG, W central SIBERIA, RUSSIA, N of the OB′ RIVER, on road, 340 mi/547 km E of KHANTY-MANSIYSK, and 75 mi/121 km NW of NIZHNE-VARTOVSK; 61°10′N 75°30′E. Arose with development of oil and gas fields. Has a heritage museum. Made city in 1985.

Langerwehe (LAHNG-uhr-VAI-uh), town, North Rhine-Westphalia, W GERMANY, 5 mi/8 km WNW of DÜREN; 50°49′N 06°22′E. Manufacturing of paper, plastics; sawmilling; potteries.

Langesund (LAHNG-uh-soon), town, TELEMARK county, S NORWAY, at mouth of LANGESUNDSFJORDEN, 16 mi/26 km SSE of SKIEN; 59°00′N 09°45′E. Mechanical manufacturing; canning; fishing. Important lumber-shipping center in sixteenth century.

Langesundsfjorden (LAHNG-uh-soons-fyawr-uhn), inlet of the SKAGERRAK at LANGESUND, TELEMARK county, S NORWAY. From its island-dotted mouth, it extends inland in two branches, FRIERFJORDEN (7 mi/11.3 km long) and Erdangerfjorden (4 mi/6.4 km long). BREVIK is on it.

Langevin, CANADA: see SAINTE-JUSTINE.

Langewiesen (LAHNG-uh-VEE-zuhn), town, THURINGIA, central GERMANY, at foot of THURINGIAN FOREST, on ILM RIVER, and 3 mi/4.8 km ESE of ILMENAU; 50°40′N 10°58′E. Summer resort.

Langfang, city (□ 388 sq mi/1,008.8 sq km; 2000 population 1,078,010), E central HEBEI province, CHINA, on BEIJING-TIANJIN railroad; 37°59′N 116°40′E. Agriculture (grains, vegetables, fruits, oil crops, cotton; hogs, cattle, sheep, poultry; dairying; freshwater fish); manufacturing (food processing, textiles, oil refining and chemicals, and machinery). In 1992 the city established an economic and technological development district to stimulate overseas investment.

Langfjellene (LAHNG-fyel-luh-nuh), collective name applied to an almost continuous range of mountains in W NORWAY, including JOTUNHEIMEN, JOSTEDALSBREEN, Hardangerfjell, HARDANGERVIDDA, HAUKELIFJELL, Bykle Mountains.

Langford (LANG-fuhrd), rural municipality (□ 217 sq mi/564.2 sq km; 2001 population 784), SW MANITOBA, W central CANADA; 50°10′N 99°25′W. Agriculture. Surrounds S part of NEEPAWA.

Langford (LANG-fuhrd), village (□ 15 sq mi/39 sq km; 2001 population 18,840), SW BRITISH COLUMBIA, W CANADA, on SE VANCOUVER ISLAND, 6 mi/10 km W of VICTORIA, in CAPITAL REGIONAL DISTRICT; 48°27′N 123°30′W. Logging; mixed farming, fruit growing, vegetables, dairying. Settled 1851; incorporated 1992. Formerly Langford Station.

Langford (LANG-fuhrd), village (2001 population 2,882), E central BEDFORDSHIRE, central ENGLAND, on IVEL RIVER, and 2 mi/3.2 km S of BIGGLESWADE; 52°03′N 00°16′W. Has 14th-century church.

Langford, village (2006 population 287), MARSHALL county, NE SOUTH DAKOTA, 14 mi/23 km S of BRITTON; 45°36′N 97°49′W. Fort Sisseton State Park to E.

Langford Station, CANADA: see LANGFORD.

Langgöns (LAHNG-guhns), town, HESSE, central GERMANY, 7 mi/11.3 km S of GIESSEN; 50°30′N 08°40′E. Castle (11th-century) and 16th-century church.

Langham (LANG-uhm), town (2006 population 1,120), central SASKATCHEWAN, CANADA, near NORTH SASKATCHEWAN RIVER, 20 mi/32 km NW of SASKATOON; 52°22′N 106°58′W. Grain elevators; dairying; ranching.

Langhirano (lahn-gee-RAH-no), town, PARMA province, EMILIA-ROMAGNA, N central ITALY, on PARMA RIVER and 13 mi/21 km S of PARMA; 44°37′N 10°16′E. Sausage, canned tomatoes.

Lang Hit (LAHNG HIT), village, BAC THAI province, N VIETNAM, 14 mi/23 km N of THAI NGUYEN; 21°44′N 105°51′E. Zinc-mining center. Vietnamese with Tai minority.

Langholm (LANG-olm), town (2001 population 2,311), DUMFRIES AND GALLOWAY, S Scotland, on ESK RIVER, and 18 mi/29 km N of CARLISLE; 55°09′N 03°00′W. Previously woolen mills (tweed) and tanneries.

Langhorne, borough (2006 population 1,967), BUCKS county, SE PENNSYLVANIA, 19 mi/31 km NE of PHILADELPHIA, and 9 mi/14.5 km WSW of TRENTON, NEW JERSEY, near Neshaminu Creek. Agr. to N (livestock; dairying); manufacturing (machinery, medical instruments, sand blasting equipment, valves, steel processing, metal fabrication, power transmission gears, plastic products, printing, folding cartons). Buehl Field airport to NE. Sesame Place, theme park based upon *Sesame Street* television show, to E. Laid out 1783, incorporated 1874.

Langhorne Manor, borough (2006 population 1,076), BUCKS county, SE PENNSYLVANIA, residential suburb, 18 mi/29 km NE of downtown PHILADELPHIA and 1 mi/1.6 km S of LANGHORNE.

Langisjór (LOUNG-gis-yawr), two lakes, S ICELAND, on SW edge of vatnajokull. Larger lake (64°11′N 18°17′W) is 10 mi/16 km long, 1 mi/1.6 km wide. Drained SW by Skafta River

Langjökull (LOUNG-yuh-kuh-tuhl), extensive glacier, W ICELAND, 50 mi/80 km NE of REYKJAVIK; 40 mi/64 km long, 5 mi/8 km–15 mi/24 km wide. Rises to 4,757 ft/1,450 m at 64°36′N 20°36′W.

Langkawi Island (LAHNG-kah-we), main island of LANGKAWI group (□ 203 sq mi/527.8 sq km), in Strait of MALACCA, MALAYSIA, on THAILAND border, 20 mi/32 km off PERLIS; 06°22′N 99°50′E. Rugged, low mountains; 18 mi/29 km long, 10 mi/16 km wide; rises to 2,898 ft/883 m at Gunong Raya. Coconuts, rice, rubber; rubber plantations. Fish (silver fish, *ikan bilis*, laid out on mats to dry; used as food). Main village of KUAH is on S shore; ferry from KUALA PERLIS on mainland; administered as part of KEDAH. Popular tourist destination beginning in the 1980s.

Langkazi, CHINA: see NAGARZE.

Langkha Tuk (LUHNG-KAH TUK), peak (4,173 ft/1,272 m), S THAILAND, on MALAY PENINSULA, 50 mi/80 km WNW of SURATTHANI.

Langkloof Valley, SOUTH AFRICA: see KAREEDOUW.

Langlade (LANG-laid), county (2006 population 20,631), NE WISCONSIN; ⊙ ANTIGO; 45°15′N 89°04′W. Lumbering is chief industry; agriculture (barley, oats, wheat, corn, beans, potatoes; poultry). Drained by WOLF and EAU CLAIRE rivers. Numerous lakes, especially N and far E; part of Nicolet National Forest in extreme E; Langlade Fish Hatchery in E. Formed 1879.

Langlade (lahn-GLAHD), unincorporated village, W central QUEBEC, E CANADA, included in SENNETERRE; 48°14′N 75°59′W. Garnet mining.

Langley (LANG-lee), city (□ 4 sq mi/10.4 sq km; 2001 population 23,643), SW BRITISH COLUMBIA, W CANADA, in lower FRASER River valley, 20 mi/32 km SE of VANCOUVER, and in GREATER VANCOUVER regional district; 49°06′N 122°39′W. In dairying, fruit- and hops-growing region; manufacturing (plastic and metal products, clothing); retail trade. A campus of Kwantlen University College here (opened 1993). Incorporated 1955. Formerly called Langley Prairie.

Langley (LANG-lee), township (□ 119 sq mi/309.4 sq km; 2001 population 86,896), SW BRITISH COLUMBIA, W CANADA, in the GREATER VANCOUVER regional district; 49°05′N 122°35′W. Agriculture (mushrooms, berries; poultry, livestock; dairying); horse breeding and training; flowers. Established 1873.

Langley (LANG-lee), unincorporated town, FAIRFAX county, NE VIRGINIA, residential suburb 7 mi/11 km NW of WASHINGTON, D.C., on POTOMAC RIVER; 38°56′N 77°09′W. Federal Highway Administration Research Station and Central Intelligence Agency (CIA) headquarters are here. Little Falls Dam to E.

Langley, town (2006 population 1,026), Island county, NW WASHINGTON, on WHIDBEY ISLAND on Saratoga Passage, and 10 mi/16 km WNW of EVERETT; 48°02′N 122°25′W. Resort; trade center for agricultural area (livestock; berries; dairy products).

Langley (LANG-lee), village (2006 population 3,400), E CHESHIRE, W central ENGLAND; 53°14′N 02°05′W. Silk mills.

Langley, unincorporated village, FLOYD county, E KENTUCKY, 9 mi/14.5 km S of PRESTONSBURG, on Right Fork River. Bituminous coal, natural gas. Manufacturing (natural gas processing).

Langley, village (2006 population 688), MAYES county, NE OKLAHOMA, 13 mi/21 km SSE of VINITA; 36°27′N 95°02′W. Manufacturing (signs). PENSACOLA DAM (NEOSHO River) is E; forms LAKE OF THE CHEROKEES. Incorporated 1939.

Langley, unincorporated village, AIKEN county, SW SOUTH CAROLINA, 7 mi/11.3 km WSW of AIKEN. Manufacturing (sealants); kaolin clay processing.

Langley Air Force Base (LANG-lee), U.S. military installation, HAMPTON city, SE VIRGINIA, 4 mi/6 km N of downtown HAMPTON; on Back River estuary, arm of CHESAPEAKE BAY. Covers 5 sq mi/13 sq km (3,195 acres/1,293 ha). Named for aviation pioneer Samuel P. Langley. Oldest continuously active air force base in U.S.; headquarters of Tactical Air Command; air-defense missile units. NASA Langley Research Center. Established 1916.

Langley, Mount (14,027 ft/4,275 m), on INYO-TULARE county line, E CALIFORNIA, in the SIERRA NEVADA, c.5 mi/8 km SSE of MOUNT WHITNEY, on E boundary of SEQUOIA NATIONAL PARK. Formerly called Mount Corcoran.

Langley Park, unincorporated town (2000 population 16,214), PRINCE GEORGES county, W central MARYLAND, a suburb of WASHINGTON, D.C.; 38°59′N 76°59′W. The town has grown with the development of the greater Washington, D.C. metropolitan area. Light industries and businesses.

Langley Prairie, CANADA: see LANGLEY.

Langnau am Albis, commune, ZÜRICH canton, N SWITZERLAND, 5 mi/8 km E of ZÜRICH; 47°17′N 08°32′E. Game preserve nearby.

Langnau im Emmental or **Langnau**, commune, BERN canton, W central SWITZERLAND, in the EMMENTAL valley, on Ilfis River, and 16 mi/26 km E of BERN. Elevation 2,218 ft/676 m. Export center of Emmental cheese; textiles; metalworking; canning.

Langney Point (LANG-nee POINT), promontory on the CHANNEL between Pevensey Bay and BEACHY HEAD, SE East SUSSEX, SE ENGLAND, 2 mi/3.2 km NE of EASTBOURNE; 50°47′N 00°18′E.

Langogne (lahn-GO-nyuh), commune (□ 11 sq mi/28.6 sq km), LOZÈRE department, LANGUEDOC-ROUSSILLON region, S FRANCE, near headwaters of the ALLIER, 22 mi/35 km NE of MENDE; 44°43′N 03°51′E. Cattle market; meat processing; sawmilling. Summer resort area at nearby reservoir. The town's old section preserves medieval features.

Langon (lahn-GON), town (□ 5 sq mi/13 sq km), subprefecture of GIRONDE department, AQUITAINE region, SW FRANCE, on left bank of GARONNE RIVER, and 26 mi/42 km SE of BORDEAUX; 44°33′N 00°14′W. Center of Bordeaux region wine trade; the SAUTERNES and BARSAC vineyards extend SW of town, while the

---

Cross-references are shown in SMALL CAPITALS. The pronunciation guide is shown on page xix. The sources of population figures are shown on page xvii.

Côte de Bordeaux wines orginate in vineyards on right bank of the Garonne.

Langøya (LAHNG-uh-yah) [Norwegian=long island], island (□ 332 sq mi/863.2 sq km) in NORTH SEA, NORDLAND county, N NORWAY, in the VESTERÅLEN group, separated from HINNØYA (E) by narrow strait; 35 mi/56 km long (NNE–SSW), 25 mi/40 km wide. Deeply indented (SE) by VESTERÅLSFJORDEN. Important fisheries. Chief villages: Eidsfjord, SORTLAND.

Langport (LANG-pawt), town (2001 population 2,735), central SOMERSET, SW ENGLAND, river PARRETT, and 12 mi/19 km E of TAUNTON; 51°02′N 02°50′W. Agricultural market in dairying and flower-growing region. Has 15th-century church. Walter Bagehot born here.

Langquaid (LAHNG-kveit), village, BAVARIA, S GERMANY, in LOWER BAVARIA, on GREAT LAABER RIVER, 13 mi/21 km S of REGENSBURG; 48°50′N 12°03′E.

Langreo (lahng-GRAI-o) or **Sama de Langreo** (SAH-mah dhai lahng-GRAI-o), city, OVIEDO province, NW SPAIN, in NALÓN valley, 10 mi/16 km SE of OVIEDO. Bituminous coal.

Langres (LAHN-gruh), town (□ 8 sq mi/20.8 sq km; 2004 population 8,761), sub-prefecture of HAUTE-MARNE department, CHAMPAGNE-ARDENNE region, E central FRANCE, on a rocky spur of the Plateau of LANGRES, a limestone tableland; 47°52′N 05°20′E. An agricultural distribution center, its manufacturing includes machinery and plastics. It has an old and famous cutlery industry. An episcopal see since the 3rd century, Langres has preserved most of its ancient Roman fortifications, including towers, gates, and ramparts. The large cathedral of SAINT-MAMMÈS dates from the 12th century. Following several fires, it was restored in 18th century with a classic facade. Diderot, whose father was a cutler, was born here. Its Roman name was Andematunum.

Langres, Plateau of (LAHN-gruh), forested limestone tableland of E central FRANCE, in N CÔTE-D'OR and S HAUTE-MARNE departments; 47°41′N 05°03′E. Extends c.50 mi/80 km from DIJON (S) to beyond LANGRES (N), forming a watershed between the SEINE and SAÔNE river systems. Rises to c.1,800 ft/549 m. Continued by CÔTE D'OR (S), Côtes de Moselle (N), and Monts Faucilles (NE). Sheep raising, cheese manufacturing. Traversed by PARIS-Dijon and Paris-BELFORT railroad lines, and by MARNE-SAÔNE CANAL, and by major N-S highways. The Seine, AUBE, MARNE, and MEUSE rivers rise here within a short distance of one another.

Langrune-sur-Mer (lahn-GROON–syur–MER), commune (□ 1 sq mi/2.6 sq km; 2004 population 1,688), CALVADOS department, BASSE-NORMANDIE region, NW FRANCE, 10 mi/16 km N of CAEN; 49°19′N 00°22′W. Ancient fishing port. Has 13th-century church. Allied landings took place here in June 1944 in World War II.

Langruth (LANG-gruhth), unincorporated town, S MANITOBA, W central CANADA, 33 mi/53 km NNW of PORTAGE LA PRAIRIE; 50°23′N 98°40′W. Main center within LAKEVIEW rural municipality.

Langsa (LAHNG-sah), town, ACEH province, NE SUMATRA, INDONESIA, near Strait of MALACCA, 80 mi/129 km NW of MEDAN, on Medan–BANDA ACEH railroad; ⊙ E. Aceh district; 04°28′N 97°58′E. Trade center for rubber and palm-oil growing. Its port, KUALALANGSA (or Koealalangsa), is 4 mi/6.4 km NE on the strait; exports rubber, resin, palm oil, petroleum (from nearby E Aceh fields).

Långsele (LONG-SE-le), town, VÄSTERNORRLAND county, NE SWEDEN, on Faxälven River, near its mouth on ÅNGERMANÄLVEN RIVER, 7 mi/11.3 km W of SOLLEFTEÅ; 63°11′N 17°04′E. Railroad junction; manufacturing. Foundation of medieval church.

Langshan (LAHNG-SHAHN), town, W INNER MONGOLIA AUTONOMOUS REGION, N CHINA, 45 mi/72 km

W of WUYUAN, in HETAO oasis; 41°05′N 107°30′E. Cattle raising; grain, sugar beets, oilseeds.

Långshyttan (LONGS-HIT-tahn), village, KOPPARBERG county, central SWEDEN, 19 mi/31 km SE of FALUN; 60°27′N 16°02′E.

Langside (LANG-seid), district of GLASGOW, S central Scotland. At the battle of Langside (1568) the first earl of Moray defeated the forces of Mary, Queen of Scots led by Archibald Campbell, fifth earl of Argyll. As a result, Mary fled to England.

Lang Son (LAHNG SON), province (□ 3,160 sq mi/8,216 sq km), N VIETNAM, in NE mountain region, N border with CAO BANG province, NE border with CHINA, E border with QUANG NINH province, S border with HA BAC province, W border with BAC THAI province; ⊙ LANG SON; 21°55′N 106°30′E. Strategically located on Vietnam's NE border with China, the rugged highland landscapes still display significant tracts of ecologically complex forests despite widespread deforestation in recent generations. Diverse soils and mineral resources, shifting cultivation, opium growing, livestock raising, forest products (medicinals, honey, beeswax, foraged foods), and commercial agriculture (tea, coffee, mulberry, aromatic produce, vegetables, fruits). Lumbering, sawmilling, sericulture, food processing, brick and tile, handicrafts, carpet manufacturing. Scene of military conflict between Vietnam and China in 1979. Kinh population with Tay, Nung, Dao, San Chay, and other minorities.

Lang Son (LAHNG SON), city, ⊙ LANG SON province, N VIETNAM, near CHINA border, on HANOI-VAN LANG (formerly NA SAM) railroad, and 85 mi/137 km NE of Hanoi, on the Song Ky Cung (right headstream of LI RIVER); 21°50′N 106°47′E. Commercial center and transportation hub, point of trade with China, and military post; sericulture; shifting cultivation, tea, coffee, aromatic produce, and forest products. Copper, lead, zinc, and coal and oil-shale deposits nearby. Chinese fort (15th century). Captured 1885 by French and 1950 by the Viet Minh. Partly destroyed in 1979 by invading Chinese military forces. Tho, Nung, Man, Dao, and other minorities. Formerly spelled Langson.

Langston, town (2006 population 1,698), LOGAN county, central OKLAHOMA, 10 mi/16 km ENE of GUTHRIE, near CIMARRON RIVER. Seat of Langston University.

Langstone Harbour or **Langston Harbour** (LANG-stun), 4 mi/6.4 km long, in HAMPSHIRE, S ENGLAND, between PORTSEA and HAYLING islands; 50°47′N 01°02′W.

Lang Suan (LUHNG SU-ahn), town and district center, CHUMPHON province, S THAILAND, in ISTHMUS OF KRA, port on GULF OF THAILAND, on railroad, and 40 mi/64 km S of CHUMPHON; 09°57′N 99°04′E. Tin mining; fruit gardening, fishing. Sometimes spelled Langsuen.

Langtang (LUHNG-tahng), village, N central NEPAL, at CHINA (TIBET) border and N of LANGTANG NATIONAL PARK; 28°12′N 85°36′E. Elevation 12,001 ft/3,658 m. Airport.

Langtang Lirung (LAHNG-tahng lee-ROONG), mountain peak (23,730 ft/7,234 m), N central NEPAL; 28°15′N 85°31′E.

Langtang National Park (LAHNG-tahng) (□ 660 sq mi/1,716 sq km), N central NEPAL, in inhabited valley N of KATHMANDU; 38°06′N 85°30′E. Established 1976.

Langtoft (LANG-tahft), village (2001 population 1,976), LINCOLNSHIRE, E ENGLAND, 1 mi/1.6 km N of MARKET DEEPING; 52°42′N 00°20′W.

Langton Bay, CANADA: see FRANKLIN BAY.

Langtry (LANG-tree), unincorporated village, VAL VERDE county, SW TEXAS, on the RIO GRANDE (Mexican border), near mouth of the PECOS and c.50 mi/80 km NW of DEL RIO; 29°48′N 101°33′W. Elevation 1,315 ft/401 m. Nearby, at old town of Langtry,

Judge Roy Bean, "the law west of the Pecos," meted out justice in his frontier saloon. Amistad Reservoir and Amistad National Recreation Area downstream (SE); Seminole Canyon State Park to SE.

Languard, Piz, SWITZERLAND: see PIZ LANGUARD.

Langue (LAHN-gai), town, VALLE department, S HONDURAS, on INTER-AMERICAN HIGHWAY, and 11 mi/18 km WNW of NACAOME; 13°37′N 87°39′W. Commercial center; beverages, rope milling; henequen, livestock. Has noted church (1804).

Languedoc (lahn-guh-dok), ancient region and former province of S FRANCE, bounded by the foothills of the PYRENEES (S), the MASSIF CENTRAL and the CÉVENNES Mountains (NW), the RHÔNE valley (E), and the MEDITERRANEAN (SE). It comprises the departments of AUDE, GARD, HÉRAULT, LOZÈRE, and PYRÉNÉES-ORIENTALES as well as parts of ARIÈGE, HAUTE-GARONNE, and TARN departments. The region's name was derived from Provençal or *Langue d'Oc*, the language of its inhabitants, whose word for "yes" was "oc," in contrast to the Northerners who spoke *Langeu d'Oil*. This idiom was the language of the troubadours. Geographically, the region of name now refers to Lower Languedoc, an alluvial plain along the Mediterranean, with a warm climate; wine is the chief product, but agriculture is diversifying through irrigation. MONTPELLIER, NÎMES, SÈTE, BÉZIERS, and NARBONNE are the chief cities. Historic CARCASSONNE is also here. The Massif Central rises to NW and is less densely populated. Historically, Languedoc roughly corresponds to Narbonensis province of Roman GAUL; Lower Languedoc was later called SEPTIMANIA. Its history from the Frankish conquest (completed 8th century) to its final incorporated into the French royal domain (1271) is largely that of the counts of Toulouse. Under the old regime the parliament of Languedoc sat at TOULOUSE; that provincial assembly retained its importance until the French Revolution, when the departments were created. Historic Languedoc's E part was incorporated into the administrative region of LANGUEDOC-ROUSSILLON under the regionalization laws of the 1970s and 1980s.

Languedoc-Roussillon (lahn-guh-dok–roo-see-YON), administrative region (□ 10,570 sq mi/27,482 sq km) of S FRANCE; ⊙ MONTPELLIER; 43°40′N 03°10′E. Created in 1980s as part of regionalization laws adopted by French government. It is composed of the departments of AUDE, GARD, HÉRAULT, LOZÈRE, and PYRÉNÉES-ORIENTALES, and thus extends from the lower RHÔNE valley SW to the PYRENEES and the Spanish border, along the MEDITERRANEAN shore. It is approximately coextensive with historic Lower LANGUEDOC, but it does not include the territory of Upper Languedoc, which includes TOULOUSE (the historic capital of Langue d'oc) and the Upper Garonne valley.

Langueux (lahn-gu), town (□ 3 sq mi/7.8 sq km), CÔTES-D'ARMOR department, in BRITTANY, NW FRANCE, just SE of SAINT-BRIEUC; 48°30′N 02°43′W. Agricultural processing industry.

Languidic (lahn-gwee-deek), town, MORBIHAN department, NW FRANCE, in BRITTANY, 12 mi/19 km NE of LORIENT; 47°50′N 03°10′W. Food processing, especially fish canning and freezing.

Languilayo, Laguna (lahn-gee-LAH-yo, lah-GOO-nah), lake, CUSCO region, S PERU, in the ANDES; 14°28′S 71°13′W. Alternate spellings include: Laguna Langui Layo; Laguna de Langui y Layo; Laguna Langui.

L'Anguille River (lan-GWEEL), c.110 mi/177 km long, E ARKANSAS; rises in CRAIGHEAD county, S of JONESBORO; flows S and SE, past FORREST CITY and MARIANNA, to SAINT FRANCIS River in LEE county.

Languiñeo (lahn-geen-YAI-o), village, W CHUBUT province, ARGENTINA, 50 mi/80 km SE of ESQUEL; 43°18′S 70°25′W. Sheep, cattle.

Languiñeo, department, ARGENTINA: see TECKA.

Area is shown by the symbol □, and capital city or county seat by ⊙.

Langwedel (lahng-VAI-duhl), village, LOWER SAXONY, NW GERMANY, 4 mi/6.4 km N of VERDEN; 52°58′N 09°11′E.

Langweid am Lech (LAHNG-veid ahm LEKH), village, BAVARIA, S GERMANY, in SWABIA, on right bank of LECH RIVER, 8 mi/12.9 km N of AUGSBURG; 48°30′N 10°50′E.

Langxi (LANG-SEE), town, ☉ Langxi county, SE ANHUI province, CHINA, near JIANGSU border, 45 mi/72 km ESE of WUHU; 31°08′N 119°10′E. Rice, oilseeds.

Langzhong (LANG-JUNG), city (□ 725 sq mi/1,885 sq km), N SICHUAN province, CHINA, 55 mi/89 km N of NANCHONG, and on left bank of the JIALING River; 31°39′N 105°56′E. Agriculture and light industry are the largest sources of income for the city. Crop growing, animal husbandry (grain, oil crops, vegetables; hogs), manufacturing (food, textiles).

Lanham (LA-nuhm), village, PRINCE GEORGES county, central MARYLAND, ENE of WASHINGTON, D.C.; 38°58′N 76°50′W. Next to Washington Bible College is the private estate of Azalea Acres, where the public can view the flowers in season. Pop. figure includes SEABROOK.

Lanier (luh-NIR), county (□ 200 sq mi/520 sq km; 2006 population 7,723), S GEORGIA; ☉ LAKELAND; 31°02′N 83°04′W. Coastal plain area intersected by ALAPAHA RIVER. Agriculture (corn, tobacco, peanuts, soybeans, cotton, fruit; cattle, hogs); forestry products Formed 1919.

Lanier (lah-nee-ER), swamp, ISLA DE LA JUVENTUD, off S coast of CUBA, extending c.25 mi/40 km E-W across central part of the island; 21°34′N 82°45′W. Formerly called Liguanea.

Lanier, Lake, in DAWSON, FULTON, GWINNETT, and HALL counties, GEORGIA, NE of ATLANTA. Built in the 1950s for flood control, electrical power generation, and to assist downriver shipping, the lake has become the leading recreational facility in the state and the source of the fresh water supply for most residents of metropolitan Atlanta.

Lanigan (LAN-i-guhn), town (2006 population 1,233), S central SASKATCHEWAN, CANADA, 70 mi/113 km ESE of SASKATOON; 51°51′N 105°02′W. Grain elevators, lumbering.

Lanikai (LAH-nee-KEI), village, E OAHU, HONOLULU county, HAWAII, on NE coast, 14 mi/23 km N of HONOLULU, between Alala Point and Wailea. Mokulua Islands offshore are state seabird sanctuaries.

Lanín Volcano (lah-NEEN), Andean peak (c.12,300 ft/3,749 m) on ARGENTINA-CHILE border, S of MAMUIL-MALAL PASS. Site of national park.

Lanivtsi (lah-nif-TSEE) (Russian *Lanovtsy*) (Polish *Lanowce*), town (2004 population 7,900), NE TERNOPIL′ oblast, UKRAINE, on right tributary of HORYN′ River, 24 mi/39 km SE of KREMENETS′; 48°51′N 26°01′E. Elevation 921 ft/280 m. Raion center; food processing (flour, sugar, butter, feed); manufacturing (asphalt, reinforced concrete, rubber toys); repair shops. Vocational technical school; heritage museum. Known since 1444; town since 1956. Jewish community since the 16th century, decimated by the 1918–1920 civil war pogroms (630 remaining in 1939); eliminated by the Nazis in 1941.

Lanjarón (lahn-hah-RON), town, GRANADA province, S SPAIN, on S slope (2,300 ft/701 m) of the SIERRA NEVADA, 19 mi/31 km SSE of GRANADA; 36°55′N 03°29′W. Olive oil processing, liqueur manufacturing, bottled water. Oranges and other fruit, chestnuts, vegetables. Has famous mineral springs. Dominated by ancient castle. Gypsum quarries nearby.

Lankada, Lake, Greece: see KORONIA, LAKE.

Lankadia, Greece: see LANGADIA.

Lankao (LAN-GOU), town, ☉ Lankao county, N HENAN province, CHINA, on LONGHAI Railroad, 27 mi/43 km E of KAIFENG; 34°50′N 114°49′E. Grain, cotton, oilseeds; food industry, chemicals. Well known for its

dedicated chief between 1960s and 1970s, Jiao Yulu. Also known as Lanfeng.

Länkärän (LAHN-kahr-ahn) or **Lenkoran**, city, SE AZERBAIJAN, near the IRANIAN border, on the CASPIAN SEA; 38°45′N 48°50′E. It is a port and an important food-processing center for fish and tea. Its inhabitants are mostly Talysh, an Iranian-speaking people who are Shiah Muslims. Lenkoran, known since the 17th century, was the capital of the Talysh khanate under PERSIA in the 18th century and was ceded to RUSSIA by Persia in 1813. The Lenkoran Lowland, a coastal strip c.40 mi/60 km long, has a humid, subtropical climate. Citrus fruit, tea, and rice are grown here. Manufacturing (furniture, bricks); timber combine.

Lankatilaka, SRI LANKA: see GADALADENIYA.

Lånke (LAWNG-kuh), village, NORD-TRØNDELAG county, central NORWAY, on STJØRDALSELVA River, on railroad, and 21 mi/34 km E of TRONDHEIM. Agriculture; forestry.

Lankershim, California: see NORTH HOLLYWOOD.

Lankin, village (2006 population 118), WALSH co., NE NORTH DAKOTA, 25 mi/40 km WSW of GRAFTON, near N branch of FOREST RIVER; 48°18′N 97°55′W. Founded in 1905 and incorporated in 1908. Named for James Lankin, owner of townsite.

Lankio (LAN-kyo), town, Zanzan region, NE CÔTE D'IVOIRE, 52 mi/83 km NW of BOUNA, near BURKINA FASO border; 09°52′N 03°32′W. Agriculture (sorghum, corn, beans).

Lanlacuni Bajo (lahn-lah-KOO-nee BAH-ho), town, PUNO region, SE PERU, on affluent of INAMBARI, 102 mi/164 km E of CUSCO; 13°30′S 70°25′W.

Lanlate (lahn-LAH-te), town, OYO state, SW NIGERIA, on road, 20 mi/32 km S of ISEYIN; 07°36′N 03°27′E. Market town. Yams, millet, maize.

Lanmeur (lahn-MUHR), commune (□ 10 sq mi/26 sq km; 2004 population 2,139), FINISTÈRE department, in BRITTANY, W FRANCE, 7 mi/11.3 km NE of MORLAIX; 48°39′N 03°43′W. Market gardening for early vegetables; woodworking. Has 12th–16th-century chapel with a crypt of pre-Roman date.

Lann-Bihoué, FRANCE: see LORIENT.

Lannemezan (lah-nuh-me-zahn), town (□ 7 sq mi/18.2 sq km), HAUTES-PYRÉNÉES department, MIDI-PYRÉNÉES region, SW FRANCE, in S part of LANNEMEZAN PLATEAU, 17 mi/27 km SE of TARBES; 43°08′N 00°22′E. Commercial center (livestock and dairy market); aluminum-reduction plant. Chief distribution point for electricity generated in the PYRENEES. Has psychiatric hospital.

Lannemezan Plateau (lah-nah-me-zahn), tableland of SW FRANCE, in MIDI-PYRÉNÉES region, at N foot of the PYRENEES, occupying part of HAUTES-PYRÉNÉES, GERS, and HAUTE-GARONNE departments; 43°09′N 00°27′E. Average elevation 2,000 ft/610 m. It is bounded by ADOUR RIVER (W), NESTE RIVER and foothills of central Pyrenees (S), and upper GARONNE RIVER (SE). Sloping gently northward, this glacial outwash delta gives rise to numerous streams (BAÏSE, GERS, GIMONE, SAVE), all of which drain N and NE to the Garonne. ARMAGNAC brandy originates here. Horse breeding.

Lannilis (lahn-ee-LEES), town (□ 9 sq mi/23.4 sq km), FINISTÈRE department, in BRITTANY, NW FRANCE, 13 mi/21 km N of BREST, between two deep inlets of the ENGLISH CHANNEL; 48°34′N 04°31′W. Fruits and vegetables; pottery. Tourist center for visits to rocky coast.

Lannion (lahn-yon), town (□ 17 sq mi/44.2 sq km), sub-prefecture of CÔTES-D'ARMOR, near BRITTANY's N coast, 50 mi/80 km NE of BREST; 48°44′N 03°28′W. Commercial center with electronics industry. Just N is the National Center for Telecommunications Research. Town has many 15th–16th-century houses and a 12th–15th-century church built by the Templars.

Lannon, town (2006 population 995), WAUKESHA county, SE WISCONSIN, suburb 14 mi/23 km NW of

MILWAUKEE; 43°08′N 88°09′W. In dairy and farm area. Manufacturing (fabricated metal products, packaging).

L'Annonciation (lah-non-see-ah-SYON), former village, SW QUEBEC, E CANADA, in the LAURENTIANS, 40 mi/64 km NW of SAINTE-AGATHE-DES-MONTS; 46°24′N 74°52′W. Dairying. Merged into RIVIÈRE-ROUGE in 2002.

Lanoraie (lah-no-RAI), village (□ 39 sq mi/101.4 sq km), LANAUDIÈRE region, S QUEBEC, E CANADA, on the SAINT LAWRENCE RIVER, 35 mi/56 km NNE of MONTREAL; 45°58′N 73°13′W. Dairying; resort.

La Noria (lah no-REE-ah), village, TARAPACÁ region, N CHILE, 5 mi/8 km W of the SALAR DE PINTADOS, 22 mi/35 km SE of IQUIQUE. Former nitrate-mining center. Flourished c.1900.

Lanovtsy, UKRAINE: see LANIVTSI.

Lanowce, UKRAINE: see LANIVTSI.

Lanping (LAN-PING), town, ☉ Lanping county, NW YUNNAN province, CHINA, near MEKONG River, 65 mi/105 km NW of DALI; 26°29′N 99°16′E. Elevation 10,138 ft/3,090 m. Timber, grain; non-ferrous metal smelting and rolling, non-ferrous ore-mining.

Lanquín (lahn-KEEN), town, ALTA VERAPAZ department, GUATEMALA, 28 mi/45 km ENE of COBÁN; 15°34′N 89°58′W. Coffee; subsistence farming (corn, beans, plantains, bananas). Grutas de Lanquín National Park, a large and little-explored limestone cavern system, is 1.2 mi/1.9 km N; Kekchi-speaking population.

Lans, FRANCE: see LANS-EN-VERCORS.

Lansallos (lan-SA-luhs), village (2001 population 1,584), E CORNWALL, SW ENGLAND, on the CHANNEL, 3 mi/4.8 km E of FOWEY; 50°22′N 04°34′W. Has 15th-century church. Coast to S owned by National Trust.

Lansdale, borough (2006 population 15,720), MONTGOMERY county, SE PENNSYLVANIA, suburb 20 mi/32 km NNW of PHILADELPHIA; 40°14′N 75°16′W. It is a farm processing and industrial center. Manufacturing (food products, building materials, apparel, plumbing equipment, commercial printing, machinery). Agriculture (grain, apples; livestock; dairying). Evansburg State Park to W. The Jenkins House here dates from 1702. Incorporated 1872.

Lansdowne (LANZ-doun), rural municipality (□ 296 sq mi/769.6 sq km; 2001 population 877), SW MANITOBA, W central CANADA, 99 mi/158 km NW of WINNIPEG, and traversed by WHITEMUD RIVER; 50°19′N 99°15′W. Agriculture and related businesses. Includes ARDEN, TENBY. Incorporated 1884.

Lansdowne (LAHNS-doun), cantonment town, GARHWAL district, N UTTAR PRADESH state, N central INDIA, 32 mi/51 km SSE of HARIDWAR; 29°50′N 78°41′E. Elev. 5,026 ft/1,532 m. Wheat, barley, rice. Founded 1887. Sometimes spelled Lansdown.

Lansdowne, unincorporated town, BALTIMORE county, NE MARYLAND, suburb of BALTIMORE; 39°14′N 76°39′W. Settled by an Irishman named McGrath who named the area after the Marquis of Lansdowne, the postmaster of ENGLAND in 1782. Iron was mined here in the 18th century and the Lansdowne Christian Church has a small Civil War museum.

Lansdowne (LANZ-doun), unincorporated village, SE ONTARIO, E central CANADA, 27 mi/43 km ENE of KINGSTON, and included in township of LEEDS AND THE THOUSAND ISLANDS; 44°24′N 76°01′W. Dairying, mixed farming.

Lansdowne, borough (2006 population 10,759), DELAWARE county, SE PENNSYLVANIA, a residential suburb 6 mi/9.7 km WSW of PHILADELPHIA, on Darby Creek; 39°56′N 75°16′W. Manufacturing (paper products, printing). Incorporated 1893.

Lansdown Hill (LANZ-duhn), ridge (780 ft/238 m), Bath and North East Somerset, SW ENGLAND, extends 5 mi/8 km NW from BATH; 51°25′N 02°23′W. Noted for its breed of sheep. Site of a racecourse. Battle fought here in 1643.

L'Anse (LANZ), town, ⊙ BARAGA county, NW UPPER PENINSULA, MICHIGAN, 27 mi/43 km SSE of HOUGHTON, at head of KEWEENAW BAY; 46°45′N 88°27′W. In farm and lumber area. Fisheries; manufacturing (acoustical ceiling panels, lumber). Keweenaw Bay Indian Reservation is E; L'Anse Indian Reservation to NE and NW (both sides of bay); Ottawa National Forest to W; six-story-high copper statue on the lake bluff of Father Frederick Baraga, the Snowshoe Priest, honoring his good works in area. Baraga State Park to W. Incorporated in 1873.

Lanse (LANS), unincorporated village, Cooper township, CLEARFIELD county, central PENNSYLVANIA, 6 mi/9.7 km NE of PHILIPSBURG; 40°58′N 78°07′W. Surface bituminous coal.

L' Anse Amour, CANADA: see ANSE AMOUR, L'.

L'Anse aux Epines (LANS uh-PEEN) or **Lance aux Epines** (lans-o-PEEN), peninsula, GRENADA, WEST INDIES; 11°59′N 61°45′W. Beach hotel and resort area on S coast.

L'Anse aux Meadows National Historic Park (LAN-see MED-owz), national park and historic site, NEWFOUNDLAND AND LABRADOR, CANADA, at northernmost tip of island, 12 mi/19 km N of St. ANTHONY; 51°40′N 55°30′W. Site of Viking settlement on Sacred Bay, STRAIT OF BELLE ISLE; established c.1000. Discovered in 1960 by Norwegian archaeologists looking for Viking settlement of Vineland. Declared federal historic park, 1970; World Heritage Area, 1988. Reconstructed stone and sod-roofed village. Wooden huts protect excavations.

Lans-en-Vercors (lahn–zahn–ver-kor), commune (□ 14 sq mi/36.4 sq km; 2004 population 2,303), ISÈRE department, RHÔNE-ALPES region, SE FRANCE, in VERCORS mountain range, 8 mi/12.9 km SW of GRENOBLE; 45°07′N 05°35′E. Winter-sports resort with ski slopes at 5,000 ft/1,524 m–6,200 ft/1,890 m. Formerly called Lans.

L'Anse-Saint-Jean (LAHNS–san–ZHAHN), village (□ 204 sq mi/530.4 sq km; 2006 population 1,148), SAGUENAY—LAC-SAINT-JEAN region, S central QUEBEC, E CANADA, 10 mi/16 km from RIVIÈRE-ÉTERNITÉ; 48°14′N 70°12′W.

Lansford, village (2006 population 238), BOTTINEAU CO., N NORTH DAKOTA, 28 mi/45 km N of MINOT; 48°37′N 101°22′W. Dairy products; wheat; livestock. Founded 1903 and incorporated in 1904.

Lansford (LANS-fuhrd), borough (2006 population 4,186), CARBON county, E Pennsylvania, 8 mi/12.9 km WSW of JIM THORPE. Agriculture (hay, potatoes, dairying); manufacturing (apparel). Founded 1846, incorporated 1877.

Lanshan (LAN-SHAN), town, ⊙ Lanshan county, S HUNAN province, CHINA, near GUANGDONG border, 70 mi/113 km SW of CHEN XIAN; 25°18′N 112°06′E. Tobacco; logging, food industry.

Lansing, city (2000 population 119,128), ⊙ MICHIGAN and INGHAM, CLINTON, and EATON counties, S MICHIGAN, 75 mi/121 km NW of DETROIT, on the GRAND RIVER at its confluence with the RED CEDAR RIVER; 42°42′N 84°32′W. Lansing is a trade and processing center for its surrounding agricultural area. Manufacturing (paper products, machinery, medical equipment, fabricated metal and plastic products, building materials, printing). Railroad junction. The city grew after it was made the state capital (1847), and industrial development came with the railroads (1870s) and the automobile industry (1897). The state capitol houses a museum, Fenner Arboretum; the state office building contains the state library and the state historical office. Lansing Community College and the Michigan School for the Blind are here. American author Ray Baker was born here. The adjacent suburb of EAST LANSING is the seat of Michigan State University. Sleepy Hollow State Park to NE; Capital City Airport to N. Incorporated in 1859.

Lansing, town (2000 population 1,012), ALLAMAKEE county, extreme NE IOWA, at foot of bluffs on MISSISSIPPI RIVER (bridged here), 14 mi/23 km ENE of WAUKON; 43°21′N 91°13′W. Limestone quarries, lead and zinc deposits nearby. Large group of Indian effigy mounds and a state fish hatchery are in vicinity. Laid out 1851, incorporated 1867.

Lansing, town (2000 population 9,199), LEAVENWORTH county, NE KANSAS, suburb 4 mi/6.4 km S of LEAVENWORTH and 16 mi/26 km NW of KANSAS CITY, near MISSOURI River; 39°15′N 94°53′W. In general agricultural region. Kansas State Prison is here. St. Mary College is in N.

Lansing, village (2000 population 28,332), COOK county, NE ILLINOIS, S suburb of CHICAGO, near the INDIANA state line; 41°34′N 87°32′W. Among the industries are meat packing, food processing, and manufacturing of metal products. Incorporated 1893.

Lansing, village, MOWER county, SE MINNESOTA, on CEDAR RIVER, 6 mi/9.7 km N of AUSTIN; 43°44′N 92°58′W. Corn, soybeans; livestock; dairying; manufacturing (feeds).

Lansing (LAN-seeng), village (2006 population 149), ASHE county, NW NORTH CAROLINA, 24 mi/39 km NNE of BOONE, on North Fork of NEW RIVER; 36°30′N 81°30′W. Manufacturing (machinery); retail trade.

Lansing (LAN-sing), unincorporated village, BELMONT county, E OHIO, 5 mi/8 km NNW of BELLAIRE, on small Wheeling Creek, near WEST VIRGINIA state line; 40°04′N 80°47′W. Manufacturing. In coal-mining area.

Lansing, Lake, MICHIGAN: see HASLETT.

Lanskroun (LAHNSH-kroun), Czech *Lanškroun*, German *Landskron*, town, VYCHODOCESKY province, E BOHEMIA, CZECH REPUBLIC, 40 mi/64 km SE of HRADEC KRÁLOVÉ; 49°55′N 16°37′E. Railroad terminus; manufacturing (electronics, paper); brewery (established 1700). Renaissance town hall; folk architecture.

Lanslebourg-Mont-Cenis (lahn-luh-boor–mon-suh-nee), commune (□ 36 sq mi/93.6 sq km; 2004 population 604), SAVOIE department, RHÔNE-ALPES region, SE FRANCE, in upper MAURIENNE valley of the Savoy Alps (ALPES FRANÇAISES), on the ARC, 12 mi/19 km NE of MODANE, 13 mi/21 km NW of Susa (Italy); 45°17′N 06°52′E. Road junction for MONT CENIS pass (2 mi/3.2 km SSE) and Col de l'ISERAN (12 mi/19 km NE). Winter sports in VAL-CENIS (E); fine trails for summer hiking. VANOISE NATIONAL PARK is N of this valley.

Lanslevillard (lahn-luh-vee-lahr), village (□ 16 sq mi/41.6 sq km), SAVOIE department, RHÔNE-ALPES region, SE FRANCE, in upper MAURIENNE valley of the Savoy Alps (ALPES FRANÇAISES), on ARC RIVER, 14 mi/23 km NE of MODANE; 45°17′N 06°55′E. Together with neighboring LANSLEBOURG-MONT-CENIS (2 mi/3.2 km W), these villages are the base for VAL-CENIS ski area on N slope of Mont Cenis massif (4,500 ft/1,372 m–9,000 ft/2,743 m) created in 1967, and providing one of the most extensive winter-sport terrains in the upper Maurienne valley.

Lantadilla (lahn-tah-DHEE-lyah), town, PALENCIA province, N central SPAIN, on PISUERGA RIVER, 27 mi/43 km NE of PALENCIA; 42°20′N 04°16′W. Wheat; wine; sheep.

Lanta, Ko (LUHN-TAH, GO), island, KRABI province, S THAILAND, in GULF OF THAILAND, off coast of MALAY PENINSULA, 30 mi/48 km SSE of KRABI; 15 mi/24 km long, 5 mi/8 km wide. Town and district have same name.

Lantana (lan-TAN-nuh), town (□ 2 sq mi/5.2 sq km; 2000 population 9,437), PALM BEACH county, SE FLORIDA, 9 mi/14.5 km S of WEST PALM BEACH, on LAKE WORTH lagoon; 26°34′N 80°03′W. Large Finnish-American community.

Lan Tao, HONG KONG: see LANTAU ISLAND.

Lantau Island (LAN-TOU), NEW TERRITORIES, HONG KONG, CHINA; separated from Hong Kong island (E)

by 6-mi/9.7-km-wide West Lamma Channel; 15 mi/24 km long, 6 mi/9.7 km wide. Lead (SILVER MINE BAY) and tungsten deposits. Site of the new CHEK LAP KOK airport and a major Buddhist monastery on the mountaintop. Also spelled Lan Tao.

Lanteira (lahn-TAI-rah), town, GRANADA province, S SPAIN, on N slope of the SIERRA NEVADA, 9 mi/14.5 km S of GUADIX; 37°10′N 03°09′W. Flour mills. Chestnuts, lumber. Mineral springs. Iron mines nearby.

Lantejuela, La (lahn-tai-HWAI-lah, lah), town, SEVILLE province, SW SPAIN, in lake district, 40 mi/64 km E of SEVILLE; 37°21′N 05°13′W. Olives, cereals; livestock. Sulphur spa nearby.

Lantian (LAN-TYAN), town, ⊙ Lantian county, SE SHAANXI province, CHINA, 20 mi/32 km ESE of XI'AN; 34°09′N 109°19′E. Grain; food processing, logging, engineering. Site of Paleolithic Lantian culture.

Lantier (lahn-TYAI), village (□ 17 sq mi/44.2 sq km; 2006 population 641), LAURENTIDES region, S QUEBEC, E CANADA, 7 mi/11 km from SAINTE-AGATHE-DES-MONTS; 46°09′N 74°15′W.

Lanton (LAHN-ton), resort commune (2004 population 5,621), GIRONDE department, AQUITAINE region, SW FRANCE, on ARCACHON BASIN, 23 mi/37 km SW of BORDEAUX; 44°42′N 01°02′W.

Lantzville, district, SW BRITISH COLUMBIA, W CANADA, in NANAIMO regional district; 49°15′N 124°04′W. Formed in 1920. Traditional coal-mining area. Once known as Grant's Mine.

Lanús (lah-NOOS), city, BUENOS AIRES province, E ARGENTINA; 34°43′S 58°24′W. An administrative center in the Greater Buenos Aires area. Named for Anacarsis Lanús, a local landowner, merchant, and politician of the 19th century. Manufacturing (textiles, paper, wire; meat processing, leather products, and rubber).

Lanusei (lah-noo-ZAI), town, NUORO province, E SARDINIA, S ITALY, 33 mi/53 km SSE of NUORO; 39°53′N 09°32′E. Wine; fruit; livestock. Nuraghi.

Lanut (lah-NOOT), island, in the DANUBE, N central BULGARIA; 43°43′N 25°00′E.

Lanuvio (lah-NOO-vyo), town, ROMA province, LATIUM, central ITALY, in ALBAN HILLS, 4 mi/6 km W of VELLETRI; 41°40′N 12°42′E. Occupies site of ancient LANUVIUM and has Roman ruins. Formerly called Civita Lavinia.

Lanuvium (luh-NOO-vee-uhm), ancient city of LATIUM, ITALY, c.20 mi/32 km S of ROME, in the ALBAN HILLS near the APPIAN WAY. It was celebrated for its temple of Juno. The modern village is LANUVIO; there are ruins of a temple and Roman walls on the site.

Lanvaux, Landes de (lahn-vo, lahnd duh), low, rocky hill range, MORBIHAN department, W FRANCE, in BRITTANY, about 12 mi/19 km inland from VANNES; 47°47′N 02°36′W. It is tree-covered with some cattle pastures. Megalithic stone monuments abound.

Lanvéoc (lahn-vai-ok), commune (□ 7 sq mi/18.2 sq km), FINISTÈRE department, in BRITTANY, NW FRANCE, on the BREST ROADS (inlet of ATLANTIC OCEAN), opposite and 6 mi/9.7 km from BREST; 48°17′N 04°28′W. Site of French naval academy. Often referred to as Lanvéoc-Poulmio.

Lanvéoc-Poulmio, FRANCE: see LANVÉOC.

Lanxi (LAN-SEE), city (□ 506 sq mi/1,311 sq km; 1994 estimated urban population 90,100; 1994 estimated total population 650,200), W central ZHEJIANG province, China, on railroad spur, 12 mi/19 km NW of JINHUA, on the Fuchun River; 29°12′N 119°27′E. Light industry and agriculture are main sources of income for the city. Main agriculture (rice, oilseeds, cotton, sugarcane; livestock); manufacturing (food processing, textiles, chemicals, nonferrous metals, machinery, electrical equipment). Nearby is Zhuge Eight Diagram, a village created during Song dynasty (1129–1279) to honor Zhuge Kungming, a famous military strategist during the Three Kingdoms period (220–280). The village's layout resembles an Eight Diagram, the mystical Chinese symbol; it contains Zhuge ancestry temples and well-preserved homes built since the Ming (1368–1644) and Qing dynasties (1644–1840).

Lanxi (LAN-SEE), town, ⊙ Lanxi county, SW HEILONGJIANG province, Northeast CHINA, 35 mi/56 km NW of HARBIN; 46°15′N 126°14′E. Grain, tobacco, jute, sugar beets; linen textiles.

Lan Xian (LAN SI-AN), town, ⊙ Lan Xian county, NW SHANXI province, CHINA, 60 mi/97 km NW of TAIYUAN; 38°17′N 111°40′E. Grain, oilseeds.

Lany (LAH-ni), Czech *Lány*, German *Lana*, village, STREDOCESKY province, central BOHEMIA, CZECH REPUBLIC, on railroad, 6 mi/9.7 km WSW of KLADNO. Has 18th century hunting lodge, park, and game preserve. Summer residence of Thomas G. Masaryk, who is buried here.

Lan Yü Island (LAN YOO-I) [*Orchid Island* in Chinese], off SE TAIWAN, 45 mi/72 km E of Cape OLUANPI; 8 mi/12.9 km long, 1.5 mi/2.4 km to 4 mi/6.4 km wide; 22°03′N 121°33′E. Little Lan Yü Island is 3 mi/4.8 km S. Designated a national park, Lan Yü is inhabited by some 2,000 aborigines belonging to the dwindling Yami tribe. Most of the island is uninhabited. Formerly called Botel Tobago.

Lanywa (LAHN-ywah), village, PAKOKKU township, MAGWE division, MYANMAR, on right bank of AYEYARWADY RIVER, 30 mi/48 km SW of PAKOKKU; 20°58′N 94°49′E. Petroleum center (first well sunk in 1927) in SINGU oil fields.

Lanza (LAHN-sah), town and canton, INQUISIVI province, LA PAZ department, W BOLIVIA, on E slopes of Eastern Cordillera of the ANDES Mountains, 20 mi/32 km SSE of INQUISIVI; 17°22′S 66°57′W.

Lanzahita (lahn-thah-EE-tah), town, ÁVILA province, central SPAIN, in the SIERRA DE GREDOS, 32 mi/51 km SSW of ÁVILA; 40°12′N 04°56′W. Olives, grapes; livestock; olive oil pressing.

Lanzarote (lahn-thah-RO-tai), ancient *Capraria*, northernmost and smallest island (□ 307 sq mi/798.2 sq km) of the larger CANARY ISLANDS, in LAS PALMAS province, SPAIN, in the ATLANTIC, between FUERTEVENTURA (S) and GRACIOSA ISLAND (N), c.110 mi/177 km NE of LAS PALMAS, 80 mi/129 km NW of CAPE JUBY on coast of Spanish Morocco; 29°14′N 13°28′W to 28°52′N 13°51′W. Chief town and port is ARRECIFE. Island is 35 mi/56 km long (NE-SW) and is up to 12 mi/19 km wide. Deeply indented, Lanzarote is composed of volcanic rocks, rising in the FAMARA MASSIF to 2,215 ft/675 m. Still active is the MONTAÑA DE FUEGO or Montañas del Fuego. The climate is arid and tropical. Although water is scarce, crops are raised for export, chiefly onions, tomatoes, figs, grapes. Some livestock raising. Fish—canned or dried—is another important source of income. There are several saltworks. Minor industries are processing and quarrying. Practically all foreign trade passes through Arrecife. Among other towns are HARÍA, SAN BARTOLOMÉ, and TEGUISE. Rich in associations with early Spanish and Portuguese discoveries, Lanzarote was variously known as Torcusa, Isla del Infierno, Tierra del Fuego, Lancelot, etc. Large volcanic eruptions were recorded 1730–1736 and 1824–1825. Tourism.

Lanzhot (LAHNZH-hot), Czech *Lanžhot*, German *Landshut*, village, JIHOMORAVSKY province, SE MORAVIA, CZECH REPUBLIC, on railroad, 4 mi/6.4 km SE of BREČLAV, near AUSTRIA and SLOVAKIA borders; 48°44′N 16°58′E. Agriculture (barley, wheat); wine production. Retains colorful regional costumes and customs.

Lanzhou (LAN-JO), city (□ 632 sq mi/1,632 sq km; 1994 estimated urban population 1,295,600; 1994 estimated total population 1,612,600), ⊙ GANSU province, W CHINA, on the HUANG HE (Yellow River) at its confluence with the WEI; 36°03′N 103°41′E. It is a railroad, highway, and air hub and the junction point to remote XINJIANG in extreme NW China. Lanzhou is

linked by railroad to BEIJING and to the MONGOLIAN People's Republic and RUSSIA. It is on the highway to TIBET. A rapidly growing industrial city, it receives its power from a nearby hydroelectric facility and from the coal that is mined in the area. It has one of the largest oil refineries in the country, in addition to textile mills, petrochemical, rubber, and fertilizer plants, and machine manufacturing. It has a gas-diffusion plant for processing plutonium, and is the center of China's atomic energy industry. The city is well known for its tea ceremony. An old walled city, Lanzhou was the scene (1936) of a successful Chinese Communist revolt. It is the seat of an oil research institution, Lanzhou University, and numerous technical colleges. Formally called Gaolan or Kaolan. The name sometimes appears as Lan-chou or Lan-chow.

Lanzo Torinese (LAHN-tso to-ree-NAI-ze), resort village, TORINO province, PIEDMONT, NW ITALY, 17 mi/27 km NNW of TURIN; 45°16′N 07°28′E. Textile industry, machinery.

Laoag (lah-WAHG), city (□ 42 sq mi/109.2 sq km; 2000 population 94,466), ⊙ ILOCOS NORTE province, NW LUZON, PHILIPPINES, on small Laoag River near its mouth on SOUTH CHINA SEA, 125 mi/201 km N of BAGUIO; 18°12′N 120°36′E. Trade center for rice area. Northernmost Philippine city, founded 1586. Center of political unrest in the late 18th and early 19th centuries. Provincial capital from 1818; chartered as city 1965. Notable Spanish colonial–era structures (including a cathedral, a bell tower, and waterworks).

Laoang Island (LAH-wahng), 7 mi/11.3 km long, 3 mi/4.8 km wide, in NORTHERN SAMAR province, PHILIPPINES, in PHILIPPINE SEA, near BATAG ISLAND, nearly connected to N coast of SAMAR island; 12°35′N 125°01′E. Coconut growing, fishing. On SW coast is Laoang town (1990 population 42,048), an agricultural center (rice). Formerly called Calamutang Island.

Lao Bao (LOU BOU), town, QUANG TRI province, N central VIETNAM, along the Song Xe Pon River; 16°37′N 106°37′W. The area is an active trading center for Thai goods smuggled in from LAOS.

Lao Cai (LOU KEI), province (□ 3,107 sq mi/8,078.2 sq km), N VIETNAM, in NW mountain region, N border with CHINA, NE border with HA GIANG province, SE border with YEN BAI province, SW border with SON LA province, W border with LAI CHAU province; ⊙ LAO CAI; 22°10′N 104°00′E. Marking a sector of the Vietnam-China border, this mountainous province has a number of environmental riches, is home to many ethnic groups, and still embraces tree-covered landscapes despite continuing deforestation. Diverse soils and mineral resources (phosphate, granite), shifting cultivation, opium growing, forest products (medicinals, honey, beeswax, exotic plants, foraged foods), and some commercial agriculture (tea, cardamom, vegetables, fruits). Lumbering, sawmilling, handicrafts, domestic tourism. Kinh population with H'mong, Dao, Nung, Giay, Tay, and other minorities.

Lao Cai (LOU KEI), city, ⊙ LAO CAI province, N VIETNAM, on RED RIVER (CHINA frontier), opposite Hokow, on HANOI-KUNMING railroad, 160 mi/257 km NW of Hanoi; 22°30′N 103°57′E. Transportation and trading center. Phosphate and graphite deposits; shifting cultivation; cardamom. Tourism potential. Tay, Dao, H'mong, and other minorities. Formerly Lao Cai or LAOKAY.

Laoet, INDONESIA: see PULAU LAUT.

Laohekou (LAW-HUH-KO), city (□ 391 sq mi/1,013 sq km; 1994 estimated urban population 148,100; 1994 estimated total population 488,500), NW HUBEI province, CHINA, near HENAN border, 200 mi/322 km NW of WUHAN, on the HAN RIVER; 32°24′N 111°40′E. Major trade center serving NW Hubei, SW Henan, and SE SHAANXI. Agriculture and heavy industry are the largest sectors of the city's economy. Main agriculture products are rice, tobacco, cotton, oilseeds,

jute, and medicinal herbs. Main industries include food processing, textiles, chemicals, and machinery. A mission center originally. Also known as Guanghua.

Laoighis, Laois (both: LAI-ish) or **Leix** (LAIKS), county (□ 664 sq mi/1,726.4 sq km; 2006 population 67,059), LEINSTER province, E central IRELAND; ⊙ PORTLAOISE. Bounded by KILKENNY (S), TIPPERARY (SW), OFFALY (W and N), KILDARE (E), and CARLOW (SE) counties. Drained by NORE and BARROW rivers. Part of the central plain of Ireland, Laoighis is generally level, except for the SLIEVE BLOOM mountains (1,733 ft/528 m) in NW. Agriculture (wheat, barley, and sugar beets); dairying; manufacturing (woolens, woodworking, pharmaceuticals). Towns include ABBEYLEIX and MOUNTMELLICK.

Laokay, VIETNAM: see LAO CAI, city.

Laolong (LOU-LUNG), town, E GUANGDONG province, CHINA, on EAST RIVER (head of navigation), 5 mi/8 km NE of LONGCHUAN. Commercial center; in rice-growing region.

Laolong, CHINA: see LONGCHUAN, GUANGDONG province.

Laon (lah-on), town (□ 16 sq mi/41.6 sq km); ⊙ AISNE department, PICARDIE region, N FRANCE, on a rocky height 300 ft/91 m above the surrounding plain, 29 mi/47 km NW of REIMS, on the historic border between CHAMPAGNE and ÎLE-DE-FRANCE; 49°34′N 03°37′E. A road and railroad center, Laon has a metalworking industry (cable drawing, heating equipment) and a large sugar mill nearby. It was fortified in Roman times and was capital of France in the Carolingian era. Laon was an episcopal see from the 5th century until the French Revolution. During the Middle Ages it was torn by bitter struggles against the bishops by the burghers who ultimately succeeded (12th century) in obtaining recognition of their charter. Notable monuments include the vast Gothic Cathedral of Notre Dame, with an exquisite nave, St. Martin's Church (both built 12th–13th century), and an octagonal chapel of the Knights Templars (12th century). The old town is accessible through medieval gates that are reached by steep roads and a rack-and-pinion streetcar; the ramparts are well preserved. The citadel (E) is the administrative center of the department.

Laona (lai-O-nuh), village, FOREST county, NE WISCONSIN, 36 mi/58 km ESE of RHINELANDER, in Nicolet National Forest, in lake region. Lumbering, logging; tourism; fishing, hatchery. Logging museum; Potawatomi Indian Reservation to SE.

Lao Ngam (LOU NGAHM), town, district administrative center, SALAVAN province, S LAOS, SW of SALAVAN town; 15°10′N 106°27′E. Transport hub, market center. Also called BANG LAO NGAM.

Laongo (lah-AWN-go), village, BAS-CONGO province, W CONGO, 50 mi/80 km N of BOMA; 05°10′S 12°59′E.

La Orchila, Isla (LAH or-CHEE-lah, EES-lah) or **Orchila, Isla** in the CARIBBEAN, DEPENDENCIAS FEDERALES, VENEZUELA, 105 mi/169 km NE of CARACAS; 7 mi/11 km long, 4 mi/6 km wide; 11°48′N 66°10′W. Some goat grazing. Guano deposits. Sometimes spelled Orchilla.

Lao River (lou), 25 mi/40 km long, S ITALY; rises in the APENNINES 7 mi/11 km E of Lauria, flows SSW to TYRRHENIAN SEA 3 mi/5 km S of SCALEA. Formerly also called Laino.

La Oroya (lah o-RO-yah), town (2005 population 19,204), ⊙ YAULI province, JUNÍN region, central PERU, in Cordillera CENTRAL of the ANDES, on MANTARO RIVER at mouth of the Yauli, and 80 mi/129 km ENE of LIMA (connected by railroad and highway); 11°32′S 75°54′W. Elevation 12,178 ft/3,712 m. Railroad and road junction of lines to CERRO DE PASCO and HUANCAYO. Metallurgical center (copper, silver, zinc, lead); ore smelting, electrolytic copper refining. Hydroelectric power station (fed by Pomacocha Reservoir). Produces bismuth, arsenic, and cadmium as by-products. Ships minerals, grain; livestock. Receives ores from Cerro de Pasco, MOROCOCHA, and Casapalca mines; connected with Pachacacha power plant.

Laos (lah-OUS), republic (□ 91,428 sq mi/236,800 sq km; 2004 estimated population 6,068,117; 2007 estimated population 6,521,998), officially, the Lao People's Democratic Republic, SE ASIA; ⊙ VIENTIANE; 18°00′N 105°00′E.

Geography

The administrative capital is Vientiane, the historic royal capital, LUANG PHABANG. A landlocked country, Laos is bordered by CHINA on the N, by VIETNAM on the E, by CAMBODIA on the S, and by THAILAND and MYANMAR on the W. In general, the MEKONG RIVER, most of which flows in a broad valley, forms the boundaries with Myanmar and Thailand. For two lengthy stretches, however—one greater than 300 mi/ 480 km—the Mekong flows entirely through the territory of Laos. Except for the Mekong lowlands and three major plateaus, the terrain of Laos is rugged, mountainous, and still heavily wooded despite serious deforestation from logging and shifting cultivation; jagged crests in the N tower over 9,000 ft/2,740 m.

Population

About half the population is Lao, a people ethnically related to the Thai. Those settled along the Mekong River are termed Lao Lam, while the so-called Lao Thai, who are more "tribal" in character, have resisted the cultural influences of lowland settled peoples and generally reside in upland valleys rather than the Mekong flood plains. Upland tribes include the many groups that constitute the Lao Theung, or various Mountain Mon-Khmer groups, as well as the Meo, Mien, Black Thai, Dao, and several Tibeto-Burman speaking peoples. There are also important minorities of Vietnamese and Chinese. A majority of Laotians are Theravada Buddhists; although the 40% of the people are generally animists. Many Laotians are descendants of Thai immigrants who were pushed south from YÜNNAN, China, beginning in the 13th century and gradually infiltrated the territory of the KHMER EMPIRE.

Economy

Laos is one of the regions of SE Asia least touched by modern civilization. There are no railroads, though a railroad bridge over the Mekong connects Vientiane with Thailand's railroad system at NONG KHAI. Commercial airports are few, and roads and trails are limited. Use of the country's main communications artery, the Mekong River, is impeded by many falls and rapids. More than half the population lives along the Mekong and its tributaries. Most are subsistence farmers, who also weave their own cloth, produce most domestic utensils and tools, and still maintain diverse folk crafts. Rice is by far the chief crop; corn and vegetables are also grown. Commercial crops include coffee, tobacco, sugarcane, and cotton. Illegal opium poppies and cannabis sativa are grown by shifting cultivators in a NW region bordering Thailand and Myanmar known as the "Golden Triangle." Fish from the rivers and game from the forests supplement the diet. Forests cover about two-thirds of the country; teak is cut and lac is extracted, but poor transportation and the lack of industry limit production. Some tin is mined, and other known mineral resources include coal, iron, graphite, copper, gold, lead, salt, and zinc. None have been effectively developed. The principal exports of Laos are tin, timber, and coffee. Since almost all manufactured items have to be imported, however, there is a continuing foreign trade deficit. A growing competitive advantage of Laos is hydroelectricity. In 1992, electricity exports, primarily to Thailand, comprised nearly 47% of export value. In 2001, 400 million kw of electricity were exported–almost one-third of all electricity produced. The first dam, built along the Mekong, was NAM NGUM, and a second one, Nam Theum 2, is currently in the works. Plans for future

dam sites are underway as well, although the projects have caused some controversy because they often require displacement of residents in the area and cause damage to wildlife. The projects should give Laos a productive capacity of 20,000 MW of hydroelectricity.

History: to 19th Century

In the mid-14th century a powerful kingdom called Lan Xang was founded in Laos by Fa Ngoun (1353–1373), who is also credited with the introduction of Theravada Buddhism and much of Khmer civilization into Laos. Lan Xang waged intermittent wars with the Khmer, Burmese, Vietnamese, and Thai, and by the 17th century it held sway over sections of Yünnan, China, of S Myanmar, of the Vietnamese and Cambodian plateaus, and large stretches of N Thailand. In 1707, however, internal dissensions brought about a split of Lan Xang into two kingdoms: LUANG PHABANG in upper (N) Laos and Vientiane in lower (S) Laos. During the next century the two states, constantly quarreling, were overrun by the armies of neighboring countries.

History: 19th Century to 1955

In the early 19th century SIAM was dominant over the two Laotian kingdoms, although Siamese claims were disputed by ANNAM. After French explorations in the late 19th century, Siam was forced (1893) to recognize a French protectorate over Laos, which was incorporated into the union of INDOCHINA. During World War II, Laos was gradually occupied by the Japanese, who in 1945 persuaded the king of Luang Prabang to declare the country's independence. The French nevertheless reestablished (1946) dominion over Laos, recognizing the king as constitutional monarch of the entire country. The French granted an increasing measure of self-government, and in 1949 Laos became a semiautonomous state within the FRENCH UNION. In 1951, a communist Laotian nationalist movement, the Pathet Lao, was formed by Prince Souphanouvong in NORTH VIETNAM. In 1953, Pathet Lao guerrillas accompanied a Viet Minh invasion of Laos from Vietnam and established a government at SAM NEUA in N Laos. That year Laos attained full sovereignty; admission into the UN came in 1955.

History: 1955 to 1961

The new country faced immediate civil war as Pathet Lao forces, supported by the Viet Minh, made incursions into central Laos, soon occupying sizable portions of the country. Agreements reached at the GENEVA Conference of 1954 provided for the withdrawal of foreign troops and the establishment of the Pathet Lao in two N provinces. In 1957 an agreement was reached between the royal forces and the Pathet Lao, but in 1959 the coalition government collapsed and hostilities were renewed. A succession of coups resulted (1960) in a three-way struggle for power between neutralist, rightist, and Communist forces. The Communist Pathet Lao rebels remained under the leadership of Prince Souphanouvong in the N provinces. The right-wing government of Boun Oum, installed in Vientiane, was recognized by the U.S. and other Western countries and controlled the bulk of the royal Laotian army. The Soviet Union and its allies continued to recognize the deposed neutralist government of Souvanna Phouma, who had fled to neighboring Cambodia.

History: 1961 to 1971

In May 1961, with Pathet Lao and neutralist forces in control of about half the country, a cease-fire was arranged. A fourteen-nation conference convened in Geneva, producing (1962) another agreement providing for the neutrality of Laos under a unified government. A provisional coalition government, with all factions represented, was accordingly established under the premiership of Souvanna Phouma. Attempts to integrate the three military forces failed,

however, and the Pathet Lao began moving against neutralist troops. Open warfare resumed in 1963, and the Pathet Lao, bolstered by supplies and troops from North Vietnam, solidified control over most of N and E Laos. Disgruntled right-wing military leaders staged a coup in 1964 and attempted to force the resignation of Souvanna Phouma; the U.S. and the Soviet Union emphasized their support of the premier, however, and he remained in office with a right-wing neutralist government. Pathet Lao guerrilla activity decreased after the start (1965) of U.S. bombings of North Vietnamese military bases and communications routes. The bombings also included attacks on what came to be known as the Ho Chi Minh Trail, a lengthy and complex system of co-lateral pathways that together comprised the North Vietnamese supply route through E Laos. Communist pressure increased during 1969, and early in 1970 the Pathet Lao launched several major offensives.

History: 1971 to 1975

At the beginning of 1971, South Vietnamese troops invaded Laotian territory in an unsuccessful attempt to block the Ho Chi Minh Trail. The attack drove the North Vietnamese deeper into Laos, and Laos became another multi-frontal battleground of the Vietnam War, with heavy U.S. aerial bombardments. The U.S. extended enormous military and economic aid to the Laotian government and financed the use of Thai mercenary troops, whose numbers peaked to over 21,000 in 1972. The Pathet Lao, supported by North Vietnamese troops, scored major gains, consolidating their control over more than two-thirds of Laotian territory (mainly in the more rugged mountain sectors) but over only one-third of the population. Heavy fighting persisted until February 1973, when a cease-fire was finally declared. A final agreement between the government and the Pathet Lao, concluded in September 1973, provided for the formation of a coalition government under the premiership of Souvanna Phouma (inaugurated 1974), the stationing of an equal number of government and Pathet Lao troops in the two capitals, and the withdrawal of all foreign troops and advisers. After Communist victories in Vietnam and Cambodia, the Pathet Lao took control of the country in 1975, and officially changed its name to the Lao People's Democratic Republic. Souphanouvong became president, and Kaysone Phomvihane became premier.

History: 1975 to Present

Between 1975 and 1985, over 300,000 Laotians fled their country and sought refuge in the U.S. and elsewhere. The Laotian political economy faced enormous difficulties in the immediate post-war years, with government leadership rolling back efforts to collectivize agriculture and restrict private enterprise by the mid-1980s. The shift in "chintanakan mai" (new thinking) was further entrenched with the collapse of many socialist states in the late 1980s. Besides increasing revenues from tourism, Laos benefited from improved relations with its SE Asian neighbors. ASEAN granted Laos observer status in 1992 and the country became a full member in 1997. Kaysone became president in 1991. He died the following year and was succeeded as president by Nouhak Phoumsavan. Khamtay Siphandone became party leader and, when Nouhak retired in 1998, assumed the job of president as well. Khamtay was succeeded as party leader and president by Vice President Choummali Saignason in 2006.

Government

Laos is governed under the constitution of 1991. The president, who is the head of state, is elected by the legislature for a five-year term. The government is headed by the prime minister, or premier, who is appointed by the president. The unicameral legislature consists of the 115-seat National Assembly, whose members are popularly elected for five-year terms. The only permitted political party is the Lao People's

Revolutionary Party. The current head of state is President Choummali Saignason (since 2006). The current head of government is Prime Minister Bouasone Bouphavanh (since 2006). Administratively, the country is divided into fifteen provinces and one municipality (the capital).

Lao Shan (LOU SHAN), mountain (3,707 ft/1,130 m) in E SHANDONG province, CHINA, overlooking YELLOW SEA, 20 mi/32 km NE of QINGDAO; 36°10′N 120°37′E. Famous for its scenery, Taoist mountains, and cliff writing.

Laotie Shan (LOU-TYE SHAN), mountain in southernmost headland of LIAODONG peninsula, S Northeast CHINA, overlooking the Strait of BOHAI, 6 mi/9.7 km SW of LUSHAN (Port Arthur); 38°46′N 121°10′E.

Laotto, village, NOBLE county, NE INDIANA, 15 mi/24 km N of FORT WAYNE. Manufacturing (metal products). Corn, wheat; cattle. Laid out 1871.

Laou, Oued (LAW, wahd), river, 40 mi/64 km long, NW MOROCCO; rises in RIF mountains S of CHEFCHAOUEN; flows NNE to the MEDITERRANEAN SEA at OUAD LAOU village.

Lapa (LAH-pah), city (2007 population 41,677), SE PARANÁ, BRAZIL, on railroad, 35 mi/56 km SW of CURITIBA; 25°45′S 49°42′W. Flour milling, maté processing; rye, tobacco, cattle. Sand deposits, clay quarries.

Lapa, Brazil: see BOM JESUS DA LAPA.

Lapachito (lah-pah-CHEE-to), village, E Chaco national territory, ARGENTINA, 23 mi/37 km NW of RESISTENCIA; 27°10′S 59°24′W. Railroad junction; cotton, livestock; forest products; sawmills, cotton gins.

Lapac Island (lah-PAHK) (□ 16 sq mi/41.6 sq km), in TAPUL GROUP, SULU province, PHILIPPINES, in SULU ARCHIPELAGO, just W of SIASI, 25 mi/40 km SW of JOLO ISLAND; 05°32′N 120°47′E.

Lapa, La (LAH-pah, lah), village, BADAJOZ province, W SPAIN, 38 mi/61 km SE of BADAJOZ. Olives, cereals, vegetables; livestock; tiles, pottery.

La Palca (lah PAHL-kah), canton, OROPEZA province, CHUQUISACA department, SE BOLIVIA, 22 mi/35 km N of SUCRE; 19°10′S 65°16′W. Elevation 8,212 ft/2,503 m. Copper-bearing lode, iron (Mina Okekhasa and Mina Virgén de Copacabana) and manganese (Mina Lourdes) mining, and gypsum deposits. Agriculture (potatoes, yucca, bananas, corn, wheat, oats, rye, peanuts, barley); cattle.

Lapalisse (lah-pah-lees), commune (□ 12 sq mi/31.2 sq km), ALLIER department, central FRANCE in AUVERGNE, on the Besbre, 13 mi/21 km NE of VICHY; 46°15′N 03°38′E. Road junction and agriculture market center; manufacturing (leather goods, knitwear, hardware, and food preserves); flour- and sawmilling. Its 15th–16th-century Renaissance castle has a chamber with a golden ceiling and Flemish tapestries.

La Palma, city (2000 population 15,408), ORANGE county, S CALIFORNIA; suburb 19 mi/31 km SE of downtown LOS ANGELES, 5 mi/8 km W of ANAHEIM; 33°51′N 118°02′W. Knots Berry Farm, at Buena Park, to E. Manufacturing (aircraft parts, chemicals, metal, paper, and plastic products; printing).

La Palma (LAH PAHL-mah), town, ⊙ La Palma municipio, CUNDINAMARCA department, central COLOMBIA, on W slopes of Cordillera ORIENTAL, 45 mi/72 km NW of BOGOTÁ; 05°21′N 74°23′W. Elevation 4,796 ft/1,462 m. Agricultural center (sugarcane, coffee, corn, potatoes).

La Palma (lah PAHL-mah), town, PINAR DEL RÍO province, W CUBA, 25 mi/40 km NNE of PINAR DEL RÍO; 22°45′N 83°33′W. In agricultural region (sugarcane, tobacco); manufacturing of cigars; lumbering. The Manuel Sanguily sugar mill is 5 mi/8 km NE.

La Palma (lah PAHL-mah), municipality and town, CHALATENANGO department, EL SALVADOR, NW of CHALATENANGO city in NW part of department.

La Palma (lah PAHL-mah), town and township, ⊙ Chepigana district, DARIÉN province, E PA-NAMA, port on TUIRA RIVER estuary, on SAN MIGUEL GULF, 90 mi/145 km SE of PANAMA city. Agriculture (plantains, corn, rice); livestock raising; sawmilling.

La Palma (lah PAHL-mah), village, LOS SANTOS province, S central PANAMA, in PACIFIC lowland, 4 mi/6.4 km SE of LAS TABLAS. A minor civil division of Las Tablas district. Sugarcane, coffee; livestock.

La Palma (lah PAHL-mah), saddle depression (elevation 6,000 ft/1,830 m) in the Central CORDILLERA, central COSTA RICA, between BARVA (W) and IRAZÚ (E) volcanoes, NE of SAN JOSÉ; approximately 10°00′N 84°00′W. Rain-bearing CARIBBEAN air masses enter central plateau here. Crossed by San José-LIMÓN Highway. Principal access route to BRAULIO CARRILLO NATIONAL PARK.

La Paloma (lah pah-LO-mah) town, ROCHA department, SE URUGUAY, port on the ATLANTIC adjoining CAPE SANTA MARÍA, 16 mi/26 km SE of ROCHA, on railroad; 34°40′S 54°09′W. Beach resort; fishing and fish-processing center. Sometimes called Puerto La Paloma.

La Paloma (LAH pah-LO-mah), town, ZULIA state, NW VENEZUELA, in MARACAIBO lowlands, 28 mi/45 km SW of ENCONTRADOS; 08°48′N 72°32′W. Petroleum drilling in the Tarra oil field, served by local railroad and by pipeline along ESCALANTE RIVER (NE) to Lake MARACAIBO.

Lapaluoto, FINLAND: see RAAHE.

La Pampa (lah PAHM-pah), interior province (2001 population 299,294), central ARGENTINA; ⊙ SANTA ROSA. Low, grassy area (dry PAMPA), sloping gradually E from MENDOZA province border (NW). Bordered S by the RÍO COLORADO, watered by ATUEL RIVER, and the Río Salado, which reach the central marshes. Contains a number of salt marshes and salt deposits. Climate is generally dry and temperate. Mostly livestock raising (cattle, sheep, goats, horses). Agriculture in N and E: barley, wheat, alfalfa, corn, oats, rye, sunflowers. Lumbering of hardwood (cedar, carob, oak) in NE. Also salt mining, dairying. Meat packing at Santa Rosa; flour milling at Santa Rosa, GENERAL PICO, GENERAL SAN MARTÍN; petroleum extracted in SW. Established as a national territory 1884, became a province 1951.

La Panza Range, one of the COAST RANGES, SAN LUIS OBISPO county, SW CALIFORNIA, extends c.30 mi/48 km NW-SE between SANTA LUCIA RANGE (W) and TEMBLOR RANGE (E); rises to 4,054 ft/1,236 m, 24 mi/39 km E of San Luis Obispo. A N extension of the Sierra Madre; in Los Padres National Forest.

Lapão (lah-POUN), city (2007 population 25,580), N central BAHIA, BRAZIL, 11 mi/18 km SE of IRECÊ; 11°23′S 41°50′W.

La Para (lah PAH-rah), town, N CÓRDOBA province, ARGENTINA, on the RÍO PRIMERO, near the MAR CHIQUITA, and 80 mi/129 km NE of CÓRDOBA; 30°52′S 62°59′W. Railroad junction and agriculture center. Wheat, flax, alfalfa; dairying, livestock raising; meat-packing plant.

La Paragua (LAH pah-RAH-gwah), town, BOLÍVAR state, SE VENEZUELA, on PARAGUA RIVER, and 100 mi/161 km SSE of CIUDAD BOLÍVAR; 06°52′N 63°19′W. Tropical woods, rice.

La Parguera (lah pahr-GAI-rah), fishing village, SW coast of PUERTO RICO, in E branch of Boquerón State Park (also known as La Parguera Park). Tourism; on Phosphorescent Bay where trillions of sea micro-organisms breed and illuminate the water. University of Puerto Rico Marine Sciences research facilities located on Magueyes Island nearby (S).

La Parida, VENEZUELA: see BOLÍVAR, CERRO.

Lapas (LAH-pash), village, SANTARÉM district, central PORTUGAL, suburb of TORRES NOVAS, 19 mi/31 km NNE of SANTARÉM. Alcohol distilling.

Lapa, Serra da (LAH-pah, SER-rah dah), mountain range (3,100 ft/945 m) of N central PORTUGAL, in

VISEU and GUARDA districts, NW of TRANCOSO; 40°54′N 07°30′W. VOUGA RIVER rises here.

Lapas, Las, SPAIN: see FRONTERA.

LaPasse (lah-PAHS), unincorporated village, SE ONTARIO, E central CANADA, from RENFREW, and included in WHITEWATER Region township; 45°48′N 76°46′W. Tourism. Also written La Passe.

Lapataia (lah-pah-tah-EE-ah), village, SW TIERRA DEL FUEGO national territory, ARGENTINA, port on inlet of BEAGLE CHANNEL, near CHILE border, 10 mi/16 km W of USHUAIA; 54°50′S 68°34′W. Small sheep-raising settlement.

La Patrie (lah pah-TREE), village (□ 80 sq mi/208 sq km), S QUEBEC, E CANADA, on SALMON RIVER (a tributary of SAINT FRANÇOIS River), 30 mi/48 km E of SHERBROOKE, at foot of MEGANTIC MOUNTAIN; 45°24′N 71°15′W. Dairying; livestock raising.

La Paz, department, ARGENTINA: see SAN ANTONIO, CATAMARCA province.

La Paz, department (□ 51,732 sq mi/134,503.2 sq km; 2005 population 2,630,381), W BOLIVIA; ⊙ LA PAZ; 15°30′S 68°00′W. Bordered W by PERU, in extreme SW by CHILE. Includes E part of Lake TITICACA. The E Cordillera of the ANDES mountains (with its peaks ILLIMANI, ANCOHUMA, and ILLAMPU) crosses department NW-SE, separating the Altiplano, or high plateau (S; 13,000 ft/3,962 m–14,500 ft/4,420 m), from the YUNGAS and tropical lowlands (N). The Altiplano is drained by DESAGUADERO RIVER, tropical lowlands by BENI RIVER and its affluents, La Paz, KAKA, TUICHI, and MADIDI rivers. Grain, potatoes, sheep, and llama raised on plateau, tropical agriculture (coffee, cacao, cinchona, coca, fruit) in lowlands. Rubber plantations along Beni and Madre de Dios rivers (N). Mining centers include COROCORO (copper; closed 1986), Cordillera de Tres Cruces (tin, tungsten, antimony), TIPUANI RIVER, GUANAY, MAPIRI, Huayti, Teoporto (gold). Main industrial centers are La Paz and VIACHA (railroad junction), with railroad connection to ARICA (Chile). Roads cross E CORDILLERA into lowlands where river transportation prevails. Archaeological sites at TIAHUANACO and TITICACA and Coati islands Dept. created January 23, 1826, by Antonio José de Sucre.

La Paz, department (□ 909 sq mi/2,363.4 sq km), S EL SALVADOR, on the PACIFIC; ⊙ ZACATECOLUCA. Slopes from coastal range S to the Pacific; drained by JIBOA RIVER. Includes part of LAKE ILOPANGO (NW) and SW slopes of volcano SAN VICENTE (NE). Grain; livestock raising. Salt extraction (at JALTEPEQUE LAGOON) is important. Main centers: Zacatecoluca, SAN JUAN NONUALCO, SANTIAGO NONUALCO; linked by road with SAN SALVADOR. Scene of Nonualco Indian revolt, 1843. Formed 1833.

La Paz (lah PAHS), department (□ 1,247 sq mi/3,242.2 sq km; 2001 population 156,560), SW HONDURAS, on EL SALVADOR border; ⊙ LA PAZ; 14°05′N 87°55′W. Astride Continental Divide; includes MONTAÑAS LA SIERRA (W), upper Goascorán River valley (SE), upper COMAYAGUA RIVER valley (NE). Coffee, wheat in highlands (MARCALA, OPATORO), henequen (La Paz); cattle raising. Ropemaking, palm-hat manufacturing, ceramics are local industries. Main centers include La Paz (oriented toward Comayagua) and Marcala (linked by road with LA ESPERANZA). Formed 1869.

La Paz, county (□ 4,513 sq mi/11,689 sq km; 1990 population 13,844; 2000 population 19,715), W ARIZONA; ⊙ PARKER; 33°43′N 113°58′W. COLORADO RIVER (CALIFORNIA state line and Pacific/Mountain time zone boundary) bounds county on W. BILL WILLIAMS and SANTA MARIA rivers form N county line. Cattle; alfalfa, hay. Mountain ranges include TRIGO MOUNTAINS (SW), DOME ROCK MOUNTAINS (W), HARCUVAR MOUNTAINS (E), and part of KOFA MOUNTAINS (S). Part of Colorado Indian Reservation in W; parts of Cibola and Imperial National Wildlife

Areas in SW; part of Kofa National Wildlife Refuge in S; part of Havasu National Wildlife Refuge in NW; part of Eagletail Mountains Wilderness Area in SE; part of Harquahala Mountains Wilderness Area on E boundary. Alamo State Park in NE. Part of large Yuma Proving Ground (U.S. Army) in SW. County formed in 1983 from YUMA county.

La Paz, city (2005 population 839,169), MURILLO province, ⊙ LA PAZ department, administrative ⊙ BOLIVIA (since 1898), W Boliva, and largest city of Bolivia; 16°30′S 68°09′W. Elevation 11,897 ft/3,626 m (making it the highest capital city in the world). La Paz is crowded into a long, narrow valley cut by the LA PAZ RIVER. It is an agricultural market and has light manufacturing industries. LA PAZ-EL ALTO INTERNATIONAL AIRPORT to W. The site, where there was an Inca village, was chosen by Alonso de Mendoza in 1548 because it offered a modicum of protection in winter from the wind and cold of the barren high plateau c.1,400 ft/430 m above. Because of the narrowness of the valley, the city could not be laid out in the customary Spanish grid pattern. The Plaza Murillo, named after the independence leader Pedro Domingo Murillo, with the national palace, cathedral, and other buildings, is small; there are only a few broad, long avenues, and the streets ascend steeply on either side. Climate is generally cool and extreme variations in temperature are common; agriculture is difficult. La Paz's location on colonial trade routes made it the commercial and political focus of colonial life; some of the colonial architecture remains. Its University of San Andrés was founded in 1830, and a Catholic university in 1966. There are extraordinary tourist attractions in the region, notably the ANDEAN peaks ILLIMANI, ANCOHUMA, and ILLAMPÚ, LAKE TITICACA, the ruins of TIAHUANACO, and the adjacent tropical YUNGAS. The city's full name is La Paz de Ayacucho, after a Bolivian victory at AYACUCHO (PERU) in the war for independence (1809–1825); originally called Nuestra Señora de la Paz.

La Paz (lah PAHS), city (2001 poulation 16,947), ⊙ LA PAZ department, SW HONDURAS, in COMAYAGUA RIVER valley, 10 mi/16 km SSW of COMAYAGUA; 14°19′N 87°41′W. Elevation 2,297 ft/700 m. Commercial center; ropemaking; henequen, coffee, livestock. Founded 1792; known as La Paz since 1851.

La Paz (lah PAHS), city (2005 population 189,176) and township, largest city and ⊙ BAJA CALIFORNIA SUR state, W MEXICO; 24°10′N 110°17′W. A tourist spot and transportation hub for the S BAJA CALIFORNIA peninsula, La Paz was first settled in 1811. The city was known for its pearl fishing until the middle of the 20th century when the oyster beds were destroyed by disease. It is known for its water sports, as well as being an entry point for the LOS CABOS resort area to the S. La Paz is linked to MAZATLÁN and LOS MOCHIS by ferry, and to TIJUANA and MEXICALI by Mexico Highway 1. International airport to SE. Marinas.

La Paz, town (1991 population16,659), ⊙ La Paz department (□ 2,685 sq mi/6,981 sq km), NW ENTRE RÍOS province, ARGENTINA, on PARANÁ RIVER and 80 mi/129 km NE of PARANÁ. Railroad terminus, port, and agriculture center (corn, oats, alfalfa, rice, olives, fruit; livestock). Lime factory, flour mills. Has museum, golf, amusement park.

La Paz, town (1991 population 5,618), ⊙ La Paz department (□ 2,650 sq mi/6,890 sq km), E MENDOZA province, ARGENTINA, on TUNUYÁN RIVER (irrigation) and 85 mi/137 km SE of MENDOZA. Railroad junction, lumbering and farming center (fruit, wine, alfalfa; livestock); wine making.

La Paz (LAH PAHS), town, ⊙ La Paz municipio, SANTANDER department, N central COLOMBIA, 63 mi/101 km SW of BUCARAMANGA; 06°11′N 73°34′W. Coffee, corn, sugarcane; livestock.

Lapaz or **La Paz**, town (2000 population 489), MARSHALL county, N INDIANA, 8 mi/12.9 km N of PLYMOUTH. In agricultural area.

La Paz (lah PAHS) (2004 population 19,832), town, CANELONES department, S URUGUAY, on railroad, and 10 mi/16 km N of MONTEVIDEO; 34°46′S 56°15′W. Winegrowing; granite quarrying.

La Paz, town, COLONIA department, SW URUGUAY, urban nucleus of the COLONIA VALDENSE agricultural settlement, 32 mi/51 km ENE of Colonia; 34°21′S 57°18′W. Sometimes called Colonia Piamontesa, a name also applied to the Colonia Valdense.

La Paz, MEXICO: see LOS REYES ACAQUILPAN.

La Paz, MEXICO: see VILLA DE LA PAZ.

La Paz, NICARAGUA: see LA PAZ DE ORIENTE.

La Paz, PHILIPPINES: see PAZ, LA.

La Paz Bay (lah PAHS), large sheltered, deep-water inlet of Gulf of CALIFORNIA, on SE coast of BAJA CALIFORNIA SUR, NW MEXICO, bordered E by ESPÍRITU SANTO Island; 50 mi/80 km long NW-SE, c.20 mi/32 km wide. City of LA PAZ is at its head.

La Paz Centro (lah PAHS SEN-tro), town, (2005 population 19,010), LEÓN department, ⊙ La Paz Centro municipio, W NICARAGUA, near LAKE MANAGUA, 16 mi/26 km SE of LEÓN; 12°20′N 86°41′W. Brick-and tile-making center; manufacturing (pottery); agriculture (corn, sesame). Old town of La Paz Viejo is on railroad, 2 mi/3.2 km NW.

La Paz, Cordillera de (lah PAHS, kor-di-YAI-rah dai), highest range in the E Cordillera of the ANDES mountains, LA PAZ department, W BOLIVIA; extends 170 mi/274 km SE from Nudo de Apolobamba on PERU border to LA PAZ RIVER. Rises to 21,490 ft/6,550 m in the ANCOHUMA, a peak of the ILLAMPÚ. Tungsten deposits at SE foot of peak ILLIMANI (large canyon at base of this peak is 11,483 ft/3,500 m deep). Also called Cordillera Real.

La Paz de Oriente (lah PAHS dai o-ree-AIN-tai), town, CARAZO department, SW NICARAGUA, 5 mi/8 km ESE of JINOTEPE; 11°49′N 86°08′W. In sugarcane zone. Also called La Paz.

La Paz-El Alto International Airport (lah PAHZ–el AHL-to), LA PAZ department, W BOLIVIA; 8 mi/14 km X of LA PAZ; elevation 13,313 ft/4,058 m; 16°30′S 68°11′W. Reputed to be the highest airport in the world. Airport code LPB.

La Paz River, c. 100 mi/161 km long, LA PAZ department, W BOLIVIA; rises in 2 branches (CHOQUEYAPU and CHUQUIAGUILLO rivers) on the CHACALTAYA, 10 mi/16 km N of LA PAZ; flows SE and NE through Eastern Cordillera of the Andes Mountains, past La Paz, Obrajes, and Mecapeca, joining TAMAMPAYA RIVER 20 mi/32 km ENE of CHULUMANI to form BOPI RIVER; 15°41′S 67°15′W. Receives Palca River (left), LURIBAY River (right). Considered chief headstream of BENI RIVER.

La Pe, town, S central OAXACA, MEXICO, 34 mi/55 km S of OAXACA de JUÁREZ, and 7 mi/11 km NW of Ejotga de Erespo. Elevation 4,856 ft/1,480 m. In the upper Miahuatlán River valley. Temperate climate. Agriculture.

La Pêche (lah PESH), village (□ 231 sq mi/600.6 sq km; 2006 population 6,585), OUTAOUAIS region, SW QUEBEC, E CANADA, 10 mi/16 km from PONTIAC; 45°41′N 75°59′W.

Lapeer (luh-PIR), county (□ 663 sq mi/1,723.8 sq km; 2006 population 93,761), E MICHIGAN; 43°05′N 83°13′W; ⊙ LAPEER. Drained by FLINT (Holloway Reservoir on W boundary) and BELLE rivers and by short Mill Creek. Agriculture (cattle, hogs, sheep, poultry; dairying; apples, corn, wheat, oats, soybeans, barley, potatoes, carrots, sugar beets, beans, celery, onions). Manufacturing at Lapeer (motor vehicle parts, fabricated metal products, machining, plumbing equipment, plastic products). Numerous small lakes, especially in W. Organized 1833.

Lapeer (luh-PIR), city (2000 population 9,072), ⊙ LAPEER county, E MICHIGAN, 20 mi/32 km E of FLINT, on South Branch of FLINT RIVER; 43°02′N 83°19′W. In dairying and grain-growing area; manufacturing (metal products, transportation equipment, plastics products, furniture). Has state home and school for mentally ill. Airport to E. Metamora-Hadley State Recreation Area to S. Settled 1831, incorporated as city 1869.

Lapeer Heights (luh-PIR), village, GENESEE county, SE central MICHIGAN, suburb of FLINT, 3 mi/4.8 km E; 43°00′N 83°35′W. Now part of city of BURTON.

Lapel, town (2000 population 1,855), MADISON county, E central INDIANA, 8 mi/12.9 km WSW of ANDERSON; 40°04′N 85°51′W. In livestock and grain area; glass products. Laid out 1876.

La Pelada (lah pe-LAH-dah), village, central SANTA FE province, ARGENTINA, 55 mi/89 km NNW of SANTA FE; 30°52′S 60°59′W. Railroad junction, agriculture center (corn, soybeans, flax, wheat; livestock, poultry).

La Peña (LAH PAI-nyah), town, ⊙ La Peña municipio, CUNDINAMARCA department, central COLOMBIA, 33 mi/53 km NW of BOGOTÁ. Coffee, corn, sugarcane; livestock.

La Peña (lah PAIN-yah), village, VERAGUAS province, W central PANAMA, near INTER-AMERICAN HIGHWAY, 4 mi/6.4 km NW of SANTIAGO. Sugarcane; livestock. A minor civil division of Santiago district.

La Peña, Embalse de, reservoir, HUESCA province, NE SPAIN, c.22 mi/35 km NNW of HUESCA; 42°23′N 00°43′W. Collects waters of Asabón and GALLEGO rivers.

La Pérade, CANADA: see SAINTE-ANNE-DE-LA-PÉRADE.

La Perla (lah PER-lah), town, VERACRUZ, E MEXICO, at SE foot of PICO DE ORIZABA, 6 mi/10 km N of ORIZABA. Fruit.

La Perla (LAH PER-lah), beach suburb, Callao province, CALLAO region, W PERU; 11°22′S 76°48′W.

La Perla, S suburb of CALLAO city, Callao constitutional province, PERU; 12°05′S 77°08′W.

La Perouse, Mount (per-ROOS) (10,728 ft/3,270 m), SE ALASKA, in FAIRWEATHER RANGE, near Gulf of ALASKA, 100 mi/161 km NW of JUNEAU, in GLACIER BAY National Monument; 58°34′N 137°05′W.

La Perouse Pinnacle, HAWAII: see FRENCH FRIGATE SHOAL.

La Pérouse Strait (LAH pai-ROOZ) or **soya strait**, Japanese *Soya-kaikyo* (SO-yah-KAH-ee-kyo), channel, 25 mi/40 km wide, separating N HOKKAIDO island, extreme N JAPAN, from S SAKHALIN island, RUSSIAN FAR EAST, and connecting the SEA OF JAPAN on the W with the SEA OF OKHOTSK on the E.

Lapeza (lah-PAI-thah), town, GRANADA province, S SPAIN, 18 mi/29 km ENE of GRANADA; 37°17′N 03°17′W. Olive oil, cereals, fruit, sugar beets.

Lapi, FINLAND: see LAPIN.

La Piedad de Cabadas (lah pee-ai-DAHD dai kah-BAH-dahs), city, ⊙ La Piedad municipio, MICHOACÁN, central MEXICO, on central plateau, on LERMA River (GUANAJUATO border), and 90 mi/145 km ESE of GUADALAJARA; 20°20′N 102°01′W. Manufacturing and agricultural center (cereals, sugarcane, fruit, vegetables; livestock); tanneries, rayon mills; manufacturing (native shawls, sweets). Formerly called La Piedad de Cavadas.

Lapin (LAH-pin), province (□ 39,575 sq mi/102,895 sq km), N FINLAND; ⊙ ROVANIEMI. Bordered in N by NORWAY, E by RUSSIA, S by OULUN province, W by Sweden. Contains thirty percent of country's land but only four percent of its population. Much of Lapin is above ARCTIC CIRCLE and was formed by glaciation, producing moors and fells; KEMIJOKI is major river in region. Open-pit quartz mining in TORNIO; chromite and ferrochrome mining and production, integrated with Tornio stainless steel plant in KEMI. Settled by Europeans during 1860s gold prospecting. Tourism. Commonly known as Lapland.

Lapinin Island (lah-PEE-neen), BOHOL province, PHILIPPINES, in CANIGAO CHANNEL, just off NE coast of

Area is shown by the symbol □, and capital city or county seat by ⊙.

Bohol island; c.3 mi/4.8 km long; 10°06′N 124°33′E. Flat and fringed with mangrove trees. Coconut growing, fishing. Chief town, Presidente Carlos P. Garcia (1990 population 21,173).

La Pintada (peen-TAH-dah), town, ⊙ La Pintada district, COCLÉ province, central PANAMA, in Coclé Mountains, 6 mi/9.7 km NNW of PENONOMÉ. Road junction point. Rice, corn, beans; livestock.

Lapithos (LAH-pee–[th]-os) [Turkish= LAPTA], town, KYRENIA district, N CYPRUS, 1 mi/1.6 km S of MEDITERRANEAN coast, 15 mi/24 km NW of NICOSIA; 35°20′N 33°10′E. KYRENIA MOUNTAINS to S. Citrus, almonds, carobs, olives, vegetables, potatoes; sheep, goats.

La Place (luh PLAHS), unincorporated city, SAINT JOHN THE BAPTIST parish, SE LOUISIANA, on E bank (levee) of the MISSISSIPPI RIVER, and 25 mi/40 km WNW of NEW ORLEANS; 30°21′N 91°51′W. In sugarcane and farming (vegetables; cattle) area; fishing (catfish, crawfish, alligators); manufacturing (structural steel, synthetic rubber; printing and publishing). Lake PONTCHARTRAIN shore, 4 mi/6 km NE. Manchac State Wildlife Area to N. BONNET CARRE SPILLWAY to SE.

La Plaine, village, SE DOMINICA, BRITISH WEST INDIES, 10 mi/16 km ENE of ROSEAU; 15°20′N 61°15′W. Limes. Agricultural demonstration center.

Lapland, Finnish *Lappi*, Swedish *Lappland*, vast region of N EUROPE, largely within the ARCTIC CIRCLE. It includes the Norwegian provinces of FINNMARK and TROMS and part of NORDLAND; Swedish historic province of Lappland; LAPIN province, N FINLAND; and the KOLA PENINSULA of RUSSIA. Swedish Lappland is now included in NORRBOTTEN and VÄSTERBOTTEN counties.

Lapland is mountainous in N NORWAY and SWEDEN, reaching its highest point (6,965 ft/2,123 m) in KEBNEKAISE (Sweden), and consists largely of tundra in the NE. There are also extensive forests and many lakes and rivers. The climate is arctic and the vegetation is generally sparse, except in the forested S zone. Lapland is very rich in mineral resources, particularly in high-grade iron ore at GÄLLIVARE and KIRUNA (Sweden), in copper at SULITJELMA (Norway), and in nickel and apatite in Russia. KIRKENES and NARVIK (both in Norway) are the chief maritime outlets for Scandinavian Lapland, and MURMANSK is the port for Russian Lapland. The region abounds in sea and river fisheries and in aquatic and land fowl. Reindeer are essential to the economy; there is a growing tourist industry in the region.

The Lapps (or Laplanders), who constitute the indigenous population, number about 30,000 and are concentrated mainly in Norway (about 21,000), where they are called Samme or Finns (hence the Finnmark). They speak a Finno-Ugric language. Largely nomadic, many Lapps follow their reindeer herds, wintering in the lowlands and summering in the W mountains. Their movements today are more restricted than in former times. Other Lapps depend on sea and river fishing and hunting.

Little is known of their early history; it is believed that they came from central ASIA and were pushed to the N extremity of Europe by the migrations of the Finns, Goths, and Slavs. They may have assumed their Finnic language in the last millennium B.C.E. Though mainly conquered by Sweden and Norway in the Middle Ages, the Lapps resisted Christianization until the 18th century, when Russian and Scandinavian missionaries came to the region.

Laplandskiy Preserve (lah-PLAHND-skeeyee), wildlife refuge (□ 1,048 sq mi/2,724.8 sq km), western KOLA PENINSULA, MURMANSK oblast, extreme NW European Russia, N of ARCTIC CIRCLE. Tayga and tundra; mountainous with many valleys and lakes (Chunozero is the largest, at 514 acres/208 ha). Fauna include wild reindeer, bear, beaver, and fox. Established in 1930.

La Plata, county (□ 1,699 sq mi/4,417.4 sq km; 2006 population 47,936), SW COLORADO; ⊙ DURANGO; 37°17′N 107°50′W. Cattle, sheep, grazing area, bordering on NEW MEXICO (S); bounded W by LA PLATA MOUNTAINS; drained by ANIMAS, Florida (forms LEMON RESERVOIR) and Los Pinas (forms VALLECITO RESERVOIR in E) rivers. Gold, silver, lead, and coal mines near Durango. Includes part of San Juan National Forest in much of northern half. Part of Southern Ute Indian Reservation (S); Purgatory Ski Area (N). Formed 1874.

La Plata (lah PLAH-tah), city, ⊙ of BUENOS AIRES province, E central ARGENTINA, 5 mi/8 km inland from ENSENADA, its port on the RÍO DE LA PLATA; 34°55′S 57°56′W. La Plata's chief function is that of provincial capital, but industrial growth has been steady, and meat packing and large quantities of processed food, chemicals, steel, and petroleum are produced. Although the proximity of BUENOS AIRES has to some extent checked its development, La Plata is also a major cultural center, with fine museum and colleges and a national university. The national naval academy is located nearby (Ensenada). The city was founded in 1882, after Buenos Aires was federalized as the national capital. During the dictatorship of Juan Perón (1946–1955) both city and province were renamed Eva Perón, in honor of his wife. The name La Plata was restored when Perón's regime was overthrown (1955).

La Plata (lah-PLAI-tuh), city (2000 population 1,486), MACON county, N central MISSOURI, 20 mi/32 km N of MACON; 40°01′N 92°29′W. Railroad junction. Agriculture (corn, wheat, soybeans; cattle, hogs); manufacturing (wood cabinets). Laid out 1855.

La Plata (LAH PLAH-tah), town, ⊙ La Plata municipio, HUILA department, S central COLOMBIA, on affluent of upper MAGDALENA River, in E foothills of Cordillera CENTRAL, and 55 mi/89 km SW of NEIVA; 02°23′N 75°53′W. Elevation 4,324 ft/1,317 m. Cacao, coffee, rice, plantains; livestock. Silver mines, worked since pre-Spanish days. Waterfall nearby. Founded 1551.

La Plata (luh PLAH-tah), town (2000 population 6,551), ⊙ CHARLES county (since 1895), S MARYLAND, 25 mi/40 km S of WASHINGTON, D.C.; 38°32′N 76°58′W. Tobacco market; large warehouses. Arose in 1872 as a result of the coming of the railroad, and replaced Port Tobacco as county seat in 1895. U.S. Army Radio Receiving Station on 500 acres/202 ha is (1943) 2 mi/3.2 km NE.

La Plata, Isla de (lah PLAH-tah, EES-lah dai), island, off PACIFIC coast of MANABÍ province, W ECUADOR, 20 mi/32 km SW of Cape SAN LORENZO; 01°20′S 81°05′W.

La Plata Mountains, SW COLORADO and NW NEW MEXICO, spur of SAN JUAN MOUNTAINS extending N-S between LA PLATA and ANIMAS rivers. Highest point is Hesperus Peak (13,232 ft/4,033 m). Gold, silver, coal mined.

La Plata Peak (14,361 ft/4,377 m), CHAFFEE county, central COLORADO, in SAWATCH MOUNTAINS, 18 mi/29 km SW of LEADVILLE, in San Isabel National Forest.

La Plata River, c.70 mi/113 km long, in SW COLORADO and NW NEW MEXICO; rises in LA PLATA MOUNTAINS, W LA PLATA county, Colorado; flows S to SAN JUAN RIVER just W of FARMINGTON, New Mexico.

La Plata River, ARGENTINA and URUGUAY: see PLATA, RÍO DE LA.

La Playa (LAH PLEI-yah), town, ⊙ La Playa municipio, NORTE DE SANTANDER department, N COLOMBIA, in the ANDES, 50 mi/80 km NW of CÚCUTA; 08°12′N 73°14′W. Elevation 5,164 ft/1,573 m. Coffee, corn; livestock.

Lapleau (lah-PLO), commune (□ 6 sq mi/15.6 sq km), CORRÈZE department, LIMOUSIN administrative region, S central FRANCE, 9 mi/14.5 km WNW of MAURIAC, in mountainous Auvergne region; 45°30′N

02°08′E. Agriculture. Dam (295 ft/90 m high, 900 ft/274 m long) and hydroelectric plant of L'Aigle are 4 mi/6.4 km SE, on the DORDOGNE.

La Plume, township, LACKAWANNA county, NE PENNSYLVANIA, 12 mi/19 km NNW of SCRANTON; 41°33′N 75°45′W. Agriculture includes livestock, vegetables; dairying. Seat of Keystone Junior College; Lackawanna State Park to NE.

La Pocatière (lah po-kah-TYER), city (□ 9 sq mi/23.4 sq km; 2006 population 4,508), BAS-SAINT-LAURENT region, S QUEBEC, E CANADA, 16 mi/26 km from KAMOURASKA; 47°22′N 70°02′W. Manufacturing of railway cars.

La Pointe, village, SW Madeline Island, in APOSTLE ISLANDS, ASHLAND county, N WISCONSIN, on narrow channel of LAKE SUPERIOR, on W side of entrance to CHEQUAMEGON BAY; 46°55′N 90°38′W. Fishing; mustard, coffee. A French fortified trading post was built here in 1693, evacuated in 1698, and reoccupied 1718–1759. In early-19th century, site of an American Fur Company post. Ferry from BAYFIELD; tourism and historical museum. APOSTLE ISLANDS NATIONAL LAKESHORE on adjacent island; Big Bay State Park to E.

La Poma (lah PO-mah), village, ⊙ La Poma department (□ 1,770 sq mi/4,602 sq km), and central SALTA province, ARGENTINA, on CALCHAQUÍ RIVER, and 50 mi/80 km WNW of SALTA; 24°43′S 66°13′W. In grain and sheep area.

La Porte, county (□ 613 sq mi/1,593.8 sq km; 2000 population 110,106), NW INDIANA, bounded NW by Lake MICHIGAN, N by MICHIGAN state line, partly S by KANKAKEE RIVER; ⊙ LA PORTE; 41°33′N 86°44′W. Resorts on Lake Michigan. Manufacturing, especially at MICHIGAN CITY and La Porte. Fruit, corn, soybeans; hogs, cattle; lake shipping; fisheries; timber. Natural lakes, glacially formed. Kingsbury State Fish and Wildlife Area in SE, part of Kankakee State Fish and Wildlife Area in S. Formed 1832.

La Porte, city (2000 population 21,621), ⊙ LA PORTE county, NW INDIANA; 41°37′N 86°43′W. It is a manufacturing center in a fertile farmland on the edge of the CALUMET industrial region. Manufacturing (machinery, plastic containers, agriculture, paper, wood, and rubber products, furniture, chemicals, metals, transportation equipment, fabricated metal products, and baked goods). Pine Lake is on the NW edge of the city. Laid out 1833; incorporated 1835.

La Porte (luh PORT), city (2006 population 33,886), HARRIS county, SE TEXAS, suburb 18 mi/29 km ESE of downtown HOUSTON on GALVESTON BAY near HOUSTON SHIP CHANNEL (SAN JACINTO RIVER); 29°40′N 95°02′W. Oil wells, refineries; manufacturing (chemicals, plastics, agriculture products, marine cargo containers). La Porte Municipal Airport in city. Bridge and tunnel to BAYTOWN across ship channel; Sylvan Beach County Park; San Jacinto Battleground and Battleship Texas State Historic Site to N. Settled 1889, incorporated 1892.

Laporte, village, LARIMER county, N COLORADO, E of FRONT RANGE, 4 mi/6.4 km NW of FORT COLLINS, on CACHE LA POUDRE RIVER. Elevation 5,061 ft/1,543 m. Supply point; manufacturing (cement). Lory State Park and HORSETOOTH RESERVOIR to SW.

Laporte, village (2000 population 145), HUBBARD county, N central MINNESOTA, 19 mi/31 km SSE of BEMIDJI; 47°12′N 94°45′W. Manufacturing (stained glass). LEECH LAKE reservoir and Leech Lake Indian Reservation to E, Paul Bunyan State Forest to SW; several small natural lakes in area, including Garfield Lake to NE.

Laporte (LA-port), borough (2006 population 271), ⊙ SULLIVAN county, NE PENNSYLVANIA, 32 mi/51 km ENE of WILLIAMSPORT, on Lake Mokoma reservoir. Agriculture (hay, dairying); timber. Worlds End State Park to NW; Wyoming State Forest to W.

La Porte City, town (2000 population 2,275), BLACK HAWK county, E central IOWA, on WOLF CREEK near its mouth on CEDAR RIVER, and 15 mi/24 km SSE of WATERLOO; 42°18′N 92°11′W. Manufacturing (meat processing, dairy products, feed, concrete blocks); limestone quarry nearby. Incorporated 1871.

Láposbánya, ROMANIA: see BAIȚA.

Lapoutroie (lah-poo-trwah), summer commune (□ 8 sq mi/20.8 sq km), HAUT-RHIN department, in AL-SACE, E FRANCE, 10 mi/16 km NW of COLMAR, on road to Col du BONHOMME (W), near crest of the VOSGES MOUNTAINS; 48°09′N 07°10′E. Cheese manufacturing. Formerly spelled La Poutroye.

Lapovo (LAH-po-vo), village (2002 population 8,228), central SERBIA, 15 mi/24 km NE of KRAGUJEVAC, near the MORAVA RIVER; 44°11′N 21°06′E. Railroad junction.

Lappa (LAH-PAH), town, S GUANGDONG province, CHINA, port on LAPPA island, just W of MACAO (across channel of PEARL RIVER delta); commercial center. Opened to foreign trade in 1871.

Lappeenranta (LAHP-pain-RAHN-tah), Swedish *Villmanstrand*, city, KYMEN province, SE FINLAND; 61°04′N 28°11′E. Elevation 330 ft/100 m. On LAKE SAIMAA at N end of SAIMAA CANAL, near Russian border. An important trade and industrial center, with chemical works, lumber mills, and cement factories. The city was chartered in 1649 and became an important border fortress after the Treaty of Nystad (1721); was ceded to RUSSIA 1743–1811. Site of oldest Orthodox church in Finland (1785). Distribution point for VYBORG-FINLAND natural-gas pipeline. Spa (established 1824). Airport.

Lappersdorf (LAHP-puhrs-dorf), suburb of REGENS-BURG, BAVARIA, S GERMANY, in UPPER PALATINATE, just N of city; 49°03′N 12°05′E.

Lappo, FINLAND: see LAPUA.

Laprairie (lah-PRE-ree), former county (□ 170 sq mi/442 sq km), S QUEBEC, E CANADA, on the SAINT LAWRENCE RIVER, just S of MONTREAL; the county seat was LAPRAIRIE; 45°20′N 73°35′W. Also written La Prairie.

La Prairie (lah pre-REE), city (□ 17 sq mi/44.2 sq km; 2006 population 20,901), MONTÉRÉGIE region, S QUEBEC, E CANADA, 10 mi/16 km from MONTREAL; 45°25′N 73°30′W. Part of the Metropolitan Community of Montreal (*Communauté Métropolitaine de Montréal*).

Laprairie (lah-PRE-ree), town (□ 17 sq mi/44.2 sq km; 2006 population 20,901), MONTÉRÉGIE region, S QUEBEC, E CANADA, on the SAINT LAWRENCE RIVER, near E end of the LACHINE RAPIDS, opposite MON-TREAL; 45°25′N 73°29′W. Food processing, manufacturing (metalworking machinery, electrical components, baked goods). Suburb of Montreal. Settled 1673, was site of fort built by Frontenac; attacked (1691) by New Englanders under Peter Schuyler. The first railroad in Canada was built (1836) between here and Saint-Jean. Was seat of historic LAPRAIRIE county. Also written La Prairie.

La Prairie, town (2000 population 60), ADAMS county, W ILLINOIS, 26 mi/42 km ENE of QUINCY; 40°08′N 91°00′W. In agricultural area.

La Prairie, village (2000 population 605), ITASCA county, N central MINNESOTA, 2 mi/3.2 km E of GRAND RAPIDS on MISSISSIPPI RIVER, at mouth of Prairie River; 47°13′N 93°29′W. Agriculture (alfalfa; cattle; dairying); timber. Golden Anniversary State Forest to SE.

La Prenza, unincorporated town, SAN DIEGO county, S CALIFORNIA; residential suburb 8 mi/12.9 km E of downtown SAN DIEGO. Sweetwater Reservoir to S.

La Présentation (lah prai-zahn-tah-SYON), parish (□ 41 sq mi/106.6 sq km; 2006 population 1,963), MONTÉRÉGIE region, S QUEBEC, E CANADA, 17 mi/27 km from VERCHÈRES; 45°40′N 73°03′W.

Laprida (lah-PREE-dah), town (1991 population 7,341), ⊙ Laprida district (□ 1,334 sq mi/3,468.4 sq km), S

central BUENOS AIRES province, ARGENTINA, 50 mi/80 km SSW of OLAVARRÍA; 37°30′S 60°45′W. Agricultural center (oats, wheat; sheep; cattle); flour milling.

Laprida, ARGENTINA: see LA HUERTA.

La Protección, HONDURAS: see PROTECCIÓN.

La Providencia (lah pro-vee-DAIN-see-ah), village, central BUENOS AIRES province, ARGENTINA, 9 mi/14.5 km SSE of OLAVARRÍA. Railhead; limestone-quarrying center.

La Pryor, (luh PRAI-yor), town (2000 population 1,491), ZAVALA county, SW TEXAS, 18 mi/29 km N of CRYSTAL CITY, near NUECES RIVER; 28°56′N 99°50′W. Irrigated area (spinach, vegetables, corn, pecans; cotton).

Lapsaki, TURKEY: see LAPSEKI.

Lapseki, Turkish *Lâpseki*, ancient *Lampsacus*, village, NW TURKEY in ASIA, on E shore of DARDANELLES, opposite GELIBOLU, 20 mi/32 km NE of ÇANAKKALE; 40°22′N 26°42′E. Lignite deposits; cereals, lentils. Formerly also Lapsaki.

Lapstone (LAP-ston), town, NEW SOUTH WALES, SE AUSTRALIA, 37 mi/60 km from SYDNEY, on edge of BLUE MOUNTAINS. Walks, lookouts. Named for the area's stones worn by water.

Lapta, CYPRUS: see LAPITHOS.

Laptevo, RUSSIA: see YASNOGORSK.

Laptev Sea (LAHP-tyef), section (□ 250,000 sq mi/650,000 sq km) of the ARCTIC OCEAN, N Siberian Russia, between the TAYMYR PENINSULA and the NEW SIBERIAN ISLANDS. It is a shallow sea and is frozen for most of the year. The LENA RIVER empties into it through an extensive delta; the sea also receives the Khatanga and Yana rivers. The LAPTEV Sea, part of the NORTHERN SEA ROUTE, is navigable only during August and September; TIKSI and NORDVIK are the chief ports. Formerly called the Nordenskjöld (or Nor-denskiöld) Sea for the Swedish explorer Nils Adolf Nordenskjöld, it was renamed in honor of Khariton and Dmitri Laptev, two Russian arctic explorers of the second expedition organized and commanded by Vitus Bering.

Laptev Strait, RUSSIA: see DMITRIY LAPTEV STRAIT.

Lapua (LAH-poo-ah), Swedish *Lappo*, village, VAASAN province, W FINLAND, 45 mi/72 km E of VAASA; 62°57′N 23°00′E. On 100 mi/161 km long Lapuanjoki (river); metalworking, lumbering. Scene of 1808 Finnish victory over RUSSIA is nearby.

Lapuente, village, RIVERA department, NE URUGUAY, on the ARROYO YAGUARÍ, near BRAZIL border, 55 mi/89 km SE of RIVERA; 31°31′S 54°58′W. Grain, vegetables; cattle, sheep.

La Puente, city (2000 population 41,063), LOS ANGELES county, S CALIFORNIA; growing suburb 16 mi/26 km E of downtown LOS ANGELES; 34°02′N 117°57′W. Primarily residential, it has some diverse light manufacturing (hardware, electronics, paper products, metal plating, scew-machine products). Laid out 1841, incorporated 1956.

La Puerta (LAH PWER-tah), town, TRUJILLO state, W VENEZUELA, in ANDEAN spur, on highway, 16 mi/26 km SSW of VALERA; 09°07′N 70°44′W. Elevation 5,768 ft/1,758 m. Wheat, corn, potatoes.

La Puerta (lah PWER-tah), village (1991 population 558), ⊙ Ambato department (□ 933 sq mi/2,425.8 sq km), SE CATAMARCA province, ARGENTINA, on the RÍO DEL VALLE, and 2 mi/3.2 km N of CATAMARCA. Mixed farming; forest industry.

La Puerta, village, N CÓRDOBA province, ARGENTINA, 75 mi/121 km NE of CÓRDOBA. Wheat, flax, corn; livestock raising; dairying.

Lapu-Lapu (LAH-poo–LAH-poo), city (□ 22 sq mi/57.2 sq km; 2000 population 217,019) of MACTAN Island, CEBU province, PHILIPPINES, on W coast of island, 3 mi/4.8 km E of CEBU across narrow channel; 10°19′N 123°E. Coconut-growing and fishing center. Nearby is a monument dedicated to Magellan, killed here in 1521. Formerly called Opon.

La Punta (lah POON-tah), village, SW SANTIAGO DEL ESTERO province, ARGENTINA, at S foot of SIERRA DE GUASAYÁN, on railroad, and 50 mi/80 km SW of SANTIAGO DEL ESTERO; 28°22′S 64°47′W. Livestock-raising and lumbering center; gypsum works.

La Punta (LAH POON-tah), W suburb of CALLAO, Callao province, Callao region, W central PERU, situated on spit of land stretching out into the PACIFIC, 2 mi/3 km W of Callao (linked by electric ferries); 12°05′S 77°11′W. Bathing resort.

La Punta, PERU: see PUNTA DE BOMBÓN.

La Purísima (lah poo-REE-see-mah), settlement, BAJA CALIFORNIA SUR, COMONDÚ municipio, NW MEXICO, in valley, 180 mi/290 km NW of LA PAZ; 26°10′N 112°05′W. Some agriculture (grain, livestock).

La Purísima Concepción, Mission, California: see LOMPOC.

La Push, unincorporated village, CLALLAM county, NW WASHINGTON, on Pacific coast, at mouth of QUIL-LAYUTE RIVER, and 25 mi/40 km S of Cape Flattery. Headquarters of Quillayute (Quilleute) Indian Reservation; located in reservation. Fish; timber. Flattery Rocks National Wildlife Refuge to N; Quillayute National Wildlife Refuge to S, both offshore. Reservation bounded N, E, and S by coastal unit of OLYMPIC NATIONAL PARK, including Lake Ozette to N. Quillayute State Airport to NE.

Lapushna, MOLDOVA: see LAPUȘNA.

Lapușna (lah-POOSH-nah), village, W MOLDOVA, 22 mi/35 km WSW of CHISINAU (Kishinev); 46°53′N 28°24′E. Wine, corn. Formerly spelled Lapushna.

Lapwai, town (2000 population 1,134), NEZ PERCE county, W IDAHO, 10 mi/16 km E of LEWISTON, in W part of Nez Perce Indian Reservation; 46°24′N 116°48′W. Trade center; logging; cattle; alfalfa, wheat, barley; potatoes, vegetables; manufacturing (agriculture lime, machinery, logs) and agency headquarters for Coeur d'Alene and Nez Perce Indian reservations in Idaho, and for Kalispel Indian Reservation in WA-SHINGTON. Spalding Mission (1836) to N, now Spalding Area Unit of NEZ PERCE NATIONAL HIS-TORICAL PARK.

Lapy (LAH-pee), Polish *Łapy*, town, BIAŁYSTOK province, NE POLAND, on Narew River, 15 mi/24 km SW of BIAŁYSTOK; 52°59′N 22°52′E. Railroad junction (repair shops), food processing (sugar).

La Quebrada (LAH kai-BRAH-dah), town, ⊙ Urdaneta municipio, TRUJILLO state, W VENEZUELA, in ANDEAN spur, 10 mi/16 km SSE of VALERA; 09°09′N 70°34′W. Elevation 4,659 ft/1,420 m. Wheat, corn, potatoes.

La Quemada, MEXICO: see QUEMADA.

La Quiaca (lah kee-AH-sah), town (1991 population 11,576), ⊙ Yaví department (□ 1,330 sq mi/3,458 sq km), N JUJUY province, ARGENTINA, on BOLIVIA border opposite VILLAZÓN, 155 mi/249 km NNW of JUJUY; 22°06′S 65°37′W. Elevation 11,300 ft/3,444 m. Customhouse. Trade and mining center, with oil and silver deposits. Meteorological station.

L'Aquila (LAH-kwee-lah), city (2001 population 68,503), ⊙ L'AQUILA province and of ABRUZZI, central ITALY, on the PESCARA RIVER; 42°22′N 13°22′E. It is an agricultural and industrial center, and a summer resort. A motorway that connects it with ROME has been extended to the ADRIATIC coast. Strong industrial growth 1970–1994, with highly diversified consumer industries. Manufacturing includes building materials, construction equipment, machinery, and electronic equipment. L'Aquila is situated at the foot of the GRAN SASSO D'ITALIA mountain group and is a popular base for mountain climbing. Built around a castle (13th–16th century), it rose to importance in the 13th century and later became the second city of the kingdom of NAPLES. However, the city's influence declined during the wars of the 16th century. Despite several devastating earthquakes L'Aquila has retained its medieval fortifications and a number of impressive old buildings, including Saint Bernardino's

Basilica (15th–16th century). There is a university in the city.

Laquinamaya (lah-kee-nah-MAH-yah), canton, INGAVI province, LA PAZ department, W BOLIVIA, 22 mi/35 km W of SAN ANDRES, on the Peruvian border, 12 mi/20 km NW of SANTO DOMINGO DE MACHACA and the nearest road; 16°55′S 69°25′W. Elevation 12,641 ft/3,853 m. Abundant gas wells in area. Lead-bearing lode; limestone and gypsum deposits. Agriculture (potatoes, yucca, bananas, barley, rye); cattle.

La Quinta, city (2000 population 23,694), RIVERSIDE county, S CALIFORNIA; suburb 17 mi/27 km SE of PALM SPRINGS, 5 mi/8 km SW of INDIO, in N foothills of SANTA ROSA MOUNTAINS; 33°40′N 116°17′W. San Bernardino National Forest to SW. Rapidly growing urban fringe in irrigated agricultural area (citrus, dates, cotton, grain; nursery livestock, poultry).

Lar, town, DEORIA district, E UTTAR PRADESH state, N central INDIA, 50 mi/80 km SSE of GORAKHPUR; 26°13′N 83°58′E. Trades in rice, wheat, barley, oilseeds, sugarcane.

Lar (LAHR), town (2006 population 54,688), FĀRS province, S IRAN, 170 mi/274 km SE of SHIRAZ on Shiraz–Bandar Abbas road; 27°40′N 54°20′E. Main town of LARISTAN. Tobacco, cotton, dates, mustard seed. Airport.

Lara (LAH-rah), inland state (☐ 7,640 sq mi/19,864 sq km; 2001 population 1,556,415), N VENEZUELA; ⊙ BARQUISIMETO; 10°10′N 69°50′W. Mountainous region, containing NE section of great ANDEAN spur (S), outliers of CORDILLERA DE LA COSTA (E), and arid SEGOVIA HIGHLANDS. Drained by TOCUYO RIVER. Climate dry and tropical, with rains generally in fall. Predominantly agricultural region, producing coffee, cacao, sugarcane, corn, cotton, tobacco, vegetables, bananas, sisal; wheat, barley, potatoes at higher elevations. Cattle grazing on well-watered plains around CARORA, goat grazing in arid uplands. Other principal cities include CARORA, EL TOCUYO, QUÍBOR, CABUDARE, DUACA. Includes PARQUES NACIONALES YACAMBÚ, TEREPAIMA, and CERRO SAROCHE, as well as part of DINIRA.

Lara (LAH-ruh), town, VICTORIA, SE AUSTRALIA, N of GEELONG; 38°02′S 144°25′E. Wetlands, reserves in area. Heavily damaged by 1969 bushfires.

Larache (la-RAHSH), city, Larache province, Tanger-Tétouan administrative region, N MOROCCO, on road to TANGER, on the ATLANTIC OCEAN; 35°11′N 06°09′W. Vegetables, cork, wool, and timber are exported. Fishing. The Phoenicians founded a trading post across the river in 1100 B.C.E., which was later captured by the Romans and called Lixus. Today, remains of both civilizations, including a Roman amphitheater, can be found. SPAIN held the city twice (1610–1691 and 1911–1956).

Laragne-Montéglin (lah-rah-nyuh–mon-tai-glan), commune (☐ 9 sq mi/23.4 sq km), HAUTES-ALPES department, PROVENCE-ALPES-CÔTE D'AZUR region, SE FRANCE, on the BUËCH, 10 mi/16 km NW of SISTERON; 44°19′N 05°49′E. Brickworks; flour milling, lavender essence distilling; orchards.

Larak (lah-RAHK), island, Hormozgān province, SE IRAN, in Strait of HORMUZ, 20 mi/32 km S of BANDAR Abbas, off NE tip of Qishm island; 26°51′N 56°21′E. Island is 7 mi/11.3 km long, 5 mi/8 km wide. Fishing, pearling. Also spelled LARK.

La Ramada, town and canton, TAPACARI province, COCHABAMBA department, central BOLIVIA, on S slopes of Cordillera de COCHABAMBA, on road, and 8 mi/12.9 km E of TAPACARÍ; 17°34′S 66°28′W. Barley; livestock.

La Ramada (lah rah-MAH-dah), village, NE TUCUMÁN province, ARGENTINA, on railroad, and 20 mi/32 km NE of TUCUMÁN. Lumbering; livestock.

Laramie (LA-ruh-mee), county (☐ 2,687 sq mi/6,986.2 sq km; 2006 population 85,384), SE WYOMING; ⊙CHEYENNE; 41°19′N 104°41′W. Agricultural area bordering on COLORADO (S) and NEBRASKA (E); watered by CHUGWATER, HORSE, North Bear, Little Bear, and LODGEPOLE creeks. Agriculture (wheat, sugar beets, hay, alfalfa, corn, beans, oats; cattle, sheep); dairying; timber. Foothills of LARAMIE MOUNTAINS in W; Curt Gowdy State Park in SW. Formed 1867.

Laramie (LA-ruh-mee), city (2006 population 25,688), ⊙ ALBANY county, SE WYOMING, on the LARAMIE RIVER, 40 mi/64 km WNW of CHEYENNE, 20 mi/32 km N of COLORADO state line; 41°18′N 105°34′W. Elevation 7,165 ft/2,184 m. It is a commercial, railroad, and industrial center for a livestock and timber region. Laramie has railroad yards, sawmills, a cement factory, and meat storage facilities. Manufacturing also includes beverages; cabinetry; printing and publishing; and computer software. Tourism is an important economic activity; the city is surrounded by mountain ranges and many nearby ski, hunting, and fishing areas. The city is the seat of the University of Wyoming. Fairgrounds to S; airport to W. Laramie was settled in 1868 with the arrival of the railroad and grew with the development of the surrounding ranch country and local mining enterprises. Historic Territorial Prison at Wyoming Territorial State Park, in W part of city. It is headquarters for the Medicine Bow National Forest. Nearby is the site of Fort Sanders, established in 1866 to protect the OVERLAND TRAIL and workers on the Union Pacific railroad. LARAMIE MOUNTAINS to E; part of Medicine Bow National Forest to E, includes Vedauwoo Recreation Area to SE, rock formations; Curt Gowdy State Park also to SE. Hutton Lake National Wildlife Refuge to SW. Incorporated 1874.

Laramie Basin, WYOMING: see LARAMIE PLAINS.

Laramie, Fort, WYOMING: see FORT LARAMIE.

Laramie Mountains, range of the ROCKY MOUNTAINS, N extension of FRONT RANGE in COLORADO, c.130 mi/209 km long and c.20 mi/32 km wide; mainly in SE WYOMING, from N of FORT COLLINS, Colorado, to CASPER and North Platte River (N). LARAMIE PLAINS are to W. Highest point is LARAMIE PEAK (10,274 ft/3,132 m), on boundary between CONVERSE county and ALBANY county, Wyoming. N part of range in Medicine Bow National Forest.

Laramie Peak (LA-ruh-mee) (10,272 ft/3,131 m), highest point in LARAMIE MOUNTAINS, ALBANY and CONVERSE counties, SE WYOMING, 65 mi/105 km N of LARAMIE, in Medicine Bow National Forest.

Laramie Plains (LA-ruh-mee) or **Laramie Basin** (7,000 ft/2,134 m–8,000 ft/2,438 m), ALBANY and CARBON counties, SE WYOMING; high basin drained by MEDICINE BOW and LARAMIE rivers; bounded E and NE by LARAMIE MOUNTAINS, W by MEDICINE BOW MOUNTAINS; cattle-grazing area. City of LARAMIE is on E edge of plains.

Laramie River, 216 mi/348 km long, in N COLORADO and SE WYOMING; rises in FRONT RANGE, W larimer county, N Colorado, near CONTINENTAL DIVIDE; flows N and NE, past LARAMIE, Wyoming, through the LARAMIE PLAINS and Wheatland Reservoir, through LARAMIE MOUNTAINS and Grayrocks Reservoir, to North Platte River opposite FORT LARAMIE. Supplies water to CACHE LA POUDRE RIVER in N Colorado through Laramie-Poudre, or Greeley-Poudre, Tunnel (c.2 mi/3.2 km long; finished 1911), unit in irrigation project.

Laranda, TURKEY: see KARAMAN.

Laranja da Terra (LAH-rahn-zhah dah TE-rah), city (2007 population 10,740), E central ESPÍRITO SANTO, BRAZIL, 27 mi/43 km NNE of AFONSO CLAUDIO; 19°50′S 41°08′W.

Laranjal (LAH-rahn-zhahl), town, W PARANÁ state, BRAZIL, 64 mi/103 km E of CASCAVEL; 24°52′S 52°37′W.

Laranjal do Jari (LAH-rahn-ZAHL do zhah-REE), city (2007 population 35,680), AMAPÁ, BRAZIL.

Laranjal Paulista (LAH-rahn-zhal POU-lee-stah), city (2007 population 24,454), S central SÃO PAULO, BRA-ZIL, on railroad, 21 mi/34 km SSW of PIRACICABA; 23°04′S 47°51′W. Manufacturing of pottery and agricultural implements, distilling; agriculture (coffee, sugar, fruit). Until 1944, called Laranjal.

Laranjeiras (LAH-rahn-zhai-rahs), city (2007 population 23,923), E SERGIPE, NE BRAZIL, on Cotingüiba River, on railroad, 12 mi/19 km NW of ARACAJU; 10°48′S 37°10′W. Sugar-growing and -milling center. Ships sugar, coffee, cotton, salt.

Laranjeiras, Brazil: see ALAGOA NOVA.

Laranjeiras do Sul (LAH-rahn-zhai-rahs do sool), city (2007 population 30,466), SW central PARANÁ, BRA-ZIL, 50 mi/80 km W of GUARAPUAVA; 25°25′S 52°25′W. Maté collecting, cattle raising. Until 1944, called Laranjeiras; and 1944–1948, called Iguaçu. Was capital of Iguaçu territory (1943–1946).

Larantoeka, INDONESIA: see FLORES.

Larantuka, INDONESIA: see FLORES.

Laraos (lah-RAH-os), town, HUAROCHIRÍ province, LIMA region, W central PERU, in CORDILLERA OCCIDENTAL, 12 mi/19 km ENE of YAUYOS; 11°40′S 76°34′W. Potatoes, cereals; livestock.

Larap (LAH-rahp), town, in JOSE PAÑGANIBAN municipality, CAMARINES NORTE province, SE LUZON, PHILIPPINES, just W of Jose Pañganiban, 55 mi/89 km NW of NAGA, at base of small peninsula in PHILIPPINE SEA; iron-mining center (with mine shaft and open pit).

Laraquete (lah-rah-KAI-tai), village, CONCEPCIÓN province, BÍO-BÍO region, S central CHILE, minor port on PACIFIC coast, on railroad, 26 mi/42 km SSW of CONCEPCIÓN; 37°10′S 73°11′W. Swimming and fishing resort.

Larat, INDONESIA: see TANIMBAR ISLANDS.

Larbba Naït Irathen (lahr-BAH NAH-tee-rah-TEN), town, TIZI OUZOU wilaya, N coastal ALGERIA; 36°33′N 03°12′E. Craft center. Was first set up as a French fort to pacify Greater KABYLIA. Formerly called Fort National.

Larbert (LAHR-buhrt), town (2001 population 6,425), FALKIRK, central Scotland, 3 mi/4.8 km NW of Falkirk; 56°01′N 03°50′W. Previous industries included iron founding and manufacturing of stoves, cookers, and castings.

Larchmont, upper-income suburban residential village (☐ 1 sq mi/2.6 sq km; 2006 population 6,530), part of town of MAMARONECK, WESTCHESTER county, SE NEW YORK, on harbor on LONG ISLAND SOUND, between NEW ROCHELLE (SW) and Mamaroneck; 40°55′N 73°45′W. A few small light industries. Yachting center (annual regattas). A number of prominent Americans have lived here. Developed c.1845, incorporated 1891.

Larch River (LAHRCH), 270 mi/435 km long, N QUEBEC, E CANADA; issues from Lower Seal Lakes, just N of Clearwater Lake; flows NE to confluence with CANIAPISCAU RIVER, 50 mi/80 km SW of KUUJJUAQ, here forming KOKSOAK RIVER, which flows NE to UNGAVA BAY. Numerous rapids.

Larchwood, town (2000 population 788), LYON county, extreme NW IOWA, near MINNESOTA state line, 13 mi/21 km W of ROCK RAPIDS; 43°27′N 96°26′W. State park and site of large Indian village are nearby.

Lårdal (LAWR-dahl), village, TELEMARK county, S NORWAY, on BANDAK lake, 55 mi/89 km WNW of SKIEN. Cattle breeding. Hotels.

Larderello (lahr-de-REL-lo), village, PISA province, TUSCANY, central ITALY, on small affluent of CECINA RIVER, 12 mi/19 km S of VOLTERRA; 43°14′N 10°53′E. Has *Soffioni* once used to produce boric acid and now only electricity. Factories, badly damaged in World War II, have been repaired. Named for Francesco de Larderel; who initiated extraction of boric acid from thermal mists.

Larder Lake (LAHR-duhr), township (☐ 88 sq mi/228.8 sq km; 2001 population 790), E ONTARIO, E central CANADA, on Lander Lake, 45 mi/72 km N of HAI-

LEYBURY, and c.9 mi/15 km from QUEBEC border; 48°07′N 79°41′W. Gold mining, forestry. Incorporated 1938.

Lardero (lahr-DHAI-ro), town, LA RIOJA province, N SPAIN, 3 mi/4.8 km S of LOGROÑO; 42°26′N 02°28′W. Fruit, olive oil, wine, vegetables; cattle, sheep.

Lardier, Cape (lahr-DYAI), VAR department, PROVENCE-ALPES-CÔTE D'AZUR region, SE FRANCE, on the MEDITERRANEAN, formed by a rugged spur of the MAURES massif, 8 mi/12.9 km S of SAINT-TROPEZ; 43°10′N 06°37′E. It forms the S tip of the Saint-Tropez peninsula.

Lardin-Saint-Lazare, Le (lahr-dan-san-lah-ZAHR, luh), commune (□ 4 sq mi/10.4 sq km), DORDOGNE department, AQUITAINE region, SW FRANCE, on VÉZÈRE RIVER, 15 mi/24 km W of BRIVE-LA-GAILLARDE. Paper mill.

L'Ardoise (lahrd-WAHZ), village, E NOVA SCOTIA, CANADA, on S coast of CAPE BRETON ISLAND 30 mi/48 km E of PORT HAWKESBURY. In mid-18th century it was important center of Acadian fishing and slate quarrying region, with active fur trade. More recently tourism has become its primary business.

Lardosa (lahr-DO-zah), village, CASTELO BRANCO district, central PORTUGAL, on railroad, 12 mi/19 km NNE of CASTELO BRANCO. Grain, corn, olives, beans.

Lardy (lahr-DEE), commune (□ 2 sq mi/5.2 sq km), ESSONNE department, ÎLE-DE-FRANCE region, N central FRANCE, on Juine River, 7 mi/11 km ENE of ÉTAMPES; 48°31′N 02°16′E. On railroad.

Lare (LAH-reh), town, Meru district, EASTERN province, KENYA, on NYAMBENI range 25 mi/40 km E of ISIOLO; 00°20′S 37°38′E. Market center; livestock; crafts.

Larecaja, province, LA PAZ department, W BOLIVIA, ⊙ SORATA; 15°30′S 68°20′W.

La Rédemption (lah rai-dahn-SYON), village (□ 45 sq mi/117 sq km; 2006 population 531), BAS-SAINT-LAURENT region, S QUEBEC, E CANADA, 5 mi/8 km from SAINT-MOÏSE; 48°27′N 67°53′W.

Laredo, city, CANTABRIA province, N SPAIN, on BAY OF BISCAY, 20 mi/32 km E of SANTANDER; 39°29′N 02°54′E. Fishing and ore-shipping port; fish processing (salmon, sardines, anchovies), boatbuilding, sawmilling; chemical works. Corn, potatoes, lemons, cattle in area. Bathing resort.

Laredo (luh-RAI-do), city (2000 population 250), GRUNDY county, N MISSOURI, 10 mi/16 km ESE of TRENTON; 40°01′N 93°27′W. Corn, wheat; cattle.

Laredo (luh-RAI-do), city (2006 population 215,484), ⊙ WEBB county, S TEXAS, 130 mi/209 km WSW of CORPUS CHRISTI, on the RIO GRANDE; 27°31′N 99°29′W. It is a port of entry on the U.S.-Mexico border, with a thriving export-import trade and a tourist industry. During the 1980s and early 1990s, Laredo became one of the fastest growing U.S. cities. It is a wholesale and retail center for a large area on both sides of the Rio Grande. Important to its economy are cattle ranching, irrigated farming (vegetables; cotton; sorghum), oil production, and mining and smelting. A wide variety of products are manufactured, such as clothing, military supplies, candles, fabricated steel products, and leather goods. Laredo has close economic ties with its large sister city in Mexican, NUEVO LAREDO ("New Laredo," 1990 population 350,000; founded 1775), with which it is linked by two international bridges. Laredo, a blend of Spanish, Mexican, and American frontier influences, grew as a post on the road to San Antonio and other Texas cities. After the Texas Revolution its ownership remained in doubt until the S boundary of Texas was established by the Mexican-American War; during that period the city was the capital of the "Republic of the Rio Grande" (the capitol building, erected in 1755, still stands). Laredo's growth was aided by the arrival of the railroad (1880s), the development of irrigated farming, the discovery of oil and natural gas, and the

opening (1936) of a highway to Mexico City. Laredo International Airport to E. The former army post Fort McIntosh was founded in 1849 and intermittently rebuilt and used until 1946. Texas A and M International University and Laredo Junior College are on grounds of fort. An extension center of the Texas Arts and Industries University is also there. Casa Blanca Lake and Lake Casa Blanca State Park to E. Founded 1755, incorporated 1852.

Laredo (lah-RAI-do), village (2005 population 23,954), LA LIBERTAD region, NW PERU, on irrigated coastal plain, 4 mi/6 km ENE of TRUJILLO; 08°06′S 78°57′W. Sugar, rice.

Laredo (luh-RAI-do), village, HILL county, N MONTANA, near Big Sandy Creek, 12 mi/19 km SW of HAVRE, at N edge of Rocky Boy's Indian Reservation. Grain, livestock region. Fort Assiniboine to NE, BEAVER CREEK Park and Reservoir to E.

La Reforma (lah re-FOR-mah), town, SAN MARCOS department, SW GUATEMALA, in Pacific piedmont, 10 mi/16 km S of SAN MARCOS; 14°48′N 91°49′W. Elevation 3,740 ft/1,140 m. Coffee, sugarcane, grain; livestock.

La Reforma (lah re-FOR-mah), town in far SW OAXACA, MEXICO, 24 mi/39 km NE of SANTIAGO PINOTEPA NACIONAL. A mountainous region; little fertile soil, hot climate. Agriculture (cereals, fruits); woods. No roads. Connected to Mexico Highway 125 by unpaved road (12 mi/20 km). Elevation 3,117 ft/950 m.

La Reforma, MEXICO: see REFORMA, town.

La Reid (lah RED), wildlife park near THEUX, LIÈGE province, E BELGIUM, 5 mi/8 km SSW of VERVIERS.

La Reina (lah RAI-nah), municipality and town, CHALATENANGO department, EL SALVADOR, WNW of CHALATENANGO city in W part of department.

La Reina, NICARAGUA: see SAN RAMÓN.

La Reine (lah REN), village (□ 39 sq mi/101.4 sq km), ABITIBI-TÉMISCAMINGUE region, W QUEBEC, E CANADA, on ONTARIO border, 50 mi/80 km NNW of ROUYN-NORANDA; 48°52′N 79°30′W. Gold, copper, zinc mining.

Laren (LAH-ruhn), town (2001 population 23,285), NORTH HOLLAND province, W central NETHERLANDS, 3 mi/4.8 km NE of HILVERSUM, 17 mi/27 km ESE of AMSTERDAM; 52°15′N 05°14′E. GOOIMEER 3 mi/4.8 km to N. Dairying; cattle, poultry; vegetables, fruit, nursery stock; manufacturing (carpeting).

Lares (LAH-res), town, CALCA province, CUSCO region, S central PERU, on upper URUBAMBA RIVER, 15 mi/24 km NNW of CUSCO; 13°08′S 72°10′W. In agricultural region (coca, cacao, cereals, potatoes, sugarcane); thermal spring. Mostly copper mines, and some silver mines nearby. Lares Pass (14,711 ft/4,484 m) to N.

Lares (LAH-res), town (2006 population 37,164), W PUERTO RICO, W of GUAJATACA RIVER, 15 mi/24 km SW of ARECIBO. Elevation 1,125 ft/343 m. Agriculture (coffee, oranges, plantains, bananas); varied light manufacturing. Copper deposits in area.

La Resolana, MEXICO: see CASIMIRO CASTILLO.

Lares Pass, PERU: see LARES.

Largeau, CHAD: see FAYA.

Largentière (lahr-zhahn-tyer), commune (□ 2 sq mi/5.2 sq km), sub-prefecture of ARDÈCHE department, RHÔNE-ALPES region, S FRANCE, 7 mi/11.3 km SW of AUBENAS, in foothills of the CÉVENNES Mountains; 44°32′N 04°18′E. Commercial center in vineyard area. Old center has medieval appearance, with 15th-century castle (now a hospital) named for the argent [French=silver] mined here in the 10th–15th centuries by the counts of TOULOUSE.

Largo (LAHR-go), city (□ 14 sq mi/36.4 sq km; 2000 population 69,371), PINELLAS county, W central FLORIDA, on the Pinellas peninsula and the Gulf coast, 15 mi/24 km W of TAMPA; 27°54′N 82°46′W. It is a packing, canning, and shipping center in a citrus fruit–growing and fishing area. Its beaches and many

recreational facilities make it a popular resort spot. Settled 1853, incorporated 1905.

Largo (LAHR-go) or **Kirkton of Largo**, village, FIFE, E Scotland, consists of adjacent Lower Largo and Upper Largo, on Largo Bay of the Firth of Forth, and 3 mi/4.8 km NE of LEVEN; 56°13′N 02°55′W. Golfing resort.

Largo, Cayo (LAHR-go, KEI-yo), narrow islet, largest key at E part of LOS CANARREOS Archipelago, within ISLA DE LA JUVENTUD Special Municipality, off S coast of CUBA, on Jardines Bank, 65 mi/105 km WSW of CIENFUEGOS; 21°41′N 81°30′W. Islet is 15 mi/24 km long NE-SW; its 6 hotels and 13,000-ft/3,962-m airplane runway cater to tourists. Also has 3 small fishing villages.

Largs (LAHRGZ), community, SW MANITOBA, W central CANADA, 4 mi/7 km from MINNEDOSA, in MINTO rural municipality; 50°16′N 99°55′W.

Largs (LAHRGZ), township, NEW SOUTH WALES, SE AUSTRALIA, 26 mi/42 km NW of NEWCASTLE, 2.8 mi/4.5 km N of MORPETH; 32°42′S 151°36′E.

Largs (LAHRGZ), town (2001 population 11,241), North Ayrshire, SW Scotland, on Firth of Clyde, 11 mi/18 km SSW of GREENOCK; 55°48′N 04°53′W. Port, seaside resort. In Firth of Clyde, 2 mi/3.2 km W, is GREAT CUMBRAE island. At battle of Largs (1263) Alexander III of Scotland defeated attempted invasion of Scotland by Haakon IV of Norway. Sir Thomas Brisbane born in nearby Brisbane House.

Lari (LAH-ree), town, PISA province, TUSCANY, central ITALY, 14 mi/23 km SE of PISA; 43°34′N 10°35′E. Highly diversified secondary manufacturing center; pasta manufacturing; marble quarrying.

Larimer, county (□ 2,633 sq mi/6,845.8 sq km; 2006 population 276,253), N COLORADO, ⊙ FORT COLLINS; 40°38′N 105°27′W. Irrigated agricultural area, bordering on WYOMING (N); drained by CACHE LA POUDRE and LARAMIE rivers. CONTINENTAL DIVIDE forms county boundary in SW. Limestone, timber; cattle, sheep; fruit, vegetables, wheat, hay, beans, barley, corn, sugar beets. Manufacturing at Fort Collins and LOVELAND. Feature of irrigation system is Laramie-Poudre (or Greeley-Poudre) Tunnel, connecting Laramie and Cache la Poudre rivers; used to irrigate 125,000 acres/50,588 ha in Larimer and WELD counties. Includes part of ROCKY MOUNTAIN NATIONAL PARK in SW and of FRONT RANGE in W; part of Roosevelt National Forest in W half of county; Picnic Rock, Lory, and Boyd Lake state parks in E; small part of Colorado State Forest on W boundary. Formed 1861.

Larimer (LER-i-muhr), unincorporated town, NORTH HUNTINGDON township, WESTMORELAND county, SW PENNSYLVANIA, residential suburb 15 mi/24 km SE of PITTSBURGH; 40°20′N 79°43′W.

Larimna (LAH-reem-nah), village, FTHIOTIDA prefecture, E central GREECE department, on Gulf of ÉVVIA, 16 mi/26 km N of THEBES; 38°34′N 23°17′E. Limonite and nickel mining. In ancient times, a major port at old mouth of Kifissos River. Also spelled Larymna.

Larimore, town (2006 population 1,289), GRAND FORKS co., E NORTH DAKOTA, 28 mi/45 km W of GRAND FORKS; 47°54′N 97°37′W. Dairy; livestock; grain, potatoes. Turtle River State Park to NE. Founded in 1881 and named for Newell Green Larimore who developed a 15,000 acre farm in the area.

Laringovi, Greece: see ARNAIA.

Larino (lah-REE-no), town, CAMPOBASSO province, MOLISE, S central ITALY, 21 mi/34 km NNE of CAMPOBASSO; 41°48′N 14°54′E. Highly diversified secondary manufacturing center includes pasta, flour, olive oil, wine. Bishopric. Has cathedral (completed 1319) and several palaces. Nearby are ruins (amphitheater) of ancient Larinum.

Larino, UKRAINE: see LARYNE.

Larinskiy, RUSSIA: see NEVER.

La Rioja (lah ree-O-hah), province (□ 35,691 sq mi/ 92,796.6 sq km; 2001 population 289,983), W ARGENTINA; ⊙ LA RIOJA. Bounded NW by the ANDES where it borders CHILE. Drained by headwaters of BERMEJO RIVER. On E border is the SALINAS GRANDES. Contains the SIERRA PUNILLA (W), SIERRA DE FAMATINA (N), SIERRA DE VELASCO (center), and SIERRA DE LOS LLANOS (SE). The fertile, inhabited valleys enjoy Mediterranean climate. Major resort is CHILECITO in S FAMATINA VALLEY. Game reserve for protection of vicuna near LAGUNA BRAVA. Separated from CÓRDOBA province in 1816.

La Rioja (lah ree-O-hah), autonomous region and province (□ 2,044 sq mi/5,314.4 sq km; 2001 population 276,702), N SPAIN; ⊙ LOGROÑO. Comprises fertile agricultural region sloping from central plateau to EBRO RIVER, between BURGOS and NAVARRE provinces. Important administrative, agricultural, and commercial center. Copper, tin, lead, and coal deposits only partly exploited. Noted for its wine and fruit; also cereals, vegetables, olive oil, sugar beets, sheep, and some cattle and hogs. Wine production is chief industry; other agricultural industries include canning (vegetables, fruit), meat processing, distilling (alcohol, brandy); manufacturing (packaging containers, wine-making machinery, textiles, tobacco, and chemicals). Chief cities/towns are Logroño, CALAHORRA, HARO, and SANTO DOMINGO DE LA CALZADA. Formerly known as Logroño province (until 1980).

La Rioja (lah ree-O-hah), city (□ 5,770 sq mi/15,002 sq km; 2001 population 289,983), ⊙ LA RIOJA province and La Rioja department (□ 5,770 sq mi/14,944 sq km; 1991 population 220,729), W ARGENTINA, on the small Rioja River at E foot of SIERRA DE VELASCO, on railroad, and 90 mi/145 km SW of CATAMARCA, 600 mi/ 966 km NW of BUENOS AIRES; 29°24′S 66°53′W. Elevation 1,620 ft/494 m. Farming, lumbering, and trading center; sawmills, textile mills, wine making; agriculture (olives, fruit, grain; livestock). An old colonial city, founded 1591, it has modern administrative buildings, national college, archaeological museum, seismographic station. Severely damaged by earthquake (1894). The completed dam (1930) on the Rioja irrigates and supplies power to the area.

Lárissa (LAH-ree-sah), prefecture (□ 2,080 sq mi/5,408 sq km), E THESSALY department, N GREECE, on NW AEGEAN SEA; ⊙ LÁRISSA; 39°30′N 22°30′E. Bounded N by West and CENTRAL MACEDONIA departments, SE by MAGNESIA prefecture, SW by CENTRAL GREECE department, and W by TRIKKALA and KARDITSA prefectures. OLYMPOS and Kamvounia mountains in N; Orthys mountains in S. Contains the fertile Lárissa lowland drained by PENEIOS River. Its 30-mi/48-km-long Aegean coastline, astride Mount OSSA, is devoid of ports. Agriculture includes wheat, vegetables, tobacco, olives, almonds, citrus fruit; livestock raising. Chromite mining E of PHARSALA. Traversed by THESSALONÍKI-ATHENS railroad; main centers are Lárissa, TIRNAVOS, Pharsala, and AYIA. Scene of heavy fighting between Greeks and Germans in World War II. Also spelled Larisa.

Lárissa (LAH-ree-sah), city (2001 population 124,394), ⊙ LÁRISSA prefecture, E THESSALY department, N GREECE, on the PINEIOS RIVER; 39°38′N 22°25′W. Agricultural trade center and transportation hub, linked by railroad with the port of VÓLOS and with THESSALONÍKI and ATHENS. Textiles. The chief city of ancient THESSALY, it was annexed (4th century B.C.E.) by Philip II of Macedon (MACEDONIA) and in 196 B.C.E. became an ally of ROME. Taken from the BYZANTINE EMPIRE by BULGARIA and later held by SERBIA, with which it passed (15th century) under the rule of the Ottoman Turks. In the Greek War of Independence the city was (1821) the headquarters of Ali Pasha. Ceded by TURKEY to Greece in 1881. Formerly called Larisa.

Laristan (lahr-ee-STAHN), former province or region, S IRAN. Situated in central ZAGROS Mountain ranges, and once independent, it became part of IRAN under Abbas I in early 17th century, and flourished in early 18th century with the prosperity of nearby port of BANDAR ABBAS. Now largely a part of Fārs province.

La Rivera (lah ree-VE-rah), canton, ATAHUALLPA province, ORURO department, W central BOLIVIA, N of Salar de COIPASA; 19°01′S 68°39′W. Elevation 12,113 ft/3,692 m. Tungsten-bearing lode. Limestone and gypsum deposits. Agriculture (potatoes, yucca, bananas); cattle.

La Riviera, unincorporated city (2000 population 10,273), SACRAMENTO county, central CALIFORNIA; residential suburb 6 mi/9.7 km E of downtown SACRAMENTO, on AMERICAN RIVER; 38°34′N 121°22′W. California State University, Sacramento and Goethe Arboretum to W.

La Rivière (lah ree-VYER), community, S MANITOBA, W central CANADA, 6 mi/10 km W of MANITOU, in PEMBINA rural municipality; 49°14′N 98°40′W.

La Rivière Anglaise, district, NE MAHÉ ISLAND, SEYCHELLES, on INDIAN OCEAN (to E); 04°35′S 55°27′E. Borders SAINT LOUIS (S), MONT BUXTON (W), and ANSE ETOILE (N) districts. Formed c.1979. Also called ENGLISH RIVER.

Lark, IRAN: see LARAK.

Larkana (luhr-kah-NAH), district (□ 2,857 sq mi/ 7,428.2 sq km), NW SIND province, S central PAKISTAN; ⊙ LARKANA. Bounded W by N KIRTHAR RANGE, E by INDUS River; irrigated by right-bank canals of SUKKUR BARRAGE system. Prehistoric remains at MOHENJO-DARO. Sometimes spelled Larkhana.

Larkana (luhr-kah-NAH), city, ⊙ LARKANA district, SIND province, S central PAKISTAN, 20 mi/32 km NNE of KARACHI, on the Ghar Canal; 27°33′N 68°13′E. Famous for the quality of its rice; it is an important grain market and a trading center for silk and cotton goods. Brass and other metalware are manufactured. Mangoes and dates grown nearby. A sugar factory and agricultural research station. Has two colleges affiliated with Sind University. Named for the Larak tribe that inhabited the neighboring area. Former Pakistani prime minister Zulfikar Ali Bhutto was born here. Remains of the ancient city of MOHENJO-DARO have been uncovered c.15 mi/8 km to S.

Larkfield–Wikiup, unincorporated town, SONOMA county, W CALIFORNIA; residential suburb 7 mi/11.3 km NW of SANTA ROSA, near RUSSIAN RIVER. Fruit, grain; dairying; poultry. Airport.

Larkhall (LAHRK-hawl), town (2001 population 15,549), South Lanarkshire, S Scotland, on the AVON WATER, near the CLYDE RIVER, and 3 mi/4.8 km SE of HAMILTON; 55°44′N 03°58′W. Previously light-metals production.

Lark River (LARK), 26 mi/42 km long, SUFFOLK, E ENGLAND; rises in several branches 4 mi/6.4 km S of BURY ST. EDMUNDS; flows NW, past Bury St. Edmunds and MILDENHALL, to OUSE RIVER 3 mi/4.8 km NE of ELY in Cambridgeshire.

Larkspur, city (2000 population 12,014), MARIN county, W CALIFORNIA; prestigious residential suburb 12 mi/ 19 km NNW of SAN FRANCISCO, and 2 mi/3.2 km W of SAN FRANCISCO BAY, near MOUNT TAMALPAIS; 37°57′N 122°32′W. The region's scenic beauty and excellent beaches attract many visitors. Nearby Larkspur Canyon has a redwood grove. Sharpe Army Depot to NE; Mount Tamalpais State Park, MUIR WOODS NATIONAL MONUMENT to W. Incorporated 1908.

Larkspur (LAHRK-spuhr), summer village (2001 population 21), central ALBERTA, W CANADA, 19 mi/30 km from WESTLOCK, in WESTLOCK COUNTY; 54°25′N 113°46′W. Incorporated 1985.

Larkspur, village (2000 population 234), DOUGLAS county, central COLORADO, 33 mi/53 km S of DENVER, 27 mi/43 km N of COLORADO SPRINGS, on PLUM CREEK, at E edge of FRONT RANGE; 39°10′N 104°54′W. Elevation 6,680 ft/2,036 m. Cattle; fruit, nuts, wheat; timber; manufacturing of wood products. Pike National Forest to W.

Larksville, borough (2006 population 4,500), LUZERNE county, NE central PENNSYLVANIA, residential suburb 2 mi/3.2 km WNW of WILKES-BARRE, on SUSQUEHANNA RIVER. Light manufacturing. Larksville Mountain to NW. Incorporated 1909.

Larmor-Plage (lahr-mor–plazh), town (□ 2 sq mi/5.2 sq km) S suburb of LORIENT, MORBIHAN department, W FRANCE, in BRITTANY, on Bay of BISCAY and at entrance to Lorient harbor; 47°42′N 03°23′W. Popular bathing resort and residential area experiencing rapid growth. Has 15th–17th-century church containing fine works of art.

Larnaca (lahr-NAH-kah), district (□ 435 sq mi/1,131 sq km; 2001 population 117,124), SE CYPRUS, on the MEDITERRANEAN; ⊙ LARNACA. Attila Line crosses NE corner; c. 10% of land in Turkish occupied zone. Part of DHEKELIA U.K. SOVEREIGN BASE AREA; Cape PYLA in E; Cape KITI in S; beach resort area in NE, on LARNACA BAY. Among other towns are ATHIENOU and LEFKARA. Mostly lowland, with foothills of the TROODOS MOUNTAINS in N. Predominantly agricultural (wheat, oats, vetches, citrus, almonds, carobs, grapes, olives; sheep, goats, poultry, hogs). Fishing. Pyrite mines near Kalavaso; also gypsum, umber, bentonite. Ancient city of Citium partly occupied by LARNACA city.

Larnaca (lahr-NAH-kah), city (2001 population 46,666), ⊙ LARNACA district, SE CYPRUS, on LARNACA BAY, 24 mi/39 km SE of NICOSIA; 34°55′N 33°38′E. It is a major port and manufacturing center (chemicals, petroleum refining, food processing, textiles, footwear, building materials, tobacco products). The modern section of the town occupies the site of ancient CITIUM. There is a tradition that Saint Lazarus settled in Larnaca after his resurrection and became its first bishop. It is also the birthplace of the Greek stoic Zenon (336 B.C.E.). Church of Saint Lazarus (685–912); fort at harbor built by the Turks in 1625. Moslem holy site to SW; museum, international airport to S; DHEKELIA U.K. SOVEREIGN BASE AREA 8 mi/12.9 km to NE. Also spelled Larnaka.

Larnaca Bay (lahr-NAH-kah), arm of MEDITERRANEAN SEA, SE CYPRUS, c. 14 mi/23 km long, 7 mi/11.3 km wide; city of LARNACA on W shore. Livadhia resort area on N, DHEKELIA U.K. SOVEREIGN BASE AREA on NE; Cape PYLA to E.

Larnaca Salt Lake (lahr-NAH-kah), LARNACA district, S CYPRUS, S of LARNACA. Parallels LARNACA BAY, MEDITERRANEAN SEA; 4 mi/6.4 km long, 1 mi/1.6 km wide. Salt Lake Park on N shore. Proposed nature reserve.

Larne (LAHRN), Gaelic *Latharna*, town (2001 population 18,228), ANTRIM, NE Northern Ireland, on an inlet of the NORTH CHANNEL; 54°51′N 05°49′W. Seaport and tourist center. Car-ferry service to Stranraer, Scotland. Beef and potatoes are exported; among the industrial products are electrical equipment, paper, and linen. Pharmaceuticals. In 1315, Edward Bruce landed at Larne. In 1914 the Ulster Volunteers ran German guns ashore here.

Laroche, BELGIUM: see ROCHE-EN-ARDENNE, LA.

Larned (LAHR-ned), town (2000 population 4,236), ⊙ PAWNEE county, SW central KANSAS, on ARKANSAS RIVER, at mouth of PAWNEE RIVER, 22 mi/35 km SW of GREAT BEND; 38°10′N 99°05′W. Railroad junction. Trade center for agricultural area (wheat, alfalfa, sugar beets); dairying. Manufacturing (meat packing). Fort Larned National Historical Site 3 mi/4.8 km to W. Laid out 1873, incorporated 1886.

La Roche (LA RUHSH), town, NW SASKATCHEWAN, CANADA, on E side of Lac La Loche, 260 mi/418 km NW of SASKATOON, 25 mi/40 km E of ALBERTA border. Sport fishing and hunting area. Timber, furs.

Airstrip; highway from SE. Clearwater Provincial Park is 25 mi/40 km N.

La Roche-en-Ardenne, BELGIUM: see ROCHE-EN-ARDENNE, LA.

Laroche-Migennes, FRANCE: see MIGENNES.

Larochette (lah-ro-SHET), German *Fels*, village, LAROCHETTE commune, E central LUXEMBOURG, 7 mi/11.3 km SE of ETTELBRUCK; 49°46'N 06°14'E. Woolen textiles, brushes; market center for fruit-growing area (apples, pears). Tourism; castle ruins on cliffs above village.

Laroles (lah-RO-les), village, GRANADA province, S SPAIN, 22 mi/35 km SSE of GUADIX; 37°01'N 03°00'W. Olive-oil processing, flour milling. Ships chestnuts, nuts, figs.

La Romana (lah ro-MAH-nah), province (□ 209 sq mi/543.4 sq km; 2002 population 219,812), SE DOMINICAN REPUBLIC, along the CARIBBEAN SEA coast; ⊙ LA ROMANA; 18°30'N 68°58'W. Agricultural (sugarcane; sugarcane milling.) The province lies in the coastal plain.

La Romana (lah ro-MAH-nah), city (2002 population 191,303), SE DOMINICAN REPUBLIC, on the CARIBBEAN SEA; ⊙ of LA ROMANA province (1981 population 109,769); 18°30'N 69°00'W. Major port of La Romana province; airport; site of several resort hotels.

La Ronge (LA RANJ), town (2006 population 2,725), central SASKATCHEWAN, CANADA, W shore of LAC LA RONGE; 55°06'N 105°18'W. Former site of Indian residential school and Hudson's Bay Company posts in area. Native community doubled since 1971. Tourist, administrative center, and prospecting departure point.

Laropi (lah-RO-pee), village, ADJUMANI district, NORTHERN region, NW UGANDA, on the ALBERT NILE RIVER, 9 mi/14.5 km SSE of MOYO. Cotton, peanuts, sesame, millet. Was part of former NILE province.

Laroque, FRANCE: see GANGES.

Laroquebrou (lah-rok-uh-broo), commune (□ 6 sq mi/15.6 sq km), CANTAL department, in AUVERGNE, S central FRANCE, in the MASSIF CENTRAL, on the CÈRE, 12 mi/19 km W of AURILLAC; 44°58'N 02°12'E. Pottery workshops. Has ruins of a feudal castle and an ornate 14th–15th-century late Gothic church. Dam and hydroelectric plant 3 mi/4.8 km SE.

Laroque-d'Olmes (lah-rok-dol-muh), commune (□ 5 sq mi/13 sq km), ARIÈGE department, MIDI-PYRÉNÉES region, S FRANCE, in the MIREPOIX district, 13 mi/21 km E of FOIX; 42°58'N 01°52'E. Textiles.

Larose (luh-ROZ), unincorporated town (2000 population 7,306), LAFOURCHE parish, SE LOUISIANA, 30 mi/48 km SW of NEW ORLEANS, on Bayou LAFOURCHE, at intersection of INTRACOASTAL WATERWAY; 29°34'N 90°22'W. Manufacturing (boats, oil-field equipment, concrete). Pointe au Chien State Wildlife Area to SW.

La Rose, village (2000 population 159), MARSHALL county, N central ILLINOIS, near CROW CREEK, 10 mi/16 km ESE of LACON; 40°58'N 89°13'W. In agricultural area.

La Rosita, NICARAGUA: see ROSITA.

Larouche (lah-ROOSH), village (□ 34 sq mi/88.4 sq km; 2006 population 1,046), SAGUENAY—LAC-SAINT-JEAN region, S central QUEBEC, E CANADA, 9 mi/15 km from ALMA; 48°27'N 71°31'W.

Larrabee, town (2000 population 149), CHEROKEE county, NW IOWA, 8 mi/12.9 km N of CHEROKEE; 42°51'N 95°32'W.

Larraga (lah-RAH-gah), town, NAVARRE province, N SPAIN, near ARGA RIVER, 12 mi/19 km SE of ESTELLA; 42°34'N 01°51'W. Olive oil, wine, cereals.

Larrainzar (lah-rah-een-SAHR), town, CHIAPAS, S MEXICO, in Sierra de HUEYTEPEC, 12 mi/19 km NW of SAN CRISTÓBAL DE LAS CASAS; 16°53'N 92°44'W. Wheat, fruit. A Tzotzic Mayan community. Also known as SAN ANDRÉS LARRAINZAR.

Larreynaga (lah-rai-NAH-gah), town (2005 population 11,292), LEÓN department, W NICARAGUA, 20 mi/

32 km NE of LEÓN; 12°40'N 86°34'W. Livestock; sesame, corn.

Larrimah (LA-ri-muh), settlement, N central NORTHERN TERRITORY, AUSTRALIA, 265 mi/426 km SE of DARWIN, 110 mi/177 km S of KATHERINE; 15°31'S 133°10'E. Cattle. World War II airfield site and dirt airstrip c.6 mi/10 km N of LARRIMAH. Formerly called Birdum.

Larroque (lahr-RO-ke) or **Villa Larroque**, town, SE ENTRE RÍOS province, ARGENTINA, on railroad, 29 mi/47 km W of GUALEGUAYCHÚ; 33°02'S 59°01'W. Agriculture center (flax, grain growing, stock raising, dairying).

Larroudé, ARGENTINA: see BENJAMÍN LARROUDÉ.

Larsa (LAHR-suh), ancient city of S BABYLONIA, in modern IRAQ, 26 mi/42 km NW of AN NASIRIYAH. It was the biblical Ellasar (Genesis 14.1). When the last king of the third dynasty of UR was overthrown (c.1950 B.C.E.) by the Elamites and Amorites, the cities of ISIN and Larsa were rivals for hegemony in MESOPOTAMIA. In 1763 B.C.E. Hammurabi defeated Larsa and succeeded in uniting Babylonia under his power. The city was dedicated to the sun god, Shamash. Temple libraries and important documents have been found in the ruins.

Lars Christensen Coast, ANTARCTICA, on INDIAN OCEAN and along AMERY ICE SHELF, extends E from 67°E to CAPE DARNLEY, and then S to head of Amery Ice Shelf at 71°S 71°E. Discovered 1931 by Norwegian whalers.

Larsen Bay, S ALASKA, S arm of UYAK BAY, inlet of SHELIKOF STRAIT, W KODIAK ISLAND, 60 mi/97 km WSW of KODIAK, 20 mi/32 km long; 57°32'N 154°03'W. Salmon canning. LARSEN BAY village is on shore.

Larsen Ice Shelf, lies along most of the E coast of the ANTARCTIC PENINSULA; 67°30'S 62°30'W.

Larson, village (2000 population 17), BURKE co., NW NORTH DAKOTA, 29 mi/47 km WNW of BOWBELLS; 48°53'N 102°52'W. Founded in 1907 and incorporated in 1911. Named for the postmaster, Columbia Larson.

Larto, Lake (LAHR-to), oxbow lake, CATAHOULA parish, E central LOUISIANA, formed by a cutoff of OUACHITA RIVER, 30 mi/48 km E of ALEXANDRIA; 31°22'N 91°55'W. Lake is c. 10 mi/16 km long. Fishing. Catahoula Lake Diversion Canal runs E-W to N of Lake Larto. Saline State Wildlife Area adjacent to W.

Larue (luh-ROO), county (□ 263 sq mi/683.8 sq km; 2006 population 13,791), central KENTUCKY; ⊙ HODGENVILLE; 37°32'N 85°42'W. Bounded E and NE by ROLLING FORK; drained by NOLIN RIVER. Rolling agricultural area (burley tobacco, corn, hay, alfalfa, soybeans, wheat; hogs, cattle, poultry; dairying); limestone quarries; timber. Includes ABRAHAM LINCOLN BIRTHPLACE NATIONAL HISTORIC SITE, at center of county, near Hodgenville, Lincoln boyhood home in NE, near White City. Formed 1843.

La Rue (luh ROO), village (2006 population 724), MARION county, central OHIO, 13 mi/21 km W of MARION, on SCIOTO RIVER; 40°34'N 83°23'W.

La Ruinette, SWITZERLAND: see RUINETTE, LA.

Laruns (lah-ruhn), commune, PYRÉNÉES-ATLANTIQUES department, AQUITAINE region, SW FRANCE, on the GAVE D'OSSAU, in Ossau valley of central PYRÉNÉES, 21 mi/34 km S of PAU; 42°59'N 00°25'W. Junction of trans-Pyrenean highway to SPAIN via POURTALET pass (12 mi/19 km S) with road leading E to LOURDES via Col d'AUBISQUE. Sawmilling. Nearby are the spas of Les EAUX-CHAUDES and EAUX-BONNES.

La Russell, city (2000 population 138), JASPER county, SW MISSOURI, on SPRING RIVER, and 14 mi/23 km E of CARTHAGE; 37°08'N 94°03'W. Wheat, hay, sorghum; cattle.

Larvik (LAHR-veek), city (2007 population 41,364), VESTFOLD county, SE NORWAY, at the head of the Larviksfjorden (an arm of the SKAGERRAK); 59°04'N 10°00'E. Historically, a shipping and whaling center.

Tanum Church, a medieval stone structure, is nearby. Port with ferry service to DENMARK (Fredrikshavn). Chartered 1671.

Larwill, town (2000 population 282), WHITLEY county, NE INDIANA, 7 mi/11.3 km WNW of COLUMBIA CITY; 41°11'N 85°37'W. In agricultural area. Laid out 1854.

Lar'yak (lahr-YAHK), village, SE KHANTY-MANSI AUTONOMOUS OKRUG, TYUMEN oblast, NW SIBERIA, RUSSIA, on the VAKH RIVER (OB' RIVER basin), 370 mi/595 km E of KHANTY-MANSIYSK; 61°06'N 80°15'E. Elevation 111 ft/33 m. Fish canning, reindeer raising.

Larymna, Greece: see LARIMNA.

Laryne (LAH-ri-ne) (Russian *Larino*), town, central DONETS'K oblast, UKRAINE, in the DONBAS, 8 mi/13 km SE, and under jurisdiction, of DONETS'K; 47°53'N 37°56'E. Elevation 488 ft/148 m. Limestone quarries; brickworks; landfill serving Donets'k. Established in 1872, town since 1938.

Lasa, CHINA: see LHASA.

La-sa, CHINA: see LHASA.

La Sábana (lah SAH-bah-nah), village, ⊙ Tapenagá department, S Chaco province, ARGENTINA, on railroad, and 65 mi/105 km SW of RESISTENCIA; 27°52'S 59°57'W. In livestock area; sawmills.

Las Acequias (LAHS ah-say-KEE-ahs), town, S central CÓRDOBA province, Argentina, 24 mi/39 km SE of RÍO CUARTO; 33°17'S 63°59'W. Soybeans, wheat, flax; cattle. Variant spellings include: Las Acequitas; Las Arequias.

La Sal, unincorporated village, SAN JUAN county, SE UTAH. Near COLORADO state line, 25 mi/40 km SE of MOAB, in LA SAL MOUNTAINS, in Lisbon Valley. Wheat; oil and natural gas; lumber. Elevation 7,125 ft/2,172 m. Part of Manti-La Sal National Forest to N.

La Salina (LAH sah-LEE-nah), town, ⊙ La Salina municipio, CASANARE department, S COLOMBIA, 45 mi/72 km N of YOPAL, at the foot of the Sierra Nevada de COCUY; 06°07'N 72°20'W. Elevation 5,593 ft/1,704 m. Sugarcane, corn; livestock.

La Salle (lah SAL), community, S MANITOBA, W central CANADA, 14 mi/23 km S of WINNIPEG, and in MACDONALD rural municipality; 49°41'N 97°20'W.

La Salle (luh SAL), county (□ 1,148 sq mi/2,984.8 sq km; 2000 population 111,509), N ILLINOIS; ⊙ OTTAWA; 41°20'N 88°52'W. Drained by ILLINOIS, FOX, VERMILION, and LITTLE VERMILION rivers. Includes part of Illinois AND Michigan canal Parkway. Agriculture (corn, soybeans, wheat, cattle, hogs; dairy products); clay. Diversified manufacturing, chiefly at LA SALLE, PERU, Ottawa, and STREATOR. Includes STARVED ROCK, Buffalo Rock, Illini, and Mathieson state parks. Nuclear power plants: LA SALLE 1 (initial criticality June 21, 1982) and La Salle 2 (initial criticality March 10, 1984) are 11 mi/18 km SE of Ottawa. Use cooling water from a reservoir, and each has a maximum dependable capacity of 1,036 MWe. Morseilles National Guard Training Area near E side. Formed 1831.

La Salle (luh SAL), county (□ 1,494 sq mi/3,884.4 sq km; 2006 population 5,969), S TEXAS; ⊙ COTULLA. Drained by NUECES and FRIO rivers. Mainly cattle-ranching area; partly in irrigated WINTER GARDEN agricultural region (peanuts, watermelons, sorghum, corn). Oil and natural gas wells. Formed 1858.

LaSalle (luh-SAL) or **Ville Lasalle** (VEEL lah-SAL), former city, borough (French *arrondissement*) of MONTREAL (□ 6 sq mi/15.6 sq km), S QUEBEC, E CANADA, on S MONTREAL ISLAND, on the SAINT LAWRENCE RIVER at the head of the LACHINE RAPIDS; 45°26'N 73°40'W.

La Salle (luh SAHL), city, S central QUEBEC, CANADA, residential suburb 6 mi/10 km SSW of downtown MONTREAL, on MONTREAL ISLAND, on bend on NE side of St. Lawrence River, N of Mercier Bridge; 49°41'N 97°16'W. canal LACHINE to NW.

La Salle (luh SAL), city (2006 population 9,511), LA SALLE county, N ILLINOIS, on the ILLINOIS RIVER (bridged here); 41°20′N 89°05′W. It forms a tri-city unit with PERU and OGLESBY. Agriculture (corn, wheat, soybeans; cattle, hogs); manufacturing (chemicals, cement, circuit boards, nonferrous metal products). The city developed as an important water transportation center after the opening of the ILLINOIS AND MICHIGAN CANAL in 1848. Illinois Valley Community College across river. Settled 1830; incorporated 1852.

LaSalle (lah-SAHL), town (□ 25 sq mi/65 sq km; 2001 population 25,285), S ONTARIO, E central CANADA, 6 mi/10 km S of WINDSOR; 42°13′N 83°02′W. The DETROIT River forms the town's W border.

La Salle (luh SAHL), unincorporated residential town, S ONTARIO, E central CANADA, on DETROIT RIVER, 8 mi/12.9 km SW of WINDSOR, and included in town of LASALLE; 42°14′N 83°05′W.

La Salle (luh SAL), town (2000 population 1,849), WELD county, N COLORADO, on SOUTH PLATTE RIVER, and suburb 5 mi/8 km S of GREELEY; 40°21′N 104°42′W. Elevation 4,676 ft/1,425 m. Railroad junction. In irrigated wheat, beans, vegetable, cattle and sugar beet region. Light manufacturing (feeds). Lower Latham Reservoir to E.

La Salle (luh SAL), village (2000 population 90), WATONWAN county, S MINNESOTA, 7 mi/11.3 km NNE of ST. JAMES on WATONWAN RIVER; 44°04′N 94°34′W. Grain, soybeans; livestock; manufacturing (feeds). Lake HANSKA to N.

La Salle (luh SAL), parish (□ 638 sq mi/1,658.8 sq km; 2006 population 14,093), central LOUISIANA; ⊙ JENA; 31°42′N 92°08′W. Bounded W by LITTLE RIVER, CASTOR CREEK, and BIG SALINE BAYOU. Includes CATAHOULA LAKE in S center, and SALINE LAKE on S boundary (fishing, waterfowl hunting). Oil and natural-gas fields; agriculture (cotton, sugarcane, home gardens, soybeans, hay; cattle, hogs); logging, timber, lumber milling. CATAHOULA LAKE Diversion Canal crosses S part. Includes Catahoula National Wildlife Refuge at E end of CATAHOULA LAKE, in SE, part of Saline State Wildlife Area in S. Named after the French explorer. Formed 1908.

La Salle 1 and 2 Nuclear Power Plants, ILLINOIS: see LA SALLE, county.

La Sal Mountains, range, La Sal National Forest, SAN JUAN and GRAND counties, E UTAH, between COLORADO RIVER and COLORADO state line. Chief peaks are Mountain TOMASAKI (12,230 ft/3,728 m), Mount WAAS (12,311 ft/3,752 m), and Mount PEALE (12,721 ft/3,877 m). Copper deposits. Partly in Manti-La Sal National Forest.

La Salud (lah sah-LOOD), town, LA HABANA province, W CUBA, on railroad, and 18 mi/29 km SSW of HAVANA; 22°42′N 82°26′W. Tobacco, oranges, vegetables.

Las Animas, county (□ 4,775 sq mi/12,415 sq km; 2006 population 15,564), SE COLORADO; ⊙ TRINIDAD; 37°19′N 104°02′W. Largest county in Colorado; bordering on NEW MEXICO; drained by PURGATOIRE and APISHAPA rivers. Coal-mining and cattle-grazing area; cattle, horses; wheat, hay, sorghum. Part of SANGRE DE CRISTO MOUNTAINS and San Isabel National Forest in W; Trinidad State Park in W center; parts of Comanche National Grassland in E and NE. Pinon Canyon Military Reserve in N. Formed 1866.

Las Animas (lahs ah-NEE-mahs), town, VALDIVIA province, LOS LAGOS region, S central CHILE, 2 mi/3 km ESE of VALDIVIA. Suburb in agricultural area (cereals, potatoes, fruit; livestock); lumbering. An old Spanish fort in colonial times.

Las Animas (lahs ah-NEE-mahs), town (2000 population 2,758), ⊙ BENT county, SE COLORADO, on ARKANSAS RIVER, W of mouth of Purgatoine River, 75 mi/121 km E of PUEBLO; 38°04′N 103°13′W. Elevation 3,901 ft/1,189 m. Trade center in irrigated vegetables, wheat, barley, corn, sorghum, cattle region. Kit Carson Museum (in cabin where the scout died) and U.S. veterans' hospital. Railroad junction to E; JOHN MARTIN RESERVOIR to E; ADOBE CREEK RESERVOIR (Blue Lake) to N; HORSE CREEK RESERVOIR to NW; BENT'S OLD FORT to W. Oldest active courthouse in Colorado (1888). City founded 1869, moved to present site 1873 to be on railroad; incorporated 1886.

Las Anod (LAHS ah-NOD), township, ⊙ SOOL region, NW central SOMALIA, in HAUD plateau, 235 mi/378 km ESE of HARGEISA. Important road junction; stock-raising center. Also spelled Lasanod.

La Sarine, SWITZERLAND: see SAANE RIVER.

La Sarre (lah SAHR), town (□ 57 sq mi/148.2 sq km), ⊙ ABITIBI-OUEST county, ABITIBI-TÉMISCAMINGUE region, W QUEBEC, E CANADA, 40 mi/64 km NNW of ROUYN-NORANDA; 48°48′N 79°12′W. Gold and copper mining, lumbering, pulpwood milling, dairying.

Las Banderas (lahs bahn-DAI-rahs), village, MANAGUA department, SW NICARAGUA, 25 mi/40 km NE of MANAGUA; 12°20′N 85°57′W. Road junction. Corn, beans.

Las Barrancas (lahs bahr-RAHN-kahs), seaside resort town in greater BUENOS AIRES, ARGENTINA, on the RÍO DE LA PLATA, adjoining SAN ISIDRO, 13 mi/21 km NW of Buenos Aires.

Las Barrancas (lahs bahr-RAHN-kahs) or **Barrancas**, village, N MENDOZA province, Argentina, in MENDOZA RIVER valley (irrigation area), 15 mi/24 km SSE of MENDOZA. In agricultural area (wine, corn, alfalfa); oil fields, asphalt deposits.

Las Bela (luhs BAI-lah), district (□ 7,043 sq mi/18,311.8 sq km), KALAT division, BALUCHISTAN province, SW PAKISTAN; ⊙ LAS BELA. Bounded E by KIRTHAR RANGE, SE by HAB RIVER; crossed N-S by PAB RANGE; central lowland drained by PORALI RIVER; S coast lies along SONMIANI Bay, with long strip extending W between ARABIAN SEA and MAKRAN COAST RANGE. Crossed 325 B.C.E. by part of army of Alexander the Great on return to PERSIA, while fleet cruised offshore; visited by Arab traders in Middle Ages. Formerly under suzerainty of KALAT; acceded independently in 1948 to PAKISTAN. Sometimes spelled LUS BELA.

Las Bela (luhs BAI-lah), town, ⊙ LAS BELA district, KALAT division, SE BALUCHISTAN province, SW PAKISTAN, near PORALI RIVER, 105 mi/169 km NNW of KARACHI; 25°45′N 66°35′E. Trades in oilseeds, millet, wool, rice; crocheting, woolen-carpet weaving. An 8th century Buddhist center; on important medieval Arab trade route from PERSIA to SIND.

Lasberg (LAHS-berg), township, NE UPPER AUSTRIA, AUSTRIA, 3 mi/4.8 km SSE of FREISTADT; 48°28′N 14°32′E.

Las Bonítas (LAHS bo-NEE-tahs), town, BOLÍVAR state, SE VENEZUELA, landing on ORINOCO RIVER, 40 mi/64 km ENE of CAICARA DEL ORINOCO; 07°47′N 65°39′W. Livestock.

Las Breñas (lahs BRAIN-yahs), town, SW Chaco province, ARGENTINA, on railroad, 45 mi/72 km SW of PRESIDENCIA ROQUE SÁENZ PEÑA; 27°05′S 61°05′W. Agriculture (cotton, corn, alfalfa, flax; livestock), lumbering, cotton ginning.

Las Cabras (lahs KAH-brahs), town, ⊙ Las Cabras comuna, CACHAPOAL province, LIBERTADOR GENERAL BERNARDO O'HIGGINS region, central CHILE, on CACHAPOAL RIVER, on railroad, 38 mi/61 km WSW of RANCAGUA; 34°18′S 71°19′W. Agricultural center (cereals, vegetables, alfalfa, fruit; livestock).

Lascahobas (lah-skah-O-bahs), town, CENTRE department, S central HAITI, on ARTIBONITE PLAIN, 32 mi/51 km NE of PORT-AU-PRINCE; 18°50′N 71°56′W. Tobacco, sisal, sugarcane, and coffee.

Las Cañas (lahs KAHN-yahs), village, SE CATAMARCA province, ARGENTINA, 40 mi/64 km NE of CATAMARCA. In corn, livestock, and lumber area.

Las Cañas (lahs KAHN-yas), village, COQUIMBO province, COQUIMBO region, N central CHILE, on railroad, and 8 mi/13 km SSW of ILLAPEL; 29°54′S 70°53′W. Lead mining and smelting.

Lascano (lahs-KAH-no), city, ROCHA department, SE URUGUAY, on highway, 29 mi/47 km SSE of TREINTA Y TRES; 33°40′S 54°12′W. Trading center in stock-raising region (cattle, sheep).

Las Carabelas, ARGENTINA: see CARABELAS.

Las Casas (lahs KAH-sahs), canton, AZURDUY province, CHUQUISACA department, SE BOLIVIA, SE of AZERO RIVER, at the foot of the Khosko Toro mountains. Elevation 8,163 ft/2,488 m. Abundant gas and oil wells in area. Agriculture (potatoes, yucca, bananas, corn, rye, sweet potatoes, peanuts); cattle and pig raising.

Las Casuarinas (lahs kah-swah-REE-nahs), town, S SAN JUAN province, ARGENTINA, in SAN JUAN RIVER valley (irrigation area), on railroad, 23 mi/37 km SE of SAN JUAN; 31°49′S 68°19′W. Viticulture center.

Las Catitas (lahs kah-TEE-tahs), town, NE MENDOZA province, ARGENTINA, on railroad, on TUNUYÁN RIVER (irrigation), 55 mi/89 km SE of MENDOZA; 33°18′S 68°02′W. Lumbering and agricultural center (fruit, wine, grain; livestock). Formerly called José Néstor Lencinas.

Lascaux Cave (lahs-KO), historic site, near MONTIGNAC, DORDOGNE department, AQUITAINE region, SW central FRANCE, near VÉZÈRE RIVER; 45°03′N 01°11′E. In this cave are more than 600 paintings (believed to be c.17,000 years old) of prehistoric animals. Discovered by accident in 1940, the cave was opened to the public in 1948. The site attracted more than 1 million visitors in the first 15 years, and the resulting increase in humidity caused the growth of algae and the formation of calcite, which threatened to damage the paintings. To preserve this prehistoric treasure, the cave was closed to the public. Due to its enormous cultural interest, the departmental authorities erected a reproduction of the "hall of the bulls," a 600-ft/183-m section of the cave, in the upper galleries of the cavern. Lascaux II, as it has come to be known, has been open to the public since 1983 and continues to attract large numbers of visitors. The valley of the Vézère, between Montignac and les EYZIES-DE-TAYAC (12 mi/19 km SW), has been declared a world treasure by UNESCO, the UN cultural organization. A museum of the prehistoric discoveries in the valley and their significance for human development is located at Le Thot, 4 mi/6.4 km SW of Montignac, near the earlier discoveries of Cro-Magnon man.

Las Cejas (lahs SAI-hahs), village, E TUCUMÁN province, ARGENTINA, 30 mi/48 km E of TUCUMÁN; 26°53′S 64°44′W. Railroad junction. Livestock and lumbering center.

L'Ascension (lah-son-SYON), village (□ 132 sq mi/343.2 sq km; 2006 population 805), LAURENTIDES region, S QUEBEC, E CANADA, 19 mi/31 km from LABELLE; 46°33′N 74°50′W.

L'Ascension-de-Notre-Seigneur (lah-son-SYON–duh–NO-truh–sen-YUHR), village (□ 51 sq mi/132.6 sq km; 2006 population 1,938), SAGUENAY—LAC-SAINT-JEAN region, S central QUEBEC, E CANADA, 11 mi/17 km from ALMA; 48°42′N 71°41′W.

L'Ascension-de-Patapédia (lah-son-SYON–duh–pah-tah-PAI-dyah), village (□ 37 sq mi/96.2 sq km; 2006 population 221), GASPÉSIE—ÎLES-DE-LA-MADELEINE region, SE QUEBEC, E CANADA, 5 mi/8 km from SAINT-FRANÇOIS-D'ASSISE; 47°56′N 67°15′W.

Las Choapas (lahs cho-AH-pahs), city, SE VERACRUZ, MEXICO, 7 mi/11 km SE of the port of COATZACOALCOS, located in a plains region on the banks of the TONALÁ River, which forms the border between the states of Veracruz and TABASCO; 17°56′N 94°05′W. There is a railroad line to CAMPECHE, and roads connect with those that go from Coatzacoalcos to SALINA CRUZ and VILLAHERMOSA. There are many roads built by Petróleos Mexicanos. Hot climate. Petroleum center.

Cross-references are shown in SMALL CAPITALS. The pronunciation guide is shown on page xix. The sources of population figures are shown on page xvii.

Las Colonias, ARGENTINA: see ESPERANZA.

Las Coloradas (lahs ko-lo-RAH-dahs), village (1991 population 605), ⊙ Catán-Lil department, S NEUQUÉN province, ARGENTINA, in foothills of the ANDES, 55 mi/89 km SSW of ZAPALA; 39°33′S 70°35′W. Sheep-raising center.

Las Condes (lahs KON-dais), village, ⊙ Las Condes comuna, SANTIAGO province, METROPOLITANA DE SANTIAGO region, central CHILE, in Andean foothills, suburb 9 mi/15 km NE of SANTIAGO; 33°22′S 70°31′W.

Las Cotorras, TRINIDAD AND TOBAGO: see FIVE ISLANDS.

Las Cruces (lahs KROO-suhs) [Span.=the crosses], city (2006 population 86,268), ⊙ DOÑA ANA county, SW NEW MEXICO, 42 mi/68 km NNW of EL PASO, 225 mi/362 km S of ALBUQUERQUE, on the RIO GRANDE; 32°20′N 106°45′W. Elevation 3,896 ft/1,188 m. In a farm area irrigated by the ELEPHANT BUTTE system. The second-largest city in New Mexico and one of the country's fastest growing metropolitan areas in 1990s. Its economy is based chiefly on agriculture and the nearby WHITE SANDS MISSILE RANGE, testing grounds for the first atomic bomb and a major military and NASA testing site. Cattle, sheep; dairying; vegetables, melons, corn, cotton, pecans, triticale. Manufacturing (food processing, winery, beverages, printing and publishing, aircraft parts, concrete, consumer goods, wood products, dairy products). The city has a textile industry and canning and processing plants for various products, such as sugar beets, pecans, and cotton. The name refers to a massacre (1830) of some forty travelers by Apaches on this site. New Mexico State University is to the SE. Doña Ana Branch Community College is here. Nearby are the historic village of LA MESILLA to SW, Fort Fillmore (1851) ruins to S, the village of TORTUGAS to S. Aguirre Springs National Recreation Area to E, in ORGAN MOUNTAINS; Leasburg Dam State Park and Fort Selden (1865) State Monument to NW; Las Cruces Airport to W; White Sands Missile Range to NE. Founded 1848, incorporated 1907.

Las Cuevas, MEXICO: see MATAMOROS, town.

Las Cuevas (lahs KWAI-vahs), village, NW MENDOZA province, ARGENTINA, 70 mi/113 km W of MENDOZA, on Transandine railroad (Mendoza-VALPARAISO), at CHILE border near USPALLATA PASS. Customhouse. Copper deposits nearby.

Las Daua, SOMALIA: see LASO DAUAO.

Lasdehnen, RUSSIA: see KRASNOZNAMENSK.

Las Delicias (lahs de-LEE-see-ahs), village, W ENTRE RÍOS province, ARGENTINA, on railroad, 16 mi/26 km SSE of PARANÁ. Wheat, flax; livestock.

Las Dureh (LAHS doo-rai), village, N SOMALIA, in OGO highland, on road, 55 mi/89 km NNE of BURAO. Gums; camels, sheep, goats. Also spelled Las Duareh.

Lasem (LAH-suhm), town, N Java Tengah province, INDONESIA, c.71 mi/114 km ENE of SEMARANG, near JAVA SEA; 06°42′S 111°26′E. Trade center for agricultural area (rice, corn, cassava, peanuts); batik printing. Mount Lasem lies 5.6 mi/9 km E. Sometimes spelled Lassem.

La Serena (lah se-RAI-nah), city (2002 population 147,815), ⊙ La Serena comuna, ELQUI province, and ⊙ COQUIMBO region, N central CHILE, on the ELQUI RIVER and the PAN-AMERICAN HIGHWAY; 29°54′S 71°15′W. A commercial and agricultural center in a region of orchards and vineyards; popular resort. La Serena was founded in 1543, destroyed by Native Americans in 1549, and sacked by the English in 1680. Often damaged by earthquakes, La Serena is a city of old-world charm, noted for its cathedral, fine buildings, and gardens. Airport. Iron-ore, silver and copper mines nearby.

Las Esperanzas (lahs es-pai-RAHN-sahs), mining settlement, Múzquiz municipio, COAHUILA, N MEXICO, in SABINAS coal district, on railroad, 12 mi/19 km SE of MÚZQUIZ.

Lasethi, Greece: see LASITHI, prefecture.

Las Flores (lahs FLO-res), town (1991 population 18,385), ⊙ Las Flores district (□ 1,290 sq mi/3,354 sq km), E central BUENOS AIRES province, ARGENTINA, 110 mi/177 km SSW of BUENOS AIRES; 36°05′S 59°05′W. Railroad junction; agricultural center (corn, sunflowers, wheat; livestock). Formerly called Carmen de las Flores.

Las Flores (lahs FLO-res), town, LEMPIRA department, NW HONDURAS, 9 mi/15 km N of GRACIAS, on paved highway; 14°40′N 88°37′W. Small farming.

Las Flores or **Flores**, town, MALDONADO department, S URUGUAY, 21 mi/34 km WNW of MALDONADO; 34°44′S 55°20′W. In agricultural region (sugar beets, grain; stock). Adjoining S is a beach resort.

Las Garzas (lahs GAHR-zahs), village, NE SANTA FE province, ARGENTINA, 30 mi/48 km NNE of RECONQUISTA. Agriculture center (cotton, tobacco, flax; livestock).

Las Guabas (GWAH-bahs), town and minor civil division of Los Santos district, LOS SANTOS province, S central PANAMA, in PACIFIC lowland, 8 mi/12.9 km SW of LOS SANTOS. Bananas, sugarcane; livestock.

Las Guaranas (lahs GWAH-rah-nahs), town, 8 mi/13 km SE of SAN FRANCISCO DE MACORÍS, DUARTE province, central DOMINICAN REPUBLIC; 19°12′N 70°13′W.

Lash, NORTH CAROLINA: see WALNUT COVE.

Lashburn (LASH-buhrn), town (2006 population 914), W SASKATCHEWAN, CANADA, 20 mi/32 km SE of LLOYDMINSTER, near ALBERTA border; 53°07′N 109°37′W. Wheat.

Las Heras (lahs AI-rahs), town, ⊙ Las Heras department (□ 4,290 sq mi/11,154 sq km), N MENDOZA province, ARGENTINA, on railroad, in MENDOZA RIVER valley (irrigation area), just N of MENDOZA; 32°51′S 68°49′W. Agricultural and manufacturing center. Produces cement, pottery, dried fruit, wine; has sawmills, lime quarries. Agriculture products (wine, fruit, potatoes, alfalfa). Airport.

Las Heras, ARGENTINA: see COLONIA LAS HERAS.

Lashio (LAHSH-yo), town, SHAN STATE, E central MYANMAR; 22°56′N 97°45′E. It is an important center for trade between MYANMAR and CHINA and is the terminus of the railroad line from MANDALAY. Lashio was famous in World War II as the starting point of the BURMA ROAD.

Lash-Jawain (lahsh–jah-WAH-ein), town, NIMRUZ province, SW AFGHANISTAN, on the lower Farah Rud River, 55 mi/89 km SW of FARAH, near IRAN border; 31°43′N 61°37′E. At edge of Seistan lake depression. Consists of 2 sections that developed around ancient forts: Lash on right bank, and Jawain on left bank. Also spelled Lash Juwain or Lash-o-Jovain.

Lashkar, city, GWALIOR district, N central MADHYA PRADESH state, central INDIA, SE of GWALIOR town; 26°10′N 78°10′E. Today it is a modern commercial center and transportation hub: factories produce yarn, paint, ceramics, chemicals, and leather products. Victoria College, the palace of the Maharaja of Gwalior, and a state museum are here.

Lashkarak (lahsh-kah-RAHK), village, Tehrān province, N IRAN, 13 mi/21 km NE of TEHRAN, in ELBURZ mountains. Popular skiing resort.

Lashkargah (lahsh-kah-RAHG-ah), town, S AFGHANISTAN, ⊙ HELMAND province, on HELMAND River, c.80 mi/129 km E of KANDAHAR; 31°35′N 64°21′E. Manufacturing include carpet, marble; food processing (oils, milk products). Domestic airport. It replaced GIRISHK and BUST as province capital. Lashkargah (Persian, place of the army) is said to have been built in Ghaznawid times (11th century) and rebuilt in the 1960s.

Lashma (LAHSH-mah), town (2006 population 1,720), N RYAZAN oblast, central European Russia, on the OKA River, 8 mi/13 km W of KASIMOV; 54°56′N 41°09′E. Elevation 390 ft/118 m. Iron foundry.

Las Hortensias (lahs or-TEN-see-ahs), village, CAUTÍN province, ARAUCANIA region, S central CHILE, on railroad, 25 mi/40 km SE of TEMUCO; 38°57′S 72°10′W. In agricultural area (cereals, vegetables, potatoes; livestock).

Lashva, BOSNIA AND HERZEGOVINA: see LAŠVA.

La Sierpe (lah see-ER-pai), town, SE SANCTI SPÍRITUS province, central CUBA, 9 mi/15 km ESE of ZAZA Reservoir; 21°46′N 79°14′W.

La Sierra (LAH see-ER-rah), town, ⊙ La Sierra municipio, CAUCA department, SW COLOMBIA, 21 mi/34 km SW of POPAYÁN; 02°10′N 76°45′W. Elevation 6,000 ft/1,828 m. Coffee, corn, sugarcane; livestock.

La Sierra (lah SYE-rah), town, GUANACASTE province, NW COSTA RICA, on ABANGARES RIVER, and 3 mi/4.8 km ENE of LAS JUNTAS. Former gold-mining center in Abangares district.

La Sierra (lah SYE-rah), town, MALDONADO department, S URUGUAY, on railroad, and 40 mi/64 km ENE of MONTEVIDEO; 34°46′S 55°25′W. In sugar beet region; sugar refinery; agriculture (wheat; cattle).

La Sierra (lah see-E-ruh), suburban section of RIVERSIDE, RIVERSIDE county, S CALIFORNIA, 6 mi/9.7 km SW of downtown Riverside, at La Sierra and Magnolia avenues. LAKE MATHEWS reservoir to S. Site of Loma Linda University.

La Silla de Caracas (LAH SEE-yah dai kah-RAH-kahs), range with two peaks (PICO OCCIDENTAL, 8,163 ft/2,488 m; and PICO ORIENTAL, 8,661 ft/2,640 m), N VENEZUELA, 6 mi/10 km ENE of CARACAS; 10°33′N 66°51′W. Second-highest elevation in CORDILLERA DE LA COSTA.

Lasin (LAH-seen), Polish Łasin, German Lessen, town, Toruń province, N POLAND, 14 mi/23 km E of GRUDZIADZ; 53°32′N 19°06′E. Railroad spur terminus; manufacturing of agricultural machinery, sawmilling, flour milling. Monastery.

Lasithi (lah-SEE-thee), prefecture (□ 738 sq mi/1,918.8 sq km), CRETE department, GREECE, E of Mount DIKTI, and on MEDITERRANEAN SEA (S), Sea of CRETE and Gulf of MERABELLO (N), and Kasos Strait (E); ⊙ AYIOS NIKOLAOS, 35°05′N 25°50′E. Easternmost prefecture of the department, it is bordered W by IRÁKLION prefecture. Agriculture includes olives, carob, raisins, wheat, olive oil; stock raising (sheep, goats); fisheries. Main ports are Ayios Nikolaos, SITIA, and IERAPETRA. Also spelled Lasethi.

Lasithi (lah-SEE-thee), plain (□ 15 sq mi/39 sq km), LASITHI prefecture, CRETE department, GREECE, in Lasithi (Dikti) Mountains. Elevation 2,625 ft/800 m. Rich alluvial soil dedicated to agriculture (including potatoes, fruit trees) by villagers living on adjacent slopes. Once distinguished by numerous white-sailed windmills used for irrigation; few remain in operation.

Lasithi, Mount, Greece: see DIKTI, MOUNT.

Las Juntas, COSTA RICA: see JUNTAS.

Las Junturas (lahs hoon-TOO-rahs), village, central CÓRDOBA province, ARGENTINA, 55 mi/89 km SE of CÓRDOBA; 31°49′S 63°26′W. Soybeans, wheat, corn, flax; livestock.

Lask (lahsk), Polish Łask, town, Sieradz province, central POLAND, 18 mi/29 km SW of ŁÓDZ; 51°35′N 19°08′E. Tanning, distilling, manufacturing of apparel; textile industry.

Laskay (las-KAI), unincorporated village, YORK region, S ONTARIO, E central CANADA, 16 mi/25 km from TORONTO, and included in the township of KING; 43°54′N 79°34′W.

Läskelä, RUSSIA: see LYASKELYA.

Lasker (LAS-kuhr), village (□ 1 sq mi/2.6 sq km; 2006 population 97), NORTHAMPTON county, NE NORTH CAROLINA, 21 mi/34 km ESE of ROANOKE RAPIDS; 36°21′N 77°18′W. Manufacturing (feeds); agriculture (tobacco, peanuts, cotton, grain; livestock).

Las Khoreh, SOMALIA: see LAS KORAI.

Laško (LAHSH-ko), village, central SLOVENIA, on SAVINJA RIVER, on railroad, and 35 mi/56 km E of

LJUBLJANA; 46°09′N 15°14′E. In former lignite-mining area; brewery; textile and leather industry. Health resort with radioactive warm springs.

Las Korai (LAHS KO-rai), village, N SOMALIA, on Gulf of ADEN, 70 mi/113 km NE of ERIGAVO. Gums (frankincense, myrrh); fisheries. Formerly called LAS KHOREH.

Las Lajas (lahs LAH-hahs), town (1991 population 3,787), ⊙ Picunches department, central NEUQUÉN province, ARGENTINA, on the RÍO AGRIO, in ANDEAN foothills, 30 mi/48 km NW of ZAPALA; 38°31′S 70°22′W. Road center on route to PINO HACHADO PASS (CHILE border). Stock raising.

Las Lajas (las LAH-hahs), town and municipio (1996 est. population 8,796), Comayagua department, 30 mi/50 km NNW of Comayagua; 14°46′N 87°45′W. Located between Humaya and Sulaco arms of El Cajón Reservoir. Small farming, grain, timber. Poor road access.

Las Lajas (LAH-hahs), town, ⊙ San Felix district, CHIRIQUÍ province, W PANAMA, in PACIFIC lowland, 38 mi/61 km ESE of DAVID, on INTER-AMERICAN HIGHWAY. Hardwood lumbering; stock raising.

Las Lomas, unincorporated town (2000 population 3,078), MONTEREY county, W CALIFORNIA; residential suburb 3 mi/4.8 km SSE of WATSONVILLE, 4 mi/6.4 km E of PACIFIC OCEAN; 36°52′N 121°44′W.

Las Lomas (lahs LO-mahs), village and minor civil division of David district (2000 population 10,439), CHIRIQUÍ province, W PANAMA, 1 mi/1.6 km E of DAVID, on INTER-AMERICAN HIGHWAY. Coffee, bananas; livestock.

Las Lomas (LAHS LO-mahs), village, PIURA province, PIURA region, NW PERU, in W foothills of CORDILLERA OCCIDENTAL, 45 mi/72 km NE of PIURA; 04°40′S 80°15′W. Vineyards, orchards; sugarcane. Near Reservorio San Lorenzo.

Las Lomitas (lahs lo-MEE-tahs), town, central FORMOSA province, ARGENTINA, on railroad, 180 mi/290 km NW of FORMOSA. Stock-raising center; lumbering. Airport.

Las Maderas (lahs mah-DAI-rahs), village, MANAGUA department, SW NICARAGUA, on INTER-AMERICAN HIGHWAY, 31 mi/50 km NE of MANAGUA. Lumbering.

Las Majaguas, Embalse (LAHS mah-HAH-gwahs, em-BAHL-sai), reservoir, PORTUGUESA state, W VENEZUELA, near SAN RAFAEL DE ONOTO, and connected to the COJEDES RIVER. Opened for service in 1963. Irrigation; tourism; fishing.

Las Mañanitas, urban district of PANAMA City, PANAMA province, PANAMA.

Las Manos (lahs MAH-nos), village, NUEVA SEGOVIA department, NW NICARAGUA, 12 mi/19 km N of OCOTAL; 13°47′N 86°34′W. Frontier station on HONDURAS border, on road to Danlí (Honduras).

Las Margaritas (lahs mahr-gahr-REE-tahs), city, ⊙ Las Margaritas municipio, CHIAPAS, S MEXICO, in SIERRA MADRE, 11 mi/18 km ENE of COMITÁN DE DOMÍNGUEZ; 16°19′N 91°59′W. Elevation 4,970 ft/1,515 m. Sugarcane, cereals, fruit, stock. Tojolabal-speaking Maya Indians occupy rural areas of the municipio.

Las Marías (lahs mah-REE-yahs), town (2006 population 11,948), W PUERTO RICO, 10 mi/16 km ENE of Mayagüez. Coffee center. Agriculture (coffee, oranges, grapefruit); light manufacturing.

Las Matas (lahs MAH-tas) or **Las Matas de Farfán**, town (2002 population 21,271), SAN JUAN province, W DOMINICAN REPUBLIC, in irrigated SAN JUAN VALLEY, on highway, 20 mi/32 km W of SAN JUAN. Agricultural (coffee, bananas, vegetables, cereals.)

Las Matas de Santa Cruz (lahs MAH-tas dai SAHN-tah croos), town, 19 mi/30 km SE of MONTE CRISTI, MONTE CRISTI province, NW DOMINICAN REPUBLIC; 19°40′N 71°30′W. In banana-producing region.

Las Mercedes (lahs mer-SAI-des), canton, SUD YUNGAS province, LA PAZ department, W BOLIVIA, SW of LA PAZ. Elevation 5,686 ft/1,733 m. Copper-bearing

lode; tungsten mining at Mina Reconquista, Mina CHOJLLA, and the cooperative Mina Bolsa Negra; clay, limestone, and gypsum deposits. Agriculture (potatoes, yucca, soy, tobacco, coffee, tea, citrus fruits, bananas, rye); cattle.

Las Mercedes (lahs mer-SAI-des), town, S ENTRE RÍOS province, ARGENTINA, 25 mi/40 km SW of GUALEGUAYCHÚ; 25°57′S 64°40′W. In flax, grain, and livestock area; dairying.

Las Mercedes (LAHS mer-SAI-des), town, ⊙ Las Mercedes municipio, GUÁRICO state, central VENEZUELA, of VALLE DE LA PASCUA; 09°08′N 66°27′W. Livestock. Oil wells.

Las Minas (lahs MEE-nahs), town, VERACRUZ, E MEXICO, 19 mi/31 km NW of XALAPA ENRÍQUEZ on Bobos River. Elevation 5,755 ft/1,754 m. Hydroelectric plant. In Totomac Indian area.

Las Minas (lahs MEE-nahs), town, ⊙ LAS MINAS district, HERRERA province, S central PANAMA, 20 mi/32 km WSW of CHITRÉ. Stock raising. Former gold-mining center.

Las Minas, PANAMA: see PUERTO PILÓN.

Lasnamäe (LUHS-nah-mah-ai), suburban district within TALLINN, ESTONIA.

Lasne (LAZ-nuh), commune (2006 population 13,917), Nivelles district, BRABANT province, central BELGIUM, 12 mi/19 km SE of BRUSSELS.

Las Ninfas National Park, GUATEMALA: see AMATITLÁN.

Laso Dauao (LAH-so dah-WAH-ou), Italian *las daua*, village, NE SOMALIA, 55 mi/89 km S of BOSSASSO. Road junction. Formerly spelled LAZ DAUA.

Las Palmas (lahs PAHL-mahs), town, E Chaco province, ARGENTINA, on railroad, 30 mi/48 km NNE of RESISTENCIA. Processing and agricultural center. Its port is 4 mi/6.4 km SE on PARAGUAY RIVER. Population figure includes LA LEONESA.

Las Palmas, town, N BUENOS AIRES province, ARGENTINA, 8 mi/12.9 km W of ZÁRATE. Cattle-raising and meat-packing center; vegetable products.

Las Palmas (lahs PAHL-mahs), town, ⊙ Las Palmas district, VERAGUAS province, W central PANAMA, 30 mi/48 km W of SANTIAGO. Manganese deposits. It is situated on Las Palmas Peninsula, a S projection of Panama, between TABASARÁ and SAN PABLO rivers, and forming W side of MONTIJO GULF.

Las Palmas, SPAIN: see PALMAS, LAS, province and city.

Las Pampitas (lahs pahm-PEE-tahs), canton, MAMORÉ province, BENI department, NE BOLIVIA, 32 mi/50 km SE of PUERTO SILES; 12°55′S 64°39′W. Agriculture (potatoes, yucca, bananas, coffee, tobacco, cotton, cacao, peanuts); horses and cattle. Unpaved road 12 mi/20 km W.

Las Paredes (lahs pah-RAI-des), town, central MENDOZA province, ARGENTINA, on railroad (Capitán Montoya station), 7 mi/11.3 km W of SAN RAFAEL, near DIAMANTE RIVER (irrigation area); 34°35′S 68°27′W. Wine, potatoes, fruit; wine making, dried fruit processing.

Las Parejas (lahs pah-RAI-hahs), town, S central SANTA FE province, ARGENTINA, 55 mi/89 km WNW of ROSARIO; 32°41′S 61°32′W. Agricultural center (wheat, corn, soybeans, sunflowers, flax; livestock); dairying.

Las Peñas (lahs PAIN-yahs), village, N CÓRDOBA province, ARGENTINA, 60 mi/97 km NNE of CÓRDOBA. Corn, alfalfa, livestock; quarrying, woodcutting.

Las Perdices (lahs per-DEE-ses), town, central CÓRDOBA province, ARGENTINA, 32 mi/51 km SW of VILLA MARIA; 32°41′S 63°41′W. Soybeans, grain; livestock.

Las Petas (lahs PAI-tahs), town and canton, ANGEL SANDOVAL province, SANTA CRUZ department, E BOLIVIA, 120 mi/193 km E of SAN IGNACIO, near BRAZIL border; 16°23′S 59°11′W.

Las Piedras, city (2004 population 69,222), CANELONES department, S URUGUAY, near railroad, 9 mi/14.5 km NNW of MONTEVIDEO; 34°41′S 56°08′W. In grape-growing and ostrich-farming district. Known for its

racetrack. Has Gothic chapel of the Salesians. Site of decisive victory (1811) of Artigas over the Spanish.

Las Piedras (lahs pee-AI-drahs), town and canton, MADRE DE DIOS province, PANDO department, N BOLIVIA, on MADRE DE DIOS RIVER, 5 mi/8 km W of RIBERALTA; 11°06′S 66°10′W. Rubber.

Las Piedras (lahs pee-AI-drahs), town (2006 population 38,631), E PUERTO RICO, 3 mi/4.8 km NW of HUMACAO. Industrial center. Native artisanry (wood, thread, and clay products, hammocks).

Las Piedras (LAHS pee-AI-drahs), village, FALCÓN state, NW VENEZUELA, on W shore of PARAGUANÁ PENINSULA, 26 mi/42 km WSW of PUEBLO NUEVO, outside of AMUAY; 11°45′N 70°20′W. Petroleum-shipping point; refinery.

Las Pilas (lahs PEE-lahs), W NICARAGUA, volcano (3,514 ft/983 m) in CORDILLERA DE LOS MARIBIOS, 14 mi/23 km ENE of LEÓN.

Las Piñas, PHILIPPINES: see PIÑAS, LAS.

Las Plumas (lahs PLOO-mahs), village, ⊙ Mártires department, E central CHUBUT province, ARGENTINA, on CHUBUT RIVER, 115 mi/185 km WSW of RAWSON; 43°43′S 67°15′W. Sheep raising.

Las Quemas (lahs KE-mahs), village, LLANQUIHUE province, LOS LAGOS region, S central CHILE, on MAULLÍN RIVER, 14 mi/23 km NW of PUERTO MONTT. In agricultural area (wheat, flax, potatoes; livestock); dairying, lumbering.

Lasqueti Island (las-KE-tee) (□ 26 sq mi/67.6 sq km; 2006 population 367), SW BRITISH COLUMBIA, W CANADA, in Strait of GEORGIA just S of TEXADA ISLAND, 50 mi/80 km WNW of VANCOUVER; 10 mi/16 km long, 3 mi/5 km wide; 49°29′N 124°16′W. Copper and gold mining in N central area. Village of Lasqueti is on W shore. Part of ISLANDS TRUST regional district.

Las Rosas (lahs RO-sahs), town (1991 population 11,022), ⊙ Belgrano department (□ 940 sq mi/2,444 sq km), S central SANTA FE province, ARGENTINA, 65 mi/105 km NW of ROSARIO; 32°28′S 61°35′W. Railroad junction and agricultural center (wheat, soybeans, corn, flax, alfalfa, fruit; livestock).

Las Rosas (lah RO-sahs), town, ⊙ Las Rosas municipio, CHIAPAS, S MEXICO, at SW foot of Sierra de HUEYTEPEC, 29 mi/47 km SE of SAN CRISTÓBAL DE LAS CASAS. Agricultural center (cereals, sugarcane, coffee, fruit; livestock; timber). Called PINOLA until 1934.

Lassa, CHINA: see LHASA.

Las Sabanas (lahs sah-BAH-nahs), town, MADRIZ department, NICARAGUA, S of SOMOTO.

Lassance (LAH-sahn-se), town (2007 population 6,284), N central MINAS GERAIS, BRAZIL, on RIO DAS VELHAS, 43 mi/69 km NW of CORINTO; 17°48′S 44°35′W.

Lassay-les-Châteaux (lah-sai–lai–shah-to), commune (□ 22 sq mi/57.2 sq km), MAYENNE department, PAYS DE LA LOIRE region, W FRANCE, 11 mi/18 km NNE of MAYENNE; 48°26′N 00°29′W. Flour milling. Has 15th-century fortified castle. The regional park of Normandy-Maine is just N.

Lassem, INDONESIA: see LASEM.

Lassen (LA-suhn), county (□ 4,558 sq mi/11,850.8 sq km; 2006 population 34,715), NE CALIFORNIA; ⊙ SUSANVILLE; 40°39′N 120°35′W. On high volcanic plateau (more than 4,000 ft/1,219 m) extending E from CASCADE RANGE; the SIERRA NEVADA is along SW and S borders, bounded by NEVADA on E. The county's highest elevation (more than 8,000 ft/2,438 m) is in the WARNER MOUNTAINS (NE). Drained by Pit and Susan rivers. Densely forested (pine, fir, cedar); logging and lumber milling are chief industries. Timber; stock grazing (cattle); farms (some irrigation) chiefly in HONEY LAKE valley produce potatoes, rice, wheat, barley, oats, garlic, and strawberries. Fishing, hunting, camping, and winter sports attract vacationers. Includes part of LASSEN VOLCANIC NATIONAL PARK in SW. Much of county is in national forests: Modoc (N), Plumas (S), and Lassen (W). Eagle and Honey lakes are here. Sierra Army Depot in SE, SE of Honey Lake. Region resisted (1863) California's jurisdiction until Lassen county was formed in 1864.

Lassen Peak or **Mount Lassen** (10,457 ft/3,187 m), SE SHASTA county, N CALIFORNIA, at S end of CASCADE RANGE, N of the SIERRA NEVADA, c.50 mi/80 km E of REDDING. In W part of LASSEN VOLCANIC NATIONAL PARK. Until MOUNT ST. HELENS erupted in 1980, Lassen Peak was considered to be the only active volcano in the continental U.S. The last major eruption here occurred in 1914, and the volcano was intermittently active until 1921. Discovered (probably 1821) by Luis Argüello, pioneer and guide. Later named for Peter Lassen, pioneer and guide. The peak was a prominent landmark in the mid-1800s for westward travelers to California. Lassen Peak Ski Area to S.

Lassen Volcanic National Park (□ 166 sq mi/430 sq km), SHASTA and LASSEN counties (also extends into PLUMAS and TEHAMA counties), N CALIFORNIA, at the S end of the CASCADE RANGE; 40°28′N 121°22′W. The park contains volcanic peaks, lava flows, vents, and hot springs. LASSEN PEAK (10,457 ft/3,187 m), in W part of park, is an active volcano. Established 1916.

Lassiter Coast, on the E coast of PALMER LAND, ANTARCTIC PENINSULA, between Cape Mackintosh and Cape Adams; 73°45′S 62°00′W.

L'Assomption (lah-sonp-SYON), county (□ 247 sq mi/ 642.2 sq km), LANAUDIÈRE region, S QUEBEC, E CANADA, on the SAINT LAWRENCE RIVER, N of MONTREAL; ⊙ L'ASSOMPTION; 45°45′N 73°30′W. Composed of six municipalities. Formed in 1983.

L'Assomption (lah-sonp-SYON), town (□ 39 sq mi/ 101.4 sq km), ⊙ L'ASSOMPTION county, LANAUDIÈRE region, S QUEBEC, E Canada, on L'ASSOMPTION RIVER, and 22 mi/35 km NNE of MONTREAL; 45°49′N 73°26′W. Lumbering, woodworking, light manufacturing; market in tobacco growing, dairying region. Site of Agriculture and Agri-Food Canada Experimental Farm. Part of the Metropolitan Community of Montreal (*Communauté Metropolitaine de Montréal*).

L'Assomption River (lah-sonp-SYON), French, *Rivière L'Assomption* (ree-VYER lah-sonp-SYON), 100 mi/161 km long, S QUEBEC, E CANADA; rises in the LAURENTIANS in MONTAGNE TREMBLANTE PARC; flows SE to JOLIETTE, thence S, past L'ASSOMPTION, to the SAINT LAWRENCE RIVER at N end of MONTREAL ISLAND; 45°42′N 73°28′W.

Lasswade, Scotland: see BONNYRIGG.

Las Tablas (lahs TAH-blahs), town (2000 population 7,980), ⊙ Las Tablas district, ⊙ LOS SANTOS province, central PANAMA, in PACIFIC lowland, on branch of INTER-AMERICAN HIGHWAY 90 mi/145 km SW of PANAMA city. Commercial center; agriculture (sugarcane, corn, rice, yucca, coffee, beans; horse raising). Formerly a gold-mining town.

Lastarria (lahs-tahr-REE-ah), village, CAUTÍN province, ARAUCANIA region, S central CHILE, on railroad, 35 mi/56 km S of TEMUCO; 39°14′S 72°41′W. In agricultural area (wheat, barley, peas, potatoes; livestock).

Lastarria Volcano, ARGENTINA and CHILE: see AZUFRE VOLCANO. La Station-du-Coteau, CANADA: see COTEAU-STATION.

Las Tejerías (LAHS tai-her-EE-ahs), town, ⊙ Santos Michelena municipio, ARAGUA state, N VENEZUELA, on TUY RIVER, on railroad and highway, 25 mi/40 km SW of CARACAS; 10°17′N 67°12′W. Coffee, cacao, corn, fruit.

Las Tetas de María Guevara, Monumento Natural (LAHS TAI-tahs dai mah-REE-ah gai-VAH-rah monoo-MEN-to nah-too-RAHL), natural monument (□ 6.5 sq mi/17 sq km), NUEVA ESPARTA state, on MARGARITA ISLAND; 10°55′N 64°07′W. Two hills emerge from the coastal plain, resembling the form of a woman. There are also three coastal lagoons surrounded by mangroves. Created 1974.

Last Frontier, The: see ALASKA.

Lástimas, Cerro (LAHS-tee-mahs, SER-ro), Andean peak (10,000 ft/3,048 m), LINARES province, MAULE region, S central CHILE, 25 mi/40 km SE of LINARES; 35°59′S 71°08′W.

Last Mountain (LAST), (2,275 ft/693 m), S central SASKATCHEWAN, CANADA, 50 mi/80 km NNW of REGINA; 51°10′N 105°15′W. Nearby is LAST MOUNTAIN LAKE.

Last Mountain Lake (LAST), (□ 89 sq mi/231.4 sq km), S central SASKATCHEWAN, CANADA, 22 mi/35 km NW of REGINA; 58 mi/93 km long, 3 mi/5 km wide; 51°10′N 105°15′W. Drains S into QU'APPELLE RIVER.

Las Torres (lahs TOR-rais), national park, in Aisén del GENERAL CARLOS Ibáñez del Campo region, 35 mi/56 km E of PUERTO CISNES.

Las Toscas (lahs TOS-kahs), town, NE SANTA FE province, ARGENTINA, on railroad, 60 mi/97 km NNE of RECONQUISTA. Agriculture center (sugar, flax, corn, potatoes, rice, tobacco, cotton; livestock); sugar refineries, forest industries.

Las Toscas, village, CANELONES department, S URUGUAY, on the Río de la PLATA, 25 mi/40 km ENE of MONTEVIDEO, near ATLÁNTIDA; 34°43′S 55°43′W.

Lastourville (lahs-TOOR-veel), city OGOOUÉ-LOLO province, S GABON, on OGOOUÉ RIVER, 30 mi/48 km NE of KOULAMOUTOU; 00°47′S 12°43′E. Cacao-producing area. Airport. Gold deposits nearby. Sometimes spelled Lastoursville.

Lastovo Island, Italian *Lagosta*, Dalmatian island in ADRIATIC SEA, S CROATIA, 60 mi/97 km W of DUBROVNIK; 6 mi/9.7 km long, 3 mi/4.8 km wide. Wine growing. Village of Lastovo (42°45′N 16°54′E) is on N shore. Lastovo Channel (Croatian *Lastovski Kanal*) separates it from Korčula Island (N).

Lastra a Signa (LAH-strah ah SEE-nyah), town, FIRENZE province, TUSCANY, central ITALY, on ARNO RIVER, 8 mi/13 km W of FLORENCE; 43°46′N 11°06′E. Manufacturing (chemicals, textiles, clothing).

Las Trancas Reserve, LA PAZ department, SW HONDURAS, between OPATORO and GUAJIQUIRO; 14°06′N 87°50′W. Highland forest, quetzal habitat. Small wildlife reserve (foot access only).

Lastras de Cuéllar (LAH-strahs dhai KWE-lyahr), town, SEGOVIA province, central SPAIN, 23 mi/37 km N of SEGOVIA; 41°18′N 04°06′W. Cereals, grapes; resins; livestock. Lumbering; flour milling, pottery manufacturing.

Lastres (LAH-stres), village, OVIEDO province, NW SPAIN, fishing port on BAY OF BISCAY, 20 mi/32 km E of GIJÓN; 43°31′N 05°16′W. Fish processing, flour milling. Apple orchards in area. Coal mining.

Lastres, Cape, OVIEDO province, NW SPAIN, on BAY OF BISCAY, 18 mi/29 km E of GIJÓN; 43°34′N 05°18′W.

Las Tres Vírgenes, Volcán (lahs trais VIR-he-nes), volcanic massif in BAJA CALIFORNIA SUR, NW MEXICO, near coast of Gulf of CALIFORNIA, 25 mi/40 km NW of SANTA ROSALÍA in EL VIZCAÍNO BIOSPHERE RESERVE. Rises to 6,547 ft/1,996 m. Last erupted 1746.

Las Trincheras (LAHS trin-CHAI-rahs), town, CARABOBO state, N VENEZUELA, 9.0 mi/14.5 km NNW of VALENCIA; 10°18′N 68°05′W. Elevation 2,191 ft/667 m. Health resort with thermal springs.

Las Trojes, Honduras: see TROJES.

Lastrup (LAHS-trup), village, LOWER SAXONY, NW GERMANY, 8 mi/12.9 km WSW of CLOPPENBURG; 52°47′N 07°52′E. Livestock (hogs, chickens).

Lastrup (LAS-truhp), village (2000 population 99), MORRISON county, central MINNESOTA, 17 mi/27 km ENE of LITTLE FALLS on Little Mink Creek; 46°02′N 94°03′W. Dairying; manufacturing (feeds).

Las Tunas (lahs TOO-nahs), province (□ 2,662 sq mi/ 6,921.2 sq km; 2002 population 525,485), E CUBA; ⊙ LAS TUNAS. Created from former Camagüey and ORIENTE provinces in 1976. Bordered N by ATLANTIC OCEAN, S by GULF OF GUACANAYABO and GRANMA province, E

by HOLGUÍN province, and W by CAMAGÜEY province. Important centers are LAS TUNAS, PUERTO PADRE, Amancio, and JOBABO. The province produces c.9% of the nation's total amount of sugar each year; and processing at 3 mills (Peru, Argelia Libre, and Antonio Guiteras) is just over half of provincial total.

Las Tunas (lahs TOO-nuhz), town (2002 population NaN), ⊙ LAS TUNAS province, E CUBA, on Central Highway and railroad, 100 mi/161 km NW of SANTIAGO DE CUBA; 20°56′N 76°56′W. Linked by railroad with MANATÍ (25 mi/40 km N) on N coast. Trading and agricultural center (sugarcane, bananas, oranges, cattle, beeswax, honey). Airfield, with 7,200 ft/2,195 m runway. Steel structures and glass are manufactured there. Scene of fighting in revolutionary war of 1950s.

La Suiza (lah SWEE-sah), village, CARTAGO province, central COSTA RICA, 6 mi/9.7 km SE of TURRIALBA; 09°51′N 83°37′W. In coffee area; livestock raising.

Lašva (LAHS-vah), hamlet, central BOSNIA, BOSNIA AND HERZEGOVINA, on BOSNA RIVER, at mouth of Lašva River, and 5 mi/8 km S of ZENICA, on ZENICA-SARAJEVO railroad; 44°08′N 17°56′E. Also spelled Lashva.

Las Vacas (lahs VAH-kahs), village, COQUIMBO region, N central CHILE, on railroad, 18 mi/29 km SW of ILLAPEL. In agricultural area (fruit, grain; livestock); gold mining.

Las Varillas (lahs vah-REE-yahs), town, E CÓRDOBA province, ARGENTINA, 50 mi/80 km SW of SAN FRANCISCO; 31°52′S 62°43′W. Railroad junction and agricultural center (wheat, flax, corn; livestock); produces casein, cheese.

Lašva River (LAHS-vah), 30 mi/48 km long, central BOSNIA, BOSNIA AND HERZEGOVINA; rises NNE of DONJI VAKUF, at S foot of Vlašić Mountains; flows ESE, past TRAVNIK (hydroelectric plant), to BOSNA RIVER at Lašva. Also spelled Lashva River.

Las Vegas, city (□ 83 sq mi/215.8 sq km; 2006 population 552,539), ⊙ Clark co., SE Nevada; 36°12′N 115°13′W. Elev. 2,025 ft/617 m. Only the N ½ of physical city (much of it commercial) is incorporated The area generally S of Sahara Avenue, including the Las Vegas Boulevard gambling and entertainment district (the Strip) is unincorporated, and its population is counted as part of the unincorporated cities of Paradise and Winchester. Commonly referred to as 'Sin City,' it is the largest city in Nevada and one of the fastest growing cities in the country, which is largely caused by an influx of retirees and the growth of industries that support the retirement community. The city's population increased by more than 56% between 1980 and 1990. It is considered *the* gambling capital of the world; gambling was legalized in 1931. Revenue from hotels, gambling, entertainment, and other tourist-oriented industries forms the backbone of Las Vegas's economy. Its nightclubs, casinos, and championship boxing matches are world famous, and entertainment enterprises have led to an increasing array of music, sports, and gambling centers up and down the Strip, a.k.a "Glitter Gulch." The city is also the commercial hub of a ranching and mining area. Manufacturing (paper products, printing and publishing, apparel, chemicals, signs, machinery, gaming equipment and devices, electronics, dairy products, fabricated metal products, lumber, building materials, hardware, beverages, jewelry, cut stone and stone products, furniture, consumer goods). In the 19th century, Las Vegas was a watering place for travelers to S California In 1855–1857 the Mormons maintained a fort there, and in 1864, Fort Baker was built by the U.S. Army. In 1867, Las Vegas was detached from the Arizona Territory and joined to Nevada Its main growth began with the completion of a railroad in 1905. The University of Nevada is in unincorporated S part of city. Large Nellis Air Force Bombing and Gunnery Range to the NE of the city; Hoover Dam is

---

Area is shown by the symbol □, and capital city or county seat by ⊙.

25 mi/40 km SE, forms large Lake Mead (L. Mead National Reservoir Area). Old Las Vegas Mormon Fort State Park at Las Vegas Boulevard and Washington Avenue Floyd Lamb State Park to NW. Mt. Charleston, in Toiyabe National Forest to W. Desert View Natural Area to NW. Indian Springs Air Force Base to NW. Valley of Fire State Park to NE. Pabco Gypsum Mines 14 mi/23 km to NE. McCarran International Airport in S unincorporated part of city. Convention Center in unincorporated S part of city. Nevada State Museum; Las Vegas Natural History Museum Inc. 1911.

Las Vegas, city (2006 population 13,889), ⊙ SAN MIGUEL county, N central NEW MEXICO, on GALLINAS RIVER, in SANGRE DE CRISTO MOUNTAINS, 64 mi/103 km ESE of SANTA FE, in irrigated region; 35°36′N 105°13′W. Elevation 6,435 ft/1,961 m. Las Vegas is a mountain and health resort, as well as shipping center for livestock, lumber, wool, and hides in grain (alfalfa, peas) and livestock (cattle, sheep) area. New Mexico Highlands University is here. Hot springs 6 mi/9.7 km N; ruins of FORT UNION (1851–1891), dude ranches, and parts of Santa Fe National Forest to W and S; Las Vegas National Wildlife Refuge to SE, at McAllister Lake; Storrie Lake State Park to N; Villanueva State Park to SSW. Founded 1609.

Las Vegas, town (1988 population 7,307) and municipio; Santa Bárbara department, on road to El Mochito (formermining center), borders Lake Yojoa on E, 10 mi/16 km ESE of Santa Bábara; 14°42′N 88°03′W. Lacks direct road access. Independent smallmining, coffee.

Las Vegas, CUBA: see VEGAS.

Las Vigas de Ramírez, town, ⊙ Las Vigas de Ramírez municipio, VERACRUZ, E MEXICO, in SIERRA MADRE ORIENTAL, on railroad, 13 mi/21 km NW of XALAPA ENRÍQUEZ on Mexico Highway 140. Elevation 4,265 ft/1,300 m. Agricultural center (cereals, coffee, fruit).

Las Vueltas (lahs VWEL-tahs), municipality and town, CHALATENANGO department, EL SALVADOR, NE of CHALATENANGO city.

Laswari (luhs-WAH-ree), village, ALWAR district, E RAJASTHAN state, NW INDIA, 20 mi/32 km E of Alwar. In 1803, British defeated Marathas in nearby battle.

Las Yaras (LAHS YAH-rahs), town, TACNA region, S PERU, in irrigated zone on SAMA RIVER, 25 mi/40 km NW of TACNA; 17°51′S 70°33′W. Elevation 1,240 ft/377 m. Sugarcane, rice, cotton, corn.

Lataband Pass (lah-TAH-bahnd) (5,900 ft/1,798 m), in N outlier of the SAFED KOH, E AFGHANISTAN, 21 mi/34 km E of KABUL, on highway from Khyber Pass; 34°30′N 69°34′E.

La Tablada, ARGENTINA: see TABLADA.

Latacunga (lah-tah-KOON-gah), city (2001 population 51,689), COTOPAXI province, N central ECUADOR, c.20 mi/32 km SW from COTOPAXI volcano; 00°56′S 78°37′W. A town of the ancient Incas and Puruhá Indians (before the Incas), it is in a high mountain basin (c.9,100 ft/2,774 m) between the E and W Andean cordilleras. It lies within an agricultural and livestock region. Manufacturing include ceramics, ammunition, and furniture. Because of its proximity to the active Cotopaxi volcano, the city has suffered severe earthquakes. A gateway to the Parque Nacional COTOPAXI.

Latady Island, S of CHARCOT Island and W of ALEXANDER Island in the E BELLINGSHAUSEN Sea, 35 mi/60 km long and over 10 mi/16 km wide; 70°45′S 74°35′W. Ice-covered.

Latah (LAI-tah), county (□ 1,076 sq mi/2,797.6 sq km; 2006 population 35,029), in Palouse region, N IDAHO; ⊙ MOSCOW; 46°49′N 116°43′W. Situated between Coeur d'Alene (N) and Nez Perce (S) Indian reservations. Long, rolling hills, drained by PALOUSE and POTLATCH rivers. Borders on WASHINGTON (W).

Lumber; sheep, cattle; alfalfa; wheat, oats, barley; peas, lentils. Includes part of St. Joe National Forest in N. Formed 1888.

Latah (LAI-tah), village (2006 population 151), SPOKANE county, E WASHINGTON, 28 mi/45 km SE of SPOKANE, on Hangman Creek; 47°17′N 117°09′W. Grain; peas; sheep, hogs.

Lata, La, URUGUAY: see CARDONA.

La Tapera (lah tah-PE-rah), town, ⊙ Lago Verde comuna, COIHAIQUE province, AISÉN DEL GENERAL CARLOS IBAÑEZ DEL CAMPO, S CHILE, 70 mi/113 km NE of COIHAIQUE. Tourist resort on road to ARGENTINA.

Latchford (LACH-fuhrd), town (□ 2 sq mi/5.2 sq km; 2001 population 363), E ONTARIO, E central CANADA, on MONTREAL RIVER, 12 mi/19 km SW of HAILEYBURY; 47°20′N 73°54′W. In mining region (silver, cobalt, nickel, bismuth, arsenic); lumbering. World's shortest covered bridge (18 ft/5 m).

Latchingdon (LACH-ing-duhn), village (2006 population 1,203), ESSEX, SE ENGLAND, 4 mi/6.4 km SE of MALDON; 51°40′N 00°45′E.

Late (LAH-TAI), cultivated, but usually uninhabited, island, N PACIFIC OCEAN, W VAVA'U group; 18°48′N 174°39′E. Rises to 1,827 ft/557 m. Volcano dormant since 1854.

La Tebaida (LAH tai-BEI-dah), town, ⊙ La Tebaida municipio, QUINDÍO department, W central COLOMBIA, on railroad, 8.0 mi/12.9 km SW of ARMENIA; 04°27′N 75°47′W. Elevation 3,907 ft/1,190 m. Coffee, plantains, cassava; livestock.

Låtefoss (LAW-tuh-faws), waterfall (538 ft/164 m) in HORDALAND county, SW NORWAY, 8 mi/12.9 km S of ODDA. Tourist attraction.

Latehar (lah-teh-HAR), district (□ 3,825 sq mi/9,945 sq km; 2001 population 467,071), W central Jharkhand state, E INDIA, ⊙ Latehar. Created in 2001 out of PALAMU district. Primarily tribal population. Agricultural district: rice, maize, cereals, wheat, oil seeds. Rich in mineral deposits: coal, bauxite, laterite, dolomite, graphite, granite, quartz, fireclay, felspar. Through the 1990s and 2000s, the district has been a site of Naxalite activity.

Latehar, town (2001 population 19,067), ⊙ LATEHAR district, W Jharkhand state, E central INDIA.

Lateiki, TONGA: see METIS SHOAL.

La Tène (lah TEN-uh), ancient Celtic site on LAKE NEUCHÂTEL, NW SWITZERLAND, that gives its name to the cultures of the Late Iron Age. It is characterized by an art style that drew upon Greek, Etruscan, and Scythian motifs and translated them into abstract designs in metal, pottery, and wood. The earliest phase of La Tène culture, from the 6th to the late 5th century B.C.E. (the period of the first of the great Celtic migrations), spread from the middle RHINE region E into the DANUBE valley, S into Switzerland, and W and N into FRANCE, the LOW COUNTRIES, DENMARK, and the BRITISH ISLES. It flourished until overrun by Roman culture.

Laterrière or **La Terrière** (both: lah-te-ree-ER), former town, S central QUEBEC, E CANADA, 9 mi/14 km SSW of CHICOUTIMI; 48°18′N 71°07′W. Lumbering; dairying. Amalgamated into Saguenay in 2002.

Laterza (lah-TER-tsah), town, Taranto province, APULIA, S ITALY, 10 mi/16 km ESE of MATERA; 40°37′N 16°48′E. In agricultural region (olives, grapes, cereals). Has castle and cathedral.

Latex (LAI-teks), village, HOUSTON county, E TEXAS, 5 mi/8 km N of CROCKETT. Agricultural area; oil and natural gas. Manufacturing (chemicals, plastics).

Latgale (LAHT-gah-lai) or **Latgallia**, region and former district, LATVIA, N of the DVINA (DAUGAVA) River. Flax growing and processing. DAUGAVPILS was the chief city. The region was settled by the Latgalians, a Baltic tribe, who were closely akin to the Letts and spoke a Lettish dialect. Latgale shared the history of

LIVONIA (of which it formed the S part) until 1561, when it passed to POLAND. Unlike the rest of Latvia, however, Latgale retained Roman Catholicism. The area was ceded to Russia during the Polish partition of 1772. In 1918 it became part of newly independent Latvia.

Latham (LAI-thuhm), unincorporated community, MONITEAU county, central MISSOURI, 8 mi/12.9 km SW of CALIFORNIA.

Latham (LAI-thuhm), city (2000 population 164), BUTLER county, SE KANSAS, 23 mi/37 km SE of EL DORADO; 37°32′N 96°38′W. In cattle region. Butler State Fishing Lake to NW.

Latham (LAI-thuhm), village (2000 population 371), LOGAN county, central ILLINOIS, 13 mi/21 km NW of DECATUR; 39°58′N 89°09′W. In agricultural area.

Lathen (LAH-tuhn), village, LOWER SAXONY, NW GERMANY, on right bank of EMS RIVER, 12 mi/19 km N of MEPPEN; 52°51′N 07°18′E. Site of test track for MAGLEV (magnetic levitation train).

Lathi (LAH-tee), town, AMRELI district, GUJARAT state, W central INDIA, 50 mi/80 km W of BHAVNAGAR. Local agriculture market (millet, cotton). Was capital of former EAST KATHIAWAR state of LATHI of WEST INDIA STATES agency; state merged 1948 with SAURASHTRA and later (1960) with Gujarat state.

Lathqiya, Al-, SYRIA: see LATTAKIA, district.

Lathrop (LAI-thruhp), city (2000 population 10,445), SAN JOAQUIN county, central CALIFORNIA, 9 mi/14.5 km S of STOCKTON, near SAN JOAQUIN RIVER; 37°49′N 121°17′W. Railroad junction. Fruit, pumpkins, nuts, vegetables, grain, sugar beets; nursery products; dairying; cattle; manufacturing (concrete, glass, and rubber products; building components, pesticides, feeds, prefabricated buildings). Caswell Memorial State Park to S.

Lathrop, city (2000 population 2,092), CLINTON county, NW MISSOURI, 30 mi/48 km SE of SAINT JOSEPH; 39°32′N 94°19′W. Corn, soybeans; cattle; historic mule market (shipped 170,000 mules during Boer War in South Africa and 90,000 to Europe during World War I). Manufacturing (candy). Platted 1867.

Lathrup Village (LA-thrup), village (2000 population 4,236), OAKLAND county, SE MICHIGAN, residential suburb 15 mi/24 km NW of DETROIT; 42°29′N 83°13′W. Completely surrounded by city of SOUTHFIELD. Light manufacturing.

La Tigra National Park (lah TEE-grah), reserve and buffer zone (□ 92 sq mi/239.2 sq km), FRANCISCO MORAZÁN department, S central HONDURAS, 7 mi/11 km NE of TEGUCIGALPA. Protects part of the Tegucigalpa watershed; a former mining zone, most of the park is secondary forest.

Latimer, county (□ 729 sq mi/1,895.4 sq km; 2006 population 10,562), SE OKLAHOMA; ⊙ WILBURTON; 34°52′N 95°14′W. Drained by small Fourche Maline Creek; in OUACHITA MOUNTAINS Cattle; lumbering, coal mining (though activity has declined); manufacturing (apparel, electric motors); oil and natural-gas wells; some agriculture. Includes Robbers Cave State Park in NW. SARDIS LAKE reservoir on S boundary; Sansbois Mountains (extension of Ouachitas) on N boundary. Formed 1907.

Latimer, town, FRANKLIN county, N central IOWA, 8 mi/12.9 km W of HAMPTON. In livestock and grain area.

Latimer, unincorporated town (2000 population 4,288), JACKSON county, SE MISSISSIPPI, 9 mi/14.5 km NNE of BILOXI, near Tchoutacabouffa River, at SE edge of De Soto National Forest; 30°30′N 88°51′W. Timber.

Latimer, village (2000 population 21), MORRIS county, E central KANSAS, 20 mi/32 km W of COUNCIL GROVE; 38°44′N 96°50′W. Grazing, farming.

Latina (lah-TEE-nah), province (□ 868 sq mi/2,256.8 sq km), LATIUM, S central ITALY; ⊙ LATINA; 41°27′N

13°06′E. Borders on TYRRHENIAN SEA, with reclaimed PONTINE MARSHES on coast, surrounded by Apennine hills. Agriculture (cereals, grapes, olives, citrus fruit, sugar beets); stock raising (cattle, sheep). Fishing (GAETA, TERRACINA). Manufacturing at FORMIA, Latina, POMEZIA, and APRILIA. Tourism, bathing resorts, Circeo National Park. Formed 1934 from ROMA province. Until 1947 called Littoria.

Latina (lah-TEE-nah), city (2001 population 107,898), ⊙ LATINA province, in LATIUM, central ITALY, near the TYRRHENIAN SEA. It is an industrial, commercial, and agricultural center. Manufacturing includes fabricated metals, machinery, textiles, paper, wood products, tires, chemicals, and processed food. There is a nuclear power station in the city. It was the first community founded (1932) by Mussolini in the reclaimed PONTINE and was known as Littoria until 1947.

Latin America, term referring to the Spanish-speaking, Portuguese-speaking, and French-speaking countries (except Canada) of N. America, S. America, Central America, and the West Indies. The twenty republics are Argentina, Bolivia, Brazil, Chile, Colombia, Costa Rica, Cuba, Dominican Republic, Ecuador, El Salvador, Guatemala, Haiti, Honduras, Mexico, Nicaragua, Panama, Paraguay, Peru, Uruguay, and Venezuela. The term also includes Puerto Rico, the French West Indies, and other islands of the West Indies where a Romance language is spoken. Occasionally the term is used to include Belize, Guyana, French Guiana, and Suriname as well.

Latin Quarter (LA-tin kwor-tuhr), French, *Quartier Latin* (ka-tyai lah-tan), section of PARIS, ÎLE-DE-FRANCE region, FRANCE, on the left bank of the SEINE RIVER, administratively divided between the 5th (Panthéon) and the 6th (Luxembourg) *arrondissements*. Here the University of Paris and related educational and cultural institutions have been concentrated since the 12th century, giving the district an intellectual allure.

Latir Peak (luh-TIR) (12,708 ft/3,873 m), SANGRE DE CRISTO MOUNTAINS, NE TAOS county, N NEW MEXICO, near COLORADO state line, 9 mi/14.5 km NE of QUESTA on N boundary of Carson National Forest. Latir Lakes to NE.

Latisana (lah-tee-ZAH-nah), town, UDINE province, FRUILI-VENEZIA GIULIA, NE ITALY, on TAGLIAMENTO RIVER, 23 mi/37 km SSW of UDINE; 45°46′N 12°59′E. Agricultural center; wine making. Sustained World War II damage and has also suffered from floods.

Latium (LAI-shee-uhm), Italian *Lazio*, region (□ 6,642 sq mi/17,269.2 sq km), central ITALY, extending from the APENNINES westward to the TYRRHENIAN SEA; ⊙ ROME. The region is divided into FROSINONE, LATINA, RIETI, ROME, and VITERBO provinces (named for their capitals). The region is mostly hilly and mountainous, with a narrow coastal plain, much of which has been reclaimed in the 20th century. Agriculture forms the backbone of the regional economy; products include cereals, vegetables, grapes, olives, and fodder. Sheep and cattle are raised. Rome is Latium's main commercial, service, and industrial center. Industry in the region has been spurred (mid-20th century) by the construction of hydroelectric facilities on the ANIENE and LIRI rivers and a nuclear power plant at Latina. Manufacturing includes chemicals, cement, textiles, construction materials, and processed food. There is a large tourist industry, and fishing is pursued along the coast, especially at CIVITAVECCHIA, the region's chief port. In ancient times, Latium comprised a limited area E and S of the TIBER River that extended to the ALBAN HILLS; only after it became part of Italy in 1870 did it approximately reach its present limits. In early Roman times Latium was inhabited by the Latins, the Etruscans (N of the Tiber River), and several Italic tribes. In the 3rd century

B.C.E., Rome subdued all of Latium. The fertile coastal plain became marshy, malaria-infested, and impoverished during the late Roman Empire and early years of the republic. After the fall of Rome, Latium was invaded in turn by the Visigoths, the Vandals, and the Lombards. From the 8th century the duchy of Rome, including most of modern Latium, belonged to the popes. Their authority was not always recognized in the towns, which were ruled at times as free communes or by local feudal lords. Except for the area S of TERRACINA, which belonged to the kingdom of NAPLES, Latium remained a part of the PAPAL until 1870. In World War II, S Latium was the scene of bloody battles during the Allied drive on Rome. There are two universities in Rome, which is also the site of the VATICAN.

Lat Lum Kaeo (LAHT LUM KA-OO), village and district center, PATHUMTHANI province, S THAILAND, 20 mi/32 km NE of BANGKOK, in irrigated area; 14°02′N 100°25′E. Rice-milling center. Also spelled Lam Luk Ka.

Latnaya (LAHT-nah-yah), town (2006 population 7,200), W VORONEZH oblast, S central European Russia, on road and railroad, 14 mi/23 km W of VORONEZH; 51°39′N 38°54′E. Elevation 472 ft/143 m. Construction materials (bricks, concrete panels). Refractory clay and quartzite quarries.

La Tolita Island, ECUADOR: see TOLA, ISLA DE LA.

La Toma (lah TO-mah), town (1991 population 5,612), ⊙ Pringles or Coronel Pringles department (□ 1,625 sq mi/4,225 sq km), NE SAN LUIS province, ARGENTINA, on railroad, and 45 mi/72 km ENE of SAN LUIS. Farming and mining center. Agricultural products (flax, wheat, corn, sunflowers; livestock). Travertine, granite, onyx, quartzite, feldspar, beryllium, strontium sulphate deposits nearby. Formerly called Cuatro de Junio or 4 de Junio.

La Toma (lah TO-mah), resort, PERAVIA province, S DOMINICAN REPUBLIC, 15 mi/24 km W of SANTO DOMINGO.

Laton, unincorporated town (2000 population 1,236), FRESNO county, central CALIFORNIA, 7 mi/11.3 km NNW of HANFORD, on KINGS RIVER; 36°26′N 119°42′W. Citrus, nuts, cotton, grain; cattle; dairying.

Latorica River (lah-TO-ri-TSAH), (Hungarian *Latorca*), c.127 mi/204 km long, UKRAINE (W TRANSCARPATHIAN section) and SE SLOVAKIA (VYCHODOSLOVENSKY province); rises in Ukraine on S slope of the BESHCHADY MOUNTAINS 13 mi/21 km WNW of VERETS′KYY PASS; flows S and W past SVALYAVA and MUKACHEVE, joining ONDAVA RIVER 31 mi/50 km SE of KOŠICE (Slovakia) to form BODROG RIVER; length in Slovakia, 38 mi/61 km.

La Torre (lah TO-rai), canton, SUD CINTI province, CHUQUISACA department, SE BOLIVIA, 10 mi/16 km S of VILLA ABECIA; 21°11′S 65°30′W. Elevation 7,575 ft/2,309 m. Clay, limestone, and gypsum deposits. Agriculture (potatoes, yucca, bananas, corn, wheat, oats, rye, peanuts); cattle and pig raising.

Latorytsya River (lah-to-RI-tsyah), 62 mi/100 km long; a tributary of the TYSA River, in TRANSCARPATHIAN oblast, W UKRAINE. Floods frequently since 2002 due to increased rainfall and dike accidents in bordering HUNGARY.

Latouche Island (lah-TOOSH), S ALASKA, in Gulf of ALASKA, at entrance of PRINCE WILLIAM SOUND, off E coast of KENAI PENINSULA, 50 mi/80 km E of SEWARD; 60°00′N 147°55′W. Rises to c.2,000 ft/610 m; 12 mi/19 km long, 2 mi/3.2 km–4 mi/6.4 km wide. Named by George Vancouver in 1794. Latouche village, NW.

La Tour (luh TOOR), unincorporated town, JOHNSON county, W central MISSOURI, 12 mi/19 km E of HARRISONVILLE.

Latour-de-Carol (lah-TOOR–duh–kah-ROL), commune (□ 4 sq mi/10.4 sq km; 2004 population 84), PYRÉNÉES-ORIENTALES department, Cerdagne (CER-DAÑA) region, in LANGUEDOC-ROUSSILLON administrative region, S FRANCE, on Spanish border, 3 mi/4.8 km NW of PUIGCERDÁ (Spain); 42°28′N 01°54′E. It is the border crossing station of the trans-Pyrenean railroad c.7 mi/11.3 km SE of tunnel under PUYMORENS pass (a crossing of the principal range of the Pyrenees). Formerly spelled La Tour-de-Carol.

La Tour-de-Peilz, SWITZERLAND: see TOUR-DE-PEILZ, LA.

Látrabjarg (LOU-trah-BYAHRK), steep rocky cape, W extremity of ICELAND, at tip of VESTFJARÐA Peninsula, on DENMARK Strait; 65°30′N 24°32′W. Lighthouse; scene of many shipwrecks.

La Trinidad (lah tree-nee-DAHD), town, COMAYAGUA department, N central HONDURAS, on COMAYAGUA RIVER, 19 mi/30 km N of COMAYAGUA; 14°04′N 88°34′W. Small farming, corn, beans; livestock.

La Trinidad (lah tree-nee-DAHD), town, ESTELÍ department, W NICARAGUA, 16 mi/26 km SSE of ESTELÍ, on INTER-AMERICAN HIGHWAY. Rice, corn; livestock, poultry.

La Trinidad, PHILIPPINES: see TRINIDAD, LA.

La Trinidad Vista Hermosa (lah tree-nee-DAHD VEES-tah er-MO-sah), town, NW OAXACA, MEXICO, 20 mi/32 km E of HUAJUAPAM de LEÓN; 17°42′N 97°32′W. Elevation 7,152 ft/2,180 m. Agricultural resources. Cold climate. A Mixtec community.

La Trinitaria, MEXICO: see VILLA LA TRINITARIA.

La Trinité-des-Monts (lah tree-nee-TAI–dai–MON), parish (□ 90 sq mi/234 sq km; 2006 population 278), BAS-SAINT-LAURENT region, S QUEBEC, E CANADA, 22 mi/36 km from RIMOUSKI; 48°08′N 68°28′W.

Latrobe (luh-TROB), town, N TASMANIA, SE AUSTRALIA, 40 mi/64 km WNW of LAUNCESTON, near mouth of MERSEY RIVER; 41°14′S 146°25′E. Commuter satellite of DEVONPORT, 6 mi/9 km N. Sheep, cattle; forage crops, wheat, barley, vegetables (peas, potatoes), poppies, and orchards; dairying; wood fiber and paper mills. Significant retirement location.

Latrobe, borough (2006 population 8,561), WESTMORELAND county, SW PENNSYLVANIA, 33 mi/53 km ESE of PITTSBURGH, on LOYALHANNA CREEK, in the foothills of the ALLEGHENY MOUNTAINS; 40°18′N 79°22′W. Manufacturing (foam rubber, asphalt, lumber, printing and publishing, building materials, steel, plastic products, beer). Agriculture (corn, hay; livestock; dairying. St. Vincent College to SW. County Airport to SW. Keystone State Park to N. Incorporated 1854.

Latrobe, Mount, AUSTRALIA: see WILSON'S PROMONTORY.

La Tronosa Forest Reserve, LOS SANTOS province, PANAMA, S part of AZUERO PENINSULA.

Latrun (lah-TROON), historic site, ISRAEL, in JUDEAN FOOTHILLS, 15 mi/24 km WNW of JERUSALEM. Captured by Israel in 1967 Six Day War. Was a road center. Site of a large Trappist monastery with vineyards and olive orchards known for their wine and oil production. The monastery was first built in 1890, destroyed by the Ottomans during World War I, and rebuilt in 1927. The new highway to Jerusalem runs in the vicinity of the monastery. On the hill above the monastery are the ruins of a Crusader's fort, which was an important stronghold on the road from the coast to Jerusalem. Site of large British detention camp after 1945. In 1948 Arab stronghold athwart road from coast to Jerusalem; blowing up of pumping station here cut Jerusalem's water supply. On hill below monastery is a former British police station during mandate times. Station is now a memorial site for Israeli armored and tank corp. Tank museum.

Latta (LAT-ah), town (2006 population 1,490), DILLON county, NE SOUTH CAROLINA, 21 mi/34 km NE of FLORENCE; 34°20′N 79°25′W. Manufacturing includes plastics, apparel, ordnance. Trading and shipping center for cotton, grain, soybeans, livestock.

Lattakia (lah-tah-KI-ye), Arabic *Al-lathqiya*, district (□ 2,433 sq mi/6,325.8 sq km), W SYRIA, on the MEDITERRANEAN Sea, bounded E by Jebel Ansariya (Ansariya Range), now officially Al Jibal As-Sahiliya; ☉ LATTAKIA. A region of tobacco, cotton, cereals, olives, figs, and sericulture. Agriculture in the LATTAKIA region will be served by the Tishin Dam (now under construction) on the Kebir Shemali River After World War I and under the French mandate of Syria, the area had a semi-autonomous status as the territory of the Alawites or Alaouites until it was incorporated into the newly formed Syrian republic.

Lattakia, city, ☉ LATTAKIA district, W SYRIA, on the MEDITERRANEAN Sea; 35°31′N 35°46′E. It is Syria's leading port, exporting bitumen, asphalt, cereals, raw cotton, fruit, and the famous Lattakia tobacco (cultivated since the 17th century). Industries include sponge fishing, vegetable-oil milling, and cotton ginning. Formerly the ancient Phoenician city of Ramitha, it was rebuilt (c.290 B.C.E.) by Seleucus I and later prospered as the Roman Laodicea ad Mare. Byzantines and Arabs fought over it from the 7th to 11th century. The city was captured in 1098 by the Crusaders and flourished in the 12th century until after its capture in 1188 by Saladin. From the 16th century to World War I it was part of the Ottoman Empire. While Syria was under the French League of Nations mandate, LATTAKIA was capital of the autonomous territory of the Alawites (1920–1942). A deep-water port was completed in 1959. Air terminal. Naval base; ship repair facilities. Landmarks include ancient columns and a Roman arch. The city is the seat of the University of Lattakia.

Lattes (lah-tuh), town, HÉRAULT department, LANGUEDOC-ROUSSILLON region, S FRANCE, near the MEDITERRANEAN coast, 3 mi/4.8 km SSE of MONTPELLIER; 43°34′N 03°54′E. Archeologists have uncovered (since 1963) the once-flourishing port of Lattara (6th century B.C.E.–3rd century C.E.), which eventually sank into marshland with the rise of the sea level. Nearby museum exhibits and documents archaeological discoveries.

Lattimore (LAT-uh-mor), village (□ 1 sq mi/2.6 sq km; 2006 population 417), CLEVELAND county, S NORTH CAROLINA, 6 mi/9.7 km W of SHELBY; 35°19′N 81°39′W. Manufacturing; agriculture (cotton, grain; livestock).

Lattingtown, residential village (□ 3 sq mi/7.8 sq km; 2000 population 1,860), NASSAU county, SE NEW YORK, on N shore of W LONG ISLAND, just NE of GLEN COVE; 40°53′N 73°35′W. Near marshlands. Bailey Arboretum here.

Latty (LAT-ee), village (2006 population 188), PAULDING county, NW OHIO, 17 mi/27 km SW of DEFIANCE; 41°05′N 84°35′W. In agricultural area.

Latulipe-et-Gaboury (lah-too-LEEP-et-gah-boo-REE), village (□ 115 sq mi/299 sq km; 2006 population 319), ABITIBI-TÉMISCAMINGUE region, SW QUEBEC, E CANADA; 47°24′N 79°00′W. Formed 1924.

La Tuque (lah tuhk), town, S QUEBEC, CANADA, on the St. Maurice River, NW of QUEBEC; 47°27′N 72°47′W. La Tuque, in a lumbering and farming region, was established as a trading post in the French period; it grew after the coming of the railroad in 1908. Pulp and paper center with a hydroelectric power station.

La Tuque (lah TOOK), town (□ 10,973 sq mi/28,529.8 sq km), MAURICIE region, S QUEBEC, E CANADA, on SAINT MAURICE River, NW of QUEBEC city, and 80 mi/129 km N of TROIS-RIVIÈRES; 47°27′N 72°47′W. In lumbering, agricultural region; pulp milling center, manufacturing. Hydroelectric power station. Was established in the French period as a trading post; grew after the coming of the railroad in 1908. Restructured in 2003.

Latur (LAH-toor), district (□ 2,763 sq mi/7,183.8 sq km), MAHARASHTRA state, W central INDIA; ☉ Latur.

Latur (LAH-toor), city (2001 population 299,985), LATUR district, MAHARASHTRA state, central INDIA, 39 mi/63 km ENE of OSMANABAD; 18°24′N 76°35′E. Railroad spur terminus; cotton ginning and grain trade (chiefly wheat, millet, rice) center. Ships cotton to BARSI mills, 60 mi/97 km W, by railroad. LATUR Road station is on another railroad, 19 mi/31 km ENE. A severe 1993 earthquake killed thousands of people here.

Latvia (LAT-vee-ah), *Latvija*, republic (□ 24,590 sq mi/63,688 sq km; 2004 estimated population 2,306,306; 2007 estimated population 2,259,810); ☉ RĪGA.

Geography

It borders on ESTONIA in the N; LITHUANIA in the S; the BALTIC SEA (with the Gulf of RIGA) in the W; RUSSIA in the E; BELARUS in the SE. In addition to Riga, LIEPĀJA, DAUGAVPILS, CĒSIS, and JELGAVA are the chief cities. LATVIA falls into four historic regions: N of the DVINA River are VIDZEME and LATGALE, which were parts of LIVONIA in the N; S of the DAUGAVA (DVINA) River are KURZEME and ZEMGALE, which belonged to the former duchy of COURLAND Latvia is largely a fertile lowland, drained by the Daugava (Dvina), the VENTA, the GAUJA, and the LIELUPE rivers. There are numerous lakes and swamps, and morainic hills rise to the E.

Economy

The country has livestock (dairying, stock raising) and valuable timber resources (39% of the country is forest). It was an important industrial center of the former SOVIET UNION; manufacturing includes machinery, metals, electrical equipment, and textiles; food and dairy processing, distilling, and shipbuilding are also important to the economy. A major portion of Latvia's trade is with GREAT BRITAIN, GERMANY, SWEDEN, the U.S., ESTONIA, and Russia.

Population

About half of the population consists of the Letts and of the closely related Latgalians, both members of the Baltic language subfamily. About one-third of the people are Russians, and there are Belarussian, Ukranian, Lithuanian, Polish, and Jewish minorities. Latvian is the official language. The Latvians are mainly Lutherans, but there are large numbers of Roman Catholics in LATGATE.

History: to 1885

The Letts (after whom the country was called LETT-LAND in its early history) were conquered and Christianized by the Livonian Brothers of the Sword in the 13th century. Their country formed the S part of LIVONIA until 1561, when COURLAND became a vassal duchy under Polish suzerainty and Livonia passed to POLAND. In 1629, Sweden conquered Livonia (except Latgale), and the Russians took it in 1721. With the first (1721) and third (1795) partitions of Poland, Latgale and Courland also passed to RUSSIA. The region had been dominated since the 13th century by German merchants and by a German landowning aristocracy that reduced the Letts to servitude. Under the Russian regime these German "Baltic Barons" retained their power, and German remained the official language until 1885, when it was replaced by Russian.

History: 1885 to 1990

By the end of the 19th century there was great agricultural and industrial prosperity (with the coming of the railroad). Although Latvia was devastated in World War I, the collapse of Russia and Germany made Latvian independence possible in 1918. Soviet troops and German volunteer bands were expelled from the country. Peace with Russia followed in 1920. The Latvian constitution (1920) provided for a democratic republic, and the largest land holdings were expropriated. However, there was no political stability and in 1934 the constituent assembly and political parties were dissolved. Soviet pressure forced Latvia

to grant (1939) the USSR several naval and military bases; a subsequent Latvian-German agreement provided for the transfer of the German minority to Germany. Soviet troops occupied Latvia in 1940 and subsequently elections held under Soviet auspices resulted in the absorption of Latvia into the USSR as a constituent republic. Occupied (1941–1944) by German troops, whom the Latvians supported, in World War I, Latvia was reconquered by the Soviet Union. In the postwar years, the remaining estates were initially distributed to landless peasants, but soon Latvia's resources and industry were nationalized and a program of industrialization was pursued by the Soviet regime.

History: 1990 to Present

In May 1990 the parliament of Latvia annulled its annexation, reestablished the constitution of 1922, and adopted a declaration of independence. Latvia passed a referendum on independence in March 1991, and officially declared its independence on August 20, 1991. A.V. Gorbunous became Latvia's first head of state since 1940, and Ivars Godmanis became prime minister. Guntis Ulmanis became president in 1993. All Russian troops left Latvia by August 30, 1994. Vaira Vike-Freiberga succeeded Ulmanis as president in 1999; she was reelected in 2003. Latvia formally joined NATO (North Atlantic Treaty Organization) on March 29, 2004 and became a full member of the European Union (EU) on May 1, 2004. Latvian governments have tended to be unstable, center-right coalitions, with prime ministers in office for relatively short terms, but Prime Minister Aigars Kalvitis's coalition, which has held power since 2004, won reelection in 2006. In 2007 Valdis Zatlers was elected president.

Government

Latvia is governed under the constitution of 1922 (restored 1991), as amended. The president, who is the head of state, is elected by parliament for a four-year term; there are no term limits. The government is headed by the prime minister, who is appointed by the president. The unicameral parliament (*Saeima*) has 100 members who are popularly elected for four-year terms. Latvia has over twenty political parties and most governments are formed by coalition. The current head of state is President Valdis Zatlers; the current head of government is Prime Minister Aigars Kalvitis. Administratively, the country is divided into twenty-six counties and seven municipalities.

Latyan (laht-YAHN), village, Tehrān province, N IRAN, in ELBURZ mountains, 15 mi/24 km NE of TEHRAN, and on the river JAJ Rud. Site of irrigation and hydroelectric project.

Lau (LOU), town, TARABA state, E NIGERIA, landing on BENUE River and 21 mi/34 km NNW of JALINGO. Cassava, millet, durra; cattle raising.

Lau (LOU), archipelago of volcanic and limestone islands, FIJI, SW PACIFIC Ocean, 150 mi/241 km E of VITI LEVU; 17°00′S 178°30′W. Includes VANUA BALAVU (largest island), KANACEA, MAGO, CICIA, MOALA, TOTOYA, MATUKU, KABARA, and several smaller islands. One N Lau group was named Exploring Islands by U.S. expedition (1840).

Laubach (LOU-bahkh), town, HESSE, W GERMANY, 13 mi/21 km ESE of GIESSEN; 50°33′N 08°59′E. Metal- and woodworking; manufacturing (electronic equipment); climatic health resort. Has Renaissance castle.

Lauban, POLAND: see LUBAN.

Lauca (LOU-kah), national park, Tarapacá region, 15 mi/24 km E of PUTRE, in the Andean front range. Protects bird life near the lakes and the vicuña on the adjacent grasslands.

Laucala (lou-DHAH-lah), volcanic island (□ 5 sq mi/13 sq km), FIJI, SW PACIFIC OCEAN, 10 mi/16 km E of TAVEUNI. Comprised of eroded remnants of shield volcanoes, enclosed with Qamia in Taveuni's reef; 4 mi/6.4 km long. Copra. Formerly spelled Lauthala.

Cross-references are shown in SMALL CAPITALS. The pronunciation guide is shown on page xix. The sources of population figures are shown on page xvii.

Laucala Bay (lou-DHAH-lah), SE Viti Levu, FIJI, SW PACIFIC OCEAN, E of SUVA; 4 mi/6.4 km E-W, 2 mi/3.2 km N-S. Site of University of the South Pacific. Sometimes spelled Lauthala Bay.

Lauca River (LOU-kah), 120 mi/193 km long, ORURO department, W BOLIVIA; rises in the ANDES Mountains on Chilean border; flows SE, across high plateau, to Lake COIPASA 29 mi/47 km SSE of HUACHACALLA; 19°10′S 68°10′W. Receives Turco River. Crosses provinces of SAJAMA, LITORAL, ATAHUALLPA. A dam and hydropower project built by CHILE in 1962 caused a rift between the two countries on the grounds that the dam was reducing the flow of water into Bolivia and contributing to the salinity of Lake COIPASA, hence affecting Bolivian agriculture and cattle.

Laucha (LOU-khah), town, SAXONY-ANHALT, central GERMANY, on the UNSTRUT, 8 mi/12.9 km NW of NAUMBURG; 51°14′N 11°40′E. Noted for its bell foundry until 1914. Has 15th-century church and remains of old town walls.

Lauchhammer (LOUKH-hahm-muhr), town, SAXONY-ANHALT, central GERMANY, 9 mi/14.5 km W of SENFTENBERG; 51°29′N 13°45′E. Lignite mining; bronze casting (since 1725). Power station. Became a town in 1953 through the unification of several villages, all built in the vicinity of a foundry from 1725.

Lauchheim (LOUKH-heim), town, BADEN-WÜRTTEMBERG, S GERMANY, on the JAGST, 7 mi/11.3 km ENE of AALEN; 48°52′N 10°15′E. Has 18th-century palace.

Lauch River (lah-osh), c.20 mi/32 km long, HAUT-RHIN department, ALSACE, E FRANCE; rises in the VOSGES MOUNTAINS at foot of the Ballon de GUEBWILLER, flows E, past GUEBWILLER, then N in Alsatian lowland, to the THUR 3 mi/4.8 km above COLMAR; 48°02′N 07°21′E.

Lauchstädt, Bad, GERMANY: see BAD LAUCHSTÄDT.

Lauda-Königshofen (LOU-dah-ko-niks-HO-fuhn), town, BADEN-WÜRTTEMBERG, GERMANY, on the TAUBER, 7 mi/11.3 km NNW of MERGENTHEIM; 49°33′N 09°45′E. Center for wine making and trading. Manufacturing of measuring instruments, dyes, furniture. Has 16th-century bridge. Lauda and Königshofen, formerly two independent towns, were unified in 1950.

Laudar, YEMEN: see LODAR.

Lauder (LAW-duhr), community, SW MANITOBA, W central CANADA, 9 mi/15 km SW of HARTNEY, in CAMERON rural municipality; 49°23′N 100°40′W.

Lauder (LAW-duhr), town (2001 population 1,081), Scottish Borders, SE Scotland, in Lauderdale valley, on LEADER WATER, and 8 mi/12.9 km NNE of GALASHIELS; 55°42′N 02°45′W. Nearby is Thirlestane Castle, seat of earl of Lauderdale.

Lauderdale (LAW-duhr-dail), county (□ 718 sq mi/1,866.8 sq km; 2006 population 87,891), extreme NW Alabama; ⊙ Florence, 34°54′N 87°38′W. Borders Mississippi and Tennessee, drained in S by Pickwick Landing Reservoir, Wilson Lake, and Wheeler Lake Wilson and Wheeler dams provide hydroelectric power for industries at Florence. Cattle, poultry; corn, cotton, hay, and soybeans. Formed 1918. Originally included in Houston County, it was named in honor of John Houston, governor of GA. It was then made part of Elk County before becoming Lauderdale County after James Lauderdale, a lieutenant colonel who died during the Battle of Talladega in 1814.

Lauderdale, county (□ 715 sq mi/1,859 sq km; 2006 population 76,724), E MISSISSIPPI, bordering E on ALABAMA; ⊙ MERIDIAN; 32°30′N 88°40′W. Drained by CHUNKY, OKATIBBEE, and BUCATUNNA creeks. Agr. (cotton, corn, sweet potatoes; cattle, hogs); timber. Meridian Naval Air Station in N; Sam Dale Memorial State Historical Site in N; Okatibbee Lake reservoir in NW; Lake Tom Bailey State Lake in E. Formed 1833.

Lauderdale, county (□ 487 sq mi/1,266.2 sq km; 2006 population 26,732), W TENNESSEE; ⊙ RIPLEY; 35°46′N 89°38′W. Bounded W by the MISSISSIPPI River, N by FORKED DEER River, S by HATCHIE River. Manufacturing; agriculture. Formed 1835.

Lauderdale, town (2000 population 2,364), RAMSEY county, E MINNESOTA, residential suburb 5 mi/8 km WNW of ST. PAUL, and 3 mi/4.8 km ENE of MINNEAPOLIS; 44°59′N 93°12′W.

Lauderdale, Scotland: see LEADER WATER.

Lauderdale-by-the-Sea (LAW-duhr-dail), town (□ 1 sq mi/2.6 sq km; 2000 population 2,563), BROWARD county, SE FLORIDA, 7 mi/11.3 km N of FORT LAUDERDALE on the ATLANTIC OCEAN; 26°11′N 80°05′W.

Lauderdale Lakes (LAW-duhr-dail), city (□ 3 sq mi/7.8 sq km; 2000 population 31,705), BROWARD county, SE FLORIDA, 3 mi/4.8 km NW of FORT LAUDERDALE; 26°10′N 80°12′W. Manufacturing includes furniture, commercial glass.

Lauderhill (LAW-duhr-hil), city (□ 7 sq mi/18.2 sq km; 2000 population 57,585), BROWARD county, SE FLORIDA, 5 mi/8 km NW of FORT LAUDERDALE; 26°10′N 80°14′W. Manufacturing includes plastic products, furniture, commercial printing, shelving, and motor vehicle parts.

Laudo, Cumbre de (LAHO-do COOM-brai dai), volcanic peak, in the SIERRA NEVADA DE LAGUNAS BRAVAS, in the ANDES, NW ARGENTINA; approximately 26°31′S 68°33′W. Lies in an extremely remote region of the Andes and is largely unexplored. Elevation has been recorded variously between 20,505 ft/6,250 m to 21,083 ft/6,426m; most commonly cited as 20,997 ft/6,400 m.

Laudun (lo-duhn), town (□ 24 sq mi/62.4 sq km), GARD department, LANGUEDOC-ROUSSILLON region, S FRANCE, near RHÔNE RIVER, 13 mi/21 km NW of AVIGNON; 44°06′N 04°40′E. Vineyards, wine trade. The Phénix nuclear power facility is 4 mi/6.4 km NE on the Rhône.

Lauenburg (LOU-uhn-burg), former duchy, N central GERMANY, on the right bank of the lower ELBE. The duchy belonged to a branch of the house of SAXONY from the 12th to the late 17th century, when it passed to the house of HANOVER. Lauenburg was occupied by FRANCE from 1803 to 1813. The Congress of VIENNA awarded (1815) it to PRUSSIA and made it a member state of the German Confederation, but Prussia ceded it to the Danish crown in exchange for W Pomerania. In the Danish War of 1864 the duchy was seized by Prussia and AUSTRIA, and Austria soon afterward ceded its rights to Prussia. Lauenburg was incorporated into the province of SCHLESWIG-HOLSTEIN in 1876 and ceased to be a duchy in 1918.

Lauenburg, POLAND: see LEBORK.

Lauenburg an der Elbe (LOU-uhn-burg ahn der ELbuh), town, SCHLESWIG-HOLSTEIN, N GERMANY, harbor on right bank of ELBE RIVER, at mouth of ELBE-LÜBECK CANAL, 10 mi/16 km NE of LÜNEBURG; 53°22′N 10°34′E. Manufacturing of matches; shipbuilding. Has many 16th- and 17th-century paneled buildings. Chartered c.1260.

Lauerz, Lake, German *Lauerzer See* (□ 1.2 sq mi/3.1 sq km), SCHWYZ canton, central SWITZERLAND, WNW of SCHWYZ, at foot of the ROSSBERG; 46 ft/14 m deep; 47°00′N 08°36′E. Elevation 1,467 ft/447 m. The commune of Lauerz is on the W shore. Schwanau, an island with ruined castle, is in lake.

Lauf (LOUF), village, BADEN-WÜRTTEMBERG, SW GERMANY, on W slope of BLACK FOREST, 15 mi/24 km SSW of RASTATT; 48°39′N 08°09′E.

Laufach (LOUF-ahkh), village, BAVARIA, central GERMANY, in LOWER FRANCONIA, 7 mi/11.3 km ENE of ASCHAFFENBURG; 50°01′N 09°17′E.

Lauf an der Pegnitz (LOUF uhn der PEG-nits), city, MIDDLE FRANCONIA, N central BAVARIA, GERMANY, on the PEGNITZ, 10 mi/16 km ENE of NUREMBERG; 49°30′N 11°17′E. Old city center is on the right bank of the PEGNITZ, opposite which is new industrial center; manufacturing (ceramics, plastics, tools, motor vehicles); woodworking. On island in the PEGNITZ is early-15th-century castle.

Laufen, commune, BASEL-LAND canton, NW SWITZERLAND, on BIRS RIVER, 10 mi/16 km SSW of BASEL. Elevation 1,171 ft/357 m. Metal products, cement, tiles, textiles; woodworking. Previously (until 1994) in BERN canton.

Laufen, district, BASEL-LAND canton, W SWITZERLAND. Main town is LAUFEN; population is German-speaking and Protestant. Previously in BERN canton; plebiscite transferred this district to Basel-Land canton in 1994.

Laufen (LOU-fen), town, BAVARIA, GERMANY, in UPPER BAVARIA, on the SALZACH, 14 mi/23 km ENE of TRAUNSTEIN, on Austrian border; 47°57′N 12°57′E. Tourism; school for nature preservation. Has mid-14th-century church. Chartered c.1041.

Laufenburg, commune, AARGAU canton, N SWITZERLAND, on the RHINE (German border; bridge), 12 mi/19 km N of AARAU. Hydroelectric plant; chemicals, woodworking. Late Gothic church, ruined castle.

Laufenburg, district, N AARGAU canton, N SWITZERLAND; 47°34′N 08°03′E. Population is German-speaking and Protestant.

Laufenburg (LOU-fuhn-burg), town, BADEN-WÜRTTEMBERG, SW GERMANY, at S foot of BLACK FOREST, on the RHINE (Swiss border; bridge, lock), 5 mi/8 km E of SÄCKINGEN; 47°33′N 08°04′E. Hydroelectric plant; manufacturing (chemicals). Formerly an Austrian town; in 1801 divided into the Swiss town of Laufenburg on the left bank of the Rhine, and the German town of the same name on the right bank. German town is sometimes called Kleinlaufenburg.

Lauffen (LOU-fen), town, BADEN-WÜRTTEMBERG, GERMANY, on the NECKAR, 5 mi/8 km SW of HEILBRONN; 49°05′N 09°09′E. Railroad junction; manufacturing (machinery, furniture, Portland cement); agriculture (fruit, vegetables; vineyards). Has Gothic church and medieval castle now seat of the town council. Poet Hölderlin was born here.

Laugerie-Haute, FRANCE: see EYZIES-DE-TAYAC, LES.

Laugharne (LAHRN), village (2001 population 1,320), CARMARTHENSHIRE, SW Wales, on TAF RIVER estuary, and 9 mi/14.5 km SW of CARMARTHEN; 51°46′N 04°28′W. Model for Llareggub of Dylan Thomas. Army training area to S. Formerly in DYFED, abolished 1996.

Laughery Creek, c.50 mi/80 km long, in SE INDIANA; rises in N RIPLEY county; flows S past VERSAILLES, to Friendship, where it turns ENE to flow along the boundary between DEARBORN and OHIO counties to join the OHIO River SE of AURORA.

Laughing Bird Caye National Park, small islet, BELIZE, 12 mi/19 km ESE of Palentia; 16°26′N 88°12′W. Designated a national park in 1991 to protect nesting area of laughing gulls and other seabirds. Uninhabited.

Laughlin (LAHF-lin), unincorporated town (2000 population 7,076), CLARK county, SE NEVADA, 72 mi/116 km SSE of LAS VEGAS, 20 mi/32 km N of NEEDLES (CALIFORNIA), on COLORADO RIVER (bridged), below (S of) DAVIS DAM (forms Lake MOHAVE), opposite BULLHEAD CITY, ARIZONA; 35°08′N 114°37′W. Elev. 535 ft/163 m. The southernmost town in Nevada, Laughlin began with a couple of casinos and a fishing dock in mid-1980s; by 1996, it was 3rd in gambling revenue in Nevada with many hotels. Located at extreme S end of LAKE MEAD NATIONAL RECREATION AREA; Fort Mohave Indian Reservation to S; Grapevine Canyon to N. Laughlin/Bullhead City Airport to E in Arizona Lowest point in Nevada (469 ft/143 m) 10 mi/16 km to SSW, on Colorado River, at S tip of state.

Lauingen (LOU-ing-uhn), town, SWABIA, W BAVARIA, GERMANY, on the DANUBE, 23 mi/37 km ENE of ULM; 48°34′N 10°26′E. Manufacturing of textiles. Has remains of medieval town wall; tower (1478); late-Gothic church, and late-18th-century town hall. Chartered c.1156, it was second capital of the small principality of Pfalz-Neuburg. Albertus Magnus (1193–1280) was born here.

Laujar de Andarax (lou-HAHR dhai ahn-dah-RAHKS), town, ALMERÍA province, S SPAIN, 11 mi/18 km NNE of BERJA; 36°59′N 02°51′W. Olive oil processing, flour milling, knit items. Ships grapes. Wine, cereals, fruit, vegetables, sugar beets produced in area. Mineral springs.

Laukaa (LOU-kah), Swedish *Laukas*, village, KESKI-SUOMEN province, S central FINLAND, 14 mi/23 km NNE of JYVÄSKYLÄ; 62°25′N 25°57′E. In lake region and lumbering area. Airport.

Laun, CZECH REPUBLIC: see LOUNY.

Launaguet (lo-nah-ge), commune (□ 2 sq mi/5.2 sq km), HAUTE-GARONNE department, MIDI-PYRÉNÉES region, FRANCE, 5 mi/8 km N of TOULOUSE; 43°41′N 01°27′E.

Launay (lo-NAI), village (□ 97 sq mi/252.2 sq km), ABITIBI-TÉMISCAMINGUE region, W QUEBEC, E CANADA, 20 mi/32 km WNW of AMOS; 48°39′N 78°32′W. Gold mining.

Launceston (LON-ses-tuhn), city and port (2001 population 68,443), N TASMANIA, SE AUSTRALIA, 124 mi/199 km N of HOBART, at head of TAMAR RIVER, at confluence of NORTH ESK and SOUTH ESK rivers; 41°45′S 147°06′E. The second-largest and second-most-populous city of Tasmania, it is the main port for trade with the Australian mainland. The principal exports are dairy products, flour, and lumber. There are woolen mills and aluminum works. The city is a major regional center for tertiary education. Founded 1806.

Launceston (LAWNS-tuhn), town (2001 population 7,135), CORNWALL, SW ENGLAND, on hill above Kensey River, 20 mi/32 km NW of PLYMOUTH; 50°38′N 04°22′W. Known in ancient Celtic or Brythonic as Dunheved ("hill-head"). Formerly county capital. Has remains of Norman castle. River is crossed by two old bridges.

Launching Place (LAWN-cheng), township, suburb of MELBOURNE, VICTORIA, SE AUSTRALIA, between WARBURTON and LILYDALE, and on YARRA RIVER. Residential area; camping grounds. Formerly called Ewart's, for David Ewart's Home Hotel.

Launglon (loung-LON), township, TANINTHARYI division, MYANMAR, on peninsula, 10 mi/16 km SW of TAVOY; 13°58′N 98°07′E.

Laungowal (loun-GO-wahl), town, SANGRUR district, PUNJAB state, N INDIA, 45 mi/72 km WSW of PATIALA; 30°13′N 75°41′E. Gram, wheat. Sometimes spelled LONGOWAL.

La Unión (lah oon-YON), province (□ 1,136 sq mi/2,953.6 sq km), AREQUIPA region, ⊙ COTAHUASI, S PERU; 15°00′S 72°50′W. Reaches into the ANDES from the PACIFIC coast. Drained by Cotahuasi (see OCOÑA RIVER) and its tributaries. Cereals, potatoes; livestock.

La Unión, department (□ 883 sq mi/2,295.8 sq km), E EL SALVADOR; ⊙ LA UNIÓN. Bounded N by HONDURAS, E by NICARAGUA (along Goascorán River), SE by GULF OF FONSECA of the Pacific; mountainous (W), sloping E to coastal lowland; contains CONCHAGUA volcano. Agriculture (grain, coffee, sugarcane; livestock); fisheries, salt production, and tortoise-shell collection along coast. Main centers are LA UNIÓN and SANTA ROSA DE LIMA, served by railroad and INTER-AMERICAN HIGHWAY. Formed 1865.

La Union, community, DOÑA ANA county, NEW MEXICO, 17 mi/27 km NNW of EL PASO central business district. One of the older communities in the Mesilla Valley. It was founded in 1853 near the banks of the RIO GRANDE. It was constantly in danger of being in a flood until the construction of ELEPHANT BUTTE DAM.

La Unión (lah oon-YON), city, ⊙ La Unión comuna (2002 population 25,615), VALDIVIA province, LOS LAGOS region, S CHILE, 38 mi/61 km SSE of VALDIVIA, on railroad, on Llollelhue River, in lake district; 40°15′S 73°02′W. Railroad junction; com-

mercial and agricultural center (cereals, sugar beets; livestock); sawmills, food processing, textiles. Founded by German colonists.

La Unión (lah oon-YON), city and municipality, ⊙ LA UNIÓN department, E EL SALVADOR, port on LA UNIÓN BAY (NW inlet of GULF OF FONSECA), on spur of INTER-AMERICAN HIGHWAY and railroad, and 95 mi/153 km ESE of SAN SALVADOR, at N foot of CONCHAGUA volcano; 13°20′N 87°51′W. Main PACIFIC port of Salvador, handling c.50% of country's ocean trade; exports coffee, sugar, cotton, henequen. Tortoise-shell industry, fisheries, livestock raising. Most harbor facilities at port of CUTUCO (1.5 mi/2.4 km. E); developed beach resorts nearby.

La Unión (LAH oon-YON), city, ⊙ DOS DE MAYO province, HUÁNUCO region, central PERU, on E slopes of Cordillera BLANCA of the ANDES, on the Vizcarra River, 40 mi/64 km WNW of HUÁNUCO; 09°46′S 76°48′W. Cereals, potatoes; livestock.

La Unión (LAH oon-YON), town, ⊙ La Unión municipio, ANTIOQUIA department, NW central COLOMBIA, in the Cordillera CENTRAL, 21 mi/34 km SE of MEDELLÍN; 05°58′N 75°21′W. Coffee, plantains, cacao; livestock.

La Unión (LAH oon-YON), town, ⊙ La Unión municipio, NARIÑO department, SW COLOMBIA, in Cordillera CENTRAL, 28 mi/45 km NNE of PASTO; 01°36′N 77°08′W. Elevation 5,856 ft/1,785 m. Cereals, cacao, coffee, sugarcane, potatoes, fruit; livestock.

La Unión (LAH oon-YON), town, ⊙ La Unión municipio, SUCRE department, N central COLOMBIA, 27 mi/43 km S of SINCELEJO; 08°51′N 75°16′W. Corn, rice; livestock.

La Unión (LAH oon-YON), town, ⊙ La Unión municipio, VALLE DEL CAUCA department, W COLOMBIA, 155 mi/249 km NNE of CALI near the CAUCA River; 04°32′N 76°06′W. Corn, soybeans, sorghum, sugarcane; livestock.

La Unión (lah oon-YON), town, ZACAPA department, E GUATEMALA, in highlands, 11 mi/18 km SSE of GUALÁN; 14°58′N 89°17′W. Elevation 3,609 ft/1,100 m. Coffee, sugarcane; livestock. Until c.1920 called Monte Oscuro.

La Unión (lah oon-YON), town, COPÁN department, W HONDURAS, 7 mi/11.3 km SW of SANTA ROSA DE COPÁN; 14°49′N 88°24′W. In tobacco area; cigar manufacturing; agriculture (sugarcane; livestock).

La Unión, town, LEMPIRA department, W HONDURAS, 17 mi/27 km NE of GRACIAS; 14°40′N 88°54′W. Tobacco, grain.

La Unión, town, OLANCHO department, central HONDURAS, 11 mi/18 km NNW of SALAMÁ; 13°57′N 87°02′W. Sugar milling; coffee, sugar, rice. Airfield.

La Unión (lah oon-YON), town, GUERRERO, SW MEXICO, in PACIFIC lowland, on La Unión River, 31 mi/50 km NW of ZIHUATANEJO, 4 mi/6km N of Mexico Highway 200; 18°00′N 101°48′W. Rice, sugarcane, fruit. Silver deposits nearby.

La Unión (LAH oon-YON), town, PIURA province, PIURA region, NW PERU, on coastal plain, on PIURA RIVER, and 15 mi/24 km SSW of PIURA; 05°24′S 80°45′W. In irrigated rice and cotton area.

La Unión (LAH oon-YON), town, BARINAS state, W VENEZUELA, landing on PORTUGUESA River, at mouth of GUANARE River, and 55 mi/89 km SSW of CALABOZO (GUÁRICO state); 08°10′N 67°44′W. Livestock; sugarcane.

La Unión, COSTA RICA: see TRES RÍOS, canton.

La Unión, GUATEMALA: see SAN MARCOS, city.

La Union, PHILIPPINES: see UNION, LA.

La Unión Bay, EL SALVADOR, bay in GULF OF FONSECA, NE of LA UNIÓN. Large inlet, forms the harbor for El Salvador's busiest port.

Laupahoehoe (LOU-pah-HOI-HOI), village (2000 population 473), NE HAWAII island, HAWAII county, HAWAII, 21 mi/34 km NNW of HILO, on Hamakua coast; 19°58′N 155°14′W. On leaf-shaped point is a

memorial to victims of the 1946 tidal wave; lighthouse. Hilo Forest Reserve to S; Manowaialee Forest Reserve to SW.

Laupen, commune, BERN canton, W SWITZERLAND, on SAANE RIVER, at mouth of SENSE RIVER, near FRIBOURG border, 10 mi/16 km WSW of BERN. In the Battle of Laupen (1339), Fribourg, with Burgundian help, tried to defeat Bernese hegemony, but was beaten by Rudolf von Erlach.

Laupen, district, W BERN canton, W central SWITZERLAND; 46°54′N 07°15′E. Population is German-speaking and Protestant.

Lauperswil, commune, BERN canton, W central SWITZERLAND, in the EMMENTAL valley, 14 mi/23 km E of BERN.

Laupheim (LOUP-heim), town, BADEN-WÜRTTEMBERG, S Germany, 12 mi/19 km SW of ULM; 48°14′N 09°53′E. Manufacturing (pharmaceuticals, tools). Has 17th-century church and castles from the 16th and 18th centuries.

Laur (lah-OOR), town, NUEVA ECIJA province, central LUZON, PHILIPPINES, 16 mi/26 km ENE of CABANATUAN; 15°30′N 121°09′E. Agricultural center (rice, corn). Manganese deposits.

Laura (LAW-ruh), town, S SOUTH AUSTRALIA, 17 mi/27 km E of PORT PIRIE, on slopes of FLINDERS RANGES; 33°11′S 138°18′E. Wheat, fruit, and dairy products. Childhood home of Australian poet C.J. Dennis. Annual folk fair.

Laura (LOR-uh), village (2006 population 505), MIAMI county, W OHIO, 15 mi/24 km SW of PIQUA; 39°59′N 84°24′W. In agricultural area.

Laura (LAW-ruh), settlement, NE QUEENSLAND, AUSTRALIA, 125 mi/201 km NW of CAIRNS; 15°30′N 144°30′E. Terminus of railroad from COOKTOWN. Fruit; cattle. Aboriginal artworks.

Lauragais (lo-rah-gai), small region and former district of LANGUEDOC province, S FRANCE, SE of TOULOUSE, now partly in TARN, HAUTE-GARONNE, and AUDE departments; ⊙ CASTELNAUDARY. Agriculture (wheat, oats; cattle and poultry raising). Preparation of down and feathers for bedding.

La Urbana (LAH oor-BAH-nah), town, BOLÍVAR state, SE VENEZUELA, landing on ORINOCO RIVER (APURE state border), and 65 mi/105 km SW of CAICARA DEL ORINOCO; 07°03′N 66°54′W. Forest products (rubber, balata gum); livestock. Airport.

Laurel, county (□ 443 sq mi/1,151.8 sq km; 2006 population 56,979), SE KENTUCKY, ⊙ LONDON; 37°06′N 84°07′W. Bounded W by ROCKCASTLE RIVER, NE in part by South Fork of Rockcastle River, S by LAUREL RIVER (forms LAUREL RIVER LAKE reservoir); drained by several creeks. Mountain agricultural area, in CUMBERLAND foothills (cattle, hogs, poultry; dairying; burley tobacco, corn, hay, alfalfa); coal mines, timber. Includes Levi Jackson Wilderness Road State Park in center and part of Daniel Boone National Forest in W half. Formed 1825.

Laurel, city (2000 population 19,960), PRINCE GEORGES county, central MARYLAND, about halfway between WASHINGTON, D.C., and BALTIMORE; 39°06′N 76°52′W. Originally a small industrial center known as Laurel Factory, now primarily residential, Laurel has light manufacturing and a growing number of businesses. The Washington, D.C. Children's Center and the famous Laurel racetrack (opened in 1911) are here. Montpelier, one of Maryland's famous mansions, completed about 1783 and now owned by the state, is one of several in the area. Fort George G. Meade is also here. Patented in the late 1600s, incorporated 1870.

Laurel, city (2000 population 18,393), ⊙ JONES county, SE MISSISSIPPI, on TALLAHALA CREEK; 31°41′N 89°08′W. Agriculture (cotton, corn; cattle, poultry; dairying); timber; manufacturing (automotive parts, wood products, apparel, chemicals, furniture, machinery, concrete, electrical equipment, fabricated metal products, food processing); oil and natural gas. The city was founded as the site of a sawmill in 1882. Oil was discovered in the vicinity in 1944. Southeastern Baptist College is in Laurel. Lake Bogue Homa State Lake to E; De Soto National Forest to SE; Hesler-Noble Municipal Airport in SW. Incorporated 1892.

Laurel (LAW-ruhl), unincorporated city, HENRICO county, E central VIRGINIA, residential suburb 8 mi/13 km NNW of RICHMOND; 37°37′N 77°30′W. V.E. Randolph Museum and Meadow Farm Museum are here. Henrico County Government Center in S.

Laurel, town (2000 population 3,668), SUSSEX county, SW DELAWARE, 15 mi/24 km N of SALISBURY, Maryland, on BROAD CREEK; 38°32′N 75°34′W. Elevation 19 ft/5 m. Shipping center for fruit, vegetables. Manufacturing (wood products). Robert Lake Graham Wildlife Area to W. Mason-Dixon Marker (1768), in SW corner of Delaware, 8 mi/12.9 km to SW. Numerous ponds in the area make Laurel a popular boating, fishing, and recreation center. Laid out 1802, incorporated 1883.

Laurel (LAW-ruhl), unincorporated town (□ 6 sq mi/15.6 sq km; 2000 population 8,393), SARASOTA county, W central FLORIDA, 15 mi/24 km S of SARASOTA; 27°08′N 82°27′W.

Laurel, town (2000 population 579), FRANKLIN county, SE INDIANA, 13 mi/21 km NW of BROOKVILLE; 39°30′N 85°11′W. In agricultural area. On West Fork of WHITEWATER RIVER. Laid out 1836.

Laurel, town (2000 population 266), MARSHALL county, central IOWA, 12 mi/19 km S of MARSHALLTOWN; 41°52′N 92°55′W. In agricultural area.

Laurel, town (2000 population 6,255), YELLOWSTONE county, S MONTANA, on YELLOWSTONE RIVER, near mouth of CLARKS FORK, 16 mi/26 km SW of BILLINGS; 45°40′N 108°46′W. Oil refinery; shipping point for copper, zinc, lumber, grain, hay, livestock, wool. Dairying; sugar beets, honey. Manufacturing (petroleum products, meat processing). Herbsfest. Laurel Municipal Airport and Canyon Creek Battlefield–Nez Perce National Historical Park to N; large Crow Indian Reservation to SE. Inc. 1908. Formerly called Carlton.

Laurel, town (2006 population 905), CEDAR county, NE NEBRASKA, 17 mi/27 km SSE of HARTINGTON, on LOGAN CREEK; 42°25′N 97°05′W. Livestock; grain, alfalfa mill.

Laurel (LAW-ruhl), hamlet, S ONTARIO, E central CANADA, included in the township of AMARANTH; 43°57′N 80°12′W.

Laurel Bay, unincorporated town (2000 population 6,625), BEAUFORT county, S SOUTH CAROLINA, 6 mi/9.7 km W of BEAUFORT, on BROAD RIVER. U.S. Marine Base to E; 32°27′N 80°47′W. Industry in fish and oysters. Agricultural interests in vegetables.

Laureldale, borough (2006 population 3,780), BERKS county, SE central PENNSYLVANIA, suburb 3 mi/4.8 km N of READING. Manufacturing (fabricated steel products, textiles, commercial printing, batteries). Settled 1902, incorporated 1930.

Laureles (lou-RE-les), town, Ñeembucú department, S PARAGUAY, near Paraná River, on road, 65 mi/105 km ESE of PILAR; 27°14′S 57°29′W. Stock raising.

Laureles, MEXICO: see BENITO JUÁREZ, MICHOACÁN state.

Laureles, Los, CHILE: see LOS LAURELES.

Laureles, Poblado, village, TACUAREMBÓ department, N URUGUAY, on railroad, 25 mi/40 km NNE of TACUAREMBÓ; 31°22′S 55°51′W.

Laurel Fork, river, c.50 mi/80 km long, E WEST VIRGINIA; rises in E RANDOLPH county; flows NNE, receives Gandy Creek from S and Glady Creek from SW, turns WNW, joins SHAVERS FORK at PARSONS to form CHEAT RIVER.

Laurel Hill, unincorporated village, SCOTLAND co., S NORTH CAROLINA, 5 mi/8 km WNW of LAURINBURG.

Agr. area (cotton, grain, tobacco, soybeans; chickens, hogs). Manufacturing (textiles).

Laurel Hill (LAH-ruhl HIL), ridge (2,400 ft/732 m–2,900 ft/884 m) in the ALLEGHENY MOUNTAINS, in SW PENNSYLVANIA, on FAYETTE/SOMERSET county line, runs 55 mi/89 km NE from S Fayette county to just W of NANTY GLO; 40°13′N 79°06′W–39°49′N 79°26′W. YOUGHIOGHENY RIVER cuts through just below CONFLUENCE, CONEMAUGH RIVER cuts through W of JOHNSTOWN. Much of ridge in Forbes State Forest. Bituminous coal, limestone, sandstone, clay, shale.

Laurel Hill, WEST VIRGINIA: see LAUREL RIDGE.

Laurel Hollow, village (□ 3 sq mi/7.8 sq km; 2006 population 1,989), NASSAU county, SE NEW YORK, on N shore of LONG ISLAND, 3 mi/4.8 km W of HUNTINGTON; 40°51′N 73°28′W. In summer-resort area. JFK INTERNATIONAL AIRPORT to the SW. Until 1935, called Laurelton.

Laurel Mountain, borough (2006 population 174), SOMERSET county, SW PENNSYLVANIA, 15 mi/24 km SE of LATROBE on LAUREL HILL Ridge; 40°12′N 79°10′W. Ski resort to W. Several state parks and forests in area.

Laurel Park (LAH-ruhl PAHRK), town (□ 2 sq mi/5.2 sq km; 2006 population 2,110), HENDERSON county, W NORTH CAROLINA, residential suburb 2 mi/3.2 km W of HENDERSONVILLE, 26 mi/41.8 km SE of ASHEVILLE, 44 mi/70.8 km NW of SPARTANBURG, at S edge of BLUE RIDGE MOUNTAINS; 35°18′N 82°30′W. Service industries; manufacturing. Incorporated 1925.

Laurel Ridge or **Laurel Hill** (c.3,300 ft/1,006 m), N WEST VIRGINIA, in the ALLEGHENY MOUNTAINS. Extends SSW from CHEAT RIVER W of ROWLESBURG to TYGART RIVER W of ELKINS; 32 mi/51 km long. Rich Mountain is its S continuation. Scene (July 8, 1861) of Civil War engagement (Battle of Laurel Hill) E of BELINGTON, in which Confederate troops were forced to retreat.

Laurel River, 38 mi/61 km long, SE KENTUCKY; rises in the CUMBERLANDS in E LAUREL county; flows generally SW past CORBIN and through LAUREL RIVER LAKE reservoir, to CUMBERLAND RIVER 11 mi/18 km W of CORBIN.

Laurel River Lake, reservoir, on the border between LAUREL and WHITLEY counties, SE KENTUCKY, on LAUREL RIVER, in Daniel Boone National Forest, 18 mi/29 km SW of LONDON; c.13 mi/21 km long; 36°57′N 84°17′W. Maximum capacity 435,600 acre-ft. Formed by Laurel River Dam (282 ft/86 m high), built by the Army Corps of Engineers for flood control and power generation.

Laurel Run, borough (2006 population 698), LUZERNE county, NE central PENNSYLVANIA, residential suburb 2 mi/3.2 km SE of WILKES-BARRE. Wilkes-Barre Mountain in SE.

Laurel Springs (law-RUHL), borough (2006 population 1,923), CAMDEN county, SW NEW JERSEY, 10 mi/16 km SE of CAMDEN; 39°49′N 75°00′W. Largely residential. Incorporated 1913.

Laurelton, residential section of SE borough of QUEENS, NEW YORK city, SE NEW YORK, NE of JFK INTERNATIONAL AIRPORT, 2.5 mi/4 km W of VALLEY STREAM. Predominantly African-American population, many from the Caribbean.

Laurelton, NEW YORK: see LAUREL HOLLOW.

Laurelville (LOR-uhl-vil), village (2006 population 549), HOCKING county, S central OHIO, 16 mi/26 km NE of CHILLICOTHE; 39°28′N 82°44′W. In agricultural area; grain products, lumber.

Laurencekirk (LAW-ruhns-kuhrk), town (2001 population 1,808), Aberdeenshire, NE Scotland, 13 mi/21 km SW of STONEHAVEN; 56°49′N 02°27′W. Previously agricultural machinery, linen manufacturing. Formerly in Grampian, abolished 1996.

Laurens (LOR-uhns), county (□ 818 sq km; 2006 population 47,316), central GEORGIA, ⊙ DUBLIN; 32°28′N 82°56′W. Coastal plain agriculture

(tobacco, wheat, cotton, corn, peanuts; cattle, hogs) and timber area intersected by OCONEE River. Formed 1807.

Laurens (LOR-enz), county (□ 722 sq mi/1,877.2 sq km; 2006 population 70,374), NW central SOUTH CAROLINA, ⊙ LAURENS; 34°28′N 82°00′W. Bounded SW by SALUDA RIVER, NE by ENOREE RIVER; part of Lake Greenwood in S. Includes part of Sumter National Forest. Manufacturing includes granite, sand, vermiculite, textiles. Agriculture includes chickens, hogs, cattle; eggs; dairying; corn, wheat, rye, soybeans, sorghum, hay. Formed 1785.

Laurens (LOR-enz), city (2006 population 9,849), ⊙ LAURENS county, NW SOUTH CAROLINA; 34°30′N 82°01′W. Manufacturing includes printing and publishing, ceramics, motor vehicle parts, glass and fiberglass products, machining, textiles, paper products. Agriculture includes livestock; dairying, grain, soybeans, sorghum. Incorporated 1875.

Laurens, town (2000 population 1,476), POCAHONTAS county, N central IOWA, 11 mi/18 km NW of POCAHONTAS; 42°51′N 94°50′W. Incorporated 1890.

Laurens, village (2006 population 261), OTSEGO county, central NEW YORK, 4 mi/6.4 km N of ONEONTA; 42°31′N 75°05′W. In dairying area. Nearby is Gilbert Lake State Park.

Laurentian Hills (law-REN-chuhn HILZ), town (□ 247 sq mi/642.2 sq km; 2001 population 2,750), SE ONTARIO, E central CANADA, 22 mi/36 km from PEMBROKE; 45°58′N 77°30′W. Formed in 2000 from the merger of the United Townships of ROLPH, BUCHANAN, WYLIE, AND MCKAY, with CHALK RIVER village.

Laurentian Mountains (lo-REN-shuhn) or **Laurentides** (lo-rahn-TEED), S QUEBEC, E CANADA, N of the SAINT LAWRENCE and OTTAWA rivers, rising to 3,150 ft/960 m in Mont TREMBLANT; 48°00′N 71°00′W. The GATINEAU, L'ASSOMPTION, LIÈVRE, MONTMORENCY, and SAINT MAURICE rivers rise in lakes in this region, which is a popular year-round recreational area, especially for MONTREAL and OTTAWA. MONTAGNE TREMBLANTE Provincial Park is here.

Laurentian Plateau, CANADA: see CANADIAN SHIELD.

Laurentian Valley (law-REN-chuhn VA-lee), town (□ 213 sq mi/553.8 sq km; 2001 population 8,733), SE ONTARIO, E central CANADA, 7 mi/11 km from PEMBROKE; 45°46′N 77°13′W. Formed in 2000 from Stafford-Pembroke, ALICE AND FRASER.

Laurentides (LO-ruhn-teidz), region (□ 7,938 sq mi/20,638.8 sq km; 2005 population 509,459), S central QUEBEC, E CANADA, c.90 mi/145 km NW of MONTREAL; 46°26′N 74°59′W. Composed of eighty eight municipalities, centered on SAINT-JÉRÔME. High-technology industries; tourism (MONT-TREMBLANT ski center).

Laurentides (LO-ruhn-teidz), former town, SW QUEBEC, E CANADA, 25 mi/40 km NNW of MONTREAL; 45°51′N 73°46′W. Woodworking; vegetable growing; dairying; tobacco growing and processing. Sir Wilfrid Laurier born here. Town was called Saint-Lin until it was incorporated 1883. Amalgamated into Saint-Lin–Laurentides in 2000.

Laurentides, CANADA: see LAURENTIAN MOUNTAINS.

Laurentides, Les (LO-ruhn-teidz, lai), county (□ 46 sq mi/119.6 sq km; 2006 population 40,860), LAURENTIDES region, S QUEBEC, E CANADA; ⊙ SAINT-FAUSTIN–LAC-CARRÉ; 46°07′N 74°35′W. Composed of twenty municipalities. Formed in 1983.

Laurentides Park (LO-ruhn-teidz), provincial park (□ 4,000 sq mi/10,400 sq km), S central QUEBEC, E CANADA, on the LAURENTIAN PLATEAU, N of QUEBEC city, and S of Lake SAINT JOHN; 80 mi/129 km long, 60 mi/97 km wide. Rises to 3,800 ft/1,158 m (S). Public recreation ground and game reserve, park contains c.1,600 lakes and vast network of streams. CHICOUTIMI River rises in center of park. Established 1895.

Area is shown by the symbol □, and capital city or county seat by ⊙.

Laurentien (lo-rahn-TYEN), central borough of QUE-BEC city, CAPITALE-NATIONALE region, S QUEBEC, E CANADA; 46°47′N 71°27′W.

Lauricocha, Lake (lou-ree-KO-chah), Spanish *Lago Lauricocha*, small ANDEAN lake (c.4 mi/6 km long), HUÁNUCO region, central PERU, 38 mi/61 km NE of CERRO DE PASCO; 10°19′S 76°41′W. Elevation 14,270 ft/4,349 m. Feeding upper MARAÑÓN RIVER, it is considered by some to be the source of the AMAZON.

Laurie Island, SOUTH ORKNEY ISLANDS, in the South ATLANTIC, E of CORONATION ISLAND; 60°45′S 44°35′W. Island is 145 mi/233 km long, 1.5 mi/2.4 km–5 mi/8 km wide. Discovered 1821. Claimed by BRITAIN as part of British Antarctic Territory, but ARGENTINA maintains a meteorological and radio station here. Uninhabited.

Laurier (lor-YAI), community, S MANITOBA, W central CANADA, 12 mi/19 km S of SAINTE ROSE DU LAC, in Ste. Rose rural municipality; 50°53′N 99°33′W.

Laurier (lor-ee-AI), unincorporated village, FERRY county, NE WASHINGTON, port of entry at BRITISH COLUMBIA (CANADA) border, opposite CASCADE (British Columbia), 11 mi/18 km ESE of GRAND FORKS (British Columbia), 32 mi/51 km NE of Republic. Christian Lake (British Columbia) to NE; parts of Colville National Forest to E and W.

Laurier-Station (LO-ryai–stah-SYON), town (□ 5 sq mi/13 sq km; 2006 population 2,424), CHAUDIÈRE-APPALACHES region, S QUEBEC, E CANADA, 15 mi/25 km from LOTBINIÈRE; 46°32′N 71°38′W.

Laurierville (LO-ree-ai-vil, French, LO-ree-ai-VEEL), village (□ 43 sq mi/111.8 sq km), CENTRE-DU-QUÉBEC region, S QUEBEC, E CANADA, 22 mi/35 km NW of THETFORD MINES; 46°18′N 71°39′W. Dairying.

Laurieton (LAW-ree-tuhn), town, NEW SOUTH WALES, SE AUSTRALIA, 236 mi/380 km N of SYDNEY; 31°39′S 152°48′E. Holiday destination.

Laurin (lah-RAI), village, MADISON county, SW MONTANA, on RUBY RIVER, 48 mi/77 km SSE of BUTTE. In livestock region. Parts of Beaverhead National Forest to NE and SE. Robbers Roost to NW. Formerly called Cicero.

Laurinburg (LAW-ruhn-buhrg), city (□ 12 sq mi/31.2 sq km; 2006 population 15,766), ⊙ SCOTLAND county, S NORTH CAROLINA, 37 mi/60 km SW of FAYETTEVILLE, near SOUTH CAROLINA state line; 34°46′N 79°28′W. Railroad junction. Manufacturing (pharmaceuticals, metal fabricating, building materials, cobalt powder, wooden products, textiles, feeds, glass products); service industries; agriculture (cotton, tobacco, grain, soybeans; poultry, hogs). St. Andrews Presbyterian College, Indian Museum of the Carolinas, Native American Library. Incorporated 1877.

Laurion, Greece: see LÁVRION.

Laurium (LAW-ree-uhm), town (2000 population 2,126), HOUGHTON county, NW UPPER PENINSULA, MICHIGAN, 11 mi/18 km NE of HOUGHTON; 47°14′N 88°26′W. Former copper-mining region. Incorporated in 1889.

Lauriya Nandangarh, INDIA: see BETTIAH.

Lauro de Freitas (LOU-ro dee FRAI-tahs), city (2007 population 142,307), BAHIA state, E BRAZIL, on ATLANTIC OCEAN coast, 6 mi/10 km NE of SALVADOR; 12°54′S 38°20′W.

Lauro, Monte (LOU-ro, MON-te), highest peak (3,231 ft/985 m) in Monti IBLEI and in SIRACUSA province, SE SICILY, ITALY; 37°07′N 14°49′E. Source of IRMINIO, ACATE, and ANAPO rivers.

Lauro Müller (LOU-ro MEE-ler), city (2007 population 13,700), SE SANTA CATARINA, BRAZIL, railroad terminus 40 mi/64 km W of LAGUNA; 28°24′S 49°23′W. Coal mining. Named for first governor of state after establishment of republic (1889).

Laurys Station, unincorporated town, LEHIGH county, E PENNSYLVANIA, 9 mi/14.5 km NNW of ALLENTOWN, on LEHIGH RIVER; 40°43′N 76°31′W. Manufacturing (wood products). Agriculture includes dairying, livestock; grain; timber.

Lausanne, district, S central VAUD canton, W SWITZERLAND, on LAKE GENEVA; 46°32′N 06°40′E. Main cities/towns are LAUSANNE, PRILLY, PULLY, and RENENS; population is French-speaking and Protestant.

Lausanne, city (2000 population 124,914), ⊙ VAUD canton, W SWITZERLAND, on LAKE GENEVA, c.50 mi/80 km SW of BERN; 46°32′N 06°40′E. Elevation ranges from 1,220 ft/372 m at lakeshore to over 1,600 ft/488 m in upper town. An important railroad junction and lake port, it is the trade and commercial center of a rich agricultural region. The construction of the Simplon Tunnel in 1906 gave Lausanne much greater commercial significance, putting it on the route between PARIS and MILAN. Food and tobacco products are manufactured, as well as precision instruments, clothing, metal products, and leather goods. Lausanne is also a well-known resort city. It is the headquarters of the International Olympic Committee and the seat of the Swiss federal court of appeal. The old town, with its steep streets, clusters around the cathedral, the finest Gothic church in Switzerland, the recently built Palais de Rumine, the Chateau (built for the prince-bishops, later used by Bernese bailiffs, and now used by the cantonal government), and several museums.

A neolithic, and later a Celtic, settlement, Lausanne became a Roman military camp (*Lousanna*) with Roman roads under present highways. An episcopal see since the late 6th century, it was ruled by prince-bishops until 1536, when it was conquered by Bern and accepted the Reformation. Bernese rule ended in 1798, and Lausanne became (1803) the capital of the newly formed canton of Vaud. The scene of brilliant social life in the 18th century, Lausanne was the residence of Gibbon, Rousseau, and Voltaire. Lausanne has been the site of several international conferences, including the one in 1928 that led to a treaty fixing TURKEY's borders after World War I. The University of Lausanne was founded as a Protestant school of theology in 1537 and became famous as a center of Calvinism. It was made a university in 1890, and is currently partly housed in the Palais de Rumine (its main campus is in Chavannes, a W suburb of Lausanne).

Lauscha (LOU-shah), town, THURINGIA, central GERMANY, in THURINGIAN FOREST, 8 mi/12.9 km N of SONNEBERG; 50°29′N 11°10′E. Glass-manufacturing center (industry was introduced in the late 16th century). Health and winter-sports resort. Has glass-trade school.

Lausick, Bad, GERMANY: see BAD LAUSICK.

Lausitz, GERMANY and POLAND: see LUSATIA.

Lausitzer Gebirge, CZECH REPUBLIC: see LUSATIAN MOUNTAINS.

Lausitzer Neisse, POLAND: see NEISSE, river.

Laussedat, Mount (los-DAH), (10,035 ft/3,059 m), SE BRITISH COLUMBIA, W CANADA, near ALBERTA border, in ROCKY MOUNTAINS, in HAMBER PROVINCIAL PARK, 65 mi/105 km WNW of BANFF (Alberta); 51°35′N 116°58′W.

Laut, INDONESIA: see PULAU LAUT.

Lauta (LOU-tah), town, BRANDENBURG, E GERMANY, in LOWER LUSATIA, 5 mi/8 km SE of SENFTENBERG; 51°28′N 14°04′E. Manufacturing of mining equipment; aluminum smelting and refining. Power station.

Lautaret, Col du (lo-tah-re, kol dyoo), Alpine pass (6,752 ft/2,058 m), HAUTES-ALPES department, PROVENCE-ALPES-CÔTE D'AZUR region, SE FRANCE, connecting upper valley of the ROMANCHE RIVER (W) with valley of the GUISANE (which flows SE to the DURANCE); 45°02′N 06°24′E. On GRENOBLE-BRIANÇON road at junction for Col du GALIBIER road. Many winter-sport stations are located at the approaches to Lautaret. La MEIJE, a prominent peak (13,065 ft/3,982 m) of the Massif des ÉCRINS, rises 6

mi/9.7 km SW of the pass, in the ÉCRINS NATIONAL PARK. Lautaret Pass forms a natural divide between the N and S French ALPS. Panoramic views of many Alpine peaks.

Lautaro (lou-TAH-ro), town, ⊙ Lautaro comuna (2002 population 18,808), CAUTÍN province, ARAUCANIA region, S central CHILE, on CAUTÍN RIVER, on railroad, 16 mi/26 km NNE of TEMUCO; 38°31′S 72°27′W. Agricultural center (cereals, vegetables; livestock; food processing, lumbering. Founded 1881.

Lautem (LOU-tuhm), town, EAST TIMOR, on Wetar Strait, 80 mi/129 km ENE of DILI; 08°22′S 126°54′E. Fishing.

Lautem, EAST TIMOR: see LOS PALOS.

Lautenbach (lo-tahn-BAHK), village (□ 5 sq mi/13 sq km), HAUT-RHIN department, in ALSACE, E FRANCE, in SE VOSGES MOUNTAINS, in narrow valley of the LAUCH RIVER, 4 mi/6.4 km NW of GUEBWILLER; 47°57′N 07°09′E. Thread manufacturing, sawmilling. The Ballon de GUEBWILLER is 4 mi/6.4 km SW. Lautenbach dates from 8th century and developed around Benedictine abbey, of which only the 13th-century church survives.

Lautenburg, POLAND: see LIDZBARK.

Lautenthal (LOU-tuhn-tahl), suburb of LANGELSHEIM, LOWER SAXONY, N GERMANY, in the upper HARZ, on INNERSTE RIVER, 4 mi/6.4 km NNW of CLAUSTHAL-ZELLERFELD; 51°56′N 10°20′E. Summer resort with mineral baths. Incorporated 1975 into LANGELSHEIM.

Lauter (LOU-tahr), town, SAXONY, E central GERMANY, in the ERZGEBIRGE, 3 mi/4.8 km SE of AUE, near Czech border; 50°34′N 12°45′E. Manufacturing of machinery, paper, enamelware; woodworking. Chartered 1962.

Lauteraarhorn or **Gross Lauteraarhorn**, peak (13,261 ft/4,042 m) in BERNESE ALPS, BERN canton, S central SWITZERLAND, 5 mi/8 km SE of GRINDELWALD.

Lauterach (LOU-ter-ahkh), township, VORARLBERG, W Austria, 2 mi/3.2 km S of BREGENZ on the BREGEN-ZERACH; 47°28′N 09°44′E. Market and transportation center.

Lauterbach (LOU-ter-bahkh), town, HESSE, central GERMANY, 13 mi/21 km WNW of FULDA; 50°38′N 09°23′E. Health resort; manufacturing of paper; woodworking; food processing. Has two castles, a late Gothic church, and many half-timbered houses.

Lauterberg im Harz, Bad, GERMANY: see BAD LAU-TERBERG IM HARZ.

Lauterbourg (lo-ter-BOOR), commune (□ 4 sq mi/10.4 sq km; 2004 population 2,247), BAS-RHIN department, in ALSACE, NE FRANCE, on the LAUTER (German border) near its mouth on the RHINE, 48°58′N 08°11′E. Chemical works. Formerly fortified. Easternmost community in France.

Lauterbrunnen, commune, BERN canton, S central SWITZERLAND, SE of INTERLAKEN, in the BERNESE ALPS. Elevation 2,612 ft/796 m. It is famous for its springs and waterfalls, such as the Staubbach, which falls nearly 1,000 ft/305 m from MÜRREN, and the Trummelbach, which descends in a series of cascades. Gateway to Mürren and Wengen resorts.

Lauterique (lou-te-REE-kai), town, LA PAZ department, SW HONDURAS, 21 mi/34 km NW of NACAOME; 13°48′N 87°37′W. Subsistence farming, grain, beans; livestock.

Lauter River (LOU-tuhr), 35 mi/56 km long, in W GERMANY and E FRANCE; formed by several mountain streams 6 mi/9.7 km ESE of PIRMASENS, flows generally SE to the RHINE 2 mi/3.2 km E of LAUTERBOURG. Source at 49°11′N 48°59′E; forms French-German border below WISSEMBOURG.

Lauter River, 25 mi/40 km long, RHINELAND-PALATINATE, W GERMANY; rises 4 mi/6.4 km SE of KAISERSLAUTERN; flows NNW, past KAISERSLAUTERN, to the GLAN 6.3 mi/10.1 km SSW of Alsenz; source at 49°25′N 07°50′E.

Lauter River (LO-tuhr), border stream, 50 mi/80 km long, between FRANCE and GERMANY, in northernmost ALSACE, between WISSEMBOURG and LAUTERBOURG where it joins the RHINE.

Lauthala, FIJI: see LAUCALA.

Lautiosaari (LOU-tee-o-SAH-ree), village, LAPIN province, NW FINLAND, on KEMIJOKI RIVER, 4 mi/6.4 km NNW of KEMI; 65°43′N 24°37′E. Railroad and road bridge; site of Kemi airport.

Laut, Nusa (LOUT, NOO-sah), island (□ 24 sq mi/62.4 sq km), ULIASER Islands, INDONESIA, in BANDA SEA, just SE of SAPARUA, near SW coast of SERAM; 03°40′S 128°47′E. Roughly circular, c.5 mi/8 km in diameter. Coconuts, cloves, sago. Also spelled Noesa Laoet.

Lautoka (lou-TO-kah), town, NW Viti Levu, FIJI, SW PACIFIC OCEAN; 17°37′S 177°28′E. Entry port (Second largest port in country); center of sugar industry and shipment. Agriculture school. Airfield.

Lautrec (lo-TREK), commune (□ 21 sq mi/54.6 sq km), TARN department, MIDI-PYRÉNÉES region, S FRANCE, 9 mi/14.5 km NW of CASTRES; 43°43′N 02°09′E. Pink garlic grown here. Has 15th–18th-century churches and medieval houses on center square.

La Uvita (LAH oo-VEE-tah), town, ⊙ La Uvita municipio, BOYACÁ department, central COLOMBIA, 80 mi/129 km NE of TUNJA; 06°19′N 72°33′W. Elevation 8,093 ft/2,466 m. Agriculture includes, coffee, sugarcane, corn; livestock.

La Uvita (lah oo-VEE-tah), village, PUNTARENAS province, SW COSTA RICA, small PACIFIC port 16 mi/26 km NW of PUERTO CORTÉS. Minor beach resort.

Lauvsnes (LOUFS-nais), village, NORD-TRØNDELAG county, central NORWAY, on the NORTH SEA, 18 mi/29 km W of NAMSOS; 64°30′N 10°55′E. Wood-pulp production; fishing. Administrative center for the region.

Lauwe (LOU-wuh), village, WEVELGEM commune, Kortrijk district, WEST FLANDERS province, W BELGIUM, on Leie River, 3 mi/4.8 km WSW of KORTRIJK; 50°48′N 03°11′E. Flax-growing area; brick manufacturing.

Lauwersmeer (LOU-uhrz-mair), lake, N NETHERLANDS, 8 mi/12.9 km ENE of DOKKUM, former inlet of WADDENZEE, arm of NORTH SEA, enclosed by barrier dam (1969). Receives Reitdorp River from SE; forms part of boundary between FRIESLAND and GRONINGEN provinces; 6 mi/9.7 km long, 1 mi/1.6 km wide. Expozee exposition at LAUWERSOOG. Formerly called Lauwers Zee.

Lauwersoog (LOU-uhr-zawkh), town, GRONINGEN province, N NETHERLANDS, 19 mi/31 km NW of GRONINGEN, between WADDENZEE (N) and LAUWERSMEER (S; former inlet of Waddenzee); 53°24′N 06°14′E. Ferry to SCHIERMONNIKOOG island (national park) to W. Fish; dairying; cattle, sheep, vegetables, grain. Expozee exhibition is held here.

Lauwers Zee, NETHERLANDS: see LAUWERSMEER.

Lauzet-Ubaye, Le (lo-ze–u-bei, luh), Alpine village (□ 26 sq mi/67.6 sq km), ALPES-DE-HAUTE-PROVENCE department, PROVENCE-ALPES-D'AZUR region, SE FRANCE, in the UBAYE valley, 11 mi/18 km WNW of BARCELONNETTE; 44°25′N 06°26′E. Stock raising, lumbering. The large reservoir and hydroelectric plant of SERRE-PONÇON are just NW.

Lauzon (lo-ZON) or **Lévis-Lauzon** (lai-VEE–lo-ZON), former city, S central QUEBEC, E CANADA, suburb 1 mi/2 km SSE of QUEBEC city, on SE shore of SAINT LAWRENCE RIVER; 46°50′N 71°10′W. Connected to Quebec by ferry. Shipbuilding, manufacturing (industrial machinery, plastic products); general farming.

Lauzoua (lo-ZWAH), village, Sud-Bandama region, S CÔTE D'IVOIRE, on NW shore of a coastal lagoon, 20 mi/32 km W of GRAND-LAHOU; 05°12′N 05°19′W. Agriculture (palm kernels, manioc, bananas, rubber). Also spelled Lozoua.

Lava Beds National Monument (□ 73 sq mi/189 sq km), SISKIYOU and MODOC counties, N CALIFORNIA, 12 mi/19 km S of OREGON state line. Bounded by Tule Lake National Wildlife Refuge on N; Modoc National Forest W, S, and E; and Klamath National Forest to SW. Examples of rugged volcanic landscape. Used as a fortress by Native Americans during Modoc Indian War of 1872–1873. Canby's Massacre Site in N; ice caves in SW. Authorized 1925.

Lavaca (lah-VAH-kah), county (□ 970 sq mi/2,522 sq km; 2006 population 18,970), S TEXAS; ⊙ HALLETTSVILLE; 29°22′N 96°56′W. In coastal plains region; drained by LAVACA and NAVIDAD rivers. Cattle; agriculture (corn, milo, rice, hay). Some oil and natural gas fields. Formed 1846.

Lavaca (luh-VAK-uh), town (2000 population 1,825), SEBASTIAN county, W ARKANSAS, 13 mi/21 km ESE of FORT SMITH, near ARKANSAS RIVER; 35°19′N 94°10′W. Light manufacturing. Fort Chaffee Military Reservation to S.

Lavaca River (lah-VAH-kah), 100 mi/161 km long, S TEXAS; rises in N LAVACA county, N of MOULTON; flows generally SSE to Lavaca Bay, the NW arm of MATAGORDA BAY. Receives NAVIDAD RIVER 11 mi/18 km above mouth.

Lavadores (lah-vah-DHO-res), SE suburb of VIGO, PONTEVEDRA province, GALICIA region, NW SPAIN; 42°14′N 08°41′W. Fish processing, boat building, metal stamping.

Lavagna (lah-VAH-nyah), town, GENOVA province, LIGURIA, N ITALY, port on Gulf of Rapallo; separated from CHIAVARI by small stream; 44°18′N 09°20′E. Cotton and silk mills, shipyards; resort. Noted for slate quarries. Pope Innocent IV was born here.

Lava Hot Springs, village (2000 population 521), BANNOCK county, SE IDAHO, 25 mi/40 km SE of POCATELLO, on PORTNEUF RIVER; 42°37′N 112°01′W. Elevation 5,072 ft/1,546 m. Part of Caribou National Forest to NW; Portneuf Reservoir partially in SE corner of Fort Hall Indian Reservation to N.

Lavak, Pristan, TURKMENISTAN: see PRISTAN LAVAK.

Laval (lah-VAHL), administrative region (□ 95 sq mi/247 sq km; 2005 population 370,368), S QUEBEC, E CANADA; 45°35′N 73°45′W. Composed of the city of LAVAL.

Laval (lah-VAHL), county (□ 95 sq mi/247 sq km; 2006 population 364,756), Laval administrative region, S QUEBEC, E CANADA; ⊙ LAVAL; 45°35′N 73°45′W. Coextensive with Île Jésus (Jesus Island). Composed of one municipality, Laval. Formed in 1980.

Laval (lah-VAHL), city (1991 population 314,398), coextensive with Île Jésus (□ 94 sq mi/244.4 sq km), ⊙ LAVAL county, Laval administrative region, S QUEBEC, E CANADA, between the Rivière des mille Îles and the Rivière des PRAIRIES, just NW of MONTREAL; 45°36′N 73°44′W. The second-largest city in QUEBEC, Laval was created in 1965, when fourteen small communities on the island were amalgamated. It is a largely residential suburb of Montreal, with summer tourism facilities, and is part of the Metropolitan Community of Montreal (*Communauté Metropolitaine de Montréal*). The island was known as Montmagny Island until 1699, when it was granted to the Jesuits of Quebec and began to be settled.

Laval (lah-VAHL), island, in SE PERSIAN GULF, off the coast of IRAN, W of LENGEH. Border post with oil installation.

Lava Lava (LAH-vah LAH-vah), canton, CHAPARÉ province, COCHABAMBA department, central BOLIVIA, 3 mi/5 km SE of SACABA; 17°25′S 65°59′W. Elevation 8,940 ft/2,725 m. Clay, limestone, and gypsum. Agriculture (potatoes, yucca, bananas, corn, rice, rye, sweet potatoes, soy, tobacco, coffee, coca, tea, citrus fruits); beef and dairy cattle.

Lavalle (lah-VAH-ye), village, W SANTIAGO DEL ESTERO province, ARGENTINA, at W foot of SIERRA DE GUASAYÁN, on CATAMARCA province border, on railroad, 32 mi/51 km N of FRÍAS. In stock-raising and lumbering area; sulphur springs; sawmills.

Lavalle, ARGENTINA: see GENERAL LAVALLE, village.

La Valle (luh VAL), village (2006 population 306), SAUK county, S central WISCONSIN, on BARABOO RIVER, and 21 mi/34 km WNW of BARABOO; 43°34′N 90°07′W. In dairy and livestock region, on a mill pond. Cheese. Dutch Hollow Lake to W; Redstone Lake to NE.

Lavalle, ARGENTINA: see SANTA LUCÍA, CORRIENTES province.

Lavalle, ARGENTINA: see PUERTO LAVALLE.

LaVallee (lah-vah-LAI), township (□ 92 sq mi/239.2 sq km; 2001 population 1,073), NW ONTARIO, E central CANADA, 12 mi/19 km from FORT FRANCES, and bounded for 6 mi/10 km by Rainy River, which forms the U.S.-Canada border; 48°39′N 93°39′W. Composed of DEVLIN, BURRISS, and WOODYATT townships. Also written La Vallee.

Lavalleja (lah-vah-YAI-hah), department (□ 3,867 sq mi/10,054.2 sq km; 2004 population 60,925), SE URUGUAY; ⊙ MINAS; 34°00′S 55°00′W. Bounded by the CUCHILLA GRANDE PRINCIPAL (W); drained (N) by CEBOLLATÍ RIVER. Principally a cattle- and sheep-raising area; agricultural products include wheat, corn, oats. Lead deposits and quarries near Minas. Main centers are Minas, JOSÉ BATLLE Y ORDÓÑEZ, SOLÍS. Dept. was formed 1816. Formerly called Minas.

Lavalleja or **Colonia Lavalleja**, village, SALTO department, NW URUGUAY, on road, 55 mi/89 km ENE of SALTO; 31°06′S 57°01′W. Agricultural center (cereals; cattle, sheep, horses).

Lavallette (lah-vuh-LET), resort borough (2006 population 2,752), OCEAN county, E NEW JERSEY, on peninsula between BARNEGAT BAY and the ATLANTIC OCEAN, 7 mi/11.3 km ENE of TOMS RIVER; 39°58′N 74°04′W.

Lavaltrie (lah-vahl-TREE), village (□ 27 sq mi/70.2 sq km), LANAUDIÈRE region, SW QUEBEC, E CANADA, on the SAINT LAWRENCE RIVER, 9 mi/14 km SW of SOREL-TRACY; 45°53′N 73°17′W. Dairying. Just E, in the Saint Lawrence, is islet of Lavaltrie. Restructured in 2001.

Lavamünd (lah-vah-MYOONT), township, CARINTHIA, S Austria, on the DRAU, at mouth of the LAVANT, near Slovenian border, 15 mi/24 km S of WOLFSBERG; 46°39′N 14°56′E. Railroad terminus; customs station; hydroelectric plant.

Lavandou, Le (lah-vahn-DOO, luh), town (□ 11 sq mi/28.6 sq km), VAR department, PROVENCE-ALPES-CÔTE D'AZUR region, SE FRANCE, 22 mi/35 km E of TOULON, at foot of the Monts des MAURES; 43°13′N 06°21′E. Bathing resort and sheltered fishing port on the MEDITERRANEAN; beach and sailboat harbor at La Favière (S). Sister resort of BORMES-LES-MIMOSAS is 2 mi/3.2 km NW amidst subtropical vegetation. Scenic coastal highway (known as Corniche des Maures) links town to SAINT-TROPEZ, 16 mi/26 km NE along the French RIVIERA. Le Lavandou is embarkation point for ferries to the HYÈRES islands, 8 mi/12.9 km offshore.

Lavan, Jazire-ye (lah-VAHN, jah-ZIR–YE), island (□ 29 sq mi/75.4 sq km), Hormozgān province, S IRAN, in PERSIAN GULF, 95 mi/153 km WNW of Bandar-e Lengeh; 26°50′N 53°15′E. Island is c.14 mi/23 km long, 4 mi/6.4 km wide. An important oil refinery. Airport. Formerly called Sheykh Sho'eyb or Shaikh Shu'aib.

Lavant River (LAH-vahnt), 50 mi/80 km long, STYRIA, CARINTHIA, S Austria; rises S of JUDENBURG in the Sectal Alps; flows S, past WOLFSBERG, to DRAU RIVER at Lavamünd. Crossed by Lavant Motorway Bridge, one of Europe's highest (541 ft/165 m). Excellent cattle (Lavanttaler Rind) bred in upper valley; fruit cultivation below Wolfsberg.

Area is shown by the symbol □, and capital city or county seat by ⊙.

Lavapié Point (lah-vah-PEE-ai), PACIFIC cape at entrance to ARAUCO GULF, S central CHILE, 37 mi/60 km SW of CONCEPCIÓN.

Lavara (LAH-vah-rah), town, Évros prefecture, EAST MACEDONIA AND THRACE department, extreme NE GREECE, on railroad, 30 mi/48 km SSW of EDIRNE (TURKEY), on MARITSA River (Turkish border); 41°16′N 26°23′E.

Lavardac (lah-vahr-DAHK), commune (□ 5 sq mi/13 sq km), LOT-ET-GARONNE department, AQUITAINE region, SW FRANCE, on the BÄISE RIVER, 4 mi/6.4 km NNW of NÉRAC, on BORDEAUX-TOULOUSE highway; 44°11′N 00°18′E. Metalworking; ARMAGNAC brandy distilling and shipping.

Lavassaare, town, on island LAVASSAARE in GULF OF FINLAND, ESTONIA, 80 mi/129 km W of SAINT PETERSBURG and 40 mi/64 km NNW of NARVA. Naval fortifications. Beach resort. Under Finnish rule (until 1940).

Lavaur (lah-VOR), town (□ 24 sq mi/62.4 sq km), TARN department, MIDI-PYRÉNÉES region, S FRANCE, on AGOUT RIVER, 20 mi/32 km ENE of TOULOUSE; 43°42′N 01°49′E. Commercial center. Metalworking, printing, flour milling. Has 13th–15th-century Gothic church. An early stronghold of the Albigenses, it was captured (1211) by Simon de Montfort. Episcopal see (1317–1790).

Lavaux (lah-VO), district, S central VAUD canton, W SWITZERLAND, on LAKE GENEVA. Main town is LUTRY; vineyards; Corniche road along the lake. Population is French-speaking and Protestant.

Lavayén River (lah-vah-YAIN), left headstream of SAN FRANCISCO RIVER, c.75 mi/121 km long, in SALTA and JUJUY provinces, ARGENTINA; flows NE, joining the RÍO GRANDE DE JUJUY near SAN PEDRO to form the San Francisco.

Lavedan (lah-ve-dahn), region of BÉARN, in HAUTES-PYRÉNÉES department, MIDI-PYRÉNÉES region, SW FRANCE. Upper valley of the GAVE DE PAU.

Laveen, unincorporated town, MARICOPA county, central ARIZONA, residential suburb 7 mi/11.3 km SW of PHOENIX, S of SALT RIVER. Manufacturing (cotton milling, ironworks). Phoenix South Mountain Park to SE. Gila River Indian Reservation to SW, including Saint Johns Museum.

La Vega (lah VAI-gah), province (□ 916 sq mi/2,381.6 sq km; 2002 population 385,101), central DOMINICAN REPUBLIC; ⊙ LA VEGA; 19°07′N 70°37′W. A mountainous interior province, bounded by Cordillera CENTRAL (S) and Cordillera SEPTENTRIONAL (N); drained by the YAQUE DEL NORTE and CAMÚ RIVER, which here forms LA VEGA REAL valley, part of the republic's most fertile and densely populated CIBAO region. Main crops include tobacco, cacao; also coffee, rice, corn; cattle; wheat in the uplands. CONSTANZA and JARABACOA, with nearby JIMENOA Falls, are mountain resorts. The region was visited in 1492 by Columbus, who built a fort. Became a province 1845.

La Vega (lah VAI-gah), city (2002 population 98,386), central DOMINICAN REPUBLIC, on the CAMÚ RIVER; ⊙ LA VEGA province; 19°15′N 70°30′W. La Vega is the commercial and processing center of a rich agricultural region. A religious sanctuary erected on the site of an important battle in the colonial period is nearby. The city was founded in 1495.

La Vega (LAH VAI-gah), town, ⊙ La Vega municipio, CUNDINAMARCA department, central COLOMBIA, on W slopes of Cordillera ORIENTAL, 30 mi/48 km NW of BOGOTÁ; 05°00′N 74°20′W. Elevation 3,986 ft/1,215 m. In agricultural region (sugarcane, coffee, corn, potatoes; livestock).

La Vega (LAH VAI-gah), town, ⊙ La Vega municipio, CAUCA department, SW COLOMBIA, 30 mi/48 km SW of POPAYÁN; 02°00′N 76°46′W. Agriculture includes coffee, corn, sugarcane; livestock.

La Vega (LAH VAI-gah), town, DISTRITO CAPITAL, N VENEZUELA, on GUAIRE RIVER, and on railroad, SW suburb of CARACAS; 10°28′N 66°57′W. Lime-quarrying and cement-milling center.

La Vega Real (lah VAI-gah re-AHL), valley, NE DOMINICAN REPUBLIC, fertile lowland, E section of the CIBAO along CAMÚ and YUNA rivers, extending c. 60 mi/97 km E from LA VEGA city to SÁNCHEZ. Main crops are cacao, coffee, rice, corn, tropical fruit.

La Vela de Coro (LAH VAI-lah dai KO-ro), town, ⊙ Colina municipio, FALCÓN state, NW VENEZUELA, port on the CARIBBEAN, at base of Isthmus of MÉDANOS leading N to PARAGUANÁ PENINSULA, 8.0 mi/12.9 km ENE of CORO (linked by highway, and by railroad built in 1986–1990); 11°27′N 69°34′W.

Lavelanet (lah-vuh-lah-NAI), town (□ 4 sq mi/10.4 sq km), ARIÈGE department, MIDI-PYRÉNÉES region, S FRANCE, 13 mi/21 km E of FOIX; 42°56′N 01°51′E. Textile milling, manufacturing of hosiery. Bauxite deposits nearby.

Lavello (lah-VEL-lo), town, POTENZA province, BASILICATA, S ITALY, 8 mi/13 km NE of MELFI; 41°03′N 14°48′E. Pasta, wine, olive oil, cement. Roman ruins, medieval church, castle.

Lavena, gorge and hydroelectric plant, S LIECHTENSTEIN, near Swiss border, 5 mi/8 km SSE of VADUZ.

La Venada Island, NICARAGUA: see SOLENTINAME ISLANDS.

Lavendon, village (2001 population 1,230), Milton Keynes, central ENGLAND, on OUSE RIVER, 2 mi/3.2 km NW of OLNEY; 52°11′N 00°40′W.

Lavenham (LAV-uhn-uhm), town (2001 population 1,738), W SUFFOLK, E ENGLAND, 6 mi/9.7 km NE of SUDBURY; 52°06′N 00°47′E. Agricultural market. Has 15th–16th-century church with tall tower, 16th-century guildhall, and ancient market cross. Lavenham was a center of medieval East Anglia cloth industry.

L'Avenir (lahv-NIR), village (□ 37 sq mi/96.2 sq km), CENTRE-DU-QUÉBEC region, S QUEBEC, E CANADA, near SAINT FRANÇOIS River, 12 mi/19 km SE of DRUMMONDVILLE; 45°46′N 72°18′W. Dairying; cattle, pigs.

L'Avenir, village (2000 population 2,374), MOKA district, MAURITIUS, 10 mi/16 km SE of PORT LOUIS, and 1.25 mi/2 km E of SAINT PIERRE. Sugarcane, vegetables, livestock; apparel and textile manufacturing.

Laveno-Mombello (lah-VAI-no–mom-BEL-lo), town, VARESE province, LOMBARDY, N ITALY, on E shore of LAGO MAGGIORE, 12 mi/19 km NW of VARESE; 45°55′N 08°37′E. Resort, port; ceramic, pottery works, alcohol distillery. Feldspar quarries nearby. Ferry to Sasso del Ferro.

La Venta (lah VEN-tah), town, FRANCISCO MORAZÁN department, central HONDURAS, 17 mi/27 km NE of TEGUCIGALPA, on secondary road, N of CHOLUTECA RIVER; 14°18′N 87°10′W. Small farming, grain, beans; livestock.

La Venta (lah VEN-tah), village, Huimanguillo municipio, W TABASCO, MEXICO, in coastal mangrove swamps, 30 mi/48 km E of COATZACOALCOS, near TONALÁ RIVER (VERACRUZ border); 18°07′N 94°03′W. One of the most important Olmec archaeological sites in Mexico; colossal stone heads discovered here have been moved to VILLAHERMOSA, but largest pyramid in Olmec region remains. Site is now adjacent to a petroleum complex.

Laventie (lah-vahn-TEE), town (□ 7 sq mi/18.2 sq km), PAS-DE-CALAIS department, NORD-PAS-DE-CALAIS region, N FRANCE, 9 mi/14.5 km NE of BÉTHUNE; 50°38′N 02°46′E. Woodworking and footwear manufacturing. Potatoes and sugar beets.

Laventure, village (2000 population 5,654), FLACQ district, MAURITIUS, 19 mi/30.4 km E of PORT LOUIS. Sugarcane; livestock.

La Verde (lah VER-dai), town, E Chaco province, ARGENTINA, 35 mi/56 km NW of RESISTENCIA. Agricultural center (sorghum, cotton; livestock); sawmills, cotton gins.

La Vérendrye (lah VAI-ruhn-drei), waterway provincial park (□ 71 sq mi/184.6 sq km), ONTARIO, E CANADA, 80 km SW of THUNDER BAY, on U.S. (MINNESOTA) border; 47°19′N 77°00′W. Rare flora. Camping permitted. No facilities.

La Vergne, city (2006 population 27,255), RUTHERFORD county, central TENNESSEE, 15 mi/24 km SE of NASHVILLE, near J. PERCY PRIEST Reservoir (STONES RIVER), 36°00′N 86°34′W. Manufacturing. One of the fastest growing communities in the state. Incorporated 1972.

La Verkin, town (2006 population 4,142), WASHINGTON county, SW UTAH, 20 mi/32 km NE of SAINT GEORGE, and on VIRGIN RIVER; 37°12′N 113°16′W. Fruit; cattle. ZION NATIONAL PARK to E; Dixie National Forest to NW. Quail Creek reservoir and State Park to W.

Laverlochère (lah-ver-lo-SHER), village (□ 41 sq mi/106.6 sq km), ABITIBI-TÉMISCAMINGUE region, W QUEBEC, E CANADA, 16 mi/26 km W of HAILEYBURY; 47°25′N 79°17′W. Agriculture; dairy products; gold mining.

La Verne, city (2000 population 31,638), LOS ANGELES county, S CALIFORNIA; suburb 26 mi/42 km ENE of downtown LOS ANGELES; 34°07′N 117°46′W. La Verne has expanded from a citrus-growing town to a city with significant residential developments and light manufacturing (electronic components, apparel, hand tools, ceramics, stone products, plastics products, synthetic rubber). Oil refining, high-technology computers, and aerospace and defense-related indutries are in adjacent areas. University of La Verne and a water-filtration plant that serves much of the Southern California region are in the city. Marshall Canyon Regional Park to NE, Bonelli Regional Park to SW; Angeles National Forest to N. Incorporated 1906.

Laverne, town (2006 population 1,025), HARPER county, NW OKLAHOMA, 33 mi/53 km NW of WOODWARD, near NORTH CANADIAN (Beaver) River; 36°42′N 99°54′W. In agricultural area (wheat; cattle); manufacturing (meat processing).

La Vernia (luh VER-nee-ah), village (2006 population 1,168), WILSON county, S TEXAS, 24 mi/39 km ESE of SAN ANTONIO, on CIBOLO CREEK; 29°21′N 98°06′W. Agriculture (poultry; peanuts, vegetables, melons); manufacturing (steel storage tanks).

Lavers Hill, resort village, VICTORIA, SE AUSTRALIA, 144 mi/232 km SW of MELBOURNE, and on GREAT OCEAN ROAD; 38°40′S 143°24′E.

Laverton (LA-vuhr-tuhn), town, S central WESTERN AUSTRALIA, 155 mi/249 km NNE of KALGOORLIE; 28°49′S 122°25′E. Terminus of railroad from Kalgoorlie. Former gold-mining center; nickel mining.

La Veta, village (2000 population 924), HUERFANO county, S COLORADO, on CUCHARAS RIVER, in E foothills of SANGRE DE CRISTO MOUNTAINS, 15 mi/24 km WSW of WALSENBURG; 37°30′N 105°00′W. Elevation c.7,013 ft/2,138 m. Resort and trading point in agriculture and coal-mining region; livestock and grain products La Veta Pass (9,382 ft/2,860 m), W of La Veta, crossed by railroad, in Sangre de Cristo Mountains, is crossed by U.S. Highway 160. Cuchara Valley Ski Area to SW. Lathrop State Park to NE; part of San Isabel National Forest to SW; part of San Isabel National Forest with the Great Dikes of the Spanish Peaks rock formations nearby.

Lavezares (lah-ve-SAH-res), town, NORTHERN SAMAR province, NW SAMAR island, PHILIPPINES, on SAN BERNARDINO STRAIT, 65 mi/105 km NW of CATBALOGAN; 12°32′N 124°21′E. Agricultural center (corn, rice, coconuts).

Lavi (lah-VEE), kibbutz, Israel, 6 mi/10 km W of TIBERIAS in LOWER GALILEE; 32°47′N 35°26′E. Elevation 593 ft/180 m. Mixed farming (orchards; dairy; livestock), some industry (synagogue furniture), and a guest house. Founded in 1948 by orthodox Jews.

La Victoria (lah veek-TO-ree-ah), canton, MÉNDEZ province, TARIJA department, S central BOLIVIA. Elevation 6,555 ft/1,998 m. Clay, limestone, phosphates, and gypsum deposits. Agriculture (potatoes, yucca, bananas, corn, wheat, barley, sweet potatoes); cattle.

La Victoria (LAH veek-TO-ree-ah), town, ⊙ La Victoria municipio, VALLE DEL CAUCA department, W COLOMBIA, in CAUCA valley, on railroad, 16 mi/26 km NNE of CALI. Agriculture includes tobacco, sugarcane, coffee, sorghum, and soybeans. Silver, gold, and platinum mined nearby.

La Victoria (LAH veek-TO-ree-ah), town, ⊙ La Victoria municipio, BOYACÁ department, central COLOMBIA, 60 mi/97 km W of TUNJA. Agriculture includes corn, coffee, sugarcane; livestock.

La Victoria (LAH veek-TO-ree-ah), town, ⊙ José Félix Ribas municipio, ARAGUA state, N VENEZUELA, on ARAGUA RIVER, in valley of CORDILLERA DE LA COSTA, on PAN-AMERICAN HIGHWAY, on railroad, and 36 mi/58 km WSW of CARACAS; 10°13′N 67°19′W. Elevation 2,007 ft/611 m. Trading and agricultural center (coffee, cacao, sugarcane, tobacco, vegetables, cereals, grapes); manufacturing.

La Victoria (LAH veek-TO-ree-ah), E suburb of LIMA, LIMA province, W central PERU; 11°28′S 77°14′W. Industrial suburb (manufacturing; food processing).

La Victoria, NICARAGUA: see NIQUINOHOMO.

La Villa, town (2006 population 1,458), HIDALGO county, S TEXAS, 14 mi/23 km E of EDINBURG; 26°18′N 97°55′W. Irrigated agricultural area in Rio Grande Valley (cotton, citrus, vegetables); manufacturing (pickles).

La Villa del Rosario (LAH VEE-yah del ro-SAH-ree-o), town, ⊙ Rosario de Perijá municipio, ZULIA state, NW VENEZUELA, in MARACAIBO lowlands, 25 mi/40 km NE of MACHIQUES; 10°20′N 72°18′W. Sugarcane; livestock.

La Villa River, 74 mi/119 km long; forms part of border between HERRERA and LOS SANTOS provinces, PANAMA; rises in AZUERO highlands; enters Parita Bay W of CHITRÉ.

La Villita, MEXICO: see NOMBRE DE DIOS.

La Viña (lah VEE-nyah), canton, ESTEBAN ARCE province, COCHABAMBA department, central BOLIVIA, 13 mi/20 km SE of CAPINOTA, on an unpaved road; 17°58′S 65°50′W. Elevation 9,026 ft/2,751 m. Clay, limestone, and gypsum. Agriculture (potatoes, yucca, bananas, corn, rye, soy, coffee); cattle.

La Viña (lah VEEN-yah), village (1991 population 1,241), ⊙ La Viña department (□ 840 sq mi/2,184 sq km), S SALTA province, ARGENTINA, on railroad (Castañares station), and 50 mi/80 km S of SALTA. Agriculture center (alfalfa, wheat, corn, oats, wine; livestock); flour milling.

Lavina (luh-VEE-nuh), village (2000 population 209), GOLDEN VALLEY county, S central MONTANA, on MUSSELSHELL RIVER opposite mouth of Big Coulee Creek, 40 mi/64 km NW of BILLINGS; 46°18′N 108°57′W. Wheat; sheep, cattle, sheepdogs.

Lavínia (LAH-vee-nee-ah), town (2007 population 7,984), NW SÃO PAULO, BRAZIL, on railroad and 38 mi/61 km W of ARAÇATUBA; 21°09′S 51°02′W. Cattle raising.

La Virgen (lah VIR-hen), village, HEREDIA province, N COSTA RICA, on SARAPIQUÍ RIVER, and 27 mi/43 km N of HEREDIA; 10°25′N 84°09′W. Bananas, fodder crops; livestock.

La Virginia, (LAH veer-HEE-nee-ah) town, ⊙ La Virginia municipio, RISARALDA department, W central COLOMBIA, on the CAUCA River, 12 mi/19 km NW of PEREIRA; 04°53′N 75°52′W. Agriculture includes sugarcane, plantains, coffee; livestock.

La Virtud (lah vir-TOOD), town, LEMPIRA department, W HONDURAS, near Lempa River (El Salvador border), 37 mi/60 km S of GRACIAS; 14°03′N 88°42′W. Grain; livestock.

Lavis (LAH-vees), town, TRENTO province, TRENTINO–ALTO ADIGE, N ITALY, 5 mi/8 km N of TRENT, on AVISIO RIVER, near its confluence with the ADIGE; 46°08′N 11°07′E. Wine, food processing; wood products.

La Visitation-de-l'Île-Dupas (lah vee-zee-tah-SYON–duh–LEEL–dyoo-PAH), village (□ 10 sq mi/26 sq km; 2006 population 597), LANAUDIÈRE region, S QUEBEC, E CANADA, 2 mi/3 km from BERTHIERVILLE; 46°05′N 73°09′W.

La Visitation-de-Yamaska (lah vee-zee-tah-SYON–duh–yuh-MA-skuh), village (□ 16 sq mi/41.6 sq km; 2006 population 379), CENTRE-DU-QUÉBEC region, S QUEBEC, E CANADA; 46°08′N 72°36′W.

Lavizzara, Val (lah-vee-ZAHR-rah), valley, TICINO canton, S SWITZERLAND; upper valley of the MAGGIA RIVER.

Lavon (LAH-vuhn), village (2006 population 420), COLLIN county, N TEXAS, 27 mi/43 km ENE of DALLAS, on urban fringe of Dallas–FORT WORTH area; 33°01′N 96°26′W. Agricultural area. LAVON LAKE reservoir (East Fork of TRINITY RIVER) to NW; LAKE RAY HUBBARD reservoir to SW.

Lavonia (luh-VO-nee-uh), town (2000 population 1,827), FRANKLIN county, NE GEORGIA, near SOUTH CAROLINA state line, 16 mi/26 km SE of TOCCOA; 34°26′N 83°07′W. Manufacturing includes apparel, textiles, feeds, train parts, industrial machinery.

Lavon Lake (LAH-vuhn), reservoir (□ 33 sq mi/85.8 sq km), COLLIN county, NE TEXAS, on East Fork of TRINITY RIVER, 25 mi/40 km NE of DALLAS; 33°02′N 96°29′W. Maximum capacity 921,100 acre-ft. Formed by Lavon Dam (81 ft/25 m high), built (1974) by Army Corps of Engineers for water supply; also used for flood control and recreation.

Lavoy (luh-VOI), former village (2001 population 108), central ALBERTA, W CANADA, 9 mi/14 km ESE of VEGREVILLE, in MINBURN COUNTY NO. 27; 53°28′N 111°52′W. Grain, dairying. Dissolved 1999.

Lavra, Cape, Greece: see AKRATHOS.

Lavradio (liv-RAH-dee-o), town, SETÚBAL district, S central PORTUGAL, across the TAGUS RIVER from LISBON, 2 mi/3.2 km SE of BARREIRO; 38°40′N 09°03′W.

Lavras (lah-vrahs), city (2007 population 87,421), S MINAS GERAIS, BRAZIL, on railroad, 45 mi/72 km W of SÃO JOÃO DEL REI; 21°15′S 44°59′W. Textile milling, distilling; exports (coffee, tobacco; cattle). Agricultural colony established 1908 by American Presbyterian Church.

Lavras, Brazil: see LAVRAS DO SUL.

Lavras da Mangabeira (lah-vrahs dah mahn-gah-bai-rah), city (2007 population 29,736), SE CEARÁ, BRAZIL, on FORTALEZA-CRATO railroad, 45 mi/72 km NE of Crato, in semiarid cattle-raising area; 06°45′S 38°57′W. Until 1944, called Lavras.

Lavras do Sul (LAH-vrahs do sool), town (2007 population 8,116), S central RIO GRANDE DO SUL, BRAZIL, 35 mi/56 km NNE of BAGÉ; 30°49′S 53°55′W. Rice, wheat. Gold mining. Copper deposits. Until 1944, called Lavras.

Lavrentiya (lah-VRYEN-tee-yah), village (2005 population 1,215), NE CHUKCHI AUTONOMOUS OKRUG, NE RUSSIAN FAR EAST, on Lavrentiya [Russian=(St.) Lawrence] Bay of the BERING SEA, 350 mi/563 km ENE of ANADYR'; 65°35′N 171°00′W. Government arctic station, trading post; airfield. Formerly known as Chukotskaya Kul'tbaza (Russian=Chukchi cultural base).

Lávrion, Latin *Laurium*, city, ATTICA prefecture, ATTICA department, E central GREECE, port on Gulf of PETALION of AEGEAN SEA, opposite MAKRONISOS island, on railroad (terminus), and 26 mi/42 km SE of ATHENS; 37°43′N 24°03′E. Major smelting and shipping center for complex ores of manganese, cadmium, and lead—zinc, iron, and silver mined at Kamariza (2.5 mi/4 km W; included in Lávrion municipality). Ancient Laurium was important for its silver mines (from 6th century B.C.E.) which were one of the chief sources for Athenian revenue in the 5th century B.C.E.

Mining ceased 400 B.C.E.; resumed 1864 by French, and 1875 by Greeks and the present town (formerly Ergasteria) dates from this period. Nearby are ruins of a temple of Poseidon. Also called Laurion.

Lavumisa (lah-VOO-mee-suh), town, SHISELWENI district, SE SWAZILAND, 83 mi/134 km SE of MBABANE, on railroad, at SOUTH AFRICA border, opposite Golela Road junction; MOZAMBIQUE border 5 mi/8 km to E; 27°20′S 31°53′E. Jozini Pongolapoost Dam (on PONGOLA RIVER, South Africa) to SE (N arm enters Swaziland to E). Cattle, goats, sheep, hogs; corn, pineapples, sugarcane, cotton. Customs office. Airstrip.

Law (LAW), village (2001 population 2,883), South Lanarkshire, S Scotland, 2 mi/3.2 km SE of WISHAW; 55°45′N 03°53′W. Formerly in Strathclyde, abolished 1996.

Lawa (LAH-wah), village, TONK district, central RAJASTHAN state, NW INDIA, 18 mi/29 km NW of Tonk. Local agricultural market (millet, cotton, gram). Was capital of former petty state of Lawa; in 1948, state merged with union of Rajasthan.

Lawai (LAH-WAH-ee), town (2000 population 1,984), KAUAI island, KAUAI county, HAWAII, 2 mi/3.2 km inland from S coast, on Lawai Stream, 9 mi/14.5 km WSW of LIHUE, on Kaumualii Highway; 21°55′N 159°30′W. Sugarcane, fruit. Aepoalua Reservoir and several smaller reservoirs to SE. National Tropical Botanical Garden to S. Lihue-Koloa Forest Reserve to N.

Lawar (LUH-wahr), town, MEERUT district, NW UTTAR PRADESH state, N central INDIA, 9 mi/14.5 km NNE of Meerut. Wheat, gram, jowar, sugarcane, oilseeds.

Lawa River, SURINAME and FRENCH GUIANA: see MARONI RIVER.

Lawas (luh-WAHS), town, N SARAWAK, NW BORNEO, E MALAYSIA, near BRUNEI BAY, on Lawas River (24 mi/39 km long), about halfway between BRUNEI border on SW and Subah state boundary on NE, 32 mi/51 km E of Brunei; 04°56′N 115°22′E. Buffalo; cattle; rubber; mackerel fishing.

Lawdar, YEMEN: see LODAR.

Law Dome, near-coastal ice dome on the BUDD COAST of WILKES LAND, ANTARCTICA; 66°44′S 112°50′E. It rises to 4,580 ft/1,395 m.

Lawiczka, POLAND: see RESKO.

Lawit, Gunong (LAH-wit, GOO-nong), highest peak (4,982 ft/1,519 m), of TRENGGANU, NE MALAYSIA, 40 mi/64 km WNW of KUALA TERENGGANU; 05°25′N 102°35′E. Sometimes called Gunong Batil.

Lawksawk, township (□ 2,362 sq mi/6,141.2 sq km), SHAN STATE, MYANMAR, ⊙ Lawksawk (a village 35 mi/56 km NW of TAUNGGYI and on ZAWGYI RIVER); 21°15′N 96°52′E. Parallel mountain ranges extend S with wide plateau (elevation 3,500 ft/1,070 m) in center. Teak forests; some irrigated rice. Formerly a state (sawbwaship).

Lawler, town (2000 population 461), CHICKASAW county, NE IOWA, 8 mi/12.9 km E of NEW HAMPTON; 43°04′N 92°09′W. In corn, hog, dairy, fertilizer and concrete area.

Lawn (LAWN), village (2006 population 348), TAYLOR county, W central TEXAS, 21 mi/34 km S of ABILENE; 32°08′N 99°45′W. In cotton, cattle area.

Lawndale, city (2000 population 31,711), LOS ANGELES county, S CALIFORNIA; residential suburb 12 mi/19 km SW of LOS ANGELES, and 3 mi/4.8 km NE of PACIFIC OCEAN, in the Centinela Valley; 33°54′N 118°21′W. The population of Lawndale grew rapidly in the 1950s, but has been stable since early 1960s. Incorporated 1959.

Lawndale (LAWN-dail), village (□ 1 sq mi/2.6 sq km; 2006 population 633), CLEVELAND county, S NORTH CAROLINA, 8 mi/12.9 km N of SHELBY; 35°24′N 81°33′W. Manufacturing (textiles, furniture, motor vehicle parts); agriculture (cotton, grain; livestock).

Lawnside, borough (2006 population 2,800), CAMDEN county, SW NEW JERSEY, 7 mi/11.3 km SE of CAMDEN;

Area is shown by the symbol □, and capital city or county seat by ⊙.

39°52'N 75°01'W. Site bought by abolitionists for free blacks (1840), and first called Free Haven. Pop. is overwhelmingly African-American. Incorporated 1926.

Lawnton, unincorporated town (2000 population 3,787), DAUPHIN county, S PENNSYLVANIA, residential suburb near Spring Creek; 40°16'N 76°47'W.

Lawoe, Mount, INDONESIA: see LAWU, MOUNT.

Lawra (LOU-ruh), town (2000 population 5,763), district administrative center, UPPER WEST region, GHANA, on MOUHOUN RIVER, 50 mi/80 km NNW of Wa at CÔTE D'IVOIRE border; 10°39'N 05°52'E. Road junction; shea nuts, millet, durra, yams; cattle ranching, skins, and hides.

Lawrence (LAW-rens), rural municipality (□ 294 sq mi/764.4 sq km; 2001 population 540), S MANITOBA, W central CANADA; 51°20'N 99°30'W. Service industries. Main center is Rorketon. Incorporated 1914.

Lawrence, county (□ 718 sq mi/1,866.8 sq km; 2006 population 34,312), NW Alabama; ☉ Moulton; 34°32'N 87°20'W. Drained in N by Wheeler Reservoir (in Tennessee River). Part of William B. Bankhead National Forest in S. Cotton, corn, soybeans; poultry, cattle; deposits of coal, limestone, and asphalt. Formed 1818. Named for James Lawrence, a NJ naval commander who fought at Tripoli and in the War of 1812. The county seat was Marathon until 1820 when Moulton was chosen.

Lawrence, county (□ 592 sq mi/1,539.2 sq km; 2006 population 16,899), NE ARKANSAS; ☉ WALNUT RIDGE; 36°02'N 91°07'W. Bounded E by CACHE RIVER and on N in part by SPRING RIVER; drained by BLACK and STRAWBERRY rivers. Agriculture (cattle, hogs; rice, wheat, soybeans, sorghum). Lake Charles State Park near center; Shirey Bay–Rainey Brake Wildlife Management Area in SW. Formed 1815.

Lawrence, county (□ 373 sq mi/969.8 sq km; 2006 population 15,887), SE ILLINOIS, bounded E by WABASH RIVER; ☉ LAWRENCEVILLE; 38°42'N 87°43'W. Drained by EMBARRAS RIVER. Agricultural area, with oil and natural-gas wells; livestock; soybeans, corn, wheat. Oil refineries; also other manufacturing at Lawrenceville. Includes Red Hills State Park and Lincoln Trail State Memorial. Formed 1821.

Lawrence, county (□ 452 sq mi/1,175.2 sq km; 2006 population 46,413), S INDIANA; ☉ BEDFORD; 38°50'N 86°29'W. Drained by SALT CREEK and East Fork of WHITE RIVER. Large limestone quarries; agriculture (fruit, corn, soybeans); manufacturing at Bedford, MITCHELL. Site of Purdue University. Moses Fell Annex Farm W of Bedford. Avoca State Fishing Area in NW, Williams Dam State Fishing Area in W, Spring Mill State Park in SE. Hoosier National Forest in NE and SW. Karst topography (sinkholes) in S. Formed 1818.

Lawrence, county (□ 420 sq mi/1,092 sq km; 2006 population 16,321), NE KENTUCKY; ☉ LOUISA; 38°04'N 82°46'W. Bounded E by BIG SANDY and TUG FORK Big Sandy rivers (both form WEST VIRGINIA state line here); drained by LEVISA FORK, Big Sandy River, and Blaine Creek (forms YATESVILLE LAKE reservoir in center of county). Mountain agricultural area (corn, burley tobacco, hay, alfalfa; cattle); oil and gas wells, coal mines, fireclay, and sand pits; timber. Formed 1821.

Lawrence, county (□ 435 sq mi/1,131 sq km; 2006 population 13,457), S central MISSISSIPPI; ☉ MONTICELLO; 31°33'N 90°06'W. Drained by PEARL RIVER. Agr. (cotton, corn; poultry, cattle, hogs); timber. Lake Mary Crawford State Lake in W. Formed 1874.

Lawrence, county (□ 619 sq mi/1,609.4 sq km; 2006 population 37,400), SW MISSOURI; ☉ MOUNT VERNON; 37°07'N 93°50'W. In the OZARKS, drained by SPRING RIVER. Wheat, hay, oats, barley, corn, apples, peaches, vegetables; turkeys, cattle; dairying. Manu-

facturing at AURORA, Mount Vernon, PIERCE CITY, and MARIONVILLE (dairy and grain products); limestone. Formed 1845.

Lawrence, county (□ 456 sq mi/1,185.6 sq km; 2006 population 63,179), S OHIO; ☉ IRONTON; 38°35'N 82°33'W. Bounded S by OHIO RIVER, here forming boundary with KENTUCKY and WEST VIRGINIA; drained by SYMMES CREEK. In the Unglaciated Plain physiographic region. Agriculture (dairy products; livestock; grain, fruit, tobacco, hay); manufacturing (chemicals, steel, iron) at Ironton; limestone quarrying. Formed 1815.

Lawrence, county (□ 362 sq mi/941.2 sq km; 2006 population 91,795), W PENNSYLVANIA; ☉ NEW CASTLE; 40°59'N 80°19'W. Bounded W by OHIO state line; drained by SHENANGO, MAHONING, and BEAVER rivers, and Neshannock Creek. Agriculture (corn, oats, wheat, hay, alfalfa, apples; sheep, hogs, cattle, dairying); sand and gravel, limestone, coal. Manufacturing at New Castle. Iron center in mid-nineteenth century. McConnells Mill State Park in E. Formed 1849.

Lawrence, county (□ 800 sq mi/2,080 sq km; 2006 population 22,685), W SOUTH DAKOTA, on WYOMING state line; ☉ DEADWOOD; 44°21'N 103°47'W. Drained by Spearfish, Whitewood, and Elk creeks. Farming (in N); forest and mining region in BLACK HILLS. Gold, silver, quartz, timber; cattle, dairying; grain, honey; tourism. Development of county parallels growth of Homestake Mining Company and legalized gaming at Deadwood. The S two-thirds of the county is in Black Hills National Forest, except for pocket around LEAD and Deadwood that is set aside for mining interests; TERRY PEAK is near Lead. Roughneck Falls in W; Terry Peak and Deer Mountain ski areas in SW. Formed 1875.

Lawrence, county (2006 population 40,934), S TENNESSEE; ☉ LAWRENCEBURG; 35°13'N 87°23'W. Bounded S by ALABAMA; drained by upper BUFFALO RIVER and SHOAL CREEK. Upland agricultural region; manufacturing. Formed 1817.

Lawrence, city (2000 population 38,915), MARION county, central INDIANA, residential suburb 7 mi/11.3 km NE of INDIANAPOLIS, on Fall Creek; 39°52'N 85°59'W. Light manufacturing. Fort Benjamin Harrison is here.

Lawrence, city (2000 population 80,098), ☉ DOUGLAS county, a metropolitan statistical area in NE KANSAS, on the KANSAS River, 32 mi/51 km WSW of KANSAS CITY, 23 mi/37 km ESE of TOPEKA; 38°57'N 95°15'W. Railroad junction. Although agriculture trade and light manufacturing are economically important (especially as the headquarters of the greeting card company Hallmark), the city's major employer is the University of Kansas, which has hundreds of administrative offices, laboratories, and research facilities. Manufacturing (printing, medical equipment, construction machinery, feeds, fertilizers, chemicals, communications equipment, textiles, asphalt, paper products, concrete, pharmaceuticals, plastics products). Lawrence was founded in 1854 by the New England Emigrant Aid Company. The political center of the Free Staters, it was actually, though not legally, capital for a short time after 1857. LAWRENCE was an important stop on the Underground Railroad and the base for many Abolitionist organizations. In 1856 a proslavery raid on the town instigated the retaliatory POTTAWATOMIE killings by John Brown. In 1863 the town was again sacked and burned by William Quantrill. The Plymouth Congregational Church here was the first church built (1854) by settlers in KANSAS. Clinton Lake reservoir and Clinton State Park to W. The Haskell Indian Community College (1884), a large school for Native Americans, is here. Natural History Museum, Spencer Art Museum here. Lawrence Municipal Airport to NE. Incorporated 1858.

Lawrence, city (□ 7 sq mi/18.1 sq km; 1990 population 70,207; 2000 population 72,043), ☉ ESSEX county, NE MASSACHUSETTS, on the MERRIMACK RIVER; 42°42'N 71°10'W. It is a port of entry. Textiles, clothing, electrical equipment, athletic shoes, and rubber and paper products are manufactured. High-technology industries in the area also contribute to Lawrence's economy. Boston capitalists laid out an industrial town here in 1845 and built a granite dam on the Merrimack River. They also built mills and workers' dwellings, which were soon crowded with laborers, mainly from EUROPE, and Lawrence became one of the world's greatest centers for woolen textiles. By 1911 it was known as the "cloth-making capital of the world" (it was the headquarters for American Woolen Company, among others). Several disastrous events have occurred here—the collapse and burning of the Pemberton Mill in 1860, when over 500 trapped workers were killed or injured; the tornado of 1890; and the protracted labor strike by members of the International Workers of the World in 1912. Leonard Bernstein was born here. Has a community college. Heritage State Park. Settled 1655, set off from ANDOVER and METHUEN 1847, incorporated as a city 1853.

Lawrence, township, CLUTHA district, SE SOUTH ISLAND, NEW ZEALAND, 40 mi/64 km W of DUNEDIN; 45°55'S 169°42'E. Sheep-farming center; tourism based on gold-mining nostalgia.

Lawrence, township, MERCER county, central NEW JERSEY, 6 mi/9.7 km N of TRENTON; 40°17'N 74°43'W. Incorporated 1798.

Lawrence, unincorporated town, Peters township, WASHINGTON county, SW PENNSYLVANIA, suburb 11 mi/18 km SSW of PITTSBURGH, 12 mi/19 km NE of WASHINGTON, on Chartiers Creek; 40°18'N 80°07'W. Manufacturing (machinery, data communications equipment, commercial printing). Also known as Lawrence Hills.

Lawrence, village (2000 population 1,059), VAN BUREN county, SW MICHIGAN, 8 mi/12.9 km W of PAW PAW, on PAW PAW RIVER, in fertile area; 42°13'N 86°02'W. Agriculture (vegetables, fruit; hogs, poultry); manufacturing (food processing, walk-in coolers and freezers, canned and frozen fruits and vegetables, electroplating, molded plastic products).

Lawrence, village (2006 population 292), NUCKOLLS county, S NEBRASKA, 12 mi/19 km NW of NELSON; 40°17'N 98°15'W. Dairying; grain; livestock.

Lawrence, affluent residential village (□ 4 sq mi/10.4 sq km; 2006 population 6,442), NASSAU county, SE NEW YORK, one of "Five Towns of Long Island," S shore of W LONG ISLAND, 9 mi/14.5 km SE of JAMAICA; 40°36'N 73°42'W. A few light industries. In resort area. Incorporated 1897.

Lawrenceburg, city (2000 population 4,685), ☉ DEARBORN county, SE INDIANA, on OHIO River, 50 mi/80 km S of RICHMOND; 39°06'N 84°52'W. In agricultural area; manufacturing (whiskey, feed, machinery, lumber). Port of entry. Prehistoric fortifications found near here. Laid out 1802. Inundated by flood in 1937.

Lawrenceburg, city (2006 population 10,819), ☉ LAWRENCE county, S TENNESSEE, on SHOAL CREEK (source of hydroelectric power), 32 mi/51 km SSW of COLUMBIA; 35°14'N 87°20'W. Manufacturing. Founded c.1815. David Crockett State Park is just W.

Lawrenceburg, town (2000 population 9,014), ☉ ANDERSON county, central KENTUCKY, 22 mi/35 km W of LEXINGTON; 38°01'N 84°53'W. Trade center in BLUEGRASS agricultural region (dairying; poultry, horses; burley tobacco, corn, hay); manufacturing (paper products, motor vehicle parts, communications equipment, distilled whiskey, crushed limestone, computer equipment). Buckley Hills Wildlife Sanctuary to N; Beaver Lake reservoir 10 mi/16 km to SW. Settled 1776; incorporated 1820.

Cross-references are shown in SMALL CAPITALS. The pronunciation guide is shown on page xix. The sources of population figures are shown on page xvii.

Lawrence, Cape, NE ELLESMERE ISLAND, BAFFIN region, NUNAVUT territory, CANADA, on KENNEDY CHANNEL; 80°21'N 69°15'W.

Lawrence Hills, Pennsylvania: see LAWRENCE.

Lawrence Park, township (2000 population 4,048), ERIE county, NW PENNSYLVANIA, residential suburb 3 mi/4.8 km ENE of ERIE, on LAKE ERIE; 42°08'N 80°01'W.

Lawrencetown (LOR-unts-toun), village, W NOVA SCOTIA, CANADA, on ANNAPOLIS RIVER, 20 mi/32 km NE of ANNAPOLIS ROYAL. Apple packing, barrel manufacturing; fishing, hunting center.

Lawrenceville (LAH-rens-vil), village (□ 7 sq mi/18.2 sq km), ESTRIE region, SW QUEBEC, E CANADA, 19 mi/31 km E of GRANBY; 45°25'N 72°21'W. Dairying, lumbering, woodworking.

Lawrenceville, city (2000 population 22,397), ⊙ GWINNETT county, N central GEORGIA, 26 mi/42 km ENE of ATLANTA; 33°57'N 83°59'W. Suburb of Atlanta. Diversified economy with business and industrial parks. Manufacturing of displays, industrial machinery, consumer goods, paper products, printing and publishing, foods, plastics. New municipal government center opened in mid-1990s. Incorporated 1821.

Lawrenceville, city (2000 population 4,745), ⊙ LAWRENCE county, SE ILLINOIS, on EMBARRAS RIVER near the WABASH, 8 mi/12.9 km WNW of VINCENNES (Indiana); 38°43'N 87°41'W. In oil, natural gas, and agricultural area. Oil refineries; manufacturing (metal industries, electronic components). Livestock; soybeans, corn, wheat. Oil was discovered here in 1906. Founded 1821, incorporated 1835.

Lawrenceville (LAW-rens-vil), town (2006 population 1,155), ⊙ BRUNSWICK county, S VIRGINIA, on Great Creek, near MEHERRIN RIVER, 40 mi/64 km SW of PETERSBURG; 36°45'N 77°50'W. Manufacturing (clothing, motors, wood products, furniture, textiles, plastic and metal products, lumber); in agricultural area (tobacco, grain, cotton, peanuts; livestock; dairying). St. Paul's College; Southside Virginia Community College (Christiana campus) to NW at Cochran. Founded 1814; incorporated 1874.

Lawrenceville, village (2000 population 4,081), MERCER county, W NEW JERSEY, 5 mi/8 km NNE of TRENTON; 40°17'N 74°43'W. Turbine manufacturing. Lawrenceville School for boys (1810) and Rider University are located here. Largely residential. Settled 1692.

Lawrenceville, unincorporated village (2006 population 292), CLARK county, W central OHIO, 5 mi/8 km NNW of SPRINGFIELD; 39°59'N 83°52'W.

Lawrenceville, borough (2006 population 617), TIOGA county, N PENNSYLVANIA, 13 mi/21 km N of MANSFIELD, 17 mi/27 km WSW of ELMIRA, at NEW YORK state line, on TIOGA RIVER. Agriculture (dairying).

Lawson (LAW-suhn), township, commuter suburb 58 mi/93 km from SYDNEY, NEW SOUTH WALES, SE AUSTRALIA, in BLUE MOUNTAINS; 33°43'S 150°26'E. Walking tracks; Aboriginal rock art at Lawson's E edge.

Lawson, city (2000 population 2,336), RAY and CLAY counties, NW MISSOURI, 16 mi/26 km NW of RICHMOND; 39°26'N 94°12'W. Corn, soybeans; hogs, cattle.

Lawson Army Airfield, airport for Fort Benning located on the FORT BENNING MILITARY RESERVATION, GEORGIA. Built on site of Cusseta Town, which was the capital of the Lower Creek Nation before its removal. Floodplain of CHATTAHOOCHEE RIVER.

Lawson Heights, unincorporated town (2000 population 2,339), WESTMORELAND county, SW PENNSYLVANIA, residential suburb 2 mi/3.2 km SSW of LATROBE; 40°17'N 79°23'W. Agriculture includes dairying. St. Vincent College to W.

Lawsonia (law-SO-nee-uh), village, SOMERSET county, SE MARYLAND, on the EASTERN SHORE near TANGIER SOUND, 31 mi/50 km SSW of SALISBURY; 37°58'N 75°50'W. Nearby is Pocomoke Sound Wildlife Management Area.

Lawton, city (2006 population 87,540), ⊙ COMANCHE county, SW OKLAHOMA, 72 mi/116 km SW of OKLAHOMA CITY; drained by CACHE CREEK; 34°36'N 98°25'W. Elevation 1,117 ft/340 m. It is a commercial and trade center for the surrounding cotton, wheat, and cattle area and for FORT SILL, a military reservation to N. The fort is the largest local civilian employer; 25 mi/40 km long (E-W), and c.5 mi/8 km wide, it is practically a twin city to Lawton. Manufacturing (machining, bakery products, meat, rubber products, apparel, fabricated steel, publishing and printing, concrete). Cameron University is in the city. Comanche Tribal Headquarters.; Museum of the Great Plains; Fort Sill Museum at Fort Sill; county fairgrounds. Nearby is a large limestone quarry and the WICHITA MOUNTAINS Wildlife Refuge to NW. Incorporated 1901.

Lawton, town (2000 population 697), WOODBURY county, W IOWA, 11 mi/18 km E of SIOUX CITY; 42°28'N 96°10'W. In agricultural area, with manufacturing of construction equipment and nylon rope.

Lawton, town (2000 population 1,859), VAN BUREN county, SW MICHIGAN, 16 mi/26 km SW of KALAMAZOO; 42°10'N 85°50'W. In area of vineyards and small lakes (to S). Manufacturing (vegetables and fruit canning, food processing, juices and jellies, tools, foundry products).

Lawton, village (2006 population 39), RAMSEY CO., NE NORTH DAKOTA, 27 mi/43 km NE of DEVILS LAKE; 48°17'N 98°22'W. Founded in 1899 and incorporated in 1911. Named for General George Lawton, veteran of the Spanish-American War.

Lawtonka, Lake, reservoir, COMANCHE county, SW OKLAHOMA, on branch of CACHE CREEK, at edge of WICHITA MOUNTAINS, 12 mi/19 km NW of LAWTON; 5 mi/8 km long; 34°44'N 98°30'W. Formed by dam built for water supply. FORT SILL Military Reservation just S.

Lawu, Mount (LAH-woo) (10,712 ft/3,265 m), central JAVA, INDONESIA, 25 mi/40 km E of SURAKARTA; 07°38'S 111°11'E. Also spelled Mount Lawoe.

Laxa (LAHKS-O), town, ÖREBRO county, S central SWEDEN, 30 mi/48 km SW of ÖREBRO; 58°59'N 14°37'E. Railroad junction; manufacturing (lumber mills; mechanical industries, mineral water bottling); agriculture.

Laxapana Falls (LUHKS-shuh-PAH-nuh), waterfall in HATTON PLATEAU, CENTRAL PROVINCE, S central SRI LANKA, on left headstream of the KELANI GANGA River, and 7 mi/11.3 km W of HATTON; 377 ft/115 m high. Hydroelectric power station. Also called Laxapanagala and RAKSAPANA; also spelled LAKSAPANA FALLS.

Laxenburg (LAHK-sen-burg), township, E LOWER AUSTRIA, on SCHWECHAT RIVER, at the urban fringe of VIENNA, 10 mi/16 km S of city center; 48°04'N 16°27'E. Has several former imperial castles; one of them is a UN research center, and another is used by the International Institute for Applied System Analysis (IIASA). Trade and transportation enterprises. Summer festival; large park.

Laxey (LAKS-ee), town (2001 population 1,725), E coast of ISLE OF MAN, GREAT BRITAIN, 7 mi/11.3 km NE of DOUGLAS; 54°14'N 04°24'W. Seaside resort, with woolen milling. Lead was formerly mined here. Garwick Glen is old smugglers' cave. Nearby Laxton is England's last open-field village.

Laxmishwar, INDIA: see LAKSHMESHWAR.

Laxou (lah-SHOO), W suburb of NANCY, MEURTHE-ET-MOSELLE department, LORRAINE, NE FRANCE; located near the N-S expressway bypass of Nancy; 48°41'N 06°09'E. The recreational forest of Haye is just W.

Laya (LAH-yah), village, W SVERDLOVSK oblast, E central URALS, W Siberian Russia, on road and near railroad, 8 mi/13 km NNW and, under administrative jurisdiction, of NIZHNIY TAGIL; 58°03'N 59°55'E. Elevation 702 ft/213 m. Site of former ironworks called Laiskiy Zavod.

La Yesca (lah YES-kah), town, NAYARIT, W MEXICO, on JALISCO border, 55 mi/89 km ESE of TEPIC, in an isolated part of SIERRA MADRE OCCIDENTAL; 21°19'N 104°00'W. Agriculture; mining.

Laykovats, SERBIA: see LAJKOVAC.

Laylah, SAUDI ARABIA: see LAILA.

Lay Lake, on border of Chilton and Coosa counties, central Alabama, in Coosa River, 41 mi/66 km NNW of Montgomery; c.4.8 mi/77 km long; 32°57'N 86°31'W. Extends NNE into Talladega and Shelby counties to Logan Martin Dam. Formed by Lay Dam (104 ft/32 m high, 1,603 ft/489 m long), a privately built power dam completed 1914; crossed by road. Both the dam and lake were named for William P. Lay, an engineer and promotor of river navigation.

Laymantown (LAI-muhn-toun), unincorporated town, BOTETOURT county, W central VIRGINIA, 9 mi/15 km NE of ROANOKE; 37°22'N 79°51'W. George Washington National Forest, APPALACHIAN TRAIL to N.

Layon River (le-YON), 60 mi/97 km long, MAINE-ET-LOIRE department, PAYS DE LA LOIRE region, W FRANCE; 47°21'N 00°45'W. Rises 3 mi/4.8 km NE of CHOLET; flows NE then NW, past THOUARCÉ, to the LOIRE at CHALONNES-SUR-LOIRE.

Layou, town, W SAINT VINCENT, WEST INDIES, on Mount Wynn Bay, 4 mi/6.4 km NW of KINGSTOWN; 13°12'N 61°16'W. Cotton, arrowroot; fishing.

Layou River (luh-YOO), c.10 mi/16 km long, W central DOMINICA, BRITISH WEST INDIES, flows W, through Layou Plateau, to the coast.

Layrac (lai-rahk), commune (□ 14 sq mi/36.4 sq km), LOT-ET-GARONNE department, AQUITAINE region, SW FRANCE, on the GERS near its junction with the GARONNE, 5 mi/8 km SSE of AGEN; 44°08'N 00°40'E. Market for peaches, raisins, plums in a fertile agricultural valley; manufacturing of agricultural tools.

Lay River, 80 mi/129 km long, VENDÉE department, PAYS DE LA LOIRE region, W FRANCE; rises 4 mi/6.4 km N of La CHÂTAIGNERAIE, in the hills of the Vendée; flows SW, past MAREUIL-SUR-LAY-DISSAIS, to the Pertuis BRETON (a strait of the BAY OF BISCAY), at L'AIGUILLON-SUR-MER. In its lower course it crosses a vast coastal wetland with numerous drainage canals, all contained within the regional park of the Poitevin marshland (MARAIS POITEVIN, VAL DE SÈVRE ET DE LA VENDÉE NATURAL REGIONAL PARK).

Laysan (lai-SAHN), island, N PACIFIC, part of HONOLULU county, HAWAII, c. 890 mi/1,432 km NW of HONOLULU, c.160 mi/257 km N of TROPIC OF CANCER, 25°46'N 171°44'W. Coral with hypersaline lagoon; known for its large bird population (rookery for albatross, frigate birds, gulls). Annexed 1857 by Hawaiian kingdom; now U.S. possession. Part of Hawaiian Islands National Wildlife Refuge.

Laytapi (lai-TAH-pee), canton, NOR CHICHAS province, POTOSÍ department, W central BOLIVIA. Elevation 8,596 ft/2,620 m. Antimony mining at Mina Churquini; limestone deposits. Agriculture (potatoes, yucca, bananas); cattle.

Layton, city (2006 population 62,716), DAVIS county, N UTAH, suburb 9 mi/14.5 km S of OGDEN and 21 mi/34 km N of SALT LAKE CITY, near GREAT SALT LAKE; 41°04'N 111°57'W. Elevation 4,400 ft/1,341 m. In an irrigated farm area served by the Weber basin project. Drained by HOLMES and Hobbs creeks. Fruits and vegetables; dairying; cattle, sheep; manufacturing (bakery goods, dairy products, computer software). Causeway leads to ANTELOPE ISLAND State Park, in Great Salt Lake, to W. During the 1970s and 1980s the city profited from the prosperous commercial and financial activities of nearby Salt Lake City. Housing developments, business offices, and light manufacturing plants were constructed in Layton to accommodate its increased growth and industry. HILL AIR

FORCE BASE, with 17,000 workers, is Utah's largest employer. WASATCH RANGE and National Forest to E; Hill Air Force Base to N. Layton Heritage Museum.

Laytonsville, town (2000 population 277), MONTGOMERY county, central MARYLAND, 23 mi/37 km NNW of WASHINGTON, D.C.; 39°13′N 77°08′W. The first house here was believed to have been built by John Layton. For one year, 1848, it was called Cracklintown in honor of a nearby tavern where cracklin' bread was made.

Laytonville, unincorporated town (2000 population 1,301), MENDOCINO county, NW CALIFORNIA, 38 mi/61 km NNW of UKIAH, in COAST RANGES; 39°40′N 123°30′W. Laytonville Indian Reservation to W; Round Valley Indian Reservation to NE. Fruit; cattle; nursery stock; timber.

Laza (LAH-sah), town and canton, SUD YUNGAS province, LA PAZ department, W BOLIVIA, in the YUNGAS, 3 mi/4.8 km NNE of IRUPANA; 16°25′S 67°26′W. Subtropical agriculture (coffee, tea, quihoa); citrus fruits, bananas.

Lazareto (lah-zah-RE-to), canton, CERCADO province, TARIJA department, S central BOLIVIA. Elevation 6,883 ft/2,098 m. Extensive gas wells in area; clay deposits. Agriculture (potatoes, yucca, bananas, corn, barley, sweet potatoes; cattle).

Lazarev (LAH-zah-reef), settlement (2005 population 1,890), E KHABAROVSK TERRITORY, RUSSIAN FAR EAST, on the W shore of the narrowest (approximately 5 mi/ 8 km) S part of the TATAR STRAIT, on local road, 43 mi/69 km NE of BOGORODSKOYE; 52°13′N 141°31′E. Port facilities; local airfield.

Lazarevac (LAHZ-ah-re-vahts), town (2002 population 58,511), central SERBIA, on railroad, and 30 mi/48 km SSW of BELGRADE, in the SUMADIJA region; 44°22′N 20°15′E. Also spelled Lazarevats.

Lazarev Ice Shelf, the part of the ice shelf along the PRINCESS ASTRID COAST of QUEEN MAUD LAND, ANTARCTICA, that lies between Leningradskiy and VERBLYUD Islands, 50 mi/80 km long; 69°37′S 14°34′E.

Lazarevka, RUSSIA: see LAZAREVO, KIROV OBLAST.

Lazarevo (LAH-zah-ree-vuh), village, central TULA oblast, central European Russia, on railroad, 25 mi/40 km S of TULA; 53°51′N 37°29′E. Elevation 636 ft/193 m. In agricultural area (sugar beets, sunflowers, wheat, rye, vegetables); food processing.

Lazarevo (LAH-zah-ree-vuh), village, SE KIROV oblast, E central European Russia, in the VYATKA River basin, 8 mi/11 km SSW of SHURMA; 56°49′N 50°12′E. Elevation 429 ft/130 m. In agricultural area (wheat, flax, potatoes). Distillery. Also called Lazarevka (LAH-zah-reef-kah).

Lazarevskoye (LAH-zah-ryef-skuh-ye), health resort, S KRASNODAR TERRITORY, extreme NW CAUCASUS, RUSSIA, on the BLACK SEA coastal railroad and highway, 38 mi/61 km N, and under administrative jurisdiction, of SOCHI, and 18 mi/29 km SE of TUAPSE; 43°54′N 39°20′E. Elevation 216 ft/65 m. A fortified point in 19th century Russian wars against the Circassians.

Lazarev Trough (LA-zuh-ruhv), ANTARCTICA, depression in continental shelf off CLARIE COAST; 65°35′S 130°00′E.

Lázaro Cárdenas (LAH-zah-ro KAHR-dai-nahs), city and port, in extreme S central MICHOACÁN, MEXICO, on the PACIFIC OCEAN, 30 mi/48 km S of ARTEAGA, on Mexico Highway 200. A heavy industrial center specializing in iron, steel, and petrochemicals; also an important shipping center. Also called Ciudada Lázaro Cárdenas; sometimes spelled Lázero Cárdenas. Formerly called Los Llanitos.

Lázaro Cárdenas, MEXICO: see NANCHITAL DE LÁZARO CÁRDENAS DEL RÍO, PRESA LÁZARO CÁRDENAS, or SANCTÓRUM.

Laz Daua, SOMALIA: see LASO DAUAO.

Lazdijai (LAHZ-dee-yei), Polish Łozdzieje, city, S LITHUANIA, 23 mi/37 km SSE of MARIJAMPOLE, near Polish border; 54°14′N 23°31′E. Road junction; shoe manufacturing, flour milling. Important border crossing on VIA BALTICA HIGHWAY. Passed in 1795 to PRUSSIA, in 1815 to Russian POLAND; in Suvalki government until 1920. Also spelled Lazdiyay and Lozdzieje.

Lazi (LAH-see), town, on S Siquijor island, SIQUIJOR province, PHILIPPINES, on small inlet of MINDANAO SEA; 09°08′N 123°37′E. Rice growing, fishing.

Lazio, ITALY: see LATIUM.

Lazirky, UKRAINE: see NOVOORZHYTS′KE.

Laziska Gorne (lah-ZEES-kah GOOR-nee), Polish Łaziska Górne, town (2002 population 22,070), Katowice province, S POLAND, 11 mi/18 km SW of KATOWICE; 50°09′N 18°51′E. Foundries; large coal-fed power plant.

Lazne Jesenik, CZECH REPUBLIC: see JESENIK.

Lazne Kundratice (LAHZ-nye KUN-drah-TI-tse), Czech Lázně Kundratice, German Kunnersdorf, village, SEVEROCESKY province, N BOHEMIA, CZECH REPUBLIC, 7 mi/11.3 km SW of LIBEREC; 50°42′N 14°54′E. Health resort with mineral baths.

Lazne Kynzvart (LAHZ-nye KINZH-vahrt), Czech Lázně Kynžvart, German Bad Königswart, town, ZAPADOCESKY province, W BOHEMIA, CZECH REPUBLIC, on railroad, 4 mi/6.4 km NW of MARIANSKE LAZNE; 50°01′N 12°38′E. Health resort (elevation 2,208 ft/673 m) with carbonated mineral springs and baths. Fine castle.

Lázně Kynžvart, CZECH REPUBLIC: see LAZNE KYNZVART.

Lazne Libverda, CZECH REPUBLIC: see HEJNICE.

Lazne Tousen (LAHZ-nye TOU-shen-yuh), Czech Lázneři Toušeň, German Bad Tauschim, village, STREDOCESKY province, N central BOHEMIA, CZECH REPUBLIC, on ELBE RIVER, on railroad, 14 mi/23 km ENE of PRAGUE. Health resort with mineral springs.

Lazo (lah-ZO), village (2006 population 3,525), SE MARITIME TERRITORY, SE RUSSIAN FAR EAST, on the slopes of the S SIKHOTE-ALIN mountains, on highway junction, 80 mi/129 km NE of NAKHODKA; 45°52′N 133°39′E. Elevation 314 ft/95 m. Lumbering. Formerly known as Vangou (vahn-GO-oo).

Lazo (la-ZO), village (2006 population 3,215), W MARITIME TERRITORY, SE RUSSIAN FAR EAST, 4 mi/6 km E of the USSURI River where it creates a natural border between RUSSIA and CHINA, on the TRANS-SIBERIAN RAILROAD, 5 mi/8 km S of DAL′NERECHENSK; 45°52′N 133°39′E. Elevation 314 ft/95 m. In agricultural area (rice, soybeans).

Lazorki, UKRAINE: see NOVOORZHYTS′KE.

Lazovskiy Preserve (lah-ZOF-skeeye), nature reserve (□ 450 sq mi/1,170 sq km), in the S part of SIKHOTE-ALIN, MARITIME TERRITORY, RUSSIAN FAR EAST, extreme SE Siberian Russia. Mostly forest. Local fauna include roe deer, wild boar, lynx, musk deer, bear, and sable. Established in 1935. To ensure the area's ecological purity, no more than 1,000 are allowed in the preserve annually.

Lazurne (lah-ZOOR-ne), (Russian Lazurnoye), town, SW KHERSON oblast, UKRAINE, on BLACK SEA coast, 37 mi/60 km S of KHERSON, 9 mi/15 km W of SKADOVS′K; 46°04′N 32°32′E. Summer recreation center (tourist facilities, youth camps); forestry, Prymorske irrigation administration. Known since 1803, town since 1975.

Lazurnoye, UKRAINE: see LAZURNE.

Lea (LEE), county (□ 4,394 sq mi/11,424.4 sq km; 2006 population 57,312), extreme SE NEW MEXICO, in LLANO ESTACADO; ⊙ LOVINGTON; 32°47′N 103°25′W. Cattle, sheep; dairying; chiles, corn, sorghum, hay, alfalfa, melons, peas, spinach, pecans, peanuts, cotton, wheat, oats, barley, millet, rye. Bounded S and E by TEXAS (Central-Mountain time zone boundary on both sides; New Mexico in Mountain time zone). Petroleum, natural-gas fields. Harry McAdams State Park in E corner, at HOBBS. Formed 1917.

Leach River, ENGLAND: see THAMES RIVER.

Leachville, town, MISSISSIPPI county, NE ARKANSAS, 20 mi/32 km W of BLYTHEVILLE, near MISSOURI state line; 35°55′N 90°15′W. In cotton, rice, and soybean area. Manufacturing (consumer goods, printing).

Leacock (LEE-kahk), unincorporated town, LANCASTER county, SE PENNSYLVANIA, residential suburb 5 mi/8 km NE of LANCASTER; 40°05′N 76°12′W.

Lead (LEED), city (2006 population 2,860), LAWRENCE county, W SOUTH DAKOTA, 30 mi/48 km NW of RAPID CITY, in the BLACK HILLS. Site of the famous Homestake Mine (to E), which closed in 2001 (had been in operation since 1877). Tourism, gold and silver mining; surface tours; railroad terminus. Deer Mountain and Terry Peak ski areas to W. Laid out 1876 after discovery of gold, incorporated 1890.

Leadbetter Island (LED-be-tuhr), KNOX county, S MAINE, in PENOBSCOT BAY just W of Vinalhaven Island; 1 mi/1.6 km long, 0.5 mi/0.8 km wide.

Leader (LEE-duhr), town (2006 population 881), SW SASKATCHEWAN, CANADA, 45 mi/72 km SSW of KINDERSLEY; 50°53′N 109°33′W. Railroad junction; grain elevators, flour mills.

Leader Water (LEED-uhr), river, 21 mi/34 km long, Scottish Borders, SE Scotland; rises on Lammer Law; flows SSE, past LAUDER and EARLSTON, to the TWEED RIVER 2 mi/3.2 km E of MELROSE. Its valley, noted for scenic beauty and angling resorts, is called Lauderdale. Formerly in Borders, abolished 1996.

Lead Hill, village (2000 population 287), BOONE county, N ARKANSAS, 17 mi/27 km NE of HARRISON, in the OZARKS, near MISSOURI state line; at base of peninsula of BULL SHOALS LAKE reservoir (WHITE RIVER) to N; 36°24′N 92°54′W.

Leadhills (LED-hilz), village, South Lanarkshire, S Scotland, in LOWTHER HILLS, and 12 mi/19.2 km N of THORNHILL; 55°25′N 03°47′W. Elevation 1,350 ft/411 m. Former lead-mining town. Allan Ramsay born here. Just SW, in DUMFRIES AND GALLOWAY, is lead-mining village of WANLOCKHEAD. Formely in Strathclyde, abolished 1996.

Leadon River (LEE-duhn), 22 mi/35 km long, W ENGLAND; rises 7 mi/11.3 km WNW of Great MALVERN, Worcestershire; flows SE, past LEDBURY, to the SEVERN just NW of GLOUCESTER.

Leadore, village (2000 population 90), LEMHI county, E IDAHO, on LEMHI RIVER, 45 mi/72 km SSE of SALMON; 44°41′N 113°22′W. Sheep, cattle; alfalfa; oats. Salmon National Forest to NE and SW.

Leadville, town (2000 population 2,821), ⊙ LAKE county, central COLORADO, near the headwaters of the ARKANSAS RIVER, in the ROCKY MOUNTAINS; 39°15′N 106°17′W. Elevation c.10,152 ft/3,094 m. Some mining and smelting are still carried on (at nearby CLIMAX to NE huge deposits of molybdenum are mined), and farming, ranching, and the tourist trade have kept this famous city from becoming another ghost town. Rich placer gold deposits were discovered c.1860 in California Gulch. Oro City, the principal camp, flourished for about two years. The camps were virtually deserted until 1877, when the discovery of carbonates of lead with a high silver content again transformed Oro City into a boomtown. By 1880, two years after its incorporation, Leadville had become one of the greatest silver camps in the world. In 1893, with the repeal of the Sherman Silver Act, silver mining collapsed; but in the late 1890s, with the discovery of gold nearby, Leadville again revived. The spectacular history of Leadville is epitomized in the life of Horace Tabor. Points of interest include the restored Tabor home; the Matchless Mine, now a museum; and the Healy House–Dexter Cabin Museum. TURQUOISE LAKE reservoir to W, parts of San Isabel National Forest to N, W, and S. Pike National Forest to E (beyond county line), Leadville National Fish Hatchery to W. Incorporated 1878.

Leadwood (LED-wud), village (2000 population 1,160), SAINT FRANCOIS county, E MISSOURI, in SAINT

FRANCOIS MOUNTAINS, 3 mi/4.8 km W of PARK HILLS; 37°51′N 90°35′W. Former mining town, primarily residential; sand and gravel.

Leaf Rapids (LEEF RA-pidz), town (□ 492 sq mi/ 1,279.2 sq km; 2001 population 1,309), MANITOBA, W central CANADA, 1.9 mi/3.1 km S of CHURCHILL River, 606 mi/975 km NW of WINNIPEG; 56°28′N 99°45′W. Copper-mine community; tourism. Incorporated 1974.

Leaf River, village (2000 population 555), OGLE county, N ILLINOIS, on LEAF RIVER and 6 mi/9.7 km NW of OREGON; 42°07′N 89°24′W. In rich agricultural area.

Leaf River (LEEF), 300 mi/483 km long, N QUEBEC, E CANADA; issues from Lake MINTO; flows NE, through tidal Leaf Lake (30 mi/48 km long, 15 mi/24 km wide), to UNGAVA BAY 65 mi/105 km NNW of KUUJJUAQ.

Leaf River, c.25 mi/40 km long, N ILLINOIS; rises S of FREEPORT; flows generally SE to ROCK RIVER N of OREGON; 42°08′N 89°41′W.

Leaf River, 50 mi/80 km long, in W and W central MINNESOTA; rises in E central OTTER TAIL county, in Ground Lake, SE of OTTER TAIL LAKE; 46°24′N 95°25′W. Flows E, through West and East Leaf lakes, past BLUFFTON and to N of WADENA, to CROW WING RIVER 10 mi/16 km NNW of STAPLES.

Leaf River, c.180 mi/290 km long, in S central and SE MISSISSIPPI; rises in SCOTT county; flows S past TAYLORSVILLE and HATTIESBURG, then SE past BEAUMONT and MCLAIN, joining CHICKASAWHAY RIVER to form PASCAGOULA RIVER in N GEORGE county.

League City (LEEG), city (2006 population 65,351), GALVESTON county, SE TEXAS, suburb 13 mi/21 km NW of TEXAS CITY and 21 mi/34 km SE of HOUSTON; 29°29′N 95°06′W. Near GALVESTON BAY, bounded by Clear Lake (NE) and Clear Creek (N). The aeronautics industry is of prime importance to the area; the Lyndon B. Johnson Space Center is to N. Diversified light manufacturing. Incorporated 1961.

Leake (LEEK), county (□ 585 sq mi/1,521 sq km; 2006 population 22,769), central MISSISSIPPI; ⊙ CARTHAGE; 32°45′N 89°31′W. Drained by PEARL and YOCKANOOKANY rivers and LOBUTCHA CREEK. Includes two small Native Amer. reservations in center and SE (Choctaw). Agriculture (cotton, corn; poultry, cattle; dairying); timber. NATCHEZ TRACE PARKWAY passes through NW. Formed 1833.

Leake, East (LEEK), village (2001 population 6,236), S Nottinghamshire, central ENGLAND, 4 mi/6.4 km NNE of LOUGHBOROUGH; 52°49′N 01°10′W. Previous industries included hosiery and basket weaving. Has 13th-century church.

Leakesville (LEEKS-vil), town (2000 population 1,026), ⊙ GREENE county; 31°08′N 88°33′W, SE MISSISSIPPI, 45 mi/72 km ESE of HATTIESBURG and on CHICKASAWHAY RIVER. Agriculture (cotton, corn; cattle, poultry); timber; manufacturing (lumber, wood processing). Parts of De Soto National Forest to NW and SW.

Leakey (LEE-kee), village (2006 population 374), ⊙ REAL county, SW TEXAS, 36 mi/58 km N of UVALDE and on FRIO RIVER; 29°43′N 99°45′W. Elevation 1,609 ft/490 m. In ranching area (goats, sheep, cattle). The scenic Frio Canyon is here. Cedar timber; manufacturing (cedar wood oil, perfume). Lost Maples State Park to E.

Leaksville (LEEKS-vil), town, ROCKINGHAM county, N NORTH CAROLINA, 11 mi/18 km NNW of REIDSVILLE, on DAN RIVER. Established 1797; merged with the towns of SPRAY (1 mi/1.6 km to NE) and DRAPER in 1967 to form city of EDEN; 36°29′N 79°46′W.

Leal, village (2006 population 34), BARNES co., E central NORTH DAKOTA, 20 mi/32 km NW of VALLEY CITY; 47°06′N 98°18′W. Founded in 1892 and incorporated 1911. Named for settlers from Canada of Scottish-English descent. Leal is Scottish for faithful and true.

Leales (le-AH-les), village, ⊙ Leales department, central TUCUMÁN province, ARGENTINA, at confluence of the RÍO COLORADO and SALÍ RIVER, on railroad, 27 mi/ 43 km SSW of TUCUMÁN; 27°12′S 65°18′W. In agricultural area (corn, alfalfa, sugarcane, cotton; livestock); apiculture.

Lealman (LEEL-muhn), unincorporated city (□ 5 sq mi/13 sq km), W central FLORIDA, PINELLAS county, 3 mi/4.8 km N of ST. PETERSBURG; 27°49′N 82°41′W.

Lealui, ZAMBIA: see MONGU.

Leamington (LEE-ming-tuhn), town (□ 102 sq mi/ 265.2 sq km; 2001 population 27,138), S ONTARIO, E central CANADA, on Lake ERIE; 42°03′N 82°35′W. In a market-gardening area, it has large food-processing plants. Gateway to POINT PELEE NATIONAL PARK.

Leamington, village (2006 population 212), MILLARD county, W central UTAH, 20 mi/32 km NE of DELTA, on SEVIER RIVER; 39°31′N 112°17′W. Irrigated agricultural area. White Sand Dunes and Little Sahara Recreation Area to NW; Fishlake National Forest to SE. Fool Creek reservoir to SW.

Leamington, ENGLAND: see ROYAL LEAMINGTON SPA.

Leam River (LEM), 25 mi/40 km long, Northamptonshire and Warwickshire, central ENGLAND; rises 5 mi/8 km NW of DAVENTRY; flows W, past Leamington, to AVON River E of Warwick.

Le'an (LUH-AN), town, ⊙ LE'AN county, central JIANGXI province, CHINA, 45 mi/72 km ENE of JI'AN; 27°24′N 115°49′E. Grain; logging, food processing, coal mining.

Leander (LEE-an-duhr), town (2006 population 20,451), WILLIAMSON county, central TEXAS, 22 mi/35 km NNW of AUSTIN; 30°33′N 97°51′W. In agricultural area (cotton, corn, wheat; cattle); limestone quarries; manufacturing (concrete, plumbing equipment, power supplies).

Leandro N. Alem (le-AHN-dro AH-lem), town, S MISIONES province, ARGENTINA, 40 mi/64 km SE of POSADAS; 34°30′S 61°24′W. Agricultural center (corn, tobacco, cotton, maté, tung, potatoes, jute, tea, citrus fruit, grapes); lumbering; maté and tobacco processing.

Leandro N. Alem, ARGENTINA: see VEDIA.

Leane, Lough, IRELAND: see KILLARNEY, LAKES OF.

León River (le-AHN), c.50 mi/80 km long, N HONDURAS; rises on YORO department border in CORDILLERA DE NOMBRE DE DIOS; flows NNE to CARIBBEAN SEA 10 mi/16 km ENE of TELA; 15°47′N 87°20′W.

Le'an River (LUH-AN), 130 mi/209 km long, NE JIANGXI province, CHINA; rises on ANHUI-ZHEJIANG border; flows W, past WUYUAN and LEPING, joining CHANG River to form short BO RIVER at BOYANG, on BOYANG Lake.

Leão, Brazil: see BUTIÁ.

Lea River (LEE) or **Lee River**, 46 mi/74 km long, SE ENGLAND; rises just N of LUTON, BEDFORDSHIRE; flows SE and S, through HERTFORDSHIRE, along Essex–Greater London boundary, and past Hatfield, Hertfield, and Ware, to the Thames at Blackwall. Receives STORT RIVER just E of Hoddesdon.

Learned (LUHR-ned), village (2000 population 50), HINDS county, W MISSISSIPPI, 22 mi/35 km WSW of JACKSON; 32°12′N 90°32′W. Agricultural and timber area. NATCHEZ TRACE PARKWAY passes to NW.

Leary, town (2000 population 666), CALHOUN county, SW GEORGIA, 20 mi/32 km WSW of ALBANY; 31°29′N 84°31′W.

Leary (LEE-ree), village (2006 population 585), BOWIE county, NE TEXAS, 8 mi/12.9 km W of TEXARKANA; 33°28′N 94°12′W. Agricultural area. Oil and natural gas.

Leasburg, town (2000 population 323), CRAWFORD county, E central MISSOURI, in the OZARKS, near MERAMEC RIVER, 9 mi/14.5 km N of STEELVILLE; 38°05′N 91°17′W. Elevation 1,024 ft/312 m. Onondaga Cave State Park to SE.

Leask (LEESK), village (2006 population 418), central SASKATCHEWAN, CANADA, 40 mi/64 km WSW of PRINCE ALBERT; 53°01′N 106°45′W. Farming, dairying.

Leatherhead (LE-thuhr-hed), town (2001 population 9,685), central SURREY, SE ENGLAND, on MOLE RIVER, 17 mi/27 km SW of LONDON; 51°17′N 00°19′W. Manufacturing of electrical equipment. Has 12th-century church and several old houses.

Leatherman Peak, Idaho: see LOST RIVER RANGE.

Leaton, Fort, Texas: see PRESIDIO, county and PRESIDIO, town.

Léau, BELGIUM: see ZOUTLEEUW.

L'Eau d'Heure (LO DUHR), recreational park near BOUSSU-LEZ-WALCOURT, HAINAUT province, SW BELGIUM, 14 mi/23 km S of CHARLEROI.

Leau-Gamka, river, SOUTH AFRICA: see GAMKA RIVER.

Leavenheath (lev-uhn-HEETH), village (2001 population 1,744), SUFFOLK, E ENGLAND, 8 mi/12.9 km NW of COLCHESTER; 51°59′N 00°50′E. Box River 2 mi/3.2 km to NE.

Leavenworth (LEV-uhn-wuhrth), county (□ 468 sq mi/1,216.8 sq km; 2006 population 73,628), NE KANSAS; ⊙ LEAVENWORTH; 39°12′N 95°02′W. Gently rolling to hilly area, bounded E by MISSOURI River and MISSOURI, S by KANSAS River. Wheat, sorghum, soybeans, hay, vegetables, apples; cattle, hogs, poultry; dairying; paper products, industrial machinery. Oil. Formed 1855.

Leavenworth (LEV-uhn-wuhrth), city (2000 population 35,420), ⊙ LEAVENWORTH county, NE KANSAS, on the MISSOURI River. Satellite community of KANSAS CITY, KANSAS; 39°19′N 94°55′W. Railroad junction. It is the commercial center of a farm and livestock region, with flour mills and plants that make automobile batteries, machinery, furniture, and metal products. Nearby FORT LEAVENWORTH, with its various institutions (including U.S. Army Command and General Staff College and the federal penitentiary, which is located on the grounds although operated by the Justice Dept.), is central to the city's economy. St. Mary College is nearby. Leavenworth is the oldest city in KANSAS and was the first city in the state to be incorporated. It was settled (1854) near the fort by proslavery Missourians, but later became a Union supporter during the Civil War. It also flourished as a supply point on westward travel routes. The state's first newspaper was printed here in 1854. Flooding occurred in the area in 1993. Incorporated 1855.

Leavenworth, town (2000 population 353), CRAWFORD county, S INDIANA, near OHIO River, 11 mi/18 km SSE of ENGLISH; 38°12′N 86°21′W. In agricultural area. Manufacturing (crushed stone). Moved (1937–1938) to higher ground from flood-ravaged former site on Ohio River. Harrison-Crawford State Forest and WYANDOTTE CAVE nearby to E.

Leavenworth, town (2006 population 2,225), CHELAN county, central WASHINGTON, 20 mi/32 km NW of WENATCHEE and on Wenatchee River, near mouth of Icicle Creek; 47°35′N 120°40′W. Apples, pears; timber; manufacturing (fishing lures). Camping, fishing; winter sports; Leavenworth Ski Area to N. STEVENS PASS (4,061 ft/1,238 m), U.S. Highway 2, and Cascade Railroad Tunnel (8 mi/12.9 km long) through crest of CASCADE RANGE 20 mi/32 km to NW. Surrounded, except NE, by Wenatchee National Forest, including Alpine Lakes Wilderness Area to W. Lake Wenatchee reservoir and State Park to N. Incorporated 1906.

Leavenworth, Fort, KANSAS: see Fort Leavenworth.

Leavitt Peak (11,569 ft/3,526 m), TUOLUMNE-MONO county line, E CALIFORNIA, in the SIERRA NEVADA, 32 mi/51 km NW of MONO LAKE.

Leavittsburg (LE-vits-buhrg), unincorporated village (□ 2 sq mi/5.2 sq km), TRUMBULL county, NE OHIO, 3 mi/4.8 km W of WARREN, on MAHONING RIVER; 41°14′N 80°52′W.

Leawood (LEE-wud), city (2000 population 27,656), JOHNSON county, NE KANSAS, suburb 8 mi/12.9 km S of KANSAS CITY; 38°54'N 94°37'W. It is an agriculture trading and processing point that has undergone major suburban development as an outgrowth of Kansas City. Business offices and light manufacturing are located here. Manufacturing (meat products). State Line Airport in S. Borders MISSOURI on E. Incorporated 1948.

Leba (LE-bah), Polish Łeba, German *Leba*, town in POMERANIA, Słupsk province, N POLAND, on LEBA RIVER, near the BALTIC SEA, 18 mi/29 km NNW of LĘBORK; 54°45'N 17°33'E. Railroad spur terminus; fishing port; fish canning and smoking. Founded 14th century by Teutonic Knights; chartered 1357.

Lebach (LAI-bahkh), town, SAARLAND, SW GERMANY, on Theel River, 10 mi/16 km NE of SAARLOUIS; 49°26'N 06°55'E. Railroad junction; manufacturing of motor vehicle parts, wire. Grew in 1974 through incorporated of neighboring villages. Chartered 1979.

Lebadea, Greece: see LEVADIA.

Lébamba (lai-BAHM-buh), town, NGOUNIÉ province, SW GABON, 37 mi/60 km SE of MOUILA; 02°12'S 11°29'E. Coffee growing.

Lebane (LEB-ah-nai), village (2002 population 24,918), S SERBIA, 12 mi/19 km SW of LESKOVAC; 42°55'N 21°45'E.

Lebanon (LE-buh-nuhn), republic (□ 4,015 sq mi/10,400 sq km; 2004 estimated population 3,777,218; 2007 estimated population 3,925,502), SW ASIA, on the MEDITERRANEAN SEA; ⊙ BEIRUT; 33°50'N 35°50'E.

Geography

The country faces the E shore of the Mediterranean and is bordered on the N and E by SYRIA and on the S by ISRAEL. In addition to Beirut there are four ports: TRIPOLI in the N, SAIDA and TYRE (now Sur) in the S, and JUNIYE in central Lebanon. There are four physical regions. The fertile narrow coastal plain is the most densely populated part of the country, containing most of Lebanon's big cities and intensive agriculture. The W highlands, rising abruptly from the coastal plain, consist of two main ranges: the Lebanon Mountains (extending N of the LITANI RIVER to the Nahr el Kebir on the Syria border) and the N part of the GALILEE Highlands (stretching from the Israeli border to the Litani River). The fertile Al BEQA'A Valley (average elevation 3,280 ft/1,000 m) lies in the Syrian-African rift, an extension of the JORDAN VALLEY, and is drained in the N by the ORONTES Asi River and in the S by the Litani River; and the ANTI-LEBANON range, which is more densely inhabited on its W slopes. Much of the terrain is mountainous; the Lebanon Mountains, which run parallel to the coast, reach their highest point at QURNET ES SAUDA (elevation 10,131 ft/3,088 m); on the E border are Mount HERMON and the Anti-Lebanon range, along which the boundary of Lebanon and Syria extends. The Orontes in the N and the Litani in the S are the main rivers.

Population

About 93% of Lebanese are Arabs. Arabic is the official language; English is also spoken, as is Armenian to a limited extent. The country's last official census was in 1932; at that time, the Christian population was 53% of the total. No census has been taken since in order to maintain the fiction of a balanced Christian-Muslim population; however, most knowledgeable observers believe that about two-thirds of the population is now Muslim. The rapidly growing group of Shiites, in the N Beka and S of Beirut, now rival the Sunni in number. About one-quarter of the population is divided into Greek Orthodox, Armenians, and Maronites (an ancient Eastern Christian church that was independent until it was united with the Catholic Church). There is also a minority of Druze, whose religion combines Muslim and Christian elements.

The civil war has resulted in the extensive emigration of Maronites and Armenians. Political life is profoundly affected by the country's religious diversity; political groups that are mainly Christian, especially of the Maronite sect, generally favor an independent course for Lebanon, stressing its ties with Europe; the Muslims, however, favor closer ties with the surrounding Arab countries. Maronites, Sunni Muslims, and Shiite Muslims are represented in the legislature, cabinet, and civil service.

Economy

Lebanon is the distribution center for the Middle East, and commerce is its major industry. Lebanon itself is largely agricultural. The main crops are citrus fruits, apples, bananas, pears, olives, vegetables, and grapes. Grain is grown, but not enough for Lebanon to be self-sufficient. Sugarcane, cotton, and tobacco are also grown, and hemp (hashish) is grown and illegally exported. Livestock are raised, and poultry is a large source of revenue. The largest manufacturing industries are engaged in food processing and the production of textiles, apparel, carpeting, and tobacco products. Lebanon has few minerals. There used to be two oil pipelines terminating in Lebanon—one from Iraq to Tripoli and another from Saudi Arabia to Sidon (Zahrani)—but they have closed because they fell victim to Middle East conflicts. The refineries near Sidon and Tripoli that were destroyed are being rebuilt. Not many of the famed cedars remain, although oak and pine are exploited. Tourism and financial services, which flourished in the 1950s and 1960s, has decreased dramatically with the onset of Lebanon's political unrest. The country exports fruit, vegetables, and textiles, largely to the other Arab countries; imports include grain, flour, and manufactured goods from Italy, France, Germany, Syria, and China.

History: to 16th Century

In ancient times the area of Lebanon and Syria was occupied by the Canaanites, who founded the great cities of PHOENICIA and later established a commercial maritime empire. Lebanon's cities, as well as its forests and iron and copper mines (since exhausted), attracted the successive dominant powers in the Middle East. The Phoenician cities occupied a favored position in the Persian Empire and were conquered by Alexander the Great. The region came under Roman dominion starting in 64 B.C.E. (there are notable Roman ruins at BAALBEK) and was Christianized before the Arab conquest in the 7th century. By then the Maronites had established themselves—a cardinal fact in the history of Lebanon, which long remained predominantly Christian while Syria became Muslim. Later (11th century) the Druze settled in the S part of Mount Lebanon and in adjacent regions of Syria, and trouble between them and the Christians was to become a constant theme in regional history. The Crusaders were active in Lebanon (late 11th century) and were aided by the Lebanese Christians. After the Crusaders, Lebanon was loosely ruled by the Mameluks (c.1300). Invasions by Mongols and others contributed to the decline of trade until the reunification of the Middle East under the Ottoman Turks (early 16th century).

History: 16th Century to World War II

In the later years of Ottoman control, Lebanon had considerable autonomy, and powerful families ruled the country. Conflict among the religious communities, culminating in massacres of the Maronites by the Druze in 1860, led to intervention by FRANCE (1861), and the Ottoman sultan was forced to appoint a Christian governor for Lebanon. Mount Lebanon was set aside as a special, semi-autonomous district within the Ottoman Empire, and its boundary was drawn to ensure a Christian majority. This was called "smaller Lebanon" and controlled by the dominant Christian

group, the Maronites. The French were given the mandate of Syria after World War I by the League of Nations, and Lebanon was a part of that mandate. The French, being Catholic, separated Lebanon (home of most of the Maronite Catholics) from Syria, thus creating a new state. They added the Tripoli area to the N, the Seka in the E, and the Mutaribri area in the S, creating a "greater Lebanon." In 1926 the mandate was given a republican constitution. A treaty with France in 1936 provided for independence after a three-year transition period, but it was not ratified by France.

History: World War II to 1975

In World War II the French VICHY government controlled Lebanon until a British–Free French force conquered (June–July 1941) both Lebanon and Syria. The Free French proclaimed Lebanon an independent republic. Lebanon became independent on January 1, 1945. In that year it became a member of the UN, and soon afterward all British and French troops were evacuated. As a member of the Arab League, Lebanon declared war on Israel in 1948 but took little part in the conflict; an armistice agreement with Israel was signed in 1949. In the spring of 1958 opposition to the ruling president Camille Chamoun's pro-Western policies and his acceptance of U.S. aid under the Eisenhower doctrine erupted in rioting in Tripoli, Beirut, and elsewhere. The rioting grew into full-scale rebellion, and Chamoun called in U.S. forces (July 1958). General Fouad Chehab, a nonpolitical personality, was elected to succeed Chamoun, and the rebellion ebbed. By the fall, U.S. forces had left the country. In 1962 a military coup was attempted in Beirut but was crushed. During the 1967 Arab-Israeli War, Lebanon gave verbal support to the Arab effort against Israel but did not become involved in any military action. Israel, however, has repeatedly accused Lebanon of not doing enough to prevent the Palestinian commandos from operating against Israel from Lebanese soil. After the bloody suppression in 1970–1971 of the guerrillas in JORDAN, large numbers of Palestinians fled to S Lebanon and Beirut. Lebanon did not enter the 1973 Arab-Israeli War, nor did the Lebanese army interfere with Palestinian guerrillas operating in S Lebanon. Israel continued its attacks on Palestinian guerrilla bases in S Lebanon.

History: 1975 to 1987

Lebanon became embroiled in civil war among the Christians, Muslims, and Palestinians from early 1975 to late 1976. At the request of the president, Suleiman Franjieh, Syrian forces entered Lebanon (April 1976), halting Muslim and Palestinian advances. An estimated 50,000 Lebanese were killed and twice that number wounded. The country became devastated, the economy crippled, and tourism plummeted to the point of nonexistence. A cease-fire in October 1976, proved unstable, and hostilities resumed full scale in 1977. In response to guerrilla attacks by the Palestinian Liberation Organization (PLO), Israel occupied S Lebanon in March 1978, but withdrew in June. In 1981 fighting continued between Christian (Phalangist) and Syrian forces, as Israeli air raids on Beirut became a destructive form of reprisal. On June 6, 1982, Israeli forces invaded Lebanon full scale, primarily to eliminate Palestinian guerrilla bases. Israeli troops and their Christian allies laid siege to Beirut. Nearly 7,000 Palestinian guerrillas were forced to leave Lebanon, which was accomplished under the supervision of a Multinational Force (MNF) comprised of U.S. and European-allied troops, and Israeli forces occupied W Beirut on September 15. The next day, Phalangist militia began the massacres of Palestinians in the refugee camps of Sabra and Chatila on the SW part of Beirut. The uproar that followed, both internationally and in Israel, was a major reason for the return of the MNF and Israeli pullback from W Beirut. The U.S.

---

and MNF withdrew from Beirut after the Palestinian bombings of the U.S. embassy (April 1983; sixty-three killed) and the U.S. Marine barracks (October 1983; 241 killed). As Israeli troops slowly left the Beirut and S area, Lebanese militias fought in the wake of the Israeli withdrawal. Israel completed its withdrawal in mid-1985, but left soldiers to work in conjunction with the Christian South Lebanese Army (SLA) to maintain a security ("buffer") zone.

History: 1987 to Present

Beirut remained a major battle area, and in February 1987, Syrian troops moved into the city to suppress the attacking factions. A tentative peace accord in 1989 between Christian and Muslim representatives was aided by the Arab League to halt fighting. Throughout the 1980s, Westerners were the targets of numerous kidnappings by radical Shiite groups, and Lebanon became notorious for taking Western hostages. In early 1991, Lebanese troops organized to regain control of the S from PLO guerrillas and Israelis who controlled a 6-mi/10-km-deep security zone. By 1994 neither the Israeli nor the Syrian forces had quit the country, and clashes between Israeli troops and Palestinian units across the border and extreme militant Shiite organizations supported by the Hezbollah group in IRAN, as well as among the existing Lebanese militias, were still common. A formula on communal representation based on the 1932 census distribution of the population has traditionally provided for the president of Lebanon to be a Maronite Christian, the prime minister a Sunni Muslim, and the speaker of the Parliament a Shiite Muslim. This arrangement lasted until 1989, when a Muslim president, René Mowad (who was backed by Syria), was elected. Mowad was assassinated that November, and Elias Hrawi succeeded him. The many years of fighting ruined Lebanon's infrastructure, national economy, and tourism industry. The Ta'if Accord, a blueprint for reconciliation has provided for a more equitable political system by allowing Muslims greater input into the political process. Since 1989 there have been several successful elections, the militias have been weakened or abandoned, and the Lebanese Army has restored central government authority over about two-thirds of the country, though Hezbollah continues to retain its weapons. Hrawi's presidential term was extended for three years in 1995, as the country continued to recover from years of heavy fighting which had crippled its infrastructure and economy. General Emile Lahoud was elected president in 1998. In May 2000, Israel withdrew its troops from S Lebanon. Syria's influence in Lebanese politics became blatant in 2004 when Lahoud's term was extended by constitutional amendment at its behest; the action was denounced by the UN Security Council. The assassination of former Prime Minister Rafik Hariri in February 2005 sparked anti-Syrian demonstrations and increased international pressure on Syria, which then agreed to withdraw all its troops from Lebanon (completed in April). Parliamentary elections resulted in an anti-Syrian majority; Fouad Siniora became prime minister. A UN investigation into Hariri's killing implicated (2005) senior Lebanese and Syrian officials. In July 2006 Israel's air force attacked Lebanese targets after Hezbollah captured two Israeli soldiers. Hezbollah launched rockets against N Israel in response, and Israel sent troops into S Lebanon. A UN-mediated cease-fire (mid-August) and Israeli withdrawal (complete in October) left a UN peacekeepers and the Lebanese army deployed in S Lebanon. Hezbollah's prestige was enhanced by its resistance to Israeli forces, and it sought a re-formed government in which it and its allies would hold greater power, leading to a political stalemate that continued to 2007.

Government

The country is governed under the constitution of 1926 as amended. Under the constitution, the presi-
dent, who is the head of state and wields real power, is elected by the legislature for a six-year term and cannot serve consecutive terms. The government is headed by the prime minister, who is appointed by the president. The unicameral legislature consists of the 128-seat National Assembly, whose members are elected by popular vote on the basis of sectarian proportional representation for four-year terms. The Ta'if accord of 1989 calls for all main religious groups to be represented in the cabinet. The current head of state is President Emile Lahud (since 1998), and the current head of government is Prime Minister Fuad Siniora. Lebanon is divided into eight administrative districts or governorates: Aakar, Baalbek-Hermel, Beirut, Beqa'a, Mount Lebanon, Nabatiye, North Lebanon, and South Lebanon.

Lebanon, county (□ 362 sq mi/941.2 sq km; 2006 population 126,883), SE central PENNSYLVANIA; ⊙ LEBANON. Drained by SWATARA CREEK. Agriculture (corn, wheat, oats, barley, hay, alfalfa, soybeans; sheep, hogs, cattle, chickens, eggs, dairying) area; limestone. Manufacturing at Lebanon, MYERSTOWN, and PALMYRA. Stiegel glassware made here in 18th century. Indiantown Gap Military Reservation in NW, 10 mi/16 km NW of Lebanon, established 1935. APPALACHIAN TRAIL passes through N part of county, on Stony and BLUE MOUNTAIN ridges; Swatara State Park in N; Lebanon Valley in S; part of Middle Creek Waterfowl Management Area in SE. Settled c.1710 by Germans; formed 1813.

Lebanon (LE-buh-nuhn), city (2000 population 3,523), SAINT CLAIR county, SW ILLINOIS, suburb of SAINT LOUIS, near SILVER CREEK, 12 mi/19 km NE of BELLEVILLE; 38°36′N 89°48′W. In agricultural area (wheat, soybeans; hogs, poultry; dairy products). Seat of McKendree College. Main Street has nineteenth century buildings with shops, taverns, restaurants; an old inn (1830) is here. SCOTT AIR FORCE BASE to SW. Settled in early nineteenth century; incorporated 1857.

Lebanon, city (2000 population 14,222), ⊙ BOONE county, central INDIANA, 25 mi/40 km NW of INDIANAPOLIS; 40°03′N 86°28′W. In dairying and farming area; manufacturing (machinery, electrical equipment, animal feed, plastics, paper products, asphalt, frozen foods). Laid out 1832.

Lebanon, city (2000 population 12,155), ⊙ LACLEDE county, S central MISSOURI, in the OZARKS, 47 mi/76 km NE of SPRINGFIELD; 37°40′N 92°39′W. Shipping center for grain and dairy products; major service center on I-44. Manufacturing (wood products, machinery, fabricated metal products, electrical equipment, apparel, electric motors, food-processing plants). Bennett Spring State Park to W (DALLAS county). Founded c.1849.

Lebanon, city (2006 population 12,586), GRAFTON county, W NEW HAMPSHIRE, 43 mi/69 km NW of CONCORD; 43°38′N 72°15′W. Bounded W by CONNECTICUT RIVER (VERMONT state line), drained by MASCOMA RIVER. Includes village of West Lebanon. Manufacturing (machinery, software, structural steel products, printing and publishing, consumer goods); soapstone quarrying; agriculture (nursery crops, apples, vegetables; cattle, poultry; dairying). Railroad junction at West Lebanon. Part of Moose Mountains in NE. Founded 1761.

Lebanon (LEB-uh-nahn), city (□ 12 sq mi/31.2 sq km; 2006 population 20,346), ⊙ WARREN county, SW OHIO, 28 mi/45 km NE of CINCINNATI, on small Turtle Creek; 39°25′N 84°13′W. FORT ANCIENT STATE MEMORIAL PARK nearby. Laid out 1802.

Lebanon, city (2006 population 14,416), LINN county, W OREGON, on South SANTIAM RIVER, 12 mi/19 km SE of ALBANY; 44°31′N 122°54′W. Railroad junction. Fruits, grains; dairy products. Lumber milling. Fish hatchery to NE. Foster Reservoir, with fish hatchery and GREEN PETER RESERVOIR, to SE. Site of Lebanon Strawberry Festival. Incorporated 1878.
Lebanon, city (2006 population 24,180), ⊙ LEBANON county, SE PENNSYLVANIA, 28 mi/45 km E of HARRISBURG, on Quittapahilla Creek, in Pennsylvania Dutch farm country; 40°20′N 76°24′W. Railroad junction. Manufacturing (consumer goods, fabricated metal products, lumber, commercial printing, apparel, machinery, tools, machinery, textiles, fertilizers, wood products, food, plastic products). It has steel and steel-fabricating industries, although the industry declined significantly in the 1970s and 1980s. Agriculture (grain, soybeans, apples; livestock; dairying). Lebanon was a flourishing town before 1790, and early 18th-century German religious groups are still represented here. Lebanon Valley Airport to E. The city has a historical museum, Lebanon Historical Society, Stoy Museum, and horse shows. Also in the area are the Cornwall Furnace (operated 1742–1883) and the Union Canal tunnel, a civil engineering landmark. Swatara State Park and APPALACHIAN TRAIL to N; Fort Indiantown Military Reservation and Memorial Lake State Park to NW. Founded 1753, incorporated as a city 1868.

Lebanon, city (2006 population 23,702), ⊙ WILSON county, N central TENNESSEE, 28 mi/45 km E of NASHVILLE; 36°12′N 86°18′W. Manufacturing; recreation. Seat of Cumberland University and a community college. Sam Houston practiced law here. Fine antebellum homes nearby include the Hermitage. Cedars of Lebanon State Park is in the extensive cedar glades. Founded c.1802.

Lebanon (LE-buh-nuhn), town, NEW LONDON county, E central CONNECTICUT, 11 mi/18 km NW of NORWICH; 41°37′N 72°14′W. Egg production and farming. Revolutionary War Office (1727), Governor Trumbull house (1740; now historical museum), other 18th-century houses here. Incorporated 1700.

Lebanon, town (2000 population 5,718), ⊙ MARION county, central KENTUCKY, 28 mi/45 km WSW of DANVILLE, in outer BLUEGRASS REGION; 37°34′N 85°15′W. Agriculture (burley tobacco, grain; dairying); manufacturing (textiles, wood products, fabricated metal, metal fabrication, building materials, paper products, printing and publishing). Springfield-Lebanon Airport to N. Lebanon National Cemetery is here; Saint Mary's College to W at SAINT MARY. Established 1815.

Lebanon, town, YORK county, SW MAINE, near SALMON FALLS RIVER, 9 mi/14.5 km SW of ALFRED; 43°23′N 70°54′W. Settled 1738, incorporated 1767.

Lebanon (LE-buh-non), town (2006 population 3,201), ⊙ RUSSELL county, SW VIRGINIA, 22 mi/35 km NNE of BRISTOL, on Little Cedar Creek; 36°53′N 82°04′W. Manufacturing (building materials, apparel, motor vehicle parts); agriculture (corn, soybeans; livestock; dairying); timber.

Lebanon (LE-buh-nuhn), village (2000 population 303), SMITH county, N KANSAS, 12 mi/19 km ENE of SMITH CENTER; 39°48′N 98°33′W. Corn; livestock. Geographic center (39°50′N 98°35′W) of the continental U.S. is 2 mi/3.2 km NW of here.

Lebanon, village (2006 population 67), RED WILLOW county, S NEBRASKA, 20 mi/32 km SE of MCCOOK, on BEAVER CREEK, near KANSAS state line; 40°02′N 100°16′W.

Lebanon, village (2006 population 75), POTTER county, N central SOUTH DAKOTA, 10 mi/16 km ENE of GETTYSBURG; 45°04′N 99°46′W.

Lebanon (LE-buh-nuhn), borough (2006 population 1,830), HUNTERDON county, W NEW JERSEY, 9 mi/14.5 km N of FLEMINGTON; 40°38′N 74°50′W. Agricultural area; machinery manufacturing.

Lebanon, ancient *Libanus*, range, c.100 mi/161 km long, paralleling the MEDITERRANEAN SEA from the lower LITANI VALLEY, S LEBANON, N, rising steeply from the coast; 34°00′N 36°00′E. The valley of Nahr el Kebir, along which the Lebanese-SYRIAN boundary runs, is the natural N limit of the Lebanese Mountains.

QURNET ES SAUDA (10,131 ft/3,088 m) is the highest peak. A great fault line, forming a steep escarpment, leads in the E down to the fertile AL BEQA'A valley, which separates the Lebanon from the ANTI-LEBANON mountains to the E. The Litani River rises in the valley and flows S and then W to the Mediterranean. The mountains were famed in ancient times for the huge old cedars that extended in a narrow strip for 85 mi/ 137 km along the upper W slope of the range. However, these trees were depleted by long use as a building material and a fuel, and only 10 small isolated groves remain. (Solomon's temple in Jerusalem was built in the 10th century B.C.E. with cedars brought from Lebanon.) Apples, olives, apricots, and other fruits are grown in large orchards. Through history the Lebanon Mountains have provided refuge for persecuted minorities, such as the Druzes and the Maronites, who settled on the fertile middle slopes. Many springs, fed by the melting snow, exit from the mountainside and make intensive irrigation possible. Clusters of villages are found on the terraced slopes. The E and W slopes have become summer and ski resorts.

Lebanon Junction (LE-buh-nuhn), town (2000 population 1,801), BULLITT county, central KENTUCKY, near ROLLING FORK, 11 mi/18 km S of SHEPHERDSVILLE; 37°49'N 85°43'W. In agricultural area (burley tobacco, grain; livestock; dairying); manufacturing (machinery, printing and publishing). FORT KNOX Military Reservation to NW.

Lebanon Valley, Pennsylvania: see LEBANON, CO.

Lebaoth, ISRAEL: see BETH-LEBAOTH.

Lebap (le-BAHP), weloyat (□ 35,900 sq mi/93,340 sq km), E TURKMENISTAN; ⊙ CHARJEW. Drained by the AMU DARYA river flowing through the KARA KUM desert; intensive cotton and some wheat cultivation in its valley; sericulture. Cotton and silk processing at Charjew, KERKI, and KERKICHI. Sulphur mining and chemical industry (GAURDAK), coal (SVINTSOVII RUDNIK). Goat and karakul-sheep raising on desert. Crossed by TRANS-CASPIAN RAILROAD (N), KAGAN (UZBEKISTAN)–DUSHANBE (TAJIKISTAN) railroad (S). Population consists of Turkmens, Uzbeks, and Russians. Formed 1939.

Lebap (le-BAHP), town, LEBAP weloyat, TURKMENISTAN, alongside W bank of AMU DARYA and UZBEKISTAN border, c.149 mi/240 km NW of CHARJEW; 41°03'N 61°53'E.

Leba River (LE-bah), Polish Łeba, German Leba, 80 mi/ 129 km long, NW POLAND, in POMERANIA; rises W of KARTUZY; flows N, W, past LEBORK, and N, through Leba Lake, past LEBA town, to the BALTIC SEA just N of Leba. Leba Lake (□ 29 sq mi/75 sq km), is just W of the town, on Koszalin-Gdansk province border, separated from the Baltic Sea by a narrow spit; 10 mi/16 km long, 2 mi/3.2 km–5 mi/8 km wide.

Lebbeke (le-BAI-kuh), commune (2006 population 17,623), Dendermonde district, EAST FLANDERS province, W BELGIUM, 3 mi/4.8 km SSE of DENDERMONDE; 51°00'N 04°08'E. Wool weaving; agriculture market.

Lebda, LIBYA: see KHUMS, AL.

Lebec, unincorporated village (2000 population 1,285), KERN county, S central CALIFORNIA, in TEHACHAPI MOUNTAINS, 37 mi/60 km S of BAKERSFIELD. Oil refining. Cattle; dairying; grain. Los Padres National Forest to SW; Site of Fort Tejon State Historical Park; TEJON PASS to S.

Lebedin, UKRAINE: see LEBEDYN.

Lebedinovka (le-be-DEE-nuhv-kuh), village, CHÜY region, KYRGYZSTAN, in CHU valley, just E of BISHKEK; 42°53'N 74°43'E. Orchards, vegetables. Tertiary-level administrative center.

Lebedyachi Islands, UKRAINE: see CRIMEAN HUNTING PRESERVE.

Lebedyan' (lye-bye-DYAHN), city (2006 population 22,975), central LIPETSK oblast, S central European

Russia, on the DON River, on railroad, 38 mi/61 km NW of LIPETSK, and 36 mi/58 km NE of YELETS; 53°01'N 39°08'E. Elevation 442 ft/134 m. Highway hub; machinery, metal, and food industries. Founded in the 16th century as a fortress. Made city in 1779.

Lebedyn (le-be-DIN), (Russian Lebedin), city, S SUMY oblast, UKRAINE, near PSEL RIVER, on railroad spur, 26 mi/42 km SSW of SUMY; 50°35'N 34°29'E. Elevation 465 ft/141 m. Raion center; metalworking and machine tool plants; food processing (fruit canning, dairy, meat, flour); sewing; manufacturing (furniture, building materials). Teachers' college, medical school, vocational technical school; art museum, memorial museum; wooden church (1748). Established in 1652, city since 1655 (company center of the Sumy Cossack Regiment).

Lebel-sur-Quévillon (luh-BEL–syur–kai-vee-YON), city (□ 17 sq mi/44.2 sq km; 2006 population 3,142), NORD-DU-QUÉBEC region, W QUEBEC, E CANADA; 49°03'N 76°59'W. Forestry, pulp and paper products; mining; tourism.

Lebenstedt (LAI-buhn-shtet), N district of SALZGITTER, LOWER SAXONY, NW GERMANY. Potash mines, oil wells.

Lébény (LAI-bai-nyuh), German Leiden, village, Györ-sopron county, NW HUNGARY, 12 mi/19 km WNW of Györ; 47°44'N 17°23'E. Noted Romanesque church.

Lebern (LAI-buhrn), district, SW SOLOTHURN canton, N SWITZERLAND. Main town is GRENCHEN; population is German-speaking and mainly Roman Catholic.

Lebialem, department (2001 population 144,560), SOUTH-WEST province, CAMEROON.

Le Bic (luh BEEK), village (□ 31 sq mi/80.6 sq km; 2006 population 2,889), SE QUEBEC, E CANADA, on the SAINT LAWRENCE RIVER, and 10 mi/16 km SW of RIMOUSKI; 48°22'N 68°42'W. Resort; woodworking. Nearby is BIC ISLAND.

Le Blanc, FRANCE: see BLANC, LE.

Lebnitsa (LEB-neet-sah), river, SW BULGARIA, tributary of the STRUMA River; 41°23'N 23°10'E.

Lebo (LEE-bo), village (2000 population 961), COFFEY county, E KANSAS, 16 mi/26 km NNW of BURLINGTON; 38°24'N 95°51'W. In livestock and grain region.

Lebombo Mountains (le-BOM-bo), mountain range, SE AFRICA; runs S-N, from PONGOLA RIVER, KWAZULU-NATAL province, E SOUTH AFRICA, through E SWAZILAND, SW MOZAMBIQUE, and NE South Africa to LIMPOPO RIVER; forms low escarpment generally 1,312 ft/400 m–1,969 ft/600 m overlooking coastal plain of INDIAN OCEAN; c.250 mi/402 km long, c.10 mi/16 km wide, 96 mi/150 km along border; 26°15'S 32°00'E. Has Border Peak (2,402 ft/732 m), the highest point in SWAZILAND. Range bisected by GREAT USUTU (Lusutufu), KOMATI, and Sabie rivers. KRUGER NATIONAL PARK (South Africa) in N. Also spelled LUBOMBO. S extension called Ubombo Mountains.

Lebong (LAI-bawng), town, DARJILING district, N WEST BENGAL state, E INDIA, N suburb of Darjiling. Elevation 8,000 ft/2,438 m. Former BRITISH cantonment. Famous for its racecourse.

Lebon Régis (LE-bon RE-zhees), city (2007 population 11,735), central SANTA CATARINA state, BRAZIL, 26 mi/ 42 km SE of CAÇADOR; 26°56'S 50°42'W. Wheat, corn, potatoes.

Lebork, Polish Lebork, German Lauenburg, town (2002 population 35,252), Słupsk province, N POLAND, in POMERANIA region, on LEBA river, 30 mi/48 km ENE of SŁUPSK; 54°33'N 17°45'E. Railroad junction; linen milling, food canning, woodworking, agriculture-implement manufacturing. Founded 1341 by Teutonic Knights, who built castle. Passed 1657 to BRANDENBURG.

Lebret (luh-BRET), village (2006 population 203), SE SASKATCHEWAN, CANADA, on the FISHING LAKES, 16 mi/26 km N of INDIAN HEAD; 50°45'N 103°47'W. Mixed farming, fishing.

Lebrija, city, SEVILLE province, SW SPAIN, in ANDALUSIA, in lower basin of the GUADALQUIVIR, on railroad, 32 mi/51 km S of SEVILLE; 36°55'N 06°04'W. Processing and trading center for agricultural products of fertile region (grapes, olives, cereals, fruit; livestock). Mining of aluminum silicate, which is exported. Flour mills, potteries. Its outstanding buildings, in Mozarabic style, include 12th-century church (formerly a mosque), ruins of Moorish castle and Carthusian convent, and 18th-century tower modeled after the Giralda of Seville. Of ancient origin, Lebrija was the Roman Nerissa Veneria, where Venus was worshiped. Birthplace of the Spanish humanist Antonio de Nebrija and navigator Díaz de Solís.

Lebrija (lai-BREE-hah), town, ⊙ Lebrija municipio, SANTANDER department, N central COLOMBIA, 5 mi/8 km W of BUCARAMANGA; 07°07'N 73°13'W. Elevation 3,536 ft/1,077 m. Coffee-growing center; corn, cassava, cacao; livestock.

Lebrija River (lai-BREE-hah), c.100 mi/161 km long, N central COLOMBIA; rises in Cordillera ORIENTAL NW of BUCARAMANGA; flows N to MAGDALENA River at Bodega Central; 08°08'N 73°47'W.

Lebu (LAI-boo), town, ⊙ Lebu comuna (2002 population 20,838), ARAUCO province, BÍO-BÍO region, S central CHILE, on the PACIFIC, on Lebu River, 65 mi/ 105 km SSW of CONCEPCIÓN; 37°38'S 73°41'W. Port, railroad terminus, coal-mining center; beach resort. Fishing. Agricultural products (cereals, vegetables); livestock. Airport. Founded 1852.

Lebyazhye (lye-BYAHZH-ye), town (2005 population 3,425), S KIROV oblast, European Russia, on the VYATKA River, on highway, 18 mi/29 km SW of NOLINSK; 57°25'N 49°31'E. Elevation 252 ft/76 m. In agricultural area (wheat, sunflowers, sugar beets); grain storage, dairy products, bakery; livestock veterinary station; logging, lumbering, sawmilling.

Lebyazhye (lye-BYAHZH-ye), town (2005 population 6,685), E KURGAN oblast, SW SIBERIA, RUSSIA, on crossroads and the TRANS-SIBERIAN RAILROAD, approximately 36 mi/58 km N of the KAZAKHSTAN border, 45 mi/72 km ESE of KURGAN; 55°16'N 66°29'E. Elevation 482 ft/146 m. Dairy products; brewery.

Lebyazhye (leb-YAHZH-ye), village, SE PAVLODAR region, KAZAKHSTAN, on IRTYSH RIVER, 65 mi/105 km SSE of PAVLODAR; 51°30'N 77°45'E. Tertiary-level (raion) administrative center. In agricultural area; cattle.

Lebyazhye (lye-BYAHZH-ye), village (2005 population 5,565), W LENINGRAD oblast, NW European Russia, on the SE shore of the Gulf of FINLAND, 29 mi/47 km W of SAINT PETERSBURG; 59°57'N 29°25'E. Summer homes.

Lebyazhye (lye-BYAHZH-ye), health resort, SW ALTAI TERRITORY, S central SIBERIA, RUSSIA, on a small bitter-salt lake, on road, 40 mi/64 km N of RUBTSOVSK; 51°42'N 80°50'E. Elevation 777 ft/236 m. Sanatoria for tuberculosis, spas.

Lebyazhye (lye-BYAHZH-ye), peak of the central URAL Mountains, W SVERDLOVSK oblast, RUSSIA, near the VYSOKAYA mountain and NIZHNIY TAGIL; 67°50'N 31°04'E. Phosphorous-iron mines, limestone-crushing works, supplying metallurgical plants of Nizhniy Tagil.

Leça da Palmeira (lai-sah dah pahl-MAI-rah), town, PÔRTO district, N PORTUGAL, at mouth of small Leca River on the ATLANTIC OCEAN, 6 mi/9.7 km NW of OPORTO; 41°12'N 08°42'W. Seaside resort. With adjoining MATOZINHOS (SE), it encloses artificial harbor of LEIXÕES. Has 17th-century fortress.

Lecanto (luh-KAN-to), unincorporated town (□ 6 sq mi/15.6 sq km; 2000 population 5,161), CITRUS county, W central FLORIDA, 10 mi/16 km W of INVERNESS; 28°51'N 82°30'W. Manufacturing includes building materials, furniture.

Le Cap, HAITI: see CAP-HAÏTIEN.

Cross-references are shown in SMALL CAPITALS. The pronunciation guide is shown on page xix. The sources of population figures are shown on page xvii.

Lecce (LET-che), province (□ 1,065 sq mi/2,769 sq km), APULIA, S ITALY; ⊙ LECCE. Between the ADRIATIC SEA and GULF OF TARANTO; forms S extremity of "heel" of Italian peninsula. Plain in N; low, hilly terrain in S, rising to 659 ft/201 m. Watered by a few small streams. Leads Italy in production of olive oil and tobacco. Other major crops are grapes, figs, cereals, citrus fruit. Livestock raising; fishing. Manufacturing at Lecce and Gallipoli, primarily in agricultural products, some building materials, clothing. Area reduced to form provinces of Tarento (1923) and BRINDISI (1927).

Lecce (LET-che), city (2001 population 83,303, ⊙ LECCE province, APULIA region, S ITALY; 40°23′N 18°11′E. It is an industrial and agricultural center. Manufacturing includes fabricated metal products, machinery, ceramics, leather, paper, food products, tobacco products, and wine. A Greek and later a Roman town, Lecce was from 1053 to 1463 a semi-independent country under various lords. In the 16th and 17th centuries, culture and commerce flourished here. There are many fine churches and palaces built or restored in a characteristic baroque style. The city has a university (founded 1959).

Lecco (LEK-ko), town, COMO province, LOMBARDY, N ITALY, port on Lake of Lecco (SE arm of LAKE COMO), at efflux of ADDA River, 15 mi/24 km ENE of COMO; 45°51′N 09°23′E. Resort and highly diversified industrial center; iron- and steelworks, silk and paper mills, lime kilns, nail factories, food canneries; manufacturing of machinery, chemicals, plastics, wine. A major market and exporter of cheese, especially Gorgonzola. Has picturesque medieval bridge (14th century; partly modernized) over the Adda and a museum of natural history.

Le Center, town (2000 population 2,240), ⊙ LE SUEUR county, S MINNESOTA, 20 mi/32 km NE of MANKATO, near Le Sueur Creek; 44°23′N 93°43′W. Elevation 1,066 ft/325 m. Agricultural area (corn, oats, peas, alfalfa; livestock; dairying); manufacturing (building materials, fiberglass products, plastics, machinery). County fairgrounds. Settled 1864, incorporated 1876. Known as Le Sueur Center until 1931.

Lécera (LAI-thai-rah), town, ZARAGOZA province, NE SPAIN, 30 mi/48 km SSE of ZARAGOZA. In fertile agricultural area (cereals, wine).

Lech (LEKH), village, VORARLBERG, W Austria. Ski and tennis resort.

Lech, village, VORARLBERG, W Austria, on upper LECH RIVER, 16 mi/26 km NE of BLUDENZ; 47°13′N, 10°09′E. Elevation 4,401 ft/1,341 m. International winter-sports center; cable cars to Rüfikopf (7,199 ft/2,194 m), Oberlech (5,087 ft/1,551 m) and Zuger Hochlicht (7,227 ft/2,203 m). Settled by Walsers around year 1300.

Lechaina, Greece: see LEKHAINA.

Lechang (LUH-CHANG), town, ⊙ Lechang county, N GUANGDONG province, CHINA, on WU RIVER, 37 mi/60 km NW of SHAOGUAN, and on GUANGZHOU-WUHAN railroad; 25°08′N 113°20′E. Rice, oilseeds, jute; cotton, textiles.

Le Chasseral, SWITZERLAND: see CHASSERAL, LE.

Le Chasseron, SWITZERLAND: see CHASSERON, LE.

Leche Lagoon (LAI-chai), CAMAGÜEY province, E CUBA, 3 mi/5 km N of MORÓN, and bounded N by TURIGUANÓ Island; 22°12′N 78°37′W. Lagoon is 7 mi/11 km long, up to 5 mi/8 km wide. Linked by tidal marshes with the sea (inlets of OLD BAHAMA CHANNEL). Its milky color is due to lime sulphates.

Lechena, Greece: see LEKHAINA.

Léchère, La, FRANCE: see LÉCHÈRE-LES-BAINS, LA.

Léchère-les-Bains, La (lai-sher–lai-ban, lah), commune, spa in SAVOIE department, RHÔNE-ALPES region, SE FRANCE, on ISÈRE river, and 11 mi/18 km SE of ALBERTVILLE, in the Savoy Alps (ALPES FRANÇAISES). One of the newest thermal resorts of the ALPS whose springs were discovered in 1869 as a result of a rock slide. Also known as La Léchère.

Lechfeld (LEKH-felt), plain near AUGSBURG, BAVARIA, S GERMANY, drained by the LECH RIVER; 48°00′N 10°45′E–48°20′N 10°55′E. There in 955, King (later Emperor) Otto I defeated the Magyars and stopped their expansion into central EUROPE.

Lechiguanas Islands (le-chee-GWAH-nahs), in PARANÁ RIVER delta, E ARGENTINA. Bounded by PARANÁ PAVÓN and PARANÁ IBICUY arms (N) and main Paraná River channel (S); extend c.70 mi/113 km from VILLA CONSTITUCIÓN (SANTA FE province) to IBICUY.

Lechinkay (lye-cheen-KEI), village (2005 population 4,380), central KABARDINO-BALKAR REPUBLIC, N CAUCASUS, S European Russia, on the CHEGEM RIVER, on highway, 9 mi/14 km WNW of NAL'CHIK; 43°34′N 43°26′E. Elevation 2,165 ft/659 m. Agricultural products. Has a mosque. Archaeological digs in the vicinity.

Lechinţa (le-KEEN-tsah), German Lechnitz, Hungarian Szászlekence, village, BISTRIŢA-NĂSĂUD county, N central ROMANIA, 10 mi/16 km SW of BISTRIŢA; railroad junction; agricultural center. In HUNGARY, 1940–1945.

Lechkhumi Range (lech-KHOO-mee), S spur of the central Greater Caucasus, in NW GEORGIA; 37 mi/60 km long, forming watershed between the TSKHENISTSKALI and upper RION rivers; rises to 11,844 ft/3,610 m.

Lechlade (LECH-laid), village (2001 population 2,759), GLOUCESTERSHIRE, W ENGLAND, on THAMES RIVER, 10 mi/16 km NE of SWINDON; 51°43′N 01°41′W. Tourist site, especially for boaters traveling along the Thames.

Lechnitz, ROMANIA: see LECHINŢA.

Lech River (LEKH), c.175 mi/282 km long; rising in VORARLBERG, W Austria, and flowing NE into S GERMANY past AUGSBURG to the DANUBE RIVER. The WERTACH RIVER is its chief tributary. There are about twenty hydroelectric stations on the river, of which Rain (105,000 kw capacity) is the largest. In 1632 Gustavus II of SWEDEN defeated the Count of Tilly near the mouth of the Lech.

Lechtal Alps (LEKH-tahl), German Lechtaler Alpen, range of the NORTHERN LIMESTONE ALPS, TYROL, W Austria. They extend 35 mi/56 km NE from the ARLBERG pass, rising to 9,967 ft/3,038 m in the PARSEIERSPITZE mountains; pastures, cattle. The Letchtal, a valley of the upper LECH RIVER, parallels the range on the N; the STANZER TAL and the valley of the Inn (below LANDECK) parallel it to the S.

Lechuguilla Island (le-choo-GEE-yah) (□ 13 sq mi/33.8 sq km), narrow barrier island in SE Gulf of CALIFORNIA, off coast of SINALOA, NW MEXICO, at mouth of Río Fuerte, 9 mi/14 km W of TOPOLOBAMPO; 12 mi/19 km long, 1 mi/1.6 km–2 mi/3.2 km wide.

Leciñena (lai-thee-NYAI-nah), town, ZARAGOZA province, NE SPAIN, 16 mi/26 km NE of ZARAGOZA; 41°48′N 00°37′W. Cereals; wine; sheep; lumber.

Leck (LEK), village, SCHLESWIG-HOLSTEIN, NW GERMANY, 18 mi/29 km W of FLENSBURG, in N FRIESLAND; 54°47′N 08°58′E. Market center for cattle region.

Le Claire, town (2000 population 2,847), SCOTT county, E IOWA, on outer MISSISSIPPI RIVER, suburb 12 mi/19 km ENE of DAVENPORT; 41°36′N 90°21′W. Manufacturing (metal, wood and limestone products). Lock and Dam No. 14 is immediately downstream; Buffalo Bill Museum here. Spring flooding occurred in 1993.

Leclercville (luh-KLERK-vil), village (□ 52 sq mi/135.2 sq km), CHAUDIÈRE-APPALACHES region, S QUEBEC, E CANADA, on the SAINT LAWRENCE RIVER, 30 mi/48 km NE of TROIS-RIVIÈRES; 46°34′N 72°00′W. Lumbering; dairying.

L'Écluse, NETHERLANDS: see SLUIS.

Lecompte (luh-KAHMP), town (2000 population 1,366), RAPIDES parish, central LOUISIANA, 15 mi/24 km S of ALEXANDRIA; 31°05′N 92°24′W. In agricultural area (cotton, sugarcane; cattle, horses); timber; manufacturing (feeds, charcoal). Alexander State Forest to W. Settled c. 1855.

Lecompton (luh-KAHMP-tuhn), village (2000 population 608), DOUGLAS county, NE KANSAS, on the KANSAS River, 10 mi/16 km NW of LAWRENCE; 39°02′N 95°23′W. The pro-slavery Lecompton Constitution was formulated here September 1857, and it was rejected by Kansas voters in 1858. Clinton Lake reservoir and Clinton State Park to S.

Le Conte, Mount (luh-KAWNT) (6,593 ft/2,010 m), in GREAT SMOKY MOUNTAINS, SEVIER county, E TENNESSEE, 6 mi/10 km SE of GATLINBURG; 35°39′N 83°26′W. Tourist lodge, campsite here.

Lecques, Les, FRANCE: see SAINT-CYR-SUR-MER.

Lectoure (lek-TOOR), town (□ 32 sq mi/83.2 sq km), GERS department, MIDI-PYRÉNÉES region, SW FRANCE, on GERS RIVER, 21 mi/34 km N of AUCH; 43°56′N 00°37′E. Agriculture market (wheat; cattle, poultry; wine); ARMAGNAC brandy distilling, brewing. Has an archeological museum and a 15th–17th-century church (former cathedral). It was the seat of the counts of Armagnac until 1473, and of a bishop until 1790.

Leczna (LENCH-nah), Polish Łęczna, town (2002 population 22,166), Lublin province, E POLAND, on WIEPRZ RIVER, 14 mi/23 km ENE of LUBLIN; 51°17′N 22°52′E. Flour milling, manufacturing of soap; horse trading. Church and synagogue from 16th–17th century; museum.

Leczyca (len-CHEE-tsah), Polish Łęczyca, Russian Lenchitsa, town, PŁOCK province, central POLAND, on BZURA RIVER, 23 mi/37 km NNW of ŁÓDŹ. Flour milling, brewing, manufacturing of cement, starch. Romanesque church, built 1161, was restored 1951. During World War II, under German rule, called Lentschütz.

Ledaña (lai-DHAH-nyah), town, CUENCA province, E central SPAIN, 27 mi/43 km NNE of ALBACETE; 39°22′N 01°42′W. Saffron, cereals, grapes, olives, vegetables; livestock.

Ledang, Gunong, Malaysia: see OPHIR, MOUNT.

Ledava River (LE-dah-vah), Hungarian Lendva, c.50 mi/80 km long, in AUSTRIA and NE SLOVENIA; rises in Austria 7 mi/11.3 km SSE of FELDBACH; flows SE, through the PREKMURJE region (Slovenia), past MURSKA and LENDAVA, to MURA RIVER on Hungarian border, 9 mi/14.5 km SE of LENDAVA. Sometimes called Lendava River.

Ledbury (LED-buh-ree), town (2001 population 9,221), E Herefordshire, W ENGLAND, on LEADON RIVER, 12 mi/19 km E of HEREFORD; 52°03′N 02°25′W. Former agricultural market. Has Norman church, 17th-century market house, and several old inns. John Masefield born here.

Lede (LAI-duh), agricultural commune (2006 population 17,098), Aalst district, EAST FLANDERS province, N central BELGIUM, 4 mi/6.4 km NW of AALST; 50°58′N 03°59′E.

Ledeberg (LAI-duh-berkh), town, Ghent district, EAST FLANDERS province, NW BELGIUM, on SCHELDT RIVER, just S of GHENT; 51°02′N 03°45′E.

Ledec nad Sazavou (LE-dech NAHD SAH-zah-VO), Czech Ledeč nad Sázavou, German Ledetsch an der Sazau, town, VYCHODOCESKY province, E BOHEMIA, CZECH REPUBLIC, on SAZAVA RIVER, on railroad, 14 mi/6.4 km NW of HAVLÍČKŮV BROD; 49°42′N 15°17′E. Agriculture (barley, oats); manufacturing (machinery, footwear). Has Renaissance castle, museum of earthenware.

Ledegem (LAI-duh-khem), agricultural commune (2006 population 9,364), Roeselare district, WEST FLANDERS province, W BELGIUM, 6 mi/9.7 km W of KORTRIJK; 50°51′N 03°07′E.

Ledengskoye, RUSSIA: see BABUSHKINA, IMENI.

Ledesma, town, SALAMANCA province, W SPAIN, on TORMES RIVER, 20 mi/32 km NW of SALAMANCA.

Sawmilling, tanning; cereals; livestock. Mineral springs.

Ledesma, ARGENTINA: see LIBERTADOR GENERAL SAN MARTÍN.

Ledetsch an der Sazau, CZECH REPUBLIC: see LEDEC NAD SAZAVOU.

Lediba (le-DEE-buh), village, BANDUNDU province, W CONGO, on right bank of KWA RIVER, 30 mi/48 km W of BANDUNDU; 03°03′S 16°32′E. Elev. 515 ft/156 m.

Lednice (LED-ni-TSE), German *Eisgrub*, village, JIHO-MORAVSKY province, S MORAVIA, CZECH REPUBLIC, on Dyje River, 28 mi/45 km SSE of BRNO. Railroad terminus. Agriculture (grapes, fruit, tobacco, wheat); wine making. Has a picturesque castle; was part of the former domain of Prince Liechtenstein, with valuable collections and extensive park. Paleolithic-era archaeological site.

Lednicke Rovne (led-NYITS-kai rou-NE), Slovak *Lednické Rovne*, Hungarian *Lednicróna*, village, ZA-PADOSLOVENSKY province, W SLOVAKIA, in NE foothill of the WHITE CARPATHIAN MOUNTAINS, on VÁH RIVER, 23 mi/37 km SW of ŽILINA; 49°04′N 18°17′E. Has a railroad terminus; large glassworks, distillery; food processing.

Ledo (LEE-do), village, DIBRUGARH district, NE ASSAM state, NE INDIA, on BURHI DIHING RIVER and 55 mi/89 km ESE of DIBRUGARH. Railroad terminus; India terminus of Ledo (Stilwell) Road; brick and pottery works.

Ledong (LUH-DUNG), town, ⊙ Ledong county, SW HAINAN province, CHINA; 18°44′N 109°09′E. Rice, tropical crops; beverages, timber, textiles.

Ledrada (laidh-RAH-dhah), village, SALAMANCA province, W SPAIN, 6 mi/9.7 km NNE of BÉJAR; 40°28′N 05°43′W. Meat processing; cereals, wine.

Ledrae, CYPRUS: see NICOSIA.

Le Droit Park, small, historic residential section in NW WASHINGTON, D.C., SE of Howard University; 38°55′N 77°01′W. Many fine homes were built, beginning in 1873. By 1900, the area had become one of Washington's first "suburban" developments open to African-American residents.

Ledu, town, ⊙ Ledu county, NE QINGHAI province, CHINA, 35 mi/56 km ESE of XINING, and on XINING RIVER; 36°30′N 102°22′E. Grain, livestock; machinery, tobacco processing.

Leduc (luh-DOOK), city, (□ 14 sq mi/36.4 sq km; 2005 population 15,630), central ALBERTA, W CANADA, S of EDMONTON; 53°16′N 113°32′W. It is the center of the Leduc oil field (discovered 1947), which is now mostly depleted. The city is an oil storage and pumping station and also serves as an agricultural distribution center. First incorporated as a village in 1899; became a town in 1906 and a city in 1983.

Leduc County (luh-DOOK), municipality (□ 1,010 sq mi/2,626 sq km; 2001 population 12,528), central AL-BERTA, W CANADA; 53°14′N 113°48′W. Dairying; livestock (hogs, chickens, beef); crops (canola, cereal grains, legumes); oil production; tourism. Includes BEAUMONT, THORSBY, CALMAR, WARBURG, NEW SAR-EPTA; the summer villages of ITASKA BEACH, SUN-DANCE BEACH, GOLDEN DAYS; and the hamlets of BUFORD, NISKU, TELFORDVILLE, KAVANAGH, ROLLY VIEW, LOOMA, and SUNNYBROOK. Formed 1944.

Ledyard, town, NEW LONDON county, SE CONNECTI-CUT; 41°26′N 72°01′W. It is a farm center. The site of Fort Decatur is marked here. The Foxwoods casino complex, opened in 1991 by the Mashantucket Pequots on their reservation overlooking the THAMES River, had grown into the largest casino in the world (measured by gaming area) by 1996. The complex (located on the former site of nuclear reactor components) includes gaming rooms, hotels, and historical reference rooms that cover the history of the Mashantucket Pequots. A tribal museum was under construction in early 1996. Settled c.1653, incorporated 1836.

Ledyard, town (2000 population 147), KOSSUTH county, N IOWA, near MINNESOTA state line, 25 mi/40 km N of ALGONA; 43°25′N 94°09′W.

Ledyczek (le-DEE-cheek), Polish *Ledyczek*, German *Landeck*, town, Koszalin province, NW POLAND, 16 mi/26 km SSE of SZCZECINEK. Dairying; grain; livestock. Until 1938, in former Prussian province of Granzmark Posen–Westpreussen; was in POMERANIA until 1945.

Lee, county (□ 615 sq mi/1,599 sq km; 2006 population 125,781), E ALABAMA; ⊙ Opelika. Piedmont area leveling off to flat farm lands below Fall Line; bounded on E by Chattahoochee River and Georgia Cotton, corn; textiles. Granite, dolomite, manganese. Formed 1866. Named for Robert E. Lee, commander-in-chief of the Confederate armies during the Civil War.

Lee, county (□ 619 sq mi/1,609.4 sq km; 2006 population 11,379), E ARKANSAS; ⊙ MARIANNA; 34°46′N 90°46′W. Bounded E by the MISSISSIPPI; drained by SAINT FRANCIS and L'ANGUILLE rivers and Rig Creek. Agriculture (wheat, cotton, rice, soybeans; hogs); timber. Industries at Marianna. Part of Saint Francis National Forest in SE; Louisiana Purchase State Historical Monument at SW corner. Formed 1873.

Lee (LEE), county (□ 1,212 sq mi/3,151.2 sq km; 2006 population 571,344), SW FLORIDA, on GULF OF MEXICO (W); ⊙ FORT MYERS; 26°34′N 81°55′W. Lowland area, swampy in SE, drained by CALOOSAHATCHEE RIVER. Bordered by a chain of barrier islands (LACOSTA, CAPTIVA, SANIBEL, and Estero islands) sheltering several lagoons (Pine Island Sound, San Carlos Bay, ESTERO BAY) and PINE ISLAND. Agriculture (gladioli growing; citrus fruit, vegetables), cattle raising, fishing, and major tourist industry. Formed 1887. Named after Civil War general Robert E. Lee. Major growth around Fort Myers since 1970s has made this one of country's fastest growing counties.

Lee, county (□ 362 sq mi/941.2 sq km; 2006 population 32,495) SW central GEORGIA; ⊙ LEESBURG; 31°47′N 84°08′W. Bounded E by FLINT RIVER; drained by KINCHAFOONEE RIVER and MUCKALEE CREEK. Coastal plain agriculture (peanuts, soybeans, cotton, wheat, corn; cattle, hogs) and timber area. Formed 1826.

Lee, county (□ 729 sq mi/1,895.4 sq km; 2006 population 35,701), N ILLINOIS; ⊙ DIXON; 41°44′N 89°17′W. Agriculture (corn, soybeans; cattle; poultry; dairying). Manufacturing (food products, cement products, metal products, industrial machinery). Sand, gravel pits. Drained by ROCK, GREEN, and KYTE rivers, and BUREAU CREEK. Formed 1839.

Lee, county (□ 538 sq mi/1,398.8 sq km; 2006 population 36,338), extreme SE IOWA; ⊙ FORT MADISON and KEOKUK; 40°38′N 91°28′W. Bounded NE by SKUNK RIVER, E by MISSISSIPPI RIVER (forms ILLINOIS state line here), and S by DES MOINES RIVER (forms MIS-SOURI state line here). Prairie agricultural area (hogs, cattle, poultry, sheep; corn, oats, soybeans); limestone quarries, coal deposits. Manufacturing at Fort Madison and Keokuk. Lock and Dam No. 19 at Keokuk. Part of Shimek State Forest in W. Formed 1836.

Lee, county (□ 211 sq mi/548.6 sq km; 2006 population 7,648), E central KENTUCKY; ⊙ BEATTYVILLE; 37°35′N 83°43′W. In the CUMBERLAND MOUNTAINS; drained by KENTUCKY RIVER and its North, Middle, and South forks. Mountain agricultural area (livestock; burley tobacco, hay); coal oil; hardwood timber; limestone. Includes part of Daniel Boone National Forest in W half of county. Formed 1870.

Lee, county (□ 453 sq mi/1,177.8 sq km; 2006 population 79,714), NE MISSISSIPPI; ⊙ TUPELO; 34°17′N 88°40′W. Drained by Chiwapa and Oldtown creeks. Agriculture (cotton, corn, soybeans, wheat, honey; poultry; cattle; dairying); timber. BRICES CROSS ROADS National Battlefield Site in N (W of Baldwin); Tupelo National Battlefield Site in W part of Tupelo; Tombigbee State Park in E; Lake Lamar Bruce State Lake in W;

NATCHEZ TRACE PARKWAY passes SW-NE through county. Formed 1866.

Lee (LEE), county (□ 259 sq mi/673.4 sq km; 2006 population 56,908), central NORTH CAROLINA; ⊙ SANFORD; 35°28′N 79°10′W. In forested PIEDMONT region in N and sandhill area in S; bounded NE by CAPE FEAR RIVER and NW by DEEP RIVER. Manufacturing at Sanford (sand, stone quarrying); service industries; agriculture (especially tobacco, cotton, corn, wheat, oats, soybeans, hay; poultry), timber. Formed 1907 from Moore and Chatham counties. Named for Robert E. Lee (1807–1870), general-in-chief of the Confederate army during US Civil War.

Lee, county (□ 411 sq mi/1,068.6 sq km; 2006 population 20,559), NE central SOUTH CAROLINA; ⊙ BISH-OPVILLE; 34°09′N 80°15′W. Drained by LYNCHES and BLACK rivers. Manufacturing of clay and sand. Agricultural area (cotton, peanuts, corn, oats, soybeans, hay; hogs, cattle); timber. State Penitentiary and Lee State Park (2,839 acres/1,149 ha) is on Lynches River in E. Formed 1902.

Lee (LEE), county (□ 634 sq mi/1,648.4 sq km; 2006 population 16,573), S central TEXAS; ⊙ GIDDINGS; 30°18′N 96°57′W. Bounded NE by E yegua creek, SE in part by Yegua Creek, head of SOMERVILLE LAKE reservoir in E corner, confluence of the two creeks. Diversified agriculture, hogs, cattle; peanuts, corn, sorghum. Oil and gas, clay, Fuller's earth. Formed 1874.

Lee (LEE), county (□ 437 sq mi/1,136.2 sq km; 2006 population 23,787), extreme SW VIRGINIA, in wedge formed by KENTUCKY (NW) and TENNESSEE (S) state lines; ⊙ JONESVILLE; 36°42′N 83°07′W. Westernmost county in Virginia; mountain and valley region, with CUMBERLAND Mountains along Kentucky state line, CUMBERLAND GAP pass at SW tip, part of Cumberland Gap National Historical Park in SW (extends into Kentucky, Tennessee), part of POWELL MOUNTAIN in E; includes parts of Jefferson National Forest in NE, including Cave Springs Recreational Area. Drained by POWELL RIVER. Agriculture (tobacco, corn, hay, alfalfa; cattle); timber; extensive bituminous-coal mining, limestone quarrying. Limestone caves. Formed 1792.

Lee, town, PENOBSCOT county, E central MAINE, 45 mi/72 km NNE of BANGOR; 45°22′N 68°17′W. In hunting, fishing area.

Lee, town, BERKSHIRE county, W MASSACHUSETTS, in the BERKSHIRES, on HOUSATONIC River, 9 mi/14.5 km S of PITTSFIELD; 42°18′N 73°14′W. Includes Lee village. Resort; paper and lumber mills; marble quarries. October Mountain State Forest nearby. Villages of East Lee, South Lee, and Jacobs Pillow (resort and site of summer dance festival). Settled 1760, set off from GREAT BARRINGTON and WASHINGTON in 1777.

Lee, town, STRAFFORD county, SE NEW HAMPSHIRE, 9 mi/14.5 km SW of DOVER; 43°07′N 71°00′W. Drained by LAMPREY, North, and OYSTER rivers. Manufacturing (machinery); agriculture (nursery crops, corn, apples; cattle; dairying).

Lee, village (2000 population 313), in LEE and DE KALB counties, N ILLINOIS, 18 mi/29 km SW of SYCAMORE; 41°47′N 88°56′W. In rich agricultural area.

Leechburg, borough (2006 population 2,246), ARM-STRONG county, W central PENNSYLVANIA, 24 mi/39 km NE of PITTSBURGH, on KISKIMINETAS RIVER. Agriculture (corn, hay; dairying); manufacturing (fabricated steel, medical equipment, metal fabrication); bituminous coal. CROOKED CREEK LAKE reservoir and park to NE. Laid out 1828, incorporated 1850.

Leech Lake (□ 251 sq mi/650 sq km), CASS county (W end extends into HUBBARD county), N central MIN-NESOTA, 25 mi/40 km SE of BEMIDJI, largely in Leech Lake Indian Reservation; 47°10′N 94°25′W. The sec-ond-largest lake in Minnesota, after RED LAKE. Fed by Kabekuna and Steamboat rivers at W end, by BOY RIVER at E end; drains from NE end through Leech

River (c.30 mi/48 km long; dammed just W of FEDERAL DAM village) into MISSISSIPPI RIVER. Lake is 20 mi/32 km long, maximum width 15 mi/24 km; elevation 1,296 ft/395 m. Maximum capacity c.1,000,000 acre-ft. Bear Island (3.5 mi/5.6 km long, 1 mi/1.6 km wide) is in SE. Fishing, bathing, and boating resorts. Town of WALKER is on SW shore, on Walker Bay. Lake is surrounded by Chippewa National Forest and used as reservoir. Shoreline indented by large bays, BOY and Headquarters bays in E, Socker Bay in N, Steamboat and Walker bays in W.

Lee City, village, WOLFE county, E central KENTUCKY, in the CUMBERLANDS, 50 mi/80 km ESE of WINCHESTER. Agriculture (tobacco, corn; cattle).

Leedale (LEE-dail), hamlet, S central ALBERTA, W CANADA, 39 mi/62 km from PONOKA, in PONOKA COUNTY; 52°35′N 114°29′W.

Leedey (LEE-dee), village (2006 population 326), DEWEY county, W OKLAHOMA, 39 mi/63 km S of WOODWARD; 35°52′N 99°20′W. In agricultural area; manufacturing (furniture).

Leeds (LEEDZ), city (2001 population 443,247), ⊙ WEST YORKSHIRE, N central ENGLAND, on the AIRE RIVER; 53°48′N 01°32′W. It lies between one of England's leading manufacturing regions on the W and S and an agricultural region on the N and E. Leeds is a communications and regional government center and a junction of transportation routes, both railroad and water; canal and river connect Leeds with both E and W coasts. Airport. Manufacturing includes woolens (produced since the 14th century) and clothing, for which Leeds is a center of wholesale trade. Metal goods (locomotives, machinery, farm implements, and airplane parts), leather goods, and chemicals are also produced. Extensive slum-clearance and rehousing efforts have been undertaken since 1920. Yorkshire College, founded 1874, became a constituent college of Victoria University in 1887 and the independent University of Leeds in 1904. Leeds Metropolitan University was established in 1992. Among the other educational institutions is a 16th-century grammar school. Leeds has a classical town hall (1858) in which triennial musical festivals are held. Several sports arenas have been constructed and opened here in the 1970s and 1980s. Also of interest are St. Peter's Church, the Cathedral of St. Anne, St. John's Church, and the City Art Gallery. Kirkstall Abbey, founded in the 12th century, is near the city. Joseph Priestley was pastor at Mill Hill Chapel. The districts of the city include Armley, Beeston, Bramley, Bramhope, Farsley, Holbeck, Horsforth, Hunslet, Kirkstall, and Wortley.

Leeds, city (2000 population 10,455), on Jefferson–St. Clair co. line, N central ALABAMA, 10 mi/16 km E of Birmingham. In coal, iron, and limestone area; lumber and steel and wire products, plastics, furniture. Founded 1881. First known as 'Cedar Grove,' and then 'Oak Ridge,' it was renamed after the city in Yorkshire, England, which is also known for its iron industry. Inc. in 1887.

Leeds, town, ANDROSCOGGIN county, SW MAINE, on the ANDROSCOGGIN, 15 mi/24 km NNE of AUBURN; 44°17′N 70°07′W. Vegetables; food processing.

Leeds (LEEDZ), village (2001 population 2,224), central KENT, SE ENGLAND, 4 mi/6.4 km ESE of MAIDSTONE; 51°15′N 00°37′E. Has Norman church. Castle, dating from Saxon times, was rebuilt in Norman and Tudor eras and again in 19th century. As a princess, Elizabeth Tudor was imprisoned here.

Leeds, village (2006 population 452), BENSON co., N central NORTH DAKOTA, 30 mi/48 km WNW of DEVILS LAKE; 48°17′N 99°26′W. Hurricane Lake to NNW, Lake Ibsen to SE. Founded in 1887 and incorporated 1899. Named for LEEDS, ENGLAND.

Leeds, village (2006 population 720), WASHINGTON county, SW UTAH, 15 mi/24 km NE of SAINT GEORGE; 37°14′N 113°21′W. Fruit; cattle. Dixie National Forest, including PINE VALLEY MOUNTAINS Wilderness

Area, to NW; Quail Creek reservoir and State Park to SW. Mining boom 1870s and 1880s. Pop. has grown in 1980s and 1990s with influx of retired citizens.

Leeds, hamlet, GREENE county, SE NEW YORK, on CATSKILL CREEK 3 mi/4.8 km NW of CATSKILL; 42°15′N 73°53′W. Has 18th-century stone bridge, one of the oldest stone structures in the state.

Leeds, MASSACHUSETTS: see NORTHAMPTON.

Leeds and Grenville United Counties (LEEDZ, GREN-vil), (□ 1,294 sq mi/3,364.4 sq km; 2001 population 96,606), SE ONTARIO, E central CANADA, on the SAINT LAWRENCE RIVER, and on NEW YORK border; ⊙ BROCKVILLE; 44°50′N 75°40′W. Gateway to THOUSAND ISLANDS region; tourism; also manufacturing, agriculture. Leeds county, chief town, Brockville. Grenville county, chief town, Prescott. Composed of the separated Brockville, GANANOQUE, and PRESCOTT; and NORTH GRENVILLE, ATHENS and Rear of Yonge and Escott, AUGUSTA, EDWARDSBURGH/CARDINAL, ELIZABETHTOWN-KITLEY, LEEDS AND THE THOUSAND ISLANDS, FRONT OF YONGE, RIDEAU LAKES, MERRICKVILLE-WOLFORD, and WESTPORT.

Leeds and the Thousand Islands (LEEDZ, THOU-zuhnd EI-luhndz), township (□ 234 sq mi/608.4 sq km; 2001 population 9,069), SE ONTARIO, E central CANADA; 44°29′N 76°05′W.

Leedstown (LEEDZ-toun), unincorporated village, WESTMORELAND county, E VIRGINIA, 30 mi/48 km SE of FREDERICKSBURG, on RAPPAHANNOCK RIVER; 38°06′N 76°59′W. Agriculture (grain, soybeans; cattle). Here, in 1766, were drawn up the Leedstown Resolutions, embodying points later included in Declaration of Independence.

Lee, Fort, Virginia: see PETERSBURG.

Leegebruch (LAI-ge-brukh), village, BRANDENBURG, E GERMANY, 3 mi/4.8 km SW of ORANIENBURG; 52°44′N 13°11′E.

Leek (LAIK), city, GRONINGEN province, N NETHERLANDS, 8 mi/12.9 km WSW of GRONINGEN; 53°10′N 06°23′E. Leekstermeer Lake to NE; recreational center to E. Dairying; cattle, poultry raising; vegetables, fruit, grain; manufacturing (food processing). Site of National Carriage Museum.

Leek (LEEK), town (2001 population 18,768), N STAFFORDSHIRE, W ENGLAND, near the Churnet River, 10 mi/16 km NE of STOKE-ON-TRENT; 53°06′N 02°01′W. Manufacturing of textile machinery and machine tools. Previously a silk-milling center. Has remains of Cistercian abbey founded 1214. Parish church has 14th-century tower and four Saxon crosses in churchyard.

Lee Lake, WASHINGTON county, W MISSISSIPPI, and CHICOT county, ARKANSAS, E of MISSISSIPPI RIVER, 8 mi/12.9 km S of GREENVILLE; 33°16′N 91°02′W. Oxbow lake is c.8 mi/12.9 km long. The section inside of this former bend in Mississippi River remained part of ARKANSAS when river channel shifted W.

Leelanau (lee-LA-nou), county (□ 2,533 sq mi/6,585.8 sq km; 2006 population 22,112), NW MICHIGAN; ⊙ LELAND. A peninsula (LEELANAU PENINSULA) bounded W by LAKE MICHIGAN and E by GRAND TRAVERSE BAY and its W arm; 45°07′N 86°01′W. Area known for cherry growing; also apples, plums, strawberries, grapes, corn, wheat; cattle, hogs, poultry. Fisheries; resorts. It is a former lumber region. Lighthouse and Leelanau State Park at tip of peninsula (N). Sleeping Bear Dunes on Lake Michigan shore, including N and S MANITOU ISLANDS. Timber Line Ski Area in S, Sugar Loaf Mountain Ski Area in center. Organized 1863.

Leelanau, Lake (lee-LA-nou), LEELANAU county, NW MICHIGAN, on LEELANAU PENINSULA, just E of LELAND; c.4.5 mi/7.2 km long; 44°54′N 85°43′W. Resort. Connected to LOWER LEELANAU LAKE (c.9 mi/14.5 km long; just S) by short stream.

Leelanau Peninsula, MICHIGAN: see LEELANAU county.

Leeman (LEE-muhn), resort village, WESTERN AUSTRALIA state, W AUSTRALIA, 168 mi/270 km N of

PERTH; 29°56′S 114°58′E. Crayfishing. Commuter zone for Enneaba, 25 mi/40 km to E. Tourism; resort. Originally called Snag Island.

Leende (LAIN-duh), village, NORTH BRABANT province, S NETHERLANDS, 7 mi/11.3 km SSE of EINDHOVEN; 51°21′N 05°33′E. Belgian border 4 mi/6.4 km to SSW. Dairying; cattle, hog raising; agriculture (grain, vegetables).

Lee-on-the-Solent (LEE–on–thuh–SO-lent), seaside resort (2001 population 7,067), HAMPSHIRE, S ENGLAND, 4 mi/6.4 km W of GOSPORT; 50°48′N 01°12′W. Museum.

Lee Park, unincorporated town, LUZERNE county, NE central PENNSYLVANIA, residential suburb 3 mi/4.8 km WSW of WILKES-BARRE on Solomon Creek; 41°13′N 75°54′W.

Leeper, Mount (9,603 ft/2,927 m), S ALASKA, in CHUGACH MOUNTAINS, 20 mi/32 km NE of Cape YAKATAGA; 60°17′N 142°05′W.

Leer (LER), city, LOWER SAXONY, NW GERMANY, port on right bank of the EMS, at mouth of Leda River (W end of EMS-HUNTE CANAL), 13 mi/21 km SE of EMDEN; 53°14′N 07°27′E. Railroad junction. Foundry; manufacturing of agricultural machinery; food processing (chocolate, cacao, condensed milk, milk sugar). Trades in coffee and tea. Built around a church in EAST FRIESLAND (800); developed as a linen-weaving center and as a port; chartered 1823. Has 16th-century castle, 17th- and 18th-century churches.

Leerdam (LAIR-dahm), town (2001 population 17,156), SOUTH HOLLAND province, S central NETHERLANDS, on Linge River, 14 mi/23 km S of UTRECHT; 51°58′N 05°06′E. It is famous for its glassware and ceramics. Other industries include dairying; cattle, hogs, poultry; vegetables, sugar beets; manufacturing (cheese). Museum of Glassware.

Lee River, Gaelic An Laoi, 50 mi/80 km long, in CORK county, SW IRELAND; rises in SHEHY MOUNTAINS, flows E, past Macroom and Cork, to LOUGH MAHON, NW reach of CORK HARBOUR.

Lee River, Gaelic An Laoi, 10 mi/16 km long, KERRY county, SW IRELAND; rises ENE of TRALEE, flows SW past Tralee to Tralee Bay.

Leers (LER), town (□ 2 sq mi/5.2 sq km), NORD department, NORD-PAS-DE-CALAIS region, N FRANCE, 3 mi/4.8 km ESE of ROUBAIX, at Belgian border; 50°41′N 03°15′E. Textile milling.

Lees, village, CUMBERLAND county, S NEW JERSEY, on MAURICE RIVER, 10 mi/16 km SSE of MILLVILLE. Farming area. State prison is nearby.

Leesburg (LEEZ-buhrg), city (□ 12 sq mi/31.2 sq km; 2000 population 15,956), Lake county, central FLORIDA, in hilly and lake region, 30 mi/48 km SE of OCALA; 28°49′N 81°53′W. Retirement destination. Formerly a major citrus fruit-growing area. Incorporated 1875.

Leesburg (LEEZ-buhrg), city (2006 population 37,476), ⊙ LOUDOUN county, N VIRGINIA, 32 mi/51 km WNW of WASHINGTON, D.C., near the POTOMAC RIVER; 39°06′N 77°33′W. Manufacturing (building materials, printing and publishing, concrete blocks, wine); agriculture (grain, apples, soybeans); dairying. Trade center in region known for livestock breeding (horses, cattle). Limestone quarrying. Site of Civil War engagement of Ball's Bluff (October 1861) to NE, on Potomac River; a Confederate victory. Balls Bluff National Cemetery. Oatlands, home of George Carter; Oak Hill, home of James Monroe, to S. DULLES INTERNATIONAL AIRPORT 10 mi/16 km to SE. Settled 1749; incorporated 1758.

Leesburg, town (2000 population 2,633), ⊙ LEE county, SW central GEORGIA, 10 mi/16 km N of ALBANY, near KINCHAFOONEE RIVER; 31°44′N 84°10′W. Manufacturing includes apparel, crushed stone, machine parts.

Leesburg, town (2000 population 625), KOSCIUSKO county, N INDIANA, 6 mi/9.7 km N of WARSAW; 41°20′N 85°51′W. In agricultural area. Laid out 1835.

Leesburg (LEEZ-buhrg), village (2006 population 1,336), HIGHLAND county, SW OHIO, 10 mi/16 km NNE of HILLSBORO; 39°20′N 83°33′W. In livestock-raising and farming area.

Leesport, borough (2006 population 1,924), BERKS county, SE central PENNSYLVANIA, suburb 7 mi/11.3 km NNW of Reading on SCHUYLKILL RIVER; 40°26′N 75°58′W. Manufacturing includes crushed stone, concrete, apple cider, copper alloy, furniture, fertilizer, and medical equipment. Agriculture includes dairying, livestock; grain, apples, soybeans. Reservoirs nearby.

Lee's Summit, city (2000 population 70,700), JACKSON and CASS county, W MISSOURI, large residential and industrial suburb 16 mi/26 km SSE of KANSAS CITY. Trucking center. Manufacturing (communications equipment, appliances, pharmaceuticals, plastic containers, tool and die, metal products). Richards-Gebaur Airport nearby. James A. Reed Memorial Wildlife Area to SE; Lake Jacomo Park to E; Longview Community College. Incorporated 1868.

Lee State Park, SOUTH CAROLINA: see LEE.

Leeston, town (□ 2,531 sq mi/6,580.6 sq km), E SOUTH ISLAND, NEW ZEALAND, 23 mi/37 km SW of CHRISTCHURCH; 43°46′S 172°18′E. Grain; sheep.

Leesville, city, ⊙ VERNON parish, W LOUISIANA, 50 mi/80 km WSW of ALEXANDRIA; 31°08′N 93°16′W. Manufacturing (apparel, bottled water, lumber; printing and publishing); agriculture (livestock; corn, sweet potatoes). FORT POLK MILITARY RESERVE and Kisatchie National Forest are SE; Peason Ridge State Wildlife Area to N; Anacoco Lake and Anacoco Prairie State Game and Fish Preserve to W; Boise-Vernon State Wildlife Area to SW. W Louisiana Frontier Festival. Incorporated 1899.

Leesville (LEEZ-vil), village (2006 population 183), CARROLL county, E OHIO, 24 mi/39 km SSE of CANTON; 40°27′N 81°12′W. Nearby flood-control dam impounds Leesville Reservoir (capacity 37,400 acre-ft) in a small tributary of TUSCARAWAS RIVER.

Leesville, CONNECTICUT: see EAST HADDAM.

Leesville, SOUTH CAROLINA: see BATESBURG.

Leesville Lake (LEEZ-vil), reservoir, on PITTSYLVANIA and CAMPBELL county border, SW central VIRGINIA, on ROANOKE (Staunton) River, extends SW into BEDFORD county, 25 mi/40 km SSW of LYNCHBURG; 37°05′N 79°23′W. Maximum capacity 94,960 acre-ft; c.20 mi/32 km long. Formed by Leesville Dam (83 ft/25 m high), built (1963) for power generation.

Leeton (LEE-tuhn), town, S NEW SOUTH WALES, SE AUSTRALIA, 165 mi/266 km WNW of CANBERRA, in RIVERINA region; 34°34′S 146°24′E. Dairying center (cheeses); sun-dried tomatoes; vineyards. A range of birdlife at Fivebough Swamp (1.5 sq mi/3.9 sq km) near town.

Leeton, town (2000 population 619), JOHNSON county, W central MISSOURI, 12 mi/19 km S of WARRENSBURG; 38°34′N 93°42′W.

Leetonia (lee-TO-nee-yuh), village, ST. LOUIS county, NE MINNESOTA, in MESABI IRON RANGE 2 mi/3.2 km W of HIBBING; 47°25′N 92°59′W. Railroad spur terminus. Iron mines nearby.

Leetonia (lee-TON-ee-ah), village (□ 2 sq mi/5.2 sq km; 2006 population 2,036), COLUMBIANA county, E OHIO, 15 mi/24 km S of YOUNGSTOWN; 40°52′N 80°46′W. Laid out 1866.

Leetsdale, borough (2006 population 1,135), ALLEGHENY county, W PENNSYLVANIA, suburb 15 mi/24 km NW of PITTSBURGH, on OHIO RIVER, between mouths of Big Sewickley (NW) and Little Sewickley (SE) creeks. Agriculture to NE (corn, hay; livestock; dairying); manufacturing (fabricated metal products, office supplies, plastic products). Settled 1796, incorporated 1904.

Leeuwarden (LAI-uh-vahr-duhn), city (2001 population 86,429), ⊙ FRIESLAND province, N NETHERLANDS, 70 mi/113 km NNE of AMSTERDAM; 53°12′N 05°48′E. WADDENZEE, arm of NORTH SEA, 10 mi/16 km to NW. It is the center of an agricultural and dairying region; hub for network of canals, including the VAN HARINXMA (W), Dokkumer Ee (N), and Zwertevaart (SW). The PRINSES MARGRIET CANAL passes to SE. Railroad junction; airport to NW. Dairying; cattle, sheep, poultry; vegetables, fruit, grain; manufacturing (automobile parts, gas heating systems, acrylics, food processing, alcohol). Chartered in 1435, Leeuwarden was (sixteenth–seventeenth century) the center of a goldworking and silverworking industry. Noteworthy buildings include the Weigh House (1598), Chancellor Building (1571), and Town Hall (1715). Center of Frisian culture; Princesshof Museum, Museum of Friesland, National Museum of Ceramics are here. The notorious dancer and spy Mata Hari born here.

Leeuwin, Cape (LOO-in), SW WESTERN AUSTRALIA, in INDIAN OCEAN, at W end of FLINDERS BAY; 34°22′S 115°08′E. In area known for treacherous currents and gales. Lighthouse.

Leeuwin-Naturaliste National Park (LOO-in–NAH-chu-rah-leest) (□ 62 sq mi/161.2 sq km), SW WESTERN AUSTRALIA state, W AUSTRALIA, 170 mi/274 km SSW of PERTH, on INDIAN OCEAN. Has 90 mi/145 km of coastline from Cape NATURALISTE in N to Cape LEEUWIN in S; lighthouses at both capes. Sea cliffs, headlands, beaches, dunes, caves, swamps, and karri forest. Limestone karst topography punctuated by granite capes. Last habitat of marsupial predator, the chuditch. Grey kangaroos, possums; numerous bird species. Camping, picnicking, hiking, swimming, fishing, surfing, caving, and whale watching.

Leeuw-Saint-Pierre, BELGIUM: see SINT-PIETERS-LEEUW.

Leeville, unincorporated village, LAFOURCHE parish, extreme SE LOUISIANA, 39 mi/63 km SE of HOUMA, on navigable Bayou LAFOURCHE, in marshy region between CAMINADA and TIMBALIER bays, c. 50 mi/80 km S of NEW ORLEANS; 29°14′N 90°12′W. Center of oil field; shrimp processing. Wisner State Wildlife Area to E.

Lee Vining, unincorporated village, MONO county, E CALIFORNIA, 52 mi/84 km NNW of BISHOP, in the SIERRA NEVADA, near W shore of MONO LAKE. In mining and recreational region. Cattle. Canyon of Lee Vining Creek (W) has hydroelectric plant. Forms the E gateway to YOSEMITE NATIONAL PARK (boundary 7 mi/11.3 km to W) via TIOGA PASS (9,944 ft/3,031 m). Inyo National Forest to W, Toiyabe National Forest to NW; Saddlebag Lake to W.

Leeward Islands, N group of the LESSER ANTILLES in the WEST INDIES, extending SE from PUERTO RICO to the WINDWARD ISLANDS. The principal islands are the VIRGIN ISLANDS of the U.S.; the French island and overseas department of GUADELOUPE and its dependencies; the Dutch islands of SAINT EUSTATIUS and SABA; the jointly owned (Dutch and French) SAINT MARTIN; the islands of the independent states of SAINT KITTS AND NEVIS and ANTIGUA; and the islands of the British dependent territories of ANGUILLA, MONTSERRAT, and the BRITISH VIRGIN ISLANDS. Largely volcanic in origin, the Leeward Islands have lush, subtropical vegetation, rich soil, and abundant rainfall. The warm, delightful climate is tempered by the surrounding water so that there is little variation in temperature. Most of the islands have become popular tourist destinations. Products for the most part are agriculture—fruits, vegetables, sugar, cotton, coffee, and tobacco. Columbus first sighted the Leeward Islands in 1493, but settlement began only after the British arrived in the 17th century. Sir Thomas Warner, sent to Saint Kitts in 1623, was made governor general of the yet uncolonized neighboring islands (Nevis, Antigua, Montserrat, and BARBUDA), and in the same year the Frenchman Pierre Bélain d'Esnambuc also established a colony on Saint Kitts. By 1632, when the English had settled the neighboring islands, the sharp, 3-way colonial conflict of England, France, and Spain had begun. The Spanish were forced from the struggle, but for nearly 2 centuries the islands were pawns in the Anglo-French worldwide wars. They changed hands with each fresh attack by British or French forces and were reshuffled in ownership whenever a new treaty was signed. Their final disposition did not come until the end of the Napoleonic Wars in 1815.

Leeward Islands, CAPE VERDE: see SOTAVENTO ISLANDS.

Leeward Islands, FRENCH POLYNESIA: see SOCIETY ISLANDS.

Lefebvre (luh-FE-vruh), village (□ 26 sq mi/67.6 sq km; 2006 population 840), CENTRE-DU-QUÉBEC region, S QUEBEC, E CANADA, 12 mi/19 km from DRUMMONDVILLE; 45°43′N 72°25′W.

Lefedzha (le-FED-zhah), river, RUSE oblast, BULGARIA, tributary of the YANTRA RIVER; 43°11′N 25°52′E.

Leffe (LEF-fe), village, BERGAMO province, LOMBARDY, N ITALY, 12 mi/19 km NE of BERGAMO; 45°48′N 09°53′E. Bedcover factories, cotton mills; clothing.

Leffrinckoucke (lef-rahn-KOOK), town (□ 3 sq mi/7.8 sq km), NORD department, NORD-PAS-DE-CALAIS region, N FRANCE, 4 mi/6.4 km E of DUNKERQUE; 51°02′N 02°28′E. Metalworking.

Lefka (LEF-kah), town, LEFKOSIA district, NW CYPRUS, 3 mi/4.8 km S of MORPHOU BAY, on Setrakhos River, 30 mi/48 km W of NICOSIA; 35°07′N 32°51′E. Attila Line passes to E, S, and W; KARAVOSTASI 3 mi/4.8 km to NW. Agricultural region (citrus, olives, grain; sheep, goats). Formerly the processing and shipping point for EVRYKHOU VALLEY pyrite mines, which lie S of Turkish occupation line; pyrite and copper locally mined. Seat of University of Lefka. Also known as Lefke or LEUKA.

Lefkadia (lef-kah-[TH]YAH), village, EMATHEIA prefecture, CENTRAL MACEDONIA department, NE GREECE, 19 mi/31 km S of EDHESSA; 40°39′N 22°07′E. Site of several subterranean Macedonian tombs, dating from 4th–2nd century B.C.E. Possibly the site of ancient Mieza, where Aristotle taught.

Lefka Ori (lef-KAH O-ree) [Greek=white mountains], range, KHANIÁ prefecture, at the W end of CRETE department, GREECE; 15 mi/24 km S of Khaniá; 35°20′N 24°00′E. Highest peak, Mount PAKHNES (8,043 ft/2,452 m). Dominates this section of CRETE. Snow covers the upper reaches of the range from late autumn to early summer. Many wildflowers endemic to Crete. Referred to by foreigners as the White Mountains and by Cretans as the Madares. Site of Allied and Greek guerilla action against Germans during WW II. Also called Mount Levka.

Lefkara, CYPRUS: see PANO LEFKARA.

Lefkas (lef-KAHS), prefecture, IONIAN ISLANDS department, off W coast of GREECE; ⊙ Lefkas, on LEFKAS island; 38°45′N 20°40′E. Includes Lefkas, MEGANESI, KASTOS, and KALAMOS islands. Kefallinía prefecture to S.

Lefkas (lef-KAHS), Latin *Leucas*, town, ⊙ LEFKAS prefecture, IONIAN ISLANDS department, off W coast of GREECE, on NE end of LEFKAS island, on channel separating island from AKARNANIA prefecture on the Greek mainland; 38°50′N 20°42′E. Trade center; salines. Connected by road across lagoon with Venetian fort of Santa Maura. Seat of Greek metropolitan. Site of ancient city 2 mi/3.2 km S. Formerly spelled Leukas.

Lefkas (lef-KAHS), mountainous island (□ 115 sq mi/299 sq km), LEFKAS prefecture, IONIAN ISLANDS department, off W coast of GREECE, in the IONIAN SEA; 38°43′N 20°38′E. Lefkas, the chief town, is at the N end of the island. Olive oil, currants, wine, and tobacco are produced. Airport at Aktion. Colonized (7th century B.C.E.) by Corinthians; ally of Corinth in the Peloponnesian War. Later was capital of the Acarnanian League (3rd century B.C.E.). Captured (1697) from the Ottoman Turks by Venice, which held it until

1797. Ruins of Cyclopean walls and a temple to Apollo Leukates. Sappho is said, probably falsely, to have committed suicide by plunging into the sea from a cliff here. Also known as Santa Maura. Formerly spelled Levkás.

Lefkimmi (lef-KEE-mee), town, S Kérkira island, Kérkira prefecture, IONIAN ISLANDS department, off W coast of GREECE, 17 mi/27 km SSE of Kérkira city; 41°01′N 26°12′E. Olive oil, wine, citrus fruits. Also spelled Levkimmi and Leukimme.

Lefkoniko (lef-KO-nee-ko), [urkish= Geçitkale], town, LEFKOSIA district, NE CYPRUS, 25 mi/40 km ENE of NICOSIA; 35°15′N 33°44′E. KYRENIA MOUNTAINS to N. Agricultural region (grain, vetches, tobacco, olives, almonds; sheep, cattle, goats).

Lefkoşa, CYPRUS: see NICOSIA.

Lefkosia (lef-kuh-SEE-yah), district (□ 1,053 sq mi/ 2,737.8 sq km; 2001 population 27,945), central and NW CYPRUS; ⊙ NICOSIA. Bounded NW by MEDITERRANEAN SEA; officially covering area between TROODOS MOUNTAINS in S and KYRENIA MOUNTAINS in N. Major towns are MORPHOU, LEFKA, KARAVOSTASI, and KYTHREA. Turkish forces have occupied the N d of district since 1974; the Turkish name for this area is Lefkoşa, and it is also sometimes called North LEFKOSIA district. The Greek sector includes part of MESAORIA lowland agricultural region in N and Troodos mountain resort area in S. Citrus, deciduous fruit, melons, grain, olives, wine and table grapes, almonds, vegetables, potatoes; goats, sheep, poultry. Manufacturing at NICOSIA. Ancient ruined city Soli is NW on MORPHOU BAY.

Lefkosia, CYPRUS: see NICOSIA.

Leflore (luh-FLOR), county (□ 606 sq mi/1,575.6 sq km; 2006 population 35,752), W central MISSISSIPPI; ⊙ GREENWOOD; 33°32′N 90°01′W. TALLAHATCHIE and YALOBUSHA rivers join in E center to form YAZOO RIVER. Agriculture (cotton, rice, sorghum, wheat, soybeans; cattle); catfish. Greenwood is market, processing center. Matthews Brake National Wildlife Refuge in S; Florewood River Plantation State Park in center. Formed 1871.

Le Flore (luh FLOR), county (□ 1,608 sq mi/4,180.8 sq km; 2006 population 50,079), SE OKLAHOMA; ⊙ POTEAU; 34°53′N 94°42′W. Bounded N by ARKANSAS RIVER, E by ARKANSAS state line; drained by POTEAU and KIAMICHI rivers; in OUACHITA MOUNTAINS. Agriculture (corn, fruit, vegetables, hay, soybeans, potatoes; cattle, poultry); oil and natural gas wells; timber. Part of Ouachita National Forest is in county, mostly in S half. Kiamichi Mountains (extension of Ouachitas) in S; Spiro Mounds in N. Heavener-Runestone State Park in E; Lake Wister State Park at center; Talimena State Park in W. ROBERT S. KERR LAKE and Dam on Arkansas River in NW corner. Formed 1907.

Le Flore, village (2006 population 174), LE FLORE county, E OKLAHOMA, 20 mi/32 km WSW of POTEAU, near Fourche Maline Creek; 34°53′N 94°58′W. Agricultural and timber area. Ouachita National Forest to SE.

Leforest (luh-for-e), town (□ 2 sq mi/5.2 sq km), PASDE-CALAIS department, NORD-PAS-DE-CALAIS region, N FRANCE, 5 mi/8 km N of DOUAI; 50°26′N 03°04′E. In former coal-mining district.

Lefors (LUH-fors), village (2006 population 554), GRAY county, extreme N TEXAS, in the PANHANDLE, 65 mi/ 105 km ENE of AMARILLO and on North Fork of RED RIVER; 35°26′N 100°47′W. In oil, gas, and cattle area; wheat, sorghum, corn, forage crops; manufacturing (gas processing). MCCLELLAN CREEK National Grassland to S.

Lefroy, Lake, salt lake, AUSTRALIA: see KAMBALDA.

Lefroy, Mount (luh-FROI) (11,230 ft/3,423 m), on ALBERTA–BRITISH COLUMBIA border, W CANADA, in ROCKY MOUNTAINS, on W edge of BANFF NATIONAL PARK, 35 mi/56 km WNW of BANFF; 51°22′N 116°16′W.

Legal (luh-GAL), town (□ 1 sq mi/2.6 sq km; 2001 population 1,058), central ALBERTA, W CANADA, near Manawan Lake (4 mi/6 km long), 28 mi/45 km N of EDMONTON, in STURGEON COUNTY; 53°57′N 113°38′W. Grain elevators, mixed farming. French-Canada settlement, founded 1898; changed in status from a village to a town in 1998.

Leganés (lai-gah-NAIS), city (2001 population 173,584), MADRID province, central SPAIN, 7 mi/11.3 km SSW of MADRID; 40°19′N 03°45′W. Most workers now commute to nearby Madrid.

Leganiel (lai-gah-NYEL), village, CUENCA province, central SPAIN, 40 mi/64 km ESE of MADRID; 40°10′N 02°57′W. Cereals, grapes, olives; sheep.

Legaspi (le-GAHS-pee), city (□ 59 sq mi/153.4 sq km; 2000 population 157,010), ⊙ ALBAY province, SE LUZON, PHILIPPINES, on ALBAY GULF; 13°08′N 123°44′E. It is a large seaport and the S terminus of the MANILA railroad. Copra and hemp are shipped. Airport. Founded c.1639 as Albay, it was renamed Legaspi in 1925. In World War II it was the scene (December 12, 1941) of a large Japanese landing, part of a pincers movement on Manila. Towering directly behind the city is the spectacular active volcano MOUNT MAYON. Its eruption in 1814 severely damaged the town and killed over 1,000 people.

Legazpia (lai-GAHTH-pyah), town, GUIPÚZCOA province, N SPAIN, 14 mi/23 km WSW of TOLOSA; 43°03′N 02°20′W. Metalworking; manufacturing of paper; wine production; flour milling. Lead and zinc mines nearby.

Legden, village, North Rhine-Westphalia, NW GERMANY, 23 mi/37 km NW of MÜNSTER; 52°03′N 07°06′E.

Légé (lai-ZHAI), commune (□ 24 sq mi/62.4 sq km), LOIRE-ATLANTIQUE department, PAYS DE LA LOIRE region, W FRANCE, 23 mi/37 km S of NANTES. Agriculture market; hog raising.

Lège-Cap-Ferret (lezh–kahp–fe–re), town (□ 28 sq mi/ 72.8 sq km), bathing resort in GIRONDE department, AQUITAINE region, SW FRANCE on BAY OF BISCAY, on a sandspit at entrance to ARCACHON BASIN, directly opposite the town of ARCACHON (3 mi/4.8 km E). It is better known as Cap-Ferret. Oyster beds. Lighthouse.

Legge Peak, AUSTRALIA: see LEGGES TOR.

Legges Tor (LEG TOR), highest peak (5,160 ft/1,573 m) in TASMANIA, SE AUSTRALIA, in BEN LOMOND range; 41°33′S 147°39′E. Sometimes called Legge Peak.

Legget (LEG-et), unincorporated village, POLK county, E TEXAS, 30 mi/48 km S of LUFKIN. Timber. Oil and natural gas. Manufacturing (liquid-propane processing). Alabama and Coushatta Indian Reservation to SE.

Leggett (LEG-uht), village (2006 population 71), EDGECOMBE county, E central NORTH CAROLINA, 14 mi/23 km ENE of ROCKY MOUNT, near TAR RIVER; 35°59′N 77°34′W. Service industries; agriculture (grain, tobacco, cotton, peanuts; livestock).

Leghorn, ITALY: see LIVORNO.

Legi Oberskie, POLAND: see OBRA RIVER.

Legion Mine, village, MATABELELAND SOUTH province, SW ZIMBABWE, 27 mi/43 km SSE of ANTELOPE, near SHASHANI RIVER; 21°30′S 28°32′E. Road terminus; BOTSWANA border 18 mi/29 km to S. Gold mining. Livestock; grain.

L'église (lai-GLEEZ), commune, Neufchâteau district, LUXEMBOURG province, SE BELGIUM, 6 mi/10 km SE of NEUFCHÂTEAU; 49°48′N 05°32′E.

Legnago (len-YAH-go), town, VERONA province, VENETO, N ITALY, on Adige River, 23 mi/37 km SE of VERONA; 45°11′N 11°18′E. Railroad junction. Highly diversified small industrial center. Manufacturing (agricultural and woodworking machinery, buttons, chemicals, fertilizer, castor oil). Large beet-sugar refinery. Was SE fortress of the "Quadrilateral" in 16th century and again after 1814.

Legnano (len-YAH-no), city (2001 population 53,797), LOMBARDY, NW ITALY, near MILAN; 45°36′N 08°54′E. Manufacturing of this important industrial center includes plastics, steel, machinery, and textiles. The Lombard League defeated (1176) Emperor Frederick I near Legnano.

Legnica (leg-NEE-tsah), German Liegnitz, city (2002 population 107,100), SW POLAND, on the KACZAWA River; 51°12′N 16°12′E. The center of a vegetable-growing region, it also has manufacturing (metal goods, textiles, and foodstuffs); copper deposits nearby. Chartered in 1252, it was the capital of a duchy ruled by a branch of the Piast dynasty, until 1675. In the War of the Austrian Succession, it was acquired (1742) by PRUSSIA. The city was heavily damaged in World War II, but it has retained its 11th-century castle (rebuilt 1835), parts of its medieval walls and towers, and two churches (13th–14th century), one of which contains the tombs of the Piasts.

Legnickie Pole, POLAND: see WAHLSTATT.

Legnone, Monte (len-YO-ne, MON-te), highest summit (8,563 ft/2,610 m) in LAKE COMO region, LOMBARDY, N ITALY, 17 mi/27 km NNE of LECCO.

Legoendi, INDONESIA: see LEGUNDI.

Legon (le-GAHN), town, GREATER ACCRA region, GHANA, 5 mi/8 km NE of Accra; 05°39′N 00°11′W. Seat of University of Ghana, the nation's first co-ed university. Now merged with ACCRA.

Legostayevo (lye-guh-STAH-ee-vuh), village, E NOVOSIBIRSK oblast, SW SIBERIA, RUSSIA, in the OB′ RIVER basin, near highway, 15 mi/24 km E of ISKITIM; 54°38′N 83°49′E. Elevation 613 ft/186 m. Dairy farming.

Legrad (LE-grahd), Hungarian Légrád, village, N CROATIA, on DRAVA RIVER, at MURA RIVER mouth, 25 mi/40 km E of Varaždin, in PODRAVINA region, on Hungarian border. Trade center for agriculture (poultry, dairy products) area.

Le Grand, unincorporated town (2000 population 1,760), MERCED county, central CALIFORNIA, 12 mi/19 km ESE of MERCED; 37°13′N 120°16′W. Cattle, poultry; dairying; alfalfa, grain, sugar beets, melons, almonds.

Le Grand, town (2000 population 883), MARSHALL county, central IOWA, near IOWA RIVER, 7 mi/11.3 km ESE of MARSHALLTOWN; 42°00′N 92°46′W. Concrete, asphalt. Limestone quarries nearby.

Leg River, c.50 mi/80 km long, Rzeszow province, SE POLAND; rises 11 mi/18 km N of RZESZOW; flows N to VISTULA River 3 mi/4.8 km E of SANDOMIERZ. Another LEG RIVER, German Lyck, right tributary of BIEBRZA RIVER, is in Białystok province.

Leguan Island (le-GWAHN), ⊙ ESSEQUIBO ISLANDS–WEST DEMERARA district, N GUYANA; 06°55′N 58°25′W. Island is 8 mi/12.9 km long; located in ESSEQUIBO River estuary, on the ATLANTIC OCEAN, SE of WAKENAAM ISLAND, 18 mi/29 km WNW of GEORGETOWN. Rice-growing area.

Léguevin (lai-guh-VAN), town (□ 9 sq mi/23.4 sq km), HAUTE-GARONNE department, MIDI-PYRÉNÉES region, S FRANCE, 10 mi/16 km W of TOULOUSE; 43°36′N 01°15′E. Agriculture market for fruit, vegetables, and poultry of the Toulouse region.

Legundi (LUH-goon-dee), islet, INDONESIA, in SUNDA STRAIT, just off S coast of SUMATRA, at W entrance to LAMPUNG BAY, 25 mi/40 km S of BANDAR LAMPUNG; 05°50′S 105°16′E. Wooded, hilly, rising to 1,125 ft/343 m; 5 mi/8 km long. Coconuts. Also spelled Legoendi.

Leh (LAI), district (□ 31,917 sq mi/82,984.2 sq km), JAMMU AND KASHMIR state, extreme N INDIA; ⊙ LEH. The NE portion, known as AKSAI CHIN, has been occupied by CHINA since the 1962 Sino-Indian War, and 14,499 sq mi/37,552 sq km of the district are under disputed Chinese occupation. Formerly Ladakh district, which included KARGIL district. See also LADAKH region.

Leh (LAI), town, ⊙ LEH district, JAMMU AND KASHMIR state, extreme N INDIA. Elevation c.11,500 ft/3,505 m (this is among the world's highest continually inhabited towns). Much of the trade between India and TIBET passed through here until 1959, when a dispute broke out between India and CHINA. The uninhabited AKSAI CHIN region around Leh was contested by India and China in the 1960s and is still occupied by China. The palace of the former rulers of W Tibet and a Lamaist monastery are here. Tourist spot.

Lehavim (le-hah-VEEM), urban settlement, S ISRAEL, 8 mi/12.9 km N of BEERSHEBA in N NEGEV. Most residents commute to jobs in the region. Founded 1983.

Lehchevo (le-CHE-vo), village, MONTANA oblast, BOICHINOVTSI obshtina, NW BULGARIA, on the OGOSTA RIVER, 18 mi/29 km NE of MONTANA; 43°32′N 23°32′E. Grain, legumes; livestock.

Lehe, GERMANY: see BREMERHAVEN.

Lehesten (LAI-uhs-tuhn), town, THURINGIA, central GERMANY, in THURINGIAN FOREST, 13 mi/21 km SSE of SAALFELD; 50°29′N 11°27′E. Slate-quarrying center (since 13th century), with paper mills. Health and winter-sports resort.

Lehi (LEE-hei), town (2006 population 36,021), UTAH county, N central UTAH, 25 mi/40 km S of SALT LAKE CITY, 14 mi/23 km NW of PROVO, on Dry Creek, near JORDAN RIVER. WASATCH RANGE and Uinta National Forest to E; 40°23′N 111°50′W. Elevation 4,562 ft/1,390 m. Trading point for agricultural area; alfalfa, flour mill; dairying; manufacturing (cheese, explosives, clay products). Calcite and clay mining; limestone quarrying. Saratoga Resort to SW on lake. Surrounding region is irrigated by water from UTAH LAKE (just S) and PROVO RIVER. Formerly called Lehi City. Settled by Mormons 1850; incorporated 1852.

Lehigh (LEE-hei), county (□ 348 sq mi/904.8 sq km; 2006 population 335,544), E PENNSYLVANIA; ⊙ ALLENTOWN. Bounded in NE and drained by LEHIGH RIVER, and Jordan and Little Lehigh creeks. Rolling industrial and farm area with BLUE MOUNTAIN ridge running length of NW boundary. Agriculture (corn, wheat, oats, barley, hay, alfalfa, soybeans, potatoes, apples; chicken, sheep, hogs, cattle, dairying); limestone, slate; sand and gravel. Manufacturing at Allentown, EMMAUS, and BETHLEHEM. Trexler Lehigh County Game Preserve in N center; five covered bridges in center, on Jordan Creek. The county is urbanized in SE around twin cities of Allentown and Bethlehem. Formed 1812.

Lehigh, town (2000 population 497), WEBSTER county, central IOWA, on DES MOINES RIVER, 12 mi/19 km SSE of FORT DODGE; 42°21′N 94°02′W. Brick and tile plant. Clay, sand, gravel pits nearby. Dolliver Memorial State Park in NW; Brushy Creek State Park in NE.

Lehigh (LEE-hei), village (2000 population 215), MARION county, central KANSAS, 16 mi/26 km W of MARION; 38°22′N 97°17′W. In grain, livestock, and oil-producing region.

Lehigh, village, Stark co., W NORTH DAKOTA, 5 mi/8 km E of DICKINSON, and on HEART River; 46°52′N 102°41′W. Founded around 1890 and named for Lehigh, PENNSYLVANIA.

Lehigh, village (2006 population 295), COAL county, S central OKLAHOMA, 5 mi/8 km S of COALGATE, near MUDDY BOGGY CREEK; 34°28′N 96°13′W. In agricultural area.

Lehigh (LEE-hei), river, 103 mi/166 km long, on LACKAWANNA-WAYNE county line, NE PENNSYLVANIA; rises in group of small ponds; flows SW through Francis Walter Reservoir, turns S at White Haven, flows past JIM THORPE, LEHIGHTON, and ALLENTOWN, then E past BETHLEHEM to DELAWARE RIVER at EASTON; 41°16′N 75°24′W. Area around Allentown, Bethlehem, and Easton once noted for its anthracite-coal mining and steel mills.

Lehigh Acres (LEE-hei), unincorporated town (□ 96 sq mi/249.6 sq km; 2000 population 33,430), LEE county, SW FLORIDA, 13 mi/21 km E of FORT MYERS; 26°36′N 81°38′W. Wilderness surrounds this planned community. Manufacturing includes building materials,

printing and publishing. Population has expanded greatly since 1990s.

Lehighton (LEE-hei-tuhn), borough (2006 population 5,494), CARBON county, E PENNSYLVANIA, 20 mi/32 km NW of ALLENTOWN, on LEHIGH RIVER. Agriculture (corn, hay, potatoes; dairying; timber); manufacturing (machinery, fabricated metal products, lumber, insulation, apparel). Jake Arnerican Memorial Airport to W. Mauch Chunk Lake reservoir to W; BELTZVILLE LAKE reservoir and Beltzville State Park to E. Settled 1746, incorporated 1866.

Lehigh Valley International Airport, is located between ALLENTOWN and BETHLEHEM, PENNSYLVANIA and is accessible from both Interstate 78 and Route 22; elevation 393 ft/120 m; 40°39′N 75°27′W. The airport has two runways, one, three-level passenger terminal (some million passengers annually), and one cargo terminal (some 23,500 tonnes annually). The airport opened as a private air field in 1927 and as a commerical airport in 1935. Airport code ABE.

Lehinch (luh-HINCH) or **Lahinch**, Gaelic *Leacht Ui Chonchubháir*, town (2006 population 607), W CLARE county, W IRELAND, on Liscannor Bay, 2 mi/3.2 km W of ENNISTYMON; 52°56′N 09°21′W. Seaside and golfing resort.

Lehliu Gară (lekh-LYOO GAH-ruh), town, CĂLĂRASI county, SE ROMANIA, 32 mi/51 km NW of CĂLĂRASI; 44°26′N 26°51′E. Manufacturing of foodstuffs. Also called Lehliu.

Lehman Caves National Monument (LEE-muhn) (□ 1 sq mi/2.6 sq km), WHITE PINE county, E NEVADA, 35 mi/56 km ESE of ELY. Monument (est. 1922) was included with GREAT BASIN NATIONAL PARK when the park was est. in 1985.

Lehmann (LAI-mahn), village, central SANTA FE province, ARGENTINA, 60 mi/97 km NW of SANTA FE; 31°08′S 61°27′W. Agriculture center (soybeans, alfalfa, wheat, flax; livestock); dairying.

Lehnin (lai-NEEN), village, BRANDENBURG, E GERMANY, 10 mi/16 km SE of BRANDENBURG; 52°20′N 12°45′E. Health resort. Has remains of Cistercian monastery (founded 1180; dissolved 1542).

Le Hochet, village (2000 population 13,878), PAMPLEMOUSSES district, MAURITIUS, 3 mi/4.8 km N of PORT LOUIS. Sugarcane; garment and textile manufacturing.

Lehota pod Vtacnikom (le-HO-tah POT FTAHCHhyi-kom), Slovak *Lehota pod Vtáčnikom*, village, STREDOSLOVENSKY province, W central SLOVAKIA, 5 mi/8 km S of PRIEVIDZA; 48°19′N 18°00′E. Potatoes; manufacturing of building materials; summer resort. MOUNT VTACNIK (4,416 ft/1,346 m), Slovak *Vtáčnik*, is 5 mi/8 km SSE.

Lehr (LER), village (2006 population 101), LOGAN and MCINTOSH counties, S NORTH DAKOTA, 17 mi/27 km N of Ashley; 46°16′N 99°20′W. Green Lake and Doyle Memorial State Park to SW. Founded in 1899 and named for Andreas and Johann Lehr who came here from SOUTH DAKOTA.

Lehre (LE-re), town, LOWER SAXONY, N GERMANY, 7 mi/11.3 km NE of BRUNSWICK; 52°20′N 10°40′E.

Lehrte (LER-te), city, LOWER SAXONY, NW GERMANY, 10 mi/16 km E of HANOVER; 52°22′N 09°58′E. Connected by railroad with HANOVER. Potash mining; metalworking; sugar refining.

Lehto, RUSSIA: see LEKHTA.

Lehua (lai-HOO-ah), small island 0.5 mi/0.8 km off N tip of NIIHAU, KAUAI county, HAWAII. Highest point rises to 702 ft/214 m. Light beacon. KAULAKAHI CHANNEL to E.

Lehututu (LAI-hoo-TOO-too), village (2001 population 1,719), N KGALAGADI DISTRICT, SW BOTSWANA, in KALAHARI DESERT, 112 mi/180 km NW of GABORONE; 23°59′S 21°53′E. Elevation 3,150 ft/960 m. Road junction.

Leiah (lai-YAH), town, MUZAFFARGARH district, SW PUNJAB province, central PAKISTAN, on railroad, 65

mi/105 km NNW of MUZAFFARGARH; 30°58′N 70°56′E. Market center for wheat, millet, rice, dates; handloom weaving, textile manufacturing, sugar factory.

Leibnitz, town, STYRIA, SE AUSTRIA, near MUR RIVER, 20 mi/32 km S of GRAZ at the Leibnitzer Feld; 46°47′N 15°32′E. Market center of S Styria with a catchment area extending into SLOVENIA; manufacturing includes machines, plastics, timber, construction materials, textiles, foodstuffs; vineyards.

Leibo (LAI-BO), town, ⊙ Leibo county, SW SICHUAN province, CHINA, 65 mi/105 km WSW of YIBIN, near CHANG JIANG (Yangzi River) and YUNNAN border; 28°15′N 103°34′E. Grain; logging.

Leicester (LEST-uh), city and county (2001 population 279,921), central ENGLAND; 52°38′N 01°08′W. The city is connected by canals with the TRENT RIVER and LONDON, and it is also a railroad center. Leicester was of industrial importance as early as the 14th century; the making of hosiery, knitwear, and shoes are long-established industries. Other manufacturing includes chemicals, dyes, textiles, textile and woodworking machinery, scientific services, and light-metal products. Leicester was the Ratae Coritanorum, or Ratae, of the Romans, whose FOSSE WAY passes nearby. It was also a town of the ancient Britons and was one of the Five Boroughs of the Danes. Its antiquities include the Jewry Wall, a Roman structure 18 ft/5 m high and 70 ft/21 m long (near which extensive Roman relics have been found); remains of a Norman castle; and the ruins of an abbey founded in 1143, in which Cardinal Wolsey died in 1530. Several of the churches (St. Nicholas, St. Mary de Castro, and All Saints) show Norman work, and Trinity Hospital is a 14th-century foundation. Richard III stayed in Leicester the night before he was killed in the Battle of BOSWORTH FIELD; his body was brought back to Leicester for burial. The University College, now the University of Leicester, was founded in 1918 and chartered in 1957. De Montfort University was established in 1992. Districts of the city include Birstall, Blaby, Braunstone, Evington, and Humberstone. Formerly part of LEICESTERSHIRE.

Leicester (LES-tuhr), residential town, WORCESTER county, central MASSACHUSETTS, just W of WORCESTER; 42°14′N 71°55′W. Has a junior college. Includes villages of Leicester Center, Cherry Valley, and Rochdale. Settled 1713, incorporated 1722.

Leicester, town, ADDISON co., W VERMONT, near Lake DUNMORE, in GREEN MOUNTAINS, 11 mi/18 km S of MIDDLEBURY; 43°52′N 73°05′W.

Leicester, village, SIERRA LEONE, on SIERRA LEONE PENINSULA, at NE foot of LEICESTER PEAK, 2 mi/3.2 km SSE of FREETOWN; 08°28′N 13°13′W. Cassava, corn.

Leicester (LES-tuhr), village (2006 population 448), LIVINGSTON county, W central NEW YORK, 32 mi/51 km SSW of ROCHESTER; 42°46′N 77°54′W.

Leicester Peak (1,952 ft/595 m), SIERRA LEONE, on SIERRA LEONE PENINSULA, 2.5 mi/4 km S of FREETOWN; 08°27′N 13°13′W.

Leicestershire (LEST-uh-shir), county (□ 832 sq mi/ 2,163.2 sq km; 2006 population 915,900), central ENGLAND; ⊙ LEICESTER; 52°40′N 01°10′W. Fertile farming land exists in the uplands of the E, while the W is devoted to industry and some mining. The hilly CHARNWOOD FOREST is in the NW. The SOAR and the WREAKE are the principal rivers. Leicestershire is primarily an agricultural county (sheep, dairy cattle, wheat, and barley). Stilton cheese is a well-known dairy product of the region. Leicester is an important industrial city and the center of the boot and shoe industry. LOUGHBOROUGH and HINCKLEY AND BOSWORTH also have industrial concentrations, and Plungar has an oilfield. Melton Mowbray and Market Harborough are famous fox-hunting centers. Other towns include Ashby-de-la-Zouch, Coalville, and Kegworth. Leicestershire was part of the Anglo-Saxon kingdom of MERCIA. At Bosworth Field, in 1485,

---

Cross-references are shown in SMALL CAPITALS. The pronunciation guide is shown on page xix. The sources of population figures are shown on page xvii.

2096 **LEICHHARDT**

Richard III was slain by the forces of Henry Tudor, who ascended the throne as Henry VII. Leicestershire was reorganized as a new nonmetropolitan county in 1974.

Leichhardt (LEIK-hahrt), suburb, E NEW SOUTH WALES, AUSTRALIA, on S shore of Port JACKSON, 3 mi/5 km WSW of SYDNEY; 33°53′S 151°09′E. In metropolitan area; highly diversified light industries, commercial and retail. Italian influences.

Leichhardt Range (LEIK-hahrt), E spur of GREAT DIVIDING RANGE, E QUEENSLAND, AUSTRALIA; extends c.130 mi/209 km SSE from HOME HILL; 20°40′S 147°25′E. Rises to 4,190 ft/1,277 m (Mount Dalrymple).

Leichhardt River (LEIK-hahrt), 300 mi/483 km long, N QUEENSLAND, AUSTRALIA; rises in hills SW of DOBBYN; flows NNW to Gulf of CARPENTARIA; 17°35′S 139°48′E. Shallow. ALEXANDRA RIVER is main tributary.

Leichlingen (LEIKH-ling-uhn), town, North Rhine-Westphalia, W GERMANY, on the WUPPER, 5 mi/8 km SW of SOLINGEN; 51°06′N 07°01′E. Summer resort; metalworking; food processing; manufacturing of electrical goods. Center for the teaching of disabled persons. Site of major hospital.

Leiden (LEI-duhn), city (2001 population 251,212), SOUTH HOLLAND province, W NETHERLANDS, on the OLD RHINE RIVER, 10 mi/16 km NNE of The HAGUE; 52°09′N 04°30′E. NORTH SEA 6 mi/9.7 km to W; Rijn-Schie Canal passes to SE. Railroad junction; airport to W. Dairying; cattle; flowers, nursery stock, vegetables, fruit; manufacturing (cosmetics, boats, aircraft systems). The city dates from Roman times, and Leiden has had an important textile industry since the sixteenth century, when an influx of weavers came from YPRES. The city took a prominent part in the revolt (late sixteenth century) of the Netherlands against Spanish rule. Besieged and reduced to starvation in 1574, it was saved from surrender when William the Silent ordered the flooding of the surrounding land by cutting the dikes, thus enabling the fleet of the Beggars of the Sea to sail to the city's rescue across the countryside. It was the home of many of the Pilgrims for about ten years before they embarked (1620) for America. Leiden was the birthplace of the Anabaptist leader John of Leiden and of the painters Jan van Goyen, Jan Steen, Lucas van Leyden, and Rembrandt. The city has a tenth-century fortress; two historic churches, the Pieterskerk (fourteenth century) and the Hooglandse Kerk (fifteenth century); several museums; and many seventeenth-century houses. The famous State University of Leiden, the oldest in the Netherlands, is here (founded 1575). It was a center for the study of Protestant theology, classical and Oriental languages, science, and medicine in the seventeenth and eighteenth centuries. The university is particularly noted for its departments of Asian studies, physics, and astronomy; the Leyden jar was invented here. Also spelled Leyden.

Leiderdorp (LEI-duhr-dawrp), town, SOUTH HOLLAND province, W NETHERLANDS, on OLD RHINE RIVER, 3 mi/4.8 km E of LEIDEN; 52°09′N 04°32′E. Dairying; cattle; flowers, nursery stock, vegetables, fruit; manufacturing (furniture).

Leidschendam (LEID-skhen-dahm), town, SOUTH HOLLAND province, W NETHERLANDS, 4 mi/6.4 km NE of The HAGUE, on Vliet River; 52°05′N 04°24′E. NORTH SEA 5 mi/8 km to NW. Dairying; cattle; vegetables, flowers, fruit; manufacturing (hardware).

Leidy, Mount (LEI-dee) (10,317 ft/3,145 m), peak in ROCKY MOUNTAINS, TETON county, NW WYOMING, 25 mi/40 km NE of JACKSON. Rises above E side of JACKSON HOLE.

Leiferde (LEI-fer-te), village, LOWER SAXONY, N GERMANY, 15 mi/24 km W of WOLFSBURG; 52°26′N 10°26′E.

Leigh (LEE), town (2001 population 43,150), GREATER MANCHESTER, NW ENGLAND, 11 mi/17.6 km W of MANCHESTER; 53°30′N 02°33′W. Manufacturing of

agricultural machinery, electrical goods, and metalworks.

Leigh (LEE), village (2006 population 418), COLFAX county, E NEBRASKA, 20 mi/32 km NNW of SCHUYLER; 41°42′N 97°14′W. Livestock; grain. Manufacturing (fertilizer, feed).

Leigh Creek (LEE), village, E central SOUTH AUSTRALIA, 170 mi/274 km N of PORT PIRIE; 30°31′S 138°25′E. Wool; livestock; company coal-mining village.

Leighlin, IRELAND: see OLD LEIGHLIN.

Leighlinbridge (LAHK-luhn-brij), Gaelic *Leithghlinn an Droichid*, village (2006 population 674), W CARLOW county, SE IRELAND, on BARROW RIVER, 7 mi/11.3 km SSW of CARLOW; 52°44′N 06°58′W. Agricultural market. Has remains of Black Castle, built 1181 by Hugh de Lacy and of a Carmelite friary.

Leigh-on-Sea, ENGLAND: see SOUTHEND-ON-SEA.

Leigh River (LEE), river, VICTORIA, SE AUSTRALIA; 38°07′S 144°04′E. Meets BARWON RIVER at INVERLEIGH township.

Leigh Smith Island, in S FRANZ JOSEF LAND, ARCHANGEL oblast, extreme N European Russia, in the ARCTIC OCEAN, E of HOOKER ISLAND; 80°15′N 54°00′E. Named after the British explorer who discovered it in 1880.

Leighton (LAI-tuhn), town (2000 population 849), Colbert co., NW Alabama, 10 mi/16 km E of Tuscumbia. Known earlier as 'Cross Roads' or 'Jeffers Crossroads,' it was renamed after William Leigh, grand master of the Masonic Grand Lodge of AL in 1834. Inc. in 1891.

Leighton, town (2000 population 153), MAHASKA county, S central IOWA, 8 mi/12.9 km WNW of OSKALOOSA; 41°20′N 92°47′W. Meat processing.

Leighton Buzzard (LAI-tuhn BUZ-uhrd), town (2001 population 32,417), SW BEDFORDSHIRE, central ENGLAND, 11 mi/18 km WNW of LUTON; 51°55′N 00°40′W. Former agricultural market. Silica-sand quarries, cement works. Has 15th-century market cross and 13th-century church.

Leijun, YEMEN: see GHEIL BIN YUMEIN.

Leimebamba (lai-mai-BAHM-bah), town, CHACHAPOYAS province, AMAZONAS region, N PERU, on UTCUBAMBA RIVER, 32 mi/51 km S of CHACHAPOYAS; 06°41′S 77°47′W. Sugarcane, plantains, rice, yucca. Variant spelling: Leymebamba.

Leimen (LEI-muhn), town, BADEN-WÜRTTEMBERG, SW GERMANY, 4 mi/6.4 km S of HEIDELBERG; 49°21′N 08°42′E. Manufacturing of cement, building materials; brewing; wine making.

Leinburg, village, BAVARIA, S GERMANY, in MIDDLE FRANCONIA, 10 mi/16 km E of NUREMBERG; 49°27′N 11°18′E.

Leinefelde (LEI-ne-fel-de), town, THURINGIA, central GERMANY, on LEINE RIVER, 20 mi/32 km SE of GÖTTINGEN; 51°23′N 10°19′E. Textile industry. Chartered 1969.

Leine River (LEI-ne), c.120 mi/193 km long, N GERMANY; rises near WORBIS; 51°25′N 10°18′E. Flows W, then generally N, past GÖTTINGEN and HANOVER (head of navigation), to the ALLER 11 mi/18 km NW of WIETZE. Receives the INNERSTE (right).

Leinfelden-Echterdingen (LEIN-fel-den–EKH-tuhr-ding-uhn), suburb of STUTTGART, BADEN-WÜRTTEMBERG, SW GERMANY, 6 mi/9.7 km S of city center; 48°41′N 09°09′E. Electronics industry; metalworking. Site of Stuttgart airport.

Leingarten (LEIN-gahr-tuhn), suburb of HEILBRONN, BADEN-WÜRTTEMBERG, SW GERMANY, 5 mi/8 km W of city center; 49°08′N 09°07′E.

Leinster (LEN-stuhr), Gaelic *Cúige Laighean*, province (□ 7,634 sq mi/19,848.4 sq km; 2006 population 2,295,123), E IRELAND, comprising the counties of CARLOW, DUBLIN, KILDARE, KILKENNY, LAOIGHIS, LONGFORD, LOUTH, MEATH, OFFALY, WESTMEATH, WEXFORD, and WICKLOW. It contains the city of DUBLIN. Its wealth and accessibility made the ancient

province subject to Danish and Anglo-Norman invasions.

Leinster, Mount (LEN-stuhr), Gaelic *Cnoc Laighean* (2,610 ft/796 m), on border between WEXFORD and CARLOW counties, SE IRELAND, 12 mi/19 km NW of ENNISCORTHY; 52°37′N 06°47′W. Highest point of BLACKSTAIRS MOUNTAINS.

Leipe, POLAND: see LIPNO.

Leipheim (LEIP-heim), town, SWABIA, W BAVARIA, GERMANY, on the DANUBE, 3 mi/4.8 km W of GÜNZBURG; 48°27′N 10°13′E. Metalworking. Site of air base. Chartered c.1330.

Leipnik, CZECH REPUBLIC: see LIPNIK NAD BECVOU.

Leipsic (LIP-sik), village (2000 population 203), KENT county, central DELAWARE, 6 mi/9.7 km N of DOVER, and on Leipsic River, 5 mi/8 km W of its mouth on DELAWARE BAY; 39°14′N 75°31′W. Elevation 3 ft/0.9 m. Named for LEIPZIG, GERMANY. An early fur-trading center and port. Oystering was a major industry in the late 19th and early 20th century, and began to pick up again in the late 1990s.

Leipsic (LEIP-sik), village (□ 3 sq mi/7.8 sq km; 2006 population 2,209), PUTNAM county, NW OHIO, 18 mi/29 km WNW of FINDLAY; 41°05′N 83°59′W. In grain-growing area; food and dairy products, clay and cement products.

Leipsos, Greece: see LIPSOI.

Leipzig (LEIP-sig), city (2006 population 506,578), SAXONY, E central GERMANY, at the confluence of the PLEISSE, WHITE ELSTER, and Parthe rivers; 51°22′N 12°22′E. Historically, one of Germany's major industrial, commercial, and transportation centers, it has many railroad lines and two airports. However, since reunification in 1989, approximately eighty percent of industrial-sector jobs left Leipzig, and throughout the 1990s and the 2000s, high levels of unemployment plagued the region. In the 2000s, some multinational and domestic companies began to invest in the city, which led to a degree of revitalization; however, this investment by no means brought the city back to its former level of prosperity. Manufacturing includes textiles, electrical products, machine tools, and chemicals. The city used to harbor major industries in heavy construction, publishing, fur trade, and engineering. Most of the companies engaged in these industries have left Leipzig, long considered the birthplace and the heart of the German printing and publishing industries. There have been major attempts to improve the city's infrastructure, especially in transportation. The area is heavily polluted with sulfur dioxide from nearby coal-processing plants. Important international trade and industrial fairs have been held in the city since the Middle Ages. Originally a Slavic settlement called Lipsk, Leipzig was chartered at the end of the 12th century and rapidly developed into a commercial center located at the intersection of important trade routes. A printing industry, which later became important, was started there c.1480. Scene of the famous religious debate between Martin Luther, Carlstadt, and Johann Eck in 1519. Accepted the Reformation in 1539. During the Thirty Years War, three great battles were fought near Leipzig (two at BREITENFELD and one at LÜTZEN). One of the leading cultural centers of EUROPE in the age of the philosopher and mathematician Leibnitz (born here 1646) and of the composer Johann Sebastian Bach, who was cantor at the Church of St. Thomas from 1723 until his death. The University of Leipzig (founded 1409; renamed Karl Marx University in 1953) became one of the most important in Germany. In the 18th century Gottsched, Gellert, Schiller, and many others made Leipzig a literary center; the young Goethe studied here in 1765. The city's musical reputation reached its peak in the 19th and early 20th centuries. Felix Mendelssohn, who died here in 1847, made the Gewandhaus concerts (begun in the 18th century in a former guildhouse and still continuing)

Area is shown by the symbol □, and capital city or county seat by ⊙.

internationally famous. Robert Schumann worked in Leipzig, Richard Wagner was born here in 1813, and the Leipzig Conservatory (founded by Mendelssohn in 1842–1843) became one of the world's best-known musical acadamies. The Battle of Leipzig (October 16–19, 1813), also called the Battle of the Nations, was a decisive victory of the Austrian, Russian, and Prussian forces over Napoleon I. The battle is commemorated by a large monument in the city. Until World War II, Leipzig was the center of the German book and music publishing industries, and the center of the European trade in furs and smoked foods. The city (including the book-trade quarter) was badly damaged in World War II. Noteworthy buildings include the Church of St. Thomas (late 15th century), which has housed the tomb of Bach since 1950; the Gewandhaus, built in 1884 to replace the earlier structure; the 13th-century Pauline Church; Auerbach's Keller (16th century), an inn in which a scene of Goethe's *Faust* is set; the old city hall (1558); the old stock exchange (1682); the Church of St. John (17th century); the large main railroad station with adjoining conference and exhibition hall (now being restored to include retail and office space); the former German supreme court building (which now houses an art museum); and the opera (1960). In addition to the university, the city has institutes of applied radioactivity and stable isotopes.

Leipzig, UKRAINE: see SERPNEVE.

Leirfossen (LER-faws-suh), two waterfalls on NIDELVA River, in SØR-TRØNDELAG county, central NORWAY, 4 mi/6.4 km S of TRONDHEIM. One of the falls is 105 ft/32 m; the other is 92 ft/28 m high. Hydroelectric plants.

Leiria (lai-REE-ah), district (□ 1,307 sq mi/3,398.2 sq km; 2001 population 459,426), W central PORTUGAL, divided between BEIRA LITORAL (N) and ESTREMADURA (S) provinces; ☉ LEIRIA. Extends along ATLANTIC coast from Cape Carvoeiro (S) almost to mouth of MONDEGO RIVER at FIGUEIRA DA FOZ. Includes Portugal's largest pine forest W of Leiria. In addition to Leiria, chief towns are CALDAS DA RAINHA (spa), PENICHE (fish-processing center), ALCOBAÇA (noted for its abbey), NAZARÉ, and MARINHA GRANDE (glass manufacturing). Also here are battlefields of ALJUBARROTA and the Dominican monastery of BATALHA.

Leiria (lai-REE-ah), town, ☉ LEIRIA district, W central PORTUGAL, in BEIRA LITORAL; 39°45′N 08°48′W. Agriculture trade center producing leather goods and cement. Built on the site of Collipo, a Roman settlement. Alfonso I erected (beginning 1135) a castle on a cliff above the present city; taken and retaken in the wars with the Moors. In 1254, Alfonso III summoned to Leiria the first Cortes to have representatives of the towns. The first duke of Braganza, ancestor of the Braganza royal line, was reared here. Nearby, at MARINHA GRANDE, is the national glass factory. In 1466, one of the first printing presses in Portugal was opened here.

Lei River (LAI), Mandarin *Lei Shui* (LAI SHUAI), right tributary of XIANG River, c.200 mi/322 km long, HUNAN province, CHINA; rises near GUIDONG on JIANGXI border; flows SW and NW, past RUCHENG, YONGXING, and LEIYANG, to Xiang River at HENGYANG.

Leirvik (LER-veek), village, HORDALAND county, SW NORWAY, on SE STORD island, 26 mi/42 km NNE of HAUGESUND; 63°06′N 07°38′E. Port; herring fishing and curing; shipbuilding. Construction of offshore installations. Burial grounds from Bronze Age are here. Sometimes spelled Lervik.

Leisenring (LEI-suhn-ring), unincorporated town, Franklin township, FAYETTE county, SW PENNSYLVANIA, 2 mi/3.2 km WSW of CONNELLSVILLE; 39°59′N 79°38′W.

Leishan (LAI-SHAN), town, ☉ Leishan county, SE GUIZHOU province, CHINA, 30 mi/48 km NNW of RONGJIANG; 26°23′N 108°04′E. Tobacco, grain; logging, beverages. Also known as Danjiang.

Lei Shui, CHINA: see LEI RIVER.

Leisnig (LEIS-nik), town, SAXONY, E central GERMANY, on the FREIBERGER MULDE, 9 mi/14.5 km WNW of DÖBELN; 51°10′N 12°55′E. Manufacturing of textile machinery, electrical equipment, furniture. Old Mildenstein castle towers over town.

Leiston (LAI-stuhn), town (2001 population 6,240), E SUFFOLK, E ENGLAND, 20 mi/32 km NE of IPSWICH; 52°12′N 01°34′E. Nearby are remains of 14th-century abbey. On NORTH SEA, 2 mi/3.2 km E, is village of Sizewell, site of the Sizewell A (two units, 580 MW capacity) and Sizewell B (one unit, 1,188 MW capacity) nuclear power plants; A went on-line in 1966, B in 1995.

Leisure Village–Pine Lake Park, unincorporated area, retirement community, OCEAN county, E central NEW JERSEY, 4 mi/6.4 km E of LAKEHURST; 40°02′N 74°11′W.

Leitches Creek (LEE-chiz), village, NE NOVA SCOTIA, CANADA, NE CAPE BRETON ISLAND, on SYDNEY HARBOUR, 6 mi/10 km WNW of SYDNEY.

Leitchfield (LICH-feeld), town (2000 population 6,139), ☉ GRAYSON county, W central KENTUCKY, 36 mi/58 km NNE of BOWLING GREEN; 37°28′N 86°17′W. In agricultural area (dairying; livestock; burley tobacco, corn, hay; timber), stone-quarrying area; manufacturing (lumber, machinery, cheese, apparel, furniture, crushed limestone). Grayson County Airport to S. Jack Thomas House (c.1810). NOLIN RIVER Lake reservoir to SE; ROUGH RIVER LAKE reservoir and Rough River Dam State Resort Park to NW.

Leitersburg (LEET-erz-berg), village, WASHINGTON county, W MARYLAND, 6 mi/9.7 km NE of HAGERSTOWN. A grain mill has operated here since 1792. Nearby is a Mennonite church, in use since 1835. Small plastic works here.

Leiters Ford, village, FULTON county, N INDIANA, 10 mi/16 km WNW of ROCHESTER. In agricultural area. Manufacturing (lighting fixtures).

Leith, unincorporated town, South Union township, FAYETTE county, SW PENNSYLVANIA, suburb 1 mi/1.6 km S of UNIONTOWN; 39°53′N 79°43′W.

Leith (LEETH), village (2006 population 25), GRANT co., S NORTH DAKOTA, 52 mi/84 km SW of BISMARCK; 46°21′N 101°38′W. Founded in 1910 and incorporated in 1915. Named for Leith, Scotland.

Leith, Scotland: see EDINBURGH.

Leitha Mountains (LEI-tah), German *Leithagebirge,* E AUSTRIA, on BURGENLAND–LOWER AUSTRIA border; extends from PARNDORF c.20 mi/32 km SW, along NW shore of LAKE NEUSIEDL, rising 1,585 ft/483 m in the SONNENBERG. Large limestone quarries; vineyards, orchards (cherries, peaches) on lower slopes. EISENSTADT at S foot.

Leitha River (LEI-tah), Hungarian *Lajta,* 112 mi/180 km long; formed in E AUSTRIA by the confluence of the SCHWARZA and PITTEN rivers. It flows generally E to an arm of the DANUBE RIVER near Mosonmagyaróvár, HUNGARY. It was the historic boundary between Austria and Hungary. The terms Cisleithania and Transleithania for the Austrian and Hungarian part of the dual monarchy were derived from this river.

Leith Hill (LEETH) (965 ft/294 m), SURREY, SE ENGLAND; highest point of the NORTH DOWNS; 51°10′N 00°22′W. On the summit is a tower, from where there is a view on clear days of LONDON and the ENGLISH CHANNEL.

Leith, Water of (LEETH), river, 23 mi/37 km long, WEST LOTHIAN and EDINBURGH, E Scotland; rises 5 mi/8 km SE of WEST CALDER in PENTLAND HILLS; flows NE, past Edinburgh, to Firth of Forth at Leith.

Leitmeritz, CZECH REPUBLIC: see LITOMĚŘICE.

Leitomischl, CZECH REPUBLIC: see LITOMYSL.

Leitre (LEI-ter), village, SANDAUN province, N central NEW GUINEA island, NW PAPUA NEW GUINEA, 25 mi/40 km ESE of VANIMO, on PACIFIC OCEAN coast; 02°50′S 141°38′E. Bananas, coconuts, palm oil; fish.

Leitrim (LEE-truhm), Gaelic *Liatroim,* county (□ 613 sq mi/1,593.8 sq km; 2006 population 28,950), N central IRELAND; ☉ CARRICK-ON-SHANNON; 53°10′N 08°28′W. Borders NORTHERN IRELAND to NE, CAVAN county to E, LONGFORD county to S, ROSCOMMON and SLIGO counties to W, and DONEGAL county to N. Leitrim is divided into two parts by LOUGH ALLEN; the N part is mountainous, the S part level. Potatoes and oats are grown. Industries include textiles, electrical goods, and automotive parts. The population has declined by more than 100,000 since the 1840s.

Leiva (LAI-vah), village, ☉ Leiva municipio, BOYACÁ department, N central COLOMBIA, in Cordillera ORIENTAL, 14 mi/23 km NW of TUNJA; 01°56′N 77°18′W. Elevation 7,050 ft/2,149 m. Wheat, corn, potatoes, coffee; livestock. Emerald mines. Silver deposits nearby. An early congress (1812) was held here.

Leivadion, Greece: see LIVADION.

Leivadostra, Bay of, Greece: see LIVADOSTRA, BAY OF.

Leix, IRELAND: see LAOIGHIS.

Leixlip (LEEKS-lip), Gaelic *Léim an Bhradáin,* village (2006 population 14,676), NE KILDARE county, E IRELAND, on the LIFFEY RIVER and ROYAL CANAL, and 9 mi/14.5 km W of DUBLIN; 53°22′N 06°29′W. Agricultural market. Power station driven by waterfall in the river. Bridge (built 1308) across the Liffey and 13th-century castle.

Leixões (LAI-shesh), artificial seaport of OPORTO, PÔRTO district, NW PORTUGAL, on the ATLANTIC OCEAN, at mouth of small Leca River, 5 mi/8 km NW of city. Consists of two curved moles (0.75 mi/1.21 km and 1 mi/1.6 km long) at the base of which are the twin towns of LEÇA DA PALMEIRA and MATOZINHOS. Harbor (completed 1890) was built to accommodate vessels unable to ascend DUERO RIVER (sandbar at mouth) to Oporto. Its chief export is port wine. Also called Pôrto de Leixões.

Leiyang (LAI-YANG), city (□ 1,025 sq mi/2,656 sq km; 1994 estimated urban population 148,000; 1994 estimated total population 1,167,800), SE HUNAN province, CHINA, on LEI RIVER, on WUHAN-GUANGZHOU railroad, 38 mi/61 km SSE of HENGYANG; 26°25′N 112°51′E. Rice, tobacco, oilseeds, medicinal herbs; food industry, engineering, chemicals, building materials, iron smelting.

Lei Yue Mun Strait (LAI YOO-E MUN), strait, HONG KONG, CHINA, at E end of Hong Kong harbor, between Hong Kong Island (S) and mainland (N); ¼ mi/⅖ km–½ mi/⅘ km wide.

Leizhou, CHINA: see HAIKANG.

Leizhou Peninsula (LAI-JO), SW GUANGDONG province, CHINA, on South CHINA Sea, between Guangdong Bay (E) and Gulf of TONKIN (W); 90 mi/145 km long, 30 mi/48 km–45 mi/72 km wide. Separated from HAINAN island (S) by QIONGZHOU STRAIT. Subtropical vegetation. Principal towns are ZHANJIANG, HAIKANG, XUWEN.

Lejamaní (le-hah-mah-NEE), town, COMAYAGUA department, W central HONDURAS, 3 mi/4.8 km N of LA PAZ, on secondary road connecting with Highway 1; 14°19′N 87°42′W. Small farming, grain, beans.

Lejeune (luh-ZHUN), village (□ 104 sq mi/270.4 sq km; 2006 population 364), BAS-SAINT-LAURENT region, S QUEBEC, E CANADA, 5 mi/7 km from AUCLAIR; 47°46′N 68°34′W.

Lejía, Cerro (lai-HEE-ah, SER-ro), peak (17,585 ft/5,360 m) of N CHILE, ANTOFAGASTA region, in the PUNA DE ATACAMA; 23°33′S 67°48′W.

Lejona (lai-HO-nah) or **Lexona**, municipio, NW of BILBAO, VIZCAYA province, N SPAIN. Metal foundries. Chief village, Lamicaco, on NERVIÓN RIVER, has glassworks and distilleries.

Lejunior (lah-JOON-yuhr), village, HARLAN county, SE KENTUCKY, 11 mi/18 km ENE of HARLAN, on CLOVER FORK of CUMBERLAND RIVER. Bituminous coal. Timber. Formerly called Highsplint.

Leka (LAI-kah), island (□ 22 sq mi/57.2 sq km) in NORTH SEA, NORD-TRØNDELAG county, W NORWAY, just off mainland, 40 mi/64 km N of NAMSOS; 10 mi/16 km long, 1 mi/1.6 km–6 mi/9.7 km wide. Leka fishing village on E shore.

Lékana, town, PLATEAUX region, SW Congo Republic, NW of DJAMBALA. Administrative center.

Le Kef, province, TUNISIA: see KAF, AL, province.

Lekemti, ETHIOPIA: see NEKEMTE.

Leketi River, REPUBLIC OF THE CONGO: see ALIMA RIVER.

Lekhaina (le-khe-NAH), town, ILIA prefecture, WESTERN GREECE department, W PELOPONNESUS, GREECE, on railroad, 20 mi/32 km NW of PÍRGOS; 37°56′N 21°16′E. Livestock market (cattle, sheep); Zante currants, figs, wheat; wine. Also spelled Lechaina, Lechena, and Lekhaena.

Lekhta (LYEKH-tah), Finnish *Lehto*, village, central Republic of KARELIA, NW European Russia, on road, 23 mi/37 km WSW of BELOMORSK; 64°26′N 33°58′E. Elevation 374 ft/113 m. Dairying.

Lekie, department (2001 population 354,864), CENTRAL province, CAMEROON, ⊙ MONATÉLÉ.

Lékila (lai-KEE-luh), town, HAUT-OGOOUÉ province, SE GABON, 47 mi/76 km NNE of FRANCEVILLE; 00°58′S 13°49′E.

Lekkerkerk (LE-kuhr-kerk), town, SOUTH HOLLAND province, W NETHERLANDS, on LEK RIVER, 8 mi/12.9 km E of ROTTERDAM; 51°54′N 04°41′E. Dairying; cattle, hogs; grain, vegetables, sugar beets; manufacturing (food processing, building materials).

Lekki (LE-kee), town, LAGOS state, SW NIGERIA, on Gulf of GUINEA, 50 mi/80 km E of LAGOS. Fisheries; palm oil and kernels. Lekki Lagoon (N) is E part of coastal lagoon paralleling Gulf of Guinea.

Lékoni (LAI-ko-nee), town, HAUT-OGOOUÉ province, SE GABON, 50 mi/80 km E of FRANCEVILLE; 01°34′S 14°16′E.

Lékoumou, region (2007 population 83,216), SW Congo Republic, bordered by PLATEAUX, POOL, and BOUENZA regions to the E, Gabon to the N, and NIARI region to the W; ⊙ SIBITI; 03°00′S 13°30′E. The OGOOUÉ River flows E-W across the N part of the region. Agriculture (groundnuts, tobacco, corn, soya, sweet potatoes); oil-palm manufacturing; gold, lead, and zinc mining. Rubber plantations located c.20 mi/32 km NNW of Sibiti.

Lek River (LEK), channel of the RHINE RIVER, 40 mi/64 km long, SW central NETHERLANDS LOWER RHINE RIVER divides into Lek River (W) and CROOKED RHINE RIVER (NW) 13 mi/21 km SE of UTRECHT; flows into NEW MAAS RIVER 6 mi/9.7 km E of ROTTERDAM at KRIMPEN AAN DEN LEK. Flows past CULEMBORG, VIANEN, and SCHOONHOVEN. It is navigable for its entire length.

Leksand (LEK-SAHND), town, KOPPARBERG county, central SWEDEN, on Österdalälven River, at SE end of LAKE SILJAN, 20 mi/32 km WNW of FALUN; 60°44′N 15°01′E. In tourist area. Church (c.1300–1700); museum. Scene of Dalecarlian mid-summer festival.

Leksha, Lake, RUSSIA: see LEKSOZERO.

Leksmond, NETHERLANDS: see LEXMOND.

Leksozero (lyek-SO-zee-ruh), Finnish *Lieksajärvi*, lake (surface □ 85 sq mi/220 sq km), W Republic of KARELIA, NW European Russia, 5 mi/8 km–10 mi/16 km from the Finnish border; 25 mi/40 km long, 8 mi/13 km wide; 63°45′N 31°00′E. REBOLY village on the NW shore. Formerly spelled Leksha.

Lekst, Jbel (LE-kist, zhe-BEL), peak (7,740 ft/2,359 m), MOROCCO, in Anti-ATLAS mountains, 6 mi/10 km N of TAFRAOUT.

Leku, town (2007 population 16,459), SOUTHERN NATIONS state, S ETHIOPIA, on road, near SE shore of LAKE AWASA; 06°55′N 38°28′E.

Leland, town (2000 population 258), WINNEBAGO county, N IOWA, on LIME CREEK, 4 mi/6.4 km N of FOREST CITY; 43°19′N 93°38′W.

Leland (LEE-luhnd), town (2000 population 5,502), WASHINGTON county, W MISSISSIPPI, 10 mi/16 km E of GREENVILLE, on Deer Creek; 33°23′N 90°54′W. Railroad junction. Agriculture (cotton, grain, soybeans; cattle); timber; manufacturing (machinery, furniture, light manufacturing). "Birthplace of the Frog," boyhood home of Jim Henson, creator of the Muppets. Stoneville National Wildlife Refuge to N. Settled 1847, laid out 1884.

Leland (LEE-luhnd), town (□ 4 sq mi/10.4 sq km; 2006 population 4,616), BRUNSWICK county, SE NORTH CAROLINA, 6 mi/9.7 km WNW of WILMINGTON, near CAPE FEAR RIVER; 34°14′N 78°00′W. Railroad junction. Retail trade; manufacturing (textiles, metal fabricating, building materials); agriculture (tobacco, grain, soybeans; livestock). In late 2004, Leland doubled its geographic size with the voluntary annexation of a 4,900-acre tract of land commonly known as Brunswick Forest. Incorporated 1989.

Leland, village (2000 population 970), LA SALLE county, N ILLINOIS, 18 mi/29 km N of OTTAWA; 41°36′N 88°47′W. In agricultural area.

Leland (LEE-luhnd), village, ⊙ LEELANAU county, NW MICHIGAN, 19 mi/31 km NNW of TRAVERSE CITY, between LAKE LEELANAU (E) and LAKE MICHIGAN; 45°01′N 85°45′W. In dairy, livestock-raising, and fruit area; lumber; printing; winery; resort. Sleeping Bear Dunes National Lakeshore to SW; ferries to N and S MANITOU ISLANDS, part of national lakeshore.

Le Landeron, SWITZERLAND: see LANDERON, LE.

Leland Grove, village (2000 population 1,592), SANGAMON county, central ILLINOIS, residential suburb 2 mi/3.2 km SW of SPRINGFIELD, nearly surrounded by city of Springfield; 39°46′N 89°40′W.

Leland, Mount (LEE-luhnd) (7,810 ft/2,380 m), on ALASKA–BRITISH COLUMBIA (CANADA) border, in SAINT ELIAS MOUNTAINS, 40 mi/64 km W of SKAGWAY; 59°22′N 136°29′W.

Lelången (LE-LONG-en), lake, W SWEDEN, VÄRMLAND and ÄLVSBORG counties, near NORWEGIAN border, at BENGTSFORS. Connected N with FOXEN and STORA LE Lakes; 28 mi/45 km long, 1 mi/1.6 km wide. Drained E into LAKE VÄNERN.

Lel'chitsy (LEL CHEE-tsee), urban settlement, GOMEL oblast, BELARUS, ⊙ Lel'chitsy region, in PRIPET Marshes, on Ubort River (tributary of Pripiat), 45 mi/72 km SW of MOZYR, 51°48′N 28°20′E. Manufacturing (construction materials, fruit and vegetable drying); creamery.

Lelekivka (le-LE-keef-kah), (Russian *Lelekovka*), NNW suburb of KIROVOHRAD, central KIROVOHRAD oblast, UKRAINE, 5 mi/8 km NNW of city center; 48°34′N 32°14′E. Elevation 393 ft/119 m. Mostly residential. Site of major battles between the German and Soviet forces in January–March 1944, with the Germans successfully escaping a number of encirclement attempts.

Lelekovka, UKRAINE: see LELEKIVKA.

Leleque (le-LAI-ke), village, ⊙ Cushamén department, NW CHUBUT province, ARGENTINA, 37 mi/60 km NNE of ESQUEL; 42°23′S 71°03′W. Oats, wheat, alfalfa; sheep, cattle.

Lelija (LE-lee-yah), mountain in DINARIC ALPS, between upper and lower HERZEGOVINA, BOSNIA AND HERZEGOVINA, along right bank of upper NERETVA RIVER, 5 mi/8 km S of KALINOVIK; 43°25′N 18°29′E. Highest point, VELIKA LELIJA (6,665 ft/2,301 m). Also spelled Leliya.

Leling (LUH-LING), city (□ 453 sq mi/1,173 sq km; 1994 estimated urban population 62,000; 1994 estimated total population 631,500), N SHANDONG province, CHINA, 75 mi/121 km NNE of JINAN, near HEBEI border; 37°37′N 117°15′E. Agriculture is the largest source of income for the city. Crop growing, livestock-raising. Grains, cotton, fruits, eggs; hogs; manufacturing (food processing, textiles).

Lella Khedidja (le-LAH ke-dee-JAH), highest peak (7,572 ft/2,308 m) in the DJURDJURA range of Great KABYLIA, BOUÏRA wilaya, N central ALGERIA, 70 mi/113 km ESE of ALGIERS, and culminating point of Algeria's coastal TELL ranges. Also spelled Lalla Khadidja.

Le Locle, SWITZERLAND: see LOCLE, LE.

Lélouma, prefecture, Labé administrative region, W central GUINEA, in Moyenne-Guinée geographic region; ⊙ Lélouma. Bordered N by Mali prefecture, E by Labé prefecture, SE by Pita prefecture, S by Télimélé prefecture, and W by Gaoual prefecture. Part of FOUTA DJALLON massif here. Summit (elevation 4,085 ft/1,245 m) S of Lélouma town. Road between Labé and Kounsitél towns runs NW-SE through center of prefecture; a secondary road connects Lélouma town to this road.

Lélouma, town, ⊙ Lélouma prefecture, Labé administrative region, NW GUINEA, in Moyenne-Guinée geographic region; 11°29′ 12°39′. Connected by secondary road to main road between LABÉ town to SE and KOUNDARA and SENEGAL to NW.

Lelydorp (LIR-li-dawrp), village, ⊙ WANICA district, N SURINAME, 10 mi/16 km S of PARAMARIBO; 05°40′N 55°08′W. Coffee, rice, fruit.

Lely Mountains (LE-lee), NE spur of the GUIANA HIGHLANDS, E SURINAME, 100 mi/161 km SSE of PARAMARIBO; extend c.40 km/64 km along TAPANAHONI and MARONI rivers near BRAZIL border. Rise to 2,428 ft/740 m.

Lelystad (LAI-lee-staht), city (2001 population 62,500), ⊙ FLEVOLAND province, EASTERN FLEVOLAND polder, W central NETHERLANDS, 25 mi/40 km ENE of AMSTERDAM, at SE end of Markerwaarddijk barrier dam; 52°31′N 05°26′E. IJSSELMEER to N, MARKERMEER to W; railroad terminus to N; Oostvaardersplassen nature area to SW. Cattle; corn, vegetables, potatoes, sugar beets, wheat, flower bulbs; manufacturing (glass, paint, rubber products, sporting goods). Research station for cattle, sheep, and horse husbandry; observatory; museum. Airport to S.

Lema (LAI-mah), town, KEBBI state, NW NIGERIA, near NIGERborder, 30 mi/48 km N of BIRNIN KEBBI; 12°57′N 04°14′E. Market town. Groundnuts, maize, millet.

Lemahabang (luh-MAH-hah-bahng), town, E Java Barat province, INDONESIA, near JAVA SEA, 10 mi/16 km SE of CIREBON; 06°20′S 108°20′E. Trade center for agricultural area (sugar, rice, peanuts); textile mills.

Le Maire Strait (lah mah-EE-re), channel in the South Atlantic between SE tip of main island of TIERRA DEL FUEGO and ISLA DE LOS ESTADOS, S Argentina; 18 mi/29 km wide. Named for the Dutch navigator Le Maire, who discovered it in 1616.

Lema Islands, CHINA: see DANGAN ISLANDS.

Léman, Lac, SWITZERLAND: see GENEVA, LAKE.

Le Mars, city (2000 population 9,237), ⊙ PLYMOUTH county, NW IOWA, on FLOYD RIVER, 23 mi/37 km NNE of SIOUX CITY; 42°47′N 96°10′W. Railroad junction; agriculture-trade and manufacturing center; dairy products, apparel, detergents, printing, rendering plant, meat processing; grain mill (cereals, feed), creameries; cement work. Sand and gravel pits nearby. Westmar College (established 1890) and historical museum are here. Founded in late 1870s; incorporated 1881.

Lemba (LEM-buh), village, BAS-CONGO province, W CONGO, on road 30 mi/48 km N of BOMA; 05°36′S 13°01′E. Elev. 813 ft/247 m.

Lembach (LEM-bahk), village (□ 15 sq mi/39 sq km), BAS-RHIN department, in ALSACE, NE FRANCE, on NE slopes of N VOSGES MOUNTAINS, 9 mi/14.5 km WSW of WISSEMBOURG; 49°00′N 07°48′E. Interesting relics of the MAGINOT LINE can be seen 2 mi/3.2 km SE. The ruined castle of Fleckenstein (1 mi/1.6 km N), built in 12th century and destroyed in 1680, once commanded the lowland of N Alsace to the RHINE.

Lembang (LAIM-bahng), health resort, PREANGER region, Java Barat province, INDONESIA, 7 mi/11.3 km N of BANDUNG; 06°49′S 107°36′E. Elevation 4,091 ft/1,247 m. Hot springs.

Area is shown by the symbol □, and capital city or county seat by ⊙.

Lembeek (LEM-baik), village, HALLE commune, Halle-Vilvoorde district, BRABANT province, central BELGIUM, on CHARLEROI-BRUSSELS CANAL, 2 mi/3.2 km SSW of Halle; 50°43′N 04°13′E.

Lembeni (laim-BAI-nee), village, KILIMANJARO region, NE TANZANIA, 35 mi/56 km SE of MOSHI, on railroad; PARE MOUNTAINS to E; MKOMAZI Game Reserve to SE; 03°50′S 37°38′E. Grain; livestock; timber.

Lemberg (LEM-buhrg), town (2006 population 255), SE SASKATCHEWAN, CANADA, 24 mi/39 km NE of INDIAN HEAD; 50°44′N 103°12′W. Grain-shipping center; grain elevators; dairying; livestock raising.

Lemberg (LEM-berg), village, RHINELAND-PALATINATE, W GERMANY, 3 mi/4.8 km SE of PIRMASENS; 49°10′N 07°40′E.

Lemberg (LEM-berg), highest peak (3,330 ft/1,015 m) of SWABIAN JURA, SW GERMANY, 5 mi/8 km ESE of ROTTWEIL; 48°10′N 08°49′E.

Lemberg, UKRAINE: see L′VIV, city.

Lembongan (laim-BONG-ahn), island, INDONESIA, in BADUNG Strait, just NW of PENIDA island, near BALI; 08°41′S 115°27′E. Lighthouse. Also called Nusa Lembongan.

Lem Dam, port, KOH KONG province, W CAMBODIA; 11°15′N 103°01′E.

Leme (LE-mai), city (2007 population 84,450), E central SÃO PAULO, BRAZIL, on railroad, 50 mi/80 km NNW of CAMPINAS; 22°12′S 47°24′W. Pottery manufacturing, food processing; agriculture (coffee, cotton).

Lemem Bar (le-MAIM), village, SOMALI state, SE ETHIOPIA, in the OGADEN, near SOMALIAN border, 50 mi/80 km E of K′ELAFO; 05°34′N 44°55′E. In camel and sheep raising region. Oasis.

Lemery (le-ME-ree), town, BATANGAS province, S LUZON, PHILIPPINES, on BALAYAN BAY, 30 mi/48 km WSW of SAN PABLO; 13°58′N 120°53′E. Fishing and agricultural center (rice, sugarcane, corn, coconuts). Archaelogical sites.

Lemeshenskiy, RUSSIA: see ORGTRUD.

Lemeshkino (lee-MYESH-kee-nuh), village, N VOLGOGRAD oblast, SE European Russia, on road, 31 mi/50 km E of YELAN′; 51°01′N 44°28′E. Elevation 482 ft/146 m. In agricultural area; flour mill.

Lemesos, CYPRUS: see LIMASSOL.

Lemfu (LEM-foo), village, BAS-CONGO province, W CONGO, on left bank of INKISI RIVER, 145 mi/233 km ENE of BOMA; 05°18′S 15°13′E. Elev. 1,482 ft/451 m. Roman Catholic missionary center with small seminary and convents for nuns and monks; also center of trade (manioc, yams, bananas, plantains, palm products).

Lemgo (LEM-go), town, North Rhine-Westphalia, NW GERMANY, 11 mi/18 km SE of HERFORD; 52°02′N 08°54′E. Manufacturing of furniture, textiles; metalworking. Has two Gothic churches, 14th–17th-century town hall. Was member of HANSEATIC LEAGUE.

Lemhi, county (□ 4,569 sq mi/11,879.4 sq km; 2006 population 7,930), E IDAHO; ⊙ SALMON; 44°58′N 113°57′W. Mountain and valley area drained by SALMON RIVER and tributaries and bounded on E by BITTERROOT RANGE and MONTANA (CONTINENTAL DIVIDE). Sheep, cattle; alfalfa, oats; mining (lead, silver, copper); manganese and uranium deposits; tourism, recreation, white water rafting on Salmon River In Bitteroot Mountain Region; Salmon River and YELLOWJACKET MOUNTAINS are in NW, LEMHI RANGE in E. Parts of Salmon National Forest throughout county, especially E; small part of Targhee National Forest in SE; part of Frank Church–River of No Return Wilderness Area. Formed 1869.

Lemhi Range, range, E IDAHO, running NW-SE between town of SALMON and SNAKE RIVER PLAIN, mainly in LEMHI county, in Salmon National Forest. Flatiron Mountain (11,019 ft/3,359 m) is highest point.

Lemhi River, 70 mi/113 km long, E IDAHO, near MONTANA state line; formed by confluence of several forks in E LEMHI county; flows NNW, between BITTERROOT and LEMHI ranges, to SALMON RIVER at SALMON. River is entirely within Lemhi county.

Lemieux (luh-MYU), village (□ 29 sq mi/75.4 sq km; 2006 population 338), CENTRE-DU-QUÉBEC region, S QUEBEC, E CANADA; 46°18′N 72°07′W.

Lemington, town, ESSEX CO., NE VERMONT, on the CONNECTICUT River, 20 mi/32 km N of GUILDHALL; 44°52′N 71°36′W. In hunting, fishing region. Monadnock Mountain is here.

Lemland (LEM-lahnd), fishing village, ÅLAND province, SW FINLAND, 6 mi/9.7 km ESE of MAARIANHAMINA; 60°05′N 20°06′E. On Lemland island (□ 42 sq mi/109 sq km), one of Ahvenanmaa group.

Lemmenjoki National Park (LEM-men-YO-kee), LAPIN province, FINLAND, 155 mi/250 km N of ROVANIEMI. The country's northernmost national park, it protects vast areas of wilderness and preserves the resources on which Finland's traditional Sami culture is based. Special allowance is made for traditional Sami land uses, such as reindeer herding, fishing, and berry picking. Popular hiking area.

Lemmer (LE-muhr), city, FRIESLAND province, N NETHERLANDS, minor port on the IJSSELMEER, at entrance to LEMSTERVAART canal, 12 mi/19 km SW of HEERENVEEN; 52°51′N 05°43′E. Tjeukemmer lake to NE; pumping station for NORTHEAST POLDER to S. Dairying; cattle; grain, vegetables, fruit, sugar beets.

Lemmi (LEM-mee), town, BAUCHI state, E central NIGERIA, on road, 45 mi/72 km WNW of BAUCHI. Major tin-mining center.

Lemmon, town (2006 population 1,227), PERKINS county, NW SOUTH DAKOTA, 140 mi/225 km NW of PIERRE, on NORTH DAKOTA state line; 45°56′N 102°09′W. Trading point for large grain and livestock region in North Dakota and South Dakota; wheat, alfalfa; manufacturing (jewelry). Grand River National Grassland is to S; Petrified Wood Park to S; Standing Rock Indian Reservation to E. Llewellyn Johns Memorial and Shadehill State Recreation Area to S.

Lemmon, Mount, highest peak (9,157 ft/2,791 m) in SANTA CATALINA MOUNTAINS, SE ARIZONA, 20 mi/27 km NNE of TUCSON. On N side is the site of Biosphere Two, in which researchers lived in an enclosed self-contained environment. Mount Lemmon Ski Valley. Sometimes called Lemmon Mountain.

Lemnos, locality, N VICTORIA, SE AUSTRALIA, 5 mi/8 km N of SHEPPARTON, in GOULBURN valley; 36°21′S 145°28′E. Orchards; dairying; food processing. Named for the Greek island to which Australian soldiers were evacuated following the Gallipoli campaign of World War I.

Le Moléson, SWITZERLAND: see: MOLÉSON.

Lemona (lai-MO-nah), commune, Vizcaya province, N SPAIN, 8 mi/12.9 km ESE of BILBAO. Cement works; manufacturing of leather belts. Anaiba is seat of commune.

Lemon Fair River, c.20 mi/32 km long, W VERMONT; rises near SUDBURY; flows N to OTTER CREEK near WEYBRIDGE.

Lemon Grove, city (2000 population 24,918), SAN DIEGO county, S CALIFORNIA; residential suburb 7 mi/11.3 km E of SAN DIEGO; 32°44′N 117°02′W. In agricultural area, with some small industries (manufacturing of wire products, motor-vehicle parts, and bricks; printing). Lemon Grove has benefited from the growth of nearby San Diego.

Lemon, Lake, reservoir, MONROE and BROWN counties, S central INDIANA, on Beanblossom Creek, 10 mi/16 km NE of HUNTINGTON; 3 mi/4.8 km long; 39°16′N 86°24′W. Formed by Lake Lemon Dam.

Lemon Reservoir (□ 1 sq mi/2.6 sq km), LA PLATA county, SW COLORADO, on Florida River, in San Juan National Forest, in the ROCKY MOUNTAINS, 12 mi/19 km NE of DURANGO; 37°23′N 107°40′W. Maximum capacity 487,660 acre-ft. Formed by Lemon Dam (284 ft/87 m high), built (1963) by the Bureau of Reclamation for irrigation; also used for flood control, recreation, and power generation. Transfer State Park on N end of Reservoir.

Lemont (lah-MONT), village (2000 population 13,098), COOK county, NE ILLINOIS, SW of CHICAGO, on Des Plaines River, SANITARY AND SHIP CANAL (ILLINOIS AND MICHIGAN CANAL NATIONAL HERITAGE CORRIDOR), 10 mi/16 km NNE of JOLIET; 41°40′N 87°59′W. Oil refining; manufacturing of aluminum products; limestone quarries. Ships petroleum, stone. ARGONNE NATIONAL LABORATORY (for atomic research) and Argonne Forest Preserve (recreational area) are nearby. Incorporated 1873.

Lemonweir River, c.60 mi/97 km long, W WISCONSIN; rises in JACKSON county; flows SE, past NEW LISBON and MAUSTON, to WISCONSIN RIVER 11 mi/18 km E of Mauston.

Lemoore, city (2000 population 19,712), KINGS county, S central CALIFORNIA, in San Joaquin Valley, 30 mi/48 km S of FRESNO; 36°18′N 119°48′W. Dairying; poultry; cantaloupes, plums, peaches, nectarines, olives, pistachios, almonds, cotton, barley; manufacturing (textiles, printing and publishing). Lemoore Naval Air Station to W. TULARE LAKE irrigation reservoir to S. Incorporated 1900.

Le Morne, village (2000 population 1,143), BLACK RIVER district, MAURITIUS, 31 mi/49.6 km SSW of PORT LOUIS. Tourist resort; deer ranching; fishing village.

LeMoyne (luh-MWAHN), former municipality, S QUEBEC, E CANADA; 45°30′N 73°29′W. Part of the VIEUX-LONGUEUIL borough of LONGUEUIL, into which LeMoyne amalgamated in 2002.

Lemoyne (LUH-moin), borough (2006 population 3,968), CUMBERLAND county, S central PENNSYLVANIA, suburb 2 mi/3.2 km SW of HARRISBURG, on SUSQUEHANNA RIVER (bridged). Railroad junction. Manufacturing (meat processing, signs). Northernmost point of Confederate advance, 1863. Camp Hill State Correctional Institution to SW.

Lempdes (lahn-puh), suburb (□ 4 sq mi/10.4 sq km; 2004 population 8,579) of CLERMONT-FERRAND, PUY-DE-DÔME department, in AUVERGNE, central FRANCE; 45°46′N 03°13′E.

Lempira (lem-PEE-rah), department (□ 1,295 sq mi/3,367 sq km; 2001 population 250,067), W HONDURAS, on EL SALVADOR border; ⊙ GRACIAS; 14°30′N 88°35′W. Astride Continental Divide; bounded S by LEMPA and SUMPUL rivers; drained N by JICATUYO RIVER branches, S by MOCAL RIVER. Mainly agricultural (coffee, tobacco, rice, wheat); cattle, hogs; alcohol distilling. Rich opal deposits near ERANDIQUE were formerly exploited. In addition to Gracias, main centers include Erandique and GUARITA. Formed 1825; called Gracias until 1943.

Lempster, town, SULLIVAN county, SW NEW HAMPSHIRE, 9 mi/14.5 km S of NEWPORT; 43°13′N 72°10′W. Manufacturing (lumber, metal fabrication); timber; agriculture (nursery crops, hay; cattle, poultry). Part of Honey Brook State Forest in SW. Long Pond in SE.

Lemro River, c.180 mi/290 km long, CHIN and RAKHINE States, MYANMAR; rises in S CHIN HILLS at 22°N; flows S past MYAUNGBWE (head of navigation) and MINBYA, to the BAY OF BENGAL at MYEBON. Navigable 30 mi/48 km above mouth.

Lemsal, LATVIA: see Limbaži.

Lemstervaart (LEM-stuhr-vahrt), canal, NORTH-EAST POLDER, FLEVOLAND province, central NETHERLANDS, runs N-S through center of polder, joining URKERVAART and ZWOLSEVAART canals just E of EMMELOORD.

Lemui Island (le-MOO-ee) (□ 49 sq mi/127.4 sq km), off E coast of CHILOÉ ISLAND, CHILOÉ province, LOS LAGOS region, S CHILE, 10 mi/16 km SE of CASTRO;

9 mi/15 km long, 3 mi/5 km–5 mi/8 km wide. PU-QUELDÓN, on N shore, is chief village. Also spelled Lemuy.

Lemuria: see ATLANTIS.

Lemvig (LEM-vee), port city, RINGKJØBING county, W JUTLAND, DENMARK, on LIMFJORD, 31 mi/50 km N of RINGKØBING; 56°25′N 08°20′E. Manufacturing (windmill production, shipbuilding, machinery, wood products, furniture); fishing.

Lemwerder (LEM-VER-der), suburb of BREMEN, LOWER SAXONY, NW GERMANY, on left bank of WESER RIVER, 8 mi/12.9 km NW of city center; 53°07′N 08°38′E.

Lemyethna (lem-YET-nah), township, AYEYARWADY division, MYANMAR, on W bank of BASSEIN RIVER, 22 mi/35 km WSW of HENZADA; 17°36′N 95°09′E.

Lena (LEE-nah), village (2000 population 2,887), STE-PHENSON county, N ILLINOIS, near WISCONSIN state line, 11 mi/18 km NW of FREEPORT; 42°22′N 89°49′W. In agricultural area; processes dairy products. Incorporated 1869.

Lena (LEE-nuh), village (2000 population 167), LEAKE county, central MISSISSIPPI, 10 mi/16 km SSW of CARTHAGE; 32°35′N 89°35′W. Agriculture (cotton, corn; cattle); manufacturing (apparel). Bienville National Forest to S.

Lena, village (2006 population 500), OCONTO county, NE WISCONSIN, 30 mi/48 km N of GREEN BAY; 44°57′N 88°02′W. Dairying region; vegetables; wood products. Near Copper Culture State Park.

Lena-Angara Plateau (LYE-nah–ahn-gah-RAH), S IRKUTSK oblast, Siberian Russia, between the ANGARA and KIRENGA (tributary of the LENA RIVER) rivers; 372 mi/600 km long, 236 mi/380 km wide; elevation 3,281 ft/1,000 m. Tayga. Mineral resources include iron and copper ores and common salt. Provides raw resouces for the manufacture of building materials.

Lena Basin River Ports (LYE-nah), group of important water transport junctions linking the industrial and agricultural regions of SAKHA REPUBLIC and N IRKUTSK oblast, RUSSIAN FAR EAST, with the rest of the country. Includes ports, landings, and stopping points. General purpose ports and industrial wharves organize the cargo and passenger transport on the LENA RIVER and its tributaries, especially since the area lacks a well-developed network of railroad and highways. Cargo transported down the Lena from Osetrovo (Irkutsk oblast) to the river mouth include manufactured goods, foodstuffs, machinery, metals, hardware, cement, petroleum products, and coal. Ore and timber are moved upriver. Main ports include Osetrovo, Kirensk (Irkutsk oblast), Lensk (Sakha Republic), Olekminsk, Pokrovsk, Yakutsk, Sangary, Bodaybo (Vitim River), Khandyga, and Dzhebariki-Khaya (Aldan River). Operating during ice-free period, 125–170 days per year.

Lena Beach (LEE-nuh), fishing village, SE ALASKA, on LYNN CANAL, 15 mi/24 km NW of JUNEAU, and on GLACIER HIGHWAY; 58°24′N 134°45′W.

Lena Coal Basin (LYE-nah) (□ 289,575 sq mi/752,895 sq km), in SAKHA REPUBLIC and KRASNOYARSK TERRITORY, RUSSIAN FAR EAST, mainly within the Central Yakut Lowland, in the basin of the LENA RIVER and its tributaries (Aldan and Vilyuy rivers). In the N, the basin extends along the LAPTEV SEA from the Lena River mouth to the KHATANGA GULF. Lignite and hard coal.

Lenakel, VANUATU: see TANNA.

Lenapah (LE-nuh-pah), village (2006 population 306), NOWATA county, NE OKLAHOMA, 10 mi/16 km N of NOWATA; 36°51′N 95°37′W. In livestock area; crushed rock.

Lenape Heights (LE-nah-pee), unincorporated town (2000 population 1,212), ARMSTRONG county, W central PENNSYLVANIA, residential suburb 1 mi/1.6 km SE of FORD CITY; 40°45′N 79°31′W.

Lena River (LYE-nah), 2,648 mi/4,261 km long, easternmost and longest of great rivers of Siberian Russia; rises near Lake BAYKAL in the W BAYKAL RANGE (IRKUTSK oblast) at approximately 4,700 ft/1,433 m; flows NE, through the CENTRAL SIBERIAN PLATEAU, past UST′-KUT, KIRENSK, and VITIM, receiving (right) Kirenga and Vitim rivers. Entering SAKHA REPUBLIC, its valley widens progressively as it passes Peleduy, OLEKMINSK, POKROVSK, and YAKUTSK (here flowing N) and receives (right) Olekma, Botoma, and Aldan rivers. Here turning NW, the Lena flows parallel to and 100 mi/161 km–150 mi/241 km W of the VER-KHOYANSK RANGE, reaching its greatest width of 9 mi/14 km above ZHIGANSK; receives Vilyuy River (left). Valley narrows again as the Lena flows between BULUN and KYUSYUR, entering the LAPTEV SEA through a 150-mi/241-km-wide delta mouth; Olenek, Bykov, and Trofimov are main mouth branches. Navigation, mainly for lumber and grain, upstream to Ust′-Kut (2,135 mi/3,436 km from mouth; May-November); at high water, ships go to Kachuga, 310 mi/499 km further upstream. At Yakutsk (915 mi/1,473 km from mouth), river is ice-free June through October, at its delta, July through September. Coal, natural gas, diamonds, and gold are found along the Lena and its tributaries (Vilyuy, Vitim, Aldan). Reached by the Russians in 1630.

Lenart (LE-nahrt), village, NE SLOVENIA, NE of MAR-IBOR; 46°36′N 15°52′E. Principal trade center of SLO-VENSKE mountain region since 1332. Has 16th-centuey church.

Lenauheim (LE-nou-heim), Hungarian *Csatád*, village, TIMIŞ county, W ROMANIA, 20 mi/32 km NW of TI-MIŞOARA; 45°52′N 20°48′E. Austrian poet N. Lenau b. here.

Lenawee (LE-nuh-wee), county (□ 761 sq mi/1,978.6 sq km; 2006 population 102,191), SE MICHIGAN; 41°53′N 84°04′W; ⊙ ADRIAN. Bounded S by OHIO state line; drained by RIVER RAISIN and its branches, and by TIFFIN RIVER. Agriculture (soybeans, wheat, oats, sugar beets, corn, beans; cattle, hogs, sheep, dairy products); manufacturing at ADRIAN, HUDSON, MOR-ENCI, and TECUMSEH; hatcheries; sand and gravel; chrysanthemum raising; lake resorts. Walter J. Hayes State Park in N on WAMPLERS LAKE; Lake Hudson State Recreation Area in W; Devils Lake in NW. Organized 1826.

Lenchitsa, POLAND: see LECZYCA.

Lencloître (lahn-klwah-truh), commune (□ 7 sq mi/18.2 sq km), VIENNE department, POITOU-CHARENTES region, W central FRANCE, 10 mi/16 km W of CHÂ-TELLERAULT; 46°49′N 00°20′E. Market-gardening center (chiefly asparagus), dairying. Has 12th–15th-century abbey church.

Lençóis (len-sois), town (2007 population 9,629), central BAHIA, BRAZIL, at foot of the CHAPADA DIA-MANTINA, 17 mi/27 km NNW of ANDARAÍ; 12°30′S 41°25′W. Old diamond-mining center, highly productive until end of 19th century. Semiprecious stones also found here. Agriculture, ranching.

Lençóis, Brazil: see LENÇÓIS PAULISTA.

Lençóis Paulista (LEN-sois POU-lee-stah), city (2007 population 59,459), central SÃO PAULO, BRAZIL, on railroad, 26 mi/42 km SE of BAURU; 22°36′S 48°47′W. Cotton, manioc, and rice processing; distilling. Until 1944, called Lençóis; was called Ubirama 1944–1948.

Lend (LENDT), village, SALZBURG W central AUSTRIA, in the PINZGAU region on the SALZACH river, near mouth of the GASTEINER ACHE, 12 mi/19 km ESE of ZELL AM SEE. 47°48′N 13°03′E. Aluminum plant; important road junction. Hydroelectric station to the SE, in Gastein Valley.

Lendak (len-DAHK), Hungarian *Lándok*, village, VY-CHODOSLOVENSKY province, N SLOVAKIA, on E slope of HIGH TATRAS, and 13 mi/21 km NNE of POPRAD; 49°14′N 20°21′E. Has potatoes; cattle. Has a 14th-

century Gothic church. Known for folk costumes and architecture.

Lendava (len-DAH-vah), town, NE SLOVENIA, on LE-DAVA RIVER, on railroad, 38 mi/61 km E of MARIBOR, on Hungarian border; 46°34′N 16°27′E. Local trade center; petrochemical, pharmaceutical, and electronic industries. Petroleum deposits nearby.

Lendava River, AUSTRIA and SLOVENIA: see LEDAVA RIVER.

Lendelede (LEN-duh-lai-duh), commune (2006 population 5,372), Kortrijk district, WEST FLANDERS province, W BELGIUM, 5 mi/8 km NNW of KORTRIJK; 50°53′N 03°14′E. Textiles; agricultural market.

Lendinara (len-dee-NAH-rah), town, ROVIGO province, VENETO, N ITALY, 9 mi/14 km W of ROVIGO; 45°05′N 11°36′E. Highly diversified secondary industrial center; manufacturing (fabricated metals, agricultural machinery, fertilizer, jute products, vinegar, beet sugar, clothing).

Lendva, AUSTRIA: see LEDAVA RIVER.

Lene, Lough (LEEN, LAHK), lake (3 mi/4.8 km long, 1 mi/1.6 km wide), NE WESTMEATH county, central IRELAND, 3 mi/4.8 km E of Castlepollard; 53°39′N 07°14′W.

Lenexa (luh-NEK-suh), city (2000 population 40,238), JOHNSON county, E KANSAS, suburb 12 mi/19 km SW of KANSAS CITY; 38°57′N 94°47′W. Manufacturing (apparel, machinery, electronic equipment, plastics, medical equipment, printing, bakery products, ink, consumer goods, beverages, paper products).

L'Enfant-Jésus, CANADA: see VALLÉE-JONCTION.

Lengana River, SOUTH AFRICA: see EXCELSIOR.

Lengby (LENG-bee), village (2000 population 79), POLK county, 32 mi/51 km W of BEMIDJI, on N shore of Spring Lake, near Poplar River; 47°30′N 95°37′W. White Earth Indian Reservation to S. Dairying; grain.

Lengede (LEN-ge-de), town, LOWER SAXONY, NW GERMANY, 8 mi/12.9 km SSE of PEINE; 52°13′N 10°18′E. Primarily residential; manufacturing of agricultural machinery. Iron ore was mined until 1977.

Lengenfeld (LENG-uhn-felt), town, SAXONY, E central GERMANY, in the ERZGEBIRGE, on GÖLTZSCH RIVER, 12 mi/19 km NE of PLAUEN; 50°33′N 12°17′E. Manufacturing of textiles.

Lenger (leng-GER), city, SE SOUTH KAZAKHSTAN region, KAZAKHSTAN, on spur of TURK-SIB RAILROAD, 15 mi/24 km SE of SHYMKENT; 42°10′N 69°54′E. Lignite-mining center, supplying Shymkent lead works.

Lengerich (LENG-ge-rikh), town, North Rhine-Westphalia, NW GERMANY, in TEUTOBURG FOREST, 10 mi/16 km SW of OSNABRÜCK; 52°12′N 07°51′E. Railroad junction. Manufacturing of machinery, pharmaceuticals, plastics; cement and lime works, paper mills. Chartered 1727. Has late-Romanesque church.

Lenggries (LENG-grees), village, UPPER BAVARIA, BA-VARIA, in Bavarian Alps, on the ISAR, 5 mi/8 km S of BAD TÖLZ; 47°41′N 11°35′E. Railroad terminus; brewing; sawmilling. Summer resort (elevation 2,228 ft/679 m).

Lengnau, commune, BERN canton, NW SWITZERLAND, 6 mi/9.7 km ENE of BIEL.

Lengshuijiang (LENG-SHUAI-JIANG), city (□ 169 sq mi/438 sq km; 1994 estimated urban population 160,700; 1994 estimated total population 336,000), central W HUNAN province, CHINA; 27°40′N 111°26′E. Heavy industry is the largest source of income for the city. Crop growing, livestock raising. Grain, hogs; manufacturing (utilities, chemicals, iron and steel).

Lengshuitan (LENG-SHUAI-TAN), city (□ 471 sq mi/1,220 sq km; 1994 estimated urban population 139,600; 1994 estimated total population 450,900), HUNAN province, CHINA; 26°27′N 111°35′E. Agriculture and heavy industry are the largest sources of income for the city. Crop growing. Grain, hogs; manufacturing (food processing, machinery).

Lengua de Pájaro, CUBA: see NICARO.

Lengua de Vaca Pass (LEN-gwah de VAH-kah), on MICHOACÁN-MEXICO state border, central MEXICO, on Mexico Highway 41 from TOLUCA DE LERDO to ZITÁCUARO, 45 mi/72 km W of Toluca. Elevation 9,348 ft/2,849 m.

Lengua de Vaca, Punta (LEN-gwah dai VAH-kah, POON-tah), cape on PACIFIC coast, COQUIMBO region, N central CHILE, at W end of Tongoi Bay.

Lenguazaque (len-gwah-SAH-kai), town, ⊙ Lenguazaque municipio, CUNDINAMARCA department, central COLOMBIA, 42 mi/68 km NE of BOGOTÁ; 05°17′N 73°43′W. Elevation 8,989 ft/2,239 m. Coffee, corn, sugarcane; livestock.

Lengwethen, RUSSIA: see LUNINO, KALININGRAD oblast.

Lenham (LEN-uhm), village (2006 population 3,200), central KENT, SE ENGLAND, 9 mi/14.5 km NW of ASHFORD; 51°15′N 00°43′E. Former agricultural market. Has 14th-century church.

Lenhartsville (LEN-uhrts-vil), borough (2006 population 171), BERKS county, E central PENNSYLVANIA, 16 mi/26 km. N of READING, on Maiden Creek. Agriculture (grain; livestock; dairying); light manufacturing Pennsylvania Dutch Folk Culture Center is here.

Lenhovda (LEN-HOOV-dah), village, KRONOBERG county, S SWEDEN, 19 mi/31 km ENE of VÄXJÖ; 57°00′N 15°17′E.

Lenin (LE-nin), town, GOMEL oblast, BELARUS, on SLUCH RIVER, 30 mi/48 km ENE of LUNINETS, on former USSR-POLAND border. Flour milling; sawmilling.

Lenin, AZERBAIJAN: see CHINARLY.

Lenina (LYE-nee-nah), village (2005 population 5,285), central KRASNODAR TERRITORY, S European Russia, on road and near railroad, 48 mi/77 km NNE of KRASNODAR; 46°04′N 39°47′E. Elevation 141 ft/42 m. In agricultural area (grain, sunflowers, castor bean, fruits). Developed around a collective farm called Imeni Lenina (Russian=named after Lenin).

Lenina, RUSSIA: see GORYACHEISTOCHNENSKAYA.

Leninabad, Takikistan: see KHUDJAND.

Lenina, Imeni (LE-nee-nah, EE-me-nee), N ironmining suburb of KRYVYY RIH, DNIPROPETROVS'K oblast, UKRAINE, on the right bank of the SAKSAHAN' RIVER, and 14 mi/23 km NNE of city center; 48°05′N 33°29′E. Elevation 462 ft/140 m. Until about 1926, called Kalachevs'kyy Rudnyk (Russian *Kalachevskiy Rudnik*) or Kalachevs'ke (Russian *Kalachevskoye*).

Lenina, Imeni V. I. (LYE-nee-nah EE-mye-nee), town (2006 population 2,880), W ULYANOVSK oblast, E central European Russia, on the BARYSH River (tributary of the SURA River, VOLGA RIVER basin), on road, 12 mi/19 km SSW of BARYSH, to which it is administratively subordinate; 53°34′N 46°58′E. Elevation 830 ft/252 m. Cloth factory. Until about 1938, called Rumyantsevo.

Leninaul (lye-neen-ah-OOL), village (2005 population 7,685), W central DAGESTAN REPUBLIC, in the NE foothills of the Greater CAUCASUS Mountains, SE European Russia, on road, 50 mi/80 km W of MAKHACHKALA, and 11 mi/18 km S of KHASAVYURT; 43°05′N 46°34′E. Elevation 1,479 ft/450 m. Agriculture (grain, livestock). Until 1944, called Aktash-Aul; then, for a short time (until 1954), Stalinaul. Part of the disputed territory between CHECHNYA and Russia; briefly occupied by Chechen military units in 1999, leading to the renewal of hostilities with RUSSIA.

Leninavan (le-NEEN-ah-vahn), urban settlement, NAGORNO-KARABAKH region, AZERBAIJAN, on the right bank of TERTER River Seed-cleaning plant, winery.

Lenin-dzhol, KYRGYZSTAN: see MASSY.

Lenine (LE-nee-ne) (Russian *Lenino*), town, E Republic of CRIMEA, UKRAINE, on KERCH PENINSULA, on the NORTH CRIMEAN CANAL and railroad, 34 mi/55 km W of KERCH; 45°18′N 35°47′E. Elevation 118 ft/35 m. Raion center; wine making; grain and feed milling.

Established at end of the 19th century, named Sim Kolodyaziv (Russian *Sem' Kolodezey*); renamed and town since 1957.

Lenine (LE-nee-ne) (Russian *Lenino*), town (2005 population 8,300), E Republic of CRIMEA, UKRAINE, on road and railroad (Sem' Kolodezey station), 33 mi/53 km W of KERCH; 45°18′N 35°47′E. Elevation 118 ft/35 m. In agricultural area (wheat). Formerly called Sem' Kolodezey [Ukrainian=seven wells].

Leningori, GEORGIA: see AHALGORI.

Leningrad (le-nin-GRUHD), oblast (□ 28,708 sq mi/ 74,640.8 sq km; 2005 population 1,658,545), NW European Russia, in the NW part of the East European Plain, on the Gulf of FINLAND, an inlet of the BALTIC SEA; ⊙ SAINT PETERSBURG (though the city is an independent municipality and politically subordinate to the RUSSIAN FEDERATION); 60°00′N 32°00′E. Bordered N by FINLAND, the Republic of KARELIA, and lakes LADOGA and ONEGA, S by NOVGOROD and PSKOV oblasts, E by VOLOGDA oblast, and W by ESTONIA and the Gulf of Finland. Flat terrain, with traces of glacier activity; mostly low-lying areas, including the Baltic and Ladoga depressions and the Neva, Vuoksa and Svir' lowlands. The high Baltic-Ladoga Scarp (131 ft/40 m–197 ft/60 m) extends S of the Gulf of Finland and Lake Ladoga. Half the area is forested. Maritime to continental climate, with warm winters and cool summers; snow cover lasts 2 to 3 months and the growing season lasts for a slightly longer period. The dense river network is nearly completely in the Baltic Sea basin; major rivers include the Neva, Volkhov, Svir' (reservoir), Vuoksa, Narva (reservoir), Sias', and Luga and are used for hydroelectric power, navigation, and timber floating. The VOLGA-BALTIC WATERWAY plays an important role, as do the BALTIC-WHITE SEA and SAIMAA canals. River ports include St. Petersburg (also a seaport), PETROKREPOST', SVIRITSA, PODPOROZHYE, and Lodeynoye Pole; VYBORG is a seaport. Ladoga and Onega are the largest lakes and there are many small lakes, especially on the Karelian Isthmus. Local fauna include marten, fox, elk, beaver, and mink. Among the most economically developed oblasts in the country; St. Petersburg, with its diversified economy, is one of the largest Russian industrial centers. Main industries include machine building, metallurgy, petroleum refining, logging and processing of forestry and agricultural products, and light manufacturing; chemicals, building materials (cement, glass, bricks, stone), foodstuffs. Dairy cattle. Limestone. Industrial pollution is a significant concern. The population is mostly comprised of Russian, with Ukrainian, Finns, Byelorussians, Veps, Jews, Estonians, Tatars, and Karelians. Formed in 1927.

Leningrad, RUSSIA: see SAINT PETERSBURG.

Leningrad-300, RUSSIA: see MIRNYY, ARCHANGEL oblast.

Leningrad Health Resort Region (lye-neen-GRAHT), coastal strip of the KARELIAN ISTHMUS, LENINGRAD oblast, NW European Russia, along NE and N coasts of the Gulf of FINLAND, from SESTRORETSK to the mouth of the Ioninioka River, 22 mi/35 km NW of SAINT PETERSBURG; 25 mi/40 km long, 1 mi/1.6 km-4 mi/6 km wide. Ridge of sand dunes and wooded hills stretching along the coast. Shchuch'ye, Kayavan, Lampi, and Krasavitsa lakes are here. Mild summers; annual precipitation, 24 in/60 cm. Numerous resorts, sanatoria, boarding houses, and tourist centers. Includes SESTRORETSK, SOLNECHNOYE, REPINO, KOMAROVO, ZELENOGORSK, USHKOVO, SEROVO, Molodëzhnoye, Smolyachkovo, and Chernaya Rechka.

Leningradskaya (lye-neen-GRAHT-skah-yah), city (2005 population 38,895), N central KRASNODAR TERRITORY, S European Russia, on the SOSYKA River (left tributary of the YEYA RIVER), on road and railroad, 59 mi/95 km S of ROSTOV-NA-DONU, and 20 mi/

32 km SW of KUSHCHËVSKAYA; 46°19′N 39°23′E. Elevation 108 ft/32 m. Flour mill. Until the 1930s, an urban settlement called Umanskaya.

Leningradskii, town, E KHATLON viloyat, TAJIKISTAN, 18 mi/29 km NE of KULOB; 38°05′N 70°00′E. Wheat. Formerly called Muminabad. Also spelled Leningradskiy.

Leningradskiy (lye-neen-GRAHT-skeeyee), settlement (2005 population 695), N CHUKCHI AUTONOMOUS OKRUG, N RUSSIAN FAR EAST, N of the ARCTIC CIRCLE, in the lowlands near the coast of the CHUKCHI SEA, 178 mi/286 km E of PEVEK; 69°23′N 178°25′E. Mineral ore processing.

Leninkent (lye-neen-KYENT), urban settlement (2005 population 14,245), E central DAGESTAN REPUBLIC, SE European Russia, on road, 8 mi/13 km W, and under administrative jurisdiction, of MAKHACHKALA; 42°58′N 47°21′E. Elevation 472 ft/143 m. In oil- and gas-producing region.

Lenino (LYE-nee-nuh), former town, central MOSCOW oblast, central European Russia, now a S suburb of MOSCOW; 55°37′N 37°39′E. Elevation 488 ft/148 m. Cultural tourism center. Has ruins of palace of Catherine II, amidst English gardens. Formerly called Tsaritsyno.

Lenino (LYE-nee-nuh), N suburb of IRKUTSK, SE IRKUTSK oblast, E central SIBERIA, RUSSIA, on the left bank of the ANGARA RIVER, at the Irkutsk II station of the TRANS-SIBERIAN RAILROAD; 54°36′N 100°43′E. Elevation 1,463 ft/445 m. Railroad repair shops. Formerly called Innokentyevskoye; incorporated into Irkutsk in 1930.

Lenino, RUSSIA: see LENINSK-KUZNETSKIY.

Lenino, UKRAINE: see LENINE.

Leninobod, viloyat, NW TAJIKISTAN; ⊙ KHUDJAND. Area along ZERAVSHAN RIVER (S) and projecting across TURKESTAN RANGE NE to W FERGANA VALLEY at the Syr Darya and KURAMA RANGE; bounded S by GISSAR RANGE, N by UZBEKISTAN and KYRGYZSTAN. Agriculture chiefly in Fergana Valley, along rivers, and on lower mountain slopes (wheat, cotton, and grapes); silkworm breeding; fruit growing; sheep and goat raising. Cotton and silk processing, silk weaving, fruit canning, cottonseed oil extracting; wine making. Extensive mining area on S slope of Kurama Range; lead and zinc at KANSAI and Kara-Mazar, radioactive ores and vanadium at TABOSHAR, bismuth at ADRASMAN, arsenic at TAKELI, tungsten at CHORUKH-DAIRON; coal mining at SHURAB, oil fields at KIM and NEFTEABAD. Fergana Valley railroad runs along the SYR DARYA, past NAU, PROLETARSK, Khudjand, and KANIBADAM, the region's chief urban centers. Population consists of Tajiks, Uzbeks. Includes large Kairakkum reservoir E of Khujarid. Formed 1939. Also spelled Leninabad.

Leninogorsk (le-nee-no-GORSK), city, NE EAST KAZAKHSTAN region, KAZAKHSTAN, in NW ALTAI MOUNTAINS, near ULBA RIVER, 50 mi/80 km NE of UST-KAMENOGORSK; 50°23′N 83°32′E. Railroad terminus. Mining center in an area that is the chief source of Kazak zinc and lead. Silver, copper, and gold are also mined. The 1st mines opened in the late 18th century. Metallurgical factories produce lead, copper, zinc concentrates, and cadmium. Until 1940, called Ridder; also spelled Leninogor.

Leninogorsk (lye-nee-nuh-GORSK), city (2006 population 66,505), SE TATARSTAN Republic, E European Russia, on road and railroad (Pismyanka station), 200 mi/322 km SE of KAZAN', and 13 mi/21 km WNW of BUGUL'MA; 54°36′N 52°30′E. Elevation 652 ft/198 m. Production of petroleum and natural gas; motor vehicle parts, machinery, concrete products, garments, food processing (dairy, bakery). Arose with the discovery of the Romashkinskoye oil field in 1948. Called Novaya Pis'myanka until 1955, when it was made a city.

Lenino-Kokukshino (LYE-nee-nuh–kuh-KOOK-shi-nuh), village, NW TATARSTAN Republic, E European Russia, near the Mesha River (right tributary of the KAMA River), on road, 21 mi/34 km E of KAZAN', and 5 mi/8 km NNE of PESTRETSY, to which it is administratively subordinate; 55°49′N 49°41′E. Elevation 416 ft/126 m. Produce processing, fish farm.

Lenin Peak (23,405 ft/7,134 m), on TAJIKISTAN-KYRGYZSTAN border. Highest point in the TRANS-ALAI (Chong-Alay) range, and 2nd-highest peak in Kyrgyzstan.

Leninpol, KYRGYZSTAN: see BAKAI-ATA.

Leninsk, RUSSIA: see TALDOM.

Leninsk (LYE-neensk), city (2006 population 15,275), E VOLGOGRAD oblast, SE European Russia, on the AKHTUBA RIVER (VOLGA RIVER basin), on road and railroad, 34 mi/55 km E of VOLGOGRAD; 48°42′N 45°12′E. Fishing fleet base; fish processing (canneries), flour mill. Made city in 1963.

Leninsk (LYE-neensk), town (2004 population 4,700), W central CHELYABINSK oblast, SE URALS, RUSSIA, 19 mi/30 km SW, and under administrative jurisdiction, of MIASS; 54°53′N 59°52′E. Elevation 1,332 ft/405 m. Gold mining.

Leninsk (le-NEENSK), town, DASHHOWUZ weloyat, TURKMENISTAN, in Khiva oasis, in AMU DARYA delta region, 35 mi/56 km NW of DASHHOWUZ; 42°03′N 59°24′E. Tertiary-level administrative center. Cotton. Formerly called Ak-Tepe.

Leninsk, UZBEKISTAN: see ASSAKE.

Leninskaya Sloboda (LYE-neen-skah-yah sluh-buh-DAH), town, central NIZHEGOROD oblast, central European Russia, on the VOLGA RIVER, 25 mi/40 km SE of NIZHNIY NOVGOROD, and 9 mi/14 km ESE of KSTOVO, to which it is administratively subordinate; 56°04′N 44°28′E. Elevation 423 ft/128 m. In agricultural area (feed corn; dairy livestock).

Lenins'ke (LE-neen-ske) (Russian *Leninskoye*), town, SE LUHANS'K oblast, UKRAINE, in the DONBAS, 4 mi/6.4 km WSW of SVERDLOVS'K and subordinated to its city council; 48°04′N 39°33′E. Elevation 994 ft/302 m. Coal mine, aluminum smelter. Established in 1900 as a settlement of Mine No. 9 (until 1936), then settlement of Mine Imeni Lenina (until 1944), then Lenins'kyy (Russian *Leninskiy*).

Leninskii, town, W TAJIKISTAN, 8 mi/12.9 km S of DUSHANBE (linked by narrow-gauge railroad); 38°25′N 68°50′E. Area is under no direct viloyat administrative division, rather, it is under direct republic supervision. Wheat, horses; metalworks; tertiary level administrative center. Formerly called Koktash. Also spelled Leninskiy.

Leninskiy (LYE-neen-skeeyee), town (2006 population 9,740), central TULA oblast, central European Russia, on the Volot River (OKA RIVER basin), on railroad (Obidimo station), 8 mi/13 km NW of TULA; 53°37′N 37°03′E. Elevation 770 ft/234 m. Mining, chemical, and asphalt works. Until 1939, called Domman–Asfal'tovyy Zavod.

Leninskiy (LYE-neen-skeeyee), settlement, W MARI EL REPUBLIC, E central European Russia, on the VETLUGA RIVER (tributary of the VOLGA RIVER), near highway, 25 mi/40 km NW of KOZ'MODEM'YANSK; 56°34′N 45°56′E. Elevation 252 ft/76 m. Glassworks.

Leninskiy, RUSSIA: see LENINSKOYE, KIROV oblast.

Leninskiy, UKRAINE: see LENINS'KE.

Leninsk-Kuznetskiy (LYE-neensk–kooz-NYETS-keeyee), city (2005 population 107,990), central KEMEROVO oblast, S central SIBERIA, RUSSIA, on the INYA RIVER (OB' RIVER basin), on road and railroad, 80 mi/129 km S of KEMEROVO; 54°39′N 86°09′E. Elevation 761 ft/231 m. It is a coal center in the KUZNETSK BASIN; coal processing, firefighting equipment for coal mines, servicing of mining equipment. Also manufacturing (textiles, clothing, footwear, building materials, heavy industrial machinery, furniture) and food processing (bakery). Founded in 1864 as a mining settlement,

called Kol'chugino until 1922, and Lenino, 1922–1925. Made city in 1925, it underwent rapid development in the 1930s and during World War II. With a switch to free-market economy in 1992, poor management has caused some of the mines to shut down, resulting in a sharp population decline as miners and their families moved away.

Leninskoe (LE-nin-skaw-yuh), village, NE OSH region, KYRGYZSTAN, in FERGANA VALLEY, 23 mi/37 km NE of OSH; 40°30′N 73°09′E. Cotton, silk. Until 1937, called POKROVKA.

Leninskoye (LYE-neen-skuh-ye), town (2005 population 5,270), W KIROV oblast, E European Russia, on highway and the TRANS-SIBERIAN RAILROAD (Shabalino station), 45 mi/72 km W of KOTEL'NICH; 58°18′N 47°05′E. Elevation 492 ft/149 m. Food processing (dairy, bakery), flax processing; livestock veterinary station; agricultural supplies and machinery repair; furniture making; concrete plant. Established in 1906 in conjunction with the construction of the railroad. Formerly called Shabalino; since 1940, also called Leninskiy. Town since 1945.

Leninskoye (LYE-neen-skuh-ye), village (2005 population 6,755), S JEWISH AUTONOMOUS OBLAST, S RUSSIAN FAR EAST, on the AMUR River, 60 mi/97 km S of BIROBIDZHAN (to which it is connected by railroad); 47°56′N 132°37′E. Elevation 183 ft/55 m. In agricultural area. Formerly called Mikhaylo-Semenovskoye; then called Blyukherovo until 1939.

Leninskoye, KAZAKHSTAN: see SHARAPKHANA.

Leninskoye, UKRAINE: see LENINS'KE.

Lenins'kyy, UKRAINE: see LENINS'KE.

Lenjan (lehn-JAHN), district, Esfahān province, W central IRAN, just SW of ESFAHAN, along Zaindeh River; 32°23′N 51°33′E. One of the chief rice-producing regions of Iran; also cotton and millet.

Lenk, commune, BERN canton, SW central SWITZERLAND, on SIMME RIVER, in OBERSIMMENTAL, 22 mi/35 km SSW of THUN. Elevation 3,504 ft/1,068 m. Health resort, winter-sports center.

Lenne River (LEN-ne), c.60 mi/97 km long, W GERMANY; rises on the KAHLE ASTEN; 51°11′N 08°29′E. Flows NW, past ALTENA and HOHENLIMBURG, to the RUHR just NW of HAGEN.

Lennestadt (LEN-ne-shtaht), town, North Rhine-Westphalia, W GERMANY, on LENNE RIVER, 18 mi/29 km N of SIEGEN; 51°09′N 08°04′E. Sulphur mining; metalworking; electronics and textile industries. Open-air theater in nearby suburb of ELSPE.

Lennik (LEN-nik), commune (2006 population 8,724), Halle-Vilvoorde district, BRABANT province, central BELGIUM, 6 mi/10 km ESE of NINOVE.

Lenningen (LE-ning-uhn), village, Lenningen commune, SE LUXEMBOURG, 4 mi/6.4 km N of REMICH; 49°36′N 06°22′E. Vineyards.

Lenno, POLAND: see WLEN.

Lennonville (LE-nuhn-vil), village, W central WESTERN AUSTRALIA, AUSTRALIA, 210 mi/338 km ENE of GERALDTON; 27°58′S 117°50′E. On Geraldton-Wiluna railroad; mining center in MURCHISON GOLDFIELD.

Lennox, unincorporated city (2000 population 22,950), LOS ANGELES county, S CALIFORNIA; residential suburb 9 mi/14.5 km SW of LOS ANGELES, between INGLEWOOD (N) and HAWTHORNE (S); 33°57′N 118°22′W. Los Angeles International Airport to W, Hawthorne Municipal Airport to S. Hollywood Park racetrack to NE.

Lennox, town (2006 population 2,138), LINCOLN county, SE SOUTH DAKOTA, 15 mi/24 km SW of SIOUX FALLS; 43°21′N 96°54′W. Livestock center; grain, animal feed; meat products. Manufacturing (food processing, smoked fish). County fair takes place here. Settled 1879, incorporated 1906.

Lennox and Addington (LE-nuhks, A-deeng-tuhn), county (□ 1,072 sq mi/2,787.2 sq km; 2001 population 39,461), SE ONTARIO, E central CANADA, on Lake ONTARIO; ⊙ NAPANEE; 44°40′N 77°10′W. Manu-

facturing (tires); construction; agriculture (grapes); electricity generating station.

Lennox Head (LE-nuhks HED), village, N NEW SOUTH WALES, SE AUSTRALIA, 475 mi/765 km N of SYDNEY, near BALLINA, and at S end of Seven Mile Beach; 28°49′S 153°35′E. Holiday destination; surfing at Lennox Head point (28°50′S 153°37′E). Lake Ainsworth, N of town, has water the color of surrounding tea trees.

Lennox Hills (LEN-ahks HILZ), range, S STIRLING, central Scotland, extends 15 mi/24 km along East Dumbartonshire and North Lanarkshire border. The W part is called Campsie Fells, the E part Kilsyth Hills. Highest points are Earl's Seat (1,896 ft/578 m), 3 mi/4.8 km N of STRATHBLANE, and Meikle Bin (1,870 ft/570 m), 4 mi/6.4 km NW of KILSYTH.

Lennox Island (LEN-nuhks) (□ 51 sq mi/132.6 sq km), in TIERRA DEL FUEGO, at mouth of BEAGLE CHANNEL, just E of NAVARINO ISLAND, 20 mi/32 km S of the main island of the archipelago; c.9 mi/15 km long, 9 mi/14.5 km wide. Ownership of the island is disputed between CHILE and ARGENTINA. Both nations have an interest in the natural resources (petroleum, mineral deposits, and krill) of these islands, as well as in their strategic position. A number of the islands in the Beagle Channel have been awarded to Chile, and though Argentina has rejected the award, the two nations have begun to develop joint energy ventures in the area.

Lennox-King Glacier, outlet glacier in the TRANSANTARCTIC MOUNTAINS of EAST ANTARCTICA, flowing from the polar plateau through the QUEEN ALEXANDRA RANGE into the ROSS ICE SHELF, 40 mi/65 km long; 83°25′S 168°00′E.

Lennoxtown (LEN-ahks-TOUN), town (2001 population 3,773), East Dunbartonshire, W Scotland, at foot of Campsie Fells, 8 mi/12.9 km N of GLASGOW; 55°58′N 04°12′W. Previously textile printing. Formerly in Strathclyde, abolished 1996.

Lennoxville (LE-nuhks-vil), borough (French *arrondissement*), former town, S QUEBEC, E CANADA, at the confluence of the SAINT FRANÇOIS and Massawippi rivers; 45°22′N 71°52′W. Chiefly residential. Seat of Bishop's University (1843).

Leno (LAI-no), town, BRESCIA province, LOMBARDY, N ITALY, near MELLA RIVER, 12 mi/19 km S of BRESCIA; 45°22′N 10°13′E. Manufacturing of electrical apparatus, fabricated metals, machinery, clothing. Has Benedictine convent founded 758.

Lenoir (luh-NOR), county (□ 402 sq mi/1,045.2 sq km; 2006 population 57,662), E central NORTH CAROLINA; ⊙ KINSTON; 35°14′N 77°38′W. Bounded NE by Contentnea Creek. On coastal plain; drained by NEUSE RIVER; source of Treat River in S. Service industries; manufacturing at Kinston; agriculture (tobacco, corn, cotton, wheat, soybeans, hay, sweet potatoes; poultry, hogs). Formed 1791 when Dobbs county was divided into Glasgow and Lenoir counties. Named for Wiliam Lenoir (1751–1839), a hero of the Battle of Kings Mountain.

Lenoir (luh-NOR), city (□ 16 sq mi/41.6 sq km; 2006 population 18,018), ⊙ CALDWELL county, W central NORTH CAROLINA; 35°54′N 81°31′W. Manufacturing (furniture, medical equipment, pharmaceuticals, machinery, apparel; wood, glass, synthetic, and plastic products; printing and publishing); service industries. Tuttle Educational State Forest to SW; Pisgah National Forest to W; Lake Rhodhiss reservoir (CATAWBA RIVER) to S. Incorporated 1851.

Lenoir City (luh-NOOR), city (2006 population 7,703), LOUDON county, E TENNESSEE, on TENNESSEE RIVER, 23 mi/37 km SW of KNOXVILLE; 35°48′N 84°16′W. In timber and agricultural area; manufacturing; lumbering. Site of former cotton mill. Fort Loudon Dam and Reservoir (water power) nearby. Founded 1890.

Lenora (luh-NOR-uh), village (2000 population 306), NORTON county, NW KANSAS, on North Fork of

SOLOMON RIVER, 16 mi/26 km SSW of NORTON; 39°36′N 100°00′W. Railroad terminus. Wheat, corn; cattle.

Lenora, CZECH REPUBLIC: see VIMPERK.

Lenore (luh-NOR), community, SW MANITOBA, W central CANADA, 37 mi/59 km WNW of BRANDON, and in WOODWORTH rural municipality; 49°57′N 100°45′W.

Lenore Lake (luh-NOR), central SASKATCHEWAN, CANADA, 17 mi/27 km NNE of HUMBOLDT; 17 mi/27 km long, 3 mi/5 km–7 mi/11 km wide; 52°30′N 105°00′W.

Lenox (LEN-uhks), town (2000 population 889), COOK county, S GEORGIA, 12 mi/19 km SSE of TIFTON; 31°16′N 83°28′W.

Lenox, town (2000 population 1,401), TAYLOR county, SW IOWA, 16 mi/26 km NNW of BEDFORD; 40°52′N 94°33′W. Shipping point in livestock, grain area. Incorporated 1875.

Lenox, resort town, including Lenox village, BERKSHIRE county, W MASSACHUSETTS, in the BERKSHIRES, 7 mi/11.3 km S of PITTSFIELD; 42°22′N 73°17′W. Paper mill. Scene of annual Tanglewood Music Festival (begun 1934) at "Tanglewood," a former estate, mainly in adjoining STOCKBRIDGE town. At Edith Wharton's estate, the Mount, there is an annual performance of Shakespeare's plays. Hawthorne's cottage, burned in 1890, was rebuilt and dedicated as a shrine in 1948. Lenox is noted for its many estates. Settled c.1750, set off from RICHMOND 1767. Includes state forest and villages of New Lenox and Lenox Dale.

Lens (LENZ), commune (2006 population 4,024), Mons district, HAINAUT province, S BELGIUM, on DENDER RIVER, 7 mi/11.3 km NNW of MONS; 50°33′N 03°54′E. Agriculture. Has eighteenth-century monastery buildings.

Lens, commune, VALAIS canton, S SWITZERLAND, on N slope of RHÔNE valley, between SIERRE and SION. Resort.

Lens (LAHNS), industrial town (□ 5 sq mi/13 sq km), sub-prefecture of PAS-DE-CALAIS department, NORD-PAS-DE-CALAIS region, N FRANCE, 17 mi/27 km SW of LILLE; 50°26′N 02°50′E. Center of an urban area with a population of more than 300,000 in the midst of what had been France's most productive coal fields in the 18th century. Most mines have now declined; an adjustment to metalworking and more diversified industry is under way, with coal pellets and special coke products being delivered to the steel industry, and coal-mine gas being used for regional energy supply, including thermal power generation. Metalworks now include cable and wire drawing, copper and brass foundries. Lens figured prominently in the Thirty Years War (1648), and again in World War I when bloody battles were fought at nearby HILL 70 (NW) and VIMY Ridge (SSW). Since 19th century, Lens has attracted a large Polish population, mostly for work in the mines.

Lensahn (LEN-sahn), village, SCHLESWIG-HOLSTEIN, N GERMANY, 25 mi/40 km NNE of LÜBECK; 54°13′N 10°53′E.

Lensk (LYENSK), city (2006 population 23,405), SW SAKHA REPUBLIC, W RUSSIAN FAR EAST, port on the LENA RIVER, connected by road N to MIRNYY, 520 mi/837 km W of YAKUTSK; 60°43′N 114°55′E. Elevation 633 ft/192 m. Diamond trade, lumbering and woodworking industries. Established as a settlement of Mukhtuya; city status and current name since 1963.

Lenskoye (LYEN-skuh-ye), village, E central SVERDLOVSK oblast, W SIBERIA, RUSSIA, on the TURA RIVER (tributary of the TOBOL River), on road, 20 mi/32 km WNW of TURINSK; 58°09′N 63°12′E. Elevation 219 ft/66 m. Agricultural products.

Lent, village, GELDERLAND province, E NETHERLANDS, on WAAL RIVER, 2 mi/3.2 km N of NIJMEGEN, opposite Nijmegen; 51°52′N 05°53′E. Dairying; cattle; grain, vegetables, fruit; manufacturing (kilns).

Lentekhi (len-TE-khee), urban settlement, and center of Lentekhi region, NW GEORGIA, in SVANETIA, 35 mi/56 km N of KUTAISI; 42°47′N 42°44′E. Footwear. Until 1938, called Leksura.

Lenti (LEN-ti), city, ZALA county, W HUNGARY, 20 mi/32 km SW of ZALAEGERSZEG; 46°37′N 16°33′E. Apples, pears, rape, alfalfa, silage corn; cattle; manufacturing (furniture, milling, handicrafts).

Lentilly (lahn-tee-YEE), commune (□ 7 sq mi/18.2 sq km; 2004 population 5,013), RHÔNE department, RHÔNE-ALPES region, W central FRANCE; 45°49′N 04°40′E.

Lentini (len-TEE-nee), ancient *Leontini*, town, SIRACUSA province, E SICILY, ITALY, near LAKE LENTINI, 16 mi/26 km S of CATANIA; 37°17′N 15°00′E. In citrus fruit-growing area. Produces citrus syrups, pasta, glass, lime, cement. Exports citrus fruit. One of oldest Greek settlements in Sicily; founded 729 B.C.E. on site of ancient Siculian fortress by Chalcidians from NAXOS. Overwhelmed (498 B.C.E.) by Hippocrates; further colonized (476 B.C.E.) by Hieron of Syracuse. Stormed by Romans (C.E. 214), Saracens (C.E. 848). Largely destroyed by earthquake of 1693. Gorgias was born here.

Lentini, Lake, largest permanent lake (□ 4 sq mi/10.4 sq km) in SICILY, ITALY, in SIRACUSA province, 2 mi/3 km NW of LENTINI; 37°19′N 14°57′E. Partly artificial; waterfowl, fish. Its vapors make Lentini unhealthy in the summer. Also called Biviere di Lentini.

Lentschütz, POLAND: see LECZYCA.

Lentvaris (LENT-vuh-ris), Polish *Landwarów*, city (2001 population 11,773), SE LITHUANIA, 9 mi/14 km WSW of VILNA; 54°39′N 25°03′E. Railroad junction; manufacturing (nails, screws, carpets); stone quarries. In Russian Vilna government until it passed to POLAND in 1921, then (1939) Lithuania.

Lenwood, unincorporated town (2000 population 3,222), SAN BERNARDINO county, S CALIFORNIA, 5 mi/8 km WSW of BARSTOW, in MOJAVE DESERT, and on MOJAVE RIVER; 34°54′N 117°06′W. Cattle.

Lenzburg, commune, AARGAU canton, N SWITZERLAND, on AA RIVER, 7 mi/11.3 km E of AARAU. Elevation 1,227 ft/374 m. Food processing; paper products; metalworking; printing.

Lenzburg, district, S central AARGAU canton, N SWITZERLAND; 47°23′N 08°11′E. Main town is LENZBURG; population is German-speaking and Protestant.

Lenzburg, village (2000 population 577), SAINT CLAIR county, SW ILLINOIS, 18 mi/29 km SSW of BELLEVILLE; 38°17′N 89°49′W. In bituminous-coal and agricultural area.

Lenzerheide (LEHN-tzuhr-high-duh), Romansh *Planüra*, health resort, GRISONS canton, E SWITZERLAND, 8 mi/12.9 km S of CHUR. Elevation 4,843 ft/1,476 m. Winter sports. Near Lenzerheide Pass (5,075 ft/1,547 m) leading N to Chur.

Lenzie (len-ZEE), town (2001 population 8,873), East Dunbartonshire, W SCOTLAND, 5 mi/8 km NE of GLASGOW; 55°55′N 04°09′W. Light industry. Formerly in Strathclyde, abolished 1996.

Lenzing (LEN-tsing), township, S central UPPER AUSTRIA, on AGER RIVER, 3 mi/4.8 km SW of VÖCKLABRUCK; 47°58′N 13°37′E. Pulp and rayon milling.

Lenzkirch (LENTS-kirkh), village, BADEN-WÜRTTEMBERG, GERMANY, in BLACK FOREST, on the WUTACH, 3 mi/4.8 km SSW of NEUSTADT; 47°52′N 08°12′E. Manufacturing of precision instruments and medical equipment. Health resort, winter-sports center (elevation 2,657 ft/810 m).

Leo, town, ALLEN county, NE INDIANA, 10 mi/16 km NE of FORT WAYNE. Corn, soybeans; cattle. Manufacturing (furniture). Laid out 1849.

Léo (LAI-o), town, ⊙ SISSILI province, CENTRE-OUEST region, S BURKINA FASO, 90 mi/145 km SSW of OUAGADOUGOU; 11°07′N 02°08′W. On the lower central plateau, near GHANA border. Agriculture (shea nuts, peanuts) and livestock (cattle, sheep, goats).

Leoben (lai-O-ben), town, central STYRIA province, S central AUSTRIA, on the MUR RIVER; 47°23′N 15°06′E. An industrial center in a former mining region, it has large ironworks, lumber mills, a mining school, and breweries; railroad junction. Founded as a town by King Ottokar II of Bohemia between 1261–1280, Leoben still has some of the medieval fortifications, several notable churches and burger houses. An armistice between FRANCE and Austria, preliminary to the Treaty of Campo Formio, was signed (1797) at Leoben to conclude Napoleon I's victorious Italian campaign.

Leobersdorf (lai-O-bers-dorf), township, E LOWER AUSTRIA, on TRIESTING RIVER, 6 mi/9.7 km S of BADEN; 47°56′N 16°13′E. Railroad junction; manufacturing of industrial machinery.

Leobschütz, POLAND: see GLUBCZYCE.

Léogâne (lai-o-GAHN), town, OUEST department, S HAITI, port on NE coast of JACMEL PENINSULA, 18 mi/29 km W of PORT-AU-PRINCE; 18°31′N 72°38′W. Tobacco, coffee, and sugar growing. Has grotto.

Leogang (LAI-o-gahng), village, SALZBURG, W central AUSTRIA, in the PINZGAU 7 mi/11.3 km NNW of ZELL AM SEE; 47°26′N 12°46′E. Former mines, mining museum; winter and summer tourism.

Léognan (lai-o-nyahn), town (□ 16 sq mi/41.6 sq km), GIRONDE department, AQUITAINE region, SW FRANCE, 8 mi/12.9 km S of BORDEAUX; 44°44′N 00°36′W. Production of dry white Bordeaux wines.

Leola, city (2006 population 402), ⊙ MCPHERSON county, N SOUTH DAKOTA, 28 mi/45 km NW of ABERDEEN; 45°43′N 98°56′W. Shipping point for grain and hogs; duck hunting in vicinity.

Leola (lee-O-lah), unincorporated town, LANCASTER county, SE PENNSYLVANIA, industrial suburb 6 mi/9.7 km NE of LANCASTER; 40°05′N 76°11′W. Light manufacturing.

Leola (lee-O-luh), village (2000 population 515), GRANT county, central ARKANSAS, 33 mi/53 km W of PINE BLUFF; 34°10′N 92°35′W. Manufacturing (food processing; lumber; building materials). Jenkins Ferry Battleground State Historical Monument to NE.

Leominster (LE-min-stuhr), city (2000 population 41,303), WORCESTER county, N central MASSACHUSETTS; 42°31′N 71°46′W. Manufacturing (plastics, metal fabrication, locks, processed foods, and chemicals). Birthplace of John Chapman (Johnny Appleseed). Leominster State Forest here. Set off from LANCASTER 1740, incorporated as a city 1915.

Leominster (LEE-O-min-stuhr), town (2001 population 11,114), N Herefordshire, W ENGLAND, on LUGG RIVER, 12 mi/19 km N of HEREFORD; 52°14′N 02°44′W. Agricultural market, with agricultural-machinery works. Has 11th-century minster and many old half-timbered buildings.

Cadadón León (kahn-yah-DON lai-ON), village, ⊙ Río Chico department (1991 population 2,696), central SANTA CRUZ province, ARGENTINA, on the RÍO CHICO, at S foot of the GRAN MESETA CENTRAL, 120 mi/193 km NW of SANTA CRUZ. Alfalfa-growing and sheep-raising center. Agricultural experiment station. Airport.

León, historic region and former kingdom, NW SPAIN, E of PORTUGAL and GALICIA, now part of CASTILE-LEÓN. It includes the provinces of LEÓN, SALAMANCA, and ZAMORA, named after their chief cities. It is sparsely populated, and the climate is harsh; winters are long and cold, and the summers are extremely hot and often accompanied by drought. N León, which is crossed by the CANTABRIAN MOUNTAINS, has coal mines, forests, and mountain pastures; the rest of the region is a dry plateau drained by the DUERO RIVER and its tributaries. León has long been noted for its linen manufacturing. Early in the Christian reconquest, the kings of ASTURIAS gained control over León (8th–9th century); their territory, of which the city of LEÓN was made the capital in the 10th century, became the kingdom of Asturias and León. The power

Cross-references are shown in SMALL CAPITALS. The pronunciation guide is shown on page xix. The sources of population figures are shown on page xvii.

of the kings also extended over Galicia and part of CASTILE, NAVARRE, and the Basque Provinces, but it was too weak to prevent the rise of the independent kingdoms of Navarre and Castile. León was conquered (1037) by Ferdinand I of Castile, on whose death (1065) the kingdoms again became separate. Reunited in 1072 under Alfonso VI, León and Castile were again separated in the 12th century and remained so until Ferdinand III accomplished the final reunion in 1230.

León (lai-ON), department (□ 2,355 sq mi/6,123 sq km), W NICARAGUA, on the PACIFIC; ☉ LEÓN; 12°35′N 86°35′W. Bounded E by LAKE MANAGUA; includes volcanic range of CORDILLERA DE LOS MARABIOS (TELICA, ROTA, CERRO NEGRO, LAS PILAS, and VOLCÁN DEL HOYO volcanoes), and MOMOTOMBO volcano. Agriculture (corn, sesame, beans, cotton, coffee, sugarcane, cacao; livestock; cheese production). Sugar mills near QUEZALGUAQUE. Gold and silver mining at EL LIMÓN, LA INDIA, and VALLE DE LAS ZAPATAS. Manufacturing concentrated in LEÓN. Main centers include León, LA PAZ CENTRO, NAGAROTE, EL SAUCE.

León, province (□ 5,432 sq mi/14,123.2 sq km; 2001 population 488,751), NW SPAIN, in CASTILE-LEÓN region; ☉ LEÓN. Bounded by crest of CANTABRIAN MOUNTAINS (N) and crossed by their offshoots; rest of province is part of central high plateau. Drained by ESLA RIVER and its tributaries and by SIL RIVER. Sufficient rainfall in N and W, where there are extensive forests and pastures (cattle and sheep raising; transhumance) and fertile valleys; barren, cold areas on plateau. Essentially agricultural: cereals, potatoes, wine, vegetables, sugar beets, flax, some fruit. Leading Spanish province for anthracite (about 70% of total) and second for bituminous-coal production from mines in Cantabrian Mountains (VILLABLINO, TREMOR, TORENO, POLA DE GORDÓN); also iron (Ponferrada), zinc (PICOS DE EUROPA), barite, and some tungsten mines. Clay and limestone quarries. Industries include metallurgy, chemicals, textiles, pharmaceuticals, and food processing; also makes coal briquettes, ceramics, candy, soap. Chief cities/towns include León, ASTORGA, PONFERRADA.

Leon (LEE-on), county (□ 701 sq mi/1,822.6 sq km; 2006 population 245,625), NW FLORIDA, on GEORGIA state line (N), bounded W by OCHLOCKONEE RIVER and LAKE TALQUIN; ☉ TALLAHASSEE; 30°27′N 84°16′W. Rolling terrain in N, coastal plain in S; includes many lakes and part of Apalachicola National Forest. Agriculture (corn, peanuts, cotton, vegetables; cattle, hogs, poultry; dairy products), and forestry (lumber). Industry and two universities (Florida A&M and Florida State) at Tallahassee. Named after Juan Ponce de León. Formed 1824.

Leon (LEE-ahn), county (□ 1,080 sq mi/2,808 sq km; 2006 population 16,538), E central TEXAS; ☉ CENTERVILLE; 31°18′N 96°00′W. Bounded W by NAVASOTA RIVER, E by TRINITY RIVER. Agriculture (cattle, hogs; grain, vegetables, watermelons, hay); timber; oil and natural-gas wells; lignite; iron ore. Hunting; fishing. Formed 1846.

León (lai-ON), city (2005 population 1,137,465) and township; ☉ León municipio, GUANAJUATO state, central MEXICO; 21°07′N 101°01′W. It is located in a fertile river valley 6,184 ft/1,885 m high, but with a mild, temperate climate. Site of a famous flood, which in 1888 almost washed the city away. León, on the main railroad line between EL PASO, TEXAS, and MEXICO CITY, and on Mexico Highways 37 and 45, is a commercial, agriculture, and mining center and one of Mexico's leading leather working and shoemaking cities. The local mines yield gold, copper, silver, lead, and tin. León was officially founded in 1577. Formerly called León de los Aldama.

León (lai-ON), city (2005 population 139,433) and township, W NICARAGUA; 12°26′N 86°53′W. It is the second-largest city of the republic. León is commercial

center between CORINTO and MANAGUA. It was founded in 1524 on LAKE MANAGUA by Francisco Fernández de Córdoba and moved W to its present site in 1610 after a severe earthquake. In colonial times León was the political hub of Nicaragua. Center of the intellectuals and artisans, León became the stronghold of the liberal forces after independence from Spain (1821) and engaged in bitter rivalry with conservative GRANADA. Costly revolutions, in one of which León accepted aid from U.S. filibuster William Walker, led to the founding of Managua (1855). The city is still the liberal center of the country; it was a heavily pro-Sandinista city during the revolution against the Somoza dictatorship, and in 1990 was one of the few major cities to vote for the Sandinista party. The poet Rubén Darío is buried in the cathedral here. Damaged by ash fall from eruption of CERRO NEGRO volcano in 1992.

León, city (2001 population 130,916), ☉ LEÓN province, NW SPAIN, in CASTILE-LEÓN, at the foot of the CANTABRIAN MOUNTAINS and at the confluence of the BERNESGA and Torio rivers; 42°35′N 05°34′W. It is an agricultural and commercial center. Dating from Roman times, it was reconquered from the Moors in 882 by Alfonso III of ASTURIAS. Early in the 10th century, León replaced OVIEDO as the capital of the kingdom of Asturias, which became the kingdom of LEÓN. The city flourished in the 12th and 13th centuries as a trade center but declined after the kings of León and Castile made VALLADOLID their favored residence. It still retains a medieval atmosphere, and its many historic monuments attract tourists. Most notable is the Spanish Gothic cathedral (13th–14th century).

Leon, city (2000 population 1,983), ☉ DECATUR county, S IOWA, c.60 mi/97 km S of DES MOINES; 40°44′N 93°45′W. Manufacturing (dairy products, concrete blocks, wood, building materials, apparel). Nine Eagles State Park to S. Settled 1840, incorporated 1867.

Leon (le-YON), town, ILOILO province, SE PANAY island, PHILIPPINES, 14 mi/23 km WNW of ILOILO; 10°50′N 122°20′E. Rice growing.

León (lai-ON), village, S JUJUY province, ARGENTINA, on railroad, on the RÍO GRANDE DE JUJUY, 13 mi/21 km NW of JUJUY; 24°02′S 65°25′W. In fruit-growing, livestock-raising, and mining area. Limestone, lead, silver deposits.

Leon (LEE-ahn), village (2000 population 645), BUTLER county, SE KANSAS, 10 mi/16 km SE of EL DORADO; 37°41′N 96°46′W. In cattle and grain area. Oil wells nearby.

Leon, village (2006 population 98), LOVE county, S OKLAHOMA, near RED RIVER, 26 mi/42 km SW of ARDMORE; 33°52′N 97°25′W. In diversified farm area.

Leon (LEE-ahn), village (2006 population 127), MASON county, W WEST VIRGINIA, on KANAWHA RIVER, 12 mi/19 km SE of POINT PLEASANT; 38°45′N 81°57′W.

Leon, hamlet, CATTARAUGUS county, W NEW YORK, 17 mi/27 km NE of JAMESTOWN; 42°17′N 79°00′W. In agricultural area.

Leona (lee-O-nuh), village (2000 population 88), DONIPHAN county, NE KANSAS, 12 mi/19 km W of TROY; 39°47′N 95°19′W. Agriculture (chiefly apples). Brown State Fishing Lake to NW.

Leonard, town (2000 population 66), SHELBY county, NE MISSOURI, 10 mi/16 km NW of SHELBYVILLE; 39°53′N 92°10′W. Corn, soybeans; cattle, hogs.

Leonard (LE-nuhrd), town (2006 population 2,098), FANNIN county, NE TEXAS, 9 mi/14.5 km S of BONHAM; 33°22′N 96°16′W. Market in cotton, sorghum, soybeans, cattle. Agribusiness; manufacturing (office supplies, crushed stone). Settled c.1880.

Leonard, village (2000 population 332), OAKLAND county, SE MICHIGAN, 18 mi/29 km. NE of PONTIAC; 42°52′N 83°08′W. Manufacturing (motor vehicle parts).

Leonard, village (2000 population 29), CLEARWATER county, NW MINNESOTA, 21 mi/34 km NW of BE-

MIDJI, between Four Legged Lake (S) and Stenlund Lake (N); 47°38′N 95°16′W. Dairying.

Leonardo (lee-uh-NAR-do), village (2000 population 2,823), MONMOUTH county, E NEW JERSEY, near SANDY HOOK BAY, 15 mi/24 km NE of FREEHOLD; 40°25′N 74°03′W. Largely residential.

Leonardo Bravo, MEXICO: see CHICHIHUALCO.

Leonardtown, town (2000 population 1,896), ☉ SAINT MARYS county, S MARYLAND, c.45 mi/72 km SSE of WASHINGTON, D.C., and on navigable estuary entering the POTOMAC; 38°18′N 76°38′W. Laid out as Shepherd's Old Fields in 1708, the name was changed in honor of Benedict Leonard Calvert, the fourth Lord Baltimore, in 1728. Has been the county capital since 1710. Raided by the British in the War of 1812, its business today revolves around tobacco, seafood, and summer visitors. The many historic buildings in the town include St. Francis Xavier Church (built 1767 on site of church erected c.1654), the oldest Catholic church in Maryland; and Tudor Hall, built in the early 18th century, once owned by ancestors of Francis Scott Key and now the county public library The cannon outside the Old Jail (c.1895), now St. Mary's County Historical Society, is reputed to have come from the Ark, a ship that transported the original settlers to Maryland in 1634.

Leonardville, village (2000 population 398), RILEY county, NE KANSAS, 20 mi/32 km NW of MANHATTAN; 39°21′N 96°51′W. Trading point in livestock and grain region.

Leonard Wood, Fort, MISSOURI: see FORT LEONARD WOOD.

Leonarisso (lee-yo-NAH-ris-so) [Turkish=*Ziyamet*], village, FAMAGUSTA district, NE CYPRUS, 27 mi/43 km NNE of FAMAGUSTA, on KARPAS Peninsula; 35°28′N 34°08′E. MEDITERRANEAN SEA 3 mi/4.8 km to NW and 3 mi/4.8 km to SE. Tobacco, grain, grapes, citrus, olives; livestock. Church of Panyia Kanakaria (orig. 5th century) and ruins of 12th-century castle to NE.

Leona River (LEE-o-nuh), c.75 mi/121 km long, S central TEXAS; rises N of UVALDE, flows generally SE through livestock-ranching area, to the FRIO RIVER 10 mi/16 km S of PEARSALL.

Leonberg (LE-awn-berg), town, BADEN-WÜRTTEMBERG, GERMANY, 7 mi/11.3 km WNW of STUTTGART; 48°48′N 09°00′E. Manufacturing of machinery, tools, precision instruments, chemicals, optical instruments, rubber and leather goods. Has 13th-century church, Renaissance castle. Schelling was born here.

León Cancha (lai-ON KAHN-chah), canton, MÉNDEZ province, TARIJA department, S central BOLIVIA, 6 mi/10 km N of SELLA; 64°44′S 21°11′W. Elevation 6,555 ft/1,998 m. Extensive gas wells in area. Clay, limestone, phosphates, and some gypsum deposits. Agriculture (potatoes, yucca, bananas, corn, wheat, barley, sweet potatoes); cattle.

Leoncio Prado (lai-on-SEE-o PRAH-do), province, HUÁNUCO region, central PERU; ☉ TINGO MARÍA; 09°10′S 76°00′W. Traversed by HUALLAGA RIVER, with borders on SAN MARTÍN, LORETO, and UCAYALI regions. PARQUE NACIONAL TINGO MARÍA on W fringe.

Leoncio Rodrigues (LAI-on-SEE-o rah-DREE-ges), township, W ACRE, BRAZIL, on BR 364 Highway; 07°55′S 71°28′W.

León Cortés, COSTA RICA: see SAN PABLO, SAN JOSÉ province.

Leonding (LE-awn-ding), town, N central UPPER AUSTRIA, at the urban fringe of LINZ, 3 mi/4.8 km SW of city center; 48°17′N 14°15′E. Manufacturing of batteries and textile machinery, electrotechnical products; market center.

Leone (lai-O-nai), village, W TUTUILA, American Samoa.

Leone, Monte (le-O-ne, MON-te), highest peak (11,683 ft/3,561 m) of LEPONTINE ALPS, on Italian-Swiss border, 8 mi/12.6 km SE of BRIG. Pierced by Simplon Tunnel. SIMPLON PASS is just W.

Leones (le-O-nes), town, E CÓRDOBA province, AR-
GENTINA, 12 mi/19 km W of MARCOS JUÁREZ; 32°39′S
62°18′W. Agricultural center (wheat, flax, corn, soybeans, alfalfa, potatoes, vegetables; cattle); dairying;
tanning.

Leones, Cerro (lai-O-nais, SER-ro) or **Cerro Alto de
los Leones** (SER-ro AHL-to dai lohs lai-O-nais),
Andean peak (19,455 ft/5,930 m), VALPARAISO region,
central CHILE, just S of 33°S.

Leones Islands (le-O-nes), small archipelago in the
ATLANTIC, belonging to ARGENTINA, 2 mi/3.2 km off
coast of CHUBUT province, 20 mi/32 km SSE of CA-
MARONES. Guano deposits.

Leonessa (le-O-NES-sah), village, RIETI province, LA-
TIUM, central ITALY, 12 mi/19 km NNE of RIETI;
42°34′N 12°58′E. Woodworking. Ski, summer resort.
Medieval churches, houses of nobility.

Leongatha (lee-uhn-GA-thuh), town, S VICTORIA, SE
AUSTRALIA, 70 mi/113 km SE of MELBOURNE, in S
GIPPSLAND region; 38°29′S 145°57′E. Dairy products;
in livestock, agricultural area (potatoes).

Leonia (lee-O-nee-uh), residential borough (2006
population 8,799), BERGEN county, NE NEW JERSEY, 3
mi/4.8 km ESE of HACKENSACK, near W approach to
GEORGE WASHINGTON BRIDGE; 40°51′N 73°59′W. Incorporated 1894.

Leonidas (lee-YAW-nuh-duhs), village (2000 population 60), ST. LOUIS county, NE MINNESOTA, in MESABI
IRON RANGE, 4 mi/6.4 km SSW of VIRGINIA, 1 mi/1.6
km W of EVELETH; 47°28′N 92°34′W. Iron mines
nearby.

Leonidion (le-o-NEE–[th]-ee-on), town, ARKADIA
prefecture, SE PELOPONNESE department, SE mainland GREECE, near Gulf of ARGOLIS, 24 mi/39 km ENE
of SPARTA; 37°10′N 22°52′E. Wine, almonds. Ancient
Doric dialect still spoken here. Formerly called Leo-
nidi. Also spelled Leonidhion.

Leonidovo (lye-uh-NEE-duh-vuh), town (2006 population 1,255), S central SAKHALIN oblast, RUSSIAN FAR
EAST, on coastal railroad and highway, 12 mi/19 km
WNW of PORONAYSK; 49°17′N 142°52′E. Elevation 200
ft/60 m. Railroad junction; coal mining. Under Japanese rule (1905–1945), was called Kami-shikuka.

Leonine City (LEE-o-nein), part of ROME, ITALY, W of
the TIBER. Site of the VATICAN.

León Island, Spanish *Isla de León* (EE-slah dhai lai-
ON) or **San Fernando Island** (sahn fer-NAHN-do),
offshore island on Atlantic coast of CÁDIZ province,
SW SPAIN, separated from mainland by narrow canals; 36°27′N 06°12′W. Irregularly shaped and c.10 mi/
16 km long, it encloses (S and W) the BAY of CÁDIZ.
Cities of CÁDIZ and SAN FERNANDO are on it. Several
saltworks. Island was named for Juan Ponce de León,
to whom Henry IV of Castile ceded it.

León Mountains (lai-ON), S offshoots of the CAN-
TABRIAN MOUNTAINS, LEÓN province, N SPAIN, covering W part of province, between the Órbigo (E) and
the Sil (W) valleys; 42°30′N 06°18′W. Highest peak:
Teleno (6,950 ft/2,118 m).

León Muerto, Sierra (le-ON MWER-to, see-YER-rah),
subandean mountain range in CATAMARCA and SALTA
provinces, ARGENTINA, E of ANTOFAGASTA; extends E-
W c.20 mi/32 km. Rises to c.17,500 ft/5,334 m.

Leonora (lee-uh-NO-ruh), town, S central WESTERN
AUSTRALIA, on spur of KALGOORLIE-LAVERTON rail-
road, 518 mi/833 km E of PERTH, 130 mi/209 km N of
Kalgoorlie; 28°53′S 121°19′E. Gold mining.

Leonora, village, ESSEQUIBO ISLANDS–WEST DEMERARA
district, N GUYANA; 06°51′N 58°17′W. Located near
ATLANTIC OCEAN coast, on railroad, 10 mi/16 km
WNW of GEORGETOWN.

Leonore, village (2000 population 110), LA SALLE
county, N ILLINOIS, 12 mi/19 km SW of OTTAWA;
41°11′N 88°58′W. In agricultural area.

Léon, Pays de (lai-on, pai-yee duh), region of the NW
extremity of BRITTANY in FINISTÈRE department, W
FRANCE. Fruit and vegetable crops.

León River (lai-ON), c.75 mi/121 km long, ANTIOQUIA
department, NW COLOMBIA; rises on W slopes of
Serranía de ABIBE near CHOCÓ department border;
flows NNW to the CARIBBEAN at Golfo de URABÁ;
07°15′N 76°03′W.

Leon River (LEE-ahn), c.145 mi/233 km long, central
TEXAS; rises in EASTLAND county, W of EASTLAND;
flows generally SE, through Lake Leon (Eastland
county) and PROCTOR LAKE (COMANCHE county), past
GATESVILLE. It continues through large BELTON LAKE
reservoir, on E side of FORT HOOD Military Reservation, past city of TEMPLE, joining LAMPASAS RIVER to
form LITTLE RIVER c.9 mi/14.5 km S of Temple.

Leontes, LEBANON: see LITANI RIVER.

Leontini (lee-uhn-TEI-nei), ancient city, E SICILY,
ITALY, c.20 mi/32 km S of CATANIA. It was (729 B.C.E.)
a colony of Chalcidians from the island of NAXOS and
passed (5th century B.C.E.) under the rule of SYR-
ACUSE. It was the legendary home of the Laestrygones,
a group of giants encountered by Odysseus. The
modern town occupying the site is LENTINI.

Leon Valley (LEE-ahn VA-lee), town (2006 population
9,795), BEXAR county, S central TEXAS, residential
suburb 8 mi/12.9 km WNW of SAN ANTONIO, near
Leon Creek; 29°30′N 98°36′W.

León Viejo (lai-ON vee-AI-ho), original site of LEÓN,
NICARAGUA, abandoned after earthquake in 1609.
Located at foot of MOMOTOMBO. Burial place of
Pedrárias Davila.

Leonville (lee-AHN-vil), town (2000 population
1,007), SAINT LANDRY parish, S central LOUISIANA, 16
mi/26 km NNE of LAFAYETTE, and on Bayou TECHE;
30°28′N 91°59′W. In agricultural area (rice, vegetables,
peaches, sweet potatoes, sugarcane, cotton; cattle;
dairying); timber.

Leopold (LEE-uh-pold), residential suburb 6 mi/10 km
ESE of GEELONG, VICTORIA, SE AUSTRALIA, S of PORT
PHILLIP BAY, N of Lake Connewarre and BASS STRAIT,
and on Bellarine Peninsula; 38°11′S 144°28′E. Formerly
called Kensington.

Leopold and Astrid Coast, ANTARCTICA, on INDIAN
OCEAN, extends between 81°30′E and 88°00′E. Dis-
covered 1934 by expedition under Norwegian explorer
Lars Christensen.

Leopold Canal, GERMANY: see DREISAM RIVER.

Leopoldina (lai-o-pol-zhee-nah), city (2007 population
49,969), SE MINAS GERAIS, BRAZIL, near RIO DE JA-
NEIRO border, 50 mi/80 km NE of JUIZ DE FORA;
21°30′S 42°41′W. Railroad terminus; agricultural trade
center (coffee, sugar, dairy products, tobacco) with
processing plants.

Leopoldina, Brazil: see ARUANÁ.

Leopoldina, Brazil: see COLONIA LEOPOLDINA.

Leopoldina, Brazil: see PARNAMIRIM.

Leopold Island, NUNAVUT territory, CANADA, in DAVIS
STRAIT, off Cape MERCY, SE BAFFIN ISLAND, near
entrance of CUMBERLAND SOUND; 64°59′N 63°18′W.
Island is 7 mi/11 km long. Steep cliffs rise to c.2,000 ft/
610 m.

Leopoldo de Bulhões (LE-o-pol-do dee bool-YOUNS),
town (2007 population 8,787), central GOIÁS, BRAZIL, 37
mi/60 km SE of ANÁPOLIS; 16°38′S 48°40′W.

Leopoldov, German *Leopoldstadt,* Hungarian *Lipotrár,*
town, W SLOVAKIA, on VAH RIVER, just NW of HLO-
HOVEC. In agricultural area (wheat, corn, sugar beets).
At a railroad junction; has distilling, food processing;
and a former fortress (built 1665–1669) that was
converted into a prison in 1854.

Leopoldsburg (lai-o-POLTS-burkh), French *Bourg-
Léopold* (BOOR–LAI-o-pold), commune (2006 population 14,409), Hasselt district, LIMBURG province,
NE BELGIUM, 14 mi/23 km NNW of HASSELT; 51°07′N
05°15′E. Glass and explosives manufacturing.

Leopoldsdorf im Marchfelde (LAI-O-polds-dorf in
MAHRKH-fel-de), township, E LOWER AUSTRIA, in
the MARCHFELD, 15 mi/24 km E of VIENNA; 48°13′N
16°41′E. Sugar refinery, sugar beets, vegetables.

Leopoldshöhe (LE-aw-puhlds-HOE-e), town, North
Rhine-Westphalia, NW GERMANY, in TEUTO-
BURG FOREST, 8 mi/12.9 km E of BIELEFELD; 51°58′N
08°45′E.

Leopoldstadt (LAI-o-polds-shtaht), district (□ 7 sq mi/
18.2 sq km) of VIENNA, AUSTRIA, on island formed by
the Danube and DANUBE CANAL, just E of the city
center. Has a large amusement park, the well-known
Prater. Designated the official ghetto of Vienna in the
17th centruy. Vienna fairs take place here.

Leopoldville, former province (□ 140,017 sq mi/
364,044.2 sq km), in W and SW CONGO; the capital
was KINSHASA. Other principal centers were BOMA,
MATADI, THYSVILLE, KIKWIT, INONGO. Bordered in
the S by ANGOLA (partly along estuary of CONGO
RIVER), W by the ATLANTIC OCEAN and CABINDA (an
exclave of ANGOLA) along SHILOANGO RIVER, N by
CONGO REPUBLIC (mostly along CONGO RIVER). Now
divided into three administrative regions: KINSHASA,
BAS-ZAÏRE, and BANDUNDU. Drained by CONGO,
KASAI, and KWANGO rivers. Has both equatorial rain
forest and parklike savanna, with baobabs. In SE are
grown palm products, fibers, sesame; in NE, hardwood, copal, rice, groundnuts. Along lower CONGO,
sugar, cattle, and food staples (manioc, yams, plantains) are raised. Most productive area was the
MAYUMBE, with coffee, cacao, and banana plantations,
elaeis-palm groves, gold mines, and wild rubber and
tropical hardwood forests. BOMA-TSHELA and MA-
TADI-KINSHASA railroads in W; river traffic in E.

Leopoldville, CONGO: see KINSHASA.

Leópolis (lai-O-po-lees), town (2007 population
4,230), N PARANÁ state, BRAZIL, 31 mi/50 km NW of
LONDRINA; 23°04′S 50°45′W. Coffee, rice, cotton,
corn; livestock.

Leopolis, UKRAINE: see L′VIV, city.

Leoti (lee-O-tee), town (2000 population 1,598), ☉ WI-
CHITA county, W KANSAS, c.45 mi/72 km NW of
GARDEN CITY, N of White Women Creek; 38°28′N
101°21′W. Elevation 3,297 ft/1,005 m. Shipping point
for grain and cattle area.

Leova (LEE-O-vah), city (2004 population 10,000), SW
MOLDOVA, on PRUT RIVER (head of navigation), 45
mi/72 km SW of CHISINAU (Kishinev), on Romanian
border; 46°28′N 28°15′E. Agriculture center; flour and
oilseed milling, lumbering. Formerly spelled Leovo.

Leoville (LEE-o-vil), village (2006 population 341), W
central SASKATCHEWAN, CANADA, 70 mi/113 km NE of
NORTH BATTLEFORD; 53°38′N 107°33′W. Mixed farming; dairying.

Lepaera (le-pah-RE-rah), town, LEMPIRA department,
W HONDURAS, 14 mi/23 km N of GRACIAS; 14°47′N
88°35′W. Tobacco, coffee, corn.

Lepakshi, INDIA: see HINDUPUR.

Lepanto (leh-PAHN-to), town, PUNTARENAS province,
W COSTA RICA, small port on GULF OF NICOYA, on
NICOYA PENINSULA, 12 mi/19 km W of PUNTARENAS
(across the gulf); 09°57′N 85°02′W.

Lepanto (luh-PAN-to), town (2000 population 2,133),
POINSETT county, NE ARKANSAS, 26 mi/42 km SE of
JONESBORO, on left-hand chute of LITTLE RIVER;
35°36′N 90°19′W. In agricultural area; manufacturing
(apparel). Incorporated 1909.

Lepanto, Greece: see NAVPAKTOS.

Lepanto, Gulf of, Greece: see CORINTH, GULF OF.

Lepanto, Strait of, Greece: see RION STRAIT.

Lepar (LAI-pahr), island (c.75 sq mi/194 sq km), IN-
DONESIA, in JAVA SEA, just off SE coast of BANGKA;
02°58′S 106°41′E. Island is 13 mi/21 km long, 10 mi/16
km wide. Principal town is Tanjunglabu.

Lepaterique (le-pah-te-REE-kai), town, FRANCISCO
MORAZÁN department, S central HONDURAS, in SIERRA
DE LEPATERIQUE, 16 mi/26 km WSW of TEGUCIGALPA;
14°04′N 87°28′W. Elevation 7,359 ft/2,243 m. Pottery
making, tanning; corn, wheat, beans; livestock.

Lepaterique, Sierra de (le-pah-te-REE-kai, see-E-
rah de), section of Continental Divide in S central

HONDURAS, SW of TEGUCIGALPA; extends from LE-PATERIQUE c.20 mi/32 km SE; forms divide between upper CHOLUTECA (N) and upper NACAOME (S) rivers; 14°02′N 87°21′W. Elevation 7,359 ft/2,243 m.

Lepe (LAI-pai), town, HUELVA province, SW SPAIN, near the ATLANTIC, 14 mi/23 km W of HUELVA; 37°15′N 07°12′W. Summer resort and agricultural center (olives, cereals, fruit; livestock). Processing of figs and chicory; flour milling; lumbering; fish salting.

Lepel' (LE-pel), city, W VITEBSK oblast, BELARUS, on ULLA RIVER, on small LEPEL Lake, 65 mi/105 km WSW of VITEBSK, 54°50′N 25°40′E. Railroad terminus; manufacturing (milk canning; baked goods; fish processing).

Lepenac River (LEP-e-nahts), KOSOVO province, SW SERBIA and N MACEDONIA; rises 7 mi/11.3 km E of PRIZREN, Serbia; flows E and SSE, past KACANIK, through Kacanik defile into Macedonia, to VARDAR River 3 mi/4.8 km WNW of SKOPJE, Macedonia. Also spelled Lepenats.

Lepers Island, VANUATU: see AOBA.

Lepi (LEP-ee), town, HUAMBO province, W central ANGOLA, on Benguela railroad, 27 mi/43 km WSW of HUAMBO.

Lépine, FRANCE: see ÉPINE, L'.

Leping (LUH-PING), town, ⊙ Leping county, NE JIANGXI province, CHINA, 30 mi/48 km ESE of BOY-ANG, on LE'AN River (left headstream of BO RIVER); 28°58′N 117°07′E. Rice, cotton, oilseeds, sugarcane; pharmaceuticals; coal mining.

L'Épiphanie (lai-pee-fah-NEE), village (□ 1 sq mi/2.6 sq km), LANAUDIÈRE region, S QUEBEC, E CANADA, 4 mi/6 km NW of L'ASSOMPTION; 45°51′N 73°29′W. Woodworking; dairying.

Lepoglava (LE-po-glah-vah), village, N CROATIA, on railroad, 16 mi/26 km WSW of Varaždin, at W foot of Ivančica Mountain, in HRVATSKO ZAGORJE. In former lignite-mining area.

Lepontine Alps or **Adula Alpen**, French *Alpes Lé-pontiennes*, Italian *Alpi Lepontine*, German *Walliser Alpen*, division of Central Alps, along Italian-Swiss border and N into TICINO and GRISONS cantons of SWITZERLAND; they extend from PENNINE ALPS at SIMPLON PASS (WSW) to RHAETIAN ALPS at Splügen Pass (ENE); 46°25′N 08°40′E. Bounded by N upper RHÔNE and VORDERRHEIN valleys, S by Italian lake district (LAGO MAGGIORE; LUGANO and COMO lakes). Highest peak, MONTE LEONE (11,683 ft/3,561 m) on Italian border, just E of Simplon Pass. Range includes ADULA and ST. GOTTHARD groups. The valleys of VAL LEVENTINA and VALLE MESOLCINA cut deeply into the S slope. Important passes on N edge are Simplon, St. Gotthard, LUKMANIER, and SAN BERNARDINO.

Leppävirta (LEP-pah-VIR-tah), village, KUOPION province, S central FINLAND, 25 mi/40 km S of KUOPIO; 62°29′N 27°47′E. In SAIMAA lake region. Agriculture market in grain-growing and lumbering region; lake port.

Lepreau, Point (luh-PRO), promontory on the Bay of FUNDY, S NEW BRUNSWICK, CANADA, 24 mi/39 km SW of SAINT JOHN; 45°03′N 66°28′W. Lighthouse; fishing ports; blueberries. New Brunswick's only nuclear power plant is here.

Lepreon (le-PRE-on), ancient city in what is now ILIA prefecture, WESTERN GREECE department, Greece, W PELOPONNESUS, 22 mi/35 km SE of Pírgos; 37°26′N 21°43′E. Remains include acropolis and small temple. Originally independent, the city was ruled by ELIS (c.450–400 B.C.E.), later allied with Arcadia (ARKA-DIA), and became a member of Achaean League. Just S is modern village of Lepreon, formerly called Strovitsi. Formerly spelled Lepreum.

Lepsinsk (lep-SEENSK), town, NE TALDYKORGAN region, KAZAKHSTAN, in the DZUNGARIAN ALATAU Mountains, on LEPSY RIVER, 115 mi/185 km NE of TALDYKORGAN; 45°31′N 80°35′E. In wheat area; distillery.

Lepsy (lep-SEE), town, N TALDYKORGAN region, KA-ZAKHSTAN, on TURK-SIB RAILROAD, on LEPSY RIVER, 85 mi/137 km NNE of TALDYKORGAN; 46°14′N 78°56′E. In desert area. Tertiary-level (raion) administrative center.

Lepsy River (lep-SEE), 210 mi/338 km long, TALDY-KORGAN region, KAZAKHSTAN; rises in the DZUN-GARIAN ALATAU Mountains; flows NW past LEPSINSK, to E end of LAKE BALKASH. Used for irrigation. Formerly spelled Lepsa.

Leptis (LEP-tis), ancient city of LIBYA, just E of Al KHUMS. Founded (c.600 B.C.E.) by Phoenicians from Sidon. Annexed (46 B.C.E.) to Roman province of AFRICA; flourished as important Roman port, especially under Septimius Severus (born here). Among the world's most impressive Roman African ruins (including walls, baths, arches, temples, and forums). Also known as Lepcis; sometimes called Leptis Magna to distinguish it from another Leptis, S of HA-DRUMETUM, in present TUNISIA.

Leptis Magna, LIBYA: see LEPTIS and KHUMS, AL.

Leque (LE-ke), canton, TAPACARÍ province, COCHA-BAMBA department, central BOLIVIA, 22 mi/35 km NW of TAPACARÍ, on the INDEPENDENCIA-ORURO highway, at the Maso Cruz Mountains; 17°30′S 66°49′W. Elevation 9,521 ft/2,902 m. Iron, clay, limestone, and gypsum deposits. Agriculture (potatoes, yucca, bananas, corn, rye, soy); cattle.

Lequeitio (le-KAI-tyo), town, VIZCAYA province, N SPAIN, fishing port on BAY OF BISCAY, 22 mi/35 km ENE of BILBAO; 43°22′N 02°30′W. Fish processing, boatbuilding, sawmilling. Cereals, fruit, cattle, lumber in area. Bathing resort.

Lequepalca (le-ke-PAHL-kah), canton, CERCADO province, ORURO department, W central BOLIVIA, 18 mi/30 km E of CARACOLLO, on the COCHABAMBA border, on a branch of the ORURO–LA PAZ highway; 17°38′S 66°55′W. Elevation 12,372 ft/3,771 m. Clay and dolomite deposits. Agriculture (potatoes, yucca, bananas, oats, barley); cattle.

Léraba, province (□ 1,209 sq mi/3,143.4 sq km; 2005 population 110,907), CASCADES region, SW BURKINA FASO; ⊙ Sindou; 10°40′N 05°12′W. Bordered on NNE by KÉNÉDOUGOU province, E by COMOÉ province, S by CÔTE D'IVOIRE, and WNW by MALI. Established in 1997 with fourteen other new provinces.

L'Érable (lai-RAH-bluh), county (□ 495 sq mi/1,287 sq km; 2006 population 23,917), CENTRE-DU-QUÉBEC region, S QUEBEC, E CANADA; ⊙ PLESSISVILLE; 46°15′N 71°45′W. Composed of eleven municipalities. Formed in 1982.

Le Raysville (LUH RAIS-vil), borough (2006 population 306), BRADFORD county, NE PENNSYLVANIA, 14 mi/23 km ENE of TOWANDA. Agriculture (dairying).

Lerberget (LER-bery-et), fishing village, SKÅNE county, SW SWEDEN, on KATTEGATT strait, 10 mi/16 km NNW of HELSINGBORG; 56°11′N 12°34′E. Seaside resort.

Lerdal (LER-DAHL), village, KOPPARBERG county, central SWEDEN, on E shore of LAKE SILJAN, at RÄTTVIK.

Lerdo, MEXICO: see CIUDAD LERDO.

Lerdo de Tejada (LER-do de te-HE-dah), city and township, VERACRUZ state, SE MEXICO, at edge of Papa-loapan River flood plain, and 24 mi/38 km NW of SAN ANDRÉS TUXTLA on Mexico Highway 180; 18°37′N 95°31′W. Agricultural center (sugarcane, fruit, cattle); sugar processing center.

Léré (lai-RAI), town, MAYO-KÉBBI OUEST administrative region, SW CHAD, on MAYO-KÉBBI RIVER near CAMEROON border, and 90 mi/145 km SW of BONGOR; 09°39′N 14°13′E. Cotton ginning; millet; livestock.

Léré (LAI-rai), town, SIXTH REGION/TIMBUKTU, MALI, near border with MAURITANIA, 144 mi/240 km SW of TIMBUKTU; 15°43′N 04°55′W.

Lere, NIGERIA: see LERI.

Lerez River (lai-RETH), PONTEVEDRA province, NW SPAIN; rises in CANTABRIAN MOUNTAINS 8 mi/12.9 km

WSW of LALÍN, flows 27 mi/43 km WSW to PONTE-VEDRA BAY at PONTEVEDRA.

Leri (LAI-ree), town, BAUCHI state, E central NIGERIA, 55 mi/89 km SW of BAUCHI. Tin-mining center. Sometimes spelled Lere.

Leribe (leh-REE-beh), district (□ 1,092 sq mi/2,839.2 sq km; 2001 population 362,339), NW LESOTHO, bounded NW by CALEDON RIVER (SOUTH AFRICAN border), on S in part by Phuthiatsana and Bokong rivers; ⊙ HLOTSE (LERIBE). Drained in SE by Malibamatso River (forms Katse Reservoir; KATSE DAM in SE corner); MALUTI MOUNTAINS in E. Sheep, goats, cattle, horses; corn, sorghum, vegetables, peaches. Clay mining.

Leribe, LESOTHO: see HLOTSE.

Lerici (LAI-ree-chee), town, LA SPEZIA province, LI-GURIA, N ITALY, port on GULF OF SPEZIA, and 5 mi/8 km SE of SPEZIA; 44°04′N 09°55′E. In agricultural region (olives, grapes, citrus fruit); sea mussels, oysters; fabricated metal products, machinery. Sea bathing, winter resort. Has 12th-century Pisan castle (now a marine observatory).

Lérida, Catalan *Lleida*, province (□ 4,659 sq mi/12,113.4 sq km; 2001 population 362,206), NE SPAIN, in CAT-ALONIA; ⊙ LÉRIDA. Bounded by FRANCE (border along crest of central PYRENEES) and ANDORRA (N). Occupied by S slopes and spurs of the Pyrenees (with several peaks of c.10,000 ft/3,048 m), by the valley de Arán, by the CERDAÑA (Cerdagne), and by the Urgel plain. Drained by the SEGRE and its tributaries (numerous hydroelectric plants); province is greatest producer of hydroelectric energy in Spain. Lignite and some zinc deposits. Predominantly agricultural (cereals, olive oil, wine, sugar beets, alfalfa; livestock); manufacturing (cement, hats, soap). Major towns include BALAGUER, BORJAS BLANCAS, CERVERA, and SEO DE URGEL.

Lérida (LAI-ree-dhah) or **Lleida**, city (2001 population 112,199), ⊙ LÉRIDA province, NE SPAIN, in CATALO-NIA, on the SEGRE RIVER, and c.80 mi/128 km E of ZARAGOZA; 41°37′N 00°37′E. Lérida is the center of a fertile farm area (fruit, vegetables; livestock) and has a limited variety of manufactures (dairy, electric energy production, food processing, lumbering, construction, textile and chemical manufacturing). The ancient Ilerda, it was taken (49 B.C.E.) by Julius Caesar, who defeated Pompey's generals there. Lleida fell to the Moors in C.E. 714 and was liberated (1149) by Raymond Berengar IV of Barcelona. The university founded there (c.1300) by James II of Aragón was discontinued in 1717. Traditionally a strategic, forti-fied city, Lleida was a key defense point for BARCE-LONA in the Spanish civil war; it fell (April 1938) after a nine-month battle. The old section of the city is dominated by the castle, whose ramparts enclose a Romanesque cathedral.

Lérida (LAI-ree-dah), town, ⊙ Lérida municipio, TO-LIMA department, W central COLOMBIA, in MAGDA-LENA Valley, 33 mi/53 km NE of IBAGUÉ; 04°51′N 74°54′W. Elevation 1,076 ft/327 m. Bananas, corn, coffee, sorghum, cotton; livestock.

Lerik (le-REEK), urban settlement and administrative center of Lerik region, SE AZERBAIJAN, on N slope of TALYSH MOUNTAINS, 23 mi/37 km W of LÄNKÄRAN; 38°46′N 48°24′E. Livestock, wheat, potatoes. Manu-facturing (bricks, tile, asphalt).

Lerín (lai-REEN), town, NAVARRE province, N SPAIN, on EGA RIVER, and 13 mi/21 km SSE of ESTELLA; 42°29′N 01°58′W. Legumes, vegetables, pimientos.

Lérins (lai-ran), group of four small islands, ALPES-MARITIMES department, PROVENCE-ALPES-CÔTE D'AZUR region, SE FRANCE, in the MEDITERRANEAN SEA, S of CANNES; 43°31′N 07°03′E. Sainte-Marguerite is the largest island (2 mi/3.2 km long; 0.5 mi/0.8 km wide). On Saint-Honorat is the oldest occupied for-tified monastery in W Europe, founded by St. Hon-oratus, rebuilt in 1073, and enlarged in 19th century. Excellent views of French RIVIERA shoreline and MARITIME ALPS in background.

Area is shown by the symbol □, and capital city or county seat by ⊙.

Lerma, town, BURGOS province, N SPAIN, on ARLANZA RIVER (affluent of the DUERO), on Burgos-Madrid highway, and 20 mi/32 km S of BURGOS; 42°02'N 03°45'W. In fertile region (grapes, cereals, chick peas; livestock); tanning, liquor distilling. Historic town has superb palace (built in early 17th century by Cardinal Duke Francisco Gómez de Sandoval), college, collegiate church, and convents.

Lerma (LER-man), river, c.350 mi/563 km long; rising in MEXICO state, central MEXICO; flowing NW and W through GUANAJUATO state to Lake CHAPALA, crossing the part of the central plateau known as the ANÁHUAC. The river draining the lake to N, and flowing NW through JALISCO state to the PACIFIC OCEAN is generally called the Río GRANDE DE SANTIAGO (c.200 mi/320 km long) but it is considered a continuation of the LERMA. The river system is extensively used for irrigation and hydroelectric power.

Lerma de Villada (ler-mah dai vee-YA-dah), city and township, MEXICO state, central MEXICO, on upper LERMA River, and 28 mi/45 km WSW of MEXICO CITY, 8 mi/12.9 km E of TOLUCA DE LERDO; ⊙ Lerma Municipio; 19°17'N 99°28'W. Elevation 8,458 ft/2,578 m. On railroad and on Mexico City-Toluca Expressway (Mexico Highway 15). Cereals, livestock.

Lerma Valley (LER-mah), subandean valley in central SALTA province, ARGENTINA, along TORO RIVER (irrigation), extending c.30 mi/48 km S from SALTA. Grain, fruit, and tobacco are grown. Main centers: Salta, ROSARIO DE LERMA, CERRILLOS, CHICOANA.

Lermontov (LYER-muhn-tuhf), city (2006 population 23,515), S STAVROPOL TERRITORY, N CAUCASUS, S European Russsia, on highway branch and near railroad spur (Skachki station), 120 mi/193 km SW of STAVROPOL, and 10 mi/16 km NW of PYATIGORSK; 44°06'N 42°58'E. Elevation 2,447 ft/745 m. Metallurgical combine. Founded in 1953 for mining uranium ore. Became a city in 1956.

Lermontovka (LYER-muhn-tuhf-kah), village (2005 population 4,435), SW KHABAROVSK TERRITORY, RUSSIAN FAR EAST, 5 mi/8 km E of the Russia-CHINA border, in the USSURI River basin, on road and the TRANS-SIBERIAN RAILROAD, 69 mi/111 km SSW of KHABAROVSK; 47°09'N 134°19'E. Elevation 209 ft/63 m. Logging and lumbering; gold mining in the vicinity.

Lermontov-Yurt, RUSSIA: see KHAMBI-IRZE.

Lerna, village (2000 population 322), COLES county, E central ILLINOIS, 8 mi/12.9 km SW of CHARLESTON; 39°25'N 88°17'W. In rich agricultural area.

Lerna (LER-nah), archaeological site, ARGOLIS prefecture, PELOPONNESE department, S mainland GREECE, 7 mi/11.3 km S of ARGOS, near shore of Gulf of ARGOLIS. Occupied 4th millenium c.1200 B.C.E. Noteworty for its Early Helladic (late 3rd millenium B.C.E.) remains, especially the House of Tiles. Later, the Lernean mysteries in honor of the goddess Demeter were celebrated here.

Le Roeulx, BELGIUM: see ROEULX, LE.

Léros (LE-ros), island (⊙ c.20 sq mi/52 sq km; 1981 population 8,127), DODECANESE prefecture, SOUTH AEGEAN department, SE GREECE, in the AEGEAN SEA, N of Kalymnos. Airport. Occupied by Italy between 1912 and 1945, which built a naval base on the island and used it as a submarine base during World War II.

Le Roy, city (2000 population 3,332), MCLEAN county, central ILLINOIS, 14 mi/23 km SE of BLOOMINGTON; 40°20'N 88°45'W. Trade and processing center in rich agricultural area; food processing. Near Moraine View State Park. Incorporated 1857.

Leroy (luh-ROI), town, S central SASKATCHEWAN, CANADA, 22 mi/35 km SE of HUMBOLDT; 52°00'N 104°45'W. Mixed farming, dairying.

Le Roy, town (2000 population 13), DECATUR county, S IOWA, 13 mi/21 km NE of LEON; 40°52'N 93°35'W. In livestock area.

Le Roy, town (2000 population 925), MOWER county, SE MINNESOTA, on UPPER IOWA RIVER, near IOWA state line, 26 mi/42 km ESE of AUSTIN; 43°30'N 92°30'W. Corn, oats, soybeans, alfalfa, peas; cattle, sheep, hogs, poultry; dairying; manufacturing (feeds, fertilizers, electrical equipment, furniture). Limestone deposits nearby. Lake Louise State Park to NE.

Le Roy, village, COFFEY county, E KANSAS, on NEOSHO River, 9 mi/14.5 km SE of BURLINGTON; 38°05'N 95°37'W. In grain area. Manufacturing (fabricated metal products, meat, eggs).

Le Roy, village (2000 population 267), OSCEOLA county, central MICHIGAN, 14 mi/23 km S of CADILLAC; 44°02'N 85°27'W. In stock-raising area. Manufacturing (hardwood lumber, machining); lake resorts.

Le Roy, village (□ 2 sq mi/5.2 sq km; 2006 population 4,254), GENESEE county, W NEW YORK, 10 mi/16 km E of BATAVIA; 42°58'N 77°59'W. Diversified manufacturing; in agricultural area. In 1897, Pearl Bixby Wait, a local carpenter, perfected the formula for Jello gelatin dessert, then sold it for $450. Until 1964, General Foods had a plant that produced Jello in the village. Jello Museum is housed in Le Roy Historical Society Bldg. Settled 1793, incorporated 1834.

Leroy (LEE-roi), village (2006 population 332), MCLENNAN county, E central TEXAS, 14 mi/23 km NNE of WACO; 31°43'N 97°01'W. Cattle; dairying; grain. Feeds.

Lerum (LER-OOM), residential town, ÄLVSBORG county, SW SWEDEN, on SÄVEÅN RIVER, 13 mi/21 km ENE of GÖTEBORG; 57°47'N 12°17'E. Includes Aspenäs and Hedefors (HE-de-FORSH), villages.

Lerwick (LUHR-wik), town (2001 population 6,830), ⊙ SHETLAND ISLANDS, extreme N Scotland; 60°09'N 01°09'W. Northernmost town in Great Britain. Located on the SE coast of Mainland island, it is the central market town of the Shetlands, dealing in produce, cattle, and sheep. Previously hosiery. Terminus of passenger boat services from the mainland of Scotland. It also has become an important service base for the NORTH SEA oil industry, with the development of associated engineering works. Lerwick, a fishing town, grew up around a Dutch trading post in the 17th century. Cromwell built a fort here, named Fort Charlotte under George III, that is now a coast guard station. The Norse festival of Up Helly Aa is held annually at the end of January.

Léry (lai-REE), city (□ 4 sq mi/10.4 sq km; 2006 population 2,370), MONTÉRÉGIE region, S QUEBEC, E CANADA, 14 mi/22 km from MONTREAL; 45°21'N 73°48'W. Part of the Metropolitan Community of Montreal (*Communauté Metropolitaine de Montréal*).

Lesaca (lai-SAH-kah), town, NAVARRE province, N SPAIN, in BIDASSOA valley, 7 mi/11.3 km SSE of IRÚN; 43°15'N 01°42'W. Corn, beans, apples. Iron mines nearby.

Lesachtal (LAI-sahkh-tahl), high alpine valley of GAIL RIVER between the EAST TYROL-CARINTHIAN border and Kötschach-Mauthen, Carinthia, S AUSTRIA; 46°40'–43'N, 12°43'–13°00'E. N are the LIENZ DOLOMITES of the GAILTAL ALPS, S are the CARNIC ALPS. Gail River is deeply insected; settlements and alpine agriculture c. 900 ft/274 m above valley bottom. LIESING and Sankt Lorenzen im Lesachtal are the main settlements; Maria Luggau is a prominent pilgrim place that also attracts pilgrims from adjacent ITALY and SLOVENIA. Mountaineering and summer tourism; beautiful folk costumes.

Les Bergeronnes (lai ber-zhe-RUHN), village (2006 population 690), CÔTE-NORD region, E QUEBEC, E CANADA; 48°15'N 69°33'W. Formed in 1999 from BERGERONNES, GRANDES-BERGERONNES.

Les Bons Villers, BELGIUM: see BONS VILLERS, LES.

Lesbos (LES-vos), prefecture, NORTH AEGEAN department, Greece, in E AEGEAN SEA, near NW TURKEY, on LESBOS island; ⊙ Mitilíni, 39°10'N 26°20'E. Almost coterminous with island of Lesbos.

Lesbos (LES-vos), island (□ 630 sq mi/1,638 sq km), LESBOS prefecture, NORTH AEGEAN department, E GREECE, in the E AEGEAN SEA near TURKEY; 39°10'N 26°32'E. Sometimes included with SOUTHERN SPORADES group. A fertile island, it has vast olive groves and also produces wheat, wine, and citrus fruit. Fishing, tanning, and livestock raising are significant industries. Manganese. Frequent earthquakes. Mitilíni is the island's chief town. Airport. Center of Bronze Age civilization and later (c.1000 B.C.E.) settled by Aeolians. Brilliant cultural center from the 7th to the 6th century B.C.E., when the poets Alcaeus and Sappho and the statesman Pittacus were active here. Aristotle and Epicurus lived here, and Theophrastus was born here. Joined the Delian League and revolted unsuccessfully against ATHENS in 428–427 B.C.E. Later, passed to MACEDONIA, ROME, and the BYZANTINE EMPIRE. Taken by the Ottoman Turks in 1462; became part of Greece in 1913. Occupied by Germans from April 1941 to October 1944. Sometimes known as Mytilene, which is a variation of Mitilíni; also Lésvos.

Lescar (les-kahr), residential suburb (□ 10 sq mi/26 sq km; 2004 population 9,439), of PAU, PYRÉNÉES-ATLANTIQUES department, AQUITAINE region, SW FRANCE, 4 mi/6.4 km NW of city center; 43°20'N 00°25'E. Produces BASQUE berets, furniture, woolens. Has large 12th-century Romanesque church (cathedral until 1790). Founded by Romans as Beneharnum, it became the chief town of BÉARN district until its destruction by Normans in 9th century. Rebuilt in 12th century.

Les Cayes (lai-KAI), **Cayes** or **Aux Cayes**, town (2003 population 48,095), SW HAITI, on the CARIBBEAN SEA; ⊙ SUD department; 18°12'N 73°45'W. Haiti's chief S port and fishing port, it handles exports, mainly molasses and coffee. Tobacco; cotton; cattle; sugarcane growing and processing; cigar manufacturing.

Lesce, village, NW SLOVENIA, 9 mi/14.5 km SE of JESENICE and 3 mi/4.8 km SE of BLED, on Ljubljana-Jesenice, railroad; 46°21'N 14°09'E. Elev. 1,654 ft/504 m. Manufacturing (machinery, food); tourism. Training center for pilots here also hosts international air shows. Has one of the oldest churches in Slovenia.

Les Cèdres (lai SE-druh), village (□ 30 sq mi/78 sq km; 2006 population 5,687), MONTÉRÉGIE region, S QUEBEC, E CANADA, 25 mi/39 km W of MONTREAL; 45°18'N 74°03'W. Part of the Metropolitan Community of Montreal (*Communauté Metropolitaine de Montréal*).

Les Cheneaux Islands (LE shuh-NO), MICHIGAN, group of about 35 small wooded islands in LAKE HURON, just S of SE UPPER PENINSULA, and NE of the STRAITS of MACKINAC; 45°58'N 84°20'W. Locally called "The Snows." Resort area; annual regatta. Largest of group is MARQUETTE ISLAND, c.5 mi/8 km long.

Leschkirch, ROMANIA: see NOCHRICH.

Leschnitz, POLAND: see LESNICA.

Les Coteaux (lai ko-TO), village (□ 5 sq mi/13 sq km; 2006 population 3,599), MONTÉRÉGIE region, S QUEBEC, E CANADA; 45°17'N 74°14'W. Formed in 1994 from COTEAU-LANDING, COTEAU-STATION villages.

Les Coteaux (lai ko-TO), village, W TOBAGO, TRINIDAD AND TOBAGO, 3 mi/5 km N of SCARBOROUGH at crossroads in Courland Valley. Agriculture includes cacao- and coconut-growing.

Lescure-d'Albigeois (les-kur–dahl-bee-zhwah), commune (□ 5 sq mi/13 sq km) suburb of ALBI, TARN department, MIDI-PYRÉNÉES region, S FRANCE, on TARN RIVER, and 2 mi/3.2 km NNE of Albi; 43°57'N 02°10'E.

Les Diablerets, SWITZERLAND: see DIABLERETS, LES.

Leše, hamlet, N SLOVENIA, just SW of PREVALJE, near Meža River; 46°31'N 14°54'E. Once had the largest lignite mines in the country; closed since 1939.

Les Échelles-entre-deux-Guiers, FRANCE: see ÉCHELLES, LES.

Leseru (lai-SAI-rooh), town, RIFT VALLEY province, W KENYA, railroad junction for KITALE, 10 mi/16 km NW

Cross-references are shown in SMALL CAPITALS. The pronunciation guide is shown on page xix. The sources of population figures are shown on page xvii.

of ELDORET; elevation 6,489 ft/1,978 m; 00°35′N 35°12′E. Agriculture (coffee, wheat, corn, tea, wattle); dairy farming.

Leshan (LUH-SHAN), city (□ 971 sq mi/2,515 sq km; 1994 estimated non-agriculture population 384,300; estimated total population 1,104,700), central SI-CHUAN province, CHINA, just S of CHENGDU, on the MIN River; 29°34′N 103°42′E. Heavy industry is the largest source of income for the city; agriculture and light industry are also important. Crop growing, livestock raising. Grain, oil crops, cotton, vegetables, fruits, hogs, eggs, poultry; manufacturing (food, textiles, paper, chemicals, pharmaceuticals, iron and steel, non-ferrous metals, machinery, electrical equipment). Nearby are decorated grottoes, a huge stone Buddha, and the sacred peak Emei. Formerly called JIADING or Kiating. Sometimes spelled Loshan.

Leshara (le-SHAHR-uh), village (2006 population 107), SAUNDERS county, E NEBRASKA, 7 mi/11.3 km SE of FREMONT, and on PLATTE RIVER; 41°19′N 96°25′W.

Les Hauteurs (laiz o-TUHR), village (□ 41 sq mi/106.6 sq km; 2006 population 566), BAS-SAINT-LAURENT region, S QUEBEC, E CANADA, 5 mi/7 km from SAINT-CHARLES-GARNIER; 48°23′N 68°07′W.

Leshchëv (lee-SHCHOF), settlement, SE VOLGOGRAD oblast, SE European Russia, on the VOLGA RIVER, 14 mi/23 km SW of LENINSK, to which it is administratively subordinate; 48°29′N 44°47′E. Below sea level. Logging, lumbering, woodworking.

Leshukonskoye (lye-shoo-KON-skuh-ye), village (2005 population 4,800), S central ARCHANGEL oblast, N European Russia, on the MEZEN′ River (head of navigation), near highway, 150 mi/241 km E of ARCHANGEL; 64°54′N 45°46′E. Elevation 104 ft/31 m. Lumbering. Until 1929, called Ust′-Vashka.

Lesichevo (le-see-CHE-vo), village, PLOVDIV oblast, Lesichevo obshtina (1993 population 7,117), W central BULGARIA, on the TOPOLNITSA RIVER, 14 mi/23 km NW of PAZARDZHIK; 42°22′N 24°07′E. Vineyards, rice.

Lésigny (lai-see-nyee), town (□ 3 sq mi/7.8 sq km), outer residential suburb of PARIS, SEINE-ET-MARNE department, ÎLE-DE-FRANCE region, N central FRANCE, 15 mi/24 km SE of Notre Dame Cathedral. It lies on outer circumferential expressway of the Paris metropolitan area; 48°44′N 02°37′E. The forest of Notre Dame is just W.

Les Îles-de-la-Madeleine (laiz EEL–duh–lah–mahd-LEN), municipality (□ 64 sq mi/166.4 sq km; 2006 population 12,573), GASPÉSIE—ÎLES-DE-LA-MADE-LEINE region, QUEBEC, E CANADA; 47°24′N 61°47′W. Formed in 2001 from seven other municipalities.

Lesima, Monte (MON-te le-ZEE-mah), peak (5,656 ft/1,724 m) in LIGURIAN APENNINES, N central ITALY, 9 mi/14 km SSE of Varzi; 44°41′N 09°15′E.

Lesina, CROATIA: see HVAR.

Lesina, Lago di (LAH-go dee LAI-zee-nah), shallow lagoon (□ 20 sq mi/52 sq km) on N coast of GARGANO promontory, S ITALY. Separated from the ADRIATIC by a sand bar (two canals in E); 14 mi/23 km long, 2 mi/3 km wide. Fishing; bird hunting. On SW shore is town of LESINA (1991 population 6,415).

Lesja (LAI-shah), village, OPPLAND county, S central NORWAY, at head of the GUDBRANDSDAL, on LÅGEN River, on railroad, and 80 mi/129 km SSE of KRIS-TIANSUND. Livestock; some manufacturing; tourism. Church built 1748. Lesjaverk village, 10 mi/16 km WNW, was site (1650–1812) of iron foundry.

Lesjöfors (LE-SHUH-FORSH), village, VÄRMLAND county, W SWEDEN, in BERGSLAGEN region, 18 mi/29 km N of FILIPSTAD; 59°59′N 14°10′E. Former iron-mining region.

Lesko (les-ko), German *Lisko*, town, Krosno province, SE POLAND, on SAN RIVER, 8 mi/12.9 km SE of SANOK; 49°27′N 22°19′E.

Leskolovo (lyes-KO-luh-vuh), settlement (2005 population 4,035), N central LENINGRAD oblast, NW European Russia, on the KARELIAN ISTHMUS, on

railroad, 23 mi/37 km N of SAINT PETERSBURG; 60°16′N 30°27′E. Elevation 246 ft/74 m. Poultry processing.

Leskovac (LESK-o-vahts), city (2002 population 156,252), SE SERBIA, on railroad, and 23 mi/37 km S of NIS, near the SOUTHERN MORAVA RIVER; 43°00′N 21°57′E. Textile manufacturing. Quince and hemp growing nearby; vineyards. Dates from early 19th century; under Turkish rule until 1878. Also spelled Leskovats.

Leskovec, Bulgaria: see LYASKOVETS.

Leskovik (les-ko-VEEK), village, S ALBANIA, 34 mi/55 km SSW of KORÇE, near Greek border; 40°09′N 20°35′E. Sulphur springs nearby. Also spelled Leskoviku.

Leslie, county (□ 404 sq mi/1,050.4 sq km; 2006 population 11,973), SE KENTUCKY; in CUMBERLAND MOUNTAINS, in Daniel Boone National Forest; ⊙ HYDEN; 37°05′N 83°22′W. Drained by Middle Fork KENTUCKY RIVER (forms Buckhorn Lake reservoir, on N boundary) and Cutshin Creek. Mountain agricultural area, (livestock; burley tobacco); timber; bituminous-coal mines. Formed 1878.

Leslie (LEZ-lee), town (2001 population 2,998), FIFE, E Scotland, 7 mi/11.3 km NNW of KIRKCALDY; 56°13′N 03°13′W. Paper milling, flax spinning.

Leslie, town (2000 population 2,044), INGHAM county, S central MICHIGAN, 14 mi/23 km N of JACKSON; 42°27′N 84°25′W. In agricultural area (livestock; dairy; grain, apples, corn, soybeans); light manufacturing. Settled 1836, incorporated 1869.

Leslie, town (2000 population 87), FRANKLIN county, E central MISSOURI, in OZARK region, 15 mi/24 km SW of WASHINGTON; 38°25′N 91°13′W. Hay, cattle.

Leslie (LES-lee), village (2000 population 482), SEARCY county, N ARKANSAS, c.50 mi/80 km N of CONWAY, and near Middle Fork of LITTLE RED RIVER; 35°49′N 92°33′W.

Leslie, village (2000 population 455), SUMTER county, SW central GEORGIA, 12 mi/19 km SE of AMERICUS; 31°57′N 84°05′W.

Leslieville (LEZ-lee-vil), hamlet, S central ALBERTA, W CANADA, 13 mi/22 km E of ROCKY MOUNTAIN HOUSE, in CLEARWATER COUNTY; 52°23′N 114°36′W.

Lesmahagow (LES-muh-HA-go), town (2001 population 3,685), South Lanarkshire, S Scotland, on Nethan River, and 5 mi/8 km SW of LANARK; 55°37′N 03°54′W. Apparel. Monks of Lesmahagow introduced Clydeside fruitgrowing industry. Sometimes called Abbey Green, after abbey founded here c.1140 (no remains). Formerly in Strathclyde, abolished 1996.

Les Mamelles, district, E MAHÉ ISLAND, SEYCHELLES; 04°37′S 55°30′E. Borders ROCHE CAIMAN (N and E), CASCADE and GRAND′ ANSE (Mahé) (S), and PLAI-SANCE (W and N) districts. One of three districts formed c.1997.

Les Méchins (lai mai-SHAN), village (□ 175 sq mi/455 sq km; 2006 population 1,177), BAS-SAINT-LAURENT region, S QUEBEC, E CANADA, on the SAINTE LAWRENCE RIVER, and midway between SAINTE-ANNE-DES-MONTS and MATANE; 49°00′N 66°59′W.

Lesna (LE-shnah), Polish *Leśna*, German *Marklissa* (MAHR-kleesah), town in LOWER SILESIA, Jelenia Góra province, SW POLAND, in UPPER LUSATIA, on KWISA RIVER (irrigation dam 2 mi/3.2 km E), and 15 mi/24 km SE of GÖRLITZ. Cotton milling.

Lesna River (les-NAH), Polish *Leśna*, c.90 mi/145 km long, N BREST oblast, in BELARUS; rises in two branches in BIALOWIEZA. Forest joining NW of KAMENETS, flows generally SW past KAMENETS, to BUG RIVER 9 mi/14 km NW of BREST. Logging.

Lesnaya Volchanka, RUSSIA: see VOLCHANSK.

Lesneven (le-nuh-ven), town (□ 4 sq mi/10.4 sq km), FINISTÈRE department, W FRANCE, in BRITTANY, 15 mi/24 km NNE of BREST; 48°34′N 04°19′W. Market center; woodworking. Founded in 5th century, town preserves old stone houses. Its museum relates the history of the LÉON district.

Lesnica (le-SHNEE-tsah), Polish *Leśnica*, German *Bergstadt*, town in UPPER SILESIA, after 1945 in OPOLE province, S POLAND, on BYSTRZYCA RIVER, and 20 mi/32 km SSE of OPOLE (Oppeln). Leather manufacturing. Until 1937, called Leschnitz.

Lešnica (LESH-nee-tsah), village, W SERBIA, on railroad, and 21 mi/34 km SW of SABAC; 44°39′N 19°18′E. Also spelled Leshnitsa.

Lesnikovo (lyes-nee-KO-vuh), village (2005 population 5,820), S central KURGAN oblast, SW SIBERIA, RUSSIA, on the E bank of the TOBOL River, on road, 13 mi/21 km S of KURGAN; 55°17′N 65°19′E. Elevation 344 ft/104 m. In agricultural area; livestock breeding. Site of Kurgan state agricultural academy.

Lesnoy (lees-NO-yee), city (2006 population 52,045), W SVERDLOVSK oblast, extreme W Siberian Russia, on the TURA RIVER, on road junction and railroad, 18 mi/29 km N of VERKHNYAYA TURA, and 9 mi/14 km S of Is; 58°39′N 59°48′E. Elevation 633 ft/192 m. Storage of fissile materials; electrical equipment for chemical industry; research and development enterprises; bakery. A closed city, formerly (until 2001) home to a nuclear munitions plant and known under its secret designation of Sverdlovsk-45.

Lesnoy (lyes-NO-yee), town (2005 population 6,845), NE KIROV oblast, E European Russia, on local railroad, 10 mi/16 km NW of RUDNICHNYY; 59°47′N 52°10′E. Elevation 623 ft/189 m. Administrative center for regional disciplinary and correctional facilities.

Lesnoy (lyes-NO-yee), town (2006 population 8,665), central MOSCOW oblast, central European Russia, on road and near railroad, 3 mi/5 km NE of PUSHKINO, to which it is administratively subordinate; 56°04′N 37°55′E. Elevation 629 ft/191 m. Mechanized lifting equipment; meat processing.

Lesnoy (lyes-NO-yee), village, central PENZA oblast, S central European Russia, on road and railroad, 17 mi/27 km SW of NIZHNIY LOMOV; 53°20′N 43°20′E. Elevation 797 ft/242 m. Forestry.

Lesnoy, RUSSIA: see UMBA.

Lesnoye (lyes-NO-ye), village (2006 population 1,865), N TVER oblast, W European Russia, 22 mi/35 km SSW of PESTOVO; 58°17′N 35°31′E. Elevation 501 ft/152 m. Flax.

Lesnoye, RUSSIA: see ROSHNI-CHU.

Lesnoye Konobeyevo (lyes-NO-ye kuh-nuh-BYE-ee-vuh), village, E RYAZAN oblast, central European Russia, on the TSNA RIVER (tributary of the MOKSHA River), on highway branch and near railroad, 17 mi/27 km NE of SHATSK; 54°02′N 41°55′E. Elevation 344 ft/104 m. In agricultural area (wheat, rye, oats, sunflowers). Also called Konobeyevo-Lesnoye.

Lesnoy Gorodok (lyes-NO-yee guh-ruh-DOK), settlement (2006 population 5,045), central MOSCOW oblast, central European Russia, on road and railroad, 4 mi/6 km WSW, and under administrative jurisdiction, of ODINTSOVO; 55°38′N 37°12′E. Elevation 561 ft/170 m. Seed storage.

Lesnoy Karamysh, RUSSIA: see KAMENSKIY.

Lesnyaki, UKRAINE: see YAHOTYN.

Lesnyye Polyany (lyes-NI-ye puh-LYAH-ni), village (2006 population 3,685), central MOSCOW oblast, central European Russia, on road and railroad, 3 mi/5 km S of (and administratively subordinate to) PUSHKINO; 55°58′N 37°52′E. Elevation 524 ft/159 m. In agricultural area; pedigree livestock breeding, bakery.

Lesnyye Polyany (lees-NI-ye puh-LYAH-ni), village (2005 population 1,715), NE KIROV oblast, E European Russia, in the VYATKA River basin, on the N-bound spur of the TRANS-SIBERIAN RAILROAD, 22 mi/35 km N of (and administratively subordinate to) Omutninsk, and 2 mi/3.2 km SE of PESKOVKA; 59°01′N 52°25′E. Elevation 682 ft/207 m. Flax processing; textiles.

Lesobeng (leh-SOO-ben), village, MAFETENG district, LESOTHO, 65 mi/105 km SE of MASERU, near Lesobeng River; 29°45′S 28°24′E. CENTRAL RANGE to E. Cattle, sheep, goats, horses; peas, beans.

Lesobirzha, RUSSIA: see ZELENOBORSKIY.

Lesogorsk (lye-suh-GORSK), city, W central SAKHA-LIN oblast, RUSSIAN FAR EAST, on the TATAR STRAIT, on coastal highway and railroad, 257 mi/414 km NNW of YUZHNO-SAKHALINSK, 100 mi/161 km S of ALEK-SANDROVSK-SAKHALINSKIY, and 3 mi/5 km N of TEL'NOVSKIY; 49°27′N 142°08′E. Sawmill. Under Japanese rule (1905–1945), called Nayoshi or Kitanayoshi. Founded in 1860.

Lesogorsk (lye-suh-GORSK), town (2005 population 5,760), W IRKUTSK oblast, E central SIBERIA, RUSSIA, on road, 78 mi/126 km W of BRATSK; 56°02′N 99°31′E. Elevation 892 ft/271 m. Railroad junction on the BAYKAL-AMUR MAINLINE; railroad depots. Logging, lumbering, woodworking.

Lesogorsk (lye-suh-GORSK), urban settlement (2006 population 1,055), S NIZHEGOROD oblast, central European Russia, on local road and railroad, 18 mi/29 km S of ARZAMAS, and 8 mi/13 km SW of SHATKI, to which it is administratively subordinate; 55°06′N 43°56′E. Elevation 672 ft/204 m. Manufacturing (electrical machinery).

Lesogorskiy (lye-suh-GORS-keeyee), town (2005 population 2,985), NW LENINGRAD oblast, NW European Russia, on the VUOKSI River, approximately 12 mi/19 km SE of the Finnish border, on road, 23 mi/37 km NNE of VYBORG; 61°02′N 28°56′E. Sawmilling, wood pulp, artificial silk. Its Finnish name was Jääski (Russian *Yaski*); town passed from FINLAND to USSR in 1940 and in 1948 was renamed.

Lesopil'noye (lye-suh-PEEL-nuh-ye), town, S KHA-BAROVSK TERRITORY, RUSSIAN FAR EAST, on the TRANS-SIBERIAN RAILROAD (Zvenyevoy station), on the Bikin River (tributary of the USSURI River), less than 14 mi/23 km E of the Chinese border, 6 mi/10 km SE of BIKIN; 46°44′N 134°20′E. Elevation 187 ft/56 m. Sawmilling.

Lesosibirsk (lye-suh-see-BEERSK), city (2005 population 66,105), S central KRASNOYARSK TERRITORY, SE SIBERIA, RUSSIA, terminus of a branch of the TRANS-SIBERIAN RAILROAD, 285 mi/459 km N of KRAS-NOYARSK, and 170 mi/274 km N of ACHINSK; 58°14′N 92°29′E. Elevation 324 ft/98 m. Landing on the YENI-SEY RIVER, near its confluence with the ANGARA RIVER; lumber and wood chemical industries. Formed by merging the settlements of Maklakovo (dating from 1640) and Novomaklakovo. The project for the Middle Yenisey hydroelectric power station lies 16 mi/25 km to the S. Made city in 1975.

Lesostepnaya fiziko-geograficheskaya zona Ukrainy, UKRAINE: see FOREST-STEPPE ZONE, UKRAINIAN.

Lesotho (leh-SOO-tho), kingdom (□ 11,720 sq mi/30,355 sq km; 2004 estimated population 1,865,040; 2007 estimated population 2,125,262), S AFRICA, bounded on all sides by the REPUBLIC OF SOUTH AFRICA; ⊙ MASERU.

Geography

The DRAKENSBERG RANGE, the CENTRAL RANGE, and THABA PUTSUA RANGE occupies the central and E parts of the country, with the MALOTI Mountains in the W; elevation varies from lowest level near MO-HALES HOEK c.4265 ft/1301 m, in SW, to more than 11,000 ft/3,353 m along the E frontier. The country is generally mountainous with high rainfall, having the highest peak in S Africa, THABANA-NTLENYANA (11,425 ft/3,482 km) and is commonly called "The Kingdom in the Sky," "The Roof of Africa," and "The Switzerland of Africa."

Population

About 80% of the Basotho are Christian, and the country has one of the highest literacy rates in Africa. At ROMA, near MASERU, is the University of Lesotho. English and Sesotho (a Bantu language) are the official languages of the kingdom.

Economy

The kingdom is heavily populated, especially in W, where most of the 11% of arable land is located, the remainder being mainly pasture land. Corn and sorghum are the main crops; also wheat, beans, peas, asparagus, tomatoes, peaches. Sheep are bred for wool, Angora goats for mohair; cattle, hogs, and chickens are also raised. Breeding programs started in the 1970s have brought the sturdy, sure-footed Basotho pony into popularity for herding livestock and for tourist pony trekking. All land in Lesotho is held by the king and is allocated to Basotho residents by local chiefs; foreigners are forbidden to own land. Many Basotho are employed in South African mines and industries. Mining within Lesotho is limited to sandstone and clay quarrying. A diamond mine at Letseng-la-Terae, in NE, closed in 1982. Light manufacturing is concentrated in Maseru and most major towns and is central to Lesotho's plans for economic growth. The local currency is the loti (plural, maloti); Lesotho belongs to a common monetary agreement with South Africa, NAMIBIA, and SWAZILAND. LESOTHO HIGHLANDS WATER SCHEME at the center of the country, with one dam completed in 1996, another in 1998, and one under construction, and four more planned, will provide power for Lesotho and, eventually, to a power grid comprising all SADC countries and provide water for South Africa, to be completed by 2020. SEHLA-BATHEBE NATIONAL PARK (25 sq mi/65 sq km) is in SE, MONT AUX SOURCES NATIONAL PARK in NE; both border on Drakensberg. Wildlife in Lesotho is mainly limited to birds and small land animals.

History: to 1871

San (Bushmen), who were the region's earliest known inhabitants, were supplanted several centuries prior to colonization by various Bantu-speaking peoples. The Basotho are made up of remnants of ethnic groups that were scattered during the disturbances accompanying the rise of the Zulu (1816–1830). They were rallied c.1820 by Moshoeshoe (MOO-shwee-shwee), a commoner who founded a dynasty in what is now Lesotho. Moshoeshoe not only defended his people from Zulu raids but preserved their independence against Boer and British interlopers. During one of the many attacks upon his people, Moshoeshoe used a flat-topped mountain he named Thaba-Bosiu (Mountain of the Night) as his fortress against enemy sieges. He also welcomed Catholic and Protestant missionaries, who still play an important role in the kingdom. Following wars with the Boer-ruled OR-ANGE FREE STATE in 1858 and 1865, Moshoeshoe put the Basotho under British protection (1868).

History: 1871 to 1973

The protectorate was annexed to CAPE COLONY in 1871 without Basotho consent; but in 1884 it was placed under the direct control of Britain. A resident commissioner was established at Maseru, where he administered through Basotho chiefs. When the UNION OF SOUTH AFRICA was forged in 1910, Basutoland came under the jurisdiction of the British High Commissioner in South Africa. Provisions were made for the eventual incorporation of the territory into the Union; but Basotho opposition, especially after the rise of the Nationalist party with its apartheid policy, prevented annexation. In 1960 the British granted Basutoland a new constitution that paved the way for internal self-government. On October 4, 1966, Basutoland became independent as Lesotho. Following general elections in early 1970, which the opposition Congress party apparently won, Prime Minister Leabua Jonathan declared a state of emergency and suspended the constitution; King Moshoeshoe II went into exile but returned at the end of the year.

History: 1973 to 1990

In 1973, an interim assembly began work on a new constitution, but the Congress party, led by Ntsu Mokhehle, refused to participate. In January 1974, Chief Leabua Jonathan accused the Congress party of attempting to stage a coup d'etat, and in the months that followed hundreds of its members reportedly were killed. Jonathan refused to concede democratic reforms to the outlawed Basutoland Congress party (BCP), and his government backed attacks on BCP supporters. Armed clashes between the Lesotho Liberation Army (the militarized segment of the BCP) and the government were common throughout the 1970s and 1980s. In the late 1970s, Jonathan exploited growing popular resentment against South Africa and its policies of apartheid. South Africa responded by organizing economic blockades and military raids against Lesotho. Major-General Justinus Lekhanya led a coup in 1986 that installed King Moshoeshoe II as head of state. After prolonged disputes with Lekhanya over power, the king was deposed in 1990, and his son became king as Letsie III.

History: 1990 to Present

In 1990, Lekhanya announced that civilian rule would be restored, but he was overthrown in a bloodless coup in 1991. In 1993, an elected government assumed office and in 1995 King Moshoeshoe was restored to the throne. The king died in 1996, and Letsie III succeeded him on throne. In 1998, the army mutinied and there were violent protests by the people after a contentious election led to bloody South African military intervention. However, by 2002 peaceful parliamentary elections were held, and political stability restored.

Government

Lesotho is a parliamentary constitutional monarchy governed under the constitution of 1993. The king is head of state but has no executive or legislative powers. The government is headed by a prime minister, who is the leader of the majority party in the Assembly. There is a bicameral Parliament. The thirty-three-member Senate consists of the twenty-two principal chiefs and eleven other members appointed by the ruling party. Of the 120 members of the Assembly, eighty are elected by popular vote and forty by proportional vote, all for five-year terms. Since 1996, King Letsie III has been head of state. The current head of government is Prime Minister Pakalitha Mosisili (since 1998). Lesotho is divided into ten primary administrative divisions, called districts: BEREA, BUTHA-BUTHE, LERIBE, MAFETENG, MASERU, MOHALES HOEK, MOKHOTLONG, QACHAS NEK, QU-THING, and THABA-TSEKA.

Lesotho Highlands Water Scheme (leh-SOO-tho), central LESOTHO, consists of KATSE DAM (completed 1996) on Malibamatso River; also Mashai Dam on ORANGE (Senqu) River and Mohale Dam on SENQU-NYANE RIVER, both under construction. Plan calls for seven dams; two tunnels will divert water to SOUTH AFRICA. Project to reach full potential by 2020. The scheme will supply water to South Africa and electricity to the power grid eventually comprising all SADE countries. Project developed in four phases. Beginning with Katse Dam (completed 1996) ending with Ntoahae Dam in 2015. Project requires approximately 75 mi/120 km of tunnels and will provide hydroelectricity and water to the subregion, jobs to Lesotho citizens, and royalties to the country for all water leaving Lesotho.

Lesovka, UKRAINE: see UKRAYINS'K.

Lesozavod, RUSSIA: see MIRNYY, NIZHEGOROD oblast.

Lesozavodsk (lye-suh-zah-VOTSK), city (2006 population 41,650), W MARITIME TERRITORY, SE RUSSIAN FAR EAST, on road and the TRANS-SIBERIAN RAILROAD (Ussuri station), on the USSURI River (head of navigation), approximately 6 mi/10 km SE of the Chinese border, 215 mi/346 km NE of VLADIVOSTOK, and 135 mi/217 km NNE of USSURIYSK; 45°29′N 133°25′E. Elevation 213 ft/64 m. Sawmilling; furniture, biochemicals, clothing; food processing (meat-processing plant, dairy, bakery). Agriculture (rice, soybeans, potatoes, vegetables; meat and dairy livestock). Granite and limestone deposits in the vicinity. Railroad shops in Ruzhino, its N suburb. Shmakovka, a health resort with baths, lies SE of the city. Developed during the 1930s. Made city in 1938.

Lesozavodskiy (lye-suh-zah-VOT-skeeyee), town, SW MURMANSK oblast, NW European Russia, on an island in the KANDALAKSHA BAY of the WHITE SEA, 34 mi/55 km SSE of (and administratively subordinate to) KANDALAKSHA, and 3 mi/5 km N of KOVDA; 66°43′N 32°50′E. Sawmilling.

Lesozavodskiy, RUSSIA: see NOVOVYATSK.

Lesparre-Médoc (les-pahr–mai-dok), town (□ 18 sq mi/46.8 sq km), sub-prefecture of GIRONDE department, AQUITAINE region, SW FRANCE, 37 mi/60 km NNW of BORDEAUX at the N edge of the LANDES district; 45°18′N 00°56′W. Market center (fruits, vegetables). Has 14th-century tower.

L'Esperance (le-spai-RAWNGS), township on W coast of MAHÉ ISLAND, SEYCHELLES, on the Anse à la Mouche (inlet of INDIAN OCEAN), 8 mi/12.9 km S of VICTORIA. Tourism; fishing.

Les Pléiades, SWITZERLAND: see PLÉIADES, LES.

Lespugue (lai-pyu-guh), commune (□ 2 sq mi/5.2 sq km), HAUTE-GARONNE department, MIDI-PYRÉNÉES region, SW FRANCE on LANNEMEZAN PLATEAU, 45 mi/72 km SW of TOULOUSE; 43°14′N 00°40′E. Here a statue of a female figure carved from a mammoth's tusk (the "Venus of Lespugue") was uncovered, and found to be more than 25,000 years old. It is now exhibited in the Musée de l'Homme, in PARIS.

Lesquin (lai-kan), SE suburb (□ 3 sq mi/7.8 sq km; 2004 population 6,347) of LILLE, NORD department, NORD PAS-DE-CALAIS region, N FRANCE; 50°35′N 03°07′E. Household consumer products made here. Lille-Lesquin international airport.

Les Rangiers, SWITZERLAND: see RANGIERS, LES.

Les Rivières (lai ree-VYER), central borough (French *arrondissement*) of QUEBEC city, CAPITALE-NATIONALE region, S QUEBEC, E CANADA; 46°49′N 71°17′W.

Lessay (le-sai), commune (□ 8 sq mi/20.8 sq km), MANCHE department, BASSE-NORMANDIE region, NW FRANCE, near the ENGLISH CHANNEL, on COTENTIN PENINSULA, 13 mi/21 km NNW of COUTANCES; 49°13′N 01°32′W. Dairying. Has beautifully restored 11th–12th-century Romanesque church of an abbey founded in 1056, and severely damaged in World War II. The British Island of JERSEY is 18 mi/29 km offshore.

Lesse (LES), river, c.50 mi/80 km long; rising in the ARDENNES, SE BELGIUM, and flowing NW to join the MEUSE RIVER near DINANT. It passes in its middle course through underground limestone caves.

Lessebo (LES-se-BOO), town, KRONOBERG county, S SWEDEN, on small Lake Läen, 20 mi/32 km ESE of VÄXJÖ; 56°45′N 15°16′E.

Lessen, BELGIUM: see LESSINES.

Lessen, POLAND: see LASIN.

Lesser, in Russian names: see also MALAYA, MALO-MALOYE, MALY, MALYE.

Lesser Antilles, CARIBBEAN: see WEST INDIES.

Lesser Anyuy River, RUSSIA: see ANYUY River.

Lesser Balkan Range, TURKMENISTAN: see BALKAN.

Lesser Cheremshan River, RUSSIA: see GREATER CHEREMSHAN RIVER.

Lesser Fatra (faht-RAH), Slovak *Malá Fatra* (mah-LAH faht-RAH), GERMAN *Kleine Fatra*, HUNGARIAN *Kis Fátra*, mountain range of the CARPATHIANS, STREDOSLOVENSKY province, NW SLOVAKIA; extend c.40 mi/64 km NE-SW, along both banks of upper Váh River, SE of ŽILINA; rise to 5,607 ft/1,709 m in VELKY KRIVAN.

Lesser Himalaya, PAKISTAN, INDIA, CHINA (TIBET), and NEPAL: see HIMALAYA.

Lesser Irgiz River, RUSSIA: see IRGIZ RIVER.

Lesser Kas River, RUSSIA: see KAS RIVER.

Lesser Kemin River, KYRGYZSTAN: see KEMIN.

Lesser Kinel River, RUSSIA: see GREATER KINEL RIVER.

Lesser Kokshaga River (kuhk-shah-GUH), Russian *Malaya Kokshaga*, approximately 125 mi/201 km long, in MARI EL REPUBLIC, European Russia; rises 17 mi/27 km WNW of NOVYY TORYAL, on the border of KIROV oblast; flows S past IOSHKAR-OLA, to the VOLGA RIVER, 8 mi/12.9 km ESE of (opposite) MARIINSKIY POSAD (CHUVASH REPUBLIC); timber floating. Receives the Lesser Kundysh River (left). Also called Kokshaga.

Lesser Kundysh River (koon-DISH), Russian *Malyy Kundysh*, approximately 50 mi/80 km long, in MARI EL REPUBLIC, European Russia; rises 13 mi/21 km S of NOVYY TORYAL; flows SW to the LESSER KOKSHAGA RIVER, 20 mi/32 km S of IOSHKAR-OLA. Timber floating. Also called Lower Kundysh [Russian *Nizhniy Kundysh*] or simply Kundysh.

Lesser Northern Dvina River (dvee-NAH), Russian *Malaya Severnaya Dvina*, NW European Russia; formed by junction of the SUKHONA and YUG rivers at VELIKIY USTYUG, VOLOGDA oblast; flows N, past KRASAVINO, joining the VYCHEGDA River at KOTLAS (ARCHANGEL oblast) to become Northern Dvina River proper. The name Sukhona is sometimes applied to the Lesser Northern Dvina.

Lesser Slave Lake (LE-suhr SLAIV), central ALBERTA, W CANADA, NW of EDMONTON; 60 mi/97 km long and 3 mi/5 km–10 mi/16 km wide; 55°27′N 115°27′W. It drains E into the ATHABASCA RIVER by the Lesser Slave River. Commercial fishing; lumbering and farming on shores.

Lesser Slave River No. 124 (LE-suhr SLAIV), municipal district (□ 3,885 sq mi/10,101 sq km; 2001 population 2,825), central ALBERTA, W CANADA; 55°13′N 114°20′W. Includes the town of SLAVE LAKE and the hamlets of CANYON CREEK, Marten Beach, WIDEWATER, CHISHOLM, SMITH, FLATBUSH, and WAGNER. Formed as an improvement district in 1969; incorporated 1995.

Lesser Sundas, INDONESIA: see SUNDA ISLANDS.

Lesser Tsivil River, RUSSIA: see TSIVIL.

Lesser Usen River, RUSSIA: see USEN.

Lesser Walachia, ROMANIA: see WALACHIA.

Lesser Yenisey River, RUSSIA: see YENISEY RIVER.

Lessines (le-SEEN), Flemish *Lessen*, commune (2006 population 17,980), Soignies district, HAINAUT province, SW central BELGIUM, on DENDER RIVER, and 6 mi/9.7 km NNE of ATH; 50°43′N 03°50′E.

Lessini, Monti (MON-tee les-SEE-nee), mountain group rising to 7,424 ft/2,263 m in CIMA CAREGA, N ITALY; 45°41′N 11°13′E. Separated from MONTE BALDO group by ADIGE RIVER. Marble and limestone quarries.

Lessudden, Scotland: see ST. BOSWELLS.

Lester, town (2000 population 251), LYON county, NW IOWA, 8 mi/12.9 km W of ROCK RAPIDS; 43°26′N 96°19′W. Livestock, grain.

Lester, village (2006 population 315), RALEIGH county, S WEST VIRGINIA, 7 mi/11.3 km WSW of BECKLEY; 37°44′N 81°17′W. Semibituminous-coal area. Light manufacturing.

Lester B. Pearson International Airport, S central ONTARIO, CANADA; covers 4,428 acres/1,792 ha. Official name of Toronto International Airport, 12 mi/19 km W of downtown TORONTO at Malton, in N part of city of MISSISSAUGA, just NW of MACDONALD-CARTIER FREEWAY. Canada's largest and busiest airport. Built in 1938 and greatly expanded, it is surrounded by light manufacturing and office parks, hotels, commercial areas. Airport Code YYZ.

Lester Prairie, town (2000 population 1,377), MCLEOD county, S central MINNESOTA, 40 mi/64 km WSW of MINNEAPOLIS, on South Fork CROW River W of mouth of Otter Creek; 44°52′N 94°02′W. Manufacturing (feeds, plastic products, construction materials).

Lesterville, village (2006 population 154), YANKTON county, SE SOUTH DAKOTA, 15 mi/24 km NW of YANKTON; 43°02′N 97°35′W.

Lestock (luh-STAHK), village (2006 population 138), SE central SASKATCHEWAN, CANADA, 65 mi/105 km NNE of REGINA; 51°18′N 103°50′W. Wheat, dairying.

Lestrem (les-trem), commune (□ 8 sq mi/20.8 sq km), PAS-DE-CALAIS department, NORD-PAS-DE-CALAIS region, N FRANCE, near LYS RIVER; and 7 mi/11 km N of BÉTHUNE-Beuvry; 50°37′N 02°41′E.

L'Estrie or **Cantons de l'est**, S QUEBEC, CANADA, collective name of townships S of the St. Lawrence, centered on SHERBROOKE, 1st surveyed after 1791, when English land laws briefly replaced French system of seigneurial tenure. Name was used to distinguish region from the Western Townships, N of the St. Lawrence and on the BAY OF QUINTE, ONTARIO, surveyed 1783–1784; latter designation is no longer used.

Le Suchet, SWITZERLAND: see SUCHET, LE.

Le Sueur (luh SOOR), county (□ 473 sq mi/1,229.8 sq km; 2006 population 27,895), S MINNESOTA; ⊙ LE CENTER; 44°22′N 93°43′W. Bordered W by MINNESOTA RIVER. Agricultural area (soybeans, alfalfa, peas; sheep, hogs, cattle, poultry; dairying); silica, sand, marble, limestone. Part of Sakatah Lake State Park in SE corner; numerous small lakes in county, especially in S and E. Formed 1853.

Le Sueur (luh SOOR), town (2000 population 3,922), LE SUEUR county, S MINNESOTA, 20 mi/32 km N of MANKATO, on MINNESOTA RIVER at mouth of Le Sueur Creek; 44°27′N 93°54′W. Grain, soybeans, peas; livestock; dairying; manufacturing (whey and dairy products, baked goods, frozen food and beverages, greeting cards, injection molding, printing and publishing). Municipal Airport to S. Home of W.W. Mayo, founder of Mayo Clinic in ROCHESTER (built 1859). Settled 1852.

Le Sueur Center, MINNESOTA: see LE CENTER.

Le Sueur River (luh SOOR), 80 mi/129 km long, S MINNESOTA; rises in N FREEBORN county, c.9 mi/14.5 km NNW of ALBERT LEA; 43°46′N 93°28′W. Flows N and W, in very serpentine course, passes NE of NEW RICHLAND and SW of WASECA, past ST. CLAIR, receives Big Cobb and MAPLE rivers S of MANKATO, enters BLUE EARTH RIVER 4 mi/6.4 km SW of Mankato and 3 mi/4.8 km S of its confluence with MINNESOTA RIVER.

Lesura (LE-soo-rah), village, MONTANA oblast, KRIVODOL obshtina, BULGARIA; 43°38′N 23°34′E.

Lésvos, Greece: see LESBOS.

Leszno (LE-shno), German *Lissa*, town (2002 population 63,660), SW POLAND. A railroad junction, it is a center for metallurgy and light industry. Chartered in 1547, it passed to PRUSSIA in 1793 and again in 1815. It reverted to POLAND in 1919. Leszno was a center of the Protestant Reformation of the 16th century and the chief seat of the Moravian Brethren in Poland. John Amos Comenius was a rector of the famous Moravian school here. The town has an 18th century palace.

Letälven (LET-ELV-en), river, 100 mi/161 km long, S central SWEDEN; rises S of VANSBRO; flows generally S, through LAKE MÖCKELN, past DEGERFORS, to LAKE SKAGERN 10 mi/16 km WNW of LAXÅ.

L'Étang du Nord, CANADA: see ÉTANG-DU-NORD.

L'Étang-la-Ville (lai-tahn–lah–veel), town (□ 2 sq mi/5.2 sq km), YVELINES department, ÎLE-DE-FRANCE region, N central FRANCE, 3 mi/5 km S of SAINT-GERMAIN-EN-LAYE; 48°52′N 02°04′E.

Letcher (LECH-uhr), county (□ 339 sq mi/881.4 sq km; 2006 population 24,520), SE KENTUCKY; ⊙ WHITESBURG; 37°07′N 82°50′W. In the CUMBERLAND MOUNTAINS. Bounded E and SE by Virginia; drained by North Fork KENTUCKY RIVER and POOR FORK of CUMBERLAND RIVER and Rockhouse Creek. Important bituminous coal-mining area; clay, sand, and gravel pits, stone quarries, timber; some agriculture (livestock; burley tobacco). Includes part of PINE MOUNTAIN and part of Jefferson National Forest, both on SE boundary; Kingdom Come State Park on SW boundary. Lilley Cornett Woods preserve in SW; Pine Mountain Wildlife Management Area and Bad Branch State Nature Preserve in S. Formed 1842.

Letcher, village (2006 population 164), SANBORN county, SE central SOUTH DAKOTA, 15 mi/24 km NNW of MITCHELL; 43°53′N 98°08′W. Trade center for diversified farming region; (corn, hogs, poultry). Twin Lakes State Lakeside Use Area to NW.

Letchworth (LECH-wuhth), town (2001 population 32,932), HERTFORDSHIRE, E central ENGLAND, 5 mi/8 km N of STEVENAGE; 51°58′N 00°13′W. It was the first Garden City, founded in 1903 by Sir Ebenezer Howard. Industries focus on printing and the manufacturing of printing machinery.

Letchworth State Park, New York: see GENESEE RIVER.

Letea Island (LE-ta) or **Leti Island** (LE-tee), marshy island (c.40 mi/64 km long, 18 mi/29 km wide) in Danube Delta on BLACK SEA coast of ROMANIA, formed by Brațul Chilia arm (N) and Brațul Sulina arm (S).

Letellier (luh-tel-YAI, luh-TEL-yah), unincorporated village, SE MANITOBA, W central CANADA, on Marais River, 11 mi/18 km NW of EMERSON, and in MONTCALM rural municipality; 49°08′N 97°18′W. Grain, livestock.

Leténye (LE-te-nye), town, ZALA county, W HUNGARY, 13 mi/21 km W of NAGYKANIZSA; 46°26′N 16°44′E. Potatoes, grain, silage corn, apples; cattle, hogs. One of two road crossings and custom point to CROATIA.

Lethbridge (LETH-brij), city (□ 47 sq mi/122.2 sq km; 2006 population 78,713), S ALBERTA, W CANADA, on the OLDMAN RIVER; 49°42′N 112°50′W. Formerly a coal-mining center, Lethbridge is now a commercial and service center for an irrigated farming and ranching district. Diverse manufacturing includes food, fabricated metal products, and electronic equipment. Federal agricultural and veterinary research stations. The University of Lethbridge (1967). Incorporated as a town in 1890; became a city in 1906.

Lethbridge County (LETH-brij), municipality (□ 1,096 sq mi/2,849.6 sq km), S ALBERTA, W CANADA; 49°49′N 112°48′W. Agriculture. Includes COALDALE, BARONS, COALHURST, NOBLEFORD, PICTURE BUTTE, and seven hamlets. Established as a municipal district in 1954; became a county in 1960.

Lethem (le-TEM), village, UPPER TAKUTU-UPPER ESSEQUIBO district, GUYANA; 03°23′N 59°48′W. Located on BRAZIL border, 260 mi/418 km SSW of GEORGETOWN. Tourist services, airfield. Formerly called BON SUCCESS.

Leti (LAI-tee), island (□ 48 sq mi/124.8 sq km). LETI ISLANDS, S MALUKU province, INDONESIA, in BANDA SEA, just W of MOA, 25 mi/40 km E of TIMOR; 11 mi/18 km long, 5 mi/8 km wide; 03°11′S 127°19′E. Coconuts; fishing. Principal town is Serwaru.

Letichev, UKRAINE: see LETYCHIV.

Leticia (lai-TEE-see-ah), town, ☉ Amazonas department, SE COLOMBIA, on the upper AMAZON; 04°12′S 69°56′W. The Leticia region, a narrow strip of land extending S of the PUTUMAYO River to the Amazon, was disputed, at times violently, between Colombia and PERU (1932–1934). The region was awarded to Colombia by the League of Nations in 1934. Commercial town and tourist center for exploring indigenous tribes in the surrounding area. An isolated region of the country. Airport.

Leti Island, ROMANIA: see LETEA ISLAND.

Leti Islands (LAI-tee), group (□ c.290 sq mi/751 sq km), S MALUKU province, INDONESIA, in BANDA SEA, off E tip of TIMOR; 08°20′S 127°57′E. Comprise MOA (largest island), LAKOR, LETI, and several islets. Islands are hilly and forested. Cattle raising, agriculture (rice, tobacco), copra producing; fishing. Also spelled Letti Islands.

Leting (LUH-TING), town, NE HEBEI province, CHINA, 40 mi/64 km ESE of TANGSHAN, near LUAN RIVER, on Gulf of BOHAI; ☉ Leting county (1990 population479,438); 39°26′N 118°56′E. Grain, cotton, oilseeds; food, textiles; salt mining.

Letiny, CZECH REPUBLIC: see BLOVICE.

Letka (LYET-kah), village (2005 population 2,935), SW KOMI REPUBLIC, NE European Russia, on the LETKA River (right affluent of the upper VYATKA River), on road, 147 mi/237 km S of SYKTYVKAR; 59°36′N 49°25′E. Elevation 541 ft/164 m. Logging, lumbering, sawmilling.

Letka River (LYET-kah), approximately 99 mi/159 km long, in E European Russia. Rises in SW KOMI REPUBLIC, in the NE NORTHERN UVALS, and flows SW for about one-quarter of its course to Vorchanka, where it turns W and continues in a long, W-to-SW arch, to LETKA. At this point, the river turns SSE and continues, with many small twists and turns, for the remaining two-thirds of its course, past Sludka, into N KIROV oblast, to flow into the VYATKA River approximately 4 mi/6 km N of Shestakovo. Timber floating and non-commercial fishing along most of its course.

Letlhakeng (LAI-lah-keng), town (2001 population 6,032), KWENENG DISTRICT, BOTSWANA, 69 mi/111 km NW of GABORONE; 24°06′S 25°04′E. At end of paved road to CENTRAL KALAHARI GAME RESERVE. Commercial center for the densely population district.

Letmathe (LET-mah-te), suburb of ISERLOHN, North Rhine-Westphalia, W GERMANY, on the LENNE, 4 mi/6.4 km W of city center; 51°22′N 07°42′E. Inc. 1975 into ISERLOHN.

Letnan, CZECH REPUBLIC: see LETNANY.

Letnany (LET-nyah-NI), Czech Letňany, German Letnan, NE district of PRAGUE, PRAGUE PROVINCE, central BOHEMIA, CZECH REPUBLIC, on railroad, 5 mi/8 km from city center; 50°08′N 14°31′E. Manufacturing (machinery). Research Institute of Aviation.

Letnerechenskiy (lyet-nye-RYE-cheen-skeeyee), town, E Republic of KARELIA, NW European Russia, on highway branch and the Murmansk railroad (Letniy station), 18 mi/29 km SW of BELOMORSK; 64°17′N 34°23′E. Elevation 177 ft/53 m. Sawmilling; manufacturing of construction materials.

Letnik (LYET-neek), village (2006 population 3,140), S ROSTOV oblast, S European Russia, on the YEGORLYK RIVER, on road, 11 mi/18 km S of RAZVIL′NOYE; 46°03′N 41°16′E. Elevation 170 ft/51 m. Agricultural products.

Letnitsa (LET-neet-sah), city, PLEVEN oblast, obshtina center, N BULGARIA, on the VIT RIVER, 5 mi/8 km SW of LEVSKI; 43°18′N 25°04′E. Flour milling; agriculture (grain, vineyards, watermelons, vegetables, fodder plants; livestock, poultry). Workshops of the machine building plants from LOVECH and TROYAN.

Letnyaya Stavka (LYET-nyah-yah STAHF-kah) [Russian=summer headquarters], settlement (2006 population 5,000), central STAVROPOL TERRITORY, N CAUCASUS, S European Russia, on road junction, 67 mi/108 km S of ELISTA (Republic of KALMYKIA-KHALMG-TANGEH), and 29 mi/47 km N of BLAGODARNYY; 45°26′N 43°27′E. Elevation 511 ft/155 m. In cattle and sheep-raising area. Population largely Turkmen.

Letohrad (LE-to-HRAHT), town, VYCHODOCESKY province, NE BOHEMIA, CZECH REPUBLIC, 32 mi/51 km E of PARDUBICE; 50°02′N 16°30′E. Railroad junction. Manufacturing (electronics and textiles); wood and food processing. Has a 17th century baroque castle with town hall. Until 1950, the town was called KYSPERK, German geyersberg.

Letona (lee-TO-nuh), village (2000 population 201), WHITE county, central ARKANSAS, 9 mi/14.5 km NW of SEARCY; 35°21′N 91°49′W.

Letopolis, EGYPT: see AUSIM.

Letovice (LE-to-VI-tse), German Lettowitz, town, JIHOMORAVSKY province, W MORAVIA, CZECH REPUBLIC, on SVITAVA River, on railroad, and 24 mi/39 km N of BRNO; 49°33′N 16°35′E. Manufacturing (textiles, ceramics, chemicals).

Letpadan (let-pah-DAHN), township, BAGO division, MYANMAR, on Yangon-Mandalay railroad, and 70 mi/113 km NNW of YANGON. Head of railroad to THARRAWAW.

Letskiy Reyd (LYETS-keeyee RYAID), village, N KIROV oblast, E central European Russia, just S of the confluence of the LETKA and VYATKA rivers, on road, 18 mi/29 km N, and under administrative jurisdiction, of SLOBODSKOY; 58°27′N 50°12′E. Elevation 459 ft/139 m. Timber floating.

Letterkenny (le-tuhr-KE-nee), Gaelic Leitir Ceanainn, town (2006 population 15,062), E central DONEGAL county, N IRELAND, at head of LOUGH SWILLY, at mouth of SWILLY RIVER, and 17 mi/27 km W of LONDONDERRY, Northern Ireland; 54°57′N 07°44′W. Fishing port, agricultural market. Cathedral completed 1901.

Lettermore, Gaelic Leitir Mór, island (□ 4 sq mi/10.4 sq km) in KILKIERAN BAY (inlet of GALWAY BAY), SW GALWAY county, W IRELAND, 25 mi/40 km W of GALWAY, just N of GORUMNA island; 53°18′N 09°40′W.

Lettermullen (le-tuhr-MUH-len) or **Lettermullan**, Gaelic Leitir Meallláin, island (□ 1 sq mi/2.6 sq km) in KILKIERAN BAY (inlet of GALWAY BAY), SW GALWAY county, W IRELAND, 28 mi/45 km W of GALWAY, just W of GORUMNA island, with which it is linked by a road bridge; 53°14′N 09°44′W.

Letti Islands, INDONESIA: see LETI ISLANDS.

Lettland, LATVIA: see LATVIA.

Lettowitz, CZECH REPUBLIC: see LETOVICE.

Letts, town (2000 population 392), LOUISA county, SE IOWA, 12 mi/19 km WSW of MUSCATINE; 41°19′N 91°13′W. Livestock, grain.

Letur (lai-TOOR), town, ALBACETE province, SE central SPAIN, 25 mi/40 km WSW of HELLÍN; 38°22′N 02°06′W. Olive-oil processing, flour milling, basket manufacturing. Honey, esparto, cereals.

Letychiv (le-TI-cheef) (Russian Letichev), town, E KHMEL′NYTS′KYY oblast, UKRAINE, on the Southern BUH, and 28 mi/45 km E of KHMEL′NYTS′KYY; 49°23′N 27°37′E. Elevation 895 ft/272 m. Raion center; manufacturing (machines, ceramics, bricks, furniture); sewing; food processing. Local history museum. Walls and tower of a 16th-century castle, two 17th-century Orthodox and Roman Catholic churches. Known since 1411; under Poland since 1434; site of many battles among the Cossacks, Poles, and Turks (17th century); passed to Russia (1793); site of Karmelyuk uprising (beginning of the 19th century). Jewish community since 1765, by the beginning of the 19th century, one of the centers of Hasidic learning in Podolian Ukraine; reduced by the pogroms and subsequent emigration of the 1880s and the civil war pogroms by the Ukrainian nationalists in 1919, but still numbering over 2,400 in 1939; placed in a ghetto by the Nazis in 1941 and finally wiped out in 1942.

Letzeburg, LUXEMBOURG: see LUXEMBOURG, city.

Leubus (LEE-boos), Polish Lubiąż, village in LOWER SILESIA, after 1945 in Wroclaw province, SW POLAND, on the ODER, and 14 mi/23 km ENE of LEGNICA (Liegnitz). Has former Cistercian monastery, founded c.1175, rebuilt 18th century, secularized 1810. Sacked 1432 by Hussites and (1639), in Thirty Years' War, by Swedes.

Leucadia, Greece: see LEFKAS.

Leucadia, Cape, Greece: see DOUKATO, CAPE.

Leucallec Island, CHILE: see GUAITECAS ISLANDS.

Leucas, Greece: see LEFKAS.

Leucate (lu-kaht) or **Port-Leucate** (por–lu-kaht), village (□ 9 sq mi/23.4 sq km; 2004 population 3,392), AUDE department, LANGUEDOC-ROUSSILLON region, S FRANCE, 16 mi/26 km NNE of PERPIGNAN; on sandy coastal strip between Leucate lagoon and MEDITERRANEAN SEA; 42°55′N 03°02′E. Its waterfront forms part of the beaches of ROUSSILLON, which extend c.25 mi/40 km N-S near Perpignan. Area was formerly subject to flooding in stormy sea weather.

Leucates, Cape, Greece: see DOUKATO, CAPE.

Leucayec Island, CHILE: see GUAITECAS ISLANDS.

Leuchars (LOO-kahrs), village (2001 population 2,518), FIFE, E Scotland, near EDEN RIVER estuary, 5 mi/8 km NW of ST. ANDREWS; 56°23′N 02°53′W. Paper-milling village of Guardbridge 2 mi/3.2 km S.

Leuk, French *Loèche*, commune, VALAIS canton, S SWITZERLAND, on the RHÔNE, 5 mi/8 km ENE of SIERRE. Elevation 2,339 ft/713 m. Resort village of LEUKERBAD is 4 mi/6.4 km N.

Leuk (LUHYK), district, N central VALAIS canton, S SWITZERLAND; 46°19′N 07°38′E. Population is German-speaking and Roman Catholic.

Leuka, CYPRUS: see LEFKA.

Leukas, Greece: see LEFKAS.

Leukerbad (LUHYKER-bahd), French *Loèche-les-Bains*, village, VALAIS canton, S SWITZERLAND, on short Dala River, 4 mi/6.4 km N of LEUK; 46°23′N 07°38′E. Elevation 4,596 ft/1,401 m. Mountain health resort noted for thermal springs.

Leukimme, Greece: see LEFKIMMI.

Leuktra (LEFK-trah), village of ancient GREECE, in what is now BOEOTIA prefecture, CENTRAL GREECE department, 7 mi/11.3 km SW of THEBES; 38°15′N 23°11′E. Here the Spartans were defeated (371 B.C.E.) by the Thebans under Epaminondas. A brilliant tactical success, the battle also dealt a severe blow to Spartan hegemony. Formerly Leuctra.

Leun (LOIN), town, HESSE, central GERMANY, on LAHN RIVER, 14 mi/23 km W of GIESSEN; 50°33′N 08°21′E.

Leuna (LOI-nah), town, SAXONY-ANHALT, S central GERMANY, 2.5 mi/4 km S of MERSEBURG; 51°20′N 12°00′E. Here, in 1916, the first synthetic nitrogen plant began to operate after the invention of the Haber process. The city was badly damaged in World War II because of its extensive chemical installations. Nevertheless, the city grew as a center of the German synthetic chemical industry, and until reunification in 1990 it was an oil-refining and petrochemical center for E Germany. The city suffered from considerable chemical pollution and has experienced a severe economic depression. The population declined by almost one third in the four years following unification.

Leupp, unincorporated town (2000 population 970), COCONINO county, N central ARIZONA, 39 mi/63 km E of FLAGSTAFF, in SW part of Navajo Indian Reservation, on LITTLE COLORADO RIVER. Sheep, cattle. Crafts.

Leura (LOO-ruh), town, NEW SOUTH WALES, SE AUSTRALIA, 65 mi/105 km W of SYDNEY, near BLUE MOUNTAINS NATIONAL PARK; 33°43′S 150°20′E. Noted for its gardens and main street.

Leusden (LUHS-duhn), town (2001 population 23,121), UTRECHT province, W central NETHERLANDS, 3 mi/4.8 km SE of AMERSFOORT, on Eem River; 52°08′N 05°26′E. Dairying; cattle, poultry; fruit, grain, vegetables.

Leuser, Mount (LAI-suhr) (11,092 ft/3,381 m), SW ACEH province, N SUMATRA, INDONESIA, in N BARISAN MOUNTAINS, near W coast, 110 mi/177 km W of MEDAN; 03°45′N 97°11′E. MOUNT LEUSER NATIONAL PARK encompasses E slopes. Formerly Mount Loser.

Leutensdorf, CZECH REPUBLIC: see LITVÍNOV.

Leutershausen (loi-tuhrs-HOU-suhn), town, MIDDLE FRANCONIA, W BAVARIA, GERMANY, on the ALTMÜHL, 7 mi/11.3 km W of ANSBACH; 49°18′N 10°25′E. Manufacturing of fabricated metal products, paper, textiles. Granite quarries.

Leuthen or **Lutynia**, village in LOWER SILESIA, after 1945 in Wroclaw province, SW POLAND, 11 mi/18 km W of WROCŁAW (Breslau). Here, in 1757, Austrians defeated by Prussians under Frederick the Great. After 1945, briefly called Litom.

Leutkirch im Allgäu (LOIT-kirkh-im AHL-goi), town, BADEN-WÜRTTEMBERG, S GERMANY, 19 mi/31 km ENE of RAVENSBURG; 47°50′N 10°01′E. Manufacturing of plastics, wood products, fabricated metal products, food. Has rococo palace from 17th century.

Leutschau, SLOVAKIA: see LEVOCA.

Leuven (LUH-vuhn), French *Louvain* (loo-vainh) city (□ 449 sq mi/1,167.4 sq km; 2006 population 90,854), ⊙ Leuven district, BRABANT province, central BELGIUM, on the DIJLE RIVER; 50°53′N 04°42′E. It is a cultural, commercial, and industrial center, as well as a railroad junction. Mentioned in the ninth century, Leuven was a center of the wool trade and of the cloth industry in the Middle Ages. For a time it was the capital of the duchy of BRABANT, and in 1356 the *Joyeuse Entrée*, a charter of liberties, was granted here. In the fourteenth century, strife between the nobles and the weavers was prevalent; after the nobles gained authority most of the weavers emigrated to HOLLAND and ENGLAND, and the city declined. In 1426, Duke John IV of Brabant founded a university. Its library was destroyed by the order of the German military in World War I and damaged in World War II, but was rebuilt after each war. In 1968, as a result of a long-standing dispute between Flemish- and French-speaking sections of the university, it was divided into two autonomous units. The Flemish-speaking Universiteit de Leuven remained in Leuven, and the French-speaking Université Catholique de Louvain was established at OTTIGNIES. Among the noted buildings of Leuven are the Gothic city hall (fifteenth century; damaged in both World Wars); the fourteenth-century. Cloth Workers' Hall, and several medieval churches. Brewery.

Leuze-en-Hainaut (LUHZ–AW–ai-NO), commune (2006 population 13,235), Tournai district, HAINAUT province, W BELGIUM, on branch of DENDER RIVER, and 10 mi/16 km E of TOURNAI. Textiles.

Léva, SLOVAKIA: see LEVICE.

Levack (luh-VAK), unincorporated town, SE central ONTARIO, E central CANADA; included in Greater SUDBURY; 46°38′N 81°23′W. Nickel and copper mining.

Levadia (lee-vah-[TH]YAH), Latin *Lebadea*, city (2001 population 20,061), E central GREECE department, on railroad, and 60 mi/97 km NW of ATHENS; ⊙ BOEOTIA prefecture; 38°25′N 22°54′E. Cotton and woolen milling center; trades in wheat, cotton, tobacco, wine. Ancient Lebadea was the site of the oracle of Trophonios, a Boeotian god. Flourished in Middle Ages and was an administrative center of central Greece under Turkish rule. Sometimes spelled Livadia, Levadeia, or Levadhia.

Levallois-Perret (luh-vahl-lwah–pe-re), NW residential and industrial suburb of PARIS, HAUTS-DE-SEINE department, ÎLE-DE-FRANCE region, N central FRANCE, on right bank of the SEINE RIVER, and 4 mi/6.4 km from Notre Dame Cathedral; 48°54′N 02°17′E. Electrical equipment, perfume. Town is reached by Paris metro subway system.

Levan (LUH-van), village (2006 population 834), JUAB county, central UTAH, 10 mi/16 km SSW of NEPHI; 39°33′N 111°52′W. Elevation 5,314 ft/1,620 m. Wheat, barley, alfalfa; cattle. Chicken Creek reservoir to SW; SEVIER BRIDGE RESERVOIR and Painted Rock State Park to S. San Pitch Mountains and part of Uinta National Forest to E.

Levanger (lai-VAHNG-uhr), town, NORD-TRØNDELAG county, central NORWAY, port on E shore of TRONDHEIMSFJORDEN, on railroad, and 35 mi/56 km NE of TRONDHEIM; 63°45′N 11°18′E. Agriculture, livestock; forestry; some light manufacturing; exports lumber. Alstadhaug church (4 mi/6.4 km SW) antedates 1250. Mentioned in 1247 as Lifangr; incorporated 1836; rebuilt after fire in 1897.

Levant [Italian=east], collective name for the countries of the E shore of the MEDITERRANEAN from EGYPT to, and including, TURKEY. The divisions of the French mandate over SYRIA and LEBANON were called the Levant States, and the term is still sometimes applied to those two nations.

Levant (lev-ANT), agricultural town, PENOBSCOT county, S MAINE, 8 mi/12.9 km NW of BANGOR; 44°53′N 68°59′W.

Levant, Île du (le-vahn, eel dyoo), island (5 mi/8 km long, 1 mi/1.6 km wide), easternmost of HYÈRES islands, in MEDITERRANEAN off S coast of FRANCE, in VAR department, PROVENCE-ALPES-CÔTE D'AZUR region, 18 mi/29 km ESE of HYÈRES; 43°03′N 06°28′E. A rocky and barren hogback; nature center. Reached from Le LAVANDOU (9 mi/14.5 km NNW).

Levanto (LAI-vahn-to), town, LA SPEZIA province, LIGURIA, N ITALY, port on GULF OF GENOA and 12 mi/19 km NW of SPEZIA. In olive- and grape-growing region; resort. Has Gothic church (1232), Franciscan convent (1449). Marble, sandstone, serpentine quarries nearby.

Levanzo (LAI-vahn-tso), ancient *Phorbantia* (□ 2 sq mi/5.2 sq km), one of EGADI Islands, in MEDITERRANEAN SEA off W SICILY, ITALY 9 mi/14 km NW of TRAPANI; 3 mi/5 km long; rises to 912 ft/278 m; 37°59′N 12°20′E. Major blue fin tunny fisheries. Chief port, Levanzo, on S coast. Along with island of Favigana and Marèttimo, forms one municipality. 10,000-year-old cave paintings in Grotto del Genovese.

Levashi (lye-vah-SHI), village (2005 population 8,005), E central DAGESTAN REPUBLIC, NE CAUCASUS, S European Russia, in the GIMRY RANGE, on road, 30 mi/48 km SSE of BUYNAKSK, and 27 mi/43 km S of MAKHACHKALA; 42°26′N 47°19′E. Elevation 3,946 ft/1,202 m. Population mostly Darghin and Avar, with a Chechen minority, largely recent refugees from war zones.

Levashovo (lee-VAH-shuh-vuh), village (2005 population 3,980), N central LENINGRAD oblast, NW European Russia, suburb of SAINT PETERSBURG, 12 mi/19 km NNW of the city center (connected by highway and railroad); 60°06′N 30°13′E. Mostly residential. A military airfield in the vicinity.

Levasy (LE-vuh-see), town, JACKSON county, W MISSOURI, 25 mi/40 km E of downtown KANSAS CITY, on MISSOURI River; 39°07′N 94°07′W. Soybeans, wheat, corn; cattle. Fort Osage National Historic Landmark to N.

Levaya Rossosh (LYE-vah-yah ROS-suhsh), village, central VORONEZH oblast, S central European Russia, on road and near railroad (Anoshkino station), 25 mi/40 km SSE of VORONEZH; 51°20′N 39°24′E. Elevation 423 ft/128 m. In agricultural area (grain, hemp, sunflowers).

Level Island, CHILE: see CHONOS ARCHIPELAGO.

Levelland (LE-vuhl-land), city (2006 population 12,674), NW TEXAS, 30 mi/48 km W of LUBBOCK on the LLANO ESTACADO; ⊙ HOCKLEY county; 33°34′N 102°21′W. Elevation 3,523 ft/1,074 m. The economy is based chiefly on oil and natural gas, agriculture (cotton; cattle), and manufacturing (food, transportation equipment). South Plains College (two-year) is in LEVELLAND. Incorporated 1926.

Levelock (LE-ve-lawk), village (2000 population 122), SW ALASKA, 50 mi/80 km E of DILLINGHAM; 59°06′N 156°53′W.

Level Park, village, CALHOUN county, S MICHIGAN, suburb 7 mi/11.3 km NW of BATTLE CREEK, on KALAMAZOO river; 42°21′N 85°16′W.

Level Plains, town (2000 population 1,544), Dale co., SE ALABAMA, 4 mi/6.4 km W of Daleville; 31°18′N 85°46′W. Inc. in 1965.

Levelwood (LE-vel-wood), settlement and local administrative district (2006 population 260), E central island of SAINT HELENA, in the British dependency of St. Helena, in the ATLANTIC ocean, under 3 mi/5 km SE of JAMESTOWN; 15°58′S 05°41′W. Elevation 1,069 ft/325 m. Developed around a military fort built in the 19th century to protect the island's E approaches.

Leven (LEV-uhn), town (2001 population 8,051), FIFE, E Scotland, at the mouth of the LEVEN RIVER on the Largo Bay of the Firth of Forth, and 8 mi/12.8 km NE of KIRKCALDY; 56°12′N 03°00′W. Manufacturing of paper, consumer goods. Previously iron, steel, sawmilling, and coal mining. Summer resort, famous for its golf links and beaches. Nearby docks at METHIL.

Leven, Loch (LEV-uhn), sea inlet in LOCHABER district of HIGHLAND, N Scotland, extending 9 mi/14.5 km E from LOCH LINNHE at BALLACHULISH; up to 1 mi/1.6 km wide. Receives Coe River. At head of loch, the canalized Leven River (16 mi/26 km long), previously supplied hydroelectric power station at KINLO-CHLEVEN.

Leven, Loch (LEV-uhn), lake (□ 8 sq mi/20.8 sq km), SE PERTH AND KINROSS, E Scotland; 56°13′N 03°23′W. Among its islands are Castle Island, 2 mi/3.2 km E of KINROSS, with remains of 15th-century Loch Leven Castle, scene of imprisonment of Mary, Queen of Scots 1567–1568, and St. Serf's Island (ruins of old priory). The loch is noted for its pink trout. Leven River National Nature Reserve here. Formerly in Tayside, abolished 1996.

Leven River (LE-ven), 50 mi/80 km long, N TASMANIA, SE AUSTRALIA; rises SE of WARATAH; flows NE to BASS STRAIT at ULVERSTONE, on N coast.

Leven River (LEE-vuhn), 11 mi/18 km long, in CUMBRIA, NW ENGLAND; rises in WINDERMERE LAKE; flows SW and S to MORECAMBE BAY just E of ULVERSTON.

Leven River (LEV-uhn), 7 mi/11.3 km long, in West Dunbartonshire, W Scotland; originates from S end of LOCH LOMOND at Balloch; flows S past Jamestown, Alexandria, and Bonhill, to the CLYDE RIVER at DUMBARTON. Its valley is center of textile printing and dyeing industry.

Leven River (LEV-uhn), 15 mi/24 km long, FIFE, E Scotland; rises at SE end of LOCH LEVEN, flows E to Largo Bay of Firth of Forth at LEVEN.

Levens (LEE-vuhnz), village (2001 population 1,654), CUMBRIA, NW ENGLAND, on KENT RIVER, and 4 mi/6.4 km SSW of KENDAL; 54°16′N 02°47′W. Sheep raising, agriculture. Tudor house with formal gardens.

Levenshulme, ENGLAND: see MANCHESTER.

Leventina, district NW TICINO canton, S SWITZERLAND. Population is Italian-speaking and Roman Catholic.

Leventina, valley, TICINO canton, S SWITZERLAND. The Valle Leventina (German *Livinental*) follows TICINO RIVER from AIROLO to BIASCA. Main town, Faido.

Leveque, Cape (luh-VEK), N WESTERN AUSTRALIA, in INDIAN OCEAN, at W end of KING SOUND; 16°23′S 122°55′E. Lighthouse.

Leverano (le-ve-RAH-no), town, LECCE province, APULIA, S ITALY, 10 mi/16 km WSW of LECCE; 40°17′N 18°00′E. Agricultural center (wine, olive oil, cheese, wheat, tobacco).

Leverburgh (LEV-uhr-buhrg), village, EILEAN SIAR, NW Scotland, on SW coast of Harris, LEWIS AND HARRIS island; 57°46′N 07°00′W. Village and harbor were part of industrial development of island by Lord Leverhulme in 1920s. Formerly called Obbe. Lighthouse.

Leverett (LEV-ret), town, FRANKLIN county, N central MASSACHUSETTS, 10 mi/16 km SSE of GREENFIELD; 42°28′N 72°29′W.

Leverger, Brazil: see SANTO ANTÔNIO DO LEVERGER.

Levering, village, EMMET county, NW MICHIGAN, 11 mi/18 km SSW of MACKINAW CITY, in lake region; 45°38′N 84°47′W. Ships farm produce. LAKE PARADISE to NE.

Leverington (LEV-uh-ring-tuhn), village (2001 population 3,978), N CAMBRIDGESHIRE, E ENGLAND, just NW of WISBECH; 52°40′N 00°08′W. Former agricultural market in fruit-growing region. Has 13th-century church.

Leverkusen (lai-ver-KOO-suhn), city, North Rhine-Westphalia, W GERMANY, port on the RHINE RIVER; 51°03′N 07°00′E. Industrial center and a road and railroad junction; manufacturing includes chemicals, pharmaceuticals, photographic film, machinery, and textiles. There is a noted chemical research library in the city. In castle Morsbroich is a museum of modern art.

Lever, Little (LEE-vuh), town (2001 population 11,505), GREATER MANCHESTER, W ENGLAND, 3 mi/4.8 km ESE of BOLTON; 53°33′N 02°22′W. Manufacturing of paper, plastics, and chemicals.

Leverstad, CONGO: see LUSANGA.

Leverville, CONGO: see LUSANGA.

Levetsova, Greece: see KROKEAI.

Levezou, FRANCE: see LÉVÉZOU.

Lévézou (lai-vai-zoo) or **Lévezou** (lai-ve-zoo), mountain range, AVEYRON department, MIDI-PYRÉNÉES region, S FRANCE, in MASSIF CENTRAL, extends c.15 mi/24 km S-N between TARN and AVEYRON rivers just W of MILLAU; 44°09′N 02°53′E. Rises to 3,800 ft/1,158 m. Sheep raising. Also spelled Levezou (le-ve-zoo).

Levice (le-VI-tse), German *Lewentz*, Hungarian *Léva*, city (2000 population 36,538), ZAPADOSLOVENSKY province, S SLOVAKIA; 48°13′N 18°36′E. Has a railroad junction; manufacturing (textiles, furniture, cosmetics, food). Has picturesque 18th-century castle ruins and museum of TEKOV region. Founded in the 12th century. Neolithic-era archaeological site. Several villages in vicinity noted for colorful regional costumes. A part of Hungary from 1938–1945.

Levico Terme (LAI-vee-ko TER-me), village, TRENTO province, TRENTINO–ALTO ADIGE, N ITALY, 10 mi/16 km SE of TRENT, in VALSUGANA, near LAGO DI LEVICO (1.5 mi/2.4 km long; a source of BRENTA RIVER); 46°01′N 11°18′E. Resort (elevation 1,660 ft/506 m) with mineral springs, thermal baths.

Levidi (le-VEE-[th]-ee), ancient *Elymia*, town, ARKADIA prefecture, central PELOPONNESE department, S mainland GREECE, 13 mi/21 km NNW of Trípolis; 37°41′N 22°18′E. Wheat, tobacco; livestock (sheep, goats). Scene of Greek victory over Turks (1821). Sometimes Levidhion, Levidion, or Levidhi.

Levie (luh-vee), commune (□ 33 sq mi/85.8 sq km), S central CORSICA, FRANCE, in CORSE-DU-SUD department, 10 mi/16 km NE of SARTÈNE; 41°42′N 09°07′E. Museum exhibits archeological relics excavated on Levie Plateau.

Levier (le-vye), commune (□ 14 sq mi/36.4 sq km), DOUBS department, FRANCHE-COMTÉ region, E FRANCE, in the JURA Mountains, 12 mi/19 km WNW of PONTARLIER; 46°57′N 06°08′E. Forestry, cheese making. The forest of Joux, which extends c.20 mi/32 km N-S from Levier to CHAMPAGNOLE, is a well-managed spruce forest with many old specimen trees. It is traversed by a scenic road.

Levi Iskur (LE-vee EES-kuhr), river, SW BULGARIA; 42°13′N 23°31′E. Tributary of the ISKUR River.

Levikha (LYE-vee-hah), town (2006 population 3,435), central SVERDLOVSK oblast, E URALS, extreme W Siberian Russia, on crossroads, 58 mi/93 km N of YE-KATERINBURG, and 12 mi/19 km NNW, and under jurisdiction, of KIROVGRAD; 57°35′N 59°54′E. Elevation 826 ft/251 m. Railroad spur terminus; copper- and zinc-mining, supplying the Kirovgrad refinery; sawmilling.

Levin (luh-VIN), town (□ 411 sq mi/1,068.6 sq km; 2006 population 19,134), WANGANUI-MANAWATU region, S NORTH ISLAND, NEW ZEALAND, 59 mi/95 km N of WELLINGTON; HOROWHENUA district (□ 411 sq mi/1,063 sq km). On narrow but productive coastal plain, with good road, rail links; servicing of intensive dairying, vegetable, fruit research area.

Levin, CZECH REPUBLIC: see USTEK.

Lévis (lai-VEE), former county (□ 272 sq mi/707.2 sq km), S QUEBEC, E CANADA, on the SAINT LAWRENCE RIVER, just S of QUEBEC city; 46°40′N 71°10′W. The city of LÉVIS was its county seat.

Lévis (lai-VEE), city (□ 171 sq mi/444.6 sq km; 2006 population 127,352), CHAUDIÈRE-APPALACHES region, S QUEBEC, E CANADA, 2 mi/3 km from QUEBEC city; 46°48′N 71°11′W. Large refinery; financial services; foods; building of ships and heavy industrial products; tourism. Was seat of historic LÉVIS county. Restructured in 2002. Part of the Metropolitan Community of Quebec (*Communauté Métropolitaine de Québec*).

Levisa Bay (lai-VEE-sah), small inlet, of the ATLANTIC OCEAN, in HOLGUÍN province, E CUBA, on N coast, 7 mi/11 km E of MAYARÍ; 8 mi/13 km long, 4 mi/6 km wide. Connects with sea through narrows; 20°43′N 75°31′W. Boca Carenerito Nicaro is on S shore.

Levisa Fork (luh-VAL-zuh), river, 164 mi/264 km long, SW VIRGINIA and E KENTUCKY; rises in E BUCHANAN county, Virginia; flows NW past GRUNDY, into PIKE county, Kentucky, through FISHTRAP LAKE reservoir, past PIKEVILLE and PRESTONSBURG, and N past PAINTSVILLE, joins TUG FORK at LOUISA to form BIG SANDY RIVER; 38°07′N 82°36′W. Partially navigable; has locks, dams. Receives RUSSELL FORK RIVER from SE 3 mi/5 km W of Fishtrap Lake.

Levitha (LE-ree-thah), Italian *Levita*, uninhabited island (□ 4 sq mi/10.4 sq km), DODECANESE prefecture, SOUTH AEGEAN department, GREECE; 37°00′N 26°30′E. Considered part of Greece during Italian rule of the DODECANESE.

Levittown, unincorporated city (2000 population 53,966), Bristol township, BUCKS county, E PENNSYLVANIA, suburb 20 mi/32 km NE of downtown PHILADELPHIA, and 6 mi/9.7 km SW of TRENTON, NEW JERSEY, near DELAWARE RIVER; 40°08′N 74°50′W. Manufacturing (fabricated metal products, electronic equipment, printing and publishing, plastic products). It was the second housing development built (1951–1955) by Levitt and Sons, Incorporated, who repeated the low-cost residence plan of their Levittown, New York, development. The very name itself, Levittown, has come to symbolize the U.S. post-World War II suburban phenomenon, which first gave middle-class families the option of inexpensive, single-unit housing outside the urban sector. Pennsbury Manor State Park to E; Historic FALLSINGTON to NE. Van Sliver Lake (backwater lake) to E.

Levittown, unincorporated suburban town (□ 6 sq mi/15.6 sq km; 2000 population 53,067), NASSAU county, SE NEW YORK, on LONG ISLAND; 40°43′N 73°30′W. It was originally developed by Levitt and Sons, Inc. as a mass-produced area of private, low-cost housing, and became the propotoype for many postwar housing developments throughout the country. Founded 1947.

Levka, Mount, Greece: see LEFKA ORI.

Levkimmi, Greece: see LEFKIMMI.

Levkosia, CYPRUS: see NICOSIA.

Levoca (le-VO-chah), Slovak *Levoča*, German *Leutschau*, Hungarian *Löcse*, town, VYCHODOSLOVENSKY province, N central SLOVAKIA, 36 mi/58 km NW of KOŠICE; 49°02′N 20°36′E. Railroad terminus; manufacturing (textiles, machinery; wood products). Has an old cultural center famous for its late-Gothic and Renaissance architecture: St. James's cathedral, began in the 13th century, with its remarkable woodwork and fifteen altars; a 14th-century Franciscan monastery and church; a 16th-century town hall-turned-museum; medieval burghers' houses; remains of fortifications. Wood carving blossomed here in the Middle Ages; printing in the 16th century.

Levokumskoye (lye-vuh-KOOM-skuh-ye), village (2006 population 11,450), E STAVROPOL TERRITORY, N CAUCASUS, S European Russia, on the left bank of the KUMA RIVER, on road, 25 mi/40 km E of BU-DËNNOVSK; 44°49′N 44°39′E. Elevation 200 ft/60 m. Wine making, flour milling.

Levroux (le-vroo), commune (□ 21 sq mi/54.6 sq km), INDRE department, CENTRE administrative region, central FRANCE, 12 mi/19 km NNW of CHÂTEAUROUX; 46°59′N 01°37′E. Noted for its goat cheese. Tanning, small-scale manufacturing. Has fine 13th-century church of Saint-Sylvain built atop ruins of a Gallo-Roman temple.

Levshino, RUSSIA: see PERM, city.

Levski (LEV-skee), city (1993 population 13,222), LOVECH oblast, Levski obshtina, N BULGARIA, near the OSUM RIVER, 26 mi/42 km ESE of PLEVEN; 43°22′N 25°10′E. Railroad junction; grain trading, poultry raising center; manufacturing (machinery, foodstuffs, hemp). Formerly known as Kara Agach.

Levski (LEV-skee), peak (7,104 ft/2,166 m) in the Troyan Mountains, N central BULGARIA, 5 mi/8 km NNW of KARLOVO; 42°43′N 24°47′E. Also called Ambaritsa.

Lev Tolstoy (LYEF tuhl-STO-yee), town (2006 population 8,930), NE LIPETSK oblast, S central Russia, on road, 22 mi/35 km W of CHAPLYGIN; 53°12′N 39°27′E. Elevation 692 ft/210 m. Railroad junction; railroad shops. Agricultural chemicals; bakery. Station room, where the renowned Russian novelist Lev N. Tolstoy died (1910), is now a museum. Until 1927, called Astapovo.

Lev Tolstoy, RUSSIA: see L'VA TOLSTOGO.

Levuka (le-VOO-kah), entry port town, E OVALAU, FIJI, SW PACIFIC OCEAN; colony (1874–1882). Site of early white settlement in Fiji. Bananas, pineapples. Former copra and sandalwood center. Now focus of Japanese and Fijian commercial fishing industry.

Levunovo (le-VOO-no-vo), village, SOFIA oblast, SANDANSKI obshtina, SW BULGARIA, on the Struma River, 8 mi/13 km NE of PETRICH; 41°29′N 23°28′E. Trades in tobacco, rice, cotton.

Levy (LE-vee), county (□ 1,412 sq mi/3,671.2 sq km; 2006 population 39,076), N central FLORIDA, bounded by GULF OF MEXICO (S, W) and by SUWANNEE (W) and WITHLACOOCHEE (S) rivers; ☉ BRONSON; 29°16′N 82°47′W. Flatwoods area, with many small lakes and some swamps. Livestock raising (hogs, cattle), farming (corn, vegetables, peanuts), lumbering, fishing, and some quarrying (limestone, dolomite). Formed 1845.

Levyye Lamki, Vtoryye, RUSSIA: see VTORYYE LEVYYE LAMKI.

Lewarae (LOO-uh-rai), unincorporated village, RICHMOND county, S NORTH CAROLINA, 4 mi/6.4 km W of ROCKINGHAM.

Lewe (LE-we), southwesternmost township, MANDALAY division, MYANMAR, 10 mi/16 km SW of PYINMANA, and served by two railroad lines, one to YANGON and one to CHAUK oil fields; 19°38′N 96°07′E.

Lewellen, village (2006 population 243), GARDEN county, W NEBRASKA, 10 mi/16 km ESE of OSHKOSH, and on NORTH PLATTE RIVER; 41°19′N 102°08′W. Livestock, grain, sunflower seeds. Ash Hollow State Historical Park to SE.

Lewes (LYOO-wis), town (2001 population 15,988) and district, SE ENGLAND; ☉ East SUSSEX, 8 mi/12.8 km SSW of UCKFIELD; 50°52′N 00°01′E. Farm market; light manufacturing. St. Pancras priory was founded here in the 11th century; its ruins remain. In 1264, Lewes was the scene of a victory by Simon de Montfort, Earl of Leicester, over Henry III. T. Paine lived here briefly.

Lewes (LOO-is), resort town (2000 population 2,932), SUSSEX county, SE DELAWARE, 15 mi/24 km NE of GEORGETOWN, just W of CAPE HENLOPEN, on LEWES AND REHOBOTH CANAL; shore of DELAWARE BAY 2 mi/3.2 km to NE; 38°46′N 75°09′W. Elevation 9 ft/2 m. Railroad and car (U.S. Highway 9) ferry to CAPE MAY, NEW JERSEY. Deep-sea fishing; sand and gravel pits. Port of entry. Former Fort Miles Military Reservation to E, is now Cape Henlopen State Park. Prime Hook National Wildlife Reserve to NW. Settled by Dutch in 1631 as first white settlement along the Delaware River. Many historic homes and buildings still stand. Historic and maritime exhibits are at Zwaanendael Museum. Incorporated 1857.

Lewes and Rehoboth Canal, SE DELAWARE, waterway connecting DELAWARE BAY and N end of REHOBOTH BAY; c.15 mi/24 km long; 38°45′N 75°06′W. Starts in N at BROADKILL CREEK, extends SE parallel to DELAWARE BAY, past LEWES and CAPE HENLOPEN (to NE) turns S parallel to ATLANTIC coast, to N end of Rehoboth Bay. For small craft. N entrance 2 mi/3.2 km NW of Lewes; S entrance to ocean via INDIAN RIVER BAY.

Lewes River (LOO-is), 338 mi/544 km long, in S YUKON, CANADA, on the upper course of YUKON RIVER; issues from TAGISH LAKE on BRITISH COLUMBIA border, flows N, through Marsh Lake, past WHITEHORSE (head of navigation), through LAKE LABERGE, and NW past CARMACKS, to confluence with PELLY RIVER at FORT SELKIRK, forming Yukon River proper. Receives Teslin River.

Lewin Brzeski (LE-veen BREE-skee), German *Löwen*, town in LOWER SILESIA, after 1945 in OPOLE province, SW POLAND, on the GLATZER NEISSE (head of navigation), and 10 mi/16 km SE of BRZEG (Brieg); 50°45′N 17°37′E. Sugar refining, tile manufacturing, sawmilling. After 1945, briefly called Lubien, Polish *Lubień*.

Lewin Kłodzki (LE-veen KLOTS-kee), German *Hummelstadt*, town in LOWER SILESIA, Wałbrzych province, SW POLAND, near CZECH border, at SW foot of HEUSCHEUER MOUNTAINS, 16 mi/26 km WSW of KLODZKO (Glatz); 50°24′N 16°18′E. Health resort. Until 1938, German name was Lewin.

Lewis, county (□ 479 sq mi/1,245.4 sq km; 2006 population 3,756), W IDAHO; ☉ NEZPERCE; 46°14′N 116°26′W. Agricultural area bounded on E by CLEARWATER RIVER; on, in part, SALMON RIVER (extreme S), and Lawyer's Creek. Cattle; alfalfa; barley; wheat; beans; lumber; wholesale and retail trade. Includes part of Nez Perce Indian Reservation, covers all but S end of county. Winchester Lake State Park on W boundary. Formed 1911.

Lewis, county (□ 495 sq mi/1,287 sq km; 2006 population 14,012), NE KENTUCKY; ☉ VANCEBURG; 38°32′N 83°29′W. Bounded N by OHIO RIVER (OHIO state line) SW by North Fork LICKING RIVER; drained by Grassy Fork and Kinniconick creeks. Rolling agricultural area (burley tobacco, hay, alfalfa, soybeans, wheat, corn; cattle, poultry; dairying); some manufacturing at Vanceburg. Formed 1806.

Lewis, county (□ 505 sq mi/1,313 sq km; 2006 population 10,152), NE MISSOURI; ☉ MONTICELLO; 40°06′N 91°43′W. Bounded E by MISSISSIPPI RIVER, drained by WYACONDA RIVER and NORTH and MIDDLE Fabius rivers. Corn, wheat, soy beans; hogs; cattle; lumber; manufacturing at LA GRANGE. Formed 1832.

Lewis, county (2006 population 26,685), N central NEW YORK; ☉ LOWVILLE; 43°47′N 75°27′W. Rises to foothills of the ADIRONDACK MOUNTAINS in E; drained by BLACK and MOOSE rivers (water power). Diversified light manufacturing and agriculture. Named for Morgan Lewis, New York governor at time of county's creation. Historically part of romantic scheme to settle refugees from French Revolution. Lake BONAPARTE named for Napoleon's brother Joseph, once King of Spain, who owned large tract of land here. Building of Black River canal from UTICA to CARTHAGE begun in 1836; it encompasses large part of Tug Hill Plateau, a winter recreational region and source of pulpwood for paper and lumber mills. Formed 1805.

Lewis, county (2006 population 11,588), central TENNESSEE; ☉ HOHENWALD; 35°31′N 87°30′W. Drained by BUFFALO RIVER and small Swan Creek. Diverse manufacturing. Meriwether Lewis National Monument located here on NATCHEZ TRACE PARKWAY. Formed 1843.

Lewis, county (□ 2,436 sq mi/6,333.6 sq km; 2006 population 73,585), SW WASHINGTON; ☉ CHEHALIS; 46°35′N 122°24′W. Drained by COWLITZ and CHEHALIS rivers. Peas, alfalfa, hay, oats, vegetables; dairying; poultry; quicksilver; food processing. Riffa Lake (MOSSYROCK DAM) and Mayfield Lake (Mayfield Dam) in S center; part of MOUNT SAINT HELENS NATIONAL VOLCANIC MONUMENT in SE. Includes parts of Mount Baker–Snoqualmie (NE) and Gifford Pinchot (E) national forests and S part of MOUNT RAINIER NATIONAL PARK (NE); includes Tatoosh Wilderness Area and parts of Goat Rocks and William O. Douglas

Wilderness areas, in E; LEWIS AND CLARK and Ike Kinswa state parks in S center; Rainbow Falls State Park in W. Formed 1845.

Lewis, county (□ 389 sq mi/1,011.4 sq km; 2006 population 17,129), central WEST VIRGINIA; ☉ WESTON; 39°00′N 80°30′W. On ALLEGHENY PLATEAU; drained by the West Fork River (a headstream of the MONONGAHELA RIVER). Agriculture (corn, potatoes, alfalfa, hay); cattle, sheep. Manufacturing at WESTON. Natural-gas and oil wells. STONEWALL JACKSON LAKE RESERVOIR (State Park and Wildlife Management Area) in S; part of Stonecoal Lake reservoir (Wildlife Management Area) in SE; Jackson's Mill historic site and state 4-H camp in N. Formed 1816.

Lewis, town (2000 population 438), CASS county, SW IOWA, on EAST NISHNABOTNA RIVER, and 8 mi/12.9 km SSW of ATLANTIC, near Cold Springs State Park; 41°18′N 95°04′W. In rich agricultural region; limestone quarry.

Lewis, village (2000 population 486), EDWARDS county, S central KANSAS, 9 mi/14.5 km E of KINSLEY; 37°56′N 99°15′W. In grain area. Light manufacturing.

Lewis, lumbering village, ESSEX county, NE NEW YORK, in the ADIRONDACK MOUNTAINS, 30 mi/48 km SSW of PLATTSBURGH; 44°17′N 73°34′W.

Lewis, Scotland: see LEWIS AND HARRIS.

Lewis and Clark, county (□ 3,497 sq mi/9,057 sq km; 1990 population 47,495; 2000 population 55,716), W central MONTANA; ☉ HELENA; 47°07′N 112°23′W. Mountain region crossed by CONTINENTAL DIVIDE that passes through W half of county; drained by MISSOURI, Dearborn, and BLACKFOOT rivers. N boundary formed by North Fork SUN and SUN rivers (forms Gibson Reservoir; NW). Wheat, barley, oats, hay, cattle, sheep. Sand and gravel, gold. Missouri River forms 3 reservoirs: HOLTER LAKE in E, HAUSER LAKE (and Black Sandy State Park) in SE, and part of large Canyon Ferry Lake (and state park) in SE corner. Gates of the Mountains (Missouri River Canyon) in E. Part of Lewis and Clark National Forest in NW, all three forests include parts of Bob Marshall and Scapegoat wilderness areas, in NW. Lincoln State Forest and Great Divide Ski Area in SW; sections of Helena National Forest in SE and W; Willow Creek Wildlife Management Area and Reservoir in N. Initially formed 1865; present boundaries established 1941. First called Edgerton county, then Lewis and Clarke county (1870); the "e" was later dropped.

Lewis and Clark Affiliated Area, MISSOURI, NEBRASKA, SOUTH DAKOTA, NORTH DAKOTA, MONTANA, IDAHO, OREGON, historic trail (4,500 mi/7,242 km long), commemorates the Lewis and Clark expedition. Authorized 1978.

Lewis and Clark Caverns State Park, SE JEFFERSON county, SW MONTANA, recreational area in canyon of JEFFERSON RIVER, 34 mi/55 km ESE of BUTTE. Includes large limestone cave consisting of underground passages and chambers fretted with stalactites and stalagmites. Formerly known as Morrison Cave and Lewis and Clark Caverns National Monument. Covers 2,735 acres/1,107 ha.

Lewis and Clark Lake, reservoir (□ 50 sq mi/130 sq km), on NE NEBRASKA (KNOX county) and extreme SE SOUTH DAKOTA (YANKTON county) border, on MISSOURI RIVER, 7 mi/11 km E of YANKTON; 42°51′N 97°29′W. Maximum capacity 540,000 acre-ft. Formed by Gavins Point Dam (74 ft/23 m high), built (1958) by Army Corps of Engineers for flood control; also used for power generation, irrigation, navigation, and recreation.

Lewis and Harris (LOO-is and HER-is), **Lewis with Harris**, **Lewis** or **The Lews**, island (□ 825 sq mi/2,145 sq km), largest and northernmost of the OUTER HEBRIDES, Eilean Siar, NW Scotland, 24 mi/39 km from the Scottish mainland across THE MINCH; 58°08′N 06°41′W. Harris has hilly terrain. Central Lewis is a vast, wet moor, uninhabited and unproductive. All the towns lie on the coast, and the bulk of the island's population is in STORNOWAY, which has a significant port, and the N parish of Ness. Crofting, fishing, and livestock raising are the main occupations. The thriving Harris tweed industry is centered in Stornoway, but utilizes home looms throughout the island. Gaelic is spoken. A prehistoric monument with large stones stands at CALLANISH in Lewis.

Lewisberry, borough (2006 population 389), YORK county, S PENNSYLVANIA, 9 mi/14.5 km S of HARRISBURG. Agriculture (grain; livestock; dairying); light manufacturing Gifford Pinchot State Park to S.

Lewisburg, city (2006 population 10,834), central TENNESSEE, 15 mi/24 km SE of COLUMBIA; ⊙ MARSHALL county; 35°27′N 86°48′W. Former trade, shipping, processing center for prosperous livestock-raising; now home to several nationally known industries and world headquarters of the Tennessee Walking Horse Breeders' and Exhibitors' Association. Incorporated 1837.

Lewisburg, town (2000 population 903), LOGAN county, S KENTUCKY, near MUD RIVER, 10 mi/16 km NNW of RUSSELLVILLE; 36°59′N 86°57′W. In agricultural and timber area; manufacturing (lumber, apparel). Lake Malone reservoir and State Park to NW.

Lewisburg, town (2006 population 3,561), SE WEST VIRGINIA, near GREENBRIER RIVER, 25 mi/40 km ENE of HINTON; ⊙ GREENBRIER county; 37°48′N 80°25′W. Elevation c.2,300 ft/701 m. Resort area. Grain, apples; tobacco; livestock; dairying. Limestone quarrying. Manufacturing (construction equipment, transportation equipment, food, printing and publishing). Greenbrier Valley Airport to NE. Greenbrier Military School. Old stone church (1796), and site of Fort Savannah (later Fort Union). Carnegie Hall (1902). Lost World Caverns to N; West Virginia State Fairgrounds to S; Greenbrier State Forest to SE; S terminus of Greenbrier River State Trail to E. Incorporated 1782.

Lewisburg (LOO-uhs-buhrg), village (□ 1 sq mi/2.6 sq km; 2006 population 1,774), PREBLE county, W OHIO, 21 mi/34 km WNW of DAYTON, on small Twin Creek; 39°51′N 84°32′W. Trade center for farm and orchard area; nurseries.

Lewisburg, borough (2006 population 5,578), central PENNSYLVANIA, 20 mi/32 km SSE of WILLIAMSPORT, on West Branch of SUSQUEHANNA RIVER; ⊙ UNION county. Agriculture area (grain, soybeans; poultry, livestock, dairying); manufacturing (food products, fabricated metal products, lumber, wood products, printing and publishing; gypsum. Bucknell University here. Lewisburg Federal Penitentiary to N. Tiadaghton State Forest to NW; Shamokin Mountain ridge to S. Laid out 1785, incorporated 1813.

Lewis Cass, Mount (6,864 ft/2,092 m), on ALASKA-BRITISH COLUMBIA border, in COAST RANGE, 50 mi/80 km E of Wrangell; 56°24′N 131°05′W.

Lewis, Fort, WASHINGTON: see Tacoma.

Lewisham (LYOO-wish-uhm), inner borough (□ 14 sq mi/36.4 sq km; 2001 population 248,922) of GREATER LONDON, SE ENGLAND, on the THAMES; 51°27′N 00°01′E. Mainly residential with a large shopping center; some light manufacturing. Lewisham, which was noted in Elizabethan times for its cattle market and royal dockyard, trades in timber. The writer Christopher Marlowe was killed in a brawl at Deptford in 1593. Goldsmith's College, a faculty of the University of London, is in the borough. Lewisham includes New Cross, an industrial district, and Sydenham, site of the Crystal Palace until 1936 (it burned down).

Lewis Hills, section of the LONG RANGE rising to 2,672 ft/814 m, SW NEWFOUNDLAND AND LABRADOR, S of the BAY OF ISLANDS. Highest point on the island of Newfoundland.

Lewis Pass, level pass (2,835 ft/864 m), between CANTERBURY and WEST COAST regions, NEW ZEALAND, in N central SOUTH ISLAND, where the SOUTHERN ALPS

meet the SPENSER MOUNTAINS to the N, the Spenser-Wairau river system to the E, and the Maruia-Buller system to the W. Open country for E-W traffic hot springs, small lakes, now Lewis Pass National Reserve.

Lewisport, town (2000 population 1,639), HANCOCK county, NW KENTUCKY, 17 mi/27 km NE of OWENSBORO, on the OHIO RIVER; 37°55′N 86°54′W. In agricultural area (burley and dark tobacco, grain; livestock). Squire Pate House (1822) and Emmick Plantation House (1850) to E.

Lewis Range, E Front Range of ROCKY MOUNTAINS in NW MONTANA, extends c.160 mi/257 km SSE from near Waterton Lake on U.S.-CANADA border, through GLACIER NATIONAL PARK, to BLACKFOOT RIVER NW of HELENA. Forms part of CONTINENTAL DIVIDE, follows length of range. Within park is highest portion of range, including some of the most spectacular summits of the Rockies; chief peaks are Mount Cleveland (10,466 ft/3,185 m, highest in range and park), Mount Stimson (10,142 ft/3,091 m), Kintla Peak (10,110 ft/3,079 m), Mount Jackson (10,052 ft/3,064 m), Mount Siyeh (10,014 ft/3,052 m), Going-to-the-Sun Mountain (9,642 ft/2,939 m). Within park, range is crossed by Marias and LOGAN passes.

Lewis River, c.95 mi/153 km long, W of Mount Adams, in NE SKAMANIA county, SW WASHINGTON; rising in the CASCADE RANGE, and flowing SW to S of MOUNT SAINT HELENS through to the COLUMBIA River c.15 mi/24 km N of VANCOUVER, Washington. Swift Reservoir, YALE LAKE reservoir, and Lake Mervin reservoir furnish hydroelectric power and form a string of lakes along the river's middle course.

Lewis River, c. 30 mi/48 km long, in NW WYOMING; rises in geyser region of S central YELLOWSTONE NATIONAL PARK; flows S through Shoshone Lake and Lewis Lake, to SNAKE RIVER near S boundary of park.

Lewis River, Idaho: see SNAKE RIVER.

Lewis Run, borough (2006 population 570), MCKEAN county, N PENNSYLVANIA, 5 mi/8 km S of BRADFORD, on East Branch of Tunungant Creek. Agriculture (grain; livestock, dairying); light manufacturing Bradford Regional Airport 4 mi/6.4 km to S. Allegheny National Forest to W.

Lewis Smith Lake, on border of Cullman and Walker counties, N central ALABAMA, on Sipsey Fork River, 34 mi/55 km NW of Birmingham; 33°57′N 87°05′W. Sprawling reservoir has 3 major arms; NW arm extends c.35 mi/56 km into William B. Bankhead National Forest. Max. capacity of 2,203,000 acre-ft. Formed by Lewis Smith Dam (265 ft/81 m high), built (1961) by the Alabama Power Co. for hydroelectric power generation and flood control. Major recreation area. Both the dam and lake were named after Lews M. Smith, an official of Alabama Power Co.

Lewiston, city (2000 population 30,904), NW IDAHO, at the WASHINGTON state line, and on SNAKE RIVER at mouth of CLEARWATER RIVER, opposite CLARKSTON, Washington, 185 mi/298 km NNW of BOISE, and 85 mi/137 km SSE of SPOKANE; ⊙ NEZ PERCE county; 46°23′N 117°00′W. It is the commercial and industrial center of a timber, grain, and livestock region (cattle; wheat, barley; alfalfa; vegetables, potatoes) that also has lime, clay, and silica deposits. The city has food processing plants, a large pulp and paper mill, a small arm ammunitions factory, and lumber factories. Lewis and Clark camped there in 1805. At nearby LAPWAI, Henry H. Spalding established (1836) a mission and operated the first printing press in the Pacific Northwest. Lewiston grew as a supply and shipping center after gold was discovered on the Clearwater River. It was the first capital (1863–1864) of Idaho Territory and had the first newspaper, the *Golden Age* (1862), in Idaho. Lewiston-Nez Perce Airport on S part of city. Lewis-Clark State College is in the city, Nez Perce County Historical Museum. Lowest point in Idaho 710 ft/216 m, 2 mi/3.2 km N, where Snake River turns W into Washington. Uma-

tilla National Forest to SW (Washington); Nez Perce Indian Reservation to E; Hells Gate State Park to S. Founded 1861.

Lewiston, industrial city (2000 population 35,690), ANDROSCOGGIN county, SW MAINE, on the E band of the ANDROSCOGGIN RIVER opposite AUBURN; 44°05′N 70°10′W. A 50 ft/15 m waterfall supplied power for early textile mills; diversified industry. Bates College (1855) and the Memorial Armoury (1927), with its large auditoriums, are in Lewiston. Nearby is a bird sanctuary. Incorporated 1795.

Lewiston, unincorporated town (2000 population 1,305), TRINITY county, NW CALIFORNIA, 7 mi/11.3 km ESE of WEAVERVILLE, on TRINITY RIVER (forms LEWISTON LAKE reservoir to N and the larger CLAIR ENGLE LAKE reservoir); 40°42′N 122°48′W. Trinity Dam, 7 mi/11.3 km NNE, and both lakes are in Trinity Unit of WHISKEYTOWN-SHASTA-TRINITY NATIONAL RECREATION AREA. Shasta-Trinity National Forest to NE and NW. Timber, hay, cattle, sheep, lambs.

Lewiston, town (2000 population 1,484), WINONA county, SE MINNESOTA, 13 mi/21 km WSW of WINONA; 43°58′N 91°52′W. Grain; livestock; poultry; dairying; manufacturing (feeds, fertilizers, electronics, printing and publishing). Richard J. Dorer Hardwood State Forest to S, N, and E. Whitewater Wildlife Area to NW.

Lewiston, town (2006 population 1,652), CACHE county, N UTAH, at IDAHO state line, 17 mi/27 km N of LOGAN, near BEAR RIVER, in Cache Valley; 41°57′N 111°52′W. Elevation 4,506 ft/1,373 m. Dairying and irrigated agricultural area; wheat, barley, alfalfa, sugar beets, vegetables; dairying (fifty-six dairy farms within city limits); manufacturing (plastic products). Settled 1870 by Mormons; incorporated 1904.

Lewiston, village, MONTMORENCY county, N MICHIGAN, c.45 mi/72 km SW of ALPENA, near Twin Lakes (each c.2 mi/3.2 km long); 44°53′N 84°18′W. In resort and agricultural region. Light manufacturing.

Lewiston, village (2006 population 79), PAWNEE county, SE NEBRASKA, 15 mi/24 km NW of PAWNEE CITY; 40°14′N 96°24′W.

Lewiston, village (□ 1 sq mi/2.6 sq km; 2006 population 2,662) and port of entry, NIAGARA county, W NEW YORK, on NIAGARA River (bridged here to QUEENSTON, ONTARIO), just N of NIAGARA FALLS, 20 mi/32 km NNW of BUFFALO; 43°10′N 79°02′W. Some manufacturing. Massive 2.4-million kw hydroelectric pumped storage project S and E of village. As with LEWISTON, named after Governor Morgan Lewis. Artpark, a 200-acre/81-ha theater and arts complex here. Settled c.1796, incorporated 1822.

Lewiston Lake, reservoir (□ 1 sq mi/2.6 sq km), TRINITY county, N central CALIFORNIA, on TRINITY RIVER, at S end of CLAIR ENGLE LAKE reservoir, in WHISKEYTOWN-SHASTA-TRINITY NATIONAL RECREATION AREA within Trinity National Forest, and 23 mi/37 km WNW of REDDING; 40°44′N 122°48′W. Maximum capacity 14,660 acre-ft. Formed by Lewiston Dam (91 ft/28 m high), built (1963) by the Bureau of Reclamation for irrigation.

Lewiston-Woodville (LOO-uhs-tuhn-WUD-vil), town (□ 2 sq mi/5.2 sq km; 2006 population 584), BERTIE county, NE NORTH CAROLINA, 15 mi/24 km NW of WINDSOR; 36°06′N 77°10′W. Manufacturing; retail trade; agricultural area (cotton, tobacco, grain, peanuts; poultry, livestock). Peanut Belt State Research Station nearby.

Lewistown, city (2000 population 2,522), central ILLINOIS, 36 mi/58 km WSW of PEORIA; ⊙ FULTON county; 40°23′N 90°09′W. In agricultural and bituminous-coal-mining area; corn, wheat, livestock. City was the home of Edgar Lee Masters. The territory, its people, and legends are reflected in his *Spoon River Anthology*. Lincoln and Douglas delivered speeches here. DICKSON MOUNDS MUSEUM is nearby. Settled 1821, incorporated 1857.

Lewistown, town (2000 population 595), LEWIS county, NE MISSOURI, near MIDDLE FABIUS RIVER, 6 mi/9.7 km W of MONTICELLO; 40°04′N 91°48′W. Agriculture.

Lewistown, town (2000 population 5,813), central MONTANA, on Big Spring Creek, and 90 mi/145 km ESE of GREAT FALLS; ⊙ FERGUS county; 47°04′N 109°26′W. Trade center for agriculture and mining region; coal, gypsum, kaolin clay (whiteware clay) gold mines; cattle, sheep, hogs, wheat, barley, oats, dairying; some manufacturing Judith Mountains to NE; Big Snowy Mountains in part of Lewis and Clark National Forest to S; Big Springs Trout Hatchery to SE. Laid out as Reed's Fort 1882, incorporated as Lewistown 1899.

Lewistown, village, FREDERICK county, N MARYLAND, 9 mi/14.5 km N of FREDERICK. The town was laid out by Daniel Fundenburg in 1815. A fish hatchery here, constructed by the WPA in the 1930s on Fishing Creek, was recently consolidated with the one on BEAVER CREEK near SOUTH MOUNTAIN because of a dwindling water supply.

Lewistown, borough (2006 population 8,582), central PENNSYLVANIA, 42 mi/68 km NW of HARRISBURG, on the JUNIATA RIVER; ⊙ MIFFLIN county; 40°36′N 77°34′W. Manufacturing (paper products, electronic equipment, machinery, printing and publishing, construction materials, fabricated metal products, apparel, wood products, plastic products). In a lush farm and dairy area (grain, soybeans; livestock, poultry, dairying). Many Amish live and farm in the surrounding area. Lewiston Airport to SW. McCoy House (late 1700s). BLUE MOUNTAIN ridge to SE, JACKS MOUNTAIN ridge to NW; Reeds Gap State Park to NE; Tuscarora State Forest to S. Incorporated 1795.

Lewistown Junction, unincorporated village, Granville township, MIFFLIN county, central PENNSYLVANIA, 1 mi/1.6 km SW of Lewiston, on Junction River; 40°35′N 77°34′W.

Lewisville (LOO-wis-vil), city (2006 population 94,589), DENTON county, N TEXAS, suburb 22 mi/35 km NNW of DALLAS–FORT WORTH; 33°02′N 96°58′W. Elevation 490 ft/149 m. Rapidly growing urban margin of Dallas in farm area (cattle; wheat; peanuts; cotton). Manufacturing (fabricated metal products, machinery, furniture, concrete, chemicals). LEWISVILLE LAKE reservoir and Lake Lewisville State Park to N; 5 mi/8 km-long dam (on Elm Fork TRINITY RIVER) on N edge of city.

Lewisville, town (2000 population 1,285), SW ARKANSAS, 27 mi/43 km E of TEXARKANA, near RED RIVER; ⊙ LAFAYETTE county; 33°21′N 93°34′W. Railroad junction. In diversified agricultural area; manufacturing (furniture, feeds), oil and gas field. Lake ERLING reservoir to S.

Lewisville, town (2000 population 395), HENRY county, E INDIANA, near FLATROCK RIVER, 9 mi/14.5 km S of NEW CASTLE; 39°49′N 85°21′W. In agricultural area. Laid out 1829.

Lewisville (LOO-uhs-vil), unincorporated town (□ 11 sq mi/28.6 sq km; 2006 population 12,444), FORSYTH county, N central NORTH CAROLINA, residential suburb 10 mi/16 km W of WINSTON-SALEM, near Yadkin (PEE DEE) RIVER; 36°05′N 80°24′W. Service industries; manufacturing; in agricultural area (tobacco, grain, cattle). Incorporated 1991.

Lewisville, village (2000 population 467), JEFFERSON county, SE IDAHO, 5 mi/8 km WNW of RIGBY, inside bend of SNAKE RIVER; 43°42′N 112°01′W. Elevation 4,790 ft/1,460 m. Railroad terminus of railroad spur from UCON. Grain; sugar beets; potatoes; alfalfa; cattle, sheep; dairying; manufacturing (food).

Lewisville, village (2000 population 274), WATONWAN county, S MINNESOTA, 11 mi/18 km ESE of ST. JAMES; 43°55′N 94°26′W. Grain, soybeans; livestock.

Lewisville (LOO-uhs-vil), village (2006 population 230), MONROE county, central OHIO, 15 mi/24 km S of BARNESVILLE; 39°46′N 81°13′W. In agricultural area.

Lewisville Lake (LOO-wis-vil), reservoir (15 mi/24 km long), at LEWISVILLE, DENTON county, N TEXAS, on Elm Fork of the TRINITY RIVER, 23 mi/37 km NW of DALLAS; 38°04′N 96°59′W. Maximum capacity 2,329,900 acre-ft. Has 8 mi/12.9 km NE arm. Formed by Lewisville Dam (118 ft/36 m high), built (1954) by the Army Corps of Engineers for water supply and flood control. Residential community of LAKE DALLAS on W shore. Lake Lewisville State Park on E shore.

Lewis with Harris, Scotland: see LEWIS AND HARRIS.

Lews, The, Scotland: see LEWIS AND HARRIS.

Lexington, county (□ 759 sq mi/1,973.4 sq km; 2006 population 240,160), central SOUTH CAROLINA; ⊙ LEXINGTON; 33°53′N 81°16′W. Bounded NE by CONGAREE RIVER, SW by North Fork of the EDISTO; drained by SALUDA RIVER (dammed to form Lake MURRAY) in N. In Sand Hill belt. Sand, clay, shale, granite kaolin; some agriculture (chickens, hogs; dairying; vegetables, corn, wheat, rye, oats, soybeans, hay, peaches). Residential and manufacturing suburbs of COLUMBIA in E. Formed 1785.

Lexington, city (2000 population 1,912), MCLEAN county, central ILLINOIS, 15 mi/24 km NE of BLOOMINGTON; 40°38′N 88°46′W. Trade and shipping center in rich agricultural area; corn, wheat, soybeans, livestock, dairy products. Settled 1828, incorporated 1867.

Lexington, city, central KENTUCKY, in the heart of the BLUEGRASS REGION; ⊙ FAYETTE county; 38°02′N 84°27′W. Elevation 983 ft/300 m. The outstanding center in the U.S. for the raising of thoroughbred horses (several hundred horse farms in 35 mi/56 km radius), it is also an important market for burley tobacco, cattle, hogs, and bluegrass seed as well as a railroad shipping point for E Kentucky's oil, coal, farm produce, and quarry products. Lexington has railroad junction and center, manufacturing (fixtures, printing and publishing, fabricated metal products, consumer goods, food, machinery, transportation equipment, electronic equipment). The city was named in 1775 by a group of hunters who were encamped on the site when they heard the news of the battle of Lexington. Lexington Airport (Blue Grass Field) in W; Civic Center (Rupp Arena); Bluegrass Army Depot (chemical weapons). The city is the seat of the University of Kentucky, Transylvania University Places of interest include "Ashland," the home of Henry Clay (designed by Latrobe in 1806 and rebuilt with the original materials in the 1850s); "Hopemont," the home of John Hunt Morgan (1814); the Thomas Hart house (1794); the home of Mary Todd Lincoln; and the library, which has a file of the *Kentucky Gazette,* founded by John Bradford in 1787. Headquarters of American Thoroughbred Breeding Center. Aviation Museum of Kentucky; Whitney Museum; Boone Station State Historic Site in SE; Waveland State Historic Site in S; Kentucky Horse Park (State Park) in N, includes International Museum of the Horse; Red Mile Harness Track (1875), Keeneland Racetrack in W. Lexington cemetery contains the graves of Clay, Morgan, J. C. Breckinridge, and the author James Lane Allen. A number of hospitals and a federal narcotics facility are located in the city. Lexington National Cemetery in N. Incorporated 1832.

Lexington, city (2000 population 4,453), W central MISSOURI, on MISSOURI River and 35 mi/56 km E of KANSAS CITY; ⊙ LAFAYETTE county; 39°10′N 93°52′W. Corn, sorghum, soybeans, wheat; cattle, hogs; manufacturing (apparel, wood products); former coal mines, limestone, rock quarries. Wentworth Military College (junior) here. Civil War battle, September 18–20, 1861. Historic court house (1847–1849). Numerous historic buildings. Laid out 1822, incorporated 1845.

Lexington, city (2006 population 10,251), ⊙ DAWSON county, S central NEBRASKA, 35 mi/56 km W of KEARNEY, and on PLATTE RIVER; 40°46′N 99°44′W.

Flour; grain, livestock, dairy and poultry products. Manufacturing (food, machinery). Johnson Reservoir and Johnson Lake State Recreation Area to S. Laid out as Plum Creek (on the OREGON TRAIL) 1872, changed to Lexington 1889.

Lexington (LEK-seeng-tuhn), city (□ 17 sq mi/44.2 sq km; 2006 population 20,382), central NORTH CAROLINA, 22 mi/35 km S of WINSTON-SALEM; ⊙ DAVIDSON county; 35°48′N 80°15′W. HIGH ROCK LAKE reservoir (Yadkin River) to SW. Manufacturing (paper products, food, machinery; printing and publishing, lumber, furniture, textiles); service industries. Davidson County Community College. Boone's Cave State Park to W. Incorporated 1827.

Lexington, city (2006 population 7,780), W TENNESSEE, 24 mi/39 km E of JACKSON; ⊙ HENDERSON county; 35°39′N 88°24′W. Trade center; retail and heavy manufacturing. Natchez Trace Forest State Park is nearby.

Lexington (LEKS-eeng-tuhn), independent city (□ 11 sq mi/28.6 sq km; 2006 population 6,739), NW VIRGINIA, in SHENANDOAH VALLEY; ⊙ surrounding ROCKBRIDGE county; 37°46′N 79°26′W. Manufacturing (printing and publishing, fixtures); agriculture in lush farm area (dairying; livestock, horses; grain, soybeans, apples, peaches). Bombarded and partially burned by General David Hunter in 1864. Virginia Military Institute (V.M.I.); Washington and Lee University (including burial site of Robert E. Lee and George C. Marshall Museum). Stonewall Jackson also buried in city; Stonewall Jackson Home. NATURAL BRIDGE to S, Virginia Horse Center to N. Laid out 1777, incorporated 1841.

Lexington, town (2000 population 30,355), MIDDLESEX county, E MASSACHUSETTS, a residential suburb 7 mi/11.3 km NW of BOSTON; 42°27′N 71°14′W. Manufacturing (printing and publishing, computer software). Corporate headquarters for major defense company. On April 19, 1775, the first battle of the Revolution was fought here. The site is marked by a monument on the triangular green, around which are several 17th-century buildings that include Buckman Tavern (1710), where the minutemen assembled. Other attractions are Monroe Tavern (1695), British headquarters during the battle; and the Hancock-Clarke House (1698), where John Hancock and Samuel Adams were awakened by Paul Revere's alarm. The first state normal school in the country, now Farmington State College, was established here in 1839. The theologian and reformer Theodore Parker was born in Lexington. Museum of Our National Heritage. Minutemen National Park. Settled c.1640, incorporated 1713.

Lexington, town (2000 population 2,214), ANOKA county, E MINNESOTA, 11 mi/18 km NNE of downtown MINNEAPOLIS; 45°08′N 93°10′W. Light manufacturing. Small natural lakes to E.

Lexington, town (2000 population 2,025), central MISSISSIPPI, 29 mi/47 km SSE of GREENWOOD; ⊙ HOLMES county; 33°07′N 90°02′W. Agriculture (cotton, corn, soybeans, sorghum; cattle); timber; manufacturing (textiles, apparel); sand and gravel. Booker-Thomas Museum; Hillside National Wildlife Refuge to W. Incorporated 1836.

Lexington, town (2006 population 2,073), CLEVELAND county, central OKLAHOMA, 33 mi/53 km SSE of OKLAHOMA CITY, on CANADIAN River, opposite PURCELL; 35°01′N 97°20′W. Trading point in agricultural area (wheat, oats, livestock; dairying); light manufacturing. Settled 1889.

Lexington, town (2006 population 14,110), central SOUTH CAROLINA, 12 mi/19 km W of COLUMBIA, near Lake MURRAY; ⊙ LEXINGTON county; 33°59′N 81°13′W. Manufacturing includes food, plastics, electronic equipment, printing and publishing, apparel; agricultural area includes chickens, eggs, hogs, dairying, vegetables, apples, and peaches.

Lexington (LEK-sing-tuhn), town (2006 population 1,255), LEE county, S central TEXAS, c.40 mi/64 km ENE of AUSTIN; 30°24′N 97°00′W. Elevation 456 ft/139 m. In agricultural area (cotton, cattle, corn); manufacturing (fixtures). Established 1850s.

Lexington, unincorporated town, COWLITZ county, SW WASHINGTON, residential suburb 3 mi/4.8 km N of KELSO, on COWLITZ RIVER, opposite mouth of Ostrander Creek. Mud flows from 1980 eruption of MOUNT ST. HELENS, which surged down Toutle-Cowlitz river system and reached as far as Lexington. Dairying, vegetables.

Lexington, village (2000 population 239), NE GEORGIA, 17 mi/27 km ESE of ATHENS; ⊙ OGLETHORPE county; 33°52′N 83°07′W.

Lexington, village, SCOTT county, S INDIANA, 7 mi/11.3 km ESE of SCOTTSBURG. In agricultural area. Lumber, crushed stone, agricultural lime. Founded c.1811.

Lexington, village (2000 population 1,104), SANILAC county, E MICHIGAN, 21 mi/34 km NNW of PORT HURON, on LAKE HURON; 43°16′N 82°31′W. Fisheries; lumber; tourism, resorts; light manufacturing.

Lexington (LEK-sing-tuhn), village (□ 4 sq mi/10.4 sq km; 2006 population 4,185), RICHLAND county, N central OHIO, 6 mi/10 km SW of MANSFIELD, on Clear Fork of MOHICAN RIVER; 40°40′N 82°35′W.

Lexington, village (2006 population 275), MORROW county, N OREGON, 9 mi/14.5 km NW of HEPPNER, on WILLOW CREEK; 45°27′N 119°41′W. Agriculture (wheat, alfalfa, potatoes, onions; sheep, cattle).

Lexington, hamlet, GREENE county, SE NEW YORK, in the CATSKILL MOUNTAINS, on SCHOHARIE CREEK, 24 mi/39 km NW of SAUGERTIES; 42°12′N 74°21′W. In summer-recreation area.

Lexington Hills, unincorporated town (2000 population 2,454), SANTA CLARA county, W CALIFORNIA; residential suburb 11 mi/18 km SSW of downtown SAN JOSE; 37°10′N 121°58′W. Surrounds Lexington Reservoir (Los Gatos Creek) in SANTA CRUZ MOUNTAINS; Sierra Azul to E.

Lexington Park, village (2000 population 11,021), SAINT MARYS county, S MARYLAND, 10 mi/16 km E of LEONARDTOWN; 38°16′N 76°27′W. Patuxent Naval Air Test Center is nearby. On the grounds of the center, Mattapany, built early in the 18th century, is the official residence of the commanding officer. First established as a mission by Jesuits, it was reclaimed by Lord Baltimore in 1641 to prevent the priests from establishing independent communities. It was called Jarboesville, for an early postmaster, until 1950.

Lexmond (LEKS-mawnt), village, SOUTH HOLLAND province, W central NETHERLANDS, near LEK RIVER, 8 mi/12.9 km S of UTRECHT; 51°58′N 05°02′E. Dairying; cattle raising; vegetables, sugar beets; some manufacturing.

Lexouríon, Greece: see LIXOURI.

Leyburn (LAI-buhrn), town, QUEENSLAND, NE AUSTRALIA, 136 mi/219 km SW of BRISBANE; 28°01′S 151°34′E. Sawmilling; former gold-mining center.

Leyburn (LAI-buhn), town (2001 population 2,208), NORTH YORKSHIRE, N ENGLAND, near URE RIVER, 7 mi/11.3 km SSW of RICHMOND; 54°18′N 01°49′W.

Leyden (LEI-den), agricultural town, FRANKLIN county, NW MASSACHUSETTS, on GREEN RIVER, and 8 mi/12.9 km N of GREENFIELD; 42°42′N 72°38′W.

Leyden, NETHERLANDS: see LEIDEN.

Leye, FRANCE: see LYS (Flemish *Leie*).

Leye (LUH-YE), town, W GUANGXI ZHUANG AUTONOMOUS REGION, CHINA, 60 mi/97 km N of BOSE, near HONGSHUI River (GUIZHOU border); ⊙ Leye county (1990 population 132,204); 24°48′N 106°34′E. Grain; food industry, construction materials.

Leyland (LEE-luhnd), town (2001 population 37,103), W central LANCASHIRE, NW ENGLAND, 5 mi/8 km S of PRESTON; 53°41′N 02°42′W. Steel industry; manufacturing of heavy vehicles (trucks and buses), chemicals,

health care products; textile industry. Has Elizabethan mansion, 17th-century inn.

Leymebamba, PERU: see LEIMEBAMBA.

Leyptsig, UKRAINE: see SERPNEVE.

Leyptsigskaya, UKRAINE: see SERPNEVE.

Leyre River (LER), 50 mi/80 km long in NW LANDES and GIRONDE departments, AQUITAINE region, SW FRANCE; rises in the LANDES region, flows to the ARCACHON BASIN of the Bay of BISCAY; 44°35′N 00°52′W. It traverses the Regional Park of the Landes of Gascony (LANDES DE GASCOGNE NATURAL REGIONAL PARK).

Leysdown (LAIZ-doun), village (2006 population 1,700), KENT, SE ENGLAND, E end of Isle of SHEPPEY, on THAMES estuary, 6 mi/9.7 km SW of MINSTER; 51°24′N 00°55′E. Sailing, fishing.

Leysin (leh-SAHN), commune, VAUD canton, SW SWITZERLAND, 5 mi/8 km NE of AIGLE. Elevation 4,111 ft/1,253 m. Year-round health resort overlooking RHÔNE valley and DENTS DU MIDI with numerous sanatoria for the tubercular.

Leyte (LAI-te), province (□ 2,206 sq mi/5,735.6 sq km), in EASTERN VISAYAS region, N LEYTE, PHILIPPINES; ⊙ TACLOBAN; 10°55′N 124°40′E. Population is 33.3% urban, 66.7% rural; in 1991, 100% of urban and 66% of rural settlements had electricity. Leyte Valley on E side of island (populated by Waray) and Ormoc Valley on W (populated by Cebuano) separated by range of extinct volcanoes. In typhoon belt, with year-round rain; heaviest in November–January. Agricultural region (coconuts, rice, abaca, sugarcane, tobacco as cash crops; corn as food crop). Mining (pyrites, limestone, phosphates); logging; fishing. Food processing. Wood, metal, ceramic handicrafts. Industrial complex at ISABEL (copper smelting, phosphate fertilizers) created in part to stem Isabel outmigration; power supplied to area from Tongonan Geothermal Plant NW of ORMOC. Tacloban is commercial center for entire region, with airport and international port. Secondary port and airport at Ormoc. Airstrip at HILONGOS. Spanish settlement from 1596; province created 1768. Site of initial landings and intensive battles during American liberation of Philippines (1944).

Leyte (LAI-te), town, LEYTE province, N LEYTE, PHILIPPINES, at head of Leyte Bay (small inlet of Biliran Strait; 7 mi/11.3 km long, 2 mi/3.2 km wide), 36 mi WNW of TACLOBAN; 11°22′N 124°30′E. Agricultural center (rice, sugarcane).

Leyte (LAI-te), island (□ 2,785 sq mi/7,241 sq km), one of the VISAYAN ISLANDS, PHILIPPINES, between LUZON and MINDANAO; 10°50′N 124°50′E. Area includes 116 islands, sixty-four of which are named. A fertile agricultural land, it is the nation's leading producer of sweet potatoes and bananas and a major producer of corn and peanuts. It has commercial coconut plantations and extensive forest reserves; lumbering is an important industry. In World War II, Leyte was occupied by the Japanese in early 1942. It was the scene of the first main American landing (October 20, 1944) in the campaign to recover the Philippines. That landing was followed by the battle of LEYTE GULF, the greatest naval engagement of all time, in which American naval forces destroyed the Japanese fleet.

Leyte Gulf (LAI-te), large inlet of PHILIPPINE SEA between LEYTE and SAMAR, PHILIPPINES. HOMONHON ISLAND guards the entrance, and SURIGAO STRAIT opens S to MINDANAO SEA; a narrow channel in N, between Leyte and Samar, leads to SAMAR SEA and the inland waters. In World War II, here was fought the battle of LEYTE GULF (also called second battle of the Philippine Sea) on October 23–26, 1944. The battle, between almost all of the Japanese fleet and the U.S. Third and Seventh Fleets, resulted in a great U.S. victory, which definitively destroyed Japan's sea power. The action was fought in three general areas: in Surigao Strait; off Samar; and off Cape Engaño.

Leyton, ENGLAND: see WALTHAM FOREST.

Lezajsk (LE-zah-yuhsk), Polish *Leż ajsk*, town, Rzeszów province, SE POLAND, on railroad, and 24 mi/39 km NE of RZESZÓW, near SAN RIVER; 50°16′N 22°25′E. Brick and cement-works; tanning, distilling. Cloister.

Lézardrieux (lai-zahr-dree-u), commune (□ 4 sq mi/10.4 sq km), CÔTES-D'ARMOR department, W FRANCE, in BRITTANY, on estuary of TRIEUX RIVER, on ENGLISH CHANNEL, and 22 mi/35 km NW of SAINT-BRIEUC; 48°46′N 03°06′W. Agriculture (early potatoes), fishing, cattle raising. Has 18th-century church with a striking bell tower.

Lezay (luh-zai), commune (□ 17 sq mi/44.2 sq km), DEUX-SÈVRES department, POITOU-CHARENTES region, W FRANCE, 22 mi/35 km ESE of NIORT; 46°16′N 00°01′W. Dairying.

Lezha (lee-ZHAH), town, S VOLOGDA oblast, central European Russia, on the Lezha River (right affluent of the SUKHONA River), on railroad, 38 mi/61 km SE of VOLOGDA; 58°56′N 40°45′E. Elevation 459 ft/139 m. Flax processing, dairying.

Lezhë (LE-ZHE), ancient *Lissus*, town (□ 190 sq km; 2001 population 14,495), N ALBANIA, on left and near mouth of old DRIN River arm, 20 mi/32 km SSE of SHKODËR, on main Tiranë-Shkodër road; ⊙ Lezhë district (□ 190 sq mi/480 sq km; 2001 population 159,182); 41°48′N 19°40′E. Linked by road with its ADRIATIC SEA port, SHËNGJIN. Copper mining at nearby new town of RUBIK (7 mi/11.3 km E). Has ruins of 15th-century Venetian citadel with tomb of Scanderbeg (died here 1468). Founded (4th century B.C.E.) as a Syracusan colony; held (1393–1478) by Venetians. Also Leshi.

Lezhi (LUH-JI), town, central SICHUAN province, CHINA, 60 mi/97 km ESE of CHENGDU; ⊙ Lezhi county (1990 population 838,826); 30°18′N 105°00′E. Rice, sugarcane, oilseeds; food and beverages, textiles.

Lezhnëvo (leezh-NYO-vuh), town (2005 population 7,940), S IVANOVO oblast, central European Russia, on the Ukhtoma River (KLYAZ'MA RIVER basin), 16 mi/26 km S of IVANOVO; 56°46′N 40°53′E. Elevation 351 ft/106 m. Cotton textiles; agricultural machinery repair. Has a heritage museum and theater. Known since the early 13th century. Throughout the 17th century, owned by the ancestral family of the renowned Russian poet and writer Alexander S. Pushkin.

Lézignan-Corbières (lai-zee-nyahn–kor-byer), town (□ 14 sq mi/36.4 sq km), AUDE department, LANGUEDOC-ROUSSILLON region, S FRANCE, 12 mi/19 km W of NARBONNE; 43°12′N 02°46′E. Wine-trading center. Manufacturing of casks and vineyard equipment.

Lezo (LAI-tho), town, GUIPÚZCOA province, N SPAIN, 4 mi/6.4 km E of SAN SEBASTIÁN; 43°19′N 01°54′W. Chemical works; rubber processing.

Le Zoute, BELGIUM: see KNOKKE-HEIST.

Lezoux (luh-zoo), town (□ 13 sq mi/33.8 sq km), PUY-DE-DÔME department, AUVERGNE region, central FRANCE, in the LIMAGNE, lowland (within the MASSIF CENTRAL), 15 mi/24 km ENE of CLERMONT-FERRAND; 45°50′N 03°23′E. Agriculture market. Ceramic objects with distinctive seals, produced in Gallo-Roman times, are on exhibit.

Lezuza (lai-THOO-thah), town, ALBACETE province, SE central SPAIN, 27 mi/43 km W of ALBACETE; 38°57′N 02°21′W. Cereals, saffron, sheep, goats.

Lgarara, Lake (ee-gah-RAH-rah), in N TANZANIA, SHINYANGA, and ARUSHA regions, 115 mi/185 km WNW of ARUSHA, on E boundary of SERENGETI NATIONAL PARK; 2 mi/3.2 km wide; 02°59′S 35°01′E. OLDUVAI GORGE to E.

L'gov (LGOF), city (2005 population 23,100), W KURSK oblast, SW European Russia, on the SEYM RIVER (tributary of the DESNA River), on crossroads and railroad, 50 mi/80 km W of KURSK; 51°41′N 35°17′E. Elevation 538 ft/163 m. Railroad junction. Center of agricultural area; food products, wine. The city grew

---

Cross-references are shown in SMALL CAPITALS. The pronunciation guide is shown on page xix. The sources of population figures are shown on page xvii.

around a monastery built here in 1669. Made city in 1779. Site of an experimental federal penal colony.

Lhakhang (LAH-KANG), town, S TIBET, SW CHINA, in central ASSAM HIMALAYAS, near BHUTAN border, on tributary of MANAS RIVER, and 110 mi/177 km S of LHASA. Elevation 9,850 ft/3,002 m.

Lhanbryde (LAHN-breid) or **Lhanbryd**, agricultural village (2001 population 1,845), MORAY, NE Scotland, 4 mi/6.4 km ESE of ELGIN; 57°37′N 03°13′W. Formerly in Grampian, abolished 1996.

Lhari, CHINA: see LHORONG.

Lharigo, CHINA: see LHORONG.

Lhasa (LAH-SAH), city (□ 214 sq mi/556.4 sq km; 2000 population 139,822), SW CHINA, on a tributary of the Yarlung Zangbo (BRAHMAPUTRA) River; ⊙ TIBET Autonomous Region; 29°39′N 91°06′E. Elevation c.11,800 ft/3,597 m. Lhasa is the chief Tibetan trade center, connected by road with the Chinese provinces of QINGHAI, SICHUAN, and XINJIANG UYGUR AUTONOMOUS REGION, and with INDIA, KASHMIR, and Nepal. Railway (2006), highest in the world, connects to Chinese cities. Airport. Manufacturing includes chemicals, machinery, and wool and leather products. Because of the remoteness of the city and the traditional hostility of the Tibetan clergy toward foreigners, Lhasa has long been called the Forbidden City. Prior to 1951 when Tibet came under the Chinese government, Lhasa was the center of Lamaism, and about one-half of its population were Lamaist monks. Lhasa has little noteworthy architecture, but there are impressive religious edifices. On a nearby hill, there stands the magnificent Potala, the former palace of the Dalai Lama, a gigantic block of buildings nine stories high, home for thousands of monks. A smaller palace of the Dalai Lama is set in the beautifully wooded grounds of Jewel Park. Near the city is the Drepung monastery, one of the largest in the world. The holiest temple in Lhasa, unimpressive from the outside, is the Jokang, which contains a jeweled image of the young Buddha. Several of the religious edifices were damaged during China's imposition of direct political control over Tibet (1959–1960), during which the Dalai Lama and other Tibetans fled to India. Increased protests and uprisings in the late 1980s against Chinese control of Tibet led China to impose (March 1989) martial law on the region. A modern highway bridge, made of reinforced concrete (c.2,400 ft/730 m long), crosses the river at Lhasa. The city's name also appears as Lassa, Lasa, or La-sa.

Lhasa He (LAH-SAH HUH), river, 195 mi/314 km long, SE TIBET, SW CHINA; chief left tributary of upper BRAHMAPUTRA RIVER; rises in E central NYENCHEN TANGLHA range in several headstreams joining 40 mi/64 km NNE of LHASA; flows SE and SW, past Gandaingoin, DAGZE (Dechen), and LHASA, to the Brahmaputra 4 mi/6.4 km N of GONGGAR. Used as transportation (mainly yak-hide coracles) artery for Lhasa. Also called Kyi Chu, Kyi Qu, or Ky Qu.

Lhaze (LAH-ZI), town, S TIBET, SW CHINA, on the BRAHMAPUTRA RIVER, on main LEH-LHASA trade route, and 75 mi/121 km W of XIGAZE; 29°08′N 87°43′E. Elevation 13,010 ft/3,965 m. Lamasery.

L'Hers, FRANCE: see HERS RIVER.

L'hirondelle (LI-rawn-DEL), amusement park near Oteppe, LIÈGE province, E BELGIUM, 7 mi/11.3 km NW of HUY.

Lhokseumawe (lawk-suh-MAH-wai), town, N coast of ACEH province, NE SUMATRA, INDONESIA, port on Strait of MALACCA, 130 mi/209 km ESE of BANDA ACEH; ⊙ North Aceh district; 05°10′N 97°08′E. Trade center for agricultural area, shipping resin, copra, spices, coffee, tea. Port shipping liquified natural gas from nearby ARUN natural-gas field. Also Lhoseumawe; formerly Lho Sumawe or Lho Somawe.

Lhorong (LO-RONG), town, ⊙ Lhorong county, E TIBET, SW CHINA, in KHAM region, 250 mi/402 km NE of LHASA, and on road to QAMDO; 30°45′N 95°49′E.

Farming, livestock raising. Hot springs nearby. Also appears as Lhari, Lharigo, or Luolong.

Lho Sumawe, INDONESIA: see LHOKSEUMAWE.

Lhotse (LUHT-sai), mountain peak (27,940 ft/8,516 m), E NEPAL, on Nepal-TIBET (CHINA) border, just S of MOUNT EVEREST; 27°58′N 86°56′E. World's forth-highest mt. First climbed by Swiss party in 1956. Also called E1.

Lhotse Shar (LUHT-sai SHAHR), peak (27,560 ft/8,400 m), E NEPAL, on S ridge of LHOTSE; 27°57′N 86°57′E. Eighth-highest peak in the world.

Lhuntse (LUHN-tsee), village, NE BHUTAN, 48 mi/77 km N of MONGAR; 27°45′N 91°11′E. Area is well known for producing high-quality weavings. Village of Dungkar, N, is original home of the Wangchuk royal family. Also spelled Lhuntshi.

Lhuntshi, Bhutan: see LHUNTSE.

Liampo, CHINA: see NINGBO.

Lian (lee-AHN), town, BATANGAS province, S LUZON, PHILIPPINES, near NASUGBU, 45 mi/72 km SW of MANILA; 14°00′N 120°39′E. Agricultural center (rice, sugarcane, corn, coconuts).

Liancheng (LIAN-CHENG), town, SW FUJIAN province, CHINA, 25 mi/40 km E of CHANGTING; ⊙ Liancheng county (1990 population 302,734); 25°47′N 116°48′E. Rice, tobacco, oilseeds; logging, chemicals, paper; iron-ore mining, non-ferrous ore mining, coal mining.

Liancourt (lee-ahn-koor), town (□ 1 sq mi/2.6 sq km), OISE department, PICARDIE region, N FRANCE, 10 mi/16 km NW of SENLIS; 49°20′N 02°28′E. Agricultural market (beans); manufacturing (consumer goods, furniture). Has 15th–17th century church.

Lianga (lee-AHN-gah), town, SURIGAO DEL SUR province, PHILIPPINES, on E coast of MINDANAO, at head of Lianga Bay (deep inlet of PHILIPPINE SEA); 08°39′N 126°05′E. Port.

Liangcheng (LIANG-CHENG), town, ⊙ Liangcheng county, S INNER MONGOLIA AUTONOMOUS REGION, N CHINA, 45 mi/72 km SE of HOHHOT; 40°25′N 112°20′E. Cattle raising; millet, grain, oilseeds.

Liangdang (LIANG-DANG), town, SE GANSU province, CHINA, 45 mi/72 km SE of TIANSHUI, near SHAANXI border; ⊙ Liangdang county (1990 population 50,070); 33°56′N 106°12′E. Rice, wheat, millet; furniture; coal mining.

Lianghe (LIANG-HUH), town, W YUNNAN province, CHINA, 18 mi/29 km SW of TENGCHONG; ⊙ Lianghe county (1990 population 141,557); 24°50′N 98°20′E. Sugar refining; pharmaceuticals; silk, rice, millet, beans, sugarcane; non-ferrous ore mining.

Liang-hsiang, CHINA: see LIANGXIANG.

Liang Mountains (LIANG), Mandarin *Liang Shan* (LIANG-SHAN), mountain range, W SHANDONG province, CHINA, N of JINING; 35°46′N 116°06′E. Originally called Shouliang Mountains, the mountains were the royal hunting ground during the Han dynasty. From the 5th dynasty to the Song dynasty, the Huang He (Yellow River) broke its levee three times and flooded the Liang mountain area, forming LIANGSHAN Lake. The mountains gained fame due to the classic novel *By the Lake*, which tells the story of rebellious peasants who established their power base here. The martial arts have become a tradition in the area.

Liangping (LIANG-PING), town, E SICHUAN province, CHINA, 35 mi/56 km WSW of WANXIAN; ⊙ Liangping county (1990 population 818,902). Manufacturing of food; rice, sweet potatoes, millet, beans, wheat, tobacco, oilseeds.

Liang Shan, CHINA: see LIANG MOUNTAINS.

Liangshan Bo, CHINA: see LIANGSHAN LAKE.

Liangshan Lake (LIANG-SHAN), Mandarin *Liangshan Bo* (LIANG-SHAN BU-uh), lake in W SHANDONG province, CHINA. In 959 C.E. (during the Period of the Five Dynasties and Ten Kingdoms), HUANG HE (Yellow River) broke its levee and flooded areas surrounding the nearby Liangshan Mountains, thus

forming this lake. Numerous subsequent river floodings sustained the lake over the years, but eventually Liangshan Lake lost its stable water supply due to the shifting channel of the Huang He and the implementation of river control projects. Much of the lake basin has been converted into agricultural land, and today, Dongping Lake (in DONGPING county) is the only remnant of Liangshan Lake. During the Northern Song dynasty (960–1127), the Liangshan Mountains and Liangshan Lake became a stronghold of rebel peasants; this history formed the basis for the classic Chinese novel *By the Lake*.

Liangxiang (LIANG-SIANG), town, CHINA, 15 mi/24 km SW of BEIJING and on Beijing-WUHAN railroad. An administrative unit of Beijing municipality; grain; construction materials. Sometimes appears as Liang-hsiang.

Lianhua (LIAN-HUAH), town, W JIANGXI province, CHINA, near HUNAN border, 65 mi/105 km W of JI'AN; ⊙ Lianhua county (1990 population 217,075); 27°03′N 113°54′E. Rice, tea; paper, printing and publishing; coal mining.

Lianjiang (LIAN-JIANG), town, E FUJIAN province, CHINA, 15 mi/24 km ENE of FUZHOU, on East CHINA SEA; ⊙ Lianjiang county (1990 population 583,418); 26°12′N 119°32′E. Rice, sugarcane; food and beverages, chemicals; timber.

Lianjiang, town, SW GUANGDONG province, CHINA, 30 mi/48 km NNW of ZHANJIANG; ⊙ Lianjiang county; 21°36′N 110°16′E. Sugarcane, rice, fruits; sugar refining.

Lianjiangkou (LIAN-JIANG-KO), town, N HEILONGJIANG province, NORTHEAST, CHINA, 10 mi/16 km NW of JIAMUSI on N bank of SONGHUA River; 46°53′N 130°16′E. On railroad. Grain, sugar beets.

Liannan (LIAN-NAN), town, NW GUANGDONG province, CHINA, 10 mi/16 km SW of LIAN XIAN; ⊙ Liannan county (1990 population 137,113); 24°43′N 112°17′E. Rice, tobacco, jute, medicinal herbs; chemicals, food; non-ferrous ore mining; logging.

Lianping (LIAN-PING), town, N central GUANGDONG province, CHINA, 70 mi/113 km SE of SHAOGUAN, near JIANGXI border; ⊙ Lianping county (1990 population 313,720); 24°22′N 114°30′E. Rice, fruit, sugarcane, oilseeds; chemicals, pharmaceuticals, food; logging; non-ferrous ore mining.

Lian River (LIAN), 135 mi/217 km long, SE GUANGXI ZHUANG AUTONOMOUS REGION, CHINA; rises in Yunkai Dashan Mountains; flows SW past YULIN and BOBAI, and HEPU, to Gulf of TONKIN, forming delta N of BEIHAI.

Lianshan (LIAN-SHAN), town, NW GUANGDONG province, CHINA, 30 mi/48 km SW of LIAN XIAN; ⊙ Lianshan county (1990 population 100,315); 24°35′N 112°06′E. Lumbering; bamboo; rice; food, chemicals.

Lianshui (LIAN-SHUAI), town, N JIANGSU province, CHINA, 18 mi/29 km NE of HUAIYIN and on Yun River; ⊙ Lianshui county (1990 population 953,345); 33°46′N 119°18′E. Rice, oilseeds, cotton; food and beverages, chemicals, textiles, pharmaceuticals; iron smelting.

Lian Xian (LIAN SIAN), town, N GUANGDONG province, CHINA, 70 mi/113 km W of SHAOGUAN; ⊙ Lian Xian county (1990 population 460,075); 24°48′N 112°26′E. Rice, medicinal herbs, tobacco; construction materials, paper, chemicals.

Lianyuan (LIAN-YUAN), city (□ 732 sq mi/1,896 sq km; 1994 estimated urban population 116,600; 1994 estimated total population 1,034,800), central HUNAN province, CHINA, on railroad, 10 mi/16 km E of Lengshuijinag; 27°42′N 111°41′E. Agriculture and heavy industry are the largest sources of income for the city. Crop growing, livestock raising. Grain, vegetables, fruits, hogs; manufacturing (food and beverages, chemicals, machinery); coal mining.

Lianyungang (LIAN-YOOIN-GANG), city (□ 320 sq mi/829 sq km; 1994 estimated urban population 401,100; 1994 estimated total population 559,200), N JIANGSU province, CHINA, on the LONGHAI railroad and near YELLOW SEA; 34°37′N 119°10′E. It developed significantly with the construction of the Longhai railroad and its deepwater port in the 1930s. Light industry, commerce, and transportation are the largest sectors of the city's economy. Grains, oil crops, cotton, vegetables, fruits, aquatic products, hogs, eggs, poultry; manufacturing (salt, food, textiles, paper, chemicals, pharmaceuticals, plastics, machinery). Artifacts of the Longshan culture and Yunshi culture were found in the city in 1996. Also known as Xinpu.

Lianzhou River (LIAN-JO), 100 mi/161 km long, N GUANGDONG province, CHINA; rises in Nan Ling Mountains on HUNAN-GUANGDONG border N of LIAN XIAN (Lianzhou); flows SE past Lian Xian and YANGSHAN, to NORTH RIVER S of YINGDE.

Liao (LIOU), principal river of NORTHEAST, CHINA, c.900 mi/1,448 km long; rising in INNER Mongolia and flowing E then S through the fertile Liao alluvial plain to the Gulf of LIAODONG. The E branch, its main tributary, joins it at the JILIN-LIAONING province border. The shallow, silt-laden Liao is navigable for light junks c.400 mi/644 km upstream. It was the main route into S Northeast until the construction of the railroad in the early 20th century. Through sedimentation, the Liao delta has steadily grown. Also appears as Liao He, or Liao River.

Liaobei (LIOU-BAI), former province (□ 47,000 sq mi/122,200 sq km), NORTHEAST (former MANCHURIA), CHINA. The ☉ was Liaoyuan. It was one of nine provinces created in Manchuria in 1945 by the Chinese Nationalist government. Since the Nationalists never gained effective control of Manchuria after World War II, the province existed only on paper. It was later divided between the INNER MONGOLIA AUTONOMOUS REGION and the province of Liaoning.

Liaocheng (LIOU-CHENG), district, W SHANDONG province, CHINA. There are six counties and two cities (Liaocheng and LINQING) in the district's administrative area. It has rich reserves of coal, crude oil, and natural gas. The newly completed BEIJING-JIULONG Railroad passes through the region. In ancient times, the district was one of nine prosperous trading areas along the GRAND CANAL. A significant proportion of tributes to the imperial courts came from here. Among the numerous historic sites is the Haiyangge, one of four major book depositories during the Qing Dynasty (1644–1911).

Liaocheng (LIOU-CHENG), city (□ 481 sq mi/1,250.6 sq km; 2000 population 838,309), W SHANDONG province, CHINA, on the GRAND CANAL and 60 mi/97 km WSW of JINAN; 36°29′N 115°55′E. A city for 2,500 years, Liaocheng was named after the Liao, an ancient river that no longer exists. It is rich in coal, crude oil, natural gas, and thermal resources. Agriculture and light industry are the largest sectors of the city's economy. Crop growing, animal husbandry. Grains, cotton, vegetables, eggs; manufacturing (food, textiles, chemicals, machinery, transportation equipment). The city prospered during the Ming (1368–1644) and Qing (1644–1911) dynasties as a shipping stop on the Grand Canal. Historic sites include the Guangyue Pavilion, a wooden structure built during the Ming period; the Shanshan Hall, built during the Qing period; the Haiyuan Building, a Qing private library; and a cast iron tower built during the Song dynasty (960–1279). Surrounding the old city, the Dongchang Lake is the largest artificial freshwater lake N of CHANG JIANG (Yangzi River). In 1995 excavations led to the discovery of the remains of a city built 5,800 years ago in the area.

Liaodong, Gulf of (LIOU-DUNG), N arm of the Gulf of BOHAI, in SW NORTHEAST, CHINA, between SHANHAIGUAN corridor of LIAONING province (W) and Liaodong peninsula (E); 100 mi/161 km long, 60 mi/97 km wide. Receives LIAO River (N). Its main ports are YINGKOU and HULUDAO.

Liao He, CHINA: see LIAO.

Liaoning (LIOU), province (□ 58,400 sq mi/151,840 sq km; 2000 population 41,824,412), NE CHINA, on the BOHAI and W KOREA BAY; ☉ SHENYANG; 41°00′N 123°00′E. A part of the NORTHEAST, it encompasses the Liaodong peninsula and the plain of the LIAO RIVER. Rainfall is adequate, but long, severe winters permit only one harvest annually. Soybeans are the major crop, and millet, kaoliang, wheat, rice, sweet potatoes, beans, cotton, tobacco, fruit, and oakleaf silk (pongee silk) are also produced. Liaoning is China's largest producer of heavy industrial products, and it supplies one-fifth of China's electrical power. It is a major coal-producing area and contains a large percentage of China's iron-ore reserves; there are large deposits of oil and magnesite and smaller ones of copper, lead, zinc, and molybdenum. Shenyang is the center of a vast heavy-industrial complex (metallurgy, machinery, chemicals, petroleum, and coal) that also embraces ANSHAN, a major city for iron and steel; FUSHUN, a coal and a shale-oil producing center; and DALIAN, the chief commercial port of the Northeast. Important manufacturing includes locomotives, tractors, and a wide range of heavy equipment. Liaoning is also a leading producer of machine-made paper, and it has numerous brick and tile factories that utilize waste ash and slag. Textiles and food are also produced. Along the coast, salt production and fishing are key. The Liao River, which crosses Liaoning, is navigable in its lower reaches, and an extensive railroad net, including sections of the Southern Northeast railroad, connects the interior with the ports along the coast.

The growth of railroads after 1900 spurred the development of the province; the Japanese concentrated heavy industry here, especially after 1931. The SUPUNG dam on the YALU RIVER, built by the Japanese, supplies power to Liaoning and NORTH KOREA. Liaoning's fine harbors were long coveted by RUSSIA and JAPAN for their strategic positions. Japan acquired (1895) the Liaodong peninsula after the first Sino-Japanese War, but was forced by Russia, GERMANY, and FRANCE to return it to China that same year. Russia received in 1898 the S portion of the Liaodong peninsula as a twenty-five-year leasehold. After the Russo-Japanese War (1904–1905), Japan took this territory and held it until the end of World War II. The E part of REHE province became part of Liaoning in 1956, and more than 30,000 sq mi/77,700 sq km of territory from the INNER MONGOLIA AUTONOMOUS REGION was added to W Liaoning in 1970. This territory was returned to Inner Mongolia in 1979.

Liaoning (LIOU-NING), short name.

Liao River, CHINA: see LIAO.

Liaoyang (LIOU-YANG), city (□ 216 sq mi/561.6 sq km; 2000 population 639,553), central LIAONING province, Northeast, CHINA, on a tributary of the Hun River; 41°17′N 123°11′E. It is an industrial city with manufacturing of chemicals as its major industry. Other manufacturing includes machinery, textiles, and paper. It has evolved into a supply center for neighboring ANSHAN, providing it with food and other support services. One of the oldest cities of the Northeast, Liaoyang contains several Buddhist temples built in the 11th century. In the Russo-Japanese War it was the site of a battle (August 23–September 3, 1904) in which the RUSSIANS were forced to retreat.

Liaoyuan (LIOU-YUAN), city (□ 80 sq mi/208 sq km; 2000 population 411,073), SW JILIN province, CHINA, on railroad connecting CHANGCHUN and DALIAN; 42°53′N 125°10′E. The city is an industrial center with sizable light and heavy industries. Crop growing accounts for most of the city's agricultural activity, followed by livestock raising. Grains, vegetables, hogs, eggs, beef, lamb; manufacturing (food, textiles, chemicals, pharmaceuticals, synthetic fibers, plastics, machinery); coal mining.

Liaozhong (LIOU-JUNG), town, central LIAONING province, NORTHEAST, CHINA, on Liao plain, 45 mi/72 km SW of SHENYANG, and on left bank of lower LIAO RIVER; ☉ Liaozhong county (1990 population 492,010); 41°30′N 122°42′E. Grain, soybeans, oilseeds; food, textiles, apparel; iron smelting.

Liard River (LEE-ahrd), river, 755 mi/1,215 km long, SE YUKON TERRITORY, CANADA; rising in the PELLY MOUNTAINS and flowing SE into N BRITISH COLUMBIA, passes through the main range of the ROCKY MOUNTAINS, thence NE through densely wooded country to the MACKENZIE RIVER at FORT SIMPSON, FORT SMITH region, NORTHWEST TERRITORIES. Navigable to FORT LIARD, an old Hudson's Bay Company trading post, c.165 mi/270 km from its mouth. The SOUTH NAHANNI RIVER and FORT NELSON rivers are its chief tributaries. Part of its course is followed by the Alaska HIGHWAY.

Liari (LYAH-ree), village, LAS BELA district, SE BALUCHISTAN, SW PAKISTAN, 39 mi/63 km SSE of LAS BELA; 25°41′N 66°29′E.

Liat (LEE-aht), island, South Sumatra province, INDONESIA, in JAVA SEA, 35 mi/56 km E of SE coast of BANGKA; 8 mi/12.9 km long, 6 mi/9.7 km wide; 02°53′S 107°05′E. Community port of Pongeh.

Lib (LEEB), coral island, Ralik Chain, Marshall Islands, W central Pacific, c.60 mi/97 km SW of Kwajalein; 08°19′N 167°25′E.

Liban (LI-bahn-yuh), Czech *Libáň*, village, VYCHODOCESKY province, N BOHEMIA, CZECH REPUBLIC, on railroad, 28 mi/45 km SSE of LIBEREC; 50°47′N 15°03′E. In a sugar-beet area. Manufacturing (plastics); food production.

Libang (LEE-bahng), town, W NEPAL; ☉ ROLPA district; 28°16′N 82°46′E. Elevation 4,100 ft/1,250 m.

Líbano (LEE-bah-no), town, ☉ Líbano municipio, TOLIMA department, W central COLOMBIA, at E foot of Nevado del RUIZ in Cordillera CENTRAL, 33 mi/53 km N of IBAGUÉ; 04°55′N 75°02′W. Elevation 5,200 ft/1,585 m. Mining (silver, gold); agricultural center (coffee, corn, bananas, sorghum; livestock); food processing.

Líbano (LEE-bah-no), village, GUANACASTE province, NW COSTA RICA, 4 mi/6.4 km SSE of TILARÁN; 10°25′N 85°00′W. Coffee, grain, sugarcane; livestock.

Libanovon, Greece: see AIGINION.

Libanus, LEBANON: see LEBANON, mountain range.

Libau (lee-BO), community, SE MANITOBA, W central CANADA, 33 mi/53 km NE of WINNIPEG, and in ST. CLEMENTS rural municipality; 50°16′N 96°43′W.

Libau, LATVIA: see LIEPĀJA.

Libava, CZECH REPUBLIC: see MESTO LIBAVA.

Libava, LATVIA: see LIEPĀJA.

Libby, town (2000 population 2,626), NW MONTANA, 56 mi/90 km WNW of KALISPELL, and on KOOTENAI River, at mouth of Libby Creek, just NE of CABINET MOUNTAINS; ☉ LINCOLN county; 48°23′N 115°34′W. Visitors center; trade center for timber area; agriculture (hay, some livestock). Gold, silver, lead, zinc, vermiculite mines nearby. Area surrounded by Kootenai National Forest; Cabinet Mountains Wilderness Area to SW; Kootenai Falls on Kootenai River to W; Turner Mountain Ski Area to N. Libby Dam on Kootenai River, 10 mi/16 km to E (upstream), forms Lake KOOCANUSA, which extends 85 mi/137 km, including 40 mi/64 km into CANADA (BRITISH COLUMBIA). Founded as mining village in 1860s, incorporated 1910. Heritage Museum; Montana City Old Town; Logger Days (second week of July), Nordicfest (mid-September).

Libbyville, Alaska: see NAKNEK.

Libcice nad Vltavou (LIP-chi-TSE NAHD VUHL-tah-VOU), Czech *Libčice nad Vltavou*, German *Libeschitz an der Moldau*, town, STREDOCESKY province, N central BOHEMIA, CZECH REPUBLIC, on VLTAVA RIVER, on railroad, and 8 mi/12.9 km NNW of PRAGUE; 50°12′N 14°21′E. Metallurgy. Has an 18th century Gothic church.

Liben (LI-ben-yuh), Czech *Libeň*, German *Lieben*, N district of PRAGUE, PRAGUE-CITY province, CZECH REPUBLIC, 2 mi/3.2 km from city center, on right bank of VLTAVA RIVER. Manufacturing (machinery, chemicals).

Libenge (lee-BENG-gai), town, Équateur province, NW CONGO, on UBANGI RIVER (CENTRAL AFRICAN REPUBLIC border) opposite MONGOUMBA, and 225 mi/362 km NW of LISALA; 03°39′N 18°38′E. Elev. 1,072 ft/326 m. Customs station, river port, and trading center; cotton ginning. Airport. Coffee and rubber plantations; extensive farming of food staples (manioc, yams, plantains) in vicinity. Libenge has Capuchin and Augustinian missions, trade schools, hospital.

Liberal, city (2000 population 19,666), SW KANSAS; ⊙ SEWARD county; 37°02′N 100°55′W. Elevation 2,836 ft/864 m. Railroad junction. It is the trade center for a grazing and farm area. Manufacturing (food, helium). Sand and gravel; oil and natural gas are extracted. The traditional International Pancake Race between the housewives of Liberal and OLNEY, ENGLAND, is held annually on Shrove Tuesday. Seward County Community College is in Liberal. County Museum here. Founded 1888, incorporated 1945.

Liberal, city (2000 population 779), BARTON county, SW MISSOURI, 15 mi/24 km NE of PITTSBURG, Kansas; 37°33′N 94°31′W. Corn, wheat, sorghum; cattle. Prairie State Park to SW.

Liberalitas Julia, PORTUGAL: see ÉVORA.

Liberato Salzano (LEE-be-rah-to SAHL-sah-no), town, N RIO GRANDE DO SUL state, BRAZIL, 25 mi/40 km NE of PALMEIRA DAS MISSÕES; 27°35′S 53°04′W. Wheat, soybeans, corn, potatoes, manioc; livestock.

Libercourt (lee-ber-koor), town (□ 2 sq mi/5.2 sq km), PAS-DE-CALAIS department, NORD-PAS-DE-CALAIS region, N FRANCE, 8 mi/12.9 km ENE of LENS, in former coal-mining district; 50°29′N 03°01′E.

Liberdade (lee-ber-dah-zhee), town (2007 population 5,395), S MINAS GERAIS, BRAZIL, in the SERRA DA MANTIQUEIRA, on railroad, and 45 mi/72 km NW of BARRA DO PIRAÍ (RIO DE JANEIRO); 22°03′S 44°28′W. Nickel mining; electrometallurgy.

Liberec (LI-be-RETS), German *Reichenberg*, city (2001 population 99,102), SEVEROCESKY province, CZECH REPUBLIC, N BOHEMIA, on the LAUSITZER NEISSE RIVER, and near the German and Polish borders; 50°47′N 15°03′E. Manufacturing (textiles, known especially for its woolens; machinery; glassware). Founded c.1350, Liberec has enjoyed prosperity since the 16th century, when cloth making was introduced; the first textile factories were built in the 18th century. It later became the center of Nazi movement among ethnic Germans in CZECHOSLOVAKIA; when German troops occupied the city in 1938, Czech residents were expelled. After World War II, Germans were forced out in turn.

Liberia [Latin=place of freedom], republic (□ 43,000 sq mi/111,370 sq km; 2004 estimated population 3,390,635; 2007 estimated population 3,195,931), W AFRICA; ⊙ MONROVIA.

Geography

Liberia fronts on the ATLANTIC OCEAN for some 350 mi/563 km and is bordered on the NW by SIERRA LEONE, on the N by GUINEA, and on the E by CÔTE D'IVOIRE. Monrovia is the main port and administrative and commercial center. Other important towns include BUCHANAN and HARPER, both ports. Liberia is divided into fifteen first-order administrative divisions called counties—BOMI, BONG, GBARPOLU, GRAND BASSA, GRAND CAPE MOUNT, GRAND GEDEH, GRAND KRU, LOFA, MARGIBI, MARYLAND, MONTSERRADO, NIMBA, RIVERCESS, RIVER GEE, and SINOE—and sixty-six second-order administrative divisions called districts. Liberia can be divided into three distinct topographical areas. First, a flat coastal plain of some 10 mi/16 km to 50 mi/80 km, with creeks, lagoons, and mangrove swamps; second, an area of broken, forested hills with elevations from 600 ft/185 m to 1,200 ft/365 m, which covers most of the country; and third, an area of mountains in the N

highlands, with elevations reaching 4,540 ft/1,384 m in the NIMBA MOUNTAINS and 4,528 ft/1,380 m in the Wutivi Mountains. The six main rivers, which flow into the Atlantic, divide the country at right angles to the coast. Vegetation in much of the country is dense forest growth. The climate is tropical and humid, with a heavy rainfall, averaging 183 in/465 cm on the coast and some 88 in/224 cm in the southeastern interior. There are two rainy seasons and a dry, harmattan season in December and January.

Population

The majority of the population are members of some sixteen ethnic groups. These include the Kpelle, the Mano, the Bassa, the Grebo, the Kru, and the Vai. Decentralized political organizations are common, with government appointed chiefs directing most local affairs. Poro, a men's organization with educational, legal, and religious functions, continues to be of importance, particularly among the Vai, Kpelle, and Gola peoples. The official religion of Liberia is Christianity, but traditional religions and Islam are practiced—the population is 40% animist, 20% Muslim, and 40% Christian. English is the official language, and African languages are used extensively. Far less numerous, but of great political importance, are the descendants of American freed slaves who migrated to Liberia in the 19th century. These people, formerly called Americo-Liberians, are concentrated in the towns, where they provide the country's Westernized leadership and, for the most part, are adherents of various Protestant sects.

Economy

Until the 1950s, Liberia's economy was almost totally dependent upon subsistence farming and the production of rubber. The American-owned Firestone plantation was the country's largest employer and held a concession on some 1,563 sq mi/4,047 sq km of land. With the discovery of high-grade iron ore, first at BOMI HILLS, and then at Bong and Nimba, the production and export of minerals became the country's major cash-earning economic activity. Other important minerals include gold, diamonds, barite, and kyanite. Some three-quarters of the population remain in the agricultural economy, producing such crops as rice, cassava, yams, and okra. Much rice, the main staple, is imported, but efforts have been made to develop intensive rice production and to establish fish farms. Rubber and timber, produced mainly on foreign concessions, are the main nonagricultural exports, while coffee and cacao are also exported. Iron ore, rubber, and diamonds provide the bulk of the export earnings. Much of the country's industry is concentrated around Monrovia and is directed toward the production of iron ore. The lack of skilled and technical labor has slowed the growth of the manufacturing sector. Mineral processing plants are located near Buchanan and Bong. The government derives a sizable income from registering ships; low fees and lack of control over shipping operations have made the Liberian merchant marine one of the world's largest. Internal communications are poor, with few paved roads and only a few short, freight-carrying railroad lines. However civil war and poor government have destroyed much of Liberia's economic infrastructure, and the current National Transition Government of Liberia will have to implement sound economic policies to revitalize the economy.

History: to 1909

Liberia was founded in 1821, when officials of the American Colonization Society were granted possession of Cape MESURADO by local chiefs. African-American settlers were landed in 1822, the first of some 15,000 to settle in Liberia. The survival of the colony during its early years was due primarily to the work of Jehudi Ashmun, one of the society's agents. In 1847, primarily due to British pressures, the colony was declared an independent republic. The Americo-

Liberian minority controlled the country's politics and new immigration virtually came to an end with the American Civil War. Liberia was involved in efforts to end the W African slave trade. Attempts to modernize the economy led to a rising foreign debt in 1871, which the republic had serious difficulty repaying. The debt problem and constitutional issues led to the overthrow of the government in 1871. Conflicts over territorial claims resulted in the loss of large areas of claimed, but uncontrolled, lands to Britain and France in 1885, 1892, and 1919, but rivalries between the Europeans colonizing W Africa and the interest of the U.S. helped preserve Liberian independence during this period. Nevertheless, the decline of Liberia's exports and its inability to pay its debts resulted in a large measure of foreign interference.

History: 1909 to 1990

In 1909 the government was bankrupt, and a series of international loans were floated. Firestone leased large areas for rubber production in 1926. In 1930 scandals broke out over the exportation of forced labor from Liberia, and a League of Nations investigation upheld the charges that slave trading had gone on with the connivance of the government. President King and his associates resigned, and international control of the republic was proposed. Under the leadership of presidents Edwin Barclay (1930–1944) and William V. S. Tubman (1944–1971), however, Liberia avoided such control. Under Tubman, new policies to open the country to international investment and to allow the indigenous peoples a greater say in Liberian affairs were undertaken. Upon Tubman's death in 1971, Vice President W. River Tolbert took charge, and in 1972 he was elected to the presidency. Although Tolbert cultivated a democratic climate and favorable relations abroad, an organized opposition emerged early in his regime, some of it from Liberian students living in the U.S. In 1980 Tolbert was assassinated in a coup led by Master-Sergeant Samuel K. Doe. Doe became Liberia's first indigenous president by a fraudulent election in 1985, banning observation by opposition parties and destroying ballots. The Doe government was infamous for corruption and human-rights abuses; it also became the target of numerous coup attempts. Thousands of refugees fled to Guinea and Côte d'Ivoire during this period.

History: 1990 to Present In 1990

Liberia was invaded by rebel forces of the National Patriotic Front of Liberia (NPFL), led by Charles Taylor, who proclaimed himself president. The U.S. sent troops to the area when the leader of the NPFL threatened to take foreign hostages. Doe was assassinated in 1990 by another group of rebels led by Prince Yormie Johnson, who also sought the presidency. The Economic Community of West African States (ECOWAS) intervened to negotiate a peace settlement between the two rebel groups and the government. ECOWAS also sent a Nigerian-led peacekeeping force to Monrovia and installed an interim government led by Amos Sawyer. Taylor's forces, with military aid from LIBYA and BURKINA FASO, began a siege of Monrovia in 1992 in an attempt to seize control and engaged in fighting with ECOWAS forces. The warring parties agreed to a peace settlement in August 1995, signing the ABUJA peace accords. A transitional coalition government was formed in September 1995 with Wilton Sankawulo as head of a Council of State. Fighting resumed in April 1996, causing further damage in Monrovia; thousands of foreign residents sought refuge or evacuated the city. President Taylor, who won the presidency in 1997 after eight years of civil war, was unable to completely eliminate rebel groups. Because of this and the UN-imposed sanction for Taylor's meddling in Sierra Leone's civil war, Taylor abdicated power in August 2003. A transitional government formed with rebel, government, and civil groups, assumed control in October 2003. In 2005,

Ellen Johnson-Sirleaf, of the Unity Party, was elected president after a run-off vote, defeating popular football star George Weah to become the first elected female president of an African country.

Government

Liberia is governed under the constitution of 1986. The executive branch is headed by a president, who is popularly elected for a renewable six-year term. The president is both the head of state and the head of government. The bicameral legislature, the National Assembly, consists of the 30-seat Senate, whose members are popularly elected for nine-year terms, and the 64-seat House of Representatives, whose members are popularly elected for six-year terms. The current head of state is President Ellen Johnson-Sirleaf (since 2006).

Liberia (lee-BAI-ree-ah), city (2000 population 34,469), NW COSTA RICA, on INTER-AMERICAN HIGHWAY, on Liberia River (left affluent of TEMPISQUE RIVER), and 105 mi/169 km NW of SAN JOSÉ; ⊙ GUANACASTE province and of Liberia canton (1995 estimated population 40,021); 10°45′N 85°30′W. Agriculture center (livestock, fruit, grain, sugarcane). Site of Tomás Guardia International Airport (also known as Llano Grande), with direct access to developing coastal resorts.

Liberta, village (2001 population 2,560), S ANTIGUA, ANTIGUA AND BARBUDA Republic, WEST INDIES, 6 mi/ 9.7 km SSE of SAINT JOHN'S. Sugarcane, sea-island cotton. Founded after emancipation of slaves (1834).

Libertad, city, SAN JOSÉ department, S URUGUAY, on highway, and 30 mi/48 km NW of MONTEVIDEO; 34°38′S 56°36′W. In agricultural region (cereals, livestock); flour milling.

Libertad (lee-ber-TAHD), town, NE BUENOS AIRES province, ARGENTINA, 3 mi/4.8 km SE of MERLO; 27°20′S 59°20′W. Cattle raising, dairying, fruit; plant nurseries.

Libertad (lee-ber-TAHD), town, ⊙ Rojas municipio, BARINAS state, W VENEZUELA, on MASPARRO RIVER, and 45 mi/72 km ESE of BARINAS; 08°19′N 69°37′W. Livestock, vegetables.

Libertad (lee-ber-TAHD), town, ⊙ Ricaurte municipio, COJEDES state, N VENEZUELA, in LLANOS, 25 mi/40 km S of SAN CARLOS; 09°23′N 68°44′W. Livestock.

Libertad (lee-ber-TAHD), town, TÁCHIRA state, W VENEZUELA, in ANDEAN spur, 8 mi/13 km NW of SAN CRISTÓBAL; 07°35′N 72°19′W. Elevation 4,416 ft/1,346 m. Coffee, vegetables.

Libertad, PERU: see LA LIBERTAD.

Libertad, La, MEXICO: see LA LIBERTAD.

Libertador, province, DOMINICAN REPUBLIC: see DAJABÓN.

Libertador, village, DOMINICAN REPUBLIC: see PEPILLO SALCEDO.

Libertador, El, moutain, Argentina: see CACHI, NEVADO DE.

Libertador General Bernardo O'Higgins (lee-ber-tah-DOR he-ne-RAHL ber-NAHR-do o-EE-geens) or **VI region** (□ 2,746 sq mi/7,112 sq km; 2002 population 775,883; 2005 estimated population 840,555), central CHILE; ⊙ RANCAGUA; 34°30′S 71°00′W. The ANDES (E) give way to the fertile central valley (W), watered by CACHAPOAL RIVER; on SW is RAPEL RIVER, to the PACIFIC OCEAN on the W. Has mild climate. Predominantly agricultural (cereals, potatoes, fruit, grapes, marigolds, sugar beets; livestock). Food processing, metalworking, dairying, lumbering. Composed of CACHAPOAL, CARDENAL CARO, and COLCHAGUA provinces. Main cities are Rancagua, Pichilemu, and SAN FERNANDO. Large lake, Embalse de Rapel, on RAPEL RIVER. Site of Parque Nacional Palmas de Cocalán. At EL TENIENTE are some of largest copper mines in Chile. Textiles and wickerware are artisanal industries.

Libertador General San Martín (lee-ber-tah-DOR he-ne-RAHL sahn mahr-TEEN), town (□ 1,122 sq mi/ 2,917.2 sq km), ⊙ Ledesma department (□ 1,122 sq mi/

2,906 sq km; 1991 population 69,215), SE JUJUY province, ARGENTINA, on railroad and 40 mi/64 km NE of JUJUY. Lumbering and agricultural center (sugarcane, alfalfa, fruit, vegetables; livestock). Sugar refineries, alcohol distilleries, sawmills. Lead mines nearby. Formerly Ledesma.

Liberton, Scotland: see EDINBURGH.

Liberty (LI-buhr-TEE), county (□ 843 sq mi/2,191.8 sq km; 2006 population 7,782), NW FLORIDA, bounded by OCHLOCKONEE (E) and APALACHICOLA (W) rivers; ⊙ BRISTOL; 30°13′N 84°53′W. Lumber; agriculture (livestock, corn, peanuts). The S half of county included in Apalachicola National Forest. Formed 1855.

Liberty, county (□ 602 sq mi/1,565.2 sq km; 2006 population 62,571), SE GEORGIA; ⊙ HINESVILLE; 31°48′N 81°28′W. Bounded SE by the ATLANTIC OCEAN, NE by CANOOCHEE RIVER. Includes SAINT CATHERINES ISLAND. Coastal plain agriculture (corn, sugarcane, rice, tobacco; cattle, hogs, poultry); lumber, logging and wood products; fishing area. FORT STEWART MILITARY RESERVATION in NW part of county. Formed 1777.

Liberty, county (1,447 sq mi/3,748 sq km; 1990 population 2,295; 2000 population 2,158), N MONTANA; ⊙ CHESTER; 48°33′N 111°02′W. Agricultural area bordering on CANADA (ALBERTA) in N; drained in S by MARIAS River, source of Sage Creek in NE. Wheat, barley, oats, hogs. Light manufacturing. Tiber Dam forms Lake ELWELL reservoir in W. Formed 1920.

Liberty (LI-buhr-tee), county (□ 1,176 sq mi/3,057.6 sq km; 2006 population 75,685), SE TEXAS; ⊙ LIBERTY; 30°08′N 94°48′W. S part is on GULF OF MEXICO coastal plain; N is rolling, wooded. Bounded SW by Cedar Bayou; drained by TRINITY RIVER (shallow-draft navigation to Liberty) and San Jacinto River (NW). Agriculture (rice, soybeans, sorghum, corn); livestock (especially cattle). Timber (chiefly pine), lumber milling; oil and natural-gas wells, oil refining; sulphur production; sand and gravel. Parts of BIG THICKET NATIONAL PRESERVE in NE. Formed 1836.

Liberty, city (2006 population 29,581), satellite city 13 mi/21 km NW of downtown KANSAS CITY; W central MISSOURI; ⊙ CLAY county; 39°14′N 94°25′W. Shipping point and regional service center; grain elevators. Corn, wheat, soybeans; hogs, cattle; manufacturing (printing and publishing, fabricated metal products, food). William Jewell College is here. Laid out 1822.

Liberty, town (2000 population 2,061), E INDIANA, 13 mi/21 km S of RICHMOND; ⊙ UNION county; 39°38′N 84°56′W. In agricultural area (livestock; dairying; grain). Laid out 1822.

Liberty, town (2000 population 1,850), central KENTUCKY, on GREEN RIVER, and 22 mi/35 km SSW of DANVILLE; ⊙ CASEY county; 37°19′N 84°55′W. Agriculture (burley tobacco, corn; livestock, poultry; dairying); timber, limestone; manufacturing (apparel, furniture, lumber). Settled 1791. One of five places so named in Kentucky.

Liberty, resort town, WALDO county, S MAINE, 14 mi/23 km WSW of BELFAST; 44°21′N 69°20′W. Wood products, fabricated metal products. Includes part of Lake Saint George State Park.

Liberty (LI-buhr-tee), town (□ 2 sq mi/5.2 sq km; 2006 population 2,735), RANDOLPH county, central NORTH CAROLINA, 18 mi/29 km SE of GREENSBORO near source of ROCKY RIVER; 35°51′N 79°34′W. Manufacturing (feeds, construction materials, fixtures, textiles, apparel); service industries; agricultural area (tobacco, soybeans, grain; dairying; poultry, livestock).

Liberty, town (2000 population 3,031), PICKENS county, NW SOUTH CAROLINA, 17 mi/27 km WSW of GREENVILLE; 34°47′N 82°42′W. Manufacturing includes transportation equipment, food, apparel; agriculture includes dairying, poultry, hogs, corn.

Liberty, town (2006 population 385), DE KALB county, central TENNESSEE, 8 mi/13 km NW of SMITHVILLE; 36°00′N 85°58′W.

Liberty (LI-buhr-tee), town (2006 population 8,443), SE TEXAS, on TRINITY RIVER at head of shallow-draft navigation, and 40 mi/64 km NE of HOUSTON; ⊙ LIBERTY county; 30°03′N 94°47′W. Elevation 30 ft/9 m. Trade center in agriculture (rice, sorghum, soybeans, corn, cattle), timber, oil and natural-gas producing area. Manufacturing (chemicals, machinery). Founded c.1830.

Liberty, village, MONTGOMERY county, SE KANSAS, on VERDIGRIS RIVER, and 9 mi/14.5 km N of COFFEYVILLE; 37°09′N 95°35′W. Livestock raising. Oil and gas fields nearby. Montgomery State Fishing Lake to W.

Liberty (2000 population 633), SW MISSISSIPPI, 42 mi/68 km SE of NATCHEZ, between East and West forks AMITE RIVER; ⊙ AMITE county; 31°09′N 90°47′W. Agriculture (cotton, corn; cattle; dairying); catfish; some manufacturing.

Liberty, village (2006 population 86), GAGE county, SE NEBRASKA, 18 mi/29 km SE of BEATRICE, and on branch of BIG BLUE RIVER, near KANSAS state line; 40°05′N 96°28′W. Light manufacturing.

Liberty, village (□ 2 sq mi/5.2 sq km; 2006 population 3,976), SULLIVAN county, SE NEW YORK, in the CATSKILLS, 30 mi/48 km NW of MIDDLETOWN; 41°47′N 74°45′W. NEVERSINK and RONDOUT reservoirs, both part of New York city's water supply, nearby. Settled 1793, incorporated 1870.

Liberty, village (2006 population 178), TULSA county, NE OKLAHOMA, 18 mi/29 km SSE of downtown TULSA; 35°51′N 95°58′W. Agricultural area located in margin of urban growth.

Liberty, borough (2000 population 2,670), ALLEGHENY county, SW PENNSYLVANIA, residential suburb 11 mi/ 18 km SE of downtown PITTSBURGH, and 2 mi/3.2 km SSW of MCKEESPORT, near YOUGHIOGHENY RIVER. Incorporated c.1912.

Liberty, borough (2000 population 230), TIOGA county, N PENNSYLVANIA, 23 mi/37 km N of WILLIAMSPORT. Agriculture (dairying; livestock; grain); light manufacturing Tioga State Forest to N, Tidaghton State Forest to SE.

Liberty Cap (LI-buhr-tee kap), mountain peak (14,112 ft/4,301 m), PIERCE county, W central WASHINGTON; 46°52′N 121°46′W. Subsidiary volcanic peak on the Mount RAINIER massiff.

Liberty Center (LIB-uhr-tee SEN-tuhr), village (2006 population 1,113), HENRY county, NW OHIO, 7 mi/11 km ENE of NAPOLEON, near MAUMEE RIVER; 41°26′N 84°00′W.

Liberty Hill (LI-buhr-tee), unincorporated village (2006 population 1,510), WILLIAMSON county, central TEXAS, 29 mi/47 km NNW of AUSTIN, on S Fork SAN GABRIEL RIVER. Light manufacturing. Lake Georgetown reservoir to NE; 30°39′N 97°55′W.

Liberty Island, NEW YORK: see STATUE OF LIBERTY NATIONAL MONUMENT.

Liberty Lake, unincorporated town, SPOKANE county, E WASHINGTON, industrial suburb 15 mi/24 km E of SPOKANE, 12 mi/19 km WSW of COEUR D'ALENE, IDAHO, 2 mi/3.2 km W of Idaho state line, on LIBERTY LAKE reservoir; 47°39′N 117°05′W. Manufacturing (plastic products, electronic equipment, computers, machinery, printing and publishing).

Libertytown, village, FREDERICK county, N MARYLAND, 10 mi/16 km ENE of FREDERICK. Highway junction in agricultural area.

Libertyville, town (2000 population 325), JEFFERSON county, SE IOWA, 6 mi/9.7 km SW of FAIRFIELD, in bituminous-coal-mining and livestock area; 40°57′N 92°02′W.

Libertyville, village (2000 population 20,742), Lake county, NE ILLINOIS, in a lake area; 42°16′N 87°58′W. Remnant agriculture: manufacturing (paper products, electronic equipment). A naval training station is nearby. Incorporated 1882.

LibEschitz an der Moldau, CZECH REPUBLIC: see LIBCICE NAD VLTAVOU.

Libia: see LIBYA.

Libice nad Cidlinou (LI-bi-TSE NAHT TSI-dli-NOU), German *Libitz an der Cidlina*, village, STREDOCESKY province, E central BOHEMIA, CZECH REPUBLIC, on railroad, on Cidlina River, and 6 mi/9.7 km NNW of KOLIN; 50°08′N 15°12′E. Manufacturing (machinery); agriculture (wheat; sugar-beets). Archaeological site with museum of 8th century Slavonic castle.

Libin (lee-BE), commune (2006 population 4,611), Neufchâteau district, LUXEMBOURG province, SE BELGIUM, 7 mi/11.3 km NW of LIBRAMONT, in the ARDENNES. Kaolin-earth quarrying.

Libina (LI-bi-NAH), German *Liebau*, village, Severocesky province, W MORAVIA, CZECH REPUBLIC, on railroad, and 7 mi/11.3 km SE of Šumperk. Manufacturing (textiles); fruit growing. KAMENNA, (Czech *Kamenná*), foundry is 2 mi/3.2 km SW.

Libitz an der Cidlina, CZECH REPUBLIC: see LIBICE NAD CIDLINOU.

Libmanan (leeb-MAH-nahn), town, CAMARINES SUR province, SE LUZON, PHILIPPINES, on Sipocot River (tributary of BICOL RIVER), on railroad, and 10 mi/16 km NW of NAGA 13°40′N 122°59′E. Trade center for agricultural area (rice, abaca, corn).

Libo (LEE-BO), town, S GUIZHOU province, CHINA, 55 mi/89 km SW of RONGJIANG; ⊙ Libo county; 25°25′N 107°53′E. Elevation 1,453 ft/443 m. Grain; logging, forage processing, coal mining; manufacturing of food.

Libochovice (LI-bo-KHO-vi-TSE), German *Libochowitz*, town, SEVEROCESKY province, W central BOHEMIA, CZECH REPUBLIC, on OHRE RIVER, and 28 mi/45 km NW of PRAGUE; 50°24′N 14°02′E. Railroad junction. Manufacturing (glass, food). Has a baroque castle with park (tropical flora).

Libochowitz, CZECH REPUBLIC: see LIBOCHOVICE.

Libohovë (lee-bo-HOV), town, S ALBANIA, near Greek border, 7 mi/11.3 km SE of GJIROKASTËR; 40°02′N 20°16′E. Agricultural center. Has early 19th-century castle built by Ali Pasha. Also spelled Libohova.

Libon (lee-BON), town, ALBAY province, SE LUZON, PHILIPPINES, near LAKE BATO, 23 mi/37 km WNW of LEGASPI; 13°16′N 123°23′E. Agricultural center (abaca, rice, coconuts).

Libono (lee-BOH-noh), town, BUTHA-BUTHE district, N LESOTHO, 85 mi/137 km NE of MASERU; 28°37′S 29°35′E. CALEDON RIVER (SOUTH AFRICAN border) 1 mi/1.6 km to N, Monontsa Pass border crossing 5 mi/8 km to NE. Cattle, goats, sheep, horses; corn, beans, peas. Postal center with some shops.

Liborina (lee-bo-REE-nah), town, ⊙ Liborina municipio, ANTIOQUIA department, NW central COLOMBIA, in the CAUCA Valley, 30 mi/48 km NW of MEDELLÍN; 06°41′N 75°48′W. Elevation 2,847 ft/867 m. Coffee, cacao, plantains; livestock.

Libourne (lee-boor-nuh), town (□ 7 sq mi/18.2 sq km), sub-prefecture of GIRONDE department, AQUITAINE region, SW FRANCE, river port on the DORDOGNE at mouth of ISLE RIVER (head of ocean navigation), and 17 mi/27 km ENE of BORDEAUX; 44°55′N 00°14′W. Commercial and shipping center for Bordeaux wines, especially from SAINT-ÉMILION vineyards (just E); glassworks. The 16th-century town hall is situated on a fine town square, along with 15th-century church. Founded 1286 as a bastide (fortified village) by Edward I of England. Residence of Edward the Black Prince and birthplace of Richard II. There is a research center for development of electronic mail.

Libramont-Chevigny (lee-brah-MON–shuh-vi-NYI), commune (2006 population 9,922), Roeselare district, LUXEMBOURG province, SE central BELGIUM, 16 mi/26 km WSW of BASTOGNE, in the ARDENNES. Railroad junction; highest point (1,570 ft/479 m) of BRUSSELS-LUXEMBOURG railroad; agriculture and cattle market. Recogne, 1 mi/1.6 km W, was refuge of Napoleon III after battle of SEDAN (1871).

Library, unincorporated town, SOUTH PARK township, ALLEGHENY county, W PENNSYLVANIA, residential suburb 10 mi/16 km SSW of downtown PITTSBURGH; 40°17′N 80°00′W. Some manufacturing South Regional Park to N.

Library of Congress, U.S. government institution, Independence Avenue and 1st St., SE WASHINGTON, D.C. A complex of three buildings containing the world's largest collection of books, maps, manuscripts, media items, etc. The Jefferson Building (1897) contains the impressive Main Reading Room. The Adams Building (1939) and Madison Building (1980) have added much-needed shelf space for the more than one million items added to the collection each year.

Librazhd (lee-BRAHZHD), town, central ALBANIA, on Shkumbin River, and 14 mi/23 km ENE of ELBASAN; ⊙ Librazhd district (□ 390 sq mi/1,014 sq km); 41°10′N 20°20′E. Road center. Built for iron mining and processing. Also spelled Librazhdi.

Libres (LEE-bres), town, ⊙ Libres municipio, PUEBLA state, central MEXICO, 45 mi/72 km NE of PUEBLA on railroad, on Mexico Highway 129; 19°27′N 97°41′W. Elevation 7,808 ft/2,380 m. Cereals, maguey; livestock. Formerly called San Juan de los Llanos.

Libres, ARGENTINA: see PASO DE LOS LIBRES.

Libreville (LEE-bruh-veel), city (2003 population 661,600), a port on the GABON RIVER estuary, near the GULF OF GUINEA; ⊙ of GABON and of ESTUAIRE province; 00°23′N 09°26′E. Primarily an administrative center; is also a trade center for a lumbering region. A FRENCH fort was built here in 1843. Freed slaves were sent there, and in 1849 it was named Libreville [=freetown]. It was the chief port of French Equatorial Africa before the development (1934–1946) of POINTE NOIRE, in the CONGO REPUBLIC. Gabon's school of administration and school of law are in Libreville. An international airport is nearby.

Librilla (lee-VREE-lyah), town, MURCIA province, SE SPAIN, 15 mi/24 km SW of MURCIA; 37°53′N 01°21′W. Grapes, oranges, lemons.

Libus (LI-bush), Czech *Libuš*, S district of PRAGUE, PRAGUE-CITY province, central BOHEMIA, CZECH REPUBLIC; 50°01′N 14°28′E. Poultry; food processing.

Libverda, CZECH REPUBLIC: see HEJNICE.

Libya or **Libia,** officially Socialist People's Libyan Arab Jamahirya [Arabic=state of the masses], Arabic *Jamahiriya Al Arabiya Al Libiya Al Shabiya Al Ishtirakiya Al Uzma*, republic (□ 679,358 sq mi/1,759,540 sq km; 2004 estimated population 5,631,585; 2007 estimated population 6,036,914), N AFRICA, bordering on ALGERIA (SW), TUNISIA (NW), MEDITERRANEAN SEA (N), EGYPT (E), SUDAN (SE), and CHAD and NIGER (S); ⊙ TRIPOLI (Arabic *Tarabulus*); 25°00′N 17°00′E.

Geography

Other cities include AJDABIYAH, AL BAYDA, AL MARJ, BANGHAZI, MISRATAH, and TOBRUK. The country is divided into three main geographical regions: CYRENAICA (350,000 sq mi/905,000 sq km), also known as Eastern Provinces; FAZZAN (220,000 sq mi/570,000 sq km); and TRIPOLITANIA (110,000 sq mi/285,000 sq km), also called Western Provinces. Administratively, the country is divided into thirty-four municipalities (*shabiyat*): Ajdābiyā, Al-Butnān, Al-Hizam al-Akhdar, Al-Jabal al-Akhdar, Al-Jifarah, Al-Jufrah, Al-Kufrah, Al-Marj, Al-Marqab, Al-Qatrūn, Al-Qubbah, Al-Wāhah, An-Nuqāt al-Khams, Ash-Shāti', Az-Zāwiyah, Banghāzī, Banī Walīd, Darnah, Ghadāmis, Gharyān, Ghāt, Jaghbūb, Marzūq, Misrātah, Mizdah, Nālūt, Sabhā, Sabrātah Surmān, Surt, Tājūrā'wa an-Nawahi al-Arba', Tarābulus, Tarhūnah-Masallātah, Wādī al-Hayāt, and Yafran-Jādū.

Desert accounts for 97% of the land; 1.4% of it is arable, 0.1% irrigated. Libya is part of the vast North African Plateau. In NW Tripolitania, the JIFARAH coastal plain rises slowly to meet the N-facing limestone escarpment of JABAL NAFUSAH, which extends SSW from Al KHUMS to NALUT, near the Tunisian border. To the S, the upland plateau of GHUDAMIS BASIN and desert landscape of HAMMADA AL HAMRA

gradually give way to the MARZUQ BASIN of FAZZAN. Here, small groups of agricultural oases are located, where water resources and trade routes meet. In N Fazzan, oases are located along wadis in a series of long E-W depressions. Elsewhere are vast sand deserts, such as Sahra AWBARI and Sahra MARZUQ, as well as Jabal As SAWDA and AL HARUJ AL ASWAD, mountains of volcanic origin. In NE Cyrenaica, a narrow coastal plain (featuring the cities of Banghazi and Darnah) borders the Mediterranean Sea. Jebel Al AKHDAR, a hilly limestone plateau (maximum elevation 2,860 ft/872 m) rises sharply from the coast (extending inland c.20 mi/32 km) and then drops sharply in the W toward the E coast of Gulf of SURT on the Mediterranean and the SURT BASIN (135,135 sq mi/ 350,000 sq km), where most of the country's major oil fields are located. In the E the plateau slopes gradually toward Egyptian border as a series of low ridges parallel to the coast known as Marmarica.

South of Jebel Al Akhdar is a vast desert lowland—part of AS SAHRA AL LIBIYAH (Libyan Desert), which extends to the Egyptian border—where SARIR KALANSHIYU, a large "sea" of fine mobile sands and dunes, As Sarir, and RAMLAT RABYANAH are located. Settlement is concentrated in a few widely separated oases: WAHAT JALU and JAGHBUB in N and Az ZIGHAN and Al JAWF (the largest oases in Wahat Al KUFRA) in the S. In SW Cyrenaica, SARIR TIBASTI and a narrow arm of the central Saharan Tibasti Mountains cross the border into Chad. Picco Bette, Libya's highest point (elevation 7,438 ft/2,267 m) is here.

The climate in most of the country is Saharan, with little rainfall and marked seasonal variations in temperature. There are narrow bands of Mediterranean-type climate along Jifarah and the N Cyrenaica coastal plain, with semiarid steppe weather conditions (long, hot summers and short, cool winters) inland, on Jebel Nafusah and Jebel Al Akhdar. Summer temperatures in Tripoli average 72°F/22°C–86°F/ 30°C; the average annual rainfall here is 15 in/38 cm. AL AZIZIYAH, S of Tripoli, has the world's highest recorded shade temperature (c.141°F/60°C). Lengthy droughts (sometimes lasting for years) and long periods of desiccation are common. Ghiblis (hot, dry, and sand-laden S Saharan winds that can last several days) exaggerate arid conditions during spring and autumn and often pose agricultural and public health hazards.

Population

The population is unevenly distributed. Approximately 66% live in Tripolitania (mostly in Tripoli, Misratah, and regional administrative and commercial centers in Jifarah and Jebel Nafusah). About 20% live in Cyrenaica, primarily in Banghazi, Tobruk, Al Bayda, Darnah, and smaller settlements on Jebel Al Akhdar plateau. The rest of the population is concentrated in oases, mostly in Fazzan. An estimated 86% of residents now live in urban centers, due to heavy migration from rural areas in recent years. Labor shortages in the agriculture and petroleum industries have attracted many foreign workers, mostly from Egypt, Tunisia, and TURKEY. In 1995, 30,000 Palestinian workers were expelled because of the Israel-PLO peace agreement; some Egyptian workers were also expelled. Berbers once constituted the chief ethnic group, but now they have been largely assimilated into Arab culture. In Fazzan there are many persons are of mixed Berber and black African descent; there are also smaller groups of Italians and Greeks. About 5% of population live as pastoral nomads, mostly in Cyrenaica. Arabic is the official language, and the greater majority of people are Sunni Muslim.

Economy

Libya was a very poor, agriculturally-oriented country with bleak economic prospects until 1958, when petroleum was discovered 200 mi/320 km–300 mi/480 km S and SE of Gulf of Surt; crude petroleum was exported on an increasingly significant scale between 1961 and 1981. Income increased markedly when government nationalized (with compensation) 51% ownership in subsidiaries of foreign petroleum firms operating in the country (1972–1973). The remaining subsidiaries were completely nationalized. At the same time, the price of petroleum rose dramatically, further enhancing Libya's wealth. Since then, the economy has been almost inextricably linked to world oil prices. Much of the income from petroleum was used to create and improve cities, to modernize transportation and agriculture, and to build up the military. The rise of the petroleum industry resulted in the migration of many Libyans to urban areas and the growth of unemployment; to solve this problem, the government invested more money in agricultural development in an attempt to make farming more attractive. Although petroleum production has dropped since the 1970s, petroleum exports continue to generate a substantial percentage of the country's GDP. The country is also a major exporter of natural gas and has several large gas liquefication plants. In addition, gypsum, salt, and limestone are produced in significant quantities. Although farming is severely limited by the small amount of fertile soil and the lack of rainfall, agriculture still employs about 17% of the labor force (2004 estimate). To increase the amount of land that can be cultivated, Libya has initiated several reclamation projects on the Jifarah plain, at TAWURGHA, and in the Saharan oases of Al Kufra, As Sarir, and Wadi Ajarif. One of the largest is the GREAT MANMADE RIVER (GMR) project (begun 1984) designed to carry fossil water from Saharan artesian wells through a 2,400 mi/3,862 km pipeline system to irrigate 313 sq mi/811 sq km in the coastal region. The largest agricultural irrigation project in the Middle East, the GMR is expected to take 25 years to complete and cost an estimated $25 billion. By 1997, the system was connected to the cities of Tripoli, Cirt, and Benghazi and also provided thousands of acres of farmland with irrigation water. Chief agricultural products are grain, olives, tobacco, fruits, vegetables, and nuts; sheep and goats are raised. In addition to petroleum and natural-gas production, manufacturing includes construction materials (especially cement), chemicals, processed food, beverages, clothing, footwear, soap, and cigarettes. Libya has a large military industry. Libya's annual earnings from exports usually far exceed the cost of imports; it has the highest per capita GDP in Africa (estimated at $6,400 in 2003). Crude petroleum is the leading export; the main imports are machinery, foodstuffs, transport equipment, and consumer goods. Principal trade partners are Italy, Germany, Great Britain, France, and Spain.

History to 1815

Throughout most of its history the territory that constitutes modern Libya has been held by foreign powers. Tripolitania and Cyrenaica had divergent histories up to Ottoman conquest (mid-16th century). Fazzan was captured by Ottomans only in 1842. Ottomans gained control of most of N Africa (16th century), dividing it into three regencies—Algeria, Tunisia, and Tripoli (which included Cyrenaica). The Janissaries, professional soldiers of slave origins, wielded considerable influence over the Ottoman governor. From the early 1600s, Janissaries chose a leader called the Dey, who sometimes had as much power as the Ottoman governor sent from Constantinople. Pirates based in Tripoli preyed on shipping of Christian nations in the Mediterranean Sea. Janissary Ahmad Karamanli, who became Dey in 1711, killed the Ottoman governor and prevailed upon the Ottomans to appoint him to the post. The position of governor remained hereditary in the Karamanli family until 1835. In the 18th century and during the Napoleonic Wars, the Dey took in a large amount of

revenue from piracy and extended central government control into much of the country's interior. U.S. and Tripoli went to war (1801–1805) over the amount of tribute to be paid to the Dey for protection from pirate raids.

History: 1815 to 1943

After 1815, England, France, and the Kingdom of the Two Sicilies undertook a successful campaign against the pirates; this undermined the Dey's finances and facilitated reestablishment of direct Ottoman rule in Tripoli (1835). Italy conquered N Tripoli in the Turko-Italian War (1911–1912); Tripoli and N Libya received autonomy from Turkey by the Treaty of Ouchy. Libyans continued to fight the Italians, but by 1914 Italy had occupied much of the country. However, Italy was forced to undertake a long series of wars of pacification against government leaders and their allies. During the 1930s, while the country was still under Italian rule, roads, civic buildings, schools, and hospitals were built. In 1934 Tripolitania (which included Fazzan at the time) and Cyrenaica formally united to form the colony of Libya. It was made an integral part of Italy (1939) and was one of the main battlegrounds of N Africa after Italy entered World War II (June 1940).

History: 1943 to 1973

After the Allied victory in N Africa (1943), Libya was placed under an Anglo-French military government. The UN was given jurisdiction over the country in 1949, and it decided that Libya should be independent. On December 24, 1951, it became the United Kingdom of Libya, ruled by King Idris I. Italian colonists began leaving in droves just after World War II, and by 1964 most of them were gone. In 1970, Libya confiscated all Italian and Jewish property. The country joined the Arab League in 1953 and was admitted to the UN in 1955. The 1950s were characterized by great poverty, though the government maintained a balanced budget; a minimal amount of economic development was made possible only by subsidies and loans received from various Western nations. But the discovery of petroleum vastly improved Libya's prospects. By the early 1960s the country was receiving a large amount of revenue from oil sales. A 1953 Anglo-Libyan treaty that had allowed Britain to establish military bases in Libya in return for economic subsidies was terminated by Libya in 1964; most British troops were withdrawn in early 1966. In September 1969, a group of army officers led by 27-year-old Muammar al-Qaddafi ousted King Idris in a coup d'etat. The constitution of 1951 was abrogated, and the government was placed in the hands of a twelve-member Revolutionary Command Council (RCC) headed by Qaddafi, who appointed himself prime minister. In mid-1972, Qaddafi turned the post of prime minister over to Abdul Salam Jallud, but he remained the RCC's president, the country's most important political and military office. The regime pursued a policy of Arab nationalism and strict adherence to Islamic law; though Qaddafi espoused socialist principles, he was strongly anticommunist. He was especially concerned with reducing Western influences; as part of that effort, Great Britain was forced to evacuate its remaining bases in Libya, and the U.S. abandoned its military base at Wheelus Field, near Tripoli.

History: 1973 to 1988

An implacable foe of ISRAEL, Libya contributed men and materials (especially aircraft) to the Arab side in the Arab-Israeli war of October 1973. After the war, Libya was a strong advocate of reducing sales of petroleum to nations that had supported Israel and was also a leading force in increasing the price of crude petroleum. Relations with Egypt declined steadily after 1973 when Qaddafi failed to push through a merger with Egypt. Qaddafi survived a coup attempt (1975) and began ordering the assassination of dissi-

dents who were living in exile in Europe (1980). In 1981, two Libyan fighter planes attacked U.S. forces on maneuvers in Gulf of Surt (which Libya claims as national waters), and the U.S. shot them down. Libya's relations with the U.S. worsened when it became the alleged base for international terrorism and periodically offered financial assistance to terrorist organizations. The U.S. banned Libyan oil imports (1982). In 1982, the U.S. accused Libya of sponsoring a West Berlin nightclub bombing that killed two American servicemen; as a reprisal, President Ronald Reagan ordered a bombing raid against Tripoli and Banghazi in an apparent attempt to kill Qaddafi (1986).

History: 1988 to Present

In December 1988, a bomb blew up a Pan Am commercial airplane over LOCKERBIE, SCOTLAND, killing 270 people. International warrants were issued for the arrest and extradition to Great Britain of two Libyans, but the government refused to surrender the suspects. In 1989, it was discovered that a West German company was selling Libya equipment for the construction of a chemical weapons plant at RABTA. These actions, as well as the widespread belief in the U.S. and Europe that Qaddafi's regime was responsible for terrorist activities, led to American and UN sanctions against Libya (1992). In 1990 the dispute over the AOUZOU STRIP was submitted to the International Court of Justice, which ruled in Chad's favor, and the strip was returned to Chad in 1994. In April 1999, Libya handed over the suspects in the Lockerbie crash to the UN, which suspended its sanctions; they were lifted in 2003 after settlements were reached with the families of the victims of the two airline bombings. Also in 2003 Libya pledged to renounce the production and use of weapons of mass destruction. In 2004 it identified chemical weapons stockpiles to be destroyed, leading to an end of the U.S. trade embargo (1986–2004) and the subsequent restoration of U.S.-Libyan diplomatic relations.

Government

Libya, which professes to have a government in which the people rule directly, has no formal constitution. The highest official organ is the General People's Congress, consisting of some 2,700 representatives from local peoples' committees. In practice, Libya is a military regime, with power vested in the revolutionary leader, Muammar al-Qaddafi, who holds no official title but is the de facto head of state. The head of government is the secretary of the General People's Committee (the cabinet); al-Baghdadi Ali al-Mahmudi has held the post since March 2006.

Libyan Desert, Arabic *As Sahrā Al Libiyah* (es SAH-rah ahl lib-EE-yeh), NE part of the Sahara desert, NE Africa, in W Egypt, E Libya, and NW Sudan. It is a region of sand dunes, stony plains, and rocky plateaus, with few inhabitants and little traffic across it; Al Kufrah is the chief oasis in Libya; KHARGA and FARAFRA are the main oases in Egypt. Sometimes includes Sarīr Kalanshiyū, As Sarīr, and Ramlat Rabyānah. Known in Egypt as *As Sahrā Al Gharbīyah* [Arabic=western desert].

Libyssa, ancient town of BITHYNIA, NW ASIA MINOR, whose ruins are on N shore of Gulf of IZMIT, just E of GEBZE, 35 mi/56 km ESE of ISTANBUL, TURKEY. Scene of Hannibal's suicide.

Licab (LEE-kahb), town, NUEVA ECIJA province, central LUZON, PHILIPPINES, 12 mi/19 km ENE of TARLAC; 15°33′N 120°46′E. Rice, corn.

Licancábur Volcano (lee-kahn-KAH-boor), Andean peak (19,455 ft/5,930 m) in ANTOFAGASTA region, N CHILE, at BOLIVIA border; 22°50′S.

Licanray (lee-KAHN-rei), village, VALDIVIA province, LOS LAGOS region, S central CHILE, on N shore of LAKE CALAFQUÉN, in Chilean lake district, 65 mi/105 km NE of VALDIVIA; 39°28′S 72°07′W. Resort; fishing.

Licantén (lee-kahn-TAIN), village, ⊙ Licantén comuna, CURICÓ province, MAULE region, central CHILE,

---

on MATAQUITO RIVER, on railroad, and 45 mi/72 km W of CURICÓ; 34°59′S 72°00′W. Cereal; livestock.

Licata (lee-KAH-tah), town, S SICILY, ITALY, on the MEDITERRANEAN SEA at the mouth of the SALSO River; 37°06′N 13°56′E. Licata is a seaport, seaside resort, and commercial and industrial center (textiles, food processing). Sulfur and asphalt are shipped through its port. It was founded in the early 3rd century B.C.E. as a refuge for the inhabitants of GELA after that city's destruction and was called Phintias after the tyrant of Acragas. Off nearby Cape Economus (now Poggio di Sant'Angelo), the Roman consul Regulus won (256 B.C.E.) a decisive battle in the first of the Punic Wars.

Lice, town, E TURKEY, between TIGRIS and BATMAN rivers, 45 mi/72 km NNE of DIYARBAKIR; 38°29′N 40°39′E. Wheat.

Lich (LIKH), town, HESSE, central GERMANY, 7 mi/11.3 km SE of GIESSEN; 50°31′N 08°49′E. Brewing; pharmaceuticals. Has 16th-century church, 18th-century castle, and many half-timbered houses from 16th–17th centuries.

Licheng (LEE-CHENG), town, NW SHANDONG province, CHINA, 10 mi/16 km ENE of JINAN, and on QINGDAO-JINAN railroad; an administrative district of Jinan; 36°41′N 117°05′E. Wheat, millet, peanuts, oilseeds.

Licheng, town, ⊙ Licheng county, SE SHANXI province, CHINA, on railroad, and 36 mi/58 km NE of CHANGZHI, near HEBEI border; 36°30′N 113°23′E. Grain; engineering, papermaking, iron-ore mining and smelting.

Lichfield (LICH-feeld), city (2001 population 28,435) and district, STAFFORDSHIRE, W central ENGLAND, about 14 mi/22.4 km NNE of BIRMINGHAM; 52°41′N 01°49′W. Lichfield is a market town with light industries, famous for its three-spired cathedral and its close associations with Dr. Samuel Johnson, born here in 1709. The cathedral, dating from the 13th and 14th centuries, replaced the original church built by St. Chad, who founded it in the 7th century. Johnson's house was turned into a museum that contains relics of his life and works, and a statue of him rests in the market square. Lichfield has a very old grammar school (founded 1497). In the 18th century, a literary circle that included Erasmus Darwin, Thomas Day, and Anna Seward was known as the Lichfield group.

Lichinga (li-CHEEN-guh), town (1997 population 85,758), ⊙ LAGO district (□ 46,224 sq mi/120,182.4 sq km) and NIASSA province, N MOZAMBIQUE, near LAKE MALAWI, 140 mi/225 km N of ZOMBA (MALAWI); 13°25′S 35°20′E. Cotton, corn, beans. Projected W terminus of railroad from LUMBO (opposite Mozambique city), completed (1950) to Lúrio River W of Melawa. Airport. Formerly known as VILA CABRAL.

Lichinga District, MOZAMBIQUE: see NIASSA.

Lichnice, CZECH REPUBLIC: see PRACHOVICE.

Lichtenau (LIKH-tuhn-ou), town, BADEN-WÜRTTEMBERG, SW GERMANY, near RHINE RIVER, 12 mi/19 km SW of RASTATT; 48°44′N 08°00′E.

Lichtenau (LIKH-tuhn-ou), town, North Rhine-Westphalia, central GERMANY, 9 mi/14.5 km SE of PADERBORN; 51°37′N 08°54′E. Manufacturing of steel; woodworking; tourism.

Lichtenau (LIKH-tuhn-ou), village, BAVARIA, S GERMANY, in MIDDLE FRANCONIA, on FRANCONIAN REZAT RIVER, 23 mi/37 km SW of NUREMBERG; 49°16′N 10°40′E. Has 18th-century church, 16th-century fortress, and many half-timbered houses.

Lichtenau, Hessisch, GERMANY: see HESSISCH LICHTENAU.

Lichtenberg (LIKH-tuhn-berg), district, BERLIN, GERMANY, on the SPREE, 5 mi/8 km E of city center; 52°30′N 13°24′E. In Soviet sector, 1945–1990.

Lichtenberg, NW district of SALZGITTER, LOWER SAXONY, NW GERMANY. Potash mining.

Lichtenberg, FRANCE: see INGWILLER.

Lichtenburg, town, NORTH-WEST province, SOUTH AFRICA, 120 mi/193 km W of JOHANNESBURG on the

HARTZ RIVER; 26°10′S 26°11′E. Elevation 4,887 ft/1,490 m. Center of mining (diamonds, gold) and agricultural (corn) region; cement manufacturing. Named by president Thomas Francois Burgers as "Town of Light" in 1873. Airfield and railroad junction; on N14 highway.

Lichtenfels (LIKH-tuhn-fels), town, UPPER FRANCONIA, BAVARIA, central GERMANY, on the MAIN, 19 mi/31 km NNE of BAMBERG; 50°09′N 11°03′E. Railroad junction. Manufacturing of clothing, metal goods, toys; brewing. A former basket-weaving center. Has 15th-century church and 18th-century town hall. Chartered 1206.

Lichtenfels, town, HESSE, central GERMANY, 27 mi/43 km N of MARBURG; 51°12′N 08°48′E.

Lichtenfels (LIKH-tuhn-FELS), Greenlandic *Akunnaaq*, abandoned settlement, Nuuk (Godthåb) commune, SW GREENLAND, on small Irkens Havn inlet of the ATLANTIC, 2 mi/3.2 km SW of FISKENAESSET, Greenlandic *Qeqertarsuatsiaat*. Site (1758–1900) of Moravian mission. Church extant.

Lichtenstein (LIKH-tuhn-shtein), village, BADEN-WÜRTTEMBERG, SW GERMANY, 6 mi/9.7 km S of REUTLINGEN; 48°25′N 09°15′E.

Lichtenstein in Sachsen (LIKH-tuhn-shtein in SAHK-suhn), town, SAXONY, E central GERMANY, 7 mi/11.3 km ENE of ZWICKAU; 50°47′N 12°37′E. Textile industry; manufacturing of transformers. Includes former commune of CALLNBERG (S). Formerly called Lichtenstein-Callnberg.

Lichtental (LIKH-tuhn-tahl), E suburb of BADEN-BADEN, BADEN-WÜRTTEMBERG, SW GERMANY, on the OOS; 48°45′N 08°13′E. Site of Cistercian nunnery, founded 1245.

Lichtenvoorde (LIKH-tuhn-vawr-duh), town, GELDERLAND province, E NETHERLANDS, 27 mi/43 km ENE of ARNHEM; 51°59′N 06°34′E. German border 6 mi/9.7 km to NE and SE. Dairying; cattle, hogs, poultry; grain, vegetables; manufacturing (food processing, boilers).

Lichtervelde (LISH-tuhr-vel-duh), commune (2006 population 8,449), WEST FLANDERS province, NW BELGIUM, 12 mi/20 km SSW of BRUGES; 51°02′N 03°09′E. Town in an agriculture and dairy area.

Lichtervelde (LIKH-tuhr-vel-duh), agricultural village, WEST FLANDERS province, W BELGIUM, 3 mi/4.8 km SSE of TORHOUT.

Lichuan (LEE-CHUAN), city (□ 1,779 sq mi/4,625.4 sq km; 2000 population 764,267), SW HUBEI province, CHINA, 30 mi/48 km W of ENSHI, near SICHUAN border, and on the upper stream of the QING JIANG River; 30°20′N 108°57′E. Agriculture and light industry are the largest sectors of the city's economy. Crop growing (grain, oilseeds, tobacco); animal husbandry.

Lichuan (LEE-CHUAN), town, ⊙ Lichuan county, E JIANGXI province, CHINA, 55 mi/89 km SE of LICHUAN city, in the Bohea Hills; 27°14′N 116°51′E. Rice, sugarcane, oilseeds, jute; cotton textiles, engineering, food processing.

Licínio de Almeida (lee-SEEN-yo dee ahl-MAI-dah), city (2007 population 12,656), S central BAHIA, BRAZIL, on SALVADOR-BELO HORIZONTE railroad, 23 mi/37 km SW of CACULÉ; 14°40′S 42°29′W.

Licking (LIK-ing), county (□ 686 sq mi/1,783.6 sq km; 2006 population 156,287), central OHIO; ⊙ NEWARK; 40°05′N 82°30′W. Drained by LICKING RIVER and RACCOON CREEK. Includes part of BUCKEYE LAKE reservoir (recreation). In the Glaciated Plain physiographic region. Agricultural area (hogs; dairy products; corn); manufacturing (stone, clay, and glass products; commercial printing; motor vehicle parts); sand and gravel pits. Formed 1808.

Licking (LIK-ing), town, TEXAS county, S central MISSOURI, in the Ozarks, 14 mi/23 km N of HOUSTON; 37°30′N 91°51′W. Elevation 1,259 ft/384 m. Grain; cattle; lumbering; manufacturing (apparel). Has center for a recreational area. Montauk State Park to E.

Licking River, E and N KENTUCKY, c.320 mi/515 km long; rises in SE MAGOFFIN county, E Kentucky; flows NW past SALYERSVILLE and WEST LIBERTY, through CAVE RUN LAKE reservoir, receives North Fork from E, then South Fork from S at FALMOUTH, enters OHIO RIVER at COVINGTON and NEWPORT, opposite CINCINNATI, OHIO. The Licking was an important means of travel for Native Americans and pioneers and later a busy trade route. In 1780, at the river's mouth, George Rogers Clark's frontiersmen gathered for their march up the Little Miami; the battle of Blue Licks (1782) occurred in the Licking valley.

Licking River (LIK-ing), in central OHIO, formed by North and South forks and RACCOON CREEK at NEWARK, flows c.40 mi/64 km E and SE to the MUSKINGUM RIVER at ZANESVILLE; 39°56′N 82°00′W. North Fork rises in KNOX county, flows c.35 mi/56 km E and S. South Fork rises in LICKING county, flows SE and NE for c.30 mi/48 km. Raccoon Creek (c.25 mi/40 km long), entirely within Licking county, flows SE to junction with the other headstreams.

Lick Observatory, astronomical observatory on MOUNT HAMILTON, SANTA CLARA county, W CALIFORNIA, 15 mi/24 km E of SAN JOSE, in DIABLO RANGE. The first mountaintop observatory in the world, it was founded through gifts made by James Lick in 1874–1875 and came under the direction of the University of California in 1888. The original telescope at the observatory is a 36-in/91-cm refracting telescope, second largest in the world after the 40-in/102-cm refractor at YERKES OBSERVATORY (Wisconsin). The principal research instrument is now a 120-in/305-cm reflecting telescope that went into operation in 1959. Other equipment includes 36-in/91-cm and 22-in/56-cm reflectors, a 12-in/30-cm refractor, and a 20-in/51-cm twin-astrographic telescope.

Licola, village, VICTORIA, SE AUSTRALIA, 158 mi/254 km E of MELBOURNE, and beside Macalister River; 37°38′S 146°37′E. Village owned by Lions Club of Victoria, which runs Licola Wilderness Village, providing facilities and activities for disadvantaged young people.

Licomapampa (lee-ko-mah-PAHM-pah), canton, INQUISIVI province, LA PAZ department, W BOLIVIA. Elevation 8,986 ft/2,739 m. Tin, lead, zinc, and silver mining at Mina COLQUIRI; tungsten mining at Mina Chambilaya and Mina Chicote Grande; clay, limestone, and gypsum deposits. Agriculture (potatoes, yucca, bananas, rye, coca); cattle.

Licosa, Punta (POON-tah lee-KO-zah), promotory, S ITALY, at SE end of GULF OF SALERNO, 20 mi/32 km W of VALLO DELLA LUCANIA 40°15′N 14°54′E.

Lida (LEE-duh), city, NE GRODNO oblast, BELARUS, 55 mi/89 km S of VILNA, 60 mi/97 km ENE of GRODNO; 53°50′N 25°19′E. Railroad junction (repair shops); manufacturing center (varnishes and paints, electrical devices, furniture, building supplies); meat and milk canning; food concentrate combines. Has ruins of old castle. Founded 1323 by Lithuanian duke Gedymin; sacked (1392) by Teutonic Knights. Developed as flax- and livestock-trading center in sixteenth-century. Passed (1795) from POLAND to RUSSIA; reverted (1919) to Poland; ceded to USSR in 1945.

Lida, Lake (LEI-duh) (□ 10 sq mi/26 sq km), OTTER TAIL county, W MINNESOTA, 19 mi/31 km NNE of FERGUS FALLS, separated by narrow land bridges from Crystal Lake (N) and Lake LIZZIE (NW), throughout which it drains to PELICAN RIVER; 4 mi/6.4 km long, 2.5 mi/4 km wide. Causeway separates South Arm (2 mi/3.2 km long) from remainder of lake; 46°34′N 95°58′W. Elevation 1,313 ft/400 m. Resorts. Maplewood State Park at S end.

Lidcombe (LID-kuhm), suburb, E NEW SOUTH WALES, AUSTRALIA, 11 mi/18 km W of SYDNEY; 33°52′S 151°02′E. In metropolitan area; railroad junction; brewery; manufacturing (air conditioners, refrigerators).

Liddel Water (LID-uhl), river, Scottish Borders, SE Scotland; rises 11 mi/18 km SE of Hawick in CHEVIOT

HILLS; flows 27 mi/43 km SW, past Newcastleton, forming the border between Scotland and England, to ESK RIVER, 2 mi/3.2 km S of Canonbie. Formerly in Borders, abolished 1996.

Lidderdale, town (2000 population 186), CARROLL county, W central IOWA, 6 mi/9.7 km NE of CARROLL; 42°07′N 94°46′W. Feed milling.

Lidgerwood (LIJ-uhr-wud), town (2006 population 729), RICHLAND CO., SE NORTH DAKOTA, 29 mi/47 km SW of WAHPETON; 46°04′N 97°09′W. Dairy products; poultry; grain. (Sisseton) Wahpeton Indian Reservation to S. Lake Tewaukon and Tewaukon National Wildlife Refuge to SW. Founded in 1887; incorporated in 1895. Named for George I. Lidgerwood, a railroad agent.

Lidhorikion, Greece: see LIDORIKI.

Lidhult (LEED-HULT), village, KRONOBERG county, S SWEDEN, near LAKE BOLMEN, 19 mi/31 km W of LJUNGBY; 56°50′N 13°27′E.

Lídice (LEE-dee-see), city, extreme W RIO DE JANEIRO state, BRAZIL, 22 mi/35 km S of BARRA MANSA, on railroad; 22°51′S 44°12′W.

Lidice (LI-dyi-TSE), village, STREDOCESKY province, central BOHEMIA, CZECH REPUBLIC, 3 mi/5 km E of KLADNO; 50°08′N 14°12′E. In reprisal for the assassination of Reinhard Heydrich, the Germans in 1942 "liquidated" Lidice by killing all the men, deporting all women and children, and razing the village. After World War II a new village was built near the site of old Lidice, which is now a national park and memorial.

Lidingö (LEED-eeng-UH), town, STOCKHOLM county, SE SWEDEN, residential suburb 5 mi/8 km E of STOCKHOLM, on Lidingön Island in BALTIC SEA; 59°22′N 18°10′E. Resort. Chartered 1926.

Lidke Ice Stream, glacier, 25 mi/40 km long, WEST ANTARCTICA; flows into the STANGE ICE SHELF, 30 mi/50 km E of BERG ICE STREAM, on ENGLISH COAST of ELLSWORTH LAND; 73°30′S 76°30′W.

Lidköping (LEED-SHUHP-eeng), town, SKARABORG county, S SWEDEN, port on LAKE VÄNERN; 58°30′N 13°10′E. Manufacturing includes porcelain; machine shops. Läckö castle (thirteenth–seventeenth century) nearby. Chartered 1446.

Lido di Roma (LEE-do dee RO-mah), town, ROMA province, LATIUM, central ITALY, on TYRRHENIAN SEA, near mouth of the TIBER, 15 mi/24 km SW of ROME; 41°44′N 12°16′E. Beach resort.

Lido Isle, residential island in NEWPORT BAY, CALIFORNIA.

Lidoriki (lee-[th]-o-ree-KEE), village, PHOCIS prefecture, W CENTRAL GREECE department, 9 mi/14.5 km W of AMPHISSA; 38°32′N 22°12′E. Sheep and goat raising. Also called Lidorikion or Lidhorikion.

Lidzbark (LEETS-bahrk), German Lautenburg, town, Ciechanów province, N POLAND, on railroad, and 15 mi/24 km W of DZIAŁDOWO; 53°16′N 19°49′E. Manufacturing of chemicals, mineral water; tanning, distilling, lumbering.

Lidzbark Warminski (LEETS-bahrk vahr-MEEN-skee), Polish Lidzbark Warmiński, German Heilsberg, town, EAST PRUSSIA; after 1945 in OLSZTYN province, NE POLAND, on ŁYNA RIVER, 25 mi/40 km N of OLSZTYN (Allenstein); 54°08′N 20°35′E. Grain and cattle market; sawmilling. Seat (1350–1772) of bishops of ERMLAND. Has remains of 15th century town walls. In World War II, c.50% destroyed.

Liebau, CZECH REPUBLIC: see LIBAVA.

Lieben, CZECH REPUBLIC: see LIBEN.

Liebenau (LEE-ben-ou), township, NE UPPER AUSTRIA, in the Mühlviertel, 14 mi/23 km E of FREISTADT, near Lower Austrian line; 48°32′N 14°49′E. Dairying.

Liebenau (LEE-be-nou), village, LOWER SAXONY, N GERMANY, on left bank of WESER RIVER, 6 mi/9.7 km SW of NIENBURG; 52°36′N 09°05′E.

Liebenau an der Diemel (LEE-be-nou ahn der DEE-mel), town, HESSE, W GERMANY, on the DIEMEL, 4 mi/

6.4 km W of HOFGEISMAR; 51°30′N 09°16′E. Agricultural center.

Liebenau an der Mohelka, CZECH REPUBLIC: see HODKOVICE NAD MOHELKOU.

Liebenau bei Schwiebus, POLAND: see LUBRZA.

Liebenburg (LEE-ben-burg), village, LOWER SAXONY, NW GERMANY, 8 mi/12.9 km N of GOSLAR; 52°02′N 10°26′E. Has 18th-century castle.

Liebenfelde, RUSSIA: see ZALES'YE.

Liebenstein, Bad, GERMANY: see BAD LIEBENSTEIN.

Liebenthal (LEE-buhn-thahl), village (2000 population 111), RUSH county, W central KANSAS, 8 mi/12.9 km N of LA CROSSE; 38°38′N 99°19′W. In wheat and cattle area.

Liebenthal, POLAND: see LUBOMIERZ.

Liebenwerda, Bad, GERMANY: see BAD LIEBENWERDA.

Liebenzell, Bad, GERMANY: see BAD LIEBENZELL.

Lieberwolkwitz (LEE-ber-vuhlk-vits), village, SAXONY, E central GERMANY, 5 mi/8 km SE of LEIPZIG; 51°18′N 12°27′E. Scene (October 1813) of opening engagement of battle of Leipzig.

Liebwerda, CZECH REPUBLIC: see HEJNICE.

Liechtenstein (LEEK-ten-shtein), principality (□ 62 sq mi/161.2 sq km; 2007 population 34,247), W central EUROPE, in the ALPS between AUSTRIA and SWITZERLAND; ⊙ VADUZ.

Economy

Bounded in the W by the RHINE RIVER, Liechtenstein had traditionally been an agricultural country, but it has been industrialized in recent years. Only a fraction of the population still engages in agriculture (dairying and the raising of livestock and cereals). The leading manufactured products are machinery and other metal goods, construction materials, pharmaceuticals, optical lenses, electronic equipment, and consumer goods. A large part of the production is exported. Tourism and the sale of postage stamps are major sources of revenue. The country has no army and only a small police force. Much of the country's high income is derived from the minimal taxes imposed on international corporations, which, because of the low taxes, are headquartered here. The stable political environment and the secrecy of its banks contribute to its reputation as a banking and financial service center.

History

Catholicism is the dominant religion, and German is the national language. The Liechtenstein ruling house is an old Austrian family. The principality was created in 1719 by uniting the county of Vaduz with the barony of Schellenburg. The princes, vassals of the Holy Roman emperors, also owned huge estates (many times larger than their principality) in Austria and adjacent territories; they rarely visited their country but were active in the service of the Hapsburg monarchy. Liechtenstein became independent in 1866, after having been a member of the GERMAN CONFEDERATION from 1815 to 1866. The principality escaped the major upheavals of the 19th and 20th century. A parliament-approved proposal granting women the right to vote was decisively defeated in two referendums (1971, 1973), thus making Liechtenstein the only Western European country to deny women suffrage. In 1984 women were granted suffrage on national, but not local, issues. Liechtenstein is a member of EFTA, the UN, and the WTO; it is not a member of the EU, but it is in a customs union with Switzerland.

Government

Liechtenstein is a constitutional monarchy governed under the constitution of 1921 as amended. The hereditary monarch is the head of state. The head of government is appointed by the monarch, and the cabinet is elected by the legislature. Members of the 25-seat unicameral Parliament or Landtag are elected by popular vote for four-year terms. Administratively, Liechtenstein is divided into 11 communes. Since 1919,

Liechtenstein has been represented abroad through Switzerland. It adopted Swiss currency in 1921 and formed a customs union with Switzerland in 1924. Prince Hans Adam II is the ruling prince and head of state. Otmar Hasler has been the head of government since 2001.

Liedekerke (LEE-duh-ker-kuh), commune (2006 population 11,993), Halle-Vilvoorde district, BRABANT province, central BELGIUM, 13 mi/21 km W of BRUSSELS. Agricultural market. Has ruins of ancient Carmelite monastery.

Liège (lyezh), Flemish Luik, German Lüttich, province (□ 1,491 sq mi/3,876.6 sq km; 2006 population 1,042,840), E BELGIUM, bordering on GERMANY in the E; ⊙ LIÈGE; 50°38′N 05°35′E. The chief cities are Liège, VERVIERS, HERSTAL, HUY, and SERAING. The province is French-speaking except in the E districts of Eupen and Malmédy, located near the German border, where the German language prevails. The province was part of the prince-bishopric of Liège, and thus part of the Holy Roman Empire, until 1792. Liège is part of the industrial MEUSE Valley and of the agricultural ARDENNES Plateau. Some farming and lumbering are in the province. Manufacturing includes machinery, ordnance, and textiles.

Liège (lyezh), Flemish Luik, German Lüttich, city (□ 308 sq mi/800.8 sq km; 2006 population 187,432), ⊙ LIÈGE district and province, E BELGIUM, at the confluence of the MEUSE and OURTHE rivers, near the DUTCH and GERMAN borders; 50°38′N 05°34′E. Greater Liège includes the suburbs of HERSTAL, OUGRÉE, and GRIVEGNÉE. The commercial center of the industrial Meuse valley, Liège is also an important transportation hub. It is located on the ALBERT CANAL and on the LIÈGE-MAASTRICHT CANAL and is the center of a road and railroad network connecting Belgium and Germany. Manufacturing includes fabricated metal products, ordnance, motor vehicles, electronic equipment, chemicals, glass, furniture. A growing trade center by the tenth century, Liège became the capital of the extensive prince-bishopric of Liège, which included most of Liège province and parts of LIMBURG and NAMUR provinces. This ecclesiastical state, part of the Holy Roman Empire, lasted until 1792. In the late Middle Ages it was torn by bitter social strife. In the fourteenth century the workers (organized in guilds) won far-reaching concessions from the nobles and the wealthy bourgeoisie, but Charles the Bold, duke of BURGUNDY, abolished the citizens' communal liberties in 1467. The citizens of Liège, encouraged by Louis XI of FRANCE, rose in rebellion, but Charles forced Louis to assist him in suppressing the revolt and then sacked the city (1468). After Liège came under HAPSBURG rule (1477), the prince-bishops, though technically sovereign members of the Holy Roman Empire, were dependent on the Spanish kings and, after 1714, the emperors. In 1792 the French under Dumouriez entered the city and annexed it to the republic. In the nineteenth century, Liège was a center of Walloon particularism, rapid industrial growth as one of the earliest modern steelmaking centers, and social unrest. It was liberated (May 1944) from its second German occupation in thirty years by U.S. forces, but during the Battle of the Bulge (December 1944–January 1945) it suffered considerable destruction from German rockets. In the 1950s and 1960s, the decline of the steel industry led to massive unemployment, and Liège was again a center of social and political unrest. The city is modern yet retains some historic buildings, including a cathedral (founded 971), the Church of the Holy Cross (tenth century), the Church of St. Denis (tenth–eleventh century), and the sixteenth-century palace of justice (the former residence of the prince-bishops). The city is the cultural center of French-speaking Belgium. It has a university (founded 1816) and a music conservatory. The composer César Franck was born here.

Liège-Maastricht Canal (lyezh–mah-STRIKHT), 16 mi/26 km long, E BELGIUM; section of ALBERT CANAL between LIÈGE (Belgium) and MAASTRICHT (NETHERLANDS).

Liége-Saint-Lambert, CONGO: see KATANA.

Liegnitz, POLAND: see LEGNICA.

Lieksa (LEE-ek-sah), town, POHJOIS-KARJALAN province, E FINLAND, 50 mi/80 km N of JOENSUU; 63°19′N 30°01′E. Elevation 330 ft/100 m. Near Russian border, on E shore of LAKE PIELINEN. Sawmilling, woodworking; popular water-sports center.

Lieksajärvi, RUSSIA: see LEKSOZERO.

Lielahti (LEE-eh-lah-tee), suburb of TAMPERE, HÄMEEN province, SW FINLAND, 2 mi/3.2 km W of city center; 61°31′N 23°41′E. Elevation 330 ft/100 m. Located between lakes NÄSIJÄRVI (N) and PYHÄJÄRVI (S). Lumber, pulp, cellulose, paper mills.

Lielupe River (LYE-loo-pai), German *aa* or *kurische* [=COURLAND Aa] also BOLDERAA, in W central LATVIA; formed at BAUSKA by union of MEMELE and MUSA rivers; flows NW past JELGAVA, N and ENE to Gulf of RIGA, joining mouth of DAUGAVA River between DAUGAVGRIVA and BOLDERAJA; 74 mi/119 km long; including either headstream, c.190 mi/306 km long. Parallels coast of Gulf of RIGA in lower course, forming a sandspit on which is JÜRMALA. Navigable for entire course.

Liénart (lee-en-AHRT), village, ORIENTALE province, N CONGO, 50 mi/80 km ENE of BUTA; 03°04′N 25°31′E. Elev. 1998 ft/608 m. Railroad junction and trading post in cotton-growing area.

Lienen (LEE-nen), village, North Rhine-Westphalia, NW GERMANY, in TEUTOBURG FOREST, 9 mi/14.5 km SSW of OSNABRÜCK; 52°09′N 07°58′E.

Lienz (LEE-ents), town, EAST TYROL, AUSTRIA, on the ISEL RIVER, and 75 mi/121 km W of KLAGENFURT. Market center; important road junction, hydroelectric station on DRAU RIVER; 46°50′N 12°46′E. Has two Gothic churches and 16th century Bruck castle. LIENZ is at the N foot of the LIENZ DOLOMITES (W group of GAILTAL ALPS). Manufacturing of machinery; brewing. Summer and winter tourism. Cable cars to the Zellersfeld at an elevation of 6,700 ft/2,042 m.

Lienz Dolomites (LEE-ents DO-luh-meits), subrange of the GAILTAL ALPS, S AUSTRIA (highest peak, SANDSPITZE, 8,449 ft/2,575 m), along EAST TYROL–CARINTHIA line, S of DRAU RIVER. At the N foot lies LIENZ. Noted for mountaineering and rock climbing.

Liepāja (LYE-pah-yuh) or **liepaya**, German *libau*, city (2000 population 89,448), W LATVIA; 56°31′N 21°01′E. An ice-free port on the BALTIC SEA, it is located at the end of an isthmus separating the Baltic Sea from Lake Liepaja. The city has a naval base as well as a commercial harbor exporting timber, metals, and oil (28% of cargo handled in 1993), with ferry service to ROSTOCK and COPENHAGEN. Liepaja is second only to Rīga in size and industrial development among Latvian cities. Metallurgy is the leading industry (iron and steel); others include shipbuilding, food and fish processing, and sugar refining. Founded by the Teutonic Knights in 1263, it had a population of 1,500 by 1625, the city was part of LIVONIA and later of the duchy of COURLAND, with which it passed to RUSSIA in 1795. In the late 19th and early 20th century Liepaja acquired great commercial importance and became one of the main Russian emigration ports with a direct shipping line to the U.S. The city was under German occupation during most of World War I. It was briefly the site of the provisional Latvian government when Bolshevik forces attacked Rīga in 1918. Held by the Germans 1941–1945, Liepaja suffered heavy damage. After World War II it was annexed by the USSR along with the rest of Latvia. City landmarks include a residence of Peter the Great and the 18th-century Church of the Trinity.

Liepaya, LATVIA: see LIEPĀJA.

Lier (LYAWR), French *Lierre* (lee-air), commune (2006 population 33,282), Mechelen district, ANTWERPEN province, N BELGIUM, at confluence of GROTE NETE and KLEINE NETE rivers (here forming NETE RIVER), 9 mi/14.5 km SE of ANTWERP. Manufacturing: fabricated metal prods, lace making. Has fifteenth-century church of St. Gommarius, seventeenth–eighteenth-century town hall with fourteenth-century belfry.

Lier (LEE-uhr), village (2007 population 22,257), BUSKERUD county, SE NORWAY, on LIERELVA River, on railroad, and 3 mi/4.8 km NNE of DRAMMEN; 59°35′N 05°29′E. Manufacturing (fiberglass, chemicals); lumber milling; market gardening. Gardener's College here. Commuting to OSLO and Drammen is common.

Lierde (LYAWR-duh), commune (2006 population 6,442), Oudenaarde district, EAST FLANDERS province, W central BELGIUM, 12 mi/19 km E of OUDENAARDE.

Lierelva (LEE-uhr-EL-vah), river, BUSKERUD county, SE NORWAY; rises near S side of TYRIFJORDEN; flows c.25 mi/40 km S to head of DRAMMENSFJORDEN 3 mi/4.8 km E of DRAMMEN. Valley is one of SE Norway's most fertile regions.

Lierneux (lyawr-NUH), commune (2006 population 3,342), Verviers district, LIÈGE province, E BELGIUM, 25 mi/40 km SSE of LIÈGE. Agriculture, lumbering.

Lierre, BELGIUM: see LIER.

Liesbeeck, river, SOUTH AFRICA: see KIRTENBOSCH NATIONAL BOTANICAL GARDEN.

Lieshout (LEES-hout), village, NORTH BRABANT province, SE NETHERLANDS, 7 mi/11.3 km NE of EINDHOVEN, on WILHELMINA CANAL; 51°31′N 05°36′E. Dairying; cattle, hogs; grain, vegetables, potatoes, fruits, grain, vegetables.

Liesing (LEES-ing), outer SW district (□ 12 sq mi/31.2 sq km) of VIENNA, AUSTRIA, 6 mi/9.7 km SW of city center; one of Vienna's industrial districts. Manufacturing (motor vehicles, chemicals, pharmaceuticals, electronics, processed foods, brewing). Intersdorf, with its market for fruits and vegetables and its motorway junction, is part of the district. Watruba church.

Liesse-Notre-Dame (lye-suh–no-truh–dahm), commune (□ 3 sq mi/7.8 sq km), AISNE department, PICARDIE region, N FRANCE, 9 mi/14.5 km ENE of LAON. Its 14th–15th-century church of Notre-Dame-de-Liesse (with 12th-century image of the Virgin) has been goal of pilgrimages since the Crusades.

Liestal (LEASE-tal), district, N BASEL-LAND half-canton, N SWITZERLAND; 47°28′N 07°44′E. Main towns are LIESTAL and PRATTELN; population is German-speaking and Protestant.

Liestal (LEASE-tal), town (2000 population 10,429), ⊙ BASEL-LAND half-canton, N SWITZERLAND, on ERGOLZ RIVER, and 8 mi/12.9 km SE of BASEL; 47°28′N 07°44′E. Elevation 1,079 ft/329 m. Textiles, consumer goods, chemicals, metal products; printing. Has old town hall and cantonal museum.

Liétor (LYAI-tor), town, ALBACETE province, SE central SPAIN, on MUNDO RIVER, and 15 mi/24 km W of HELLÍN; 38°32′N 01°57′W. Esparto processing; olive oil, honey, fruit. Large reservoir nearby.

Lieusaint (lyu-sen), town (□ 4 sq mi/10.4 sq km), SEINE-ET-MARNE department, ÎLE-DE-FRANCE region, N central FRANCE, near border of ESSONNE department, 21 mi/33 km SE of PARIS; 48°38′N 02°33′E.

Lieuvin (lyu-van), region of NORMANDY in CÔTES-D'ARMOR and SEINE-MARITIME departments, W FRANCE. Pasture and cereal cropland between RISLE (E) and TOUQUES (W) rivers.

Lievenhof: see LITHUANIA.

Lievenhof, LATVIA: see LĪVĀNI.

Lievestuore (LEE-e-ve-STOO-o-rah), village, KESKISUOMEN province, S central FINLAND, 14 mi/23 km E of JYVÄSKYLÄ; 62°16′N 26°12′E. Elevation 495 ft/150 m. On small lake near N end of Lake PÄIJÄNNE. Pulp mill, wood-alcohol plant.

Liévin (lyai-van), town (□ 5 sq mi/13 sq km), PAS-DE-CALAIS department, NORD-PAS-DE-CALAIS region, N FRANCE, 3 mi/4.8 km WSW of LENS; 50°25′N 02°46′E. Fertilizer manufacturing. A former coal-mining center with miners' row housing, Liévin is retooling for economic diversification. Just S rises VIMY RIDGE of World War I fame.

Lièvre (lee-AI-vruh), river, c.200 mi/320 km long, S QUEBEC, E CANADA; rising in KEMPT LAKE; flowing generally SW into the OTTAWA River near BUCKINGHAM; 45°31′N 75°25′W. Parts of it are navigable. There are five hydroelectric plants along its course; two of the most important are at MASSON and High Falls.

Lièvres, Île aux, CANADA: see HARE ISLAND.

Liezen (LEE-tsen), town, NW STYRIA, central AUSTRIA, near ENNS RIVER, 42 mi/68 km WNW of LEOBEN; 47°34′N 14°14′E. Market center of the Styrian Enns valley, important road junction. Manufacturing of machines.

Lifake, CONGO REPUBLIC: see DJOLU.

Liffey River (LI-fee), Gaelic *An Life*, c.50 mi/80 km long; rises in the Wicklow Mountains, WICKLOW county, E IRELAND; flows W, NE, and then E through DUBLIN to DUBLIN BAY. Three electric power stations on river.

Liffol-le-Grand (lee-fol–luh–grahn), commune (2004 population 2,410), VOSGES department, E FRANCE, in LORRAINE, 6 mi/9.7 km SW of NEUFCHÂTEAU; 48°19′N 05°35′E. Furniture manufacturing.

Lifford (LI-fuhrd), Gaelic *Leifear*, town (2006 population 1,448), ⊙ DONEGAL county, N IRELAND, at confluence of Mourne and FINN rivers, here forming FOYLE RIVER, opposite STRABANE (Northern Ireland), and 110 mi/177 km NNW of DUBLIN; 54°50′N 07°29′W. Manufacturing of shirts, furniture. It was scene (1600) of battle between Hugh Roe O'Donnell and English garrison of Derry.

Liffré (lee-fre), town (□ 28 sq mi/72.8 sq km), ILLE-ET-VILAINE department, in BRITTANY, W FRANCE, 10 mi/16 km NE of RENNES; 48°13′N 01°30′W. Electronics industry, woodworking.

Lifou (LEE-FOO), raised coral island, largest of LOYALTY ISLANDS, SW PACIFIC OCEAN, 60 mi/97 km E of NEW CALEDONIA; c.40 mi/64 km long, 10 mi/16 km wide; 20°55′S 167°15′E. Copra. CHÉPÉNÉHÉ is chief town.

Lifudzin, RUSSIA: see RUDNYY.

Lifuka (lee-FOO-kah), low, shoaly, coral island, HA'APAI group, central TONGA, S PACIFIC OCEAN; c.5 mi/8 km long; 19°48′N 174°21′E. Pangai ⊙ HA'APAI group is here; a port of entry. The name "Friendly Islands" originally applied by Capt. Cook to Lifuka, later generally used for TONGA.

Liganga (lee-GAN-gah), village, IRINGA region, S TANZANIA, 40 mi/64 km SSE of NJOMBE, NE of Lake NYASA; 10°08′S 35°07′E. Iron reserve.

Ligao (LEE-gou), town (2000 population 90,603), ALBAY province, SE LUZON, PHILIPPINES, on railroad, and 15 mi/24 km WNW of LEGASPI; 13°14′N 123°33′E. Trade center for agricultural area.

Ligatne (LEE-gaht-nai), town, N central LATVIA, on railroad, and 6 mi/10 km ENE of SIGULDA. Large paper and pulp mill (present factory first built in 1858) on GAUJA RIVER.

Liggett (LIG-et), unincorporated village, HARLAN county, SE KENTUCKY, in the CUMBERLAND MOUNTAINS, 8 mi/12.9 km SSW of HARLAN. Bituminous coal.

Lighthouse Point (LEIT-hous), city (□ 2 sq mi/5.2 sq km; 2000 population 10,767), BROWARD county, SE FLORIDA, 3 mi/4.8 km S of DEERFIELD BEACH; 26°16′N 80°05′W. Manufacturing includes flow pumps.

Lighthouse Reef, BELIZE, E of TURNEFFE ISLANDS, and 50 mi/80 km offshore; 30 mi/48 km long, 8 mi/12.9 km wide; 17°20′N 87°32′W. Site of Half Moon Cay natural monument and Blue Hole undersea sink hole.

Lightning Ridge (LEIT-neeng RIJ), village, N NEW SOUTH WALES, AUSTRALIA, 335 mi/539 km NW of

NEWCASTLE; 29°26′S 147°59′E. Opal mines; noted for its black opal, a dark stone with blue, green, and red flecks. Mineral springs spa just E.

Light Oak (LEIT OK), unincorporated town (□ 1 sq mi/ 2.6 sq km; 2000 population 779), CLEVELAND county, S NORTH CAROLINA, residential suburb 4 mi/6.4 km ESE of SHELBY; 35°16′N 81°28′W. Manufacturing; service industries.

Light River (LEIT), 100 mi/161 km long, SE SOUTH AUSTRALIA; rises in MOUNT LOFTY RANGES; flows SW, past KAPUNDA and HAMLEY BRIDGE, to Gulf SAINT VINCENT 20 mi/32 km N of PORT ADELAIDE; 34°23′S 138°33′E.

Ligné (lee-nyai), commune (□ 17 sq mi/44.2 sq km), LOIRE-ATLANTIQUE department, PAYS DE LA LOIRE region, W FRANCE, 16 mi/26 km NE of NANTES; 47°25′N 01°23′W. Coal deposits nearby.

Lignières (lee-NYAIR), commune, NEUCHÂTEL canton, W SWITZERLAND; northeasternmost commune in the canton on border of BERN canton. Elevation 2,635 ft/ 803 m. Birthplace of Frédéric Chiffelle.

Lignite (LIG-neit), village, BURKE co., NW NORTH DAKOTA, 15 mi/24 km WNW of BOWBELLS, near lignite-mining center; 48°52′N 102°33′W. Founded in 1907 and incorporated in 1915. Named for the coal in the area.

Lignite, locality, S central ALASKA, NE of MOUNT MCKINLEY NATIONAL PARK, 5 mi/8 km N of HEALY. On Alaska railroad; sub-bituminous-coal deposits.

Lignon du Velay, FRANCE: see LIGNON RIVER.

Lignon River (lee-nyon), tributary of the LOIRE RIVER, 60 mi/97 km long, in HAUTE-LOIRE department, AUVERGNE region, central FRANCE; rises in S MASSIF CENTRAL, near Mont MÉZENC, 17 mi/27 km SE of Le PUY-EN-VELAY; flows N, past TENCE, in deep gorges to the Loire above MONISTROL-SUR-LOIRE; 45°15′N 04°08′E. Dammed for hydroelectric power below Tence. Sometimes called Lignon du Velay.

Ligny (leen-YEE), village in commune of SOMBREFFE, Namur district, NAMUR province, central BELGIUM, near NAMUR. At Ligny, on June 16, 1815, Napoleon I defeated the Prussians under Blücher early in the Waterloo campaign.

Ligny-en-Barrois (lee-NYEE–ahn–bahr-WAH), town (□ 13 sq mi/33.8 sq km), MEUSE department, LORRAINE region, NE FRANCE, on the ORNAIN and MARNE-RHINE CANAL, and 9 mi/15 km SE of BAR-LE-DUC; 48°41′N 05°20′E. Optical industry; woodworking (furniture, toys, and measuring devices).

Ligny-en-Cambrésis, FRANCE: see CAUDRY.

Ligon (LIG-uhn), village, FLOYD county, E KENTUCKY, in CUMBERLAND foothills, 14 mi/23 km SW of PIKEVILLE. Bituminous coal.

Ligonier (lig-uh-NEER), city (2000 population 4,357), NOBLE county, NE INDIANA, on ELKHART RIVER, and 9 mi/14.5 km NW of ALBION; 41°28′N 85°35′W. Trade center in dairying and poultry-raising area; manufacturing (building materials, gaskets, hydraulic equipment, marshmallows, plastic products, flour and feed milling, motor vehicle parts, molded products, sample books). Laid out 1835.

Ligonier (LI-guh-nir), borough (2006 population 1,640), WESTMORELAND county, SW PENNSYLVANIA, on LOYALHANNA CREEK, 17 mi/27 km ESE of GREENSBURG. Agriculture (corn, hay; dairying); manufacturing (wood products); bituminous coal; timber. Fort Ligonier built c.1758. Linn Run State Park and Forbes Satte Forest to S, LAUREL RIDGE State Park to SE and E; CHESTNUT RIDGE to NW. Laid out 1817; incorporated 1834.

Ligovo (LEE-guh-vuh), former town, central LENINGRAD oblast, NW European Russia, adjoins URITSK (0.5 mi/1 km to the W), which was formerly (until 1925) called Ligovo; 59°50′N 30°12′E. Now part of SAINT PETERSBURG, less than 7 mi/11 km SW of the city center, conected by road and local railway. Birthplace of a number of members of the Russian royal family in the late 19th century.

Ligua, CHILE: see LA LIGUA.

Liguanea, CUBA: see LANIER.

Liguanea Plain, fertile lowland in SE JAMAICA, includes all KINGSTON area and lower SAINT ANDREW; 17°58′N–18°01′N 76°45′W–76°52′W. VERE PLAIN continues it W.

Ligua River (LEE-gwah), Spanish, *Río de la Ligua* (REE-o de lah LEE-gwah), c.70 mi/113 km long, VALPARAISO region, central CHILE; rises in the ANDES at c.32°15′N; flows W past CABILDO and LA LIGUA, to the PACIFIC N of PAPUDO.

Liguasan Marsh (leeg-WAH-sahn), extensive swamp region, c.25 mi/40 km long and 20 mi/32 km wide, along the PULANGI RIVER, S central MINDANAO, PHILIPPINES. Fertile rice-growing areas and mangrove forests; game refuge and bird sanctuary (□ c.170 sq mi/440 sq km) was established here in 1941.

Ligueil (lee-guh-yuh), commune (□ 11 sq mi/28.6 sq km; 2004 population 2,180), INDRE-ET-LOIRE department, CENTRE administrative region, W central FRANCE, 20 mi/32 km NE of CHÂTELLERAULT; 47°02′N 00°49′E. Dairying.

Ligugé (lee-gu-zhe), commune (□ 9 sq mi/23.4 sq km), VIENNE department, POITOU-CHARENTES region, W central FRANCE, 5 mi/8 km S of POITIERS; 46°31′N 00°20′E. Paper mill. Has ancient monastery still occupied by Benedictine monks. Pre-Roman and Gallo-Roman buildings were discovered in 1953 under church of Saint-Martin.

Liguria (li-GOOR-ee-uh), region (□ 2,098 sq mi/5,454.8 sq km), NW ITALY, extending along the LIGURIAN SEA and bordering FRANCE on the W; 44°30′N 08°50′E. The generally mountainous region has a steep, narrow coastal strip that includes the beautiful ITALIAN RIVIERA. In the interior, the LIGURIAN ALPS rise in the W and the LIGURIAN APENNINES in the E. GENOA is ⊙ of Liguria, which is divided into Genoa, IMPERIA, LA SPEZIA, and SAVONA provinces (named for their capitals, all of which are seaports). Flowers (mostly for use in making perfume), olives, wine grapes, citrus fruits, mushrooms, and cereals are grown. Chestnuts are gathered in the mountains, where there are extensive pastures, timberland, and marble, slate, quartz, and limestone quarries. Fishing is pursued along the coast. Manufacturing of the region includes iron and steel, ships, machinery, textiles, chemicals, processed foods, and forest products. Liguria derives its name from the ancient Ligurii, who occupied the MEDITERRANEAN coast from the RHÔNE River to the ARNO River. In the 4th century the Ligurii were driven from the Alpine regions by Celtic immigrants, while Phoenicians, Greeks, and Carthaginians colonized the coast. In the 2nd century B.C.E. the entire region was subdued by the Romans. Throughout the Middle Ages, Genoa struggled with local feudal lords (and at times with Venice) for control of the area. By the 16th century it controlled virtually all of present-day Liguria, and from that time until its annexation (1815) by the kingdom of Sardinia, Liguria shared the history of Genoa. There is a university at Genoa.

Ligurian Alps (li-GYOOR-ee-uhn), Italian *Alpi Liguri*, E division of MARITIME ALPS, NW ITALY; extend 60 mi/97 km W from LIGURIAN APENNINES at CADIBONA Pass to TENDA Pass; 44°10′N 08°05′E; rise to 8,697 ft/ 2,651 m in MONTE MARGUAREIS, and to 7,218 ft/2,200 m in MONTE SACCARELLO. W end forms part of French-Italian border.

Ligurian Apennines (A-pe-neinz), ITALY, W division of N APENNINES, extend 100 mi/161 km between MARITIME ALPS at CADIBONA Pass (W) and ETRUSCAN APENNINES at LA CISA Pass (E). Chief peaks: Monte MAGGIORASCA (5,915 ft/1,803 m), Monte BUE (5,840 ft/ 1,780 m), Monte ANTOLA (5,243 ft/1,598 m), and Monte BEIGUA (4,222 ft/1,287 m). Passes include GIOVI Pass, Passo del TURCHINO, and Colle della SCOFFERA. In N is source of Orba, SCRIVIA, TARO, and TREBBIA rivers; in S, of Lavagna and VARA rivers.

Ligurian Sea (lig-YOOR-ee-uhn), arm of the MEDITERRANEAN Sea, between the Ligurian coast (Italian Riviera) and the islands of CORSICA and ELBA; the Gulf of GENOA is its northernmost part. The sea receives the ARNO River from the E. The ports of GENOA, La SPEZIA, and LIVORNO are on its rocky coast. The sea's NW coast is noted for its favorable climate and scenic beauty.

Ligwera (lee-GWAI-rah), village, LINDI region, S TANZANIA, 50 mi/80 km S of LIWALE, near Lumesule River; 10°18′S 37°38′E. Cashews, bananas, corn, sweet potatoes; goats; sheep; timber.

Lihir Islands, small volcanic group (□ 70 sq mi/182 sq km), NEW IRELAND district, BISMARCK ARCHIPELAGO, PAPUA NEW GUINEA, SW PACIFIC OCEAN, 30 mi/48 km NE of New Ireland; comprise circa 5 islands. Largest island, LIHIR, is circa 12 mi/19 km long, rises to 1,640 ft/500 m; 03°05′S 152°35′E. Gold deposits.

Lihou (lee-HO-oo), island, one of the CHANNEL ISLANDS, just off the W coast of GUERNSEY (connected at low tide); 49°27′N 02°40′W. Covers 38 acres/15 ha; has remains of a 12th-century priory of St. Mary. Had an iodine factory before World War II, which was destroyed during German occupation (1940–1945), when the island was used for target practice by heavy artillery.

Lihue (LEE-HOO-ai), town (2000 population 5,674), ⊙ KAUAI county, SE KAUAI, HAWAII, near NAWILIWILI HARBOR, 110 mi/177 km WNW of HONOLULU; 21°58′N 159°20′W. Junction of Kaumualii Highway, to S coast, and Kuhio Highway, to N coast. Manufacturing (coatings, printing and publishing, lumber, soaps, sugarcane processing, construction machinery). Lihue Plantation to W. Kauai Community College to W at PUHI. Grove Farm Homestead Museum, Kauai Museum, Kilohana Crater (1,133 ft/345 m) to NW; Lihue-Koloa Forest Reserve to NW; Haleia National Wildlife Refuge to SW; Wailua River State Park (including Wailua Falls) to N; Ahukini State Recreational Park to NE.

Lihuel Calel, ARGENTINA: see CUCHILLO-CÓ.

Lija (LEE-jah), village, central MALTA, 3.5 mi/5.6 km W of VALLETTA; 35°54′N 14°26′E. Lija and nearby BALZAN and ATTARD are called the "Three Villages."

Lijiang (LEE-JIANG), town, ⊙ Lijiang county, NW YUNNAN province, CHINA, near CHANG JIANG (Yangzi River), 75 mi/121 km N of DALI; 26°48′N 100°16′E. Elevation 8,136 ft/2,480 m. Rice, medicinal herbs; logging, timber processing, papermaking, food industry. Lijiang is an ancient town with its main part built in late Song dynasty (1129–1279) and early Yuan dynasty (1206–1368). The town has a dense network of canals and arched bridges. Residential houses have white exterior walls and gray roofs. The majority of the town residents are the Naxi ethnic group.

Lijin (LEE-JIN), town, ⊙ Lijin county, N SHANDONG province, CHINA, 75 mi/121 km NW of WEIFANG, and on HUANG HE (Yellow River) at head of its delta; 37°29′N 118°16′E. Grain, cotton, oilseeds.

Lik (LEEK), village, MONTANA oblast, MEZDRA obshtina, BULGARIA; 43°06′N 23°46′E.

Lika (LEE-kah), karst region in W CROATIA, between ADRIATIC SEA and BOSNIA AND HERZEGOVINA border. Large areas forested. Livestock raising (dairy cattle, sheep); agr. (rye, barley, potatoes). Lumber and wood processing. Chief town, GOSPIC. Earliest settlement dates back to Illyrian and Celtic tribes. Part of medieval Croatian state. Occupied by Turks (1525–1699); part of Hapsburg Vojna Krajina (military frontier) until 1881. Scene of operations by Yugoslav partisans in World War II, when it was devastated. Part of Serb (Srpska) Krajina (1991–1995); sustained heavy damage in 1991 war. Plitvice National Park is here.

Lika River (LEE-kah), c.35 mi/56 km long, W CROATIA, in LIKA region; rises 12 mi/19 km ESE of GOSPIC; flows NW, disappearing into the karst 18 mi/29 km NNW of Gospić. Dammed N of Gospić, forming Krušičko Lake Hydroelectric plant (built 1970) at Sklope.

Cross-references are shown in SMALL CAPITALS. The pronunciation guide is shown on page xix. The sources of population figures are shown on page xvii.

Likasi (lee-KAH-see), city, KATANGA province, SE CONGO REPUBLIC; 10°59′S 26°44′E. Elev. 4,327 ft/1,318 m. Major industrial, mining, and transportation center; copper and cobalt mining and refining (though mineral production has greatly declined in recent years); manufacturing: cement, chemicals, beverages. Archaeological and mineral museum. Formerly called JADOTVILLE or JADOTSVILLE.

Likati (leee-KAH-tee), village, ORIENTALE province, N CONGO, on a tributary of ITIMBIRI RIVER, on railroad and 65 mi/105 km NW of BUTA; 03°21′N 23°53′E. Elev. 1,400 ft/426 m. Center of rice and cotton area; cotton ginning. Has Roman Catholic mission. Coffee and aleurite plantations in vicinity.

Likavka (li-KAU-ka), village STREDOSLOVENSKY province, N SLOVAKIA, on railroad, on VAH RIVER, and just N of RUŽOMBEROK. Machinery, woodworking; wheat, potatoes. Likava Castle ruins just N.

Likely (LEIK-lee), unincorporated village, E central BRITISH COLUMBIA, W CANADA, 41 mi/66 km NNE of WILLIAMS LAKE, in CARIBOO REGIONAL DISTRICT; 52°37′N 121°33′W. Old gold-rush town. The QUESNEL RIVER, a chief tributary of the FRASER, begins here; location of Quesnel River Research Centre (University of Northern British Columbia). Formerly called Quesnel Dam.

Likėnai (li-KEN-ei), town, N LITHUANIA, 9 mi/14 km W of Birzėnai. Mineral springs and health resort. Also spelled Likenay.

Likeri, Lake, Greece: see ILIKI, LAKE.

Likhapani (li-kuh-PAH-nee), village, DIBRUGARH district, NE ASSAM state, NE INDIA, on BURHI DIHING RIVER, and 60 mi/97 km ESE of DIBRUGARH. Railroad spur terminus; coal-mining center. Also spelled Lakhapani.

Likhaya, RUSSIA: see LIKHOVSKOY.

Likhma (LIK-muh), village, RAIPUR district, central CHHATTISGARH state, central INDIA, 50 mi/80 km SSE of DHAMTARI. Terminus of lumber railroad; sawmilling.

Likhoslavl' (lee-huh-SLAHVL), city (2006 population 12,220), central TVER oblast, W European Russia, on crossroads and railroad, 25 mi/40 km NW of TVER; 57°08′N 35°28′E. Elevation 501 ft/152 m. Railroad junction; lighting, radiators, artistic handicrafts; dairying. Peat extraction in the vicinity. Was the capital of the former Karelian National Okrug (1937–1939). Made city in 1925.

Likhovka, UKRAINE: see LYKHIVKA.

Likhovskoy (lee-huhf-SKO-yee), town (2006 population 13,615), W ROSTOV oblast, S European Russia, on highway, 55 mi/88 km N of ROSTOV-NA-DONU, and 12 mi/19 km S of KAMENSK-SHAKHTINSKIY; 48°09′N 40°11′E. Elevation 613 ft/186 m. Railroad junction; railroad shops; local industries. Coal mining in the area. Until 1930, called Likhaya.

Likhu Kola (lee-KOO ko-LAH), river, central NEPAL; joins the SUN KOSI RIVER at 27°16′N 86°13′E.

Likhvin, RUSSIA: see CHEKALIN.

Liki, INDONESIA: see KUMAMBA ISLANDS.

Likiep (LEE-kee-ep), atoll (□ 4 sq mi/10.4 sq km; 1999 population 527), RATAK CHAIN, MAJURO district, MARSHALL ISLANDS, W central PACIFIC, 125 mi/201 km NE of KWAJALEIN; sixty-four islets on lagoon 26 mi/42 km long; 09°49′N 168°18′E. Sometimes spelled Likieb.

Likimi (lee-KEE-mee), village, Équateur province, NW CONGO, on MONGALA RIVER, and 65 mi/105 km NW of LISALA; 02°50′N 20°45′E. Elev. 1,299 ft/395 m. Agriculture (cotton, sesame) and trading center.

Likino (LEE-kee-nuh), village, central MOSCOW oblast, central European Russia, on road and near railroad, 5 mi/8 km W of ODINTSOVO, to which it is administratively subordinate, 55°37′N 37°08′E. Elevation 662 ft/201 m. Poultry factory.

Likino-Dulëvo (LEE-kee-nuh–doo-LYO-vuh), city (2006 population 30,640), E MOSCOW oblast, central European Russia, on road and railroad, 60 mi/97 km E of MOSCOW, and 5 mi/8 km S of OREKHOVO-ZUYEVO, to which it is administratively subordinate; 55°42′N 38°57′E. Elevation 446 ft/135 m. Formed in 1930 by amalgamation of Likino (N; cotton textiles [from 1870], lumber machines) and Dulëvo (S; porcelain works [from 1832], paint making). Bus factory from 1959. Made city in 1937.

Likisia (li-KEE-sah), town, ⊙ Likisa district, Indonesian-controlled E TIMOR, in central TIMOR, on OMBAI Strait, 17 mi/27 km W of DILI. Fishing; copra. Sometimes spelled Liquissa or Liquiçá.

Likodhimos, Greece: see LYKODIMO.

Likoma, district (2007 population 11,094), Northern region, N MALAWI. District is composed of a group of islands in LAKE NYASA and is named after LIKOMA ISLAND, the largest island of the group.

Likoma Island (lik-ko-mah), Likoma district, Northern region, E MALAWI, in Lake Malawi (LAKE NYASA), off E shore, 40 mi/64 km ESE of CHINTHECHE; 5 mi/8 km long, 4 mi/6.4 km wide. The village of Likoma is on SE shore of island. Fishing; rice, corn.

Likorema, Greece: see PERISTERA.

Likouala (lee-kwah-LAH), region (2007 population 89,347), NE Congo Republic, with SANGHA region to the W and S, CUVETTE region to the extreme SE, CENTRAL AFRICAN REPUBLIC to the N, and CONGO to the E; ⊙ IMPFONDO; 02°00′S 17°30′E. The LIKOUALA-AUX-HERBES RIVER rises near EPÉNA and flows S into Cuvette region, and the UBANGI RIVER flows S past Impfondo (on the W bank) and along the border with Congo.

Likouala-aux-herbes River (lee-kwah-LAH–o–ZERB), circa 300 mi/483 km long, Congo Republic; rises circa 60 mi/97 km WNW of Dongou; flows S, past EPÉNA and through a region of swamps, to SANGHA RIVER 10 mi/16 km W of LIRANGA. Navigable circa 150 mi/241 km in middle and lower course. Also called Likouala-Essoubi.

Likouala-Essoubi, Congo Republic; see LIKOUALA-AUX-HERBES RIVER.

Likouala River (lee-kwah-LAH), circa 325 mi/523 km long, Congo Republic; formed by several headstreams 40 mi/64 km W of MAKOUA; flows E, past Makoua, and SSE to CONGO RIVER at Mossaka (also point of confluence of SANGHA RIVER). Forms extensive swamps in lower course and is connected by a network of side streams with the lower Sangha. Navigable for circa 250 mi/402 km below Makoua. Was explored (1878) by de Brazza.

Likuyu (lee-KOO-yoo), village, RUVUMA region, S TANZANIA, 10 mi/16 km W of SONGEA, 1 mi/1.6 km S of PERAMIHO; 10°44′S 35°29′E. Timber; tobacco, corn; goats, sheep.

Lila Lake, HAMILTON county, NE central NEW YORK, on branch of BEAVER RIVER, in the ADIRONDACK MOUNTAINS, in Adirondack Park, 30 mi/48 km SW of SARANAC LAKE town; c.2.5 mi/4 km long, max. 1.5 mi/2.4 km wide; 44°00′N 74°45′W.

Lilas, Les (lee-lah, lai), inner NE suburb of PARIS, SEINE-SAINT-DENIS department, ÎLE-DE-FRANCE region, N central FRANCE, 3.5 mi/5.6 km from Notre Dame Cathedral and just outside Paris circumferential highway; 48°53′N 02°25′E. Diverse manufacturing.

Lilayi (li-LAY-ee), township, LUSAKA PROVINCE, S central ZAMBIA, 7 mi/11.3 km S of LUSAKA; 15°32′S 28°18′E. On railroad. Agriculture (tobacco, corn, potatoes, cotton, soybeans, peanuts, vegetables); cattle. Talc mining. Lusaka International Airport to NE.

Lilbourn (LIL-buhrn), city (2000 population 1,303), NEW MADRID county, extreme SE MISSOURI, near MISSISSIPPI RIVER, 5 mi/8 km N of NEW MADRID; 36°35′N 89°36′W. Cotton, rice, wheat, soybeans; lumber mills. Incorporated as city 1910.

Lilburn, suburb (2000 population 11,307) of ATLANTA, GWINNETT county, GEORGIA, 5 mi/8 km E of DORAVILLE; 33°53′N 84°08′W. Bedroom community for Atlanta region as well as an emerging business hub.

Manufacturing (glass, concrete, printing plates, steel fabricating, signs).

L'Île-Bizard–Sainte-Geneviève (LEEL-bee-ZAHRD–sent-zhe-ne-VYEV), island, borough (French *arrondissement*) of MONTREAL, S QUEBEC, E CANADA, off NW coast of MONTREAL ISLAND; 45°29′N 73°52′W. Île-Bizard is connected to Sainte-Geneviève by Jacques-Bizard bridge.

L'Île-Cadieux (LEEL-kah-DYU), town (2006 population 137), MONTÉRÉGIE region, S QUEBEC, E CANADA, 19 mi/30 km from MONTREAL; 45°26′N 74°01′W. Incorporated 1922. Part of the Metropolitan Community of Montreal (*Communauté Metropolitaine de Montréal*).

L'Île-d'Anticosti (LEEL–DAHN-tuh-KOS-tee), village (□ 3,059 sq mi/7,953.4 sq km; 2006 population 270), CÔTE-NORD region, E QUEBEC, E CANADA; 49°30′N 63°00′W.

L'Île-d'Orleans (LEEL–dor-lai-AHN), county (□ 74 sq mi/192.4 sq km; 2006 population 6,834), CAPITALE-NATIONALE region, S central QUEBEC, E CANADA; ⊙ SAINTE-FAMILLE; 46°55′N 70°54′W. Composed of six municipalities. Formed in 1982.

L'Île-Dorval (LEEL–dor-VAHL), town, island (2006 population 2), MONTRÉAL administrative region, S QUEBEC, E Canada, off S coast of MONTREAL ISLAND; 45°26′N 73°45′W. Part of the Metropolitan Community of Montreal (*Communauté Metropolitaine de Montréal*).

Lilesville (LEILZ-vil), village (□ 1 sq mi/2.6 sq km; 2006 population 432), ANSON county, S NORTH CAROLINA, 2 mi/3.2 km E of WADESBORO; 34°58′N 79°59′W. Manufacturing (sand and gravel processing, fabric goods, lumber); service industries; in agricultural (cotton, grain; poultry, livestock) and timber area. BLEWETT FALLS LAKE reservoir to E.

Lilienfeld (LIL-een-feld), town, S central LOWER AUSTRIA, on TRAISEN RIVER, and 13 mi/21 km S of Sankt Pölten; 48°01′N 15°36′E. Cistercian abbey founded 1202–1206 with a late Romanesque/early Gothic church. Birthplace of alpine skiing. In 1905 Malthias Zdarsky organized the first slalom at the nearby Muckenkogel.

Lilienthal (LIL-e-uhn-tahl), town, LOWER SAXONY, NW GERMANY, 5 mi/8 km NNE of BREMEN; 53°08′N 08°54′E. Primarily residential. Has church of former Cistercian nunnery.

Lilikse, GHANA: see LILIXIA.

Liling (LEE-LING), city (□ 833 sq mi/2,165.8 sq km; 2000 population 936,626), E HUNAN province, CHINA, near JIANGXI border, on railroad, 45 mi/72 km SE of CHANGSHA; 27°42′N 113°29′E. Agriculture is the largest sector of the city's economy; light and heavy industries are also important. Main industries include crafts, porcelain manufacturing, kaolin quarrying, and construction materials.

Liliw (LEE-lee-uh), town, LAGUNA province, S LUZON, PHILIPPINES, 8 mi/12.9 km ENE of SAN PABLO; 14°07′N 121°27′E. Agricultural center (rice, coconuts, sugarcane).

Lilixia (li-liks-EE-uh), town, UPPER WEST REGION, GHANA, 15 mi/24 km SW of TUMU; 10°47′N 02°05′W. Livestock, shea-nut butter.

Liliya Mota, town, AMRELI district, GUJARAT state, W central INDIA, 55 mi/89 km WSW of BHAVNAGAR. Millet, cotton, oilseeds, sugarcane.

Lilla Edet (LIL-lah ED-et), town, ÄLVSBORG county, SW SWEDEN, on GÖTA ÄLV RIVER (falls; navigation lock), 12 mi/19 km SW of TROLLHÄTTAN; 58°08′N 12°07′E.

Lillafüred (LEL-lah-fue-red), year-round resort, BORSOD-ABAÚJ-ZEMPLÉN county, NE HUNGARY, on Lake HÁMOR, 7 mi/11 km W of MISKOLC, in Bükk Mountains; 48°06′N 20°38′E. Large tourist trade.

Lille (LEEL), commune (2006 population 15,574), Turnhout district, ANTWERPEN province, N BELGIUM, 8 mi/13 km SW of TURNHOUT.

Area is shown by the symbol □, and capital city or county seat by ⊙.

Lille (LEEL), Flemish *Ryssel* or *Rijssel*, city (□ 9 sq mi/23.4 sq km), ⊙ NORD department, N FRANCE, in Flanders, on canalized DEÛLE RIVER, and 130 mi/209 km NNE of PARIS; 50°38′N 03°04′E. Forms a major industrial urban complex, together with ROUBAIX (6 mi/9.7 km NE), TOURCOING (8 mi/12.9 km NNE), and a ring of suburban communities. Lille has also become the administrative headquarters of NORD-PAS-DE-CALAIS region, established in the 1980s. Lille-Lesquin international airport 4 mi/6.4 km SE of the city. A leading textile center, historically known for its lisle, it also has an important metallurgical industry, chemical works, breweries, beet-sugar refineries, and print shops. Commercial activity culminates in well-known annual (June) fair. Lille's economic strength dates from the mid-19th century, when coal provided the energy for the industrial revolution. The extensive coal-mining area of N France supported France's growth for 100 years, but since World War II its output has declined drastically due to the exhaustion of economically recoverable coal and the aging infrastructure. As a result, the region turned into a "rust belt," which only since the 1980s is experiencing a revival as new industries are attracted by Lille's strategic position as NW Europe's key railroad center. With the completion in 1994 of the CHANNEL TUNNEL (Chunnel) under the ENGLISH CHANNEL, Lille lies at the junction of passenger and freight railroad traffic reaching France from ENGLAND and fanning out toward Paris, E France, BELGIUM, and GERMANY from a new Lille railroad terminal with extensive rail yards. The TGV (high-speed passenger trains) railroad network also focuses on Lille as the major urban center closest to the Channel ports and tunnel. Lille was the site of the historic agreement in the early 1980s between France and the United Kingdom that led to the construction of the Chunnel project. Lille is regularly laid out, in the shape of a triangle. Of its 17th-century fortifications (built by Vauban) a large citadel remains upon a commanding site. The city has a 17th-century stock exchange; several churches containing paintings of the Flemish school; and an unfinished cathedral (begun 1854). Its famed art museum has outstanding collection of Flemish, Dutch, French, and Spanish masters. Seat of a university (since 1807), a Catholic seminary, and bishopric (since 1913). Lille became the chief city of the medieval county of Flanders; the residence of 15th-century dukes of BURGUNDY, and (after 1668) capital of FRENCH FLANDERS. Taken (1708) by Eugene of SAVOY and the duke of MARL-BOROUGH after a costly siege; restored to France by treaty of UTRECHT (1713); and fought off a siege by Austrian forces in 1792. Occupied by Germans, 1914–1918, and 1940–1944, it was heavily damaged by Allied air raids in World War II. General Charles de Gaulle was born here in 1890. Formerly spelled L'Isle and Lisle.

Lille Bælt (LI-luh belt) [Dan.=little belt], strait, S DENMARK, connects the KATTEGAT strait with the BALTIC SEA; 30 mi/48 km long and 0.5 mi/0.8 km–18 mi/29 km wide; separates FYN island and JUTLAND; crossed by road and railroad bridge.

Lillebonne (lee-luh-buhn), ancient *Juliobona*, town (□ 5 sq mi/13 sq km; 2004 population 9,769), SEINE-MARITIME department, HAUTE-NORMANDIE region, N FRANCE, near SEINE RIVER estuary, 19 mi/31 km E of Le HAVRE; 49°31′N 00°33′E. Cotton mills and chemical plants. Has ruins of a Roman amphitheater (C.E. c.120) and of a castle tower built by William the Conqueror. Castle of TANCARVILLE is 4 mi/6.4 km SW.

Lillehammer (LIL-luh-HAHM-muhr), city (2007 population 25,537), ⊙ OPPLAND county, S NORWAY, at the N end of Lake MJØSA; 61°08′N 10°30′E. It is a commercial center for the fertile GUDBRANDSDALEN valley and is a popular summer and winter resort. Its open-air museum, Maihaugen (founded 1887), fea-tures complete farms, peasant cottages, workshops, and handicrafts of the region. It was the site of the 1994 Winter Olympic Games.

Lille Koldewey, GREENLAND: see GREAT KOLDEWEY.

Lillerød (LI-luh-rudh), town, FREDERIKSBORG county, SJÆLLAND, DENMARK, 5 mi/8 km SSE of HILLERØD; 55°52′N 12°22′E. Orchards (apples); dairying.

Lillers (lee-ler), town (□ 10 sq mi/26 sq km), PAS-DE-CALAIS department, NORD-PAS-DE-CALAIS region, N FRANCE, 8 mi/12.9 km WNW of BÉTHUNE; 50°34′N 02°29′E. In fertile agricultural district (wheat, sugar beets, and tobacco). Has major sugar refinery. Also produces agricultural equipment and fertilizer. Has restored 12th-century Romanesque church. First artesian well (so named for ARTOIS historic region) is said to have been sunk here.

Lillesand (LIL-luh-sahn), town, AUST-AGDER county, S NORWAY, on an inlet of the SKAGERRAK, on railroad, and 15 mi/24 km ENE of KRISTIANSAND; 58°15′N 08°22′E. Shipping and boatbuilding; canning; manufacturing of glassware; exports timber, feldspar, quartz.

Lille Sotra, NORWAY: see SOTRA.

Lillestrøm (LIL-luh-struhm), village, AKERSHUS county, SE NORWAY, on N shore of ØYEREN Lake, and 10 mi/16 km ENE of OSLO; 59°57′N 11°05′E. Railroad junction. This is a regional trade center with lumber milling; manufacturing (chemicals; plastics; pre-fabricated houses; knit goods).

Lillhagen (LIL-HAH-gen), residential suburb, GÖTE-BORG OCH BOHUS county, SW SWEDEN, 4 mi/6.4 km N of GÖTEBORG, near GÖTA ÄLV RIVER.

Lillie Glacier, in E VICTORIA LAND, EAST ANTARCTICA; 70°45′S 163°55′E. Flows to the coast at Ob' Bay on the OATES COAST, 100 mi/160 km long and 10 mi/15 km wide.

Lillie Glacier Tongue, the floating extension of LILLIE GLACIER in Ob' Bay on the OATES COAST of VICTORIA LAND, EAST ANTARCTICA; 70°34′S 163°48′E.

Lillington (LIL-eeng-tuhn), town (□ 4 sq mi/10.4 sq km; 2006 population 3,171), ⊙ HARNETT county, central NORTH CAROLINA, 28 mi/45 km SSW of RALEIGH, on CAPE FEAR RIVER; 35°23′N 78°49′W. Service industries; manufacturing (vanity tops and bathtubs, mobile homes, signs, gun holsters, apparel, turnstile and crowd control equipment, ordnance, sand and gravel, machining, garden supplies); agriculture (grain, tobacco, cotton, sweet potatoes; poultry, livestock). Raven Rock State Park to NW.

Lillinonah, Lake, reservoir (□ 2 sq mi/5.2 sq km), FAIRFIELD county, SW CONNECTICUT, on HOUSA-TONIC River, 12 mi/20 km E of DANBURY; 41°23′N 73°10′W. Maximum capacity 37,200 acre-ft. Formed by Stevenson Dam (83 ft/25 m high), built (1919) for power generation. Paugusett State Forest (W) and state parks here.

Lillo (LEE-lyo), town, TOLEDO province, central SPAIN; road center in upper LA MANCHA, 27 mi/43 km SE of ARANJUEZ. In fertile region (cereals, potatoes, sugar beets, saffron, olives, grapes, truck produce, fruits; sheep). Cheese processing, olive-oil pressing.

Lillooah, INDIA: see LILUAH.

Lillooet (LI-luh-wet), village (□ 10 sq mi/26 sq km; 2001 population 2,741), S BRITISH COLUMBIA, W CANADA, on FRASER River, near SETON Lake, 70 mi/113 km W of KAMLOOPS, in SQUAMISH-LILLOOET regional district; elevation 820 ft/250 m; 50°41′N 121°56′W. In gold-mining and mixed farming region; hydroelectric power. Semi-arid climate. Originally called Cayoosh Flat.

Lillooet Lake (LI-luh-wet), SW BRITISH COLUMBIA, W CANADA, expansion of LILLOOET RIVER, in COAST MOUNTAINS, 40 mi/64 km SW of LILLOOET; 21 mi/34 km long, 1 mi/2 km–2 mi/3 km wide; 50°15′N 122°30′W.

Lillooet River (LI-luh-wet), 130 mi/209 km long, SW BRITISH COLUMBIA, W CANADA; rises in COAST MOUNTAINS at foot of Mount Dalgleish, W of LIL-LOOET; flows SE, through LILLOOET LAKE, to NW end of HARRISON LAKE, which drains into FRASER River; 49°45′N 122°09′W.

Lilly, town (2000 population 221), DOOLY county, central GEORGIA, 6 mi/9.7 km WNW of VIENNA; 32°09′N 83°53′W.

Lilly, borough (2006 population 885), CAMBRIA county, W central PENNSYLVANIA, 21 mi/34 km ENE of JOHNSTOWN, on Little Conemaugh River. Agriculture (corn, hay; dairying); manufacturing (dresses); bituminous coal.

Lilly Grove, unincorporated town, MERCER county, S WEST VIRGINIA, 2 mi/3.2 km NW of PRINCETON; 37°22′N 81°04′W.

Liloan (LEEL-wahn), town, central Cebu island, PHI-LIPPINES, on CAMOTES SEA, 10 mi/16 km NE of CEBU; 10°25′N 123°58′E. Agricultural center (corn, coconuts).

Lilongwe, district (2007 population 1,951,278), Central region, central MALAWI; ⊙ LILONGWE. Bordered by MCHINJI (NW), KASUNGU (N), DOWA (NEE), and DEDZA (ESE) districts and MOZAMBIQUE (S and W).

Lilongwe (li-lo-ngweh), city (2007 population 744,436), MALAWI; ⊙ Malawi since 1973 and Lilongwe district and Central region; 13°59′S 33°25′E. An administrative and commercial center in a fertile agricultural area. The city was founded in 1947 as an agricultural marketing center. Largely a planned city, government programs in the 1970s and 1980s expanded the city's transportation facilities (which included the construction of an international airport to N and railroad connections to neighboring cities) and its agricultural potential. Banda College of Agriculture is in Lilongwe.

Liloy (LEE-loi), town, ZAMBOANGA DEL NORTE province, W MINDANAO, PHILIPPINES, on SULU SEA, 55 mi/89 km WNW of PAGADIAN; 08°06′N 122°40′E. Fishing; coconuts, rice, corn. Airstrip.

Liluah (LIL-wah), N suburb of HAORA, HAORA district, S W. BENGAL state, E INDIA, near HUGLI River, 2.5 mi/4 km N of Haora city, and 3.5 mi/5.6 km NNW of CALCUTTA city center. Large railroad shops; extensive iron- and steelworks; manufacturing of chemicals, brass and copper sheets, silk and cotton cloth, hosiery, rubber goods, pottery, cement, soap, paint, and motor vehicles; oilseed milling. Also spelled LILLOOAH.

Lil, Wadi al (WAH-dee el LEEL), river, N TUNISIA, 30 mi/48 km long; rises in the Majardah Mountains; flows SE to the MAJARDAH RIVER near SUQ AL KHAMIS. Irrigation and flood-control dam N of SUQ AL KHA-MIS.

Lily, unincorporated town, LAUREL county, SE KEN-TUCKY, 5 mi/8 km N of CORBIN, in CUMBERLAND foothills, on LAUREL RIVER. Bituminous coal; manufacturing (machining, mining lights). LONDON-Corbin Airport to N. Levi Jackson Wilderness Road State Park to NE.

Lily, village (2006 population 19), DAY county, NE SOUTH DAKOTA, 13 mi/21 km SW of WEBSTER; 45°10′N 97°40′W. Waubay Lake and National Wildlife Refuge to SE; Pickerel Lake State Recreation Area to E.

Lilyak (LEEL-yahk), village, RUSE oblast, Targovishte obshtina, BULGARIA; 43°14′N 26°28′E.

Lilyaksko (LEEL-yahk-sko), plateau, DANUBE Plain, NE BULGARIA; 43°15′N 26°26′E.

Lilybaeum (li-li-BEE-uhm), ancient city of SICILY, on the extreme W coast. It is the modern MARSALA. It was founded (396 B.C.E.) by CARTHAGE and became a stronghold. In the First Punic War it resisted a long Roman siege (250–242 B.C.E.). Rome finally won (241 B.C.E.) the city and used it as a base for the African campaign of Scipio Africanus Major. The city was famous for its harbor.

Lily Bay, bay, PISCATAQUIS county, central MAINE, on MOOSEHEAD LAKE, and 31 mi/50 km NW of DOVER-FOXCROFT. Lily Bay State Park.

Lilydale (LI-lee-dail), town, N TASMANIA, SE AUS-TRALIA, 13 mi/21 km NNE of LAUNCESTON; 41°14′S

147°14′E. Agricultural center; lavender farms, orchards. Originally called Germantown.

Lilydale (LI-lee-dail), town, S VICTORIA, AUSTRALIA, 20 mi/32 km ENE of MELBOURNE; 37°45′S 145°21′E. Railroad junction; fruit-growing center; wineries. Limestone quarries. Annual agricultural and horticultural show.

Lilydale, village (2000 population 552), DAKOTA county, SE MINNESOTA, residential suburb 3 mi/4.8 km SW of downtown ST. PAUL, 8 mi/12.9 km SE of downtown MINNEAPOLIS, on MISSISSIPPI RIVER (bridged), immediately below (E of) mouth of MINNESOTA RIVER; 44°54′N 93°07′W.

Lily Dale, resort village, CHAUTAUQUA county, extreme W NEW YORK, 9 mi/14.5 km S of DUNKIRK, on CASSADAGA LAKES; 42°21′N 79°20′W. In agricultural area. This small, gated community has been a center of spiritualism since organization of Lily Dale Assembly (1879), attracting thousands of summer tourists. Technically part of Pomfret township.

Lilyfield, community, S MANITOBA, W central CANADA, 10 mi/16 km from WINNIPEG, and in ROSSER rural municipality; 50°00′N 97°16′W.

Lily Lake, village, KANE county, NE ILLINOIS, 10 mi/16 km WNW of GENEVA; 41°56′N 88°28′W.

Lilypons (LIL-ee-ponz), village, FREDERICK county, central MARYLAND, 10 mi/16 km S of FREDERICK. Named after the opera singer. Tropical fish ponds here are described as the largest commercial operation of this kind in the country.

Lima (LEE-mah), region (□ 12,408 sq mi/32,260.8 sq km), W central PERU, on the PACIFIC; ⊙ HUACHO. Includes Don Martín, SAN LORENZO, and HORMIGAS DE AFUERA islands. The CORDILLERA OCCIDENTAL of the ANDES crosses the region N-S. Drained by CHANCAY, HUAURA, MALA, RÍMAC, and CAÑETE rivers. Sugar and cotton plantations on irrigated coastal plain; cereals, corn, potatoes, coffee; sheep raising in mountains. Fisheries (Callao, Huacho). Mining of copper at Casapalca; salt at Huacho and CHILCA; coal fields at OYÓN. Manufacturing of chemicals, electronics, textiles in Lima, Callao, and suburbs. Beach resorts in Lima metropolitan area, thermal baths at CHURÍN and Chilca. Ruins of pre-Incan and Incan times at PACHACAMAC and Cajamarquilla. Served by railroad and international airport, and crossed by PAN-AMERICAN HIGHWAY. Main centers: Lima, Callao, Huacho, Barranca, Cajatambo, Huaral, Canta, Matucana, Yavyos, San Vicente de Cañete. As part of the 2002 regionalization law, the former Lima department split into Lima province and Lima region, and Callao Constitutional Province, already forming a separate administrative unit, changed to Callao region.

Lima (LEE-mah), province, W central PERU, ⊙ LIMA. Bordering on the PACIFIC OCEAN. In 2002, Lima department split into Lima province and LIMA region. Today, Lima is the only province that does not belong to one of the twenty-five regions.

Lima (LEE-mah), city (2005 population 7,753,439), W PERU, capital and largest city of Peru and ⊙ Lima department. Its port is CALLAO. The Lima urban area is Peru's economic center and the site of oil-refining and diversified manufacturing industries. The city was founded on January 18, 1535, by Francisco Pizarro and is the second oldest capital city in South America. It was named the City of Kings by Pizarro. It was the capital of Spain's New World empire until the 19th century. Its cultural supremacy on the continent was contested in colonial times only by Bogotá, Colombia, and Lima's only rival in magnificence and political prestige was MEXICO CITY. A sharp rise in its population in the 20th century has resulted in serious overcrowding and a wide gap between rich and poor. Rebuilt several times, Lima reflects the architectural styles prevalent in various periods; much of the city is characterized by modern steel and concrete buildings.

Although many streets are narrow and preserve a colonial atmosphere, spacious boulevards traverse the entire metropolitan area. Small squares, statues of national heroes, parks, and gardens are common. The focal point of the city's life is the central square, the Plaza de las Armas. It is dominated by the huge national palace and cathedral. The cathedral, begun by Pizarro and containing what are claimed to be his remains, was almost totally destroyed by earthquakes in 1687 and 1746, along with much of the city. Besides the palace, the cathedral, and numerous churches, including the monastery of Santa Rosa with the relics of St. Rose of Lima, notable public buildings include the National Library, founded in 1821 by José de San Martín, and the University of San Marcos, founded in 1551, and the Catholic University. The city has many museums, several devoted to anthropology and the archaeology of the many ancient cultures of Peru. One fashionable suburb is Miraflores, on the bluffs above the Pacific. Lima has a uniformly cool climate and during the winter is subject to the fogs (*garva*) and heavy mists peculiar to Peru's S desert coast. It almost never rains. Not far from the city are the pre-Incan ruins at PACHACAMAC. Jorge Chávez International Airport and railroad service to HUANCAYO.

Lima (LEI-muh), city (□ 13 sq mi/33.8 sq km); 2006 population 38,219), ⊙ ALLEN county, NW OHIO, 80 mi/128 km NW of COLUMBUS; 40°45′N 84°07′W. Located in a fertile farm area, it is a processing and marketing center for grain, dairy, and meat products. Manufacturing included auto engines, school buses, electric signs and motors, cranes and power shovels, petroleum products (refinery), machine tools, plastics, chemicals, engines, and fertilizers. Lima, formerly a large oil producer (1885–1910), houses a symphony orchestra and a branch of Ohio State University. Settled 1831; incorporated 1842. Many industries left Lima in the 1970s and 1980s during the Rust Belt decline. Population has decreased since the late 20th century.

Lima (LEE-mah), town, N BUENOS AIRES province, ARGENTINA, 11 mi/18 km WNW of ZÁRATE; 34°03′S 59°12′W. Agriculture center: vegetables; seed production; cattle raising; dairying.

Lima (LEE-mah), town, San Pedro department, central PARAGUAY, on a tributary of the Jejuí-Guazú River and 45 mi/72 km NE of San Pedro; 23°52′S 56°30′W. Livestock-raising center in maté and lumber area.

Lima (LEI-muh), village (2000 population 159), ADAMS county, W ILLINOIS, 16 mi/26 km N of QUINCY; 40°10′N 91°22′W. In agricultural area.

Lima, village (2000 population 242), BEAVERHEAD county, extreme SW MONTANA, 40 mi/64 km S of DILLON, near IDAHO state line, and on RED ROCK RIVER at mouth of Junction Creek; 44°38′N 112°35′W. Sheep, cattle; hay, potatoes. CONTINENTAL DIVIDE to S. Parts of Beaverhead National Forest to S, W, and NE. LIMA RESERVOIR used for irrigation and recreation to E on Red Rock River. Originally called Allerdice, then Spring Hill.

Lima, village (□ 1 sq mi/2.6 sq km; 2006 population 2,417), LIVINGSTON county, W central NEW YORK, 18 mi/29 km S of ROCHESTER; 42°54′N 77°36′W. Makes concrete; agricultural products. Incorporated 1797.

Lima, village (2006 population 73), SEMINOLE county, central OKLAHOMA, 6 mi/9.7 km WNW of WEWOKA; 35°10′N 96°35′W. In agricultural area.

Lima (LEE-muh), unincorporated village, GREENVILLE county, NW SOUTH CAROLINA, in the BLUE RIDGE, on North SALUDA RIVER and 17 mi/27 km N of GREENVILLE.

Lima Campos (LEE-mah KAHM-pos), city (2007 population 11,411), N central MARANHÃO state, BRAZIL, 12 mi/19 km E of PEDREIRAS; 04°29′S 44°29′W.

Limache (lee-MAH-chai), town (2002 population 34,948), VALPARAISO province, VALPARAISO region,

central CHILE, in ACONCAGUA RIVER valley (on Limache River), 20 mi/32 km E of VALPARAISO; 33°01′S 71°16′W. Agricultural center (cereals, fruits, wine, beans; cattle) at QUILPUE nearby. Just NW is San Francisco de Limache.

Lima Duarte (lee-mah doo-ahr-tai), city (2007 population 15,819), S MINAS GERAIS state, BRAZIL, in the SERRA DA MANTIQUEIRA, 30 mi/48 km W of JUIZ DE FORA; 21°48′S 43°46′W. Railroad terminus; low-grade rutile deposits.

Limagne (lee-mahn-yuh), fertile lowland of PUY-DE-DÔME and ALLIER departments, in AUVERGNE region, central FRANCE; bounded by volcanic AUVERGNE MOUNTAINS (W) and granitic Monts du Forez (E). Traversed lengthwise by valley of ALLIER RIVER; c.60 mi/97 km long, 10 mi/16 km–25 mi/40 km wide; occupies an ancient lake bed. On its alluvial soil (enriched by volcanic ash and old lava flows) wheat, sugar beets, fruit trees, vegetables, and tobacco are intensively cultivated. Wine grapes are grown on W slopes. Limagne also contains an important industrial district centered on CLERMONT-FERRAND and THIERS. Other towns: VICHY, GANNAT, RIOM, ISSOIRE. Sometimes spelled Limagnes.

Limagnes, FRANCE: see LIMAGNE.

Lima, La, HONDURAS: see LA LIMA.

Liman (lee-MAHN), town (2005 population 8,850), SE ASTRAKHAN oblast, SE European Russia, near the indented CASPIAN SEA coast, on road, 55 mi/89 km SW of ASTRAKHAN; 45°47′N 47°14′E. Below sea level. Until 1944, called Dolban.

Liman, UKRAINE: see KRASNYY LYMAN.

Liman, UKRAINE: see LYMAN, Kharkiv oblast.

Liman, Mount, INDONESIA: see WILIS MOUNTAINS.

Limanowa (lee-mah-NO-vah), town, Nowy Sącz province, S POLAND, on railroad, and 14 mi/23 km WNW of NOWY SACZ; 49°42′N 20°26′E. Tanning. In World War II, under German rule, called Ilmenau.

Limanskoye, UKRAINE: see LYMANS'KE.

Lima Reservoir (LEI-muh), SW MONTANA, in BEAVERHEAD county, just S of SNOWCREST Range, near IDAHO state line, formed by Lima Dam in RED ROCK RIVER; 8 mi/12.9 km long, 1 mi/1.6 km wide. Irrigates livestock region. Town of LIMA 12 mi/19 km W. Red Rock Lakes National Wildlife Refuge, important nesting site of trumpeter swan, upstream (E) from lake.

Limari (lee-MAH-ree), N province of COQUIMBO region, N CHILE; ⊙ OVALLE; 30°45′S 71°00′W. Dispersed mining or farming settlements except along the coast, where economic opportunities in agriculture are greater due to more water and better transportation. Fishing; livestock; beach resorts.

Limarí River (lee-mah-REE), 125 mi/201 km long, COQUIMBO region, N central CHILE; rises in the ANDES near ARGENTINA border; flows W, past SOTAQUÍ and OVALLE, to the PACIFIC 30 mi/48 km WSW of Ovalle. Irrigates fruit-growing area. Not navigable.

Lima River (LEE-muh), 20 mi/32 km long, TUSCANY, central ITALY; rises in ETRUSCAN APENNINES E of Passo dell'Abetone; flows S, past La Lima, and W, to Serchio River 1 mi/1.6 km SW of Bagni di Lucca.

Lima River (LEE-muh), Spanish *Limia* (LEE-myah), 70 mi/113 km long, in NW SPAIN and N PORTUGAL; rises in LAKE ANTELA (ORENSE province); flows SW, past PONTE DA BARCA (head of navigation), to the Atlantic Ocean at VIANA DO CASTELO. Hydroelectric plant W of LINDOSO.

Limas (lee-mah), commune (□ 2 sq mi/5.2 sq km), RHÔNE department, RHÔNE-ALPES region, W central FRANCE; 45°58′N 04°42′E.

Limassol (lee-mah-SO), Greek *Lemesos* (le-ME-sos), district (□ 537 sq mi/1,396.2 sq km; 2001 population 201,057), SW CYPRUS; ⊙ LIMASSOL. Bounded on S by MEDITERRANEAN Sea, which includes AKROTIRI and EPISKOPI bays and AKROTIRI Peninsula. District comprised of coastal plain in S, foothills of TROODOS MOUNTAINS in N. LIMASSOL city is the largest port in

CYPRUS. AKROTIRI U.K. SOVEREIGN BASE AREA is W of city, on EPISKOPI BAY. Agricultural area. Grain, carob, wine grapes, olives, almonds, deciduous fruits, citrus, vegetables, potatoes; sheep, goats, cattle, poultry. Fishing. Copper, gypsum, and umber are mined. Resorts in the forested mountains of the N; beach resorts E of LIMASSOL city. District contains ruins of ancient cities AMATHOUS and KOURION.

Limassol (lee-mah-SO), city (2001 population 94,250), S CYPRUS, on AKROTIRI BAY, ⊙ LIMASSOL District. Located on AKROTIRI BAY, MEDITERRANEAN SEA, 40 mi/64 km SSW of NICOSIA; 34°40′N 33°02′E. Cape GATA and Lady's Mile Beach to S; beach resort area to E; AKROTIRI SALT LAKE and AKROTIRI PENINSULA to SW; AKROTIRI U.K. SOVEREIGN BASE AREA to W and SW. The city is served by the international airports at LARNACA (E) and PAPHOS (W); it is linked by motorway to NICOSIA and LARNACA. Ferries connect Limassol with Latakia, SYRIA; HAIFA, ISRAEL; and RHODES and Pireas, GREECE. Manufacturing (cement, textiles, food processing, cigarettes, wine); wine and agricultural goods are exported. Asbestos, lumber, copper, and gypsum are mined in region. Limassol was established on the site of ancient city of KOURION, and gained importance during medieval times. Seat of University College of Arts and Sciences. Byzantine castle, museum, zoo, public gardens. AMATHOUS ruins to E. Also spelled Lemesos.

Limavady (li-muh-VA-dee), Gaelic *Léim an Mhadaidh*, town (2001 population 12,135), N Londonderry, NW Northern Ireland, on Roe River and 15 mi/24 km ENE of Londonderry; 55°03′N 06°58′W. Market.

Limaville (LEI-muh-vil), village (2006 population 185), STARK county, E central OHIO, 17 mi/27 km NE of CANTON; 40°59′N 81°08′W.

Limay (lee-mai), town (☐ 4 sq mi/10.4 sq km), YVELINES department, ÎLE-DE-FRANCE region, N central FRANCE, on right bank of the SEINE opposite MANTES-LA-JOLIE, and 27 mi/43 km NW of PARIS; 48°59′N 01°44′E. Cement works.

Limay (lee-MEI), town, Bataan province, S LUZON, PHILIPPINES, on W BATAAN peninsula, on MANILA BAY, 27 mi/43 km W of MANILA; 14°32′N 120°33′E. Sugarcane, rice. Petroleum refinery.

Limay, NICARAGUA: see SAN JUAN DE LIMAY.

Limay Mahuida (lee-MEI), village, ⊙ Limay Mahuida department, W central LA PAMPA province, ARGENTINA, on the Río Salado, and 120 mi/193 km W of GENERAL ACHA; 37°12′S 66°42′W. In livestock area.

Limay River (lee-MEI), c.260 mi/418 km long, in NEUQUÉN and RÍO NEGRO provinces, ARGENTINA; rises in Lake NAHUEL HUAPÍ at NAHUEL HUAPÍ in Argentinian lake district; flows NE, past PIEDRA DEL AGUILA, PICÚN-LEUFÚ, and PLOTTIER, to join NEUQUÉN RIVER at NEUQUÉN, forming the Río Negro. Its lower course is used for irrigation (fruit growing, viticulture) and water power. Receives COLLÓN CURÁ RIVER and ARROYO PICÚN-LEUFÚ.

Limbach (LIM-bahkh), town, SAARLAND, SW GERMANY, on BLIES RIVER, 3 mi/4.8 km W of HOMBURG; 49°19′N 07°16′E.

Limbach-Oberfrohna (LIM-bahkh–O-buhr-FRAW-nah), town, SAXONY, E central GERMANY, 8 mi/12.9 km WNW of CHEMNITZ; 50°53′N 12°45′E. Hosiery-knitting center; manufacturing of machinery, gloves, cardboard, pharmaceuticals; textile dyeing. Was first center of Saxon knitting industry in 18th century. In 1950 the formerly independent towns of LIMBACH and Oberfrohna were unified.

Limbang (LIM-bahng), town, N SARAWAK, NW BORNEO, E MALAYSIA, on Limbang River (122 mi/196 km long), near its mouth on BRUNEI BAY, 10 mi/16 km SSE of BRUNEI, on narrow strip separating two sections of sultanate of Brunei; 04°49′N 115°02′E. Trade center: rubber, rice; buffalo, cattle; timber; fisheries. GUNUNG MULU NATIONAL PARK 30 mi/48 km S; airport.

Limbara, Monte (MON-te leem-BAH-rah), highest point (4,468 ft/1,362 m) in SASSARI province, N SARDINIA, S ITALY, 3 mi/53 km ENE of SASSARI; 40°51′N 09°10′E.

Limbazhi, LATVIA: see LIMBAŽI.

Limbaži (LIM-bah-zhee) or **limbazhi**, German *lemsal*, city (2000 population 9,237), N LATVIA, in VIDZEME, 45 mi/72 km NNE of Rīga; 57°31′N 24°42′E. In timber and potato district; manufacturing (leather goods, woolens, flour, metalworks). Castle ruins. Was Hanseatic town in 14th century.

Limbdi (LIM-buh-dee), town, SURENDRANAGAR district, GUJARAT state, W central INDIA, on KATHIAWAR PENINSULA, 12 mi/19 km SE of WADHWAN; 22°34′N 71°48′E. Agriculture market (cotton, millet, wheat); cotton ginning, handicrafts (cloth fabrics, ivory bangles, metalware). Was capital of the former princely state of Limbdi of WESTERN INDIA STATES agency; state merged 1948 with SAURASHTRA, becoming part of Gujarat state in 1960.

Limbe (LIM-bai), city (2001 population 84,500), ⊙ FAKO department, South-West province, CAMEROON, port on AMBAS Bay of Gulf of Guinea, at S foot of Cameroon Mt., 10 mi/16 km S of BUEA; 04°03′N 09°12′E. Exports bananas, cacao, palm oil and kernels, hardwood, and rubber; tourist center. Petroleum refineries at nearby Cape Limbo. Has famous botanical gardens; hospital founded 1858 by Baptist missionaries. Formerly called Victoria.

Limbé (lang-BAI), town, NORD department, N HAITI, 12 mi/19 km WSW of CAP-HAÏTIEN; 19°42′N 72°24′W. Agriculture (cacao, coffee, cotton, oranges, limes, sugarcane); copper deposits nearby. Old colonial town. Also called Le Limbé.

Limbe (lih-mbeh), town, Southern region, S MALAWI, in SHIRE HIGHLANDS, on railroad and 5 mi/8 km SE of BLANTYRE; 15°49′S 35°03′E. Elevation 3,800 ft/1,158 m. Major commercial and transportation center; manufacturing (cigarettes, soap); tea, tobacco, tung. Anglican and Roman Catholic cathedrals. Became township in 1909.

Limberg Dam, AUSTRIA: see KAPRUN.

Limbiate (leem-BYAH-te), town, MILANO province, LOMBARDY, N ITALY, 10 mi/16 km N of MILAN; 45°36′N 09°07′E. Highly diversified secondary industrial center; machinery, fabricated metals, wood products.

Limboto (lim-BO-to), town, ⊙ Limboto District, Sulawesi Utara province, INDONESIA, 36 mi/58 km NNW of GORONTALO.

Limbourg (lim-BOOR), Flemish *Limburg*, commune (2006 population 5,669), Verviers district, LIÈGE province, E BELGIUM, on VESDRE RIVER, and 4 mi/6.4 km ENE of VERVIERS. Wool spinning and weaving. Has church with sixteenth-century apse. Until 1648, capital of the duchy of Limburg, now divided between Belgium and the NETHERLANDS.

Limbourg, Belgian: see LIMBURG, province.

Limburg (LIM-berg; Flemish LIM-berkh), French *Limbourg* (lahm-BOOR), province (☐ 930 sq mi/2,418 sq km; 2006 population 817,206), NE BELGIUM, bordering on the NETHERLANDS in the N; ⊙ HASSELT; 50°56′N 05°20′E. The chief cities are Hasselt, TONGEREN, and SINT-TRUIDEN. The province is bordered in the E by the MEUSE RIVER and is crossed by the ALBERT CANAL. It is largely agricultural: fruits, grains, and sugar beets are cultivated; dairy products. Market gardening is also an activity, and coal is mined, but in decreasing quantities, in the CAMPINE (Flemish *Kempenland*) region in the N. Most of Limburg was included in the prince-bishopric of Liège until 1792. It became (1815) part of the Dutch province of Limburg, which was divided between Belgium and the Netherlands in 1839. Limburg's population is largely Flemish-speaking.

Limburg (LIM-buhrkh), province (☐ c.850 sq mi/2,200 sq km; 2007 estimated population 1,127,805), SE

NETHERLANDS, ⊙ MAASTRICHT. Bounded by BELGIUM (W, S), by GERMANY (E). Includes S land extension of the Netherlands 30 mi/48 km long between Belgium and Germany; drained by MAAS RIVER, forms part of border with Belgium, paralleled on E by JULIANA CANAL. Major cities include VENLO, HEERLEN, ROERMOND, WEERT, and SITTARD. As opposed to the rest of the country, which is generally level and at or below sea level, the province has low rolling hills, especially in far SE. The highest point in the Netherlands, Vaalserberg (1,035 ft/321 m) is located at the common intersection of the Netherlands, Belgium, and Germany. The land is generally agricultural, with a densely populated industrial region in SE corner. Dairying; cattle, hogs, poultry; wheat, rye, vegetables, sugar beets, fruit, wine grapes. The province was once a prominent coal-mining area with modern mines developing in the twentieth century, and the seams presumably are continuations of the German coal fields in AACHEN and extending W into Belgian KEMPENLAND; production ceased in 1974. Rich in historic antiquities, the province takes its name from the former duchy of Limburg, which comprised the S part of the modern province, including Maastricht, and an E portion of modern LIÈGE province in Belgium. Founded in the eleventh century, the duchy was divided in the Peace of Westphalia (1648) between the United Netherlands (which received Maastricht) and the Spanish Netherlands. The duchy was united (1815) under the Kingdom of the Netherlands. Limburg province, as established in 1815, did not correspond to the borders of the old duchy. It was contested after the establishment (1831) of an independent Belgium. The Dutch-Belgian treaty of 1839 divided the territory, which was incorporated into the Dutch and Belgian provinces of Limburg. Belgian separatist feeling existed in Limburg province in the nineteenth century; the area was not fully integrated into the Dutch national structure until the early twentieth century.

Limburg, Belgian: see LIMBOURG, commune.

Limburg an der Lahn (LIM-burg ahn der LAHN), town, HESSE, W central GERMANY, on the LAHN, 22 mi/35 km NNW of WIESBADEN; 50°23′N 08°04′E. Manufacturing of road-building machinery, metal goods; pottery works. Roman Catholic bishopric since 1827. Has 13th-century cathedral, built on site of 10th-century church. Belonged to electors of TRIER 1420–1803.

Limburgerhof (LIM-bur-guhr-huhf), suburb of MANNHEIM, BADEN-WÜRTTEMBERG, SW GERMANY, just SW of city; 49°25′N 08°24′E.

Lime Creek, 78 mi/126 km long, S MINNESOTA and N IOWA; rises in Bear Lake, FREEBORN county, Minnesota, 7 mi/11.3 km SW of ALBERT LEA, 43°31′N 93°29′W; flows SSW into WINNEBAGO county, N Iowa to Ford city, then ESE, through MASON CITY, to SHELL ROCK RIVER 2 mi/3.2 km S of ROCKFORD. From FOREST CITY to its mouth the stream is called Winnebago River.

Limedale, village, PUTNAM county, W central INDIANA, 2 mi/3.2 km SSW of GREENCASTLE. Limestone quarries. Laid out 1864.

Lime Hall, town, SAINT ANN parish, N JAMAICA, 2 mi/3.2 km S of SAINT ANN'S BAY; 18°24′N 77°11′W. In agricultural region (citrus fruits, corn, pimentos, coffee; cattle).

Limehouse, ENGLAND: see TOWER HAMLETS.

Limeil-Brévannes (lee-mai–brai-vahn), town (☐ 2 sq mi/5.2 sq km), VAL-DE-MARNE department, ÎLE-DE-FRANCE region, N central FRANCE, a SE suburb of PARIS, 10 mi/16 km from Notre Dame Cathedral; 48°44′N 02°29′E. Several medical facilities and a nuclear studies center are located here.

Limeira (LEE-mai-rah), city (2007 population 272,734), E central SÃO PAULO, BRAZIL, on railroad, 32 mi/51 km NW of CAMPINAS; 22°34′S 47°24′W. Orange-growing center; produces cotton gins and coffee-processing machinery, apiary supplies, hats; has meat-packing plant. Sericulture and apiculture in area.

Limeira, Brazil: see JOAÇABA.

Lime Kiln, village, FREDERICK county, N MARYLAND, near MONOCACY River, 4 mi/6.4 km S of FREDERICK. Always associated with limestone operations, it was once called Slabtown because many houses and walls were made of limestone slabs. The BALTIMORE AND OHIO RAILROAD still carries out lime-based portland cement.

Limekilns (LEIM-kilnz), fishing village (2001 population 1,411), FIFE, E Scotland, on the Firth of Forth, 3 mi/4.8 km S of DUNFERMLINE; 56°02′N 03°28′W.

Lime Lake, CATTARAUGUS county, W NEW YORK, resort lake (c.1 mi/1.6 km long), 25 mi/40 km N of OLEAN; 42°26′N 78°29′W. Lime Lake village on N shore; MACHIAS just SW.

Limena (lee-MAI-nah), village, PADOVA province, VE-NETO, N ITALY, on BRENTA river, and 5 mi/8 km N of PADUA; 45°29′N 11°50′E. Fabricated metals, machinery, clothing, wood products, wine machinery.

Limenda (LEE-meen-dah), village, S ARCHANGEL oblast, N European Russia, near road and railroad, 25 mi/40 km S of KOTLAS; 61°05′N 46°48′E. Elevation 314 ft/95 m. Pulp, paper; building of river tugs.

Limerick (LIM-rik), Gaelic *Luimneach*, county (□ 1,065 sq mi/2,769 sq km; 2006 population 184,055), SW IRELAND; ☉ LIMERICK; 52°40′N 08°37′W. Borders KERRY county to W, CORK county to S, TIPPERARY county to E, and CLARE county to N. The region is an agricultural plain lying S of the SHANNON estuary. The Golden Vale in the E part of the county and the Shannon bank are especially fertile. Dairy farming and salmon fishing are the chief occupations. On the Shannon River above Limerick is an important hydroelectric plant. Main manufactures include aluminum castings, automotive parts, concrete pipes, and office equpiment. After the Anglo-Norman invasion and the organization of Limerick as a shire (c.1200), the district was controlled for many centuries by the earls of Desmond. The chief towns include KILMALLOCK, ADARE, NEWCASTLE WEST, FOYNES, and AB-BEYFEALE.

Limerick (LIM-rik), Gaelic *Luimneach*, city (2006 population 52,539), ☉ LIMERICK county, SW IRELAND, at the head of the SHANNON estuary, and 20 mi/32 km SE of ENNIS; 52°40′N 08°37′W. The city has a port with two docks. The primary imports are grain, timber, and coal; exports include produce and fish. Former industries included salmon fishing, food processing, flour milling, and lacemaking. Limerick was occupied by the Norsemen in the 9th century, became the capital of MUNSTER under Brian Boru (c.1000), was taken by the English toward the end of the 12th century, and was James II's last stronghold in Ireland after the Glorious Revolution.

The city has three sections—English Town, the oldest, on King's Island; Irish Town to the S; and Newtown Pery, S of Irish Town, founded in 1769. Preserved in Limerick is the Treaty Stone on which was signed (1691) the treaty granting the Irish Catholics certain rights, chiefly the guarantee of political and religious liberty. The repeated violations of this treaty during the reigns of William III and Queen Anne caused Limerick to be called City of the Violated Treaty. Of notable interest are a Protestant cathedral (12th century), a Roman Catholic cathedral, King John's Castle (begun 1210), and the Walled City in the urban center. The Dockside area (occupied by breweries, foundries, bacon factories, and warehouses) along the river is now being rehabilitated with a promenade, parks, arts centers, museum, and restaurants. Limerick is the site of a teacher's college and the National Institute for Higher Education, a branch of the National University of Ireland. The city is the setting for Frank McCourt's memoir *Angela's Ashes*.

Limerick (LI-muh-rik), township (□ 78 sq mi/202.8 sq km; 2001 population 362), SE ONTARIO, E central CANADA, 50 mi/80 km N of BELLEVILLE; 44°55′N 77°42′W. Tourism based on lake activities.

Limerick (LIM-uhr-ik), town, YORK county, SW MAINE, 15 mi/24 km NNW of ALFRED; 43°41′N 70°47′W. Yarn mills. Settled c.1775; incorporated 1787.

Limerick (LIM-uhr-ick), village (2006 population 130), S SASKATCHEWAN, CANADA, 13 mi/21 km W of ASSI-NIBOIA; 49°39′N 106°16′W. Wheat.

Limerick 1 and 2 Nuclear Power Plants, Pennsylvania: see MONTGOMERY county.

Lime Ridge, unincorporated town (2000 population 951), COLUMBIA county, E central PENNSYLVANIA, residential suburb 5 mi/8 km W of BERWICK on SUS-QUEHANNA RIVER; 41°01′N 76°21′W.

Lime Ridge, village (2006 population 158), SAUK county, S central WISCONSIN, 21 mi/34 km W of BARABOO; 43°28′N 90°09′W. In dairy and livestock region.

Lime Rock, CONNECTICUT: see SALISBURY.

Limerock, Rhode Island: see LINCOLN.

Lime Springs, town (2000 population 496), HOWARD county, NE IOWA, near UPPER IOWA RIVER, 10 mi/16 km NW of CRESCO; 43°27′N 92°16′W. Feed milling. Limestone quarries, sand and gravel pits nearby.

Limestone, county (□ 607 sq mi/1,578.2 sq km; 2006 population 72,446), N Alabama; ☉ Athens. Bounded N by Tennessee, drained by Wheeler Lake (on Tennessee River) and Elk River. Cotton, timber, corn, soybeans. First Alabama co. invaded in Civil War, 1862. Nuclear power plants: Browns Ferry 1 (initial criticality, August 17, 1973), Browns Ferry 2 (initial criticality, July 20, 1974), and Browns Ferry 3 (initial criticality, August 8, 1976) are 10 mi/16 km NW of Decatur. Use cooling water from the Tennessee River, and each has a max. dependable capacity of 1,065 MWe. Formed 1818. Originally part of Elk County, it was named for the creek flowing through it.

Limestone (LEIM-ston), county (□ 933 sq mi/2,425.8 sq km; 2006 population 22,720), E central TEXAS; ☉ GROESBECK; 31°32′N 96°34′W. Drained by NAVA-SOTA RIVER. Agriculture (cotton, corn, peaches, hay, vegetables, pecans, nursery crops); dairying; livestock (cattle, horses, hogs, sheep, goats). Oil, natural gas; clay, stone, lignite, sand and gravel. Manufacturing at Groesbeck, Mexia. Includes Fort Parker State Park and Old Fort Parker State Historic Site, near Lake Mexia, in N center; Confederate Reunion Grounds State Historic Site in N; part of LAKE LIMESTONE in SE. Formed 1846.

Limestone, town, AROOSTOOK county, NE MAINE, 17 mi/27 km NNE of PRESQUE ISLE, at NEW BRUNSWICK border; 46°54′N 67°51′W. In potato and pea country. Port of entry; terminus of Bangor and Aroostook railroad. Loring Air Force Base (closed) here. Settled 1849; incorporated 1869.

Limestone, village (□ 1 sq mi/2.6 sq km; 2006 population 397), CATTARAUGUS county, W NEW YORK, on ALLEGHENY RIVER, 10 mi/16 km WSW of OLEAN; 42°01′N 78°37′W. Incorporated 1877.

Limestone, Lake (LEIM-ston), reservoir (□ 21 sq mi/54.6 sq km), ROBERTSON county, E central TEXAS, on NAVASOTA RIVER, 50 mi/80 km ESE of WACO; 31°20′N 96°19′W. Maximum capacity 557,878 acre-ft. Formed by Sterling C. Robertson Dam (65 ft/16 m high), built (1978) for water supply; also used for irrigation and recreation.

Limfjord (LEEM-FYUHRD), waterway, c.110 mi/180 km long, cutting across N JUTLAND, DENMARK, and connecting the NORTH SEA with the KATTEGAT strait. It is very irregular in shape, forming LØGSTØR, a lagoon 15 mi/24 km wide in its middle section; its max. depth is c.50 ft/15 m. There are several islands, notably MORS. Before 1825, when the fjord cut through to the North Sea, its W part consisted of several freshwater lakes that drained E into the Kattegat strait. The Thyborøn

Canal keeps the W entrance of Limfjord open. ÅL-BORG is the chief port on the waterway.

Limia River, SPAIN and PORTUGAL: see LIMA RIVER.

Limingen (LEE-ming-uhn), lake (□ 36 sq mi/93.6 sq km), NORD-TRØNDELAG county, central NORWAY, extends 15 mi/24 km SE from Gjersvika; 4 mi/6.4 km wide.

Limington (LI-ming-ton), town, YORK county, SW MAINE, on SACO RIVER, and 18 mi/29 km N of ALFRED; 43°43′N 70°42′W. Wood products.

Limin Vatheos, Greece: see SÁMOS.

Limmared (LIM-mah-RED), village, ÄLVSBORG county, SW SWEDEN, 20 mi/32 km SE of BORÅS; 57°32′N 13°21′E. Railroad junction. Manufacturing (glassworks founded 1740). Ruins of fourteenth-century fortress.

Limmat River, SWITZERLAND: see LINTH RIVER.

Limmen Bight (LI-muhn BEIT), in Gulf of CARPEN-TARIA, indentation of NE coast of NORTHERN TERRI-TORY, AUSTRALIA, between GROOTE EYLANDT (N) and SIR EDWARD PELLEW ISLANDS (S); 85 mi/137 km NW-SE. Receives ROPER and LIMMEN BIGHT rivers; 14°45′S 135°40′E.

Limmen Bight River (LI-muhn BEIT), 140 mi/225 km long, N NORTHERN TERRITORY, AUSTRALIA; rises in N hills of BARKLY TABLELAND; flows NNE to LIMMEN BIGHT of Gulf of CARPENTARIA; 15°07′S 135°44′E. Variant name: Parsons Creek.

Limne, Greece: see LIMNI.

Limni (LEEM-nee), ancient *Aegae*, town, NW ÉVVIA island and prefecture, CENTRAL GREECE department, port on N Gulf of Évvia, 25 mi/40 km NNW of KHALKÍS; 38°46′N 23°19′E. Nickel and magnesite deposits (S); fisheries. Some tourism. Also spelled Limne.

Limnia (LIM-yah), town, FAMAGUSTA district, E CY-PRUS, 7 mi/11.3 km NNW of FAMAGUSTA; 35°12′N 33°51′E. MEDITERRANEAN SEA to E. Grain, grapes, citrus, olives, vegetables; livestock. Turkish name Mermenekse.

Límnos (LEEM-nos), island (□ 186 sq mi/483.6 sq km), LESBOS prefecture, NORTH AEGEAN department, NE GREECE, in the AEGEAN SEA near TURKEY; 2.5 mi/4 km–25 mi/40 km (SW-NE; at central isthmus) wide; 39°54′N 25°21′E. Rises to 1,411 ft/430 m. Irregular coastline; deeply indented N and S. Largely mountainous, in deforested volcanic formations, whose lava flows have created fertile soils. Agriculture includes grain (wheat), fruit (figs, grapes), almonds, cotton, silk, and tobacco; sheep and goats; fish. A medicinal earth (Lemnian Bole), used as an astringent, has been produced here since ancient times. Popular tourist center. Airport. KÁSTRON and MOUDROS, a port with a strategic gulf, are the island's chief towns. In ancient Greece, the island, because of its volcanic origin, was sacred to Hephaestus, who was hurled here from Olympos by Zeus. Also associated with the stories of Hypsipyle and Philoctetes. Conquered (510 B.C.E.) by Athenian general Miltiades; remained an Athenian possession throughout antiquity. After the fall (1204) of the BYZANTINE EMPIRE, Límnos was captured by the Genoese, who held it until 1464, when it passed to Venice. Seized by the Ottoman Turks in 1479 and became part of Greece in 1913. Nearby are remains of a Bronze Age settlement. Also called Lemnos.

Limnytsya River (LEEM-ni-tsyah) (Russian *Lomnitsa*) (Polish *Lomnica*), 75 mi/121 km long, IVANO-FRAN-KIVS'K oblast, UKRAINE; formed 13 mi/21 km SSW of PEREHINS'KE by the confluence of headstreams rising in the GORGANY; flows NNE, past Perehins'ke, and NE to the DNIESTER (Ukrainian *Dnister*) River just NW of HALYCH.

Limoeiro (LEE-mo-AI-ro), city (2007 population 55,870), E PERNAMBUCO state, NE BRAZIL, on railroad, and 40 mi/64 km WNW of RECIFE; 07°52′S 35°27′W. Agriculture (fruits, manioc); manufacturing (sugar milling, alcohol distilling); ships cotton and livestock.

Limoeiro de Anadia (LEE-mo-AI-ro dee ah-HAN-zhe-ah), city (2007 population 25,487), central ALAGOAS state, NE BRAZIL, 20 mi/32 km S of PALMEIRA DOS ÍNDIOS; 09°45′S 36°31′W. Cotton, sugar, tropical fruits. Until 1944, called Limoeiro.

Limoeiro do Ajuru (LEE-mo-ai-ro do AH-zhoo-roo), city (2007 population 23,284), N central PARÁ state, BRAZIL, on RIO TOCANTINS near confluence with RIO PARÁ; 01°55′S 49°24′W.

Limoeiro do Norte (lee-mo-AI-ro do nor-chee), city (2007 population 53,599), E CEARÁ state, BRAZIL, on RIO JAGUARIBE, near RIO GRANDE DO NORTE border, and 45 mi/72 km SSW of ARACATI; 05°13′S 38°09′W. Cotton; cattle; ships carnauba wax. Until 1944, called Limoeiro.

Limoges (lee-mozh), city (□ 30 sq mi/78 sq km); ⊙ HAUTE-VIENNE department, W central FRANCE, on the VIENNE RIVER and 220 mi/354 km SSW of PARIS; 45°51′N 01°15′E. Chief center of French porcelain industry, which uses the abundant kaolin found in the area; the city's porcelain workshops are its major employers. The shoe industry is also important to Limoges. Uranium is mined nearby, and automobile parts and electrical equipment are manufactured. Of Roman origin (*Limonum*), the city became (11th century) the seat of the viscounts of Limoges and (1589) the capital of Limousin province. It was often visited by war, pestilence, and famine. Richard I of England was killed in battle near Limoges (1199). In 1370, Edward the Black Prince burned the city and massacred its inhabitants. The famous Limoges enamel industry (dating from Merovingian times) was fully developed by the 13th century, but declined when Limoges was once more devastated in the Wars of Religion. Turgot, who was intendant of Limousin from 1761 to 1764, brought back prosperity by reintroducing (1771) the china manufactures. Limoges has a cathedral (chiefly 13th–16th century), a notable national ceramics and porcelain museum, and an art gallery containing many works by Renoir, born here. University. Following France's regionalization act in the 1980s, which created regional entities, Limoges became the administrative seat of the LIMOUSIN region, embracing three departments of W central France. The metropolitan area's population is c.170,000. Limoges-Bellegarde international airport 5 mi/8 km NW of city.

Limoges (lee-MOZH), unincorporated township (2006 population 2,000), SE ONTARIO, E central CANADA, 16 mi/26 km from OTTAWA, and included in RUSSELL township; 45°19′N 75°15′W. Built around a church, established 1884. Primarily French-speaking population. Formerly called South Indian; once part of Embrun village.

Limoilou (lee-mwah-LOO), E borough (French *arrondissement*) of QUEBEC city, CAPITALE-NATIONALE region, S QUEBEC, E CANADA; 46°50′N 71°13′W.

Limón (lee-MON), province (□ 3,549 sq mi/9,227.4 sq km; 2000 population 432,923) of E COSTA RICA, on CARIBBEAN SEA; ⊙ LIMÓN; 10°00′N 83°13′W. Extends from SAN JUAN River (NICARAGUA border) S to SIXAOLA RIVER (PANAMA border), and is drained by TORTUGUERO, PARISMINA, and REVENTAZÓN rivers. Largely swampy, malarial lowland along N coast, it becomes mountainous near the Continental Divide (Cordillera de TALAMANCA; SW). Agriculture: cacao, coconuts, and bananas. Main food crops are beans, corn, fruits. Fodder crops support some livestock. N sector of province is served by coastal shipping (ports of Limón, BARRA DE COLORADO, Tortuguero), paved highways. Main centers are Limón, SIQUIRRES, MATINA, and GUÁPILES. S areas were badly damaged by earthquake in April 1991.

Limón (lee-MON), city, ⊙ LIMÓN canton and province, COSTA RICA, on the CARIBBEAN SEA; 09°46′N 83°10′W. Once the leading port of Costa Rica, it has been superseded by nearby MOÍN. Limón gained importance with the construction of the railroad to SAN JOSÉ in the late 1800s and has been the point of export for Costa Rica's large banana industry.

Columbus may have visited the site on his 1502 voyage.

Limón (lee-MON), town, COLÓN department, N HONDURAS, in Mosquitia, on CARIBBEAN SEA, at mouth of small Limón River, 30 mi/48 km E of TRUJILLO; 15°52′N 85°33′W. Coconuts, corn, beans; livestock. Sometimes called Barra de Limón.

Limon, town (2000 population 2,071), LINCOLN county, E central COLORADO, on BIG SANDY CREEK, 80 mi/129 km SE of DENVER, and 65 mi/105 km NE of COLORADO SPRINGS; 39°16′N 103°41′W. Agricultural trade region: wheat, sunflowers; cattle; manufacturing (concrete). Founded 1888; incorporated 1909.

Limonade (lee-mo-NAHD), town, NORD department, N HAITI, near the ATLANTIC OCEAN, 8 mi/12.9 km SE of CAP-HAÏTIEN; 19°40′N 72°07′W. Agriculture (coffee, tobacco, sugarcane, bananas, citrus fruits); essential oils distillery; copper deposits nearby.

Limón, Altos de, part of CONTINENTAL DIVIDE, DARIÉN province, SE PANAMA, on COLOMBIA border. Elevation 4,642 ft/1,415 m.

Limonar (lee-mo-NAHR), town, MATANZAS province, W CUBA, on Central Highway, on railroad, and 12 mi/19 km ESE of MATANZAS; 22°46′N 81°25′W. In sugar-growing region. Also apiculture; cattle raising. Nearby are refineries and centrals of Horacio (NE) and Fructoso Rodríguez (S).

Limón Bay (lee-MON), inlet of CARIBBEAN SEA, in PANAMA CANAL AREA, at N end of PANAMA CANAL; c.4.5 mi/7.2 km long, c.2.5 mi/4 km wide. Entrance protected by breakwaters. On E shore are COLÓN, CRISTÓBAL, and Telefers Island. At NW gate are TORO POINT and FORT SHERMAN.

Limón, El, MEXICO: see EL LIMÓN.

Limón, El, NICARAGUA: see EL LIMÓN.

Limone Piemonte (lee-MO-ne pye-MON-te), village, CUNEO province, PIEDMONT, NW ITALY, in MARITIME ALPS, N of TENDA Pass, 13 mi/21 km S of CUNEO; 44°12′N 07°34′E. Elevation 3,248 ft/990 m. Customs station. Tourist resort, especially for winter sports.

Limones (lee-MO-nes), town, ESMERALDAS province, N ECUADOR, minor PACIFIC port on offshore island in SANTIAGO RIVER estuary, 45 mi/72 km ENE of ESMERALDAS. In region of tropical lowland forests (mangrove, tagua nuts, bananas, coconuts); sawmills. Near Archaeological National Park on Manta de Oro Island. Formerly called Valdez.

Limonest (lee-mo-NAI), commune (□ 3 sq mi/7.8 sq km), RHÔNE department, RHÔNE-ALPES region, E central FRANCE, 7 mi/11.3 km NNW of LYON; 45°50′N 04°46′E. Tileworks.

Limonum, FRANCE: see POITIERS.

Limoquije (lee-mo-KEE-hai), canton, MARBÁN province, BENI department, NE BOLIVIA, 18 mi/30 km SW of LORETO, on the MAMORÉ RIVER, 10 mi/15 km E of unpaved road; 15°26′S 64°48′W. Elevation 525 ft/160 m. Agriculture (potatoes, yucca, bananas, coffee, tobacco, cotton, cacao, peanuts); cattle and horse raising.

Limours (lee-moor), town (□ 5 sq mi/13 sq km; 2004 population 6,414), ESSONNE department, ÎLE-DE-FRANCE region, N central FRANCE, 11 mi/18 km SSW of VERSAILLES; 48°39′N 02°05′E. Has 16th-century church with noted stained-glass windows.

Limousin (lee-mu-zan), region and former province, SW central FRANCE, in the hilly country W of the MASSIF CENTRAL, divided into Haut Limousin (around LIMOGES, N of the Monts du LIMOUSIN), and Bas Limousin (around TULLE, S of the mountains); 45°30′N 01°50′E. Limoges, the historic capital, is the center of the porcelain industry, for which the abundant kaolin of the region is used; both Limoges and Tulle are important markets for cattle raised in most of Limousin and for their leather products. BRIVE-LA-GAILLARDE is an agricultural market center with diverse industries. In 918, Limousin became a fiefdom of the duchy of AQUITAINE, and much of its history is essentially that of Aquitaine. Ravaged by Edward the Black Prince in the

Hundred Years War, Limousin was reconquered for France (1370–1374) by the constable Bertrand du Guesclin, but it was not permanently attached to the French crown until the reign of Henry IV (1607). It remained a depressed area until Turgot became intendant (1761–1764) and introduced notable economic reforms, including the reestablishment of the porcelain china industry. Under France's regionalization law of the 1980s Limousin became one of twenty-two administrative regions, comprising the departments of HAUTE-VIENNE, CREUSE, and CORRÈZE (□ 6,541 sq mi/16,941 sq km), and headquarters at Limoges.

Limousin, Monts du (lee-mu-zan, mon dyoo), W offshoots of AUVERGNE MOUNTAINS, in MASSIF CENTRAL, W central FRANCE, extending c.60 mi/97 km SW-NE across parts of HAUTE-VIENNE, CORRÈZE, and CREUSE departments, LIMOUSIN region. Rise to almost 3,200 ft/975 m. Important cattle-raising district.

Limoux (lee-moo), town (□ 12 sq mi/31.2 sq km), subprefecture of AUDE department, LANGUEDOC-ROUSSILLON administrative region, S FRANCE, on AUDE RIVER and 13 mi/21 km SSW of CARCASSONNE; 43°03′N 02°13′E. Known for its sparkling white wines (*blanquette de Limoux*); honey and nougat manufacturing, woodworking. Has 12th–16th-century church and a 14th-century bridge. Noted for its annual carnival.

Limpias (LEEM-pyahs), town, CANTABRIA province, N SPAIN, 20 mi/32 km ESE of SANTANDER; 43°22′N 03°25′W. Corn, potatoes, fruits. Shrine here is place of pilgrimage.

Limpio (LIM-pee-o), town (2002 population 73,158) Central department, S PARAGUAY, on Paraguay River 12 mi/19 km NE of Asunción (suburb, part of metropolitan area); 25°10′S 57°32′W. Livestock and meat-packing center; apiculture. Founded 1785.

Limpopo, river, c.650 mi/1,046 km long, SOUTH AFRICA, BOTSWANA, ZIMBABWE, and MOZAMBIQUE; formed in South Africa by joining of MARICO and Ngotwane rivers 190 mi/306 km NNW of JOHANNESBURG, in W part of NORTHERN CAPE province. It flows in a great arc, first N (forming part of the South Africa–Botswana border), then E (forming the South Africa–Zimbabwe border), and finally SE through Mozambique to the INDIAN OCEAN, 80 mi/129 km ENE of MAPUTO. The upper Limpopo is also known as the Krokodil, or CROCODILE RIVER. The river's main tributary, the OLIFANTS, enters the Limpopo c.130 mi/210 km from its mouth; below this point the Limpopo is permanently navigable. The lower Limpopo waters a fertile and heavily populated region in Mozambique.

Limpopo Province (□ 47,598 sq mi/123,280 sq km; 2004 estimated population 5,300,139; 2007 estimated population 5,402,900), SOUTH AFRICA; ⊙ POLOKWANE; 24°00′S 29°30′E. Most N province, includes original N TRANSVAAL. Enclosed to NW and N by LIMPOPO River (ZIMBABWE and BOTSWANA border), to E by MOZAMBIQUE, to S by MPUMALANGA, GAUTENG, and NE edge of NORTH-WEST provinces. Largely subtropical and dry savanna with scattered isolated rock outcrops. Rich in minerals and extensive farming of crops and livestock. Sparsely populated in W, denser in E and S. Its area is 10.1% of the republic's total land area. Also called Northern province.

Limpsfield (LIMPS-feeld), village (2001 population 3,680), E SURREY, SE ENGLAND, at foot of NORTH DOWNS, 9 mi/14.5 km E of REIGATE; 51°16′N 00°01′E. Has 12th-century church.

Lim River (LEEM), 136 mi/219 km long, in E MONTENEGRO, W SERBIA, and S BOSNIA-HERZEGOVINA; longest (right) affluent of the DRINA River; rises in PLAV Lake (Montenegro) in NORTH ALBANIAN ALPS; flows NNW, past ANDRIJEVICA, IVANGRAD, BIJELO POLJE, PRIJEPOLJE (Serbia), and PRIBOJ, to Drina River 5 mi/8 km SW of VISEGRAD (Bosnia-Herzegovina). Navigable for 104 mi/167 km. The POLIMLJE Valley extends along its upper course.

Limski Zaljev, Croatian *zaljev* [=gulf], gulf, W CROA-TIA, on W coast of ISTRIA, N of ROVINJ. Popularly called a "fjord" because of its appearance, it represents the submerged mouth of a former river. Protected as a special marine reserve and "distinctive landscape." Bird sanctuary. Fish hatchery (shellfish).

Limuru (lee-MOO-roo), town (1999 population 4,141), CENTRAL province, S central KENYA, on railroad, and 18 mi/29 km NW of NAIROBI; 01°07′S 36°38′E. Elevation 7,340 ft/2,237 m. Agricultural center; coffee and tea plantations, vegetable farms. One of the first of Kenya's districts settled by Europeans.

Linakeng (lee-nah-KEHNG), village, MOKHOTLONG district, E central LESOTHO, 82 mi/132 km ESE of MA-SERU, on Mashai Reservoir (under construction; formed on Lineheng River tributary of ORANGE RIVER), on N shore of Linaheng River arm; 30 mi/48 km W of Sani Pass and border crossing; 29°31′S 28°48′E. Horses, sheep, goats, cattle; beans, peas. Also called Linakino.

Lin'an (LIN-AN), town, ⊙ Lin'an county, NW ZHE-JIANG province, CHINA, 28 mi/45 km W of HANG-ZHOU, S of TIANMU MOUNTAINS; 30°14′N 119°43′E. Rice, wheat, tea, silk, bamboo.

Linao (lee-NAH-o), village, CHILOÉ province, LOS LAGOS region, S CHILE, on NE coast of CHILOÉ ISLAND, 15 mi/39 km SE of ANCUD; 41°58′S 73°33′W. Minor port in agricultural area (potatoes, wheat; livestock); fishing; lumbering.

Linapacan Island (lee-nah-PAH-kahn), one of the CALAMIAN ISLANDS, PALAWAN province, PHILIPPINES, 15 mi/24 km off N tip of PALAWAN; 10 mi/16 km long; 13°53′N 124°24′E. Separated from CULION ISLAND by 12 mi/19 km wide Linapacan Strait leading to SOUTH CHINA SEA. Chief town is Linapacan (1990 population 5,835). Coconuts. Fishing.

Linard, Piz, SWITZERLAND: see SILVRETTA.

Linares (lee-NAH-rais), Southeastern province of MAULE region, S central CHILE, bordering the ANDES; ⊙ LINARES; 36°00′S 71°45′W. Fertile agricultural area in the central valley, watered by LONGAVÍ, LONCO-MILLA, and Melado rivers. Cereals, vegetables, grapes, fruits; livestock. Lumbering, food processing. Main centers: Linares, VILLA ALEGRE, YERBAS BUENAS, COLBÚN, RETIRO, LONGAVÍ, PARRAL, SAN JAVIER.

Linares (lee-NAH-res), city and township, NUEVO LEÓN, N MEXICO, at foot of SIERRA MADRE ORIENTAL, at junction Mexico Highways 60 and 85, railroad, on INTER-AMERICAN HIGHWAY, and 75 mi/121 km SE of MONTERREY; 24°51′N 99°34′W. Elevation 2,244 ft/684 m. Trading center in rich farming (sugarcane, oranges, cotton, cereals) and grazing district. Old colonial churches.

Linares (lee-NAH-rais), city, JAÉN province, S SPAIN, in ANDALUSIA. The rich silver and lead mines nearby have brought prosperity to the city, which now has many metallurgical industries. Powder and dynamite are chief products.

Linares (lee-NAH-rais), town, ⊙ Linares comuna (2002 population 65,133) and province, MAULE region, S central CHILE, in the central valley, 175 mi/282 km SW of SANTIAGO; 35°51′S 71°36′W. On railroad. Commercial and agricultural center (cereals, grapes, sugar beets, fruits, vegetables; livestock; dairying); food processing. Airport. Founded 1794.

Linares (lee-NAH-res), town, ⊙ Linares municipio, NARIÑO department, SW COLOMBIA, in S CORDILLERA OCCIDENTAL, 23 mi/37 km WNW of PASTO; 01°21′N 77°31′W. Elevation 5,036 ft/1,535 m. Corn, wheat, cacao, coffee, sugarcane; livestock.

Linares de la Sierra (lee-NAH-rais dhai lah SYE-rah), village, HUELVA province, SW SPAIN, c.4 mi/6.4 km WSW of ARACENA; 37°53′N 06°37′W. Olives, acorns, corn, fruits, cereals; sheep, hogs; timber.

Linares de Riofrío (lee-NAH-rais dhai REE-o-FREE-o), town, SALAMANCA province, W SPAIN, 16 mi/26 km NW of BÉJAR; 40°35′N 05°55′W. Flour- and saw-

milling; livestock; cereals, nuts, vegetables. Limestone quarries and anthracite deposits nearby.

Linas (lee-nahs), town (□ 2 sq mi/5.2 sq km; 2004 population 5,854), ESSONNE department, ÎLE-DE-FRANCE region, N central FRANCE; 48°38′N 02°16′E. Just S of MONTLHÉRY on Orge River, and 15 mi/24 km SSW of PARIS. Speedway, road-testing laboratory. Has 13th–16th century church.

Linas, Monte (LEE-nahs, MON-te), highest point (4,045 ft/1,233 m) in CAGLIARI province, SW SARDI-NIA, ITALY 11 mi/18 km NNE of IGLESIAS; 39°27′N 08°37′E.

Lincang (LIN-ZANG), town, ⊙ Lincang county, SW central YUNNAN province, CHINA, 100 mi/161 km SE of BAOSHAN; 23°54′N 100°01′E. Elevation 4,987 ft/1,520 m. Food and beverages, metal smelting, papermaking; rice, millet, beans, oilseeds; nonferrous ore mining.

Lincent (LIN-sah), commune (2006 population 3,015), Waremme district, LIÈGE province, E BELGIUM, 10 mi/16 km W of WAREMME.

Lincheng (LIN-CHENG), town, ⊙ Lincheng county, SW HEBEI province, CHINA, 40 mi/64 km S of SHI-JIAZHUANG, and on spur of BEIJING-WUHAN railroad; 37°26′N 114°34′E. Coal-mining center; agriculture (grain, oilseeds, cotton); manufacturing (engineering, textiles).

Linchuan (LIN-CHUAN), city (□ 819 sq mi/2,121 sq km; 1994 estimated non-agricultural population 192,800; estimated total population 893,600), N JIANGXI province, CHINA, on railroad, at road junction, on the XU RIVER; 27°55′N 116°16′E. Agriculture and light industry are the largest sectors of the city's economy. Agriculture (grains, oilseeds, vegetables; hogs, poultry); manufacturing (food, textiles, chemicals, transportation equipment). Formerly called Fuzhou or Fuchow.

Lincoln (LEEN-kuhn), former county (□ 332 sq mi/863.2 sq km), S ONTARIO, E central CANADA, on Lake ONTARIO, and on NIAGARA River, on U.S. (NEW YORK) border; 43°05′N 79°34′W. SAINT CATHARINES was its county seat. LINCOLN and WELLAND counties were succeeded by the Niagara regional government.

Lincoln, county (□ 572 sq mi/1,487.2 sq km; 2006 population 14,125), SE ARKANSAS; ⊙ STAR CITY; 33°58′N 91°44′W. Bounded NE by ARKANSAS RIVER; drained by Bayou BARTHOLOMEW. Agriculture (cotton, soybeans, wheat, rice; cattle, hogs, poultry). Joe Harden Lock and Dam in N; Cane Creek State Park at center. Formed 1871.

Lincoln, county (□ 2,586 sq mi/6,723.6 sq km; 2006 population 5,458), E central COLORADO; ⊙ HUGO; 39°03′N 103°27′W. Agricultural area, drained by BIG SANDY, HORSE, Rush, and Hell creeks. Wheat, sunflowers; cattle. Source of South and North Forks AR-IKAREE RIVER and South Fork REPUBLICAN RIVER in N. Formed 1889.

Lincoln, county (□ 257 sq mi/668.2 sq km; 2006 population 8,257), NE GEORGIA; ⊙ LINCOLNTON; 33°47′N 82°27′W. Bounded E by SOUTH CAROLINA state line, formed here by SAVANNAH River, and S by LITTLE RIVER. Piedmont agricultural area (cotton, corn, hay, agr., fruits); cattle, hogs, poultry; manufacturing (apparel, textiles); sawmilling. Formed 1796.

Lincoln, county (□ 1,205 sq mi/3,133 sq km; 2006 population 4,522), S IDAHO; ⊙ SHOSHONE; 42°59′N 114°08′W. In SNAKE RIVER PLAIN; watered by BIG WOOD and LITTLE WOOD rivers. Irrigated region in SW, around Shoshone, produces cattle; alfalfa, oats, barley, corn, sugar beets, potatoes, dry beans. Shoshone Ice Caves in NW. Formed 1895.

Lincoln, county (□ 719 sq mi/1,869.4 sq km; 2006 population 3,396), central KANSAS; ⊙ LINCOLN; 39°03′N 98°12′W. Smoky Hills region, drained by SALINE RIVER and Salt Creek. WILSON LAKE and Dam on W border, with Wilson State Park on S shore (in RUSSELL county). Cattle, sheep; wheat, sorghum, alfalfa, hay. Formed 1870.

Lincoln, county (□ 336 sq mi/873.6 sq km; 2006 population 25,361), central KENTUCKY; ⊙ STANFORD; 37°27′N 84°40′W. Drained by DIX and GREEN rivers and Fishing and Buck creeks (source of latter three streams). Rolling upland agricultural area, partly in outer BLUEGRASS REGION (burley tobacco, hay, alfalfa, soybeans, wheat, corn; hogs, cattle, poultry; dairying; timber). Isaac Shelby State Historic Site in NW; William Whitley House State Historic Site in E. Formed 1780 from Kentucky county, VIRGINIA, one of three original counties of Kentucky district of Virginia.

Lincoln, coastal county (□ 699 sq mi/1,817.4 sq km; 2006 population 35,234), S MAINE; ⊙ WISCASSET; 44°03′N 69°32′W. Fishing and resort area, with some agriculture in N part; boatbuilding, shipping, and canning of seafood. Resorts dot its rugged coast and islands. SHEEPSCOT and EASTERN rivers, DAMARISCOTTA and MEDOMAK inlets. Formed 1760. Maine Yankee Nuclear Power Plant, initial criticality October 23, 1972, is 10 mi/16 km N of BATH, uses cooling water from the Back River, and has a max. dependable capacity of 860 MWe; was temporarily closed for repair in 1995.

Lincoln, county (□ 548 sq mi/1,424.8 sq km; 2006 population 5,963), SW MINNESOTA; ⊙ IVANHOE; 44°25′N 96°16′W. Bounded by SOUTH DAKOTA on W. Agricultural area (oats, wheat, soybeans, hay, alfalfa; sheep, hogs, cattle, poultry); dairying. Includes part of COTEAU DES PRAIRIES; several lakes scattered through county, including Lake BENTON in S and Lake Hendricks on W boundary. Formed 1873.

Lincoln, county (□ 588 sq mi/1,528.8 sq km; 2006 population 34,404), SW MISSISSIPPI; ⊙ BROOKHAVEN; 31°32′N 90°27′W. Drained by BOGUE CHITTO and East Fork AMITE rivers. Agriculture (cotton, corn; cattle, poultry; dairying); timber. Part of Homochitto National Forest in NW. Formed 1870.

Lincoln, county (□ 629 sq mi/1,635.4 sq km; 2006 population 50,123), E MISSOURI; ⊙ TROY; 39°04′N 90°58′W. Bounded E by MISSISSIPPI RIVER, drained by the CUIVRE. Soybeans, corn, apples; cattle, hogs; limestone; manufacturing at Troy. Cuivre River State Park NE of Troy. Formed 1818.

Lincoln, county (□ 3,675 sq mi/9,518 sq km; 1990 population 17,481; 2000 population 18,837), extreme NW MONTANA; ⊙ LIBBY; 48°32′N 115°25′W. Mountain region bordering on CANADA (BRITISH COLUMBIA, N) and IDAHO (W); drained by KOOTENAI River (Lake KOOCANUSA formed by Libby Dam, E center). W state line forms Mountain/Pacific time zone boundary. Lowest point in Montana (c.1,800 ft/549 m) at point where Kootenai River flows W into IDAHO. Kootenai National Forest covers most of county, including part of CABINET MOUNTAINS Wilderness Area in S, parts of Kaniken National Forest along W boundary, part of Stillwater State Forest in E, Turner Mountains Ski Area in NW center. Thompson Lakes source of THOMPSON RIVER, in SE corner, source of STILLWATER RIVER in E. Kootenai National Forest and Yaak River, Cabinet Mountains in SW, PURCELL MOUNTAINS in NW, WHITEFISH RANGE in NE, Salish Mountains in E. Hay; lumber; some cattle, sheep; lead, silver, gold. Formed 1909.

Lincoln, county (□ 2,575 sq mi/6,695 sq km; 2006 population 35,865), SW central NEBRASKA; ⊙ NORTH PLATTE; 41°03′N 100°46′W. The SOUTH PLATTE and NORTH PLATTE rivers join here to form the PLATTE. Grazing and agricultural area: cattle, hogs; dairying; corn, wheat, alfalfa, wild hay, sunflower seeds. Central/Mountain time zone boundary follows W boundary. Fort McPherson National Cemetery in E, S of Maxwell; Lake Maloney State Recreation Area at center, S of North Platte; Sutherland State Recreation Area in W. Old OREGON TRAIL follows S side of Platte/South Platte River. Formed 1860.

Lincoln, county (□ 10,637 sq mi/27,656.2 sq km; 2006 population 4,738), SE NEVADA; ⊙ PIOCHE; 37°40′N 114°40′W. Seventh largest U.S. co. in land ara.

Mountain region bordering on ARIZONA and UTAH, both on E; E boundary is also Pacific/Mountain time zone boundary; watered by MEADOW VALLEY WASH. Part of Desert National Wildlife Range in SW, also part of Nellis Air Force Bombing and Gunnery Range in SW. Pahranagat National Wildlife Refuge in SW center. Beaver Dam, Kershaw-Ryan, Spring Valley, and Cathedral Gorge state parks in E; Echo Canyon State Recreational Area in E. White River Petroglyphs Archaeological Site, Leviathan Cave Natural Area, and small part of Humboldt National Forest in W. Mining (perlite, sand and gravel), ranching. Part of EGAN RANGE is in N. Formed 1866.

Lincoln, county (□ 4,831 sq mi/12,560.6 sq km; 2006 population 21,223), S central NEW MEXICO; ⊙ CARRIZOZO; 33°44′N 105°27′W. Parts of Lincoln National Forest in center and N; ranges of SACRAMENTO MOUNTAINS extend N-S. Lincoln State Monument and Smokey Bear State Historical Park in S center; Valley of Fires National Recreation Area, in the MALPAIS volcanic flow area, in W; Ski Apache ski area in S; part of Cibola National Forest in N. Sheep, cattle; hay, alfalfa, some wheat, oats; watered by RIO HONDO. Formed 1869.

Lincoln (LEEN-cuhn), county (□ 307 sq mi/798.2 sq km; 2006 population 71,894), W central NORTH CAROLINA; ⊙ LINCOLNTON; 35°28′N 81°13′W. In PIEDMONT region; bounded E by CATAWBA RIVER (Cowans Ford Dam forms Lake NORMAN reservoir in upper reach of Mountains Island Lake reservoir below [S] dam). Drained by South Fork Catawba River. Manufacturing at Lincolnton; service industries; agriculture (corn, wheat, hay, barley, soybeans; poultry, cattle; dairying). Formed 1779 when Tryon County was divided to form Lincoln and Rutherford counties. Named for Benjamin Lincoln (1733–1810), general of the American Revolution.

Lincoln, county (□ 965 sq mi/2,509 sq km; 2006 population 32,645), central OKLAHOMA; ⊙ CHANDLER; 35°42′N 96°52′W. Intersected by the DEEP FORK of the Canadian River (touches SW corner). Diversified agriculture (cotton, sorghum, castor beans, vegetables, pecans); dairying; livestock and poultry raising; beekeeping; manufacturing (dairy products, apparel, electronic components) at Chandler. Oil (at Chandler and Stroud) and natural-gas wells. Formed 1890.

Lincoln, county (□ 1,193 sq mi/3,101.8 sq km; 2006 population 46,199), W OREGON; ⊙ TOLEDO; 44°38′N 123°54′W. Bounded W by PACIFIC OCEAN, drained by ALSEA, Siletz, and YAQUINA rivers. Small portion of county in S transferred (1949) to BENTON county (E). Home of Oregon's leading fishing company (salmon, tuna, halibut, crabs, shrimp, clams). Paper products; timber; dairy products; poultry, sheep, cattle. COAST RANGE to E. Part of Siuslaw National Forest in N and S; Ellmaker State Park in E. More than twenty state parks and waysides on or near coast, notably Devils Lake State Park in NW, Yaquina Bay State Park in W, and Beachside State Park in SW. Formed 1893.

Lincoln, county (□ 578 sq mi/1,502.8 sq km; 2006 population 35,239), SE SOUTH DAKOTA; ⊙ CANTON; 43°18′N 96°46′W. Rolling prairie region bounded E by BIG SIOUX RIVER (IOWA state line). Corn, soybeans; cattle, hogs. Urbanized in N central part with SIOUX FALLS and other area towns. Drained in W by East Fork of VERMILLION RIVER. Newton Hills State Park in E, S of Canton. Formed 1862.

Lincoln, county (2006 population 32,728), S TENNESSEE; ⊙ FAYETTEVILLE; 35°08′N 86°36′W. Bounded S by ALABAMA; crossed by ELK RIVER. Rich agricultural area; manufacturing. Formed 1809.

Lincoln, county (□ 2,339 sq mi/6,081.4 sq km; 2006 population 10,376), E WASHINGTON; ⊙ DAVENPORT; 47°34′N 118°25′W. Bounded on N by COLUMBIA and SPOKANE rivers; GRAND COULEE DAM (just beyond NW corner of county) forms FRANKLIN D. ROOSEVELT LAKE, which also creates SPOKANE RIVER Arm in NE;

Little Falls and Long Lake dams on Spokane River in NE; Coulee Dam National Recreation Area surrounds lake and area. Potatoes, wheat, alfalfa, hay, barley, oats, rye. Formed 1883.

Lincoln, county (□ 439 sq mi/1,141.4 sq km; 2006 population 22,357), W WEST VIRGINIA; ⊙ HAMLIN; 38°10′N 82°04′W. On ALLEGHENY PLATEAU; drained by GUYANDOTTE, MUD, and COAL rivers. Agriculture (corn, potatoes, alfalfa, hay, nursery crops, tobacco); cattle; honey. Oil and natural-gas wells; bituminous coal. Big Ugly Wildlife Management Area in S; Hilbert Wildlife Management Area in NE. Formed 1867.

Lincoln, county (□ 906 sq mi/2,355.6 sq km; 2006 population 30,151), N central WISCONSIN; ⊙ MERRILL; 45°19′N 89°45′W. Wooded lake region in N is resort area; dairying and farming (barley) in S part. Paper milling and manufacturing at Merrill and TOMAHAWK. Drained by WISCONSIN RIVER. Contains Council Grounds State Park in S. Camp and Ten and Harrison Hills ski areas in NE. Formed 1874.

Lincoln, county (□ 4,089 sq mi/10,631.4 sq km; 2006 population 16,383), W WYOMING; ⊙ KEMMERER; 42°15′N 110°42′W. Grain (barley, hay, alfalfa); livestock (sheep, cattle); dairying area bordering UTAH and IDAHO; drained by Green, SNAKE, GREYS, and SALT rivers; GREEN RIVER forms FONTENELLE RESERVOIR in E. Coal, oil, and phosphate found here. SALT RIVER RANGE and Bridger-Teton National Forest in N part; GRAND CANYON OF THE SNAKE RIVER in far N. Pine Creek Ski Area in W; FOSSIL BUTTE NATIONAL MONUMENT in SW. Formed 1911.

Lincoln (LING-kuhn), city (2001 population 85,595) and district, ⊙ LINCOLNSHIRE, E ENGLAND, on the WITHAM RIVER; 53°14′N 00°32′W. Located at the junction of the Roman FOSSE WAY and ERMINE STREET, the city is a center of road and railroad transportation. Manufacturing includes heavy machinery, light-metal products, motor vehicle and electronic parts, and food products. Lincoln was an ancient British settlement, the Roman Lindum or Lindum Colonia, and was one of the Five Boroughs of the Danes. Lincoln Castle, begun by William I in 1068, was contested in the civil war between Matilda and Stephen (12th century). The town was burned in the 12th century; three parliaments were held in Lincoln in the 14th century. Parliamentarians captured Lincoln in 1644. For centuries horse races and fairs have been held here. The Lincoln Cathedral, first built from 1075 to 1501, has a central tower 271 ft/83 m high, containing the famous bell "Great Tom of Lincoln." One of the few extant copies of the Magna Carta is in the cathedral. In Lincoln are teacher-training, theological, art, and technical colleges.

Lincoln, city (2000 population 4,577), Talladega co., E central Alabama, 18 mi/29 km W of Anniston, and 13 mi/21 km N of Talladega, near Coosa River; 33°36′N 86°08′W. Originally known as 'Kingsville,' it was renamed in 1856 and incorporated in 1911.

Lincoln, city (2000 population 11,205), PLACER county, central CALIFORNIA, 25 mi/40 km NNE of SACRAMENTO, in Sacramento Valley; 38°54′N 121°19′W. Ships fruits, grain, vegetables; manufacturing (plastics and clay products, sawmilling, cabinets). Incorporated 1890.

Lincoln, city (2000 population 15,369), ⊙ LOGAN county, central ILLINOIS; 40°08′N 89°22′W. The city was plotted and promoted (1853) with the aid of Abraham Lincoln and named for him when he was still an unknown country lawyer. Lincoln practiced law there from 1847 to 1859, and buildings and places associated with him have been preserved or reconstructed, including Postville Courthouse Historic Site. Lincoln College, Lincoln Christian College, and Lincoln and Logan Correctional Centers are here. Edward A. Madigan State Park, formerly Railsplitter State Park, to S. It is a shipping and industrial center. In agricultural area (corn, soybeans; cattle, hogs); manufacturing (glass containers, electrical controls,

corrugated containers, cleaning products, and shampoos). Incorporated 1865.

Lincoln, city (2006 population 241,167), ⊙ NEBRASKA and LANCASTER county, SE Nebraska, 50 mi/80 km SW of OMAHA; 40°49′N 96°41′W. Major railroad junction and center. It is the trade and industrial center for a large grain and livestock area. Manufacturing (printing and publishing, processed and prepared foods, beverages, construction materials, electronics, motorcycles, sports equipment, valves and cylinders, asphalt, automotive parts). Cattle are slaughtered and processed. Many insurance companies have their home offices here. It is the seat of the University of Nebraska (two campuses), Union College, Nebraska Wesleyan University, and Southeast Community College (Lincoln campus). Lied Center for Performing Arts, which attracts musicians and other performers from around the world, is located on University of Nebraska campus. Sheldon Memorial Art Gallery, State Historical Society Museum and State Museum, American Historical Society of Germans from Russia Museum, planetarium, sculpture garden, and several parks. Several state recreation areas to S and W. State fairgrounds, veterans hospital, municipal airport. State penitentiary S of city. The state capitol, designed by B. G. Goodhue, with sculptures by Lee Lawrie, was completed in 1934. William Jennings Bryan lived in Lincoln from 1887 to 1916; his home is preserved. Founded in 1864 as Lancaster, the city was chosen as the site of the capital in 1867 and renamed. Incorporated 1869.

Lincoln, town (1991 population 24,486), ⊙ Lincoln district (□ 2,381 sq mi/6,190.6 sq km), N BUENOS AIRES province, ARGENTINA, 38 mi/61 km SW of JUNÍN; 35°00′S 61°45′W. Railroad junction and agriculture center (grain; cattle and sheep raising; dairying); meat-packing.

Lincoln (LEEN-kuhn), town (□ 63 sq mi/163.8 sq km; 2001 population 20,612), NIAGARA region, S ONTARIO, E central CANADA, on Niagara Peninsula, immediately W of ST. CATHARINES; 43°09′N 79°25′W. Fruits; nurseries; wineries; industrial distribution center. Suburban.

Lincoln, township, SELWYN district, CANTERBURY region, SOUTH ISLAND, NEW ZEALAND, 11 mi/18 km SW of CHRISTCHURCH; 43°39′S 172°29′E. Agricultural center. Site of Lincoln University, specializing in agriculture and horticulture.

Lincoln, town (2000 population 1,752), WASHINGTON county, NW ARKANSAS, 17 mi/27 km WSW of FAYETTEVILLE, in the OZARKS, near OKLAHOMA state line; 35°57′N 94°25′W. Poultry; manufacturing (egg processing, compost turners, poultry houses, printing).

Lincoln, town (2000 population 182), TAMA county, central IOWA, 20 mi/32 km NNW of TOLEDO; 42°15′N 92°41′W. In agricultural area.

Lincoln or **Lincoln Center,** town, ⊙ LINCOLN county, N central KANSAS, on SALINE RIVER and 32 mi/51 km WNW of SALINA. Railroad junction. Shipping center for grain and cattle area. Manufacturing (acrylics, crushed stone). Founded 1871; incorporated 1879.

Lincoln, town, including Lincoln village, PENOBSCOT county, central MAINE, on the PENOBSCOT and c.45 mi/72 km N of BANGOR, on Mattanawook Pond; 45°21′N 68°27′W. Trade center, with wood products, pulp and paper mill. Settled c.1825; incorporated 1829.

Lincoln, residential town, MIDDLESEX county, E MASSACHUSETTS, 14 mi/23 km WNW of BOSTON; 42°26′N 71°19′W. Agricultural, in suburbanizing area. Gropius, the German architectural founder of the Bauhaus School, lived here. Cambridge Reservoir to the E. Site of Decordova Museum and park. Drumlin Farm headquarters of Massachusetts Audobon Society. Settled c.1650; incorporated 1754.

Lincoln, town (2000 population 1,026), BENTON county, central MISSOURI, 9 mi/14.5 km N of WARSAW; 38°23′N 93°19′W. Corn, wheat; cattle.

Lincoln, town, GRAFTON county, N central NEW HAMPSHIRE, 18 mi/29 km SSE of LITTLETON; 44°05′N 71°34′W. Drained by PEMIGEWASSET RIVER and its East Branch. Railroad terminus. Agriculture (cattle; dairying; timber); manufacturing (electrical connections, printing and publishing). Parts of White Mountain National Forest in E and W; Hancock and Big Rock state campgrounds in E; Kancamagus Pass (2,860 ft/872 m) on Kancamagus Highway on E boundary; Loon Mountain Ski Area in SE; APPALACHIAN TRAIL crosses in NE. Incorporated 1764.

Lincoln, town (2006 population 2,431), BURLEIGH co., S central NORTH DAKOTA, suburb 4 mi/6.4 km SE of BISMARCK, on Apple Creek, near Bismarck Municipal Airport; 46°46′N 100°42′W. Residential. A rural subdivision started in the 1970s as Fort Lincoln Estates that incorporated in 1977 as Lincoln.

Lincoln, town (□ 18 sq mi/46.8 sq km), PROVIDENCE county, NE RHODE ISLAND. Set off from SMITHFIELD, named for Abraham Lincoln; 41°55′N 71°27′W. Includes villages of MANVILLE, SAYLESVILLE, Limerock, and part of LONSDALE. Once a textile town, its manufacturing includes wire, tubing, metal parts, and thread. Corporate headquarters for Cross Pens. Limestone has been quarried here since colonial times. Many pre-Revolutionary houses, a state park, and Lincoln Downs (racetrack) are in the town. Incorporated 1871.

Lincoln, town, ADDISON co., W central VERMONT, on NEW HAVEN RIVER and 11 mi/18 km NE of MIDDLEBURY, in GREEN MOUNTAINS; wood products; 44°05′N 72°58′W. Settled 1795 by Quakers. Named for Benjamin Lincoln (1733–1840), major general in the American Revolutionary War.

Lincoln, unincorporated village, BONNEVILLE county, SE IDAHO, residential suburb 4 mi/6.4 km E of IDAHO FALLS. Agricultural area.

Lincoln, village (2000 population 364), ALCONA county, NE MICHIGAN, 7 mi/11.3 km WNW of HARRISVILLE, near LAKE HURON; 44°41′N 83°24′W. Manufacturing (cutting tools). Small lakes (resorts) nearby. Mount Maria Ski Area and HUBBARD LAKE to NW.

Lincoln, village, LEWIS AND CLARK county, W central MONTANA, 35 mi/56 km NW of HELENA, between BLACKFOOT RIVER and Keep Cool Creek, at center of Lincoln State Forest, which is surrounded by Helena National Forest. One of the country's largest open-pit gold and silver mines. Lead, copper, molybdenum are also mined. Cattle, hay, timber; sport fishing. Manufacturing (wood fuel pellets, beef snacks). CONTINENTAL DIVIDE to E. Scapegoat Wilderness Area to N. Setting for novel *A River Runs Through It* by Norman Maclean.

Lincoln, village, LINCOLN county, central NEW MEXICO, on Rio Bonito, just S of CAPITAN MOUNTAINS, 31 mi/50 km SSE of CARRIZOZO; elevation c.5,715 ft/1,742 m. Agriculture; livestock. Center of Lincoln county cattle war, 1877–1878. State museum, formerly county courthouse, in Lincoln State Monument, contains historical and archaeological collection. Billy the Kid imprisoned here 1881. Parts of Lincoln National Forest to N and SW. EL CAPITAN MOUNTAIN (10,083 ft/3,073 m), in Capitan Mountains, to NE.

Lincoln, unincorporated village, Ephrata township, LANCASTER county, SE PENNSYLVANIA, suburb 2 mi/3.2 km NW of EPHRATA; 40°11′N 76°12′W. In rich agricultural area.

Lincoln, borough (2006 population 1,136), ALLEGHENY county, W PENNSYLVANIA, 12 mi/19 km SSW of PITTSBURGH and 4 mi/6.4 km SSW of MCKEESPORT on MONONGAHELA RIVER; 40°17′N 79°50′W. Agriculture includes dairying; livestock; corn, hay.

Lincoln (LING-kuhn), parish (□ 469 sq mi/1,219.4 sq km; 2006 population 41,857), N LOUISIANA; ⊙ RUSTON; 32°32′N 92°39′W. Drained by Middle Fork of Bayou D'ARBONNE. Agriculture (peaches, blueberries, home gardens; hay; cattle, poultry); logging; manu-

facturing (lumber, wood products, glass containers, metal products, lighting fixtures); oil and natural gas. Small part of Jackson Bienville State Wildlife Area on S boundary. Named after President Abraham Lincoln. Formed 1873.

Lincoln, plantation, OXFORD county, W MAINE, on MAGALLOWAY RIVER, and c.35 mi/56 km NW of RUMFORD; 44°57′N 71°00′W.

Lincoln Boyhood National Memorial, SPENCER county, SW INDIANA near LINCOLN CITY and Lincoln State Park. Site of the farm where Abraham Lincoln was raised and the burial place of his mother, Mary Hanks Lincoln. Authorized 1962.

Lincoln Center for the Performing Arts, arts center, on Upper W Side of MANHATTAN, SE NEW YORK, occupying a triangular area NW of Columbus Circle and bounded by BROADWAY on NE, by Columbus Avenue on E, by Amsterdam Avenue on W, by West 66th Street on N, and by 62nd Street on S. Perhaps the foremost performing arts center of its kind in the world, it is the realization of the 1930s dream of Charles Spafford, one of the "Dukes of New York," a group of anonymous, powerful people serving on boards of most of the city's major institutions. John D. Rockefeller III was chairman of the Exploratory Committee that ultimately led to the Center's construction, and he was instrumental in raising its $185 million cost from corporations, foundations, foreign countries, state and federal funds, and private contributions. The site of the project was an area W of CENTRAL PARK known as Lincoln Square, which had become one of the city's most congested slums by the late 1940s. Built between 1959 and 1972, the complex has seven main buildings: Avery Fisher Hall, New York State Theater, Metropolitan Opera House, New York Public Library for the Performing Arts, Vivian Beaumont Theater, Samuel B. and David Rose Building, and Julliard Building, which houses Alice Tully Hall. Another facility, Frederick P. Rose Hall, was added in the adjacent Time Warner Center in 2004. Lincoln Center tenants include the Metropolitan Opera, New York Philharmonic, New York City Ballet, New York City Opera, Chamber Music Society of Lincoln Center, School of American Ballet, Julliard School of Music, New York Public Library at Lincoln Center, Lincoln Center Theater, Jazz at Lincoln Center, and Film Society of Lincoln Center. Damrosch Park, with its band shell, grove of trees, public plaza, and other features, is also part of the Center.

Lincoln City, town (2006 population 7,944), LINCOLN county, W OREGON, 46 mi/74 km W of SALEM, on PACIFIC OCEAN, N of Siletz Bay and mouth of Siletz River; 44°58′N 124°00′W. Fish, timber, tourism. Manufacturing (textile products). Part of Siuslaw National Forest to E; Devils Lake State Park to E and S; several state waysides along coast.

Lincoln City, village, SPENCER county, SW INDIANA, 34 mi/55 km ENE of EVANSVILLE. Laid out 1872 on site of farm of Thomas Lincoln (Abraham Lincoln's father). Lincoln State Park (just S). Nearby LINCOLN BOYHOOD NATIONAL MEMORIAL has grave of Nancy Hanks Lincoln and site of Lincoln cabin built in 1816.

Lincoln Heights, unincorporated town, WESTMORELAND county, SW PENNSYLVANIA, residential suburb 2 mi/3.2 km S of JEANNETTE; 40°18′N 79°37′W. Agriculture includes dairying; corn, hay.

Lincoln Heights (LINK-uhn HEITZ), village (2006 population 3,747), HAMILTON county, extreme SW OHIO; suburb 10 mi/16 km N of downtown CINCINNATI; 39°14′N 84°27′W. Incorporated 1946.

Lincoln Highway, historic road extending for more than 3,300 mi/5,311 km from NEW YORK City to SAN FRANCISCO; built 1913–1927. Now part of U.S. highway system.

Lincoln Homestead, National Historic Site, central ILLINOIS, located in SPRINGFIELD. Authorized 1971.

Only private home owned by Abraham Lincoln; he was living there when he was elected president.

Lincoln Homestead State Park, KENTUCKY: see SPRINGFIELD.

Lincolnia, unincorporated city, FAIRFAX county, NE VIRGINIA, residential suburb 5 mi/8 km W of ALEXANDRIA, 9 mi/15 km SW of WASHINGTON, D.C.; 38°49′N 77°09′W.

Lincolnia Heights, unincorporated town, FAIRFAX county, NE VIRGINIA, residential suburb 5 mi/8 km WNW of ALEXANDRIA, 8 mi/13 km SW of WASHINGTON, D.C.; 38°49′N 77°08′W. Lake Barcroft reservoir to N.

Lincoln Island, CHINA: see DONG DAO ISLAND.

Lincoln Log Cabin State Historic Site, E central ILLINOIS, 8 mi/12.9 km S of CHARLESTON, on site of last Lincoln family homestead in Illinois; contains reconstruction of Lincoln cabin built in 1837; 39°22′N 88°12′W. Covers 86 acres/35 ha.

Lincoln Memorial, monument, 164 acres/66 ha, in Potomac Park, WASHINGTON, D.C. Authorized 1911; built 1914–1917. The building, designed by Henry Bacon and styled after a Greek temple, has thirty-six Doric columns representing the states of the Union at the time of Lincoln's death. Inside the building is a heroic statue of Lincoln by Daniel Chester French and two murals by Jules Guerin.

Lincoln, Mount (14,286 ft/4,354 m), in PARK RANGE, NW PARK county, central COLORADO, 12 mi/19 km NE of LEADVILLE in Pike National Forest. Highest peak in range. CONTINENTAL DIVIDE passes to N, source of SOUTH PLATTE RIVER 5 mi/8 km SW of Summit.

Lincoln, Mount, NEW HAMPSHIRE: see FRANCONIA MOUNTAINS.

Lincoln Mountain, W central VERMONT, in GREEN MOUNTAINS, 15 mi/24 km NE of MIDDLEBURY; one of its summits, Mt. Ellen (4,135 ft/1,260 m), is third highest in range. Sugarbush Valley and Mad River Glen ski areas are nearby.

Lincoln Park, city (2000 population 40,008), WAYNE county, SE MICHIGAN, a suburb 7 mi/11.3 km SW of downtown DETROIT; 42°14′N 83°10′W. Borders North Branch of Ecorse River on N and E, drained by South Branch. ECORSE in S, 0.5 mi/0.8 km W of Detroit River. Light manufacturing. It is a residential community in an area marked by a significant decline in industry. Incorporated in 1921.

Lincoln Park, unincorporated town (2000 population 3,904), FREMONT county, S central COLORADO, suburb 2 mi/3.2 km SE of CAÑON CITY, near ARKANSAS RIVER; 38°25′N 105°12′W. In coal-mining area. Glass, machinery.

Lincoln Park, unincorporated town, BERKS county, SE central PENNSYLVANIA, residential suburb 3 mi/4.8 km WSW of READING; 40°18′N 75°59′W.

Lincoln Park, village, UPSON county, W central GEORGIA, just S of THOMASTON; 32°52′N 84°19′W.

Lincoln Park, village (□ 1 sq mi/2.6 sq km; 2000 population 2,337), ULSTER county, SE NEW YORK; 41°57′N 74°00′W.

Lincoln Park, borough (2006 population 10,856), MORRIS county, N NEW JERSEY, 7 mi/11.3 km W of PATERSON; 40°55′N 74°17′W. Agriculture and manufacturing. Incorporated 1922.

Lincoln Sea, part of the ARCTIC OCEAN off NE ELLESMERE ISLAND (CANADA) and NW GREENLAND.

Lincolnshire (LING-kuhn-shir) or **Lincs**, county (□ 2,662 sq mi/6,921.2 sq km; 2001 population 646,645), E ENGLAND, on the HUMBER estuary, the NORTH SEA, and The WASH; ⊙ LINCOLN; 53°10′N 00°20′W. Within Lincolnshire are seven districts: North Kesteven, South Kesteven, East Lindsey, West Lindsey, South Holland, Boston (borough), and Lincoln (city). The county is generally low and flat, with extensive marshes along the coast. It is crossed by many dikes and canals, some of which, notably the Foss Dyke, date back to Roman times. Lincolnshire, the second largest county in England, is an important agricultural area (potatoes, vegetables, sugar beets). The area also profits from tourism. Great GRIMSBY is a fishing port, and the county's industries include equipment manufacturing and some steelmaking. Significant towns include Boston, Gainsborough, Grantham, Louth, Mablethorpe, Market Rasen, Skegness, Sleaford, Spalding, and Stamford. In Anglo-Saxon times, Lincolnshire was variously under the control of MERCIA and NORTHUMBERLAND. Relics from a number of medieval churches remain. Lincolnshire was reorganized as a nonmetropolitan county in 1974.

Lincolnshire, village (2000 population 6,108), Lake county, NE ILLINOIS, suburb 30 mi/48 km NNW of downtown CHICAGO, 6 mi/9.7 km W of HIGHLAND PARK; 42°12′N 87°55′W. Drained by Des Plaines River. Manufacturing (construction and mining equipment, educational aids, toxic-gas–monitoring equipment, storage tanks).

Lincoln's New Salem Historic Site, restored historic village (a state historical site), SANGAMON county, central ILLINOIS, on SANGAMON RIVER, and 16 mi/26 km NW of SPRINGFIELD; 39°58′N 89°50′W. Here was home of Abraham Lincoln, 1831–1837; buildings, which were standing in Lincoln's day (including the Rutledge Tavern, Danton Offut's store, and the Lincoln-Berry store) have been restored. A small museum houses pioneer relics. Settled 1828; decline and abandonment came after 1839. Official title is Lincoln's New Salem Historic Site.

Lincolnton, city (2006 population 10,599), ☉ LINCOLN co., W central NORTH CAROLINA, 28 mi/45 km NW of CHARLOTTE, on South Fork CATAWBA RIVER; 35°28′N 81°15′W. Railroad junction. Agricultural area (cotton, grain, chickens, hogs, cattle; dairying). Manufacturing (textiles, apparel, pharmaceuticals, furniture, machinery, printing and publishing, food). Incorporated 1785.

Lincolnton, town (2000 population 1,595), ☉ LINCOLN county, NE GEORGIA, 34 mi/55 km NW of AUGUSTA, near SAVANNAH River; 33°47′N 82°29′W. Manufacturing (lumber, apparel, yarns, wood products food processing).

Lincoln Tunnel, three-tube vehicular tunnel under HUDSON RIVER between midtown MANHATTAN borough of NEW YORK city and WEEHAWKEN, NEW JERSEY; 40°44′N 74°01′W. The tunnel is 8,215 ft/2,504 m long (portal to portal), nearly 100 ft/30 m below surface of the river. Opened 1937.

Lincoln University, unincorporated village, CHESTER county, SE PENNSYLVANIA, 3 mi/4.8 km ENE of OXFORD; 39°48′N 75°55′W. Seat of Lincoln University Founded 1854 as one of first U.S. universities for African-Americans.

Lincoln Village, unincorporated town (2000 population 4,216), SAN JOAQUIN county, central CALIFORNIA; residential suburb 4 mi/6.4 km NNW of downtown STOCKTON; 38°00′N 121°20′W.

Lincolnville, resort town, WALDO county, S MAINE, on PENOBSCOT BAY, and 10 mi/16 km S of BELFAST; 44°17′N 69°04′W.

Lincolnville, village (2000 population 225), MARION county, E central KANSAS, 10 mi/16 km NNE of MARION; 38°29′N 96°57′W. In grain and livestock region.

Lincolnville, village (2006 population 850), CHARLESTON county, SE SOUTH CAROLINA, 20 mi/32 km NW of CHARLESTON, and 1 mi/1.6 km E of SUMMERVILLE; 33°00′N 80°09′W. Agriculture (poultry, hogs; corn, cotton).

Lincolnwood, village (2000 population 12,359), COOK county, NE ILLINOIS, residential suburb 9 mi/14.5 km NW of downtown CHICAGO; 42°00′N 87°43′W. Bounded by SKOKIE on N, by Chicago all other sides. Manufacturing (machinery, bags, food products, audio visual equipment, computer equipment, lighting systems). Until 1935 called Tessville.

Lincs, ENGLAND: see LINCOLNSHIRE.

Lincura (leen-KOO-rah), town, ÑUBLE province, BÍO-BÍO region, S central CHILE, 18 mi/29 km SSW of BULNES, on Hata River; 36°52′S 72°23′W.

Lind, village (2006 population 575), ADAMS county, SE WASHINGTON, 15 mi/24 km SW of RITZVILLE, on Lind Coulee; 46°58′N 118°37′W. In COLUMBIA BASIN agricultural region (wheat; cattle, poultry).

Linda, unincorporated city (2000 population 13,474), YUBA county, N central CALIFORNIA; residential suburb 3 mi/4.8 km ESE of MARYSVILLE, near YUBA RIVER; 39°07′N 121°33′W. Beals Air Force Base to E. Seat of Yuba College. Agriculture (grain, walnuts, peaches, prunes, olives; dairying, cattle).

Linda (LEEN-dah), village, W NIZHEGOROD oblast, central European Russia, on the Linda River (left tributary of the VOLGA RIVER), on highway branch and railroad, 17 mi/27 km N of (and administratively subordinate to) BOR, and 13 mi/21 km SW of SEMËNOV; 56°37′N 44°06′E. Elevation 295 ft/89 m. Poultry farm.

Linda-a-Velha (lin-dah–ah–VEL-yah) [Spanish=beautiful old woman], suburb of LISBON, LISBOA district, PORTUGAL, 6 mi/9.7 km W of city center, 38°43′N 09°15′W. Commercial center.

Lindale (LIN-dal), town (2006 population 4,290), SMITH county, E TEXAS, 13 mi/21 km NNW of TYLER, near Sabine River; 32°30′N 95°24′W. Canning, shipping center. Agriculture (fruit, vegetables, horticultural crops, rose bushes). Manufacturing (food products, printing and publishing).

Lindale, village (2000 population 4,088), FLOYD county, NW GEORGIA, 4 mi/6.4 km S of ROME; 34°11′N 85°10′W. Textile manufacturing.

Lindale (LIN-dail), hamlet, central ALBERTA, W CANADA, 14 mi/22 km from DRAYTON VALLEY, in BRAZEAU county; 53°15′N 114°39′W.

Lindås (LIND-OS), village, KALMAR county, SE SWEDEN, agglomerated with EMMABODA.

Lindau (LIN-dou), town, BAVARIA, S GERMANY, in SWABIA, on an island in LAKE CONSTANCE; 47°33′N 09°41′E. Connected by bridges with the mainland, it is a picturesque summer resort and tourist center and a base for lake steamer service to AUSTRIA and SWITZERLAND. Lindau was an imperial city from 1275 to 1803 and passed to BAVARIA in 1805.

Linda Vista, suburban section of SAN DIEGO, SAN DIEGO county, S CALIFORNIA, 5 mi/8 km N of downtown San Diego, N of SAN DIEGO RIVER. Miramar Naval Air Station to NE; Tecolote Canyon Natural Park to NW; Mission Bay to W. Residential area.

Lindeman Island (LIN-de-muhn), island (□ 3 sq mi/7.8 sq km), at S end of Whitsunday group, off QUEENSLAND coast, in CORAL SEA, NE AUSTRALIA; 20°27′S 149°02′E. Resort (Club Med). Walking tracks.

Linden, city (2006 population 39,874), UNION county, NE NEW JERSEY; 40°37′N 74°14′W. During the first half of the 20th century, Linden changed from an agricultural district to a city of diverse manufacturing. The city, named for the linden trees in the vicinity, was part of ELIZABETH until 1861. Incorporated 1925.

Linden (LIN-duhn), town, HESSE, central GERMANY, 4 mi/6.4 km S of GIESSEN; 50°32′N 08°40′E. Manufacturing (furniture).

Linden, town (2002 population 29,502), UPPER DEMERARA–BERBICE district, GUYANA, on E bank of DEMERARA RIVER, 70 mi/113 km S of GEORGETOWN; 06°00′N 58°18′W. Bauxite mining. The second-largest town in Guyana; has airfield.

Linden, town (2000 population 2,424), ☉ Marengo co., W Alabama, 15 mi/24 km S of DEMOPOLIS; 32°17′N 87°47′W. Corn area. Manufacturing (apparel); lumber milling. Founded 1823. Originally known as 'Town of Marengo' for the county, it was renamed for a shortened version of 'Hohenlinden' in honor of the victory of the French general Moreau over the Austrian forces of Archduke John at Hohenlinden in Bavaria in 1800. Inc. in 1839

Linden, unincorporated town (2000 population 1,103), SAN JOAQUIN county, central CALIFORNIA, 12 mi/19 km ENE of STOCKTON, near Mormon Slough, and S of CALAVERAS RIVER; 38°01′N 121°06′W. Agriculture (dairying, cattle; fruit, nuts, vegetables, sugar beets, beans, nursery products, grain). Manufacturing (farm machinery, printing and publishing).

Linden, town (2000 population 700), MONTGOMERY county, W INDIANA, 17 mi/27 km S of LAFAYETTE; 40°11′N 86°54′W. In agricultural area.

Linden, town (2000 population 226), DALLAS county, central IOWA, 9 mi/14.5 km N of CRAWFORDSVILLE; 41°38′N 94°16′W. In agricultural area. Manufacturing (feed).

Linden, town (2000 population 2,861), GENESEE county, SE central MICHIGAN, 14 mi/23 km SSW of FLINT; 42°49′N 83°46′W. In lake and farm area. Food processing. Several lakes to E and SW.

Linden, town (2006 population 990), ☉ PERRY county, W central TENNESSEE, on BUFFALO RIVER, and 46 mi/74 km W of COLUMBIA; 35°37′N 87°50′W. Trade center for lumbering and agricultural area; manufacturing; recreation.

Linden (LIN-duhn), town (2006 population 2,190), ☉ CASS county, NE TEXAS, 33 mi/53 km N of MARSHALL; 33°00′N 94°21′W. Elevation 410 ft/125 m. Agriculture (vegetables, fruits; chickens); timber. Manufacturing (oil and gas equipment). Founded c.1850.

Linden (LIN-den), village, NEW SOUTH WALES, SE AUSTRALIA, 50 mi/81 km from SYDNEY, in BLUE MOUNTAINS. Nearby is Caley's Repulse, a pyramid-shaped mound of rocks thought to have been created by Indigenous Australians. Formerly called Seventeen Mile Hollow due to its distance 17 mi/11 km from Nepean River.

Linden (LIN-den), village (□ 1 sq mi/2.6 sq km; 2001 population 636), S ALBERTA, W CANADA, 46 mi/73 km from CALGARY, in KNEEHILL county; 51°35′N 113°29′W. Manufacturing (animal feed, truck parts); dairying; construction. First settled by Mennonite farmers in 1902; incorporated 1964.

Linden (LIN-duhn), village (2006 population 126), CUMBERLAND county, S central NORTH CAROLINA, 15 mi/24 km NNE of FAYETTEVILLE on Little River (mouth in CAPE FEAR RIVER 3 mi/4.8 km to E); 35°15′N 78°45′W. Manufacturing; service industries.

Linden (LIN-den), unincorporated village, WARREN county, N VIRGINIA, 7 mi/11 km E of FRONT ROYAL, at MANASSAS GAP; 38°54′N 78°04′W. Manufacturing (sparkling cider, wine).

Linden, village (2006 population 591), IOWA county, SW WISCONSIN, 8 mi/12.9 km SW of DODGEVILLE; 42°55′N 90°16′W. Cheese; in dairying and hog-raising area.

Lindenberg im Allgäu (LIN-den-berg im AHL-goi), town, BAVARIA, GERMANY, in the ALLGÄU, 10 mi/16 km NE of LINDAU, near Austrian border; 47°36′N 09°54′E. Railroad junction. Manufacturing (apparel; food processing [cheese], glider planes). Tourism; summer resort.

Lindenfels (LIN-den-fels), town, HESSE, central GERMANY, 7 mi/11.3 km E of BENSHEIM, in the ODENWALD; 49°41′N 08°47′E. Manufacturing (jewelry). Tourist center. Has ruined castle.

Lindenhurst, village (2000 population 12,539), Lake county NE ILLINOIS, residential suburb, 35 mi/56 km NW of downtown CHICAGO, 10 mi/16 km WNW of WAUKEGAN, on Hastings Creek; 42°25′N 88°01′W. Small lakes in area; recreation.

Lindenhurst, village (□ 3 sq mi/7.8 sq km; 2006 population 27,937), SUFFOLK county, SE NEW YORK, on S LONG ISLAND; 40°41′N 73°22′W. Primarily residential. Light manufacturing and services. Incorporated 1923.

Lindenow Fjord (LIN-duh-no), Greenlandic *Kangerlussuatsiaq*, inlet of the ATLANTIC, SE GREENLAND, 60 mi/97 km E of JULIANEHAAB; 35 mi/56 km long, 1 mi/

1.6 km–3 mi/4.8 km wide; 60°32′N 43°45′W. Extends inland to edge of ice cap, which rises steeply to 7,500 ft/2,286 m on N shore. Ruins of medieval Scandinavian settlement found here.

Lindenwold, borough (2006 population 17,160), CAMDEN county, SW NEW JERSEY, 9 mi/14.5 km SE of CAMDEN; 39°49′N 74°59′W. Manufacturing (plastics, nails, machinery, food). Terminus for a light railroad line from PHILADELPHIA. Settled 1742, incorporated 1929.

Lindern (LIN-dern), village, LOWER SAXONY, NW GERMANY, 28 mi/45 km SW of OLDENBURG; 52°50′N 07°46′E. In peat region.

Lindesay, Mount (LIND-zee) (elevation 4,064 ft/1,239 m), E AUSTRALIA, in MCPHERSON RANGE, on QUEENSLAND–NEW SOUTH WALES border, 60 mi/97 km SSW of BRISBANE; 28°21′S 152°43′E.

Lindesberg (LIN-duhs-BER-yuh), town, ÖREBRO county, S central SWEDEN, on Hörksälven River, 20 mi/32 km N of ÖREBRO; 59°36′N 15°13′E. Manufacturing (machinery; pulp and paper mill). Trade center since Middle Ages. Medieval church. Chartered 1643.

Lindesnes (LIN-nuhs-nais), cape, in VEST-AGDER county, southernmost point of the Norwegian mainland, projecting into the NORTH SEA at the entrance to the SKAGERRAK. Norway's first lighthouse (1655) is here.

Lindewiese-Bad, CZECH REPUBLIC: see LIPOVA-LAZNE.

Lindhorst (LIND-horst), village, LOWER SAXONY, N GERMANY, 19 mi/31 km W of HANOVER; 52°22′N 09°17′E.

Lindhos, Greece: see LINDOS.

Lindi (LEEN-dee), region (2006 population 852,000), SE TANZANIA, ⊙ LINDI, bounded E by INDIAN OCEAN. Part of SELOUS GAME RESERVE in W; drained by MATANDU and Mbwemburu rivers. KILWA KISIWANI historical site on Kilwa Island in INDIAN OCEAN, in NE. Agriculture (cashews, bananas, copra, manioc; sheep, goats); fish. Part of former MTWARA province.

Lindi (LEEN-dee), town, ⊙ LINDI region (□ 55,223 sq mi/143,579.8 sq km), SE TANZANIA, 225 mi/362 km SSE of DAR ES SALAAM, on Lindi Bay, INDIAN OCEAN; at mouth of navigable LUKULEDI RIVER; 09°59′S 39°42′E. Agriculture (cashews, peanuts, copra, bananas; goats; sheep); fish. Manufacturing (edible oil milling). Airport to N. Complete dinosaur remains discovered at Tendunguru, to SW, in 1912.

Lindian (LIN-DIAN), town, ⊙ Lindian county, SW HEILONGJIANG province, NORTHEAST, CHINA, 45 mi/72 km ESE of QIQIHAR; 47°15′N 124°51′E. Agriculture (grain, soybeans, sugar beets); sugar refining.

Lindi River (LEEN-dee), c.375 mi/603 km long, E CONGO; rises 35 mi/56 km SW of LUBERO; flows NW, W, and SW, past MAKALA and BAFWASENDE, through dense equatorial forest, to CONGO RIVER 5 mi/8 km WNW of KISANGANI. Numerous rapids in lower course. Rugged terrain along most of its course, combined with the lack of funding, has prevented detailed exploration of LINDI RIVER.

Lindisfarne, ENGLAND: see HOLY ISLAND.

Lind Island, CANADA: see JENNY LIND ISLAND.

Lindlar (LIND-lahr), town, North Rhine-Westphalia, W GERMANY, 8 mi/12.9 km W of GUMMERSBACH; 51°01′N 07°22′E. Metal- and woodworking. Manufacturing (paper, plastics). Tourism.

Lindley, town, E FREE STATE province, SOUTH AFRICA, on Vals River, and 35 mi/56 km NW of BETHLEHEM; 27°52′S 27°55′E. Elevation 5,232 ft/1,595 m. Agricultural center (grain; livestock, dairying). On railroad link to VEREENIGING. Weir on river provides water and recreation area. Airfield. Established in 1875, named for American missionary Daniel Lindley.

Lindley, ENGLAND: see HUDDERSFIELD.

Lindóia (LEEN-doi-ah), town (2007 population 5,657), E SÃO PAULO, BRAZIL, 36 mi/58 km NE of CAMPINAS, near MINAS GERAIS border; 22°31′S 46°39′W. Elevation

3,100 ft/945 m. Resort with hot springs (developed after 1920).

Lindon, town (2006 population 9,758), UTAH county, N central UTAH, suburb 7 mi/11.3 km NNW of PROVO, and 43 mi/69 km SSE of SALT LAKE CITY; 40°20′N 111°43′W. Elevation 4,700 ft/1,433 m. Agriculture (berries, vegetables, alfalfa; dairying; cattle, sheep). Manufacturing (metal galvanizing, wood trusses). Served by PROVO RIVER irrigation project. Geneva Steel Plant nearby. Mount Timpanagos (11,750 ft/3,581 m) to NE. Uinta National Forest to E; UTAH LAKE to SW. Originally called Stringtown.

Lindong, CHINA: see BAIRIN ZUOQI.

Lindos (LEEN–[th]-os), Italian *Lindo*, Latin *Lindus*, village, on E shore of RHODES island, DODECANESE prefecture, SOUTH AEGEAN department, GREECE, 26 mi/42 km S of RHODES. Popular tourist area. One of the leading city-states of ancient Rhodes and a member of the Dorian Hexapolis; noted for its shrine to Athena. Has small Crusaders' church and collections of 17th-cent. oriental faience. The apostle Paul landed here on his voyage to Rome. Also spelled Lindhos.

Lindoso (leen-DO-soo), town, VIANA DO CASTELO district, N PORTUGAL, 26 mi/42 km NE of BRAGA, on Spanish border; 41°52′N 08°12′W. Hydroelectric plant 3 mi/4.8 km W on LIMA RIVER.

Lindsay, city (2000 population 10,297), TULARE county, S central CALIFORNIA, 12 mi/19 km SE of VISALIA, in SIERRA NEVADA foothills; 36°13′N 119°05′W. Packs and ships oranges, olives, grapes, pistachios, almonds, walnuts. Cattle; poultry. Manufacturing (citrus pulp, food and beverage machinery, plastic products). Incorporated 1910.

Lindsay (LIND-zee), former town (□ 6 sq mi/15.6 sq km; 2001 population 16,930), SE ONTARIO, E central CANADA, on the Scugog River, 53 mi/85 km NE of TORONTO; 44°21′N 78°44′W. Industrial town, with woolen, flour, and lumber mills, in an agricultural and scenic lake district (KAWARTHA LAKES). Amalgamated into Kawartha Lakes when that city was created in 2001 to replace Victoria county, of which Lindsay was the seat.

Lindsay (LIND-zee), former township (□ 105 sq mi/273 sq km; 2001 population 599), SW ONTARIO, E central CANADA; 45°07′N 81°22′W. Amalgamated into NORTHERN BRUCE PENINSULA municipality in 1999.

Lindsay (LIN-zee), town (2006 population 2,915), GARVIN county, S central OKLAHOMA, 23 mi/37 km WNW of PAULS VALLEY, and 23 mi/37 km SE of CHICKASHA, and on WASHITA River; 34°50′N 97°36′W. In agricultural area. Manufacturing (machinery, fishing lures). Historic Murray-Lindsay Mansion.

Lindsay, village (2006 population 278), PLATTE county, E central NEBRASKA, 23 mi/37 km NW of COLUMBUS, and on branch of PLATTE RIVER; 41°42′N 97°41′W. Agriculture (grain, livestock). Manufacturing (irrigation systems).

Lindsay (LIN-zee), village (2006 population 1,003), COOKE county, N TEXAS, 3 mi/4.8 km W of GAINESVILLE, near Elm Fork TRINITY RIVER; 33°38′N 97°13′W. Agricultural area (cattle, sheep, goats; grain).

Lindsborg (LINZ-buhrg), town (2000 population 3,321), MCPHERSON county, central KANSAS, on SMOKY HILL RIVER, and 13 mi/21 km N of MCPHERSON; 38°34′N 97°40′W. Railroad junction. Trade center for wheat and livestock area; flour. Manufacturing (signs, metal products). Seat of Bethany College (Lutheran; founded 1881). Founded 1868, incorporated 1879.

Lindsey (LIN-zee), village (□ 2 sq mi/5.2 sq km; 2006 population 482), SANDUSKY county, N OHIO, 7 mi/11 km NW of FREMONT; 41°25′N 83°13′W. In agricultural area. Manufacturing (meat products).

Lindsey, ENGLAND: see KERSEY.

Lindstrom, town (2000 population 3,015), CHISAGO county, E MINNESOTA, 33 mi/53 km NNE of ST. PAUL; 45°23′N 92°50′W. Agricultural area (grain; cattle, poultry, dairying). Manufacturing (food products,

plastic products, consumer goods, printing and publishing). Between four lakes: North Lindstrom (NW), South Lindstrom (SW), South Center (SE), North Center (NE). St. Croix Wild River State Park to NE; Carlos Avery Wildlife Area to W. Early Swedish settlement.

Lindum or **Lindum Colonia**, ENGLAND: see LINCOLN.

Lindus, Greece: see LINDOS.

Línea, La (LEE-nai-ah, lah) or **La Línea de la Concepción**, city, CÁDIZ province, S SPAIN, on the STRAIT OF GIBRALTAR, and 35 mi/56 km SW of MARBELLA; 36°10′N 05°21′W. Cereals, fruit, vegetables; manufacturing (apparel). Situated on the Spanish border N of the neutral zone that separates the city from the British colony.

Line Islands, coral group (□ 43 sq mi/111.8 sq km), central and S PACIFIC; 00°00′N 157°00′W. Once valued for their guano deposits, the islands now have coconut groves, airfields, and meteorological stations. Of the eleven coral islands and atolls in the archipelago, eight are part of the Republic of KIRIBATI: TERAINA, TABUAERAN, KIRITIMATI (formerly Washington, Fanning, and Christmas islands), MALDEN, STARBUCK, CAROLINE, VOSTOK, and FLINT. KINGMAN REEF, PALMYRA, and JARVIS Island are dependencies of the UNITED STATES. The islands were uninhabited when discovered by American sailors in 1798, although a few show evidence of ancient Polynesian contacts. The British government once conducted hydrogen bomb tests on Malden. Also called Equatorial Islands.

Linekin Neck (LIN-uh-kin), peninsula, LINCOLN county, S MAINE, E of BOOTHBAY Harbor, terminating in Ocean Point. Resort villages.

Lines of Torres Vedres, defensive redoubt prepared by British General Arthur Wellington to protect his outnumbered British-Portuguese force from French Napoleonic armies, IBERIAN PENINSULA, 1810. Three fortified lines, running 30 mi/48 km between the TAGUS River and the ATLANTIC N of LISBON. The fortifications, many cut directly into the natural landscape, were armed with more than 500 guns, a phenomenal total for the time. The French probed the position in October–Nov. 1810, but withdrew, deciding correctly that it was impregnable.

Linesville, borough (2006 population 1,116), CRAWFORD county, NW PENNSYLVANIA, 14 mi/23 km W of MEADVILLE, on Lineville Creek, at its entrance on NE shore of PYMATUNING RESERVOIR (causeway). Agriculture area (dairying, livestock; potatoes, corn). Manufacturing (aluminum processing, fiberglass boxes and units, machinery, lumber). Pymatuning Airport to NW. Incorporated 1862.

Lineville, town (2000 population 2,401), Clay co., E Alabama, just NE of Ashland. Manufacturing (apparel; food products, plastic products); lumber. First known as 'Lundies Crossroads' for William Y. and Thomas Lundie, it was later known as 'County Line.' It became Lineville when the town became the temporary county seat for Clay County in 1866. Inc. in 1898.

Lineville, town (2000 population 273), WAYNE county, S IOWA, at MISSOURI state line, 15 mi/24 km SW of CORYDON; 40°35′N 93°31′W. Livestock; grain.

Linëvo (lee-NYO-vuh), town (2006 population 21,450), E NOVOSIBIRSK oblast, SW SIBERIA, RUSSIA, on highway and railroad, 37 mi/60 km SSE of NOVOSIBIRSK, and 11 mi/18 km S of ISKITIM, to which it is administratively subordinate; 54°26′N 83°22′E. Elevation 708 ft/215 m. Gas pipeline construction and repair; manufacturing (electrical components and wiring). Made town in 1974.

Linëvo (lee-NYO-vuh), town (2006 population 6,720), N VOLGOGRAD oblast, SE European Russia, near the MEDVEDITSA River (DON River basin), on road and railroad, 39 mi/63 km WSW of KRASNOARMEYSK; 50°53′N 44°49′E. Elevation 449 ft/136 m. Flour and feed milling. Established in German VOLGA ASSR as Gussenbakh; German population removed and the town renamed Medveditskoye in 1941; current name since 1996.

Linfen (LIN-FEN), city (□ 503 sq mi/1,307.8 sq km; 2000 population 582,690), S SHANXI province, CHINA, on FEN River, on railroad, and 135 mi/217 km SSW of TAIYUAN; 36°08′N 111°34′E. Agriculture, market, and industrial center in irrigated farming district of the Fen valley. Heavy and light industry, agriculture (especially crop growing). Main agriculture products include grains, eggs, and hogs. Industries include textiles and iron and steel production. Traditionally the residence of legendary emperor Yao (2357–2255 B.C.E.).

Linfield, unincorporated village, Limerick township, MONTGOMERY county, SE PENNSYLVANIA, 5 mi/8 km SE of POTTSTOWN, on SCHUYLKILL RIVER; 40°12′N 75°34′W.

Linford (LIN-fuhd), locality (2006 population 6,200), THURROCK, SE ENGLAND, on THAMES estuary, 4 mi/6.4 km NE of TILBURY; 51°29′N 00°25′E.

Linford, Great (LIN-fuhd), town (2001 population 16,912), Milton Keynes, S central ENGLAND, on Grand Union Canal, 3 mi/4.8 km NE of MILTON KEYNES; 52°04′N 00°45′W. Has 12th-century church. Tudor rectory.

Lingao (LIN-GOU), town, ⊙ Lingao county, N HAINAN province, CHINA, 40 mi/64 km W of QIONGSHAN; 19°54′N 109°40′E. In sugar-growing area; sugar refining.

Linga, West (LIN-gah), uninhabited islet (1.5 mi/2.4 km long) of the SHETLAND ISLANDS, extreme N Scotland, just W of WHALSAY island across 1-mi/1.6-km-wide Linga Channel; 60°21′N 01°02′W.

Lingayen (leeng-gah-YEN), town, ⊙ PANGASINAN province, central LUZON, PHILIPPINES, port on S shore of LINGAYEN GULF, on AGNO delta, 7 mi/11.3 km W of DAGUPAN; 16°01′N 120°14′E. Agricultural center (rice, corn; copra). Airstrip.

Lingayen Gulf (leeng-gah-YEN), large inlet of SOUTH CHINA SEA, central LUZON, PHILIPPINES, between Santiago Islands (W) and SAN FERNANDO POINT (E); 26 mi/42 km wide at entrance, extends c.35 mi/56 km inland. Contains CABARRUYAN and SANTIAGO islands. In World War II, the Japanese landed here in December 1941, and U.S. forces in January 1945.

Lingbao (LING-BOU), town, ⊙ Lingbao county, W HENAN province, CHINA, near LONGHAI Railroad, and HUANG HE (Yellow River), 24 mi/39 km E of TONGGUAN; 34°34′N 110°42′E. Agriculture (grain, oilseeds, tobacco). Nonferrous ore mining.

Lingbi (LING-BEE), town, ⊙ Lingbi county, N Anhui province, CHINA, 45 mi/72 km NNE of BENGBU; 33°33′N 117°33′E. Grain, oilseeds; cotton, jute.

Lingbo (LING-BOO), village, GÄVLEBORG county, E SWEDEN, 20 mi/32 km SW of SÖDERHAMN; 61°03′N 16°41′E.

Lingchuan (LING-CHUAN), town, ⊙ Lingchuan county, NE GUANGXI ZHUANG AUTONOMOUS REGION, CHINA, on upper GUI RIVER, on railroad, and 12 mi/19 km NNE of GUILIN; 25°25′N 110°20′E. Agriculture (rice, tea). Manufacturing (chemicals, building materials, food processing, machinery and equipment manufacturing); iron smelting.

Lingchuan, town, ⊙ Lingchuan county, SE SHANXI province, CHINA, 35 mi/56 km SSE of CHANGZHI, in TAIHANG MOUNTAINS; 35°46′N 113°16′E. Agriculture (grain). Manufacturing (chemicals, machinery and equipment manufacturing); coal mining, iron smelting.

Lingding Island (LING-DING), Mandarin *Lingting*, in PEARL RIVER estuary, S GUANGDONG province, CHINA, 22 mi/35 km NE of MACAO, near HONG KONG border. Noted for conical peak rising to 833 ft/254 m; in the Lingding Sea, where General Wen Tianxiang of late Ming dynasty (1368–1644) composed his famous patriotic poem.

Lingen (LING-uhn), city, LOWER SAXONY, NW GERMANY, on EMS RIVER, and DORTMUND-EMS CANAL, 17 mi/27 km NNW of RHEINE; 52°32′N 07°20′E. Port and industrial center; oil refining; chemical industry; metalworking (steel); livestock. Nearby is a nuclear power plant (built 1988). Noteworthy buildings are the baroque town hall and a 17th-century palace. City first mentioned in 975; chartered 13th century; developed as a trading center on the EMS.

Lingenfeld (LING-uhn-felt), village, RHINELAND-PALATINATE, W GERMANY, on an arm of the RHINE, 6 mi/9.7 km SW of SPEYER; 49°15′N 08°21′E.

Lingfield (LING-feeld), village (2001 population 4,214), SE SURREY, SE ENGLAND, on Eden River, and 3 mi/4.8 km N of East GRINSTEAD; 51°10′N 00°01′E. Former agricultural market. Has 15th-century church. Site of racecourse.

Lingga (LEENG-gah), island (□ c.360 sq mi/932 sq km), LINGGA ARCHIPELAGO, Riau province, INDONESIA, in SOUTH CHINA SEA, off E coast of SUMATRA, and S of RIAU ARCHIPELAGO, just N of SINGKEP, 100 mi/161 km SSE of SINGAPORE; 40 mi/64 km long, up to 20 mi/32 km wide; 00°09′S 104°39′E. Generally low, with hills rising to 3,266 ft/995 m in SW. Agriculture and forest products (sago, copra, gambier, pepper; rattan, timber). Fishing. Chief town and port is Daik or Kotadaik on S coast of island.

Lingga Archipelago (LEENG-gah), island group (□ 842 sq mi/2,189.2 sq km), RIAU province, INDONESIA, in SOUTH CHINA SEA, off E coast of SUMATRA, and S of RIAU ARCHIPELAGO; 00°09′S 104°39′E. Comprises numerous islands, largest being LINGGA and SINGKEP. Between Lingga and Singkep is small but important islet of Penuba or Penoeba (7 mi/11.3 km long, 3 mi/4.8 km wide) and small port of Penuba. Daik, the chief town and port of Lingga group, is on Lingga Island N of Lingga are smaller islands of Sebangka (20 mi/32 km long, 4 mi/6.4 km wide), Bakung or Bakoeng (10 mi/16 km long, 3 mi/4.8 km wide), and Temiang (8 mi/12.9 km long, 2 mi/3.2 km wide). Islands are generally low and of coral formation.

Linggajati, INDONESIA: see CIREBON, city.

Linggi River (LING-gee), 40 mi/64 km long, W NEGRI SEMBILAN, MALAYSIA; rises in central Malayan range NW of SEREMBAN; flows S, past SEREMBAN, to Strait of MALACCA on MALACCA border below Pengkalan Kempas.

Linghed (LING-HED), village, KOPPARBERG county, central SWEDEN, 15 mi/24 km NE of FALUN; 60°47′N 15°53′E.

Lingle, village (2006 population 488), GOSHEN county, SE WYOMING, on North Platte River, and 9 mi/14.5 km NW of TORRINGTON; 42°08′N 104°20′W. Elevation 4,165 ft/1,269 m. Shipping point for sugar beets, beans, and cattle; in irrigated region.

Linglestown, unincorporated town (2000 population 6,414), DAUPHIN county, S PENNSYLVANIA, residential suburb 7 mi/11.3 km NE of HARRISBURG; 40°20′N 76°47′W. BLUE RIDGE MOUNTAINS to N.

Lingling, CHINA: see YONGZHOU.

Lingolsheim (lan-gol-ZEM), German (LING-ols-heim), SW residential suburb (□ 2 sq mi/5.2 sq km), of STRASBOURG, BAS-RHIN department, E FRANCE, in ALSACE; 48°34′N 07°41′E.

Lingqiu (LING-CHIU), town, ⊙ Lingqiu county, NE SHANXI province, CHINA, in WUTAI Mountains, 65 mi/105 km SE of DATONG, near SHANSI-HEBEI border; 39°26′N 114°14′E. Cattle raising; agriculture (wheat, kaoliang, beans). Manufacturing (building materials, machinery and equipment manufacturing, chemicals, food processing).

Ling River (LING), 35 mi/56 km long, SE ZHEJIANG province, CHINA; rises in KUOCANG MOUNTAINS, flows NE and E, turning SE at LINHAI, from which it flows SE, past HAIMEN, to Taizhou Bay of EAST CHINA SEA.

Lingshan (LIN-SHAN), town, ⊙ Lingshan county, S GUANGXI ZHUANG AUTONOMOUS REGION, CHINA, on

QIN RIVER, and 65 mi/105 km ESE of NANNING; 22°26′N 109°17′E. Agriculture (rice, sugarcane, oilseeds). Manufacturing (food processing, crafts); fur processing.

Lingshi (LING-SHI), town, ⊙ Lingshi county, S central SHANXI province, CHINA, on FEN River, and 30 mi/48 km S of FENYANG, and on railroad; 36°51′N 111°46′E. Agriculture (wheat, millet, beans). Manufacturing (chemicals); coal mining, iron smelting.

Lingshi (LING-SHI), pass (16,118 ft/4,913 m), in main range of W ASSAM Himalayas, 2.5 mi/4 km NNW of LINGSHI (NW BHUTAN).

Lingshou (LING-SHO), town, ⊙ Lingshou county, SW HEBEI province, CHINA, 20 mi/32 km NNW of SHIJIAZHUANG, and on HUTUO RIVER; 38°18′N 114°22′E. Agriculture (grain; tobacco, cotton, oilseeds). Manufacturing (textiles, apparel, machinery and equipment manufacturing, electronics, food processing, chemicals).

Lingshui (LING-SHUAI), town, ⊙ Lingshui county, SE HAINAN province, CHINA, 110 mi/177 km SSW of HAIKOU; 18°31′N 110°01′E. Agriculture (rice, animal husbandry); fisheries. Manufacturing (beverages).

Lingsugur (ling-suh-GOOR), town, tahsil headquarters, RAICHUR district, KARNATAKA state, SW INDIA, 55 mi/89 km W of RAICHUR; 16°10′N 76°31′E. Cotton ginning; millet, oilseeds. Was capital of former Lingsugur district (divided 1905 between Raichur and GULBARGA districts).

Lingtai (LING-TEI), town, ⊙ Lingtai county, SE GANSU province, CHINA, 35 mi/56 km SSE of JINGCHUAN, near SHAANXI border; 35°04′N 107°37′E. Grain, oilseeds; tobacco.

Linguaglossa (LEENG-gwah-GLOS-sah), village, CATANIA province, E SICILY, ITALY, on NE slope of MOUNT ETNA, 16 mi/26 km N of ACIREALE; 37°50′N 15°08′E. Woodworking center; wine. Winter ski resort.

Linguère (LEEN-ger), town (2002 population 11,667), LOUGA administrative region, N central SENEGAL, terminus of abandoned Louga-Linguère branch of Dakar–Saint-Louis railroad, 102 mi/164 km SE of SAINT-LOUIS; 15°24′N 15°07′W. Exports peanuts and gum arabic. Corn growing, livestock raising. Airfield.

Lingwick, village (□ 94 sq mi/244.4 sq km; 2006 population 427), ESTRIE region, S QUEBEC, E CANADA; 45°35′N 71°20′W.

Lingwu (LING-WU), town, ⊙ Lingwu county, N central NINGXIA HUI AUTONOMOUS REGION, CHINA, 25 mi/40 km S of YINCHUAN, across HUANG HE (Yellow River); 38°05′N 106°20′E. Cattle raising; grain.

Lingxi, CHINA: see YONGSHUN.

Ling Xian (LING SIAN), town, ⊙ Ling Xian county, NW SHANDONG province, CHINA, 20 mi/32 km ESE of DEZHOU; 37°21′N 116°31′E. Manufacturing (textiles). Agriculture (grain, peanuts, melons; cotton).

Ling Xian, town, ⊙ Ling Xian county, SE HUNAN province, CHINA, near JIANGXI border, 75 mi/121 km ESE of HENGYANG; 26°26′N 113°45′E. Tea, rice; logging, timber processing.

Lingyuan (LING-YUAN), city (□ 1,266 sq mi/3,279 sq km; 1994 estimated urban population 129,700; estimated total population 623,800), W LIAONING province, CHINA, on railroad, 75 mi/121 km ENE of CHENGDE; 41°12′N 119°16′E. Agriculture and heavy industry are the main sources of income for the city. Manufacturing (chemicals, iron and steel, transportation equipment).

Lingyun (LING-YOOIN), town, ⊙ Lingyun county, W GUANGXI ZHUANG AUTONOMOUS REGION, CHINA, 30 mi/48 km N of BOSE; 24°24′N 106°31′E. Agriculture (grain); logging. Manufacturing (furniture, food products).

Linhai (LIN-HEI), city (□ 838 sq mi/2,178.8 sq km; 2000 population 980,541), E ZHEJIANG province, CHINA, on small LING RIVER, and 65 mi/105 km NNE of WENZHOU, 25 mi/40 km from the Taizhou Bay of EAST

CHINA SEA; 28°54′N 121°08′E. Agriculture and light industry are the largest sectors of the city's economy. Agriculture (grains, fruits, hogs). Manufacturing (food processing, textiles, crafts, chemicals, machinery). Also known as Taizhou.

Linhares (leen-yah-res), city (2007 population 124,581), N central ESPÍRITO SANTO, BRAZIL, on the swampy lower RIO DOCE, and 65 mi/105 km N of VITÓRIA; 19°25′S 40°02′W. Cacao, coffee. Animal reserve established nearby in 1943.

Linhares (leen-YAHR-ish), village, GUARDA district, N central PORTUGAL, 10 mi/16 km W of GUARDA. Its 12th-century castle played part in Moorish wars.

Linh Cam, VIETNAM: see DUC THO.

Linhe (LIN-HUH), city (□ 899 sq mi/2,337.4 sq km; 2000 population 425,973), W INNER MONGOLIA Autonomous Region, CHINA, near the HUANG HE (Yellow River), in HETAO oasis, on railroad, and 135 mi/217 km W of BATOU; 40°45′N 107°26′E. Agriculture (especially crop growing and animal husbandry) is the largest source of income for the city. Main agriculture products include grains, vegetables, oil crops, fruits, and hogs. Manufacturing (food, textiles).

Lin-hsia, CHINA: see LINXIA.

Linhuaiguan (LIN-HUEI-GUAN), town, N ANHUI province, CHINA, 20 mi/32 km E of BENGBU, on HUAI River, and TIANJIN-PUKOU railroad; 32°53′N 117°45′E. Agriculture (rice, oilseeds).

Liniers (lee-nee-ERS), W industrial section of BUENOS AIRES, ARGENTINA. Textile mills, railroad shops.

Linjan, IRAN: see LENJAN.

Linjiang (LIN-JIANG), town, SE JILIN province, NORTHEAST, CHINA, 50 mi/80 km E of TONGHUA, and on upper YALU RIVER (NORTH KOREA border); an administrative unit of HUNJIANG; 41°45′N 126°56′E. Agriculture (medicinal herbs); logging. Manufacturing (papermaking).

Linjin (LIN-JIN), town, SW SHANXI province, CHINA, 22 mi/35 km NE of YONGJI, near HUANG HE (Yellow River). Agriculture (grain, oilseeds; cotton).

Linkebeek (LING-kuh-baik), commune (2006 population 4,729), Halle-Vilvoorde district, BRABANT province, central BELGIUM, 5 mi/8 km S of BRUSSELS.

Linkenheim-Hochstätten (LING-kuhn-heim–HOKH-shtet-tuhn), town, BADEN-WÜRTTEMBERG, GERMANY, 8 mi/12.9 km N of KARLSRUHE; 49°08′N 08°25′E. Agriculture (asparagus, strawberries).

Linkinhorne (LING-kin-hawn), agricultural village (2001 population 1,471), E CORNWALL, SW ENGLAND, 7 mi/11.3 km S of LAUNCESTON; 50°32′N 04°22′W. Has Elizabethan manor house and church.

Linköping (LIN-SHUHP-eeng), town, ⊙ ÖSTERGÖTLAND county, S SWEDEN, near LAKE ROXEN; 58°25′N 15°38′E. Commercial, industrial, and educational center. Manufacturing (transportation equipment, electrical equipment, processed food). University and teaching hospital. An episcopal see since 1120; flourishing medieval intellectual and religious center. Defeat of Swedish king Sigismund III (shortly before he was formally deposed) by future Charles IX at nearby Stångebro (1598). Romanesque cathedral (twelfth century; rebuilt 1230s), with 344-ft/105-m spire; thirteenth-century castle (restored 1931–1932). Large library; outdoor museum; Gamla (Old) Linköping. International airport to S.

Linkou (LIN-KO), town, ⊙ Linkou county, SE HEILONGJIANG province, NORTHEAST, CHINA, 60 mi/97 km NE of MUDANJIANG; 45°18′N 130°17′E. Railroad junction. Agriculture (tobacco, medicinal herbs).

Link River, small stream 2 mi/3.2 km long, KLAMATH county, S OREGON, at city of KLAMATH FALLS; connects UPPER KLAMATH LAKE with Lake Ewauna, which is drained by KLAMATH RIVER. Site of small dam.

Linli (LIN-LEE), town, ⊙ Linli county, N HUNAN province, CHINA, 25 mi/40 km N of CHANGDE; 29°27′N 111°39′E. Agriculture (rice, sugarcane, oilseeds; cot-

ton). Manufacturing (food processing, construction materials). Also called Anfu.

Linlin Island (LEEN-leen) (□ 3 sq mi/7.8 sq km), just off E coast of CHILOÉ ISLAND, S CHILE; 42°23′S 73°26′W.

Linlithgow (LIN-lith-gou), town (2001 population 13,370), WEST LOTHIAN, central Scotland, 16 mi/26 km W of EDINBURGH; 55°58′N 03°37′W. Manufacturing (electronics, pharmaceuticals, whiskey). Linlithgow Palace, now a ruin, was a seat of Stuart kings and the birthplace of James V and Mary, Queen of Scots. Begun in the 15th century by James I, it was occupied (1651–1659) by Cromwell and his forces and burned in 1746. The first earl of Murray, regent of Scotland, was murdered here in 1570. Formerly in Lothian, abolished 1996.

Linn, county (□ 724 sq mi/1,882.4 sq km; 2006 population 201,853), E IOWA; ⊙ CEDAR RAPIDS; 42°04′N 91°35′W. Prairie agricultural area (hogs, cattle, poultry; corn, oats) drained by CEDAR and WAPSIPINICON rivers and BUFFALO CREEK. Many limestone quarries, sand and gravel pits. Manufacturing at Cedar Rapids and MARION. Duane Arnold nuclear power plant, initial criticality March 23, 1974, is 8 mi/12.9 km NW of Cedar Rapids, uses cooling water from the Cedar Rapids River, and has a max. dependable capacity of 515 MW. Palisades Kepler State Park in SE; Pleasant Creek State Park in NW. Flooding occurred along rivers in 1993. Formed 1837.

Linn, county (□ 606 sq mi/1,575.6 sq km; 2006 population 9,962), E KANSAS; ⊙ MOUND CITY; 38°12′N 94°50′W. Prairie region, bordering E on MISSOURI; drained (NE) by MARAIS DES CYGNES River. Agriculture (cattle, hogs; wheat, sorghum, soybeans; hay). Formed 1855.

Linn, county (□ 624 sq mi/1,622.4 sq km; 2006 population 12,865), N central MISSOURI; ⊙ LINNEUS; 39°52′N 93°06′W. Agriculture (corn, wheat, hay, soybeans; sheep, cattle, hogs). Coal; manufacturing at BROOKFIELD and MARCELINE. Pershing State Park SW of LACLEDE; Fountain Grove Conservation Area in SW corner. General John Joseph Pershing born here 1860. Formed 1837.

Linn, county (□ 2,309 sq mi/6,003.4 sq km; 2006 population 111,489), W OREGON; ⊙ ALBANY; 44°29′N 122°31′W. Level farm land rising E to CASCADE RANGE, bounded W by WILLAMETTE RIVER. Most of N boundary formed by North SANTIAM River, forms DETROIT LAKE Reservoir in NE. Manufacturing (paper products, food processing, metal industries); logging. Agriculture (corn, beans, wheat, oats, barley; poultry, hogs, sheep, cattle); dairy products; nurseries, wineries. Part of Willamette National Forest in E, including part of Mount Washington Wilderness Area in SE and part of Mount Jefferson Wilderness Area in NE. Summit of MOUNT JEFFERSON (10,495 ft/3,199 m) marks NE corner of county. Foster and GREEN PETER RESERVOIRS, Cascadia State Park in E. Formed 1847.

Linn, city (2000 population 1,354), ⊙ OSAGE county, central MISSOURI, 19 mi/31 km ESE of JEFFERSON CITY; 38°28′N 91°50′W. Agriculture (corn, wheat, vegetables; cattle, dairying; poultry). Manufacturing (transportation equipment, plastic products); clay pits. Linn State Technical College.

Linn, village (2000 population 425), WASHINGTON county, N KANSAS, 10 mi/16 km S of WASHINGTON; 39°40′N 97°05′W. Shipping point in grain and livestock region; dairying; poultry and produce packing.

Linn (LIN), unincorporated village, HIDALGO county, S TEXAS, 26 mi/42 km NNE of MCALLEN, N of Rio Grande Valley; 26°33′N 98°07′W. Cattle; cotton. Oil and natural gas.

Linn Creek, town (2000 population 280), CAMDEN county, central MISSOURI, in the OZARKS, on LAKE OF THE OZARKS, just N of CAMDENTON; 38°02′N 92°42′W. Manufacturing (model railroad accessories). Tour-

ism; resorts. Original town of Linn Creek was submerged by the Lake of the Ozarks in 1931.

Linndale (LIN-dail), village (2006 population 91), CUYAHOGA county, N OHIO; SW suburb of CLEVELAND; 41°26′N 81°46′W.

Linne (LI-nuh), village, LIMBURG province, SE NETHERLANDS, on MAAS RIVER, and 4 mi/6.4 km SW of ROERMOND; 51°09′N 05°56′E. Dairying; cattle; vegetables, grain.

Linné, Cape (li-NAI), W West Spitsbergen, SPITSBERGEN group, on ARCTIC OCEAN, on S side of mouth of Is Fjord, 30 mi/48 km WSW of Longyear City; 78°03′N 13°35′E. Meteorological observatory (est. 1933), radio station, lighthouse, radio beacon.

Linneus (LIN-nee-uhs), city (2000 population 369), ⊙ LINN county, N central MISSOURI, 8 mi/12.9 km NW of BROOKFIELD; 39°52′N 93°11′W. Corn, wheat, soybeans; sheep, cattle, hogs.

Linneus (LIN-ee-uhs), town, AROOSTOOK county, E MAINE, 8 mi/12.9 km SW of HOULTON; 46°02′N 67°58′W. In agricultural, hunting, fishing area.

Linn Grove, town (2000 population 211), BUENA VISTA county, NW IOWA, on Little Sioux River, and 18 mi/29 km N of STORM LAKE; 42°53′N 95°14′W. Manufacturing (hydraulic cylinder components). Sand and gravel pits nearby. Wanata State Park to NW.

Linnhe, Loch (LIN), inlet (22 mi/35 km long, 1 mi/1.6 km–5 mi/8 km wide) on W coast of Scotland, between HIGHLAND and Argyll and Bute; 55°58′N 03°37′W. At its head is FORT WILLIAM, where it joins the 7-mi/11.3-km-long Loch Eil at right angles, and where the CALEDONIAN CANAL begins. At mouth of Loch Linnhe is LISMORE island.

Linnich (LIN-nikh), town, North Rhine-Westphalia, W GERMANY, on the ROER RIVER, 6 mi/9.7 km NW of JÜLICH; 50°59′N 06°16′E. Glass painting; packaging technology.

Linntown, unincorporated town (2000 population 1,542), East Buffalo township, UNION county, central PENNSYLVANIA, residential suburb 1 mi/1.6 km SW of LEWISBURG; 40°57′N 76°54′W.

Liñola (lee-NYO-lah), town, LÉRIDA province, NE SPAIN, 16 mi/26 km ENE of LÉRIDA; 41°42′N 00°55′E. In irrigated agricultural area (olive oil, cereals, alfalfa); sheep raising; gravel.

Lino Lakes (LEI-no), town (2000 population 16,791), ANOKA county, E MINNESOTA, residential suburb 14 mi/23 km NNE of downtown MINNEAPOLIS, and 14 mi/23 km N of downtown ST. PAUL; 45°09′N 93°04′W. Chain of small natural lakes through center of community along Rice Creek; lakes also in SE corner. Manufacturing (plastic products, machining, building materials, corrosion inhibiting products, electrical products). Minnesota Correctional Facility to W.

Linosa (lee-NAW-zah), ancient *Aegusa*, island (□ 2 sq mi/5.2 sq km), one of PELAGIE Islands, SICILY, ITALY, in MEDITERRANEAN Sea between MALTA and TUNIS, 105 mi/169 km SW of LICATA, Sicily; 2 mi/3.2 km long. Rises to 640 ft/195 m. Fisheries; vineyards, orchards. Chief port, Linosa (S).

Linovitsa, UKRAINE: see LYNOVYTSYA.

Linping (LIN-PING), town, N ZHEJIANG province, CHINA, 13 mi/21 km NE of HANGZHOU, and on railroad to SHANGHAI.

Linpu (LIN-PU), town, N ZHEJIANG province, CHINA, 8 mi/12.9 km S of XIAOSHAN, and on ZHEJIANG-JIANGXI railroad.

Linqing (LIN-CHING), city (□ 369 sq mi/959.4 sq km; 2000 population 672,759), W SHANDONG province, CHINA, on the GRAND CANAL, at mouth of the WEI RIVER, and 70 mi/113 km W of JINAN; 36°51′N 115°42′E. Agriculture and light industry are the largest sectors of the city's economy. Agriculture (animal husbandry; grains, fruits, eggs; cotton). Manufacturing (food processing, textiles). The city is called the "land of the Beijing Opera" because of its long tradition of love for that opera. Numerous famous Beijing Opera artists have come here to perform.

Linqu (LIN-CHYOOI), town, ⊙ Linqu county, central SHANDONG province, CHINA, 13 mi/21 km SSE of YIDU; 36°30′N 118°32′E. Tobacco-growing center; grain, oilseeds. Well known for its scenic setting (the YI MOUNTAINS and YI RIVER are nearby).

Lins (LEENS), city (2007 population 69,240), W central SÃO PAULO, BRAZIL, on railroad, 60 mi/97 km NW of BAURU; 21°40′S 49°45′W. Coffee growing and processing. Manufacturing (furniture, pottery); sawmilling. Ships coffee, manioc, rice, dairy produce. Bishopric. Has business school (established 1943). Formerly called Albuquerque Lins.

Linsan (LEEN-san), town, Kindia prefecture, Kindia administrative region, SW GUINEA, in Guinée-Maritime geographic region, 40 mi/64 km E of KINDIA; 10°17′N 12°26′W. Bananas.

Linselles (lin-sel), town (□ 4 sq mi/10.4 sq km), NORD department, NORD-PAS-DE-CALAIS region, N FRANCE, 6 mi/9.7 km N of LILLE; 50°44′N 03°05′E. Manufacturing (textiles).

Linshui (LIN-SHUAI), town, ⊙ Linshui county, E central SICHUAN province, CHINA, 50 mi/80 km NNE of CHONGQING city; 30°18′N 106°55′E. Agriculture (rice, oilseeds; jute, tobacco). Manufacturing (machinery and equipment manufacturing, food processing, chemicals); iron smelting, coal mining.

Linsia, CHINA: see LINXIA.

Linslade (LIN-slaid), town (2001 population 4,488), E BEDFORDSHIRE, central ENGLAND, on Ouzel River, and just W of LEIGHTON BUZZARD; 51°55′N 00°40′W. Former agricultural market. Has 15th-century church.

Linstead, town, SAINT CATHERINE parish, central JAMAICA, on railroad, and 12 mi/19 km NNW of SPANISH TOWN; 18°09′N 77°01′W. Annatto, tropical fruit, coffee; livestock.

Lint (LINT), commune (2006 population 8,007), Antwerp district, ANTWERPEN province, N BELGIUM, 3 mi/5 km W of LIER.

Lintaca (leen-TAH-kah), canton, NOR CINTI province, CHUQUISACA department, SE BOLIVIA, 12 mi/20 km SW of CAMARGO, at the foot of the Lique Mountains; 20°46′S 65°20′W. Elevation 7,894 ft/2,406 m. Extensive gas wells in area; also, limestone deposits. Agriculture (potatoes, yucca, bananas, corn, wheat, oats, rye, peanuts); cattle and hog raising.

Lintan (LIN-TAN), town, ⊙ Lintan county, SE GANSU province, CHINA, 90 mi/145 km S of LANZHOU; 34°40′N 103°23′E. Cattle and sheep raising; medicinal herbs, oilseeds.

Lintao (LIN-TOU), town, ⊙ Lintao county, SE GANSU province, CHINA, 45 mi/72 km S of LANZHOU, and on TAO RIVER; 35°20′N 104°00′E. Manufacturing (food and forage processing, leather and fur processing, crafts, machinery and equipment manufacturing, chemicals). Agriculture (grain, oilseeds, sugar beets, medicinal herbs). Population is largely Muslim.

Linter (LIN-tuhr), commune (2006 population 7,034), Leuven district, BRABANT province, central BELGIUM, 4 mi/6 km NW of TIENEN, near GEET RIVER.

Lintfort, GERMANY: see KAMP-LINTFORT.

Lintgen (LINT-guhn), town, LINTGEN commune, S central LUXEMBOURG, on ALZETTE RIVER, and 7 mi/11.3 km N of LUXEMBOURG city; 49°43′N 06°08′E. Manufacturing Just SW is agricultural village of PRETTINGEN or PRETTEN.

Linthal (LIN-tahl), commune, GLARUS canton, E central SWITZERLAND, on LINTH RIVER, and 9 mi/14.5 km SSW of GLARUS. Cotton textiles. Road leads to KLAUSEN PASS and ALTDORF.

Linth Canal, NE SWITZERLAND, 11 mi/18 km long; canalizes LINTH RIVER from LAKE WALEN to LAKE ZÜRICH.

Linthicum (LIN-thi-cum), suburban village (2000 population 7,539), ANNE ARUNDEL county, central MARYLAND, 7 mi/11.3 km SSW of downtown BALTIMORE;

39°13′N 76°40′W. Abner Linthicum bought the land here in 1801, and he and his descendants farmed it until 1908, when the estate was broken up. Baltimore-Washington International Airport is nearby.

Linth River, 87 mi/140 km long, N SWITZERLAND; formed by two headstreams rising on KLAUSEN PASS in Uri canton and slopes of TÖDI in GLARUS canton; flows N, past GLARUS, through ESCHER CANAL, to LAKE WALEN (its old bed bypassed the lake), thence through LINTH CANAL to LAKE ZÜRICH; emerges from Lake Zürich at Zürich as LIMMAT RIVER; flows NW, past BADEN, to AARE RIVER. Drains 933 sq mi/2,416 sq km.

Linton, city (2000 population 5,774), GREENE county, SW INDIANA, 12 mi/19 km W of BLOOMFIELD; 39°02′N 87°10′W. In agricultural area. Manufacturing (machinery, coal industry equipment, furniture); bituminous-coal mines. Settled 1816. Laid out 1850; incorporated as town in 1886, as city in 1900.

Linton, town (2006 population 1,097), ⊙ EMMONS CO., S NORTH DAKOTA, 46 mi/74 km SSE of BISMARCK, and on Beaver Creek, 15 mi/24 km E of MISSOURI River (Lake Oahe Reservoir); 46°16′N 100°13′W. Farming center (grain; livestock; dairying). Founded in 1899 and named for attorney George W. Lynn.

Linton (LIN-tuhn), village (2001 population 4,614), S CAMBRIDGESHIRE, E ENGLAND, on CAM RIVER, and 10 mi/16 km SE of CAMBRIDGE; 52°06′N 00°17′E. Former agricultural market. Has 15th-century church.

Linton (LIN-tuhn), village (2001 population 4,700), S DERBYSHIRE, central ENGLAND, 4 mi/6.4 km SSE of BURTON UPON TRENT; 52°44′N 01°35′W.

Linton, East (LIN-tuhn), town (2001 population 1,744), East Lothian, E Scotland, on TYNE RIVER, and 5 mi/8 km ENE of HADDINGTON; 55°59′N 02°39′W. Nearby are remains of Hailes Castle, one-time residence of Mary, Queen of Scots.

Lintong (LIN-TUNG), town, ⊙ Lintong county, SE central SHAANXI province, CHINA, 18 mi/29 km ENE of XI'AN, and on LONGHAI RAILROAD; 34°22′N 109°12′E. Agriculture (grain; cotton, oilseeds). Manufacturing (food processing, consumer goods, machinery).

Linton, West (LIN-tuhn), town (2001 population 1,459), Scottish Borders, SE Scotland, 10 mi/16 km NW of PEEBLES, on E flank of PENTLAND HILLS; 55°45′N 03°21′W. Resort. Formerly in Borders, abolished 1996.

Lintzford, ENGLAND: see WHICKHAM.

Linville (LIN-vil), unincorporated village, AVERY county, NW NORTH CAROLINA, 24 mi/39 km NNW of MORGANTON, and on LINVILLE RIVER at edge of Pisgah National Forest (N and E); 36°03′N 81°52′W. Manufacturing (gravel processing). Nearby are Linville Dam and Linville Falls (S) and Linville Caverns (12 mi/19 km SSW), with stalactite and stalagmite formations. BLUE RIDGE PARKWAY passes to SE.

Linville River (LIN-vil RI-vuhr), c.30 mi/48 km long, NW NORTH CAROLINA; rises in the BLUE RIDGE MOUNTAINS in central AVERY county; flows S, past LINVILLE, CATAWBA RIVER in Lake JAMES reservoir, forms NW arm extending 4 mi/6.4 km from dam; 36°07′N 81°50′W. Linville Falls, two cascades, and Linville Caverns, S of Linville. On upper course is Linville Dam (160 ft/49 m high, 1,326 ft/404 m long; for hydroelectric power; completed 1919).

Linwood, city (2000 population 374), LEAVENWORTH county, NE KANSAS, on KANSAS River, 11 mi/18 km ENE of LAWRENCE; 39°00′N 95°01′W. General agriculture.

Linwood, city (2006 population 7,354), ATLANTIC county, SE NEW JERSEY, 7 mi/11.3 km W of ATLANTIC CITY; 39°20′N 74°34′W. Primarily residential; some floriculture. Incorporated as borough 1889, as city 1931.

Linwood (LIN-wuhd), town (2001 population 9,058), Renfrewshire, W Scotland, adjacent to JOHNSTONE,

on BLACK CART WATER, and 3 mi/4.8 km W of PAISLEY; 55°51′N 04°29′W. Previously site of motor vehicle assembly plant. Formerly in Strathclyde, abolished 1996.

Linwood, town, WALKER county, NW GEORGIA, just NW of LA FAYETTE; 34°43′N 85°17′W.

Linwood, unincorporated town, DELAWARE county, SE PENNSYLVANIA, residential suburb 15 mi/24 km SW of downtown PHILADELPHIA, and 2 mi/3.2 km WSW of CHESTER, near DELAWARE state line; DELAWARE RIVER 1 mi/1.6 km to SE; 39°49′N 75°25′W. Manufacturing (sheet-metal fabrication).

Linwood (LIN-wud), unincorporated village, WATERLOO region, S ONTARIO, E central CANADA, 12 mi/19 km NW of KITCHENER, and included in WELLESLEY township; 43°34′N 80°43′W. Dairying, mixed farming.

Linwood, village (2006 population 114), BUTLER county, E NEBRASKA, 15 mi/24 km NE of DAVID CITY, near PLATTE RIVER; 41°24′N 96°55′W.

Linwood, MASSACHUSETTS: see NORTHBRIDGE.

Linwu (LIN-WU), town, ⊙ Linwu county, SE HUNAN province, CHINA, on BEI RIVER, and 45 mi/72 km SW of CHEN XIAN, near GUANGDONG border; 25°17′N 112°33′E. Agriculture (rice, medicinal herbs). Manufacturing (chemicals); nonferrous ore mining.

Linxi (LIN-SEE), town, ⊙ Linxi county, central INNER MONGOLIA AUTONOMOUS REGION, N CHINA, 95 mi/153 km NNW of CHIFENG; 43°31′N 118°02′E. Agriculture (grain, oilseeds, sugar beets). Manufacturing (food industry).

Linxia (LIN-SIAH), city (□ 34 sq mi/88.4 sq km; 2000 population 168,714), SE GANSU province, CHINA, 50 mi/80 km SW of LANZHOU; 35°36′N 103°05′E. The city is a gateway to the Tibetan areas near QINGHAI and SICHUAN borders. Agriculture is the largest source of income for the city; crop growing, animal husbandry. Manufacturing (food, leather, machinery). The name may appear as Lin-hsia or Linsia. Also called Hanjiaji.

Lin Xian (LIN SIAN), town, ⊙ Lin Xian county, N HENAN province, CHINA, 30 mi/48 km W of ANYANG, at E foot of TAIHANG Mountains; 36°01′N 113°51′E. Grain, cotton; iron-ore mining and iron smelting; building materials.

Lin Xian, town, ⊙ Lin Xian county, W SHANXI province, CHINA, 32 mi/51 km NNW of LISHI; 37°57′N 110°59′E. Grain, oilseeds; chemicals, furniture, food processing, iron-ore mining, coal mining.

Linxiang (LIN-SIANG), town, ⊙ Linxiang county, northeasternmost HUNAN province, CHINA, on GUANGZHOU-WUHAN railroad, and 20 mi/32 km ENE of YUEYANG, near HUBEI border; 29°29′N 113°27′E. Agriculture (tea, rice); fish. Manufacturing (food, chemicals, building materials); lead-zinc ore mining.

Linyanti (lin-YAHN-tee), village, S CAPRIVI REGION, NAMIBIA, 60 mi/97 km SSE of KATIMA MULILO, on the LINYANTI RIVER; 18°04′S 24°01′E. Tourist haven for game watching. Flooded during rainy season.

Linyanti River, ANGOLA, ZAMBIA, NAMIBIA, and BOTSWANA: see CUANDO RIVER.

Linyi (LIN-YEE), city (□ 675 sq mi/1,748 sq km; 1994 estimated urban population 520,800; estimated total population 1,704,300), NW SHANDONG province, CHINA, 35 mi/56 km N of JINAN; 35°10′N 118°18′E. Agriculture and light industry are the largest sectors of the city's economy. Heavy industry is also important. Agriculture (grains, fruits; cotton, oil crops; eggs, hogs, poultry). Manufacturing (food processing, beverages, textiles, chemicals, machinery); coal mining.

Linyi (LIN-YEE), town, ⊙ Linyi county, S SHANDONG province, CHINA, on Beng River, and 135 mi/217 km SE of JINAN; 37°12′N 116°54′E. Commercial and road center. Manufacturing (cotton textiles). Agriculture (peanuts, grain; oilseeds, tobacco). Has noted monasteries.

Linyi, town, ⊙ Linyin county, SW SHANXI province, CHINA, 35 mi/56 km NE of YONGJI near SHAANXI border; 35°11′N 110°47′E. Manufacturing (food processing, textiles, machinery and equipment manufacturing, chemicals). Agriculture (grain; cotton, oilseeds).

Linying (LIN-YING), town, ⊙ Linying county, central HENAN province, CHINA, on YING RIVER, on BEIJING-WUHAN railroad, and 15 mi/24 km NNW of YANCHENG; 33°48′N 113°59′E. Grain; oilseeds, tobacco, cotton.

Linyou (LIN-YO), town, ⊙ Linyou county, W SHAANXI province, CHINA, 45 mi/72 km NE of BAOJI, in mountain region; 34°41′N 107°48′E. Agriculture (grain, oilseeds). Manufacturing (food); coal mining.

Lin-yu, CHINA: see SHANHAIGUAN.

Linyu, CHINA: see SHANHAIGUAN.

Linyüan (LIN-YOOI-AN), town, S TAIWAN, near W coast, 8 mi/12.9 km SSE of FENGSHAN; 22°30′N 120°23′E. Sugarcane, pineapples.

Linz (LINTS), city, UPPER AUSTRIA, NW Austria, major port on DANUBE RIVER; 48°18′N 14°17′E. Commercial and industrial center and railroad junction. International airport to SW; third-largest city in Austria. See of a bishop. Manufacturing (iron and steel, machinery, shipyards, electrical equipment and electrical chemicals, metal products, glass, furniture, textiles). Two thermoelectric power stations. Originally a Roman settlement called Lentia, Linz was made provincial capital of Holy Roman Empire in late 15th century. City has numerous historic structures, including Romanesque church of St. Martin (8th century); baroque cathedral (18th century) where composer Anton Bruckner was an organist (1855–1868); city hall (17th century); baroque bishop's palace (1721–1726); new neo-Gothic cathedral (19th–20th century). The Provincial Museum in Linz contains paintings, folk art, and Roman artifacts. Annual Bruckner festival including arts and electronica, a festival of electronic music. Pilgrim church at Pöstlingberg. University Abbey of St. Florian 7 mi/11.3 km SE.

Linz am Rhein (LINTS ahm REIN), town, RHINELAND-PALATINATE, W GERMANY, on right bank of the RHINE (landing), and 14 mi/23 km SE of BONN; 50°34′N 07°17′E. Railroad junction; basalt quarrying and trade; tourism. Has late-Romanesque-Gothic church. Grapes for red wine growing in vicinity.

Linze (LIN-ZUH), town, ⊙ Linze county, N GANSU province, CHINA, on SILK ROAD to XINJIANG, 30 mi/48 km NW of ZHANGYE, and on HEI RIVER, at the GREAT WALL; 39°10′N 100°21′E. Grain, sugar beets, oilseeds.

Linzee, Cape (LIN-zee) westernmost point of CAPE BRETON ISLAND, NE NOVA SCOTIA, CANADA, on GULF OF ST. LAWRENCE, 3 mi/5 km NNW of PORT HOOD; 46°03′N 61°32′W.

Linzi (LIN-ZI), town, N central SHANDONG province, CHINA, 20 mi/32 km NNE of ZIBO, near QINGDAO-JINAN railroad; 36°55′N 118°20′E. Agriculture (grain; cotton, tobacco). Petroleum refining, electric-power generation.

Lion-d'Angers, Le (lee-on–dahn-zhe, luh), commune (□ 15 sq mi/39 sq km), MAINE-ET-LOIRE department, PAYS DE LA LOIRE region, W FRANCE, on OUDON RIVER, and 13 mi/21 km NW of ANGERS; 47°37′N 00°42′W. Horse-breeding center and site of annual international riding event. Has a partly Romanesque 11th–15th-century church.

Lionel Town, town, CLARENDON parish, S central JAMAICA, 9 mi/15 km S of MAY PEN; 17°48′N 77°14′W. Market town in sugarcane growing area. Sugar factory.

Lion, Gulf of (LEI-uhn), French *Golfe du Lion* (golf dyoo lee-on), large embayment of the MEDITERRANEAN SEA, off coast of S FRANCE, extending from the French-Spanish border to TOULON; 43°00′N 04°00′E. Its coastline includes many lagoons, saltworks, and inlets, as well as the RHÔNE delta. MARSEILLE is the chief port with a heavy waterfront industrial district extending W to the Rhône delta. Toulon and SÈTE are also important ports on the gulf.

Lion's Bay (LEI-uhnz), village (□ 1 sq mi/2.6 sq km; 2001 population 1,379), SW BRITISH COLUMBIA, W CANADA, 7 mi/10 km N of WEST VANCOUVER, and in GREATER VANCOUVER regional district; 49°27′N 123°14′W. Incorporated 1971.

Lion's Den, town, Mashonaland North province, N central ZIMBABWE, 14 mi/23 km NW of CHINHOYI, on Angwe River, on railroad; 17°16′S 30°01′E. Road junction. Copper mining. Agriculture (cattle, sheep, goats; cotton, corn, wheat, tobacco). CHINHOYI CAVES RECREATIONAL PARK to SE.

Lions Head (LEI-uhnz HED), former village (□ 1 sq mi/2.6 sq km; 2001 population 500), SW ONTARIO, E central CANADA, on SAUGEEN PENINSULA, on GEORGIAN BAY, 33 mi/53 km NNW of OWEN SOUND; 44°59′N 81°15′W. Dairying, mixed farming. Amalgamated into NORTHERN BRUCE PENINSULA municipality in 1999.

Lion-sur-Mer (lee-on–syur–mer), commune (□ 1 sq mi/2.6 sq km; 2004 population 2,508), beach resort in CALVADOS department, BASSE-NORMANDIE region, NW FRANCE, on the ENGLISH CHANNEL, 9 mi/14.5 km N of CAEN; 49°18′N 00°19′W. Its waterfront was the scene of Allied landing (June 1944) in NORMANDY during World War II.

Lionville (LAI-uhn-vil), unincorporated town, CHESTER county, SE PENNSYLVANIA, 9 mi/14.5 km SW of PHOENIXVILLE; 40°03′N 75°39′W. Manufacturing (sump pumps, electronic equipment). Agriculture (dairying, livestock, poultry; grain, apples). Marsh Creek State Park to W. Thomas Newcomer Library and Museum to E.

Liopesi (lee-o-PE-see), town, ATTICA prefecture, ATTICA department, E central GREECE, on railroad, and 7 mi/11.3 km E of ATHENS. Wine center; wheat, olive oil; livestock raising (sheep, goats). Formerly called Paiania.

Lioran, Le (lee-o-rahn, luh), resort village, CANTAL department, S central FRANCE, on N slope of Plomb du Cantal, in the MASSIF CENTRAL, 17 mi/27 km WNW of SAINT-FLOUR; winter sports at SUPERLIORAN (elevation 4,000 ft/1,219 m–6,000 ft/1,829 m) amid spruce forests. Road and railroad tunnels under Lioran Pass (4,270 ft/1,301 m) were built in mid-19th century to avoid snow on the passes in the AUVERGNE MOUNTAINS. Plomb du Cantal (6,120 ft/1,865 m) and Puy Griou (5,600 ft/1,707 m), both volcanic summits a few miles S of Le Lioran, provide panoramic views of the chain of CANTAL mountains that forms part of the volcanic AUVERGNE region.

Liozno (li-OZ-no), urban settlement, E VITEBSK oblast, BELARUS, ⊙ LIOZNO region, 28 mi/45 km SE of VITEBSK. Manufacturing (flax mill; food processing).

Lipa (LEE-pah), city (□ 81 sq mi/210.6 sq km; 2000 population 218,447), BATANGAS province, S LUZON, PHILIPPINES, 14 mi/23 km SW of SAN PABLO, near railroad; 13°57′N 121°10′E. Trade center for agricultural area (rice, sugarcane, citrus fruits, coffee, coconuts). Manufacturing (cutlery, handicrafts, apparel).

Lipanos, Bolsón de los (lee-PAH-nos, bol-SON dai los), arid depression in N outliers of SIERRA MADRE ORIENTAL of COAHUILA, N MEXICO, on CHIHUAHUA border, NE of Bolsón de MAPIMÍ. Elevation c.3,000 ft/914 m.

Lipany (li-PAH-ni), Hungarian *Héthárs*, town, VYCHODOSLOVENSKY province, NE SLOVAKIA, on railroad, on the TORYSA RIVER, and 16 mi/26 km NW of PRESOV; 49°09′N 20°58′E. Manufacturing (apparel). Has 14th-century church. Kamenicky Castle is 2 mi/3.2 km N and HANIGOVSKY is 3 mi/4.8 km NE.

Lipany, CZECH REPUBLIC: see CESKY BROD.

Lipari (LI-puh-ree, It. LEE-pah-ree), ancient *Lipara*, island (□ 14 sq mi/36.4 sq km), largest of LIPARI ISLANDS, in TYRRHENIAN SEA off NE SICILY, ITALY, between SALINA (NW) and VULCANO (S), 22 mi/35 km NW of MILAZZO, in Messina province; 6 mi/10 km long, 4 mi/6 km wide; 38°29′N 14°56′E. Rises to 1,978 ft/603 m in Monte Chirica (N). Pumice industry centered at Canneto, on E coast; quarries on Campo Bianco, a nearby mountain slope. Agriculture (capers, grapes, figs, olives); lobster fisheries. Chief port, Lipari (1991 population 10,382), on SE coast; commercial center of islands; manufacturing of pumice soap; exports pumice, currants, malmsey wine. Resort with sulphur baths. Bishopric since 1400; has Norman cathedral, restored 1654. Under the Roman Empire and during the Fascist regime in Italy (20th century) the island group served as a place of exile for political prisoners.

Lipari Islands (LIP-uh-ree), Italian *Isole Eolie*, volcanic island group (□ 44 sq mi/114.4 sq km), Messina province, NE SICILY, ITALY, in the TYRRHENIAN SEA. The group includes LIPARI (14.5 sq mi/37.6 sq km), an exporter of pumice and the site of Lipari, the group's main town; SALINA, where malmsey wine and currants are produced; VULCANO, the site in former times of the worship of the mythical fire god, with a high volcano that emits hot sulfurous vapors; STROMBOLI, with an active volcano (3,040 ft/927 m) that has several craters; Panarea, FILICUDI, ALICUDI; and 11 other islands. Fishing is an important occupation, and there is a growing tourist industry. The mythical residence of Aeolus, the wind god, for whom islands were formerly named, they were colonized by the Greeks in the 6th century B.C.E. Under the Roman Empire and during the Fascist regime in Italy (20th century) the island group served as a place of exile for political prisoners.

Lipcani (lip-KAHN), town, N MOLDOVA, on PRUT RIVER, on railroad, and 40 mi/64 km E of CHERNOVTSY, UKRAINE, on Romania-Ukraine border; 48°15′N 26°47′E. Agricultural center; flour milling; gypsum quarry. Until World War II, population largely Jewish. Formerly spelled Lipkany.

Lipen (LEE-pen), village, MONTANA oblast, MONTANA obshtina, BULGARIA; 43°24′N 23°25′E.

Lipetrén, Sierra (lee-pe-TRAIN, see-YER-rah), low pre-Andean range in S RÍO NEGRO province, ARGENTINA, S of LAKE CARRI LAUFQUÉN; extends to Chubut border in arid Patagonian plateau.

Lipetsk (LEE-pyetsk), oblast (□ 9,282 sq mi/24,133.2 sq km; 2006 population 1,199,540), in the center of European Russia and the E European Plain, in the basin of the upper DON River; ⊙ LIPETSK. Much of the area is part of the CENTRAL RUSSIAN UPLAND, with rolling plains dissected by ravines and gullies; elevation in W reaches 820 ft/250 m. Drained by the Don River and its tributaries, including the VORONEZH, SOSNA, and Krasnaya Mecha rivers. Moderate climate, with a three-month growing season; annual precipitation reaches 20 in/500 mm. Many karst caverns, seasonal small rivers, and karst springs. Mostly fertile blackearth (chernozem) soil, with some forest and mixedgrass steppe. Local fauna include rodents, foxes, and wolves. Because of its central location, the oblast is a major crossroads for the national road and railroad networks. Ferrous metallurgy in the area dates back to Peter the Great (late 17th century). Other industries include machines, building materials, appliances, foodstuffs, furniture, tractors, machine tools, consumer goods, chemicals, and medical equipment. Agriculture includes grain (wheat, rye, barley, oats, millet, buckwheat), sugar beets, sunflowers, potatoes, and fodder crops; also livestock raising. Some ore and limestone mined near Lipetsk. Major cities include Lipetsk, YELETS, DANKOV, and GRYAZI. Population mostly Russian. Formed in 1954. Consists of 8 cities, 5 towns, and 299 villages.

Lipetsk (LEE-pyetsk), city (2006 population 519,025), ⊙ LIPETSK oblast, S central European Russia, on the VORONEZH RIVER, on railroad, 316 mi/509 km SSE of MOSCOW; 52°37′N 39°34′E. Elevation 515 ft/156 m. Railroad junction and highway hub. It is the center of an iron ore-mining area and of ferrous metallurgy.

Industrial products include iron and steel, pipes, machinery, chemicals (nitrogen fertilizers, plastic goods), cement, handicrafts, food products. The city has mineral springs and has been a major health resort since 1803. It was founded in the 13th century, destroyed by the Tatars, and rebuilt (1707) by Peter the Great as a metallurgical center. Made a city in 1779.

Lípez, Cordillera de (kor-dee-YAI-rah dai LEE-pes), range in the ANDES Mountains, southernmost part of the Eastern Cordillera, in POTOSÍ department, SW BOLIVIA; extends c.160 mi/257 km SW from PORTUGALETE (at S end of Cordillera de CHICHAS) to Cerro ZAPALERI, where Bolivia, CHILE, and ARGENTINA meet; rises to 19,225 ft/5,860 m; 22°15′S 67°15′W. Forms watershed between Río GRANDE DE LÍPEZ (W) and SAN JUAN RIVER (E).

Liphook, ENGLAND: see BRAMSHOTT.

Lipiany (lee-PYAH-nee), German *Lippehne* (leep-ehne), town in BRANDENBURG, after 1945 in Szczecin province, NW POLAND, on small lake, 25 mi/40 km S of STARGARD SZCZECINSKI. Agricultural market (grain, sugar beets, potatoes; livestock). Manufacturing (food processing, furniture). After 1945, briefly called Lipiny.

Lipik (LEE-peek), town, E CROATIA, on PAKRA RIVER, on railroad, and 2 mi/3.2 km SW of PAKRAC, at W foot of PSUNJ Mountain, in SLAVONIA. Spa and health resort, with waters rich in iodine. Petroleum and natural gas in vicinity. Sustained heavy Serbian damage in 1991.

Lipin Bor (LEE-peen BOR), village (2006 population 3,745), W VOLOGDA oblast, N central European Russia, on the E shore of BELOYE OZERO (lake), on road, 32 mi/51 km NNW of KIRILLOV; 60°16′N 37°57′E. Elevation 413 ft/125 m. Lumbering.

Liping (LEE-PING), town, ⊙ Liping county, SE GUIZHOU province, CHINA, 150 mi/241 km ESE of GUIYANG, near HUNAN-GUANGXI border; 26°16′N 109°08′E. Grain, oilseeds, tobacco, timber.

Lipiny, POLAND: see LIPIANY.

Lipitsy (LEE-pee-tsi), village, S TULA oblast, central European Russia, on road, 30 mi/48 km ENE of MTSENSK; 53°21′N 37°16′E. Elevation 938 ft/285 m. In agricultural area (flax, grain, potatoes, sunflowers).

Lipiya (LEE-pee-yah), former town, SW NIZHEGOROD oblast, E central European Russia, near the OKA River, 7 mi/11 km ESE of MUROM. Shipyards nearby. Now incorporated into NAVASHINO.

Lipkany, MOLDOVA: see LIPCANI.

Lipki (LEEP-kee), city (2006 population 9,570), central TULA oblast, W central European Russia, on the UPA RIVER (tributary of the OKA River), near railroad, 25 mi/40 km S of TULA; 53°58′N 37°42′E. Elevation 738 ft/224 m. Coal mining (in the MOSCOW BASIN). Formerly known as Lipovskiy, or Lipkovskiy; current name and city status since 1955.

Lipkovskiy, RUSSIA: see LIPKI.

Lipljan, town and municipality (□ 163 sq mi/423.8 sq km; 2000 population 76,143), KOSOVO province, SW SERBIA, 10 mi/16 km SSW of PRIŠTINA; 42°31′N 21°07′E. On road and railroad; railroad station. Agriculture and mining. Widespread unemployment since mid-1990s.

Lipnik nad Becvou (LIP-nyeek NAHD BECH-vou), Czech *Lipník nad Bečvou*, German *Leipnik*, town, SEVEROMORAVSKY province, NE central MORAVIA, CZECH REPUBLIC, on BECVA RIVER, on railroad, and 15 mi/24 km ESE of OLOMOUC; 49°32′N 17°36′E. Manufacturing (machinery, consumer goods); cheese making. The town has a 13th century church, which had been rebuilt in 18th century, a 16th century castle, a Jewish cemetery, a synagogue, and a 17th century Renaissance belfry. Paleolithic-era archaeological site.

Lipno (LEEP-no), town, Włocławek province, central POLAND, on railroad, and 14 mi/23 km NNE of WŁOCŁAWEK. Cement manufacturing, tanning, flour milling. During World War II, German administrative headquarters, called Leipe.

Lipno Dam (LIP-no), JIHOCESKY province, S BOHEMIA, CZECH REPUBLIC, in Sumava Mountains, on VLTAVA RIVER, and 24 mi/39 km SW of ČESKÉ BUDEJOVICE. The largest Czech dam, it was built between 1950–1959. It impounds a reservoir (□ 18.8 sq mi/48.7 sq km, max. depth 72 ft/22 m); elevation 2,362 ft/720 m. Power plant; trout breeding; noted summer resort (water sports).

Lipnyaki, UKRAINE: see BARYSHIVKA.

Lipon (LIP-uhn), village, HOOD county, N central TEXAS, 43 mi/69 km WSW of FORT WORTH. Agricultural area (cattle; peanuts, pecans).

Lipova (lee-PO-vah), Hungarian *Lippa*, town, ARAD county, W ROMANIA, in BANAT, on left bank of MURES RIVER opposite RADNA, on railroad, and 20 mi/32 km ESE of ARAD. Health resort with mineral springs; also trading (notably in livestock) and processing center (brewing, flour milling). Manufacturing (carbonated waters, tiles, and bricks); granite quarrying. Has 13th-century castle remains and a fine Orthodox church built originally in Byzantine style (14th century) and repeatedly restored.

Lipova (LI-po-VAH), Czech *Lipová*, village, SEVEROMORAVSKY province, N BOHEMIA, CZECH REPUBLIC, on railroad, and 28 mi/45 km NE of ÚSTÍ NAD LABEM. Manufacturing (cutlery). Until 1947, the town was known in Czech as *Hanspach*, German *Hainspach*.

Lipova-lazne (LI-po-VAH–LAHZ-nye), Czech *Lipová-lázně*, German *Lindewiese-Bad*, village, SEVEROMORAVSKY province, W SILESIA, CZECH REPUBLIC, 3 mi/4.8 km W of JESENIK. Elevation 1,640 ft/500 m. Railroad junction; health resort in the JESENIKS.

Lipovaya Dolina, UKRAINE: see LYPOVA DOLYNA.

Lipovchik, RUSSIA: see KSHENSKIY.

Lipovets, UKRAINE: see LYPOVETS'.

Lipovskiy, RUSSIA: see LIPKI.

Lipovtsy (LEE-puhf-tsi), town (2006 population 6,530), SW MARITIME TERRITORY, SE RUSSIAN FAR EAST, on road and China-bound branch line connecting the TRANS-SIBERIAN RAILROAD to HARBIN, 26 mi/42 km NNW of USSURIYSK; 44°12′N 131°44′E. Elevation 475 ft/144 m. Railroad depots. In agricultural area (rice, soybeans, potatoes, vegetables). Formerly called Krasnopol'ye.

Lippa, ROMANIA: see LIPOVA.

Lippe (LIP-pe), former state, N central GERMANY, between the TEUTOBURG FOREST and the WESER RIVER; DETMOLD was capital. Incorporated 1947 into the state of North Rhine-Westphalia.

Lippe (LIP-pe), river, W GERMANY; rises in the TEUTOBURG FOREST, near BAD LIPPSPRINGE; flows W into the RHINE RIVER; 51°47′N 08°49′E. It is canalized to permit barge navigation. Water from the Lippe is used in the RUHR canal system.

Lippehne, POLAND: see LIPIANY.

Lippe Lateral Canal, German *Lippe-Seitenkanal* (LIP-pe SEI-tuhn-kah-nahl), 66 mi/106 km long, W GERMANY; a major transportation artery of the RUHR connecting RHINE RIVER and DORTMUND-EMS CANAL (at DATTELN); extends between WESEL (W) and HAMM (E); 51°38′N 06°23′E–51°39′N 07°20′E. Navigable for vessels up to 1,000 tons; 8 locks. Parallels lower course of LIPPE RIVER. Also called WESEL-DATTELN CANAL.

Lippe-Seitenkanal, GERMANY: see LIPPE LATERAL CANAL.

Lippetal (LIP-pe-tahl), town, North Rhine-Westphalia, W GERMANY, on LIPPE RIVER, 9 mi/14.5 km E of HAMM; 51°40′N 08°03′E.

Lippspringe, Bad, GERMANY: see BAD LIPPSPRINGE.

Lippstadt (LIP-shtaht), town, North Rhine-Westphalia, W GERMANY, on the LIPPE (head of navigation), 17 mi/27 km W of PADERBORN; 51°40′N 08°21′E. Railroad junction. Manufacturing (automobile parts; metalworking). Has 12th- and 13th-century churches, and a 17th-century palace. Founded 1170. Joined HANSEATIC LEAGUE in 1280. In World War II, U.S. troops met here (April 1945), completing encirclement of the RUHR.

Lipscomb (LIPS-com), county (□ 932 sq mi/2,423.2 sq km; 2006 population 3,114), extreme N TEXAS; ⊙ LIPSCOMB; 36°16′N 100°16′W. Elevation 2,500 ft/762 m–2,800 ft/853 m. In NE corner of the PANHANDLE. In W, high plains broken by deep valley of WOLF CREEK, also drained by Kiowa Creek in NW; in E, rolling hills. Grain (wheat, grain sorghum, milo, alfalfa); cattle. Oil and gas. Hunting for quail, wild turkey, deer. Formed 1876. Acquired part of ELLIS county, OKLAHOMA, in resurvey of 100th meridian (1930).

Lipscomb (LIPS-kuhm), city (2000 population 2,458), Jefferson co., N central Alabama, a suburb 8 mi/12.9 km SW of Birmingham. Settled c.1890, incorporated 1910. Named for a local family whose house was located at a frequently used streetcar stop.

Lipscomb (LIPS-com), unincorporated village, ⊙ LIPSCOMB county, extreme N TEXAS, on edge of high plains of the PANHANDLE, c.55 mi/89 km WSW of WOODWARD, OKLAHOMA, and on WOLF CREEK. Elevation 2,430 ft/741 m. In farm, livestock area.

Lipsoi (leep-SEE), Italian *Lisso*, island (□ 6 sq mi/15.6 sq km), DODECANESE prefecture, SOUTH AEGEAN department, GREECE, in the AEGEAN SEA, E of PÁTMOS; 4 mi/6.4 km long, 1 mi/1.6 km wide; 37°18′N 26°45′E. Rises to 900 ft/274 m. Wheat, figs, raisins; sponge fisheries. Also Leipsos and Lipsos.

Liptau, SLOVAKIA: see LIPTOV.

Liptó, SLOVAKIA: see LIPTOV.

Liptólúzsna, SLOVAKIA: see LIPTOVSKA LUZNA.

Lipton (LIP-tuhn), village (2006 population 342), SE central SASKATCHEWAN, CANADA, 45 mi/72 km NE of REGINA; 50°54′N 103°51′W. Grain elevators, lumbering, mixed farming.

Liptov (lip-TAWF), GERMAN *Liptau*, HUNGARIAN *Liptó*, historical region (□ 867 sq mi/2,254.2 sq km), STREDOSLOVENSKY and VYCHODOSLOVENSKY provinces, N SLOVAKIA, by POLISH border, in VÁH RIVER valley, and between HIGH and LOW TATRAS. Settled since Neolithic era; known as a region since early 14th century. Its original center was LIPTOVSKY HRADOK (castle) and from 1677, LIPTOVSKY MIKULÁŠ. Agriculture (barley, potatoes; sheep) and lumbering were the main economic activities until the late 19th century. Main centers include RUŽOMBEROK (paper and textiles), Liptovský Mikuláš (by LIPTOVSKA MARA DAM; footwear), and LIPTOVSKY HRADOK (machinery).

Liptovska Luzna (lip-TAWF-skah loozh-NAH), Slovak *Liptovská Lúžna*, Hungarian *Liptólúzsna*, village, STREDOSLOVENSKY province, central SLOVAKIA, in LOW TATRAS, and 14 mi/23 km S of RUŽOMBEROK; 46°13′N 15°23′E. Cattle breeding. Noted folk costumes and traditional log cabins. Health resort of KORYTNICA (ko-RIT-nyi-TSAH) (elevation 2,792 ft/851 m) is 4 mi/6.4 km SSW.

Liptovska Mara Dam (lip-TAWF-skah mah-RAH), SLOVAK *Liptovská Mara*, STREDOSLOVENSKY province, N Slovakia, on VÁH RIVER, and just WNW of LIPTOVSKÝ MIKULÁŠ. Impounds reservoir (□ 8.3 sq mi/21.5 sq km, maximum depth 141 ft/43 m). Two power plants (total output 198 MW); important for maintaining water level of river (because there are another fifteen power plants downstream). Summer resorts on N banks; boat transport.

Liptovsky Hradok (lip-TAWF-skee hrah-DOK), Slovak *Liptovský Hrádok*, Hungarian *Liptóújvár*, town, ZAPADOSLOVENSKY province, N Slovakia, on VÁH RIVER, on railroad, and 19 mi/31 km ESE of RUŽOMBEROK; 49°02′N 19°44′E. Manufacturing (machinery, food processing, woodworking); large sawmill. Ruins of castle; museum.

Liptovský Mikuláš (lip-TAWF-skee mi-KU-lahsh), city (2000 population 33,007), STREDOSLOVENSKY province, N SLOVAKIA, on right bank of VÁH RIVER, on railroad; 49°05′N 19°37′E. Manufacturing (machinery, footwear), tanning, food and wood processing. Has 13th-century church, restored in 18th century, and museum of SLOVAK KARST. Former cultural center of

Slovak Protestants; first popular demonstration for Slovak independence held here in 1848. Center of LIPTOV region since 17th-century military base. Until 1952, was called Liptovsky Sväty Mikulaš, Slovak *Liptovsky Sväty Mikuláš*, GERMAN *Liptau Sankt Niklas*, HUNGARIAN *Liptószentmiklós*.

Liptrap, Cape (LIP-trap), S VICTORIA, AUSTRALIA, in BASS STRAIT, just W of WARATAH BAY; 38°55′S 145°55′E. Its face is nearly perpendicular, rising to 297 ft/91 m; lighthouse.

Liptsy, UKRAINE: see LYPTSI.

Lipu (LEE-PU), town, ⊙ Lipu county, NE GUANGXI ZHUANG AUTONOMOUS REGION, CHINA, 60 mi/97 km ENE of LIUZHOU; 24°30′N 110°24′E. Agriculture (grain, jute). Manufacturing (food and beverages, papermaking).

Lipu La (LEE-PU-LAH), pass (c.17,000 ft/5,182 m) in SE ZASKAR RANGE of Kumaun HIMALAYAS, SW TIBET, CHINA, 9 mi/14.5 km WSW of TAKLAKOT, on pilgrimage route to MANASAROWAR LAKE; 30°14′N 81°02′E. Also called Lipu Lekh; sometimes written Lipulekh.

Lipu Lekh, CHINA: see LIPU LA.

Lipulekh, CHINA: see LIPU LA.

Lipumba (lee-POOM-bah), village, RUVUMA region, S TANZANIA, 40 mi/64 km WSW of SONGEA; 10°50′S 35°05′E. Timber; tobacco, corn; goats, sheep.

Liquan (LEE-QUAN), town, ⊙ Liquan county, S central SHAANXI province, CHINA, 30 mi/48 km NW of XI'AN. Grain, oilseeds.

Lira (LEE-ruh), former administrative district (□ 2,800 sq mi/7,280 sq km; 2005 population 832,600), NORTHERN region, central UGANDA, NE of LAKE KYOGA; capital was LIRA; 02°05′N 33°10′E. Average elevation 3,839 ft/1,170 m. As of Uganda's division into fifty-six districts, was bordered by PADER (N), KOTIDO (NE), MOROTO, KATAKWI, and KABERAMAIDO (E), KAMULI, KAYUNGA, and NAKASONGOLA (S, on opposite shore of Lake Kyoga), and APAC (W) districts. The native inhabitants and primary ethnic group were the Langi people. Vegetation was primarily savannah, with some tropical savannah woodlands; there were also wetlands. Highest point was Otuke Peak (5,214 ft/1,589 m). LAKE KWANIA in SW (on border with Apac). An agricultural region (large producer of oil seeds [including shea butter, simsim, and sunflower]; also millet, cassava, beans, maize, sorghum, sweet potatoes); cattle raising was important (also poultry, pigs, and sheep). Lira was only large town. In 2005 S portion of district was carved out to form AMOLATAR district; in 2006 central portion was carved out to form DOKOLO district and N portion was formed into current LIRA district.

Lira, administrative district, NORTHERN region, central UGANDA; ⊙ LIRA. As of Uganda's division into eighty districts, borders PADER (N), ABIM (NE), MOROTO and AMURIA (E), KABERAMAIDO (SE), DOKOLO (S), APAC (W), and OYAM and GULU (NW) districts. The Langi people live here. Agricultural area. Secondary railroad travels through Lira town, running NW to GULU town and SE to join main railroad between KASESE town (W Uganda) and MOMBASA (SE KENYA). Several roads branch out of Lira town, including a main road traveling SE to TORORO town. Formed in 2005–2006 from N portion of former LIRA district (AMOLATAR district formed from S portion in 2005 and Dokolo district from central portion in 2006).

Lira (LEE-ruh), town (2002 population 80,879), ⊙ LIRA district, NORTHERN region, central UGANDA, 50 mi/80 km SE of GULU; 03°34′N 31°51′E. Agricultural trade center (cotton, peanuts, sesame). Inhabited by Nilotic Lango people. Was capital of former NORTHERN province.

Liranga (lee-rahng-GAH), village, CUVETTE region, E central Congo Republic, on CONGO RIVER (CONGO border), just below its confluence with UBANGI RIVER, and 60 mi/97 km ESE of OWANDO; 00°39′S 17°36′E. Banana plantations, palm groves.

Lircay (leer-KEI), city, ⊙ ANGARAES province, HUANCAVELICA region, S central PERU, in CORDILLERA OCCIDENTAL of the ANDES, 26 mi/42 km ESE of HUANCAVELICA by highway; 13°00′S 74°43′W. Elevation 10,732 ft/3,271 m. Agriculture (wheat, corn, alfalfa; livestock). Manufacturing.

Liré (lee-rai), commune (□ 12 sq mi/31.2 sq km), MAINE-ET-LOIRE department, PAYS DE LA LOIRE region, W FRANCE, near left bank of the lower LOIRE RIVER, 18 mi/29 km NE of NANTES; 47°21′N 01°10′W. Wine grape vineyards. Birthplace of Joachim du Bellay, a French classical poet of 16th century, who founded a literary group called La Pléiade.

Liri (LEE-ree), river, 98 mi/158 km long, LATIUM, central ITALY; rising in the APENNINES, flowing generally SE to the TYRRHENIAN Sea. Below its junction with the RAPIDO River near CASSINO it is called the GARIGLIANO. There are hydroelectric stations along its course. In World War II the area around the river was the scene of heavy fighting between Allied and German troops.

Liria (LEE-ryah) or **Llíria**, town, VALENCIA province, E SPAIN, 17 mi/27 km NW of VALENCIA; 39°38′N 00°36′W. Manufacturing (flour milling, wine production, furniture, construction materials); sawmilling. Cereals. Has 14th-century church and Renaissance palace of duke of Berwick, who was invested (18th century) with town by Philip V.

Li River (LEE), Mandarin *Li Shui* (LEE SHUAI), river, 250 mi/402 km long, NW HUNAN province, CHINA; rises on HUBEI border; flows SE and E, past CILI, LI XIAN, and JINSHI, to DONGTING Lake near ANXIANG.

Liro River (LEE-ro), 15 mi/24 km long, SONDRIO province, N ITALY; rises in small lake 1 mi/1.6 km W of SPLÜGEN Pass; flows S to MERA RIVER 1 mi/1.6 km SW of CHIAVENNA. Dammed 1 mi/1.6 km S of pass to form reservoir (1 mi/1.6 km long), which supplies large Mese hydroelectric plant near Chiavenna.

Lirquén (leer-KEN), town, CONCEPCIÓN province, BÍO-BÍO region, S central CHILE, on CONCEPCIÓN BAY, on railroad, and 8 mi/13 km NNE of CONCEPCIÓN; 36°43′S 72°58′W. Industrial zone.

Lisac (lee-SHAHK), mountain (6,344 ft/1,934 m), MACEDONIA, 14 mi/23 km WNW of TITOV VELES. Also spelled Lisats.

Lisafa (lee-SAH-fuh), village, Équateur province, NW CONGO, on navigable route to BOENDE, and 85 mi/137 km E of MBANDAKA; 01°07′N 19°45′E. Elev. 1,253 ft/381 m. Fishing, oil palms.

Lisakovsk (lee-sah-KOVSK), city, KOSTANAI region, KAZAKHSTAN, 66 mi/110 km SW of KOSTANAI; 52°32′N 62°32′E. Gold mining. Construction materials; food processing.

Lisala (lee-SAH-lah), town, Équateur province, NW CONGO, on right bank of CONGO RIVER, and 270 mi/435 km NE of MBANDAKA; 02°09′N 21°31′E. Elev. 1,345 ft/409 m. Commercial center, notably for produce (copal, cotton, palm kernels, raphia); palm-oil milling. Has medical and accounting schools, hospital, Roman Catholic missions, and airfield. Seat of vicar apostolic is 3 mi/4.8 km SW. UPOTO Baptist mission is 3 mi/4.8 km SW. UMANGI Roman Catholic mission is 12 mi/19 km W. Birthplace of late Pres. Mobutu Sese Seku.

Lisa Mountains (LEE-sah), E central BULGARIA, N spur of the E STARA PLANINA Mountains; extend N from the KOTEL MOUNTAINS; rise to over 3,000 ft/914 m; 43°02′N 26°20′E. Forested. Form a watershed between the GOLYAMA KAMCHIYA and YANTRA rivers.

Lisan, Al, JORDAN: see LISAN, EL.

Lisan, El (lee-SAN, el), peninsula on DEAD SEA, S central JORDAN, WNW of AL KARAK; 6 mi/9.7 km wide, 10 mi/16 km long. Airfield; large Catholic monastery with vineyards famous for wine production. Ruins of important Crusader fort nearby. The small Arab village near the monastery was totally destroyed and abandoned during the 1967 Six-Day War. The new highway

to JERUSALEM runs in the vicinity. Agriculture, to a small extent, is practiced only on the E fringes of the peninsula. The peninsula is now limited to its narrow N part. The fall in the level of the Dead Sea by over 34 ft/10 m during the last 30 years resulted in the drying up of the shallow S basin of the sea and of the strait that separated the Lisan from the W coast of the sea. Also spelled Al-Lisan.

Lisats, MACEDONIA: see LISAC.

Lisboa (leezh-BO-ah), district (□ 1,061 sq mi/2,758.6 sq km; 2001 population 2,136,013), W central PORTUGAL, in ESTREMADURA and W part of RIBATEJO provinces; ⊙ LISBON; 39°00′N 09°08′W. Bounded by the ATLANTIC OCEAN (W and S), by lower TAGUS RIVER and its estuary (SE). Contains CABO DA ROCA, westernmost point of continental Europe. Agriculture, fishing. Tourist centers include Lisbon, SINTRA, and the seaside resorts of ESTORIL and CASCAIS.

Lisbon (LIZ-bin), Portuguese *Lisboa* (leezh-BO-ah) ancient *Olisipo* (oo-LIS-ah-poo), city (2007 population 564,657), ⊙ PORTUGAL, ESTREMADURA province, and LISBOA district, W PORTUGAL, on the TAGUS R. where it broadens to enter the ATLANTIC OCEAN, and set on seven terraced hills; 38°43′N 09°08′W. Portugal's largest city and its cultural, administrative, commercial, and industrial hub. International airport to N. Has one of the best harbors in Europe, handling a large trade. Agricultural and forest products and fish are exported. The city's industries include the production of textiles, machinery, chemicals, electronics, cement, and steel; oil and sugar refining; food processing; and shipbuilding. A large transient and tourist trade is drawn here.

The Castelo de São Jorge, a fort that dominates the city, may have been built by the Romans on the site of the citadel of the early inhabitants, who traded with Phoenician and Carthaginian navigators. The Romans occupied the town in 205 B.C.E., and it was conquered by the Moors in C.E. 714. The city's true importance dates, however, from 1147, when King Alfonso I, with the help of Crusaders, drove out the Moors. Alfonso III transferred (c.1260) his court here from Coimbra, and the city rose to great prosperity in the 16th century with the establishment of Portugal's empire in Africa and India. Although many of the old buildings were destroyed by earthquakes, particularly the disastrous earthquake of 1755, some of the medieval ones remain. The old quarter, the picturesque and crowded Alfama, surrounds the 12th-century cathedral (rebuilt later). The new quarter, built by the Marqués de Pombal after the great earthquake, centers about a large square, the Terreiro do Paço. Some of the well-known buildings in and near Lisbon are the Renaissance Monastery of São Vicente de Fora, with the tombs of the Braganza kings; the Church of St. Roque, with the fine Chapel of St. John (built by John V in the 18th century); and the magnificent monastery at Belém, on the N bank of the Tagus facing the sea, built by Manuel I to commemorate the discovery of the route to India by Vasco da Gama. Camões was born here. The University of Lisbon (originally founded 1292, but transferred to Coimbra in 1537), was reestablished here in 1911. In 1966 the Ponte 25 de Abril (known as Salazar Bridge until 1975), one of the world's longest (3,323 ft/1,013 m) suspension bridges, was completed across the Tagus, linking Lisbon with the Setubal Peninsula.

Lisbon (LIZ-bin), town, NEW LONDON county, SE CONNECTICUT, on QUINEBAUG RIVER, and 7 mi/11 km NE of NORWICH; 41°36′N 72°00′W. Garbage incinerator plant, capable of burning 500 tons/551 metric tons of trash per day to generate electricity. Also an agricultural center. Incorporated 1786.

Lisbon, town (2000 population 1,898), LINN county, E IOWA, near CEDAR RIVER, 15 mi/24 km ESE of CEDAR RAPIDS; 41°55′N 91°23′W. Manufacturing (feed; chiropractic equipment). Limestone quarries nearby.

Lisbon (LIZ-buhn), town, ANDROSCOGGIN county, SW MAINE, on the ANDROSCOGGIN, and 7 mi/11.3 km SE of LEWISTON; 44°01′N 70°05′W. Light manufacturing at Lisbon Falls (1990 population 4,674) and Lisbon Center villages. In Bowdoin until incorporated 1799.

Lisbon, town, GRAFTON county, NW NEW HAMPSHIRE, 10 mi/16 km SSW of LITTLETON; 44°14′N 71°52′W. Drained by AMMONOOSUC RIVER. Manufacturing (women's shoes, wire products, furniture, maple products); timber. Agriculture (vegetables, nursery crops, sugar maples; poultry, cattle; dairying). Iron, gold, other minerals formerly mined. Settled 1753, was named Concord until 1824.

Lisbon, town (2006 population 2,197), ⊙ RANSOM county, SE NORTH DAKOTA, 52 mi/84 km SW of FARGO, and on SHEYENNE RIVER; 46°26′N 97°40′W. Agriculture, dairy products. Site of state soldiers' home. Settled 1878, incorporated 1883. Fort Ransom State Park and Historic Site to NW. Fish hatchery and sunflower products. Founded in 1880.

Lisbon, village (2000 population 227), KENDALL county, NE ILLINOIS, 20 mi/32 km SSW of AURORA; 41°28′N 88°28′W. In rich agricultural area.

Lisbon (LIZ-buhn), village (2000 population 162), CLAIBORNE parish, N LOUISIANA, 43 mi/69 km WNW of MONROE; 32°48′N 92°52′W. Lake Claiborne State Park to SW. In agricultural area; oil and natural gas; gasoline and diesel fuel processing. Two small units of Kisatchie National Forest to NE and NW.

Lisbon, village, on KENT/OTTAWA county line, SW MICHIGAN, 14 mi/23 km NNW of GRAND RAPIDS. In farm area.

Lisbon (LIZ-buhn), village (□ 1 sq mi/2.6 sq km; 2006 population 3,063), ⊙ COLUMBIANA county, E OHIO, 22 mi/35 km SSW of YOUNGSTOWN; 40°46′N 80°46′W. In coal, clay, and limestone area. Founded as New Lisbon 1802.

Lisbon Bay (LIZ-bin), Portuguese *Mar da Palha*, upper section of TAGUS RIVER estuary, LISBOA and SETÚBAL districts, W central PORTUGAL, forming a lake (7 mi/11.3 km wide, 12 mi/19 km long) just above city of LISBON. Its E shore is low and marshy, while the higher W shore is lined with Lisbon's N suburbs and part of its harbor installation. The bay is linked with the ATLANTIC OCEAN by a channel (c.2 mi/3.2 km wide) extending 8 mi/12.9 km W from Lisbon.

Lisbon Falls, MAINE: see LISBON.

Lisburn (LIZ-buhrn), Gaelic *Lios na gCearrbhach*, town (2001 population 71,465), ANTRIM, NE Northern Ireland, on the LAGAN RIVER; 54°31′N 06°04′W. The chief industry was historically linen manufacturing, introduced by the Huguenots after the revocation of the Edict of Nantes (1685). The Lambeg Industrial Research Association, a major fiber research laboratory, is here. Other manufacturing includes motor vehicle parts and sheet metal. Major retail and outlet shopping center. Lisburn is the seat of the Roman Catholic bishop of Down and Connor and of the Protestant bishop of Connor. A technical school is located in the former home of Sir William Wallace of the Wallace Collection in LONDON.

Lisburne, Cape (LIZ-buhrn), NW ALASKA, on CHUKCHI SEA, 40 mi/64 km NNE of Point HOPE; 68°53′N 166°04′W. Here are large rookeries. Coal deposits nearby.

Liscannor (lis-KA-nuhr), Gaelic *Lios Ceanúir*, village, W Clare co., Ireland, on Liscannor Bay (4 mi/6.4 km wide, 5 mi/8 km deep) of the Atlantic, 4 mi/6.4 km W of Ennistymon; 52°56′N 09°23′W. Fishing port. Has remains of anc. castle of the O'Briens. John P. Holland, inventor of the submarine, was b. here. The 668 ft/204 m sheer Cliffs of Moher are 3 mi/4.8 km NW.

Lischanna, Piz, SWITZERLAND: see PIZ LISCHANNA.

Lischau, CZECH REPUBLIC: see LISOV.

Liscomb, town (2000 population 272), MARSHALL county, central IOWA, near IOWA RIVER, 11 mi/18 km NNW of MARSHALLTOWN; 42°11′N 93°00′W. Feed milling.

Lisdoonvarna (lis-DOON-VAHR-nuh), Gaelic *Lios Dúin Bhearna*, spa town (2006 population 767), CLARE county, W IRELAND, 8 mi/12.9 km N of ENNISTYMON; 53°01′N 09°17′W. Sulfur and chalybeate springs; adjoins The Burren, a limestone plateau rich in geological interest. Noted for annual gathering of farmers seeking wives.

Lisen (LEE-shen-yuh), Czech *Líšeň*, German *Lösch*, E district of BRNO, JIHOMORAVSKY province, S MORAVIA, CZECH REPUBLIC; 49°12′N 16°42′E. Manufacturing (tractors).

Lisets, UKRAINE: see LYSETS'.

Lishan (LEE-SHAN), town, N HUBEI province, CHINA, 100 mi/161 km NW of WUHAN; 31°53′N 113°19′E. Agriculture (rice; cotton, oilseeds, tobacco).

Lishan (LEE-SHAN) or **Pear Mountain**, TAIWAN. Fertile plateau (8,530 ft/2,600 m) in high CENTRAL MOUNTAIN RANGE. CENTRAL CROSS-ISLAND HIGHWAY (E-W) here, constructed (1950s) by retired Chinese servicemen, many of whom returned as fruit growers. Area known for its orchids; pears, apples, peaches, plums, cherries, and tropical fruits are also grown. Highway branch from N connects Lishan with NORTHERN CROSS-ISLAND HIGHWAY at Chilan.

Lishi (LEE-SHI), town, ⊙ Lishi county, W SHANXI province, CHINA, 45 mi/72 km WNW of FENYANG, and on main road to HUANG HE (Yellow River); 37°31′N 111°08′E. Agriculture (wheat). Manufacturing (machinery and equipment); coal mining, lumbering.

Lishu (LEE-SHU), town, ⊙ Lishu county, SW JILIN province, NORTHEAST, CHINA, 10 mi/16 km N of SIPING; 43°18′N 124°19′E. Agriculture (soybeans, kaoliang, millet, wheat, sugar beets). An old NE town, it was largely eclipsed by the newer railroad city of Siping.

Lishui (LEE-SHUAI), city (□ 580 sq mi/1,508 sq km; 2000 population 323,933), S central ZHEJIANG province, CHINA, 55 mi/89 km NW of WENZHOU, and on headstream of the WU RIVER; 28°30′N 119°59′E. Agriculture (grains, fruits, hogs). Manufacturing (textiles, apparel, food processing, chemicals, machinery).

Lishui (LEE-SHUAI), town, ⊙ Lishui county, S JIANGSU province, CHINA, 35 mi/56 km SSE of NANJING, and on Qinhua River; 31°38′N 119°02′E. Rice; oilseeds, jute, cotton.

Li Shui, CHINA: see LI RIVER.

Lisianski Inlet (li-see-YAN-skee), SE ALASKA, long narrow fjord on NW coast of CHICHAGOF ISLAND; 25 mi/40 km long, opening into CROSS SOUND at 57°06′N 136°28′W.

Lisianski Island, N PACIFIC, part of HONOLULU county, HAWAII, c. 1,080 mi/1,738 km NW of HONOLULU, 170 mi/274 km N of TROPIC OF CANCER; 26°04′N 173°58′W. Level; coral and coral sand. Annexed 1857 by Hawaiian kingdom. Part of Hawaiian Islands National Wildlife Refuge. Sometimes written Lisyanski.

Lisichansk, UKRAINE: see LYSYCHANS'K.

Lisieux (lee-syu), town (□ 4 sq mi/10.4 sq km), subprefecture of CALVADOS department, BASSE-NORMANDIE region, NW FRANCE, 27 mi/43 km E of CAEN; 49°09′N 00°14′E. It is one of the oldest towns in NORMANDY. Its modern importance dates from the canonization (1925) of St. Theresa, whose shrine there (a basilica built 1929–1975) attracts many pilgrims. Lisieux has some small industries (metalworking, food processing, and woodworking). Most of the town was destroyed in World War II, but the Cathedral of Saint-Pierre survived.

Lisimachia, Lake, Greece: see LYSIMACHIA, LAKE.

Lisina Mountain (LEE-shee-nah), in DINARIC ALPS, central BOSNIA, BOSNIA AND HERZEGOVINA. Highest point, LISINA (4,812 ft/1,467 m) is 9 mi/14.5 km SW of Mrkonjić GRAD.

Lisiy Nos (LEE-seeyee–NOS) [Russian=Fox Cape], town (2005 population 2,490), W LENINGRAD oblast, NW European Russia, on the Gulf of FINLAND, opposite KRONSHTADT, 12 mi/19 km NW of SAINT PETERSBURG, to which it is administratively subordinate; 60°01′N 30°01′E. Vacation homes; amusement park. Formerly known as Vladimirovka.

Liska (LEESH-kah), mountain (6,262 ft/1,909 m) in MACEDONIA, 12 mi/19 km S of Kicevo; 41°19′N 20°58′E.

Liskamm, (Lis-kam), peak (14,852 ft/4,527 m), in the Southern ALPS, border of ITALY and SWITZERLAND, highest peak in LISKAMM GROUP; 45°55′N 07°50′E.

Liskamm Group (Lis-kam Groop), mountain (14,852 ft/4,527 m), in the Southern ALPS; 45°55′N 07°50′E. Liskamm Group's notable peaks include LISKAMM (14,852 ft/4,527 m), CASTOR (13,871 ft/4,228 m), BREITHORN (13,661 ft/4,164m), and POLLUX (13,425 ft/4,092).

Liskeard (lis-KAHRD), town (2001 population 8,656), E CORNWALL, SW ENGLAND, between Looe and Seaton rivers, 11 mi/18 km E of BODMIN; 50°27′N 04°27′W. Agricultural market town; former mining center. Has 15th-century church. Just S is Well of St. Keyne, celebrated in poem by Southey.

Liskhimbud, UKRAINE: see LYSYCHANS'K.

Liskhimstroy, UKRAINE: see LYSYCHANS'K.

Liski (LEE-skee), city (2006 population 55,975), central VORONEZH oblast, S central European Russia, port on the left bank of the DON River, on railroad and highway, 60 mi/97 km SSE of VORONEZH; 50°58′N 39°29′E. Elevation 374 ft/113 m. Railroad junction (repair shops); mechanical and assembly shops; meat packing; flour milling, sugar refining. Called Svoboda, 1918–1943. Became city in 1937. Liski village, across the Don River, was absorbed and gave the new city its name in 1943. Names of earlier settlements: Novaya Pokrovka (Bobrovsk), Petropavlovskoye, Zaluzhnoye. Called Georgiu Dezh, 1965–1991.

Lisko, POLAND: see LESKO.

Lisle, village (2000 population 21,182), DU PAGE county, NE ILLINOIS, residential suburb 23 mi/37 km W of downtown CHICAGO; 41°47′N 88°05′W. East DU PAGE RIVER runs through village. Remnant agriculture (oats, corn). Diverse manufacturing. Illinois Benedictine College here. Morton Arboretum to NE.

Lisle, village (2006 population 288), Broome county, S NEW YORK, on TIOUGHNIOGA RIVER, 18 mi/29 km NNW of BINGHAMTON; 42°21′N 76°00′W. In dairying area. Incorporated 1876.

L'Isle, FRANCE: see LILLE.

L'Isle, FRANCE: see LILLE.

L'Isle-aux-Allumettes (LEEL–o–ah-lyoo-MET), former municipality (□ 34 sq mi/88.4 sq km; 2006 population 552), S QUEBEC, E CANADA. Amalgamated into L'Isle-aux-Allumettes in 1998.

L'Isle-aux-Allumettes (LEEL–o–zah-lyoo-MET), village (□ 73 sq mi/189.8 sq km; 2006 population 1,400), OUTAOUAIS region, S QUEBEC, E CANADA; 45°55′N 77°04′W. Restructured in 1998.

L'Isle-aux-Allumettes-Partie-Est (leel–o–zah-lyoo-MET–pahr-tee–EST), former municipality (□ 38 sq mi/98.8 sq km; 2001 population 422), S QUEBEC, E CANADA. Amalgamated into L'Isle-aux-Allumettes in 1998.

L'Isle-aux-Coudres (leel-o-KOO-druh), village (□ 12 sq mi/31.2 sq km; 2006 population 1,333), CAPITALE-NATIONALE region, SE QUEBEC, E CANADA, at S end of Île aux COUDRES; 47°23′N 70°25′W.

Lisle-sur-Tarn (leel–syur–tahrn), commune (□ 33 sq mi/85.8 sq km), TARN department, MIDI-PYRÉNÉES region, S FRANCE, on TARN RIVER, and 17 mi/27 km WSW of ALBI; 43°51′N 01°48′E. Wine trade.

L'Islet (lee-LAI), county (□ 808 sq mi/2,100.8 sq km; 2006 population 19,259), CHAUDIÈRE-APPALACHES region, S QUEBEC, E CANADA, on the SAINT LAWRENCE and on U.S. (MAINE) border; ⊙ SAINT-JEAN-PORT-JOLI; 47°00′N 70°00′W. Composed of fourteen municipalities. Formed in 1982.

L'Islet (lee-LAI) or **Bon Secours** (bon suh-KOOR), town (□ 46 sq mi/119.6 sq km), CHAUDIÈRE-APPAL-

ACHES region, SE QUEBEC, E CANADA, on the SAINT LAWRENCE RIVER, and 14 mi/23 km NE of MONTMAGNY; 47°06′N 70°21′W. Lumbering, woodworking, dairying, metal casting.

L'Isle-Verte (leel–VERT), village (□ 43 sq mi/111.8 sq km; 2006 population 1,498), BAS-SAINT-LAURENT region, S QUEBEC, E CANADA, on the SAINT LAWRENCE RIVER and 16 mi/26 km NE of RIVIÈRE-DU-LOUP; 48°01′N 69°20′W. Agriculture (dairying, potatoes), lumbering, peat moss manufacturing. Opposite is Île VERTE.

Lislique (lees-LEE-kai), municipality and town, LA UNIÓN department, EL SALVADOR, N of SANTA ROSA DE LIMA.

Lisman (LIS-muhn), town (2000 population 653), Choctaw co., SW Alabama, near Mississippi state line, 7 mi/11.3 km NW of Butler; 32°10′N 88°17′W. Lumber. Named for a supporter of the Tuscaloosa Belt Line RR. Inc. in 1971.

Lismore (LIZ-mor), city (2001 population 27,358), N NEW SOUTH WALES, E AUSTRALIA, 482 mi/776 km NE of SYDNEY, on the North Arm of the WILSONS RIVER; 28°48′S 153°16′E. An important industrial city, Lismore is a leading producer of butter. Dairying; tropical fruits. Tourism. Higher education. Its port is Ballina.

Lismore (liz-MOR), Gaelic *Lios Mór*, town (2006 population 790), WATERFORD county, S IRELAND, on the BLACKWATER RIVER, and 18 mi/29 km SW of CLONMEL; 52°08′N 07°55′W. It is a market town with a salmon-fishing industry. St. Carthagh founded a monastery here in the 7th century. It was a famed center of learning by the 8th century. Lismore Castle, which has been restored, was built under King John in 1185. Robert Boyle born here.

Lismore (LIZ-mor), village (2000 population 238), NOBLES county, SW MINNESOTA, 21 mi/34 km NW of WORTHINGTON; 43°45′N 95°57′W. Grain, soybeans; livestock; dairying.

Lismore (LIZ-mor), island (9.5 mi/15.3 km long, 1.5 mi/2.4 km wide), Argyll and Bute, W Scotland, in LOCH LINNHE; 56°30′N 05°33′W. Fertile soils. There are ruins of several old castles, one of which was a 9th-century Viking fortress, another the residence of the bishops of Argyll. The 16th-century Book of the Dean of Lismore is a volume of Scottish Gaelic and other verse, compiled by Dean James MacGregor and his brother Duncan. It is one of the oldest Scottish Gaelic collections. Formerly in Strathclyde, abolished 1996.

Lisnaskea (lis-nuh-SKEE), Gaelic *Lios na Scéithe*, village (2001 population 2,739), S FERMANAGH, SW Northern Ireland, near Upper Lough Erne, 10 mi/16 km SE of ENNISKILLEN; 54°15′N 07°27′W. Market. Originally inauguration place of The Maguire, it has ruins of a planter castle.

Lisnyaky, UKRAINE: see YAHOTYN.

Lisostepova fizyko-heohrafichna zona Ukrayiny, UKRAINE: see FOREST-STEPPE ZONE, UKRAINIAN.

Lisov (LI-shof), Czech *Lišov*, German *Lischau*, town, JIHOCESKY province, S BOHEMIA, CZECH REPUBLIC, 6 mi/9.7 km NE of ČESKÉ BUDEJOVICE; 49°01′N 14°36′E. Agriculture (barley, oats, rye); manufacturing (furniture, bricks).

Lisovka, UKRAINE: see UKRAYINS'K.

Lispeszentadorján (LISH-pe-sant-ah-dor-yahn), village, ZALA county, W HUNGARY, 15 mi/24 km WNW of NAGYKANIZSA; 46°32′N 16°42′E. Extraction of petroleum and natural gas nearby begun 1937. Gas pipeline to BUDAPEST completed 1949. Includes former hamlet of Lispe. Gas reserves exhausted, oil reserves near exhaustion.

Liss (LIS), village (2001 population 6,051), E HAMPSHIRE, S ENGLAND, 4 mi/6.4 km NNE of PETERSFIELD; 51°02′N 00°54′W. Has 14th-century church.

Lissa, CROATIA: see VIS.

Lissa, POLAND: see LESZNO.

Lissa an der Elbe, CZECH REPUBLIC: see LYSA NAD LABEM.

Lisse (LI-suh), town (2001 population 22,401), SOUTH HOLLAND province, W NETHERLANDS, 9 mi/14.5 km S of HAARLEM, on the RINGVAART canal; 52°16′N 04°33′E. NORTH SEA 5 mi/8 km to W. Agriculture (flower-bulb–growing center; dairying; cattle; vegetables, fruit, nursery stock). Manufacturing (prepacked flowers, dairy products). Site of well-known Keukenhof Flower Park.

Lisses (LEES), town (□ 4 sq mi/10.4 sq km), Essonne department, ÎLE-DE-FRANCE region, N central FRANCE. 2 mi/4 km W of CORBEIL-ESSONNES; 48°36′N 02°25′E.

Lissone (lees-SO-ne), town, MILANO province, LOMBARDY, N ITALY, 2 mi/3 km NW of MONZA; 45°37′N 09°14′E. Manufacturing (fabricated metal, machinery, wood products, plastics; furniture, cosmetics, sausage); cotton mills, glass factories, foundry.

Lissoy, IRELAND: see AUBURN.

Lissus, ALBANIA: see LEZHË.

List (LIST), village, SCHLESWIG-HOLSTEIN, NW GERMANY, on SYLT island, 9 mi/14.5 km NNE of WESTERLAND; 55°01′N 08°21′E. NORTH SEA resort. Formerly noted for its oyster culture. Was site of two schools for fliers until 1945.

Lista (LIS-tah), peninsula (□ c.54 sq mi/140 sq km) in VEST-AGDER county, S NORWAY, on the NORTH SEA, and flanked by two fjords; c.12 mi/19 km long, 10 mi/16 km wide. Flat, partly forested area. Many farmers here are repatriated American immigrants. Archaeological findings from Stone-Age period.

Lister, GERMANY: see OLPE.

Listerlandet (LIS-ter-lahnd-et), peninsula, SKÅNE county, S SWEDEN, on BALTIC SEA at N end of Hanöbukten, 20 mi/32 km E of KRISTIANSTAD; 7 mi/11.3 km long, 6 mi/9.7 km wide; 56°02′N 14°40′E. SÖLVESBORG town at base; HÄLLEVIK and HÖRVIK fishing ports.

Lister, Mount (LI-stuh) (13,200 ft/4,023 m), highest peak in ROYAL SOCIETY RANGE, ANTARCTICA, W of MCMURDO SOUND; 78°04′S 162°41′E. Discovered 1902 by Robert F. Scott, British explorer.

Listopadovka (lee-stuh-PAH-duhf-kah), village, E VORONEZH oblast, S central European Russia, on crossroads, 25 mi/40 km W of BORISOGLEBSK; 51°25′N 41°26′E. Elevation 485 ft/147 m. In agricultural area (grain, flax, sugar beets).

Listowel (lis-TO-uhl), former town (□ 2 sq mi/5.2 sq km; 2001 population 5,905), S ONTARIO, E central CANADA, on Middle Maitland River, and 24 mi/39 km N of STRATFORD; 43°44′N 80°58′W. Dairying. Textile milling, furniture manufacturing. Amalgamated into NORTH PERTH in 1998.

Listowel (li-STO-wuhl), Gaelic *Lios Tuathail*, town (2006 population 3,901), NE KERRY county, SW IRELAND, on FEALE RIVER, and 16 mi/26 km NE of TRALEE; 52°27′N 09°29′W. Agricultural market. There are slight remains of ancient castle of the Desmonds.

Listvenichnoye, RUSSIA: see LISTVYANKA.

Listvyagi (leest-VYAH-gee), industrial settlement (2005 population 4,590), W KEMEROVO oblast, S central SIBERIA, RUSSIA, 4 mi/6 km SW, and under administrative jurisdiction, of NOVOKUZNETSK; 53°41′N 86°59′E. Elevation 1,364 ft/415 m. Open-pit coal mining.

Listvyanka (leest-VYAHN-kah), town (2005 population 1,685), S IRKUTSK oblast, E central SIBERIA, RUSSIA, port on BAYKAL LAKE, near the ANGARA RIVER outlet, on coastal highway, 40 mi/64 km SE of IRKUTSK; 51°51′N 104°52′E. Elevation 1,669 ft/508 m. Terminus of a railroad spur; shipyard. Formerly called Listvenichnoye.

Listvyanskiy (leest-VYAHN-skeeye), town (2006 population 2,660), SE NOVOSIBIRSK oblast, SW SIBERIA, RUSSIA, near highway, 15 mi/24 km ESE of CHEREPANOVO, and 11 mi/18 km SE of ISKITIM, to which it is administratively subordinate; 54°27′N 83°29′E. Elevation 593 ft/180 m. Coal mining (Gorlov anthracite basin).

Lisyanka, UKRAINE: see LYSYANKA.

Lisyanski Island, HAWAII: see LISIANSKI ISLAND.

Litakovo (le-TAHK-o-vo), village, SOFIA oblast, BOTEVGRAD obshtina, W BULGARIA, in the Botevgrad Basin, 7 mi/11 km WNW of Botevgrad; 42°57′N 23°41′E. Dairying, hog raising; fruit, vegetables.

Litang (LEE-TANG), town, ⊙ Litang county, W SICHUAN province, CHINA, 105 mi/169 km W of KANGDING; 30°00′N 100°16′E. Livestock; logging; timber.

Litani River (lee-TAH-nee), ancient *Leontes*, 90 mi/145 km long, in LEBANON; rises near BAALBEK and flows SSW through the fertile AL BEQA'A valley, between LEBANON and ANTI-LEBANON ranges, turning abruptly W 6 mi/10 km SW of MARJ 'UYUN to enter the MEDITERRANEAN 5 mi/8 km N of TYRE. The short W section of its lower course is called the Qasimiye or Qasimiyah. The Lower Litaini has a dam and two hydroelectric stations but, because of the political instability, the irrigation potential for the S Beqa'a is largely unrealized.

Litchfield, county (□ 944 sq mi/2,454.4 sq km; 2006 population 190,119), NW CONNECTICUT, on MASSACHUSETTS and NEW YORK state lines; ⊙ WINSTED; 41°47′N 73°15′W. Diversified manufacturing in TORRINGTON, THOMASTON, and Winsted (metal products, electrical equipment, consumer goods, machinery, glass and plastic products, textiles, apparel, machinery, furniture). Agriculture (dairy products, vegetables, fruit; tobacco; poultry). Many resorts on lakes and in LITCHFIELD HILLS; Mount Frissell (2,380 ft/725 m), highest point in Connecticut, is near SALISBURY. Includes part of Lake CANDLEWOOD, and WARAMAUG, WONONSKOPOMUC, HIGHLAND, and BANTAM lakes, several state parks and forests; winter sports center at Mohawk Mountain State Park. Drained by HOUSATONIC, NAUGATUCK, SHEPAUG, FARMINGTON, POMPERAUG, and STILL rivers. Constituted 1751.

Litchfield, city (2000 population 6,815), MONTGOMERY county, SW central ILLINOIS, 9 mi/14.5 km W of HILLSBORO; 39°10′N 89°39′W. Trade center for bituminous coal–mining and agricultural area (corn, wheat, soybeans; livestock). Manufacturing (metal products, dairy products). Oil was first commercially produced in Illinois here in 1880s. Incorporated 1859.

Litchfield, town, a ⊙ LITCHFIELD county, W CONNECTICUT, just SW of TORRINGTON; 41°44′N 73°11′W. Includes boroughs of Litchfield and Bantam (1990 population 757). Agriculture (poultry; fruit; dairy products). Resorts on BANTAM LAKE (S); has state parks. First UNITED STATES school exclusively for law students here, 1784–1833. Ethan Allen, Henry Ward Beecher, Harriet Beecher Stowe born here. Incorporated 1719.

Litchfield (LICH-feeld), town, KENNEBEC county, S MAINE, 10 mi/16 km SW of AUGUSTA; 44°09′N 69°56′W. In farming, resort region.

Litchfield, town (2000 population 1,458), HILLSDALE county, S MICHIGAN, 10 mi/16 km NW of HILLSDALE, and on SAINT JOSEPH RIVER; 42°02′N 84°45′W. Agriculture (poultry; farming; dairy). Manufacturing (apparel, transportation equipment, metal products).

Litchfield, town (2000 population 6,562), ⊙ MEEKER county, S central MINNESOTA, 63 mi/101 km W of MINNEAPOLIS; 45°07′N 94°31′W. Elevation 1,134 ft/346 m. Trading point in agricultural area (grain, soybeans, peas, beans; livestock, poultry; dairying). Manufacturing (manufacturing equipment, leather products, food, wood products, machinery, apparel, textiles, transportation equipment). Lake Ripley to S, several small lakes in area. Settled 1856, plotted 1869, incorporated 1872.

Litchfield, town, HILLSBOROUGH county, S NEW HAMPSHIRE, 10 mi/16 km S of MANCHESTER, and 5 mi/8 km N of NASHUA; 42°50′N 71°27′W. Bounded W

Area is shown by the symbol □, and capital city or county seat by ⊙.

by MERRIMACK RIVER. Agriculture (fruit, vegetables, nursery products; livestock, poultry; dairying).

Litchfield (LICH-feeld), village (□ 69 sq mi/179.4 sq km; 2006 population 525), OUTAOUAIS region, SW QUEBEC, E CANADA, 5 mi/8 km from BRYSON; 45°45′N 76°35′W.

Litchfield, village (2006 population 257), SHERMAN county, central NEBRASKA, 12 mi/19 km SW of LOUP CITY, and on MUD CREEK; 41°09′N 99°09′W. Grain; livestock.

Litchfield Hills, NW CONNECTICUT, S extension of the Berkshires running E of HOUSATONIC River, in NW LITCHFIELD county.

Litchfield National Park (LICH-feeld) (□ 564 sq mi/1,466.4 sq km), N central NORTHERN TERRITORY, N central AUSTRALIA, 60 mi/97 km S of DARWIN; 40 mi/64 km long, 32 mi/51 km wide. Administered by Northern Territory Conservation Commission. Waterfalls descend from sandstone escarpment of Tabletop Range. Pockets of rainforest. Hot springs. "Magnetic" termite mounds, fan-shaped structures on N-S axis. Camping, picnicking, hiking. Established 1988.

Litchfield Park, town (2000 population 3,810), MARICOPA county, S central ARIZONA, near AGUA FRIA RIVER, suburbs 19 mi/31 km W of downtown PHOENIX; 33°30′N 112°21′W. Irrigated agricultural area, especially to SW (fruit, vegetables, grain; cattle, sheep). Luke Air Force Base to N.

Litchville, village (2006 population 171), BARNES county, SE NORTH DAKOTA, 21 mi/34 km SSW of VALLEY CITY; 46°39′N 98°11′W. Cattle, poultry; dairy products. Founded in 1900 and incorporated in 1903.

Lith, town, S HEJAZ, MAKKA province, SAUDI ARABIA, RED SEA port 120 mi/193 km SE of JIDDA. Agriculture (sorghum, dates, rice, vegetables, fruit).

Lithada, Cape, Greece: see KINAION, CAPE.

Litherland (LITH-uh-luhnd), town (1991 population 12,088; 2001 population 12,074), MERSEYSIDE, NW ENGLAND, 4 mi/6.4 km N of LIVERPOOL; 53°28′N 03°00′W. Leather-tanning center; also tin smelting, manufacturing (chemicals, soap).

Lithgow (LITH-go), municipality (2001 population 11,441), E NEW SOUTH WALES, SE AUSTRALIA, in BLUE MOUNTAINS, and 70 mi/113 km WNW of SYDNEY; 33°29′S 150°09′E. Coal-mining center; ironworks, steel, woolen, textile, and flour mills; small arms factory; brewery; bricks, pottery. Power station.

Lithia Springs (LITH-ee-uh), suburb (2000 population 2,072), DOUGLAS county, GEORGIA, 8 mi/12.9 km NE of DOUGLASVILLE, and 16 mi/26 km W of ATLANTA; 33°47′N 84°40′W. Bedroom community for Atlanta and a growing commercial center. Manufacturing (plastic products, printing, glass products, consumer goods, crushed stone, concrete).

Lithinon, Cape (LEE-thee-non), headland of IRÁKLION prefecture, S CRETE department, GREECE, on MEDITERRANEAN SEA; 34°55′N 24°44′E. Southernmost point of CRETE.

Lithonia (lith-O-nee-uh), town (2000 population 2,187), DEKALB county, NW central GEORGIA, 16 mi/26 km ESE of ATLANTA; 33°43′N 84°07′W. Emerging suburban residential and commercial area in metropolitan Atlanta. Manufacturing (plastics, crushed stone, concrete, processed foods, computer equipment, metal products, transportation equipment); granite processing.

Lithopolis (lith-AH-puh-lis), village (2006 population 910), FAIRFIELD county, central OHIO, 15 mi/24 km SE of COLUMBUS; 39°48′N 82°49′W.

Lithou, U.K.: see CHANNEL ISLANDS.

Lithuania, Lithuanian *Lietuva*, republic (□ 25,174 sq mi/65,201 sq km; 2004 estimated population 3,607,899; 2007 estimated population 3,575,439); ⊙ VILNIUS. Other important cities include KAUNAS, KLAIPṪDA, SIAULIAI, and PANEVĖŽYS.

Geography
Lithuania borders on the BALTIC SEA in the W, LATVIA in the N, BELARUS in the E, POLAND in the S, and KALININGRAD region (formerly EAST PRUSSIA) in the SW. Lithuania is a flatland. The NEMANUS RIVER receives 72% of the country's run-off.

Population
About 80% of the population is Lithuanian and largely Roman Catholic; there are Russians (9%), Poles (7%), Belarussian, Lettish, and Jewish minorities. The natives speak Lithuanian, a Baltic language. The republic's educational and cultural institutions include universities at VILNIUS and KAUNAS and the Lithuanian Academy of Sciences.

Economy
Primarily agricultural before 1940, Lithuania has since developed considerable industry, including food processing, and shipbuilding and manufacturing (textiles, machinery, metal products, chemicals, electrical equipment). Agriculture (grains, flax, sugar beets, potatoes, vegetables), dairy farming, and livestock raising are still carried on extensively.

History: to 1386
The pagan Liths, or Lithuanians, may have settled along the Nemen as early as 1500 B.C.E. In the 13th century the Livonian Brothers of the Sword and the Teutonic Knights conquered the region now comprising ESTONIA, Latvia, and parts of Lithuania. To protect themselves against the Knights, who pressed them from the N and the S, the Lithuanians formed (13th century) a strong unified state, expanding the territories at the expense of the neighboring Russian principalities. Lithuania became one of the largest states of medieval Europe, including all Belorussia, a large part of UKRAINE, and sections of Great RUSSIA; at its farthest extent it touched the BLACK SEA.

History: 1386 to 1860
Jogaila, the son of the grand duke Algirdas, became king of Poland (1386) as Ladislaus II through his marriage with Jadwiga (daughter of Louis I of Poland), both accepting and introducing Christianity to the new expanding nation. The union between Lithuania and Poland had at first the character of an alliance between independent nations. Vytautas, a cousin of Ladislaus II, ruled Lithuania independently (1392–1430) and brought it to the height of its power and expansion. In 1410 the Polish-Lithuanian forces severely defeated the Teutonic Knights at TANNENBERG and NOVGOROD; however, with Vytautas's death decline set in. The Belorussians, who had retained their Greek Orthodox faith, inclined toward the rising grand duchy of MOSCOW. In 1569, hard pressed by the Russians under Ivan IV, Lithuania was joined with Poland by the Union of LUBLIN to form a commonwealth. The Lithuanian aristocracy and burghers became thoroughly Polonized. By the three successive partitions of Poland (1772, 1793, 1795) Lithuania disappeared as a national unit and passed to Russia. A Lithuanian linguistic and cultural revival began in the 19th century, inspired largely by the Roman Catholic clergy and accompanied by frequent anti-Russian uprisings.

History: 1860 to 1940
By 1860 it had a population of two million. World War I and the consequent collapse of Russia and GERMANY made Lithuanian independence possible. Proclaimed (February 1918) an independent kingdom under German protection, Lithuania became (November 1918) an independent republic. It resisted attacks by Bolshevik troops and by volunteer bands of German adventurers, but in 1920 Vilnius was seized by Poland. Lithuania remained technically at war with Poland until 1927. In 1923, Lithuania seized the MEMEL TERRITORY. Internal politics were unstable. The virtual dictatorship (1926–1929) of Augustine Voldemaras was succeeded (1929–1939) by that of Antanas Smetona, and an authoritarian constitution on cor-

porative (fascist) lines became effective in 1938. Poland forced the official cession of Vilnius (1938), which, however, passed to Lithuania after the Soviet-German partition of Poland in 1939. During that same year, a German ultimatum forced the restitution of Memel.

History: 1940 to Present
In 1940 the USSR, which had obtained military bases in Lithuania, occupied the country. After a Soviet-sponsored "election," Lithuania became a constituent republic of the USSR. During the German occupation (1941–1944) of Lithuania in World War II, during which Lithuanians fought alongside the Germans, a considerable Jewish minority of Lithuanians was largely exterminated. In 1944 the Communist government returned. Mass deportations of intellectuals and farmers occurred from 1944 to 1949. In March 1990 the Lithuanian parliament declared independence from the USSR. The Soviet Union responded with an oil embargo and troop actions. A referendum on independence passed in February 1991, and Lithuania's independence was recognized by the USSR on September 6, 1991. Vytautas Landsbergis, of the pro-independence Sajudis party, became Lithuania's first post-Soviet head of state. Russian troops were not completely withdrawn, however, until 1993. In 1992 the Democratic Labor (former Communist) party defeated Sajudis, and Algirdas Brazauskas, a former Communist, was elected president (1993). Lithuania signed a free-trade agreement with Estonia and Latvia in 1993. A Lithuanian-American, Valdas Adamkus, was elected president in 1998. In 2002 Adamkus was defeated for reelection by Liberal Democrat Rolandas Paksas, but Paksas was impeached and removed from office in 2004 on charges of corruption and having links to Russian organized crime. Parliament speaker Arturas Paulauskas became acting president, and subsequently Adamkus was elected to the office. Lithuania formally joined NATO (North Atlantic Treaty Organization) on March 29, 2004 and became a full member of the European Union (EU) on May 1, 2004.

Government
Lithuania is governed under the constitution of 1992. The president, who is the head of state, is elected by popular vote for a five-year term and is eligible for a second term. The prime minister, who is the head of government, is appointed by the president, as is the cabinet. The unicameral Parliament (*Seimas*) has 141 members; seventy-one are elected by popular vote and seventy by proportional representation, all for four-year terms. The current president is Valdas Adamkus (since July 2004); the current head of government is Prime Minister Gediminas Kirkilas (since July 2006). Administratively, the country is divided into ten counties.

Lithuanian-Belorussian Upland, Russian *Litovsko-Belorusskaya Vozvyshennost*, moraine region in POLAND, the BALTIC STATES, and BELARUS; extends c.300 mi/483 km from Masurian Lakes in Poland NE to DVINA River; forms divide between BALTIC SEA and BLACK SEA drainage areas. Rises to 1,140 ft/347 m; gives rise to VILIYA RIVER (N), BEREZINA, PTICH, and SLUCH rivers (S); crossed by Nemunas River.

Litija (lee-TEE-yah), town (2002 population 6,357), central SLOVENIA, on Sava River, on railroad, and 16 mi/26 km E of LJUBLJANA; 46°03′N 14°49′E. Lead mine (worked from Roman times to mid-1960s); lead smelter closed in 1966. Manufacturing (woolen and cotton textiles); timber industry.

Litin, UKRAINE: see LITYN.

Lititz (LI-tits), borough (2006 population 9,029), LANCASTER county, SE PENNSYLVANIA, 8 mi/12.9 km N of LANCASTER; 40°08′N 78°18′W. Agriculture (grain, soybeans, apples; eggs; livestock; dairying). Manufacturing (footwear, machinery, commercial printing, consumer goods, millwork, food processing, crushed stone, transportation equipment, pharmaceuticals, furniture). Lancaster Airport to S. Speedwell Forge Lake reservoir to N. Settled c.1740 by Moravians, laid out 1757, incorporated 1759.

Litlabø (LIT-lah-buh), village, HORDALAND county, SW NORWAY, on SW shore of STORD island, 26 mi/42 km NNE of HAUGESUND; 59°48′N 05°27′E.

Litokhoron (lee-TO-kho-ron), town, THESSALONÍKI prefecture, CENTRAL MACEDONIA department, NE GREECE, 32 mi/51 km N of LÁRISSA, at foot of Mount Olympos (ascended from here); 40°06′N 22°30′E. Tourist trade. Formerly Litochoron.

Litom, POLAND: see LEUTHEN.

Litoměřice (LI-to-MNYE-rzhi-TSE), German *Leitmeritz*, city (2001 population 24,879), SEVEROCESKY province, N BOHEMIA, CZECH REPUBLIC, on the ELBE RIVER, opposite OHRE RIVER mouth, on railroad; 50°32′N 14°08′E. River port. Noted for tanning industry; manufacturing (metals, food processing); brewery (established 1720); extensive vineyards and orchards (peaches, apricots) in vicinity. Has a 16th century town hall, and a 16th century building with cup-shaped tower (now a museum). It has been the seat of the Roman Catholic bishop since 1635. Annual agricultural exhibition called "Garden of Bohemia" (*Zahrada Čech*).

Litomysl (LI-to-MISHL), Czech *Litomyšl*, German *Leitomischl*, town, VYCHODOCESKY province, E BOHEMIA, CZECH REPUBLIC, 27 mi/43 km SE of PARDUBICE; 49°52′N 16°19′E. Railroad terminus. Manufacturing (glass, machinery, synthetic fibers); agriculture (fruit and ornamental nurseries). Has a 14th century church, a 16th century Renaissance castle with art gallery, an 18th century Piarist church, a town hall with an extensive library, a Moorish-style synagogue, and the remains of Medieval forts. Composer Bedřich Smetana was born here in 1824. Scene of annual opera festival.

Litoral (lee-TOR-uhl), province (□ 2,573 sq mi/6,689.8 sq km; 2001 population 298,414), W EQUATORIAL GUINEA; BATA is ⊙, only city, and chief port; 01°31′N 09°49′E. Bounded on S by GABON, E by CENTRO-SUR province, N by CAMEROON, W by GULF OF GUINEA. The most developed portion of the country, growing timber, oil palms, and coffee.

Litoral, Cadena del (lee-to-RAHL, kah-DAI-nah del), mountain range, N VENEZUELA; 10°30′N 67°00′W. Makes up the N half of the W portion of CORDILLERA DE LA COSTA, with Cadena del INTERIOR making up the S half. Pico NAIGUATÁ is the highest peak in Cadena del Litoral and in all of Cordillera de la Costa. Variant name: Cordillera del Litoral.

Litoral, Cordillera del, VENEZUELA: see LITORAL, CADENA DEL.

Litoral, Serranía del, VENEZUELA: see LITORAL, CADENA DEL.

Litovel (LI-to-VEL), German *Littau*, town, SEVEROMORAVSKY province, N central MORAVIA, CZECH REPUBLIC, on MORAVA river, and 11 mi/18 km NW of OLOMOUC; 49°43′N 17°05′E. Agriculture (sugar beets, oats, barley); manufacturing (paper, machinery); brewing; distilling; sugar refining; food processing.

Litovko (lee-TOF-kuh), town (2005 population 2,645), S KHABAROVSK TERRITORY, RUSSIAN FAR EAST, on branch of the TRANS-SIBERIAN RAILROAD, 115 mi/185 km SSW of KOMSOMOL'SK-NA-AMURE; 49°14′N 135°10′E. Elevation 180 ft/54 m. Sawmilling, plywood.

Litschau (LIT-chou), town, NW LOWER AUSTRIA, N Austria, in the WALDVIERTEL near the CZECH border, 7 mi/11.3 km NNE of Gmünd; 48°57′N 15°03′E. Elevation 1,634 ft/498 m. Railroad terminus. Manufacturing of metals; summer tourism; old glass and textile industry; castles, parts of town walls conserved. Town created before 1386.

Littau (LIT-ow), town (2000 population 15,929), LUCERNE canton, central SWITZERLAND, 2 mi/3.2 km W

of LUCERNE; 47°03′N 08°15′E. Elevation 1,677 ft/511 m. Textiles, chemicals, metalworks, printing. Includes part of EMMENBRÜCKE.

Littau, CZECH REPUBLIC: see LITOVEL.

Little Abaco Island, Abaco district, N BAHAMAS, just NW of GREAT ABACO ISLAND, NE of GRAND BAHAMA Island, 125 mi/201 km N of NASSAU; 26°53′N 77°43′W.

Little Aden (A-den), volcanic peninsula (□ 15 sq mi/39 sq km), W of ADEN, separated from Aden peninsula (E) by ADEN Bay; 12°44′N 44°52′E. Rocky and barren; rises to 1,218 ft/371 m. Has two fishing villages: FUKUM on W shore and BUREIKA on E shore. Adjoining the isthmus (N) is the settlement of Little Aden saltworks. Peninsula was incorporated (c.1868) into ADEN. Large refinery was built in 1954. There are now some associated industries.

Little Alföld (AHL-fold), HUNGARIAN *Kis Alföld*, CZECH *Podunajská nížina*, fertile plain (□ 3,860 sq mi/10,036 sq km), SLOVAKIA and NW Hungary; bounded S and E by BAKONY and VÉRTES Mountains, W by AUSTRIAN border, N by spurs of the CARPATHIANS. Drained by DANUBE, RÁBA, and VAH rivers; HANSÁG swamps are SW. Agriculture (wheat, rye, barley, corn, sugar beets, potatoes, vegetables, fodder crops); large-scale dairy farming; breeding of riding horses; orchards, vineyards, tobacco. Main cities are GYÖR, KOMARNO (in Slovakia), and KOMÁROM. Historically, the Little Alföld has been the most developed region of Hungary after the BUDAPEST era. It suffered less than other areas from OTTOMAN warfare and devastation and has benefited from proximity to Austria and BOHEMIA. It has been restructuring its economy quite successfully and, combined with immediately adjoining regions of TRANSDANUBIA, received more than 30 percent of all direct foreign investment in 1990–1995.

Little America, village, SWEETWATER county, SW WYOMING, 20 mi/32 km W of Green River town, near Blacks Fork River, on N side of Interstate Highway 80. Home base of chain of Little America truck stops located across Wyoming, other W states. Located 150 miles east of SALT LAKE CITY, UTAH, 300 miles west of CHEYENNE and a few hours from YELLOWSTONE and Grand Teton National Parks.

Little America, ANTARCTICA. Name of five former U.S. stations on ROSS ICE SHELF from 1929–1958. I–IV were on the BAY OF WHALES, but V was built several miles away near KAINAN BAY (at 78°11′S 162°10′W). Base for Richard E. Byrd expeditions.

Little America Basin, basin in SE ROSS SEA, ANTARCTICA, S PACIFIC OCEAN, W of PRESTRUD BANK, and N of ROOSEVELT ISLAND.

Little Andaman Island, in ANDAMAN AND NICOBAR ISLANDS UNION TERRITORY, INDIA, in BAY OF BENGAL, southernmost of ANDAMAN ISLANDS, separated from main group by DUNCAN PASSAGE, from CAR NICOBAR ISLAND (S) by TEN DEGREE CHANNEL; 26 mi/42 km long N-S, 16 mi/26 km wide; 10°45′N 92°30′E.

Little Arkansas River (ahr-KAN-zez), 90 mi/145 km long, S central KANSAS; rises in N RICE county N of LYONS; flows SE past BUTLER and HALSTEAD to ARKANSAS RIVER at WICHITA.

Little Auglaize River (o-GLAIZ), c.45 mi/72 km long, W OHIO; rises W of LIMA; flows generally N to the AUGLAIZE RIVER just N of MELROSE; 41°06′N 84°24′W.

Little Bahama Bank, shoal, NW BAHAMAS, N of GRAND BAHAMA Island, 60 mi/97 km E of WEST PALM BEACH (across Straits of FLORIDA); c.150 mi/241 km long NW-SE, c.50 mi/80 km wide. Surrounded by many cays and islands.

Little Barren River, c.45 mi/72 km long, S central KENTUCKY; rises in SE METCALFE county, SE of EDMONTON; flows N past Edmonton, enters GREEN RIVER 9 mi/14.5 km W of GREENSBURG. BARREN RIVER is to SW.

Little Barrier Island, NNE of AUCKLAND, NEW ZEALAND, formed by extinct cone-shaped volcano; 36°12′S 175°05′E. Rises to 2,369 ft/722 m, 50 mi/80 km. Since

1890s, a floral and faunal reserve, now part of HAURAKI GULF MARITIME PARK.

Little Bassa, LIBERIA: see MARSHALL.

Little Basses, Sinhalese *Kuda Ravana Kotuwa* [=little Ravana's rocks], small group of rocks off SOUTHERN PROVINCE, SE coast of SRI LANKA; 06°24′N 81°44′E. Lighthouse.

Little Bay, SE NEW JERSEY, small inlet just S of GREAT BAY; protected from the ATLANTIC OCEAN by islands; 39°22′N 74°28′W; traversed by INTRACOASTAL WATERWAY channel.

Little Bay De Noc (duh NAHK), a N arm of GREEN BAY indenting SW shore of DELTA county, UPPER PENINSULA, MICHIGAN; c.16 mi/26 km long N-S, 1 mi/1.6 km–4 mi/6.4 km wide; 45°46′N 87°00′W. ESCANABA is on W shore. Peninsula separates it from BIG BAY DE NOC (E).

Little Bay Island (□ 2 sq mi/5.2 sq km; 1991 population 261), SE NEWFOUNDLAND AND LABRADOR, CANADA, in NOTRE DAME BAY, 25 mi/40 km SW of Cape St. JOHN; 49°39′N 55°50′W. Fishing.

Little Bear Creek Reservoir, FRANKLIN county, NW ALABAMA, on Little Bear Creek, 30 mi/48 km SW of FLORENCE; 34°26′N 87°58′W. Formed by Little Bear Creek Dam (69 ft/21 m high), built (1974) by the Tennessee Valley Authority for flood control and recreation.

Little Bear Peak (14,037 ft/4,278 m), in ROCKY MOUNTAINS, ALAMOSA and COSTILLA counties, S COLORADO, 10 mi/16 km NNW of FORT GARLAND, and 1 mi/1.6 km SW of BLANCA PEAK.

Little Bear River, 50 mi/80 km long, N UTAH; rises in WASATCH RANGE NNE of OGDEN; flows generally N, through HYRUM RESERVOIR past HYRUM and WELLSVILLE, to BEAR RIVER 8 mi/12.9 km NW of LOGAN. Hyrum Dam (98 ft/30 m high, 540 ft/165 m long; completed 1935), just SW of Hyrum, is used for irrigation. Hyrum Reservoir State Park in CACHE county is a state recreation area.

Little Beaver River (BEE-vuhr), c.7 mi/11 km long, E OHIO and W PENNSYLVANIA; formed by Middle, West, and North forks in SE COLUMBIANA county, Ohio; flows SE to the OHIO RIVER just across Pennsylvania state line. Main headstream (Middle Fork, c.35 mi/56 km long) rises in MAHONING county, flows SE.

Little Belt Mountains, range of ROCKY MOUNTAINS in central MONTANA, rise SE of GREAT FALLS, extend c.68 mi/109 km SE to MUSSELSHELL RIVER. Lie within Lewis and Clark National Forest. JUDITH RIVER Wildlife Management Area on E side, has sapphire mine; Showdown Ski Area at center of range. Highest point, Big Baldy Mountain (9,175 ft/2,797 m). Silver, lead, gold, zinc, sapphires, coal mined.

Little Bighorn, river, c. 90 mi/145 km long, N WYOMING and S MONTANA; rising in the BIGHORN MOUNTAINS in NW SHERIDAN county, N Wyoming, and flowing N past CROW AGENCY to join the BIGHORN RIVER at HARDIN in S Montana. On June 25–26, 1876, Sioux and Cheyenne warriors defeated the forces of College George Custer in the Little Bighorn valley. Little Bighorn Battlefield National Monument occupies the site of the battle at Crow Agency.

Little Bighorn Battlefield, national monument (765 acres/310 ha), BIG HORN county, SE MONTANA, 14 mi/23 km SE of HARDIN, 2 mi/3.2 km SE of CROW AGENCY. Authorized 1879. Site of the battle between the Seventh Cavalry, in NE part Crow Indian Reservation, commanded by George Armstrong Custer, and the Sioux and Northern Cheyenne. Battle of Little Bighorn June 25–26, 1876. Formerly Custer Battlefield National Monument, the name was changed in 1993. Includes Custer National Cemetery, within Monument park.

Little Bitter Lake, EGYPT: see BITTER LAKES.

Little Blue River, 450 mi/724 km long, in S NEBRASKA and N KANSAS; rises in KEARNEY county, Nebraska;

flows ESE, past Hebron and Fairbury, Nebraska, then SSE to BIG BLUE RIVER at BLUE RAPIDS, Kansas.

Little Boars Head, NEW HAMPSHIRE: see GREAT BOARS HEAD.

Little Bonaire (bawn-ER), Dutch *Klein Bonaire*, islet, just off W coast of BONAIRE island, NETHERLANDS ANTILLES, 40 mi/6 km E of WILLEMSTAD; 2 mi/3.2 km long, 1.5 mi/2.4 km wide; 12°10′N 68°19′W.

Littleborough (LIT-uhl-buh-ruh), town (2001 population 12,275), NE GREATER MANCHESTER, NW ENGLAND, on ROCH RIVER, and 14 mi/23 km NNE of MANCHESTER; 53°39′N 02°05′W. Manufacturing (chemicals, pharmaceuticals); light industry. Previously cotton mills. Stubley Old Hall, one of oldest (16th century) mansions of the region, is here.

Little Brewster Island, MASSACHUSETTS: see BREWSTER ISLANDS.

Little Buffalo (LI-tuhl BUH-fuh-lo), First Nations settlement (□ 5 sq mi/13 sq km), central ALBERTA, W CANADA, 47 mi/75 km from PEACE RIVER town, in NORTHERN SUNRISE COUNTY; 56°26′N 116°06′W.

Little Caledon River, c.60 mi/97 km long, N LESOTHO; rises c.30 mi/48 km E of MASERU in MALUTI MOUNTAINS; flows WSW, joins CALEDON RIVER 5 mi/8 km SW of Maseru. Also called Thupa Kuba.

Little Calumet River, ILLINOIS and INDIANA: see CALUMET RIVER.

Little Canada, town (2000 population 9,771), RAMSEY county, E MINNESOTA, suburb 5 mi/8 km N of downtown ST. PAUL, and 9 mi/14.5 km ENE of downtown MINNEAPOLIS; 45°01′N 93°04′W. Railroad junction. Manufacturing (plastic products, store fixtures, printing, medical supplies). Numerous small lakes in area; Gervais Lake in E.

Little Cape Mount River, LIBERIA: see LOFA RIVER.

Little Captain Island, SW CONNECTICUT, two small islands joined by reefs, in LONG ISLAND SOUND, 2 mi/3.2 km offshore, S of GREENWICH. Public recreation center here.

Little Carpathian Mountains, Slovak *Malé Karpaty* (mah-LAI kahr-PAH-ti), GERMAN *Kleine Karpaten*, HUNGARIAN *Kis Karpatok*, range in ZAPADOSLOVENSKY province, SW SLOVAKIA; extends c.55 mi/89 km NE–SW, between NOVÉ MESTO NAD (N) and DEVIN (S). Rise to 2,520 ft/768 m. Best-known mountain is BRADLO, 1,781 ft/543 m. Broad-leaved forests, extensive vineyards; limestone and granite quarries.

Little Cayman, island (□ 9 sq mi/23.4 sq km; 1999 population 115) of CAYMAN ISLANDS, British crown colony, WEST INDIES, separated by narrow channel from CAYMAN BRAC (E), and 60 mi/97 km ENE of GRAND CAYMAN; c.9 mi/14.5 km long, 1 mi/1.6 km wide; 19°40′N 80°05′W. Tourism at small lodges, scuba diving, sport fishing.

Little Cedar River c.40 mi/64 km long, in SW UPPER PENINSULA, MICHIGAN; rises NW of HERMANSVILLE in MENOMINEE county; flows S, past Hermansville and DAGGETT, to MENOMINEE RIVER 8 mi/12.9 km S of STEPHENSON; 45°45′N 87°39′W.

Little Cedar River, 60 mi/97 km long, in MINNESOTA and IOWA; rises in MOWER county, SE Minnesota, c.25 mi/40 km SSW of ROCHESTER; flows S, into N Iowa, past STACYVILLE, to CEDAR RIVER at NASHUA; 43°38′N 92°45′W.

Little Chazy River (shai-ZEE), c.20 mi/32 km long, NE NEW YORK; rises in E central CLINTON county; flows E and NE, past CHAZY, to Lake CHAMPLAIN 6 mi/9.7 km S of ROUSES POINT.

Little Chebeague Island (shuh-BEEG), SW MAINE, in CASCO BAY off PORTLAND; c.0.75 mi/1.2 km long.

Little Chenier, ridge, CAMERON parish, SW LOUISIANA; 29°50′N 92°59′W. One of the Cheniers (oak tree-covered ridges) paralleling the GULF coast of SW LOUISIANA. There are numerous Native American sites.

Little Choptank River, c.15 mi/24 km long, E MARYLAND; tidal arm of CHESAPEAKE BAY, penetrating the EASTERN SHORE in DORCHESTER county, just N of TAYLORS ISLAND.

Little Chuckwalla Mountains, California: see CHUCKWALLA MOUNTAINS.

Little Chute, city (2006 population 11,035), OUTAGAMIE county, E WISCONSIN, opposite KIMBERLY, on FOX RIVER, and a suburb 5 mi/8 km ENE of APPLETON; 44°17′N 88°18′W. Manufacturing (printing, building materials, prepared food, cheese, concrete). A dam is here. Settled 1850, incorporated 1899.

Little City, village, MARSHALL county, S OKLAHOMA, 9 mi/14.5 km E of MADILL. In agricultural area. Lake TEXOMA to E and S.

Little Colinet Island, CANADA: see GREAT COLINET ISLAND.

Little Colorado River, 315 mi/507 km long, largely in ARIZONA; rises in SE APACHE county, E Arizona; flows generally NW through Arizona, past SAINT JOHNS, HOLBROOK, and WINSLOW, along PAINTED DESERT, to COLORADO RIVER in the GRAND CANYON, in Grand Canyon National Park, 70 mi/113 km N of FLAGSTAFF. Stream is dammed, forms Lyman Lake (State Park), for irrigation 10 mi/16 km S of Saint Johns.

Little Compton, town (2000 population 3,593), NEWPORT county, SE RHODE ISLAND, between Sakonnet River and MASSACHUSETTS state line, and bounded S by the Atlantic; 41°31′N 71°10′W. Agricultural, fishing, resort area. Includes villages of ADAMSVILLE, LITTLE COMPTON, and SAKONNET. John and Priscilla Alden's daughter Elizabeth lived and is buried here; Benjamin Church also lived here. Rhode Island Red fowl originated in the town. Incorporated as a Plymouth Colony town in 1682, passed to Rhode Island 1747.

Little Courland Bay, TOBAGO: see MOUNT IRVINE BAY.

Little Creek, village (2000 population 195), KENT county, E DELAWARE, 4 mi/6.4 km E of DOVER, on Little Creek; 39°10′N 75°27′W. Elevation 3 ft/o m. Oysters. Little Creek Wildlife Area to SE.

Little Curaçao (kyoo-ruh-SOU), Dutch *Klein Curaçao*, islet in NETHERLANDS ANTILLES, 20 mi/32 km ESE of WILLEMSTAD, CURAÇAO; 12°00′N 68°40′W. Lighthouse.

Little Current (LI-tuhl KUHR-ent), unincorporated town (□ 2 sq mi/5.2 sq km; 2001 population 1,526), ONTARIO, E central CANADA, on N Manitoulin Island, on NORTH CHANNEL of Lake HURON, and included in the town of Northeastern Manitoulin and The Islands; 45°58′N 81°56′W. A port and a popular yachting resort; railroad connections with the mainland.

Little Cypress Bayou (SEI-pruhs), c.50 mi/80 km long, NE TEXAS; rises E of WINNSBORO in SW CAMP county, near source of BIG CYPRESS CREEK in several streams N of Gilmer; flows generally SE, then NE to join Cypress Bayou at W end of CADDO LAKE c.6 mi/9.7 km E of JEFFERSON; 32°39′N 94°42′W.

Little Danube River, SLOVAK *Malý Dunaj* (mah-LEE du-NAI), GERMAN *Kleine Donau*, HUNGARIAN *Kis Duna*, arm of the DANUBE RIVER, c.98 mi/158 km long, ZAPADOSLOVENSKY province, SW Slovakia; branches off main stream 2 mi/3.2 km SE of BRATISLAVA; meanders SE, joining VÁH RIVER at KOLAROVO; encloses, together with the Váh and the Danube, at GREAT RYE ISLAND.

Little Desert National Park (LI-tuhl DE-zuhrt) (□ 510 sq mi/1,326 sq km), NW VICTORIA, SE AUSTRALIA, 250 mi/402 km NW of MELBOURNE; 2 mi/35 km long, 10 mi/16 km wide. Semi-desert area. Mallee and heath grow among windblown sands; clay-pan areas support yellow gums, stringybarks. Parrots, wrens, currawongs, honey eaters, mallee fowl; possums, kangaroos, bats; bearded dragons. Camping, picnicking, and hiking. Established 1968.

Little Diomede Island (DEI-yo-meed), DIOMEDE ISLANDS, NW ALASKA, in BERING STRAIT, 20 mi/32 km WNW of Cape PRINCE OF WALES (SEWARD PENINSULA), 27 mi/43 km SE of CAPE DEZHNEV (SIBERIA), and 4.5 mi/7.2 km E of Russian island of RATMANOV (Big Diomede); 2 mi/3.2 km long, 1 mi/1.6 km wide; 65°45′N 168°57′W. Rises to 1,200 ft/366 m. On W coast is Inuit settlement of Diomede village (1990 population 178); school. Inuit inhabitants noted for skill as seamen. Just W is international boundary between U.S. and RUSSIAN FAR EAST. Discovered by Bering, 1728. Also known as Kruzenshtern (Krusenstern) and Ignaluk (IG-nuh-luhk).

Little Downham, ENGLAND: see DOWNHAM.

Little Dragoon Mountains, ARIZONA: see DRAGOON MOUNTAINS.

Little Dunmow, ENGLAND: see DUNMOW, GREAT.

Little Eaton (EE-tuhn), village (2001 population 2,557), S DERBYSHIRE, central ENGLAND, on DERWENT River, and 4 mi/6.4 km N of DERBY; 52°58′N 01°27′W. Former paper-milling site.

Little Eau Pleine River (oh PLAIN), c.35 mi/56 km long, central WISCONSIN; rises NW of MARSHFIELD near CLARK/MARATHON county line; flows generally SE through RICE LAKE to WISCONSIN RIVER (DuBay Reservoir) 12 mi/19 km NNW of STEVENS POINT.

Little Edisto Island, SOUTH CAROLINA: see EDISTO ISLAND.

Little Egg Harbor, township, OCEAN county, E central NEW JERSEY, on LITTLE EGG HARBOR, 15 mi/24 km NE of ATLANTIC CITY; 39°36′N 74°20′W. Incorporated 1798.

Little Egg Harbor, SE NEW JERSEY; inlet of the ATLANTIC OCEAN E of TUCKERTON; sheltered from ocean by S end of LONG BEACH island; c.6 mi/9.7 km long, 4 mi/6.4 km wide; 39°35′N 74°16′W. Link in INTRACOASTAL WATERWAY, entering from Manahawkin Bay (N) and continuing into GREAT BAY (S). Beach Haven Inlet, S of BEACH HAVEN, and Little Egg Inlet are entrances from the Atlantic Ocean.

Little Elm (ELM), town (2006 population 21,287), DENTON county, N TEXAS, residential suburb 25 mi/40 km NNW of downtown DALLAS, on LEWISVILLE LAKE reservoir (bridged); 33°09′N 96°55′W. Agriculture and recreation area. Manufacturing (printing press blankets). Lake Lewisville State Park to S.

Little Exuma Island, central Bahamas, southernmost (apart from small Hog Cay) of the Exuma islands, and adjoining GREAT EXUMA ISLAND, 150 mi/241 km SE of NASSAU; 13 mi/21 km long, c.1 mi/1.6 km wide; 23°27′N 75°37′W. It is crossed by the TROPIC OF CANCER. Main settlement is WILLIAMS TOWN.

Little Falls, city (2000 population 7,719), ⊙ MORRISON county, central MINNESOTA, on MISSISSIPPI RIVER, and 30 mi/48 km NNW of ST. CLOUD, on both sides of Mississippi River (four bridges); 45°58′N 94°21′W. Elevation 1,118 ft/341 m. Resort and trade center for agricultural area (grain, sunflowers, potatoes; livestock; dairying). Manufacturing (metal products, hardwood products, beverages, paper, concrete, transportation equipment, plastic products, printing and publishing). Granite quarry nearby. Grew with establishment of mills that used falls in river as source of water power. Little Falls–Morrison County Airport to SE. Point of interest is Charles Lindbergh State Park to SW, surrounding aviator's childhood home. Camp Ripley Military Reservation to N. Settled 1855, incorporated as village 1879, as city 1889.

Little Falls, city (□ 3 sq mi/7.8 sq km; 2006 population 4,980), HERKIMER county, central NEW YORK, 18 mi/29 km ESE of UTICA; 43°02′N 74°51′W. At falls of MOHAWK RIVER (water power) and on NEW YORK STATE BARGE CANAL (locks here). Manufacturing (paper, fiberglass products, printed packaging materials) and services. Home of General Nicholas Herkimer, hero of the Battle of Oriskany. Settled c.1725; incorporated as city 1895.

Little Falls, township (2000 population 10,855), PASSAIC county, NE NEW JERSEY, on PASSAIC River, and 3 mi/4.8 km SW of PATERSON; 40°52′N 74°13′W. Large laundry plant; manufacturing (metal products, athletic goods, concrete products). Includes SINGAC village. Settled 1711.

Little Ferry, borough (2006 population 10,715), BERGEN county, NE NEW JERSEY, on HACKENSACK River, and 4 mi/6.4 km ESE of PATERSON; 40°51′N 74°02′W. Manufacturing; residential. Settled 1636, incorporated 1894.

Littlefield (LI-tuhl-feeld), town (2006 population 6,246), ☉ LAMB county, NW TEXAS, on the LLANO ESTACADO, 35 mi/56 km NW of LUBBOCK; 33°55′N 102°19′W. Elevation 3,556 ft/1,084 m. Trade, shipping, processing center for agricultural/ and livestock (grain sorghum, cotton). Manufacturing (fibers, turbine pumps). Lakes nearby (hunting, fishing). Settled 1912, incorporated 1925; became county seat 1946.

Littlefield, unincorporated village, MOHAVE county, NW ARIZONA, 23 mi/37 km WSW of SAINT GEORGE, UTAH, and 9 mi/14.5 km ENE of MESQUITE, NEVADA, on VIRGIN River, at mouth of Beaver Dam Wash. Elevation 1,858 ft/566 m. Cattle. VIRGIN MOUNTAINS to S; Paiute Wilderness Area, in Black Rock Mountains, to SE; Beaver Dam Mountains Wilderness Area to E. Service center on Interstate Highway 15 the only viable community in this isolated NW corner of Arizona.

Little Fish River, river, SOUTH AFRICA: see GREAT FISH RIVER.

Little Fogo Islands (FO-go), group of ten islets at entrance of NOTRE DAME BAY, SE NEWFOUNDLAND AND LABRADOR, CANADA, 5 mi/8 km N of Fogo. Northernmost island has lighthouse (49°49′N 54°07′W). Once had fishing communities, now uninhabited.

Littlefork, town (2000 population 680), KOOCHICHING county, N MINNESOTA, on LITTLE FORK RIVER 6 mi/9.7 km SW of INTERNATIONAL FALLS; 48°23′N 93°33′W. Agriculture (cattle; alfalfa). Manufacturing (lumber; fishing lures); timber. Nett Lake Indian Reservation to SE; Koochiching State Forest to SW; Smokey Bear State Forest to NW. Often spelled Little Fork.

Little Fork River, 150 mi/241 km long, N MINNESOTA; rises in Swampy area of ST. LOUIS county, 9 mi/14.5 km E of COOK and S of VERMILION LAKE; flows first W past Cook, then NW through Koochiching State Forest, past LITTLEFORK (Little Fork) village, to RAINY RIVER on CANADA (ONTARIO) border, 10 mi/16 km SW of INTERNATIONAL FALLS; 47°49′N 92°27′W. Mouth of BIG FORK RIVER, on Rainy River, 6 mi/9.7 km to W.

Little Gombrani, island, S of RODRIGUEZ Island, dependency of MAURITIUS; 328 mi/528 km W of GOMBRANI ISLAND; 19°47′S 63°25′E. Small coralline island (1.85 acres/0.75 ha) with scattered vegetation. Also called Cesseli Island.

Little Goose Creek, c. 30 mi/48 km long, N WYOMING; rises in BIGHORN MOUNTAINS near CLOUD PEAK in NW JOHNSON county; flows N to GOOSE CREEK at SHERIDAN.

Little Goose Lock and Dam, WASHINGTON: see BRYAN, LAKE.

Little Grand Lake, SW NEWFOUNDLAND AND LABRADOR, CANADA, 23 mi/37 km S of CORNER BROOK; 10 mi/16 km long, 1 mi/2 km wide. Drains into GRAND LAKE.

Little Gull Island, NEW YORK: see GULL ISLANDS.

Little Gunpowder Falls, stream, c.25 mi/40 km long, N MARYLAND; flows SE, forming part of BALTIMORE-HARFORD county line, to GUNPOWDER RIVER (estuary), c.15 mi/24 km ENE of BALTIMORE.

Littlehampton (LIT-uhl-hamp-tuhn), town (2001 population 25,593), West SUSSEX, SE ENGLAND, on the CHANNEL at mouth of ARUN RIVER, 8 mi/12.9 km W of WORTHING; 50°48′N 00°32′W. Resort. In Middle Ages it was port for nearby ARUNDEL. Nearby (N) is Wick.

Little Horton, ENGLAND: see BRADFORD.

Little Humboldt River (HUHM-bolt), c.60 mi/97 km long, N NEVADA; intermittent stream formed in E HUMBOLDT county by joining of North and South Forks in Chimney Dam Reservoir 70 mi/113 km NE of

WINNEMUCCA; flows W and SW to HUMBOLDT River 4 mi/6.4 km NE of Winnemucca. North Fork, c.45 mi/72 km long, rises in NE Humboldt county; South Fork, c.40 mi/64 km long, rises in W ELKO county, flows NNW, through SW.

Little Inagua Island, S BAHAMAS, just NE of GREAT INAGUA ISLAND, SW of CAICOS ISLANDS, 360 mi/579 km SE of NASSAU; roughly 10 mi/16 km long, up to 10 mi/16 km wide; 21°30′N 73°00′W. Practically uninhabited. Little Inagua Land and Sea Park.

Little Irchester, ENGLAND: see IRCHESTER.

Little Italy, neighborhood, S central borough of MANHATTAN, NEW YORK city, SE NEW YORK, bounded approximately by Canal Street on S, East Houston Street on N, the Bowery on E, and Cleveland Place and Lafayette Street on W; 40°43′N 73°59′W. The district's ethnic flavor was established 1890–1924 with the arrival of a flood of Italian immigrants. In 1932, 98% of the population was Italian, but now the district is more ethnically mixed due to the encroachment of CHINATOWN from the S and other demographic changes.

Littlejohn Island, resort island, SW MAINE, in CASCO BAY off Yarmouth; 1 mi/1.6 km long. Bridge connects with COUSINS ISLAND.

Little Juniata River, Pennsylvania: see JUNIATA RIVER.

Little Kai, INDONESIA: see KAI ISLANDS.

Little Kanawha River (kuh-NAW-uh), c.160 mi/257 km long, central and NW WEST VIRGINIA; rises in S UPSHUR county; flows generally W through Burnsville Lake reservoir (Wildlife Management Area), past BURNSVILLE, GLENVILLE, and GRANTSVILLE, then NW to OHIO RIVER at PARKERSBURG; receives HUGHES RIVER from E, 12 mi/19 km SE of Parkersburg.

Little Karoo, SOUTH AFRICA: see SOUTHERN KAROO.

Little Kentucky River, c.35 mi/56 km long, N KENTUCKY; rises in SW HENRY county, flows generally N, enters OHIO RIVER 2 mi/3.2 km W of CARROLLTON, and 1 mi/1.6 km W of mouth of KENTUCKY RIVER.

Little Koldewey, GREENLAND: see GREAT KOLDEWEY.

Little Lake, in extreme SE LOUISIANA, inlet of Gulf of MEXICO, c. 25 mi/40 km SSW of NEW ORLEANS; c. 10 mi/16 km long. Joined directly to BARATARIA BAY (SE) and Lake SALVADOR and the INTRACOASTAL WATERWAY (N). Indirect connection to BARATARIA BAY through Bayou Saint Denis (ESE).

Little Lake, NEW YORK: see WANETA LAKE.

Little London, town, WESTMORELAND parish, SW JAMAICA, in coastal lowland, 5 mi/8 km WNW of SAVANNA-LA-MAR; 18°18′N 78°12′W. Sugar, rice, breadfruit; livestock.

Little Lost River, c.60 mi/97 km long, E IDAHO; rises in LEMHI RANGE, CUSTER/LEMHI county line; flows SSE, terminating in a depression ENE of ARCO in E part of BUTTE county, in tract of Idaho Engineering Laboratory. Same termination area as BIG LOST RIVER.

Little Machipongo Inlet, VIRGINIA: see HOG ISLAND; PARRAMORE ISLAND.

Little Madawaska River (mad-uh-WAHS-kuh), c.35 mi/56 km long, NE MAINE; rises in NE AROOSTOOK county; flows NE and SE to the AROOSTOOK near CARIBOU.

Little Makin (MA-kin), atoll (☐ 3 sq mi/7.8 sq km), northernmost of GILBERT ISLANDS, KIRIBATI, W central PACIFIC OCEAN; 03°17′N 172°58′E. Copra. Also called Makin Meang (Little Makin).

Little Malad River, c.45 mi/72 km long, SE IDAHO; formed by confluence of two forks in ONEIDA county; flows S, through Daniels Reservoir, joining Deep Creek and Devil Creek near MALAD CITY to form MALAD RIVER.

Little Manistee River (MAN-is-TEE), c.50 mi/80 km long, W MICHIGAN; rises near LUTHER in Lake county; flows NW to MANISTEE LAKE and MANISTEE RIVER, near LAKE MICHIGAN, at MANISTEE; 44°00′N 85°36′W.

Little Marlow (MAH-lo), agricultural village (2001 population 1,331), S BUCKINGHAMSHIRE, S central

ENGLAND, on the THAMES, and 2 mi/3.2 km ENE of MARLOW; 51°35′N 00°44′W. Has Norman church.

Little Martinique, SAINT VINCENT AND THE GRENADINES: see PETITE MARTINIQUE.

Little Meadows, borough (2006 population 289), SUSQUEHANNA county, NE PENNSYLVANIA, 17 mi/27 km NW of MONTROSE, at NEW YORK state line, on Appalachian Creek; 41°59′N 76°07′W. Agriculture (corn, hay; dairying).

Little Missenden (MIS-uhn-duhn), village (2001 population 6,510), S BUCKINGHAMSHIRE, S central ENGLAND, 5 mi/8 km NE of HIGH WYCOMBE; 51°40′N 00°39′W. Has Norman church and Elizabethan manor home.

Little Missouri, river, c. 560 mi/901 km long, WYOMING, MONTANA, and SOUTH DAKOTA; rising in W CROOK county, NE Wyoming, and flowing NE through SE corner of Montana, NW corner of South Dakota, where it flows through Little Missouri National Grassland and N and S units of THEODORE ROOSEVELT NATIONAL PARK, and past Elkhorn Ranch Site into Garrison Reservoir on the Missouri River 13 mi/21 km N of Kildere, where it forms Little Missouri Bay (Little Missouri Bay State Park on S shore), an arm of the reservoir c. 30 mi/48 km long. Little Missouri River channel c. 25 mi/40 km NE of Kildere.

Little Missouri River, c.145 mi/233 km long, SW ARKANSAS; rising in the OUACHITA MOUNTAINS, SE POLK county, and flowing generally SE to join the OUACHITA RIVER N of CAMDEN. North of MURFREESBORO is Narrows Dam (1950), which impounds Lake GREESON.

Little Moose Mountain (3,630 ft/1,106 m), HAMILTON county, NE central NEW YORK, in ADIRONDACK MOUNTAINS, 15 mi/24 km NW of SPECULATOR; 43°40′N 74°35′W. Little Moose Lake (c.1 mi/1.6 km long) just NE.

Littlemore, ENGLAND: see OXFORD.

Little Mountain, village (2006 population 262), NEWBERRY county, NW central SOUTH CAROLINA, 24 mi/39 km WNW of COLUMBIA; 34°12′N 81°24′W. Manufacturing (marine transmissions). Agriculture (livestock, poultry; grain); timber.

Little Muddy River, c.60 mi/97 km long, S ILLINOIS; rises in SE WASHINGTON county; flows generally S, into BIG MUDDY RIVER W of HURST; 38°21′N 89°12′W.

Little Muskegon River (mus-KEE-guhn), c.35 mi/56 km long, central MICHIGAN; rises in small lakes in MECOSTA county; flows SW, past MORLEY, to MUSKEGON RIVER, 7 mi/11.3 km E of NEWAYGO; 43°34′N 85°16′W.

Little Muskingum River (muhs-KING-guhm, muhs-KING-uhm), c.65 mi/105 km long, SE OHIO; rises in MONROE county; flows generally SW, through WASHINGTON county, to OHIO RIVER 3 mi/5 km SE of MARIETTA; 39°22′N 81°24′W.

Little Namaqualand, SOUTH AFRICA: see NAMAQUALAND.

Little Narragansett Bay, on RHODE ISLAND-CONNECTICUT state line, inlet of the Atlantic estuary of PAWCATUCK RIVER. Sheltered by curving peninsula, site of WATCH HILL resort village. Yatching, fishing, harbor.

Little Neck, residential section of NE QUEENS borough of NEW YORK city, SE NEW YORK, on LITTLE NECK BAY; 40°46′N 72°45′W.

Little Neck Bay, inlet of LONG ISLAND SOUND indenting N shore of W LONG ISLAND, NASSAU county; SE NEW YORK, between QUEENS borough (W) and GREAT NECK peninsula (E); c.1.5 mi/2.4 km wide at entrance, 2.5 mi/4 km long; 40°48′N 72°46′W. At the NE corner of the bay entrance is Willet's Point, on which the Civil War–era fortification later known as Fort Totten was built.

Little Nemaha River (nee-MAH-hah), c.75 mi/121 km long, SE NEBRASKA; rises near LINCOLN; flows SE, past Syracuse and Auburn, to MISSOURI RIVER near Nemaha. Channel straightened in parts.

Little Niangua River (nei-ANG-gwuh), c.40 mi/64 km long, central MISSOURI; rises in the OZARKS in DALLAS county; flows NE to NIANGUA arm of LAKE OF THE OZARKS in CAMDEN county.

Little Nicobar Island, INDIA: see ANDAMAN AND NICOBAR ISLANDS.

Little Ocmulgee River (OK-muhl-gee), c.70 mi/113 km long, S central GEORGIA; rises in S TWIGGS county; flows SE, past MCRAE, to OCMULGEE River just SE of LUMBER CITY; 32°07′N 82°54′W.

Little Osage River (O-saij), 68 mi/109 km long, in W MISSOURI and E KANSAS; rises near MORAN, Kansas; flows E, joining MARAIS DES CYGNES River to form OSAGE RIVER SE of RICH HILL, Missouri.

Little Ossipee Pond (AHS-uh-pee), SW MAINE, center of WATERBORO resort area; 2.5 mi/4 km long; drains N into Little Ossipee River (17 mi/27 km long), which enters the SACO at LIMINGTON.

Little Pamir, AFGHANISTAN: see WAKHAN.

Little Paternoster Islands, INDONESIA: see BALABALAGAN ISLANDS.

Little Patuxent River (puh-TUX-ent), c.35 mi/56 km long, central MARYLAND; rises in N HOWARD county; flows SE, past Fort George G. Meade, to the PATUXENT 4 mi/6.4 km ESE of BOWIE.

Little Peconic Bay, SE NEW YORK, between N and S peninsulas of E LONG ISLAND, E of GREAT PECONIC BAY; c.6 mi/9.7 km long, 4.5 mi/7.2 km wide; 41°00′N 72°24′W.

Little Pee Dee River, c.90 mi/145 km long, E SOUTH CAROLINA; rises in E MARLBORO county near NORTH CAROLINA state line; flows SE past DILLON to GREAT PEE DEE RIVER, 18 mi/29 km W of MYRTLE BEACH.

Little Pendulum Island, GREENLAND: see PENDULUM ISLAND.

Little Pigeon Creek, c.27 mi/43 km long, SW INDIANA; rises in E WARRICK county; flows SW along most of the boundary between Warrick and SPENCER counties to join the OHIO River c.5 mi/8 km SE (upstream) of NEWBURGH.

Little Platte River, c.170 mi/274 km long, SW IOWA and NW MISSOURI; rises near CRESTON, Iowa; flows generally S to MISSOURI RIVER, below LEAVENWORTH, KANSAS. Sometimes called PLATTE RIVER.

Little Point, village, MORGAN county, central INDIANA, 14 mi/23 km NW of MARTINSVILLE. In agricultural area. Founded 1829.

Little Polissya, UKRAINE: see VOLHYNIAN UPLAND.

Little Popo, TOGO: see ANÉHO.

Little Popo Agie River, WYOMING: see POPO AGIE RIVER.

Littleport, town (2000 population 26), CLAYTON county, NE IOWA, on VOLGA RIVER, and 7 mi/11.3 km S of ELKADER; 42°45′N 91°22′W. In corn, hog, dairy region.

Littleport (LIT-uhl-pawt), agricultural village (2001 population 7,521), N CAMBRIDGESHIRE, E ENGLAND, on OUSE RIVER, and 4 mi/6.4 km NNE of ELY; 52°28′N 00°19′E.

Little Powder River, c. 100 mi/161 km long, in WYOMING and MONTANA; rises in NE CAMPBELL county, NE Wyoming, 25 mi/40 km NNE of GILLETTE at joining of Cottonwood and Rawhide creeks; flows N to POWDER RIVER 5 mi/8 km NE of BROADUS, POWDER RIVER county, SE Montana.

Little Prairie, CANADA: see CHETWYND.

Little Prespa Lake (PRES-pah), Albanian *Liqen i Prespes e vogël*, Greek *Limni Mikre* (or *Mikri*) *Prespa* (□ 19 sq mi/49 sq km), on Albanian-Greek border, near Macedonia border, just S of PRESPA LAKE (separated by 1.5-mi/2.4-km-long sandspit); 12 mi/19 km long, 4 mi/6.4 km wide; c.16 ft/5 m deep. Fisheries. A former inlet of Prespa Lake, it is gradually drying out.

Little Quarter, Czech MALÁ STRANA (MAH-lah STRAH-nah), German *Kleinseite*, W central district of PRAGUE, PRAGUE-CITY province, central BOHEMIA, CZECH REPUBLIC, on left bank of VLTAVA RIVER. After Stare Mesto, Prague's second town, founded by Premysl Otakar II in 1257; it is the residence of the Czech Parliament and many foreign embassies.

Little Red River, 105 mi/169 km long, NW ARKANSAS; rising in the BOSTON MOUNTAINS, and flowing SE to the WHITE RIVER. Archeys Fork (30 mi/48 km long) joins South Fork at CLINTON (40 mi/64 km long); both rise in W VAN BUREN county. Middle Fork (60 mi/97 km long) rises in S SEARCY county, and joins South Fork at EDGEMONT, CLEBURNE county, in GREERS FERRY LAKE. Main river continues another 75 mi/121 km to White River, part SEARCY. Greers Ferry Dam and reservoir (completed 1964) provide flood control and hydroelectric power.

Little Rich Mountain (3,320 ft/1,012 m), in the BLUE RIDGE, GREENVILLE county, NW SOUTH CAROLINA, c.25 mi/40 km NNW of GREENVILLE, near NORTH CAROLINA state line.

Little River, county (□ 564 sq mi/1,466.4 sq km; 2006 population 13,074), extreme SW ARKANSAS; ⊙ ASHDOWN; 33°42′N 94°14′W. Bounded W by TEXAS, S by RED RIVER, N and E by LITTLE RIVER. Agriculture (wheat, soybeans; cattle, hogs). Manufacturing (wood products); lumber milling. Timber; sand, gravel, limestone. MILLWOOD LAKE (Little River) and State Park in E. Formed 1867.

Little River (LI-tuhl RI-vuhr), township, VICTORIA, SE AUSTRALIA, 27 mi/44 km WSW of MELBOURNE, and N of GEELONG. Railway station.

Little River, township, Banks Peninsula district, SOUTH ISLAND, NEW ZEALAND, on SW BANKS PENINSULA, near LAKE ELLESMERE, 17 mi/27 km S of CHRISTCHURCH; 43°46′S 172°47′E. Dairy and sheep farming.

Little River, town (2000 population 7,027), HORRY county, E SOUTH CAROLINA, 25 mi/40 km E of CONWAY, and on tidal LITTLE RIVER near NORTH CAROLINA state line; 33°52′N 78°37′W. Manufacturing (pulpwood). Agriculture (timber; livestock; watermelons, vegetables, grain).

Little River, village (2000 population 536), RICE county, central KANSAS, on LITTLE ARKANSAS RIVER, and 11 mi/18 km ENE of LYONS; 38°23′N 98°00′W. In wheat area. Oil wells nearby.

Little River, c.30 mi/48 km long, in NE ALABAMA; formed by confluence of two headstreams in Lookout Mountain, NE Alabama; flows SW to WEISS LAKE, 5 mi/8 km NNE of CENTRE.

Little River, c. 25 mi/40 km long, in E CONNECTICUT; rises near HAMPTON; flows S to SHETUCKET RIVER, 5 mi/8 km NNE of NORWICH.

Little River, c.75 mi/121 km long, E GEORGIA; rises near MAXEYS; meanders E to SAVANNAH River, 20 mi/32 km NW of AUGUSTA; 33°37′N 82°52′W.

Little River, c.70 mi/113 km long, in SW KENTUCKY; formed in CHRISTIAN county S of HOPKINSVILLE by junction of its North and South forks (North Fork rises in Hopkinsville, flows S c.10 mi/16 km; South Fork rises in E Christian county, flows WSW c.20 mi/32 km.); flows SW, then WNW, past CADIZ, to CUMBERLAND RIVER. 8 mi/12.9 km W of Cadiz, receives Muddy Fork (rises in NW Christian county, flows WSW c.30 mi/48 km). 3 mi/4.8 km W of Cadiz; river below Cadiz forms E arm of LAKE BARKLEY reservoir.

Little River, c. 90 mi/145 km long, in central LOUISIANA; formed N of GEORGETOWN by junction of DUGDEMONA RIVER and CASTOR CREEK; flows SE to CATAHOULA LAKE, and ENE to OUACHITA RIVER (BLACK RIVER) at JONESVILLE; 31°37′N 91°48′W. Little River State Wildlife Area on W bank.

Little River, c.70 mi/113 km long, SE MISSOURI; formerly the outflow of swamps and wetlands of SE Missouri in NEW MADRID, PEMISCOT, and DUNKLIN counties, into which the CASTOR and WHITEWATER rivers emptied. Now completely channelized to the ARKANSAS state line and referred to as the Little River drainage channels. Remnants of the original channel remain. Continues in Arkansas as Little River and joins the SAINT FRANCIS River at MARKED TREE, Arkansas. Swamps and wetlands were affected by NEW MADRID earthquakes 1811–1812.

Little River, 90 mi/145 km long, central OKLAHOMA; rises SE of OKLAHOMA CITY; flows generally SE, through city of NORMAN and part of MACOMB, to CANADIAN River, in HUGHES county, c.5 mi/8 km S of HOLDENVILLE. Dammed in CLEVELAND county, forming Lake Thunderbird. Little River State Park on Lake Thunderbird, at Norman.

Little River, c.220 mi/354 km long, OKLAHOMA and ARKANSAS; rises S of PINE VALLEY in OUACHITA MOUNTAINS (Kiamichi Mountains) in SW LE FLORE county, Oklahoma; flows SW, then SE, past WRIGHT CITY, into Arkansas, through MILLWOOD LAKE reservoir, where it receives SALINE RIVER from N; joins RED RIVER just W of Fulton, Arkansas Mountain Fork, c.40 mi/64 km long, rises in W Arkansas, enters MCCURTAIN county, Oklahoma, through BROKEN BOW LAKE reservoir, entering main stream 10 mi/16 km SE of Broken Bow. PINE CREEK LAKE formed in McCurtain county. Millwood Reservoir (for flood control) is c.10 mi/16 km NW of Fulton. Little River crosses into Arkansas at elevation 287 ft/87 m, the lowest in Oklahoma.

Little River, c.50 mi/80 km long, in E TENNESSEE; rises in Great Smoky Mountains National Park, on CLINGMANS DOME, near NORTH CAROLINA state line; flows NW past ELKMONT, TOWNSEND, and WALLAND, to FORT LOUDOUN Reservoir (TENNESSEE RIVER) 8 mi/13 km S of KNOXVILLE; 35°52′N 83°59′W.

Little River (RIV-uhr), c.75 mi/121 km long, in central TEXAS; formed by LEON and LAMPASAS rivers 9 mi/14.5 km S of TEMPLE; flows generally SE and E past CAMERON, to BRAZOS river, c.5 mi/8 km W of HEARNE.

Little River (LI-tuhl), c.30 mi/48 km long, N VIRGINIA; rises in NE FAUQUIER county; flows NE to Goose Creek (a tributary of POTOMAC RIVER), in LOUDOUN county, 3 mi/5 km SE of LEESBURG.

Little River (LI-tuhl), c.50 mi/80 km long, SW VIRGINIA; rises in NE FLOYD county; flows S and WNW, to NEW RIVER, 3 mi/5 km S of RADFORD.

Little River (LI-tuhl), c.40 mi/64 km long, central VIRGINIA; rises in E LOUISA county; flows SE to NORTH ANNA RIVER, 5 mi/8 km W of HANOVER.

Little River, GEORGIA: see WITHLACOOCHEE RIVER.

Little River–Academy (uh-KAD-uh-mee), town, Bell county, central TEXAS, suburb 7 mi/11.3 km S of TEMPLE, NE of point where LEON and LAMPASAS rivers form LITTLE RIVER; 30°59′N 97°20′W. Agricultural area. Manufacturing (feeds, fertilizers).

Little River Dam, SOUTH CAROLINA: see KEOWEE, LAKE.

Little Rocher (LI-tuhl rahsh), village, SE NEW BRUNSWICK, CANADA, on CHIGNECTO BAY, 18 mi/29 km SW of HOPEWELL CAPE. Gypsum quarrying.

Little Rock, city (2000 population 183,133), ⊙ of state and of PULASKI county, central ARKANSAS, on the ARKANSAS RIVER; 34°43′N 92°20′W. Drained by Fourche Creek. MURRAY LOCK AND DAM to NW; Lake MAUMELLE reservoir and Pinnacles State Park to NW. It is a river port and the administrative, commercial, transportation, and cultural center of the state. Diversified manufacturing and consumer goods. The city's industries process agricultural products, fish, beef, poultry, and bauxite and timber. Its manufacturing industries are closely related with those of NORTH LITTLE ROCK across the river. The settlement was a well-known river crossing when Arkansas Territory was established in 1819. It became territorial capital in 1821 and state capital when Arkansas entered the Union in 1836. In the Civil War the battle of Little Rock (1863) was fought there. The city became a center of world attention in 1957, when Federal troops were sent there to enforce a 1954 U.S. Supreme Court ruling against segregation in the public schools. Little Rock is the seat of Philander Smith College, Arkansas Baptist College, the University of Arkansas at Little Rock, and several other branches of the University, including the law and medical schools. Of interest are the beautiful old statehouse (which served as capitol from 1836 to 1910) and several museums; the present capitol building was built in 1911. The city also contains several state institutions. Little Rock Air Force Base is to NE in JACKSONVILLE. Arkansas School for Blind and Deaf here. Camp Robinson National Guard Training Area to N; Metropolitan Vocational Technical Educational Center. Livestock Showgrounds and Barton Coliseum; Adams Field Municipal Airport and Little Rock Post Industrial Park (on Arkansas River) in E end. Incorporated 1831.

Littlerock, unincorporated town (2000 population 1,402), LOS ANGELES county, S CALIFORNIA, 14 mi/23 km SE of LANCASTER; 34°32′N 117°59′W. California Aqueduct passes to S. SAN GABRIEL MOUNTAINS to S. Pears, grain; dairying; poultry, cattle.

Little Rock, town (2000 population 489), LYON county, NW IOWA, on LITTLE ROCK RIVER, and 15 mi/24 km E of ROCK RAPIDS; 43°26′N 95°52′W. In livestock and grain area.

Little Rock River, 40 mi/64 km long, SW MINNESOTA and NW IOWA; rises in central NOBLES county, SW Minnesota, c.10 mi/16 km W of Washington, flows SW into NW Iowa, past LITTLE ROCK town, to ROCK RIVER 4 mi/6.4 km NE of ROCK VALLEY; 43°36′N 95°40′W.

Little Ruaha River (roo-AH-hah), c. 140 mi/225 km long, IRINGA region, central TANZANIA, in SOUTHERN HIGHLANDS; rises near Ihende, 60 mi/97 km SW of IRINGA; flows NE, then NW, past Iringa; joins GREAT RUAHA RIVER 35 mi/56 km NW of Iringa.

Little Rye Island, SLOVAKIA and HUNGARY: see GREAT RYE ISLAND.

Little Sac River (SAK), c.45 mi/72 km long, SW central MISSOURI; rises in the OZARKS N of SPRINGFIELD; flows NW to SAC RIVER in CEDAR county, as part of STOCKTON LAKE.

Little Saint Bernard Pass, French *Petit-Saint-Bernard, Col du* (pe-tee–san–ber-nahrd, kol dyoo), Italian *Piccolo-San-Bernardo*, Alpine pass (7,178 ft/2,188 m) between the TARENTAISE (upper valley of ISÈRE RIVER), SE FRANCE, and the Val d'AOSTA in NW Italy; 45°40′N 06°53′E. It is crossed by scenic road from BOURG-SAINT-MAURICE (7 mi/11.3 km SW, in SAVOIE department, RHÔNE-ALPES region) to Morgex (10 mi/16 km NE) in Italy. Steep curvy ascents, especially on French side. La ROSIÈRE ski resort (elevation 6,070 ft/1,850 m) lies on French approach to pass. Hospice (just SW of pass), founded in 10th century by St. Bernard of Menthon, lies in ruins. The pass has witnessed the crossing of many armies over the centuries; now it is of tourist interest as a leg in the automobile route surrounding the MONT BLANC massif. The Italian side of Mont Blanc (10 mi/16 km N) can be seen from the pass.

Little Salkehatchie River, SOUTH CAROLINA: see COMBAHEE RIVER.

Little Salmon River, 40 mi/64 km long, W IDAHO; rises in mountain region S of NEW MEADOWS, E ADAMS county; flows N, through deep canyon, to SALMON RIVER at RIGGINS, in SW IDAHO county.

Little San Bernardino Mountains, SE continuation of SAN BERNARDINO MOUNTAINS, mainly RIVERSIDE county, S CALIFORNIA, extend NW into SAN BERNARDINO county and c.40 mi/64 km NW-SE along E side of COACHELLA VALLEY. Elevation c.4,000 ft/1,219 m–5,500 ft/1,676 m. Partly (E slope) within SW part of JOSHUA TREE NATIONAL MONUMENT. COLORADO RIVER AQUEDUCT passes along SW base of range.

Little Sandy River, c.90 mi/145 km long, NE KENTUCKY; rises in SW ELLIOTT county; flows generally NNE, past SANDY HOOK, through GRAYSON LAKE reservoir, and past GRAYSON, to OHIO RIVER at Greenup.

Little San Salvador, islet, central BAHAMAS, just W of N CAT ISLAND, 95 mi/153 km ESE of NASSAU; a narrow, bifurcated bar, c.6 mi/9.7 km long W-E; 24°35′N 75°55′W. San Salvador or Watling Island, where Columbus made his first landfall, is 95 mi/153 km ESE, on the other side of Cat Island.

Little Satilla River (suh-TIL-uh), c.30 mi/48 km long, in SE GEORGIA; rises in E BRANTLEY county; flows ESE forming the county border between GLYNN and CAMDEN counties, emptying into Saint Andrews Sound between the sea islands of JEKYLL and CUMBERLAND; 31°07′N 81°40′W.

Little Scarcies River (SKAHR-seez), c.170 mi/274 km long, largely in SIERRA LEONE; rises SW of DABOLA, S central GUINEA; flows SW past MANGE, and forms common estuary with the GREAT SCARCIES 25 mi/40 km N of FREETOWN. Navigable for 22 mi/35 km below Mange. Also called Kaba River.

Little Scioto River (sei-O-tuh), c.40 mi/64 km long, S OHIO; rises in JACKSON county; flows S through SCIOTO county to the OHIO RIVER 6 mi/10 km E of PORTSMOUTH; 38°45′N 82°53′W.

Little Sea, POLAND: see PUCK BAY.

Little Sebago Lake (suh-BAI-go), SW MAINE, in central CUMBERLAND county, E of SEBAGO LAKE; 6 mi/9.7 km long, 0.5 mi/0.8 km–1 mi/1.6 km wide. Drains SSW into PRESUMPSCOT RIVER.

Little Silver, borough (2006 population 6,089), MONMOUTH county, E NEW JERSEY, just SE of RED BANK, and 13 mi/21 km NE of FREEHOLD; 40°19′N 74°01′W. Fort Monmouth nearby. Incorporated 1923.

Little Sioux, town (2000 population 217), HARRISON county, W IOWA, on Little Sioux River near its mouth on MISSOURI RIVER, and 17 mi/27 km NW of LOGAN; 41°48′N 96°01′W.

Little Skellig, IRELAND: see SKELLIGS.

Little Smoky (LI-tuhl SMO-kee), hamlet, W central ALBERTA, W CANADA, in GREENVIEW NO. 16 municipal district; 54°44′N 117°11′W.

Little Smoky River (LI-tuhl SMO-kee), 185 mi/298 km long, ALBERTA, W CANADA; rises in ROCKY MOUNTAINS N of JASPER NATIONAL PARK; flows E and N to SMOKY RIVER 60 mi/97 km NE of GRANDE PRAIRIE; 55°40′N 117°38′W.

Little Snake River, c.150 mi/241 km long, in NW COLORADO and S WYOMING; rises in N tip of PARK RANGE, N ROUTT county, near CONTINENTAL DIVIDE, Routt National Forest Colorado; flows W, along Wyoming/Colorado state line, meanders into CARBON county, Wyoming; past DIXON and BAGGS, Wyoming, then SW into Colorado, to YAMPA RIVER c.45 mi/72 km W of CRAIG, E of Dinosaur National Monument.

Little Sodbury, ENGLAND: see OLD SODBURY.

Little Sodus Bay (SO-duhs), inlet of Lake ONTARIO, CAYUGA county, W central NEW YORK, 12 mi/19 km E of SODUS BAY; c.2 mi/3.2 km long, 0.5 mi/0.8 km–0.75 mi/1.21 km wide; 43°20′N 76°43′W. FAIR HAVEN (resort); Fair Haven Beach State Park.

Little Sound, BERMUDA: see PORT ROYAL BAY.

Little Squam Lake, NEW HAMPSHIRE: see SQUAM LAKE.

Little St. Lawrence, village, SE NEWFOUNDLAND AND LABRADOR, CANADA, on SW side of PLACENTIA BAY, on BURIN PENINSULA, 22 mi/35 km ESE of GRAND BANK; 46°55′N 55°21′W.

Little Stour River, ENGLAND: see GREAT STOUR RIVER.

Littlestown, borough (2006 population 4,157), ADAMS county, S PENNSYLVANIA, 10 mi/16 km SE of GETTYSBURG, near MARYLAND state line; 39°44′N 77°05′W. Agricultural area (grain, soybeans; poultry, livestock; dairying). Manufacturing (furniture, consumer goods, machinery, metal products, food products). Long Arm Reservoir to E. Laid out 1765, incorporated 1864.

Little Switzerland (LIT-uhl SWITS-uhr-luhnd), unincorporated village, MCDOWELL county, W NORTH CAROLINA, 12 mi/19 km NNW of MARION, in the BLUE RIDGE MOUNTAINS, in Pisgah National Forest. BLUE RIDGE PARKWAY passes to NW; 35°50′N 82°05′W. Resort area. Gems. Emerald Village, North Carolina Mining Museum.

Little Tallahatchie River, MISSISSIPPI: see TALLAHATCHIE RIVER.

Little Tallapoosa River (tal-uh-POO-suh), c.90 mi/145 km long, in W GEORGIA and E ALABAMA; rises in N CARROLL county; flows SW, past CARROLLTON, into ALABAMA to TALLAPOOSA River, 6 mi/9.7 km W of WEDOWEE; 33°46′N 84°57′W.

Little Tancook Island (tan-kook), islet in MAHONE BAY, S NOVA SCOTIA, CANADA, 8 mi/13 km SE of CHESTER; 44°28′N 64°08′W.

Little Tennessee River, c.135 mi/217 km long, GEORGIA, NORTH CAROLINA, and TENNESSEE; rising in the BLUE RIDGE, NE Georgia, and flowing generally NW across SW North Carolina and through E Tennessee to the TENNESSEE RIVER opposite LENOIR CITY; 34°55′N 83°26′W. On the river in North Carolina, near the Tennessee state line, is Fontana Dam (480 ft/146 m high; 2,365 ft/721 m long; completed 1945), impounding FONTANA LAKE. It is part of the Tennessee Valley Authority and is the highest dam E of the ROCKY MOUNTAINS. The dam provides flood control, river regulation, and hydroelectricity. CHEOAH DAM in North Carolina, and CALDERWOOD and Chilhowee dams in Tennessee, are also part of the Tennessee Valley Authority.

Little Tibet, PAKISTAN and INDIA: see BALTISTAN LADAKH.

Little Tobago Island (to-BAI-go), islet off NE TOBAGO, 18 mi/29 km NE of SCARBOROUGH; 11°13′N 60°30′W. Noted as reserve for birds of paradise, introduced from DUTCH NEW GUINEA in 1909. The island (c.500 acres/202 ha) was presented to government of TRINIDAD AND TOBAGO in 1929. Sometimes called BIRD OF PARADISE ISLAND.

Littleton, city (2000 population 40,340), ⊙ ARAPAHOE county, N central COLORADO, suburb 7 mi/11.3 km S of downtown DENVER, on SOUTH PLATTE RIVER; 39°35′N 105°00′W. Located in an irrigated farm area rapidly being displaced by urbanization. Manufacturing (construction materials, electronic games, medical supplies, metal products, consumer goods, furniture). Arapahoe Community College, Littleton Historical Museum, Arapaho County Fairgrounds, and a thoroughbred racing track, Continental Racetrack, are here. CHATFIELD LAKE and State Park to SW. Plotted 1812, incorporated 1890. Site of 1999 Columbine High School massacre.

Littleton, town, AROOSTOOK county, E MAINE, 8 mi/12.9 km N of HOULTON; 46°13′N 67°50′W. In potato-growing area. Incorporated 1856.

Littleton, rural town, MIDDLESEX county, NE central MASSACHUSETTS, 13 mi/21 km SW of LOWELL; 42°32′N 71°29′W. Agriculture (poultry; dairying; apples). Manufacturing (consumer goods, medical supplies, computer equipment). Skiing at Nashoba Valley. Includes village of Littleton Common. Settled on site of "praying Indian" village of Nashoba, established c.1656; incorporated 1715.

Littleton, town, GRAFTON county, NW NEW HAMPSHIRE; bounded NW by CONNECTICUT RIVER (VERMONT state line); 44°19′N 71°48′W. Drained by AMMONOOSUC RIVER. Agriculture (cattle, poultry; dairying; vegetables; nursery crops, sugar maples; timber). Manufacturing (electrical equipment, food products, machinery, lumber, pharmaceuticals, printing and publishing, metal products). Moore Dam on CONNECTICUT RIVER forms MOORE RESERVOIR in N. Settled 1769, incorporated 1784.

Littleton, village (2000 population 197), SCHUYLER county, W ILLINOIS, 8 mi/12.9 km NNW of RUSHVILLE; 40°13′N 90°37′W. In agricultural area.

Littleton (LI-tuhl-tuhn), village (□ 1 sq mi/2.6 sq km; 2006 population 655), WARREN and HALIFAX counties, N NORTH CAROLINA, 15 mi/24 km W of ROANOKE RAPIDS; 36°25′N 77°54′W. Service industries; manufacturing (wood products, apparel); agriculture (tobacco, cotton, peanuts, grain, sweet potatoes;

Area is shown by the symbol □, and capital city or county seat by ⊙.

livestock). Lake GASTON reservoir (ROANOKE RIVER) to N. Founded before 1775.

Littleton, village (2006 population 198), WETZEL county, N WEST VIRGINIA, 19 mi/31 km ENE of NEW MARTINSVILLE, on Fish Creek; 39°42'N 80°30'W. SW corner of PENNSYLVANIA (W end of MASON-DIXON LINE) 2 mi/3.2 km to NNE.

Little Traverse Bay, NW MICHIGAN, inlet of Lake MICHIGAN, c.15 mi/24 km NE of GRAND TRAVERSE BAY; c.10 mi/16 km long, 5 mi/8 km wide; 45°24'N 85°00'W. BAY VIEW, PETOSKEY, HARBOR SPRINGS, and resort villages are on its shores.

Little Truckee River, c.30 mi/48 km long, E CALIFORNIA; rises in Weber Lake in the SIERRA NEVADA; flows E and S past Sierraville, and through Boca Reservoir to TRUCKEE RIVER, 6 mi/9.7 km NE of Truckee, near Nevada state line. Boca Dam (1,629 ft/497 m long, 116 ft/35 m high; completed 1939 by Bureau of Reclamation) is on lower course near mouth. Forms small reservoir (capacity 40,900 acre-ft) and is chief unit in Truckee storage project. Water from reservoir is released into Truckee River and used for irrigation of 47 sq mi/122 sq km in Washoe and Storey counties, W Nevada, and to supplement CARSON and Truckee rivers in supplying Newlands irrigation project in vicinity of Fallon, W Nevada.

Little Tupper Lake, NEW YORK: see TUPPER LAKE.

Little Unadilla Lake, NEW YORK: see MILLERS MILLS.

Little Usutu River, SWAZILAND: see LUSUSHWANA RIVER.

Little Valley, village (□ 1 sq mi/2.6 sq km; 2006 population 1,057), ⊙ CATTARAUGUS county, W NEW YORK, 7 mi/11.3 km NNW of SALAMANCA; 42°15'N 78°47'W. Light manufacturing, agriculture, and services. Incorporated 1876.

Little Vermilion River, c.55 mi/89 km long, in E ILLINOIS and W INDIANA; rises in SE CHAMPAIGN county, Illinois; flows generally E to the WABASH just E of NEWPORT, Indiana; 39°58'N 88°02'W.

Littleville Lake, reservoir, on HAMPDEN/HAMPSHIRE county border, W MASSACHUSETTS, on Middle Branch WESTFIELD RIVER, 12 mi/19 km WSW of NORTHAMPTON; 4 mi/6.4 km long; 42°15'N 72°50'W. Maximum capacity 40,500 acre-ft. Formed by Littleville Dam (earth construction; 159 ft/48 m high), built (1965) by the Army Corps of Engineers for flood control, water supply, and recreation.

Little Wabash River, c.200 mi/322 km long, E central and SE ILLINOIS; rises near MATTOON; flows S and SE to the WABASH near NEW HAVEN; 39°28'N 88°27'W. Dam impounds Lake Mattoon near Mattoon.

Little Wabash River (WAH-bash), c.30 mi/48 km long, in NE INDIANA; rises in W ALLEN county at FORT WAYNE; flows SW, past HUNTINGTON, to the WABASH RIVER c.2 mi/3.2 km W of Huntington.

Little Walsingham (WAWL-sing-uhm), village (2001 population 773), N NORFOLK, E ENGLAND, 5 mi/8 km N of FAKENHAM; 52°53'N 00°52'E. Site of Walsingham Abbey, one of the great shrines of medieval England. There are ruins of an Augustinian priory founded 1149. The sacred shrine of Our Lady of Walsingham, built 1061, became a center of medieval pilgrimages; there are no remains, but the priory chapel was restored in 1921 and the wayside Slipper Chapel (where pilgrims left their shoes) in 1934. There is also a 15th-century church here. Just NE is agricultural village of Great Walsingham, with 14th-century church.

Little Washita River (WAHSH-uh-tah), c.30 mi/48 km long, S OKLAHOMA; rises in SE CADDO county; flows SE and then NE, through GRADY county, to WASHITA River just SE of CHICKASHA.

Little Watts Island, Virginia: see WATTS ISLAND.

Little White River, c.135 mi/217 km long, S SOUTH DAKOTA; rises in SE SHANNON county; flows E past MARTIN, then NNE past ROSEBUD and village of WHITE RIVER to WHITE RIVER 12 mi/19 km SSE of MURDO;

43°11'N 102°09'W. Formerly South Fork of White River.

Little Wichita River (WICH-i-tah), c.50 mi/80 km long, N TEXAS; rises in ARCHER county; flows generally NE to RED RIVER 14 mi/23 km ENE of HENRIETTA. Dam impounds Lake Kickapoo (capacity 105,000 acre-ft), 26 mi/42 km SW of WICHITA FALLS. ARROWHEAD LAKE 13 mi/21 km SE of Wichita Falls.

Little Wood River, 90 mi/145 km long, S central IDAHO; rises in PIONEER MOUNTAINS, N BLAINE county; flows S, through Little Wood River reservoir, and past RICHFIELD, then W, past SHOSHONE, to BIG WOOD RIVER just W of GOODING.

Little York, town (2000 population 185), WASHINGTON county, S INDIANA, 13 mi/21 km NE of SALEM; 38°42'N 85°54'W. Agricultural area. Laid out 1831.

Little York, village (2000 population 269), WARREN county, NW ILLINOIS, 20 mi/32 km W of GALESBURG; 41°00'N 90°45'W. In agricultural area. On Cedar Creek.

Little Zab, IRAQ and IRAN: see ZAB, LITTLE.

Little Zelenchuk River, RUSSIA: see ZELENCHUK River.

Littoinen (LIT-toi-nen), Swedish *Littois*, suburb of TURKU, TURUN JA PORIN province, SW FINLAND, 4 mi/6.4 km ENE of city center; 60°27'N 22°24'E. Elevation 66 ft/20 m. Woolen mills.

Littoral, department (□ 31 sq mi/80.6 sq km; 2002 population 665,100), extreme S BENIN; ⊙ COTONOU; 06°22'N 02°25'E. Bordered on W and N by ATLANTIQUE department, E by OUÉMÉ department, and S by BIGHT OF BENIN. Established in 1999 out of the SE corner of Atlantique department. Includes Cotonou, Benin's largest city.

Littoral, province (□ 7,814 sq mi/20,316.4 sq km; 2004 population 2,380,000), W CAMEROON; ⊙ and largest city is DOUALA; 04°15'N 10°00'E. Bounded on S by South province, E by Central province, N by South-West and West provinces, W by Gulf of Guinea. Lowland area where bananas, palm oil, and cocoa are important. Douala has many industries and port facilities. Nkongsamba and Edéa are other important cities. Includes the departments of MOUNGO, NKAM, SANAGA-MARITIME, and WOURI.

Littoral Range, RUSSIA: see BAYKAL RANGE.

Litueche (lee-TWAI-chai), town, ⊙ Litueche comuna, CARDENAL CARO province, LIBERTADOR GENERAL BERNARDO O'HIGGINS region, central CHILE, 25 mi/40 km NE of PICHILEMU. Ancient Indian community. Fruit, vegetables, cereals.

Lituya Bay (li-TOO-yuh), SE ALASKA, inlet of Gulf of ALASKA, 100 mi/161 km SE of YAKUTAT, SSW of Mount FAIRWEATHER; 9 mi/14.5 km long, 2 mi/3.2 km wide; 58°38'N 137°34'W. At head of bay, mountains rise to 11,924 ft/3,634 m. Discovered 1786 by Count de la Pérouse, who named it Port des Français. Famous for extraordinary high tides.

Litvino, RUSSIA: see SOSNOVOBORSK, PENZA oblast.

Litvínov (LIT-vee-NOF), German *Leutensdorf*, city (2001 population 27,397), SEVEROCESKY province, NW BOHEMIA, CZECH REPUBLIC, in NE foothills of the ORE MOUNTAINS, on railroad; 50°36'N 13°37'E. Manufacturing (textiles); gas, electrical, and chemical works; power plant; synthetic fuel plant; intensive lignite mining in vicinity. Formed, c.1948, by union of HORNI LITVINOV, Czech *Horní Litvínov*, German OBERLEUTENSDORF, and DOLNI LITVINOV, Czech *Dolní Litvínov*.

Litvinovka (leet-VEE-nuhf-kah), village, W central ROSTOV oblast, S European RUSSIA, on the KALITVA RIVER, on road, 15 mi/24 km NNE of BELAYA KALITVA; 48°23'N 40°52'E. In agricultural area (sunflowers, sugar beets, feed corn, wheat); food processing.

Litvintsi, UKRAINE: see NOVA USHYTSYA.

Lityn (LEE-tin), (Russian *Litin*), town, W VINNYTSYA oblast, UKRAINE, 18 mi/29 km WNW of VINNYTSYA; 49°20'N 28°04'E. Elevation 836 ft/254 m. Raion center. Manufacturing (building material, peat); food pro-

cessing (cannery, dairy); feed milling; granite quarrying. Regional museum. Known since 1431, when a castle was built there; damaged during the Cossack-Polish War (1648–1657); involved in popular rebellions (1687, 1702–1703, 1750, 1768); passed to Russia (1793); involved in Yuriy Karmaliuk's rebellion (1820s–1830s); site of battles during struggle for Ukraine's independence (1919–1920).

Litzmannstadt, POLAND: see ŁÓDŹ, city.

Litz Manor, village, SULLIVAN county, NE TENNESSEE, suburb of KINGSPORT.

Liu'an (LIU-AN), city (□ 1,380 sq mi/3,574 sq km; 1994 estimated urban population 260,700; estimated total population 1,666,200), W ANHUI province, CHINA; 45 mi/72 km W of HEFEI, on PI RIVER; 31°48'N 116°30'E. Industry is the largest sector of the city's economy. Manufacturing (food processing, textiles, paper, chemicals, machinery). Agriculture (crop growing, animal husbandry). Sometimes appears as Lu'an.

Liuba (LIU-BAH), town, ⊙ Liuba county, SW SHAANXI province, CHINA, 40 mi/64 km N of Nanzheng; 33°37'N 106°55'E. Logging, timber processing, furniture.

Liucheng (LIU-CHENG), town, ⊙ Liucheng county, N central GUANGXI ZHUANG AUTONOMOUS REGION, CHINA, 10 mi/16 km NNW of LIUZHOU, and on LIU RIVER; 24°39'N 109°14'E. Agriculture (rice, wheat, beans, sugarcane). Manufacturing (chemicals, pharmaceuticals); sugar refining.

Liu-cheng, CHINA: see WILLOW PALISADE.

Liuchiu Island (LIU-CHIU) or **Hsiao Liuchiu** (SHIAO LIU-CHIU), island, TAIWAN, c.19 mi/30 km SSE of KAOHSIUNG; 22°37'N 120°24'E. Tourism has mostly replaced fishing as main industry. Known for seafood, coves, coral formations, and a sandy beach.

Liu-chou, CHINA: see LIUZHOU.

Liuhe (LIU-HUH), town, S JIANGSU province, CHINA, 25 mi/40 km NW of SHANGHAI, and on CHANG JIANG (Yangzi River). Commercial center; cotton, rice, beans; fisheries.

Liuhe, town, ⊙ Liuhe county, S JILIN province, NORTHEAST, CHINA, 40 mi/64 km NNW of TONGHUA, and on railroad; 42°15'N 125°42'E. Manufacturing (machinery, pharmaceuticals, food products, building materials). Agriculture (soybeans, grain; tobacco, medicinal plants).

Liuhekou (LIU-HUH-KO), town, NW HENAN province, CHINA, at E foot of TAIHANG MOUNTAINS, 20 mi/32 km NW of ANYANG. Coal mining.

Liu-kiu Islands, JAPAN: see RYUKYU ISLANDS.

Liukuei (LIU-GUEI) [means *Six Turtles* in Chinese], town, SW TAIWAN, 35 mi/56 km NE of KAOHSIUNG, on Laonung River. In a valley surrounded by high mountains.

Liuli (lee-OO-lee), village, RUVUMA region, S TANZANIA, 70 mi/113 km WSW of SONGEA, near Lake NYASA; 10°03'N 34°35'E. Timber; corn; goats, sheep; fish.

Liupan Mountains (LIU-PAN), SE GANSU province and S NINGXIA HUI Autonomous Region, CHINA, extends in N-S direction; rise to c.10,000 ft/3,048 m, 10 mi/16 km NNE of LONGDE. As an E edge of the central GANSU plateau, the LIUPAN Mountains is part of the HELAN-LIUPAN-Longmen-Ilao Mountain series that form the boundaries between W and E China. In 1935, the Communist Red Army climbed over the Liupan Mountains during the famous Long March. The mountains gained fame from Mao Zedong's poem recording the event. The region is among the least developed in China.

Liupanshui (LIU-PAN-SHUAI), city (□ 2,422 sq mi/6,273 sq km; 1994 estimated urban population 408,000; estimated total population 1,909,900), W GUIZHOU province, CHINA; 26°42'N 104°49'E. Agriculture (grain, vegetables, hogs). Manufacturing (iron and steel); major coal-mining center.

Liure (lee-OO-rai), town, EL PARAÍSO department, SE HONDURAS, 19 mi/30 km N of CHOLUTECA; 13°30'N

87°05′W. Elevation 3,701 ft/1,128 m. Subsistence farming; livestock.

Liu River (LIU), 300 mi/483 km long, in SE GUIZHOU and NE GUANGXI ZHUANG AUTONOMOUS REGION, CHINA; rises along Guizhou-Guangxi border; flows to HONGSHUI, two main tributaries are Rong River, which flows S past RONG'AN and LIUZHOU, and Hongshui River, which flows S and SE past DU'AN and Laibin. The Liu joins the Yu River at GUIPING to form Xi River.

Liuwa Plain National Park (□ 1,412 sq mi/3,671.2 sq km), WESTERN province, W ZAMBIA, 50 mi/80 km NW of MONGU, near ANGOLA border; between Luambimba River (forms E park boundary) and Luanginga River in Liuwa Plain. Abundant wildlife. Access from KALABO to S.

Liuyang (LIU-YANG), town, ⊙ Liuyang county, NE HUNAN province, CHINA, 38 mi/61 km E of CHANG-SHA; 28°12′N 113°36′E. Agriculture (oilseeds, tobacco). Manufacturing (crafts, papermaking, chemicals, food). Site of the Liuyang Lake, an artificial lake on the Liuyang River dammed in 1985.

Liuzhou (LIU-JO), city (□ 251 sq mi/652.6 sq km; 2000 population 751,311), N central GUANGXI ZHUANG Autonomous Region, S CHINA, on the LIU RIVER at the intersection of highways and three railroads; 24°17′N 109°15′E. Manufacturing (paper products, wood products, chemicals, textiles, food processing); iron and steel. Sometimes spelled Liu-chou.

Livadhion, Greece: see LIVADION.

Livadhostra, Bay of, Greece: see LIVADOSTRA, BAY OF.

Livadia, Greece: see LEVADIA.

Livadica (lee-VAHD-ee-tsah), peak (8,170 ft/2,490 m) in SAR Mountains, SERBIA, on MACEDONIA border, 13 mi/21 km NNE of TETOVO; 42°11′N 21°04′E. Also spelled Livaditsa.

Livadion (lee-VAH-[th]-ee-on), town, LÁRISSA prefecture, N THESSALY department, N GREECE, 36 mi/58 km NNW of LÁRISSA, at W foot of Olympos; 40°30′N 22°26′E. Olives; livestock products Passed to Greece in 1913. Also spelled Leivadion and Livadhion.

Livadiya (lee-VAH-dee-yah), town (2006 population 12,555), SW MARITIME TERRITORY, SE RUSSIAN FAR EAST, on the Vostok Bay of the Sea of JAPAN, on coastal highway extension, 32 mi/51 km SE of VLADI-VOSTOK, and 12 mi/19 km WNW of NAKHODKA; 42°52′N 132°40′E. Fish canning, ship repair.

Livadiya (lee-WAH-dee-yah), town, S Republic of CRIMEA, UKRAINE, on the BLACK SEA, 2 mi/3.2 km SW of YALTA city center; 44°28′N 34°09′E. Subordinated to Yalta city council. It produces wine and is a noted health resort. Dating from medieval times, Livadiya became a summer residence of the Russian czars in 1861. The Livadiya palace, built in 1910–1911, is now a sanatorium. It was the meeting place of the Yalta Conference in 1945.

Livadostra, Bay of (lee-vah-[TH]O-strah), inlet of Gulf of CORINTH, in E central GREECE, on N side of Megara Peninsula; 10 mi/16 km wide, 15 mi/24 km long. Also Bay of Leivadostra, or Livadhostra.

Līvāni (LEE-vah-nee) or **Livany**, German *lievenhof*, city (2000 population 10,368), SE LATVIA, in LATGALE, on right bank of the DVINA (DAUGAVA) River, and 35 mi/56 km NNW of DAUGAVPILS. Wool processing, weaving. In Russian VITEBSK government until 1920.

Livanjsko Polje (lee-VAHN-sko POL-ye) or **Livno Plain,** historical region, W BOSNIA, BOSNIA AND HERZEGOVINA, in karst. Some agriculture. Principal town is LIVNO.

Livany, LATVIA: see LĪVĀNI.

Livarot (lee-vah-ro), commune (□ 4 sq mi/10.4 sq km), CALVADOS department, NW central FRANCE, in BASSE-NORMANDIE region, 10 mi/16 km SSW of LI-SIEUX; 49°01′N 00°09′E. Noted cheese-manufacturing center; cider distilling.

Livengood (LEI-ven-gud), village, central ALASKA, 50 mi/80 km NNW of FAIRBANKS, at N end of ELLIOTT

HIGHWAY from Fairbanks. Placer gold mining; out-fitting center for prospectors. Airfield. Near Alaska Pipeline and DALTON HIGHWAY.

Livenka (LEE-veen-kah), village (2004 population 2,900), S central BELGOROD oblast, SW European RUSSIA; on road and railroad, 31 mi/50 km NNE of URAZOVO; 50°27′N 38°16′E. Elevation 446 ft/135 m. Machine building; winery.

Livenza River (lee-VEN-tsah), 70 mi/113 km long, NE ITALY; rises 4 mi/6 km N of SACILE; flows SE, across Venetian plain, past Sacile and MOTTA DI LIVENZA, to the ADRIATIC near CAORLE. Navigable for 30 mi/48 km.

Live Oak (LEIV OK), county (□ 1,078 sq mi/2,802.8 sq km; 2006 population 11,522), S TEXAS; ⊙ George West; 28°21′N 98°07′W. Drained by FRIO, ATASCOSA, and NUECES rivers. Cattle ranching, hogs, agriculture (grain sorghum, corn, cotton; hay). Oil, natural gas wells; sand and gravel. Tips State Recreation Park in center; part of CHOKE CANYON LAKE reservoir (Frio R) and State Park on W boundary; part of Lake Corpus Christi (Nueces River) in SE corner. Formed 1856.

Live Oak, city (2000 population 6,229), SUTTER county, N central CALIFORNIA, in Sacramento Valley, near FEATHER RIVER, 10 mi/16 km N of YUBA CITY; 39°16′N 121°40′W. Trade and shipping center for agricultural area (vegetables, melons, walnuts, prunes, peaches, pears); millwork. Sutter Butte to SW. Incorporated 1947.

Live Oak (LEIV OK), city (□ 7 sq mi/18.2 sq km; 2000 population 6,480), ⊙ SUWANNEE county, N central FLORIDA, c.80 mi/129 km E of TALLAHASSEE; 30°18′N 82°59′W. Railroad junction; chief bright-leaf tobacco market of the state; lumber milling; manufacturing (naval stores, apparel). Former site of Florida Memorial College, which acquired university status and relocated to the greater MIAMI area in the 1960s.

Live Oak (LIEV OK), city (2006 population 11,704), BEXAR county, S central TEXAS, suburb 14 mi/23 km NE of downtown SAN ANTONIO; 29°33′N 98°20′W. Agricultural area; oil and natural gas. Manufacturing (dog and fish food).

Live Oak, unincorporated town (2000 population 16,628), SANTA CRUZ county, W CALIFORNIA; residential suburb 1 mi/1.6 km NE of SANTA CRUZ, 2 mi/3.2 km N of MONTEREY BAY; 36°59′N 122°01′W. New Brighton State Beach to S. Apples, berries, plums, nursery stock.

Live Oak Manor, unincorporated town, JEFFERSON parish, SE LOUISIANA, residential suburb 7 mi/11 km W of downtown NEW ORLEANS, on the MISSISSIPPI RIVER. Elevated 9 ft/3 m.

Liverdun (lee-ver-duhn), town (□ 9 sq mi/23.4 sq km), MEURTHE-ET-MOSELLE department, in LORRAINE, NE FRANCE, on a meander of the MOSELLE, and on MARNE-RHINE CANAL, 7 mi/11.3 km NW of NANCY; 48°45′N 06°03′E. Metalworks (pumps), fruit preserving. Has 12th-century church.

Livermore, city (2000 population 73,345), ALAMEDA county, W central CALIFORNIA; suburb 28 mi/45 km ESE of downtown OAKLAND, in Livermore Valley; 37°42′N 121°46′W. Agriculture (grapes, oats; nursery products, roses; cattle); wineries. Manufacturing (construction materials, household items, metal and plastic products, communications equipment). Lawrence Radiation Laboratory of the University of California conducts nuclear research. Los Positas College (two-year). Livermore Municipal Airport to SW. Del Valle State Recreation Area to E. HETCH HETCHY AQUEDUCT passes E-W to S. Incorporated 1876.

Livermore, town (2000 population 431), HUMBOLDT county, N central IOWA, near East DES MOINES RIVER, 10 mi/16 km N of DAKOTA CITY; 42°52′N 94°10′W. Feed milling.

Livermore, town (2000 population 1,482), MCLEAN county, W KENTUCKY, 18 mi/29 km S of OWENSBORO, on GREEN RIVER, mouth of ROUGH RIVER to SE; 37°29′N 87°07′W. In agricultural area (tobacco, grain;

---

livestock). Manufacturing (metal fabrication, plastic products, furniture).

Livermore, town, ANDROSCOGGIN county, SW MAINE, 20 mi/32 km N of AUBURN; 44°24′N 70°12′W. Farming. Includes villages of Livermore and North Livermore. Incorporated 1795.

Livermore, town, GRAFTON county, N central NEW HAMPSHIRE, 20 mi/32 km NNE of PLYMOUTH, in White Mountain National Forest; 44°01′N 71°27′W. Drained by Sawyer River. Crossed by Kancamagus Highway. Timber.

Livermore Falls, town, ANDROSCOGGIN county, SW MAINE, 25 mi/40 km N of LEWISTON, and on the ANDROSCOGGIN RIVER; 44°25′N 70°09′W. Paper mills; shoes. Called East Livermore until 1930. Settled 1786, incorporated 1843.

Livermore, Mount (LI-vuhr-mor), (8,381 ft/2,555 m), extreme W TEXAS, 24 mi/39 km NNW of MARFA; 30°37′N 104°10′W. Highest peak in DAVIS MOUNTAINS and second highest in state. Sometimes called Baldy Peak or Old Baldy.

Liverpool (LIV-uhr-pool), city (2001 population 439,473), ⊙ MERSEYSIDE, NW ENGLAND, on the MERSEY River near its mouth; 53°25′N 03°00′W. One of Britain's largest cities. A large center for food processing (especially flour and sugar), Liverpool has a variety of industries, including the manufacturing of electrical equipment, chemicals, and rubber. Its first wet dock was completed by 1715; today, Liverpool's docks are more than 7 mi/11.3 km long. Once Britain's greatest port, Liverpool suffered extreme setbacks with the advent of container ships, which it could not handle, and the shift in Great Britain's trade focus from the U.S. to the EU. The city is connected by tunnels, with BIRKENHEAD across the Mersey. Liverpool was once famous for its pottery, and its textile industry was also prosperous; however, since World War II its cotton market has declined considerably. In the mid-1980s, unemployment rose in the metropolitan area, especially among people under the age of 27.

In 1207, King John granted Liverpool its first charter. In 1644, during the English Civil War, Liverpool surrendered to the royalists under Prince Rupert after several sieges. Air raids during World War II caused heavy damage and casualties. Liverpool Cathedral, the largest in England, designed by Sir Giles Gilbert Scott, was begun in 1904 and completed in 1978. A Roman Catholic cathedral was consecrated in 1967. St. George's Hall is an imposing building in a group that includes libraries and art galleries. The Walker Gallery has a fine collection of Italian and Flemish paintings, as well as more modern works. The University of Liverpool was incorporated in 1903. There is a separate school of tropical medicine. John Moores University was incorporated in 1992. The statesman William Gladstone, the artist George Stubbs, and the members of the musical group the Beatles were born here. The district includes Garston, Kirkdale, Speke, Walton, Wavertree, and West Derby.

Liverpool (LI-vur-pool), town (2001 population 2,866), ⊙ QUEENS county, SW NOVA SCOTIA, CANADA, at head of Liverpool Bay (5 mi/8 km long) of the ATLANTIC OCEAN, at mouth of Mersey River, 70 mi/113 km NW of HALIFAX; 44°02′N 64°43′W. Elevation 98 ft/29 m. Shipbuilding and lumbering were important industries. Recent focus on newsprint and paper products production, fish processing, and tourism.

Liverpool, village (2006 population 2,396), ONONDAGA county, central NEW YORK, on ONONDAGA LAKE, just N of SYRACUSE; 43°06′N 76°12′W. Formerly known for its salt-mining and willow-weaving industries, today the village supports a variety of light manufacturing and service industries. Part of town of Salina. Incorporated 1830.

Liverpool (LI-vuhr-pool), village (2006 population 423), BRAZORIA county, SE TEXAS, residential suburb 32 mi/51 km S of downtown HOUSTON, and 22 mi/

Area is shown by the symbol □, and capital city or county seat by ⊙.

35 km WSW of TEXAS CITY, on Chocolate Bayou; 29°18′N 95°16′W. Agricultural area. Oil and natural gas.

Liverpool, borough (2006 population 887), PERRY county, central PENNSYLVANIA, 22 mi/35 km NNW of HARRISBURG, on SUSQUEHANNA RIVER (ferry to MILLERSBURG 2 mi/3.2 km to S); 40°34′N 76°59′W. Agricultural area (corn, hay; dairying). Manufacturing (modular homes, molded rubber products). Buffalo Mountain ridge to SW, end of Mahantango Mountain ridge to E.

Liverpool (LI-vuhr-pool), residential suburb, E NEW SOUTH WALES, AUSTRALIA, 17 mi/27 km WSW of SYDNEY; 33°54′S 150°56′E. Some agriculture.

Liverpool Bay, NOVA SCOTIA, CANADA: see LIVERPOOL.

Liverpool, Cape, N BYLOT ISLAND, NUNAVUT territory, CANADA, on BAFFIN BAY, at E end of LANCASTER SOUND; 73°45′N 77°45′W.

Liverpool Coast, Danish *Liverpool Kyst*, region, E GREENLAND, on GREENLAND SEA, extends 80 mi/129 km N from mouth of SCORESBY SOUND; 71°00′N 21°30′W. Rugged and indented, noted for its dangerous currents, and scene of numerous shipwrecks.

Liverpool, Curiche (koo-REE-chai) [Spanish= Liverpool lagoon], marshy lake in SANTA CRUZ department, NE BOLIVIA, 120 mi/193 km N of CONCEPCIÓN; 15 mi/24 km long, 13 mi/21 km wide; 14°21′S 62°6′W. Affluent and outlet: SAN MARTÍN RIVER. Formerly called Lake Rey.

Liverpool Range (LI-vuhr-pool), E central NEW SOUTH WALES, AUSTRALIA, part of GREAT DIVIDING RANGE, extending E-W between COONABARABRAN and MURRURUNDI; 31°45′S 150°45′E. Rises to 4,500 ft/1,372 m (Oxley's Peak).

Livet-et-Gavet (lee-ve–e–gah-ve), rural township (□ 18 sq mi/46.8 sq km), ISÈRE department, RHÔNE-ALPES region, SE FRANCE, in Oisans valley (along deep Alpine gorge of ROMANCHE RIVER), 11 mi/18 km SE of GRENOBLE; 45°06′N 05°56′E. Between Livet and Gavet (4 mi/6.4 km apart) are electrochemical and electrometallurgical works powered by local hydroelectric plants. The nearest town is VIZILLE near junction of Romanche and DRAC rivers.

Livets River, POLAND: see LIWIEC RIVER.

Livet Water, Scotland: see GLENLIVET.

Liv Glacier, outlet glacier in E ANTARCTICA, flowing from the polar plateau through the QUEEN MAUD MOUNTAINS, into SE ROSS ICE SHELF; 40 mi/65 km long; 84°55′S 168°00′W.

Livindo River, CAMEROON: see IVINDO RIVER.

Livingston, county (□ 1,045 sq mi/2,717 sq km; 2006 population 38,658), E central ILLINOIS; ⊙ PONTIAC; 40°53′N 88°33′W. Agriculture (corn, wheat, soybeans; livestock; dairy products). Limestone, clay. Diversified manufacturing. Drained by VERMILION RIVER. Site of maximum security Pontiac Correctional Center. Formed 1837.

Livingston, county (□ 342 sq mi/889.2 sq km; 2006 population 9,797), W KENTUCKY; ⊙ SMITHLAND; 37°12′N 88°20′W. Bounded W and N by OHIO RIVER (ILLINOIS state line), S by TENNESSEE RIVER (joins Ohio River at SW tip of county); crossed by CUMBERLAND RIVER (forms part of E boundary in SE). Kentucky Dam and part of KENTUCKY LAKE reservoir are on S boundary (Tennessee River), Barkley Dam and part of LAKE BARKLEY reservoir on SE boundary. Gently rolling agricultural area (tobacco; corn, wheat, soybeans; hay, alfalfa; hogs, cattle, poultry; dairying); catfish. Limestone quarries. Tourism. Formed 1798.

Livingston, county (□ 585 sq mi/1,521 sq km; 2006 population 184,511), SE MICHIGAN; ⊙ HOWELL; 42°36′N 83°54′W. Drained by RED CEDAR, HURON, and SHIAWASSEE rivers. Agriculture (cattle, hogs, sheep; poultry; corn, oats, soybeans, apples; dairy products). Manufacturing at Howell. Summer resorts; numerous small lakes in SE ¼ and extreme NE. Pinckney State Recreation Area on S boundary, Island

Lake State Recreation Area, Brighton State Recreation Area, and Mount Brighton Ski Area, all in SE. Organized 1836.

Livingston, county (□ 533 sq mi/1,385.8 sq km; 2006 population 14,291), N central MISSOURI; ⊙ CHILLICOTHE; 39°46′N 93°32′W. Drained by GRAND RIVER. Agriculture (corn, wheat, soybeans; cattle). Manufacturing at Chillicothe. Fountain Grove Conservation Area in SE. Severe flooding on Grand River in 1993. Formed 1837.

Livingston, county (□ 640 sq mi/1,664 sq km; 2006 population 64,173), W central NEW YORK; ⊙ GENESEO; 42°43′N 77°46′W. In FINGER LAKES region; bisected S-N by broad Genesee River valley; drained also by Canaseraga and Honeoye creeks. CONESUS and HEMLOCK lakes in county. Some manufacturing at AVON, GENESEO, DANSVILLE, CALEDONIA. Rich agricultural area (dairying; grain, vegetables, hay; poultry). Salt mines; gypsum and limestone quarries. Horse farms. James and William Wadsworth of Connecticut, most prominent early settlers, purchased 55 sq mi/142 sq km of land here and encouraged liberal settlement in the early 19th century. Named for Robert Livingston, Continental Congress delegate and drafter of Declaration of Independence. Well-known Genesee Valley Hunt founded 1876; nation's second oldest. Includes N part of LETCHWORTH STATE PARK. Formed 1821.

Livingston, parish (□ 665 sq mi/1,729 sq km; 2006 population 114,805), SE LOUISIANA; ⊙ LIVINGSTON; 32°30′N 90°45′W. Bounded W by AMITE RIVER and S by Petite Amite and Blind rivers, partly E by NATALBANY RIVER, SE by Lake MAUREPAS; drained by TICKFAW RIVER. Agriculture (home gardens, nursery crops, cucumbers, peppers; cattle, horses, poultry, exotic fowl, hogs; dairying); alligators; logging. Hunting, fishing. Named after Edward Livingston, who formulated LOUISIANA's code of law. Located in the "FLORIDA parishes" of Louisiana, part of former British colony of W FLORIDA. Formed 1832.

Livingston, city (2000 population 10,473), MERCED county, central CALIFORNIA, in San Joaquin Valley, near MERCED RIVER, 14 mi/23 km NW of MERCED; 37°23′N 120°43′W. Agriculture (dairying; poultry; grain, vegetables, fruit, almonds; cotton, alfalfa). Manufacturing (farm machinery); poultry processing, dehydrated fruits and vegetables.

Livingston (LEE-veen-ston), town, IZABAL department, E GUATEMALA, minor port on Bay of AMATIQUE (inlet of CARIBBEAN SEA), at mouth of Río DULCE, and 11 mi/18 km WNW of PUERTO BARRIOS; 15°50′N 88°45′W. Supply point for Lake IZABAL region; boatbuilding, mahogany working; customhouse. Exports bananas, rubber, sarsaparilla, lumber. Until rise of Puerto Barrios, leading ATLANTIC OCEAN port of Guatemala; until 1920, the capital of Izabal department. Population largely of Garífuna and W Indian origin.

Livingston (LIV-ing-stuhn), town (2001 population 50,826), ⊙ WEST LOTHIAN, central Scotland, 4 mi/6.4 km E of BATHGATE; 55°53′N 03°32′W. Designated a New Town in 1962; variety of light industries. Computer manufacturing. Formerly in Lothian, abolished 1996.

Livingston, town (2000 population 3,297), ⊙ Sumter co., W Alabama, 22 mi/35 km W of Demopolis, between Tombigbee River and Mississippi state line. Manufacturing (paper products apparel). Livingston University here. Founded c.1833. Named for Edward Livingston, secretary of state under Andrew Jackson. Inc. in 1867.

Livingston, town (2000 population 1,342), ⊙ LIVINGSTON parish, SE LOUISIANA, 25 mi/40 km E of BATON ROUGE; 30°30′N 90°45′W. In lumbering area. Manufacturing (wood products, apparel); oil deposits. Founded 1918.

Livingston, town (2000 population 6,851), ⊙ Park county, S MONTANA, 24 mi/39 km E of BOZEMAN, on

YELLOWSTONE RIVER, N of YELLOWSTONE NATIONAL PARK, and just NW of ABSAROKA RANGE; 45°40′N 110°34′W. Railroad; tourism; trade center for mining and agricultural area. Coal, arsenic, silver, gold mines; marble, granite. Agriculture (dairying; cattle; vegetables, barley, oats, alfalfa). Manufacturing (pottery, fishing flies, printing and publishing, wood products). Parts of Gallatin National Forest to SE, SW, NW, and NE; ABSAROKA-BEARTOOTH Wilderness Area to SE; Paradise Valley of Yellowstone River to SW, scenic area leading to Yellowstone National Park (U.S. Highway 89). Calamity Jane lived here. Livingston Professional Cowboys Association Rodeo (July 4 weekend). Park County Museum, Depot Center, Sleeping Giant Wildlife Museum to S. Originally called Clark City. Founded 1882, incorporated 1889.

Livingston, township (2000 population 27,391), ESSEX county, NE NEW JERSEY, near PASSAIC River, 8 mi/12.9 km NW of NEWARK; 40°47′N 74°19′W. Largely residential.

Livingston, town (2006 population 3,517), ⊙ OVERTON county, N TENNESSEE, 18 mi/29 km NNE of COOKEVILLE; 36°23′N 85°19′W. Residential; small business enterprises; recreation. Standing Stone State Park and DALE HOLLOW Reservoir are nearby.

Livingston (LI-ving-stuhn), town (2006 population 6,430), ⊙ POLK county, E TEXAS, 45 mi/72 km S of LUFKIN; 30°42′N 94°56′W. Elevation 194 ft/59 m. Trade, shipping center in oil, timber. Manufacturing (lumber, machinery, printing); wood processing. Alabama-Coushatta Indian Reservation is c.15 mi/24 km E. Oil discovered 1940s. Part of BIG THICKET NATIONAL PRESERVE to SE; LAKE LIVINGSTON reservoir and State Park to SW.

Livingston, village (2000 population 825), MADISON county, SW ILLINOIS, 14 mi/23 km NE of EDWARDSVILLE; 38°58′N 89°45′W. In agricultural area. Incorporated 1905.

Livingston, village (2000 population 228), ROCKCASTLE county, E KENTUCKY, on ROCKCASTLE RIVER, and 27 mi/43 km NE of SOMERSET, in Daniel Boone National Forest; 37°17′N 84°13′W. Coal mining, agriculture. Hunting and fishing in vicinity. Great Saltpetre Cave to NE; Camp Wildcat Battle Monument to S.

Livingston, village (2006 population 143), ORANGEBURG county, W central SOUTH CAROLINA, 15 mi/24 km WNW of ORANGEBURG; 33°32′N 81°07′W.

Livingston, village, GRANT and IOWA counties, SW WISCONSIN, 11 mi/18 km N of PLATTEVILLE; 42°53′N 90°25′W. In dairy and diversified-farming area.

Livingstone, ZAMBIA: see MARAMBA.

Livingstone Falls, 32 cataracts of lower CONGO RIVER, partly in W CONGO, partly along CONGO REPUBLIC border, extending between MATADI and KINSHASA. Here CONGO RIVER cuts a narrow gorge through CRYSTAL MOUNTAINS and falls a total of 850 ft/259 m over a distance of c.220 mi/354 km. An 80-mi/129-km placid stretch exists in the rapids between MANYANGA and ISANGILA. Unsuccessful ascent of the cataracts was attempted (1816) by British explorer Capt. J. K. Tuckey. Henry M. Stanley conquered the falls (1877) at the end of his journey down CONGO RIVER. In 1890–1898, MATADI-KINSHASA railroad was built to circumvent the falls.

Livingstone, Fort (LI-veeng-ston), post of North West Mounted Police, W MANITOBA, W central CANADA, near town of SWAN RIVER. From 1875 to 1877, was capital of NORTHWEST TERRITORIES and headquarters of Mounted Police.

Livingstone Memorial National Monument, ZAMBIA: see CHITAMBO.

Livingstone Mountains, range, IRINGA region, S TANZANIA, on NE shore of Lake NYASA, N of MANDA. From lake surface, range rises sharply more than 3,281 ft/1,000; crest of range 6,565 ft/2,001 m.

Livingston, Fort, LOUISIANA: see GRAND TERRE ISLAND.

Livingstonia (li-ving-stoh-neh-ha), village, Northern region, NE MALAWI, on E edge of NYIKA PLATEAU, on road, and 15 mi/24 km SW of CHILUMBA BAY; 10°36'S 34°07'E. Elevation 4,500 ft/1,372 m. Headquarters of Church of SCOTLAND mission (established 1894), with technical, vocational, and teachers training schools, hospital. Coal deposits nearby. Formerly called Kondowe. Its port is Florence Bay, 5 mi/8 km NE, on Lake Malawi (LAKE NYASA).

Livingston Island, SOUTH SHETLAND ISLANDS, off GRAHAM LAND, ANTARCTICA; 38 mi/61 km long, 23 mi/37 km wide; 62°36'S 60°30'W. Also known as Friesland Island, Smiths Island, or Smolensk Island.

Livingston, Lake (LI-ving-stuhn), reservoir, on SAN JACINTO/POLK county border, E TEXAS, on TRINITY RIVER, 6 mi/9.7 km SSW of LIVINGSTON, extends NW into TRINITY and WALKER counties; c.45 mi/72 km long; 30°38'N 95°10'W. Maximum capacity 2,040,000 acre-ft. Formed by Livingston Dam (89 ft/27 m high), built (1968) by the city of HOUSTON for water supply.

Livingston Manor, village (□ 3 sq mi/7.8 sq km; 2000 population 1,355), SULLIVAN county, SE NEW YORK, in CATSKILL MOUNTAINS, on small Willowemoc Creek, 8 mi/12.9 km NW of LIBERTY; 41°53'N 74°49'W. Manufacturing (corrugated pipe); lumber milling. State brown trout hatchery 7 mi/11.3 km NE. Summer and winter (skiing) resort; heart of DELAWARE-Sullivan counties trout production and fly-fishing region.

Livitaca (lee-vee-TAH-kah), town, CHUMBIVILCAS province, CUSCO region, S PERU, in the ANDES, 60 mi/97 km SSE of CUSCO; 14°18'S 71°41'W. Elevation 11,975 ft/3,649 m. Cereals, potatoes. Variant spelling: Livitica.

Livitica, PERU: see LIVITACA.

Livland, LATVIA and ESTONIA: see LIVONIA.

Livländische Aa, LATVIA: see GAUJA RIVER.

Livno (LEEV-no), town, W BOSNIA, BOSNIA AND HERZEGOVINA, 36 mi/58 km NE of SPLIT (CROATIA); 43°49'N 17°00'E. Handicraft center. Lignite mine nearby. Town has been known since 9th century.

Livny (LEEV-ni), city (2006 population 52,950), S OREL oblast, SW European Russia, on the SOSNA RIVER (DON River basin), on railroad, 95 mi/153 km SE of ORËL; 52°25'N 37°36'E. Elevation 426 ft/129 m. Highway junction; regional transshipment center. Manufacturing (machinery, plastics); flour milling, distilling. Known since the end of the 12th century. Chartered in 1586.

Livonia (li-VO-nee-ah), region and former Russian province, comprising present ESTONIA and parts of LATVIA (VIDZEME and LATGALE). It borders on the BALTIC SEA and its arms, the GULF OF RIGA and the GULF OF FINLAND, in the W and the N and extends E to LAKE PEIPSI and the NARVA RIVER. Livonia, also known as LIVLAND, was named for the Livs (a Finno-Ugric tribe in the 13th century inhabiting the coast when the Livonian Brothers of the Sword conquered the entire region). The knights formed a strong state and threatened LITHUANIA and NOVGOROD in the 13th and 14th century. The chief cities—notably RĪGA, TARTU, and TALLINN—were Germanic in culture and were members of the HANSEATIC LEAGUE. After the dissolution (1561) of the Livonian Order, Livonia was contested by POLAND, RUSSIA, and SWEDEN. COURLAND, in the SW, became a duchy under Polish suzerainty, and LATGALE, in the SE, became part of Poland. VIDZEME, in the center, passed first to Poland, then (1629) to Sweden, which also held the N part (ESTONIA). The Swedish share was conquered (1710) in the Northern War by Peter I of Russia, who kept it at the Peace of NYSTAD (1721). Latgale passed to Russia in 1772. In 1783, Livonia was constituted a Russian province, and in 1918 it was divided between Estonia and Latvia.

Livonia (li-VON-yah), city (2000 population 100,545), WAYNE county, SE MICHIGAN, a suburb of DETROIT;

42°23'N 83°22'W. Drained in S by Middle RIVER ROUGE (parkway), in NE by Upper River Rouge. Manufacturing (transportation equipment, plastic and steel products, textiles); food processing. The city is the seat of Madonna University and Schoolcraft Junior College. The Wolverine Harness Raceway is here. Nankin Mills Nature Center on S boundary with WESTLAND. Founded 1835, incorporated 1950.

Livonia (li-VON-yuh), town (2000 population 112), WASHINGTON county, S INDIANA, 10 mi/16 km WSW of SALEM; 38°34'N 86°17'W. Agricultural area. Laid out 1819.

Livonia, town, POINTE COUPEE parish, SE central LOUISIANA, 22 mi/35 km WNW of BATON ROUGE, on Bayou Maringouin; 30°34'N 91°33'W. Railroad junction. Agricultural area (cotton, rice, sugarcane, vegetables; cattle); timber; crawfish. Manufacturing (concrete, lumber). Atchafalaya National Wildlife Refuge to SW.

Livonia, town (2000 population 114), PUTNAM county, N MISSOURI, on CHARITON RIVER, and 15 mi/24 km E of UNIONVILLE; 40°29'N 92°42'W.

Livonia, village (2006 population 1,612), Livingston county, W central NEW YORK, near CONESUS LAKE, 23 mi/37 km S of ROCHESTER; 42°49'N 77°40'W. In agricultural area. Summer residences and recreation.

Livonian Switzerland, scenic resort district in central LATVIA, NE of Rīga, in VIDZEME, on middle GAUJA River Has wooded hills, caves, grottoes. Its center is SIGULDA.

Livorno (lee-VOR-no), province (□ 471 sq mi/1,224.6 sq km), TUSCANY, central ITALY; ⊙ LIVORNO; 43°14'N 10°35'E. Borders on LIGURIAN SEA; comprises narrow, 50-mi/80-km-long coastal strip enclosed by Apennine hills. Includes islands of ELBA, CAPRAIA, PIANOSA, GORGONA, and MONTE CRISTO. Watered by CECINA River and small streams. Agriculture (grapes, olives, cereals, fruit), livestock raising; fishing. Chief producer of Italy's iron ore, on Elba. Other mines (tin, copper, iron) at CAMPIGLIA MARITTIMA. Manufacturing at Livorno, piombino, portoferraio, and cecina includes shipbuilding, chemicals and petrochemicals, glassworks, fish canning and processing, apparel, cement. Area increased by addition of territory from PISA province in 1925.

Livorno (lee-VOR-no), English *Leghorn*, city (2001 population 156,274), ⊙ LIVORNO province, TUSCANY, central ITALY, on the LIGURIAN SEA, and on the Aurelian Way; 43°33'N 10°19'E. It is a busy commercial, industrial, and tourist center and is one of the most important ports of Italy. Manufacturing (iron, steel, aluminum, copper, metal minerals, chemicals, transportation equipment, machinery, electrical equipment); petroleum refining. The city has major shipyards and a fishing industry. A fortified castle in the Middle Ages, Livorno was developed (16th century) into a flourishing city by the Medici. In 1590, Ferdinand I, grand duke of Tuscany, made it a free port and opened it to all religious and political refugees. The city was badly damaged in World War II. Points of interest include the cathedral (16th century, restored after 1945) and the remains of the 17th-century city wall. The Italian naval academy is there.

Livorno (LUH-vawr-no), village, WANICA district, N SURINAME, on SURINAME river, and 3 mi/4.8 km S of PARAMARIBO; 05°44'N 55°12'W. Rice, coffee, fruit.

Livradois-Forez Natural Regional Park (lee-vrah-dwah–for-AI) (□ 1,158 sq mi/3,010.8 sq km), HAUTE-LOIRE and PUY-DE-DÔME departments, central FRANCE; mountains in AUVERGNE region (DORE RIVER drainage).

Livradois, Massif du (lee-vrah-dwah, mah-seef dyoo), granitic mountain range and plateau of the MASSIF CENTRAL, central FRANCE, between ALLIER and DORE river valleys in PUY-DE-DÔME and HAUTE-LOIRE departments, AUVERGNE region; 45°30'N 03°33'E. Average elevation 3,000 ft/915 m. There are several extinct

volcanoes within LIVRADOIS-FOREZ NATURAL REGIONAL PARK.

Livramento (LEE-vrah-men-to), town (2007 population 7,101), central PARAÍBA, BRAZIL, 9 mi/14.5 km W of SAO JOSÉ DOS CORDEIROS; 07°23'S 36°57'W.

Livramento, Brazil: see JOSÉ DE FREITAS.

Livramento, Brazil: see NOSSA SENHORA DO LIVRAMENTO.

Livramento, Brazil: see OLIVEIRA FORTES.

Livramento, Brazil: see SANTANA DO LIVRAMENTO.

Livramento do Brumado (lee-vrah-men-to do broo-mah-do), city, central BAHIA, BRAZIL, 65 mi/105 km SSW of ANDARAÍ; 13°35'S 41°50'W. Gold mining. Until 1944, called Livramento.

Livron-sur-Drôme (lee-vron–syur–dro-muh), town (□ 15 sq mi/39 sq km), DRÔME department, RHÔNE-ALPES region, SE FRANCE, on the DRÔME RIVER near its mouth on the RHÔNE, and 11 mi/18 km S of VALENCE; 44°46'N 04°51'E. Manufacturing (chemicals, pharmaceuticals); vineyards nearby.

Livry-Gargan (lee-vree–gahr-gahn), town (□ 2 sq mi/5.2 sq km), SEINE-SAINT-DENIS department, ÎLE-DE-FRANCE region, N central FRANCE, an outer ENE suburb of PARIS, 10 mi/16 km from Notre Dame Cathedral; 48°55'N 02°33'E. Metalworks. Its abbey, founded 1186 and destroyed in the French Revolution, was rebuilt in 19th century.

Liwa (LEE-wuh), township (2003 population 25,776), N OMAN, GULF OF OMAN, 14 mi/23 km NW of SOHAR.

Liwale (lee-WAH-lai), town, LINDI region, SE central TANZANIA, 115 mi/185 km W of LINDI, on Liwale Makubwa River, at junction of several roads; 09°48'S 37°59'E. SELOUS GAME RESERVE to N and W. Wheat, corn; cattle, sheep, goats.

Liwang (LEE-wahng), town, ROLPA district, SW NEPAL; 28°16'N 82°46'E. Elevation 4,101 ft/1,250 m.

Liwiec River (LEE-vee-ets), Russian *Livets*, 72 mi/116 km long, in E POLAND; rises 3 mi/4.8 km S of MORDY; flows NW past WEGROW to BUG RIVER 4 mi/6.4 km E of WYSZKOW.

Liwonde (lih-woh-deh), village, Southern region, S MALAWI, on left bank of SHIRE RIVER, on road, and 20 mi/32 km NNW of ZOMBA. Tobacco, cotton, tung, corn, rice; livestock.

Liwonde National Park, Southern region, MALAWI, on E bank of SHIRE RIVER, and bordering LAKE MALOMBE in the N, 40 mi/64 km N of ZOMBA.

Li Xian (LEE SIAN), town, ⊙ Li Xian county, SE GANSU province, CHINA, 45 mi/72 km WSW of TIANSHUI; 34°11'N 105°02'E. Grain, oilseeds, medicinal herbs; food and forage processing; chemicals.

Li Xian, town, ⊙ Li Xian county, W central HEBEI province, CHINA, 25 mi/40 km S of BAODING; 38°29'N 115°34'E. Grain, oilseeds, cotton; textiles.

Li Xian, town, ⊙ Li Xian county, N HUNAN province, CHINA, on GULF OF OMAN, and 40 mi/64 km NNE of CHANGDE, and on road to Yangzi port of SHASHI; 29°38'N 111°45'E. Rice, oilseeds, cotton, jute. Coal mining nearby.

Li Xian, town, ⊙ Li Xian county, W central SICHUAN province, CHINA, on tributary of MIN River, and 70 mi/113 km NW of CHENGDU; 31°28'N 103°17'E. In mountain region; grain, logging.

Lixouri (lee-KSOO-ree), town, on SW peninsula of KEFALLINÍA island, KEFALLINÍA prefecture, IONIAN ISLANDS department, W coast of GREECE, 2 mi/3.2 km W of ARGOSTOLI, across Argostoli Bay; 38°12'N 20°26'E. Trades in currants, olive oil, wine. Fisheries. Also called Lexuri, Lixuri, Lexourion, or Lixourion. The ancient *Pale* (just N) was a leading city of Kefallinía.

Lixuri, Greece: see LIXOURI.

Lixus, MOROCCO: see LARACHE.

Liyang (LEE-YANG), city (□ 593 sq mi/1,541.8 sq km), SW JIANGSU province, CHINA, near ANHUI-ZHEJIANG border, 50 mi/80 km S of ZHENJIANG; 31°26'N 119°27'E. Agriculture and heavy industry predominate. Rice, oilseeds, cotton, jute; food processing, textiles, apparel, chemicals, iron and steel, machinery, electrical equipment.

Liyü-t'an (LEE-YOOI–TAN) or **Grass Carp Lake**, deep mountain lake in a scenic area SW of HUALIEN, TAIWAN, popular for boating and fishing (carp). Guava, tangerines, lemons, bananas, and papaya are grown in the area.

Lizarda (LEE-sahr-dah), town (2007 population 3,635), E TOCANTINS state, BRAZIL, 195 mi/314 km SE of PALMAS, on border with MARANHÃO; 09°36′S 46°41′W.

Lizard Island (LI-zuhrd) (□ 4 sq mi/10.4 sq km), mountainous island with coral reef perimeter, off NE QUEENSLAND coast, AUSTRALIA, 60 mi/97 km NE of COOKTOWN; 14°40′S 145°28′E. Lizard Island National Park covers entire island; surrounded by GREAT BARRIER REEF MARINE PARK. Highest point, Cook's Lookout (1,178 ft/359 m), climbed by Captain James Cook, August 1770. Named for monitor lizards (giant goannas) found on island. Resort with limited accommodations. Access by private boat or scheduled flights from Cooktown and CAIRNS. Sport fishing, swimming, snorkeling, scuba diving, hiking. Established 1939.

Lizard, The (LIZ-uhd), promontory, SW CORNWALL, SW ENGLAND, on the ENGLISH CHANNEL, 15 mi/24 km SSW of FALMOUTH. The whole peninsula S of Helford River is sometimes called The Lizard, and its S extremity (southernmost point of Great Britain; 49°57′N 05°13′W), Lizard Point or Lizard Head, is the site of two lighthouses. Coast has colored serpentine rocks, inlets, caves, islets, and dangerous reefs (hence the promontory's nickname, "Graveyard of Ships").

Lizella (luh-ZEL-uh), town, BIBB county, GEORGIA, 11 mi/18 km SW of MACON; 32°48′N 83°49′W. Pottery.

Lizton, town (2000 population 372), HENDRICKS county, central INDIANA, 8 mi/12.9 km N of DANVILLE; 39°53′N 86°32′W. In agricultural area. Manufacturing (lawn mowers).

Lizy-sur-Ourcq (lee-zee–syur–oor-kuh), commune (□ 4 sq mi/10.4 sq km), SEINE-ET-MARNE department, ÎLE-DE-FRANCE region, N central FRANCE, on the OURCQ RIVER near its influx into the MARNE, and 8 mi/12.9 km NE of MEAUX; 49°01′N 03°02′E. Metalworking, sugar milling.

Lizzano in Belvedere (leet-SAH-no een bel-ve-DAI-re), village, BOLOGNA province, EMILIA-ROMAGNA, N central ITALY, in ETRUSCAN APENNINES, 4 mi/6 km W of PORRETTA TERME; 44°10′N 10°53′E. Resort (elevation 2,100 ft/640 m); tourism.

Lizzie, Lake (□ 8 sq mi/20.7 sq km), OTTER TAIL county, W MINNESOTA, 22 mi/35 km N of FERGUS FALLS; 4 mi/6.4 km long, 3 mi/4.8 km wide; 46°37′N 96°00′W. Flows W through short stream to Prairie Lake. Lake LIDA to SE drains into Lake Lizzie; Crystal Lake to E; PELICAN LAKE to N flows through short stream to Lake Lizzie. Resorts.

Ljeskovec, Bulgaria: see LYASKOVETS.

Ljósafoss (LYO-sah-FAWS), waterfall, SW ICELAND, on SOG River, 25 mi/40 km E of REYKJAVIK. Power station here.

Ljubelj Pass, AUSTRIA and SLOVENIA: see LOIBL PASS.

Ljubija (LYOO-bee-yah), village, N BOSNIA, BOSNIA AND HERZEGOVINA, 11 mi/16 km SW of PRIJEDOR; 44°54′N 16°36′E. In Serb Republic of Bosnia and Herzegovina. Center of iron-mining area. Mines opened in 1916; closed in 1992. Also spelled Lyubiya.

Ljubinje (lyoo-BEEN-ye), village, S HERZEGOVINA, BOSNIA AND HERZEGOVINA, on MOSTAR-TREBINJE road, and 30 mi/48 km SSE of MOSTAR; 42°56′N 18°05′E. Local trade center. Also spelled Lyubinye.

Ljubisnja (LYOO-bees-nyah), Serbian *Ljubišnja* (LYOO-beesh-nyah), mountain in DINARIC ALPS, N MONTENEGRO, between TARA and COTINA rivers; highest point (7,341 ft/2,238 m) is 12 mi/19 km N of ZABLJAK; 43°20′N 19°07′E. Also spelled Lyubishnya.

Ljubljana (lyoo-BLYAH-nah), German *Laibach*, city (2002 population 257,338), ⊙ SLOVENIA, on the SAVA River; 46°03′N 14°32′E. An industrial and transportation center, manufacturing chemicals, electronics, machinery, textiles, and food products. Roman Catholic archiepiscopal see. Seat of the Slovene Academy of Arts and Sciences and the University of Ljubljana (founded 1919). International airport to N. Known as Emona in Roman times. Passed (1277) to the Hapsburgs and became the chief city of the Austrian province of Carniola. Held briefly by the French during the Napoleonic wars. After Slovenia joined the former Yugoslavia in 1918, it was made capital of Slovenia; after 1991, of the Republic of Slovenia. Center of the Slovene national movement in the 19th century. Has a medieval fortress and several fine palaces and churches.

Ljubljanica River (lyoo-BLYAH-nee-tsah), c.20 mi/32 km long, in W SLOVENIA; rises near VRHNIKA; flows ENE past Vrhnika and LJUBLJANA to SAVA River just ENE of Ljubljana.

Ljubostinja (LYOO-bo-stin-yah), monastery, central SERBIA, 17 mi/27 km WNW of KRUSEVAC, near the WESTERN MORAVA RIVER. Also spelled Lyubostinya.

Ljuboten (lyoo-bo-TEN), peak (8,187 ft/2,495 m) in SAR MOUNTAINS, on Macedonian-Serbian border, 16 mi/26 km NNE of TETOVO, MACEDONIA, above KACANIK defile. Nearby Ljuboten mines produce chromium processed at RADUSA (Macedonia). Also spelled Lyuboten.

Ljubovija (lyoo-BO-vee-yah), village (2002 population 17,052), W SERBIA, on DRINA River (BOSNIA-HERZE-GOVINA border), and 26 mi/42 km W of VALJEVO; 44°11′N 19°22′E. Also spelled Lyuboviya.

Ljubuša Mountains (LYOO-boo-shah), Serbo-Croatian *Ljubuša Planina*, in DINARIC ALPS, W BOSNIA, BOSNIA AND HERZEGOVINA; c. 10 mi/16 km long; 43°44′N 17°23′E. Highest peak, Velika Ljubuša (5,894 ft/1,796 m), is 12 mi/19 km WSW of PROZOR. Also spelled Lyubusha Mountains.

Ljubuški (LYOO-boosh-kee), town, lower HERZEGOVINA, BOSNIA AND HERZEGOVINA, 16 mi/26 km SW of MOSTAR, near CROATIA (DALMATIA) border; 43°11′N 17°32′E. Chief town of Ljubuško polje. Center of tobacco-growing region. Also spelled Lyubushki.

Ljubuško Polje (lyoo-BOOSH-ko POL-ye) or **Ljubuški Plain**, plain in karst topography, and historical region, lower HERZEGOVINA, BOSNIA AND HERZEGOVINA. Principal town is Ljubuški.

Ljugarn (YOO-GAHRN), village, GOTLAND county, SE SWEDEN, on E coast of Gotland island, 25 mi/40 km SE of VISBY; 57°20′N 18°42′E. Seaside resort. Archaeological museum.

Ljungan (YOONG-ahn), river, 200 mi/322 km long, in N SWEDEN; rises near NORWEGIAN border SE of TRONDHEIM; flows generally ESE, through several small lakes, past Ånge, LJUNGAVERK (power station), and MATFORS, to GULF OF BOTHNIA 4 mi/6.4 km SE of SUNDSVALL. Salmon fishing.

Ljungaverk (YUNG-ah-verk), village VÄSTERNORRLAND county, NE SWEDEN, on LJUNGAN RIVER, 40 mi/64 km W of SUNDSVALL; 62°30′N 16°03′E. Manufacturing (chemicals). Large hydroelectric station.

Ljungby (YUNG-BEE), town, KRONOBERG county, S SWEDEN, on LAGAN RIVER, 30 mi/48 km W of VÄXJÖ; 56°50′N 13°56′E. Railroad junction; light manufacturing; paper-based products. Art museum.

Ljungbyhed (YUNG-BEE-HED), town, SKÅNE county, S SWEDEN, 20 mi/32 km E of HELSINGBORG; 56°04′N 13°14′E. Military flight school.

Ljungsbro (YUNGS-BROO), town, ÖSTERGÖTLAND county, SE SWEDEN, on MOTALA STRÖM RIVER (falls), and on GÖTA CANAL, 7 mi/11.3 km NW of LINKÖPING; 58°30′N 15°31′E. Food processing.

Ljungskile (YUNGS-SHEE-le), town, GÖTEBORG OCH BOHUS county, SW SWEDEN, on narrow channel of SKAGERRAK strait, opposite ORUST island, 8 mi/12.9 km S of UDDEVALLA; 58°13′N 11°55′E. Seaside resort.

Ljusdal (YOOS-DAHL), town, GÄVLEBORG county, E SWEDEN, on LJUSNAN RIVER, 50 mi/80 km W of HUDIKSVALL; 61°50′N 16°5′E. Railroad junction. Light manufacturing in agricultural and forest region. Many historic wooden manor houses.

Ljusnan (YOOS-nahn), river, 270 mi/435 km long, in N central SWEDEN; rises in NORWEGIAN border mountains W of RÖROS; flows in winding course generally SE, past SVEG, LJUSDAL, and BOLLNÄS, to GULF OF BOTHNIA at LJUSNE. Hydroelectric stations. Receives VOXNAN River.

Ljusne (YOOS-ne), town, GÄVLEBORG county, E SWEDEN, on GULF OF BOTHNIA at mouth of LJUSNAN River, 6 mi/9.7 km SE of SÖDERHAMN; 61°13′N 17°08′E. Lumber milling. Hydroelectric station. Includes Ala village.

Ljutomer (LYOO-to-mer), village, NE SLOVENIA, 26 mi/42 km E of MARIBOR, at SE foot of the SLOVENSKE GORICE mountains; 46°31′N 16°11′E. Railroad junction. Winery; trade in wine and horses. Has castle. First mentioned in 1265.

Llagostera (lyah-go-STAI-rah), town, GERONA province, NE SPAIN, 11 mi/18 km SSE of GERONA; 41°49′N 02°54′E. Road junction. Cork processing; cereals, fruit.

Llaillai, CHILE: see LLAY-LLAY.

Llaima (YEI-mah), village, CAUTÍN province, ARAUCANIA region, S central CHILE, on railroad, and 32 mi/51 km ENE of TEMUCO. Cereals; livestock.

Llaima Volcano (YEI-mah), Andean peak (10,040 ft/3,060 m), CAUTÍN province, ARAUCANIA region, S central CHILE, 45 mi/72 km ENE of TEMUCO. Active volcano. Winter sports.

Llajta Mauca (YAH-tah MOU-kah), town, E SANTIAGO DEL ESTERO province, ARGENTINA, 20 mi/32 km NW of AÑATUYA; 28°12′S 63°05′W. Cotton; livestock. Formerly called Kilómetro 511.

Llallagua (yah-YAH-gwah), town (2001 population 20,065) and canton, RAFAEL BUSTILLO province, POTOSÍ department, W central BOLIVIA, on E slopes of Cordillera de AZANAQUES, 3 mi/4.8 km NNW of UNCÍA, and on MACHACAMARCA-Uncía railroad; 18°25′S 66°38′W. Elevation 12,733 ft/3,881 m. Formerly chief tin-mining center of Bolivia; gradually being replaced by CATAVI mines, 2 mi/3.2 km NE. Originally owned by Chilean Llallagua Company. Siglo XX, most productive mine, is largest tin mine in Bolivia, with 496 mi/800 km of passages.

Llama (YAH-mah), town, CHOTA province, CAJAMARCA region, NW PERU, in CORDILLERA OCCIDENTAL, 32 mi/51 km W of CHOTA; 06°31′S 79°08′W. Cereals, corn, potatoes.

Llambrión, Torre de (lyahm-bree-ON, TO-rai dhai), peak (8,586 ft/2,617 m) of CANTABRIAN MOUNTAINS, N SPAIN, in PICOS DE EUROPA (massif), 12 mi/19 km NE of RIAÑO.

Llamellín (yah-mai-YEEN), town, ⊙ ANTONIO RAYMONDI province, ANCASH region, W central PERU, on E slopes of Cordillera BLANCA, near MARAÑÓN RIVER, 25 mi/40 km NE of HUARI; 09°08′S 77°01′W. Cereals, corn; livestock.

Llanaber, Wales: see BARMOUTH.

Llanaelhaearn (hla-neil-HEI-uhrn), village (2001 population 1,067), GWYNEDD, NW Wales, on LLEYN PENINSULA, 6 mi/9.7 km N of PWLLHELI; 52°59′N 04°24′W.

Llanarmon-yn-Ial (hla-NAHR-mon–in–EI-ahl), village (2001 population 1,069), DENBIGHSHIRE, NE Wales, 3 mi/4.8 km SE of RUTHIN; 53°05′N 03°15′W. Church (18th century) with 13th-century effigies. Formerly in CLWYD, abolished 1996.

Llanarth (HLA-nahrth), village (2001 population 1,564), CEREDIGION, W Wales, 3 mi/4.8 km SE of NEW QUAY; 52°11′N 04°18′W. Formerly in DYFED, abolished 1996.

Llanasa (hla-NA-suh), village (2001 population 4,820), FLINTSHIRE, NE Wales, 3 mi/4.8 km ESE of PRESTATYN; 53°19′N 03°21′W. Formerly in CLWYD, abolished 1996.

Llanbadrig (hlan-BA-drig), village (2001 population 1,392), N ISLE OF ANGLESEY, NW Wales, on IRISH SEA, 11 mi/18 km NE of HOLYHEAD; 53°25′N 04°26′W. Just ENE is Llanlleiana Head promontory, northernmost point of Wales; 53°25′N 04°25′W.

Llanbedrog (hlan-BED-rog), village (2001 population 1,020), GWYNEDD, NW Wales, on LLEYN PENINSULA, 4 mi/6.4 km SW of PWLLHELI; 52°51′N 04°29′W. Previously granite quarrying.

Llanberis (hlan-BE-ris), town (2001 population 2,018), GWYNEDD, NW Wales, 8 mi/12.9 km ESE of CAERNARVON, at S end of Llyn Padarn (2 mi/3.2 km long); 53°07′N 04°08′W. Previously slate quarrying. Terminal of railroad to SNOWDON summit. Nearby is 13th-century Dolbadarn Castle. At foot of Pass of Llanberis. Just ENE is former slate-quarrying village of DINORWIG; also Dinorwig hydroelectric pump; storage scheme of Marchlyn Mawr Reservoir.

Llanboidy (hlan-BOI-dee), village (2001 population 988), CARMARTHENSHIRE, SW Wales, 12 mi/19 km WNW of CARMARTHEN; 51°52′N 04°35′W. Previously woolen milling. Formerly in DYFED, abolished 1996.

Llanbradach (hlan-BRA-duhkh), village (2001 population 4,622), CAERPHILLY, SE Wales, on RHYMNEY RIVER, 2 mi/3.2 km N of Caerphilly; 51°36′N 03°14′W. Formerly in MID GLAMORGAN, abolished 1996.

Llanbrynmair (hlan-brin-MEIR), village (2001 population 958), POWYS, E Wales, 9 mi/14.5 km ENE of MACHYNLLETH; 52°36′N 03°38′W. Has 15th-century church.

Llancanelo, Lake (yahn-kah-NAI-lo) or **Lake Llancanello**, salt lake (□ 185 sq mi/481 sq km), S MENDOZA province, ARGENTINA, 75 mi/121 km SW of SAN RAFAEL, in N part of a salt desert.

Llancynfelyn, Wales: see BORTH.

Llandaff, Wales: see CARDIFF.

Llandarcy, Wales: see SWANSEA.

Llanddarog (hlan-[TH]A-rog), village (1991 population 1,103; 2001 population 1,095), CARMARTHENSHIRE, SW Wales, 6 mi/9.7 km SE of CARMARTHEN; 51°49′N 04°10′W. Formerly in DYFED, abolished 1996.

Llanddeiniolen (hlan-thein-YO-luhn), village (2001 population 4,885), GWYNEDD, NW Wales, 5 mi/8 km NE of CAERNARVON; 53°11′N 04°13′W. Just NW, on MENAI STRAIT, is former slate-shipping port of Port Dinorwic, linked by railroad with LLANBERIS.

Llandegai (hlan-de-GEI), village, GWYNEDD, NW Wales, just SE of BANGOR; 53°13′N 04°06′W. Had model cottage housing for workers when the slate-industry was active.

Llandeilo (hlan-DEI-lo), town (2001 population 1,731), CARMARTHENSHIRE, SW Wales, on TOWY RIVER, and 13 mi/21 km N of CARMARTHEN; 51°53′N 04°00′W. Resort. Just W is Dynevor Park, with ruins of 13th-century Dynevor Castle. Formerly in DYFED, abolished 1996.

Llandinam (hlan-DEE-nam), village (2001 population 942), POWYS, E Wales, on SEVERN RIVER, 5 mi/8 km NE of LLANIDLOES; 52°29′N 03°26′W.

Llandovery (hlan-DUHV-ree), town (2001 population 2,235), E CARMARTHENSHIRE, SW Wales, on TOWY RIVER, and 12 mi/19 km NE of LLANDEILO; 51°59′N 03°47′W. Agricultural market; light manufacturing. Has ruins of Norman castle. Site of Roman fort. Formerly in DYFED, abolished 1996.

Llandrindod Wells (hlan-DRIN-dod), spa town (2001 population 5,024), ⊙ POWYS, E Wales, 17 mi/27 km WSW of KNIGHTON; 52°14′N 03°23′W. Resort with mineral springs. Previously woolen milling.

Llandudno (hlan-DID-no), resort (2001 population 20,090), CONWY, NW Wales, at base of GREAT ORMES HEAD, and at mouth of CONWY RIVER; 53°20′N 03°50′W.

Llandudoch, Wales: see ST. DOGMAELS.

Llandwrog (hlan-DOO-rog), village (2001 population 2,466), GWYNEDD, NW Wales, 5 mi/8 km SSW of CAERNARVON; 53°04′N 04°19′W.

Llandybie (hlan-DUH-bee) or **Llandebie**, village (2001 population 9,634), CARMARTHENSHIRE, SW Wales, 4 mi/6.4 km SSW of LLANDEILO; 51°49′N 04°00′W. Formerly in DYFED, abolished 1996.

Llandysul (hlan-DI-sil), town (2001 population 1,521), CEREDIGION, W Wales, on TEIFI RIVER, and 12 mi/19 km SW of LAMPETER; 52°02′N 04°18′W. Previously woolen milling. Formerly in DYFED, abolished 1996.

Llanegwad (hla-NEG-wuhd), village (2001 population 1,388), CARMARTHENSHIRE, SW Wales, on TOWY RIVER, and 7 mi/11.3 km E of CARMARTHEN; 51°52′N 04°09′W. Formerly in DYFED, abolished 1996.

Llanelli (hla-NE-hlee), town and former district (2001 population 23,422), CARMARTHENSHIRE, S Wales, on the LOUGHOR RIVER estuary, and 14 mi/22.4 km SE of CARMARTHEN; 51°41′N 04°09′W. Docks closed in 1950s. Previously tinplate and steel works; glass and chemicals. Was also a district of DYFED, abolished 1996.

Llanelly (hla-NE-hlee), village (2001 population 3,810), MONMOUTHSHIRE, SE Wales, in BRECON BEACONS NATIONAL PARK, and 4 mi/6.4 km ENE of BRYNMAWR; 51°49′N 03°06′W. Previously paper milling. Formerly in GWENT, abolished 1996.

Llanelwy, Wales: see ST. ASAPH.

Llanerch, Pennsylvania: see HAVERFORD.

Llanes (LYAH-nes), city, OVIEDO province, NW SPAIN, in ASTURIAS, fishing port on BAY OF BISCAY, 50 mi/80 km W of SANTANDER. Fish salting; boat-building; dairy products. Has 15th-century Gothic church and remains of medieval walls. Mineral springs.

Llanfaelog (hlan-VEI-log), village (2001 population 1,679), W ISLE OF ANGLESEY, NW Wales, 8 mi/12.9 km SE of HOLYHEAD; 53°14′N 04°28′W.

Llanfair-ar-y-Bryn (HLAN-veir–ar–uh–BRIN), village (2001 population 635), CARMARTHENSHIRE, SW Wales, on Crychan River, 2 mi/3.2 km NE of LLANDOVERY; 52°02′N 03°44′W. Has 13th–15th century church, built within Roman walls. Formerly in DYFED, abolished 1996.

Llanfair Caereinion (HLAN-veir kah-REIN-yon), town (2001 population 1,616), POWYS, E Wales, on Einion River, and 8 mi/12.9 km W of WELSHPOOL; 52°40′N 03°19′W.

Llanfair Clydogau (HLAN-veir kli-DO-gei), village (2001 population 770), CEREDIGION, W Wales, on TEIFI RIVER, and 4 mi/6.4 km NE of LAMPETER; 52°08′N 04°01′W. Site of Roman silver mines. Formerly in DYFED, abolished 1996.

Llanfair-Dyffryn-Clwyd (HLAN-veir–DUH-frin–KLOO-wid), village (2001 population 1,070), DENBIGHSHIRE, NE Wales, 2 mi/3.2 km NE of RUTHIN; 53°05′N 03°18′W. Church (15th century) with 17th-century communion table. Formerly in CLWYD, abolished 1996.

Llanfairfechan (hlan-feir-VE-khuhn), resort (2001 population 3,755), CONWY, N Wales, on CONWY BAY of Irish Sea, 7 mi/11.3 km ENE of BANGOR, at base of PENMAENMAWR; 53°15′N 03°59′W.

Llanfairpwllgwyngyll (HLAN-veir-poohl-GWIN-gihl) or **Llanfairpwllgwyngyllgogerychwyrndrobwllllantysiliogogogoch**, village (2001 population 3,040), SE ISLE OF ANGLESEY, NW Wales, near MENAI STRAIT, 2 mi/3.2 km W of MENAI BRIDGE; 53°13′N 04°12′W. Longest place name (fifty-eight letters in English, fifty-one in Welsh) in Northern Hemisphere.

Llanfair Talhaiarn (HLAN-veir tal-HEI-uhrn), village (2001 population 979), CONWY, N Wales, on Elwy River, 6 mi/9.7 km SW of ABERGELE; 53°13′N 03°36′W. Burial place of poet John Jones (1810–1869). Formerly in CLWYD, abolished 1996.

Llanfair-ym-Muallt, Wales: see BUILTH WELLS.

Llanfihangel ar Arth (hlan-vi-HANG-uhl ar AHRTH) or **Llanfihangel ararth**, village (2001 population 2,051), CARMARTHENSHIRE, SW Wales, on TEIFI RIVER, and 9 mi/14.5 km E of NEWCASTLE EMLYN; 52°01′N 04°15′W. Previously woolen milling. Just ENE is former

woolen-milling village of Maesycrugiau. Formerly in DYFED, abolished 1996.

Llanfor (HLAN-vuhr), village, GWYNEDD, NW Wales, on the DEE RIVER just NE of BALA; 52°55′N 03°35′W.

Llanfyllin (hlan-VUH-hlin), town (2001 population 1,407), POWYS, E Wales, 9 mi/14.5 km NNW of WELSHPOOL; 52°45′N 03°15′W.

Llanfynydd (hlan-VUH-ni-[th]-), village (1991 population 1,699; 2001 population 1,752), FLINTSHIRE, NE Wales, 6 mi/9.7 km NW of WREXHAM, on line of OFFA'S DYKE; 51°55′N 04°06′W. Formerly in CLWYD, abolished 1996.

Llanga Belén (YAHN-gah be-LEN), canton, AROMA province, LA PAZ department, W BOLIVIA, 6 mi/10 km SE of Sicasica, on a branch of the ORURO–LA PAZ highway; 17°24′S 67°44′W. Elevation 12,851 ft/3,917 m. Clay, limestone, gypsum deposits. Potatoes, yucca, bananas, rye; cattle.

Llangadog (hlan-GA-dog), town (2001 population 1,303), CARMARTHENSHIRE, SW Wales, on TOWY RIVER, and 6 mi/9.7 km NE of LLANDEILO; 51°56′N 03°53′W. Nearby is prehistoric camp of Carn Goch. Formerly in DYFED, abolished 1996.

Llanganates, Cordillera de los (yahn-gah-NAH-tes, kor-dee-YER-rah dai los), E Andean massif, central ECUADOR, E of AMBATO; includes several volcanic peaks, highest of which is Cerro HERMOSO (15,216 ft/4,638 m). Region rich in minerals (vanadium, molybdenum, chromium, iron, phosphorus, arsenic).

Llangattock (hlan-GA-tuhk), village (2001 population 1,006), POWYS, E Wales, on USK RIVER, adjacent to CRICKHOWELL; 51°51′N 03°08′W. Eastern extremity of BRECON BEACONS.

Llangefni (hlan-GEV-nee), town (2001 population 4,662), ⊙ ISLE OF ANGLESEY, NW Wales, 8 mi/12.9 km NNW of CAERNARVON; 53°16′N 04°20′W.

Llangeler (hlan-GE-luhr), village (2001 population 3,222), CARMARTHENSHIRE, SW Wales, on TEIFI RIVER, and 4 mi/6.4 km E of NEWCASTLE EMLYN; 52°01′N 04°23′W. Former woolen-milling center. Nearby are former woolen-milling villages of Drefach, Velindre, and Pentrecwrt. Formerly in DYFED, abolished 1996.

Llangennech (hlan-GE-nekh), village (2001 population 4,510), CARMARTHENSHIRE, SW Wales, on LOUGHOR RIVER, and 4 mi/6.4 km ENE of LLANELLI; 51°41′N 04°05′W. Previously tinplating. Formerly in DYFED, abolished 1996.

Llangernyw (hlan-GER-nyoo), village (2001 population 982), CONWY, N Wales, on Elwy River, 8 mi/12.9 km SE of CONWY; 53°11′N 03°41′W. Tourism. Formerly in CLWYD, abolished 1996.

Llangoedmor (hlan-GOID-mor), village (2001 population 1,174), CEREDIGION, W Wales, near TEIFI RIVER, just E of CARDIGAN; 52°05′N 04°38′W. Previously woolen milling. Formerly in DYFED, abolished 1996.

Llangollen (hlan-GO-hlen), town (2001 population 3,412), DENBIGHSHIRE, NE Wales, at the head of the Vale of Llangollen on the DEE RIVER, and 8 mi/12.8 km E of CORWEN; 52°58′N 03°10′W. Tourist center; attractions include the Dinas Bran castle (13th century), Eliseg's Pillar, Valle Crucis Abbey (1200), and the Shropshire Union Canal. The International Musical Eisteddfod is held here. Formerly in CLWYD, abolished 1996.

Llangors (hlan-GORS), village (2001 population 1,045), POWYS, E Wales, 4 mi/6.4 km S of TALGARTH; 51°56′N 03°16′W. On LLANGORSE LAKE, the second largest natural lake in Wales; also in BRECON BEACONS NATIONAL PARK. Has 14th–15th century church.

Llangorse Lake (hlan-GORS), Welsh *Llyn Syfaddan*, POWYS, E Wales, 5 mi/8 km ESE of BRECON; 1 mi/1.6 km in diameter; 51°56′N 03°16′W. Drained by short tributary of WYE RIVER. At N end is village of LLANGORS. Second largest natural lake in Wales.

Llangynwyd (hlan-GUH-noo-id), village (2001 population 3,310), BRIDGEND, S Wales, 2 mi/3.2 km S of

MAESTEG; 51°35′N 03°39′W. Formerly in MID GLAMORGAN, abolished 1996.

Llanharan (hlan-HA-ruhn), village (2001 population 3,421), RHONDDA CYNON TAFF, S Wales, 7 mi/11.3 km ENE of BRIDGEND; 51°32′N 03°26′W. Mineral spring. Formerly in MID GLAMORGAN, abolished 1996.

Llanharry (hlan-HA-ree), village (2001 population 2,919), RHONDDA CYNON TAFF, S Wales, 4 mi/6.4 km NE of BRIDGEND; 51°30′N 03°26′W. Farming. Formerly in MID GLAMORGAN, abolished 1996.

Llanhilleth, Wales: see ABERTILLERY.

Llanidloes (hlan-NID-lois), town (2001 population 2,807), POWYS, E Wales, on the SEVERN RIVER, and 11 mi/18 km SW of NEWTOWN; 52°28′N 03°32′W. Previously leather manufacturing and woolen (flannel) milling. Museum. Annual quilt festival.

Llanilltud Fardre, Wales: see LLANTWIT FARDRE.

Llanilltud Fawr, Wales: see LLANTWIT MAJOR.

Llanlleiana Head, Wales: see LLANBADRIG.

Llanllwchaiarn, Wales: see NEWTOWN.

Llanllwni (hlan-HLOO-nee), village (2001 population 676), CARMARTHENSHIRE, SW Wales, on TEIFI RIVER, 7 mi/11.3 km SW of LAMPETER; 52°02′N 04°12′W. In agricultural area. Formerly in DYFED, abolished 1996.

Llanllyfni (hlan-HLUHV-nee), village (2001 population 1,206), GWYNEDD, NW Wales, 6 mi/9.7 km S of CAERNARVON; 53°03′N 04°17′W. SNOWDONIA NATIONAL PARK to E, CAERNARVON BAY 3 mi/4.8 km to W.

Llanllyfni (hlan-HLUHV-nee), village, Gwynedd, Wales, 7 mi/11.3 km S of Caernarvon; 53°02′N 04°17′W. Woolen milling.

Llano (LAN-o), county (□ 966 sq mi/2,511.6 sq km; 2006 population 18,269), central TEXAS; ⊙ LLANO; 30°42′N 98°40′W. Elevation 650 ft/198 m–1,800 ft/549 m. Hilly area on E Edwards Plateau, bounded E by COLORADO RIVER (forms LAKE BUCHANAN, INKS LAKE, and Lake Lyndon B. Johnson); drained by LLANO RIVER and tributaries. Ranching (cattle, sheep, goats, hogs) and agriculture (peanuts, pecans, peaches, grapes, grain; hay). Granite, stone, vermiculite quarrying. Scenery, hunting, fishing attract tourists. Formed 1856.

Llano (LAN-o), town (2006 population 3,327), ⊙ LLANO county, central TEXAS, on EDWARDS PLATEAU, c.65 mi/105 km NW of AUSTIN, and on LLANO RIVER railroad terminus in ranching (cattle and sheep) and agriculture (grain, peanuts, peaches, grapes) area; 30°45′N 98°40′W. Granite quarrying and cutting; light manufacturing. Tourism (hunting, fishing). LAKE BUCHANAN is c.12 mi/19 km E; Enchanted Rock State Natural Area to SW (GILLESPIE county). Founded 1855; incorporated 1901.

Llano, El, PANAMA: see EL LLANO.

Llano Estacado (YAH-no es-tah-KAH-do), level, semiarid, plateau like region of the S GREAT PLAINS (□ c.40,000 sq mi/103,600 sq km), E NEW MEXICO and W TEXAS, between the PECOS R and the CAPROCK ESCARPMENT. Most irrigation is drawn from subsurface water tables, leading to concern over groundwater depletion. The High Plains (c.4,000 ft/1,220 m) of the Texas Panhandle, centered around AMARILLO, are usually distinguished from the somewhat lower South Plains (c.2,500 ft/760 m), centered around LUBBOCK. Both are wind-swept grasslands. Formerly used for cattle ranching, the plains are dotted with dry-land and irrigated farms as well as oil and natural-gas fields. Also called the Staked Plain.

Llano River (LAN-o), c.105 mi/169 km long, in central TEXAS; formed at town of JUNCTION, KIMBLE county, by N Llano and S Llano rivers, both rising on EDWARDS PLATEAU; flows generally E to the COLORADO RIVER 50 mi/80 km NW of AUSTIN. N Llano rises in SUTTON county, flows c.60 mi/97 km E to Junction. S Llano rises NW EDWARDS county, flows c.80 mi/129 km E and NE to Junction.

Llanos de Aridane, Los, town, SANTA CRUZ DE TENERIFE province, PALMA island, CANARY ISLANDS,

SPAIN, 9 mi/14.5 km WSW of SANTA CRUZ DE LA PALMA; 28°39′N 17°54′W. In agricultural region (cereals, sugarcane, fruit, grapes; livestock). Wine making, silk manufacturing.

Llanos, Los, DOMINICAN REPUBLIC: see LOS LLANOS.

Llanos, Sierra de los (YAH-nos, see-YER-rah dai los), pampean mountain range in SE LA RIOJA province, ARGENTINA, SE of PATQUIA; c.25 mi/40 km long (N-S).

Llanquera (yahn-KE-rah), canton, CARANGAS province, ORURO department, W central BOLIVIA, 19 mi/30 km NE of CHOQUECOTA, on a branch of the ORURO-SABAYA road; 18°06′S 67°47′W. Elevation 12,448 ft/3,794 m. Lead-bearing lode. Gypsum, copper deposits in the Huayllamarca Mountains. Potatoes, yucca, bananas; cattle.

Llanquihue (yahn-KEE-wai), province (□ 7,107 sq mi/18,478.2 sq km) of LOS LAGOS region, S central CHILE; ⊙ PUERTO MONTT; 41°30′S 73°05′W. Located between the PACIFIC (W), the ANDES (E), and GULF OF ANCUD (S), just N and NE of CHILOÉ ISLAND, it embraces part of the Chilean lake district, notably LAKE LLANQUIHUE and LAKE TODOS LOS SANTOS, over which tower OSORNO and CALBUCO volcanoes. Numerous islands in Gulf of Ancud and RELONCAVÍ SOUND include: MAILLÉN, GUAR, Calbuco, PULUQUI, QUENU. Has temperate, humid climate. Lumbering, dairying, and agricultural region (cereals, vegetables, fruit; livestock), with fisheries in the gulfs and in the freshwater lakes. Fish canneries concentrated at Calbuco; at Puerto Montt are breweries, lumber mills, dry docks. A noted tourist country with fine scenery, major resorts are: PUERTO VARAS, ENSENADA, PETROHUÉ, PEULLA.

Llanquihue (yahn-KEE-wai), village, ⊙ Llanquihue comuna, LLANQUIHUE province, LOS LAGOS region, S central CHILE, on SW shore of LAKE LLANQUIHUE, on railroad, and 16 mi/26 km N of PUERTO MONTT; 41°15′S 73°01′W. Dairying; lumbering.

Llanquihue, Lake (yahn-KEE-hwai) (□ 300 sq mi/780 sq km), in OSORNO and LLANQUIHUE provinces, LOS LAGOS region, S central CHILE, in Chilean lake district, 10 mi/16 km N of PUERTO MONTT; 22 mi/35 km long, 25 mi/40 km wide; 41°08′S 72°48′W. Bounded by the volcanoes OSORNO (E) and CALBUCO (SE) and by wooded subandean hills, it is a well-known resort, bordered by villages of PUERTO OCTAY (N), FRUTILLAR (W), PUERTO VARAS (S), ENSENADA (E). Lumbering, fishing; sports. Depths of almost 5,000 ft/1,524 m have been measured. Outlet: MAULLÍN RIVER.

Llanrhaeadr (hlan-REI-uh-duhr), village (2001 population 1,080), DENBIGHSHIRE, NE Wales, 2 mi/3.2 km SE of DENBIGH; 53°09′N 03°23′W. Has 15th-century church with 13th-century tower. Formerly in CLWYD, abolished 1996.

Llanrhidian (hlan-RID-yuhn), village (2001 population 5,675), SWANSEA, S Wales, on GOWER peninsula, 10 mi/16 km W of Swansea; 51°36′N 04°10′W. Extensive salt marshes to N. Formerly in WEST GLAMORGAN, abolished 1996.

Llanrug (hlan-REEG), village (2001 population 2,755), GWYNEDD, NW Wales, 3 mi/4.8 km E of CAERNARVON; 53°08′N 04°12′W.

Llanrwst (hlan-ROOST), town (2001 population 3,037), CONWY, N Wales, on CONWY RIVER, and 11 mi/18 km SSW of COLWYN BAY; 53°09′N 03°48′W. Agricultural market. Previously leather manufacturing. Has 15th-century church, reputed to contain tomb of Llewellyn the Great.

Llansá (lyahn-SAH), town, GERONA province, NE SPAIN, 12 mi/19 km NE of FIGUERAS, near the MEDITERRANEAN SEA; 42°22′N 03°09′E. Olives, grapes, vegetables. Feldspar mines. Small port of Puerto de Llansá is 1 mi/1.6 km NNE.

Llansantffraid (hlan-sant-FREID), village (2001 population 2,482), CEREDIGION, W Wales, on CARDIGAN BAY, 5 mi/8 km NE of ABERAERON; 52°18′N 04°10′W.

Llansantffraid (hlan-sant-FREID), village, S POWYS, E Wales, 6 mi/9.7 km SW of BRECON; 51°53′N 03°27′W.

Llansantffraid Deythur, Wales: see LLANSANTFFRAID-YM-MECHAIN.

Llansantffraid-ym-Mechain (hlan-sant-FREID–uhm–ME-hein), village (2001 population 1,215), N POWYS, E Wales, on VYRNWY RIVER, and 8 mi/12.9 km N of WELSHPOOL; 52°46′N 03°09′W. Just S is village of Llansantffraid Deythur.

Llansawel, Wales: see BRITON FERRY.

Llansteffan (hlan-STE-fuhn) or **Llanstephen**, village (2001 population 1,076), CARMARTHENSHIRE, SW Wales, on TOWY RIVER estuary, and 7 mi/11.3 km SW of CARMARTHEN; 51°46′N 04°23′W. Previously woolen mills. Site of St. Anthony's wishing well. Formerly in DYFED, abolished 1996.

Llanta (YAHN-tah), copper-mining settlement, ATACAMA region, N CHILE, on railroad, and 55 mi/89 km E of CHAÑARAL, on the QUEBRADA DEL SALADO RIVER.

Llantilio Pertholey (thlan-TIL-i-o puh-THO-lee), Welsh *Llandeilo Bertholau*, village (2001 population 3,965), MONMOUTHSHIRE, SE Wales, just N of ABERGAVENNY; 51°51′N 03°01′W. Has 14th-century church. Formerly in GWENT, abolished 1996.

Llantrisant (hlan-TRI-sant), town (2001 population 14,915), RHONDDA CYNON TAFF, S Wales, 10 mi/16 km WNW of CARDIFF; 51°33′N 03°23′W. Remains of 13th-century castle. Formerly in MID GLAMORGAN, abolished 1996.

Llantwit Fardre (HLAN-twit VAHR-drai), Welsh *Llanilltud Fardre*, village (2001 population 6,214), CAERPHILLY, SE Wales, 3 mi/4.8 km NE of LLANTRISANT; 51°34′N 03°19′W. Formerly in MID GLAMORGAN, abolished 1996.

Llantwit Major (LAN-twit), Welsh *Llanilltud Fawr*, town (2001 population 9,687), VALE OF GLAMORGAN, S Wales, near BRISTOL CHANNEL, 8 mi/12.9 km SE of BRIDGEND; 51°24′N 03°29′W. Llandow industrial estate 2 mi/3.2 km N. Site of Roman villa and Bedford Castle. Formerly in SOUTH GLAMORGAN, abolished 1996.

Llanura de Guacanayabo (yah-NOOR-ah dai gwah-kahn-nah-YAH-bo), flood plain, GRANMA province, CUBA, straddling the island's longest river. Part of 3,463 sq mi/8,969 sq km watershed in E Cuba. Rich agricultural lands. Also known as Llanura del Coato.

Llanwenog (hlan-WE-nog), village (2001 population 1,883), CEREDIGION, W Wales, 5 mi/8 km WSW of LAMPETER; 52°05′N 04°12′W. Formerly in DYFED, abolished 1996.

Llanwrda (hla-NOOR-duh), village (2001 population 513), CARMARTHENSHIRE, SW Wales, on TOWY RIVER, and 4 mi/6.4 km SW of LLANDOVERY; 51°58′N 03°52′W. Previously woolen milling.

Llanwrtyd Wells (hla-NOOR-tid), resort (2001 population 762), POWYS, E Wales, 15 mi/24 km NW of BRECON; 52°06′N 03°38′W. Previously sulfur spring and tweed mill.

Llanybydder (hla-nuh-BUH-duh) or **Llanybyther**, village (2001 population 1,420), CARMARTHENSHIRE, SW Wales, on TEIFI RIVER, 4 mi/6.4 km SW of LAMPETER; 52°04′N 04°10′W. Previously woolen milling. Formerly in DYFED, abolished 1996.

Llanyre (hla-NIR), village (2001 population 1,061), POWYS, E Wales, 1 mi/1.6 km W of LLANDRINDOD WELLS; 52°15′N 03°24′W. Remains of Roman fort Castel Collen (C.E. 75–78) to N.

Llao-Llao (YOU–YOU), village, SW RÍO NEGRO province, ARGENTINA, on LAKE NAHUEL HUAPÍ, 15 mi/24 km WNW of SAN CARLOS DE BARILOCHE; 41°03′S 71°32′W. Resort with numerous hotels.

Llardecáns (lyahr-dai-KAHNS), town, LÉRIDA province, NE SPAIN, 17 mi/27 km SSW of LÉRIDA; 41°22′N 00°33′E. Sheep raising. Soap manufacturing, olive-oil processing.

Llaretas, Cordón de las (yah-RAI-tahs, kor-DON dai lahs), Andean range in W central MENDOZA province, ARGENTINA; rises to over 16,000 ft/4,877 m.

Cross-references are shown in SMALL CAPITALS. The pronunciation guide is shown on page xix. The sources of population figures are shown on page xvii.

Llata (YAH-tah), city, ⊙ HUAMALÍES province, HUÁNUCO region, central PERU, on E slopes of Cordillera BLANCA of the ANDES, near MARAÑON RIVER, 50 mi/80 km NW of HUÁNUCO; 09°25′S 76°47′W. Elevation 11,246 ft/3,428 m. Weaving of native textiles; cereals, corn, potatoes; livestock.

Llaurí (lyou-REE), town, VALENCIA province, E SPAIN, 6 mi/9.7 km E of ALZIRA; 39°09′N 00°20′W. Rice, oranges.

Llavalol (yah-vah-LOL), residential district in greater BUENOS AIRES, ARGENTINA, SSW of BUENOS AIRES; 34°47′S 58°26′W.

Llavica (yah-VEE-kah), canton, NOR LÍPEZ province, POTOSÍ department, W central BOLIVIA, SSW of SANTIAGO DE AGENCHA; 20°37′S 67°42′W. Elevation 12,402 ft/3,780 m. Gas deposits to E. Tungsten-bearing lode; lead and zinc mining at Mina Toldos, owned by the Yana Mallcu Mining Company, which has been mining since the Colonial era; salt extraction. Potatoes, yucca, bananas; cattle.

Llay (HLEI), town (2001 population 4,905), WREXHAM, NE Wales, on ALYN RIVER, and 3 mi/4.8 km N of Wrexham; 53°06′N 03°00′W. Formerly in CLWYD, abolished 1996.

Llay-Llay (YAI–YAI), town, VALPARAISO province, central CHILE, on ACONCAGUA RIVER, and 37 mi/60 km NE of VALPARAISO. Sometimes Llaillai.

Lleida, SPAIN: see LÉRIDA.

Llera (LYAI-rah), town, BADAJOZ province, W SPAIN, 15 mi/24 km N of LLERENA; 38°27′N 06°03′W. Mining (lead, pitchblende, zinc) and agriculture (wheat, barley, sheep).

Llera, MEXICO: see LLERA DE CANALES.

Llera de Canales (YE-rah dau kah-NAH-les), town, ⊙ Llera township, TAMAULIPAS, NE MEXICO, in E outliers of SIERRA MADRE ORIENTAL, 30 mi/48 km SSE of CIUDAD VICTORIA; 23°19′N 99°01′W. In agricultural area.

Llerena (lyai-RAI-nah), town, BADAJOZ province, W SPAIN, in EXTREMADURA, in NW SIERRA MORENA, on railroad to SEVILLE, and 70 mi/113 km SE of BADAJOZ; 38°14′N 06°01′W. Livestock raising and processing (sheep, hogs), and agricultural center (cereals, chick peas, grapes, olives; honey). Liquor distilling, wine making, olive oil extracting, textiles. Area is rich in minerals (silver mines, coal deposits). Noted plateresque church with Mudejar tower. Ancient city was the Roman *Degina Turdulorum*. Heavily disputed by Moors, it was finally taken (1241) by the Knights of the Order of Santiago, who made it their headquarters. Sacked by French during Peninsular War.

Lleulleu, Lake (YAI-oo-YAI-oo) (☐ 27 sq mi/70.2 sq km), ARAUCO province, BÍO-BÍO region, S central CHILE, within 5 mi/8 km of the PACIFIC, at W foot of CORDILLERA DE NAHUELBUTA, 40 mi/64 km SE of LEBU; 38°09′S 73°20′W. Has several long, narrow arms; c.10 mi/16 km long. Tourist site.

Llewellyn (loo-WEL-lin), unincorporated town, Branch township, SCHUYLKILL county, E central PENNSYLVANIA, 4 mi/6.4 km NW of POTTSVILLE, on West Creek; 40°40′N 76°16′W. In anthracite-coal region.

Lleyn Peninsula (HLEEN) or **The Lleyn**, GWYNEDD, NW Wales, between CAERNARVON BAY (N) and CARDIGAN BAY (S) of Irish Sea; c.30 mi/48 km long, 5 mi/8 km–15 mi/24 km wide; 52°55′N 04°30′W. Chief town, PWLLHELI. At SW extremity is promontory of BRAICH-Y-PWLL.

Llica (YEE-kah), town and canton, ⊙ DANIEL CAMPOS province, POTOSÍ department, SW BOLIVIA, in Cordillera de LLICA, on NW shore of Salar de UYUNI salt flat, and 40 mi/64 km WSW of SALINAS DE GARCI MENDOZA; 19°50′S 68°18′W. Elevation 12,051 ft/3,673 m. Quinoa, potatoes.

Llica, Cordillera de, spur of the western Cordillera of the ANDES Mountains, ORURO department, SW BOLIVIA; extends 60 mi/97 km E from Cordillera de SILLAJHUAY to SALINAS DE GARCI MENDOZA; separates Salar de COIPASA (N) and Salar de UYUNI (S); 20°46′S 67°42′W.

Llico (YEE-ko), village, ARAUCO province, BÍO-BÍO region, S central CHILE, on the PACIFIC (ARAUCO GULF) 32 mi/52 km N of LEBU. Grain, beans; livestock; fishing.

Llico (YEE-ko), village, CURICÓ province, MAULE region, central CHILE, on the coast, near Vichuquén Lagoon, 55 mi/89 km NW of CURICÓ. Resort.

Llifén (yee-FEN), village, VALDIVIA province, LOS LAGOS region, S central CHILE, on E bank of LAKE RANCO, in Chilean lake district, 60 mi/97 km SE of VALDIVIA. Resort; thermal springs. Lumbering.

Llinás (lyee-NAHS), town, BARCELONA province, NE SPAIN, 7 mi/11.3 km NNW of MATARÓ; 41°38′N 02°24′E. Knitwear. Cork, wine, lumber in area. Summer resort.

Llingua Island (YEEN-gwah) (☐ 2 sq mi/5.2 sq km), just off E coast of CHILOÉ ISLAND, S CHILE; 42°25′S 73°27′W.

Llipi Llipi (YEE-pee YEE-pee), canton, LOAYZA province, LA PAZ department, W BOLIVIA, 9 mi/15 km SW of CAXATA; 17°11′S 67°23′W. Elevation 8,333 ft/2,540 m. Tin mining at Mina Viloco; clay, limestone deposits. Potatoes, yucca, bananas, rye; cattle.

Lliuco (yee-OO-ko), village, CHILOÉ province, LOS LAGOS region, S CHILE, on NE coast of CHILOÉ ISLAND, 22 mi/35 km SE of ANCUD; 42°02′S 73°29′W. In agricultural area (potatoes, wheat; livestock); lumbering.

Llivia (LYEE-vyah), village (☐ 5 sq mi/13 sq km); exclave of NE SPAIN (GERONA province), within PYRÉNÉES-ORIENTALES department, SW FRANCE, 4 mi/6.4 km NE of PUIGCERDÁ (Spain); 42°28′N 01°59′E. Tourism; dairying. Has 15th-century stone church and tower. En Louces Catalan Museum. Was capital of CERDAÑA (Cerdagne) until 12th century. A Spanish exclave within French territory, created by Treaty of the Pyrenees (1659), which also provided for a neutral road through French territory from Puigcerdá.

Lloa (YO-ah), village, PICHINCHA province, N central ECUADOR, in the ANDES, 7 mi/11.3 km WSW of QUITO; 00°15′S 78°35′W. Starting point for ascent of Cerro PICHINCHA.

Llobregat River (lyo-vrai-GAHT), 105 mi/169 km long, in BARCELONA province, NE SPAIN; rises in SIERRA DE CADÍ of the E PYRENEES, flows S and SE to the MEDITERRANEAN SEA S of BARCELONA. Receives CARDONER RIVER. Irrigates fertile Llobregat coastal plain. Vineyards and several hydroelectric plants along its course.

Llocllapampa (yok-yah-PAHM-pah), town, JAUJA province, JUNÍN region, central PERU, on LA OROYA–HUANCAVELICA railroad, on MANTARO RIVER, and 8 mi/13 km WSW of JAUJA; 11°49′S 75°40′W. Cereals, potatoes.

Llodio (LYO-dhyo), village, ÁLAVA province, N SPAIN, on NERVIÓN RIVER, and 8 mi/12.9 km SSW of BILBAO; 43°09′N 02°58′W. Metalworks; glass manufacturing, flour- and sawmilling. Cereals, potatoes; cattle, hogs. Locally called Plaza.

Llogora Pass (lo-GO-rah) (3,461 ft/1,055 m), in RRZEZA E KANALIT, S ALBANIA, on VLORË-HIMARË road, and 9 mi/14.5 km NW of Himarë, at foot of Mt. Çikës.

Llolleo (yo-YE-o), town, SANTIAGO province, METROPOLITANA DE SANTIAGO region, central CHILE, on the PACIFIC at mouth of MAIPO RIVER, just S of SAN ANTONIO; 33°37′S 71°36′W. Beach resort.

Llombay (lyom-BEI), town, VALENCIA province, E SPAIN, 17 mi/27 km SW of VALENCIA. Olive-oil processing, flour milling; wine, raisins, oranges. Gypsum quarries. Marquisate belonged to Osuna family.

Llorente (lyo-REN-te), town, EASTERN SAMAR province, SE SAMAR island, PHILIPPINES, on PHILIPPINE SEA, 50 mi/80 km SE of CATBALOGAN; 11°21′N 125°27′E. Agricultural center (coconuts, hemp).

Lloret de Mar (lyo-RET dhai MAHR), city, GERONA province, NE SPAIN, 19 mi/31 km SSE of GERONA; 41°42′N 02°51′E. Beach resort on the MEDITERRANEAN SEA; cork, wine, olive oil.

Lloret de Vista Alegre (VEE-stah ah-LE-grai), town, MAJORCA, BALEARIC ISLANDS, 18 mi/29 km E of PALMA; 39°37′N 02°58′E. Cereals, vegetables, figs, apricots; poultry, livestock; lumbering.

Lloró (yo-RO), town, ⊙ Lloró municipio, CHOCÓ department, W COLOMBIA, 12 mi/19 km SW of QUIBDÓ; 05°30′N 76°32′W. Port on the ATRATO RIVER. Agriculture includes plantains, cassava, and cacao.

Llosa de Ranes (LYO-sah dhai RAH-nes), town, VALENCIA province, E SPAIN, 3 mi/4.8 km N of JÁTIVA; 39°01′N 00°32′W. Olive oil, cereals, vegetables.

Lloseta (lyo-SAI-tah), town, MAJORCA island, BALEARIC ISLANDS, on railroad, and 15 mi/24 km NE of PALMA; 39°43′N 02°52′E. Olives, almonds, carobs, grapes. Food processing, manufacturing (shoes, tiles), gravel quarrying.

Lloyd Barrage, PAKISTAN: see SUKKUR BARRAGE.

Lloyd Dam, PUNE district, MAHARASHTRA state, W central INDIA, on upper NIRA RIVER, and 22 mi/35 km S of Pune. Dam (190 ft/58 m high) impounds reservoir (☐ 14.5 sq mi/37.6 sq km) powering hydroelectric plant and supplying Nira River canal irrigation system (headworks 16 mi/26 km E). In operation since 1928.

Lloydell (LOI-del), unincorporated village, Adams township, CAMBRIA county, W central PENNSYLVANIA, 12 mi/19 km E of JOHNSTOWN, on South Fork of Little Conemaugh River; 40°18′N 78°41′W. Beaverdam Run Reservoir to E.

Lloyd George, Mount (LOID JORJ) (10,000 ft/3,048 m), N central BRITISH COLUMBIA, W CANADA, in ROCKY MOUNTAINS; 57°51′N 124°57′W.

Lloyd Harbor, village (☐ 10 sq mi/26 sq km; 2006 population 3,654), SUFFOLK county, SE NEW YORK, on LONG ISLAND N shore, on LLOYD NECK near LLOYD HARBOR (arm of LONG ISLAND SOUND), 2 mi/3.2 km NW of HUNTINGTON; 40°55′N 73°27′W. Incorporated 1926; part of Long Island's affluent Gold Coast.

Lloyd Harbor, arm of LONG ISLAND SOUND indenting E shore of LLOYD NECK in N LONG ISLAND, SE NEW YORK; extends c.2.5 mi/4 km W from its mouth on HUNTINGTON BAY N of HUNTINGTON. 0.25 mi/0.40 km–1 mi/1.6 km wide; 40°55′N 73°26′W.

Lloydminster (LOID-min-stuhr), city (total ☐ 17 sq mi/43 sq km; 2001 population in ALBERTA 13,148; in SASKATCHEWAN 7,241 in 1991, 7,840 in 2001), on the Alberta-Saskatchewan border, CANADA; 53°17′N 110°00′W. The city is chartered by both provinces. Farming and ranching; oil, natural gas, coal, and salt deposits. The area was first settled in 1903; incorporated as a city in 1958.

Lloyd Neck, peninsula on LONG ISLAND N shore, SE NEW YORK, between OYSTER BAY (W) and HUNTINGTON BAY (E); c.5 mi/8 km long, 1.5 mi/2.4 km–3.5 mi/5.6 km wide; 40°56′N 73°28′W. E shore deeply indented by LLOYD HARBOR. Lloyd Point at tip. Caumsett State Park; Target Rock National Wildlife Refuge.

Lloyd Point, NEW YORK: see LLOYD NECK.

Lloyd Shoals Reservoir, GEORGIA: see JACKSON LAKE.

Lloyds Lake, SW NEWFOUNDLAND AND LABRADOR, CANADA, on LLOYDS RIVER, and 45 mi/72 km SSE of CORNER BROOK; 12 mi/19 km long, 1 mi/2 km wide; 48°25′N 57°35′W. On S shore are ANNIEOPSQUOTCH MOUNTAINS.

Lloyds River, SW NEWFOUNDLAND AND LABRADOR, CANADA, upper course of EXPLOITS RIVER, flows 60 mi/97 km ENE, through King George IV Lake, LLOYDS LAKE, and RED INDIAN LAKE, where it becomes EXPLOITS RIVER proper.

Llubí (lyoo-VEE), town, MAJORCA island, BALEARIC ISLANDS, 21 mi/34 km ENE of PALMA; 39°42′N 03°00′E. Capers, apricots, almonds, grapes; livestock. Liquor distilling, wine making.

Lluchmayor (lyooch-mei-OR) or **Llummayor** (lyoo-mei-OR), inland city, MAJORCA, BALEARIC ISLANDS, on railroad, and 14 mi/23 km ESE of PALMA. Industrial center in agricultural region (cereals, grapes, carobs, almonds, fruit; sheep, hogs). Liquor distilling, wine making, textile milling; manufacturing of shoes, felt, lace, paper. Gravel quarries.

Lluidas Vale (loo-EI-dus), town, SAINT CATHERINE parish, central JAMAICA, 15 mi/24 km NW of SPANISH TOWN; 18°07′N 77°09′W. Sugar-growing center in lush green valley. In 1600s, rebel slaves formed one of the first free black settlements in the New World here.

Llullaillaco (yoo-yei-YAH-ko), mountain and extinct volcano (22,057 ft/6,723 m), on the border of CHILE and ARGENTINA; 24°42′S; 68°33′W. One of the highest peaks in the ANDES and perpetually snowcapped, it overlooks a pass used for railroad and highway traffic between Chile and Argentina.

Llummayor, SPAIN: see LLUCHMAYOR.

Llusco (YOOS-ko), town, CHUMBIVILCAS province, CUSCO region, S PERU, in ANDEAN valley, 55 mi/89 km S of CUSCO; 14°19′S 72°06′W. Elevation 11,066 ft/3,372 m. Cereals, potatoes; copper and silver production; woolen goods.

Lluta River (YOO-tah), c.100 mi/161 km long, in N CHILE, rises in the ANDES at N foot of Cerro de Tacora on PERU border, flows S and WSW, through arid subandean plateaus to the PACIFIC 5 mi/8 km N of ARICA. In its lower irrigated reaches cotton and fruit are grown.

Llwchwr, Afon, Wales: see LOUGHOR RIVER.

Llyn Alaw (HLIN A-lou), large man-made reservoir (3 mi/4.8 km long, 0.5 mi/0.8 km wide), ISLE OF ANGLESEY, NW Wales, 4 mi/6.4 km SE of AMLWCH; 53°21′N 04°26′W. Supplies drinking water to N half of island. Operational since 1966.

Llyn Brenig (HLIN BRE-nig), large reservoir (3 mi/4.8 km long) on CONWY-DENBIGHSHIRE border, N Wales, 8 mi/12.9 km SSW of DENBIGH; 53°05′N 03°32′W. Nature reserve. Regulates flow of the DEE RIVER. Archaeological trail of sites from the Stone Age to the 16th century. Operational since 1979.

Llyn Brianne (HLIN BREE-an), large reservoir on CEREDIGION-POWYS border, central Wales, 10 mi/16 km N of LLANDOVERY; 52°09′N 03°46′W. Dam (c.3 mi/4.8 km long) at S end. Constructed in early 1970s to regulate flow of TOWY RIVER.

Llyn Celyn (HLIN KE-lin), large reservoir, GWYNEDD, NW Wales, 4 mi/6.4 km NW of BALA; 52°57′N 03°42′W. Dam at E end; fish stocks. Regulates flow of DEE RIVER. Operational since 1965.

Llyn Cwellyn (HLIN KWE-hlin) or **Lake Quellyn**, reservoir (□ 1 sq mi/2.6 sq km), GWYNEDD, NW Wales, at foot of SNOWDON mountain, and 7 mi/11.3 km SE of CAERNARVON; 53°04′N 04°09′W. Dam at N end. Supplies drinking water to parts of Gwynedd and ISLE OF ANGLESEY.

Llyn Gwynant (HLIN GWI-nuhnt), lake in GWYNEDD, NW Wales, 3 mi/4.8 km SE of SNOWDON summit; 1 mi/1.6 km long; 53°02′N 04°01′W.

Llyn Llydaw (HLIN HLUH-dou), lake in GWYNEDD, NW Wales, just E of SNOWDON summit; 1 mi/1.6 km long; 53°04′N 04°02′W.

Llyn Stwlan (HLIN STOO-luhn), reservoir, GWYNEDD, NW Wales, 1 mi/1.6 km SW of Blaenau Ffestiniog; 52°59′N 04°00′W. Upper lake site of Tan-y-Grisiau power station, first pumped water storage scheme (1968).

Llyn Syfaddan, Wales: see LLANGORSE LAKE.

Llyn Tegid, Wales: see BALA LAKE.

Llyn Trawsfynydd (HLIN trans-FUH-nuh-[th]-), reservoir, GWYNEDD, NW Wales, 2 mi/3.2 km S of FFESTINIOG; 52°54′N 03°55′W. In operation since 1930. Nuclear power station at N end (1964).

Loa, village (2006 population 515), ⊙ WAYNE county, S central UTAH, 35 mi/56 km SE of RICHFIELD, and on FREMONT RIVER; 38°23′N 111°38′W. Elevation 7,020 ft/2,140 m. Alfalfa, barley; dairying; cattle; manufacturing (cheese). FISH LAKE Reservoir to NW; Fishlake

National Forest to E and N. Loa Fish Hatchery to NE; Bicknell Bottoms Fish Hatchery to SE. Named for MAUNA LOA Volcano by Mormon missionary to HAWAII. Settled 1870s.

Loa (LO-ah), longest river of CHILE, 275 mi/443 km long, flowing S from the ANDES, N Chile, then W and N through the ATACAMA DESERT, before turning W to the PACIFIC OCEAN. It is not navigable but affords some water supply and hydroelectric power for copper and nitrate-mining communities in its vicinity.

Lo Aguirre (lo ah-GEER-re), village, SANTIAGO province, METROPOLITANA DE SANTIAGO region, central CHILE, 19 mi/31 km W of SANTIAGO. Copper mining.

Loami (loe-AM-ih), village (2000 population 804), SANGAMON county, central ILLINOIS, 13 mi/21 km SW of SPRINGFIELD; 39°40′N 89°50′W. In agricultural area.

Loandjili, town, KOUILOU region, SW Congo Republic, on Loango Bay, 10 mi/16 km NNW of POINTE-NOIRE (of which it is now a part); 04°45′S 11°51′E. Former port and caravan terminus noted throughout 19th and early 20th century for its trade in tropical commodities. Site of gas-fueled power plant (45.3 MW capacity). Seat of Roman Catholic vicar apostolic. French factories in area existed as early as 17th century.

Loanhead (LON-hed), town (2001 population 6,384), MIDLOTHIAN, E Scotland, on North Esk River, and 6 mi/9.7 km SSE of EDINBURGH; 55°52′N 03°09′W. Light manufacturing. Previously coal mining and limestone quarrying. Formerly in Lothian, abolished 1996.

Loano (lo-AH-no), town, SAVONA province, LIGURIA, NW ITALY, port on GULF OF GENOA, and 6 mi/10 km N of ALBENGA; 44°08′N 08°15′E. In agricultural region; fish canning; shoe manufacturing. Resort on Riviera di Ponente. Has palace (1578), now town hall.

Loay (LO-wei), town, BOHOL province, SW Bohol island, PHILIPPINES, on MINDANAO SEA, 11 mi/18 km ESE of TAGBILARAN; 09°36′N 124°02′E. Agricultural center (rice, coconuts).

Loayza, province, LA PAZ department, W BOLIVIA, ⊙ LURIBAY; 17°00′S 67°40′W.

Lobamba (LOU-bahm-bah), village, royal and parliamentary capital of SWAZILAND, HHOHHO district, W central Swaziland, 9 mi/14.5 km SE of MBABANE; 26°26′S 31°11′E. Embo State Palace is here, also National Museum, and Somhlolo National Stadium. Lozitha State House, royal residence, 7 mi/11.3 km to ENE. Located in Ezulwini (Royal) Valley. Also spelled Lubombo.

Lobatera (lo-bah-TER-ah), town, ⊙ Lobatera municipio, TÁCHIRA state, W VENEZUELA, in Andean outliers, 12 mi/19 km N of SAN CRISTÓBAL; 07°55′N 72°14′W. Elevation 3,700 ft/1,127 m. Coffee, vegetables.

Lobato (LO-bah-to), city (2007 population 4,219), NW PARANÁ state, BRAZIL, 28 mi/45 km N of MARINGÁ; 22°59′S 51°57′W. Coffee, rice, cotton, corn; livestock.

Lobatse (LO-BAH-tsai), town (2001 population 29,689), ⊙ SOUTH-EAST DISTRICT, SE BOTSWANA, on railroad, near Republic of SOUTH AFRICA border, 40 mi/64 km S of GABORONE; 25°15′S 25°40′E. Elevation 3,988 ft/1,216 m. Town Council administrative area. Dairying. Headquarters of Baralong tribe; hospital. Area includes Lobatsi block of farms.

Löbau (LOB-ou), town, SAXONY, E central GERMANY, in UPPER LUSATIA, at NE foot of LUSATIAN MOUNTAINS, 14 mi/23 km WSW of GÖRLITZ; 51°06′N 14°40′E. Railroad junction. Textile industry; manufacturing of pianos, shoes. Has several late-gothic churches. Town first mentioned 1221. Was member of Lusatian League, founded here in 1346.

Lobau (LO-bou), flood plain on the left bank of the Danube, SE of VIENNA and adjacent to parts of LOWER AUSTRIA. Scene of noted battle between French and Austrian troops in 1809. Part of Danube flood plains national park.

Löbau, POLAND: see LUBAWA.

Lobaye (lo-bah-YAI), prefecture (□ 7,425 sq mi/19,305 sq km; 2003 population 246,875), SW CENTRAL AFRICAN REPUBLIC; ⊙ M'BAÏKI. Bordered N by OMBELLA M'POKO prefecture, E and S by CONGO, SW by SANGHA-MBAÉRÉ economic prefecture, and NW by MAMBÉRÉ-KADÉÏ prefecture. Drained by LOBAYE and Pama rivers. Agriculture (coffee, cotton, cocoa, rubber, palm products); timber; mineral and ore mining (gold, diamond, copper); sawmilling. Main centers are M'Baïki, BODA, and Ngoto.

Lobaye River (lo-bah-YAI), 275 mi/443 km long, in W and SW CENTRAL AFRICAN REPUBLIC; rises in NANA-MAMBÉRÉ prefecture just E of BOUAR; flows SE and E through HAUTE-SANGHA and LOBAYE prefectures to UBANGI RIVER just N of MONGOUMBA. Navigable c.40 mi/64 km upstream.

Lobbes (LOB), commune (2006 population 5,531), Thuir district, HAINAUT province, S BELGIUM, on SAMBRE RIVER, and 9 mi/14.5 km WSW of CHARLEROI.

Loben, POLAND: see LUBLINIEC.

Lobendava (LO-ben-DAH-vah), German *Lobendau*, village, SEVEROCESKY province, N BOHEMIA, northernmost village in CZECH REPUBLIC, 28 mi/45 km NNE of ÚSTÍ NAD LABEM, near German border; 51°02′N 14°20′E. Produces artificial flowers, feather ornaments.

Lobenstein (LO-buhn-shtein), town, THURINGIA, central GERMANY, in FRANCONIAN FOREST, near the SAALE RIVER, 15 mi/24 km NW of HOF; 50°27′N 11°38′E. Electronics. Spa with mud baths. Has baroque palace.

Lobería (lo-be-REE-ah), town (1991 population 10,046), ⊙ Lobería district (□ 2,010 sq mi/5,226 sq km), S BUENOS AIRES province, ARGENTINA, 27 mi/43 km N of NECOCHEA; 38°05′S 58°45′W. Agriculture center (grain; cattle, sheep); flour milling.

Löberöd (LUH-ber-RUHD), village, SKÅNE county, S SWEDEN, 18 mi/29 km ENE of LUND; 55°26′N 14°07′E.

Lobethal (LU-buh-thuhl), village, SOUTH AUSTRALIA state, S central AUSTRALIA, 21 mi/33 km from ADELAIDE, and on W slopes of MOUNT LOFTY RANGES; 34°54′S 138°52′E. Orchards (stone fruits, apples, and pears); dairying; beef cattle. Settled c.1840 by Germans. Briefly known as Tweedvale.

Lobez (WO-beez), Polish *Łobez*, German *Labes*, town in POMERANIA, after 1945 in Szczecin province, NW POLAND, on the REGA RIVER, and 30 mi/48 km NE of STARGARD SZCZECINSKI. Metalworking, limestone quarrying, flour milling, potato processing, manufacturing of agricultural implements. In World War II, c.50% destroyed.

Lobith (LAW-bit), village, GELDERLAND province, E NETHERLANDS, 12 mi/19 km SE of ARNHEM; 51°52′N 06°07′E. German border 1 mi/1.6 km to N and S, 2 mi/3.2 km to E; RHINE RIVER passes to S. Dairying; cattle, hogs; vegetables, grain.

Lobito (lo-BEE-to), city (2004 population 137,400), BENGUELA province, W central ANGOLA, on the ATLANTIC OCEAN. Angola's largest and busiest port, it is also a road hub, and the W terminus of the trans-African Benguela railroad. The harbor, protected by a sand bar, is among the best on Africa's W coast. There are bulk loading facilities for ores and grain; other exports include coffee, sisal, sugar, fish, salt, and beans. The port ships minerals from ZAMBIA and the SHABA (Katanga) region of the Democratic Republic of the CONGO. Shipbuilding, food processing, manufacturing (textiles, cement, building materials). The completion of the railroad from BENGUELA (1929) made it an important commercial center. It is built mainly on reclaimed land. Grains, fruits, sisal, coconuts grown in area. Founded by the Portuguese in 1843.

Lobitos (lo-BEE-tos), town, TALARA province, PIURA region, NW PERU; port on the PACIFIC, 8 mi/13 km N of TALARA; 04°26′S 81°17′W. In petroleum area; oil-shipping center; petroleum storage. Power plant, machine shops; refrigerated fish plants; manufacturing.

Mining of calcium, graphite, carbon, salt, phosphates. Administrative center for surrounding oil fields of EL ALTO, CABO BLANCO.

Lob Nor, CHINA: see LUOBU BO.

Lobnya (LOB-nyah), city (2006 population 61,850), N central MOSCOW oblast, central European Russia, on railroad, 18 mi/29 km NNW of MOSCOW; 56°06′N 37°29′E. Elevation 659 ft/200 m. Construction materials, electrical equipment, ceramics, cotton milling, vegetable oil. Made city in 1961. Incorporated Krasnaya Polyana in 1975.

Lobo (LO-bo), former township (□ 76 sq mi/197.6 sq km; 2001 population 6,084), SW ONTARIO, E central CANADA, 10 mi/16 km NW of LONDON; 43°00′N 81°25′W. Amalgamated into MIDDLESEX CENTRE township in 1998.

Lobón (lo-VON), town, BADAJOZ province, W SPAIN, on the GUADIANA RIVER, and 16 mi/26 km WSW of MÉRIDA; 38°51′N 06°37′W. Olives, cereals, grapes, asparagus; livestock.

Lobos (LO-bos), town (1991 population 23,112), ⊙ Lobos district (□ 666 sq mi/1,731.6 sq km), NE BUENOS AIRES province, ARGENTINA, 55 mi/89 km SW of BUENOS AIRES; 35°15′S 59°10′W. Railroad hub and agriculture center (grain; livestock, poultry); light manufacturing; food processing.

Lobos Cay, BAHAMA Islands: see CAY LOBOS.

Lobos, Isla de (LO-vos, EE-slah dai), volcanic islet (□ 2 sq mi/5.2 sq km), CANARY ISLANDS, in LA BOCAYNA channel, just off NE FUERTEVENTURA; c.2 mi/3.2 km long; 28°45′N 13°49′W. Supports a few fishermen. Lighthouse.

Lobos Island (□ 5.5 sq mi/14.2 sq km), in Gulf of CALIFORNIA, off coast of SONORA, NW MEXICO, 45 mi/72 km SSE of GUAYMAS; low and sandy, 12 mi/19 km long, c.1 mi/1.6 km wide; 27°28′N 97°13′W. Lighthouse on W coast, at Point Lobos (27°22′N 110°38′W). Separated from mainland by Estero de Lobos.

Lobos Island, off coast of MALDONADO department, S URUGUAY, at mouth of the Río de la PLATA, 10 mi/16 km SSE of MALDONADO; 35°02′S 54°50′W. Said to have most powerful lighthouse in S AMERICA.

Lobos Islands (LO-bos), two groups of guano islands off PACIFIC coast of LAMBAYEQUE region, NW PERU, 35 mi/56 km apart. Largest and northernmost is *Isla Lobos de Tierra* [Spanish=landward seal island], 75 mi/121 km NW of PUERTO ETEN and 10 mi/16 km offshore; 2 mi/3 km wide, 6 mi/10 km long; 06°28′S 80°52′W. To S are the *Islas Lobos de Afuera* [Spanish=seaward seal islands], consisting of two islands (each c.2 mi/3 km long, 1 mi/2 km wide) separated by a narrow channel, 55 mi/89 km NW of Puerto Eten, and 40 mi/64 km offshore; 06°57′S 80°43′W.

Lobositz, CZECH REPUBLIC: see LOVOSICE.

Lobos, Point, promontory, SAN FRANCISCO county, W CALIFORNIA, on S side of entrance to GOLDEN GATE strait (entrance to SAN FRANCISCO BAY), 3 mi/4.8 km WSW of GOLDEN GATE BRIDGE, and 6 mi/9.7 km W of SAN FRANCISCO. Within GOLDEN GATE NATIONAL RECREATION AREA.

Lobos, Point, promontory on S shore of CARMEL BAY, MONTEREY county, W CALIFORNIA, 6 mi/9.7 km SW of MONTEREY, S of MONTEREY BAY. Point Lobos State Reserve.

Lobos, Punta de, cape, MONTEVIDEO department, S URUGUAY, on N bank of the Río de la PLATA, guarding W gate of the port of MONTEVIDEO; 34°54′S 56°15′W.

Lobsens, POLAND: see LOBZENICA.

Lobster House, mountain (1,916 ft/584 m), SW NEWFOUNDLAND AND LABRADOR, CANADA, 14 mi/23 km NNW of BUCHANS.

Lobster Lake, PISCATAQUIS county, W central MAINE, 27 mi/43 km N of GREENVILLE, just E of MOOSEHEAD LAKE; irregularly shaped; 4 mi/6.4 km long, 2 mi/3.2 km wide. In recreational area.

Loburg (LO-burg), town, SAXONY-ANHALT, central GERMANY, 14 mi/23 km SE of BURG; 52°07′N 12°05′E.

Has old town gate, eighteenth-century town hall, remains of early-Gothic castle.

Loburn, township, Waimakariri district (□ 856 sq mi/2,225.6 sq km), CANTERBURY region, NEW ZEALAND, 21 mi/34 km NNW of CHRISTCHURCH; 43°15′S 172°32′E. Sheep; fruit.

Lobutcha Creek (lo-BOO-chuh), c.50 mi/80 km long, in central MISSISSIPPI, rises in NW WINSTON county, flows SW into PEARL RIVER 3 mi/4.8 km E of CARTHAGE.

Lobva (LOB-vah), town (2006 population 8,360), W SVERDLOVSK oblast, W SIBERIA, RUSSIA, on the Lobva River (left tributary of the LYALYA RIVER, in the OB′ RIVER basin), on road and railroad, 10 mi/16 km NNW (and under jurisdiction) of NOVAYA LYALYA; 59°11′N 60°29′E. Elevation 265 ft/80 m. Woodworking, sawmilling; wood chemicals.

Lobzenica (wob-zee-NEET-sah), Polish *Łobż enica*, German *Lobsens*, town, Piła province, NW POLAND, 32 mi/51 km WNW of BYDGOSZCZ; 53°16′N 17°16′E. Brewing, cement manufacturing.

Locarno, district, SW TICINO canton, S SWITZERLAND; 46°10′N 08°48′E. Main town is LOCARNO; population is Italian-speaking and Roman Catholic.

Locarno, town (2000 population 14,561), TICINO canton, S SWITZERLAND, at N end of LAGO MAGGIORE; 46°10′N 08°48′E. Elevation 673 ft/205 m. In a beautiful resort region with a Mediterranean climate, Locarno attracts a great number of tourists. Machinery and electro-chemical products are made. The Swiss cantons took the town from MILAN in 1512, and it was incorporated in Ticino canton in 1803. Has noted pilgrimage church, Madonna del Sasso (first built 1480), which has a painting by Bramantino. Site of conference (1925) leading to Locarno Pact, which guaranteed the common boundaries of BELGIUM, FRANCE, and GERMANY and permitted Germany to join the League of Nations.

Locate di Triulzi (lo-KAH-te dee tree-OOL-tsee), town, MILANO province, LOMBARDY, N ITALY, 7 mi/11 km S of MILAN; 45°21′N 09°13′E. Food processing; paper.

Loc Binh (LAHK BIN), town, LANG SON province, N VIETNAM, 12 mi/19 km SE of LANG SON, on the Song Ky Cung (right headstream of LI RIVER); 21°45′N 106°55′E. Market and transportation center. Lignite mining. Livestock; rice, tea, coffee; forest products. Tho, Dao, and other minority peoples. Formerly Locbinh.

Loccum, GERMANY: see REHBURG-LOCCUM.

Loch, for names of Scottish lakes and inlets beginning with "Loch," see second part of element; e.g., for Loch Awe, see AWE, LOCH.

Lochaber (lah-KAH-ber), canton (□ 24 sq mi/62.4 sq km; 2006 population 477), OUTAOUAIS region, SW QUEBEC, E CANADA, 3 mi/5 km from THURSO; 45°38′N 75°13′W.

Lochaber (LAWK-ah-buhr), mountainous administrative district, SW HIGHLAND, W Scotland. It includes BEN NEVIS (4,406 ft/1,343 m), Great Britain's highest peak. The Lochaber power development provides electricity for FORT WILLIAM's aluminum works. A ferry runs from MALLAIG, a herring fishery, to the ISLE OF SKYE.

Lochaber-Partie-Ouest (lah-KAH-ber–pahr-TEE–WEST), canton (□ 24 sq mi/62.4 sq km; 2006 population 448), OUTAOUAIS region, SW QUEBEC, E CANADA, 3 mi/4 km from THURSO; 45°37′N 75°18′W.

Lo Chacón (lo chah-KON), town, SANTIAGO province, METROPOLITANA DE SANTIAGO region, central CHILE, 3 mi/5 km SW of EL MONTE, near MAIPO RIVER.

Lochalsh, Scotland: see KYLE OF LOCHALSH.

Locharbriggs (LAWK-ahr-brigs), suburb (2001 population 6,096), DUMFRIES AND GALLOWAY, S Scotland, on Lochar Water, and 3 mi/4.8 km NE of DUMFRIES; 55°06′N 03°35′W.

Lochard, village, MATABELELAND NORTH province, SW central ZIMBABWE, 30 mi/48 km ENE of BULAWAYO, on railroad; 19°56′S 29°01′E. Livestock; grain.

Lochau (LAWKH-ou), village, VORARLBERG, W AUSTRIA, on LAKE CONSTANCE, near German border, 2 mi/3.2 km N of BREGENZ; 47°32′N 09°45′E. Elev. 1,265 ft/386 m. Manufacturing of cheese, food processing. Has Renaissance castle, Neu-Hofen.

Lochboisdale (LAWK-boiz-dail), fishing port, chief center of SOUTH UIST island, Eilean Siar, NW Scotland, at head of Loch Boisdale (4 mi/6.4 km long, up to 2 mi/3.2 km wide), on SE coast of island; 57°10′N 07°19′W.

Lochbuie, town (2000 population 2,049), WELD county, N central COLORADO, 25 mi/40 km NW of DENVER; 40°00′N 104°43′W. Elevation 4,980 ft/1,518 m. Cattle; grain, sugar beets, beans, fruit. Denver International Airport to S. BARR LAKE State Park to SW; HORSE CREEK RESERVOIR to E.

Lochee, Scotland: see DUNDEE.

Lochem (LAW-khuhm), city, GELDERLAND province, E NETHERLANDS, on BERKEL RIVER, and 9 mi/14.5 km E of ZUTPHEN; 52°09′N 06°25′E. TWENTE CANAL passes to N; Lochemse Berg hill to SE. Dairying; cattle, hogs, poultry; grain, vegetables; light manufacturing Castle Ampsen to N.

Loches (lo-shuh), town (□ 10 sq mi/26 km), subprefecture of INDRE-ET-LOIRE department, CENTRE administrative region, W central FRANCE, in TOURAINE region, on the INDRE RIVER, and 23 mi/37 km SE of TOURS; 47°08′N 01°00′E. Manufacturing of electronic equipment. A picturesque site famous for its medieval "cities," especially the 14th-century castles, ancient dungeon, and fortifications that dominate the river valley. Established (10th century) by the counts of ANJOU, it later became (13th century) a royal residence and then a state prison. The royal lodge, built by Charles VII, contains the tomb of Agnès Sorel and the oratory of Anne of Brittany.

Loch Garman, IRELAND: see WEXFORD, county and town.

Lochgelly (LAWK-ge-lee), town (2001 population 6,749), FIFE, E Scotland, 6 mi/9.7 km W of KIRKCALDY; 56°08′N 03°19′W.

Lochgilphead (LAWK-gilp-heed), town (2001 population 2,326), ⊙ Argyll and Bute, W Scotland, at head of Loch Gilp (inlet of LOCH FYNE, 1 mi/1.6 km long), on Crinan Canal, and 20 mi/32 km WNW of DUNOON; 56°03′N 05°26′W. In sheep-raising region. Tourist and shopping center. Formerly in Strathclyde, abolished 1996.

Lochiel, unincorporated village, SE ONTARIO, E central CANADA, 25 mi/40 km from CORNWALL, and included in NORTH GLENGARRY township; 45°22′N 74°37′W.

Lochinvar, town, MASHONALAND EAST province, NE central ZIMBABWE, suburb 5 mi/8 km SW of HARARE; 17°53′S 31°00′E.

Lochinver (lawk-IN-vuhr), village, N HIGHLAND, N Scotland, on Loch Inver (3 mi/4.8 km long sea inlet), and 18 mi/29 km N of ULLAPOOL; 58°09′N 05°15′W. Tourist and angling resort.

Loch Lomond (LAHK LO-muhnd), unincorporated town, PRINCE WILLIAM county, NE VIRGINIA, residential suburb 2 mi/3 km N of MANASSAS, near BULL RUN creek; 38°46′N 77°28′W. Manassas National Battlefield Park to NW.

Loch Lynn, town, GARRETT county, W MARYLAND, in the ALLEGHENIES, just SE of OAKLAND.

Loch Lynn Heights, town (2000 population 469), GARRETT county, W MARYLAND, 2 mi/3.2 km E of OAKLAND; 39°23′N 79°22′W. It originated as a rival resort to MOUNTAIN LAKE PARK. Now mainly residential. Incorporated 1896.

Lochmaben (lawk-MAH-ben), small town (2001 population 1,952), DUMFRIES AND GALLOWAY, S Scotland, 8 mi/12.9 km NE of DUMFRIES; 55°07′N 03°27′W. Has ruins of 14th-century castle of Robert the Bruce. Nearby are several small lakes forming nature reserve.

Loch Maddy, Scotland: see NORTH UIST.

Lochmoor, MICHIGAN: see GROSSE POINTE WOODS.

Area is shown by the symbol □, and capital city or county seat by ⊙.

Lochnagar (lawk-NAH-guhr), basin and ridge (2,198 ft/ 670 m) of GRAMPIAN MOUNTAINS, Aberdeenshire, NE Scotland, 7 mi/11.3 km SE of BRAEMAR; 56°58′N 03°14′W. Distillery at foot of mountain at Balmoral estate.

Loch Raven Reservoir (LOK RAI-ven), BALTIMORE county, N MARYLAND, on Gunpowder Falls River, in Loch Raven Park, 10 mi/16 km NNE of BALTIMORE; c.10 mi/16 km long; 39°25′N 76°32′W. Formed by Loch Raven Dam (75 ft/23 m high, 650 ft/198 m long), built (1922) for Baltimore water supply.

Lochristi (lo-KRIS-tee), commune (2006 population 20,238), Ghent district, EAST FLANDERS province, NW BELGIUM, 6 mi/9.7 km NE of GHENT. Vegetables, flowers; tree nurseries.

Lochsa River (lock-sah), c.65 mi/105 km long, in N IDAHO; rises in NE IDAHO county in BITTERROOT RANGE, near LOLO PASS; flows WSW past Lochsa Historic Ranger Station, joining SELWAY RIVER at Lowell to form Middle Fork CLEARWATER RIVER. Nez Perce National Forest to S.

Loch Sheldrake, hamlet, SULLIVAN county, SE NEW YORK, on small Loch Sheldrake, 5 mi/8 km SE of LIBERTY; 41°46′N 74°39′W.

Lochwinnoch (lawk-WI-nuhk), town (2001 population 2,570), Renfrewshire, W Scotland, at SW end of Castle Semple Loch (2 mi/3.2 km long), and 6 mi/9.7 km SW of JOHNSTONE; 55°47′N 04°38′W. Printing, furniture manufacturing. Previously silk milling. Bronze Age artifacts have been found here. BLACK CART WATER rises in Castle Semple Loch. Formerly in Strathclyde, abolished 1996.

Lochwiza, UKRAINE: see LOKHVYTSYA.

Lochy, Loch (LAW-kee), lake (10 mi/16 km long, 1.5 mi/ 2.4 km wide; 531 ft/162 m deep), HIGHLAND, N Scotland, between LOCH OICH (NE) and LOCH LINNHE (SW); 56°58′N 04°55′W. Forms part of CALEDONIAN CANAL.

Lock (LAHK), town, SOUTH AUSTRALIA state, S central AUSTRALIA, 186 mi/300 km SW of PORT AUGUSTA and at EYRE PENINSULA center; 33°33′S 135°45′E. Wheat, sheep. Arid area; conservation parks nearby. Heritage museum housed in old police station.

Lock and Dam Number 13, ARKANSAS: see JAMES W. TRIMBLE LOCK AND DAM.

Lock and Dam Number 5, JEFFERSON county, central ARKANSAS, on ARKANSAS RIVER, 23 mi/37 km SSE of LITTLE ROCK; 34°25′N 92°06′W. Dam (54 ft/16 m high) built (1968) by Army Corps of Engineers for flood control, recreation, and as a fish and wildlife pond. Also known as Pool Number Five.

Lockbourne (LAHK-buhrn), village (2006 population 264), FRANKLIN county, central OHIO, 10 mi/16 km S of COLUMBUS; 39°48′N 82°58′W.

Locke, town, CAYUGA county, W central NEW YORK, in FINGER LAKES region, 20 mi/32 km SSE of AUBURN; 42°39′N 76°25′W. President Millard Fillmore born here, in division that later became town of Summerhill.

Lockeford, unincorporated town (2000 population 3,179), SAN JOAQUIN county, central CALIFORNIA, 17 mi/27 km NE of STOCKTON, on CALAVERAS RIVER; 38°09′N 121°10′W. Nuts, fruits, grapes, pumpkins, vegetables, grain, nursery products, sugar beets; dairying; cattle. Manufacturing (prefabricated metal buildings, wood-fiber products). Mokelumne Aqueduct passes to SE; CAMANCHE RESERVOIR to NE.

Locke Mills, MAINE: see GREENWOOD.

Locke, Mount, Texas: see DAVIS MOUNTAINS.

Lockenhaus (LOK-ken-hous), township, central BURGENLAND, E AUSTRIA, at the foot of Güns Mountains, 15 mi/24 km NNW of SZOMBATHELY, HUNGARY; 47°24′N 16°25′E. Castle; music festival.

Lockeport (LAHK-port), town, SW NOVA SCOTIA, CANADA, on Lockeport Harbour (14 mi/23 km long) of the ATLANTIC OCEAN, 50 mi/80 km ESE of YARMOUTH; 43°42′N 65°7′W. Elevation 3 ft/0.9 m. Fishing.

Lockerbie (LAW-kuhr-bee), town (2001 population 4,009), DUMFRIES AND GALLOWAY, S Scotland, 11 mi/ 17.6 km ENE of DUMFRIES; 55°07′N 03°21′W. Livestock fairs. Industries include manufacturing of cheese and cream. Site of Pan Am airplane crash (1988) as a result of bomb attributed to Middle East terrorists.

Lockesburg, town (2000 population 711), SEVIER county, SW ARKANSAS, 10 mi/16 km ESE of DE QUEEN; 33°58′N 94°10′W. Manufacturing (industrial control systems).

Lockhart (LOK-hahrt), town, S NEW SOUTH WALES, SE AUSTRALIA, 135 mi/217 km W of CANBERRA, in RIVERINA region; 35°14′S 146°43′E. Wheat; also sheep, wool, cattle, and poultry.

Lockhart, town (2000 population 548), Covington co., S Alabama, 22 mi/35 km SSE of Andalusia, near Florida state line. Founded c. 1910 when the Jackson Lumber Co. built a mill in the area, the town was named for Aaron Lockhart, a county commissioner in 1821. Inc. in 1931.

Lockhart (LAHK-hahrt), unincorporated town (□ 5 sq mi/13 sq km; 2000 population 12,944), ORANGE county, central FLORIDA, 5 mi/8 km NW of ORLANDO; 28°37′N 81°26′W. Light manufacturing.

Lockhart (LAHK-hahrt), town (2006 population 13,642), ⊙ CALDWELL county, S central TEXAS, 27 mi/ 43 km S of AUSTIN; 29°52′N 97°40′W. Agricultural area (cotton, sorghum, corn; turkeys; eggs); manufacturing (sausage processing, laboratory equipment, apparel, printing). Oil fields nearby. Lockhart State Park to SE is on site of battle of Plum Creek (1840), Texan defeat of Comanche Indians after they swept through the area's settlements. Founded 1848, incorporated 1870.

Lockhart, village (2006 population 502), UNION county, NW SOUTH CAROLINA, on BROAD RIVER, and 18 mi/29 km ESE of SPARTANBURG; 34°47′N 81°27′W. Textiles. Sumter National Forest to S.

Lock Haven, city (2006 population 8,652), ⊙ CLINTON county, N central PENNSYLVANIA, 28 mi/45 km WSW of WILLIAMSPORT, on West Branch of SUSQUEHANNA RIVER, 2 mi/3.2 km W of mouth of BALD EAGLE CREEK; 41°08′N 77°27′W. Manufacturing (prefabricated housing, concrete, paper, printing, aircraft). Agriculture (grain; livestock; dairying). Seat of Lock Haven University (University of Pennsylvania). Piper Memorial Airport to E. Bald Eagle State Park to SW; Seroul State Forest to NW, Bald Eagle State Forest to SE, Tiadaghton State Forest and Ravensburg State Park to E. Settled 1769, incorporated as a city 1870.

Lockington (LAHK-ing-tuhn), village (2006 population 202), SHELBY county, W OHIO, on GREAT MIAMI RIVER, 4 mi/6 km N of PIQUA; 40°12′N 84°14′W. Lockington dam is just S.

Lockland (LAHK-luhnd), village (□ 1 sq mi/2.6 sq km; 2006 population 3,321), HAMILTON county, extreme SW OHIO, 10 mi/16 km NNE of CINCINNATI; 39°13′N 84°27′W. Plotted 1828, incorporated 1865.

Lockney (LAHK-nee), town (2006 population 1,831), FLOYD county, NW TEXAS, on the LLANO ESTACADO, c.45 mi/72 km NE of LUBBOCK, near WHITE RIVER; 34°07′N 101°26′W. Cattle; wheat, sorghum, cotton, soybeans, pumpkins, sunflowers. Beef processing, farm equipment. Settled 1894, incorporated 1907.

Löcknitz (LUHK-nits), village, MECKLENBURG–WESTERN POMERANIA, NE GERMANY, 13 mi/21 km W of SZCZECIN, POLAND; 53°27′N 14°12′E.

Lockport (LAHK-port), community, SE MANITOBA, W central CANADA, 15 mi/25 km N of PETERSFIELD, 16 mi/ 26 km from WINNIPEG, and in ST. ANDREWS rural municipality; 50°04′N 96°56′W.

Lockport, city (2000 population 15,191), WILL county, NE ILLINOIS, 5 mi/8 km N of JOLIET, at locks connecting SANITARY AND SHIP CANAL with Des Plaines River; 41°35′N 88°02′W. In agricultural (corn, soybeans; dairying) area; manufacturing (aluminum powders and pigments, plastic pipe, auto parts, farm equipment). Historic locks of old Illinois and Michigan Canal and Visitor Center. Lewis University campus; airport on border with Romeoville. Stateville state prison across river. Laid out 1837, incorporated 1853.

Lockport, industrial city (□ 8 sq mi/20.8 sq km; 2006 population 21,035), ⊙ NIAGARA county, W NEW YORK, on NEW YORK STATE BARGE CANAL; 43°10′N 78°42′W. Manufacturing includes metal and paper products, chemicals, plastics. In a rich fruit and dairy region. Built around a series of locks on the old ERIE CANAL. Settled 1821, incorporated 1865. Writer Joyce Carol Oates born here.

Lockport, town (2000 population 2,624), LAFOURCHE parish, SE LOUISIANA, 34 mi/55 km SW of NEW ORLEANS, and on Bayou LAFOURCHE; 29°39′N 90°33′W. Offshore oil exploration and outfitting center; manufacturing (drilling platforms, paper, phenolic resin, shipbuilding). Also called Longville.

Lockport, village, HENRY county, N KENTUCKY, 18 mi/29 km NNW of FRANKFORT, on KENTUCKY RIVER. Zinc and lead deposits in area. Tobacco; concrete and asphalt mixers.

Lockridge, town (2000 population 275), JEFFERSON county, SE IOWA, 11 mi/18 km E of FAIRFIELD; 40°59′N 91°45′W.

Locks Heath (LAHKS HEETH), residential suburb (2001 population 7,035), HAMPSHIRE, S ENGLAND, 1 mi/1.6 km N of SOUTHAMPTON WATER, and 1 mi/1.6 km S of Park Gate; 50°52′N 01°15′W.

Lock Springs, town (2000 population 64), DAVIESS county, NW MISSOURI, near GRAND RIVER, 10 mi/16 km SE of GALLATIN; 39°51′N 93°46′W. Damaged by flooding in 1993.

Lockwood, city (2000 population 989), DADE county, SW MISSOURI, 38 mi/61 km NE of JOPLIN; 37°23′N 93°57′W. Corn, sorghum, wheat, cattle; light manufacturing; limestone quarries.

Lockwood, Cape, W ELLESMERE ISLAND, BAFFIN region, NUNAVUT territory, CANADA, on GREELY FJORD; 80°29′N 82°55′W.

Lockwood Island, in ARCTIC OCEAN, just off PEARY LAND region, N GREENLAND; 6 mi/9.7 km long, 5 mi/8 km wide, rises to 2,461 ft/750 m; 83°20′N 40°08′W. Discovered (May 1882) by James B. Lockwood and David Legg Brainard of Greeley expedition (1881–1884) under the command of Adolphus W. Greeley.

Locle, Le (LOCK-luh, luh), district, NW NEUCHÂTEL canton, NW SWITZERLAND. Main town is Le Locle; population is French-speaking and Protestant.

Locle, Le (LOCK-luh, luh), town (2000 population 10,529), NEUCHÂTEL canton, NW SWITZERLAND, in the Jura mountains near the French border, and 5 mi/ 8 km SE of LA CHAUX-DE-FONDS. Elevation 3,005 ft/ 916 m. It has been a watchmaking center since the 17th century.

Locmaria-Plouzané (lok-mah-ree-ah–ploo-zah-nai), agricultural commune (□ 8 sq mi/20.8 sq km), FINISTÈRE department, in BRITTANY, W FRANCE; 7 mi/ 12 km W of BREST.

Locmariaquer (lok-mah-ree-ah-ker), commune, MORBIHAN department, W FRANCE, in brittany, 11 mi/18 km SW of VANNES, on the Gulf of MORBIHAN (ATLANTIC OCEAN); 47°34′N 02°57′W. Oyster beds. Numerous megalithic monuments (including a menhir over 65 ft/20 m high) attract tourists. Point of departure for boat trips around the Gulf of Morbihan.

Locminé (lok-me-nai), commune (□ 1 sq mi/2.6 sq km), MORBIHAN department, W FRANCE, in BRITTANY, 16 mi/26 km NNW of VANNES; 47°53′N 02°50′W. Cattle market and food-processing center. Has remains of an abbey founded in 6th century.

Loc Ninh (LAHK NIN), town, SONG BE province, S VIETNAM, railroad terminus 75 mi/121 km N of HO CHI MINH CITY, near CAMBODIA border; 11°51′N 106°36′E. Transportation and trading center. Rubber plantations; forestry and wood products. Tuy, Khmer, Kihn, and other peoples. Formerly Locninh.

Loco, village (2006 population 152), STEPHENS county, S OKLAHOMA, 20 mi/32 km SE of DUNCAN; 34°19′N 97°40′W. Agricultural area in oil region.

Locorotondo (LO-ko-ro-TON-do), town, BARI province, APULIA, S ITALY, 14 mi/23 km SSE of MONOPOLI; 40°45′N 17°20′E. Wine making, olive oil; almonds.

Locri (LO-kree), town, REGGIO DI CALABRIA province, CALABRIA, S ITALY, port on IONIAN Sea, 3 mi/5 km SSW of SIDERNO MARINA; 38°14′N 16°16′E. Bathing resort; olive-oil refining, wine making, manufacturing (furniture, bricks). Hot mineral baths nearby. Town badly damaged by earthquakes (1907–1908). Called Gerace Marina until 1934. Archaeological museum has finds from ancient Locri Epizephyrii, a colony of Greek Locris, founded in 7th century B.C.E. Its ruins, 2 mi/3 km SW, on coast, include temples, town walls, extensive necropolis.

Locris (lok-REES), ancient region and state in what are now FTHIOTIDA and PHOCIS prefectures, CENTRAL GREECE department; 38°40′N 22°50′E. The state was probably in existence before the arrival of the Phocians. The rise of DORIS and Phocis split the original region into western and eastern portions. Eastern Locris, along the MALIAKOS Gulf and Gulf of Euboea (now Gulf of ÉVVIA) between Thermopylae and Larimna, was again split (6th century B.C.E.) by Phocis into Epicnemedian in the W and Opuntian in the E. Western, or Ozolaean, Locris was on the N coast of the Gulf of Corinth and had for its principal towns Amphissa and Naupactus (now Navpaktos). Largely hemmed in by stronger states, the Locrians played a minor role in Greek history. However, they founded (c.700 B.C.E.) one of the earliest Greek colonies in S Italy, Epizephyrian Locris, near the promontory of Zephyrium in the toe of the peninsula. The earliest written legal code in Europe, attributed to Zaleucus (7th century B.C.E.), was used here.

Locroja (lo-KRO-hah), town, CHURCAMPA province, HUANCAVELICA region, S central PERU, in Cordillera CENTRAL, 10 mi/16 km W of CHURCAMPA; 12°40′S 74°25′W. Elevation 12,654 ft/3,856 m. Cereals, sugarcane; livestock. Variant spelling: Lucruja.

Locronan (lok-ro-nahn), commune (□ 3 sq mi/7.8 sq km), FINISTÈRE department, NW FRANCE, in BRITTANY, 21 mi/34 km SSE of BREST, near ATLANTIC coastline; 48°06′N 04°12′W. Has picturesque church square lined with stone houses, 15th-century church, and chapel. A former canvas-making center, village retains several handicraft textile and glass-blowing shops. Scene of a major religious pilgrimage to the former Benedictine priory of Locronan (atop nearby hill), which takes place once every 6 years in honor of St. Ronan, a 5th-century Irish monk.

Loctudy (lok-tu-dee), commune (□ 4 sq mi/10.4 sq km), FINISTÈRE department, in BRITTANY, W FRANCE, fishing port at inlet of BAY OF BISCAY, 13 mi/21 km SSW of QUIMPER; 47°50′N 04°10′W. It is also a bathing resort. The manor house of Kérazan is nearby.

Locumba (lo-KOOM-bah), town, ☉ JORGE BASADRE province, TACNA region, S PERU, on LOCUMBA RIVER, and 40 mi/64 km NW of TACNA; 17°36′S 70°45′W. Elevation 2,221 ft/676 m. Cotton, rice, vineyards; livestock. Archaeological site nearby.

Locumba River (lo-KOOM-bah), TACNA region, S PERU; rises in the ANDES; flows SW past LOCUMBA to the PACIFIC OCEAN; 17°57′S 70°55′W.

Locust (LO-cuhst), town (□ 5 sq mi/13 sq km; 2006 population 2,562), CABARRUS and STANLY counties, S central NORTH CAROLINA, 15 mi/24 km SSW of ALBEMARLE and 23 mi/37 km E of CHARLOTTE; 35°15′N 80°25′W. Light manufacturing; service industries; agriculture (cotton, grain, soybeans; livestock; dairying). Read Gold Mine State Historical Site to N, first gold mine in U.S. (1799). Established in W Stanly county in the late 1860s by German, Scotch-Irish, and English immigrants. Formerly Locust Level.

Locust Creek, c.85 mi/137 km long, in S IOWA and N MISSOURI, rises 10 mi/16 km SE of CORYDON in WAYNE county (Iowa); flows S, into Missouri, to GRAND RIVER 2 mi/3.2 km W of SUMNER.

Locust Fork, stream, c.110 mi/177 km long; rises near Boaz, NE Alabama; flows SW to Mulberry Fork c.20 mi/32 km W of Birmingham, forming Black Warrior River.

Locust Gap, unincorporated village, Mount Carmel township, NORTHUMBERLAND county, E central PENNSYLVANIA, 14 mi/23 km NW of POTTSVILLE; 40°46′N 76°26′W.

Locust Grove, town (2000 population 2,322), HENRY county, N central GEORGIA, 11 mi/18 km NE of GRIFFIN; 33°21′N 84°07′W. Exurban railroad community S of ATLANTA. Light manufacturing

Locust Grove, town (2006 population 1,587), MAYES county, NE OKLAHOMA, 12 mi/19 km SE of PRYOR; 36°12′N 95°10′W. In livestock-raising and agricultural area; manufacturing (chemical plant equipment, boilers). Lake Hudson (Markham Ferry Dam) to N, FORT GIBSON LAKE to SW, both on NEOSHO River.

Locust Level, NORTH CAROLINA: see LOCUST.

Locust Valley, village (□ 1 sq mi/2.6 sq km; 2000 population 3,521), NASSAU county, SE NEW YORK, near N shore of W LONG ISLAND, 3 mi/4.8 km E of GLEN COVE; 40°52′N 73°35′W. Affluent residential area. Some light manufacturing.

Lod (LOD), city (2006 population 66,800), central ISRAEL, 9 mi/15 km SE of TEL AVIV; 31°57′N 34°53′E. Elevation 291 ft/88 m. The city is a railroad and road junction. Manufacturing includes paper products, chemicals, oil products, electronic equipment, processed food, and cigarettes. Nearby is Ben Gurion Airport, the nation's chief international airport and the center of its large aircraft industry. Lod is mentioned in the Bible. It was destroyed (C.E. 66–70) by the Romans and, after the destruction of the Temple in JERUSALEM (C.E. 70), became the temporary seat of many famous Jewish teachers. Hadrian rebuilt the city and named it Diospolis. It is the traditional home and place of burial of St. George, England's patron saint, and has a church in his honor. In the 5th century it was the seat of a bishop; a synod of bishops met there in 415. Lod was occupied by the Crusaders in 1099, destroyed by Saladin in 1191, and rebuilt by King Richard I (Richard Lion Heart) of ENGLAND. During the Arab-Israeli War of 1948 most Arabs left the city, which was then settled by Jewish immigrants. Also known as Lydda.

Loda (LOW-dah), village (2000 population 419), IROQUOIS county, E ILLINOIS, 4 mi/6.4 km NNE of PAXTON; 40°31′N 88°04′W. Corn, soybeans; cattle, hogs; egg and poultry processing; fertilizer.

Lodar (LO-dahr), S YEMEN, on road, and 40 mi/64 km NNE of SHUQRA. A market center; has airfield. Former chief town of AUDHALI sultanate. Also spelled Laudar or Lawdar.

Löddeköpinge (LUHD-de-SHUHP-eeng-e), residential town, SKÅNE county, S SWEDEN, 10 mi/16 km NW of LUND; 55°45′N 13°02′E.

Loddon (LUD-uhn), market town (2001 population 2,622), SE NORFOLK, E ENGLAND, 11 mi/18 km SE of NORWICH; 52°32′N 01°29′E. Has flint church, built c.1480.

Loddon River (LAH-duhn), 155 mi/249 km long, in central VICTORIA, AUSTRALIA; rises in GREAT DIVIDING RANGE, near CRESWICK; flows N, past BRIDGEWATER and KERANG, to MURRAY RIVER 20 mi/32 km E of SWAN HILL; 35°32′S 143°52′E. Used for irrigation; frequently dry.

Loddon River (LUD-uhn), 30 mi/48 km long, in HAMPSHIRE and Wokingham, S ENGLAND; rises at BASINGSTOKE; flows NE to the THAMES at WARGRAVE. Receives BLACKWATER RIVER.

Lodelinsart (lo-duh-len-SAHR), town in commune of CHARLEROI, Charleroi district, HAINAUT province, S central BELGIUM, 2 mi/3.2 km N of Charleroi.

Lodenice (LO-dye-NYI-tse), Czech *Loděnice*, German *Lodenitz*, village, STREDOCESKY province, central BOHEMIA, CZECH REPUBLIC, on railroad, and 5 mi/8 km NE of BEROUN. Manufacturing of sound recordings; limestone quarry in vicinity; vineyard. Has an 18th century church.

Lodenitz, CZECH REPUBLIC: see LODENICE.

Löderup (LUH-de-ROOP), village, SKÅNE county, S SWEDEN, near BALTIC SEA, 12 mi/19 km SW of SIMRISHAMN; 55°26′N 14°07′E.

Lodève (lo-de-vuh), town (□ 8 sq mi/20.8 sq km), sub-prefecture of HÉRAULT department, LANGUEDOC-ROUSSILLON region, S FRANCE, at foot of the CÉVENNES MOUNTAINS, 27 mi/43 km N of BÉZIERS; 43°43′N 03°19′E. Woodworking, textiles. Has 14th-century church. Roman Catholic episcopal see from 4th century to 1970. Uranium ore mined and processed nearby since 1980s; it is one of France's major sources of uranium.

Lodeynoye Pole (luh-DYAI-nuh-ye PO-lye), city (2005 population 21,945), NE LENINGRAD oblast, NW European Russia, on the SVIR' River, on road and railroad, 150 mi/241 km ENE of SAINT PETERSBURG; 60°44′N 33°33′E. Railroad shops, machinery (transport hoisting equipment), sawmills. Russian military air defense base in the vicinity. Site of the Olonets shipyards (founded in 1702 by Peter the Great), in operation until the 1830s. Chartered in 1785.

Lodge, village (2006 population 114), COLLETON county, S central SOUTH CAROLINA, 30 mi/48 km S of ORANGEBURG; 33°04′N 80°57′W. Timber; livestock; grain; watermelons.

Lodge Grass, village, BIG HORN county, S Montana, on LITTLE BIGHORN River at mouth of Lodge Grass Creek, and 30 mi/48 km SSE of HARDIN, in E part of Crow Indian Reservation; 45°18′N 107°22′W. Cattle. Lodge Grass Storage Reservoir to SW; Rosebud Battlefield State Park to E; BIGHORN CANYON NATIONAL RECREATION AREA to W, including Old Fort C. F. Smith and YELLOWTAIL DAM.

Lodgepole, village (2006 population 353), CHEYENNE county, W NEBRASKA, 20 mi/32 km E of SIDNEY, and on LODGEPOLE CREEK; 41°08′N 102°38′W. Dairying; livestock; grain.

Lodgepole (LAHJ-pol), hamlet (2005 population 179), central ALBERTA, W CANADA, 16 mi/26 km from DRAYTON VALLEY, in BRAZEAU COUNTY; 53°06′N 115°19′W. Near Brazeau Dam; recreation area.

Lodgepole Creek, 212 mi/341 km long, in WYOMING, NEBRASKA, and COLORADO; rises at elevation of 8,000 ft/2,438 m in LARAMIE MOUNTAINS, SE ALBANY county, SE Wyoming, c. 10 mi/16 km E of LARAMIE; flows E past PINE BLUFFS, Wyoming, and KIMBALL, SIDNEY, and CHAPPELL, W Nebraska, then SE to join SOUTH PLATTE RIVER 5 mi/8 km WSW of JULESBURG, NE Colorado.

Lodhran (LOD-huhr-ahn), town, MULTAN district, S PUNJAB province, central PAKISTAN, 45 mi/72 km SSE of MULTAN; 29°32′N 71°38′E.

Lodi, city (2000 population 56,999), SAN JOAQUIN county, central CALIFORNIA, 8 mi/12.9 km NNE of STOCKTON, 30 mi/48 km S of SACRAMENTO, on MOKELUMNE RIVER; 38°07′N 121°17′W. In a rich farm area (nursery products; sugar beets, fruit, nuts, grapes, pumpkins, vegetables, grain; dairying; cattle). Diversified manufacturing; food processing, wineries. San Joaquin County Historical Museum to S. Founded in 1869 and settled by wheat farmers from the Dakotas, mostly of German descent. Incorporated 1906.

Lodi (LO-dei), city (2006 population 2,940), COLUMBIA county, S central WISCONSIN, on small Spring Creek, and 18 mi/29 km NNW of MADISON; 43°19′N 89°31′W. In diversified farming area. Food processing, machinery and equipment manufacturing. Incorporated as village in 1872, as city in 1941.

Lodi (LO-dee), town, LOMBARDY, N ITALY, on the ADDA River, near MILAN; 45°19′N 09°30′E. A center for dairy products, food processing, and light manufacturing. The city is located near the site of ancient Laus Pompeia, which was destroyed by Milan in C.E. 1111. At Lodi on May 10, 1796, Napoleon defeated the Austrians after personally leading his troops across the bitterly contested bridge over the Adda. Has Romanesque cathedral (12th century) and Renaissance-style Church of the Incoronata.

Lodi (LO-dee), village (□ 2 sq mi/5.2 sq km; 2006 population 3,344), MEDINA county, N OHIO, 27 mi/43 km WSW of AKRON, on East Branch of BLACK RIVER; 41°02′N 82°01′W.

Lodi (LO-dei), industrial borough (2006 population 24,310), BERGEN county, NE NEW JERSEY; 40°52′N 74°04′W. Light manufacturing. Incorporated 1894.

Lodja (LO-jah), town, KASAI-ORIENTAL province, central CONGO, on right bank of LUKENIE RIVER, and 100 mi/161 km N of LUSAMBO; 03°29′S 23°26′E. Elev. 1,479 ft/450 m. Terminus of steamboat navigation. Trading center; cotton ginning, rice milling. Has Roman Catholic mission, hospital. Also spelled Loja.

Lodomeria, UKRAINE: Latinized name of the duchy of Volodymyr. See VOLODYMYR-VOLYNS'KYY.

Lodore, Falls of, ENGLAND: see DERWENT WATER.

Lodosa (lo-DHO-sah), town, NAVARRE province, N SPAIN, on the EBRO RIVER, and 20 mi/32 km E of LOGROÑO; 42°25′N 02°05′W. In irrigated agricultural area (cereals, fruit, pepper). Wine. Sawmilling, olive oil processing, tanning; soap, fertilizers, brandy, plaster. Hydroelectric plant nearby. Cave dwellings in area.

Lodosa Canal, in NAVARRE and ZARAGOZA province, N SPAIN, parallels the EBRO RIVER for c.50 mi/80 km. Used for irrigation.

Lodwar (lo-DWAHR), town (1999 population 16,981), ⊙ Turkana district, RIFT VALLEY province, NW KENYA, on TURKWEL RIVER, on road, and 177 mi/285 km NNE of KITALE; 03°08′N 35°37′E. Livestock; peanuts, sesame, corn. Airfield.

Lodz (LOODZH), Polish *Łódź* (LOD-zee), province (□ 6,503 sq mi/16,907.8 sq km), central POLAND; ⊙ Łódź. Hilly plateau in S, sloping to undulating plain (N); drained by WARTA, PILICA, PROSNA, and BZURA rivers. Cotton milling, principal industry of province, is concentrated in larger cities (ŁÓDŹ, Piotrkow, PABIANICE, TOMASZOW, ZGIERZ). Principal crops are rye, potatoes, oats, barley, rapeseed; livestock raising is important. Boundaries of pre–World War II province (□ 7,349 sq mi/19,034 sq km) were changed by transfer of territory to Kielce, Poznan, and Warszawa provinces. Includes greater part of former PETROKOV (Piotrkow) and Kalish (KALISZ) governments of Russian Poland. Łódź city has status of autonomous province.

Łódź (LOODZH), city (2002 population 789,318), central POLAND. The second-largest city of Poland and an important industrial center, Łódź is the center of the Polish textile industry. Other manufacturing includes chemicals, textiles, machinery, food processing, and radios. Chartered in 1423, the city passed to PRUSSIA in 1793 and to RUSSIA in 1815. It reverted to Poland in 1919. The first textile mills were established in the city c.1830 by German weavers, but the industry grew only after 1870, when local Jewish community, centered in nearby village of Balut and which had provided contracted labor, expanded. Most of the industry was cotton, but there was some wool and linen also. Łódź was a center for Jewish trade unionism. By the outbreak of World War II, Jews represented over 30% of the total population and were then exterminated. The city was also the center of the Polish labor and socialist movements. In World War II it was incorporated into GERMANY, renamed Litzmannstadt, and subjected to ruthless Germanization. The city has a university (founded in 1945).

Loe Band (LO BAND), village, ZHOB district, NE BALUCHISTAN province, SW PAKISTAN, 110 mi/177 km W of ZHOB, on Afghan border; 31°15′N 67°37′E.

Loèche, SWITZERLAND: see LEUK.

Loèche-les-Bains, SWITZERLAND: see LEUKERBAD.

Loeches (lo-E-ches), town, MADRID province, central SPAIN, 15 mi/24 km E of MADRID; 40°23′N 03°24′W. Noted spa with several mineral springs. In agricultural area.

Loe Dakka, AFGHANISTAN: see DAKKA.

Loei (LUH-ee), province (□ 4,344 sq mi/11,294.4 sq km), NE THAILAND, along the border with LAOS; ⊙ Loei; 17°25′N 101°30′E. Rice, cotton, cotton weaving, gold mining at CHIANG KHAN. Has annual Cotton Blossom Festival in Loei. Also famous for the Phi Ta Khon Festival, in Dan Sai, part of the rocket and rain festivals held throughout the NE. National parks here include PHU KRADUNG NATIONAL PARK, highest point in the province (4,921 ft/1,500 m).

Loei (LUHRY), town, ⊙ LOEI province, N central THAILAND, on NW edge of KORAT PLATEAU, on Loei River (minor affluent of the MEKONG) and 100 mi/161 km NW of KHON KAEN; 17°29′N 101°43′E. Rice, cotton. Gold mining at CHIANG KHAN (N).

Loemadjang, INDONESIA: see LUMAJANG.

Loenen (LOO-nuhn), village, UTRECHT province, W central NETHERLANDS, on VECHT RIVER, and 9 mi/14.5 km NNW of UTRECHT; 52°13′N 05°01′E. Dairying; livestock; flowers, nursery stock, vegetables, fruit. Winter sports resort, lake fishing. AMSTERDAM-RHINE CANAL passes to W; Loosdrechtse Plassen lake area to E.

Loenvatn (LAW-uhn-VAH-tuhn), lake (□ 4 sq mi/10.4 sq km) in a fissure of the JOSTEDALSBREEN glacier, SOGN OG FJORDANE county, W NORWAY. Extends c.7 mi/11.3 km SE from Loen village at head of NORDFJORD, 65 mi/105 km ENE of FLORØ. Tourist area. Landslide in 1905 caused waves that threw a steamer 300 ft/91 m onto a mountainside, where it still reposes. Nearby, 8 mi/12.9 km SE, is highest point of the Jostedalsbreen (6,700 ft/2,042 m).

Loeriesfontein, town, NORTHERN CAPE province, SOUTH AFRICA, on NE edge of GREAT KAROO, 40 mi/64 km NW of CALVINIA. Serves as center for sheep farming in what is called the Hantam area.

Lo Espejo (lo es-PAI-ho), town and comuna, suburb 6 mi/10 km S of SANTIAGO, METROPOLITANA DE SANTIAGO region, central CHILE, on railroad.

Loess Plateau, Mandarin *Huangtu Gaoyuan*, central NW CHINA, a vast highland (□ 115,830 sq mi/301,158 sq km) covered with 328 ft/100 m–656 ft/200 m of loess and various erosional landforms. It is the middle section of the massive loess deposits of N China; bounded by the GREAT WALL (N), QINLING Mountains (S), the FEN valley (E), and the HELAN Mountains and QINGHAI Lake (W). The plateau extends over SE GANSU, S NINGXIA, central SHAANXI, and W SHANXI provinces. Its origin has been attributed to deposits of airborne and eastward-floating glacial loess caused by the interception of moisture brought westward by ancient monsoons beginning c.250,000 years ago. The HUANG HE (Yellow River) flows N through the W plateau and, after forming a bend in HETAO area, runs S across the E plateau. The fine-grain loess washes off along the valley, giving the river its name. Agriculture is characterized by dry land farming (sorghum, millet, corn, sweet potatoes, beans). In irrigated lands, rice and wheat are also produced. Deforestation, overfarming, and the soil's loose texture, compounded by periods of drought, have caused severe erosion, low agricultural productivity, and ecological imbalance. Ancient Chinese civilization originated here. The ancient Hua and Change peoples were believed to have lived in the area; the two groups eventually fused to form the Hua-Xia, the forerunners of the Han Chinese. Here were the political centers of the Western Zhou (1126–771 B.C.E.), Qin (221–206 B.C.E.), and many other ancient Chinese states. The Chinese communists established their political center in the Loess Plateau (Yan'an) during the 1930s and 1940s.

Lofa (LO-fah), county (□ 4,493 sq mi/11,681.8 sq km; 1999 population 351,492), NW LIBERIA; ⊙ VOINJAMA; 07°45′N 10°00′W. Borders SIERRA LEONE (W), GUINEA (N, E), BONG county (SE), and Gbarpolu county (S). Forested in E. Agriculture (rice, kola, cassava). Main centers include KOLAHUN. The S portion of the county was separated and formed into Gbarpolu county c.2002. Lofa is divided into six districts: Vahun (SW), Kolahun (W), Foya (NW), Voinjama (N central), Zorzor (E), and Salayea (SE).

Lofa River (LO-fah), c.200 mi/322 km long, in W LIBERIA, rises on GUINEA border; flows SW to the ATLANTIC OCEAN 30 mi/48 km WNW of MONROVIA. Sometimes spelled Loffa; also called Little Cape Mount River.

Lofer (LO-fer), township, SALZBURG, W AUSTRIA, in the PINZGAU, on SAALACH RIVER, and 22 mi/35 km SW of SALZBURG, near TYROL border; 47°35′N 12°42′E. Road junction. Summer resort; winter sports nearby at Loferer Alm (4,343 ft/1,324 m).

Loffa River, LIBERIA: see LOFA RIVER.

Löffingen (LOF-fing-uhn), village, BADEN-WÜRTTEMBERG, SW GERMANY, in BLACK FOREST, 6 mi/9.7 km ESE of NEUSTADT; 47°53′N 08°21′E. Elevation 2,638 ft/804 m. Summer resort and pilgrimage center.

Lofley, Cape (LOF-lyai), S point of ALEXANDRA LAND, part of FRANZ JOSEF LAND, ARCHANGEL oblast, extreme N European Russia, in the ARCTIC OCEAN; 80°27′N 45°30′E.

Lofoten and Vesterålen (LOO-foo-tuhn, VES-tuhr-aw-luhn), two island groups, NORDLAND and TROMS counties, NW NORWAY, in the NORWEGIAN SEA. Situated within the ARCTIC CIRCLE, the islands extend c.150 mi/240 km from NE to SW and are from 1 mi/1.6 km to 50 mi/80 km off the mainland. The North Atlantic Drift gives these northern islands a temperate climate. The chief islands of the Lofoten group are Røstøya, VÆRØY, MOSKENESØYA, VESTVÅGØY, and AUSTVÅGØY; the celebrated MOSKENSTRAUMEN is S of Moskenesøya. The Vesterålen group, separated from the Lofoten by the narrow Raftsundet, includes the islands of HINNØYA (the largest island of Norway), LANGØYA, and Andøya. SVOLVÆR (on Austvågøya) and HARSTAD (on Hinnøya) are main trading centers. Cattle and sheep raising; coal and magnetite are mined on Andøya and Langøya. The cod and herring fisheries here are among the richest in the world. The codfish shoal on the E coast of the islands from February to April, the herrings on the W coast from August to November. During these seasons thousands of fishing craft come to the fish banks, but treacherous tidal currents make operations difficult.

Lofthouse and Stanley (LAHFT-uhs and STAN-lee), villages (2001 population 19,136), WEST YORKSHIRE, N ENGLAND, 5 mi/8 km S of LEEDS; 53°43′N 01°29′W. Previously brick manufacturing. Aire-Calder Canal to E.

Loftus (LAHFT-us), town (2001 population 7,075), Redcar and Cleveland, NE ENGLAND, 8 mi/12.9 km ESE of REDCAR; 54°33′N 00°53′W. Iron and steel mills.

Lofty, Mount, AUSTRALIA: see MOUNT LOFTY RANGES.

Log (LOG), town (2006 population 3,700), central VOLGOGRAD oblast, SE European Russia, on road and railroad, 20 mi/32 km SSE of FROLOVO, and 12 mi/19 km NNW of ILOVLYA, to which it is administratively subordinate; 49°29′N 43°51′E. Elevation 367 ft/111 m. Food processing. Sometimes called Logovskiy.

Loga (LO-guh), town, DOSSO province, NIGER, 75 mi/121 km E of NIAMEY; 13°37′N 03°14′E. Administrative center.

Loga (LAW-gah), village, VEST-AGDER county, S NORWAY, on railroad, and 2 mi/3.2 km N of (and incorporated into) FLEKKEFJORD; 58°19′N 06°40′E. Textile milling.

Logan, county (□ 731 sq mi/1,900.6 sq km; 2006 population 22,903), W ARKANSAS; ⊙ BOONEVILLE and PARIS; 35°13′N 93°42′W. Bounded N by ARKANSAS

RIVER; drained by PETIT JEAN RIVER. Soybeans; chickens, cattle, hogs; timber. Coal mining, sawmilling, cotton ginning. Hunting, fishing. Part of Ozark National Forest in SE; part of Ouachita National Forest in SW; part of Blue Mountain reservoir (Petit Jean River) in S. MAGAZINE MOUNTAIN is in county. Formed 1873 as Sarber county, renamed 1874.

Logan, county (□ 1,844 sq mi/4,794.4 sq km; 2006 population 20,780), NE COLORADO; ⊙ STERLING; 40°43′N 103°06′W. Irrigated agricultural area, bordering on N NEBRASKA, drained by SOUTH PLATTE RIVER. Sugar beets, beans, wheat, hay, sunflowers, corn; cattle. STERLING RESERVOIR and North Sterling State Park in W center; Julesburg Reservoir in NE on E boundary. Formed 1887.

Logan, county (□ 619 sq mi/1,609.4 sq km; 2006 population 30,302), central ILLINOIS; ⊙ LINCOLN; 40°07′N 89°21′W. Agriculture (corn, wheat, soybeans, cattle, hogs). Some manufacturing. Drained by SALT and KICKAPOO creeks and small Sugar Creek. Edward River Madigan (formerly Railsplitter) State Park is S of Lincoln. Formed 1839.

Logan, county (□ 1,073 sq mi/2,789.8 sq km; 2006 population 2,675), W KANSAS; ⊙ OAKLEY (in far NE corner); 38°55′N 101°09′W. Farming and grazing area, drained by SMOKY HILL RIVER. Wheat, sorghum, cattle. Logan State Fishing Lake at center. In Central time zone: Mountain/Central time zone boundary follows W county border. Formed 1887.

Logan, county (□ 557 sq mi/1,448.2 sq km; 2006 population 27,363), S KENTUCKY; ⊙ RUSSELLVILLE; 36°51′N 86°52′W. Bounded S by TENNESSEE; drained by MUD, RED, and Gasper rivers and Whippoorwill, Wolf Lick, and Elk Lick creeks. Rolling agricultural area (dark and burley tobacco, corn, wheat, barley, soybeans, alfalfa, hay; cattle, hogs, poultry; dairying; timber). Bituminous-coal and asphalt mines, limestone quarries. Some manufacturing at Russellville. Part of Lake Malone reservoir in NW corner, Lake Herndon reservoir in center. Formed 1792.

Logan, county (□ 571 sq mi/1,484.6 sq km; 2006 population 749), central NEBRASKA; ⊙ STAPLETON; 41°33′N 100°28′W. Located in Sand Hills region. Drained by SOUTH LOUP RIVER. Grazing area; cattle, hogs, corn. Formed 1885.

Logan, county (□ 1,000 sq mi/2,600 sq km; 2006 population 1,999), S NORTH DAKOTA; ⊙ NAPOLEON; 46°27′N 99°28′W. Agricultural area watered by Beaver Creek. Cattle; sunflowers, wheat, barley; dairy products. Rush Lake in NW. ALKALI LAKE on N boundary. Beaver Lake State Park near center. Formed 1873 and government organized in 1884. Named for John A. Logan (1826–1886), a Civil War general.

Logan (LO-guhn), county (□ 461 sq mi/1,198.6 sq km; 2006 population 46,189), W central OHIO; ⊙ BELLEFONTAINE; 40°22′N 83°46′W. Drained by GREAT MIAMI and MAD rivers and small Mill and Rush creeks. Includes Indian Lake State Park (resort) and CAMPBELL HILL (1,549 ft/472 m), state's highest point. In Till Plains physiographic region. Agricultural area (livestock, dairy products, grain); manufacturing at Bellefontaine. Formed 1817.

Logan, county (□ 748 sq mi/1,944.8 sq km; 2006 population 36,971), central OKLAHOMA; ⊙ GUTHRIE; 35°54′N 97°27′W. Intersected by CIMARRON RIVER (forms part of NE boundary) and by small Ephraim and Cottonwood creeks. Diversified agriculture (wheat, fruit); cattle and poultry; dairying. Manufacturing at Guthrie. Formed 1890.

Logan, county (□ 456 sq mi/1,185.6 sq km; 2006 population 36,218), SW WEST VIRGINIA; ⊙ LOGAN; 37°49′N 81°56′W. On ALLEGHENY PLATEAU; drained by GUYANDOTTE RIVER. Bituminous-coal-mining region; natural-gas fields. Manufacturing at Logan. Some agriculture (tobacco) and retail trade. Chief Logan State Park in N center. Formed 1824.

Logan (LO-guhn), city (□ 3 sq mi/7.8 sq km; 2006 population 7,368), ⊙ HOCKING county, S central OHIO, 34 mi/55 km SW of ZANESVILLE, on HOCKING RIVER; 39°32′N 82°24′W. In rich agricultural area. Manufacturing (rubber and plastic products). Oil and gas wells. Founded 1816, incorporated 1839.

Logan, city (2006 population 47,660), ⊙ CACHE county, N UTAH, on the Logan River; 41°44′N 111°50′W. It is the center of an irrigated dairy and farm area, with huge cheese plants, other food-processing facilities, and diverse light manufacturing. A Latter-Day Saints tabernacle, Logan Temple, and Utah State University are located there. Utah Festival Opera. Wasatch National Forest to E. LOGAN PEAK is visible from the city. Mount Naomi Wilderness Area to NE. Founded by Mormons and incorporated 1859.

Logan, town (2000 population 1,545), ⊙ HARRISON county, W IOWA, on BOYER RIVER, and 26 mi/42 km N of COUNCIL BLUFFS; 41°38′N 95°47′W. Railroad junction. Manufacturing (textbook printing; limestone products; conveyors). Settled as Boyer Falls; renamed Logan 1864; incorporated 1876.

Logan, town (2000 population 603), PHILLIPS county, N KANSAS, on North Fork SOLOMON RIVER, and 15 mi/24 km WSW of PHILLIPSBURG; 39°39′N 99°34′W. In corn, wheat, cattle area. Manufacturing (bakery products). Hansen Memorial Museum.

Logan, township, GLOUCESTER county, S NEW JERSEY, 15 mi/24 km W of WOODBURY; 39°47′N 75°21′W. Incorporated 1877.

Logan, town (2006 population 978), QUAY county, E NEW MEXICO, on CANADIAN RIVER, near TEXAS border, and 24 mi/39 km NE of TUCUMCARI; 35°21′N 103°26′W. Cattle- and sheep-ranching region; cotton, grain. Ute Lake reservoir and State Park nearby.

Logan, town (2006 population 1,537), ⊙ LOGAN county, SW WEST VIRGINIA, on GUYANDOTTE RIVER, 38 mi/61 km SW of CHARLESTON; 37°51′N 81°59′W. Coal mining; gas wells. Some agriculture (tobacco). Trade center. Manufacturing (mining equipment, hydraulic cylinders, concrete, printing and publishing). South West Virginia Community College. Chief Logan State Park to NW.

Logan, village, GALLATIN county, SW MONTANA, on Gallatin River, 5 mi/8 km SE of MISSOURI River headwaters, and 25 mi/40 km NW of BOZEMAN. In grain and livestock region. Madison Buffalo Jump State Park to S; Missouri Headwaters State Park to NW. Originally called Cannon House.

Logan, ILLINOIS: see HANAFORD.

Logan', RUSSIA: see LAGAN'.

Logan City (LO-guhn), city, suburban area (□ 97 sq mi/252.2 sq km), QUEENSLAND, NE AUSTRALIA, 12 mi/19 km S of BRISBANE, and 32 mi/51 km N of GOLD COAST. Established 1981.

Logan Creek, 85 mi/137 km long, in E NEBRASKA; rises in CEDAR county; flows SSE and S past Wakefield, Pender, and Lyons, to ELKHORN RIVER at Winslow.

Logan Glacier, in SAINT ELIAS MOUNTAINS glacier system, S ALASKA, 3 mi/4.8 km wide, extends 65 mi/105 km WNW from Mount LOGAN, near 60°50′N 141°00′W. Flows into CHITINA RIVER.

Logan International Airport, BOSTON, MASSACHUSETTS. Opened September 1923 and operated by Massachusetts Port Authority. Has five runways and five passenger terminals serving 22 million passengers and 364,000 metric tons annually. In 1994, Logan launched a $4.4 billion modernization program. On September 11, 2001, terrorists highjacked two flights originating from Logan and flew them into the twin towers of the WORLD TRADE CENTER in NEW YORK city. Airport Code BOS.

Logan Lake (LO-guhn), district municipality (□ 125 sq mi/325 sq km; 2001 population 2,185), S BRITISH COLUMBIA, W CANADA, 37 mi/60 km S of KAMLOOPS,

in THOMPSON-NICOLA regional district; 50°30′N 120°48′W. Incorporated 1983.

Logan Martin Lake, reservoir, on the border of St. Clair and Talladega counties, NE central Alabama, on the Coosa River, 27 mi/43 km ESE of Birmingham; c.40 mi/64 km long; 33°24′N 86°20′W. Extends NNE to H. Neely Henry Dam. Max. capacity 642,000 acre-ft. Formed by Logan Martin Dam (75 ft/23 m high), built in 1964 by the Alabama Power Co. for power generation and flood control. Both the dam and lake were named for Logan Matin, an official of Alabama Power Co.

Logan, Mount (19,850 ft/6,050 m), extreme SW YUKON TERRITORY, CANADA, just E of ALASKA; highest mountain in Canada and second highest in NORTH AMERICA; 60°33′N 140°24′W. It caps an immense tableland and is the center of the greatest glacial expanse in North America. The first ascent was made in 1925.

Logan, Mount (LO-guhn) (3,700 ft/1,128 m), E QUEBEC, E CANADA, on NW GASPÉ PENINSULA, in SHICKSHOCK MOUNTAINS, 40 mi/64 km E of MATANE; 48°52′N 66°47′W.

Logan Pass (6,646 ft/2,026 m), in LEWIS RANGE, NW MONTANA, near center of GLACIER NATIONAL PARK, on CONTINENTAL DIVIDE and GLACIER/FLATHEAD county line; crossed by Going-to-the-Sun Highway. Nearby points of interest are Logan Pass Visitor Center, Logan Pass Boardwalk Wildlife Viewing Area, Hidden Lake to SW, and Hanging Gardens (with colorful displays of wildflowers).

Logan Peak (9,710 ft/2,960 m), CACHE county, N UTAH, 5 mi/8 km E of LOGAN, in Wasatch National Forest.

Logansport, city (2000 population 19,684), ⊙ CASS county, N central INDIANA, at the confluence of WABASH and EEL rivers; 40°45′N 86°22′W. In a fertile farm area. Diversified manufacturing (pork products, power equipment, lumber, electronic controls, fibercoating, stampings, cement and masonry powders, motor vehicle parts, animal feeds). GRISSOM AIR BASE is nearby. Laid out 1828. Incorporated 1838.

Logansport, town, DE SOTO parish, NW LOUISIANA, 40 mi/64 km SSW of SHREVEPORT, and on SABINE River (here forming TOLEDO BEND RESERVOIR and TEXAS state line; bridged to JOAQUIN, TEXAS); 31°58′N 93°57′W. Cotton, cattle, timber; dairying; manufacturing of plaster molds, lumber, plywood. Gas field nearby. Founded in 1830s.

Loganton (LO-guhn-tuhn), borough (2006 population 436), CLINTON county, central PENNSYLVANIA, 10 mi/16 km SE of LOCK HAVEN. Agriculture (grain; livestock, dairying); light manufacturing. Ravensburg State Park to NE, part of Bald Eagle State Forest to N and S.

Loganville, town (2000 population 5,435), WALTON and GWINNETT counties, N central GEORGIA, 29 mi/47 km ENE of ATLANTA; 33°50′N 83°54′W. Exurban community on fringe of Atlanta region. Light manufacturing.

Loganville, village (2006 population 272), SAUK county, S central WISCONSIN, on small Narrows Creek, and 15 mi/24 km W of BARABOO; 43°26′N 90°02′W. In dairy and livestock region. Natural Bridge State Park to SE.

Loganville (LO-guhn-vil), borough (2006 population 1,027), YORK county, S PENNSYLVANIA, 8 mi/12.9 km SSE of YORK. Agriculture (grain, apples; livestock, dairying); manufacturing (electrical connectors).

Logar (lo-GAHR), province (2005 population 326,100), E AFGHANISTAN; ⊙ PUL-I-ALAM. Bounded by KABUL province (N), NANGARHAR province (NE), PAKTIA province (S and E), GHAZNI province (S), and WARDAK province (W). Watered by LOGAR RIVER; Koshar and Sultan mountains are highest peaks. Population largely Pashtun, with some Tajiks. Called the "granary of Kabul"; agriculture (grapes, apples, vegetables). Livestock. Province was largely depopulated during 1980s, but most of population returned after the fall of the Marxist regime in 1992. Prov. capital (1964–1975) was Baraki Barak.

Logaros Lagoon (lo-yah-ROS) (□ 12 sq mi/31.2 sq km), in Árta prefecture, S EPIRUS department, NW GREECE, on N shore of Gulf of Árta 7 mi/11.3 km SW of Árta; 5 mi/8 km long, 3 mi/4.8 km wide. Fisheries.

Logar River (lo-GAHR), 150 mi/241 km long, in E AFGHANISTAN; rises in SW outliers of the Hindu Kush near 34°00′N 68°00′E; flows E past SHAIKHABAD and BARAKI RAJAN, and N past KULANGAR to KABUL RIVER E of KABUL; 34°33′N 69°17′E. Used for timber floating and for irrigation through canal. Called WARDAK RIVER in middle course. Chrome mining in valley. Also spelled Lohgar and Lohgard.

Logarska Dolina (lo-GAHR-skah do-LEE-nah), Alpine glacial valley, N SLOVENIA, in KAMNIK-SAVINJA ALPS, NW of GRINTAVEC peak; 5.6 mi/9 km long, 1,640 ft/500 m wide. SAVINJA RIVER rises here.

Logatec (lo-GAH-tek), ancient *Longaticum*, town (2002 population 7,630), central SLOVENIA, 15 mi/24 km SW of LJUBLJANA, on Ljubljana-Postojna railroad; 45°54′N 14°13′E. Elev. 1,562 ft/476 m. Local trade center; wood processing, paper and textile manufacturing. Has castle (1600). Formed after World War II by merging Gorenji Logatec and Dolenji Logatec with several villages.

Lögdeå (LUHG-de-O), village, VÄSTERBOTTEN county, N SWEDEN, on islet in GULF OF BOTHNIA, 20 mi/32 km SW of UMEÅ; 63°33′N 19°23′E.

Loge (LOZH-ai), river, 180 mi/290 km long, in NW ANGOLA; rises in Malanje Plateau region and flows W entering ATLANTIC OCEAN at port of AMBRIZ.

Loggieville, former village, NE NEW BRUNSWICK, CANADA, on MIRAMICHI RIVER estuary, and 5 mi/8 km NE of CHATHAM. Lumber port. Amalgamated into the City of MIRAMICHI in 1995.

Loggiovano, CROATIA: see LOZOVAC.

Logierait (LO-gee-rait), agricultural village, PERTH AND KINROSS, E Scotland, on the TAY RIVER, and 8 mi/12.9 km ENE of ABERFELDY; 56°38′N 03°41′W. It was seat of dukes of ATHOLL, who held quasi-regal court here.

Logishin (lo-GI-shin), Polish *Lohiszyn*, urban settlement, BREST oblast, BELARUS, 17 mi/27 km NNW of PINSK. Limestone plant.

Logis-Neuf, Le (lo-zhee–nuf, luh), section of community of Coucourde, DRÔME department, RHÔNE-ALPES region, S FRANCE, on left bank of RHÔNE RIVER, and 8 mi/12.9 km N of MONTÉLIMAR. Here, on a diversion canal, is the large hydroelectric project known as Baix-le-Logis-Neuf, part of a power-generating system built along the mid- and lower course of the Rhône since World War II.

Lognes (lon-yuh), town (□ 1 sq mi/2.6 sq km), SEINE-ET-MARNE department, ÎLE-DE-FRANCE region, on regional railroad, N central FRANCE; 48°50′N 02°37′E.

Logoisk (lo-GOISK), urban settlement, ⊙ LOGOISK region, W central MINSK oblast, BELARUS, on Gaina River (tributary of BEREZINA), and 25 mi/40 km NNE of MINSK; 54°08′N 27°42′E. Manufacturing (woodworking, starch making, bricks). Dairying.

Logone Birni, town, Far-North province, CAMEROON, 25 mi/40 km S of N'Djamena (CHAD); 11°48′N 15°03′E.

Logone-et-Chari, department (2001 population 405,035), Far-North province, CAMEROON, ⊙ Kousséri.

Logone Occidental (lo-GON aw-ksee-dawng-TAHL), administrative region (2000 population 592,468), SW CHAD; ⊙ MOUNDOU; 08°30′N 16°00′E. Borders MAYO-KEBBI OUEST (WNW), TANDJILÉ (N), and LOGONE ORIENTAL (E and SSW) administrative regions. Drained by LOGONE RIVER tributaries. Logone Occidental was formerly a prefecture (with the same capital) until, following Chad's administrative division reorganization from fourteen prefectures to twenty-eight departments, it became a department. It became a region following a decree in October 2002 that re-

organized Chad's administrative divisions into eighteen regions.

Logone Oriental (lo-GON o-ree-awng-TAHL), administrative region (2000 population 452,566), extreme S CHAD; ⊙ DOBA; 08°30′N 16°30′E. Borders LOGONE OCCIDENTAL (LOGONE RIVER headstream) and TANDJILÉ administrative regions (N), MANDOUL administrative region (E), CENTRAL AFRICAN REPUBLIC (S), and CAMEROON (W). Drained by headstreams of Logone and M'BÉRÉ rivers. Main centers include Doba, GORÉ, Baïbokourn. Logone Oriental was a prefecture (with the same capital) prior to Chad's administrative division reorganization from fourteen prefectures to twenty-eight departments. It was recreated as a region following a decree in October 2002 that reorganized Chad's administrative divisions into eighteen regions.

Logone River (lo-GON), 240 mi/386 km long, main tributary of CHARI River in W CHAD and along CAMEROON border. Formed by M'BÉRÉ River (W branch of Logone) and PENDÉ River (E branch of Logone) 28 mi/45 km SSE of LAÏ. Flows NW and N, past Laï and BONGOR, to join with the Chari at N'DJAMENA to form a wide delta entering Lake CHAD. Its total length with M'Béré River (sometimes considered as main headstream) is c.500 mi/805 km. Forms extensive swamps through most of its course and at times of high water is linked with BENOUÉ River (Benue) through FIANGA and TIKEM swamps and MAYO-KÉBBI River. Navigable for small power boats part of the year below Bongor. Fishing.

Logoualé (lo-gwah-LAI), town, Dix-Huit Montagnes region, W CÔTE D'IVOIRE, 28 mi/45 km NW of DUÉKOUÉ; 07°07′N 07°33′W. Agriculture (manioc, bananas, rice, palm oil).

Logovskiy, RUSSIA: see LOG.

Logroño (lo-GRO-nyo), city (2001 population 133,058), ⊙ LA RIOJA province, N SPAIN, in La Rioja, on the EBRO RIVER, and 48 mi/77 km SW of PAMPLONA; 42°28′N 02°27′W. It is a farm-processing center noted for its Rioja wine. Wood and metal products; textiles. The kings of Navarre and Castile fought over Logroño from the 10th century until its annexation (1173) to Castile.

Logrosán (lo-gro-SAHN), town, CÁCERES province, W SPAIN, 23 mi/37 km SE of TRUJILLO; 39°20′N 05°29′W. Chemical-fertilizer processing, flour milling; agriculture trade (cereals, livestock, lumber). Mineral springs. Phosphate mines and limestone quarries.

Løgstør (luk-STUR), town, NORDJYLLAND county, N JUTLAND, DENMARK, on LIMFJORD, and 25 mi/40 km W of ÅLBORG; 56°55′N 09°20′E. Fisheries (eels); meat cannery; lime kilns. Limestone quarrying in area.

Logtak Lake, INDIA: see LOKTAK LAKE.

Løgumkloster (LU-khuhm-KLO-stuhr), town, SØNDERJYLLAND county, S JUTLAND, DENMARK, 12 mi/19 km NW of TØNDER; 55°03′N 09°04′E. Site of twelfth-century Lygum Monastery.

Loh, VANUATU: see TORRES ISLANDS.

Lohame Hageta'ot (lo-khah-ME hah-ge-tah-OT), kibbutz, N ISRAEL, 2.5 mi/4 km N of AKKO (Acre); 32°57′N 35°05′E. Elevation 49 ft/14 m. Mixed farming, manufacturing (electronic equipment, processed-meat substitutes). Holocaust museum and an institute researching the Holocaust. Parts of a 19th century Turkish aqueduct through which water was transported from KABRI to Akko. Founded in 1949 by former partisans and ghetto fighters from POLAND and LITHUANIA.

Lohardagga (lo-HAHR-dah-gah) or **Lohardaga**, district (□ 576 sq mi/1,497.6 sq km), SW Jharkhand state, E INDIA; ⊙ LOHARDAGGA. Formed in 1983 in BIHAR state; joined twenty-one other districts in 2000 to form Jharkhand. Primarily agricultural district; main crops include rice, maize, millet, wheat, oilseeds, vegetables. Many residents engaged in small-scale indus-

tries, such as manufacturing bricks, soap, oil, candles, aluminum goods, wooden furniture, earthern pots, bamboo baskets, and weaving cloth and carpets. Over half the population is tribal (Oraon, Asur, Birijia, Munda, Kharia).

Lohardagga (lo-HAHR-dah-gah) or **Lohardaga**, town (2001 population 46,204), Lohardagga district, W central Jharkhand state, E INDIA, on CHOTA NAGPUR PLATEAU, 41 mi/66 km W of RANCHI; 23°27′N 84°42′E. Railroad terminus. Trades in rice, oilseeds, corn, cotton; manufacturing (brass utensils).

Loharu (lo-HAH-roo), town, BHIWANI district, HARYANA state, N INDIA, 50 mi/80 km S of HISAR; 28°27′N 75°49′E. Local market center for millet, cotton, salt, oilseeds. Was capital of former princely state of Loharu of PUNJAB STATES; incorporated 1948 into HISAR district and later into Bhiwani district.

Lohawat (lo-HAH-wuht), town, JODHPUR district, W central RAJASTHAN state, NW India, 55 mi/89 km NNW of Jodhpur. Road and railroad junction. Local market for hides, salt, woolen handicrafts.

Loheia, YEMEN: see LUHAIYA.

Lohfelden (LO-fel-den), town, HESSE, central GERMANY, 4 mi/6.4 km SSE of KASSEL; 51°17′N 09°32′E.

Lohgar River, AFGHANISTAN: see LOGAR RIVER.

Lohit, district (□ 4,402 sq mi/11,445.2 sq km), ARUNACHAL PRADESH state, extreme NE INDIA; ⊙ TEZU.

Lohja (LO-yah), Swedish *Lojo*, town, UUDENMAAN province, S FINLAND, 30 mi/48 km W of HELSINKI; 60°15′N 24°05′E. In lake region. Pulp, lumber, cellulose, and plywood mills; cement works; limestone quarries. Has fourteenth-century church.

Lohman, town (2000 population 168), COLE county, central MISSOURI, 11 mi/18 km W of JEFFERSON CITY. In agricultural area (grain, soybeans; hogs, cattle).

Lohmar (LO-mahr), town, NORTH RHINE–WESTPHALIA, W GERMANY, 8 mi/12.9 km NW of BONN; 50°50′N 07°40′E. Metalworking. Has eighteenth-century palace (rebuilt in twentieth century), which is now a hotel.

Lohmühle, LUXEMBOURG: see VIANDEN.

Löhnberg (LON-berg), village, HESSE, central GERMANY, on LAHN River, 19 mi/31 km WSW of GIESSEN; 50°31′N 08°16′E.

Löhne (LO-ne), city, NORTH RHINE–WESTPHALIA, NW GERMANY, on the WERRE, 5 mi/8 km N of HERFORD; 52°11′N 08°42′E. Railroad junction. Manufacturing of textiles, furniture, plastics. Chartered 1969.

Lohne (LO-ne), town, LOWER SAXONY, NW GERMANY, 27 mi/43 km NNE of OSNABRÜCK; 52°40′N 08°14′E. Plastics; metalworking; meat processing (ham), distilling (brandy).

Lohner (LOAN-uhr) or **Gross Lohner**, peak (10,003 ft/3,049 m) in BERNESE ALPS, BERN canton, SW central SWITZERLAND, 3 mi/4.8 km SE of ADELBODEN; 46°27′N 07°36′E.

Lohr (LOR), town, LOWER FRANCONIA, BAVARIA, central GERMANY, on the MAIN River, 19 mi/31 km E of ASCHAFFENBURG; 49°59′N 09°34′E. Hydroelectric plant; manufacturing of metal products, glass, electronics. Has Gothic church (thirteenth century) and castle (fifteenth to sixteenth century). Chartered 1333.

Lohra (LO-rah), village, HESSE, central GERMANY, 9 mi/14.5 km SW of MARBURG; 50°45′N 08°38′E. Has thirteenth-century church and many half-timbered houses from seventeenth century.

Lohrville, town (2000 population 431), CALHOUN county, central IOWA, 10 mi/16 km SSE of ROCKWELL CITY; 42°16′N 94°32′W. Metal products; feed; fertilizer.

Lohrville (LOR-vil), village (2006 population 413), WAUSHARA county, central WISCONSIN, 29 mi/47 km W of OSHKOSH; 44°02′N 89°07′W. In dairy and farm area.

Lohumbo (lo-HOOM-bo), village, SHINYANGA region, NW TANZANIA, 28 mi/45 km WSW of SHINYANGA, and on railroad; 03°49'S 33°04'E. Corn, wheat, millet; livestock. Gold and limestone deposits.

Loibl Pass (LOIBL), Slovenian *Ljubelj* (4,170 ft/1,271 m), in the KARAWANKEN ALPS, on Austro-Slovenian border, 5 mi/8 km N of TRŽIČ (SLOVENIA); 46°26'N 16°16'E. KLAGENFURT (Austria)–Ljubljana (Slovenia) road tunnel here (1963) is 1 mi/1.6 km long; elevation 3,255 ft/992 m.

Loikaw (LOI-KAW), town, ⊙ KAYAH STATE, MYANMAR, on the NAM PILU RIVER, on road, and 70 mi/113 km NE of TOUNGOO; 19°40'N 97°13'E. Timber center in region rich in rice and corn.

Loilem (LOI-LEM), township, SHAN STATE, MYANMAR, on Thazi-Kentung road, and 35 mi/56 km ENE of TAUNGGYI. Head of road (N) to HSIPAW. Rice production.

Loilong (loi-LONG), former SW state (myosaship), now township (□ 1,098 sq mi/2,854.8 sq km), S SHAN STATE, MYANMAR, on the NAM PILU; ⊙ PINLAUNG, a village 45 mi/72 km SW of TAUNGGYI. Hilly (highest point, 6,124 ft/1,867 m). Forested.

Loimaa (LOI-mah), town, TURUN JA PORIN province, SW FINLAND, 40 mi/64 km NE of TURKU; 60°51'N 23°03'E. Metalworking; leather goods.

Loimwe (LOI-mwai), village, KENGTUNG township, SHAN STATE, MYANMAR, 13 mi/21 km SE of KENGTUNG, on road to LAMPANG (THAILAND).

Loing Canal (LWAN), French, *Canal du Loing* (kahnahl dyoo lwan), in LOIRET and SEINE-ET-MARNE departments, N central FRANCE, formed by junction of ORLÉANS CANAL with BRIARE CANAL just below MONTARGIS, paralleling LOING RIVER to its mouth on the SEINE below MORET-SUR-LOING, near FONTAINEBLEAU; 48°22'N 02°50'E.

Loing River (LWAN), c.100 mi/161 km long, in YONNE, LOIRET, and SEINE-ET-MARNE departments, N central FRANCE; rises near Saint-Sauveur; flows N, past MONTARGIS and NEMOURS, to the SEINE below MORET-SUR-LOING. Paralleled by BRIARE CANAL (above Montargis), and by LOING CANAL to its mouth.

Loire (LWAHR), department (□ 1,846 sq mi/4,799.6 sq km), in LYONNAIS, E central FRANCE; ⊙ SAINT-ÉTIENNE; 45°30'N 04°00'E. It lies wholly within the N MASSIF CENTRAL, bounded by the Monts du Forez (W), Monts du BEAUJOLAIS (NE), Monts du LYONNAIS (E), Mont PILAT and RHÔNE RIVER (SE). It is drained (S-N) by the upper LOIRE RIVER, which alternately flows through narrow gorges and level basins (FOREZ plain, Roanne basin). Agriculture (rye, wheat, potatoes, vegetables) in the lowlands; sheep, cattle, and hogs in the uplands. Formerly one of France's leading industrial departments, it has suffered from decline in the coal-mining and steel-manufacturing complex in the district surrounding Saint-Étienne, the largest city. The ROANNE district remains a cotton-textile manufacturing center. Former capitals of the department are FEURS (1793–1801) and MONTBRISON (1801–1856). Administratively, this department falls into the RHÔNE-ALPES region of France.

Loire-Atlantique (lwahr–aht-lahn-teek), department (□ 2,631 sq mi/6,840.6 sq km), S BRITTANY, W FRANCE; ⊙ NANTES; 47°15'N 01°50'W. Touching the Bay of BISCAY (W) and bisected by the widening LOIRE RIVER estuary, this mainly level region is drained by the lower Loire River and its tributaries, the ERDRE (right) and the SÈVRE NANTAISE (left). Numerous depressions are filled with salt marshes, peat bogs (GRANDE-BRIÈRE), and shallow lagoons, generally near the sea. Extensive livestock raising and dairying; vineyards (muscadet), apple orchards, vegetable farms (near Nantes). Chief crops are wheat, buckwheat, potatoes, and forage. Large industrial complex (shipbuilding, oil refining, chemical installations, wood- and leatherworking, and food processing) is centered on Nantes and SAINT-NAZAIRE, which also carry on export trade. Densely populated

N bank of Loire estuary has several popular bathing resorts (especially La BAULE-ESCOUBLAC, Le CROISIC). Administratively, the department forms part of the PAYS DE LA LOIRE region. It was previously named Loire-Inférieure, Loire-Maritime.

Loire Châteaux (LWAHR shah-to), French *Châteaux de la Loire (shah-to duh lah lwahr)*, grouping of royal and nobles' mansions and castles in the LOIRE RIVER valley of W FRANCE. Most were built in 15th and 16th centuries in the old provinces of ANJOU, TOURAINE, BLÉSOIS (region of Blois), and ORLÉANAIS. A partial list of the best-known châteaux includes LANGEAIS, AZAY-LE-RIDEAU, VILLANDRY, AMBOISE, CHENONCEAUX, CHAUMONT, BLOIS, CHAMBORD, and VALENCAY. The gardens of these châteaux are said to reproduce an idealized French landscape in abstract.

Loire, Haute-, FRANCE: see HAUTE-LOIRE.

Loire-Inférieur, FRANCE: see LOIRE-ATLANTIQUE.

Loire Lateral Canal (LWAHR LA-tuhr-uhl kuh-NAL), French, *Canal latéral à la Loire (kah-nahl lah-tai-rahl ah lah LWAHR)*, c.150 mi/241 km long, parallels the middle course of LOIRE RIVER between ROANNE (S terminus) and BRIARE (LOIRET department). Built 1822–1856, it spans the ALLIER on an aqueduct 5 mi/8 km SW of NEVERS. Chief towns on canal are DIGOIN (junction with Canal du CENTRE), DECIZE (junction with NIVERNAIS CANAL), NEVERS, and SANCERRE. At Marseille-les-Aubigny it is joined by BERRY CANAL. At BRIARE (N terminus) navigation continues on the Loire proper and on Briare Canal toward the SEINE. Building materials, lumber, and wine were traditionally shipped via canals, which have become a route for pleasure craft and tourism.

Loire-Maritime, FRANCE: see LOIRE-ATLANTIQUE.

Loire, Pays de la, FRANCE: see PAYS DE LA LOIRE.

Loire River (LWAHR), ancient *Liger*, longest river of FRANCE, c.630 mi/1,014 km long; rises in the VIVARAIS mountains of the MASSIF CENTRAL, SE central France, at foot of Mont GERBIER DE JONC and flows N in a great arc through central and W France to the ATLANTIC OCEAN at SAINT-NAZAIRE; 47°16'N 02°11'W. The upper LOIRE flows swiftly northwestward through numerous gorges and ancient lake beds in the Massif Central. Its major tributary is the ALLIER. At ORLÉANS it swings westward and enters a wide valley, known as the PAYS DE LA LOIRE (the Loire country). TOURS, BLOIS, SAUMUR, and ANGERS are river towns. In the Loire basin lie the rich fields, gardens, and vineyards of ORLÉANAIS, TOURAINE, and ANJOU, all former provinces of France. At the head of the Loire estuary, c.35 mi/56 km from the sea, is the large city of NANTES. The Loire's chief tributaries are the Allier, CHER, INDRE, CREUSE, and VIENNE (from S and W), and the SARTHE, with its tributary, the LOIR RIVER from the N. Silting, shallow reaches, and seasonal volume fluctuations limit the use of the Loire for navigation. Because it is subject to heavy flooding, its banks in some areas are lined with dikes. The LOIRE LATERAL CANAL parallels the river from ROANNE to BRIARE. Other canals connect the river with the SEINE and RHÔNE river systems. The LOIRE CHÂTEAUX are embodiments of French history and civilization. The valley is also noted for its wines (Sancerre, Pouilly, Vouvray). There are 4 nuclear power plants on the Loire: at BELLEVILLE-SUR-LOIRE (Cher department), at DAMPIERRE-EN-BURLY (Loiret department), at SAINT-LAURENT-NOUAN (Loir-et-Cher department), and at Avoine-Chinon (Indre-et-Loire department), above the influx of Vienne River into the Loire.

Loire-sur-Rhône (lwahr–syur–ro-nuh), commune (□ 6 sq mi/15.6 sq km), RHÔNE department, RHÔNE-ALPES region, E central FRANCE, on right bank of RHÔNE RIVER, and 4 mi/6.4 km NW of VIENNE. Markets fresh fruit. Here is a large steam-driven electric power plant.

Loiret (lwah-re), department (□ 2,631 sq mi/6,840.6 sq km), in former ORLÉANAIS province, and historically

part of the earliest nucleus of the French state, N central FRANCE; ⊙ ORLÉANS; 47°55'N 02°20'E. Generally flat, it is traversed E-W by the mid-course of the LOIRE RIVER. Also drained by LOING and ESSONNE rivers, tributaries of the SEINE. The ORLÉANS, BRIARE, and LOING canals connect the Seine with the Loire. Primarily agricultural, the department contains S part of the wheat-growing BEAUCE (N and NW); the GÂTINAIS region (NE), known for its honey; part of the poorly drained, lake-studded SOLOGNE department; (S of the Loire); and the vast Forest of ORLÉANS. There are apple and cherry orchards, tree nurseries, and vineyards in the river valleys. Food processing (vinegar, cider, honey, sugar, candies, and biscuits). Wool is made into blankets at Orléans, which now has a diversified economic base. There is a nuclear power plant on the Loire at DAMPIERRE-EN-BURLY near GIEN. Administratively, this department forms part of the CENTRE region. The department's population is growing due to proximity of the PARIS urban region (N), to which it is linked by high-speed rail (from Orléans) and N-S expressways.

Loir-et-Cher (lwahr–e–sher), department (□ 2,449 sq mi/6,367.4 sq km), part of ORLÉANAIS province, N central FRANCE; ⊙ BLOIS; 47°30'N 01°35'E. Generally level terrain, well drained by the LOIRE RIVER in its central portion (fruit and vegetable growing) and by two important tributaries, the LOIR (N) and the CHER (S), which give department its name. Cattle raising and wheat growing in N (edge of BEAUCE district); poorer SOLOGNE district (SE) produces rye, corn, vegetables, and poultry. Light manufacturing is concentrated in the principal towns: Blois, ROMORANTIN-LANTHENAY (blankets, cottons), VENDÔME (gloves). The LOIRE CHÂTEAUX (Blois, Chambord, Chaumont, Cheverny) attracts tourists. A nuclear power plant operates at SAINT-LAURENT-NOUAN halfway between ORLÉANS and Blois, on left bank of the Loire. Administratively, this department forms part of the CENTRE region.

Loir River (LWAHR), 193 mi/311 km long, in NW central FRANCE; rises SW of CHARTRES (EURE-ET-LOIR department); flows generally SW, past CHÂTEAUDUN, VENDÔME, CHÂTEAU-DU-LOIR (head of navigation), and La FLÈCHE, to the SARTHE RIVER just above ANGERS (MAINE-ET-LOIRE department). Traverses wheat-growing Little Beauce district in its upper course, then a deep and fertile alluvial valley.

Loisach River (LOI-sakh), 62 mi/100 km long, in BAVARIA, GERMANY; rises W of EHRWALD (AUSTRIA) near 47°24'N 10°57'E; flows NNE, past GARMISCH-PARTENKIRCHEN, through the KOCHELSEE, to the ISAR River 1 mi/1.6 km N of WOLFRATSHAUSEN.

Loison-sous-Lens (lwah-zon–soo–lahns), town (□ 6 sq mi/15.6 sq km), PAS-DE-CALAIS department, NORD-PAS-DE-CALAIS region, N FRANCE, suburb E of LENS, on Canal de la Souchez; 50°26'N 02°52'E.

Loita Hills (lo-EE-tah), Narok district, RIFT VALLEY province, S KENYA, N of Mount ELGON; 01°45'N 34°43'E. Rises to 8,078 ft/2,462 m.

Loitz (LOITS), town, MECKLENBURG–WESTERN POMERANIA N GERMANY, on the PEENE River, 14 mi/23 km SW of GREIFSWALD; 53°58'N 13°07'E. Woodworking; electronics, high-technology industries.

Loíza (loo-EE-zah), town (2006 population 33,634), NE PUERTO RICO, E of LOÍZA RIVER, 15 mi/24 km ESE of SAN JUAN. Industrial and commercial area; light manufacturing; tourism. Has one of highest percentages of African descendants (Yoruban slaves) of all island towns; the center of African-Hispanic culture. Known for traditional masks. San Patricio Church (begun 1645), one of the oldest on island, still active. Settled 1511 at site of a pre-Columbian village. Frequently attacked during 16th century by Carib Indians. Formerly LOÍZA ALDEA. Piñones Forest Reserve is W.

Loíza Aldea, PUERTO RICO: see LOÍZA.

Area is shown by the symbol □, and capital city or county seat by ⊙.

Loíza River (lo-EE-zah) or **Río Grande de Loíza**, c. 40 mi/64 km long, in E PUERTO RICO; rises in the Sierra de CAYEY S of SAN LORENZO; flows N and NE through fertile Caguas valley, past San Lorenzo, CAGUAS, TRUJILLO ALTO, and CAROLINA, to the ATLANTIC at LOÍZA. Used for irrigation.

Loja, province (2001 population 404,835), S ECUADOR, bordering on PERU; ⊙ LOJA. Situated in the ANDES, intersected by the Río Grande or CATAMAYO RIVER. Climate is semitropical; main rainy season December–April; there are large arid sections. Primarily agricultural (cereals, potatoes, sugarcane, coffee, subtropical fruit, fiber plants). Considerable cattle and sheep raising. The province was in colonial times the chief source for cinchona bark. Woolen goods (blankets, carpets) are made at most of its trading centers, such as Loja and GONZANAMÁ.

Loja (YO-ah), city (2001 population 118,532), ⊙ LOJA province, S ECUADOR, on the ZAMORA RIVER, at the terminus of the Ecuadorian section of the PAN-AMERICAN HIGHWAY, and 270 mi/435 km SSW of QUITO; 04°01′S 79°12′W. Elev. c.7,300 ft/2,225 m. With a humid, though pleasant, subtropical climate, it is a trading center in fertile agricultural region (sugarcane, coffee, tobacco, cereals, potatoes, pomegranates, cinchona; livestock). Tanning; manufacturing of textile goods (woolens, blankets, carpets). An old colonial city with two universities and a law school, it was formerly the world's center for cinchona production: the bark of four species of this genus was processed to produce quinine and quinidine to treat malaria. Nearby are coal, copper, iron, kaolin, and marble deposits, little exploited. Its airport, La Toma, is 22 mi/35 km W in Catamayo, near entrance to Parque Nacional PODOCARPUS, created in 1982 to protect a wide range of mountain habitats. Founded 1553 by Alonso de Mercadillo.

Lojo, FINLAND: see LOHJA.

Loka (LO-kuh), village, Équateur province, NW CONGO, on left bank of UBANGI RIVER, and 18 mi/29 km NW of MBANDAKA; 00°20′N 17°57′E. Elev. 1,099 ft/334 m. Located adjacent to border of CONGO REPUBLIC.

Lokachi (lo-kah-CHEE) (Polish *Lokacze*), town (2004 population 3,500), S VOLYN' oblast, UKRAINE, on the Luh River (right tributary of the Western BUH), 30 mi/48 km W of LUTS'K; 50°44′N 24°39′E. Elevation 649 ft/197 m. Raion center. Flour- and feed milling. Heritage museum; church (1609). Known since 1542, town since 1940.

Lokacze, UKRAINE: see LOKACHI.

Lokandu (lo-KAHN-doo), village, MANIEMA province, central CONGO, on LUALABA RIVER, and 150 mi/241 km NNW of KASONGO; 02°31′S 25°47′E. Elev. 1,564 ft/476 m. Steamboat landing, trading center in rice-growing area. Also center of training for troops. Has Roman Catholic mission, hospital. Built on former site of RIBA-RIBA, a noted Arab slave-traders' camp razed in 1893.

Lokbatan (lawk-bah-TAHN), town, Karadagsky district of greater BAKY, AZERBAIJAN, on CASPIAN SEA, 8 mi/13 km SW of Baky; 40°19′N 49°43′E. Machinery manufacturing; food processing. Oil wells.

Lokeren (LO-kuh-ruhn), commune (2006 population 38,133), Sint-Niklaas district, EAST FLANDERS province, NW BELGIUM, 12 mi/19 km ENE of GHENT. Light manufacturing. Has church with carved eighteenth-century pulpit.

Loket (LO-ket), German *Elbogen*, town, ZAPADOCESKY province, W BOHEMIA, CZECH REPUBLIC, on OHRE RIVER, on railroad, and 5 mi/8 km E of SOKOLOV; 50°11′N 12°45′E. Poultry; porcelain manufacturing; lignite mining. Has a 12th century castle, church (built 1701–1734) with Gothic Madonna, and museum.

Lokha (LO-kah), canton, MANCO KÁPAC province, LA PAZ department, W BOLIVIA, 3 mi/5 km E of COPACABANA, on peninsula; 16°15′S 69°10′W. Elevation 12,602 ft/3,841 m. Potatoes, yucca, bananas, rye.

Lokhvitsa, UKRAINE: see LOKHVYTSYA.

Lokhvytsya (LOKH-vi-tsyah) (Russian *Lokhvitsa*), (Polish *Lochwiza*), city, N POLTAVA oblast, UKRAINE, near SULA RIVER, 26 mi/42 km NNE of LUBNY; 50°22′N 33°16′E. Elevation 347 ft/105 m. Textiles; food processing. Heritage museum. Known since 1320 in Pereyaslav principality; granted city status (1628) when it belonged to the Vyshnevetskyy princely family; Cossack company center (1648–1649, 1658–1781).

Loki, Mount (LO-kee) (9,090 ft/2,771 m), SE BRITISH COLUMBIA, W CANADA, in SELKIRK MOUNTAINS, 35 mi/56 km NE of NELSON; 49°51′N 116°45′W.

Lokitaung (lo-kee-TAHNG), village, RIFT VALLEY province, KENYA, near borders of SUDAN and ETHIOPIA, on road, and 80 mi/129 km N of LODWAR; 04°15′N 35°45′E. Livestock; peanuts, sesame, corn. Near Lake TURKANA.

Løkken (LU-kuhn), town, NORDJYLLAND county, N JUTLAND, DENMARK, on the SKAGERRAK, and 12 mi/19 km SW of HJØRRING; 57°22′N 09°50′E. Fisheries.

Løkken, NORWAY: see MELDAL.

Loknya (LOK-nyah), town (2006 population 4,720), N PSKOV oblast, W European Russia, on road and railroad, 35 mi/56 km NNW of VELIKIYE LUKI; 56°49′N 30°08′E. Elevation 374 ft/113 m. Dairy products; sawmilling.

Loko (LO-ko), town, BENUE state, central NIGERIA, on BENUE River, and 38 mi/61 km SSE of NASARAWA. Shea nuts, cassava, durra.

Lokoja (lo-ko-JAH), town, ⊙ KOGI state, central NIGERIA, at the junction of the NIGER and BENUE rivers; 07°48′N 06°44′E. Trade and distribution center for an agricultural (chiefly cotton) region. Food processing. Iron-ore deposits serve nearby AJAOKUTA mill. In 1859 a British trading and missionary settlement was founded here. In 1900, Lokoja served as the staging point for the British conquest of N Nigeria and became the temporary capital of the protectorate of North Nigeria.

Lokolenge (lo-ko-LEN-gai), village, NW CONGO, 19 mi/30.6 km W of BAMBU; 01°11′N 22°40′E. Elev. 1,377 ft/419 m. Subsistence farming.

Lokosovo (luh-KO-suh-vuh), industrial settlement (2005 population 3,945), central KHANTY-MANSI AUTONOMOUS OKRUG, central SIBERIA, RUSSIA, on the left bank of the OB' RIVER, 41 mi/66 km E of SURGUT; 61°08′N 74°49′E. Elevation 144 ft/43 m. Oil and gas processing.

Lokossa, town, ⊙ MONO department, SW BENIN, 54 mi/87 km NW of COTONOU, on road to TOGO border; 06°38′N 01°43′E. Oil palm.

Lokot' (LO-kut), town (2005 population 11,780), SE BRYANSK oblast, W central European Russia, on road and railroad, approximately 30 mi/48 km NW of the RUSSIA-UKRAINE border, 45 mi/72 km SSE of BRYANSK; 52°34′N 34°34′E. Elevation 757 ft/230 m. Distilling; machine tools; furniture.

Lokot' (LO-kuht), village, S ALTAI TERRITORY, S SIBERIA, RUSSIA, near the ALEY RIVER, on the border with KAZAKHSTAN, on road and near junction of the TURKISTAN-SIBERIA RAILROAD and branch line to LENINOGORSK (Kazakhstan), 20 mi/32 km S of RUBTSOVSK; 51°11′N 81°11′E. Elevation 751 ft/228 m.

Lokrum Island Special Reserve (☐ 310 sq km), S CROATIA, in ADRIATIC SEA, across from DUBROVNIK. Entire island protected as a special reserve for its lush Mediterranean vegetation and transplanted subtropical flora.

Loksa (LOK-suh), town, N ESTONIA, port on GULF OF FINLAND, 35 mi/56 km ENE of TALLINN, in LAHEMAA NATIONAL PARK; 59°34′N 25°43′E. Ship repair dock, sawmill, tile works; bathing beach.

Loks Land, island, NUNAVUT territory, CANADA, in DAVIS STRAIT, off SE BAFFIN ISLAND, at entrance of FROBISHER BAY; 20 mi/32 km long, 15 mi/24 km wide; 62°27′N 64°35′W.

Loktak Lake (LOK-tuhk) or **Logtak**, large marshy lake (c.25 mi/40 km long N-S; 4 mi/6.4 km–13 mi/21 km wide), S central MANIPUR state, extreme NE INDIA, S of IMPHAL; 24°33′N 93°50′E. Drained by MANIPUR (or Imphal) River.

Lokva, mountain, MACEDONIA, near right bank of VARDAR River; highest peak (4,238 ft/1,292 m) is 13 mi/21 km NNW of DJEVDJELIJA; 41°25′N 20°40′E.

Lola, prefecture, N'ZÉRÉKORÉ administrative region, extreme SE GUINEA, in Guinée-Forestière geographic region; ⊙ LOLA. Bordered N by Beyla prefecture, E and S by CÔTE D'IVOIRE, SW by LIBERIA, and W by N'ZÉRÉKORÉ prefecture. Gouan River in N. Mount Tétini (elevation 4,124 ft/1,257 m) in N. Towns include Guéassou, Lola, and Nzoo. Main road runs through Lola town, connecting it to N'ZÉRÉKORÉ town to W and Nzoo and Côte d'Ivoire to SE; secondary roads connect Guéassou to BOOLA to NW and Côte d'Ivoire to E.

Lola, town, ⊙ Lola prefecture, N'ZÉRÉKORÉ administrative region, SE GUINEA, in Guinée-Forestière geographic region, near LIBERIA border, and 20 mi/32 km E of N'ZÉRÉKORÉ; 70°52′N 08°29′W. Coffee, rice, kola nuts; livestock. Iron ore deposits.

Lola, Mount, peak (9,143 ft/2,787 m) of the SIERRA NEVADA, NEVADA county, E CALIFORNIA, N of DONNER PASS, W of Independence Lake, and 30 mi/48 km W of RENO, NEVADA. In Tahoe National Forest.

Lolgorien (lol-GO-ree-en), market center, RIFT VALLEY province, KENYA, 70 mi/113 km SSW of KERICHO; 01°14′S 34°49′E.

Loliondo (lo-lee-ON-do), village, ARUSHA region, N TANZANIA, 120 mi/193 km NW of ARUSHA, near KENYA border; 02°04′S 35°40′E. Airstrip. Gold deposits. SERENGETI NATIONAL PARK to W.

Lolita (LO-lee-tuh), unincorporated village, JACKSON county, S TEXAS, 25 mi/40 km E of VICTORIA, near confluence of LAVACA and NAVIDAD rivers; 28°50′N 96°32′W. Agricultural area (rice, cotton, cattle). Manufacturing (plastic bags). LAKE TEXANA reservoir to N.

Lolland or **Laaland** (both: LO-lahn), island (☐ 479 sq mi/1,245.4 sq km), SE DENMARK, in the BALTIC SEA, E of LANGELAND, S of SJÆLLAND, and W of FALSTER; 54°46′N 11°30′E. The island is low-lying and agricultural; sugar beets are the main crop. The chief cities are MARIBO, SAKSKØBING, and NAKSKOV. There are numerous summer resorts on the SW coast.

Lollar (LUHL-lahr), town, HESSE, central GERMANY, on LAHN River, and 4 mi/6.4 km N of GIESSEN; 50°39′N 08°43′E. Metal- and ironworking.

Lolo, town (2000 population 3,388), MISSOULA county, W MONTANA, 10 mi/16 km SSW of MISSOULA, on BITTERROOT RIVER; 46°46′N 114°07′W. Cattle, horses; hay. Lolo National Forest to E and W; SAPPHIRE MOUNTAINS to E, BITTERROOT RANGE to W. Selway-Bitterroot Wilderness Area to SW; Lolo Hot Springs and LOLO PASS to W.

Lolodorf (LO-lo-dorf), village, South province, SW CAMEROON, 60 mi/97 km ENE of Kribi; 03°11′N 10°44′E. Road junction; cacao plantations.

Lolog, Lake (lo-LOG) (☐ 13 sq mi/33.8 sq km), in the ANDES, SW NEUQUÉN province, ARGENTINA, in lake district, 15 mi/24 km NW of JUNÍN DE LOS ANDES; c.15 mi/24 km long (E-W), 1 mi/1.6 km–2 mi/3.2 km wide, from a point 4 mi/6.4 km E of CHILE border. Elevation 3,146 ft/959 m.

Lolol (lo-LOL), town, ⊙ Lolol comuna, COLCHAGUA province, LIBERTADOR GENERAL BERNARDO O'HIGGINS region, central CHILE, 37 mi/60 km SSW of SAN FERNANDO. Livestock; grapes, fruit, vegetables.

Lolo Pass (5,235 ft/1,596 m), between IDAHO and MONTANA, in BITTERROOT RANGE, IDAHO county (Idaho) and MISSOULA county (Montana), 32 mi/51 km WSW of MISSOULA. LOLO PASS Visitors Center on Idaho side.

Cross-references are shown in SMALL CAPITALS. The pronunciation guide is shown on page xix. The sources of population figures are shown on page xvii.

Lolotique (lo-lo-TEE-kai), municipality and town, W SAN MIGUEL department, EL SALVADOR, WNW of SAN MIGUEL.

Lolotiquillo, municipality and town, MORAZÁN department, EL SALVADOR, N of SAN FRANCISCO GOTERA.

Lolotla (lo-LOT-lah), town, HIDALGO, central MEXICO, 50 mi/80 km N of PACHUCA DE SOTO, on Mexico Highway 105; 20°49′N 98°36′W. Elevation 5,577 ft/ 1,700 m. Cereals, beans, fruit, livestock.

Lom (LOM), city (1993 population 31,189), MONTANA oblast, Lom obshtina, NW BULGARIA, port on the right bank of the DANUBE (Romanian border) at the mouth of the LOM RIVER, 45 mi/72 km NNW of VRATSA; 43°50′N 23°15′E. Railroad terminus and commercial center. Machinery manufacturing, food processing, consumer goods. Vineyards and vegetable farms nearby. Lignite deposits just S, at Momin Brod (suburb). Has ruins of a Roman camp and Turkish fortress. Founded in 1st century C.E. as a Roman fortress of Almus. Was a small commercial town under Turkish rule (15th–19th century); developed as a major river port following the construction of the Sofia–Vratsa–Lom railroad. Formerly called Lom Palanka.

Lom (LOM), German *Bruch*, town, SEVEROČESKY province, NW BOHEMIA, CZECH REPUBLIC, on railroad, and 5 mi/8 km N of MOST. Deep-shaft coal mining.

Lom (LAWM), village, OPPLAND county, S central NORWAY, on OTTA RIVER, at N foot of the JOTUNHEIMEN Mountains, 80 mi/129 km NW of LILLEHAMMER; 61°50′N 08°33′E. Farming, livestock raising, lumbering. Tourism is becoming increasingly important. Has eleventh-century stave church and several notable old wooden buildings. Knut Hamsun born here.

Loma (LOM-uh), village, MESA county, W COLORADO, near COLORADO RIVER and UTAH state line, 15 mi/24 km NW of GRAND JUNCTION. Elevation 4,511 ft/1,375 m. In irrigated agriculture and livestock region. Oil and gas. Highline State Recreation Area to N.

Loma, village (2006 population 18), CAVALIER county, NE NORTH DAKOTA, 12 mi/19 km SSW of LANGDON; 48°38′N 98°31′W. Founded in 1905 as Irene and renamed Loma in 1906. Post office closed in 1965.

Loma Alta (LO-mah AHL-tah), town and canton, GENERAL FEDERICO ROMÁN province, PANDO department, N BOLIVIA, on BENI RIVER, and 16 mi/26 km NNE of RIBERALTA; 10°48′S 66°09′W.

Loma Bonita (LO-mah bo-NEE-tah), city and township, OAXACA, S MEXICO, on railroad, on Mexico Highway 145, in Papaloapan lowlands, 16 mi/26 km E of TUXTEPEC; 18°00′N 95°58′W. Pineapple-growing center, developed in 1930s, most important central producer for state; exports to UNITED STATES and CANADA. Cattle raising.

Loma de Cabrera (LO-mah dai kah-BRAI-rah), town, DAJABÓN province, NW DOMINICAN REPUBLIC, on MASSACRE RIVER, near HAITI border, and 6 mi/9.7 km SSE of DAJABÓN; 19°25′N 71°35′W. Agricultural center (coffee, rice; goats; hides; beeswax, honey). Formerly LA LOMA.

Loma de Tierra (LO-mah dai tee-ER-rah), town, LA HABANA province, W CUBA, on Central Highway, and 10 mi/16 km SE of HAVANA; 23°02′N 82°15′W. In sugar-growing and dairying region.

Loma Larga (LO-mah LAR-gah), canton, VALLEGRANDE province, SANTA CRUZ department, E central BOLIVIA, between the Altares and El Bosque mountains; 18°45′S 63°54′W. Elevation 6,660 ft/2,030 m. Abundant gas resources in area. Limestone deposits. Potatoes, yucca, bananas, corn, sweet potatoes, peanuts, tobacco; cattle.

Loma Linda, city (2000 population 18,681), SAN BERNARDINO county, S CALIFORNIA; suburb 5 mi/8 km SSE of SAN BERNARDINO, and 56 mi/90 km E of LOS ANGELES; 34°02′N 117°15′W. Drained by Santa Ana Wash. Citrus and vegetable region being rapidly displaced by urban development. Loma Linda University Hospital. Tri-City Airport in NW. Norton Air Force Base to NE. San Timoteo Canyon to SE.

Lomami River (lo-MAH-mee), c.900 mi/1,448 km long, left tributary of CONGO RIVER in SE and central CONGO; rises in KATANGA highlands 15 mi/24 km W of KAMINA; flows N in wide curves parallel to the LUALABA, past TSHOFA and OPALA, to CONGO RIVER at ISANGI. Navigable for barges in lower course for 245 mi/394 km below BENA KAMBA. Upper LOMAMI region is cattle-raising area.

Loma Mountains (LO-mah), NE SIERRA LEONE, 40 mi/64 km N of SEFADU; 09°10′N 11°07′W. The range, c.20 mi/32 km long, includes Bintimane peak (6,390 ft/1,948 m), highest in Sierra Leone. The NIGER RIVER rises just E.

Loma Negra (LO-mah NAI-grah), town, central BUENOS AIRES province, ARGENTINA, 7 mi/11.3 km SSE of OLAVARRÍA; 36°58′S 60°15′W. Railroad terminus. Cement-milling center; clay and limestone quarries.

Loma, Point, S tip of high rugged peninsula sheltering SAN DIEGO BAY from the PACIFIC OCEAN, SAN DIEGO county, S CALIFORNIA, opposite North Island (peninsula) and Coronado, W of SAN DIEGO. Old lighthouse (1855) on crest near tip is included in CABRILLO NATIONAL MONUMENT (established 1913, includes former U.S. Fort Rosecrans Military Reservation), set aside in memory of Portuguese explorer Juan Rodríquez Cabrillo, who discovered the bay (1542). Point Loma residential district of San Diego is on peninsula. Point Loma Nazarene College and U.S. International University.

Loma Rica, unincorporated town (2000 population 2,075), YUBA county, N central CALIFORNIA, 15 mi/24 km NE of MARYSVILLE, in W foothills of SIERRA NEVADA; 39°19′N 121°24′W. Grain, nuts, fruit; cattle; dairying; timber.

Lomas (LO-mahs), town, CARAVELÍ province, AREQUIPA region, S PERU, landing on the PACIFIC, near PAN-AMERICAN HIGHWAY, and 55 mi/89 km S of NAZCA (ICA region); 15°34′S 74°50′W. Port for cattle and cotton-growing region.

Lomas Barbudal Biological Reserve (LO-mas bahr-BOO-dahl) (□ 9 sq mi/23.4 sq km), GUANACASTE province, COSTA RICA, E of BAGACES, in zone of semideciduous forest with highly varied insect and bird life, including sixty species of butterflies.

Lomas de Chapultepec (LO-mahs dai chah-POOL-te-pek), neighborhood, Federal Distrito, MEXICO. An exclusive residential area of MIGUEL HIDALGO Delegación, MEXICO CITY.

Lomas de Zamora (LO-mahs dai zah-MO-rah), city and district (□ 38 sq mi/98.8 sq km), in greater BUENOS AIRES, ARGENTINA, 9 mi/14.5 km SSW of BUENOS AIRES; 34°46′S 58°24′W. Suburb and major industrial center (electrical manufacturing, chemicals, cement). Has agricultural school, national university, art gallery.

Lomas, Las, PANAMA: see LAS LOMAS.

Loma Tina, DOMINICAN REPUBLIC: see TINA, MONTE.

Lomati River, c.75 mi/121 km long, in SOUTH AFRICA and SWAZILAND; rises c.5 mi/8 km S of BARBERTON, MPUMALANGA province, NE South Africa; flows ENE, through extreme N Swaziland; reenters South Africa at Jeppe's Reef town, joins KOMATI RIVER 17 mi/27 km SW of KOMATIPOORT. Also called Mlumati River.

Lomaviti Group, consisting of 9 main volcanic islands, FIJI, SW PACIFIC OCEAN, in KORO SEA E of Viti Levu and S of VANUA LEVU. The largest islands are GAU, KORO, and OVALAU.

Lomax (LO-maks), unincorporated town, HOWARD county, W central TEXAS, 12 mi/19 km SW of BIG SPRING, near BEALS CREEK. Oil and natural gas. Cattle; cotton, vegetables, sesame.

Lomax (LOW-max), village (2000 population 477), HENDERSON county, W ILLINOIS, on the MISSISSIPPI RIVER, and 9 mi/14.5 km S of BURLINGTON, IOWA; 40°40′N 91°04′W. Grain; livestock.

Lomazzo (lo-MAH-tso), town, COMO province, LOMBARDY, N ITALY, 8 mi/13 km SSW of COMO; 45°42′N 09°02′E. Diversified manufacturing (shoes, textile machinery, candy).

Lombard, village (2000 population 42,322), DU PAGE county, NE ILLINOIS, a residential suburb of CHICAGO; 41°52′N 88°00′W. Manufacturing (plastics). The village is known for its lilacs. Incorporated 1869.

Lombardía (lom-bahr-DEE-ah), town, ⊙ municipality of GABRIEL ZAMORA, central MICHOACÁN, MEXICO, on N bank of Cupatitzio River, on railroad, on road, and 23 mi/37 km N of URUAPAN; 19°09′N 102°02′W. Semidry climate. Corn, beans, rice, avocado, oranges, lemons; livestock. Formerly GABRIEL ZAMORA.

Lombard Street, ENGLAND: see LONDON.

Lombardy (LOM-buhr-dee), Italian *Lombardia*, region (□ 9,200 sq mi/23,920 sq km), N ITALY, bordering on SWITZERLAND in the N; ⊙ MILAN; 45°40′N 09°30′E. The region is divided into the provinces of BERGAMO, BRESCIA, COMO, CREMONA, Mantua, MILAN, PAVIA, SONDRIO, and VARESE (named for their capitals). There are Alpine peaks and glaciers in the N, the VALTELLINA valley in the NE, several picturesque lakes, and upland pastures that slope to the rich, irrigated PO valley in the S. Rice, cereals, forage, flax, sugar beets, and mulberry are the main crops. Lombardy is the country's leading industrial region, and Milan is the chief commercial, industrial, and financial center in Italy. Manufacturing (textiles, apparel, iron and steel, machinery, motor vehicles, chemicals, furniture, and wine). The Lombard plain, located in the central part of Lombardy at the confluence of several Alpine passes, has been a battlefield in many wars. First inhabited by a Gallic people, the region became (3rd century B.C.E.) part of the Roman province of CISALPINE GAUL. In C.E. 569 the region was made the center of the kingdom of the Lombards, for whom it was named. Lombardy was united in 774 with the empire of Charlemagne. After a period of confusion (10th century), power gradually passed (11th century) from feudal lords to autonomous communes. Trade between N EUROPE and the E MEDITERRANEAN was largely carried on via the Po valley, and Lombard merchants and bankers did business throughout Europe. In the 12th century several cities united in the Lombard League and defeated Emperor Frederick I at Legnano (1176). The 13th century was marked by struggles between Guelphs (pro-papal) and Ghibellines (pro-imperial). Except for Mantua (ruled by the Gonzaga family), Lombardy fell (14th–15th century) under the sway of the Visconti family and the Sforza dukes of Milan; however, Bergamo and Brescia (1428) and Cremona (1529) were lost to Venice and the Valtellina valley was taken by the Grisons (1512). After the end (mid-16th century) of the Italian Wars, the rest of Lombardy followed the fortunes of Milan. Spanish rule (1535–1713) was followed by that of AUSTRIA (1713–1796) and of France (1796–1814). The Lombardo-Venetian kingdom was est. (1815) under the Austrians; in 1859 they were permanently removed and the kingdom was dissolved. In the 11th–12th century there was a characteristic Lombard Romanesque architecture, and during the Renaissance, Lombardy had a flourishing school of painting whose leading figures were Bernardino Luini and Gaudenzio Ferrari. There are universities at Milan and Pavia.

Lombe (LOM-bai), village, MALANJE province, NW ANGOLA, on railroad, and 14 mi/23 km WNW of MALANJE. Rice, beans.

Lomblen (LAWM-blen), largest island (□ 499 sq mi/1,297.4 sq km) of SOLOR ISLANDS, Lesser SUNDA IS-

LANDS, INDONESIA, between FLORES SEA (N) and SAVU SEA (S), 25 mi/40 km E of FLORES, just W of ADONARA; 50 mi/80 km long, 20 mi/32 km wide; 08°25′S 123°30′E. Irregular in shape and mountainous, rising to 5,394 ft/1,644 m. Agriculture; fishing. Also called Kawula or Kawoela.

Lombok (LAWM-bawk), island (□ c.1,825 sq mi/4,727 sq km; 1990 population 2,403,025), E INDONESIA, one of the Lesser SUNDA ISLANDS, separated from BALI by the Strait of LOMBOK; 08°45′S 116°30′E. MATARAM, with the port of AMPENAN to the W, is the capital of Nusa Tenggara Barat province and the chief town on the island. The volcanic and mountainous terrain rises to 12,224 ft/3,726 m at Mount RINJANI. Its S area is a fertile plain producing maize, rice, coffee, cotton, textiles, and tobacco. The population is mainly Sasaks, Muslims of Malay descent; there are also Balinese and ethnic Chinese. First visited by the Dutch in 1674, Lombok became part of the Netherlands East Indies in 1894. The English naturalist Alfred River Wallace posited that Lombok was the meeting point for Asian and Australian fauna. Many zoogeographers, however, have come to doubt the validity of this theory. A state university is in Mataram.

Lombo Kangra (LUHM-BO KAHNG-GRAH), highest peak (23,165 ft/7,061 m) in KAILAS RANGE (Gangdisê Shan), S TIBET, SW CHINA, 40 mi/64 km NW of SAGA (Gya'gya).

Lombok, Peak of, INDONESIA: see RINJANI, MOUNT.

Lombok Strait, channel (50 mi/80 km long, 20 mi/32 km–40 mi/64 km wide) connecting INDIAN OCEAN (S) and JAVA SEA (N), between BALI (W) and LOMBOK (E) in W Lesser SUNDA Islands; contains NUSA PENIDA island. Its W arm (between Bali and Nusa Penida) is BADUNG PASSAGE; 08°30′S 115°50′E.

Lomé (LO-mai), city (2003 population 839,000) ⊙ GOLFE prefecture and MARITIME region and TOGO, on Gulf of GUINEA, at border with GHANA; 06°07′N 01°13′E. It is the country's administrative, communications, and industrial center, and its chief port handles coffee, cacao, copra, and palm nuts. In 1978, the city's oil refinery began production. Railroads connect the city with the agricultural interior and with BENIN. University. Airport. Lomé was a small village until 1897, when it became the capital of the German colony of Togo. The city is sometimes categorized as a commune forming its own administrative division.

Lomela (lo-ME-lah), village, KASAI-ORIENTAL province, central CONGO, on right bank of LOMELA RIVER, and 180 mi/290 km N of LUSAMBO; 02°18′S 23°17′E. Elev. 1,509 ft/459 m. Terminus of steam navigation; trading center; cotton ginning, rice fields.

Lomela River (lo-ME-lah), c.500 mi/805 km long, in central CONGO; rises 10 mi/16 km N of KATAKO KOMBE; flows NW and WNW past LOMELA to join TSHUAPA RIVER just W of BOENDE, forming the BUSIRA RIVER. Navigable for 400 mi/644 km below the village of LOMELA.

Lometa (LO-me-tuh), village (2006 population 890), LAMPASAS county, central TEXAS, near the COLORADO RIVER, c.50 mi/80 km SE of BROWNWOOD; 31°13′N 98°23′W. Railroad junction in agriculture (cattle; peanuts, watermelons) area; manufacturing (farm gates and equipment). Colorado Bend State Park to SW.

Lom-et-Djérem (lawm–ai–JAI-rim), department (2001 population 228,691), East province, CAMEROON; ⊙ BERTOUA. Pangar Djérem National Game Reserve in NW.

Lomié (LO-mee-ai), village, East province, S CAMEROON, 65 mi/105 km SSE of Abong-M'Bang; 03°12′N 13°39′E. Coffee plantations.

Lomintsevskiy (LO-meen-tsif-skeeyee), settlement (2006 population 4,625), central TULA oblast, W central European Russia, near railroad spur, 10 mi/16 km SE of SHCHËKINO, to which it is administratively

subordinate; 53°59′N 37°40′E. Elevation 570 ft/173 m. Ceramics manufacturing.

Lomira (lo-MEI-ruh), town (2006 population 2,511), DODGE county, E WISCONSIN, 13 mi/21 km S of FOND DU LAC; 43°35′N 88°26′W. In dairying region. Light manufacturing.

Lo Miranda (lo mee-RAHN-dah), town, LIBERTADOR GENERAL BERNARDO O'HIGGINS region, central CHILE, on CACHAPOAL RIVER, on railroad, and 7 mi/11 km WSW of RANCAGUA; 34°11′S 70°54′W.

Lomita, city (2000 population 20,046), LOS ANGELES county, S CALIFORNIA; residential suburb 18 mi/29 km SSW of LOS ANGELES, S of Torrance, W of San Pedro section of Los Angeles; 33°48′N 118°19′W. Torrance Municipal Airport to W. South Coast Botanical Gardens to SW.

Lomitas, Las, ARGENTINA: see LAS LOMITAS.

Lomma (LOM-mah), town, SKÅNE county, S SWEDEN, suburb 5 mi/8 km NE of MALMÖ, on ÖRESUND; 55°40′N 13°05′E. Seat of agricultural university.

Lommatzsch (LUHM-mahtsh), town, SAXONY, E central GERMANY, 8 mi/12.9 km WNW of MEISSEN; 51°12′N 13°18′E. Center of vegetable- and fruit-growing region; manufacturing (agricultural machinery, glass, mirrors); food processing. Has late-Gothic church.

Lomme (lo-muh), outer W suburb (□ 3 sq mi/7.8 sq km) of LILLE, NORD department, NORD-PAS-DE-CALAIS region, N FRANCE; 50°39′N 02°59′E. Textile milling, metalworking.

Lommel (LAH-muhl), commune (2006 population 32,031), Maaseik district, LIMBURG province, NE BELGIUM, 20 mi/32 km N of HASSELT.

Lomnica Peak (lom-NYITS-kee), SLOVAK Lomnický štít (lom-NYITS-kee SHTYEET), GERMAN Lomnitzer Spitze, HUNGARIAN Lomnici Csúcs, second-highest peak (8,635 ft/2,632 m) of the HIGH TATRAS, VYCHO-DOSLOVENSKY province, N Slovakia, 10 mi/16 km NNW of POPRAD; 49°12′N 20°13′E. Has cable railroad. Meteorological observatory on top. TATRANSKA LOMNICA is at SE foot, KEZMARSKE ZLABY at E foot.

Lomnica River, UKRAINE: see LIMNYTSYA RIVER.

Lomnice (LOM-nyi-TSE), German Lomnitz, village, JIHOMORAVSKY province, W central MORAVIA, CZECH REPUBLIC, 17 mi/27 km NW of BRNO. Agriculture (potatoes, flax); manufacturing (textiles). Has a 13th century castle, and a baroque town hall and church. The 16th century castle of Lysice is 6 mi/9.7 km NE.

Lomnice nad Luznici (LOM-nyi-TSE NAHD LUZH-nyi-SEE), Czech Lomnice nad Lužnicí, German Lomnitz an der Linde, town, JIHOCESKY province, S BOHEMIA, CZECH REPUBLIC, on LUZNICE RIVER, on railroad, and 13 mi/21 km NE of ČESKÉ BUDĚJOVICE; 49°05′N 14°43′E. In region of large fish ponds; carp breeding and fishing, mostly for export; also tench, pike, and perch. Wood processing.

Lomnice nad Popelkou (LOM-nyi-TSE NAHD PO-pel-KOU), German Lomnitz an der Popelka, town, VYCHODOCESKY province, E BOHEMIA, CZECH REPUBLIC, on railroad, and 21 mi/34 km SE of LIBEREC; 50°32′N 15°23′E. Food processing; textiles. Has a baroque castle, museum.

Lomnitsa River, UKRAINE: see LIMNYTSYA RIVER.

Lomnitz, CZECH REPUBLIC: see LOMNICE.

Lomnitz an der Linde, CZECH REPUBLIC: see LOMNICE NAD LUZNICI.

Lomnitz an der Popelka, CZECH REPUBLIC: see LOMNICE NAD POPELKOU.

Lomnitzer Spitze, SLOVAKIA: see LOMNICA PEAK.

Lomo de Arico, CANARY ISLANDS: see ARICO.

Lomond (LO-muhnd), village (2001 population 171), S ALBERTA, W CANADA, 50 mi/80 km N of LETHBRIDGE, in VULCAN COUNTY; 50°21′N 112°39′W. Wheat, flax; cattle. Incorporated 1916.

Lomond Hills (LAH-muhnd HILZ), upland on border of FIFE and PERTH AND KINROSS, E Scotland, extends

6 mi/9.7 km E-W, NE of LOCH LEVEN. Highest points are East Lomond (1,471 ft/448 m), just SSW of FALK-LAND, and West Lomond (1,713 ft/522 m), 4 mi/6.4 km W of Falkland.

Lomond, Loch (LAH-muhnd), largest freshwater lake (23 mi/37 km long, 1 mi/1.6 km–5 mi/8 km wide) in Great Britain, forming border of Argyll and Bute, West Dunbartonshire, and STIRLING, W Scotland; 56°05′N 04°36′W. The LEVEN RIVER drains the lake into the CLYDE RIVER. At the S end of the lake, near its outlet, are numerous wooded islands. The N end is overlooked by BEN LOMOND (3,192 ft/973 m high). The hydroelectric plant at the NW end of the lake is fed by water from LOCH SLOY. The lake has become a popular tourist attraction. It has numerous associations with Rob Roy, and a cave here was once used as a refuge by Robert the Bruce. Formerly in Central and Strathclyde regions, abolished 1996.

Lomond, Loch, lake, S NEW BRUNSWICK, CANADA, 10 mi/16 km NE of SAINT JOHN; 4 mi/6 km long, 1 mi/2 km wide.

Lomond, West, Scotland: see LOMOND HILLS.

Lomonosov (luh-muh-NO-suhf), city (2005 population 36,460), W central LENINGRAD oblast, NW European Russia, on the S shore of the Gulf of FINLAND, 25 mi/40 km W of, and administratively subordinate to, SAINT PETERSBURG; 59°54′N 29°44′E. Railroad terminus, summer resort, and tourist attraction. Foundries and light manufacturing. Palace built (1710–1725) by Peter the Great and the "Chinese Palace," built (1762–1768) by Catherine the Great. Until 1948, called Oranienbaum.

Lomont (lo-mon), narrow range forming NW escarpment of the JURA mountains, in DOUBS department, FRANCHE-COMTÉ region, E FRANCE, extending c.25 mi/40 km between BAUME-LES-DAMES (W) and Swiss border (E); crossed by the DOUBS RIVER in deep valley near PONT-DE-ROIDE; 47°21′N 06°36′E. Rises to 2,750 ft/838 m.

Lomovka (LO-muhf-kah), village, S NIZHEGOROD oblast, central European Russia, on local road and near railroad, 10 mi/16 km N of (and administratively subordinate to) ARZAMAS; 55°32′N 43°50′E. Elevation 465 ft/141 m. Agricultural products; distillery.

Lomovka (LO-muhf-kah), village (2005 population 3,180), E BASHKORTOSTAN Republic, extreme W Siberian Russia, in the SE URALS, on road and railroad, 34 mi/55 km NW of MAGNITOGORSK, and 3 mi/5 km S, and under administrative jurisdiction of BELORETSK; 53°55′N 58°21′E. Elevation 1,584 ft/482 m. Distillery.

Lom Palanka, Bulgaria: see LOM.

Lomphat (LAHM-PAHT), town, E CAMBODIA, ⊙ RA-TANAKIRI province; 13°30′N 106°59′E. Transportation hub in forested area; rubber plantations, shifting cultivation; diverse manufacturing. Jarai, Budong, and other minorities.

Lompobatang, Mount (LOM-po-bah-tang) (9,419 ft/2,871 m), SULAWESI, INDONESIA, in Sulawesi Selatan province, 40 mi/64 km ESE of UJUNG PANDANG; 05°20′S 119°55′E. Also called Bonthain Peak. Sometimes spelled Lompobattang.

Lompoc (LAHM-pahk), city (□ 12 sq mi/31.2 sq km; 2000 population 41,103), SANTA BARBARA county, S CALIFORNIA, 44 mi/71 km WNW of SANTA BARBARA, on SANTA YNEZ RIVER, and 8 mi/12.9 km E of PACIFIC OCEAN; 34°40′N 120°28′W. Railroad terminus. In oil and agricultural (fruit, avocados, grain; cattle) area. It has a huge flower-seed industry and two large silica-earth mines. Petroleum processing; light manufacturing; food processing. Lompoc Museum and Lompoc Valley Historical Society. La Purísima Mission (1791) State Historic Park to NE. Vandenberg Air Force Base, large missile-testing base with launch site for military satellites, to NW. POINT ARGUELLO to SW. Incorporated 1888.

Cross-references are shown in SMALL CAPITALS. The pronunciation guide is shown on page xix. The sources of population figures are shown on page xvii.

Lom River (LAWM), 58 mi/93 km long, in NW BULGARIA, formed by a confluence of streams rising in the Sveti Nikola Mountains near the Sveti Nikola Pass, flows NE to the DANUBE at LOM; 43°37′N 22°48′E.

Lom River, left headstream of SANAGA River, in E CAMEROON; rises near Central Afr. Republic border 40 mi/64 km ENE of MEIGANGA; flows generally SW, joining another headstream 90 mi/145 km NW of BATOURI to form the Sanaga.

Lom Sak (LOM SUHG), town and district center, Petchabun province, N THAILAND, on PA SAK RIVER, and 25 mi/40 km N of Petchabun; 16°47′N 101°15′E. Gold mining. Airport.

Łomża (WOM-shah), town (2002 population 63,936), NE POLAND, on the Narew River railroad terminus. Manufacturing (food processing, paper, and textile milling). Dates from c.1000; it passed to PRUSSIA in 1795 and to RUSSIA in 1815 and reverted to Poland in 1921.

Lonaconing (lon-ah-KON-ing), mining town (2000 population 1,205), ALLEGANY county, W MARYLAND, in the ALLEGHENIES, and 14 mi/23 km WSW of CUMBERLAND; 39°34′N 78°59′W. In bituminous-coal area. An old iron furnace said to have been built in 1837 and the first in the nation to use coke is on the Register of Historic Places. The name is either derived from several Indian words with different meanings or from an Indian scout called Nemacolin, but the inhabitants refer to it as "Coney." BIG SAVAGE MOUNTAINS and Savage River State Forest are just W. Settled 1835.

Lonar, INDIA: see MEHEKAR.

Lonate Pozzolo (lo-NAH-te pot-SO-lo), town, VARESE province, LOMBARDY, N ITALY, near TICINO River, 5 mi/8 km WSW of BUSTO ARSIZIO; 45°36′N 08°45′E. Fabricated metals, machinery, textiles, apparel.

Lonato (lo-NAH-to), village, BRESCIA province, LOMBARDY, N ITALY, near SW shore of Lake GARDA, 14 mi/23 km ESE of BRESCIA; 45°27′N 10°29′E. Silk industry center. Machinery, chemicals, fabricated metals. Peat fields nearby. Noted for victory (1796) of Napoleon over Austrians.

Lonavale, town, PUNE district, MAHARASHTRA state, W central INDIA, in WESTERN GHATS, 36 mi/58 km NW of Pune; 18°45′N 73°25′E. Hill station; has railroad station and workshops. Just NW are two reservoirs supplying Khopoli hydroelectric plant.

Loncoche (lon-KO-chai), town, ⊙ Loncoche comuna, CAUTÍN province, ARAUCANIA region, S central CHILE, on railroad, and PAN-AMERICAN HIGHWAY, 45 mi/72 km S of TEMUCO; 39°22′S 72°38′W. Agricultural and commercial center. Cereals, vegetables; livestock. Food processing, dairying, lumbering.

Loncomilla River (lon-ko-MEE-yah), c.20 mi/32 km long, in LINARES province, MAULE region, S central CHILE, formed by union of LONGAVÍ and PERQUILAUQUÉN rivers 10 mi/16 km WNW of LINARES, flows N, past SAN JAVIER, to MAULE RIVER. Used for irrigation.

Loncopué (lon-ko-POO-e), village (1991 population 3,059), ⊙ Loncopué department, W central NEUQUÉN province, ARGENTINA, on the RÍO AGRIO, and 65 mi/105 km NW of ZAPALA; 38°04′S 70°37′W. Livestock-raising center.

Londe-les-Maures, La (lon-duh–lai–mor, lah), town (□ 30 sq mi/78 sq km), VAR department, PROVENCE-ALPES-CÔTE D'AZUR region, SE FRANCE, near the bay of HYÈRES on the MEDITERRANEAN, 5 mi/8 km ENE of HYÈRES, at foot of the MAURES massif; 43°08′N 06°14′E. Cork manufacturing; vineyards. Saltworks nearby. Has exotic bird park.

Londerzeel (LAHN-duhr-zail), agricultural commune (2006 population 17,412), Halle-Vilvoorde district, BRABANT province, central BELGIUM, 11 mi/18 km NW of BRUSSELS. Has thirteenth-century church.

Londiani (lo-ndi-AH-ni), town (1999 population 3,996), RIFT VALLEY province, W central KENYA, on railroad, and 60 mi/97 km E of KISUMU; 00°10′S 35°36′E. Elevation 7,533 ft/2,296 m. Agriculture and trade center. Coffee, tea, wheat, corn; dairying.

Londinium, ENGLAND: see LONDON.

Londoko (luhn-duh-KO), town (2005 population 1,285), NW JEWISH AUTONOMOUS OBLAST, S RUSSIAN FAR EAST, on road and the TRANS-SIBERIAN RAILROAD, near the BIRA RIVER (AMUR River basin), 45 mi/72 km W of BIROBIDZHAN; 49°02′N 131°59′E. Elevation 862 ft/262 m. In the protected old-growth forest region. Lime works. Formerly known as Londoko-Kamenushka.

Londoko-Kamenushka, RUSSIA: see LONDOKO.

London (LUHN-duhn), city (□ 163 sq mi/423.8 sq km; 2001 population 336,539), ⊙ MIDDLESEX county, SW ONTARIO, E central CANADA, on the THAMES River, and 100 mi/161 km SW of TORONTO; 42°58′N 81°13′W. In one of Canada's richest agricultural districts, it is an industrial, commercial, service, and financial center. Manufacturing includes electrical products, locomotive and motor vehicle parts. The University of Western Ontario (1878) and the affiliated Ursuline and Huron colleges are here. The site was chosen in 1792 by Government Simcoe to be capital of Upper Canada, but York was made capital instead. Its streets and bridges are named for those of London in England. Settled 1826.

London (LUHN-duhn), ancient *Londinium*, city; ⊙ Great Britain; SE ENGLAND, on both sides of the THAMES River; 51°30′N 00°09′W.

Geography

Since 1965, officially Greater London (□ c.620 sq mi/1,606 sq km; 1991 population 6,378,600; 2001 population 7,172,091), it consists of the Corporation of the City of London (□ 1 sq mi/2.6 sq km; 1991 population 4,000) and 32 boroughs, each a separate local authority: CAMDEN, HACKNEY, HAMMERSMITH AND FULHAM, HARINGEY, ISLINGTON, KENSINGTON AND CHELSEA, LAMBETH, LEWISHAM, NEWHAM, SOUTHWARK, TOWER HAMLETS, WANDSWORTH, City of WESTMINSTER (the inner boroughs); BARKING AND DAGENHAM, BARNET, BEXLEY, BRENT, BROMLEY, CROYDON, EALING, ENFIELD, GREENWICH, HARROW, HAVERING, HILLINGDON, HOUNSLOW, KINGSTON UPON THAMES, MERTON, REDBRIDGE, RICHMOND UPON THAMES, SUTTON, and WALTHAM FOREST (the outer boroughs).

Greater London includes the area of the former county of London, most of the former county of Middlesex, and areas that were formerly in Surrey, Kent, Essex, and Hertfordshire. Each of the boroughs elects a council. The corporation of the City, the core of London historically and commercially, elects a lord mayor, aldermen, and councilmen.

Economy

London is one of the world's foremost financial, commercial, industrial, cultural, and tourist centers. The Bank of England, Lloyds, and numerous banks and investment companies have their headquarters here, primarily in the City. It is a center for international finance, especially for large investment houses looking for a strong foothold in the EU. London is one of the world's greatest ports, handling 50,500,000 tons/45,803,500 metric tons in 1993. Oil and oil products have made up for a loss of containerized freight. Since 1991 the government has encouraged the privatization of port facilities and new piers have been built to accommodate imports of steel, lead, other metals, lumber, and newsprint. Diverse manufacturing includes printing, publishing, brewing, and food processing. It exports manufacturing goods and imports petroleum, tea, wool, raw sugar, timber, butter, metals, and meat. Many London area workers are employed in manufacturing clothing, furniture, precision instruments, jewelry, cement, chemicals, and stationery. Engineering and scientific research are also important. The city is served by five airports: HEATHROW to W, London Gatwick to S, Stansted and London LUTON to the N, and London City Airport in the Docklands.

Culture, Education, and Notable Sights

London is rich in artistic and cultural activity, with numerous theaters, cinemas, museums, galleries, and opera and concert halls. London has an ethnically and culturally diverse population, with large groups of immigrants from Commonwealth nations. Besides the British Museum, the art galleries and museums of London include the Victoria and Albert Museum, the National Gallery, and the Tate Gallery. Notable places include Old Bailey, or Central Criminal Court, on the site of Newgate Prison; Smithfield, site of many political executions as well as London's chief horse and cattle market; and Cheapside, a short and busy thoroughfare known for its bargain shopping. The University of London is the largest in Great Britain, and includes the London School of Economics. Inner London also houses City University, Guildhall University, University of East London, Middlesex University, and University of North London. The Lloyds building opened in 1986. Among the more recent developments is the Canary Wharf office complex, on Isle of Dogs. The city is host of the 2012 Summer Olympics.

The best-known streets of London are Fleet Street, the Strand, PICCADILLY, Whitehall, Pall Mall, Downing Street, Lombard Street, and Bond and Regent streets (noted for their shops). Municipal parks include HYDE PARK, Kensington Gardens, and Regent's Park.

Early History

Little is known of London prior to A.D. 61, when followers of Queen Boadicea are said to have slaughtered the inhabitants of the Roman fort here. Roman authority was restored and first city walls (of which there are remains) were built. London emerged as an important town in 886 under the firm control of King Alfred, who gave the city a government. London put up some resistance to William I in 1066, but he subsequently treated the city well. During his reign the White Tower, the nucleus of the TOWER OF LONDON, was built just E of the city wall. Under the Normans and Plantagenets, the city grew commercially and politically and, during the reign of Richard I (1189–1199), obtained a form of municipal government from which the modern City Corporation developed. Medieval London saw the foundation of the Inns of Court and the construction of WESTMINSTER ABBEY.

History: Rise to Empire

By the 14th century London had become the capital of England. The reign of Elizabeth I brought London to a level of great wealth, power, and influence—the undisputed center of England's Renaissance culture. With the advent (1603) of the Stuarts to the throne, the city became involved in struggles with the crown (particularly King Charles I) on behalf of its democratic privileges, culminating in the English Civil War, the execution of Charles, the Interregnum (1649–1660), and the restoration of Charles II to the throne. The Whig party rose from political factions, largely supported by London merchants and several powerful aristocrats, that opposed Charles II and his efforts to promote Catholicism in Britain and impose his will upon Parliament. In 1665 the great plague took some 75,000 lives. A great fire in September 1666 lasted five days and virtually destroyed London. Sir Christopher Wren played a large role in rebuilding the city. He designed more than 51 churches, notably the rebuilt St. Paul's Cathedral.

History: Empire and World War II

In the 19th century London began a period of extraordinary growth. The area of present-day Greater London had about 1.1 million people in 1801; by 1851 the population had increased to 2.7 million, and by 1901 to

6.6 million. During the Victorian era London acquired tremendous prestige as the capital of the British Empire and as a cultural and intellectual center. Britain's free political institutions and intellectual atmosphere continued to make London a haven for persons unsafe in their own countries. The Italian Giuseppe Mazzini, the Russian Alexander Herzen, and the German Karl Marx were among many politically controversial figures who lived for long periods in London. Many buildings of central London were completely destroyed or partially damaged in air raids during World War II. These include the Guildhall (scene of the lord mayor's banquets and other public functions); No. 10 Downing Street, the British Prime Minister's residence; the Inns of Court; Westminster Hall and the Houses of Parliament; St. George's Cathedral; and many of the great halls of the ancient livery companies.

Modern Growth

The growth of London in the 20th century has been extensively planned. One notable feature is the concept of a "Green Belt" to save certain areas from intensive urban development.

London (LUHN-duhn), city (□ 9 sq mi/23.4 sq km; 2006 population 9,496), ⊙ MADISON county, central OHIO, 24 mi/39 km WSW of COLUMBUS; 39°53′N 83°26′W. In livestock and grain area; fish hatchery. London Correctional Institute. Founded 1811, incorporated 1831.

London (LUHN-duhn), former township (□ 114 sq mi/296.4 sq km; 2001 population 5,637), SW ONTARIO, E central CANADA, 9 mi/14 km from the city of LONDON; 43°05′N 81°17′W. Amalgamated into MIDDLESEX CENTRE in 1998.

London, town (2000 population 925), POPE county, N central ARKANSAS, 8 mi/12.9 km WNW of RUSSELLVILLE, and on Lake DARDANELLE (ARKANSAS RIVER); 35°19′N 93°14′W.

London, unincorporated town (2000 population 1,848), TULARE county, S central CALIFORNIA, 13 mi/21 km NW of VISALIA; 36°29′N 119°27′W. Citrus, peaches, plums, olives, nuts, nursery products; cattle, hogs, poultry.

London, town (2000 population 5,692), ⊙ LAUREL county, SE KENTUCKY, in CUMBERLAND foothills, 29 mi/47 km E of SOMERSET; 37°07′N 84°04′W. Trade center for agriculture (poultry; dairy; tobacco, corn, wheat), coal, and timber area. Light manufacturing. Seat of Sue Bennet College (1897). Colonel Harlan Sanders' original chicken restaurant 10 mi/16 km to S at CORBIN. Annual World Chicken Festival (September). Camp Wildcat Battle Monument to NW. Levi Jackson Wilderness Road State Park to SE; Daniel Boone National Forest to W and E; Holly Bay Recreation Area on LAUREL RIVER LAKE reservoir to SW; Wood Creek Lake reservoir to NW.

London, unincorporated village, KANAWHA county, W central WEST VIRGINIA, on KANAWHA RIVER, c.17 mi/27 km SE of CHARLESTON; 38°11′N 81°22′W.

London Airport, ENGLAND, United Kingdom: see HEATHROW AIRPORT.

London Colney (LUHN-duhn KAHL-nee), village (2001 population 7,742), HERTFORDSHIRE, E ENGLAND, on COLNE RIVER, 4 mi/6.4 km N of BOREHAMWOOD; 51°44′N 01°18′W. Aircraft museum 1 mi/1.6 km to SE.

Londonderry (LUHN-duhn-dE-ree) or **Derry**, Gaelic *Doire*, city (□ 148 sq mi/384.8 sq km; 2001 population 83,699), ⊙ LONDONDERRY, NW Northern Ireland, on the FOYLE RIVER, and 15 mi/24 km WSW of LIMAVADY; 55°00′N 07°20′W. Originally called Derry, the area was dominated for many centuries by the O'Neill family, whose confiscated estates were granted in 1609 to the city companies of LONDON, from which the name Londonderry came. It is a seaport with extensive exporting of livestock. Industries include food processing and chemicals. The city grew up around an

abbey founded in 546 by St. Columba. It was burned by the Danes in 812. In 1311 it was granted to Richard de Burgh, earl of ULSTER. The old town walls are well preserved. In the siege of Londonderry by the forces of James II (beginning in April 1689), it was held for 105 days under the leadership of George Walker; a triumphal arch, a column, and one of the town gates commemorate the siege. The city contains a Protestant cathedral (built 1628–1633; restored 1886–1887), a Roman Catholic cathedral, and a monastery church (founded 1164). Magee College is affiliated with The Queen's University of Belfast. To the W of the city center are the Catholic districts of Creggan and the Bogside. The E suburbs are mostly Protestant. Site of bombings between Catholics and Protestants. The nationalists of Northern Ireland prefer the city be called Derry, while the unionists use Londonderry, the city's official name, as stated in the city's royal charter and confirmed by a high court in January 2007.

Londonderry (LUHN-duhn-de-ree), town, ROCKINGHAM county, SE NEW HAMPSHIRE, 10 mi/16 km SSE of MANCHESTER, and 7 mi/11.3 km NE of NASHUA; 42°52′N 71°23′W. Drained by Beaver Brook. Diversified manufacturing. Agriculture (vegetables, beans, apples; cattle; dairying). Former NW section now in DERRY. Settled by Scots Irish 1719, chartered 1722.

Londonderry, town, WINDHAM co., S central VERMONT, on West River, and 20 mi/32 km NNW of Newfane; 43°12′N 72°47′W. Partly in Green Mountain National Forest. Ski resorts. Named for LONDONDERRY, NEW HAMPSHIRE.

Londonderry, Northern Ireland: see DERRY.

Londonderry, Cape (LUHN-duhn-de-ree), northernmost point of WESTERN AUSTRALIA, NE coast of TIMOR SEA, at W end of JOSEPH BONAPARTE GULF; 13°45′S 126°55′E.

Londonderry Island (LUHN-duhn-de-ree) (27 mi/43 km long), TIERRA DEL FUEGO, CHILE, on the PACIFIC, SW of main island of the archipelago; 55°03′S 70°35′W. Just off SW coast is small Treble Island.

London Mills, village (2000 population 447), FULTON county, W central ILLINOIS, on SPOON RIVER (bridged here), 16 mi/26 km NW of CANTON, and 24 mi/39 km NNW of Lewiston; 40°42′N 90°16′W. In agricultural (corn, wheat, sorghum; cattle) and bituminous-coal-mining area. Feed milling.

London Mountain, CANADA: see WHISTLER MOUNTAIN.

Londres (LON-dres), town, S central CATAMARCA province, ARGENTINA, 6 mi/9.7 km SW of BELÉN; 27°43′S 67°07′W. Cereals, alfalfa, walnuts, olives; livestock; wine. Tungsten, coal, and tin mines nearby.

Londrina (LON-dree-nah), city (2007 population 497,833), N PARANÁ, BRAZIL, on railroad, and 190 mi/306 km NW of CURITIBA; 23°18′S 51°09′W. Chief trade center of N Paraná; coffee, cotton, rice, fruit, and livestock processing; manufacturing (furniture), paper milling, distilling. Has large German and Slavic population. Founded 1932.

Londža River, c.30 mi/48 km long, in SLAVONIA, E CROATIA; rises 5 mi/8 km SSW of Našice; flows SW to ORLJAVA River just S of PLETERNICA.

Lone Cone, peak (12,613 ft/3,844 m), in SAN JUAN MOUNTAINS, SAN MIGUEL and DOLORES counties, SW COLORADO, 23 mi/37 km W of TELLURIDE, in Uncompahre National Forest.

Lone Elm, village (2000 population 27), ANDERSON county, E KANSAS, 13 mi/21 km S of GARNETT; 38°04′N 95°14′W. Livestock; grain; dairying

Lone Grove, town (2006 population 5,156), CARTER county, S OKLAHOMA, suburb 8 mi/12.9 km W of ARDMORE; 34°11′N 97°16′W. Agricultural and oil production area. Manufacturing (glass and fiberglass products).

Lonely Island, RUSSIA: see UYEDINENIYE ISLAND.

Lonely Mine, town, MATABELELAND NORTH province, central ZIMBABWE, 45 mi/72 km NNE of BULAWAYO;

19°31′S 28°44′E. Livestock; grain. Former gold-mining center.

Lone Oak, town (2000 population 104), MERIWETHER county, W GEORGIA, 14 mi/23 km NE of LA GRANGE; 33°10′N 84°49′W.

Lone Oak, village (2000 population 454), MCCRACKEN county, W KENTUCKY, 5 mi/8 km SW of PADUCAH; 37°02′N 88°40′W. Tobacco, sorghum, grain; livestock.

Lone Oak (LON OK), village (2006 population 570), HUNT county, NE TEXAS, near Sabine River, 14 mi/23 km SE of GREENVILLE; 33°00′N 95°56′W. In agricultural area. LAKE TAWAKONI reservoir to S.

Lone Pine, unincorporated town (2000 population 1,655), INYO county, E CALIFORNIA, in Owens Valley, 65 mi/105 km ENE of VISALIA, on OWENS RIVER; 36°34′N 118°05′W. Lead and silver mining; cattle. Los Angeles Aqueduct passes to W. MOUNT WHITNEY, highest point in continental U.S., 10 mi/16 km W; DEATH VALLEY NATIONAL MONUMENT to E; SEQUOIA NATIONAL PARK to W; part of Inyo National Forest to W; Lake Owens (dry) to SE. Lone Pine Indian Reservation to S.

Lone Rock, town (2000 population 157), KOSSUTH county, N IOWA, 11 mi/18 km NNW of ALGONA; 43°13′N 94°19′W.

Lonerock, village (2006 population 22), GILLIAM county, N OREGON, 24 mi/39 km SW of HEPPNER, on Lone Rock Creek; 45°05′N 119°52′W. Part of Umatilla National Forest to S.

Lone Rock, village (2006 population 876), RICHLAND county, SW WISCONSIN, on WISCONSIN RIVER, on railroad, and 14 mi/23 km SE of RICHLAND CENTER; 43°10′N 90°12′W. In timber and agricultural area. Lumber; cheese; feeds; hunting.

Lone Star (LON STAHR), town (2006 population 1,601), MORRIS county, NE TEXAS, 30 mi/48 km N of LONGVIEW; 32°56′N 94°42′W. In agricultural area (poultry; peanuts, watermelons). Manufacturing (construction materials; coal tar and slag processing). Ellison Creek Reservoir to W; Lake O' the Pines (Big Cypress Creek) to S.

Lone Star, village, CALHOUN county, central SOUTH CAROLINA, 18 mi/29 km ENE of ORANGEBURG. In agricultural area (cotton, soybeans). Lake MARION to E.

Lone Star State: see TEXAS.

Lone Tree, town (2000 population 1,151), JOHNSON county, E IOWA, 13 mi/21 km SSE of IOWA CITY; 41°29′N 91°25′W. Feed milling.

Lonevåg (LAW-nuh-vawg), village, HORDALAND county, SW NORWAY, on W OSTERØY, 11 mi/18 km NNE of BERGEN; 60°32′N 05°30′E. Leather; metal goods.

Lone Wolf, village (2006 population 470), KIOWA county, SW OKLAHOMA, 8 mi/12.9 km WSW of HOBART; 34°59′N 99°15′W. In cotton, grain, livestock area. ALTUS LAKE reservoir to SW.

Long, county (□ 404 sq mi/1,050.4 sq km; 2006 population 11,452), SE GEORGIA; ⊙ LUDOWICI; 31°46′N 81°45′W. Bounded SW by ALTAMAHA RIVER. Coastal plain agriculture (corn, soybeans, tobacco); livestock (hogs, cattle), and forestry area. Formed 1920.

Long, village, W ALASKA, 24 mi/39 km S of RUBY. Placer gold mining. Airfield.

Long, CHINA: see GANSU.

Longa (LON-guh), town, CUANDO CUBANGO province, ANGOLA, on main road, and 55 mi/89 km E of Menogue; 14°44′S 18°36′E. Market center.

Longa Island (LAHN-gah), uninhabited island (1 mi/1.6 km long, 0.5 mi/0.8 km wide) at mouth of Gair Loch, W HIGHLAND, N Scotland; 57°43′N 05°47′W. Rises to 229 ft/70 m.

Long An, province (□ 1,675 sq mi/4,355 sq km), S VIETNAM, in MEKONG Delta; ⊙ Tan An; 10°40′N 106°25′E. Borders on CAMBODIA and TAY NINH province (N), HO CHI MINH urban region (E), TIEN

Cross-references are shown in SMALL CAPITALS. The pronunciation guide is shown on page xix. The sources of population figures are shown on page xvii.

GIANG province (S), and DONG THAP province (W). One of three provinces lying in extensive marshlands known as the PLAIN OF REEDS (DONG THAP MUOI), it is famous for its lakes and ponds, forests, diverse bird life, and wet rice farming. Landscape is braided by drainage ditches and irrigation canals. Fertile alluvial soils sustain a thriving agricultural economy (vegetables, fruits, coconuts, sugarcane, cassava). Aquaculture and riverine fisheries, food processing, light manufacturing. Population is Kinh, with small Khmer minority.

Long'an (LUNG-AN), town, ⊙ Long'an county, SW GUANGXI ZHUANG AUTONOMOUS REGION, CHINA, 50 mi/80 km NW of NANNING, and on right bank of YOU RIVER; 23°11′N 107°41′E. Grain, sugarcane.

Longares (long-GAH-res), town, ZARAGOZA province, NE SPAIN, 24 mi/39 km SW of ZARAGOZA; 41°24′N 01°11′W. Wine; cereals; sheep.

Longá River (LON-gah), 150 mi/241 km long, in PIAUÍ, NE BRAZIL; rises S of CAMPO MAIOR; flows N to PARNAÍBA RIVER 25 mi/40 km above PARNAÍBA. Not navigable.

Long Ashton (LAHNG ASH-tuhn), residential town (2001 population 4,981), North Somerset, SW ENGLAND, 3 mi/4.8 km WSW of BRISTOL; 51°26′N 02°39′W. Large orthopedic hospital and 15th-century Ashton Court.

Longaví (lon-gah-VEE), town, ⊙ Longaví comuna, LINARES province, MAULE region, S central CHILE, on railroad, on PAN-AMERICAN HIGHWAY, and 10 mi/16 km SW of LINARES; 35°58′S 71°41′W. Agricultural area (cereals, vegetables, fruit, grapes; livestock).

Longaví, Nevado (lon-gah-VEE, nai-VAH-do), Andean peak (10,600 ft/3,231 m), LINARES province, MAULE region, S central CHILE, 35 mi/56 km SE of LINARES.

Longaví River (lon-gah-VEE), c.60 mi/97 km long, in LINARES province, MAULE region, S central CHILE; rises in the ANDES 12 mi/19 km WNW of DIAL PASS (ARGENTINA border); flows NW to join PERQUILAUQUÉN RIVER 10 mi/16 km WNW of LINARES, to form LONCOMILLA RIVER.

Longay (LAHN-gai), uninhabited island (0.5 mi/0.8 km long, 0.5 mi/0.8 km wide), INNER HEBRIDES, HIGHLAND, N Scotland, just E of SCALPAY island; 57°19′N 05°55′W. Rises to 221 ft/67 m.

Long Barn, unincorporated village, TUOLUMNE county, central CALIFORNIA, in the SIERRA NEVADA, 15 mi/24 km NE of SONORA. Resort area. In Stanislaus National Forest.

Long Beach, city (2000 population 461,522), LOS ANGELES county, S CALIFORNIA; suburb 18 mi/29 km S of LOS ANGELES, on SAN PEDRO BAY; 33°48′N 118°10′W. Drained in W by LOS ANGELES RIVER, in E by SAN GABRIEL RIVER. Having an excellent harbor, it serves as Los Angeles's port and a year-round resort noted for its long, wide beaches and active marina. The city has a large oil industry; oil (discovered 1921) is found both underground and offshore. Manufacturing includes aircraft, automobile parts, electronic equipment, audio/visual equipment, oil-well services, and home furnishings. The city grew rapidly as a result of the high-technology and aerospace industries in the area, and its population increased significantly 1940–1960 and continued to increase 1960–1990. It has the largest Cambodian community in the U.S. Points of interest include an adobe ranch house (1844) that has become a museum; four man-made oil islands in the harbor; and the ocean liner *Queen Mary*, which was purchased 1967 and converted into a museum, hotel, and tourist center. California State University, Long Beach and Long Beach City College. Long Beach Marina at ALAMITOS BAY in SE corner. Long Beach Airport (Daugherty Field) in N. Long Beach Naval Shipyard (TERMINAL ISLAND), closed in 1997. Ferries to SANTA CATALINA ISLAND. Incorporated 1888.

Long Beach, city (2000 population 17,320), HARRISON county, SE MISSISSIPPI, 4 mi/6.4 km WSW of GULFPORT, on MISSISSIPPI SOUND; 30°21′N 89°10′W. Light manufacturing. Seat of Gulf Park College. Gulfport Naval Center in NE; University of Southern Mississippi (Gulfport campus). Beach resort area. Suffered widespread damage during Hurricane Katrina, August 2005.

Long Beach, suburban city and beach community (□ 3 sq mi/7.8 sq km; 2006 population 35,111), NASSAU county, SE NEW YORK, on barrier island off S shore of W LONG ISLAND (railroad, highway connections), 8 mi/12.9 km SE of JAMAICA; 40°35′N 73°40′W. Light manufacturing and services; railroad terminus. Former resort. Popular beach. Incorporated 1922.

Long Beach, town (2000 population 1,559), LA PORTE county, NW INDIANA, on Lake MICHIGAN, 11 mi/18 km NW of LA PORTE; 41°45′N 86°51′W.

Long Beach, township, OCEAN county, SE NEW JERSEY; 39°38′N 74°12′W. Resort and summer art colony communities on coastal LONG BEACH island. Incorporated 1899.

Long Beach (LAHNG BEECH), town, BRUNSWICK county, SE NORTH CAROLINA, 26 mi/42 km SSW of WILMINGTON, on Oak Island, on ATLANTIC OCEAN (S); 33°55′N 78°09′W. Together with YAUPON BEACH, forms Oak Island, which incorporated 1999. Beach resort area. INTRACOASTAL WATERWAY canal passes to N. Lockwood Folly Inlet to W, mouth of CAPE FEAR RIVER estuary to E. Originally incorporated 1955.

Long Beach, town (2006 population 1,394), PACIFIC county, SW WASHINGTON, on coast, 5 mi/8 km N of mouth of the COLUMBIA River, and 15 mi/24 km NW of Astoria, OREGON; 46°22′N 124°03′W. Manufacturing (printing and publishing); beach resorts. Willapa Bay and National Wildlife Refuge to NE.

Long Beach, village (2000 population 271), POPE county, W MINNESOTA, 2 mi/3.2 km W of GLENWOOD, on N shore of Lake MINNEWASKA; 45°38′N 95°25′W. PELICAN LAKE to W.

Long Beach, narrow island (c.19 mi/31 km long), E NEW JERSEY, sheltering LITTLE EGG HARBOR, Manahawkin Bay, and S end of BARNEGAT BAY from the ATLANTIC OCEAN. Barnegat City is at BARNEGAT LIGHT at N end; BEACH HAVEN and Beach Haven Inlet are at S end. Other resorts: HARVEY CEDARS, SURF CITY, SHIP BOTTOM (bridge to mainland here).

Longbenton and Killingworth (lahng-BEN-tuhn and KIL-ing-wuhth), town (2001 population 15,070), TYNE AND WEAR, NE ENGLAND, 4 mi/6.4 km NE of NEWCASTLE UPON TYNE; 55°00′N 01°34′W. Light industry and steel milling. Gosforth Park (site of racecourse). Burradon located 3 mi/4.8 km N.

Longboat Key (LAHNG-bot), town (2000 population 7,603), SARASOTA and MANATEE counties, W central FLORIDA, 12 mi/19 km NW of SARASOTA; 27°23′N 82°38′W. Manufacturing (printing and publishing).

Longboat Key, narrow barrier island (□ 17 sq mi/44.2 sq km), MANATEE and SARASOTA counties, W central FLORIDA, sheltering Sarasota Bay, in GULF OF MEXICO; c.10 mi/16 km long; 27°23′N 82°38′W. Entire island located within boundaries of town of LONGBOAT KEY. Exclusive resort located here.

Long Branch, residential city (2006 population 32,314), MONMOUTH county, E central NEW JERSEY, on the ATLANTIC coast; 40°17′N 73°59′W. Residential but remains popular resort area. Presidents Grant, Hayes, Garfield, and Arthur summered here, and President Wilson's summer house (now part of Monmouth College) was at WEST LONG BRANCH. President Garfield died in LONG BRANCH in 1881. Historical museum, art center. Monmouth Park Racetrack is nearby. Settled 1740, incorporated 1904.

Long Branch, borough (2006 population 514), WASHINGTON county, SW PENNSYLVANIA, 3 mi/4.8 km S of CHARLEROI, near MONONGAHELA RIVER. Corn, hay; dairying.

Long Buckby (LAHNG BUK-bee), village (2001 population 4,224), W NORTHAMPTONSHIRE, central ENGLAND, 10 mi/16 km WNW of NORTHAMPTON; 52°18′N 01°04′W. Has 13th-century church.

Long Cay, islet (10 mi/16 km long NE-SW, c.1 mi/1.6 km wide) and district (□ 8 sq mi/20.8 sq km), S BAHAMAS, just W of CROOKED ISLAND, 250 mi/402 km SE of NASSAU. Main settlement, ALBERT TOWN (center); 22°37′N 74°20′W. Also known as Fortune Island.

Longchang (LUNG-CHANG), town, ⊙ Longchang county, S central SICHUAN province, CHINA, on railroad to CHENGDU, and 80 mi/129 km W of CHONGQING; 29°21′N 105°18′E. Rice, sugarcane, oilseeds.

Long Chau Islands (LOUNG CHOU), group of islets in Gulf of BAC BO, N VIETNAM, 10 mi/16 km SE of CAT BA ISLAND; 20°37′N 107°09′E. Formerly NORWAY ISLANDS.

Longchuan (LUNG-CHUAN), town, ⊙ Longchuan county, E GUANGDONG province, CHINA, on EAST RIVER, and 65 mi/105 km WSW of MEIZHOU; 24°06′N 115°15′E. Rice. Also known as Laolong.

Longchuan, town, ⊙ Longchuan county, westernmost YUNNAN province, CHINA, on MYANMAR border, near SHWELI RIVER, 75 mi/121 km SW of TENGCHONG; 24°21′N 97°56′E. Rice, sugarcane, medicinal herbs; sugar refining. Formerly Changfengkai.

Longchuan River, CHINA: see SHWELI RIVER.

Long Clawson (LAHNG CLAW-suhn), agricultural village (2001 population 2,428), NE LEICESTERSHIRE, central ENGLAND, 6 mi/9.7 km NNW of MELTON MOWBRAY; 52°50′N 00°56′W. Cheese making. Has 13th-century church.

Long Creek, village (2000 population 1,364), MACON county, central ILLINOIS, suburb 7 mi/11.3 km ESE of DECATUR; 39°47′N 88°50′W. Corn, soybeans. Spitler Woods State Natural Area to S.

Long Creek, village (2006 population 199), GRANT county, NE central OREGON, 70 mi/113 km SSW of PENDLETON; 44°42′N 119°05′W. Elevation 3,754 ft/1,144 m. Grain; livestock. Malheur National Forest in SE; Umatilla National Forest to NE.

Longdale, village (2006 population 311), BLAINE county, W central OKLAHOMA, 21 mi/34 km NNW of WATONGA; 36°07′N 98°32′W. In agricultural area (grain; livestock). Oil and natural-gas deposits. CANTON LAKE to SW (NORTH CANADIAN River).

Longde (LUNG-DUH), town, ⊙ Longde county, S NINGXIA HUI AUTONOMOUS REGION, CHINA, 35 mi/56 km W of PINGLIANG; 35°38′N 106°06′E. Grain, oilseeds.

Longdendale (LAHNG-uhn-dail), valley of Etherow River, PEAK DISTRICT, DERBYSHIRE, central ENGLAND, W of MANCHESTER; 53°28′N 01°56′W.

Long Eaton (LAHNG EE-tuhn), town (2001 population 44,826), DERBYSHIRE, central ENGLAND, 6 mi/9.7 km SW of NOTTINGHAM; 52°54′N 01°16′W. Synthetics, electrical equipment, railroad carriages.

Longeville-lès-Metz (lon-zhuh-veel—le-mets), town (□ 1 sq mi/2.6 sq km), W suburb of METZ, MOSELLE department, in LORRAINE, NE FRANCE, on left bank of MOSELLE RIVER; 49°07′N 06°08′E.

Longeville-lès-Saint-Avold (lon-zhuh-veel—le-santah-vohl), commune (□ 9 sq mi/23.4 sq km), suburb of SAINT-AVOLD, MOSELLE department, in LORRAINE, NE FRANCE, 16 mi/25 km west of SARREGUEMINES; 49°07′N 06°38′E.

Longfellow-Evangeline Memorial State Commemorative Area (□ 0.2 sq mi/0.6 sq km), ST. MARTIN parish, S central LOUISIANA, along BAYOU TECHE, just N of ST. MARTINVILLE; 30°08′N 91°49′W. Established 1934 to commemorate supposed real-life heroine of Longfellow's poem *Evangeline*. Commemorates French history of S Louisiana. Depicts plantation life in 1800s.

Area is shown by the symbol □, and capital city or county seat by ⊙.

Longfellow Historic Site, CAMBRIDGE, E MASSACHU-SETTS, home of Henry Wadsworth Longfellow (1837–1882); also George Washington's headquarters during the siege of BOSTON (1775–1776). Authorized 1972.

Longfield and New Ash Green (LAHNG-feeld and NYOO ASH GREEN), suburbs (2001 population 12,762), NW KENT, SE ENGLAND; 51°22′N 00°19′E. Longfield is just outside of HARTLEY; New Ash Green is 2 mi/3.2 km S of Longfield. Large housing development. London "overspill" resides here.

Longford (LAWNG-fuhrd), Gaelic *An Longfort*, county (□ 421 sq mi/1,094.6 sq km; 2006 population 34,391), N central IRELAND; ⊙ LONGFORD. Borders CAVAN and LEITRIM counties to N, ROSCOMMON county to W, and WESTMEATH county to SE. A part of the central plain of Ireland, it has level land with numerous small lakes, bogs, and marshes. The SHANNON RIVER and LOUGH REE form its western border. Raising beef cattle is the principal occupation; oats and potatoes are the chief crops.

Longford (LAHNG-fuhrd), town, NE central TASMANIA, SE AUSTRALIA, 12 mi/19 km S of LAUNCESTON, and on SOUTH ESK RIVER; 41°36′S 147°07′E. Livestock; dairy products; wool. Waterfowl sanctuary near town.

Longford (LAWNG-fuhrd), Gaelic *An Longfort*, town (2006 population 7,622), ⊙ LONGFORD county, N central IRELAND, on the Camlin River, and 24 mi/38 km NW of MULLINGAR; 53°44′N 07°48′W. Primarily a farm market. The Roman Catholic bishopric of Ardagh and Clonmacnois is here.

Longford (LAHNG-fuhd), village (2006 population 1,300), GLOUCESTERSHIRE, W central ENGLAND, 2 mi/3.2 km N of GLOUCESTER; 51°55′N 02°14′W. Farming.

Longford, village (2000 population 94), CLAY county, N central KANSAS, 17 mi/27 km SW of CLAY CENTER; 39°10′N 97°19′W. Grain; livestock.

Longford, ENGLAND: see STOKE-ON-TRENT.

Longframlington (lahng-FRAM-ling-tuhn), village (2001 population 1,179), E central NORTHUMBERLAND, NE ENGLAND, 8 mi/12.9 km SSW of ALNWICK; 55°17′N 01°47′W. Former coal-mining community.

Long Grove, town (2000 population 597), SCOTT county, E IOWA, 11 mi/18 km N of DAVENPORT; 41°41′N 90°34′W. In agricultural area.

Long Grove, village (2000 population 6,735), Lake county, NE ILLINOIS, suburb 29 mi/47 km NW of CHICAGO, 10 mi/16 km W of HIGHLAND PARK; 42°12′N 88°00′W.

Longguan (LUNG-GUAN), town, NE HEBEI province, CHINA, 40 mi/64 km E of ZHANGJIAKOU, near the GREAT WALL; 40°45′N 115°43′E. Livestock, oilseeds.

Longgun (LUNG-GUN), town, E HAINAN province, CHINA, at mouth of Wanchuan River, 60 mi/97 km S of QIONGSHAN; 19°12′N 110°31′E. Rattan cane, fruits, tropical crops; hogs, cattle; fisheries.

Longhai (LUNG-HEI), town, ⊙ Longhai county, S FUJIAN province, CHINA, on left bank of LONG RIVER, and 25 mi/40 km WNW of XIAMEN; 24°27′N 117°49′E. Commercial center; rice, sugarcane; food processing, chemicals, engineering, building materials. Long a leading city on the Fujian coast, it declined (19th century) and was supplanted by XIAMEN (Amoy). Had Portuguese settlement (1547–1549). Also called Shima.

Longhai Railroad (LUNG-HEI), chief E-W railroad of central CHINA, extending from YELLOW SEA port of LIANYUNGANG, W past XUZHOU (crossing of TIANJIN-PUKOU railroad), ZHENGZHOU (crossing of BEIJING-GUANGZHOU railroad), XI'AN, TIANSHUI, and LANZHOU.

Longhorn Caverns State Park (LAWNG-horn), limestone cave, BURNET county, central TEXAS, 8 mi/12.9 km SW of BURNET; area of 1 sq mi/3 sq km; 30°41′N 98°21′W. On the N edge of the EDWARDS PLATEAU, the cave lies beneath a triangular ridge rising above the valley of the COLORADO RIVER; 2 mi/3.2

km long. Nature trails on surface. Home of prehistoric cave dwellers. Gunpowder once manufactured here for Confederate Army. Museum.

Longhua (LUNG-HUAH), town, ⊙ Longhua county, NE HEBEI province, CHINA, 27 mi/43 km NNW of CHENGDE; 41°17′N 117°37′E. Grain; livestock.

Longhui (LUNG-HUAI), town, ⊙ Longhui county, central HUNAN province, CHINA, 20 mi/32 km SW of SHAOYANG city; 27°07′N 111°02′E. Rice, medicinal herbs; food and beverages, tobacco processing, papermaking. Also called Taohuaping.

Longhurst (LAHNG-huhrst), unincorporated village, PERSON county, N NORTH CAROLINA, 2 mi/3.2 km NNE of ROXBORO; 36°25′N 78°58′W. Agriculture (tobacco, grain; livestock).

Longido (lon-GEE-do), village, ARUSHA region, N TANZANIA, 43 mi/69 km N of ARUSHA, near KENYA border, crossing 12 mi/19 km to NE; LONGIDO MOUNTAIN to NE; 02°44′S 36°40′E. Corn, wheat; cattle, sheep, goats.

Longido Mountain (lon-GEE-do), extinct volcano, ARUSHA region, N TANZANIA, 45 mi/72 km N of ARUSHA, and 2 mi/3.2 km NE of LONGIDO, near KENYA border; 02°43′S 36°41′E. Elevation 8,625 ft/2,629 m.

Longing, Cape, a rock cape at the end of a large, ice-covered promontory on the E coast of GRAHAM LAND, ANTARCTIC PENINSULA; 64°33′S 58°50′W. It marks the junction of TRINITY PENINSULA and Nordenskjöld Coast.

Long Island, village (2000 population 155), PHILLIPS county, N KANSAS, on PRAIRIE DOG CREEK, near NEBRASKA state line, and 17 mi/27 km NW of PHILLIPSBURG; 39°57′N 99°31′W. Corn; livestock.

Long Island, village, SULLIVAN county, NE TENNESSEE, suburb just S of KINGSPORT; 36°31′N 82°33′W.

Long Island (LAHNG), island, Whitsunday group, in CORAL SEA, off QUEENSLAND coast, NE AUSTRALIA, part of GREAT BARRIER REEF MARINE PARK, and 6 mi/9 km long, 1 mi/2 km wide. Resorts.

Long Island (□ 13 sq mi/34 sq km), SE NEWFOUNDLAND AND LABRADOR, CANADA, in PLACENTIA BAY, 65 mi/105 km W of ST. JOHN'S; 15 mi/24 km long, 2 mi/3 km wide; 47°35′N 54°00′W. On SE coast is fishing settlement of HARBOUR BUFFET. Tourism.

Long Island, SE NEWFOUNDLAND AND LABRADOR, Canada, in NOTRE DAME BAY, 25 mi/40 km SSW of CAPE ST. JOHN; 6 mi/10 km long, up to 4 mi/6 km wide; 49°35′N 55°40′W.

Long Island, in the Atlantic, W NOVA SCOTIA, CANADA, at entrance to BAY OF FUNDY, forming SW shore of ST. MARY BAY, 30 mi/48 km SW of DIGBY; 12 mi/19 km long, 2 mi/3 km wide. At its extremities are Boar Head (N) and Dartmouth Point (S).

Long Island, SE ALASKA, in Cordova Bay, N arm of DIXON ENTRANCE, 55 mi/89 km SW of KETCHIKAN; 15 mi/24 km long, 2 mi/3.2 km–6 mi/9.7 km wide; 54°51′N 132°42′W. Fishing at Howkan village (W) and in Elbow Bay region (NE). Sometimes called Howkan Island.

Long Island (c.1 mi/1.6 km long), E MASSACHUSETTS, in QUINCY BAY section of Boston Harbor, SE of downtown BOSTON. Connected to mainland by bridge (3 mi/4.8 km long) opened 1951, it is site of city hospital and a lighthouse. Largest island in Boston Harbor. Linked to MOON ISLAND by bridge and from Moon via causeway to SQUANTUM on mainland. Restricted to use by the city of Boston.

Long Island (□ 1,723 sq mi/4,463 sq km; 1990 population 6,861,454), 118 mi/190 km long, and 12 mi/19 km–20 mi/32 km wide, SE NEW YORK; fourth-largest island of the U.S. and the largest outside Alaska and Hawaii. Separated from STATEN ISLAND by The Narrows, from MANHATTAN and the BRONX by the EAST RIVER, and from CONNECTICUT by the LONG ISLAND SOUND; on the S is the ATLANTIC OCEAN. Comprises four counties—Kings, QUEENS, NASSAU, and SUFFOLK; Kings

(BROOKLYN) and Queens are boroughs of NEW YORK city. E Long Island has two flukelike peninsulas separated by GREAT and LITTLE PECONIC BAYS. The N fluke, terminating in Orient Point, follows part of the Harbor Hill moraine, a hilly ridge that extends W along N Long Island to The Narrows and was deposited by melting ice during the last stage of the Pleistocene epoch. The S fluke, terminating in MONTAUK POINT, follows the Ronkonkoma moraine, a somewhat older morainal ridge that extends W to join the Harbor Hill moraine at Lake Success. Low, wooded hills, capped by glacial deposits, lie N of the moraines and contrast with a broad, low-lying outwash plain to S; the highest point on the island is c.400 ft/122 m above sea level. Long beaches, backed by dunes and shallow lagoons, fringe the S shore; the N shore has low cliffs, deeply indented by bays. With no large streams, water supply is limited and is obtained from groundwater. Large recharge basins catch surplus rainwater to replenish underground supplies, and strict conservation measures have been imposed to prevent further contamination of groundwater from sewage disposal and detergents and from encroachment by seawater. Both the Dutch and the English established farming, whaling, and fishing settlements on Long Island, but it remained sparsely settled until railroad, bridges, and highways provided easy access to New York city. The LONG ISLAND EXPRESSWAY is particularly high-trafficked. Fashionable summer estates started being built on the N shore—known as "The Gold Coast"—in early 20th century. Industrial and residential growth, largely concentrated in Nassau county and the boroughs, occurred rapidly after World War II. During the same period, central Long Island communities and those along both shores experienced major residential growth. Despite the decline in farming in E Long Island, the region still retains a rural character, quite in contrast with island to the W. Sand and gravel are quarried from the island's glacial deposits. Sport and commercial fishing is important on the S and E coasts. The S shore is a popular recreational area and includes FIRE ISLAND National Seashore, Robert Moses, and JONES BEACH state parks, CONEY ISLAND, and parts of Gateway National Recreation Area. The Long Island Pine Barrens in Suffolk county, with pitch pine and scrub oak, overlie the aquifers. The Barrens have been set aside as a preserve where development is either precluded or highly regulated. Also here is an affluent residential and beach community called the Hamptons (EAST HAMPTON, SOUTHAMPTON, and WESTHAMPTON). LAGUARDIA and JFK INTERNATIONAL airports are on W Long Island; MacArthur Airport is in HOLBROOK, in central Long Island. BROOKHAVEN NATIONAL LABORATORY is in the E. State universities of New York at STONY BROOK, OLD WESTBURY, and FARMINGDALE; Long Island University, Hofstra University, New York Institute of Technology, Adelphi University, branches of City University of New York, and Suffolk Community College.

Long Island, plantation, HANCOCK county, S MAINE, on LONG ISLAND (c.2 mi/3.2 km diameter), which is site of Frenchboro village (1990 population 44), and on smaller Placentia, Black, Great Duck (lighthouse), and Little Duck (bird sanctuary) islands; c.8 mi/12.9 km S of MOUNT DESERT ISLAND, near entrance to BLUE HILL BAY.

Long Island, resort, fishing, and farming island (□ c.1.4 sq mi/3.6 sq km) off PORTLAND, MAINE, in CASCO BAY.

Long Island, volcanic island (□ 160 sq mi/416 sq km), MADANG province, N central PAPUA NEW GUINEA, circa 40 mi/64 km N of HUON PENINSULA, E NEW GUINEA island, across Vitiaz Strait; 05°20′S 147°06′E. Main town is Malala on NE side. Copra, coconuts; fish. Doughnut-shaped island has large lake-filled crater, Lake Wiscton, in center, and another crater on

Reumur Peak (4,278 ft/1,304 m) at N end. Also called AROP Island.

Long Island, BAHAMAS: see BAHAMAS.

Long Island, CANADA: see GAULTOIS ISLAND.

Long Island, NEW ZEALAND: see DUSKY SOUND.

Long Island City, area of NEW YORK city, in SW QUEENS county, SE NEW YORK, on LONG ISLAND; 40°45′N 73°55′W. An industrial and residential district, it has a waterfront on the EAST RIVER, and is connected with MANHATTAN by the Queensborough Bridge. Manufacturing, distribution, and industrial services; film and television studios. The Isamu Naguchi Garden Museum (sculpture) and P.S. 1 Contemporary Art Center are here.

Long Island Expressway (L.I.E.), SE NEW YORK, one of three major E-W highways connecting MANHATTAN to the boroughs of BROOKLYN and QUEENS, and NASSAU and SUFFOLK counties; Interstate Route 495 (I-495) from Manhattan via the Queens Midtown Tunnel to CALVERTON, 5 mi/8 km W of RIVERHEAD in E LONG ISLAND. By mid-1950s New York state recognized need for thoroughfare through central Long Island for commercial vehicles and trucks. Of monumental size for the era, the project completed the first section (from EAST RIVER to Shelter Rock Road) in 1958, and by 1962 the six-lane highway stretched through Nassau county to the Suffolk county line. Its construction spurred industrial development throughout the area, particularly around ROOSEVELT FIELD, JERICHO, FARMINGDALE, and HICKSVILLE. The system, 71 mi/114 km in length, is barely able to handle the congestion of rush-hour traffic, particularly in the western half; it has been nicknamed "the world's longest parking lot."

Long Island Sound, arm of the ATLANTIC OCEAN, c.90 mi/145 km long, and 3 mi/4.8 km–20 mi/32 km wide, separating LONG ISLAND, NEW YORK, from the SE New York mainland and CONNECTICUT; 41°00′N 73°15′W. On the W EAST RIVER joins it with NEW YORK BAY. The sound is fed from the N by the HOUSATONIC, CONNECTICUT, and THAMES rivers. It existed as a water basin before the last glacial era c.18,000 years ago, but its current dimensions were formed largely as a result of glacial advance and retreat. Historically the sound served as a marine fisheries center, including operations involving shellfish. By the last decades of the 20th century, however, pollution had become a serious problem. The sound remains a popular leisure-boating center. NEW HAVEN, NEW LONDON, and BRIDGEPORT, Connecticut, are the largest port cities; ferries run between PORT JEFFERSON, New York, and Bridgeport, and between ORIENT Point, New York, and New London. Many residential communities, some of them among the wealthiest in the nation, line the sound.

Longjing (LUNG-JING), city (□ 1,232 sq mi/3,191 sq km; 1994 estimated urban population 142,500; estimated total population 273,100), E JILIN province, CHINA, 10 mi/16 km SW of Yangji, and on railroad; 42°45′N 129°25′E. Light industry and agriculture are the main sources of income for the city. Agriculture (grains; hogs, poultry); manufacturing (synthetic fibers, paper).

Longjumeau (lon-zhu-mo), town (□ 1 sq mi/2.6 sq km), ESSONNE department, ÎLE-DE-FRANCE region, N central FRANCE, on small Yvette River, and 11 mi/18 km S of PARIS; 48°42′N 02°18′E. Glassworks; manufacturing of pharmaceuticals. In 1568, a treaty between Catholics and Protestants was signed here.

Long Key, narrow barrier island (c.6 mi/9.7 km long) and resort area, PINELLAS county, W central FLORIDA, in GULF OF MEXICO, near mouth of TAMPA BAY, 7 mi/11.3 km WSW of ST. PETERSBURG; 27°42′N 82°44′W. Connected by causeway with nearby mainland.

Longkou (LUNG-KO), city (□ 324 sq mi/842.4 sq km; 2000 population 599,386), NE SHANDONG province, CHINA, port on N coast of the Shandong peninsula, on BOHAI Bay, 110 mi/177 km N of QINGDAO; 37°41′N

120°18′E. Heavy and light manufacturing are the largest sectors of the city's economy; agriculture is also important. Main industries include coal mining, textiles, utilities, chemicals, machinery, and electrical equipment.

Longlac (LONG-lak), unincorporated town (□ 9 sq mi/ 23.4 sq km), central ONTARIO, E central CANADA, on LONG LAKE, 150 mi/241 km NE of PORT ARTHUR; 49°46′N 86°32′W. Elevation 1,035 ft/315 m. Gold mining.

Long Lake, town (2000 population 1,842), HENNEPIN county, E MINNESOTA, suburb 15 mi/24 km W of MINNEAPOLIS on S shore of Long Lake, surrounded by ORONO; 44°59′N 93°34′W. Light manufacturing. Morris T. Baker Park Reserve to NW; Lake MINNETONKA to S.

Long Lake, unincorporated village (2000 population 3,356), Lake county, NE ILLINOIS, suburb 38 mi/61 km NW of CHICAGO, 15 mi/24 km W of WAUKEGAN, on S side of Long Lake; 42°22′N 88°07′W. In agriculture area.

Long Lake, village (2006 population 51), MCPHERSON county, N SOUTH DAKOTA, 18 mi/29 km NW of LEOLA, at E end of Long Lake; 45°51′N 99°12′W.

Long Lake, hamlet, HAMILTON county, NE central NEW YORK, in the ADIRONDACKS, on E shore of Long Lake (14 mi/23 km long, c.1 mi/1.6 km wide), c.75 mi/121 km NE of UTICA; 43°57′N 74°35′W. Lake receives and discharges RAQUETTE RIVER.

Long Lake (□ 3 sq mi/8 sq km), N central NEW BRUNSWICK, CANADA, 10 mi/16 km E of GRAND FALLS; 6 mi/10 km long, 1 mi/2 km wide.

Long Lake (LAHNG) (□ 75 sq mi/195 sq km), central ONTARIO, E central CANADA, 150 mi/241 km NE of PORT ARTHUR; 45 mi/72 km long, 2 mi/3 km wide. Drains S into Lake SUPERIOR. At N end is LONGLAC.

Long Lake, SIERRA county, NE CALIFORNIA, in the SIERRA NEVADA, at base of MOUNT ELWELL, 9 mi/14.5 km NE of DOWNIEVILLE, near Gold Lake Campgrounds; fishing. In Tahoe National Forest.

Long Lake, AROOSTOOK county, N MAINE, lake in course of the ALLAGASH RIVER, in recreational area 68 mi/109 km W of PRESQUE ISLE; 4 mi/6.4 km long.

Long Lake, AROOSTOOK county, N MAINE, S of CANADIAN border, 10 mi/16 km W of VAN BUREN, and c.25 mi/40 km NNW of PRESQUE ISLE; 11 mi/18 km long; easternmost of FISH RIVER LAKES.

Long Lake, CUMBERLAND county, SW MAINE, center of resort area; 13.5 mi/23 km long, 0.5 mi/0.8 km–1 mi/1.6 km wide. Discharges S through SONGO RIVER into SEBAGO LAKE.

Long Lake, ALPENA and PRESQUE ISLE counties, NE MICHIGAN, 7 mi/11.3 km N of ALPENA, near LAKE HURON; c.8 mi/12.9 km long, 1 mi/1.6 km wide; 45°12′N 83°29′W. Resort.

Long Lake, GRAND TRAVERSE county, NW MICHIGAN, c.6 mi/9.7 km SW of TRAVERSE CITY, in forested resort area; c.4 mi/6.4 km long, 1 mi/1.6 km wide; 44°42′N 85°44′W.

Long Lake, HUBBARD county, W MINNESOTA, 4 mi/6.4 km E of PARK RAPIDS; 6 mi/9.7 km long, 0.5 mi/0.8 km wide; 46°54′N 94°59′W. Drains S past small dam through small stream into FISH HOOK RIVER, which flows short distance (0.5 mi/0.8 km) S to SHELL RIVER. Resort area. Village of Hubbard at S end.

Long Lake, BURLEIGH and KIDDER counties, S central NORTH DAKOTA, 30 mi/48 km ESE of BISMARCK; 20 mi/32 km long, 2 mi/3.2 km wide; 46°41′N 100°17′W. Used as migratory waterfowl refuge. Slade National Wildlife Refuge to NE.

Long Lake, reservoir, WASHBURN county, NW WISCONSIN, on N branch of RED CEDAR RIVER, 11 mi/18 km SE of SPOONER; c.13 mi/21 km long, with 4 mi/6.4 km NE arm at S end; 45°40′N 91°41′W. Narrow, fishhook-shaped. Lumber; resort area, fishing (pike and walleye).

Long Lake, reservoir, LINCOLN county, NE WASHINGTON, on SPOKANE RIVER, 23 mi/37 km NW of SPOKANE; 47°50′N 117°52′W. Maximum capacity 105,080 acre-ft. Formed by LONG LAKE DAM (213 ft/ 65 m high), built (1915) for power generation; also used for recreation. Spokane Indian Reservation just W.

Long Lake, CANADA: see THORHILD COUNTY NO. 7.

Long Lake Dam, WASHINGTON: see SPOKANE RIVER.

Longlaville (lon-glah-veel), commune (□ 1 sq mi/2.6 sq km) of Longwy, MEURTHE-ET-MOSELLE department, in LORRAINE, NE FRANCE, at LUXEMBOURG border, in LONGWY iron-mining and metallurgical district; 49°32′N 05°47′E.

Longli (LUNG-LEE), town, ⊙ Longli county, S central GUIZHOU province, CHINA, on railroad, and 18 mi/29 km SE of GUIYANG; 26°27′N 106°58′E. Grain, oilseeds, tobacco; rubber products.

Longling (LUNG-LING), town, ⊙ Longling county, W YUNNAN province, CHINA, on road to MYANMAR, and 50 mi/80 km SW of BAOSHAN; 24°35′N 98°41′E. Elevation 1,585 ft/483 m. Rice, sugarcane; food industry.

Long, Loch (LAHNG), inlet of the Firth of Clyde, extending NE-SW in Argyll and Bute, W Scotland; 56°05′N 04°50′W. Oil is piped 57 mi/92 km to the GRANGEMOUTH refinery.

Long Marston, ENGLAND: see MARSTON MOOR.

Longmeadow, town (2000 population 15,633), HAMPDEN county, SW MASSACHUSETTS, a residential suburb adjoining SPRINGFIELD, on the CONNECTICUT River; 42°03′N 72°34′W. Settled 1644; set off and incorporated 1783; Bay Path College is here.

Long Melford (LAHNG MEL-fuhd), town (2001 population 3,675), S SUFFOLK, E ENGLAND, on STOUR RIVER, and 3 mi/4.8 km N of SUDBURY; 52°04′N 00°43′E. Former agricultural market. Pharmaceutical manufacturing. Has 15th-century inn, Perpendicular church, and two moated Elizabethan mansions.

Longmen (LUNG-MEN), town, ⊙ Longmen county, central GUANGDONG province, CHINA, on the ZENG River, and 45 mi/72 km SSW of LIANPING; 23°44′N 114°15′E. Rice, sugarcane, oilseeds.

Longmen (LUNG-MEN), caves in the Song Mountains, NW HENAN province, CHINA, 12 mi/19 km SW of LUOYANG. Contain large rock carvings of the Buddha and rock temples, dating from C.E. 500.

Longming (LUNG-MING), town, SW GUANGXI ZHUANG AUTONOMOUS, CHINA, 45 mi/72 km NW of Zhongdong. Rice, wheat, timber, bamboo, sugarcane.

Longmont, city (2000 population 71,093), BOULDER county, N COLORADO 28 mi/45 km NNW of DENVER, on SAINT VRAIN CREEK; 40°10′N 105°06′W. Elevation 4,979 ft/1,518 m. Railroad junction. Growing trade and processing center for a rich farm area of the Saint Vrain Valley, irrigated by the Colorado–Big Thompson project. Manufacturing (vitamins, primary metal products, consumer products, manufacturing equipment, publishing and printing, building materials). Dickens Opera House; Pioneer Museum here. Roosevelt National Forest and ROCKY MOUNTAIN NATIONAL PARK to W; Barbour Ponds State Park to E. Incorporated 1885.

Long Mountain, village (2000 population 7,100), PAMPLEMOUSSES district, MAURITIUS, on hillside, 6 mi/9.6 km ENE of PORT LOUIS. Sugarcane, vegetables, fruit, ginger.

Longnan (LUNG-NAN), town, ⊙ Longnan county, S JIANGXI province, CHINA, 80 mi/129 km SSW of GANZHOU, in JIULIAN MOUNTAINS, near GUANGDONG border; 24°54′N 114°47′E. Rice; nonferrous-ore mining, logging, timber processing.

Longniddry (LAHNG-nid-dree), village (2001 population 2,613), East Lothian, E Scotland, on Firth of Forth, 10 mi/16 km E of EDINBURGH; 55°58′N 02°53′W. Good beaches; golf. Formerly in Lothian, abolished 1996.

Area is shown by the symbol □, and capital city or county seat by ⊙.

Longobucco (long-go-BOOK-ko), town, COSENZA province, CALABRIA, S ITALY, near Trionto River, 9 mi/14 km S of ROSSANO; 39°27′N 16°37′E. In fruit-growing region; domestic weaving of silk, cotton, wool.

Longonjo (lon-GON-jo), town, HUAMBO province, W ANGOLA, on Benguela railroad, and 35 mi/56 km WSW of HUAMBO; 12°50′S 15°15′E. Elev. 4,650 ft/1,417 m.

Longonot (lon-GO-not), village, RIFT VALLEY province, W central KENYA, in GREAT RIFT VALLEY at foot of Longonot volcano (elevation 9,350 ft/2,850 m), and 10 mi/16 km SSE of NAIVASHA; 00°53′S 36°30′E. On railroad. Wheat-growing center; coffee, sisal, corn, fruits.

Longoro (lon-GO-ro), town, BRONG-AHAFO region, GHANA; 08°11′N 01°53′W. Located 15 mi/24 km NW of KINTAMPO, 35 mi/56 km downstream of BUI DAM on MOUHOUN RIVER.

Longos, Greece: see SITHONIA.

Longotoma (lon-go-TO-mah), village, SAN FELIPE DE ACONCAGUA province, VALPARAISO region, central CHILE, 9 mi/15 km NW of LA LIGUA; 32°23′S 71°22′W. On railroad. Agriculture (potatoes, oats; livestock).

Longowal, INDIA: see LAUNGOWAL.

Longoza, Bulgaria: see KAMCHIYA RIVER.

Long Phu (LOUNG POO), town, HAU GIANG province, S VIETNAM, in MEKONG Delta, on BASSAC RIVER, and 10 mi/16 km E of SOC TRANG; 09°37′N 106°10′E. Rice, fishery. Khmer minority. Formerly BANGLONG or BANG LONG.

Long Pine, village (2006 population 342), BROWN county, N NEBRASKA, 8 mi/12.9 km E of AINSWORTH; 42°31′N 99°42′W. Resort. Livestock; vegetables; lumber. Long Pine State Recreation Area and dam nearby.

Long Point (LAHNG), peninsula, S ONTARIO, E central CANADA, extends 20 mi/32 km E into Lake ERIE, ESE of PORT ROWAN; 1 mi/2 km–4 mi/6 km wide; 42°34′N 80°15′W.

Long Point, village (2000 population 247), LIVINGSTON county, N central ILLINOIS, 16 mi/26 km WNW of PONTIAC; 41°00′N 88°53′W. In agricultural area.

Long Point, SE MASSACHUSETTS, sandspit on N tip of CAPE COD; curves SE to shelter PROVINCETOWN harbor. Has lighthouse at 42°02′N 70°12′W.

Long Pond, village, SOMERSET county, W MAINE, on Long Pond (8 mi/12.9 km long, 1 mi/1.6 km wide), and 27 mi/43 km NW of GREENVILLE. In lumbering, hunting, fishing area.

Long Pond, reservoir, PLYMOUTH and BRISTOL counties, SE MASSACHUSETTS, on short stream draining N to ASSAWOMPSETT POND, 9 mi/14.5 km SE of TAUNTON; c.3.5 mi/5.6 km long; 41°48′N 70°56′W. Heaven Heights village on W shore.

Long Pond, NEW YORK: see WILLSBORO.

Longpont-sur-Orge (lon-pon–syoor–orzh), town (1 sq mi/2.6 sq km), ESSONNE department, ÎLE-DE-FRANCE region, N central FRANCE; 48°38′N 02°17′E.

Longport, borough (2006 population 1,088), ATLANTIC county, SE NEW JERSEY, on the ATLANTIC coast, 5 mi/8 km SW of ATLANTIC CITY; 39°18′N 74°31′W.

Long Prairie, town (2000 population 3,040), ⊙ TODD county, central MINNESOTA, on LONG PRAIRIE RIVER, 25 mi/40 km W of LITTLE FALLS; 45°58′N 94°51′W. Elevation 1,295 ft/395 m. Trading point in agricultural area (grain, potatoes, beans; livestock, poultry; dairying); manufacturing (animal protein products, food processing, dump trailers, printing and publishing). Winnebago Indian Agency here 1848–1855. Plotted 1867, incorporated 1883.

Long Prairie River, 120 mi/193 km long; rises in chain of lakes, E DOUGLAS county, W MINNESOTA (Lake CARLOS is source of main stream; 45°58′N 95°19′W); flows generally E in very serpentine course, passes Lake OSAKIS on N (man-made channel connects river with lake), past LONG PRAIRIE, then N past Brownesville,

and NE to CROW WING RIVER 2 mi/3.2 km SE of MOTLEY.

Longquan (LUNG-CHUAN), city (1,180 sq mi/3,068 sq km), SW ZHEJIANG province, CHINA, near FUJIAN and JIANGXI borders, 55 mi/89 km WSW of LISHUI, in mountain area, and on headstream of the WU RIVER; 28°09′N 119°14′E. Agriculture and light industry are the largest sectors of the city's economy. Main agriculture (grains, tea; hogs); manufacturing (timber). Formerly a town, Longquan was redesignated as a city in 1990.

Long Range, mountain range, c.300 mi/480 km long, SW NEWFOUNDLAND AND LABRADOR, CANADA, rising to 2,672 ft/814 m in the LEWIS HILLS. It forms the GREAT NORTHERN PENINSULA of NW Newfoundland. Part of the Appalachian system, the range consists of parallel ridges that rise steeply from the coast and slope gently E. A depression, of which GRAND LAKE and St. George's Bay are part, divides Long Range into two sections. Timber. GROS MORNE NATIONAL PARK is here.

Longreach (LAHNG-REECH), town, central QUEENSLAND, AUSTRALIA, 445 mi/716 km W of ROCKHAMPTON on Landsborough Highway; 23°27′S 144°15′E. Opal-mining district; mines at Opalton and Mayneside to W. Beef cattle; wool. Stockman's Hall of Fame; Outback Heritage Center, housed in the original Qantas airlines hanger; Qantas' first commercial flight departed from Longreach, February 1921. Airport.

Longridge (LAHNG-rij), town (2001 population 7,490), N central LANCASHIRE, NW ENGLAND, 6 mi/9.7 km NE of PRESTON; 53°50′N 02°35′W. Cheese processing. Previously stone quarrying. Site of Preston reservoirs.

Long River (LUNG), 60 mi/97 km long, in S FUJIAN province, CHINA; rises in several branches in the coastal ranges; flows SE past NANJING and LONGHAI, forming common estuary with Jiulong River (left) on Xiamen Bay of TAIWAN STRAIT.

Long Sault Island (SOO), ST. LAWRENCE county, New York, in the SAINT LAWRENCE River (Long Sault rapids here), at ONTARIO border, 3 mi/4.8 km N of MASSENA; c.4.5 mi/7.2 km long, 0.25 mi/0.4–1 mi/1.6 km wide; 44°58′N 74°55′W.

Long Sault Rapids (SOO), in the SAINT LAWRENCE RIVER, SE ONTARIO, E central CANADA, 12 mi/19 km WSW of CORNWALL; 9 mi/14 km long; 48°38′N 94°04′W. Bypassed by Cornwall Canal (11 mi/18 km long).

Longshan (LUNG-SHAN), town, ⊙ Longshan county, northwesternmost HUNAN province, CHINA, on HUBEI border, 85 mi/137 km NW of YUANLING; 29°27′N 109°23′E. Tea, tobacco, medicinal herbs; tobacco processing, nonferrous-ore mining, metal smelting.

Longsheng (LUNG-SHENG), town, ⊙ Longsheng county, NE GUANGXI ZHUANG AUTONOMOUS REGION, CHINA, 30 mi/48 km NNW of GUILIN, near HUNAN border; 25°48′N 110°00′E. Grain; building materials, logging, timber processing. Population mostly of Miao and Yao tribes.

Longs Peak (14,255 ft/4,345 m), LARIMER and BOULDER counties, N COLORADO, in the FRONT RANGE of the ROCKY MOUNTAINS, just E of CONTINENTAL DIVIDE, in SE ROCKY MOUNTAIN NATIONAL PARK. On E side of its snow-capped peak is a 2,000 ft/610 m drop to Chasm Lake. Peak named for explorer Stephen H. Long.

Longstone Island, ENGLAND: see FARNE ISLANDS.

Long Stop Hill, N Tunisia, near left bank of MAJARDAH R., 6 mi/9.7 km N of Madjez al-Bab. Elev. 951 ft/290 m. Scene of bitter fighting in last phase of Tunisian campaign (April–May 1943) in World War II.

Long Strait, joins EAST SIBERIAN and CHUKCHI seas at 70°00′N 177°00′E, separating WRANGEL ISLAND from mainland SAKHA REPUBLIC, RUSSIAN FAR EAST; 85 mi/137 km wide. Named for the U.S. explorer George De Long, who first navigated here in 1867. Also known as De Long Strait.

Longstreet, village (2000 population 163), DE SOTO parish, NW LOUISIANA, 31 mi/50 km SSW of SHREVEPORT; 32°06′N 93°57′W. In agricultural area.

Long Sutton (LAHNG SUHT-uhn), town (2001 population 6,461), LINCOLNSHIRE, E ENGLAND, 5 mi/8 km ESE of HOLBEACH; 52°47′N 00°08′E. Agricultural market in fruit-growing region. Has church of Norman origin with 15th-century additions.

Long Thanh (LOUNG TAHN), town, DONG NAI province, S VIETNAM, 17 mi/27 km E of HO CHI MINH CITY; 10°47′N 106°59′E. Market and transportation center; light manufacturing, rubber plantations and processing. Airport. Formerly Longthanh.

Long Tom River, 50 mi/80 km long, in LANE county, W OREGON; rises 10 mi/16 km W of Beneta, on E slopes of COAST RANGE; flows E past VENETA and into FERN RIDGE RESERVOIR, then N past Munroe to WILLAMETTE RIVER 12 mi/19 km S of CORVALLIS.

Longton, village (2000 population 394), ELK county, SE KANSAS, on ELK RIVER, and 12 mi/19 km SE of HOWARD; 37°22′N 96°04′W. Shipping and trading point in grain and cattle area.

Longton, ENGLAND: see STOKE-ON-TRENT.

Longtown (LAHNG-town), town (2006 population 3,000), CUMBRIA, NW ENGLAND, on ESK RIVER, on Scottish border, and 8 mi/12.9 km N of CARLISLE; 55°01′N 02°58′W. The E end of Scot's Dyke, a wooded fortified line 3 mi/4.8 km long, built in 1552, is 3 mi/4.8 km N of the town.

Longtown, town (2000 population 76), PERRY county, E MISSOURI, 6 mi/9.7 km SE of PERRYVILLE; 37°40′N 89°46′W.

Long Trail: see GREEN MOUNTAINS.

Longueau (lon-zhuh), outer SE suburb (1 sq mi/2.6 sq km; 2004 population 5,179) of AMIENS, SOMME department, PICARDIE region, N FRANCE, on the AVRE near its mouth into the SOMME; 49°52′N 02°21′E. Railroad junction and railroad yards. Airport.

Longue, Île (lon-guh, eel), peninsula extending N into BREST ROADS (inlet of ATLANTIC OCEAN), FINISTÈRE department, BRITTANY, NW FRANCE, c.5 mi/8 km N of CROZON, on CROZON PENINSULA; 43°02′N 06°05′E. A nuclear submarine base was established here in 1970s.

Longue Island (LONG-guh) or **Long Island,** one of the SEYCHELLES, in MAHÉ GROUP, off NE coast of MAHÉ ISLAND, 4 mi/6.4 km E of VICTORIA; 04°37′S 55°31′E; 0.5 mi/0.8 km long, 0.25 mi/0.4 km wide. Closed to the public.

Longué-Jumelles (lon-gai–zhou-MEL), town (37 sq mi/96.2 sq km), MAINE-ET-LOIRE department, PAYS DE LA LOIRE region, FRANCE, 22 mi/35 km E of ANGERS.

Longue-Pointe-de-Mingan (long–pwant–duh–MIN-guhn), village (161 sq mi/418.6 sq km; 2006 population 518), CÔTE-NORD region, E QUEBEC, E CANADA, 8 mi/14 km from RIVIÈRE-SAINT-JEAN; 50°16′N 64°09′W.

Longue-Rive (LONG–REEV), village (114 sq mi/296.4 sq km; 2006 population 1,328), CÔTE-NORD region, E QUEBEC, E CANADA; 48°33′N 69°15′W. Name changed from Saint-Paul-du-Nord–Sault-au-Mouton in 1998.

Longueuil (lon-GUH-ee), city (43 sq mi/111.8 sq km), MONTÉRÉGIE region, S QUEBEC, E CANADA, on the SAINT LAWRENCE RIVER opposite MONTREAL; 45°32′N 73°31′W. Residential and industrial suburb of Montreal. It annexed Montreal South in 1961, and it merged with the city of JACQUES-CARTIER in 1969. In 2002 it restructured and is composed of the boroughs of GREENFIELD PARK, SAINT-HUBERT, and VIEUX-LONGUEIL. BOUCHERVILLE, BROSSARD, SAINT-BRUNO-DE-MONTARVILLE, and SAINT-LAMBERT voted to regain their independence as of 2006, but along with Longueuil, form the agglomeration of Longueuil. The city is part of the Metropolitan Community of Montreal (*Communauté Métropolitaine de Montréal*).

Longuyon (lon-gwee-yon), town (□ 11 sq mi/28.6 sq km; 2004 population 5,782), MEURTHE-ET-MOSELLE department, LORRAINE region, NE FRANCE, on CHIERS RIVER, and 9 mi/14.5 km SW of LONGWY; 49°26′N 05°36′E. Railroad junction. Metalworking.

Longvic (lon-VEEK), town (□ 4 sq mi/10.4 sq km), CÔTE-D'OR department, in BURGUNDY, NW FRANCE, S suburb of DIJON on OUCHE RIVER; 47°17′N 05°04′E. Military airbase. Electrical appliances. Dijon-Longvic international airport.

Longview (LONG-vyoo), city (2006 population 76,524), ⊙ GREGG county, extends E into HARRIS county, E TEXAS, 22 mi/35 km W of MARSHALL, and 34 mi/55 km NE of TYLER, near Sabine River; 32°31′N 94°45′W. Elevation 339 ft/103 m. A growing manufacturing, business, and distributing center for the rich E Texas oil field. Highly industrialized city located in major oil-producing area. Cattle, race horses; timber. Manufacturing (RR cars, oil field equipment, travel trailers, natural gas processing, printing, machinery). Also a livestock center. City boomed with the discovery of the oil field in 1930. It is the seat of LeTourneau University. Gregg county airport to SE. Annual horse and cattle shows are held in Longview. Lake Cherokee reservoir to SE; Lone Star Speedway. Settled early 1800s; incorporated 1872.

Longview, city (2006 population 36,767), Cowlitz county, SW WASHINGTON, a port of entry on the CO-LUMBIA River, at mouth of COWLITZ RIVER; 46°09′N 122°57′W. It is a railroad junction and transportation center, with the Lewis and Clark Bridge (highway) across the Columbia to OREGON. Its manufacturing includes plastic, paper, and wood products; aluminum; steel foundry; logging. The city was founded in 1922 as a lumber town on the site of the historic settlement Monticello, which had been swept away by a flood in 1867. Twin city with KELSO, adjoins LONG-VIEW to NE. The Lower Columbia College (2-year) is here; Lake Sacajawea at center of city; Seaquest State Park, on Silver River reservoir to NE. Incorporated 1924.

Long View (LAHNG VYOO), town (□ 3 sq mi/7.8 sq km; 2006 population 4,889), BURKE and CATAWBA counties, W central NORTH CAROLINA, residential suburb 2 mi/3.2 km W of HICKORY; 35°43′N 81°23′W. Hickory Regional Airport to NW. Incorporated 1907.

Longview (LAHNG-vyoo), village (2001 population 300), SW ALBERTA, W CANADA, 12 mi/19 km S of BLACK DIAMOND, in FOOTHILLS NO. 31 municipal district; 50°32′N 114°14′W. Incorporated 1964.

Longview, village, CHAMPAIGN county, E ILLINOIS, 18 mi/29 km SSE of CHAMPAIGN; 39°53′N 88°03′W. In agricultural area.

Longview, village, OKTIBBEHA county, E MISSISSIPPI, 7 mi/11.3 km WSW of STARKVILLE. Noxubee National Wildlife Refuge to S.

Longville, village (2000 population 180), CASS county, N central MINNESOTA, 42 mi/68 km N of BRAINERD, at NE end of WOMAN LAKE, S of boundary of Leech Lake Indian Reservation, and at S edge of Chippewa National Forest; 46°59′N 94°13′W. Dairying; livestock; oats; timber. LEECH LAKE to NW; numerous small lakes in area.

Longwood (LAHNG-wuhd), city (□ 6 sq mi/15.6 sq km; 2000 population 13,745), SEMINOLE county, central FLORIDA; suburb 11 mi/18 km N of ORLANDO; 28°42′N 81°21′W. Has experienced major growth since 1975.

Longwood (LONG-wud), settlement (2006 population 665), central SAINT HELENA island, in S ATLANTIC, on the interior plateau, on highway, 2 mi/3.2 km SE of JAMESTOWN; 15°57′S 05°41′W. Elevation 1,305 ft/397 m. Famed as residence (1815–1821) of Napoleon, who was confined here by the British until his death. His body was removed (1840) to France. Site of government-owned flax-fiber mill.

Longwood Park (LAHNG-wud PAHRK), unincorporated village, RICHMOND county, S NORTH CAROLINA, 3 mi/4.8 km N of HAMLET.

Longwu (LUNG-WU), town, S YUNNAN province, CHINA, 35 mi/56 km NW of JIANSHUI, in mountain region; 24°01′N 102°24′E. Timber, rice, tobacco.

Longwy (lon-gwee), town (□ 2 sq mi/5.2 sq km), MEURTHE-ET-MOSELLE department, LORRAINE region, NE FRANCE, near the Belgian and LUXEMBOURG borders; 49°31′N 05°46′E. The center of the Lorraine iron and steel industry, Longwy has been diversifying its economic base into ceramics and steel products. Longwy-Haut, c.500 ft/152 m above the lower, newer business section of town, has kept the 17th-century fortifications built by Vauban.

Longxi (LUNG-SEE), town, ⊙ Longxi county, SE GANSU province, CHINA, 90 mi/145 km SE of LANZ-HOU; 35°03′N 104°38′E. Agriculture (grain, medicinal herbs, oilseeds; sheep); manufacturing (food and beverages, engineering, papermaking, nonferrous-metal smelting).

Long Xian (LUNG SIAN), town, ⊙ Long Xian county, SW SHAANXI province, CHINA, 30 mi/48 km NW of BAOJI, and on road to LANZHOU; 34°54′N 106°51′E. Grain, oilseeds, tobacco, medicinal herbs; food processing, building materials, engineering.

Long Xian, Wengyuan county, CHINA: see WENGYUAN.

Long Xuyen (LOUNG soo-YEN), city, ⊙ AN GIANG province, S VIETNAM, in MEKONG Delta, on right bank of Song Hau (Hau River), and 90 mi/145 km WSW of HO CHI MINH CITY; 10°23′N 105°28′E. Market hub; administrative complex; transportation center. Highly irrigated area; major rice and agriculture (sugarcane, corn, beans, peanuts, tobacco, fruit) center; sericulture. Former Khmer territory; passed (18th century) to Vietnamese. Kinh population with Khmer minority. Formerly Longxuyen.

Longyan (LUNG-YAN), city (□ 1,034 sq mi/2,678 sq km; 1994 estimated urban population 217,900; estimated total population 433,000), SW FUJIAN province, CHINA, 60 mi/97 km SE of CHANGTING; 25°10′N 117°00′E. Light and heavy industries form the economic core of the city. Crop growing and animal husbandry are also important. Grains, sugar; hogs; manufacturing (food, tobacco, textiles, iron and steel, machinery, transportation equipment); coal mining.

Longyao (LUNG-YOU), town, ⊙ Longyao county, SW HEBEI province, CHINA, 25 mi/40 km NE of XINGTAI, near BEIJING-WUHAN railroad; 37°21′N 114°46′E. Grain, cotton; textiles.

Longyearbyen (LAWNG-yir-BU-uhn), town and administrative center of SVALBARD, on ISFJORDEN, SPITSBERGEN island, NORWAY. A coal-mining settlement founded (1905) by an American company and named after the American miner J. M. Longyear. Its coal mines were transferred to a Norwegian company in 1916. Destroyed (September 1943) by German battleships but was quickly rebuilt.

Longyou (LUNG-YO), town, ⊙ Longyou county, SW ZHEJIANG province, CHINA, on tributary of QIANTANG RIVER, and 28 mi/45 km W of JINHUA; 29°00′N 119°10′E. On railroad. Rice, oilseeds; papermaking, food and beverages, chemicals, building materials.

Longzhen (LUNG-JEN), town, W HEILONGJIANG province, NE CHINA, 50 mi/80 km NNE of BEI'AN; 48°43′N 126°45′E. Oilseeds, sugar beets; lumbering.

Longzhou (LUNG-JO), town, Longzhou county, SW Guangxi Uygur Autonomous Region, CHINA, port on LI RIVER, and 80 mi/129 km WSW of NANNING, near VIETNAM border; 22°24′N 106°50′E. Major transit center for trade with Vietnam; rice, oilseeds, sugarcane; sugar refining. Opened to foreign trade in 1887.

Loni (LO-nee), town, GHAZIABAD district, NW UTTAR PRADESH state, N central INDIA, 7 mi/11.3 km NNE of DELHI. Wheat, gram, jowar, sugarcane, oilseeds.

Lonigo (lo-NEE-go), town, VICENZA province, VENETO, N ITALY, at SW foot of MONTI BERICI, 13 mi/21 km SW of VICENZA; 45°23′N 11°23′E. Agricultural center; manufacturing (metalworking machinery, cable, electric drills, fabricated products, chemicals, leather goods, silk and cotton textiles). Noted for its annual horse fair.

Löningen (LO-ning-uhn) town, LOWER SAXONY, NW GERMANY, 14 mi/23 km SW of CLOPPENBURG; 52°44′N 07°45′E. Chemical industry; printing; food processing. Chartered 1982.

Lonja River (LO-nyah), c.100 mi/161 km long, central CROATIA; rises 9 mi/14.5 km SSW of Varaždin; flows S, past IVANIC Grad, to SAVA River 9 mi/14.5 km S of KUTINA. Called Trebeš in lower course. Receives Česma, ILOVA, and PAKRA rivers. LONJSKO POLJE NATURE PARK on the floodplain between LONJA and Sava rivers.

Lonjsko Polje Nature Park [Croatian=Lonja field] (□ 196 sq mi/509.6 sq km), in central and E CROATIA, in POSAVINA region, on floodplain along SAVA River, S of KUTINA and between SISAK and Nova Gradiška. Wetlands; winter bird sanctuary.

Lonkin (long-KIN), village, KAMAING township, KA-CHIN STATE, MYANMAR, on UYU RIVER (left affluent of CHINDWIN RIVER), and 60 mi/97 km WNW of MYITKYINA; 25°39′N 96°22′E. Jade-mining center, linked by road with MOGAUNG.

Lonneker (LAW-nuh-kuhr), town, OVERIJSSEL province, E NETHERLANDS, 2 mi/3.2 km ENE of ENSCHEDE; 52°15′N 06°55′E. Lonneker Berg hill to N; airport to NW. Dairying; livestock; vegetables, grain; light manufacturing.

Lonoke (LO-nok), county (□ 802 sq mi/2,085.2 sq km; 2006 population 62,902), central ARKANSAS; ⊙ LO-NOKE; 34°45′N 91°52′W. Bounded N by Cypress Bayou; drained by Wabbaseka, Meto, and Two Prairie (forms part of E boundary) bayous. Agriculture (rice, cotton, wheat, strawberries, soybeans; livestock). Manufacturing of wood products; rice milling. Toltec Mounds State Park (on Mound Lake) in SW; AR-KANSAS RIVER to W. Formed 1873.

Lonoke (LO-nok), city (2000 population 4,287), ⊙ LO-NOKE county, central ARKANSAS, 21 mi/34 km E of LITTLE ROCK, near Bayou Meto; 34°47′N 91°54′W. Trade center and shipping point for area producing rice, cotton, pecans, strawberries. Manufacturing (wooden cabinets, apparel, ordnance, safety gates).

Lonoli (lo-NO-lee), village, Équateur province, central CONGO, near TSHUAPA RIVER, 50 mi/80 km ESE of BOENDE; 00°28′S 21°38′E. Elev. 1,368 ft/416 m. Palm-oil milling, hardwood lumbering.

Lonquimay (lon-kee-MEI), village, E LA PAMPA province, ARGENTINA, on railroad and 40 mi/64 km ENE of SANTA ROSA; 36°28′S 63°37′W. Grain and livestock center.

Lonquimay (lon-KEE-mei), village, ⊙ Lonquimay co-muna, MALLECO province, ARAUCANIA region, S central CHILE, 95 mi/153 km SE of ANGOL, on trans-Andean road to ZAPALA, ARGENTINA; 38°26′S 71°14′W. Health resort; livestock. Lonquimay Volcano (9,480 ft/2,890 m) is 20 mi/32 km NW.

Lons (LON), town (□ 5 sq mi/13 sq km), PYRÉNÉES-ATLANTIQUES department, AQUITAINE region, SW FRANCE, NW suburb of PAU.

Lönsboda (LUHNS-BOO-dah), town, SKÅNE county, S SWEDEN, 25 mi/40 km N of KRISTIANSTAD; 56°24′N 14°19′E.

Lønsdal (LUHNS-dahl), village, NORDLAND county, N central NORWAY, in the LØNSDAL (valley), within ARCTIC CIRCLE, on tributary of SALTELVA River, and 40 mi/64 km SE of BODØ. Tourist resort; on railroad.

Lonsdale, town (2000 population 1,491), RICE county, S MINNESOTA, 35 mi/56 km SW of ST. PAUL; 44°28′N 93°25′W. Grain; livestock; dairying; manufacturing (die-cutting and laminating, light manufacturing). Small lakes in area.

Area is shown by the symbol □, and capital city or county seat by ⊙.

Lonsdale (LAHNZ-dail), village (2000 population 118), GARLAND county, central ARKANSAS, 14 mi/23 km E of HOT SPRINGS; 34°32′N 92°48′W.

Lonsdale, industrial village in CUMBERLAND and LINCOLN towns, PROVIDENCE county, NE RHODE ISLAND, on BLACKSTONE RIVER (bridged here), and 6 mi/9.7 km N of PROVIDENCE. Former textile center.

Lonsdale, Point (LAHNZ-dail), S VICTORIA, SE AUSTRALIA, in BASS STRAIT; forms W side of entrance to PORT PHILLIP BAY; 38°18′S 144°37′E. Surrounded by reefs; lighthouse overlooking. The Rip, a notoriously turbulent stretch of water. On rocky outcrop forming W head of Port Phillip Bay is Point Lonsdale township, resort, 63 mi/101 km S of MELBOURNE.

Lons-le-Saunier (lon–luh–so–nyai), town (□ 3 sq mi/7.8 sq km), ⊙ JURA department, FRANCHE-COMTÉ region, E FRANCE, at W foot of the JURA MOUNTAINS, 70 mi/113 km NNE of LYON; 46°40′N 05°33′E. A saltwater spa since Roman times, the town has food (cheese, sparkling wine) and textile industries, and is a trade and administrative center, with extensive hospital facilities. Parts of its Romanesque church date from the 11th century. Rouget de Lisle, poet and composer of the *Marseillaise*, was born here in 1760. In the mountains, 4 mi/6.4 km E, the Cirque of Baume offers a spectacular panorama onto the BURGUNDY hills (W) and intervening lowlands. The picturesque village of BAUME-LES-MESSIEURS lies in a deeply incised valley.

Lontra (LON-trah), town (2007 population 7,946), N central MINAS GERAIS state, BRAZIL, 42 mi/68 km S of JANUÁRIA; 15°52′S 44°15′W.

Lontras (LON-trahs), town (2007 population 9,180), E SANTA CATARINA state, BRAZIL, 38 mi/61 km SW of BLUMENAU; 27°10′S 49°33′W. Rice, corn, manioc; livestock.

Löntsch, SWITZERLAND: see NETSTAL.

Lontué (lon-too-AI), town, TALCA province, MAULE region, central CHILE, near LONTUÉ RIVER, 4 mi/6 km NNE of MOLINA; 35°04′S 71°16′W. On railroad. Wine, wheat, barley, tobacco. Hydroelectric station nearby.

Lontué River (lon-too-AI), c.75 mi/121 km long, central CHILE; rises in the ANDES near ARGENTINA border; flows along CURICÓ and TALCA province borders of MAULE region to join TENO RIVER 5 mi/8 km WNW of CURICÓ, forming MATAQUITO RIVER. The upper course is called Río Colorado. Used for hydroelectric power and irrigation. Known for grapes grown along its valley.

Lontzen (LAHNT-zuhn), commune (2006 population 5,109), Verviers district, LIÈGE province, E BELGIUM, 8 mi/12.9 km S of AACHEN, near German border.

Looc (lo-OK), town, SW TABLAS ISLAND, ROMBLON province, PHILIPPINES, on Looc Bay (inlet of TABLAS STRAIT); 12°19′N 122°01′E. Agricultural center (rice, coconuts).

Loo-choo Islands, JAPAN: see RYUKYU ISLANDS.

Looe (LOO), town (2001 population 5,280), E CORNWALL, SW ENGLAND, on Looe Bay of the ENGLISH CHANNEL, at mouth of Looe River, and 14 mi/23 km SE of BODMIN; 50°21′N 04°27′W. Includes fishing ports and tourist resorts of East Looe, on E bank of Looe River, and West Looe, opposite. West Looe has 16th-century inn, former pirates' and smugglers' resort, and 14th-century church. Just offshore is Looe Island.

Looe, East, ENGLAND: see LOOE.

Loogootee (lo-GO-dee), city (2000 population 2,741), MARTIN county, SW INDIANA, near East Fork of WHITE RIVER, 7 mi/11.3 km W of SHOALS; 38°41′N 86°55′W. In agricultural area (wheat, corn, hay; livestock); lumber milling; manufacturing (animal feed, textiles). Bituminous-coal mines to W.

Lookeba (LOOK-uh-bah), village (2006 population 131), CADDO county, W central OKLAHOMA, 25 mi/40 km WSW of EL RENO, on Sugar Creek; 35°21′N 98°22′W. In agricultural area. Red Rock Canyon State Park to N.

Looking Glass River, c.65 mi/105 km long, S central MICHIGAN; rises in NW LIVINGSTON county; flows N and W, past DE WITT, to GRAND RIVER at PORTLAND; 42°46′N 84°08′W.

Lookout, village, PIKE county, E KENTUCKY, in the CUMBERLAND MOUNTAINS, 11 mi/18 km S of PIKEVILLE. Bituminous coal.

Lookout, Cape (LUK-out), N ONTARIO, E central CANADA, on HUDSON BAY, 65 mi/105 km W of entrance of JAMES BAY; 55°18′N 83°55′W.

Lookout, Cape, TILLAMOOK county, NW OREGON, coastal promontory c.10 mi/16 km SW of TILLAMOOK. Site of Cape Lookout State Park.

Lookout, Cape (LUK-out, CAIP), promontory, CARTERET county, E NORTH CAROLINA, at SW end of HATTERAS ISLAND, in OUTER BANKS (sand barrier islands on ATLANTIC OCEAN coast); 34°35′N 76°32′W. MOREHEAD CITY 14 mi/23 km SE of CAPE LOOKOUT NATIONAL SEASHORE. Cape Lookout Lighthouse (1859); ferry to HARKERS ISLAND, near mainland.

Lookout Heights, KENTUCKY: see FORT WRIGHT.

Lookout, Mount, AUSTRALIA: see ABERFELDY.

Lookout Mountain, residential town (2006 population 1,881), HAMILTON county, SE TENNESSEE, along N ridge of LOOKOUT MOUNTAIN at GEORGIA state line, suburb 3 mi/5 km SW of CHATTANOOGA; reached by road and cable railroad from Chattanooga; 34°59′N 85°21′W. Elevation 2,126 ft/648 m. Has limestone caves, interesting rock formations, Adolph S. Ochs Observatory and Museum (dedicated 1940). Ruby Falls, an underground waterfall, and Rock City Gardens, were made famous by barn-roof advertising. Civil War "Battle above the Clouds" fought here. Part of surrounding area is in CHICKAMAUGA AND CHATTANOOGA NATIONAL MILITARY PARK.

Lookout Mountain, suburb (2000 population 1,617), WALKER county, GEORGIA, a suburb of CHATTANOOGA, TENNESSEE, 3 mi/4.8 km S across Tennessee-Ga. state line from LOOKOUT MOUNTAIN, Tennessee; 34°58′N 85°22′W. Named for the mountain, this village has become a tourist destination due to the beautiful vistas from the mountain. The world's first miniature golf course was built in Lookout Mountain in the 1920s. Manufacturing of display boards, lamps.

Lookout Mountain, narrow ridge (c.2,000 ft/610 m) of the CUMBERLAND PLATEAU in TENNESSEE, GEORGIA, and ALABAMA, parallel to SAND MOUNTAIN; from Moccasin Bend of TENNESSEE RIVER near CHATTANOOGA, Tennessee, extends c.75 mi. SSW, across NW corner of Georgia, to GADSDEN, Alabama; summit, 35°00′N 85°20′W. Cable railroad and road ascend NE end (elevation 2,392 ft/729 m); a popular tourist area with a magnificent view, interesting limestone caverns, notable Rock City Gardens, and Adolph S. Ochs Observatory and Museum This portion of ridge contains residential town of LOOKOUT MOUNTAIN, was site (1863) of Civil War Battle of Lookout Mountain ("Battle above the Clouds"), and is partly included in CHICKAMAUGA AND CHATTANOOGA NATIONAL MILITARY PARK.

Lookout Mountain, peak (6,505 ft/1,983 m) in the SIERRA ANCHA, GILA county, central ARIZONA, c.60 mi/97 km NE of PHOENIX.

Lookout Mountain, peak (7,375 ft/2,248 m) in FRONT RANGE, JEFFERSON county, N central COLORADO, 2 mi/3.2 km SW of GOLDEN. Reached by Lariat Loop road from N, by Lookout Mountain road from S. On summit is grave of William F. Cody (Buffalo Bill).

Lookout Mountain, peak (9,128 ft/2,782 m) in W NEW MEXICO, in ZUÑI MOUNTAINS, 34 mi/55 km SE of GALLUP, NE of CONTINENTAL DIVIDE, in Cibola National Forest.

Lookout, Point, low headland, SAINT MARYS county, S MARYLAND, at N side of mouth of the POTOMAC on CHESAPEAKE BAY; lighthouse (built 1830). Visitors can picnic, swim, and fish in 700 acres/283 ha Point Lookout State Park, where a notorious prison housed 20,000 Confederate soldiers during the Civil War, of whom 3,384 died. One survivor was Sidney Lanier, the poet.

Lookout Point Lake, reservoir (□ 7 sq mi/18.2 sq km), LANE county, W central OREGON, on Middle Fork WILLAMETTE RIVER, on western edge of Willamette National Forest, 19 mi/31 km SE of EUGENE; 43°55′N 122°45′W. Maximum capacity 477,700 acre-ft. Formed by Lookout Point Dam (276 ft/84 m high), built (1953) by Army Corps of Engineers and U.S. Department of Agriculture Forest Service for flood control.

Lookout Shoals Lake (LUK-out SHOLZ LAIK), reservoir, on IREDELL-CATAWBA county border and in ALEXANDER county, W central NORTH CAROLINA, on CATAWBA RIVER, 15 mi/24 km E of HICKORY, and 3 mi/4.8 km N of CATAWBA; c.10 mi/16 km long; 35°44′N 81°04′W. Maximum capacity 366,840 acre-ft; rises to 838.1 ft/255.5 m; has 37 mi/59.5 km of shoreline. Formed by OXFORD DAM (97 ft/30 m high), built by Duke Power Company in 1915 for power generation. Two public boating access areas. Lake NORMAN reservoir below dam.

Loolmalasin, Mount (lo-ol-mah-lah-SEEN), extinct volcano, ARUSHA region, N TANZANIA, 62 mi/100 km WNW of ARUSHA, in Ngorongoro Conservation Area; 03°41′S 35°48′E. NGORONGORO CRATER to SW; Embului and Embagai craters to N.

Looma (LOO-muh), hamlet, central ALBERTA, W CANADA, 14 mi/23 km from LEDUC, in LEDUC COUNTY; 53°22′N 113°15′W.

Loomis, city (2000 population 6,260), PLACER county, central CALIFORNIA, in Sacramento Valley, 23 mi/37 km NE of SACRAMENTO; 38°49′N 121°11′W. Ships plums, kiwi fruit, vegetables, walnuts; sheep, cattle; manufacturing (lumber, sheet-metal work, screw-machine products). Folsom Lake State Recreation Area (AMERICAN RIVER) to SE.

Loomis, village (2006 population 375), PHELPS county, S NEBRASKA, 6 mi/9.7 km WNW of HOLDREGE; 40°28′N 99°30′W. Manufacturing (clothing).

Loon (lo-ON), town, BOHOL province, W Bohol island, PHILIPPINES, on BOHOL STRAIT, 11 mi/18 km NNW of TAGBILARAN; 09°48′N 123°50′E. Agricultural center (rice, coconuts).

Loon Lake, hamlet, FRANKLIN county, NE NEW YORK, on Loon Lake (c.2 mi/3.2 km long), 31 mi/50 km WSW of PLATTSBURGH; 44°33′N 74°04′W.

Loon Lake, PISCATAQUIS county, NW MAINE, 47 mi/76 km N of GREENVILLE, in wilderness recreational area; 4.5 mi/7.2 km long.

Loon Lake, 4 mi/6.4 km long, EL DORADO county, central CALIFORNIA, on Gerle Creek, 21 mi/34 km WNW of SOUTH LAKE TAHOE, in Eldorado National Forest; 39°00′N 120°20′W. Elevation 6,378 ft/1,944 m. Maximum capacity 76,500 acre-ft. Formed by Loon Lake Dam (100 ft/30 m high), built (1963) for SACRAMENTO water supply and debris control.

Loon Lake (LOON), settlement, N ALBERTA, W CANADA, 88 mi/141 km from HIGH PRAIRIE, in OPPORTUNITY NO. 17 municipal district; 56°33′N 115°24′W.

Loon op Zand (LAWN awp SAHNT), town, NORTH BRABANT province, S NETHERLANDS, 5 mi/8 km N of TILBURG; 51°38′N 05°05′E. Loonse en Drunense Duinen National Park to NE. Dairying; cattle, hogs; vegetables, potatoes. Castle to N.

Loon-Plage, town (□ 14 sq mi/36.4 sq km), NORD department, NORD-PAS-DE-CALAIS region, N FRANCE; 50°59′N 02°13′E. Seaside resort on NORTH SEA 15 mi/25 km NNE of CALAIS.

Loop Head, Gaelic *Ceann Léime*, promontory on the ATLANTIC, SW CLARE county, W IRELAND, on N shore of mouth of the SHANNON, 20 mi/32 km WSW of KILRUSH. Lighthouse (52°34′N 09°56′W).

Loos (LOS), WSW suburb (□ 3 sq mi/7.8 sq km), of LILLE, NORD department, NORD-PAS-DE-CALAIS region, N FRANCE; 50°37′N 03°01′E. Chemical works, printshops, and textile mills. Its old Cistercian abbey is now a prison.

Loos, FRANCE: see LOOS-EN-GOHELLE.

Loosahatchie River (loo-suh-HA-chee), c.65 mi/105 km long, SW TENNESSEE; rises SW of HARDEMAN; flows W past SOMERVILLE, to MISSISSIPPI RIVER near MEMPHIS; 35°12′N 90°03′W. Upper course canalized.

Loosdorf (LOS-dorf), township, central LOWER AUSTRIA, on PIELACH RIVER, 10 mi/16 km W of SANKT POLTEN; 48°12′N 15°24′E. Agriculture (grain, orchards; cattle, poultry). Located 2 mi/3.2 km SW of Schallaburg, a Renaissance castle with a fine arcade courtyard.

Loosdrechtse Plassen (LAWS-drekht-suh PLAH-suhn), group of shallow lakes in UTRECHT province, W central NETHERLANDS, 6 mi/9.7 km NNW of UTRECHT; 4 mi/6.4 km long, 3 mi/4.8 km wide. Came into being as a result of peat-digging in former centuries. Fishing; water sports.

Loosduinen (LAWS-doi-nuhn), suburb of The HAGUE, SOUTH HOLLAND province, W NETHERLANDS, 4 mi/6.4 km SSW of city center; 52°16′N 04°14′E. NORTH SEA 1 mi/1.6 km to NW. Vegetables, flowers, nursery stock.

Loose (LOOZ), village (2001 population 2,207), central KENT, SE ENGLAND, 2 mi/3.2 km S of MAIDSTONE; 51°14′N 00°31′E.

Loos-en-Gohelle (los–ahn–go-el), town (□ 4 sq mi/10.4 sq km), PAS-DE-CALAIS department, NORD-PAS-DE-CALAIS region, N FRANCE, 3 mi/4.8 km NW of LENS, in coal-mining district; 50°27′N 02°47′E. Recaptured by British in costly battle of Loos (1915) in World War I. Nearby Double Crassier Hill and HILL 70 were scene of desperate fighting. Several British cemeteries in area. Formerly known as Loos.

Loos, Iles de, GUINEA: see LOS ISLANDS.

Lop (LO-PU), town and oasis, ☉ Lop county, SW XINJIANG UYGUR AUTONOMOUS REGION, CHINA, 20 mi/32 km SE of HOTAN, and on highway; 37°03′N 81°05′E; grain, cotton; jade. Also spelled Luopu.

Lopandino (luh-PAHN-dee-nuh), town (2004 population 2,140), SE BRYANSK oblast, central European Russia, on railroad, 32 mi/51 km NE of the RUSSIA-UKRAINE border, 50 mi/80 km S of BRYANSK, and 4 mi/6.4 km NNE of KOMARICHI; 52°28′N 34°49′E. Elevation 659 ft/200 m. Highway junction. Sugar refining.

Lopar, town, W CROATIA, in KVARNER region, on RAB Island. Seaside resort.

Lopasnya, RUSSIA: see CHEKHOV, MOSCOW oblast.

Lopatcong, township, WARREN county, NW NEW JERSEY, 3 mi/4.8 km E of PHILLIPSBURG; 40°42′N 75°09′W. Incorporated 1851.

Lopatin (luh-PAH-teen), settlement, N DAGESTAN REPUBLIC, NE CAUCASUS, RUSSIA, fishing port on CASPIAN SEA, on the N tip of the AGRAKHAN Peninsula, 60 mi/97 km N (under jurisdiction) of MAKHACHKALA. Fish processing.

Lopatin, UKRAINE: see LOPATYN.

Lopatino (luh-PAH-tee-nuh), village, SE PENZA oblast, E European Russia, near highway, 19 mi/30 km SE of SURSK; 53°02′N 46°04′E. Elevation 606 ft/184 m. In agricultural area (buckwheat, sunflowers, wheat, rye; livestock raising); food processing (dairy, bakery, vegetable oil).

Lopatino, RUSSIA: see VOLZHSK.

Lopatinskiy (luh-PAH-teen-skeeye), town (2006 population 14,185), SE MOSCOW oblast, central European Russia, on road and railroad, 30 mi/48 km SE of MOSCOW, and 3 mi/5 km NE of VOSKRESENSK, to which it is administratively subordinate; 55°20′N 38°44′E. Elevation 436 ft/132 m. Railroad establishments. Processing of phosphate for the Voskresensk chemical combine; paper and cardboard products manufacturing. Originally a village of Dvoriki, until 1962, when a mineral-chemical combine was founded,

and the village was expanded into town and renamed Lopatinskiy Rudnik; current name since the 1980s.

Lopatinskiy Rudnik, RUSSIA: see LOPATINSKIY.

Lopatka, Cape (luh-PAHT-kah), S extremity of KAMCHATKA Peninsula, KAMCHATKA oblast, extreme NE SIBERIA, RUSSIAN FAR EAST; 50°52′N 156°40′E. Separated by First Kuril Strait from SHUMSHU ISLAND, northernmost of the KURIL ISLANDS.

Lopatki (luh-PAHT-kee), village, SE KURGAN oblast, SW SIBERIA, RUSSIA, on road, approximately 17 mi/27 km N of the Russia-KAZAKHSTAN border, on road, 20 mi/32 km S of LEBYAZHYE; 54°59′N 66°37′E. Elevation 492 ft/149 m. In agricultural area (buckwheat, corn, rye, oats, sunflowers).

Lopatyn (lo-PAH-tin), (Russian *Lopatin*), (Polish *Lopatyn*), town (2004 population 4,600), NE L'VIV oblast, UKRAINE, on crossroads, 42 mi/ E of L'VIV and 15 mi/24 km NW of BRODY; 50°13′N 24°51′E. Elevation 708 ft/215 m. Agricultural processing (grain, vegetables, hops); peat bricketing; forestry. Known since 1366; town since 1956.

Lop Buri (LOP BU-REE), province, central Thailand; ☉ Lop Buri; 14°56′N 100°45′E. Rice, corn, millet, beans; animal husbandry. Most famous for the Khmer temples in the provincial capital that were built duing the period from the 10th–13th cent. when Lop Buri was a part of the Angkor Empire. A large number of monkeys live at the San Phra Kan (Kala Shrine) in Lop Buri town, and a local businessman has been holding an elaborate annual feast for the monkeys of Kala Shrine. Somdet Phra Narai National Museum is here. Also spelled Lopburi.

Lop Buri (LOP BU-REE), town, ☉ Lop Buri prov., central Thailand; 14°48′N 100°37′E. Originally called Lavo, it was ruled by the Mons in the 7th and 8th cents. and by the Khmers from the 10th to the 13th cent. While Ayutthaya was capital of Thailand, Lop Buri served as an alternate capital and the former summer capital of Siam (Lop Buri Palace). Well-known Hindu and Khmer ruins here. Also spelled Lopburi.

Lopburi Palace (LOP-BU-REE), Thai *Phra Narai Ratchaniwet*, historic site, Lopburi province, THAILAND. In 1665, King Narai began construction on and completed the palace in 1677. Reflects the influence of the French architects who contributed to its design, as well as the influence of traditional Khmer architectural style. Among the buildings in the palace compound are the remains of the royal elephant stables, a royal reception hall, a pavilion, an audience hall, residence quarters for the king's harem, and the Lopburi National Museum.

Lopburi River (LOP-BU-REE), central THAILAND, a left arm of CHAO PHRAYA RIVER 75 mi/121 km; branches off below SINGBURI; flows S, past Lopburi and AYUTTHAYA (where it receives PA SAK RIVER), and rejoins CHAO PHRAYA RIVER at RATCHAKHRAM.

Lopé-Okanda Reserve, park and wildlife refuge in central GABON, OGOOUÉ-IVINDO, and OGOOUÉ-LOLO provinces, 15 mi/24 km SW of BOOUÉ; 00°15′S 11°35′E.

Lopera (lo-PAI-rah), town, JAÉN province, S SPAIN, 10 mi/16 km SW of ANDÚJAR; 37°57′N 04°12′W. Olive-oil processing, soap manufacturing. Cereals, melons, wine, and tobacco in area. Has ruins of Moorish castle. Birthplace of Bernabe Cobo, colonial chronicler of the New World.

Lopevi (lo-PAI-vee), active volcanic island (c.4 mi/6.4 km long; rises to 4,636 ft/1,413 m), VANUATU, SW PACIFIC OCEAN, 10 mi/16 km SSE of AMBRYM; 16°30′S 168°21′E. Also called Ulveah.

López (LO-pes), town, ☉ López municipio, CAUCA department, SW COLOMBIA, 42 mi/68 km NW of POPAYÁN, on E edge of Pacific coastal plain; 02°51′N 77°15′W. Sugarcane, corn, coffee, rice, plantains; livestock. Sometimes called Micay or San Miguel.

López (LO-pes), town, CHIHUAHUA, N MEXICO, on FLORIDO River, and 40 mi/64 km E of HIDALGO DEL PARRAL. Elevation 3,707 ft/1,130 m. Corn, wheat,

cotton, tobacco, livestock. Also known as VILLA LÓPEZ.

Lopez (LO-pes), town, QUEZON province, S LUZON, PHILIPPINES, near LOPEZ BAY, 45 mi/72 km E of LUCENA; 13°50′N 122°17′E. Copra center. Its port is Hondagua, just N, on SE shore of Lopez Bay.

Lopez, unincorporated village, Colley township, SULLIVAN county, NE PENNSYLVANIA, 34 mi/55 km W of SCRANTON, on LOYALSOCK CREEK; 41°27′N 76°20′W.

Lopez, unincorporated village, SAN JUAN county, NW WASHINGTON, on W side of LOPEZ ISLAND, SAN JUAN ISLANDS, 5 mi/8 km E of FRIDAY HARBOR, on San Juan Channel. Fishing; tourism. Lopez Island Airport to S. Terminus of ferries from ANACORTES and Friday Harbor, landing 4 mi/6.4 km NNE at Upright Head. Spencer Spit State Park to NE.

Lopez Bay (LO-pes), S arm of LAMON BAY, PHILIPPINES, between ALABAT ISLAND (E) and S coast of LUZON (W and S); 36 mi/58 km long, 3 mi/4.8 km–11 mi/18 km) wide. CABALETE ISLAND is at entrance; ATIMONAN is on SW shore.

Lopez, Cape (LO-pez), low, forested headland in W GABON, OGOOUÉ-MARITIME province, on the GULF OF GUINEA at the extremity of an island formed by two mouths of OGOOUÉ RIVER, just NNW of PORT-GENTIL; 00°38′S 08°42′E. Mangroves; guano deposits. It is the southernmost limit of BIGHT OF BIAFRA.

López, Cerro (LO-pez, SER-ro), ANDEAN peak (6,890 ft/2,100 m) in SW RÍO NEGRO province, ARGENTINA, on S shore of LAKE NAHUEL HUAPÍ, 15 mi/24 km W of SAN CARLOS DE BARILOCHE. Popular skiing ground.

López de Filippis, PARAGUAY: see MARISCAL ESTIGARRIBIA.

Lopez Island (12 mi/19 km long; 5 mi/8 km wide), SAN JUAN county, NW WASHINGTON, in E part of SAN JUAN ISLANDS. Bounded E by Lopez Sound, N by Upright Channel, W by San Juan Channel, S by STRAIT OF JUAN DE FUCA. Village of LOPEZ in NW; Spencer Spit State Park in NE. Ferry landing at N end. Dairying; cattle; hay. Fishing. Tourism.

Lopez Point, coastal promontory, MONTEREY county, W CALIFORNIA, c.40 mi/64 km SSE of MONTEREY.

Lopik (LAW-pik), village, UTRECHT province, W central NETHERLANDS, 11 mi/18 km SSW of UTRECHT, near LEK RIVER; 51°58′N 04°57′E. Dairying; livestock; vegetables, sugar beets; manufacturing (food, awnings, aluminum dies).

Lopnur, CHINA: see YULI.

Lopori River (lo-PO-ree), c.380 mi/612 km long, W CONGO; rises 70 mi/113 km S of Yakuma; flows NW and W, past SIMBA, LOKOLENGE, and BONGANDANGA, to join MARINGA RIVER at BASANKUSU, forming the LULONGA. Navigable for 280 mi/451 km downstream from SIMBA.

Lopud (LO-pud), Italian *Mezzo*, ancient *Delaphodia*, Dalmatian island in ADRIATIC SEA, S CROATIA; 2 mi/3.2 km long. Chief village, Lopud, is 8 mi/12.9 km WNW of DUBROVNIK, in tourist area. Ruins of chapels and Venetian forts.

Lo-pu-po, CHINA: see LUOBU BO.

Lora de Estepa (LO-rah dhai es-TAI-pah), town, SEVILLE province, SW SPAIN, in outliers of the CORDILLERA PENIBÉTICA, 16 mi/26 km E of OSUNA; 37°16′N 04°50′W. Olives, cereals, esparto, onions.

Lora del Río (dhel REE-o), city, SEVILLE province, SW SPAIN, on the GUADALQUIVIR RIVER, on railroad, and 31 mi/50 km NE of SEVILLE; 37°39′N 05°32′W. Trading and processing center in agricultural region (olives, cereals, fruits; livestock). Vegetable canning, flour milling, olive-oil pressing. Has graphite, sand, and granite quarries. Ruins of old fortress are nearby. The ancient town is of Iberian origin and has been identified with the Roman *Axatiana*, mentioned by Pliny. The town was taken (1243) from the Moors by Ferdinand III. Ruins of Setefillas castle are 3 mi/4.8 km NE.

Lorado (lor-AID-o), unincorporated village, LOGAN county, SW WEST VIRGINIA, 15 mi/24 km ESE of LOGAN; 37°47′N 81°42′W. Coal-mining region.

Lorain (lo-RAIN), county (□ 495 sq mi/1,287 sq km; 2006 population 301,993), N OHIO; ⊙ ELYRIA; 41°20′N 82°09′W. Bounded N by LAKE ERIE; drained by BLACK and VERMILION rivers. Agricultural area (grain, fruit; poultry, livestock; dairy products); manufacturing at LORAIN, Elyria, WELLINGTON. Commercial fishing; sandstone quarries; lake resorts. Formed 1822.

Lorain (lo-RAIN), city (□ 24 sq mi/62.4 sq km; 2006 population 70,592), LORAIN county, N OHIO, 25 mi/40 km W of CLEVELAND, on LAKE ERIE at the mouth of the BLACK RIVER; 41°26′N 82°11′W. Once an important ore-shipping point, Lorain has shipyards, steel works, automobile-assembly plants, and commercial fisheries. Port activities, once integral to the city's economy, have declined in bulk and importance over the years. Power equipment, automotive and building materials, navigation equipment, and toys are among the manufactures. Lorain also has numerous boating facilities. Incorporated 1834.

Lorain (LUH-rain), borough (2006 population 698), CAMBRIA county, SW central PENNSYLVANIA, residential suburb 2 mi/3.2 km SE of JOHNSTOWN; 40°17′N 78°54′W.

Loraine, village (2000 population 363), ADAMS county, W ILLINOIS, 18 mi/29 km NNE of QUINCY; 40°08′N 91°13′W. In agricultural area.

Loraine, village (2006 population 18), RENVILLE county, N NORTH DAKOTA, 8 mi/12.9 km N of MOHALL; 48°52′N 101°34′W.

Loraine (LOR-ain), village (2006 population 609), MITCHELL county, W TEXAS, 8 mi/12.9 km E of COLORADO CITY; 32°24′N 100°42′W. Oil and gas. Cattle, sheep; grains, cotton, alfalfa. Incorporated 1907.

Lora, La (lah LO-rah), plateau, NW BURGOS province, N SPAIN, along CANTABRIA province border. SEDANO is its agricultural center.

Loralai (LO-rah-lei), district (□ 7,375 sq mi/19,175 sq km), BALUCHISTAN province, SW PAKISTAN; ⊙ LORALAI. Bordered E by SULAIMAN Range, S by CENTRAL BRAHUI RANGE; drained by several seasonal streams and (W) by upper course of NARI RIVER. Some irrigation; wheat and millet grown in valleys, olives in SULAIMAN Range (E). Limestone deposits in hills (center, NE); some gypsum (SE). Handicrafts (felts, felt coats, mats, saddlebags); cattle raising (Barkhan noted for its horses).

Loralai (LO-rah-lei), town, ⊙ LORALAI district, NE BALUCHISTAN province, SW PAKISTAN, near NARI (LORALAI) River, 95 mi/153 km E of QUETTA; 30°22′N 68°36′E.

Loralai River, PAKISTAN: see NARI RIVER.

Loramie Creek (LO-ruh-mee), c.40 mi/64 km long, W OHIO; rises in N SHELBY county, flows SW to Lake Loramie or Loramie Reservoir (c.7 mi/11 km long) just SE of MINSTER, then SE to GREAT MIAMI RIVER just N of PIQUA; 40°11′N 84°14′W.

Loramie, Lake, Ohio: see LORAMIE CREEK.

Loranca del Campo (lo-RAHNG-ko dhel KAHM-po), town, CUENCA province, E central SPAIN, 31 mi/50 km W of CUENCA; 40°05′N 02°41′W. Cereals, grapes, olives, saffron, chick peas, honey; sheep.

Lorane (LUH-rain), unincorporated town (2000 population 2,994), BERKS county, SE central PENNSYLVANIA, residential suburb 5 mi/8 km SE of READING on SCHUYLKILL RIVER; 40°17′N 75°50′W. Agriculture includes dairying, livestock, poultry; apples, plums, grain. Daniel Boone Homestead State Historical Site to E.

Lorca (LOR-kah), city (2001 population 77,477), MURCIA province, SE SPAIN, in Murcia, on the Guadalentín River, and c.40 mi/64 km W of CARTAGENA; 37°40′N 01°41′W. A market center for a fertile, irrigated basin producing cereals, fruits, and vegetables. Hemp sandals and woolen products are made in Lorca. Gypsum

quarries and sulfur and iron mines nearby. Taken by the Moors in the 8th century, the city was liberated in 1243. It has a Moorish castle, a 17th-century collegiate church, and several old mansions.

Lorch (LORKH), town, BADEN-WÜRTTEMBERG, S GERMANY, on the REMS, 5 mi/8 km W of SCHWÄBISCH GMÜND; 48°48′N 09°42′E. Manufacturing of chassis; metalworking. Has Benedictine monastery (founded c.1100, partly destroyed 1525), with Romanesque church. On site of Roman castrum.

Lorch, town, HESSE, central GERMANY, on right bank of the RHINE, and 19 mi/31 km W of WIESBADEN; 50°03′N 07°49′E. Wine; tourism. Has Gothic church.

Lorch, AUSTRIA: see ENNS.

Lorcha (LOR-chah), town, ALICANTE province, E SPAIN, 11 mi/18 km SW of GANDÍA; 38°51′N 00°18′W. Olive oil processing; cereals, wine; sheep.

Lordegan (luhr-de-GAHN), town, Chahār Mahāll va Bākhtīarī province, SW IRAN, 60 mi/97 km S of SHAHR-e Kord; 31°30′N 50°50′E. Some wheat cultivation.

Lord Hood Island, S TUAMOTU ARCHIPELAGO: see MARUTEA SOUTH.

Lord Howe Island (LORD HOU), volcanic island (□ 5 sq mi/13 sq km), S PACIFIC, a dependency of NEW SOUTH WALES, SE AUSTRALIA; 31°33′S 159°05′E. It is a resort c.300 mi/480 km E of the Australian coast. The island was explored in 1788 by the British and was settled in 1834. Palm seeds are the only export.

Lord Howe Island, SOLOMON ISLANDS: see ONTONG JAVA ATOLL.

Lord Howe Island Group (LORD HOU), NEW SOUTH WALES, SE AUSTRALIA, S PACIFIC OCEAN, 380 mi/612 km E of PORT MACQUARIE. Main island, LORD HOWE ISLAND, 31°35′S 159°10′E; 6 mi/10 km long, 1.8 mi/2.9 km wide. Also, ADMIRALTY ISLANDS 1 mi/1.6 km N, Mutton Bird Island 1 mi/1.6 km NE of Rocky Point, Balls Pyramid 20 mi/32 km SSE. Highest point, Mount Gower (2,871 ft/875 m) at S end; second-highest, Mount Lidgfird (2,549 ft/777 m), volcanic origin. Wooded hills, palm groves, swamps, grass areas. Southernmost coral reefs in world; sand beaches. Preserve and home to more than 120 native and introduced bird species, including boobies, muttonbirds, tropicbirds, terns, gannets, and the once nearly-extinct Lord Howe Island woodhen. Resort community near N end. Landing strip. Road 2 mi/3 km S to Kings Beach. Walking tracks N to Fishy Point and S to Mount Gower. Tourism. Declared UNESCO World Heritage Area 1982.

Lordsburg, town (2006 population 2,762), ⊙ HIDALGO county, SW NEW MEXICO, 44 mi/71 km SW of SILVER CITY, near ARIZONA state line; 32°20′N 108°42′W. Elevation 4,258 ft/1,298 m. Railroad junction. Trade center, resort. Cattle; vegetables, cotton, chiles, jalapeños, wheat, oats, barley, alfalfa, Christmas trees. Copper, silver, gold, and lead mines in vicinity. Southwest New Mexico fair and livestock show take place here. Part of Gila National Forest, in BURRO MOUNTAINS, to NE; Shakespeare Ghost Town to S; Stein's Ghost Town to SW.

Loreauville (luh-RO-vil), village (2000 population 938), IBERIA parish, S LOUISIANA, 6 mi/10 km NE of NEW IBERIA, and on Bayou TECHE; 30°04′N 91°45′W. In agricultural area (sugar, rice; cattle); catfish, crawfish; manufacturing (aluminum boats, harvesting equipment).

Lore City (LOR), village (2006 population 302), GUERNSEY county, E OHIO, 7 mi/11 km ESE of CAMBRIDGE; 39°59′N 81°27′W.

Lorelei (lor-e-LEI), cliff (433 ft/132 m) on the right bank of the RHINE River, near St. Goarshausen, W GERMANY, about midway between KOBLENZ and BINGEN; 50°08′N 07°44′E. The Rhine forms a dangerous narrows here, and in German legend a fairy similar to the Greek Sirens lived on the rock and by her singing lured sailors to their death. The rock has sometimes been identified as the place where the hoard of the Nibelungs is hidden under the Rhine.

Lorena (LO-re-nah), city (2007 population 79,394), SE SÃO PAULO, BRAZIL, on PARAÍBA River, and 110 mi/177 km NE of São Paulo; 22°44′S 45°08′W. Railroad junction on RIO DE JANEIRO–SÃO PAULO railroad (spur to MINAS GERAIS). Sugar milling, alcohol distilling, meat drying, dairying; plant nursery. Trades in coffee, rice, corn. Talc quarries.

Lorena (LOR-ai-nuh), town (2006 population 1,628), MCLENNAN county, E central TEXAS, 10 mi/16 km SSW of WACO; 31°22′N 97°12′W. Agricultural area (cattle; dairying; cotton, grain). Light manufacturing.

Lorengau (law-reng-OU), town, ⊙ MANUS province, N PAPUA NEW GUINEA, on NE coast of MANUS island, ADMIRALTY ISLANDS, on Seeadler Harbor, arm of PACIFIC OCEAN; 02°05′S 147°20′E. Connected by bridge to LOS NEGROS ISLAND, to E. After World War II, briefly replaced as administrative center by Inrim plantation, 10 mi/16 km W.

Lo-Reninge (LO–RAI-ni-nyuh), commune (2006 population 3,323), Diksmuide district, WEST FLANDERS province, NW BELGIUM, 6 mi/9.7 km SW of DIKSMUIDE. Farming, cattle raising, dairying. Fortified in 1167; had important textile industry in Middle Ages.

Lorentzweiler (LO-rents-WEI-luhr), village, Lorentzweiler commune, S central LUXEMBOURG, on ALZETTE RIVER, and 6 mi/9.7 km N of LUXEMBOURG city; 49°42′N 06°09′E.

Lorenzo (law-REN-zo), town (2006 population 1,256), CROSBY county, NW TEXAS, on the LLANO ESTACADO, 20 mi/32 km ENE of LUBBOCK; 33°40′N 101°31′W. Wheat; cotton; cattle; oil and gas; manufacturing (textile yarns).

Lorenzo Geyres, village, PAYSANDÚ department, NW URUGUAY, near QUEGUAY RIVER, on railroad, and 17 mi/27 km NNE of PAYSANDÚ; 32°05′S 57°55′W. Wheat; cattle.

Lorestan (luhr-ei-STAHN), province (□ 11,121 sq mi/28,914.6 sq km), W IRAN; 33°30′N 48°30′E. The chief cities are the capital, KHORRAMABAD, and BORUJERD. The region consists mainly of forested and pastured mountain ranges in the ZAGROS Mountains; the highest point is Oshtorankuh (c.12,960 ft/4,050 m). Though it has large petroleum deposits, agriculture is the chief industry. Crops include grain, cotton, fruit, vegetables, and oilseed; also cotton-ginning and food-processing industries; sheep and goat herding is important. The inhabitants are mainly Lurs and Bakhtiari. The Kassite conquerors of BABYLONIA came from Lorestan (18th century B.C.E.). The noted Lorestan bronzes, found in the province beginning c.1930, include cups, horse bits, daggers, and shields, ornamented with animal motifs, checkerboards, wavy lines, and crosses. They were probably made in the 8th and 7th century B.C.E. by local metalworkers for Scythian, Cimmerian, or Median nomads.

Loreto (lo-RAI-to), region (□ 142,415 sq mi/370,279 sq km), NE and E PERU, just S of the equator; ⊙ IQUITOS; 05°00′S 75°00′W. Borders NW on ECUADOR, NE (PUTUMAYO River) on COLOMBIA, and E (JAVARI RIVER) on BRAZIL. The largest administrative division of Peru, covering approximately one-third of the country. Bounded on the W by E outliers of the ANDES, it is entirely within the AMAZON basin, here frequently called Montaña. A vast network of streams, tributaries to the Amazon, intersect the region; among them are the UCAYALI and MARAÑÓN, which form the Amazon S of Iquitos. The climate is humid and tropical with little variation; main rainy season November–May. In spite of its rich vegetation, the region is opening to colonization, with crops including cotton, sugarcane, rice, cacao, coca, yucca, beans, corn, and tropical fruit of all kinds thriving. Once predominantly a rubber-producing region, its virgin forests yield rubber, balata, chicle, cascarilla, Brazil nuts, tanning barks, aromatic and medicinal plants, and a variety of fine timber. There are gold, petroleum, salt, and gypsum deposits. Its large navigable rivers

---

Cross-references are shown in SMALL CAPITALS. The pronunciation guide is shown on page xix. The sources of population figures are shown on page xvii.

serve as main lines of communication, and its trade via Amazon River is oriented toward the ATLANTIC. Iquitos, its leading inland port and trading and processing center, can be reached by ocean-going vessels. Site of RESERVA NACIONAL PACAYA-SAMIRIA. In addition to Iquitos, main cities are CABALLOCOCHA, NAUTA, YURIMAGUAS, REQUENA, and CONTAMANA.

Loreto (lo-RAI-to), province, LORETO region, NE PERU; ⊙ NAUTA; 04°00′S 75°10′W. N central province of Loreto region and N border on ECUADOR.

Loreto (lo-re-to), city (2007 population 10,243), S central MARANHÃO, BRAZIL, on RIO BALSAS, and 50 mi/80 km NE of BALSAS (connected by road); 07°06′S 45°14′W. Hides, rubber, resins; cattle raising. Airfield.

Loreto (lo-RAI-to), Loreto department (□ 1,115 sq mi/2,899 sq km), SW SANTIAGO DEL ESTERO province, ARGENTINA, on railroad, and 55 mi/89 km S of SANTIAGO DEL ESTERO; 27°46′S 57°17′W. Agricultural center (wheat, vegetables, grapes; livestock). Formerly Ciudad de Loreto.

Loreto, town and canton, ⊙ MARBÁN province, BENI department, NE BOLIVIA, in the LLANOS, 34 mi/55 km SSE of TRINIDAD; 15°16′S 64°40′W. Elevation 525 ft/ 160 m. Cattle; tropical agricultural products (coffee, mangoes, cacao, coca, bananas); potatoes.

Loreto (lo-RE-to), town, in the MARCHE, central ITALY, on a hill overlooking the ADRIATIC Sea. Highly diversified secondary industrial center, including fabricated metals, wood products. It is a famous place of pilgrimage. According to legend, the Holy House of the Virgin in Nazareth was brought to Loreto through the air by angels in 1294. Around the Holy House (a small brick building) there is a church—the Santuario della Santa Casa—begun in 1468 by Pope Paul II; Bramante contributed to its construction. It has fine bronze doors (16th–17th century) and frescoes by Melozzo da Forli and Luca Signorelli. Our Lady of Loreto is a patron of aviators. The Loretto (or Loreto) order of nuns, named for the town, was founded in Ireland in 1822.

Loreto (lo-RE-to), town, ZACATECAS, N central MEXICO, on railroad, and 50 mi/80 km SE of ZACATECAS; 22°18′N 102°00′W. Grain; livestock.

Loreto (lo-RE-to), town, Concepción department, central PARAGUAY, 13 mi/21 km NE of Concepción; 23°15′S 57°11′W. Livestock-raising and agricultural center (corn, alfalfa, vegetables, maté).

Loreto (lo-RAI-to), village, N CORRIENTES province, ARGENTINA, 100 mi/161 km ESE of CORRIENTES; 27°19′S 55°32′W. Agriculture center (tobacco, sugarcane, cotton; livestock).

Loreto (lo-RE-to), village, MAGALLANES province, MAGALLANES Y LA ANTARTICA region, S CHILE, on N BRUNSWICK PENINSULA, 4 mi/6 km NW of PUNTA ARENAS (linked by short railroad line). Coal mining.

Loreto (lo-RE-to), village, LA PAZ municipio, BAJA CALIFORNIA SUR, NW MEXICO, port on Gulf of CALIFORNIA, on Mexico Highway 1, and 145 mi/233 km N of La Paz; 26°00′N 111°20′W. Sugarcane, dates, figs. Formerly a Jesuit mission, founded 1697. The oldest of the Californian missions; site of planned tourist resort development begun in the early 1990s.

Lorette (lo-RET), community, SE MANITOBA, W central CANADA, 16 mi/25 km SE of WINNIPEG, and in TACHÉ rural municipality; 49°44′N 96°52′W.

Lorette (lo-ret), town (□ 1 sq mi/2.6 sq km), LOIRE department, RHÔNE-ALPES region, SE central FRANCE, on GIER RIVER, and 11 mi/18 km ENE of SAINT-ÉTIENNE, in metallurgical industry district; 45°31′N 04°35′E. Quartzite quarries nearby.

Loretteville (lo-REHT-vil), former town, S central QUEBEC, E CANADA, on west Nelson River, and 8 mi/ 13 km NW of QUEBEC; 46°51′N 71°20′W. Manufacturing of gloves, skis, canoes. Huron settlement began here 1697; village was known as Indian Lorette or Jeune Lorette. Nearby is L'ANCIENNE-LORETTE. Amalgamated into Quebec city in 2002.

Loretto, city (2006 population 1,701), LAWRENCE county, S TENNESSEE, 13 mi/21 km SSW of LAWRENCEBURG; 35°05′N 87°26′W.

Loretto (luh-RET-o), town (2000 population 623), MARION county, central KENTUCKY, 27 mi/43 km E of ELIZABETHTOWN; 37°38′N 85°24′W. Manufacturing (whiskey). Headquarters of Sisters of Loretto; Holy Cross Church (1820s).

Loretto, village (2000 population 570), HENNEPIN county, E MINNESOTA, suburb 18 mi/29 km W of downtown MINNEAPOLIS, surrounded by MEDINA; 45°03′N 93°38′W. Manufacturing (wedding cakes, metal stampings).

Loretto (luh-RE-to), borough (2006 population 1,259), CAMBRIA county, SW central PENNSYLVANIA, 5 mi/8 km ENE of EBENSBURG; 40°30′N 78°38′W. Agricultural area (corn, hay, potatoes; dairying); manufacturing (dairy products). St. Francis College here.

Lorgues (lor-guh), town (□ 25 sq mi/65 sq km), VAR department, PROVENCE-ALPES-CÔTE D'AZUR region, SE FRANCE, in Provence (MARITIME) Alps 6 mi/9.7 km SW of DRAGUIGNAN; 43°29′N 06°22′E. Surrounded by olive groves and vineyards, Lorgues has long been a major producer of olive oil and raisins. Abbey of Thoronet, built by Cistercian order in 12th century, is 5 mi/8 km SW.

Lorica (lo-REE-kah), town, ⊙ Lorica municipio, CÓRDOBA department, N COLOMBIA, in CARIBBEAN LOWLANDS, located around navigable SINÚ RIVER and streams from CIÉNAGA GRANDE, and 33 mi/53 km N of MONTERÍA; 09°14′N 75°49′W. Agricultural center (rice, corn, bananas, sugarcane; livestock); fishing.

Lorient (lo-ryahn), port city (□ 6 sq mi/15.6 sq km), sub-prefecture of MORBIHAN department, BRITTANY, NW FRANCE, naval station on an inlet of the ATLANTIC OCEAN, 65 mi/105 km SE of BREST; 47°45′N 03°22′W. A shipbuilding center and submarine base, Lorient also is home port to a fishing fleet and has extensive fish-processing and -storage facilities. Its industries provide navigational equipment and other naval support materials. Established (17th century) as a port to serve the French East India Company, it was developed as a naval base by Napoleon I and became the country's chief naval station under Napoleon III, in mid-19th century. In World War II it was the Germans' major submarine base on the Atlantic. Almost totally destroyed by Allied bombs in 1942–1944, it has been completely rebuilt. W of Lorient is naval air station of Lann-Bihoué.

Loriga (lo-REE-gah), town, GUARDA district, N central PORTUGAL, in SERRA DA ESTRÊLA, 28 mi/45 km SW of GUARDA; 40°19′N 07°42′W. Wool; dairying.

L'Orignal (lo-reen-YAHL), unincorporated village (□ 2 sq mi/5.2 sq km), 2001 population 2,033), ⊙ PRESCOTT and RUSSELL counties, SE ONTARIO, E central CANADA, on OTTAWA River, 50 mi/80 km ENE of OTTAWA, and included in CHAMPLAIN township; 45°37′N 74°41′W. Lumbering center, dairying; resort.

Lorimor, town (2000 population 427), UNION county, S IOWA, 17 mi/27 km ENE of CRESTON; 41°07′N 94°03′W. Metal products.

Lőrinci (LUH-rin-tsi), Hungarian *Lőrinci*, town, HEVES county, N HUNGARY, on ZAGYVA RIVER, and 32 mi/51 km NE of BUDAPEST; 47°44′N 19°41′E. Sugar refineries, flour mills. Limestone quarry nearby.

Loring, fishing village, on W side of REVILLAGIGEDO ISLAND, 20 mi/32 km N of KETCHIKAN.

Loriol-sur-Drôme (lor-ee-ol-syur-drom), town (□ 11 sq mi/28.6 sq km), DRÔME department, RHÔNE-ALPES region, SE FRANCE, near the DRÔME and RHÔNE rivers, 13 mi/21 km SSW of VALENCE; 44°45′N 04°49′E. Diverse manufacturing; peach and pear orchards in area. Dam on Rhône at head of bypass canal, SW of town.

Lori-Pambak, The, region encompassing the basins of the DEBED, PAMBAK, and Dzoraget rivers, ARMENIA. Preponderance of steppe and forest landscapes. The region is rich in copper and polymetalic deposits.

Loris (LOR-is), town (2006 population 2,290), HORRY county, E SOUTH CAROLINA, 17 mi/27 km NNE of CONWAY, near NORTH CAROLINA state line; 34°03′N 78°53′W. Diversified manufacturing. Agriculture includes watermelons, tobacco, grain, strawberries; livestock.

Lorman, unincorporated village, JEFFERSON county, SW MISSISSIPPI, 26 mi/42 km NE of NATCHEZ. Cotton, rice, corn; cattle; catfish. Old Country Store, in continuous operation since 1890s. NATCHEZ TRACE PARKWAY passes to W; Canemount (c.1855), and Rosewood (c.1857) plantations to NW. Alcorn State University 4 mi/6.4 km NW.

Lormont (lor-mon), NE suburb (□ 3 sq mi/7.8 sq km) of BORDEAUX, GIRONDE department, AQUITAINE region, SW FRANCE, on right bank of GARONNE RIVER; forms part of port of Bordeaux; 44°52′N 00°32′W. Warehousing, cement manufacturing, and woodworking.

Lorn (LORN), mountainous district (c.25 mi/40 km long, 20 mi/32 km wide) of Argyll and Bute, W Scotland, bounded W by the Firth of Lorn, N by LOCH LEVEN, NW by LOCH LINNHE, E by PERTH AND KINROSS and NW STIRLING, and S by LOCH AWE. Chief town is OBAN. Formerly in Strathclyde, abolished 1996.

Lorne (LORN), rural municipality (□ 350 sq mi/910 sq km; 2001 population 2,033), S MANITOBA, W central CANADA; 49°25′N 98°44′W. Farming (canola; flax, hemp; sunflowers; livestock); dairying. Includes SWAN LAKE community. Established 1880.

Lorne (LORN), community, S central ONTARIO, E central CANADA; included BALDWIN township; 46°17′N 81°41′W.

Lorne (LORN), village, S VICTORIA, SE AUSTRALIA, 75 mi/121 km SW of MELBOURNE, on BASS STRAIT, and on GREAT OCEAN ROAD; 38°34′S 144°01′E. Seaside resort.

Loros, CHILE: see LOS LOROS.

Loroum, province (□ 1,385 sq mi/3,601 sq km; 2005 population 141,904), NORD region, N central BURKINA FASO; ⊙ Titao; 13°55′N 02°10′W. Bordered on N by MALI, ENE by SOUM province, SE by BAM province, and S and W by YATENGA province. Established in 1997 with fourteen other new provinces.

Loroux-Bottereau, Le (lo-roo–bo-ter-o, luh), town (□ 17 sq mi/44.2 sq km), LOIRE-ATLANTIQUE department, PAYS DE LA LOIRE region, W FRANCE, 10 mi/16 km E of NANTES. Agriculture market and wine center.

Lorquí (lor-KEE), town, MURCIA province, SE SPAIN, on SEGURA RIVER, and 10 mi/16 km NW of MURCIA; 38°05′N 01°15′W. Fruit-conserve manufacturing, pepper processing, sawmilling. Agricultural trade (fruit, saffron, cereals, peanuts). The Roman generals Publius and Gnaeus Scipio were defeated (212 B.C.E) by the Numidian Masinissa near here.

Lörrach (LOR-rahkh), city, BADEN-WÜRTTEMBERG, SW GERMANY, at SW foot of BLACK FOREST, 4 mi/6.4 km NNE of BASEL, on Swiss border; 47°37′N 07°40′E. Railroad junction. Manufacturing (machinery textiles); metal- and woodworking, food processing (chocolate). Trades in wine, fruit, timber. Has fifteenth-century church and nineteenth-century chalet and church. Obtained market rights in 1403; chartered 1682.

Lorrain (law-RANG), town, NE MARTINIQUE, French WEST INDIES, on the ATLANTIC, and 15 mi/24 km N of FORT-DE-FRANCE; 14°49′N 61°04′W. Rum distilling. Sometimes called Le Lorrain.

Lorrain (lo-RAIN), unincorporated village, E ONTARIO, E central CANADA, on MONTREAL RIVER, 5 mi/8 km WSW of COBALT, and included in COLEMAN township; 47°21′N 79°46′W. Silver and cobalt mining.

Lorraine (lo-rain), German *Lothringen*, region and former province (□ 9,092 sq mi/23,639.2 sq km), NE FRANCE; 49°00′N 06°00′E. Bordering on Belgium, Luxembourg, and Germany (N), on ALSACE (E), FRANCHE-COMTÉ (S), and on CHAMPAGNE (W). It is now an administrative region comprising 4 departments (MEUSE, MEURTHE-ET-MOSELLE, MOSELLE, and VOSGES), dominated by 2 cities—NANCY, which is the historic capital of the old province and METZ, which was chosen as the capital of the new administrative region. In the W of Moselle department, a Germanic dialect is spoken along with French. The rest of Lorraine is French-speaking. Historically, Lorraine extended from the W slopes of the VOSGES MOUNTAINS (a vast forested and lake-strewn plateau) to the E edges of the PARIS BASIN, where 2 major cuestas, the CÔTES DE MEUSE (E) and the Côte de Moselle (W) overlook lower-lying ground (WOËVRE, BASSIGNY, and Valley of the Moselle). Much of the rural population has migrated to urban areas. Agriculture (cereals, mostly wheat), cattle raising; dairying. Forestry is important; a third of the land is in managed forests. Coal and low-grade-iron-ore mining has declined markedly, as has local metallurgy. Manufacturing (chemicals, cotton textiles). Lorraine was, in the 9th century, part of the kingdom of Lotharingia; it became a duchy under the Holy Roman Empire. Several fiefs that escaped the control of the dukes, notably the duchy of Bar and the 3 bishoprics of Metz, TOUL, and VERDUN. The bishoprics were finally annexed by France in 1552. In the 16th century a cadet branch of the house of Lorraine, the Guise family, gained tremendous influence in France, while Lorraine itself, under Duke Charles II (1559–1608), enjoyed a period of relative order and prosperity amidst a Europe torn by religious strife. By an arrangement (1735) with Louis XV, Francis I, emperor of Austria and founder of the house of Hapsburg-Lorraine, exchanged the duchies of Lorraine and Bar for Tuscany; Lorraine and Bar were given to Louis XV's father-in-law, Stanislaus I, ex-king of Poland, upon whose death (1766) they passed to France. In 1871, the NE quarter of Lorraine was ceded to Germany and united with Alsace as the imperial land (Reichsland) of ALSACE-LORRAINE. Those parts of Lorraine remaining French were organized into the present departments of Meurthe-et-Moselle, Meuse, and Vosges. After World War I, Alsace-Lorraine was returned to France, but it was again annexed (1940–1944) by Germany during World War II. During both World Wars Lorraine suffered heavy damage and many of its citizens were deported.

Lorraine (lo-REN), city (□ 2 sq mi/5.2 sq km; 2006 population 9,943), LAURENTIDES region, S QUEBEC, E CANADA, 13 mi/21 km from MONTREAL; 45°41′N 73°47′W. Part of the Metropolitan Community of Montreal (*Communauté Métropolitaine de Montréal*).

Lorraine (lo-RAIN), village (2000 population 136), ELLSWORTH county, central KANSAS, 11 mi/18 km SSW of ELLSWORTH; 38°34′N 98°19′W. Railroad junction. Wheat; cattle.

Lorraine Belge (LAH-rain BEL-zhuh), southeasternmost part of BELGIUM in LUXEMBOURG province, bordering on French LORRAINE. It is also known as the *Gaume*; in German-speaking parts of the province and in the grand duchy of LUXEMBOURG, it is known as the *Gutland* because of its good soils.

Lorraine Natural Regional Park (lo-rain), MEUSE and MEURTHE-ET-MOSELLE departments (□ 892 sq mi/2,319.2 sq km), LORRAINE region, NE FRANCE; WOËVRE district.

Lorrainville (lo-RAIN-vil, French, lo-ran-VEEL), village (□ 33 sq mi/85.8 sq km), ABITIBI-TÉMISCAMINGUE region, W QUEBEC, E CANADA, 16 mi/26 km ESE of HAILEYBURY; 47°21′N 79°21′W. Dairying, cattle raising; gold-mining region. Restructured in 1994.

Lorris (lo-ree), commune (□ 17 sq mi/44.2 sq km), LOIRET department, CENTRE administrative region, N central FRANCE, just N of LOIRE RIVER valley, and 27 mi/43 km E of ORLÉANS; 47°53′N 02°31′E. Livestock market. Has Renaissance town hall, and a fine 12th–13th-century church.

Lorsch (LORSH), town, HESSE, central GERMANY, 3 mi/4.8 km SW of BENSHEIM; 49°39′N 08°29′E. Agricultural center; manufacturing of machinery, furniture. Has ruins of one of most powerful abbeys of Germany, founded in eighth century; passed to archbishops of MAINZ in 1232. Louis the German and Louis the Younger buried here. Mentioned in Nibelungenlied as burial place of Siegfried.

Lorton (LOR-tuhn), unincorporated city, FAIRFAX county, NE VIRGINIA, suburb 15 mi/24 km SE of WASHINGTON, D.C.; 38°42′N 77°14′W. Manufacturing (metal finishing, printing and publishing, lumber, concrete). Site of District of Columbia Correctional Facility (closed 2001). Gunston Cove of POTOMAC RIVER, FORT BELVOIR Military Reservation to E, Mason Neck State Park and National Wildlife Refuge, Pohick Bay Regional Park to SE, Occoquan Regional Park to W, Gunston Hall historic site to SE.

Lorton, village (2006 population 39), OTOE county, SE NEBRASKA, 10 mi/16 km WSW of NEBRASKA CITY, and on branch of LITTLE NEMAHA RIVER; 40°36′N 96°01′W.

Losai National Reserve (□ 697 sq mi/1,805 sq km), EASTERN province, E KENYA; 01°27′N 37°37′E. Designated a national reserve in 1976.

Los Alamitos, city (2000 population 11,536), ORANGE county, S CALIFORNIA; suburb 18 mi/29 km SSE of downtown LOS ANGELES, 7 mi/11.3 km E of LONG BEACH, and on SAN GABRIEL RIVER; 33°48′N 118°04′W. Diversified manufacturing. Los Alamitos Racetrack to N; Los Alamitos Naval Air Station in SE; U.S. Naval Weapons Center to S, at Seal Beach; ALAMITOS BAY and Long Beach Marina 5 mi/8 km SW. Incorporated 1960.

Los Alamos (LOS AH-lah-mos), county (□ 109 sq mi/283.4 sq km; 2006 population 19,022), N NEW MEXICO; ⊙ LOS ALAMOS; 35°52′N 106°18′W. Smallest county in NEW MEXICO. Includes unincorporated community of Los Alamos and part of WHITE ROCK. All government is at county level. High plateau area largely within VALLE GRANDE MOUNTAINS. Atomic research at Los Alamos. Formed 1949 from parts of SANDOVAL and SANTA FE counties. Part of Jemez Mountain in W; small part of Santa Fe National Forest along N and W boundaries; Pajarito Ski Area in W; bounded by BANDELIER NATIONAL MONUMENT on S; part of Bandelier National Monument in SE; bounded by RIO GRANDE in SE.

Los Alamos, unincorporated city (2000 population 11,909), ⊙ LOS ALAMOS county, N central NEW MEXICO, 23 mi/37 km NW of SANTA FE; 35°53′N 106°16′W. Elevation 7,410 ft/2,259 m. It is on a long mesa extending E from JEMEZ MOUNTAIN. The U.S. government chose the site in 1942 for atomic research, and the first atomic bombs were produced and tested here. In 1947 the Atomic Energy Commission took over the town. In 1962 government control ended; the county was incorporated in 1969, and all government is at the county level. Light, high-technology manufacturing and incipient tourism. Has the only fully operational plutonium plant in country. Nuclear weapons design ended in 1988, and bombs now being dismantled. Bradbury Science Museum. The Los Alamos Scientific Laboratory, operated by the University of California, is a national historic landmark. Santa Clara Indian Reservation to N; BANDELIER NATIONAL MONUMENT to S; Santa Fe National Forest to N and SW; Pajarito Ski Area to W.

Los Alamos (los AH-lah-mos), village, ⊙ Los Alamos comuna, ARAUCO province, BÍO-BÍO region, S central CHILE, 11 mi/18 km E of LEBU; 37°37′S 73°28′W. Coal mining; also agriculture (cereals, vegetables; livestock).

Los Alamos, unincorporated village, SANTA BARBARA county, SW CALIFORNIA, 16 mi/26 km SSE of SANTA MARIA. Oil field nearby.

Los Aldamas (los ahl-DAH-mahs), town, NUEVO LEÓN, N MEXICO, in lowland, on SAN JUAN RIVER near PRESA

MARTE R. GÓMEZ, and 75 mi/121 km ENE of MONTERREY; 26°02′N 99°12′W. Cotton, corn, sugarcane.

Los Alerces (LOS ah-LAIR-ses), Argentinian national park in W CHUBUT province, Argentina, in the Patagonian lake district; comprises lakes of EPUYÉN, PUELO, MENÉNDEZ, and FUTALAUFQUÉN. Tourist resort; lumbering and fishing district. Established 1937.

Los Altos (los AHL-tos), region, W sector of the volcanic highlands, especially QUEZALTENANGO, SAN MARCOS, and HUEHUETENANGO departments, GUATEMALA; 14°42′N 90°28′W–14°55′N 90°38′W. Historically important because these three departments separated from Guatemala and formed the Republic of Los Altos in 1838. Incorporated as the sixth province in the Central American Federation, but annexed by Guatemala in 1840. A popular uprising in 1848 failed.

Los Altos, city (2000 population 27,693), SANTA CLARA county, W CALIFORNIA; residential suburb 9 mi/14.5 km WNW of downtown SAN JOSE, and 30 mi/48 km SSE of downtown SAN FRANCISCO; 37°22′N 122°06′W. Drained by Adobe Creek. Diversified light manufacturing. Junior college in nearby LOS ALTOS HILLS. Moffett Field Naval Air Station to NE. HETCH HETCHY AQUEDUCT runs E-W to N. Portola and Castle Rock state parks and SANTA CRUZ MOUNTAINS to SW. Incorporated 1952.

Los Altos (LOS AHL-tos), town, SUCRE state, NE VENEZUELA, near CARIBBEAN coast, 21 mi/34 km SW of CUMANÁ. Cacao, coffee, sugarcane.

Los Altos Hills, city (2000 population 7,902), SANTA CLARA county, W CALIFORNIA; residential suburb 14 mi/23 km W of downtown SAN JOSE, and 30 mi/48 km SSE of downtown SAN FRANCISCO, just WSW of LOS ALTOS, in NE foothills of SANTA CRUZ MOUNTAINS; 37°22′N 122°08′W. Stanford University to N. Portola and Castle Rock state parks to S. Seat of Foothill College (two-year).

Los Amates (los ah-MAH-tes), town, IZABAL department, E GUATEMALA, on MOTAGUA River, on railroad, and 45 mi/72 km SW of PUERTO BARRIOS; 15°16′N 89°06′W. In banana area; grain; livestock; lumbering. QUIRIGUÁ archaeological site is in Los Amates municipio, 5.6 mi/9 km from the town. The town marks a climatic transition from the dry middle Motagua River valley (ZACAPA, EL PROGRESO) and the humid lower valley (Izabal department), which is Guatemala's principal banana production center.

Los Andes, province, LA PAZ department, W BOLIVIA; ⊙ PUCARANI; 16°20′S 68°30′W.

Los Andes (los AHN-dais), easternmost province of VALPARAISO region, N central CHILE; ⊙ LOS ANDES; 32°45′S 70°20′W. Mountainous area. Mining; agriculture in broader valleys and foothills of the ANDES.

Los Andes (los AHN-dais), town, ⊙ Los Andes comuna (2002 population 55,127) and province, VALPARAISO region, central CHILE, in the Andes, on ACONCAGUA RIVER, and 45 mi/72 km N of SANTIAGO, on TRANSANDINE RAILWAY and highway; 32°50′S 70°37′W. Elevation 2,700 ft/823 m. Agriculture (cereals, grapes; livestock) and industrial center, with copper mining and marble quarrying. Food processing. Artisan center. Formerly Santa Rosa de los Andes.

Los Andes (LOS AHN-dais), town, ⊙ Los Andes municipio, NARIÑO department, SW COLOMBIA, 25 mi/40 km NW of PASTO in the Pacual River valley; 01°30′N 77°29′W. Elevation 2,857 ft/870 m. Sugarcane, wheat; livestock. Sometimes called Sotomayor.

Los Angeles, county (□ 4,060 sq mi/10,556 sq km; 2006 population 9,948,081), S CALIFORNIA; ⊙ LOS ANGELES; 34°11′N 118°16′W. The fertile Los Angeles basin, a plain reaching to the PACIFIC OCEAN on the W, is almost surrounded by mountains covering c.50 percent of county's surface; in the basin and in tributary of SAN FERNANDO VALLEY are the scores of cities of the metropolitan area, many of them virtually part of Los Angeles, and all including large suburban and semirural areas. Even areas at edge of the desert, N of SAN

GABRIEL, had experienced rapid development by the mid-1990s. Many cities near Los Angeles have now declined economically. Among the largest (in population) are LONG BEACH, PASADENA, GLENDALE, SANTA CLARITA, POMONA, and TORRANCE.

SAN GABRIEL MOUNTAINS (NE wall of Los Angeles basin) have peaks over 10,000 ft/3,048 m; coastal ranges (including SANTA MONICA MOUNTAINS, SANTA SUSANA MOUNTAINS) are lower. Antelope Valley (part of MOJAVE DESERT; irrigated agriculture) is in N. Off coast, which is indented by SAN PEDRO BAY (LOS ANGELES HARBOR here) and Santa Monica Bay, are CHANNEL ISLANDS, including SANTA CATALINA (also called Catalina), noted resort, and SAN CLEMENTE ISLAND. Intermittent LOS ANGELES and SAN GABRIEL rivers and their tributaries have flood-control works. California Aqueduct crosses N part of county NW-SE, Los Angeles Aqueduct crosses county N-S.

County's mediterranean climate and its resorts have long attracted winter residents and year-round tourists. Has some of nation's most valuable farmland (irrigation required, as c.50 percent of them are very small), but much of it has been replaced by urbanization. Important industries (motion pictures at HOLLYWOOD, CULVER CITY, and BURBANK; oil refining, automobile assembling; manufacturing of aircraft, tires and tubes, steel, foundry products, clothing, furniture, food products, computer equipment and software), rich oil and natural-gas fields. Chief farm products are dairy products; cattle, poultry; honey, strawberries, peaches, onions, barley; ornamentals and bedding plants. Part of EDWARDS AIR FORCE BASE on N boundary in NE; part of SANTA MONICA MOUNTAINS NATIONAL RECREATION AREA in W; parts of Angeles National Forest in center (San Gabriel Mountains) and NW; Saddleback Butte State Park in NE; part of Mojave Desert in NE; part of Antelope Valley in NW; numerous state beaches on coast. Formed 1850.

Los Ángeles (los AHN-hai-lais), city, ⊙ Los Ángeles comuna (2002 population 117,972) and BÍO-BÍO province, BÍO-BÍO region, S central CHILE, on railroad and the PAN-AMERICAN HIGHWAY, 60 mi/97 km SE of CONCEPCIÓN, 300 mi/483 km SSW of SANTIAGO. Airport. Agricultural center (cereals, fruit, grapes, vegetables); food processing, wood products, lumbering. Popular Salta de Laja recreation area is nearby. Founded 1739 as a fort on the Indian frontier.

Los Angeles, city (2000 population 3,694,820), ⊙ LOS ANGELES county, S CALIFORNIA, bounding PACIFIC OCEAN in three places, at PACIFIC PALISADES c.15 mi/24 km W of downtown, at VENICE 12 mi/19 km SW of downtown, and at SAN PEDRO (Long Beach area) 20 mi/32 km S of downtown; 34°07′N 118°25′W. A port of entry on the Pacific coast, with a fine harbor at San Pedro Bay, it is the second largest U.S. city in population and seventh largest in area. Often referred to as the City of Angels.

Drained by LOS ANGELES RIVER, a large concrete flood-control channel that can swell into a raging torrent during Pacific storms. Two low mountain ranges, the Santa Monica and Verdugo, cut across the center of the city. Los Angeles's warm climate is tempered by cool ocean currents that flow S along the coast from the N Pacific.

Los Angeles is a shipping, industrial, communication, financial, fashion, and distribution center for the W U.S. and much of the the PACIFIC RIM. It is also the motion-picture, television, radio, and recording capital of the U.S., if not the world, with numerous studios. Once an agricultural-distribution center, Los Angeles is one of the country's largest center of the clothing and textile industries. It is also a leading producer of aircraft, computers, paper, toys, glass, furniture, wire products, biomedical-industry products, electrical and electronic machinery, pharmaceuticals, petrochemicals, and fabricated metal. Tourism,

multimedia and cybernet-entertainment software technology, printing and publishing, food processing, and oil refining are also major industries. Growth in many of these areas has offset a decline in financial, real estate, and insurance sectors that has occurred in the mid-1990s, as well as the relocation of several large corporate headquarters. Oil and natural-gas fields in part of city and environs. LOS ANGELES HARBOR and Long Beach Harbor, together, are Los Angeles's port, part in City of LONG BEACH and part in Los Angeles (the two cities adjoin each other). Los Angeles has one of the busiest ports in the U.S.; roughly half of its commerce is foreign. There is a smaller boat harbor in MARINA DEL REY, Venice area. It is the principal financial center for the West Coast and the E Pacific Basin.

The vast Los Angeles metropolitan area covers five counties (Los Angeles, Orange, Riverside, San Bernardino, and Ventura) and encompasses 34,000 sq mi/88,060 sq km with over 14.5 million people. Los Angeles's metro area has practically merged with San Diego's metro area to S. The Los Angeles metropolitan area is connected by a freeway system that is increasingly unable to accommodate the area's traffic. The enormous number of motor vehicles, combined with the city's valley location, creates dangerously high levels of smog. A light-rail system (Los Angeles County Metropolitan Transportation Authority; opened in 1990) and bus service alleviate only a small percentage of freeway congestion. Los Angeles's only subway line began operation in 1993 and was competed in 2000; other light-rail lines connected with the system were completed in 2003, linking the downtown area to Long Beach in the S and LOS ANGELES INTERNATIONAL AIRPORT in the N.

Maintaining an adequate water supply for the city has long been a problem; the metropolitan area obtains its water from the OWENS RIVER rising in the Sierra Nevada (via Los Angeles Aqueduct), the COLORADO RIVER (via COLORADO RIVER AQUEDUCT) to E, and from N California (via California Aqueduct) through California's CENTRAL VALLEY to the N.

As Los Angeles rapidly expanded throughout the 20th century, it absorbed numerous communities and enclosed independent municipalities. Among the communities now part of Los Angeles are Century City, Brentwood, HOLLYWOOD, San Pedro, Sylmar, Watts, Westwood Hills, Bel Air, and Boyle Heights, and several large suburbs in the SAN FERNANDO VALLEY, such as Northridge, Sherman Oaks, and Van Nuys. Two moderate earthquakes (1971 and 1994), each claiming more than sixty lives, were epicentered in the San Fernando Valley. Independent municipalities surrounded by Los Angeles include SANTA MONICA (2000 population 84,084), BEVERLY HILLS (2000 population 33,784), and SAN FERNANDO (2000 population 23,564). Incorporated cities in the broader metropolitan region with populations of 80,000 or more include Alhambra, ANAHEIM, Burbank, Downey, El Monte, FULLERTON, Garden Grove, Glendale, HUNTINGTON BEACH, Irvine, Inglewood, Long Beach, Moreno Valley, Norwalk, Oceanside, Ontario, Orange, Oxnard, PASADENA, Pomona, Rancho Cucamonga, Riverside, San Bernardino, Santa Ana, Santa Clarita, Santa Monica, Simi Valley, Thousand Oaks, and Torrance.

Spanish explorer Gaspar de Portola visited the site of the city in 1769, and in 1781 El Pueblo de Nuestra Señora de los Angeles de Porciuncula (The Town of Our Lady the Queen of the Angels of Porciuncula) was founded. The city served several times as the capital of the Spanish colonial province of ALTA CALIFORNIA and was a cattle-ranching center. U.S. forces captured Los Angeles from the Mexicans in 1846. Los Angeles incorporated four years later. The arrival of the railroad (Southern Pacific in 1876; Santa Fe in 1885) and the discovery of oil in the early 1890s stimulated ex-

pansion, as did the development of the motion-picture industry in the early 20th century. During World War II, Los Angeles boomed as a center for the production of war supplies and munitions and thousands of African-Americans migrated to Los Angeles to fill factory jobs. After the war, massive suburban growth made the city enormously prosperous, but also created or exacerbated a variety of urban problems. In 1965, the African-American community of WATTS was the site of six days of rioting that left thirty-four people dead and caused over $200 million in property damage.

Los Angeles experienced dramatic growth through immigration in the 1970s and 1980s. Today, Los Angeles is one of the most racially and ethnically diverse cities in the U.S. In 1990, the Hispanic population of metropolitan Los Angeles was almost five million (almost 40% of the population) and the area's Asian population was over 1.3 million. In addition to an already well established Japanese-American community, recent immigration has come from China, South Korea, Vietnam, Cambodia, the Philippines, and other nations. Chinatown is N of downtown, Little Tokyo lies SE of downtown. In the 1980s, violent gang warfare over the illegal-drug (especially crack) trade became a serious problem for law-enforcement officials. In April 1992, four white Los Angeles police officers were found not guilty of police brutality after they had been videotaped beating Rodney King, a black motorist they were pursuing; the acquittal touched off riots in S central Los Angeles and many other areas that resulted in over fifty deaths, thousands of arrests, and approximately $1 billion in property damage.

In Los Angeles are botanical gardens; the Los Angeles County Museum of Art; the Museum of Contemporary Art; history, movie, industrial, and science museums; and many parks, including Griffith Park, one of the largest urban parks in the world, with a zoo and planetarium. The LA BREA Tar Pits, near Beverly Hills, are famous for Ice Age fossils. The Los Angeles Philharmonic is internationally famous as is the city's opera company. In 1982, Los Angeles gained its second National Football League team (the Rams were the first) when the Oakland Raiders moved to the city, but the Rams moved to St. Louis and the Raiders back to Oakland (both in 1995), leaving the city without a National Football League franchise. In baseball, the National League's Los Angeles Dodgers and the American League's Los Angeles Angels of Anaheim represent the area. The city also has two National Basketball Association teams (the Lakers and the Clippers) and a National Hockey League team (the Kings). Los Angeles hosted the summer Olympic Games twice (1932 and 1984).

The motion-picture and television industries, the proximity of many resorts and theme parks (Six Flags Magic Mountain and Six Flags Hurricane Harbor), the fine beaches, and a climate that encourages year-round outdoor recreation attract millions of tourists annually to Los Angeles. Other attractions in the region include the Santa Anita and Hollywood Park racetracks, Knott's Berry Farm, and Disneyland (at Anaheim). Among the city's many educational institutions are the University of Southern California; the University of California, Los Angeles (UCLA); California State University, Los Angeles, in NE; California State University, Northridge, in NW; Occidental College; Loyola Marymount University; Mount St Mary's College; Fashion Institute of Design and Merchandising (two-year), and numerous other two-year colleges. Will Rogers State Historical Park and State Beach in W; Topanga State Park and State Beach in W; Los Encinos State Historical Park in W; El Pueblo de Los Angeles State Historical Park in downtown, across the street from Union Station; Venice City and Dockweiler state beaches in W; Royal Palms State Beach in S; Santa Monica Mountains National Recreation Area in W, extends W into Ventura county; Angeles National Forest, in San Gabriel Mountains, to N; Busch Gardens in San Fernando Valley.

Los Angeles Aqueduct, California; see OWENS RIVER.

Los Angeles Harbor, man-made port (□ 12 sq mi/31.2 sq km) of LOS ANGELES, LOS ANGELES COUNTY, S CALIFORNIA, on SAN PEDRO BAY, 20 mi/32 km S of city's center, at SAN PEDRO and WILMINGTON. Adjoins Long Beach Harbor to E; both share same breakwater-system shelter and both function as Los Angeles's port. Consists of two major parts: outer harbor (U.S. Navy's chief Pacific coast anchorage) in SAN PEDRO BAY, sheltered by breakwater (in two sections; 4.5 mi/ 7.2 km long), extending E from POINT FERMIN; inner harbor consists of channels and turning basins dredged in former mudflats and of TERMINAL ISLAND (largely artificial) lying between inner and outer harbors. Cerritos Channel connects Inner Harbor with Long Beach Harbor (E) at U.S. Naval Shipyard. Port generally ranks high in U.S., and first on Pacific coast, in volume of cargo, most of which is outgoing (foreign and coastwise); twelfth largest in U.S., Long Beach tenth largest, together fourth largest in U.S.

Ships coal, oil and petroleum products, manufactured goods, citrus and other fruit, canned fish, vegetables, cotton, borax, potash, soda ash, cement; receives iron, steel, lumber, bananas, copra, jute, hemp, fertilizers, rubber, sugar, spices, coffee, tea, newsprint, wool. Harbor district (Wilmington, San Pedro, HARBOR CITY) is served by 120 mi/193 km of railroad, connecting with three transcontinental lines; container port transfers cargo directly from trains and trucks to container ships. Port has c.25 mi/40 km of dock frontage, oil terminals and refineries, shipyards, dry docks, a foreign trade zone (free port) opened in 1949, a naval-operating base and other naval and coast guard installations, and a fishing port which generally leads California in catches (tuna, mackerel, sardines). Passenger terminals (at Wilmington) and automobile-assembling and other industrial plants.

Harbor's modern development began in 1899, when breakwater was begun in exposed anchorage that had been in use as port since 1850s, and which had been connected (1869) by railroad to Los Angeles; Los Angeles annexed San Pedro and Wilmington in 1909, and formed a harbor district to coordinate development; dredging of inner harbor was done 1912–1914. Opening (1914) of PANAMA CANAL, and the accelerating agricultural and commercial growth of Southern California, as well as development of area's great oil fields, stimulated trade of port and led to subsequent improvements.

Los Angeles International Airport, (□ 5 sq mi/13 sq km) LOS ANGELES COUNTY, S CALIFORNIA, serving the greater LOS ANGELES metropolitan region, 10 mi/16 km from the city center (road, rail links); 33°56′N 118°24′W. Elevation 125 ft/38 m. The airport is the fifth-busiest in the world with four parallel E-W runways, three of which are over 10,000 ft/3,050 m. The airport handles over sixty million passengers and more than two million tons of cargo annually. Opened to general aviation as Mines Field in 1928, used for military aviation during World War II, it opened to commercial aviation in 1946. Vice President Lyndon B. Johnson dedicated the current terminal complex at its opening in 1961. A U.S. Coast Guard Air Station is also part of the complex. The airport is probably better known by its airport code: LAX.

Los Angeles River, channeled intermittent stream, c.50 mi/80 km long, LOS ANGELES COUNTY, S CALIFORNIA; rises c.25 mi/40 km NW of downtown LOS ANGELES, in SANTA SUSANA MOUNTAINS, enters Los Angeles, drains SAN FERNANDO VALLEY, flows E along N base of SANTA MONICA MOUNTAINS, through Sepulveda Flood Control Dam, where it receives LOS ANGELES AQUEDUCT from N, then S, past E end of range and through downtown Los Angeles, S through its S in-

dustrial district and SE suburbs, to SAN PEDRO BAY at Long Beach. Torrential rainy-season flows are controlled by masonry embankments and huge catchment basins on upper river and tributaries (Pacoima River, TUJUNGA CREEK, short Rio Hondo). Dry concrete culverts become raging torrents during storms.

Los Ánjeles, CHILE: see LOS ÁNGELES.

Los Antiguos (los ahn-TEE-gwos), village, SANTA CRUZ province, ARGENTINA, on S shore of LAKE BUENOS AIRES, on Chile border, 130 mi/209 km W of COLONIA LAS HERAS; 46°33′S 71°37′W. Livestock-raising (sheep, goats) and lumbering area. Customhouse, airport.

Losantville, town (2000 population 280), RANDOLPH county, E INDIANA, 14 mi/23 km SW of WINCHESTER; 40°01′N 85°11′W. Agricultural area. Also called Bronson.

Losap (LO-sahp), atoll, State of CHUUK, E CAROLINE ISLANDS, Federated States of MICRONESIA, W PACIFIC, 75 mi/121 km SE of CHUUK ISLANDS; 5 mi/8 km long; 17 islets on triangular reef.

Los Arabos (los ah-RAH-bos), town, MATANZAS province, W CUBA, on Central Highway and 60 mi/97 km ESE of MATANZAS; 22°44′N 80°44′W. Railroad junction in agricultural region (sugarcane, bananas). The refinery and central of México is NE.

Losarcos, SPAIN: see ARCOS, LOS.

Losar de la Vera (lo-SAHR dhai lah VAI-rah), town, CÁCERES province, W SPAIN, 27 mi/43 km ENE of PLASENCIA; 40°07′N 05°36′W. Olive-oil processing; livestock raising; ships pepper and fruit.

Los Asientos (an-see-YEN-tos), village and minor civil division of Pedasí district, LOS SANTOS province, S central PANAMA, in PACIFIC lowland, 3 mi/4.8 km W of PEDASÍ. Sugarcane, coffee, corn; livestock.

Los Ausoles (los au-SO-les), EL SALVADOR, near AHUACHAPÁN. Site of large geothermal power generation project.

Los Bajos (los BAH-hos), village, LLANQUIHUE province, LOS LAGOS region, S central CHILE, on headland at W shore of LAKE LLANQUIHUE, 28 mi/45 km N of PUERTO MONTT. Dairying, lumbering.

Los Banos, city (2000 population 25,869), MERCED county, central CALIFORNIA, in San Joaquin Valley, 22 mi/35 km SW of MERCED; 37°04′N 120°51′W. Dairying; poultry; irrigated farming (beans, tomatoes, sweet potatoes, almonds, rice, wheat, corn, cotton); manufacturing (electrical machinery; printing and publishing). Fremont Ford State Recreation Area to N, San Luis State Recreation Area to W; San Luis and Merced national wildlife refuges to NE. DELTA-MENDOTA CANAL and California Aqueduct to SW.

Los Cabos, MEXICO: see SAN JOSÉ DEL CABOS.

Los Cerrillos (LOS se-REE-os), village, SANTA FE county, N central NEW MEXICO, on Gallisteo Creek, and 27 mi/43 km SSW of SANTA FE. Elevation 5,888 ft/ 1,795 m. Cattle, sheep; corn, grain, alfalfa. Several Pueblo Indian villages in vicinity. Santo Domingo Indian Reservation to W.

Los Cerrillos, URUGUAY: see CERRILLOS.

Lösch, CZECH REPUBLIC: see LISEN.

Los Chacos (los CHA-kos), canton, WARNES province, SANTA CRUZ department, E central BOLIVIA NE of AZUZAQUÍ; 14°33′S 62°11′W. Elevation 1,089 ft/332 m. Potatoes, yucca, bananas, corn, rice, cotton, peanuts; cattle.

Los Chaves (LOS CHAH-ves), unincorporated village (2000 population 5,033), VALENCIA county, central NEW MEXICO, 27 mi/43 km S of ALBUQUERQUE, on RIO GRANDE; 34°43′N 106°45′W. Agricultural area (dairying; cattle, sheep; grain, alfalfa, grapes). Senator William M. Chaves State Park located here.

Los Chiles (lohs CHEE-lais), town, ⊙ Los Chiles canton, N ALAJUELA province, COSTA RICA, 2 mi/3.2 km S of NICARAGUA border, and on Palos River (with drainage to SAN JUAN RIVER); 10°52′N 84°40′W. Livestock; bananas, plantains, manioc, corn; lumbering. Principal center of GUATUSO plain.

Loschtitz, CZECH REPUBLIC: see LOSTICE.

Los Coconucos (LOS ko-ko-NOO-kos), volcanic massif, SW COLOMBIA, in Cordillera CENTRAL, 20 mi/ 32 km SE of POPAYÁN; 02°15′N 76°23′W. Has 2 major peaks, rising to 14,908 ft/4,544 m.

Los Cóndores, village, N SAN LUIS province, ARGENTINA, 60 mi/97 km NE of SAN LUIS; 32°20′S 64°16′W. Tungsten.

Los Córdobas (LOS KOR-do-bahs), town, ⊙ Los Córdobas municipio, CÓRDOBA department, N COLOMBIA, 25 mi/40 km WNW of MONTERÍA; 08°53′N 76°21′W. Agriculture includes cassava, rice; livestock.

Los Corralitos (los kor-rah-LEE-tos), town, N MENDOZA province, ARGENTINA, in MENDOZA RIVER valley (irrigation area), 10 mi/16 km ESE of MENDOZA. In fruit-growing area.

Los Duranes (LOS du-RAHN-es), village, BERNALILLO county, central NEW MEXICO, on RIO GRANDE, part of city of ALBUQUERQUE, near Old Albuquerque (Old Town).

Los Encuentros (los en-KWEN-tros), village, SOLOLÁ department, SW central GUATEMALA, road center 4.5 mi/7.2 km N of SOLOLÁ; 14°51′N 91°09′W. Elevation 8,860 ft/2,701 m. Important stopping point for vehicle traffic on INTER-AMERICAN HIGHWAY and an intersection, where CHICHICASTENANGO–QUICHÉ highway branches from the Inter-American Highway.

Loser (LO-ser), peak (5,602 ft/1,707 m) of the TOTES GEBIRGE, NW STYRIA, W central AUSTRIA, overlooking ALTAUSSEE in the SALZKAMMERGUT; 47°40′N 13°46′E. Excellent view.

Loser, Mount, INDONESIA: see LEUSER, MOUNT.

Losevo (LO-see-vuh), village (2006 population 4,455), S central VORONEZH oblast, S central European Russia, on the BITYUG RIVER, on highway branch, 38 mi/61 km NNE of ROSSOSH; 51°44′N 38°50′E. Elevation 416 ft/126 m. In agricultural area (grains, hemp, sugar beets; livestock raising); food processing.

Los Fresnos (LOS FRES-nos), town (2006 population 5,345), CAMERON county, extreme S TEXAS, 12 mi/19 km N of BROWNSVILLE; 26°04′N 97°28′W. Rich irrigated farming area (cotton, vegetables, sugarcane) of lower Rio Grande valley; manufacturing (building materials, meat rendering). Palo Alto Battlefield National Historic Site to S; Laguna Atacosca National Wildlife Refuge to NE. Incorporated after 1940.

Los Gatos, city (2000 population 28,592), SANTA CLARA county, W CALIFORNIA; suburb 9 mi/14.5 km SW of SAN JOSE, at S edge of Bay Area urbanized area; 37°14′N 121°58′W. Drained by Los Gatos Creek. It is an affluent residential community. Agriculture to S (cherries, vegetables, nursery products; poultry; dairying); manufacturing (electronic components, computer peripherals). Los Gatos, Spanish for "the cats," got its name from the wildcats that abounded in the SANTA CLARA VALLEY at the time the city was founded. SANTA CRUZ MOUNTAINS to SW; Forest of Nisene Marks State Park to S; Big Basin Redwoods State Park to W; Lexington Reservoir to S. Incorporated 1887.

Loshan, CHINA: see LESHAN.

Losheim (LOS-heim), town, SAARLAND, SW GERMANY, 7 mi/11.3 km NE of MERZIG; 49°31′N 06°44′E. Woodworking; tourism.

Los Herreras (los he-RE-rahs), town, NUEVO LEÓN, MEXICO, on railroad, on PESQUERÍA RIVER, and 60 mi/ 97 km ENE of MONTERREY. Elevation 656 ft/200 m. Cotton, sugarcane, cereals, fruit.

Los Huemules (los wai-MOO-lais), national park, in AISÉN DEL GENERAL CARLOS IBAÑEZ DEL CAMPO region, CHILE, 35 mi/56 km SW of PUERTO AISÉN.

Losice (lo-SEE-tse), Polish Łosice, Russian Lositsy, town, Biała Podlaska province, E POLAND, 19 mi/31 km E of SIEDLCE; 52°13′N 22°43′E.

Losimingur, Mount (lo-see-meen-GOOR) (7,546 ft/ 2,300 m), ARUSHA region, N TANZANIA, 40 mi/64 km W of Arusha, 12 mi/19 km NE of Lake MANYARA; 03°26′S 36°04′E. Olkerii Escarpment to N.

Lošinj Island (LO-sheen-ye), Ital. *Lussino*, (□ 29 sq mi/
75.4 sq km) in ADRIATIC SEA, W CROATIA, in KVARNER
GULF; 19 mi/31 km long N-S; rises (N) to c.1,930 ft/588
m; linked with CRES Isl. by bridge. Southernmost
point is 40 mi/64 km NW of ZADAR. Tourism. Livestock, agr. (olives and grapes), fishing; shipbuilding.
Passed (1918) to Italy and (1947) to the former Yugoslavia. Chief town, MALI LOSINJ, (Ital. *Lussinpiccolo*)
(1991 population 6,566), is on S part of isl.; resort. First
mentioned in 14th cent. as Malo Selo [=small village].
Village of VELI LOSINJ nearby.

Losinoostrovskaya, RUSSIA: see BABUSHKIN, MOSCOW
oblast.

Losino-Petrovskiy (luh-SEE-nuh–peet-ROF-skeeyee),
city (2006 population 22,050), E central MOSCOW
oblast, central European Russia, on the KLYAZ'MA
RIVER, on road and railroad, 32 mi/51 km ENE of
MOSCOW, and 9 mi/14 km W of NOGINSK; 55°52′N
38°12′E. Elevation 456 ft/138 m. Textile industry (initial processing of wool, worsteds, silk cloth); agricultural chemicals. Made city in 1951. Until 1928, called
Petrovskaya Sloboda.

Losinovka, UKRAINE: see LOSYNIVKA.

Losinyy (luh-SEE-niyee), town (2006 population
2,180), S SVERDLOVSK oblast, W SIBERIA, RUSSIA, on
road and railroad (Aduy station), 17 mi/27 km NNE
(under jurisdiction) of BERËZOVSKIY; 57°08′N 61°04′E.
Elevation 800 ft/243 m. Peat briquettes, sport shoes.

Los Islands, French *Îles de Los* or *Îles de Loos*, group of
ATLANTIC OCEAN islets just off CONAKRY, GUINEA;
09°30′N 13°50′W. There are five larger islands, among
them TAMARA and FACTORY islands. Bauxite deposits.
Beaches, tourism, fisheries. Principal villages are FO-
TOBA and Kassa. Ceded 1904 by GREAT BRITAIN to
FRANCE.

Lositsy, POLAND: see LOSICE.

Loskop Dam, MPUMALANGA, SOUTH AFRICA, 15 mi/24
km NE of TSHWANE (formerly Pretoria) on WILGE
RIVER. Supplies water to mining areas around MARBLE
HALL and Groblersdal.

Loskutovo (luhs-KOO-tuh-vuh), village (2006 population 3,125), SE TOMSK oblast, S central Siberian
Russia, near highway and railroad, 7 mi/11 km SE, and
under administrative jurisdiction, of TOMSK; 56°24′N
85°08′E. Elevation 521 ft/158 m. Agricultural chemicals.

Los Lagos (los LAH-gos) or **X region** (2002 population
1,066,310), 2005 estimated population 1,156,304), S
CHILE, extending from the ARAUCANIA region in the N
to the AISÉN DEL GENERAL CARLOS IBAÑEZ DEL CAMPO
region in the S; ⊙ PUERTO MONTT; 41°45′S 73°00′W. A
land of lakes, fjords, and forest, Los Lagos stretches
from the ANDES in the E to the PACIFIC OCEAN in
the W. Its provinces are VALDIVIA, OSORNO, PUERTO
MONTT, CASTRO, and CHAITÉN. Major occupations
include extensive agriculture (livestock, dairying),
lumbering, fishing. Food processing in larger cities.
Tourism. National parks include PIRIHUEICO, PEREZ
ROSALES, ALERCE ANDINO, CHILOÉ.

Los Lagos (los LAH-gos), city, ⊙ Los Lagos comuna,
VALDIVIA province, LOS LAGOS region, S central
CHILE, on CALLE-CALLE RIVER, and on railroad and
PAN-AMERICAN HIGHWAY, 23 mi/37 km E of VALDI-
VIA; 39°51′S 72°50′W. Agricultural area (cereals, veg-
etables; livestock); food processing, lumbering.

Los Lagos, ARGENTINA: see NAHUEL HUAPÍ, village.

Loslau, POLAND: see WODZISLAW.

Los Laureles (los lou-RAI-lais), town, CAUTÍN province, ARAUCANIA region, S central CHILE, on railroad,
and 18 mi/29 km ESE of TEMUCO. Agricultural center
(wheat, barley, peas, potatoes; livestock); flour milling, lumbering.

Los Libertadores-Wari (LOS lee-ber-tah-DO-res–
WAH-ree), former planning region, SW PERU. Consisted of Ica, Ayacucho, and Huancavelica departments
(today ICA, AYACUCHO, and HUANCAVELICA regions).
Rich in mineral resources (lead, zinc, gold, silver, salt,
copper, iron, mercury, limestone); hot springs. Agri-

culture (rice, corn, beans, potatoes, wheat, cassava,
coffee, cacao, barley, plantains, citrus fruits); meat and
meat products (sheep, fowl, pork, beef); dairying;
wool. Created as part of Peru's 1988 regionalization
program. These regions never quite caught on and
were abandoned.

Los Llanos (los YAH-nos), officially Villa de San José
de Los Llanos, town, SAN PEDRO DE MACORÍS province, SE DOMINICAN REPUBLIC, 28 mi/45 km ENE
of SANTO DOMINGO; 18°38′N 69°30′W. Sugar; cattle.
Sugar mill to SE.

Los Loros (los LO-ros), village, ATACAMA region, N
central CHILE, on COPIAPÓ RIVER (irrigation), on
railroad, and 40 mi/64 km SE of COPIAPÓ; 27°50′S
70°06′W. Alfalfa, clover, corn, subtropical fruit; goats.

Los Lunas (LOS LOON-uhs), town (2006 population
11,803), ⊙ VALENCIA county, central NEW MEXICO, on
RIO GRANDE, and 32 mi/51 km S of ALBUQUERQUE;
34°48′N 106°44′W. Elevation 4,852 ft/1,479 m. Trade
center, dairying and livestock-shipping point (cattle,
sheep) in irrigated grain- and vegetable-farming
(corn, alfalfa) area. Manufacturing (trailers). University of New Mexico, Valencia campus (two year).
ISLETA Pueblo Indian village to N in Isleta Indian
Reservation; MANZANO MOUNTAINS to E in part of
Cibola National Forest to E; Senator Willie M. Chavez
State Park to S.

Los Mármoles (los MAR-mo-les), national park (□ 75
sq mi/195 sq km), in NW HIDALGO, MEXICO, on
Mexico Highway 85 N of ZIMAPÁN. Has natural
springs and a large canyon (San Vicente). The forest
terrain includes oak, pine, fir, and walnut trees. The
park is underdeveloped and has no visitor facilities.
Named for large marble outcrops.

Los Menucos (los me-NOO-kos), town, S central RÍO
NEGRO province, ARGENTINA, on railroad, and 45 mi/
72 km NE of MAQUINCHAO; 40°50′S 68°08′W. Sheep,
goats.

Los Millanes (LOS mee-YAH-nes), town, on MAR-
GARITA ISLAND, NUEVA ESPARTA state, NE VENE-
ZUELA; 7 mi/11 km W of LA ASUNCIÓN; 11°04′N
63°57′W. Corn, sugarcane, fruit.

Los Mochis (los MO-chees), city (2005 population
231,977) and township, ⊙ Ahome Municipio, SINALOA
state, W MEXICO, on railroad, and on Mexico Highway
15, 140 mi/226 km SW of CIUDAD OBREGON; 25°48′N
109°00′E. It is the commercial and processing center of
the rich agricultural area (Río Fuerte Irrigation district) irrigated by the FUERTE River. Produts include
grains, sugarcane, and tomatoes, as well as cattle and
pigs. Los Mochis is additionally a tourist center.

Los Molinos, unincorporated town (2000 population
1,952), TEHAMA county, N CALIFORNIA, 12 mi/19 km
SSE of RED BLUFF, on SACRAMENTO RIVER, at mouth
of Mill River; 40°02′N 122°06′W. Woodson Bridge
State Recreation Area to S. Part of Lassen National
Forest to E; Tehama State Game Refuge to E. Olives,
walnuts, prunes; dairying; poultry.

Los Molles (los MO-yes), village, SW MENDOZA
province, ARGENTINA, on tributary of ATUEL RIVER,
and 100 mi/161 km WSW of SAN RAFAEL. Health resort with hot springs.

Los Monjes Islands (LOS MON-hais), VENEZUELA,
NE of LA GUAJIRA Peninsula (whose SE edge is in
Venezuela), consisting of Northern el Monje (12°29′N
70°55′W); Eastern el Monje (12°23′N 70°54′W); and
Monje del Sur (12°21′N 70°54′W). Venezuela's sovereignty over the islands was recognized by COLOMBIA
in 1976, but a dispute remains over who owns the
waters and oil resources around them, as well as who
has jurisdiction over the waters of the mouth of the
Gulf of VENEZUELA.

Los Muermos (los MWER-mos), village, ⊙ Los
Muermos comuna, LLANQUIHUE province, LOS LAGOS
region, S central CHILE, 27 mi/43 km WNW of PUERTO
MONTT; 41°24′S 73°29′W. Cereals, vegetables; livestock; dairying, lumbering.

Losna (luhs-NAH), settlement, central SMOLENSK oblast, W European Russia, near highway, 13 mi/21 km
NNW of POCHINOK, to which it is administratively
subordinate; 54°36′N 32°18′E. Elevation 695 ft/211 m.
Stone quarries. Sometimes spelled Losnaya.

Los Negros Island, BISMARCK ARCHIPELAGO, MANUS
province, N PAPUA NEW GUINEA, 1 mi/1.6 km E of
MANUS island (connected by bridge across Loniu
Passage), in PACIFIC OCEAN; 02°00′S 147°25′E. Includes townships of Lombrum and Momote. Fruits,
vegetables, coconuts. BISMARCK SEA to S. Variant
name: ROSU NEGUROSU JIMA.

Los Nietos, unincorporated town, LOS ANGELES county,
S CALIFORNIA; suburb 10 mi/16 km SE of downtown
LOS ANGELES, near WHITTIER, on SAN GABRIEL RIVER.
Oil wells. Pio Pico State Historical Park to NE.

Los Olivos, unincorporated village, SANTA BARBARA
county, SW CALIFORNIA, 26 mi/42 km NW of SANTA
BARBARA. Fruit, grapes, grain; cattle. Winery. Los
Padres National Forest and Sierra Madre to NE.

Losombo (lo-SAWM-bo), village, Équateur province,
NW CONGO, on LULONGA RIVER, and 80 mi/129 km
NNE of MBANDAKA; 00°12′N 18°53′E. Coffee; fishing.

Losonc, SLOVAKIA: see LUČENEC.

Losone, commune, TICINO canton, S SWITZERLAND, 2
mi/3.2 km W of LOCARNO; 46°08′N 08°45′E.

Lo, Song (LO, SOUNG), [Vietnamese=clear river],
Chinese *P'an-lung Chiang*, French *Rivière Claire*,
river, 250 mi/402 km long, rises in CHINA's YUNNAN
province N of WENSHAN at elevation of c.4,000 ft/1,219
m, flows SE into RED RIVER at VIET TRI, N VIETNAM;
22°00′N 105°09′E. Navigable for 5 ft/1.5 m-draught
boats up to TUYEN QUANG.

Lososina (luh-SAW-see-nuh) [Russian=salmon], village (2005 population 3,120), SE KHABAROVSK TER-
RITORY, RUSSIAN FAR EAST, on the TATAR STRAIT,
4 mi/6 km NNE of SOVETSKAYA GAVAN', to which it
is administratively subordinate; 49°01′N 140°21′E.
Fisheries.

Los Osos, unincorporated city, SAN LUIS OBISPO
county, SW CALIFORNIA, 10 mi/16 km WNW of SAN
LUIS OBISPO, 1 mi/1.6 km SE of BAYWOOD PARK, 2 mi/
3.2 km E of PACIFIC OCEAN. Statistically reported as
Baywood–Los Osos (1990 population 14,733). Flowers,
nursery stock, strawberries, apples, avocados, vegetables, grain; cattle. Morro Bay State Park and Atascadero State Beach to NW; Montana de Oro State Park
to S.

Los Palacios (los pah-LAH-see-os), town, PINAR DEL
RÍO province, W CUBA, on Los Palacios River, on
railroad, and 30 mi/48 km ENE of PINAR DEL RÍO;
22°36′N 83°15′W. In agricultural region (sugarcane,
tobacco, pineapples; cattle). Manufacturing of cigars.

Los Palmitos (los pahl-MEE-tos), town, ⊙ Los Palmitos municipio, SUCRE department, N central CO-
LOMBIA, 7 mi/11.3 km ENE of SINCELEJO 09°22′N
75°16′W. Corn, yams, rice; livestock.

Los Palos (los PAH-laws), town, ⊙ LAUTEM district
(□ 1,266 sq mi/3,291.6 sq km), INDONESIAN-controlled
EAST TIMOR, 95 mi/153 km E of DILI; 08°28′S 127°01′E.
Copra, tobacco, hardwood.

Los Paraguas, Cordillera de, Colombia: see LOS
PARAGUAS, SERRANÍA DE.

Los Paraguas, Serranía de (LOS pah-RAH-gwahs,
ser-rah-NEE-ah dai), ANDEAN range, W COLOMBIA,
forming part of CORDILLERA OCCIDENTAL, a territorial
extension of 136 sq mi/352 sq km; 04°40′N 76°20′W.
High levels of bio-diversity. Also known as Cordillera
de Los Paraguas.

Los Pinales, CUBA: see NIPE, SIERRA DE.

Los Pinos (los PEE-nos), town, LA HABANA province,
W CUBA, on railroad, W of HAVANA, and midway
between Havana and MATANZAS. In dairying region.

Los Pinos (los PEE-nos), village, SE BUENOS AIRES
province, ARGENTINA, 8 mi/12.9 km SW of BALCARCE.
In pine woods supplying timber for MAR DEL PLATA
paper mills; quartzite quarries.

Los Pinos River, c.75 mi/121 km long, in SW COLORADO and NW NEW MEXICO; rises at CONTINENTAL DIVIDE, in SAN JUAN MOUNTAINS in S central HINSDALE county, SW Colorado, source within 2 mi/3.2 km of RIO GRANDE, across Divide; flows SSW through Valecito Reservoir, past BAYFIELD and IGNACIO (Colorado), to SAN JUAN RIVER (NAVAJO Reservoir) in SAN JUAN county, New Mexico. Sometimes called PINE RIVER. VALLECITO DAM (162 ft/49 m high, 4,010 ft/1,222 m long; completed 1941) is 10 mi/16 km N of BAYFIELD; creates VALLECITO RESERVOIR (capacity 129,700 ft/39,533 m), used to irrigate 33,100 acres/13,396 ha. Lower 8 mi/12.9 km forms N arm of NAVAJO RESERVOIR.

Los Planes (los PLAH-nes) or **Planes de Renderos**, village, SAN SALVADOR department, S central SALVADOR, in coastal range, 5 mi/8 km S of San Salvador (linked by road). Residential mountain resort; tuberculosis sanatorium.

Los Pozos (PO-sos), town, ⊙ Los Pozos district, HERRERA province, S central PANAMA, 20 mi/32 km SW of CHITRÉ. Livestock raising.

Los Puertos de Altagracia (LOS PWER-tos dai ahl-tah-GRAH-see-ah) or **Altagracia**, town, ⊙ Miranda municipio, ZULIA state, NW VENEZUELA, port on narrows of Lake MARACAIBO, across from and 7 mi/11.3 km NE of MARACAIBO; 10°44′N 71°31′W. Terminus of railroad (built 1986–1990) and pipeline from El Mene oil fields, 35 mi/56 km E. Fishing and agricultural center (corn, beans, bananas, coconuts, cotton).

Los Queltehues, CHILE: see QUELTEHUES.

Los Quirquinchos (los kir-KEEN-chos), town, SW SANTA FE province, ARGENTINA, 70 mi/113 km SW of ROSARIO; 33°22′S 61°43′W. Agriculture center (wheat, corn, soybeans, alfalfa, flax; livestock, poultry; apiculture); flour milling.

Los Ralos (los RAH-los), town, central TUCUMÁN province, ARGENTINA, on railroad, and 14 mi/23 km ESE of TUCUMÁN. Agricultural center (sugarcane, corn, alfalfa, fruit; livestock); sugar refinery.

Los Ramones (los rah-MO-nes), town, NUEVO LEÓN, N MEXICO, on railroad, on PESQUERÍA RIVER, and 40 mi/64 km E of MONTERREY; 25°42′N 99°34′W. Cotton, sugarcane, grain.

Los Ranchos de Albuquerque (LOS RAHN-chos de Al-buh-kuhr-kee), town (2006 population 5,416), BERNALILLO county, N central NEW MEXICO, residential suburb 4 mi/6.4 km N of downtown ALBUQUERQUE, on RIO GRANDE; 35°10′N 106°39′W.

Los Rastrojos (LOS ruhs-TRO-hos), town, LARA state, NW VENEZUELA, on ACARIGUA-BARQUISIMETO highway, and 4 mi/6 km SE of Barquisimeto; 10°01′N 69°14′W. Elevation 1,492 ft/454 m. Sugarcane, corn, fruit; livestock.

Los Reyes (los RE-yes), residential town, part of Coyoacán delegación, Federal Distrito, central MEXICO, 6 mi/9.7 km S of MEXICO CITY.

Los Reyes, town, VERACRUZ, E MEXICO, in SIERRA MADRE ORIENTAL, 17 mi/27 km SSW of CÓRDOBA. Elevation 4,134 ft/1,260 m. Fruit.

Los Reyes Acaquilpan (los RE-yes ah-kah-KEEL-pahn), city (2005 population 232,211), LA PAZ municipio, MEXICO state, central MEXICO, 12 mi/19 km SE of MEXICO CITY. railroad junction. Part of the AREA METROPOLITANA DE LA CIUDAD DE MÉXICO. Formerly an agricultural community, now rapidly urbanizing. Also known as Los Reyes la Paz.

Los Reyes de Juárez (los RE-yes dai HWAH-res), town, PUEBLA, central MEXICO, on railroad, and 28 mi/45 km ESE of PUEBLA. Agricultural center (cereals, maguey, fruit).

Los Reyes de Salgado (los RE-yes dai sahl-GAH-do), town, MICHOACÁN, central MEXICO, in TRANSVERSE VOLCANIC AXIS, 30 mi/48 km SSW of ZAMORA DE HIDALGO on Mexico Highway 40; 19°35′N 102°29′W. Agricultural center (corn, alfalfa, tobacco, fruit; livestock).

Los Reyes la Paz, MEXICO: see LOS REYES ACAQUILPAN.

Los Ríos (los REE-os), province (□ c.2,600 sq mi/6,734 sq km; 2001 population 650,178), W central ECUADOR; ⊙ BABAHOYO. Inland, bounded E by the ANDES, it consists almost entirely of densely forested lowlands traversed by numerous rivers tributary to the GUAYAS. Has a humid, tropical climate with considerable rainfall December–June. A leading cacao-growing region, it also produces coffee, sugarcane, rice, coconuts, melons, oranges, bananas, cereals, oil palm, yucca. The luxuriant forests yield balsa wood, cedar, laurel, tagua nuts, *toquilla* straw, rubber, etc. Some cattle raising. Industries include wood processing and food processing. The region's products are shipped via Guayas River and its affluents to GUAYAQUIL, VINCES, and BABAHOYO, which serve as main river ports.

Los Robles, VENEZUELA: see EL PILAR.

Los Roques, Archipiélago, VENEZUELA: see LOS ROQUES, ISLAS.

Los Roques, Islas (LOS RO-kais, EES-lahs), group of small CARIBBEAN islands, Dependencias Federales, VENEZUELA, 85 mi/137 km N of CARACAS; 11°50′N 66°45′W. Consisting of 2 larger, narrow islets and a great number of smaller ones. Sparsely inhabited; have salt and guano deposits; some sponge fishing. Also referred to as Archipiélago Los Roques.

Los Santos (SAHN-tos), province (□ 1,411 sq mi/3,668.6 sq km; 2000 population 83,495) central PANAMA, on PACIFIC coast; ⊙ LAS TABLAS; 07°35′N 80°20′E. Occupies SE part of AZUERO PENINSULA. Agriculture (plantains, coffee, sugarcane, corn, rice, beans, yucca); poultry raising is important. There are deposits of gold. Province is served by a branch of INTER-AMERICAN HIGHWAY on the E coast. Main centers are Las Tablas and LOS SANTOS. Formed 1855 out of old Azuero province.

Los Santos (LOS SAHN-tos), town, ⊙ Los Santos municipio, SANTANDER department, N central COLOMBIA, 18 mi/29 km S of BUCARAMANGA; 06°46′N 73°06′W. Coffee, corn, cassava, sugarcane; livestock.

Los Santos, town, ⊙ Los Santos district, LOS SANTOS province, S central PANAMA, in PACIFIC lowland, on branch of INTER-AMERICAN HIGHWAY, and 14 mi/23 km NW of LAS TABLAS; 07°56′N 80°25′W. Commercial and industrial center; distilling, saltworking. Agriculture (bananas, corn, rice, beans, coffee; livestock). Founded 1555.

Los Santos Peninsula, PANAMA: see AZUERO PENINSULA.

Los Sauces (los SOU-sais), town, ⊙ Los Sauces comuna, MALLECO province, ARAUCANIA region, S central CHILE, 14 mi/23 km SSW of ANGOL, on railroad and Rehaé River; 37°58′S 72°50′W. Agricultural center (cereals, fruit, grapes, vegetables; livestock); food processing, lumbering.

Los Sauces, ARGENTINA: see SAN BLAS.

Lossburg (LUHS-burg), village, BADEN-WÜRTTEMBERG, SW GERMANY, in BLACK FOREST, 5 mi/8 km S of FREUDENSTADT; 48°25′N 08°27′E. Summer resort.

Losser (LAW-suhr), town, OVERIJSSEL province, E NETHERLANDS, 6 mi/9.7 km NE of ENSCHEDE, near Dinkel River; 52°16′N 07°01′E. German border 2 mi/3.2 km to E and 3 mi/4.8 km to S. Dairying; cattle, hogs, poultry; grain, vegetables; food processing, farming machines.

Los Serranos, unincorporated town, SAN BERNARDINO county, S CALIFORNIA; residential suburb 32 mi/51 km ESE of downtown LOS ANGELES, and 7 mi/11.3 km SW of ONTARIO; 33°58′N 117°42′W. Prado Flood Control Basin (Santa Ana River) to SE. Chino Hills State Park to S.

Lossiemouth (LO-see-mouth), town (2001 population 6,803), MORAY, NE Scotland, on MORAY FIRTH, at mouth of LOSSIE RIVER, and 5 mi/8 km N of ELGIN; 57°42′N 17°10′W. Fishing port, seaside resort. Town includes fishing port of Branderburgh. Formerly in Grampian, abolished 1996.

Lossie River (LO-see), 31 mi/50 km long, MORAY, NE Scotland; rises 10 mi/16 km W of Aberlour; flows NE, past Elgin, to MORAY FIRTH at LOSSIEMOUTH. Formerly in Grampian, abolished 1996.

Lössnitz (LOS-nits), town, SAXONY, E central GERMANY, in the ERZGEBIRGE, 3 mi/4.8 km NNE of AUE; 50°38′N 12°43′E. Textile and paper industry. Chartered 1284.

Los Surgentes (los soor-HAIN-tes), town, E CÓRDOBA province, ARGENTINA, on the RÍO TERCERO, and 20 mi/32 km S of MARCOS JUÁREZ; 32°58′S 62°01′W. Wheat, corn, alfalfa, soybeans; horticulture; livestock, poultry.

Lostant, village (2000 population 486), LA SALLE county, N ILLINOIS, 13 mi/21 km SW of OTTAWA; 41°08′N 89°03′W. In agricultural area.

Lost Cabin, village, FREMONT county, central WYOMING, on Badwater Creek on branch of BIGHORN RIVER, and 75 mi/121 km WNW of CASPER.

Lost Creek, unincorporated town, SCHUYLKILL county, E central PENNSYLVANIA, 2 mi/3.2 km NW of FRACKVILLE, on Shenandoah Creek; 40°48′N 76°14′W. In anthracite-coal region.

Lost Creek, village (2006 population 507), HARRISON county, N WEST VIRGINIA, 8 mi/12.9 km S of CLARKSBURG; 39°09′N 80°20′W. Corn; cattle, poultry. Manufacturing (meat processing, limestone processing). Watters Smith Memorial State Park to W.

Los Telares (los te-LAH-res), village, SANTIAGO DEL ESTERO province, ARGENTINA, 30 mi/48 km NNE of OJO DE AGUA. In agricultural (sunflowers), sheep, cattle, and lumbering (quebracho, carob) area; charcoal burning.

Los Telares, village (1991 population 1,336), ⊙ Salavina department (□ 1,086 sq mi/2,823.6 sq km), S SANTIAGO DEL ESTERO province, ARGENTINA, on W bank of the RÍO DULCE, and 70 mi/113 km SE of LORETO; 28°59′S 63°26′W. Grain and livestock.

Los Teques (LOS TAI-kes), city, ⊙ MIRANDA state, N VENEZUELA, in CORDILLERA DE LA COSTA, on highway and railroad to CARACAS, and 13 mi/21 km SW of Caracas; 10°20′N 67°02′W. Elevation 3,864 ft/1,178 m. Trading center in agricultural region (coffee, sugarcane, cacao, cereals). Ceramics and textiles. Popular residential area and tourist resort for citizens of Caracas; noted for beautiful parks.

Los Testigos (LOS tes-TEE-gos), small CARIBBEAN island group, Dependencias Federales, VENEZUELA, 50 mi/80 km NE of MARGARITA ISLAND; 11°23′N 63°07′W.

Lost Hills, unincorporated town (2000 population 1,938), KERN county, S central CALIFORNIA, 42 mi/68 km NW of BAKERSFIELD, on California Aqueduct; 35°37′N 119°42′W. Agriculture (cotton, grain, sugar beets, fruits, nuts; cattle; dairying). Kern National Wildlife Refuge to N; TEMBLOR RANGE to SW.

Lostice (LOSH-tyi-TSE), Czech *Loštice*, German *Loschtitz*, town, SEVEROMORAVSKY province, NW central MORAVIA, CZECH REPUBLIC, on railroad, and 18 mi/29 km NW of OLOMOUC; 49°45′N 16°56′E. Noted for its cheese pastry; clothing; woodworking. Has museum of pottery. BOUZOV castle museum is 3 mi/4.8 km SW.

Lostine (LAHS-teen), village (2006 population 243), WALLOWA county, NE OREGON, 9 mi/14.5 km NW of ENTERPRISE, on Lostine River; 45°29′N 117°25′W. Grain; livestock. Wallowa River 2 mi/3.2 km to E. Part of Wallowa-Whitman National Forest to SW, including Eagle Cap Wilderness Area.

Lost Mine Peak, Texas: see CHISOS MOUNTAINS.

Lost Nation, town (2000 population 497), CLINTON county, E IOWA, 34 mi/55 km WNW of CLINTON; 41°58′N 90°49′W. Feed milling; fertilizers.

Los Toldos (los TOL-dos), town (1991 population 12,115), ⊙ General Viamonte district (□ 828 sq mi/2,152.8 sq km), N central BUENOS AIRES province, ARGENTINA, 30 mi/48 km S of JUNÍN. In agricultural zone (grain; livestock); dairying (butter, cheese). Birthplace of Eva Duarte de Peron.

Cross-references are shown in SMALL CAPITALS. The pronunciation guide is shown on page xix. The sources of population figures are shown on page xvii.

Lost River, village (2000 population 26), CUSTER county, central IDAHO, 13 mi/21 km WNW of ARCO; 43°43′N 113°32′W. Challis National Forest to W.

Lost River, c.70 mi/113 km long, in CALIFORNIA and OREGON; begins at joining of North and South forks, 30 mi/48 km NW of Alturas, Modoc National Forest, N MODOC county; flows W 5 mi/8 km to CLEAR LAKE RESERVOIR, then NW (inflow and outflow are both in NE corner of reservoir), into Oregon, W past Bonanza, to 5 mi/8 km SE of KLAMATH FALLS, then SSE, re-entering California and into TULE LAKE (National Wildlife Refuge), 4 mi/6.4 km S of Oregon state line, and N of LAVA BEDS NATIONAL MONUMENT. Dam at reservoir and system of small diversion dams and canals along middle course are units in Klamath irrigation project, supplying water to agricultural area in Siskiyou county, California, and Klamath county, Oregon.

Lost River, c.75 mi/121 km long, S INDIANA; rises in W WASHINGTON county; flows W (partly in subterranean channel), past WEST BADEN SPRINGS, to East Fork of WHITE RIVER in S MARTIN county.

Lost River, NEW HAMPSHIRE: see KINSMAN NOTCH.

Lost River, West Virginia: see CACAPON RIVER.

Lost River Range, in CUSTER and BUTTE counties, E IDAHO, between BIG LOST and LITTLE LOST rivers, in part of Challis National Forest. Chief peaks are BORAH PEAK (12,662 ft/3,859 m; highest point in state), Dorion Peak (12,016 ft/3,662 m), Leatherman Peak (12,228 ft/3,727 m), Invisible Mountain (11,330 ft/3,453 m), Dickey Peak (11,141 ft/3,396 m).

Lost Springs, village (2000 population 71), MARION county, central KANSAS, 16 mi/26 km N of MARION; 38°34′N 96°57′W. In grain, livestock, and oil-producing area.

Lost Springs, village (2006 population 1), CONVERSE county, E WYOMING, 23 mi/37 km E of DOUGLAS; 42°46′N 104°55′W. Elevation 4,996 ft/1,523 m. Coal mines and oil field nearby. Formerly Lost Spring.

Los Tuxtlas, MEXICO: see SAN ANDRÉS TUXTLA.

Lostwithiel (luhst-WI[TH]-i-uhl), town (1991 population 2,402; 2001 population 2,739), E CORNWALL, SW ENGLAND, on FOWEY RIVER (with 14th-century bridge), and 5 mi/8 km SSE of BODMIN; 50°24′N 04°40′W. Fishing port. Has 13th-century church and remains of 14th-century Duchy Palace. Just N is 12th-century Restormel Castle.

Losuia (lo-SOO-yuh), town, MILNE BAY province, TROBRIAND ISLANDS, SE PAPUA NEW GUINEA, port on W coast of KIRIWINA island; 08°30′S 151°05′E. Yams, coconuts; timber; fish, prawns, lobsters, trepang. Chief town of island group, still ruled by powerful chieftains, matrilineal inheritances. Annual Yam Festival, July and August.

Los Vilos (los VEE-los), village, ⊙ Los Vilos comuna, CHOAPA province, COQUIMBO region, N central CHILE, minor port on the PACIFIC, 30 mi/48 km SW of ILLAPEL; 31°54′S 71°30′W. On railroad and PAN-AMERICAN HIGHWAY. Beach resort, agricultural center (cereals; fruit; livestock). Silver and manganese deposits nearby. Fishing.

Losynivka (lo-SI-neef-kah) (Russian *Losinovka*), town, S CHERNIHIV oblast, UKRAINE, 14 mi/23 km S of NIZHYN; 50°50′N 31°55′E. Elevation 380 ft/115 m. Food processing (flour, cheese); brickworks. Known since 1627, town since 1958.

Lot (LO), department (□ 2,018 sq mi/5,246.8 sq km), in old QUERCY province, SW central FRANCE; ⊙ CAHORS; 44°30′N 01°30′E. It slopes SW from the MASSIF CENTRAL to the GARONNE valley. The center and SE of department is occupied by the arid limestone upland known as Causses du Quercy. Drained (E-W) by the DORDOGNE, LOT, and CÉLÉ rivers. Agriculture (wine, fruits, vegetables, tobacco, nuts; goat cheese) thrives in river valleys; sheep graze in the Causses. Truffle cultivation; hog, turkey, and goose raising; and liqueur distilling underpin the department's economy.

Little industry aside from food processing, goose-liver pate, canning. Chief towns are CAHORS (metalworks) and FIGEAC (agricultural trade). Tourism flourishes in Dordogne River valley and at PADIRAC and ROCAMADOUR. Administratively, the department forms part of the MIDI-PYRÉNÉES region. It is losing population due to its relative isolation and absence of major urban centers.

Lot, FRANCE: see LOT RIVER.

Lota (LO-tah), city (2002 population 48,975), ⊙ Lota comuna, CONCEPCIÓN province, BÍO-BÍO region, S central CHILE, a port on the ARAUCO GULF, an inlet of the PACIFIC OCEAN; 37°05′S 73°10′W. Founded 1841, the city grew rapidly after coal was discovered in the region. Most of the mines are now closed. Home to Chile's oldest hydroelectric power station.

Lotbinière (lo-bin-YER), county (□ 641 sq mi/1,666.6 sq km), CHAUDIÈRE-APPALACHES region, S QUEBEC, E CANADA, on the SAINT LAWRENCE RIVER; ⊙ SAINTE-CROIX; 46°31′N 71°36′W. Composed of eighteen municipalities. Formed in 1982.

Lotbinière (lo-bin-YER), village (□ 30 sq mi/78 sq km), CHAUDIÈRE-APPALACHES region, S QUEBEC, E CANADA, on the SAINT LAWRENCE RIVER, and 35 mi/56 km WSW of QUEBEC city; 46°37′N 71°56′W. Lumbering, dairying.

Lot-et-Garonne (lo–eh–gah-ron), department (□ 2,070 sq mi/5,382 sq km), SW FRANCE, in former GUIENNE and GASCONY provinces; ⊙ AGEN; 44°20′N 00°30′E. Lies in AQUITAINE basin, bounded by the LANDES district (SW); drained SE-NW by the middle GARONNE RIVER and its tributaries (LOT, GERS, BAÏSE). Rich agriculture river valleys (wheat, tobacco, plums, raisins, wine, and vegetables); cattle and poultry raising. Armagnac brandy distilling (Nérac area), lumbering, and resin extraction (Landes). Industry limited to food processing. Chief towns are Agen (fruit-preserving center), VILLENEUVE-SUR-LOT (canning), MARMANDE (distilling, tomato shipping), and NÉRAC. Main Bordeaux-Toulouse transportation route crosses department parallel to Garonne River. Administratively, the department forms part of the Aquitaine region.

Lotfabad (luht-fah-BAHD), village, Khorāsān province, NE IRAN, 55 mi/89 km ENE of QUCHAN, near TRANS-CASPIAN railroad. Frontier station on TURKMENISTAN boundary in DARAGAZ agricultural district; grain, cotton, raisins. In earthquake zone.

Lothair (LO-thur), village, PERRY county, SE KENTUCKY, 2 mi/3.2 km SE of HAZARD, in CUMBERLAND foothills, near North Fork of KENTUCKY RIVER. Bituminous coal.

Lothair (LAHT-HER), village,, LIBERTY county, N MONTANA, 12 mi/19 km W of CHESTER. Wheat, barley, oats, rye, hay; cattle, hogs, sheep; gas and oil. Lake ELWELL reservoir (formed by Tiber Dam on MARIAS River) to S.

Lothian (LO-thee-uhn), former administrative region of Scotland, S of Firth of Forth and N of Scottish Borders; 55°56′N 03°05′W. Lothian was abolished in 1996 and divided into East Lothian, WEST LOTHIAN, MIDLOTHIAN, and EDINBURGH. Includes the PENTLAND, MOORFOOT and LAMMERMUIR Hills, and the rich farmlands around HADDINGTON. Wheat, oats, potatoes, and whiskey are produced here.

Lothringen, FRANCE: see LORRAINE.

Lotikovo, UKRAINE: see LOTYKOVE.

Lotikovskiy Rudnik, UKRAINE: see LOTYKOVE.

Lotoshino (luh-tuh-shi-NO), town (2006 population 5,610), NW MOSCOW oblast, central European Russia, on the Lob River (VOLGA RIVER basin), on road junction, 36 mi/58 km S of TVER, and 17 mi/27 km NW of VOLOKOLAMSK; 56°13′N 35°38′E. Elevation 528 ft/160 m. Flax processing.

Lot River (LO), ancient *Oltis*, c.300 mi/483 km long, in S central FRANCE, rises in the CÉVENNES MOUNTAINS, near Mont LOZÈRE, and flows W past MENDE and CAHORS to join the GARONNE RIVER below AGEN in

AQUITAINE. The limestone uplands through which the Lot winds are intersected by some fertile basins. Vineyards and orchards line the lower valley. The TRUYÈRE and the CÉLÉ are its right tributaries.

Lötschberg, railroad tunnel (9 mi/15 km long) in BERNESE ALPS, central SWITZERLAND, between KANDERSTEG (BERN canton) and RHÔNE valley in Valais canton; opened in 1913. Elevation 4,068 ft/1,240 m. It created, in conjunction with the Simplon tunnel, the second railroad route through the Swiss Alps.

Lötschen Pass (8,825 ft/2,690 m), in BERNESE ALPS, S SWITZERLAND, 1 mi/1.6 km ESE of the BALMHORN. Leads from S BERN canton into VALAIS canton. No road. LÖTSCHBERG tunnel is below the pass.

Lötschental, valley, VALAIS canton, S SWITZERLAND, extending SW along SE flank of PETERSGRAT range to the RHÔNE valley; watered by Langgletscher and by Lonza River, which joins the Rhône at Gampel.

Lotsmano-Kamenka, UKRAINE: see POBIDA.

Lotsmano-Kamyanka, UKRAINE: see POBIDA.

Lott (LAHT), town (2006 population 684), FALLS county, E central TEXAS, near BRAZOS RIVER, 25 mi/40 km S of WACO; 31°12′N 97°01′W. In cotton, corn, cattle area; light manufacturing.

Lotte (LUHT-te), town, NORTH RHINE-WESTPHALIA, NW GERMANY, 6 mi/9.7 km W of OSNABRÜCK; 52°16′N 07°54′E.

Lottefors (LOT-te-FORSH), village, GÄVLEBORG county, E SWEDEN, on LJUSNAN RIVER, 5 mi/8 km N of BOLLNÄS; 61°25′N 16°24′E.

Lottsburg (LAHTS-buhrg), unincorporated village, NORTHUMBERLAND county, E VIRGINIA, 58 mi/93 km SE of FREDERICKSBURG, near Coan River, arm of POTOMAC RIVER estuary; 37°57′N 76°31′W. Manufacturing (canned seafoods); agriculture (tomatoes, grain; poultry, cattle); fish, crabs, oysters.

Lotung (LUH-DUNG), town, N TAIWAN, 5 mi/8 km S of ILAN; 24°41′N 121°46′E. Railroad junction; industrial center; lumber milling; ferroalloy metallurgy (carbide, manganese, ferrosilicon, steel alloys); camphor and camphor oil processing, sugar milling. Hydroelectric plant nearby.

Lotus Island, ST. LAWRENCE county, N NEW YORK, in the SAINT LAWRENCE River, at ONTARIO border, 2.5 mi/4 km NE of ALEXANDRIA BAY; c.1 mi/1.6 km in diameter; 44°23′N 75°54′W.

Lotykivs'kyy Rudnyk, UKRAINE: see LOTYKOVE.

Lotykove (LO-ti-ko-ve), (Russian *Lotikovo*), town (2004 population 8,033), S central LUHANS'K oblast, UKRAINE, in the DONBAS, 7 mi/11 km NE of ALCHEVS'K; 48°32′N 38°55′E. Elevation 623 ft/189 m. Coal mines; dry goods factory. Established at the beginning of the 20th century; until 1912 called Ivanivs'kyy Rudnyk (Russian *Ivanovskiy Rudnik*); then, until 1923, Hustav (Russian *Gustav*); and then, until 1938 Lotykivs'kyy Rudnyk (Russian *Lotikovskiy Rudnik*).

Lötzen, POLAND: see GIŻYCKO.

Loualidia (lwah-li-DEE-yuh), village, MOROCCO, SW of EL JADIDA, on lagoon of ATLANTIC OCEAN. Oyster farms; seaside resort.

Louann (loo-AN), village (2000 population 195), OUACHITA county, S ARKANSAS, 15 mi/24 km NW of EL DORADO; 33°23′N 92°47′W.

Loubet Coast, WEST ANTARCTICA, on the W coast of GRAHAM LAND, ANTARCTIC PENINSULA, between Cape Bellue and the head of Bourgeois Fjord; 67°00′S 66°00′W.

Loubomo, town (2007 population 110,128), ⊙ NIARI region, SW REPUBLIC OF THE CONGO, on railroad, and 180 mi/290 km W of BRAZZAVILLE, 70 mi/113 km NE of POINTE-NOIRE; 04°11′S 12°40′E. Founded 1934, it is a growing center of a mining (gold, lead) and agricultural region; tanning, fiber processing, livestock raising; sisal plantations. Airfield. Also called Dolisie.

Loucna nad Desnou (LOUCH-nah NAHD DESnou), Czech *Loučná nad Desnou*, formerly *Vízmberk* (VEEZM-berk), German *Wiesenberg*, village, SEVER-

OMORAVSKY province, N MORAVIA, CZECH REPUBLIC, in the JESENIKS, on railroad, and 9 mi/14.5 km NNE of Šumperk; 50°04′N 17°06′E. Manufacturing of bicycles in nearby SOBOTIN, Czech *Sobotín* (SO-bo-TYEEN), 4 mi/6.4 km S; lumbering.

Loucovice (LOU-cho-VI-tse), Czech *Loučovice*, German *Kienberg*, village, JIHOCESKY province, S BOHEMIA, CZECH REPUBLIC, on VLTAVA RIVER, on railroad, and 26 mi/42 km SSW of ČESKÉ BUDEJOVICE; 48°38′N 14°15′E. Paper mill. Village has two 14th century Gothic churches. A hill (huge rock wall) called Devil's Wall, Czech *Čertova stěna*, which rises from a rocky field, is just NE.

Loudéac (loo-dai-ak), town (□ 31 sq mi/80.6 sq km), CÔTES-D'ARMOR department, W FRANCE, in BRITTANY, 48 mi/77 km W of RENNES; 48°10′N 02°45′W. A market center for poultry and pigs. Storage silos, food-processing plants. Noted for its fairs and horse-racing events.

Loudi (LO-DEE), city (□ 166 sq mi/431.6 sq km; 2000 population 300,428), central HUNAN province, CHINA; 27°45′N 111°59′E. Heavy industry is the largest source of income for the city. Crop growing (grain), animal husbandry (hogs). Manufacturing (utilities, iron and steel, coal mining).

Loudias River (loo-&hardTH;EE-ahs), 25 mi/40 km long, in CENTRAL MACEDONIA department, NE GREECE; rises in the PAIKON; flows SE, through drained Lake Yiannitsa, to the Gulf of Thessaloníki in delta 15 mi/24 km WSW of THESSALONÍKI. Also Loudhias River

Loudima (loo-dee-MAH), town, LÉKOUMOU region, SW Congo Republic, on NIARI RIVER, on railroad, and 150 mi/241 km W of Brazzaville; 04°06′S 13°02′E. Agricultural center (groundnuts, tobacco, corn, soya, sweet potatoes), site of experimental station for mechanized tropical farming.

Loudon (LOU-duhn), county (□ 240 sq mi/624 sq km; 2006 population 44,566), E TENNESSEE; ⊙ LOUDON; 35°44′N 84°19′W. In GREAT APPALACHIAN VALLEY; crossed by TENNESSEE and LITTLE TENNESSEE rivers; bounded NW by CLINCH River. Includes part of FORT LOUDOUN LAKE (reservoir). Agriculture; manufacturing. Formed 1870.

Loudon, town, MERRIMACK county, S central NEW HAMPSHIRE, 7 mi/11.3 km NE of CONCORD; 43°19′N 71°26′W. Drained by SOUCOOK RIVER and Bee Hole Brook. Agriculture (poultry, livestock; corn, vegetables, nursery crops; dairying); light manufacturing. Site of New Hampshire International Speedway.

Loudon, town (2006 population 4,872), ⊙ LOUDON county, E TENNESSEE, on TENNESSEE RIVER, and 28 mi/45 km SW of KNOXVILLE; 35°45′N 84°20′W. Manufacturing. Nearby is site of Fort Loudon (built 1756), which fell to the Cherokee in 1760 after a long siege. FORT LOUDOUN Reservoir and Dam are NE. Settled 1828; incorporated 1927.

Loudon, Mount (LOU-duhn), (10,550 ft/3,216 m), SW ALBERTA, W CANADA, near BRITISH COLUMBIA border, in ROCKY MOUNTAINS, near BANFF NATIONAL PARK, 65 mi/105 km NW of BANFF; 51°55′N 116°26′W.

Loudonville (LOO-duhn-vil), village (□ 3 sq mi/7.8 sq km; 2006 population 2,989), ASHLAND county, N central OHIO, 17 mi/27 km ESE of MANSFIELD, on Black Fork of MOHICAN RIVER; 40°38′N 82°14′W. Manufacturing (foodstuffs, oils). Mohican State Forest and Pleasant Hill Dam are nearby. Laid out 1814.

Loudonville, residential suburb (□ 5 sq mi/13 sq km) of ALBANY, ALBANY county, E NEW YORK, 3 mi/4.8 km N of downtown; 42°42′N 73°46′W. Seat of St. Bernardine of Siena College.

Loudoun (LOUD-uhn), county (□ 521 sq mi/1,354.6 sq km; 2006 population 268,817), N VIRGINIA; ⊙ LEESBURG; 39°05′N 77°38′W. In rolling PIEDMONT region, rising to the BLUE RIDGE in W; bounded NW by WEST VIRGINIA state line, N and NE by POTOMAC RIVER

(MARYLAND state line); drained by LITTLE RIVER and short Goose Creek. Agriculture and country-estate area (known for horse and cattle breeding, fox hunting; also wheat, corn, barley, hay, soybeans, alfalfa, tobacco, apples; cattle, sheep, poultry; dairying); timber; limestone. Part of Harpers Ferry National Historical Park in W. Part of DULLES INTERNATIONAL AIRPORT in SE. APPALACHIAN TRAIL parallels W county boundary. Formed 1757.

Loudoun Hill, Scotland: see DARVEL.

Louds Island, S MAINE, in MUSCONGUS BAY, just E of BRISTOL; 3 mi/4.8 km long, c.0.5 mi/0.8 km wide. Sometimes called Muscongus.

Loudun (loo-duhn), town (□ 15 sq mi/39 sq km), VIENNE department, POITOU-CHARENTES region, W central FRANCE, 19 mi/31 km SSE of SAUMUR; 47°00′N 00°04′E. Road center and cattle market; manufacturing of agricultural implements. Town prospered as an intellectual center in 17th-century, but suffered from revocation (1685) of Edict of NANTES, resulting in Protestant outmigration. Has medieval churches and a square keep.

Louellen (loo-EL-en), village, HARLAN county, SE KENTUCKY, on CLOVER FORK of CUMBERLAND RIVER, and 13 mi/21 km ENE of HARLAN. Bituminous coal.

Loue River (LOO), c.80 mi/129 km long, in DOUBS and JURA departments, FRANCHE-COMTÉ region, E FRANCE; rises in the central JURA mountains, 7 mi/11.3 km NNW of PONTARLIER (its source results from underground connection with upper DOUBS RIVER); 47°00′N 05°27′E. Flows in a picturesque valley generally W, past ORNANS to the Doubs 5 mi/8 km SSW of DOLE. Used for canoeing and hydroelectric power. Formerly called Louve.

Louga (LOO-guh), administrative region (□ 11,269 sq mi/29,299.4 sq km; 2004 population 714,732), NW SENEGAL; ⊙ LOUGA; 15°15′N 15°45′W. Bounded on N by SAINT-LOUIS administrative region, E by MATAM administrative region, S by TAMBACOUNDA, KAOLACK, FATICK, DIOURBEL, and THIÈS administrative regions, W by ATLANTIC OCEAN. Peanut farming and livestock production. Formerly part of a larger Diourbel province. The rural areas of Lougré Thioly and Vélingara (in SE portion of the region) became part of Matam administrative region when it was created in 2002.

Louga (LOO-guh), town (2002 population 73,662), ⊙ LOUGA administrative region, W SENEGAL, 34 mi/55 km SE of SAINT-LOUIS; 15°37′N 16°13′W. Located in a region where peanuts, cassava, and gum arabic are produced, Louga is a road junction and station on the Dakar–Saint-Louis railroad. Administrative center.

Lough, for names of Irish lakes and inlets beginning with "Lough," see second part of element; e.g., for Lough Erne, see ERNE, LOUGH. See also LAKE.

Loughborough (LUHF-buh-ro), former township (□ 82 sq mi/213.2 sq km; 2001 population 5,250), SE ONTARIO, E central CANADA; 44°26′N 76°31′W. Amalgamated into SOUTH FRONTENAC.

Loughborough (LUHF-buh-ruh), town (2001 population 57,626), LEICESTERSHIRE, central ENGLAND, on the SOAR RIVER; 52°46′N 01°12′W. Market town with machinery and equipment manufacturing works; other manufacturing includes hosiery, shoes, pharmaceuticals, boilers, and pottery. Bell foundries were built in 1840; the great bell of St. Paul's Cathedral in London was cast here in 1881. Loughborough's war memorial, with a carillon of 47 bells, was built in 1923. In the town are the Loughborough University and a grammar school (founded in 1495).

Loughbrickland (lahk-BRIK-luhnd), village (2001 population 681), W DOWN, Northern Ireland, 10 mi/16 km SW of DROMORE; 54°19′N 06°18′W. Founded for English Protestants in 1585. Just S is small Lough Brickland. At Emdale, 4 mi/6.4 km SE, the Reverand Patrick Brontë, (Ó Pronntaigh), father of writers Emily, Anne, and Charlotte, was born.

Lougheed (lah-HEED, LAH-heed), village (2001 population 228), E ALBERTA, W CANADA, 30 mi/48 km WSW of WAINWRIGHT, in FLAGSTAFF COUNTY; 52°44′N 111°33′W. Grain elevators, stockyards. Incorporated 1911.

Lougheed Island, largest (□ 504 sq mi/1,310.4 sq km) of the FINDLAY ISLANDS, NUNAVUT territory, CANADA, in the ARCTIC OCEAN, N of BATHURST ISLAND, and separated from BORDEN ISLANDS (W) by PRINCE GUSTAF ADOLF SEA and from ELLEF RINGNES ISLAND (NE) by Maclean Strait; 77°30′N 105°00′W. Island is 50 mi/80 km long, 12 mi/19 km–15 mi/24 km wide.

Loughman (LOU-muhn), unincorporated town (□ 4 sq mi/10.4 sq km; 2000 population 1,385), POLK county, central FLORIDA, 4 mi/6.4 km SW of KISSIMMEE; 28°15′N 81°34′W. Increasing population due to its proximity to Walt Disney World.

Loughor (LUH-khuhr), Welsh *Casllwchwr*, town (2001 population 4,991), SWANSEA, S WALES, on LOUGHOR RIVER estuary, 6 mi/9.7 km WNW of Swansea; 51°40′N 04°03′W. Former steel mill and tin-smelting center. Commuters to Swansea. Has remains of Norman castle. Formerly in WEST GLAMORGAN, abolished 1996.

Loughor River (LUH-khuhr), Welsh *Afon Llwchwr*, 15 mi/24 km long, CARMARTHENSHIRE, SW Wales; rises 3 mi/4.8 km NE of AMMANFORD; flows SW, past Ammanford, Pontardulais, Llwchwr, forming the border between Carmarthenshrie and SWANSEA, and past LLANELLI to CARMARTHEN BAY 2 mi/3.2 km W of BURRY PORT. Formerly in DYFED, abolished 1996.

Loughrea (lahk-RAI), Gaelic *Baile Locha Riach*, town (2006 population 4,532), S GALWAY county, W IRELAND, on Lough Rea (2 mi/3.2 km long), and 21 mi/34 km ESE of GALWAY; 53°12′N 08°34′W. Agricultural market; cotton spinning. There are remains of Burke Castle and Carmelite Friary, both founded c.1300.

Loughrin (LAHF-rin), former township, S central ONTARIO, E central CANADA; 46°34′N 80°29′W. Amalgamated into town of MARKSTAY-WARREN in 1999.

Loughton (LOU-tuhn), town (2001 population 30,340), SW ESSEX, SE ENGLAND, 3 mi/4.8 km NE of CHINGFORD, at E end of Epping Forest; 51°39′N 00°00′E. Electrical equipment, printing works. Just N are remains of ancient British earthworks.

Louhans (loo-ahn), town (□ 8 sq mi/20.8 sq km; 2004 population 6,422), sub-prefecture of SAÔNE-ET-LOIRE department, in BURGUNDY, E central FRANCE, on the SEILLE RIVER, 16 mi/26 km WSW of LONS-LE-SAUNIER; 46°38′N 05°13′E. Market for butter, eggs, poultry from the BRESSE agricultural region. Has important cattle and hog fairs.

Louhi, Russia: see LOUKHI.

Louin (LOO-in), village (2000 population 339), JASPER county, E central MISSISSIPPI, 38 mi/61 km WSW of MERIDIAN; 32°04′N 89°15′W. Agriculture (cotton, corn; cattle, poultry; dairying); timber. Bienville National Forest to N.

Louisa, county (□ 417 sq mi/1,084.2 sq km; 2006 population 11,858), SE IOWA, on ILLINOIS state line (to E; here formed by MISSISSIPPI RIVER); ⊙ WAPELLO; 41°13′N 91°15′W. Prairie agricultural area (cattle, hogs, poultry; corn, oats, wheat, soybeans) drained by IOWA RIVER. Fertile E section (between the Iowa and the Mississippi) is artificially drained (pumping stations, ditches). Gypsum and limestone quarries. Lock and Dam No. 17 above Toolesburg. Mark Twain State Wildlife Refuge in E; Toolesburg Indian Mounds in SE near Mississippi River. Widespread flooding in 1993. Formed 1836.

Louisa (loo-EE-zuh), county (□ 510 sq mi/1,326 sq km; 2006 population 31,226), central VIRGINIA; ⊙ LOUISA; 37°59′N 77°57′W. Bounded N and NE by NORTH ANNA RIVER; drained by SOUTH ANNA RIVER. Agriculture (tobacco, barley, wheat, corn, soybeans, hay; cattle); some timber. North Anna 1 and North Anna 2 nuclear power plants are in NE on LAKE ANNA reservoir in town of MINERAL. Formed 1742.

Cross-references are shown in SMALL CAPITALS. The pronunciation guide is shown on page xix. The sources of population figures are shown on page xvii.

Louisa, town (2000 population 2,018), ⊙ LAWRENCE county, NE KENTUCKY, 26 mi/42 km S of ASHLAND, on BIG SANDY RIVER (here formed by junction of Levisa and Tug forks); 38°06′N 82°35′W. Trade and shipping center for mountain agricultural area (livestock; corn, alfalfa, tobacco); manufacturing (apparel, beverages); also coal mines and fireclay and sand pits. Birthplace of U.S. Supreme Court Justice Fred Vinson. YATESVILLE LAKE reservoir, on Blaine Creek, to W. Settled 1789; established 1822.

Louisa (loo-EE-zuh), town (2006 population 1,537), ⊙ LOUISA county, central VIRGINIA, 26 mi/42 km E of CHARLOTTESVILLE; 38°01′N 78°00′W. Manufacturing (plastic products, printing and publishing, tobacco processing, paper products, lumber, clothing); in agricultural area (tobacco, grain, soybeans; cattle, poultry; dairying); timber. GREEN SPRINGS National Historic Landmark District.

Louisa (luh-WEE-suh), unincorporated village, SAINT MARY parish, S LOUISIANA, 33 mi/53 km SSE of LAFAYETTE, near W COTE BLANCHE BAY, Gulf of MEXICO, on COTE BLANCHE ISLAND salt dome. Shallow-draft port; manufacturing (boat parts, shrimp processing). Cypremort Point State Park to SW.

Louisburg (LOO-is-buhrg), town, E CAPE BRETON ISLAND, NOVA SCOTIA, CANADA; 45°55′ N 59°58′W. Elevation 3 ft/0.9 m. The town, an ice-free port, is near the site of the fortress of Louisbourg, built by France in the early 18th century. The site is now a national historic park.

Louisburg, town (2000 population 2,576), MIAMI county, E KANSAS, near MISSOURI state line, 10 mi/16 km ENE of PAOLA; 38°37′N 94°40′W. Shipping point for livestock. Gas transmission and distribution.

Louisburg (LOO-ee-buhrg), town (□ 2 sq mi/5.2 sq km; 2006 population 3,726), ⊙ FRANKLIN county, N central NORTH CAROLINA, on TAR RIVER, 28 mi/45 km NE of RALEIGH; 36°06′N 78°17′W. Railroad spur terminus. Service industries; manufacturing (food processing, apparel, steel products, furniture, wooden products, lumber). Trade center for agricultural area (tobacco). Seat of Louisburg College Laurel Mills (1769) to NE. Settled 1758.

Louisburg (LOO-wis-buhrg), village (2000 population 26), LAC QUI PARLE county, SW MINNESOTA, near MINNESOTA RIVER, 11 mi/18 km N of MADISON; 45°10′N 96°10′W. Grain; dairying; livestock. MARSH LAKE (Minnesota River) to NE, in Lac qui Parle Wildlife Area.

Louisburgh (LOO-wis-buh-ruh), Gaelic *Cluain Chearbán*, village (2006 population 314), SW MAYO county, NW IRELAND, on S shore of CLEW BAY, 12 mi/19 km WSW of WESTPORT; 53°46′N 09°49′W. Fishing port. Overlooked by CROAGH PATRICK mountain.

Louisburgh, Scotland: see WICK.

Louise (loo-EEZ), rural municipality (□ 360 sq mi/936 sq km; 2001 population 989), S MANITOBA, W central CANADA, 67 mi/107 km from BRANDON, on U.S. (NORTH DAKOTA) border; 49°09′N 98°54′W. Agriculture (grain; livestock); tourism. Printing museum. Encompasses CRYSTAL CITY, CLEARWATER, PILOT MOUND communities. Incorporated 1880.

Louise (loo-EEZ), village (2000 population 315), HUMPHREYS county, W MISSISSIPPI, 15 mi/24 km SSW of BELZONI; 32°58′N 90°35′W. In agricultural area (cotton, grain; cattle). Delta National Forest to SW.

Louise Island (loo-EEZ) (□ 105 sq mi/273 sq km), W BRITISH COLUMBIA, W CANADA, one of the QUEEN CHARLOTTE ISLANDS, in HECATE STRAIT, E of N MORESBY ISLAND; 15 mi/24 km long, 2 mi/3 km wide; 52°59′N 131°47′W. Rises to 3,550 ft/1,082 m.

Louise, Lake (loo-EEZ), SW ALBERTA, W CANADA, in the ROCKY MOUNTAINS, in BANFF NATIONAL PARK; 51°25′N 116°14′W. Lake is 2 mi/3 km long, elevation 5,680 ft/1,731 m. Noted for its scenic beauty, especially its sunrises, it is surrounded by high peaks, glaciers, and snow fields, which are reflected in its waters. The lake was explored in 1882 and later named for Princess Louise. It has become a popular year-round tourist and mountain-climbing center. The lake drains E into BOW RIVER.

Louiseville (loo-EEZ-vil), town (□ 39 sq mi/101.4 sq km), ⊙ MASKINONGÉ county, MAURICIE region, S QUEBEC, E CANADA, on Rivière du Loup, near its mouth on the SAINT LAWRENCE RIVER, 20 mi/32 km WSW of TROIS-RIVIÈRES; 46°15′N 72°57′W. Milling, lumbering, food processing; manufacturing.

Louisiade Archipelago (loo-EE-zee-ad), SW PACIFIC, SE MILNE BAY province, SE PAPUA NEW GUINEA; 11°12′S 153°00′E. The archipelago comprises circa ten volcanic islands and numerous coral reefs, with elevations rising to 2,800 ft/853 m. The major islands are TAGULA (the largest), ROSSEL, MISIMA, and Panaete (latter two are in DEBOYNE Islands). CORAL SEA to S, SOLOMON SEA to N. The inhabitants are Papuans. BWAGAOIA, on MISIMA island, is the chief village of the group. Most of the islands have gold reserves, but mining largely ceased after World War II; there is a major new gold mine on Misima. The archipelago was explored by the French navigator J. A. B. d'Entrecasteaux and was named in honor of the king of FRANCE.

Louisiana (luh-wee-zee-AN-uh), state (□ 51,843 sq mi/ 134,791.8 sq km; 2006 population 4,287,768), S central UNITED STATES, admitted to the Union in 1812 as the eighteenth state; ⊙ BATON ROUGE; 31°04′N 92°00′W. BATON ROUGE is the state's second-largest city; the largest city is NEW ORLEANS, whose seaport is the busiest in the nation. Other major cities are SHREVEPORT, LAKE CHARLES, KENNER, and LAFAYETTE.

Geography
Louisiana is bounded on the N by ARKANSAS, on the E by MISSISSIPPI (the state line is formed by the MISSISSIPPI RIVER in the N, by the PEARL RIVER in the extreme SE), on the S by the GULF OF MEXICO, and on the W by TEXAS (the SABINE RIVER marks most of the boundary). A low country on the Gulf coastal plain and the Mississippi alluvial plain, Louisiana rises gradually into low hills in the NW; the highest point is DRISKILL MOUNTAIN (535 ft/163 m) in BIENVILLE parish near ARKANSAS. The rainy coast country contains marshes and fertile delta lands; inland are rolling pine hills and prairies. The MISSISSIPPI RIVER dominates the many waterways, but there are other rivers (e.g., the RED RIVER, the OUACHITA, the ATCHAFALAYA, and the CALCASIEU). The "Pelican State," especially the coast, is threaded by many slow-moving bayous (e.g., the TECHE, the MACON, and the LAFOURCHE), all former channels of the MISSISSIPPI RIVER and other rivers that continue to provide outlets during floods. The marshy coast is indented by bays and saltwater lakes, the largest being LAKE PONTCHARTRAIN, N of NEW ORLEANS. There are numerous backwater lakes throughout the delta and coastal areas, and oxbow and horseshoe lakes abound along river margins. Manmade canals criss-cross the state, constructed for shipping and floodwater diversion. The state's canal system is dominated by the Gulf branch of the INTRACOASTAL WATERWAY, which provides a sheltered barge canal from FLORIDA to TEXAS.

Economy
Major Louisiana shipping outlets include the MISSISSIPPI RIVER GULF OUTLET SE of NEW ORLEANS, HOUMA NAVIGATION CANAL, and PORT ALLEN—MORGAN CITY ALTERNATE ROUTE. The system allows for direct ship-to-barge transfer of goods. The climate (subtropical in the S and temperate in the N), together with the rich alluvial soil, makes the state one of the nation's leading producers of sweet potatoes, rice, and sugarcane. Other major agricultural commodities are soybeans, cotton; cattle; and dairy products. Strawberries, vegetables, corn, hay, pecans, and watermelons are also produced in great quantity.

Home gardens and nursery crops are major agricultural income sources. Fishing is a major industry; shrimp, menhaden, crawfish, crabs, and oysters are the principal catches, as well as catfish on inland waters. Louisiana is a leading fur-trapping state (nutria is the primary source of fur); its marshes (□ 7,409 sq mi/19,189 sq km of the state's area is underwater) supply most of the country's muskrat furs. Pelts are also obtained from mink, coypus, opossums, otter, and raccoon. The raising of ratite birds (emus, ostriches) for their meat has become a widespread activity. Alligators are farmed throughout the marshy S part of the state for their meat and hides. The state has great mineral wealth. It leads the nation in the production of salt and sulfur, and it ranks second in the production of crude petroleum (of which many deposits are offshore), natural gas, and natural-gas liquids. Timber is plentiful; forests cover almost 50% of the land area. There is one national forest. The state rapidly industrialized in the 1960s and 1970s. It has giant oil refineries, petrochemical plants, metal foundries, sawmills, and paper mills. Other industries produce foods, transportation equipment, and electronic equipment.

Education
Among the state's more prominent institutions of higher learning are Tulane University, University of New Orleans, Dillard University, Southern University, Loyola University, and Newcomb College, all at NEW ORLEANS; Louisiana State Universtiy and Agricultural and Mechanical (A&M) College, main campus at BATON ROUGE; Southern Universtiy and Agricultural and Mechanical (A&M) College (the largest predominantly African-American institution in the country), main campus in LAFAYETTE; Grambling State University at GRAMBLING; Manesse State University in LAKE CHARLES; and Louisiana Tech University, at RUSTON.

Tourism
Cajun and Creole foods, music, and traditions became popular throughout the U.S. during the 1980s and 1990s, resulting in increased tourism to the state. NEW ORLEANS, with its exciting nightlife and Old World charm, is the major attraction. It is especially noted for its picturesque French Quarter, which has many celebrated restaurants, and for the colorful Mardi Gras—perhaps the most famous festival in the U.S.—held annually, on the day leading up to Ash Wednesday, since 1838. Another yearly attraction is the Sugar Bowl football game, staged in the Superdome on New Year's Day. Elsewhere a variety of recreational facilities makes the state an excellent vacationland; some of its lakes (e.g., PONTCHARTRAIN) have been highly developed as resort areas, and there is superb hunting and fishing throughout much of the region. Louisiana is rich in tradition and legend, and three different groups have contributed to its unique heritage: the Creoles, descendants of the original French colonists; the Cajuns (the term derived from Acadian), whose French ancestors were expelled from the ACADIA region of NOVA SCOTIA, CANADA, by the British in the 18th century; and the American cotton planters. Along the rivers, bayous, and cypress swamps overhung with Spanish moss, some of the old mansions remain, recalling the elegance and splendor of Southern antebellum days. Plantation tours from BATON ROUGE, NATCHITOCHES, and many other cities are very popular, while the Cajun country in the Mississippi delta land also attracts visitors—most particularly the LONGFELLOW-EVANGELINE MEMORIAL STATE COMMEMORATIVE AREA, at SAINT MARTINVILLE; and JEAN LAFITTE NATIONAL HISTORICAL PARK, with four units in and around NEW ORLEANS and LAFAYETTE.

History: to 1712
Louisiana has a long and colorful history. The region was possibly visited by Cabeza de Vaca and his fellow survivors of a Spanish expedition of 1528, and it was

certainly seen by some of Hernando De Soto's men (1541–1542). In 1682, René-Robert Cavelier, sieur de La Salle reached the mouth of the MISSISSIPPI and claimed for FRANCE all of the land drained by that river and its tributaries, naming it Louisiana after Louis XIV. Europeans did not permanently settle here until 1699, when Pierre le Moyne, sieur d'Iberville, founded a settlement near BILOXI. This settlement became the seat of government for Louisiana, an enormous territory embracing the entire Mississippi drainage basin. In 1702, Iberville's brother, the sieur de Bienville, was appointed governor and moved the territorial government to Fort Louis on the Mobile River. This colony was later moved (1710) to the present site of MOBILE (ALABAMA), and Mobile became capital of Louisiana. French missionaries and fur traders explored some of the vast territory, and NATCHITOCHES (the oldest settlement within the present boundaries of the state of Louisiana) grew from a French military and trading post established (c. 1714) to protect the RED RIVER area from the Spanish.

History: 1712 to 1776
To increase the value of the colony, FRANCE granted (1712) a monopoly of commercial privileges, which in 1717 passed to a company organized by John Law. The promise of riches under Law's Mississippi Scheme brought many settlers to Louisiana, and a large number of them remained even after his scheme had collapsed. NEW ORLEANS was founded in 1718 and in 1723 became the capital. Large numbers of blacks were brought in as slaves, and the *Code Noir*, adopted in 1724, provided for the rigid control of their lives and the protection of the whites. After the French lost the last conflict of the French and Indian Wars (1754–1763), they secretly ceded (by the Treaty of FONTAINEBLEAU in 1762) the area W of the MISSISSIPPI and the "Isle of Orleans" to SPAIN to keep the entire Louisiana territory from falling into the hands of the British. By the Treaty of PARIS (1763), GREAT BRITAIN gained control of all Louisiana E of the MISSISSIPPI except the Isle of Orleans; these changes were announced in 1764. The French colonists resisted the new Spanish rule but were subdued, and finally a Spanish mercantilistic monopoly of trade was instituted. During the Spanish years, agriculture flourished with the cultivation of rice and sugarcane, and NEW ORLEANS grew as a major port and trading center. The Spanish government welcomed thousands of Acadians to the area from CANADA, and they settled what came to be known as the Cajun country.

History: 1776 to 1815
During the American Revolution, NEW ORLEANS was a center for Spanish aid to the colonies. After Spain declared war on GREAT BRITAIN in 1779, Bernardo de Gálvez, governor of Louisiana, became an active ally of the revolutionists, capturing BATON ROUGE and NATCHEZ (1779), Mobile (1780), and PENSACOLA (1781). After the war Louisiana's control of the great inland trade route, the MISSISSIPPI, led to heated controversy with the Americans. In the secret Treaty of SAN ILDEFONSO (1800), Napoleon I forced the retrocession of the territory to FRANCE. Revelation of this treaty caused profound concern in the UNITED STATES. President Jefferson attempted to purchase the Isle of Orleans from FRANCE. To the surprise of the American representatives in France, Napoleon decided to sell all of Louisiana to the U.S. The U.S. took possession in 1803, and in 1804 the territory was divided into two parts. The S part, which was called the Territory of Orleans, was admitted to the Union in 1812 as the state of Louisiana. Settlement (1819) of the W FLORIDA Controversy gave Louisiana the area between the Mississippi and PEARL rivers, which formerly had been part of Florida. After statehood French and Spanish influence remained, not only in the Creole and Cajun societies but also in the civil law

(based on French and Spanish codes) and in the division of the state into parishes rather than counties.

History: 1815 to 1862
In the early years of the 19th century the diverse people of Louisiana—the French, the Spanish, the Germans, and Isleños brought by Gálvez from the CANARY ISLANDS—united behind Andrew Jackson to defeat (1815) the British at the Battle of New Orleans during the War of 1812. This victory brought Jackson to national prominence even though the war was officially over by the time the battle was fought. (The battle site is contained in CHALMETTE National Historical Park.) With settlers pouring in from other Southern states, great sugar and cotton plantations developed rapidly in the fertile lowlands, and the less productive uplands also were settled. The state capital was moved several times, finally ending up at Baton Rouge in 1849. The advent of steam propulsion on the Mississippi (the first steamboat to navigate the river arrived in New Orleans in 1812) was a boon to the state's economy; by 1840, New Orleans was the nation's second-largest port. Plantation owners, with their large landholdings and many slaves (more than 50% of the population) dominated politics and largely controlled the state. On January 26, 1861, Louisiana seceded from the Union and six weeks later joined the Confederacy.

History: 1862 to 1882
The fall of New Orleans to David G. Farragut in 1862 prefaced the detested military occupation under General B. F. Butler. Occupied Louisiana was a proving ground for Lincoln's moderate restoration program, but after Lincoln's assassination radical Republicans seized control, and Louisiana suffered greatly during Reconstruction. The Ku Klux Klan was particularly active from 1866 to 1871. In the election of 1872 the radical Republican candidate for governor lost but was installed with the help of Federal troops. Reconstruction in Louisiana finally ended with the disputed presidential election of 1876, when Louisiana's electoral votes were "traded" to the Republicans (whose candidate was Rutherford B. Hayes) in exchange for the withdrawal of Federal troops from the state. Francis River T. Nicholls, a Democrat, became governor, and white control of the state was reestablished. Economic recovery was slow. The disrupted plantation system was largely replaced by farm tenancy and sharecropping. The decline of steamboat traffic was offset somewhat by new railroad building and the opening of the Mississippi River for oceangoing vessels from New Orleans to the sea (a feat accomplished by James B. Eads) with the construction of jetties at the mouth of the river.

History: 1882 to 1935
Mississippi floods constituted a serious problem, and levee building increased after the flood of 1882; it was only after the flood of 1927 (at the time called America's greatest peacetime disaster), however, that the Federal government undertook a vast control system. The water resources development program encompasses flood control, navigation, drainage, and irrigation. The pattern of Louisiana economy was changed by the discovery of oil and natural gas in 1901, and industries began to grow on the basis of cheap fuel and cheap labor. Medical advances helped to curb the yellow-fever epidemics that had periodically disrupted the state. Industrial growth and the continuing woes of the tenant farmers did not alter control of the state by "Bourbon" Democrats, but in 1928 a virtual revolution occurred when Huey P. Long was elected governor. His almost dictatorial rule, detested by liberals across the nation, brought material progress at the cost of widespread official corruption. Long withstood all outside pressures, including the opposition of President Franklin D. Roosevelt's administration. Long resigned the governorship in 1931 to become a U.S. senator, but he retained control over the state. After his assassina-

tion in 1935, his political heirs made their peace with the New Deal, and Federal funds, withheld during Long's last years, were poured into the state. The BONNET-CARRE SPILLWAY, built to divert Mississippi floodwaters into Lake Pontchartrain and protect New Orleans and other downstream communities from flood damage, was completed in 1935.

History: 1935 to Present
The state has a large African-American population, and the issue of civil rights has long been a bitter one. The process of integration following the 1954 Supreme Court ruling against racial segregation in the public schools was difficult. In 1965, Hurricane Betsy struck Louisiana, killing seventy-four and causing property damage in excess of $1 billion. Hurricane Camille (1969) was much more destructive, ravaging Louisiana and a number of other states and killing 256 people. In April 1973, the Mississippi River rose to its highest level recorded in Louisiana. Its floodwaters, together with those from its tributaries and distributaries, covered more than 10% of the state and caused millions of dollars of damage to crops and property. In 1992, Hurricane Andrew, after bringing severe damage to S Florida, struck S central Louisiana, killing eleven. Louisiana enjoyed an oil boom in the early 1980s, but when oil prices collapsed in 1986, the state economy did as well. The state's unemployment rate rose to the highest in the nation, and it suffered the effects of massive outmigration. This placed an ever-greater economic burden on the tourist industry and led to increased efforts to diversify the state economy. Environmental problems also plagued Louisiana during the 1980s and early 1990s. It was discovered that natural erosion, oil exploitation, and river control projects had severely degraded Louisiana's freshwater marshlands and its wildlife and plant communities, especially the Mississippi delta land. New Orleans and the Mississippi delta region (along with the Gulf Coast regions of Mississippi and ALABAMA) suffered catastrophic damage from Hurricane Katrina in 2005. New Orleans was particularly damaged when several levees failed and the city flooded stranding tens of thousands of residents and causing widespread property damage. In the aftermath, the city was almost completely evacuated and is still trying to recover.

Government
The state has had eleven constitutions since it was admitted to the Union in 1812. Its present constitution (1975) replaced the constitution of 1921, which had been amended more than 500 times. The state's executive branch is headed by a governor elected for a four-year term and allowed one reelection. The current governor is Kathleen Babineaux Blanco. Louisiana's bicameral legislature has a senate with thirty-nine members and a house of representatives with 105 members, all elected for four-year terms. The state elects two senators and seven representatives to the U.S. Congress.

Louisiana has sixty-four parishes: ACADIA, ALLEN, ASCENSION, ASSUMPTION, AVOYELLES, BEAUREGARD, BIENVILLE, BOSSIER, CADDO, CALCASIEU, CALDWELL, CAMERON, CATAHOULA, CLAIBORNE, CONCORDIA, DE SOTO, EAST BATON ROUGE, EAST CARROLL, EAST FELICIANA, EVANGELINE, FRANKLIN, GRANT, IBERIA, IBERVILLE, JACKSON, JEFFERSON, JEFFERSON DAVIS, LAFAYETTE, LAFOURCHE, LA SALLE, LINCOLN, LIVINGSTON, MADISON, MOREHOUSE, NATCHITOCHES, ORLEANS, OUACHITA, PLAQUEMINES, POINTE COUPEE, RAPIDES, RED RIVER, RICHLAND, SABINE, SAINT BERNARD, SAINT CHARLES, SAINT HELENA, SAINT JAMES, SAINT JOHN THE BAPTIST, SAINT LANDRY, SAINT MARTIN, SAINT MARY, SAINT TAMMANY, TANGIPAHOA, TENSAS, TERREBONNE, UNION, VERMILION, VERNON, WASHINGTON, WEBSTER, WEST BATON ROUGE, WEST CARROLL, WEST FELICIANA, and WINN.

Louisiana (loo-WEE-zee-a-nuh), city (2000 population 3,863), PIKE county, E MISSOURI, on MISSISSIPPI RIVER

Cross-references are shown in SMALL CAPITALS. The pronunciation guide is shown on page xix. The sources of population figures are shown on page xvii.

(here spanned by Champ Clark Bridge), 25 mi/40 km SE of HANNIBAL; 39°26′N 91°03′W. Agriculture (grain, soybeans, corn; hogs; nursery products); manufacturing (apparel, chemicals, machinery, metal fabrication, tools). Ammonia plant has been converted for study and production of synthetic fuels. Laid out 1818.

Louisiana Point, extreme SW point of LOUISIANA, on GULF coast at E side of SABINE PASS entrance, 14 mi/23 km SSE of PORT ARTHUR (TEXAS). SABINE PASS lighthouse (29°43′N 93°51′W) is 2 mi/3.2 km N.

Louis Napoleon, Cape, E ELLESMERE ISLAND, BAFFIN region, NUNAVUT territory, CANADA, on KANE BASIN, at end of Darling Peninsula (20 mi/32 km long); 79°38′N 72°17′W. Shoreline was explored by Shackleton, 1935.

Louis Philippe Peninsula (loo-EES fee-LEEP-pai), Antartica, in South PACIFIC, on the extreme N tip of Palmer (Antarctic) Peninsula, extending c.93 mi/150 km NE from Charcot Bay and Sjögren Fiord. Discovered 1838 by Dumont d'Urville, French navigator. Chile established in 1949 a military base on W coast at 63°20′S 57°54′W. Sometimes called Trinity (Spanish *Trinidad*) Peninsula.

Louis Trichardt, town, SOUTH AFRICA: see MAKHADO.

Louisville (LOO-is-vil), city (2000 population 18,937), BOULDER county, N central COLORADO, just E of FRONT RANGE, suburb 15 mi/24 km NNW of DENVER; 39°58′N 105°08′W. Elevation 5,337 ft/1,627 m. Coal-mining point in grain, sugar-beet region. Manufacturing (medical equipment, communications equipment, information storage systems). Incorporated 1892.

Louisville, city (2000 population 209), POTTAWATOMIE county, NE KANSAS, on small affluent of KANSAS River, 14 mi/23 km ENE of MANHATTAN, 3 mi/4.8 km N of WAMEGO; 39°15′N 96°19′W. Livestock; grain.

Louisville (LOO-uh-vuhl), city (2000 population 256,231), ⊙ JEFFERSON county, N KENTUCKY, on the OHIO RIVER, at the Falls of the Ohio, opposite JEFFERSONVILLE and NEW ALBANY (INDIANA); 38°13′N 85°44′W. Commonly referred to as Falls City and "gateway to the South." It is the largest city in Kentucky, a major river port, and one of the important industrial, financial, marketing, and shipping centers for the South and Midwest. A diverse manufacturing base, including motor vehicles, consumer goods, naval ordnance, and whiskey; other manufacturing (wood and paper products, tobacco products, food processing, computers and software, chemical processing, aluminum processing, concrete, printing and publishing). Louisville Slugger baseball bat factory and baseball museum moved here from Jeffersonville, Indiana, in 1996. Since the 1970s many industries have relocated to the city's suburbs, following the trend of many U.S. cities. Naval Surface Warfare Center scheduled to close. A settlement grew around a fort built by George Rogers Clark (1778) a base of operations against the British and the Native Americans. The city was chartered by the VIRGINIA legislature in 1780 (when Kentucky was part of Virginia) and named for Louis XVI of FRANCE. Louisville developed at the falls, as a portaging place (a canal was built in 1830; the McAlpine Locks currently allow modern barge traffic to pass the falls). After the arrival of the railroads in the mid-nineteenth century, its role as a shipping center became even more important. During the Civil War it was a center of pro-Union activity in the state and a military and supply base for Federal forces. Louisville International Airport (Standiford Field) in S part of city, Bowman Field airport is E of downtown. The University of Louisville (established 1798), Bellarmine College, Spalding University, Jefferson Community College (University of Kentucky), and Presbyterian and Southern Baptist Theological Seminaries are here. Commonwealth Convention Center. Churchill Downs, a noted horse racetrack and scene of the Kentucky Derby (held annually in May

since 1875), is S of downtown; also includes Kentucky Derby Museum. The city has many parks and is the site of the Kentucky Fair and Exhibition Grounds. Among the many points of interest are the American Printing House for the Blind; the J. B. Speed Art Museum; Howard Steamboat Museum (in Jeffersonville, Indiana), Sons of American Revolution Museum, Colonel Harland Sanders Museum at Kentucky Fried Chicken headquarters; Farmington historic home built 1810; the Filson Club, with a historical library and museum; Louisville Zoo; Bernheim Arboretum and Research Forest; Whitehall Mansion (1855); the Jefferson County Courthouse (1850); and old Cave Hill Cemetery (E of downtown), where George Rogers Clark is buried. Nearby is "Locust Grove," built 1790, the last home of Clark; and the Zachary Taylor National Cemetery, in the E suburbs, where President Taylor and his wife are buried. The FORT KNOX gold bullion depository is 20 mi/32 km SSW (downstream on Ohio River). E. P. "Tom" Sawyer State Park to E. Birthplace of boxer Muhammad Ali and writer Hunter S. Thompson. Incorporated 1780.

Louisville (LOO-is-vil), city (□ 5 sq mi/13 sq km; 2006 population 9,442), STARK county, E central OHIO, 6 mi/10 km NE of CANTON; 40°50′N 81°16′W.

Louisville (LOO-ee-vil), town (2000 population 612), Barbour co., SE Alabama, 10 mi/16 km SW of Clayton. Lumber; pecans. Originally named 'Lewisville' for Daniel Lewis, an early settler in the region, the spelling was soon changed to Louisville. Inc. in 1834.

Louisville (LOO-is-vil), town (2000 population 2,712), ⊙ JEFFERSON county, E GEORGIA, on OGEECHEE RIVER, c.40 mi/64 km SW of AUGUSTA; 33°00′N 82°24′W. Manufacturing of textiles, clothing, electrical transformers. Has old slave market (built before 1800) referred to locally as Mauat House in the center of Broad Street, Several late-eighteenth-century houses. Laid out 1786 as the prospective capital of Georgia. State buildings completed 1795; seat of government until 1804, when MILLEDGEVILLE became capital. The Georgia legislature held its first session here in 1776. Town square.

Louisville (LOO-is-vuhl), town (2000 population 7,006), ⊙ WINSTON county; 33°07′N 89°02′W, E central MISSISSIPPI, 45 mi/72 km SW of COLUMBUS. In agricultural area (cotton, corn; dairying); timber; manufacturing (apparel, furniture, bricks, lumber, consumer goods, motor vehicle parts, construction equipment). Legion State Park to N; Noxubee National Wildlife Refuge to NE; part of Tombigbee National Forest to N: Nanih Waiya Historical Site, ancient Choctaw burial mounds, to SE. Incorporated 1836.

Louisville, town (2006 population 1,070), CASS county, E NEBRASKA, 20 mi/32 km SW of OMAHA, and on PLATTE RIVER; 41°00′N 96°09′W. Railroad junction. Grain; manufacturing (portland cement). Louisville Lakes State Recreation Area nearby.

Louisville (LOU-is-vil), village (2000 population 1,242), ⊙ CLAY county, S central ILLINOIS, 22 mi/35 km W of OLNEY; 38°46′N 88°30′W. In agricultural, oil, and natural-gas area; corn, wheat, fruit; poultry. On LITTLE WABASH RIVER.

Loukhi (LO-oo-hee), Finnish *Louhi*, town (2005 population 5,715), N Republic of KARELIA, NW European Russia, on highway branch and the Murmansk railroad, 75 mi/121 km S of KANDALAKSHA; 66°04′N 33°02′E. Elevation 314 ft/95 m. Junction of a railroad spur to Kestenga; railroad enterprises.

Loukoléla (loo-ko-lai-LAH), village, CUVETTE region, E central Congo Republic, on CONGO RIVER opposite LUKOLELA (Congo), 250 mi/402 km NNE of Brazzaville; 01°02′S 17°06′E. In 1977, site of UN camp for Rwanda Itutu refugees who fled from the Tutsi-led rebel forces during Congoan civil war, after they had concentrated in camps on the E side of the Congo River.

Loukos River (wahd LOO-kos), 85 mi/137 km long, W MOROCCO; rises on W slope of RIF mountains, flows W and NW, past KSAR EL KEBIR, to the ATLANTIC OCEAN at LARACHE. Formed border between Spanish and French protectorates in upper course. Two dams, Oued Makhazine (built 1979), near Larache, and the Garde (built 1981), help irrigate valley. Fertile agricultural region (vegetables, citrus fruit). Also spelled Oued Loukkos.

Loulé (loo-LAI), town, FARO district, S PORTUGAL, 10 mi/16 km NW of FARO; 37°08′N 08°02′W. Pottery, copper work, baskets. Agricultural market town; figs, almonds, vegetables. Ruins of Moorish fortifications.

Loulombo, town, BOUENZA region, S Congo Republic, 10 mi/16 km W of MINDOULI; 04°15′S 14°05′E.

Loulouni (LOO-loon-ee), town, THIRD REGION/SIKASSO, MALI, 31 mi/50 km S of SIKASSO.

Loum (loom), city (1998 estimated population 115,781; 2001 estimated population 141,400), LITTORAL province, CAMEROON, 45 mi/72 km N of DOUALA; 04°43′N 09°44′E.

Louny (LOU-ni), German *Laun*, city, SEVEROCESKY province, NW BOHEMIA, CZECH REPUBLIC, on OHRE RIVER; 50°21′N 13°48′E. Railroad junction. Manufacturing (machinery, porcelain goods); food processing; brewery (established 1892). Quality hops grown in vicinity. Once a royal seat, it has a 16th century church and castle ruins. Poet Jaroslav Vrchlický was born here in 1853. Paleolithic-era archaeological site.

Loup (LOOP), county (□ 571 sq mi/1,484.6 sq km; 2006 population 656), central NEBRASKA, ⊙ TAYLOR; 41°55′N 99°27′W. Sand Hills grazing region drained by CALAMUS and NORTH LOUP rivers. Cattle, hogs; corn. Formed 1883.

Loup City (LOOP), city (2006 population 914), ⊙ SHERMAN county, central NEBRASKA, 40 mi/64 km NW of GRAND ISLAND, and on MIDDLE LOUP RIVER; 41°16′N 98°58′W. Grain. Bowman Lake State Recreation Area to SW; Sherman Reservoir and State Recreation Area to NE. Settled 1873.

Loupe, La (loo-puh, lah), commune (□ 2 sq mi/5.2 sq km), EURE-ET-LOIR department, CENTRE administrative region, NW central FRANCE, 21 mi/34 km W of CHARTRES; 48°29′N 01°01′E. Agricultural trade center in PERCHE district noted for horse breeding.

Loup River (LOO), 30 mi/48 km long, ALPES-MARITIMES department, PROVENCE-ALPES-CÔTE D'AZUR region, SE FRANCE; rises in Provence (MARITIME) Alps, flows SE through narrow gorge to the MEDITERRANEAN S of CAGNES-SUR-MER. Hydroelectric plant.

Loup River (LOOP), 68 mi/109 km long, E central NEBRASKA; formed by North and Middle Loup rivers in HOWARD county, near St. Paul; flows E, past Fullerton and Genoa, to PLATTE RIVER at COLUMBUS. Its tributaries flow SE: North Loup, 212 mi/341 km long, rises in CHERRY county; flows past Ord to join the Middle Loup, which also rises in Cherry county and flows 221 mi/356 km, past NEBRASKA NATIONAL FOREST and Loup City. South Loup, 152 mi/245 km long, rises in LOGAN county; flows past Ravenna to join Middle Loup E of Boelus. The CALAMUS RIVER joins the North Loup, the DISMAL RIVER joins the Middle Loup, and the CEDAR RIVER joins the main stream. Diversion dam SW of Genoa is unit in Loup River power project, directing water through a canal to generators at Monroe and Columbus.

Lourches (loor-shuh), commune (□ 1 sq mi/2.6 sq km), NORD department, NORD-PAS-DE-CALAIS region, N FRANCE, on the SCHELDT RIVER (canalized), 2 mi/3.2 km SW of DENAIN; 50°19′N 03°21′E. In declining coal-mining area. Metalworks.

Lourdes (LOR-dais), town, ⊙ Lourdes municipio, NORTE DE SANTANDER department, N COLOMBIA, 19 mi/31 km WNW of CÚCUTA; 07°57′N 72°49′. Elevation 4,678 ft/1,425 m. Coffee, plantains; livestock.

Lourdes (LOOR-des), town, VILLA CLARA province, N central CUBA; 22°51′N 80°27′W. Small agricultural community in sugarcane region, on railroad. Located 6 mi/10 km W of Quntín Banderas sugar mill and 7 mi/12 km W of José River Riquelme sugar mill.

Lourdes (loor-duh), town (□ 14 sq mi/36.4 sq km), pilgrimage center in HAUTES-PYRÉNÉES department, MIDI-PYRÉNÉES region, SW FRANCE, at foot of the PYRENEES, 15 mi/24 km SE of PAU; 43°06′N 00°02′W. It is famous for its Roman Catholic shrine where Our Lady of Lourdes (whose feast day is February 11) is believed to have repeatedly appeared (1858) to a peasant girl, 14-year-old Bernadette Soubirous (now St. Bernadette). Millions of people, especially the ill and invalids, make the pilgrimage to Lourdes each year, drawn by their faith in the miraculous cures attributed to the waters of the shrine. The grotto and underground basilica containing the shrine are located in a park within a bend of the Gave de PAU River. There are hospitals and a convent in town, as well as a busy commercial center selling religious articles. A medieval castle built on a rock overlooks the shrine; it contains a museum. The Pic du GER (elevation 2,782 ft/848 m) is reached from town by aerial tramway, offering panoramic views of the central Pyrenees. Summer is the peak season for pilgrimages to Lourdes. The main summits of the Pyrenees (contained within the PYRENEES NATIONAL PARK) are less than 25 mi/40 km S of Lourdes. TARBES-OSSUN-Lourdes international airport 6 mi/9.7 km NE of Lourdes.

Lourenço (lo-REN-so), village, NC Amapá, Brazil, 61 mi/98 km W of CALÇOENE, in Serra Lombarda; 09°19′N 51°38′W. Gold mining.

Lourenço Marques, MOZAMBIQUE: see MAPUTO, city and province.

Loures (LOR-esh), town, LISBOA district, central PORTUGAL, 7 mi/11.3 km N of LISBON. Sugar refining, chemical fertilizer manufacturing, textile dyeing.

Louriçal (lor-ee-SAHL), village, LEIRIA district, W central PORTUGAL, 19 mi/31 km NNE of LEIRIA. Wine, rice, resin (from nearby pine woods). Has 17th-century convent.

Lourinhã (lor-een-YAH), town, LISBOA district, central PORTUGAL, near the ATLANTIC OCEAN, 9 mi/14.5 km SSE of PENICHE. Distilling; wine, olives, almonds.

Louros River (LOO-ros), 47 mi/76 km long, in S EPIRUS department, NW GREECE; rises in the TOMAROS mountains, flows S to the Gulf of Árta of IONIAN SEA 13 mi/21 km SW of Árta. Hydroelectric plant. Louros is also the name of a gorge along the river.

Louroujina (loo-ro-YE-nah) [Turkish=Annçilar], town, LEFKOSIA district, S central CYPRUS, 13 mi/21 km SSE of NICOSIA; 35°01′N 33°28′E. Located at S end of 5 mi/8 km-long extension of Turkish occupied zone; road access only from N. Grain, vegetables, potatoes, citrus; sheep, goats.

Lousã (lo-ZAH), town, COIMBRA district, N central PORTUGAL, on railroad, 12 mi/19 km SE of COIMBRA. At N foot of Serra da Lousã (3,950 ft/1,204 m), SW outlier of SERRA DA ESTRÊLA. Manufacturing (paper, rugs, blankets, hats); sawmilling.

Lousa, village, BRAGANÇA district, N PORTUGAL, 30 mi/48 km ESE of VILA REAL, near right bank of the DUERO RIVER. Vineyards (producing port wine).

Lousada (lo-ZAH-dah), town, PÔRTO district, N PORTUGAL, on railroad, 18 mi/29 km ENE of OPORTO; 41°18′N 08°15′W. Agricultural trade (corn, beans, olives, figs); vineyards.

Lousal (lo-ZAHL) or **Mina do Lousal**, town, SETÚBAL district, S central PORTUGAL, on railroad, 13 mi/21 km SE of GRÂNDOLA; 38°03′N 08°25′W. Copper mining.

Lousana, hamlet, S central ALBERTA, W CANADA, 29 mi/46 km from RED DEER, in RED DEER COUNTY; 52°07′N 113°10′W.

Louth (LOU[TH]), Gaelic *Lú*, county (□ 321 sq km; 2002 population 101,821; 2006 population 111,267), NE IRELAND; ⊙ DUNDALK; 53°57′N 06°32′W. Borders the IRISH SEA from the mouth of the BOYNE RIVER to CARLINGFORD LOUGH; MONAGHAN county to NW; and MEATH county to SW. The terrain is an undulating plain, except for a hilly district in the N. The principal rivers are the FANE, GLYDE, and DEE. Among the industries are cotton and linen manufacturing, brewing, and food processing. Dundalk, DROGHEDA, and GREENORE are ports. The county is associated with the exploits of the legendary Irish hero Cú Chulainn.

Louth (LOUTH), market town (2001 population 15,930), NE central LINCOLNSHIRE, E ENGLAND, on Lud River, 23 mi/37 km ENE of LINCOLN; 53°22′N 00°01′W. Food processing; malt and chemical manufacturing. Has 13th-century grammar school that Alfred Tennyson attended; parish church dating from late 12th century, with 300-ft/91-m perpendicular spire (16th century); and, nearby, ruins of Cistercian Louth Park Abbey, founded 1139. For centuries Louth was an important religious center, site (1536) of a meeting of partisans affiliated with the Pilgrimage of Grace.

Louth Bay (LOUTH), inlet of SPENCER GULF, SOUTH AUSTRALIA, on SE EYRE PENINSULA, between BOSTON POINT (S) and Point Bolingbroke (N); 12 mi/19 km long, 7 mi/11 km wide; 34°34′S 136°02′E. Sheltered by SIR JOSEPH BANKS islands.

Loutra (loo-TRAH) or **Therma**, village and hot springs, on N central coast of SAMOTHRACE island, Évros prefecture, EAST MACEDONIA AND THRACE department, extreme NE GREECE; 39°03′N 26°33′E. Popular spa and base for climbing Mount Fengari. Sanctuary remains 5 mi/8 km W.

Loutra Aidhipsou (loo-TRAH-e[th]-eep-SOO) [Greek=spa of Aidhipsou], city, NW ÉVVIA island and prefecture, CENTRAL GREECE department, GREECE, port on N Gulf of Évvia, 40 mi/64 km NW of KHALKÍS; 38°51′N 23°03′E. Health resort with sulphur springs, frequented since ancient times; fisheries. The municipality (formed c.1940) includes villages of Aidhipsos, 2 mi/3.2 km N; Ayios (or Hagios), 4 mi/6.4 km N; Gialtra (or Yialtra), 4 mi/6.4 km W across Gulf of Aidhipsou; and Gourgouvitsa, 2 mi/3.2 km ENE. Also spelled Loutra Aidepsos, Loutra Aedipsos.

Loutraki (loo-TRAH-kee), city, KORINTHIA prefecture, NE PELOPONNESE department, S mainland GREECE, port on Bay of CORINTH, 4 mi/6.4 km NE of CORINTH; 37°59′N 22°58′E. Summer resort (alkaline springs). Includes town of Perachora or Perakhora, 4 mi/6.4 km NW; and Poseidonia, 2.5 mi/4 km S. Also called Loutrakio.

Loutro (loo-TRO), harbor, SW CRETE department, GREECE; 35°12′N 24°05′E. Mentioned in the New Testament. Also called Phenice, Phoenix. An alternate spelling is Lutro.

Louvain, BELGIUM: see LEUVEN.

Louve, FRANCE: see LOUE RIVER.

Louveciennes (loo-vuh-syen), W residential suburb (□ 2 sq mi/5.2 sq km) of PARIS, YVELINES department, ÎLE-DE-FRANCE region, N central FRANCE, on S bend of SEINE RIVER, 10 mi/16 km W of Notre Dame Cathedral; 48°52′N 02°07′E. Surrounded by parkland and Forest of MARLY-LE-ROI (S). Has several small estates with châteaux and regional waterworks, including châteaux once owned by Madame du Barry, mistress of Louis XV.

Louveira (LOO-vai-rah), city (2007 population 30,038), S SÃO PAULO state, BRAZIL, 6 mi/9.7 km NW of JUNDIAÍ, on railroad; 23°04′S 46°58′W.

Louvicourt (loo-vee-KOOR), unincorporated village, W QUEBEC, E CANADA, included in VAL-D'OR; 48°04′N 77°23′W. Gold and copper mining.

Louvière, La (loo-VYER, lah), commune, Soignies district, HAINAUT province, S BELGIUM. It was an industrial center of the Bassin du Centre coal-mining region, but is having to change its economic focus since the coal mines have ceased producing.

Louviers (loo-vye), town (□ 10 sq mi/26 sq km), EURE department, HAUTE-NORMANDIE region, NW central FRANCE, on left bank of the braided EURE RIVER, on PARIS-NORMANDY expressway, 16 mi/26 km SSE of ROUEN; 49°13′N 01°11′E. Woolen-manufacturing center; woodworking, manufacturing of audiovisual equipment. It has a 13th–16th-century Flamboyant Gothic church containing significant artwork.

Louviers, unincorporated village, DOUGLAS county, central COLORADO, on PLUM CREEK, near SOUTH PLATTE RIVER, 15 mi/24 km S of DENVER, near E edge of FRONT RANGE. Elevation 5,680 ft/1,731 m. Manufacturing (prefabricated buildings). Roxborough State Park and Pike National Forest to W.

Louvigné-du-Désert (loo-vee-nyai–dyoo–dai-zer), town (□ 16 sq mi/41.6 sq km), ILLE-ET-VILAINE department, in BRITTANY, W FRANCE, 10 mi/16 km NNE of FOUGÈRES; 48°29′N 01°08′W. Granite-quarrying center. Manufacturing of ready-to-wear clothing. Megalithic monuments nearby.

Louvres (loo-vruh), industrial community (□ 4 sq mi/10.4 sq km), VAL-D'OISE department, ÎLE-DE-FRANCE region, N central FRANCE, 14 mi/23 km NNE of PARIS, and 3 mi/4.8 km N of Charles de Gaulle airport; 49°02′N 02°30′E. New TGV (high speed railroad) track to N France passes by Louvres.

Louvroil (loo-vroi), town (□ 2 sq mi/5.2 sq km) industrial suburb of MAUBEUGE, NORD department, NORD-PAS-DE-CALAIS region, N FRANCE, on the SAMBRE; 50°16′N 03°58′E. Rolling mills; manufacturing of tools and fabricated iron and steel products.

Lo Valdés (lo vahl-DAIS), ski slopes, SANTIAGO province, METROPOLITANA DE SANTIAGO region, central CHILE, in the ANDES, near ARGENTINA border, 40 mi/64 km SE of SANTIAGO. Accommodations at year-round base for mountain excursions.

Lövånger (LUHV-ONG-er), village, VÄSTERBOTTEN county, N SWEDEN, on small lake near GULF OF BOTHNIA, 30 mi/48 km SSE of SKELLEFTEÅ; 64°22′N 21°19′E. Has fifteenth-century church. Tourist and conference center in old "church town."

Lovango Cay (luh-VAING-go kee), islet, U.S. VIRGIN ISLANDS, off New St. John Island, 8 mi/12.9 km E of CHARLOTTE AMALIE; 18°22′N 64°47′W.

Lovászi (LO-vah-se), Hungarian *Lovászi*, village, ZALA county, W HUNGARY, 20 mi/32 km WNW of NAGY-KANIZSA; 46°33′N 16°34′E. Center of first natural-gas field in Hungary, but now of very minor importance; power plant.

Lovatni-Yerik, RUSSIA: see ILOVATKA.

Lovat' River (luh-VAHT), 335 mi/539 km long, in W European Russia; rises in PSKOV oblast SE of NEVEL, on the RUSSIA-Belorussian border, flows N, past VELIKIYE LUKI and KHOLM (NOVGOROD oblast), to Lake IL'MEN, forming a joint delta mouth with the POLA RIVER. Receives Kunya (right) and Polist (left) rivers. Navigable for 40 mi/64 km in its lower course. Part of the ancient Baltic-Black Sea water route.

Lovcen (LOV-tsen), Serbian *Lovčen* (LOV-chen), basaltic mountain in DINARIC ALPS, SW MONTENEGRO on Gulf of KOTOR; highest point (5,737 ft/1,749 m) is 3 mi/4.8 km SE of KOTOR; 42°23′N 18°51′E. Crossed by scenic serpentine road connecting Kotor on ADRIATIC coast with PODGORICA, via CETINJE (at E foot). National park (□ 25 sq mi/65 sq km). Was refuge of Serb nobles after battle of Kosovo (1389) and center of Montenegrin resistance against Turks. Also spelled Lovchen.

Love, county (□ 531 sq mi/1,380.6 sq km; 2006 population 9,162), S OKLAHOMA; ⊙ MARIETTA; 33°57′N 97°15′W. Bounded S by RED RIVER, here forming TEXAS state line. Includes part of Lake Murray (with state park) in NE and headquarters of Lake TEXOMA in SE (Red River). Agricultural area (grain, corn, asparagus and other vegetables, watermelons, hay, peanuts; cattle, sheep, hogs, horses); manufacturing at Marietta. Formed 1907.

Love (LUHV), village (2006 population 55), E central SASKATCHEWAN, CANADA, 10 mi/16 km NW of NIPAWIN; 53°28′N 104°09′W. Wheat; dairying.

Lovea (LO-VIR), village, W CAMBODIA, 16 mi/26 km S of SISOPHON. In THAILAND, 1941–1946.

Love Canal, section of NIAGARA FALLS, NEW YORK, and a former chemical disposal site. The empty canal was used 1942–1953 by Hooker Chemicals and Plastics Corp. to dump 22,000 tons/19,954 metric tons–25,000 tons/22,675 metric tons of toxic waste; the waste was sealed in metal drums in a manner that has since been declared illegal. The canal was then filled in and the land sold for $1 to the expanding city of Niagara Falls by the chemical company. Housing and an elementary school were built on the site. By the late 1970s several hazardous chemicals had leaked through their drums and risen to the surface. Investigations confirmed the existence of toxins in the soil and determined that they were responsible for the area's unusually high rates of birth defects, miscarriages, cancer, illness, and chromosome damage. Families were evacuated by New York state from the area in 1978, and in 1980 the Love Canal was proclaimed by President Jimmy Carter to be a Federal Disaster Area. At the same time, reacting to the resulting national outrage over hidden industrial pollution and triggered by the Love Canal situation, Congress created the Federal Superfund. In early 1980s, the government bought and razed 238 houses located next to the dump; 550 houses, farther away, were bought, and two-thirds of them were declared in 1986 to be habitable. In March 1995 New York state closed its public information office, declaring the Love Canal neighborhood to be safe. Since then, the 234 homes in the Emergency Declaration Area Habitable Zone put back on the market have been sold and the neighborhood renamed Black Creek. The original corporation's successor in December 1995 settled with the federal government, agreeing to pay $102 million to the Federal Superfund and an additional $27 million to the Federal Emergency Management Agency, which handled the early evacuation crisis. New York state settled their suit with the same company for $98 million in June 1994. Total estimated cost of the cleanup was $129 million, and more than $20 million in damages was paid by the chemical company and the city of Niagara Falls to a group of former residents.

Lovech (LO-vech), oblast (□ 5,869 sq mi/15,259.4 sq km), in central part of the DANUBE plain; 43°09′N 24°43′E. Bordering on MONTANA oblast to the W and RUSE oblast to the E, STARA PLANINA to the S (PLOVDIV and HASKOVO oblasts, small parts of SOFIA and BURGAS oblasts), Danube River and ROMANIA to the N, lying between the ISKUR and YANTRA rivers. VIT, OSUM, and ROSITSA rivers flow here. Flat to the N, hilly and mountainous to the S. Area constitutes c.14% of Bulgaria. Has 40 cities. Provides c.12.5% of the nation's industrial and c.13% of agricultural production. Oil wells near PLEVEN, coal deposits used for coking; clays and limestone for cement production; water and forest resources. Electrotechnical industry, radio, TV, automobile assembly, petrochemicals, heavy machinery, textiles; wood processing, leather. Agriculture (grain crops, fodder, viticulture, sugar beets, sunflowers, tomatoes; cattle, sheep); fruit and vegetable canning, wine making. Historic sites, folk architecture and craft in VELIKO TURNOVO (former capital of Bulgaria), ETURA, ORESHAK.

Lovech (LO-vech), city (1993 population 48,528), LOVECH oblast and obshtina, N BULGARIA, on OSUM RIVER, 19 mi/31 km S of PLEVEN. Agriculture (grain; livestock), tanning center; motor vehicle assembly; manufacturing (pig-iron parts, motorcycles and bicycles, plastics, electric tools, leather goods, furniture). Hydroelectricity. Leathercraft school. Once a Roman settlement; became an important stronghold under Turkish rule (15th–19th century).

Lovejoy, town (2000 population 2,495), CLAYTON county, N central GEORGIA, 21 mi/34 km S of ATLANTA; 33°26′N 84°19′W. Exurb of Atlanta. Manufacturing includes motor vehicle parts, meat products, food-processing equipment.

Lovejoy, ILLINOIS: see BROOKLYN.

Lovelady (LUHV-lai-dee), village (2006 population 607), HOUSTON county, E TEXAS, 13 mi/21 km S of CROCKETT; 31°07′N 95°27′W. In agricultural area; timber; manufacturing (oil and gas processing, plastic processing).

Loveland, city (2000 population 50,608), LARIMER county, N COLORADO, 10 mi/16 km S of FORT COLLINS on BIG THOMPSON RIVER; 40°25′N 105°04′W. Elevation 4,982 ft/1,519 m. Loveland lies in a fertile farm area, irrigated by the Colorado–Big Thompson project, where sugar beets and wheat are grown, also fruit and vegetables, beans, barley; cattle, sheep. The city is also a growing industrial hub, manufacturing (building materials, metal products, electrical equipment, computer equipment, chemicals, printing and publishing, concrete, medical equipment). Gateway to Rocky Mountain State Park 25 mi/40 km to W; Roosevelt National Forest to W; Boyd Lake State Park to NE; CARTER LAKE to SW. Incorporated 1881.

Loveland (LUHV-luhnd), city (□ 5 sq mi/13 sq km; 2006 population 11,154), on CLERMONT-HAMILTON-WARREN county line, SW OHIO, 18 mi/29 km NE of CINCINNATI, on Little Miami River, on edge of suburban development; 39°16′N 84°13′W. Settled 1825, incorporated 1876.

Loveland, village (2006 population 13), TILLMAN county, SW OKLAHOMA, 15 mi/24 km ESE of FREDERICK; 34°17′N 98°46′W. In agricultural area (cotton, grain, peanuts).

Loveland, Lake, California: see SWEETWATER RIVER.

Loveland Pass (11,992 ft/3,655 m), N central COLORADO, crosses CONTINENTAL DIVIDE, in FRONT RANGE, c.55 mi/89 km W of DENVER. U.S. Highway 6 passes through it. EISENHOWER MEMORIAL TUNNEL to NW, carries Interstate 70 across Divide. Winter sports area. GRAYS PEAK is just E.

Lovell (LUHV-uhl), town, OXFORD county, W MAINE, on KEZAR LAKE, 20 mi/32 km WSW of SOUTH PARIS; 44°11′N 70°53′W. In resort area; wood products.

Lovell, town (2006 population 2,288), BIG HORN county, N WYOMING, on SHOSHONE RIVER, near MONTANA state line, 70 mi/113 km W of SHERIDAN, 65 mi/105 km S of BILLINGS, Montana; 44°50′N 108°23′W. Elevation 3,814 ft/1,163 m. Supply and processing point in irrigated sugar-beet and grain region; agriculture (beet sugar, sugar-beet pellets and pulp, beans); cattle, sheep; manufacturing (concrete; gypsum products, steel fabricating). Oil wells in vicinity. Big Horn Lake reservoir (Big Horn River) and Big Horn Canyon National Recreation Area to E; Big Horn Mountains and Big Horn National Forest also to E. Pryor Mountain Wild Horse Range to NE; Medicine Wheel National Historic Site to E. Laid out 1900 by Mormons.

Lovell, village, LOGAN county, central OKLAHOMA, 17 mi/27 km NW of GUTHRIE. In agricultural area.

Lovell's Island (LUH-vuhlz), island, E MASSACHUSETTS, in QUINCY BAY section of Boston Harbor, 6 mi/9.7 km SE of BOSTON. Part of Boston Harbor Islands State Park. Former site of Fort Standish, abandoned after World War II.

Lovelock (LUHV-LAHK), town (2006 population 1,903), ⊙ PERSHING county, W central NEVADA, on HUMBOLDT River, between TRINITY (W) and HUMBOLDT (E) ranges, and c.83 mi/134 km NE of RENO; 40°10′N 118°28′W. Elevation 3,977 ft/1,212 m. Trade center for Humboldt irrigation project (barley; cattle, sheep; dairying). Gold, silver, copper, lead, and diatomite are mined in vicinity. HUMBOLDT SINK in Humboldt Wildlife Management Area to SW. RYE PATCH Reservoir and State Recreation Area to N, on Humboldt River; museum. Settled 1860, incorporated 1917.

Lovely (LUHV-lee), unincorporated town, MARTIN county, E KENTUCKY, 13 mi/21 km NW of WILLIAMSON, WEST VIRGINIA, on TUG FORK RIVER (West Virginia state line) at mouth of Wolf Creek, in CUMBERLAND MOUNTAINS. Bituminous coal. Manufacturing (coal processing).

Lovely Banks (LUHV-lee BANKS), suburb just NW of GEELONG, VICTORIA, SE AUSTRALIA; 38°04′S 144°20′E. Elevation 328 ft/100 m; has views of Geelong and CORIO BAY.

Lovendegem (lo-VEN-duh-khem), commune (2006 population 9,381), Ghent district, EAST FLANDERS province, NW BELGIUM, on BRUGES-Ghent Canal, 5 mi/8 km NW of GHENT. Agriculture.

Lovenia, Mount, peak (13,219 ft/4,029 m) of UINTA MOUNTAINS, N DUCHESNE county, NE UTAH, 38 mi/61 km N of DUCHESNE, in High Uintas Wilderness Area, Ashley National Forest.

Lövenich (LO-ve-nikh), suburb of COLOGNE, NORTH RHINE-WESTPHALIA, W GERMANY; 50°56′N 06°57′E.

Love Point, MARYLAND: see KENT ISLAND.

Loves Park, city (2000 population 20,044), WINNEBAGO county, N ILLINOIS, on the ROCK RIVER; 42°20′N 89°00′W. It is chiefly residential. Next to Rock Cut State Park. Incorporated 1947.

Lövestad (LUHV-e-STAHD), village, SKÅNE county, S SWEDEN, 15 mi/24 km NNE of YSTAD; 55°39′N 13°53′E.

Lövéte, ROMANIA: see LUETA.

Lovett (LUHV-et), town, LAURENS county, central GEORGIA, 11 mi/18 km NE of DUBLIN; 32°38′N 82°46′W.

Lovettsville (LUH-vets-vil), town (2006 population 1,204), LOUDOUN county, N VIRGINIA, 8 mi/13 km NNE of LEESBURG, near POTOMAC RIVER (MARYLAND state line), 2 mi/3 km S of BRUNSWICK (Maryland) across the river; 39°16′N 77°38′W. Light manufacturing; agriculture (dairying; livestock; grain, apples, soybeans). Harpers Ferry National Historical Park to W.

Love Valley (LUHV VAL-ee), village (2006 population 55), IREDELL county, W central NORTH CAROLINA, 17 mi/27 km NNW of STATESVILLE; 35°59′N 80°59′W. Construction; agriculture (tobacco, grain; livestock; timber). Horse-riding trails, camping.

Lovewell Mountain (LUH-vuhl), peak (2,496 ft/761 m), SW NEW HAMPSHIRE, on SULLIVAN-MERRIMACK county line, near WASHINGTON.

Lovewell Pond, OXFORD county, W MAINE, near FRYEBURG; 2 mi/3.2 km long. Monument marks site of victory over Indians, 1725.

Lovewell Reservoir (□ 12 sq mi/31 sq km), JEWELL county, N central KANSAS, on White Rock Creek, 64 mi/103 km NNW of SALINA; 39°53′N 98°02′W. Maximum capacity 186,290 acre-ft. Formed by Lovewell Dam (93 ft/28 m high), built (1957) by the Bureau of Reclamation for flood control; also used for irrigation and recreation. Lovewell State Park on N shore.

Lovich, POLAND: see LOWICZ.

Loviisa, FINLAND: see LOVISA.

Lovilia, town (2000 population 583), MONROE county, S IOWA, near CEDAR CREEK, 9 mi/14.5 km NW of ALBIA; 41°08′N 92°54′W. In bituminous-coal-mining and livestock area; manufacturing (feed, concrete blocks).

Loving (LUHV-ing), county (□ 676 sq mi/1,757.6 sq km; 2006 population 60), W TEXAS; ⊙ MENTONE; 31°50′N 103°34′W. High rolling prairies; bordered N by NEW MEXICO state line, SW by PECOS RIVER. Elevation c.2,500 ft/762 m–3,400 ft/1,036 m. Some cattle ranching. Oil and natural gas. RED BLUFF LAKE (Pecos River), used for irrigation and recreation on county line in NW. Smallest county (in terms of population) in the lower forty-eight states. Formed 1887.

Loving, town (2006 population 1,317), EDDY county, SE NEW MEXICO, near PECOS RIVER, 13 mi/21 km SSE of CARLSBAD; 32°17′N 104°05′W. In irrigated cotton and alfalfa region; potash and salt mining. CARLSBAD CAVERNS NATIONAL PARK to WSW. Junction of railroad spur to potash mines to NE.

Area is shown by the symbol □, and capital city or county seat by ⊙.

Lovingston (LUH-veengz-tuhn), unincorporated village, ⊙ NELSON county, central VIRGINIA, 29 mi/47 km NE of LYNCHBURG; 37°45′N 78°52′W. Manufacturing (clothing); agriculture (primarily apples; also corn, alfalfa; cattle).

Lovington, town (2006 population 9,693), ⊙ LEA county, SE NEW MEXICO, on LLANO ESTACADO, near TEXAS state line, 20 mi/32 km NW of HOBBS; 32°57′N 103°20′W. Railroad terminus. Trade center in livestock, grain, vegetables, melons; dairying and poultry raising, cotton. Manufacturing (food, engines). Potash mines and oil wells in vicinity. Rodeo and county fair take place here in August. Founded 1908.

Lovington, village (2000 population 1,222), MOULTRIE county, central ILLINOIS, 17 mi/27 km ESE of DECATUR; 39°42′N 88°37′W. Corn, wheat, soybeans; livestock; dairy products. Incorporated 1873.

Lovinobana (lo-VI-no-BA-nya), Slovak *Lovinobaňa* or *Lovinovaňa*, Hungarian *Lónyabánya*, village, STREDOSLOVENSKY province, S SLOVAKIA, on railroad, 7 mi/11.3 km NNW of LUČENEC. Clothing manufacturing; mining and processing of magnesite. Summer resort Ruzina (ru-ZHI-nah) is just W, and 17th-century castle of Divin (dyi-VEEN) is 2 mi/3.2 km W.

Lovisa (loo-VEE-SAH), town, UUDENMAAN province, S FINLAND, at head of small bay in GULF OF FINLAND, 50 mi/80 km ENE of HELSINKI; 60°27′N 26°14′E. Fishing port; railroad terminus; seaside resort. Population is largely Swedish-speaking. Founded in 1745 as Degerby, renamed in 1752. Nearby fortifications destroyed in 1855 by British. Major nuclear power station on Hästholm island. Station has two of the country's four nuclear reactors (went on-line in 1977 and 1978), with 890 MW combined capacity. Also has two conventional oil-fueled power plants (40 MW combined), which went on-line in 1974. Also spelled Loviisa.

Lovnidol (LOV-nee-dol), village, LOVECH oblast, SEVLIEVO obshtina, N central BULGARIA, on the N slope of the BALKAN MOUNTAINS, 8 mi/13 km ESE of SEVLIEVO; 43°00′N 25°14′E. Horse-raising center; horticulture. Formerly called Chiflik.

Lovö, SWEDEN: see DROTTNINGHOLM.

Lovosice (LO-vo-SI-tse), German *Lobositz*, town, SEVEROCESKY province, N BOHEMIA, CZECH REPUBLIC, on ELBE RIVER, 10 mi/16 km S of ÚSTÍ NAD LABEM; 50°31′N 14°04′E. Railroad junction. Manufacturing of chemicals (fertilizers, synthetic fibers); food processing (edible oils). Garnet deposits nearby. Frederick the Great defeated the Austrian army here in 1756.

Lovozero (luh-VO-zye-ruh), village (2006 population 2,905), central MURMANSK oblast, NW European Russia, on the KOLA PENINSULA, near Lake LOVOZERO, terminus of a highway branch connecting it to OLENEGORSK, 85 mi/137 km SE of MURMANSK; 68°03′N 35°00′E. Elevation 688 ft/209 m. In lumbering and reindeer-raising area.

Lovozero (luh-VO-zee-ruh), lake (surface □ 77 sq mi/200 sq km), central MURMANSK oblast, 30 mi/48 km long, 3 mi/5 km wide, max depth 20 ft/6 m; 67°40′N 35°10′E. Turned into a reservoir after the construction of the Serebryansk hydroelectric power plant.

Lovran, Italian *Laurana*, town, KVARNER region, W CROATIA, on E coast of ISTRIA, in OPATIJA Riviera, 3 mi/4.8 km S of Opatija. Popular seaside resort. Cardiovascular and orthopedic clinic. Was first mentioned in 7th century as Lauriana. Church of Sveti [=saint] Juraj dates to 14th century.

Lovushki Islands (luh-VOOSH-kee), Japanese *Mushiru-retsugan*, group of reefs in N main KURIL ISLANDS group, SAKHALIN oblast, extreme E SIBERIA, RUSSIAN FAR EAST; separated from SHIASHKOTAN ISLAND (N) by FORTUNA STRAIT, from RAIKOKE ISLAND (S) by KRUZENSHTERN STRAIT; 48°33′N 153°50′E.

Low (LO), canton (□ 100 sq mi/260 sq km; 2006 population 864), OUTAOUAIS region, SW QUEBEC, E CANADA, 27 mi/44 km from GATINEAU; 45°48′N 75°56′W. Covered bridge. Established 1858.

Lowa (LO-wah), village, Équateur province, E CONGO, on LUALABA RIVER opposite mouth of LOWA RIVER, 135 mi/217 km SSE of KISANGANI. Steamboat landing, trading center. Roman Catholic mission. Airfield.

Low and Burbanks Grant, land grant, COOS county, N central NEW HAMPSHIRE, 13 mi/21 km SW of BERLIN. Wilderness area in White Mountain National Forest in PRESIDENTIAL RANGE of WHITE MOUNTAINS.

Lowarai (LO-wah-rei), major pass between DIR and CHITRAL districts, NORTH-WEST FRONTIER PROVINCE, N PAKISTAN, near AFGHANISTAN border; 35°20′N 71°38′E. Elevation 10,499 ft/3,200 m. Closed six months of the year because of snow landslides, thereby cutting off supplies to CHITRAL. Tunnel underpass partially completed but abandoned.

Low Archipelago, FRENCH POLYNESIA: see TUAMOTU ARCHIPELAGO.

Lowa River (LO-wah), c.275 mi/443 km long, E tributary of the LUALABA, in E CONGO; formed by several headstreams rising on N and NW slopes of the KAHUSI range c.25 mi/40 km W of BOBANDANA, flows generally W, past WALIKALE and through dense equatorial forest, to LUALABA RIVER opposite LOWA village. Drains important tin-mining area.

Low, Cape, S extremity of SOUTHAMPTON ISLAND, KIVALLIQ region, NUNAVUT territory, CANADA, on HUDSON BAY; 63°07′N 85°18′W.

Low Countries, region of NW EUROPE comprising the NETHERLANDS, BELGIUM, and grand duchy of LUXEMBOURG. The N parts of the Netherlands and Belgium form a low plain bordering on NORTH SEA, but S Belgium and Luxembourg are part of the ARDENNES. The name "Low Countries" thus is a political and historic rather than a strictly geographic concept. One of the wealthiest areas of medieval and modern Europe, it also has been a chronic theater of war. For the geography and history, see articles on the individual countries and on their provinces.

Lowden, town (2000 population 794), CEDAR county, E IOWA, 12 mi/19 km NE of TIPTON; 41°51′N 90°55′W. Limestone quarries nearby.

Lowden, unincorporated village, WALLA WALLA county, SE WASHINGTON, 11 mi/18 km W of WALLA WALLA, on WALLA WALLA RIVER Grain, vegetables, grapes; livestock. Wineries.

Lowdham (LOUD-uhm), village (2001 population 2,365), central NOTTINGHAMSHIRE, central ENGLAND, 7 mi/11.3 km NE of NOTTINGHAM; 53°01′N 01°00′W. Has church dating mainly from 15th century.

Lowe Farm (LO), community, S MANITOBA, W central CANADA, 10 mi/16 km W of town of MORRIS, in MORRIS rural municipality; 49°21′N 97°35′W.

Lowell (LO-uhl), city (□ 14 sq mi/36.4 sq km; 2000 population 105,167), ⊙ MIDDLESEX county, NE MASSACHUSETTS, at the confluence of the MERRIMACK and CONCORD rivers; 42°38′N 71°19′W. High-technology computer industries have developed here; other manufacturing includes electronic, electrical, and telecommunications equipment, textiles, rubber products, chemicals, machine parts, foodstuffs, shoes, paper, and plastics. The city grew after textile mills were built at Pawtucket Falls, and it became one of the major textile centers of the country. It has rebounded somewhat after the collapse of the computer manufacturing industry in 1992. Lowell State College and Lowell Technological Institute and a branch of University of Massachusetts are here. Now home of second-largest Cambodian population in U.S. The city has several fine parks, and James Whistler's birthplace is preserved. Oldest American boat shop (1793) still in operation is here. Charles Dickens visited Lowell in 1842 and described it in *American Notes*. Site of LOWELL HISTORICAL PARK, commemorating American Industrial Revolution; Heritage State Park. Settled 1653, set off from CHELMSFORD 1826, incorporated as a city 1836.

Lowell, town (2000 population 5,013), BENTON county, extreme NW ARKANSAS, suburb 5 mi/8 km from ROGERS (W) and SPRINGDALE (S), in the OZARKS; 36°16′N 94°08′W. Manufacturing (sheet metal).

Lowell, town (2000 population 7,505), Lake county, NW INDIANA, 10 mi/16 km SSW of CROWN POINT; 41°17′N 87°25′W. Consumer goods, automotive parts, furniture. Settled 1849, laid out 1853.

Lowell, town, PENOBSCOT county, central MAINE, 30 mi/48 km NNE of BANGOR; 45°13′N 68°30′W. In hunting, fishing area.

Lowell, town (2000 population 4,013), KENT county, SW MICHIGAN, 17 mi/27 km E of GRAND RAPIDS, on GRAND RIVER at mouth of FLAT RIVER; 42°55′N 85°20′W. Railroad junction. In agricultural area (apples, cherries); manufacturing (motor vehicle parts, wire products, chemical sprayers). Resort. Settled 1821, incorporated 1859.

Lowell (LO-uhl), town (□ 2 sq mi/5.2 sq km; 2006 population 2,705), GASTON county, S NORTH CAROLINA, suburb 4 mi/6.4 km E of GASTONIA near South Fork of CATAWBA RIVER; 35°16′N 81°05′W. Manufacturing (metal fabrication, dye, textiles); service industries.

Lowell, town (2006 population 948), LANE county, W OREGON, 17 mi/27 km SE of EUGENE, on Middle Fork of WILLAMETTE RIVER, below LOOKOUT POINT DAM; 43°55′N 122°46′W. Timber; dairying; poultry. FALL CREEK RESERVOIR to N. Elijah Bristowl State Park to W; Willamette National Forest to E; Umpqua National Forest to SE.

Lowell, town, ORLEANS co., N VERMONT, on Missisquoi River, 16 mi/26 km SW of Newport; 44°47′N 72°27′W. The last asbestos mine in the eastern United States that closed in 1993.

Lowell, village, COCHISE county, SE ARIZONA, in MULE MOUNTAINS, near Mexican border, 3 mi/4.8 km SE of BISBEE. Elevation 5,250 ft/1,600 m. Copper mining.

Lowell (LO-uhl), village (2006 population 598), WASHINGTON county, SE OHIO, 8 mi/13 km NNW of MARIETTA, on MUSKINGUM RIVER; 39°31′N 81°30′W. In agricultural area.

Lowell, village, SNOHOMISH county, NW WASHINGTON, suburb 3 mi/4.8 km S of EVERETT, on SNOHOMISH RIVER.

Lowell, village (2006 population 361), DODGE county, S central WISCONSIN, 8 mi/12.9 km S of BEAVER DAM; 43°20′N 88°49′W. In dairying region. Cheese.

Lowell Historical Park (LO-uhl), NE MASSACHUSETTS, authorized 1978. Restored site of first integrated textile mill. National Historic Site maintained by National Park Service.

Lowell, Lake, reservoir, CANYON county, SW IDAHO, between SNAKE (SW) and BOISE (NE) rivers, in Deer Flat National Wildlife Refuge, 5 mi/8 km W of NAMPA; 43°34′N 116°45′W. Maximum capacity c.177,000 acre-ft; 9 mi/14.5 km long, 2 mi/3.2 km wide. Formed by three earth-fill dams; fed by irrigation canal extending W from diversion dam on Boise River 7 mi/11.3 km SE of BOISE. Supplies water for ARROW ROCK division of BOISE IRRIGATION PROJECT. Also known as Deer Flat Reservoir.

Lowell Observatory, astronomical observatory located in FLAGSTAFF, COCONINO county, N central ARIZONA, on Mars Hill, near Thorpe park, W of downtown. It was founded in 1894 by Percival Lowell, the American amateur astronomer who popularized the idea that Mars might support intelligent life. Its original telescope, still in operation, is a 24-in/61-cm refractor. A 42-in/107-cm reflector was added, and in 1929 the 13-in/33-cm A. Lawrence Lowell photographic camera began operation. Also at the observatory's original Mars Hill site and at its nearby Anderson Mesa station are 72-in/183-cm, 31-in/79-cm, 24-in/61-cm, and 21-in/53-cm reflecting telescopes. Many discoveries of fundamental importance were made at the observatory, especially by V. M. Slipher, its director from 1916 to 1954. By 1917 he had determined through spectroscopic analysis the radial velocities of most spiral nebulae then known. His discovery that nearly all these nebulae, now known as galaxies, were apparently moving away from the earth led to Hubble's work and the discovery of the expanding universe. Beginning in 1905 the observatory made a concerted search for a trans-Neptunian planet; this led to the discovery of Pluto by Clyde Tombaugh in 1930. Principal research programs involve the discovery and determination of orbits for new asteroids, a search for nearby stars, and the measurement of light and motion of close double stars, nebulae, and other galactic objects.

Lowellville (LO-uhl-vil), village (□ 1 sq mi/2.6 sq km; 2006 population 1,179), MAHONING county, NE OHIO, 8 mi/13 km SE of YOUNGSTOWN, on MAHONING RIVER, near PENNSYLVANIA state line; 41°02′N 80°32′W. Settled c.1800, incorporated 1836.

Lowe, Mount (5,603 ft/1,708 m), LOS ANGELES county, S CALIFORNIA, in SAN GABRIEL MOUNTAINS, N of PASADENA and just NW of MOUNT WILSON. Astronomical observatory (built 1894). Formerly had incline railroad to summit. In Angeles National Forest. Bear Canyon to N.

Löwen, POLAND: see LEWIN BRZESKI.

Löwenberg, POLAND: see LWOWEK SLASKI.

Löwenhagen, RUSSIA: see KOMSOMOL'SK, KALININGRAD oblast.

Löwenstein (LO-ven-shtein), town, BADEN-WÜRTTEMBERG, S GERMANY, 8 mi/12.9 km SE of HEILBRONN; 49°05′N 09°22′E. Wine; mineral springs nearby. Has ruined castle.

Löwentin Lake, POLAND: see NIEGOCIN, LAKE.

Lower, township, CAPE MAY county, S NEW JERSEY, 2 mi/3.2 km N of CAPE MAY; 38°58′N 74°54′W. Incorporated 1798.

Lower, in Russian names: see also NIZHNE-, NIZHNEYE, NIZHNI, NIZHNIYE, NIZHNYAYA.

Lower Allen, unincorporated town (2000 population 6,619), CUMBERLAND county, S PENNSYLVANIA, residential suburb 5 mi/8 km SW of HARRISBURG; 40°13′N 76°54′W. Camp Hill State Correctional Center to E.

Lower Ammonoosuc River, NEW HAMPSHIRE: see AMMONOOSUC RIVER.

Lower Arrow Lake, CANADA: see ARROW LAKES.

Lower Aulaqi, YEMEN: see AULAQI.

Lower Austria, German Niederösterreich, province (□ 7,400 sq mi/19,240 sq km; 2005 population 1,575,291), NE AUSTRIA; ⊙ SANKT PÖLTEN (since 1986); 48°20′N 15°45′E. Lower Austria is the largest Austrian province; it borders the CZECH REPUBLIC in the N and SLOVAKIA in the NE. A picturesque, hilly region, drained by DANUBE RIVER and containing peaks of Eastern Alps and WIENERWALD (Vienna Woods). Province includes roughly half of the country's arable land and is noted for grain production and wines. In the WEINVIERTEL, Lower Austria's NE part, most of Austria's crude oil and gas are produced. They are refined at Schwechat, Austria's largest refinery. There are five large hydropower stations and three large thermo power stations on the Danube, which provide for a good share of Austria's electricity supply. The valleys and basins around VIENNA, WIENER NEUSTADT and Sankt Pölten contain about 20% of all Austrian industry, which includes the manufacturing of metal, textiles, chemicals, and an oil refinery. Region also supports industries in food processing, sugar refining, brewing, saw milling. BADEN has a well-known spa, and SEMMERING region in S is tourist and health center. The most important tourist region of the province is the WACHAU, a narrow valley of the Danube between MELK and KREMS AN DER DONAU. Almost all of Lower Austria belongs to the recreation belt of Vienna with weekend houses and leisure facilities, especially the Vienna Woods and the Waldviertel during summer and the Lower Austrian Limestone Alps in the winter. Since the fall of the Iron Curtain, Lower Austria again is situated on a major

transit route between GERMANY, Western Europe, and SE EUROPE. The province has several medieval castles and abbeys. In c.1450 a permanent split was made between Upper and Lower Austria. The region became a province in 1918; it lost a smaller part in the N to CZECHOSLOVAKIA in 1919 and Vienna in 1920, which became a federal province (Bundesland) of its own. The history of Lower Austria, the historical heartland of the later Hapsburg empire and present Austria, coincides with that of Austria.

Lower Avon, ENGLAND: see AVON, river.

Lower Bari Doab Canal, India and Pakistan: see BARI DOAB CANAL, LOWER.

Lower Bavaria (buh-VA-ree-uh), German Niederbayern (NEED-uhr-bai-uhrn), administrative division (German Regierungsbezirk) (□ 4,153 sq mi/10,797.8 sq km) of E BAVARIA, GERMANY; ⊙ LANDSHUT. Bounded N by UPPER PALATINATE, W and S by UPPER BAVARIA, SE and E by AUSTRIA, NE by CZECH REPUBLIC. Includes BOHEMIAN FOREST (NE). Drained by DANUBE, INN, and ISAR rivers. Wheat (S of Danube only), barley, rye, cabbage; livestock; intensive hops growing around Landshut. Industries (metals, wood, beer) at Landshut, PASSAU, and STRAUBING. Glassworks in Bohemian Forest (Frauenau, ZWIESEL). Tourism (in Danube region and in Bohemian Forest).

Lower Beeding (LO-wuhr BEED-ing), village (2001 population 1,001), West SUSSEX, SE ENGLAND, 4 mi/6.4 km SE of HORSHAM; 51°02′N 00°17′W.

Lower Bentham (LO-wuhr BEN-thuhm), village (2006 population 3,200), NORTH YORKSHIRE, N ENGLAND, 6 mi/9.7 km SE of KIRKBY LONSDALE; 54°06′N 02°32′W.

Lower Buchanan, LIBERIA: see BUCHANAN.

Lower Burrell, city (2006 population 12,350), WESTMORELAND county, SW PENNSYLVANIA, suburb 18 mi/29 km NE of PITTSBURGH, on ALLEGHENY RIVER, Little Pucketta Creek to S; 40°34′N 79°42′W. Manufacturing (steel fabrication, machinery, fabricated metal products). Agriculture (grain, soybeans, apples; livestock; dairying). City's steel-based economy declined in the 1970s and 1980s. Incorporated 1959.

Lower California: see BAJA CALIFORNIA.

Lower Canada: see QUEBEC, province.

Lower Chateaugay Lake, NEW YORK: see UPPER CHATEAUGAY LAKE.

Lower Chindwin, former district (□ 3,676 sq mi/9,557.6 sq km) comprised of four townships in the SW portion of SAGAING division, MYANMAR; capital was at MONYWA. Irrigated agriculture (rice, cotton, beans); teak forests, gold placers. Astride lower CHINDWIN RIVER, in dry zone (annual rainfall, 31 in/79 cm). Population is nearly entirely Burmese. District is served by Mandalay-Yeu railroad and CHINDWIN RIVER navigation.

Lower Crystal Springs, reservoir (□ 2 sq mi/5.2 sq km), SAN MATEO county, central CALIFORNIA, on San Mateo Creek, in San Andreas Rift Zone and San Francisco State Fish and Game Refuge, 4 mi/9.7 km WSW of SAN MATEO; 37°32′N 122°22′W. Maximum capacity 71,570 acre-ft. Formed by Lower Crystal Springs Dam (140 ft/43 m high), built (1888) for water supply; owned by city and county of SAN FRANCISCO.

Lower Dniester Irrigation System (DHEES-tuhr), (Ukrainian Nyzhn'odnistrovs'ka Zroshuval'na Systema), irrigation system in central ODESSA oblast, UKRAINE, 16 mi/26 km WSW of ODESSA, between DNIESTER estuary (W) and BARABOY RIVER (a minor, intermittent stream, E). Water is drawn from Dniester River from a point 1 mi/1.6 km upstream from Mayaky, and conveyed by a pair of pipes (6 ft/1.8 m in diameter) 5 mi/7.8 km inland to a water reservoir, and then distributed by the mainline canal (17 mi/28 km) and distributary canals (255 mi/411 km). The system, under construction since 1964, by 1990 provided for the irrigation of 91,000 acres/37,000 ha, and the drainage of 13,800 acres/5,600 ha. Water is applied by

sprinklers to grow vegetables, potatoes, grapes (in vineyards), and grain.

Lower East Side, section of SE MANHATTAN, NEW YORK city, SE NEW YORK, bounded on E by EAST RIVER, S by Fulton and Franklin streets, W by Pearl Street and BROADWAY, and N by 14th Street; 40°43′N 73°59′W. Prototypical immigrant neighborhood of New York city, first populated by Irish immigrants in the 1840s. Its distinctive character, however, was established by the flood (beginning in the 1880s) of E European Jews, Russians, Ukrainians, Slovaks, Romanians, Hungarians, Poles, Greeks, and Italians. In 1920 the Jewish enclave alone numbered 400,000, fostering some 500 synagogues and religious schools. By the late 19th century, it was such a rundown slum that it defied even the efforts of noted reformer Jacob A. Riis to improve it. In 1936, in an attempt to replace the squalid tenements here, the city's housing authority constructed its first houses at 3rd Street and 1st Avenue. Following World War II, it became the first integrated neighborhood in the city, with an influx of African Americans and Puerto Ricans. Now it is heavily Chinese, especially in the S part; other ethnic groups living here include E and S Asians, Dominicans, and Latin and Central Americans. Home of such artists and entertainers as Al Jolson, George and Ira Gershwin, Irving Berlin, and the Marx Brothers. In recent years it has experienced modest gentrification, partly as a result of its proximity to the EAST VILLAGE.

Lower Fox River, WISCONSIN: see FOX RIVER.

Lower Franconia (frang-KO-nee-uh), German Unterfranken (UN-tuhr-frahn-ken), administrative division (German Regierungsbezirk) (□ 3,277 sq mi/8,520.2 sq km) of NW BAVARIA, Germany; ⊙ WÜRZBURG; 49°30′N 09°00′E–50°35′N 10°50′E. Bounded by HESSE (W, N), THURINGIA (N), UPPER FRANCONIA and MIDDLE FRANCONIA (E), and BADEN-WÜRTTEMBERG (S). Hilly region including the SPESSART and part of RHÖN MOUNTAINS; drained by the MAIN RIVER. Agriculture. (wheat, barley) and stock raising; wine growing in river valleys. Industrial centers at historic towns of Würzburg (machine tools), SCHWEINFURT (dyes), and ASCHAFFENBURG (textiles, paper). BAD KISSINGEN is noted resort. Part of old historic region of FRANCONIA. Between 1938 and 1945 called Mainfranken.

Lower Galilee, ISRAEL: see GALILEE.

Lower Ganga Canal, UTTAR PRADESH state, N central INDIA; leaves right bank of the GANGA RIVER 5 mi/8 km E of DIBAI (headworks at NARAURA) in Bulandshwar district, runs SSE and S, splitting (ETAH district) into FARRUKHABAD and BEWAR branches (E) and BHOGNIPUR Branch (W) distributaries. The latter branch crosses KANPUR and ETAWAH branches of UPPER GANGA CANAL; Etawah Branch continues ESE as LOWER GANGA CANAL, dividing into various distributaries in SE end of GANGA-YAMUNA DOAB. Total length of main canal and distributaries is c.5,120 mi/8,240 km. Entire system irrigates over 1,700 sq mi/4,403 sq km and also furnishes hydroelectric power. Opened 1879–1880.

Lower Glenelg National Park (LO-uhr glen-ELG) (□ 105 sq mi/273 sq km), SW VICTORIA, SE AUSTRALIA. Limestone gorge (c.164 ft/50 m deep), Princess Margaret Rose caves carved by GLENELG RIVER and rainwaters. Forest, heath, swamp. Diverse flora and fauna, including 50 orchid species and platypus, ducks, moorhens, emus, kangaroos, wallabies, possums, koalas, echidnas, wombats, and gliders. Canoeing, fishing, and walks.

Lower Granite Lake, reservoir, on GARFIELD-WHITMAN county border, SE WASHINGTON, on SNAKE RIVER, extends SE into ASOTIN and NEZ PERCE counties, 13 mi/21 km SW of PULLMAN; 46°38′N 117°11′W. Maximum capacity 483,800 acre-ft; 58 mi/93 km long. Clearwater River enters from E at LEWISTON, forming 15-mi/24-km E arm. Formed by Lower Granite Lock and Dam (105 ft/32 m high), built (1975) by the Army Corps of Engineers for power generation and navigation.

Lower Hutt, NEW ZEALAND: see HUTT CITY.

Lower Kalskag, village (2000 population 267), SW ALASKA, near KALSKAG, on KUSKOKWIM River; 61°31′N 160°20′W.

Lower Klamath Lake, intermittent lake in NE corner of SISKIYOU county, N CALIFORNIA, near OREGON state line, c.20 mi/32 km S of KLAMATH FALLS, Oregon. Variable in size and depth. Serves as catch basin for surplus irrigation water, which is pumped (NW) into KLAMATH RIVER through Klamath Strait (7 mi/11.3 km long; largely in Oregon) and used to generate power at Copco No. 1 Dam on Klamath River in California. U.S. bird refuge (□ 128 sq mi/331 sq km) occupies N half of Lower Klamath Lake and extends N into KLAMATH county, Oregon.

Lower Lake, unincorporated town (2000 population 1,755), Lake county, NW CALIFORNIA, at S end of CLEAR LAKE, 30 mi/48 km NNE of SANTA CLARA, on CACHE CREEK; 38°55′N 122°37′W. Farm center (walnuts, pears, grapes, oats; cattle). Mineral springs (resorts) nearby. Clear Lake State Park to NW.

Lower Largo, Scotland: see LARGO.

Lower Leelanau Lake, MICHIGAN: see LEELANAU, LAKE.

Lower Lusatia, GERMANY and POLAND: see LUSATIA.

Lower Macnean (mak-NEEN), lake, SW FERMANAGH, SW Northern Ireland, 8 mi/12.9 km WSW of ENNISKILLEN; 2 mi/3.2 km long, 1 mi/1.6 km wide; 54°17′N 07°50′W.

Lower Merion, township and upper-income residential suburb of PHILADELPHIA, MONTGOMERY county, SE PENNSYLVANIA 8 mi/12.9 km WNW of downtown; 39°59′N 75°16′W. Includes ARDMORE (chief center of township); Bala-Cynwyd, part of BRYN MAWR, seat of Bryn Mawr College (in W); Gladwyne; MERION (or Merion Station); PENN WYNNE; WEST MANAYUNK; WYNNEWOOD; and part of ROSEMONT, seat of Rosemont College (in W). Bounded NW by SCHUYLKILL RIVER, also in part of NW and SE by Philadelphia. Eastern Baptist Theological Seminary and St. Charles Borromeo Seminary in S. Appleford-Parsons Bank Arboretum and Henry Foundation Botanical Research Center in N.

Lower Merwede River (MER-vai-duh), Dutch *Beneden Merwede*, channel of WAAL and Maas (MEUSE) rivers, SW NETHERLANDS; formed 3 mi/4.8 km W of GORINCHEM by dividing of UPPER MERWEDE RIVER into NEW MERWEDE and Lower Merwede rivers. Flows 12 mi/19 km W, divides into NOORD RIVER (NNW) and OLD MAAS RIVER (SW) 1 mi/1.6 km N of DORDRECHT. Entire length navigable.

Lower Michigan, region (□ 40,494 sq mi/105,284.4 sq km), S MICHIGAN. The region lies S of STRAITS OF MACKINAC, which separate it from the UPPER PENINSULA (U.P.), or Upper Michigan; connected to the Upper Peninsula by Mackinac Bridge. Also referred to as the lower peninsula, it is bounded W by LAKE MICHIGAN, N by the Straits, E by LAKE HURON, SAINT CLAIR RIVER, LAKE SAINT CLAIR, Detroit River, and LAKE ERIE, S by OHIO and INDIANA. Comprises 67 of state's 83 counties. Unlike the Upper Peninsula, it is highly industrialized, especially the southern ½, and it is noted for its agricultural products, especially its fruits and vegetables. It is characterized by low moraines, sand dunes, marshes, and lakes, all remnants of past glaciation.

Lower Navarre (nah-vahr) or **French Navarre**, French *Basse Navarre* (bahs nah-vahr) or *Navarre Française* (nah-vahr frahn-sez), small historic region of SW FRANCE, in what is now PYRÉNÉES-ATLANTIQUES department, AQUITAINE administrative region, in the BASQUE country; its capital was SAINT-JEAN-PIED-DE-PORT. Consists of section of old kingdom of Navarre on N slope of the W PYRENEES. It was annexed to France by Henry IV in 1589.

Lower Neckar (NEK-ahr), German *Unterer Neckar* (UN-tuhr-uhr NEK-ahr), region (□ 943 sq mi/2,451.8 sq km), BADEN-WÜRTTEMBERG, S GERMANY. Chief town is MANNHEIM.

Lower Paia (PAH-EE-ah), village, N MAUI, MAUI county, HAWAII, 8 mi/12.9 km ENE of WAILUKU, on coast. Paia Sugar Mill at PAIA to SE. Mantokui Buddhist Mission to NE; H. P. Baldwin Beach Park to SW.

Lower Palatinate, GERMANY: see PALATINATE.

Lower Plenty (LO-wuhr PLEN-tee), locality, VICTORIA, SE AUSTRALIA, 10 mi/16 km NE of MELBOURNE, and at junction of Plenty, YARRA rivers; 37°44′S 145°07′E.

Lower Red Lake, MINNESOTA: see RED LAKE.

Lower Red Rock Lake, MONTANA: see RED ROCK LAKES.

Lower Rhine River (REIN), Dutch *Neder Rijn*, channel of the RHINE RIVER, S central NETHERLANDS; formed by forking of Rhine River into Lower Rhine and WAAL rivers 9 mi/14.5 km SE of ARNHEM and 2 mi/3.2 km W of German border. Flows 17 mi/27 km NW to Arnhem, where IJSSEL RIVER, another channel of the Rhine, divides to N. Lower Rhine continues W past OOSTERBEEK and RHENEN to WIJK BIJ DUURSTEDE, 13 mi/21 km SE of UTRECHT, where it divides to form LEK RIVER (W) and CROOKED RHINE RIVER (NW). Course between Rhine and Ijssel rivers is canalized as Pennerdens Canal; AMSTERDAM-RHINE CANAL crosses Lek River W of its dividing point from Lower Rhine.

Lower Rice Lake, CLEARWATER county, NW central MINNESOTA, 27 mi/43 km WSW of UPPER RICE LAKE, fed from S and drained NW by WILD RICE RIVER; also fed from NE by Mosquito River; 47°21′N 95°28′W. Marshy area in White Earth Indian Reservation.

Lower River, administrative division coextensive with MANSA KONKO local government area (□ 625 sq mi/1,625 sq km; 2003 population 72,546), central THE GAMBIA, along S bank of GAMBIA RIVER; ⊙ MANSA KONKO (the only large urban area); 13°24′N 15°35′W. Bordered N by NORTH BANK division (Gambia River, border), S by SENEGAL, E by CENTRAL RIVER division, and W by WESTERN division. Extends 60 mi/97 km along S bank of Gambia River. Produces groundnuts and some rice near the river. KIANG WEST NATIONAL PARK is here. Comprises six smaller administrative divisions called districts: Jarra Central, Jarra East, Jarra West, Kiang Central, Kiang East, and Kiang West.

Lower Saint Croix National Scenic Riverway (KROI) (□ 15 sq mi/39 sq km), WASHINGTON and CHISAGO counties, E MINNESOTA and PIERCE, ST. CROIX, and POLK counties, NW WISCONSIN. Scenic lower course (52 mi/84 km) of the ST. CROIX River; lower 27 mi/43 km federally owned. Part of the Wild and Scenic Rivers System. Headquarters at ST. CROIX FALLS, Wisconsin. Authorized 1972.

Lower Saint Mary Lake, MONTANA: see GLACIER NATIONAL PARK.

Lower Saint Regis Lake, NEW YORK: see SAINT REGIS RIVER.

Lower Salem (LO-uhr SAI-luhm), village (2006 population 103), WASHINGTON county, SE OHIO, 10 mi/16 km NNE of MARIETTA; 39°34′N 81°23′W.

Lower Saranac Lake, NEW YORK: see SARANAC LAKES.

Lower Savage Islands, group of three small islands, NUNAVUT territory, CANADA, off SE BAFFIN ISLAND, in Gabriel Strait (arm of HUDSON STRAIT); 61°48′N 65°48′W.

Lower Saxony (SAK-suh-nee), German *Niedersachsen*, state (□ 18,295 sq mi/47,567 sq km; 2005 population 7,993,946), NW GERMANY; ⊙ HANOVER; 52°40′N 09°00′E. The state was formed in 1946 by the merger of the former Prussian province of Hanover and the former states of BRUNSWICK, OLDENBURG, and SCHAUMBURG-LIPPE. Situated on the N German plain, it is bordered by the NETHERLANDS on the W; the states of NORTH RHINE-WESTPHALIA and HESSE on the S; the states of MECKLENBURG–WESTERN POMERANIA, BRANDENBURG, SAXONY-ANHALT, and THURINGIA on the E; and the states of BREMEN, SCHLESWIG-HOLSTEIN, and HAMBURG and the NORTH SEA on the N. The state is mountainous in the S (notably the HARZ and WESER MOUNTAINS); heaths and moors form the central belt. Drained by the WESER, EMS, ALLER, LEINE, and ELBE rivers. Farming and cattle raising are important occupations. Industry (including the manufacturing of iron and steel, textiles, machinery, food products, and chemicals) is well developed in the cities of BRUNSWICK, CELLE, GOSLAR, HANOVER, and OSNABRÜCK. There are oil wells in the EMSLAND, large iron-ore deposits at WATENSTEDT-SALZGITTER, and lignite mines near HELMSTEDT. EMDEN, WILHELM-SHAVEN, and CUXHAVEN are the chief North Sea ports. The region of Lower Saxony has had no historic unity since 1180, when Emperor Frederick I broke up the duchy of Henry the Lion of Saxony, of which it was a part. The term "Lower Saxony" continued, however, as a geographic expression. It also designated (sixteenth century to 1806) one of the imperial circles of the Holy Roman Empire; the circle included Mecklenburg, HOLSTEIN, and Bremen, in addition to present-day Lower Saxony.

Lower Silesia: see SILESIA.

Lower Subansiri (soo-BAHN-see-ree), district (□ 5,023 sq mi/13,059.8 sq km), ARUNACHAL PRADESH state, extreme NE INDIA; ⊙ ZIRO.

Lower Tallassee Dam, ALABAMA: see THURLOW DAM.

Lower Tanshui River, TAIWAN: see TANSHUI RIVER.

Lower Tunguska River (toon-GOOS-kah), Russian *Nizhnyaya Tunguska*, 1,587 mi/2,554 km long, Siberian Russia; rises on the CENTRAL SIBERIAN PLATEAU, N of UST'-KUT (IRKUTSK oblast); flows N, past YERBOGACHEN, and WNW, past TURA (KRASNOYARSK TERRITORY), to the YENISEY RIVER at TURUKHANSK. Navigable May through October for 1,100 mi/1,770 km below Tura. In its upper course, approaches within 18 mi/29 km of the LENA RIVER. Receives Kochechuma (right) and Taymura (left) rivers.

Lower Vacherie (VASH-ree), unincorporated town, SAINT JAMES parish, LOUISIANA, 9 mi/14.5 km SE of CONVENT; 29°56′N 90°40′W. Oak Alley plantation 5 mi/8 km N; gas field nearby.

Lower Yafa, YEMEN: see YAFA.

Lower Yosemite Falls, California: see YOSEMITE NATIONAL PARK.

Lower Zambezi National Park (□ 1,579 sq mi/4,105.4 sq km), LUSAKA province, SE ZAMBIA, 70 mi/113 km E of LUSAKA. Bounded on S by ZAMBEZI RIVER (Zimbabwe border), opposite MANA POOLS National Park in ZIMBABWE. Drained in W by Chongwe River. Abundant wildlife (leopards, cheetahs, lions, elephants, hippos, buffalo, zebras, baboons, impalas, and crocodiles); fish (giant catfish, vundu), and bird species. Landscape consists of acacia woodland and flood plain grasslands.

Lowes, Loch of, Scotland: see ST. MARY'S LOCH.

Lowestoft (LO-wuh-stahft), town (2001 population 57,746), SUFFOLK, E ENGLAND, on NORTH SEA coast; 52°29′N 01°45′E. Popular seaside resort (the easternmost town in England); fishing; shipbuilding; food processing; light manufacturing. The resort area is separated from Old Lowestoft by Lake Lothing and the harbor. Most of the old houses were destroyed by a fire in the 17th century. St. Margaret's Church, from the 15th century, has an ancient tower. Birthplace of the satirist Thomas Nashe.

Lowesville (LOZ-vil), unincorporated town (□ 6 sq mi/15.6 sq km; 2000 population 1,440), LINCOLN county, W central North Carolina, residential community 16 mi/26 km NNW of CHARLOTTE, near CATAWBA RIVER; 35°25′N 81°00′W. (Mountain Island Lake reservoir to SE; Cowans Ford Dam to NE forms Lake NORMAN reservoir). Manufacturing; service industries; agriculture (grain; livestock; dairying).

Loweswater (LOZ-waw-tuh), small lake in the LAKE DISTRICT, CUMBRIA, NW ENGLAND, 6 mi/9.7 km

---

Cross-references are shown in SMALL CAPITALS. The pronunciation guide is shown on page xix. The sources of population figures are shown on page xvii.

S of COCKERMOUTH; 1.5 mi/2.4 km long; 54°34′N 03°20′W.

Low Head (LO HED), headland, N TASMANIA, SE AUSTRALIA, in BASS STRAIT, at E end of Port Dalrymple (mouth of TAMAR RIVER); 41°02′S 146°46′E. Lighthouse.

Lowicz (LO-veech), Polish Łowicz, Russian Lovich, town (2002 population 30,735), Skierniewice province, central POLAND, on BZURA RIVER, 45 mi/72 km WSW of WARSAW; 52°07′N 19°56′E. Railroad junction; manufacturing of agricultural machinery, wheels, bricks, tiles, cement, carpets, vinegar, candy. Weaving and spinning; brewing; flour milling. Has two monasteries. Castle ruins nearby.

Low Island (LO), SOUTH SHETLAND ISLANDS, off PALMER PENINSULA, ANTARCTICA; 63°17′S 62°09′W. Island is 10 mi/16 km long, 6 mi/9.7 km wide. Also known as Jameson Island.

Low Moor, town (2000 population 240), CLINTON county, E IOWA, 9 mi/14.5 km WSW of CLINTON; 41°47′N 90°20′W. In agricultural area.

Low Moor (LO MOOR), unincorporated village, ALLEGHANY county, NW VIRGINIA, 5 mi/8 km E of COVINGTON, on JACKSON RIVER, in George Washington National Forest; 37°47′N 79°59′W. Light manufacturing; agriculture (grain, apples; cattle); timber.

Lowndes (LOUNDS), county (□ 725 sq mi/1,885 sq km; 2006 population 12,759), S central Alabama, ⊙ Hayneville. In the Black Belt; bounded N by Alabama River; drained by its tributaries. Cotton, corn, soybeans; dairying; poultry; lumber milling. Formed 1830. Named for William Jones Lowndes, SC legislator and U.S. representative.

Lowndes (LOUNDZ), county (□ 511 sq mi/1,328.6 sq km; 2006 population 97,844), S GEORGIA, ⊙ VALDOSTA; 30°50′N 83°16′W. Bounded S by FLORIDA state line; drained by WITHLACOOCHEE RIVER. Coastal plain agriculture (tobacco, cotton, soybeans, peanuts, corn); cattle, hogs; forestry (lumber and by-products). Formed 1825.

Lowndes, county (□ 516 sq mi/1,341.6 sq km; 2006 population 59,773), E MISSISSIPPI; ⊙ COLUMBUS; 33°28′N 88°26′W. Bordered E by ALABAMA state line; drained by TOMBIGBEE River (forms Columbus Lake reservoir on NW boundary) and Luxapalila Creek. Agriculture (cotton, corn, hay, soybeans, wheat; cattle); timber. Lake Lowndes State Park in E; Buttahatchie River forms part of N boundary. Formed 1830.

Lowndesville (LOUNZ-vil), village (2006 population 163), ABBEVILLE county, NW SOUTH CAROLINA, 19 mi/31 km S of ANDERSON; 34°12′N 82°39′W. Agricultural area.

Low Point, cape, NE NOVA SCOTIA, CANADA, on NE CAPE BRETON ISLAND, 9 mi/14 km N of SYDNEY; 46°16′N 60°07′W. Lighthouse.

Lowry (LOU-ree), village (2000 population 271), POPE county, W MINNESOTA, 7 mi/11.3 km NW of GLENWOOD; 45°42′N 95°31′W. Livestock; grain; dairying; manufacturing (printing and publishing). Lake RENO to NE.

Lowry, village (2006 population 9), WALWORTH county, N central SOUTH DAKOTA, 13 mi/21 km S of SELBY, on Swan Creek; 45°19′N 99°58′W.

Lowry Air Force Base, COLORADO: see DENVER, city.

Lowry City (LOU-ree), city (2000 population 728), SAINT CLAIR county, W MISSOURI, near Truman Lake (OSAGE RIVER), 7 mi/11.3 km N of OSCEOLA; 38°08′N 93°43′W. Corn, hay; cattle.

Lowry Crossing (LOU-ree), town (2006 population 1,825), COLLIN county, N TEXAS, residential suburb 28 mi/45 km NE of DALLAS, and 5 mi/8 km E of MCKINNEY, on East Fork of TRINITY RIVER, just N of LAVON LAKE reservoir; 33°10′N 96°32′W. Agricultural area on urban fringe.

Lowrys (LOU-reez), village (2006 population 199), CHESTER county, N SOUTH CAROLINA, 15 mi/24 km

SW of ROCK HILL; 34°47′N 81°14′W. Agriculture in area includes livestock; grain; dairying.

Low Tatras, SLOVAK Nízke Tatry, GERMAN Lage Tatra, HUNGARIAN Alacsony Tátra, POLISH Tatry Niż ne, section of the CARPATHIANS in central Slovakia, parallel to and S of the HIGH TATRAS; extend c.65 mi/105 km E-W, between the GREATER FATRA (W), Hernad and VÁH rivers (N), HRON RIVER (S), and VONDRISEL (E). Rise to 6,703 ft/2,043 m in DUMBIER, to 6,391 ft/1,948 m in KRALOVA HOLA peaks. Extensively forested; noted for picturesque scenery, underground rivers, and stalactite caverns in limestone formations to N (DEMANOVA CAVE); trout fishing in streams to S. CIERNY VÁH, HERNAD, and HRON rivers rise here.

Lowther Hills (LO-thuhr HILZ), range, on the border between South Lanarkshire and DUMFRIES AND GALLOWAY, S Scotland; extending 20 mi/32 km in semicircle and continuing E in MOFFAT HILLS; 55°19′N 03°38′W. Highest peaks are Queensberry (2,285 ft/696 m) in Dumfries and Galloway, 7 mi/11.3 km SW of MOFFAT; and Green Lowther (2,403 ft/732 m) in South Lanarkshire, 2 mi/3.2 km E of WANLOCKHEAD.

Lowther Island, NUNAVUT territory, CANADA, in BARROW STRAIT, between BATHURST and PRINCE OF WALES islands; 74°35′N 97°35′W. Island is 17 mi/27 km long, 2 mi/3 km–6 mi/10 km wide.

Low Veld, SOUTH AFRICA: see VELD.

Lowville (LOU-vil), village (□ 1 sq mi/2.6 sq km; 2006 population 3,247), ⊙ LEWIS county, N central NEW YORK, on Black River, 25 mi/40 km SE of WATERTOWN; 43°47′N 75°29′W. Dairy products, timber, light manufacturing. Settled 1798, incorporated 1854.

Loxley, town (2000 population 1,348), Baldwin co., SW Alabama, 19 mi/31 km ESE of Mobile; 30°37′N 87°45′W. Cotton ginning; meat processing and rendering; furniture manufacturing; metal fabrication; wood products. Named for an early family in the lumbering business. Inc. in 1957.

Loxstedt (LOKS-stet), commune, LOWER SAXONY, NW GERMANY, 5 mi/8 km SE of BREMERHAVEN; 53°28′N 08°39′E. Manufacturing of chemicals.

Loxton (LAHKS-tuhn), village, SE SOUTH AUSTRALIA, 120 mi/193 km ENE of ADELAIDE, and on MURRAY RIVER, near VICTORIA border; 34°27′S 140°34′E. Railroad terminus; sheep, some fruit.

Loyada (lo-YAH-dah), village, DJIBOUTI, on S coast of GULF OF TADJOURA, 12 mi/19 km SE of DJIBOUTI; 11°28′N 43°15′E. Police post on SOMALIA border; fisheries.

Loyal, town (2006 population 1,257), CLARK county, central WISCONSIN, 17 mi/27 km WNW of MARSHFIELD; 44°44′N 90°30′W. In dairying region. Cheese; agricultural machinery. Incorporated as city in 1948.

Loyal, village (2006 population 83), KINGFISHER county, central OKLAHOMA, 13 mi/21 km NW of KINGFISHER; 35°58′N 98°07′W. In agricultural area; manufacturing (steel items). Incorporated 1930.

Loyalhanna (LOI-uhl-HA-nah), unincorporated town, WESTMORELAND county, SW PENNSYLVANIA, suburb 1 mi/1.6 km ENE of LATROBE; 40°19′N 79°21′W. Manufacturing (metal and plastic fabrication, electrical equipment).

Loyalhanna Creek (LOI-uhl-HA-nah), c.50 mi/80 km long, SW PENNSYLVANIA; rises in S WESTMORELAND county, flows first NE, then NNW past LIGONIER and LATROBE, joins Conemaugh River at SALTSBURG to form KISKIMINETAS RIVER; 40°07′N 79°20′W. Loyalhanna Lake, flood-control reservoir, 4.5 mi/7.2 km above mouth.

Loyalist (LOI-yuh-list), township (□ 131 sq mi/340.6 sq km; 2001 population 14,590), SE ONTARIO, E central CANADA, on Lake ONTARIO, and 10 mi/16 km from GREATER NAPANEE; 44°14′N 76°46′W. Formed in 1998 from ERNESTOWN, AMHERST ISLAND, and BATH.

Loyall (LOI-uhl), town, Harlan county, SE KENTUCKY, 2 mi/3.2 km W of HARLAN, in the CUMBERLAND MOUNTAINS, on CUMBERLAND RIVER; 36°51′N

83°20′W. Bituminous coal; timber; manufacturing (concrete, signs). Kentenia State Forest to NE; part of Daniel Boone National Forest to N. Incorporated 1928.

Loyalsock Creek (LOI-uhl-sahk), c.60 mi/97 km long, N central PENNSYLVANIA; rises in E SULLIVAN county at WYOMING county line; 41°28′N 76°14′W. Flows WSW past Mildred, through Wyoming and Tiadaghton state forests, to West Branch of SUSQUEHANNA RIVER at MONTOURSVILLE, 3 mi/4.8 km E of WILLIAMSPORT.

Loyalton, city (2000 population 862), SIERRA county, NE CALIFORNIA, in Sierra Valley of the SIERRA NEVADA, 23 mi/37 km NW of RENO (Nevada); 39°41′N 120°14′W. Dairying; alfalfa, hay, field crops; lumber.

Loyalton, village, EDMUNDS county, N central SOUTH DAKOTA, 16 mi/26 km SW of IPSWICH; 45°17′N 99°16′W.

Loyalty Islands (LOI-yuhl-tee), French, Loyauté, Iles (lwah-yo-tai, eel) coral group (□ 800 sq mi/2,080 sq km), generally raised and tilted NW, S PACIFIC OCEAN, a part of the French overseas territory of NEW CALEDONIA; 21°00′S 167°00′E. The group consists of three islands (LIFOU, MARÉ, and OUVÉA atoll) and some islets. The chief exports are coconuts and copra. Melanesian reserve area.

Loyang (lo-YAHNG), town, E Singapore island, SINGAPORE, 11 mi/18 km NE of SINGAPORE, on Serangoon Harbour, Selat JOHOR STRAIT, 2 mi/3.2 km WSW of CHANGI, at mouth of Tampines River; 01°22′N 103°58′E.

Loyang, CHINA: see LUOYANG.

Loyev (LO-yev), urban settlement, S GOMEL oblast, ⊙ Loyev region, BELARUS, on DNIEPER River (landing), at mouth of the SOZH, and 35 mi/56 km SSW of GOMEL. Manufacturing (building materials).

Loyno (LO-yee-nuh), village, NE KIROV oblast, E central European Russia, on the KAMA River (head of navigation), on road and railroad spur, 75 mi/121 km NNE of OMUTNINSK; 59°43′N 52°39′E. Elevation 469 ft/142 m. Ships phosphorite from nearby Rudnichnyy field; logging, lumbering; food processing (bakery).

Loyno, RUSSIA: see LOINO.

Loyola, unincorporated town (2000 population 3,478), SANTA CLARA county, W CALIFORNIA; residential suburb 12 mi/19 km W of SAN JOSE, and 4 mi/6.4 km SSW of MOUNTAIN VIEW; 37°21′N 122°06′W. Castle Rock State Park to S; Portola State Park to SW. SANTA CRUZ MOUNTAINS to SW.

Loyola, SPAIN: see AZPEITIA.

Lozarevo (lo-zahr-EV-o), village, BURGAS oblast, SUNGURLARE obshtina, BULGARIA; 42°46′N 26°53′E. Railroad station. Wine making.

Lozdzieje, LITHUANIA: see LAZDIJAI.

Lozen (LO-zen), village, SOFIA oblast, Greater SOFIA, BULGARIA; 42°36′N 23°29′E.

Lozère (lo-zer), department (□ 1,995 sq mi/5,187 sq km), upper LANGUEDOC, S FRANCE; ⊙ MENDE; 44°30′N 03°30′E. Wholly within the MASSIF CENTRAL, it is traversed by Mont d'AUBRAC (NW), MARGERIDE MOUNTAINS (N), and the CÉVENNES (SE), which culminate in Mont Lozère (5,573 ft/1,699 m). The department is drained northward by ALLIER and TRUYÈRE rivers, and westward by LOT and TARN rivers. Sheep raising, dairying, and fruit growing are chief agricultural activities. Mende (textiles) is only sizeable town. Lozère is France's least populous department, and its population is still declining. Caves in the limestone hills of the CAUSSES and the remarkable Tarn River gorge are chief tourist attractions. Department forms part of administrative region of LANGUEDOC-ROUSSILLON. CÉVENNES NATIONAL PARK (headquarters at FLORAC) occupies the SE part of Lozère department.

Lozère, Mont (lo-zer, mon), granitic massif (5,573 ft/1,699 m), E LOZÈRE department, LANGUEDOC-ROUSSILLON region, S FRANCE, in the CÉVENNES MOUNTAINS, 10 mi/16 km–15 mi/24 km ESE of MENDE;

44°25′N 03°46′E. TARN, LOT, and CÈZE rivers rise here. The summit and upper elevations are bare and rocky. Cattle graze at lower elevations in summer. Mont LOZÈRE lies within CÉVENNES NATIONAL PARK, with headquarters at FLORAC.

Lozivs'kyy (lo-ZEEV-skiee) (Russian *Lozovskiy*), town, central LUHANS'K oblast, UKRAINE, on railroad spur 19 mi/30 km W of LUHANS'K city center, and 2.5 mi/4 km SW of ZYMOHIR'YA; 48°34′N 38°53′E. Elevation 544 ft/165 m. Coal mine, enrichment plant. Several burial mounds from the 11th to 13th century are nearby. Established in 1949, town since 1953.

Loznica (LOZ-nee-tsah), town (2002 population 86,413), ⊙ Jadar county, W SERBIA, on railroad, 27 mi/43 km SW of Sabar, near the DRINA River (BOSNIA-HERZEGOVINA border); 44°31′N 19°14′E. Ships plums; antimony smelter. Antimony mines nearby. Major synthetic fiber factory. Also spelled Loznitsa.

Loznitsa (LOZ-neet-sah), city (1993 population 2,959), RUSE oblast, Loznitsa obshtina, BULGARIA; 43°23′N 26°36′E. Grain, vegetables, fruit; livestock; metal processing.

Lozno-Aleksandrovka, UKRAINE: see LOZNO-OLEKSANDRIVKA.

Lozno-Oleksandrivka (LOZ-no-o-lek-SAHN-dreefkah) (Russian *Lozno-Aleksandrovka*), town (2004 population 1,700), NW LUHANS'K oblast, UKRAINE, on right tributary of Aydar River, 40 mi/64 km NNW of STAROBIL'S'K; 49°50′N 38°44′E. Elevation 357 ft/108 m. Brickworks. Known since the 17th century; until mid-1930s, called Oleksandrivka (Russian *Aleksandrovka*).

Lozova (lo-zo-VAH) (Russian *Lozovaya*), city (2001 population 64,041), S KHARKIV oblast, UKRAINE, 75 mi/121 km S of KHARKIV; 48°53′N 36°23′E. Elevation 462 ft/140 m. Raion center; railroad junction, railroad servicing; manufacturing (machinery, building materials) food processing (flour, cheese, meat); sewing; metalworks. Technical vocational school; regional history museum. Established in the 1860s, city since 1938.

Lozova-Azovs'ka, UKRAINE: see PANYUTYNE.

Lozovac (LO-zo-vahts), Italian *Loggiovano*, village, W CROATIA, in DALMATIA, in ZAGORA region, on KRKA RIVER opposite SKRADIN. Formerly center of bauxite mining and aluminum manufacturing.

Lozova Pavlivka, UKRAINE: see BRYANKA.

Lozovaya, UKRAINE: see LOZOVA.

Lozovaya-Azovskaya, UKRAINE: see PANYUTYNE.

Lozovaya Pavlovka, UKRAINE: see BRYANKA.

Lozove (lo-zo-VE) (Russian *Lozovoye*), town, E central KHMEL'NYTS'KYY oblast, W UKRAINE. Instrument manufacturing, peat excavation. Established in 1929 as Torforozrobka [Ukrainian=peat gathering]; incorporated as town 1949; renamed 1950.

Lozovik (LOZ-o-veek), village, N central SERBIA, near the MORAVA RIVER, 10 mi/16 km NE of SMEDEREVSKA PALANKA.

Lozovoye, UKRAINE: see LOZOVE.

Lozovskiy, UKRAINE: see LOZIVS'KYY.

Lozoya (lo-THOI-ah), village, MADRID province, central SPAIN, on LOZOYA RIVER, on E slopes of SIERRA DE GUADARRAMA, 37 mi/60 km N of MADRID; 40°57′N 03°47′W. Potatoes, rye; livestock; apiculture; lumbering.

Lozoya Canal, SPAIN: see ISABEL II CANAL.

Lozoya River, c.50 mi/80 km long, MADRID province, central SPAIN; rises in the SIERRA DE GUADARRAMA near SEGOVIA province border, flows NE and E, past RASCAFRÍA and BUITRAGO DEL LOZOYA, to JARAMA RIVER 6 mi/9.7 km NE of TORRELAGUNA. Near its mouth are several reservoirs that feed the ISABEL II CANAL, supplying MADRID with fresh water.

Lozoyuela (lo-thoi-WAI-lah), town, MADRID province, central SPAIN, 35 mi/56 km N of MADRID; 40°55′N 03°37′W. Rye, potatoes; cattle, sheep. Granite quarries.

Loz'va River (luhz-VAH), 265 mi/426 km long, SVERDLOVSK oblast, W Siberian Russia; rises in the central URAL Mountains at 61°50′N 59°30′E–61°50′N 62°20′E. Flows generally SE, joining the SOSVA RIVER 8 mi/13 km N of GARI to form the TAVDA RIVER. Receives the Ivdel' River (right). Navigable below Ivdel'; logging.

Ltava, UKRAINE: see POLTAVA, city.

Lu, CHINA: see SHANDONG.

Luabo (loo-AH-bo), town, Zambézia province, S central MOZAMBIQUE, on lower ZAMBEZI River, and 22 mi/35 km NW of CHINDE; 18°23′S 36°06′E. Sugar mill.

Lualaba River, CONGO: see CONGO, River.

Luali (loo-AH-lee), village, BAS-CONGO province, W CONGO, on border of CONGO REPUBLIC, and 70 mi/113 km NW of BOMA; 05°06′S 12°29′E. Elev. 137 ft/41 m.

Lu'an, CHINA: see LIU'AN.

Luana, town (2000 population 249), CLAYTON county, NE IOWA, 15 mi/24 km N of ELKADER; 43°03′N 91°27′W. In corn, livestock, dairying (cheese, cream, whey products) region.

Luana Point, cape, SAINT ELIZABETH parish, SW JAMAICA, 4 mi/6.4 km W of BLACK RIVER; 18°03′N 77°55′W.

Luancheng (LUAN-CHENG), town, ⊙ Luancheng county, SW HEBEI province, CHINA, 14 mi/23 km SE of SHIJIAZHUANG, near BEIJING-WUHAN railroad; 37°53′N 114°39′E. Cotton, wheat, millet, oilseeds.

Luanco (LWAHNG-ko), town, OVIEDO province, NW SPAIN, on BAY OF BISCAY, 8 mi/12.8 km NW of GIJÓN; 43°37′N 05°47′W. Fishing port and bathing resort; fish salting, boat building. Iron mines nearby.

Luanda, province (□ 934 sq mi/2,428.4 sq km), W ANGOLA, on ATLANTIC OCEAN (to W); ⊙ LUANDA; 08°50′S 13°20′E. Surrounded by BENGO province to N, E, and S. Small province including Luanda city and the area surrounding it.

Luanda (loo-AHN-dah), city (2004 population 2,783,000), ⊙ Luanda province and ANGOLA, W Angola, port on the ATLANTIC OCEAN. W of international airport. It is Angola's largest city and its administrative center. Because of the influx of refugees from the countryside during the civil war, the city's population may have reached 3 million in early 1990s. Manufactures include processed foods, beverages, textiles, cement and other construction materials; plastic products, metalware, cigarettes, and shoes. Petroleum, found nearby, is refined in the city. Luanda has a natural harbor, with a fine port. The chief exports are coffee, cotton, sugar, diamonds, iron, and salt. Luanda's market (Roque Santiero) stretches along the beach, attracting thousands of shoppers on peak days. Founded in 1575 by the Portuguese as São Paulo de Luanda, the city has been the administrative center of Angola since 1627 (except for 1640–1648). From c.1550 to c.1850, it was the center of a large slave trade to BRAZIL. After Angolan independence (1975), much of the city's large Portuguese population left. For a period in the late 1970s and early 1980s, numerous Cuban soldiers and civilian advisors were quartered in the city to lend support to the Marxist government. In the early 1980s, the city's oil refinery was damaged during civil war. It is the seat of a Roman Catholic archbishop. The University of Angola, the 17th-century Fort of São Miguel, and the Governor's Palace are in Luanda.

Luanda (loo-AHN-dah), town, WESTERN province, W KENYA, on railroad, and 12 mi/19 km SSE of BUTERE; 00°02′N 34°36′E. Cotton, peanuts, sesame, corn.

Luang, THAILAND: see CHOM THONG.

Luang Nam Tha (LOO-ahng NAHM TAH) province (2005 population 145,231), NW LAOS, NW border with MYANMAR, N border with CHINA, SW border with BOKEO province, SE UDOMSAI province; ⊙ NAM THA; 20°40′N 101°20′E. Formerly part of Haut-Mekong province. Shifting cultivation; forest products. Opium manufacturing along MYANMAR border. Meo, Lao, Mien, Thai, Lolo, Yao, and other minority peoples.

Luango (loo-AHN-go), town, BAS-CONGO province, W CONGO, 55 mi/89 km N of BOMA; 05°10′S 12°59′E.

Luang Phabang (LOO-ahng pah-BAHNG), province (2005 population 405,949), N LAOS, SSE border with VIENTIANE, XIENG KHUANG and HUA PHAN provinces, NNW borders with PHONG SALI, UDOMSAI, and SAYABULI, NE border with VIETNAM; ⊙ LUANG PHABANG; 20°00′N 102°30′E. Rice farming; shifting cultivation; opium growing. Lao, Meo, Thai, Khmer, and other minority peoples.

Luang Phabang (LOO-ahng pah-BAHNG), city, ⊙ LUANG PHABANG province and the historic and cultural ⊙ LAOS, NW LAOS, on the left bank of MEKONG River, slightly downstream from 2 of its major tributaries (Nam Ou and Nam Seng) and at the confluence of the NAM KHAN river; 19°52′N 102°08′E. The economic center of N LAOS, it is a thriving river port and a market for rubber, rice, teak, and fish. Most shops are Chinese-owned. Zinc is mined nearby. According to tradition, Luang Phabang was founded by Indian Buddhist missionaries. For several centuries during the Lan Xang period (14th–17th century) it was the center of a Laotian-Thai kingdom that controlled most of Laos and parts of SIAM; it is home of the former Royal Palace. Luang Phabang came under French rule in 1893. Today, the Royal Palace Museum, nearly 30 historic temples, French architecture, and the city's diverse population make Luang Phabang one of the major tourist sites in contemporary Laos. Newly expanded commercial airport. Sometimes spelled Luang Prabang.

Luang Prabang Range (lu-ahng PRUH-BAHNG), Thai *Luang Phra Bang*, on THAILAND-LAOS border, extending along (W of) MEKONG River to PHETCHABUN RANGE (S); densely forested (teak), it rises to 6,735 ft/2,053 m.

Luang River, THAILAND: see TAPI RIVER.

Luanguinga River (loo-ahn-GEEN-guh), c.250 mi/402 km long, SW central AFRICA, right tributary of the ZAMBEZI; rises on central plateau of ANGOLA; flows SE into Barotseland (ZAMBIA) and to the Zambezi below KALABO.

Luangwa (loo-AH-eng-gwah), township, LUSAKA PROVINCE, SE ZAMBIA, on ZAMBEZI RIVER 140 mi/225 km ESE of LUSAKA; 15°37′S 30°25′E. At mouth of Luangwa (Arwangua) River, opposite (W of) ZUMBO, MOZAMBIQUE, at corner of Mozambique-Zimbabwe border. Agriculture (tobacco, corn); cattle. LOWER ZAMBEZI NATIONAL PARK to W.

Luangwa River, c.500 mi/800 km long, ZAMBIA and MOZAMBIQUE, S central AFRICA. Source in NE Zambia in NE NORTHERN province, 120 mi/193 km ENE of Kasama; flows SSW past NORTH LUANGWA, Luambe, and SOUTH LUANGWA national parks. Forms part of border between Northern and EASTERN provinces; receives LUNSEMFWA RIVER from W as it turns S to form part of Zambia-Mozambique border; enters ZAMBEZI RIVER at FEIRA, Zambia.

Luan He, CHINA: see LUAN RIVER.

Luanping (LUAN-PING), town, ⊙ Luanping county, NE HEBEI province, CHINA, on LUAN RIVER, and 9 mi/14.5 km W of CHENGDE, and on railroad; 40°55′N 117°17′E. Building materials, chemicals, food industry, coal and iron-ore mining. Also known as Anjiangying.

Luan River (LUAN), Mandarin *Luan He*, 500 mi/805 km long, NE CHINA; rises as Shandian River in NE HEBEI province, flows N and then SE past LUANPING, XIFENGKOU, and LUAN XIAN, to Gulf of BOHAI. Unnavigable, it is obstructed by rapids in upper course. Sometimes spelled Lwan.

Luanshya (loo-WAH-en-shah), city (2000 population 144,009), COPPERBELT province, N central ZAMBIA, 18 mi/29 km SW of NDOLA, near CONGO border; 13°08′S 28°24′E. Terminus of railroad spur from Ndola. It is a uranium- and copper-mining center, located in the COPPERBELT REGION. Agriculture (grains, peanuts, tobacco, cotton, soybeans, flowers). Manufacturing (copper and aluminum cables, mining equipment, clothing; orchids and gladioli processing).

Luan Xian (LUAN SIAN), town, ⊙ Luan Xian county, NE HEBEI province, CHINA, on LUAN RIVER, 30 mi/48 km ENE of TANGSHAN, and on TIANJIN-SHENYANG railroad; 39°45′N 118°44′E. Coal-mining center; clay quarrying; oilseeds, medicinal herbs.

Luapula (loo-WAH-poo-lah), province (□ 19,524 sq mi/50,762.4 sq km; 2000 population 775,353), N ZAMBIA; ⊙ MANSA (formerly Fort Rosebery); 11°00′S 29°00′E. Bounded S, W, and N by CONGO, E by NORTHERN province; SE corner touches CENTRAL province; created from Central province LUAPULA RIVER forms S and W borders, drains from S end of LAKE BANGWEULU in SE, flows into Lake Mwera on NW border. Part of Bangweulu Swamps in SE. Lusenga Plain National Park and part of Mweru Wantipa National Park in N. Manganese reserves in vicinity of Mansa in S. Agriculture (rice, vegetables, bananas, cassava, corn, chilies); cattle. Fish from lakes Banghweulu and Mweru.

Luapula River (loo-WAH-poo-lah), biggest river in northern ZAMBIA. Starts at S tip of LAKE MWERU (straddling CONGO and Zambian border), and stretches c.316 mi/510 km to discharge into LAKE BANGWEULU. The river forms a natural boundary between Zambia and Congo.

Luarca (LWAHR-kah), town, OVIEDO province, NW SPAIN, on BAY OF BISCAY, 35 mi/56 km WNW of OVIEDO; 43°32′N 06°32′W. Fishing port and bathing resort; fish and meat processing, tanning, furniture manufacturing; dairy products. Coal shipping.

Lua River (LOO-ah), c.190 mi/306 km long, NW CONGO; rises 40 mi/64 km ESE of BOSOBOLO, flows SSW and SW to UBANGI RIVER at DONGO. Navigable for c.80 mi/129 km downstream from MOGALE.

Luashi (LWAH-shee), village, KATANGA province, SE CONGO, on ANGOLA border, 200 mi/322 km W of LIKASI; 10°56′S 23°37′E. Elev. 3,897 ft/1,187 m. Customs station and trading post in cotton area. Roman Catholic and Protestant missions.

Luatamba (loo-ah-TAHM-buh), town, MOXICO province, ANGOLA, on road, and 35 mi/56 km SE of LUENA; 12°06′S 20°19′E. Market center.

Luau (loo-OU), town, MOXICO province, E ANGOLA, frontier station on Benguela railroad, opposite DILOLO-GARE (Democratic Republic of the CONGO), and 180 mi/290 km ENE of Luena; 10°42′S 22°07′E. Formerly Teixeira de Sousa (te-SHEI-rah de SO-zah).

Luba (LOO-buh), town (2003 population 6,800), ⊙ BIOKO SUR province, on SW coast of BIOKO island, EQUATORIAL GUINEA, 25 mi/40 km SSW of MALABO; 03°27′N 08°33′E. Important fishing industry. Ships cacao, bananas, coffee, palm oil, kola, and coconuts. Has oil storage depot. Formerly San Carlos.

Lubaantun, archaeological site, TOLEDO district, BELIZE, 20 mi/32 km NW of PUNTA GORDA; 16°17′N 88°55′W. Important prehistoric site of S Belize, featuring late classic pyramids and buildings.

Lubaczow (loo-BAH-choov), Polish *Lubaczów*, town, Przemyśl province, SE POLAND, on LUBACZÓWKA RIVER, and 30 mi/48 km NNE of Przemyśl, near UKRAINE border; 50°10′N 23°08′E. Flour milling; lumber mill. Natural-gas deposits nearby.

Lubaczowka River (loo-bah-CHOW-kah) (Polish *Lubaczówka*) (Ukrainian *Lyubachivka*), approximately 50 mi/80 km long, W UKRAINE and SE POLAND; rises 6 mi/10 km SE of NEMYRIV (Ukraine); flows generally W, past LUBACZÓW (Poland), to San River 7 mi/11 km N of JAROSLAW. On both sides of the international border, the river basin contained largely Ukrainian rural population until Polish-Soviet population exchanges in 1945–1946 reduced the Ukrainian population on the Polish side.

Lubaga (loo-BAH-gah), village, MWANZA region, NW TANZANIA, 40 mi/64 km N of MWANZA, on N shore of UKEREWE ISLAND on Massonga Bay, Lake VICTORIA; 01°57′S 32°57′E. Cotton, peanuts, subsistence crops; livestock.

Luban (LOO-bahn), Polish *Lubań*, German *Lauban*, town (2002 population 22,852) in LOWER SILESIA, Jelenia Góra province, SW POLAND, in UPPER LUSATIA, on KWISA RIVER, and 14 mi/23 km E of GÖRLITZ. In lignite-mining region; cotton milling, metalworking. In World War II, heavily damaged (c.65% destroyed). In Middle Ages, a member of Lusatian League (founded 1346).

Lubānas Lake (LU-bah-nuhs), Latvian *Lubānas Ezers*, largest lake (□ 34 sq mi/88.4 sq km) of LATVIA, 23 mi/37 km NW of REZEKNE; 10 mi/16 km long, 5 mi/8 km wide; 56°46′N 26°53′E. AIVIEKSTERiver is outlet.

Lubang Islands (LOO-bahng), small island group, MINDORO OCCIDENTAL province, PHILIPPINES, in SOUTH CHINA SEA, SW of LUZON, just off NW coast of MINDORO island across Calavite Passage (c.6 mi/9.7 km wide); 13°45′N 120°15′E. Comprises Lubang Island (largest), AMBIL ISLAND, GOLO ISLAND, CABRA ISLAND, and several islets. Generally mountainous. Rice growing; livestock raising. Inhabited by Tagalogs. Lubang Island (□ 74 sq mi/192 sq km) is 12 mi/19 km off NW coast of Mindoro island; 18 mi/29 km long, 6 mi/9.7 km wide; rises to 1,967 ft/600 m. Lubang (1990 population 18,800), on N coast of island, is chief town of group.

Lubango (loo-BAHN-go), city, ⊙ HUÍLA province, SW ANGOLA, 160 mi/257 km S of BENGUELA, near W edge of central plateau; 14°55′S 13°30′E. Inland terminus of railroad from NAMIBE (90 mi/145 km WSW). Agricultural trade center in upland region (4,000 ft/1,219 m–6,000 ft/1,829 m) formerly a center of European settlement. Ships hides and skins, dairy produce, rice, flour. Agricultural processing, tanning. Airfield. Formerly called Sá da Bandeira (SAH dah ban-DEI-rah).

Lubao (LOO-bou), town, PAMPANGA province, central LUZON, PHILIPPINES, on railroad, and 9 mi/14.5 km SW of SAN FERNANDO; 14°55′N 120°33′E. Sugar-growing center. Sugar mill and distillery at nearby village of Del Carmen.

Lubarika, CONGO: see LUVUNGI.

Lubartow (loo-BAHR-toov), Polish *Lubartów*, Russian *Lyubartov*, town (2002 population 23,166), Lublin province, E POLAND, on railroad, and 15 mi/24 km N of LUBLIN, near WIEPRZ RIVER; 51°28′N 22°38′E. Manufacturing (glass, concrete blocks; vinegar, beer, flour). Before World War II, population 50% Jewish.

Lubawa (LOO-bah-vah), German *Löbau*, town, OLSZTYN province, N POLAND, 9 mi/14.5 km NNW of Nowe Miasto; 53°30′N 19°45′E. Railroad spur terminus; flour milling, sawmilling. Ruins of 14th-century castle.

Lubawka (LOO-bah-vkah), German *Liebau* (lee-bah), town, in LOWER SILESIA, Jelenia Góra province, SW POLAND, at E foot of the RIESENGEBIRGE, on BOBRAWA RIVER, and 13 mi/21 km WSW of WALDENBURG (Wałbrzych); 50°43′N 16°00′E. Frontier station on CZECH border; coal mining, cotton and linen milling; health and winter sports resort.

Lubban, El, Arab village, Ramallah district, of RAMALLAH, in the Samarian Highlands, WEST BANK; 32°02′N 35°02′E. Agriculture (olives, cereals). It is believed to be the site of a Talmudic period settlement.

Lubban Sharqiya, Arab village, Nablus district, 10 mi/16 km S of NABLUS, in the Samarian Highlands, WEST BANK; 32°04′N 35°14′E. Agriculture (fruit, olives, cereal).

Lübbecke (LOOB-bek-ke), town, NORTH RHINE-WESTPHALIA, N central GERMANY, on N slope of WIEHEN MOUNTAINS, 12 mi/19 km W of MINDEN; 52°18′N 08°27′E. Manufacturing of furniture, cigars, machinery, paper. Chartered 1279. Has twelfth-century church.

Lubbeek (LOO-baik), commune (2006 population 13,684), Leuven district, BRABANT province, central BELGIUM, 6 mi/10 km E of LEUVEN.

Lübben (LOOB-ben), town, BRANDENBURG, E GERMANY, in LOWER LUSATIA, on several small islands in the SPREE, 23 mi/37 km NW of COTTBUS, in SPREE FOREST; 51°51′N 13°54′E. Cardboard manufacturing; canning. Tourist resort. Has late-Gothic church, sixteenth-century palace. Town first mentioned 1007; in Middle Ages, chief town of Saxonian Lower Lusatia; passed to PRUSSIA 1815.

Lübbenau (LOOB-buhn-ou), town, BRANDENBURG, E GERMANY, in LOWER LUSATIA, on several islands in the SPREE, 18 mi/29 km WNW of COTTBUS, in SPREE FOREST; 51°53′N 13°58′E. Central town of the Spree Forest; tourist resort. Manufacturing (building material); canning. Has nineteenth-century palace with big park.

Lubbock (LUH-buhk), county (□ 900 sq mi/2,340 sq km; 2006 population 254,862), NW TEXAS; ⊙ LUBBOCK; 33°36′N 101°49′W. Elevation 3,000 ft/914 m–3,500 ft/1,067 m. Drained by intermittent North Fork of the Double Mountain Fork of BRAZOS RIVER (forms Buffalo Springs Lake in SE), Yellow House Draw and Blackwater Draw. Major railroad junction and center; one of state's leading agricultural counties (cotton, sorghum, wheat, hay, vegetables, sunflowers, soybeans; beef cattle, sheep, hogs, poultry; eggs). Oil and gas; stone, sand, and gravel. Lubbock Lake Landmark State Park at center of county, in Lubbock city. Hunting, fishing. Formed 1876.

Lubbock (LUH-buhk), city (2006 population 212,169), ⊙ LUBBOCK county, NW TEXAS, c.260 mi/418 km WNW of FORT WORTH; 33°34′N 101°52′W. Elevation 3,241 ft/988 m. In the LLANO ESTACADO region of the GREAT PLAINS, on North Fork of the Double Mountain Fork of the BRAZOS RIVER, at confluence of Yellow House Draw and Blackwater Draw. Lubbock was settled in 1879 by Quakers. It is the trade center for the cotton- and grain-growing region of Texas and E NEW MEXICO. Many residents are employed in education, retail services, and technology companies. Manufacturing (cottonseed oil, earth-moving equipment, wineries, food processing, dairy products, pumps, and irrigation equipment). Lubbock International Airport in N. In Mackenzie Park a prairie-dog town has been preserved. Lubbock Lake Landmark State Park is in N part of city. Site is an important geological formation; Buffalo Springs Lake reservoir to SE. Texas Tech University and Lubbock Christian University are in the city. Reese Air Force Base where jet pilots are trained, is at W end of city. Buddy Holly Statue and Walk of Fame honors Texas music legends. Incorporated 1909.

Lubec (loo-BEK), town, WASHINGTON county, E MAINE, on the coast, 25 mi/40 km ENE of MACHIAS; 44°49′N 67°01′W. Resort, fishing (sardine canning). WEST QUODDY HEAD, SE of Lubec village (1990 population 1,536), is easternmost point of U.S. On TREAT'S ISLAND, in COBSCOOK BAY, is North Lubec village. Settled c.1780, incorporated 1811.

Lübeck (LOO-bek), city (2005 population 211,825), SCHLESWIG-HOLSTEIN, N GERMANY, on the TRAVE RIVER near its mouth on the BALTIC SEA; 53°52′N 10°41′E. A major port and commercial and industrial center; the port is the city's primary employer. Among its industries are shipbuilding, metalworking, food processing, and manufacturing of ceramics, wood products, and medical instruments; famous for marzipan. Known in the eleventh century, Lübeck was destroyed by fire in 1138, but was refounded in 1143. Acquired and chartered by Henry the Lion c.1158; the charter, which granted far-reaching communal rights, was copied by more than 100 other cities in the Baltic area. In 1226, Frederick II made Lübeck a free imperial city. Ruled by a merchant aristocracy, it soon rose to great commercial prosperity, acquired hegemony over the Baltic trade, and headed the HANSEATIC LEAGUE. In 1630 the last of the Hanseatic diets was held there. The city escaped the ravages of the Thirty Years War (1618–1648), and, in spite of a decline in Lübeck's power, its patrician merchant families continued to prosper. In the French Revolutionary Wars, Lübeck was sacked by French troops in 1803, and, after the Prussian army under Blücher surrendered (1806) at nearby RATEKAU, the city was occupied by the French. Lübeck, governed by a senate, joined the North German Confederation and later the German Empire as a free Hanseatic city; it retained that status until 1937, when it was incorporated into Schleswig-Holstein. The opening (1900) of the ELBE-LÜBECK CANAL (formerly called the Elbe-Trave Canal) helped increase Lübeck's trade. Despite heavy damage by bombing in World War II, the inner city of Lübeck remains one of the finest examples of medieval Gothic architecture in N EUROPE. Among the buildings that have been restored are the magnificent city hall (thirteenth to fifteenth century); the churches of St. Catherine and St. Jacob (both from fourteenth century); the Hospital and Church of the Holy Ghost (thirteenth century); the Holstentor (completed 1477), an imposing city gate flanked by two round towers; the cathedral (founded in 1173); the large brick Church of St. Mary (thirteenth to fourteenth century); and many of the old patrician residences. There are also several museums in the city, as well as the Günter Grass Haus, commemorating the artist's visual and written work. Dietrich Buxtehude, the composer and organist, was active in Lübeck from 1668 to 1707. Thomas Mann and his brother Heinrich Mann born here.

Lubeck (loo-BEK), unincorporated town (2000 population 1,303), Wood county, NW WEST VIRGINIA, 5 mi/8 km SW of PARKERSBURG; 39°14′N 81°37′W. Agriculture (grain, tobacco); livestock; poultry. Blennerhassett Island Historic State Park, on BLENNERHASSETT ISLAND in OHIO RIVER, to N.

Lübeck Bay (LOO-bek), N GERMANY, SW arm of MECKLENBURG BAY of the BALTIC SEA; c.20 mi/32 km long, c.10 mi/16 km wide; 53°54′N 10°46′E–54°00′N 11°00′E. Receives TRAVE RIVER at TRAVEMÜNDE.

Lubefu (loo-BE-foo), village, KASAI-ORIENTAL province, central CONGO, on Lubefu River (a headstream of the SANKURU), and 75 mi/121 km ENE of LUSAMBO; 04°43′S 24°25′E. Elev. 1,732 ft/527 m. Cotton ginning. Has Roman Catholic mission.

Lubelenge (loo-be-LENG-gai), village, MANIEMA province, central CONGO, on LUALABA RIVER, on railroad, and 80 mi/129 km NNW of KASONGO; 03°05′S 25°51′E. Elev. 1,683 ft/512 m. Sawmilling; palm-oil milling; coffee plantations.

Lüben, POLAND: see LUBIN.

Lubenham (LUHB-uhn-uhm), village (2001 population 2,419), LEICESTERSHIRE, central ENGLAND, 1 mi/1.6 km NW of MARKET HARBOROUGH; 52°29′N 00°58′W. Fox-hunting center.

Lubenik (lu-BE-nyeek), Slovak *Lubeník*, Hungarian *Lubény*, village, VYCHODOSLOVENSKY province, SE central SLOVAKIA, on railroad, and 15 mi/24 km W of ROZNAVA; 48°40′N 20°12′E. Mining and manufacturing of magnesite.

Luben-ń Wielki, UKRAINE: see VELYKYY LYUBIN'.

Lubény, SLOVAKIA: see LUBENIK.

Lubero (loo-BE-ro), village, NORD-KIVU province, E CONGO, 80 mi/129 km N of BUKAVU; 00°06′S 29°06′E. Elev. 4,835 ft/1,473 m. Center for cultivation of wheat, coffee, and pyrethrum; large flour mills. Strawberries, European vegetables, and flowers also grown here. Tourist center. Has Roman Catholic mission, hospital. Lubero was founded 1925.

Lubéron, Montagne du (luh-bai-ron, mon-tahn-yuh dyoo), narrow scrubby or wooded range in VAUCLUSE department, PROVENCE-ALPES-CÔTE D'AZUR region, SE FRANCE, extending c.35 mi/56 km between CAVAILLON (W) and MANOSQUE (E) N of the DURANCE RIVER; 43°48′N 05°22′E. It rises to 3,700 ft/1,128 m. Many old villages are perched on its lower slopes. The limestone hills are dotted with stone shelters formerly used by shepherds. Since 1977, the Lubéron has been included in the Lubéron Natural Regional Park,

covering some 502 sq ft/47 sq m and 60 small communities, to preserve the natural and cultural character of this typically Provençal district. Panoramic views from the crestline, especially from the Mourre Nègre (highest point). APT is the nearest town and tourist base.

Lubéron Natural Regional Park, FRANCE: see LUBÉRON, MONTAGNE DU.

Lubersac (lu-ber-sahk), commune (□ 22 sq mi/57.2 sq km), CORRÈZE department, LIMOUSIN region, S central FRANCE, 20 mi/32 km NNW of BRIVE-LA-GAILLARDE; 45°27′N 01°24′E. Preserves and ships fruits and vegetables. Has Romanesque church.

Lubiaz, POLAND: see LEUBUS.

Lubien (LOO-bee-nee), Polish *Lubień Kujawski*, Russian *Lyuben* or *Lyuben'*, town, Włocławek province, central POLAND, 18 mi/29 km SSE of WŁOCŁAWEK. Distilling, flour milling; center of folk art.

Lubien, town, OPOLE province, POLAND: see LEWIN BRZESKI.

Lubieszow, UKRAINE: see LYUBESHIV.

Lubilash River, CONGO: see SANKURU RIVER.

Lubimbi, village, MATABELELAND NORTH province, W central ZIMBABWE, 52 mi/84 km ESE of HWANGE, near SHANGANI RIVER, 7 mi/11.3 km E, its mouth on GWAYI RIVER; 18°28′S 27°17′E. Site of Lubimbi Hot Springs. Livestock; grain.

Lubin (LOO-been), German *Lüben* (lee-ben), town (2002 population 78,544) in LOWER SILESIA, Legnica province, SW POLAND, 13 mi/21 km N of LEGNICA (Liegnitz). Copper mines; Manufacturing of food products, furniture, pianos; woodworking. Suffered heavy damage in World War II.

Lubisi Dam, village, SOUTH AFRICA: see GREAT KEI RIVER.

Lublin (LOOB-leen), city (2002 population 357,110), SE POLAND; 51°15′N 22°34′E. It is a railroad junction and industrial center. Manufacturing includes motor vehicles, agricultural machinery, chemicals, and foodstuffs. One of the oldest Polish towns, Lublin became the ⊙ of a province in 1474 and the seat of a tribunal in 1578. It was the meeting place of several diets (16th–18th century), one of which united (1569) Poland with LITHUANIA. Lublin passed to AUSTRIA in 1795 and to RUSSIA in 1815. It was (1918) the seat of a temporary Polish Socialist government. In 1941, Majdanek concentration camp was established by the Nazis in Lublin. In 1944, it was the seat of a provisional government rivaling the Polish government-in-exile in LONDON. At the YALTA Conference (February 1945) an agreement was reached to broaden the Lublin government by including members of the London cabinet; the Lublin government was recognized as the sole Polish authority at the Potsdam Conference (August 1945). The Catholic University at Lublin (founded 1918) and Maria Curie-Sklodowska University are here. Lublin's most notable buildings are a 14th-century city hall (rebuilt 1787), a 14th-century castle (rebuilt 1826), and a 16th-century cathedral. Jewish community dates from 14th century. Formed majority of population in 19th century and dominated the leather industry. A center for Jewish religion and trade-union activities, the community was exterminated by the Nazis.

Lublin, village (2006 population 96), TAYLOR county, central WISCONSIN, 41 mi/66 km NE of EAU CLAIRE; 45°04′N 90°43′W. In lumbering and dairying region. On the S edge of Chequamegon National Forest to NE.

Lubliniec (loob-LEE-neets), German *Lublinitz*, town (2002 population 24,601), Częstochowa province, S POLAND, 21 mi/34 km WSW of CZĘSTOCHOWA; 50°40′N 18°41′E. Railroad junction; manufacturing of textiles, machines, cement goods; flour milling; ironworks. Passed from GERMANY to Poland, 1921. During World War II, called Loben under German rule.

Lubnaig, Loch (LUB-naig), lake (4 mi/6.4 km long, 146 ft/45 m deep), STIRLING, central Scotland, 3 mi/4.8 km

NW of CALLANDER, at foot of BEN LEDI; 56°16′N 04°16′W. Outlet: Leny River, flowing to TEITH RIVER. Formerly in Central region, abolished 1996.

Lubny (loob-NEE), city (2001 population 52,572), N central POLTAVA oblast, UKRAINE, on high right bank of SULA RIVER and 75 mi/121 km WNW of POLTAVA; 50°01′N 33°00′E. Elevation 501 ft/152 m. Raion center. Manufacturing (machine tool, machine building, ceramics, building materials, furniture); pharmaceutical chemicals; food processing (flour, dairy, meat, food flavoring); sewing. Museum; art gallery. Founded in 988 by Prince Volodymyr; site of Rus' victory over the Cumans (1107); destroyed by the Mongols (1239); rebuilt in the 16th century by Prince Wiśniowiecki (Ukrainian *Vyshnevets'kyy*); regimental capital in Hetman state (1648, 1658–1781); site of botanical garden with medicinal plants and first field apothecary in Ukraine in the early 18th century; city since 1783. Also known in Ukrainian as Lubni.

Lubochna, SLOVAKIA: see KRALOVANY.

Lubokpakam, town, ⊙ DELI-SERDANG province, NE SUMATRA, INDONESIA, 11 mi/18 km E of MEDAN; 03°33′N 98°52′E.

Lubombo (LOO-bawm-boh), district (□ 2,296 sq mi/5,969.6 sq km), E SWAZILAND, bounded on N and NE by SOUTH AFRICA, on E by MOZAMBIQUE; ⊙ SITEKI (Stegi); 26°25′S 31°45′E. Drained by MBULUZI, BLACK MBULUZI, WHITE MBULUZI, and GREAT USUTU rivers. Hlane Royal National Park and Mlawula Nature Reserve in N, Mkhaya Nature Reserve in SW; Lebombo Mountains in E, Swaziland Irrigation Scheme in N and the Usutu Irrigation Scheme in SE; sugarcane, cotton, corn, pineapples, citrus; cattle, goats, sheep, hogs. Coal mining E Central; diamond mining in N.

Lubombo (loo-BO-em-bo), township, SOUTHERN province, S central ZAMBIA, 10 mi/16 km E of MAZABUKA; 15°48′S 27°45′E. On railroad. Agriculture (tobacco, corn, sugarcane, cotton); cattle.

Lubombo Mountains, mountain range, SE AFRICA (MOZAMBIQUE, SOUTH AFRICA, and SWAZILAND): see LEBOMBO MOUNTAINS.

Lubomierz (loo-BO-meez), German *Liebenthal*, town in LOWER SILESIA, Jelenia Góra province, SW POLAND, 13 mi/21 km NW of JELENIA GÓRA (Hirschberg). Agricultural market (grain, potatoes; livestock). Has church of Benedictine (later Ursuline) convent, founded 1278. After 1945, briefly called Milosna, Polish *Mitosna*.

Luboml, UKRAINE: see LYUBOML'.

Lubonga, river, CONGO. A tributary of the MWALESHI RIVER, it mostly flows through the NORTH LUANGWA NATIONAL PARK. One of the highest concentration of wildlife, both flora and fauna, is on its shores.

Lubovna, SLOVAKIA: see STARA LUBOVNA.

Lubraniec (loob-RAH-neets), Russian *Lyubranets*, town, Włocławek province, central POLAND, 13 mi/21 km SW of WŁOCŁAWEK; 52°33′N 18°51′E. Flour milling.

Lubrín (loo-VREEN), town, ALMERÍA province, S SPAIN, 14 mi/23 km SW of HUÉRCAL-OVERA; 37°13′N 02°04′W. Olive oil, cereals, almonds. Iron deposits.

Lubrza (LOOB-zah), German *Liebenau bei Schwiebus*, town in BRANDENBURG, after 1945 in Zielona Góra province, W POLAND, 6 mi/9.7 km NW of SWIEBODZIN; 52°18′N 15°26′E. Agricultural market (grain, vegetables, potatoes; livestock). Tourist resort.

Lubsko (LOOB-sko), German *Sommerfeld*, town in BRANDENBURG, after 1945 in Zielona Góra province, W POLAND, 16 mi/26 km SE of GUBEN, in LOWER LUSATIA; 51°48′N 14°58′E. Railroad junction; woolen milling, stone quarrying. Chartered 1283. Has 13th-century church, 16th-century castle, remains of 15th-century town walls.

Lübtheen (LOOBT-hain), town, MECKLENBURG–WESTERN POMERANIA, N GERMANY, 25 mi/40 km SW of SCHWERIN; 53°19′N 11°05′E. Manufacturing of automobile parts.

Lubuagan (loo-boo-AHG-ahn), town, KALINGA-APAYAO province, N LUZON, PHILIPPINES, on small tributary of CHICO RIVER, and 40 mi/64 km ESE of BANGUED; 17°18′N 121°04′E. Lumbering, rice growing.

Lubudi (loo-BOO-dee), town, KATANGA province, SE CONGO, on railroad, and 90 mi/145 km NW of LIKASI; 09°57′S 25°58′E. Elev. 4,291 ft/1,307 m. Industrial center. Tin smelting; manufacturing of cement, cast-iron, and bronze articles, calcium carbide. Has hydroelectric power plant. Roman Catholic mission.

Lubuklinggau (loo-book-LING-gou), city, ⊙ Musi Rawas district, W central South Sumatra province, INDONESIA; 03°10′S 102°57′E. Railroad terminus for line from PRABUMULIH.

Lubumbashi (loo-boom-BAH-shee), city (2005 population 1,102,000), ⊙ KATANGA province, SE CONGO, near the border with ZAMBIA; 11°40′S 27°28′E. Elevation 3,966 ft/1,208 m. The second-largest city in the country, it is a commercial and industrial center. Copper is smelted here; textiles, food products, beverages, printed materials, and bricks are manufactured. Founded in 1910, Lubumbashi prospered with the development of the region's copper-mining industry. It also serves as a distribution center for other minerals, including cobalt, zinc, tin, and coal. Lubumbashi was the capital of the secessionist state of KATANGA (1960-1963), as SHABA was then called, and was the scene of intense fighting between UN troops and Katangan forces. The city is the site of the University of Lubumbashi (founded 1964 as the state-run Université Officielle du Congo), a regional museum, and a modern airport. It is situated on a transcontinental railroad that links LUANDA on the W coast of Africa with BEIRA on the E coast. Formerly called ELISABETHVILLE or ELIZABETHSTAD. Also spelled Lumumbashi.

Lubunda (loo-BOON-duh), village, KATANGA province, E CONGO, on railroad, and 185 mi/298 km WNW of KALEMIE. Rice, cotton. Has noted Roman Catholic church and schools. Formerly called BRAINE-L'AL-LEUD SAINT-JOSEPH.

Lubutu (loo-BOO-too), village, ORIENTALE province, E CONGO, 120 mi/193 km SE of KISANGANI; 00°44′S 26°35′E. Elev. 2,198 ft/669 m. In rice-producing area; trading center. Has Roman Catholic and Protestant missions. Former Arab post. Emin Pasha was murdered here (1892) by the Arabs.

Luby (LU-bi), German Schönbach, town, ZAPADOCESKY province, W BOHEMIA, CZECH REPUBLIC, in SW part of ORE MOUNTAINS, and 12 mi/19 km NNE of CHEB; 50°15′N 12°24′E. Railroad terminus. Traditionally known for the manufacturing of musical instruments (strings) since 16th century. Town has a 17th-century Renaissance castle, museum.

Lubyany (loo-BYAH-ni), town (2006 population 1,900), N TATARSTAN Republic, E European Russia, on the VYATKA River, on local highway branch, 36 mi/58 km N of NIZHNEKAMSK, and 22 mi/35 km N of MAMADYSH; 56°02′N 51°24′E. Elevation 305 ft/92 m. Sawmilling. Has a forestry industry vocational school; mosque.

Lübz (LOOBTS), town, MECKLENBURG–WESTERN POMERANIA, N GERMANY, on the regulated ELDE, 8 mi/12.9 km ENE of PARCHIM; 53°28′N 12°01′E. Sugar refining; brewing.

Luc (LOOK), town (□ 10 sq mi/26 sq km), AVEYRON department, MIDI-PYRÉNÉES region, S FRANCE. S suburb of RODEZ.

Luca, MALTA: see LUQA.

Lucainena de las Torres (loo-kei-NAI-nah dhai lahs TO-res), town, ALMERÍA province, S SPAIN, 20 mi/32 km NE of ALMERÍA; 37°03′N 02°12′W. Terminus of mining railroad from Mediterranean port of AGUA AMARGA. Iron-mining center with iron foundries. Olive oil, cereals, esparto.

Lucala (loo-KAH-luh), town, CUANZA NORTE province, ANGOLA, on road and railroad junction, 30 mi/48 km E of N'dalatando; 09°16′S 15°17′E. Market center.

Lucama (LOOK-uh-mah), town (2006 population 861), WILSON county, E central NORTH CAROLINA, 7 mi/11.3 km SW of WILSON; 35°38′N 78°00′W. Retail trade;

manufacturing (apparel, pallets); agriculture (tobacco, cotton, sweet potatoes, grain; livestock). Incorporated 1889.

Lucan (LOO-kuhn), Gaelic Leamhcán, town (2006 population 37,424), DUBLIN county, E IRELAND; suburb 7 mi/11.3 km W of DUBLIN, on the LIFFEY RIVER; 53°21′N 06°27′W. Spa resort and hunting center.

Lucan (LOO-kuhn), former village (□ 1 sq mi/2.6 sq km; 2001 population 2,010), S ONTARIO, E central CANADA, 16 mi/26 km NW of LONDON; 43°11′N 81°24′W. Lumbering, fruit growing. Merged with Biddulph in 1999 to form Lucan Biddulph township.

Lucan (loo-KAN), village (2000 population 226), REDWOOD county, SW MINNESOTA, 17 mi/27 km SW of REDWOOD FALLS, near Sleepy Eye Creek; 44°24′N 95°24′W. Corn, oats, soybeans; livestock; dairying; manufacturing (chassis liners, millwork).

Lucanas (loo-KAH-nahs), province (□ 6,818 sq mi/17,726.8 sq km), AYACUCHO region, S PERU; ⊙ PUQUÍO; 14°30′S 74°20′W. Largest province of Ayacucho; drained by Chincha River (E) and Palpa River (W). Cereals, potatoes; livestock.

Lucan Biddulph (LOO-kuhn BI-duhlf), township (□ 65 sq mi/169 sq km; 2001 population 4,201), SW ONTARIO, E central CANADA, 19 mi/31 km from LONDON; 43°13′N 81°23′W. Farming. Museum of local history. Formed in 1999 from LUCAN, Biddulph.

Lucania (loo-KAI-nee-uh), ancient region of S ITALY. It was bounded on the E by the Gulf of Tarentum (now TARANTO) and by APULIA, on the N by SAMNIUM and CAMPANIA, on the W by the TYRRHENIAN Sea, and on the S by BRUTTIUM. Italic tribes and Greek colonists lived there before the Roman conquest in the 3rd century B.C.E. (see MAGNA GRAECIA). Their chief cities were HERACLEA and METAPONTUM on the Gulf of Tarentum and PAESTUM and Buxentum on the Tyrrhenian coast. The non-Greek Lucanians were Samnites. The W portion of ancient Lucania is now in Campania; the larger E part is in BASILICATA.

Lucania, Mount, 17,147 ft/5,226 m high, in the ST. ELIAS MOUNTAINS, SW YUKON TERRITORY, CANADA, near the ALASKA border; 61°01′N 140°28′W. Canada's third-tallest peak; first climbed in 1937.

Lucapa (loo-KAH-puh), city, ⊙ of LUNDA NORTE province, ANGOLA, on major road 90 mi/145 km NNE of SAURIMO; 08°25′S 20°45′E. Administrative center. Rice, manioc, corn; diamonds.

Lucas, county (□ 434 sq mi/1,128.4 sq km; 2006 population 9,543), S IOWA; ⊙ CHARITON; 41°01′N 93°27′W. Prairie agricultural area (hogs, cattle, poultry, corn, hay) drained by CHARITON RIVER and WHITEBREAST CREEK. Bituminous-coal deposits mined in E. Has state parks. Formed 1846.

Lucas (LOO-kuhs), county (□ 343 sq mi/891.8 sq km; 2006 population 445,281), NW OHIO; ⊙ TOLEDO; 41°39′N 83°40′W. Bounded N by MICHIGAN state line, SE by MAUMEE RIVER, and NE by W end of LAKE ERIE. Chief agricultural products are corn, vegetables, wheat; agricultural bedding plants. Manufacturing (paper products, glass, motor vehicles and associated products). Includes FALLEN TIMBERS STATE PARK and site of old Fort Meigs. Formed 1835.

Lucas, town (2000 population 243), LUCAS county, S IOWA, on WHITEBREAST CREEK, and 7 mi/11.3 km W of CHARITON; 41°01′N 93°27′W. In livestock and grain area. John Lake Lewis born here in 1880.

Lucas (LOO-kuhs), town (2006 population 4,297), COLLIN county, N TEXAS, residential suburb 25 mi/40 km NE of DALLAS, and 10 mi/16 km ENE of PLANO, near W shore of LAVON LAKE reservoir, in rapidly growing urban fringe; 33°06′N 96°34′W. Recreation.

Lucas (LOO-kuhs), village (2000 population 436), RUSSELL county, central KANSAS, 18 mi/29 km NE of RUSSELL; 39°03′N 98°32′W. Livestock, grain. Manufacturing (farm machinery, feeds). Concrete sculpture of the Garden of Eden is here. Wilson Lake Reservoir and dam; Wilson State Park to S.

Lucas (LOO-kuhs), village (2006 population 602), RICHLAND county, N central OHIO, 6 mi/10 km SE of MANSFIELD; 40°42′N 82°25′W.

Lucas do Rio Verde (LOO-kas do REE-o VER-zhee), city (2007 population 30,781), MATO GROSSO, BRAZIL.

Lucas González (LOO-kahs gon-ZAH-lez), town, S central ENTRE RÍOS province, ARGENTINA, on railroad, and 15 mi/24 km E of NOGOYÁ; 32°24′S 59°33′W. Agriculture (fruit, grain, livestock, poultry); dairy products; flour mills.

Lucasville (LOO-kuhs-vil), unincorporated village (□ 3 sq mi/7.8 sq km; 2000 population 1,588), SCIOTO county, S OHIO, 9 mi/14 km N of PORTSMOUTH; 38°52′N 82°59′W. Location of a federal maximum-security prison.

Lucaya (loo-KIE-yuh), locality, just E of FREEPORT, on GRAND BAHAMA Island, N BAHAMAS; 26°32′N 78°40′W. A resort and residential community with marinas, golf courses, and a Museum of Underwater Exploration.

Lucban (look-BAHN), town, QUEZON province, S LUZON, PHILIPPINES, 16 mi/26 km ENE of SAN PABLO, at foot of MOUNT BANAHAO in agricultural area (coconuts, rice).

Lucca (LOOK-kah), province (□ 684 sq mi/1,778.4 sq km), TUSCANY, central ITALY; ⊙ LUCCA; 44°02′N 10°27′E. Borders on LIGURIAN Sea; mountainous terrain (E APUANE Alps, GARFAGNANA) rising from narrow coastal plain. Drained by SERCHIO River and its tributaries. Agriculture (grapes, olives, corn, fruit); livestock raising (cattle, sheep); forestry. Marble quarries (SERAVEZZA, PIETRASANTA, Querceta) in Apuane Alps. Bathing resorts (VIAREGGIO, FORTE DEI MARMI) along coast. Manufacturing at Lucca and Viareggio. Area decreased in 1927 to help form PISTOIA province.

Lucca (LOOK-kah), city (2001 population 81,862), ⊙ LUCCA province, TUSCANY, N central ITALY, near the LIGURIAN Sea; 43°50′N 10°29′E. It is a commercial and industrial center and an agricultural market (olive oil, wine, and tobacco). Manufactures include machinery, fabricated metals, chemicals, plastics, textiles (especially silk), clothing, wood, paper, and food products. A Ligurian settlement, later a Roman town, Lucca became (6th century) the capital of a Lombard duchy and (12th century) a free commune, which soon developed into a republic. In spite of ruthless strife and frequent wars, the city prospered. Its bankers and merchants were noted throughout Europe, as were its velvets and damasks. Lucchese sculpture flourished in the 15th century with Matteo Civitali, whose fine works adorn the cathedral. Numerous churches, showing Pisan influence, were built from the 12th to the 14th century. Lucca remained an independent republic until Napoleon I made it a principality (1805) for his brother-in-law, Felice Baciocchi, and his sister Elisa. In 1817, Lucca became part of the duchy of Parma and in 1847 of the grand duchy of Tuscany; in 1860 it was annexed to the kingdom of Sardinia. The cathedral (11th–15th century) and the churches of San Frediano (begun in the 6th century) and San Michele (12th century) have fine marble facades. The city's ramparts (16th–17th century) are also notable.

Luce (LOOS), county (□ 1,911 sq mi/4,968.6 sq km; 2006 population 6,684), NE UPPER PENINSULA, MICHIGAN; ⊙ NEWBERRY; 46°45′N 85°35′W. Bounded N by LAKE SUPERIOR; drained by TAHQUAMENON RIVER and small Two Hearted River. Forest and farm area (forage, potatoes, hay, oats; cattle, livestock); lumbering; recreation. Some manufacturing at Newberry. Includes part of MANISTIQUE LAKE and NORTH MANISTIQUE LAKE. Part of Tahquamenon Falls State Park in NE; Muskallonge Lake State Park in N; Big Village Ski Area in S. Formed and organized 1887.

Lucé (lu-sai), SW town (□ 2 sq mi/5.2 sq km) suburb of CHARTRES, EURE-ET-LOIR department, CENTRE administrative region, N central FRANCE; 48°26′N

01°28′E. Metalworking. Has museum of regional agriculture and agricultural machinery built in this area.

Lucea, town, ⊙ HANOVER parish, NW JAMAICA, minor port 17 mi/27 km W of MONTEGO BAY, 95 mi/153 km WNW of KINGSTON; 18°27′N 78°10′W. Has fine, almost landlocked, harbor. Exports bananas and yams. Has old churches. Phosphate deposits nearby.

Luce Bay (LOOS), inlet (20 mi/32 km long; 19 mi/31 km wide) of IRISH SEA, DUMFRIES AND GALLOWAY, S Scotland, at mouth between MULL OF GALLOWAY and BURROW HEAD; 54°49′N 04°50′W.

Lucedale (LOOS-dail), town (2000 population 2,458), ⊙ GEORGE county, 30°55′N 88°35′W, SE MISSISSIPPI, 36 mi/58 km WNW of MOBILE, ALABAMA, and 38 mi/61 km N of PASCAGOULA. Agricultural (cotton, corn, vegetables) and timber area; manufacturing (storage tanks, jewelry, concrete, wood products, knitting kits). PASCAGOULA RIVER Wildlife Management Area to SW; Palestine Gardens to N, 20-acre/8-ha scale model of the Holy Land.

Lucélia (LOO-se-lee-ah), city (2007 population 19,212), W SÃO PAULO, BRAZIL, 75 mi/121 km WNW of MARÍLIA, on railroad; 21°44′S 51°01′W. Coffee, cotton, grain; livestock.

Lucena (loo-SE-nah), city (□ 26 sq mi/67.6 sq km; 2000 population 196,075), ⊙ QUEZON province, S LUZON, PHILIPPINES, near TAYABAS BAY, 22 mi/35 km SE of SAN PABLO, on railroad; 13°56′N 121°37′E. Trade center for fishing and agricultural area (rice, corn).

Lucena (loo-THAI-nah), city, CÓRDOBA province, S SPAIN, in ANDALUSIA, 37 mi/60 km SE of CÓRDOBA; 37°24′N 04°29′W. Industrial and agricultural center. Olive-oil processing, wine making; cereals. Medicinal springs. Has some notable churches. Scene of Christian victory (1483) over the Moors led by Boabdil, last king of Granada, who was captured.

Lucena (LOO-se-nah), town (2007 population 10,943), central coast coast of PARAÍBA, BRAZIL, 29 mi/47 km N of JOÃO PESSOA; 06°54′S 34°52′W.

Lucena del Cid (LOO-se-nah dhel THEEDH), town, Castellón de la Plana province, E SPAIN, 17 mi/27 km NW of CASTELLÓN DE LA PLANA; 40°08′N 00°17′W. Meat processing; livestock market; cereals, wine. Anthracite, iron, and lead mining; marble quarries. Has Roman remains and medieval castle.

Lucena del Puerto (LOO-se-nah dhel PWER-to), town, HUELVA province, SW SPAIN, on the RÍO TINTO, and 8 mi/12.9 km ENE of HUELVA; 37°18′N 06°43′W. Cereals, olives, white wine; timber, livestock.

Lučenec (lu-CHE-nets), Slovak *Lučenec*, Hungarian *Losonc*, city (2000 population 28,332), STREDOSLOVENSKY province, S SLOVAKIA; 48°20′N 19°40′E. In fertile agricultural region (wheat, potatoes, sugar beets; vineyards). Contains a railroad junction. Manufacturing (machinery, textiles; food processing; woodworking; distilling; wine making). Magnesite mining, tobacco growing in vicinity. Mostly Hungarian population. Held by Hungary between 1938–1945. Founded in the 13th century.

Luceni (loo-THAI-nee), village, ZARAGOZA province, NE SPAIN, near EBRO RIVER, 24 mi/39 km NW of ZARAGOZA; 41°50′N 01°14′W. Cereals, beets, alfalfa.

Lucens, commune, VAUD canton, SW SWITZERLAND, on BROYE RIVER, and 16 mi/25 km NE of LAUSANNE. Elevation 1,647 ft/502 m. Imposing 13th- and 16th-century castle, former residence of the bishops of Lausanne. Location of A. Conan Doyle Museum (founded 1966).

Lucera (loo-CHE-rah), town, APULIA, S ITALY; 41°30′N 15°20′E. It is an agricultural and diversified secondary industrial center. Already important in the 4th century B.C.E., the town was destroyed by the Byzantines in the 7th century C.E. It was revived (13th century) by Emperor Frederick II, who built a castle (now in ruins) that was the most important fortress in Apulia. Lucera also has a 14th-century cathedral.

Lucéram (lu-sai-rahm), picturesque village (□ 25 sq mi/65 sq km), ALPES-MARITIMES department, PROVENCE-ALPES-CÔTE D'AZUR region, SE FRANCE, 13 mi/21 km NNE of NICE; 43°53′N 07°22′E. Medieval streets and alleys; partly Gothic church restored in 18th century. Popular summer and winter sports resort of Peïra-Cava (elevation c.5,000 ft/1,524 m) is 4 mi/6.4 km N in MARITIME ALPS. MERCANTOUR NATIONAL PARK entrance is 6 mi/9.7 km NE.

Lucerna (loo-CER-nah), town, OCOTEPEQUE department, W HONDURAS, on major highway, 17 mi/27 km NE of NUEVA OCOTEPEQUE; 14°33′N 88°56′W. Small farming.

Lucerne (loo-SUHRN), German *Luzern*, canton (□ 576 sq mi/1,497.6 sq km), central SWITZERLAND; ⊙ LUCERNE. Drained by the REUSS and KLEINE EMME rivers, Lucerne is mainly an agricultural and pastoral region, with orchards and large forested areas. It contains LAKE SEMPACH and borders on LAKE LUCERNE. Several resort areas are here, notably along the NW shores of Lake Lucerne. The population is mainly German-speaking and Roman Catholic. Manufacturing includes machinery, textiles, metallurgic goods, electrical equipment, paper, and wood products. Boat building and automobile assembly are also important. One of the FOUR FOREST CANTONS, its history is that of its chief city, Lucerne. Lucerne canton joined the Swiss Confederation in 1332.

Lucerne (loo-SUHRN), German *Luzern*, city, ⊙ LUCERNE canton, central SWITZERLAND, on LAKE LUCERNE, and c.15 mi/24 km SW of ZUG; 47°04′N 08°16′E. Elevation, 1,440 ft/439 m. Situated on both banks of the REUSS RIVER that flows out of Lake Lucerne, it is one of the largest resorts (mainly summer) in Switzerland. A direct narrow-gauge trail links Lucerne to the winter-sports center of ENGELBERG in Obwalden.

The city grew around the monastery of St. Leodegar, founded in the 8th century. An important trade center on the St. Gotthard route, it became a Hapsburg possession in 1291. Lucerne joined the Swiss Confederation in 1332 and gained full freedom after the battle of Sempach (1386). It became the capital of the HELVETIC REPUBLIC in 1798. Lucerne was one of the chief towns of the League of Catholic cantons, the Sonderbund (1845–1847). The noted monument, the Lion of Lucerne (designed by A. B. Thorvaldsen), was erected (1820–1821) in memory of the Swiss guards killed in PARIS in 1792. Other points of interest are the Kapellbrücke, an old covered bridge across the Reuss, a mainly 17th-century church (Hofkirche), the Glacier Garden, the cantonal buildings, and several museums. An international music festival is held in Lucerne every summer.

Lucerne (loo-SUHRN), unincorporated town (2000 population 2,870), Lake county, NW CALIFORNIA, 21 mi/34 km ESE of UKIAH, on N shore of CLEAR LAKE; 39°05′N 122°49′W. Mendocino National Forest to N. Walnuts, pears, oats, cattle. Tourism.

Lucerne, town (2000 population 92), PUTNAM county, N MISSOURI, 16 mi/26 km W of UNIONVILLE; 40°27′N 93°17′W.

Lucerne, village, CASS county, N central INDIANA, 8 mi/12.9 km NNW of LOGANSPORT. In agricultural area (corn, soybeans, hogs); manufacturing (food grinding, fertilizer blending).

Lucerne, FRANCE: see GRANVILLE.

Lucerne-in-Maine, MAINE: see DEDHAM.

Lucerne, Lake or **Lake of the Four Mountain Cantons**, German *Vierwaldstätter See*, French *Lac de Lucerne*, irregular-shaped lake (□ 44 sq mi/114.4 sq km), central SWITZERLAND. Elevation 1,424 ft/434 m. It has a maximum depth of c.700 ft/213 m. The lake is fed and drained by the REUSS RIVER. The four cantons of its German name are Unterwalden, Uri, SCHWYZ, and LUCERNE. Surrounded by mountains, Lake Lucerne is noted for its scenic beauty; many resort towns are

along its shores. LUCERNE (German *Luzern*), the principal lakeside city, is located at its N outlet. The northern arm of Lake Lucerne is sometimes called Lake Küssnacht, the SW arm is known as Lake Alpnacht, and the SE arm is Lake Uri (German *Urner See*).

Lucerne Lake, intermittently dry bed (c.6 mi/9.7 km long), SAN BERNARDINO county, S CALIFORNIA, in MOJAVE DESERT, 17 mi/27 km E of VICTORVILLE. Unincorporated town of LUCERNE VALLEY to S.

Lucerne Mines (LOO-suhrn) or **Lucerne**, unincorporated town, Center township, INDIANA county, W central PENNSYLVANIA, 4 mi/6.4 km S of INDIANA, on Yellow Creek, just NE of its mouth on Two Lick Creek; 40°32′N 79°09′W. Agriculture (corn, hay, dairying); manufacturing (voltage suppressors, hydraulic equipment).

Lucerne Valley, unincorporated town, SAN BERNARDINO county, S CALIFORNIA, 30 mi/48 km NE of SAN BERNARDINO, at N edge of SAN BERNARDINO MOUNTAINS, and SW edge of MOJAVE DESERT. Cattle, fruit, alfalfa, grain. Manufacturing (cement, industrial chemicals). Limestone quarrying. San Bernardino National Forest to S.

Luceville (LOOS-vil, French, loos-VEEL), former village, SE QUEBEC, E CANADA, near the SAINT LAWRENCE RIVER, 11 mi/18 km NE of RIMOUSKI; 48°31′N 68°20′W. Dairying, pig raising. Amalgamated into SAINTE-LUCE in 2001.

Luchegorsk (loo-chye-GORSK), town (2006 population 21,660), W MARITIME TERRITORY, SE RUSSIAN FAR EAST, in the E part of the USSURI River basin, approximately 15 mi/24 km E of the Russia-China border, near road and railroad on the Khabarovsk-Vladivostok line, 6 mi/10 km from station; 46°29′N 134°12′E. Elevation 291 ft/88 m. Maritime regional electric power-generating station.

Lucheng (LU-CHENG), town, ⊙ Lucheng county, SE SHANXI province, CHINA, on road, and 16 mi/26 km NE of CHANGZHI; 36°14′N 113°20′E. Grain, medicinal herbs; coal mining, building materials, food processing, non-ferrous metal smelting.

Lucheng, town, W GUANGXI Zhuang Autonomous Region, CHINA, 45 mi/72 km NW of BOSE. Grain.

Luchente (loo-CHEN-tai), town, VALENCIA province, E SPAIN, 9 mi/14.5 km ESE of JÁTIVA; 38°56′N 00°21′W. Olive-oil processing; wine, almonds, cereals.

Luchenza (lah-chen-zah), town and railroad station, Southern region, S MALAWI, 24 mi/39 km SE of BLANTYRE; 16°01′S 35°18′E. In agricultural area; cotton, tobacco, tung, tea.

Luches'k, UKRAINE: see LUTS'K, city.

Luches'k Velykyy, UKRAINE: see LUTS'K, city.

Lu-chiang, TAIWAN: see Lukang.

Luchinskoye (LOO-cheen-skuh-ye), settlement, W central MOSCOW oblast, central European Russia, on the ISTRA RIVER, on road and near railroad, 2 mi/3.2 km S, and under administrative jurisdiction, of ISTRA; 55°53′N 36°50′E. Elevation 541 ft/164 m. Brick works.

Luchki, RUSSIA: see YAROSLAVSKIY.

Luchon, FRANCE: see BAGNÈRES-DE-LUCHON.

Luchon Valley (lu-shon), in central PYRENEES, HAUTE-GARONNE department, MIDI-PYRÉNÉES region, S FRANCE, extending 10 mi/16 km N-S from Marignac to BAGNÈRES-DE-LUCHON; it is drained by a small tributary of GARONNE RIVER; 42°51′N 00°36′E. Active resort area centered on Bagnères-de-Luchon and SUPERBAGNÈRES.

Luchou (LU-JOE), township, N TAIWAN, across TANSHUI RIVER from TAIPEI.

Lüchow (LOO-khou), town, LOWER SAXONY, N GERMANY, 8 mi/12.9 km N of SALZWEDEL (E Germany); 52°58′N 11°10′E. Manufacturing of ball bearings; metalworking; food processing. Chartered 1293.

Luchow, CHINA: see HEFEI.

Luchuan (LU-CHUAN), town, ⊙ Luchuan county, SE GUANGXI Zhuang Autonomous Region, CHINA,

20 mi/32 km SE of YULIN; 22°21′N 110°15′E. Rice, wheat, sugarcane; building materials, chemicals, textiles.

Luciana (loo-THYAH-nah), town, CIUDAD REAL province, S central SPAIN, on GUADIANA RIVER at mouth of BULLAQUE RIVER, and 20 mi/32 km of CIUDAD REAL; 38°59′N 04°17′W. Cereals, olives; livestock.

Lucianópolis (LOO-see-ah-NO-po-lees), town (2007 population 2,299), SW SÃO PAULO state, BRAZIL, 30 mi/48 km SW of BAURU; 22°27′S 49°31′W.

Luciára (LOO-see-ah-rah), town (2007 population 2,419), NE MatoGrosso on Rio Araguaia, on Goiás border; 11°08′S 50°40′W.

Lucien (LOO-see-en), village, NOBLE county, N OKLAHOMA, 9 mi/14.5 km W of PERRY.

Lucie River, river, SIPALIWINI district, SURINAME, tributary of the CORANTIJN RIVER; rises in central SURINAME in the WILHELMINA MOUNTAINS; feeds TOEKOMSTIG RESERVOIR, which flows into CORANTIJN RIVER; waterfalls and rapids.

Lucignano (loo-chee-NYAH-no), village, AREZZO province, TUSCANY, central ITALY, 15 mi/24 km SSW of AREZZO. Has 13th-century church and palace with museum containing pictures by Luca Signorelli.

Lucija (LOO-tsee-yah), Italian *Lucia*, town (2002 population 5,579), SW SLOVENIA, near Portorož, in KOPRSKO PRIMORJE (Koper Riviera) seaside resort area. Old saltworks have been transformed into a marina.

Lucillos (loo-THEE-lyos), village, TOLEDO province, central SPAIN, 11 mi/18 km E of TALAVERA DE LA REINA; 39°59′N 04°37′W. Cereals, grapes, olives; sheep.

Lucindale (LOO-sin-dail), village, SE SOUTH AUSTRALIA, 175 mi/282 km SSE of ADELAIDE; 36°59′S 140°22′E. On NARACOORTE-KINGSTON railroad; dairy products, livestock.

Lucinda Point (loo-SIN-duh), village and small port, E QUEENSLAND, AUSTRALIA, on headland forming N end of entrance to HALIFAX BAY, 60 mi/97 km NW of TOWNSVILLE; 18°31′S 146°20′E. Sugar port for INGHAM, 13 mi/21 km inland, with which it is connected by electric railroad. Sometimes called Dungeness.

Lucio Vicente López (LOO-see-o vee-SAIN-te LO-pez), village, S SANTA FE province, ARGENTINA, on CARCARAÑA RIVER, and 26 mi/42 km NW of ROSARIO; 32°43′S 61°02′W. Hydroelectric station in agricultural area (wheat, corn, flax, livestock, poultry); dairying.

Lucivna, SLOVAKIA: see STRBA.

Luck, town (2006 population 1,186), POLK county, NW WISCONSIN, 37 mi/60 km WNW of RICE LAKE, in wooded lake area; 45°34′N 92°28′W. Manufacturing (food products, wood moldings and picture frames; wire screen cloth, lumber, furniture).

Luck, UKRAINE: see LUTS′K.

Lucka (LUHK-kah), town, THURINGIA, central GERMANY, 17 mi/27 km S of LEIPZIG; 51°06′N 12°20′E. Paper industry; foundry. Lignite was formerly mined in the area.

Luckau (LUHK-kou), town, BRANDENBURG, E GERMANY, in LOWER LUSATIA, near SPREE FOREST, 15 mi/24 km N of FINSTERWALDE; 51°53′N 13°43′E. Spa. Town retains much of its medieval character; has sixteenth-century church, many half-timbered houses, and remains of thirteenth-century town wall.

Luckeesarai (LUHK-ee-suh-rei) or **Lakhisarai**, town (2001 population 77,840), ⊙ LAKHISARAI district, SE BIHAR state, E INDIA, on tributary of the GANGA River and 28 mi/45 km WSW of MUNGER; 25°11′N 86°05′E. Railroad and road junction; rice, corn, wheat, grain, barley. Extensive Buddhist ruins nearby at Rajauna. Village of Barahiya, 9 mi/15 km N, has Maharani Asthan, the highest temple in Bihar.

Luckenbach (LOO-ken-bahk), unincorporated village, GILLESPIE county, S central TEXAS, 55 mi/89 km NNW of SAN ANTONIO; 30°10′N 98°45′W. Elevation 1,561 ft/476 m. Cattle-ranching area; tourism. German hamlet established 1850; purchased in 1970s by Texas humorist Hondo Crouch.

Luckenwalde (LUHK-kuhn-vahl-de), town, BRANDENBURG, E GERMANY, 30 mi/48 km SSW of BERLIN; 52°05′N 13°10′E. Metalworking; manufacturing of machinery, plastics, electronics. Site of concentration camp under Hitler regime. Has sixteenth-century church, old fortified tower. Chartered 1808.

Luckey (LUHK-ee), village (2006 population 983), Wood county, NW OHIO, 10 mi/16 km ENE of BOWLING GREEN; 41°27′N 83°29′W. Limestone quarries.

Lucknow (luhk-NOU), district (□ 976 sq mi/2,537.6 sq km), central UTTAR PRADESH state, N central INDIA; ⊙ Lucknow. On GANGA plain; drained by the GOMATI RIVER; irrigated by Sarda Canal system. Agriculture (wheat, rice, gram, millet, oilseeds, barley, sugarcane, corn); mango, orange, and ber groves. Main centers: Lucknow (mfg.), MALIHABAD, KAKORI, AMETHI. Formerly also spelled LAKHNAU.

Lucknow (luhk-NOU), city (2001 population 2,245,509), ⊙ LUCKNOW district and UTTAR PRADESH state, N central INDIA, on the GOMATI RIVER; 26°51′N 80°55′E. Educational and cultural center; a major transportation hub and agriculture market. Industries: food processing, railroad shops, and handicrafts. It was capital of the kingdom of OUDH (1775–1856) and then of Oudh province. It became the capital of the UNITED PROVINCES when AGRA and Oudh merged in 1877. Notable architectural structures include the Imambara (mausoleum) of Asuf-ad-daula, Oudh's greatest king; the Pearl Palace. During the Indian Mutiny, the BRITISH garrison in Lucknow suffered heavy casualties during a siege (June–November 1857). Although the siege was broken, the British evacuated the city, returning a year later, when they regained control of India. Seat of a well-known university and of several scientific laboratories. Famous for chikan, a type of embroidery used on fabrics. Also famous for unique culture, which includes qawwali music, Urdu poetry, courtly mannerisms, and rich cuisine.

Lucknow, village, E central NEW SOUTH WALES, SE AUSTRALIA, 155 mi/250 km NW of SYDNEY, 6 mi/9 km SE of ORANGE; 33°23′S 149°12′E. Former gold-mining center.

Lucknow (LUHK-no), former village (□ 1 sq mi/2.6 sq km; 2001 population 1,136), SW ONTARIO, E central CANADA, 18 mi/29 km NE of GODERICH; 43°57′N 81°31′W. Manufacturing; dairying. Amalgamated into HURON-KINLOSS township in 1999.

Lucky, town (2000 population 355), BIENVILLE parish, LOUISIANA, 16 mi/26 km W of JONESBORO; 32°15′N 93°00′W.

Lucky (looch-KI), Slovak *Lúčky*, Hungarian *Lucski*, village, STREDOSLOVENSKY province, N SLOVAKIA, 6 mi/9.7 km NE of RUŽOMBEROK; 49°08′N 19°24′E. Elevation of 1,903 ft/580 m. Health resort with ferruginous thermal springs.

Lucky Hill, village, SAINT MARY parish, N JAMAICA, just S of GAYLE, and 8 mi/12.9 km SW of PORT MARIA; 18°18′N 77°01′W. A community project. Principal crops are bananas, citrus fruit, corn, peas, cacao.

Lucky Lake (LUHK-ee), village (2006 population 295), SW SASKATCHEWAN, CANADA, near LUCKY LAKE (4 mi/6 km long, 3 mi/5 km wide), 55 mi/89 km NE of SWIFT CURRENT; 50°59′N 107°08′W. Magnesium-sulphate production; grain elevators, lumbering.

Lucky Peak Reservoir, SW IDAHO, on BOISE RIVER, 10 mi/16 km SE of BOISE; c.10 mi/16 km long; 43°30′N 116°04′W. Maximum capacity 307,000 acre-ft. Formed by Lucky Peak Dam (230 ft/70 m high), built (1955) by the Army Corps of Engineers for flood control.

Lucky Shot, hamlet, Lucky Shot Landing, S ALASKA, 40 mi/64 km NNE of ANCHORAGE. Airstrip.

Luc, Le (LOOK, luh), town (□ 17 sq mi/44.2 sq km), VAR department, PROVENCE-ALPES-CÔTE D'AZUR region, SE FRANCE, 13 mi/21 km SW of DRAGUIGNAN, on expressway from RHÔNE valley and MARSEILLE to the French RIVIERA; 43°24′N 06°19′E. Trade center for wine and olive oil. Medieval town center is dominated by six-sided 16th-century church steeple.

Lucma (LOOK-mah), town, LA CONVENCIÓN province, CUSCO region, S central PERU, on affluent of URUBAMBA RIVER, in Cordillera VILCABAMBA, and 15 mi/24 km SW of QUILLABAMBA; 13°03′S 72°56′W. Elevation 9,632 ft/2,935 m. Sugarcane, coca, cereals. Silver, copper deposits nearby.

Lucmagn, Culom, SWITZERLAND: see LUKMANIER PASS.

Lucomagno, Paso del, SWITZERLAND: see LUKMANIER PASS.

Luçon (lu-son), town (□ 12 sq mi/31.2 sq km), VENDÉE department, PAYS DE LA LOIRE region, W FRANCE, 18 mi/29 km SE of LA ROCHE-SUR-YON; 46°28′N 01°10′W. Its partially silted small port is connected by 9 mi/14.5 km canal across Marais POITEVIN (i.e., the marshland of old POITOU province) with Bay of BISCAY; agricultural distribution and horticultural center. Seat of a bishop. Has 12th–16th-century cathedral with a 280-ft/85-m high spire; its episcopal palace encloses a 15th-century cloister. As a young man, Richelieu was bishop of Luçon, and led the town's growth. The regional park of the MARAIS POITEVIN extends from Luçon to the ATLANTIC (SW) and to the poorly drained delta (S) of the SÈVRE NIORTAISE RIVER. For many centuries that marsh was insalubrious, preventing the region's development. It now attracts fishermen and hunters.

Lucre (LOO-krai), town, CUSCO region, S central PERU, 15 mi/24 km ESE of CUSCO; 13°38′S 71°44′W. Elevation 10,675 ft/3,253 m. In agricultural region (cereals, potatoes).

Lucrécia (LOO-kre-see-ah), town (2007 population 3,423), SW RIO GRANDE DO NORTE state, BRAZIL 66 mi/106 km SSW of MOSSORÓ; 06°07′S 37°49′W.

Lucrecia Cape (loo-KRAI-see-ah), on ATLANTIC coast of E CUBA, HOLGUÍN province, 45 mi/72 km ENE of HOLGUÍN; 21°05′N 75°37′W.

Lucrino, Lago (LAH-go loo-KREE-no), small coastal lake in CAMPANIA, S ITALY, 2 mi/3 km WNW of POZZUOLI. Oyster culture; bathing. An ancient embankment, Via Herculea, can still be traced under the water.

Lucroja, PERU: see LOCROJA.

Luc-sur-Mer (look–syur–mer), village (□ 1 sq mi/2.6 sq km), CALVADOS department, BASSE-NORMANDIE region, NW FRANCE, beach resort on the ENGLISH CHANNEL, 9 mi/14.5 km N of CAEN; 49°18′N 00°21′W. Has marine zoological laboratory. Located near Allied landing beaches of NORMANDY invasion (1944) in World War II.

Lucun (LU-ZUN), locality, salt pan in SW SHANXI province, CHINA, 35 mi/56 km E of YONGJI, along railroad; 18 mi/29 km long, 3 mi/4.8 km wide. Major salt-production source, with centers in towns of XIEXIAN and ANYI.

Lucusse (loo-KOO-sai), town, MOXICO province, ANGOLA, on road junction, and 85 mi/137 km SE of LUENA; 12°38′S 20°52′E. Market center.

Lüda (LOO-DAH), city, S LIAONING province, CHINA, at the tip of the Liaodong peninsula. A former city that comprised the important and historic municipalities of LÜSHUN (Port Arthur) and DALIAN (Dairen). The combined area is also known as Dalian. Sometimes spelled Lü-ta.

Ludajana Rive, Bulgaria: see LUDA YANA RIVER.

Luda Kamchiya River (LOO-dah KAM-chee-yah), c.75 mi/121 km long, E BULGARIA; rises SW of KOTEL in the KOTEL MOUNTAINS; flows E and ENE to TSONEVO, here joining the GOLYAMA to form the KAMCHIYA RIVER; 42°53′N 27°11′E. Tsonevo Reservoir is on the Luda Kamchiya River, c.10 mi/16 km S of PROVADIYA. Also called Luda Ticha River.

Lüdao Island, TAIWAN: see Lütao Island.

Luda Ticha River, Bulgaria: see LUDA KAMCHIYA RIVER.

Luda Yana River (LOO-dah YAH-nah), 49 mi/79 km long, W central BULGARIA; rises in the central SREDNA

GORA 7 mi/11 km S of PIRDOP; flows generally S, past PANAGYURISHTE, to the MARITSA RIVER 3 mi/5 km ESE of PAZARDZHIK; 42°15′N 24°23′E. Has gold-carrying sand. Sometimes spelled Ludajana River.

Ludbreg (LUD-breg), village, central CROATIA, on railroad, and 14 mi/23 km E of Varaždin; 46°15′N 16°38′E. Flour milling, brick manufacturing Former lignite-mining area.

Ludden (LUHD-uhn), village (2006 population 27), DICKEY county, SE NORTH DAKOTA, 9 mi/14.5 km S of OAKES, near JAMES RIVER; 46°00′N 98°07′W. Founded in 1884 and incorporated in 1909. Named for Mr. and Mrs. John D. Ludden of St. Paul, MINNESOTA. Post office closed in 1986.

Luddenden Foot (LUH-duhn-duhn FU-it), village (2001 population 10,899), WEST YORKSHIRE, N ENGLAND, at foot of CALDER RIVER valley, 3 mi/4.8 km W of HALIFAX; 53°43′N 01°56′W. Previously woolen milling.

Lude, Le (lu-duh, luh), town (□ 17 sq mi/44.2 sq km), SARTHE department, PAYS DE LA LOIRE region, NW FRANCE, on LOIR RIVER, and 11 mi/18 km ESE of La FLÈCHE; 47°39′N 00°09′E. Market center; noted for its 15th-century Gothic and Renaissance castle (built on older foundations) in the shape of a quadrilateral with machicolated towers. Nighttime "light and sound" spectacle for visitors.

Lüdenscheid (LOO-den-sheid), city, NORTH RHINE–WESTPHALIA, W GERMANY; 51°13′N 07°28′E. Industrial center; manufacturing includes aluminum, metal products, plastics, and synthetics. Chartered in 1287 and later was a member of the HANSEATIC LEAGUE. Noteworthy buildings include Neuenhof (1694), a moated castle.

Lüderitz (LOO-duh-rits) or **Angra Pequena**, town, SW NAMIBIA, on Angra Pequena or Lüderitz Bay (5 mi/8 km long, 25 mi/40 km wide) of the ATLANTIC OCEAN, 180 mi/290 km W of KEETMANSHOOP; 26°39′S 15°09′E. Chief seaport of S part of country; serves surrounding diamond-mining region. Important fishing port and tourist destination. Important crayfish fisheries and several canneries. railroad terminus; airfield. Bartholomew Diaz landed here in 1486 and erected a cross. Site was acquired 1883 by F.A.E. Lüderitz, a German merchant, and was taken under German protection (April 24, 1884).

Ludgerovice (LUD-ge-RZHO-vi-TSE), Czech *Ludgeřovice*, German *Ludgerstal*, village, SEVEROMORAVSKY province, E central SILESIA, CZECH REPUBLIC, 5 mi/8 km NW of OSTRAVA; 49°54′N 18°15′E. In a coal-mining area. Gothic church. Part of GERMANY until 1920.

Ludgershall (LUD-guhr-shuhl), town (2001 population 3,775), E WILTSHIRE, S ENGLAND, 7 mi/11.3 km NW of ANDOVER; 51°16′N 01°37′W. Military residential area. Has remains of Norman castle. Norman church dating back to 13th century.

Ludgerstal, CZECH REPUBLIC: see LUDGEROVICE.

Ludgvan (LUJ-vahn), village (2001 population 5,304), W CORNWALL, SW ENGLAND, 3 mi/4.8 km NE of PENZANCE; 50°08′N 05°29′W. Former tin- and copper-mining community; market town in fruit- and vegetable-growing region. Has 15th-century church.

Ludhiana (loo-DYAH-nuh), district (□ 1,452 sq mi/3,775.2 sq km), central PUNJAB state, N INDIA; ⊙ Ludhiana. Bounded N by SUTLEJ RIVER, S by SANGRUR and PATIALA districts. Irrigated by SIRHIND CANAL system; agriculture (wheat, grain, corn, cotton, oilseeds); hand-loom weaving. Chief towns are Ludhiana, JAGRAON, KHANNA. Invaded by Ranjit Singh, 1806–1809; under BRITISH control soon after.

Ludhiana (loo-DYAH-nuh), city (2001 population 1,398,467), ⊙ LUDHIANA district, PUNJAB state, N INDIA; 30°54′N 75°51′E. Founded in the late 15th century, it lies on the Grand Trunk Road, connecting DELHI with AMRITSAR. Hosiery, cotton textiles, bicycle parts, and sewing machines. Seat of Punjab Agri-

culture University (PAU), one of the country's important agriculture universities, recognized for wheat research.

Ludian (LU-DIAN), town, ⊙ Ludian county, northeasternmost YUNNAN province, CHINA, 14 mi/23 km SW of ZHAOTONG; 27°11′N 103°33′E. Tobacco; tobacco processing.

Luding (LU-DING), town, ⊙ Luding county, W SICHUAN province, CHINA, on DADU RIVER, and 20 mi/32 km SE of KANGDING, and on highway; 29°56′N 102°12′E. Grain, livestock; logging, leather and fur processing.

Lüdinghausen (LOO-ding-HOU-shun), town, NORTH RHINE–WESTPHALIA, GERMANY, 15 mi/24 km SW of MÜNSTER; 51°47′N 07°27′E. Foundries; distilling; woodworking; food processing. Has late-Gothic church and moated castle (sixteenth century).

Ludington, town (2000 population 8,357), ⊙ MASON county, W MICHIGAN, on LAKE MICHIGAN at mouth of PERE MARQUETTE RIVER; 43°57′N 86°26′W. Port for GREAT LAKES shipping; railroad terminus. Manufacturing (furniture, industrial chemicals, industrial machinery, styrofoam, wire products, railroad maintenance equipment, aluminum fabrication). Agriculture (apples, cherries, peaches, vegetables). Mason County Airport to E. Historical museum. Resort; fishing in many nearby streams and lakes. Has monument on site of first burial place (1675) of Father Marquette. Car ferry to MANITOWOC, WISCONSIN, carries U.S. Highway 10 across Lake Michigan (4 hour crossing). Ludington State Park, with its dunes, to N, between Hankin Lake and Lake Michigan. Manistee National Forest to N. Incorporated as city 1873.

Luditz, CZECH REPUBLIC: see ZLUTICE.

Ludlam Bay (LUHD-luhm), SE NEW JERSEY, inlet of the ATLANTIC OCEAN (c.2.5 mi/4 km long) NW of SEA ISLE CITY; 39°10′N 74°41′W. Entered from ocean by CORSONS INLET (NE); crossed by INTRACOASTAL WATERWAY. Between bay and ocean is Ludlam Beach, barrier island (c.7 mi/11.3 km long) between Corsons Inlet (N) and TOWNSENDS INLET (S); site of Sea Isle City (bridge to mainland here).

Ludlow (LUHD-lo), town (2001 population 9,548), S SHROPSHIRE, W ENGLAND, on TEME RIVER at mouth of Corve River, and 23 mi/37 km S of SHREWSBURY; 52°22′N 02°44′W. Former agricultural market; previously sand quarrying. Has ruins of 11th-century castle and an important fortress on the Welsh border. There are several half-timbered houses, one of the original seven town gates, a 12th-century chapel, and a large 13th–15th-century parish church. The first performance of Milton's *Comus* took place in Ludlow in 1634. Butler wrote *Hudibras* here.

Ludlow (LUHD-lo), town (2000 population 4,409), KENTON county, N KENTUCKY, residential suburb 2 mi/3.2 km WSW of CINCINNATI, OHIO, 1 mi/1.6 km W of COVINGTON, Kentucky, on the OHIO RIVER (railroad bridge); 39°05′N 84°32′W. Diverse light manufacturing. Settled c.1790; incorporated as village 1864, as city 1925.

Ludlow, agricultural town, AROOSTOOK county, E MAINE, 10 mi/16 km W of HOULTON; 46°09′N 67°58′W.

Ludlow, town, HAMPDEN county, SW MASSACHUSETTS, on the CHICOPEE river; 42°11′N 72°28′W. Residential suburb of SPRINGFIELD and CHICOPEE. Manufactures include industrial molds, plastic products, twines. Ludlow State Park is within the town. Settled c.1750, set off from Springfield 1774, incorporated 1775.

Ludlow, town (2000 population 204), LIVINGSTON county, N central MISSOURI on Shoal Creek, 12 mi/19 km SW of CHILLICOTHE; 39°38′N 93°42′W.

Ludlow, unincorporated town, MCKEAN county, N PENNSYLVANIA, 12 mi/19 km SE of WARREN, on Twomile Run, in Allegheny National Forest; 41°43′N 78°56′W. Timber; manufacturing (lumber).

Ludlow, town (2006 population 1,053), including Ludlow village, WINDSOR CO., S central VERMONT, on

BLACK RIVER, and 20 mi/32 km SSW of Woodstock, in GREEN MOUNTAINS; 43°23′N 72°42′W. Okemo State Forest and Okemo Ski Area are nearby. Calvin Coolidge (1872–1933), 30th President of the United States, attended Black River Academy here.

Ludlow, village (2000 population 324), CHAMPAIGN county, E ILLINOIS, 19 mi/31 km NNE of CHAMPAIGN; 40°23′N 88°07′W. In agricultural area.

Ludlow Falls (LUHD-lo FAHLZ), village (2006 population 214), MIAMI county, W OHIO, 12 mi/19 km SSW of PIQUA; 39°59′N 84°20′W.

Ludogoriye (loo-do-GOR-ee-yah), hilly upland in NE BULGARIA, between the BLACK SEA and the DANUBE, S of the Dobrudzha; rises to over 1,600 ft/488 m, 15 mi/24 km E of RAZGRAD; 43°32′N 27°15′E. Consisting largely of Cretaceous marls, the dry, sparsely populated region has oak forests (partly cleared) and some agriculture (grain, tobacco). Formerly called Deliorman.

Ludogorsko (loo-do-GOR-sko), plateau, DANUBE Plain, NE BULGARIA; 43°30′N 27°00′E.

Ludowici (loo-duh-WEE-see), town (2000 population 1,440), ⊙ LONG county, SE GEORGIA, c.45 mi/72 km SW of SAVANNAH, near the ALTAMAHA RIVER; 31°43′N 81°45′W.

Ludres (lu-druh), town (□ 3 sq mi/7.8 sq km), suburb of NANCY in MEURTHE-ET-MOSELLE department, LORRAINE region, NE FRANCE, 5 mi/8 km S of Nancy in iron-mining area; 48°37′N 06°10′E. Distribution of milk and dairy products.

Ludsen, LATVIA: see LUDZA.

Ludus (LOO-doosh), Hungarian *Marosludas*, town, MURES county, NW central ROMANIA, in TRANSYLVANIA, on MURES RIVER, and 15 mi/24 km SE of TURDA; 46°29′N 24°06′E. Railroad junction, agricultural market (notably for grain and livestock).

Ludvika (LOOD-VEE-kah), town, KOPPARBERG county, W central SWEDEN, in BERGSLAGEN region, on 10-mi/16-km-long Lake Väsman, 35 mi/56 km SW of FALUN; 60°09′N 15°11′E. Railroad junction; manufacturing (heavy electrical equipment); beverages). Large ironworks (sixteenth–early twentieth century) now closed. Mining museum GRÄNGESBERG, with important iron mines, 9 mi/14.5 km SW. Incorporated 1919 as city.

Ludwig Canal, GERMANY: see RHINE-MAIN-DANUBE CANAL.

Ludwigsau (LOOD-viks-ou), village, HESSE, central GERMANY, on FULDA River, 24 mi/39 km N of FULDA; 50°54′N 09°45′E.

Ludwigsburg (LOOD-viks-burg), city, BADEN-WÜRTTEMBERG, SW GERMANY, near the NECKAR River; 48°54′N 09°12′E. Transportation and industrial center. Manufacturing (machine tools, iron and wire goods, porcelain, and organs). Mineral springs here) in the suburb of HOHENECK. Foundation on a large baroque castle built (1704–1733) in imitation of VERSAILLES by Duke Eberhard Ludwig of Württemberg. Schiller lived in Ludwigsburg (1768–1773; 1793–1794); attended a nearby military school. Birthplace of the poets J. Kerner and E. Mörike; the theologian D. F. Strauss, and the philosopher F. T. Vischer.

Ludwigsfelde (LOOD-viks-fel-de), town, BRANDENBURG, E GERMANY, 16 mi/26 km SSW of BERLIN; 52°19′N 13°15′E. Manufacturing (automobiles, diesel engines, pipes).

Ludwigshafen am Rhein (LOOD-viks-hah-fuhn ahm REIN), city, RHINELAND-PALATINATE, W GERMANY, a port on the left bank of the RHINE RIVER, connected by bridge with MANNHEIM; 49°29′N 08°26′E. A major trans-shipment point and leading center of the German chemical industry; other manufacturing includes machinery and motor vehicles. Founded as a small fortress in the seventeenth century, Ludwigshafen was named and developed by King Louis I of Bavaria in the mid-nineteenth century. Badly damaged in World War II; was the scene (1948) of a disastrous explosion of several chemical plants. The city has since been rebuilt.

Ludwigskanal, GERMANY: see RHINE-MAIN-DANUBE CANAL.

Ludwigslust (LOOD-viks-luhst), town, MECKLEN-BURG–WESTERN POMERANIA, N GERMANY, on Ludwigslust Canal (linking ELBE and ELDE rivers), and 21 mi/34 km S of SCHWERIN; 53°20′N 11°30′E. Road junction. Canning; manufacturing of agricultural machinery; meat processing. Has eighteenth-century former grand-ducal palace. Founded 1756; until 1837 residence of grand dukes of MECKLENBURG-SCHWERIN. Chartered 1876.

Ludwigsort, RUSSIA: see LADUSHKIN.

Ludwigsstadt (LOOD-viks-shtaht), town, UPPER FRANCONIA, N BAVARIA, GERMANY, in THURINGIAN FOREST, 17 mi/27 km NNE of KRONACH, 3 mi/4.8 km S of Probstzella; 50°29′N 11°23′E. Manufacturing of metal products, medical equipment, plastics. Has early-Romanesque chapel. Chartered 1377.

Ludwikowo, POLAND: see MOSINA.

Ludwipol, UKRAINE: see SOSNOVE.

Ludza (LOOD-zuh), German *Ludsen*, Russian (until 1917) *lyutsin*, city (2000 population 10,822), LATGALE region, E LATVIA, 15 mi/24 km E of REZEKNE, on small Ludza Lake; 56°33′N 27°43′E. Tanning, sawmilling (timber trade), flour milling; rye, flax. Castle ruins. In VITEBSK government until 1920.

Luebo (LWE-bo), town, ⊙ KASAI district, KASAI-OCCIDENTAL province, S CONGO, on left bank of LULUA RIVER, and 140 mi/225 km WSW of LUSAMBO; 05°21′S 21°25′E. Elev. 1,571 ft/478 m. In fiber-growing area; terminus of steam navigation, trading center. Airfield. Has Roman Catholic and Protestant missions, hospital. Noted diamond-mining region extends S and E.

Lueders, village (2006 population 274), JONES county, W central TEXAS, 25 mi/40 km N of ABILENE, and on Clear Fork of the BRAZOS; 32°47′N 99°37′W. In cotton, cattle area; limestone deposits; oil refinery.

Luella, village, GRAYSON county, N TEXAS, 5 mi/8 km SE of SHERMAN; 33°34′N 96°32′W. Cattle, wheat, peanuts. Oil and natural gas.

Luembe River (loo-EM-bai), c.350 mi/563 km long, left tributary of KASAI River, in NE ANGOLA and S Democratic Republic of the CONGO; rises above Mulonda (Lunda Sul); flows N to the Kasai 15 mi/24 km above TSHIKAPA village. Receives (left) the Chiumbe.

Luena (loo-AI-nuh), town, ⊙ MOXICO province, BIÉ province, E central ANGOLA, on Benguela railroad, and 200 mi/322 km ENE of Cuito; 11°48′S 19°52′E. Elev. 4,300 ft/1,311 m. Agricultural trade center (rice, manioc, corn). Brick manufacturing. Road to CHICAPA RIVER diamond washings (N). Formerly Vila Luso (VEE-lah LOO-soo) and Moxico. Airport.

Luena (LWE-nah), village, KATANGA province, SE CONGO, on railroad, and 100 mi/161 km ESE of KAMINA; 09°27′S 25°47′E. Elev. 2,319 ft/706 m. Center of coal-mining area supplying, notably, the copper works at LIKASI.

Luena (loo-AI-nuh), river, E ANGOLA; rises in highlands in MOXICO province; flows 330 mi/531 km ESE joining the Zambesi River in ZAMBIA.

Luepa (loo-AI-pah), village, BOLÍVAR state, SE VENEZUELA, in GUIANA HIGHLANDS, on the Aponguao River, 110 mi/177 km SSE of EL CALLAO; 05°44′N 61°30′W. Diamond-bearing area. Airfield.

Luesia (LWAI-syah), village, ZARAGOZA province, N SPAIN, 36 mi/58 km NW of HUESCA; 42°22′N 01°01′W. Tanning; cereals, wine; livestock, lumber.

Lueta (loo-E-tah), Hungarian *Lövéte*, village, HARGHITA county, central ROMANIA, 10 mi/16 km ESE of ODORHEIU-SECUIESC; 46°16′N 25°29′E. Iron mining; ironworking. In HUNGARY, 1940–1945.

Lueyang (LOOI-E-YANG), town, ⊙ Lueyang county, SW SHAANXI province, CHINA, on upper JIALING River, and 65 mi/105 km WNW of Nanzheng, in

mountain region; 33°20′N 106°09′E. Electric power generation, iron smelting.

Lufeng (LU-FENG), town, ⊙ Lufeng county, SE GUANGDONG province, CHINA, on coast, 70 mi/113 km WSW of SHANTOU; 22°57′N 115°38′E. Rice, wheat, sugarcane, oilseeds; food industry, crafts, plastics; machinery and equipment manufacturing.

Lufeng, town, ⊙ Lufeng county, N central YUNNAN province, CHINA, 35 mi/56 km W of KUNMING; 25°08′N 102°05′E. Elevation 5,418 ft/1,651 m. Rice, wheat, millet, tobacco; chemicals; salt and coal mining; iron smelting.

Lufira River (loo-FEE-rah), c.300 mi/483 km long, SE CONGO; rises on KATANGA highlands near ANGOLA border 70 mi/113 km WNW of LUBUMBASHI; flows NE and NNW to join LUALABA RIVER in S end of LAKE KISALE, 70 mi/113 km NE of BUKAMA. Many rapids throughout its course. A major water reservoir has been formed in its lower middle valley c.20 mi/32 km E of LIKASI; CORNET FALLS are the site of the largest hydroelectric plant in CONGO.

Lufkin (LUHF-kin), city (2006 population 33,863), ⊙ ANGELINA county, E TEXAS, 115 mi/185 km NNE of HOUSTON; 31°19′N 94°43′W. Elevation 328 ft/100 m. Forest industries with many sawmills; the first plant to make newsprint from native pine. Railroad junction; manufacturing (pumping units, iron castings, trailer parts, carbonated beverages, electronics for missiles, industrial chrome plating); agriculture (cattle, poultry; hay). Fuller's earth is found in the region. Angelina College (two-year), and a state school for the mentally retarded are in Lufkin. Angelina National Forest to E, David Crockett National Forest to W, and SAM RAYBURN RESERVOIR. Incorporated 1890.

Luga (LOO-gah), city (2005 population 40,015), SW LENINGRAD oblast, NW European Russia, on the LUGA RIVER, 86 mi/138 km SSW of SAINT PETERSBURG; 58°44′N 29°50′E. Elevation 137 ft/41 m. Railroad and highway junction; foundry shop, abrasives, chemicals, and microbiological products; knitwear. Has an agricultural institute. Chartered in 1777.

Luganka River, UKRAINE: see LUHAN' RIVER.

Lugano, district, S TICINO canton, S SWITZERLAND; 46°00′N 08°58′E. Main town is LUGANO; Italian-speaking and Roman Catholic.

Lugano, town (2000 population 48,478), TICINO canton, S SWITZERLAND, near the Italian border, and c.15 mi/24 km SSW of BELLINZONA; 46°01′N 08°58′E. Elevation 886 ft/270 m. A commercial center in the Middle Ages; popular scenic resort. Tourism; international banking. Other industries are machinery, textiles, tobacco products, and chocolate. The Swiss cantons took Lugano from MILAN in 1512, but the town has retained its Italian character. Notable buildings are the Romanesque Cathedral of San Lorenzo and the 15th-century Monastery of Santa Maria degli Angioli, with its frescoes by Bernardino Luini. A fine art museum is here. The town is situated on LAKE LUGANO (Italian *Lago di Lugano* or *Ceresio*), narrow and irregular in shape, which lies between Switzerland and ITALY.

Lugano, Lake of, Italian *Lago di Lugano* (LAH-go dee loo-GAH-no) or *Lago Ceresio* (chai-RAI-zyo) (□ 19 sq mi/49.4 sq km), between ITALY and SWITZERLAND, and between LAGO MAGGIORE and LAKE COMO; 45°58′N 09°00′E. Narrow, very irregular in shape; c.20 mi/32 km long, elevation 789 ft/240 m, maximum depth 944 ft/288 m. Numerous mountain streams fall into lake; drained by short Tresa River into Lago Maggiore. Fine scenery on shores near Lugano; NE arm is bounded by steep, rocky mountains. Bridge between Melide (W) and Bissone (E) connects two banks. Main town on lake, LUGANO.

Lugan' River, E UKRAINE: see LUHAN' RIVER.

Lugansk, Ukrane: see LUHANS'K, OBLAST.

Lugansk, UKRAINE: see LUHANS'K, city.

Lugansk, UKRAINE: see STANYCHNO-LUHANS'KE.

Luganskaya oblast, UKRAINE: see LUHANS'K oblast.

Luganskiy zapovednik, UKRAINE: see LUHANS'K NATURE PRESERVE.

Luganskoye, UKRAINE: see LUHANS'KE, Donets'k oblast.

Luganskoye, UKRAINE: see PAVLOHRAD.

Luganville (LOO-gahn-vil), town, VANUATU, SW PACIFIC OCEAN. Second largest town and port of Vanuatu (after PORT VILA), in SE ESPÍRITU SANTO; 15°32′S 167°10′E. Many World War II structures and relics, as in nearby Million Dollar Point. Located on N shore of SEGOND CANAL, facing Aore Island Airfield. Also called Santo or Santo Town.

Lugareño (loo-gahr-AIN-yo), sugar-mill village of SIERRA DE CUBITAS, CAMAGÜEY province, E CUBA, on railroad, and 28 mi/45 km ENE of CAMAGÜEY; 21°34′N 77°28′W.

Lugari (loo-GAH-rai), village, WESTERN province, W KENYA, on railroad, and 30 mi/48 km WNW of ELDORET; 00°39′N 34°54′E. Coffee, tea, wheat, sisal, corn.

Luga River (LOO-gah), 215 mi/346 km long, NW European Russia; rises in marshes NW of NOVGOROD, on the border of NOVGOROD and LENINGRAD oblasts; flows S into the NW corner of Novgorod oblast, and then NW across Leningrad oblast, past LUGA and KINGISEPP, to the Luga Bay of the Gulf of FINLAND at UST'-LUGA. Forms rapids; navigable (April–December) for 85 mi/137 km above its mouth. Receives the Oredezh River (right).

Lugar Water (LOO-gahr), river (15 mi/24 km long), East Ayrshire, S Scotland; rises 3 mi/4.8 km WSW of MUIRKIRK; flows generally W, past Lugar, Cumnock, and Ochiltree, to Ayr River just SSW of Mauchline. Formerly in Strathclyde, abolished 1996.

Lugasi (loo-GAH-see), town, CHHATARPUR district, NE MADHYA PRADESH state, central INDIA 8 mi/12.9 km E of NOWGONG; 25°05′N 79°35′E. Former capital of petty state of Lugasi of Central India agency; since 1948, state merged with VINDHYA PRADESH and later with Madhya Pradesh state.

Lugau (LOO-gou), town, SAXONY, E central GERMANY, at N foot of the ERZGEBIRGE, 11 mi/18 km SW of CHEMNITZ; 50°45′N 12°45′E. Manufacturing of steel, machinery. Coal was mined until 1960.

Lugazi (loo-GAH-zee), town (2002 population 27,979), MUKONO district, CENTRAL region, S UGANDA, 26 mi/42 km E of KAMPALA.; 00°23′N 32°57′E. Sugar mill; cotton, coffee, sugarcane. Was part of former Buganda province.

Lügde (LOOG-de), town, NORTH RHINE-WESTPHALIA, NW GERMANY, 2 mi/3.2 km S of BAD PYRMONT; 51°57′N 09°14′E. Forestry; manufacturing of furniture, electrical goods; canning. Has twelfth-century church.

Lugdunum, FRANCE: see LYON.

Lugela (loo-GE-luh), village, Zambézia province, central MOZAMBIQUE, 95 mi/153 km N of QUELIMANE; 17°30′S 37°16′E.

Lugela District, one of 15 districts of Zambézia, MOZAMBIQUE: see ZAMBÉZIA, province.

Lugenda River (loo-GEN-duh), 300 mi/483 km long, N MOZAMBIQUE; rises in Lake CHIUTA (MALAWI border); flows NE to the ROVUMA (TANZANIA border), at 11°24′S 38°30′E. Not navigable.

Lugg River (LUHG), E Wales and England; rises in POWYS 7 mi/11.3 km W of KNIGHTON; flows 40 mi/64 km SE into Herefordshire, past LEOMINSTER. Receives ARROW RIVER just SSE of Leominster, and FROME RIVER 3 mi/4.8 km E of HEREFORD, before joining the WYE, 4 mi/6.4 km ESE of Hereford.

Lugh Ferrandi, SOMALIA: see LUK.

Lugia, ETHIOPIA: see LOGIYA.

Lugin, UKRAINE: see LUHYNY.

Lugino, UKRAINE: see LUHYNY.

Luginy, UKRAINE: see LUHYNY.

Lugnaquilla (luhg-nuh-KWI-luh), mountain (3,039 ft/926 m), central WICKLOW county, E IRELAND, 18 mi/29 km W of WICKLOW; 52°58′N 06°27′W.

Lugnvik (LUGN-VEEK), village, JÄMTLAND county, N central SWEDEN, on E shore of STORSJÖN LAKE, 2 mi/3.2 km N of ÖSTERSUND; 63°12′N 14°38′E.

Lugo (LOO-go), province (□ 3,815 sq mi/9,919 sq km; 2001 population 357,648), NW SPAIN, in GALICIA, on BAY OF BISCAY; ⊙ LUGO. Crossed by GALICIAN MOUNTAINS; has indented, rocky coast line. Drained by MINHO RIVER and its tributaries flowing S, and by several short rivers (including DEVA) flowing N to Bay of Biscay. Temperate climate with abundant rainfall near coast; cold with less rain in interior. Widely scattered population, with few towns; poor communications. Of mineral deposits (iron, antimony, copper, coal), only iron mines near VILLAODRID and VIVEIRO are exploited. Chief resources: livestock raising (cattle, horses, hogs) and lumbering; important fisheries along coast. Agriculture (potatoes, rye, corn, nuts, vegetables, and honey); vineyards in Minho valley. Chief cities/towns: Lugo, MONFORTE, Viveiro.

Lugo (LOO-go), city (2001 population 88,414), ⊙ LUGO province, NW SPAIN, in GALICIA, on the MINHO RIVER, and c.60 mi/96 km NW of PONFERRADA; 43°00′N 07°34′W. The city is the processing and trade center for a fertile farm area. Slaughterhouse, refrigerator factory, livestock derivatives industry; lumbering. It has well-preserved Roman walls (3rd century B.C.E.) and a 12th-century cathedral (restored).

Lugo (LOO-go), town, RAVENNA province, EMILIA-ROMAGNA, N central ITALY, near SENIO River, 14 mi/23 km W of RAVENNA. Railroad junction; commercial center; manufacturing (fabricated metals, machinery, alcohol, barrels, clothing, shoes, soap, wax, explosives); food processing. Has ancient castle.

Lugoba (loo-GO-bah), village, PWANI region, E TANZANIA, 65 mi/105 km WNW of DAR ES SALAAM; 06°30′S 38°14′E. Highway junction. Sisal, corn; goats; sheep; timber.

Lugo Bridge, CHINA: see FENGTAI.

Lugoff (LOOG-ahf), unincorporated town (2000 population 6,278), KERSHAW county, N central SOUTH CAROLINA, 5 mi/8 km W of CAMDEN, near WATEREE River; 34°13′N 80°40′W. Manufacturing includes tools, chassis, synthetic fibers, contract embroidery, laundry bags, machining, industrial gases, women's sportswear; vessels and tanks; silica sand processing. Agriculture (poultry, cattle, grain, and cotton). Wateree Dam and reservoir to NW; Lake Wateree State Park in SW shore.

Lugogo (LOO-go-go), river, rises from highlands of S central UGANDA; flows c.80 mi/129 km NW, and joins KAFU RIVER. Drains CENTRAL region.

Lugoj (loo-GOZH), Hungarian *Lugos*, city, TIMIȘ county, W ROMANIA, in BANAT, on TIMIȘ RIVER, 33 mi/53 km ESE of TIMIȘOARA; 45°40′N 21°54′E. Railroad hub and commercial center (trade in grain, fruit, and livestock). Produces textiles (notably silk), leather goods, spraying equipment, bricks, tiles, and paper; flour milling, distilling. Experimental sericulture station; extensive vineyards in vicinity. Has several old churches, a baroque 18th-century Minorite church, remains of old monastery; museum with Roman mementos. Originally Roman fortress, it was a royal city in 14th century; chartered 1428; occupied by Turks (1658–1695). Was the political and cultural center of Romanians in Banat throughout 19th century. Orthodox and Uniate bishoprics.

Lugones (loo-GO-nes), outer NE suburb of OVIEDO, OVIEDO province, NW SPAIN; 43°24′N 05°48′W. Copper and tin processing; chemical works.

Lugos, ROMANIA: see LUGOJ.

Lugovaya (loo-guh-VAH-yah), village (2006 population 3,080), central MOSCOW oblast, central European Russia, on railroad, 8 mi/13 km N of DOLGOPRUDNYY, and 3 mi/5 km S of LOBNYA; 56°03′N 37°29′E. Elevation 643 ft/195 m. In agricultural area. Research institute of the Russian Academy of Agrosciences is located here.

Lugo-Vodyanskoye, RUSSIA: see PRIMORSK, VOLGOGRAD oblast.

Lugovoi (loo-guh-VOI), town, SW ZHAMBYL region, KAZAKHSTAN, on TURK–SIB RAILROAD, 70 mi/113 km E of ZHAMBYL; 42°54′N 72°45′E. Tertiary-level (raion) administrative center. Irrigated agricultural area (wheat); metalworks. Junction of railroad branch line to BISHKEK and Issyk-Kol (KYRGYZSTAN).

Lugovoy (loo-guh-VO-yee), settlement (2005 population 2,010), SW KHANTY-MANSI AUTONOMOUS OKRUG, W central Siberian Russia, near Tuman Lake, on road, 41 mi/66 km Se of URAY; 59°44′N 65°50′E. Elevation 137 ft/41 m. Logging, lumbering.

Lugovskiy (LOO-guhf-skeeyee), settlement (2005 population 611), NE IRKUTSK oblast, E central SIBERIA, RUSSIA, on road, 15 mi/24 km S of MAMA; 58°03′N 112°53′E. Elevation 1,466 ft/466 m. Logging, timbering. Formerly known as Ust'-Lugovka.

Lugovskoy (loo-guhfs-KO-yee), settlement (2006 population 2,955), SE SVERDLOVSK oblast, W Siberian Russia, in the PYSHMA RIVER basin, on road and railroad, 7 mi/11 km S of (and administratively subordinate to) TUGULYM; 56°57′N 64°31′E. Elevation 193 ft/58 m. Logging, lumbering.

Lugufu (loo-GOO-foo), village, KIGOMA region, W TANZANIA, 35 mi/56 km ESE of KIGOMA, on railroad; 05°01′S 30°05′E. Corn, wheat; sheep, goats.

Lugufu River (loo-GOO-foo), c. 125 mi/201 km long, W TANZANIA; rises in NW RUKWA region, c.100 mi/161 km SE of KIGOMA; flows generally NW SSE of Kigoma.

Luguru (loo-goo-ROO), village, MWANZA region, NW TANZANIA, 75 mi/121 km ESE of MWANZA, near SIMIYU RIVER; 03°14′S 33°55′E. Cotton, corn, wheat; cattle, sheep, goats.

Lugus Island (LOO-goos), □ 14.8 sq mi/38.3 sq km, in TAPUL GROUP, SULU province, PHILIPPINES, in SULU ARCHIPELAGO; 05°41′N 120°50′E.

Luhacovice (LU-hah-CHO-vi-TSE), Czech *Luhačovice*, German *Luhatschowitz*, town, JIHOMORAVSKY province, E MORAVIA, CZECH REPUBLIC, in W foothills of the White Carpathians, 9 mi/14.5 km SSE of Zlín; 49°06′N 17°46′E. Railroad terminus. Popular health resort with alkaline muriatic springs; sulphur and peat baths, among extensive forests. Known since 16th century; bathing facilities were established in 1790.

Luhaiya (loo-HAI- yuh), town, W YEMEN, minor port on RED SEA, 65 mi/105 km N of HODEIDA; trade in coffee and agricultural products, fishing; rock-salt deposits. Also spelled Loheia and Luhayyah.

Luhanka River, UKRAINE: see LUHAN' RIVER.

Luhan' River (loo-HAHN) (Russian *Lugan'*), river, E UKRAINE; rises 11 mi/18 km NE of HORLIVKA (DONETS'K oblast); flows 123 mi/198 km, first NE past PERVOMAYS'K and KIROVS'K, then E past LUHANS'K, to DONETS River, 7 mi/11 km E of Luhans'k. Also called Luhanka (Russian *Luganka*).

Luhans'k (loo-HAHNSK) (Russian *Lugansk*), city (2001 population 463,097), ⊙ LUHANS'K oblast, E UKRAINE, at the confluence of the LUHAN' and Vilkhivka rivers, in the DONETS BASIN; 48°34′N 39°20′E. Elevation 347 ft/105 m. A major industrial center, it produces locomotives, processed coal, chemicals, steel pipes, mining equipment, machine tools, textiles, and foodstuffs. Has fourty scientific research and project construction institutes, four post-secondary institutes, seventeen vocational technical schools, two theaters, a circus, and four museums. Founded in 1795 around a cannon foundry, and is the oldest center of the Donets Basin. A city since 1882, Luhans'k was called Voroshylovhrad (Russian *Voroshilovgrad*) from 1935 to 1958 and Luhans'k from 1958 to 1970. Renamed Voroshylovhrad in 1970, it was once again renamed Luhans'k in 1990. Also called, in Ukrainian, until 1935, Luhans'ke. Sizeable Jewish community since 1878, including one of the largest Hasidic communities in Eastern Ukraine; numbered over 7,100 overall in 1939;

the majority survived World War II through evacuation, and at least half returned to the city throughout the late 1940s; close to 10,000 Jews in the city in 2005.

Luhans'k, UKRAINE: see STANYCHNO-LUHANS'KE.

Luhans'ka oblast, UKRAINE: see LUHANS'K oblast.

Luhans'ke (loo-HAHN-ske) (Russian *Luganskoye*), town (2004 population 6,100), E DONETS'K oblast, UKRAINE, in the DONBAS, on the Myronovs'kyy Reservoir of the LUHAN' RIVER, right tributary of the DONETS, 15 mi/24 km N of YENAKIYEVE; 48°25′N 38°16′E. Elevation 554 ft/168 m. Metalworks. Established in 1701, town since 1938. Until 1922, called Pyatnadtsyata Rota (Ukrainian=Fifteenth Company).

Luhans'ke, UKRAINE: see LUHANS'K, city.

Luhans'k Nature Preserve (loo-HAHNSK) (Ukrainian *Luhans'kyy zapovidnyk*) (Russian *Luganskiy zapovednik*), preserve (□ 4,000 acres/1,600 ha), in three separate locations of LUHANS'K oblast, UKRAINE. Established in 1968. Comprises three divisions: Stril'tsivs'kyy Step, [Ukrainian=Riflemen's steppe], Proval's'kyy Step [Ukrainian=Precipice steppe], and Stanychno-Luhans'ke division. The Stril'tsivs'kyy Step, in NE Luhans'k oblast, near the border with the Russian Federation, 6 mi/9.7 km SW of MILOVE, on S outliers of the CENTRAL RUSSIAN UPLAND, is representative of feathergrass steppe. The Proval's'kyy Step, on the DONETS RIDGE in SE Luhans'k oblast, 8 mi/13 km, and 12 mi/19 km NE of SVERDLOVS'K, contains two parcels of grass-clad upland with ravines supporting indigenous forests. The Stanychno-Luhans'ke plot, on the left bank of the DONETS River floodplain, 6 mi/10 km NW of Stanychno-Luhans'ke, consists of a riverine complex with wood, meadows, marsh, aquatic, and sandy-steppe vegetation. Studies involve researching and analyzing changes in the steppe ecosystems as impacted by human activities.

Luhans'k oblast (loo-HAHNSK) (Ukrainian *Luhans'ka*) (Russian *Luganskaya*), oblast (□ 10,303 sq mi/26,787.8 sq km; 2001 population 2,546,178), E UKRAINE; ⊙ LUHANS'K. Includes SE parts of CENTRAL RUSSIAN UPLAND (N) and DONETS RIDGE (S); drained by DONETS River and its affluent, Aydar River. Population is predominantly Ukrainian (51.9%) and Russian (44.8%), with Belorussian (1.2%), Tatar (0.4%), Jewish (0.3%), Moldavian (0.2%), and other minorities. Rich agricultural area (N) with barley, wheat, corn for grain and silage, sunflowers; vegetable produce and orchards, suburban dairy, broilers (S). S area forms part of highly industrialized and urbanized DONETS BASIN, with chief coal-mining centers at ANTRATSYT, KRASNODON, LYSYCHANS'K, ROVEN'KY, STAKHANOV, and SVERDLOVS'K; steel, metallurgy and manufacturing industries at Luhans'k, Stakhanov, and ALCHEVS'K. Also chemicals, glass, cement, food processing. Dense railroad network(s). Formed 1938 as Voroshylovhrad (Ukrainian *Voroshylovhrads'ka*) oblast (Russian *Voroshilovgradskaya*); re-named Luhans'k in 1958; reverted to Voroshylovhrad in 1970; again reverted to Luhans'k in 1990. In 1993 had thirty seven cities, 109 towns, and eighteen rural raions.

Luhans'kyy zapovidnyk, UKRAINE: see LUHANS'K NATURE PRESERVE.

Luhatschovitz, CZECH REPUBLIC: see LUHACOVICE.

Luhayyah, YEMEN: see LUHAIYA.

Luhe (LU-HUH), town, ⊙ Luhe county, N JIANGSU province, CHINA, 18 mi/29 km NNE of NANJING, across CHANG JIANG (Yangzi River); 32°20′N 118°52′E. Rice, medicinal herbs, oilseeds.

Luhuo (LU-HO), town, ⊙ Luhuo county, W SICHUAN province, CHINA, 125 mi/201 km NW of KANGDING, and on highway; 31°24′N 100°38′E. Logging.

Luhyn, UKRAINE: see LUHYNY.

Luhyne, UKRAINE: see LUHYNY.

Luhyny (loo-HI-nee) (Russian *Luginy*), town, N ZHYTOMYR oblast, UKRAINE, 12 mi/19 km NW of Korosten'; 51°05′N 28°24′E. Elevation 570 ft/173 m. Raion center;

flour milling; sawmilling; sewing. Formerly called Luhyn (Russian *Lugin*) or Luhyne (Russian *Lugino*). First mentioned in 1606, town since 1967.

Lui (loo-EE), river, 150 mi/241 km long, N ANGOLA; rises in northern highlands; flows N, joining the CUANGO RIVER near Democratic Republic of the CONGO border.

Luiana (loo-EE-ah-nuh), town, CUANDO CUBANGO province, ANGOLA, along CUANDO RIVER, close to ZAMBIA border, 380 mi/612 km SE of Menogue; 17°15′S 22°50′E. Market center.

Luia River, ZIMBABWE: see RUYA RIVER.

Luichart, Loch (LOO-wee-kahrt, LAWK), lake (5 mi/8 km long, 1 mi/1.6 km wide, 164 ft/50 m deep), HIGHLAND, N Scotland, 17 mi/27 km W of DINGWALL; 57°36′N 04°45′W. Fed and drained by CONON RIVER, reservoir of North of Scotland Hydroelectric Board. Power station below dam.

Luigi Razza, LIBYA: see MESSA.

Luik, BELGIUM: see LIÉGE.

Luilaka River, CONGO: see MOMBOYO RIVER.

Luina, settlement, ghost town, NW TASMANIA, SE AUSTRALIA, 81 mi/131 km N of QUEENSTOWN, 40 mi/64 km SW of BURNIE, 18 mi/29 km W of Murchison Highway, and on Whyte River. Area of tin-mining boom late 1800s. Mining revival during 1980s here and in Savage River and North Valley near WARATAH.

Luing (LOO-eeng), island (6 mi/9.7 km long, 2 mi/3.2 km wide; rises to 306 ft/93 m) of the INNER HEBRIDES, Argyll and Bute, W Scotland, in Firth of Lorn, just S of SEIL island, and 12 mi/19 km SSW of OBAN; 56°12′N 05°40′W. Cattle breeding. Just off W coast is islet with lighthouse. Formerly in Strathclyde, abolished 1996.

Luino (LWEE-no), town, VARESE province, LOMBARDY, N ITALY, port on E shore of LAGO MAGGIORE, 13 mi/21 km NNW of VARESE; 46°00′N 08°44′E. Resort; secondary diversified industrial center. Silk, rayon, and cotton mills; manufacturing of sewing machines, fabricated metals, shoes, pharmaceuticals. Fish hatchery.

Luís Alves (LOO-ees AHL-ves), town, NE SANTA CATARINA state, BRAZIL, 28 mi/45 km SSW of JOINVILLE; 26°44′S 48°57′W. Rice, manioc; livestock. Also spelled Luiz Alves.

Luisant (lwee-zahn), S suburb (□ 1 sq mi/2.6 sq km) of CHARTRES, EURE-ET-LOIR department, CENTRE administrative region, NW central FRANCE, in the BEAUCE wheat-growing district; 48°25′N 01°29′E.

Luís Antônio (LOO-ees ahn-TO-nee-o), town (2007 population 10,272), N SÃO PAULO state, BRAZIL, 28 mi/45 km S of Riberão Preto; 21°30′S 47°41′W.

Luis Calvo, province, CHUQUISACA department, SE BOLIVIA; ⊙ Villa Vaca Guzmán (Muyupampa); 20°40′S 63°30′W.

Luís Correia (LOO-ees ko-rai-ah), city (2007 population 26,178), extreme N PIAUÍ, BRAZIL, on the Atlantic, 9 mi/14.5 km ENE of PARNAÍBA; 02°53′S 41°40′W. Limited modern dock facilities; hydroplane landing. Saltworks. Until 1939, called Amarração. Formerly also spelled Luiz Correia.

Luís Domingues (loo-EES do-meen-ges), town (2007 population 6,675), N MARANHÃO, BRAZIL, near the Atlantic and near PARÁ border, 130 mi/209 km NW of SÃO LUÍS, in gold-mining area; 01°17′S 45°46′W.

Luís Gomes (LOO-ees GO-mes), city (2007 population 9,734), SW RIO GRANDE DO NORTE, NE BRAZIL, 35 mi/56 km NNE of CAJAZEIRAS (PARAÍBA); 06°25′S 38°23′W. Cotton, corn, rice, manioc, hides. Formerly spelled Luiz Gomes.

Luishia (LWEESH-yah), village, KATANGA province, SE CONGO, on railroad, and 20 mi/32 km SE of LIKASI; 11°10′S 27°02′E. Elev. 4,232 ft/1,289 m. Copper- and cobalt-mining center. Also copper and cobalt mines at KAMVALI, 10 mi/16 km NNE; copper mines at KANSONGWE, 5 mi/8 km NNW, and SHANDWE, 6 mi/9.7 km ENE.

Luisiana (lwees-YAH-nah), town, LAGUNA province, S LUZON, PHILIPPINES, 15 mi/24 km NE of SAN PABLO;

14°11′N 121°31′E. Agricultural center (rice, coconuts, sugarcane).

Luisiana, La (lwees-YAH-nah, lah), town, SEVILLE province, SW SPAIN, on railroad, and 40 mi/64 km ENE of SEVILLE; 37°32′N 05°15′W. Olives, cereals, livestock (hogs, goats, sheep); vegetable-oil extracting.

Luisiânia (LOO-ee-see-ah-nee-ah), town, W SÃO PAULO state, BRAZIL, 42 mi/68 km NW of MARÍLIA; 21°41′S 50°17′W. Coffee growing. Also spelled Luiziânia.

Luislândia (loo-ees-LAHN-zhee-ah), town (2007 population 6,432), N central MINAS GERAIS, BRAZIL, 19 mi/31 km SE of SÃO FRANCISCO; 16°12′S 44°33′W.

Luis Moya (LOO-ees MO-yah), town, ZACATECAS, N central MEXICO, 30 mi/48 km SE of ZACATECAS, on Mexico Highway 45; 22°28′N 102°17′W. Maguey, grain, beans, livestock. Formerly San Francisco de los Adame.

Luiswishi (lwee-SWEE-shee), village, KATANGA province, SE CONGO, on railroad, and 10 mi/16 km N of Lubambashi; 11°31′S 28°12′E. Copper and cobalt mining.

Luitpold Coast (LOOT-puhld), ANTARCTICA, forms part of COATS LAND, on WEDDELL SEA; extends between 28° and 36°W. Discovered 1912 by Wilhelm Filchner, German explorer.

Luiza (lwee-ZAH), village, KASAI-OCCIDENTAL province, S CONGO, on LULUA RIVER, and 150 mi/241 km SW of KABINDA, near ANGOLA border; 07°12′S 22°25′E. Elev. 2,857 ft/870 m. Cotton ginning, cottonseed-oil milling. Also spelled Luisa.

Luján (loo-HAHN), city (1991 population 66,226), ⊙ Luján district (□ 300 sq mi/780 sq km), NE BUENOS AIRES province, ARGENTINA, on LUJÁN RIVER, and 40 mi/64 km W of BUENOS AIRES; 34°34′S 59°06′W. Railroad junction and agricultural center (grain; livestock). A noted pilgrimage city, it has Gothic basilica of Our Lady of Luján, and museum of colonial history, several libraries and higher schools (including agricultural college). Founded 1630.

Luján (loo-HAHN) or **Luján de Cuyo**, town, ⊙ Luján de Cuyo department (□ 1,965 sq mi/5,109 sq km), N MENDOZA province, ARGENTINA, on MENDOZA RIVER (irrigation), and 11 mi/18 km S of MENDOZA. Railroad junction. Agricultural and manufacturing center. Agriculture (wine, alfalfa, fruit, potatoes, grain). Irrigation dam and hydroelectric plant nearby.

Luján, town, N SAN LUIS province, ARGENTINA, at foot of SIERRA DE SAN LUIS, 70 mi/113 km NNE of SAN LUIS. Agriculture center (grain, goats, cattle).

Luján de Cuyo, ARGENTINA: see LUJÁN, MENDOZA province.

Luján River (loo-HAHN), c.75 mi/121 km long, NE BUENOS AIRES province, ARGENTINA; rises near SUIPACHA 15 mi/24 km SW of MERCEDES; flows ENE, past Mercedes and LUJÁN, to the Río de la PLATA at PARANÁ delta at TIGRE, 16 mi/26 km NNW of BUENOS AIRES. Navigable for small steamers.

Lujeni, UKRAINE: see LUZHANY.

Lujiang (LU-JIANG), town, ⊙ Lujiang county, N ANHUI province, CHINA, 45 mi/72 km S of HEFEI, SW of CHAO Lake; 31°14′N 117°17′E. Rice, cotton, wheat, oilseeds, jute; food processing; machinery and equipment manufacturing, chemicals.

Luk or **Lugh**, township, SW SOMALIA, on the JUBBA River near ETHIOPIA; 03°48′N 42°34′E. Road junction. Agricultural region (sorghum, maize). Trade center. Formerly LUGH FERRANDI.

Lukachëk (loo-kah-CHOK), settlement, E AMUR oblast, RUSSIAN FAR EAST, near the SELEMDZHA RIVER, on road, 10 mi/16 km WNW of EKIMCHAN; 53°03′N 132°16′E. Elevation 1,535 ft/467 m. Gold mines. Developed during World War II.

Lukachukai (loo-kuh-choo-kei), unincorporated village, APACHE county, NE ARIZONA, 117 mi/188 km NNE of HOLBROOK, Arizona, and 36 mi/58 km SW of SHIPROCK, NEW MEXICO, on Lukachukai Creek, in

Navajo Indian Reservation. Elevation 6,450 ft/1,966 m. Cattle, sheep; crafts. CHUSKA MOUNTAINS to NE. CANYON DE CHELLY NATIONAL MONUMENT to S.

Lukala (loo-KAH-lah), village, BAS-CONGO province, W CONGO, on railroad, and 100 mi/161 km ENE of BOMA; 05°31′S 14°32′E. Elev. 1,761 ft/536 m. Cement works.

Lukang (LOO-GANG), town, W central TAIWAN, minor port on W coast, 7 mi/11.3 km WSW of CHANGHUA; 24°03′N 120°25′E. Saltworks; incense manufacturing, sugar refining, fish processing. Formerly a leading port of Taiwan (17th century to late 19th century); closed during Japanese occupation. It still has many original buildings of historical interest. Manufacturing of traditional Chinese furniture and religious artifacts. Also spelled Lu-chiang.

Lukashëvka (loo-kah-SHOF-kah), village, W central KURSK oblast, SW European Russia, on road and railroad, 25 mi/40 km W of KURSK; 51°38′N 35°36′E. Elevation 666 ft/202 m. Distilling.

Lukavac (LOO-kah-vahts), town, N BOSNIA, BOSNIA AND HERZEGOVINA, in Tuzla valley.

Lukavytsya, UKRAINE: see OBUKHIV.

Lukaya (LOO-kei-yah), town (2002 population 14,147), MASAKA district, CENTRAL region, S UGANDA, 15 mi/25 km NE of MASAKA; 00°08′S 31°53′E. Trade center.

Luke, town (2000 population 80), ALLEGANY county, W MARYLAND, on North Branch of the POTOMAC at mouth of SAVAGE RIVER, and 20 mi/32 km SW of CUMBERLAND; 39°29′N 79°04′W. The town is named for the Luke family, which began a paper mill here in 1888 that is now one of the largest in the world. The mill turns out 1,000 tons/1,102 metric tons of high quality paper using 60 million gals/227.1 million liters of water from the Potomac every day, 24 hours a day.

Lukenie River, CONGO: see FIMI RIVER.

Lukeville, unincorporated village, PIMA county, SW ARIZONA, 34 mi/55 km S of AJO, on Mexican border, opposite SONOITA, SONORA, in S part of ORGAN PIPE CACTUS NATIONAL MONUMENT. Elevation 1,814 ft/553 m. Gulf of CALIFORNIA (PUERTO PEÑASCO) 45 mi/72 km to SW.

Lukh (LOOKH), town (2005 population 3,120), E central IVANOVO oblast, central European Russia, on the Lukh River (tributary of the KLYAZ'MA RIVER), 58 mi/93 km E of IVANOVO, and 19 mi/31 km SE of VICHUGA; 57°05′N 42°15′E. Elevation 364 ft/110 m. Food processing (vegetable drying); dairying; textiles; furniture making; brick making. Has a heritage museum. Known from the 14th century, when in Nizhegorod-Suzdal' principality until joining the Moscow principality in 1392. Sacked by the Tatar-Mongols in 1428; attacked during the combined Polish-Lithuanian force in 1608, during a series of short Russian wars of succession. Made city in 1778 as part of Kostroma guberniia; reduced to status of a village, 1925–1959; made town in 1959. Historical landmarks include the Svyato-Troitskiy (Holy Trinity) Cathedral and a number of 16th- and 17th-century private residences and public buildings.

Lukhovitsy (loo-huh-VEE-tsi), city (2006 population 32,385), SE MOSCOW oblast, central European Russia, on road and railroad, 84 mi/135 km SE of MOSCOW, and 15 mi/24 km SE of KOLOMNA; 54°56′N 39°01′E. Elevation 498 ft/151 m. Railroad junction; stone quarries; clothing, dairy products, flour milling. Made city in 1957.

Lukhovka (LOO-huhf-kah), town (2006 population 8,825), central MORDVA REPUBLIC, central European Russia, on road, 3 mi/5 km SE of SARANSK, to which it is administratively subordinate; 54°09′N 45°14′E. Elevation 524 ft/159 m. Engine repair facilities for passenger and industrial vehicles. Site of an experimental farm of Mordva State University.

Luki (LOO-kee), city (1993 population 3,528), PLOVDIV oblast, Luki obshtina, BULGARIA, at Dzhurkovska and Manastirska rivers; 41°50′N 24°50′E. Hunting resort; lead and zinc nearby.

Area is shown by the symbol □, and capital city or county seat by ⊙.

Lukino (LOO-kee-nuh), village (2006 population 3,600), W NIZHEGOROD oblast, central European Russia, on the VOLGA River, on railroad, 14 mi/23 km NW of NIZHNIY NOVGOROD, and 3 mi/5 km S of BALAKHNA, to which it is administratively subordinate; 56°26′N 43°38′E. Elevation 265 ft/80 m. Agricultural products.

Lukino (LOO-kee-nuh), urban settlement, central MOSCOW oblast, central European Russia, on highway and near railroad, 2 mi/3.2 km N, and under administrative jurisdiction, of BALASHIKHA; 55°50′N 37°56′E. Elevation 482 ft/146 m. Paints; tin works.

Lukiv (LOO-keef) (Russian *Lukov*), town (2004 population 3,900), W central VOLYN′ oblast, UKRAINE, on railroad, 16 mi/26 km W of KOVEL′; 51°03′N 25°24′E. Elevation 547 ft/166 m. Chalk- and kaolin-quarrying center; tanning, household goods; brick manufacturing. Has old monastery, churches. Until 1946, known as Matseyiv (Polish *Maciejów*, Russian *Matseyevo*). Known since 1430; town since 1940.

Lukla (LUHK-lah), village, central NEPAL; 27°43′N 86°43′E. Elevation 9,100 ft/2,774 m. Site of the STOL airport serving the Everest (KHUMBU) region, a major tourist destination.

Lukmanier Pass (LOOK-mahn-yuhr), Romansch *Cuolm Lucmagn*, Italian *Paso del Lucomagno* (6,280 ft/1,914 m), in the LEPONTINE ALPS, SE central SWITZERLAND, between Val Medel in GRISONS canton and Val Blenio in TICINO canton, 12 mi/19 km S of DISENTIS.

Luknovo (LOOK-nuh-vuh), town (2006 population 2,595), NE VLADIMIR oblast, central European Russia, on road and near railroad, 7 mi/11 km SSW of VYAZNIKI; 56°12′N 42°02′E. Elevation 465 ft/141 m. Linen spinning and weaving.

Lukolela (loo-ko-LE-lah), village, Équateur province, W CONGO, on left bank of CONGO RIVER opposite LOUKOLELA, and 115 mi/185 km SSW of MBANDAKA; 01°03′S 17°12′E. Elev. 938 ft/285 m. Trading and agricultural center; steamboat landing; cacao and rubber plantations, hardwood lumbering. Has Roman Catholic and Baptist missions. Established in 1883 by Stanley, it was one of the first trading posts in Central Africa.

Lukoma Bay (loo-KO-mah), inlet, Lake NYASA, RUVUMA region, S TANZANIA, just N of MOZAMBIQUE border; c. 20 mi/32 km wide. Main village is CHIWANDA.

Lukonde (loo-KAWN-dai), village, KATANGA province, SE CONGO, near MANIKA PLATEAU (UPEMBA NATIONAL PARK); 09°54′S 27°28′E. Elev. 2,893 ft/881 m.

Lukongole (loo-kon-GO-lai) or **Lukongole Camp**, village, LINDI region, S central TANZANIA, 140 mi/225 km W of KILWA MASOKO, on Njenje River, in SELOUS GAME RESERVE; 09°06′S 37°29′E. Livestock.

Lukonzolwa (loo-kon-ZOL-wah), village, KATANGA province, SE CONGO, on W bank of LAKE MWERU, and 210 mi/338 km NNE of LUBUMBASHI; 08°47′S 28°39′E. Elev. 3,015 ft/918 m. Cattle raising. Roman Catholic mission.

Lukosi, village, MATABELELAND NORTH province, W ZIMBABWE, 10 mi/16 km S of HWANGE, on Lukosi River, on railroad; 18°30′S 26°30′E. HWANGE NATIONAL PARK to S. Livestock; grain.

Lukov, POLAND: see LUKOW.

Lukov, UKRAINE: see LUKIV.

Lukovë, new town, SW ALBANIA, on ADRIATIC SEA coast, N of SARANDË; 39°59′N 19°54′E. Citrus orchards.

Lukovit (LOO-ko-veet), city, LOVECH oblast, LUKOVIT obshtina (1993 population 23,405), N BULGARIA, on the PANEGA RIVER (hydroelectric station), 27 mi/43 km SW of PLEVEN; 43°12′N 24°11′E. Agricultural center (grain, vineyards, vegetables, fruit, sheep; dairying); flour milling, canning, tanning, manufacturing (auto parts, wrapping materials, carpets, bricks). Karstlike rock formations and grottoes nearby.

Lukovnikovo (loo-KOV-nee-kuh-vuh), village, SW TVER oblast, W European Russia, on road, 28 mi/45 km N of RZHEV; 56°39′N 34°21′E. Elevation 725 ft/220 m. Flax processing.

Lukovskaya (loo-kuhf-SKAH-yah), village (2006 population 5,335), N NORTH OSSETIAN REPUBLIC, SE European Russia, on the TEREK RIVER, near highway and on railroad, 3 mi/5 km E of MOZDOK; 43°43′N 44°35′E. Elevation 446 ft/135 m. Railroad depots. Produce processing.

Lukow (WOO-koof), Polish *Łuków*, Russian *Lukov*, town (2002 population 30,819), Siedlce province, E POLAND, on KRZNA RIVER, and 17 mi/27 km S of SIEDLCE. Railroad junction; manufacturing of soap, vinegar, food products; weaving, brewing, distilling; brickworks. Before World War II, population over 50% Jewish.

Lukoyanov (loo-kuh-YAH-nuhf), city (2006 population 12,935), SE NIZHEGOROD oblast, central European Russia, on the TESHA River (tributary of the OKA River), on railroad, 107 mi/172 km SSE of NIZHNIY NOVGOROD; 55°02′N 44°29′E. Elevation 610 ft/185 m. Highway junction; local transshipment point. Ceramic-brick industry; ball-bearing factory; meat-packing plant. Chartered in 1779.

Lükqün (LU-KUH-CHIN), town, E central XINJIANG Uygur Autonomous Region, CHINA, in the TURFAN depression, 30 mi/48 km ESE of Turfan; 42°44′N 89°42′E. The depression is sometimes called Lukqün.

Lukqün, CHINA: see LÜKQÜN.

Lukuga River (loo-KOO-gah), c.200 mi/322 km long, E CONGO; issues from W shore of LAKE TANGANYIKA at KALEMIE; flows W to LUALABA RIVER, 25 mi/40 km N of KABALO. There are low-grade coal deposits along its tributaries, N of KALEMIE and at GREINERVILLE. It is LAKE TANGANYIKA's only outlet.

Lukula (loo-KOO-lah), town, BAS-CONGO province, W CONGO, on railroad, and 40 mi/64 km NNW of BOMA; 05°23′S 12°57′E. Elev. 646 ft/196 m. Sawmilling and agricultural center; palm-oil milling, rubber, palm, and cacao plantations.

Lukula, town, WESTERN province, W ZAMBIA, 60 mi/97 km N of MONGU, on ZAMBEZI RIVER; 14°23′S 23°15′E. At S edge of Luena Flats swamp; Kabompo River enters Zambezi River to N. LIUWA PLAIN NATIONAL PARK to W. Airfield. Agriculture (honey, beeswax); cattle. Timber.

Lukuledi River (loo-koo-LAI-dee), river, c. 110 mi/177 km long, SE TANZANIA; rises c. 20 mi/32 km WNW of MASASI; flows ENE to LINDI Bay, INDIAN OCEAN. Navigable to Mwakya, c. 10 mi/16 km. MAKONDE PLATEAU to S.

Lukull, Cape, UKRAINE: see KALAMIT BAY.

Lukumburu (loo-koom-BOO-roo), village, IRINGA region, S central TANZANIA, 70 mi/113 km NNW of SONGEA; 09°42′S 35°09′E. Sheep, goats; corn, wheat.

Lukuni (loo-KOO-nee), village, KATANGA province, SE CONGO, on railroad, and 12 mi/19 km NNW of LUBUMBASHI; 11°30′S 27°25′E. Elev. 4,704 ft/1,433 m. Copper mining. Also spelled Lukumi.

Lukunor (LOOK-ah-nor), atoll (□ 1 sq mi/2.6 sq km), MORTLOCK or NOMOI Islands, State of CHUUK, E CAROLINE ISLANDS, MICRONESIA, W PACIFIC, c.6 mi/9.7 km NE of SATAWAN; 05°31′N 153°46′E; 7 mi/11.3 km long, 4 mi/6.4 km wide; 18 low islets on triangular reef, largest island being Lukunor (c.2 mi/3.2 km long).

Lukwika (loo-KWEE-kah), village, MTWARA region, SE TANZANIA, 37 mi/60 km SW of Masai; 11°09′S 36°35′E. Timber; cashews, bananas, manioc; sheep, goats.

Lukyanovo (look-YAH-nuh-vuh), urban settlement, S MOSCOW oblast, central European Russia, on the OKA River, just N of the administrative border with TULA oblast, on highway and railroad, 3 mi/5 km N of (and administratively subordinate to) SERPUKHOV; 54°52′N 37°25′E. Elevation 505 ft/153 m. Brick works.

Luky Lake, UKRAINE: see SHATS'K LAKES.

Lula (LOO-luh), town (2000 population 1,438), HALL county, NE GEORGIA, 10 mi/16 km NE of GAINESVILLE, near source of OCONEE River; 34°23′N 83°40′W.

Industry includes poultry processing equipment and light manufacturing.

Lula (LOO-lah), village, ORIENTALE province, E CONGO, on railroad, and 3 mi/4.8 km SSE of KISANGANI; 00°26′N 25°12′E. Elev. 1,433 ft/436 m. Has agricultural research station with rubber and coffee plantations.

Lula (LOO-luh), village (2000 population 370), COAHOMA county, NW MISSISSIPPI, 18 mi/29 km NNE of CLARKSDALE; 34°27′N 90°28′W. Timber. Many antebellum homes in the vicinity. Moon Lake, natural oxbow lake near MISSISSIPPI RIVER and source of SUNFLOWER RIVER (c.4 mi/6.4 km long; resort) is to W.

Luleå (LOO-le-O), town, ⊙ NORRBOTTEN county, NE SWEDEN, port on GULF OF BOTHNIA at mouth of LULEÄLVEN RIVER; 65°35′N 22°10′E. Manufacturing (steel works). Airport. Despite icebound harbor most of winter, iron ore is important export. Chartered by Gustavus II (1621). Burned down (1887), rebuilt in modern style. Gammelstad church village (c. 1400). Seat of Lutheran bishopric.

Luleälven (LOO-le-ELV-en), river, c. 275 mi/443 km long; rises near NORWEGIAN border, NORRBOTTEN county, N SWEDEN; flows SE to GULF OF BOTHNIA at LULEÅ. Falls at STORA SJÖFALLET, Porjus, and Harsprånget. Power plants at Porjus (1910–1914) and at Harsprånget (1945–1952), the latter, with a 940-MW capacity, one of Europe's largest. Power from Porjus operates many industries in N Sweden, especially iron mines at KIRUNA and GÄLLIVARE. Power transported by high tension wires to central Sweden.

Lule Burgas, TURKEY: see LULEBURGAZ.

Luleburgaz, Turkish *Lüleburgaz* (loo-luh-bur-GAHZ), ancient *Bergulae*, town (2000 population 79,002), European TURKEY, 23 mi/37 km SSE of KIRKLARELI; 41°25′N 27°22′E. Agricultural trade center (grain, sugar beets, beans, potatoes). Also spelled Lule Burgas.

Lulenga, CONGO: see RUGARI.

Lules (LOO-les), town, central TUCUMÁN province, ARGENTINA, on LULES RIVER, and on railroad, 13 mi/21 km SW of TUCUMÁN; 26°56′S 65°21′W. Lumbering and agricultural center (sugarcane, tomatoes, corn, rice; livestock); sugar refinery. Hydroelectric station nearby.

Lules River (LOO-les), W central TUCUMÁN province, ARGENTINA; rises in N outliers of CUMBRE DE POTRERILLO; flows 45 mi/72 km ESE, past LULES, to the SALÍ RIVER 4 mi/6.4 km NE of BELLA VISTA. Irrigates sugarcane area; hydroelectric stations along its course.

Luliang (LU-LIANG), town, ⊙ Luliang county, E YUNNAN province, CHINA, 60 mi/97 km E of KUNMING; 25°03′N 103°39′E. Elevation 6,102 ft/1,860 m. Road junction; pears, rice, wheat, millet, beans, tobacco; machinery and equipment manufacturing; chemicals.

Lüliang Mountains (LOO-LYAHNG), c.9,000 ft/2,743 m, W SHANXI province, CHINA, form divide between FEN River and HUANG HE (Yellow River); 35 mi/56 km NE of LISHI; 37°45′N 111°25′E.

Luling (LOO-ling), unincorporated town (2000 population 11,512), SAINT CHARLES parish, SE LOUISIANA, on W bank (levee) of the MISSISSIPPI RIVER, and 16 mi/26 km W of NEW ORLEANS; 29°55′N 90°22′W. Elevated 9 ft/3 m. Sugarcane, vegetables (especially okra); catfish, alligators, crabs; manufacturing (safety valves, chemicals). Highway bridge to DESTREHAN.

Luling (LOO-ling), town (2006 population 5,398), CALDWELL county, S central TEXAS, on SAN MARCOS RIVER, and c.40 mi/64 km S of AUSTIN; 29°40′N 97°39′W. Elevation 418 ft/127 m. Oil field supply and shipping center; manufacturing (oil field storage tanks, concrete, rubber products, machining). Palmetto State Park to SE. Founded 1874; long a cow town, it boomed after oil discovery, 1922.

Lulong (LU-LUNG), town, ⊙ Lulong county, NE HEBEI province, CHINA, 40 mi/64 km NE of TANGSHAN, and on LUAN RIVER; 39°52′N 118°52′E. Fruits, oilseeds, grain; building materials, machinery and equipment manufacturing; chemicals.

Lulonga (loo-LAWNG-gah), village, Équateur province, NW CONGO, on left bank of CONGO RIVER at mouth of LULONGA RIVER, and 45 mi/72 km N of MBANDAKA; 00°37′N 18°23′E. Elev. 1,128 ft/343 m. Steamboat landing and mission center (Roman Catholic and Protestant), palm-oil milling, copal treating.

Lulonga River (loo-LAWN-guh), c.130 mi/209 km long, W CONGO; formed by union of the LOPORI and the MARINGA at BASANKUSU; flows W and SW to CONGO RIVER at LULONGA. Navigable for steamboats along entire course.

Luluabourg, CONGO: see KANANGA.

Lulua River (LOOL-wah), 550 mi/885 km long, right tributary of the KASAI, S CONGO; rises at ANGOLA border 25 mi/40 km S of MALONGA; flows in wide curves N and NW, past SANDOA, LUIZA, KANANGA, and LUEBO, to KASAI RIVER 28 mi/45 km N of CHARLESVILLE. Navigable for steamers for c.35 mi/56 km in lower course (below LUEBO).

Lulworth Cove (LUHL-wuhth), circular bay on the DORSET coast, S ENGLAND, about 0.5 mi/0.8 km S of WEST LULWORTH; 50°37′N 02°15′W. The cove is nearly enclosed by limestone cliffs. Tourist site.

Lumaco (loo-MAH-ko), town, ⊙ Lumaco comuna, MALLECO province, ARAUCANIA region, S central CHILE, on railroad, on Lumaco River, and 28 mi/45 km SSW of ANGOL; 38°09′S 72°55′W. Agriculture (cereals, vegetables, fruit, livestock); lumbering, food processing.

Lumadjang, INDONESIA: see LUMAJANG.

Lumajang (LOO-mah-jahng), town, ⊙ Lumajang district, Java Timur province, INDONESIA, near INDIAN OCEAN, 70 mi/113 km SE of SURABAYA; 08°08′S 113°13′E. Trade center for agricultural area (sugar, tobacco, coffee, rubber, tea, cinchona bark). Nearby are sugar mills. Formerly Lumadjang or Loemadjang.

Lumaku, Gunung (loo-MAH-koo GOO nong) (8,200 ft/2,500 m), SW SABAH, E MALAYSIA, on SARAWAK border, 25 mi/40 km S of BEAUFORT; 04°54′N 115°32′E.

Lumame, town (2007 population 10,060), AMHARA state, central ETHIOPIA; 10°15′N 37°56′E. Roadside town 12 mi/19 km ESE of DEBRE MARKOS, in livestock raising area.

Lumbala Kaquengue (loom-BAH-luh ka-KAIN-gai), town, MOXICO province, SE ANGOLA, near ZAMBIA border, on road, and 190 mi/306 km SE of LUENA. Formerly Gago Coutinho (GAH-goo koo-TEEN-yo).

Lumban (lum-BAHN), town, LAGUNA province, S LUZON, PHILIPPINES, near LAGUNA DE BAY, 18 mi/29 km NNE of SAN PABLO; 14°18′N 121°29′E. Agricultural center (rice, coconuts, sugar).

Lumber Bridge (LUHM-buhr BRIJ), village (2006 population 120), ROBESON county, S NORTH CAROLINA, 16 mi/26 km SW of FAYETTEVILLE near Big Swamp River; 34°53′N 79°04′W. Retail trade; manufacturing (construction materials, poultry processing); agriculture (grain, tobacco, poultry, livestock). Incorporated 1891.

Lumber City, town (2000 population 1,247), TELFAIR county, S central GEORGIA, 22 mi/35 km WNW of BAXLEY, near junction of OCMULGEE and LITTLE OCMULGEE rivers; 31°56′N 82°41′W. Shipping point for hardwood and pine lumber. Manufacturing (steel tire cords, animal feeds). Incorporated 1889.

Lumber City, borough (2006 population 82), CLEARFIELD county, W central PENNSYLVANIA, 4 mi/6.4 km SW of CURWENSVILLE, on West Branch of SUSQUEHANNA RIVER; 40°55′N 78°34′W. Agriculture (corn, hay, dairying). Curwensville Lake reservoir to NE.

Lumberport, town (2006 population 970), HARRISON county, N WEST VIRGINIA, on West Fork River (headstream of MONONGAHELA RIVER), 6 mi/9.7 km NNW of CLARKSBURG; 39°22′N 80°20′W. Coal; oil. Agriculture (corn); cattle; poultry; dairying. Incorporated 1901.

Lumber River (LUHM-buhr RI-vuhr), c.125 mi/201 km long, in NORTH CAROLINA and SOUTH CAROLINA; rises as Drowning Creek on MONTGOMERY-MOORE county line E of BISCOE in central North Carolina; flows SE, past LUMBERTON, approximate point of name change ot Lumber River, receives Big Swamp River from N as it turns to SSW into South Carolina, to LITTLE PEE DEE RIVER 3 mi/4.8 km SSW of NICHOLS; c.50 mi/80 km as Lumber River; c.65 mi/105 km as Drowning Creek; 34°57′N 79°21′W.

Lumberton (LUHM-buhr-tuhn), city (□ 15 sq mi/39 sq km; 2006 population 21,894), ⊙ ROBESON county, SE NORTH CAROLINA, 32 mi/51 km S of FAYETTEVILLE, on the LUMBER RIVER; 34°37′N 79°00′W. (Drowning Creek changes name to Lumber River about here). Railroad junction. Service industries; manufacturing (transformers, textiles, textile dyeing and finishing, polypropylene, apparel, sheet metal fabricating, electrical products, concrete, paper packaging); agriculture (tobacco, grain, soybeans; cattle, dairying, hogs); timber. Seat of Pembroke State University to W at Pembroke. Robeson County Planetarium. Lumber River State Park to SE. Founded 1787, incorporated 1859.

Lumberton, town (2000 population 2,228), LAMAR county, SE MISSISSIPPI, on RED RIVER, 24 mi/39 km SSW of HATTIESBURG, on Red River and RED CREEK; 31°00′N 89°27′W. In agricultural (Cotton, corn, pecans; cattle, poultry) and pine-timber area. Manufacturing (electrical transformers, concrete, feeds, apparel, petroleum refining); oil and natural gas. Large pecan nursery. De Soto National Forest to E. Settled in 1880s.

Lumberton, township, BURLINGTON county, S NEW JERSEY, 2 mi/3.2 km S of MOUNT HOLLY; 39°57′N 74°47′W. Boat manufacturing, printing. Incorporated 1860.

Lumberton (LUHM-buhr-tuhn), town (2006 population 9,728), HARDIN county, SE TEXAS, 13 mi/21 km NNW of BEAUMONT; 30°15′N 94°12′W. Railroad junction. Timber; oil and natural gas; cattle. Manufacturing (heat exchangers and pressure valves, refractory anchors, industrial flanges). Main unit of BIG THICKET NATIONAL PRESERVE to E and S. Village Creek State Park is here.

Lumbier (loom-BYER) or **Irunberri**, town, NAVARRE province, N SPAIN, 20 mi/32 km SE of PAMPLONA; 42°39′N 01°18′W. Cereals, livestock.

Lumbini (loom-BEE-nee), administrative zone (2001 population 2,526,868), central NEPAL. Includes the districts of ARGHAKHANCHI, GULMI, KAPILAVASTU, NAWALPARASI, PALPA, and RUPANDEHI.

Lumbini (loom-BEE-nee), religious site, S NEPAL, in the TERAI, 15 mi/24 km SSW of BHAIRAWA; 27°27′N 83°16′E. Elevation 312 ft/95 m. In ancient Lumbini (or Rummindei) Garden, here, a 3rd century B.C.E. Asokan pillar was discovered (1895), erected on site of birthplace (c. 563 B.C.E.) of Gautama Siddhartha of the Sakya clan, who later became the Buddha; one of eight great ancient Buddhist pilgrimage centers. Nearby (W) is TAULIHAWA, capital of ancient Sakya kingdom of KAPILAVASTU.

Lumbo (LOOM-bo), town, NE MOZAMBIQUE, on MOZAMBIQUE CHANNEL of INDIAN OCEAN opposite Mozambique Island (2 mi/3.2 km E); 15°01′S 40°40′E. Ocean terminus of railroad to NAMPULA (95 mi/153 km W). Has Mozambique city's airport.

Lumbrales (loom-BRAH-les), town, SALAMANCA province, W SPAIN, 26 mi/42 km NNW of CIUDAD RODRIGO; 40°56′N 06°43′W. Agricultural trade center (cereals, vegetables, wine, olive oil); livestock raising. Lead mines nearby.

Lumbreras, town, MURCIA province, SE SPAIN, 10 mi/16 km SW of LORCA. Almonds and other fruit, cereals, wine. Its railroad station is 7 mi/11.3 km SSE.

Lumbreras (loom-BRAI-rahs), village, S central SALTA province, ARGENTINA, at W foot of SIERRA LUM-

BRERA, on Pasaje or Juramento River, on railroad, and 45 mi/72 km SE of SALTA; 25°12′S 64°55′W. Lumbering and livestock-raising center; sawmills.

Lumbrera, Sierra (loom-BRAI-rah, see-YER-rah), subandean mountain range in central SALTA province, ARGENTINA, E of LUMBRERAS; extends c.40 mi/64 km ENE-WSW; rises to c.3,500 ft/1,067 m.

Lumbres (luhn-bruh), town (□ 3 sq mi/7.8 sq km), PAS-DE-CALAIS department, NORD-PAS-DE-CALAIS region, N FRANCE, on the AA RIVER, and 7 mi/11.3 km SW of SAINT-OMER; 50°42′N 02°00′E. Paper milling, portland cement manufacturing.

Lumby (LUHM-bee), village (□ 2 sq mi/5.2 sq km; 2001 population 1,618), S BRITISH COLUMBIA, W CANADA, 14 mi/23 km E of VERNON, in NORTH OKANAGAN regional district; 50°15′N 118°58′W. Fruit, vegetables; manufacturing (wood products).

Lumding (LOOM-ding), town, NAGAON district, central ASSAM state, NE INDIA, 50 mi/80 km SSE of NAGAON; 25°45′N 93°10′E. Railroad junction; rice, jute; rape and mustard; tea.

Lumezzane (loo-me-TSAH-ne), commune, BRESCIA province, LOMBARDY, N ITALY, 8 mi/13 km N of BRESCIA; 45°39′N 10°15′E. Chief towns: San Sebastiano, Sant'Apollonio, Pieve. Numerous factories (fabricated metal products, machinery, firearms, cutlery).

Lumiar (LOO-mee-ahr), city, E RIO DE JANEIRO state, BRAZIL, 13 mi/21 km SE of NOVA FRIBURGO; 22°22′S 42°12′W.

Lumière, Cape (loom-YER), in NORTHUMBERLAND STRAIT, E NEW BRUNSWICK, CANADA, 40 mi/64 km N of MONCTON; 46°40′N 64°43′W.

Lumley, resort suburb, 4 mi/6.4 km SE of FREETOWN, SIERRA LEONE; 08°27′N 13°16′N. Road terminus for taxis from downtown FREETOWN. Tourism (beach, hotels).

Lummen (LOOM-men), commune (2006 population 13,739), Hasselt district, LIMBURG province, NE BELGIUM, 8 mi/12.9 km WNW of HASSELT. Agriculture, cattle raising.

Lummi Island (LUHM-mee), WHATCOM county, NW WASHINGTON, 8 mi/12.9 km SW of BELLINGHAM; 9 mi/14.5 km long; 0.5 mi/0.8 km–1 mi/1.6 km wide. ROSARIO STRAIT to S, Strait of Georgia to NW, Bellingham Bay to E. Hale Passage to NE separates island from mainland. Village of Lummi on narrow part of island in NW. Ferry from Lummi Peninsula, to N. Fish; oysters, crabs. Highest point, Lummi Peak, 1,509 ft/460 m. SAN JUAN ISLANDS to W.

Lumparland (LUM-pahr-lahnd), fishing village (□ 14 sq mi/36.4 sq km), ÅLAND province, SW FINLAND; 60°07′N 20°15′E. Elevation 33 ft/10 m. On Lumparland Island (□ 13.5 sq mi/35 sq km), in AHVENANMAA group, 11 mi/18 km E of MAARIANHAMINA.

Lumphanan (luhm-PAH-nuhn), agricultural village, Aberdeenshire, NE Scotland, 7 mi/11.3 km S of ALFORD; 57°08′N 02°42′W. Just NW is Macbeth's Cairn, reputed to mark site where Macbeth was killed by Macduff. Formerly in Grampian, abolished 1996.

Lumpiaque (loom-PYAH-kai), town, ZARAGOZA province, NE SPAIN, near JALÓN RIVER, 27 mi/43 km W of ZARAGOZA; 41°38′N 01°18′W. Produces sugar beets and wine.

Lumpkin, county (□ 285 sq mi/741 sq km; 2006 population 25,462), N GEORGIA, ⊙ DAHLONEGA; 34°34′N 84°00′W. In blue ridge area, drained by CHESTATEE and ETOWAH rivers. Agriculture (corn, hay, potatoes); poultry, cattle, hogs; timber. CHATTAHOOCHEE National Forest occupies N half of county. Formed 1832.

Lumpkin, town (2000 population 1,369), ⊙ STEWART county, SW GEORGIA, 30 mi/48 km SSE of COLUMBUS; 32°03′N 84°48′W. Manufacturing of dyes and paints. Incorporated 1831.

Lumsden (LUHMZ-duhn), town (2006 population 1,523), S SASKATCHEWAN, CANADA, on Q'APPELLE RIVER, and 17 mi/27 km NW of REGINA; 50°38′N 104°60′W. Grain elevators, livestock.

Lumsden, township, SOUTHLAND district, S SOUTH ISLAND, NEW ZEALAND, 53 mi/85 km N of INVERCARGILL, and on WAIMEA plains, ORETI RIVER; 45°44′S 168°27′E. Sheep raising, dairying.

Lumut (LOO-moot), town (2000 population 31,882), W PERAK, MALAYSIA, port on Dindings River off Strait of MALACCA, 45 mi/72 km SSW of TAIPING; 04°14′N 100°38′E. Chief town of the DINDINGS, on Dinding Straits (between mainland and PANGKOR ISLAND). Ferry to Pangkor Island; fisheries.

Luna (LOO-nuh), county (□ 2,965 sq mi/7,709 sq km; 2006 population 27,205), SW NEW MEXICO; ⊙ DEMING; 32°10′N 107°45′W. Livestock and grain (cattle, some sheep; melons, grapes, fruit, nuts, jalapeños, chiles, onions, green beans, cabbage, some lettuce; pecans, cotton, corn, sorghum, hay, alfalfa, wheat, oats, barley) area bordering on MEXICO (CHIHUAHUA state) in S. FLORIDA MOUNTAINS in SE, part of MIMBRES MOUNTAINS, including Cooke's Peak, in N. Formed 1901.

Luna, town, LA UNION province, N central LUZON, PHILIPPINES, on W coast, 16 mi/26 km NNE of SAN FERNANDO; 16°51′N 120°23′E. Rice-growing center.

Luna, town, ZARAGOZA province, N SPAIN, 28 mi/45 km W of HUESCA. Cereals, livestock.

Lunacharskoye, UZBEKISTAN: see KIBRAI.

Lunahuaná (loo-nah-hwah-NAH), town, CAÑETE province, LIMA region, W central PERU, in irrigated CAÑETE RIVER valley, on SAN VICENTE DE CAÑETE–HUANCAYO road, and 19 mi/31 km ENE of San Vicente de Cañete; 12°58′S 76°08′W. Major viticultural center; fruit.

Luna, Isla de la, BOLIVIA: see COATI ISLAND.

Luna, Laguna de (LOO-nah, lah-GOO-nah dai), lake (□ 75 sq mi/195 sq km) in ESTEROS DEL IBERÁ (swamps), N CORRIENTES province, ARGENTINA, 3 mi/4.8 km N of LAKE IBERÁ; 12 mi/19 km long.

Lunan (LU-NAN), town, ⊙ Lunan county, E YUNNAN province, CHINA, 40 mi/64 km ESE of KUNMING; 24°47′N 103°16′E. Elevation 5,433 ft/1,656 m. Rice, wheat, millet, tobacco; food processing, building materials.

Lunan Water (LOO-nuhn), river (14 mi/23 km long), ANGUS, E Scotland; rises E of FORFAR; flows E, past Friockheim and Inverkeilor, to NORTH SEA 4 mi/6.4 km SSW of MONTROSE. Formerly in Tayside, abolished 1996.

Lunavada (loo-nah-VAH-dah), town, tahsil headquarters, PANCH MAHALS district, GUJARAT state, W central INDIA, 25 mi/40 km N of GODHRA; 23°08′N 73°37′E. Railroad spur terminus; trades in corn, rice, millet, timber; match manufacturing, rice husking; leather goods. Was the capital of former princely state of Lunavada in Gujarat States, BOMBAY; state incorporated 1949 into PANCH MAHALS district. Sometimes spelled Lunawada.

Lunawa, SRI LANKA: see MORATUWA.

Luncarty (luhn-KAR-tee), village (2001 population 1,265), PERTH AND KINROSS, E Scotland, on the TAY RIVER, and 4 mi/6.4 km N of PERTH; 56°27′N 03°28′W. Salmon hatchery. Formerly in Tayside, abolished 1996.

Lund (LUND), town, SKÅNE county, S SWEDEN; 55°42′N 13°12′E. Educational, commercial, and industrial center; railroad junction. Manufacturing (packaging, medical equipment, medicine; research industries). Mentioned (c.920) in sagas as Lunda; Catholic archiepiscopal see for SCANDINAVIA (1103–1104), flourishing medieval trade center. Declined after becoming Lutheran bishopric (1536), devastated during seventeenth-century Danish-Swedish wars. Passed to Sweden with Skåne province (1658). Seat of University of LUND, dedicated by Charles XI (1668), where poet Esaias Tegnér (1782–1846) taught; well-known theological faculty in nineteenth century. Seat of technical university. Romanesque cathedral (eleventh century); museum of folk customs.

Lund (LUHND), unincorporated village, SW BRITISH COLUMBIA, W CANADA, 13 mi/21 km from town of POWELL RIVER, in POWELL RIVER regional district; 49°58′N 124°46′W. Tourism. Founded 1889.

Lund (LUHNT), unincorporated village, WHITE PINE county, E NEVADA, 27 mi/43 km SSW of ELY, on White River. Cattle, sheep. EGAN RANGE to E; parts of Humboldt National Forest to W and N; Wayne E. Kirch Wildlife Management Area to S.

Lundale, unincorporated village, LOGAN county, SW WEST VIRGINIA, 12 mi/19 km ESE of LOGAN.

Lunda Norte (LOON-duh nor-TAI), province (□ 39,674 sq mi/103,152.4 sq km), NE ANGOLA; ⊙ LUCAPA. Bordered by Congo, E by KASAI RIVER and Congo, S by LUNDA SUL province, W by MALANJE province Drained by CUILO, Luangue, CHICAPA, Chiumbe, LUIA, CUANGO rivers. Agriculture includes rice, manioc, corn, palms. Minerals include diamonds. Main centers are LUCAPA, CUANGO, LUREMO, Luachimo, CHITATO, and CAUNGULA.

Lundar (LUN-dahr), community, S MANITOBA, W central CANADA, 39 mi/63 km E of Lake MANITOBA, 69 mi/110 km NW of WINNIPEG, in COLDWELL rural municipality; 50°41′N 98°01′W. Service center for agricultural, fishing, and recreational activity, and tourism. Founded 1887; first known as Swan Lake Settlement.

Lunda Sul (LOON-duh sool), province (□ 17,621 sq mi/45,814.6 sq km), NE ANGOLA; ⊙ SAURIMO. Bordered N by LUNDA NORTE, E by Congo, S by Rio Cassai and MOXICO province, W by Rio Cuango and MALANJE province. Drained by CUILO, Luachimo, CHICAPA, Chiumbe, Mombo rivers. Agriculture includes rice, manioc, corn. Minerals include diamonds, manganese, iron. Main centers are Suarimo, MUCONDA, MURIEGE, and ALTO CHICAPA.

Lundazi, town, EASTERN province, E ZAMBIA, 100 mi/161 km NNE of Chipat, near MALAWI border; 12°18′S 33°11′E. Road junction. Agriculture (tobacco, corn, cotton, peanuts). Gem mining (aquamarine, tourmaline). NORTH LUANGWA NATIONAL to W, Luambe National Park to SW.

Lundbreck (LUND-brek), hamlet, ALBERTA, W CANADA, in PINCHER CREEK NO. 9; 49°35′N 114°10′W.

Lunde (LUN-duh), village, TELEMARK county, S NORWAY, on railroad, and 19 mi/31 km WNW of SKIEN; 59°05′N 09°42′E. Sawmill, brickworks.

Lunde (LUND-e), village, VÄSTERNORRLAND county, NE SWEDEN, on ÅNGERMANÄLVEN RIVER estuary, 16 mi/26 km N of HÄRNÖSAND; 62°53′N 17°52′E.

Lundenburg, CZECH REPUBLIC: see BŘECLAV.

Lunderskov (LOO-nuhr-skou), town, VEJLE county, S JUTLAND, DENMARK, 33 mi/53 km E of ESBJERG; 55°30′N 09°20′E. Railroad junction. Manufacturing (boats; fabricated metal products); dairying.

Lundevatn (LUN-duh-VAH-tuhn), lake (□ 10 sq mi/26 sq km), ROGALAND county, SW NORWAY, on SIRA River, 4 mi/6.4 km NW of FLEKKEFJORD; 1,017 ft/310 m deep.

Lundin Links (LUHN-din LINKZ), village (2001 population 2,090), FIFE, E Scotland, on Largo Bay, and 2 mi/3.2 km NE of LEVEN; 56°13′N 02°57′W. Firth of Forth to S. Three large standing stones (dating from the second millennium B.C.E.) can be seen on one of the holes on the golf links.

Lundi River, ZIMBABWE: see RUNDE RIVER.

Lundsbrunn (LUNDS-BRUN), village, SKARABORG county, SW SWEDEN, 6 mi/9.7 km N of SKARA; 58°28′N 13°27′E. Health resort; conference site.

Lundy Isle (LUHN-dee), 3 mi/4.8 km long, off DEVON, SW ENGLAND, at the mouth of the BRISTOL CHANNEL; 51°11′N 04°40′W. Granite was quarried here for centuries. Inhabited in prehistoric times, Lundy Isle was a stronghold of pirates and smugglers from the Middle Ages until the 18th century. Ruins of a 13th-century castle remain. National Trust property.

Lundy's Lane (LUHN-deez), locality in S ONTARIO, E central CANADA, just W of NIAGARA FALLS, scene of a stubborn engagement of the War of 1812, fought July 25, 1814; 43°05′N 79°06′W. The American forces commanded by General Winfield Scott and led by General Jacob J. Brown, pushing into Canada, encountered British troops posted along Lundy's Lane. After prolonged fighting, the Americans fell back to FORT ERIE, their former position.

Lüneburg (LOO-ne-burg), city, LOWER SAXONY, N GERMANY, on the ILMENAU River; 53°15′N 10°25′E. Railroad junction and river port. Industries include saltworks, chemicals, and textiles; also trade in foodstuffs, metal, and coal. Famous hot salt springs and mud baths. Dating from the tenth century, was the capital of the dukes of Brunswick-Lüneburg and was important member of the HANSEATIC LEAGUE. Predominately built in the late-Gothic and Renaissance styles, the city has several fine churches, a large city hall (begun thirteenth century, additions as late as the eighteenth century), and many gabled houses in the characteristic N German style. The LÜNEBURGER HEIDE, a vast heath, is SW of the city.

Lüneburger Heide (LOO-nuh-BUR-guhr HEI-duh), large heath, NW GERMANY, SW of LÜNEBURG between the ELBE and ALLER rivers; 52°40′N 09°30′E–53°15′N 10°30′E. Much of the heath has been reforested with pine; the sandy regions that remain are mostly protected as a nature preserve. There is some agriculture and tourism.

Lunel (lyu-nel), town (□ 9 sq mi/23.4 sq km), HÉRAULT department, LANGUEDOC-ROUSSILLON region, S FRANCE, near VIDOURLE River, 14 mi/23 km ENE of MONTPELLIER; 43°40′N 04°08′E. Vineyards; wine-shipping center of LANGUEDOC; distilling, fruit and vegetable preserving. Town has a bullfighting tradition. Also called Lunel-Viel.

Lünen (LOO-nen), city, NORTH RHINE-WESTPHALIA, W GERMANY, on the LIPPE River; 51°36′N 07°31′E. Important port and railroad junction; manufacturing includes motor vehicles, metal and fabricated metal products, and textiles. Coal is mined and, in part, feeds large coal power plant. Lünen, first mentioned in 1195, was chartered in 1265 and passed to the counts of Mark in 1302.

Lunenburg (LOO-nuhn-buhrg), county (□ 1,169 sq mi/3,039.4 sq km; 2001 population 47,591), SW NOVA SCOTIA, CANADA, on the ATLANTIC OCEAN; ⊙ LUNENBURG; 44°22′N 64°19′W. Elevation 3 ft/0.9 m. Split from HALIFAX COUNTY in 1759.

Lunenburg (LOO-nuhn-buhrg), county (□ 432 sq mi/1,123.2 sq km; 2006 population 13,219), S VIRGINIA; ⊙ LUNENBURG; 36°57′N 78°14′W. Rolling agricultural region; bounded S by MEHERRIN RIVER, N by NOTTOWAY RIVER. Manufacturing at KENBRIDGE; lumber milling; agriculture (tobacco; also hay, wheat, barley, soybeans; cattle); timber (pine, oak). Formed 1746.

Lunenburg, (LOO-nuhn-buhrg) town (2001 population 2,568), cap. LUNENBURG county, SW NOVA SCOTIA, CANADA, on Lunenburg Bay, S of HALIFAX; 44°22′N 64°19′W. Elevation 0 ft/0 m. Fishing, shipbuilding; fisheries museum. Town founded in 1753 by Col. Charles Lawrence and settled by about 1500 "foreign"Protestants from GERMANY, FRANCE, and SWITZERLAND. Name comes form royal house of Brunswick-Lunenburg. Designated a UNESCO World Heritage Site in 1995.

Lunenburg (LOO-nen-buhrg), agricultural town, WORCESTER county, N MASSACHUSETTS, 22 mi/35 km N of WORCESTER; 42°35′N 71°43′W. Settled 1721, incorporated 1728. Large community of Finnish origin. Amusement park.

Lunenburg (LOON-uhn-buhrg), town, ESSEX CO., NE VERMONT, on CONNECTICUT River, and 16 mi/26 km E of St. Johnsbury; 44°28′N 71°42′W. Includes village of Gilman. Chartered 1763.

Lunenburg (LOO-nuhn-buhrg), unincorporated village, ⊙ LUNENBURG county, S VIRGINIA, 25 mi/40 km SSE of FARMVILLE, 3 mi/5 km SW of VICTORIA;

36°57′N 78°15′W. Agriculture (cattle; tobacco, grain, soybeans).

Lune River (LOON), CUMBRIA and LANCASHIRE, NW ENGLAND; rises near Ravenstonedale; flows 45 mi/72 km S and SW, past Kirkby Lonsdale, Tunstall, and Lancaster, to MORECAMBE BAY 6 mi/9.7 km SW of Lancaster. Navigable below Lancaster.

Lunéville (lyu-nai-veel), town (□ 6 sq mi/15.6 sq km), sub-prefecture of MEURTHE-ET-MOSELLE department, NE FRANCE, on the MEURTHE RIVER, in LORRAINE, 16 mi/26 km ESE of NANCY; 48°36′N 06°30′E. It is known for its fine faïence crockery. Railroad and electrical equipment; textiles and toys are also made here. One of the 3 independent bishoprics annexed to France in 1552. Its 18th-century palace served as the residence of Leopold, duke of Lorraine, and later of Stanislaus I, exiled king of Poland. A formal garden adjoins the palace, in imitation of VERSAILLES. A treaty signed there in 1801 between France and Austria confirmed and supplemented the terms of the treaty of Campo Formio under which France's growing influence over Italy was established.

Lunga (LUHN-gah), uninhabited island (2.5 mi/4 km long, up to 1 mi/1.6 km wide; rises to 323 ft/98 m) of the INNER HEBRIDES, Argyll and Bute, W Scotland, just N of SCARBA, from which it is separated by strait noted for violent current; 56°13′N 05°42′W. Formerly in Strathclyde, abolished 1996.

Lunga, Isola, CROATIA: see DUGI OTOK.

Lungarës (luhng-GAHRS), a prong of RRZEZA E KA-NALIT Mountains of S ALBANIA, slopes c.30 mi/48 km from Mt. ÇIKËS N to the lower VIJOSË RIVER. Also Lungara.

Lungau (LUN-gou), catchment area of the MUR RIVER, in the province of SALZBURG, W central AUSTRIA; it coincides with the administrative district of TAMS-WEG. It is crossed by the Tauern motorway and has a mixed economy of dairy, farming, and tourism.

Lungern (LOON-guhrn), commune, Obwalden half-canton, central SWITZERLAND, at S end of LAKE LUNGERN, 8 mi/12.9 km SSW of SARNEN, near BRÜNIG PASS. Elevation 2,467 ft/752 m. Resort; woodworking. Lungernase hydroelectric station is N.

Lungern, Lake, German *Lungernsee* (□ 1 sq mi/2.6 sq km), Obwalden half-canton, central SWITZERLAND; 223 ft/68 m deep; 46°48′N 08°11′E. Elevation 2,257 ft/688 m. Supplies Lungernsee hydroelectric plant (N).

Lungga Point, SOLOMON ISLANDS: see GUADALCANAL.

Lungi (LUHNG-gee), town, NORTHERN province, W SIERRA LEONE, on ATLANTIC coast, 10 mi/16 km N of FREETOWN (across SIERRA LEONE RIVER mouth); 08°08′N 12°24′W. Sierra Leone's chief airport, formerly at WATERLOO, is now 3 mi/4.8 km SSE of LUNGI (reached by ferryboat from Freetown, and road through PORT LOKO).

Lunging Island, NEW HAMPSHIRE: see ISLES OF SHOALS.

Lunglei, district (□ 1,751 sq mi/4,552.6 sq km), MIZORAM state, NE INDIA; ⊙ Lunglei.

Lunglei, town (2001 population 47,355), ⊙ LUNGLEI district, MIZORAM state, NE INDIA, 58 mi/93 km S of AIZAWL, in LUSHAI HILLS (extensive bamboo tracts); 22°53′N 92°44′E. Rice, cotton.

Lungro (LOONG-gro), town, COSENZA province, CA-LABRIA, S ITALY, 7 mi/11 km SW of CASTROVILLARI; 39°44′N 16°07′E. Salt refining, cheese making; wine, olive oil. Bishopric.

Lungtan (LONG-TAN), town, NW TAIWAN, 16 mi/26 km ENE of HSINCHU; center for tea production and processing; peanuts, sweet potatoes, tea oil.

Lungué-Bungo River (loon-GOO-ai-BOON-go), c.400 mi/644 km long, SW central AFRICA, right tributary of the ZAMBEZI; rises in central plateau of ANGOLA; flows SE into BAROTSELAND (ZAMBIA), and to the Zambezi 60 mi/97 km N of MONGU. Formerly Lungwebungu.

Luni (LOO-nee), ancient Etruscan town, N ITALY, on the Macra (MAGRA), at its mouth on GULF OF GENOA,

and 4 mi/6 km SE of SARZANA; 44°04′N 09°59′E. Made a Roman colony 177 B.C.E. Famed for its wine and marble (from CARRARA). Destroyed by Saracens 1016. Scanty ruins include amphitheater, tower, Christian church. Latin name, Luna.

Luni (LOO-nee), village, JODHPUR district, W central RAJASTHAN, NW INDIA, 19 mi/31 km S of Jodhpur, near LUNI RIVER railroad junction; millet, wheat, oilseeds.

Lunigiana (loo-nee-JAH-nah), mountain district (□ 375 sq mi/975 sq km), in MASSA CARRARA province, TUSCANY, central ITALY. Watered by upper and middle course of MAGRA RIVER. Has mild climate and abundant rainfall. Agriculture (cereals, chestnuts, grapes, vegetables); livestock (cattle, sheep). Chief center, Pontremoli. Colonized by Romans, who founded Luna (177 B.C.E.).

Luninets (loo-ni-NETS), Polish *Luniniec*, city, BREST oblast, BELARUS, ⊙ LUNINETS region, in PRIPET Marshes, 32 mi/51 km ENE of PINSK, 52°18′N 26°50′E. Railroad junction; railroad transport service and maintenance shops; manufacturing (baked goods, dry milk); logging and timber distribution. Passed (1793) from POLAND to RUSSIA; reverted (1921) to Poland; ceded to USSR in 1945.

Luning (LU-NING), town, SW SICHUAN province, CHINA, 50 mi/80 km NW of XICHANG, and on YALONG River bend; 28°22′N 101°56′E; grain.

Lunino (LOO-nee-nuh), town (2005 population 8,200), N PENZA oblast, E European Russia, near the confluence of the Shuksha and SURA rivers, on crossroads and railroad spur, 30 mi/48 km NNE of PENZA; 53°35′N 45°14′E. Elevation 469 ft/142 m. Hemp mill; grain elevator; bakery; livestock feed.

Lunino (LOO-nee-nuh), village, N KALININGRAD oblast, W European Russia, on crossroads, 13 mi/21 km SE of SOVETSK; 54°55′N 22°04′E. Elevation 187 ft/56 m. Until 1945, in EAST PRUSSIA and called Lengwethen until 1938, then Hohensalzburg; renamed after the territory was ceded to the USSR following World War II.

Luni River (LOO-nee), seasonal stream, c.320 mi/515 km long, W RAJASTHAN state, NW INDIA; rises in Aravalli Range, in two main headstreams joining W of AJMER city; flows WSW past BALOTRA, and SSW into NE end of RANN OF CUTCH. Receives drainage from SW slopes of Aravalli Range.

Lunkaransar (LOON-duh-ruhn-suhr), town, tahsil headquarters, BIKANER district, N RAJASTHAN state, NW INDIA, 45 mi/72 km NE of Bikaner; 28°29′N 73°45′E. Exports salt (natural deposits just N). Selenite deposits nearby; red sandstone quarried (S).

Lunlunta (loon-LOON-tah), village, N MENDOZA province, ARGENTINA, in MENDOZA River valley, on railroad, and 10 mi/16 km S of MENDOZA; 33°02′S 68°49′W. Wine producing, livestock raising; oil wells.

Lunna (loo-NAH), Polish *Lunna*, town, central GRODNO oblast, BELARUS, on NEMAN River (landing), and 13 mi/21 km W of MOSTY. Flour milling, brick manufacturing. Has old church.

Lunsar, town, NORTHERN province, SIERRA LEONE, 20 mi/32 km E of PORT LOKO, 2 mi/3.2 km W of MAR-AMPA; 08°41′N 12°32′W. Important chiefdom town, trade center, and road junction in route to MAKENI and KABALA. Has eye clinic that serves people from the N province and other parts of the country.

Lunsemfwa River, ZAMBIA: see LUANGWA RIVER.

Luntai (LUN-TEI), town and oasis, ⊙ Luntai county, central XINJIANG UYGUR AUTONOMOUS REGION, CHINA, 95 mi/153 km W of KORLA, and on highway S of the TIANSHAN; 41°46′N 84°10′E. Sericulture; grain; cattle, sheep; food processing, coal mining. The name sometimes appears as Bügür.

Lunteren (LUHN-tuh-ruhn), town, GELDERLAND province, central NETHERLANDS, 17 mi/27 km SW of APELDOORN, at W edge of VELUWE nature area; 52°05′N 05°37′E. Dairying; cattle, poultry; fruit, vegetables, potatoes.

Lunugala, town, UVA PROVINCE, SE SRI LANKA, in E SRI LANKA HILL COUNTRY, 10 mi/16 km ENE of BADULLA; 07°02′N 81°12′E. Tea, rice, rubber, vegetables.

Lunugala Ridge, UVA PROVINCE, E outlier of SRI LANKA HILL COUNTRY; 16 mi/26 km long N-S, up to 3 mi/4.8 km wide. Highest point, DOREPOTAGALA PEAK (4,964 ft/1,513 m), is 6 mi/9.7 km SW of BIBILE. Rubber plantations on lower, tea gardens on upper slopes.

Lunuwila, village, NORTH WESTERN PROVINCE, SRI LANKA, 10 mi/16 km N of NEGOMBO; 07°21′N 79°51′E. Trades in coconuts and rice. Seat of Coconut Research Institute (1929).

Lun'yevka (LOON-yeef-kah), town (2006 population 350), E central PERM oblast, W URALS, E European Russia, 5 mi/8 km NNE of KIZEL; 59°09′N 57°40′E. Elevation 987 ft/300 m. Mining center in Kizel bituminous-coal basin; declined in importance following the post-Soviet switch to market economy. Developed prior to World War I. Until about 1928, called Lun'yevskiye Kopi.

Lun'yevskiye Kopi, RUSSIA: see LUN'YEVKA.

Lunz am See (LUNTS ahm SAI), township, SW LOWER AUSTRIA, picturesque Lunzersee (small Alpine lake) nearby, and 20 mi/32 km SE of AMSTETTEN; 47°52′N 15°02′E. Railroad terminus; summer resort, and winter sports regions of Lackenhof and Hochkar are nearby.

Lunzenau (LUHN-tsuhn-ou), town, SAXONY, E central GERMANY, on the ZWICKAUER MULDE, 12 mi/19 km NW of CHEMNITZ; 50°58′N 12°45′E.

Luobei (LU-uh-BAI), town, ⊙ Luobei county, NE HEILONGJIANG province, NE CHINA, 25 mi/40 km NE of JIAMUSI, and on right bank of AMUR River (RUS-SIAN border); 47°35′N 130°50′E. Grain, sugar beets; food processing, logging.

Luobu bo (LU-uh-BU BO), salt basin, SE XINJIANG UYGUR AUTONOMOUS REGION, CHINA, in the TARIM River basin. Since 1964, Luobu bo has been used by the Chinese Communist government for its nuclear test explosions. Once a large salt lake (as mapped by ancient Chinese geographers), it is now largely dried up, with marshes and small, shifting lakes receiving the channels of the Tarim River. The region was explored by N. M. Przhevalsky and Sven Hedin. In 1928, at the time of the last expedition, the lake covered c.1,200 sq mi/3,108 sq km. The name sometimes appears as Lo-pu-po. Formerly Lob Norwegian.

Luocheng (LU-CHENG), town, ⊙ Luocheng county, N GUANGXI ZHUANG AUTONOMOUS REGION, CHINA, 40 mi/64 km NW of LIUZHOU; 24°47′N 108°54′E. Rice; chemicals, lumbering, building materials, coal mining, iron-ore and non-ferrous-ore mining.

Luochuan (LU-uh-CHUAN), town, ⊙ Luochuan county, N central SHAANXI province, CHINA, 50 mi/80 km S of YAN'AN, near LUO RIVER; 35°46′N 109°25′E. Grain, tobacco; food processing.

Luoci (LU-uh-ZI), town, NE central YUNNAN province, CHINA, 28 mi/45 km NW of KUNMING; 25°15′N 102°12′E. Elevation 6,403 ft/1,952 m. Timber, rice, wheat, millet, beans; tobacco.

Luodian (LU-uh-DIAN), town, ⊙ Luodian county, S GUIZHOU province, CHINA, 80 mi/129 km S of GUIYANG; 25°29′N 106°39′E. Wheat, rice, millet; food processing, papermaking; logging.

Luoding (LU-uh-DING), town, ⊙ Luoding county, W GUANGDONG province, CHINA, 55 mi/89 km S of WUZHOU; 22°46′N 111°34′E. Grain, oilseeds, jute; machinery and equipment manufacturing; logging, chemicals; papermaking, crafts, printing.

Luofu Mountains (LU-uh-FU), S GUANGDONG province, CHINA, extend 60 mi/97 km between ZENG RIVER and EAST RIVER. Rise to over 1,300 ft/396 m at Mount Luofu (SW); 23°40′N 114°36′E. Site of numerous temples.

Luohe (LU-uh-HUH), city (□ 23 sq mi/59.8 sq km; 2000 population 187,792), S central HENAN province, CHINA, on BEIJING-WUHAN railroad; 33°33′N 114°00′E.

An industrial center. Designated as an open city in 1992, it has established a high-tech industrial park, a trade-commercial district, and a number of trading markets to attract investment from foreign countries, and other provinces. Light industry is the largest sector of the city's economy. Crop growing, animal husbandry, and commercial agriculture. Grains, oil crops, cotton, vegetables, hogs, eggs, beef, lamb. Manufacturing (food processing, textiles, leather, paper).

Luojiang (LU-uh-JIANG), town, ⊙ Luojiang county, central N SICHUAN province, CHINA, 50 mi/80 km NE of CHENGDU; 31°20′N 104°31′E; rice, sugarcane, jute, oilseeds.

Luolong, CHINA: see LHORONG.

Luonan (LU-uh-NAN), town, ⊙ Luonan county, SE SHAANXI province, CHINA, near HENAN border, 65 mi/105 km E of XI'AN, in mountain region; 34°09′N 110°04′E; grain, medicinal herbs; machinery and equipment manufacturing; electronics.

Luoning (LU-uh-NING), town, ⊙ Luoning county, NW HENAN province, CHINA, on LUO RIVER, and 55 mi/89 km SW of LUOYANG; 34°22′N 111°38′E. Grain, tobacco.

Luoping (LU-uh-PING), town, ⊙ Luoping county, E YUNNAN province, CHINA, near GUIZHOU border, 100 mi/161 km ESE of KUNMING; 24°58′N 104°20′E. Elevation 6,561 ft/2,000 m. Rice, wheat, millet, beans, oilseeds; tobacco.

Luopu, CHINA: see LOP.

Luo River (LU-uh), Mandarin *Luo Shui* (LWAW SHUAI), 200 mi/322 km long, in NW HENAN province, CHINA; rises in mountains on SHAANXI border; flows ENE, past LUSHI, LUONING, YIYANG, and LUOYANG, to HUANG HE (Yellow River) below GONG Xian. Valley is center of earliest recorded Chinese civilization.

Luo River, 250 mi/402 km long, in SHAANXI province, CHINA; rises in Baiyu Mountains on northwesternmost Shaanxi-NINGXIA border; flows SSE across the LOESS PLATEAU, and past GANQUAN, FU XIAN, and DALI, to HUANG HE (Yellow River), near Huang He and WEI RIVER confluence.

Luorong (LU-uh-RUNG), town, NE central GUANGXI ZHUANG Autonomous Region, CHINA, 20 mi/32 km NE of LIUZHOU, and on railroad; 24°26′N 109°32′E. Rice, sugarcane, oilseeds.

Luoshan (LU-uh-SHAN), town, ⊙ Luoshan county, S HENAN province, CHINA, 23 mi/37 km E of XINYANG; 32°12′N 114°32′E. Rice, wheat, kaoliang, beans, jute, oilseeds.

Luo Shui, CHINA: see LUO RIVER.

Luotian (LU-uh-TIAN), town, ⊙ Luotian county, E HUBEI province, CHINA, 75 mi/121 km ENE of WUHAN; 30°47′N 115°20′E. Rice, oilseeds.

Luoyang (LU-uh-YANG), city (□ 210 sq mi/546 sq km; 2000 population 1,202,192), NW HENAN province, CHINA, on the LUO RIVER; 34°41′N 112°28′E. The city is the hub of several highways and is located on the LONGHAI RAILROAD. A new industrial center with a variety of heavy and light industries, it has quadrupled in size since 1949. LUOYANG, a major Chinese cultural center, was the center of several ancient dynasties, particularly that of the Eastern Zhou (770–256 B.C.E.) and the Tang dynasty (C.E. 618–906). Under the Zhou, it was the seat of several schools of philosophy. The nearby LONGMEN grottoes, embellished in the 6th century C.E., contain colossal carvings of Buddha; Baima Si, or White Horse Temple, built in C.E. 68, is the oldest temple in China. Luoyang was formerly called Henanfu or Honanfu. Sometimes spelled Loyang.

Luoyuan (LU-uh-YUAN), town, ⊙ Luoyuan county, NE FUJIAN province, CHINA, 30 mi/48 km NNE of FUZHOU, on EAST CHINA SEA coast; 26°30′N 119°33′E. Rice, wheat, sugarcane; food industry, tobacco processing.

Luozi (LWO-zee), village, BAS-CONGO province, W CONGO, on right bank of CONGO RIVER, and 90 mi/145 km NE of BOMA; 04°57′S 14°08′E. Elev. 321 ft/97 m. Palm-oil milling, fiber growing.

Lupac (loo-PAHK), Hungarian *Lupák*, village, CARAȘ-SEVERIN county, SW ROMANIA, 5 mi/8 km SW of REȘIȚA; 45°17′N 21°49′E. Coal mining.

Lupa Goldfields, TANZANIA: see CHUNYA.

Lupák, ROMANIA: see LUPAC.

Lupa Market (LOO-pah), village, MBEYA region, SW TANZANIA, 20 mi/32 km NW of MBEYA, on Zira River; MBEYA RANGE to S; 08°39′S 33°16′E. Coffee, corn, sweet potatoes, pyrethrum; cattle, sheep, goats. Former gold-mining region.

Lupane, town, MATABELELAND NORTH province, W central ZIMBABWE, 95 mi/153 km NNW of BULAWAYO, on LUPANE RIVER, opposite mouth of BUBI RIVER, and 2 mi/3.2 km SE of mouth of Lupane River on GWAYI RIVER; 18°56′S 27°46′E. Road junction. Livestock; grain.

Lupane River, c. 70 mi/113 km long, MATABELELAND NORTH province, W central ZIMBABWE; rises in Zenka Pan 10 mi/16 km SW of NKAYI; flows W past LUPANE, entering GWAYI RIVER 2 mi/3.2 km W of Lupane. BUBI RIVER enters from SE at Lupane. Also spelled Luapne River.

Lupao (LOO-pou), town, NUEVA ECIJA province, central LUZON, PHILIPPINES, 27 mi/43 km NNW of CABANATUAN; 15°51′N 120°55′E. Rice, corn.

Lupa River (LOO-pah), c. 85 mi/137 km long, MBEYA region, SW central TANZANIA; rises c. 85 mi/137 km N of MBEYA; flows S; enters Zira River 25 mi/40 km NW of Mbeya near village of LUPA MARKET.

Lupar River (LOO-pahr), 142 mi/229 km long; S SARAWAK, in W BORNEO, MALAYSIA; rises in KAPUAS Mountains on Indonesian border; flows SW to Lubok Antu, then WNW past SIMANGGANG to SOUTH CHINA SEA 40 mi/64 km E of KUCHING. Navigable by small craft.

Lupeni (loo-PEN), Hungarian *Lupény*, town, HUNEDOARA county, W central ROMANIA, on headstream of JIU river in the TRANSYLVANIAN ALPS, and 9 mi/14 km SW of PETROȘANI; railroad terminus, coal-mining center. Manufacturing of cellulose, rayon, and soap; coke ovens and flour mills.

Lupény, ROMANIA: see LUPENI.

Lupérico (LOO-pe-ree-ko), town, central SÃO PAULO state, BRAZIL, 13 mi/21 km SE of MARÍLIA; 22°25′S 49°48′W.

Luperón (loo-pe-RON), town, PUERTO PLATA province, N DOMINICAN REPUBLIC, on small inlet of the ATLANTIC OCEAN, and 17 mi/27 km WNW of PUERTO PLATA; 19°50′N 71°03′W. In agricultural region (coffee, cacao, tobacco, corn.) Until 1927, called BLANCO. The ruins of ISABELA are 9 mi/14.5 km W.

Lupionopolis (LOO-po-NO-po-lees), town (2007 population 4,377), N PARANÁ state, BRAZIL, on PARANÁPANEMA River (border with SÃO PAULO state); 22°44′S 51°40′W. Coffee, rice, cotton; livestock.

Lupiro (loo-PEE-ro), village, MOROGORO region, central TANZANIA, 75 mi/121 km SE of IRINGA; Mount Chikweta to S; 08°25′S 36°37′E. Road junction. Cattle, goats, sheep; corn, wheat.

Lupkow Pass (LOOP-kov), Polish *Przełęcz Łupkowska* (pshe-wench loop-kov-ska), pass (1,917 ft/584 m) between E BESKIDS and Beschady Mountains, on POLAND-SLOVAKIA border, 22 mi/35 km SSW of SANOK, Poland; used by railroad.

Lupombelo (loo-pom-BAI-lo), village, MBEYA region, SW central TANZANIA, 85 mi/137 km NNW of MBEYA; 07°51′S 33°00′E. Sheep, goats; corn, wheat.

Lupton City, village, HAMILTON county, SE TENNESSEE, N suburb of CHATTANOOGA, N of the TENNESSEE RIVER; 35°06′N 85°15′W.

Luputa (lo-POO-tah), village, KASAI-OCCIDENTAL province, S CONGO, on railroad, and 85 mi/137 km SW of KABINDA; 07°10′S 23°42′E. Elev. 2,762 ft/841 m.

Agricultural center (manioc, yams, groundnuts, livestock), with veterinary laboratory.

Luqa (LOO-kah), town (2005 population 6,072), E central MALTA, 3 mi/4.8 km SSW of VALLETTA; 35°51′N 14°29′E. In agricultural region (vegetables, livestock). International airport. Suffered severely in World War II air raids. Also spelled Luca.

Luquan (LU-CHUAN), town, ⊙ Luquan county, NE central YUNNAN province, CHINA, 38 mi/61 km NNW of KUNMING, in mountain region; 25°35′N 102°30′E. Grain, tobacco, timber; food processing.

Luque (LOO-kai), city, (2002 population 170,986), Central department, S PARAGUAY, on railroad and road junction, and 8 mi/13 km E of ASUNCIÓN; 25°15′S 57°29′W. Manufacturing and agricultural center (fruit, sugarcane, tobacco, cotton; livestock); soap factories, liquor distilleries, brick- and tileworks, maté mills. Founded 1635; temporary ⊙ during War of the Triple Alliance (1865–1870). The city is now part of the Asunción metropolitan area.

Luque (LOO-kai), town, central CÓRDOBA province, ARGENTINA, 55 mi/89 km ESE of CÓRDOBA; 31°39′S 63°22′W. Grain, soybeans, fruit, cattle.

Luque, town, CÓRDOBA province, S SPAIN, 5 mi/8 km SE of BAENA; 37°33′N 04°16′W. Olive-oil processing. Agricultural trade (cereals, almonds, vegetables); lumber; livestock raising. Marble, stone, and gypsum quarries.

Luquillo (loo-KEE-yo), town (2006 population 20,452), NE PUERTO RICO, on the ATLANTIC, 27 mi/43 km ESE of SAN JUAN. Industrial, commercial, and tourism area. Luquillo Beach bathing resort (one of island's most beautiful beaches), towered over by Sierra de LUQUILLO.

Luquillo, Sierra de (loo-KEE-yo, see-YER-rah dai), mountain range in NE PUERTO RICO, 20 mi/32 km SE of SAN JUAN. The CARIBBEAN NATIONAL FOREST is located here; extends c. 15 mi/24 km SW-NE, rising to c. 3,494 ft/1,065 m. Its highest and best-known peak is EL YUNQUE. Largely forested; timber is used for charcoal. Resort area. El Portal del Yunque, a visitor center, with ecological exhibits on the tropical forest.

Luquisani (loo-kee-SAH-nee), canton, MUÑECAS province, LA PAZ department, W BOLIVIA, SE of TUILUNI; 15°24′S 68°45′W. Elevation 12,303 ft/3,750 m. Limestone and gypsum deposits. Agriculture (potatoes, yucca, bananas, rye); cattle.

Lurate Caccivio (loo-RAH-te kaht-CHEE-vyo), commune, COMO province, LOMBARDY, N ITALY, 5 mi/8 km WSW of COMO; 45°46′N 09°01′E. Chief town: Caccivio; commune seat, nearby Lurate; secondary diversified industrial center, including silk-milling centers.

Luray, town (2000 population 102), CLARK county, extreme NE MISSOURI, near North WYACONDA RIVER, 9 mi/14.5 km WNW of KAHOKA; 40°27′N 91°52′W.

Luray (LOO-rai), town (2006 population 4,878), ⊙ PAGE county, N VIRGINIA, 27 mi/43 km NE of HARRISONBURG, on Hawksbill Creek, in SHENANDOAH VALLEY; 38°39′N 78°27′W. Manufacturing (automotive door panels, canoes, printing and publishing; clothing, leather finishing); in agricultural area (apples, peaches, grain; poultry, livestock; dairying). Headquarters of SHENANDOAH NATIONAL PARK. George Washington National Forest is N and W. The Luray Caverns to W discovered in 1878, noted for their large stalagmite and stalactite formations, and water pools. Incorporated 1812.

Luray (lor-AI), village (2000 population 203), RUSSELL county, central KANSAS, 16 mi/26 km NNE of RUSSELL; 39°07′N 98°41′W. Livestock, grain.

Luray (luh-RAI), village (2006 population 113), HAMPTON county, SW SOUTH CAROLINA, 15 mi/24 km SSE of ALLENDALE; 32°48′N 81°14′W. Cotton, soybeans, watermelons; livestock.

Lurcy-Lévis (luhr-see-lai-vee), commune (□ 27 sq mi/70.2 sq km), ALLIER department, AUVERGNE region, central FRANCE, 22 mi/35 km NW of MOULINS; 46°43′N 02°56′E. Manufacturing of woodworking equipment. Has a Romanesque church. Gypsum and kaolin quarries nearby.

Lure (LYOOR), town (□ 9 sq mi/23.4 sq km), subprefecture of HAUTE-SAÔNE department, FRANCHE-COMTÉ region, E FRANCE, near the OGNON RIVER, 17 mi/27 km WNW of BELFORT; 47°42′N 06°30′E. A regional market with metal and chemical works; woodworking. A 7th-century Benedictine abbey, rebuilt in 18th century, now houses administrative offices.

Lure, Lake, NORTH CAROLINA: see LAKE LURE TOWN.

Luremo (loo-RAI-mo), town, LUNDA NORTE province, ANGOLA, road, and 180 mi/290 km W of LUCAPA; 08°31′S 17°50′E. Market center.

Lure Mountain (LYOOR), a low range of the MARITIME ALPS, in ALPES-DE-HAUTE-PROVENCE department, PROVENCE-ALPES-CÔTE D'AZUR region, SE FRANCE, extending c.20 mi/32 km from the DURANCE valley below SISTERON westward to Mont VENTOUX; 44°07′N 05°47′E. Rises to 6,000 ft/1,829 m. Vegetation consists of oaks, cedars, and lavender fields below barren crestline. Excellent views of high ALPS to NE. Lure Mountain figures in the writings of Jean Giono.

Luretha, KENTUCKY: see FERGUSON.

Lurgan (LUHR-guhn), town (2001 population 24,000), ARMAGH, SW Northern Ireland, 3 mi/4.8 km NE of CRAIGAVON, 2 mi/3.2 km S of Lough Neagh; 54°28′N 06°20′W. Textiles. Linen Hall is the hub of the town. Center for trout fishing.

Luribay (loo-ri-BEI), town and canton, ⊙ LOAYZA (until 1930s, Loaiza) province, LA PAZ department, W BOLIVIA, on Luribay River (branch of LA PAZ RIVER), and 55 mi/89 km SE of LA PAZ; 17°14′S 67°44′W. Elevation 8,331 ft/2,539 m. Vineyards, orchards (bananas, mangoes, apples).

Luribay Valley, in LA PAZ department, W BOLIVIA, between Serranía de SICASICA (SW) and Cordillera de TRES CRUCES (NE); watered by Luribay River and its affluent, the Caracato River.

Lurigancho-Chosica (loo-ree-GAHN-cho–cho-SEE-kah), town, LIMA province, W central PERU, in ANDEAN foothills, at confluence of RÍMAC and SANTA EULALIA rivers, on LIMA–LA OROYA railroad, on highway, and 27 mi/43 km ENE of Lima; 11°54′S 76°42′W. Resort, situated above coastal mists; elevation c.2,800 ft/853 m. Sugarcane, fruit. Has hydroelectric plant serving Lima. Paper mill. Pre-Columbian ruins nearby.

Lurín (loo-REEN), town, LIMA province, W central PERU, on coastal plain, near the PACIFIC, on PAN-AMERICAN HIGHWAY, and 13 mi/21 km SE of LIMA; 12°19′S 76°15′W. Sugarcane, cotton, vegetables.

Lúrio River (LO-ree-o), 335 mi/539 km long, N MOZAMBIQUE; rises W of NAMULI MOUNTAINS near the MALAWI border; flows NE to the MOZAMBIQUE CHANNEL 40 mi/64 km S of PEMBA. Not navigable. Reached by railroad from LUMBO (opposite Mozambique city).

Luristan, IRAN: see LORESTAN.

Lurnfeld (LURN-feld), region in CARINTHIA, S AUSTRIA, wide section (basin) of the Drau Valley between confluence of Möll River (NW) and Lieser River (SE), just NW of SPITTAL AN DER DRAU; 46°48′–51′N, 13°21′–28′E. Since ancient times, a focus on settlement and traffic; the Roman town, Teurnia, was located here (near Sankt Peter im Holz). Main settlements are Spittal an der Drau and Möllbrücke.

Luruaco (lor-WAH-ko), town, ⊙ Luruaco municipio, ATLANTICO department, N COLOMBIA, 30 mi/48 km SW of BARRANQUILLA; 10°37′N 75°09′W. Cotton, corn, soybeans, plantains; livestock.

Lurzut (luhr-ZOOT), village, QABILI province, SE TUNISIA, 47 mi/76 km S of Ramada. Road links village to border post at Sinawan, c.75 mi/121 km SE.

Lusahunga (loo-sah-HOON-gah), village, KAGERA region, NW TANZANIA, 18 mi/29 km S of Biharmulo;

02°53′S 31°13′E. Road junction. Tobacco, corn; goats, sheep. Also spelled Lusahanga.

Lusaka (loo-SAH-ka), province (□ 8,454 sq mi/21,980.4 sq km; 2000 population 1,391,329), S central ZAMBIA; ⊙ LUSAKA; 15°25′S 29°00′E. Bounded on W and N by CENTRAL province, on E by EASTERN province, (LUNSEMFWA RIVER) and MOZAMBIQUE (LUANGWA RIVER), on S by ZIMBABWE (ZAMBEZI RIVER) and SOUTHERN province, (KAFUE RIVER); created from Central province. LOWER ZAMBEZI NATIONAL PARK in E. Zambia's main railroad line crosses province in W. Gold deposits in center and E. W part in Zambia's main agricultural belt (cotton, tobacco, soybeans, corn, peanuts, sorghum, coffee, vegetables, fruit, flowers); cattle; poultry; ostriches. Manufacturing at Lusaka.

Lusaka (loo-SAH-ka), city (2000 population 1,084,703), CENTRAL province, ⊙ ZAMBIA, S central Zambia; 15°25′S 28°07′E. Elevation 4,200 ft/1,280 m. A sprawling city located in a productive farm area, Lusaka is an administrative, financial, and commercial center. Agriculture (tobacco, cotton, peanuts, grain, soybeans, vegetables, coffee, flowers, chilies, sorghum); poultry; cattle; ostriches. Manufacturing (food processing, diversified metal manufacturing; construction products, zinc oxide, agrichemicals, textiles, leather and marble products, gems; glass and mirrors, tobacco, alcoholic spirits; petroleum oils, motor vehicles). The city is at the junction of the Great North Road (to TANZANIA) and the Great East Road (to MALAWI), and is on Zambia's main railroad—the Tazara, or Tan Zam, railroad—connecting through KAPIRI MPOSHI (100 mi/161 km to N) with Dar-es-Salaam, Tanzania, providing Zambia with alternative railroad access to the ocean. Lusaka International Airport 20 mi/32 km to SE. University of Zambia (1965); National Archives and Library; Art Center; Military Museum; Lusaka Stock Exchange (LUSE; established February 1994); Luburma Market; Freedom Statue. Munda Wanga Botanical Gardens and Lusaka Zoo to S. Founded in 1905 by Europeans and named after the headman of a nearby African village, the city's main growth occurred after 1935, when it replaced Livingstone (now MARAMBA) as the capital of the British colony of Northern Rhodesia (Zambia). The city and its surrounding area became a province in 1976.

Lusaka-Saint-Jacques (loo-SAH-kuh—SAIN—ZHAHK), village, KATANGA province, SE CONGO, 85 mi/137 km SSE of KALEMIE; 07°10′S 29°27′E. Elev. 4,212 ft/1,283 m. Cattle raising, vegetable farming. Roman Catholic mission with small seminary and convent.

Lusakert, urban-type settlement, ARMENIA; 40°23′N 44°36′E. Tool-making plant. Arose in 1949 in connection with the construction of the Giumush Hydroelectric Power Plant.

Lusambo (loo-SAHM-bo), town, KASAI-ORIENTAL province, central CONGO, on right bank of SANKURU RIVER, and 550 mi/885 km E of KINSHASA; 04°58′S 23°26′E. Elev. 1,305 ft/397 m. Commercial, cotton, and communications center. Saw-milling (notably railroad-sleeper manufacturing), cotton ginning, cottonseed-oil milling, rice processing. Coffee, cacao, and rubber plantations in vicinity. Has Roman Catholic and Protestant missions, various schools, including business school for sons of chiefs, hospitals, and an airport. Lusambo post was founded in 1889 and was the headquarters of CONGO FREE STATE troops during the campaign against Arab slave traders (1892–1894).

Lusanga (loo-SAHN-guh), village, BANDUNDU province, SW CONGO, on left bank of Kwenge River, and 13 mi/21 km NNW of KIKWIT; 04°50′S 18°44′E. Elev. 1,010 ft/307 m. Major palm products center; palm-oil milling. Has Roman Catholic missions, hospitals, mechanical and other trade schools. Formerly known as LEVERVILLE or LEVERSTAD.

Lusangania (loo-suhn-GAHN-nee-yuh), village, Équateur province, NW CONGO, 72 mi/116 km E of MBANDAKA; 00°10′N 24°34′E. Elev. 1,486 ft/452 m.

Lusatia (LOO-sai-shuh), German *Lausitz* (LOU-sits), Polish Łużyce, region of E GERMANY and SW POLAND; extends N from the LUSATIAN MOUNTAINS, at the CZECH border, and W from the ODER River. The hilly and fertile S section is known as UPPER LUSATIA, the sandy and forested N part as LOWER LUSATIA. The LUSATIAN NEISSE separates E Germany and SW Poland. Forestry, farming, and livestock raising are the chief occupations. There are lignite mines, textile mills, and glass-making factories. BAUTZEN, COTTBUS, GÖRLITZ, ZAGAN, and ZITTAU are the main towns. The Lusatians are descended from the Slavic Wends, and part of the population, especially in the SPREE FOREST, still speaks Wendish and has preserved traditional dress and customs. The region was colonized by the Germans beginning in the tenth century and was constituted into the margraviates of Upper and Lower Lusatia. Both margraviates changed hands frequently among SAXONY, BOHEMIA, and BRANDENBURG. In 1346 several towns of the region formed the Lusatian League and preserved considerable independence. Under the Treaty of Prague (1635), all of Lusatia passed to Saxony. The Congress of Vienna awarded (1815) Lower Lusatia and a large part of Upper Lusatia to PRUSSIA. After World War II, the Lusatian Wends (or Sorbs, as they are also called) sought unsuccessfully to obtain national recognition.

Lusatian Mountains (loo-SAI-shyuhn), Czech *Lužické hory* (LU-zhits-KEE HO-ri), German *Lausitzer Gebirge*, Polish *Góry Łużyckie*, westernmost range of the SUDETES, in UPPER LUSATIA, along CZECH borders with GERMANY and POLAND; extend c.60 mi/97 km between the ELBE (W) and upper LUSATIAN NEISSE (E) rivers. Highest peaks are JESTED (3,320 ft/1,012 m), 3 mi/4.8 km SW of LIBEREC, and Luž (2,060 ft/793 m), 12 mi/19 km. NNE of CESKA LIPA. Semiprecious stones (Bohemian garnets, agates, amethysts) quarried on SE slope (gem industry at TURNOV).

Lusatian Neisse, river, POLAND: see NEISSE.

Lusavan, ARMENIA: see CHARENTSAVAN.

Lus Bela, PAKISTAN: see LAS BELA.

Lusby (LUZ-bee), village, CALVERT county, S MARYLAND, near CHESAPEAKE BAY, 12 mi/19 km SE of PRINCE FREDERICK. Nearby is "Charlesgift," formerly called Preston-at-Patuxent (built 1650), one of oldest Maryland houses. Nearby is the Calvert Cliffs Nuclear Power Station, and the nearby beach area famous for marine fossils.

Luscar (LUH-skuhr), unincorporated village, W ALBERTA, in ROCKY MOUNTAINS at foot of Luscar Mountain (8,534 ft/2,601 m), near E side of JASPER NATIONAL PARK, 30 mi/48 km ENE of JASPER, in YELLOWHEAD COUNTY; 53°04′N 117°24′W. Coal mining; cattle.

Luseland (LOOS-land), town (2006 population 571), W SASKATCHEWAN, CANADA, 65 mi/105 km SW of NORTH BATTLEFORD; 52°05′N 109°26′W. Wheat.

Lusengo (loo-SENG-go), village, Équateur province, NW CONGO, on right bank of CONGO RIVER, and 20 mi/32 km ENE of NOUVELLE-ANVERS; 01°46′N 19°29′E. Elev. 1,272 ft/387 m. Steamboat landing and trading post; palm groves.

Luserna San Giovanni (loo-ZER-nah sahn jo-VAHN-nee), commune, TORINO province, PIEDMONT, NW ITALY, in MONTE VISO region of COTTIAN ALPS; 44°48′N 07°15′E. Includes Airali, 6 mi/10 km SW of PINEROLO, and adjacent villages of Luserna and San Giovanni. Secondary diversified industrial center. Produces woolen textiles, chocolate.

Lushai Hills (LOO-shei), in MIZORAM state, extreme NE INDIA, between MYANMAR (E) along TYAO and BOINU rivers (headstreams of KALADAN RIVER) and BANGLADESH (W). Coextensive with LUSHAI HILLS (S continuation of MANIPUR HILLS), rising to over 5,500 ft/1,676 m. Drained by tributaries of BARAK (SURMA; N), KALADAN (S), and KARNAPHULI (W) rivers. Agriculture (rice, cotton, sesame, sugarcane, tobacco); orange groves (center at SAIRANG). Extensive bamboo tracts. Main villages: AIZAWL, LUNGLEI. Formerly an autonomous district, received special status (1950) in accordance with Indian Constitution; now part of Mizoram state.

Lushan (LU-SHAN), town, ⊙ Lushan county, W HENAN province, CHINA, on SHA RIVER, and 50 mi/80 km NNE of NANYANG; 33°44′N 112°55′E. Grain, tobacco, oilseeds; building materials, machinery and equipment manufacturing; chemicals, food processing, coal mining.

Lushan, town, ⊙ Lushan county, W SICHUAN province, CHINA, 55 mi/89 km ENE of KANGDING; 30°10′N 102°59′E. Grain, tea, oilseeds, medicinal herbs; papermaking.

Lushan (LU-SHAN), mountain (4,836 ft/1,474 m), N JIANGXI province, CHINA; formed in a tectonic movement some 20,000,000 years ago. The most recent ice age left various erosional and depositional landforms. Due to its humid subtropical climate and various microclimatic conditions, a variety of plant species grow in the Lushan. Forest covers over 76% of the area. The Lushan botanical garden (established 1934), the oldest in China, contains more than 2,400 plant species. Its natural beauty has attracted visitors from across the nation since ancient times. Site of numerous Buddhist and Taoist temples. The Pure Land sect, established during the Eastern Jin dynasty (316–450), was the Buddhist center in S China. The White Deer Cave Academy was revitalized by Zhu Xi in 1179 and became one of the four most important higher learning centers in China at the time. Many historic figures, writers, and painters have praised the area's beauty in their work. In the late 19th century, European and Chinese elite came to build resort homes. Popular for conventions and recreation.

Lushi (LU-SHI), town, ⊙ Lushi county, NW HENAN province, CHINA, on upper LUO RIVER, and 100 mi/161 km WSW of LUOYANG, in mountain region; 34°04′N 111°02′E. Grain, tobacco; food processing, non-ferrous metal smelting.

Lushnje (LOOSH-nye), town (2001 population 37,872), ⊙ Lushnje district (□ 270 sq mi/702 sq km; 2001 population 37,872), W central ALBANIA, on road, and 21 mi/34 km NW of BERAT at N edge of MYZEQE plain; 40°55′N 19°39′E. Agricultural center with agriculture school and tree nursery. Also spelled Lushnja.

Lushoto (loo-SHO-toh), town, TANGA region, NE TANZANIA, in USAMBARA Mountains, 60 mi/97 km WNW of TANGA; 04°48′S 38°16′E. Elevation 4,579 ft/1,396 m. Mountain resort and market center. Seat of Urdambara Trade School and Silviculture Research Institute. Known as Wilhelmstal under German rule.

Lushton, village (2006 population 33), YORK county, SE NEBRASKA, 12 mi/19 km SW of YORK, and on West Fork of BIG BLUE RIVER; 40°43′N 97°43′W.

Lushui (LU-SHUAI), town, ⊙ Lushui county, NW YUNNAN province, CHINA, 60 mi/97 km NNW of BAOSHAN, and on right bank of SALWEEN RIVER, near HPIMAW (MYANMAR border town); 25°58′N 98°50′E. Grain, timber; food industry, non-ferrous-ore mining.

Lüshun (LOO-SHUN), city, SW LIAONING province, CHINA, at the tip of the LIAODONG peninsula; 38°46′N 121°15′E. It was formerly combined with DALIAN (Dairen) into the joint municipality of LÜDA. Now, it is an administrative unit of Dalian. Lüshun is an important naval base dominating the entrance to the BOHAI; it is also a S terminus of the South Liaoning railroad. The city was the administrative center of the Liaodong leasehold from 1898 to 1945 (see LIAONING). It was near the marine battlefield during the Sino-Japanese War (1895). As a RUSSIAN base (1898–1905), it was the site, on February 8, 1904, of the surprise Japanese naval attack that precipitated the Russo-

Japanese War. The city passed to Japan by the Treaty of Portsmouth (1905). In 1945 it became the headquarters of the Port Arthur Naval Base District under joint Sino-Soviet administration. China regained exclusive control in 1955. The naval base has opened nearly half of its territory to tourism and investment since 1996.

Lusignan (lyu-zee-nyahn), commune (□ 14 sq mi/36.4 sq km), VIENNE department, POITOU-CHARENTES region, W central FRANCE, 14 mi/23 km SW of POITIERS; 46°26′N 00°07′E. Cattle market. Limekilns. Has ruins of a castle inhabited by the Lusignan family, rulers of CYPRUS during the Middle Ages.

Lusigny-sur-Barse (lyu-zee-nyee–syur–bahr-suh), village (□ 14 sq mi/36.4 sq km), AUBE department, CHAMPAGNE-ARDENNE region, NE central FRANCE, 9 mi/14.5 km ESE of TROYES; 48°15′N 04°16′E. Sawmilling. Lies within the FORÊT D'ORIENT NATURAL REGIONAL PARK, with three interconnected lakes and a wildlife preserve.

Lusikisiki, town, EASTERN CAPE province, SOUTH AFRICA, 15 mi/24 km inland from INDIAN OCEAN coast (called the Wild Coast), 25 mi/40 km N of PORT ST. JOHNS. Serves as a commercial and transportation center for farming community and a number of large tea estates.

Lusitania (LOO-sah-TAI-nee-yah), Roman province in the IBERIAN PENINSULA and historic name for PORTUGAL. As constituted (C.E. c.5) by Augustus, the province included all of modern central Portugal as well as much of W SPAIN. It took its name from the Lusitani, a group of warlike tribes who, despite defeats, resisted Roman domination until their great leader, Viriatus, was killed (139 B.C.E.) by treachery. In the first century B.C.E. they joined in supporting Sertorius, who set up an independent state in Spain. The old identification of Portugal with Lusitania and of the ancestors of the Portuguese with the Lusitanians (hence Camões's great epic was entitled *Os Lusíadas*) is now largely ignored, but the creation of Lusitania may have resonated in the setting up of the separate kingdom of Portugal many centuries later.

Lusk (LUHSK), Gaelic *Lusca*, town (2006 population 5,236), NE DUBLIN county, E IRELAND, 14 mi/23 km NE of DUBLIN; 53°31′N 06°10′W. Agricultural market. Remnants of monastery founded in 5th century, including round tower.

Lusk (LUHSK), town (2006 population 1,330), ⊙ NIOBRARA county, E WYOMING, on Niobrara River (intermittent), and 100 mi/161 km E of CASPER; 42°45′N 104°27′W. Elevation 5,015 ft/1,529 m. Trading point in dry-farming, cattle, and sheep region; wheat, oats. LANCE CREEK oil field to NW. Silver and radium once mined here. Town refers to itself as seat of least populated county in least populated state (in numbers, Wyoming is 50th); oil boom of 1918 briefly swelled population to 10,000. Incorporated 1898.

Luso (LOO-zoo), village, AVEIRO district, N central PORTUGAL, on railroad, and 13 mi/21 km NNE of COIMBRA; 40°23′N 08°23′W. Spa with alkaline springs.

Lussac (lyu-sahk), commune (□ 9 sq mi/23.4 sq km), GIRONDE department, AQUITAINE region, SW FRANCE, 8 mi/12.9 km ENE of LIBOURNE; 44°57′N 00°06′W. SAINT-ÉMILION vineyards surround the village.

Lussac-les-Châteaux (lyu-sahk–lai–shah-to), commune (□ 10 sq mi/26 sq km), VIENNE department, POITOU-CHARENTES region, W central FRANCE, near VIENNE river, 21 mi/34 km SE of POITIERS; 46°24′N 00°43′E. Flour milling. Has museum of archeology displaying a large number of prehistoric relics, including tools, animal drawings, and human profiles discovered in local caves.

Lussino, CROATIA: see LOŠINJ ISLAND.

Lustenau (LUS-ten-ou), township, VORARLBERG, W AUSTRIA, on the RHINE (Swiss border), 4 mi/6.4 km W of DORNBIRN; 47°26′N 09°40′E. Customs station; mar-

ket center; embroidery center, textiles. Manufacturing of metals and timber goods.

Lustrafjorden (LOOST-rah-fyawr-uhn), N arm of SOGNEFJORDEN, in SOGN OG FJORDANE county, W NORWAY; c.30 mi/48 km long, 2 mi/3.2 km wide. Terminates at SKJOLDEN, where glacier streams from the JOSTEDALSBREEN give fjord water a milky appearance.

Lusushwane River, c.75 mi/121 km long, SE Africa and SWAZILAND; rises at Lochiel, MPUMALANGA province, E SOUTH AFRICA; flows in serpentine course, enters Swaziland 12 mi/19 km W of MBABANE; passes SW of Mbabane, through MLILWANE WILDLIFE SANCTUARY, joins GREAT USUTU RIVER at SIDVOKODVO. Called Little Usutu River in Swaziland.

Lusutufu River, E SOUTH AFRICA: see GREAT USUTU RIVER.

Luswaka (loos-WAH-kuh), village, KATANGA province, SE CONGO, close to MANIKA PLATEAU (UPEMBA NATIONAL PARK); 09°51′S 26°30′E. Elev. 3,415 ft/1,040 m.

Lütao Island or **Green Island**, also called *Huo-shao Tao* in Chinese, volcanic island (□ 11 sq mi/28.6 sq km), off SE TAIWAN coast, 20 mi/32 km SE of TAITUNG; 23°09′N 120°29′E. Sweet potatoes, rice, peanuts, livestock; fisheries. Road on perimeter of island; two lighthouses. Nanliao is main village. Formerly called Huo-shao Island. Also spelled Lüdao.

Lutcher (luhch-uhr), town (2000 population 3,735), SAINT JAMES parish, SE central LOUISIANA, on E bank (levee) of the MISSISSIPPI RIVER (toll ferry here), and 36 mi/58 km W of NEW ORLEANS; 30°04′N 90°43′W. In agricultural area (okra, sugarcane, cattle); crawfish; chemicals. Manufacturing (concrete). Founded c. 1890. Chemical plants in area.

Lutécia (LOO-te-see-ah), town (2007 population 2,788), W SÃO PAULO, BRAZIL, 28 mi/45 km WSW of MARÍLIA; 22°20′S 50°23′W. Agriculture (coffee, cotton).

Lutesville, former town, BOLLINGER county, SE MISSOURI, 17 mi/27 km SW of JACKSON; annexed by MARBLE HILL in 1986.

Lutetia, FRANCE: see PARIS.

Lutfabad, IRAN: see LOTFABAD.

Luther, town (2000 population 158), BOONE county, central IOWA, 8 mi/12.9 km SSE of BOONE; 41°58′N 93°49′W. In agricultural area.

Luther, town (2006 population 1,096), OKLAHOMA county, central OKLAHOMA, residential suburb 22 mi/35 km NE of downtown OKLAHOMA CITY, near the DEEP FORK of Canadian River; 35°39′N 97°10′W.

Luther, village (2000 population 339), Lake county, W central MICHIGAN, 20 mi/32 km SW of CADILLAC; 44°02′N 85°40′W. In farm area. Part of Manistee National Forest to NE.

Lutherstadt Wittenberg, GERMANY: see WITTENBERG.

Luthersville (LOO-thuhrs-vil), town (2000 population 783), MERIWETHER county, W GEORGIA, 19 mi/31 km NE of LA GRANGE; 33°13′N 84°44′W. Lumber.

Lutherville, suburban village, BALTIMORE county, N MARYLAND, between LOCH RAVEN RESERVOIR and Lake Roland, and 9 mi/14.5 km N of downtown BALTIMORE; 39°26′N 76°37′W. Named for Martin Luther by its founder, John C. Morris, it is the site of former Maryland College for Women (seminary chartered 1853). The Fire Museum of Maryland is located here. Timonium Fairgrounds here, one of the 3.5-mi/5.6-km race tracks in Maryland, is the site every year of the 10-day state fair held by the Baltimore State Fair and Agricultural Society of Baltimore County.

Lutin (LU-tyeen), Czech *Lutdn*, village (1991 population 3,331), SEVEROMORAVSKÝ province, central MORAVIA, CZECH REPUBLIC, 6 mi/9.7 km WSW of OLOMOUC; 49°34′N 17°08′E. Manufacturing (machinery, primarily pumps; plastic); agriculture (wheat, sugar beets, vegetables); cattle.

Lütjenburg (LOOT-yen-boorg), town, SCHLESWIG-HOLSTEIN, NW GERMANY, 18 mi/29 km E of KIEL; 54°18′N 10°36′E. Distilling; manufacturing of medical equipment. Has late-Romanesque church; chartered 1275.

Lutkun (loot-KOON), village (2005 population 3,340), S DAGESTAN REPUBLIC, E CAUCASUS, extreme SE European Russia, on mountain highway, 82 mi/132 km S of MAKHACHKALA, and less than 3 mi/5 km W of AKHTY; 41°29′N 47°41′E. Elevation 3,562 ft/1,085 m. Ecotourism base; indigenous handicrafts. Site of sporadic fighting between Chechen and Russian forces at the beginning of the second Russian-Chechen conflict in 1999.

Luton (LOO-tuhn), town and county (□ 17 sq mi/44.2 sq km; 2001 population 184,371), S central ENGLAND on the LEA RIVER; 51°53′N 00°25′W. Manufacturing of hats, motor vehicles, ball bearings, and aircraft parts. The English millinery industry was established here during the time of James I. International airport 2 mi/3.2 km E. Formerly in N BUCKINGHAMSHIRE.

Lutope River, c. 90 mi/145 km long, NW central ZIMBABWE; rises in MAFUNGABUSI PLATEAU, 17 mi/27 km S of GOKWE; flows W, then N, entering SENGWA RIVER 48 mi/77 km WNW of Gokwe. Forms part of boundary between MATABELELAND NORTH and MIDLANDS provinces.

Lutro, Greece: see LOUTRO.

Lutry, residential commune, VAUD canton, W SWITZERLAND, on LAKE GENEVA, 3 mi/4.8 km E of LAUSANNE. Church (13th–16th century); fine views (across lake to SE) of CHABLAIS mountains (French Alps).

Lütschine, 17 mi/27 km long (including the White Lütschine), S central SWITZERLAND; rises in BERNESE ALPS in two headstreams, the White and the Black Lütschine, which join 4 mi/6.4 km S of INTERLAKEN; flows N to Lake Brienz.

Lutsen (LOOT-suhn), unincorporated village, COOK county, NE MINNESOTA, 15 mi/24 km WSW of GRAND MARAIS; 47°38′N 90°40′W. Resort community on Lake SUPERIOR, in Superior National Forest. Manufacturing of draperies and upholstery. Cascade River State Park to NE; Ray Berglund State Wayside to SW. Lutsen Ski Area to N. Lutsen Mountains Resort to SW.

Lutshwadi River (loot-SHWAH-dee), tributary of the KASAI RIVER, KASAI-OCCIDENTAL province, SW CONGO; 04°20′S 20°34′E. Merges with the KASAI RIVER at ILEBO (PORT-FRANCQUI).

Luts'k (LOOTSK) (Russian *Lutsk*) (Polish *Łuck*), city (2001 population 208,816), ⊙ VOLYN' oblast, UKRAINE, on the STYR RIVER; 50°45′N 25°20′E. Elevation 574 ft/174 m. A river port, it has industries producing automobiles, bearings, electric appliances, scientific instruments; manufacturing (plastics, building materials, footwear, clothing, food products, textiles). It houses a pedagogical institute, a branch of the L'viv Polytechnical Institute, five special schools, six vocational technical schools; two theaters, philharmonic orchestra; Volhynia heritage museum. First mentioned in 1085 as Luches'k, it is one of the oldest cities of Volhynia. It was the main fortress of the Luchan tribe and was called Luches'k Velykyy. Luts'k, together with all of Volhynia, was part of Kievan Russia until 1154, when it became the capital of the Luts'k independent principality. It was included in the Halych-Volyn' principality in the 13th century, was taken by Lithuania in the 14th century, and was an important trade city from the 14th to the 16th century Luts'k was part of Poland from the second half of the 16th century, was taken by Russia in 1791, was Polish again from 1919 to 1939, and was annexed to the Ukrainian SSR in 1939. Jewish community dated back to the 10th century. In 13th century, it grew rapidly to become a majority of the population in the early 1920s, and nearly half of the population before World War II. The community was liquidated by the Nazis during World War II. Architectural monuments include the walls and turrets of a castle (13th–16th century) and several churches (14th–17th century). Also called, in Ukrainian in the 1920s, Luts'ke.

Lutsk, UKRAINE: see LUTS'K.

Luts'ke, UKRAINE: see LUTS'K.

Luttelgeest (LUH-tuhl-khaist), village, NORTH-EAST POLDER, FLEVOLAND province, central NETHERLANDS, c.10 mi/16 km SE of LEMMER, and 5 mi/8 km NE of EMMELOORD.

Lutten (LUH-tuhn), village, OVERIJSSEL province, E NETHERLANDS, 9 mi/14.5 km SSE of HOOGEVEEN; 52°37′N 06°35′E. Shetland Ponypark to NW. Dairying; cattle, hogs; vegetables, grain.

Lutter am Bärenberge (LUHT-ter ahm BER-en-berge), village, LOWER SAXONY, N GERMANY, 9 mi/14.5 km NW of GOSLAR; 51°39′N 10°17′E. Scene (1626) of Tilly's victory over Christian IV of Denmark.

Lutterbach (lu-ter-BAHK), German (LOO-tuhr-bahk), outer WNW suburb (□ 3 sq mi/7.8 sq km; 2004 population 6,070), of MULHOUSE, HAUT-RHIN department, in ALSACE, E FRANCE; 47°46′N 07°17′E. Diverse manufacturing.

Lutterworth (LUHT-uh-wuhth), town (2001 population 8,293), S LEICESTERSHIRE, central ENGLAND, 6 mi/9.7 km NNE of RUGBY; 52°27′N 01°12′W. Clothing; light machinery and equipment manufacturing. Has 12th-century church at which John Wycliffe was rector in his last years.

Luttrell (LUH-truhl), town (2006 population 949), UNION county, NE TENNESSEE, 17 mi/27 km NNE of KNOXVILLE; 36°13′N 83°45′W.

Lutugino, UKRAINE: see LUTUHYNE.

Lutuhyne (loo-TOO-hi-ne) (Russian *Lutugino*), city, S central LUHANS'K oblast, UKRAINE, in the DONBAS, at railroad junction, 11 mi/18 km SSW of LUHANS'K; 48°24′N 39°12′E. Elevation 465 ft/141 m. Raion center; machine construction, metalworking, coal mining. Established in 1896 as Shmidtivka (Russian *Shmidtovka*); renamed in 1925 after Leonid I. Lutugin, Russian geologist of the Donets Basin; city since 1960.

Lutunguru (loo-tung-GOO-roo), village, NORD-KIVU province, E CONGO, 70 mi/113 km N of BUKAVU; 00°28′S 28°49′E. Elev. 5,823 ft/1,774 m. Gold-mining center. Also gold mining at BILATI, 3 mi/4.8 km SE, and MOHANGA, 11 mi/18 km ENE. Vicinity of LUTUNGURU is known for production of food staples (manioc, corn).

Lutynia, POLAND: see LEUTHEN.

Lutz (LUHTZ), unincorporated town (□ 24 sq mi/62.4 sq km; 2000 population 17,081), HILLSBOROUGH county, W central FLORIDA, 10 mi/16 km N of TAMPA; 28°08′N 82°28′W. Manufacturing (neon signs, metal roofing; printing and publishing; hydraulic units).

Lützelflüh, commune, BERN canton, NW central SWITZERLAND, on EMME RIVER, 12 mi/19 km ENE of BERN. Textiles.

Lützen (LOO-tsuhn), town, SAXONY, S central GERMANY, 10 mi/16 km SW of LEIPZIG; 51°17′N 12°08′E. There, in the Thirty Years War, Gustavus II of Sweden defeated (1632) General Albrecht Wallenstein, but was killed in the battle. In 1813, Napoleon I defeated the Russian and Prussian forces at nearby Grossgörschen (also spelled Gross Görschen).

Lutzmannsburg (LUTS-mahns-burg), township, central BURGENLAND, E AUSTRIA, near RABNITZ RIVER, and on the Hungarian border, 16 mi/26 km N of SZOMBATHELY, Hungary; 47°28′N, 16°38′E. Elevation 628 ft/191 m. Thermal bath, wine.

Lützow-Holm Bay (LOOT-sou-HOWM), large bay in ANTARCTICA, on INDIAN OCEAN; 120 mi/193 km wide, 60 mi/97 km deep; 67°10′S 37°30′E. Bordered by PRINCE OLAV and PRINCE HARALD coasts. Discovered 1931 by the Norwegian explorer Hjalmar Riiser-Larsen.

Luverne (LUH-vuhrn), city (2000 population 4,617), ⊙ ROCK county, extreme SW MINNESOTA, 27 mi/43 km ENE of SIOUX FALLS, SOUTH DAKOTA, and 24 mi/39 km SSE of PIPESTONE, on ROCK RIVER; 43°38′N 96°12′W. Elevation 1,454 ft/443 m. Trading point in agricultural area (grain, milk, soybeans; livestock, poultry; dairying). Manufacturing (water tanks, truck mounts, and

farm equipment, printing); granite quarries. Blue Mounds State Park to N. Settled 1867, plotted 1870, incorporated as village 1877, as city 1904.

Luverne, town (2000 population 2,635), ⊙ Crenshaw co., S Alabama, on Patsaliga Creek, and 45 mi/72 km S of Montgomery; 31°42′N 86°15′W. Processing and shipping point in corn, peanut, and poultry area; clothing; soft drink bottling; lumber. Inc. 1891. Named for the wife of M. P. LeGrand of Montgomery, owner of a real estate business. County seat since 1893.

Luverne, town, on HUMBOLDT-KOSSUTH county line, N central IOWA, 15 mi/24 km NNE of DAKOTA CITY. Livestock, grain.

Luverne (loo-VUHRN), village (2006 population 38), STEELE county, E NORTH DAKOTA, 23 mi/37 km NE of VALLEY CITY, near SHEYENNE RIVER; 47°15′N 97°55′W. LAKE ASHTABULA Reservoir to W. Founded in 1912. Many original setters were from DENMARK and it is known as Little Denmark.

Luvua River (LOOV-wah), c.215 mi/346 km long, SE CONGO; issues from N end of LAKE MWERU; flows NW, past KIAMBI, to LUALABA RIVER opposite ANKORO. Navigable for shallow-draught boats for c.100 mi/161 km in its lower course below KIAMBI. Rapids in its middle course. Large hydroelectric plant at PIANA-MWANGA. LUVUA RIVER is sometimes considered to be a continuation of LUAPULA RIVER (which enters LAKE MWERU from the S), so that the whole LUVUA-MWERU-LUAPULA-BANGWEULU-CHAMBEZI system represents an E headstream of the CONGO.

Luvungi (loo-VUNG-gee), village, SUD-KIVU province, E CONGO, near RUZIZI RIVER, on railroad, and 28 mi/45 km SSE of BUKAVU; 02°52′S 29°02′E. Elev. 2,913 ft/887 m. Trading center in cotton-growing area; cotton ginning, palm-oil milling. Airfield. LUBARIKA, 4 mi/6.4 km WSW, is a center of agricultural research, notably on cotton cultivation.

Luwegu River (loo-WAI-goo), c. 210 mi/338 km long, S central TANZANIA; rises in central RUVUMA region c. 25 mi/40 km E of SONGEA; flows generally NE through SW part of SELOUS GAME RESERVE, past SIGURI FALLS; joins Luhombero River below falls, 115 mi/185 km S of MOROGORO.

Luwero (loo-WAI-roo), former administrative district (□ 3,551 sq mi/9,232.6 sq km), central UGANDA, SW of LAKE KYOGA and N of KAMPALA; capital was LUWERO; 01°00′N 32°20′E. As of Uganda's division into thirty-nine districts, was bordered by APAC (N), LIRA (NE), MUKONO (E), MPIGI (S), MUBENDE (SW), KIBOGA (W), and MASINDI (NW) districts. Agricultural area (bananas, beans). Luwero and BOMBO were important towns. In 1997 S and W portions of district became (now former) new LUWERO district and NE portion became NAKASONGOLA district.

Luwero, former administrative district (□ 2,152 sq mi/5,595.2 sq km; 2005 population 508,300), CENTRAL region, central UGANDA; capital was LUWERO; 01°00′N 32°20′E. As of Uganda's division into fifty-six districts, was bordered by NAKASONGOLA (N), KAYUNGA (E), MUKONO (SE), WAKISO (S), MUBENDE (SW), KIBOGA (W), and MASINDI (NW) districts. Area's original inhabitants were the Baganda people, also living there were the Banyankole, Banyarwanda, and Nubian peoples and Luo speakers. Vegetation was primarily savannah grasslands with some hills; marsh area W of Luwero town. Majority of population was agricultural, particularly in S (including bananas, cassava, fruit, honey, maize, and potatoes); livestock raised in N (including dairy products and beef). Towns included Luwero and BOMBO. Created in 1997 from S and W portions of former LUWERO district (Nakasongola district was created from NE portion). In 2005 W portion of district was carved out to form NAKASEKE district and E portion was formed into current LUWERO district.

Luwero, administrative district, CENTRAL region, central UGANDA; ⊙ LUWERO. As of Uganda's division into eighty districts, borders NAKASONGOLA (N), KAYUNGA (E), MUKONO (SE), WAKISO (S), and NAKASEKE (W) districts. Some marsh area W of Luwero town. Primarily agricultural area. Towns include Luwero and BOMBO. Luwero and Bombo towns on road between KAMPALA city and MASINDI town. Formed in 2005 from E portion of former LUWERO district created in 1997 (Nakaseke district formed from W portion).

Luwero (loo-WAI-roo), town (2002 population 23,497), ⊙ LUWERO district, CENTRAL region, central UGANDA, on road, and 40 mi/64 km N of KAMPALA, E of LU-GOGO RIVER; 00°51′N 32°30′E. Was part of former NORTH BUGANDA province.

Luwingu (loo-WI-eng-goo), township, NORTHERN province, N ZAMBIA, 80 mi/129 km W of KASAMA, on Kasama-Munanga road; 10°16′S 29°54′E. Corn; livestock. LAKE BANGWEULU and Bangweulu Swamps to S.

Luwuk (LOO-wook), town, ⊙ Luwuk district, on PE-LENG STRAIT, SULAWESI, INDONESIA, 180 mi/290 km ENE of POSO.

Luxembourg (LUK-sahm-boor), grand duchy (□ 998 sq mi/2,594.8 sq km; 2007 population 480,222), W EUROPE.

Geography

Roughly triangular, it borders on BELGIUM in the W and N, GERMANY in the E, and FRANCE in the S; ⊙ LUXEMBOURG (city). The country is divided into the following three administrative districts DIEKIRCH, GREVENMACHER,LUXEMBOURG. These three districts are then further divided into the following twelve cantons: CLERVAUX, WILTZ, VIANDEN, DIEKIRCH, RE-DANGE, ECHTERNACH, GREVENMACHER, REMICH, MERSCH, CAPELLEN, ESCH-SUR-ALZETTE, and LUX-EMBOURG. These cantons are then further divided into communes, which are then divided into towns and villages. The three largest towns are Wilz, Vianden, and Clervaux.

Population

The languages of French, German, and Letzebuergesh, the Middle German dialect that prevails in Luxembourg, have nearly equal status. Hence the country has three official names, Grand Duché de Luxembourg, Grossherzogtum Luxemburg, and Groussherzogtum Letzeburg. Most Luxembourgers are bilingual and many are trilingual. Over 95% are Roman Catholic.

Economy

The N part of the country, the *Oesling*, is part of the ARDENNES massif, strongly dissected by the SÛRE and its tributaries. Agriculture remains one of the mainstays of this region, while traditional industries involving woolen textiles and leather tanning have practically disappeared. Tourism has become important to the economy. The S part of the country, the *Gutland*, is the country's leading agricultural, manufacturing, and tourist region; its capital is also a major banking and financial service center. The Gutland consists of three cuestas, NW facing escarpments that cross the country from SW to NE, forming a gentle rolling landscape. The soils and climate more easily produce crops than those of the *Oesling*, hence the name *Gutland*, the good land. Luxembourg City lies on the extreme N cuesta, whose sandstone formation bears the name of the city. Agriculture is a dominant activity, but iron ore in the bottom lands of the extreme S escarpment made the fortune of modern Luxembourg. A series of small towns, of which ESCH-SUR-ALZETTE is the most important, developed as the core of one of the most productive iron and steel manufacturing regions in the world. Luxembourg's ores are depleted, but the steel industry continues, using iron imported from France. In addition to its farming and manufacturing resources, the Gutland has picturesque towns and recreational facilities, such as campgrounds, that attract large numbers of Belgian, Dutch, and German tourists. Luxembourg's GNP is among the highest in the world and it has an enviable standard of living, but the decline of the steel industry and Luxembourg's low birthrate have created economic and demographic problems, exemplified by the fact that almost 30% of the country's population is foreign-born workers.

History

Luxembourg, an early Frankish county, became an important fief of the Holy Roman Empire and was raised to a duchy in 1354. No fewer than three of its rulers were elected Holy Roman Emperors. In 1443, Luxembourg was seized by Philip the Good of BUR-GUNDY, and as a result passed, by marriage, into Hapsburg hands in 1482. For the next three centuries, Luxembourg shared the history of the S Netherlands, as a Spanish province until 1714 (except for a brief French interlude) and as an Austrian possession until the French Revolution, when it briefly became the *Département des Forêts*. At the Congress of Vienna (1814–1815) it was made a grand duchy in personal union with the Netherlands through common allegiance to William I. Luxembourg joined the Belgians in their revolt against William. After the revolt, in 1839, Luxembourg was divided, the W half going to BELGIUM, and the E part retaining its status as a grand duchy ruled from The NETHERLANDS. At the London Conference of 1867, Luxembourg was declared neutral territory, Prussian troops were withdrawn, and the fortress was dismantled. The neutrality of the country was violated by Germany in both World Wars and occupied as enemy territory. After regaining its independence in 1945, Luxembourg became an effective voice for European cooperation. In 1946 it entered the UN, joining NATO in 1949. It joined the European Economic Community (the predecessor of the European Union) as a founding member in 1957. In 1958, it joined Belgium and The Netherlands in a full economic union.

Government

Luxembourg is a constitutional monarchy governed under the constitution of 1868 as revised. The hereditary monarch is the titular head of state. The government is headed by the prime minister, who is appointed by the monarch with the approval of the legislature. The sixty members of the unicameral legislature, the Chamber of Deputies, are elected by popular vote to five-year terms. The twenty-one members of the Council of State, an advisory body to the legislature, are appointed by the monarch on the advice of the prime minister. The current head of state is Grand Duke Henri (since October 2000). Prime Minister Jean-Claude Juncker has been head of government since 1995.

Luxembourg (LEWKS-ahn-boor), Flemish *Luxemburg,* province (□ 1,714 sq mi/4,456.4 sq km; 2006 population 259,698), SE BELGIUM, in the ARDENNES, bordering on the GRAND DUCHY OF LUXEMBOURG in the E and on FRANCE in the S; ⊙ ARLON; 49°41′N 05°49′E. The chief towns are Arlon, BASTOGNE, and MARCHE-EN-FAMENNE. The province is drained by the OURTHE, SEMOIS, and LESSE rivers. It is mainly agriculture (grain, clover, tobacco, potatoes; hogs, cattle; dairying). Iron is mined, and timber is exported. The population is largely French-speaking, although Letzeburgisch, a Middle German dialect, is spoken in the E. The province was detached from the Grand Duchy of Luxembourg and given to Belgium in 1839. In World War II it was a major battleground in the Battle of the Bulge (December 1944–January 1945). Tourism is extensive.

Luxembourg (LUK-suhm-burk), administrative district (□ 349 sq mi/907.4 sq km; 2001 population 320,137), SW LUXEMBOURG, ⊙ Luxembourg; 49°35′N 06°10′E. Bordered to the N and E by the neighboring districts of DIEKIRCH and GREVENMACHER, respectively, to the S by FRANCE, and to the W by BELGIUM. Contains the cantons of CAPELLEN, ESCH-SUR-ALZ-ETTE, Luxembourg, and MERSCH.

Luxembourg (LUK-suhm-burk) or **Luxemburg** city (*Letzeburg* or *Letzebuerg* in the local dialect); ⊙ Grand Duchy of Luxembourg and Luxembourg canton (□ 92 sq mi/238 sq km), S Luxembourg, at the confluence of the ALZETTE and Pétrusse rivers; 49°36′N 06°10′E. The entrenched meanders of the two rivers are crossed by spectacular bridges, including Adolphus Bridge and the new bridge of EUROPE that spans the Alzette. The upper town is a substantial melding of both the ancient and the modern. The original nucleus of the city is built on an old defensive site (inside the rivers' meanders). It consists of numerous medieval houses and churches, the most notable of which is the grand-ducal palace and cathedral (both 16th century). Newer features such as the city hall and the Chamber of Deputies are also located there. The modern upper town is located on the W: a busy commercial center bordered by a complex of parks replacing the old fortifications. S of the Pétrusse, close to the railroad terminal, hotels, theaters, and financial and corporate headquarters (ARBED) are grouped. Kirchberg stretches NW toward the airport of FINDEL, where the new institutions of the EU, including the Court of Justice, the European Parliament, and the European Investment Bank, as well as Radio-Television-Luxembourg, are found. The lower town, in the winding valley bottoms, is mostly industrial—GRUND, CLAU-SEN, Pfaffental, and Mühlenbach. The town shares the history of the Grand-Duchy (q.v.). First established by the Romans, it was known by the Saxons as *Luci-linburhuc* [Engish=*the little fortress*]. It was successively controlled and elaborated by Frankish counts, vassals of the Holy Roman Emperor, Spaniards, Austrians, and Dutch, until it was dismantled in 1867. Luxembourg's present importance stems from its role as a financial service center, as well as its selection, after World War II, as the seat of the European Coal and Steel Community and later of the European Parliament and other institutions of the EEC and its role as a major financial and telecommunications center.

Luxembourg Palace (luhks-em-boorg), large Renaissance palace in PARIS, ÎLE-DE-FRANCE, FRANCE, on the left bank of the SEINE near the Sorbonne. It was built (1612–1620) for Marie de' Medici by Salomon de Brosse on the site of a former palace belonging to the duke of Piney-Luxembourg (hence its name). The 24 panels painted by Rubens are now at the Louvre. The palace was used for the Paris Peace Conference of 1946. It is now the seat of the Senate of the French Republic. It contains valuable paintings, notably those by Delacroix. The beautiful Luxembourg Gardens, with meticulously maintained flowerbeds, are a major Left Bank attraction.

Luxemburg, town, DUBUQUE county, E IOWA, 21 mi/34 km WNW of DUBUQUE. State park nearby.

Luxemburg, town (2006 population 2,241), KEWAUNEE county, E WISCONSIN, on DOOR PENINSULA, 15 mi/24 km E of GREEN BAY; 44°32′N 87°42′W. Dairying; manufacturing (wood cabinets, cheese, plastic bottles). Railroad junction to E. Speedway here.

Luxemburg, Luxembourg: see GRAND DUCHY, province, and city.

Luxemburg, UKRAINE: see ROZIVKA.

Luxeuil, FRANCE: see LUXEUIL-LES-BAINS.

Luxeuil-les-Bains (luk-SUH-ee–lai–ban), town (□ 8 sq mi/20.8 sq km), thermal spa in HAUTE-SAÔNE department, FRANCHE-COMTÉ region, E FRANCE, at W foot of VOSGES MOUNTAINS, 17 mi/27 km NNE of VESOUL; 47°49′N 06°23′E. Noted watering place with mineral springs. Also noted for its textile and lace industry. Has 14th-century abbatial church, 16th-century palace, and several remarkable lay buildings in central quarter. Known to Romans for the curative powers of its springs, town was destroyed (451) by the Huns, but rebuilt in 6th century around the now famed Abbey of Luxeuil (founded by St. Columban).

Devastated by the Saracens c.732, the abbey was restored by Charlemagne and became a great center of Christian learning in the Middle Ages. It was secularized during the French Revolution.

Luxhattipet, INDIA: see LAKSHETTIPET.

Luxi (LU-SEE), town, ⊙ Luxi county, NW HUNAN province, CHINA, on YUAN RIVER, and 60 mi/97 km NE of ZHIJIANG; 28°17′N 110°10′E. River crossing. Agriculture (grain, oilseeds, tobacco); manufacturing (food processing, engineering, chemicals, coal mining).

Luxi, town, ⊙ Luxi county, E YUNNAN province, CHINA, 75 mi/121 km SE of KUNMING, in mountain region; 24°31′N 103°46′E. Rice, millet, beans, oilseeds, tobacco; food processing, chemicals, machinery, coal mining.

Luxi, town, ⊙ Luxi county, westernmost YUNNAN province, CHINA, 50 mi/80 km S of TENGCHONG, near road to MYANMAR, and near Myanmar border; 24°27′N 98°36′E. Rice, sugarcane; sugar refining. Also called Mangshi.

Lu Xian (LU-SIAN), town, ⊙ Lu Xian county, S central SICHUAN province, CHINA, on left bank of CHANG JIANG (Yangtze River), at mouth of TUO RIVER, 50 mi/80 km ENE of YIBIN; 28°58′N 105°26′E. Rice, sugarcane; food and beverages, engineering; building materials; coal mining.

Luxikou (LU-SEE-KO), town, SE HUBEI province, CHINA, on YANGZI RIVER, and 55 mi/89 km SW of WUHAN, on HUNAN border and E shore of Heng Lake, former treaty port of call.

Luxor, city and urban governate (2004 population 414,389), QENA province, Upper EGYPT, on the E bank of the NILE RIVER; 25°41′N 32°39′E. It is 1 mi/1.6 km SW of KARNAK and occupies part of the site of ancient THEBES. The temple of Luxor, the greatest monument of antiquity in the city, was built in the reign of Amenhotep III (1414 B.C.E.–1397 B.C.E.) as a temple to Amon. The temple, 780 ft/230 m long, was much altered by succeeding pharaohs, especially by Ramses II, who had colossal statues of himself erected on the grounds. In early Christian times the temple was made into a church, and later a shrine to a Muslim saint was built in the great hall. The temple was restored, beginning in 1883. Numerous temples and burial grounds, including the Valley of the Kings, are nearby on the W side of the Nile River.

Luxora (luhks-OR-uh), town, MISSISSIPPI county, NE ARKANSAS, 10 mi/16 km S of BLYTHEVILLE, near the MISSISSIPPI RIVER; 35°45′N 89°55′W. In cotton-growing and rice area. Manufacturing (asphalt, metal stampings). Founded 1882.

Luxulyan (luh-SOOL-yuhn) or **Luxulian**, village (2001 population 1,371), central CORNWALL, SW ENGLAND, 6 mi/9.7 km S of BODMIN; 50°22′N 04°44′W. Previously granite quarrying and paper milling. Treffry's Viaduct (100 ft/30 m high, 660 ft/201 m long) spans the valley. Has 15th-century church.

Luya (LOO-yah), province (□ 3,289 sq mi/8,551.4 sq km), AMAZONAS region, PERU; ⊙ LAMUD; 06°25′S 78°00′W. On edge of the ANDES and AMAZON. Coffee, coca, potatoes, rice, cereals, sugarcane; livestock.

Luya (LOO-yah), city, LUYA province, AMAZONAS region, N PERU, on E ANDEAN slopes, 5 mi/8 km NNW of CHACHAPOYAS; 06°11′S 77°54′W. Corn, potatoes, cereals, alfalfa, sugarcane.

Luya Mountains, Mandarin *Luya Shan* (LU-YAH-SHAN), NW SHANXI province, CHINA, a branch of the LÜLIANG MOUNTAINS, rise to over 4,500 ft/1,372 m 25 mi/40 km SW of NINGWU.

Luyengo (lou-yain-GHO), town, MANZINI district, W central SWAZILAND, 18 mi/29 km S of MBABANE, near Great Usutu River; 26°35′S 31°10′E. Road junction. Seat of branch campus of University of Swaziland. Timber; citrus, corn, vegetables; cattle, goats, sheep, hogs. Also spelled Loyengo.

Luyi (LU-YEE), town, ⊙ Luyi county, E HENAN province, CHINA, 95 mi/153 km SE of KAIFENG, on ANHUI

border; 33°52′N 115°28′E. Grain, cotton, oilseeds, tobacco; food and beverages.

Luynes (LWEEN), town (□ 13 sq mi/33.8 sq km; 2004 population 4,945), INDRE-ET-LOIRE department, CENTRE administrative region, W central FRANCE, in LOIRE RIVER valley, and 5 mi/8 km W of TOURS; 47°23′N 00°34′E. Wine-producing center. Has cave dwellings and a 13th-century feudal castle.

Luz (loos), city (2007 population 17,170), W central MINAS GERAIS state, BRAZIL, near upper SÃO FRANCISCO River, 50 mi/80 km NW of DIVINÓPOLIS; 19°45′S 45°42′W. Dairying. Bishopric.

Luz (LOOSH) or **Luz de Lagos** (LOOSH dai LAH-goosh), village, FARO district, S PORTUGAL, on the ATLANTIC OCEAN (S coast), 3 mi/4.8 km WSW of LAGOS. Seaside resort.

Luz, ISRAEL: see BETHEL.

Luza (LOO-zah), city (2005 population 11,815), NW KIROV oblast, E European Russia, on road and railroad, on the LUZA RIVER, 310 mi/499 km NW of KIROV, and 45 mi/72 km SSE of KOTLAS; 60°37′N 47°16′E. Elevation 318 ft/96 m. Sawmilling, logging, lumbering, woodworking, construction materials; gas pipeline service station; mechanical repair plant; seed inspection, grain storage, food processing (bakery). Has a sanatorium. Became city in 1944.

Luzarches (lyu-zahr-shuh), commune (□ 7 sq mi/18.2 sq km), VAL-D'OISE department, ÎLE-DE-FRANCE region, N central FRANCE, 18 mi/29 km NNE of Paris; 49°07′N 02°25′E. Light manufacturing. Has 12th–16th-century church.

Luza River (loo-ZAH) approximately 225 mi/362 km long, N central European Russia; rises SE of OPARINO, KIROV oblast; flows NNE, into the KOMI REPUBLIC, past OBYACHEVO, and generally W, past LAL'SK (Kirov oblast) and LUZA, into the YUG RIVER S of VELIKIY US-TYUG (VOLOGDA oblast). Navigable for 50 mi/80 km above its mouth, below Lal'sk.

Luze (LU-zhe), Czech *Luže*, village, VYCHODOCESKY province, E BOHEMIA, CZECH REPUBLIC, 7 mi/11.3 km SW of VYSOKE MYTO; 49°54′N 16°02′E. Summer resort noted for trout fishing. Manufacturing (machinery). Ruins of a 14th century KOSUMBERK castle are 7 mi/11.3 km SE.

Luzech (lyu-ZESH), village (□ 9 sq mi/23.4 sq km), LOT department, MIDI-PYRÉNÉES region, SW FRANCE, within a bend of LOT RIVER, 8 mi/12.9 km W of CA-HORS; 44°29′N 01°17′E. Furniture manufacturing; trade in wine, fruits, and tobacco. Has ruins of a Gallic fort known as an *oppidum*.

Luzern, district, LUCERNE canton, central SWITZER-LAND. Main cities/towns are LUCERNE, HORW, KRIENS, and LITTAU; 47°05′N 08°16′E. Population is German-speaking and Roman Catholic.

Luzerna (LOO-ser-nah), town (2007 population 5,391), central SANTA CATARINA state, BRAZIL, 3 mi/4.8 km NE of JOAÇABA, on railroad; 27°07′S 51°28′W. Wheat, corn; livestock.

Luzerne (LOO-suhrn), county (□ 907 sq mi/2,358.2 sq km; 2006 population 313,020), E central PENNSYLVA-NIA; ⊙ WILKES-BARRE. Bounded SE by LEHIGH RIVER; drained by SUSQUEHANNA RIVER (forms part of NE boundary); Nescopeck Mountain ridge crosses E-W in center; WYOMING VALLEY in NE center. Agriculture (corn, wheat, oats, hay, alfalfa, vegetables, potatoes, apples; hogs, cattle; dairying); some anthracite coal; sandstone. Manufacturing at Wilkes-Barre, MOUN-TAIN TOP, HAZLETON, WEST HAZLETON, and KINGS-TON. First permanent settlements (1753) by people from Connecticut Part of Ricketts Glen State Park in NW, Frances Slocum State Park in NE; part of Lackawanna State Forest in N center. Nuclear power plants: Susquehanna 1 (initial criticality September 10, 1982; max. dependable capacity of 1040 MWe) and Susquehanna 2 (initial criticality May 8, 1984; maximum dependable capacity of 1044 MWe) are 7 mi/11.3

km NE of BERWICK.; both use cooling water from the Susquehanna River Formed 1786.

Luzerne, town (2000 population 105), BENTON county, E central IOWA, 19 mi/31 km SSW of VINTON; 41°54′N 92°10′W. In agricultural area.

Luzerne, borough (2006 population 2,792), LUZERNE county, NE central PENNSYLVANIA, suburb 3 mi/4.8 km N of WILKES-BARRE, on Huntsville Creek; 41°17′N 75°54′W. Manufacturing (sheet-metal components, wood products); some anthracite coal in area. Frances Slocum State Park to N. Incorporated 1882.

Luzerne, NEW YORK: see LAKE LUZERNE.

Luzhany (loo-ZHAH-ni) (Russian *Luzheny*) (Romanian *Lujeni*), town (2004 population 12,000), N CHERNIVTSI oblast, UKRAINE, in N BUKOVYNA, on railroad junction, and 8 mi/13 km NW of CHERNIVTSI, on left bank of PRUT RIVER; 48°22′N 25°47′E. Elevation 583 ft/177 m. Household goods manufacturing. Has the Church of Assumption (1453–1455), 19th-century park. Known since 1453; town since 1968.

Luzhene, Bulgaria: see VELINGRAD.

Luzheny, UKRAINE: see LUZHANY.

Luzhou (LU-JO), city (□ 83 sq mi/215 sq km; 1994 estimated non-agricultural population 284,100; estimated total population 415,900), S SICHUAN province, CHINA, on CHANG JIANG (Yangzi River), and 50 mi/80 km E of YIBIN; 28°55′N 105°25′E. An industrial and commercial center; manufacturing (oil and natural-gas extraction, food processing, chemicals, machinery). Crop growing, animal husbandry; grains, oil crops, vegetables, fruits; hogs, poultry.

Luzhou, CHINA: see HEFEI.

Luziânia (loo-see-AH-nee-ah), city (2007 population 196,046), E GOIÁS state, central BRAZIL, 65 mi/105 km E of ANÁPOLIS; 16°14′S 47°55′W. Cheese manufacturing, fruit preserving, tobacco and coffee shipping. Nitrate and rock-crystal deposits in vicinity. Until 1944, called Santa Luzia.

Lužická Nisa River, POLAND and CZECH REPUBLIC: see NEISSE, river.

Lužické hory, CZECH REPUBLIC: see LUSATIAN MOUN-TAINS.

Luzilândia (LOO-see-lahn-zhee-ah), city (2007 population 24,257), N PIAUÍ state, BRAZIL, on right bank of PARNAÍBA River (MARANHÃO border), and 50 mi/80 km SW of PARNAÍBA city; 03°28′S 42°22′W. Rice, tobacco. Until 1944, called Pôrto Alegre.

Luzinga (loo-ZEEN-guh), EASTERN region, SE UGANDA, on railroad, and 18 mi/29 km N of Jinja. Cotton, tobacco, coffee, bananas, millet. Was part of former BUSOGA province.

Luzino (LOO-zee-nuh), urban settlement (2006 population 9,860), S OMSK oblast, SW Siberian Russia, on highway and railroad, 12 mi/19 km W, and under administrative jurisdiction, of OMSK; 54°57′N 73°02′E. Elevation 331 ft/100 m. Chemicals; pork processing.

Luz Island, CHILE: see CHONOS ARCHIPELAGO.

Luz, La, MEXICO: see LA LUZ.

Luz, La, NICARAGUA: see SIUNA.

Luznice River (LUZH-ni-TSE), Czech *Lužnice*, German *Lainsitz*, 129 mi/208 km long, CZECH REPUBLIC (S BOHEMIA), and N AUSTRIA; rises at Czech-Austrian border 12 mi/19 km SE of KAPLICE; flows c.16 mi/26 km E and NNE in AUSTRIA, past ČESKE VELENICE (Czech Republic), then NNW, through fish-pond area (around TREBON), past SOBESLAV and TÁBOR, and SW to VLTAVA RIVER 1 mi/1.6 km WNW of TYN NAD VLTAVOU.

Luzon (loo-ZON), island (□ 40,420 sq mi/104,688 sq km; 1990 population 30,797,458), largest, most populous, and most important of the PHILIPPINE Islands; 16°00′N 121°00′E. The irregular coastline includes 2,016 islands (589 that are named) and provides several fine bays, most notably MANILA BAY, which is considered the best natural harbor in the Orient and one of the finest in the world. N Luzon, which is drained by the CAGAYAN RIVER, is very mountainous; the highest peak, MOUNT PULOG, rises to 9,606 ft/2,928 m. In the E, the great SIERRA MADRE range so closely parallels the shore that almost no coastal plain exists. Mountains extend generally along the entire length of the island, into the irregular Bicol peninsula to the SE, and feature several volcanoes, including Mountains PINATUBO, Mayou, and Taal. In the W, the Zambales range runs from LINGAYEN GULF S to BATAAN peninsula. The island has two large lakes, LAGUNA DE BAY and TAAL. Between the rugged coastal mountains, in central Luzon, lies the Central Plain, watered by the PAMPANGA and AGNO rivers. Barely above sea level, c.100 mi/161 km long and 40 mi/64 km wide, it is the most important agricultural land in all the Philippines. It supplies food for almost the entire MANILA area and is the nation's major rice-producing region and its second sugarcane-producing area (after NEGROS island). Elsewhere, the Bicol peninsula is known for its extensive coconut plantations; the Cagayan River valley, for its tobacco and corn. Other major crops are fruits, vegetables, and cacao. Luzon has important lumbering and mining industries; there are gold, chromite, nickel, copper, and iron deposits, and the bamboo on Bataan peninsula has many commercial uses. Manufacturing is centered in the Manila metropolitan area, where the major industries produce textiles, chemicals, and metal products. Scattered throughout the island are fertilizer plants, an occasional oil refinery, cement factories, and plywood mills and wood product plants. There is a well-defined highway network, strong maritime trade and shipping activity, two airports at Manila, and a single railroad line from Manila S to LEGASPI. Outside of the NATIONAL CAPITAL REGION (where the population is 100% urban), the population is 46.7% urban, 53.3% rural. The inhabitants are almost all Christian and are principally Tagalogs and Ilocanos. Indigenous peoples include the Negritos and Igorots (the latter's famous rice terraces on steep mountain slopes are considered one of the agricultural wonders of the world). As the major island, Luzon has played the leading role in the nation's history. Manila harbor has been important since the arrival of the Spanish in the late 16th century. It was on Luzon that the Filipino revolt against Spanish rule began (1896), that U.S. forces wrested control of the islands from SPAIN (1898), and that the Philippine insurrection against U.S. rule broke out (1899). The island was invaded by Japanese forces in several places on December 10, 1941, and in early 1942 the Allied forces made their last stand on Bataan peninsula and CORREGIDOR. Luzon was recovered (1945) after a major landing from Lingayen Gulf (January), a bloody fight for Manila (February), and protracted clean-up operations, which were not completed until June. In mid-1972, Luzon experienced the worst flooding in the nation's history; more than 400 lives were lost, and there was extensive property damage. Luzon's several U.S. military bases were closed down between 1971 and 1992, adversely affecting the island's economy. Also devastating were several natural disasters, including typhoons, the July 1990 earthquake in BAGUIO (registering 7.7 on the Richter scale), and the eruption of Mount Pinatubo in June 1991. The government has responded to the crisis with a series of regional development plans for the areas N and S of Manila, which are designed to foster industrialization and expand on the existing infrastructure, including using former U.S. bases. But the resulting urban growth and environmental degradation may pose an even greater threat to the island's welfare.

Luzon Strait (loo-ZON), channel connecting SOUTH CHINA SEA and PHILIPPINE SEA, Philippines, between TAIWAN (N) and LUZON (S). Divided into 3 channels: Babuyan Channel (bah-BOO-yahn), between BABUYAN Islands and Luzon; Balintang Channel (bah-leen-TAHNG), between Babuyan Islands and BATAN Islands; and Bashi Channel (BAH-shee), between Batan Islands and Taiwan.

Luz-Saint-Sauveur (lu-zuh–san–so-vuhr), commune (□ 9 sq mi/23.4 sq km; 2004 population 1,077), HAUTES-PYRÉNÉES department, MIDI-PYRÉNÉES region, SW FRANCE, in central PYRENEES, in valley of the Gave de Pau, and 10 mi/16 km SSE of ARGELÈS-GAZOST; 42°52′N 00°00′E. Active tourist resort; manufacturing of woolen textiles. Has 12th–14th-century fortified church. Saint-Sauveur quarter (across the Gave de Pau stream) has sulfur springs.

Luzuriaga (loo-zoo-ree-AH-gah), town, N MENDOZA province, ARGENTINA, in MENDOZA RIVER valley (irrigation area), on railroad, and 5 mi/8 km SSE of MAIPÚ; 32°57′S 68°49′W. Wine making.

Luzy (lyu-zee), town (□ 16 sq mi/41.6 sq km), NIÈVRE department, in BURGUNDY, central FRANCE, in the S MORVAN massif, 19 mi/31 km SW of AUTUN; 46°47′N 03°58′E. Road junction and livestock market. Nearby is the S entrance to MORVAN NATURAL REGIONAL PARK (created 1970), which extends 40 mi/64 km northward to vicinity of AVALLON.

Luzyca, POLAND: see LUSATIA.

Luzyckie, Gory, POLAND: see LUSATIAN MOUNTAINS.

Luzzara (loo-TSAH-rah), town, REGGIO NELL'EMILIA province, EMILIA-ROMAGNA, N central ITALY, near PO RIVER, 3 mi/5 km NE of GUASTALLA; 44°58′N 10°41′E. Machinery, fabricated metals, wood products; wine, sausage, straw hats. Scene of defeat (1702) of Imperialists under Prince Eugene by French and Spanish under Vendôme.

Luzzi (LOO-tsee), town, COSENZA province, CALABRIA, S ITALY, 11 mi/18 km NNE of COSENZA; 39°27′N 16°17′E. Dried figs, wine, olive oil; light manufacturing.

L'va Tolstogo (LVAH tuhl-STO-vuh), village (2005 population 3,540), central KALUGA oblast, central European Russia, in a wooded area near the UGRA RIVER, on road, 12 mi/19 km NW of KALUGA; 54°36′N 36°02′E. Elevation 570 ft/173 m. Agricultural market; poultry plant; furniture manufacturing. Alternatively called Lev Tolstoy.

L'viv (LVEEV) (Latin *Leopolis*) (Polish *Lwów*) (German *Lemberg*) (Russian *L'vov*), city (2001 population 732,818), oblast center, historical capital of Western UKRAINE and, after KIEV, the second cultural, political, and religious center of Ukraine. It is a major transportation node (hub of railroads, highways; international airport) and industrial complex (68% of industrial output of the oblast). Industries include transportation equipment, electronic equipment, chemicals, pharmaceuticals, textiles, footwear, food and beverage processing, musical instruments, glass-blowing and ceramics.

Research and higher learning institutions include those in astronomy, applied mechanics and mathematics, geological exploration, geology and geochemistry of fuel resources, measurement and control devices. Among the institutions of higher learning are the Ivan Franko L'viv University and ten specialized institutes, a branch of the Academy of Sciences of Ukraine, and the main office of the Shevchenko Scientific Society. There are many theatres, operas, philharmonic orchestras, and many museums and galleries.

L'viv was established in the 13th century by Prince Danylo Romanovych of Halych-Volhynia State; captured by Casimir III the Great of POLAND (1349); remained capital of Regnum Russiae, a separate country united with Poland; capital of the Polish province. Rus voivodeship (1434–1772); capital of the Austrian Crownland of GALICIA (1772–1918); capital of the short-lived Western Ukrainian National Republic (October 1918); part of Poland (1919–1939); annexed by the USSR to Soviet Ukraine (1939–1941); under German occupation, General Government (1941–1944); reclaimed by the USSR (1945–1991); center of Ukrainian nationalism; since 1991, part of independent Ukraine.

Jewish community, mainly mercantile class, dated back to traders from BYZANTIUM and later in the 14th century, from BOHEMIA and GERMANY. In 1939, it represented one-third of the total population, with a vigorous religious and secular cultural life, and included merchants, artisans and professionals. The community was exterminated by the Nazis during World War II.

The only Ukrainian city with Renaissance architecture, L'viv also has baroque-style churches, and some rococo buildings, which include St. George's Cathedral, seat of the Ukrainian Catholic (Uniate) church.

L'viv, (Ukrainian *L'vivs'ka*) (Russian *L'vovskaya*), oblast (□ 8,429 sq mi/21,915.4 sq km; 2001 population 2,626,543), W UKRAINE; ⊙ L'VIV. In highlands on Polish border (W and NW); includes SE ROZTOCHCHYA mountains, and HOLOHORY mountains in the middle and a section of the N slope of CARPATHIAN Mountains in S. Drained by upper BUH and STYR rivers, by upper DNIESTER River (Ukrainian *Dnister*) with its tributaries, and the tributaries of the San River (Ukrainian *Syan*); 50°00′N 24°00′E. Dark prairie soils (on upland) and brown forest (in mountains). Humid continental climate (short summers). Population mostly Ukrainian (90.4%), with Russian (7.2%), Polish (1%), and Jewish (0.5%) minorities. Principal extraction of mineral resources includes natural gas and petroleum in the Subcarpathian hydrocarbon-bearing zone (BORYSLAV, DASHAVA), bituminous coal in the L'VIV-VOLYN' COAL BASIN (CHERVONOHRAD), some lignite near ZOLOCHIV, ZHOVKVA, and RAVA-RUS'KA, potassium and common salt in the Subcarpathian salt-bearing basin (STEBNYK, MORSHYN), native sulphur in the Subcarpathian sulphur-bearing basin (NOVYY ROZDOL), and ozocerite at Boryslav. Agriculture (wheat, rye, oats; livestock, flax); flour milling, distilling, brewing, meat and vegetable preserving. Lumbering, stone quarrying, petroleum refining. L'viv is main industrial center; light manufacturing at Zolochiv, Zhovkva, and BRODY. Metalworking and light industries in DROHOBYCH, Boryslav, STRYY, SAMBIR. Health resorts with mineral springs (TRUSKAVETS', Morshyn). Formed (1939) out of parts of Polish Lwow and Tarnopol provinces, following Soviet occupation of E Poland; held by Germany (1941–1944); ceded to USSR in 1945. In 1959, expanded S by incorporating Drohobych oblast, which was also formed (1939) out of parts of Polish Lwow and Stanislawow provinces, following Soviet occupation of East Poland. Has (1993) fourty two cities, thirty five towns, and twenty rural raions.

L'vivs'ka, UKRAINE: see L'VIV, oblast.

L'viv-Volhynia Coal Basin (Ukrainian *L'vivs'ko-Volyns'kyy Kam'yanovuhil'nyy Baseyn*) (□ approximately 3,800 sq mi/10,000 sq km), in N L'viv oblast and SW VOLYN' oblast, W UKRAINE, extends NW into Poland. Explored in detail, 1948–1950, exploitation began in 1954, resulting in the formation of two mining districts, at CHERVONOHRAD (N L'viv oblast) and NOVOVOLYNS'K (SW Volyn' oblast), with twenty one coal mines, producing about 9 million tons/10 million metric tonnes/year. Coal is mined (at depths 1,000 ft/300 m–3,000 ft/900 m) in seams 2 ft/0.6 m–9.2 ft/2.8 m thick, ash content is high (5–35%), some sulfur (1.5–4%); used for electric power generation, heating. Also called L'viv-Volyn' Coal Basin.

L'viv-Volyn' Coal Basin, UKRAINE: see L'VIV-VOLHYNIA COAL BASIN.

L'vov, UKRAINE: see L'VIV, city.

L'vovskaya, UKRAINE: see L'VIV, oblast.

L'vovskiy (LVOF-skeeyee), town (2006 population 11,780), central MOSCOW oblast, central European

Russia, on road and railroad, 35 mi/56 km S of Moscow, and 7 mi/11 km S of PODOL'SK, to which it is administratively subordinate; 55°18′N 37°31′E. Elevation 600 ft/182 m. Non-ferrous metal works.

L'vovskoye (LVOF-skuh-ye), settlement (2005 population 4,795), S central KRASNODAR TERRITORY, S European Russia, on the left bank of the KUBAN' River, on road, 23 mi/37 km W of KRASNODAR; 45°01′N 38°36′E. In oil- and gas-producing region.

Lwan, CHINA: see LUAN RIVER.

Lwów, UKRAINE: see L'VIV, city.

Lwowek (LVOO-vek), Polish *Lwówek*, German *Neustadt bei Pinne* (new-stad bee peene), town, Poznań province, W POLAND, 32 mi/51 km W of POZNAŃ. Railroad spur terminus; machine manufacturing, sawmilling.

Lwowek Slaski (LVOO-vek SLON-skee), Polish *Lwówek Śląski*, German *Löwenberg* (leven-berg), town in LOWER SILESIA, after 1945 in Wrocław province, SW POLAND, on BOBRAWA RIVER, and 25 mi/40 km WSW of LEGNICA (Liegnitz). Agricultural market (grain, malt, potatoes; livestock); gypsum quarrying. Has 16th century church, remains of medieval town walls. Chartered 1217. Had important woolen trade until Thirty Years War.

Lyady (LYA-dee), town, SE VITEBSK oblast, BELARUS, 31 mi/50 km ENE of ORSHA. Linen milling.

Lyady (LYAH-di), village, N PSKOV oblast, W European Russia, on the PLYUSSA RIVER, on road, 52 mi/84 km N of PSKOV, and 40 mi/64 km WSW of LUGA; 58°37′N 28°47′E. Dairying.

Lyakhi (LYAH-hee), village, SE VLADIMIR oblast, central European Russia, on the OKA River, on highway, 18 mi/29 km SSW of MUROM; 55°20′N 41°55′E. Elevation 328 ft/99 m. In agricultural region (sunflowers, grains, hemp); native handicrafts.

Lyakhivtsi, UKRAINE: see BILOHIRYA.

Lyakhovichi (LYA-ho-vi-chi), Polish *Lachowicze*, city, S BREST oblast, Lyath Raion, BELARUS, 12 mi/19 km SE of BARANOVICHI. Has seventeenth-century castle. Old Polish fortress, assaulted (1660) by Russians. Passed (1795) from POLAND to RUSSIA; reverted (1921) to Poland; ceded to USSR in 1945.

Lyakhovichi (LYA-ho-vi-chi), town, BREST oblast, BELARUS; ⊙ Lyath Rayon.

Lyakhov Islands (LYAH-huhf) (□ 2,700 sq mi/7,020 sq km), southern group of the NEW SIBERIAN ISLANDS, SAKHA REPUBLIC, RUSSIAN FAR EAST, between the LAPTEV SEA and the EAST SIBERIAN SEA; 73°30′N 141°00′E. They include Bol'shoy LYAKHOV, Malyy LYAKHOV, and STOLBOVOY islands and are separated from the Anjou group of the New Siberian Islands by the Sannikov Strait and from the mainland by the Dmitriy Laptev Strait. They were discovered (1770) by Ivan Lyakhov, a Russian merchant. In 1928, the Soviet government established a geophysical station here.

Lyakhovtsy, UKRAINE: see BILOHIRYA.

Lyaki (LYAH-kee), town, central AZERBAIJAN, railroad station 6 mi/10 km S of AGDASH. Cotton ginning.

Lyallpur, PAKISTAN: see FAISALABAD, city.

Lyalta (LEI-uhl-tuh), hamlet, S ALBERTA, W CANADA, 11 mi/17 km from STRATHMORE, in WHEATLAND COUNTY; 51°07′N 113°36′W.

Lyalya River (LYAH-lyah), 110 mi/177 km long, SVERDLOVSK oblast, W Siberian Russia; rises in the central URAL Mountains, approximately 15 mi/24 km WSW of PAVDA; flows generally E, past Pavda, STARAYA LYALYA, and NOVAYA LYALYA, to the SOSVA RIVER, 12 mi/19 km E of SOSVA; timber floating. Receives the LOBVA River (left).

Lyambir' (LYAHM-beer), village (2006 population 8,210), central MORDVA REPUBLIC, central European Russia, on road, 7 mi/11 km NNW of SARANSK; 54°17′N 45°07′E. Elevation 636 ft/193 m. In agricultural area; livestock breeding, food processing.

Lyamino (LYA-mee-nuh), town (2006 population 4,960), E central PERM oblast, at the foot of the W

URALS, E European Russia, near the CHUSOVAYA River (KAMA River basin), near railroad, 4 mi/6 km W, and under administrative jurisdiction of CHUSOVOY; 58°17′N 57°43′E. Elevation 377 ft/114 m. House-building materials, sawmilling.

Lyangar, town, NAWOIY wiloyat, UZBEKISTAN, in the AK-TAU mountains, 40 mi/64 km NW of KATTA-KURGAN; 40°25′N 65°59′E. Molybdenum and tungsten mining.

Lyangar (lee-ahn-GAHR), village, FERGANA wiloyat, NE UZBEKISTAN. Tertiary-level administrative center.

Lyangasovo (lyahn-GAH-suh-vuh), town (2005 population 12,500), central KIROV oblast, E European Russia, on railroad, 7 mi/11 km WSW of KIROV; 58°31′N 49°27′E. Elevation 498 ft/151 m. Railroad depots; toys; sawmilling, woodworking; food processing (bakery, confectionery).

Lyanozovo, RUSSIA: see VAGONOREMONT.

Lyantonde, administrative district, CENTRAL region, S UGANDA. LAKE VICTORIA to E and TANZANIA to S. Agricultural and dairy area. LYANTONDE town on road between MBARARA and MASAKA towns. Formed in 2006 from part of former RAKAI district (current RAKAI district also formed from part of former Rakai district).

Lyantonde (lee-an-TON-dai), town (2002 population 7,508), LYANTONDE district, CENTRAL region, SW UGANDA, 40 mi/65 km WSW of MASAKA; 00°24′S 31°09′E. Dairy and agricultural center (bananas, coffee).

Lyantor (lyahn-TOR), city (2005 population 36,755), central KHANTY-MANSI AUTONOMOUS OKRUG, W central SIBERIA, RUSSIA, 25 mi/40 km W of SURGUT; 61°25′N 72°31′E. Elevation 177 ft/53 m. Arose as a workers settlement of Lyantorskiy, with the development of the Lyantor oil and gas field. City status and current name since 1992.

Lyaskelya (LYAHS-kee-lyah), Finnish *Läskelä*, town (2005 population 3,540), SW Republic of KARELIA, NW European Russia, on road and railroad, on the Yanisioki River (Lake LADOGA basin), 12 mi/19 km ENE of SORTAVALA; 61°46′N 31°01′E. Wood pulp and paper. In FINLAND until 1940.

Lyaskovets (LYAS-ko-vets), city, LOVECH oblast, Lyaskovets obshtina (1993 population 17,185), N BULGARIA, adjoining GORNA ORYAHOVITSA; 43°07′N 25°43′E. On railroad spur; market center in vegetable and fruit-growing district; wine and sparkling wine making; seed-producing center; manufacturing (motor vehicle parts, scales). Ruins of an ancient monastery nearby. Sometimes spelled Leskovec and Ljeskovec.

Lyasomin Island (lyah-SO-meen), in ARCHANGEL oblast, NW European Russia, in the W section of the NORTHERN DVINA River delta, 10 mi/16 km NW of ARCHANGEL; 64°40′N 40°04′E. Forested and swampy.

Lycaonia (lei-kai-O-nee-uh), ancient country of S central ASIA MINOR (now in TURKEY), W and SW of Lake Tuzbet. GALATIA and CILICIA on the N and S, PHRYGIA and CAPPADOCIA on the W and E. Passing successively to the Persians, Syrians, and Romans, it was divided by the Romans between Galatia and CAPPADOCIA. It was visited by Paul and Barnabas (Acts 14:6). Its chief city was ICONIUM.

Lychen (LI-khen), town, BRANDENBURG, E GERMANY, on small Lychen Lake (connected by short canal with the HAVEL), 15 mi/24 km SE of NEUSTRELITZ; 53°13′N 13°19′E. Health resort. Has early Gothic church, and remains of old town walls. Founded 1248.

Lychkovo (lich-KO-vuh), town, S central NOVGOROD oblast, NW European Russia, on the Polomet' River (right tributary of the POLA River), on highway branch and railroad, 38 mi/61 km E of STARAYA RUSSA, and 18 mi/29 km N of DEMYANSK, to which it is administratively subordinate; 57°55′N 32°24′E. Elevation 141 ft/42 m. Flax processing, grain storage, furniture.

Lycia (LI-shuh), ancient country, SW ASIA MINOR, located in modern TURKEY. Egyptian sources ally the Lycians to the Hittites at the time of Ramses II. Lycia was frequently mentioned by Homer in Greek mythology. In historic times it was held by the Persians, the Seleucids, and the Romans (from 189 B.C.E.). Its chief towns, PATARA and MYRA, were visited by St. Paul (Acts 27:5). Ruins include rock-cut tombs and Grecian sculptures dating from the 5th century B.C.E. The modern-day coastline from BODRUM to ANTALYA is a major center for tourism.

Lyck, POLAND: see ELK.

Lyckeby (LIK-ke-BEE), village, BLEKINGE county, S SWEDEN, 3 mi/4.8 km NNE of KARLSKRONA; 56°13′N 15°37′E.

Lyck River, POLAND: see LEG RIVER.

Lycksele (LIK-SE-le), town, VÄSTERBOTTEN county, N SWEDEN, on UMEÄLVEN RIVER, 70 mi/113 km NW of UMEÅ; 64°36′N 18°42′E. Small industrial companies; forestry center. Airport. Outdoor animal park. Incorporated 1946.

Lycoming (lei-KO-ming), county (□ 1,243 sq mi/3,231.8 sq km; 2006 population 117,668), N central PENNSYLVANIA; ⊙ WILLIAMSPORT. Hilly agricultural and forested region; drained by West Branch of SUSQUEHANNA RIVER and by LYCOMING (forms part of N boundary), LOYALSOCK, and PINE creeks; 41°21′ 77°03′. Agriculture (corn, wheat, oats, barley, hay, alfalfa, soybeans, potatoes, apples; sheep, hogs, cattle; dairying). Manufacturing at WILLIAMSPORT, MONTGOMERY, MONTOURSVILLE, and MUNCY. Largest county in land area in Pennsylvania. Part of Allenwood Federal Prison Camp in SE. Little Pine and Upper Pine Bottom state parks in W, Susquehanna State Park in S center, at Williamsport; parts of Tiadaghton State Forest throughout county, especially in N and S. Formed 1795.

Lycopolis, EGYPT: see ASYUT, city.

Lycus, IRAQ: see ZAB, GREAT.

Lycus River, in W ASIA MINOR, a tributary of the Maeander (modern BÜYÜK MENDERES).

Lycus River, TURKEY: see KELKIT RIVER.

Lydbury North (LID-buh-ree), village (2001 population 1,204), SHROPSHIRE, W ENGLAND, 3 mi/4.8 km S of Lydham; 52°28′N 02°57′W. Large Norman church. Plowden Hall 2 mi/3.2 km to E. WALES border 5 mi/8 km to W.

Lydd (LID), town (2001 population 5,782), SE KENT, SE ENGLAND, 14 mi/22.4 km S of ASHFORD; 50°57′N 00°55′E. A military training center, Lydd gave its name to *lyddite* (picric acid), an explosive that was tested at the military camp here in 1888. Lydd was a member of the CINQUE PORTS but is no longer a seaport because of changes in the shoreline. Airport to E.

Lydda, ISRAEL: see LOD.

Lyddan Island, an island at the SW extremity of the RIISER-LARSEN ICE SHELF, 20 mi/30 km off the PRINCESS MARTHA COAST, W QUEEN MAUD LAND, EAST ANTARCTICA; 45 mi/73 km long; 74°25′S 20°45′W.

Lydenburg, town (□ 3 sq mi/7.8 sq km), MPUMALANGA province, SOUTH AFRICA, on Dorps River, a tributary of the OLIFANTS RIVER, and 150 mi/241 km ENE of TSHWANE (formerly Pretoria); 25°06′S 30°28′E. Elevation 5,740 ft/1,750 m. Platinum-mining and agricultural center (fruits, cotton, soya beans, wheat, tobacco; sheep; dairying). Founded 1849, it was capital of an independent republic from 1857 to 1860. Surrounded by Mydenburg Nature Reserve (□ 3 sq mi/7.8 sq km). Railroad and airfield. Oldest school building in Transvaal, dating back to 1851, is located here.

Lydia, ancient country, W ASIA MINOR, N of CARIA and S of MYSIA (now NW TURKEY). The tyrant Gyges was the founder of the Mermnadae dynasty, which lasted from c.700 B.C.E. to 550 B.C.E. The little kingdom grew to an empire in the chaos that had been left after the fall of the Neo-Hittite kingdom. Lydia was proverbially golden with wealth, and its capital, SARDIS, was magnificent. To Lydian rulers is ascribed the first use of coined money (7th century B.C.E.). Lydia had close ties with the Greek cities of Asia, which were for a time within the Lydian empire. Cyrus the Great of PERSIA defeated (c.546 B.C.E.) Croesus, Lydia's last ruler, and Lydia was absorbed into the Persian Empire.

Lydia (LID-ee-uh), unincorporated town (2000 population 1,079), IBERIA parish, LOUISIANA, 6 mi/9.7 km SE of NEW IBERIA; 31°55′N 91°48′W.

Lydia, village, DARLINGTON county, central SOUTH CAROLINA, 10 mi/16 km SW of HARTSVILLE. Agriculture includes soybeans, wheat, cotton, corn.

Lydia Mills, textile village, LAURENS county, NW SOUTH CAROLINA, adjacent to CLINTON.

Lydia, Mount, TURKEY: see MYCALE.

Lydick, village, SAINT JOSEPH county, N INDIANA, 7 mi/11.3 km WNW of SOUTH BEND.

Lydney (LID-nee), town (2001 population 8,960), W GLOUCESTERSHIRE, W ENGLAND, near SEVERN estuary, 19 mi/31 km N of BRISTOL; 51°43′N 02°32′W. Light industry. Church dates from 13th century; has 14th-century cross. Site of Roman settlement.

Lye (LEI), town (2001 population 12,402), WEST MIDLANDS, central ENGLAND, on STOUR RIVER, and 10 mi/16 km W of BIRMINGHAM; 52°27′N 02°07′W. Just to E is Wollescote, former site of sheet-metal industry.

Lyeksa, Lake, RUSSIA: see LEKSOZERO.

Lyell Land (luh-EL), Danish *Lyell Land*, peninsula (35 mi/56 km long, 16 mi/24 km–28 mi/45 km wide), E GREENLAND, on KING OSCAR FJORD; 72°35′N 25°30′W. Rises to 7,216 ft/2,199 m near its base.

Lyell, Mount (LEI-uhl) (11,495 ft/3,504 m), on ALBERTA–BRITISH COLUMBIA border, W CANADA, in ROCKY MOUNTAINS, on W edge of BANFF NATIONAL PARK, 75 mi/121 km SE of JASPER; 51°57′N 117°06′W.

Lyell, Mount, peak (13,114 ft/3,997 m) of the SIERRA NEVADA, NE MADERA county, E CALIFORNIA, in YOSEMITE NATIONAL PARK, near its E boundary, and c.20 mi/32 km SW of MONO LAKE. Has small glacier.

Lyerly (LEI-uhr-lee), village (2006 population 488), CHATTOOGA county, NW GEORGIA, 17 mi/27 km NW of ROME, and on CHATTOOGA RIVER; 34°24′N 85°24′W. Manufacturing of work gloves and carpets.

Lyford (LEI-fuhrd), town (2006 population 2,456), WILLACY county, extreme S TEXAS, 15 mi/24 km NNW of HARLINGEN; 26°24′N 97°47′W. In irrigated farm area (sugarcane, vegetables, cotton; cattle); oil and gas.

Lygoudista, Greece: see KHORA.

Lygudista, Greece: see KHORA.

Lykens (LEI-kens), borough (2006 population 1,846), DAUPHIN county, E central PENNSYLVANIA, 22 mi/35 km NNE of HARRISBURG, on Wiconisco Creek. Manufacturing (heat exchangers, work clothes). Part of Weiser State Forest to E and SW. Settled c.1740; laid out 1848; incorporated 1872.

Lykhivka (LI-kheef-kah) (Russian *Likhovka*), town (2004 population 4,500), W DNIPROPETROVS'K oblast, UKRAINE, on Omel'nyk River, right tributary of the DNIEPER (Ukrainian *Dnipro*) River, 18 mi/29 km WNW of VERKHN'ODNIPROVS'K; 48°41′N 33°55′E. Elevation 282 ft/85 m. Food processing (flour, dairy, poultry). Established in 1740; town since 1957.

Lykodimo (lee-KO-[th]ee-mo), mountain, MESSENIA prefecture, SW PELOPONNESE department, extreme SW Greece, in MESSENIA Peninsula, rises to 3,146 ft/959 m, 16 mi/26 km WSW of KALAMATA; 36°56′N 21°51′W. Also called Mathia. Also spelled Likodhimos.

Lyle, unincorporated town, KLICKITAT county, S WASHINGTON, 8 mi/12.9 km NW of THE DALLES, OREGON, on COLUMBIA River (Bonneville reservoir), at mouth of KLICKITAT RIVER railroad junction. Wheat, alfalfa; cattle.

Lyle, village (2000 population 566), MOWER county, SE MINNESOTA, on IOWA state line, near CEDAR RIVER, and 11 mi/18 km S of AUSTIN; 43°30′N 92°56′W.

Railroad junction. Alfalfa, soybeans, grain; dairying; poultry; manufacturing (feeds, grain processing).

Lyman, county (□ 1,707 sq mi/4,438.2 sq km; 2006 population 3,929), S central SOUTH DAKOTA; ⊙ KENNEBEC; 43°53′N 99°50′W. Agricultural and cattle-raising region bounded S by WHITE RIVER, E by MISSOURI RIVER; Lower Brule Indian Reservation in N. Corn, wheat; cattle, hogs, poultry; dairy products. Part of Fort Pierre National Grassland in NW. Formed 1890.

Lyman (li-MAHN) (Russian *Liman*), town (2004 population 6,300), central KHARKIV oblast, UKRAINE, on NW end of Lake Lyman, on left bank of DONETS River, on road and railroad spur 7 mi/12 km SE of ZMIYIV; 49°35′N 36°28′E. Elevation 321 ft/97 m. Known since 1682; town since 1958.

Lyman (LEI-muhn), unincorporated town (2000 population 1,081), HARRISON county, SE MISSISSIPPI, 8 mi/12.9 km N of GULFPORT. Timber; corn, cotton, pecans, citrus. De Soto National Forest to NE.

Lyman, town, GRAFTON county, NW NEW HAMPSHIRE, 8 mi/12.9 km WSW of LITTLETON; 44°16′N 71°56′W. Agriculture (cattle, poultry; vegetables; dairying; nursery crops, sugar maples); manufacturing (briefcases and portfolios); recreational area. Gardner Mountain (2,330 ft/710 m) on NW boundary.

Lyman, town (2006 population 2,798), SPARTANBURG county, NW SOUTH CAROLINA, near TYGER RIVER, 11 mi/18 km W of SPARTANBURG; 34°57′N 82°07′W. Manufacturing includes gears, copper film negatives, apparel fabrics. Agriculture includes dairying; livestock, poultry; grain, soybeans, peaches, apples.

Lyman, town (2006 population 1,962), UINTA county, SW WYOMING, 35 mi/56 km E of EVANSTON near Blacks Fork River; 41°19′N 110°17′W. Elevation 6,695 ft/2,041 m.

Lyman (LEI-men), village (2006 population 404), SCOTTS BLUFF county, W NEBRASKA, 20 mi/32 km WNW of SCOTTSBLUFF, near NORTH PLATTE RIVER, and at WYOMING state line; 41°55′N 104°02′W. Livestock; beet sugar, grain, potatoes. Manufacturing (screw-machine products).

Lyman (LEI-muhn), village (2006 population 230), WAYNE county, S central UTAH, 2 mi/3.2 km E of LOA; 38°23′N 111°35′W. Elevation c.7,200 ft/2,195 m. Cattle; alfalfa, barley. Fishlake National Forest to E; Loa State Fish Hatchery to N.

Lyman, village (2006 population 423), SKAGIT county, NW WASHINGTON, 15 mi/24 km NE of MOUNT VERNON, and on SKAGIT RIVER; 48°31′N 122°04′W. Lumber; vegetables, berries; poultry; dairying. Mount Baker–Snoqualmie National Forest to NE and SE, Mount Baker (10,775 ft/3,284 m) to NE.

Lyman, UKRAINE: see KRASNYY LYMAN.

Lymans'ke (li-MAHN-ske) (Russian *Limanskoye*), town, W central ODESSA oblast, UKRAINE, near the border with Moldova, on the E bank of Kuchurhan Liman 37 mi/60 km WNW of ODESSA, and 13 mi/21 km SSW of ROZDIL'NA; 46°40′N 29°58′E. Elevation 114 ft/34 m. Furniture manufacturing, two vineyards with wineries; vocational school. Established at end of the 18th century as Zel'tsi (Russian *Zel'tsy*); renamed in 1944; town since 1957.

Lymbia (LIM-byah), town, LEFKOSIA district, E central CYPRUS, 14 mi/23 km SSE of NICOSIA; 34°59′N 33°38′E. Grain, citrus, vegetables, olives; livestock. Attila Line passes to N. Also spelled Lympia.

Lyme, town, NEW LONDON county, SE CONNECTICUT, on the CONNECTICUT River, and 12 mi/19 km WNW of NEW LONDON; 41°23′N 72°20′W. Agriculture; dairying; resorts. Includes HADLYME village and Hamburg (surrounds Hamburg Cove, finger offshoot of the Connecticut River), with yacht harbor at mouth of Eight Mile River; 3 state parks. Formerly included towns of EAST LYME and OLD LYME. Motor vehicle sales and repairs.

Lyme, town, GRAFTON county, W NEW HAMPSHIRE, 12 mi/19 km NNE of LEBANON; 43°47′N 72°07′W.

Bounded on W by the CONNECTICUT RIVER (VERMONT state line). Agriculture (cattle, poultry; vegetables; dairying; nursery crops; timber); manufacturing (software). Dartmouth Skiway Ski Area in E; APPALACHIAN TRAIL crosses town in E.

Lyme Regis (LEIM REE-jis), town (2001 population 3,513), DORSET, SW ENGLAND, 20 mi/32 km SW of YEOVIL; 50°44′N 02°56′W. Tourist resort. Paleontological discoveries have been made in the Blue Lias rocks quarried nearby.

Lyminge (LIM-inj), village (2001 population 2,688), E KENT, SE ENGLAND, 4 mi/6.4 km N of HYTHE; 51°08′N 01°05′E. Former agricultural market. The church was first built in 965.

Lymington (LI-ming-tuhn), market town (2001 population 5,923), HAMPSHIRE, S ENGLAND, on the SOLENT channel at the mouth of the Lymington River; 50°45′N 00°33′W. Resort, port; coast trading and yacht building are pursued, and piston rings are produced. A Roman camp was in the vicinity.

Lymm (LIM), town (2001 population 10,552), Warrington, W ENGLAND, on Bridgewater Canal, and 5 mi/8 km E of WARRINGTON; 53°22′N 02°27′W. Salt and tool manufacturing. Previously stone quarrying.

Lympne (LIMP-nee), village (2001 population 1,516), KENT, SE ENGLAND, 3 mi/4.8 km W of HYTHE; 51°08′N 01°05′E. Castle (13th–15th century), church (11th century), and early 20th-century country house nearby.

Lympstone (LIMP-stuhn), village (2001 population 1,754), S DEVON, SW ENGLAND, on EXE RIVER estuary, and 7 mi/11.3 km SE of EXETER; Fishing port. Its 15th-century church contains personal relics of Sir Francis Drake.

Lyna River (LEE-nah), Polish *Łyna*, German *Alle*, 137 mi/220 km long, in EAST PRUSSIA, after 1945 in NE POLAND and KALININGRAD oblast, RUSSIA; rises in Poland in small lake S of OLSZTYN (Allenstein); flows generally N, past Olsztyn, LIDZBARK WARMINSKI, and BARTOSZYCE, into Russia, past PRAVDINSK and DRUZHBA (head of navigation; junction with MASURIAN CANAL), to PREGEL RIVER at ZNAMENSK.

Lynas, Point, Wales: see AMLWCH.

Lynbrook, village (□ 1 sq mi/2.6 sq km; 2006 population 19,457), NASSAU county, SE NEW YORK; 40°39′N 73°40′W. It is a suburb of NEW YORK city on S shore of LONG ISLAND, located inside the township of HEMPSTEAD. The area was settled in 1785 and was called Pearsalls. Old Church dates from 1800. The name *Lynbrook* (formed by reversing the syllables in *Brooklyn*) was adopted in 1895. Incorporated 1911.

Lynch (LINCH), **Villa Lynch**, or **Villa Linch**, town in W greater BUENOS AIRES, ARGENTINA, on federal district line; 34°36′S 58°31′W.

Lynch (LINCH), town (2000 population 900), HARLAN county, SE KENTUCKY, 5 mi/8 km E of CUMBERLAND, in the CUMBERLAND MOUNTAINS near VIRGINIA state line; 36°57′N 82°54′W. Bituminous coal; coal processing. Black Mountain (4,139 ft/1,262 m), highest peak in Kentucky, is to S. Portal thirty-one Mine Tour. Jefferson National Forest to E. Founded 1917.

Lynch, village (2006 population 230), BOYD county, N NEBRASKA, 20 mi/32 km ESE of BUTTE, and on PONCA CREEK; 42°49′N 98°28′W. Livestock; grain. Excavation of prehistoric settlements was begun nearby in 1936.

Lynchburg (LINCH-buhrg), independent city (□ 50 sq mi/130 sq km; 2006 population 67,720), central VIRGINIA, 100 mi/161 km WSW of RICHMOND, on JAMES RIVER; separate from adjoining AMHERST, BEDFORD, and CAMPBELL counties; 37°23′N 79°11′W. Railroad junction; trade center and tobacco market in the foothills of the BLUE RIDGE MOUNTAINS; manufacturing (machining, shelving, heating equipment, sheet metal fabrication, conveyor systems, nuclear power products, crushed limestone, lumber, power transformers, printing and publishing, iron pipe, communications equipment, automatic teller machines, pharmaceuticals, bakery products, metal plating, electronic capacitors, corrugated containers, furniture, prepared foods, gravure ink, cosmetics, dairy products, metal cylinders, motor vehicle parts, footwear, tools, lumber, and wood products). Confederate supply base in the Civil War; Union attack repulsed (1864). Randolph College, Lynchburg College, Liberty University, Virginia Theological Seminary, Lynchburg College, Central Virginia Community College, Maier Museum of Art (at Randolph College); Pest House Medical Museum. Notable historic houses include Poplar Forest to SW, built by Thomas Jefferson. Lynchburg Regional Airport to S. Settled 1757; incorporated as a city 1852.

Lynchburg, unincorporated town (2000 population 2,959), DE SOTO county, NW MISSISSIPPI, residential suburb 14 mi/23 km SSW of downtown MEMPHIS, TENNESSEE, and 3 mi/4.8 km E of Walls; 34°58′N 90°05′W.

Lynchburg, town, ⊙ MOORE county, S TENNESSEE, 11 mi/18 km SW of TULLAHOMA; 35°17′N 86°22′W. In timber and agrarian area; Jack Daniel Distillery is a major tourist attraction. Community college here.

Lynchburg (LINCH-buhrg), village (2006 population 1,418), HIGHLAND county, SW OHIO, 14 mi/23 km S of WILMINGTON, on East Fork of Little Miami River; 39°14′N 83°47′W.

Lynchburg, village (2006 population 580), LEE county, NE central SOUTH CAROLINA, 18 mi/29 km NE of SUMTER, near LYNCHES RIVER, and 20 mi/32 km SW of FLORENCE; 34°03′N 80°04′W. Fabric finishing, draperies, house furnishings.

Lynches River (LIN-chez), c.140 mi/225 km long, S. NORTH CAROLINA and NE SOUTH CAROLINA, in south UNION co.; rises just over state line in North Carolina; flows SE Johnsonville, South Carolina Enters GREAT PEE DEE RIVER 4 mi/6.4 km ENE of Johnsonville. Lee State Park is along river near Bishopville.

Lynd (LIND), village (2000 population 346), LYON county, SW MINNESOTA, 5 mi/8 km SW of MARSHALL, on REDWOOD RIVER; 44°23′N 95°54′W. Agriculture (dairying; poultry; livestock; grain, soybeans). Camden State Park to SW.

Lyndeborough (LIND-buh-ro), town, HILLSBOROUGH county, S NEW HAMPSHIRE, 15 mi/24 km WSW of MANCHESTER; 42°53′N 71°46′W. Drained by Purgatory Brook. Agriculture (livestock, poultry; fruits, vegetables; dairying; timber). Wapack National Wildlife Refuge in W.

Lynden, town (2006 population 10,912), WHATCOM county, NW WASHINGTON, 15 mi/24 km N of Bellingham, and on NOOKSACK RIVER; 48°57′N 122°28′W. Dairying; poultry; berries; tulips. Port of entry 3 mi/4.8 km S of Canadian (BRITISH COLUMBIA) border; ALDERGROVE, British Columbia, 7 mi/11.3 km N. Manufacturing (iron furniture, farm machinery, structural wood products, millwork, printing and publishing, fertilizer). Pioneer Museum Settled c.1860; incorporated 1891.

Lynden (LIN-den), unincorporated village, S ONTARIO, E central CANADA, included in city of HAMILTON; 43°14′N 80°08′W. Dairying; fruits.

Lyndhurst (LIND-huhrst), city (□ 4 sq mi/10.4 sq km; 2006 population 14,195), CUYAHOGA county, NE OHIO, 41°31′N 81°29′W. Residential suburb of CLEVELAND. Incorporated 1917.

Lyndhurst (LIND-huhst), town (2001 population 2,976), SW HAMPSHIRE, S ENGLAND, in center of the NEW FOREST, 8 mi/12.9 km W of SOUTHAMPTON; 50°52′N 01°35′W. Tourist resort. Site of 17th-century "King's House," residence of deputy-surveyor of the New Forest.

Lyndhurst (LIND-huhrst), township (2000 population 19,383), BERGEN county, NE NEW JERSEY, near PASSAIC River, 5 mi/8 km NNW of JERSEY CITY; 40°47′N 74°06′W. Manufacturing and residential. Incorporated 1852.

Lyndhurst (LIND-hurst), unincorporated village, SE ONTARIO, E central CANADA, on GANANOQUE RIVER, 30 mi/48 km NE of KINGSTON, and included in LEEDS AND THE THOUSAND ISLANDS township; 44°32′N 76°07′W. Dairying; mixed farming.

Lyndhurst (LIND-huhrst), locality, virtual ghost town, NE SOUTH AUSTRALIA state, S central AUSTRALIA, at intersection of Strzelecki and Oodnadatta tracks; 30°19′S 138°24′E. "Ochre Cliffs" quarry N of town contains red, brown, and yellow ores. Established 1878 as a railway siding.

Lyndoch (LIN-dahk), township, SOUTH AUSTRALIA state, S central AUSTRALIA, 36 mi/58 km from ADELAIDE, at S end of BAROSSA VALLEY; 34°37′S 138°53′E. Vineyards. Australian Museum of Mechanical Music.

Lyndon (LIN-duhn), town (2000 population 1,038), ⊙ OSAGE county, E KANSAS, 30 mi/48 km S of TOPEKA; 38°36′N 95°41′W. Elevation 1,030 ft/314 m. Trade center in livestock and grain region; paper products, metal products Melvin Lake and State Park to SW; Pomona Lake and State Park to NE.

Lyndon, town (2000 population 9,369), JEFFERSON county, N KENTUCKY, residential suburb, 9 mi/14.5 km E of downtown LOUISVILLE; 38°15′N 85°35′W. Herr House here (built 1789) is one of earliest brick houses in Kentucky.

Lyndon, town, CALEDONIA CO., NE VERMONT, on PASSUMPSIC RIVER, and 7 mi/11.3 km N of St. Johnsbury; 44°32′N 72°00′W. Includes villages of Lyndon Center, seat of teachers college, and Lyndonville. Settled 1788. Named for Josiah Lyndon Arnold.

Lyndon, village (2000 population 566), WHITESIDE county, NW ILLINOIS, on ROCK RIVER, and 7 mi/11.3 km SSE of MORRISON; 41°43′N 89°55′W. In agricultural area.

Lyndon Baines Johnson Memorial Grove on the Potomac, memorial, WASHINGTON, D.C. Grove (17 acres/7 ha) of 500 white pines overlooking POTOMAC River; vista of the Capitol. Authorized 1973.

Lyndon B. Johnson National Historic Site (LIN-duhn) (□ 2 sq mi/5.2 sq km), SE TEXAS, 50 mi/80 km W of AUSTIN; 30°15′N 98°26′W. Sites of the birthplace, boyhood home, and ranch (and grandparents' ranch) of President Lyndon B. Johnson. Authorized 1969; redesignated 1980.

Lyndon Station, village (2006 population 457), JUNEAU county, S central WISCONSIN, near the DELLS OF THE WISCONSIN, 10 mi/16 km SE of MAUSTON; 43°42′N 89°53′W. Lumber; manufacturing (wood products).

Lyndonville (LIN-duhn-vil), village (□ 1 sq mi/2.6 sq km; 2006 population 844), ORLEANS county, W NEW YORK, 35 mi/56 km NE of BUFFALO, near Lake ONTARIO; 43°19′N 78°23′W. Light maufacturing and farming. Incorporated 1903.

Lyndonville, village (2006 population 1,236), in Lyndon, CALEDONIA CO., NE VERMONT; 44°31′N 72°00′W. Veterinary medicines, dairy products, maple sugar.

Lyndora (lin-DOR-ah), unincorporated town, Butler township, BUTLER county, W PENNSYLVANIA, residential suburb 2 mi/3.2 km SW of BUTLER, on Connoquinessing Creek; 40°51′N 79°55′W. Manufacturing (industrial chemicals).

Lyneham (LEI-nuhm), village (2001 population 5,822), N WILTSHIRE, S ENGLAND, 9 mi/14.5 km WSW of SWINDON; 51°31′N 01°58′W. Has medieval church. Air force base to W.

Lynemouth (LEIN-mouth), town (2001 population 1,832), E NORTHUMBERLAND, NE ENGLAND, on NORTH SEA, 7 mi/11.3 km NE of MORPETH; 55°12′N 01°32′W. Power station to N.

Lyngdal (LUNG-dahl), village, VEST-AGDER county, S NORWAY, at head of Lyngdalsfjorden (10 mi/16 km inlet of the NORTH SEA), at mouth of the LYNGDAL-selva River, 18 mi/29 km SE of FLEKKEFJORD; 58°08′N 07°05′E. Lumber mills; light manufacturing; tourism.

Lyngdalselva (LUNG-dahls-EL-vah), river, c.40 mi/64 km long, VEST-AGDER county, S NORWAY; rises E of

Knaben NE of FLEKKEFJORD; flows S to an inlet of NORTH SEA.

Lyngen (LUNG-uhn), inlet (60 mi/97 km long, 2 mi/3.2 km–5 mi/8 km wide) of NORWEGIAN SEA, TROMS county, N NORWAY, 30 mi/48 km E of TROMSØ. Noted for spectacular scenery; mountains on W shore rise abruptly to c.4,700 ft/1,433 m. Further inland, 25 mi/40 km E of Tromsø, the JÆGERVASSTIND, surrounded by glaciers, rises to 6,283 ft/1,915 m.

Lyngør (LUNG-uhr), village, AUST-AGDER county, S NORWAY, port on a tiny island in the SKAGERRAK, 18 mi/29 km NE of ARENDAL; 58°37′N 09°08′E. Lighthouse. Scene of naval skirmish (1812) with the English.

Lynhurst, village, MARION county, central INDIANA, a W suburb 4 mi/6.4 km W of downtown INDIANAPOLIS, just S of the Indianapolis Speedway. Became part of Indianapolis 1970.

Lynkerdem, village, E. KHASI Hills district, MEGHALAYA state, NE INDIA, in Khasi Hills, 15 mi/24 km S of SHILLONG. Rice, cotton. Coal deposits nearby.

Lynmouth (LIN-muhth), village (2001 population 2,064), N DEVON, SW ENGLAND, on BRISTOL CHANNEL, and 13 mi/21 km E of ILFRACOMBE; 51°15′N 03°50′W. Tourist resort. Shelley's cottage is here. Promontory of Foreland Point, 2 mi/3.2 km NE.

Lynn (LIN), county (□ 893 sq mi/2,321.8 sq km; 2006 population 6,212), NW TEXAS; ⊙ TAHOKA; 33°10′N 101°49′W. On the LLANO ESTACADO; elevation c.3,000 ft/914 m. Irrigated agricultural area, with large crops of grain sorghum, cotton, wheat; livestock (beef cattle, hogs, sheep). Oil and gas; stone. Includes Tahoka Lake, in NE corner, and other intermittently dry lakes. Formed 1876.

Lynn, city (2000 population 89,050), ESSEX county, E MASSACHUSETTS, on MASSACHUSETTS BAY; 42°28′N 70°58′W. Lynn is an old industrial center. The first ironworks (1643) and the first fire engine (1654) in the country were built here. Formerly the shoe industry was important, but jet engines, marine turbines, dairy products, plastics, and electrical instruments have become the major products. The home of Mary Baker Eddy, the founder of Christian Science, is in Lynn. Lynn Beach and Lynn Woods Reservation are here. Incorporated as a town 1631; as a city 1850.

Lynn, town (2000 population 1,143), RANDOLPH county, E INDIANA, 9 mi/14.5 km SSE of WINCHESTER; 40°03′N 84°56′W. Corn, oats, soybeans; poultry, livestock; manufacturing (burial caskets, foundry filters). Five mi/8 km SE is state's highest point (1,257 ft/383 m) near Bethel, in WAYNE county.

Lynn or **Lynn Regis**, ENGLAND: see KING'S LYNN.

Lynn Canal, natural inlet, SE ALASKA, c.90 mi/145 km long, 7 mi/11.3 km–12 mi/19 km wide. It connects in the S with CHATHAM STRAIT and STEPHENS PASSAGE and thrusts N between mountains to break finally into the inlets of the CHILKOOT and CHILKAT rivers. Navigable to its head, Lynn Canal connects SKAGWAY with JUNEAU and is an important shipping lane. During the Alaska gold rush (1896) it was a major route to the gold fields.

Lynndyl, village (2006 population 125), MILLARD county, W UTAH, on SEVIER RIVER, 17 mi/27 km NE of DELTA; 39°30′N 112°23′W. Irrigated agricultural area: alfalfa, barley, wheat, sugar beets; dairying; cattle. Little Sahara Recreational Area to N; Fool Creek Reservoir to S. Fishlake National Forest to SE. Railroad junction.

Lynnfield, town (2000 population 11,542), ESSEX county, NE MASSACHUSETTS. Primarily residential, Lynnfield is a suburb 8 mi/12.9 km N of BOSTON; 42°32′N 71°02′W. Incorporated 1814.

Lynn Garden, village, SULLIVAN county, NE TENNESSEE, suburb just N of KINGSPORT; 36°35′N 82°34′W.

Lynn Haven (LIN HAI-vuhn), city (□ 10 sq mi/26 sq km; 2000 population 12,451), BAY county, NW FLORIDA, 5 mi/8 km N of PANAMA CITY; 30°14′N 85°39′W. Light manufacturing.

Lynnhaven (LIN-hai-ven) or **Lynnhaven Shores**, urbanized area, part of VIRGINIA BEACH city, SE VIRGINIA, 15 mi/24 km E of NORFOLK, on 2-branched Lynnhaven Bay, an arm (c.5 mi/8 km long) of CHESAPEAKE BAY, which it joins via narrow Lynnhaven Inlet (bridged); 36°54′N 76°04′W. Oysters.

Lynn Lake (LIN), town (□ 351 sq mi/912.6 sq km; 2001 population 699), NW MANITOBA, W central CANADA, 200 mi/322 km N of THE PAS; 56°51′N 101°03′W. Nickel- and copper-mining center; manufacturing of ammonium sulphate fertilizer; tourism. Succeeded (early 1950s) exhausted copper property of SHERRIDON, 120 mi/193 km S, from which railroad was extended to Lynn Lake.

Lynnview, town (2000 population 965), JEFFERSON county, N KENTUCKY, residential suburb 5 mi/8 km SSE of downtown LOUISVILLE; 38°10′N 85°42′W. Louisville International Airport (Standiford Field) to W.

Lynnville, town (2000 population 781), WARRICK county, SW INDIANA, 10 mi/16 km N of BOONVILLE; 38°12′N 87°19′W. In agricultural and bituminous-coal area.

Lynnville, town (2000 population 366), JASPER county, central IOWA, on NORTH SKUNK RIVER, and 15 mi/24 km SE of NEWTON; 41°34′N 92°47′W. Livestock; grain. Lynnville Mill upstream.

Lynnville, town (2006 population 339), GILES county, S TENNESSEE, 12 mi/19 km N of PULASKI; 35°22′N 87°00′W. One of most historical areas in county; small business enterprises; residential.

Lynnville, village (2000 population 137), MORGAN county, W central ILLINOIS, 7 mi/11.3 km WSW of JACKSONVILLE; 39°41′N 90°20′W. In agricultural area.

Lynnwood, city (2006 population 33,685), SNOHOMISH county, W central WASHINGTON, a suburb 15 mi/24 km N of downtown SEATTLE, and 12 mi/19 km SSW of EVERETT, near PUGET SOUND (2 mi/3.2 km NW of city center); 47°50′N 122°18′W. Manufacturing (aerospace parts, communications equipment, electrical equipment and electronic components, metal stampings, precious metal jewelry, scales and balances, textile bags). The city has rapidly developed in the 1980s and early 1990s along with the Seattle area. Edmonds Community College is located here. Incorporated 1959.

Lynnwood, unincorporated town, FAYETTE county, SW PENNSYLVANIA, 3 mi/4.8 km SE of MONESSEN; 40°07′N 79°50′W. Agriculture includes dairying.

Lynovytsya (li-NO-vi-tsya) (Russian *Linovitsa*), town (2004 population 2,950), SE CHERNIHIV oblast, UKRAINE, on highway and railroad 9 mi/14 km S of PRYLUKY; 50°28′N 32°24′E. Elevation 393 ft/119 m. Sugar refinery, bakery; two 18th-century parks. Known since 1629; town since 1960.

Lynton (LIN-tuhn), town (2006 population 1,513), N DEVON, SW ENGLAND, 13 mi/21 km E of ILFRACOMBE, near BRISTOL CHANNEL, on steep cliff 400 ft/122 m above Lynmouth harbor; 51°15′N 03°51′W. Tourist resort. Has 13th-century church.

Lyntupy (lyn-TOO-pee), urban settlement, VITEBSK oblast, BELARUS. Railroad station; alcoholic beverage combine.

Lynwood, city (2000 population 69,845), LOS ANGELES county, S CALIFORNIA; suburb 10 mi/16 km S of downtown LOS ANGELES; 33°55′N 118°12′W. Although primarily residential, Lynwood has printing presses and varied light manufacturing, such as truck equipment, furniture, metal products, die casting, and gears. LOS ANGELES RIVER to E. Founded 1896; incorporated 1921.

Lynwood, village (2000 population 7,377), COOK county, NE ILLINOIS, suburb 25 mi/40 km SSE of downtown CHICAGO, borders INDIANA on E; 41°31′N 87°32′W. Manufacturing: screw machine products, light manufacturing. Lansing Municipal Airport to N.

Lynxville, village (2006 population 171), CRAWFORD county, SW WISCONSIN, on the MISSISSIPPI (IOWA

state line), and 14 mi/23 km NNE of PRAIRIE DU CHIEN, in livestock and dairy region; 43°15′N 91°02′W. A U.S. fish hatchery is here. Lock and Dam No. 9 to S.

Lyon (LEI-uhn), county (2006 population 11,636), extreme NW IOWA; ⊙ ROCK RAPIDS; 43°22′N 96°12′W. Prairie agricultural area (sheep, hogs, cattle, poultry; corn, oats), drained by ROCK and LITTLE ROCK rivers, and bounded N by MINNESOTA and SOUTH DAKOTA (5 mi/8 km of N boundary borders South Dakota) and W by BIG SIOUX RIVER (forms South Dakota state line here). Formed 1851.

Lyon, county (□ 855 sq mi/2,223 sq km; 2006 population 35,369), E central KANSAS; ⊙ EMPORIA; 38°27′N 96°09′W. Located in FLINT HILLS region, level to hilly area, drained by NEOSHO and COTTONWOOD rivers. Poultry, cattle, hogs; corn, wheat, sorghum, hay. Lyon State Fishing Lake in NE. Formed 1860.

Lyon, county (□ 256 sq mi/665.6 sq km; 2006 population 8,273), W KENTUCKY; ⊙ EDDYVILLE; 37°01′N 88°04′W. Bounded SW by TENNESSEE RIVER (KENTUCKY LAKE reservoir), NW by CUMBERLAND RIVER. Includes part of Kentucky Woodlands Wildlife Refuge; part of Land Between the Lakes Recreation Area in S; Barkley Dam, on Cumberland River on W county boundary, forms LAKE BARKLEY reservoir, which crosses county; Mineral Mound State Park, on Lake Barkley, in E center. Gently rolling agricultural area (burley and dark tobacco, corn, wheat, soybeans, hay, alfalfa; hogs, cattle; catfish); limestone quarries, hardwood timber. Formed 1854.

Lyon, county (□ 721 sq mi/1,874.6 sq km; 2006 population 24,640), SW MINNESOTA; ⊙ MARSHALL; 44°24′N 95°50′W. Drained by YELLOW MEDICINE, COTTONWOOD, and REDWOOD rivers. Agricultural area (corn, oats, wheat, hay, alfalfa, soybeans; hogs, sheep, cattle, poultry; dairying; honey). Camden State Park in SW corner; small lakes in SW and NE corner. Formed 1868.

Lyon, county (□ 2,016 sq mi/5,241.6 sq km; 2006 population 51,231), W NEVADA; ⊙ YERINGTON; 39°00′N 119°11′W. East and WEST WALKER rivers form the WALKER RIVER above Yerington at center of county. Drained by CARSON RIVER in N; forms LAHONTAN RESERVOIR, Lahontan State Recreational Area; reservoir supplies water for irrigation. Part of Toiyabe National Forest in S, in SIERRA NEVADA. Mason Valley Wildlife Management Area in N center. Part of Walker River Indian Reservation in NE. Fort Churchill State Historical Park in N. Fernley Wildlife Mangement Area in N. Artesia Lake Wildlife Management Area in W. Borders CALIFORNIA on SW. Cattle, sheep, poultry; dairying; hay, vegetables; cement, sand and gravel, gypsum; copper, gold, silver, diatomite. Formed 1861.

Lyon (lee-on), traditional English spelling *Lyons*, city (□ 18 sq mi/46.8 sq km); ⊙ RHÔNE department and RHÔNE-ALPES administrative region, E central FRANCE, at confluence of the RHÔNE and SAÔNE rivers; 45°45′N 04°51′E. The population of the metropolitan area exceeds 1,250,000, making this the second-largest urban agglomeration in France. Lyon is a major transportation hub between PARIS and the MEDITERRANEAN, SWITZERLAND, and ITALY; it lies 240 mi/386 km SSE of Paris and 170 mi/274 km N of MARSEILLE. For centuries a leading commercial and textile center, Lyon now has a highly diversified economic base (chemical, electrical and electronics, mechanical equipment, and high-technology industries) in the urban areas and suburbs. Because of its accessibility and diversity, Lyon plays host to national and international trade fairs. It has a river port, a stock exchange (founded 1506, the oldest in France), a university (founded 1808), and many cultural attractions. As a year-round tourist center, it is linked to Paris and Marseille via a TGV (high-speed electric train) service; its centrally located Perrache station is a railroad, in-

terurban bus, and local transport hub with good access to the expressways radiating from the city. Lyon has long been known for its excellent cuisine and sophisticated retail district. The city has grown and spread out in a valley, dominated by the LYONNAIS MOUNTAINS (outliers of the MASSIF CENTRAL) on the W, occupied by two major S-flowing rivers (the RHÔNE and its delayed tributary the SAÔNE), which flow in parallel until they join just S of the central business district. As a result, downtown Lyon is nowhere wider than 0.5 mi/0.8 km. Within that narrow area, known as the *Presqu'ile* [French=peninsula], are the administrative buildings of the city and region, the museum (for fine arts, natural history, decorative arts, and printing history of fine textiles, especially silk manufacturing), and the spacious Place Bellecour, the city's hub. The opera house and financial district (stock exchange) are also there. The old town is dominated by Fourvière Height (elevation 960 ft/293 m), crowned by a semi-oriental basilica (built 1872–1896). The newer parts of Lyon consist of residential districts (N), university area (S), and the sprawling industrial suburbs, of which VILLEURBANNE is the largest. Lyon has wide riverside quays (now partly overshadowed by expressways) and some two dozen bridges spanning the Saône and Rhône. Among the many historic buildings are the Gothic 12th–15th-century cathedral and several medieval churches, Renaissance dwellings in the old town, the museum of the Gallo-Roman civilization built into the Fourvière hillside, and the entire quarter of La Croix-Rousse. Founded 43 B.C.E. as a Roman colony, ancient *Lugdunum* soon became the principal city of Gaul. Christianity was first introduced into Gaul here. One of the earliest archiepiscopal sees in France, Lyon was ruled by its archbishops until c.1307, when Philip IV incorporated the city and Lyonnais province into the French crownlands. Lyon became a silk center in the 15th century; at first the silkworms raised in SE France met the needs of the industry, but soon Lyon became dependent on imports from the Far East. In 1793, Lyon was devastated by French Revolutionary troops after an insurrection by the local citizenry. During the German occupation in World War II (1940–1944), Lyon was the headquarters of the French Resistance movement. In 1987, Klaus Barbie ("the butcher of Lyon"), who had been head of the Gestapo in the city during the war, was sentenced to life imprisonment for "crimes against humanity." Lyon is one of the first French cities to provide urban services and regional planning for the entire Urban Community of Lyon, embracing some 55 suburbs and towns in the metropolitan area. This Community is governed by a council of delegates. The urban region's vitality has been further strengthened by the completion of Lyon-Satolas international airport (9 mi/15 km E), the development of the Rhône valley industrial plants, and the continuing expansion of hydroelectric and nuclear power facilities along the Rhône. The older international airport, Lyon-BRON, is 6 mi/10 km ESE of the city. Research parks are found in the outlying communities, and the city itself has outstanding medical facilities, including a well-known cancer research center. For its long history of urban development, Lyon was designated a UNESCO World Heritage site in 1998.

Lyon, village (2000 population 418), COAHOMA county, NW MISSISSIPPI, 2 mi/3.2 km NE of CLARKSDALE, on SUNFLOWER RIVER; 34°13′N 90°32′W. In agricultural area (cotton, corn, soybeans, rice; cattle); manufacturing (cotton processing).

Lyon, Fort, Colorado: see FORT LYON.

Lyon, Loch (LEI-uhn), lake (2 mi/3.2 km long; elevation 1,052 ft/321 m; maximum depth 100 ft/30 m), W PERTH AND KINROSS, E Scotland, 12 mi/19 km WNW of KILLIN; 56°32′N 04°36′W. Outlet: LYON RIVER reservoir for North of Scotland Hydroelectric Board. Formerly in Central region, abolished 1996.

Lyon Mountain, village, CLINTON county, extreme NE NEW YORK, in the ADIRONDACK MOUNTAINS, 22 mi/35 km W of PLATTSBURGH. Lyon Mountain (3,810 ft/1,161 m) is just SE; 44°43′N 73°55′W. Lyon Mountain Correctional Facility.

Lyonnais (LEE-o-nai), region and former province, E central FRANCE, now divided between RHÔNE and LOIRE departments, RHÔNE-ALPES administrative region. It included Lyonnais proper (the region around LYON, its capital), which Philip IV acquired c.1307; also the former counties of Forez and Beaujolais, annexed in 1531; and the tiny dependency of Franc-Lyonnais. It is largely an industrial region whose centers are Lyon, SAINT-ÉTIENNE, and ROANNE.

Lyonnais, Monts du (LEE-o-nai, mon dyoo), mountain chain in E central FRANCE, forming part of the E escarpment of the MASSIF CENTRAL, in LOIRE and RHÔNE departments, RHÔNE-ALPES region; 45°40′N 04°30′E. Bounded by the Forez Plain (SW), the RHÔNE valley (E), and the JAREZ region (S), it rises to c.3,000 ft/914 m. Cattle and sheep raising. The mountain range overlooks LYON (NE) and SAINT-ÉTIENNE (SSW).

Lyonnesse (LEI-uhn-es), once a region W of CORNWALL, now beneath the sea more than forty fathoms deep. The Lyonnesse of Celtic legend and the home of Tristram and of the Lady of Lyones has been identified with LOTHIAN, a former district in Scotland.

Lyon River (LEI-uhn), 34 mi/55 km long, W PERTH AND KINROSS, E Scotland; rises 4 mi/6.4 km NE of TYNDRUM; flows E, through GLEN LYON, to the TAY RIVER 4 mi/6.4 km W of ABERFELDY. On upper course of river is LOCH LYON.

Lyons (LEI-uhnz), town (2000 population 1,585), BOULDER county, N COLORADO, at junction of N and S SAINT VRAIN creeks, just E of FRONT RANGE, 15 mi/24 km N of BOULDER; 40°13′N 105°16′W. Elevation c.5,374 ft/1,638 m. Mining, building stone, and lumbering point. Rocky Mountain National Park to W; Roosevelt National Forest to W.

Lyons, town (2000 population 4,169), ⊙ TOOMBS county, E central GEORGIA, 5 mi/8 km ESE of VIDALIA; 32°12′N 82°19′W. In agricultural and timber area. Manufacturing includes storage tanks, clothing, textiles, concrete products, veneers. Incorporated 1897.

Lyons, town (2000 population 748), GREENE county, SW INDIANA, 9 mi/14.5 km WSW of BLOOMFIELD; 38°59′N 87°05′W. In agricultural and bituminous-coal area.

Lyons, town (2000 population 3,732), ⊙ RICE county, central KANSAS, 25 mi/40 km NW of HUTCHINSON; 38°21′N 98°12′W. Railroad junction. Salt mining and processing. Oil and natural-gas fields. County Museum here. Quivira Relics archaeological site to W. Laid out 1876 on SANTA FE TRAIL; incorporated 1880.

Lyons, town (2006 population 896), BURT county, E NEBRASKA, 17 mi/27 km NW of TEKAMAH, and on LOGAN CREEK, near MISSOURI RIVER; 41°56′N 96°28′W. Grain; livestock; manufacturing (metal fabrication, alfalfa products). Incorporated 1869.

Lyons, town (2006 population 1,102), LINN county, W OREGON, 20 mi/32 km SE of SALEM, on North SANTIAM RIVER; 44°46′N 122°36′W. Timber; fruits, vegetables. Manufacturing (veneer). North Santiam State Park and Willamette National Forest to E.

Lyons, village (2000 population 10,255), COOK county, NE ILLINOIS, a residential suburb of CHICAGO, on the Des Plaines River; 41°48′N 87°49′W. Lyons was settled at the edge of an early travel route, the portage between the CHICAGO and the Des Plaines rivers. Incorporated 1888.

Lyons, village (2000 population 726), IONIA county, S central MICHIGAN, 6 mi/9.7 km E of IONIA, and on GRAND RIVER; 42°58′N 84°57′W. In farm area; furniture manufacturing.

Lyons, village (2006 population 3,500), ⊙ WAYNE county, W NEW YORK, on the NEW YORK STATE BARGE CANAL and CLYDE RIVER, and 32 mi/51 km ESE of ROCHESTER; 43°05′N 77°00′W. Settled 1800 and known for its peppermint crop; incorporated 1831. The Erie Canal once went through the village but was rerouted when it was expanded in the 1850s. Summer lake resort.

Lyons (LEI-uhns), village (2006 population 558), FULTON county, NW OHIO, 27 mi/43 km W of TOLEDO, near MICHIGAN state line; 41°42′N 84°04′W.

Lyons, village, MINNEHAHA county, SE SOUTH DAKOTA, 13 mi/21 km NW of SIOUX FALLS; 43°43′N 96°52′W. Agricultural area (corn, soybeans; cattle, hogs). Manufacturing (fire trucks, printing). Clear Lake to SW.

Lyons, village, WALWORTH county, SE WISCONSIN, 6 mi/9.7 km SW of BURLINGTON. Vegetables; livestock. Manufacturing (thermofoam clamshells, light aircraft components).

Lyons (LEI-uhns) or **Lyon Station**, borough (2006 population 514), BERKS county, E central PENNSYLVANIA, 12 mi/19 km NE of READING; 40°28′N 75°45′W. Agriculture (grain; livestock, poultry; dairying); manufacturing (electrical equipment).

Lyons, FRANCE: see LYON.

Lyons Falls, village (□ 1 sq mi/2.6 sq km; 2006 population 552), LEWIS county, N central NEW YORK, on Black River (falls here) at junction with MOOSE RIVER, and 14 mi/23 km SSE of LOWVILLE; 43°37′N 75°21′W. Formerly, paper milling here and at Lyonsdale, 2 mi/3.2 km E. on Moose River. Milling begun in 1838, completed in 1858, ceased operation in 1922. Never successfully competed with railroad. Completely abandoned in 1926 when its final function as a feeder of water to the NEW YORK STATE BARGE CANAL was assumed by Delta Lake (NE of ROME) and Hinckley Reservoir, N of UTICA.

Lyons-la-Forêt (lee-on–lah-fo-re), village (□ 40 sq mi/104 sq km), EURE department, HAUTE-NORMANDIE region, NW FRANCE, 16 mi/26 km E of ROUEN; 49°24′N 01°28′E. Resort surrounded by domanial forest of LYON (□ 40 sq mi/104 sq km). Attracts visitors from lower SEINE RIVER valley towns.

Lyons River, AUSTRALIA: see GASCOYNE RIVER.

Lyon Station, Pennsylvania: see LYONS.

Lypiatt Park, ENGLAND: see BISLEY.

Lypova Dolyna (LI-po-vah do-LI-nah) (Russian *Lipovaya Dolina*), town, SW SUMY oblast, UKRAINE, on KHOROL RIVER, and 18 mi/29 km SE of ROMNY; 50°34′N 33°48′E. Elevation 492 ft/149 m. Raion center; food processing; feed mill. Established in the first half of the 17th century; town since 1961.

Lypovets' (li-po-VETS) (Russian *Lipovets*), town, E VINNYTSYA oblast, UKRAINE, 25 mi/40 km E of VINNYTSYA; 49°14′N 29°03′E. Elevation 734 ft/223 m. Raion center; farm machinery repair; food processing (meat, dairy, flour); brickworks. Known since the 14th century, passed to Poland in 1569; castle built in the 17th century played an important role in the Cossack-Polish War (1648–1657), during which its early Jewish communities were destroyed, and in popular uprisings (1702–1704, 1737); passed to Russia in 1793; site of battles between the army of Ukrainian National Republic and the Red Army; town since 1925. Jews again reestablished a community constituting nearly half of the population from the mid-19th to the mid-20th century, numbering 3,600 in 1926, to be destroyed again in World War II—fewer than 100 Jews remaining in 2005.

Lyptsi (LIP-tsee) (Russian *Liptsy*), town, N KHARKIV oblast, UKRAINE, on Kharkiv River, 15 mi/24 km NNE of KHARKIV; 50°12′N 36°26′E. Elevation 508 ft/154 m. Vegetable and fruit farming. Some residents work in nearby Kharkiv.

Lys (LEES), river, 135 mi/217 km long, Flemish *Leie*, in N France and Belgium; rises in the hills of ARTOIS (N FRANCE); flows NE, forming Franco-Belgian border between ARMENTIÈRES and MENEN (French *Menin*);

50°45′N 03°00′E. It continues into Belgium past KORTRIJK (French *Courtrai*) emptying into the SCHELDT RIVER at GHENT. The Lys is canalized from AIRE-SUR-LA-LYS to Ghent. Its valley is known for textile spinning and weaving. It was the scene of severe fighting at end of World War I. Formerly spelled Leye, in Belgium.

Lysa Hora (LI-sah ho-RAH) (Russian *Lysaya Gora*), village (2004 population 3,300), NW MYKOLAYIV oblast, UKRAINE, 14 mi/23 km NE of PERVOMAYS'K; 48°10′N 31°05′E. Elevation 364 ft/110 m. Wheat, corn, sunflowers.

Lysa Hora (LI-sah HO-rah), Czech *Lysá Hora*, highest mountain (4,340 ft/1,323 m) of Czech part of the MORAVIAN-SILESIAN BESKIDS, SEVEROMORAVSKY province, E SILESIA, CZECH REPUBLIC, 5 mi/8 km SE of FRYDLANT NAD OSTRAVICI; 49°33′N 18°27′E. Wintersports area.

Lysaker (LU-sah-kuhr), village, AKERSHUS county, SE NORWAY, port at head of OSLOFJORDEN, on railroad, and 4 mi/6.4 km W of OSLO city center, next FORNEBU AIRPORT; 59°54′N 10°36′E. Varied businesses and industry; site of several corporate headquarters. Fridtjof Nansen died at his estate here.

Lysa nad Labem (LI-sah NAHD LAH-bem), Czech *Lysá nad Labem*, German *Lissa an der Elbe*, town, STREDOCESKY province, central BOHEMIA, CZECH REPUBLIC, near ELBE RIVER, 20 mi/32 km NW of KOLÍN; 50°12′N 14°50′E. Railroad junction. Agriculture (sugar-beet and vegetables); manufacturing (metal furniture, textiles, food processing). Has a picturesque castle.

Lysander, residential village, in township of Lysander, ONONDAGA county, central NEW YORK, 12 mi/19 km NW of SYRACUSE, 1 mi/1.6 km E of BALDWINSVILLE; 43°10′N 76°22′W. Also known as Radisson Community. Created in 1969 as one of many residential, commercial, and industrial projects by New York state's Urban Development Corp. that were meant to accommodate population growth in the area. Housing was to be provided for people of all income levels (planned capacity 18,000 residents), and 795 acres/322 ha were set aside for industrial uses. Despite fears of local residents of increased taxes, the loss of rural character, stifling of industrial competition, and the possibility of slums developing due to subsidized low-income housing, the community has been very successful.

Lysa Pass, Czech *Lyský průsmyk* (LIS-kee PROOS-mik) (elevation 1,732 ft/528 m), E MORAVIA, CZECH REPUBLIC, on Slovak border, 5 mi/8 km NE of VALASSKE KLOBOUKY, at NE edge of the White CARPATHIANS. Railroad corridor.

Lysaya Gora, UKRAINE: see LYSA HORA.

Lysefjorden (LU-suh-fyawr-uhn), long, narrow SE arm of BOKNAFJORDEN, ROGALAND county, SW NORWAY, extends 25 mi/40 km NE from HØGSFJORDEN. Popular tourist area. Prekestolen (the Pulpit), a 2,274 ft/693 m rock promontory, on N shore.

Lysekil (LI-se-SHEEL), town, GÖTEBORG OCH BOHUS county, SW SWEDEN, on SKAGERRAK at mouth of Gullmarn, 15 mi/24 km fjord, 45 mi/72 km NNW of GÖTEBORG; 58°17′N 11°27′E. Manufacturing (fish canning, packaging); fishing. Museum, fishing laboratory, sea aquarium. Resort. Incorporated 1903.

Lysets' (li-SETS) (Russian *Lisets*) (Polish *Lysiec*), town, central IVANO-FRANKIVS'K oblast, UKRAINE, 7 mi/11.3 km SW of IVANO-FRANKIVS'K; 48°52′N 24°36′E. Elevation 908 ft/276 m. Embroidery; land drainage administration. Known since 1491; town since 1940.

Lysi (LEE-see) [Turkish=*Akdoğan*], town, FAMAGUSTA district, E CYPRUS, 18 mi/29 km ESE of NICOSIA; 35°06′N 33°41′E. Attila Line passes to S. Citrus, grain, olives, vegetables, potatoes; livestock.

Lysica, POLAND: see SWIETOKRZYSKIE, GORY,

Lysice, CZECH REPUBLIC: see LOMNICE.

Lysiec, UKRAINE: see LYSETS'.

Lysimachia, Lake (lee-see-mah-KHEE-ah) (□ 5 sq mi/13 sq km), in AKARNANIA prefecture, WESTERN

GREECE department, on outlet of Lake TRIKHONIS, 10 mi/16 km N of MESOLONGI; 4 mi/6.4 km long, 2 mi/3.2 km wide; 38°34′N 21°23′E. Fisheries. Also spelled Lisimakhia; formerly called Angelokastro.

Lyskamm, SWITZERLAND: see MONTE ROSA.

Lyskovo (LIS-kuh-vuh), city (2006 population 23,465), E central NIZHEGOROD oblast, central European Russia, on the VOLGA RIVER, on road junction, 56 mi/90 km ESE of NIZHNIY NOVGOROD; 56°01′N 45°02′E. Elevation 341 ft/103 m. Highway junction. Electrical instruments for cars and tractors, accessories, food products (beer, canned goods, bread, dairy products). Architectural landmarks include the Church of Saint Georgiy the Victorious, the Ascension Cathedral, and the Church of the Transfiguration. Has a heritage museum. Became city in 1922.

Lysky Prusmyk, CZECH REPUBLIC: see LYSA PASS.

Lys-lez-Lannoy (lees–lai–lah-nwah), outer SE suburb (□ 1 sq mi/2.6 sq km; 2004 population 1,707) of ROUBAIX, NORD department, NORD-PAS-DE-CALAIS region, N FRANCE, at Belgian border; 50°40′N 03°13′E. Pharmaceutical industry, textiles.

Ly Son Island (LEE SON) or **cu lao re island**, VIETNAM island in SOUTH CHINA SEA, off Cape BA LANG AN (BATANGAN), central VIETNAM; 3 mi/4.8 km long, 1 mi/1.6 km wide; 15°23′N 109°07′E; lighthouse. Formerly Poulo Canton.

Lysøysund (LUS-uh-u-SOON-duh), village, SØR-TRØNDELAG county, central NORWAY, on NORTH SEA, 35 mi/56 km NNW of TRONDHEIM. Fishing.

Lys River (lees), 22 mi/35 km long, VAL D'AOSTA region, NW ITALY; rises in glaciers on S slope of Monte ROSA; flows S, through GRESSONEY valley, to DORA BALTEA River near PONT-SAINT-MARTIN. Descends over 6,000 ft/1,829 m. Used for hydroelectric power at several stations, including Gressoney-la-Trinité and Pont-St.-Martin.

Lyss (LEASE), commune (2000 population 10,659), BERN canton, NW SWITZERLAND, on AARE RIVER, 11 mi/18 km NW of BERN; 47°04′N 07°18′E. Elevation 1,457 ft/444 m. Metal products, cement, tiles, foodstuffs, watches.

Lyster (LI-stuhr), village (□ 63 sq mi/163.8 sq km), CENTRE-DU-QUÉBEC region, S QUEBEC, E CANADA, on BÉCANCOUR RIVER, and 35 mi/56 km SW of QUEBEC city; 46°22′N 71°37′W. Lumbering; dairying; livestock.

Lysterfield (LIST-uhr-feeld), suburb 22 mi/35 km SE of MELBOURNE, VICTORIA, SE AUSTRALIA; 37°56′S 145°18′E. Vegetables. Recreational facilities at Lysterfield Lake Park.

Lystra (LIS-truh), ancient city of LYCAONIA, S central ASIA MINOR, in present TURKEY, S of KONYA; 37°36′N 32°17′E. The Acts of the Apostles reports that it was visited by Paul and Barnabas. An ancient altar found there mentioned the city and helped to identify the site.

Lys'va (lis-VAH), city (2006 population 69,150), E PERM oblast, W URALS, RUSSIA, on the Lys'va River (left tributary of the CHUSOVAYA), on road and railroad, 128 mi/206 km E of PERM; 58°06′N 57°48′E. Elevation 623 ft/189 m. Ferrous metallurgy, producing quality steels and sheet metal for automobiles and aircraft; turbogenerators, metal goods, knitwear; meat processing. Founded in 1785 as an ironworking plant; became city in 1926.

Lysyanka (LI-syahn-kah) (Russian *Lisyanka*), town, central NW CHERKASY oblast, UKRAINE, 56 mi/90 km W of CHERKASY, and 45 mi/72 km NE of UMAN'; 49°15′N 30°50′E. Elevation 613 ft/186 m. Raion center; flour and feed mill; dairying; forest melioration; fishery stations. Vocational technical school; heritage museum. Known since 1593; town since 1965.

Lysychans'k (li-si-CHAHNSK) (Russian *Lisichansk*), city (2001 population 115,229), W LUHAN'S'K oblast, UKRAINE, on the DONETS, in the DONBAS, and 7 mi/11 km SSE of RUBIZHNE; 48°55′N 38°25′E. Elevation 613 ft/186 m. Coal-mining center of the Lysychans'k-

Rubizhne industrial node; first coal mines in the Donbas, six still operating; chemical and petroleum plants (soda, rubber, vulcanizing, petroleum refining); glass; food processing; building materials. Liskhimbud (Russian *Liskhimstroy*), a chemical center, is part of Lysychans'k. Established in 1795; city since 1938, incorporated a number of adjacent settlements in 1965, including Proletars'k (formerly called Nesvitovych, Russian *Nesvetevich*) with its coal mine and glassworks, and Verkhnyy (Russian *Verkhneye*) with soda and glassworks.

Lysyna-Kosmats'ka, Mount, UKRAINE: see POKUTIAN-BUKOVINIAN CARPATHIANS.

Lysyye Gory (LI-si-ye GO-ri) [Russian=bold mountains], town (2006 population 7,485), S central SARATOV oblast, SE European Russia, on the MEDVEDITSA River (DON River basin), on road and railroad spur, 15 mi/24 km E of BALANDA; 51°31′N 44°48′E. Elevation 456 ft/138 m. Flour mill, cannery.

Lytchett Minster (LIT-shuht MIN-stuh) village (2001 population 7,573), SE DORSET, SW ENGLAND, on Lytchett Bay (extension of POOLE HARBOUR), and 4 mi/6.4 km WNW of POOLE; 50°45′N 02°04′W.

Lytham St. Anne's (LITH-uhm ANZ), town (2001 population 41,327), LANCASHIRE, NW ENGLAND, on the N shore of the RIBBLE estuary; 53°44′N 02°59′W. Seaside resort. Lytham St. Anne's was founded in the 12th century by Benedictine monks. Major golf course.

Lythrodonda (li[th]-RO-don-dah), town, LEFKOSIA district, central CYPRUS, 16 mi/26 km SSW of NICOSIA, in E foothills of TROODOS MOUNTAINS; 34°57′N 33°17′E. Grain, citrus, olives, vegetables; livestock. Also spelled Lythrodhonda and Lythrodontas.

Lytkarino (lit-KAH-ree-nuh), city (2006 population 50,570), central MOSCOW oblast, central European Russia, on the MOSKVA River, on road and railroad, 20 mi/32 km SE of MOSCOW; 55°35′N 37°54′E. Elevation 554 ft/168 m. Machine building, elevator repair, manufacturing (optical glass, textiles), food processing. Made city in 1957.

Lytle (LI-tuhl), town (2006 population 2,689), ATASCOSA county, SW TEXAS, 23 mi/37 km SW of SAN ANTONIO, near source of ATASCOSA RIVER; 29°13′N 98°47′W. Agricultural area (cotton, corn, peanuts, vegetables); manufacturing (feeds).

Lytle Creek, unincorporated village, SAN BERNARDINO county, S CALIFORNIA, 14 mi/23 km NW of SAN BERNARDINO, on Lytle Creek, in SAN BERNARDINO MOUNTAINS and San Bernardino National Forest. SAN GABRIEL MOUNTAINS to W. Tourism. Manufacturing (transportation equipment).

Lytle, Lake (LI-tuhl), W central TEXAS, impounded by dam in small Lytle Creek (a S tributary of Clear Fork of Brazos River), in city of ABILENE; 1 mi/1.6 km long; 32°26′N 99°42′W.

Lyttelton, town, port of CHRISTCHURCH, CANTERBURY region, SOUTH ISLAND, NEW ZEALAND, BANKS PENINSULA district (□ 447 sq mi/1,157 sq km); 43°36′S 172°43′E. Connected with Christchurch by railroad and road tunnels (c.1.5 mi/2.4 km long); on N shore of Lyttelton harbor (1.5 mi/2.4 km wide across mouth, 8 mi/12.9 km long). Exports wool, frozen meat, hides, timber. Sometimes called PORT LYTTELTON.

Lytton, town (2000 population 305), on CALHOUN-SAC county line, central IOWA, 7 mi/11.3 km E of SAC CITY; 42°25′N 94°51′W. In agricultural area.

Lytton (lee-TON), former village, SW QUEBEC, E CANADA; 46°39′N 76°02′W. Amalgamated into MONTCERF-LYTTON in 2001.

Lytton (LI-tuhn), village (□ 3 sq mi/7.8 sq km; 2001 population 319), S BRITISH COLUMBIA, W CANADA, on FRASER River at mouth of THOMPSON River, 65 mi/105 km SW of KAMLOOPS, and in THOMPSON-NICOLA regional district; 50°14′N 121°34′W. Fruit and vegetable growing; gold mining; lumbering. Commercial apple growing in British Columbia began in this area.

Lyubachivka River, UKRAINE: see LUBACZOWKA RIVER.

Lyubachovka River, POLAND and UKRAINE: see LUBACZOWKA RIVER.

Lyuban' (lyoo-BAHN), city (2005 population 4,455), S LENINGRAD oblast, NW European Russia, on the Tigoda River (left tributary of the VOLKHOV RIVER), on road and railroad, 53 mi/85 km SE of SAINT PETERSBURG; 59°21′N 31°15′E. Logging, woodworking. Chartered in 1912.

Lyuban' (LYU-buhn), town, MINSK oblast, BELARUS, ⊙ Lyuban' region, 25 mi/40 km SE of SLUTSK. Food and flax processing.

Lyubar (LYOO-bahr), town (2004 population 4,250), SW ZHYTOMYR oblast, UKRAINE, 37 mi/60 km W of BERDYCHIV; 49°55′N 27°45′E. Elevation 738 ft/224 m. Raion center; food processing (flour, dairy, feed, meat); brickworks. Vocational technical school; heritage museum. Established in the mid-14th century; town since 1924. Jewish community since the late 15th century, mostly craftsmen and professionals; by the early 20th century, the town was almost entirely Jewish, one of the region's centers of Talmudic learning, with nine synagogues, a Jewish theater, hospital, and school. Most of the community of 7,000 eliminated by the Nazis in 1941. As of July 1999, only two Jewish families remained in Lyubar.

Lyubartov, POLAND: see LUBARTOW.

Lyubashevka, UKRAINE: see LYUBASHIVKA.

Lyubashivka (lyoo-bah-SHIF-kah) (Russian *Lyubashevka*), town, N ODESSA oblast, UKRAINE, on highway and railroad, 93 mi/150 km NNW of ODESSA, and 17 mi/27 km ENE of ANAN'YIV; 47°51′N 30°15′E. Elevation 541 ft/164 m. Raion center; elevator; feed and flour mill; dairying. Heritage museum. Established toward the end of the 18th century; town since 1957. Jewish community since the beginning of the 19th century, numbering 2,500 by 1940; destroyed during World War II—fewer than 100 Jews remaining in 2005.

Lyubazh, RUSSIA: see VERKHNIY LYUBAZH.

Lyubcha (LYUB-chuh), Polish *Lubez*, urban settlement, GRODNO oblast, BELARUS, on NEMAN River, and 15 mi/24 km NE of NOVOGRUDOK, 53°46′N 26°01′E. Railroad spur terminus; lumbering; rye, oats, potatoes. Noted residence of Polish gentry in fifteenth-century; declined in seventeenth-century.

Lyubech (LYOO-bech) (Russian *Lyubych*) (Belorussian *Lyubyach*), town, W CHERNIHIV oblast, UKRAINE, on left bank of the DNIEPER (Ukrainian *Dnipro*) River, on the border with BELARUS, and 30 mi/48 km NW of CHERNIHIV; 51°42′N 30°39′E. Elevation 465 ft/141 m. Fruit drying, cheese; souvenirs. Small lakes ecological preserve nearby. Site of the Lyubech congress of Rus' princes (1097, 1135); passed to Lithuania (1356), to Poland (1569), to Cossack state (1648–1657), to Muscovy-Russia (1657). Known since the 9th century.

Lyuben, POLAND: see LUBIEN.

Lyubertsy (LYOO-beer-tsi), city (2006 population 153,995), central MOSCOW oblast, central European Russia, on road, 12 mi/19 km ESE of MOSCOW; 55°40′N 37°53′E. Elevation 492 ft/149 m. Railroad junction. Essentially a suburb of Moscow and a weekend retreat for many of the residents of the capital; has two small natural lakes with sandy beaches. Machinery, metalworking, electrical instruments, plastics, carpets, woodworking, industrial waste processing, food industries (dairy, bakery). Agricultural machinery manufacturing at adjoining Ukhtomskaya. Known since 1621. Became city in 1925.

Lyubeshiv (LYOO-be-shif) (Russian *Lyubeshov*) (Polish *Lubieszów*), town, NE VOLYN' oblast, UKRAINE, in PRIPYAT MARSHES, on STOKHID river (head of navigation), and 25 mi/40 km NE of KAMIN'-KASHYRS'KYY; 51°46′N 25°31′E. Elevation 462 ft/140 m. Raion center; manufacturing (silicate bricks, flax); food processing (flour, feed); lumbering. Vocational technical school. Has a 17th-century monastery and an 18th-century basilica. Known since 1484, it passed from Lithuania to Poland (1569), to Russia (1795); part of autonomous or independent Ukraine (1917–1921); reverted to Poland (1921); part of Ukrainian SSR since 1939; independent Ukraine since 1991. Center for Jews; Hasidic dynasty from mid-19th to mid-20th century; destroyed during World War II.

Lyubeshov, UKRAINE: see LYUBESHIV.

Lyubilki (lyoo-BEEL-kee), settlement, S YAROSLAVL oblast, central European Russia, on road and near railroad, 14 mi/23 km SSW of ROSTOV, to which it is administratively subordinate; 56°58′N 39°13′E. Elevation 570 ft/173 m. Rock and sand quarrying; concrete products manufacturing. Formerly known as Lyubilovo.

Lyubim (lyoo-BEEM), city (2006 population 5,995), NE YAROSLAVL oblast, central European Russia, at a confluence of the Ucha and Obnora rivers (KOSTROMA RIVER basin), on crossroads and railroad, 78 mi/126 km NE of YAROSLAVL, and 20 mi/32 km NE of DANILOV; 58°21′N 40°41′E. Elevation 364 ft/110 m. Sawmilling, timbering; food industries (bakery, dairy). Known since 1546. Made city in 1777.

Lyubimets (lyoo-BEE-mets), city, HASKOVO oblast, LYUBIMETS obshtina (1993 population 12,318), SE BULGARIA, on the MARITSA RIVER, 8 mi/13 km NW of SVILENGRAD; 41°50′N 26°05′E. Railroad station. Sericulture center; cotton yarn; tobacco, grain, and grape growing; milling, canning. Formerly called Khebibchevo or Hebibchevo.

Lyubinskiy (LYOO-been-skeeye), town (2006 population 10,420), SW OMSK oblast, SW SIBERIA, RUSSIA, on crossroads and the TRANS-SIBERIAN RAILROAD, 30 mi/48 km NW of OMSK; 55°09′N 72°42′E. Elevation 344 ft/104 m. In agricultural area; brewery, dairy canning. Railroad station since 1913. Until 1947, when granted town status, called Novo-Lyubino.

Lyublin, POLAND: see LUBLIN.

Lyublinets, UKRAINE: see LYUBLYNETS'.

Lyublino (LYOO-blee-nuh), former city, central MOSCOW oblast, central European Russia, on the MOSKVA River, adjoining MOSCOW on the SE and on the E bank of its S port; 55°41′N 37°44′E. Elevation 465 ft/141 m. Railway car manufacturing center; foundries. Includes Pererva, just S; site of the Moskva River locks and dam with hydroelectric plant. Became city in 1925; incorporated into Moscow in 1960.

Lyublino, RUSSIA: see SERGEYEVKA, settlement, MARITIME TERRITORY.

Lyublinskiy (LYOO-bleen-skeeye), former town, now a SE suburb of MOSCOW, central MOSCOW oblast, central European Russia, on the E bank of the MOSKVA River; 55°40′N 37°43′E. Elevation 465 ft/141 m. Waterworks.

Lyublynets' (LYOO-bli-nets) (Russian *Lyublinets*), town, central VOLYN' oblast, NW UKRAINE, near railroad and highway, 4 mi/7 km SW of KOVEL'; 51°11′N 24°37′E. Elevation 629 ft/191 m. Building materials manufacturing, including silicate brick, reinforced concrete, fiberglass, and asphalt. Known since the second half of the 18th century; town since 1987.

Lyubokhna (lyoo-bukh-NAH), town (2005 population 5,150), NE BRYANSK oblast, central European Russia, on the Bol'va River (tributary of the DESNA River, DNIEPER River basin), on railroad and near highway, 19 mi/31 km N of BRYANSK, and 7 mi/11 km S of DYATKOVO; 53°30′N 34°23′E. Elevation 505 ft/153 m. Foundry (heating radiators).

Lyubomirovka, UKRAINE: see VERKHIVTSEVE.

Lyuboml' (LYOO-boml) (Polish *Luboml*), city, W VOLYN' oblast, UKRAINE, near BUH River (Polish border), 30 mi/48 km W of KOVEL'; 51°14′N 24°02′E. Elevation 600 ft/182 m. Known since 1287 as part of the Volodymyr-Volyns'kyy principality, it passed to Poland in the 14th century, and then to Russia (1795); part of autonomous or independent Ukraine (1917–1921), it reverted to Poland (1921); part of Ukrainian SSR since 1939 and independent Ukraine since 1991. Nearly completely Jewish city in 1921 (94% of the population); Jewish community reduced by the 1926 pogroms but still numbering over 3,100 in 1939; eliminated during World War II. Raion center; food processing (flour, dairy); manufacturing (cotton wad, blankets, machines); metal- and woodworking; sawmilling. Vocational technical school; museum; St. Gregory Church (1264), Catholic church (1412), castle (15th–16th centuries).

Lyubomyrivka, UKRAINE: see VERKHIVTSEVE.

Lyubotin, UKRAINE: see LYUBOTYN.

Lyubotyn (lyoo-BO-tin) (Russian *Lyubotin*), city, N central KHARKIV oblast, UKRAINE, 10 mi/16 km W of KHARKIV; 49°57′N 35°58′E. Elevation 662 ft/201 m. Railroad junction; railroad shops; machine works; cotton textile mill; food processing (flour, yeast, food flavoring), distillery; bricks. Vocational technical

school. Established in the mid-17th century, company center of the Kharkiv regiment (17th–18th century); city since 1938.

Lyubranets, POLAND: see LUBRANIEC.

Lyubuchany (lyoo-boo-CHAH-ni), urban settlement (2006 population 3,530), S MOSCOW oblast, central European Russia, near highway and railroad, 5 mi/8 km SSE of L'VOVSKIY, and 5 mi/8 km NNE of CHEKHOV, to which it is administratively subordinate; 55°14′N 37°33′E. Elevation 623 ft/189 m. Manufacturing (plastic products).

Lyubyach, UKRAINE: see LYUBECH.

Lyubych, UKRAINE: see LYUBECH.

Lyubytino (lyoo-BI-tee-nuh), town (2006 population 3,200), central NOVGOROD oblast, NW European Russia, on the MSTA River (Lake IL'MEN basin), on railroad, 35 mi/56 km NW of BOROVICHI, and 5 mi/8 km S of KHOLM; 58°49′N 33°23′E. Elevation 183 ft/55 m. Highway junction; local transshipment point. Mineral pigments; sawmilling, timbering. Formerly known as Beloye.

Lyudinovo (lyoo-DEE-nuh-vuh), city (2005 population 41,255), SW KALUGA oblast, central European Russia, on Lake Lompad' (formed by a dam on the Nepolot' River), on railroad, 120 mi/193 km SW of KALUGA, and 40 mi/64 km N of BRYANSK; 53°52′N 34°27′E. Elevation 574 ft/174 m. Diesel engines; iron foundry (radiators, heating equipment); clothing; food processing; lumber. Agriculture includes rye, wheat, oats, flax, and corn for livestock feed. Meat and dairy livestock raising. Phosphate, sand, and ceramic clay quarries in the vicinity. Historical landmarks include the churches of Sergiy Radonezhskiy (late 19th century), Troitskaya (1836), and Paraskeva Pyatnitsa (mid-19th century). Known since 1626; became city in 1938.

Lyudvipol', UKRAINE: see SOSNOVE.

Lyudvypil', UKRAINE: see SOSNOVE.

Lyuksemburg, UKRAINE: see ROZIVKA.

Lyulin (LYOO-lin), mountain range, W BULGARIA; 42°38′N 23°11′E.

Lyulyak (LYOO-yahk), island in the DANUBE, NE BULGARIA; 43°48′N 25°54′E.

Lyulyakovo (LYOOL-yak-o-vo), village, BURGAS oblast, RUEN obshtina, BULGARIA; 42°52′N 27°04′E.

Lyuta, Bulgaria: see VLADIMIROVO.

Lyutsin, LATVIA: see LUDZA.

Lyutsymyr Lake, UKRAINE: see SHATS'K LAKES.

M

Ma'Abar (muh-AHB-bahr), township, S central YEMEN, on central plateau, 40 mi/64 km S of SANA, on road from Sana to TAIZ. Commercial center for rural area.

Ma'abarot (mah-bah-ROT) or **Mabarot**, kibbutz, ISRAEL, 3 mi/4.8 km NNE of NETANYA, on SHARON PLAIN; 32°22'N 34°54'E. Elevation 121 ft/36 m. Fish hatchery and fish farming; mixed farming. Manufacturing of medical and veterinarian products. Burial caves and other ruins indicate prehistoric settlement. Local archaeological museum. To the N is Tel Hefer, where archaeological findings show settlement from early Bronze Age to 7th century C.E. Founded 1933.

Ma'abda, El (MAH-AHB-duh, el), village, ASYUT province, central Upper EGYPT, on the E bank of the NILE River, 15 mi/24 km NW of ASYUT. Cereals, dates, sugarcane.

Maachah (MAH-khah), ancient city-state of SYRIA, S of Mount HERMON; 33°15'N 35°34'E. Held by Arameans in biblical times. The inhabitants were Maachathi or Maachathites. Mentioned several times in the Hebrew Bible. The town of Abel-Beth-Maachah may have been here. Also spelled Maacah.

Maakel (mah-AH-kul), administrative region (2005 population 675,700), central ERITREA; ⊙ ASMARA; 15°20'N 38°56'E. Largely consists of the fertile plateau area surrounding city of Asmara. Borders regions of ANSEBA (N), SEMENAWI KAYIH BAHRI (E), DEBUB (S), GASH-BARKA (W). Including city of Asmara, it is the center of Eritrean industry and commerce. Agricultural products include cereals, fruits, vegetables.

Maala (muh-AHL-uh), urban division of ADEN town, S YEMEN, on N shore of Aden peninsula, 2 mi/3.2 km NW of Crater. Harbor for traditional boats on Aden Bay; coffee and other warehouses. Docks for ship repair, sailmaking, manufacturing a variety of products (mainly consumer goods). Includes Somali quarter known as Somalipura. Also spelled Maalla and Ma'alla.

Maalaea Bay (MAH-ah-LEI-ah), arm of PACIFIC OCEAN, on S coast of isthmus, MAUI, MAUI county, HAWAII, 6 mi/9.7 km S of WAILUKU.

Ma'ale Adumim (mah-ah-LAI ah-doo-MEEM), town, E of JERUSALEM, in the Judaean Wilderness, WEST BANK. The name derives from the reddish (Hebrew= *adumim*) local rocks. A growing commuter suburb of JERUSALEM. Nearby ruins include an ancient inn, which Christian tradition reveres as the Good Samaritan's Inn. Above the inn are the remains of a Crusader fort, which in turn was built on Roman foundations.

Ma'ale Ephraim (mah-ah-LAI ef-RAH-eem), Jewish settlement, NW of JERICHO, WEST BANK. Small industrial zone and commercial center. Founded in 1978, the settlement provides services for other nearby Jewish settlements in the JORDAN VALLEY. Also spelled Ma'ale Efrayim.

Maale Hahamisha (mah-ah-LE hah khah-mee-SHAH), kibbutz, E ISRAEL, in JUDAEAN HIGHLANDS, 7 mi/11.3 km WNW of JERUSALEM; 31°49'N 35°07'E. Elevation 2,135 ft/650 m. Manufacturing of sweets; fruit, vegetables; dairying; poultry. Summer resort. Founded 1938.

Maalhosmadulu Atoll, N group of MALDIVES, in INDIAN OCEAN, 05°01'N 72°46'E–06°00'N 73°12'E. Coconuts. Also spelled Malosmadulu.

Ma'alot-Tarshiha (mah-ah-LOT–tahr-SHEE-khah), town (2006 population 21,100), ISRAEL, 12 mi/20 km E of NAHARIYA in UPPER GALILEE; 33°00'N 35°16'E. Elevation 1,912 ft/582 m. Textiles; light manufacturing. Ma'alot began as a development township for immigrants in 1957 and eventually merged with the adjacent Arab village of Tarshiha (built on the foundations of an ancient Jewish settlement). Township had been a base for the army of Fawzi al Kaukji during both the 1936 Arab riots and Israel's 1948 War of Independence. Only joint Jewish-Arab-run municipality in Israel; about one-quarter of the town's inhabitants are Arab, and another quarter are recent immigrants from the former Soviet Union. Scene of heavy fighting between Christians and Muslims in Crusader times. Remains of a Crusader fort nearby. A school in Ma'alot was the target of a terrorist attack in the 1970s.

Maamba (mah-AH-em-bah), mining district, SOUTHERN province, S ZAMBIA, 100 mi/161 km SSW of LUSAKA, and W of SINAZONGWE; 17°22'S 27°09'E. Near LAKE KARIBA reservoir (ZAMBEZI RIVER). Zambia's only coal-mining area; also limestone and amethyst.

Ma'ameltin (mah-AH-mil-teen), coastal village, central LEBANON, on MEDITERRANEAN SEA, 12 mi/19 km NNE of BEIRUT. Summer resort. Also spelled Mu'amaltayn.

Ma'an (muh-AN), town (2004 population 26,461), S JORDAN, on main railroad line to DAMASCUS (SYRIA); 30°12'N 35°44'E. Important administrative and economic center for S Jordan. Carries on trade in agriculture, produce. Phosphate production to SW at Al-Shaziya. Was a point on an early Middle Eastern caravan route and on the pilgrimage road to MECCA. One of main stops on road between AMMAN and AQABA. The ancient Nabatean trading town of PETRA is in Ma'an's hinterland.

Ma'an Islands (MAH-AHN), northeasternmost group of ZHOUSHAN Archipelago, in EAST CHINA SEA, ZHEJIANG province, CHINA, SE of Shengsi Islands. Largest and northernmost is North Ma'an Island; 2 mi/3.2 km long, 1 mi/1.6 km wide; 30°52'N 122°40'E. Lighthouse. Also known as Saddle Islands.

Ma'anit (mah-ah-NEET), kibbutz, ISRAEL, 6 mi/10 km ENE of HADERA, on the NW edge of the Samarian foothills; 32°27'N 35°01'E. Elevation 370 ft/112 m. Mixed farming and light industry; glucose plant. Founded in 1935 at KARKUR; moved to current location in 1942 near the remains of ancient Jewish settlement and the city of Narbata, an important urban center during Roman times.

Ma'anshan (MAH-AHN-SHAHN), city (□ 110 sq mi/286 sq km), E ANHUI province, CHINA; 31°49'N 118°32'E. One of China's major centers for the iron and steel industry and heavy industry (textiles, chemicals, iron and steel, machinery; utilities). Agriculture (grains, oil crops, vegetables); animal husbandry (hogs).

Maardu (MAHR-doo), town, N ESTONIA, 10 mi/16 km E of TALLINN; 59°25'N 25°01'E. Phosphorite deposits, largely depleted, connected with Tallinn by 4-lane highway.

Ma'aret Misrin (MAH-ah-ret mis-REEN), township, ALEPPO district, NW SYRIA, 25 mi/40 km SW of ALEPPO.

Maarheeze (MAHR-hai-zuh), village, NORTH BRABANT province, SE NETHERLANDS, 11 mi/18 km SSE of EINDHOVEN; 51°18'N 05°37'E. Dairying; livestock; vegetables, grain; manufacturing (food, lighting fixtures, hardware).

Maarianhamina (MAH-ree-ahn-HAH-mi-nah), Swedish *Mariehamn*, city, ⊙ ÅLAND province, SW FINLAND, S ÅLAND Island; 60°06'N 19°57'E. Elevation 33 ft/10 m. An active trade center and a popular summer resort on Åland island. Founded in 1861 by Czar Alexander II. International airport to W.

Maarkedal (MAHR-kuh-dahl), commune (2006 population 6,490), Oudenaarde district, EAST FLANDERS province, W central BELGIUM, 6 mi/10 km SE of OUDENAARDE.

Maarmorilik (mahr-MO-ri-lik), settlement, Uummannaq commune, W GREENLAND, at head of E arm of UUMMANNAQ FJORD, near edge of inland ice, 35 mi/56 km NE of Umanak; 71°08'N 51°17'W. Marble quarries; production suspended during World War II. "Black Angel" lead-zinc ore deposits nearby were mined, 1973–1990.

Ma'arret el Nu'man, SYRIA: see MA'ARRET EN NU'MAN.

Ma'arret en Nu'man (MAH-ah-ret en NOO-mahn), town, IDLIB district, NW SYRIA, 45 mi/72 km SW of ALEPPO, on Aleppo-Hama road; 35°38'N 36°40'E. Cotton, cereals. Birthplace of noted Arabic free-thinking poet Abu-l-Ala al-Maarri, or Abul 'Ala al-Ma'ari. Also spelled Ma'arret el Nu'man.

Maarssen (MAHR-suhn), town, UTRECHT province, central NETHERLANDS, on VECHT RIVER, and 5 mi/8 km NW of UTRECHT; 52°08'N 05°08'E. AMSTERDAM-RHINE CANAL passes to SW; LOOSDRECHTSE PLASSEN lakes to NE; recreational center to E. Cattle, poultry; flowers, vegetables, nursery stock, fruit; manufacturing (pharmaceuticals, cleaning equipment; food processing).

Maas, FRANCE, BELGIUM, and NETHERLANDS: see MEUSE RIVER.

Maasai Mara Game Reserve (MAH-ah-sah-ee MAH-rah), national park (□ 649 sq mi/1,687.4 sq km), S KENYA, NW section of SERENGETI plains. Open savanna vegetation; has lion, wildebeest, giraffe, baboon, hyena, impala, buffalo, and hippopotamus.

Maasdriel, NETHERLANDS: see KERKDRIEL.

Maaseik (mah-ZEIK), commune (□ 341 sq mi/886.6 sq km; 2006 population 23,761), ⊙ Maaseik district, NE BELGIUM, on MEUSE RIVER, and 22 mi/35 km NE of HASSELT, near NETHERLANDS border; 51°06'N 05°48'E. Market center. Painters Hubert and Jan van Eyck born here.

Maaselkä, RUSSIA: see MASEL'SKAYA.

Maasin (mah-AH-sin), town, ILOILO province, S central PANAY island, PHILIPPINES, 16 mi/26 km NW of ILOILO; 10°55'N 122°23'E. Rice-growing center.

Maasin, town (2000 population 71,163), ⊙ SOUTHERN LEYTE province, PHILIPPINES, on CANIGAO CHANNEL, 80 mi/129 km SSW of TACLOBAN; 10°12'N 124°51'E. Agricultural trading center (coconuts, rice, hemp) and port.

Maasland (MAHS-lahnt), village, SOUTH HOLLAND province, SW NETHERLANDS, 11 mi/18 km WNW of ROTTERDAM, and 1.5 mi/2.4 km of MAASSLUIS; 51°56'N 04°17'E. Cattle; vegetables, flowers, nursery stock; manufacturing (farm equipment).

Maasmechelen (mahs-ME-chuh-luhn), commune (2006 population 36,397), Tongeren district, LIMBURG province, NE BELGIUM, 9 mi/14 km E of GENK, on ALBERT CANAL, and near MAAS RIVER; 50°49'N 05°40'E.

Maas River, BELGIUM, FRANCE, and NETHERLANDS: see MEUSE RIVER.

Ma'asser esh Shuf (mah-AH-sir esh SHOOF), village, central LEBANON, 21 mi/34 km SE of BEIRUT; 33°39'N 35°40'E. Known for its vineyards and the production of wine and arrack. Nearby is one of the few remaining cedar woods. Summer resort.

Maassluis (MAHS-LOIS), town, SOUTH HOLLAND province, SW NETHERLANDS, on the NEW WATERWAY, and 10 mi/16 km W of ROTTERDAM; 51°55'N 04°16'E. Entrance to Rotterdam harbor from NORTH SEA (8 mi/12.9 km to NW), opposite ROZENBURG. Large oil storage, refining, and port facilities to W. Dairying; cattle, poultry; vegetables, flowers, nursery stock; manufacturing (diamond tools).

Maastricht (MAH-strikht), city, ⊙ LIMBURG province, SE NETHERLANDS, on the MEUSE RIVER, 122 mi/196 km SE of AMSTERDAM; 50°50'N 05°42'E. Meuse River flows N through city; Belgian border 3 mi/4.8 km to NW, W, and SW; German border 15 mi/24 km to E. ZUID-WILLEMS Canal terminates 1 mi/1.6 km to N; JULIANA CANAL terminates 2 mi/3.2 km to N, ALBERT CANAL (Belgium) passes to W. Railroad junction; airport to NNE. Agricultural area (dairying; cattle, hogs, poultry; grain, vegetables, fruit, sugar beets); manufacturing (plastic ware, ceramics, building materials, fabricated metal products, consumer goods;

food processing). Formerly a center for the Dutch industry of S Limburg, which closed down during the 1970s. The Maas was forded in Roman times; the city derives its name from *Mosae Trajectum* [Latin=Maas ford]. An episcopal center 382–721, Maastricht has the oldest church in the Netherlands, the Cathedral of St. Servatius (founded in the sixth century). In 1284 the city came under the dual domination of the dukes of BRABANT and the prince-bishop of LIÈGE. Was a strategic fortress and suffered many sieges. The Spanish under Alessandro Farnese captured it (1579) from the Dutch rebels during the revolt of the Netherlands and massacred a large part of the population. In 1632 the Dutch under Prince Frederick Henry recovered the city. It later fell into French hands during the wars of the seventeenth and eighteenth century, notably in 1673 and 1794. Has many old structures including the Romanesque Church of Our Lady (eleventh century), a thirteenth-century bridge across the Maas, and the town hall (seventeenth century). Caves of Mount Sint-Peter in SW part of city, with over 20,000 passages excavated over the centuries for marble. In 1992, the Treaty of the European Union (known as the Maastricht Treaty) was signed at a summit conference here. The treaty was an important step in the integration of the EC. Cultural center, with many museum; seat of Academy of Architecture, European Institute of Public Administration, and Maastricht School of Management. Convention center here is the site of the world's largest annual art fair. The district of Wijk occupies the E bank of the Maas.

Maasvlakte (MAHS-vlahk-tuh), region, SOUTH HOLLAND province, W NETHERLANDS, reclaimed land area, 20 mi/32 km WNW of ROTTERDAM, NW extension of VOORNE region; 51°58′N 04°04′E. On S side of entrance to NEW WATERWAY; includes EUROPOORT, large port and industrial facility built in 1960s. Lighthouse.

Maas-Waal Canal (MAHS–VAHL), GELDERLAND province, E NETHERLANDS, extends 9 mi/14.5 km N-S, from WAAL RIVER (3 mi/4.8 km NW of NIJMEGEN) to MEUSE RIVER (5 mi/8 km S of Nijmegen). Completed 1927.

Ma'ayan Zvi (ma-ah-YAHN ts-VEE), kibbutz, ISRAEL, on S outskirts of ZICHRON YAACOV, 9 mi/14.5 km N of HADERA; 32°34′N 34°55′E. Elevation 436 ft/132 m. Mixed farming and optical lens production. Recreational park for fishing. Founded 1938.

Mabalacat (mah-bah-LAH-kaht), town (2000 population 171,045), PAMPANGA province, central LUZON, PHILIPPINES, 15 mi/24 km NW of SAN FERNANDO; 15°13′N 120°33′E. Sugar milling. N terminal of expressway from MANILA.

Mabalane, MOZAMBIQUE: see GAZA.

Mabamba (mah-BAHM-bah), village, KIGOMA region, NW TANZANIA, 15 mi/24 km W of KIBONDO, near BURUNDI border; 03°35′S 30°31′E. Tobacco, corn; goats, sheep.

Mabanda (muh-BAHN-duh), town, NYANGA province, SW GABON, near CONGO border, 70 mi/113 km SE of MOUILA; 02°48′S 11°32′E.

Mabang (mah-BAHNG), village, SOUTHERN province, W SIERRA LEONE, 29 mi/47 km ESE of FREETOWN.

Mabank (MAH-bank), town (2006 population 2,821), KAUFMAN county, NE TEXAS, c.50 mi/80 km SE of DALLAS; 32°22′N 96°05′W. In agricultural (cattle; cotton, sorghum, wheat), timber area; manufacturing (turbines); diversified light manufacturing); oil wells. Purtis Creek State Park to E; CEDAR CREEK RESERVOIR to SW.

Mabarot, ISRAEL: see MA'ABAROT.

Mabaruma (mah-bah-ROO-mah), village, BARIMA-WAINI district, NW GUYANA, in hills S of MORAWHANNA, 10 mi/16 km from the coast, 150 mi/241 km NW of GEORGETOWN; 08°12′N 59°47′W. In fruit-

growing and coffee region (citrus, coconuts, bananas, cacao).

Mabel, village (2000 population 766), FILLMORE county, SE MINNESOTA, near IOWA state line, 19 mi/31 km SE of PRESTON, on Riceford Creek; 43°31′N 91°46′W. Grain, soybeans; livestock, poultry; dairying. Richard J. Dorer Memorial Hardwood State Forest to N.

Mabel Lake (MAI-buhl), S BRITISH COLUMBIA, W CANADA, 27 mi/43 km SW of REVELSTOKE; 22 mi/35 km long, 1 mi/2 km–2 mi/3 km wide; 50°31′N 118°44′W. Drained W by SHUSWAP RIVER into SHUSWAP LAKE.

Maben (MAI-buhn), town (2000 population 803), OKTIBBEHA and WEBSTER counties, E MISSISSIPPI, 17 mi/27 km WNW of STARKVILLE; 33°32′N 89°04′W. Agriculture (cotton, corn, soybeans; cattle); manufacturing (wood products, apparel; meat processing). NATCHEZ TRACE PARKWAY passes to NW.

Mabesi, Lake (mah-BE-see), SE SIERRA LEONE, near ATLANTIC coast, 10 mi/16 km S of PUJEHUN; 6 mi/9.7 km long, 3 mi/4.8 km wide; 07°11′N 11°42′W. Connected with WAANJE RIVER (W). Also spelled Mabessi.

Mabi (MAH-bee), town, Kibi county, OKAYAMA prefecture, SW HONSHU, W JAPAN, 12 mi/20 km W of OKAYAMA; 34°37′N 133°41′E. Bamboo shoots.

Mabian (MAH-BYEN), town, ⊙ Mabian county, SW SICHUAN province, CHINA, 65 mi/105 km SW of LESHAN; 28°48′N 103°39′E. Grain, medicinal herbs; food processing, logging.

Mabira (mah-bee-rah), village, KAGERA region, NW TANZANIA, 60 mi/97 km WNW of BUKOBA, near KAGERA RIVER; 01°15′S 30°59′E. Timber; tin mining.

Mabla Mountains (MAH-blah), basalt range, DJIBOUTI, on N coast of GULF OF TADJOURA, 25 mi/40 km NNW of DJIBOUTI; 11°55′N 43°02′E. Rise to about 4,000 ft/1,219 m.

Mablethorpe (MAI-buhl-thawp), town (2001 population 7,000), E LINCOLNSHIRE, E ENGLAND, on NORTH SEA, 11 mi/18 km E of LOUTH; 53°20′N 00°15′E. Seaside resort. Nearby are Sutton, 2 mi/3.2 km SE, and Trusthorpe.

Mableton, growing suburb (2000 population 29,733), COBB county, GEORGIA, 5 mi/8 km NW of ATLANTA; 33°30′N 84°35′W. Manufacturing of lumber, aerospace parts, signs, coil coatings, plastic products, wood products, forestry equipment, vegetable oil, machining.

Mably (mah-BLEE), town (□ 12 sq mi/31.2 sq km), LOIRE department, RHÔNE-ALPES region, central FRANCE, on the canal paralleling the upper LOIRE RIVER, in the Roanne basin, and 4 mi/6.4 km N of ROANNE; 46°05′N 04°04′E. Textiles; food processing.

Mabole (MAH-bo-lai), town and suburb, WESTERN PROVINCE, SRI LANKA, 6 mi/9.7 km NNE of COLOMBO city center; 07°29′N 79°53′E. Administered by urban council jointly with WATTALA (1.5 mi/2.4 km SSW) and PELIYAGODA (1 mi/4.8 km S).

Mabonto (mah-BON-to), town, NORTHERN province, central SIERRA LEONE, 13 mi/21 km NE of MAGBURAKA; 08°52′N 11°49′W. Alluvial gold-mining center.

Mabote, MOZAMBIQUE: see INHAMBANE.

Mabou (MA-boo), village, NE NOVA SCOTIA, E CANADA, W CAPE BRETON ISLAND, on MABOU RIVER, near its mouth, on the GULF OF ST. LAWRENCE, 12 mi/19 km SSW of INVERNESS; 46°04′N 61°22′W. Elevation 351 ft/106 m.

Mabou River, (MA-boo) 15 mi/24 km long, NE NOVA SCOTIA, E CANADA, on W CAPE BRETON ISLAND; rises in the CRAIGNISH HILLS 4 mi/6 km SE of MABOU; flows WNW, past Mabou, to GULF OF ST. LAWRENCE 5 mi/8 km W of Mabou.

Mabrouk, village, EIGHTH REGION/KIDAL, N MALI, in the SAHARA DESERT, on desert trail, and 210 mi/338 km NE of TIMBUKTU. Oasis.

Mabscott, town (2006 population 1,351), RALEIGH county, S WEST VIRGINIA, suburb, 3 mi/4.8 km SW of

BECKLEY; 37°46′N 81°12′W. Semibituminous-coal region. Manufacturing (concrete). Incorporated 1906.

Mabton, town (2006 population 2,047), YAKIMA county, S WASHINGTON, near YAKIMA River, and 35 mi/56 km SE of YAKIMA; 46°13′N 120°00′W. At E edge of Yakima Indian Reservation. Feeds Horse Heaven Hills to S.

Mabuki (mah-BOO-kee), village, MWANZA region, NW TANZANIA, 35 mi/56 km SSE of MWANZA, near Moame River; 03°00′S 33°09′E. Road junction. Cattle, goats, sheep; cotton, corn, wheat, millet.

Mabwe (MAHB-wai), village, KATANGA province, SE CONGO, on the right bank of LAKE UPEMBA; 08°39′S 26°13′E. Elev. 1,886 ft/574 m.

Macabebe (mah-kah-BE-be), town (2000 population 65,346), PAMPANGA province, central LUZON, PHILIPPINES, 8 mi/12.9 km S of SAN FERNANDO; 14°55′N 120°41′E. Agricultural center (sugarcane, rice).

Macabí Island (mah-kah-BEE), small rocky island, 6 mi/10 km off PACIFIC coast of LA LIBERTAD region, NW PERU, 36 mi/58 km WNW of TRUJILLO; 07°49′S 79°29′W. Guano deposits. Variant names include: Nacabí; Macavi.

Macabu, Brazil: see CONCEIÇÃO DE MACABU.

Macachín (mah-kah-CHEEN), town (1991 population 3,852), ⊙ Atreucó department, E LA PAMPA province, ARGENTINA, on railroad, and 50 mi/80 km SE of SANTA ROSA; 37°09′S 63°39′W. Wheat, oats, alfalfa; livestock.

Macacu River (MAH-kah-koo), 60 mi/97 km long, central RIO DE JANEIRO state, BRAZIL; rises in the Serra do MAR S of NOVA FRIBURGO; flows SW to GUANABARA BAY 12 mi/19 km NNE of NITERÓI.

Macaé (MAH-kah-AI), city (2007 population 169,229), E RIO DE JANEIRO state, BRAZIL, port on the ATLANTIC OCEAN, on railroad, and 50 mi/80 km SW of CAMPOS; 22°23′S 41°47′W. Textile and paper milling, coffee and rice processing, brandy distilling, match manufacturing. Outport and swimming beach at Imbetiba (S). Formerly spelled Macahé.

Macael (mah-kah-EL), town, ALMERÍA province, S SPAIN, near ALMANZORA RIVER, 20 mi/32 km WSW of HUÉRCAL-OVERA; 37°20′N 02°18′W. Olive-oil processing; carob beans. Iron, copper, ochre mines; marble quarrying.

Macahé, Brazil: see MACAÉ.

Macahubas, Brazil: see MACAÚBAS.

Macahubas, Brazil: see MACAUBAL.

Macaíba (MAH-kah-ee-bah), city (2007 population 63,344), E RIO GRANDE DO NORTE state, NE BRAZIL, head of navigation on Jundiaí River, and 10 mi/16 km WSW of NATAL; 05°51′S 35°21′W. Cattle-raising center; ships cotton, corn, manioc. Formerly spelled Macahyba.

Macaira, VENEZUELA: see SAN FRANCISCO DE MACAIRA.

Macajalar Bay (mah-kah-hah-LAHR), inlet of MINDANAO SEA in N MINDANAO, PHILIPPINES; 17 mi/27 km long, 20 mi/32 km wide at mouth. Receives CAGAYAN RIVER.

Macajuba (MAH-kah-ZHOO-bah), city (2007 population 11,255), E central BAHIA state, BRAZIL, 109 mi/175 km W of Feira de Santana; 12°08′S 40°22′W.

Macambará (MAH-kahm-bah-rah), city, W RIO GRANDE DO SUL state, BRAZIL, 31 mi/50 km E of ITAQUI, on railroad; 29°08′S 56°03′W. Sheep.

Macambira (MAH-kahm-bee-rah), town, W SERGIPE state, BRAZIL, 38 mi/61 km NW of ARACAJU; 10°40′S 37°32′W. Manioc; livestock.

Macamic (mah-kah-MEEK), parish, QUEBEC, E CANADA; 45°48′N 79°00′W. Amalgamated into MACAMIC village in 2001.

Macamic (mah-kah-MEEK), village (□ 74 sq mi/192.4 sq km; 2006 population 2,838), ABITIBI-TÉMISCAMINGUE region, W QUEBEC, E CANADA, on Macamic Lake, 35 mi/56 km N of ROUYN-NORANDA; 48°46′N 79°00′W. Lumbering; dairying; mining region. Also spelled Makamik. Restructured in 2002.

Area is shown by the symbol □, and capital city or county seat by ⊙.

Macá, Monte (mah-KAH, MON-tai), Andean peak (9,510 ft/2,899 m), AISÉN DEL GENERAL CARLOS IBAÑEZ DEL CAMPO region, S CHILE, on MORALEDA CHANNEL, 25 mi/40 km NW of PUERTO AISÉN.

Macanal (mah-kah-NAHL), town, ⊙ Macanal municipio, BOYACÁ department, central COLOMBIA, 38 mi/61 km S of TUNJA 04°58′N 73°19′W. Corn, sugarcane, coffee; livestock.

Macanga, MOZAMBIQUE: see TETE.

Macao (mah-KOU), Mandarin *Aomen* (OU-muhn), former Portuguese overseas province (□ 9 sq mi/23.4 sq km), adjoining GUANGDONG province, SE CHINA, on the PEARL RIVER estuary, 40 mi/64 km W of HONG KONG, and 65 mi/105 km S of GUANGZHOU (daily ferry and bus service); 22°10′N 113°33′E. It consists of a rocky, hilly peninsula (□ c.2 sq mi/5.2 sq km), connected by a sandy 700-ft/213-m-wide isthmus to China's Zhongshan (Tangjiahuan) island and the two small islands of TAIPA and COLOANE (linked by causeway; finished 1967). The city of Macao is approximately coextensive with the peninsula and contains almost the entire population of the former province, which is overwhelmingly Chinese. Under the terms of a 1987 agreement between China and PORTUGAL, Macao became a special administrative region under Chinese sovereignty in December 1999. Its formal name is thus the Macao Special Administrative Region of China. Macao has been promised fifty years of noninterference in its economic and social systems.

A free port, Macao is a leading trade, tourist, and fishing center, with gambling casinos and textile, plastics, and food-processing industries. Transportation is mostly by water. Most of Macao's transit trade with China is by way of its shallow harbor on the W side of the peninsula. Much of the fresh meat, vegetables, and fruits in Macao is supplied by the cities of ZHUHAI and ZHONGSHAN (Guangdong province). Much of the beer sold in Macao comes from QINGDAO (SHANDONG province). Tourism is extremely important, accounting for up to 20% of Macao's economic output; more than 80% of the tourists come from nearby Hong Kong (linked by ferry, hydrofoil, and helicopter). Finance is also significant.

The colony's name is derived from the Ma Kwok temple, built here in the 14th century. The oldest permanent European settlement in E ASIA, it was a parched and desolate spot before a trading post was established here in 1557 by the Portuguese. For nearly 300 years, the Portuguese paid China an annual tribute for the use of the peninsula, but Portugal proclaimed it a free port in 1849; this action was confirmed by China in the Protocol of Lisbon in 1887. With the gradual silting up of its harbor and the rise (late 19th century) of Hong Kong, Macao lost its preeminent position and became identified to a large extent with smuggling and gambling interests. Since 1949 the population has been greatly swelled by an influx of Chinese refugees from the mainland. Communist-organized riots shook the area in the winter of 1966–1967, resulting in a capitulation by the Portuguese authorities to the Chinese demands to bar entry to refugees and prohibit anti-Communist activities.

Macao was established in 1974 as a Chinese territory under Portuguese administration. It was ruled by a governor, appointed by the Portuguese president, and a legislative assembly. Since 1999, Macao has its own chief executive; the current one is Edmund Ho Hau Wah. Historic structures include the remaining facade of St. Paul's Basilica (built 1635 by Roman Catholic Japanese artisans; burned 1835), a fascinating example of late Italian Renaissance architecture, with mixed Western and Asian motifs; St. Domingo's church and convent (founded c.1670); the fort and chapel of Guia (1626); the fort of São Paulo do Monte (16th century); and statues to Gama and Luís de Camões, who wrote (1558–1559) part of *The Lusiads* here.

Macao is separated from China by a barrier gate (built 1849, replacing one erected by the Chinese in 1573). Formerly spelled Macau.

Mação (mah-SOU), town, SANTARÉM district, central PORTUGAL, 12 mi/19 km NE of ABRANTES; 39°33′N 08°00′W. Manufacturing of carpets.

Macão, Brazil: see MACAU.

Macapá (mah-kah-PAH), city (2007 population 344,194), ⊙ AMAPÁ state, extreme N BRAZIL, on the AMAZON River; 00°05′N 51°45′W. Mining is central to its economy; exports tin iron, gold, and manganese; lumber and fish. Manufacturing (rubber products for motor vehicles). Airport and maritime port. Founded (1688) by military men in the vicinity of a fortress (now regional museum), Macapá grew very slowly until it became the capital of the federal territory (created 1944; made a state in 1988).

Macapá, Brazil: see PERI MIRIM.

Macaparana (MAH-kah-pah-rah-nah), city (2007 population 23,144), NE PERNAMBUCO state, BRAZIL, 53 mi/85 km NW of RECIFE; 07°35′S 35°27′W. Aloe, corn, fruit.

Macará (mah-kah-RAH), town, LOJA province, S ECUADOR, on Andean slopes, at PERU border, on PAN-AMERICAN HIGHWAY, 60 mi/97 km WSW of LOJA; 04°23′S 79°57′W. Trade center in livestock-raising area. Border crossing to Peru.

Macaracas (mah-kah-RAH-kahs), town, ⊙ Macaracas district, LOS SANTOS province, S central PANAMA, in PACIFIC lowland, 18 mi/29 km WSW of LAS TABLAS; 07°44′N 80°33′W. Bananas; livestock.

Macarani (MAH-kah-rah-NEE), city (2007 population 15,728), SE BAHIA state, BRAZIL, 23 mi/37 km SW of ITAPETINGA; 15°30′S 40°27′W.

Macarao (mah-kah-RAH-o), town, DISTRITO CAPITAL, N VENEZUELA, 10 mi/16 km SW of and suburb of CARACAS; 10°26′N 67°02′W. Coffee, corn, fruit.

Macará River (mah-kah-RAH), Andean stream, c.75 mi/121 km long, on border of ECUADOR and PERU; flows along border to the CATAMAYO RIVER to form the CHIRA RIVER 20 mi/32 km SW of CELICA.

Macaravita (mah-kah-rah-VEE-tuh), town, ⊙ Macaravita municipio, SANTANDER department, N central COLOMBIA, 48 mi/77 km SE of BUCARAMANGA, in the CHICAMOCHA RIVER valley; 06°30′N 72°35′W. Elevation 7,545 ft/2,299 m. Agriculture includes coffee, sugarcane; livestock.

Macareo, Caño (mah-kah-RAI-o, KAHN-yo), central arm of ORINOCO RIVER delta, c.100 mi/161 km long, DELTA AMACURO state, NE VENEZUELA; branches off 4 mi/6 km S of COPORITO; flows NNE to the ATLANTIC OCEAN into SERPENT'S MOUTH at 09°48′N 61°35′W. Arm most frequently used for navigation, providing link with TRINIDAD.

Macaroca (MAH-kah-RO-kah), village, N central BAHIA state, BRAZIL, on SALVADOR-JUAZEIRO railroad, 36 mi/58 km SE of Juazeiro; 09°51′S 40°17′W.

Macarsca, CROATIA: see MAKARSKA.

Macarthur (muh-KAHR-thur), town, VICTORIA, SE AUSTRALIA, 196 mi/315 km W of MELBOURNE; 38°02′S 142°00′E. Memorial Rose Garden.

MacArthur, unincorporated town (2000 population 1,693), RALEIGH county, S WEST VIRGINIA, suburb, 2 mi/3.2 km SSW of BECKLEY; 37°45′N 81°12′W. Coal region. Manufacturing (building materials, wood products).

MacArthur, Fort, California: see SAN PEDRO.

Macas (MAH-kahs), town (2001 population 13,602), ⊙ MORONA-SANTIAGO province, SE ECUADOR, on E slopes of the ANDES Mountains, on the UPANO RIVER, 60 mi/97 km SE of RIOBAMBA, 150 mi/241 km S of QUITO; 02°20′S 78°06′W. Elev. 3,445 ft/1,050 m. Livestock raising; tropical woods, lumbering, and agricultural center (yuca, tropical fruits, corn, cereals, sugarcane, bananas, papaya, coffee, cacao, curare drug). Airport.

Macassar, INDONESIA: see UJUNG PANDANG.

Macassar Strait, INDONESIA: see MAKASAR STRAIT.

Macatuba (MAH-kah-too-bah), city (2007 population 16,177), central SÃO PAULO state, BRAZIL, 15 mi/24 km SW of JAÚ; 22°31′S 48°41′W. Distilling; coffee, rice, beans. Until 1944, Bocaiúva.

Macau (mah-KO), commune (□ 7 sq mi/18.2 sq km; 2004 population 3,154), in GIRONDE department, AQUITAINE region, SW FRANCE, on left bank of GARONNE RIVER, opposite Bec D'Ambès industrial district, at the confluence of the Garonne and the DORDOGNE rivers, and 12 mi/19 km NNW of BORDEAUX; 45°01′N 00°37′W. Vineyards; vegetables. Suburban development along the railroad line from Bordeaux to its outport, VERDON-SUR-MER.

Macau (MAH-kou), city (2007 population 27,129), N RIO GRANDE DO NORTE state, NE BRAZIL, port on the ATLANTIC OCEAN, at mouth of PIRANHAS (or Açu) River, and 45 mi/72 km E of MOSSORÓ; 05°07′S 36°38′W. Connected by railroad to NATAL. Important saltworking and shipping center; also exports cotton, carnauba wax. Manganese deposits in area. Formerly spelled Macão.

Macau, CHINA: see MACAO.

Macaubal (MAH-kou-bahl), town (2007 population 7,407), NW SÃO PAULO state, BRAZIL, 50 mi/80 km W of SÃO JOSÉ DO RIO PRÊTO; 20°48′S 49°57′W. Coffee, cotton, fruit; cattle. Until 1944, called Macaúbas (formerly spelled Macahubas).

Macaúbas (mah-kah-OO-bahs), city, W central BAHIA state, BRAZIL, 35 mi/56 km SE of PARATINGA; 13°02′S 42°45′W. Deposits of semiprecious stones. Formerly spelled Macahubas.

Macaúbas, Brazil: see MACAUBAL.

Macauley Island, NEW ZEALAND: see KERMADEC ISLANDS.

Macavi, PERU: see MACABÍ ISLAND.

Macaya (mah-KAH-yah), canton, SAJAMA province, ORURO department, W central BOLIVIA, 18 mi/30 km S of SAJAMA, on the TOTORA-Pisiga road, and S of CHACHACOMANI; 18°36′S 69°02′W. Elevation 12,789 ft/3,898 m. Agriculture (potatoes, yucca, bananas).

Maccabim-Re'ut (mah-kah-BEEM—re-OOT), Jewish township, central Israel, W of RAMALLAH, in the JUDAEAN FOOTHILLS. The two townships of Makkabim and Re'ut were integrated in 1990. Most residents commute to jobs in JERUSALEM and TEL AVIV, as the township is equidistant from the two cities. What is regarded as the birthplace of the Maccabee clan (1st–2nd century B.C.E.) is adjacent to the town.

MacCarthy Island, administrative division, THE GAMBIA: see CENTRAL RIVER.

MacCarthy Island, island (□ 3 sq mi/7.8 sq km) in GAMBIA RIVER, CENTRAL RIVER division, central port of THE GAMBIA, 120 mi/193 km E of BANJUL; 3 mi/4.8 km long, 1 mi/1.6 km wide; 13°31′N 14°47′W. Site of GEORGETOWN. Acquired by BRITISH traders in 1785; ceded 1823 to the crown and renamed MacCarthy Island; became settlement for liberated African slaves and headquarters of Wesleyan mission. Part of Gambia colony land, it was placed (1896) under the protectorate for administration. Formerly called Lemain Island.

Macclenny (muh-KLEN-nee), city (□ 3 sq mi/7.8 sq km; 2005 population 5,186), ⊙ BAKER county, NE FLORIDA, 28 mi/45 km W of JACKSONVILLE; 30°16′N 82°07′W. In lumbering and farming area.

Macclesfield (MAK-uhlz-feeld), town (2001 population 50,688), CHESHIRE, W ENGLAND; 53°15′N 02°07′W. Formerly principal center of silk manufacturing in England; other manufacturing includes clothing, shoes, electrical appliances, and paper. The Church of St. Michael dates from 1278 and the grammar school from the beginning of the 16th century. Macclesfield Forest is a moorland E of the town.

Macclesfield (MAK-uhlz-feeld), village (2006 population 421), EDGECOMBE county, E central NORTH CAROLINA, 13 mi/21 km E of WILSON; 35°45′N 77°40′W.

Manufacturing (consumer products, rugs, plastic products, furniture); agriculture (tobacco, cotton, soybeans, peanuts, grain, sweet potatoes, strawberries; poultry, livestock).

Macclesfield (MAK-uhlz-feeld), suburb 29 mi/47 km N of MELBOURNE, VICTORIA, SE AUSTRALIA, just N of EMERALD; 37°53′S 145°29′E.

Macclesfield Bank, CHINA: see ZHONGSHA ISLANDS.

Maccles Lake (□ 12 sq mi/31 sq km), SE NEWFOUND-LAND AND LABRADOR, E CANADA, 30 mi/48 km SE of GANDER; 8 mi/13 km long, 5 mi/8 km wide.

MacClintock Island, in S FRANZ JOSEF LAND, ARCH-ANGEL oblast, extreme N European Russia, in the ARCTIC OCEAN, W of HALL ISLAND; 20 mi/32 km long, 15 mi/24 km wide; 80°15′N 56°30′E. Rises to 1,624 ft/495 m.

Macdonald (muhk-DAH-nuhld), rural municipality (□ 447 sq mi/1,162.2 sq km; 2001 population 5,320), S MANITOBA, W central CANADA, adjacent WINNIPEG; 49°40′N 97°30′W. Agriculture; wood, building products. Includes communities of OAK BLUFF, LA SALLE, DOMAIN, BRUNKILD, SANFORD, STARBUCK, and OS-BORNE. Incorporated 1881.

MacDonald-Cartier Freeway (muhk-DAH-nuhld–KAHR-tyai), (Highway 401), S ONTARIO, E central CANADA, 600 mi/966 km divided expressway from WINDSOR to MONTREAL (QUEBEC). Commonly re-ferred to as "The 401," it is considered Canada's "Main Street" and is central to Ontario's 400-series freeway system. Built during the late 1960s and early 1970s, it links LONDON, KITCHENER/WATERLOO, TOR-ONTO, KINGSTON, and CORNWALL. Follows N shore of Lake ONTARIO and SAINT LAWRENCE RIVER. It enters Quebec 40 mi/64 km SW of Montreal, where it be-comes Highway 20 and connects with Montreal's Boulevard Metropolitain. One section in suburban Toronto has sixteen lanes.

Macdonald Group, AUSTRALIA: see MCDONALD IS-LANDS.

Macdonald, Lake (muhk-DAH-nuhld), E WESTERN AUSTRALIA, 540 mi/869 km S of WYNDHAM, near W border of NORTHERN TERRITORY; 20 mi/32 km long, 12 mi/19 km wide; 23°35′S 128°55′E. Usually dry; sur-rounded by swamps.

Macdonald, Meredith and Aberdeen Additional (muhk-DAH-nuhld, ME-re-dith, a-buhr-DEEN), township, central ONTARIO, E central CANADA; 46°28′N 83°57′W. Agriculture. Macdonald was created in 1863; Macdonald and Meredith were incorpo-rated as a municipality in 1892; and in 1899 Aberdeen joined to form the existing township.

MacDonald Pass (6,320 ft/1,926 m), on CONTINENTAL DIVIDE, on border between POWELL and LEWIS AND CLARK counties, W central MONTANA, 14 mi/23 km W of HELENA. Named for Alexander MacDonald, who built and maintained first road (1870–1875) through pass; now U.S. Highway 12.

Macdonnell Ranges (muhk-DAH-nel), S NORTHERN TERRITORY, AUSTRALIA, consist of two parallel ranges extending 200 mi/322 km W from ARLTUNGA; highest peak, Mount Zeil (4,955 ft/1,510 m); 23°45′S 133°20′E. Quartzite, eucalyptus. Site of towns of ALICE SPRINGS and HERMANNSBURG.

Macdougall, Lake (□ 265 sq mi/686 sq km), NUNAVUT territory, N CANADA, NE of Lake GARRY; 66°00′N 99°00′W; 37 mi/60 km long, 1 mi/2 km–10 mi/16 km wide. Drained SE by BACK RIVER.

MacDowell Colony, NEW HAMPSHIRE: see PETERBOR-OUGH.

Macduff (MAK-duhf), town (2001 population 3,767), Aberdeenshire, NE Scotland, on MORAY FIRTH, at mouth of DEVERON RIVER, and just E of BANFF; 57°40′N 02°29′W. Fishing center; seaside resort. For-merly in Grampian, abolished 1996.

Maceda (mah-THAI-dah), town, ORENSE province, NW SPAIN, 11 mi/18 km ESE of ORENSE. Ceramics factory; trades in rye, potatoes, cattle.

Macedo de Cavaleiros (mah-SAI-doo dai kah-vah-LAI-roosh), town, BRAGANÇA district, N PORTUGAL, on railroad, and 21 mi/34 km SSW of BRAGANÇA; 41°32′N 06°58′W. Rye, potatoes; olive oil, wine.

Macedon (MA-si-duhn), town, S central VICTORIA, AUSTRALIA, 40 mi/64 km NW of MELBOURNE, 0.5 mi/0.8 km E of Calder Highway; 37°25′S 144°34′E. Area once known for its stately homes, historic buildings, and commercial nurseries; suffered severe damage from bushfires on Ash Wednesday of 1983. Additional damage at village of Mount Macedon 3 mi/5 km NE. Cross at summit of Mount Macedon (3,323 ft/1,013 m) honors World War I dead.

Macedon, village (□ 1 sq mi/2.6 sq km; 2006 population 1,515), WAYNE county, W NEW YORK, on the BARGE CANAL, and 12 mi/19 km SE of ROCHESTER; 43°04′N 77°17′W. Manufacturing of food products. Incorpo-rated 1856.

Macedonia (ma-se-DO-nee-yah) or **the former yu-goslav republic of macedonia**, independent republic (9,930 sq mi/25,719 sq km; 2004 estimated population 2,071,210; 2007 estimated population 2,055,915), SE EUROPE, in the BALKAN PENINSULA; was a constituent republic of the former YUGOSLAVIA; ⊙ SKOPJE.

Geography

A predominantly mountainous and landlocked country, it is bordered N by SERBIA, S by GREECE, E by BULGARIA, and W by ALBANIA. Drained by the VAR-DAR RIVER, which runs right through the center of the country, and its tributaries, including the BREGAL-NICA, the CRNA REKA, and the TRESKA rivers. The main cities are Skopje, BITOLA, and PRILEP.

Population

Most of the ethnic group recognized as Macedonian reside here; Albanian minority in NW is one-fifth of the total population.

Economy

The poorest of the former Yugoslavian republics; mostly agricultural economy (grains, tobacco, cot-ton); some livestock (sheep, goats); iron, copper, and lead mining; manufacture includes chemicals, steel, machinery, textiles.

History

For history prior to 1990, see MACEDONIA region. In 1990, Macedonia (then still part of the former Yu-goslavia) elected its first non-Communist govern-ment. As the Yugoslavian federation began to disintegrate, Macedonia, led by President Kiro Gli-gorov, declared its independence in September 1991. Its sovereignty was not, at first, internationally rec-ognized, largely due to Greek protests over the name Macedonia. GREECE, fearing future territorial claims, wanted to further the distinction between indepen-dent Macedonia and Greek MACEDONIA; it later im-posed a trade embargo. The U.S. and UN recognized Macedonia in 1994; it is now a member of the UN. In 1995 Macedonia granted concessions to Greece in return for an end to the embargo. Gligorov was se-riously injured in an assassination attempt in 1995. Relations between the government and the Albanian minority, who desired greater autonomy, were tense at times in the 1990s. In 1999 the center-right candidate, Boris Trajkovski, won the presidency, but his win was marred by fraud in some areas. Fighting broke out with Albanian rebels in 2001, but an accord was signed later in the year. NATO peacekeepers disarmed the rebels, and the constitution was amended to protect ethnic Albanian rights. EU forces replaced NATO troops in 2003. In February 2004 Trajkovski was killed in a plane crash; Prime Minister Branko Crvenkovski was elected to succeed him in April.

Government

Macedonia is governed under the constitution of 1991 as amended. The president, who is the head of state, is popularly elected for a five-year term and is eligible for a second term. The government is headed by the prime minister, who is elected by the Assembly, as is the cabinet. The 120 members of the unicameral As-sembly (*Sobranie*) are elected from party lists by popular vote to serve four-year terms. The current chief of state is President Branko Crvenkovski (since May 2004), and the head of government is Prime Minister Nikola Gruevski (since August 2006). Administratively, the country is divided into eighty-five municipalities.

Macedonia (ma-se-DO-nee-yah), Macedonian *Make-doniya*, region, SE EUROPE, on the BALKAN PE-NINSULA, divided among GREECE, BULGARIA, and the Former YUGOSLAV REPUBLIC OF MACEDONIA, corre-sponding roughly with ancient Macedon, a kingdom with its capital at PELLA (c.400 B.C.–167 B.C.E.). The region extends from the AEGEAN SEA N between EPIRUS (W) and THRACE (E) and includes the VAR-DAR, STRUMA, and MESTA (in Greece, the Axios, Strimon, and Nestos) river valleys. The region is predominately mountainous, encompassing parts of the PINDOS and RODOPI mountains. Tobacco is the main crop; grains and cotton are also grown, and sheep and goats are raised. The mining of iron, cop-per, lead, and chromite is important. Greek, or Ae-gean, Macedonia (c.13,000 sq mi/33,670 sq km) includes the prefectures of EAST MACEDONIA AND THRACE, CENTRAL MACEDONIA, and WEST MACEDO-NIA, with the KHALKIDHIKÍ (Chalcidice) Peninsula, site of THESSALONÍKI (Salonica), a major industrial and shipping center. As a result of population movements after World War I, Greek Macedonia has a largely homogeneous Greek population; Bulgarian or Pirin, Macedonia is largely coextensive with the Blagoyevgrad (formerly Gorna Dzhumaya) province of BULGARIA (c.2,500 sq mi/6,475 sq km) and is largely populated by Macedonians, as is the Former Yugoslav Republic of Macedonia. Bulgaria, however, with its historic claims on Macedonia, does not acknowledge a distinct Macedonian ethnicity. The population of the region was complex when first known and in-cluded Anatolian peoples as well as several Hellenic groups. The first influence of Greek culture in Ma-cedon came from the somewhat politically isolated coastal colonies founded in the 8th century B.C.E. By the 7th century B.C.E. there was developing in W Macedon a political unit led by a Greek-speaking family, which assumed a royal status. Macedon was a Persian tributary in 500 B.C.E., but took no real part in the Persian Wars. Alexander I (d. 450 B.C.E.) was the first Macedonian king to enter into Greek politics; he began a policy of imitating features of Greek civ-ilization. By c.400 B.C.E., Macedon had adopted the Greek language and had begun to build a kingdom that was greatly enlarged by the conquests of Philip II (359 B.C.–336 B.C.E.) and Alexander the Great (336 B.C.–323 B.C.E.). Internecine strife and Roman ag-gression, however, led to the kingdom's downfall. In 168 B.C.E. Macedon was divided into four republics, and then in 146 B.C.E. annexed outright as a province of ROME. With the division (C.E. 395) of the Roman Empire, Macedonia came under BYZANTINE rule. Devastated by the Goths and Huns, it was settled (6th century) by the Slavs, who quickly made most of Macedonia a Slavic land. However, it continued under intermittent Byzantine domination until the 9th century, when most of Macedonia was wrested from the Byzantine Empire by Bulgaria. Emperor Basil II recovered it (1014–1018) for Byzantium, but after the temporary breakup (1204) of the Byzantine Empire during the Fourth Crusade, Macedonia was bitterly contested among the Latin Empire of CON-STANTINOPLE, the Bulgars under Ivan II, the despots of EPIRUS, and the emperors of Nicaea (modern IZNIK). It again became part of the Byzantine Empire, which was restored in 1261, but in the 14th century Stephen Dušan of SERBIA conquered all Macedonia except for present-day THESSALONÍKI. The fall of the Serb empire in the late 14th century brought Macedonia under the rule of the Ottoman Turks, which lasted for five centuries. In the 19th century the nationalist revival in the Balkans began; national and religious antagonism flared, and conflict was heightened by the Ottoman policy of playing 1 group against the other. Meanwhile the Ottoman Empire lost control over the major sects. of Greece, Serbia, and Bulgaria, each of which claimed Macedonia on historical or ethnic grounds. In the Treaty of San Stefano (1878), which terminated the Russo-Turk. War of 1877–1878, Bulgaria was awarded the lion's share of Macedonia. However, the settlement was nullified by the European powers at the Congress of Berlin (1878), and Macedonia was left under direct Ottoman control. A secret terrorist organization working for Macedonian independence sprang up in the late 19th century and soon wielded great power. The *komitadjis,* as the terrorist bands were called, were generally supported by Bulgaria, which gained a major share of Macedonia in the First Balkan War (1912–1913). Greece and Serbia turned against Bulgaria in the Second Balkan War, and the Treaty of Bucharest (1913) left Bulgaria only a small share of Macedonia, the rest of which was divided roughly along the present lines. Thousands of Macedonians fled to Bulgaria. In World War I the Salonica (present-day Thessaloníki) campaigns took place here. After the war, Macedonia became a hotbed of agitation and terrorism, directed largely from Bulgaria. The population exchange among Greece, Turkey, and Bulgaria after 1923 resulted in the replacement by Greek refugees from Asia Minor of most of the Slavic and Turkish elements in Greek Macedonia. Charging that the Greek minority in Bulgarian Macedonia was being mistreated, Greece in 1925 invaded Bulgaria. The League of Nations, however, forced a cession of hostilities and awarded (1926) a decision favorable to Bulgaria. Bulgarian relations with the former Yugoslavia (before 1929 the kingdom of the Serbs, Croats, and Slovenes) remained strained over the Macedonian question. Frontier incidents were frequent, as were Yugoslav charges against Bulgaria for fostering the Internal Macedonian Revolutionary Organization (IMRO), a nationalist group that used violence, in Yugoslavia. Macedonian agitation against Serb rule culminated (1934) in the assassination of King Alexander of Yugoslavia by a Macedonian nationalist at Marseilles. In World War II all Macedonia was occupied (1941–1944) by Bulgaria, which sided with the Axis against Yugoslavia and Greece, but the Bulgarian armistice treaty of 1944 restored the prewar borders, which were confirmed in the peace treaty of 1947. The Yugoslav constitution of 1946 made Yugoslav Macedonia an autonomous unit in a federal state, and the Macedonian people were recognized as a separate nationality. Tension over Macedonia continued in the early postwar years. During the Greek civil war there was much conflict between Greece and Yugoslavia over Macedonia, and the breach between Yugoslavia and Bulgaria after 1948 helped to make the Macedonian question explosive. However, with the settlement of the civil war and with the easing of Yugoslav-Bulg. relations after 1962, tension over Macedonia was reduced. For later history, see MACEDONIA (republic) and YUGOSLAVIA.

Macedonia (ma-se-DO-nee-yah), city (□ 10 sq mi/26 sq km; 2006 population 10,418), SUMMIT county, NE OHIO, 14 mi/23 km N of AKRON, just off U.S. Interstate 480, in a primarily urban area that stretches from S CLEVELAND to Akron; 41°19′N 81°29′W.

Macedônia (MAH-se-do-nee-ah), town (2007 population 3,411), NW SÃO PAULO state, BRAZIL, 10 mi/16 km N of FERNANDÓPOLIS; 20°07′S 40°13′W. Coffee growing.

Macedonia, town (2000 population 325), POTTAWATTAMIE county, SW IOWA, on WEST NISHNABOTNA RIVER, and 24 mi/39 km ESE of COUNCIL BLUFFS; 41°11′N 95°25′W. In agricultural region.

Macedonia, village (2000 population 51), HAMILTON county, S ILLINOIS, 8 mi/12.9 km WSW of MCLEANSBORO; 38°02′N 88°42′W. In agricultural area.

Macedonia, Central, Greece: see CENTRAL MACEDONIA.

Macedonia, East, Greece: see EAST MACEDONIA AND THRACE.

Macedonia, West, Greece: see WEST MACEDONIA.

Maceió (mah-sai-o), city, (2007 population 874,014), ⊙ ALAGOAS state, E BRAZIL, on a narrow strip of land between a lagoon and the Atlantic Ocean; 09°42′S 35°42′W. Industries include sugar refining, alcohol and textile production, metallurgy, and chemical processing. On the outskirts are coconut plantations. Airport with connections to all Brazilian cities; port at Jaraguá. The city grew around a sugar mill, following the early 17th-cent. Dutch occupation. In 1654, the Portuguese gained control. By the early 19th century, it had developed as an important sugar-export center, becoming the provincial capital in 1839. An important cultural center, it contains a federal university, historical institute, and academy of letters. The most outstanding landmark, a lighthouse, is in the center of the city. The Church of Bom Jesús dos Martirios is a notable example of Maceió's colonial architecture.

Macenta (mah-SEN-tah), town, ⊙ Macenta prefecture, N'Zérékoré administrative region, SE GUINEA, in Guineé-Forestière geographic region, near LIBERIA border, 125 mi/201 km S of KANKAN; 08°33′N 09°28′W. Rice, coffee, tea, palm kernels, kola nuts, manioc, millet; tea-processing industry. Bouro diamond mines nearby.

Macenta, prefecture, N'Zérékoré administrative region, SE GUINEA, in Guinée-Forestière geographic region; ⊙ MACENTA. Bordered NW tip by Kissidougou prefecture, N by Kérouané prefecture, E by Beyla prefecture, ESE by N'Zérékoré prefecture, SES tip by Yomou prefecture, and S and W by LIBERIA. LOFA RIVER (also called Lawa River here) originates in SW and flows into Liberia; Via River also in SW. Part of Fon Going ridge in NE. Ziama mountain (elevation 4,551 ft/1,387 m) in W. Towns include Koyama, Macenta, NZEBELA, SÉRÉDOU, and Yirié. Main road runs through Macenta town, connecting it to GUÉKÉDOU town to NW and Sérédou, Yirié, Nzebela, and N'ZÉRÉKORÉ towns to SE; secondary road also runs NE out of Macenta town to Kérouané prefecture.

Maceo (mah-SAI-o), town, ⊙ Maceo municipio, ANTIOQUIA department, NW central COLOMBIA, on an affluent of the MAGDALENA River, 50 mi/80 km NE of MEDELLÍN; 06°33′N 74°47′W. Coffee, cacao, plantains; livestock.

Macequece (mah-se-KAI-sai), village, MANICA province, W central MOZAMBIQUE, near Zimbabwean border, 140 mi/225 km WNW of BEIRA. Gold mines; copper and tin deposits.

Macerata (mah-chai-RAH-tah), province (□ 1,071 sq mi/2,784.6 sq km), the MARCHES, central ITALY, on the ADRIATIC Sea; ⊙ MACERATA. Crossed by the APENNINES; watered by CHIENTI, POTENZA, and MUSONE rivers. Agriculture (cereals, grapes, raw silk, olives, fruit); livestock raising (cattle, swine). Manufacturing (shoes, paper goods; leather tannery, marble works) at Macerata, Porto Civitanova, RECANATI, TOLENTINO, Pioraco, CASTELRAIMONDO, SAN SEVERINO MARCHE.

Macerata (mah-chai-RAH-tah), town, ⊙ MACERATA province, in The MARCHES, central ITALY; 43°18′N 13°27′E. Agriculture and diversified secondary industrial center. Ruled by the papacy from the mid-15th century to 1797. It retains its medieval walls. Seat of a university (founded 1290).

Maces Bay (MAY-sis), (2007 population 365) fishing village, SW NEW BRUNSWICK, E CANADA, on MACES (or Mace) Bay (inlet of the Bay of Fundy), 24 mi/39 km SW of SAINT JOHN; 45°06′N 66°29′W. Tourism.

Macestus River, TURKEY: see SIMAV RIVER.

Macgillycuddy's Reeks (muh-GI-lee-KUH-deez), Gaelic *Na Cruacha Dubha,* highest mountain range of

IRELAND, KERRY county, SW IRELAND; 52°00′N 09°42′W. It includes CARRAUNTUOHIL and other peaks more than 3,000 ft/914 m high.

MacGregor (muhk-GRE-guhr), village (□ 1 sq mi/2.6 sq km; 2001 population 882), S MANITOBA, W central CANADA, 22 mi/35 km W of PORTAGE LA PRAIRIE, in NORTH NORFOLK rural municipality; 49°58′N 98°46′W. Grain elevators, livestock; light manufacturing. Incorporated 1948.

Mach (MUHCH), town, KALAT district, KALAT division, BALUCHISTAN province, SW PAKISTAN, 27 mi/43 km SSE of QUETTA, in BOLAN PASS; 29°52′N 67°20′E. Formerly BOLAN subdivision capital. Wheat. Nearby coal and limestone deposits worked. Sometimes spelled MACHH.

Macha (MAH-chah), town and canton, CHAYANTA province, POTOSÍ department, W central BOLIVIA, 10 mi/16 km SSW of COLQUECHACA; 18°49′S 66°05′W. Road center. Potatoes.

Machaca (mah-CHAH-kah), town, AYOPAYA province, COCHABAMBA department, central BOLIVIA, 9 mi/15 km SW of INDEPENDENCIA, on the Independencia-ORURO road; 17°18′S 66°49′W. Elevation 8,602 ft/2,622 m. Tungsten mining at Mina KAMI; clay, limestone, and gypsum deposits. Agriculture (potatoes, yucca, bananas, corn, barley, rye, coffee); cattle raising for meat and dairy products.

Machacamarca (mah-chah-kah-MAHR-kah), canton, AROMA province, LA PAZ department, W BOLIVIA, SE of Sicasica; 17°37′S 67°23′W. Elevation 12,841 ft/3,914 m. Gas wells in area; clay and limestone deposits. Agriculture (potatoes, yucca, bananas, rye); cattle.

Machacamarca, canton, TOMÁS BARRÓN province, ORURO department, W central BOLIVIA, NW of ORURO; 17°37′S 67°23′W. Elevation 12,795 ft/3,900 m. Clay, limestone, and gypsum deposits. Agriculture (potatoes, yucca, bananas); cattle.

Machacamarca, canton, CORNELIO SAAVEDRA province, POTOSÍ department, W central BOLIVIA; 19°24′S 65°33′W. Elevation 10,886 ft/3,318 m. Old silver deposits S of Calavi; tin, clay, limestone, phosphate (abundant) deposits. Agriculture (potatoes, yucca, bananas, wheat, barley); cattle.

Machacamarca (mah-chah-kah-MAHR-kah), town and canton, PANTALEÓN DALENCE province, ORURO department, W BOLIVIA, in the ALTIPLANO, 17 mi/27 km SSE of ORURO, and on Oruro-uyuni railroad; 18°10′S 67°02′W. Elevation 12,165 ft/3,708 m. Clay. Junction for railroad branch to UNCÍA.

Machachi (mah-CHAH-chee), town, PICHINCHA province, N central ECUADOR, in high valley, on PAN-AMERICAN HIGHWAY, and 22 km S of QUITO; 00°30′S 78°34′W. Elev. 9,632 ft/2,936 m. Health resort and livestock-raising center in fertile region, where cereals and potatoes are grown. Manufacturing of dairy products and woolen goods; bottling of mineral water.

Machac-Marca (MAH-chahk–MAHR-kah), canton, QUILLACOLLO province, COCHABAMBA department, central BOLIVIA. Elevation 8,343 ft/2,543 m. Clay, limestone, and gypsum deposits. Agriculture (potatoes, yucca, bananas, corn, rye, soy, coffee); cattle. Also known as Machajmarca.

Machadinho (MAH-shah-zheen-yo), town (2007 population 5,503), N RIO GRANDE DO SUL state, BRAZIL, 41 mi/66 km E of ERECHIM, on PELOTAS RIVER (SANTA CATARINA state border); 27°34′S 51°40′W. Wheat, corn, potatoes, manioc; livestock.

Machado (mah-shah-do), city (2007 population 37,571), SW MINAS GERAIS state, BRAZIL, 45 mi/72 km ENE of POÇOS DE CALDAS; 21°37′S 45°55′W. Coffee-growing center.

Machododorp, town, Nkangala District Municipality, MPUMALANGA, SOUTH AFRICA. Settled 1894 as railroad stop; named for Joachim Machado, governor of MOZAMBIQUE. Cattle, sheep; manganese processing; tourism, trout fishing.

Machado River, Brazil: see GI-PARANÁ RIVER.

Machagay (mah-chah-GEI), town, S central Chaco province, ARGENTINA, on railroad, and 28 mi/45 km ESE of PRESIDENCIA ROQUE SÁENZ PEÑA; 26°56′S 60°03′W. Agricultural (cotton, corn; livestock) and lumbering center; sawmills, cotton gins. Near entrance to El Chaco National Park. Also spelled Machagai.

Machakos (mah-CHAH-kos), town (1999 population 28,891), district administrative center, EASTERN province, S central KENYA, 40 mi/64 km SE of NAIROBI; 01°38′S 37°12′E. Agricultural trade center; deciduous fruits, corn; dairy farming.

Machala (mah-CHAH-lah), town (2001 population 204,578), ☉ EL ORO province, S ECUADOR, in lowlands, 75 mi/121 km S of GUAYAQUIL, 235 mi/378 km SSW of QUITO; 03°16′S 79°57′W. Commercial center in agricultural region, trading in cacao, bananas, citrus fruit, tobacco, coffee, shrimp. Its port, PUERTO BOLÍVAR, is 2 mi/3.2 km W. Airport.

Machalí (mah-chah-LEE), town, ☉ Machalí comuna (2002 population 23,920), CACHAPOAL province, LIBERTADOR GENERAL BERNARDO O'HIGGINS region, central CHILE, in Andean foothills, 5 mi/8 km E of RANCAGUA, on railroad; 34°11′S 70°40′W. Picturesque colonial town in agricultural area (alfalfa, cereals, potatoes, fruit, grapes; livestock). Mining. Originated as a Native settlement.

Machalilla (mah-chah-LEE-yah), village, MANABÍ province, W ECUADOR, small port on the PACIFIC OCEAN, 18 mi/29 km SW of JIPIJAPA; 01°29′S 80°46′W. In agricultural region (cacao, rice, sugarcane, tagua nuts).

Machalilla National Park (mah-chah-LEE-yah), Spanish *Parque Nacional Machalilla* (180 sq mi/466 sq km), MANABÍ province, ECUADOR, S of MANTA. Ecuador's only coastal national park, it preserves beaches, dry forest, cloud forest, and archaeological sites, as well as 77 sq mi/199 sq km of ocean to protect the only coral reefs in mainland Ecuador. Also included are two small offshore islands. Headquarters at small Puerto López village. Established 1979.

Machanga, MOZAMBIQUE: see SOFALA.

Machaquilá Reserve (mah-chah-kee-LAH) (☐ 105 sq mi/273 sq km), S central PETÉN department, GUATEMALA, 30 mi/48 km W of POPTÚN; 16°28′N 89°28′W. One of several protected archaeological and forest reserves designated during the early 1990s. Centered on the Machaquilá archaeological site. No access by road.

Machar, township (☐ 71 sq mi/184.6 sq km; 2001 population 849), S ONTARIO, E central CANADA, 24 mi/38 km from KEARNEY; 45°52′N 79°27′W.

Machara, town (2007 population 8,122), OROMIYA state, central ETHIOPIA; 08°34′N 40°22′E. A busy road junction town 50 mi/80 km SW of ASEBE TEFERI. Also spelled Mechara.

Macharavialla (mah-chah-rah-VYAH-lyah), village, MÁLAGA province, S SPAIN, 12 mi/19 km E of MÁLAGA; 36°46′N 04°13′W. Grapes, raisins, lemons, almonds. Has fine church.

Macharetí (mah-chah-rai-TEE), town and canton, LUIS CALVO province, CHUQUISACA department, SE BOLIVIA, on E slopes of Serranía de AGUARAGÜE, and 32 mi/51 km NNE of VILLA MONTES; 20°49′S 63°24′W. Oil fields (37.54 percent of national oil reserves); oil pipelines to Villa Montes from La Vertiente and CAMIRI.

Machattie, Lake (muh-CHA-tee) (☐ 120 sq mi/312 sq km), SW QUEENSLAND, AUSTRALIA, 410 mi/660 km WNW of CHARLEVILLE; 17 mi/27 km long, 10 mi/16 km wide; 24°50′S 139°48′E. Dry.

Machaxalis (MAH-shah-kah-lees), town, NE MINAS GERAIS state, BRAZIL, near border with S BAHIA state, 19 mi/31 km E of Aguas Formosas; 17°10′S 40°48′W.

Machaze, MOZAMBIQUE: see MANICA.

Machecoul (mah-shuh-KOOL), town (☐ 25 sq mi/65 sq km), LOIRE-ATLANTIQUE department, PAYS DE LA LOIRE region, W FRANCE, 20 mi/32 km SW of NANTES; 47°00′N 01°49′W. Bicycle manufacturing; muscatel wine market. Has ruins of a 14th-century castle. Here in 1793 began the Vendéan royalist insurrection led by Charette. Was capital of historic district of RETZ.

Macheke, township, MASHONALAND EAST province, E ZIMBABWE, 20 mi/32 km ENE of MARONDERA, on Mucheki River (tributary of SAVE RIVER), on railroad; 18°06′S 31°50′E. Elevation 5,042 ft/1,537 m. Dairying; cattle, sheep, goats, hogs; tobacco, peanuts, citrus fruit, macadamia nuts, corn.

Machekha (MAH-chee-hah) [Russian=stepmother], village, NW VOLGOGRAD oblast, SE European Russia, on the BUZULUK RIVER, on road, 22 mi/35 km SW of YELAN; 50°48′N 43°17′E. Elevation 288 ft/87 m. Metalworks; wheat, sunflowers. Until about 1938, spelled Machikha.

Machelen (MAH-kuh-luhn), commune (2006 population 12,684), Halle-Vilvoorde district, BRABANT province, central BELGIUM, NE suburb of BRUSSELS, adjoining industrial installations along WILLEBROEK CANAL; 50°55′N 04°26′E. Has sixteenth-century church; site of Beaulieu Castle (built 1653), headquarters of William III of England in 1695.

Machena, NIGERIA: see MATSENA.

Macheng (MAH-CHUNG), city (☐ 1,392 sq mi/3,605 sq km; 1994 estimated urban population 126,900; estimated total population 1,113,600), NE HUBEI province, CHINA, 65 mi/105 km NE of WUHAN, near HENAN province border, in the DABIE MOUNTAINS; 31°13′N 115°06′E. Agriculture predominates in the local economy; light (food processing; textiles) and heavy (machinery manufacturing) industries are also important. Agriculture (grain, oil crops, vegetables); animal husbandry (hogs); eggs. Derived its name from Ma Qiu, an ancient general who built it. A township for 1,300 years, Macheng was designated a city in 1986.

Macherio (mah-KE-ree-o), town, MILANO province, LOMBARDY, N ITALY, adjacent to SOVICO, 4 mi/6 km N of MONZA. Fabricated metals, machinery, wood products; cotton-milling center.

Macherla (MAH-cher-luh), town, GUNTUR district, ANDHRA PRADESH state, S INDIA, 70 mi/113 km WNW of GUNTUR; 16°29′N 79°26′E. Railroad spur terminus. Steatite mines.

Machesney Park (ma-CHEZ-nee), city (2000 population 20,759), WINNEBAGO county, N central ILLINOIS, suburb, 6 mi/9.7 km NNE of downtown ROCKFORD, on ROCK RIVER; 42°21′N 89°02′W. Diversified light manufacturing. Rock Cut State Park to E.

Machetá (mah-chai-TAH), town, ☉ Machetá municipio, CUNDINAMARCA department, central COLOMBIA, 45 mi/72 km NE of BOGOTÁ; 05°05′N 73°37′W. Elevation 7,004 ft/2,135 m. Cereals, potatoes; livestock.

Machh, PAKISTAN: see MACH.

Machha Bhawan, INDIA: see BAWAN.

Machhapuchhare (MAH-chah-poo-chuh-rai) [Nepali=fish tail], mountain peak (22,955 ft/6,997 m), central NEPAL; 29°30′N 83°58′E. Unclimbed peak that dominates skyline N of POKHARA.

Machhlishahr (muhch-LEE-shuh-huhr), town, JAUNPUR district, SE UTTAR PRADESH state, NE INDIA, 18 mi/29 km WSW of Jaunpur; 25°41′N 82°25′E. Barley, rice, corn, wheat, sugarcane. Ancient fort, extensive mosque ruins.

Machias (muh-CHEI-uhs), town, including Machias village, ☉ WASHINGTON county, E MAINE, 65 mi/105 km ESE of BANGOR, near mouth of the MACHIAS RIVER; 44°40′N 67°27′W. Lumbering center. Site of University of Maine campus. English trading post here (1633) destroyed shortly thereafter by the French. Burnham Tavern (1770), historical museum, has mementos of early naval battle of MACHIASPORT. Settled 1763, incorporated 1784.

Machias, village, CATTARAUGUS county, W NEW YORK, 25 mi/40 km N of OLEAN; 42°23′N 78°31′W. Small lake resort (LIME LAKE nearby). Dairy and maple-sugar products; poultry. Lumber, wood products.

Machias Bay (muh-CHEI-uhs), WASHINGTON county, E MAINE, at mouths of MACHIAS and EAST MACHIAS rivers, just SE of MACHIAS; 7 mi/11.3 km long, 4 mi/6.4 km wide.

Machias Lakes (muh-CHEI-uhs), WASHINGTON county, E MAINE, five lakes (First to Fifth Machias lakes) in upper course of MACHIAS RIVER; 1 mi/1.6 km–5 mi/8 km long.

Machiasport (muh-CHEI-uhs-port), town, WASHINGTON county, E MAINE, at head of MACHIAS BAY, E of MACHIAS; 44°37′N 67°22′W. Earthworks of Revolutionary War fort taken (1814) by British in state park. Offshore, in June 1775, a British ship was captured in what became known as "first naval battle of the Revolution." Set off from MACHIAS 1862.

Machias River (muh-CHEI-uhs), c.65 mi/105 km long, in E MAINE; rises in W WASHINGTON county, forms MACHIAS LAKES in its upper course; flows c.40 mi/64 km S and SE to MACHIAS BAY at MACHIASPORT.

Machias River, c.35 mi/56 km long, in N MAINE; rises in Big Machias Lake in AROOSTOOK county; flows SE and E to the AROOSTOOK RIVER near ASHLAND.

Machias Seal Islands (muh-CHEI-uhs), group of islets, WASHINGTON county, off MAINE and NEW BRUNSWICK coasts, 24 mi/39 km SE of MACHIAS, MAINE. Canadian-operated lighthouse (44°39′N 67°06′W) on Machias Seal, the southernmost island.

Machichaco, Cape (mah-chee-CHAH-ko), Vizcaya province, N SPAIN, on BAY OF BISCAY, 17 mi/27 km NE of BILBAO; 43°28′N 02°47′W. Lighthouse.

Machico (mah-SHEE-koo), township, Madeira, PORTUGAL, on E coast of MADEIRA ISLAND, 10 mi/16 km NE of FUNCHAL; 32°43′N 16°46′W. Fishing port, resort. Mineral springs. João Goncalves Zarco landed here, 1420.

Machida (mah-CHEE-dah), city (2005 population 405,534), Tokyo prefecture, E central HONSHU, E central JAPAN, on the Tsurumi River, 34 mi/55 km W of SHINJUKU; 35°32′N 139°26′E. Industrial and residential suburb of TOKYO, and an important transportation hub.

Machilipatnam (muhch-LEE-puht-nuhm), city, ☉ KRISHNA district, ANDHRA PRADESH state, S INDIA, a port on the BAY OF BENGAL; 16°10′N 81°08′E. Railroad terminus. Educational center. It has a carpet-weaving industry; other products include scientific instruments; rice, oilseeds. In the 17th century it was a center of European trade.

Machin, township (☐ 112 sq mi/291.2 sq km; 2001 population 1,143), NW ONTARIO, E central CANADA, 40 mi/64 km W of DRYDEN, and on EAGLE LAKE; 49°46′N 93°09′W. Tourism, forestry, mining. Composed of the communities of EAGLE RIVER, MINNITAKI, and VERMILION BAY.

Machine, La (mah-SHEEN, lah), town (☐ 6 sq mi/15.6 sq km), NIÈVRE department, in BURGUNDY, central FRANCE, 16 mi/26 km ESE of NEVERS. Former coal-mining center named for novel mining machinery initially installed here in 1670. Coal mining continued until 1974, when seams were no longer worth working.

Machinery City, GEORGIA: see FAIR OAKS.

Machinga (mah-ching-gah), administrative center and district (2007 population 440,492), Southern region, MALAWI, 20 mi/32 km N of ZOMBA city; 14°58′S 35°31′E.

Machiques (mah-CHEE-kes), town, ☉ Machiques de Perijá municipio, ZULIA state, NW VENEZUELA, at E foot of Sierra de PERIJÁ, 75 mi/121 km SW of MARACAIBO; 10°03′N 72°32′W. Livestock.

Machkund River, c.190 mi/306 km long, in NE ANDHRA PRADESH and SW ORISSA states, E central INDIA; rises in EASTERN GHATS in VISHAKHAPATNAM district (Andhra Pradesh), NW of MADUGULA; flows NW and SSW, mainly along Andhra Pradesh–Orissa state border, to SABARI RIVER opposite KONTA, 60 mi/97 km

NNW of RAJAHMUNDRY. Dam at JALAPUT; hydroelectric plant planned to power industries in NE Andhra Pradesh and S Orissa. In lower course, called SILERU RIVER.

Machland (MAHKH-lahnd), flat flood plain region in NE UPPER AUSTRIA, N central AUSTRIA, on the left bank of the DANUBE RIVER, between MAUTHAUSEN (W) and SAXEN (E), E of LINZ; 48°10'–16'N, 14°32'–49'E. Fertile agricultural area with some industry; large hydropower station Wallsee on Danube. PERG is the urban center.

Machohoca (mah-cho-HO-kah), canton, SEBASTIÁN PAGADOR province, ORURO department, W central BOLIVIA. Elevation 12,139 ft/3,700 m. Gas wells in area. Lead-bearing lode, limestone deposits. Agriculture (potatoes, yucca, bananas); cattle.

Machovo Pond, CZECH REPUBLIC: see DOKSY.

Machtum (MAHK-tum), village, WORMELDANGE commune, SE LUXEMBOURG, on MOSELLE RIVER, just S of GREVENMACHER, on German border; 49°39'N 06°26'E. Vineyards; fruit growing. Small hydroelectric station.

Machu Picchu (mah-choo PEE-choo) or **Machucpicchu**, spectacular archaeological site, ceremonial center of the ancient Incas, PERU, c.50 mi/80 km NW of CUSCO, within SANTUARIO HISTORICO MACHU PICCHU; 13°09'S 72°31'W. Perched high upon a rock in a narrow saddle between two sharp mountain peaks; overlooks the URUBAMBA RIVER 2,000 ft/610 m below; it was unknown to Spanish explorers. Discovered in 1911 by the American explorer Hiram Bingham, the imposing city is one of the few urban centers of pre-Columbian America found virtually intact. Perhaps the most extraordinary ruin in the Americas, Machu Picchu contains 5 sq mi/13 sq km of terrace and construction, with more than 3,000 steps linking it to many levels. It shows admirable architectural design and execution, although the stonework is not always as refined as in other Inca sites. Accessible by bus or by train from Cusco; trekker access via the INCA TRAIL.

Machupicchu (mah-choo-PEE-choo) or **Machu Picchu** (MAH-choo PEE-choo), town, CUZCO region, S central PERU; 13°07'S 72°31'W. Located near the ancient city of MACHU PICCHU.

Machupo River (mah-CHOO-po), 150 mi/241 km long, BENI department, NE BOLIVIA; rises in marshy area near San Pedro, c.25 mi/40 km N of TRINIDAD; flows N and NE, past SAN RAMÓN and SAN JOAQUÍN, to the ITONAMAS River just above its confluence with the GUAPORÉ River; 12°34'S 64°25'W. Navigable for c.50 mi/80 km below San Ramón. Called Cocharca River in its upper course, below San Ramón.

Machva, SERBIA: see MACVA.

Machynlleth (muh-KHUHN-hleth), town (2001 population 2,147), POWYS, E Wales, on DOVEY RIVER, and 10 mi/16 km E of TYWYN; 52°36'N 03°51'W. Market and tourist center.

Macia (mah-SEE-uh), village, S MOZAMBIQUE, on road, and 35 mi/56 km W of XAIXAI. Cotton, mafura, rice, beans.

Maciejów, UKRAINE: see LUKIV.

Maciel (mah-see-AIL), town, central SANTA FE province, ARGENTINA, 37 mi/60 km NNW of ROSARIO; 32°28'S 60°53'W. Railroad junction and agriculture center (corn, wheat, flax, soybeans, potatoes; livestock).

Maciel (mah-see-EL), town, Caazapá department, S PARAGUAY, on railroad, at road junction 18 mi/29 km NNW of Yebros, and 95 mi/153 km SE of Asunción, W of Caazapá; 26°11'S 56°28'W. Lumbering; orange growing; cattle raising.

Măcin (muh-CHEEN), town, TULCEA county, SE ROMANIA, in DOBRUJA, on the Dunărea Veche arm of the DANUBE River, and 9 mi/14 km E of BRĂILA; 45°15'N 28°09'E. Trade in fish, grain, and wine. Stone and granite quarrying. Has old mosque.

Macina (ma-SEE-nah), depression along the middle NIGER RIVER, in MALI, a vast lacustrine region extending for c.300 mi/483 km NE-SW (up to c.60 mi/97 km wide) from TIMBUKTU to SÉGOU and covered by a network of lakes, swamps, and channels. Flooded during rainy period. One of AFRICA's most fertile regions, it has been utilized for irrigation (cotton, rice). A large dam built (completed 1946) at SANSANDING supplies irrigation canals. Principal lakes are DÉBO (SW) and FAGUIBINE (NE).

MacIntyre, Mount (5,112 ft/1,558 m), ESSEX county, NE NEW YORK, 10 mi/16 km S of LAKE PLACID village; 44°08'N 74°01'W. Highest peak of MacIntyre Mountains, a short range of the ADIRONDACK MOUNTAINS. The peak faces WALLFACE MOUNTAIN (W) across scenic INDIAN PASS.

Macintyre River (MA-kin-teir), NE NEW SOUTH WALES, AUSTRALIA; rises in NEW ENGLAND RANGE near TINGHA; 30°00'S 148°05'E. A head-stream of DARLING RIVER; used for irrigation. Also called Barwon River.

Mack, village, MESA county, W COLORADO, near UTAH state line (to W) and COLORADO RIVER, 20 mi/32 km NW of GRAND JUNCTION. Elevation 4,523 ft/1,379 m. In cattle and agricultural region producing wheat, potatoes, beans, oats, barley. Nearby are deposits of coal and petroleum. Highline State Park to NE.

Macka, village, NE TURKEY, 12 mi/19 km SSW of TRABZON; 40°50'N 39°39'E. Antimony. Formerly Cevizlik.

Mackay (muh-KEI), city (2001 population 57,649), QUEENSLAND, NE AUSTRALIA, 209 mi/336 km N of ROCKHAMPTON, on the Pioneer River; 21°09'S 149°12'E. Railroad junction. A major port city, Mackay exports sugar, beef, and coal. Coal ship loading terminal at Hay Point. Sugar mills; dairying; hogs; tropical fruit; timber. Tourism.

Mackay (ma-kee), village (2000 population 566), CUSTER county, S central IDAHO, 25 mi/40 km NW of ARCO, on BIG LOST RIVER; 43°55'N 113°37'W. Elevation 5,897 ft/1,797 m. Railroad terminus. Gold, silver, zinc; agriculture. Mackay Reservoir to NW; BORAH PEAK (12,662 ft/3,859 m; highest point in Idaho) to NNW, in LOST RIVER RANGE; parts of Challis National Forest to NE and SW.

MacKay (muh-KEI), hamlet, central ALBERTA, W CANADA, 35 mi/57 km from EDSON, in YELLOWHEAD COUNTY; 53°39'N 115°35'W.

Mackay Glacier, a major outlet glacier in VICTORIA LAND, EAST ANTARCTICA; flowing from the polar plateau into the ROSS SEA at Granite Harbor, SCOTT COAST; 76°58'S 162°00'E.

Mackay Glacier Tongue, the floating extension of MACKAY GLACIER in Granite Harbor, on the SCOTT COAST of VICTORIA LAND, EAST ANTARCTICA; 78°58'S 162°20'E.

Mackay, Lake (muh-KEI), W central AUSTRALIA, on WESTERN AUSTRALIA–NORTHERN TERRITORY border, 320 mi/515 km WNW of ALICE SPRINGS; 65 mi/105 km long, 40 mi/64 km wide; 22°30'S 129°00'E. Usually dry.

Mackeim, POLAND: see MAKOW.

Mackenna, ARGENTINA: see VICUÑA MACKENNA.

Mackensie River, AUSTRALIA: see MACKENZIE RIVER.

Mackenzie (muh-KEN-zee), district municipality (□ 85 sq mi/221 sq km; 2006 population 5,206), central BRITISH COLUMBIA, W CANADA, in the Northern Rocky Mountain Trench, 98 mi/156 km N of PRINCE GEORGE; 55°18'N 123°10'W. Primarily forestry.

Mackenzie, former district, provisional administrative division (□ 527,490 sq mi/1,371,474 sq km) of NORTHWEST TERRITORIES, N CANADA, comprising W mainland part of territory, bounded S by SASKATCHEWAN, ALBERTA, and BRITISH COLUMBIA, W by the YUKON, N by several arms of the ARCTIC OCEAN (BEAUFORT SEA, AMUNDSEN GULF, CORONATION GULF, DEASE STRAIT, and QUEEN MAUD GULF), and E by Keewatin (now KIVALLIQ region); 65°00'N 115°00'W. In W are MACKENZIE MOUNTAINS, N range of the ROCKY MOUNTAINS, here rising to 9,049 ft/2,758 m at

Mount SIR JAMES MCBRIEN. E of this range extends the MACKENZIE RIVER valley, widest part of which lies between GREAT BEAR and GREAT SLAVE lakes. E of lakes is plateau, c.350 mi/563 km wide (E-W), followed by plain E of line between DUBAWNT LAKE and BATHURST INLET. District is drained by Mackenzie River and its tributaries (HAY, SLAVE, LIARD, ARCTIC RED, and Great Bear rivers), and by COPPERMINE, ANDERSON, and THELON rivers. N coastline is irregular and indented by several large bays.

Gold mining centered on YELLOWKNIFE, largest town of the Northwest Territories, site of discovery of important deposits in 1934. Uranium and pitchblende mined on S shore of Great Bear Lake; operations centered on PORT RADIUM. Oil found near NORMAN WELLS; during World War II, terminal of Canol project pipeline. Copper, found near COPPERMINE, not exploited because of transportation difficulties. Norman Wells is the chief oil-producing town. In the early 1970s large natural-gas fields were discovered in the Mackenzie delta region. A plan to construct the Mackenzie Valley Pipeline from the Arctic Ocean to Alberta, which would have been the greatest construction project ever undertaken, was shelved in 1977 after a federal royal commission concluded that, though feasible, the project involved serious legal, political, and environmental problems. Tungsten-copper mined at Tungsten, on Yukon border. Lead-zinc mining at PINE POINT, 1962–1987. Fur trapping and crafts are major occupations of native (Inuit and other population).

Chief towns are Fort Smith, INUVIK, and HAY RIVER; other important trading posts are AKLAVIK, TULITA, FORT SIMPSON, FORT PROVIDENCE, Fort Reliance, Hay River, Coppermine, FORT LIARD, and ARCTIC RED RIVER. Extensive game preserves, including the Reindeer Grazing Reserve and MACKENZIE MOUNTAINS, Yellowknife, and Slave River preserves. Thelon Game Sanctuary has largest herd of musk ox on North American mainland. Transportation is by river and lake ships during summer navigation season (June–October), winter road system during freeze-up. Nearly all localities served by scheduled air service; private planes are principal means of transportation. Created 1895. NE part included in territory of NUNAVUT (1999).

Mackenzie, town, SAINT LOUIS county, E MISSOURI, residential suburb of ST. LOUIS, 8 mi/12.9 km SW of downtown.

Mackenzie, locality, UPPER DEMERARA–BERBICE district, N central GUYANA, on port on right bank of DEMERARA RIVER, just SE of WISMAR, and 55 mi/89 km S of GEORGETOWN; 06°41'N 57°55'W. Once a major bauxite-mining and shipping area, it is now part of the town of LINDEN.

MacKenzie Bay, in ANTARCTICA, on INDIAN OCEAN, along LARS CHRISTENSEN COAST, between CAPE DARNLEY and W shore of AMERY ICE SHELF; 69°30'–72°25'E. Bay is of variable size because of episodic breakouts from the Amery Ice Shelf. Discovered and named by Sir Douglas Mawson in 1931.

Mackenzie Bay, NORTHWEST TERRITORIES, N CANADA, inlet of BEAUFORT SEA of the ARCTIC OCEAN, at mouth of MACKENZIE RIVER delta; 100 mi/161 km long, 120 mi/193 km wide at mouth; 68°40'–69°45'N 134°30'–139°00'W. On E side of bay are RICHARDS, ELLICE, and several smaller islands; on W side is HERSCHEL ISLAND.

MacKenzie Bay, indentation into the W extremity of the AMERY ICE SHELF, EAST ANTARCTICA; 68°38'S 70°35'E.

Mackenzie Country, Upper WAITAKI river, Mackenzie district (□ 2,872 sq mi/7,467.2 sq km), East SOUTH ISLAND, NEW ZEALAND. A major intermontane basin in the lee of the central SOUTHERN ALPS, containing the glacially shaped lakes OHAU, PUKAKI, and TEKAPO, with outlets controlled and channeled to enhance hydroelectricity generation. Also spelled MacKenzie Country.

Mackenzie Highway (muh-KEN-zee), NW ALBERTA and SW NORTHWEST TERRITORIES, W CANADA; begins in S at GRIMSHAW (Alberta), 14 mi/23 km W of PEACE RIVER; continues N through MANNING, HIGH LEVEL, MEANDER RIVER to Enterprise (Northwest Territories), S of GREAT SLAVE LAKE, where it turns WNW and continues to FORT SIMPSON and WRIGLEY. A winter road extension beyond Wrigley goes to TULITA, NORMAN WELLS, and FORT GOOD HOPE. Major river crossings include bridge over HAY RIVER at Meander River ferry crossing of LIARD RIVER E of Fort Simpson, ferry crossing of MACKENZIE RIVER 40 mi/64 km NW of Fort Simpson. The highway and its branches to YELLOWKNIFE, Hay River, Fort Smith, and FORT RESOLUTION are collectively referred to as the Mackenzie Route. The system is used to transfer goods and services N and raw materials, mainly from mining, S. Total distance, GRIMSHAW to Wrigley, 720 mi/1,159 km. Original highway and branches constructed during World War II; section to Fort Simpson, 1970s; to Wrigley, 1994.

Mackenzie Mountains, N range of ROCKY MOUNTAINS in E YUKON and NORTHWEST TERRITORIES, NW CANADA; extends c.500 mi/805 km SE-NW between BRITISH COLUMBIA border to PEEL RIVER valley, forming S part of Yukon-Northwest Territories border. Highest peak, Mount SIR JAMES MCBRIEN (9,049 ft/2,758 m); other peaks higher than 8,000 ft/2,438 m are Mount HUNT, KEELE PEAK, DOME PEAK, Mount Sidney Dobson, and Mount IDA. In S part of range is Mackenzie Mountains Preserve, game reserve (□ 69 sq mi/179 sq km–440 sq mi/1,140 sq km) established 1938.

Mackenzie No. 23, Municipal District of (muh-KEN-zee), specialized municipality (□ 31,092 sq mi/80,839.2 sq km; 2005 population 9,687), NW ALBERTA, W CANADA; 58°41′N 117°02′W. Rural area comprising 12% of Alberta's land mass. Agriculture, forestry, oil and gas, tourism. Includes the towns of RAINBOW LAKE and HIGH LEVEL, and the hamlets of FORT VERMILION, La Crete, and Zama. Formed 1995; changed status in 1999 from a municipal district to a specialized municipality (which provides for the needs of a municipality that includes a large, urban area along with rural territory).

Mackenzie River, c.1,120 mi/1,802 km long, NORTHWEST TERRITORIES, N CANADA; issues from GREAT SLAVE LAKE, FORT SMITH region; flows generally NW to the ARCTIC OCEAN through a great delta; c.69°20′N 133°54′W. Between GREAT SLAVE LAKE and LAKE ATHABASCA it is known as the SLAVE RIVER. At Lake Athabasca, the FINLAY-PEACE river system and the ATHABASCA RIVER join the Mackenzie. The Finlay-Peace-Mackenzie system (c.2,600 mi/4,180 km long) is the second-longest continuous stream in North America. The LIARD RIVER is the largest tributary flowing directly into the Mackenzie. The river is navigable from the Arctic Ocean to Great Slave Lake between June and October. Between Great Slave Lake and Lake Athabasca there are rapids (14 mi/23 km) that must be portaged; above the rapids are more than 400 mi/644 km of navigable waters. The Liard River affords transportation between FORT NELSON (British Columbia), and the Arctic; the Athabasca-Mackenzie system is followed by a major shipping route between EDMONTON (Alberta), and the Arctic. Numerous lakes in the Mackenzie basin act as reservoirs and natural flood controls. The basin, flanked by the ROCKY MOUNTAINS and the CANADIAN SHIELD, is the N portion of the GREAT PLAINS of North America; Arctic air masses follow the valley S into the interior of the continent. Much of the Mackenzie valley is heavily forested and, where climate permits, its deep soil is well suited to agriculture. Numerous trading posts were established along the Mackenzie in the early part of the 19th century and fur trapping is still an important activity here; the chief trading posts are FORT SIMPSON, FORT PROVIDENCE, and AKLAVIK. The region was the domain of fur traders until the 1930s when vast oil fields and other mineral resources were discovered. Peter Pond was possibly the first European to enter (1777) the Mackenzie drainage area, but Sir Alexander Mackenzie, the 19th-century Canadian explorer, was the first (1789) to descend the river to the Arctic Ocean.

Mackenzie River (muh-KEN-zee), 170 mi/274 km long, E QUEENSLAND, AUSTRALIA; formed by junction of NOGOA and COMET rivers; flows NE and SSE, joining DAWSON river to form Fitzroy River; 23°38′S 149°46′E. Drains mining area (gold, copper, and coal). Also spelled Mackensie River.

Mackey (MA-kee), community, ONTARIO, E central CANADA, 41 mi/66 km from PEMBROKE, and included in HEAD, CLARA AND MARIA township; 46°10′N 77°47′W.

Mackillop, Lake, AUSTRALIA: see YAMMA YAMMA, LAKE.

Mackinac (MA-ki-naw), historic region of the Old Northwest (former NORTHWEST TERRITORY), a shortening of Michilimackinac. The name, in the past, was variously applied to different areas: to MACKINAC ISLAND, to MICHIGAN, to the whole fur-trading region supplied from the island, to the N mainland shore (SAINT IGNACE, MICHIGAN, has been sometimes called Anc. Michilimackinac), and to the S mainland shore, where MACKINAW CITY is located and where a fort called Old Mackinac once stood. The STRAITS OF MACKINAC, a passage between the UPPER and LOWER peninsulas of Michigan, connecting LAKE MICHIGAN and LAKE HURON, served for many years as an important Native American gathering place. In 1634 the French explorer Jean Nicolet was the first Europeans to pass through the straits. The French Jesuit Claude Allouez, in 1665, was the first missionary to come here; he was followed by Father Jacques Marquette, who established a mission at Saint Ignace in 1671. A fort was later built here, and it became the headquarters of French trade operations in New France and an important military post in the Old Northwest; its importance declined when DETROIT was founded in 1701. The region passed into British hands in 1761 during the last conflict of the French and Indian Wars. In 1763 members of the British garrison at Old Mackinac were attacked and killed by the Ottawa during Pontiac's Rebellion. During the American Revolution, the fort and town at Old Mackinac, threatened by the exploits of the American general George Rogers Clark, were moved to Mackinac Island. The island and the straits were awarded to the U.S. in 1783 by the Treaty of Paris, but they remained in British hands until 1794. One of the first events of the War of 1812 was the British capture of Mackinac; it was returned to U.S. control by the Treaty of Ghent in 1814. After the war, Mackinac Island became the center of operations of John Jacob Astor's American Fur Company, which thrived until the 1830s, when fur trading declined. After the 1840s the straits area changed from an important crossroads to an out-of-the-way shipping point, and the U.S. army post on the island was abandoned in 1894. Mackinac Island became a Michigan state park and, along with BOIS BLANC ISLAND, a popular summer resort. Iron-ore mining revitalized the area in the early 20th century, but the mineral was soon depleted. The Mackinac Straits Bridge (3,800 ft/1,158 m long; opened 1957) spans the straits and links Saint Ignace with Mackinaw City. The connection has stimulated the economy of the Upper Peninsula as a result of the added transportation route for tourists, vacationers and sports enthusiasts. The straits are an important link in the GREAT LAKES-SAINT LAWRENCE waterway.

Mackinac (MA-ki-naw), county (□ 2,099 sq mi/5,457.4 sq km; 2006 population 11,050), SE UPPER PENINSULA, N MICHIGAN; ⊙ SAINT IGNACE; 46°00′N 85°00′W. Bounded S by lakes MICHIGAN and HURON and by their connection, the STRAITS OF MACKINAC; drained by CARP and small PINE rivers. Part of historic MACKINAC region; includes MACKINAC and BOIS BLANC islands. Forest, resort, and agricultural area (forage crops; oats; cattle; dairy products). Several lakes (part of MANISTIQUE; and SOUTH MANISTIQUE, BREVOORT, MILAKOKIA, MILLECOQUINS) are in county. Hiawatha National Forest in E center; Straits State Park; Mackinac Island State Park in SE. Formed 1818.

Mackinac Island (MA-ki-naw), village (2000 population 523), MACKINAC county, SE UPPER PENINSULA, N MICHIGAN, 5 mi/8 km E of SAINT IGNACE, on S end of MACKINAC ISLAND (c.3 mi/4.8 km long, 2 mi/3.2 km wide; a state park since 1895), in the STRAITS OF MACKINAC; 45°51′N 84°37′W. Summer resort, connected by passenger ferry with MACKINAW CITY and Saint Ignace. No motorized vehicles on island. Has airport; only accessible by air in winter. Village maintains original 1890s buildings; transportation by bicycle, horseback, and carriage. The Astor House (built c.1817 by American Fur Company), and restored Fort Mackinac (established 1780), with 14 restored buildings, are reminders of island's military, strategic role in history of MACKINAC region. Grand Hotel has the longest porch in world. Also called Mackinac.

Mackinac, Straits of (MA-ki-naw), N MICHIGAN, channel in heart of historic MACKINAC region separating UPPER and LOWER peninsulas and forming important waterway between lakes HURON (E) and MICHIGAN (W); c.4 mi/6.4 km wide. To E and SE of SAINT IGNACE (on N shore) are MACKINAC, ROUND, and BOIS BLANC islands in Lake Huron. Saint Helena Island, to W, near N shore of Lake Michigan. South Channel of straits lies between Bois Blanc Island and the Lower Peninsula; 45°49′N 84°45′W. Spanned by the Mackinac Straits Bridge (3,800 ft/1,158 m long; opened 1957).

Mackinaw (MA-ki-naw), village (2000 population 1,452), TAZEWELL county, central ILLINOIS, on MACKINAW RIVER (bridged here), and 16 mi/26 km E of PEKIN; 40°32′N 89°21′W. In agricultural area; sand, gravel pits.

Mackinaw City (MA-ki-naw), village (2000 population 859), CHEBOYGAN and EMMET counties, N MICHIGAN, 15 mi/24 km NW of CHEBOYGAN, on the S shore of the STRAITS OF MACKINAC; 45°46′N 84°45′W. Manufacturing (candy). The region was well traveled by traders, missionaries, and explorers during the 17th and 18th century. French troops, sent to garrison Fort Michilimackinac in 1715, remained for several years until the fort was occupied by British forces. Fort Michilimackinac W of town near base of bridge in Michilimackinac State Park. MACKINAW CITY was formerly linked with the UPPER PENINSULA only by car ferry. The completion of the Mackinac Straits Bridge (1957) now links the village with SAINT IGNACE, on Upper Peninsula, to N. Passenger ferry to MACKINAC ISLAND to NE (summer). Old Mill Creek State Historical Park to SE; Wilderness State Park to W. Settled 1681, incorporated 1882.

Mackinaw River, c.130 mi/209 km long, central ILLINOIS; rises in W FORD county; flows W, SW, and N, to ILLINOIS RIVER below PEKIN; 40°35′N 88°21′W.

Mackinnon Road, village, COAST province, SE KENYA, on railroad, and 50 mi/80 km WNW of MOMBASA; 03°44′S 39°03′E. British military base developed after World War II.

Macklin (MAK-lin), town (2006 population 1,290), W SASKATCHEWAN, W CANADA, on ALBERTA border, 75 mi/121 km WSW of NORTH BATTLEFORD; 52°20′N 109°57′W. Railroad junction, with grain elevators, stockyard; dairying.

Mackmyra (MAHK-MEE-rah), village, GÄVLEBORG county, E SWEDEN, on GAVLEÅN RIVER, 6 mi/9.7 km WSW of GÄVLE; 60°37′N 16°58′E.

Macksburg, town (2000 population 142), MADISON county, S central IOWA, 12 mi/19 km SW of WINTERSET; 41°12′N 94°11′W. In agricultural area.

Macksburg (MAKS-buhrg), village (2006 population 204), WASHINGTON county, SE OHIO, 14 mi/23 km N of MARIETTA; 39°38′N 81°27′W.

Macks Creek, town (2000 population 267), CAMDEN county, central MISSOURI, in OZARKS, near LAKE OF THE OZARKS, c.14 mi/23 km WSW of CAMDENTON; 37°58′N 92°58′W. Tunnel Dam hydroelectric power station on NIANGUA RIVER to E. Timber; cattle. Recreational area.

Macksville (MAKS-vil), town, E NEW SOUTH WALES, SE AUSTRALIA, on Nambucca River, 170 mi/274 km NNE of NEWCASTLE; 30°43′S 152°54′E. Fishing, oysters; bananas and other tropical fruits; vegetables; dairy products; timber.

Macksville, village (2000 population 514), STAFFORD county, S central KANSAS, 11 mi/18 km WSW of SAINT JOHN; 37°57′N 98°58′W. In wheat and livestock region. Antique Car Museum.

Mackworth Island, SW MAINE, in CASCO BAY just NE of PORTLAND, near FALMOUTH, to which a bridge leads, and just NE of Portland; c. 0.5 mi/0.8 km in diameter.

Maclean (muh-KLAIN), municipality, NE NEW SOUTH WALES, SE AUSTRALIA, on CLARENCE RIVER, and 145 mi/233 km S of BRISBANE; 29°27′S 153°14′E. Dairying center; bananas; timber; mixed farming; tourism.

Maclean Strait, CANADA: see PRINCE GUSTAF ADOLF SEA.

Maclear, town, EASTERN CAPE province, SOUTH AFRICA, at foot of DRAKENSBERG range, on Tsitsa River, and 44 mi/70 km NE of UMTATA; 31°05′S 28°24′E. Elevation 5,182 ft/1,580 m. Railroad terminus; agricultural center (grain; livestock; dairying). Ski resort. Airfield. Named for Sir Thomas Maclear, official astronomer of CAPE COLONY, 1833–1879.

Maclear, Cape, MANGOCHE district, Southern region, central MALAWI, on S shore of Lake Malawi (LAKE NYASA), 40 mi/64 km NNW of Mangoche town. Tourist resort (hotel); airfield; landing.

Macleay River (muh-KLAI), 250 mi/402 km long, E NEW SOUTH WALES, SE AUSTRALIA; rises in NEW ENGLAND RANGE; flows generally ESE, then ENE, past KEMPSEY, to the PACIFIC OCEAN near SMOKY CAPE; 30°52′S 153°01′E. Navigable 30 mi/48 km below Kempsey by small craft.

Macleod (muh-KLOUD), suburb 9 mi/14 km NE of MELBOURNE, VICTORIA, SE AUSTRALIA, between HEIDELBERG and BUNDOORA; 37°44′S 145°04′E.

Macmerry (MAK-muh-ree), village (2001 population 1,113), East Lothian, E Scotland, just E of TRANENT; 55°56′N 02°54′W. Industrial estate nearby. Formerly in Lothian, abolished 1996.

Macmillan, river, c.200 mi/320 km long, E YUKON TERRITORY, NW CANADA; rises in two main forks in the Selwyn Mountains; flows generally W to the PELLY RIVER. Important route to the goldfields, c.1890–1900.

Macnean, Upper Lake (mak-NEEN), on border between IRELAND and NORTHERN IRELAND, between LEITRIM (W), CAVAN (S), and FERMANAGH (N) counties, 10 mi/16 km WSW of ENNISKILLEN; 5 mi/8 km long, 1 mi/1.6 km wide; 54°18′N 07°56′W. LOWER LAKE MACNEAN is just SE.

MacNider (muhk-NEI-duhr), borough (French *arrondissement*) of MÉTIS-SUR-MER, BAS-SAINT-LAURENT region, SE QUEBEC, E CANADA; 48°40′N 67°59′N.

MacNutt (mak-NUT), village (2006 population 80), SE SASKATCHEWAN, W CANADA, on MANITOBA border, 40 mi/64 km ESE of YORKTON; 51°04′N 101°36′W. Mixed farming.

Macocha, CZECH REPUBLIC: see MORAVIAN KARST.

Macolin (mah-ko-LIN), German *Magglingen*, commune, BERN canton, W SWITZERLAND, 1 mi/1.6 km ENE of BIEL. Elevation c.2,625 ft/800 m. Federal school of sports and gymnastics. Access via funicular from Biel.

Macomb (muh-KOM), county (□ 569 sq mi/1,479.4 sq km; 2006 population 832,861), SE MICHIGAN, in DETROIT metropolitan area, just N of Detroit; 42°40′N 82°54′W; ⊙ MOUNT CLEMENS. Bounded SE by LAKE SAINT CLAIR and ANCHOR BAY; drained by CLINTON RIVER and its affluents. Farm area (corn, wheat, oats, soybeans, apples, peaches; dairy products). Manufacturing (fabricated metal products, paper products, transportation equipment, machinery). Highly urbanized in S; mainly residential and commercial in W; industrial and office parks. Major suburban cities include WARREN, STERLING HEIGHTS, SAINT CLAIR SHORES. Selfridge Air National Guard Base in Lake Saint Clair in E. Dodge Brothers State Park Number Eight at Sterling Heights; Rochester-Utica State Recreation Area in SW. Formed and organized 1818.

Macomb, city (2000 population 18,558), ⊙ MCDONOUGH county, W ILLINOIS, c.59 mi/95 km WSW of PEORIA; 40°28′N 90°40′W. A trade and manufacturing center in a rich farm, clay, and coal region, the city is known for its artistic clay products. Other manufacturing includes insulated containers and roller bearings. Seat of Western Illinois University. State park nearby. Incorporated as a city 1841.

Macomb, Fort (muh-KOM), old UNITED STATES fortification, ORLEANS parish, within city of NEW ORLEANS, SE LOUISIANA, on W bank of CHEF MENTEUR PASS between lakes PONTCHARTRAIN and BORGNE, c. 20 mi/32 km ENE of downtown NEW ORLEANS; 30°03′N 89°48′W. Built c. 1828. Partly restored; in State Commemorative area (16 acres/6.5 ha).

Macomer (mah-ko-MER), town, NUORO province, W central SARDINIA, ITALY, in CATENA DEL MARGHINE, 29 mi/47 km W of NUORO; 40°16′N 08°47′E. Road and railroad junction. Cheese center; woolen mill. Nuraghi nearby.

Macomia (mah-KO-mee-uh), village, N MOZAMBIQUE, on road, and 60 mi/97 km NNW of PEMBA; 12°14′S 40°07′E. Cotton.

Macomia District, MOZAMBIQUE: see CABO DELGADO.

Macon (MAI-kuhn), county (□ 613 sq mi/1,593.8 sq km; 2006 population 22,594), E Alabama; ⊙ Tuskegee, 32°24′N 85°49′W. In the Black Belt; drained by Tallapoosa River and its branches. Cotton, corn, soybeans, peanuts, sweet potatoes; cattle; dairying. Contains Tuskegee National Forest. Formed 1832. Named for Nathaniel Macon, U.S. senator from NC.

Macon, county (□ 406 sq mi/1,055.6 sq km; 2006 population 13,817), central GEORGIA; ⊙ OGLETHORPE; 32°21′N 84°02′W. Drained by FLINT RIVER. Coastal plain agricultural area (cotton, corn, soybeans, wheat, peanuts); timber. Formed 1837.

Macon, county (□ 585 sq mi/1,521 sq km; 2006 population 109,309), central ILLINOIS; ⊙ DECATUR; 39°52′N 88°58′W. Agriculture (corn, soybeans; livestock). Diversified manufacturing Decatur is industrial and commercial center. Bituminous-coal mining. Drained by SANGAMON RIVER, dammed to form Lake Decatur (recreation area). Includes Spitler Woods Natural Area and Lincoln Trail Homestead State Park. Formed 1829.

Macon, county (□ 814 sq mi/2,116.4 sq km; 2006 population 15,651), N central MISSOURI; ⊙ MACON; 39°50′N 92°33′W. Drained by CHARITON and SALT rivers. Agriculture (corn, wheat, soybeans, hay); cattle, sheep, hogs; manufacturing at Macon. Former coal-mining area. Long Branch Lake and State Park to NW, Thomas Hill Reservoir in SW. Formed c.1838.

Macon (MAI-kuhn), county (□ 519 sq mi/1,349.4 sq km; 2006 population 32,395), W NORTH CAROLINA, on GEORGIA (S) state line; ⊙ FRANKLIN; 35°08′N 83°25′W. SE corner near common corner at Georgia, SOUTH CAROLINA, and North Carolina. Part of BLUE RIDGE MOUNTAINS in SE; crossed N-S by NANTAHALA MOUNTAINS; drained by NANTAHALA RIVER (forms Nantahala Lake reservoir in W) and the LITTLE TENNESSEE RIVER. Included in Nantahala National Forest.

Service industries; retail trade; agriculture (vegetables, apples, hay, tobacco; cattle); timber, mica mining; resort area. Lake Sequayah reservoir in SE. APPALACHIAN TRAIL (APPALACHIAN NATIONAL SCENIC TRAIL) passes N-S through W part of county. Formed 1828 from Haywood County. Named for Nathaniel Macon (1758–1837), speaker of House of Representatives and US senator.

Macon, county (□ 304 sq mi/790.4 sq km; 2006 population 21,726), N TENNESSEE, on KENTUCKY (N) state line; ⊙ LAFAYETTE; 36°32′N 86°01′W. Drained by affluents of BARREN and CUMBERLAND rivers. Agriculture; light manufacturing. Formed 1842.

Macon, city (2000 population 97,255), ⊙ BIBB county, central GEORGIA, at the head of navigation on the OCMULGEE River; 32°50′N 83°40′W. It is the industrial, processing, and shipping center for an extensive agricultural area. Many antebellum mansions remain (the city was spared from Sherman's March during the Civil War). Known for its cherry blossom festival in spring. Manufacturing includes processing of agricultural products (cotton, peanuts, soybeans; poultry; dairy products), chemicals, wood products, fabricated metal products, building materials, waste systems, transportation equipment, shipping materials; printing and publishing. Established 1806 on the E side of the Ocmulgee and renamed Newtown in 1821. A second Macon (for Nathaniel Macon) was laid out on the W side in 1823; Newtown was annexed in 1829. Seat of Wesleyan College, Mercer University, a state school for the blind, and Macon College (2-year). Also in Macon are the birthplace of Sidney Lanier, a restored grand-opera house (1884), Fort Hawkins (1806; partially restored), a museum of arts and sciences, and a planetarium. Nearby are Robins Air Force Base and the OCMULGEE NATIONAL MONUMENT. Home of Georgia Music Hall of Fame. Incorporated 1823.

Macon, city (2000 population 1,213), MACON county, central ILLINOIS, 8 mi/12.9 km S of DECATUR; 39°42′N 89°00′W. In agricultural (corn, soybeans, oats) and bituminous-coal area.

Macon, city (2000 population 5,538), ⊙ MACON county, N central MISSOURI, 25 mi/40 km S of KIRKSVILLE; 39°44′N 92°28′W. Ships agricultural products (corn, wheat, soybeans) and livestock (cattle, hogs); former coal mines; manufacturing (frozen foods, appliances). Long Branch Lake and State Park on NW side of city. Incorporated 1859.

Mâcon (mah-kon), town (□ 10 sq mi/26 sq km); ⊙ SAÔNE-ET-LOIRE department, E central FRANCE, in BURGUNDY, port on the SAÔNE RIVER, and 40 mi/64 km N of LYON; 46°18′N 04°50′E. A commercial center noted for its annual wine fair, Mâcon and suburbs have plants manufacturing motorcycles, electrical equipment, and clothing. Acquired by the French crown in 1238, passed to Burgundy by the Treaty of Arras (1435), and was recovered by France in 1477. In 16th century it was a Huguenot stronghold. Lamartine was born here. Today Mâcon is an important junction of transportation routes between PARIS, S France, and GENEVA.

Macon, town (2000 population 2,461), ⊙ NOXUBEE county, E MISSISSIPPI, 28 mi/45 km SSW of COLUMBUS, and on NOXUBEE RIVER; 33°06′N 88°33′W. Agriculture (cotton, corn, wheat, soybeans; cattle; dairying); timber; manufacturing (lumber, bricks, apparel, vinyl products, feeds, transportation equipment). County Historical Society Museum. Incorporated 1836.

Macon (MAI-kuhn), village (2006 population 107), WARREN county, N NORTH CAROLINA, 4 mi/6.4 km NE of WARRENTON; 36°26′N 78°04′W. Service industries; agriculture (tobacco, grain; livestock).

Macon, Bayou (MAI-kuhn, BAH-yoo), c.175 mi/282 km long; rising in DESHA county, SE ARKANSAS, S of DUMAS; flows S into NE LOUISIANA to the TENSAS RIVER 18 mi/29 km SSE of WINNSBORO (Louisiana).

Extremely serpentine especially in N; closely parallels MISSISSIPPI RIVER (to E) for most of its length. It was used as a rendezvous by the bandits Frank and Jesse James.

Mâconnais (mah-ko-NAI), hill district of E central FRANCE, in old BURGUNDY province, now part of SAÔNE-ET-LOIRE department Historical capital was MÂCON. Bounded by SAÔNE RIVER (E) and CHAROLAIS cattle-raising region (W); 46°15′N 04°51′E. It is noted for the vineyards first established by the monks of CLUNY. Today, the area W of Mâcon produces a variety of white wines, of which the Pouilly-Fuissé is best known. Red wines of the "gamay noir" are also shipped from here.

Macossa District, MOZAMBIQUE: see MANICA.

Macotera (mah-ko-TAI-rah), town, SALAMANCA province, W SPAIN, 22 mi/35 km ESE of SALAMANCA; 40°50′N 05°17′W. Agricultural-trade center (cereals, vegetables, wine); cattle.

Macouba (mah-koo-BAH), town, N MARTINIQUE, French WEST INDIES, 18 mi/29 km NNW of FORT-DE-FRANCE; 14°52′N 61°09′W. Cacao growing; rum distilling. Sometimes called La Macouba.

Macoupin (ma-KOO-pin), county (□ 867 sq mi; 2006 population 48,841), SW central ILLINOIS; ⊙ CARLINVILLE; 39°15′N 89°55′W. Agriculture (corn, wheat, soybeans, sorghum; cattle, hogs, poultry; dairy products). Bituminous-coal mining; clay pits. Some manufacturing (transportation equipment, food products, wood products). Drained by MACOUPIN, CAHOKIA, and small Otter creeks. Formed 1829.

Macoupin Creek, c.100 mi/161 km long, SW ILLINOIS; rises in NW MONTGOMERY county; flows SW and W to ILLINOIS RIVER SE of HARDIN; 39°24′N 89°34′W.

Macouria (mah-koor-ee-AH), town, N FRENCH GUIANA, on ATLANTIC coast, at mouth of Cayenne River, 3 mi/4.8 km SW of CAYENNE; 04°55′N 52°22′W. Coffee, manioc, fruit.

MacPherson Range, AUSTRALIA: see MCPHERSON RANGE.

Macquarie Harbour (muh-KWAW-ree), inlet of INDIAN OCEAN, on W coast of TASMANIA, AUSTRALIA; 19 mi/31 km long, 8 mi/13 km wide; c.0.5 mi/0.8 km wide at mouth, near which is Cape SORELL; 41°44′S 147°08′E. Numerous islets, including Settlement Island (5 mi/8 km long), once a penal colony; STRAHAN is on NW shore. Receives GORDON RIVER.

Macquarie Island (muh-KWAW-ree), uninhabited volcanic island in S PACIFIC OCEAN, 850 mi/1,368 km SE of TASMANIA, AUSTRALIA, to which it belongs; 21 mi/34 km long, 3 mi/5 km wide; 54°30′S 158°40′E. Rises to 1,421 ft/433 m. Rocky (pillow basalts and other extrusive rocks), with small glacial lakes, the island is the exposed crest of the MACQUARIE RIDGE, and the only place where rocks from below Earth's mantle (4 mi/6 km below ocean floor) are being exposed above sea level. It is also an exposure of a meeting point of the Pacific and Indo-Australian tectonic plates. Declared a UNESCO World Heritage Area 1997. Meteorological station. With the 2 small uninhabited groups (Bishop and Clerk, and Judge and Clerk), the group is called Macquarie Islands.

Macquarie Ridge (muh-KWAW-ree), ANTARCTICA, submarine ridge extending S from Macquarie Island toward BALLENY ISLANDS; separates SOUTH INDIAN from Southwest PACIFIC basins, centered around 57°00′S 159°00′E.

Macquarie River (muh-KWAW-ree), 590 mi/950 km long, E NEW SOUTH WALES, SE AUSTRALIA; rises in the BLUE MOUNTAINS; 30°07′S 147°24′E. Flows NW to the DARLING RIVER through an important sheep- and wheat-raising area.

Macquarie River, AUSTRALIA: see SOUTH ESK RIVER.

Mac Robertson Land, ANTARCTICA, portion of EAST ANTARCTICA S of MAWSON COAST, between ENDERBY LAND (W) and AMERY ICE SHELF and LAMBERT GLA-

CIER (E). Discovered 1930 by Sir Douglas Mawson, and named after Sir MacPherson Robertson.

Macroom (mak-KROOM), Gaelic *Maigh Chromtha,* town (2006 population 3,407), W CORK county, SW IRELAND, on Sullane River, near its mouth on LEE RIVER, and 21 mi/34 km W of CORK; 51°54′N 08°57′W. Agricultural market. MacCarthy's castle reputedly dates from 12th century. Cromwell granted manor here to father of William Penn.

Macta Marshes, crossed by Oued Habra, MOSTAGANEM wilaya, NW ALGERIA, 16 mi/25 km SW of MOSTAGANEM. In 1835, this was the site of a great defeat of the French colonial army by Algerian troops led by Emir Abdelkader. Has great ecological value, as it is a place of transition for scores of rare bird species in winter, including wild duck, partridge, bustard, and large colonies of pink swans. The marshes are threatened by both industrial expansion and industrial waste from the complexes of SIG and MOHAMMADIA.

Mactan (mahk-TAHN), coral island (□ 24 sq mi/62.4 sq km), CEBU province, the PHILIPPINES, just off the coast of Cebu island; 10°18′N 124°00′E. Highly commercialized and industrialized mixed-use island. Industry includes a port, shipyard, several musical-instrument factories, and an oil storage depot. The town of Cordoba, a fishing port, is the site of Mactan Island Export Processing Zone. Agriculture (corn, coconuts). Upscale beach and resorts on E shore. Also, site of MetroCebu's airport (country's second-largest), main transfer point between central and S islands and key element in the development of the tourist industry. Magellan was killed by Chief Lapulapu here in 1521; the spot is marked by a monument near the city of LAPU-LAPU.

Macuchi (mah-KOO-chee), village, COTOPAXI province, N central ECUADOR, on W slopes of the ANDES Mountains, 40 mi/64 km W of LATACUNGA; 00°59′S 79°04′W. Silver, gold, copper deposits.

Macuelizo (mah-ke-LEE-zo), town, SANTA BÁRBARA department, NW HONDURAS, 34 mi/55 km NW of SANTA BÁRBARA; 13°35′N 87°04′W. Livestock; corn, beans.

Macuelizo (mah-kwai-LEE-zo), town, NUEVA SEGOVIA department, NW NICARAGUA, 8 mi/12.9 km W of OCOTAL; 13°39′N 86°37′W. Sugar mill; sugarcane; livestock. Silver deposits.

Macugnaga (mah-koo-NYAH-gah), village, NOVARA province, PIEDMONT, N ITALY, near MONTE ROSA, on ANZA RIVER, and 19 mi/31 km SW of DOMODOSSOLA, near Swiss border; 45°58′N 07°58′E. Year-round ski, alpine resort.

Macumba River (muh-KUM-buh) or **Treuer River,** 145 mi/233 km long, N central SOUTH AUSTRALIA; formed by junction of two streams; flows SE to Lake Eyre; 27°52′S 137°12′E. Usually dry.

Macungie (mah-KUHNG-gee), borough (2006 population 3,122), LEHIGH county, E PENNSYLVANIA, 7 mi/11.3 km SSW of ALLENTOWN, near Swabia Creek; 40°31′N 75°32′W. Agriculture area (apples, grain, potatoes, soybeans; livestock, poultry; dairying); manufacturing (consumer goods, apparel, fabricated metal products, wood products).

Macuripe (mah-koo-REE-pai), canton, GENERAL FEDERICO ROMÁN province, PANDO department, NW BOLIVIA. Elevation 367 ft/112 m. Agriculture (rubber, rice, cacao, bananas, coffee, tobacco, cotton, peanuts); cattle and horse raising.

Macuro (mah-KOO-ro), town, Valdez municipio, SUCRE state, NE VENEZUELA, minor port on Gulf of PARIA, at E tip of PARIA peninsula, 25 mi/40 km ENE of GÜIRIA, and 29 mi/47 km W of PORT OF SPAIN; 10°41′N 61°56′W. Easternmost port of Venezuela, it ships cacao, coffee, and corn to TRINIDAD. Customhouse. Gypsum deposits nearby. Here, Christopher Columbus landed August 1, 1498.

Macurure (MAH-koo-roo-RAI), town (2007 population 7,752), NE BAHIA state, BRAZIL, 74 mi/120 km NW of PAULO AFONSO; 09°25′S 39°08′W.

Macusani (mah-koo-SAH-nee), town (□ 2,617 sq mi/6,804.2 sq km), ⊙ CARABAYA province, PUNO region, SE PERU, in Cordillera ORIENTAL of the ANDES Mountains, 125 mi/201 km NNW of PUNO; 14°05′S 70°26′W. Elevation 14,225 ft/4,336 m. Cereals, potatoes, vegetables; livestock.

Macuspana (mah-koos-PAH-nah), city and township, TABASCO, SE MEXICO, on Puxcatan River (affluent of GRIJALVA River), 27 mi/43 km SE of VILLAHERMOSA, 2 mi/3 km NE of Mexico Highway 186; 18°36′N 92°36′W. Tropical agriculture; petroleum production.

Macuto (mah-KOO-to), city, VARGAS state, N central VENEZUELA; 10°36′N 66°54′W. Popular beach resort near CARACAS and adjoining LA GUAIRA.

Mačva (MAHCH-vah), county, W SERBIA, NW of SABAC; ⊙ BOGATIC. Bounded partly by SAVA (N) and DRINA (W) rivers. Densely populated. Sugar beets, hogs. Also spelled Machva.

Mačva, Serbian *Mačvanski Okrug,* district (□ 1,261 sq mi/3,278.6 sq km; 2002 population 329,625), ⊙ ŠABAC, W central SERBIA, on BOSNIA and HERZEGOVINA border; 44°46′N 19°30′E. Includes municipalities (*opštinas*) of BOGATIC, Koceljeva, KRUPANJ, LJUBOVIJA, LOZNICA, Mali Zvornik, Sabac, and VLADIMIRCI. Industry (chemicals, plastics; food processing).

Macwahoc (muhk-WAH-hahk), plantation, AROOSTOOK county, N MAINE, 40 mi/64 km SSW of HOULTON; 45°38′N 68°14′W. Hunting, fishing.

Macy, town (2000 population 248), MIAMI county, N central INDIANA, 15 mi/24 km NNW of PERU; 40°58′N 86°08′W. In agricultural area. Laid out 1860.

Macy, village (2000 population 956), THURSTON county, NE NEBRASKA, 50 mi/80 km N of OMAHA, on OMAHA RESERVATION. Tribal agency here. Annual Omaha tribal powwow held here in late August.

Mád (MAHD), village, BORSOD-ABAÚJ-ZEMPLÉN county, NE HUNGARY, in the HEGYALJA, 23 mi/37 km ENE of MISKOLC; 48°12′N 21°17′E. Excellent wine. Szilvásfürdö, with mineral springs, is nearby.

Madaba (MAH-dah-buh), town (2004 population 70,338), S central JORDAN, E of the DEAD SEA, 19 mi/31 km SW of AMMAN; 31°43′N 35°48′E. On site of ancient town of MEDEBA.

Madaba (mah-DAH-bah), village, LINDI region, SE central TANZANIA, 120 mi/193 km WNW of KILWA MASOKO, in SELOUS GAME RESERVE; 08°41′S 37°48′E. Livestock.

Madadzi, town, MASHONALAND WEST province, N ZIMBABWE, 8 mi/ 13 km SE of KAROI; 16°55′S 29°46′E. Tobacco, cotton, wheat, corn; cattle, goats, sheep.

Madagali (mah-dah-GAH-lee), town, NE ADAMAWA state, NIGERIA, near CAMEROON border, 45 mi/72 km S of BAMA; 10°53′N 13°38′E. Peanuts, pepper, hemp, rice, cotton; cattle, skins.

Madagascar (MA-duh-GAS-kahr), Malagasy *Madagasikara* (mah-dah-gahs-KAHR), officially Democratic Republic of Madagascar, republic (□ 226,658 sq mi/587,045 sq km; 2004 estimated population 17,501,871; 2007 estimated population 19,448,815), in INDIAN OCEAN, separated from E AFRICA by the MOZAMBIQUE CHANNEL; ⊙ ANTANANARIVO.

Geography

Comprises Madagascar, the world's fourth-largest island and nearby small islands ceded by FRANCE in 1990, including JUAN DE NOVA, EUROPA, the GLORIOSO ISLANDS, and BASSAS-DE-INDIA. The country also claims the island of Tromelin to the NE, currently under France administration. Main cities include Antananarivo, ANTSIRABE, ANTSIRANANA (formerly Diégo-Suarez), FIANARANTSOA, MAHAJANGA (formerly Majunga), TOAMASINA (formerly Tamatave), and TOLIARY (formerly Tuléar). Madagascar is made up of a highland plateau fringed by a lowland coastal strip, narrow (c.30 mi/50 km) in the E and considerably wider (c.60 mi/100 km–125 mi/200 km) in the W. The plateau attains greater heights in the N, where Mount Maromokotro (9,450 ft/2,880 m), the loftiest point in the country, is located, and in the center, where the ANKARATRA Mountains reach c.8,670 ft/2,640 m. Once a mosaic of forest, brush, and grassland, the plateau is now largely deforested. A series of lagoons along much of the E coast is connected in part by the PANGALANES CANAL, which runs (c.400 mi/640 km) between FARAFANGANA and MAHAVELONA and can accommodate small boats. The island has several rivers, including the Sofia, BETSIBOKA, MANAMBOLO, MANGORO, TSIRIBIHINA, MANGOKY, MANANARA, and ONILAHY. Madagascar is divided into six provinces: ANTANANARIVO, ANTSIRANANA, FIANARANTSOA, MAHAJANGA, TOAMASINA, and TOLIARY.

Population

The inhabitants are a mix of largely Malayo-Indonesian descent and African influences, divided into roughly eighteen ethnic groups. The main Indonesian ethnic groups are the Merina, who live near Antananarivo and the Bétsiléo, who live around Fianarantsoa. The principal African groups are the Betsimisáraka, who live near Toamasina; the Tsimihety, based in the N highlands; the Sakalava and the Antandroy, who live in the W; and the Antaisaka, who live in the SE. All the people speak Malagasy, a language of Indonesian origin; it and French are official languages. About 50% of the people are Christian (equally divided between Roman Catholics and Protestants), 5% are Muslim, and the rest follow traditional beliefs.

Economy

The economy of Madagascar is overwhelmingly agricultural, largely of a subsistence type; the best farmland is in the E and NW. The principal crops are rice, cassava, corn, pulses, sugarcane, coffee, raffia, tobacco, cloves, vanilla, and litchi. In addition, large numbers of poultry, cattle, goats, sheep, and hogs are raised. Manufacturing is mostly confined to food products, beverages, and basic consumer goods such as clothing. The country's mineral production declined in the 1970s; the chief minerals extracted are chromite, graphite, phosphates, ilmenite, phlogopite, mica, marble, bauxite, zircon, and industrial beryl and garnets and precious stones. The state controls much of the economy. There is an extensive but degraded road system and only a very limited railroad network. Toamasina and Mahajanga are the chief ports. Madagascar carries on a relatively small foreign trade, and the annual value of imports is usually considerably higher than the value of exports. The main imports are metals, machinery, transport equipment, textiles, and food products; the leading exports are coffee, vanilla, rice, sugar, and cloves. The principal trade partners are the U.S., France, and GERMANY. Madagascar is an ACP (African, Caribbean, and Pacific) member of the EU and relies heavily upon assistance from its members.

History: to 1810

The earliest history of Madagascar is unclear. Africans and Indonesians reached the island no later than the 5th century C.E., the Indonesian immigration continuing until the 15th century. From the 9th century, Muslim traders (including some Arabs) from E Africa and the COMORO Islands settled in NW and SE Madagascar. Probably the first European to see Madagascar was Diogo Dias, a Portuguese navigator, in 1500. Between 1600 and 1619, Portuguese Roman Catholic missionaries tried unsuccessfully to convert the Malagasy. From 1642 until the late 18th century the French maintained footholds, first at TAOLAGNARO (formerly Fort-Dauphin) in the SE and finally on SAINTE MARIE ISLAND off the E coast. By the beginning of the 17th century there were a number of small Malagasy kingdoms, including those of the Antemoro, Antaisaka, Bétsiléo, and Merina. Later in the century the Sakalava under Andriandahifotsi conquered W and N Madagascar, but the kingdom disintegrated in the 18th century. At the end of the 18th century the Merina people of the interior were united under King Andrianampoinimerina (reigned 1787–1810), who also subjected the Bétsiléo.

History: 1810 to 1883

Radama I (reigned 1810–1828), in return for agreeing to end the slave trade, received British aid in modernizing and equipping his army, which helped him to conquer the Betsimisáraka kingdom and much of the island. The Protestant London Missionary Society was welcomed, and it gained many converts, opened schools, and helped to transcribe the Merina language. Merina culture began to spread over Madagascar. Radama was succeeded by his wife Ranavalona I (reigned 1828–1861), who, suspicious of foreigners, declared (1835) Christianity illegal and halted most foreign trade. During her rule the Merina kingdom was wracked by intermittent civil war. Under Radama II (reigned 1861–1863) and his widow and successor Rasoherina (reigned 1863–1868) the anti-European policy was reversed and missionaries (including Roman Catholics) and traders were welcomed again. Rainilaiarivony, the prime minister, controlled the government during the reigns of Ranavalona II (1868–1883) and Ranavalona III (1883–1896); by then the Merina kingdom included all Madagascar except the S and part of the W. Ranavalona II publicly recognized Christianity, and she and her husband were baptized.

History: 1883 to 1947

In 1883 the French bombarded and occupied Toamasina (then Tamatave), and in 1885 they established a protectorate over Madagascar, which was recognized by GREAT BRITAIN in 1890. Rainilaiarivony organized resistance to the French, and there was heavy fighting from 1894 to 1896. In 1896, French troops under J. S. Gallieni defeated the Merina and abolished the monarchy. By 1904 the French fully controlled the island. Under the French, who governed the Malagasy through a divide-and-rule policy, development was concentrated in the Antananarivo region, and thus the Merina benefited most from colonial rule. Merina nationalism developed early in the 20th century, and in 1916 (during World War I) a Merina secret society was suppressed by the French after a plot against the colonialists was discovered. During World War II, Madagascar was aligned with VICHY France until 1942, when it was conquered by the British; in 1943 the Free French regime assumed control.

History: 1947 to 1975

From 1947 to 1948 there was a major uprising against the French, who crushed the rebellion, killing between 11,000 and 80,000 (estimates vary) Malagasy in the process. As in other French colonies, indigenous political activity increased in 1956, and the Social Democratic Party (PSD), led by Philibert Tsiranana (a Tsimihety), gained predominance in Madagascar. On October 14, 1958, the country—renamed the Malagasy Republic—became autonomous within the French Community and Tsiranana was elected president. On June 26, 1960, it became fully independent. Under Tsiranana (reelected in 1965 and 1972), an autocratic ruler whose PSD controlled parliament, government was centralized, the coastal peoples (côtiers) were favored over those of the interior (especially the Merina), and French economic and cultural influence remained strong. In a controversial move beginning in 1967, Tsiranana cultivated economic relations with white-ruled SOUTH AFRICA. After Tsiranana was reelected in 1972, students and workers, discontented with the president's policies and with the deteriorating economic situation, staged a wave of protest demonstrations. At the height of the crisis Tsiranana handed over power to General Gabriel Ramanantsoa, who became prime minister. In October 1972, a national referendum overwhelmingly approved Ramanantsoa's plan to rule without parliament for five years; Tsiranana, who opposed the plan, resigned the presidency shortly after the vote. Ramanantsoa freed political prisoners jailed by Tsiranana, began to reduce French influence in the country, broke off relations with South Africa, and generally followed a moderately leftist course.

History: 1975 to 1986

In 1975, a new constitution was approved that renamed the Malagasy Republic the Democratic Republic of Madagascar. That same year, Ramanantsoa dissolved his government in response to mounting unrest in the military and internal disagreements regarding economic policy. Colonel Ratsimandrava assumed power but was assassinated a month later, and Lieutenant Commander Didier Ratsiraka was elected president in a referendum. The military-backed Supreme Revolutionary Council (CSR), with Ratsiraka as its head, comprised the government's executive branch. Ratsiraka's Marxist-socialist government nationalized most of the economy (industry, banking, and agriculture marketing), practiced a policy of isolation from external influences, and in 1978 borrowed widely to pay for major investments in development, industry, and the military. The nation fell into a crippling debt crisis and since 1980 has followed International Monetary Fund and World Bank structural adjustment programs. Ratsiraka's policies of censorship, regional divisiveness, and repression led to several coup attempts in the 1980s, all of which were put down.

History: 1986 to Present

Food shortages and price increases caused further social unrest in 1986 as economic conditions generally deteriorated. In foreign affairs, Madagascar under Ratsiraka later strengthened ties with the U.S. and EUROPE and continued to distance itself from South Africa. Ratsiraka was reelected in 1989 under suspicious circumstances and rioting ensued. Later in 1989, Ratsiraka survived another coup attempt. Madagascar's political and economic upheaval prompted the government to establish a multiparty system and move toward the privatization of industry in the 1990s. In 1991–1992 the nation was paralyzed as Dr. Albert Zafy led the opposition in organizing general strikes. Ratsiraka agreed to elections and was replaced by Zafy in 1993. The new government continued programs of privatization, austerity, and liberalization but saw a revolving door of prime ministers and government members. Dissatisfaction with Zafy led to his impeachment by the National Assembly in 1996. Elections that year featured fifteen candidates, including many former heads of state. Ratsiraka narrowly defeated Zafy in the run-off. In the 2001 presidential race, opposition leader Marc Ravalomanana claimed victory, but the government announced that he had won only 46% of the vote, forcing a run-off. His supporters seized control of the capital, but Ratsiraka retained the support of the much of the army and of many outside the capital. Ratsiraka rejected a recount that declared him the loser, but by early July 2002 Ravalomanana's supporters controlled much of the island, and Ratsiraka had fled the country. Ravalomanana was reelected in December 2006, but a number of candidates were barred from running.

Government

Madagascar is governed under the constitution of 1992. The president, who is head of state, is elected by popular vote for a five-year term and is eligible for a second term. The government is headed by a prime minister, who is appointed by the president. There is a

---

bicameral legislature. The Senate has 100 members, two thirds of which are selected by regional assemblies; the rest are appointed by the president. Members of the 160-seat National Assembly are popularly elected. All legislators serve four-year terms. President Marc Ravalomanana has been head of state since 2002; Prime Minister Charles Rabemananjara has been head of government since 2007.

Madail Saleh, ancient city, in what is now SAUDI ARABIA, SE of TABUK; 26°48′N 37°57′E. Was capital of the biblical region of MIDIAN, during the Nabataean occupation. Thamodic and Lihyanite settlements. Architectural monuments are the site's most significant feature; carved rock tombs and imposing templelike facades date back to the 1st century B.C.E.

Madain, Al, IRAQ: see MAIDAN, AL.

Madakasira (muh-duh-kuh-SEE-ree), town, ANANTAPUR district, ANDHRA PRADESH state, S INDIA, 55 mi/89 km SSW of ANANTAPUR; 13°56′N 77°16′E. Betel and coconut palms. Corundum mines nearby.

Madakwe Hill, ZIMBABWE: see DAWA MADAKWE.

Madalena (MAH-dah-LE-nah), city (2007 population 17,022), central CEARÁ, BRAZIL, 23 mi/37 km NE of BOA VIAGEM; 04°42′S 39°30′W.

Madalena (mah-dah-LAI-nah), town, HORTA district, central AZORES, PORTUGAL, on W coast of PICO ISLAND, 5 mi/8 km E of HORTA (on FAIAL ISLAND), across Faial Channel; 38°32′N 28°32′W. Alcohol distilling, dairying. Vineyards on slopes of Pico volcano (just SE). Formerly spelled Magdalena.

Madame, Île (mah-DAHM, eel), small island in the Pertuis d'ANTIOCHE, an inlet of Bay of BISCAY, CHARENTE-MARITIME department, POITOU-CHARENTES region, W FRANCE, at mouth of CHARENTE RIVER, 7 mi/11.3 km W of ROCHEFORT; 45°58′N 01°07′W. Oyster beds. Fortified by Vauban in 17th century as part of Rochefort's defenses.

Madame Island or **Isle Madame** (eel mah-DAHM), in the ATLANTIC OCEAN, E NOVA SCOTIA, E CANADA, just S of CAPE BRETON ISLAND, 8 mi/13 km N of CANSO, near entrance of the STRAIT OF CANSO; 12 mi/19 km long, 9 mi/14 km wide. On S coast is ARICHAT.

Madampe (MAH-thuhm-pai), town, NORTH WESTERN PROVINCE, SRI LANKA, on irrigation tank, 6.5 mi/10.5 km SSE of CHILAW; 07°13′N 79°50′E. Coconut Research Institute.

Madampe (MAH-thuhm-pai), village, SABARAGAMUWA PROVINCE, S central SRI LANKA, in SABARAGAMUWA HILL COUNTRY, 17 mi/27 km SE of RATNAPURA; 06°32′N 80°35′E. Rice, vegetables, tea, rubber.

Madan (mah-DAHN), city (1993 population 8,735), PLOVDIV oblast, Madan obshtina, S BULGARIA, in the SE RODOPI Mountains, 14 mi/23 km ESE of SMOLYAN; 41°30′N 24°57′E. Tobacco. Manufacturing of synthetic rubber. Lead, zinc mines. Mining administration center, related to PLOVDIV and KURDZHALI lead-zinc nonferrous industry.

Madan, village, BOICHINOVTSI munipality, Montana oblast, Bulgaria; 43°35′N 23°27′E.

Madanapalle (MUH-duh-nuh-puh-le), town, CHITTOOR district, ANDHRA PRADESH state, S INDIA, 45 mi/72 km NW of CHITTOOR; 13°33′N 78°30′E. Road center; rice milling; sugarcane, millet. Tuberculosis santorium. Seat of Theosophical College Dyewood (red sanders) in nearby forested hills.

Madang, province (2000 population 365,106), N central PAPUA NEW GUINEA, NE NEW GUINEA island; ⊙ Madang. Bounded on W by EAST SEPIK province, S by ENGA, WESTERN HIGHLANDS, CHIMBU, EASTERN HIGHLANDS, and MOROBE provinces, NE by BISMARCK SEA. Includes ADELBERT and FINISTERRE Ranges near coast, the RAMU RIVER basin in NW, the N slopes of BISMARCK MOUNTAINS in S. Sago, copra, coconuts, palm oil, sugarcane, bananas, rice; lobster, prawns, trepang, tuna; cattle; timber. Also includes LONG, Bogabag, KARKAR and MANAM islands, all volcanic.

Madang, town, ⊙ MADANG province, N central PAPUA NEW GUINEA, on NE NEW GUINEA island, seaport on ASTROLABE BAY, arm of BISMARCK SEA, PACIFIC OCEAN; 05°13′S 145°48′E. Exports copra, coconuts, coffee, fish and palm oil. Connected by road to LAE (SE) and along coast to Awar (NW). It was an important Japanese air base during World War II. Formerly known as Friedrich-Wilhelmshafen. Scheduled air service. Known as the "prettiest town in the South Pacific" for its parks (Elizabeth Soweby Orchid Gardens) and waterways. Mandang Cultural Center and Museum Balek Wildlife Sanctuary to SE, ADELBERT RANGE to NW.

Madanganj (maw-dawn-gawnj), village, DHAKA district, E central EAST BENGAL, BANGLADESH, on DHALESWARI RIVER, and 12 mi/19 km SE of DHAKA. Rice, jute, oilseeds; rice milling; jute press.

Madanganj, INDIA: see KISHANGARH.

Madanin or Madanīn, province, TUNISIA: see MADANIYINA.

Madaniyina (me-dah-nee-YEE-nah), Arabic *Madanīn*, province (□ 3,316 sq mi/8,621.6 sq km; 2006 population 440,100), SE TUNISIA, on MEDITERRANEAN SEA (E) and LIBYA (S) border; ⊙ MADINIYINA; 33°20′N 11°00′E. Also Médenine.

Madanpur, INDIA: see MAHRONI.

Madaoua (MAH-duh-oo-uh), town, TAHOUA province, S NIGER, near NIGERIA border, 80 mi/129 km NE of SOKOTO (NIGERIA); 14°06′N 06°26′E. Administrative center. Peanuts, millet; livestock.

Madapollam, INDIA: see NARSAPUR.

Madara (MAH-dah-rah), village, VARNA oblast, KASPICHAN obshtina, E BULGARIA, on a headstream of the PROVADIISKA RIVER, 9 mi/15 km E of SHUMEN; 43°17′N 27°06′E. Grain; livestock. Has excavations of ancient fortresses, grottoes, 10th-century bas-relief commemorating Bulgarian king Han Krum.

Madares, Mount, Greece: see LEFKA ORI.

Madarihat (MUH-dah-ree-HUHT), town, JALPAIGURI district, N WEST BENGAL state, E INDIA, 36 mi/58 km ENE of Jalpaiguri; 26°42′N 89°17′E. Railroad terminus. Tea processing, rice milling; trades in rice, tea, mustard, tobacco.

Madaripur (mah-dah-ree-poor), town (2001 population 53,688), FARIDPUR district, S central EAST BENGAL, BANGLADESH, on Arial Khan River, and 36 mi/58 km SE of FARIDPUR; 23°10′N 89°13′E. Jute trade center; agriculture (rice, oilseeds, sugarcane).

Madarounfa (mah-duh-ROON-fuh), town, MARADI province, NIGER, 40 mi/64 km S of MARADI. Administrative center.

Madaura (mah-dou-RAH), historic site, SOUK AHRAS wilaya, E ALGERIA. St. Augustine received his education here. Many ruins and Byzantine walls.

Madaus, GERMANY: see RADEBEUL.

Madawaska (ma-duh-WAH-skuh), county (□ 1,262 sq mi/3,281.2 sq km; 2001 population 35,611), NW NEW BRUNSWICK, E CANADA, on U.S. (MAINE) and QUEBEC borders; ⊙ EDMUNDSTON.

Madawaska (mad-uh-WAHS-kuh), town, AROOSTOOK county, N MAINE, on SAINT JOHN RIVER, opposite EDMUNDSTON (NEW BRUNSWICK), and 15 mi/24 km NE of FORT KENT; 47°17′N 68°15′W. Port of entry; paper mills. Includes Madawaska village and Saint David, agricultural village. Settled 1785 by Acadians, incorporated 1869.

Madawaska River (ma-duh-WAH-skuh), 250 mi/402 km long, SE ONTARIO, E central CANADA; rises in ALGONQUIN PROVINCIAL PARK; flows E, through several lakes, to CHATS Lake of OTTAWA River at ARNPRIOR; 45°26′N 76°20′W.

Madawaska River (ma-duh-WAH-skuh), 30 mi/48 km long, SE QUEBEC and NW NEW BRUNSWICK, E CANADA; issues from SE end of Lake TÉMISCOUATA; flows SE to SAINT JOHN River at EDMUNDSTON (New Brunswick); 47°22′N 68°19′W.

Madawaska Valley (ma-duh-WAH-skuh VA-lee), township (□ 259 sq mi/673.4 sq km; 2001 population

4,406), SE ONTARIO, E central CANADA, 35 mi/56 km from PEMBROKE; 45°31′N 77°41′W.

Madaya (mah-dah-YAH), township, MANDALAY district, MYANMAR, 15 mi/24 km N of MANDALAY (linked by railroad); 22°13′N 96°07′E. In irrigated rice area.

Madbury, town, STRAFFORD county, SE NEW HAMPSHIRE, 3 mi/4.8 km SW of DOVER; 43°10′N 70°56′W. Drained by BELLAMY RIVER (forms Bellamy Reservoir in W). Agriculture (cattle; vegetables, nursery crops; dairying); manufacturing (wood products; commercial printing).

Maddagiri, INDIA: see MADHUGIRI.

Maddalena (med-DAH-lee-nuh), village, CYRENAICA region, NE LIBYA, 7 mi/11.3 km NE of Al MARJ, on plateau. Agricultural settlement (cereals, olives, fruit). Founded by Italians (1936), who left after World War II, replaced by Lybian population.

Maddalena Island (□ 8 sq mi/20.8 sq km), 1 mi/1.6 km off NE SARDINIA, SASSARI province, ITALY, in TYRRHENIAN Sea; 4 mi/6 km long; 41°14′N 09°25′E. Rises to 515 ft/157 m. Lobster fisheries; granite quarries; corn, barley, vineyards. Chief port, LA MADDALENA. Causeway to CAPRERA Island (E).

Maddalena, La (mahd-dah-LAI-nah, lah), town, SASSARI province, ITALY, port on MADDALENA ISLAND, just off NE SARDINIA, 55 mi/89 km NE of SASSARI. Lobster fisheries; granite quarries; sawmills. Formerly an important naval base.

Maddalena Pass (mahd-uh-le-nah), Italian *Colle della Maddalena* or *Colle dell'Argentera*, French *Col de Larche* (kol duh lahrsh) or *Col de l'Argentière* (kol duh lahr-zhan-tyer), pass (elev. 6,532 ft/1,991 m) between MARITIME ALPS (S) and COTTIAN ALPS (N), on French-Italian border, 12 mi/19 km ENE of BARCELONNETTE (FRANCE); crossed by road (completed 1870) between CUNEO (Italy) and Barcelonnette along the upper UBAYE RIVER valley; 44°25′N 06°53′E. An army under Francis I crossed Alps here in 1515. Pass lies at N end of MERCANTOUR NATIONAL PARK, (established 1979), which extends SE along border to vicinity of TENDE (France). On Italian side there is a smaller preserve, the nature park of Argentera.

Maddaloni (mahd-dah-LO-nee), town, CASERTA province, CAMPANIA, S ITALY, 15 mi/24 km NNE of NAPLES; 41°02′N 14°23′E. Agricultural center (grapes, citrus fruit); sausage manufacturing, food canning; machinery.

Madden Dam, concrete dam on CHAGRES RIVER, PANAMA; 3,674 ft/1,120 m long, 223 ft/68 m high. Completed 1935; created ALAJUELA LAKE (formerly Madden Lake) (□ c.22 sq mi/57 sq km; c.12 mi/19 km long). Used for navigation, power, and flood control.

Maddikera (muh-dee-KAI-rah), town, KURNOOL district, TAMIL NADU state, S INDIA, 55 mi/89 km SW of Kurnool. Cotton ginning, oilseed (peanut) milling. Bamboo, dyewood in nearby forests.

Maddington (MA-deeng-tuhn), canton (□ 9 sq mi/23.4 sq km; 2006 population 450), CENTRE-DU-QUÉBEC region, S QUEBEC, E CANADA, 23 mi/37 km from NICOLET; 46°13′N 72°08′W.

Maddock (MA-dahk), village (2006 population 482), BENSON county, central NORTH DAKOTA, 33 mi/53 km WSW of DEVILS LAKE; 47°57′N 99°31′W. Dairy products; grain; livestock, poultry. Manufacturing (machinery). Founded in 1901 and incorporated in 1908.

Maddur (MUH-door), town, MANDYA district, KARNATAKA state, S INDIA, on Shimsha River, and 11 mi/18 km ENE of MYSORE. Road center in sugarcane and silk-growing area; hand-loom silk weaving.

Maddy, Loch, Scotland: see NORTH UIST.

Made en Drimmelen (MAH-duh uhn DRI-muh-luhn), town, NORTH BRABANT province, S NETHERLANDS, 7 mi/11.3 km N of BREDA; 51°41′N 04°48′E. AMER RIVER channel to N. Dairying; cattle, hogs; sugar beets, vegetables, potatoes, grain; manufacturing.

Madeira (muh-DEE-ruh), city (□ 3 sq mi/7.8 sq km; 2006 population 8,153), HAMILTON county, extreme

SW OHIO; NE suburb of CINCINNATI; 39°11′N 84°22′W.

Madeira, PORTUGAL: see MADEIRA ISLANDS.

Madeira Beach (muh-DIR-ruh), city (□ 3 sq mi/7.8 sq km; 2005 population 4,464), PINELLAS county, W central FLORIDA, 10 mi/16 km WNW of ST. PETERSBURG, on the GULF OF MEXICO; 27°48′N 82°47′W.

Madeira Islands (mah-DER-ruh), archipelago (□ 308 sq mi/800.8 sq km; 2001 population 245,011), autonomous region of PORTUGAL, in the ATLANTIC OCEAN, c.350 mi/563 km W of NW MOROCCO; 32°40′N 16°45′W. Madeira, the largest island (35 mi/56 km long and 13 mi/21 km wide), and PORTO SANTO are inhabited; two island groups, the DESERTAS and the SELVAGENS, are uninhabited. The chief town is FUNCHAL, on Madeira island. Sugarcane, Madeira wine, bananas, embroidery, and reed furniture are produced, and fishing is important. Madeira is a year-round resort. Though there have been efforts toward economic development, the islands are still heavily dependent on remittances from residents who have emigrated. Mountain peaks, which descend steeply into deep, green valleys and advance to the sea as precipitous basalt cliffs, give the island unusual scenic beauty. The delightful climate is marred only by the occasional *leste*, a hot Saharan wind. Known to the Romans as the Purple Islands; rediscovered (1418–1420) by João Gonçalves Zarco and Tristão Vas Teixeira. Settlement took place rapidly under the orders of Prince Henry the Navigator. Temporarily occupied by the British in the early 19th century.

Madeira River (muh-DAI-ruh), c.2,100 mi/3,380 km long, on the BOLIVIA-BRAZIL border; formed by the junction of the BENI and MAMORÉ rivers at 03°22′S 58°45′W; flows N along the border for c.60 mi/100 km; then NE in a winding course through the RONDÔNIA and AMAZONAS states of NW Brazil into the AMAZON River. At its mouth is Ilha Tupinambaranas, an extensive marshy region formed by the Madeira's distributaries. Receives numerous tributaries from the SE and is navigable by ocean vessels to the falls and rapids near PÔRTO VELHO (Brazil), from where the Madeira-Mamoré railroad begins a 227-mi/365-km run around the unnavigable section to GUAJARÁ MIRIM on the Mamoré River.

Madeleine, FRENCH POLYNESIA: see FATU HIVA.

Madeleine, Îles-de-la-, CANADA: see MAGDALEN ISLANDS.

Madeleine Island National Park (mahd-LEN), French *Parc National de l'Ile des Madeleines*, game reserve (□ 2 sq mi/5.2 sq km), DAKAR administrative region, W SENEGAL, on island in ATLANTIC OCEAN off CAPE VERDE peninsula, 2 mi/3.2 km W of DAKAR; 14°40′N 17°26′W. Migratory bird park.

Madeleine, La (mahd-LEN, lah), N industrial suburb (□ 1 sq mi/2.6 sq km) of LILLE, NORD department, NORD-PAS-DE-CALAIS region, N FRANCE; 50°39′N 03°04′E. Textiles (cotton, linen); metalworking (chains, auto chassis), chemicals. Also called La Madeleine-les-Lille.

Madeleine, La, FRANCE: see EYZIES-DE-TAYAC, LES.

Madeleine-les-Lille, La, FRANCE: see MADELEINE, LA.

Madeleine, Monts de la (mahd-LEN, MON duh lah), wooded upland (3,500 ft/1,067 m) of MASSIF CENTRAL, ALLIER department, AUVERGNE region, central FRANCE, on border of ALLIER and LOIRE departments, extends c.20 mi/32 km S from town of LAPALISSE (N), and overlooks the Roanne basin. Forms N spur of Monts du Forez, separating the valleys of the ALLIER (W) and the LOIRE (E) rivers. Cheese making.

Madeley (MAID-lee), town (2001 population 17,935), Telford and Wrekin, W ENGLAND, 3 mi/4.8 km S of TELFORD; 52°39′N 02°28′W. Church built by Thomas Telford. Madeley Court (Tudor mansion) now ruined.

Madeley (MAID-lee), village (2001 population 4,386), NW STAFFORDSHIRE, W ENGLAND, 7 mi/11.3 km W of STOKE-ON-TRENT; 52°59′N 02°20′W. Tile and pottery works; former agricultural market. Former site of Heighley Castle, destroyed in Civil War. Has church with 14th-century tower.

Madelia (muh-DEE-lyuh), town (2000 population 2,340), WATONWAN county, S MINNESOTA, 23 mi/37 km WSW of MANKATO, on WATONWAN RIVER; 44°02′N 94°25′W. Agricultural area (grain, soybeans; livestock, poultry); manufacturing (poultry products, building materials, foods; printing and publishing). Feji Lake to NE. Settled 1855, plotted 1857, incorporated 1873.

Madeline Island, WISCONSIN: see APOSTLE ISLANDS.

Maden, village, E central TURKEY, on DIYARBAKIR-ELAZIG railroad, on TIGRIS RIVER, and 45 mi/72 km NW of Diyarbakir; 38°24′N 39°42′E. Copper mines; also chrome; wheat. Formerly Erganimadeni.

Maden, TURKEY: see KESKIN.

Madera (mah-DAI-rah), county (□ 2,138 sq mi/5,558.8 sq km; 2006 population 146,345), central CALIFORNIA; ⊙ MADERA; 37°13′N 119°46′W. From level San Joaquin Valley in W, county stretches NE to crest of the SIERRA NEVADA; MOUNT RITTER rises to 13,156 ft/4,010 m. SAN JOAQUIN RIVER forms most of SW, S, and SE border; CHOWCHILLA RIVER forms part of NW border. Watered by FRESNO RIVER and MADERA CANAL (unit of Central Valley project). Includes part of YOSEMITE NATIONAL PARK (N), part of Sierra National Forest (N and NE), and DEVILS POSTPILE NATIONAL MONUMENT (NE). Rich irrigated valley lands produce cotton, oats, corn, wheat, barley, beans; pistachios, almonds, apples, figs, oranges, raisins, vegetables, honey; dairying; cattle, turkeys, poultry. Lumbering (chiefly pine); mining of pumice, gold, copper, sand and gravel; granite quarries. Manufacturing at Madera and Chowchilla. Pacific Coast Trail crosses county near NE border; part of Millerton Lake State Recreation Area in SE; MAMMOTH POOL RESERVOIR (NE). Formed 1893.

Madera (mah-DAI-rah), city (2000 population 43,207), ⊙ MADERA county, central CALIFORNIA, 20 mi/32 km NW of FRESNO, on FRESNO RIVER, in San Joaquin Valley; 36°58′N 120°05′W. Winery. Cattle, poultry; dairying; nuts, fruit, grain, grapes, honey. Manufacturing (machinery, consumer goods, plastic products, ornamental metal work). Granite quarry. YOSEMITE NATIONAL PARK and Sierra National Forest to NE; Millerton Lake reservoir and State Recreation Area to E; Adobe Hill (566 ft/173 m) to NE. Rapidly growing area. Incorporated 1907.

Madera (mah-DUHR-ah), unincorporated town, Bigler township, CLEARFIELD county, central PENNSYLVANIA, 22 mi/35 km N of ALTOONA, on Clearfield Creek; 40°49′N 78°26′W. Agriculture (dairying); manufacturing of clothing. Surface bituminous coal.

Madera (mah-DAI-rah), volcano (4,495 ft/1394 m), SW NICARAGUA, on S OMETEPE ISLAND in LAKE NICARAGUA, 14 mi/23 km SE of ALTAGRACIA.

Madera Acres, unincorporated town (2000 population 7,741), MADERA county, central CALIFORNIA; residential suburb 2 mi/3.2 km NNE of MADERA, on FRESNO RIVER; 37°01′N 120°04′W.

Madera Canal, irrigation unit of CENTRAL VALLEY project, MADERA county, S central CALIFORNIA. Conducts SAN JOAQUIN RIVER water (by gravity) 37 mi/60 km NW from Friant Dam (MILLERTON LAKE reservoir) to supply Madera county farmlands. Completed 1945.

Maderas, Las, NICARAGUA: see LAS MADERAS.

Maderno (mah-DER-no), village, BRESCIA province, LOMBARDY, N ITALY, port on W shore of LAGO DI GARDA, 20 mi/32 km NE of BRESCIA; 45°38′N 10°35′E. In olive-, grape-, and citrus-fruit-growing region; winter resort; manufacturing (paper, artificial silk). Has 12th-century church (restored 1580), 17th-century palace.

Madero, MEXICO: see VILLA MADERO.

Madero, Ciudad, MEXICO: see CIUDAD MADERO.

Maderuelo (mah-der-WAI-lo), village, SEGOVIA province, central SPAIN, 16 mi/26 km SE of ARANDA DE DUERO; 41°29′N 03°31′W. Flour milling; livestock raising; lumbering.

Madha (MAH-dah), town, SOLAPUR district, MAHARASHTRA state, W central INDIA, 37 mi/60 km NW of Solapur; 18°01′N 75°31′E. Agricultural market (millet, cotton, oilseeds).

Madhavpur (MAH-dahv-puhr), village, JUNAGADH district, GUJARAT state, W central INDIA, on ARABIAN SEA, 37 mi/60 km SW of JUNAGADH. Coconuts; rice; fishing. Here, according to Hindu legend, Krishna was married, probably to his main wife Rukmini.

Madhepura (muh-DAI-poor-ah), district (□ 690 sq mi/1,794 sq km), BIHAR state, E INDIA; ⊙ MADHEPURA.

Madhepura (muh-DAI-poor-ah), town (2001 population 45,015), ⊙ MADHEPURA district, BIHAR state, E INDIA; 25°55′N 86°47′E. Became own district in 1981, out of W SAHARSA district.

Madher, El (mah-DER, el), village, BATNA wilaya, NE ALGERIA, 12 mi/19 km NE of BATNA, in N outlier of the AURÈS Mountains. Sheep raising.

Madhira (muh-DEE-rah), village, WARANGAL district, ANDHRA PRADESH state, S INDIA, 27 mi/43 km SSE of KHAMMAM. Rice and oilseed milling; tobacco market. Also spelled Madira.

Madhoganj (MUH-do-guhnj), town, HARDOI district, central UTTAR PRADESH state, N central INDIA, on branch of Sarda Canal system, and 19 mi/31 km S of Hardoi; 27°07′N 80°09′E. Railroad and road junction. Trades in wheat, gram, barley, oilseeds, sugarcane.

Madhogarh (MUH-do-guhr), town, JALAUN district, S UTTAR PRADESH state, N central INDIA, 13 mi/21 km NW of Jalaun. Gram, wheat, oilseeds, jowar.

Madhopur (MUH-do-poor), village, GURDASPUR district, NW PUNJAB state, N INDIA, on RAVI River, and 26 mi/42 km NNE of Gurdaspur. Headworks and repair shops of UPPER BARI DOAB CANAL here.

Madhubani (MUHD-oo-bah-nee), district (□ 1,352 sq mi/3,515.2 sq km; 2001 population 3,570,651), N central BIHAR state, E INDIA; ⊙ MADHUBANI. Noted for distinctive artwork, referred to as either MADHUBANI or Mithila style of painting, traditionally done by women.

Madhubani (MUHD-oo-bah-nee), town (2001 population 66,285), MADHUBANI district, N central BIHAR state, E INDIA, on GANGA PLAIN, on railroad spur, and 18 mi/29 km NE of DARBHANGA; 26°22′N 86°05′E. Trades in rice, corn, wheat, barley, sugarcane; jute.

Madhugiri (MUH-duh-gee-ree), town, TUMKUR district, KARNATAKA state, S INDIA, 23 mi/37 km NNE of TUMKUR; 13°40′N 77°12′E. Hand-loom weaving; rice, millet, oilseeds, tobacco; livestock raising. Fortified in 17th century, town and hill (S) were alternately held by Mysore sultans and Marathas during 18th century. Granite and corundum quarrying in nearby hills (sandalwood, lac). Formerly spelled Maddagiri.

Madhukhali (maw-doo-krah-lee), village, FARIDPUR district, S central EAST BENGAL, BANGLADESH, on distributary of MADHUMATI RIVER, and 13 mi/21 km WSW of FARIDPUR; 23°30′N 90°40′E. Railroad junction, with spur to KAMARKHALI GHAT, 5 mi/8 km WSW.

Madhumati River (maw-du-mo-tee), main distributary of GANGA DELTA, 190 mi/306 km long, EAST BENGAL, BANGLADESH; leaves PADMA RIVER 3 mi/4.8 km N of KUSHTIA; flows SSE, past KUSHTIA, KUMARKHALI, GOPALGANJ, and PIROJPUR, through the SUNDARBANS, to BAY OF BENGAL. In upper course, called the GARAI RIVER; in Bakarganj and KHULNA districts called BA-

LESWAR RIVER; estuary mouth called HARINGHATA RIVER. Navigable for entire course by river steamers.

Madhupur (MUHD-uh-puhr), town (2001 population 47,349), DEOGHAR district, NE JHARKHAND state, E INDIA, 38 mi/61 km W of DUMKA. Railroad junction (spur to GIRIDIH coalfields); health resort; rice, corn, barley, oilseeds, rape, and mustard.

Madhupur Jungle (maw-du-poor) (□ 420 sq mi/1,092 sq km), densely wooded area, S MYMENSINGH and N DHAKA districts, EAST BENGAL, BANGLADESH; c.45 mi/ 8 km long, 6 mi/9.7 km–16 mi/26 km wide. Sal timber. Also called Garh Gazali.

Madhu Road (MAH-doo), village, NORTHERN PROVINCE, SRI LANKA, near the ARUVI ARU River, 25 mi/40 km SE of MANNAR; 08°46'N 80°09'E. Rice. Sri Lankan Christian pilgrimage center. Giant's Tank, one of the country's largest (8 sq mi/20.7 sq km) irrigation tanks, is 8 mi/12.9 km NW, on tributary of the Aruvi Aru; restored in late 19th century.

Madhya Bharat (MUH-dyuh BAH-ruht), or **Madhyabharat**, former state, W central INDIA. Comprised twenty-five former princely states. Incorporated in 1956 into MADHYA PRADESH state.

Madhya Pradesh (MUH-dyuh pruh-DAISH), state (□ 58,195 sq mi/151,307 sq km; 2001 population 60,348,023), central INDIA, between the DECCAN PLATEAU and the GANGA PLAIN; ⊙ BHOPAL. The second largest state of India, Madhya Pradesh consists (N-S) of upland zones separated by plains. Adequate rainfall and plentiful good soil permit a prosperous, predominantly agricultural economy. Grains, especially wheat, are the main crops of the N. In the SE, rice is the largest crop. The abundant cotton of the SW makes this state second only to GUJARAT state in cotton production. Other products include oilseeds, sugarcane, and maize. Spinning and weaving are the chief industries; there is a huge iron and steel mill at BHILAINAGAR and chemical and electrical industries at Bhopal. Rich in minerals; manganese, bauxite, iron ore, and coal are exploited. An aboriginal population, principally Gonds, inhabits the forested regions. Nominally within the Mogul empire, the area was ruled during the 16th and 17th century by the Gonds and in the 18th century by the Marathas. The BRITISH occupied it in 1820. Berar, originally belonging to the domain of the Nizam of HYDERABAD, was absorbed in 1903; from then until 1950 the province was called CENTRAL PROVINCES AND BERAR; renamed Madhya Pradesh in 1947. As part of the Reorganisation of States in 1956, it greatly increased its area with the addition of MADHYA BHARAT, VINDHYA PRADESH, and Bhopal, while losing Berar in the W. In 2000, sixteen SE districts of Madhya Pradesh broke off to form CHHATTISGARH state, with the result that Madhya Pradesh lost approximately half of its total area and most of its resounrce-rich land (pre-2000, total area was 171,215 sq mi/443,446 sq km). The majority of the inhabitants are Hindi-speaking Hindus, but Urdu and other languages are spoken. There are four major universities and numerous colleges in the state. Pench National Park in SE near SEONI is a tiger sanctuary. Governed by a chief minister and cabinet responsible to a bicameral legislature with one elected house and by a governor appointed by the president of India.

Madhya Saurashtra (MUH-dyuh sou-RAHSH-truh), former district, in former SAURASHTRA state, W central INDIA, now part of RAJKOT district, GUJARAT state. Also known as Central Saurashtra.

Madian (muhd-YAHN), northernmost coastal district of HEJAZ, SAUDI ARABIA, on RED SEA and its GULF OF AQABA, at JORDAN border. A wild mountainous district, also known as Jebel Esh-Shifa, with narrow TIHAMAH coastal plain (up to 15 mi/24 km wide), it forms the uptilted and buckled W edge of the Arabian shelf. Its pre-Cambrian crystalline rocks, covered with Tertiary lava flows, rise to 8,460 ft/2,579 m and are cut

by deeply incised wadis. Bedouin population is sparse, with Haql, Dhuba, and Muwailih the chief coastal settlements, and Tebuk the principal inland oasis. Believed to be the biblical Midian. Also spelled Median or Midyan.

Madibira (mah-DEE-bee-rah), village, MBEYA region, S central TANZANIA, 65 mi/105 km SW of IRINGA, near Ndembera River; 08°15'S 34°45'E. Tobacco, pyrethrum, corn, wheat; cattle, goats, sheep. Also spelled Madibura.

Madidi River (mah-DEE-dee), c.180 mi/290 km long, ITURRALDE province, LA PAZ department, NW BOLIVIA; rises on PERU border c.20 mi/32 km NNE of San Fermín; flows NE to BENI River just W of CAVINAS; 12°32'S 66°52'W. Navigable for c.125 mi/201 km, until Santiago de Pacaguaras. Dense forests along its lower course; rubber and quinine bark found here.

Madikeri, town, ⊙ KODAGU district, KARNATAKA state, S INDIA; 12°25'N 75°44'E.

Madilah, Al (mah-dee-LAH, el), mining village (2004 population 12,383), QAFSAH province, S central TUNISIA, 10 mi/16 km S of QAFSAH; 34°15'N 08°45'E. On railroad spur. Important phosphate mines; processing plant.

Madill (muh-DIL), town (2006 population 3,698), ⊙ MARSHALL county, S OKLAHOMA, 21 mi/34 km ESE of ARDMORE, near Lake TEXOMA to S and E; 34°05'N 96°46'W. Manufacturing (peanut processing; transportation equipment, apparel, fabricated metal products). Lake Texoma State Park to SE; Tishomingo National Wildlife Refuge to NE. A recreation center for nearby Lake Texoma. Settled c.1900, incorporated 1905.

Madimba (mah-DEEM-bah), village, BAS-CONGO province, W CONGO, on railroad, and 45 mi/72 km S of KINSHASA; 04°58'S 15°08'E. Elev. 1,440 ft/438 m. Agricultural center (manioc, plantains, yams, groundnuts, sweet potatoes), supplying KINSHASA.

Madina, Al (muh-DEE-nuh, ahl), township, BASRA province, SE IRAQ, on right bank of the EUPHRATES RIVER, and 45 mi/72 km NW of BASRA; 30°57'N 47°18'E. Dates, rice, corn, millet. Surrounded by large swampy area.

Madinat al 'Abid (mah-din-AHT el AH-bid), village, central YEMEN, 50 mi/80 km SSW of SANA, on motor road to HODEIDA.

Madinat ash Shab (mah-din-AHT esh shi-BAHB), town, SW YEMEN, suburb, 7 mi/11.3 km WNW of ADEN; 12°50'N 44°55'E. Formerly called al-Ittihad, it was built in the 1960s as the federal capital of the Federation of SOUTH ARABIA. From 1967 to 1970 it was, together with Aden, SOUTH YEMENI capital.

Madinat Ash Shamal (town (2004 population 2,951), Al Shamal municipality, QATAR, SE ARABIA, 107 mi/172 km NE of DOHA; 26°07'N 51°13'E. A new town serving as the administrative center of the the coastal villages in the municipality.

Madinat Esa, BAHRAIN: see ESA.

Madinat Hamad, BAHRAIN: see HAMAD.

Madingo, village, KOUILOU region, SW Congo Republic, on the Atlantic Ocean, 30 mi/48 km NW of MADINGO-KAYES. Palm mills; oil deposits nearby.

Madingo-Kayes, village, KOUILOU region, SW Congo Republic, on the Atlantic Ocean, at mouth of KOUILOU RIVER, 25 mi/40 km NNW of POINTE-NOIRE; 04°25'S 11°41'E. Manufacturing of starch and tapioca.

Madingou (mah-deeng-GOO), town, ⊙ BOUENZA region, S Congo Republic, on railroad, and 120 mi/193 km W of BRAZZAVILLE.

Madiniyina (me-de-nee-YEE-nah), town (2004 population 61,705), ⊙ MADANIYINA province, SE TUNISIA, 45 mi/72 km SSE of QABIS; 33°21'N 10°30'E. Regional market; sheep trading; date palms, olive trees, cereals, esparto. Staging center for trans-Saharan caravans and base of confederation of Berber tribes since 18th century. Leveled in 1960s, it was completely rebuilt. Its

cavelike storehouses and habitations are known as *ahorfas*.

Madioen, INDONESIA: see MADIUN.

Madira, INDIA: see MADHIRA.

Madiri (mah-DEE-ree), town, BORNO state, extreme NE NIGERIA, on road, 15 mi/24 km NE of MAIDUGURI. Market center. Groundnuts; cotton; livestock.

Madirovalo (mah-DEE-roo-VAH-loo), town, MAHAJANGA province, N central MADAGASCAR, on BETSIBOKA RIVER, and 10 mi/16 km W of AMBATO-Boény; 16°27'S 46°32'E. Rice, tobacco, cotton; cattle. Airfield.

Madison, county (□ 812 sq mi/2,111.2 sq km; 2006 population 304,307), N ALABAMA, on Tennessee (N) state line; ⊙ HUNTSVILLE, 34°44'N 86°34'W. Drained in SW by Wheeler Lake (in Tennessee River), crossed (N-S) by Flint River Cotton, corn, wheat, soybeans; mules, cattle, poultry; hay. Textiles. Formed 1808. Named for James Madison, then U.S. secretary of state.

Madison, county (□ 837 sq mi/2,176.2 sq km; 2006 population 15,361), NW ARKANSAS, in the OZARK region; ⊙ HUNTSVILLE, 36°00'N 93°42'W. Drained by WHITE and KINGS rivers and small War Eagle Creek. Agriculture (cattle, hogs, chicken, turkeys); timber. Madison County Wildlife Management Area and Withrow State Park in N; part of Hobbs Wildlife Management Area in NW corner; part of Ozark National Forest in S. Formed 1836.

Madison (MA-di-SUHN), county (□ 715 sq mi/1,859 sq km; 2006 population 19,210), N central FLORIDA, on GEORGIA state line (N); ⊙ MADISON, 30°26'N 83°28'W. Flatwoods area with swamps and many small lakes; bounded by AUCILLA (W), WITHLACOOCHEE (NE), and SUWANNEE (SE) rivers. Agriculture (corn, peanuts, cotton, tobacco; hogs, poultry) and some forestry. Formed 1827.

Madison, county (□ 286 sq mi/743.6 sq km; 2006 population 27,837), NE GEORGIA; ⊙ DANIELSVILLE, 34°08'N 83°13'W. Drained by BROAD RIVER. Manufacturing includes apparel, textiles, wood products; printing and publishing. Piedmont agricultural area (cotton, wheat, soybeans, sweet potatoes, hay); cattle, poultry, hogs. Formed 1811.

Madison, county (□ 473 sq mi/1,229.8 sq km; 2006 population 31,393), E IDAHO; ⊙ REXBURG; 43°48'N 111°40'W. Irrigated agricultural region in SNAKE RIVER PLAIN. Teton River joins HENRYS FORK RIVER in N which joins SNAKE RIVER in SW. Potatoes, sugar beets; wheat, barley, alfalfa; sheep, cattle; dairying. Part of Targhee National Forest in SE; Kelly Canyon Ski Area in SE. Formed 1913.

Madison, county (□ 740 sq mi/1,924 sq km; 2006 population 265,303), SW ILLINOIS, bounded W by MISSISSIPPI RIVER; ⊙ EDWARDSVILLE; 38°50'N 89°54'W. Part of SAINT LOUIS metropolitan area. Agriculture (wheat, soybeans, sorghum; cattle, poultry; dairy products) in E. Oil. Manufacturing at major cities of Edwardsville, GRANITE CITY, ALTON, COLLINSVILLE, and adjacent communities. Drained by CAHOKIA and SILVER creeks. Campus of Southern Illinois University at Edwardsville. Horseshoe Lake State Park in SW (large oxbow lake). Formed 1812.

Madison, county (□ 452 sq mi/1,175.2 sq km; 2006 population 130,575), E central INDIANA; ⊙ ANDERSON, 40°10'N 85°43'W. Drained by West Fork of WHITE RIVER and by small Pipe, Kilbuck, Fall, Duck, and Lick creeks. Rich agricultural area (hogs, cattle; corn, tomatoes, soybeans, vegetables; poultry). Diversified manufacturing: includes processing of farm and dairy products. Limestone quarrying. Indiana Reformatory in S at PENDLETON. Mounds State Park E of Anderson. Formed 1823.

Madison, county (□ 562 sq mi/1,461.2 sq km; 2006 population 15,547), S central IOWA; ⊙ WINTERSET, 41°20'N 94°01'W. Prairie agricultural area (hogs, cattle; corn, soybeans) drained by NORTH and MIDDLE rivers. Bituminous-coal deposits, clay, limestone quarries. Pammel State Park SW of Winterset; Badger Creek State Park in NE. Formed 1846.

Madison, county (□ 443 sq mi/1,151.8 sq km; 2006 population 79,015), central KENTUCKY; ⊙ RICHMOND, 37°44′N 84°18′W. Bounded NW, N, and NE by KENTUCKY RIVER, W in part by Paint Lick creek; drained by Silver, Muddy, and Red Lick creeks. Rolling agricultural area, in BLUEGRASS REGION (burley tobacco, corn, hay, alfalfa; poultry, cattle, horses; dairying); bituminous coal mines; clay pits. Manufacturing at BEREA and Richmond. Small parts of Daniel Boone National Forest in SE; White Hall State Historic Site in NW; Lexington-Bluegrass Army Depot (chemical weapons) in center. Formed 1785.

Madison, county (□ 742 sq mi/1,929.2 sq km; 2006 population 87,419), central MISSISSIPPI; ⊙ CANTON; 32°37′N 90°01′W. Bounded by PEARL (SE) and BIG BLACK (NW) rivers. Agriculture (cotton, corn, hay, vegetables, potatoes; poultry, cattle; dairying); timber. Manufacturing at Canton, MADISON, and RIDGELAND. S part of county is urbanized, including part of city of JACKSON and its environs. Mississippi Petrified Forest in SW; NATCHEZ TRACE PARKWAY passes through SE, on Pearl River; Cypress Swamp natural area to NE. Formed 1828.

Madison, county (□ 496 sq mi/1,289.6 sq km; 2006 population 12,109), SE MISSOURI; ⊙ FREDERICKTOWN; 37°29′N 90°21′W. Partly in SAINT FRANCOIS MOUNTAINS, drained by SAINT FRANCIS and CASTOR rivers. Farming (wheat, corn, hay; livestock) and mining (tungsten, manganese, lead, zinc, iron, cobalt, copper, antimony, nickel, granite); manufacturing at Fredericktown. Part of Mark Twain National Forest extends across co. E-W. Saint Francis River is popular rafting, canoeing, kayaking stream. Mining began in 1720s. Formed 1818.

Madison, county (□ 3,602 sq mi/9,329 sq km; 1990 population 5,989; 2000 population 6,851), SW MONTANA, ⊙ VIRGINIA CITY; 45°18′N 111°55′W. Agricultural and mining region, drained by MADISON, RUBY, BEAVERHEAD, and JEFFERSON rivers. SE corner borders on Idaho, forms CONTINENTAL DIVIDE. Several small mountain lakes in S and center; Beaverhead Rock State Park on border; ENNIS LAKE reservoir in E; part of EARTHQUAKE LAKE, formed (1959) by earthquakes in SE corner. Cattle, sheep, horses; hay, wheat, barley, potatoes. Talc mining, marble, gold, silver, lead, lignite. Parts of Beaverhead National Forest in S, SE, and center; parts of Deerlodge National Forest in N and NW; part of Gallatin National Forest in E, forest includes parts of Lee Metcalf Wilderness Area; TOBACCO ROOT MOUNTAINS in N; GRAVELLY Mountains in S; part of MADISON RANGE in E and RUBY RANGE in W. Formed 1865.

Madison, county (□ 575 sq mi/1,495 sq km; 2006 population 35,279), NE central NEBRASKA; ⊙ MADISON; 41°55′N 97°36′W. Agricultural area drained by ELKHORN RIVER. Cattle, hogs; dairying; poultry products; corn, soybeans. Manufacturing at Norfolk. Formed 1867.

Madison, county (□ 661 sq mi/1,718.6 sq km; 2006 population 70,197), central NEW YORK; ⊙ WAMPSVILLE; 42°56′N 75°39′W. Named for James Madison, who was Secretary of State in 1806, when county was formed. Almost entire county was part of the Twenty Town tract purchased by Governor George Clinton from Oneidas in 1788. Colgate University founded (1817) at HAMILTON as the Baptist Education Society. The "Perfectionists" sect, led by John H. Noyes, established the communistic Oneida Community, which turned capitalist by producing Oneida silverware and sharing stock in Oneida Community, Ltd. Lorenzo, an estate at S end of CAZENOVIA LAKE, is a State Historic Site; has a Federal-period house. Drained by CHENANGO and UNADILLA rivers and

several creeks. Includes Cazenovia Lake, and many small lakes and several reservoirs; resorts. Dairying; field crops (onions, cabbage); grain; poultry. Manufacturing concentrated at ONEIDA and CANASTOTA.

Madison (MA-duh-suhn), county (□ 451 sq mi/1,172.6 sq km; 2006 population 20,355), W NORTH CAROLINA; ⊙ MARSHALL; 35°51′N 82°42′W. APPALACHIAN MOUNTAIN region, bounded N by TENNESSEE state line; BALD MOUNTAINS along N border; drained by FRENCH BROAD RIVER. Service industries; some manufacturing at Marshall; agriculture (tobacco, corn, hay; cattle); timber; resort area. Part of Pisgah National Forest. Rural Life Museum. Named for President James Madison. Formed 1851 from Buncombe and Yancey counties. Named for James Madison, fourth president of the U.S.

Madison (MAD-i-suhn), county (□ 464 sq mi/1,206.4 sq km; 2006 population 41,496), central OHIO; ⊙ LONDON; 39°52′N 83°26′W. Drained by DEER, PAINT, and DARBY creeks. Agricultural area (livestock; corn, wheat, soybeans, fruit); manufacturing at London and MOUNT STERLING (metal products, transportation equipment). Formed 1810.

Madison, county (□ 561 sq mi/1,458.6 sq km; 2006 population 95,894), W TENNESSEE; ⊙ JACKSON; 35°37′N 88°51′W. Drained by Middle and South forks of FORKED DEER RIVER. Agriculture; diversified manufacturing. Extensive Native American mounds at Pinson Mounds Archaeological State Park, SE of Jackson. Formed 1821.

Madison (MA-di-suhn), county (□ 472 sq mi/1,227.2 sq km; 2006 population 13,310), E central TEXAS; ⊙ MADISONVILLE; 30°58′N 95°55′W. Bounded W by NAVASOTA RIVER, E by TRINITY RIVER, SE by Bedias Creek. Livestock (cattle, hogs, horses); forage crops; timber. Formed 1853.

Madison (MA-di-suhn), county (□ 321 sq mi/834.6 sq km; 2006 population 13,613), N VIRGINIA; ⊙ MADISON; 38°24′N 78°16′W. In N PIEDMONT region; rises in NW to BLUE RIDGE; bounded SW and S by RAPIDAN RIVER; drained by short Robinson River. Agriculture (barley, wheat, corn, alfalfa, soybeans, hay, apples, peaches; cattle, sheep, hogs; dairying); timber. Trout fishing. Includes part of SHENANDOAH NATIONAL PARK in NW. BLUE RIDGE (NATIONAL) PARKWAY follows NW border. Formed 1792.

Madison, parish (□ 662 sq mi/1,721.2 sq km; 2006 population 12,328), NE LOUISIANA; ⊙ TALLULAH; 32°25′N 91°11′W. Bounded W by Bayou MACON, E by MISSISSIPPI RIVER (MISSISSIPPI state line); intersected by TENSAS RIVER Fertile delta lowland agricultural area (cotton, corn, sorghum, rice, soybeans, wheat; cattle, hogs; home gardens); crawfish. Fishing and waterfowl hunting on oxbow lakes formed by the MISSISSIPPI. Possesses many Native American mounds. Named after President Madison. Formed 1838. Part of Big Lake State Wildlife Area in SW, part of TENSAS RIVER National Wildlife Refuge in S.

Madison, city (2000 population 29,329), Madison co., N ALABAMA, 9 mi/14.5 km WSW of Huntsville. Manufacturing (machinetools, machinery, fiber optics, paper products).

Madison (MA-di-SUHN), city (□ 2 sq mi/5.2 sq km; 2005 population 3,190), ⊙ MADISON county, N central FLORIDA, near GEORGIA state line, c.50 mi/80 km E of TALLAHASSEE. 30°28′N 83°25′W. Trade center for farming region.

Madison, city (2000 population 4,545), MADISON and SAINT CLAIR counties, SW ILLINOIS, suburb of SAINT LOUIS, on the MISSISSIPPI RIVER, and 5 mi/8 km N of EAST SAINT LOUIS, within Saint Louis metropolitan area; 38°40′N 90°09′W. Manufacturing (fabricated metal products, wood products; aluminum and magnesium extrusion and production). Grew with establishment of steel industry in early 1890s. Gateway

National Raceway to S. Part of Granite City Ordnance Plant near river. Incorporated 1891.

Madison, city (2000 population 12,004), ⊙ JEFFERSON county, SE INDIANA, on the OHIO River; 38°46′N 85°24′W. Port of entry and a tobacco-marketing center. Manufacturing (transportation equipment, tool and die, shoes, fabricated metal products, machinery, chemicals). Has fine examples of Georgian, Federal, Classical Revival, Gothic, Italianate, and Victorian architecture. An annual regatta is held on the Ohio River. Hanover College and Clifty Falls State Park are to W; branch of Indiana Vocational and Technical College, Jefferson Proving Grounds to N. Settled 1805, incorporated 1838.

Madison, city (2000 population 586), MONROE county, NE central MISSOURI, 12 mi/19 km ENE of MOBERLY; 39°28′N 92°12′W. Corn, wheat, soybeans; cattle, saddle horses.

Madison, city (2006 population 2,287), ⊙ MADISON county, NE central NEBRASKA, 14 mi/23 km S of NORFOLK, and on branch of ELKHORN RIVER; 41°49′N 97°27′W. Grain; dairy and poultry products; watermelons. Manufacturing (wood products; hog slaughtering). Incorporated 1873.

Madison, city (2006 population 6,258), ⊙ Lake county, E SOUTH DAKOTA, 38 mi/61 km NW of SIOUX FALLS; 44°00′N 97°06′W. Trade center for agricultural region; resort. Railroad terminus. Dairy products; livestock, poultry; grain; flour. Manufacturing (machinery, wood products; egg processing). Seat of Dakota State University, first in South Dakota (1881). Hospital and airport. Lake Madison and Lake Herman State Park to SW. Plotted 1875.

Madison, city (2006 population 223,389), ⊙ WISCONSIN and DANE county, S central WISCONSIN, city center and state capitol on an isthmus between lakes MONONA (SE) and MENDOTA (NW), in the FOUR LAKES group; 43°04′N 89°23′W. Metropolitan area; trading and manufacturing center in a fertile agricultural region. Manufacturing (printing and publishing; dairy products, chemicals, foods and beverages, machinery, lumber, packaging, paper goods, medical and medical research supplies, wood products, fabricated metal products, chemicals and chemical supplies, consumer goods); the university and the government are also large employers. Seat of the University of Wisconsin, Edgewood College, and Madison Area Technical College. Many parks that dot the wooded lake shores make it an attractive residential city. Among its points of interest are the elaborate capitol, which houses the legislative library organized by Charles McCarthy; a Unitarian church designed by Frank Lloyd Wright; University of Wisconsin Arboretum; and Vilas Park, which contains a zoo. Madison Art Center, a U.S. Forest Service Forest Products Laboratory (on University of Wisconsin campus), Mendota Mental Health Institute, and Central Wisconsin Center are also here. Governor Nelson State Park to NW, on opposite side of Lake Mendota. Dane county Regional Airport (Truax Field) in NE. Founded in 1836, and chosen (through the efforts of James Duane Doty) territorial capital before it was settled; incorporated 1856.

Madison, town (2000 population 987), SAINT FRANCIS county, E ARKANSAS, 4 mi/6.4 km E of FORREST CITY, and on SAINT FRANCIS River; 35°01′N 90°43′W. In agricultural area; manufacturing (chemicals).

Madison, resort town, NEW HAVEN county, S CONNECTICUT, on Hammonasset River and LONG ISLAND SOUND, and 17 mi/27 km E of NEW HAVEN; 41°20′N 72°37′W. Mostly residential with summer family resort at Hammonasset Beach State Park. Many fine 17th- and 18-century houses; once a shipbuilding, shipping center. Includes Madison and East River (on small East River) villages and Hammonasset

Beach State Park, at Hammonasset Point. Set off from GUILFORD 1826.

Madison, town (2000 population 3,636), ⊙ MORGAN county, N central GEORGIA, 26 mi/42 km SSW of ATHENS; 33°35′N 83°29′W. Agricultural trade center; manufacturing (clothes, furniture, lumber, consumer goods, plastics, plywood, rope). Has antebellum houses restored by residents attracted here from Atlanta and elsewhere since the 1970s. Incorporated 1809.

Madison, town (2000 population 857), GREENWOOD county, SE KANSAS, on VERDIGRIS RIVER, and 18 mi/29 km S of EMPORIA; 38°07′N 96°08′W. Manufacturing point in oil-producing and agricultural region; produces gasoline. Oil wells nearby. Laid out 1879, incorporated 1885.

Madison, town, including Madison village, SOMERSET county, central MAINE, on the KENNEBEC RIVER, near SKOWHEGAN; 44°49′N 69°47′W. Its water power developed paper and textile mills. Lakewood resort is E of village. Incorporated 1804.

Madison, town (2000 population 1,768), ⊙ LAC QUI PARLE county, SW MINNESOTA, near SOUTH DAKOTA state line, 23 mi/37 km WNW of MONTEVIDEO; 45°00′N 96°11′W. Elevation 1,091 ft/333 m. Railroad terminus. Agricultural trade center (grain, soybeans; livestock, poultry; dairying); manufacturing (fertilizers, bakery products, fabricated metal products, feeds). Lac qui Parle Wildlife Area, surrounds LAC QUI PARLE and MARSH lakes; MINNESOTA RIVER to NE. Settled c.1875, plotted 1884, incorporated as city 1902.

Madison, town (2000 population 14,692), MADISON county, central MISSISSIPPI, suburb 13 mi/21 km NNE of downtown JACKSON, on ROSS BARNETT RESERVOIR (PEARL RIVER); 32°27′N 90°06′W. Agriculture to N and W (cotton, corn, vegetables; poultry; dairying); manufacturing (liquid carbon dioxide, wood products, fabricated metal products). Bruce Campbell Field airport is here. NATCHEZ TRACE PARKWAY passes through SE, along reservoir.

Madison, town, CARROLL county, E NEW HAMPSHIRE, 29 mi/47 km NE of LACONIA; 43°53′N 71°09′W. Drained by Silver River and Forrest Brook. Agriculture (livestock, poultry; vegetables; dairying; timber); manufacturing (machine parts, asphalt). Resort area. Silver Lake in SW (includes village of Silver Lake; manufacturing wooden products, safety equipment); Madison Boulder State Reservation in NW; King Pine Ski Area in E.

Madison (MA-duh-suhn), town (□ 3 sq mi/7.8 sq km; 2006 population 2,287), ROCKINGHAM county, N NORTH CAROLINA, on DAN RIVER, and 27 mi/43 km NE of WINSTON-SALEM, and 2 mi/3.2 km S of MAYODAN; 36°23′N 79°58′W. Elevation 574 ft/175 m. Manufacturing (leather products, embroidery, building materials, textiles; commercial printing); service industries; agriculture (tobacco, grain, soybeans; livestock; dairying). BELEWS LAKE reservoir to SW. Laid out 1818, incorporated 1873.

Madison (MA-di-suhn), town (2006 population 215), ⊙ MADISON county, N VIRGINIA, near E foot of BLUE RIDGE, 26 mi/42 km NNE of CHARLOTTESVILLE, near Robinson River; 38°22′N 78°15′W. Manufacturing (wood products, lumber, clothing); agriculture (grain, apples, soybeans; livestock; dairying); timber.

Madison, town (2006 population 2,618), ⊙ BOONE county, W central WEST VIRGINIA, on Little Coal River (left tributary of BIG COAL RIVER), 23 mi/37 km SSW of CHARLESTON; 38°03′N 81°47′W. Railroad junction. Bituminous-coal region. Timber. Manufacturing (lumber, coal processing). Agriculture (tobacco, honey). Incorporated 1906.

Madison, fishing village, DORCHESTER county, E MARYLAND, 9 mi/14.5 km WSW of CAMBRIDGE. Agricultural area; vegetable cannery. Originally called Little Tobacco Stick when laid out in 1760, it was renamed in honor of James Madison, the fourth President. In local lore, it was the home of a legendary white mule

which refused every effort to harness him, was driven into the swamps by a posse, and returned drunk to harrass the town until he sobered up and became a parson. In 1814, men from here captured Captain Phipps and the eighteen-member crew of a British tender from HMS *Dauntless*, which had gone around on James Point.

Madison, village (2006 population 310), MADISON county, central NEW YORK, 20 mi/32 km SW of UTICA; 42°53′N 75°30′W. In dairying area.

Madison (MAD-i-suhn), village (□ 5 sq mi/13 sq km; 2000 population 2,921), Lake county, NE OHIO, 15 mi/24 km WSW of ASHTABULA, near LAKE ERIE; 41°46′N 81°03′W. Makes fabricated metal products, wood products, wine, mats; nurseries. Resort.

Madison, residential village, DAVIDSON county, N central TENNESSEE, suburb, 7 mi/11 km NE of NASHVILLE, on CUMBERLAND RIVER; 36°16′N 86°42′W. Manufacturing.

Madison, borough (2006 population 16,016), MORRIS county, NE NEW JERSEY; 40°45′N 74°25′W. Residential area. Corporate headquarters. Seat of Drew University and part of Fairleigh Dickinson University. Originally called Bottle Hill, it was renamed in 1834. Sayre House (1745) in Madison was General Anthony Wayne's headquarters. Once noted for its roses, this industry is commemorated in its nickname, "The Rose City"; now a suburban community with minimal industry. Settled 1685, incorporated 1889.

Madison, borough (2006 population 490), WESTMORELAND county, SW PENNSYLVANIA, 8 mi/12.9 km SW of GREENSBURG; 40°15′N 79°40′W. Agriculture (grain; livestock; dairying); manufacturing (fabricated metal products).

Madison, river, 183 mi/295 km long, in WYOMING and MONTANA; rises in YELLOWSTONE NATIONAL PARK in Park county, NW Wyoming; flows W then N through SW Montana to join the JEFFERSON and Gallatin rivers at the Three Forks of the Missouri at THREE FORKS, Montana. Impounded by Hebgen Dam (HEBGEN LAKE), GALLATIN county (Montana), in its upper course and by Madison Dam (ENNIS LAKE) MADISON county (Montana), a power facility, at midcourse. The river is used for irrigation. EARTHQUAKE or QUAKE, Lake, just downstream from Hebgen Dam was formed by an earthquake in 1959.

Madison Avenue, celebrated street of MANHATTAN, borough of NEW YORK city, SE NEW YORK. Runs from Madison Square (23rd Street) to the Madison Bridge (to the BRONX) over the HARLEM RIVER (138th Street). In the 1940s and 1950s, some of the major U.S. advertising agencies had headquarters in its Midtown section, and the name of the avenue became synonymous with the advertising industry. Now lined by expensive shops between 42nd and 96th streets.

Madison Heights, city (2000 population 31,101), OAKLAND county, SE MICHIGAN, a suburb, 12 mi/19 km NNW of downtown DETROIT; 42°30′N 83°05′W. Manufacturing (tool design and prototypes, encoder products, displays, machinery, fabricated metal products, apparel). Incorporated in 1955.

Madison Heights (MA-di-suhn HEITS), unincorporated city, AMHERST county, central VIRGINIA, suburb, 1 mi/2 km NE of LYNCHBURG, on JAMES RIVER; 37°26′N 79°05′W. Manufacturing (machinery, paper products, lumber, furniture; machining); in agricultural area (corn, apples, peaches; cattle); timber.

Madison Lake, village (2000 population 837), BLUE EARTH county, S MINNESOTA, 9 mi/14.5 km ENE of MANKATO; 44°12′N 93°48′W. Grain; livestock; light manufacturing. Madison Lake to SE, Eagle Lake to W, numerous small natural lakes in area.

Madison, Lake, Lake county, E SOUTH DAKOTA, 4 mi/6.4 km SE of MADISON; 4 mi/6.4 km long, 1.5 mi/2.4 km wide; 43°57′N 97°01′W. Popular resort.

Madison, Mount, NEW HAMPSHIRE: see PRESIDENTIAL RANGE.

Madison Range, in ROCKY MOUNTAINS of SW MONTANA, GALLATIN and MADISON counties, rises SW of BOZEMAN, extends c. 44 mi/71 km S, between GALLATIN and MADISON rivers, to HEBGEN LAKE. Partly within Gallatin and Beaverhead National Forests, includes units of Lee Metcalf Wilderness Area. Highest points are Hilgard Peak (11,316 ft/3,449 m), Koch Peak (11,286 ft/3,440 m), and Lone Mountain (11,166 ft/3,403 m).

Madison Square Garden, sports and entertainment complex, mid-MANHATTAN, NEW YORK CITY, SE NEW YORK, bounded E by 7th Avenue, W by 8th Avenue, N by 33rd Street, and S by 31st Street. The 820,000 sq ft/76,178 sq m of space includes the 20,000-seat Arena, the 5,600-seat Theater, and the 36,000-sq-ft/3,344-sq-m Exposition Rotunda, as well as an office tower and shops. Located directly above the Pennsylvania Station (MTA subways, Long Island railroad, New Jersey Transit, and Amtrak). Completed in 1967, the present complex is the fourth to bear this name; the first was an 1879 conversion of Barnum's Madison Square Hippodrome, formerly the Grand Central Depot. Each successive incarnation of the "Garden" has moved farther uptown. Home stadium of the NHL New York Rangers, NBA New York Knicks, and WNBA Liberty.

Madisonville, city (2000 population 19,307), ⊙ HOPKINS county, W KENTUCKY, 36 mi/58 km SW of OWENSBORO; 37°20′N 87°30′W. Major railroad junction. Agriculture (tobacco, grain, soybeans; livestock); manufacturing (coal processing; machinery, building materials, dairy products, ordnance, soft drinks); coal is mined here, both surface and subsurface. Madisonville Municipal Airport to E. Hopkins County Historical Society Library and Museum Seat of Madisonville Community College (University of Kentucky). Lake Pee Wee reservoir to W; White City Wildlife Management Area to E. Incorporated 1807.

Madisonville, town, SAINT TAMMANY parish, SE LOUISIANA, 8 mi/13 km SW of COVINGTON, and on TCHEFUNCTE RIVER, 2 mi/3 km N of its entrance to Lake PONTCHARTRAIN; 30°24′N 90°10′W. Barge building, manufacturing (treated wood). Hosts Wooden Boat Festival. Fairview-Riverside State Park is here.

Madisonville, town (2006 population 4,464), ⊙ MONROE county, SE TENNESSEE, 14 mi/23 km NE of ATHENS; 35°31′N 84°22′W. Agricultural area; some tourism. Seat of Hiwassee College.

Madisonville (MA-di-suhn-vil), town (2006 population 4,326), ⊙ MADISON county, E central TEXAS, c.85 mi/137 km NNW of HOUSTON; 30°56′N 95°54′W. Shipping, trade center in agriculture area (cattle, horses, hogs); timber; manufacturing (mushroom processing; feeds, transportation equipment).

Madiun (mah-dee-YOON), municipality (2000 population 164,048), ⊙ MADIUN district, Java Timur province, INDONESIA, 90 mi/145 km WSW of SURABAYA; 07°37′S 111°30′E. Trade center for agricultural area (sugar, tobacco, cinchona bark, rice, coffee, cassava, corn); textile and lumber mills, railroad workshops. Site of Communist uprising in 1948. Formerly spelled Madioen.

Madjalengka, INDONESIA: see MAJALENGKA.

Madjaz al Bab (ME-jez el BAHB), town (2004 population 20,308), BAJAH province, N central TUNISIA, on MAJARDAH RIVER, and 36 mi/58 km WSW of TUNIS; 36°39′N 09°37′E. Road center in TUNISIA's N wheat growing region.

Madjaz al Bab (ME-jez el BAHB), village, BAJAH province, NE TUNISIA, 36 mi/58 km SE of TUNIS. Founded by Andalusian immigrants in 1611 on site of former Roman settlement of Membressa (some remains have been found). National War Cemetery nearby.

Madoc (MAI-dahk), township (□ 104 sq mi/270.4 sq km; 2001 population 2,044), SE ONTARIO, E central

CANADA, 24 mi/38 km from BELLEVILLE; 44°35′N 77°29′W. Logging; marble mill; beef and dairy cattle, hogs. Historic sawmill. Incorporated in 1850.

Madoc (MAI-dahk), former village (□ 1 sq mi/2.6 sq km; 2001 population 1,375), SE ONTARIO, E central CANADA, on Deer Creek, and 24 mi/39 km N of BELLEVILLE; 44°31′N 77°29′W. Milling, dairying, mixed farming. Resort area. Amalgamated into Centre Hastings.

Madoera, INDONESIA: see MADURA.

Madona (MAH-do-nuh), German *modohn*, city (2000 population 9,553), E central LATVIA, in VIDZEME, 80 mi/129 km E of Rīga; 56°51′N 26°13′E. Railroad junction. Cement, dairy products, leather, foodstuffs; flax.

Madonie Mountains (mah-DO-nye), N SICILY, ITALY, range extending 30 mi/48 km from TORTO River E to NEBRODI Mountains; rise to 6,480 ft/1,975 m in Pizzo Antenna (center). Snow-covered nine months of year. Source of SALSO, TORTO, GRANDE rivers.

Madraka, Ras (muhd-RAH-kah, RAHS), cape on SE OMAN coast, at S end of Masira Bay of ARABIAN SEA; 19°00′N 57°52′E. Has 450-ft/137-m-high limestone cliff.

Madras (MAD-ruhs), town (2006 population 5,296), ⊙ JEFFERSON county, N central OREGON, 40 mi/64 km N of BEND; 44°37′N 121°07′W. Elevation 2,242 ft/683 m. Manufacturing (millwork; wood products, building materials, fertilizer). Grain, potatoes; sheep, cattle. LAKE CHINOOK (Round Butte Dam) to SW and Lake Simtustus (Pelton Dam) to W, both on DESCHUTES RIVER. Cove Palisades State Park surrounds Lake Chinook. Warm Springs Indian Reservation to W; Deschutes National Forest to SW; CROOKED RIVER National Grassland to S.

Madras, INDIA: see CHENNAI (city) or CHENNAI (district).

Madras River, INDIA: see COOUM RIVER.

Madras States (mud-RAHS), former agency, S INDIA, comprised of BANGANAPALLE, COCHIN, PUDUKKOTTAI, SANDUR, and TRAVANCORE. In 1939, Banganapalle and Sandur were transferred from Madras States agency to political charge of MYSORE. In 1948–1949, Banganapalle, Pudukkottai, and Sandur were incorporated into districts of MADRAS (later TAMIL NADU and ANDHRA PRADESH states), and Cochin and Travancore merged to form the union of TRAVANCORE-COCHIN (later KERALA state).

Madre de Deus (MAH-drai dee DAI-oos), town (2007 population 15,432), E central BAHIA state, BRAZIL, on TODOS OS SANTOS Bay, 22 mi/35 km NW of SALVADOR; 12°40′S 38°28′W.

Madre de Deus, Brazil: see BREJO DA MADRE DE DEUS.

Madre de Dios (MAH-drai de dee-OS), province (□ 4,200 sq mi/10,920 sq km), PANDO department, NW BOLIVIA; ⊙ PUERTO GONZALO MORENO; 11°02′S 66°13′W. Agriculture (rice, yucca, bananas, rubber).

Madre de Dios (MAH-drai dai dee-OS), region (□ 32,889 sq mi/85,511.4 sq km), SE PERU, on BRAZIL (ACRE RIVER) and BOLIVIA (HEATH RIVER) borders; ⊙ PUERTO MALDONADO; 12°00′S 70°15′W. Almost entirely in AMAZON basin, it is drained by MADRE DE DIOS RIVER and its affluents. Has a humid, tropical climate. Rainy season from November–April. Largely covered by virgin forests; rubber is still its main source of income; also balata, chicle, Brazil nuts, and timber. In the settled regions cotton, rice, sugarcane, coffee, cacao, yucca, beans, and fruit are grown. Site of PARQUE NACIONAL MANÚ.

Madre de Dios Archipelago (MAH-drai dai dee-OS), uninhabited island group off coast of S CHILE, just SW of WELLINGTON ISLAND and separated from CHATHAM and HANOVER islands (SE) by CONCEPCIÓN STRAIT; between 50°00′ and 51°00′S at 75°00′W. Main islands are Madre de Dios (N) and DUKE OF YORK (S).

Madre de Dios River (MAH-drai dai dee-OS), c.700 mi/1,127 km long, in AMAZON basin of PERU and BOLIVIA; rises in Cordillera de CARABAYA near PAU-

CARTAMBO (CUSCO region, S Peru); flows N and NE, past MANÚ, PUERTO MALDONADO (Peru), and PUERTO HEATH (Peru-Bolivia border), to BENI RIVER at RIBERALTA; 11°00′S 67°00′W. Receives MANÚ, INAMBARI, TAMBOPATA, RÍO DE LAS PIEDRAS, and Sena rivers. Generally navigable for small craft on upper course; traffic is interrupted by rapids just below Puerto Heath. It serves as main communication line in NW Bolivia. Rubber is collected along its course.

Madre, Laguna, Texas and Mexico: see LAGUNA MADRE.

Madre Vieja River (MAH-drau vee-AI-hah), c.90 mi/145 km long, SW GUATEMALA; rises in highlands W of TECPÁN GUATEMALA; flows SSW, past PATULUL, to the PACIFIC OCEAN at Barra de Madre Vieja, 19 mi/31 km S of TIQUISATE; 14°01′N 91°26′W. Forms CHIMALTENANGO-SOLOLÁ department border in upper course.

Madrid (muh-DRID), province (□ 3,089 sq mi/8,031.4 sq km; 2001 population 5,423,384), central SPAIN, in the autonomous region of Madrid; ⊙ MADRID. Situated on the monotonous central plateau (Meseta), it is bounded NW by the rugged, frequently snow-capped SIERRA DE GUADARRAMA (crossed by several passes and rising at PEÑALARA mountain to 7,972 ft/2,430 m), separating Madrid from Old Castile (SEGOVIA, ÁVILA, and LA RIOJA provinces). Bordered E by GUADALAJARA province, SE by CUENCA province, S by TOLEDO province. Watered by JARAMA, HENARES, TAJUÑA, and MANZANARES rivers, tributaries of the TAGUS RIVER. The climate varies considerably with elevation, but is generally of the rigorous continental type (long, cold winters and hot summers). Agriculture is generally backward because of soil erosion and large landholdings, but fertile sections in the S and SE grow cereals, potatoes, beets, asparagus, tomatoes, and grapes. Lime, gypsum, and stone quarries. Several medicinal springs (LOECHES, EL MOLAR). Industry, once predominantly based on agriculture, is now highly diverse (construction, metallurgy, lumbering; motor vehicles; manufacturing of chemicals, pharmaceuticals, ceramics. Other leading cities/towns include ALCALÁ DE HENARES, ARANJUEZ, GETAFE, COLMENAR VIEJO, and CHAMARTÍN. Aranjuez, ESCORIAL, and EL PARDO are known for their fine royal palaces. Apart from the SW and W outskirts of the capital, most of the province remained in Loyalist hands until the end of Spanish civil war (1936–1939), but some of the decisive battles were fought here.

Madrid (muh-DRID), city (2001 population 1,938,723), ⊙ SPAIN and of MADRID province, central SPAIN, and the focus of its own autonomous region, on the MANZANARES RIVER; 40°25′N 03°43′W. The newest of the great Spanish cities, it lacks the traditions of the ancient Castilian and Andalusian towns. Lying on a vast open plateau, it is subject to extremes of temperature; the daily variation is sometimes 40°F/22°C. Almost in the exact geographic center of Spain; the country's chief transportation and administrative center. Its commercial and industrial life developed very rapidly after the 1890s and is rivaled in Spain only by that of BARCELONA. Besides its many manufacturing industries, Madrid is an important commercial and financial center, and is foremost as a banking, distribution, education, printing, publishing, tourism, and motion-picture center. Many corporate headquarters are located here. An archepiscopal see, Madrid also has two universities, one of which was transferred from ALCALÁ DE HENARES in 1836.

Madrid was first mentioned in the 10th century as a Moorish fortress. Alfonso VI of Castile drove out the Moors in 1083. The Cortes of Castile met here several times, and Ferdinand and Isabella as well as Emperor Charles V often resided here, but Madrid became the capital of Spain only in 1561, in the reign of Philip II. The city developed slowly at first, but it expanded rapidly in the 18th century under the Bourbon kings (especially Charles III). At the beginning of the Peninsular War a popular uprising against the French

took place here on May 2, 1808, and a fierce battle was fought in the Puerta del Sol, the city's central square. In reprisal, hundreds of citizens were shot at night along the Prado promenade. The events of that day were immortalized by two of Goya's most celebrated paintings, both in the Museo del Prado. Madrid again played a heroic role in the Spanish civil war (1936–1939), when, under the command of General José Miaja, it resisted twenty-nine months of siege by the Insurgents, suffering several bombardments and air attacks and surrendering, thus ending the war, only late in March 1939.

The general aspect of Madrid is modern, with tree-lined boulevards and fashionable shopping areas, but the old quarters have picturesque winding streets. Among the many landmarks are the huge royal palace; the Buen Retiro park, opened in 1631; and the imposing 19th-century building containing the national library (founded 1712), the national archives, a museum of Spanish modern art, and an archaeological museum. The Museo del Prado houses one of the finest art collections in the world. Also noteworthy is the modern Ciudad Universitaria [University City]. Barajas, the largest airport in Spain (most air traffic), is nearby.

Madrid (MAD-rid), city (2000 population 2,264), BOONE county, central IOWA, near DES MOINES RIVER, 14 mi/23 km SSE of BOONE; 41°52′N 93°49′W. In agricultural and manufacturing area. Plotted 1852, incorporated 1883.

Madrid (mah-DREED), town, ⊙ Madrid municipio, CUNDINAMARCA department, central COLOMBIA, in Cordillera ORIENTAL, on highway and railroad, 15 mi/24 km NW of BOGOTÁ; 04°44′N 74°16′W. Elevation 7,759 ft/2,364 m. In agricultural region (cereals, potatoes; fruit; livestock). Sometimes called Serrezuela.

Madrid, town, Houston co., SE Alabama, 14 mi/23 km S of Dothan, near Florida state line. Named by J. B. Dell for the capital of Spain. Inc. in 1911.

Madrid, town, FRANKLIN county, W central MAINE, on branch of SANDY RIVER, and 21 mi/34 km NW of FARMINGTON; 44°53′N 70°25′W. Settled c.1807, incorporated 1836.

Madrid, village (2006 population 250), PERKINS county, SW central NEBRASKA, 10 mi/16 km E of GRANT; 40°51′N 101°32′W. Grain, beans, sunflower seeds.

Madrid, unincorporated village, SANTA FE county, N central NEW MEXICO, 30 mi/48 km SSW of SANTA FE. Elevation 6,020 ft/1,835 m. Cattle, sheep; corn, grain, alfalfa. Several Pueblo villages, ruins of Paako Pueblo nearby.

Madrid, village, ST. LAWRENCE county, N NEW YORK, on GRASS RIVER, and 17 mi/27 km ENE of OGDENSBURG; 44°46′N 75°07′W.

Madridejos (mah-dree-DAI-hos), town, TOLEDO province, central SPAIN, in CASTILE-LA MANCHA, on small Amarguillo River, 40 mi/64 km S of ARANJUEZ; 39°28′N 03°32′W. Communications center. Town exports and processes agricultural produce of fertile region (olives, grapes, grain, potatoes, cereals, saffron); sheep raising. Alcohol and liquor distilling, wine making, olive-oil extracting, cheese processing; manufacturing of tiles and plaster.

Madrigal de la Vera (mah-dree-GAHL dai lah VAI-rah), town, CÁCERES province, W SPAIN, 31 mi/50 km NW of TALAVERA DE LA REINA; 40°09′N 05°22′W. Olive oil, cherries, chestnuts, pepper.

Madrigalejo (mah-dree-gah-LAI-ho), town, CÁCERES province, W SPAIN, 30 mi/48 km SSE of TRUJILLO; 39°09′N 05°37′W. Olive-oil processing, flour milling; cereals, vegetables, flax. Ferdinand V died here (1516).

Madrigueras (mah-dree-GAI-rahs), town, ALBACETE province, SE central SPAIN, 17 mi/27 km NNE of ALBACETE; 39°14′N 01°48′W. Manufacturing of scales and pruning shears; alcohol and brandy distilling, olive-oil and peanut-oil processing. Wine, saffron, cereals.

Mad River, c.95 mi/153 km long, N CALIFORNIA; rises in SE TRINITY county c.45 mi/72 km SW of REDDING; flows NW to the PACIFIC OCEAN 10 mi/16 km N of Eureka and NW of Arcata.

Mad River, c.15 mi/24 km long, GRAFTON county, central NEW HAMPSHIRE; rises in WHITE MOUNTAINS E of WOODSTOCK; flows SW to PEMIGEWASSET RIVER near CAMPTON.

Mad River, c.60 mi/97 km long, W OHIO; rises in LOGAN county; flows S and SW, past SPRINGFIELD, to GREAT MIAMI RIVER at DAYTON; 39°45'N 84°11'W.

Mad River, c.25 mi/40 km long, in W central VERMONT; rises S of Warren, in GREEN MOUNTAINS; flows NE to WINOOSKI River W of MONTPELIER. Winter-sports area (skiing at Mad River Glen) near WAITSFIELD.

Mad River Glen, ski area, WASHINGTON CO., central VERMONT, in FAYSTON, and 10 mi/16 km WSW of MONTPELIER.

Madríz (mah-DREES), department (□ 530 sq mi/1,378 sq km), NW NICARAGUA, on HONDURAS border; ⊙ SOMOTO; 13°30'N 86°25'W. Drained by COCO, ESTELÍ, and Yali rivers. Varied agriculture: coffee on mountain slopes (TELPANECA), grain and livestock (SW), tobacco, sugarcane, rice (S). Manufacturing (hats, mats, hammocks). Exports dairy products. Served by INTER-AMERICAN HIGHWAY. Main centers are SOMOTO, Telpaneca. Formed 1935 out of NUEVA SEGOVIA.

Madron (MAD-ruhn), village (2001 population 1,533), W CORNWALL, SW ENGLAND, 2 mi/3.2 km NW of PENZANCE; 50°08'N 05°33'W. Former agricultural market. Church dates from 13th century. Site of closed tin mine. Prehistoric monuments nearby.

Madroñera (mah-dron-YAI-rah), town, CÁCERES province, W SPAIN, 7 mi/11.3 km ESE of TRUJILLO; 39°26'N 05°46'W. Tanning, meat and cheese processing; agricultural trade (olive oil, cereals, wine; livestock).

Madruga (mah-DROO-gah), town, LA HABANA province, W CUBA, on Central Highway, on railroad, and 34 mi/55 km ESE of HAVANA; 22°53'N 81°52'W. Spa and agricultural center (sugarcane, tobacco, coffee). The Boris Luís Santa Coloma sugar central is just SW.

Madrushkent, TAJIKISTAN: see MATCHA.

Madryn, ARGENTINA: see PUERTO MADRYN.

Madsen (MAD-sen), community, NW ONTARIO, E central CANADA, included in town of RED LAKE; 50°58'N 93°55'W.

Maduda (muh-DOO-duh), village, BAS-CONGO province, W CONGO, 72 mi/116 km NNE of BOMA; 04°55'S 13°06'E. Elev. 1,059 ft/322 m.

Madugula (MUH-doo-goo-lah), town, VISHAKHAPATNAM district, ANDHRA PRADESH state, S INDIA, 20 mi/32 km NW of ANAKAPALLE; 17°55'N 82°48'E. Oilseeds, rice, tobacco. Timber (sal, teak), myrobalan, lac, coffee in EASTERN GHATS (W).

Madukani (mah-doo-KAH-nee), village, N TANZANIA, 65 mi/105 km NW of ARUSHA, SE of Lake MANYARA, W of Lake Burungi; 03°53'S 35°48'E. TARANGIRE NATIONAL PARK to SE. Corn, wheat; cattle, goats, sheep.

Madü Lake, POLAND: see MIEDWIE LAKE.

Madumabisa (mah-DOO-mah-BEE-sah), town, MATABELELAND NORTH province, W ZIMBABWE, 7 mi/11.3 km W of HWANGE, on DEKA RIVER, on railroad; 18°21'S 26°23'E. Goba Sulphur Spring to SW. Coal mining to S. Livestock; grain.

Maduo (MAH-DWO), county (□ 9,750 sq mi/25,350 sq km), QINGHAI province, CHINA, at the head of the HUANG HE (Yellow River) basin (the first 186 mi/300 km of the Huang He is here). Elevation 13,123 ft/4,000 m. In addition to the Huang He and its tributary system, there are more than 4,000 lakes here. Among them are the only two natural lakes on the Huang He main stream: the Zhalinghu and Erlinghu. The annual average temperature is 24.8°F/−4°C. Pas- and constitutes more than 60% of the country's total area, and animal herding accounts for 66% of the

country's total income. Huangheyan on the Huang He main stream was a major trading stop and port station linking TIBET and China proper in ancient times. Today a highway bridge has replaced ferries for transportation across the river.

Madura (mah-DOO-rah), island (□ 2,113 sq mi/5,473 sq km; 1990 population 3,015,124), part of East Java province, INDONESIA, near the NE coast of JAVA, from which it is separated by Madura Strait; 07°25'S 113°25'E. Divided (along with offshore islets) into four districts; the principal cities and district capitals are PAMEKASAN, SUMENEP, SAMPANG, and BANGKALAN. Principal products are salt, obtained from pans along the coast, and fish. Cattle are extensively raised, and bull races are held annually. Densely populated. The generally chalky soil limits agriculture, and much food has to be imported. Dominated by the rulers of Java in the eleventh–eighteenth centuries. Madurese formed an important auxiliary police force under the Dutch colonial government. Made a part of Indonesia in 1949. Also spelled Madoera.

Madura, town, WESTERN AUSTRALIA state, W AUSTRALIA, 780 mi/1,255 km from PERTH; 31°58'S 127°00'E. Roadhouse, other facilities on road crossing Nullarbor Plain. Mullamullang Caves to NW.

Madurai (muh-duh-REI), district (□ 2,532 sq mi/6,583.2 sq km), TAMIL NADU state, S INDIA; ⊙ Madurai. Bordered W by WESTERN GHATS; mainly lowland (drained by VAIGAI RIVER) except for SIRUMALAI HILLS and spurs of Western Ghats (N) and Palni and VARUSHANAD hills (W), which enclose KAMBAM VALLEY. Includes black-soil cotton tract (S) and alluvial soils along tributaries of AMARAVATI RIVER (N). Mainly agriculture (cotton, grain, sesame); coffee, tea, cardamom plantations in hills; rice, tobacco, coconut and date plantations in hills; rice, tobacco, sugarcane in SE area (irrigated by channels of PERIYAR LAKE project). Teak in the hills. Cotton milling (Madurai), tobacco and cigar manufacturing (DINDIGUL), silk-weaving and dyeing industries, tanning. Health resort and solar observatory at KODAIKANAL. Architectural landmarks at Madurai. Dindigul area passed from Mysore sultans to English in 1792, rest of district from nawabs of ARCOT in 1801. Also Mathurai.

Madurai (muh-duh-REI), city (2001 population 1,203,095), ⊙ Madurai district, TAMIL NADU state, S INDIA, on the VAIGAI RIVER; 09°56'N 78°07'E. Known as the "city of festivals and temples" and is the second largest city in Tamil Nadu state. The Meenakshi temple (rebuilt 16th–17th centuries), which has 1,000 carved pillars, is especially famous. Educational and cultural center (a university town), and a market for tea, coffee, and cardamom. Important industries are the weaving and dyeing of silk and muslin cloth; also the center of S Indian cotton manufacturing. Was Pandya kingdom capital (5th century B.C.E.–11th century C.E.). In the 14th century captured by Muslim invaders, who held it until 1378, when it became part of the Hindu VIJAYANAGAR kingdom. From c.1550 until 1736, it was capital of the Nayak kingdom. The Carnatic Nawabs then gained control and in 1801 ceded it to the BRITISH. (For later history, see India.) THE NAYAK PALACE (17TH CENTURY) IS A NOTABLE BUILDING HERE.

Madurantakam (muh-doo-RAHN-tuhk-uhm), town, CHENGALPATTU-M.G.R. district, TAMIL NADU state, S INDIA, 40 mi/64 km SW of CHENNAI (Madras); 12°31'N 79°54'E. Agricultural trade (rice, oilseeds, coconuts). Megalithic remains nearby.

Maduru Oya, reservoir, EASTERN PROVINCE, SRI LANKA. Formed by 140-ft/43-m-high dam across the Maduru Oya River. Much downstream development and settlement activity has been postponed due to continuing civil unrest. Created 1982.

Madwar, El (muhd-WAHR, el), village, N JORDAN, 3.5 mi/5.6 km E of GERASA. Grain (wheat), fruit. Also spelled El Medwar and El Midwar.

Madzhalis (mah-jah-LEES), village (2005 population 6,320), SE DAGESTAN REPUBLIC, NE CAUCASUS, SE European Russia, on highway, 24 mi/39 km WNW of DERBENT; 42°07'N 47°50'E. Elevation 1,811 ft/551 m. Has a mosque. Population mostly Kaitaks and Kumyks, two of the mountain tribes of Dagestan.

Madzharovo (mahd-ZHAH-ro-vo), city, HASKOVO oblast, MADZHAROVO obshtina, BULGARIA; 41°38'N 25°52'E. Lead and zinc.

Madzia, town, POOL region, S Congo Republic, 35 mi/56 km W of BRAZZAVILLE; 04°11'S 14°48'E.

Madziwa, village, MASHONALAND CENTRAL province, NE central ZIMBABWE, 60 mi/97 km NE of HARARE, on Mufurudzi River; 17°05'S 31°37'E. Copper and nickel mining. Livestock; grain, coffee, tea.

Madziwadzido, village, MASHONALAND WEST province, NW ZIMBABWE, on UME RIVER, NE of CHIZARIRA NATIONAL PARK, 48 mi/77 km W of GOKWE; 17°38'S 28°31'E. Cattle, sheep, goats; grain, cotton.

Maeander, TURKEY: see BÜYÜK MENDERES.

Maeatinga (MEIN-ah-CHEEN-gah), city, S central BAHIA state, BRAZIL, 93 mi/150 km WNW of VITÓRIA DA CONQUISTA; 14°44'S 41°38'W.

Maebaru (mah-e-BAH-roo), city, FUKUOKA prefecture, N KYUSHU, SW JAPAN, 7 mi/11 km W of FUKUOKA; 33°33'N 130°11'E. Strawberries.

Maebashi (mah-E-bah-shee), city (2005 population 318,584), ⊙ GUMMA prefecture, central HONSHU, N central JAPAN, on the TONE RIVER; 36°23'N 139°03'E. Cucumbers; pigs; dairying. Silk cocoons. Formerly the castle town of the Matsudaira clan.

Mae Chan (MAH JUHN), village and district center, CHIANG RAI province, N THAILAND, on LAMPANG-KENGTUNG (MYANMAR) highway, and 16 mi/26 km NNE of CHIANG RAI; 20°09'N 99°52'E. Road junction; major trading center for Akha and Yao hill people. Adjoining village of Ban Kasa was known as CHIANG SAEN in 1930s.

Máe d'Água (MEI DAH-gwah), town, central PARAÍBA state, BRAZIL, 15 mi/24 km SW of SÃO JOSÉ DO BONFIM; 07°16'S 37°26'W.

Mãe do Rio (MEIN do REE-o), city (2007 population 27,619), NE PARÁ state, BRAZIL, 18 mi/29 km N of Vila Aurora; 01°44'S 47°35'W.

Mae Hong Son (MA HAWNG SAWN), province (□ 5,906 sq mi/15,355.6 sq km), NW THAILAND, on MYANMAR (KAYAH, KARENNI, and SHAN states) border; ⊙ Mae Hae Song; 18°45'N 97°55'E. A number of hill tribes live here. The provincial capital still bears traces of Burmese influence in some of its temple architecture.

Mae Hong Son (MA HAWNG SAWN), town, ⊙ MAE HONG SON province, northwesternmost THAILAND, near MYANMAR border, in DAEN LAO RANGE, on major border smuggling routes, 80 mi/129 km NW of CHIANG MAI, and 573 mi/922 km NNW of BANGKOK; 19°16'N 97°56'E. Opium trading; teak-lumbering center; airport. Lead and tungsten deposits. Sometimes spelled Meh Hongsuen.

Mae Klong, THAILAND: see SAMUT SONGKHRAM.

Mae Klong River (MA GLAWNG), 250 mi/402 km long, W central THAILAND; rises near MYANMAR border in TANEN TAUNGGYI RANGE at 16°00'N; flows S and SE in forested valley past KANCHANABURI (confluence with KHWAE NOI RIVER) and RATCHABURI, to GULF OF THAILAND at SAMUT SONGKHRAM. Navigable by shallow-draught vessels below Kanchanaburi. Spanned by railroad bridge near Kanchanaburi. Two important dams create multipurpose reservoirs. Linked by irrigation and navigation canals with THA CHIN and CHAO PHRAYA rivers. Sometimes called Khwae Yai [Thai=greater Khwae] above Kanchanaburi; also spelled Meklong.

Maella (mah-E-lyah), town, ZARAGOZA province, NE SPAIN, 12 mi/19 km SE of CASPE; 41°08'N 00°09'E. Olive-oil center; sawmilling; cereals; wine; almonds.

Maello (mah-E-lyo), town, ÁVILA province, central SPAIN, near SEGOVIA province border, 14 mi/23 km NE of ÁVILA; 40°49′N 04°30′W. Cereals, grapes; potteries.

Mae Nam, [Thai=river], name commonly applied to CHAO PHRAYA, THAILAND. For other Thai rivers beginning thus, see under following part of the name.

Mae Nam Chao Phraya, THAILAND: see CHAO PHRAYA.

Maen Du, Wales: see DINAS MAWDDWY.

Maenorbŷr, Wales: see MANORBIER.

Maentwrog (mein-TOO-rog), village (2001 population 585), GWYNEDD, NW Wales, on DWYRYD RIVER, and 4 mi/6.4 km SW of BLAENAU-FFESTINIOG; 52°56′N 03°59′W. Site of hydroelectric station.

Maeotis, Palus, RUSSIA: see AZOV, SEA OF.

Mae Sai (MA SEI), village and district center, CHIANG RAI province, N THAILAND, customs post on MYANMAR border, 35 mi/56 km N of CHIANG RAI, on LAMPANG-KENGTUNG highway; 20°26′N 99°53′E. Northernmost village in THAILAND.

Mae Sariang (MA suh-ree-AHNG), village and district center, MAE HONG SON province, NW THAILAND, near MYANMAR border, in valley E of THANON THONG CHAI RANGE, 85 mi/137 km SW of CHIANG MAI; 18°09′N 97°56′E. Teak center. Airport. Sometimes spelled Mae Sarieng.

Maesawa (mah-e-SAH-wah), town, Isawa county, IWATE prefecture, N HONSHU, NE JAPAN, on the KI-TAKAMI RIVER, 43 mi/70 km S of MORIOKA; 39°02′N 141°07′E. Beef cattle.

Maeser, village (2000 population 2,855), UINTAH county, NE UTAH, 10 mi/16 km NW of VERNAL, in S foothills of UINTA MOUNTAINS (in Ashley National Forest); 40°28′N 109°34′W. Oil and natural gas. Uintah and Ouray Indian Reservation to W.

Mae Sot (MA SAWD), village, TAK province, W THAILAND, on THAUNGYIN RIVER (MYANMAR border), opposite MYAWADDY (MYANMAR); 40 mi/64 km WSW of TAK, and 60 mi/97 km ENE of MOULMEIN (MYANMAR); 16°43′N 98°34′E. Rice. Airport. Sometimes spelled MEH SOT.

Maesteg (meis-TAIG), town (2001 population 17,859), BRIDGEND, S Wales, 6 mi/9.7 km E of PORT TALBOT; 51°37′N 03°40′W. Was a coal-mining town; now manufacturing of clothing, cosmetics, insulation, and rubber.

Maestra, Sierra, CUBA: see SIERRA MAESTRA.

Maestrazgo (mah-es-TRAHTH-go), mountainous district of Castellón de la Plana and TERUEL provinces, E SPAIN. Has forests and pastures (cattle raising) in higher W part and fertile valleys (cereals and wine) in lower E part. The passes through these mountains were ancient routes between the MEDITERRANEAN SEA and the heart of CASTILE. Served on Christian-held frontier between Christian and Muslim Spain. Was important in Middle Ages under lordship of military order of Montesa. Fortified villages on rocky heights were organized by warrior-priest knights led by El Maestre, for whom the region was named. Chief town, MORELLA.

Maestre de Campo Island (mah-ES-trah de KAHM-po), ROMBLON province, PHILIPPINES, in TABLAS STRAIT, just off E coast of MINDORO island; 3.5 mi/5.6 km in diameter; 12°56′N 121°42′E. Rice, coconuts.

Maesycrugiau, Wales: see LLANFIHANGEL AR ARTH.

Maetsue (mah-ETS-e), village, Hita county, OITA prefecture, E KYUSHU, SW JAPAN, 40 mi/65 km W of OITA; 33°12′N 130°55′E.

Maevatanana (mah-AI-vah-tah-NAHN), town, MA-HAJANGA province, N central MADAGASCAR, on highway, on IKOPA RIVER (main tributary of BETSI-BOKA RIVER), 90 mi/145 km SE of MAHAJANGA; 16°57′N 46°49′E. Rice center. Cattle raising. Hospital.

Maewo (MEI-WO), long volcanic island (□ 135 sq mi/351 sq km), VANUATU, SW PACIFIC OCEAN, 65 mi/105 km E of ESPÍRITU SANTO; c.29 mi/46 km long, 4 mi/6 km wide; 15°10′S 168°10′E. Central mountain range. Formerly Aurora Island.

Mafamude (mah-fah-MOOD), town, PÔRTO district, N PORTUGAL, S suburb of OPORTO, on left bank of DUERO RIVER, adjoining VILA NOVA DE GAIA (W); 38°56′N 09°20′W. Textile milling (carpets, knitwear); manufacturing of shoes and soap.

Mafeteng (mahf-TENG), [Lesotho=place of plump, unmarried women], district (□ 818 sq mi/2,126.8 sq km; 2001 population 238,946), W LESOTHO, ⊙ Mafeteng, on SOUTH AFRICAN (W and NW; CALEDON RIVER on NW) border. THABA PUTSUA range in E. Sheep, goats, cattle, hogs, chickens; corn, sorghum, vegetables, peaches. Clay mining.

Mafeteng (mahf-TENG) [Lesotho=place of plump, unmarried women], town (2004 population 36,000), ⊙ MAFETENG district, SW LESOTHO, 38 mi/61 km SSE of MASERU, on main N-S road; 29°49′S 27°15′E. Elevation 6,450 ft/1,966 m. SOUTH AFRICA border 5 mi/8 km to W; Tsa-Kholo Lake Reservoir 10 mi/16 km to NNW. Rock paintings archaeological site to E. Mafeteng Industrial Estate is here; manufacturing (pharmaceuticals, handicrafts). Trading and administrative center with an aerodome. Sheep, goats, cattle, hogs, chickens; corn, sorghum, vegetables, wheat, peaches.

Maffersdorf, CZECH REPUBLIC: see VRATISLAVICE NAD NISOU.

Maffo (MAH-fo), town, SANTIAGO DE CUBA province, E CUBA, at N slopes of the SIERRA MAESTRA, 32 mi/51 km WNW of SANTIAGO DE CUBA; 21°18′N 76°15′W. Fruit and sugar region.

Maffra (MA-fruh), town, S VICTORIA, SE AUSTRALIA, 110 mi/177 km E of MELBOURNE; 37°58′S 146°59′E. Railroad junction in livestock and irrigated agricultural region; dairy products, beet sugar; timber.

Mafia Channel (mah-FEE-ah), strait, INDIAN OCEAN, PWANI region, E TANZANIA, between AFRICA mainland (W) and MAFIA ISLAND (E); delta of RUFIJI RIVER narrows channel to 10 mi/16 km in S, widens to 40 mi/64 km in N. Kama and Kwale islands in NW.

Mafia Island (mah-FEE-ah), PWANI region, E TANZANIA, in INDIAN OCEAN, across MAFIA CHANNEL from mainland, 70 mi/113 km SSW of DAR ES SALAAM; 30 mi/48 km long, 10 mi/16 km wide. S end of island 10 mi/16 km E of RUFIJI RIVER delta. Town of KILINDONI is on SW shore (airport). Sport fishing (blue marlin, sailfish, yellow fin tuna). Copra, bananas; fish. Limestone deposits. Transferred from ZANZIBAR protectorate in 1922.

Mafikeng, town, ⊙ NORTH-WEST province, N central SOUTH AFRICA, in what was the South Africa homeland of BOPHUTHATSWANA, near the border of BOTSWANA, on the MOLOPO RIVER; 25°52′S 25°37′E. Elevation 4,198 ft/1,280 m. Twin town to much larger MMABATHO which was the Bophuthatswana capital. Airport and railroad junction (important depot). Market for the surrounding cattle-raising and dairy-farming area. Founded in 1885 on the site of an African settlement. In the South African War (1899–1902) the British garrison here, under Lord Baden-Powell, withstood a Boer (Afrikaner) siege for 217 days; the fort is now a national monument. During this siege Baden-Powell conceived the idea of the Boy Scouts. Was the extraterritorial capital of the Bechuanaland protectorate until it became independent as Botswana in 1965.

Máfil (MAH-feel), village, ⊙ Máfil comuna, VALDIVIA province, LOS LAGOS region, S central CHILE, on railroad, and PAN-AMERICAN HIGHWAY, 20 mi/32 km NE of VALDIVIA; 39°39′S 72°57′W. Coal-mining center.

Mafingi (mah-FEEN-gee), village, MOROGORO region, central TANZANIA, 70 mi/113 km SSE of IRINGA, near Kilimbero River; 08°35′S 36°21′E. Corn, wheat; goats, sheep.

Mafolie (mah-fo-LEE), village, St. Thomas Island, U.S. VIRGIN Islands, 75 mi/1.2 km N of CHARLOTTE AMA-LIE. Has a celebrated panorama over the whole island. Brazilian astronomers established here (1882) a station to observe transit of Venus.

Mafra (MAH-frah), city (2007 population 51,014), N SANTA CATARINA state, BRAZIL, on the RIO NEGRO (PARANÁ border) opposite city of RIO NEGRO, and 65 mi/105 km W of JOINVILLE; 26°07′S 49°49′W. Railroad junction (spur to CURITIBA); maté processing, flour milling, lumbering. Ships livestock (especially hogs).

Mafra (MAH-frah), town, LISBOA district, W central PORTUGAL, in ESTREMADURA province; 38°56′N 09°20′W. Noted for pottery manufacturing and its huge 18th-century palace and monastery, built by John V in imitation of Spain's Escorial. Rectangular in shape, the edifice is surmounted by two towers and by a central dome. It contains a church, a fine library, and extensive royal quarters. Game reserve on the outskirts of town.

Mafraq, El (MUHF-ruhk, el), town (2004 population 47,764), NW JORDAN, on HEJAZ RAILROAD, 30 mi/48 km NE of AMMAN. Elevation 2,280 ft/695 m. Important road junction; on main route from Jordan to BAGHDAD (IRAQ), and from Amman to DAMASCUS (SYRIA). Major air base. Wheat, barley. The former pipeline from N Iraq to HAIFA (ISRAEL) passed on the outskirts; now terminates here (the section from Mafraq to Haifa stopped functioning in 1948 and was later dismantled).

Mafungabusi Plateau (mah-foo-eng-gah-boo-see), MATABELELAND NORTH and MIDLANDS provinces, NW central ZIMBABWE, 120 mi/193 km NNW of BU-LAWAYO, and S of GOKWE. Average elevation higher than 4,000 ft/1,219 m. Includes the Mafungabusi Forest land. An outlier of Zimbabwe's central plateau, or high veld. The smaller Charama Plateau lies to NW, also exceeds 4,000 ft/1,219 m.

Maga (MAH-gah), town (2001 population 14,000), Far North province, CAMEROON, 43 mi/69 km NE of MAROUA; 10°50′N 14°55′E.

Magacela (mah-gah-THAI-lah), town, BADAJOZ province, SW SPAIN, 33 mi/53 km E of MÉRIDA; 38°54′N 05°44′W. Cereals, vegetables; sheep. Limekilns; pottery, tiles.

Magadalena Mixtepec (mahg-dah-LAI-nah MEESH-te-pek), town, in central OAXACA, MEXICO, 16 mi/25 km SW of OAXACA DE JUÁREZ; 16°54′N 96°54′W. Zapotec community. Farming (corn, beans). No access by road.

Magadan (muh-gah-DAHN), oblast (□ 179,249 sq mi/466,047.4 sq km; 2006 population 165,175), E RUSSIAN FAR EAST, along the N shore of the Sea of OKHOTSK, extending from the TAYGONOS PENINSULA on the SHELEKHOV GULF N to the OMOLON RIVER, and including the upper KOLYMA River basin; ⊙ MAGADAN. Kolyma Highlands in SW and along E border; also CHERSKIY Mountains (W). Mostly tundra on permafrost soil in rugged mountains broken by vast lowlands, especially in the S; swampy tayga along the coast. Cold marine climate on the coast; harsher inland. Mild summers; winter can last 8 months. Growing season barely 2 months. Aside from the Kolyma, most rivers are mountain streams that flow unevenly and flood rapidly; significant energy source. Local fauna include squirrel, blue hare, fox, bear, wolverine, weasel, reindeer, elk, walrus, seal, whale. No railroad; river (Kolyma) and air (local and international) transport networks well developed. Main road runs between Magadan and YAKUTSK (Sakha Republic) via Kolyma Gold Fields and SUSUMAN; also, Kolyma Highway between Gerba and Viliga-Kushka via Omsukchan. Foundation of regional economy is mining, especially of gold, tin, tungsten, and mercury, and nonferrous metallurgy, mainly around the upper Kolyma (also a logging area). Coal mining at Arkagala and Omsukchan. Manufacturing of machines, building materials. Agriculture includes reindeer and livestock and vegetables along Sea of Okhotsk coast and Kolyma basin headwaters, but is dominated by fishing, which contributes 18% of the oblast's total economic output. Population mostly Russian (more than 80% of the total) and Ukrainian

2246 MAGADAN

(close to 10% of the total), with indigenous Koryak, Evenki (or Lamut; largest indigenous group), Yukagir, and Chukchi. About three-quarters of population lives in urban centers in SW; main centers are Magadan and Susuman. Formed in 1953 and is currently divided into seven districts. Chukchi autonomous okrug (to the NE) was part of the oblast until 1992.

Magadan (mah-gah-DAHN), city (2006 population 93,905), ⊙ MAGADAN oblast, E RUSSIAN FAR EAST, on a branch of the Kolyma highway, 4,400 mi/7,081 km E of MOSCOW, and 1,367 mi/2,200 km SE of YAKUTSK; 59°34′N 150°48′E. Elevation 232 ft/70 m. Port on the Nagayeva Bay on the Sea of OKHOTSK, navigable from May to December. Has shipyards (fishing fleet base), machine (for mining, energy generation, ships) works, canning (fish) factories; metalworking. A major international airport (Sokol) is 20 mi/32 km N of the city center. The Kolyma highway leads from here N to the gold-mining region on the upper KOLYMA River. During the Stalin era, it was a major transit center for prisoners being sent to labor camps, and was officially a closed city until the 1980s, for propaganda purposes. Founded in 1933; made city in 1939.

Magadan Preserve (mah-gah-DAHN), wildlife refuge (□ 344 sq mi/894.4 sq km), MAGADAN oblast, extreme NE SIBERIA, RUSSIAN FAR EAST, near MAGADAN city. Established in 1982.

Magadha (MUH-gah-dah), ancient Indian kingdom, situated within the area of the modern state of BIHAR, E INDIA. Its capital was PATALIPUTRA (now PATNA). Rose to prominence in the mid-7th century B.C.E. and rapidly extended its frontiers, especially under the rule of Bimbisara (c.540 B.C.E.–c.490 B.C.E.). Fell (c.325 B.C.E.) to Chandragupta, who made it the nucleus of the Mauryan empire. After a period of obscurity, it recovered importance in the 4th century C.E. as the power base of the Gupta dynasty. Buddhism and Jainism first developed here, and the Buddha used the Magadhi dialect of Sanskrit.

Magadi (MUH-gah-dee), town, BANGALORE district, KARNATAKA state, S INDIA, 25 mi/40 km W of BANGALORE; 12°58′N 77°14′E. Trades in millet, rice; handicraft lacquerware and brass ware. Bamboo, sandalwood in nearby hills.

Magadi (mah-GAH-dee), town, ⊙ Kajiado district, RIFT VALLEY province, S KENYA, in GREAT RIFT VALLEY, on E shore of Lake MAGADI, 50 mi/80 km SW of NAIROBI; 01°55′S 36°17′E. Elevation 1,978 ft/603 m. Soda- and salt-mining center. Railroad, airfield.

Magadi, Lake (mah-GAH-dee), RIFT VALLEY province, S KENYA, in the GREAT RIFT VALLEY; c.20 mi/32 km long and 2 mi/3.2 km wide. Formed and constantly resupplied by volcanic springs, the lake has a thick crust of carbonate of soda. A floating dredge removes the crust and then pumps to refineries, where it is processed into soda ash (used in glassmaking).

Magaga (mah-GAH-gah), village, DODOMA region, central TANZANIA, 35 mi/56 km SW of DODOMA; 06°29′S 35°26′E. Sheep, goats; corn, wheat.

Magaguadavic River (ma-guh-DAI-vik), 80 mi/129 km long, SW NEW BRUNSWICK, E CANADA; rises WSW of FREDERICTON; flows S, through Lake Magaguadavic (□ 11 sq mi/28 sq km), past ST. GEORGE (hydroelectric station), to Passamaquoddy Bay 5 mi/8 km W of St. George.

Magal (mah-GAHL), kibbutz, ISRAEL, 8 mi/12.9 km SE of HADERA; 32°23′N 35°02′E. Elevation 157 ft/47 m. Mixed farming and production of irrigation systems. Founded 1953.

Magalang (mah-GAH-lahng), town, PAMPANGA province, central LUZON, PHILIPPINES, 13 mi/21 km NNW of SAN FERNANDO; 15°14′N 120°41′E. Terminus of spur of MANILA–San Fernando railroad. Agricultural center (sugarcane, rice).

Magalhães Baraba (MAH-gahl-yeins BAH-rah-bah), town, N central PARÁ state, BRAZIL, near Atlantic coast, 16 mi/26 km W of MARACANÁ; 00°47′S 47°36′W.

Magalhães de Almeida (mah-gahl-yeins di ahl-mai-dah), city (2007 population 14,272), NE MARANHÃO state, BRAZIL, on left bank of Rio PARNAÍBA (PIAUÍ state border), opposite LUZILÂNDIA, and 40 mi/64 km SW of PARNAÍBA; 03°20′S 42°17′W. Rice and tobacco growing.

Magaliesberg, Afrikaans *Magaliesberge*, mountain range, GAUTENG province, SOUTH AFRICA, extends c.100 mi/161 km E to NE part of WITWATERSRAND; rises to 6,078 ft/1,853 m on NOOITGEDACHT, 20 mi/32 km NW of KRUGERSDORP; part of the N-facing cuestas making up the rim of the Witwatersrand.

Magaliesburg, town, GAUTENG province, SOUTH AFRICA, on W WITWATERSRAND, 30 mi/48 km NW of KRUGERSDORP; 26°00′S 27°31′E. Elevation 4,682 ft/1,427 m. Railroad junction; tobacco-growing and processing center.

Ma'galim (mah-gah-LEEM), Jewish religious community, S ISRAEL, NW of BEERSHEBA, in NEGEV; 31°24′N 34°36′E. Elevation 462 ft/140 m. Center for several rural areas.

Magallanes (mah-gah-YAH-nais), province (□ 52,285 sq mi/135,941 sq km) of MAGALLANES Y LA ANTARTICA CHILENA region, S CHILE, extending to CAPE HORN; ⊙ PUNTA ARENAS; 52°55′S 71°05′W. Includes a narrow strip of mainland between ARGENTINA and the PACIFIC OCEAN, and most of the islands of TIERRA DEL FUEGO (S of the STRAIT OF MAGELLAN), as well as, N of the strait, numerous archipelagoes and islands (e.g., WELLINGTON, MADRE DE DIOS, HANOVER, ADELAIDE, RIESCO) of the partly submerged ANDES Mountains, which form a labyrinth of channels and fjords. Among its numerous peninsulas are BRUNSWICK and MUÑOZ GAMERO. Largely covered by snowcapped mountains and glaciers, it is a bleak area and has a cold, foggy climate, with heavy rainfall (up to 200 in/508 cm annually) and frequent snowstorms. Inhabited largely by small Native American tribes, who engage mostly in sheep raising and lumbering. Some agriculture (cereals, vegetables) on small scale; fishing. At Punta Arenas are food processing, sawmilling. Mineral resources include coal (on Brunswick Peninsula near Punta Arenas, and on Riesco Island near ELENA), petroleum (at SPRINGHILL), gold (mostly in stream beds).

Magallanes (mah-gah-LYAH-nes), town, SORSOGON province, extreme SE LUZON, PHILIPPINES, at S side of entrance to SORSOGON BAY, opposite Bagatao Island (2.5 mi/4 km long, 1 mi/1.6 km wide), 15 mi/24 km SW of SORSOGON; 12°49′N 123°54′E. Agricultural center (abaca, coconuts; rice); fishing.

Magallanes, ARGENTINA: see SAN JULIÁN.

Magallanes, Estrecho de, CHILE and ARGENTINA: see MAGELLAN, STRAIT OF.

Magallanes y la Antartica Chilena (mah-gah-YAH-nais ee lah ahnt-AHR-tee-kah chee-LAI-nah) or **XII region**, administrative division (2005 population 155,962), S CHILE, ⊙ PUNTA ARENAS; 69°00′S 70°00′W. Encompasses the S tip of the island of TIERRA DEL FUEGO, which Chile shares with ARGENTINA. Extreme climate conditions keep this region sparsely populated. Some mineral wealth, fishing, lumbering in a few scattered settlements. Provinces include ÚLTIMA ESPERANZA, MAGALLANES, TIERRA DEL FUEGO, and ANTÁRTICA CHILENA. (Chile claims a portion of the Antarctic continent, which is included in this region). Principal towns are PUERTO NATALES, Punta Arenas, PORVENIR, and PUERTO WILLIAMS. National parks include BERNARDO O'HIGGINS, TORRES DEL PAINE, HERNAUDO DE MAGALLANES, ALBERTO M. DE AGOSTINI, and CABO DE HORNOS.

Magallón (mah-gah-LYON), town, ZARAGOZA province, NE SPAIN, 15 mi/24 km ESE of TARAZONA; 41°50′N 01°27′W. Olive-oil processing, manufacturing of brandy and alcohol. Wine; cereals; sheep in area.

Magalloway (muh-GAL-uh-wai), plantation, OXFORD county, W MAINE, c.30 mi/48 km NW of RUMFORD, near Rangeley Lakes; 44°51′N 70°58′W.

Magalloway Mountain (muh-GA-luh-wai), peak (3,360 ft/1,024 m), COOS county, NEW HAMPSHIRE, in wilderness area, 20 mi/32 km NE of COLEBROOK, 4 mi/6.4 km W of MAINE state line. FIRST CONNECTICUT LAKE to NW. Source of Dead Diamond River.

Magalloway River (muh-GAL-uh-wai), c.15 mi/24 km long, NW MAINE and NE NEW HAMPSHIRE; rises in N OXFORD county (Maine); flows S, through PARMACHENEE LAKE, to AZISCOHOS LAKE (formed by dam), thence generally SW into New Hampshire where, with outlet of UMBAGOG LAKE, it forms ANDROSCOGGIN RIVER.

Magán (mah-GAHN), town, TOLEDO province, central SPAIN, on canal of the TAGUS RIVER, and 9 mi/14.5 km NE of TOLEDO; 39°57′N 03°55′W. Cereals, grapes, olives; olive-oil pressing, cheese processing.

Magangué (mah-gahn-GAI), town, ⊙ Magangué municipio, BOLÍVAR department, N COLOMBIA, river port on the BRAZO DE LOBA, and 120 mi/193 km S of BARRANQUILLA; 09°14′N 74°45′W. Dairying and agricultural center (corn, rice, sugarcane; livestock).

Maganik (MAH-gah-neek), mountain in DINARIC ALPS, central MONTENEGRO, between ZETA and MORACA rivers; highest point, Medjedji Vrh (7,016 ft/2,138 m), is 18 mi/29 km ESE of NIKSIC; 42°44′N 19°16′E.

Maganja Costa District, MOZAMBIQUE: see ZAMBÉZIA.

Maganja da Costa (mah-GAHN-yuh dah KOSH-tuh), village, Zambézia province, central MOZAMBIQUE, 60 mi/97 km NE of the provincial city of QUELIMANE; 17°18′S 37°30′E. Sisal, rice, copra. Also called MAGANJA.

Maganoy, town, ⊙ MAGUINDANAO province, S central MINDANAO, PHILIPPINES, 75 mi/120 km WSW of DAVAO; 06°55′N 124°16′E. Agricultural center (rice).

Magansk (mah-GAHNSK), settlement, S KRASNOYARSK TERRITORY, SE SIBERIA, Russia, in the YENISEY RIVER basin, on local railroad spur, 27 mi/43 km SE of KRASNOYARSK; 55°52′N 93°16′E. Elevation 938 ft/285 m. Sawmilling, timbering.

Magaramkent (mah-gah-rahm-KYENT), village (2005 population 6,870), SE DAGESTAN REPUBLIC, E CAUCASUS, extreme SE European Russia, on the SAMUR RIVER (irrigation), approximately 3 mi/5 km NW of the RUSSIA-AZERBAIJAN border, on road, 32 mi/51 km S of DERBENT; 41°37′N 48°21′E. Elevation 1,102 ft/335 m. Population largely Lezghian, with a sizeable Chechen minority because of influx of refugees from military hostilities in Chechnya.

Magaria (muh-GAHR-ee-uh), town, ZINDER province, S NIGER, near NIGERIA border, 55 mi/89 km S of ZINDER; 13°00′N 08°54′E. Administrative center. Peanuts, millet, and corn; livestock; food-processing industry.

Magaro, RUSSIA: see EL'BRUSSKIY.

Magar Pir, PAKISTAN: see PIR MANGHO.

Magarwara, INDIA: see UNNAO.

Magas Tatra, SLOVAKIA and POLAND: see HIGH TATRAS.

Magatá Lahjar (mah-gah-TAH lah-JAHR), village (2000 population 12,117), BRAKNA administrative region, SW MAURITANIA, on NOUAKCHOTT-NÉMA HIGHWAY, 31 mi/50 km NE of SAINT-LOUIS (SENEGAL). Livestock raising; millet, sorghum. Also spelled Magta' Lahjar.

Magat High Dam, on MAGAT RIVER, ISABELA province, N LUZON, PHILIPPINES, 62 mi/100 km NNE of BAGUIO; 16°48′N 121°30′E. The dam (2.5 mi/4 km long) is key to a major hydropower and irrigation project that will provide power for all of N and part of central Luzon, and irrigate ⅔ of available land in CAGAYAN VALLEY. Began operations in 1982, irrigation now being extended.

Magat River (MAH-gaht), c.120 mi/193 km long, N LUZON, PHILIPPINES; rises in mountains in NUEVA VIZCAYA province 30 mi/48 km SE of BAGUIO; flows generally NE, past BAYOMBONG and SOLANO, to the CAGAYAN RIVER near Hagan.

Magazine, town (2000 population 915), LOGAN county, W ARKANSAS, 37 mi/60 km ESE of FORT SMITH, 7 mi/11.3 km E of BOONEVILLE; 35°08′N 93°48′W. In rich farmland of Petit Jean valley. Manufacturing (apparel). Ozark National Forest to E. MAGAZINE MOUNTAIN is nearby.

Magazine Mountain, W central ARKANSAS, the highest peak (2,753 ft/839 m) in state, c.45 mi/72 km SE of FORT SMITH, in OUACHITA MOUNTAINS, Ozark National Forest, and 10 mi/16 km SSE of PARIS.

Magazzolo River (mah-gah-TSO-lo), 22 mi/35 km long, SW SICILY, ITALY; rises SE of SANTO STEFANO QUISQUINA; flows SW to MEDITERRANEAN Sea 3 mi/5 km NW of CAPE BIANCO.

Magba (MAHG-bah), town (2001 population 14,600), West province, CAMEROON, 65 mi/105 km NE of BAFOUSSAM; 05°58′N 11°15′E.

Magburaka (mahg-boor-AH-kah), town, ⊙ TONKOLILI district, NORTHERN province, central SIERRA LEONE, on ROKEL RIVER, and 12 mi/19 km SSE of MAKENI; 08°43′N 11°57′W. Has two secondary schools. Formerly called MAKUMP.

Magda (MAHG-dah), town (2007 population 3,146), NW SÃO PAULO state, BRAZIL, 19 mi/31 km NNE of ARAÇATUBA; 20°37′S 50°14′W. Coffee growing.

Magdagachi (mahg-dah-GAH-chee), town (2005 population 11,600), SW AMUR oblast, SE SIBERIA, RUSSIAN FAR EAST, approximately 27 mi/43 km N of the Chinese border, on road and the TRANS-SIBERIAN RAILROAD, 80 mi/129 km ESE of SKOVORODINO; 53°27′N 125°48′E. Elevation 1,171 ft/356 m. In lumbering area; railroad shops.

Magdala, ETHIOPIA: see ĀMBA MARIAM.

Magdalena (mahg-dah-LAI-nah), department (□ 20,819 sq mi/54,129.4 sq km), N COLOMBIA, on CARIBBEAN SEA; ⊙ SANTA MARTA; 10°00′N 74°30′W. Located between MAGDALENA River (W) and Cordillera ORIENTAL (E), the latter forming its border with VENEZUELA. It consists of marshy lowlands in W (Magdalena basin), dominated N by great massif Sierra Nevada de SANTA MARTA (N), rising abruptly from the coast to highest peak (18,950 ft/5,776 m) in Colombia, the Pico CRISTÓBAL COLÓN. CÉSAR and RANCHERÍA rivers water its E part. Has generally dry, tropical climate, with rainy season (May–September); temperate, alpine conditions in undeveloped high plateaus of Sierra Nevada. Mineral resources include petroleum from EL DIFÍCIL, alluvial gold in VALLEDUPAR region (E), gold, iron, and coal near RÍOHACHA, marble in CIÉNAGA vicinity. But its main production is its large crop of bananas grown in valley S of Santa Marta, and shipped from that port. Other agricultural products are cotton, corn, rice, beans, sorghum, tobacco, and yucca; coffee in uplands. The dense forests on coastal plains yield fine wood (mahogany, cedar) and medicinal plants. Magdalena has 29 municipios (2004), and its main cities include Ciénaga, FUNDACIÓN, PLATO, and EL BANCO.

Magdalena (mahg-dah-LAI-nah), town (1991 population 6,601), ⊙ Magdalena district (□ 1,288 sq mi/3,348.8 sq km), NE BUENOS AIRES province, ARGENTINA, near the RÍO DE LA PLATA, 26 mi/42 km ESE of LA PLATA; 35°04′S 57°32′W. Cattle- and sheep-raising center; meat-salting industry. Founded 1730. ATALAYA, on coast 3 mi/4.8 km N, is popular beach resort.

Magdalena, town and canton, ⊙ ITÉNEZ province, BENI department, NE BOLIVIA, port on ITONAMAS River, and 105 mi/169 km NNE of TRINIDAD; 13°22′S 64°07′W. Elevation 463 ft/141 m. Rubber-collecting center; cattle. Airport.

Magdalena (mahg-dah-LAI-nah), town, INTIBUCÁ department, SW HONDURAS, 28 mi/46 km S of LA ESPERANZA; 13°52′N 88°18′W. Subsistence agriculture, grain; livestock.

Magdalena (MAHG-dah-LAI-nah), town, JALISCO, W MEXICO, on Lake Magdalena, on railroad, and 45 mi/72 km WNW of GUADALAJARA, on Mexico Highway 15; 20°55′N 103°57′W. Elevation 4,610 ft/1,405 m. Agricultural center (grain, maguey, sugarcane, cotton, vegetables, fruit; livestock).

Magdalena, town, VERACRUZ, E MEXICO, in SIERRA MADRE ORIENTAL, 7 mi/11.3 km SSE of ORIZABA. Coffee.

Magdalena (mahg-dah-LAI-nah), town, CAJAMARCA province, AMAZONAS region, N PERU, near UTCUBAMBA River, and 9 mi/15 km SSE of CHACHAPOYAS; 06°21′S 77°49′W. Sugarcane, fruit.

Magdalena (MAG-duh-le-nuh), town (2006 population 874), SOCORRO county, W central NEW MEXICO, just NW of MAGDALENA MOUNTAINS, 27 mi/43 km WNW of SOCORRO; 34°06′N 107°13′W. Elevation 6,573 ft/2,003 m. Trade and shipping point in livestock area (cattle, sheep); alfalfa, chilies. Lead, zinc deposits nearby. National Radio Observatory and Very Large Array (V.L.A.) radiotelescope facillity to W, at NE end of Plains of San Agustin on three-pronged track system. Parts of Cibola National Forest NW, SE and SW; South Baldy peak (10,283 ft/3,134 m) 9 mi/14.5 km SSE; GALLINAS MOUNTAINS to NW; Alamo Band Navajo Tribe Reservation to NW. Founded 1884.

Magdalena (mahg-dah-LAI-nah), river, c.1,000 mi/1,609 km long, in SW COLOMBIA; rises in the Cordillera CENTRAL; flows N to the CARIBBEAN SEA near BARRANQUILLA; 11°04′N 74°50′W. It flows in a fault-block valley (c.50 mi/80 km wide) through the ANDES Mountains to a broad, swampy, alluvial plain where the CAUCA River, the chief tributary, joins its lower course. A natural and important avenue of communication, linking the interior highlands with the coastal lowlands. Its navigability is hampered by sandbars, rapids, and fluctuating water levels. LA DORADA, c.600 mi/966 km upstream, is the head of navigation. Railroads connect navigable sects. The tropical valley of the Magdalena produces bananas and plantains among other tropical crops, and oil is found here. Coffee is the chief crop along the river's upper course. Rodrigo de Bastidas, the Spanish explorer, discovered (1501) the Magdalena, and since the time of exploration (1536) by Gonzalo Jiménez de Quesada, the Spanish conquistador, the river has profoundly influenced the economic and political life of Colombia.

Magdalena, FRENCH POLYNESIA: see FATU HIVA.

Magdalena, PORTUGAL: see MADALENA.

Magdalena Apasco (mahg-dah-LAI-nah ah-PAHS-ko), town, in central OAXACA, MEXICO, 19 mi/30 km NW of OAXACA DE JUÁREZ, on railroad, and on Mexico Highway 190; 17°14′N 96°48′W. Elevation 5,446 ft/1,660 m. In Etla arm of OAXACA VALLEY irrigated by the Atoyac River. Temperate climate. Agriculture (corn, beans, fruit), woods; cattle raising. Also known as Magdalena Apasco Etla and Magdalena Apazco.

Magdalena Bay (MAHG-dah-LAI-nah), inlet of PACIFIC OCEAN, in SW coast of BAJA CALIFORNIA, NW MEXICO, SSW of CIUDAD CONSTITUCIÓN, sheltered by several islands, including SANTA MARGARITA (S); 17 mi/27 km long NW-SE, 12 mi/19 km wide. Well known as good harbor and fishing ground.

Magdalena Bay (mahg-dah-LAI-nah), Norwegian *Magdalenefjorden*, inlet of the ARCTIC OCEAN, NW West SPITSBERGEN, SVALBARD, NORWAY, 110 mi/177 km NNW of LONGYEAR City, at foot of extensive glacier region; 7 mi/11.3 km long, 2 mi/3.2 km–3 mi/4.8 km wide; 79°33′N 11°00′E. Noted for its magnificent scenery. Here are remains of twelfth-century graves of whaling crews.

Magdalena Contreras, La, MEXICO: see LA MAGDALENA CONTRERAS.

Magdalena de Kino (MAHG-dah-LAI-nah dai KEE-no), city, ⊙ Magdalena municipio, SONORA, NW MEXICO, on MAGDALENA RIVER, and 110 mi/177 km N of HERMOSILLO, 50 mi/80 km S of NOGALES, on railroad, and on Mexico Highways 2 and 15; 30°37′N 111°03′W. Elevation 2,464 ft/751 m. Agricultural center (wheat, fruit, chickpeas, vegetables, cotton) in rich silver- and copper-mining area. Yearly Native American festivals (October) in honor of Saint Francis Xavier draw many pilgrims. Nearby Gold Placers were known to Aztecs.

Magdalena Island (mahg-dah-LAI-nah), off coast of AISÉN DEL GENERAL CARLOS IBAÑEZ DEL CAMPO region, S CHILE, across MORALEDA CHANNEL from CHONOS ARCHIPELAGO; circular, c.40 mi/64 km in diameter. Mountainous, uninhabited area; elevation 5,450 ft/1,661 m.

Magdalena Jaltepec (mahg-dah-LAI-nah HAHL-tai-pek), town, in W OAXACA, MEXICO, 28 mi/45 km NW of OAXACA DE JUÁREZ, 8 mi/12 km S off Mexico Highway 190, on unpaved road; 17°18′N 97°12′W. Elevation 2,625 ft/800 m. Temperate humid climate. Agriculture (corn, beans, wheat, fruits), woods; cattle. Pottery.

Magdalena Jicotlán, MEXICO: see SANTA MAGDALENA JICOTLÁN.

Magdalena, La, MEXICO: see LA MAGDALENA TLATLAUQUITEPEC.

Magdalena Milpas Altas (mahg-dah-LAI-nah MEEL-pahs AHL-tahs), town, SACATEPÉQUEZ department, S central GUATEMALA, 5 mi/8 km SE of ANTIGUA GUATEMALA; 14°33′N 90°41′W. Elevation 6,158 ft/1,877 m. Corn, beans, vegetables, fodder grasses. Cakchiquel-speaking village.

Magdalena Mountains (MAG-duh-le-nuh), W central NEW MEXICO, in SOCORRO county, W of SOCORRO and the RIO GRANDE, largely within part of Cibola National Forest. Prominent peaks include North Baldy (9,858 ft/3,005 m) and SOUTH BALDY (10,783 ft/3,287 m; highest point). Zinc is mined.

Magdalena Ocotlán (mahg-dah-LAI-nah o-kot-LAHN), town, in S central OAXACA, MEXICO, 4 mi/6 km SSW of OCOTLÁN DE MORELOS, in the Valle Grande Arm of the OAXACA VALLEY, just W of Mexico Highway 175; 16°43′N 96°42′W. Elevation 5,085 ft/1,550 m. The total population of the municipality live here in its capital city. Temperate climate. Agriculture (cereals and fruits); woods and mezcal.

Magdalena Peñasco (mahg-dah-LAI-nah pen-YAHS-ko), town, in W OAXACA, MEXICO, 56 mi/90 km W of OAXACA DE JUÁREZ, 9 mi/14 km E of TLAXIACO, on unpaved road; 17°14′N 97°34′W. Elevation 6,201 ft/1,890 m. In the Mixteca Alberta. Temperate climate. Agriculture (corn, beans, wheat, fruits), woods, mezcal; cattle. Woven textiles.

Magdalena River, c.200 mi/322 km long, SONORA, NW MEXICO; rises SE of NOGALES near U.S. border; flows SW and W, past IMURIS, MAGDALENA DE KINO, SANTA ANA, and CABORCA, to Gulf of California 22 mi/35 km NW of Cape Tepoca. Receives Altar River, whose name sometimes designates its lower course; lower course also called Asunción River, the section above its mouth Concepción River. Used for irrigation in an otherwise arid region; chickpeas, fruit, cereals, vegetables are produced. Intermittent stream.

Magdalena Teitipac (mahg-dah-LAI-nah tai-EE-tee-PAHK), town, in S central OAXACA, MEXICO, 16 mi/25 km SE of OAXACA DE JUÁREZ, 6 mi/10 km SW of TLACOLULA; 16°55′N 96°36′W. Elevation 5,440 ft/1,658 m. In Tlacolula arm of OAXACA VALLEY. Temperate climate. Agriculture; woven textiles; mezcal processing.

Magdalena Tequisistlán (mahg-dah-LAI-nah tai-kwee-seest-LAHN), town, in SE OAXACA, MEXICO, 31 mi/50 km NW of the port of SALINA CRUZ, on Mexico Highway 190; 16°22′N 95°15′W. Agriculture (corn, coffee, sugarcane, fruits), woods. Textiles from local artisans.

Magdalena Tlacotepec (mahg-dah-LAI-nah tlah-KO-tai-pek), town, in SE OAXACA, MEXICO, 8 mi/13.5 km SW of CIUDAD IXTEPEC; 16°30′N 95°12′W. Hot climate.

Magdalena Vieja (mahg-dah-LAI-nah vee-AI-hah), SW residential section of LIMA, LIMA province, W central PERU, just N of Magdalena del Mar; 12°05′S 77°04′W. Museum of anthropology; national museum

---

Cross-references are shown in SMALL CAPITALS. The pronunciation guide is shown on page xix. The sources of population figures are shown on page xvii.

of the republic. Here resided Bolívar and San Martín after their meeting at GUAYAQUIL (1822). Incorporated 1940 into Lima proper. Also known as Pueblo Libre.

Magdalena Yodocono de Porfirio Díaz (mahg-dah-LAI-nah yo-do-KO-no dai por-FEE-ree-o DEE-ahs), town, NW OAXACA, MEXICO, 7 mi/12 km SW of ASUNCIÓN NOCHIXTLÁN; 17°23′N 97°22′W. Elevation 6,332 ft/1,930 m. Resources are livestock and poultry breeding. Formerly Yodocono de Porfirio Diaz.

Magdalena Zahuatlán (mahg-dah-LAI-nah zah-waht-LAHN), town, in NW OAXACA, MEXICO, 43 mi/70 km NW of OAXACA DE JUÁREZ, and 4.3 mi/7 km S of NOCHIXTLÁN. Elevation 7,087 ft/2,160 m. A Mixtec community. Agriculture.

Magdalen Channel, 25 mi/40 km long, S arm of STRAIT OF MAGELLAN in TIERRA DEL FUEGO, S CHILE, between CLARENCE ISLAND and W coast of main island of Tierra del Fuego.

Magdalen Islands (MAG-duh-luhn) or **Îles-de-la-Madeleine** (EEL–duh–lah–mahd-LEN), group of nine main islands and numerous islets, QUEBEC, E CANADA, in the Gulf of SAINT LAWRENCE N of PRINCE EDWARD ISLAND; 47°24′N 61°47′W. Discovered (1534) by Jacques Cartier. The main islands are ALRIGHT, AMHERST, BRION, COFFIN, East, Entry, Grindstone, GROSSE, and Wolf. Fishing and sealing are the chief occupations of the islanders, most of whom are of French descent.

Magdalinovka, UKRAINE: see MAHDALYNIVKA.

Magdapio Falls, PHILIPPINES: see PAGSANJAN FALLS.

Magdeburg (MAHG-de-boorg), city (2005 population 229,126), ⊙ SAXONY-ANHALT, central GERMANY, 48 mi/77 km ESE of BRUNSWICK on the ELBE River, with connection to MIDLAND CANAL and ELBE-HAVEL CANAL; 52°08′N 11°24′E. Large inland port, industrial center, and railroad and road junction. Manufacturing includes metal products, steel-working machinery, and chemicals. Food-processing center, primarily in sugar refining and flour milling. There are lignite and potash mines nearby. Ship elevator at E end of WESER-ELBE CANAL, and power station in ROTHENSEE district (N); airport in suburb of FRIEDRICHSTADT (E). Heavily bombed in World War II (destruction c.65 percent); entire old city destroyed; cathedral (1209–1263) damaged. Was first mentioned (805) as Saxon outpost; became center for colonization of Slav territories under Otto I. Archbishopric established 968. Town received charter from archbishops; charter became model of numerous medieval town charters in Holy Roman Empire. Was leading member of HANSEATIC LEAGUE. Accepted (1524) the Reformation; joined (1531) Schmalkaldic League and continued resistance against emperor until captured (1551) by Maurice of Saxony. The archbishops, members of the house of BRANDENBURG, were converted to Protestantism, and the family consequently ruled archbishopric as administrators. In Thirty Years War, Magdeburg held off seige (1629) by Wallenstein but was stormed and burned (May 1631) by imperial forces under Tilly. Under Treaty of Westphalia (1648), secularized archbishopic passed to Brandenburg. Was fortress until 1912. In World War II, captured by American troops (April 18, 1945); later occupied by Soviet forces. The composer Telemann, the physicist von Guericke, and Baron von Steuben (Prussian officer who served as a general in the American Revolutionary War) were born here.

Magdiwang, PHILIPPINES: see SIBUYAN ISLAND.

Magé (MAH-zhai), city (2007 population 232,171), S RIO DE JANEIRO state, BRAZIL, near N end of GUANABARA BAY, 20 mi/32 km NNE of RIO; 22°39′S 43°02′W. Railroad junction (spur to GUAPIMIRIM). Cotton mill (6 mi/9.7 km NW), match factory, pottery works. Also spelled Majé.

Magee (muh-GEE), town (2000 population 4,200), SIMPSON county, S central MISSISSIPPI, 38 mi/61 km SE of JACKSON, near Okatoma Creek; 31°52′N

89°43′W. Railroad junction to SE. Agriculture (cotton, corn; poultry, cattle); timber; manufacturing (apparel, wood products, building materials; millwork). State sanitorium to NW. Simpson County Legion Lake (state lake) to NW.

Magee, Island (ma-GEE), peninsula on SE coast of ANTRIM, NE Northern Ireland, on the NORTH CHANNEL, forming E side of Lough Larne, extending 8 mi/12.9 km NNW from BLACK HEAD; 54°50′N 05°45′W. Notorious for massacre of its Roman Catholic inhabitants in 1642.

Magelang (MAH-guh-lahng), municipality (2000 population 117,715), Java Tengah province, INDONESIA, 25 mi/40 km NNW of YOGYAKARTA, in highlands between Mounts SUMBING and MERAPI; 07°28′S 110°13′E. Elevation 1,312 ft/400 m. Trade center for agricultural area (sugar, rice, corn, tobacco, cassava); textile mills. Until 1949, site of important Dutch military establishment. BOROBUDUR (famous monument) 8 mi/12.9 km S.

Magellan, Strait of, Spanish *Estrecho de Magallanes*, c.330 mi/531 km long and 2.5 mi/4 km to 15 mi/24 km wide, separating SOUTH AMERICA from TIERRA DEL FUEGO and other islands to S. Except for a few miles at its E end in ARGENTINA, it passes through CHILE. Discovered by Ferdinand Magellan in 1520, it was important in the days of sailing ships, especially before the building of the PANAMA CANAL, and is still used by ships rounding South America. One of the most scenic waterways in the world, it affords an inland passage protected from almost continuous ocean storms. However, the strait is often foggy. The only city on the strait is PUNTA ARENAS (Chile) BRUNSWICK PENINSULA.

Magens Bay (MAI-ginz), bay of N ST. THOMAS Island, U.S. VIRGIN ISLANDS, c.1.5 mi/2.4 km N of CHARLOTTE AMALIE; 18°22′N 64°56′W. Fine beach and public park; popular tourist site.

Magenta (mah-JAIN-tah), town, MILANO province, LOMBARDY, N ITALY; 45°28′N 08°53′E. Manufacturing includes fabricated metals, machinery, matches, textiles, clothing, and paper. At the TICINO River nearby, the French and the Sardinians won a decisive victory (1859) over the Austrians, which opened the way for their march on MILAN. General MacMahon was made duke of Magenta by Napoleon III for his leading role in the battle.

Magerov, UKRAINE: see MAHERIV.

Mageroy (MAH-guhr-uh-oo), Norwegian *Magerøy*, island (□ 106 sq mi/275.6 sq km), in BARENTS SEA of the ARCTIC OCEAN, FINNMARK co., N NORWAY, 45 mi/72 km ENE of HAMMERFEST, on W side of mouth of Porsang Fjord; 22 mi/35 km long, 17 mi/27 km wide. Rises to 1,368 ft/417 m (W). On N coast are KNIVSKJELLODDEN, northernmost point of EUROPE, and NORTH CAPE. HONNINGSVÅG (SE) is chief village.

Magersfontein, locality, NORTHERN CAPE province, SOUTH AFRICA, near FREE STATE province border, 14 mi/22 km SSW of KIMBERLEY; 28°59′S 24°42′E. Scene, on December 11, 1899, of battle in South African War; Boers here checked Lord Methuen's advance on Kimberley.

Magetan (MAH-guh-tahn), town, ⊙ Magetan district, Java Timur province, INDONESIA, 35 mi/56 km E of SURAKARTA, at E foot of Mount LAWU; 07°39′S 111°20′E. Elevation 1,181 ft/360 m. Trade center for agricultural area (corn, cassava, coffee); sugar mills.

Maggia River (MAH-jyah), 35 mi/56 km long, S SWITZERLAND; rises SW of AIROLO; flows SSE, through VALLE MÁGGIA, to LAGO MAGGIORE at LOCARNO. Drains an area of 290 sq mi/751 sq km.

Maggie Lake Walker (MA-gee, WAH-kuhr), national historic site, downtown RICHMOND, E central VIRGINIA, at 110 1/2 East Lehigh Street. Home of African-American ex-slave's daughter who became bank president, early leader in the women's movement. Covers area of 1.3 acres/0.5 ha. Authorized 1978.

Maggie Valley (MAG-ee VAL-ee), village (□ 1 sq mi/2.6 sq km; 2006 population 810), HAYWOOD county, W NORTH CAROLINA, 6 mi/9.7 km WNW of WAYNESVILLE; 35°31′N 83°05′W. GREAT SMOKY MOUNTAINS National Park to N; parts of Pisgah National Forest to S and NE; BLUE RIDGE PARKWAY passes to SW. Retail trade; agriculture (corn, tobacco; livestock; dairying).

Maggiolo (mahg-gee-O-lo), town, SW SANTA FE province, ARGENTINA, 105 mi/169 km SW of ROSARIO; 33°44′S 62°16′W. Agriculture center (corn, soybeans, wheat, flax, alfalfa; livestock).

Maggiorasca, Monte (mahd-jo-RAH-skah, MON-te), highest peak (5,915 ft/1,803 m) in LIGURIAN APENNINES, N ITALY, 19 mi/31 km NNE of CHIAVARI; 44°33′N 09°29′E. Nearby is MONTE BUE.

Maggiore, Lake (mahd-JO-re), lake (□ 82 sq mi/212 sq km), partly (□ 82 sq mi/213.2 sq km) in NOVARA (W) and VARESE (E) provinces, ITALY, and partly (□ 17 sq mi/44 sq km) in TICINO canton (N), SWITZERLAND; 40 mi/64 km long, average width of 2 mi/3.2 km, maximum depth 1,220 ft/372 m; elevation 636 ft/194 m; 45°57′N 08°39′E. Second largest lake in Italy, it is traversed by the TICINO River, and Lake VARESE. Fishing (trout, perch, shad) and hatcheries; tourism. The Italian towns of PALLANZA, INTRA, ARONA, CANNOBIO, LUINO, and LAVENO-MOMBELLO and the Swiss town of LOCARNO are on its shores. Also called Lago Verbano, Lago Maggiore.

Maggiore, Monte, CROATIA: see UČKA.

Magglingen, SWITZERLAND: see MACOLIN.

Maggotty, town, SAINT ELIZABETH parish, W JAMAICA, on BLACK RIVER, on Jamaica railroad, and 23 mi/37 km SE of MONTEGO BAY; 18°09′N 77°46′W. In agricultural region (corn, spices; livestock). Known for nearby Maggotty Falls.

Maghagha (mah-GAH-guh), town, MINYA province, Upper EGYPT, on W bank of the NILE River, on IBRAHIMIYA CANAL, on railroad, and 38 mi/61 km NNE of Minya; 28°39′N 30°50′E. Cotton ginning, woolen and sugar milling; cotton, cereals, sugarcane.

Maghama (mah-GAH-mah), village (2000 population 11,367), Gorgol administrative region, S MAURITANIA, near SÉNÉGAL RIVER, 55 mi/89 km SE of KAÉDI; 15°31′N 12°51′W. Gum arabic, millet; livestock.

Maghar (muh-guhr), town, BASTI district, NE UTTAR PRADESH state, N central INDIA, 14 mi/23 km W of GORAKHPUR; 26°45′N 83°08′E. Hand-loom cotton-weaving center; rice, wheat, barley. Tomb of noted Indian saint, Kabir, acclaimed by both Hindu and Muslim disciples. Important military outpost of OUDH in 18th century.

Magherafelt (MA-huh-rah-FELT), Gaelic *Machaire Rátha*, town (2001 population 8,372), SE Londonderry, NW Northern Ireland, 15 mi/24 km WSW of BALLYMENA; 54°46′N 06°36′E. Market; trout fishing. Charles Thompson, who helped write the Declaration of Independence, was born here.

Maghnia (mahg-NYAH), town, TLEMCEN wilaya, NW ALGERIA, near MOROCCO border, in fertile lowland irrigated by the Oued TAFNA, on railroad, and 24 mi/39 km W of TLEMCEN; 34°50′N 01°45′W. Agricultural trade center (cereals, olive oil; livestock). Lead mine at GHAR ROUBAN (18 mi/29 km S) in TLEMCEN MOUNTAINS. Formerly Marnia.

Maghreb or **Magrib** (both: MUH-grib) [Arabic=the West], Arabic term for NW Africa between EGYPT (E) and the ATLANTIC OCEAN (W) and the MEDITERRANEAN SEA (N) and the SAHARA (S), specifically the ATLAS MOUNTAINS; generally applied to MOROCCO, ALGERIA, and TUNISIA. SPAIN, during the period of Muslim domination, was also included in the region.

Maghull (muh-GUHL), town (2001 population 28,848), MERSEYSIDE, NW ENGLAND, on Leeds-Liverpool Canal, and 8 mi/12.9 km N of LIVERPOOL; 53°51′N 02°57′W. Vegetables.

Magic Dam, Idaho: see BIG WOOD RIVER.

Area is shown by the symbol □, and capital city or county seat by ⊙.

Magierów, UKRAINE: see MAHERIV.

Magil, IRAQ: see MA'QIL.

Maginot Line (MAH-zhee-no), system of fortifications along NE border of FRANCE, extending from the Swiss border N and NW to BELGIUM. Named for André Maginot, French minister of war (1929–1936) who directed its construction. Although considered impregnable, the line was still not complete at the outbreak (1939) of World War II. Its actual strength was never tested, for the line was outflanked by the Germans in their invasion of 1940 (via the Low Countries). Like fortified lines since the Great Wall of CHINA, the chief effect it had was to create a false sense of security. Remnants, underground chambers and command posts of the Maginot Line are now open to vistors in various locations along its entire length.

Magione (mah-JO-ne), town, PERUGIA province, UMBRIA, central ITALY, near Lake TRASIMENO, 9 mi/14 km WNW of PERUGIA. Manufacturing (metal furniture, wire, nets).

Magistral'nyy (mah-gee-STRAHL-niyee), town, N central IRKUTSK oblast, E central Siberian Russia, on road near the BAYKAL-AMUR MAINLINE, 164 mi/264 km E of BRATSK; 56°10′N 107°26′E. Elevation 1,368 ft/416 m. Logging, lumbering.

Maglaj (MAH-glei), town, N BOSNIA, BOSNIA AND HERZEGOVINA, on BOSNA RIVER, on railroad, and 16 mi/26 km S of DOBOJ; 44°32′N 18°05′E. Center of prune- and cereal-growing area. Also spelled Maglay.

Magland, FRANCE: see CLUSES.

Maglia, unincorporated town, BUTTE county, N central CALIFORNIA, 13 mi/21 km ENE of CHICO, near Butte River. Plumas National Forest to E. LAKE OROVILLE reservoir to SE. Cattle; fruit, nuts, grain; timber.

Magliano Alpi (mah-LYAH-no AHL-pee), village, CUNEO province, PIEDMONT, NW ITALY, 15 mi/24 km NE of CUNEO; 44°27′N 07°47′E.

Magliano in Toscana (mah-LYAH-no een to-SKAH-nah), village, GROSSETO province, TUSCANY, central ITALY, 15 mi/24 km SSE of GROSSETO; 42°36′N 11°17′E. Medieval ruins.

Maglic (MAHG-leets), Serbian *Maglić*, peak (7,829 ft/2,386 m) in DINARIC ALPS, on MONTENEGRO (SE) and BOSNIA-HERZEGOVINA (NW) border, W of PIVA RIVER, 15 mi/24 km SSW of FOCA (Bosnia-Herzegovina). Also spelled Maglich.

Maglie (MAH-lye), town, LECCE province, APULIA, S ITALY, 17 mi/27 km SSE of LECCE; 40°07′N 18°18′E. Railroad junction. Secondary diversified industrial center. Agricultural trade center (olive oil, wine, figs, cereals).

Maglizh, Bulgaria: see MUGLIZH.

Maglód (MAHG-lod), Hungarian *Maglód*, village, PEST county, N central HUNGARY, 12 mi/19 km E of BUDAPEST; 47°27′N 19°22′E. Grain; dairy farming. Aluminum foundry. Residential community for Budapest workers.

Magna, village (2000 population 22,770), SALT LAKE county, N UTAH, suburb 10 mi/16 km W of downtown SALT LAKE CITY; 40°42′N 112°05′W. Elevation 4,261 ft/1,299 m. Railroad junction. Copper, silver, gold, lead processed. Diversified farming (grain, fruit, sugar beets) in vicinity; manufacturing (solid propulsion products, chemicals). Great Salt Lake (S shore) State Park to W. Large Kennecott Tailings Pond is here.

Magna Carta Island, ENGLAND: see EGHAM.

Magnac-Laval (mah-NYAHK–lah-VAHL) commune (□ 22 sq mi/70.2 sq km), HAUTE-VIENNE department, W central FRANCE, in LIMOUSIN region, 8 mi/12.9 km NE of BELLAC; 46°13′N 01°10′E. Sugar beet distilling. Site of annual religious festival in honor of St. Maximin.

Magnac-sur-Touvre (mah-NYAHK–syoor–TOOV-ruh), commune (□ 3 sq mi/7.8 sq km) of ANGOULÊME, CHARENTE department, POITOU-CHARENTES region, W FRANCE, on the TOUVRE RIVER, and 4 mi/6.4 km from city center; 45°40′N 00°14′E. Paper milling.

Magnago (mah-NYAH-go), town, MILANO province, LOMBARDY, N ITALY, 3 mi/5 km SW of BUSTO ARSIZIO. Cotton-milling center; manufacturing of textile machinery, fabricated metals, textiles, clothing.

Magna Graecia (MAG-nuh GREE-shuh), [Latin= greater Greece], ancient Greek colonies of S ITALY. The Greek overseas expansion of the 8th century B.C.E. founded a number of towns that became the centers of a new, thriving Greek territory. They were on both Italian coasts from the Bay of NAPLES and the Gulf of TARANTO S. Unlike Greek SICILY, Magna Graecia began to decline by 500 B.C.E., probably because of malaria and endless warfare among the colonies. Only Tarentum (now TARANTO) and CUMAE remained individually very significant. Magna Graecia was the center of two philosophical groups in the 6th century B.C.E., that of Parmenides (at ELEA) and that of Pythagoras (at Crotona). Through Cumae especially, the Etruscans of Capua and the Romans came into early contact with Greek civilization. The following are the chief cities of Magna Graecia (those colonized from Greece, except THURII and Elea, go back to the 8th or early 7th century B.C.E.; those colonized locally are perhaps a century younger)—on the E coast (N to S), Tarentum (colonized from SPARTA), METAPONTUM (from Achaea), HERACLEA (from Tarentum), Siris (from Colophon), SYBARIS (from Achaea), Thurii (from Athens, replacing Sybaris), Crotona (from Achaea), CAULONIA (from Crotona), Epizephyrian Locris (from Locris); on the W coast (N to S), Cumae (from Chalcis), Neapolis (now NAPLES; from Cumae), PAESTUM, or Posidonia (from Sybaris), Elea (from Phocaea in Ionia), Laos (from Sybaris), Hipponium (from Epizephyrian Locris), and Rhegium (now Reggio de Calabria; from Chalcis).

Magnesia (mah-nee-SEE-ah), prefecture (□ 1,563 sq mi/4,063.8 sq km), SE THESSALY department, N GREECE, on the AEGEAN SEA and the Gulf of VÓLOS; ⊙ VÓLOS; 39°15′N 22°45′E. Bounded NW by LÁRISSA prefecture and S by CENTRAL GREECE department Orthrys massif in the SW. The Gulf of Vólos is nearly closed off by hook-shaped Magnesia Peninsula, which rises to the PELION. Includes the Northern Sporades (except Skíros group). Agriculture includes wheat, corn, tobacco, citrus fruit, almonds, olives. Livestock raising; fisheries. Vólos is the economic center. In ancient times, Magnesia applied to the entire coastal district of Thessaly S of the Vale of Tempe. Also spelled Magnisia.

Magnesia, two ancient cities of LYDIA, W ASIA MINOR (now W TURKEY). They were colonies of the Magnetes, a tribe of E THESSALY. One city (Magnesia ad Maeandrum), SE of Smyrna (now IZMIR), was later colonized by Ionians and given by Artaxerxes I to Themistocles, who died here. Important ruins on the site include the celebrated temple of Artemis Leucophryene, built in the 2nd century B.C.E. Magnesia ad Sipylum, on the HERMUS RIVER at the foot of Mount Sipylus, NE of Smyrna, was (190 B.C.E.) the scene of the defeat of Antiochus III (Antiochus the Great) by the Romans. The modern MANISA is nearby.

Magness (MAG-nes), village (2000 population 191), INDEPENDENCE county, NE central ARKANSAS, 10 mi/16 km ESE of BATESVILLE, near WHITE RIVER; 35°42′N 91°28′W.

Magnet, village (2006 population 73), CEDAR county, NE NEBRASKA, 15 mi/24 km SW of HARTINGTON; 42°27′N 97°28′W.

Magnetawan (mag-NE-tuh-wahn), township (□ 202 sq mi/525.2 sq km; 2001 population 1,342), S ONTARIO, E central CANADA, 29 mi/46 km from city of PARRY SOUND; 45°41′N 79°40′W. Created in 1999.

Magnetawan (mag-NE-tuh-wahn), unincorporated village (2001 population 244), SE central ONTARIO, E central CANADA, on MAGNETAWAN RIVER, 30 mi/48 km NE of PARRY SOUND, and included in Magnetawan township; 45°40′N 79°38′W. Lumbering.

Magnetawan River (mag-NE-tuh-wahn), 100 mi/161 km long, SE central ONTARIO, central CANADA; rises in NW part of ALGONQUIN PROVINCIAL PARK; flows SW and W, past MAGNETAWAN, to GEORGIAN BAY of Lake HURON, through the Bying Inlet, 40 mi/64 km NW of PARRY SOUND; 45°46′N 80°30′W.

Magnetic Island (mag-NE-tik) (□ 19 sq mi/49.4 sq km), in CORAL SEA just off E coast of QUEENSLAND, AUSTRALIA, between CLEVELAND and HALIFAX bays; 19°08′S 146°50′E. Roughly triangular, 6 mi/10 km long, 5 mi/8 km wide; rises to 1,628 ft/496 m. Tourist resort. Papayas, pineapples.

Magnetic Peak (10,008 ft/3,050 m), SE MAUI island, MAUI county, HAWAII, at SW rim of Haleakala Crater, on boundary of HALEAKALA NATIONAL PARK. Maui's second-highest point; ½ mi/⁸⁄₁₀ km SE of PUU ULAULA (Red Hill), Maui's highest point.

Magnetic Springs, village (2006 population 310), UNION county, central OHIO, 17 mi/27 km SSW of MARION; 40°21′N 83°16′W.

Magnisia, Greece: see MAGNESIA.

Magnitka (mahg-NEET-kah), town, W CHELYABINSK oblast, SE URAL Mountains, RUSSIA, on a left tributary of the Kusa River (Kama River basin), on road, 11 mi/18 km N of ZLATOUST', and 9 mi/14 km E of KUSA; 55°21′N 59°41′E. Elevation 1,220 ft/371 m. On a railroad spur (Titan station); mining of iron ore and titanium. Originally called Titanogorsk.

Magnitnaya (mahg-NEET-nah-yah), mountain (2,020 ft/616 m) in S URAL Mountains, CHELYABINSK oblast, RUSSIA, just W of the upper URAL River; 53°24′N 59°10′E. Site of extensive magnetite deposits, whose exploitation led to the founding of MAGNITOGORSK (at the W foot), the largest iron and steel plant in Russia.

Magnitogorsk (mahg-nee-tuh-GORSK), city (2005 population 411,700), SW CHELYABINSK oblast, SW Siberian Russia, on the slopes of Mount Magnitnaya in the SE URAL Mountains, on the URAL River, on railroad, 285 mi/459 km SW of CHELYABINSK; 53°25′N 59°03′E. Elevation 1,145 ft/348 m. Major railroad and highway hub; local transshipment center. Built (1929–1931) under the First Five-Year Plan for ferrous metallurgy on the site of iron ore deposits, the city became an important symbol of Soviet industrial growth. Industrial cranes; cement and refractory clay works; brick making; food processing (bakery, dairy, soft-drink plant); prosthetic-orthopedic equipment; granite quarries. Coking coal comes from the Kuznetsk, Karaganda, and other distant coalfields. Numerous coke and chemical plants. As a consequence, Magnitogorsk is one of Russia's most polluted cities, which is one of the major reasons for a recent population drop. A leading steel manufacturer during World War II, it is still a major metallurgical center. Made city in 1931.

Magnolia, city (2000 population 10,858), ⊙ COLUMBIA county, SW ARKANSAS; 33°16′N 93°14′W. Manufacturing (fabricated metal products, apparel, chemicals, lumber, building materials). Its oil industry has been important since 1938. Seat of Southern Arkansas University. Incorporated 1855.

Magnolia, town (2000 population 200), HARRISON county, W IOWA, 6 mi/9.7 km NW of LOGAN; 41°41′N 95°52′W. In agricultural area.

Magnolia, town (2000 population 2,071), ⊙ PIKE county, SW MISSISSIPPI, 7 mi/11.3 km S of MCCOMB, near TANGIPAHOA RIVER; 31°08′N 90°27′W. Agriculture (cotton, corn; cattle; dairying); manufacturing (building materials, apparel, wood products, chemicals, lab equipment). Percy Quin State Park to NW.

Magnolia (mag-NOL-yuh), town (□ 1 sq mi/2.6 sq km; 2006 population 981), DUPLIN county, SE NORTH CAROLINA, 34 mi/55 km S of GOLDSBORO; 34°53′N 78°03′W. Manufacturing; agriculture (tobacco, grain, cotton, sweet potatoes; poultry, livestock). Chartered 1855.

Cross-references are shown in SMALL CAPITALS. The pronunciation guide is shown on page xix. The sources of population figures are shown on page xvii.

Magnolia (mag-NO-lee-uh), town (2006 population 1,263), MONTGOMERY county, SE TEXAS, 40 mi/64 km NW of HOUSTON; 30°12′N 95°45′W. Agricultural area (cattle, horses, ostriches; nursery products). Timber. Oil and natural gas. Manufacturing (machinery, oil field linings).

Magnolia, village (2000 population 226), KENT county, E DELAWARE, 6 mi/9.7 km SSE of DOVER; 39°04′N 75°28′W. Elevation 3 ft/0.9 m. In agricultural region. Harvey Conservation Area to E.

Magnolia, village (2000 population 279), PUTNAM county, N central ILLINOIS, 12 mi/19 km SE of HENNEPIN; 41°06′N 89°12′W. In agricultural area.

Magnolia, village (2000 population 221), ROCK county, extreme SW MINNESOTA, 7 mi/11.3 km E of LUVERNE, near IOWA state line; 43°38′N 96°04′W. Grain; livestock; dairying; manufacturing (feeds, protein blending).

Magnolia (mag-NO-lee-ah), village (2006 population 932), on border between STARK and CARROLL counties, E OHIO, 11 mi/18 km SSE of CANTON; 40°39′N 81°17′W.

Magnolia, residential borough (2006 population 4,379), CAMDEN county, SW NEW JERSEY, 8 mi/12.9 km SE of CAMDEN; 39°51′N 75°02′W. Incorporated 1915.

Magnolia, MASSACHUSETTS: see GLOUCESTER.

Magnolia Beach (mag-NO-lee-uh), village, CALHOUN county, S TEXAS, on Lavaca Bay, and 7 mi/11.3 km SE of PORT LAVACA; 28°33′N 96°32′W.

Magnolia State; see MISSISSIPPI.

Magnor (MAHNG-nawr), village, HEDMARK county, SE NORWAY, on railroad, 20 mi/32 km SE of KONGSVINGER, frontier station on Swedish border, 5 mi/8 km NW of CHARLOTTENBERG (SWEDEN), on OSLO-STOCKHOLM (Sweden) main railroad line; 59°57′N 12°12′E. Has glassworks and mechanical manufacturing. Monument here commemorates peaceful separation (1905) of Norway and Sweden.

Magny-Cours (MAH-nyee–KOOR), commune (□ 12 sq mi/31.2 sq km), NIÈVRE department, in BURGUNDY, central FRANCE, 8 mi/12.9 km S of NEVERS; 46°53′N 03°09′E. Has international auto races on a well-known road-race circuit.

Magny-en-Vexin (mah-NYEE–ahn–vek-SAN), town (□ 5 sq mi/13 sq km; 2004 population 5,470), VAL-D'OISE department, ÎLE-DE-FRANCE region, N central FRANCE, 12 mi/19 km NNE of MANTES-LA-JOLIE; 49°09′N 01°47′E. Agriculture market, furniture factory. Old houses and 15th–16th-century church.

Magny-les-Hameaux (mah-NYEE–laiz–ah-MO), town (□ 6 sq mi/15.6 sq km), YVELINES department, ÎLE-DE-FRANCE region, N central FRANCE, 5 mi/8 km SSW of VERSAILLES; 48°44′N 02°04′E. Forms part of the "new town" of SAINT-QUENTIN-EN-YVELINES. Aeronautical industry. National institute of meteorological studies. Has 12th-century church. International airport.

Mago (MAH-guh), settlement (2005 population 2,185), SE KHABAROVSK TERRITORY, RUSSIAN FAR EAST, fishing port on the lower AMUR River, 20 mi/32 km W of NIKOLAYEVSK-NA-AMURE; 53°15′N 140°11′E. Fish-processing center; can manufacturing, barrel making.

Mago (MAHNG-o), composite volcanic and raised limestone island (□ 8 sq mi/20.8 sq km), N LAU group, FIJI, SW PACIFIC OCEAN; 3 mi/4.8 km long. Copra. Sometimes spelled Mango.

Mágocs (MAH-goch), Hungarian *Mágocs*, village, BARANYA county, S HUNGARY, 5 mi/8 km E of DOMBÓVÁR; 46°21′N 18°14′E. Corn, wheat, nuts, grapes; hogs.

Magodes, UGANDA: see MOLO.

Magoffin (muh-GAHF-uhn), county (□ 309 sq mi/803.4 sq km; 2006 population 13,449), E KENTUCKY, in the CUMBERLAND MOUNTAINS; ⊙ SALYERSVILLE; 37°42′N 83°03′W. Drained by LICKING RIVER and by several creeks. Agricultural area (burley tobacco, corn, hay; cattle). Bituminous-coal mines, oil wells; timber. Formed 1860.

Magog (MAI-gahg), city (□ 56 sq mi/145.6 sq km), ⊙ MEMPHRÉMAGOG county, ESTRIE region, S QUEBEC, E CANADA, on Lake MEMPHRÉMAGOG, SW of SHERBROOKE; 45°16′N 72°09′W. Founded by Loyalist emigrants from the U.S. after 1776. Resort and trade center, with textile mills, food processing, and dairying. Restructured in 2002.

Magog (MAI-gahg), former canton, S QUEBEC, E CANADA; 45°15′N 72°10′W. Amalgamated into city of MAGOG in 2002.

Mago National Park (MAH-go), park, SOUTHERN NATIONS state, SW ETHIOPIA, 100 mi/161 km SW of ARBA MINCH; 05°30′N 36°15′E. Park protects large African game.

Magothy River (MAG-ah-thee), irregular arm of CHESAPEAKE BAY, c.12 mi/19 km long, central MARYLAND; penetrates ANNE ARUNDEL county just N of SANDY POINT.

Magoye, township, SOUTHERN province, S central ZAMBIA, on railroad, 15 mi/24 km SSW of MAZABUKA; 16°02′S 27°37′E. Agriculture (tobacco, corn, cotton, sugarcane); cattle. Kafue Flats Swamp to NW; Lochinvar National Park to NW.

Magra, INDIA: see BANSBARIA.

Magra River (MAH-grah), ancient *Macra*, 35 mi/56 km long, N ITALY; rises 4 mi/6 km SE of LA CISA pass; flows S, through the LUNIGIANA, past PONTREMOLI, AULLA, and Vezzano Ligure, to Gulf of GENOA 2 mi/3 km E of Gulf of SPEZIA. Chief affluents, VARA (right) and Auella (left) rivers. Used for irrigation and hydroelectric power.

Magrath (muh-GRATH), town (□ 2 sq mi/5.2 sq km; 2001 population 1,993), S ALBERTA, W CANADA, 19 mi/31 km S of LETHBRIDGE, in CARDSTON COUNTY; 49°25′N 112°53′W. Coal mining; oil and gas; wheat, flax, sugar beets; cattle. Mormon settlement, founded 1899.

Magrinah (mai-GREE-nah), SE suburb of TUNIS, TUNIS province, N TUNISIA, on LAKE OF TUNIS, 1 mi/1.6 km SE of city center. Lead smelter, metalworks; olive oil refinery and soap manufacturing plant. Silos. Agricultural industry includes salt working.

Magruder Mountain, NEVADA: see SILVER PEAK RANGE.

Magsingal (mahg-SEENG-ahl), town, ILOCOS SUR province, N LUZON, PHILIPPINES, 8 mi/12.9 km NNE of VIGAN, near W coast; 17°41′N 120°26′E. Rice-growing center.

Magstadt (MAHG-shtuht), village, region of STUTTGART, BADEN-WÜRTTEMBERG, GERMANY, 5 mi/8 km NNW of BÖBLINGEN; 48°44′N 08°56′E. Cattle.

Maguan (MAH-GWAHN), town, ⊙ Maguan county, SE YUNNAN province, CHINA, 25 mi/40 km S of WENSHAN, near VIETNAM border; 23°02′N 104°24′E. Grain, sugarcane, medicinal herbs; food industry. Nonferrous-ore and coal mining.

Maguarichi, mining settlement and township, CHIHUAHUA, N MEXICO, in SIERRA MADRE OCCIDENTAL, on railroad, 23 mi/37 km WNW of CREEL, and 130 mi/209 km SW of CHIHUAHUA, in Sierra Tarahumara. Silver and gold mining. Formerly MAGUARICHIC.

Maguarichic, MEXICO: see MAGUARICHI.

Magude (mah-GOO-dai), village, MAPUTO province, S MOZAMBIQUE, on INCOMATI River, on railroad, and 65 mi/105 km N of MAPUTO city; 25°02′S 32°40′E. Sugar, citrus, cotton, castor beans, corn; cattle raising.

Magude District, MOZAMBIQUE: see MAPUTO.

Mague District, MOZAMBIQUE: see TETE.

Máguez (MAH-geth), village, LANZAROTE, LAS PALMAS province, CANARY ISLANDS, SPAIN, just N of HARÍA, 13 mi/21 km N of ARRECIFE; 29°09′N 13°29′W. Cereals, fruit (especially grapes), vegetables.

Magüi (mah-GWEE), town, ⊙ Magüi municipio, NARIÑO department, SW COLOMBIA, 98 mi/158 km NW of PASTO, near the confluence of the PATÍA and the TELEMBÍ rivers in the Pacific coastal lowlands; 01°56′N 78°12′W. Rice, sugarcane, plantains; livestock. Sometimes called Payán.

Maguilla (mah-GEE-lyah), town, BADAJOZ province, W SPAIN, 13 mi/21 km NE of LLERA; 38°22′N 05°50′W. Cereals, grapes, olives, vegetables, fruit; livestock.

Maguindanao, province (□ 3,137 sq mi/8,156.2 sq km), in AUTONOMOUS REGION IN MUSLIM MINDANAO, SW MINDANAO, PHILIPPINES, on E shore of MORO GULF; ⊙ MAGANOY; 07°08′N 124°18′E. Population 1/3 urban (as of 1991); 100% of urban and 33% of rural settlements had electricity. Mountainous W and interior descends to Cotabato Plain and valley of the RIO GRANDE DE MINDANAO. Agricultural province (rice, corn as food crops; coconut as cash crop), with some fishing and aquaculture. Some agricultural processing. Timber reserves. Regional industrial center at PARANG, major port at POLLOC. Population is 60% Muslim Maguindanao, with Christian Ilocano and Cebuano and indigenous Tiruray, T'boli, and Manobo.

Magura (mah-goo-rah), town (2001 population 86,445), JESSORE district, W EAST BENGAL, BANGLADESH, on distributary of the MADHUMATI RIVER, and 26 mi/42 km NNE of JESSORE; 26°20′N 88°26′E. Rice, jute, oilseeds.

Magura-Limnyans'ka, Mount, UKRAINE: see UPPER DNIESTER BESKYDS.

Magura, Mount, UKRAINE: see SKOLE BESKYDS.

Maguse Lake (□ 540 sq mi/1,399 sq km), NUNAVUT territory, N CANADA, near HUDSON BAY; 45 mi/72 km long, 1 mi/2 km–7 mi/11 km wide; 61°30′N 95°00′W. Drained E into HUDSON BAY by Maguse River (35 mi/56 km long); at mouth of river is trading post.

Magwe (mah-GWAI), administrative division (□ 17,302 sq mi/44,985.2 sq km) of MYANMAR; ⊙ MAGWE. Located astride AYEYARWADY RIVER, between CHIN HILLS and ARAKAN YOMA (W) and PEGU YOMA (E), it includes twenty-five townships. In the dry zone (annual rainfall, 25 in/64 cm–45 in/114 cm), it has irrigated agriculture (rice, cotton, sesame, peanuts, catechu); extensive teak forests. Leading petroleum-producing region of Burma, with chief fields at YENANGYAUNG, CHAUK, YENANGYAT, and LANYWA. Served by Ayeyarwady steamers and Pyinmana-Kyaukpadaung railroad.

Magwe (mah-GWAI), former district (□ 3,724 sq mi/9,682.4 sq km), MAGWE division, MYANMAR; ⊙ was YENANGYAUNG. Between AYEYARWADY RIVER and PEGU YOMA, in dry zone (annual rainfall, 31 in/79 cm). Agriculture (nation's chief sesame producer, rice in TAUNGDWINGYI plain; also corn, peanuts, beans). Oil fields at YENANGYAUNG and CHAUK. Served by Ayeyarwady steamers and Kyaukpadaung-Pyinmana railroad.

Magwe (mah-GWAI), town and township, MAGWE division, MYANMAR, on E bank of AYEYARWADY RIVER (opposite MINBU), and 95 mi/153 km N of PROME; 20°10′N 94°55′E. Airfield.

Magyarcséke, ROMANIA: see CEICA.

Magyarlápos, ROMANIA: see TÎRGU LĂPUŞ.

Magyarország: see HUNGARY.

Magyaróvár, HUNGARY: see MOSONMAGYARÓVÁR.

Magyar Pass, UKRAINE: see YABLUNYTSYA PASS.

Mahabad (mah-ah-BAHD), town (2004 population 135,780), in ĀZERBĀYJĀN-E Gharbi province, NW IRAN, on road, and 65 mi/105 km SSE of URMIA, and S of Lake URMIA; 36°46′N 45°43′E. Elev. 4,800 ft/1,463 m. Grapes, tobacco; sheep raising. Was headquarters of Kurdish uprising of 1946. Until 1930s called SAUJBULAGH or SAVAJBOLAGH.

Mahabaleshwar (muh-HAH-buh-LAI-shwuhr), town, SATARA district, MAHARASHTRA state. W central INDIA, 27 mi/43 km NW of Satara; 17°55′N 73°40′E. Fruit and vegetable gardening. Noted health resort (sanitarium) on scenic plateau of WESTERN GHATS, rising to 4,719 ft/1,438 m in peak 1 mi/1.6 km E. Bauxite deposits nearby. Annual rainfall averages 280 in/711 cm. Source of KRISHNA River is 4 mi/6.4 km N. Other spellings are Mahabaleshvar, Mahableshwar; also called Malcompeth after its founder, Sir John Malcolm.

Mahabalipuram (muh-HAH-buh-lee-puhr-uhm), or **Mamallapuram**, village, CHENGALPATTU-M.G.R. district, TAMIL NADU state, S INDIA, on the COROMANDEL COAST. Ancient port; archaeological remains here represent some of the earliest-known examples of Dravidian architecture (c.7th century C.E.) in India. Under the patronage of the Pallava dynasty, numerous temples, hewn from granite hillocks and caves, were carved. The site is often called the Seven Pagodas because of the high pinnacles of seven of its temples, which are made of huge single blocks of stone. Tourist center (beach).

Mahaban (muh-HAH-buhn), town, MATHURA district, W UTTAR PRADESH state, N central INDIA, near the YAMUNA RIVER, 5 mi/8 km SSE of Mathura; 27°26′N 77°45′E. Pilgrimage center. Sacked by Mahmud of Ghazni in 1018. Has palace of Nanda (covered court consisting of 80 pillars; rebuilt by Muslims in 17th century).

Mahabharat Lekh (mah-hah-bah-RUHT lek), mountain range, S NEPAL; extends WNW-ESE across Nepal, between S foothills of the NEPAL HIMALAYA and SHIWALIK RANGE; 28°45′N 83°30′E. Range is 8 mi/12.9 km–10 mi/16 km wide. Highest point, 9,710 ft/2,960 m, is 18 mi/29 km SE of KATHMANDU. KATHMANDU VALLEY lies on N slope.

Mahableshwar, INDIA: see MAHABALESHWAR.

Mahabo (mah-HAH-boo), town, TOLIARY province, W MADAGASCAR, 27 mi/43 km E of MORONDAVA; 20°23′S 44°40′E. Lima beans, sugarcane, rice, peanuts. Food-processing industry.

Mahad (muh-HAHD), town, RAIGAD district, MAHARASHTRA state, W central INDIA, 70 mi/113 km SE of MUMBAI (BOMBAY); 18°05′N 73°25′E. Port (27 mi/43 km inland) on short Savitri River; trades in rice, onions, potatoes, jaggery; tanning; copper and brass products Buddhist caves (1st century C.E.) 2 mi/3.2 km NW.

Mahaddei Ueine (mah-HAH-dee YOO-en), town, E central SOMALIA, on the E bank of the Webē Shebelē River, and 15 mi/24 km N of JOUHAR; 02°58′N 45°32′E. Woodworking, leather working, rope making.

Mahadha (muh-HAHD-huh), village and oasis, NW OMAN, 22 mi/35 km NE of BURAIMI, on the W fringes of the WESTERN HAJAR highlands; 24°23′N 55°58′E. Date groves. Sometimes spelled Mahdha, Mahadhah, or Mahdah.

Mahaffey (ma-HA-fee), borough (2006 population 381), CLEARFIELD county, central PENNSYLVANIA, 22 mi/35 km WSW of CLEARFIELD, on West Branch of SUSQUEHANNA RIVER, at mouth of Chest Creek; 40°52′N 78°43′W. Agriculture (grain; livestock; dairying); lumber.

Mahagi (mah-HAH-gee), village, ORIENTALE province, NE CONGO, on UGANDA border, 95 mi/153 km NE of IRUMU; 02°18′N 30°59′E. Elev. 5,485 ft/1,671 m. Tourist center in LAKE ALBERT region. MAHAGI PORT (17 mi/27 km ESE), on the lake, is a steamboat landing for navigation on ALBERT-Nile and excursions to MURCHISON FALLS (on VICTORIA NILE).

Mahaica (mah-HEI-kah), village, MAHAICA-BERBICE district, N GUYANA, near the ATLANTIC OCEAN, on railroad, 18 mi/29 km ESE of GEORGETOWN; 06°41′N 57°55′W. Rice-growing area.

Mahaica-Berbice (mah-HEI-kah-buhr-BEES), administrative district, GUYANA, on ATLANTIC OCEAN (N). Bordered in S by UPPER DEMERARA–BERBICE, E by EAST BERBICE–CORENTYNE, and W by DEMERARA-MAHAICA district. Rice-growing area.

Mahaicony (mei-KO-nee), village, MAHAICA-BERBICE district, N GUYANA, on railroad, and 28 mi/45 km SE of GEORGETOWN; 06°33′N 57°48′W. In fertile Atlantic lowland. Rice mill nearby.

Mahajamba River (mah-ZAHM-bah), 200 mi/322 km long, MAHAJANGA province, central and NW MADAGASCAR; rises 45 mi/72 km NE of ANKAZOBE (18°02′S 47°43′E); flows NW to Mahajamba Bay of MOZAMBI-QUE CHANNEL, in a large delta. Navigable only in lower course, for c.50 mi/80 km. Rice growing in its valley. Formerly spelled Majamba.

Mahajanga (mah-ZAHNG-gah), province (□ 59,000 sq mi/153,400 sq km; 2001 population 1,733,917), NW MADAGASCAR. Drained by the BETSIBOKA and MAHAVAVY rivers. E part approaches central highlands of Madagascar, rising to 8,205 ft/2,501 m in TSARATANANA MASSIF; coffee, cattle here. The swampy coast is studded with rice fields, and cotton, tobacco, sugarcane plantations. Some gold and copper deposits. Mahajanga is an important port and chief financial and industrial center. Province formerly included what is now ANTSIRANANA province. Formerly Majunga.

Mahajanga (mah-ZAHNG-gah), city (2001 population 135,660), ⊙ MAHAJANGA province, NW MADAGASCAR, on the MOZAMBIQUE CHANNEL; 15°41′S 46°18′E. Despite its shallow harbor, Mahajanga is one of the nation's chief ports. The BETSIBOKA RIVER valley provides access to the interior. Food processing; cement, textile, and sisal processing; fishing industry. Coffee, rice, sugar, cotton, tobacco, and sisal are grown. International airport. Was capital of the Sakalava kingdom, which flourished in the 18th century FRANCE occupied MAHAJANGA 1883–1885 and retook it in 1894. The city experienced major immigration from the Comoros until riots (1976–1977) resulted in the immigrants' deportation. Formerly Majunga.

Mahakali (mah-HAH-kah-LEE), administrative zone (2001 population 860,475), W NEPAL. Includes districts of BAITADI, DADELDHURA, DARCHULA, and KANCHANPUR.

Mahakali River, NEPAL: see SARDA RIVER.

Mahakam River (MAH-hah-kahm), c.450 mi/724 km long, Indonesian BORNEO (KALIMANTAN); rises in mountains in central part of island, on SARAWAK (E MALAYSIA) border c.200 mi/322 km ENE of SINTANG; flows generally SE past SAMARINDA to MAKASAR Strait 30 mi/48 km E of Samarinda; 00°35′S 117°17′E. Has wide delta. Also called KUTAI, or Koetai, River.

Mahalapye (mah-hah-LAH-pyai), town (2001 population 39,719), Ngwato District, E BOTSWANA, 114 mi/183 km NNE of GABORONE, and on railroad; 23°34′S 26°48′E. Airfield, agricultural research center.

Mahale Mountains National Park (mah-HA-lai), KIGOMA region, W TANZANIA, 80 mi/129 km S of KIGOMA, on peninsula (35 mi/56 km wide) on E shore of Lake TANGANYIKA. Makari Hills, also called Mahale Mountains, are dominant feature of park; Mount KUNGWE (8,250 ft/2,515 m) is highest point.

Mahalingpur (mah-HAH-ling-poor), town, BIJAPUR district, KARNATAKA state, S INDIA, 50 mi/80 km SW of BIJAPUR; 16°23′N 75°07′E. Trade center for cotton, millet, peanuts, wheat; handicraft cloth weaving. Annual fair.

Mahalla el Kubra, El (MAH-HAHL-luh el KOB-ruh, el), city, GHARBIYA province, Lower EGYPT, on railroad, 4 mi/6.4 km W of DUMYAT branch of the NILE River, 16 mi/26 km NE of TANTA, 65 mi/105 km N of CAIRO; 30°58′N 31°10′E. A major textile center of Egypt: cotton ginning, cotton and silk weaving; also cigarette manufacturing, rice milling. Agriculture (cotton, cereals, rice, fruits). Sometimes spelled Mehalletel Kubra and Mehallet el Kebir. In Egypt, popularly called Mahalla.

Mahallat (mah-ahl-LAHT), town, Markazī province, N central IRAN, 60 mi/97 km SSW of QOM, on the Qom River; 33°55′N 50°27′E. In fertile region (grain; fruit). Rug making. Hot spring. Home of Agha Khan family. Was capital of former Mahallat province.

Mahallet Damana (MAH-HAHL-let dah-MAH-nuh), village, DAQAHLIYA province, Lower EGYPT, on El BAHR ES SAGHIR Canal opposite MINYET MAHALLET DAMANA, and 7 mi/11.3 km ENE of MANSURA. Cotton, cereals.

Mahallet Marhum (MAH-HAHL-let mahr-HOOM), village, GHARBIYA province, Lower EGYPT, 3 mi/4.8 km WNW of TANTA; 30°48′N 30°57′E. Cotton.

Mahallet Minuf (MAH-HAHL-let me-NOOF), village, GHARBIYA province, Lower EGYPT, 7 mi/11.3 km NNW of TANTA; 30°53′N 30°58′E. Cotton.

Mahallet Zaiyad (MAH-HAHL-let zah-YAHD), village, GHARBIYA province, Lower EGYPT, 5 mi/8 km NE of El MAHALLA EL KUBRA; 31°02′N 31°14′E. Cotton.

Maham (muh-HUHM), town, ROHTAK district, PUNJAB state, N INDIA, 17 mi/27 km WNW of Rohtak; 28°59′N 76°18′E. Millet, gram, wheat, cotton. Sometimes spelled Meham and Mahm.

Mahambo (mah-HAHM-boo), town, TOAMASINA province, E MADAGASCAR; 17°29′S 49°26′E. Tourism (beach hotels).

Mahameroe, Mount, INDONESIA: see SEMERU, MOUNT.

Mahameru, Mount, INDONESIA: see SEMERU, MOUNT.

Mahamuni (mah-hah-MOO-nee), village, KYAUKTAW township, RAKHINE STATE, MYANMAR, 5 mi/8 km E of KYAUKTAW, on arm of KALADAN RIVER. Pilgrimage center with noted pagoda; head of navigation. Also known as Thayettabin.

Mahan (mah-AHN), town, Kermān province, SE IRAN, 20 mi/32 km SE of KERMAN, and on road to ZAHEDAN; 30°03′N 57°17′E. Grain, cotton, fruit. Rug making (Kerman rugs). Has tomb of 15th-century religious leader of the Sufi sect; center of pilgrimage. Also spelled MAHUN.

Mahanadi (mah-ha-NUHN-dee), river, c.550 mi/885 km long, central and E INDIA; rises in central CHHATTISGARH state, central India; flows N then E through a gorge in the Eastern GHATS, across ORISSA state, forming a delta before entering the BAY OF BENGAL near CUTTACK. The Tel and Hasdo rivers are its main tributaries. The HIRAKUD Dam, at SAMBALPUR, is the world's largest earthen dam (2.4 mi/3.8 km long) and supports several hydroelectric plants. There is also a concrete dam (0.75 mi/1.2 km long). The Mahanadi irrigates a fertile valley whose chief crops are rice, oilseeds, and sugarcane.

Mahananda River (mah-HAH-nuhn-dah), c.225 mi/362 km long, in E INDIA (WEST BENGAL and BIHAR states) and NW BANGLADESH; formed in SE NEPAL HIMALAYA by headstreams joining NW of KURSEONG; flows SSW past Shiliguri, and SSE through rich agriculture area (rice, jute, barley, corn, mustard), past Old MALDAH (West Bengal state), ENGLISH BAZAR, and NAWABGANJ (Bangladesh), to PADMA RIVER at GODAGARI.

Mahanayim (mah-khah-NAH-yeem) or **Mahanaim**, kibbutz, NE ISRAEL, in Upper Jordan Valley, 4 mi/6.4 km ENE of ZEFAT; 32°58′N 35°34′E. Elevation 882 ft/268 m. Tile manufacturing; mixed farming. Founded 1898, but repeatedly abandoned until the present kibbutz settled here in 1939. Withstood heavy Syrian attacks in the 1948 war.

Mahanoro (mah-NOOR), town, TOAMASINA province, E MADAGASCAR, small port on coast and Canal des PANGALANES, just N of mouth of MANGORO RIVER, 125 mi/201 km SSW of TOAMASINA; 19°53′S 48°48′E. Coffee-shipping center; also vanilla, cloves.

Mahanoy City (mah-HA-noi), borough (2006 population 4,389), SCHUYLKILL county, E central PENNSYLVANIA, 10 mi/16 km NNE of POTTSVILLE, on MAHANOY CREEK; 40°48′N 76°08′W. Agriculture (corn, hay; poultry); dairying. Anthracite coal. In area are Locust Lake (S) and Tuscarora (E) state parks. Settled 1859, incorporated 1863.

Mahanoy Creek (mah-HA-noi), c.60 mi/97 km long, E central PENNSYLVANIA; rises in NE SCHUYLKILL county; flows W, past Mahony City, Fruckville, and ASHLAND, through anthracite coal-mining area, to Susquehanna River 9 mi/14.5 km S of Sunbury; 40°49′N 76°05′W.

Mahantango Creek (mai-uhn-TANG-go), c.35 mi/56 km long, E central PENNSYLVANIA; rises in W

SCHUYLKILL county; flows WSW to SUSQUEHANNA RIVER 5 mi/8 km N of MILLERSBURG; 40°43′N 76°27′W. Forms border between NORTHUMBERLAND and DAUPHIN counties.

Maha Oya (MUH-huh O-yuh), town, EASTERN PROVINCE, SRI LANKA, 28 mi/45 km SW of BATTICALOA; 07°32′N 81°21′E. Hot springs nearby.

Mahara, town, WESTERN PROVINCE, SRI LANKA; 06°59′N 79°56′E. Trades in coconuts, betel leaves, rice, and cattle.

Maharagama, town, WESTERN PROVINCE, SRI LANKA; 06°51′N 79°55′E. Residential suburb of COLOMBO. Garment factories. Cancer hospital.

Maharajganj (mah-hah-RAHJ-guhnj), district (□ 1,138 sq mi/2,958.8 sq km), UTTAR PRADESH state, N central INDIA; ⊙ Maharajganj.

Maharajganj (mah-hah-RAHJ-guhnj), town, tahsil headquarters, GORAKHPUR district, E UTTAR PRADESH state, N central INDIA, 30 mi/48 km NNE of Gorakhpur. Rice, wheat, barley.

Maharajganj, town, ⊙ MAHARAJGANJ district, central UTTAR PRADESH state, N central INDIA, 11 mi/18 km NNE of RAE BARELI. Rice, wheat, barley, gram. Sometimes spelled Maharajgunj.

Maharajnagar, INDIA: see CHARKHARI.

Maharajpur (mah-HAH-rahj-puhr), town (2001 population 21,532), CHHATARPUR district, MADHYA PRADESH state, central INDIA, 11 mi/18 km NE of CHHATARPUR; 25°01′N 79°44′E. Millet; gram, oilseeds; betel gardens.

Maharas (mah-HAH-res), town (2004 population 14,499), SAFAQIS province, E TUNISIA, fishing port on GULF OF QABIS, 20 mi/32 km SW of SAFAQIS. Oil processing and marketing center.

Maharas (mah-HAH-res), oil field, SAFAQIS province, SE TUNISIA, 14 mi/23 km SW of SAFAQIS, on coast of Gulf of Safaqis. Pipeline SW to AS-SAKHIRA.

Maharashtra (muh-hah-RAHSH-truh), state (□ 118,808 sq mi/308,900.8 sq km; 2001 population 96,878,627), W central INDIA, on the ARABIAN SEA; ⊙ MUMBAI (Bombay). Formed in 1960, when the former bilingual state of BOMBAY was split along linguistic lines into two new states, MAHARASHTRA and GUJARAT. Marathi is the official language and Hinduism is the predominant religion. The mountains of the WESTERN GHATS run parallel to the coast, leaving a narrow strip known as the KONKAN between the ARABIAN SEA and the interior plateau. There is a series of small ports along the coast in addition to Mumbai. Beyond the Western Ghats is a vast plateau drained W by the TAPI, GODAVARI, BHIMA, and KRISHNA Rivers and E by the WARDHA and Vainganga Rivers. Cotton is cultivated in the fertile belts of the Tapi valley. The heaviest rainfall is along the coastal area, where it averages 80 in/203 cm–120 in/305 cm a year. The climate in general is tropical. The plateau areas enjoy only 25 in/64 cm–80 in/203 cm of rainfall annually, creating a semiarid climatic zone. Rice, grown in the coastal area, is the primary food crop, but it is supplemented by the production of grain, sorghum, small millet, and sugarcane. The state is rich in minerals; manganese, iron ore, bauxite, coal, and salt are mined. Industry, including the manufacturing of textiles, electrical products, chemicals, and pharmaceuticals is mainly concentrated in Mumbai, PUNE, AURANGABAD, THANE, and NAGPUR. The Muslim rulers of India controlled the area of what is now Maharashtra from the early 14th century to the mid-17th century, when the great Maratha leader Shivaji founded a Maratha empire which functioned as a confederacy. In the 16th century, PORTUGAL was the leading foreign power in the region, but GREAT BRITAIN gradually gained influence and by the early 19th century had incorporated the area into the BOMBAY presidency, which later became a province of British India. Governed by a chief minister and cabinet responsible to a bicameral legislature with one elected house and by a governor appointed by the president of India.

Mahasamund (mah-HAH-suh-moond), district (□ 1,505 sq mi/3,913 sq km; 2001 population 860,176); ⊙ MAHASAMUND; E central CHHATTISGARH state, central INDIA. Bordered in W, NW, and S by RAIPUR district; N by RAIGARH district; E by ORISSA state. Formed in 1998 out of E RAIPUR district in MADHYA PRADESH; joined other districts to form Chhattisgarh state in 2000. Mineral deposits include gold, tin ore, lead ore, fluorite, beryl, granite, shale, sandstone, and limestone. Resource exploration has been limited; the economy is largely agricultural: rice, wheat, peanuts.

Mahasamund (mah-HAH-suh-moond), town (2001 population 47,203), Mahasamund district, E central CHHATTISGARH state, central INDIA, 32 mi/51 km ESE of RAIPUR; 21°06′N 82°06′E. Rice milling. Sal, bamboo, myrobalan in nearby forests (sawmilling).

Maha Sarakham (MUH-HAH SAH-ruh-KAHM), province (□ 3,463 sq mi/9,003.8 sq km), Northeastern region, THAILAND, near the center of the KORAT PLATEAU; ⊙ MAHA SARAKHAM; 16°20′N 103°40′E. Cotton, rice, corn, tobacco; hog raising.

Maha Sarakham (MUH-HAH SAH-ruh-KAHM), town, ⊙ MAHA SARAKHAM province, NE THAILAND, in KORAT PLATEAU, on CHI RIVER, and 55 mi/89 km ESE of KHON KAEN; 16°11′N 103°18′E. Road center; cotton, rice, corn, tobacco; hog raising. Sometimes spelled Mahasaragam. N section of province was separated in 1940s to form KALASIN province.

Mahaska, county (□ 573 sq mi/1,489.8 sq km; 2006 population 22,298), S central IOWA; ⊙ OSKALOOSA; 41°19′N 92°38′W. Rolling prairie agricultural area (hogs, cattle, sheep, poultry; corn, oats, hay) drained by DES MOINES, SKUNK, and NORTH SKUNK rivers. Bituminous-coal deposits mined in SW; limestone quarries. Includes Lake Keomah State Park in E. River flooding in 1993. Formed 1843.

Mahaska (muh-HAS-kuh), village (2000 population 107), WASHINGTON county, N KANSAS, near NEBRASKA state line, 20 mi/32 km NW of WASHINGTON; 39°59′N 97°20′W. In grain and livestock area.

Mahasu (muh-HAH-soo), an important former district, HIMACHAL PRADESH state, N INDIA, now part of SHIMLA district.

Mahates (mah-HAH-tais), town, ⊙ Mahates municipio, BOLÍVAR department, N COLOMBIA, in CARIBBEAN LOWLANDS, near Canal del DIQUE, 28 mi/45 km SE of CARTAGENA; 10°14′N 75°12′W. Sugarcane, corn, plantains; livestock.

Mahatsinjo (mah-TSEEN-zoo), village, MAHAJANGA province, N central MADAGASCAR, on highway, and 60 mi/97 km SSE of MAEVATANANA; 17°44′S 47°01′E. Rice; cattle.

Mahatwar, INDIA: see SAHATWAR.

Mahavavy River (mah-VAHV), c.100 mi/161 km long, in ANTSIRANANA province, N MADAGASCAR; rises in TSARATANANA MASSIF (14°08′S 49°05′E); flows N past AMBILOBE and NW, forming a large estuary, to MOZAMBIQUE CHANNEL.

Mahavavy River, c.200 mi/322 km long, in W central and NW MADAGASCAR; rises in ANTANANARIVO province 200 mi/322 km WNW of ANTANANARIVO (18°25′S 45°45′E); flows N into MAHAJANGA province, past MITSINJO and NAMAKIA to MOZAMBIQUE CHANNEL.

Mahavelona (mah-hah-VAI-loon), town, TOAMASINA province, E MADAGASCAR, on coast, 30 mi/48 km N of TOAMASINA, on Canal des PANGALANES (N terminus); 17°40′S 49°25′E. Tourist center. Formerly Foulpointe (FOOL-pwant).

Mahavinyaka (muh-HAH-vin-yuh-kuh), sacred hill (1,294 ft/394 m) in CUTTACK district, E ORISSA state, E central INDIA, 20 mi/32 km NE of Cuttack. Hindu shrines.

Mahaweli Ganga (MUH-HUH-wai-lee GAHN-gah), longest (208 mi/335 km) river of SRI LANKA; rises in CENTRAL PROVINCE, on HATTON PLATEAU just NW of HATTON; flows N through extensive tea and rubber plantations on Hatton and KANDY plateaus, past NA-

WALAPITIYA, GAMPOLA, KADUGANNAWA, and KANDY, E past MINIPE, and N past ALUTNUWARA through NORTH CENTRAL PROVINCE and across EASTERN PROVINCE to KODDIYAR BAY 7 mi/11.3 km S of TRINCOMALEE. Main tributary is the AMBAN GANGA River (left). In middle course, forms border between Central and Uva provinces. A multipurpose river-development scheme launched in the late 1970s has resulted in the construction of reservoirs at Ulhitiya (1982), Ratkinda (1983), VICTORIA (1984), MADURU OYA (1982), KOTMALE (1984), RANDENIGALA (1986), and Rantambe (1989). Together they provide 500 MW of hydropower and water to irrigate 500 sq mi/1,295 sq km of new land.

Mahaxay (mah-HAHK-sai), town, ⊙ MAHAXAY district, KHAMMUAN province, central LAOS, E of Thaket town; 17°25′N 105°10′E. Market center.

Mahbubabad (mah-BOO-bah-bahd), town, WARANGAL district, ANDHRA PRADESH state, S INDIA, 40 mi/64 km SE of WARANGAL; 17°37′N 80°01′E. Rice and oilseed milling; matches. Sometimes spelled Mahboobabad.

Mahbubnagar (mah-BOOB-nuh-gahr), district (□ 7,117 sq mi/18,504.2 sq km), ANDHRA PRADESH state, S INDIA, on DECCAN PLATEAU; ⊙ MAHBUBNAGAR. Bounded S by KRISHNA RIVER; mainly lowland except for forested hills in SE (teak, ebony, gum arabic). Largely sandy red soil; millet, oilseeds (chiefly castor beans, peanuts), rice. Oilseed and rice milling, biri manufacturing, tanning; livestock raising. Main towns are Mahbubnagar, NARAYANPET. Became part of HYDERABAD during state's formation in 18th century; later part of Andhra Pradesh state after independence.

Mahbubnagar (mah-BOOB-nuh-gahr), city, ⊙ MAHBUBNAGAR district, ANDHRA PRADESH state, S INDIA, 55 mi/89 km SW of HYDERABAD; 16°44′N 77°59′E. Road center. Millet, tobacco; oilseed and rice milling here and in surrounding villages.

Mahdah, OMAN: see MAHADHA.

Mahdalynivka (mah-hdah-LI-neef-kah), (Russian *Magdalinovka*), town, N DNIPROPETROVS'K oblast, UKRAINE, on Chaplynka River, left tributary of the ORIL' RIVER, 30 mi/48 km N of DNIPROPETROVS'K; 48°53′N 34°56′E. Elevation 364 ft/110 m. Raion center; dairy, feed and flour mill; brickworks. Heritage museum. Established in the 18th century, town since 1958.

Mahd edh Dhahab (MAH-hid ed de-HAHB), village, central HEJAZ, MEDINA province, SAUDI ARABIA, 100 mi/161 km SE of Medina; 23°30′N 40°52′E. Small quantities of gold have been mined here. Also spelled Mahad Dhahab.

Mähdelegabel (MAI-de-le-gah-bel), peak (8,059 ft/2,456 m) of Allgäu ALPS, on Austro-German border, 7 mi/11.3 km S of Oberstdorf; 47°18′N 10°18′E.

Mahdere Maryam (mah-HAH-der MER-ee-ahm), village, AMHARA state, NW ETHIOPIA, 13 mi/21 km SW of DEBRE TABOR; 11°43′N 37°55′E. In cereal growing and stock raising (horses, mules, sheep) region; trade center. Has church and monastery.

Mahdha, OMAN: see MAHADHA.

Mahdia, province, TUNISIA: see MAHDIYAH, province.

Mahdia (mah-DYAH), town, TIARET wilaya, N central ALGERIA, on the SERSOU PLATEAU, 25 mi/40 km ENE of TIARET; 35°26′N 01°37′E. Formerly called Burdeau.

Mahdia (MAH-dyah), village, POTARO-SIPARUNI district, central GUYANA, along small river (affluent of the POTARO River) of the same name, 18 mi/29 km SW of TUMATUMARI; 05°16′N 59°09′W. Near gold and diamond mines.

Mahdia, TUNISIA: see MAHDIYAH.

Mahdiyah, province (□ 1,145 sq mi/2,977 sq km; 2006 population 384,300), NE central TUNISIA, on MEDITERRANEAN SEA (E); ⊙ MAHDIYAH; 35°20′N 10°35′E. Also Mahdia and Al Mahdiyah.

Mahdiyah (mah-DEE-yah), town (2004 population 45,977), ⊙ MAHDIYAH province, E TUNISIA, fishing port on the central MEDITERRANEAN SEA, 44 mi/71 km SE of SUSAH; 33°30′N 11°04′E. Railroad spur terminus. Sardine canning, fish salting and drying, olive oil processing, soap manufacturing, textile spinning. Saltworks. Situated on a narrow promontory (1 mi/1.6 km long) called Cape Africa, Mahdiyah preserves a casbah (16th century) and old fortifications. The port (S) was cut out of sheer rock. Founded C.E. 912 on the site of a Phoenician and Roman settlement, it became the Fatimite dynasty capital. A stronghold of Barbary pirates in the 16th–17th century, the town was visited by several European punitive expeditions. MAHDIYAH province was carved out of the SUSAH province in 1941. Also spelled Mahedia, Mahdia.

Mahe, district (□ 4 sq mi/10.4 sq km), PONDICHERRY Union Territory, on SW coast of INDIA, on ARABIAN SEA; ⊙ Mahe. Consists of town proper (on coast) and a detached rural section (just NE). Formerly part of FRENCH INDIA; acceded to India in 1962.

Mahe, town, ⊙ Mahe district, PONDICHERRY Union Territory, on the SW coast of INDIA, on the ARABIAN SEA, 35 mi/56 km NW of KOZHIKODE; 11°42′N 75°32′E. Trade and administrative center. Also a tourist center (beach). Trades in coconuts, mangoes, areca nuts, and in pepper from forested hills of the WAYANAD; fishing. Occupied c.1725 by FRENCH; taken twice (1761, 1779) by ENGLISH; finally restored 1817 to French. Acceded to India 1962.

Mahé: see SEYCHELLES.

Mahébourg, town (2000 population 15,753), ⊙ GRAND PORT district, SE MAURITIUS, port on the Grand Port (inlet of INDIAN OCEAN), 20 mi/32 km SE of PORT LOUIS; divided in half by Rivière La Chaux, opposite Black City (joined by bridge). Rich sugar-growing region and tourist area. Fishing. Sugar milling and alcohol distilling at BEAU VALLON, just S. Was capital of Mauritius in 18th century. Developed as auxiliary naval base during World War II. International airport at Plaisance, 3 mi/4.8 km SW. Pointe Canon monument commemorating abolition of slavery; naval museum (built c.1771). Domaine du Chasseur, 2,000-acre/809-ha park. Town named for former governor Mahé de Labourdonnais.

Mahedia, TUNISIA: see MAHDIYAH.

Mahee (ma-HEE), island, in Strangfold Lough, NE DOWN, SE Northern Ireland, 7 mi/11.3 km SSE of NEWTOWNARDS; 1.5 mi/2.4 km long; 54°30′N 05°38′W. Golf course. Site of remains of Nendrum Abbey, probably founded in 5th century, restored 1922.

Mahé Group (mah-AI), archipelago of the SEYCHELLES. Granite group whose principal island is MAHÉ ISLAND; also includes SILHOUETTE, NORTH, PRASLIN, and LA DIGUE islands.

Mahé Island (mah-AI), main island (□ 87 sq mi/226.2 sq km; 2002 population 68,476) of SEYCHELLES, main island of MAHÉ GROUP, c.600 mi/966 km NE of MADAGASCAR; 4°42′S 55°28′E. Country's political and economic center, with 90% of Seychelles population. VICTORIA, the capital, is on NE coast. Tourism; fishing; cinnamon, tea, vanilla; coconut products.

Mahendraganj (muh-HAIN-druh-guhnj), town, WEST GARO HILLS district, MEGHALAYA state, NE INDIA, near BRAHMAPUTRA River, 27 mi/43 km WSW of Tura; 25°18′N 89°51′E. Rice, cotton, mustard, jute.

Mahendragarh (muh-HAIN-druh-gahr), town, NARNAUL district, HARYANA state, N INDIA, 29 mi/47 km W of REWARI; 28°17′N 76°09′E. Local market for millet, gram; metal handicrafts. Formerly called Kanaud.

Mahendragarh, INDIA: see NARNAUL.

Mahendragiri (muh-HAIN-druh-gir-ee), peak (4,923 ft/1,501 m) of EASTERN GHATS, in GANJAM district, SE ORISSA state, E central INDIA, 36 mi/58 km SW of BRAHMAPUR. Temples at summit include one built (11th century) by Chola dynasty. Sometimes written Mahendra Giri.

Mahendragiri, peak (5,427 ft/1,654 m) near S end of WESTERN GHATS, KANNIYAKUMARI district, TAMIL

NADU state, extreme S INDIA, 45 mi/72 km ESE of Tiruvanthapuram; 08°23′N 77°30′E.

Mahendranagar (muh-HEN-druh-nuh-guhr), city (2001 population 80,839), ⊙ KANCHANPUR district, SW NEPAL; 28°57′N 80°09′E. Elevation 650 ft/198 m. Airport.

Mahendra Rajmarg (muh-HEN-druh RAHJ-mahrg), road, E-W highway of NEPAL, extending 635 mi/1,024 km between Kakarbhita on the E (INDIA) border and 4 mi/6km W of Magendranagar on the W (also India) border.

Mahenge (mah-HAIN-gai), town, MOROGORO region, central TANZANIA, 140 mi/225 km SSW of MOROGORO; 08°42′S 36°41′E. Corn, wheat, beans, sweet potatoes, tobacco, pyrethrum; cattle, sheep, goats. Mount Chikweta to SW.

Maheriv (mah-HE-reef), (Russian *Magerov*), (Polish *Magierów*), town (2004 population 8,100), NW L'VIV oblast, UKRAINE, on road, 24 mi/38 km NW of L'VIV, and 9 mi/14 km SSE of RAVA-RUS'KA; 50°07′N 23°43′E. Elevation 793 ft/241 m. Machine testing, habedashery manufacturing. Known (and part of Poland) since the end of the 14th century; under Austrian rule (1772–1918); part of or contested by West Ukrainian Republic (1918–1919); reverted to Poland in 1919; occupied by the Soviets (1939–1941), by the Germans (1941–1944); ceded to USSR (Ukrainian SSR) in 1945; part of independent Ukraine since 1991. Town since 1990.

Mahesana, district (□ 3,485 sq mi/9,061 sq km), GUJARAT state, W central INDIA; ⊙ MAHESANA. Bounded W by RANN of Cutch, E by SABARMATI RIVER. Agriculture (millet, oilseeds, cotton, wheat); hand-loom weaving, tanning. PATAN, SIDDHAPUR, KALOL, and KADI are cotton-milling centers. District formed 1949 by merger of most of Mehsana division of former BARODA state and parts of former WESTERN INDIA STATES, including RADHANPUR and IDAR.

Mahesana, town, ⊙ MAHESANA district, GUJARAT state, W central INDIA, 40 mi/64 km NNW of AHMADABAD; 23°36′N 72°24′E. Major railroad junction; trades in millet, wheat, oilseeds, cotton fabrics; handicraft cloth weaving, chemical manufacturing.

Maheshpur, BANGLADESH: see MAHESPUR.

Maheshwar (muh-HAISH-wuhr), town (2001 population 19,646), WEST NIMAR district, central MADHYA PRADESH state, SW INDIA, on NARMADA River, and 40 mi/64 km SSW of INDORE; 22°11′N 75°40′E. Markets cotton, millet, oilseeds; handicraft cloth weaving and dyeing (saris, dhotis). An ancient site in Hindu epics Mahabharatha and Ramayana; capital of Holkar family, Maratha rulers of Indore, c.1724–1947.

Mahespur (mo-hes-poor), town (2001 population 23,100), JESSORE district, W EAST BENGAL, BANGLADESH, on river arm of GANGA DELTA, and 23 mi/37 km NW of JESSORE; 25°59′N 90°12′E. Rice, jute, oilseeds, tobacco. Also spelled Maheshpur.

Mahia Peninsula (MAH-hyah), WAIROA district, E NORTH ISLAND, NEW ZEALAND, marks N end of HAWKE BAY, 55 mi/88 km N of NAPIER; 12 mi/20 km long, 7 mi/12 km wide. Joined to mainland by narrow sandy isthmus.

Mahidasht (mah-ee-DAHSHT), village, Kermānshahān province, W IRAN, 15 mi/24 km WSW of KERMANSHAH, and on road to QASR-E-SHIRIN; 28°56′N 58°50′E. Grain, tobacco; sheep raising (wool); rug making.

Mahidpu, INDIA: see MEHIDPUR.

Mahidpur, INDIA: see MEHIDPUR.

Mahim (MAH-heem), village, THANE district, MAHARASHTRA state, W central INDIA, on ARABIAN SEA, 45 mi/72 km N of MUMBAI (BOMBAY). Trades in sugarcane, plantains, rice; betel farming, fishing (pomfrets), bone fertilizer manufacturing. Also called KELVE.

Mahim (MAH-heem), neighborhood, in NW MUMBAI (BOMBAY), MAHARASHTRA state, W INDIA, on Mahim Bay; 19°39′N 72°44′E.

Mahin (mah-HEEN), town, ONDO state, SW NIGERIA, on coast of Bight of BENIN, 100 mi/161 km ESE of LAGOS; 06°10′N 04°48′E. Market town. Kola nuts, maize, cocoyams.

Mahina (mah-HEE-nah), township and railroad station for BAFOULABÉ, FIRST REGION/KAYES, SW MALI, on BAFING RIVER, on DAKAR (SENEGAL)-NIGER railroad, and 3.5 mi/5.6 km S of Bafoulabé; 13°46′N 10°51′W. Vocational school.

Mahi River, c.360 mi/580 km long, W central INDIA; rises in W VINDHYA RANGE just S of SARDARPUR, W MADHYA PRADESH state; flows N into S RAJASTHAN state, and SW through N GUJARAT state to head of GULF OF KHAMBAT.

Mahiyangana, town, UVA PROVINCE, central SRI LANKA, on the MAHAWELI GANGA River, and 25 mi/40 km E of KANDY; 07°19′N 80°59′E. Rice, vegetables. Buddhist pilgrimage center; site of ancient stupa. Of strategic importance in ancient times. Also known as ALUTNUWARA.

Mahjaba (mah-HAHZ-uh-buh), township, S YEMEN, 70 mi/113 km NNE of ADEN; 13°49′N 45°13′E. Was capital of the former Upper YAFA sultanate. Also spelled Majaba.

Mahlaing (mah-LEIN), township, MANDALAY division, MYANMAR, on railroad, and 20 mi/32 km NW of MEIKTILA; 21°06′N 95°39′E.

Mahlata, Bulgaria: see PELOVO.

Mahlberg (MAHL-berg), town, S UPPER RHINE, BADEN-WÜRTTEMBERG, GERMANY, 5 mi/8 km SSW of LAHR; 48°16′N 07°48′E. Tobacco industry. Has castle and tobacco museum.

Mahlow (MAH-lou), village, BRANDENBURG, E GERMANY, 12 mi/19 km S of BERLIN; 52°22′N 13°24′E. Railroad junction. Fruit.

Mahm, INDIA: see MAHAM.

Mahmudabad (MAH-moo-dah-bahd), town, SITAPUR district, central UTTAR PRADESH state, N central INDIA, 32 mi/51 km SE of Sitapur; 27°18′N 81°07′E. Brass ware manufacturing; trades in wheat, rice, gram, barley, sugarcane.

Mahmudiya, Al (MAH-moo-DEE-yeh, ahl), town, BAGHDAD province, central IRAQ, on railroad, and 20 mi/32 km S of BAGHDAD, between the TIGRIS and the EUPHRATES rivers; 33°03′N 44°21′E. Market gardening; agricultural products processing. Dates, sesame, millet.

Mahmudiya Canal (mah-moo-DEE-yuh), navigable canal, Lower EGYPT, extends 50 mi/80 km from RASHID branch of the NILE River at El ATF, W through BEHEIRA province, to MEDITERRANEAN SEA at ALEXANDRIA; average width, 100 ft/30 m. Constructed 1819, it diverted most of the Nile trade to Alexandria.

Mahmudiya, El (mah-moo-DEE-yuh, el), town, BEHEIRA province, Lower EGYPT, at E end of MAHMUDIYA CANAL, on RASHID branch of the NILE River, 10 mi/16 km NNE of DAMANHUR; 31°11′N 30°32′E. Wool weaving; cotton.

Mahnomen (muh-NO-men), county (□ 583 sq mi/1,515.8 sq km; 2006 population 5,072), NW MINNESOTA; ⊙ MAHNOMEN; 47°19′N 95°48′W. Drained by WILD RICE and White Earth rivers. Agricultural area (wheat, hay, alfalfa, oats, barley, sunflowers, wild rice). Nearly entire county is within White Earth Indian Reservation which extends into CLEARWATER (E) and BECKER (S) counties; small section of White Earth State Forest in E and SE. Formed 1906.

Mahnomen, town (2000 population 1,202), ⊙ MAHNOMEN county, NW MINNESOTA, 48 mi/77 km WSW of BEMIDJI, on WILD RICE RIVER, at mouth of White Earth River, in White Earth Indian Reservation; 47°18′N 95°58′W. Elevation 1,212 ft/369 m. Trading point for agricultural area (livestock; wild rice, grain, sunflowers, alfalfa); manufacturing (building materials, chemicals, apparel). White Earth State Forest to E.

Mahnuwara, SRI LANKA: see KANDY.

Maho, town, NORTH WESTERN PROVINCE, SRI LANKA, 24 mi/39 km NNW of KURUNEGALA; 07°49′N 80°16′E. Railroad junction. Rice, vegetables, coconuts.

Mahoba (muh-HO-bah), town, HAMIRPUR district, S UTTAR PRADESH state, N central INDIA, on railroad, and 80 mi/129 km ESE of JHANSI; 25°17′N 79°52′E. Road junction; trade center (gram, jowar, oilseeds, wheat, pearl millet). Founded C.E. c.800 by a noted CHANDEL Rajput; civil capital of Chandel Rajputs until succeeded (c.1182) by KALINJAR. Captured by Kutbud-din Aibak in 1202. Has 14th-century mosque. Extensive Chandel Rajput, Jain, and Buddhist remains surround nearby artificial lakes, including 24 rock-hewn images dated 1149.

Mahoda (muh-ho-dah), district (□ 1,185 sq mi/3,081 sq km), UTTAR PRADESH state, N central INDIA; ⊙ Mahoda.

Mahoda (muh-ho-dah), town, ⊙ Mahoda district, UTTAR PRADESH state, N central INDIA.

Mahomet (mah-HOM-it), village (2000 population 4,877), CHAMPAIGN county, E central ILLINOIS, on SANGAMON RIVER, and 9 mi/14.5 km WNW of CHAMPAIGN; 40°11′N 88°24′W. In agricultural area.

Mahón (mah-ON), Catalan *Maó*, city, ⊙ and chief town of MINORCA island, Baleares province, in the W MEDITERRANEAN SEA; 39°53′N 04°15′E. A port with an excellent natural harbor defended by two fortresses, it is also an important air and naval base. Center of costume jewelry production; shoes, cheese. Named for the Carthaginian general Mago.

Mahonda (mah-HON-dah), village, N ZANZIBAR, Tanzania, 13 mi/21 km NNE of ZANZIBAR city, in N part of Zanzibar island, 3 mi/4.8 km inland (E of) ZANZIBAR CHANNEL, INDIAN OCEAN; 05°59′S 39°14′E. Cloves, copra, citrus fruit.

Mahone Bay (muh-HON), town (2001 pop. 991), S NOVA SCOTIA, E CANADA, on W shore of MAHONE BAY, 40 mi/64 km WSW of HALIFAX; 44°26′N 64°22′W. Shipbuilding, tourism.

Mahone Bay (muh-HON), inlet of the ATLANTIC OCEAN, off S NOVA SCOTIA, E CANADA, 30 mi/48 km WSW of HALIFAX; 12 mi/19 km long, 9 mi/14 km wide at entrance. On bay are towns of CHESTER and MAHONE BAY. There are numerous islands, including OAK ISLAND, reputed site of Captain Kidd's hidden treasure. TANCOOK ISLAND (3 mi/5 km long) is the largest.

Mahoning (muh-HO-ning), county (□ 419 sq mi/1,089.4 sq km; 2006 population 251,026), NE OHIO, on PENNSYLVANIA state line; ⊙ YOUNGSTOWN; 41°02′N 80°46′W. Intersected by MAHONING and LITTLE BEAVER rivers. Agriculture (livestock, poultry; grains; dairy products). Manufacturing of industrial machinery. Coal mining; limestone quarries. Includes LAKE MILTON reservoir and part of Berlin Reservoir. Formed 1846.

Mahoning Creek (ma-HO-ning), c.60 mi/97 km long, W central PENNSYLVANIA; rises in W CLEARFIELD county; flows generally W, past PUNXSUTAWNEY, through Mahoning Creek Lake (flood-control reservoir), to ALLEGHENY RIVER, just N of TEMPLETON, 9 mi/14.5 km N of KITTANNING; 41°02′N 78°42′W. Flood-control dam is c.22 mi/35 km above mouth. Another Mahoning Creek in E Pennsylvania; rises in E SCHUYLKILL county; flows E c.15 mi/24 km to LEHIGH RIVER at LEHIGHTON.

Mahoning River (muh-HON-ing), c.90 mi/145 km long, OHIO and PENNSYLVANIA; rises in NE Ohio, E of CANTON; 40°57′N 80°22′W; flows NW to ALLIANCE, then NE past WARREN, where it turns SE to flow past YOUNGSTOWN into NW Pennsylvania and joins the SHENANGO RIVER to form the BEAVER RIVER. The river drains a fertile valley. Berlin Dam (completed 1943), on the upper Mahoning, provides flood control and water supply.

Mahon, Lough (MA-huhn, LAHK), NW reach of CORK HARBOUR and estuary of LEE RIVER, CORK county, SW IRELAND; 4 mi/6.4 km long.

Mahopac (Mah-HO-pak; MAY-oh-pak), residential village (□ 6 sq mi/15.6 sq km; 2000 population 8,478), PUTNAM county, SE NEW YORK, on E shore of LAKE MAHOPAC (c.1.5 mi/2.4 km in diameter), 12 mi/19 km NE of PEEKSKILL; 41°22′N 73°44′W. Former grist mill at nearby Mahopac Falls.

Mahora (mah-O-rah), town, ALBACETE province, SE central SPAIN, 17 mi/27 km NE of ALBACETE; 39°13′N 01°44′W. Brandy distilling; livestock raising, lumbering. Wine, cereals, saffron, fruit.

Mahora, PAKISTAN: see MAHURA.

Mahoré, Comoros Islands: see MAYOTTE.

Mahot (mah-HOT), settlement of SE OMAN, on Mahot island (1 mi/1.6 km across) in the Ghubbah Hashish (an inlet of ARABIAN SEA); 20°32′N 58°10′E. Formerly trade center of SE Oman coast. Sometimes spelled Mahwat.

Mahottari (muh-HO-tuh-ree), district, SE NEPAL, in JANAKPUR zone; ⊙ JALESWAR.

Mahou, town, THIRD REGION/SIKASSO, MALI, 90 mi/150 km NE of SIKASSO, on border with BURKINA FASO. Agriculture and livestock.

Mahra, YEMEN: see MAHRI.

Mahrah, YEMEN: see MAHRI.

Mahras (mah-RAHS), town, SAFAQIS province, SE TUNISIA, 13 mi/21 km SSW of SAFAQIS. Large agricultural town surrounded by olive groves.

Mahras, El (MAH-rahs, el), village, ASYUT province, central Upper EGYPT, on IBRAHIMIYA CANAL, on railroad, 19 mi/31 km S of MINYA; 27°49′N 30°48′E. Cereals, dates, sugarcane.

Mahrauli (mah-ROU-lee), or **Meherauli** or **Mehrauli**, neighborhood of S Delhi, DELHI state, N INDIA, 8 mi/12.9 km SSW of NEW DELHI city center; 28°31′N 77°11′E. Handicraft glass-bangle making, cotton weaving, pottery, and leather work. Just N, within ruined walls of a 12th century. Chauhan Rajput citadel, is Lal Kot (=red fort) (1052). Here are several mosques and tombs (late 12th–late 14th century)—remarkable examples of synthesis of Islamic and earlier architectural styles. Most notable is the Kutab (or Qutb) MINAR, a freestanding stone tower (erected 1225), rising 238 ft/73 m from base diameter of 47 ft/14 m; its lower three stores are of fluted red sandstone, upper two (rebuilt 1368) are faced with white marble. Nearby stands the famous Iron Pillar—a shaft of purely wrought iron (c.24 ft/7 m high; c.1 ft/0.3 m in diameter; has early Gupta inscriptions). Believed forged at least as early as 5th century C.E., this remarkable example of ancient metallurgical skill has remained rustproof through the centuries; brought here from an unknown site in 11th century. Sometimes called Qutb or OLD DELHI (for the ruins of several cities in vicinity which figured in early history of Delhi state).

Mähren, CZECH REPUBLIC: see MORAVIA.

Mahri (MUH-ree), tribal area of extreme SE YEMEN, which formed the mainland section of the former MAHRI sultanate of Qishn and Socotra; its capital was QISHN. Situated on ARABIAN SEA at entrance to GULF OF ADEN, it extends from c.50°30′E to the cape RAS DHARBAT 'ALI on OMAN border. Area has fisheries and some trade in frankincense. Chief centers are Qishn, SAIHUT, and HAFAT. British protection was extended to SOCOTRA in 1886, but the British did not exercise control over the Mahri mainland. Population speaks South Arabic dialect, similar to that of Socotra and differing from the other dialects spoken in S Yemen. Now province (Mahrah) in the extreme SW part of Yemen (formerly SOUTH YEMEN). Also spelled Mahra or Mahrah.

Mährisch-Altstadt, CZECH REPUBLIC: see STARE MESTO.

Mährisch-Budwitz, CZECH REPUBLIC: see MORAVSKE BUDEJOVICE.

Mährische Slowakei, CZECH REPUBLIC: see MORAVIAN SLOVAKIA.

Mährisch-Kromau, CZECH REPUBLIC: see MORAVSKY KRUMLOV.

Mährisch-Neustadt, CZECH REPUBLIC: see UNICOV.

Mährisch Pisek, CZECH REPUBLIC: see MORAVSKY PISEK.

Mährisch-Schönberg, CZECH REPUBLIC: see ŠUMPERK.

Mährisch-Trübau, CZECH REPUBLIC: see MORAVSKA TREBOVA.

Mährisch Weisskirchen, CZECH REPUBLIC: see HRANICE.

Mahroni (mah-RO-nee), town, LALITPUR district, S UTTAR PRADESH state, N central INDIA, 21 mi/34 km ESE of Lalitpur; 24°35′N 78°43′E. Jowar, oilseeds, wheat, gram. Sandstone quarry and extensive CHANDEL Rajput remains 23 mi/37 km S, at village of MADANPUR.

Mahtomedi (mah-duh-MEE-dei), town (2000 population 7,563), WASHINGTON county, E MINNESOTA, 9 mi/14.5 km NNE of downtown ST. PAUL, on E and SE shore of WHITE BEAR LAKE; 45°03′N 92°57′W. Agricultural area (corn, oats, soybeans, alfalfa; cattle, sheep); light manufacturing. Northport Airport to NE, Lakewood Communtiy College to W, in WHITE BEAR LAKE city; Northeast Metropolitan Technical College is here. Numerous small natural lakes in area. Incorporated 1931.

Mahun, IRAN: see MAHAN.

Mahura (mah-muh-RAH), village, MUZAFFARABAD district, AZAD KASHMIR, NE PAKISTAN, on JHELUM RIVER, and 4 mi/6.4 km NE of Uri. Hydroelectric station completed 1907. Ancient Hindu temple ruins nearby. Also spelled MAHORA.

Mahuva (MUH-hoo-vah), town, BHAVNAGAR district, GUJARAT state, W central INDIA, near GULF OF KHAMBAT, 55 mi/89 km SSW of BHAVNAGAR. Market center (grain, cotton, oilseeds, timber, cloth fabrics); cotton, oilseed, and flour milling, hand-loom weaving, wood carving; sawmills. Coconuts, betel vine, mangoes grown nearby. Several annual festival fairs celebrated here. Port is 2 mi/3.2 km S; exports cotton and grain. Town formerly in BHAUNAGAR state.

Mahwah (MAH-wah), village, BERGEN county, extreme N NEW JERSEY, near Ramapo River and NEW YORK state line, 12 mi/19 km N of PATERSON; 41°04′N 74°11′W. Manufacturing of transportation equipment and electronics; former auto plant. Largely residential. Seat of Ramapo College of New Jersey. Center for the 3,000 Ramapough mountain people, who also live in nearby RINGWOOD (N.J.) and Hillbury (N.Y.); subjects of an ongoing dispute regarding their status as a Native American tribe or as descendants of late 16th-century African and Dutch farmers.

Mahwat, OMAN: see MAHOT.

Mahya Dag, TURKEY: see BUYUK MAHYA.

Maia (MEI-ah), town, PÔRTO district, N PORTUGAL, 6 mi/9.7 km N of OPORTO. Distilling.

Maiamai, IRAN: see MEYAMEY.

Maiana (MEI-AH-nah), atoll (□ 10 sq mi/26 sq km; 2005 population 1,908), N GILBERT ISLANDS, KIRIBATI, W central PACIFIC OCEAN; 9 mi/14.5 km long; 55′N 173°E. Copra. Formerly Hall Island.

Maiao (MEI-OU), island (□ 3 sq mi/7.8 sq km) with volcanic core and coral flats, Windward group, SOCIETY ISLANDS, FRENCH POLYNESIA, S PACIFIC, 60 mi/97 km W of TAHITI, of which it is a dependency; 17°39′S 150°37′W. Coconuts, large crabs for Tahiti markets. Sometimes called Tubuai-Manu or Tupuaemanu.

Maibang (MEI-bahng), town, NORTH CACHAR HILLS district, central ASSAM state, NE INDIA, in NW foothills of BARAIL RANGE, 40 mi/64 km NNE of SILCHAR; 25°18′N 93°10′E. Rice, cotton, sugarcane. Was capital of 16th century Kachari kingdom. Rock carvings nearby.

Măicăneşti (MUH-ee-kuh-NESHT), village, GALAŢI county, E central ROMANIA, 22 mi/35 km NE of RÎMNICU SĂRAT; 45°30′N 27°30′E.

Maicao (mei-KAH-o), town, ⊙ Maicao municipio, LA GUAJIRA department, NE COLOMBIA, near VENEZUELA border, 46 mi/74 km E of ríohacha; 11°22′N 72°14′W. Reportedly a center of contraband trade with Venezuela. Airport.

Mai Ceu, ETHIOPIA: see MAYCH'EW.

Maîche (mah-EESH), town (□ 6 sq mi/15.6 sq km), DOUBS department, FRANCHE-COMTÉ region, E FRANCE, in the central JURA MOUNTAINS, 18 mi/29 km S of MONTBÉLIARD; 47°15′N 06°48′E. Dairying center. Also precision instruments.

Maida (MAI-duh), village, CAVALIER county, NE NORTH DAKOTA, port of entry at Canadian (MANITOBA) border, 15 mi/24 km N of LANGDON; 48°59′N 98°21′W. Founded in 1884 and the name is the Anglo Saxon word for maiden.

Maidan, township, DIYALA province, E IRAQ, near IRAN border, on E bank of DIYALA RIVER, c.125 mi/201 km NE of BAGHDAD.

Maidan, Al (MAI-din, ahl) [Arabic=the cities], name applied by the Arabs (C.E. c.641) to a group of cities in MESOPOTAMIA (modern IRAQ), including principally CTESIPHON and SELEUCIA. Also spelled Al Madain, Al Medain, or Al-Maydan.

Maidanek, POLAND: see MAJDANEK.

Maidanpek, SERBIA: see MAJDANPEK.

Maidan Shahr, city, ⊙ WARDAK province, E central AFGHANISTAN, 22 mi/35 km W of KABUL, at the junction of the Kabul-Kandahar highway and the road leading through central Afghanistan; 34°27′N 68°48′E.

Maiden (MAI-duhn), town (□ 4 sq mi/10.4 sq km; 2006 population 3,348), CATAWBA and LINCOLN counties, W central NORTH CAROLINA, 6 mi/9.7 km S of NEWTON; 35°34′N 81°12′W. Manufacturing (textiles, security alarms, apparel, wood products, furniture, machine components); service industries; agriculture (grain, soybeans, hay; poultry, cattle, hogs; dairying).

Maiden Castle (MAY-duhn), prehistoric fortress, DORSET, S ENGLAND, near DORCHESTER; 50°43′N 02°26′W. The finest earthwork in the British Isles (□ 0.2 sq mi/0.5 sq km) is here. Excavations in 1934–1937 revealed evidences of a Neolithic village, with a two-ditch irrigation system, indicating occupation around 2000 B.C.E. On the same site are remains of an Iron-Age fortified village (300 B.C.E.), which was eventually taken by the Romans. The inhabitants ceased to occupy the castle about A.D. 70 when they moved to a town in the nearby valley.

Maidenhead (MAY-duhn-hed), town (2001 population 58,848), Windsor and Maidenhead (previously in BERKSHIRE), S central ENGLAND, on the THAMES RIVER; 51°31′N 00°43′W. Residential town with brewing and milling industries; resort. The 13th-century stone bridge was rebuilt in the 1770s.

Maiden Rock, village (2006 population 119), PIERCE county, W WISCONSIN, on LAKE PEPIN (MISSISSIPPI RIVER), 11 mi/18 km E of RED WING (MINNESOTA), surrounded by bluffs; 44°34′N 92°18′W. Manufacturing (concrete products). Silicate-rock mine is nearby.

Maidens (MAI-duhnz), unincorporated village, GOOCHLAND county, central VIRGINIA, 1 mi/1.6 km S of GOOCHLAND, 24 mi/39 km WNW of RICHMOND, on JAMES RIVER; 37°40′N 77°52′W.

Maidens, Scotland: see KIRKOSWALD.

Maidens, The or **Hulin Rocks**, group of islets in the NORTH CHANNEL off E coast of ANTRIM, Northern Ireland, 7 mi/11.3 km NE of LARNE; 54°56′N 05°44′W. Has two lighthouses.

Maidi (MAH-i-dee), town, NW YEMEN, minor RED SEA port, 105 mi/169 km N of HODEIDA, near border with SAUDI ARABIA; 16°19′N 42°49′E. Considered part of ASIR until Yemen–Saudi Arabia treaty (1934). Sometimes spelled Maydi, Medi, or Midi.

Maidos, TURKEY: see ECEABAT.

Maidstone (MAID-ston), former township (□ 70 sq mi/182 sq km; 2001 population 14,042), S ONTARIO, E central CANADA; 42°12′N 82°53′W. Amalgamated into the town of LAKESHORE in 1999.

Maidstone (MAID-ston), town (2006 population 1,037), W SASKATCHEWAN, W CANADA, 50 mi/80 km WNW of NORTH BATTLEFORD; 53°05′N 109°18′W. Grain elevators.

Maidstone (MAYD-stohn), town (2001 population 89,684), ⊙ KENT, SE ENGLAND, on the MEDWAY RIVER, 8 mi/12.8 km S of CHATHAM; 51°16′N 00°31′E. Market city with paper, printing, brewing, and machinery-manufacturing industries. Evidence of a Roman station. Chillington Manor (Elizabethan) contains the Maidstone Museum, the public library, and the headquarters of the Kent Archaeological Society. The grammar school dates from 1549. Noteworthy are the Church of All Saints, founded in the 14th century; the palace of the archbishops; and Penenden Heath, a recreation ground. Has technical, art, and adult-education schools. William Hazlitt born here.

Maidstone, town, ESSEX co., NE VERMONT, on the CONNECTICUT River, just above GUILDHALL. Maidstone Lake (c.3 mi/4.8 km long; fishing), with state park.

Maidsville, unincorporated village, MONONGALIA county, N WEST VIRGINIA, 4 mi/6.4 km N of MORGANTOWN, on MONONGAHELA RIVER. Agriculture (grain, strawberries); livestock, poultry. Bituminous coal. Manufacturing (coal processing).

Maiduguri (mei-DOO-guh-ree), town (2004 population 1,020,000), ⊙ BORNO state, extreme NE NIGERIA. Railroad, road, and air transportation center serving NE Nigeria and parts of NIGER and CHAD. Important industrial center engaged in food processing and aluminum, steel, asbestos, and cement production. Leather goods made from the hides of crocodiles caught in Lake CHAD are a leading local product. Groundnuts, cotton, and hides and skins produced in the area are exported. Important fish market. University town. Founded near Yerwa in 1907 as a British military post.

Maidum, EGYPT: see MEDUM.

Mai, El (MEI, el), village, MINUFIYA province, Lower EGYPT, 2 mi/3.2 km SW of SHIBIN EL KOM; 30°32′N 30°58′E. Cereals, cotton, flax.

Maiella (mah-YEL-lah), second-highest mountain group of the APENNINES, S central ITALY, SSE of GRAN SASSO D'ITALIA; 42°05′N 14°07′E. Extends c.20 mi/32 km N-S along border between CHIETI and AQUILA provinces; rises to 9,170 ft/2,795 m in Monte AMARO. Formerly also Majella.

Maienfeld (MEI-uhn-feld), commune, GRISONS canton, E SWITZERLAND, on the RHINE RIVER opposite BAD RAGAZ, and 11 mi/18 km N of CHUR; 47°01′N 09°32′E. Aspermont is a ruined castle on slope of Vilan mountain.

Maifa, YEMEN: see MEIFA.

Maifa'a, Wadi, YEMEN: see MEIFA'A, WADI.

Maigaiti, CHINA: see MARKIT.

Maignelay-Montigny (mai-nyuh-LAI—mon-tee-NYEE), commune (□ 7 sq mi/18.2 sq km), OISE department, PICARDIE region, N FRANCE, 20 mi/32 km NE of BEAUVAIS; 49°33′N 02°32′E. Agricultural processing. Has Gothic 15th–16th-century church.

Maigualida, Cordillera, VENEZUELA: see MAIGUALIDA, SIERRA.

Maigualida, Sierra (mei-gwah-LEE-dah, see-ER-rah), range in S VENEZUELA, on border between BOLÍVAR and AMAZONAS states, spur of GUIANA HIGHLANDS, 150 mi/241 km E of PUERTO AYACUCHO; extends c.70 mi/113 km N–S; rises over 5,000 ft/1,524 m. Also known as Cordillera Maigualida or Sierra Maigualide.

Maiguálide, Sierra, VENEZUELA: see MAIGUALIDA, SIERRA.

Mai Gudo (mei GOO-do), peak (c.10,168 ft/3,100 m), SW ETHIOPIA, 30 mi/48 km ESE of JIMMA, in mountains between OMO, GIBE, and GOJEB rivers; 07°29′N 37°12′E. Iron deposits nearby.

Maihar (MEI-huhr), town (2001 population 34,347), SATNA district, NE MADHYA PRADESH state, central INDIA, 22 mi/35 km S of SATNA; 24°16′N 80°45′E. Trades in wheat, millet, gram; limestone works. The Holy Ma Sharda Temple here is a pilgrimage site; visitors must climb 1001 stairs to get to the shrine at the top of a mountain. Former (until the withdrawal of princely support) great center of music whose alumni include Ravi Shankar and Ali Akhbar. Was capital of former princely state of Maihar of CENTRAL INDIA agency; in 1948, state merged with VINDHYA PRADESH before becoming part of Madhya Pradesh state.

Maihara (MAH-ee-hah-rah), town, Sakata county, SHIGA prefecture, S HONSHU, central JAPAN, on E shore of LAKE BIWA, 31 mi/50 km N of OTSU; 35°18′N 136°17′E. Railroad junction.

Maihingen (MEI-hing-uhn), village, SWABIA, W BAVARIA, GERMANY, 5 mi/8 km N of NÖRDLINGEN; 48°55′N 10°30′E. The former Franciscan monastery contains, since 1840, the large library of the princes of Öttingen-Wallerstein, including 1,500 incunabula; 1,500 manuscripts; 33,000 copper engravings; a collection of coins; and eleventh through sixteenth-century ivory carvings.

Maikain (mei-kei-IN), town, SW PAVLODAR region, KAZAKHSTAN, near SOUTH SIBERIAN RAILROAD, 75 mi/121 km SW of PAVLODAR; 51°28′N 75°46′E. Gold-mining center; copper deposits. Also spelled Maykain.

Maikammer (MEI-kahm-muhr), village, RHINELAND-PALATINATE, W GERMANY, on E slope of Forest of Palatinate (Pfälzes Wald), 3 mi/4.8 km S of NEUSTADT; 49°18′N 08°07′E. Wine.

Maikoor, INDONESIA: see ARU ISLANDS.

Maikop, RUSSIA: see MAYKOP.

Maikur, INDONESIA: see ARU ISLANDS.

Mailberg (MEIL-berg), township, NE LOWER AUSTRIA, NE AUSTRIA, in the WEINVIERTEL, 14 mi/23 km SE of ZNOJMO (CZECH REPUBLIC); 48°40′N, 16°11′E. Elevation: 661 ft/201 m. Noted for its wine. Famous Gothic carved altar, castle, streets with wine cellars.

Maili (MAH-EE-lee), town (2000 population 5,943), OAHU island, HONOLULU county, HAWAII, 20 mi/32 km WNW on, W coast, near mouth of Mailiili Stream; 21°25′N 158°10′W. Poultry, cattle. Lualualei Naval Reservation to E. WAIANAE Mountains to E; Maili Point to S.

Maili-Sai, KYRGYZSTAN: see MAYLUU-SUU.

Maillane (mei-YAHN), commune (□ 7 sq mi/18.2 sq km; 2004 population 2,013), BOUCHES-DU-RHÔNE department, PROVENCE-ALPES-CÔTE D'AZUR region, SE FRANCE, 12 mi/19 km SSE of AVIGNON; 43°50′N 04°47′E. Fruits, olives. Poet Mistral born nearby. Museum.

Maillén Island (mei-YEN), (□ 7 sq mi/18.2 sq km), LLANQUIHUE province, LOS LAGOS region, S central CHILE, in RELONCAVÍ SOUND, 7 mi/11 km S of PUERTO MONTT; 41°34′S 73°00′W. Livestock raising; dairying.

Maillezais (mah-yuh-ZAI), commune (□ 7 sq mi/18.2 sq km), VENDÉE department, PAYS DE LA LOIRE region, W FRANCE, 7 mi/11.3 km SSE of FONTENAY-LE-COMTE; 46°22′N 00°44′W. Has ruins of an 11th–14th-century abbey surrounded by 16th–17th-century fortifications. Just S is the vast marshland known as Marais Poitevin.

Mailly-le-Camp (mah-YEE-luh–KAHNP), commune (□ 16 sq mi/41.6 sq km), AUBE department, CHAMPAGNE-ARDENNE region, NE central FRANCE, in the chalk hills of CHAMPAGNE, 27 mi/43 km N of TROYES; 48°40′N 04°13′E. Large military camp nearby.

Mailog (MEI-log), former princely state of PUNJAB HILL STATES, N INDIA. Since 1948, merged with HIMACHAL PRADESH state.

Mailpatti, INDIA: see AMBUR.

Mailsi (MEIL-see), town, MULTAN district, S PUNJAB province, central PAKISTAN, 50 mi/80 km SE of MULTAN, near SUTLEJ River; 29°48′N 72°11′E. Local market.

Maimachen, MONGOLIA: see ALTAN BULAG.

Maimana (ma-EE-ma-nah), city, ⊙ FARYAB province, N AFGHANISTAN, near the UZBEKISTAN border, on QAISAR RIVER, 140 mi/225 km WSW of MAZAR-I-SHARIF; 35°55′N 64°47′E. Elev. 2,860 ft/872 m. A walled city inhabited mainly by Uzbeks; market for agriculture (fruit, sesame, beans); manufacturing (weaving and carpeting; woolen and cotton cloth). Important market for karakul pelts.

Maimansingh, BANGLADESH: see MYMENSINGH.

Maimará (mei-mah-RAH), town, central JUJUY province, ARGENTINA, on railroad, on the RÍO GRANDE DE JUJUY, and 40 mi/64 km NNW of JUJUY; 23°37′S 65°24′W. Corn; livestock. Nonmetallic mineral deposits nearby.

Maimon (mei-MON), town, 10 mi/16 km SE of BONAO, MONSEÑOR NOUEL province, central DOMINICAN REPUBLIC; 18°54′N 70°18′W. Nickel-mining center.

Mai Munene (MEI moo-NE-nai), village, KASAI-OCCIDENTAL province, S CONGO, landing on KASAI RIVER, and 90 mi/145 km SSW of LUEBO; 06°32′S 20°57′E. Elev. 1,624 ft/494 m. Has Roman Catholic mission, agricultural school.

Ma'in (muh-EEN), town, E central JORDAN, E of DEAD SEA, 4.5 mi/7.2 km SW of MEDEBA; 31°40′N 35°43′E. In ancient times known as Beth-ball-meon, a city of MOAB.

Main (MEIN), river, c.328 mi/528 km long, GERMANY; formed near KULMBACH, E central GERMANY, by the confluence of the ROTER MAIN and the WEISSER MAIN rivers, both of which rise in the FICHTELGEBIRGE; winds generally W through the rich farmland of central Germany and past the industrial areas of SCHWEINFURT, WÜRZBURG, ASCHAFFENBURG, and FRANKFURT to the RHINE RIVER at MAINZ. Navigable from its junction with the REGNITZ River, its chief tributary, the Main is an important E-W route. The RHINE-MAIN-DANUBE CANAL connects it with the DANUBE RIVER. The canal links the Danube with the Rhine, allowing barge traffic from the NORTH SEA to the BLACK SEA, a distance over 2,000 mi/3,219 km.

Ma'in, YEMEN: see JAUF.

Maina, Greece: see MANI.

Main à Dieu (man uh DOO), village, NE NOVA SCOTIA, E CANADA, E CAPE BRETON ISLAND, on the ATLANTIC OCEAN, 19 mi/31 km SE of SYDNEY; 46°00′N 59°51′W. Elevation 3 ft/0 m. Fishing.

Mainalon (MAI-nah-lon), Latin *Maenalus*, mountains in ARKADIA prefecture, central PELOPONNESE department, S mainland GREECE; rise to 6,496 ft/1,980 m 11 mi/18 km NNW of TRÍPOLIS; 37°32′N 22°19′E.

Mainaschaff (MEI-nah-shahf), town, LOWER FRANCONIA, BAVARIA, GERMANY, on the right bank of the RHINE RIVER opposite mouth of GERSPRENZ RIVER, 2,5 mi/4 km W of ASCHAFFENBURG; 49°50′N 09°06′E. Grain.

Mainau (MEI-nou), small island, in the Überlinger See (a branch of LAKE CONSTANCE), GERMANY, 3 mi/4.8 km N of CONSTANCE; covers 96 acres/39 ha; 47°42′N 09°12′E. Linked by bridges to mainland. Belonged (1272–1805) to Teutonic Knights. Castle (eighteenth century) is property of Swedish royal house. Botanical gardens.

Mainburg (MEIN-boorg), town, LOWER BAVARIA, GERMANY, on small Abens River, and 18 mi/29 km NW of LANDSHUT; 48°38′N 11°16′E. Brewing and hops growing; woodworking.

Main Camp, locality, MATABELELAND NORTH province, W ZIMBABWE, 40 mi/64 km SE of HWANGE, near railroad, in NE edge of HWANGE NATIONAL PARK; 18°44′S 26°57′E. Main safari camp for park. Lodge, airstrip.

Maindample, village, VICTORIA, SE AUSTRALIA, c.75 mi/120 km NE of MELBOURNE, between Bonnie Doon and MANSFIELD; 37°02′S 145°56′E.

Maindargi (MEIN-duhr-gee), town, SOLAPUR district, MAHARASHTRA state, W central INDIA, 28 mi/45 km SE of SOLAPUR; 17°28′N 76°18′E. Agriculture (millet,

cotton, wheat); handicraft cloth weaving. Was capital of former DECCAN state of KURANDVAD JUNIOR; state incorporated 1949 into Solapur and BELGAUM districts.

Mai-Ndombe, Lake (□ 900 sq mi/2,331 sq km), W CONGO, E of CONGO RIVER, and SSE of TUMBA LAKE; 90 mi/145 km long, up to 25 mi/40 km wide; mean depth 17 ft/5 m, 33 ft/10 m at the deepest point; 02°08′S 18°19′E. Empties S into the KASAI by FIMI RIVER. Shallow and of irregular shape, with low, forested shores, it increases in size by two or three times in rainy season. Main port is INONGO. Discovered for Europeans in 1882 by Stanley.

Main-Donau Kanal, GERMANY: see RHINE-MAIN-DANUBE CANAL.

Maindu, INDIA: see MENDU.

Maine (MEN), historic region and former province, NW FRANCE, S of NORMANDY and E of BRITTANY, now comprising chiefly the departments of MAYENNE and SARTHE, in PAYS DE LA LOIRE region; 48°15′N 00°00′W. Le MANS, the historic capital, is an important industrial and commercial center. Other towns are LAVAL and MAYENNE. Maine is primarily agricultural, with important livestock raising in the hilly PERCHE district; it is well watered by the MAYENNE, LOIR, and SARTHE rivers. Important during Roman times, Maine was Christianized 4th–6th century. Made a county in 10th century, it passed (1126) to ANJOU and was held for long periods by ENGLAND between 12th and 15th centuries. Finally united with the crown in 1481, and given in appanage to several Valois and Bourbon princes. Divided into present departments in 1790.

Maine, state (□ 35,387 sq mi/91,683 sq km; 2005 estimated population 1,318,220; 2000 population 1,274,923), in the extreme NE corner of the UNITED STATES; ⊙ AUGUSTA; 45°50′N 69°17′W. Maine is the largest of the NEW ENGLAND states. Admitted as the twenty-third state of the Union in 1820.

Geography
The Canadian provinces of QUEBEC and NEW BRUNSWICK border Maine from the NW around to the SE coast, with the SAINT JOHN and SAINT CROIX rivers forming part of the international border with New Brunswick. To the S is the ATLANTIC OCEAN (the BAY OF FUNDY lies off to the E). NEW HAMPSHIRE (to the W) is the only state bordering Maine. Geologic action laid down a bedrock of sandstone, shale, and limestone. Much of the soft rock eroded into tableland valleys, while the more resistant rock remained, forming the generally mountainous W, the mountains of MOUNT DESERT ISLAND in the E, and isolated peaks including KATAHDIN (5,268 ft/1,606 m), the highest point in the state. Receding glaciers deposited long drift ridges across the countryside and dammed the valleys to form more than 2,200 lakes (MOOSEHEAD LAKE is the largest) and to establish new, rugged watercourses for more than 5,000 streams and rivers. The major rivers are the ST. JOHN, PENOBSCOT, KENNEBEC, ANDROSCOGGIN, and SACO. The sea has encroached on the low coastal valleys, leaving a jigsawed, fjorded, coastline of 2,500 mi/3,219 km and numerous irregular and rocky islands offshore. E of the Kennebec the coast of Maine is rugged, but W of the river the shoreline has sandy beaches and marshy lowlands. More than 80% of Maine is forested with great stands of white pine (Maine is known as the "Pine Tree State"), hemlock, spruce, fir, and hardwoods. In the shelter of lakes and woods, particularly in the N counties, wildlife has found refuge. Moose, deer, black bear, and smaller game are still found; fish and fowl are plentiful.

Economy
Much of Maine's natural and industrial resources remain undeveloped. Many varieties of granite, including some superior ornamental types, have been used for construction throughout the nation. Sand

and gravel, zinc, and peat are found in addition to stone. The population of Maine is centered on the cleared land along the coast and major rivers. Due to a variety of economic factors—generally poor soil and a short growing season, geographical remoteness from production centers, an inadequate distribution system, lack of coal and steel, and a reluctance to adopt modern methods of production and merchandising—Maine had a very low population increase in 1860–1970 after which it experienced unprecedented growth. The economic revival experienced in port and factory towns during World War II did not continue after the war. In 1949–1950 the textile industry, after enjoying a brief expansion, suffered severely from competitive markets, and many of the old mills and plants were closed. A sharp decline in the shoe- and food-processing industries followed in the 1980s.

However, in the 1980s, Maine successfully transformed a major portion of its economy into trade, service, and finance industries, the greatest growth occurring in and near PORTLAND and, in addition, the picturesque coastal and island resorts of Maine hold a strong appeal for visitors and tourists and, combined with abundant wildlife to attract sportsmen, make the tourist trade a most important feature of Maine's economy. Many of Maine's traditional economic activities have experienced difficult times.

Fishing, one of the state's earliest industries, has declined considerably (with the exception of lobsters, still caught in abundance). Lumbering—the first sawmill in America was built in 1623 on the PISCATAQUA RIVER—dominated industry and the export trade from the days when the straight white pines provided masts for the British navy. However, since the virgin timber has been largely cut off, the timber trade has declined. Maine is still a leading producer of paper and paper products, which remain the most valuable of all manufactures in the state. The proximity of harbors and forests encouraged extensive shipbuilding, which reached its peak in the 19th century. After the decline of shipbuilding and timber trading, commercial activity slackened.

Portland, the largest port, operates far below its capacity. However, Portland, BANGOR, and ROCKLAND (with LEWISTON, the state's largest cities) are still important cities and during the summer months serve a vast resort region. Portland has also attracted commercial development for the banking, insurance, and real estate industries. Manufacturing is the largest economic sector, accounting for a fifth of all production. Limited amounts of leather goods (especially shoes), and food products are still produced in addition to transportation equipment, paper, and wood products. Printing, publishing and electronic components are also important.

Agriculture has been developed despite adverse soil and climatic conditions. Since the opening of the prairie and grasslands of the W, Maine has tended to concentrate on dairying, poultry raising, and market gardening to serve local and New England markets. The state is a leading producer of brown eggs. The growing of potatoes, particularly in AROOSTOOK county, was stimulated by the completion of the Aroostook Railroad in 1894 and in recent years maintained by introduction of freezing process. Blueberries, hay, apples, and oats are the other chief crops.

Tourism and Higher Educatiion
Places of interest in Maine include ACADIA NATIONAL PARK on Mount Desert Island; Baxter State Park, which includes the beginning of the APPALACHIAN NATIONAL SCENIC TRAIL at MOUNT KATAHDIN in the N Maine wilderness; and the Old York Gaol (1653), one of the oldest public buildings in New England. Among the state's leading educational institutions are Bowdoin College, at BRUNSWICK; Colby College, at WATERVILLE; Bates College, at Lewiston; and the University of Maine, at ORONO.

Area is shown by the symbol □, and capital city or county seat by ⊙.

History to 1652

The earliest human habitation in what is now Maine extends back to prehistoric times, as evidenced by the burial mounds of the Red Paint people found in the S central part of the state. The Native Americans who came later left enormous shell heaps, variously estimated to be 1,000–5,000 years old. At the time of settlement by Europeans the Abnaki were scattered along the coast and in some inland areas. The coast of Maine may have been visited by the Norsemen and was included in the grant that James I of ENGLAND awarded to the Plymouth Company, and colonists set out under George Popham in 1607. This settlement, Fort Saint George, on the present site of PHIPPSBURG, at the mouth of the Kennebec (then called the Sagadahoc) River, did not prosper, and the colonists returned to England in 1608. The French came to the area in 1613 and established a new colony and a Jesuit mission on Mount Desert Island; however, the English under Sir Samuel Argall expelled them. In 1620 the Council for New England (successor to the Plymouth Company) granted Ferdinando Gorges and Captain John Mason the territory between the Kennebec and MERRIMACK rivers extending 60 mi/97 km inland. At this time, the region became known as Maine. Neglected after Gorges' death in 1647, Maine settlers came under the jurisdiction of the Massachusetts Bay Colony in 1652.

History: 1652 to 1807

King Philip's War (1675–1676) was the first of many struggles between the British on one side and the French and Native Americans on the other. In 1691 MASSACHUSETTS received a new charter that confirmed its hold on Maine, and with Sir William Phips, a Maine native, as governor and the territorial question settled, local government and institutions in the Massachusetts tradition really took root here. Maine soon had prosperous fishing, lumbering, and shipbuilding industries. Dissatisfaction with British rule was first expressed openly after Parliament passed the Stamp Act in 1765. During the American Revolution, FALMOUTH was devastated by a British fleet (1775). Benedict Arnold, in 1775, led his grueling, unsuccessful expedition against Quebec N through here. During the war supplies were cut off and conflict with Native Americans was frequent, but with American independence won, economic development was rapid in what was then called the District of Maine.

History: 1807 to 1842

However, the Embargo Act of 1807 and the War of 1812 interrupted the thriving commerce and turned the district to industrial development. Agitation for statehood, which had been growing since the Revolution, now became widespread. Dissatisfaction with Massachusetts was aroused by the inadequate military protection provided during the War of 1812; by the land policy, which encouraged absentee ownership; and by the political differences between conservative Massachusetts and democratic Maine. The imminent admission of MISSOURI into the Union as a slave state hastened the separation of Maine from Massachusetts, and equality of power between North and South was preserved by admitting Maine as a free state in 1820, as part of the Missouri Compromise. With Portland as its capital (moved to Augusta in 1832) the new state entered a prosperous period. During the first half of the 19th century Maine enjoyed its greatest population increase until the 1970s. A highly profitable timber trade was carried on with the WEST INDIES, EUROPE, and ASIA, and towns such as BATH became America's leaders in shipbuilding. The long-standing Northeast Boundary Dispute almost precipitated border warfare between Maine and New Brunswick in the so-called Aroostook War of 1839; the controversy was settled by the Webster-Ashburton Treaty with GREAT BRITAIN in 1842.

History: 1842 to 1954

Political life was vigorous, particularly in the 1850s, when the reluctance of the Democrats, who had been dominant since 1820, to take a firm antislavery stand swept the new Republican party into power. Hannibal Hamlin was a leading Republican politician and was Vice President during Abraham Lincoln's first administration. Antislavery sentiment was strong, and Maine made sizable contributions of men and money to the Union in the Civil War. Generals Oliver O. Howard and Joshua Lake Chamberlain were from Maine. For decades regulation of the liquor traffic was the chief political issue in Maine, and the state was the first to adopt (1851) a prohibition law. It was incorporated into the constitution in 1884 and was not repealed until 1934. State politics entered a hectic stage in 1878 when the newly organized Greenback party combined with the Democrats to carry the election, ending more than twenty years of Republican rule. The following year the coalition was accused of manipulating election returns, a charge sustained by the state supreme court, which seated a rival legislature elected by the Republicans. In 1880 the fusionists were again successful, but from that time until the 1950s the state was generally Republican, providing that party with such national leaders as James G. Blaine, Thomas B. Reed, and Margaret Chase Smith, who in 1948 became the first female Republican U.S. senator.

History: 1954 to Present

Former U.S. Secretary of State Edmund S. Muskie was elected governor in 1954. In 1964 and 1968 (when Muskie, then a U.S. senator, ran unsuccessfully for Vice President) the state voted Democratic in the presidential election for the first time since 1912. In 1969 personal and corporate income taxes were added to the sales tax within the state. Maine's population grew 13.2% during the 1970s and 9.2% during the 1980s, its largest increases since the 1840s. Maine has a long tradition of concern for environmental issues. In the 1970s the state prohibited the use of rivers to transport logs in order to protect the salmon, and dams which produce hydroelectric energy may be removed for the same reason. The need for energy, however, led Maine voters to narrowly defeat a 1988 anti-nuclear-power referendum.

Government

Maine is governed under the 1820 constitution as amended. There is a two-house legislature of thirty-five senators and 151 representatives, all elected for two-year terms; the governor is elected for a four-year term and may succeed himself once. The current governor is John Baldacci. Maine elects two Representatives and two Senators to the U.S. Congress and has four electoral votes. Localities are classified as cities, towns, townships, and "plantations," a minor civil division for remote and sparsely populated places.

Maine has sixteen counties: ANDROSCOGGIN, AROOSTOOK, CUMBERLAND, FRANKLIN, HANCOCK, KENNEBEC, KNOX, LINCOLN, OXFORD, PENOBSCOT, PISCATAQUIS, SAGADAHOC, SOMERSET, WALDO, WASHINGTON, and YORK.

Maine, Collines du (MEN, ko-LEEN dyoo), chain of hills on the border between NORMANDY (N) and BRITTANY (S), N part of ILLE-ET-VILAINE and MAYENNE departments, W FRANCE. Highest point, Les AVALOIRS (1,368 ft/417 m).

Maine-et-Loire (MEN–ai–LWAHR), department (□ 2,767 sq mi/7,194.2 sq km), in ANJOU, W FRANCE; ⊙ ANGERS; 47°30′N 00°20′W. Generally level region traversed by the LOIRE RIVER and drained by its tributaries, the MAYENNE, SARTHE, LOIR, and THOUET rivers. Very fertile area with noted vineyards in Loire valley, orchards, and diversified agriculture, including livestock. Has slate quarries (near Angers and SEGRÉ). Industry is concentrated in Angers (food processing, metalworking), SAUMUR (sparkling wines),

and CHOLET (textiles). Department forms part of administrative region of PAYS DE LA LOIRE (French= Loire country).

Maine, Gulf of, part of the ATLANTIC OCEAN, between SE MAINE and SW NOVA SCOTIA, at the entrance of the BAY OF FUNDY. The area is noted for its scenery and fishing.

Maine River (MEN), 7 mi/11.3 km long, in MAINE-ET-LOIRE department, PAYS DE LA LOIRE region, W FRANCE; formed by confluence of the MAYENNE and the SARTHE rivers above ANGERS; empties into LOIRE RIVER 5 mi/8 km SSW of Angers; 47°25′N 00°37′W. Navigable.

Maine River, 24 mi/39 km long, SW IRELAND; rises on E border of KERRY county; flows W, past CASTLEISLAND and CASTLEMAINE, to Castlemaine Harbour of DINGLE BAY.

Mainero, MEXICO: see VILLA MAINERO.

Maïné-Soroa (mah-EEN-ai–SOR-o-uh), town, DIFFA province, S NIGER, on NIGERIA border, 210 mi/338 km E of ZINDER; 13°12′N 12°02′E. Administrative center.

Maine Turnpike, part of MAINE highway and interstate (also designated I-95) system extending NE from near PORTSMOUTH (New Hampshire) to AUGUSTA. Toll road.

Maineville (MAIN-vil), village (2006 population 1,019), WARREN county, SW OHIO, 20 mi/32 km NE of CINCINNATI; 39°19′N 84°13′W. In agricultural area.

Maine Yankee Nuclear Power Plant, MAINE: see LINCOLN.

Mainfranken, GERMANY: see LOWER FRANCONIA.

Maing (MANG), town (□ 4 sq mi/10.4 sq km), NORD department, NORD-PAS-DE-CALAIS region, N FRANCE, suburb, 3 mi/5 km SW of VALENCIENNES; 50°18′N 03°29′E.

Maingay Island, MYANMAR: see MAINGY ISLAND.

Maingkwan (meing-KWAHN), village, TANAI township, KACHIN STATE, MYANMAR, in HUKAWNG VALLEY, on road, and 80 mi/129 km NW of MYITKYINA; 26°20′N 96°36′E. Amber mines nearby.

Maingy Island (MIN-jai), in MERGUI ARCHIPELAGO, TANINTHARYI division, MYANMAR, in ANDAMAN SEA, 3 mi/4.8 km W of KING ISLAND; 5 mi/8 km long, 3 mi/4.8 km wide; 12°32′N 98°15′E. Mountainous, surrounded by mangrove swamps. Large galena deposits. Also called Maingay Island.

Mainhardt (MEIN-hahrt), village, FRANCONIA, BADEN-WÜRTTEMBERG, S GERMANY, in the Mainhardter Wald [German=Mainhardt Forest], 9 mi/14.5 km WSW of SCHWÄBISCH HALL; 49°05′N 09°34′E. Forestry.

Mainhausen (mein-HOU-suhn), village, S HESSE, central GERMANY, near the MAIN River, 16 mi/26 km ESE of FRANKFURT; 50°01′N 09°01′E. Fruit.

Mainit (mah-EE-net), town (□ 67 sq mi/174.2 sq km), SURIGAO DEL NORTE province, PHILIPPINES, NE MINDANAO, 17 mi/27 km S of SURIGAO, and on N shore of Lake Mainit (□ 67 sq mi/174 sq km; c.20 mi/32 km long); 09°33′N 125°30′E. Coconuts. Gold mining at nearby Siana village. Hot springs; possible geothermal power field.

Mainland (MAIN-land), island (□ 178 sq mi/462.8 sq km; 2001 population 15,315), ORKNEY ISLANDS, N Scotland, largest island of the archipelago; 59°00′N 03°10′W. KIRKWALL, the county capital of Orkney, is on the island. Kirkwall Bay and SCAPA FLOW deeply indent its shores. The interior has hills, moors, several lakes, and fertile valleys. Cattle, sheep, and poultry are raised. Whiskey is produced. Most famous Pictish remains are Maeshowe and the Standing Stones of Stenness. SKARA BRAE is an excavated prehistoric village. Also called Pomona.

Mainleus (MEIN-lois), village, UPPER FRANCONIA, N BAVARIA, GERMANY, at confluence of the ROTER and WHITE MAIN rivers where they form MAIN River, 3 mi/4.8 km W of KULMBACH; 50°06′N 11°22′E. Woodworking.

Cross-references are shown in SMALL CAPITALS. The pronunciation guide is shown on page xix. The sources of population figures are shown on page xvii.

Mainpuri (MEIN-puhr-ee), district (□ 1,066 sq mi/ 2,771.6 sq km), W UTTAR PRADESH state, N central INDIA; ⊙ MAINPURI. On W GANGA PLAIN and GANGA-YAMUNA DOAB; irrigated by UPPER and LOWER Ganga canals and their distributaries. Agriculture (wheat, gram, pearl millet, corn, barley, jowar, oilseeds, cotton, rice, sugarcane); mango, sissoo, and dhak groves. Main towns are Mainpuri, Skikohabad, BHONGAON, KARHAL.

Mainpuri (MEIN-puhr-ee), town, ⊙ MAINPURI district, W UTTAR PRADESH state, N central INDIA, on tributary of the GANGA RIVER, and 60 mi/97 km E of AGRA; 27°14′N 79°01′E. Road center. Oilseed milling; trades in grains, cotton, sugarcane, hides.

Main River (MAIN), Gaelic *Mean*, 30 mi/48 km long, ANTRIM, NE Northern Ireland; rises SE of BALLYMO-NEY; flows S, past RANDALSTOWN, to LOUGH NEAGH.

Maintal (MEIN-tahl), town, S HESSE, central GERMANY, on the right bank of the Main River, 7 mi/11,2 km ENE of FRANKFURT; 50°09′N 08°51′E. Manufacturing (precision instruments, synthetic fiber, machinery); metalworking; food processing; agriculture (fruit). Formed 1974 by union of Dörnigheim (first mentioned 793), Hochstadt, Wachenbuchen, and Bischofsheim.

Main-Taunus, GERMANY: see FRANKFURT.

Maintenon (man-tuh-non), town (□ 4 sq mi/10.4 sq km; 2004 population 4,472), EURE-ET-LOIR department, CENTRE administrative region, NW central FRANCE, on the EURE RIVER, and 10 mi/16 km NNE of CHARTRES; 48°35′N 01°35′E. Agriculture market. Has 13th–17th-century castle (given in 1674 by Louis XIV to Françoise d'Aubigné, later Marquise de Maintenon), and ruins of 17th-century aqueduct intended to supply VERSAILLES with water from upper Eure River, but never completed despite enormous expenditures.

Maintirano (mein-tee-RAH-noo), town, MAHAJANGA province, W MADAGASCAR, port of MOZAMBIQUE CHANNEL, 160 mi/257 km N of MORONDAVA; 18°03′S 44°02′E. Rice, cassava, beans. Airport.

Main Topsail, mountain (1,822 ft/555 m), SW NEWFOUNDLAND AND LABRADOR, E CANADA, 30 mi/48 km E of NE end of GRAND LAKE; 49°08′N 56°33′W.

Mainvilliers (man-veel-YAI), W suburb (□ 4 sq mi/ 10.4 sq km) of CHARTRES, EURE-ET-LOIR department, CENTRE administrative region, NW central FRANCE; 48°27′N 01°28′E.

Mainz (MEINTS), French *Mayence* (mah-YAWNS), city (2005 population 194,372), ⊙ RHINELAND-PALATINATE (since 1950), W GERMANY, 6 mi/9.7 km S of WIESBADEN, a port on the left bank of the RHINE RIVER, opposite the mouth of the MAIN RIVER; 49°58′N 08°14′E. Industrial, commercial (especially Rhine wines), and transportation center. Manufacturing includes chemicals, pharmaceuticals, computers, beer, wine, cement, paper, food products, furniture, machinery, and glassware. One of the great historical cities of Germany. Developed on the site of the Roman camp of Maguntiacum, or Mogontiacum (founded first century B.C.E.). St. Boniface was made (745) first archbishop of Mainz. Later archbishops acquired considerable territory around Mainz and in FRANCONIA which they ruled as princes, and from fourteenth century archchancellors and electors, of the Holy Roman Empire. City became autonomous in twelfth century, but in 1492 it again came under domination of the archbishops. Was flourshing trade and cultural center (birthplace and principal residence of Gutenberg) until French Revolutionary Wars, when electoral state collapsed. Under French rule, 1798–1814. Archbishopric secularized and see degraded to bishopric in 1803. Passed to HESSE-DARMSTADT in 1816 and became capital of newly created RHENISH HESSE province. Was fortress (1873–1918) of German Empire. Headquarters of French army of occupation (1918–1930) in the S RHINELAND.

Severely damaged during World War II (inner city 80 percent destroyed), but largely restored and rebuilt after 1945. Noteworthy structures in the old inner city include the six-towered Romanesque cathedral (consecrated 1009; restored nineteenth century); the Renaissance-style electoral (archiepiscopal) palace (seventeenth through eighteenth century), which houses an art gallery and a museum of Roman and Germanic antiquities; and the Church of St. Peter (eighteenth century). The University of Mainz was founded in 1477, was discontinued in 1816, and was reestablished in 1946 as the Johannes Gutenberg University. In 1945 the city's suburbs on the right bank of the Rhine were transferred to the state of Hesse. ZDF (German television) headquarters here.

Maio (MEI-oo), municipality and island (□ 104 sq mi/ 270.4 sq km; 2005 population 7,506), one of Cape Verde Islands, easternmost of the SOTAVENTO ISLANDS group, in the Atlantic Ocean, between Boa Vista (50 mi/80 km NNE) and Santiago (15 mi/24 km SW) islands; 14 mi/23 km long, 9 mi/14.5 km wide; ⊙ VILA DO MAIO; 15°15′N 23°10′W. Vila do Maio (also called Pôrto Inglês), its chief town, is on SW coast. Hilly interior (Monte Penoso, 1,430 ft/436 m), has a low, sandy N and W coast. Extensive salt pans. Some agriculture (corn, beans, potatoes), horse raising. Occupied by British until end of 18th century. Formerly spelled Mayo.

Maiorga (mai-OR-gah), village, LEIRIA district, central PORTUGAL, 14 mi/23 km SSW of LEIRIA; 39°35′N 08°59′W. Wine, olives. Battlefield of ALJUBARROTA just E.

Maiori (mah-YO-ree), town, Salerno province, CAMPANIA, S ITALY, port on Gulf of SALERNO, 6 mi/10 km WSW of SALERNO; 40°39′N 14°38′E. Swimming resort; fishing center. Citrus fruits.

Maipo (MEI-po), province, METROPOLITANA DE SANTIAGO region, S central CHILE. Agricultural and industrial region, with fruit, vegetables, cereals, grapes; livestock; mining (gypsum) and food processing, metal fabrication, textile production. In the greater SANTIAGO conurbation.

Maipo, CHILE: see MAIPÚ.

Mai Po, marsh, the NEW TERRITORIES, HONG KONG, SE CHINA, along the coastal edge of DEEP BAY. Marsh area established as a bird sanctuary by the World Wildlife Fund as a preserve for endangered migratory species.

Maipo Pass (MEI-po) (11,230 ft/3,423 m), in the ANDES MOUNTAINS, on ARGENTINA-CHILE border, at S foot of MAIPO VOLCANO, on road between SAN RAFAEL (Argentina) and EL VOLCÁN (Chile); 34°14′S 69°50′W.

Maipo River (MEI-po), c.155 mi/249 km long, METROPOLITANA DE SANTIAGO region, central CHILE; rises in the ANDES Mountains at foot of MAIPO VOLCANO near ARGENTINA border; flows NW and W, past QUELTEHUES (hydroelectric plant), SAN JOSÉ DE MAIPO, BUIN, MELIPILLA, and LLOLLEO, to the PACIFIC OCEAN at LA BOCA 2 mi/3 km SW of SAN ANTONIO. It receives the MAPOCHO RIVER and irrigates the fertile central valley S of SANTIAGO. Just SW of Santiago and c.10 mi/16 km from the river was fought (1818) the decisive battle of Maipú or Maipo in the war for Chilean independence. Formerly also spelled Maipú.

Maipo Volcano (MAI-po) (17,355 ft/5,290 m), in the ANDES MOUNTAINS, on ARGENTINA-CHILE border, 65 mi/105 km SE of SANTIAGO (Chile); 34°10′S 69°52′W.

Maipú (mei-POO), town (□ 1,004 sq mi/2,610.4 sq km), ⊙ Maipú district (□ 1,004 sq mi/2,600 sq km; 1991 population 10,073), E BUENOS AIRES province, ARGENTINA, 80 mi/129 km NNW of MAR DEL PLATA. Railroad junction and agricultural center (oats, sunflowers; sheep, cattle).

Maipú or **Villa Maipú**, town, ⊙ Maipú department (□ 260 sq mi/676 sq km), N MENDOZA province, ARGENTINA, in MENDOZA RIVER valley (irrigation area), on railroad, and 8 mi/12.9 km SSE of MENDOZA. Agriculture (wine, alfalfa, fruit, corn, potatoes).

Maipú (mei-POO), suburb, ⊙ Maipú comuna, METROPOLITANA DE SANTIAGO region, central CHILE; suburb of SANTIAGO; 33°31′S 70°46′W. Nearby was fought the battle of Maipú or Maipo.

Maipú (mei-POO), battlefield, METROPOLITANA DE SANTIAGO region, central CHILE, S of SANTIAGO. On April 5, 1818, San Martín routed the Spanish royalist army here and assured Chilean independence. The victory made possible his expedition to liberate PERU.

Maipures Rapids (mei-POO-rais), on COLOMBIA-VENEZUELA border, an obstacle to navigation on upper ORINOCO RIVER, c.35 mi/56 km S of ATURES RAPIDS; 05°15′N 67°30′W. The 2 sets of rapids are circumvented by a road.

Maiquetía (mei-kai-TEE-ah), town, VARGAS state, N VENEZUELA, on the CARIBBEAN SEA, at foot of CORDILLERA DE LA COSTA, just W of LA GUAIRA, 6 mi/ 10 km N of CARACAS (linked by multilane highway); 10°37′N 66°58′W. Site of SIMÓN BOLÍVAR INTERNATIONAL AIRPORT, Venezuela's principal airport, as well as the National Airport. Founded in 1670.

Maiquinique (MAI-kai-NEE-kai), town (2007 population 8,538), SE BAHIA state, BRAZIL, on Rio Maiquinique, ENE of MINAS GERAIS state border; 15°38′S 40°16′W.

Mairana (mei-RAH-nah), town and canton, FLORIDA province, SANTA CRUZ department, central BOLIVIA, in E foothills of Cordillera de COCHABAMBA, on COCHABAMBA–SANTA CRUZ road, and 9 mi/14.5 km NW of SAMAIPATA; 18°07′S 63°56′W. Corn, potatoes, tobacco, coffee, bananas, citrus fruits.

Maira River (MEI-rah), 65 mi/105 km long, NW ITALY; rises in COTTIAN ALPS 7 mi/11 km NW of ACCEGLIO; flows E, past DRONERO, and NNE, past SAVIGLIANO and RACCONIGI, to the PO RIVER 3 mi/5 km E of Pancalieri. Used for hydroelectric power at Acceglio.

Mairena del Alcor (mei-RAI-nah del ahl-KOR), town, SEVILLE province, SW SPAIN, on railroad, and 13 mi/21 km E of SEVILLE; 37°22′N 05°45′W. Agricultural center (cereals, olives, cotton; timber; livestock). Museum.

Mairena del Aljarafe (mei-RAI-nah del ahl-hah-RAH-fai), city, SEVILLE province, SW SPAIN, 5 mi/8 km SW of SEVILLE; 37°20′N 06°04′W. Olives, grapes.

Mairi (MEI-ree), city (2007 population 19,319), E central BAHIA state, BRAZIL, 112 mi /180 km NW of Feira de Santana; 11°42′S 40°08′W.

Mairiporã (MEi-ree-po-ruh), city (2007 population 71,868), SE SÃO PAULO state, BRAZIL, 15 mi/24 km N of SÃO PAULO; 23°19′S 46°35′W. Kaolin quarrying; thread manufacturing, distilling. Until 1948, called Juqueri.

Maisach (MEI-sahkh), town, UPPER BAVARIA, BAVARIA, S GERMANY, 15 mi/24 km WNW of MUNICH; 48°13′N 11°15′E. Largely residential. Town first mentioned 806.

Maisaka (MAH-ee-sah-kah), town, Hamana county, SHIZUOKA prefecture, central HONSHU, E central JAPAN, on SE shore of LAKE HAMANA, near its outlet, 50 mi/80 km S of SHIZUOKA; 34°41′N 137°36′E. Oysters, eels, soft-shelled turtles; nori. Hot springs.

Maisí (mei-SEE), scenic beach resort in GUANTÁNAMO province, at E end of CUBA, at foot of Meseta with same name; 20°15′N 74°09′W.

Maisí, Cape (mei-SEE) or **Maisí Point**, easternmost headland of CUBA, GUANTÁNAMO province, on WINDWARD PASSAGE, 110 mi/177 km E of SANTIAGO DE CUBA, 50 mi/80 km WNW of Cape SAINT NICOLAS (HAITI); 20°15′N 74°08′W. Has lighthouse. Old spelling, Maysí.

Maiskhal Island (mo-ish-kahl), just off the coast of the CHITTAGONG district, extreme SE EAST BENGAL, BANGLADESH, N of COX'S BAZAR; 16 mi/26 km long, N-S; 6 mi/9.7 km wide.

Maisome Island (mah-ee-SO-mai), MWANZA region, NW TANZANIA, in Lake Victoria, 60 mi/97 km WNW of MWANZA, at entrance to EMIN PASHA GULF, NW of mainland and E of RUBONDO ISLAND (RU-

BONDO NATIONAL PARK); 12 mi/19 km long, 8 mi/12.9 km wide; 02°18′S 32°00′E. Fish; subsistence crops.

Maisonette Point (mai-zuhn-EHT), cape on CHALEUR BAY, NE NEW BRUNSWICK, E CANADA, 35 mi/56 km NE of BATHURST; 47°50′N 65°00′W. Lighthouse.

Maisons-Alfort (mai-ZON–ahl-FOR), SE suburb (□ 2 sq mi/5.2 sq km) of PARIS, VAL-DE-MARNE department, ÎLE-DE-FRANCE region, N central FRANCE, near SEINE RIVER, and 5.5 mi/8.9 km SE of Notre Dame Cathedral; 48°48′N 02°26′E. Industrial town with chemical, cement, and metalworking establishments. A 13th-century church and a veterinary school are here.

Maisons-Laffitte (mai-ZON–lah-FEET), town (□ 2 sq mi/5.2 sq km), YVELINES department, ÎLE-DE-FRANCE region, N central FRANCE, a fine residential NW suburb of PARIS, 11 mi/18 km from Notre Dame Cathedral, on left bank of the SEINE RIVER, at E edge of Saint-Germain forest; 48°57′N 02°09′E. Manufacturing of electronics and electrical equipment. Has noted racetrack. Its 17th-century château, built by Mansart, contains museum of paintings and tapestries. Formerly named Maisons-sur-Seine.

Maisons-sur-Seine, FRANCE: see MAISONS-LAFFITTE.

Maïssade (mah-ee-SAHD), town, CENTRE department, central HAITI, 40 mi/64 km S of CAP-HAÏTIEN; 19°10′N 72°08′W. Agriculture (fruit, sugarcane, coffee, cotton); cattle; beekeeping. Lignite deposits nearby.

Maissau (MEIS-sou), town, N central LOWER AUSTRIA, NE Austria, in the WEINVIERTEL, at E slope of MANHARTSBERG mountain, 10 mi/16 km NE of KREMS AN DER DONAU; 48°34′N, 15°50′E. Elevation 1,036 ft/316 m. Wine. Founded as a town in 13th century; some fortifications conserved; medieval castle, reconstructed around 1870.

Maitén, ARGENTINA: see EL MAITÉN.

Maitencillo (mei-ten-SEE-yo), village, COQUIMBO province, N central CHILE, 18 mi/29 km SE of LA SERENA. High-grade iron ore deposits.

Maitenes (mei-TE-nes), mining settlement, SANTIAGO province, METROPOLITANA DE SANTIAGO region, central CHILE, in the ANDES Mountains, 20 mi/32 km NE of SANTIAGO. Hydroelectric station; copper mining.

Maitland (MAIT-luhnd), city (2001 population 53,470), NEW SOUTH WALES, SE AUSTRALIA, 101 mi/163 km N of SYDNEY, on the HUNTER RIVER; 32°44′S 151°33′E. Railroad junction and agriculture center with light manufacturing; also brickworks, an open-cut mine, and tourism. Began as a convict settlement in 1824. The river flooded in 1893, 1949, and 1955; spillways, levees, and channels have been built in response.

Maitland (MAIT-luhnd), city (□ 4 sq mi/10.4 sq km; 2005 population 14,125), ORANGE county, central FLORIDA, 5 mi/8 km N of ORLANDO; 28°37′N 81°22′W. Heart of major office complex along I-4. One of largest suburban business centers in SE U.S.

Maitland (MAIT-luhnd), city (2000 population 342), HOLT county, NW MISSOURI, on NODAWAY RIVER, and 33 mi/53 km NNW of SAINT JOSEPH; 40°12′N 95°04′W. Corn; cattle; hogs.

Maitland (MAIT-luhnd), town, S SOUTH AUSTRALIA state, S central AUSTRALIA, on central YORKE PENINSULA, 104 mi/168 km W of ADELAIDE, 85 mi/137 km SSW of PORT PIRIE; 34°23′S 137°40′E. Limestone soils; wheat; sheep, wool; tourism.

Maitland, (MAIT-lund) village, central NOVA SCOTIA, E CANADA, on CHIGNECTO BAY, at mouth of SHUBENACADIE RIVER, 12 mi/19 km WSW of TRURO; 45°19′N 63°31′W. Elevation 98 ft/29 m. Dairying, farming; former shipbuilding center.

Maitland, unincorporated village, MCDOWELL county, S WEST VIRGINIA, 2 mi/3.2 km ESE of WELCH; 37°25′N 81°33′W.

Maitland (MAIT-luhnd), hamlet, SE ONTARIO, E central CANADA, 7 mi/11 km from PRESCOTT, and included in AUGUSTA township; 44°38′N 75°36′W.

Maitland, suburb, WESTERN CAPE province, SOUTH AFRICA, E suburb of CAPE TOWN, 5 mi/8 km from

TABLE BAY, on N1 highway and railroad; 33°55′S 18°31′E. Just E is WINGFIELD airport.

Maitri Station (MEI-tree), ANTARCTICA, Indian station on PRINCESS ASTRID COAST; 70°46′S 11°44′E.

Maiwand (ma-EE-wahnd), village, KANDAHAR province, S AFGHANISTAN, 33 mi/53 km W of KANDAHAR, and 10 mi/16 km NE of KUSHK-I-NAKHUD; 31°44′N 65°08′E. Famous for Battle of Maiwand (1880) when Sardar Akbar defeated British General Burrows.

Maíz, MEXICO: see CIUDAD DEL MAÍZ.

Maize (MAIZ), town (2000 population 1,868), SEDGWICK county, S KANSAS, suburb, 10 mi/16 km NW of WICHITA; 37°46′N 97°27′W. In wheat region.

Maíz Gordo, Sierra del (mah-EEZ GOR-do, see-YER-rah del), subandean mountain range, on JUJUY-SALTA province border, NW ARGENTINA, N of PIQUETE; extends c.50 mi/80 km NW-SE; rises to c.6,500 ft/1,981 m. Covered with subtropical forests.

Maizières-lès-Metz (mai-ZYER–lai–METS), town (□ 3 sq mi/7.8 sq km), MOSELLE department, LORRAINE region, NE FRANCE, 7 mi/11.3 km N of METZ; 49°13′N 06°09′E. Manufacturing of electrical equipment. Amusement park.

Maizuru (mai-EEZ-roo), city, KYOTO prefecture, S HONSHU, W central JAPAN, on Maizuru Bay, 37 mi/60 km N of KYOTO; 35°28′N 135°23′E. Important port (shipbuilding) and naval base with the best natural harbor on the JAPAN SEA coast. Has textile mills and food-processing (fish paste products, *wakame* seaweed) and glass industries; plywood. Peanuts; mandarin oranges.

Majaba, YEMEN: see MAHJABA.

Majaceite River (mah-hah-THAI-tai), c.40 mi/64 km long, CÁDIZ province, SW SPAIN; rises at MÁLAGA province border, flows N, through subtropical LAKE GUADALCACÍN, to the GUADALETE RIVER 4 mi/6.4 km SSW of ARCOS DE LA FRONTERA. Used for irrigation.

Majadahonda (mah-HAH-dah-ON-dah), town, MADRID province, central SPAIN, 11 mi/18 km W of MADRID; 40°28′N 03°52′W. Agriculture.

Majadas, Las (mah-HAH-dahs, lahs), village, CUENCA province, E central SPAIN, in the Serranía de Cuenca, 17 mi/27 km NNE of CUENCA; 39°57′N 05°45′W. Cereals; goats, cattle; agriculture. Flour milling, lumbering. Unexploited coal, copper, lead deposits.

Majagua (mah-HAH-gwah), town, CIEGO DE ÁVILA province, E CUBA, on railroad, and 15 mi/24 km WNW of CIEGO DE ÁVILA; 21°55′N 79°00′W. Sugarcane; fruit; cattle. Orlando González sugar mill is 6 mi/10 km S.

Majagual (mah-hah-GWAHL), town, ⊙ Majagual municipio, SUCRE department, N COLOMBIA, in savannas, 40 mi/64 km SE of SINCELEJO; 08°32′N 74°37′W. Rice, sorghum; livestock.

Majalengka (mah-JAH-luhng-kah), town, ⊙ Majalengka district, Java Barat province, INDONESIA, 25 mi/40 km WSW of Ceribon, at foot of Mount CIREMAY; 06°50′S 108°13′E. Trade center for agricultural area (rice, sugar, peanuts); textile mills. Also spelled Madjalengka.

Majano (mah-JAH-no), town, UDINE province, FRIULI-VENEZIA GIULIA, NE ITALY, 11 mi/18 km NW of UDINE. Fabricated metals, machinery, wood products, clothing, shoes.

Majardah River (me-jer-DAH), ancient *Bagradas*, chief river of TUNISIA, c.230 mi/370 km long; rises in the Majardah Mountains of NE ALGERIA, WSW of SOUK AHRAS; flows ENE past GHARDIMAU (where it enters TUNISIA), SUQ AL ARBA, and MADJAZ AL BAB, to the GULF OF TUNIS 20 mi/32 km N of TUNIS. Only perennial river in country. Receives the MALLAGU and SILYANAH rivers (right) and the WADI AL LIL River (left). It waters TUNISIA's principal wheat growing region, between SUQ AL ARBA and MADJAZ AL BAB, and provides a chief route of access (road, railroad) to TUNIS from the W. Formerly entered the MEDITERRANEAN SEA through the LAKE OF TUNIS. Also spelled Medjerda.

Majdaha (mahzh-DAH-hah), village, S YEMEN, on the cape Ras Majdaha, on Gulf of ADEN, 10 mi/16 km E of BIR ALI. Was in the former WAHIDI sultanate of Bir Ali. Also spelled Maqdaha and Maqdahah.

Majdal, ISRAEL: see ASHKELON.

Majdal Bani Fadel, Arab village, Nablus district, 10.6 mi/17 km SE of NABLUS, on the E slopes of the SAMARIAN Highlands, WEST BANK; 32°04′N 35°21′E. Agriculture (cereal and olives); sheep raising.

Majdal Shams (MAHJ-dahl SHAHMS), Druze township, ISRAEL, 12 mi/19 km ENE of KIRYAT SHMONA, in N GOLAN HEIGHTS. Situated on the SE slopes of MOUNT HERMON (at height of 3,773 ft/1,150 m), the N section of the town is built on the foundations of a 2nd–3rd century settlement (ruins). Israel offered citizenship to the township's residents in the early 1980s; a large number refused. Many residents have been arrested for involvement in terrorist attacks on Israel.

Majdan, UKRAINE: see MAYDAN.

Majdanek (mei-DAH-nek), village, SE POLAND, a suburb of LUBLIN. The Germans established and operated a concentration camp here in World War II. About 1,500,000 persons of twenty-seven nationalities (chiefly Jews, Russians, and Poles) were annihilated here in gas chambers. Museum and Memorial. Also spelled Maidanek.

Majdanpek (MEI-dahn-PEK), town (2002 population 23,703), E SERBIA, 40 mi/64 km ESE of POZAREVAC; 44°25′N 21°56′E. Iron-pyrite mine here, connected with DONJI MILANOVAC (ENE) by 11-mi/18-km-long aerial ropeway. Former copper-mining area. Gold mining nearby. Also spelled Majdanpek, Maidanpek, or Maydanpek.

Majé, Brazil: see MAGÉ.

Majene (MAH-je-nai), town, ⊙ Majene district, Sulawesi Seletan province, INDONESIA, on Makassar Sea, 144 mi/232 km WNW of UJUNG PANDANG.

Majeri (muh-je-ree), town, MALAPPURAM district, KERALA state, S INDIA, 25 mi/40 km ESE of KOZHIKODE. Road center; rice, cassava, pepper.

Majé, Serranía de (ma-HAI, se-rah-NEE-ah dai), mountain range, in PANAMÁ and DARIÉN provinces, PANAMA, along PACIFIC Coast between Lake Bayano and SAN MIGUEL GULF; 08°55′N 78°30′W. Highest elevation is Cerro Chucantí (4,721 ft/1,439 m).

Majes River, PERU: see COLCA RIVER.

Majestic, unincorporated village, PIKE county, E KENTUCKY, in the CUMBERLAND MOUNTAINS near TUG FORK, 13 mi/21 km SE of WILLIAMSON, WEST VIRGINIA. Bituminous coal. States of West Virginia, VIRGINIA, and Kentucky converge 7 mi/11.3 km to E (easternmost point in Kentucky).

Majestic Mountain (muh-JE-stik), (10,125 ft/3,086 m), W ALBERTA, W CANADA, near BRITISH COLUMBIA border, in ROCKY MOUNTAINS, in JASPER NATIONAL PARK, 11 mi/18 km SE of JASPER; 52°53′N 118°13′W.

Majete, town (2007 population 11,374), AMHARA state, central ETHIOPIA, 50 mi/80 km SW of DESSIE; 10°27′N 39°51′E.

Majevica (mei-ye-VEET-sah), mountain range in DINARIC ALPS, NE BOSNIA, BOSNIA AND HERZEGOVINA; extends c. 30 mi/48 km NW-SE; highest point (3,094 ft/943 m) is 6 mi/9.7 km E of TUZLA; 44°35′N 18°47′E. Contains petroleum. Bituminous coal at Majevica mine, 6 mi/9.7 km NE of Tuzla. Also spelled Mayevitsa.

Maji (MAH-jee), Italian *Magi*, town (2007 population 3,071), SOUTHERN NATIONS state, SW ETHIOPIA, on plateau, 135 mi/215 km SW of JIMMA; 06°12′N 35°35′E. Elevation is 7,970 ft/2,429 m. Trade center (hides, ivory); has airfield. Maji's former slave trade market considerably depopulated the surrounding region.

Majiang (MAH-JYAHNG), town, ⊙ Majiang county, SE central GUIZHOU province, CHINA, 55 mi/89 km ESE of GUIYANG, near sources of YUAN RIVER; 26°30′N 107°35′E. Grain, tobacco, tea; food industry; logging.

Maji Moto (mah-jee MO-to), village, MARA region, N TANZANIA, 40 mi/64 km ESE of MUSOMA, SE of Masurura Swamp; 01°40′S 34°22′E. TARIME Gold Fields to N. Cattle, goats, sheep; corn, wheat, millet. One of two villages so named in Tanzania.

Maji Moto, village, RUKWA region, W TANZANIA, 47 mi/76 km NNW of SUMBAWANGA, E of Mount Yamba (6,742 ft/2,055 m); 07°15′S 31°26′E. Road junction. Corn, wheat; goats, sheep. One of two villages so named in Tanzania.

Majitha (muh-JEE-tah), town, AMRITSAR district, W PUNJAB state, N INDIA, 10 mi/16 km NNE of AMRITSAR; 31°46′N 74°57′E. Wheat, cotton, gram, rice; hand-loom woolen weaving (carpets).

Maji-ya-Chumvi (mah-jee–yah–CHOOM-vee), village, COAST province, SE KENYA, on railroad, and 25 mi/40 km NW of MOMBASA; 03°49′S 22°39′E. Sugarcane, fruits, copra.

Majma'a (muhj-MAH-ah), town (2004 population 39,875), N central NAJ'D, SAUDI ARABIA, in N Riyadh province, 110 mi/177 km NW of RIYADH; 25°55′N 45°25′E. Trading center; grain, dates, vegetables, fruit; livestock raising.

Major, county (□ 957 sq mi/2,488.2 sq km; 2006 population 7,329), NW OKLAHOMA; ⊙ FAIRVIEW; 36°18′N 98°32′W. Drained by CIMARRON RIVER (forms W part of N border) and Eagle Chief Creek. NORTH CANADIAN River drains SW corner. Agriculture (wheat, oats, soybeans, corn; cattle). Oil and natural-gas deposits. Formed 1907.

Majorca (muh-YOR-kuh), Spanish *Mallorca* (mah-LYOR-kah), island (□ 1,405 sq mi/3,653 sq km), SPAIN, largest of the BALEARIC ISLANDS, in the W MEDITERRANEAN; 39°30′N 03°00′E. PALMA is the chief city. Mountainous in the NW, rising to 4,739 ft/1,444 m in the Puig Major; the S and E form a gently rolling, fertile region. Its mild climate and beautiful scenery have long made Majorca a popular resort; tourism is its major industry. Cereals, flax, grapes, and olives are grown, a light wine is produced, hogs and sheep are raised, and lead, marble, and copper are mined. For the history of Majorca before 1276, see BALEARIC ISLANDS.

The kingdom of Majorca was formed in 1276 from the inheritance of James I of Majorca. It comprised the Balearic Islands, Roussillon and Cerdagne (between France and Spain), and several fiefs in S France. Perpignan, in Roussillon, was the capital. Peter IV of Aragón took the kingdom from James II in 1343 and reunited it with the crown of Aragón. The island's flourishing commerce declined, partly because of the warfare between the native peasantry and the Aragonese nobles and Catalan traders, but mainly because of the change in trade routes after the discovery of America. Known for its stalagmite caves and for its architectural treasures and prehistoric monuments. The abandoned old monastery where Chopin and George Sand lived is an island landmark. Inhabitants speak their own dialect of Catalan.

Majorenhof, LATVIA: see JŪRMALA.

Majori, LATVIA: see JŪRMALA.

Major Isidoro (MAH-zhor EE-see-DO-ro), city, (2007 population 18,811), central ALAGOAS state, NE BRAZIL, 25 mi/40 km WSW of PALMEIRA DOS ÍNDIOS; 09°32′S 36°58′W. Cotton. Until 1944, called Sertãozinho.

Major Viera (MAH-zhor VEE-ai-rah), town, N SANTA CATARINA state, BRAZIL, 26 mi/42 km S of CANOINHAS; 26°19′S 50°20′W. Rice, potatoes, corn, manioc.

Majszin, ROMANIA: see MOISEI.

Majuba Hill, E KWAZULU-NATAL, SOUTH AFRICA, in the DRAKENSBERG RANGE, on N11 highway and railroad to VOLKSRUST (10 mi/16 km N), in Amajuba Forest; 27°29′S 29°52′E. On February 21, 1881, a British force of 400 was routed here by 150 Boer (Afrikaner) troops under the command of P. J. Joubert.

Majulid (mah-joo-lid), village, SIBSAGAR district, E central ASSAM state, NE INDIA, on MAJULI ISLAND, 13 mi/21 km NNW of JORHAT. Surrounding area grows calamus palms.

Majuli Island (muh-JOO-lee), SIBSAGAR district, E central ASSAM state, NE INDIA, between BRAHMAPUTRA (S, E), SUBANSIRI (W) rivers, and a branch of the Brahmaputra (N) flowing to the Subansiri; 48 mi/77 km long NE-SW, 5 mi/8 km–12 mi/19 km wide; 26°58′N 94°14′E. Calamus palms grown extensively. Ferry service.

Majune District, MOZAMBIQUE: see NIASSA.

Majunga, MADAGASCAR: see MAHAJANGA.

Majuro (mah-JOO-ro), atoll and town (1999 population 23,676), RATAK CHAIN, MARSHALL ISLANDS, W central PACIFIC, ⊙ Republic of the Marshall Islands.

Maka (MAH-kah), town, BOMI county, LIBERIA, 30 mi/48 km N of MONROVIA, on road; 06°58′N 10°59′W.

Makaba (mah-KAH-bah), village, ORIENTALE province, NE CONGO, 10 mi/16 km E of IRUMU; 05°25′S 26°58′E. Elev. 1,998 ft/608 m. Livestock farms here and at N'DELE or DELE, 4 mi/6.4 km NNE.

Makabe (MAH-kah-be), town, Makabe county, IBARAKI prefecture, central HONSHU, E central JAPAN, near Mount Tsukuba, 22 mi/35 km S of MITO; 36°16′N 140°05′E. Stone.

Makado, village, MATABELELAND SOUTH province, S central ZIMBABWE, 35 mi/56 km SE of WEST NICHOLSON, near Msano River; 21°27′S 29°44′E. Road junction. Livestock; grain.

Makaha (MAH-KAH-hah), town (2000 population 7,753), OAHU island, HONOLULU county, HAWAII, 23 mi/37 km WNW of HONOLULU, on W coast, near mouth of Makaha Stream. Cattle; fish. Kaneaki Heiau (Temple) to NE. Makaha and Keeau beach parks to NW. Keeau Point State Park to NW. WAIANAE Mountains and Waianae Kai Forest Reserve to NE.

Makaha, town, MASHONALAND EAST province, NE ZIMBABWE, 108 mi/174 km ENE of HARARE, near RUENYA RIVER; 17°17′S 32°37′E. Gold mining to S. Cattle, sheep, goats.

Makaha Valley (MAH-KAH-hah), town, OAHU island, HONOLULU county, HAWAII, 23 mi/37 km WNW of HONOLULU, 1 mi/1.6 km NE of MAKAHA and W coast; 21°28′N 158°11′W. Site of Kaneaki Heiau (Temple). WAIANAE Mountains to NE.

Makahuena Point (MAH-kah-HOO-EI-nah), SE coast, KAUAI island, KAUAI county, HAWAII, 8 mi/12.9 km SW of LIHUE, SE of POIPU; 21°52′N 159°26′W.

Makak (mah-KAHK), town, Central province, CAMEROON, 42 mi/68 km SW of Yaoundé; 03°33′N 11°03′E.

Makakilo City (MAH-kah-KEE-lo), city (2000 population 13,156), OAHU island, HONOLULU county, HAWAII, 15 mi/24 km WNW of HONOLULU, 3 mi/4.8 km inland from W coast, and 4 mi/6.4 km from S coast, at S end of WAIANAE Mountains; 21°21′N 158°05′W. Dairying; poultry. Barbers Point Naval Air Station to S; Kahe Point Beach Park to W.

Makala (muh-KAH-luh), village, ORIENTALE province, E CONGO, on LINDI RIVER, and 165 mi/266 km E of KISANGANI. Noted as site of defeat of last of the Arab slave traders, marking the end of the Arab campaign (1890–1894).

Makala, CONGO: see GREINERVILLE.

Makale (mah-KAH-luh), town, ⊙ Tanah Toraja district, Sulawesi Selatan province, INDONESIA, 90 mi/145 km NW of PARE PARE.

Makale, ETHIOPIA: see MEK'ELĒ.

Makaling-Banahaw (MAKBAN) Geothermal Power Project, PHILIPPINES: see ALAMINOS.

Makalla, YEMEN: see MUKALLA.

Makallé (mah-kah-YAI), town, SE Chaco province, ARGENTINA, on railroad, and 25 mi/40 km NW of RESISTENCIA; 27°13′S 59°17′W. Cotton center; lumbering; livestock raising; citriculture.

Makalu (muh-KAH-loo), peak (27,766 ft/8,463 m), on NEPAL-CHINA (TIBET) border, in NE NEPAL HIMALAYA; 27°53′N 87°05′E. Climbed in 1955 by Jean Franco of FRANCE.

Makalu Barun National Park (muh-KAH-loo buh-ROON), (□ 579 sq mi/1,505.4 sq km), E extension of SAGARMATHA NATIONAL PARK, NE NEPAL; 27°45′N 87°06′E. Surrounded by 320-sq-mi/830-sq-km conservation area. Established 1991.

Makamba, province (□ 757 sq mi/1,968.2 sq km; 1999 population 357,492), extreme S BURUNDI, on TANZANIA border (to E and S, part of border formed by MALAGARASI RIVER) and Lake TANGANYIKA (to W); ⊙ Makamba; 04°10′S 29°45′E. Borders BURURI (NWN) and RUTANA (NNE) provinces. Rukoziri River flows through province. Airport at NYANZA LAC village.

Makamik, CANADA: see MACAMIC.

Makanalua Peninsula, HAWAII: see KALAUPAPA.

Makanchi (mah-kahn-CHEE), city, SE SEMEY region, KAZAKHSTAN, near CHINA border, 110 mi/177 km SE of AYAGUZ (joined by highway); 46°47′N 82°00′E. Tertiary-level administrative center. In irrigated agricultural area (wheat, poppies, medicinal plants). Also Mukanshy.

Makanda (mah-KAHN-dah), village, LINDI region, SE central TANZANIA, 140 mi/225 km S of MOROGORO, in SELOUS GAME RESERVE; 08°56′S 37°54′E. Livestock.

Makanda (mah-KAHN-dah), village (2000 population 419), JACKSON county, SW ILLINOIS, 12 mi/19 km SSE of MURPHYSBORO; 37°37′N 89°14′W. In agricultural region. Giant City State Park nearby.

Makangwa (mah-KAHNG-gwah), village, DODOMA region, central TANZANIA, 15 mi/24 km SSE of DODOMA; 06°57′S 35°52′E. Road junction. Cattle, sheep, goats; corn, wheat.

Makanpur, INDIA: see BILHAUR.

Makanru Island (mah-kahn-ROO), uninhabited island of N main KURIL ISLANDS group, E SAKHALIN oblast, extreme E SIBERIA, RUSSIAN FAR EAST; separated from ONEKOTAN ISLAND (SE) by Fifth Kurile Strait; 6 mi/10 km long, 4 mi/6.4 km wide; 49°47′N 154°26′E. Rises to 3,835 ft/1,169 m in an extinct volcano. Also known as Makanrushi Island.

Makanya (mah-KAH-nyah), village, KILIMANJARO region, NE TANZANIA, 75 mi/121 km SSE of MOSHI, on E edge of MASAI STEPPE, on railroad; 04°22′S 37°49′E. PARE MOUNTAINS to E; MKOMAZI Game Reserve to NE. Timber; cattle; corn, sisal.

Makanza (muh-KAHN-zuh), village, Équateur province, NW CONGO, along the CONGO RIVER, and 160 mi/257 km WSW of LISALA; 01°42′N 19°12′E. Elev. 1,154 ft/351 m. Steamboat landing and trading center, with Roman Catholic missions and schools. Formerly known as NIEUW-ANTWERPEN or NOUVELLE-ANVERS.

Makapala (MAH-kah-PAH-lah), village, N HAWAII, HAWAII county, HAWAII, on KOHALA PENINSULA, 52 mi/84 km NW of HILO. Kohala Forest Reserve to SE; Keokea Beach Park to E, near Akoakoa Point.

Makapuu Point (MAH-kah-POO-oo), SE OAHU island, HONOLULU county, HAWAII, 12 mi/19 km E of HONOLULU; 21°18′N 157°39′W. Easternmost point on Oahu. Site of lighthouse built 1909. Manana (Rabbit) Island to N.

Makarak, RUSSIA: see MAKARAKSKIY.

Makarakskiy (mah-kah-RAHK-skeeye), town (2005 population 440), NE KEMEROVO oblast, S central SIBERIA, RUSSIA, on the KIYA RIVER (OB' RIVER basin), on road, 45 mi/72 km SSW of TYAZHINSKIY; 55°35′N 88°01′E. Elevation 1,181 ft/359 m. Gold mining. Also known as Makarak.

Makarevka, RUSSIA: see DUE.

Makariv (mah-KAH-reef), (Russian *Makarov*), town, W KIEV oblast, UKRAINE, on the Zdvyzh River, right tributary of the TETERIV RIVER, and 30 mi/48 km W of KIEV; 50°28′N 29°49′E. Elevation 472 ft/143 m. Raion center. Dairy products; flour, flax; brickworks; sewing and knitting. Medical school. Known in the 16th century as the village Voronine, renamed Makariv at the end of the century; castle built by Polish landlords in the 17th century; captured by the Cossacks in 1648, became center of a Cossack company; participated in popular uprisings (1664–1665, 1694, 1702–1704). Town since 1956.

Makarov (mah-KAH-ruhf), city (2006 population 6,470), SW SAKHALIN oblast, on the SE side of SA-KHALIN Island, RUSSIAN FAR EAST, port on the TER-PENIYE GULF of the Sea of OKHOTSK, on the E coastal highway and railroad, 145 mi/233 km N of YUZHNO-SAKHALINSK; 48°38′N 142°48′E. Coal mining, wood-pulp and paper milling, wood products. Under Japanese rule (1905–1945), called Shiritori or Shir-utoru. City since 1946.

Makarov, UKRAINE: see MAKARIV.

Makarovo (mah-KAH-ruh-vuh), village, NW SARATOV oblast, SE European Russia, on the KHOPER River, on road, 20 mi/32 km W of RTISHCHEVO; 52°16′N 43°20′E. Elevation 462 ft/140 m. In agricultural area (sugar beets, grains, potatoes).

Makarpura (MUHK-uhr-puhr-uh), village, VADODARA district, GUJARAT state, W central INDIA, 4 mi/6.4 km SSW of VADODARA. Cotton, millet, tobacco. Noted stepped well nearby.

Makarska (MAH-kahr-skah), Italian *Macarsca* (mah-KAHR-skah), town, S CROATIA, small port on ADRIATIC SEA, 32 mi/51 km ESE of SPLIT, at W foot of BIOKOVO Mountain, in DALMATIA; 43°18′N 17°02′E. Seaside resort noted for its long beach; wine making (muscatel). Old churches and monastery (11th–16th centuries). First mentioned in 9th century.

Makarwal Kheji (MUK-uhr-wahl KAI-jee), village, MIANWALI district, NW PUNJAB province, central PAKISTAN, on railroad spur, and 30 mi/48 km NW of MIANWALI; 32°53′N 71°09′E. Sometimes called Makerwal.

Makary (MAH-kar-ee), town, Far-North province, CAMEROON, 7 mi/11.3 km S of Lake CHAD; 12°34′N 14°26′E. Also spelled Makari.

Makar'ye (mah-KAHR-ye), village, W KIROV oblast, E European Russia, on road and near a railroad spur terminus, 18 mi/29 km NNW of KOTEL'NICH; 57°26′N 47°08′E. Elevation 492 ft/149 m. Flax processing.

Makar'yev (mah-KAHR-yef), city (2005 population 7,505), S KOSTROMA oblast, central European Russia, on the UNZHA RIVER, on road and near railroad spur, 115 mi/185 km E of KOSTROMA, and 70 mi/113 km NE of KINESHMA; 57°53′N 43°48′E. Elevation 354 ft/107 m. Logging, timber floating, woodworking. Founded in 1439; chartered in 1778.

Makar'yev, RUSSIA: see MAKAR'YEVO.

Makar'yevo (mah-KAHR-ye-vuh), town (2006 population 217), E central NIZHEGOROD oblast, central European Russia, on the VOLGA RIVER, 55 mi/89 km ESE of NIZHNIY NOVGOROD, and 4 mi/6 km N of LYSKOVO, to which it is administratively subordinate; 56°05′N 45°03′E. Elevation 219 ft/66 m. Timber raft-ing, sawmilling. Has a 14th century monastery. Scene of large trade fairs, following their transfer (1524) from area of Kazan', until removal, in turn (1817), to Nizhniy Novgorod. Made city in 1779 and known as Makar'yev until 1925, when reduced to village status; made town in 1958.

Makasan, THAILAND: see MAKKASAN.

Makasar, INDONESIA: see UJUNG PANDANG.

Makasar Strait (MAH-kah-sahr), wide channel con-necting CELEBES SEA (N) with JAVA and FLORES seas (S), between BORNEO (W) and SULAWESI (E); 600 mi/966 km long and 80 mi/129 km–230 mi/370 km wide; 02°00′S 117°30′E. BALIKPAPAN and KOTABARU are on W shore; UJUNG PANDANG on E shore. Contains nu-merous islands, largest being PULAU LAUT and SE-BUKU off SE coast of Borneo. In World War II, the Allies inflicted (Jan. 1942) heavy losses on Japa-nese fleet in battle here, but failed to prevent Japanese landing on Balikpapan. Sometimes spelled Macassar or Makassar.

Makassar Strait, INDONESIA: see MAKASAR STRAIT.

Makat (mah-KAHT), oil town, E ATYRAU region, KA-ZAKHSTAN, on railroad, 70 mi/113 km NE of ATYRAU, in EMBA oil field; 47°38′N 53°16′E. Junction of railroad branch line to KOSCHAGYL.

Makatapora (mah-koo-to-PO-rah), village, DODOMA region, central TANZANIA, 60 mi/97 km SSE of DO-DOMA, on GREAT RUAHA RIVER; 07°05′S 35°56′E. Cattle, sheep, goats; corn, wheat.

Makatapura (mah-kah-ta-POO-rah), village, SINGIDA region, central TANZANIA, 20 mi/32 km E of MANYONI, on railroad; 05°44′S 35°09′E. Goats, sheep; grain.

Makatea (MAH-kah-TAI-ah), island, FRENCH POLY-NESIA, S PACIFIC, one of the most northwesterly of the TUAMOTU ARCHIPELAGO; 15°50′S 148°15′W. The center of the island, an uplifted atoll, was once a solid mass of phosphate that was mined jointly by British and French phosphate corporations until 1966, when the phosphate reserves were depleted. Administered as part of the Windward group of the SOCIETY ISLANDS and is no longer inhabited. Also called Aurora.

Makati (mah-KAH-tee), town (2000 population 444,867), NATIONAL CAPITAL REGION, S LUZON, PHI-LIPPINES, just SE of MANILA, on railroad; 14°34′N 121°02′E. Business and technology center (it is known as "the high-tech capital of the Philippines"). The town is also an affluent residential community with Western-style homes and shopping malls. Hub of international hotel, banking, and transportation ac-tivity.

Makato (mah-KAH-to), town, AKLAN province, NW PANAY island, PHILIPPINES, 33 mi/53 km WNW of ROXAS; 11°41′N 122°14′E. Agricultural center (tobacco, rice).

Makawao (MAH-kah-WOU), town (2000 population 6,327), NE MAUI, MAUI county, HAWAII, in interior, 6 mi/9.7 km S of N coast; 20°51′N 156°19′W. Holds annual 4th of July Rodeo. University of Hawaii Agricultural Station to SE. Olinda Prison Camp to SE. Koolau Forest Reserve to E; pine forest region to NE; eucalyptus forests to S; Makawao Forest Reserve to SE.

Makaza Pass (mah-kah-ZAH) (2,130 ft/649 m), in the SE RODOPI Mountains, on Bulgarian-Greek border, on a road between MOMCHILGRAD (BULGARIA) and Komotine (GREECE), 10 mi/16 km N of Komotine; 41°16′N 25°26′E.

MAKBAN, PHILIPPINES: see ALAMINOS.

Makedoniya, SE EUROPE: see MACEDONIA.

Makemo (mah-KAI-mo), atoll, central TUAMOTU AR-CHIPELAGO, FRENCH POLYNESIA, S PACIFIC, 125 mi/201 km E of FAKARAVA; 16°30′S 143°45′W. Formerly Phillips Island.

Makena Bay (MAH-KAI-nah), village, SW MAUI, HAWAII. MOLOKINI island 3.5 mi/5.6 km WSW, in Alalakeiki Channel. Makena Golf Course overlooks bay.

Makenge, CONGO: see KAKENGE.

Makengo, CONGO: see BOMBOMA.

Makeni (mah-KE-nee), town (□ 13,875 sq mi/36,075 sq km; 2004 population 82,840), ⊙ BOMBALI district (1985 population 277,102; 2004 population 408,390) and NORTHERN province, central SIERRA LEONE, 85 mi/137 km ENE of FREETOWN; 08°53′N 12°03′W. Trade center. Primary and secondary schools. Government livestock farm at nearby TEKO. Hydroelectric power station at BUMBUNA Falls.

Makere (mah-KAI-rai), village, KIGOMA region, W TANZANIA, 65 mi/105 km NE of KIGOMA, on Makere River; 04°19′S 30°26′E. Timber; goats, sheep; corn.

Makete (mah-KAI-tai), village, IRINGA region, SW TANZANIA, 35 mi/56 km W of NJOMBE, in KIPENGERE RANGE; 08°21′S 34°14′E. Wheat, corn; livestock.

Makeyevka, UKRAINE: see MAKIYIVKA.

Makgadikgadi Pans National Park (MAH-kgah-dee-kgah-dee) (□ 1,891 sq mi/4,916.6 sq km), CENTRAL DISTRICT, NE central BOTSWANA, on W of pan system, which includes NTWETWE and SUA Pans. Comprises pan (in SE) and open grassland. Home to herds of game.

Makhachkala (mah-hahch-kah-LAH), city (2005 population 510,775), ⊙ DAGESTAN REPUBLIC, NE CAUCASUS, SE European Russia, a port on the CAS-PIAN SEA (especially for petroleum), on road and railroad, 1,346 mi/2,166 km SE of MOSCOW; 42°58′N 47°30′E. Below sea level. Important commercial and industrial center; oil refineries linked by pipeline with the GROZNYY fields. Manufacturing includes ma-chinery and metal goods (electric welding equipment, chain bucket excavators, instruments, equipment for the food industry), textiles, chemicals; canning and other food industries. Home to the regionally famous physics institute. Founded in 1844 as a Russian stronghold of Petrovskoye. Made city and renamed Petrovsk-Port in 1857. Current name since 1921. Sharp increase in population is partially due to influx of refugees from fighting in CHECHEN REPUBLIC.

Makhadar (muh-KUHD-er), township, S YEMEN, 12 mi/19 km N of IBB, and near the main road to Yarim. Also spelled Mekhadir.

Makhado, town, LIMPOPO province, SOUTH AFRICA, at foot of SOUTPANSBERG mountains, 65 mi/105 km NE of POLOKWANE (Pietersburg); 23°03′S 29°54′E. Eleva-tion 3,280 ft/1,000 m. Stock, grain. Airfield. Popular tourist venue for hiking on Sontpanberg trails and forestry reserves E of town. The second-to-last town on railroad; also on N1 highway to ZIMBABWE border (56 mi/90 km). Main access to densely populated Vendaland to E and SE. Established 1898. Previously called Louis Trichardt after Afrikaner leader of same name. Renamed in 2003 after 19th century tribal king.

Makhaleng River (mah-KHA-leng), c.85 mi/137 km long, W LESOTHO; rises c.30 mi/48 km ESE of MASERU in MALUTI MOUNTAINS; flows SW, joining Kolo-la-Pere River at SOUTH AFRICAN border, 60 mi/97 km S of Maseru, to form Kornetspruit river, which flows S to ORANGE, or Senqu, River.

Makhambet (mah-kahm-BYET), village, ATYRAU re-gion, KAZAKHSTAN, on URAL river, c.40 mi/64 km N of ATYRAU; 47°40′N 51°35′E. Tertiary-level adminis-trative center. Wood products; melons.

Makhanpur, INDIA: see SHIKOHABAD.

Makhchesk (MAHKH-chyesk), village, SW NORTH OSSETIAN REPUBLIC, RUSSIA, in the N central Greater CAUCASUS, on the upper URUKH River (left affluent of the TEREK RIVER), approximately 15 mi/24 km N of the Russia-GEORGIA border, on mountain highway, 24 mi/39 km WSW of ALAGIR; 42°58′N 43°47′E. Elevation 4,452 ft/1,356 m. Hardy grain; livestock.

Makhdumi (muhk-DOO-mee), former petty sheikdom of SUBEIHI tribal area, SW YEMEN; MARASA was its main center. Under British protection 1897–1967. Abolished after South Yemen became independent.

Makheka (mah-KHE-kah), peak (11,360 ft/3,463 m) in the DRAKENSBERG RANGE, LESOTHO, 65 mi/105 km E of MASERU, 7 mi/11.3 km SW of SOUTH AFRICAN border; 29°14′S 29°18′E. Also called Mareka.

Makhili, Al (muh-KEE-lee, el), village, CYRENAICA re-gion, NE LIBYA, 45 mi/72 km SW of Darnah. Road junction on Tobruk-Banghazi road. Former caravan center. Formerly Makili, Mekili or El Mechili.

Makhindzhauri (mah-khin-jah-OO-ree), urban settle-ment, GEORGIA, 3.7 mi/6 km from BATUMI; 41°40′N 41°43′E. Railroad station; MAKHINDZHAURI seaside climatic and balneological health resort.

Makhir Coast (mah-KHIR), N SOMALIA, section of coast on Gulf of ADEN.

Makhmur (mahk-MOOR), township, ERBIL province, N IRAQ, in KURDISTAN, 37 mi/60 km SW of ERBIL, 45 mi/72 km SSE of MOSUL; 35°46′N 43°34′E. Sesame, millet, corn; livestock.

Makhnèvo (mahkh-NYO-vuh), town (2006 population 3,670), central SVERDLOVSK oblast, W SIBERIA, RUS-SIA, on the TAGIL River (OB' River basin), on road and railroad spur, 40 mi/64 km N of ALAPAYEVSK;

Cross-references are shown in SMALL CAPITALS. The pronunciation guide is shown on page xix. The sources of population figures are shown on page xvii.

58°27′N 61°42′E. Elevation 285 ft/86 m. Gravel pits; dairy.

Makhnivka, UKRAINE: see KOMSOMOL'S'KE, Vinnytsya oblast.

Makhnovka (mahkh-NOF-kah), village, S KURSK oblast, SW European Russia, on the Sudzha River (PSEL RIVER basin), near the border with UKRAINE, 51 mi/82 km SW of KURSK, and 2 mi/3.2 km S of SUDZHA, to which it is administratively subordinate; 51°10′N 35°18′E. Elevation 505 ft/153 m. Agricultural construction; gas pipeline equipment.

Makhnovka, UKRAINE: see KOMSOMOL'S'KE, Vinnytsya oblast.

Makhorivka, UKRAINE: see MOSPYNE.

Makhorovka, UKRAINE: see MOSPYNE.

Makhsusabad, INDIA: see MURSHIDABAD.

Makhtal (MUHK-tahl), town, MAHBUBNAGAR district, ANDHRA PRADESH state, S INDIA, 35 mi/56 km WSW of MAHBUBNAGAR; 16°30′N 77°31′E. Cotton and rice milling; biri manufacturing.

Makhtesh Ramon, crater, ISRAEL: see MIZPE RAMON.

Maki (MAH-kee), town, West Kanbara county, NIIGATA prefecture, central HONSHU, N central JAPAN, 12 mi/20 km S of NIIGATA; 37°45′N 138°53′E. Watermelons, persimmons, citrons; rice. Sado Yahiko Yoneyama quasi-national park (Kamo Lake, Donden Highlands) is nearby.

Maki (MAH-kee), village, East Kubiki county, NIIGATA prefecture, central HONSHU, N central JAPAN, 65 mi/105 km S of NIIGATA; 37°04′N 138°23′E.

Makian (MAH-kyahn), volcanic island (□ 33 sq mi/85.8 sq km), N Maluku province, INDONESIA, in MALUKU SEA, just W of HALMAHERA, 30 mi/48 km S of TERNATE; 7 mi/11.3 km in diameter; 00°19′N 127°23′E. Roughly circular. Has an active volcanic peak, rising to 4,452 ft/1,357 m. Agriculture (tobacco, sago, coconuts), fishing. First Dutch settlement 1608; Dutch sovereignty established 1683. Sometimes spelled Makjan.

Makili, LIBYA: see MAKHILI, AL.

Makimguwira (mah-keem-goo-WEE-rah), village, LINDI region, S TANZANIA, 35 mi/56 km SW of LIWALE, on S border of SELOUS GAME RESERVE; 09°59′S 37°49′E. Goats, sheep; cashews, corn; timber.

Makin, GILBERT ISLANDS, KIRIBATI: see BUTARITARI.

Makinak (MA-ki-nak), community, S MANITOBA, W central CANADA, 20 mi/33 km SE of DAUPHIN, E of RIDING MOUNTAIN NATIONAL PARK, and in OCHRE RIVER rural municipality; 50°58′N 99°39′W.

Makindu (mah-KEEN-doo), town, EASTERN province, S central KENYA, on railroad, and 100 mi/161 km SE of NAIROBI; 02°17′S 37°50′E. Elevation 3,277 ft/999 m. Sisal and rubber center. Airfield; mosque. Game reserve.

Makin Meang, GILBERT ISLANDS, KIRIBATI: see LITTLE MAKIN.

Makino (MAH-kee-no), town, Takashima county, SHIGA prefecture, S HONSHU, central JAPAN, 34 mi/55 km N of OTSU; 35°27′N 136°02′E. Shiitake mushrooms, chestnuts. Known for *tsukudani* (fish boiled in soy) and sushi.

Makinsk (mah-KEENSK), city, N AKMOLA region, KAZAKHSTAN, on railroad, 60 mi/97 km SE of KÖKSHETAU; 52°40′N 70°28′E. In wooded area. Tertiary-level administrative center. Manufacturing (agricultural machines). Until 1944, Makinka.

Makioka (mah-kee-O-kah), town, East Yamanashi county, YAMANASHI prefecture, central HONSHU, central JAPAN, 9 mi/15 km N of KOFU; 35°44′N 138°43′E.

Makivka, Mount, UKRAINE: see SKOLE BESKYDS.

Makiyivka (mah-KEE-yif-kah), (Russian *Makeyevka*), city (2001 population 389,589), central DONETS'K oblast, UKRAINE, in the DONETS BASIN; 48°02′N 37°58′E. Elevation 511 ft/155 m. Leading metallurgical and coal-mining center in the Donets'k-Makiyivka industrial node. In addition to eighteen mines and four enrichment plants, it has coking, metallurgical and pipe-making plants, light industry (cotton, textiles,

footwear); food processing and manufacturing building materials. Scientific research institutes of mining safety, engineering. Regional museum and metallurgical museum. Known since 1777; in 1870s several mines were opened here; in 1920 Makiyivka and Dmytriyevs'ke villages were merged to form city Dmytriyevs'ke (Russian *Dmitriyevsk*), renamed Makiyivka in 1931.

Makizono (mah-KEE-zo-no), town, Aira county, KAGOSHIMA prefecture, S KYUSHU, SW JAPAN, 22 mi/35 km N of KAGOSHIMA; 31°51′N 130°45′E.

Makjan, INDONESIA: see MAKIAN.

Makka (MEK-kah), city (2004 population 1,294,168), ⊙ HEJAZ and Makka province, W SAUDI ARABIA. The birthplace (C.E. c.570) of Muhammad the Prophet, it is the holiest city of Islam, and the goal of the annual Muslim hajj. It is c.45 mi/70 km from its port, JIDDA, and is in a narrow valley overlooked by hills crowned with castles. Unlike those of most Middle Eastern cities, many of the buildings, constructed of stone, are more than three stories high. The city was an ancient center of commerce and a place of great sanctity for pre-Muslim Arab sects before the rise of Muhammad. Muhammad's flight (the hegira) from here in 622 is the beginning of the rise of Islam. He captured the city shortly thereafter. Although Makka never lost its sanctity, it declined rapidly in commercial importance after its capture by the Umayyads in 692. It was sacked in 930 by the Karmathians and taken by the Ottoman Turks in 1517. The Wahhabis held it 1803–1813. Here, in 1916, Husayn ibn Ali proclaimed his independence from TURKEY and maintained himself as king of Hejaz until Makka fell to Ibn Saud in 1924. At the center of Makka is the Great Mosque, the Haram, which encloses the Kaaba, the focus of Muslim worship. Next to the Kaaba is Zamzam, a holy well used solely for religious and medicinal purposes. The bazaar outside the mosque is noted for its silks, beadwork, and perfumes. The commerce of the city depends heavily on the more than two million pilgrims who visit Makka during the annual hajj. Muslims are the only people allowed to reside here. Linked by road with many other cities in Saudi Arabia, such as MEDINA and Jidda. Has little arable land and must import most of its food. The oil boom in Saudi Arabia has significantly improved services, resulting in greater numbers of pilgrims each year. In November 1979, Muslim fundamentalists occupied the Great Mosque; after a two-week siege, more than 100 rebels were killed. Iranian pilgrims later rioted in July 1987, during the hajj, clashing with Saudi troops and ending with the death of more than 400 people. The hajj continues to be well-monitored by Saudi Arabia, yet remains a turbulent religious and increasingly political event. Seat of two colleges and the Umm al-Qura University (1979). Also spelled Mecca.

Makkasan (MUHK-GUH-SUHN), E suburb of BANGKOK, THAILAND, 3 mi/4.8 km from city center. Extensive railroad workshops. Also spelled MAKASAN.

Makkaveyevo (mahk-kah-VYE-ee-vuh), village (2005 population 3,865), central CHITA oblast, S Siberian Russia, on the INGODA River, on road and the TRANS-SIBERIAN RAILROAD, 30 mi/48 km ESE of CHITA; 51°44′N 113°58′E. Elevation 2,053 ft/625 m. Mineral mines in the vicinity.

Mak Khaeng, THAILAND: see UDON THANI.

Makkhanpur, INDIA: see SHIKOHABAD.

Makkovik (muh-KOO-vik), village (2006 population 362), E NEWFOUNDLAND AND LABRADOR, E CANADA, on Makkovik Bay of the ATLANTIC OCEAN, 124 mi/200 km No of HAPPY VALLEY-GOOSE BAY; 55°05′N 59°06′W. Fishing port focused on snow crab. Established in 1860.

Makkovik, Cape (muh-KOO-vik), on the ATLANTIC OCEAN, E NEWFOUNDLAND AND LABRADOR, E CANADA; 55°15′N 59°04′W.

Makkum (MAH-kum), town, FRIESLAND province, N NETHERLANDS, 11 mi/18 km W of SNEEK; 53°04′N 05°25′E. IJSSELMEER 1 mi/1.6 km to W; NE end of AFSLUITDIJK (Lorentz Locks) 3 mi/4.8 km to NW. Town noted for its pottery. Dairying; cattle, sheep; grain, vegetables, fruit; manufacturing (delftware, pottery, yachts). Tichelaar's Royal Makkumer Potteries here since 1660; pottery museum.

Maklakovo, RUSSIA: see LESOSIBIRSK.

Makli Hills, SIND province, SE PAKISTAN, near TATTA, off the right bank of the INDUS River. Occupied by humans since the 3rd century B.C.E. Mogul mosques and tombs.

Maknassi (mahk-NAH-see), village (2004 population 13,742), SIDI BU ZID province, S central TUNISIA, on Safaqis-Tawzar railroad, and 53 mi/85 km ENE of QAFSAH. Administrative center. Esparto trade; olive trees. Phosphate mine at Mahari Zabbūs (4 mi/6.4 km N; railroad spur).

Makó (MAH-ko), city (2001 population 25,802), S HUNGARY, on the MAROS RIVER (Romanian *Mureşul*), near the Romanian border; 46°13′N 20°29′E. Administrative and trade center in a fertile agricultural region. The center of the Hungarian onion industry; manufacturing (textiles, refrigerators, medical equipment, rubber, leather, footwear). Has large SLOVAK population. American journalist Joseph Pulitzer was born here.

Makogai (MAH-KONG-ei), volcanic island (□ 3 sq mi/7.8 sq km), FIJI, SW PACIFIC OCEAN, c.40 mi/60 km E of Viti Levu; 2.5 mi/4 km long. Was leper station until 1969, now agricultural quarantine station, and has sheep-breeding project. Sometimes spelled Makongai.

Makokou (MUH-ko-koo), city, ⊙ OGOOUÉ-IVINDO province, NE GABON, on IVINDO RIVER; 00°33′N 12°51′E. Cacao, coffee; iron deposits.

Makola, town, KOUILOU region, SW Congo Republic, 20 mi/32 km NE of POINTE-NOIRE; 04°34′S 12°03′E.

Makonde Plateau (mah-KON-dai), MTWARA region, SE TANZANIA, 60 mi/97 km SW of MTWARA, between LUKULEDI (N) and RUVUMA (MOZAMBIQUE border, S) rivers. Forested tableland exceeding 1,640 ft/500 m. Also spelled Makondi.

Makongai, FIJI: see MAKOGAI.

Makongo (mah-KON-go), town, NORTHERN REGION, GHANA, on Lake VOLTA, 25 mi/40 km SSE of SALAGA; 08°19′N 00°37′W. Ferry points 5 mi/8 km across from Yeji; port and transshipment center. Also spelled MAKONO.

Makongolosi (mah-kon-GO-lo-see), village, MBEYA region, SW TANZANIA, 35 mi/56 km NNW of MBEYA, E of Lake RUKWA; 08°27′S 33°10′E. Road junction. Corn, wheat; cattle, goats, sheep.

Makono, GHANA: see MAKONGO.

Makor, INDONESIA: see ARU ISLANDS.

Makoro (muh-KO-ro), town, ORIENTALE province, NE CONGO, on road, 30 mi/48 km W of WATSA; 03°08′N 29°53′E. Elev. 3,008 ft/916 m.

Makoshino, UKRAINE: see MAKOSHYNE.

Makoshyne (mah-KO-shi-ne), (Russian *Makoshino*), town (2004 population 3,430), E central CHERNIHIV oblast, UKRAINE, on the right bank of the DESNA River, on railroad, 6 mi/10 km SE of MENA; 51°27′N 32°02′E. Elevation 374 ft/113 m. Research-experimental reinforced concrete manufacturing plant, sand extraction; river dock. Sanatorium for children, heritage museum, hydrological preserve nearby. Established in the beginning of the 17th century; town since 1964.

Makoti (mah-KO-tee), village (2006 population 137), WARD county, NW central NORTH DAKOTA, 31 mi/50 km SW of MINOT; 47°57′N 101°47′W. Fort Berthold Indian Reservation just W. Founded in 1911 and incorporated in 1916. Name comes from maakoti, the Mandan Indian word meaning the largest of the earthen lodges.

Makotsevo (MAH-ko-tse-vo), village, SOFIA oblast, GORNA MALINA obshtina, W BULGARIA, on railroad,

25 mi/40 km E of SOFIA; 42°42′N 23°49′E. Grain; livestock. Site of mineral springs.

Makoua (mah-KWAH), town, CUVETTE region, central Congo Republic, on LIKOUALA RIVER, and 40 mi/64 km N of OWANDO; 00°00′S 15°38′E. Has Roman Catholic mission and center for treatment of leprosy and trypanosomiasis. Airport.

Makovytsya, Mount, UKRAINE: see VOLCANIC UKRAINIAN CARPATHIANS.

Makovytsya Ridge, UKRAINE: see VOLCANIC UKRAINIAN CARPATHIANS.

Makow (MAH-koov), Polish *Maków Podhalański*, town, Bielsko-Biaala province, S POLAND, on SKAWA RIVER, and 25 mi/40 km SSW of CRACOW. Sawmilling; lumber trade; health resort.

Makow, Polish *Maków Mazowiecki*, Russian *Makov*, town, Ostrołęka province, E central POLAND, on ORZYC RIVER, and 45 mi/72 km N of WARSAW. Cement manufacturing, tanning, flour milling. During World War II, under administration of German EAST PRUSSIA, called Mackeim.

Makrai (muhk-REI), village, S central HARDA district, S central MADHYA PRADESH state, central INDIA, 60 mi/97 km SW of HOSHANGABAD; 22°04′N 77°08′E. Wheat, millet, oilseeds; surrounded by dense teak and sal forests. Was capital of former princely state of MAKRAI, one of the CENTRAL INDIA states; incorporated 1948 into Hoshangabad district; into Harda district in 1998.

Makran (muk-RAHN), district (□ 23,196 sq mi/60,309.6 sq km), KALAT division, BALUCHISTAN province, SW PAKISTAN, on ARABIAN SEA (S); ⊙ TURBAT. Bounded W by IRAN, E by KHUZDAR district; bordered N by SIAHAN RANGE; crossed by CENTRAL MAKRAN and MAKRAN COAST (S) ranges. Rainfall very scanty. Chief rivers are DASHT, RAKHSHAN. Consists of barren hills and hot, dry valleys. Its dates are famous; some wheat and barley grown.

Makran (mahk-RAHN), coastal region of BALUCHESTAN, in SE IRAN and W PAKISTAN, on ARABIAN SEA. A desert plateau and mountainous area with a few cultivated valleys (dates), mainly that of DASHT (KECH or KEJ) River Came (643) under Arab control, and was visited in 1294 by Marco Polo, who called it Kes (Kej) Macoran. Its later history is that of BALUCHISTAN. Since the boundary demarcation of 1895–1896, Makran has been divided between Iran and Pakistan (then British BALUCHISTAN). Iranian Makran (also MOKRAN or MUKRAN), is now part of Hormozgān and Sīstān va Balūchestān province, is served by the ports of Bandar-e Jask, CHAHBAHAR, and GAVATER.

Makrana (muhk-RAH-nah), town, NAGAUR district, central RAJASTHAN state, NW INDIA, 120 mi/193 km NE of JODHPUR; 27°03′N 74°43′E. Trades in marble, salt, millet, hides. Noted marble quarries just W; material used in construction of TAJ MAHAL in AGRA.

Makran Coast Range (muk-RAHN), mountain range, Makran dist., Baluchistan prov., SW PAKISTAN, extends c.250 mi/402 km E-W, parallel to Arabian Sea. bet. DASHT R. (W) and W hills of PAB RANGE (E); 15 mi/24 km–40 mi/64 km wide. Several peaks higher than 4,000 ft/1,219 m; rises to 5,185 ft/1,580 m in E peak. Watered by tributaries of lower HINGOL and Dasht rivers, with many narrow gorges. Generally dry and barren.

Makran Range, Central (muk-RAHN), MAKRAN district, SW BALUCHISTAN province, SW PAKISTAN, extends c.250 mi/402 km NE-SW in arc, between IRAN border (W) and E end of SIAHAN RANGE (NE); 15 mi/24 km–40 mi/64 km wide. Rises to 7,522 ft/2,293 m in N peak. SW section watered by tributaries of DASHT RIVER.

Makresh (mah-KRESH), village and obshtina center, MONTANA oblast, VIDIN obshtina, BULGARIA; 43°47′N 22°40′E.

Makri (MAH-kree), village, Évros prefecture, EAST MACEDONIA AND THRACE department, extreme NE GREECE, 7 mi/11.3 km W of ALEXANDROÚPOLIS; 40°51′N 25°45′E. Just W, on top of a cliff overlooking the shore, is a cave that the locals call the Cave of Polyphemus, named for the Cyclops who captured Odysseus.

Makri, TURKEY: see FETHIYE.

Makrikambos, Mount (mah-kree-KAHM-bos) (5,485 ft/1,672 m), on Albanian-Greek border, in Epirus region, 14 mi/23 km ESE of GJIROKASTËR (ALBANIA); 40°00′N 20°23′E. Also spelled Makrykambos.

Makrinitsa (mah-kree-NEE-tsah), village, MAGNESIA prefecture, THESSALY department, N GREECE, 10 mi/16 km E of VÓLOS, on slopes of Pelion range; 39°24′N 22°59′E. Known for its concentration of traditional Pelion timber-framed houses.

Makronisos (mah-KRO-nee-sos), ancient *Helena*, northwesternmost island (□ 6 sq mi/15.6 sq km) of the CYCLADES, in AEGEAN SEA, in ATTICA prefecture, ATTICA department, off E central GREECE, 2 mi/3.2 km off Attica Peninsula, opposite LÁVRION, across Kéa Channel from KÉA; 8 mi/12.9 km long, 1 mi/1.6 km wide; 37°42′N 24°07′E. Mountainous, rugged terrain; fisheries. Notorious prison camp during the junta (1967–1974). Also called Makronisi or Makronisi, Helene or Eleni. Also spelled Makronesos.

Makrykambos, Mount, ALBANIA and GREECE: see MAKRIKAMBOS, MOUNT.

Makryplagi, Greece: see YERANIA.

Maksatikha (mahk-SAH-tee-hah), town (2006 population 9,340), SILYANAH N central TVER oblast, W European Russia, on the MOLOGA RIVER (VOLGA RIVER basin), on road and railroad, 58 mi/93 km N of TVER, and 30 mi/48 km W of BEZHETSK; 57°48′N 35°52′E. Elevation 413 ft/125 m. Highway junction. Oil tank farm; dairying.

Maksegnit (mak-SAI-nyet) [=Tuesday market], town (2007 population 10,416), AMHARA state, NW ETHIOPIA, 25 mi/40 km SE of GONDAR; 12°19′N 37°34′E. Market town.

Maksudabad, INDIA: see MURSHIDABAD.

Makthar (mahk-TAHR), town (2004 population 12,942), SILYANAH province, central TUNISIA, 43 mi/69 km SE of AL KAF; 35°51′N 09°12′E. Agricultural settlement; livestock; olives, esparto grass. Established by Phoenician refugees from CARTHAGE, became a regional capital and a Roman colony. Destroyed by Bani Hillali in 1050. Ancient remains include a triumphal arch (C.E. 116), temples, a Christian chapel, and twelve arches of a Roman aqueduct.

Maku (MAH-KOO), town, in Āzərbāyjān-e GHARBI province, NW IRAN, 125 mi/201 km NW of TABRIZ, near Turkish border; 27°52′N 52°26′E. Elev. 5,361 ft/1,634 m. Grain, cotton; rug weaving.

Makubetsu (mah-koo-BETS), town, S central Hokkaido prefecture, N JAPAN, 102 mi/165 km E of SAPPORO; 42°54′N 143°21′E. Agriculture (potatoes, sugar beets). Also spelled Makumbetsu.

Makuhari Mesa (mah-koo-HAH-ree), town, CHIBA prefecture, E central HONSHU, E central JAPAN, at NW base of BOSO PENINSULA, on TOKYO. Also spelled Makuwari.

Makumbako (mah-koom-BAH-ko), village, IRINGA region, SW central TANZANIA, 90 mi/145 km SW of IRINGA, on railroad; 08°52′S 34°47′E. Highway junction. Tobacco, pyrethrum, corn; wheat; cattle, goats, sheep.

Makumbi (mah-KOOM-bee), village, KASAI-OCCIDENTAL province, S CONGO, on right bank of KASAI RIVER, and 60 mi/97 km SW of LUEBO; 05°51′S 20°41′E. Elev. 1,492 ft/454 m. Regional trans-shipment point at head of railroad to CHARLESVILLE, circumnavigating the rapids; cottonseed-oil milling, soap manufacturing, sawmilling.

Makump, SIERRA LEONE: see MAGBURAKA.

Makunduchi (mah-koon-DOO-chee), town, ZANZIBAR SOUTH region, E TANZANIA, near SE coast of ZANZIBAR island, 30 mi/48 km SE of ZANZIBAR city; 06°25′S 39°31′E. Cloves, copra, bananas; fish.

Makung (mah-GUNG), chief town, of the PENGHU Archipelago (Pescadores), on W shore of Penghu Island, TAIWAN; 23°34′N 119°34′E. Naval base and fishing port; has tomb of French admiral Courbet, who occupied town during Franco-China War (1884–1885). Also appears as Makyu. Also known as Penghu.

Makungu (mah-KUNG-goo), village, SUD-KIVU province, E CONGO, near W shore of LAKE TANGANYIKA, 55 mi/89 km N of KALEMIE; 04°59′S 28°50′E. Elev. 3,986 ft/1,214 m. Gold-mining center.

Makurazaki (mah-koo-rah-ZAH-kee), city, KAGOSHIMA prefecture, SW KYUSHU, SW JAPAN, port on EAST CHINA SEA, on S SATSUMA Peninsula, 28 mi/45 km S of KAGOSHIMA; 31°16′N 130°17′E. Railroad terminus. Tea.

Makurdi (mah-KOOR-dee), town, ⊙ BENUE state, central NIGERIA, a port on the BENUE River; 07°44′N 08°32′E. Sesame seeds and cotton grown in the region are collected here for transshipment. Has an airport and is a railroad and road center and the terminus of a bridge across the Benue. Considerable limestone and marble reserves in the vicinity. Seat of a university of agriculture and Benue State University. Developed by the British in the early 20th century as a transportation and local administration center.

Makushin (MA-koo-shin), native fishing village, N UNALASKA Island, ALEUTIAN ISLANDS, SW ALASKA.

Makushino (mah-KOO-shi-nuh), city (2005 population 9,600), E KURGAN oblast, SW SIBERIA, RUSSIA, on road junction and the TRANS-SIBERIAN RAILROAD, 80 mi/129 km ESE of KURGAN; 55°12′N 67°14′E. Elevation 459 ft/139 m. Mechanical repair shops, feed mill, dairy. Founded in 1797. Made city in 1963.

Makushin Volcano (6,680 ft/2,036 m), NE UNALASKA Island, ALEUTIAN ISLANDS, SW ALASKA, 15 mi/24 km W of DUTCH HARBOR; 53°53′N 166°55′W. Flat-topped, snow-covered, active volcano.

Makutapora (mah-koo-tah-PO-rah), village, DODOMA region, central TANZANIA, 18 mi/29 km N of DODOMA, W of Lake Hombolo; 05°57′S 35°42′E. Sheep, goats; subsistence crops.

Makuti, village, MASHONALAND WEST province, N ZIMBABWE, 95 mi/153 km NW of CHINHOYI, on border between Charara and Hurungwe safari areas, at junction of road to KARIBA DAM (35 mi/56 km to WSW); 16°18′S 29°15′E. Airport. Tourism. Livestock.

Makuyuni (mah-koo-YOO-nee), village, TANGA region, NE TANZANIA, 15 mi/24 km NW of KOROGWE, on railroad, SW of USAMBARA Mountains; 05°01′S 38°15′E. Sisal, rice.

Makwanpur (MUHK-wahn-poor), district, central NEPAL, in NARAYANI zone; ⊙ HETAUDA.

Makwassie, town, NORTH-WEST province, SOUTH AFRICA, 31 mi/50 km NE of Bloemhof, and 15 mi/24 km S of WOLMARANSTAAD; 27°19′S 25°59′E. Serves as railroad junction for main railroad line between KIMBERLEY and WITWATERSRAND, and main spur to LICHTENBURG. Agriculture center for maize, groundnuts, and dairy products.

Makwiro, town, MASHONALAND WEST province, central ZIMBABWE, 22 mi/35 km NE of CHEGUTU, on Makwiro River; 17°58′S 30°25′E. Elevation 4,307 ft/1,313 m. Chromite mining. Livestock; grain, tobacco.

Mala (MAH-lah), town (2005 population 22,895), CAÑETE province, LIMA region, W central PERU, on coastal plain, on MALA RIVER, near PAN-AMERICAN HIGHWAY, and 34 mi/55 km NNW of SAN VICENTE DE CAÑETE; 12°39′S 76°38′W. In cotton- and rice-growing region.

Malå (MAHL-O), town, VÄSTERBOTTEN county, N SWEDEN, on 6-mi/9.7-km-long Lake Malåträsk, 40 mi/64 km N of LYCKSELE; 65°11′N 18°45′E.

Malaba (mah-LAH-bah), town, EASTERN region, SE UGANDA, on KENYA border, 6 mi/10 km SE of TORORO; 00°38′N 34°18′E. Border-crossing point.

Malabar (MA-luh-bahr), town (□ 13 sq mi/33.8 sq km; 2005 population 2,772), BREVARD county, E central FLORIDA, 4 mi/6.4 km S of MELBOURNE; 28°00′N 80°34′W. Manufacturing includes machinery; commercial graphics.

Malabar Coast (MAH-lah-bahr), SW coast of INDIA, stretching c.525 mi/845 km from GOA state (N) to the S tip of the peninsula at Kanniyakumri (CAPE COMORIN), primarily in KERALA state and N KARNATAKA state. Narrow coastal plain bounded by the WESTERN GHATS. Monsoon rains make the coast a fertile rice-growing region. Scene of trade struggles in the sixteenth and early seventeenth centuries between the Portuguese and their European and Indian rivals. In the late seventeenth century the British gained control of the region.

Malabo (muh-LAH-bo), city (2003 population 92,900); ⊙ BIOKO NORTE province and of EQUATORIAL GUINEA, on BIOKO island, in the GULF OF GUINEA; 03°46′N 08°46′E. Chief port and commercial center of Bioko island. Fish processing is the city's main industry, and cacao and coffee are the leading exports. Founded in 1827 by the British as a base for the suppression of the slave trade; was called Port Clarence, or Clarencetown, and then called Santa Isabel. An international airport is on the city's outskirts. Much of the city's large European population left after rioting in the late 1960s; in the 1970s, the population declined again as Nigerian workers returned to their own country.

Malaboh (MAH-lah-bo), port, IRIAN JAYA, INDONESIA; 00°49′S 131°44′E. Seaport, airport.

Malabon (mah-lah-BON), town (2000 population 338,855), NATIONAL CAPITAL REGION, S LUZON, PHILIPPINES, on MANILA BAY, just NNW of MANILA; 14°39′N 120°57′E. Agricultural center (rice, sugarcane, fruit); fishing port; sugar milling. Also called Tanong.

Malabrigo (mah-lah-BREE-go) or **Colonia Ella**, town, N SANTA FE province, ARGENTINA, on railroad, and 24 mi/39 km SW of RECONQUISTA; 29°21′S 59°59′W. In soybeans, oats, corn, livestock, and lumbering area; flour milling, sawmilling.

Mal Abrigo (MAHL ah-BREE-go), town, SAN JOSÉ department, S URUGUAY, railroad junction, 19 mi/31 km NW of SAN JOSÉ; 34°09′S 56°57′W. In livestock region.

Malabuyoc (mah-lah-boo-YOK), town, CEBU province, S Cebu island, PHILIPPINES, 15 mi/24 km NE of TANJAY across TAÑON STRAIT; 09°40′N 123°21′E. Agricultural center (corn, coconuts).

Mala Bytca, SLOVAKIA: see BYTCA.

Malaca, SPAIN: see MÁLAGA, city.

Malacacheta (mah-lah-kah-she-tah), city (2007 population 17,991), NE MINAS GERAIS state, BRAZIL, 40 mi/64 km WSW of TEÓFILO OTONI; 17°51′S 42°09′W. Coffee.

Malacatán (mah-lah-kah-TAHN), town (2002 population 11,200), SAN MARCOS department, SW GUATEMALA, in Pacific piedmont, on Pacific Coast Highway, near MEXICO border, and 17 mi/27 km WSW of SAN MARCOS; 14°54′N 92°03′W. Elevation 1,283 ft/391 m. Market center; coffee, sugarcane, grain; livestock.

Malacatancito (mah-lah-kah-tahn-SEE-to), town, HUEHUETENANGO department, W GUATEMALA, on headstream of Río Negro (CHIXOY RIVER), and 5 mi/8 km SSW of HUEHUETENANGO; 15°13′N 91°31′W. Elevation 5,607 ft/1,709 m. Corn, wheat, beans.

Malacca, MALAYSIA: see MELAKA.

Malacca, Strait of (mah-LAH-kah), between INDONESIA (SUMATRA) and MALAYSIA (the MALAY PENINSULA); c. 440 mi/708 km long and c.30 mi/48 km–200 mi/322 km wide; 02°30′N 101°20′E. Linking the INDIAN OCEAN with the SOUTH CHINA SEA, it is one of the world's most important sea passages. Chief ports include BELAWAN (Sumatra Utara province), Duma (Riau province), and Palembarg (Sumatra Selatan province), in Indonesia and MELAKA and PINANG in W Malaysia; SINGAPORE is at the S end of the strait. In recent years oil and its products and liquified natural gas from N Sumatra have been major sea cargoes. The Strait of Malacca has been controlled by the Arabs, the Dutch, the Portuguese, and the British.

In the mid-nineteenth century it was a haven for pirates who menaced Dutch and British traders.

Malacky (mah-LAHTS-ki), Hungarian *Malacka*, town, ZAPADOSLOVENSKY province, W SLOVAKIA, on railroad, and 20 mi/32 km NNW of BRATISLAVA; 48°26′N 17°01′E. Natural gas field; manufacturing (agriculture; machinery; food and wood processing; distilling). Three 17th-century churches, museum; traditional folk architecture in vicinity.

Mala Čvrsnica, BOSNIA AND HERZEGOVINA: see ČVRSNICA.

Malad (MUHL-ahd), town, MUMBAI (BOMBAY) Suburban district, MAHARASHTRA state, W central INDIA, on SALSETTE ISLAND, 17 mi/27 km N of Mumbai (Bombay) city center; 19°11′N 72°50′E. Film studios. Rice. Glass bangle factory just N.

Mala Danylivka (mah-LAH dah-NI-leef-kah), (Russian *Malaya Danilovka*), town, N KHARKIV oblast, UKRAINE, on railroad (Lozovenka station), and 3 mi/5 km S of DERHACHI, in KHARKIV metropolitan area; 50°04′N 36°09′E. Elevation 426 ft/129 m. Veterinary institute. Established in 1714, town since 1938.

Malad City, town (2000 population 2,158), ⊙ ONEIDA county, SE IDAHO, on Deep Creek, near LITTLE MALAD RIVER, near UTAH state line, and 18 mi/29 km WNW of PRESTON; 42°11′N 112°15′W. Elevation 4,700 ft/1,433 m. Railroad terminus; shipping point for agricultural area (sugar beets, wheat, barley, potatoes); dairying; cattle; manufacturing (pumice aggregate, glue flour). Caribou National Forest to E, Curlew National Grassland to W; Devil Creek Reservoir to NE; Malad Summit (pass; 5,976 ft/1,821 m) to N. Settled by Mormons 1864.

Maladetta Mountains (mah-lah-DE-tah), Spanish *Montes Malditos* (MON-tes mahl-DEE-tos), French *Monts Maudits* [=cursed mountains], massif of the central PYRENEES, NE SPAIN, near the French border. Its highest point, PICO DE ANETO (11,168 ft/3,404 m), is also the highest in the Pyrenees. The GARONNE RIVER rises in the group.

Mala Divytsya (mah-LAH dee-VI-tsyah), (Russian *Malaya Devitsa*), town (2004 population 2,900), S CHERNIHIV oblast, UKRAINE, on railroad and road, 12 mi/19 km NW of PRYLUKY; 50°41′N 32°10′E. Elevation 396 ft/120 m. Feed and flour mills. Heritage museum. Known since 1628, town since 1960.

Malad River, c.50 mi/80 km long, in SE IDAHO and N UTAH; formed by confluence of LITTLE MALAD RIVER with Deep and Devil creeks in ONEIDA county, Idaho; flows S, into BOX ELDER county, Utah, entering BEAR RIVER NW of BRIGHAM CITY.

Malá Fatra, SLOVAKIA: see LESSER FATRA.

Málaga (MA-luh-guh), province (□ 2,813 sq mi/7,313.8 sq km; 2001 population 1,287,017), S SPAIN, in ANDALUSIA, on the MEDITERRANEAN SEA; ⊙ MÁLAGA. Bounded by GRANADA province (E), CÓRDOBA and SEVILLE provinces (N), and CÁDIZ province (W). Crossed by spurs of the CORDILLERA PENIBÉTICA, among them SIERRA DE RONDA (W) and SIERRA DE YEGUAS (N). Narrow strip of lowland (*vega*) along coast. The N section belongs to the interior Andalusian plain of the GUADALQUIVIR RIVER. Drained by GUADALHORCE RIVER, GUADIARO RIVER, and tributaries of the Guadalquivir. Climate is very mild in winter (especially near the coast), cooler in the uplands; summers are hot. The fertile intramontane basins and coastal lowlands produce mostly olives, grapes, almonds, figs, citrus fruit, cereals, vegetables; livestock. Olive oil, raisins, and wine (the renowned sweet Malaga) are its chief exports. Sugarcane has nearly disappeared. Beside agriculture, mineral resources have some importance. Tourism, including retirement living, is now the major economic element. The main industries process agricultural products. Besides important fisheries and fish-salting plants, there are also tanneries, textile mills, and foundries, chiefly centered at Málaga. Other cities/towns are

ÁLORA, ANTEQUERA, RONDA, VÉLEZ-MÁLAGA, COÍN, and ESTEPONA.

Málaga (MAH-lah-gah), city (2001 population 524,414), ⊙ MÁLAGA province, S SPAIN, in ANDALUSIA, on the GUADALMEDINA RIVER and the COSTA DEL SOL, 30 mi/48 km NE of MARBELLA; 36°43′N 04°25′W. Picturesquely situated on the Bay of Málaga, it is one of the best Spanish Mediterranean ports. Olives, almonds, dried fruits, Málaga wine and iron ore are exported. Textiles and construction materials are produced. Málaga's mild climate and luxurious vegetation, as well as the beautiful beaches nearby, make it a popular resort. Founded (12th century B.C.E.) by the Phoenicians, the city passed to the Carthaginians, the Romans, the Visigoths, and finally (711) the Moors. It flourished from the 13th century as a seaport of the Moorish kingdom of Granada, until it fell to Ferdinand and Isabella in 1487. Although largely modern in aspect, the city has several historic buildings, including a cathedral begun in the 16th century, the ruins of a Moorish alcazar, and an imposing citadel called the Gibralfaro. Picasso was born here. Formerly also spelled Malaca.

Málaga (MAH-lah-gah), town, ⊙ Málaga municipio, SANTANDER department, N central COLOMBIA, in valley of Cordillera ORIENTAL, 40 mi/64 km SE of BUCARAMANGA; 06°42′N 72°44′W. Elevation 8,093 ft/2,466 m. Trading and agricultural center (coffee, cacao, yucca, corn; livestock); food processing.

Malaga (MA-luh-guh), village, GLOUCESTER county, S NEW JERSEY, on MAURICE RIVER, and 5 mi/8 km NNW of VINELAND; 39°34′N 75°02′W. In agricultural area.

Malaga, unincorporated village, CHELAN county, central WASHINGTON, 7 mi/11.3 km ESE of WENATCHEE, on the COLUMBIA River (Rock Island Lake reservoir). Apples, pears; cattle, sheep. Manufacturing (primarily aluminum production). Rock Island Dam to SE.

Malaga Lake, (mah-LAH-gah) W NOVA SCOTIA, E CANADA, 13 mi/21 km W of BRIDGEWATER; 8 mi/13 km long, 3 mi/5 km wide. Fed and drained by MEDWAY RIVER.

Malagarasi (mah-lah-ga-RAH-see), village, KIGOMA region, W TANZANIA, 82 mi/132 km ESE of KIGOMA, on MALAGARASI RIVER, on railroad; 04°49′S 30°49′E. Lake SAGARA to S. Rice, corn; goats, sheep; timber.

Malagarasi River (mah-lah-ga-RAH-see), river, S BURUNDI and W TANZANIA; rises c.35 mi/56 km NNE of KIGOMA, near Lake TANGANYIKA on Burundi-Tanzania border; flows in a circular fashion: first 250 mi/402 km NE (forms border c.60 mi/97 km), receives Lumpungu River from Burundi as it turns SSE into Tanzania, receives GOMBE River from E 130 mi/209 km E of KIGOMA, then turns W and receives UGALLA River 75 mi/121 km ESE of Kigoma, and enters Lake Tanganyika 25 mi/40 km SSE of Kigoma via a swampy delta.

Malagash (MA-luh-gash), village, N NOVA SCOTIA, E CANADA, at head of Tatamagouche Bay, 40 mi/64 km WNW of NEW GLASGOW; 45°46′N 63°23′W. Salt-production center; fishing.

Malagasy Republic: see MADAGASCAR.

Malagón (mah-lah-GON), town, CIUDAD REAL province, S central SPAIN, on railroad to MADRID, and 13 mi/21 km N of CIUDAD REAL. Processing and agricultural center (cereals, olives, vegetables, fruit, grapes; livestock); apiculture. Olive-oil pressing, liquor distilling, wine making, lumbering, wool processing; manufacturing of fertilizer. Taken from the Moors in 1212.

Malagueño (mah-lah-GAIN-yo), town, W central CÓRDOBA province, ARGENTINA, 10 mi/16 km WSW of CÓRDOBA; 31°28′S 64°22′W. Railroad terminus. Quarrying center with lime kilns, granite quarries. Agriculture (olives, flax, corn); livestock.

Malahide (ma-luh-HEID), township (□ 153 sq mi/397.8 sq km; 2001 population 8,809), SW ONTARIO, E central CANADA, 13 mi/20 km from SAINT THOMAS;

42°47′N 80°56′W. Ontario Police College. Incorporated in 1998 from the merger of Malahide, SPRINGFIELD, and SOUTH DORCHESTER. Named after a castle in IRELAND.

Malahide (ma-luh-HEID), former township (□ 101 sq mi/262.6 sq km; 2001 population 6,290), SW ONTARIO, E central CANADA, 15 mi/24 km from SAINT THOMAS; 42°43′N 80°54′W. Amalgamated into new township of Malahide, formed 1998.

Malahide (ma-luh-HEID), Gaelic *Mullach Ide*, town (2006 population 14,937), NE DUBLIN county, E IRELAND, on inlet of the IRISH SEA, 10 mi/16 km NE of DUBLIN; 53°27′N 06°09′W. Fishing port and seaside resort. Malahide Castle has been seat of lords Talbot since time of Henry II and contains noted art collection. Remains of 14th-century abbey.

Malaita (mah-LAI-tah), basaltic volcanic island (□ 1,483 sq mi/3,855.8 sq km), most productive and populous of SOLOMON ISLANDS, SW PACIFIC, 30 mi/48 km NE of GUADALCANAL; 103 mi/165 km long, 15 mi/24 km wide. Continues across narrow channel, in Maramasike. Auki, on NW coast, is a modernizing chief town. Some live on artificial islands. Produces copra, rice, cacao. Local name Mala.

Malakal (mah-lah-KAHL), town, ⊙ UPPER NILE state, S central SUDAN, on right bank of the WHITE NILE River, 430 mi/692 km S of KHARTOUM; 09°32′N 31°39′E. Road junction. Commercial and livestock-raising center (cattle, sheep, and goats); some agriculture (cotton, peanuts, sesame, corn, and durra). Veterinary laboratory. Airport. N terminus of proposed JONGLEI CANAL. Malakal was capital of former UPPER NILE province.

Malakand (MUH-lah-kuhnd), division, NORTH-WEST FRONTIER PROVINCE, N PAKISTAN; headquarters at MALAKAND. Comprises present districts of DIR, SWAT, CHITRAL, and MALAKAND.

Malakand, district, MALAKAND division, N central NORTH-WEST FRONTIER PROVINCE, N PAKISTAN, ⊙ MALAKAND. Under tribal and Islamic administration law.

Malakand (MUH-lah-kuhnd), town, ⊙ MALAKAND district, MALAKAND division, N central NORTH-WEST FRONTIER PROVINCE, N PAKISTAN, in hill pass, 44 mi/71 km NNE of PESHAWAR; 34°34′N 71°50′E. Here Upper SWAT Canal runs through hills via 2-mi/3.2-km-long tunnel; hydroelectric plant nearby. Scene of fighting during tribal disturbances in 1895 and 1897.

Mala Kapela, CROATIA: see VELIKA KAPELA.

Malakhiv Hill, UKRAINE: see MALAKHOV, hill.

Malakhov (mah-LAH-kof), (Ukrainian *Malakhiv*), hill just E of and overlooking SEVASTOPOL′, Autonomous Republic of CRIMEA, SE UKRAINE. A major fortified point in the Crimean War, it was stormed (1855) by the French after an 11-month siege. Now a memorial park, with eternal flame. Often spelled Malakoff. Also referred to by the locals as Malakhov Kurgan (Tatar= mound).

Malakhovka (mah-LAH-huhf-kah), town (2006 population 18,355), central MOSCOW oblast, central European Russia, near the MOSKVA River, on road and railroad, 18 mi/29 km SE of MOSCOW, and 5 mi/8 km E of LYUBERTSY, to which it is administratively subordinate; 55°39′N 38°01′E. Elevation 456 ft/138 m. Experimental mining equipment; food processing. A holiday retreat spot for residents of Moscow and adjacent cities; sanatoria, summer homes.

Malakoff (mah-lah-KOF), town, HAUTS-DE-SEINE department, ÎLE-DE-FRANCE region, N central FRANCE, just SSW of PARIS, on subway line 3 mi/4.8 km from Notre Dame Cathedral, between VANVES (W) and MONTROUGE (E); 48°49′N 02°18′E. Manufacturing (machine tools, electrical equipment, pharmaceuticals). Branch of University of Paris is here.

Malakoff (MAH-luh-kawf), town (2006 population 2,360), HENDERSON county, E TEXAS, 27 mi/43 km ENE of CORSICANA; 32°10′N 96°00′W. Agriculture

(vegetables, nursery crops; cattle, horses); clay; oil and gas; lignite mines; manufacturing (powdered aluminum, transportation equipment, building materials). CEDAR CREEK RESERVOIR to NW. Incorporated after 1940.

Mala Krsna (MAH-lah KUHRS-nah), village, E SERBIA, 7 mi/11.3 km SSE of SMEDEREVO; 44°35′N 21°01′E. Railroad junction.

Malakula (MAH-lah-KOO-lah), continental-type and andesitic volcanic island, 2nd-largest island of VANUATU, SW PACIFIC OCEAN, 20 mi/32 km S of ESPÍRITU SANTO, and across Bougainville Channel; 55 mi/88 km long, 14 mi/24 km wide; 16°15′S 167°30′E. Copra, coffee, cocoa project, bêches-de-mer; sandalwood. Prior to independence (1980) for Vanuatu, French administrative headquarters at Port Sandwich on SE coast, British headquarters at Bushman's Bay on NE coast. Highest peak is Mt. Penot (2,925 ft/892 m). Also Malekula or Mallicolo.

Malakwal (MUH-luk-wahl), town, GUJRAT district, NE PUNJAB province, central PAKISTAN, near JHELUM RIVER, 50 mi/80 km W of GUJRAT; 32°34′N 73°13′E.

Malali (mah-LAH-lee), village, UPPER DEMERARA–BERBICE district, central GUYANA, on right bank of DEMERARA River, and 85 mi/137 km WSW of GEORGETOWN; 05°37′N 58°21′W. Also spelled MALLALI.

Malambo (mah-LAHM-bo), town, ⊙ Malambo municipio, ATLÁNTICO department, N COLOMBIA, on MAGDALENA River, in CARIBBEAN LOWLANDS, and 8.0 mi/12.9 km S of BARRANQUILLA; 10°51′N 74°46′W. Cotton, sugarcane, corn, bananas, plantains; livestock.

Malamir, IRAN: see IZEH.

Malamocco (mah-lah-MOK-ko), village, VENEZIA province, VENETO, N ITALY, on chief island (Lido or Lido di Venezia) of chain separating Lagoon of VENICE from the ADRIATIC Sea, 3 mi/5 km S of Lido; 45°22′N 12°20′E. Resort.

Malampaka (mah-lahm-PAH-kah), village, SHINYANGA region, NW TANZANIA, 38 mi/61 km N of SHINYANGA, on railroad; 03°08′S 33°30′E. Cotton, corn, wheat, millet; cattle, sheep, goats. Airstrip at Malya, 10 mi/16 km to N.

Malanda (muh-LAN-duh), town, QUEENSLAND, NE AUSTRALIA, 53 mi/85 km from CAIRNS, 9 mi/14 km SE of ATHERTON, on ATHERTON TABLELAND; 17°21′S 145°36′E. Dairying center.

Malang (MAH-lahng), municipality (2000 population 757,383), ⊙ Malang district, Java Timur province, INDONESIA, 50 mi/80 km S of SURABAYA, at foot of Mount SEMERU; 07°59′S 112°38′E. Trade center for agricultural area (coffee, corn, rice, sugar, tea, cinchona bark, peanuts, cassava, flowers). Has textile and lumber mills, railroad workshops, factories (cigars, cigarettes, ceramics, soap). Until 1949 was important Dutch military center, and is now headquarters for an Indonesian army division and air force base. Was temporary site of Indonesian parliament during independence struggle.

Malangali (ma-leen-GAH-lee), village, IRINGA region, S central TANZANIA, 75 mi/121 km SSW of IRINGA; 08°34′S 35°01′E. Wheat, corn, pyrethrum, tobacco; cattle, sheep, goats; timber.

Malangas (mah-LAHN-gahs), town, ZAMBOANGA DEL SUR province, W MINDANAO, PHILIPPINES, 31 mi/50 km WSW of PAGADIAN, on DUMANQUILAS BAY; 07°37′N 123°00′E. Coal-mining center.

Malangawa (muh-LUHNG-wah), village, ⊙ SARLAHI district, SE NEPAL, in the TERAI, 60 mi/97 km SSE of KATHMANDU, on road to SITAMARHI (INDIA); 26°52′N 85°34′E. Rice, wheat, oilseeds, barley, millet.

Malange, ANGOLA: see MALANJE.

Malangen (MAH-lahng-uhn), fishing village, TROMS county, N NORWAY, on Malangenfjord (30-mi/48-km-long inlet of NORWEGIAN SEA), 20 mi/32 km SSW of TROMSØ; 60°34′N 11°30′E.

Malanje (mah-LAHN-jai), province (□ 37,674 sq mi/97,952.4 sq km), N ANGOLA, on ZAIRE (N) border;

⊙ MALANJE. Bordered NW by UIGE province, E by LUNDA NORTE, S by Rio Cuanza and BIE province, W by CUANZA NORTE and CUANZA SUL. Drained by Uamba and Jombo rivers. Includes Cangandala National Park. Agriculture includes cotton, mandioc, corn, peanuts, beans, rice, sisal, sunflower. Minerals include iron ore, diamonds, and copper. Main centers are MALANGE, BEMBO, CAOMBO, MASSANGO, MANGANDO, CALANDULA.

Malanje (mah-LAHN-jai), town, ⊙ MALANJE province, N central ANGOLA, on central plateau, 220 mi/354 km ESE of LUANDA; 09°33′S 16°21′E. Elev. 3,800 ft/1,158 m. Inland terminus of railroad from Luanda. Agriculture center (cotton, sisal, rice, beans, corn, manioc). Food-processing industries. Airfield. Also spelled Malange.

Malantouen (mah-LAN-too-en), town, West province, CAMEROON, 50 mi/80 km NE of BAFOUSSAM; 05°42′N 11°08′E.

Malanville (mah-lahn-VEEL), town, ALIBORI department, NE BENIN, on right (W) bank of the NIGER RIVER (NIGER border), and 65 mi/105 km NE of KANDI; 11°52′N 03°23′E. Cotton, kapok, corn, millet, shea-nut butter; livestock. Iron ore deposits to W near Guéné, extends 12 mi/20 km to SW.

Malanzán (mah-lahn-ZAHN), village (1991 population 806), General Juan F. Quiroga department, S LA RIOJA province, ARGENTINA, in Sierra de Malanzán (a short pampean mountain range), 100 mi/161 km SSE of LA RIOJA; 30°48′S 66°37′W. Agricultural center (corn, alfalfa, wine; goats, cattle). Dam nearby.

Mala Panew River (MAH-lah PAH-nev), Polish *Mała Panew*, German *Malapane*, 75 mi/121 km long, in UPPER SILESIA, OPOLE province, S POLAND; rises N of TARNOWSKIE GÓRY; flows WNW to the ODER River, just below OPOLE (Oppeln). Irrigation dam at TURAWA.

Mala Plaženica, BOSNIA AND HERZEGOVINA: see VELIKA PLAZENICA.

Mala Point (MAH-lah), SE extremity of AZUERO PENINSULA, PANAMA, on GULF OF PANAMA of the PACIFIC OCEAN; 07°28′N 80°01′W.

Malappuram (muh-lah-PUHR-ruhm), district (□ 1,371 sq mi/3,564.6 sq km), KERALA state, S INDIA; ⊙ MALAPPURAM. Lies between WESTERN GHATS (E) and MALABAR COAST of ARABIAN SEA (W). Agriculture includes rice (mainly terrace farming), coconuts, areca nuts, jackfruit, mangoes, cashew nuts (on coastal plain). Extensive tea, rubber, spice (pepper, ginger), and coffee plantations in the GHATS (mainly in the Wynaad; teak and blackwood forests). Main coconut-producing area of Kerala state; manufacturing of copra, coir rope and mats. Also spelled MALLAPURAM.

Malappuram (muh-lah-PUHR-ruhm), town, ⊙ MALAPPURAM district, KERALA state, S INDIA; 11°04′N 76°04′E. Also spelled MALLAPURAM.

Malappuram, village, BELLARY district, KARNATAKA state, S INDIA.

Malapuram, INDIA: see HOSPET.

Malaqa, El (mah-LAH-kuh, el), village, DAQAHLIYA province, Lower EGYPT, between Lake MANZALA and DUMYAT branch of the NILE River, 6 mi/9.7 km NNE of Dumyat; 31°30′N 31°50′E. Fisheries. A small settlement.

Mälaren (MEL-ah-ren), lake (□ 440 sq mi/1,144 sq km), E central SWEDEN. The country's third-largest lake, it extends c.70 mi/113 km W from STOCKHOLM, connected E to BALTIC SEA via strait through Stockholm area. Scenic shores and over 1,000 islands, notably SKOKLOSTER and GRIPSHOLM. Many important small industrial towns on lake.

Malargüe (mah-lahr-GAI), town, W MENDOZA province, ARGENTINA, 100 mi/161 km SW of SAN RAFAEL; 35°28′S 69°35′W. Railroad terminus. Coal-mining and agricultural center (alfalfa, potatoes, wheat; livestock).

Oil was discovered here 1891. Uranium, iron ore, and nonferrous metal deposits nearby.

Mala River (MAH-lah), 70 mi/113 km long, LIMA region, W central PERU; rises in CORDILLERA OCCIDENTAL of the ANDES Mountains 14 mi/23 km ESE of MATUCANA; flows S and SW, past HUAROCHIRÍ and MALA, to the PACIFIC OCEAN 5 mi/8 km WSW of Mala; 12°08′S 76°13′W. Irrigation in lower course.

Malartic (mah-lahr-TEEK), town (□ 61 sq mi/158.6 sq km), ABITIBI-TÉMISCAMINGUE region, W QUEBEC, E CANADA, 40 mi/64 km ESE of ROUYN-NORANDA; 48°08′N 78°07′W. Mining center (gold, copper, molybdenum, zinc, lead). Incorporated 1939.

Malasiqui (mah-LAH-see-kee), town, PANGASINAN province, central LUZON, PHILIPPINES, 10 mi/16 km SSE of DAGUPAN, near railroad; 15°54′N 120°28′E. Agricultural center (rice, copra, corn).

Malaspina (ma-luh-SPEE-nuh), glacier (□ 1,500 sq mi/3,900 sq km), SE ALASKA, between YAKUTAT and ICY bays and flowing into the Gulf of ALASKA. Named for an Italian navigator who explored this region for Spain in 1791. Largest piedmont glacier in N AMERICA.

Malaspina, PHILIPPINES: see MOUNT CANLAON.

Malaspina Strait (ma-luhs-PEE-nuh), arm of Strait of GEORGIA in SW BRITISH COLUMBIA, SW CANADA, between TEXADA ISLAND and mainland; 40 mi/64 km long, 3 mi/5 km–5 mi/8 km wide; 49°40′N 124°15′W. JERVIS INLET branches off (NE). Near NE entrance is POWELL RIVER.

Malá Strana, CZECH REPUBLIC: see LITTLE QUARTER.

Malato Dag, TURKEY: see SASON DAG.

Malatya, city (2000 population 381,081), E central TURKEY, in the E TAURUS MOUNTAINS, on W shore of ATATÜRK DAM AND RESERVOIR, near Karakaya Dam; 38°22′N 38°18′E. Commercial center for a farm region that produces apricots, grapes, and grains. Manufacturing includes cement, cotton textiles, and sugar. Chrome, lead, and copper deposits. Airport. Situated at a strategic crossroads in ancient times, the city was the capital of a small Hittite kingdom c.1100 B.C.; it was then known as Milidia. In Roman times it was called Melitene and was a military headquarters. An important city of CAPPADOCIA, it became a metropolitan see in early Christian times. The city frequently suffered from attack and changed hands many times. In 1516 it was annexed by the OTTOMAN EMPIRE. In 1895, Christians here were massacred. Turkish Prime Minister (1983–1989) and President (1989–1993) Turgut Özal born here. İnönü University founded here in 1975.

Malatya Mountains, S central TURKEY, part of TAURUS MOUNTAINS, extend 65 mi/105 km SW from the EUPHRATES RIVER, S of MALATYA. Includes BUZ DAGI (9,091 ft/2,771 m) and NURUHAK DAG (9,850 ft/3,002 m).

Malaucène (mah-lo-SEN), commune (□ 17 sq mi/44.2 sq km), VAUCLUSE department, SE FRANCE, in PROVENCE, at NW foot of Mont VENTOUX, 9 mi/14.5 km NNE of CARPENTRAS; 44°10′N 05°08′E. Vineyards. The village center has retained its medieval appearance, including a Romanesque and Gothic church.

Malaunay (mah-lo-NAI), town (□ 3 sq mi/7.8 sq km; 2004 population 5,948), SEINE-MARITIME department, HAUTE-NORMANDIE region, N FRANCE, 6 mi/9.7 km NNW of ROUEN, in the chalky limestone Pays de CAUX; 49°32′N 01°02′E. Railroad junction and agriculture market; manufacturing of electrical equipment.

Malavalli, INDIA: see MALVALLI.

Mala Vyska (mah-LAH VIS-kah), (Russian *Malaya Viska*), city, W KIROVOHRAD oblast, UKRAINE, 30 mi/48 km WNW of KIROVOHRAD; 48°39′N 31°37′E. Elevation 524 ft/159 m. Raion center. Sugar-refining center; distillery; powdered milk plant. Vocational technical school; museum. Established in the first half of the 18th century, city since 1957.

Malawi (mah-LAH-wi), republic (□ 47,747 sq mi/123,665 sq km, including 9,347 sq mi/24,209 sq km of inland water; 2004 estimated population 11,906,855;

2007 estimated population 13,603,181), E central AFRICA; ☉ LILONGWE.

Geography

Bordered W by ZAMBIA, N by TANZANIA, and E, S, and SW by MOZAMBIQUE. Malawi is long and narrow, and about 20% of its total area is made up of Lake NYASA (Lake Malawi). Several rivers flow into Lake Nyasa from the W, and the SHIRE RIVER (a tributary of the ZAMBEZI RIVER) drains the lake in the S. Both Lake Nyasa and the Shire River lie within the GREAT RIFT VALLEY. Much of the rest of the country is made up of a plateau that averages 2,500 ft/762 m–4,500 ft/1,372 m in elevation, but reaches c.8,000 ft/2,438 m in the N and almost 10,000 ft/3,048 m in the S. Malawi is divided into three regions—Northern, Central, and Southern—and twenty-seven administrative divisions, called districts: BALAKA, BLANTYRE, CHIKWAWA, CHIRADZULU, CHITIPA, DEDZA, DOWA, KARONGA, KASUNGU, LIKOMA, LILONGWE, MACHINGA, MANGOCHE, MCHINJI, MULANJE, MWANZA, MZIMBA, NKHATA BAY, NKHOTAKOTA, NSANJE, NTCHEU, NTCHISI, PHALOMBE, RUMPHI, SALIMA, THYOLO, and ZOMBA. The major cities are Lilongwe, BLANTYRE, MZUZU, and ZOMBA.

Population

Almost all of the country's inhabitants are Bantu speakers, of which the Tumbuka, Ngoni, and Tonga (in the N) and the Chewa, Yao, Nguru, and Nyanja (in the center and S) are the main subgroups. About 75% of Malawi is Christian (mostly Roman Catholic and Presbyterian), and roughly 20% is Muslim; the rest follow traditional beliefs. English and Chichewa are official languages.

Economy

Malawi is primarily an agricultural country, with a very low per capita income. A significant number of Malawian peasants have for generations resorted to migrant labor in SOUTH AFRICA and ZIMBABWE to cope with limited employment opportunities at home. In recent years, however, this option has disappeared, further exacerbating their misery. Most of the cultivated land is made up of small farms held under traditional terms of tenure; the principal crops raised are maize, cotton, millet, rice, groundnuts, cassava, and potatoes. The rest of the farmland is included in large estates, where tea, tobacco, sugarcane, and tung oil are produced. With the aid of foreign investment, Malawi has instituted a variety of agricultural development programs. Large numbers of poultry, goats, cattle, and pigs are raised. There is also a small fishing industry. The forestry-products industry has been expanded since the 1970s. Practically no minerals are extracted, but there are substantial deposits of bauxite and further exploration is underway for apatite and coal. Malawi's manufacture is limited to basic goods, such as processed food, beverages, clothing, footwear, construction materials, and radios. The annual value of Malawi's imports is usually considerably higher than the value of its exports, the leading imports being manufactured consumer goods, machinery, transport equipment, chemicals, and foodstuffs; the principal exports are tobacco, tea, groundnuts, and maize. The chief trade partners are South Africa, the US, Germany, Egypt, Portugal, Japan, the Netherlands, Poland, and Russia. Most of the country's foreign trade is conducted via SALIMA, a port on Lake Nyasa, which is connected by railroad with the seaports of BEIRA and NACALA in Mozambique. Malawi is an ACP (African, Caribbean, and Pacific) member of the EU. Regional membership includes SADC (Southern African Development Community) and COMESA (Community of East and Southern African States).

History: to 1859

Between the 1st and 4th centuries C.E., Bantu-speaking peoples migrated to present-day Malawi. A new wave of Bantu-speaking peoples arrived around the 14th century, and they soon coalesced into the Maravi

kingdom (late 15th–late 18th centuries), centered in the Shire River valley. In the 18th century the kingdom conquered portions of what are now Zimbabwe and Mozambique. However, shortly thereafter it declined as a result of internal rivalries and incursions by the Yao, who sold their Malawi captives as slaves to Arab and Swahili merchants living on the INDIAN OCEAN coast. In the 1840s the region was thrown into further turmoil by the arrival from S Africa of the warlike Ngoni.

History: 1859 to 1950

In 1859, David Livingstone, the Scots explorer, visited Lake Nyasa and drew European attention to the effects of the slave trade here; in 1873 two Presbyterian missionary societies established bases in the region. Missionary activity, the threat of Portuguese annexation, and the influence of Cecil Rhodes led Great Britain to send a consul to the area in 1883 and to proclaim the Shire Highlands Protectorate in 1889. In 1891 the British Central African Protectorate (or BRITISH CENTRAL AFRICA; known from 1907 until 1964 as Nyasaland), which included most of present-day Malawi, was established. During the 1890s British forces ended the slave trade in the protectorate. At the same time, Europeans established coffee-growing estates in the Shire region; they were worked by Africans who thereby earned the cash necessary to pay taxes. In 1944 the protectorate's first political movement, the moderate Nyasaland African Congress, was formed, and in 1949 the government admitted the first Africans to the legislative council. In 1953 the Federation of RHODESIA AND NYASALAND (linking Nyasaland and Northern and Southern Rhodesia) was formed, over the strong opposition of Nyasaland's African population, who feared that the more aggressively European-oriented policies of Southern Rhodesia (see Zimbabwe) would eventually be applied to them.

History: 1950 to 1990

In the mid-1950s the congress, headed by H.B.M. Chipembere and Kanyama Chiume, became more radical. In 1958, Dr. Hastings Kamuzu Banda became the leader of the movement, which was renamed the Malawi Congress Party in 1959. Banda organized protests against British rule that led to the declaration of a state of emergency in 1959–1960. The Federation of Rhodesia and Nyasaland was ended in 1963, and on July 6, 1964, Nyasaland became independent as Malawi. Banda led the country in the era of independence, first as prime minister and, after Malawi became a republic in 1966, as president; he was made president for life in 1971. He quickly alienated other leaders by governing autocratically, by allowing Europeans to retain significant influence within the country, and by refusing to oppose white-minority rule in South Africa. Arguing that the country's economic well-being depended on friendly relations with the white-run government in South Africa, Banda established diplomatic ties with the latter in 1967. This relationship drew heavy public criticism. In the succeeding decade, Malawi became a refuge for antigovernment rebels from neighboring Mozambique, causing tension between the two nations, as did the influx (in the late 1980s) of more than 600,000 civil war refugees, prompting Mozambique to close its border. The border closure forced Malawi to use South African ports at great expense.

History: 1990 to Present

The decade of the 1990s ushered a new political era in Malawi—Kamuzu Banda's grip on power weakened significantly in 1992 following unprecedented antigovernment unrest championed by an internal coalition of opposition forces led by two splinter groups—the UDF (United Democratic Front) and the AFORD (Alliance for Democracy). A referendum was called in mid-June 1993 to determine the introduction of a multiparty democracy. The combined opposition forces secured a decisive victory with 63.2% of the voters demanding an end to a single-party rule. On May 16, 1994, a provisional constitution was adopted by the National Assembly, in advance of elections in which 80% of Malawians freely and peacefully voted for the first time in thirty years for a new president and members of parliament. Bakili Muluzi won a four-candidate presidential contest to suceed the 96-year-old Kamuzu Banda and become Malawi's second president. In the parliamentary context, Muluzi's party gained eighty-four seats of the 117-member parliament. He was reelected in 1999. In 2004 Muluzi's handpicked successor, Bingu wa Mutharika, won the presidential election against a divided opposition, but when Mutharika pursued a strong anticorruption campaign he alienated Muluzi and broke with him. Mutharika's conflicts with his vice president, who remained a Muluzi supporter, culminated in the arrest of Vice President Cassim Chilumpha for treason in 2006.

Government

Malawi is governed under the constitution of 1994. The president, who is both head of state and head of government, is popularly elected for a five-year term and is eligible for a second term. The unicameral legislature consists of the 193-seat National Assembly, whose members are also elected by popular vote for five-year terms. The current head of state and government is President Bingu wa Mutharika (since May 2004).

Malawi (mah-lah-WEE), village, Loristan province, SW IRAN, on Khurramabad-Dizful road, and 35 mi/56 km SW of Khurramabad, in gorge of Kashgan River (left tributary of KARKHEH River); 33°13′N 47°46′E. Also Malavi or Tang-i-Malawi.

Malawi, Lake: see NYASA, LAKE.

Malaya: see MALAYSIA, FEDERATION OF.

Malaya [Russian=lesser, little], in Russian names: see also MALO- [Russian combining form], MALOYE, MALY, and MALYE.

Malaya Danilovka, UKRAINE: see MALA DANYLIVKA.

Malaya Devitsa, UKRAINE: see MALA DIVYTSYA.

Malayagiri (mah-LEI-yah-gee-ree), peak (3,896 ft/1,188 m) in DHENKANAL district, N ORISSA state, E central INDIA, 75 mi/121 km NW of CUTTACK, just SE of Pal Lahara; 21°23′N 85°16′E.

Malaya Ichnya, UKRAINE: see ICHNYA.

Malaya Kandala (MAH-lah-yah kahn-dah-LAH), village, NE ULYANOVSK oblast, E central European Russia, on highway branch, 20 mi/32 km NNW of DIMITROVGRAD; 54°29′N 49°23′E. Elevation 344 ft/104 m. In agricultural area (sunflowers, hemp, grains, vegetables).

Malaya Kheta (MAH-lah-yah-HYE-tah), village, W TAYMYR (DOLGAN-NENETS) AUTONOMOUS OKRUG, KRASNOYARSK TERRITORY, N Siberian Russia, N of the ARCTIC CIRCLE, on the Lesser [Russian *Malaya*] Kheta River, 45 mi/72 km W of DUDINKA; 69°34′N 84°33′E. Oil fields.

Malaya Kokshaga, RUSSIA: see LESSER KOKSHAGA RIVER.

Malayan Union: see MALAYSIA, FEDERATION OF.

Malaya Purga (MAH-lah-yah poor-GAH), village (2006 population 6,780), S UDMURT REPUBLIC, E European Russia, on road and railroad, 67 mi/108 km N of NIZHNEKAMSK, and 20 mi/32 km SSW of IZHEVSK; 56°33′N 53°01′E. Elevation 278 ft/84 m. Building combine. Weather station.

Malay Archipelago (mai-LAI), great island group of INDONESIA formerly called the East Indies. It lies between ASIA mainland (N) and AUSTRALIA (S), separating the PACIFIC OCEAN (E) from the INDIAN OCEAN (W). Also includes the PHILIPPINES Archipelago, BRUNEI, and MALAYSIA.

Malaya Serdoba (MAH-lah-yah–syer-duh-BAH), village (2005 population 4,900), S PENZA oblast, E European Russia, on the Serdoba River (left branch of the KHOPER River), on road, 46 mi/74 mi NNE of

PENZA; 52°28′N 44°56′E. Elevation 564 ft/171 m. In agricultural area; bakery.

Malaya Sopcha (MAH-lah-yah SOP-chah), former town, central MURMANSK oblast, NW European Russia, 3 mi/5 km SW of MONCHEGORSK, into which it has been incorporated. Nickel and copper mines.

Malaya Vishera (MAH-lah-yah VEE-shi-rah), city (2006 population 13,720), N NOVGOROD oblast, central European Russia, on the left headstream of the Vishera River (tributary of the VOLKHOV RIVER), on crossroads and railroad, 58 mi/93 km NE of NOVGOROD; 58°51′N 32°14′E. Elevation 226 ft/68 m. Railroad shops; glassworking; timber mill; dairy, bakery. Made city in 1921.

Malaya Viska, UKRAINE: see MALA VYSKA.

Malaybalay (mah-LEI-bah-lei), town (2000 population 123,672), ⊙ BUKIDNON province, PHILIPPINES, N central MINDANAO, 85 mi/137 km NNW of DAVAO; 08°14′N 125°10′E. In river valley in mountainous area. Abaca, coconuts, and pineapples. Airstrip. Built by Iloco migrants from LUZON.

Malayer (mah-lah-YER), town (2006 population 156,289), Hamadān province, W IRAN, NW of ARAK; 34°17′N 48°50′E.

Malay Peninsula, SE extremity (□ 70,000 sq mi/ 182,000 sq km) of Asian mainland, lying between the ANDAMAN SEA of the INDIAN OCEAN and the STRAIT OF MALACCA on the W and the GULF OF SIAM and the SOUTH CHINA SEA on the E. It stretches S for c.700 mi/ 1,127 km from the ISTHMUS OF KRA, where it is narrowest, to SINGAPORE. The N part of the peninsula forms a part of THAILAND; the S part constitutes West Malaysia, the Malayan part of the FEDERATION OF MALAYSIA. A mountain range (the highest point of which is GUNONG TAHAN, 7,186 ft/2,190 m, in Malaysia) forms the backbone of the peninsula; numerous short, swift rivers flow E and W from it. More than half of the land surface is covered with tropical rain forest; the only open areas, aside from manmade clearings, are the alluvial plains of the W central portion of the peninsula and stretches along the rivers.

The region is one of the richest in the world in the production of tin and rubber; other products include timber, copra and coconut oil, palm oil, tapioca, peanuts, pineapples, and bananas. Rice is the chief foodstuff. The peninsula forms a physical and cultural link between the mainland of Asia and the islands of INDONESIA (often included in the Malay Archipelago). The Malays, historically the dominant cultural group, probably came originally from S China (c.2,000 B.C.E.), but intermarriage with other peoples have modified their ethnic characteristics. The Chinese are now nearly as numerous as the Malays; Indians and Thais form important minority groups. Small tribes of aborigines, descendants of pre-Malay immigrants, are found in the hills and jungles. The Malay Peninsula was visited near the beginning of the Christian era by traders from India and in the succeeding centuries received, like Indonesia and Indochina, Buddhist and Brahman missionaries and Hindu colonists. Small Hinduized states sprang up, like Langkasuka in the area of modern KEDAH.

In the second half of the 8th century the peninsula fell under the domination of the Sailendra rulers of Sri Vijaya (from SUMATRA), who adopted Mahayana Buddhism. Their cities in Kedah and PATTANI rivaled the importance of their capital at PALEMBANG (on Sumatra). The peninsula was overrun in the 11th century by the Cholas from the Coromandel Coast of India; after about fifty years, the Sailendras, somewhat weakened, resumed their sway. Sailendra rule ended in the late 13th century, when Sumatra and some S areas of the Malay Peninsula fell to a Javanese invasion and when the Thai king of Sukhothai swept over the peninsula from the N. The Sumatran kingdom of Melayu next ruled over the S of the peninsula, to be followed in turn (late 14th century) by Madjapahit,

the last Hindu empire of JAVA, and by the Thai king of Ayutthaya. The fall of Madjapahit opened the way for the primacy of a Malay state, Malacca (see MELAKA).

In the 15th century, the Malays, beginning with the Malaccans, were converted to Islam (which remains the religion of most Malays). The 16th century brought the first Europeans. The Portuguese seized Malacca (1511), and soon afterward Dutch traders appeared in Malayan waters. Malacca fell to the Dutch in 1641. The important British role on the peninsula began with the founding of settlements at PINANG (1786) and Singapore (1819). The coming of the Portuguese had plunged the peninsula into anarchy. The last sultan of Malacca, in flight from the Portuguese, founded a kingdom based on the Riau Archipelago and JOHOR, but the rulers of the petty states in the S gradually achieved independence, while the rising power of Siam and an increasingly imperial Britain became rivals. The British established protectorates over several Malay states, and in 1909 the boundary between Siam and Malaya was fixed by Siam's transfer to Great Britain of suzerainty over Kedah, Perlis, Kelantan, and Terengganu. For later developments, see MALAYSIA, FEDERATION OF.

Malaysia, Federation of (muh-LAI-zhyuh), country (□ 128,430 sq mi/332,633 sq km; 2004 estimated population 23,522,482; 2007 estimated population 24,821,286), SE ASIA; ⊙ KUALA LUMPUR.

Geography

Malaysia consists of two geographic regions: West Malaysia or Peninsular Malaysia (□ 50,700 sq mi/ 131,313 sq km), on the MALAY PENINSULA, coextensive with the former British-administered Federation of Malaya, and East Malaysia (77,730 sq mi/201,320 sq km), on the island of BORNEO and the federal territory of LABUAN. The two parts are separated by c.400 mi/ 640 km of the SOUTH CHINA SEA. West Malaysia comprises the federal territories of Kuala Lumpur and Putrajaya and the following eleven states: PERLIS, KEDAH, PINANG, PERAK, KELANTAN, TERENGGANU, PAHANG, SELANGOR, NEGERI SEMBILAN, MELAKA, (Malacca), and JOHOR. East Malaysia consists of two states: SABAH (former BRITISH NORTH BORNEO) and SARAWAK (former British Northwest Borneo). West Malaysia is bordered N by THAILAND, E by the South China Sea, S by SINGAPORE (separated by the narrow JOHOR STRAIT), and W by the Strait of MALACCA and the ANDAMAN SEA. East Malaysia is bordered N by the South China and the SULU seas, E by the CELEBES SEA, and S and W by KALIMANTAN (Indonesian [S] Borneo). Along the coast within Sarawak state is the independent nation of BRUNEI. Both East and West Malaysia have mountainous interiors and coastal plains. The highest point is Mount KINABALU (13,455 ft/4,101 m) in Sabah state. The longest of the country's many rivers are the RAJANG (c.350 mi/560 km) in Sarawak, the KINABATANGAN in Sabah, and the PAHANG in West Malaysia. Lying close to the equator, Malaysia has a tropical rainy climate. Nearly three-quarters of the land area is forested.

Population

Of the total population, 58% are of Malay or indigenous descent, 24% are Chinese, and 8% are Indian or Pakistani. In West Malaysia, Malays comprise about half of the population, Chinese about one-third, and Indians and Pakistanis one-tenth. In East Malaysia, the two largest groups are the Chinese and the Ibans (Sea Dayaks), an indigenous people, who together make up about three-fifths of the total. Conflict among the ethnic groups, particularly between Malays and Chinese, has played a large role in Malaysian history. Nearly all of the Malays are Muslims, and Islam is the national religion. The majority of Chinese are Buddhists, and the majority of Indians are Hindu. The official language is Bahasa Malaysia (Malay), although English is used in the legal system. Chinese, Tamil, and regional ethnic languages and dialects are also widely spoken. Malaysia's twelve higher educational institutions include Kebangsaan University and the University of Malaya in Kuala Lumpur, the Universities Sains Malaysia in Gelugor (Pinang state), and the International Islamic University (1983).

Economy

Malaysia is a large producer of rubber and tin with significant palm-oil and crude petroleum industries. Tropical timber from Sarawak to Sabah has also become an important export item. Since the 1980s, Malaysia has become a newly industrialized country, based largely on foreign investment in manufacturing. Rice is the staple food, while fish supply most of the protein. Industry is largely concentrated in West Malaysia. Tourism is rapidly increasing parallel with political and economic stability. Genting Highlands, a plateau on the border of Selangor and Pahang, 15 mi/24 km NE of Kuala Lumpur, midway between Kuala Lumpur and BENTONG; highest point is Gunung Ulu Kali (5,814 ft/ 1,772 m); has rubber plantations and tea, and the farming of vegetables. The major cities on the Malay Peninsula are connected by state-owned railroads with Singapore, and an extensive road network covers the W coast, where most of the population is concentrated.

History: to 1786

When the Portuguese captured Malacca (1511), its sultan fled first to Pahang and then to Johor and the RIAU ARCHIPELAGO. One of his sons became the first sultan of Perak. From both Johor and ACEH in SUMATRA unsuccessful attacks were made on Malacca. Aceh and Johor also fought each other mainly over control of the Strait of Malacca. Kedah, Kelantan, and Terengganu, N of Malacca, became nominal subjects of Siam. By 1619 the Dutch had established themselves in Batavia (now JAKARTA, Indonesia), and in 1641, allied with Johor, captured Malacca after a six-month siege. In the late 17th century, the Bugis from CELEBES, a Malay people economically pressured by the Dutch, began to settle on the Malayan W coast in Selangor, where they traded in tin. The Bugis captured Johor and Riau in 1721 and maintained general control there for about a century, allowing the Johor sultanate to remain, and were also active in Perak and Kedah. Earlier, in the 15th and 16th centuries, another Malay people, the Minangkabaus from Sumatra, had peacefully settled inland from Malacca, in the area that became the state of Negeri Sembilan.

History: 1786 to 1896

The British role on the peninsula began in 1786 when Francis Light of the British East India Company, obtained the cession of the island of Pinang, for a port and naval base, from the sultan of Kedah. In 1791 the British agreed to make annual payments to the sultan. In 1800, PROVINCE WELLESLEY on the mainland was added. In 1819 the British founded Singapore, then gained formal control of Malacca from the Dutch in 1824. Pinang, Malacca, and Singapore were jointly administered as the STRAITS SETTLEMENTS. In 1816, Siam forced Kedah to invade Perak and then invaded Kedah in 1821 and exiled the sultan. The Anglo-Siamese treaty of 1821 recognized Siamese control of Kedah but left the status of Perak, Kelantan, and Terengganu ambiguous. In 1841 the sultan of Kedah was restored, but Perlis was carved out of the territory of Kedah and put under Siamese protection. Later in the century ethnic conflicts, especially between Chinese and Malays, civil war and piracy, and increasing competition from Dutch, French, and German interests led to increased involvement of GREAT BRITAIN, who at the request of Malayan merchants moved in to restore order. Treaties were made with Perak, Selangor, Pahang, and the components of what became (1895) Negeri Sembilan. In each state a British "resident" was installed to advise the sultan (who received a stipend) and to supervise administration. The Pangkor Treaty of 1874 with Perak served as a model for subsequent treaties.

History: 1896 to 1946

In 1896 the four states were grouped together as the Federated Malay States with a British resident general. Johor, which had signed a treaty of alliance with Britain in 1885, accepted a British adviser in 1914. British control of the four remaining Malayan states was acquired in 1909, when, by treaty, Siam relinquished its claims to sovereignty over Kedah, Kelantan, Perlis, and Trengganu. These four, along with Johor, became known as the Unfederated Malay States; the entire peninsula was now known as British Malaya. In the latter half of the 19th century Malaya's economy assumed many of the major aspects of its present character. The output of tin, which had been mined for centuries, increased greatly with the utilization of modern methods. Rubber trees were introduced (Indian laborers were imported to work the rubber plantations), and Malaya became a leading rubber producer. Malaya's economic character, as well as its geographic position, gave it great strategic importance, and the peninsula was quickly overrun by the Japanese at the start of World War II and held by them for the duration of the war. The British, assuming that the attack would come from sea, had built their fortifications accordingly, but a land attack quickly drove them from the island of Singapore. Malaya's Chinese population received particularly harsh treatment during the Japanese occupation.

History: 1946 to 1957

When the British returned they arranged (1946) a centralized colony, called the Malayan Union, comprising all their peninsula possessions. Influential Malays vehemently opposed the new organization; they feared that the admission of the large Chinese and Indian populations of Pinang and Malacca to Malayan citizenship would end the special position Malays had always enjoyed, and they were unwilling to surrender the political power they enjoyed within the individual sultanates. The British backed down and established in place of the Union the Federation of Malaya (1948) headed by a British high commissioner. The federation was an expansion of the former Federated Malay States. Pinang and Malacca became members in addition to the nine Malay states, but there was no common citizenship. In that same year a Communist insurrection began that was to last over a decade. The Communist guerrillas, largely recruited from among the Chinese population, employed terrorist tactics. In combating the uprising the British resettled nearly 500,000 Chinese. "The Emergency," as it was called, was declared ended in 1960.

History: 1957 to 1965

The insurrection had the positive effect of spurring the movement for Malayan independence: In 1957 the federation became an independent state within the Commonwealth of Nations and was admitted to the UN. The first prime minister was Tengku (Prince) Abdul Rahman, the leader of the Alliance Party, a loose coalition of Malay, Chinese, and Indian parties. The constitution guaranteed special privileges for Malays. In 1963 Singapore, Sabah, and Sarawak were added to the federation, creating the Federation of Malaysia. Because Singapore has a large Chinese population, the latter two states were included to maintain a non-Chinese majority. Brunei was also included in the plan but declined to join. Malaysia retained Malaya's place in the UN and the Commonwealth. The new state was immediately confronted with the hostility of Indonesia, which described the federation as a British imperialist subterfuge and waged an undeclared war against it. In the struggle Malaysia received military aid from Great Britain and other Commonwealth nations. Hostilities continued until President Sukarno's fall from power in Indonesia (1965). Nonviolent opposition came from the PHILIPPINES, which claimed ownership of Sabah until early in 1978. The merger with Singapore

did not work out satisfactorily. Friction developed between Malay leaders and Singapore's prime minister, Lee Kuan Yew, who had worked to improve the position of the Chinese minority within the Malaysian Federation.

History: 1965 to Present

In 1965, Singapore peacefully seceded from Malaysia. Intercommunal tension continued, however, between Chinese and Malays, and led in 1969 to serious violence and a temporary suspension of parliament. In order to smooth out ethnic tensions, the government instituted a New Economic Policy (NEP), an attempt to correct the economic imbalances among ethnic groups. The NEP, initiated in 1970, expired in 1990 and was replaced by the National Development Policy (NDP). The 1970s saw an increase in Malay dominance in the country's affairs, appeasing minorities, especially the crucial entrepreneurial Chinese, through government coalitions. But by 1985, Malaysia was beset by disputes within the ruling coalition, by economic recession and financial scandals. However, since 1987, the economy has boomed. In January 1993, the government moved to curb the power of the nine hereditary Malay sultans, and a constitutional crisis seemed imminent. However, public sentiment toward the monarchy had been tarnished by accusations of power abuse, and a constitutional amendment reducing its power was easily passed in May 1994. Government crackdown of Al-Arqam, a radical Islamic fundamentalist group (100,000 members), in 1994 led to disbanding of the group later that year. General elections were postponed to 1995. Construction of the proposed new capital on a former plantation at Putrajaya, SW of Kuala Lumpur, was begun in 1995, but a 1997 government austerity program called for a delay in the scheduled 2005 completion date. Like several of its neighbors, Malaysia suffered a recession in 1997–1998, but by mid-1999 the economy began to recover. Although Mahathir's coalition retained a large majority in the 1999 elections, the Islamic party of Malaysia (PAS) made significant gains. In 2003 Mahathir retired and Abdullah bin Ahmad Badawi, the deputy prime minister, succeeded him. The ruling coalition substantially increased its majority in 2003 elections, almost entirely at the expense of PAS.

Government

Malaysia is a constitutional monarchy and is governed under the constitution of 1957 as amended. The sovereign (the Yang di-Pertuan Agong) is a largely ceremonial head of state, and is elected every five years by and from the nine hereditary rulers of Perlis, Kedah, Perak, Kelantan, Terengganu, Pahang, Selangor, Negeri Sembilan, and Johor. The prime minister is head of government and must be a member and have the confidence of the House of Representatives (Dewan Ra'ayat). The cabinet is chosen by the prime minister with the consent of the sovereign. There is a bicameral Parliament. The House of Representatives consists of 219 members, all elected by popular vote in single-member districts. The House sits for a maximum of five years but may be dissolved by the sovereign. The Senate (Dewan Negara) consists of seventy members chosen for three-year terms; each state legislature elects two and the sovereign appoints the remaining forty-four. There is a high court for each half of Malaysia and a supreme court. The current sovereign (since 2006) is Sultan Mizan Zainal Abidin of Terengganu; the current head of government (since 2003) is Prime Minister Abdullah bin Ahmad Badawi.

Malay States: see MALAYSIA, FEDERATION OF.

Malazgirt (mah-lahz-KEERT), township, E TURKEY, SE of ERZURUM, near MURAT RIVER road junction; 39°09′N 42°30′E. It was an important town of ancient ARMENIA. A council held here in C.E. 726 reasserted the independence of the Armenian Church from the Orthodox Eastern Church. Here, in 1071, the Seljuk Turks under Alp Arslan routed the troops of Byzan-

---

Area is shown by the symbol □, and capital city or county seat by ⊙.

tine Emperor Romanus IV in a decisive battle that resulted in the fall of ASIA MINOR to the Seljuks.

Malbaie, La (MAL-bai, lah) or **Murray Bay**, village, E QUEBEC, E CANADA, on the St. Lawrence River, at mouth of MALBAIE RIVER, and 80 mi/129 km NE of Quebec. Lumbering, dairying; resort.

Malbaie River (mal-BAI) or **Murray River**, S QUEBEC, E CANADA, 100 mi/161 km long; rises in E part of Laurentides National Park, flows in an arc NE and then SSE to the SAINT LAWRENCE RIVER at LA MALBAIE; 47°39′N 70°08′W.

Mal Bay, inlet (6 mi/9.7 km long) of the ATLANTIC OCEAN, W CLARE county, W IRELAND; 52°51′N 09°29′W. MILTOWN MALBAY is on the bay, which contains MUTTON ISLAND, noted shipping hazard.

Malbon, village, W central QUEENSLAND, AUSTRALIA, 30 mi/48 km SSW of CLONCURRY; 21°05′S 140°18′E. Railroad junction. Livestock.

Malbork (MAHL-borg), German *Marienburg*, town (2002 population 38,545), N POLAND, on the NOGAT RIVER. Railroad junction with sugar refineries and dairies; 54°02′N 19°03′E. Originally a castle founded (1274) by the Teutonic Knights, Malbork became the seat of their grand master in 1309. It successfully withstood sieges by the Poles in 1410 and 1454, but was sold to them in 1457 by mercenaries whose pay was in arrears. Passed to PRUSSIA in 1772. GERMANY took control of the town in 1920, and it was returned to Poland in 1945. The castle (rebuilt in the 14th and 19th centuries) is one of the finest examples of German secular medieval architecture.

Malbrán (mahl-BRAHN), village, SE SANTIAGO DEL ESTERO province, ARGENTINA, on railroad, and 65 mi/105 km SE of AÑATUYA; 29°21′S 62°27′W. In agricultural area (corn, alfalfa, oats; livestock).

Mal Branch, INDIA: see UPPER GANGA CANAL.

Malbun (mul-BOON), village, SE LIECHTENSTEIN, near Austrian border, 5 mi/8 km SE of VADUZ; 47°06′N, 09°36′E. Elevation 4,877 ft/1,487 m. Winter-sports center.

Malcantone (mahl-kahn-TO-nai), hilly region in S TICINO canton, S SWITZERLAND, W of LUGANO, between W arm of LAKE LUGANO and mountains along Italian border on the W.

Mal'chevskaya (MAHL-cheef-skah-yah), village (2006 population 3,920), NW ROSTOV oblast, SW European Russia, on road and railroad, 8 mi/13 km NNW of MILLEROVO; 49°03′N 40°22′E. Elevation 662 ft/201 m. In agricultural area (wheat, sunflowers; livestock); flour mill.

Malchika (MAHL-chee-kah), village, LOVECH oblast, LEVSKI obshtina, BULGARIA; 43°25′N 25°06′E.

Malchin (MAHL-khin), town, MECKLENBURG–WESTERN POMERANIA, N GERMANY, on the PEENE RIVER, and 22 mi/35 km E of GÜSTROW, between MALCHIN and KUMMEROW lakes; 53°44′N 12°46′E. Manufacturing includes sugar refining, sawmilling; manufacturing of soap, window frames. Agricultural market (grain, sugar beets, potatoes; livestock). Has fourteenth-century church and fifteenth-century town gates. Founded in thirteenth century.

Malchin Lake (mahl-KIN), German *Malchiner See* (mahl-KIN-uhr SAI), lake (□ 5 sq mi/13 sq km), MECKLENBURG–WESTERN POMERANIA, N GERMANY, E of MALCHIN; 8 mi/12.9 km long, c.1 mi/1.6 km wide; maximum depth 36 ft/11 m, average depth 10 ft/3 m; 53°42′N 12°36′E. Connected by canal with KUMMEROW LAKE, 7 mi/11.3 km NE.

Malchow (MAHL-khou), town, MECKLENBURG–WESTERN POMERANIA, N GERMANY, on the ELDE RIVER, and 23 mi/37 km E of PARCHIM, in lake region; 53°28′N 12°53′E. Woolen- and sawmilling; brick manufacturing. Has buildings of Augustinian convent (founded in thirteenth century; secularized in sixteenth century). Town was first mentioned 1235.

Malcocinado (mahl-ko-thee-NAH-do), town, BADAJOZ province, W SPAIN, in the SIERRA MORENA, 9 mi/14.5

km S of AZUAGA; 38°07′N 05°41′W. Olives, cereals; goats, sheep. Limekilns.

Malcolm, town, POWESHIEK county, central IOWA, 9 mi/14.5 km N of MONTEZUMA; 41°42′N 92°33′W. In agricultural area; feeds.

Malcolm, village (2006 population 447), LANCASTER county, SE NEBRASKA, 10 mi/16 km NW of LINCOLN; 40°54′N 96°52′W. Branched Oak Lake (N) and Pawnee Lake (S) state recreation areas nearby.

Malcolm Island (□ 32 sq mi/83 sq km), SW BRITISH COLUMBIA, SW CANADA, in QUEEN CHARLOTTE STRAIT just off N VANCOUVER ISLAND, 3 mi/5 km N of ALERT BAY, in MOUNT WADDINGTON regional district; 15 mi/24 km long, 2 mi/3 km–4 mi/6 km wide; 50°39′N 126°59′W. Lumbering; farming; salmon fishing. SOINTULA is trade center.

Malcompeth, INDIA: see MAHABALESHWAR.

Maldah (MAHL-dah), district (□ 1,441 sq mi/3,746.6 sq km), N WEST BENGAL state, E INDIA, ⊙ ENGLISH BAZAR (also known as Ingraj Bazar). On GANGA plain (silk growing in low hill area, N); bounded W by the GANGA RIVER and BIHAR state, E by BANGLADESH. Alluvial soil; rice, wheat, oilseeds, jute, barley, corn, tobacco; noted mango-growing area; major mulberry-growing district of West Bengal. Highly malarial swamps (S). Main towns are English Bazar and OLD MALDA. Archaeological landmarks include GAUR and PANDUA. Decisive victory, here, of the Mogul Aurangzeb over his brother in 1660. Formerly called MALDAHA. Original district reduced 1947 by incorporation of SE (including NAWABGANJ town) into RAJSHAHI district, EAST BENGAL, EAST PAKISTAN (now Bangladesh), following creation of PAKISTAN.

Maldaha, INDIA: see MALDAH.

Maldegem (MAHL-duh-kuhm), commune (2006 population 22,336), Eeklo district, EAST FLANDERS province, NW BELGIUM, 16 mi/26 km NW of GHENT; 51°13′N 03°27′E. Market center for vegetable-gardening region; textile industry.

Malden (MAWL-duhn), city (2000 population 56,340), MIDDLESEX county, E MASSACHUSETTS, a residential, commercial, and industrial suburb of BOSTON, in the MYSTIC RIVER valley; 42°26′N 71°04′W. Bordered N by the cliffs of Middlesex Fells. Manufacturing includes electronic parts, fabricated metal products, paint, clothing, and footwear. A number of old historic churches are here. Michael Wigglesworth was minister here for many years. Adoniram Judson and Saul Cohen born here. Pine Barks shared with MELROSE. Settled 1640, incorporated 1881.

Malden, city (2000 population 4,782), DUNKLIN county, extreme SE MISSOURI, in MISSISSIPPI alluvial plain, 29 mi/47 km SE of POPLAR BLUFF; 36°34′N 89°58′W. Soybeans, rice, cotton; manufacturing (transportation equipment, fabricated metal products, milling tools); ships cotton, livestock. Plotted 1877.

Malden, unincorporated town, KANAWHA county, W central WEST VIRGINIA, suburb, 5 mi/8 km ESE of CHARLESTON, on KANAWHA RIVER; 38°18′N 81°33′W. Agriculture (corn, tobacco); cattle, poultry. Manufacturing (quilts).

Malden, village (2006 population 194), WHITMAN county, SE WASHINGTON, 25 mi/40 km N of COLFAX, on Pine Creek; 47°14′N 117°28′W. Agricultural region in Palouse Country.

Malden Island (MAWL-duhn), currently uninhabited coral island (□ 15 sq mi/39 sq km), LINE ISLANDS, KIRIBATI, S PACIFIC, 373 mi/600 km SE of JARVIS ISLAND; 04°03′S 154°59′W. Discovered 1825 by the British; claimed by U.S. under Guano Act (1856). Leased 1922 by the British to Australian guano firm; now abandoned. Site of ancient Polynesian temples. Formerly Independence Island.

Maldens, The, ENGLAND: see KINGSTON UPON THAMES.

Malditos, Montes, SPAIN: see MALADETTA MOUNTAINS.

Mal di Ventre Island (mahl dee VEN-tre), in MEDITERRANEAN Sea, 5 mi/8 km SW of CAPE MANNU,

W SARDINIA, ITALY; 1.5 mi/2.4 km long; 39°59′N 08°18′E.

Maldives or **Divehi Jumhuriyya**, republic (□ 115 sq mi/299 sq km; 2007 population 369,031), S ASIA, stretching c.550 mi/885 km from N to S in the N INDIAN OCEAN, SW of SRI LANKA; ⊙ MALÉ Island; 03°15′N 73°00′E.

Geography

The Maldives consists of about twenty atolls made up of approximately 2,000 coral island built up on the tops of a submarine ridge, no part of the republic is more than 8 ft/2.4 m above sea level; Mahé Island is the largest island. Maldives has a tropical monsoon climate modified by its marine location, and largely covered with tropical vegetation, particularly coconut palms. About two hundred of the islands are inhabited, and some have freshwater lagoons.

Population

Maldivians are of mixed Dravidian, Sinhalese, and Arab stock. Sunni Islam is the official religion.

Economy

Tropical fruit is raised for local consumption. Tourism; fish; coconuts and coconut products (especially copra and coir); handicrafts, apparel, and shipping are the nation's chief sources of income.

History

The country was originally settled by peoples from S Asia. In the 12th century Islam was brought here. Starting in the 16th century, with the coming of the Portuguese, the Maldives were intermittently under European influence. In 1887 they became a British protectorate and subsequently a military base but retained internal self-government. The Maldives obtained complete independence as a sultanate in 1965, but in 1968 the dynasty which had ruled the islands since the 14th century, was ended and a republic was declared. Maldives became a member of the UN in 1965 and of the South Asian Association for Regional Cooperation (SAARC) since its founding in 1985. Despite a tiny population and few resources, the Maldives enjoyed strategic importance until the British withdrawal from their base on the southernmost island of Gan in 1976. Maumoon Abdul Gayoom, who was first elected president in 1978 and has retained power since, has ruled in an authoritarian manner, although he called for democratic reform in 2005. A coup attempt (1988) by Tamil mercenaries was quashed with the help of Indian forces. In the late 1980s, the Maldives joined with a number of coral atoll nations to raise international awareness of the consequences of global warming and in 1989 hosted an international conference to discuss this issue. The December 2004 Indian Ocean tsunami caused severe damage on many of the Maldives' islands.

Government

The Maldives are governed under the constitution of 1998. The president, who is both head of state and head of government, is chosen by the legislature for a five-year term; the chosen candidate must be confirmed in a referendum. The unicameral legislature consists of the fifty-member People's Council (*Majlis*); forty-two members are elected by popular vote and eight appointed by the president. All legislators serve five-year terms. Administratively, the country is divided into nineteen atolls and the capital city.

Maldon (MAWL-duhn), town (2001 population 18,217) and district, ESSEX, SE ENGLAND, on the BLACKWATER estuary, 9 mi/14.5 km E of CHELMSFORD; 51°44′N 00°41′E. Market town with iron foundries and other small industries. The 13th-century Church of All Saints has a unique triangular tower with a hexagonal spire, and the town hall dates from the 15th century. Prehistoric traces have been found in the vicinity. A battle between Danish raiders and East Saxons was fought near here in 991.

Maldon (MOL-duhn), village, central VICTORIA, SE AUSTRALIA, 75 mi/121 km NW of MELBOURNE, 12 mi/

20 km NW of CASTLEMAINE; 36°59′S 44°05′E. In old gold-mining region; livestock; some wheat, and oats; tourism. Has well-preserved buildings from its gold-mining era; classified by the National Trust. Formerly known as Tarrangower.

Maldonado (mahl-do-NAH-do), department (□ 1,851 sq mi/4,812.6 sq km; 2004 population 140,192), S URUGUAY, on the ATLANTIC OCEAN, at mouth of the Río de la PLATA; ⊙ MALDONADO; 34°40′S 54°55′W. Smallest department after MONTEVIDEO; grew nearly 21% in population 1985–1996, making it the fastest growing department in the country. Known for its many beach resorts (e.g., PIRIÁPOLIS, PUNTA DEL ESTE, and SOLÍS), which make this hilly, partly wooded region, with its pleasant climate, a favorite playground for foreign tourists. Also important for its livestock (cattle, sheep) and agricultural crops (grain, corn, sugar beets, vegetables); wine. Maldonado has deposits of lime, granite, marble, porphyry, and other building material. Flour milling, sugar refining (LA SIERRA), wine making, quarrying. Set up 1816; formerly included Rocha.

Maldonado, city (2004 population 54,603), ⊙ MALDONADO department, S URUGUAY, near mouth of the Río de la PLATA, 65 mi/105 km E of MONTEVIDEO (connected by railroad); 34°54′S 54°57′W. Its port is protected by GORRITI ISLAND. Trade center in grain and wool. Punta del Este Intl. Airport to E. An old city (founded 1757), it still retains its colonial character. Remains of Spanish fortifications are nearby.

Malè (mah-LAI), town, TRENTO province, TRENTINO-ALTO ADIGE, N ITALY, on NOCE River, and 24 mi/39 km SW of BOLZANO; 46°21′N 10°55′E. Chief center of VAL DI SOLE.

Malé, Brazil: see MALLET.

Malea, Cape, Greece: see ARGILIOS, CAPE.

Maleas, Cape (mah-LAI-ahs), LAKONIA prefecture, on SE peninsula of PELOPONNESE department, southeasternmost point of mainland GREECE, on AEGEAN SEA, between gulfs of LAKONIA (W) and ARGOLIS (E); 36°25′N 23°12′E.

Malé Atoll (MAH-lai), central group of MALDIVES, in INDIAN OCEAN, 475 mi/764 km WSW of SRI LANKA; between 03°48′N 73°20′E and 04°42′N 73°49′E. Consists of two neighboring physical atolls (North and South Malé). Larger N section (including MALÉ ISLAND) is separated from S section by narrow channel. Agriculture (coconuts, breadfruit); fishing (bonito); palm-mat weaving; manufacturing (textiles, apparel, marine products, fiberglass boats, launches, electronic components, pipes).

Malebo Pool, lakelike expansion of the CONGO RIVER (□ 320 sq mi/832 sq km), along the CONGO-CONGO REPUBLIC border, W CENTRAL AFRICA, 350 mi/563 km from the river mouth; 22 mi/35 km long and 14 mi/23 km wide; 04°15′S 15°25′E. KINSHASA (CONGO) and BRAZZAVILLE (CONGO REPUBLIC) are important ports on its shores. BAMU ISLAND (□ 70 sq mi/180 sq km) is at its W end; island is part of CONGO REPUBLIC and also called M'BAMOU ISLAND.

Malegaon (mah-LAI-goun), city (2001 population 409,403), NASHIK district, MAHARASHTRA state, W central INDIA, at the confluence of the GIRNA and Masam rivers; 20°33′N 74°32′E. Weaving center for saris and a market for agricultural products. Formerly a military post. Captured by the British in 1818 during the war with the Pindaris, marauding tribes of landless and casteless men who were often mercenaries of Maratha leaders.

Malehra (muh-LAI-rah), or **bada malhera**, town (2001 population 15,042), CHHATARPUR district, NE MADHYA PRADESH state, N central INDIA, 10 mi/16 km NNE of CHHATARPUR; 24°34′N 79°19′E. Local market for grain, cloth fabrics.

Malé Island (MAH-lai), small island and town of MALÉ ATOLL, ⊙ MALDIVE Islands, in INDIAN OCEAN, 450 mi/724 km WSW of COLOMBO (SRI LANKA); 04°10′N 73°30′E. Malé International Airport (on adjacent

Hulule Island), seaplane base, ship anchorage; center of interisland trade (coconuts, coir, palm mats, copra, cowries); diverse manufacturing. Seat of Maldivian people's assembly and residence of sultan. Sometimes called Sultan's or King's Island.

Malé Karpaty, SLOVAKIA: see LITTLE CARPATHIAN MOUNTAINS.

Malela (mah-LE-lah), village, BAS-CONGO province, W CONGO, on N shore of CONGO RIVER estuary (ANGOLA border), and 35 mi/56 km WSW of BOMA; 05°59′S 12°37′E. Palm-oil milling, hardwood lumbering. Hydrographic station.

Malela (muh-LE-luh), village, MANIEMA province, central CONGO, near left bank of the LUALABA RIVER, on railroad, and 35 mi/56 km WNW of KASONGO; 04°22′S 26°08′E. Elev. 1,820 ft/554 m. Agricultural center (rice, cotton, coffee); palm-oil milling, rice processing, hardwood lumbering. Has Roman Catholic mission with trade schools.

Malema (mah-LAI-muh), village, NAMPULA province, N central MOZAMBIQUE, on railroad, and 130 mi/209 km W of NAMPULA; 14°56′S 37°24′E. Corn, tobacco, beans. Formerly called ENTRE RIOS.

Malema District, MOZAMBIQUE: see NAMPULA.

Maleme (mah-LE-mee), air base in KHANIÁ prefecture, W CRETE department, GREECE, on N coast of CRETE, 12 mi/19 km W of KHANIÁ; 35°31′N 23°48′E. In World War II, German parachutists landed here in a pitched battle (1941).

Malemort-sur-Corrèze (mahl-uh-MOR–syoor–ko-REZ), town (□ 6 sq mi/15.6 sq km), CORRÈZE department, LIMOUSIN region, S central FRANCE, on CORRÈZE RIVER, and 2 mi/3.2 km ENE of BRIVE-LA-GAILLARDE; 45°10′N 01°33′E. Fruit and vegetable processing; machine shops.

Male Morze, POLAND: see PUCK BAY.

Malente (muh-LEN-te) or **Malente-Gremsmühlen** (muh-LEN-te–GREMS-myool-uhn), town, SCHLESWIG-HOLSTEIN, NW GERMANY, 3 mi/4.8 km NW of EUTIN, between two lakes; 54°10′N 10°32′E. Popular summer resort and health resort. Manufacturing (precison instruments, paper products).

Malente-Gremsmühlen, GERMANY: see MALENTE.

Maleny (muh-LAI-nee), town, QUEENSLAND, NE AUSTRALIA, 56 mi/90 km N of BRISBANE, 31 mi/50 km SW of MAROOCHYDORE, and on S edge of Blackall Range; 26°45′S 152°51′E. Macademia nuts; dairying; fruit; timber; arts and crafts; tourism.

Male Polissya, UKRAINE: see VOLHYNIAN UPLAND.

Malerkotla (muh-LER-kot-lah), town, SANGRUR district, PUNJAB state, N INDIA, 33 mi/53 km NW of PATIALA; 30°31′N 75°53′E. Trades in wheat, millet, cotton, sugar, oilseeds; cotton ginning, hand-loom weaving, manufacturing of coarse paper. Was capital of former princely state of Maler Kotla of PUNJAB STATES; in 1948, merged with PATIALA AND EAST PUNJAB STATES UNION and later became part of Punjab States.

Malesherbes (mahl-ZERB), town (□ 6 sq mi/15.6 sq km), LOIRET department, CENTRE administrative region, N central FRANCE, on the ESSONNE RIVER, and 13 mi/21 km SW of FONTAINEBLEAU; 48°18′N 02°25′E. Agriculture market in rich GÂTINAIS district. Furniture manufacturing. Has 13th-century Gothic church. The 18th-century château, built on site of a 14th–15th-century feudal castle, was owned by Malesherbes, minister of Louis XVI and French Academy member, who was decapitated during the French Revolution after pleading the king's cause.

Malesia e Madhe, ALBANIA: see KOPLIK.

Maleš Mountains (MAH-lesh), Serbian *Maleške Planine*, Bulgarian *Malesheviska Planina*, mountain range, extends c.35 mi/56 km N-S, along SERBIA-BULGARIA border, E of upper BREGALNICA RIVER, between OSOGOV MOUNTAINS (N) and OGRAZDEN MOUNTAIN (S). Heavily forested. Highest peak (6,337 ft/1,932 m) is KADIITSA. Includes Vlakhina Mountains (N). Also spelled Malesh Mountains.

Malestroit (mah-les-TRWAH), commune (□ 2 sq mi/5.2 sq km), MORBIHAN department, BRITTANY, W FRANCE, on OUST RIVER (BREST–NANTES CANAL), 8 mi/12.9 km. S of PLOËRMEL; 47°49′N 02°23′W. Woodworking; horse raising. Has 12th–16th-century church of SAINT-GILLES and many old dwellings around the church.

Maletsunyane Falls (mah-let-soon-YAH-ne), on Maletsunyane River (affluent of ORANGE RIVER), MASERU district, central LESOTHO, c.50 mi/80 km SE of MASERU, and 2 mi/3.2 km S of SEMONKONG; 29°52′S 28°03′E. Drops 630 ft/192 m. Sometimes called Semon Kong Falls [Lesotho=the place that smokes], referring to the spray.

Malevon, Greece: see PARNON.

Malewa (mai-LAI-wah), river, c.67 mi/108 km long, SW KENYA; rises in W sections of ABERDARE RANGE; flows NW and then S to Lake NAIVASHA. Basin area is 639 sq mi/1,655 sq km.

Maleyeva, RUSSIA: see OKTYABR'SKIY, KRASNOYARSK Territory.

Male Zernoseky, CZECH REPUBLIC: see VELKE ZERNOSEKY.

Malgaon (MUHL-goun), town, SANGLI district, MAHARASHTRA state, W central INDIA, 10 mi/16 km E of SANGLI; 20°27′N 74°38′E. Local trade center (millet, cotton, wheat); betel farming, handicraft cloth weaving.

Malgara, TURKEY: see MALKARA.

Malgobek (mahl-guh-BYEK), city (2005 population 47,985), NW INGUSH REPUBLIC, extreme N CAUCASUS, S European Russia, in the TEREK REGION, on road, 68 mi/109 km WNW of GROZNYY (CHECHEN REPUBLIC); 43°37′N 44°27′E. Elevation 688 ft/209 m. Second-largest city in the republic. Oil- and gas-producing center; natural-gas processing. Oil is piped to MOZDOK and Groznyy, natural gas to VLADIKAVKAZ. Developed in 1933 with the opening of oil and gas fields. Became city in 1939. Recent sharp population increase due largely to influx of refugees from fighting in Chechnya.

Malgrat de Mar (mahl-GRAHT dai MAHR), town, BARCELONA province, NE SPAIN, on the MEDITERRANEAN SEA, 11 mi/18 km NE of ARENYS DE MAR; 41°39′N 02°45′E. Grapes, wheat; wine making, oil production; lace manufacturing; fishing. Tourism; summer resort. Also known as Malgrat.

Malhada (MAHL-yah-dah), city (2007 population 16,336), SW BAHIA state, BRAZIL, on Rio SÃO FRANCISCO and MINAS GERAIS state border, across river from CARINHANHA; 14°20′S 43°44′W.

Malhada de Pedras (MAHL-yah-dah dee PE-drahs), town (2007 population 8,048), S central BAHIA state, BRAZIL, on SALVADOR–BELO HORIZONTE railroad, 22 mi/35 km SW of BRUMADO; 14°22′S 41°53′W.

Malhados (MAHL-yah-dos), city, N SERGIPE state, BRAZIL, 27 mi/43 km NW of ARACAJU; 10°39′S 37°18′W. Manioc.

Malhão (mahl-YOU), highest peak (6,532 ft/1,991 m) of PORTUGAL, in SERRA DA ESTRÊLA, in SW GUARDA and NW CASTELO BRANCO districts, just NW of COVILHÃ.

Malhargarh (MUHL-hahr-guhr), town (2001 population 7,349), MANDSAUR district, NW MADHYA PRADESH state, central INDIA, 14 mi/23 km N of MANDSAUR; 24°17′N 74°59′E. Road and railroad center. Gur manufacturing, hand-loom weaving.

Malheur (muh-LOOR), county (□ 9,930 sq mi/25,818 sq km; 2006 population 31,247), SE OREGON; ⊙ VALE; 43°12′N 117°37′W. Drained by MALHEUR and OWYHEE rivers, which flow to SNAKE RIVER. Borders on NEVADA (S) and IDAHO (E). Twelfth-largest county in U.S. Agriculture (sugar beets, corn, onions, potatoes, wheat, oats, barley; hogs, sheep, cattle); dairy products. Owyhee Dam, 20 mi/32 km SW of NYSSA, creates OWYHEE RESERVOIR in E. Warm Springs Reservoir on W border. Small part of Malheur National Forest on NW border; Ontario State Park in NE; Lake Owyhee

Reservoir and State Park and Succor Creek State Recreation Area in E. Duck Pond Lake, intermittent, in SW. Lava beds in S center. Formed 1887.

Malheur (muh-LOOR), river, c. 165 mi/266 km long, SE GRANT COUNTY, OREGON; rises at confluence of Big and Lake creeks, in Malheur National Forest, in the STRAWBERRY RANGE, E Oregon; flows S then SE through Warm Springs Reservoir, receives the South Fork, then flows generally NE past JUNTURA, where it receives the North Fork, past VALE to the Snake River, N of ONTARIO. Flood-control projects; Vale project uses the Malheur for irrigation.

Malheur Lake (muh-LOOR), semidry lake, HARNEY co., SE OREGON, 20 mi/32 km SE of BURNS; c.15 mi/24 km long and up to 5 mi/8 km wide. Harney Basin in NW part of GREAT BASIN. Fed by SILVIES (from NW) and DONNER UND BLITZEN (from S) rivers. Connected by channel to HARNEY LAKE in SW. Lakes have no outlet. Separate units of Malheur National Wildlife Refuge surround Malheur and Harney lakes and Donner und Blitzen River.

Mali (MAH-lee), independent republic (☐ 478,764 sq mi/1,239,999 sq km; 2004 estimated population 11,956,788; 2007 estimated population 11,995,402), the largest country in W AFRICA; ⊙ BAMAKO.

Geography
Bordered N by ALGERIA, E and SE by NIGER, S by BURKINA FASO and CÔTE D'IVOIRE, and W by GUINEA, SENEGAL, and MAURITANIA. Mali has one district (Bamako city) and eight administrative regions (provinces) with formal/familiar appellations as follows: FIRST REGION/KAYES, SECOND REGION/KOULIKORO, THIRD REGION/SIKASSO, FOURTH REGION/SÉGOU, FIFTH REGION/MOPTI, SIXTH REGION/TIMBUKTU, SEVENTH REGION/GAO, and EIGHTH REGION/KIDAL.

Economy
In the S, traversed by the NIGER and SÉNÉGAL rivers, are fertile areas where groundnuts, rice, and cotton are grown. Elsewhere the country is arid desert or semidesert and barely supports grazing (mainly cattle, sheep, and goats). The Niger River serves as an important transportation artery and a source of fish. Groundnuts and cotton are the country's only significant cash crops, with rice, maize, sorghum, millet, and cassava being the main food crops. Live animals and preserved fish are important exports. Mali has varied light industries, including canning and preserving, cotton ginning, groundnut-oil extraction, brickmaking, and the production of textiles, cigarettes, matches, and hardware. Some salt, limestone, and gold are mined for local trade, but the country's extensive mineral resources (bauxite, manganese, iron ore, phosphates, lithium, diamonds) remain largely unexploited. The state owns much of the country's industry, but has responded to international pressure to begin privatization. Mali's chief trading partner is FRANCE. In 1968, Mali joined with Senegal, Guinea, and Mauritania in founding the Organization of Senegal River States to develop the Senegal valley.

Population
The main ethnic groups are the Bambara, Marka, Songhai, and Malinke, who are chiefly farmers and fishermen, and the Fulani and the Tuareg, who are pastoralists. The majority of the population is Muslim, while most of the remainder follow traditional religions. There are a number of wildlife refuges here, including the forest reserves of Ansongo-Ménaka, Baoulé, Faya, Mandingue, Soussan, and Tienfala.

History: to 18th Century
The Mali region has been the seat of extensive empires and kingdoms, notably those of GHANA (4th–11th century), Mali, and Gao. The medieval empire of Mali was a powerful state and one of the world's chief gold suppliers; it attained its peak in the early 14th century under Mansa (Emperor) Musa (reigned c.1312–1337), who made a famous pilgrimage to MAKKA (Mecca) in 1324 laden with gold and slaves to proclaim Mali's

prosperity and power. During his rule Muslim scholarship reached new heights here, and such cities as TIMBUKTU and DJENNÉ (Jenne) became important centers of trade, learning, and culture. The Mali empire was followed by the Songhai empire of Gao, which rose to great power in the late 15th century. In 1590 the empire, already weakened by internal divisions, was shattered by a Moroccan army. Moreover, the trans-SAHARAN trade collapsed as European ships began to ply the coast of W Africa. The Moroccans, however, could not effectively dominate the vast region, which broke up into petty states.

History: 18th Century to 1960
By the late 18th century, the area was in a semianarchic condition and was subject to incursions by the Tuareg and Fulani. The 19th century witnessed a great resurgence of Islam. The Tukolor empire of Al Hajj Umar (1794–1864) and the empire of Samory (1870–1898) emerged as Muslim states opposing French invasion of the region. By 1898 the French conquest was virtually complete; Mali, called French Sudan, became part of the Federation of FRENCH WEST AFRICA. A nationalist movement, spearheaded by trade unions, student groups, and other associations, blossomed during the period between the two world wars. The Sudanese Union, a militantly anticolonial party, became the leading political force. Its leader, Modibo Keita, was a descendant of the Mali emperors. In the French constitutional referendum of 1958, French Sudan voted to join the French Community as the autonomous Sudanese Republic. In 1959 the republic joined Senegal to form the Mali Federation, but political differences shattered the union in 1960.

History: 1960 to 1977
That same year, the Sudanese Republic, renamed the Republic of Mali, obtained full independence from France and severed ties with the French Community. Under Keita's presidency Mali became a one-party state committed to socialist policies. In 1962 the country withdrew from the franc zone and adopted a nonconvertible national currency. The resulting economic and financial difficulties forced an accommodation with France in 1967. Militant elements in the Sudanese Union opposed this rapprochement, however, and Keita formed a people's militia to destroy opposition. The arrest of several dissenting army officers by the militia in 1968 provoked a bloodless military coup that overthrew the Keita regime. The country has continued to pursue a course of nonalignment in international affairs but is also an ACP (African, Caribbean, and Pacific) member of the EU. In the early 1970s, Mali suffered from the effects of a prolonged drought that desiccated the SAHEL region of Africa. The drought, which parched the semiarid grasslands and further reduced the country's already meager water supplies, shattered Mali's agricultural economy by killing thousands of head of livestock and hindering crop production. The resulting famine, disease, and poverty contributed to the deaths of untold thousands and forced the S migration of many people. Mali received emergency aid from UN-supervised international relief programs. The drought may have permanently extended desert conditions into central and S Mali.

History: 1977 to Present
Keita died in prison in 1977, touching off a series of protests. A new constitution (1979) contained provisions for elections to be held and democratic measures were implemented in spite of an unstable political climate. President Moussa Traoré was reelected president in 1979. Traoré effectively repressed coup attempts in the late 1970s and early 1980s and was reelected in 1985. Also in 1985, a border dispute with Burkina Faso erupted into armed conflict. Neighboring nations sent troops to end the fighting, but relations between the two countries remain

strained. In 1991 Traoré was overthrown in a coup and replaced with a transitional committee. Mali was a one-party state controlled by the Democratic Union of the Malian People (UDMP) from 1974 until 1992, when a new constitution was approved providing for a multiparty democracy. That same year Alpha Oumar Konaré became Mali's first democratically elected president. A Tuareg rebellion in northern Mali during the 1990s was ended by a 1995 peace settlement after which thousands of refugees returned to Mali. In 1997 Konaré was reelected. Amadou Touré, the interim military leader in 1991–1992, was elected to succeed Konaré in 2002 and was reelected in 2007.

Government
Mali is governed under the constitution of 1992. The executive branch is headed by a president, who is the head of state and is popularly elected for a five-year term and is eligible for a second term. The prime minister, who is the head of government, is appointed by the president. The unicameral National Assembly has 147 members who are popularly elected for five-year terms. The current head of state is President Amadou Toumani Touré (since June 2002). The current head of government is Prime Minister Modibo Sidibé (since September 2007).

Mali (MAH-lee), town, ⊙ Mali prefecture, Labé administrative region, N GUINEA, in Moyenne-Guinée geographic region, in FOUTA DJALLON mountains, on road, near SENEGAL border, and 50 mi/80 km N of LABÉ; 12°05′N 12°18′W. Livestock production. Meteorological station.

Mali, prefecture, Labé administrative region, N central GUINEA, in Moyenne-Guinée geographic region; ⊙ MALI. Bordered N by SENEGAL, NE tip by MALI, E and S by Koubia prefecture, SSW by Labé prefecture, SW by Lélouma prefecture, W by Gaoual prefecture, and NW by Koundara prefecture. Gambie River in E central, Liti River in S, and Koulountou River in NW. Part of FOUTA DJALLON massif here. TAMGUÉ MASSIF (elevation 4,970 ft/1,515 m) in N central. Towns include BALAKI, Mali, and YAMBÉRING. Several roads run through the prefecture connecting it to LABÉ town, KOUNDARA town, Koubia town, and Senegal.

Malia (MAHL-yah), village, RAJKOT district, GUJARAT state, W central INDIA, 55 mi/89 km N of RAJKOT. Local trade in salt, millet, cotton. Was capital of former WESTERN KATHIAWAR state of Malia, or Maliya of WESTERN INDIA STATES agency, along SW edge of Little RANN OF CUTCH; merged (1948) with SAURASHTRA, and eventually with Gujarat state. Also spelled MALIYA.

Maliahs, The (MAHL-yahrs), broken hill ranges in W GANJAM district, S central ORISSA state, E central INDIA, within EASTERN GHATS; heights vary from 1,000 ft/305 m to 4,000 ft/1,219 m. Thickly forested (sal, bamboo); inhabited mainly by Khond tribes.

Maliakos Gulf (mah-lee-ah-KOS), Greek *Maliakos Kolpos*, Latin *Maliacus Sinus*, AEGEAN inlet in FTHIOTIDA prefecture, CENTRAL GREECE department, sheltered by ÉVVIA island (E) and connected SE with Gulf of ÉVVIA and NE with OREOS CHANNEL; 15 mi/24 km long, 5 mi/8 km wide. On N shore is Stylis, port of LAMÍA. Receives SPERCHIOS River. Also called Gulf of Lamía, Gulf of Malian, or Gulf of Malis.

Maliana, town, ⊙ BOBONARO district (☐ 565 sq mi/1,469 sq km), EAST TIMOR, in central TIMOR ISLAND, 33 mi/53 km SW of DILI; 09°02′S 125°22′E. Trade center for coffee-growing; lumbering area (sandalwood).

Malibu, city (2000 population 12,575), LOS ANGELES county, S CALIFORNIA; suburb 26 mi/42 km W of LOS ANGELES, and 11 mi/18 km W of SANTA MONICA, on PACIFIC OCEAN (faces S) at Malibu Point, at mouth of Malibu Creek. Manufacturing (electronics research; printing and publishing). Popular beach resort. Its relatively secluded area has made Malibu attractive to many entertainment and business personalities. Elaborate residences line a beach escarpment extending N from shore of Santa Monica Bay. Of interest is the

J. Paul Getty Museum, 7 mi/11.3 km E, at TOPANGA. Seat of Pepperdine University. In area are Malibu Lagoon (E) and Leo Carillo (W) state beaches; Malibu Creek State Park to N. Incorporated 1991.

Malibu Lake, reservoir (1.5 mi/2.4 km long), LOS ANGELES county, S CALIFORNIA, on Malibu Creek, at center of residential development 6 mi/9.7 km NNW of MALIBU, in SANTA MONICA MOUNTAINS NATIONAL RECREATION AREA; 34°05′N 118°42′W. Formed by small dam.

Malibu Riviera, California: see POINT DUME.

Malicorne-sur-Sarthe (mah-lee-KORN–syoor–SAHRT), commune (□ 5 sq mi/13 sq km; 2004 population 1,878), SARTHE department, PAYS DE LA LOIRE region, W FRANCE, on SARTHE RIVER, and 18 mi/29 km SW of Le MANS; 47°49′N 00°05′W. Hog market; manufacturing of ceramic art objects. Has 11th-century Romanesque church and small 17th-century castle, a favorite residence of Madame de Sévigné.

Mali Drvenik Island, CROATIA: see DRVENIK VELI ISLAND.

Mali Federation, AFRICA: see MALI.

Malignant Cove, village, E NOVA SCOTIA, E CANADA, on NORTHUMBERLAND STRAIT, 12 mi/19 km NNW of ANTIGONISH; 45°48′N 62°02′W. Resort.

Maligne Lake (muh-LEEN), W ALBERTA, W CANADA, in ROCKY MOUNTAINS, in JASPER NATIONAL PARK, at foot of Mount UNWIN, 21 mi/34 km SE of JASPER; 20 mi/32 km long, 1 mi/2 km wide; 52°40′N 117°31′W. Elevation 5,490 ft/1,673 m. Recognized for its scenic beauty. Drains NW into ATHABASCA RIVER.

Maligne Mountains (muh-LEEN), W ALBERTA, W CANADA, range of peaks in ROCKY MOUNTAINS, in JASPER NATIONAL PARK, extends 20 mi/32 km SE from JASPER; rises to 9,157 ft/2,791 m; 52°45′N 117°50′W. Overlooks upper ATHABASCA RIVER valley.

Malihabad (muhl-ee-HAH-bahd), town, LUCKNOW district, central UTTAR PRADESH state, N central INDIA, on tributary of the GOMATI RIVER, 13 mi/21 km WNW of LUCKNOW; 26°55′N 80°43′E. Wheat, rice, gram, millet. Formerly a noted Pathan residence. Extensive mango and bher orchards nearby.

Mali Hka, MYANMAR: see MALI RIVER.

Mali Idjoš, SERBIA: see KRIVAJA.

Malilipot (mah-lee-LEE-pot), town, ALBAY province, SE LUZON, PHILIPPINES, on TABACO BAY (small inlet of LAGONOY GULF), 12 mi/19 km N of LEGASPI; 13°18′N 123°43′E. Agricultural center (abaca, corn, coconuts).

Mali Lošinj, CROATIA: see LOŠINJ ISLAND.

Malimono, town, SURIGAO DEL NORTE province, NE MINDANAO, PHILIPPINES, 12 mi/20 km SSE of SURIGAO; 09°37′N 125°25′E. Gold mined from seabed by dredging.

Malin (muh-LIN), village (2006 population 635), KLAMATH county, S OREGON, 1 mi/1.6 km N of CALIFORNIA state line, 25 mi/40 km SE of KLAMATH FALLS. Agriculture (barley, wheat, potatoes; cattle); dairy products. Nearby, in California are TULE LAKE (SW) and Clear Lake (SE) national wildlife refuges.

Malin, UKRAINE: see MALYN.

Malinalco (mah-lee-NAHL-ko), town, MEXICO state, central MEXICO, 25 mi/40 km SSE of TOLUCA DE LERDO; 18°57′N 99°30′W. Sugarcane, cereals, fruit; livestock. Site of important archaeological site, famous pre-Aztec murals.

Malinaltepec (mah-lee-NAHL-te-pek), town, GUERRERO, SW MEXICO, in SIERRA MADRE DEL SUR, 35 km SSW of TLAPA DE COMONFORT; 17°03′N 98°40′W. An isolated community in the Sierra Madre del Sur. No access by road. Cereals, sugarcane, fruit, livestock.

Malinao (mah-lee-NOU), town, ALBAY province, SE LUZON, PHILIPPINES, on small TABACO BAY (inlet of LAGONOY GULF), 13 mi/21 km N of LEGASPI; 13°24′N 123°39′E. Agriculture (abaca, rice, coconuts).

Malinche (mah-LEEN-chai), dormant volcano (14,636 ft/4,461 m), central MEXICO, on PUEBLA-TLAXCALA border, 16 mi/26 km NE of Puebla. Has several extinct craters; snowcapped during winter. Aztec name is MATLALCUEYATL. Also written Malintzi or Malinzi.

Malindi (mah-LEEN-dee), town, COAST province, SE KENYA, on the INDIAN OCEAN; 03°14′S 40°05′E. Beach resort and commercial center. Probably founded in the 10th century by Arab traders, Malindi became an important city-state and a major port. The Portuguese navigator Vasco da Gama landed here in 1498 and erected a monument that still stands. Nearby are the ruins of Gedi, an ancient walled city. S of the town lies a national park established to protect coral reefs and aquatic wildlife. Airport.

Malindi, village, MATABELELAND NORTH province, W central ZIMBABWE, 43 mi/69 km SE of Hwange, on railroad, at NE edge of HWANGE NATIONAL PARK; 18°45′S 27°01′E. Livestock.

Malindidzimu Hill or **World's View,** MATABELELAND SOUTH province, SW ZIMBABWE, promontory on N side of Matopos Hills, at N edge of MATOBO NATIONAL PARK, 25 mi/40 km S of BULAWAYO; 20°29′S 28°31′E. Here are buried Cecil J. Rhodes, founder of Rhodesia, and his officer Sir Leander Starr Jameson.

Malines, BELGIUM: see MECHELEN.

Malinga (muh-LEEN-guh), town NGOUNIÉ province, S GABON, at the CONGO border; 02°25′S 12°11′E.

Malingara (ma-leen-GAH-rah), village, KIGOMA region, NW TANZANIA, 25 mi/40 km SSE of KIBONDO; 03°59′S 31°03′E. Timber; grain; livestock.

Malin Head (MA-luhn), cape on the ATLANTIC OCEAN, NE DONEGAL county, N IRELAND, 17 mi/27 km N of BUNCRANA; 55°23′N 07°24′W. Northernmost point of IRELAND.

Mălini (muh-LEEN), agricultural village, SUCEAVA county, NE ROMANIA, on MOLDOVA RIVER, and 9 mi/14 km NW of FĂLTICENI.

Malinniki (mah-LEEN-nee-kee), town, SW KALUGA oblast, central European Russia, on highway, 7 mi/11 km WSW of KHVASTOVICHI; 53°25′N 34°53′E. Elevation 633 ft/192 m. Meat packing.

Malino (MAH-lee-no), resort village, SW SULAWESI, INDONESIA, 30 mi/48 km ESE of UJUNG PANDANG, at foot of Mount LOMPOBATANG; 05°16′S 119°0′E. Scene (July 1946) of Dutch and Indonesian conference at which preliminary plans were made for foundation of federated Indonesian state.

Malino (MAH-lee-nuh), settlement (2006 population 4,315), S MOSCOW oblast, central European Russia, on crossroads and railroad, 22 mi/35 km W of KOLOMNA, and 14 mi/23 km NNE of STUPINO, to which it is administratively subordinate; 55°06′N 38°11′E. Elevation 528 ft/160 m. Mechanical repair plant; construction enterprises.

Malinovka (mah-LEE-nuhf-kah), town (2005 population 9,900), W central KEMEROVO oblast, S central SIBERIA, RUSSIA, on the KONDOMA RIVER (tributary of the TOM' River), on road and railroad (Novokuznetsk-Tashtagol line), 31 mi/50 km S of NOVOKUZNETSK, and 14 mi/23 km N of OSINNIKI, to which it is administratively subordinate; 53°24′N 87°17′E. Elevation 1,062 ft/323 m. Coal mining in the Kuznetsk coal basin. Has a profilactic sanatorium.

Malinovka (mah-LEE-nuhf-kah), village, SE TOMSK oblast, S central Siberian Russia, on road and railroad, 15 mi/25 km NE of (and administratively subordinate to) TOMSK; 56°41′N 85°20′E. Elevation 508 ft/154 m. In agricultural area; poultry factory.

Malinovka, UKRAINE: see MALYNIVKA.

Malinovoye Ozero (mah-LEE-nuh-vuh-ye–O-zee-ruh), settlement, SW ALTAI territory, S central SIBERIA, RUSSIA, on road and railroad spur, 9 mi/14 km S of MIKHAYLOVSKOYE; 51°40′N 79°47′E. Elevation 508 ft/154 m. Soda- and salt-extracting works.

Malinovskiy (mah-lee-NOF-skeeye), settlement (2005 population 2,690), W KHANTY-MANSI AUTONOMOUS OKRUG, W SIBERIA, RUSSIA, on road and railroad, 24

mi/39 km W of SOVETSKIY; 61°11′N 62°50′E. Elevation 259 ft/78 m. Sawmilling, lumbering.

Malinska (MAH-leen-skah), Italian *Roveredo*, village, W CROATIA, on KRK Island, on ADRIATIC SEA, in KVARNER region, 7 mi/11.3 km N of KRK. Seaside resort. Monastery nearby.

Malinta (muh-LIN-tuh), village (2006 population 283), HENRY county, NW OHIO, 7 mi/11 km SE of NAPOLEON; 41°19′N 84°02′W. Clay, concrete, and plaster products.

Malintzi, MEXICO: see MALINCHE.

Malinyi (ma-LEE-nyee), village, MOROGORO region, S central TANZANIA, 83 mi/133 km SSE of IRINGA, near Ruhudji River, NW of Mbariki Mountains; 08°56′S 36°08′E. Road junction. Corn, wheat, pyrethrum; cattle, goats, sheep.

Malinzi, MEXICO: see MALINCHE.

Maliovitsa (MAH-lee-o-veet-sah), peak (8,957 ft/2,730 m) in the NW RILA MOUNTAINS, W BULGARIA, 14 mi/23 km ESE of DUPNITSA; 42°11′N 23°22′E. Sometimes spelled Malevitsa.

Maliq (mah-LEECH), new town, SE ALBANIA, 7 mi/11.3 km NW of KORÇË; 40°43′N 20°41′E. Lake Maliq was drained in the 1950s and turned over to sugar-beet production. Sugar refineries. Also spelled Maliqi.

Malir (muhl-IR), village, KARACHI administration area, SE PAKISTAN, 12 mi/19 km E of KARACHI; 24°56′N 67°12′E.

Mali Rajinac, CROATIA: see VELEBIT MOUNTAIN.

Mali River (MAH-lee), Burmese *Mali Hka*, Chinese *Mei-li-kai*, 200 mi/320 km long, right (W) headstream of AYEYARWADY RIVER in KACHIN STATE, MYANMAR; rises in high mountains on INDIA border near PUTAO; flows S joining the NMAI RIVER 25 mi/40 km N of MYITKYINA to form AYEYARWADY RIVER.

Malis (mah-LEES), ancient region in what is now NE FTHIOTIDA prefecture, CENTRAL GREECE department, S of THESSALY, on S slopes of OTHRYS massif and on W shore of MALIAKOS Gulf. Its chief city was LAMÍA.

Malis, Gulf of, Greece: see MALIAKOS GULF.

Malissina (mah-lee-SEE-nah), village, FTHIOTIDA prefecture, E CENTRAL GREECE department, 47 mi/76 km ESE of LAMÍA; 38°37′N 23°14′E. Livestock raising; fisheries along coast. Formerly Malesina; also Malessini.

Mali Stapar (MAH-lee STAHP-ahr), hamlet, VOJVODINA, NW SERBIA, on DANUBE-TISZA CANAL, and 11 mi/18 km WSW of SOMBOR, in BACKA region. Terminus of 43-mi/69-km-long branch canal to NOVI SAD.

Mali Ston, CROATIA: see STON.

Malita (mah-LEE-tah), town, DAVAO DEL SUR province, S MINDANAO, PHILIPPINES, 45 mi/72 km S of DAVAO, on DAVAO GULF; 06°21′N 125°32′E. Fishing; coconuts, copra.

Malitbog (mah-leet-BOG), town, SOUTHERN LEYTE province, S LEYTE, PHILIPPINES, on SOGOD BAY, 75 mi/121 km S of TACLOBAN; 10°11′N 124°58′E. Agricultural center (coconuts, rice, hemp).

Maliya, INDIA: see MALIA.

Malka (MAHL-kah), village (2005 population 7,295), N central KABARDINO-BALKAR REPUBLIC, N CAUCASUS, S European Russia, on the MALKA RIVER, on road, 22 mi/35 km NNW of NAL'CHIK; 43°48′N 43°19′E. Elevation 2,050 ft/624 m. Open-pit iron ore mining. Agricultural products. Formerly called Ashabovo (ah-SHAH-buh-vuh).

Malka Mari National Park (□ 338 sq mi/875 sq km), NORTH EASTERN province, N KENYA, near ETHIOPIA border; 04°11′N 40°46′E. Designated a national park in 1989.

Malkangiri (muhl-kuhn-gi-ree), district (□ 2,361 sq mi/6,138.6 sq km), ORISSA state, E central INDIA; ⊙ MALKANGIRI.

Malkangiri (muhl-kuhn-gi-ree), town, ⊙ MALKANGIRI district, SW ORISSA state, E central INDIA, 55 mi/89 km

SW of JAYPUR; 18°21′N 81°54′E. Local rice and timber market. Also spelled Malkanagiri.

Malkapur (muhl-KAH-poor), town, KOLHAPUR district, MAHARASHTRA state, W central INDIA, 25 mi/40 km NW of KOLHAPUR, in WESTERN GHATS. Rice; timber (teak, sandalwood).

Malkapur, town, BULDANA district, MAHARASHTRA state, W central INDIA, 55 mi/89 km WNW of AKOLA. Millet, wheat; cotton ginning, oilseed milling.

Malkara, town, European TURKEY, 32 mi/51 km WSW of TEKIRDAG; 40°54′N 26°54′E. In grain area. Lignite. Also spelled Malgara.

Malka River (MAHL-kah), 135 mi/217 km long, KABARDINO-BALKAR REPUBLIC, SE European Russia; rises (at 10,700 ft/3,261 m) in the central Greater CAUCASUS Mountains, on the N slope of Mount EL-BRUS; flows NE and E, past SARMAKOVO and Kuba, to the TEREK RIVER W of PROKHLADNYY. Forms part of the border with STAVROPOL TERRITORY. Iron mining in upper reaches.

Malka Musala (MAHL-kah moo-sah-LAH), peak (9,521 ft/2,902 m) in the RILA MOUNTAINS, SW BULGARIA; 42°12′N 23°34′E.

Malkerns (MAHL-kahrnz), town, MANZINI district, W central SWAZILAND, 14 mi/23 km SSE of MBABANE; 26°32′S 31°11′E. Mlilwane Game Reserve to W. Manufacturing (canned fruits; handicrafts). Malkerns Research Station, fruit and crop research, located here. Timber; corn, (center of citrus growing), pineapples, avocados, tomatoes, potatoes; cattle, goats, sheep. International cannery in the area.

Malki Iskur River (MAHL-kee EES-kuhr), 50 mi/80 km long, NW BULGARIA; rises in the Etropole Mountains; flows generally N, past ETROPOLE, NW, and N to the ISKUR River at ROMAN 13 mi/21 km ESE of LUKOVIT; 43°04′N 23°54′E.

Małkinia (MAHL-kee-nya), Polish *Małkinia*, village, Ostrołęka province, E central POLAND, on right bank of Western Buh (BUG) River, and 50 mi/80 km NE of WARSAW. Railroad junction. Also called Małkinia Górna.

Malko Belovo, Bulgaria: see BELOVO.

Malko Sharkovo (MAHL-ko SHAHR-ko-vo), reservoir, BURGAS oblast, SE BULGARIA, on the POPOVSKA River; 42°07′N 26°52′E. TUNDZHA RIVER tributary.

Malko Turnovo (MAHL-ko TUHR-no-vo), city, BUR-GAS oblast, MALKO TURNOVO obshtina (1993 population 6,216), SE BULGARIA, at the NE foot of the STRANDZHA MOUNTAINS, 35 mi/56 km S of BURGAS, on the Turkish border; 41°58′N 27°32′E. Border crossing. Agricultural and livestock center; tanning, coppering, wood processing. Marble quarries. Wooden handicrafts and charcoal produced and copper mined nearby. Also known as Malka Trnovo; formerly called Turnovo.

Malla (MAH-yah), canton, LOAYZA province, LA PAZ department, W BOLIVIA; 17°05′S 67°29′W. Elevation 8,333 ft/2,540 m. Some gas wells in area. Lead-bearing lode; tin mining at Mina Viloco; clay, limestone, and gypsum deposits. Agriculture (potatoes, yucca, bananas, rye); cattle.

Mallacoota Inlet (MA-luh-KOO-tuh), lagoon, SE VICTORIA, AUSTRALIA, opening into TASMAN SEA, near Cape Howe; 5 mi/8 km long, 3 mi/5 km wide; 37°32′S 149°47′E. Irregularly shaped. Timbered shoreline; contains mullet, whiting, sea trout. Tourism. Also spelled Mallagoota.

Mallagoota Inlet, AUSTRALIA: see MALLACOOTA INLET.

Mallagu Dam, dam and reservoir, Al Kaf prov., NE Tunisia, 21 mi/34 km S of JANDUBA, on MALLAGU R. Built for flood control and irrigation.

Mallagü River (me-le-GOO), 90 mi/145 km long, NW TUNISIA; rises NW of TEBESSA (S ALGERIA); flows NE to the MAJARDAH RIVER below SUQ AL ARBA. Irrigation and flood control dam 21 mi/34 km N of AL KAF.

Mallah, SYRIA: see MELLAH.

Mallah, El (mah-LAH, el), village, AÏN TÉMOUCHENT wilaya, NW ALGERIA, on railroad, and 7 mi/11.3 km NNE of AÏN TÉMOUCHENT. Vineyard center; brick manufacturing, distilling, olive curing. Formerly Rio Salado.

Mallaig (MAL-aig), town (2001 population 797), HIGHLAND, N Scotland, on the SOUND OF SLEAT, at entrance to LOCH NEVIS, and 30 mi/48 km WNW of FORT WILLIAM; 57°00′N 05°50′W. Fishing port and terminal of steamers to SKYE island and the HEBRIDES.

Mallaig (MA-laig), unincorporated village, E ALBERTA, W CANADA, 65 mi/105 km NNW of VERMILION, in Saint Paul County No. 19; 54°13′N 111°22′W. Mixed farming, lumbering.

Mallakaster (mah-lah-KAW-ster), hill range in S central ALBANIA, NE of VLORË, between SEMAN and VI-JOSË rivers. Also spelled Mallakastra.

Mallala (ma-luh-LAH), township, SOUTH AUSTRALIA state, S central AUSTRALIA, 35 mi/57 km from ADE-LAIDE. Wheat. Museum.

Mallali, GUYANA: see MALALI.

Mallama (mah-YAH-mah), town, ⊙ Mallama municipio, NARIÑO department, SW COLOMBIA, 37 mi/60 km WSW of PASTO; 01°08′N 77°50′W. Elevation 7,467 ft/2,275 m. Wheat, sugarcane; livestock. Sometimes called Piedrancha.

Malla-Malla, Cordillera de (MAH-yah–MAH-yah, kor-dee-YER-rah dai), Andean range in VIII or BÍO-BÍO region, S central CHILE, 60 mi/97 km SE of LOS ÁNGELES, extends 12 mi/19 km ENE from CALLAQUI VOLCANO to Copahué Volcano on ARGENTINA border.

Mallankinar (muhl-uhn-kee-nahr), town, KAMARAJAR district, TAMIL NADU state, S INDIA, 7 mi/11.3 km NNW of ARUPPUKKOTTAI. In cotton area. Also spelled Mallanginar and Mallanjinar.

Mallanwan (muhl-uhn-wahn), town, HARDOI district, central UTTAR PRADESH state, N central INDIA, on branch of Sarda Canal system, and 14 mi/23 km S of HARDOI; 27°03′N 80°09′E. Metalware manufacturing; wheat, gram, barley, oilseeds, sugarcane.

Mallapuram, INDIA: see HOSPET.

Mallard, town (2000 population 298), PALO ALTO county, N IOWA, 12 mi/19 km S of EMMETSBURG; 42°56′N 94°40′W. Livestock; grain.

Mallawi (MAHL-LAH-wee), town, MINYA province, Upper EGYPT, on the W bank of the NILE River, on railroad, 25 mi/40 km S of Minya; 27°44′N 30°50′E. In a farm area, the town produces textiles and handicrafts.

Mallco Rancho (MAL-ko RAHN-cho), canton, QUIL-LACOLLO province, COCHABAMBA department, central BOLIVIA. Elevation 8,333 ft/2,540 m. Antimony mining at CARACOTA; clay, limestone, and gypsum deposits. Agriculture (potatoes, yucca, bananas, corn, rye); cattle raising for meat and dairy products.

Malle (MAH-luh), commune (2006 population 14,101), Antwerp district, ANTWERPEN province, N BELGIUM, 10 mi/16 km WSW of TURNHOUT.

Malleco (mah-YAI-ko), northernmost province (□ 5,512 sq mi/14,331.2 sq km) of ARAUCANIA region, S central CHILE; ⊙ ANGOL; 38°15′S 72°00′W. In S part of the central valley between the ANDES (E) and the CORDILLERA DE NAHUELBUTA (W), N of Quina River; drained by BÍO-BÍO and MALLECO rivers. Includes the volcanoes Tolhuaca and Lonquimay. Has humid, temperate climate. A predominantly agricultural area, it produces cereals, vegetables, grapes; fruit; livestock.

Malleco, Lake (mah-YAI-ko), Andean lake (□ 4 sq mi/10.4 sq km) in MALLECO province, ARAUCANIA region, S central CHILE, formed by upper MALLECO RIVER at NW foot of Tolhuaca Volcano near TOLHUACA; 38°13′S 71°49′W.

Malleco River (mah-YAI-ko), c.85 mi/137 km long, IX or ARAUCANIA region, S central CHILE; rises at NW foot of Tolhuaca Volcano; flows WNW, past COLLI-PULLI, joining the lesser Rehue River (left) near ANGOL to form the VERGARA RIVER.

Mallee Cliffs National Park (MA-lee KLIFS), (□ 224 sq mi/582.4 sq km), SW NEW SOUTH WALES, SE AUSTRALIA, 605 mi/974 km W of SYDNEY, 20 mi/32 km E of MILDURA (Victoria); 43 mi/69 km long, 20 mi/32 km wide; 34°15′S 142°40′E. Semiarid landscape with mallee vegetation. Habitat of endangered mallee fowl. No public access. Established 1977.

Mallemort (mal-MOR), village (□ 10 sq mi/26 sq km), BOUCHES-DU-RHÔNE department, SE FRANCE, on left bank of the DURANCE RIVER, and 19 mi/31 km NW of AIX-EN-PROVENCE; 43°44′N 05°11′E. Fruit and vegetable canning, wine growing. Dam and hydroelectric plant on Durance River canal. LUBÉRON regional park is just N.

Mallén (mah-LYEN), town, ZARAGOZA province, NE SPAIN, 8 mi/12.9 km NE of BORJA; 41°54′N 01°25′W. Agricultural-trade center (sugar beets, alfalfa, olive oil).

Mallersdorf-Pfaffenberg (MAHL-lers-dorf–PFAHF-fuhn-berg), village, LOWER BAVARIA, BAVARIA, S GERMANY, on the Small Laaber, 17 mi/27 km NNE of LANDSHUT; 48°47′N 12°14′E. Agriculture (grain, sugar beets); manufacturing (textile); dairying.

Mallet (MAH-lai), city (2007 population 12,476), S PARANÁ state, BRAZIL, on railroad, and 105 mi/169 km WSW of CURITIBA; 25°55′S 50°50′W. Agriculture (wheat, potatoes, corn). Furniture manufacturing; linen milling, tanning, wine making. Mineral springs. Ships lumber, maté. Until 1930s, called São Pedro de Mallet. Also written Malé or Malét.

Mallicolo, VANUATU: see MALAKULA.

Malling, East (MAL-ing), village (2001 population 4,448), central KENT, SE ENGLAND, 4 mi/6.4 km WNW of MAIDSTONE; 51°17′N 00°26′E. Former agricultural market. Pharmaceutical works. Site of fruit cold-storage experimental station.

Mallnitz (MAHL-nits), village, CARINTHIA, W central AUSTRIA, in the HOHE TAUERN mountain range, 21 mi/34 km NW of SPITTAL AN DER DRAU, highest station (3,627 ft/1,106 m) of Tauern railroad, near S exit of Tauern Tunnel; 46°59′N 13°10′E. Health resort. Cable cars to the ANKOGEL mountain (elevation 8,290 ft/2,527 m) with winter-sports facilities. Base of National Park Hohe Tauern. Large hydroelectric plant nearby powers Tauern railroad.

Malloa (mah-YO-ah), village, ⊙ Malloa comuna, CA-CHAPOAL province, LIBERTADOR GENERAL BERNARDO O'HIGGINS region, central CHILE, on railroad, and 23 mi/37 km SSW of RANCAGUA; 34°27′S 70°57′W. In agricultural area (cereals, vegetables, grapes; livestock); mining. Originally a Native American settlement, was founded as a Spanish town in 1742.

Malloco (mah-YO-ko), town, SANTIAGO province, METROPOLITANA DE SANTIAGO region, central CHILE, on railroad, and 15 mi/24 km SW of SANTIAGO; 33°37′S 70°52′W.

Mallorca, SPAIN: see MAJORCA.

Mallorquinas (mah-lyor-KEE-nahs), village, BARCE-LONA province, NE SPAIN, 7 mi/11.3 km NE of BAR-CELONA; 41°27′N 02°16′E. Produces chemicals (ammonic, acetic, and sulphuric acid; copper sulphate), brandy.

Mallory, unincorporated town (2000 population 1,143), LOGAN county, SW WEST VIRGINIA, near GUYAN-DOTTE RIVER, 11 mi/18 km SSE of LOGAN; 37°43′N 81°49′W. Coal mining. Manufacturing (rebuilt mining equipment). R. D. Bailey Lake (reservoir) and Wildlife Management Area to SE.

Mallow (MA-lo), Gaelic *Mala*, town (2006 population 7,864), central CORK county, SW IRELAND, on the BLACKWATER RIVER, and 16 mi/26 km NNW of CORK; 52°08′N 08°38′W. Railroad center and agricultural market, with sugar refinery, tanneries, and salmon fisheries; spa resort with mineral springs, where the Anglo-Irish "Rakes of Mallow" disported themselves.

Cross-references are shown in SMALL CAPITALS. The pronunciation guide is shown on page xix. The sources of population figures are shown on page xvii.

Racecourse. The modern Mallow Castle has adjoining fragments of ancient stronghold of the Desmonds.

Mallwen, RUSSIA: see MAYSKOYE.

Mallwischken, RUSSIA: see MAYSKOYE.

Malm (MAHLM), village, NORD-TRØNDELAG county, central NORWAY, on an inlet of TRONDHEIMSFJORDEN, 10 mi/16 km WNW of STEINKJER; 64°04′N 11°13′E. Iron mine nearby.

Malm, FINLAND: see MALMI.

Malmaison, FRANCE: see RUEIL-MALMAISON.

Malmbäck (MAHLM-BEK), village, JÖNKÖPING county, S SWEDEN, 10 mi/16 km WSW of NÄSSJÖ; 57°35′N 14°27′E.

Malmberget (MAHLM-BER-yet), town, NORRBOTTEN county, N SWEDEN, 50 mi/80 km SSE of KIRUNA; 67°10′N 20°40′E. Iron-mining center 2 mi/3.2 km N of GÄLLIVARE. Ore shipped by railroad to LULEÅ and NARVIK (NORWAY).

Malmédy (MAHL-mai-dee), commune (2006 population 11,845), Verviers district, LIÈGE province, E BELGIUM, near the GERMAN border; 50°25′N 06°02′E. Economic mainstays are tourism and manufacturing. The town and the surrounding district belonged to the abbey of nearby STAVELOT until they were given (1815) to PRUSSIA. Malmédy and EUPEN were returned to Belgium by the Treaty of Versailles after World War I. In World War II heavy fighting occurred here during the Battle of the Bulge (December 1944); seventy-two U.S. prisoners of war massacred here by German Waffen SS troops.

Malmesbury (MAHMZ-buh-ree), town (2001 population 4,631), WILTSHIRE, S ENGLAND, 9 mi/14.4 km N of CHIPPENHAM; 51°35′N 02°05′W. Agriculture and electrical equipment manufacturing are important to the local economy. Famous for its Benedictine abbey, founded in the 12th century, of which only the nave remains. King Athelstan of Wessex was buried here. Thomas Hobbes born nearby.

Malmesbury, town, WESTERN CAPE province, SOUTH AFRICA, on Diep River, on N7 highway, and 35 mi/56 km NNE of CAPE TOWN; 33°29′S 18°45′E. Elevation 1,050 ft/320 m. Center of wheat, tobacco, wine region; known as Swartland [Afrikaans=black country] because of humus-rich soil. Resort with medicinal hot springs. Founded 1745. Jan Christiaan Smuts born here. Airfield and railroad junction.

Malmi (MAHL-mee), Swedish *Malm*, suburb of HELSINKI, UUDENMAAN province, S FINLAND, 7 mi/11.3 km NNE of HELSINKI; 60°15′N 25°02′E. Elevation 33 ft/10 m. Helsinki International Airport. Machine shops.

Malmköping (MAHLM-SHUHP-eeng), town, SÖDERMANLAND county, E SWEDEN, 17 mi/27 km SSE of ESKILSTUNA; 59°08′N 16°44′E.

Malmö (MAHLM-UH), city, ⊙ SKÅNE county, S SWEDEN, on ÖRESUND opposite COPENHAGEN (DENMARK); 55°36′N 13°01′E. Sweden's third-largest city, commercial port and industrial center; manufacturing (metal goods, processed food; construction, printing). International airport to SE. Ferries to Copenhagen and between Limhavn and DRAGØR (Denmark); also bridge to Copenhagen. Important trade and shipping center during Hanseatic period; usually Danish possession until passed to Sweden with SKÅNE province (1658). Capital of former MALMÖHUS county until Skåne county created (1997). Seat of World Maritime University (UN). Malmöhus castle (begun 1434) museum; City Hall (1546); Saint Peter's Church (fourteenth century). Founded twelfth century.

Malmo, village (2006 population 103), SAUNDERS county, E NEBRASKA, 6 mi/9.7 km NW of WAHOO, and on branch of PLATTE RIVER; 41°16′N 96°43′W.

Malmöhus, SWEDEN: see SKÅNE.

Malmön (MAHLM-UHN), fishing village (□ 2 sq mi/5.2 km²), GÖTEBORG OCH BOHUS county, SW SWEDEN, on island (□ 2 sq mi/5.2 km²) of same name in SKAGERRAK strait, 6 mi/9.7 km NW of LYSEKIL; 58°21′N 11°20′E.

Malmsbury (MAHMZ-buh-ree), village, central VICTORIA, SE AUSTRALIA, on CAMPASPE RIVER, and 55 mi/89 km NW of MELBOURNE, near KYNETON; 37°11′S 144°23′E. In old gold-mining area. Site of reservoir; irrigation center.

Malmslätt (MAHLMS-LET), town, ÖSTERGÖTLAND county, SE SWEDEN, 3 mi/4.8 km W of LINKÖPING; 58°25′N 15°31′E.

Malmstrom Air Force Base, U.S. military installation (□ 6 sq mi/15.5 sq km), W central MONTANA, 4 mi/6.4 km E of GREAT FALLS, CASCADE county. Established 1942 during World War II as the take-off point for Soviet-bound lend-lease material; after the war it was a training base for crews in the Berlin Airlift. The Strategic Air Command (SAC) assumed command in 1954; SAC's 1st Minuteman missile wing was established here in 1961. The missile complex adjoining the base is one of the largest in the world. Scheduled for closing in late 1990s.

Malmyzh (mahl-MIZH), city (2005 population 8,885), SE KIROV oblast, E central European Russia, on the SHOSHMA River at its confluence with the VYATKA River, on road, 180 mi/290 km NE of KAZAN'; 56°31′N 50°41′E. Elevation 479 ft/145 m. Mechanical shops; grain storage, food industries (bakery, cannery, brewery); fishing farm. First mentioned in the 15th century as a Mari village. Former capital of Cheremiss domain.

Malmyzh (mahl-MIZH), village, S KHABAROVSK TERRITORY, SE SIBERIA, RUSSIAN FAR EAST, on the AMUR River, 45 mi/72 km S of KOMSOMOL'SK-NA-AMURE; 50°18′N 136°40′E. Elevation 452 ft/137 m. Fish cannery.

Malnate (mahl-NAH-te), town, VARESE province, LOMBARDY, N ITALY, on OLONA River, and 3 mi/5 km SE of VARESE; 45°48′N 08°53′E. Railroad junction. Highly diversified secondary industrial center including machinery, silverware, jewelry.

Malo (MAH-lo), town, VICENZA province, VENETO, N ITALY, 10 mi/16 km NW of VICENZA. Silk milling; fabricated metals, machinery, clothing.

Malo (MAH-lo), raised coral limestone island, VANUATU, SW PACIFIC OCEAN, 3 mi/4.8 km S of ESPÍRITU SANTO; 34 mi/55 km in circumference; 15°41′S 167°10′E. Rises to 1,135 ft/346 m. Copra, cacao. Formerly St. Bartholomew Island.

Malo-, [Russian combining form=lesser, little, small], in Russian names: see also MALAYA, MALOYE, MALY, AND MALYE.

Maloarkhangel'sk (mah-luh-ahr-HAHN-geelsk), city (2006 population 3,885), S ORËL oblast, SW European Russia, on road, 50 mi/80 km SSE of ORËL; 52°24′N 36°30′E. Elevation 725 ft/220 m. Center of agricultural area; hemp processing. Chartered in 1778.

Malo, Arroyo (MAH-lo, ah-ROI-o), river, 75 mi/121 km long, TACUAREMBÓ department, N central URUGUAY; rises in the CUCHILLA DE HAEDO S of TAMBORES; flows SE, past CURTINA, to the Río NEGRO 14 mi/23 km NE of SAN GREGORIO; 31°52′S 53°12′W.

Malobaranivka, UKRAINE: see IVANIVKA, Odessa oblast.

Malobaranovka, UKRAINE: see IVANIVKA, Odessa oblast.

Maloelap (MAH-LO-e-LAHP), atoll (□ 4 sq mi/10.4 sq km; 1999 population 856), RATAK CHAIN, MAJURO district, MARSHALL ISLANDS, W central PACIFIC, 220 mi/354 km E of KWAJALEIN; c.25 mi/40 km long; 08°45′N 171°03′E. Has seventy-one islets. Japanese air base in World War II. Sometimes spelled Maloelab.

Malo-Ilinovka, UKRAINE: see BRYANKA.

Malo-Illynivka, UKRAINE: see BRYANKA.

Maloja (mah-LOI-ah), district, SE GRISONS canton, E SWITZERLAND; 46°24′N 09°42′E. Elevation c.6,000 ft/1,830 m. Main town is ST. MORITZ; VAL BREGAGLIA is Italian-speaking and mainly Protestant. ENGADINE is Romansch-speaking and Protestant.

Maloja (mah-LOI-ah), Italian *Col di Maloggia*, pass (5,955 ft/1,815 m), GRISONS canton, SE SWITZERLAND, leading from the Upper ENGADINE to LAKE COMO in ITALY. The lowest pass into Italy from a latitude N of the ALPS.

Malokaterinovka, UKRAINE: see MALOKATERYNIVKA.

Malokaterynivka, (Russian *Malokaterinovka*), town, NW ZAPORIZHZHYA oblast, UKRAINE, on railroad, and 10 mi/16 km S of ZAPORIZHZHYA. Fish hatchery; vegetables. Est. 1775, town since 1938. Also spelled, Malo-Yekaterinovka.

Malokaterynivka (mah-lo-kah-te-RI-neef-kah), (Russian *Malokaterinovka* or *Malo-Yekaterinovka*), town (2004 population 4,240), N ZAPORIZHZHYA oblast, UKRAINE, on NE shore of the KAKHOVKA RESERVOIR just N of the mouth of the KINS'KA RIVER, on road and on railroad, 11 mi/18 km SSE of ZAPORIZHZHYA city center; 47°39′N 35°15′E. Fish hatchery, lake fishery. Established in 1775 as Krasnokutivka (Russian *Krasnokutovka*); renamed in 1780; town since 1938.

Malo Konare (MAH-lo ko-NAH-re), village, PLOVDIV oblast, PAZARDZHIK obshtina, W central BULGARIA, 5 mi/8 km E of Pazardzhik; 42°11′N 24°26′E. Rice; vineyards. Formerly called Doganovo Konare.

Malokuril'skoye (mah-luh-koo-REEL-skuh-ye), settlement, on the NW shore of SHIKOTAN ISLAND, in lesser Kurile group, SE SAKHALIN oblast, RUSSIAN FAR EAST; 43°52′N 146°50′E. Elevation 187 ft/56 m. Fish canning; whale oil factory. Under Japanese rule (until 1945), called Shakotan.

Malo-les-Bains (mah-LO–lai-BAN), NNE residential suburb and now part of DUNKERQUE, NORD department, NORD-PAS-DE-CALAIS region, N FRANCE, on North sea coast; 51°03′N 02°24′E. Beach resort with aquarium.

Malolo (mah-LO-lo), volcanic islands (□ 4 sq mi/10.4 sq km), S of MAMANUCA group, FIJI, SW PACIFIC OCEAN, 5 mi/8 km W of VITI LEVU; c.3 mi/4.8 km long. Copra. Tourism.

Malolos (mah-LO-los), town (2000 population 175,291), ⊙ BULACAN province, SW LUZON, the PHILIPPINES, N of MANILA; 14°51′N 120°49′E. Old marketing center for surrounding farms. The Spanish settled here in 1580. Was capital of the Philippine republic proclaimed (June 1898) by the insurrectionary leader Emilio Aguinaldo; U.S. forces captured the town in March 1899. Seat of the Bulacan College of Arts and Trades.

Malolotja Nature Reserve (ma-lou-LOUT-cha) (□ 28 sq mi/72.8 sq km), HHOHHO district, NW SWAZILAND, c.15 mi/24 km NNW of MBABANE, on SOUTH AFRICA (NW) border; 26°05′S 31°06′E. Includes Ntababovu and part of Ngwenya hills. Drained in N by KOMATI RIVER; Malolotja Falls (312 ft/95 m). Named after the 90-m/295-ft high waterfall on a tributary of the Komati River. Wildlife includes 280 bird species, hartebeest, wildebeest, zebra, leopard, baboon, vervet monkey. Prehistoric mine at Lion Cavern of red hematite dates to c.41,000 B.C.E., one of oldest sites in world. Also spelled Malolotsha.

Maloma (ma-LOU-mah), village, LUBOMBO district, S SWAZILAND, 38 mi/61 km SSE of MANZINI; 27°01′S 31°38′E. On edge of escarpment. Road junction. Coal mine opened here in 1993. Livestock; corn, vegetables, fruit.

Malomal'sk, RUSSIA: see GLUBOKAYA.

Malombe, Lake (mah-lo-mbeh), MANGOCHE district, Southern region, S MALAWI, 12 mi/19 km S of Lake Malawi (LAKE NYASA); 16 mi/26 km long, 12 mi/19 km wide; 6 ft/1.8 m–8 ft/2.4 m deep; 14°38′S 35°12′E. Marshy shores. Traversed by SHIRE RIVER. Fall of level of Lake Malawi caused Lake Malombe to dry up, c.1925. Sometimes called Lake Pamalombe.

Malomir (mah-lo-MIR), village, BURGAS oblast, TUNDZHANSKA obshtina, SE BULGARIA, 12 mi/19 km S of YAMBOL; 42°18′N 26°32′E. Grain, tobacco, rice. Formerly known as Karapcha.

Malomykolayivka (mah-lo-mi-ko-LAH-yif-kah), (Russian *Malonikolayevka*), town (2004 population

4,800), S LUHANS'K oblast, UKRAINE, in the DONBAS, on the DONETS RIDGE, on highway and railroad spur, 13 mi/21 km NNE of KRASNYY LUCH; 48°18′N 39°01′E. Elevation 757 ft/230 m. Bituminous-coal mine, stone-crushing works. Established in the beginning of the 18th century; town since 1964.

Malón (mah-LON), village, ZARAGOZA province, NE SPAIN, 5 mi/8 km NE of TARAZONA; 41°57′N 01°40′W. Sugar beets, olive oil, wine, cereals.

Malone, village (□ 2 sq mi/5.2 sq km; 2006 population 5,903), ⊙ FRANKLIN county, N NEW YORK, on SALMON RIVER, near Canada (Quebec) border, and 30 mi/48 km ESE of MASSENA; 44°51′N 74°17′W. Port of entry. Light manufacturing, sports tourism (golfing, skiing). In agricultural area (dairy products, potatoes, grain). Briefly occupied by British during War of 1812. Was gathering point for the Fenians, who raided Canada in 1866. Seat of North Country Community College. Settled c.1800, incorporated 1833.

Malone (muh-LON), village (2006 population 305), HILL county, central TEXAS, 15 mi/24 km SE of HILLSBORO; 31°55′N 96°53′W. In farm area. Manufacturing (meat products). NAVARRO MILLS LAKE reservoir to E.

Malong (MAH-LUNG), town, ⊙ Malong county, E YUNNAN province, CHINA, on railroad, and 10 mi/16 km WSW of QUJING; 25°24′N 103°34′E. Elevation 6,857 ft/2,090 m. Rice, tobacco; mining equipment. Also called Tongguan.

Malonga (mah-LAWNG-gah), village, KATANGA province, E CONGO, on railroad, and 220 mi/354 km WNW of LIKASI; 10°24′S 23°10′E. Elev. 3,369 ft/1,026 m. Agricultural trade (manioc, yams, beans).

Malonikolayevka, UKRAINE: see MALOMYKOLAYIVKA.

Malonton, hamlet, S MANITOBA, W central CANADA, 13 mi/21 km N of TEULON, in ARMSTRONG rural municipality; 50°34′N 97°12′W.

Malorad (MAH-lo-rahd), village, MONTANA oblast, BOROVAN obshtina, NW BULGARIA, 13 mi/21 km W of BYALA SLATINA; 43°29′N 23°42′E. Grain, legumes; livestock.

Malorita (mah-lo-RI-tah), Polish *Maloryta*, town, S BREST oblast, BELARUS, ⊙ MALORITA region, 27 mi/43 km SE of BREST, 51°50′N 24°08′E. Railroad station. Vegetable-drying plant.

Maloryazantseve (mah-lo-ryah-ZAHN-tse-ve), (Russian *Maloryazantsevo*), town, W LUHANS'K oblast, UKRAINE, in the DONBAS, 3 mi/5 km SW of LYSYCHANS'K; 48°53′N 38°23′E. Elevation 393 ft/119 m. Broiler factory. Landscape disturbed by past coal mining. Established in 1780, town since 1938.

Maloryazantsevo, UKRAINE: see MALORYAZANTSEVE.

Malo-Ryazantsevo, UKRAINE: see MALORYAZANTSEVE.

Maloshuyka (mah-luh-SHOO-ee-kah), town, NW ARCHANGEL oblast, N European Russia, on road and railroad, on an inlet of the ONEGA BAY on the WHITE SEA, 25 mi/40 km WSW of ONEGA; 63°42′N 37°27′E. Elevation 173 ft/52 m. In a protected old-growth forest region. Railroad shops; sawmilling. Developed in the early 1940s.

Malosmadulu Atoll, MALDIVES: see MAALHOSMADULU ATOLL.

Maloti Mountains, SOUTH AFRICA: see GOLDEN GATE HIGHLANDS NATIONAL PARK.

Malouchalinskiy (mah-luh-oo-CHAH-leen-skeeye), former town, E BASHKORTOSTAN Republic, SE URAL Mountains, extreme W Siberian Russia, now a suburb of UCHALY, less than 4 mi/6.4 km SSE of the city center (connected by road); 54°19′N 59°27′E. Elevation 1,804 ft/549 m. Mostly residential.

Måløv (MO-luv), town (2000 population 7,909), Copenhagen county, SJÆLLAND, DENMARK, 11 mi/18 km NW of COPENHAGEN; 55°45′N 12°20′E. Fruit processing; dairying; vegetable farming. Furniture manufacturing.

Malovsk (MAH-luhfsk), settlement, E central BURYAT REPUBLIC, S central SIBERIA, RUSSIA, on a left tributary of the AMALAT River, on highway, 3 mi/5 km S of

BAGDARIN; 54°24′N 113°33′E. Elevation 2,923 ft/890 m. Gold mining. Also known as Malovskiy.

Malovskiy, RUSSIA: see MALOVSK.

Maloy, town (2000 population 28), RINGGOLD county, S IOWA, on LITTLE PLATTE RIVER, and 10 mi/16 km WSW of MOUNT AYR; 40°40′N 94°24′W.

Maloyaroslavets (mah-luh-yah-ruh-SLAH-veets), city (2005 population 33,050), NE KALUGA oblast, central European Russia, on the Luzha River, on road and railroad, 38 mi/61 km NE of KALUGA, and 10 mi/16 km SW of OBNINSK; 55°01′N 36°28′E. Elevation 639 ft/194 m. Machinery and metal industries, industrial rubber products, furniture and garments making; logging and lumbering; food processing (dairy, bakery). Chartered in 1410. In 1812, scene of a Russian victory over the retreating army of Napoleon.

Maloyaz (mah-luh-YAHS), village (2005 population 4,780), NE BASHKORTOSTAN Republic, in the NW foothills of the S URAL Mountains, E European Russia, on the YURYUZAN' River (tributary of the UFA RIVER), on road, 60 mi/97 km W of ZLATOUST; 55°11′N 58°09′E. Elevation 813 ft/247 m. Lumbering. Formerly called Staryye Kartavly.

Maloye, [Russian=lesser, little, small], in Russian names: see also MALAYA, MALO- [RUSSIAN COMBINING FORM], MALY, AND MALYE.

Maloye Goloustnoye (MAH-luh-ye guh-luh-OOST-nuh-ye), settlement (2005 population 950), SE IRKUTSK oblast, E central SIBERIA, Russia, on highway, 34 mi/55 km E of IRKUTSK; 52°18′N 105°19′E. Elevation 2,162 ft/658 m. Timbering, woodworking.

Malo-Yekaterinovka, UKRAINE: see MALOKATERYNIVKA.

Maloye Kozino (MAH-luh-ye KO-zee-nuh), town (2006 population 820), W NIZHEGOROD oblast, central European Russia, on the W bank of the VOLGA RIVER, near railroad, 11 mi/18 km NW of NIZHNIY NOVGOROD, and 4 mi/6 km SE of BALAKHNA; 56°26′N 43°40′E. Elevation 265 ft/80 m. In agricultural area; food processing.

Maloye Soldatskoye (MAH-luh-ye suhl-DAHTS-kuh-ye), village, S KURSK oblast, SW European Russia, in the PSEL RIVER basin, on railroad and near highway, 48 mi/77 km SSW of KURSK, and 2 mi/3.2 km W of BELAYA, to which it is administratively subordinate; 51°02′N 35°41′E. Elevation 643 ft/195 m. In agricultural area; experimental fruit orchard and tree nursery.

Maloye Verevo (MAH-luh-ye VYE-ree-vuh), village (2005 population 4,085), W central LENINGRAD oblast, NW European Russia, on road and railroad, 16 mi/26 km SSW of SAINT PETERSBURG, and 4 mi/6 km NNE of GATCHINA, to which it is administratively subordinate; 59°37′N 30°11′E. Elevation 265 ft/80 m. In agricultural area; produce processing.

Malozemel'skaya Tundra (mah-luh-zee-MYEL-skah-yah) [Russian=little land], in NENETS AUTONOMOUS OKRUG, N ARCHANGEL oblast, N European Russia; extends along the BARENTS SEA coast from INDIGA 140 mi/225 km NE to the mouth of the PECHORA River. Reindeer raising.

Malpaís, MEXICO: see SAN NICOLÁS DE BUENOS AIRES.

Malpais, The (MAL-pei) [Spanish=bad lands], name given to lava fields in central and W NEW MEXICO. One area is in LINCOLN and OTERO counties, central New Mexico, c.45 mi/72 km N of ALAMOGORDO and SW of CARRIZOZO, in N part of Tularosa Valley, E of SIERRA OSCURA, W of SIERRA BLANCA, and with Valley of Fires National Recreation Area in N. The other area is in CIBOLA county, W NEW MEXICO, c.70 mi/113 km WSW of ALBUQUERQUE, S of GRANTS. Much of area covered by El Malpais National Monument. La Ventana Natural Arch in E; Bandera Volcano and Ice Cave to W.

Malpartida de Cáceres (mahl-pahr-TEE-dah dai KAH-the-res), town, CÁCERES province, W SPAIN, 7 mi/11.3 km WSW of CÁCERES; 39°27′N 06°30′W. Wool trade; cereals, olive oil; sheep.

Malpartida de la Serena (mahl-pahr-TEE-dah dai lah sai-RAI-nah), town, BADAJOZ province, W SPAIN, 5 mi/8 km WSW of CASTUERA; 38°40′N 05°38′W. Cereals, tubers, lentils, grapes; livestock.

Malpartida de Plasencia (mahl-pahr-TEE-dah dai plah-SEN-thyah), town, CÁCERES province, W SPAIN, 4 mi/6.4 km SE of PLASENCIA; 39°59′N 06°02′W. Cereals, grapes, olives; olive-oil processing, wine making.

Malpas (MAWL-puhs), town (2001 population 3,887), SW CHESHIRE, W ENGLAND, 13 mi/21 km SSE of CHESTER; 53°01′N 02°46′W. Former agricultural market in dairying region. Has church begun in 14th century, completed in Tudor times.

Malpas (MAW-puhs), town (2001 population 8,148), NEWPORT, SE Wales, just N of Newport; 53°01′N 02°46′W.

Malpas (MAWL-puhs), village (2000 population 200), SW CORNWALL, SW ENGLAND, just SE of TRURO; 50°16′N 05°03′W.

Mal Paso, Alto de (MAHL-PAH-so, AHL-to dai), highest point (4,330 ft/1,320 m) of HIERRO, CANARY ISLANDS, SPAIN, 8 mi/12.9 km SW of VALVERDE. Sometimes spelled Malpaso.

Malpe, INDIA: see UDUPI.

Malpelo Island (mahl-PAI-lo), small islet in the PACIFIC OCEAN, belonging to COLOMBIA, 310 mi/499 km W of BUENAVENTURA; 04°00′N 90°30′W.

Malpeque Bay (MOL-pek), inlet (12 mi/19 km long, 10 mi/16 km wide at entrance) of the Gulf of SAINT LAWRENCE, NW PRINCE EDWARD ISLAND, E CANADA, 4 mi/6 km N of SUMMERSIDE. Entrance of bay is protected by several islets. Oyster beds.

Malpica (mahl-PEE-kah), town, CASTELO BRANCO district, central PORTUGAL, near Spanish border, 12 mi/19 km SSE of CASTELO BRANCO; 39°41′N 07°24′W. Corn, wheat, olives, beans. Cork-oak woods.

Malpica (mahl-PEE-kah), town, LA CORUÑA province, NW SPAIN, on the ATLANTIC OCEAN, 21 mi/34 km W of LA CORUÑA. Fishing port, swimming resort.

Malpica, town, TOLEDO province, central SPAIN, on the TAGUS RIVER, and 16 mi/26 km ESE of TALAVERA DE LA REINA. Olives, cereals, grapes; livestock. Hunting (wild boar). Has old castle. Sometimes called Malpica de Tajo.

Malplaquet (mahl-plah-KAI), hamlet, NORD department, NORD-PAS-DE-CALAIS region, N FRANCE, near Belgian border, 5 mi/8 km NW of MAUBEUGE; 50°19′N 03°52′E. Here in 1709, Marlborough and Eugene of Savoy won costly victory over French under Marshal Villars.

Malpura (mahl-poo-rah), town, TONK district, E central RAJASTHAN state, NW INDIA, 50 mi/80 km SSW of JAIPUR; 23°21′N 73°27′E. Local market for millet, wheat, gram, cotton; handicrafts (felts, woolen blankets, saddle cloths). Headworks of small canal irrigation system 6 mi/9.7 km SSE.

Malsch (MAHLSH), town, MIDDLE UPPER RHINE, BADEN-WÜRTTEMBERG, GERMANY, 8 mi/12.9 km SW of ETTLINGEN; 48°53′N 08°20′E.

Malse River (MAHLCH), Czech *Malše*, German *Maltsch*, 62 mi/100 km long, N AUSTRIA and CZECH REPUBLIC (S BOHEMIA); rises in Austria 9 mi/14.5 km SE of Dolní Dvořiště (Czech Republic); flows generally NW, along Czech-Austrian border, and N into the Czech Republic, past KAPLICE, to VLTAVA RIVER at České Budějovice.

Malsfeld (MAHLS-feld), village, N HESSE, central GERMANY, on the left bank of the FULDA RIVER, 15 mi/24 km S of KASSEL; 51°05′N 09°32′E. Agriculture (grain); dairying.

Malsiras (mahl-see-rahs), town, SOLAPUR district, MAHARASHTRA state, W central INDIA, 70 mi/113 km WNW of SOLAPUR; 17°52′N 74°55′E. Agricultural market; sugar milling.

Malta (MAWL-tah), ancient *Melita*, republic (□ 122 sq mi/317.2 sq km; 2007 population 401,880), in the

MEDITERRANEAN SEA, S of SICILY; ⊙ VALLETTA; 35°55′N 14°26′E.

Geography

The country consists of Malta (□ 95 sq mi/246 sq km), GOZO (26 sq mi/67 sq km), and COMINO (1 sq mi/2.6 sq km) islands, and uninhabited islets of Comminotto, Fungus Rock, Filfla and Saint Paul's islands. The group is sometimes called the Maltese Islands. Malta has no rivers or lakes, no natural resources, and very few trees. Nevertheless it is of great strategic value and was an important British military base until 1979.

Economy

The decline in activity at the base, beginning in the late 1960s, created serious economic problems for the country, which has been trying since the 1970s to diversify its economy. Manufacturing and tourism (more than one million people visited in 1994) are the country's main industries; nearly half the workforce is employed in government. Shipbuilding and ship repair and the manufacture of electronics, textiles, processed food, paper, clothing, tobacco products, and construction materials are the main industries. Maltese are also employed in doing traditional lacework. Although the soil is poor, there is some agriculture (potatoes, cauliflower, grapes, wheat, barley, and other vegetables, and citrus fruits). Fishing in local waters has declined in importance. Imports include food, petroleum, machinery, and manufactured goods; exports include textiles, clothing, and ships. Most trade is with ITALY, FRANCE, and the U.K. International banking and financial services are growing, and the island is developing as an offshore tax haven.

Population

English and Maltese, a Semitic dialect, are the official languages, although Italian is also widely spoken. Roman Catholicism (claiming 98% of the population) is the state religion. Malta has a very high population density. Shortage of water has stimulated building of desalination plants (only sources until then were rainwater or wells which were depleting the underground aquifers). These plants now provide more than half of the freshwater needs.

History: to World War II

Governed under a 1974 revision of the 1964 constitution. Malta belonged successively to the Phoenicians, Greeks, Carthaginians, and Romans. St. Paul was shipwrecked here (C.E. 60). Arab rule began in C.E. 870. The Normans of SICILY occupied it c.1090. In 1530 the Hapsburg Charles V granted Malta to the Knights Hospitaller of St. John of Jerusalem. Notwithstanding a determined siege by the Turks in 1565, the knights held it until 1798, when it was surrendered to Napoleon. The British ousted the French in 1800, and made it a crown colony in 1814. For most of the 19th century Malta was ruled by a military governor. The opening of the SUEZ CANAL (1869) increased its strategic value, Malta becoming one of the principal coaling stations for steamships to INDIA and the FAR EAST.

History: World War II to 1979

During World War II Malta was subjected to extremely heavy bombing by Italian and German planes. Almost from the start of the period of British rule, the Maltese agitated for increased political freedom. A constitution promulgated in 1921 granted considerable self-government but was revoked in 1936. Malta reverted to the status of crown colony. A similar constitution was granted in 1947 but was revoked after civil disturbances in 1959. Malta became fully independent in 1964. In 1965 it joined the UN. When the Labour party came to power in 1971 Maltese prime minister Dom Mintoff threatened to break ties with the West if the British did not increase rental payments for the naval base.

History: 1979 to Present

A compromise agreement (effective through 1979) was reached by which the rent would be paid by the British while Malta would also receive a sum from NATO and aid from Italy. In August 1973, Malta initiated a seven-year economic development plan intended to free the country from its dependence on the rental payments, which were stopped in 1979 when the military bases were closed. The government of Nationalist Prime Minister Eddie Fenech Adami, elected 1987, tried to balance its foreign policy between neighboring LIBYA and the economically more important Western nations. Another issue involves the role of the Roman Catholic Church, a prominent landlord in a country where church and state were not separate until recently. The resulting tensions between the church and the earlier Labor government were especially great. The 1992, 1998, and 2003 elections returned the Nationalists and Fenech Adami to office; Lawrence Gonzi succeeded Fenech Adami as prime minister in 2004. The Labor party, with Alfred Sant as prime minister, was in power from 1996 to 1998. Malta sought full membership in the European Union (EU) starting in 1990; after freezing its application in 1996, it reactivated the application in 1998 and became a full member on May 1, 2004.

Government

Malta is governed under the constitution of 1964 as amended. The president, who is the head of state, is elected by the legislature for a five-year term and is eligible for a second term. The prime minister is the head of government. Members of the unicameral legislature, the sixty-five-seat House of Representatives, are popularly elected to five-year terms. The current head of state is President Edward Fenech Adami (since April 2004). The head of government is Prime Minister Lawrence Gonzi (since March 2004).

Malta, city (2000 population 2,120), ⊙ PHILLIPS county, N MONTANA, on MILK River, near Lake BOWDOIN, and 58 mi/93 km WNW of GLASGOW; 48°21′N 107°52′W. Trading point in irrigated agricultural area; cattle; wheat, barley, hay. Natural-gas wells to NE. Lake Bowdoin and Bowdoin National Wildlife Refuge (known for its waterfowl) to E. Phillips County Museum here. Incorporated 1909.

Malta, town (2007 population 5,654), central PARAÍBA state, BRAZIL, on JOÃO PESSOA–FORTALEZA railroad, 20 mi/32 km NW of PATOS; 06°54′S 37°32′W.

Malta, village, CARINTHIA, S AUSTRIA, in the Malta valley, 10 mi/16 km N of SPITTAL AN DER; 46°57′N, 13°30′E. Elevation: 2,545 ft/776 m. Summer tourism. In a scenic high alpine valley; panoramic road; nearby is a large storage power station (Malta-Oberstufe) with the highest dam in Austria (Kölnbreinsperre, 610 ft/ 186 m high).

Malta, village (2000 population 177), CASSIA county, S IDAHO, 25 mi/40 km SE of BURLEY, on RAFT RIVER; 42°19′N 113°22′W. Irrigated agricultural area. Cattle; dairying; grain, alfalfa.

Malta, village (2000 population 969), DE KALB county, N ILLINOIS, 10 mi/16 km WSW of SYCAMORE; 41°55′N 88°51′W. In rich agricultural area. Kishwaukee Community College nearby.

Malta (MAWL-tuh), village (2006 population 668), MORGAN county, SE OHIO, on MUSKINGUM RIVER, opposite MCCONNELSVILLE; 39°39′N 81°52′W.

Malta Bend, town (2000 population 249), SALINE county, central MISSOURI, near MISSOURI River, 10 mi/16 km NW of MARSHALL; 39°11′N 93°21′W. Grain; livestock. Grand Pass Conservation Area (wetlands) to N.

Malta Channel, strait, in the MEDITERRANEAN Sea between SE SICILY (Italy) and the MALTESE Islands; c.60 mi/97 km wide.

Maltahohe (mahl-tuh-HO-hai), German *Maltahöhe*, trading town, S central NAMIBIA, 140 mi/225 km NNW of KEETMANSHOOP; 24°50′S 16°59′E. Sheep raising. Formerly noted horse-breeding center. Site of Duwisib Castle (1900).

Maltby (MAWLT-bee), town (2001 population 17,247), SOUTH YORKSHIRE, N ENGLAND, 6 mi/9.7 km E of ROTHERHAM; 53°25′N 01°12′W. Coal mining. Machinery and equipment manufacturing.

Malte Brun Range (MAHL-te bruhn), SOUTH ISLAND, NEW ZEALAND, in SOUTHERN ALPS, and in MOUNT COOK National Park, S-W trending range, between MURCHISON and TASMAN glaciers; 43°34′S 170°19′E. Highest peak, Mount Malte Brun (elevation 10,351 ft/ 3,155 m).

Maltepe, TURKEY: see MANYAS.

Malters (MAHL-tuhrs), commune, LUCERNE canton, central SWITZERLAND, near KLEINE EMME RIVER, 6 mi/ 9.7 km W of LUCERNE; 47°02′N 08°11′E.

Maltese Islands: see MALTA.

Malton (MAWLT-uhn), town (2001 population 5,023), NORTH YORKSHIRE, N ENGLAND, on DERWENT RIVER, and 17 mi/27 km NE of YORK; 54°08′N 00°48′W. Foundries; manufacturing (agricultural implements, mineral water); light industry. Racing stables. Old Malton (just NW), has Norman church and ruins of Gilbertine priory.

Maltrata (mahl-TRAH-tah), town, VERACRUZ, E MEXICO, in valley at S foot of PICO DE ORIZABA, on railroad, and 12 mi/19 km WSW of ORIZABA; 18°48′N 97°16′W. Coffee-growing center.

Mal'tsevskaya, RUSSIA: see URITSKIY.

Maltz (MELTZ), village, NW TUNISIA, on the MAJARDAH RIVER, on railroad, and 34 mi/55 km SSW of AYN AD-DARAHAM. Cereal and livestock market. Iron mines; Chemtou marble quarries nearby.

Maluenda (mah-LWEN-dah), village, ZARAGOZA province, NE SPAIN, on JILOCA RIVER, and 5 mi/8 km SSE of CALATAYUD; 41°17′N 01°37′W. Sugar beets, cereals, wine, fruit; cattle.

Malujowice, POLAND: see MOLLWITZ.

Maluk Bliznak (MAH-lek bleez-NAHK), island, in the DANUBE River, NW BULGARIA; 43°52′N 22°51′E.

Maluk Bliznak (MAH-lek bleez-NAHK), peak, in the RILA MOUNTAINS, SW BULGARIA; 42°09′N 23°07′E. Elevation 9,029 ft/2,752 m.

Maluk Brushlen (MAH-lek brush-LEN), island, in the DANUBE River, NE BULGARIA; 44°01′N 26°20′E.

Maluk Polezhan (mah-LOOK po-le-ZHAHN), peak (9,258 ft/ 2,822 m), PIRIN MOUNTAINS, SW BULGARIA; 41°43′N 23°34′E.

Maluku (MAH-loo-koo) or **Moluccas,** Dutch *Molukken,* island group and province (□ 32,300 sq mi/ 83,657 sq km; 1990 population 1,856,075), E INDONESIA, between SULAWESI and NEW GUINEA; ⊙ AMBON, on AMBON Island; 02°00′S 128°00′E. The group's many islands include HALMAHERA, SERAM, BURU, BANDA ISLANDS, LETI, Ambon, TERNATE, TIDORE and the ARU and KAI island groups. Of volcanic origin, the Maluku are mountainous, fertile, and humid; highest peak is Gunnung Binaiya (9,902 ft/ 3,018 m) on Seram. Nutmeg and clove originated here. Other spices, rice, copra, and forest products are also produced. Sago is the staple food. Nickel mining and oil drilling on Seram. Crafts include wood carving, silver and gold filigree, and hand-loom weaving.The islands were explored by Magellan (1511–1512) and thereafter settled by the Portuguese, who established a trading center at Ternate. In the seventeenth century they were taken by the Dutch, who secured a monopoly in the clove trade. Twice the British gained a foothold in the islands, which passed definitively to the Dutch in the first quarter of the nineteenth century. Short-lived Republic of South Moluccas rebellion occured in S during 1950s. Also referred to as the Spice Islands.

Maluku (muh-LOO-koo), village, BAS-CONGO province, W CONGO, 100 mi/160 km S of KINSHASA; 04°03′S 15°34′E. Elev. 774 ft/235 m. MALEBO (STANLEY) POOL nearby.

Maluku Passage, wide channel, in INDONESIA, connecting MALUKU (SW) and SERAM (SE) seas with the PACIFIC OCEAN, between SULAWESI (W) and HALMAHERA (E) islands; c.150 mi/241 km wide; 00°00′S

125°00′E. A large W arm (Gulf of TOMINI) indents E coast of Sulawesi. Formerly spelled Molucca.

Maluku Sea, part of the PACIFIC OCEAN, in INDONESIA, between SULAWESI (W) and BURU (E) islands, merges with SERAM and BANDA seas (E), FLORES SEA (SW), and MALUKU PASSAGE (N); 00°00′S 125°00′E. Contains SULA ISLANDS. Formerly spelled Maluku.

Ma'lula (MAH-loo-lah), village, DAMASCUS district, SW SYRIA, 26 mi/42 km NNE of Damascus, on E slope of the ANTI-LEBANON mountains; 33°51′N 36°33′E. Elev. 5,650 ft/1,722 m. Summer resort. Orchards; cereals. Has a Greek Catholic convent, with the cupola dating from Byzantine times. Population largely Greek Catholic, speak a dialect of Aramaic.

Malung (MAHL-UNG), town, KOPPARBERG county, central SWEDEN, on VÄSTERDALÄLVEN RIVER, 60 mi/97 km W of FALUN; 60°41′N 13°43′E.

Malungsfors (MAHL-UNGS-FORSH), village, KOP-PARBERG county, central SWEDEN, on VÄSTER-DALÄLVEN RIVER, 70 mi/113 km W of FALUN; 60°44′N 13°33′E.

Malur (mah-loor), town, KOLAR district, KARNATAKA state, S INDIA, 15 mi/24 km SW of KOLAR; 13°00′N 77°55′E. Tobacco curing; handicrafts (pottery, biris; weaving). Kaolin deposits nearby.

Maluti Mountains (mah-LOO-ti), W LESOTHO, branch of DRAKENSBERG; extend NE and SW for 100 mi/160 km; rise to 11,000 ft/3,346 m at Mothae Peak, near juncture with Drakensberg range. Record snowfall of more than 1.5 ft/0.5 m caused national disaster in spring 1988. Also spelled Maloti.

Maluwe (mah-LOO-wai), town, NORTHERN REGION, GHANA, 20 mi/32 km N of BUI DAM, on road, and 20 mi/32 km SSE of BOLE; 08°40′N 02°17′W.

Malvalli (muhl-vah-LEE), town, MANDYA district, KARNATAKA state, S INDIA, 16 mi/26 km SE of MAN-DYA. Road center in sugarcane- and silk-growing area; hand-loom silk weaving. Famous KAVERI FALLS are 10 mi/16 km SE. Also spelled MALAVALLI.

Malvan (mahl-VAHN), town, SINDHUDURG district, MAHARASHTRA state, W central INDIA, port on ARA-BIAN SEA, 65 mi/105 km S of RATNAGIRI, in the KONKAN; 16°04′N 73°28′E. Fish-curing (mackerel, sardines, catfish) center; local market for rice, sugarcane, coconuts, mangoes. Manufacturing of pen nibs, coir products; cashew-nut processing, shark-oil extracting. Iron-ore deposits nearby. Lighthouse (S). A Maratha stronghold in 18th century

Malvar (mahl-VAHR), town, BATANGAS province, S LUZON, PHILIPPINES, 11 mi/18 km W of SAN PABLO; 14°02′N 121°09′E. Railroad junction. Agricultural center (rice, sugarcane, corn, coconuts).

Malvasia, Greece: see MONEMVASÍA.

Malvatu Oya, SRI LANKA: see ARUVI ARU.

Malveira (mahl-VEI-rah), village, LISBOA district, central PORTUGAL, on railroad, and 15 mi/24 km NNW of LISBON; 38°56′N 09°15′W. Hardware manufacturing; olives, honey.

Malvern (MAL-vuhrn), city (2000 population 9,021), ⊙ HOT SPRING county, central ARKANSAS, 17 mi/27 km SE of HOT SPRINGS, near OUACHITA RIVER; 34°22′N 92°49′W. In diversified agricultural area. Manufacturing (apparel, building materials, chemicals, lumber, fiberboard, electrical goods, wood products); sawmilling. Lake CATHERINE (Ouachita River) and Lake Catherine State Park are to NW. Incorporated as city 1876.

Malvern, town, SAINT ELIZABETH parish, SW JAMAICA, resort, 9 mi/14.5 km SE BLACK RIVER town; 17°58′N 77°43′W. In agricultural region (corn, tropical fruit and spices; livestock).

Malvern, town (2000 population 1,215), Geneva co., SE Alabama, 21 mi/34 km NE of Geneva. Originally called 'Eagan,' it was incorporated in 1904 as Malvern in honor of the Civil War Battle of Malvern Hill in VA.

Malvern, town (2000 population 1,256), MILLS county, SW IOWA, on SILVER CREEK, and 9 mi/14.5 km ESE of GLENWOOD; 41°00′N 95°35′W. In hay, grain, livestock area; feed milling. Founded 1869 with coming of railroad; incorporated 1870.

Malvern (MAL-vuhrn), village (2006 population 1,222), CARROLL county, E OHIO, 13 mi/21 km ESE of CANTON, on small Sandy Creek; 40°41′N 81°11′W. In dairying area; makes clay products.

Malvern (MAL-vuhrn), borough (2006 population 3,108), CHESTER county, SE PENNSYLVANIA, suburb, 19 mi/31 km WNW of downtown PHILADELPHIA; 40°01′N 75°30′W. Some agriculture (vegetables; livestock; dairying; nursery stock); manufacturing (biotechnical products, electronic equipment, chemicals, consumer goods, plastic products, restaurant equipment, transportation equipment, computers, machinery; publishing). Settled 1866, incorporated 1889.

Malvern (MOL-vuhrn), inner suburb 5 mi/8 km SE of MELBOURNE, S VICTORIA, SE AUSTRALIA; 37°52′S 145°02′E. In metropolitan area; residential.

Malvern, suburb, SE KWAZULU-NATAL province, SOUTH AFRICA, on inland railroad route, 7 mi/11.3 km WSW of Durban Central Business District (CBD). In fruit-growing region. Now part of conurbation of greater DURBAN area.

Malvern, suburb, E suburb of JOHANNESBURG, GAU-TENG province, SOUTH AFRICA, part of greater metropolitan Johannesburg conurbation.

Malverne, residential village (□ 1 sq mi/2.6 sq km; 2006 population 8,751), in HEMPSTEAD town, NASSAU county, SE NEW YORK, on LONG ISLAND; 40°40′N 73°40′W. Settled in the early 1800s, incorporated 1921.

Malvern East (MAL-vuhrn EEST), suburb 6 mi/10 km SE of MELBOURNE, VICTORIA, SE AUSTRALIA; 37°53′S 145°04′E. Super-regional shopping center here.

Malvern, Great (MAWL-vuhrn), town (2001 population 35,588), WORCESTERSHIRE, W central ENGLAND, on the E slopes of the scenic MALVERN HILLS, about 7 mi/11.2 km SW of WORCESTER; 52°07′N 02°19′W. Occupying the site of the medieval Chase of Malvern (royal forest; □ 11 sq mi/28 sq km), the town has become a health (mineral springs) and holiday resort. Malvern College, a co-educational (1992) public school, was founded in 1862. The priory church of Great Malvern dates from 1085; the Norman arches of the interior remain intact. The annual Malvern festival of dramatics, associated with the plays of G. B. Shaw, was instituted in 1928. Town includes West Malvern.

Malvern Hills (MAWL-vuhn), range of hills, c.9 mi/14.5 km long, W central ENGLAND, in Herefordshire and Worcestershire. Composed of metamorphic rocks. The highest points are the Worcestershire Beacon (1,395 ft/425 m) and the Herefordshire Beacon (1,114 ft/340 m); on the latter is an Iron Age fortress.

Malvési, FRANCE: see NARBONNE.

Malvinas, Islas, ARGENTINA: see FALKLAND ISLANDS.

Malwa (MAHL-wah), subdivision of former CENTRAL INDIA agency. Comprised former princely states of Alirajpur, Barwani, Dhar, Jaora, Jhabua, JOBAT, KA-THIWARA, Mathwar, PIPLODA, Ratlam, Sailana, Sitamau, and several petty states. Created 1925; today area is part of W MADHYA PRADESH state.

Malwa (MAHL-wah), plateau in W central INDIA, comprising large part of MADHYA PRADESH state and small section of SE RAJASTHAN state. Bounded S by W VINDHYA RANGE, W by offshoot of Vindhyas which extends N toward MANDSAUR, E by another offshoot which extends N toward GUNA, and N by irregular line of hills between CHITTAURGARH and Guna; average elevation c.1,600 ft/488 m. Fertile area of black-cotton soil, drained by CHAMBAL, SIPRA, KALI SINDH, and PARBATI rivers; millet, wheat, gram, cotton, poppy are chief crops. Important trade centers are INDORE, UJJAIN, RATLAM, and DEWAS. Name often applied to larger surrounding area; originally a tribal country, mentioned vaguely in early Hindu legends; in 6th century B.C.E. known as Avanti. Under Mauryan (3rd century B.C.E.) and Magadhan (4th–5th century A.D.) dynasties. Muslims first appeared in 1235; from 1401 to 1531 Malwa was strong independent state with capital at Mandu. Later (c.1560) fell to the Moguls and (in mid-18th century) to the Marathas.

Malwatu Oya, SRI LANKA: see ARUVI ARU.

Malxe River (MAHLK-se), 50 mi/80 km long, E GER-MANY; rises in LOWER LUSATIA just E of DÖBERN; flows N and WNW, in a wide arc, through SPREE FOREST, to the Spree River 1.5 mi/2.4 km SSE of LÜBBEN.

Maly, [Russian=LESSER LITTLE SMALL], in Russian names: see also MALAYA MALO- [Russian combining form], MALOYE, MALYE.

Malý Dunaj River, SLOVAKIA: see LITTLE DANUBE RIVER.

Malygin Strait, RUSSIA: see BELYY ISLAND.

Mályi (MAHL-ye), village, HUNGARY, 6.5 mi/10.5 km SE of MISKOLC; 48°01′N 20°50′E. One of the country's largest brick and tile factories.

Malyn (MAH-lin), (Russian *Malin*), city, E ZHYTOMYR oblast, UKRAINE, on the Irsha River (tributary of the TETERIV RIVER), and on railroad, 28 mi/45 km ESE of KOROSTEN'; 50°46′N 29°14′E. Elevation 449 ft/136 m. Raion center. Manufacturing (paper, cellulose, machines, building materials), food processing (butter, cheese, fruit drying); stone crushing. Forestry school; 19th-century park. Established in the 11th century, important in the Cossack-Polish War (1648–1657). City since 1938. Jewish community since 1775, numbering 4,600 in 1926; wiped out by the Nazis during World War II — fewer than 100 Jews remaining in 2005.

Malynivka (mah-LI-neef-kah), (Russian *Malinovka*), town, N central KHARKIV oblast, UKRAINE, 3 mi/5 km S of CHUHUYIV (across the DONETS River), in KHAR-KIV metropolitan area; 49°48′N 36°43′E. Elevation 360 ft/109 m. Established in 1652, town since 1938. Several Scythian settlements (5th–3rd century B.C.E.) have been excavated nearby.

Malysheva (MAH-li-shi-vuh), town (2006 population 9,900), central SVERDLOVSK oblast, at the foothills of the E URAL Mountains, W Siberian Russia, terminus of local highway branch, 50 mi/75 km NE of YEKA-TERINBURG, and 7 mi/11 km N of ASBEST, to which it is administratively subordinate; 57°07′N 61°23′E. Elevation 793 ft/241 m. Became a town in 1967.

Malyy Cheremshan, RUSSIA: see GREATER CHEREM-SHAN RIVER.

Malyye Chapurniki (MAH-li-ye chye-POOR-nee-kee), village, SE VOLGOGRAD oblast, SE European Russia, on the VOLGA-DON CANAL, on highway, 9 mi/14 km W of SVETLYY YAR, to which it is administratively subordinate; 48°27′N 44°34′E. Below sea level. In agricultural area (grain); bakery.

Malyye Derbety (MAH-li-ye deer-BYE-ti), village (2005 population 5,770), N Republic of KALMYKIA-KHALMG-TANGEH, SE European Russia, in the SARPA LAKES valley, on the Volgograd-Elista road, 55 mi/89 km S of VOLGOGRAD; 47°57′N 44°41′E. In agricultural area (wheat, livestock).

Malyye Karmakuly (MAH-li-ye kahr-mah-KOO-li), post on the W coast of the S island of NOVAYA ZEMLYA archipelago, N of the ARCTIC CIRCLE, on the E shore of the BARENTS SEA, in the ARCTIC OCEAN, N European Russia; 72°23′N 52°40′E. Government observation station.

Malyye Kolpany (MAH-li-ye kuhl-pah-NI), village, W central LENINGRAD oblast, NW European Russia, on highway, 2 mi/3.2 km SW, and under administrative jurisdiction, of GATCHINA; 59°32′N 30°05′E. Elevation 291 ft/88 m. Fodder plant.

Malyye Vyazëmy (MAH-li-ye vyah-ZYO-mi), urban settlement, W central MOSCOW oblast, central European Russia, near highway and on railroad, 10 mi/16 km WSW of (and administratively subordinate to) ODINTSOVO; 55°38′N 37°00′E. Elevation 606 ft/184 m. Experimental mechanical plant; bus works, heating equipment, sewing.

Cross-references are shown in SMALL CAPITALS. The pronunciation guide is shown on page xix. The sources of population figures are shown on page xvii.

Malyy Irgiz, RUSSIA: see IRGIZ RIVER.

Malyy Kinel, RUSSIA: see GREATER KINEL RIVER.

Malyy Kunaley (MAH-liyee koo-nah-LYAI), village (2004 population 1,020), SE BURYAT REPUBLIC, S Siberian Russia, on the KHILOK RIVER, just W of the border with CHITA oblast, on highway, 73 mi/117 km S of ULAN-UDE, and 8 mi/13 km E of BICHURA, to which it is administratively subordinate; 50°36′N 107°49′E. Elevation 2,080 ft/633 m. Agricultural cooperative.

Malyy Kundysh, RUSSIA: see LESSER KUNDYSH RIVER.

Malyy Kuyal'nyk River, UKRAINE: see KHADZHYBEY LAGOON.

Malyy Lyakhov Island (MAH-liyee–LYAH-huhf) [Russian=Lesser Lyakhov], in the LYAKHOV ISLANDS, between the LAPTEV and EAST SIBERIAN seas, off SAKHA REPUBLIC, RUSSIAN FAR EAST; separated from KOTEL'NYY ISLAND by the SANNIKOV STRAIT, from BOL'SHOY LYAKHOV ISLAND by the ETERIKAN STRAIT; 74°07′N 140°36′E.

Malyy Taymyr Island (MAH-liyee–tei-MIR) [Russian=Lesser Taymyr], in the LAPTEV SEA of the ARCTIC OCEAN, 25 mi/40 km off SE SEVERNAYA ZEMLYA archipelago, in KRASNOYARSK TERRITORY, N Siberian Russia; 78°10′N 107°00′E. Discovered in 1913 by the Russian sailor and explorer Boris A. Vilkitskiy. Formerly called Tsesarevich Aleksey Island.

Malyy Tsivil, RUSSIA: see TSIVIL.

Malyy Usen, RUSSIA: see USEN.

Malyy Utlyuk, UKRAINE: see UTLYUK LIMAN.

Malyy Yenisey River, RUSSIA: see YENISEY RIVER.

Malzéville (mahl-zai-VEEL), town (□ 2 sq mi/5.2 sq km), N residential suburb of NANCY, MEURTHE-ET-MOSELLE department, LORRAINE region, NE FRANCE, on right bank of MEURTHE RIVER; 48°43′N 06°12′E. Brewery.

Mama (MAH-mah), town, YUCATÁN, SE MEXICO, 15 mi/24 km NE of TICUL; 20°29′N 89°22′W. Henequen, sugarcane, fruit.

Mama (MAH-mah), town (2005 population 4,200), NE IRKUTSK oblast, E central SIBERIA, RUSSIA, on the VITIM River, at the mouth of the MAMA RIVER (LENA River basin), on road, 60 mi/97 km NW of BODAYBO; 58°18′N 112°54′E. Elevation 790 ft/240 m. Center of a mica-mining area; logging and lumbering.

Mamadysh (mah-mah-DISH), city (2006 population 13,960), N TATARSTAN Republic, E European Russia, on the VYATKA River (landing), on road, 104 mi/167 km E of KAZAN, and 25 mi/40 km W of YELABUGA; 55°43′N 51°24′E. Elevation 150 ft/45 m. Highway junction and river port; local transshipment point. Oil and gas processing; industrial alcohols, agricultural chemicals, woodworking; food processing (bakery, fish plant). Developed in the 1740s as a copper-smelting center. Chartered in 1781.

Mamahatun, TURKEY: see TERCAN.

Mamahuasi (mah-mah-WAH-see), town and canton, OROPEZA province, CHUQUISACA department, S central BOLIVIA, on left branch of upper PILCOMAYO River, and 15 mi/24 km NW of SUCRE; 18°56′S 65°29′W. Wheat, vegetables.

Mamaia (mah-MAH-yah), beach resort on BLACK SEA, CONSTANŢA county, SE ROMANIA, 5 mi/8 km N of CONSTANŢA. Includes Mamaia Sat (N) and Mamaia Băi (S).

Mamakhel (mah-ma-KHEL), village, NANGARHAR province, E AFGHANISTAN, on N slopes of the SAFED KOH, 26 mi/42 km SW of JALALABAD, just off highway to KABUL; 34°15′N 69°59′E.

Mamanguape (mah-mahn-gwah-pai), city (2007 population 40,297), E PARAÍBA state, NE BRAZIL, 25 mi/40 km NW of JOÃO PESSOA; 06°43′S 35°13′W. Cotton, sugar, and manioc processing; fruit, rice, medicinal plants.

Mamanuca (mah-mah-NOO-thah), volcanic group of islands in FIJI, SW PACIFIC OCEAN, c.10 mi/15 km W of VITI LEVU. Largest island is TAVUA, 1 mi/1.6 km long. Tourism. Sometimes spelled Mamanutha.

Mamanutha, FIJI: see MAMANUCA.

Mama River (MAH-mah), 200 mi/322 km long, SE Siberian Russia; rises in N BURYAT REPUBLIC NE of Lake BAYKAL; flows NE to the VITIM River at MAMA (NE IRKUTSK oblast). Large mica deposits along its course.

Mamaroneck (mah-MA-ruh-nek), residential village and town (2006 population 18,472), WESTCHESTER county, SE NEW YORK, a suburb of NEW YORK CITY, on LONG ISLAND SOUND; 40°55′N 73°43′W. Initially a farming community, Mamaroneck is a boating center with an excellent marina. There is some light manufacturing in addition to office and corporate activity. Mixed middle- and upper-class population Town includes LARCHMONT village. Settled 1661, incorporated 1895.

Mamawi Lake, NE ALBERTA, W CANADA, 8 mi/13 km SW of FORT CHIPEWYAN, in WOOD BUFFALO NATIONAL PARK, between Lake CLAIRE and Lake ATHABASCA; 16 mi/26 km long, 10 mi/16 km wide; 58°35′N 111°30′W.

Mambajao (mahm-BAH-hou), town, ⊙ CAMIGUIN ISLAND and province, PHILIPPINES, on NE coast of island; 09°13′N 124°43′E. Agricultural center (corn, coconuts). Airport. Just S of town is volcanic MOUNT HIBOK-HIBOK (4,265 ft/1,300 m); its eruption in 1948 caused temporary evacuation of island's inhabitants. The volcano has been inactive since 1953, but it remains under constant surveillance.

Mambali (mam-BAH-lee), village, TABORA region, NW central TANZANIA, 35 mi/56 km NNW of TABORA; 04°33′S 32°40′E. Road junction. Corn, wheat, cotton; cattle, goats, sheep.

Mambange, village, MATABELELAND NORTH province, W ZIMBABWE, 22 mi/35 km SE of HWANGE, on railroad, at N edge of HWANGE NATIONAL PARK; 18°23′S 26°46′E. Livestock.

Mambasa (muhm-BAH-suh), village, ORIENTALE province, NE CONGO; 01°21′N 29°03′E. Elev. 2,998 ft/913 m. Trading center; rice processing. Gold mining nearby. Has Protestant mission. Members of Pygmy tribes still live in vicinity.

Mambata Cave, ZIMBABWE: see MATOBO NATIONAL PARK.

Mamberamo River (MAHM-buh-rah-mo), c.500 mi/805 km long (including TARITATU RIVER headstream), largest river of IRIAN JAYA province, INDONESIA; formed by junction (in marshy area N of SUDIRMAN RANGE) of Taritatu and TARIKU rivers; flows generally NW to the PACIFIC OCEAN near Cape D'URVILLE; 01°26′S 137°53′E. Sometimes called Tarikaikea River

Mambéré (mahm-BAI-rai), river, W CENTRAL AFRICAN REPUBLIC; rises in Yade Massif in NANA-MAMBÉRÉ prefecture, near CAMEROON border; flows 150 mi/241 km SE to CARNOT, in HAUTE-SANGHA prefecture; then flows 120 mi/193 km S to NOLA, in SANGHA economic prefecture, where it joins SANGHA RIVER. Flows through agricultural regions (coffee, cocoa, rubber, cotton).

Mambéré-Kadéï, prefecture (□ 11,658 sq mi/30,310.8 sq km; 2003 population 364,795), SW CENTRAL AFRICAN REPUBLIC; ⊙ BERBÉRATI. Bordered N by NANA-MAMBÉRÉ prefecture, NE by OMBELLA-M'POKO prefecture, E by LOBAYE prefecture, S by SANGHA-MBAÉRÉ economic prefecture, and W by CAMEROON. Drained by Boumbe and Ekeia rivers. Agriculture (coffee, cotton); gold and diamond mining; cotton ginning. Main centers are Berbérati, CARNOT, and Gaza. Also called Haute-Sangha.

Mambilla Plateau (mahm-BEEL-lah) (around □ 772 sq mi/2,000 sq km), TARABASTATE, E NIGERIA. Rises to a range of 4,500 ft/1,372 m–5,500 ft/1,676 m. The Goetl Mountains, including Chappal Waddi (or Chapal Nadi), Nigeria's highest peak (higher than 7,500 ft/2,286 m) are here.

Mambolo (MAHM-bo-lo), coastal town, KAMBIA district, NORTHERN province, SIERRA LEONE; 08°55′N 13°02′W. Swamp rice cultivation. Agricultural research station.

Mambone (mahm-BO-nai), village, central MOZAMBIQUE, on MOZAMBIQUE CHANNEL, at mouth of SABI RIVER, 75 mi/121 km S of BEIRA; 20°59′S 33°39′E. Agricultural trade (corn, beans); fishing.

Mamborê (MAHM-bo-rai), city (2007 population 14,132), S PARANÁ state, BRAZIL, 28 mi/45 km NE of UNIÃO DA VITÓRIA; 25°55′S 50°50′W.

Mambray Creek (MAM-brai KREEK), locality, E central SOUTH AUSTRALIA state, S central AUSTRALIA, 163 mi/262 km N of ADELAIDE, on S edge of FLINDERS RANGES, and on W side of MOUNT REMARKABLE NATIONAL PARK; 32°48′S 137°59′E. Roadhouse, camping facilities.

Mambulao, PHILIPPINES: see JOSE PAÑGANIBAN.

Mambunga (muhm-BUNG-guh), village, ORIENTALE province, NE CONGO, 25 mi/40 km ESE of PAULIS; 02°37′N 27°54′E. Elev. 2,588 ft/788 m. Cattle-raising center; cotton ginning.

Mamburao, town (□ 541 sq mi/1,406.6 sq km), ⊙ MINDORO OCCIDENTAL province, W MINDORO, PHILIPPINES, 47 mi/75 km SE of BATANGAS; 13°15′N 120°40′E. Agricultural center (rice); fishing; aquaculture; shipbuilding. F. B. Harrison Game Refuge and Bird Sanctuary (□ 541 sq mi/1,400 sq km) nearby. Airport.

Mambusao (mahm-BOO-sou), town, CAPIZ province, N PANAY island, PHILIPPINES, 15 mi/24 km SW of ROXAS; 11°25′N 122°36′E. Rice-growing center. Polytechnical college.

Mamedkala (mah-myet-kah-LAH), village (2005 population 9,180), SE DAGESTAN REPUBLIC, in the W Caspian lowland just E of the foothills of the NE Greater CAUCASUS Mountains, on road and branch of coastal railroad, 53 mi/85 km SE of MAKHACHKALA, and 5 mi/8 km NW of DAGESTANSKIYE OGNI; 42°10′N 48°07′E. Agriculture (grain, fruits); wineries. Dagestan research institute of viticulture is located here. Formerly known as Khan-Mamed-Kala.

Mameki, TURKEY: see TUNCELI.

Mameli (MAH-me-lee), village, CYRENAICA region, NE Libya, on highest part of Jabal Al Akhdar plateau, on road, 17 mi/27 km SW of CYRENE, just NE of SULUNTAH. Agricultural settlement (cereals, olives, fruit, nuts). Established 1938–1939 by Italians, who left after World War II, were replaced by Libyan population. Formerly Mammelli.

Mamelle Island (mah-MEL), island (2002 population 2,352), SEYCHELLES, in INDIAN OCEAN, 10 mi/16 km NNE of VICTORIA; 04°29′S 55°32′E. Also spelled MAMELLES.

Mamelles, Les, district, SEYCHELLES: see Les Mamelles.

Ma-Me-O Beach, summer village (2001 population 81), S central ALBERTA, W CANADA, 23 mi/38 km from MILLET, in WETASKIWIN COUNTY NO. 10; 52°58′N 113°59′W. Established in 1948.

Mamer (MAH-muhr), village, MAMER commune (2001 population 6,753), SW LUXEMBOURG, 5 mi/8 km WNW of LUXEMBOURG city; 49°38′N 06°02′E.

Mamers (mah-MER), town (□ 1 sq mi/2.6 sq km), SARTHE department, PAYS DE LA LOIRE region, NW FRANCE, 13 mi/21 km ESE of ALENÇON; 48°20′N 00°15′E. Agricultural trade center; manufacturing of vacuum cleaners and camping equipment. Has 12th–16th-century church of Notre Dame.

Mamey, El, MEXICO: see MINATITLÁN.

Mameyes (mah-MAI-yes), village, NE PUERTO RICO, 23 mi/37 km ESE of SAN JUAN. Cultivates the mamey tree, which yields preserves and candy.

Mamfe (mahm-FAI), town (2001 population 20,300), ⊙ MANYU department, South-West province, CAMEROON, near border with NIGERIA, on CROSS River, and 80 mi/129 km NNW of KUMBA; 05°44′N 09°20′E. Trade center; coffee, cacao, bananas, palm oil and kernels; salt deposits; hardwood, and rubber. Has hospital Airport.

Mamiña (mah-MEE-nyah), village, TARAPACÁ region, N CHILE, on W slopes of the ANDES Mountains (9,005 ft/2,745 m), 60 mi/97 km E of IQUIQUE;

20°05′S 69°14′W. Hot sulphur springs with curative properties.

Mamison (mah-mee-SON), pass (9,550 ft/2,911 m), in the central Greater Caucasus, on the border between GEORGIA and SE European RUSSIA. Crossed by the OSSETIAN MILITARY ROAD, it links the cities of KUTAISI (Georgia) and ALAGIR (Russia), and the ARDON and RIONI river valleys.

Mamlyutka (mahm-LYOOT-kuh), city, N NORTH KAZAKHSTAN region, KAZAKHSTAN, on TRANS-SIBERIAN RAILROAD, 25 mi/40 km WNW of PETROPAVLOVSK; 54°54′N 68°36′E. In wheat area. Metalworks.

Mammelli, LIBYA: see MAMELI.

Mammoth, town (2000 population 1,762), PINAL county, SE ARIZONA, on SAN PEDRO RIVER, and 39 mi/63 km NNE of TUCSON; 32°43′N 110°38′W. Elevation 2,348 ft/716 m. Cattle, sheep. Molybdenum deposits. Busy mining (gold, silver) camp in 1880s. GALIURO MOUNTAINS to NE; parts of Coronado National Forest to E and SW; Aravaipa Canyons Wilderness area and Holy Joe Peak (6,145 ft/1,873 m) to NE.

Mammoth Cave, cavern near SAN ANDREAS, CALAVERAS county, central CALIFORNIA. Has many chambers and a subterranean lake.

Mammoth Cave National Park (□ 82 sq mi/212 sq km), EDMONSON, HART, and BARREN counties, S central KENTUCKY, in PENNYROYAL region. Site of Mammoth Cave, one of the largest known caves in the world, located in hilly, forested region drained by GREEN and NOLIN rivers. Composed of a series of subterranean chambers and narrow passages formed by the dissolution of limestone, the cave has five separate levels. The cave has at least 330 mi/531 km of explored passageways with a large variety of limestone formations (stalactites, stalagmites, and columns), and subterranean lakes and rivers. Echo River, c.360 ft/110 m below the surface, flows through the cave's lowest level and drains into the Green River. Also Hanson's Lost River, joins Mammoth Cave with the extensive Flint Ridge cave system; this long-sought link was discovered in 1972. The temperature (54°F/12°C) and relative humidity (87%) remain constant during the year throughout the cave. The cave contains the mummified body of a man believed to date from the pre-Columbian period. Eyeless fish, bats, and insects are also found. Mammoth Cave was a long-time Native American habitation before it was explored by Kentucky pioneers in 1799. During the War of 1812, saltpetre was mined in the cave for gunpowder. Numerous other caves in and around park. Variety of surface wildlife and flora. Authorized 1926.

Mammoth Caves, AUSTRALIA: see MARGARET RIVER.

Mammoth Hot Springs, WYOMING: see YELLOWSTONE NATIONAL PARK.

Mammoth Lakes, city (2000 population 7,093) MONO county, E CALIFORNIA; resort settlement and region of many small lakes, in the SIERRA NEVADA, and 35 mi/56 km NW of BISHOP; 37°38′N 118°59′W. Elevation c.8,900 ft/2,713 m. Fishing, boating, camping, hiking, horseback riding, winter sports. DEVILS POSTPILE NATIONAL MONUMENT is 5 mi/8 km W. Scene of small-scale gold rush, 1879–1880. Surrounded by Inyo National Forest to W, Sierra National Forest to W, beyond Sierra Crest; world-class Mammoth Mountain Ski Area to W; Lake Crowley reservoir to E.

Mammoth Mountain Ski Area, California: see MAMMOTH LAKES.

Mammoth Onyx Cave, KENTUCKY: see MUNFORDVILLE.

Mammoth Pool Reservoir, 9 mi/14.5 km long, on border of FRESNO and MADERA counties, S central CALIFORNIA, on SAN JOAQUIN RIVER, 47 mi/76 km NNE of FRESNO, in Sierra National Forest; 37°18′N 119°18′W. Elevation 3,330 ft/1,015 m. Extends NE. Maximum capacity 123,000 acre-ft. Formed by Mammoth Pool Dam (375 ft/114 m high), built by the Southern California Edison Company for power generation and water supply.

Mammoth Spring (MAM-uth), town (2000 population 1,147), FULTON county, N ARKANSAS, c.65 mi/105 km NW of JONESBORO, at MISSOURI state line; 36°29′N 91°32′W. In agricultural area. Has national fish hatchery. Named for spring (N end of town), one of largest in U.S. Mammoth Spring State Park, which feeds SPRING RIVER; site of resort, power plant. Manufacturing (catfish processing; walking canes).

Mamonovo (mah-MO-nuh-vuh), city (2005 population 7,230), W KALININGRAD oblast, W European Russia, on the Kaliningrad Bay of the BALTIC SEA, approximately 2 mi/3.2 km N of the Polish border, on road and railroad, 30 mi/48 km SW of KALININGRAD; 54°28′N 19°56′E. Elevation 154 ft/46 m. Fish canning. Founded in 1301. Until 1945, in German EAST PRUSSIA and called Heiligenbeil.

Mamonovshchina (mah-MO-nuhf-shchee-nah), settlement, SE NOVGOROD oblast, NW European Russia, in the POLA RIVER basin, near highway, 12 mi/19 km ENE, and under administrative jurisdiction, of MAREVO; 57°21′N 32°34′E. Elevation 669 ft/203 m. Sawmilling, lumbering.

Mamontovka (MAH-muhn-tuhf-kah), town (2006 population 8,795), central MOSCOW oblast, central European Russia, 3 mi/5 km S of (and administratively subordinate to) PUSHKINO; 55°59′N 37°49′E. Elevation 505 ft/153 m. Road construction machinery.

Mamontovo (MAH-muhn-tuh-vuh), town, S KHANTY-MANSI AUTONOMOUS OKRUG, TYUMEN oblast, W central SIBERIA, RUSSIA, near railroad, 43 mi/69 km SW of SURGUT; 60°40′N 72°50′E. Elevation 187 ft/56 m. In oil and gas field.

Mamontovo (MAH-muhn-tuh-vuh), village (2005 population 9,280), central ALTAI TERRITORY, S central SIBERIA, RUSSIA, on road, 50 mi/80 km WNW of ALEYSK; 52°43′N 81°37′E. Elevation 662 ft/201 m. Dairy processing.

Mamora (mah-MO-ruh), forest, NW MOROCCO, between SALÉ (SW) and KENITRA (NW), and extending inland 43 mi/69 km to Oued BETH stream (W). Morocco's largest cork-oak forest. Commercial cork production. Heavily grazed by sheep and sanctuary to many migratory birds.

Mamoré (mah-mor-AI), province, BENI department, NE BOLIVIA, ⊙ SAN JOAQUÍN; 13°00′S 64°55′W.

Mamoré (mah-mor-AI), canton, CARRASCO province, COCHABAMBA department, central BOLIVIA, 28 mi/45 km E of TODOS SANTOS, on the ICHILO RIVER (SANTA CRUZ department border), and 9 mi/15 km N of PUERTO VILLARROEL, in the Pojo Yungas; 16°42′S 64°48′W. Elevation 9,150 ft/2,789 m. Agriculture (potatoes, yucca, bananas, corn, rice, wheat, rye, sweet potatoes, soy, cacao, coffee); cattle for dairy products and meat.

Mamoré River (muh-moo-RAI), c.600 mi/965 km long, BOLIVIA; formed where SANTA CRUZ, BENI, and COCHABAMBA departments meet, by tributaries rising in the ANDES Mountains and plains of central Bolivia; 10°23′S 65°23′W; flows N across Beni department, past TRINIDAD, to the BRAZIL border. After forming part of the Bolivia-Brazil border, the Mamoré joins with the BENI RIVER to form the MADEIRA RIVER. With the Río Grande, its chief tributary, the Mamoré flows c.1,100 mi/1,770 km through the Bolivian lowlands (where it is navigable) and the Cordillera ORIENTAL.

Mamou, administrative region, S central GUINEA; ⊙ MAMOU. Bordered N by Labé administrative region, E by Faranah administrative region (part of N portion of border formed by BAFING RIVER), S by SIERRA LEONE, and W by Kindia administrative region (N portion of border formed by Kakrima River and central portion of border formed by KONKOURÉ RIVER). Téné River in E central, Konkouré River originates in SW (on border with Kindia administrative region), Kokoulo River in W central, Kakrima River in NW. Part of FOUTA DJALLON massif extends into the region. Mount Kavendou (elevation 4,662 ft/1,421 m) in W central and Mount Kadiondola (elevation 3,589 ft/1,094 m) in SE. Towns include DALABA, KONKOURÉ, and PITA. The railroad between CONAKRY and KANKAN town runs W-E through the S central part of the region. A main road runs roughly parallel to the railroad, going through Mamou and TIMBO towns; another main road runs NW-SE, coming from Labé town to the N (the road originates near the ATLANTIC OCEAN in THE GAMBIA and runs through SENEGAL and N Boké administrative region before reaching Labé administrative region) and going through Pita, Dalaba, Mamou, and Ouré Kaba towns, before entering Faranah administrative region in the SE. Includes the prefectures of Pita in the NNW, Dalaba in N central, and Mamou in S central. All of Mamou administrative region is in Moyenne-Guinée geographic region.

Mamou, prefecture, Mamou administrative region, S central GUINEA, in Moyenne-Guinée geographic region; ⊙ MAMOU. Bordered NWN by Pita prefecture, NNE by Tougué prefecture, NE tip by Dinguiraye prefecture, E by Dabola prefecture, ESE by Faranah prefecture, S by SIERRA LEONE, and W by Kindia prefecture. KONKOURÉ RIVER originates in W and flows N (forming N portion of the border with Kindia prefecture). FOUTA DJALLON massif extends into the prefecture. Mount Kadiondola (elevation 3,589 ft/1,094 m) in center of the prefecture. Towns include Kégnéko, KONKOURÉ, Mamou, Ouré Kaba, and TIMBO. The major railroad between CONAKRY (SW Guinea) and KANKAN town (E Guinea) travels SW-NE through the center of the prefecture, including Konkouré, Mamou, and Kégnéko towns. Main roads branch out of Mamou town connecting it to DALABA, PITA, and LABÉ towns to N; DABOLA, KOUROUSSA, and Kankan towns to E; Ouré Kaba, FARANAH, KISSIDOUGOU, and GUÉKÉDOU towns to SE; and KINDIA town and Conakry to SW.

Mamou (mah-MOO), town, ⊙ Mamou prefecture and Mamou administrative region, W central GUINEA, in Moyenne-Guinée geographic region, on railroad, and 125 mi/201 km ENE of CONAKRY; 10°15′N 12°00′W. Trading, cattle-raising, and agricultural center. Produces bananas, rice, peanuts, subsistence crops, rubber, indigo, beeswax, honey. Gold deposits nearby, bauxite to S. Roman Catholic and Protestant churches; mosque. Meteorological station.

Mamou (MAH-moo), town (2000 population 3,566), EVANGELINE parish, S central LOUISIANA, 8 mi/13 km ESE of VILLE PLATTE; 30°38′N 92°25′W. In cotton- and rice-producing area; light manufacturing. Oil field nearby.

Mamoudzou, city (□ 42 sq mi/109.2 sq km), MAYOTTE island, ⊙ MAYOTTE territory (France), COMOROS Islands, MOZAMBIQUE CHANNEL, INDIAN OCEAN, on NE coast of island, near mouth of Kaoueni River; 12°46′S 45°14′E. Fish; livestock; ylang-ylang, coconuts, coffee, vanilla; manufacturing (ylang-ylang oil). Ferry to DZAOUDZI, on PAMANDZI island to E; Pamandzi Airport, on Pamandzi island, 4 mi/6.4 km to SE. Mount Mtsapéré to W. Capital was moved here from Dzaoudzi in 1962.

Mampikony (mahm-pee-KOON), town, MAHAJANGA province, NW MADAGASCAR, 90 mi/145 km ESE of MAHAJANGA; 16°05′S 47°38′E. Market center; rice, cotton, tobacco; cattle. Airfield.

Mampong (mahm-PAWNG), town, ASHANTI region, GHANA, 30 mi/48 km NNE of KUMASI, on the African Plains; 07°04′N 01°24′W. Road and trade center. Cacao, kola nuts, hardwood, rubber. Teacher training college; agricultural experiment station; trade training center (opened 1949). Mampong Scarp, a section of Kwah plateau, to S.

Mampong, town, EASTERN REGION, GHANA, in AKWAPIM HILLS, 25 mi/40 km NNE of ACCRA; 05°55′N

MAMRE

00°08′W. Road junction. Cacao, palm oil and kernels, cassava.

Mamre, agricultural town, WESTERN CAPE province, SOUTH AFRICA, near Atlantic coast, 30 mi/48 km N of CAPE TOWN, 5 mi/8 km N of new industrial town ATLANTIS; 33°31′S 18°29′E. Military post established here 1697; later site of Moravian mission beginning 1808.

Mamry, Lake (MAHM-ree), German *Mauer*, second-largest (□ 40 sq mi/104 sq km) of Masurian Lakes, NE POLAND, between GIŻYCKO (S) and WĘGORZEWO (N), 35 mi/56 km S of CHERNYAKHOVSK (KALININGRAD oblast, RUSSIA); 12 mi/19 km long N-S, up to 8 mi/12.9 km wide. Irregular in shape. Drained (N) by ANGERAPP RIVER S terminus of MASURIAN CANAL. In German EAST PRUSSIA until 1945.

Mam Soul (MAM SOL), Gaelic *Mam Sodhail*, mountain (3,871 ft/1,180 m), HIGHLAND, N Scotland, 15 mi/24 km E of Dornie; 57°16′N 05°07′W.

Mamuil-Malal Pass (mah-moo-EEL—mah-LAHL) (4,100 ft/1,250 m–4,500 ft/1,372 m), in the ANDES MOUNTAINS, on ARGENTINA-CHILE border, at NE foot of LANÍN VOLCANO; 39°35′S 71°32′W.

Mamuju (mah-MOO-joo), town, ☉ Mamuju district, Sulawesi Seletan province, INDONESIA, 216 mi/348 km NW of UJUNG PANDANG.

Mamulique Pass (mah-moo-LEE-ke) (2,280 ft/695 m), in N outliers of SIERRA MADRE ORIENTAL, NUEVO LEÓN, N MEXICO, on INTER-AMERICAN HIGHWAY (85) and 40 mi/64 km NNE of MONTERREY.

Mamuras (mah-MOOR-uhs), new town, central ALBANIA, 17 mi/27 km N or TIRANË; 41°34′N 19°41′E. Railroad center just S of Laç. Agricultural processing. Also spelled Mamurras.

Mamurogawa (mah-moo-RO-gah-wah), town, Mogami county, YAMAGATA prefecture, N HONSHU, NE JAPAN, 43 mi/70 km N of YAMAGATA city; 38°51′N 140°15′E.

Mamykovo (mah-MI-kuh-vuh), village, S TATARSTAN Republic, E European Russia, on road, 17 mi/27 km NNW of NURLAT; 54°38′N 50°37′E. Elevation 298 ft/90 m. In agricultural area (sunflowers, grains).

Man (MAHN), town, ☉ Dix-Huit Montagnes region, W central CÔTE D'IVOIRE, at the foot of the Toura Mountains; 07°24′N 07°33′W. Administrative and commercial center for a region whose agricultural products include coffee, manioc, palm oil, rice, and cassava. Iron ore and tin are mined nearby. There is an airport.

Man, town (2006 population 712), LOGAN county, SW WEST VIRGINIA, 10 mi/16 km SW of LOGAN; 37°44′N 81°52′W. Railroad junction to E. Bituminous-coal area. Manufacturing (mining machine parts). R. D. Bailey Lake reservoir and Wildlife Management Area to SE. Incorporated 1918.

Mana (mah-NAH), town, NW FRENCH GUIANA, near mouth of MANA RIVER, on ATLANTIC coast, and 110 mi/177 km WNW of CAYENNE; 05°40′N 53°47′W.

Manaar, SRI LANKA and INDIA: see MANNAR.

Manabí (mah-nah-BEE), province (□ 7,289 sq mi/18,879 sq km; 2001 population 1,186,025), W ECUADOR, on the PACIFIC OCEAN, traversed by the equator; ☉ PORTOVIEJO. Apart from low Andean ridges (E), it consists of densely forested lowlands drained by DAULE, PORTOVIEJO, and CHONE rivers, among others. Has tropical climate, with rains December-June, when the climate is most trying. Its fertile soil yields coffee, cacao, tropical fruits, oil palm, rice, sugarcane, cotton, bananas, and forest products (*toquilla* straw, tagua nuts, balsa wood, and other fine tropical woods); livestock. Food processing, paper-prod. manufacturing, leather working. The province is the leading producer of Panama hats, which are exported to the U.S. BAHÍA DE CARÁQUEZ and MANTA are its seaports. Portoviejo, JIPIJAPA, MONTECRISTI, and CHONE are other centers. Site of the Proyecto Poza

Honda that provides a reservoir for drinking water and irrigation.

Manabique, Punta (mah-nee-BEE-kai, POON-tah), low wooded headland of GUATEMALA, on CARIBBEAN SEA; separates Bay of AMATIQUE (SW) and Gulf of HONDURAS (NE); 15°58′N 88°37′W. Also called Cabo de Tres Puntas.

Mana Camp, ZIMBABWE: see MANA POOLS NATIONAL PARK.

Manacapuru (MAH-nah-kah-poo-ROO), city, (2007 population 82,309), E central AMAZONAS state, BRAZIL, on left bank of the AMAZON River, and 40 mi/64 km WSW of MANAUS; 03°09′S 60°38′W. Rubber, Brazil nuts, hardwood.

Manacas (mah-NAH-kahs), town, VILLA CLARA province, central CUBA, on Central Highway, on railroad, and 27 mi/43 km NW of SANTA CLARA; 22°42′N 80°21′W. Sugarcane, tobacco, fruit; cattle. George Washington sugar mill 2 mi/3 km SE.

Manachanalloor, INDIA: see MANNACHCHANALLUR.

Manacle Point, ENGLAND: see SAINT KEVERNE.

Manacor (mah-nah-KOR), city, MAJORCA, BALEARIC ISLANDS, SPAIN, on railroad, and 30 mi/48 km E of PALMA; 39°34′N 03°12′E. Located on island's central plain, in agricultural region (cereals, grapes, almonds). Wine making, oil milling, flour milling, tanning; furniture and artificial-pearl manufacturing. Tourism. Has notable secular and religious buildings, among them the palace of kings of Majorca and fine parochial church. Nearby at Porto Cristo (7 mi/11.3 km E) are the Cuevas del Drach (Dragon's Cave), a favorite tourist site.

Manadhir, SAUDI ARABIA: see ABHA.

Manado (mah-NAH-do), city (2000 population 382,451), ☉ Sulawesi Utara province, on the NE coast of SULAWESI, INDONESIA; 01°29′N 124°51′E. Trade center and seaport on an inlet of the CELEBES SEA; exports include copra, coffee, spices, sugarcane, and lumber. Seat of University of North and Central Sulawesi; extension facility of the Islamic University of Indonesia. Samratulangi Airport. Also spelled Menado.

Manafwa, administrative district, EASTERN region, E UGANDA, on KENYA border (to E). As of Uganda's division into eighty districts, borders TORORO (S), MBALE (W), and BUDUDA (N) districts. Primarily agricultural area. Formed in 2005 from E and SE portions of former MBALE district created in 2000-2001 (Bududa district formed from NE portion in 2006 and W portion became current Mbale district).

Manage (mah-NAHZH), commune (2006 population 22,324), Charleroi district, HAINAUT province, S central BELGIUM, 12 mi/19 km WNW of CHARLEROI; 50°30′N 04°13′E. Manufacturing.

Managua, department (□ 1,330 sq mi/3,458 sq km), SW NICARAGUA; ☉ MANAGUA; 12°00′N 86°25′W. Includes coastal plain and E section of LAKE MANAGUA basin, separated by coastal range rising to 3,000 ft/914 m. Agriculture includes coffee (in coastal hills), corn, beans, plantains, cotton. Forests in N highlands are largely denuded. Livestock raising in Lake Managua basin. Manufacturing at Managua city. Served by INTER-AMERICAN HIGHWAY. TIPITAPA (thermal baths) and MASACHAPA (on PACIFIC coast) are tourist resorts. PUERTO SANDINO (formerly Puerto Somoza) is Pacific port, developed in 1940s.

Managua (mah-NAH-gwah), city (2005 population 908,892) and township, W NICARAGUA, ☉ and largest city of Nicaragua, on the S shore of LAKE MANAGUA; 12°00′N 86°25′W. Commercial and industrial center of the country. Situated on the INTER-AMERICAN HIGHWAY, the city is the hub of Nicaragua's railroads. Augusto César Sandino International Airport to E. Made permanent capital in 1855 to end the bitter feud between GRANADA and LEÓN. During periods of dis-

order (1912–1925 and 1926–1933) it was occupied by U.S. marines. Generally hot and sultry. A fairly constant wind blows from nearby Lake Managua, notable for the same marine phenomena as Lake Nicaragua and flanked by the smoking volcano MOMOTOMBO. Many residences and farms have been established on the cooler heights rising in the S outskirts of the city. Damaged by earthquake and fire in 1931 and by fire in 1936. On December 23, 1972, it was almost completely destroyed in an earthquake that took more than 10,000 lives. Because reconstruction was decentralized, the old city center went largely unrestored. The city suffered further damage in 1978–1979 during conflicts between government troops and the Sandinistas.

Managua (mah-NAH-gwah), town, LA HABANA province, W CUBA, at N foot of Managua hills, part of Alturas de Bejucal, 15 mi/24 km SSE of HAVANA; 22°57′N 82°18′W. Sugarcane; livestock.

Managua, Lake (mah-NAH-gwah), second-largest lake (□ 390 sq mi/1,014 sq km) of NICARAGUA, in LEÓN and MANAGUA departments, NW of LAKE NICARAGUA; 38 mi/61 km long, 16 mi/26 km wide, 65 ft/20 m deep; 12°20′N 86°20′W. Elevation 120 ft/37 m. Fisheries; alligator hunting; shallow-draught navigation. Now badly polluted from urban and industrial waste. MANAGUA is on SE shore. Other ports are MATEARE, MOMOTOMBO, San Francisco del Carnicero, TIPITAPA. Drains via TIPITAPA RIVER (SE) into Lake Nicaragua. Its native name is Xolotlán.

Manahawkin (ma-nuh-HUH-kin), village (2000 population 2,004), OCEAN county, E NEW JERSEY, near ATLANTIC coast, 18 mi/29 km S of TOMS RIVER village; 39°42′N 74°15′W. Manahawkin Bay is E; link in New Jersey section of INTRACOASTAL WATERWAY, which enters from BARNEGAT BAY (N) and continues S into LITTLE EGG HARBOR. Bay is crossed SE of MANAHAWKIN by highway bridge to LONG BEACH island, barrier between bay and the Atlantic.

Manaia (MAN-ai-ah), township, S TARANAKI district, W NORTH ISLAND, NEW ZEALAND, on S Taranaki Ring Plain, and 35 mi/56 km S of NEW PLYMOUTH; 39°33′S 174°08′E. Major dairy plant.

Manaíra (MAH-na-EE-rah), city (2007 population 10,962), SW PARAÍBA state, BRAZIL, in Serra das Princesa, 14 mi/23 km W of Princessa Isabela; 07°42′S 38°10′W.

Manakambahiny (mah-nah-KAHM-bah-EEN), town, TOAMASINA province, E MADAGASCAR, 15 mi/24 km WSW of AMBATONDRAZAKA; 17°52′S 48°17′E. Rice processing.

Manakara (mah-nah-KAHR), town, FIANARANTSOA province, E MADAGASCAR, on INDIAN OCEAN and Canal des PANGALANES, 75 mi/121 km SE of FIANARANTSOA; 22°09′S 48°01′E. Cabotage port; railroad line to Fianarantsoa; ships coffee, bananas, litchis, rice. Airport.

Manakha (muh-NUHK-uh), town, W YEMEN, on the plateau near its W fringes, 55 mi/89 km ENE of HODEIDA, and near the main road to SANA; 15°04′N 43°44′E. Elevation 7,500 ft/2,286 m. Market town of coffee-growing district; fruit and vegetables. Also spelled Menakha or Menakhah.

Manakin-Sabot, unincorporated village, GOOCHLAND county, central VIRGINIA, 14 mi/23 km WNW of RICHMOND, near JAMES RIVER; 37°36′N 77°44′W. Manufacturing (stone processing, millwork); agriculture (grain, tobacco, soybeans; cattle); limestone.

Manakundur (mah-nah-koon-door), town, KARIMNAGAR district, ANDHRA PRADESH state, S INDIA, 4 mi/6.4 km SE of KARIMNAGAR. Rice, millet, cotton.

Manalapan (ma-nuh-LA-puhn), township, MONMOUTH county, NE NEW JERSEY, 5 mi/8 km E of HIGHTSTOWN; 40°16′N 74°20′W. Incorporated 1848. Former farm community, now rapidly growing suburb.

Area is shown by the symbol □, and capital city or county seat by ☉.

Manali (muh-nah-lee), town, KULLU district, HIMACHAL PRADESH state, N INDIA, 50 mi/80 km E of Dharmsala, in Kullu valley; 32°16′N 77°10′E. Wheat, barley, rice, fruit (apples, pears, apricots). Important tourist resort and hill station. Medicinal hot springs, known as Manikaran, 2 mi/3.2 km NE.

Manam, volcanic island (□ 32 sq mi/83.2 sq km), MADANG province, PAPUA NEW GUINEA, SW PACIFIC OCEAN, 9 mi/14.5 km NE of NEW GUINEA; 6 mi/9.7 km long. Active crater (4,265 ft/1,300 m). Sometimes called VULCAN.

Manamadurai (muh-nah-MAH-doo-rei), town, PASUMPAN MUTHURAMALINGA THEVAR district, TAMIL NADU state, S INDIA, on VAIGAI RIVER, and 27 mi/43 km SE of MADURAI; 09°42′N 78°29′E. Railroad junction. Tile manufacturing. Also spelled Manamadura.

Manama (muh-NA-muh), city (□ 10 sq mi/26 sq km; 2001 population 143,025), ⊙ BAHRAIN, on the PERSIAN GULF; 26°13′N 50°35′E. Commercial center and largest city on PERSIAN GULF. Has light industries and is a free port. A causeway links it with the island of Al Muharraq to the NE. Large oil refineries S of the city. Also called Al Manamah.

Manambolo River (mah-nahm-BOOL), c.150 mi/241 km long, W MADAGASCAR; rises in ANTANANARIVO province 22 mi/35 km NE of TSIROANOMANDIDY at 18°33′S 46°20′E; flows W into N TOLIARY province, past ANKAVANDRA to large delta 30 mi/48 km N of BELO SUR TSIRIBIHINA. Lower course parallels border between MAHAJANGA and TOLIARY provinces. Navigable by shallow-draught boats to Bekopaka (30 mi/48 km upriver, MAHAJANGA province). Spectacular gorge through Bemaraha Plateau, along border of Tsingy de Bemaraha Nature Reserve.

Mánamo, Caño (MAH-nah-mo, KAHN-yo), westernmost arm of ORINOCO RIVER delta, c.110 mi/177 km long, NE VENEZUELA; branches off S of COPORITO; flows along DELTA AMACURO-MONAGAS state border to Gulf of PARIA at PEDERNALES; 09°55′N 62°16′W.

Mananara (mah-nah-NAHR), town, TOAMASINA province, NE MADAGASCAR, on coast, 140 mi/225 km NNE of TOAMASINA; 16°10′E 49°46′E. Trading center; rice, coffee, cloves. Hospital. MANANARA-NORD BIOSPHERE RESERVE, with national parks in rain forest and offshore reefs, is located here.

Mananara-Nord Biosphere Reserve (mah-nah-NAHR), park (□ 93 sq mi/241.8 sq km), TOAMASINA province, E MADAGASCAR, 125 mi/201 km N of TOAMASINA; 16°20′S 49°45′E. Includes Verezanantsoro National Park (□ 89 sq mi/230 sq km; lowland rain forest; lemur) and Nosy Atafana National Park (□ 4 sq mi/10 sq km; island and reefs). UNESCO Biosphere Reserve since 1987.

Manang (muh-NUHNG), district, N central NEPAL, in GANDAKI zone; ⊙ CHAME.

Manang (muh-NUHNG), village, central NEPAL, on N side of ANNAPURNA mountain; 28°38′N 84°00′E. Elevation 11,000 ft/3,353 m. Remote village. Airport.

Mananjary (man-nahn-ZAHR), town, FIANARANTSOA province, E MADAGASCAR, on INDIAN OCEAN and Canal des PANGALANES, at mouth of Mananjary River, 85 mi/137 km ENE of FIANARANTSOA; 21°13′S 48°21′E. Highway terminus and trade center. Produces coffee, cloves, rice, cassava, bananas, peppers, litchis.

Manantali, town, FIRST REGION/KAYES, MALI, 102 mi/170 km SE of KAYES; 13°15′N 10°29′W. Site of important hydroelectric project on BAFING RIVER.

Manantavadi (muh-nuhn-tah-vah-dee), town, WAYANAD district, KERALA state, S INDIA, 40 mi/64 km NNE of KOZHIKODE. In densely forested section of the Wynaad yielding valuable timber (teak, blackwood). Extensive tea, rubber, pepper, and coffee estates nearby.

Manantiales (mah-nahn-tee-AH-les), town, N CORRIENTES province, ARGENTINA, 55 mi/89 km SE of CORRIENTES. Agriculture (cotton, tobacco, fruit, rice, sugarcane, peanuts): subtropical woods.

Manaoag (mah-NAH-wahg), town (2000 population 54,743), PANGASINAN province, central LUZON, PHILIPPINES, 10 mi/16 km E of DAGUPAN, near LINGAYEN GULF; 16°02′N 120°29′E. Agricultural center (sugarcane, rice, copra, corn); sugar mill.

Manáos, Brazil: see MANAUS.

Manapad, INDIA: see KULASEKHARAPATNAM.

Mana Pass (MAH-NAH), pass (18,000 ft/5,486 m) in S ZASKAR RANGE of KUMAON HIMALAYA, SW TIBET, SW CHINA, 40 mi/67 km SW of ZANDA, NNW of KAMET Mountain; 31°05′N 79°25′E. Source of left headstream of ALAKNANDA RIVER.

Manapire River (mah-nah-PEE-rai), c.150 mi/241 km long, GUÁRICO state, central VENEZUELA; rises N of VALLE DE LA PASCUA; flows S to the ORINOCO RIVER, 5 mi/8 km NE of CAICARA DEL ORINOCO; 07°43′N 66°06′W.

Manapla (mah-NAH-plah), town, NEGROS OCCIDENTAL province, NW NEGROS island, PHILIPPINES, on GUIMARAS STRAIT, 23 mi/37 km NNE of BACOLOD; 10°55′N 123°08′E. Sugar milling.

Mana Pools National Park (□ 849 sq mi/2,199 sq km), MASHONALAND WEST province, N ZIMBABWE, 175 mi/282 km NW of HARARE. Bounded N by ZAMBEZI River (ZAMBIA border), E by Chewore Safari Area, W by Hurungwe Safari Area. MANA CAMP, tourist facilities, in N. Watering holes formed by channels of Zambezi River and its tributaries. Habitat for elephant, kudu, impala, zebra, wildebeest, buffalo, flamingo, and kingfisher. LOWER ZAMBEZI NATIONAL PARK. (Zambia) to N.

Manapouri, Lake (man-oo-POOR-ee), lake (□ 56 sq mi/145.6 sq km), in FIORDLAND NATIONAL PARK, SW SOUTH ISLAND, NEW ZEALAND, 70 mi/113 km NW of INVERCARGILL; 12 mi/19 km long, 6 mi/10 km wide, one of country's deepest lakes, at 1,421 ft/433 m; 45°30′S 167°30′E. Source of WAIAU RIVER, SOUTHLAND district. Often acclaimed the most beautiful lake in New Zealand; site of largest single hydroelectric power station in New Zealand.

Manappadu, INDIA: see KULASEKHARAPATNAM.

Manapparai (muh-nah-PUH-rei), town, TIRUCHCHIRAPPALLI district, TAMIL NADU state, S INDIA, 22 mi/35 km SW of Tiruchchirappalli; 10°36′N 78°25′E. Livestock market; saltpeter extraction. Mica, gypsum, limestone deposits nearby.

Manaquiri (MAH-nah-kee-ree), city (2007 population 19,314), E central AMAZONAS state, BRAZIL, 47 mi/75 km SW of MANAUS, on Rio SOLIMÕES, across from MANACAPURU; 03°10′S 60°30′W.

Manar, SRI LANKA and INDIA: see MANNAR.

Manara, ISRAEL: see MENARA.

Mana River (mah-NAH), c.200 mi/322 km long, W FRENCH GUIANA; rises at S foot of the CHAÎNE GRANITIQUE; flows N through tropical forests to the ATLANTIC OCEAN, 5 mi/8 km below MANA. Navigable for small craft c.30 mi/48 km upstream.

Mana River (MAH-nah), 330 mi/531 km long, S KRASNOYARSK TERRITORY, S central Siberian Russia; rises in the EASTERN SAYAN MOUNTAINS; flows NW, through wooded area, to the YENISEY RIVER, 15 mi/24 km W of KRASNOYARSK.

Manas (MAH-NAH-SUH), town and oasis, ⊙ Manas county, central XINJIANG UYGUR AUTONOMOUS REGION, NW CHINA, on the MANAS RIVER, in the JUNGGAR basin, 80 mi/129 km NW of URUMQI; 44°18′N 86°13′E. Center of a large mechanized-farm area. Wheat, millet, sugar beets, melons, and cotton are grown. Since 1952 an extensive irrigation project, directed by the Chinese army, has reclaimed much acreage for cultivation. Oil deposits are in the area. Manufacturing includes food processing, tobacco industry, manufacturing of cotton and wool textiles, and electric power generation. The name sometimes appears as Ma-na-ssu, or Manasi.

Manas, village, CHÜY region, KYRGYZSTAN, 16 mi/26 km NW of BISHKEK; 43°00′N 74°27′E. Poultry and cattle farming. Major international airport (1993) just N. Established 1974.

Manasa (muhn-AH-sah), town (2001 population 22,622), NIMACH district, NW MADHYA PRADESH state, central INDIA, 28 mi/45 km N of MANDSAUR; 24°29′N 75°09′E. Cotton, millet; sugar milling, handicraft blanket making.

Manasarowar Lake, CHINA: see MAPAM YUMCO.

Manasbal Lake (mahn-uhs-bahl), in N VALE OF KASHMIR, JAMMU AND KASHMIR state, N INDIA, in W central KASHMIR region, 13 mi/21 km NW of SRINAGAR; 3 mi/4.8 km long, 1 mi/1.6 km wide, c.50 ft/15 m deep. Connected by channel with right bank of JHELUM River; canal, with extensive terraces at E end, joins it with right tributary (SIND RIVER) of the Jhelum. Lake abounds in lotus; has ancient Hindu temple ruins (partially submerged); visited by houseboats. Limestone deposits nearby.

Manas Hu (MAH-NAH-SUH HOO), salt lake in NW XINJIANG UYGUR Autonomous Region, NW CHINA, 130 mi/209 km SE of TACHENG; 45°45′N 85°55′E. Receives MANAS RIVER (S).

Manasi, CHINA: see MANAS.

Manaslu (muh-NAHS-loo), peak (26,781 ft/8,163 m), in central NEPAL HIMALAYA, N NEPAL, 38 mi/61 km N of GORKHA; 28°33′N 84°34′E.

Manasquan (MA-nuhs-kwahn), resort borough (2006 population 6,199), MONMOUTH county, E NEW JERSEY, on coast, at N mouth of MANASQUAN River, and 15 mi/24 km SE of FREEHOLD; 40°06′N 74°02′W. Fishing; thriving business community. Incorporated 1887.

Manasquan River (MA-nuhs-kwahn), c.30 mi/48 km long, E NEW JERSEY; rises S of FREEHOLD; flows generally SE to the ATLANTIC OCEAN at MANASQUAN. Manasquan Inlet, at mouth of river, is N entrance of INTRACOASTAL WATERWAY, which continues S, through the Point Pleasant Canal, to head of BARNEGAT bay.

Manas River (MAH-NAH-SUH), c.220 mi/354 km long, in CHINA (TIBET), BHUTAN, and INDIA; rises in SE Tibet, N of CONA (Tsona); flows S and SSW through ASSAM HIMALAYAS and BHUTAN, bifurcating just inside ASSAM (INDIA); right stream (Manas proper) continues SSW to the BRAHMAPUTRA River opposite GOALPARA. Left stream (MORA MANAS RIVER) flows S and W, rejoining the Manas 10 mi/16 km. NNE of Goalpara. Chief tributary, Tongsa River (right).

Manassa, town (2000 population 1,042), CONEJOS county, S COLORADO, near CONEJOS RIVER, just E of SAN JUAN MOUNTAINS, and 20 mi/32 km S of ALAMOSA; 37°10′N 105°56′W. Elevation 7,683 ft/2,342 m. Manufacturing (jewelry). Trading point in SAN LUIS VALLEY. Jack Dempsey born here; museum

Manassas (muh-NA-suhs), city (2000 population 100), TATTNALL county, E central GEORGIA, 7 mi/11.3 km W of CLAXTON; 32°10′N 82°01′W.

Manassas (muh-NA-suhs), independent city (□ 10 sq mi/26 sq km; 2006 population 36,638), ⊙ surrounding PRINCE WILLIAM county, NE VIRGINIA, separate from surrounding Prince William county; 38°45′N 77°29′W. Manufacturing (building materials, medical equipment, food, machinery, electronic equipment; printing and publishing, steel fabrication, bindery services). Agricultural area to W and NW (grain, soybeans; livestock; dairying). Manassas has become a growing residential town with retail shopping centers and added housing; its development has been spurred by the expansion of WASHINGTON, D.C., suburbs, further into N Virginia. Key Civil War railroad junction; battles of BULL RUN fought nearby; sites are locally marked. Northern Virginia Community College (Manassas Campus) to N. George Mason University Prince William Campus to W. Manassas Museum.

Cross-references are shown in SMALL CAPITALS. The pronunciation guide is shown on page xix. The sources of population figures are shown on page xvii.

Manassas National Battlefield Park to NW. Manassas Regional Airport (Davis Field) to SW. Incorporated 1873, rechartered 1938.

Manassas (muh-NA-suhs), National Battlefield Park (□ 8 sq mi/20.8 sq km), PRINCE WILLIAM and FAIRFAX counties, N VIRGINIA, 5 mi/8 km NNW of MANASSAS; 38°49′N 77°31′W. Site of Civil War First (July 21, 1861) and Second (August 28–30, 1862) battles of Manassas (Bull Run). Authorized 1940.

Manassas Gap (muh-NA-suhs GAP), (c.950 ft/290 m), N VIRGINIA, on border between WARREN and FAUQUIER counties, lowest pass in BLUE RIDGE MOUNTAINS, 8 mi/13 km E of FRONT ROYAL. Railroad and Interstate highway Route 66 pass through. LINDEN village, just W, in Warren county, was formerly called Manassas Gap. APPALACHIAN TRAIL crosses gap N-S.

Manassas Park (muh-NA-suhs PAHRK), independent city (2006 population 11,642), N VIRGINIA, residential suburb, 2 mi/3 km NE of MANASSAS, near Bull Run creek, separate from surrounding PRINCE WILLIAM county and adjoining city of Manassas; 38°46′N 77°27′W. Manassas National Battlefield Park to NW.

Ma-na-ssu, CHINA: see MANAS.

Manastirishte (mah-nah-STIR-eesh-te), village, MONTANA oblast, HAIREDIN obshtina, BULGARIA; 43°35′N 23°39′E.

Manasuva Hydroelectric Project, electricity scheme on the Nanuka and Wailoa rivers, Nandrau Plateau, central Viti Levu, FIJI, SW PACIFIC OCEAN. A 270-ft-/82-m-high dam forms a lake 10 mi/16 km long to provide water through a 3.3-mi/5.3-km tunnel which drops at 45° to 420 MW-generating turbines at the Wailoa Power Station. Completed in 1985 at a cost of U.S. $165 million.

Manatee (MAN-uh-tee), county (□ 892 sq mi/2,319.2 sq km; 2006 population 313,298), W central FLORIDA, on GULF OF MEXICO and TAMPA BAY; ⊙ BRADENTON; 27°28′N 82°21′W. Level and rolling terrain, drained by MANATEE and MYAKKA rivers; has scattered lakes, part of Sarasota Bay, and small offshore islands, including Anna Maria Key. Farming and citrus fruit-growing area, with dairying, poultry raising, some fishing, and lumbering. Formed 1855.

Manatee (MAN-uh-tee), town, MANATEE county, W central FLORIDA, fishing port on TAMPA BAY, near its mouth on TAMPA BAY, and 3 mi/4.8 km E of BRADENTON; 27°29′N 82°39′W. Ships citrus fruit and vegetables.

Manatee Bay, small CARIBBEAN inlet, SAINT CATHERINE parish, S JAMAICA, at foot of the HELLSHIRE HILLS, 14 mi/23 km SW of KINGSTON; 17°50′N 77°00′W.

Manatee River (MAN-uh-tee), c.50 mi/80 km long, W central FLORIDA; rises in small lake in E MANATEE county; flows SW and W to TAMPA BAY near BRADENTON. Lower course dredged.

Manatí (mah-nah-TEE), town, Manatí municipio, ATLÁNTICO department, N COLOMBIA, in CARIBBEAN LOWLANDS, 40 mi/64 km SSW of BARRANQUILLA; 10°27′N 74°58′W. Agricultural center for cotton, corn, sugarcane, bananas, plantains.

Manatí (mah-nah-TEE), town (2006 population 48,996), N PUERTO RICO, 24 mi/39 km W of SAN JUAN. Pineapples; livestock; dairying (milk); manufacturing (chemical, plastic, and pharmaceutical products). Tourism at Mar Chiquita Beach, just N. Referred to as the "Athens of Puerto Rico." In the KARST region.

Manatí (mah-nah-TEE), sugar-mill village, LAS TUNAS province, E CUBA, near Puerto Manatí Bay (6 mi/10 km long, 3 mi/5 km wide) of the ATLANTIC OCEAN, on railroad, and 25 mi/40 km N of LAS TUNAS city; 21°19′N 76°55′W.

Manatí River (mah-nah-TEE) or **Río Grande de Manatí**, c. 40 mi/64 km long, central and N PUERTO RICO; rises in Cordillera Central just N of BARRANQUITAS; flows NW, past CIALES, to the ATLANTIC OCEAN 4 mi/6.4 km NW of MANATÍ.

Manatsuru (mah-NAHTS-roo), town, Ashigarashimo county, KANAGAWA prefecture, E central HONSHU, E central JAPAN, on SW shore of SAGAMI Bay, 34 mi/55 km S of YOKOHAMA; 35°09′N 139°08′E. Komatsu stone (used in construction). Kisen Ship Festival held here.

Manatuto (mah-nah-TOO-to), town, ⊙ MANATUTO district (□ 1,168 sq mi/3,036.8 sq km), EAST TIMOR, in central TIMOR, on WETAR STRAIT, 30 mi/48 km E of DILI; 08°30′S 126°01′E. In rice-growing area; ceramics.

Manaus (MAHN-ous), city (2007 population 1,612,475), ⊙ AMAZONAS state, NW BRAZIL, on the RIO NEGRO, on highway N to RORAIMA and then through VENEZUELA to the CARIBBEAN; 03°05′S 60°00′W. Chief commercial and cultural center of the upper AMAZON region and an important river port, with floating docks that can accommodate oceangoing vessels. Only major city in a c.600-mi/1000-km radius. Founded in 1669, Manaus grew slowly until the late 19th century, when the wild-rubber boom brought prosperity and short-lived splendor. In recent years, renewed interest in the Amazon basin and the discovery of oil nearby and the growth of the Amazon highway network brought new importance to Manaus. Now the seat of several organizations dealing with Amazonian problems, a free port and trade zone. Has an international airport, and is country's second-busiest air cargo terminal (motorcycles, motor vehicles; customs warehouses). Largest industrial complex between SÃO PAULO and CARACAS (Venezuela). Manufacturing includes electronics, chemical products, and soap; distilling, ship construction. Exports Brazil nuts, rubber, jute, and rosewood oil. Has a cathedral, opera house, zoological and botanical gardens, and a regional museum. Also spelled Manáos.

Manavadar (muh-NAH-vuh-dahr), former WESTERN KATHIAWAR state of WESTERN INDIA STATES agency. Merged 1948 with SAURASHTRA, now part of JUNAGADH district, GUJARAT state. Sometimes spelled Manavdar.

Manavadar (muh-NAH-vuh-dahr), town, JUNAGADH district, GUJARAT state, W central INDIA, 20 mi/32 km W of JUNAGADH; 21°30′N 70°08′E. Markets cotton, millet, ghee; hand-loom weaving.

Manavgat, township, SW TURKEY, near MEDITERRANEAN SEA, 40 mi/64 km E of ANTALYA; 36°47′N 31°28′E. Chromium; wheat, sesame.

Manawa, town (2006 population 1,298), WAUPACA county, central WISCONSIN, on Little Wolf River (tributary of WOLF RIVER), and 32 mi/51 km E of STEVENS POINT; 44°27′N 88°55′W. In dairying and farming area. Manufacturing (marine accessories, mill products, transportation equipment).

Manawar (muhn-ah-WAHR), town, JAMMU AND KASHMIR, extreme N INDIA, 22 mi/35 km NNW of SIALKOT (PAKISTAN); 32°48′N 74°26′E. Wheat, bajra, corn, pulses. Largely destroyed in 1947, during India-Pakistan struggle for control.

Manawar (muhn-AH-wuhr), town (2001 population 25,460), S DHAR district, SW MADHYA PRADESH state, central INDIA, 28 mi/45 km SSW of DHAR, on tributary of NARMADA River; 22°14′N 75°05′E. Cotton, corn, millet; cotton ginning.

Manawatu (man-uh-wah-TOO), district (□ 1,013 sq mi/2,633.8 sq km), NEW ZEALAND, in MANAWATU-WANGANUI region; 40°20′S 175°20′E.

Manawatu (man-uh-wah-TOO), plain, SW NORTH ISLAND, NEW ZEALAND, extending W from main range to TASMAN SEA, a broad triangular lowland bounded E by the Tararua region, crossed by the MANAWATU RIVER (that gives its name), grading from inland down land and riverine terraces to alluvial flats and coastal dunes. Intensively farmed for dairying and sheep with strong crop component. Focussed increasingly on PALMERSTON NORTH, with N-S and E-W road and railroad links. Grades S into HOROWHENUA.

Manawatu River (man-uh-wah-TOO), 113 mi/182 km land, S NORTH ISLAND, NEW ZEALAND; rises E of TARARUA and RUAHINE RANGES; flows through narrow Manawatu Gorge SW, past PALMERSTON NORTH, to COOK STRAIT 60 mi/97 km NNE of WELLINGTON. Drains agricultural lowlands.

Manawatu-Wanganui (man-uh-wah-TOO–wang-uh-NOO-ee), region (□ 9,775 sq mi/25,415 sq km), SW NORTH ISLAND, NEW ZEALAND. As defined in 1989, consists mainly of districts formerly belonging to Wellington province or land district, but excludes WELLINGTON region itself, and its borders elsewhere have been adjusted with the regions of TARANAKI, Auckland-Waikato, and HAWKE'S BAY. The formulation of this new region appears designed to group, administratively, not only the rivers that give it its name but also all streams, from the Wanganui in the W to the Waikawa S of LEVIN, that flow to NE COOK STRAIT; this incorporates the Tararua district, E of the Main Range ridges to the E. There are unities, however, in the patterns of erosion-prone rivers, in traffic flow along Main Trunk railroad and road routes, and in the influence of Wanganui and especially PALMERSTON NORTH.

Manawoka, INDONESIA: see GORONG ISLANDS.

Manay, Lake, TURKEY: see SOGUT, LAKE.

Manba (MAHN-bah), town, Tano county, GUMMA prefecture, central HONSHU, N central JAPAN, 25 mi/40 km S of MAEBASHI; 36°06′N 138°55′E.

Man Canal, MYANMAR: see MAN RIVER.

Mancarroncito Island, NICARAGUA: see SOLENTINAME ISLANDS.

Mancarrón Island, NICARAGUA: see SOLENTINAME ISLANDS.

Mancelona (man-se-LO-nuh), village (2000 population 1,408), ANTRIM county, NW MICHIGAN, 29 mi/47 km NE of TRAVERSE CITY, adjoining ANTRIM; 44°53′N 85°03′W. In dairy and agricultural area (livestock; cherries, apples; dairy); manufacturing (transportation equipment, metal stampings, storage tanks, wood products). Schuss Mountain and Shanty Creek ski areas to NW. Incorporated in 1889.

Mancenille, Baie de, HISPANIOLA: see MANZANILLO BAY.

Mancetter (MAN-suh-tuh), village (2001 population 4,716), N WARWICKSHIRE, central ENGLAND, just SE of ATHERSTONE; 52°07′N 02°19′W. Remains of granite quarry. Has church dating from 13th century. Site of a Roman station, Manduessedum.

Manchac, Bayou (MAN-shak, BEI-yoo), partly navigable waterway once connected with the MISSISSIPPI RIVER, c.19 mi/31 km long, SE LOUISIANA; begins S of BATON ROUGE; flows E, entering AMITE RIVER 9 mi/14 km S of DENHAM SPRINGS; 30°20′N 90°53′W. Also known as Bayou Iberville.

Manchac, Pass (MAN-shak), navigable waterway (shallow-draft vessels), c. 6 mi/10 km long, SE LOUISIANA, connecting Lake PONTCHARTRAIN (E) and Lake MAUREPAS (W), c. 27 mi/43 km NW of NEW ORLEANS; 30°17′N 90°18′W. Manchac State Wildlife Area on S side of channel.

Mancha, La (MAHN-chah, lah), historic region of central SPAIN, in historic CASTILE-LA MANCHA, comprising CIUDAD REAL province and part of TOLEDO, ALBACETE, and CUENCA provinces. This high, barren plateau, dotted with windmills, was made famous as the scene of most of the adventures of Don Quixote de la Mancha in the novel by Cervantes.

Manchao, Sierra de, ARGENTINA: see AMBATO, SIERRA DE.

Mancha Real (MAHN-chah rai-AHL), town, JAÉN province, S SPAIN, 10 mi/16 km ENE of JAÉN; 37°47′N 03°37′W. Manufacturing of linen and woolen cloth, plaster; olive-oil processing, flour milling. Cereals, fruit, livestock in area.

Manchas, Las (MAHN-chahs, lahs), village, PALMA island, CANARY ISLANDS, SPAIN, 8 mi/12.9 km SW of SANTA CRUZ DE LA PALMA. Bananas, tomatoes, tobacco, fruit.

Area is shown by the symbol □, and capital city or county seat by ⊙.

Manchaug, MASSACHUSETTS: see SUTTON.

Manchazh (mahn-CHAHSH), village, SW SVER-DLOVSK oblast, in the foothills of the W URAL Mountains, extreme E European Russia, on road, 17 mi/27 km SE of KRASNOUFIMSK; 56°28′N 58°08′E. Elevation 918 ft/279 m. In agricultural area (oats, rye; dairy cattle).

Manche (MAWNSH), department (□ 2,293 sq mi/ 5,961.8 sq km), NW FRANCE, in NORMANDY, on the ENGLISH CHANNEL; ☉ SAINT-LÔ; 49°00′N 01°10′W. COTENTIN PENINSULA occupies N half of department, while S is traversed by NORMANDY HILLS; CHERBOURG, at N tip of peninsula, is a major port city. The famed MONT-SAINT-MICHEL is just off the S coast, and the British CHANNEL ISLANDS are off to the W. Manche is largely agricultural. Numerous hedgerows break up the fields and apple orchards. Early fruits and vegetables are grown near N coast and GRANVILLE for export, especially to the U.K. Dairying is important in CARENTAN area; horses and cattle are bred and raised throughout department. Although industry is secondary, there are textile, small metalworking and food-processing plants, and shipbuilding at Cherbourg. There are two nuclear power facilities W and SW of Cherbourg. Fishing and tourism are important sectors of the economy. Forms part of the administrative region of BASSE-NORMANDIE.

Manche, La, Europe see ENGLISH CHANNEL.

Mancheng (MAHN-CHUNG), town, ☉ Mancheng county, central W HEBEI province, CHINA, 10 mi/16 km NW of BAODING, near BEIJING-WUHAN railroad; 38°57′N 115°20′E. Cotton, wheat, kaoliang, corn, oilseeds. Textiles; engineering.

Manchenki, UKRAINE: see MANCHENKY.

Manchenky (mahn-chen-KI), (Russian *Manchenki*), town, N KHARKIV oblast, UKRAINE, on railroad, 17 mi/ 27 km W of KHARKIV city center, and 5 mi/8 km WNW of LYUBOTYN; 49°59′N 35°52′E. Elevation 583 ft/ 177 m. Fowl processing. Established in the 17th century; town since 1957.

Manchester (MAN-ches-tuh), city and district (□ 497 sq mi/1,292.2 sq km; 2001 population 392,819), ☉ GREATER MANCHESTER, NW ENGLAND, on the IRWELL, Medlock, Irk, and Tib rivers; 53°30′N 02°13′W. Remains the center of the most densely populated area of England, despite the tremendous amount of outmigration between 1961 and 1981. New towns and complexes have been built since the 1970s. Long the leading textile city (its textile industry dates back to the 14th century) of England, the late 20th century has seen a sharp drop in Manchester's textile-based economy. Other industries, especially chemical and pharmaceutical production, financial services, and research industries, have moved to fill the void. Also the center of printing and publishing in N England. Manchester Airport (international) to S of city. The first application of steam to machinery for spinning cotton was made here in 1789, and a terminus of the first English passenger railroad (to LIVERPOOL) was constructed here by George Stephenson in 1830. The Manchester Ship Canal, opened in 1894, gave the city access to the sea. After World War I the artificial-silk industry tended to balance losses in the cotton market. A Celtic settlement is believed to have existed on the site. The Romans called the town MANCUNIUM, and there are remains of their occupation. First charter was granted in 1301. The Peterloo massacre, which occurred here in 1819, played a prominent role in liberal reform movements. Center of the Manchester school of economics and the Anti-Corn Law League, led by Richard Cobden and John Bright. The influential liberal daily, the Manchester *Guardian*, was founded in 1821. The borough has several libraries, including the John Rylands Library (founded 1899) and the Chetham Library (founded 1653), one of Europe's first free public libraries. Manchester has been an important center for scientific research. John Dalton, Lord Rutherford, and Niels Bohr, among others,

did significant work in nuclear physics here. At Jodrell Bank, nearby, is a large radio telescope, once the world's largest. Manchester has several art galleries and a symphony orchestra of international repute, the Hallé Orchestra, founded in 1858 by Sir Charles Hallé. The first municipal airport in Britain was established here in 1929. Robert Peel, the statesman, and authors Thomas de Quincey and Anthony Burgess born here. Seat of the Victoria University of Manchester (1851; formerly Owens College), the University of Manchester, Manchester Metro University, and the University of Salford. Included in the city are the districts of Ardwick, Blackley, Cheetham, Chorlton-on-Medlock (with its Royal Infirmary and University), Clayton, Crumpsall, Gorton, Levenshulme, Miles Platting, Moston, Newton Heath, Openshaw, Withington, and Wythenshawe.

Manchester, city, ☉ Delaware co., E IOWA, on Maquoketa River, and 38 mi/61 km W of Dubuque; 42°29′N 91°27′W. Agr. trade and processing center (dairy products); manufacturing (paper goods., fabricated metal products, electronic products, chemicals, machinery). A U.S. fish hatchery to SE. Settled 1850, incorporated 1886.

Manchester, city (2006 population 109,497), HILLSBOROUGH county, S NEW HAMPSHIRE, on both sides of the MERRIMACK RIVER; 42°58′N 71°26′W. Also drained by PISCATAQUOG RIVER and Cohas Brook. Largest city in NEW HAMPSHIRE. Manufacturing (computer equipment and accessories, electronic products and equipment, machinery, lobster holding systems, foods and beverages, clothing, hats, concrete products, industrial brushes, fabricated metal products, plastics, building materials, medical supplies, paper products, chemicals, sterilization systems, tool and die, machinery). The Amoskeag Falls on the MERRIMACK provided power for the first textile mills. In 1838 textile interests founded the city and established a huge textile-manufacturing company. Until the depression of the 1930s and the moving of much of the textile industry to the South, Manchester was heavily dependent on this industry. Seat of St. Anselm College and the Currier Gallery of Art. Also University of New Hampshire at Manchester, Notre Dame College, New Hampshire Technical College (2 year), Institute of Arts and Sciences, Hesser College, New Hampshire College (to N). John Stark lived and is buried here. Municipal Airport to S. MASSABESIC LAKE is on E border. McIntyre Ski Area to NE. Settled 1722, incorporated as a city 1846.

Manchester, city (2006 population 9,671), ☉ COFFEE county, central TENNESSEE, near DUCK RIVER, 12 mi/19 km NE of TULLAHOMA; 35°29′N 86°05′W. Manufacturing. Settled 1836; incorporated 1905. Old Stone Fort State Archaeological Park is nearby.

Manchester, town, HARTFORD county, central CONNECTICUT; 41°46′N 72°31′W. Pre-revolutionary sawmills and paper mills. Was also known for its production of grandfather clocks and cheney silks. Manufacturing today includes transportation equipment, tools, and dairy and paper products. A major retail shopping area has developed along the Interstate 84 corridor here. Seat of Manchester Community Technical College. Settled c.1672, incorporated 1823.

Manchester, town (2000 population 3,988), MERIWETHER and TALBOT counties, W GEORGIA, 33 mi/53 km NE of COLUMBUS, on N face of PINE MOUNTAIN (S extremity of the APPALACHIAN MOUNTAINS); 32°51′N 84°37′W. In agricultural and livestock area. Manufacturing includes building materials, chemicals, consumer goods; various light manufacturing FRANKLIN D. ROOSEVELT STATE PARK, WARM SPRINGS, and Callaway Gardens nearby. Settled 1905; incorporated 1909.

Manchester, town (2000 population 1,738), ☉ CLAY county, SE KENTUCKY, 18 mi/29 km E of LONDON, on Goose Creek, in CUMBERLAND foothills, in Daniel Boone National Forest; 37°08′N 83°46′W. In coal-

mining, timber, and agricultural (corn, tobacco, hay; livestock) area; manufacturing (lumber, tool and die, consumer goods, building materials; coal processing; window treatments). Has airport. Seat of Eastern Kentucky University–Manchester Campus and Oneida Baptist Institute. Beech Creek Wildlife Management Area to E. Established 1798; incorporated 1932.

Manchester, town, KENNEBEC county, S MAINE, just W of HALLOWELL and AUGUSTA, and on LAKE COBBOSSEECONTEE; 44°19′N 69°51′W.

Manchester, town (2000 population 3,329), CARROLL county, N MARYLAND, 29 mi/47 km NW of BALTIMORE; 39°40′N 76°53′W. Clothing and meat factories. Near site of Native Susquehannock town, whose members were massacred in battles with Seneca tribes and Colonel John Washington in 1675. Subsequent raids on European settlers contributed to Bacon's Rebellion. Laid out in 18th century by Captain Richard Richards, and named after his home city of MANCHESTER (ENGLAND).

Manchester, suburban town, ESSEX county, NE MASSACHUSETTS, on MASSACHUSETTS BAY, and 8 mi/12.9 km NE of SALEM; 42°34′N 70°46′W. Includes village of West Manchester. Former upper-class resort. Known for Singing Beach. Offshore is NORMAN'S WOE Rock. Settled 1626, incorporated 1645.

Manchester, town (2000 population 2,160), WASHTE-NAW county, SE MICHIGAN, 18 mi/29 km SW of ANN ARBOR; 42°08′N 84°02′W. In diversified agricultural area; feed milling. Manufacturing (machinery, fabricated metal products, transportation equipment). Incorporated in 1867.

Manchester, town, SAINT LOUIS county, E MISSOURI, residential suburb, 17 mi/27 km W of downtown ST. LOUIS, and 2 mi/3.2 km E of BALLWIN, near MERAMEC RIVER. Castlewood State Park to SW. Manufacturing (fabricated metal products, tool and die; printing).

Manchester, township, OCEAN county, E central NEW JERSEY, 2 mi/3.2 km SW of LAKEHURST; 39°57′N 74°22′W. Incorporated 1865.

Manchester, resort town (2006 population 716), including Manchester village, ☉ BENNINGTON co. (along with Bennington); 43°09′N 73°04′W. SW VERMONT, on BATTEN KILL, and 20 mi/32 km N of Bennington, between the TACONIC (W) and GREEN mountains; also includes Manchester Center and Manchester Depot villages. Seat of Burr and Burton Seminary (1829). Major retail center (factory outlet shops); manufacturing (consumer goods; printing); lumber, marble. Summer, winter resort; skiing at nearby BROMLEY MOUNTAIN (3,816 ft.; also known as Big Bromley), ski touring at Hildere. Southern Vermont Art Center; Mt. Equinox is just W. Settled c.1764, laid out 1784. Named for MANCHESTER, MASSACHUSETTS.

Manchester, unincorporated town, KITSAP county, W WASHINGTON, 10 mi/16 km WSW of downtown SEATTLE, on Kitsap Peninsula, on PUGET SOUND. Fishing. Manchester State Park and U.S. Naval Station to N; Blake Island State Park to SE.

Manchester, unincorporated village, MENDOCINO county, NW CALIFORNIA, near the PACIFIC OCEAN, 25 mi/40 km WSW of UKIAH. Farming (cattle; apples, pears, grapes); dairying; fish, urchins. Manchester State Beach at Point Arena, to SW.

Manchester, village (2000 population 354), SCOTT county, W central ILLINOIS, 9 mi/14.5 km SE of WINCHESTER; 39°32′N 90°19′W. In agricultural area.

Manchester, village (2000 population 102), DICKINSON county, N central KANSAS, 13 mi/21 km NNW of ABILENE; 39°05′N 97°19′W. Wheat; cattle.

Manchester, village (2000 population 81), FREEBORN county, S MINNESOTA, 7 mi/11.3 km NW of ALBERT LEA; 43°43′N 93°27′W. Dairying; manufacturing (fabricated metal products).

Manchester, village (□ 1 sq mi/2.6 sq km; 2006 population 1,439), ONTARIO county, W central NEW YORK, on Canandaigua Outlet (the stream draining CANANDAI-

GUA LAKE), and 23 mi/37 km SE of ROCHESTER; 42°58′N 77°13′W. Freight distribution point; light manufacturing; agricultural products. Incorporated 1892.

Manchester (MAN-ches-tuhr), village (□ 1 sq mi/2.6 sq km; 2006 population 2,118), ADAMS county, S OHIO, on the OHIO RIVER, 35 mi/56 km W of PORTSMOUTH; 38°42′N 83°36′W. Manufacturing. One of Ohio's earliest towns, founded 1791.

Manchester, village (2006 population 95), GRANT county, N OKLAHOMA, 40 mi/64 km N of ENID, near KANSAS state line; 36°59′N 98°02′W. In agricultural area (grain; livestock); manufacturing (machinery).

Manchester, borough (2006 population 2,467), YORK county, S PENNSYLVANIA, 6 mi/9.7 km N of YORK; 40°03′N 76°43′W. Agriculture (dairying; livestock; poultry; grain, soybeans, apples); manufacturing (plastic products, foods, fabricated metal products, transportation equipment, electronic products; machining). Laid out c.1815, incorporated c.1869.

Manchester, parish (2001 population 185,801), MIDDLESEX county, W central and S JAMAICA; ⊙ MANDEVILLE; 18°15′N–17°51′N 77°38′W–77°21′W. Bordered E by CLARENDON, N by TRELAWNY, and W by SAINT ELIZABETH parishes. This predominantly mountainous region of no rivers and scarce water supply has a salubrious climate of 55°F/13°C–88°F/31°C. Its leading settlements, such as Mandeville, CHRISTIANA, and WILLIAMSFIELD, are popular resorts. Tropical products (grapefruit, oranges, annatto, ginger, pimento) are grown widely. Traversed by Jamaica railroad.

Manchhar Lake (MUHN-chuhr), in DADU district, W SIND province, SE PAKISTAN, 8 mi/12.9 km W of SEHWAN; c.10 mi/16 km long, 2 mi/3.2 km–6 mi/9.7 km wide (considerably larger when flooded); 26°25′N 67°39′E. Formed by seasonal drainage at outliers (W and S) of KIRTHAR RANGE. Fish.

Manchinabad, PAKISTAN: see MINCHINABAD.

Manching (MAN-khing), town, UPPER BAVARIA, BAVARIA, GERMANY, on the PAAR RIVER, 5 mi/8 km SE of INGOLSTADT; 48°43′N 11°29′E. Manufacturing of airplanes. Hops, sugar beets. Air base. Town first mentioned 844.

Manchioneal, town, PORTLAND parish, NE JAMAICA, on JAMAICA CHANNEL, 14 mi/23 km SE of PORT ANTONIO; 18°03′N 76°17′W. Bananas, coconuts.

Manchita (mahn-CHEE-tah), town, BADAJOZ province, W SPAIN, 19 mi/31 km ESE of MÉRIDA; 38°49′N 06°01′W. Wheat, olives, grapes.

Mancho (MAHN-cho), peak (9,091 ft/2,771 m) in the RILA MOUNTAINS, SW BULGARIA; 42°09′N 23°36′E.

Manchoukuo, CHINA: see MANCHUKUO.

Manchouli, CHINA: see MANZHOULI.

Manchukuo (mahn-CHOO-kwaw), Chinese *Manzhouguo*, (mahn-JO-gwaw) former country, comprising the three provinces of the NORTHEAST, CHINA. The Japanese invaded the Northeast in 1931 and founded Manzhouguo in 1932, despite international protests in the League of Nations (Japan simply withdrew) and elsewhere. CHANGCHUN, the capital, was renamed Xinjing [Chinese=new capital]. Henry Pu Yi, last emperor of the Qing (Ching) dynasty (1644–1911) of China, ruled as regent and emperor. Manzhouguo, ostensibly an independent Manchu state, was actually a Japanese puppet-state. Of the major countries only JAPAN, ITALY, and GERMANY extended diplomatic recognition; few foreigners were allowed access within its borders. The Japanese military kept strict control of the administration and fought a continuing guerrilla war with native resistance groups. To develop Manzhouguo as a war base, the Japanese expanded industry and railroad system. This excellent infrastructure made the area eagerly contested by the Nationalists and Communists after World War II, when Chinese sovereignty was reasserted. Also spelled Manchoukuo.

Manchuli, CHINA: see MANZHOULI.

Manchuria, CHINA: see NORTHEAST.

Manciano (mahn-CHYAH-no), town, GROSSETO province, TUSCANY, central ITALY, 24 mi/39 km SE of GROSSETO; 42°35′N 11°31′E. In grape- and olive-growing region. Has 14th-century citadel. Travertine quarries, deposits of antimony and mercury nearby. Restructured thermal springs at Saturnia.

Mâncio Lima (MAHN-see-o LEE-mah), town (2007 population 13,753), extreme W ACRE state, BRAZIL, 12 mi/20 km W of CRUZEIRO DO SUL, near terminus of BR 364 Highway; 07°33′S 72°50′W. Extractive activities and expanding agriculture and livestock raising.

Manco Kapac (MAHN-ko kah-PAHK), province (□ 142 sq mi/369.2 sq km), LA PAZ department, W BOLIVIA, on S Lake TITICACA, 22 mi/35 km W of TIQUINA; ⊙ COPACABANA; 16°10′S 69°05′W. A mountainous (12,602 ft/3,841 m) peninsula. Climate is sunny with most rainfall in December and January. Nights are very cold with wind from the lake. Site of a religious pilgrimage for centuries. Founded June 1951 by Governor Hugh Ballivián.

Máncora (MAHN-ko-rah), village, TALARA province, PIURA region, NW PERU; minor port on the PACIFIC OCEAN, on PAN-AMERICAN HIGHWAY, 35 mi/56 km NNE of TALARA; 04°06′S 81°01′W. Fisheries. Airport.

Mancor del Valle (mahng-KOR del VAH-lyai), town, MAJORCA, BALEARIC ISLANDS, SPAIN, 17 mi/27 km NE of PALMA; 39°45′N 02°52′E. Olives, fruit, grapes; timber; livestock.

Mancos (MAHN-kos), town, YUNGAY province, ANCASH region, W central PERU, in the Callejón de HUAYLAS, on SANTA RIVER, and 3 mi/5 km SE of YUNGAY, on road to CARHUÁZ; 09°11′S 77°43′W. Corn, cereals; livestock; eggs.

Mancos, town (2000 population 1,119), MONTEZUMA county, SW COLORADO, in LA PLATA MOUNTAINS, NE Montezuma county, on MANCOS RIVER, near mouth of Chicken Creek, 23 mi/37 km W of DURANGO; 37°21′N 108°17′W. Elevation 7,030 ft/2,143 m. Cattle, sheep-shipping and outfitting point. Wheat, oats, barley, beans. Manufacturing (wood fiber). Coal and silver mines in vicinity. MESA VERDE NATIONAL PARK to SW; Ute Mountain Indian Reservation to SW; Mancos State Park and Lake to N; San Juan National Forest to NE.

Mancos River, c.60 mi/97 km long, in COLORADO and NEW MEXICO; formed in LA PLATA MOUNTAINS, NE MONTEZUMA county, SW Colorado, past E tip of MESA VERDE NATIONAL PARK, by confluence of WEST MANCOS (20 mi/32 km long) and East Mancos (15 mi/24 km long) rivers, 2 mi/3.2 km ENE of MANCOS; Middle Mancos River (c.10 mi/16 km long) joins East Mancos 3 mi/4.8 km ENE of Mancos. All three rise in NE Montezuma county. Mancos River flows intermittently SW and WSW, through part of Ute Mountain Indian Reservation, to SAN JUAN RIVER just across state line in extreme NW New Mexico, in Navajo Indian Reservation, 20 mi/32 km NW of SHIPROCK (New Mexico). Jackson Gulch Reservoir (181 ft/55 m high, 1,900 ft/579 m long; completed 1948), on diversion canal W of West Mancos River; canal continues into Chicken Creek, branch of West Mancos, 4 mi/6.4 km N of Mancos.

Mancunium, ENGLAND: see MANCHESTER.

Manda (man-DAH), town, IRINGA region, S TANZANIA, minor port 80 mi/129 km S of NJOMBE, on Lake NYASA, at mouth of RUHUHU RIVER; 10°27′S 34°33′E. Sheep, goats; subsistence crops; timber. Coal deposits along Ruhuhu River to E.

Mandaguaçu (MAHN-dah-gwah-soo), city (2007 population 18,226), NW PARANÁ state, BRAZIL, 11 mi/18 km NW of MARINGÁ; 23°20′S 52°05′W. Coffee, corn, cotton, rice; livestock.

Mandaguari (MAHN-dah-gwah-ree), city (2007 population 31,900), N PARANÁ state, BRAZIL, 17 mi/27 km SE of MARINGÁ, on railroad; 23°32′S 51°42′W. Coffee, cotton, corn, rice, potatoes; livestock.

Manda Island (MAHN-dah), island, COAST province, KENYA, off E coast, in INDIAN OCEAN, just E of Lamu Island, 40 mi/64 km NE of KIPINI; 5 mi/8 km long; 5 mi/8 km wide. Copra center; fisheries.

Mandaitivu, island in NORTHERN PROVINCE, SRI LANKA, between W JAFFNA LAGOON (N) and PALK STRAIT (S), opposite JAFFNA (ferry); 3 mi/4.8 km long, 1 mi/1.6 km wide; 09°36′N 80°00′E. Rice and coconut- and palmyra-palm plantations.

Mandal (MUHN-duhl), town, AHMADABAD district, GUJARAT state, W central INDIA, 45 mi/72 km WNW of AHMADABAD. Agriculture market (cotton, wheat, millet); handicraft cloth weaving, cotton ginning.

Mandal (MAHN-dahl), town, VEST-AGDER county, extreme S NORWAY, on the SKAGERRAK strait; 58°02′N 07°27′E. A renowned seaport in the seventeenth century, it is now a small trading and shipping center, known for its remarkable beach and for its eighteenth-century patrician houses.

Mandal (MAHN-dahl), agricultural village, SELENGE province, N MONGOLIA, on the HARAA GOL (river), and 55 mi/90 km N of ULAANBAATAR; 48°30′N 106°45′E. Fireproof clay deposits nearby.

Mandala Peak, mountain peak (15,420 ft/4,700 m), Indonesian *Pucak Mandala*, MAOKE MOUNTAINS, IRIAN JAYA province, INDONESIA; 118 mi/190 km ESE of TRIKORA PEAK; 04°44′S 140°20′E.

Mandalay (MAHN-dah-LAI), administrative division (□ 12,494 sq mi/32,484.4 sq km) of MYANMAR; ⊙ MANDALAY. Between middle AYEYARWADY RIVER (W) and SHAN PLATEAU (E). Includes twenty-six townships. Alluvial plain along AYEYARWADY RIVER; plateau N of PEGU YOMA (with MOUNT POPA); hilly region along border of SHAN STATE (E). Main rivers are AYEYARWADY, MYITNGE, SAMON, ZAWGYI. In dry zone (annual rainfall, 26 in/66 cm–30 in/76 cm) irrigated by canals and tanks; mostly agriculture (rice, sesame, cotton, sugarcane) in plain and plateau; teak forests in hills. Served by railroads and Ayeyarwady steamers.

Mandalay (MAHN-dah-LAI), city, MANDALAY division, central MYANMAR, on the AYEYARWADY RIVER; 22°00′N 96°05′E. The second-largest city in MYANMAR, it is the terminus of the main railroad line from YANGON and the starting point of branch lines to LASHIO and MYITKYINA, an important focus of trade between MYANMAR and CHINA. As a city, it dates from c.1850. It was the capital of the Burman kingdom, replacing AMARAPURA, from 1860 to 1885, when it was annexed to British BURMA. A center of Burmese Buddhism, the city is noted for the Arakan pagoda, which is built around an ancient shrine. The group of sacred buildings known as the Seven Hundred and Thirty Pagodas was erected in the reign of King Mindon (1853–1878). Heavily damaged in World War II.

Mandalay Hill (MAHN-dah-LAI) (775 ft/236 m), at the NE edge of MANDALAY city, MYANMAR. Has 1,279 stairs leading to the pagoda at the summit. Site of Japanese stronghold stormed, at great loss, by Indian and British troops during World War II.

Mandalgovd or Mandalgovi (MAHN-dahl-GO-bee), town (2000 population 14,517), ⊙ DUNDGOVI province, central MONGOLIA, 200 mi/320 km SSW of ULAANBAATAR, on Ulaanbaatar-DALANDZADGAD highway; 45°45′N 106°16′E. Elevation 4,570 ft/1,393 m. Automotive repair shops. Formerly called Sharangad; also written Mandal Gobi.

Mandali (MUHN-del-ee), town, DIYALA province, E IRAQ, 5 mi/8 km from IRAN border, 75 mi/121 km ENE of BAGHDAD; 33°44′N 45°32′E. Fruit; livestock. Sometimes spelled Mendeli and Mendali.

Mandalselva (MAHN-dahls-EL-vah), c.45 mi/72 km long, river, VEST-AGDER county, S NORWAY; rises in lake near ÅSERAL; flows S to the SKAGERRAK strait at MANDAL. Salmon fishing. Several hydroelectric power plants on the river and its tributaries.

Mandaluyong, suburb (2000 population 278,474) of MANILA, NATIONAL CAPITAL REGION, S LUZON, PHILIPPINES, on N bank of PASIG RIVER, 3 mi/4.8 km from city center; 14°35′N 121°02′E. Sugar refining, pulp and paper milling, match manufacturing.

Mandalya, Gulf of, inlet of AEGEAN SEA in SW TURKEY, 45 mi/72 km W of MUGLA; 21 mi/34 km wide, 23 mi/37 km long. Also Gulluk Gulf.

Mandamados (mahn-dah-MAH-[th]os), town, on N LESBOS island, LESBOS prefecture, NORTH AEGEAN department, GREECE, 17 mi/27 km NW of MITILÍNI; 39°19′N 26°20′E. Olive oil, wine, wheat. Also spelled Mandamadhos.

Mandan, city (2006 population 17,449), ⊙ MORTON county, S NORTH DAKOTA, on the MISSOURI RIVER, at the mouth of HEART RIVER, opposite BISMARCK; 46°49′N 100°53′W. A railroad division point, it is the distributing center for a grain, livestock, and dairy region; manufacturing (printing and publishing; building materials, wood products). Has a large cattle market and food-processing plants. Industry includes iron, fabricated metal products, tile. Lewis and Clark wintered here (1804–1805) in the Mandan Native American villages. A state industrial school is in the city, and a U.S. agricultural experiment station is nearby. Fort Lincoln State Park to S. In Central time zone; Mountain/Central time zone border (Missouri River) skirts around Mandan and NE corner of Morton county. Founded in 1878 and named for the Indian tribe of the same name.

Mandapam (MUHN-duh-puhm), town, Ramnathapuram district, TAMIL NADU state, S INDIA, on PAMBAN CHANNEL, 85 mi/137 km SE of MADURAI; 09°17′N 79°07′E. Fishing center. Fisheries Research Institute. Religious center.

Mandapeta (MUHN-dah-pe-tah), town, EAST GODAVARI district, ANDHRA PRADESH state, S INDIA, in GODAVARI RIVER delta, 21 mi/34 km WSW of KAKINADA; 16°52′N 81°56′E. Rice milling; metalware; sugarcane, tobacco, coconuts, oilseeds.

Mandara Mountains (MAN-der-uh), North province, CAMEROON and along Nigerian border, extend c.120 mi/193 km from BENUE River (NW of GAROUA) to N of MOKOLO, rising to over 3,500 ft/1,067 m. Savanna vegetation. Millet, sorghum, and some cotton; intensive terraced agr. Tourism.

Mandara Mountains (mahn-DAH-rah), long, narrow granitic range (more than 6,000 ft/1,829 m), on border of NE NIGERIA (ADAMAWA state) and N CAMEROON. Overlooks Dikwa plains S of Lake CHAD.

Mandasa (MUHN-dah-sah), town, SRIKAKULAM district, ANDHRA PRADESH state, S INDIA, 85 mi/137 km NE of VIZIANAGARAM; 18°52′N 84°28′E. Trades in coir, copra, and in products (bamboo, tanning bark, lac) of forested hills (W); aerated water manufacturing.

Mandaue (mahn-DAH-wai), city (□ 5 sq mi/13 sq km; 2000 population 259,728), CEBU province, central Cebu island, PHILIPPINES, opposite MACTAN Island, 4 mi/6.4 km NE of CEBU; 10°20′N 123°56′E. Commercial and industrial center (brewery) in MetroCebu.

Mandawa (muhn-DAH-wah), town, JHUNJHUNUN district, NE RAJASTHAN state, NW INDIA, 16 mi/26 km WSW of Jhunjhunun; 28°03′N 75°09′E. Market center for livestock, millet, hides.

Mandawa (mahn-DAH-wah), village, LINDI region, SE TANZANIA, 35 mi/56 km WNW of LINDI, near Mbwemburu River; 09°52′S 39°11′E. Timber; livestock; grain. One of two villages so named in Lindi region.

Mandawa, village, LINDI region, SE TANZANIA, 50 mi/80 km NNW of LINDI, near INDIAN OCEAN; 09°21′S 39°27′E. One of two villages so named in Lindi region. Cashews, peanuts, bananas, copra; goats, sheep.

Mandawar (muhn-DAH-wuhr), town, BIJNOR district, N UTTAR PRADESH state, N central INDIA, 8 mi/12.9 km N of BIJNOR. Rice, wheat, gram, barley, sugarcane.

Visited by Chinese Buddhist pilgrim Hsüan-Tsang in 7th century A.D. Captured by Tamerlane in 1399.

Mandayona (mahn-dei-O-nah), town, GUADALAJARA province, central SPAIN, 32 mi/51 km NE of GUADALAJARA; 40°57′N 02°45′W. Cereals, grapes, vegetables. Paper milling.

Mandelbachtal (MAHN-duhl-bahkh-tahl), commune, SAARLAND, W GERMANY, near the BLIES RIVER, 10 mi/16 km ESE of SAARBRÜCKEN; 49°10′N 07°13′E. Airport of Saarbrücken is N. Food processing. Formed by unification of several smaller villages.

Mandello del Lario (mahn-DEL-lo del LAH-ree-o), town, COMO province, LOMBARDY, N ITALY, port on E shore of Lake of Lecco, 6 mi/10 km NNW of LECCO; 45°54′N 09°19′E. Silk and paper mills; manufacturing (motorcycles, bicycles, textile machinery, fabricated metals, chemicals).

Mandelo, Cape, Greece: see MANDILI, CAPE.

Mandera (mahn-DAI-rah), town (1999 population 30,433), NORTH EASTERN province, extreme NE KENYA, where Kenya, ETHIOPIA, and SOMALIA meet, on road, and 390 mi/628 km NE of ISIOLO; 03°57′N 41°52′E. Livestock raising. Airfield.

Mandera (man-DAI-rah), village, PWANI region, E TANZANIA, 70 mi/113 km NW of DAR ES SALAAM, on WAMI RIVER; 09°12′S 38°23′E. Corn, manioc, sweet potatoes, sisal; sheep, goats.

Manderson, village (2006 population 102), BIG HORN county, N WYOMING, on Big Horn River, just W of Big Horn Mountains, and 8 mi/12.9 km SSE of BASIN; 44°16′N 107°57′W. Elevation 3,890 ft/1,186 m. Agriculture (sugar beets, beans, alfalfa; cattle, sheep); manufacturing (utility poles). Medicine Lodge State Archaeological Site to E, with prehistoric glyphs (petroglyphs, pictographs).

Mandeville (MAN-di-vil), city, SAINT TAMMANY parish, SE LOUISIANA, on Lake PONTCHARTRAIN, 23 mi/37 km N of NEW ORLEANS; 30°22′N 90°05′W. Manufacturing (wood products, electronic equipment, cultured marble, caskets, meat products, signs). N end of 30-mi/48-km PONTCHARTRAIN Causeway from NEW ORLEANS is 1 mi/2 km W. Fontainebleau State Park and SAINT TAMMANY State Refuge to SE. Hosts Mandeville Seafood Festival. Founded c. 1830, incorporated 1840.

Mandeville, town, ⊙ MANCHESTER parish, central JAMAICA; 18°02′N 77°30′W. Elevation c.2,000 ft/610 m. An inland resort town known for its cool climate and quiet, "English-village" character.

Mandeville (MAN-duh-vil, mahnd-VEEL), village (□ 128 sq mi/332.8 sq km; 2006 population 2,027), LANAUDIÈRE region, QUEBEC, E CANADA, 21 mi/34 km from BERTHIERVILLE; 46°22′N 73°21′W.

Mandhata, INDIA: see GODARPURA.

Mandhili, Cape, Greece: see MANDILI, CAPE.

Mandi (MUHN-dee), district (□ 1,525 sq mi/3,965 sq km), central HIMACHAL PRADESH state, N INDIA; ⊙ MANDI.

Mandi (MUHN-dee), town, ⊙ MANDI district, N central HIMACHAL PRADESH state, N INDIA, on BEAS River, and 45 mi/72 km NNW of SHIMLA; 31°43′N 76°55′E. Trades in grain, wool, salt, timber; hand-loom weaving; handicrafts (brass ware, woodwork). Was capital of former princely state of Mandi of PUNJAB STATES; since 1948, merged with HIMACHAL PRADESH state.

Mandiana, prefecture, Kankan administrative region, E GUINEA, in Haute-Guinée geographic region; ⊙ MANDIANA. Bordered NW by Siguiri prefecture, N and E by MALI, SE tip by CÔTE D'IVOIRE, and S and W by Kankan prefecture. Fié River in N and Sankarani River in S and center of prefecture. Towns include Mandiana, Niantanina, Saladou, Sidikila, and Tindila. Main road runs through Mandiana town, connecting it to KANKAN town to SE and Mali (country) to NE; secondary road runs N-S through Mandiana town.

Mandiana, town, ⊙ Mandiana prefecture, Kankan administrative region, E GUINEA, in Haute-Guinée geo-

graphic region, on road to MALI, and 55 mi/89 km NE of KANKAN.

Mandiankui (mahn-JEE-ahn-kwee), town, FOURTH REGION/SÉGOU, MALI, 117 mi/195 km ESE of SÉGOU. Also spelled Mandiakui.

Mandi Bahauddin (MUHN-dee buh-HOU-deen), town, GUJRAT district, NE PUNJAB province, central PAKISTAN, 34 mi/55 km W of GUJRAT; 32°35′N 73°30′E. Sometimes called BAHAUDDIN.

Mandi Burewala, PAKISTAN: see BUREWALA.

Mandi Giddarbaha, INDIA: see GIDDARBAHA.

Mandi Guru Har Sahae, INDIA: see GURU HAR SAHAI.

Mandili, Cape (mahn-DEE-lee), S extremity of Évvia island, Évvia prefecture, CENTRAL GREECE department, on AEGEAN SEA, at SW entrance to KAFIREOS CHANNEL; 37°56′N 24°30′E. Formerly called Cape Mandelo. Also spelled Cape Mandhili.

Mandimba District, MOZAMBIQUE: see NIASSA.

Mandinga (mahn-DING-gah), village, SAN BLAS territory, E PANAMA, port on Río Cangandí, near SAN BLAS GULF of CARIBBEAN SEA, 55 mi/89 km ENE of COLÓN. Coastal trade. Airfield.

Mandini, town, KWAZULU-NATAL province, SOUTH AFRICA, 10 mi/16 km inland from INDIAN OCEAN coast, 30 mi/48 km NNE of STANGER, and 50 mi/80 km SW of RICHARD'S BAY. Residential area for workers of sugar industry on KwaZulu-Natal N coast. Railroad link between DURBAN and Richard's Bay.

Mandioré, Lake (mahn-dyo-RAI), on BOLIVIA (SANTA CRUZ department)-BRAZIL (Matto Grosso do Sul state) border, 55 mi/89 km NNE of PUERTO SUÁREZ (Bolivia); 13 mi/21 km long, 5 mi/8 km wide; 18°08′S 57°33′W. Connected with PARAGUÁ River by SE outlet.

Mandi Pattoki, PAKISTAN: see PATTOKI.

Mandirituba (MAHN-dee-ree-too-bah), city (2007 population 20,408), SE PARANÁ state, BRAZIL, 23 mi/37 km S of CURITIBA; 25°46′S 49°19′W. Potatoes, manioc; livestock.

Mandi Sadiqganj (MUHN-dee suhd-EEK-gunj), town, BAHAWALNAGAR district, BAHAWALPUR division, PUNJAB province, central PAKISTAN, 135 mi/217 km NE of BAHAWALPUR; 30°10′N 73°44′E. Sometimes called Mandi Sadiq.

Mandji (MAHN-jee), town, NGOUNIÉ province, W central GABON, 46 mi/75 km NW of MOUILA; 01°41′S 10°22′E.

Mandla (MAHND-lah), district (□ 3,370 sq mi/8,762 sq km; 2001 population 894,236), SE MADHYA PRADESH state, central INDIA, on DECCAN Plateau; ⊙ MANDLA. On hilly plateau of central SATPURA RANGE; drained by NARMADA RIVER and its numerous tributaries. Rice, wheat, oilseeds in river valleys. Hills are covered with dense sal, bamboo, teak, myrobalan forests; lac growing. Sawmilling, betel farming, manufacturing of sunn hemp products (mats, cordage), mainly at Mandla; livestock raising in E hills. Kanha National Park, primarily a tiger reserve but also containing several other rare birds and mammals, is main tourist attraction. DINDORI district formed out of E part of Mandla in 1998.

Mandla (MAHND-lah), town (2001 population 45,907), ⊙ MANDLA district, SE MADHYA PRADESH state, central INDIA, on NARMADA RIVER, 45 mi/72 km SE of JABALPUR; 22°36′N 80°23′E. Terminus of railroad spur from junction of NAINPUR (20 mi/32 km SW). Sawmilling, manufacturing of sunn hemp products (mats, cordage), betel farming; rice, wheat, oilseeds. Has ruins of 17th-century Gond fortress. Lac produced in nearby dense sal forests.

Mandø (MAHN-duh), island (□ 2 sq mi/5.2 sq km) of NORTH FRISIAN group, DENMARK, in NORTH SEA, 4 mi/6.4 km off SW JUTLAND; 55°16′N 08°34′E. Mandø By, town, is on W coast.

Mandoc, EQUATORIAL GUINEA: see MBINI RIVER.

Mandoto (mahn-DOOT), town, ANTANANARIVO province, in MOYEN-OUEST region of MADAGASCAR, 55 mi/89 km WNW of ANTSIRABE; 19°32′S 46°18′E.

Important cattle-trading center on Antsirabe-Morondava road. Zone of agricultural settlement; soybean cultivation. Gold mining nearby.

Mandoudhi (mahn-DOO–[th]ee), town, NW Évvia island, Évvia prefecture, CENTRAL GREECE department, 25 mi/40 km NNW of KHALKÍS; 38°48′N 23°29′E. Wheat; wines; livestock raising (sheep, goats); magnesite deposits (S). Also called Mantoudi or Mantudi. Formerly Mantoudion or Mandoudhion.

Mandoudhion, Greece: see MANDOUDHI.

Mandoul, administrative region, S CHAD; ⊙ KOUMRA. Borders LOGONE ORIENTAL (W), TANDJILE (N), and MOYEN-CHARI (E) administrative regions and CENTRAL AFRICAN REPUBLIC (S). Drained by Ouham River (BAHR SARA river). Major centers include Koumra and MOÏSSALA. Formed following a decree in October 2002 that reorganized Chads administrative divisions from twenty-eight departments to eighteen regions. This area made up the W portion of former MOYEN-CHARI prefecture.

Mandouri (mahn-DOO-ree), town ⊙ KPENDJAL prefecture, SAVANES region, NE corner of TOGO, on OTI RIVER, near BENIN border; 10°51′N 00°49′E. Cattle, sheep, and goats.

Mandra (MAHN–[th]rah), town, ATTICA prefecture, ATTICA department, E central GREECE, 14 mi/23 km WNW of ATHENS, on road to THEBES; 38°04′N 23°30′E. Wheat, vegetables; olive oil, wine; cattle and sheep raising.

Mandra Dam (MAHN-drah), E BULGARIA, on the BLACK SEA shore; 42°25′N 27°23′E.

Mandrael (muhn-DREIL), town, SAWAI MADHOPUR district, E RAJASTHAN state, NW INDIA, 18 mi/29 km SE of KARAULI; 26°18′N 77°14′E. Wheat, millet, gram. Sometimes spelled Mandrail or Mandril.

Mandra Lake (MAHN-drah), salt lake (□ 6 sq mi/15.6 sq km) in E BULGARIA, just SW of BURGAS. Connected with the Gulf of BURGAS by a 23-ft/7-m-deep channel. Receives SREDETSKA and FAKIISKArivers.

Mandra River, Bulgaria: see SREDETSKA RIVER.

Mandritsara (mahn-dree-TSAH-rah), town, MAHAJANGA province, N central MADAGASCAR, 170 mi/274 km E of MAHAJANGA; 15°49′S 48°49′E. Rice cultivation. Airfield.

Mandsaur (muhnd-SOR), district, (□ 2,125 sq mi/5,525 sq km; 2001 population 1,183,274), NW MADHYA PRADESH state, central INDIA; ⊙ MANDSAUR. Agriculture: wheat, sorghum, maize, gram; also is a large producer of opium. NE section of district separated in 1998 to form district of NIMACH.

Mandsaur (muhnd-SOR), town (2001 population 116,483), ⊙ MANDSAUR district, MADHYA PRADESH state, central INDIA, 105 mi/169 km NNW of INDORE; 24°04′N 75°04′E. Trade center (grain, cotton, opium, cloth fabrics); cotton ginning and milling, sugar milling, food processing, hand-loom weaving. Technical institute. Two monolithic pillars, 2 mi/3.2 km SE, mark spot where Yasodharman, king of MALWA, defeated White Huns in 6th century A.D. Nearby, Humayan won victory over Bahadur Shah of Gujarat in 1535. Treaty signed here in 1818 between British and Marathas. Site of Pashupatinath Temple, containing a five ton statue of Shiva, Hindu god of destruction. Formerly in Gwalior state.

Mandu (muhn-DOO), historic fort town (□ 12 sq mi/31.2 sq km) in VINDHYA RANGE (c.2,000 ft/610 m), DHAR district, SW MADHYA PRADESH state, central INDIA, 17 mi/27 km SSE of DHAR; c.4 mi/6.4 km long, 3 mi/4.8 km wide. Religious and archaeological site; wildlife sanctuary. In 14th and 15th centuries, noted capital of Muslim state of Maiwa. Has many remains of bathing tanks, palaces, and mosques in Afghan architectural style: Jami Masjid (built 1454); Hoshang Shah's tomb, which provided inspiration to builders of the TAJ MAHAL centuries later; and Roopmati's Pavillion, built by Prince Baz Bahadur for his consort, Rani Roopmati.

Visited by Akbar and Jahangir while under Moguls (late 16th and 17th centuries); passed in 1732 to Marathas; abandoned and fell into ruins soon after.

Manduel (mawn-DUL), E suburb (□ 10 sq mi/26 sq km) of NÎMES, GARD department, LANGUEDOC-ROUSSILLON region, S FRANCE; 43°49′N 04°28′E. In vineyard area.

Mandur (MUHN-DOOR), village, EASTERN PROVINCE, SRI LANKA, 16 mi/26 km S of BATTICALOA; 07°29′N 81°45′E. Ancient temple ruins; Hindu pilgrimage center.

Mandurah (MAN-doo-rah), town (2001 population 46,697), SW WESTERN AUSTRALIA state, W AUSTRALIA, 40 mi/64 km S of PERTH, and on INDIAN OCEAN, at mouth of MURRAY RIVER; 32°33′S 115°42′E. Commuter town; summer resort.

Manduri (MAHN-doo-ree), town (2007 population 8,651), S SÃO PAULO state, BRAZIL, 34 mi/55 km E of OURINHOS, on railroad; 23°01′S 49°19′W.

Mandvi (MAHND-vee), town, SURAT district, GUJARAT state, W central INDIA, on TAPI RIVER, and 30 mi/48 km E of SURAT. Market center for rice, cotton, millet; handicraft cloth weaving.

Mandvi, town, KACHCHH district, GUJARAT state, W central INDIA, port on GULF OF KACHCHH, 35 mi/56 km SSW of BHUJ. Trades in wheat, barley, salt, cotton fabrics, gypsum; cotton ginning, oilseed milling, match manufacturing; metalworks (handicraft silver products). Airport. Lighthouse 1 mi/1.6 km W.

Mandya (MUHND-yuh), district (□ 1,915 sq mi/4,979 sq km), KARNATAKA state, S INDIA; ⊙ MANDYA. On DECCAN PLATEAU (undulating tableland); drained mainly by KAVERI River, with Krishnaraja Sagara (reservoir) on SW border and Sivasamudram island (major hydroelectric works) on SE border. SE half of district (irrigated by IRWIN CANAL) is a major sugarcane area; silk growing in SE section. Other crops include millet, rice, tobacco, cotton. Dispersed chromite, asbestos, and feldspar mining. Hand-loom weaving of silk, woolen, and cotton. Chief towns are MALVALLI, Mandya, SHRIRANGAPATTANA. Created 1939 out of NE portion of MYSORE district.

Mandya (MUHND-yuh), city, ⊙ MANDYA district, KARNATAKA state, S INDIA, 22 mi/35 km NE of MYSORE; 12°33′N 76°54′E. Sugar-milling center; syrup, candy, alcohol; tobacco and vegetable-oil processing, hand-loom weaving, handicraft cloth dyeing. Asbestos works nearby.

Måne (MAW-nuh), river, 18 mi/29 km long, in TELEMARK county, S NORWAY; flows from MØSVATN Lake E to Tinnsjøen Lake. On it are RJUKAN Falls (983 ft/300 m), which supply power for the industry of Rjukan.

Manea (MAI-nee), agricultural village (2001 population 1,579), N CAMBRIDGESHIRE, E ENGLAND, 7 mi/11.3 km NW of ELY; 52°28′N 00°12′E.

Manele (MAH-NAI-lai), village, LANAI island, MAUI county, HAWAII, on Manele Bay, S coast, 6 mi/9.7 km S of LANAI CITY (connected by road). Tourism. Hulopoe Beach Park to W.

Manendragarh (muh-NAIN-drah-guhr), town (2001 population 30,755), KORIYA district, N CHHATTISGARH state, central INDIA, 65 mi/105 km W of AMBIKAPUR. Coal mining. Lac grown in surrounding dense sal forests (bamboo, khair).

Manerbio (mah-NER-byo), town, BRESCIA province, LOMBARDY, N ITALY, on MELLA River, and 13 mi/21 km SSW of BRESCIA; 45°21′N 10°08′E. In cereal-growing, livestock-raising region; silk mills, cheese factories; manufacturing fabricated metals, clothing.

Manermiut (mah-NER-myoot), abandoned settlement, Aasiaat (Egedesminde) commune, W GREENLAND, at W end of Sarqarllit Island (sahkh-KAHR-slit) (27 mi/43 km long, 2 mi/3.2 km–7 mi/11.3 km wide), on DAVIS STRAIT, on S side of mouth of DISKO BAY, 10 mi/16 km SW of Aasiaat (EGEDESMINDE); 68°35′N 53°04′W.

Maneromango (ma-nai-ro-MAN-go), village, PWANI region, E TANZANIA, 40 mi/64 km SW of DAR ES SALAAM; 08°12′S 38°49′E. Road junction. Cashews, bananas, sisal; poultry, goats, sheep.

Manerplaw, town, HPA-AN township, KAYIN STATE, MYANMAR. A fortified hilltop location which served as the headquarters of the Karen insurgent forces until its capture in January 1995. Home to both fugitive government officials elected in 1990 but prevented from military from assuming office and to activists from Burma's prodemocracy party, the National League for Democracy.

Manetin (MAH-nye-TYEEN), Czech Manětín, town, ZAPADOCESKY province, W BOHEMIA, CZECH REPUBLIC, 18 mi/29 km NNW of PLZEŇ; 49°59′N 13°14′E. Former Czech border bastion. Baroque castle.

Manevichi, UKRAINE: see MANEVYCHI.

Manevychi (mah-NE-vi-chee), (Russian Manevichi), (Polish Maniewicze), town, NE VOLYN' oblast, UKRAINE, 36 mi/58 km ENE of KOVEL'; 51°17′N 25°32′E. Elevation 620 ft/188 m. In woodland. Sawmilling, parquetry furniture manufacturing; peat bricketing; flax making; food processing. Established in 1892 during the construction of the Kovel'-Sarny railroad; town since 1940. Small Jewish community since the town's founding, numbering 500 in 1939; completely wiped out during World War II.

Manfalut (mahn-fah-LOOT), town, ASYUT province, central Upper EGYPT, on W bank of the NILE River, on railroad, and 18 mi/29 km NW of ASYUT; 27°19′N 30°58′E. Cotton ginning, wool spinning and weaving, pottery making, wood carving; cereals, dates, sugarcane.

Manfredonia (mahn-fre-DO-nyah), city (2001 population 57,704), FOGGIA province, APULIA, S ITALY, port on GULF OF MANFREDONIA, at S foot of GARGANO promontory, 23 mi/37 km NE of FOGGIA; 41°38′N 15°55′E. Railroad terminus. Fishing. Highly diversified secondary industrial center (leather goods, stearine, lye, cement), food processing. Archbishopric. Founded by King Manfred in 13th century; has medieval fortifications. Nearby is site of ancient Sipontum (abandoned in 13th century); its cathedral (consecrated 1117) remains.

Manfredonia, Gulf of, inlet of ADRIATIC Sea, S ITALY, S of GARGANO promontory. Extends c.30 mi/48 km N-S; c.15 mi/24 km wide. Receives CANDELARO and CARAPELLE rivers. Chief port, MANFREDONIA.

Manga (MAHN-guh), region, DIFFA province, SE NIGER, in W part of the LAKE CHAD basin; 14°00′N 12°00′E. Large expanse of savanna grasslands.

Manga (mahn-gah), city (2007 population 20,907), N MINAS GERAIS state, BRAZIL, on left bank of SÃO FRANCISCO River (navigable), near BAHIA state border, and 60 mi/97 km NNE of JANUÁRIA; 14°42′S 43°49′W. Cotton, rice, brandy, hides.

Mangai (mahng-GEI), village, BANDUNDU province, central CONGO, on left bank of KASAI RIVER, and 90 mi/145 km NNE of KIKWIT; 04°03′S 19°32′E. Elev. 912 ft/277 m. Center of trade, steamboat landing; palm-oil milling. Ipamu, Roman Catholic mission and residence of prefect apostolic, is 8 mi/12.9 km SSW.

Mangaia (mahng-EI-ah), composite volcanic and limestone island (□ 25 sq mi/65 sq km), COOK ISLANDS, S PACIFIC, 110 mi/177 km SE of RAROTONGA; c.6 mi/9.7 km in diameter; 21°55′S 157°55′W. Low volcanic hills are surrounded by swamps and karstic makatea, or plateau of coral limestone, forming upland in center of island. Exports fruits, copra. Oneroa is chief settlement.

Mangaizé (MAHN-gai-zai), town, TILLABÉRY province, NIGER, 75 mi/121 km N of NIAMEY.

Mangal (MAHN-guhl), former princely state of PUNJAB HILL STATES, INDIA. Since 1948, merged with HIMACHAL PRADESH state.

Mangala (man-GAH-lah), village, MOROGORO region, E central TANZANIA, 67 mi/108 km SSW of MOROGORO, near GREAT RUAHA RIVER (bridged); 07°40′S 37°02′E. SELOUS GAME RESERVE to SE. Corn, wheat, sisal; cattle, sheep, goats.

Mangalagiri (MUHN-guh-lah-gir-ree), town, GUNTUR district, ANDHRA PRADESH state, S INDIA, in KRISHNA RIVER delta, 12 mi/19 km NE of GUNTUR; 16°26′N 80°33′E. Cement works (limestone quarries nearby); rice and oilseed milling; tobacco. Annual Hindu temple fair on hill just N.

Mangaldai (MUHN-guhl-dei), town, ⊙ DARRANG district, NW ASSAM state, NE INDIA, in Brahmaputra valley, near the BRAHMAPUTRA RIVER, 49 mi/79 km WSW of TEZPUR; 26°26′N 92°02′E. Tea, rice, rape and mustard, sugarcane, jute; tea processing.

Mangaldan (mahng-AHL-dahn), town (2000 population 82,142), PANGASINAN province, central LUZON, PHILIPPINES, near LINGAYEN GULF, 5 mi/8 km ENE of DAGUPAN; 16°04′N 120°24′E. Rice, copra, corn.

Mangalia (mahn-GAH-lyah), ancient *Callatis* or *Kallatis*, town, CONSTANŢA county, SE ROMANIA, in DOBRUJA, on the BLACK SEA, and 27 mi/43 km S of CONSTANŢA; 43°48′N 28°35′E. Maritime port, railroad terminus, and popular health resort (beach, sanatorium, and sulphurous springs). Manufacturing (textiles, foodstuffs). Ancient Greek colony established in 504 B.C.E. was restored by the Romans (29 B.C.E.) and by Constantine the Great (C.E. 4th century). Mosque (1590), archaeology museum.

Mangalmé (mahng-gahl-MAI), town, NE GUÉRA administrative region, central CHAD, 75 mi/121 km ENE of MONGO; 12°21′N 19°37′E.

Mangalore (MUHN-guh-lor), city (2001 population 539,387), DAKSHINA KANNADA district, KARNATAKA state, S INDIA, on the ARABIAN SEA; 12°52′N 74°53′E. A port, it trades in spices, rice, fish, nuts, and cardamom. Manufacturing includes roofing tiles, pottery, brick kilns, and ships. Has an airport, major road and railroad connections, and a diesel power station. Was capital (thirteenth century) of the Alupa kingdom. In the late eighteenth century, Mangalore was an important shipbuilding center. The British occupied the city in 1799.

Mangalore (MAN-guh-lor), village, N central VICTORIA, SE AUSTRALIA, 7 mi/12 km N of SEYMOUR; 36°56′S 145°11′E.

Mangalsen (muhn-GUHL-sen), town, ⊙ ACHHAM district, W NEPAL; 29°09′N 81°16′E.

Mangalvedha (muhn-guhl-VAI-dah), town, SOLAPUR district, MAHARASHTRA state, W central INDIA, 32 mi/51 km WSW of SOLAPUR; 17°31′N 75°28′E. Trades in agricultural products (millet, wheat, cotton); handicraft cloth weaving. Sometimes spelled MANGLAWEDHE.

Mangan (muhn-guhn), town, ⊙ NORTH district, SIKKIM state, NE INDIA; 27°31′N 88°32′E.

Mangando (mahn-GAHN-do), town, MALANJE province, ANGOLA, near confluence of Uamba and Cambo rivers, 130 mi/209 km NE of MALANJE; 08°02′S 17°08′E. Market center.

Manganese (MAIN-guh-neez), village, CROW WING county, central MINNESOTA, 15 mi/24 km NE of BRAINERD, and 4 mi/6.4 km NW of CROSBY, surrounded by Crow Wing State Forest; 40°31′N 94°00′W. In lake and forest area. Iron mines nearby in CUYUNA IRON RANGE. Cole Lake to W.

Manganeses de la Lampreana (mahng-gah-NAI-ses dai lah lahm-prai-AH-nah), town, ZAMORA province, NW SPAIN, 17 mi/27 km NNE of ZAMORA; 41°45′N 05°42′W. Wine, cereals; livestock.

Manganeses de la Polvorosa (mahng-gah-NAI-ses dai lah pol-vo-RO-sah), town, ZAMORA province, NW SPAIN, 5 mi/8 km NW of BENAVENTE; 42°02′N 05°45′W. Lumbering; livestock raising; cereals, wine.

Mangaon (MAHN-goun), town, RAIGAD district, MAHARASHTRA state, W central INDIA, 60 mi/97 km SSE of MUMBAI (BOMBAY). Rice; copper and brass products. Soap manufacturing 6 mi/9.7 km S, at GOREGAON.

Mangapwani (man-gah-PWAH-nee), historic site, ZANZIBAR NORTH region, E TANZANIA, on W coast of ZANZIBAR island, 12 mi/19 km N of ZANZIBAR city; 05°59′S 39°10′E. Site of Mangapwani Slave Caves, where people from interior of Africa were held, to be sold as slaves to European and Arab traders through late 19th century.

Mangaratiba (MAHN-gah-rah-chee-bah), city, SW RIO DE JANEIRO state, BRAZIL, fishing port on SEPETIBA BAY of the Atlantic Ocean, on rocky peninsula, and 50 mi/80 km W of Rio; 22°57′S 44°02′W. Railroad terminus. Fish processing, banana growing. Since 1948, coal for VOLTA REDONDA steel mill unloaded here. Surrounded by small fishing villages. Settled 16th century by Jesuits.

Mangareva, FRENCH POLYNESIA: see GAMBIER ISLANDS.

Mangarin, PHILIPPINES: see SAN JOSE.

Mangar-tepe, Bulgaria: see POLEZHAN.

Mangart, Mount (MAHNG-gahrt) (8,786 ft/2,678 m), in JULIAN ALPS, on Italo-Slovenian border, 6 mi/10 km SE of TARVISIO (ITALY).

Mangatarem (mahng-ah-TAH-rem), town, PANGASINAN province, central LUZON, PHILIPPINES, 17 mi/27 km S of DAGUPAN; 15°44′N 120°16′E. Agricultural center (rice, copra, corn).

Mangaung [Isotho=place of leopards], municipality (2001 population 645,438), FREE STATE province, SOUTH AFRICA. Created in 2000 through the merger of the judicial capital BLOEMFONTEIN, the towns of Botshabelo and THABA NCHU, and a rural area.

Mangazeya (mahn-gah-ZYE-yah), former 17th century fur-trading center in NW SIBERIA, RUSSIA, at the mouth of the TAZ River, on the TAZ Bay, on the KARA SEA of the ARCTIC OCEAN, near the site of modern TAZOVSKIY; approximately 80°00E. Named for a local Samoyed tribe and founded in 1601 by Russian fur traders; town reached its greatest development in the mid-17th century. Opening of new routes further S and local extinction of fur animals caused transfer (around 1670) of trade to Turukhansk (originally called Novaya Mangazeya [Russian=new Mangazeya]) and decline of original Mangazeya.

Mangde Chu (MAHNG-dee CHOO), river, c.100 mi/161 km long, central BHUTAN; rises in High HIMALAYA, flows N-S past TONGSA, and SSE to DANGME CHU at 26°52′N 90°57′E.

Mange (MAHNG-gai), town, NORTHERN province, W SIERRA LEONE, on LITTLE SCARCIES RIVER, and 10 mi 16 km NNW of PORT LOKO; 08°55′N 12°51′W. Trade center; palm oil and kernels, kola nuts.

Mangerton Mountain (MANG-guhr-tuhn) (2,756 ft/840 m), KERRY county, SW IRELAND, 6 mi/9.7 km S of KILLARNEY; 51°58′N 09°30′W. Near summit is the DEVIL'S PUNCH BOWL.

Mangham (MANG-guhm), town (2000 population 595), RICHLAND parish, NE LOUISIANA, 24 mi/39 km SE of MONROE, on BIG CREEK; 32°19′N 91°47′W. Cotton; manufacturing (apparel).

Mangit, city, KARAKALPAK REPUBLIC, W UZBEKISTAN, on railroad, and 50 mi/80 km NNW of URGANCH; 42°08′N 60°04′E. Cotton ginning. Severely affected by the retreat of the ARAL SEA and its ensuing ecological disaster, this city has lost the fishing-related basis of much of its economy. Formerly spelled Mangyt.

Mangla (muhng-LAH), village, MIRPUR district, AZAD KASHMIR, NE PAKISTAN, in SW KASHMIR region, on JHELUM RIVER, and 9 mi/14.5 km SW of MIRPUR; 33°07′N 73°39′E. Railroad spur terminus; headworks of Upper JHELUM CANAL. Hydroelectric station (10,000 kw) and dam nearby.

Mangla Dam, on the JHELUM RIVER, on the border between AZAD KASHMIR and PUNJAB province, NE PAKISTAN, at the juncture of the frontal HIMALAYA hills with the plains of the PUNJAB; 33°09′N 73°39′E. Has a storage capacity of 6.5 x 19m. Silting is an ongoing problem, reducing capacity. Completed in 1968.

Manglaralto (mahn-glah-RAHL-to), village, GUAYAS province, W ECUADOR, on the PACIFIC OCEAN, 65 mi/105 km WNW of GUAYAQUIL; 01°50′S 80°44′W. Minor port with good harbor facilities. In fertile agricultural region (cacao, coffee, sugarcane, rice, tagua nuts, balsa wood). Fishing. Surfing at Montañita beach.

Manglares, Punta (mahn-GLAH-res, POON-tah), PACIFIC headland on coast of NARIÑO department, SW COLOMBIA, at mouth of main arm of MIRA RIVER delta, 26 mi/42 km SW of TUMACO; 01°37′N 79°03′W. Westernmost point of Colombia.

Manglar Zapoton Ecological Reserve (mahn-glahr zah-po-TON), nature reserve, in S CHIAPAS, MEXICO, 5 mi/8 km W of HUIXTLA, on the PACIFIC OCEAN coast.

Manglaur (MUHN-glour), town, SAHARANPUR district, N UTTAR PRADESH state, N central INDIA, on UPPER GANGA CANAL, and 23 mi/37 km SE of Saharanpur; 29°48′N 77°52′E. Furniture manufacturing; wheat, rice, rape and mustard, gram. Mosque built 1825 by Balban.

Manglawedhe, INDIA: see MANGALVEDHA.

Mangles Islands (MAHN-gles) [=mangrove], group of keys off SW CUBA in Gulf of BATABANÓ, 8 mi/13 km N of ISLA DE LA JUVENTUD, forming NW islands of LOS CANARREOS archipelago; 22°04′N 82°53′W.

Manglisi (mahn-GLEE-see), town, S GEORGIA, on road, and 20 mi/32 km W of TBILISI; 41°41′N 44°22′E. Summer resort in wooded mountain district. Has cathedral (built 4th century; restored in 1850s). Until 1936, called Manglis.

Manglun (MEING-lahn), former SE state (sawbwaship), now known as Pan-Yong township (□ 3,360 sq mi/8,736 sq km), SHAN STATE, MYANMAR, on CHINA (E; YUNNAN province) border; ⊙ Pangyang, a village 35 mi/56 km SE of TANGYAN. Astride THANLWIN RIVER, which divided it into East Manglun (□ 2,482 sq mi/6,428 sq km) and West Manglun (□ 878 sq mi/2,274 sq km); bordered N by former WA STATES. Agriculture (hill rice, opium; cattle). Large Wa population.

Mango (MAIN-go), unincorporated town (□ 4 sq mi/10.4 sq km; 2000 population 8,842), HILLSBOROUGH county, W central FLORIDA, 8 mi/12.9 km E of TAMPA; 27°59′N 82°18′W.

Mango, FIJI: see MAGO.

Mango, TOGO: see SANSANNÉ-MANGO.

Mangoase (mahn-GWAH-sai), town, EASTERN REGION, GHANA, on railroad, and 10 mi/16 km NNE of NSAWAM; 05°57′N 00°18′W. Cacao, palm oil and kernels, cassava, corn. Also spelled MANGOASI.

Mangoasi, GHANA: see MANGOASE.

Mangoche (mah-ngoh-che), administrative center and district (2007 population 778,338), Southern region, MALAWI, on SHIRE RIVER, near its efflux from Lake Malawi (LAKE NYASA), and 65 mi/105 km N of ZOMBA; 14°28′S 35°16′E. Road center; tobacco, cotton, corn, peanuts. Formerly called Fort Johnston. Also spelled Mangochi.

Mangochi, MALAWI: see MANGOCHE.

Mango City, city (2001 population 166,091), EAST SINGHBHUM district, SE Jharkhand state, E INDIA.

Mangoky River (mahn-GOOK), c.350 mi/563 km long, S central and W MADAGASCAR; rises in FIANARANTSOA province as the MATSIATRA (mah-TSEE-ahch) River 30 mi/48 km S of FIANARANTSOA (21°44′S 47°03′E); flows S and SW into Toliara province to BEROROHA, and, thereafter known as Mangoky, flows W and NW to MOZAMBIQUE CHANNEL, forming a wide estuary. Navigable c.160 mi/257 km for shallow-draught boats below BEROROHA.

Mangole (mahng-GO-luh), island, SULA ISLANDS, N Moluku province, INDONESIA, in MALUKU SEA, just E of TALIABU; 72 mi/116 km long, 10 mi/16 km wide;

01°55′S 125°50′E. Mountainous; agriculture (cocounts, corn, sago); fishing. MANGOLE STRAIT to S. Chief town is Mangole. Also spelled Mangoli.

Mangole Strait (mahng-GO-luh), Maluku province, INDONESIA, between MANGOLE (N) and SANANA (S) islands; 01°56′S 125°55′E.

Mangoloma (man-GO-lo-mah), village, DODOMA region, N central TANZANIA, 45 mi/72 km E of SINGIDA; 04°51′S 35°35′E. Cattle, sheep, goats; corn, wheat.

Mangonui (MANG-guh-noo-ee), township, N NORTH ISLAND, NEW ZEALAND, 24 mi/39 km NE of KAITAIA, Far North district; 34°59′S 173°32′E. Recreational and deep-sea big game fishing.

Mangoro River (mahn-GOOR), 130 mi/209 km long, ANTANANARIVO province, E MADAGASCAR; rises S of Lake ALAOTRA, 40 mi/64 km NNW of MORAMANGA (21°44′S 47°03′E); flows S and E to INDIAN OCEAN just S of MAHANORO. Navigable for small vessels in lower course, for 65 mi/105 km.

Mangotsfield (MANG-guhts-feeld), town (2001 population 9,732), South Gloucestershire, SW ENGLAND, 5 mi/8 km ENE of BRISTOL; 51°29′N 02°28′W. Nearby are Downend (W), and Soundwell (SW), with shoe manufacturing.

Mangrol (MAHN-grol), town, SURAT district, GUJARAT state, W central INDIA, 27 mi/43 km NE of SURAT. Local market for cotton and millet; cotton ginning.

Mangrol, town, JUNAGADH district, GUJARAT state, W central INDIA, near ARABIAN SEA, 36 mi/58 km SSW of JUNAGADH. Market center for millet, cotton, fish, coconuts, rice; handicrafts (cloth fabrics, ivory and sandalwood products, metalware). Betel vine and melons grown nearby. Has fine fourteenth-century mosque. Harbor lies c.1 mi/1.6 km SW; lighthouse. Captured 1531 by Portuguese.

Mangrol, town, KOTA district, SE RAJASTHAN state, NW INDIA, 45 mi/72 km ENE of KOTA; 26°15′N 77°20′E. Agricultural market (millet, wheat, gram); hand-loom weaving.

Mangrove Bay, inlet of SOMERSET ISLAND, BERMUDA.

Mangrove Cay, settlement, W BAHAMAS, on E ANDROS Island, 60 mi/97 km SSW of NASSAU.

Mangrullo, Rincón del (mahng-GROO-yo, reeng-KON del), marshy area in CERRO LARGO department, NE URUGUAY, on Uruguay-BRAZIL border, formed by YAGUARÓN RIVER (NE), Lake MIRIM (E), and TACUARÍ RIVER (S); 15 mi/24 km wide, 20 mi/32 km long; 32°46′S 53°48′W.

Mangrul Pir (MUHN-grool pir), town, AKOLA district, MAHARASHTRA state, W central INDIA, in AJANTA HILLS, 36 mi/58 km SE of AKOLA; 20°19′N 77°21′E. Road center; millet, wheat; cotton ginning.

Mangshi, CHINA: see LUXI.

Manguaba, Brazil: see PILAR.

Manguaba Lagoon (MAHN-gwah-bah), shallow salt lake, ALAGOAS state, NE BRAZIL, separated from the ATLANTIC OCEAN by sandy coastal strip; c.12 mi/19 km long, 3 mi/4.8 km wide. Has canalized outlet to MACEIÓ (8 mi/12.9 km E). On its shore are MARECHAL DEODORO and PILAR amidst sugarcane fields.

Mangualde (mahn-GWAHLD), town, VISEU district, N central PORTUGAL, on railroad, and 9 mi/14.5 km SE of VISEU; 41°26′N 08°36′W. Flour milling, sawmilling, wool spinning, and weaving. Surrounded by vineyards and orchards.

Mangueira, Lagoa da (MAHN-gai-rah, LAH-go-ah dah), shallow lagoon in S RIO GRANDE DO SUL state, BRAZIL, between MIRIM LAKE (W) and the open ATLANTIC (from which it is separated by a narrow sandbar) just E; 60 mi/97 km long, 8 mi/12.9 km wide; 33°06′S 52°48′W. Has marshy, uninhabited shores.

Mangueirinha (MAHN-gai-reen-yah), city (2007 population 17,119), S PARANÁ state, BRAZIL; 25°04′S 52°09′W. Corn, wheat.

Mangue Seco (MAHN-zhe SE-ko), village, BAHIA state, BRAZIL, NE coast at SERGIPE state border; 11°29′S 37°23′W.

Manguito (mahn-GEE-to), town, MATANZAS province, W CUBA, on railroad, and 35 mi/56 km SE of CÁRDENAS; 22°41′N 80°55′W. In agricultural region (sugarcane, fruit; poultry, cattle). Nearby sugar mills include Seis de Agosto (NE), Jesús Rabí (SE), and Reynold García (SW).

Mangula Mine, ZIMBABWE: see MHANGURA.

Mangulile (mahn-goo-LEE-lai), town, OLANCHO department, E central HONDURAS, 22 mi/36 km E of YORO; 15°03′N 86°49′W. No road. Small farms.

Mangum (MANG-guhm), town (2006 population 2,721), ⊙ GREER county, SW OKLAHOMA, near SALT FORK of RED RIVER, and 20 mi/32 km NNW of ALTUS; 34°52′N 99°30′W. Elevation 1,608 ft/490 m. Trade center for irrigated agriculture area (peanuts; sheep); manufacturing (bricks). Seat of Mangum Jr. College Quartz Mountain State Park and Lodge (recreation) on ALTUS LAKE to E; Old Greer County Museum Laid out 1883, incorporated 1900.

Manguredjipa, CONGO: see MOTOKOLEA.

Mangush, UKRAINE: see PERSHOTRAVNEVE, Donets'k oblast.

Mangwe, town, MATABELELAND NORTH province, SW ZIMBABWE, 52 mi/84 km SW of BULAWAYO, on MANGWE RIVER; 20°42′S 28°03′E. Site of Old Mangwe Fort to S. Livestock.

Mangwendi (mah-eng-GWEE-en-dee), village, MASHONALAND EAST province, E ZIMBABWE, 20 mi/32 km SW of Arondera; 18°23′S 31°22′E. Dairying; cattle, sheep, goats; tobacco, peanuts, corn, soybeans.

Mangwe River, c.60 mi/97 km long, SW ZIMBABWE; rises at SYRINGA, 45 mi/72 km SW of BULAWAYO; flows S, past MANGWE to SIMUKWE RIVER 50 mi/80 km SSE of SYRINGA. Forms part of border between MATABELELAND NORTH and MATABELELAND SOUTH provinces.

Mangyshlak (mahng-guhsh-LAHK), city, MANGYSTAU region, KAZAKHSTAN, E of CASPIAN SEA, and 6 mi/10 km N of AKTAU; 43°42′N 51°15′E. Railroad junction in oil-producing area; local industries support oil and gas concerns.

Mangyshlak Peninsula (mahng-guhsh-LAHK), W MANGYSTAU region, KAZAKHSTAN, extending into the NE CASPIAN SEA. Major source of oil, manganese, and coal. Together with BUZACHI PENINSULA (to N), the largest output oil-production area in the country. Major oil finds here in 1950s led to subsequent oil boom in 1960s–1970s. FORT SHEVCHENKO and BAUTINO in NW corner.

Mangystau (mahng-guh-STOU), region (☐ 63,750 sq mi/165,750 sq km), extreme SW KAZAKHSTAN, E of CASPIAN SEA, and on UZBEKISTAN (E) and TURKMENISTAN (S) borders; ⊙ AKTAU. Bordered by ATYRAU (W) and AKTÖBE (NE) regions. Includes BUZACHI and MANGYSHLAK peninsulas and Mangystau Mountains. Dry, sharply continental climate. Vast oil and gas deposits drive major extracting and refining industry. Pipelines under development with nearby countries. Fishing industry now declining. Formed (1973) as Mangyshlak oblast of Kazakh SSR with territory from former Guryev oblast. Current name in use since Kazak independence (1991). Also spelled Mangghystau.

Manhao (MAHN-HOU), town, SE YUNNAN province, CHINA, 22 mi/35 km SSW of MENGZI, and on RED RIVER (head of junk navigation); 23°02′N 103°14′E. Was trading center until opening (1910) of KUNMING-HANOI (VIETNAM) railroad.

Manhartsberg (MAHN-hahrts-berg), low mountain in N LOWER AUSTRIA, NE AUSTRIA, dividing WALDVIERTEL from WEINVIERTEL; 48°33′N, 15°45′E. Elevation: 1,634 ft/498 m.

Manhasset (man-HA-suht), affluent residential village (☐ 2 sq mi/5.2 sq km; 2000 population 8,362), NASSAU county, SE NEW YORK, on N shore of W LONG ISLAND, near head of MANHASSET BAY, 4 mi/6.4 km NNW of MINEOLA; 40°47′N 73°41′W. In estates area. Some light manufacturing and services. Retailing center for area nicknamed "Miracle Mile."

Manhasset Bay, inlet of LONG ISLAND SOUND indenting N shore of W LONG ISLAND, SE NEW YORK, just E of GREAT NECK; c.3.5 mi/5.6 km long; 40°50′N 73°44′W. Is 1 mi/1.6 km wide at entrance, which lies between Barker Point on MANHASSET NECK (E) and HEWLETT POINT (W). Yachting. On or near its shores are PORT WASHINGTON, PLANDOME, MANHASSET, Great Neck.

Manhasset Neck, peninsula on N shore of W LONG ISLAND, SE NEW YORK, between MANHASSET BAY (W) and HEMPSTEAD HARBOR (E); c.5 mi/8 km long, 1.5 mi/2.4 km–3 mi/4.8 km wide; 40°50′N 73°41′W. At its blunt tip are (SW to NE) Barker Point, SANDS POINT (lighthouse), Prospect Point. Residential area.

Manhattan (man-HAT-uhn), city (2000 population 44,831), ⊙ RILEY county, NE KANSAS, at the confluence of the BIG BLUE and KANSAS rivers; 39°11′N 96°35′W. Trade and processing center of a farm area. Railroad junction. Manufacturing (printing and publishing, food industry and meat processing, research; building materials, apparel). Much of the economy is dependent upon Kansas State University and nearby Fort Riley. Author Damon Runyon born here. Pottawatonic State Fishing Lake No. 2 to NE; Tuttle Creek Dam and Reservoir and State Park to N. Incorporated 1857.

Manhattan (man-HAT-tuhn), town (2000 population 1,396), GALLATIN county, SW MONTANA, on GALLATIN River, and 20 mi/32 km NW of BOZEMAN; 45°52′N 111°20′W. Manufacturing (wood products); dairying; cattle, sheep, hogs; wheat, barley, oats, potatoes. Madison Buffalo Jump State Park to SW. Formerly called Moreland and Hamilton.

Manhattan, village (2000 population 3,330), WILL county, NE ILLINOIS, 8 mi/12.9 km SSE of JOLIET; 41°25′N 87°59′W. In agricultural area.

Manhattan, borough (☐ 33 sq mi/85.8 sq km), NEW YORK city, SE NEW YORK, coextensive with NEW YORK county; 40°46′N 73°58′W. Composed chiefly of Manhattan Island, and is bounded W by the HUDSON RIVER, S by NEW YORK BAY, E by the EAST RIVER, NE and N by the HARLEM RIVER and Spuyten Duyvil Creek. Many bridges, tunnels, and ferries link it to the other boroughs and to New Jersey. Several adjacent small islands (Randall's, Ward's, Governors, and others) belong to it. A large number of workers commute to the borough every day. Manhattan is the cultural and commercial heart of the city, and its dramatic skyline symbolizes New York city. It began as a Dutch colonial town built at the S tip of the island called NEW AMSTERDAM and served as capital of the colony of NEW NETHERLAND during the Dutch period. In 1664 the English captured New Netherland and renamed it New York. The city line first extended beyond Manhattan Island when some WESTCHESTER county towns were annexed in 1874. In the consolidation of 1898, Manhattan became one of the city's five boroughs. It is commonly divided into three regions: downtown (or, more broadly, Lower Manhattan), midtown, and uptown. Dozens of well-known neighborhoods (e.g., GREENWICH VILLAGE, SOHO, TRIBECA), parks (CENTRAL PARK), and other areas (e.g., TIMES SQUARE, BROADWAY) are located here. (For its history, cultural, educational, and religious institutions, and other points of interest, see NEW YORK, city.) See also HARLEM; LOWER EAST SIDE; UPPER EAST SIDE.

Manhattan, locality, NYE county, central NEVADA, 65 mi/105 km S of AUSTIN, in TOQUIMA RANGE. Former gold-mining town. Great gold producer (1906–1915), still actively mined. Surrounded, except W, by Toiyabe National Forest.

Manhattan Beach, city (2000 population 33,852), LOS ANGELES county, S CALIFORNIA; suburb 13 mi/21 km SW of downtown LOS ANGELES, on Santa Monica Bay; 33°54′N 118°25′W. Residential and beach community with an oil refinery and factories in adjacent communities that produce transportation equipment, electrical equipment, computers, and pottery. Manhattan State Beach is here. Incorporated 1912.

Manhattan Beach, village (2000 population 50), CROW WING county, central MINNESOTA, 26 mi/42 km N of BRAINERD, at N end of (Lower) WHITEFISH LAKE; 46°43′N 94°08′W. Crow Wing State Forest to S.

Manhattan Beach, NEW YORK: see CONEY ISLAND.

Manhay, commune (2006 population 3,204), Marche-en-Famenne district, LUXEMBOURG province, SE BELGIUM, 26 mi/42 km S of LIÈGE, in the ARDENNES; 50°18′N 05°40′E.

Manheim (MAN-heim), borough (2006 population 4,645), LANCASTER county, SE PENNSYLVANIA, 10 mi/16 km NW of LANCASTER, on Chickies Creek; 40°09′N 76°24′W. Agricultural area (dairying; livestock, poultry; grain, potatoes, soybeans, apples); manufacturing (foods and beverages, machinery, tool and die, consumer goods, apparel, plastic products; commercial printing); stone quarries. Mount Hope Estate and Winery to N; covered bridges in area. Settled 1716, laid out c.1760 by H. W. Stiegel, who probably produced the first flint glass in America here. Incorporated 1848.

Manhiça (mahn-EE-kuh), village, MAPUTO province, S MOZAMBIQUE, on INCOMATI River, and 40 mi/64 km N of MAPUTO city; 23°06′S 34°39′E. Sugarcane-growing center; rice, almonds, beans; cattle raising.

Manhiça District, MOZAMBIQUE: see MAPUTO.

Manhuaçu (mahn-wah-soo), city (2007 population 74,277), SE MINAS GERAIS state, BRAZIL, on Rio Manhuassu (right tributary of the RIO DOCE), and 110 mi/177 km W of VITÓRIA (ESPÍRITO SANTO); 20°16′S 42°01′W. Coffee center; beryl deposits. Railroad terminus; road from Vitória joins RIO DE JANEIRO-BAHIA highway.

Manhumirim (man-hoom-ee-rim), city (2007 population 20,214), SE MINAS GERAIS state, BRAZIL, at W base of the Pico da Bandeira, on railroad, and 100 mi/161 km W of VITÓRIA (ESPÍRITO SANTO); 20°21′S 41°56′W. Rice, coffee, corn.

Manhush, UKRAINE: see PERSHOTRAVNEVE, Donets'k oblast.

Mani (MAH-nee), region in the S TAIYETOS, LAKONIA prefecture, S PELOPONNESE department, extreme SE mainland GREECE. The Maniotes were known for their resistance against Turkish rule; played active role in Greek War of Independence. Formerly Maina.

Maní (mah-NEE), town, ☉ Maní municipio, CASANARE department, central COLOMBIA, on the CUSIANA River, 38 mi/61 km SSE of YOPAL, in the Oriente; 04°49′N 72°17′W. Sugarcane, plantains, cassava; livestock.

Maní (mah-NEE), town, YUCATÁN, SE MEXICO, 12 mi/19 km E of TICUL; 20°23′N 89°24′W. Henequen, sugarcane, fruit. Maya ruins nearby.

Mani (mah-NEE), village, HADJER-LAMIS administrative region, W CHAD, 55 mi/89 km NNW of N'DJAMENA; 12°44′N 14°41′E.

Maniaçu (MAH-nee-ah-SOO), village, S central BAHIA state, BRAZIL, 19 mi/30 km NNE of CAETITÉ; 13°50′S 42°24′W.

Maniago (mah-NYAH-go), town, Pordenone province, FRIULI-VENEZIA GIULIA, NE ITALY, 15 mi/24 km NNE of PORDENONE; 46°10′N 12°42′E. Manufacturing (agricultural machinery, refrigerators, artificial slate). Noted for its cutlery since medieval times. Has ancient loggia and cathedral reconstructed 1468.

Maniamba (mahn-YAHM-buh), village, N MOZAMBIQUE, near Lake MALAWI, on road, and 40 mi/64 km NNW of LICHINGA; 12°43′S 35°25′E. Cotton, corn, beans.

Manianga, CONGO: see MONIANGA.

Maniar (MUHN-yahr), town, BALLIA district, E UTTAR PRADESH state, N central INDIA, on the GHAGHARA RIVER, 16 mi/26 km N of BALLIA; 25°59′N 84°10′E. Rice, gram, oilseeds, sugarcane.

Manica (mahn-EE-kuh), province (2004 population 1,280,829), ☉ CHIMOIO, MOZAMBIQUE, following ZIMBABWE border along the central W midriff. Bounded N by TETE, W by MANICA, S by GAZA and INHAMBANE (SAVE RIVER, border), and W by SOFALA provinces railroad crosses from adjacent Sofala province at BEIRA to enter ZIMBABWE on the other side. Agriculture (sugar and cotton). Inhabited predominantly by the Shona to N and Ndau-speaking people to S. Territory administered after 1891 by Mozambique Company. Under Portuguese Government Charter, reverted to Mozambique in 1942 and became province in 1943 subdivided into Tete and Manica e Sofala provinces. Further subdivisions occured splitting Manica from Sofala province Includes nine major districts: BARUE, GURO, MACHAZE, MACOSSA, MANICA, MOSSURIZE, SUSSUNDENGA, TAMBARA, and GONDOLA.

Manica District, MOZAMBIQUE: see MANICA, province.

Manicaland, province (□ 14,077 sq mi/36,600.2 sq km; 2002 population 1,566,889), E ZIMBABWE; ☉ MUTARE. Other major towns include RUSAPE and CHIMANIMANI. Bordered E by MOZAMBIQUE; NYANGA, VUMBA, and CHIMANIMANI mountains, part of E highlands, near Mozambique border. Drained by SAVE, ODZI, and Nygadzi rivers. Nyanga and MTARAZI FALLS national parks in NE, CHIMANIMANI NATIONAL PARK and Chipinge Safari Area in SE. Agriculture (coffee, tea, citrus, wheat, corn, macadamia nuts, sorghum, tobacco); cattle, sheep, goats, hogs, poultry; dairying; timber. Gold and copper mining; limestone quarrying. Tourism.

Manicani Island (mah-nee-KAH-nee) (□ 4 sq mi/10.4 sq km), EASTERN SAMAR province, PHILIPPINES, in LEYTE GULF, just off narrow SE peninsula of SAMAR island; 2.5 mi/4 km long, 2 mi/3.2 km wide. Coconut growing. Iron ore deposits.

Manicaragua (mah-nee-kahr-RAH-gwah), town (2002 population 22,851), VILLA CLARA province, central CUBA, on upper ARIMAO RIVER, and 17 mi/27 km S of SANTA CLARA; 80°07′N 79°58′W. Tobacco-growing and -processing center. Copper deposits to ESE.

Manicoré (MAH-nee-ko-RAI), city (2007 population 44,327), S central AMAZONAS state, BRAZIL, steamer and hydroplane landing on right bank of Rio MADEIRA, 200 mi/322 km SSW of MANAUS; 03°25′S 61°12′W. Shipping center for tobacco, manioc, rubber, cacao.

Manicouagan (man-i-KWAHG-uhn), county (□ 15,236 sq mi/39,613.6 sq km; 2006 population 30,933), CÔTE-NORD region, E QUEBEC, E CANADA; ☉ BAIE-COMEAU; 49°13′N 68°09′W. Composed of nine municipalities. Formed in 1981.

Manicouagan River (man-i-KWAHG-uhn), 310 mi/499 km long, QUEBEC, E CANADA; rises in E central Quebec; flows S to the SAINT LAWRENCE RIVER near BAIE COMEAU; 49°10′N 68°11′W. The river is an important source of hydroelectricity. Also spelled Manikuagan River.

Manicuare (mah-nee-KWAH-ree), town, SUCRE state, NE venezuela, on ARAYA PENINSULA (SW), on Gulf of CARIACO, 7 mi/11 km N of CUMANÁ; 10°33′N 63°21′W. Coconuts; livestock. Saltworks nearby.

Maniema, province (2004 population 1,621,000), E central DEMOCRATIC REPUBLIC OF THE CONGO; ☉ KINDU. Bordered N by ORIENTALE province (small part of central portion of border formed by LOWA RIVER and E portion of border formed by Maiko River), NE by NORD-KIVU province, E by SUD-KIVU province (small part of N portion of border formed by Lugulu River and central portion of border formed by ELILA RIVER), S by KATANGA province, and SWW by KASAÏ-ORIENTAL (central and S portion of border formed by LOMAMI RIVER). CONGO RIVER flows N, from SE through central part of the province. LOMAMI, ULINDI, Lugulu, Lowa, and LUBUTU rivers in N; Lueki and ELILA in central part of province. Marsh area in W central part of the province around LOMAMI RIVER, on border with KASAÏ-OCCIDENTAL. A Réserve de Faune

des Hippopotemes is located here. A Réserve de Faune des Hippopotames also located in S central along CONGO RIVER and in SE. Part of Maiko National Park in NE (also in Orientale). Railroad runs from center of province (KINDU) S into KATANGA. Regional airport at KINDU. MANIEMA, NORD-KIVU, and SUD-KIVU form the region of KIVU.

Maniema, CONGO: see KIVU.

Maniewicze, UKRAINE: see MANEVYCHI.

Manifold Heights (MA-ni-fold HEITS), suburb c.1 mi/2 km WNW of GEELONG, VICTORIA, SE AUSTRALIA.

Manihi (mah-NEE-hee), atoll, TUAMOTU ARCHIPELAGO, FRENCH POLYNESIA. Productive coconut and pearl-shell resources. Its companion atoll is AHE. Discovered 1765 by John Byron, who named them Prince of Wales Islands. Local name for these twin atolls (Ahe and Manihi) is Rununga.

Manihiki (mah-nee-HEE-kee), atoll (□ 2 sq mi/5.2 sq km), N PACIFIC, in the N COOK ISLANDS. Comprises twelve islets; the whole group that includes Manihiki, RAKAHANGA, and PENRHYN is also often designated Manihiki. Manihiki was discovered in 1822 by Americans, became a British protectorate in 1889, became the New Zealand Cook Islands administration, 1901–1965, and since 1965 has been part of the Cook Islands. Copra, pearl shell, and cultured pearls are the main exports. Manihiki is also known as HUMPHREY ISLAND.

Manika (muh-NEE-kuh), plateau, SE CONGO. Part of the fertile KATANGA PLATEAU, with extensive farming and ranching areas.

Manikganj (mah-neek-gahnj), town (2001 population 52,826), DHAKA district, E central EAST BENGAL, BANGLADESH, near DHALESWARI RIVER, 27 mi/43 km WNW of DHAKA; 23°52′N 90°00′E. Rice, jute, oilseeds.

Manikiala (MUH-ni-ki-YAH-lah), village, RAWALPINDI district, N PUNJAB province, central PAKISTAN, 18 mi/29 km SE of RAWALPINDI; 33°28′N 73°15′E. Wheat, millet. Coins of Kushan period (c.1st–2nd century C.E.) found in nearby noted Buddhist stupas. Sometimes spelled MANKIALA and MANIKYALA.

Manikonde (mah-nee-KO-ndai), village, MBEYA region, W central TANZANIA, 60 mi/97 km NE of SUMBAWANGA, NE of Lake RUKWA; 07°26′S 32°27′E. Wheat; sheep, goats.

Manikpur (MUHN-ik-poor), town, Partabgarh district, SE UTTAR PRADESH state, N central INDIA, on the GANGA RIVER, 34 mi/55 km WNW of ALLAHABAD. Rice, barley, wheat, gram, mustard. Marathas defeated here by Rajputs 1761.

Manikuagan River, CANADA: see MANICOUAGAN RIVER.

Manikyala, PAKISTAN: see MANIKIALA.

Manila (mah-NI-lah), city (□ 115 sq mi/299 sq km; 2000 population 1,581,082), ☉ the PHILIPPINES, NATIONAL CAPITAL REGION, SW LUZON, on MANILA BAY; 14°35′N 121°00′E. Center of the country's largest metropolitan area, its chief port, and the focus of all governmental, commercial, industrial, and cultural activities. In addition to its extensive and superb port facilities, Manila has Ninoy Aquino International Airport, as well as Manila Domestic Airport in PASAY (6 mi/9.7 km S of city center). Also the nexus of an extensive highway network and the N terminus of a railroad to S Luzon. The manufacturing center of the Philippines, with large metal fabrication, motor vehicle assembly, and textile and garment industries. It also has food- and hemp-processing plants, cigarette factories, and establishments making toilet articles, pharmaceuticals, and other chemical products. The navigable PASIG RIVER flows through the city, dividing it into two sectors, with Intramuros (the old Spanish walled city) and Ermita (the site of most government buildings and tourist hotels) on the S bank, and the "newer" section (which includes the commercial district, many congested slum areas, and the Chinese quarter in Binondo) on the N bank. Malacañang Palace, the presidential mansion, is on the Pasig. The fortified walled colony was established here in 1571 by López de Legaspi and developed mainly by Spanish missionaries. Except for two years (1762–1764) when the city was in British hands, it remained under Spanish control until the Spanish-American War (1898), when it was seized by U.S. forces three months after the battle of Manila Bay. Filipino uprisings occurred for several years, and not until 1901 was a civil government definitely established. In World War II the city was occupied by the Japanese (from January 2, 1942). Its recovery (February 1945) involved fierce house-to-house fighting, which reduced the old walled city to rubble, destroying many fine examples of 17th-century Spanish architecture. Only the Church of San Agustin (1606) survived. Reconstruction of the Manila Cathedral began in 1958. Today Manila is one of the most cosmopolitan cities in ASIA. It has many daily newspapers and periodicals, radio and television stations, a symphony orchestra, and over twenty universities and colleges. These include the University of Santo Tomás (1611), which during World War II served as an internment camp for thousands of U.S., British, and Dutch civilian prisoners; the Ateneo de Manila (1859); the University of Manila; the University of the East; and Manila Central University. The oval-shaped Luneta, the country's national park on Manila Bay, contains a monument to José Rizal, who was executed here by a Spanish firing squad. In 1968, Manila was shaken by a severe earthquake, which killed more than 300 people and caused extensive property damage. In 1972 the city was damaged by floodwaters resulting from more than three weeks of torrential rains.

Manila (muh-NIL-uh), town, MISSISSIPPI county, NE ARKANSAS, 15 mi/24 km WSW of BLYTHEVILLE, near Right Hand Chute; 35°52′N 90°09′W. BIG LAKE NATIONAL WILDLIFE Reservoir to E; Big Lake Wildlife Management Area beyond it. Incorporated 1901.

Manila, village (2006 population 303), ⊙ DAGGETT county, NE UTAH, 35 mi/56 km NNW of VERNAL, near WYOMING state line and UINTA MOUNTAINS; 40°59′N 109°43′W. Elevation 6,375 ft/1,943 m. Sheep; timber; tourism. Uinta Mountains and Ashley National Forest to S. FLAMING GORGE RESERVOIR and National Recreation Area to E. Sheep Rock Canyon to SW. Surveyed in 1898 during capture of MANILA (Philippines) by U.S.

Manila Bay, nearly landlocked inlet of the SOUTH CHINA SEA, SW LUZON, the PHILIPPINES; c.35 mi/56 km wide at its broadest point and 30 mi/48 km long. Best natural harbor in East ASIA and one of the finest in the world. The city of MANILA is on the E shore of the bay, and on the SE is the city of CAVITE and a historic naval base. The entrance to Manila Bay (c.11 mi/18 km wide) is divided by the island of CORREGIDOR into two channels; the N channel, between Corregidor and BATAAN peninsula, is only c.2 mi/3.2 km wide. During the Spanish-American War, in the battle of Manila Bay (May 1, 1898), a U.S. squadron under Commodore George Dewey destroyed the Spanish fleet off Cavite within a few hours. The Manila Bay area was the focus, during the early phase of World War II, of a desperate attempt to save the Philippines from Japanese conquest. In the Allied recovery of the Philippines (1944–1945), many Japanese ships were sunk in the bay. The large U.S. naval base on Sangley Point was closed down in the early 1980s.

Manilla (muh-NI-luh), municipality, E central NEW SOUTH WALES, SE AUSTRALIA, 160 mi/257 km NNW of NEWCASTLE, and at junction of NAMOI, MANILLA rivers; 30°45′S 150°43′E. Sheep and agriculture center; honey and mead (liquor made from honey and water). Keepit Reservoir nearby. Laid out early 1860s.

Manilla, town (2000 population 839), CRAWFORD county, W IOWA, on branch of WEST NISHNABOTNA RIVER, and 10 mi/16 km SSE of DENISON; 41°53′N 95°14′W. Wood products; seeds; fertilizers. Incorporated 1887.

Manilva (mah-NEEL-vah), town, MÁLAGA province, S SPAIN, near the MEDITERRANEAN SEA, 17 mi/27 km NNE of GIBRALTAR; 36°23′N 05°15′W. Cereals, grapes; livestock; wine making.

Manin, SYRIA: see MENIN.

Maningory River, MADAGASCAR: see ALAOTRA, LAKE.

Maninjau, Lake (mah-NIN-jou), Sumatra Barat province, INDONESIA, 24 mi/38 km E of BUKITTINGGI; 00°20′S 100°11′E. Crater lake (elev. 1,540 ft/470 m) in one of the largest craters in the world. Also spelled Lake Manindjau.

Manion Creek, village, N central BRITISH COLUMBIA, W CANADA, 150 mi/241 km NW of PRINCE GEORGE, 20 mi/32 km SW of Williston Lake, on Manson Creek, on dirt logging road from Fort SAINT JAMES. Lodge; no other services.

Maniototo, intermontane plain, Central OTAGO, upper TAIERI RIVER, NEW ZEALAND. Upper basin of the Taieri River, containing extensive sheep farming, the former gold-mining centers of NASEBY and Ranfurly as agricultural service centers, and adjacent planted forests.

Maniow (MAH-nyoof), Polish *Maniów*, German *Mohnau*, village, Wrocław province, SW POLAND, 12 mi/19 km NE of SWIDNICA (Schweidnitz). Storage reservoir on nearby BYSTRZYCA RIVER.

Manipa (mah-NEE-pah), island (□ c.30 sq mi/78 sq km), in S Maluku province, INDONESIA, in strait between SERAM and BURU, INDONESIA; near middle of SERAM; 12 mi/19 km long and 5 mi/8 km wide; 03°17′S 127°35′E. Terrain is wooded and hilly, rising to 2,165 ft/660 m.

Manipa Strait (mah-NEE-pah), Indonesian *Selat Manipa*, channel separating BURU (W) and MANIPA (E) islands, E INDONESIA; 03°20′S 127°23′E.

Manipur (MUHN-ee-poor), state (□ 8,620 sq mi/22,412 sq km; 2001 population 2,166,788), extreme NE INDIA, on MYANMAR border (S and E); ⊙ IMPHAL. Bordered N by NAGALAND, W by ASSAM, and SW by MIZORAM states. The terrain, mostly jungle, is on a high plateau, about 2,600 ft/792 m above sea level. The MANIPUR HILLS have peaks rising to 8,500 ft/2,591 m and are mostly covered in jungle. The inhabitants are mainly of Mongoloid stock and speak Tibeto-Burmese languages. The majority are Hindu Methei, with groups of Naga and Kuki tribesmen making up the remainder of the population. Agriculture and forestry are the major sources of income; handicrafts: pottery, cane and bamboo basketry, stoneware; hand loom weaving: especially silks and tribal shawls. Famous for Manipuri dance. Game of polo originated here. The raja of Manipur signed (1762) a treaty of protection with the British, who provided forces against invading Burmese. The area was administered from Assam state until 1947, when it became a union territory under the direct control of the central government of India. Manipur became a state in 1972. It is governed by a chief minister and cabinet responsible to an elected unicameral legislature and by a governor appointed by the president of India. Known for its scenic, pristine beauty, Manipur means "Jeweled Land"; it contains several unique species of flora and fauna, especially the brow-antlered deer, called *sangai*, or "dancing deer," siroi lily, and hundreds of distinct orchid species; birdwatching.

Manipur Hills (MUHN-i-poor), name applied to hills in MANIPUR state, NE INDIA, on India-MYANMAR border; continued S by CHIN and LUSHAI hills, N by Naga Hills; consist of parallel ranges rising to 8,427 ft/2,569 m in Siruhi Kashong or Sirohifara Peak, 38 mi/61 km NE of IMPHAL. Extensive bamboo tracts; teak; silk growing. Inhabited mainly by Naga and Kuki tribes.

Manipur River (MUHN-i-poor), 200 mi/322 km long, in NE INDIA (MANIPUR state) and Upper MYANMAR (North CHIN HILLS district); rises in N MANIPUR HILLS; flows S, past IMPHAL (India), through CHIN HILLS, past FALAM (Myanmar), to MYITTHA RIVER 24 mi/39 km E of Falam. Also called Imphal or ACHAUBA RIVER.

Manipur Road, INDIA: see DIMAPUR.

Manisa, city (2000 population 214,345), W TURKEY; 38°36′N 27°29′E. Railroad junction; market center of a rich agricultural region (cotton). Mineral deposits are nearby. Has many fine buildings, notable among them the Muradiye mosque, built 1583–1588. Was the residence of Ottoman sultans Murad II and Murad III. Ruins of ancient MAGNESIA ad Sipylum are nearby.

Manisa Dag (Turkish=*Manisa Dağ*), peak (4,977 ft/1,517 m), W TURKEY, 6 mi/9.7 km S of MANISA.

Manises (mah-NEE-ses), city, VALENCIA province, E SPAIN, 5 mi/8 km W of VALENCIA, and on TURIA RIVER; 39°29′N 00°27′W. Center of ceramics. Valencia's reservoir is nearby.

Man, Isle of (MAN), island (□ 227 sq mi/588 sq km; 32 mi/48 km long, 8 mi/13 km–15 mi/24 km wide; 1991 population 69,788; 2006 estimated population 75,441), off Great Britain, in the IRISH SEA; ⊙ DOUGLAS; 54°15′N 04°30′W. The island is an internally self-governing dependent territory of the British Crown. The coast is rocky with precipitous cliffs; the Calf of Man is a detached rocky islet off the SW coast. The rounded hills in the center of the island rise to 2,034 ft/620 m at SNAEFELL. The beautiful scenery and extremely mild climate (subtropical plants are grown without protection) make the island a popular resort. Oats, barley, turnips, and potatoes are grown; sheep are raised. Dairying and fishing are important, and Manx tweeds are made from locally produced wool. Some light industry has developed, and offshore banking has become more important. Traces of occupants from Neolithic times exist. Of interest are ancient crosses and other stone monuments, a round tower, an old fort, and castles. Occupied in the 9th century by Vikings, the island was a dependency of NORWAY until 1266, when it passed to SCOTLAND, but from the 14th to the 18th centuries (except for brief periods when it reverted to the English crown) it belonged to the earls of SALISBURY and of DERBY. Since 1765, when the British Parliament purchased it from the Duke of Atholl, it has been a dependency of the crown, but is not subject to acts of Parliament. The traditional open-air assembly of the Tynwald (one of the world's oldest legislative bodies) on July 5 was attended for the first time by the British king in 1945. The island's towns include Douglas, PEEL, RAMSEY, and CASTLETOWN.

Manissobal, Brazil: see SÃO JOSÉ DO BELMONTE.

Manistee (MAN-is-TEE), county (□ 1,280 sq mi/3,328 sq km; 2006 population 25,067), NW MICHIGAN; 44°17′N 86°18′W; ⊙ MANISTEE. Bounded W by LAKE MICHIGAN; drained by MANISTEE and LITTLE MANISTEE rivers and small Bear Creek. Fruit growing (apples, cherries, peaches, berries); also cattle, hogs, poultry; potatoes, forage crops; manufacturing (paper products, chemicals, mineral products) at Manistee; salt mines; fisheries; resorts. Includes part of Manistee National Forest to S and SE; Orchard Beach State Park in SW. PORTAGE, BEAR, and MANISTEE lakes are in W. Tippy Dam Pond in SE. Organized 1855.

Manistee (MAN-is-TEE), city (2000 population 6,586), ⊙ MANISTEE county, NW MICHIGAN, c.45 mi/72 km SW of TRAVERSE CITY, port on LAKE MICHIGAN, at mouth of MANISTEE RIVER (here draining MANISTEE LAKE, just E); 44°14′N 86°19′W. Resort, shipping, and industrial center; salt mining and processing; manufacturing (salt, salt products, bulk bags, chemicals, paper, furniture, wood products, boats, textiles); fisheries. Agricultural products (fruit, potatoes; cattle, hogs, poultry). Annual forest festival held here. County Museum. Manistee-Blacker Airport NE of town. Orchard Beach State Park to N; Manistee National Forest to E. Incorporated as a city in 1869.

Manistee Lake, MICHIGAN: see MANISTEE RIVER.

Manistee River (MAN-is-TEE), c.170 mi/274 km long, NW MICHIGAN; rises in lakes near border between OTSEGO and ANTRIM counties; flows generally SW, past MESICK through Hodenpyl Dam and Tippy Dam ponds, and Manistee National Forest, to LAKE MICHIGAN at MANISTEE. Widens into MANISTEE LAKE (c.5 mi/8 km long, 1 mi/1.6 km wide) before entering Lake Michigan; 44°56′N 84°52′W.

Manistique (MAN-is-TEEK), town (2000 population 3,583), ⊙ SCHOOLCRAFT county, S UPPER PENINSULA, N MICHIGAN, c.40 mi/64 km ENE of ESCANABA, at mouth of MANISTIQUE RIVER, on LAKE MICHIGAN; 45°57′N 86°15′W. Manufacturing (wood products, paper, building materials). Resort, industrial, and shipping center; lumber and paper milling, processing of hardwood products. Limestone quarrying nearby. Indian Lake (E and W units) and Palm Book state parks, is just NW. Senny National Wildlife Refuge to NE; Hiawatha National Forest to W, including Thunder Bowl Ski Area in NW. Incorporated as village 1885, as city 1901.

Manistique Lake (MAN-is-TEEK), SE UPPER PENINSULA, N MICHIGAN, on border between LUCE (N) and MACKINAC (S) counties, 13 mi/21 km SW of NEWBERRY; c.7 mi/11.3 km long, 3 mi/4.8 km wide. Drained by MANISTIQUE RIVER; joined by streams to SOUTH MANISTIQUE LAKE (c.4.5 mi/7.2 km long, 2 mi/3.2 km wide), just S, and to NORTH MANISTIQUE LAKE (c.2 mi/3.2 km long, 1.5 mi/2.4 km wide), just NE. Village of Curtis at SE end, Helmer on NE end; 46°14′N 85°47′W.

Manistique River (MAN-is-TEEK), c.35 mi/56 km long, S UPPER PENINSULA, N MICHIGAN; rises in MANISTIQUE LAKE; flows SW to LAKE MICHIGAN just below MANISTIQUE; 46°14′N 85°51′W. Receives many small tributaries from N and NW.

Manito (MAN-i-to), village (2000 population 1,733), MASON county, central ILLINOIS, 20 mi/32 km SSW of PEORIA, near Sand Ridge State Forest; 40°25′N 89°46′W. Corn, wheat, watermelons, vegetables.

Manitoba (man-i-TO-buh), province (□ 250,934 sq mi/ 650,930 sq km, including 39,215 sq mi/101,580 sq km of water surface; 1991 population 1,119,583; 2006 population 1,148,401), W central CANADA; ⊙ WINNIPEG, largest city, accounting for more than half the province's population in its metropolitan area; 55°00′N 97°00′W. Other important cities are BRANDON, THOMPSON, PORTAGE LA PRAIRIE, and SELKIRK.

Easternmost of the Prairie Provinces, Manitoba is bounded N by the KIVALLIQ region of NUNAVUT territory (with a NE shoreline on HUDSON BAY), E by ONTARIO, S by U.S. (MINNESOTA and NORTH DAKOTA), and W by SASKATCHEWAN. Due to its central location it is a major national transportation center. The S and central part of Manitoba was once covered by Lake Agassiz (see AGASSIZ, LAKE). As its waters receded into Hudson Bay, it left behind numerous lakes (WINNIPEG, MANITOBA, and WINNIPEGOSIS) and rivers (NELSON, CHURCHILL, and HAYES), which flow NE into the bay. In some places rock formations were swept bare, and in others they were covered with rich deposits of black loam. Miles of almost uninhabited treeless tundra surround the port of CHURCHILL. Extending S from Churchill and E from Lake Winnipeg, the topography is that of the CANADIAN SHIELD; limited areas have been cleared for general farming and dairying, and the mineral and timber resources have been partly developed.

The S part of Manitoba is dominated by lakes, with Lake Winnipeg paralleled in the W by Lake Winnipegosis and Lake Manitoba. To the W and N of the RED RIVER valley, the land rises in an escarpment extending into the plateaus of the PEMBINA, TURTLE, Riding, DUCK, and PORCUPINE mountains. Much of this heavily forested area has been set off as reserves, and the RIDING MOUNTAIN area is a national park. To the S, where most of the population is concentrated,

are fields of wheat, barley, oats, rye, canola, and flax; also sunflower seeds, field peas. The well-settled Souris plains in the SW are especially famous for their wheat fields. Canada's wheat industry originated here, and Manitoba's bread wheat has set the standards for the world. Grain is shipped in quantity from Churchill (the only port in the Prairie Provinces) during the three ice-free months of the year.

Manitoba has warm summers and cold winters, and is moderately dry; most of its annual rain falls in the summer. Its heaviest snowfalls occur in the Duck and Riding mountains, in the NE part of the province. Although agriculture has been continually extended—especially in mixed farming, dairying, and poultry and livestock raising—manufacturing has nevertheless displaced it as the leading industry in the province. Foods, minerals, clothing, electrical items, chemicals, furniture, leather, fabricated metals, and transportation equipment are major products. Continuing developments in mining, pulp and paper manufacturing, and extensive hydroelectric production promise to preserve Manitoba's industrial growth. In the SW, near BRANDON, are large oil reserves, and the municipal districts of FLIN FLON and THE PAS, on the SASKATCHEWAN RIVER, are gateways to the rich mineral deposits (chiefly nickel, copper, and zinc) and timberlands of the central W Manitoba rank second only to Ontario in the production of nickel; the mines at Thompson provide most of Manitoba's nickel. Beluga whales are still caught by native fishermen at Churchill, and fur farming in the N places Manitoba third of all the provinces in the production of fur.

The modern history of Manitoba began along Hudson Bay. The search for the elusive NORTHWEST PASSAGE to the PACIFIC OCEAN drew such explorers as Henry Hudson, Thomas Button, Pierre Radisson, and Médard Chouart des Groseilliers, some of whom returned to ENGLAND laden with beaver furs. To exploit this fur wealth, King Charles II granted (1670) the Hudson's Bay Company propriety over all the lands draining into Hudson Bay. This vast area included the present-day province of Manitoba, then occupied by the Assiniboin, the Ojibwa, and the Cree peoples. The company established a trading post at PORT NELSON and soon extended its operations S to the strategic Red River valley. In 1717, Fort Prince of Wales was built at the mouth of the Churchill River (rebuilt in stone 1732–1771, it is now in Fort Prince of Wales National Historic Park). Manitoba was explored and posts were established by the French as well as the British; their rival claims were resolved when England's conquest of Canada in the French and Indian Wars was confirmed by the Treaty of Paris in 1763. Scotsmen took over much of the French fur trade, organized the North West Company, and challenged the monopoly of the Hudson's Bay Company. A crisis came when the earl of Selkirk established the Red River Settlement at present-day Assiniboine in North West Company territory. The resulting violence deterred colonization until the merger of the two companies in 1821. From then until 1870, when the Hudson's Bay Company sold its vast domain to the newly created confederation of Canada, that company was in sole control, and settlement of the area increased.

Prearrangements for the transfer of the land to the new dominion government led to conflict between government representatives and Métis (people of mixed European and Native American ancestry), who had long enjoyed almost total autonomy under the Hudson's Bay Company's rule. Fearing political persecution and the loss of their land, they staged (1869) the Red River Rebellion under the leadership of Louis Riel. The rebellion was nominally successful and the Métis were granted land and cultural rights, but after Manitoba was organized as a province in 1870, most of the Métis were harassed into moving further W. Agricultural settlement in Manitoba proceeded

slowly, but when the railroads came (1870 and 1881), they provided access to and from the grain markets on the GREAT LAKES, and the population doubled during the 1880s.

Manitoba's area was enlarged in 1881, and in 1912 it was given its present extension to Hudson Bay. The depression of 1913 and the opening of the PANAMA CANAL in 1914 ended this period of prosperity, during which Winnipeg had served as a great transportation center. With the completion of the Hudson Bay railroad to Churchill in 1929, the province was in a position to use the shorter sea route E. During the last part of the 19th century and the first part of the 20th, the Canadian government advertised for immigrants to settle the prairies, and huge numbers of Russians, Poles, Estonians, Scandinavians, and Hungarians came from EUROPE. Manitoba remains Canada's most ethnically diverse province. The largest single immigrant group was the Ukrainians, who now constitute more than 11% of the population and are an important part of Manitoba culture. A national Ukrainian festival is held each year, and there is a Ukrainian culture museum in Winnipeg. The province provided a multilingual school system from 1897 to 1916, but abolished it when the number of ethnic groups requesting such facilities grew too large. Further immigration came with World War I when American pacifist sects (e.g., the Mennonites and Hutterites), seeking to avoid military service, set up colonies of their own here. Manitoba still has problems amalgamating its many ethnic groups; they include the Métis, who have settlements (such as SAINT BONIFACE) in the province. Manitoba sends six senators (appointed) and fourteen representatives (elected) to the national Parliament. The University of Manitoba and the University of Winnipeg are at Winnipeg.

Manitoba, Lake (man-i-TO-buh) (□ 1,817 sq mi/ 4,724.2 sq km), SW MANITOBA, central CANADA; one of the largest lakes of NORTH AMERICA; 50°59′N 98°48′W. A remnant of glacial Lake AGASSIZ, it is fed by Lake WINNIPEGOSIS and drains into Lake WINNIPEG. Its shores are marshy. Commercial fisheries.

Manito Lake (MA-ni-too) (□ 67 sq mi/174.2 sq km), W SASKATCHEWAN, W CANADA, near ALBERTA border, 55 mi/89 km W of NORTH BATTLEFORD; 12 mi/19 km long, 7 mi/11 km wide; 52°04′N 109°29′W.

Manitou (MA-ni-too), village (□ 1 sq mi/2.6 sq km; 2001 population 775), S MANITOBA, W central CANADA, in PEMBINA MOUNTAINS, 50 mi/80 km S of PORTAGE LA PRAIRIE, and in PEMBINA rural municipality; 49°14′N 98°32′W. Elevation 1,590 ft/485 m. Grain elevators; dairying, mixed farming.

Manitou (MAN-uh-too), village (2006 population 275), TILLMAN county, SW OKLAHOMA, 9 mi/14.5 km NNE of FREDERICK; 34°30′N 98°58′W. In cotton and grain area. Lake Frederick reservoir to E.

Manitou Beach (MAN-i-too), town, LENAWEE county, SE MICHIGAN, on DEVILS LAKE, and 15 mi/24 km WNW of ADRIAN; 41°58′N 85°18′W.

Manitou Beach (MA-ni-too), village (2006 population 233), S SASKATCHEWAN, W CANADA, on Little Manitou Lake, 4 mi/6 km N of WATROUS; 51°26′N 105°15′W. Resort.

Manitou Falls, Big, Wisconsin: see BLACK RIVER.

Manitou Island (MAN-i-too), KEWEENAW county, off UPPER PENINSULA, N MICHIGAN, in LAKE SUPERIOR, 3 mi/4.8 km E of Keweenaw Point; c.3 mi/4.8 km long, 1 mi/1.6 km wide; 47°25′N 87°37′W. MANITOU ISLANDS are in LAKE MICHIGAN.

Manitou Islands (MAN-i-too), NW MICHIGAN, two islands in LAKE MICHIGAN, c.14 mi/23 km W of LEELANAU PENINSULA; southernmost islands of BEAVER ISLANDS archipelago. NORTH MANITOU ISLAND is 8 mi/ 12.9 km long, 4 mi/6.4 km wide; SOUTH MANITOU ISLAND is c.4 mi/6.4 km long, 3 mi/4.8 km wide. Passenger ferry from LELAND connects both islands to each other and to mainland. Resorts; sand dunes, small lakes. Part of Sleeping Bear Dunes National Lakeshore. MANITOU ISLAND is NW, in LAKE SUPERIOR.

Manitou Lake (MA-nuh-too), (□ 60 sq mi/156 sq km), W ONTARIO, E central CANADA, 50 mi/80 km E of WHITEFISH BAY (LAKE OF THE WOODS); 27 mi/43 km long, 5 mi/8 km wide; 45°46′N 81°59′W. Divided into Upper Manitou and Lower Manitou lakes by narrow strait. Drains S into RAINY LAKE.

Manitou, Lake, INDIANA: see ROCHESTER.

Manitoulin (ma-ni-TOO-lin), district (□ 1,838 sq mi/4,778.8 sq km; 2001 population 12,679), ONTARIO, E central CANADA; ⊙ GORE BAY; 45°45′N 82°30′W. Composed of MANITOULIN ISLAND, in Lake HURON, and surrounding islands.

Manitoulin Islands (man-uh-TOO-lin), archipelago consisting of three large islands and several smaller ones, in N Lake HURON, NW of GEORGIAN BAY, ONTARIO, central CANADA. The islands, in a noted fishing region, are popular resorts. The permanent population is mainly Native American. Dairying, lumbering, mixed farming, and tourism are the major activities. Manitoulin, c.80 mi/130 km long and from 2 mi/3 km–30 mi/48 km wide, is the world's largest lake island. It encloses more than 100 lakes and has a much-indented, rugged coast. Cockburn and Drummond islands are also rocky and forested. Drummond Island belongs to MICHIGAN, and the others of the group to Ontario.

Manitou Springs, town (2000 population 4,980), EL PASO county, central COLORADO, on FOUNTAIN CREEK, at foot of PIKES PEAK, and 5 mi/8 km W of COLORADO SPRINGS; 38°51′N 104°54′W. Elevation 6,412 ft/1,954 m. Tourist and health resort with mineral springs and sanitarium. Light manufacturing. Cliff Dwellings Museum and Historic Miramont Castle are here. Road and cog railroad to summit of Pikes Peak (W). Nearby are Pike National Forest (N, W, and S), Cave of the Winds (N), and Garden of the Gods (N). Founded 1872, incorporated 1888.

Manitouwadge (ma-ni-TOO-wuhj), township (□ 136 sq mi/353.6 sq km; 2001 population 2,949), NW ONTARIO, E central CANADA, 247 mi/397 km NE of THUNDER BAY; 49°08′N 85°48′W. Mining copper, lead, silver, and zinc. Ontario government-designed model town.

Manitowaning, unincorporated village, S central ONTARIO, E central CANADA, on NE Manitoulin Island, on Manitowaning Bay, 18 mi/29 km SSE of LITTLE CURRENT, and included in ASSIGINACK; 45°44′N 81°48′W. Lumbering, dairying, mixed farming.

Manitowash Waters, village, VILAS county, N WISCONSIN, 27 mi/43 km SE of IRONWOOD (MICHIGAN), in Northern Highland State Forest. Numerous lakes in area. Turtle Flambeau Flowage reservoir to W. Manufacturing (archery and craft feathers).

Manitowik Lake (ma-ni-TOU-ik), central ONTARIO, E central CANADA, 100 mi/161 km N of SAULT SAINTE MARIE; 12 mi/19 km long, 2 mi/3 km wide; 48°10′N 84°24′W. Drained S into Lake SUPERIOR by MICHIPICOTEN RIVER.

Manitowish River, c.45 mi/72 km long, N WISCONSIN; rises in lake region in VILAS county; flows W and SW through wooded lake area to FLAMBEAU RIVER, immediately E (upstream) of TURTLE-FLAMBEAU RESERVOIR. Fishing, whitewater canoeing.

Manitowoc, county (□ 1,493 sq mi/3,881.8 sq km; 2006 population 81,911) E WISCONSIN; ⊙ MANITOWOC; 44°08′N 87°32′W. Agriculture (barley, oats, wheat, corn, soybeans, peas, brans, alfalfa, hay; poultry, cattle; dairying); marble and stone. Manufacturing at Manitowoc, TWO RIVERS, and KIEL. Bounded E by LAKE MICHIGAN; drained by MANITOWOC RIVER. Point Beach State Forest in NE, in Lake Michigan; Hidden Valley Ski Area in N. Nuclear power plants, Point Beach 1 (initial criticality November 2, 1970) and Point Beach 2 (initial criticality May 30, 1972), are 13 mi/21 km NNW of Manitowoc; use cooling water

from Lake Michigan, and each has a maximum dependable capacity of 485 MWe. Formed 1836.

Manitowoc, city (2006 population 33,635), ⊙ MANITOWOC county, E WISCONSIN, 75 mi/121 km NNE of MILWAUKEE, a port of entry on LAKE MICHIGAN at the mouth of the MANITOWOC RIVER; 44°06′N 87°40′W. Manufacturing (electric equipment, malt, foods, toys, printing, fabricated metal products, consumer goods, textiles, transportation equipment, furniture, plastics products, lubrication equipment, machinery, bakery products, yachts, building materials). Railroad junction and a ship and railroad transfer point. Its shipbuilding industry dates from 1847; submarines were made here in World War II. The North West Company established a trading post on the site in 1795. Manitowoc and its twin city, TWO RIVERS (6 mi/9.7 km NE), were founded in 1836. Silver Lake College, University of Wisconsin-Manitowoc campus, and a maritime museum and submarine memorial are located here. Point Beach State Forest to NE; Lake Michigan car ferry to LUDINGTON (MICHIGAN) carries U.S. Highway 10 (last of eight ferry crossings of Lake Michigan). Incorporated 1870.

Manitowoc River, c.40 mi/64 km long, E WISCONSIN; formed by several branches rising in W CALUMET county near LAKE WINNEBAGO; flows generally E to LAKE MICHIGAN at MANITOWOC.

Maniwaki (ma-ni-WO-kee), town (□ 2 sq mi/5.2 sq km), OUTAOUAIS region, SW QUEBEC, E CANADA, on GATINEAU River, at mouth of Desert River, and 70 mi/113 km NNW of OTTAWA; 46°22′N 75°58′W. Lumbering, pulp milling, dairying. Was seat of historic Gatineau county.

Maniyachi (muhn-ee-YAH-chee), village, CHIDAMBARANAR district, TAMIL NADU state, S INDIA, 9 mi/14.5 km WNW of TUTICORIN. Railroad junction. Formerly spelled Maniyachchi.

Manizales (mah-nee-ZAH-lais), city, ⊙ CALDAS department, W central COLOMBIA, on the slopes of the Cordillera CENTRAL; 05°05′N 75°30′W. Elevation 7,063 ft/2,153 m. Commercial and agricultural center for coffee. Industries include food processing and consumer goods. Some gold and silver mines nearby. Founded in 1847 by gold prospectors from ANTIOQUIA. Destroyed by an earthquake in 1878 and by fire in 1925. Airport.

Manja (MAHNZ), town, TOLIARY province, W MADAGASCAR, 135 mi/217 km NNE of TOLIARY; 21°26′S 44°18′E. Rice, cassava, beans; cattle market.

Manjača (mahn-YAH-chah), mountain (3,982 ft/1,214 m) in DINARIC ALPS, N BOSNIA, BOSNIA AND HERZEGOVINA, along left bank of the VRBAS RIVER, near Žepče; highest point is 15 mi/24 km SSW of Banja Luka; 44°25′N 17°52′E. Also spelled Manyacha.

Manjacaze (mahn-YAH-kah-zai), village, GAZA province, S MOZAMBIQUE, on railroad, and 30 mi/48 km NNE of XAIXAI; 24°43′S 33°50′E. Cashew nuts, mafura, corn.

Manjacaze, MOZAMBIQUE: see GAZA.

Manjakandriana (mahn-zahk-ahn-DREEN), town, ANTANANARIVO province, central MADAGASCAR, on railroad, and 20 mi/32 km E of ANTANANARIVO; 18°55′S 47°49′E. Rice, market vegetables; dairy production; regional market.

Manjhand (MAHN-jund), town, DADU district, W SIND province, SE PAKISTAN, on INDUS River, and 65 mi/105 km SSE of DADU; 25°55′N 68°14′E.

Manjhanpur (MUHN-juhn-poor), town, tahsil headquarters, ALLAHABAD district, SE UTTAR PRADESH state, N central INDIA, 29 mi/47 km W of ALLAHABAD; 25°32′N 81°23′E. Gram, rice, wheat, barley, jowar. Oilseed milling 8 mi/12.9 km ENE, at BHARWARI village.

Manjil (mahn-JEEL), village, Gilān province, N IRAN, in ELBURZ range, 40 mi/64 km SSW of RASHT, on TEHRAN-RASHT road, and near the SEFID Rud; 36°43′N 49°25′E. Dam. Olive-growing center. Coal, iron, copper deposits in vicinity.

Manjimup (MAN-ji-muhp), town, SW WESTERN AUSTRALIA state, 160 mi/257 km S of PERTH; 34°14′S 116°09′E. Jarrah, karri trees in area; sawmills; woodchipping industry strongly opposed by environmentalists. Butter, dairy products, fruit (especially apples), vegetables, tobacco, lambs, wool, and grain.

Manjira River, INDIA: see MANJRA RIVER.

Manjirenji, Lake (mahn-jee-REE-en-jee), reservoir, MASVINGO province, SE central ZIMBABWE, formed by Manjirenji Dam, on Chiredzi River, 60 mi/97 km SE of MASVINGO; 9 mi/14.5 km long, 2 mi/3.2 km wide; 20°35′S 31°35′E. Surrounded by LAKE MANJIRENJI RECREATIONAL PARK.

Manjlegaon (MAHNJ-lai-goun), town, BID district, MAHARASHTRA state, W central INDIA, on tributary of the GODAVARI RIVER, and 32 mi/51 km ENE of BID; 19°09′N 76°14′E. Millet, cotton, wheat, oilseeds. Formerly MAZALGAON.

Manjo (MAHN-jo), city (2001 population 27,200), Littoral province, CAMEROON, 50 mi/80 km NNE of DOUALA; 04°50′N 09°50′E.

Manjra River (MAHNJ-rah), c.385 mi/620 km long, central MAHARASHTRA state, central INDIA; rises in isolated W hills of DECCAN PLATEAU SW of BID; flows ESE and N to GODAVARI RIVER 17 mi/27 km NW of NIZAMABAD (ANDHRA PRADESH state). Forms NIZAM SAGAR (reservoir) NW of HYDERABAD. Sometimes spelled MANJIRA.

Manjuyod (mah-NOO-yod), town, NEGROS ORIENTAL province, E NEGROS island, PHILIPPINES, on TAÑON STRAIT, 12 mi/19 km N of TANJAY; 09°45′N 123°02′E. Agricultural center (corn, coconuts, sugarcane).

Mankailpa (mahn-kai-EEL-pah), canton, VALLEGRANDE province, SANTA CRUZ department, E central BOLIVIA; 18°34′S 64°09′W. Elevation 2,030 m. Agriculture (potatoes, yucca, bananas, corn, rye, soy); cattle.

Mankato (man-KAI-do), city (2000 population 32,427), ⊙ BLUE EARTH county, extends NW into NICOLLET county, S MINNESOTA, 63 mi/101 km SSE of MINNEAPOLIS, on MINNESOTA RIVER, E of mouth of BLUE EARTH RIVER; 44°10′N 93°59′W. Elevation 793 ft/242 m. Trade and processing center for a farm region (grain, alfalfa; livestock; dairying [especially to N and E]). Manufacturing (soybean processing, dairy products, oilseed processing, printing and publishing; flour, feeds, building materials, computer equipment, machinery, paper products, fabricated metal products, consumer goods, lumber, electronic equipment, plastic products). Mankato stone has been quarried here for more than 100 years. Mankato Municipal Airport. Seat of Mankato State University and Bethany College. Sibley Park in Mankato was the site of Camp Lincoln, where more than 300 Sioux were held and thirty-eight of them hanged, after their revolt in 1862. Minneopa State Park, with waterfalls to W; Mount Kato Ski Area to S; numerous small natural lakes to NE. Incorporated 1865.

Mankato (man-KAI-to), town (2000 population 976), ⊙ JEWELL county, N KANSAS, 33 mi/53 km NW of CONCORDIA; 39°47′N 98°12′W. Railroad junction. Trading center for grain and livestock area. Lovewell Reservoir and State Park to NE; Jewall State Fishing Lake to SW. Founded 1872, incorporated 1880.

Mankayan (mahng-KAH-yahn), town, BENGUET province, N LUZON, PHILIPPINES, 35 mi/56 km NNE of BAGUIO; 16°51′N 120°48′E. Copper and gold mining.

Mankayane (man-kai-YAH-ne), town, MANZINI district, W SWAZILAND, 25 mi/40 km SSW of MBABANE, near source of NGWEMPISI RIVER; 26°41′S 31°04′E. Road junction. Timber; commercial logging. Corn, vegetables, citrus, tobacco; cattle, goats, sheep, hogs. Also spelled Makaiana.

Mankera (muhn-KAI-rah), village, MIANWALI district, W PUNJAB province, central PAKISTAN, 85 mi/137 km S of MIANWALI, in THAL region; 31°23′N 71°26′E. Sheep and goat grazing. Was 17th-century Baluchi stronghold.

Mankiala, PAKISTAN: see MANIKIALA.

Man'kivka (mahn-KEEF-kah) (Russian *Mankovka*), town, SW CHERKASY oblast, UKRAINE, 17 mi/27 km NNE of UMAN'; 48°58'N 30°20'E. Elevation 649 ft/197 m. Raion center. Manufacturing (majolica, bricks); dairy. Known since the 16th century, company center of Uman regiment (1648–1657); uprisings against the Poles (1664–1665, 1768); destroyed by the Turks and Tatars (1672). Town since 1965.

Mankono (mahng-KO-no), town, Worodougou region, central CÔTE D'IVOIRE, 75 mi/121 km WNW of BOUAKÉ; 08°03'N 06°11'W. Agricultural center (cotton, palm kernels, rice, corn, beans, tobacco, vegetables, manioc, potatoes); cotton ginning.

Mankota (man-KO-tuh), village (2006 population 238), SW SASKATCHEWAN, W CANADA, 50 mi/80 km WSW of ASSINIBOIA; 49°25'N 107°04'W. Wheat. In former coal-mining region.

Mankovka, UKRAINE: see MAN'KIVKA.

Mankoya, ZAMBIA: see KAOMA.

Mankrong (mahn-KRAWNG), village, EASTERN REGION, GHANA, on AFRAM RIVER (head of canoe traffic), 28 mi/45 km ENE of MPRAESO; 05°40'N 00°37'W. Also spelled MANKRONO.

Mankulam (MAHN-kuh-LUHM), town, NORTHERN PROVINCE, SRI LANKA, 46 mi/74 km SE of JAFFNA; 09°04'N 80°27'E. Timber center; trades in dried fish from MULLAITIVU; rice.

Mankuma (mahn-koo-muh), town, NORTHERN REGION, GHANA, 10 mi/16 km N of BOLE; 09°11'N 02°29'W.

Mankusa, town (2007 population 5,199), AMHARA state, NW ETHIOPIA, 45 mi/72 km NW of DEBRE MARKOS; 10°41'N 37°11'E.

Manlio Fabio Altamirano (mahn-lee-o FAH-dee-o ahl-tah-mee-RAh-no), town, ⊙ Manlio Rabio Altamirando municipio, VERACRUZ, E MEXICO, 14 mi/23 km SW of VERACRUZ; 19°06'N 96°20'W. Corn, fruit. Sometimes called, after its railroad station, Purga or La Purga.

Manlius, village (2000 population 355), BUREAU county, N ILLINOIS, 12 mi/19 km WNW of PRINCETON; 41°27'N 89°40'W. In agricultural area.

Manlius, residential village (□ 1 sq mi/2.6 sq km; 2006 population 4,658), ONONDAGA county, central NEW YORK, 9 mi/14.5 km ESE of SYRACUSE; 43°00'N 75°58'W. Settled 1789 incorporated 1813.

Manlleu (mahn-LYE-oo), Catalan *Manlléu*, town, BARCELONA province, NE SPAIN, in CATALONIA, on TER RIVER, and 5 mi/8 km NNE of VICH; 42°00'N 02°17'E. Cereals, legumes; textile manufacturing. Sulphur springs.

Manly, town (2000 population 1,342), WORTH county, N IOWA, 11 mi/18 km S of NORTHWOOD; 43°17'N 93°12'W. Soybean products. Incorporated 1898.

Manly (MAN-lee), unincorporated village, MOORE county, central NORTH CAROLINA, 1 mi/1.6 km NE of SOUTHERN PINES. FORT BRAGG MILITARY RESERVATION to SE; 35°11'N 79°22'W. Also called Manly Station.

Manly (MAN-lee), summer resort 10 mi/16 km E of BRISBANE, SE QUEENSLAND, NE AUSTRALIA, and on MORETON BAY; 27°28'S 153°11'E.

Manly (MAN-lee), residential suburb, E NEW SOUTH WALES, SE AUSTRALIA, on N shore of PORT JACKSON, 8 mi/13 km NNE of SYDNEY; 33°47'S 151°17'E. Direct ferry to downtown Sydney. Residential, commercial, and retail area. Seaside resort; tourism.

Manly Station, NORTH CAROLINA: see MANLY.

Manma (mahn-MAH), town, ⊙ KALIKOT district, W NEPAL; 27°09'N 81°06'E. Elevation 2,420 ft/738 m.

Manmad (MUHN-mahd), town, NASHIK district, MAHARASHTRA state, W central INDIA, 45 mi/72 km ENE of NASHIK; 20°15'N 74°27'E. Railroad junction (workshops); market center for cotton, peanuts, millet; handicraft cloth weaving.

Mannachchanellur (muh-NAH-chuh-ne-LOOR), town, TIRUCHCHIRAPPALLI district, TAMIL NADU state, S INDIA, 5 mi/8 km N of Tiruchchirappalli. Trades in agricultural products (plantains, coconuts, millet) of KAVERI RIVER valley; rice milling, castor-oil extraction. Seasonal temple-festival livestock fair at village of SAMAYAPURAM, 3 mi/4.8 km ENE. Also spelled MANACHANALLOOR and Mannachanallur.

Mannadipet (muh-NAH-dee-pet), French *Mannadipeth*, town, PONDICHERRY district, Pondicherry Union Territory, S INDIA, 12 mi/19 km W of Pondicherry. Rice, peanuts, millet. Tirubhuvane, French *Tiroubouvané*, 1.5 mi/2.4 km NE, was former commune seat. Also Mandagadipattu.

Mannadipeth, INDIA: see MANNADIPET.

Mannahill (MA-nuh-hil), village, E SOUTH AUSTRALIA, S central AUSTRALIA, 125 mi/201 km ENE of PORT PIRIE, on Port Pirie–BROKEN HILL railroad; 32°27'S 139°58'E. Wool. Sometimes spelled Manna Hill.

Manna Hill, AUSTRALIA: see MANNAHILL.

Mannar, district (□ 943 sq mi/2,451.8 sq km), NORTHERN PROVINCE, SRI LANKA, on MANNAR ISLAND; ⊙ MANNAR; 08°52'N 80°05'E.

Mannar (MAN-nahr), town, ⊙ MANNAR district, NORTHERN PROVINCE, SRI LANKA, on SE shore of MANNAR ISLAND, 50 mi/80 km S of JAFFNA; 08°58'N 79°54'E. Fishing port; rice and coconut-palm plantations. Meteorological observatory. Portuguese fort (1560) strengthened by the Dutch. Also spelled MANAAR and MANAR.

Mannargudi (muh-NAHR-goo-dee), city, Thanjavur district, TAMIL NADU state, S INDIA, on arm of VENNAR RIVER delta, and 23 mi/37 km ESE of THANJAVUR; 10°40'N 79°26'E. Railroad spur terminus (junction at NIDAMANGALAM, 7 mi/11.3 km NNW). Rice milling, silk and cotton weaving; brass and copper vessels, silverware. Hindu temples.

Mannargudi (muh-NAHR-goo-dee), town, tahsil headquarters of KATTUMANNARKOIL tahsil, SOUTH ARCOT VALLALUR district, TAMIL NADU state, S INDIA, 35 mi/56 km SSW of CUDDALORE; 11°17'N 79°33'E. Rice, cassava, sesame; cotton and silk weaving. Also KATTUMANNARKOIL.

Mannar, Gulf of (muhn-NAHR), or **Manaar**, or **Manar**, inlet of INDIAN OCEAN, between S INDIA (TAMIL NADU state) and W SRI LANKA; bounded N by RAMESWARAM ISLAND (India), ADAM'S BRIDGE [a chain of islands between PAMBAN ISLAND (India) and Sri Lanka attributed to legendary Hindu Lord Rama], and MANNAR ISLAND (Sri Lanka); 80 mi/129 km–170 mi/274 km wide (E-W), c.100 mi/161 km long. Receives TAMBRAPARNI RIVER (India) and the ARUVI ARU (Sri Lanka). TUTICORIN (Tamil Nadu state) on W shore. Noted pearl banks off Sri Lanka.

Mannar Island (MAN-nahr) (□ 50 sq mi/130 sq km), in NORTHERN PROVINCE, SRI LANKA, separating PALK STRAIT (N) from Gulf of MANNAR (S), and ADAM'S Bridge (W) from Sri Lanka mainland; 16 mi/26 km long, 1 mi/6.4 km wide; 08°00'N 79°50'E. Coconut- and palmyra-palms, rice; extensive fishing. Main ports are MANNAR, TALAIMANNAR. Connected with Sri Lanka proper by road causeway and railroad, with S INDIA by ferry. Pearl banks S, in Gulf of Mannar, famous in medieval times and under colonial rule; no longer fished regularly. Also spelled MANAAR and MANAR.

Männedorf, commune, ZÜRICH canton, N SWITZERLAND, on NE shore of LAKE ZÜRICH, and 11 mi/18 km SE of ZÜRICH; 47°15'N 08°41'E. Textiles, foodstuffs.

Mannerheim Line, former line of defensive military fortifications across the KARELIAN ISTHMUS, LENINGRAD oblast, N European Russia; named for Finnish General C.G. Mannerheim. Constructed in stages during the 1920s and 1930s. Temporarily stopped a Soviet Army advance from December 1939 to February 1940.

Mannersdorf am Leithagebirge (MAHN-ners-dorf ahm LEI-tah-ge-bir-ge), town, E LOWER AUSTRIA, 20 mi/32 km SE of VIENNA; 47°58'N 16°36'E. Railroad terminus. Cement factory, limestone quarry nearby. Castle (1600); ruins of a hermitage (1644–1783) nearby.

Mannford, town (2006 population 2,771), CREEK county, central OKLAHOMA, 22 mi/35 km W of TULSA, near CIMARRON RIVER; 36°07'N 96°19'W. In agricultural, oil-producing area; manufacturing (fabricated metal products). KEYSTONE LAKE reservoir and Keystone State Park to NE.

Mannheim (MAHN-heim), city (2004 population 325,349), LOWER NECKAR, BADEN-WÜRTTEMBERG, W central GERMANY, 3 mi/4.8 km ENE of LUDWIGSHAFEN, on the right bank of the RHINE RIVER, and at the mouth of the NECKAR RIVER; 49°29'N 08°26'E. A bridge connects it with Ludwigshafen, on the opposite bank of the Rhine; it is located where the three Federal States of RHINELAND-PALATINATE, HESSEN, and BADEN-WÜRTTEMBERG converge. Inner city is laid out in rectangular pattern. It is a major inland port and an industrial center with an important trade in coal and iron. Manufacturing and industry include electrical products, chemicals, machinery, optics, and precision mechanics. Mentioned in the eighth century as a small fishing village. Fortified and chartered 1606–1607. In 1720 the city became the residence of the electors palatine, who built (1720–1760) a large palace. Awarded to BADEN in 1802. Active trade begun on Rhine in 1834 (now extends along Neckar and canals connecting two rivers). Although many of the historic buildings were heavily damaged in World War II, the city has, since 1945, restored the château and eighteenth-century baroque buildings of the inner city, including the Jesuit church (1733–1760) and the city hall (1700–1723). The first vehicle with an internal-combustion engine (built by Carl Benz in 1885) was driven here in 1885. Historically, one of the great musical and theatrical centers of Europe. Seat of a university.

Manning (MA-neeng), town (□ 1 sq mi/2.6 sq km; 2001 population 1,293), NW ALBERTA, W CANADA, 48 mi/77 km NNW of PEACE River, on MACKENZIE HIGHWAY, and railroad line to HAY RIVER (NORTHWEST TERRITORIES), on S bank of Notikewin River, in Northern Lights No. 22 municipal district; 56°55'N 117°37'W. Established 1947 as service center for transportation, mining, agriculture; forestry; oil and gas. Outfitting for sports fishing, hunting. First formed as a village; became a town in 1957.

Manning, town (2000 population 1,490), CARROLL county, W central IOWA, on WEST NISHNABOTNA RIVER (hydroelectric plant), and 15 mi/24 km SSW of CARROLL; 41°54'N 95°03'W. Manufacturing (feed, fertilizer). Incorporated 1880.

Manning, town (2006 population 4,017), ⊙ CLARENDON county, E central SOUTH CAROLINA, 17 mi/27 km SSE of SUMTER, near Pocotaligo River; 33°41'N 80°13'W. Manufacturing includes clothing, wood products, lumber, fabricated metal products, textiles, building materials. Agriculture includes dairying; tobacco, cotton, sweet potatoes, soybeans, grain.

Manning, village, ⊙ DUNN county, W central NORTH DAKOTA, 25 mi/40 km N of DICKINSON, and on KNIFE RIVER; 47°13'N 102°46'W. Lake Ilo National Wildlife Refuge to NNE. Founded in 1908 and named for local rancher, Daniel Manning. It became the county seat in 1910.

Manningham, ENGLAND: see BRADFORD.

Manning River (MA-neeng), 139 mi/224 km long, E NEW SOUTH WALES, SE AUSTRALIA; rises in GREAT DIVIDING RANGE; flows E past WINGHAM and TAREE, to the PACIFIC OCEAN SW of CROWDY HEAD; 31°52'S 152°38'E. Navigable 27 mi/43 km below Wingham by small craft carrying dairy products.

Mannington, town (2006 population 2,085), MARION county, N WEST VIRGINIA, 12 mi/19 km WNW of FAIRMONT; 39°31'N 80°20'W. Railroad terminus and processing center in oil, natural-gas, and bituminous-coal region. Manufacturing (machining; lumber). Agriculture (corn, apples); livestock, poultry. Incorporated 1871.

Mannington, village, CHRISTIAN county, SW KENTUCKY, 18 mi/29 km N of HOPKINSVILLE. Bituminous coal; agriculture (tobacco, soybeans, grain; livestock; dairying).

Manningtree (MAN-ning-tree), town (2001 population 900), NE ESSEX, SE ENGLAND, on STOUR RIVER estuary, and 8 mi/12.9 km ENE of COLCHESTER; 51°57′N 01°04′E. Former agricultural market. Plastics works.

Manno (MAHN-no), town, Nakatado county, KAGAWA prefecture, NE SHIKOKU, W JAPAN, 16 mi/25 km S of TAKAMATSU; 34°11′N 133°50′E. Chinese quinces and quince processing. Nearby Manno Pond (□ 544.2 million cu ft/15.4 million cu m) is the oldest irrigation pond in Japan, dating back to 701–704.

Manns Choice, borough (2006 population 281), BEDFORD county, S PENNSYLVANIA, 5 mi/8 km WSW of BEDFORD, on Raystown Branch of JUNIATA RIVER; 40°00′N 78°35′W. Agriculture (corn, oats, hay; livestock; dairying); manufacturing (lumber). Shawnee Lake reservoir and Shawnee State Park to NW.

Manns Harbor (MANZ HAHR-buhr), village, DARE county, NE NORTH CAROLINA, on CROATAN SOUND (bridged) opposite ROANOKE ISLAND; 35°53′N 75°45′W. Fishing.

Mannsville, village (2006 population 403), JEFFERSON county, N NEW YORK, 20 mi/32 km SSW of WATERTOWN; 43°43′N 76°04′W. Settled 1801, incorporated 1879.

Mannsville, village (2006 population 566), JOHNSTON county, S OKLAHOMA, 15 mi/24 km E of ARDMORE, near WASHITA River; 34°11′N 96°52′W. In farm area.

Mannu, Cape (MAHN-noo), point on W coast of SARDINIA, 15 mi/24 km NW of ORISTANO; 40°02′N 08°22′E. Fisheries (tunny, lobster, coral).

Mannu d'Oschiri River (MAHN-noo DO-skee-ree), c.32 mi/51 km long, SASSARI province, N SARDINIA, ITALY; rises in MONTI DI ALÀ; flows S, cutting through Monti di Alà and CATENA DEL GOCEANO, then N to Lake COGHINAS.

Mannu d'Ozieri River (MAHN-noo dot-SYAI-ree), c.40 mi/64 km long, SASSARI province, N SARDINIA, ITALY; rises in several branches in Catena del MARGHINE; flows NNE, past Chilivani, to Lake COGHINAS.

Mannum (MA-nuhm), town, SE SOUTH AUSTRALIA state, S central AUSTRALIA, 40 mi/64 km E of ADELAIDE, and on MURRAY RIVER; 34°55′S 139°19′E. Agriculture, dairying center; citrus, dried fruit. Bird sanctuary.

Mannu River (MAHN-noo), c.35 mi/56 km long, SASSARI province, N SARDINIA, ITALY; rises near Thiesi; flows NW to Gulf of ASINARA at Porto Torres. Other Sardinian streams, including MANNU D'OSCHIRI and MANNU D'OZIERI rivers, are sometimes called simply Mannu.

Mannville (MAN-vil), village (□ 1 sq mi/2.6 sq km; 2001 population 722), E ALBERTA, W CANADA, near VERMILION RIVER, 14 mi/23 km W of VERMILION, in MINBURN COUNTY NO. 27; 53°20′N 111°10′W. Dairying; grain growing.

Mano (MAH-no), town, on SW section of SADO island, Sado county, NIIGATA prefecture, N central JAPAN, on Mano Bay, 37 mi/60 km W of NIIGATA; 37°57′N 138°20′E. Flatfish farming.

Mano (MAH-no), town, SOUTHERN province, S central SIERRA LEONE, on JONG RIVER, and 25 mi/40 km ESE of MOYAMBA, at head of road to TAIAMA; 08°02′N 12°07′W. Trade center; palm oil and kernels, rice.

Manoa (mah-NO-ah), town and canton, GENERAL FEDERICO ROMÁN province, northernmost point of BOLIVIA, PANDO department, on MADEIRA RIVER, at mouth of ABUNÁ RIVER, and 100 mi/161 km NNE of RIBERALTA, on BRAZIL border; 09°40′S 65°27′W. Rubber. ABUNÁ (Brazil), on Madeira-Mamoré railroad, is 6 mi/9.7 km ESE across the Madeira.

Manoa, Pennsylvania: see HAVERFORD.

Manoel Duarte (MAH-noo-el DOO-ahr-tai), city, N RIO DE JANEIRO state, BRAZIL, 5 mi/8 km SW of TRÊS RIOS, on railroad; 23°08′S 43°15′W.

Manoel Emidio (MAH-noo-el E-mee-zhee-o), town (2007 population 5,361), SW PIAUÍ state, BRAZIL, 95 mi/153 km SW of FLORIANO; 07°59′S 43°57′W.

Manoel Victorino (MAH-no-el vee-TOR-een-o), city, E central BAHIA, BRAZIL, 24 mi/38 km SW of Jequié; 14°08′S 40°14′W.

Man of War Bay, deep inlet of NW TOBAGO, TRINIDAD AND TOBAGO, 14 mi/23 km NE of SCARBOROUGH; c.2 mi/3.2 km long, 2 mi/3.2 km wide; 11°19′N 60°34′W. On it is CHARLOTTEVILLE. High cliffs. Sometimes spelled Man-of-War Bay.

Man-of-War Cays, Spanish *Cayos Guerrero*, islands, NICARAGUA, 5 mi/8 km NW of Matagalpa River mouth. Small islands and reefs.

Manoharpur (muh-NO-huhr-poor), town, WEST SINGHBHUM district, S Jharkhand state, E India, 70 mi/113 km WSW of JAMSHEDPUR. Rice, oilseeds, corn. Major hematite mining in nearby hills; extensive sal lumbering; bhabur; limestone quarries.

Manoharpur, town, JAIPUR district, E RAJASTHAN state, NW INDIA, 25 mi/40 km NNE of JAIPUR. Agriculture (millet, gram).

Manoir Lake, NORTHWEST TERRITORIES: see MAUNOIR, LAC.

Manokin River (mah-NOK-in), c.25 mi/40 km long, SE MARYLAND; rises in NE SOMERSET county; flows SW, past Princess Anne (head of navigation), to TANGIER SOUND just S of DEAL ISLAND. Its mouth is 5 mi/8 km wide.

Manokotak, village (2000 population 399), SW ALASKA, on small IGUSHIK River, and 22 mi/35 km WSW of DILLINGHAM; 58°59′N 159°03′W.

Manokwari (mah-NOK-wah-ree), town, ⊙ Manokwari district, IRIAN JAYA province, INDONESIA, on E DOBERAI peninsula, port on NW side of entrance to CENDERAWASIH BAY; 02°29′S 134°36′E. Trade center; exports copra, resin. Formerly capital of Netherlands New Guinea, supplanted by HOLLANDIA. Rendani Airport. Center of Roman Catholic missionary activity; agriculture college.

Manola (muh-NO-luh), hamlet, central ALBERTA, W CANADA, 7 mi/11 km E of BARRHEAD, in BARRHEAD COUNTY NO. 11; 54°06′N 114°14′W.

Manole (MAHN-o-le), village, PLOVDIV oblast, MARISHKA obshtina, BULGARIA, 42°11′N 24°56′E.

Manomet, MASSACHUSETTS: see PLYMOUTH.

Manonga River (mah-NO-ngah), N central TANZANIA; rises c. 55 mi/89 km WNW of SHINYANGA, in Siga Hills, S of Lake VICTORIA; flows S, then E into SERENGETI PLAIN, passing Shinyanga to S, joins WEMBERE RIVER 60 mi/97 km ESE of Shinyanga at upper end of Lake Kitangiri to form SIBITI RIVER Riverbed known for its fossil finds dating 5–6 million years.

Manono (mah-NO-no), village, KATANGA province, SE CONGO, 155 mi/249 km W of KALEMIE; 07°18′S 27°25′E. Elev. 2,181 ft/664 m. Center of major tin-mining area; terminus of railroad from MUYUMBA. Tin concentrating and smelting; also tantalite mining. Airport. Has Roman Catholic and Protestant missions, hospitals, extensive recreation facilities for industries personnel.

Manono (mah-NO-no), volcanic island (□ 1 sq mi/2.6 sq km), SAMOA, S PACIFIC, in 10-mi/16-km Apolima Strait; 13°50′S 172°05′W. Strait separated 'UPOLU from SAVAI'I within barrier reef surrounding 'Upolu; rises to 197 ft/60 m.

Manoomukh, BANGLADESH: see MANUMUKH.

Manor (MA-nor), town, WARE county, GEORGIA, 15 mi/24 km SW of WAYCROSS. Manufacturing of wood products.

Manor (MAI-nor), town (2006 population 2,657), TRAVIS county, S central TEXAS, 12 mi/19 km ENE of AUSTIN; 30°21′N 97°33′W. In cotton, grain area.

Manufacturing (electronic products, computer equipment).

Manor (MA-nor), village (2006 population 312), SE SASKATCHEWAN, W CANADA, at foot of MOOSE MOUNTAIN, 40 mi/64 km SSW of MOOSOMIN; 49°36′N 102°05′W. Mixed farming.

Manor (MA-nor), borough (2006 population 2,848), WESTMORELAND county, SW PENNSYLVANIA, suburb, 18 mi/29 km SE of PITTSBURGH, and 2 mi/3.2 km W of JEANNETTE, on Brush Creek; 40°21′N 79°40′W. Manufacturing (wood products, fabricated metal products). Laid out 1873, incorporated 1890.

Manorbier (ma-nuhr-BIR), Welsh *Maenorbŷr* village (2001 population 1,288), Pembrokeshire, SW Wales, on CARMARTHEN BAY of BRISTOL CHANNEL, 5 mi/8 km ESE of PEMBROKE; 51°39′N 04°48′W. Has remains of castle.

Manorhamilton (ma-nuhr-HA-muhl-tuhn), Gaelic *Cluain in Uí Ruairc*, town (2006 population 1,158), N LEITRIM county, N central IRELAND, 12 mi/19 km E of SLIGO; 54°18′N 08°11′W. Agricultural market. Has ruins of fortified mansion (1638).

Manorhaven, village (2006 population 6,307), NASSAU co., SE NEW YORK, on N shore of W LONG ISLAND, just NE of PORT WASHINGTON; 40°50′N 73°42′W. In marine-recreation area. Incorporated 1930.

Mano River (MAH-no), c.200 mi/322 km long, LIBERIA and SIERRA LEONE; rises near VOINJAMA (Liberia), on GUINEA border; flows SW and along Sierra Leone–Liberia border to the ATLANTIC OCEAN at MANO SALIJA (Sierra Leone). Morro River, its right tributary, forms upper section of border.

Manor Park, ENGLAND: see NEWHAM.

Manor Ridge, unincorporated town, LANCASTER county, SE PENNSYLVANIA, residential suburb, 4 mi/6.4 km W of LANCASTER; 40°02′N 76°21′W. Garden Spot Airport to W.

Manorville, borough (2006 population 375), ARMSTRONG county, W central PENNSYLVANIA, 2 mi/3.2 km S of KITTANNING, and 1 mi/1.6 km NNE of FORD CITY, on ALLEGHENY RIVER; 40°47′N 79°31′W. Agriculture (corn, hay; dairying).

Mano Salija (MAH-no sah-LEE-jah), village, SOUTHERN province, SE SIERRA LEONE, minor port on ATLANTIC OCEAN, at mouth of MANO RIVER (LIBERIA border), and 32 mi/51 km SE of PUJEHUN; 06°56′N 11°30′W. Palm oil and kernels, rice.

Manos, Las, NICARAGUA: see LAS MANOS.

Manosque (mah-NOSK), town (□ 21 sq mi/54.6 sq km), ALPES-DE-HAUTE-PROVENCE department, PROVENCE-ALPES-CÔTE D'AZUR region, SE FRANCE, in fertile alluvial DURANCE RIVER valley, 27 mi/43 km NE of AIX-EN-PROVENCE; 43°50′N 05°47′E. Agricultural trade center (olives, almonds, truffles, lavender, fruits, and wine), along with high-technology complex centered at nearby CADARACHE nuclear research and testing facility, 8 mi/12.9 km S at junction of Durance and VERDON rivers. Several hydroelectric generating stations are also in this reach of the Durance River. Has two restored Romanesque churches and 14th-century fortified gates, a picturesque main street, and an elegant 17th-century town hall; has a colorful outdoor market.

Manotick (MA-nuh-tik), unincorporated village, SE ONTARIO, E central CANADA, on Rideau River and RIDEAU CANAL, and included in OTTAWA; 45°15′N 75°37′W. Commuter suburb, on S edge of city of NEPEAN. Dairying; mixed farming.

Manouba, Arabic *Manūbah*, province (□ 409 sq mi/1,063.4 sq km; 2006 population 346,900), N TUNISIA; ⊙ AL MANUBA; 36°45′N 10°00′E. Created c.2003.

Manpo (MAHN-PO), city, CHAGANG province, NORTH KOREA, on railroad, and 115 mi/185 km NNE of SINUIJU, on YALU RIVER (CHINA border), opposite Jian (China); 41°10′N 126°20′E.

Manqabad (mahn-kah-BAHD), village, ASYUT province, central Upper EGYPT on W bank of the NILE

River, on railroad, 5 mi/8 km WNW of ASYUT; 27°12′N 31°07′E. Pottery making, wood carving; cereals, dates, sugar.

Manra (MAHN-rah), triangular coral island (□ 2 sq mi/5.2 sq km), PHOENIX, KIRIBATI, S PACIFIC, 110 mi/ 177 km SE of KANTON Island, c.2-mi/3.2-km-long base of triangle of islands; 04°27′S 171°15′W. Discovered 1823 by Americans, included 1937 in British GILBERT AND ELLICE ISLANDS Colony, becoming part of independent Kiribati in 1979. Populated by Gilbertese 1938–1958, when settlement removed to GHIZO (SOLOMON ISLANDS). Sometimes called Sydney Island.

Manresa (mahn-RAI-sah), city, BARCELONA province, NE SPAIN, in CATALONIA, on the CARDONER RIVER; 41°44′N 01°50′E. Industrial center with textile, metallurgical, and glass industries. Of ancient origin, Manresa has a Roman bridge and a Gothic collegiate church. Place of pilgrimage because St. Ignatius of Loyola prayed here (1522) on his way back from MONTSERRAT. Seat of a conservatory of music.

Manrique (mahn-REE-kai), town, COJEDES state, N VENEZUELA, 12 mi/19 km NNE of SAN CARLOS; 09°48′N 68°30′W. Livestock; vegetables.

Man River, Burmese *Man Chaung*, more than 50 mi/80 km long, in MINBU township, MAGWE division, MYANMAR; rises in ARAKAN YOMA; flows NE, past SAGU, to AYEYARWADY RIVER N of MINBU. Used for irrigation of S MINBU district; the left-bank MAN CANAL takes off at headworks 18 mi/29 km W of MINBU.

Mansa (mahn-SAH), district (□ 839 sq mi/2,181.4 sq km), Punjab state, N India; ⊙ Mansa.

Mansa (mahn-SAH), town, MAHESANA district, GUJARAT state, W central INDIA, 19 mi/31 km SE of MAHESANA; 23°26′N 72°40′E. Cotton, wheat, millet. Was capital of former princely state of Mansa of WESTERN INDIA STATES; incorporated 1949 into newly created Mahesana district, later made part of Gujarat state.

Mansa, town, ⊙ MANSA district, PUNJAB state, N INDIA, 65 mi/105 km WSW of PATIALA; 29°59′N 75°23′E. Market (wheat, millet, cotton).

Mansa, township, LUAPULA province, N ZAMBIA, near LUAPULA RIVER (CONGO border), 100 mi/161 km NE of ELISABETHVILLE; 11°12′S 28°53′E. Agriculture (rice, vegetables, chilies, bananas, cassava); fish. Manufacturing (batteries; rice processing). Manganese mining. Airfield. LAKE BANGWEULU to E; Mambilima Falls on Luapula River to NW. Formerly called Fort Rosebery. Transferred 1947 from NORTHERN province. Was capital of former Mweru-Luapula province.

Mansabá (mahn-suh-BUH), town, OIO province, GUINEA-BISSAU, 41 mi/66 km NE of BISSAU; 12°18′N 15°15′W. Crossroads.

Mansafis (mahn-sah-FEES), village, MINYA province, Upper EGYPT, on railroad, 5 mi/8 km NNW of ABU QURQAS; 28°00′N 30°49′E. Cotton, cereals, sugarcane.

Mansahra, district, NORTH-WEST FRONTIER PROVINCE, N PAKISTAN, at the confluence of the Kaghan valley with the Pakhli plain; ⊙ MANSAHRA. Hilly. Traditional crops of potatoes, rice, wheat, and vegetables now supplemented by tea; opium poppies on the hilltops in the Black mountain region adjacent to the TARBELA reservoir. Agriculture production has developed with the construction of the KARAKORAM HIGHWAY. Mauryan edict from the 3rd century B.C.E. (attributed to Asoka) is located along the KARAKORAM HIGHWAY here. Afghan refugees settled here from 1980 and contributed to large-scale deforestation of pine and cedar trees. The village of Khaki, NW of MANSAHRA, gave its name to the sand-colored fabric worn by villagers opposing British occupation. Cantonment town of ABBOTTABAD, immediately S of MANSAHRA, dates to the British period.

Mansahra (muhn-SAH-rah), town, ⊙ MANSAHRA district, NE NORTH-WEST FRONTIER PROVINCE, N PAKISTAN, 12 mi/19 km N of ABBOTTABAD; 34°20′N 73°12′E. Trades in maize, wheat, rice, fruit. Extensive lumbering has greatly reduced tree cover. Entrance to

Kaghan valley now a popular destination for tourists. Transportation distribution town located on KARAKORAM HIGHWAY. Rock edict of Asoka (3rd century B.C.E.) is NW. Also spelled MANSEHRA.

Mansa Konko (muhn-sah KON-ko) [=chief's hill], town, ⊙ LOWER RIVER division, THE GAMBIA, 60 mi/ 97 km E of BANJUL; 13°28′N 15°33′W. Also spelled Mansakonko.

Mansavillagra (mahn-sah-vi-YAH-grah), village, FLORIDA department, S central URUGUAY, in the CUCHILLA GRANDE INFERIOR, near the ARROYO MANSAVILLAGRA, on railroad, and 25 mi/40 km WSW of JOSÉ BATLLE Y ORDÓÑEZ; 33°38′S 55°30′W. Wheat, corn, linseed; livestock.

Mansavillagra, Arroyo, 50 mi/80 km long, river, FLORIDA department, S central URUGUAY; rises in the CUCHILLA GRANDE PRINCIPAL 10 mi/16 km SSW of ILLESCAS; flows NW to YÍ RIVER 14 mi/23 km WSW of SARANDÍ DEL YÍ; 33°26′S 55°54′W.

Manseau (mahn-SO), village (□ 40 sq mi/104 sq km; 2006 population 922), CENTRE-DU-QUÉBEC region, S QUEBEC, E CANADA, 21 mi/33 km from BÉCANCOUR; 46°22′N 72°00′W.

Mansehra, PAKISTAN: see MANSAHRA.

Mansel Island (1,317 sq mi/3,411 sq km), NUNAVUT territory, N CANADA, in HUDSON BAY, off N Ungava Peninsula; 62 mi/100 km long, 4 mi/6 km–30 mi/48 km wide. Created reindeer reserve in 1920. On N coast is trading post (62°25′N 79°36′W).

Manseriche, Pongo de (mahn-sai-REE-chai, PON-go dai), N PERU, gorge of MARAÑÓN RIVER, on border between AMAZONAS and LORETO regions, near its junction with SANTIAGO RIVER in Cerros de CAMPANQUIZ; 04°25′S 77°35′W. Marañón navigable for large boats to this point. Variant spelling: Manserriche.

Manserriche, PERU: see MANSERICHE, PONGO DE.

Mansfeld (MAHNS-felt), town, SAXONY-ANHALT, central GERMANY, at E foot of the lower HARZ, near the WIPPER RIVER, 6 mi/9.7 km NW of EISLEBEN; 51°37′N 11°27′E. Was a center (since twelfth century) of major copper-slate-mining region. Manufacturing furniture. Luther spent his childhood here. Declined in importance in sixteenth century, revived after 1671 when monopoly mining rights were abolished. Neo-Gothic castle from 1509–1518; rebuilt in 1860–1862. Has remains of eleventh-century castle.

Mansfield (MANS-feeld), city, ⊙ DE SOTO parish, NW LOUISIANA, 33 mi/53 km S of SHREVEPORT; 32°02′N 93°42′W. Trading and shipping center for fertile agricultural area; oil and natural-gas wells; timber. Lumber milling; manufacturing (fabricated metal products, transportation equipment). CLEAR and Smithport lakes to NE. Mansfield State Commemorative Area, to SE, marks the site of the Civil War battle of Sabine Crossroads (April 8, 1864), a Confederate victory. Incorporated 1847.

Mansfield (MANZ-feeld), city (□ 30 sq mi/78 sq km; 2006 population 50,201), ⊙ RICHLAND county, N central OHIO, 60 mi/96 km NNE of COLUMBUS; 40°46′N 82°31′W. In a hilly region surrounded by fertile farmlands. Commercial and insurance center. Formerly major manufacturing (steel, automotive body assembly). Among its many other products were tires, electrical appliances, sports vehicles, and brass goods. The city's economy has become more service-centered, with retail, education, and healthcare sectors, since the decline of manufacturing in the 1970s. A branch of Ohio State University is here. The home of Louis Bromfield is used as an ecological center and experimental farm. Also of interest are South Park, with a reconstructed blockhouse of the War of 1812, and Kingwood Center, with gardens, landscaped floral displays, and a pre-Civil War French-provincial mansion. Incorporated 1828.

Mansfield (MANZ-feeld), city (2006 population 41,564), TARRANT and JOHNSON counties, N TEXAS, suburb 16 mi/26 km SE of FORT WORTH; 32°34′N

97°07′W. Former agricultural area to S (cotton, corn; dairying), rapid urban growth is supplanting agriculture. Manufacturing (machinery, building materials, transportation equipment, electronic equipment, modular buildings, chemicals). Lake Joe Pool reservoir and Cedar Hill State Park to W.

Mansfield (MANZ-feeld), town, central VICTORIA, SE AUSTRALIA, 80 mi/129 km NE of MELBOURNE; 37°03′S 146°05′E. Railroad terminus in grazing and agricultural area. Former gold-mining town. Tourism.

Mansfield (MANZ-feeld), town (2001 population 69,987) and district, NOTTINGHAMSHIRE, central ENGLAND, on the W border of SHERWOOD FOREST; 53°08′N 01°12′W. In a coal district. Manufacturing of hosiery, shoes, electrical controls, and metal products. Limestone and red and white sandstone quarried nearby. Prehistoric cave dwellings in vicinity. Medieval church. The grammar school here was founded in 1561. MANSFIELD WOODHOUSE nearby.

Mansfield, town (2000 population 1,097), on border between SEBASTIAN and SCOTT counties, W ARKANSAS, 24 mi/39 km SSE of FORT SMITH; 35°03′N 94°15′W. In farm area; lumber. Ouachita National Forest to S.

Mansfield, town, TOLLAND county, NE CONNECTICUT; 41°47′N 72°13′W. Agricultural and manufacturing town, The University of Connecticut is in STORRS, which is included within Mansfield. The town also includes Mansfield Hollow, the site of a large flood-control project. Mansfield Training School for Mentally Retarded, which was the oldest such institution in the state, closed in 1993. Settled c.1692, incorporated 1702.

Mansfield, town, including Mansfield village, BRISTOL county, SW MASSACHUSETTS, 15 mi/24 km NE of PROVIDENCE (RHODE ISLAND); 42°01′N 71°13′W. Machine parts, metal products, chocolate, medical and scientific instruments and equipment, chemicals and inks. Settled 1659, set off from NORTON 1770. Great Woods Outdoor Performing Arts Center is here.

Mansfield, township, WARREN county, NW NEW JERSEY, 3 mi/4.8 km SW of HACKETTSTOWN; 40°48′N 74°54′W. Incorporated 1798.

Mansfield, village (2000 population 392), NEWTON county, N central GEORGIA, 9 mi/14.5 km SE of COVINGTON; 33°31′N 83°44′W. Manufacturing (textiles, machinery).

Mansfield, village (2000 population 949), PIATT county, central ILLINOIS, 14 mi/23 km NNE of MONTICELLO; 40°12′N 88°30′W. In grain-growing area.

Mansfield, village, PARKE county, W INDIANA, 9 mi/ 14.5 km SE of ROCKVILLE, and on Big RACCOON CREEK. Agricultural area (corn, soybeans; hogs, cattle). Covered bridges in area. Seasonal tourist site, especially in autumn.

Mansfield, village (2006 population 336), DOUGLAS county, central WASHINGTON, 15 mi/24 km NE of WENATCHEE; 47°49′N 119°38′W. Railroad terminus. In COLUMBIA basin agricultural region (wheat, barley, oats, alfalfa; cattle, sheep).

Mansfield, borough (2006 population 3,266), TIOGA county, N PENNSYLVANIA, 24 mi/39 km SW of ELMIRA, NEW YORK, on TIOGA RIVER; 41°48′N 77°04′W. Agricultural area (grain, soybeans; livestock; dairying); manufacturing (wood products); natural gas. Seat of Mansfield University of Pennsylvania Hills Creek State Park to W; Hammond and Tioga reservoirs (joined by channel, to NW). Laid out 1824, incorporated 1857.

Mansfield Dam, Texas: see TRAVIS, LAKE.

Mansfield-et-Pontefract (MANZ-feeld–ai–pont-uh-FRAHKT), village (□ 163 sq mi/423.8 sq km; 2006 population 2,099), OUTAOUAIS region, SW QUEBEC, E CANADA, 12 mi/19 km from WALTHAM; 46°02′N 76°44′W.

Mansfield Hollow Dam, CONNECTICUT: see NATCHAUG RIVER.

Mansfield Hollow Lake, reservoir (☐ 1 sq mi/2.6 sq km), TOLLAND and WINDHAM counties, W central CONNECTICUT, on NATCHAUG RIVER, 3 mi/5 km NNE of WILLIMANTIC; 41°46′N 72°11′W. Maximum capacity 79,000 acre-ft. Formed by Mansfield Hollow Dam (78 ft/24 m high), built (1952) by Army Corps of Engineers for flood control. Mansfield Hollow State Park at N end of reservoir.

Mansfield, Mount, peak (4,393 ft/1,339 m), N central VERMONT, highest peak in the GREEN MOUNTAINS and in Vermont. Most of the mountain is in Mt. Mansfield State Forest. At the foot of the mountain is a deep gorge called Smugglers Notch. With major ski area, Mt. Mansfield is a winter-sports center offering some of the finest skiing in New England.

Mansfield Woodhouse (MANZ-feeld WUHD-hous), town (2001 population 17,931), W NOTTINGHAMSHIRE, central ENGLAND, 2 mi/3.2 km N of MANSFIELD; 53°10′N 01°12′W. Stone quarrying; coal mining. Machinery, electronics, textile manufacturing. Has 14th-century church. Was once a forest post for guarding against wolves.

Manshah, El (men-SHAH, el), town, SOHAG province, central Upper EGYPT, on W bank of the NILE River, on railroad, 11 mi/18 km NW of GIRGA; 26°28′N 31°48′E. Cotton ginning, pottery making; dairying; cotton, cereals, dates, sugarcane. Site of a town built by Ptolemy I Soter, first king on the Acedonian dynasty of ancient Egypt (c.300 B.C.E.). Also spelled El Menshah.

Mansidão (MAHN-see-DOUN), city (2007 population 11,697), NW BAHIA state, BRAZIL, in Serra da TABATINGA near PIAUÍ state border; 10°44′S 44°02′W.

Mansilla (mahn-SEE-yah) or **Gobernador Mansilla**, town, S central ENTRE RÍOS province, ARGENTINA, on railroad, and 40 mi/64 km N of GUALEGUAY. Flax, wheat, oats; livestock.

Mansilla de las Mulas (mahn-SEE-lyah dai lahs MOO-lahs), town, LEÓN province, NW SPAIN, on ESLA RIVER, and 11 mi/18 km SE of LEÓN; 42°30′N 05°25′W. Tanning, candy manufacturing. Cereals; livestock; lumber. Has medieval castle.

Mansinha (MAHN-seen-yah), city, E TOCANTINS state, BRAZIL, 166 mi/267 km SE of PALMAS; 09°28′S 47°42′W.

Mansión, La, COSTA RICA: see LA MANSIÓN.

Mans, Le (MAWN, luh), city (☐ 20 sq mi/52 sq km); ⊙ SARTHE department, PAYS DE LA LOIRE region, NW FRANCE, on the SARTHE RIVER, and 115 mi/185 km SW of PARIS; 48°00′N 00°12′E. The historical capital of MAINE province, it is a regional manufacturing (motor vehicles, agricultural machinery), commercial, educational, and communications center, with a metropolitan area population of c.190,000. Its service industries, especially insurance and finance, are important. Le Mans, which dates from pre-Roman (Celtic) times, has witnessed frequent sieges and battles, especially in the time of William the Conqueror and King John, and again in the Hundred Years War, Huguenot uprising (1562), and VENDÉE rebellion (1793). The Cathedral of St. Julien (11th–13th century), which contains the tomb of Berengaria, queen of Richard the Lionhearted (Richard I of England), is partly Romanesque; its Gothic part (fine stained-glass windows) has perhaps the most daring system of flying buttresses of any Gothic cathedral. The old quarter, known as Vieux Mans, near the cathedral, is a quaint retail and restaurant district. There are several museums (including an automobile museum) and smaller churches as well as the university and a bishop's see. Henry II of England and John II of France born here. Today, Le Mans is famous for its annual 24-hour international auto race, which is run on local roads, near the fairgrounds.

Manso (mahn-SO), town, WESTERN REGION, GHANA, 20 mi/32 km SE of TARKWA; 05°05′N 01°50′W. Railroad stop. Rubber, manganese, palm oil.

Mansôa (MAHNSH-o-uh), town, OIO province, W GUINEA-BISSAU, 23 mi/37 km NE of BISSAU; 12°10′N 14°36′W. Rice, cashew nuts, coconuts, peanuts, palm oil.

Manson (MAN-suhn), community, SW MANITOBA, W central CANADA, 15 mi/24 km W of MINIOTA, in ARCHIE rural municipality; 50°08′N 101°22′W.

Manson, town (2000 population 1,893), CALHOUN county, central IOWA, 10 mi/16 km NE of ROCKWELL CITY; 42°31′N 94°32′W. Twin Lakes State Park nearby. Founded 1872, incorporated 1877.

Mansonville (MAN-suhn-vil), unincorporated village, S QUEBEC, E canada, on Missisquoi River, near U.S. (VERMONT) border, 35 mi/56 km SW of SHERBROOKE, and included in POTTON; 45°03′N 72°23′W. Dairying.

Manso, Rio, Brazil: see RIO DAS MORTES.

Mansourah, El (mahn-soo-RAH, el) [Arab.=the victorious], site of historic town, near the walls of TLEMCEN, TLEMCEN wilaya, NW ALGERIA. Founded in the late 13th century by Abou Yakoub, built by Merinids, and surrounded by a massive rampart, it was the base for a great siege (1299–1307) against their rivals, the Zianids. After the fall of Tlemcen, it was made the capital of the Merinid government of central Maghreb, and a vast palace was built for Abou El Hassan, the leader. When the Zianids reconquered Tlemcen, they destroyed El Mansourah. The many defensive towers still stand amidst olive groves. The site of the Great Mosque is used as a quarry. The site is the source of many of the marble and onyx columns supporting the mosques of Tlemcen and some private residences.

Mansura (man-SOO-ruh), town (2000 population 1,573), AVOYELLES parish, E central LOUISIANA, 29 mi/47 km SE of ALEXANDRIA; 31°04′N 92°03′W. In sugarcane and cotton area; timber; manufacturing (custom millwork, valves). Spring Bayou State Wildlife Area to NE, Grand Cote National Wildlife Refuge to NW.

Mansurabad, IRAN: see MEHRAN.

Mansurah, Al (mahn-SOO-ruh, el), city, ⊙ DAQAHLIYA province, N EGYPT, a port in the DUMYAT branch of the NILE River delta. Agricultural market and industrial center. Manufacturing includes ginned cotton, cottonseed oil, and textiles. Founded in 1221 to replace Dumyat (Damietta), then occupied by Crusaders. In 1250, Crusaders under Louis IX of FRANCE suffered a crushing defeat here at the hands of the Mamluks. Seat of a branch of the University of Cairo, the Inst. of Al Mansurah (affiliated with Al Azhar University in CAIRO), and Al Mansurah Polytechnic Inst.

Mansuri (muhn-SOO-ree), former petty sheikdom of SUBEIHI tribal area, SW YEMEN. Am Masharij was its center. Under British protection, 1871–1967. Abolished when SOUTH YEMEN became independent.

Mansuriya, Al (MUHN-soor-EE-yeh, ahl), township, DIYALA province, central IRAQ, near E bank of the TIGRIS River, and 30 mi/48 km N of BAGHDAD. Dates, fruits; livestock.

Mansuriya, El (mahn-SOO-RAI-yuh, el), township, GIZA province, Upper EGYPT, 12 mi/19 km NW of CAIRO city center.

Manta, town, ATAKORA department, NW BENIN, 20 mi/32 km W of NATITINGOU; 10°21′N 01°07′E. Cotton; livestock; shea-nut butter.

Manta (MAHN-tah), town, ⊙ Manta municipio, CUNDINAMARCA department, central COLOMBIA, 33 mi/53 km NE of BOGOTÁ; 05°01′N 73°32′W. Elevation 5,190 ft/1,581 m. Agriculture includes coffee, corn; livestock.

Manta (MAHN-tah), town (2001 population 183,105), MANABÍ province, W ECUADOR, PACIFIC port on Manta Bay, 105 mi/169 km NW of GUAYAQUIL, 22 mi/35 km WNW of PORTOVIEJO (linked by railroad); 00°57′S 80°44′W. Fishing-boat harbor. Exports products of fertile agricultural hinterlands, including coffee, bananas, cacao, tagua nuts, balsa wood, fish, hides. Has sawmills, cotton textile factories, rice mills,

tanneries. Airport, customhouse. Center for Ecuadorian tourism. Site of ancient Manta culture.

Mantachie (man-TACH-ee), village (2000 population 1,107), ITAWAMBA county, NE MISSISSIPPI, 5 mi/8 km NW of FULTON; 34°19′N 88°29′W. Agriculture (cotton, corn, soybeans; cattle, poultry); manufacturing (furniture).

Mantador (MANT-uh-dor), village (2006 population 67), RICHLAND county, extreme SE NORTH DAKOTA, 19 mi/31 km WSW of WAHPETON, and on WILD RICE RIVER; 46°10′N 96°58′W. Founded in 1893 and incorporated 1947.

Mantady National Park (mahn-tah-DEE), TOAMASINA province, E MADAGASCAR, 20 mi/32 km NE of MORAMANGA; 18°10′S 48°30′E. Elevation 2,500 ft/762 m–3,500 ft/1,067 m. New park of upland rain forest near ANDASIBE. Lemur.

Mantai (MUHN-thai), ancient *Mantota* or *Mahatittha*, village, NORTHERN PROVINCE, SRI LANKA, 6 mi/9.7 km E of MANNAR; 08°57′N 79°58′E. Coconut-palm plantations, vegetable gardens. Chief ancient landing place of Sri Lanka.

Mantakari (MAHN-tuh-kah-ree), town, TAHOUA province, NIGER, 200 mi/322 km ENE of NIAMEY.

Mantare (mahn-TAH-rai), town, MWANZA region, NW TANZANIA, 25 mi/40 km SE of MWANZA, on railroad; 02°44′S 33°10′E. Sugarcane, cotton, corn, wheat; cattle, sheep, goats. Diamond deposits at MISUNGWI, 10 mi/16 km SW.

Mantaro River (mahn-TAH-ro), c.360 mi/579 km long, central PERU; rises near CERRO DE PASCO (PASCO region) in CORDILLERA OCCIDENTAL; flows SE almost to HUANTA, where it turns sharply NW and finally ENE to the APURÍMAC (Apurímac River is from here on called Ene River); 12°16′S 73°57′W. On it are LA OROYA, JAUJA, and HUANCAYO. Used for irrigation in ANDEAN valleys; not navigable. Its upper course is linked through small affluent with Lake JUNÍN.

Mantasoa (mahn-tah-SOO), town, ANTANANARIVO province, central MADAGASCAR, 20 mi/32 km W of ANTANANARIVO; 19°01′S 47°50′E. Town and resort complex on Mantasoa reservoir. Forest plantations (pine, eucalyptus); charcoal production. Popular weekend retreat for Antananarivo elite. Pioneering industrial site of Jean Laborde in mid-19th century.

Mantasoa Reservoir, MADAGASCAR: see IKOPA RIVER.

Manteca, city (2000 population 49,258), SAN JOAQUIN county, central CALIFORNIA, 10 mi/16 km SSE of STOCKTON; 37°48′N 121°13′W. Railroad junction. Diverse agriculture (grapes, sugar beets, fruit, nuts, beans, vegetables, pumpkins; nursery products); manufacturing (frozen fruits and vegetables, cut stone, electrical machinery; sugar processing). Caswell Memorial State Park to S; Sharpe Army Depot to NW. Founded 1870, incorporated 1918.

Mante, Ciudad, MEXICO: see CIUDAD MANTE.

Mantee (man-TEE), village (2000 population 169), WEBSTER county, central MISSISSIPPI, 22 mi/35 km NW of STARKVILLE; 33°43′N 89°03′W. Agriculture and timber; manufacturing (wood products). NATCHEZ TRACE PARKWAY passes to E.

Manteigas (mahn-TAI-gash), town, GUARDA district, N central PORTUGAL, in SERRA DA ESTRÊLA, near headwaters of ZÊZERE RIVER, 18 mi/29 km SW of GUARDA. Dairy; textiles. Sulphur springs just S, at Caldas de Manteigas.

Mantena (MAHN-te-nah), city (2007 population 26,716), E central MINAS GERAIS state, BRAZIL, near border with ESPÍRITO SANTO state, 68 mi/109 km E of GOVERNADOR VALADARES; 18°47′S 41°05′W.

Mantena, Brazil: see BARRA DE SÃO FRANCISCO.

Manteno (man-TEE-no), village (2000 population 6,414), KANKAKEE county, NE ILLINOIS, 9 mi/14.5 km N of KANKAKEE; 41°15′N 87°50′W. In agricultural area. Incorporated 1878.

Mantenópolis (MAHN-ten-O-po-lees), city (2007 population 11,454), NW ESPÍRITO SANTO state, BRA-

ZIL, 25 mi/40 km SW of BARRA DE SÃO FRANCISCO; 18°52′S 41°13′W.

Manteo (man-TEE-o), town (□ 1 sq mi/2.6 sq km; 2006 population 1,290), ⊙ DARE county, NE NORTH CAROLINA, near N end of ROANOKE ISLAND, 40 mi/64 km SE of ELIZABETH CITY, on ROANOKE SOUND; 35°53′N 75°40′W. FORT RALEIGH NATIONAL HISTORIC SITE to NW. ATLANTIC OCEAN 5 mi/8 km to E. Bridges to mainland (across CROATAN SOUND to W, and to OUTER BANKS, to E Roanoke Sound). Service industries; retail trade; manufacturing (wood products; printing and publishing). North Carolina State Aquarium is here. A replica of original masted ship that brought English colonists to America moored here, at the Elizabeth II State Historical Site.

Manter (MAN-tuhr), village (2000 population 178), STANTON county, SW KANSAS, 7 mi/11.3 km SW of JOHNSON, near COLORADO state line; 37°31′N 101°52′W. Grain.

Manternach (MAHN-tuhr-nahk), town, MANTERNACH commune, SE LUXEMBOURG, on SYRE RIVER, and 2 mi/3.2 km NNW of GREVENMACHER; 49°43′N 06°26′E.

Mantero, ARGENTINA: see VILLA MANTERO.

Mantes, FRANCE: see MANTES-LA-JOLIE.

Mantes-la-Jolie (MAWNT–lah–zho-LEE), city (□ 3 sq mi/7.8 sq km), YVELINES department, ÎLE-DE-FRANCE region, N central FRANCE, port on left bank of the SEINE RIVER, opposite LIMAY, and 30 mi/48 km WNW of PARIS; 48°59′N 01°43′E. Industrial and trade center with metalworking establishments (railroad and electrical equipment, cast steel); motor vehicle manufacturing and diverse manufacturing (musical instruments, chemicals, beer, hosiery) here and in adjacent MANTES-LA-VILLE. Has noteworthy 12th–14th-century Gothic church overlooking the Seine, and a small adjacent old quarter. Formerly called Mantes or Mantes-sur-Seine.

Mantes-la-Ville (MAWNT–lah–VEEL), S suburb (□ 2 sq mi/5.2 sq km) of MANTES-LA-JOLIE, YVELINES department, ÎLE-DE-FRANCE region, N central FRANCE, c. 30 mi/48 km WNW of PARIS; 48°58′N 01°43′E. Manufacturing (musical instruments, cement, and paints).

Mantes-sur-Seine, FRANCE: see MANTES-LA-JOLIE.

Manthani (muhn-TAH-nee), city, KARIMNAGAR district, ANDHRA PRADESH state, S INDIA, on GODAVARI RIVER, and 37 mi/60 km ENE of KARIMNAGAR; 18°39′N 79°40′E. Rice, mangoes, tamarind. Bamboo (used in paper manufacturing) in nearby forests.

Manti (MAN-tei), town (2006 population 3,180), ⊙ SANPETE county, central UTAH, near SAN PITCH RIVER, in irrigated Sanpete Valley, 70 mi/113 km S of PROVO; 39°16′N 111°38′W. Elevation c.5,530 ft/1,686 m. Processing point in cattle, sheep, hogs, poultry, and dairying area; manufacturing (apparel). Oolite quarries nearby. Mormon temple (built 1877–1888) here. Founded by Mormons 1849. WASATCH PLATEAU is just E, in Manti National Forest. Gunnison Reservoir to SW. Palisade State Park to S. Settled 1849, incorporated 1851.

Mantinea (mahn-dee-NAI-ah), city of ancient GREECE, in E central Arcadia, in what is now N ARKADIA prefecture, central PELOPONNESE department, S mainland Greece, near border of ARGOLIS prefecture. In the Peloponnesian War a coalition led by Mantinea and Argos and urged on by ATHENS was defeated (418 B.C.E.) here by SPARTA. Also the scene of the victory of THEBES over Sparta in which Epaminondas was killed (362 B.C.E.).

Manto (MAHN-to), town, OLANCHO department, central HONDURAS, 21 mi/34 km NNW of JUTICALPA; 14°55′N 86°23′W. Sugarcane, rice; livestock. Airfield.

Mantoloking (man-tuh-LO-king), resort borough (2006 population 451), OCEAN county, E NEW JERSEY, on peninsula between BARNEGAT BAY (bridged here) and the ATLANTIC OCEAN, 10 mi/16 km NE of TOMS RIVER; 40°02′N 74°02′W.

Manton, town (2000 population 1,221), WEXFORD county, NW MICHIGAN, 11 mi/18 km N of CADILLAC; 44°24′N 85°24′W. Livestock; fruit, potatoes, beans. Manufacturing (machinery, building materials, lumber products). Settled 1871 as lumber town; incorporated as village 1877, as city 1924.

Mantorville, town (2000 population 1,054), ⊙ DODGE county, SE MINNESOTA, 15 mi/24 km WNW of ROCHESTER, and 2 mi/3.2 km N of KASSON, on South Branch Middle ZUMBRO RIVER; 44°04′N 92°45′W. Elevation 836 ft/255 m. Grain, soybeans, peas; livestock, poultry; dairying.

Mantova (MAHN-to-vah), province (□ 903 sq mi/2,347.8 sq km), LOMBARDY, N ITALY; ⊙ MANTOVA (Mantua); 45°10′N 10°47′E. Consists of fertile, irrigated PO plain, S of Lago di GARDA, with small area of glacial moraine hills in N; watered by Po, MINCIO, OGLIO, and SECCHIA rivers. Agriculture (wheat, corn, sugar beets, raw silk, grapes) and livestock raising (cattle, swine, horses) predominate. Produces 25% of Italy's cheese, including Parmesan. Peat digging. Secondary manufacturing principally petrochemicals, machinery, paper mills, clothing, toys.

Mantova, city, ⊙ MANTOVA province, LOMBARDY, N ITALY, bordered on three sides by lakes formed by the MINCIO River; 45°09′N 10°48′E. Agricultural, industrial, and tourist center. Manufacturing includes machinery, metals, furniture, textiles, clothing, chemicals, and refined petroleum; food processing. Originally an Etruscan settlement, Mantova was later a Roman town and afterwards a free commune (12th–13th century). It flourished under the Gonzaga family (1328–1708), who were magnificent patrons of the arts. Passed to Austria in 1708, taken by Napoleon I in 1797, retaken by Austria in 1815, and became part of Italy in 1866. The Gonzaga palace (13th–18th century), among the largest and finest in Europe, has frescoes by Mantegna and Giulio Romano and numerous other works of art. Other landmarks include the Palazzo del Te (1525–1535); the Church of Sant' Andrea (15th–18th century), designed by Alberti, where Mantegna is buried; and the law courts (13th century). Also known as Mantua.

Mänttä (MANT-tah), town, HÄMEEN province, S central FINLAND, 45 mi/72 km NE of TAMPERE; 62°02′N 24°38′E. Elevation 330 ft/100 m. In lake region; woodworking; pulp, cellulose, and paper mills.

Mantua (mahn-TOO-ah), town, PINAR DEL RÍO province, W CUBA, on small Mantua River, and 40 mi/64 km WSW of PINAR DEL RÍO; 22°18′N 84°18′W. Tobacco, fruit; cattle; lumbering.

Mantua, unincorporated town, FAIRFAX county, N VIRGINIA, residential suburb 12 mi/19 km WSW of WASHINGTON, D.C., 3 mi/5 km E of FAIRFAX; 38°51′N 77°15′W. Northern Virginia Community College (Annandale Campus) to SE.

Mantua (MAN-choo-ah), village, GLOUCESTER county, SW NEW JERSEY, on MANTUA CREEK, and 10 mi/16 km S of CAMDEN, in suburbanizing area; 39°45′N 75°10′W.

Mantua (MAN-tuh-wai), village (□ 1 sq mi/2.6 sq km; 2006 population 1,016), PORTAGE county, NE OHIO, 20 mi/32 km NE of AKRON, on CUYAHOGA river; 41°17′N 81°13′W. In dairy, poultry, and vegetable area.

Mantua (MAN-oo-ai), village (2006 population 769), BOX ELDER county, N UTAH, 5 mi/8 km NE of Brigham City, on Mantua Reservoir; 41°30′N 111°55′W. Agriculture. Elevation 5,175 ft/1,577 m. In WASATCH RANGE and National Forest.

Mantua Creek (MAN-chuh), c.16 mi/26 km long, GLOUCESTER county, SW NEW JERSEY; rises N of GLASSBORO; flows generally NW, past WENONAH and PAULSBORO, to DELAWARE RIVER opposite S PHILADELPHIA. Navigable for c.9 mi/14.5 km above mouth.

Mantua River (mahn-TOO-ah), 41 mi/66 km long, PINAR DEL RÍO province, W CUBA; rises in the SIERRA DE LOS ÓRGANOS; flows SW, past Mantua, to the ENSENADA DE GUADIANA, GULF OF MEXICO.

Mantudi, Greece: see MANDOUDHI.

Manturovo (MAHN-too-ruh-vuh), city (2005 population 18,610), S KOSTROMA oblast, central European Russia, on the UNZHA RIVER (tributary of the VOLGA RIVER), on road and railroad, 162 mi/261 km NE of KOSTROMA, and 28 mi/45 km W of SHARYA; 58°20′N 44°45′E. Elevation 410 ft/124 m. Biochemical, lumbering, woodworking, and foundry industries; bakery. Made city in 1958.

Manturovo (mahn-TOO-ruh-vuh), village (2005 population 2,735), E central KURSK oblast, SW European Russia, on road, 32 mi/51 km WNW of STARYY OSKOL; 51°27′N 37°07′E. Elevation 620 ft/188 m. In agricultural area; food processing.

Mäntyharju (MAN-tuh-HAHR-yoo), village, MIKKELIN province, S FINLAND, 20 mi/32 km SW of MIKKELI; 61°25′N 26°53′E. Elevation 330 ft/100 m. In lake region; granite quarry.

Mäntyluoto (MAN-tuh-LOO-o-to), outport of PORI, TURUN JA PORIN province, SW FINLAND, on small island (bridge to mainland) in GULF OF BOTHNIA, 9 mi/14.5 km NW of Pori city center; 61°35′N 21°29′E. Elevation 33 ft/10 m.

Manú (mah-NOO), province (□ 12,538 sq mi/32,598.8 sq km), MADRE DE DIOS region, SE PERU, in AMAZON basin; ⊙ MANÚ; 12°15′S 71°00′W. Drained by MADRE DE DIOS RIVER. Site of PARQUE NACIONAL MANÚ.

Manú (mah-NOO), town, ⊙ MANÚ province, MADRE DE DIOS region, SE Peru, landing at junction of MADRE DE DIOS and MANÚ rivers, 115 mi/185 km WNW of PUERTO MALDONADO; 12°15′S 70°54′W. In rubber region. Near PARQUE NACIONAL MANÚ. Airport.

Manu'a, volcanic island group and district (total □ 22 sq mi/57 sq km; 1980 pop. 1,732) of Amer. Samoa, comprising the islands of Ta'u, Ofu, and Olosega. According to Samoan tradition, the Manu'a group is the cradle of therace. The main settlement is Luma, on Ta'u Island, which rises to a central mount of 3,280 ft/1,000 m.

Manuae (mah-noo-AH-e), atoll (□ 2 sq mi/5.2 sq km), COOK ISLANDS, S PACIFIC, 124 mi/200 km NE of RAROTONGA. Consists of two islets joined by coral reef. W island (c.3 mi/5 km in circumference) is named Manuae (sometimes referred to as Hervey Island); E island (2 mi/3.2 km long, 1 mi/1.6 km wide) is named Te-Au-O-Tu. Copra plantation. Manuae proper, TE-AU-O-TU, and TAKUTEA are sometimes collectively called the Hervey Islands.

Manuae (mah-noo-AH-e), uninhabited atoll, one of most westerly of SOCIETY ISLANDS, FRENCH POLYNESIA, S PACIFIC; 03°52′N 159°22′W. Owned by French copra company. Also called Fenuaura and Scilly Island.

Manūba, Al (me-noo-BAH, el), town (2004 population 26,666), ⊙ MANUBA province, N TUNISIA, W residential suburb of TUNIS; 36°48′N 10°06′E. Manufacturing of explosives; tanning; flour milling, olive preserving. Orange groves, vineyards.

Manubah or **Manūbah**, province, TUNISIA: see MANOUBA, province.

Manucho (mah-NOO-cho), town, E central SANTA FE province, ARGENTINA, 26 mi/42 km NNW of SANTA FE; 31°16′S 60°48′W. In agricultural area (flax, corn, soybeans, wheat; livestock; apiculture; dairying.

Manuel (mahn-WEL), town, VALENCIA province, E SPAIN, 5 mi/8 km NNW of JÁTIVA; 39°03′N 00°30′W. Rice milling, sawmilling; cereals, oranges.

Manuel Antonio National Park, COSTA RICA: see QUEPOS.

Manuel Benavides (mahn-WEL bai-nah-VEE-des), town, CHIHUAHUA, N MEXICO, near RIO GRANDE, 43 mi/69 km ESE of OJINAGA; 29°06′N 103°54′W. On unpaved road. Lead mining; cattle raising. Also SAN CARLOS.

Manuel Derqui (mahn-WAIL DAIR-kee), town, NW CORRIENTES province, ARGENTINA, on PARANÁ RIVER, on railroad, and 27 mi/43 km S of CORRIENTES;

27°50′S 58°48′W. Agricultural center (corn, rice, cotton, sugarcane, peanuts, oranges; livestock).

Manuel Doblado, MEXICO: see CIUDAD MANUEL DOBLADO.

Manuel Florenco Mantilla, ARGENTINA: see PEDRO RIVER FERNÁNDEZ.

Manuel María Caballero (mahn-WEL mah-REE-ah kah-bah-YAI-ro), province (□ 892 sq mi/2,319.2 sq km), SANTA CRUZ department, E central BOLIVIA; ⊙ COMARAPA; 17°55′S 64°33′W. Elevation 5,955 ft/1,815 m.

Manuel M. Diéguez, town, JALISCO, central MEXICO, on central plateau, 55 mi/89 km SE of SAYULA; 19°34′N 102°55′W. Grain, beans, fruit; livestock. Also SANTA MARÍA DEL ORO.

Manuel Ribas (MAHN-oo-el REE-bahs), city, W central PARANÁ state, BRAZIL, 62 mi/100 km NNW of GUARAPUAVA; 24°31′S 51°39′W. Yerba-maté tea, timber.

Manuel Urbano (MAHN-oo-el OOR-bah-no), town, ACRE state, extreme W BRAZIL, on Purus River, near AMAZONAS state border, and 45 mi/72 km W of SENA MADUREIRA; 08°50′S 69°18′W. Until 1944, called Castelo.

Manukan (mah-NOOK-ahn), town, ZAMBOANGA DEL NORTE province, NW MINDANAO, PHILIPPINES, 50 mi/80 km NW of PAGADIAN, on SULU SEA; 08°26′N 123°07′E. Coconuts, rice, corn.

Manukau City (MAN-oo-kow), former S suburbs of AUCKLAND urban area (□ 1,065 sq mi/2,769 sq km), NEW ZEALAND, extending from MANUKAU HARBOUR to HAURAKI GULF, including Mangere (with bridge and international airport), PAPATOETOE, Howick, and others; 36°58′S 174°48′E. Manufacturing (motor vehicles).

Manukau Harbour (MAN-oo-kou), N NORTH ISLAND, NEW ZEALAND, SW harbor of AUCKLAND, 5 mi/8 km S of WAITEMATA HARBOUR on E coast; 15 mi/24 km long, 10 mi/16 km wide. Connected with TASMAN SEA by passage 2 mi/3.2 km wide. Coastal port ONEHUNGA is on NE shore, PAPAKURA on SE shore, WAIUKU on small S arm of harbor, with GLENBROOK steel mills. Auckland International Airport on E shore.

Manuk Manka Island (mah-NOOK MAHNG-kou) (□ 6 sq mi/15.6 sq km), TAWI-TAWI province, PHILIPPINES, at S end of SULU ARCHIPELAGO, 1.9 mi/3 km S of SIMUNUL ISLAND; 04°51′N 119°49′E. Southernmost in KINAPUSAN ISLAND group.

Manumukh (mah-noo-mook), village, SYLHET district, E EAST BENGAL, BANGLADESH, near the KUSIYARA RIVER, 23 mi/37 km SSW of SYLHET; 24°39′N 91°39′E. Rice, tea, oilseeds; sawmilling. Also spelled Manoomukh.

Manurewa (MAN-you-REE-wah), suburb within MANUKAU CITY, N NORTH ISLAND, NEW ZEALAND, 15 mi/24 km SE of AUCKLAND. Residential, industrial, and agricultural center. Incorporated 1989 into Manukau City.

Manuripi, province, PANDO department, NW BOLIVIA; ⊙ PUERTO RICO; 11°20′S 67°30′W.

Manuripi-Heath (mah-noo-REE-pee–HEETH), national reserve, MANURIPI and MADRE DE DIOS provinces, PANDO department and CAMACHO, BAUTISTA SAAVEDRA, FRANZ TAMAYO provinces, LA PAZ department, NE BOLIVIA. Intended to preserve the Amazón rain forest between the HEATH and MANURIPI rivers. There are a number of rubber and brazil nut trees and a variety of jungle wildlife.

Manuripi River (mah-noo-REE-pee), c.250 mi/402 km long, PERU and BOLIVIA; rises in Peru; flows NE, crossing into Bolivia, past SAN MIGUELITO, joining TAHUAMANU RIVER at PUERTO RICO to form ORTON RIVER; 11°30′S 68°00′W. Navigable in middle and lower course, c.90 mi/145 km long. Rubber is exploited in tropical forests along its banks in Bolivia.

Manú River (mah-NOO), c.120 mi/193 km long, MADRE DE DIOS region, SE PERU; rises in E ANDEAN outliers at 12°15′S 71°25′W. Flows SE, through tropical lowlands, to MADRE DE DIOS RIVER at MANÚ. Navigable for small craft. Linked through small affluent with a tributary of URUBAMBA RIVER.

Manus, province (2000 population 43,387), N central PAPUA NEW GUINEA, between BISMARCK SEA (S) and PACIFIC OCEAN (N and W); ⊙ LORENGAU. Includes main island of MANUS and Rambutyo, Tong, and Lou islands, numerous small islands, atolls, reefs mainly to SE and NW. Volcanic. Some of the smaller treeless islands around Manus are known for their canoes made from logs that have floated out to sea from SEPIK RIVER circa 150 mi/241 km to SW. Vegetables, coconuts, palm oil; fish; timber.

Manus, volcanic island (□ 633 sq mi/1,645.8 sq km), largest of ADMIRALTY ISLANDS, MANUS district, BISMARCK ARCHIPELAGO, PAPUA NEW GUINEA, SW PACIFIC OCEAN, 230 mi/370 km NW of NEW BRITAIN; 50 mi/80 km long, 20 mi/32 km wide; 02°04′S 147°E. Rises to 3,000 ft/914 m. Coconut plantations. Sometimes called ADMIRALTY ISLAND.

Manvel (MAN-vuhl), town (2006 population 4,600), BRAZORIA county, SE TEXAS, suburb 18 mi/29 km S of downtown HOUSTON, on Mustang Bayou; 29°28′N 95°21′W. Urban growth area in agricultural region. Oil and natural gas. Manufacturing (plastic products, electronic equipment).

Manvel (MAN-vuhl), village (2006 population 331), GRAND FORKS county, E NORTH DAKOTA, 13 mi/21 km NNW of GRAND FORKS, and on Turtle River; 48°04′N 97°10′W. Founded in 1882 and incorporated 1930. Named for Allen A. Manvel, a railroad agent.

Manvi (MAHN-vee), town, RAICHUR district, KARNATAKA state, S INDIA, 26 mi/42 km SW of RAICHUR; 15°59′N 77°03′E. Millet, oilseeds. Cotton ginning nearby.

Manville, village in LINCOLN town, PROVIDENCE county, NE RHODE ISLAND, on BLACKSTONE RIVER, and 11 mi/18 km N of PROVIDENCE.

Manville, village (2006 population 97), NIOBRARA county, E WYOMING, 10 mi/16 km W of LUSK; 42°46′N 104°37′W. Elevation c. 5,245 ft/1,599 m. Trading point in ranching and wheat region.

Manville, borough (2006 population 10,481), SOMERSET county, central NEW JERSEY; 40°32′N 74°35′W. Laid out 1906, incorporated 1929.

Manwat (mahn-wuht), town, PARBHANI district, MAHARASHTRA state, W central INDIA, 18 mi/29 km W of PARBHANI; 19°18′N 76°30′E. Cotton ginning, rice and oilseed milling. Manwat Road, railroad station, is 5 mi/8 km NE.

Many (ME-nee), town (2000 population 2,889), ⊙ SABINE parish, W LOUISIANA, c. 70 mi/113 km S of SHREVEPORT; 31°34′N 93°28′W. Commercial center for lumber, oil, and natural gas, and agricultural area (cattle, poultry, exotic fowl; dairying); timber. Manufacturing (transportation equipment, poultry feed, consumer goods; publishing and printing). Settled in early nineteenth century. Recreation. Fort JESUP and Los Adaes state commemorative areas to NE. Rebel State Commemorative Areas to N. Sabine State Wildlife Area to W.

Manyacha, BOSNIA AND HERZEGOVINA: see MANJAČA.

Manyame, Lake, reservoir, MIDLANDS province, N central ZIMBABWE, formed by Manyame Dam on MANYAME RIVER 40 mi/64 km W of HARARE; 20 mi/32 km long, 4 mi/6.4 km wide; 17°49′S 30°28′E. Surrounded by LAKE MANYAME RECREATIONAL PARK; LAKE CHIVERO reservoir (recreatonal park) to SE, upstream. Formerly called Lake Robertson.

Manyame River, c. 275 mi/443 km long, ZIMBABWE and MOZAMBIQUE; rises c. 35 mi/56 km SE of HARARE (Zimbabwe); flows WNW, passing S of Harare, and continuing through lakes CHIVERO and MANYAME reservoirs (both recreational parks); then N, past CHINHOYI, entering extreme W Mozambique; receives ANGWA (Duangua) River from SW before entering Lago de Cabora Bassa reservoir on the ZAMBEZI River, 15 mi/24 km E of ZUMBO. Called Panhame River in Mozambique; formerly called Hunyani River in Zimbabwe.

Manyanga (mahn-YAHNG-gah), village, BAS-CONGO province, W CONGO, on both banks of CONGO RIVER, on CONGO REPUBLIC border, and 105 mi/169 km NE of BOMA. Trading center. Founded by Henry M. Stanley. Though this section of CONGO RIVER is part of the noted LIVINGSTONE FALLS, the river is navigable between here and ISANGILA.

Manyangau, mountain (4,639 ft/1,414 m), MASHONALAND WEST province, N central ZIMBABWE, 75 mi/121 km NNW of CHINHOYI; 16°26′S 29°36′E.

Manyara, region (2006 population 1,198,000), NE TANZANIA; ⊙ BABATI; 04°45′S 36°40′E. Bordered by ARUSHA (N), KILIMANJARO (NE), TANGA (E), MOROGORO (SE point), DODOMA (S), SINGIDA (W), and SHINYANGA (NW) regions. PANGANI RIVER forms most of border with Kilimanjaro region. MASAI STEPPE plateau (3,937 ft/1,200 m) and Kitwei Plain in E. Part of TARANGIRE NATIONAL PARK in center (also in Arusha and Dodoma regions). Main road between ARUSHA city and SINGIDA town travels through W part of region. Created c. 2004 out of S portion of Arusha region.

Manyara, Lake (mah-NYA-rah), ARUSHA region, N TANZANIA, 60 mi/97 km WSW of ARUSHA; 30 mi/48 km long, 10 mi/16 km wide. Brackish lake, no outlet; salt and phosphate deposits along margins. LAKE MANYARA NATIONAL PARK on NW shore and N half of lake.

Manyas, village, NW TURKEY, near KOCA RIVER and Lake MANYAS, 28 mi/45 km N of BALIKESIR; 40°02′N 27°58′E. Also called Maltepe.

Manyas, Lake (□ 69 sq mi/179 sq km), NW TURKEY, 8 mi/12.9 km S of BANDIRMA; 12 mi/19 km long, 10 mi/16 km wide. KOCA RIVER enters and leaves it.

Manyberries (ME-nee-be-reez), hamlet, SE ALBERTA, W CANADA, 44 mi/71 km from BOW ISLAND, in FORTY MILE COUNTY NO. 8; 49°24′N 110°42′W.

Manych (MAH-nich), two rivers, STAVROPOL TERRITORY, SE European Russia. The Western Manych, approximately 200 mi/322 km long; rises near STAVROPOL in the N CAUCASUS Mountains; flows NW through Lake MANYCH-GUDILO into the lower DON River. The Eastern Manych rises in a marshy area; flows approximately 100 mi/161 km E to a system of salt lakes and marshes approximately 75 mi/121 km W of the CASPIAN SEA, but it reaches the sea only in rare spring floods. In spring, the two join in the center of the Manych Depression, a broad, valleylike lowland extending approximately 350 mi/563 km SE from the lower Don to the Caspian Sea. During the Pleistocene period, it was a glacial spillway from the Caspian Sea to the Sea of AZOV whenever the Caspian's waters rose above sea level. A variant spelling is Manich.

Manych-Gudilo, Lake (MUH-nich–goo-DEE-luh), Russian *Ozero Bol'shoy Manych* [=Great Manich Lake] (□ 135 sq mi/351 sq km), in central section of the Manych Depression, N CAUCASUS Mountains, in the W extension of Republic of KALMYKIA-KHALMG-TANGEH, at the borders of ROSTOV and STAVROPOL oblasts, SE European Russia, NW of DIVNOYE; 35 mi/56 km long, 5 mi/8 km wide; 46°20′N 42°40′E. Traversed by the canalized Western Manych River. The W half of the lake is part of the Chernye Zemli [Russian=black grounds] Nature Reserve, protecting wildlife in the local wetlands.

Many Farms, unincorporated town (2000 population 1,548), APACHE county, NE ARIZONA, 107 mi/172 km NNE of HOLBROOK, near Chinle Creek. Cattle, sheep; hay. Crafts.

Manyoni (mah-NYO-nee), town, SINGIDA region, central TANZANIA, 65 mi/105 km WNW of DODOMA, junction of railroad spur to SINGIDA; 05°47′S 34°43′E. Airstrip. Livestock; grain, subsistence crops.

Area is shown by the symbol □, and capital city or county seat by ⊙.

Manyovu (mah-NYO-voo), village, KIGOMA region, NW TANZANIA, 30 mi/48 km NNW of KIGOMA, near BURUNDI border; 04°29′S 29°51′E. Tobacco, corn; sheep, goats; timber.

Manyu, department (2001 population 177,389), SOUTHWEST province, CAMEROON; ⊙ MAMFE.

Manzala (MAHN-zah-luh, el), town, DAQAHLIYA province, NE Lower EGYPT, on El BAHR ES SAGHIR (a delta canal), and 34 mi/55 km ENE of MANSURA, near Lake MANZALA, 20 mi/32 km SSE of DUMYAT. Fisheries.

Manzala, Lake (MAHN-zah-luh), salt water lagoon (□ 660 sq mi/1,716 sq km), NE EGYPT, near PORT SAID, partly separated from the MEDITERRANEAN SEA by a narrow spit through which there are several passages. The SUEZ CANAL cuts through the E part of the lake's basin. Also spelled Menzaleh.

Manzanal (mahn-zah-NAHL), canton, MANUEL MARÍA CABALLERO province, SANTA CRUZ department, E central BOLIVIA, 50 mi/80 km N of COMARAPA; 21°06′S 65°35′W. Elevation 5,922 ft/1,805 m. Agriculture (potatoes, yucca, bananas, corn, sweet potatoes, cotton, peanuts, citrus fruits); cattle.

Manzanar, historic site, INYO county, E CALIFORNIA, W of OWENS RIVER, in Owens Valley, between LONE PINE and INDEPENDENCE. Site during World War II of a relocation camp for interned Pacific Coast residents of Japanese descent.

Manzanares (mahn-thah-NAH-res), city, CIUDAD REAL province, S central SPAIN, in CASTILE-LA MANCHA, on small AZUER RIVER, and 30 mi/48 km E of CIUDAD REAL. Railroad and road junction. Agricultural center (chiefly viticultural) on LA MANCHA plain (also grapes, olives, saffron, potatoes, vegetables; sheep). Industries include liquor and alcohol distilling, wine making, fruit canning, sawmilling, lime quarrying; manufacturing of tartaric acid, sulphur, plaster, soap, firearms, knives, chocolate, jam, textile goods. Has old castle.

Manzanares (mahn-zah-NAHR-ais), town, ⊙ Manzanares municipio, CALDAS department, central COLOMBIA, in Cordillera CENTRAL, 30 mi/48 km ENE of MANIZALES; 05°19′N 75°09′W. Elevation 8,080 ft/2,462 m. Coffee growing; livestock. Gold mines nearby.

Manzanares (mahn-thah-NAH-rais), river, c.55 mi/90 km long, central SPAIN; rises in the SIERRA DE GUADARRAMA; flows S past MADRID (where it is canalized) into the JARAMA RIVER. Used for irrigation and hydroelectric-power generation.

Manzanares el Real (mahn-thah-NAH-rais el rai-AHL), town, MADRID province, central SPAIN, 21 mi/34 km NNW of MADRID; 40°44′N 03°52′W. Lumbering; stone quarrying; livestock raising; dairying. Nearby is the Santillana dam and hydroelectric plant, adjoined by fine medieval castle.

Manzanares River (mahn-sah-NAH-rais), c.40 mi/64 km long, SUCRE state, NE VENEZUELA; rises at NW foot of Cerro TURIMIQUIRE; flows NW, past CUMANACOA and CUMANÁ, to the CARIBBEAN SEA, 1.5 mi/2.4 km NW of Cumaná; 10°28′N 64°11′W.

Manzaneda, Cabeza de (mahn-thah-NAI-dah, kah-VAI-thah dai), highest peak (5,833 ft/1,778 m) of GALICIAN MOUNTAINS, ORENSE province, NW SPAIN, 30 mi/48 km WSW of ORENSE.

Manzaneque (mahn-thah-NAI-kai), town, TOLEDO province, central SPAIN, on railroad, and 20 mi/32 km SE of TOLEDO; 39°38′N 03°47′W. Cereals, grapes, olives; sheep, goats.

Manzanera (mahn-thah-NAI-rah), town, TERUEL province, E SPAIN, 23 mi/37 km SE of TERUEL; 40°03′N 00°50′W. Cereals, potatoes; sheep.

Manzanilla (mahn-thah-NEE-lyah), town, HUELVA province, SW SPAIN, 24 mi/39 km W of SEVILLE; 37°23′N 06°25′W. Viticulture; olive growing.

Manzanilla (MAHN-zah-nil-ah), village, E TRINIDAD, TRINIDAD AND TOBAGO, 33 mi/53 km ESE of PORT OF SPAIN, near Manzanilla Bay; 10°37′N 61°08′W. Manzanilla Beach, lined by coconut palms, is a popular tourist site. Manzanilla Point is 3 mi/4.8 km E.

Manzanilla, La, MEXICO: see LA MANZANILLA DE LA PAZ.

Manzanillo (mahn-sah-NEE-yo), city, (2002 population 98,038), GRANMA province, SE CUBA, a port on the GUACANAYABO GULF of the CARIBBEAN SEA; 20°21′N 77°08′W. A leading city on Cuba's S coast, Manzanillo is a commercial center and the exportation point for the agricultural produce (sugarcane, rice, tobacco) of the Cauto plain. Founded in 1784; long a smuggling center involving British merchants from Jamaica. An attack by Great Britain in 1792 destroyed numerous Spanish ships in the harbor and led to the fortification of the city.

Manzanillo (mahn-sah-NEE-yo), city and township, COLIMA state, SW MEXICO. One of Mexico's chief PACIFIC ports, Manzanillo has a fine harbor and modern railroad and highway connections with MEXICO CITY. It handles many imports and ships out minerals, fruit, and lumber. Excellent beaches, a tropical climate, and resources for hunting and fishing have made Manzanillo a popular international resort.

Manzanillo (mahn-sahn-NEE-yo), village, PUNTARENAS province, W COSTA RICA, port on GULF OF NICOYA of the PACIFIC OCEAN, 16 mi/26 km NW of PUNTARENAS. In past exported gold from Abangares mines. Livestock raising. Also called Puerto Iglesias.

Manzanillo Bay (mahn-zahn-NEE-yo), French *Baie de mancenille*, inlet of ATLANTIC OCEAN on N coast of HISPANIOLA, at border between HAITI and DOMINICAN REPUBLIC; 19°45′N 71°45′E. Sheltered by MANZANILLO Point (N). Port of PEPILLO SALCEDO is on S shore, at mouth of MASSACRE (DAJABÓN) River.

Manzanillo Bay, inlet of CARIBBEAN SEA, in PANAMA CANAL AREA, E of MANZANILLO Isl. (site of COLÓN city), which separates it from LIMÓN bay; c.3 mi/4.8 km long, 1 mi/1.6 km wide. On E shore are FORT RANDOLPH and FRANCE FIELD, transferred from U.S. to Panama in 2000.

Manzanillo Island, indented peninsula guarding ATLANTIC entrance to the PANAMA CANAL, situated between LIMÓN (W) and MANZANILLO (E) bays; c.1 mi/1.6 km long, 0.75 mi/1.2 km wide. On it is COLÓN city.

Manzanillo Keys (mahn-sah-NEE-yo), tiny coral reefs in the Gulf of GUACANAYABO, GRANMA province, SE CUBA, just outside MANZANILLO, the harbor of which they protect; 20°23′N 77°11′W. Covered by mangroves. On Perla Key (W) is a lighthouse.

Manzanillo Point, northernmost point of PANAMA, on CARIBBEAN SEA, 31 mi/50 km NE of COLÓN; 09°38′N 79°33′W.

Manzanillo Point (mahn-sah-NEE-yo), on CARIBBEAN coast of FALCÓN state, NW VENEZUELA, 29 mi/47 km ENE of CORO; 11°32′N 69°16′W.

Manzanita (man-zuh-NEET-uh), town (2006 population 630), TILLAMOOK county, NW OREGON, on the PACIFIC OCEAN, 32 mi/51 km SSW of ASTORIA; 45°43′N 123°55′W. Dairy products; timber; tourism. Oswald West State Park to NW; Tillamook State Forest to E.

Manzano (mahn-TSAH-no), town, UDINE province, FRIULI-VENEZIA GIULIA, NE ITALY, 9 mi/14 km SE of UDINE. Wood products.

Manzano Island, CHILE: see GUAITECAS ISLANDS.

Manzanola, village (2000 population 525), OTERO county, SE COLORADO, on ARKANSAS RIVER, near mouth of APISHAPA RIVER, and 40 mi/64 km ESE of PUEBLO; 38°06′N 103°52′W. Elevation 4,252 ft/1,296 m. Trading point in region producing cantelopes, melons, vegetables, sugar beets, wheat, beans, corn; cattle. Lake Meredith Reservoir to NE.

Manzano Mountains (mahn-ZAHN-o), central NEW MEXICO, E of RIO GRANDE; extends c.40 mi/64 km N from MOUNTAINAIR; largely within parts of Cibola National Forest and part of Isleta Indian Reservation. Prominent points are Mosca (9,509 ft/2,898 m), Manzano (10,098 ft/3,078 m) peaks.

Manzhouli (mahn-ZHO-LEE), city (□ 269 sq mi/699.4 sq km; 2000 population 137,790), NE INNER MONGOLIA Autonomous Region, CHINA, on RUSSIAN border; 49°36′N 117°28′E. Heavy industry is the most important economic activity; main industries include coal mining and food processing. Animal husbandry and crop growing account for most agricultural income (milk, eggs). Developed after the 1903 construction of the Chinese Eastern railroad (now the Northeast China railroad) and was an important customs station and trade point between RUSSIA and China. Many Russian immigrants settled here after the Bolshevik Revolution. Also spelled Manchouli, Manchuli.

Manzil Abu Ruqaybah (men-ZEL ah-boo roo-kai-BAH), town, BIZERTE province, N TUNISIA, on S shore of LAKE BIZERTE, 9 mi/14.5 km SSW of BIZERTE; 37°10′N 09°48′E. Administrative center. Manufacturing includes steel mill; iron ore; tire manufacturing. Site of former French naval base, arsenal, and other military installations. Formerly called Ferryville, the town was renamed in honor of Habib Abu Ruqaybah, Tunisia's leader, after independence (1956).

Manzil, El (MEN-zil, el), village, S central JORDAN, on HEJAZ RAILROAD, and 19 mi/31 km ESE of AL KARAK. Barley; camel raising.

Manzil Tamīmah (men-ZEL tai-MEE-mah), town (2004 population 34,528), NABUL province, NE TUNISIA, on E coast of CAPE BON peninsula, 55 mi/89 km E of TUNIS. Cereals, olives; fishing, handicraft manufacturing. Lignite mines nearby.

Manzini (mahn-ZEE-nee), district (□ 1,571 sq mi/4,084.6 sq km), W and central SWAZILAND, on SOUTH AFRICA (W) border; ⊙ MANZINI; 26°15′S 31°30′E. Part of N border formed by BLACK MBULUZI RIVER to Majoli Dam; source of WHITE MBULUZI RIVER in NE. Bordered S by Mkhonduo River and Mahlangatja Hills. Drained in S by Great Usutu and Ngwempisi rivers. Timber; citrus, tea, corn, pineapples, cotton; cattle, goats, sheep, hogs.

Manzini (mahn-ZEE-nee), city, ⊙ MANZINI district, central SWAZILAND, 18 mi/29 km SE of MBABANE; 26°30′S 31°22′E. Largest city in Swaziland. Industrial center (food processing; handicrafts, beverages); most activity at MATSAPA to W. Trade center in agricultural area (corn, vegetables, citrus, pineapples; cattle, goats, sheep, hogs; timber). Matsapa International Airport to W, University of Swaziland at Kwaluseni, 4 mi/6.4 km to WNW. Was capital of Swaziland protectorate 1894–1902. Formerly Bremersdorp. Named for local chief in 1885, seat of Roman Catholic Bishop of Swaziland, and site of Cathedral since 1961. Founded 1890.

Manzovka, RUSSIA: see SIBIRTSEVO.

Mao (MOU), town, ⊙ KANEM administrative region, W CHAD, 85 mi/137 km WNW of MOUSSORO; 14°07′N 15°19′E. Market for cotton, millet, camels, livestock; vegetable growing; date palms. Was capital of former KANEM prefecture. Former Kanem state capital.

Mao, town (2002 population 47,828), SANTIAGO province, NW DOMINICAN REPUBLIC, in fertile CIBAO region, 25 mi/40 km WNW of SANTIAGO. Rice-growing and -milling center. Lumbering and gold washing in vicinity. Also called VALVERDE.

Maó, SPAIN: see MAHÓN.

Maobisse, EAST TIMOR: see MAUBISSE.

Maoemere, INDONESIA: see FLORES.

Maoflang, India: see MAWPHLANG.

Maoka, RUSSIA: see KHOLMSK.

Maoke Mountains (MOU-kai), Dutch *Sneeuw Gebergte*, collective name for mountain ranges extending c.400 mi/644 km E-W in IRIAN JAYA, INDONESIA, N central NEW GUINEA island; main sections are SUDIRMAN RANGE (W) and Jayawijaya (E) ranges; 04°00′S 138°00′E. Formerly called Snow Mountains.

Mao Khe (MOU KAI), town, QUANG NINH province, N VIETNAM, 4 mi/6.4 km ESE of DONG TRIEU; 21°03′N 106°37′E. Market hub and transportation complex. Anthracite-mining center. Formerly Maokhe.

Cross-references are shown in SMALL CAPITALS. The pronunciation guide is shown on page xix. The sources of population figures are shown on page xvii.

Maolang, India: see MAWSYNRAM.

Maoming (MOU-MING), city (□ 188 sq mi/488.8 sq km; 2000 population 532,715), SW GUANGDONG province, CHINA, on the FOSHAN River, and 50 mi/80 km NE of ZHANJIANG; 21°39′N 110°54′E. A center for heavy industry (utilities, chemicals; oil refining); also textiles, crafts. Grain, oil crops, vegetables, fruits; hogs, poultry; eggs.

Maon Shan, mountain, the New Territories, HONG KONG, SE CHINA, NE of Shatin, 7 mi/11.3 km NNE of KOWLOON; 22°25′N 114°15′E. Rises to 2,297 ft/700 m. A recently created town of the same name lies at the base of the mountains along Shatin harbor.

Maore, Comoros: see MAYOTTE.

Maowen (MOU-WUN), town, ⊙ Maowen county, NW SICHUAN province, CHINA, on left bank of MIN River, and 70 mi/113 km N of CHENGDU, in mountain region; 31°41′N 103°52′E. Logging; papermaking. Also called Fengyizhen or Maoxian.

Mao Xian, CHINA: see MAOWEN.

Ma'oz Hayim (mah-OZ khah-YEEM) or **Ma'oz Haim**, kibbutz, N ISRAEL, 3 mi/4.8 km E of BEIT SHE'AN, near JORDAN RIVER; 32°28′N 35°32′E. Below sea level 757 ft/230 m. Manufacturing of plastic packaging materials. Dairying; fish breeding; grain, vegetables. Migratory birds can be found around the fish ponds here during their flights between Europe and Africa. Signs of ancient settlement found to E, including 3rd-century (or early 4th-century) synagogue that remained in use for more than three centuries. SHEIKH HUSSEIN BRIDGE, crossing the Jordan River, is 1.2 mi/2 km E. Zakum nature reserve just SE. Founded 1930s.

Ma'oz Zion, ISRAEL: see MEVASSERET ZION.

Mapam Yumco (MAH-PUNG YUNG-ZO), lake (□ 200 sq mi/520 sq km) in W HIMALAYA, SW TIBET, SW CHINA, between KAILAS (N) and GURLA MANDHATA (S) peaks; 12 mi/19 km long, 14 mi/23 km wide; 30°40′N 81°25′E. Elevation 14,950 ft/4,557 m. Linked with Lake La'nga Co (W) by its N outlet, the Ganga Chu River or Ganga Qu (6 mi/9.7 km long, 40 ft/12 m–100 ft/30 m wide; hot springs near banks). A Hindu pilgrimage center, lake is encircled by a path 54 mi/87 km in circumference; traveled annually by large bands of pilgrims. In Hindu mythology, formed by Brahma's soul. Borax deposits nearby. Also called Manasarowar Lake.

Mapangu (muh-PAHN-goo), town, KASAI-OCCIDENTAL province, SW CONGO, on road, and 18 mi/29 km SW of ILEBO; 04°25′S 20°20′E. Elev. 1,102 ft/335 m. Oil palms.

Mapararí (mah-pahr-ah-REE), town, FALCÓN state, NW VENEZUELA, 45 mi/72 km SSE of CORO; 10°48′N 69°26′W. Elevation 2,473 ft/753 m. Coffee, cacao, corn; livestock.

Mapastepec (mah-PAHS-te-pek), town, ⊙ Mapastepec municipio, CHIAPAS, S MEXICO, in PACIFIC OCEAN lowland, on railroad, and 55 mi/89 km NW of TAPACHULA, on Mexico Highway 200; 15°25′N 92°54′W. Elevation 105 ft/32 m. Cacao, tobacco, sugarcane, fruit, livestock.

Mapavri, TURKEY: see CAYBASI.

Mape, Lake (MAH-pai), lagoon, in SE SIERRA LEONE, on N side of TURNER'S PENINSULA, 15 mi/24 km SSW of PUJEHUN; 9 mi/14.5 km long, 3 mi/4.8 km wide. Formerly called Lake Kasse.

Maphisa, town, MATABELELAND SOUTH province, SSW ZIMBABWE, 75 mi/121 km S of BULAWAYO; 21°03′S 28°27′E. Irrigated agriculture; corn, subsistence crops; cattle, goats, sheep.

Mapimí (mah-pee-MEE), town, ⊙ Mapimí municipio, DURANGO, N MEXICO, at S edge of Bolsón de MAPIMÍ, 29 mi/46 km NW of GÓMEZ PALACIO, on Mexico Highway 30; 25°51′N 103°50′W. Elevation 4,485 ft/1,367 m. Silver, gold, lead mining.

Mapimí, Bolsón de (mah-pee-MEE, bo-SON dai), arid depression in plateau of N MEXICO, in states of CHIHUAHUA, COAHUILA, and DURANGO, N of MAPIMÍ

(Durango). Desert region. Potentially fertile; has been irrigated in S, where cotton, wheat, alfalfa are grown. Average elevation c.4,485 ft/1,367 m.

Mapire (mah-PEE-rai), town, ⊙ José Gregorio Monagas municipio, ANZOÁTEGUI state, E central VENEZUELA, landing on ORINOCO RIVER (BOLÍVAR state border), and 85 mi/137 km WSW of CIUDAD BOLÍVAR; 07°46′N 64°40′W. Livestock. Airport.

Mapiri (mah-PEE-ree), town and canton, LARECAJA province, LA PAZ department, W BOLIVIA, port on MAPIRI RIVER, on SORATA-APOLO road, and 50 mi/80 km NE of SORATA; 15°15′S 68°10′W. Tropical agricultural center (bananas, tea, quinine).

Mapiri River, c.40 mi/64 km long, LA PAZ department, W BOLIVIA; rises in two main branches (Consata and Camata rivers) in Cordillera de LA PAZ; flows ESE past MAPIRI and GUANAY, joining COROICO River at Puerto Ballivián to form KAKA River; 10°11′S 66°34′W. Receives TIPUANI and Challana rivers near its mouth. Name sometimes applied also to Consata River below Consata and to Kaka River.

Maple (MAI-puhl), unincorporated village, in city of VAUGHAN, S ONTARIO, E central CANADA, 15 mi/24 km N of TORONTO; 43°51′N 79°31′W. Suburban community. Dairying; vegetable gardening. Manufacturing (concrete products, electronic goods, plastics, structural wood). Site of Canada's Wonderland amusement park.

Maple Bluff, village (2006 population 1,295), DANE county, S WISCONSIN, at E end of Lake Mendota, a suburb, 2 mi/3.2 km N of downtown MADISON; 43°06′N 89°22′W. Executive Mansion is here. Incorporated 1930 as Lakewood Bluff; renamed 1931.

Maple Creek (MAI-puhl), town (2006 population 2,198), SW SASKATCHEWAN, W CANADA, on MAPLE CREEK, and 55 mi/89 km ESE of MEDICINE HAT, at foot of the CYPRESS HILLS; 49°55′N 109°28′W. Grain elevators; dairying, lumbering; livestock raising.

Maplecrest, hamlet, GREENE CO., SE NEW YORK, in the CATSKILL MOUNTAINS, 17 mi/27 km WNW of CATSKILL; 42°16′N 74°11′W.

Maple Glen, unincorporated town (2000 population 7,042), MONTGOMERY county, SE PENNSYLVANIA, residential suburb, 12 mi/19 km N of downtown PHILADELPHIA; 40°10′N 75°10′W. Willow Grove Naval Air Station to NE.

Maple Grove, city, HENNEPIN county, E MINNESOTA, suburb, 12 mi/19 km NW of downtown MINNEAPOLIS; 45°06′N 93°27′W. Manufacturing (machining, gun drilling, sheet metal fabricating; ink, furniture, concrete products, machinery, aerospace material, building materials, foam products, wood products, burglar alarms, tool and die products, transportation equipment, medical equipment). Several small lakes in vicinity. Elm Creek Park Reserve on N border.

Maple Grove (MAI-puhl GROV), former village, S QUEBEC, E CANADA; 45°19′N 73°50′W. Amalgamated into BEAUHARNOIS in 2002.

Maple Heights (MAI-puhl HEITZ), city (□ 5 sq mi/13 sq km; 2006 population 24,293), CUYAHOGA county, NE OHIO; residential suburb 5 mi/8 km SE of CLEVELAND; 41°24′N 81°34′W. Major shopping centers; miscellaneous manufacturing. Incorporated 1932. Population has declined since late 20th century.

Maple Hill, city (2000 population 469), WABAUNSEE county, NE KANSAS, on Mill Creek, near its confluence with KANSAS River, 14 mi/23 km ENE of ALMA; 39°05′N 96°01′W. In cattle, poultry, and grain area.

Maple Island (MAI-puhl), community, S ONTARIO, E central CANADA, 25 mi/40 km from PARRY SOUND, and included in WHITESTONE township; 45°41′N 79°52′W.

Maple Lake, town (2000 population 1,633), WRIGHT county, S central MINNESOTA, 7 mi/11.3 km WNW of BUFFALO; 45°13′N 94°00′W. Grain, soybeans; livestock, poultry; dairying. Manufacturing (metal fabrication; frozen foods, consumer goods, feeds and fertilizers, boats). Several small natural lakes in area,

including Maple (E) and Mink (N) lakes; Lake Marion State Park to N.

Maple Park, village (2000 population 765), KANE county, NE ILLINOIS, 15 mi/24 km W of GENEVA; 41°54′N 88°35′W. In agricultural area (dairy products; livestock).

Maple Plain, town (2000 population 2,088), HENNEPIN county, E MINNESOTA, suburb 18 mi/29 km W of downtown MINNEAPOLIS; 45°00′N 93°39′W. Bounded on N, W, and S by INDEPENDENCE, on E by MEDINA. Dairying in area; manufacturing (concrete products, industrial patterns, wood products, machinery, molding; machining). Morris T. Baker Park Reserve to E; Lake Independence to NE; Lake MINNETONKA to SE.

Maple Rapids, village (2000 population 643), CLINTON county, S central MICHIGAN, 9 mi/14.5 km NW of SAINT JOHNS, and on MAPLE River; 43°06′N 84°41′W. In farm area.

Maple Ridge (MAI-puhl RIJ), city (□ 103 sq mi/267.8 sq km; 2001 population 63,169), SW BRITISH COLUMBIA, SW CANADA, suburb 27 mi/43 km ESE of downtown VANCOUVER, on N bank of FRASER River, and in GREATER VANCOUVER regional district; 49°13′N 122°36′W. Includes locality of Haney. River port. Logging, sawmilling, shipbuilding; dairying; fruit (berries); greenhouse crops; fisheries; mining; manufacturing (batteries, plastic and metal products); printing and publishing. Malcolm Knapp University of British Columbia Research Forest, a training facility for forestry students, is here, as is an entrance to Golden Ears Provincial Park. Incorporated 1874.

Maple River, 90 mi/145 km long, W IOWA; rises in BUENA VISTA county, flows S, past IDA GROVE, and SW to Little Sioux River 7 mi/11.3 km ESE of ONAWA.

Maple River, c.65 mi/105 km long, S central MICHIGAN; rises S of CORUNNA in SHIAWASSEE county; flows NW, past OVID and ELSIE, then W and SW, past MAPLE RAPIDS, to GRAND RIVER at MUIR; 42°56′N 84°03′W.

Maple River, S MINNESOTA; rises in Penny Lake, on border between FARIBAULT and FREEBORN counties; flows W past WELLS, then NW, receives waters from Minnesota Lake from two outflows, then N 5 mi/8 km W of MAPLETON, past GOOD THUNDER, to LE SUEUR RIVER 5 mi/8 km S of MANKATO; 43°47′N 93°38′W.

Maple River, 100 mi/161 km long, E NORTH DAKOTA; rises in STEELE county; flows S largely through CASS county to ENDERLIN, then NE to SHEYENNE RIVER, 10 mi/16 km NW of FARGO, near its confluence with RED RIVER OF THE NORTH; 47°27′N 97°50′W.

Maple Shade, village, BURLINGTON county, SW NEW JERSEY, 5 mi/8 km E of CAMDEN; 39°57′N 75°00′W. Light industry and commerce. Largely residential.

Maple Springs, hamlet, CHAUTAUQUA county, extreme W New York, on CHAUTAUQUA LAKE, 11 mi/18 km NW of JAMESTOWN; 42°12′N 79°25′W. Trolley park.

Maplesville, town (2000 population 672), Chilton co., central Alabama, on Mulberry River, and 15 mi/24 km WSW of Clanton. Paper, lumber and veneer. Named for Steven W. Maples, a settler and storeowner. Inc. in 1914.

Mapleton (MAI-puhl-tuhn), township (□ 207 sq mi/538.2 sq km; 2001 population 9,303), S ONTARIO, E central CANADA, 28 mi/44 km from GUELPH; 43°45′N 80°41′W. Farming and related industrties; Mennonite furniture, woodworking shops. Area first settled in the mid-1800s. The township was formed in 1999 with the amalgamation of PEEL, MARYBOROUGH, and DRAYTON.

Mapleton, town (2000 population 1,416), MONONA county, W IOWA, on MAPLE RIVER, and 40 mi/64 km SE of SIOUX CITY; 42°10′N 95°47′W. Concrete. Incorporated 1878.

Mapleton, town, AROOSTOOK county, NE MAINE, just W of PRESQUE ISLE; 46°42′N 68°07′W. Potato-producing area. Incorporated 1880.

Area is shown by the symbol □, and capital city or county seat by ⊙.

Mapleton, town (2000 population 1,678), BLUE EARTH county, S MINNESOTA, 17 mi/27 km S of MANKATO, near MAPLE RIVER; 43°55′N 93°57′W. Grain; livestock; light manufacturing. Laid out and incorporated 1878.

Mapleton, town (2006 population 7,157), UTAH county, N central UTAH, suburb, 8 mi/12.9 km SE of PROVO, on Hobble Creek; 40°07′N 111°34′W. Elevation 4,537 ft/1,383 m. Fruit, vegetables; dairying; cattle, sheep. Spanish Fork Peak to SE; UTAH LAKE to NW. Unita National Forest and WASATCH RANGE to E. Originally named Union Beach, name changed in 1901.

Mapleton, village (2000 population 98), BOURBON county, SE KANSAS, 15 mi/24 km NW of FORT SCOTT; 38°01′N 94°52′W. Dairying, general agriculture.

Mapleton (MAI-puhl-tuhn), unincorporated village, HERTFORD county, NE NORTH CAROLINA, 37 mi/60 km E of ROANOKE RAPIDS, near MEHERRIN RIVER; 36°25′N 77°02′W. Agriculture (tobacco, cotton, peanuts, grain; poultry, livestock).

Mapleton, village (2006 population 598), CASS county, E NORTH DAKOTA, 12 mi/19 km W of FARGO, and on MAPLE RIVER; 46°53′N 97°02′W. Founded in 1875 and incorporated in 1884. Named for the Maple River.

Mapleton Depot, borough (2006 population 443), HUNTINGDON county, S central PENNSYLVANIA, 3 mi/4.8 km W of MOUNT UNION, on JUNIATA RIVER; 40°23′N 77°56′W. Agriculture (alfalfa; livestock, poultry); manufacturing (feeds, glass sand). Parts of Rothrock State Forest to S, NE, and W. Also known as Mapleton.

Maple Valley, unincorporated town (2006 population 16,440), KING county, W WASHINGTON, suburb 20 mi/32 km SE of downtown SEATTLE, on CEDAR RIVER; 47°24′N 122°02′W. Railroad junction. Manufacturing (wood products). Lake Youngs reservoir to W.

Mapleview, village (2000 population 189), MOWER county, SE MINNESOTA, 2 mi/3.2 km N of AUSTIN, on CEDAR RIVER; 43°41′N 92°58′W. Grain; dairying.

Maplewood, city (2000 population 34,947), RAMSEY county, SE MINNESOTA, residential suburb, 5 mi/8 km NE of downtown ST. PAUL, immediately SW of city of St. Paul, with E extension adjoining the E border of St. Paul; 44°59′N 93°01′W. Manufacturing (dairy products, signs). MISSISSIPPI RIVER to SW; numerous small lakes in area. Incorporated 1957.

Maplewood, city (2000 population 9,228), SAINT LOUIS county, E MISSOURI, a suburb, 7 mi/11.3 km SW of downtown ST. LOUIS; 38°36′N 90°19′W. Manufacturing (fabricated metal products, tools, transportation equipment, machinery). Settled 1825, incorporated 1908.

Maplewood, township (2000 population 23,868), ESSEX county, NE NEW JERSEY, just W of NEWARK; 40°43′N 74°16′W. Primarily residential area.

Maplewood, unincorporated village, CALCASIEU parish, SW LOUISIANA, 6 mi/10 km W of LAKE CHARLES, and 4 mi/6 km E of SULPHUR; 30°12′N 93°18′W. Railroad junction; oil and natural gas.

Maplewood, NEW HAMPSHIRE: see BETHLEHEM.

Maplewood Heights, unincorporated town, KING county, W WASHINGTON, residential suburb, 12 mi/19 km SE of downtown SEATTLE, 4 mi/6.4 km E of RENTON, on CEDAR RIVER. Lake Youngs reservoir to S.

Mapocho River (mah-PO-cho), 75 mi/121 km long, METROPOLITANA DE SANTIAGO region, central CHILE; rises in the ANDES near ARGENTINA border; flows W and S, past LAS CONDES, SANTIAGO, PEÑAFLOR, and TALAGANTE, to MAIPO RIVER 10 mi/16 km E of MELIPILLA. Irrigates fertile central valley near Santiago. In its valley are copper and lead deposits.

Mapou, village (2000 population 1,187), RIVIÈRE DU REMPART district, N MAURITIUS, 6 mi/9.7 km WNW of RIVIÈRE DU REMPART. Sugarcane, Anjalay station, district court.

Mapperley, ENGLAND: see NOTTINGHAM.

Mappsville (MAPS-vil), unincorporated village, ACCOMACK county, E VIRGINIA, 10 mi/16 km NNE of ACCOMAC, in EASTERN SHORE area, W of ATLANTIC OCEAN, and E of CHESAPEAKE BAY; 37°50′N 75°34′W. Manufacturing (clam processing; bronze sculptures); agriculture (dairying; livestock; vegetables); fish, oysters, clams.

Mapulaca (mah-poo-LAH-kah), town, LEMPIRA department, W HONDURAS, near RÍO LEMPA (EL SALVADOR border), 39 mi/63 km S of GRACIAS; 14°02′N 88°37′W. Corn, beans.

Mapusa (muh-poo-SAH), town, NORTH GOA district, GOA state, W central INDIA, 6 mi/9.7 km N of PANAJI; 15°36′N 73°49′E. Trades in timber, fish, rice, copra, cashew nuts; livestock raising.

Maputo (mah-POO-to), province (2004 population 1,074,793), ⊙ MAPUTO, southernmost tip of MOZAMBIQUE, on MAPUTO BAY (E) and INDIAN OCEAN (SE), and on SOUTH AFRICA (W) and SWAZILAND (SW) borders. Bordered N and NE by GAZA province. High coastal plain containing spectacular beaches near Ponta de Oro and natural reserves of the country. Commercial hub of the country; agriculture (sugar, sisal, copra). Prinicipal port outlet for neighboring South Africa. A short railroad along the LIMPOPO River links Maputo to commercial centers in South Africa. Has seven populous districts (BOANE, MAGUDE, Manhiça, MARRAQUENE, MATUTUINE, MOAMBA, and NAMAACHA) that spread along 66 habitats of the predominately Tsonga- and Ronga-speaking peoples. Maputo city and the environs has the country's largest Indo-Pakistani population of traders and professionals. Formerly called Lourenço Marques.

Maputo (mah-POO-to), city (2004 population 1,744,300), ⊙ MOZAMBIQUE, a port on the INDIAN OCEAN; 14°50′S 40°43′E. Mozambique's largest city and its administrative, communications, and commercial center. The economy is dominated by the modern port, on Malputo Bay; coal, cotton, sugar, chrome, ore, sisal, copra, and hardwood are the chief exports. Main manufacturing includes food products, beverages, cement, pottery, furniture, shoes, and rubber. People of Indo-Pakistani background play an important role in retail trade. Prior to Mozambique's independence in 1975, tourists from SOUTH AFRICA and ZIMBABWE (formerly RHODESIA) frequented the city and its excellent beaches. Since then, tourism has declined. Linked by railroad with South Africa, SWAZILAND, and ZIMBABWE and by an all-weather road with JOHANNESBURG (South Africa). International airport to N. Founded in the late 18th century, the city is named for the Portuguese trader who first explored the area in 1544. Its main growth dates from 1895, when a railroad to PRETORIA (South Africa) was completed. In 1907, Lourenço Marques became the capital of Mozambique. After independence, most of the city's large Portuguese population left and its named was changed to Maputo. Local economy suffered as Mozambique broke ties with South Africa in the 1970s and 1980s. Seat of the University of Eduardo Mondlane (1962). Museum on Mozambique's history, a military museum, and the Roman Catholic Cathedral of Our Lady of Fatima.

Maputo Bay (mah-POO-to), inlet of the INDIAN OCEAN, S MOZAMBIQUE, SE AFRICA; c.55 mi/89 km long, 20 mi/32 km wide. MAPUTO, the capital and chief port of MOZAMBIQUE, is on the bay. Maputo Bay is a large deepwater harbor, with numerous quays to handle oceangoing vessels; railroads lead into the interior. The first Westerner to visit (1502) the bay was António do Campo, one of Vasco da Gama's captains; the area was explored in 1544 by Lourenço Marques, the Portuguese trader. In the 1700s, Dutch and Austrian trading companies tried to establish posts on the bay; both were driven out by malaria and the Portuguese. In 1787, PORTUGAL built a fort here, around which the town of Lourenço Marques (now named Maputo, the capital) grew. In the mid-1800s, Portugal's claim to the area was challenged by GREAT BRITAIN and by the TRANSVAAL when it was realized that the bay provided a major access route to the Kimberley diamond mines. The Transvaal recognized Portugal's sovereignty in 1869, and in 1875 FRANCE, acting as arbiter, awarded the area to Portugal. Formerly DELAGOA BAY.

Maputo Elephant Reserve (mah-POO-to), MAPUTO province, MOZAMBIQUE, 15 mi/24 km N of Zitundo. A wilderness of savannas and lakes bordered by 60 mi/97 km of crystalline sands along the INDIAN OCEAN. A World Biodiversity site. Ecotourism is proposed for 400 sq mi/1,036 sq km of areas S and W of reserve.

Maputo River, MOZAMBIQUE: see PONGOLA RIVER.

Maputsoe (mah-POOT-so), town (2004 population 36,200), LERIBE district, NW LESOTHO, 38 mi/61 km NNE of MASERU, on CALEDON RIVER (South African border), opposite FICKSBURG (South Africa; crossing at Ficksburg Bridge); 28°53′S 27°54′E. Ficksburg Airport to N, across river. Maputsoe and Ha Nyenye industrial estates; manufacturing (clothing, footwear, cornmeal, umbrellas). Agricultural area (corn, sorghum, vegetables; cattle, sheep, goats, hogs).

Maqatin (muh-KAH-tin), village, S YEMEN, on Gulf of ADEN, 50 mi/80 km E of SHUQRA. Fisheries. Was in former FADHLI sultanate.

Maqdaha, YEMEN: see MAJDAHA.

Ma'qil (muh-AH-kil), port for BASRA, SE IRAQ, at head of large-ship navigation on the SHATT AL ARAB, just N of BASRA, and 80 mi/129 km from the PERSIAN GULF; here, too, is a railroad station for BASRA. Ma'qil, a modern port, was built up by the British after World War I. Port lost much of its importance with decline of ship traffic up the Shatt al Arab. Almost paralyzed since the beginning of the IRAN-Iraq War, and it was partly devastated during the fighting. Sometimes spelled Magil.

Maqteir (mahk-te-IR), dunes, N MAURITANIA, range of dune formation c.124 mi/200 km NE of ATAR; c.21°25′N 12°00′W.

Maquapit Lake (MA-kwuh-pit) (□ 7 sq mi/18.2 sq km), S NEW BRUNSWICK, E CANADA, 20 mi/32 km E of FREDERICTON, and 2 mi/3 km W of GRAND LAKE; 5 mi/8 km long, 2 mi/3 km wide.

Maqueda Bay (mah-KAI-dah), inlet of SAMAR SEA, in W SAMAR island, PHILIPPINES; 9 mi/14.5 km long, 7 mi/11.3 km wide. BUAD ISLAND is near entrance. WRIGHT is at head of bay, and CATBALOGAN near N side of entrance.

Maqueda Channel (mah-KAI-dah), strait in PHILIPPINES connecting LAGONOY GULF with PHILIPPINE SEA, between RUNGUS POINT of SE LUZON and CATANDUANES island; c.5 mi/8 km wide.

Maquela do Zombo (mah-KAI-luh do ZOM-bo), town, UÍGE province, NW ANGOLA, near Zaire border, 120 mi/193 km S of KINSHASA (Democratic Republic of the CONGO). Cashews, almonds, manioc, sesame, corn; pottery manufacturing. Airfield.

Maqueripe Bay (MAH-kah-reep), swimming beach, NW TRINIDAD, TRINIDAD AND TOBAGO, 8 mi/12.9 km NW of PORT OF SPAIN; 10°44′N 61°37′W.

Maquinchao (mah-keen-CHOU), town (1991 population 1,930), ⊙ Veinticinco [=twenty-five] de Mayo department, S RÍO NEGRO province, ARGENTINA, on railroad, 125 mi/201 km E of SAN CARLOS DE BARILOCHE; 41°15′S 68°44′W. Sheep-raising and lead-mining center.

Maquiné (MAH-kee-ne), city, E RIO GRANDE DO SUL state, BRAZIL, 59 mi/95 km NE of PÔRTO ALEGRE; 29°41′S 51°11′W.

Maquiné, Brazil: see CORDISBURGO.

Maquoit Bay (muh-KOIT), SW MAINE, indentation of CASCO BAY, just SW of BRUNSWICK, and E of FREEPORT; 4 mi/6.4 km long.

Maquoketa (ma-KO-kid-uh), city (2000 population 6,112), ⊙ JACKSON county, E IOWA, on MAQUOKETA RIVER, near mouth of North Fork Maquoketa River, and 32 mi/51 km S of DUBUQUE; 42°04′N 90°40′W.

Manufacturing (consumer goods, machinery, fabricated metal products, feeds, concrete products, textiles, transportation equipment; printing). Maquoketa Caves State Park to W. Flooding occurred in 1993. Incorporated 1853.

Maquoketa River, E IOWA, c.130 mi/209 km long; rises in SE FAYETTE county; flows SE, past MANCHESTER, MONTICELLO, and MAQUOKETA, to MISSISSIPPI RIVER 7 mi/11.3 km SE of BELLEVUE. Receives North Fork (c.75 mi/121 km long) near Maquoketa.

Maquon, village (2000 population 318), KNOX county, W central ILLINOIS, on SPOON RIVER, and 15 mi/24 km SE of GALESBURG; 40°47′N 90°09′W. In agricultural area.

Mara (MAH-rah), region (2006 population 1,572,000), N TANZANIA, on UGANDA (on Lake VICTORIA) and KENYA (both N) borders, ⊙ MUSOMA. Lake Victoria to E (excluding UKEREWE and other islands). Masuru Swamp in N; part of SERENGETI PLAIN and SERENGETI NATIONAL PARK in SE. Ikorongo (E) and Gruneti (S) game reserves. Cotton, grain; cattle, sheep, goats; fish. Gold mining in center and N. Part of former Lake province.

Mara (MAH-rah), town, COTABAMBAS province, APURÍMAC region, S central PERU, in the ANDES Mountains, near Santo Tomás River (an affluent of APURÍMAC RIVER), and 45 mi/72 km SSW of CUSCO; 14°06′S 72°07′W. Potatoes, cereals; livestock.

Mara (MAH-rah), village, Kajiado district, RIFT VALLEY province, S KENYA, on left bank of MARA RIVER, and 55 mi/89 km WSW of NAROK. Agriculture (peanuts, sesame, corn) and livestock. Also called Mara Bridge. Lolgorien, or Longorien, goldfields (SW) were discovered 1920.

Mara (MAH-rah), river, KENYA and TANZANIA, c.180 mi/290 km in length; rises in the MAU ESCARPMENT in the SW portion of Kenya; flows due S and W through Tanzania to enter Lake VICTORIA at town of MUSOMA. Basin drainage area is 3,000 sq mi/7,770 sq km.

Maraã (mah-rah-AHN), city (2007 population 17,517), W central AMAZONAS state, BRAZIL, 99 mi/160 km NW of TEFÉ, on Rio Japurá; 01°50′S 65°21′W.

Maraada, LIBYA: see MARADAH.

Marabá (mah-rah-BAH), city (2007 population 196,468), SE PARÁ state, BRAZIL, on left bank of TOCANTINS River, near GOIÁS and MARANHÃO state borders, and 270 mi/435 km S of BELÉM; 05°20′S 49°05′W. Rubber, Brazil nuts; fishing. Important node on Trans-Amazon Highway. Near TUCURUÍ dam and hydroelectric station on Tocantins. Diamonds found nearby. Airfield.

Marabadyasa (mah-rah-bah-DYAH-sah), town, central CÔTE D'IVOIRE, 24 mi/38 km W of KATIOLA; 08°06′N 05°26′W. Agriculture (corn, beans, tobacco, peanuts, cotton). Also spelled Marabadiassa.

Marabahan (MAH-rah-BAH-hahn), town, ⊙ Baritokuala district, Kalimantan Seletan province, SE BORNEO, INDONESIA, 36 mi/58 km NE of BANJARMASIN.

Marabá Paulista (MEE-rah-bah POU-lee-stah), city, W SÃO PAULO state, BRAZIL, 36 mi/58 km W of PRESIDENTE PRUDENTE; 22°16′S 51°56′W.

Marabella (MAH-rah-bel-ah), village, W TRINIDAD, TRINIDAD AND TOBAGO, on the Gulf of PARIA, on railroad, and 2 mi/3.2 km N of SAN FERNANDO. Coconuts; manjak deposits. Thermal springs nearby.

Marabios, Cordillera de los, NICARAGUA: see MARIBIOS, CORDILLERA DE LOS.

Maracaçumé (MAH-rah-kah-soo-MAI), city (2007 population 17,633), N MARANHÃO state, BRAZIL, on BR 316 Highway, near PARÁ state border; 02°06′S 45°57′W.

Maracaí (MAH-rah-kah-EE), city (2007 population 13,166), W SÃO PAULO state, BRAZIL, 55 mi/89 km ESE of PRESIDENTE PRUDENTE; 22°36′S 50°39′W. Dairy products; lard processing, sawmilling. Formerly Maracahy.

Maracaibo (mah-rah-KEI-bo), city, ⊙ ZULIA state, NW VENEZUELA, at the outlet of Lake MARACAIBO; 10°42′N 71°37′W. Venezuela's second-largest city, a commercial and industrial center, and the oil capital of SOUTH AMERICA. Besides oil, exports include lumber, processed textiles, and soap. Founded in 1574. In the 17th century it was sacked 5 times, notably by Sir Henry Morgan in 1669. Until the establishment of the oil industry after 1918, Maracaibo was extremely underdeveloped, but exploitation by foreign interests of the vast petroleum resources of the MARACAIBO BASIN resulted in a rapid expansion and modernization of the city. The dredging of the lake also increased Maracaibo's importance as a shipping point for inland products. The General Rafael Urdaneta Bridge (c.5-mi/8-km-long) is S of the city. Native Guajiro craft markets.

Maracaibo Basin or **Maracaibo Lowlands** (mah-rah-KEI-bo), region, NW VENEZUELA; 10°00′N 71°40′W. Bounded by mountains on three sides and opened to the CARIBBEAN SEA in the N. Lake MARACAIBO occupies much of the hot and humid low-lying region, yielding rich petroleum deposits. The basin is semiarid in the N, but has an average annual rainfall of 50 in/127 cm in the S.

Maracaibo, Lake (mah-rah-KEI-bo), largest lake (□ 5,100 sq mi/13,260 sq km) of SOUTH AMERICA, NW VENEZUELA, extending c.110 mi/177 km inland; 09°40′N 71°30′W. A strait, 34 mi/55 km long, connects it with the Gulf of VENEZUELA. Discovered in 1499 by Alonso de Ojeda, the Spanish explorer, the lake lies in the extremely hot and humid lowlands of the Maracaibo basin, a region which, almost enclosed by mountains, is semiarid in the N but has an average annual rainfall of 50 in/127 cm in the S. Sugarcane, cacao, and some livestock is raised. By far the most vital activity is production of petroleum. Developed since 1918 by foreign concerns, and later by the Venezuela government, the region is one of the greatest oil-producing areas in the world. Lake Maracaibo, with the CATATUMBO RIVER, its chief tributary, is a major artery of communication for products of the adjacent region and those of the COLOMBIAN-Venezuelan highlands. A dredged channel gives oceangoing vessels access to the lake. CABIMAS and the port of MARACAIBO are the principal cities on the lake. General Rafael Urdaneta Bridge (c.5 mi/8 km long; completed 1962), spanning the lake's outlet, is one of the longest bridges in the world.

Maracajá, Brazil: see GOVERNADOR ISLAND.

Maracaju (mah-rah-kah-ZHOO), city (2007 population 30,924), S Mato Grosso do Sul state, BRAZIL, 85 mi/137 km SSW of CAMPO GRANDE, on railroad between Campo Grande and PONTA PORÃ; 21°35′S 55°14′W. Cattle- and maté-shipping center. Was de jure capital of former Ponta Porã territory (1943–1946).

Maracaju, Serra de (MAH-rah-kah-zhoo, SER-rah zhe), Spanish *Cordillera de Mbaracayú*, diabase range in PARANÁ and MATO GROSSO states, S BRAZIL, extending c.50 mi/80 km E-W on either side of GUAÍRA FALLS, where it is crossed by PARANÁ River Its W part forms Brazil-PARAGUAY border and sends spurs into Paraguay. Continued (W) by Serra de AMAMBAÍ. Rises to c.1,500 ft/457 m.

Maracanã (mah-rah-kah-nuh), city (2007 population 28,299), extreme E PARÁ state, BRAZIL, near the Atlantic Ocean, 85 mi/137 km NE of BELÉM; 00°40′S 47°25′W. Fishing; cattle raising; lumber shipping.

Maracanaú (mar-rah-kah-nah-OO), city (2007 population 197,301), N CEARÁ state, BRAZIL, on FORTALEZA-CRATO railroad (junction of spur to MARANGUAPE), and 12 mi/19 km SW of Fortaleza; 03°46′S 38°34′W.

Maracanda, UZBEKISTAN: see SAMARKAND.

Maracás (mah-rah-KAHS), city (2007 population 34,180), E central BAHIA state, BRAZIL, in the Serra do SINCORÁ, 40 mi/64 km NW of JIQUIÉ; 13°30′S 40°29′W. Coffee growing; kaolin deposits.

Maracas (MAH-rah-kas), village, NW TRINIDAD, TRINIDAD AND TOBAGO, at S foot of El Tucuche Mountain, 7 mi/11.3 km E of PORT OF SPAIN. The Maracas Falls are 2 mi/3.2 km N. Public beach, destination for day visitors; MARACAS BAY village resort scheduled for development.

Maracas Bay, inlet of N TRINIDAD, TRINIDAD AND TOBAGO, towered over by the forest-clad El Tucuche Mountain, 16 mi/26 km NE of PORT OF SPAIN; 10°45′N 61°26′W. Resort.

Maracay (mah-rah-KEI), city, ⊙ ARAGUA state, N VENEZUELA, near NE shore of Lago de VALENCIA, in basin of CORDILLERA DE LA COSTA, on PAN-AMERICAN HIGHWAY, on railroad, and 50 mi/80 km WSW of CARACAS; 10°15′N 67°35′W. Trading and industrial center in fertile country producing coffee, cacao, indigo, sugarcane, tobacco; livestock; timber. Textile and paper milling, meat canning; manufacturing. Air force base.

Maracena (mah-rah-THAI-nah), NW suburb of GRANADA, GRANADA province, S SPAIN; 37°12′N 03°38′W. In fertile district yielding cereals, olive oil, sugar beets, tobacco. Large meat-processing plants.

Maracha-Terego, administrative district, NORTHERN region, NW UGANDA, on DEMOCRATIC REPUBLIC OF THE CONGO border (W). As of Uganda's division into eighty districts, borders KOBOKO (N), YUMBE (N and E), and ARUA (S) districts. Primarily agricultural. Formed in 2006 from N central portion of former ARUA district created in 2000 (Koboko district formed from extreme N portion in 2005 and current Arua district from central and S portions). Also called Nyadri.

Maracó, ARGENTINA: see GENERAL PICO.

Maradah (muh-RAH-de), village, CYRENAICA region, N central LIBYA, in oasis on TRIPOLITANIA border, on road, 205 mi/330 km SSW of BANGHAZI, c. 70 mi/113 km S of Gulf of SIRTE; 29°14′N 19°13′E. Carnallite deposits discovered (1936). Oil and gas fields nearby. Also spelled Maraada.

Maradana (MUH-ruh-thah-nuh), section of COLOMBO, WESTERN PROVINCE, SRI LANKA, 2 mi/3.2 km E of city center; 06°15′N 79°52′E. Railroad junction.

Maradi (MUH-rah-dee), province (2005 population 2,552,800), S central NIGER, on NIGERIA (S) border; ⊙ MARADI; 14°00′N 07°00′E. Bounded N by AGADEZ, E by ZINDER, and W by TAHOUA provinces.

Maradi (MUH-rah-dee), city (2001 population 148,017), ⊙ MARADI province, S NIGER, near the border with NIGERIA; 13°29′N 07°06′E. Administrative and commercial center for an agricultural region that specializes in peanut growing and goat raising. A major road connects Maradi with KANO (Nigeria). Has a technical college, a center for research on poultry and goat breeding, an airport, and a hospital.

Maragha, IRAN: see MARAGHEH.

Maragha, El (MAH-RAH-guh, el), village, SOHAG province, central Upper EGYPT, on W bank of the NILE River, on railroad, and 8 mi/12.9 km SE of TAHTA; 26°42′N 31°36′E. Cotton, cereals, dates, sugar.

Maragheh (mahr-RAH-GE), city (2006 population 149,929), Āżerbāyjān-e sharqi province, NW IRAN, on S slopes of Mount SAHAND; 37°22′N 46°15′E. The trade and transportation center of a fertile fruit-growing region; dried fruits are exported. After Arab conquest in the 7th century, Maragheh developed rapidly as a provincial capital. In 1029 it was seized by the Oghuz Turks, but they were driven out by a subsequent Kurdish dynasty. Destroyed by the Mongols in 1221, but Hulagu Khan held court here until the establishment of a fixed capital at TABRIZ. Temporarily occupied by Russia in 1828. Celebrated observatory (13th century) now in ruins. Also known as Maragha.

Maragini (MAR-a-JEE-nee), village, SANDAUN (West Sepik) province, N central NEW GUINEA island, NW PAPUA NEW GUINEA, 65 mi/105 km SSE of VANIMO, 40 mi/64 km E of INDONESIAN (IRIAN JAYA) border, on S edge of Bewani Mountains, in SEPIK RIVER basin.

Marshy area. Boat access only. Coconuts, sago, palm oil; timber.

Maragogipe (MAH-rah-go-ZHEE-pe), city (2007 population 42,086), E central BAHIA state, BRAZIL, on right bank of Rio PARAGUAÇU at its influx into TODOS OS SANTOS Bay, 32 mi/51 km NW of SALVADOR; 12°43′S 38°54′W. Port shipping fruit, coffee. Cigar-manufacturing center. Fishing.

Maragoji (MAH-rah-GO-zhe), city, E ALAGOAS state, NE BRAZIL, on the Atlantic Ocean, 55 mi/89 km NE of MACEIÓ; 09°01′S 35°12′W. Sugar, copra, fruit. Formerly spelled Maragogy.

Maragua (mah-RAH-gwah), canton, OROPEZA province, CHUQUISACA department, SE BOLIVIA; 19°02′S 65°25′W. Elevation 8,212 ft/2,503 m. Iron (Mina Okekhasa and Mina Virgén de Copacabana) and manganese (Mina Lourdes) mining; limestone and gypsum deposits. Agriculture (potatoes, yucca, bananas, corn, barley, oats, rye, peanuts); cattle.

Maragua, canton, CHAYANTA province, POTOSÍ department, W central BOLIVIA, 25 mi/40 km SE of COLQUECHACA, 5 mi/8 km S of OCURI; 18°56′S 65°48′W. Elevation 13,642 ft/4,158 m. Insignificant iron deposits. Agriculture (potatoes, cassava, bananas, barley); cattle.

Maragua (mah-RAH-goo-ah), town, CENTRAL province, S central KENYA, on railroad, and 7 mi/11.3 km S of MURANG'A; 00°48′S 37°08′E. Coffee, wheat, corn.

Marahoué, region (□ 3,280 sq mi/8,528 sq km; 2002 population 651,700), central CÔTE D'IVOIRE; ⊙ BOUAFLÉ; 07°10′N 05°50′W. Bordered N by Worodougo region, NE by Vallée du Bandama region (S portion of border formed by Lake Kossou), E by Lacs region (N portion of border formed by Lake Kossou), S by Fromager region, and WNW by Haut-Sassandra region. Part of Lake Kossou in E; Béré River in N. Maraoué National Park in central part of the region. Towns include Bouaflé, SINFRA, and ZUÉNOULA.

Marahra (MAH-rah-rah), town, ETAH district, W UTTAR PRADESH state, N central INDIA, 7 mi/11.3 km SW of KASGANJ; 27°44′N 78°35′E. Glass-bangle manufacturing; wheat, pearl millet, barley, corn, jowar, oilseeds. Has seventeenth-century tombs, eighteenth-century mosque. Also spelled Marehra.

Marahú, Brazil: see MARAÚ.

Marahuaca, Cerro (mah-rah-HWAH-kah, SER-ro), peak (8,461 ft/2,579 m) in AMAZONAS state, S VENEZUELA, just NE of Cerro DUIDA, 200 mi/322 km SE of PUERTO AYACUCHO; 03°34′N 65°27′W.

Marais Breton, FRANCE: see BRETON, MARAIS.

Maraisburg, SOUTH AFRICA: see ROODEPOORT.

Marais des Cygnes (muhr-ee dah SEEN) river, c.140 mi/225 km long, KANSAS and MISSOURI; rises in S WABAUNSEE county c.25 mi/40 km SW of TOPEKA, E central Kansas; flows SE through Melvern Lake Reservoir, past OTTAWA and OSAWATOMIE, then into W MISSOURI to join the LITTLE OSAGE RIVER 6 mi/9.7 km W of Scheli City to form the OSAGE RIVER. Subject to heavy flooding, the river has many flood control projects.

Marais du Cotentin et du Bessin Natural Regional Park (mah-rai dyoo ko-tawn-TAN ai dyoo be-SAN) (□ 463 sq mi/1,203.8 sq km), MANCHE and CALVADOS departments, BASSE-NORMANDIE region, NW FRANCE. Poorly drained marshland.

Marais, Le (mah-RAI, luh) [French=the swamp], old quarter of PARIS, ÎLE-DE-FRANCE, FRANCE, on right bank of the SEINE RIVER, now comprising the 3rd and 4th *arrondissements* of the city. Until 18th century it was the most aristocratic section of Paris. The Hôtel des Tournelles, long the residence of the kings of France (Henry II was killed in its court during a joust), was replaced by the Place des Vosges. The Marais park, surrounded by uniform houses in pink brick and gray slate, remains a perfect ensemble of 17th-century architecture. Nearby is the Musée Carnavalet, once the home of Madame de Sévigné, which

now houses the municipal museum of Paris. During 19th century it became a ghetto area for Jewish refugees from E Europe. Since 1969, a major restoration program has been underway, including renovation of several museums, mansions, and hotels, such as the 17th-century Hôtel Sully.

Marais Poitevin, FRANCE: see POITEVIN, MARAIS.

Marais Poitevin, Val de Sèvre et de la Vendée Natural Regional Park (mah-rai pwah-tuh-VAN, vahl duh SEV-ruh ai duh lah vawn-DAI), S VENDÉE (□ 772 sq mi/2,007.2 sq km), N CHARENTE-MARITIME, W central DEUX-SÈVRES departments, W FRANCE. Marshland stretching inland from the ATLANTIC coast along the SÈVRE NIORTAISE RIVER.

Maraita (mah-rah-EE-tah), town, FRANCISCO MORAZÁN department, S HONDURAS, 20 mi/32 km SE of TEGUCIGALPA; 13°53′N 87°02′W. Corn, wheat; livestock.

Marajó (MAH-rah-ZHO), island, N Brazil, at the mouth of the AMAZON River; c.150 mi/241 km long, c.100 mi/161 km wide. The island divides the river into the Amazon proper and the PARÁ. Cattle are raised on the extensive E grasslands, and water buffaloes are bred in the low, swampy W. The island is famous for its prehistoric mounds, which yield pottery.

Marakabei (mah-RAH-ke-bai), town, MASERU district, central LESOTHO, 42 mi/68 km ESE of MASERU, on SENQUNYANE RIVER (Mohale Dam site 6 mi/9.7 km to NNW), on N end of the THABA PUTSUA Range; 29°33′S 28°09′E. Horses, sheep, goats, cattle; peas, beans. Tourism.

Marakei (mah-RAH-kai), atoll (□ 4 sq mi/10.4 sq km; 2005 population 2,741), N GILBERT ISLANDS, KIRIBATI, W central PACIFIC OCEAN; 02°N 173°20′E. Sponges.

Marakkanam (muh-rah-kuhn-NUHM), town, SOUTH ARCOT VALLALUR district, TAMIL NADU state, S INDIA, on lagoon on COROMANDEL COAST of BAY OF BENGAL, 21 mi/34 km E of TINDIVANAM; 12°12′N 79°57′E. S terminus of BUCKINGHAM CANAL. Ships casuarina and cashew plants. Saltworks. Also spelled Merkanam, Markanam.

Marala (ma-rah-LAH), village, SIALKOT district, E PUNJAB province, central PAKISTAN, near CHENAB River, 12 mi/19 km NNW of SIALKOT; 32°39′N 74°30′E. Headworks of Upper CHENAB canal are just W. Sometimes spelled MERALA.

Maralal (mah-rah-LAL), town (1999 population 16,281), Samburu district, RIFT VALLEY province, KENYA, on road, and 106 mi/170 km NE of NAKURU; 01°05′N 36°42′E. Market and trading center.

Marale (mah-RAH-lai), town, FRANCISCO MORAZÁN department, central HONDURAS, 20 mi/32 km S of YORO; 14°53′N 87°09′W. Corn, wheat, coffee.

Maralik (muhr-ah-LYEK), urban settlement and administrative center of the Ari region, NW ARMENIA, at W foot of MOUNT ARAGATS, on the Kumayri-Yerevan Highway, 15 mi/24 km S of KUMAYRI; 40°34′N 43°52′E. Has railroad station. Wheat; livestock. Manufacturing of magnetic amplifiers, building materials, cheese; dairy.

Máramarossziget, ROMANIA: see SIGHETU MARMAȚIEI.

Maramba (mah-LAH-em-bah), city, SOUTHERN province, S ZAMBIA, on the ZAMBEZI RIVER (ZIMBABWE border; bridged to VICTORIA FALLS township, Zimbabwe); 17°51′S 25°52′E. railroad-spur junction to MULOBEZI to NW. Airport. Industrial, commercial, transportation, and tourist center. Manufacturing (locks and latches, clothing and textiles, railroad sleepers, lumber). Cattle; timber. Headwaters of LAKE KARIBA reservoir c.70 mi/113 km to E; MOSI-OA-TUNYA NATIONAL PARK and Victoria Falls National Park (Zimbabwe), protect VICTORIA FALLS immediately upstream (to W). Eastern Cataract Field Museum, Steam Railroad Museum, and Rainbow Lodge at Victoria Falls. The National Museum contains archaeological, ethnological, and historical materials, including letters and relics of Livingstone. Founded in

1905, the city was initially named for David Livingstone, the Scots explorer. From 1911 to 1935 it served as capital of the British protectorate of Northern Rhodesia. Formerly called Livingstone.

Marambaia Island (MAH-rahm-bei-ah), off coast of RIO DE JANEIRO state, BRAZIL, 8 mi/12.9 km SSE of MANGARATIBA. Rises to more than 2,000 ft/610 m. Fishing trade school. With the narrow, sandy tombolo (25 mi/40 km long) linking it to mainland (25 mi/40 km W of RIO), it encloses SEPETIBA Bay.

Marambio Station (mah-RAHM-bee-yo), ANTARCTICA, Argentinian station on SEYMOUR ISLAND; 64°14′S 56°37′W.

Maramec (MAR-uh-mek), village (2006 population 106), PAWNEE county, N OKLAHOMA, 9 mi/14.5 km SE of PAWNEE; 36°14′N 96°40′W. In agricultural area.

Maramjhiri, INDIA: see BETUL.

Maramoros'ka Solotvyna, UKRAINE: see SOLOTVYNA.

Marampa (mah-RAHM-pah), town, NORTHERN province, W SIERRA LEONE, near ROKEL RIVER, 22 mi/35 km ESE of PORT LOKO; 08°45′N 10°28′W. Trade center of Marampa chiefdom. Just W are important open-pit iron (hematite) mines. Was once linked by private railroad with ocean-shipping point of PEPEL (35 mi/56 km WSW).

Maramureş (mah-rah-MOO-resh), county, N ROMANIA, in Crasna-Mures, on UKRAINE border, in CARPATHIAN Mountains; ⊙ BAIA MARE; 47°40′N 24°00′E. Mining, forestry, agriculture.

Maramureş, ROMANIA: see CRIŞANA-MARAMUREŞ.

Marana, town (2000 population 13,556), PIMA county, S ARIZONA, near SANTA CRUZ RIVER, 21 mi/34 km NW of TUCSON; 32°24′N 111°10′W. Elevation 2,055 ft/626 m. Trade center in cotton-growing area; cattle; manufacturing (machinery parts). Tortolita Mountains to NE; TUCSON MOUNTAIN unit of SAGUARO NATIONAL MONUMENT to S.

Maranacook, Lake (muh-RAN-uh-kuk), lake, KENNEBEC county, S MAINE, just N of WINTHROP; 5.5 mi/8.9 km long.

Maranboy, settlement, NW central NORTHERN TERRITORY, AUSTRALIA, 190 mi/306 km SE of DARWIN; 14°32′S 132°47′E. Tin and wolfram mines. Police station here; entrance to Beswick Aboriginal Reserve.

Maranchón (mah-rahn-CHON), village, GUADALAJARA province, central SPAIN, 55 mi/89 km NE of GUADALAJARA; 41°03′N 02°12′W. Grain growing, sheep raising.

Marand (mah-RAHND), town (2006 population 114,841), ÁŻERBĀYJĀN-E SHARQI province, NW IRAN, on railroad, and 35 mi/56 km NW of TABRIZ; 38°25′N 45°46′E. Elev. 4,377 ft/1,334 m. Road junction for KHOI, JULFA and URMIA. Grain, cotton; dry fruit. Has many gardens and fruit orchards. Also spelled MERAND or MEREND.

Marandellas, ZIMBABWE: see MARONDERA.

Maranello (mah-rah-NEL-lo), town, MODENA province, EMILIA-ROMAGNA, N central ITALY, 9 mi/14 km SSW of MODENA; 44°32′N 10°52′E. Agricultural machinery, metals, machinery, textiles. Has ruined castle.

Marang (MAH-rahng), town, E central TERENGGANU, E MALAYSIA, on SOUTH CHINA Sea, at mouth of small Marang River, 10 mi/16 km SSE of KUALA TERENGGANU; 05°12′N 103°13′E. Coconuts; fisheries.

Maranganí (mah-rahn-gah-NEE), town, CANCHIS province, CUSCO region, S PERU, on Vilcanota River (see URUBAMBA RIVER), on CUSCO-PUNO railroad, and 8 mi/13 km SSE of SICUANI; 14°22′S 71°10′W. Elevation 11,532 ft/3,515 m. Cereals, potatoes.

Marange-Silvange (mah-RAWNZH–seel-VAWNZH), town (□ 5 sq mi/13 sq km; 2004 population 5,650), MOSELLE department, LORRAINE region, NE FRANCE, 8 mi/12.9 km NNW of METZ; 49°13′N 06°06′E. In declining iron-mining district.

Maranguape (mah-rahn-gwah-pai), city (2007 population 103,181), N CEARÁ state, BRAZIL, on railroad spur, and 15 mi/24 km SW of FORTALEZA; 03°47′S

38°35′W. Hill resort; ships coffee, sugar, cotton, and fruit to Fortaleza. Carnauba-wax mill.

Maranhão (mah-rahn-YOUN), state (□ 126,897 sq mi/ 329,932.2 sq km; 2007 population 6,117,966), NE BRAZIL, on the ATLANTIC OCEAN; ⊙ SÃO LUÍS. Site of major agricultural and industrial colonization projects since the 1970s; agriculture and livestock remain central to the region's economy. Industries now include steel manufacturing and food processing. Major iron pellatizing and export facility at Ponta da Madeira near São Luis. Long been inhabited by the Tupinambás when the Portuguese established the first European settlement here in 1534.

Maranhão (mahr-ahn-YOU), reservoir and lake, SW PORTALEGRE district, E central PORTUGAL, on Seda River; 39°05′N 07°55′W. Receives Aviz River Seda at N tip, Benavila and Aviz on E shore.

Marañón (mah-rahn-YON), province (□ 2,142 sq mi/ 5,569.2 sq km), HUÁNUCO region, central PERU; ⊙ HUACRACHUCO; 08°45′S 76°40′W. Northernmost province of Huánuco region, bordered by ANCASH, LA LIBERTAD, and SAN MARTÍN regions.

Marañón River (mah-rahn-YON), c.1,000 mi/1,609 km long, PERU; rises in Lake LAURICOCHA in the CORDILLERA OCCIDENTAL, W central Peru; flows generally NW, then E across the ANDES Mountains to join the UCAYALI River in NE Peru where it forms the AMAZON River; some consider the Marañón to be the authentic headwater of the Amazon; 04°27′S 73°30′W. It is navigable to the Pongo de MANSERICHE, the gorge in NW Peru through which it flows before reaching the Amazon basin. The HUALLAGA RIVER is its chief tributary. Pedro de Ursúa, the Spanish explorer, descended the Marañón in 1560.

Marano Vicentino (mah-RAH-no vee-chen-TEE-no), town, VICENZA province, VENETO, N ITALY, 12 mi/19 km NNW of VICENZA; 45°41′N 11°25′E. Foundries, machine shops, woolen mill.

Marans (mah-RAWN), town (□ 150 sq mi/390 sq km), CHARENTE-MARITIME department, POITOU-CHARENTES region, W FRANCE, in Marais POITEVIN, formerly an active port on the SÈVRE NIORTAISE connected by canal with La ROCHELLE (12 mi/19 km SW); 46°19′N 01°00′W. Grain and poultry market; pharmaceutical laboratories, frozen foods manufacturing, and boat maintenance. Surrounding marshland (□ 150 sq mi/389 sq km), reclaimed since 17th century, now in agriculture.

Marão, Serra do (mahr-OU, SER-rah doo), short range in SE PÔRTO and SW VILA REAL districts, N PORTUGAL, N of DUERO RIVER, and just WSW of VILA REAL; 41°15′N 07°55′W. Rises to 4,642 ft/1,415 m. Vineyards on S slope.

Marapanim (mah-rah-pah-neem), city (2007 population 26,651), extreme E PARÁ state, BRAZIL, on the Atlantic Ocean, and 80 mi/129 km NE of BELÉM; 00°20′S 47°40′W. Fishing; lumbering; port.

Marapa River (mah-RAH-pah), 110 mi/177 km long, S TUCUMÁN province, ARGENTINA; rises in S outliers of NEVADO DEL ACONQUIJA; flows E past GRANEROS and LA MADRID, and joins the río CHICO to form the río HONDO, which enters the Salí River (upper course of the río DULCE). Irrigates sugarcane area; dam at Escaba.

Marapi, Mount (MAH-rah-pee), volcanic peak (9,485 ft/2,891 m), W Sumatra Barat province, INDONESIA, in PADANG HIGHLANDS of BARISAN MOUNTAINS, 40 mi/ 64 km NNE of PADANG; 00°23′S 100°28′E. In some eruptions, it has produced tremendous amounts of hot clouds of gases, ashes, and lava. Also called Mount Merapi. See also MERAPI, MOUNT, Java.

Maraqué River, CÔTE D'IVOIRE: see BANDAMA ROUGE RIVER.

Mara River (MAH-rah), c. 145 mi/233 km long, TANZANIA and KENYA; formed by joining of Amala and Kanunda rivers (both c. 50 mi/80 km long; rise in MAU ESCARPMENT, SW Kenya); flows SSW through Masai Mara National Reserve (Kenya), enters Tanzania,

continues W through Masaruru Swamp, enters Mara Bay, Lake VICTORIA, 10 mi/16 km E of MUSOMA.

Mara Rosa (MAH-rah RO-sah), city (2007 population 9,933), NW GOIÁS state, BRAZIL, 22 mi/35 km SSW of ESTRELA DO NORTE; 14°02′S 49°17′W.

Maras (MAH-rahs), town, URUBAMBA province, CUSCO region, S central PERU, 22 mi/35 km NW of CUSCO; 13°19′S 72°09′W. Elevation 10,505 ft/3,201 m. In agricultural region (cereals, vegetables).

Marasa (MAH-rah-sah), village, SUBEIHI tribal area, SW YEMEN, 26 mi/42 km WNW of LAHEJ, near former border between NORTH YEMEN and SOUTH YEMEN; 13°11′N 44°31′E. Was capital of former MAKHDUMI sheikdom. Also called Dar Murshid. Also spelled Wadi Marasa and Wadi Marasah.

Marasalcakmak, TURKEY: see OVACIK.

Mărăşeşti (muh-ruh-SHESHT), town, VRANCEA CO., E ROMANIA, 12 mi/19 km N of FOCŞANI. Important railroad junction; metallurgy, chemical works (glue, gelatin, sulphuric and hydrochloric acids). Romanian army here repulsed (1917) a great German offensive under Mackensen.

Marashki Trustenik, Bulgaria: see TRUSTENIK, LOVECH oblast.

Marataizes (MAH-rah-tah-EE-ses), city (2007 population 31,230), S coast of ESPÍRITO SANTO state, BRAZIL, 4 mi/7 km S of ITAPEMIRIM; 21°05′S 40°49′W.

Maratea (mah-rah-TAI-ah), village, POTENZA province, BASILICATA, S ITALY, on Gulf of Policatro, 7 mi/ 11 km SE of SAPRI; 39°59′N 15°43′E. Swimming resort; textiles.

Marathokambos (mah-rah-THO-kahm-bos), town, on W SÁMOS island, SÁMOS prefecture, NORTH AEGEAN department, GREECE, 20 mi/32 km W of SÁMOS town; 37°43′N 26°42′E. Wine, olive oil, tobacco.

Marathon, county (□ 1,576 sq km/4,097.6 sq km; 2006 population 130,223), central WISCONSIN; ⊙ WAUSAU; 44°53′N 89°45′W. Primarily a dairying area, county is a major producer of cheese; lumbering; agriculture (oats, barley, corn, potatoes, alfalfa, hay, ginseng, cranberries; cattle, sheep, hogs, poultry); diversified manufacturing at Wausau. Drained by WISCONSIN RIVER and its tributaries (EAU CLAIRE, BIG and LITTLE EAU PLEINE), Big RIB and Placer rivers and Big Eau Pleine and Du Bay reservoirs (both in S). RIB MOUNTAIN State Park, SW of Wausau, at center, including Rib Mountain (1,940 ft/591 m), winter sports center. N fringe of city of MARSHFIELD in SW. Formed 1850.

Marathon (MER-uh-thahn), city (□ 6 sq mi/15.6 sq km; 2005 population 9,822), ⊙ MONROE county, FLORIDA, in middle FLORIDA KEYS, halfway between KEY WEST and KEY LARGO; 24°42′N 81°04′W. Center of major resort area.

Marathon (MA-ruh-thahn), town (□ 66 sq mi/171.6 sq km; 2001 population 4,416), central ONTARIO, E central CANADA, on Lake SUPERIOR, 130 mi/209 km E of THUNDER BAY; 49°46′N 86°26′W. Gold mining; paper mill. Secondary airport. Formerly Peninsula.

Marathon, town (2000 population 302), BUENA VISTA county, NW IOWA, near source of RACCOON River, 18 mi/29 km NE of STORM LAKE; 42°51′N 94°58′W. In livestock and grain area.

Marathon, town, MARATHON county, central WISCONSIN, on Big RIB River, and 10 mi/16 km W of WAUSAU; 44°55′N 89°50′W. In dairying and lumbering area. Manufacturing (dairy products, lumber, beverages, building materials, wood products, feeds and fertilizers).

Marathon, village (2006 population 1,030), CORTLAND county, central NEW YORK, on TIOUGHNIOGA RIVER, and 25 mi/40 km NNW of BINGHAMTON; 42°26′N 76°02′W. Manufacturing (boats, transportation equipment), retail, and commercial services. In agricultural area. Incorporated 1861.

Marathon (MA-ruh-thahn), unincorporated village, BREWSTER county, extreme W TEXAS, 28 mi/45 km SE

of ALPINE; 30°12′N 103°14′W. Elevation 4,043 ft/1,232 m. In ranch region (sheep, cattle, goats); manufacturing (machinery). Headquarters for BIG BEND NATIONAL PARK (c.35 mi/56 km S).

Marathon (MAHR-a-thon), Greek *Marathón*, ancient Greek village and plain, in what is now E ATTICA prefecture, ATTICA department, E central GREECE, 20 mi/32 km NE of ATHENS; 38°09′N 23°58′E. Here the Athenians and Plataeans under Miltiades defeated a Persian army in 490 B.C.E. in the Persian Wars. Inspiration for the modern 26.2-mi/42-km race, based on the story of a messenger running this distance to give the news of victory, and promptly dying from the exertion.

Marathonisi (mah-rah-tho-NEE-see), islet in Gulf of LAKONIA, in GYTHION town, LAKONIA prefecture, SE PELOPONNESE department, extreme SE mainland GREECE; 37°41′N 20°53′E. Also, former name of Gythion. Supposed Cranae of ancient Greece, where Paris brought Helen after her abduction.

Marathonisi, Gulf of, Greece: see LAKONIA, GULF OF.

Marathovouno (mah-rah-THO-voo-no), Turkish *ulukisla*, town, FAMAGUSTA district, E CYPRUS, 15 mi/24 km ENE of NICOSIA; 35°13′N 33°37′E. PEDHIEOS RIVER passes to S, KYRENIA MOUNTAINS are to N. Grain, vetch, citrus, olives, vegetables, potatoes; goats, sheep, cattle, poultry.

Marathus, SYRIA: see AMRIT.

Marathwara (muh-RAHT-wah-rah), W division of former HYDERABAD state, W central INDIA. Comprises Marathi-speaking districts of AURANGABAD, BID, NANDED, OSMANABAD, and PARBHANI (now part of MAHARASHTRA state). Also Marathwada.

Maraú (mah-rah-OO), city (2007 population 17,104), E BAHIA state, BRAZIL, on the Atlantic Ocean, 50 mi/80 km N of ILHÉUS; 14°09′S 39°01′W. Oil-shale and asphalt deposits. Formerly spelled Marahú.

Marau (MAH-rah-oo), city (2007 population 33,783), N RIO GRANDE DO SUL state, BRAZIL, 17 mi/27 km SE of PASSO FUNDO; 28°27′S 52°12′W. Grapes, wheat, corn, potatoes; livestock.

Maraval (MAH-rah-val), N residential suburb of PORT OF SPAIN, N TRINIDAD, TRINIDAD AND TOBAGO.

Maravatío de Ocampo (mah-rah-vah-TEE-o dai o-KAHM-po), city and township, ⊙ Maravatío municipio, MICHOACÁN, central MEXICO, on central plateau, 45 mi/72 km ENE of MORELIA; 19°53′N 100°26′W. Elevation 6,824 ft/2,080 m. Railroad junction; processing and agricultural center (cereals, fruit, vegetables; livestock); flour milling, tanning, lumbering; manufacturing (shoes, textiles).

Maravia District, MOZAMBIQUE: see TETE.

Maravilha (MAH-rah-VEEL-yah), city (2007 population 10,229), central W ALAGOAS state, BRAZIL, 16 mi/ 26 km NW of SANTANA DO IPANEMA; 09°13′S 37°18′W.

Maravilha, city (2007 population 21,684), far W SANTA CATARINA state, BRAZIL; 26°47′S 53°09′W. Wheat; livestock.

Maravilla (mah-rah-VEE-yah), town and canton, MANURIPI province, PANDO department, NW BOLIVIA, near MADRE DE DIOS RIVER, 50 mi/80 km ESE of PUERTO RICO; 11°15′S 66°54′W.

Maravilla, Lake, CHILE: see TORO, LAKE.

Mara Vista, MASSACHUSETTS: see FALMOUTH.

Marawah (muh-ROU-wuh), township, CYRENAICA region, NE LIBYA, 30 mi/48 km E of Al MARJ, on Jabal Al AKHDAR plateau; 32°29′N 21°25′E. Road junction. Scene of fighting (1942) between Germans and British in World War II. Formerly Maraua.

Marawi (mah-rah-WEE), ancient city in the Nile state of SUDAN, on the right bank of the NILE River, 125 mi/ 201 km NNE of what is now KHARTOUM. In the mid-6th century B.C.E., Marawi replaced NAPATA as the central city of the Cushite dynasty and from 530 B.C.E. until C.E. 350 served as the capital. By the 1st century B.C.E., it was a major center for iron smelting. It is believed that knowledge of iron casting was carried

(7th–10th century C.E.) from the middle Nile to the middle Niger by a great African overland route. Among Marawi's extensive ruins are royal palaces (6th century B.C.E.) and a temple of Amon. Nearby are cemeteries and three groups of pyramids. Modern Marawi located further down NILE RIVER in separate state.

Marawi (mah-rah-WEE), town, Northern state, SUDAN, port on left bank of the NILE River (in great bend), opposite KARIMA railhead, and 150 mi/241 km NW of AD-DAMER; 18°29′N 31°49′E. Cotton center; also trade in wheat, barley, corn, fruits, and livestock. Has small museum of antiquities. Opposite the modern town, on right NILE bank, is site of ancient NAPATA. Also called Merowe.

Marawi City (mah-RAH-wee), city (□ 9 sq mi/23.4 sq km; 2000 population 131,090), ☉ LANAO DEL SUR province, W central MINDANAO, PHILIPPINES, 40 mi/64 km SW of CAGAYAN, near N shore of LANAO, LAKE; 08°00′N 124°17′E. Rice, corn; fishing. Formerly known as Dansalan.

Marayes (mah-RAH-yes), village, SE SAN JUAN province, ARGENTINA, at S foot of SIERRA DE LA HUERTA, on railroad, and 70 mi/113 km ENE of SAN JUAN; 31°29′S 67°20′W.

Maraza (mah-RAH-zah), urban settlement, E AZERBAIJAN, at SE end of the Greater CAUCASUS, on road, and 50 mi/80 km W of BAKY; 40°31′N 48°55′E. Livestock; dairying; winery.

Marazion (MAR-uh-ZEI-uhn), village (2001 population 1,495), W CORNWALL, SW ENGLAND, on MOUNT'S BAY of the CHANNEL, and 3 mi/4.8 km E of PENZANCE; 50°07′N 05°27′W. Fishing port and tourist resort for ST. MICHAEL'S MOUNT, to SW.

Marbach (MAHR-bahkh) or **Marbach am Neckar**, town, region of STUTTGART, BADEN-WÜRTTEMBERG, GERMANY, on the NECKAR River, and 7 mi/11.2 km NE of LUDWIGSBURG; 48°56′N 09°16′E. Railroad junction. Wine. Has 15th-century church; small local museum Poet Schiller was born here. Important library here has archives of famous German writers.

Mårbacka (MOR-BAHK-kah), village, VÄRMLAND county, W SWEDEN, on LAKE FRYKEN, 25 mi/40 km NW of KARLSTAD. Here was residence of Selma Lagerlöf.

Marbán, province, BENI department, BOLIVIA, ☉ LORETO; 15°40′S 64°20′W.

Marbat (MAHR-bet), village, Southern (Janubiah) region, S OMAN, on ARABIAN SEA, 40 mi/64 km E of SALALA, on the cape Ras Murbat, at foot of the JABAL SAMHAN. Sheltered anchorage and chief port of DHOFAR country; airfield. Sometimes Murbat.

Marbel, PHILIPPINES: see KORONADAL.

Marbella (mahr-BAI-lyah), city (2001 population 100,036), MÁLAGA province, S SPAIN, in ANDALUSIA, on the MEDITERRANEAN SEA; 36°31′N 04°53′W. City is a noted resort for wealthy tourists. Many of its permanent residents are foreigners.

Marbial (bahr-BYAHL), OUEST department, S HAITI, rural region just N of JACMEL; 18°20′N 72°28′W. Coffee and fruit growing.

Marble, village (2000 population 105), GUNNISON county, W central COLORADO, on CRYSTAL RIVER, in foothills of ELK MOUNTAINS, and 40 mi/64 km NNW of GUNNISON; 39°04′N 107°11′W. Elevation 7,950 ft/2,423 m. Quarries produce high-grade marble. Paonia State Park to SW, surrounded by Gunnison National Forest; White River National Forest to N.

Marble, village (2000 population 695), ITASCA county, N central MINNESOTA, 14 mi/23 km ENE of GRAND RAPIDS, in MESABI IRON RANGE; 47°19′N 93°17′W. Open-pit iron mines in area. SWAN LAKE 5 mi/8 km E.

Marble (MAHR-buhl), unincorporated village, CHEROKEE county, extreme W NORTH CAROLINA, 8 mi/12.9 km NE of MURPHY in Nantahala National Forest; 35°10′N 83°55′W. Timber; marble quarrying. Manufacturing (textiles, apparel, wood chips).

Marble Arch, LIBYA, name given by British troops in 1943, during World War II battles, to Italian-built arch marking 1937 completion of Tripoli-Banghazi road (connecting both parts of Italian colony). Nickname alludes to LONDON's famous Marble Arch at NE entrance to HYDE PARK.

Marble Bar (MAHR-buhl BAHR), town, N WESTERN AUSTRALIA state, W AUSTRALIA, 90 mi/145 km SE of PORT HEDLAND; 21°11′S 119°44′E. Terminus of railroad from Port Hedland; old mining center of PILBARA goldfield; quartz, granite; grazing area. Hot climate. Named for a local deposit first thought to be marble, but later discovered to be jasper, a variety of quartz.

Marble Canyon, COCONINO county, N ARIZONA, extends c.60 mi/97 km S along COLORADO RIVER from mouth of PARIA RIVER, near UTAH state line, to mouth of LITTLE COLORADO RIVER, at E end of Grand Canyon National Park. Also known as Marble Gorge. Sometimes defined as upper part of GRAND CANYON. Established 1969 as Marble Canyon National Monument; added to Grand Canyon National Park 1975 (in NE extension of park).

Marble City, village (2006 population 251), SEQUOYAH county, E OKLAHOMA, 8 mi/12.9 km N of SALLISAW; 35°34′N 94°49′W. In agricultural area; manufacturing (limestone processing). Crystal Caves and Sallisaw State Park (to S) are nearby.

Marble Cliff (MAHR-buhl KLIF), village (2006 population 607), FRANKLIN county, central OHIO, just W of COLUMBUS, on SCIOTO RIVER; 39°59′N 83°04′W. Generally suburban. Named after limestone quarry here.

Marble Falls (MAHR-buhl), town (2006 population 7,186), BURNET county, central TEXAS, on COLORADO RIVER (bridged), and 39 mi/63 km NW of AUSTIN; 30°34′N 98°16′W. Elevation 764 ft/233 m. Railroad terminus. Agriculture (cattle, sheep; pecans, fruit); cedar timber; granite quarries. Manufacturing (crushed stone, oil field tanks, electronic components). Lake Marble Falls (Colorado River) impounds here, Lake Lyndon B. Johnson dammed to W; LONGHORN CAVERNS STATE PARK to NW. Settled 1887, incorporated 1908.

Marble Hall, town, LIMPOPO province, SOUTH AFRICA, 90 mi/145 km NE of TSHWANE, near the confluence of the Elands and OLIFANTS rivers; 24°59′S 29°16′E. Elevation 3159 ft/963 m. Railroad terminus. Originally a marble-quarrying center, now center for vanadium mining; mines to E of town. Established 1929.

Marblehead (MAHR-buhl-hed), suburban town (2000 population 20,377), ESSEX county, NE MASSACHUSETTS, on the ATLANTIC coast; 42°29′N 70°51′W. A fishing village for many years, Marblehead became a resort in the nineteenth century; it is a picturesque town especially famous for yachting and antiques. Has many eighteenth-century buildings, including politician Elbridge Gerry's birthplace. Abbot Hall contains Archibald Willard's painting *Spirit of '76*. In Burial Hill cemetery are the graves of hundreds of American Revolutionary soldiers and a monument to the sixty-five Marblehead residents who died in a gale in 1846. The Revolutionary War Fort Sewall is in a seaside park. Includes village of Clifton. Site of Marblehead Regatta held every July. Incorporated 1649.

Marblehead (MAHR-buhl-hed), village (□ 4 sq mi/10.4 sq km; 2006 population 848), OTTAWA county, N OHIO, 6 mi/10 km N of SANDUSKY, at tip of Marblehead Peninsula, which extends c.15 mi/24 km E into LAKE ERIE and shelters SANDUSKY BAY on N; 41°32′N 82°43′W. Resort, fishing center; has lighthouse. Limestone quarries; fruit orchards.

Marble Hill, a residential district of MANHATTAN borough of NEW YORK city, SE NEW YORK, across HARLEM RIVER from N end of Manhattan Island and surrounded W, N, and E by the Bronx. Was originally part of Manhattan Island until Harlem River was redirected S of it in 1895, cutting it off; in 1913, it was physically joined to the Bronx by landfill.

Marble Hill, town (2000 population 1,502), ☉ BOLLINGER county, SE MISSOURI, 24 mi/39 km W of CAPE GIRARDEAU; 37°17′N 89°58′W. Corn, wheat; livestock; timber; manufacturing (headwear). Annexed adjacent Lutesville in 1986.

Marble Rock, town (2000 population 326), FLOYD county, N IOWA, on SHELL ROCK RIVER, and 11 mi/18 km WSW of CHARLES CITY; 42°57′N 92°52′W. In livestock area; limestone quarries.

Marble Rocks, gorge of NARMADA RIVER, JABALPUR district, N Madhya Pradesh state, central India, WSW of JABALPUR; c.2 mi/3.2 km long. Lies between magnesian limestone cliffs which rise higher than 100 ft/30 m (several inscriptions and sculptures date from 12th century). Chaunsat Yogini, nearby 10th-century Hindu temple, is scene of annual pilgrimage. Steatite quarries in vicinity. Locally called Bheraghat. Tourist center.

Marbleton, village (2006 population 862), SUBLETTE county, W WYOMING, near GREEN RIVER, 25 mi/40 km SW of PINEDALE, and 1 mi/1.6 km N of BIG PINEY; 42°33′N 110°05′W. Elevation 6,850 ft/2,088 m.

Marbleton (MAHR-buhl-tuhn), locality, unincorporated village, ESTRIE region, S QUEBEC, E CANADA; 45°37′N 71°35′W. Included in DUDSWELL since 1995.

Marburg (MAHR-buhrg), town, QUEENSLAND, NE AUSTRALIA, 34 mi/55 km W of BRISBANE, between Brisbane and TOOWOOMBA; 27°34′S 152°35′E.

Marburg, SLOVENIA: see MARIBOR.

Marburg an der Lahn (MAHR-boorg ahn der LAHN), city, HESSE, GERMANY, 15 mi/24 km N of GIESSEN, on the LAHN RIVER; 50°49′N 08°46′E. Chiefly known for its Protestant university, founded in 1527 by Philip of HESSE. Tourism is its largest industry; manufacturing includes woodworking and leather. Marburg grew in the 12th century around a castle; it was chartered in c.1220 and, at intervals during the 13th–17th century, served as the residence of the landgraves of HESSE; was known as HESSEN-MARBURG. Became part of the Prussian province of HESSE-NASSAU in 1866. Gothic church of St. Elizabeth. Knights' Hall of 11th–16th-cent. castle was scene (1529) of religious disputation between Luther and Zwingli. Was residence of Hessian rulers from 13th to 17th centuries. Part (1567–1604) of independent Hessen-Darmstadt and HESSE KASSEL. Undamaged in World War II.

Marbury, village, CHARLES county, S MARYLAND, near MATTAWOMAN CREEK, 24 mi/39 km SSW of WASHINGTON, D.C.; 38°34′N 77°10′W. Nearby is Smallwood State Park and Indian Head Naval Ordnance Station.

Marca (MAHR-kah), city, RECUAY province, ANCASH region, W central PERU, in Cordillera NEGRA of the ANDES Mountains, 40 mi/64 km S of HUARÁZ; 10°05′S 77°30′W. Cereals, alfalfa; livestock.

Marca, SOMALIA: see MARKA.

Marcaconga (mahr-kah-KON-gah), town, CUSCO region, S PERU, in the ANDES Mountains, 40 mi/64 km SE of CUSCO; 13°58′S 71°33′W. Elevation 13,018 ft/3,967 m. Cereals, potatoes, fruit.

Marcala (mahr-KAH-lah), city (2001 population 18,489), LA PAZ department, SW HONDURAS, near Continental Divide, 50 mi/80 km W of TEGUCIGALPA, 25 mi/40 km ESE of LA ESPERANZA (linked by road); 14°09′N 88°02′W. Elevation 5,249 ft/1,600 m. Commercial center; coffee, wheat.

Marcali (MAHR-tsah-le), city, SOMOGY county, SW HUNGARY, 22 mi/35 km NE of NAGYKANIZSA; 46°35′N 17°25′E. Manufacturing (leather goods, telecommunication components, food-processing machinery, baking and pasta, dairy products). Before 1918, Marczali.

Mar Cantábrico, SPAIN: see BISCAY, BAY OF.

Marcapata (mahr-kah-PAH-tah), town, QUISPICANCHI province, CUSCO region, S central PERU, in Cordillera de CARABAYA, 75 mi/121 km E of CUSCO, on URCOS–PUERTO MALDONADO road; 13°30′S 70°55′W. Elevation

10,335 ft/3,150 m. In agricultural region (potatoes, cereals, coca, coffee).

Marcavi (mar-KAH-vee), canton, W GENERAL B. BIL-BAO province, POTOSÍ department, W central BOLIVIA, near ESTEBAN ARCE province. Elevation 9,974 ft/3,040 m. Agriculture (potatoes, yucca, bananas, barley); cattle.

Marcelândia (mahr-se-LAHN-zhee-ah), city (2007 population 14,080), N central MATO GROSSO state, BRAZIL, near Serra Formosa; 11°02′S 54°32′W.

Marcelháza, SLOVAKIA: see MARCELOVA.

Marcelin (MAHRS-lin), village (2006 population 169), central SASKATCHEWAN, W CANADA, 45 mi/72 km WSW of PRINCE ALBERT; 52°56′N 106°47′W. Mixed farming, dairying.

Marceline (mahr-suh-LEEN), city (2000 population 2,558), LINN and CHARITON counties, N central MIS-SOURI, 8 mi/12.9 km SE of BROOKFIELD; 39°43′N 92°57′W. Shipping center in grain, livestock (esp. dairy), cattle, hogs; manufacturing (printing, publishing, steel fabrication). Coal area but no longer mined. Plotted 1887. Boyhood home of Walt Disney.

Marcelino Ramos (MAHR-se-lee-no RAH-mos), town (2007 population 5,372), N RIO GRANDE DO SUL state, BRAZIL, on URUGUAY RIVER (SANTA CATARINA state border) at mouth of PEIXE River, on railroad, and 30 mi/48 km NE of ERECHIM; 27°28′S 51°54′W. Industrial center (meat processing; lumbering).

Marcelino Ugarte, ARGENTINA: see SALTO.

Marcelino Vieira (MAHR-se-lee-no VEE-ai-rah), town (2007 population 8,184), extreme SW RIO GRANDE DO NORTE state, BRAZIL, 95 mi/153 km SW of MOSSORÓ; 06°18′S 38°10′W.

Marcellina, Mount, peak (11,349 ft/3,459 m) in ROCKY MOUNTAINS, GUNNISON county, W COLORADO, 14 mi/23 km WNW of CRESTED BUTTE, in the Raggeds Wilderness.

Marcellus (mahr-SEL-uhs), town (2000 population 1,162), CASS county, SW MICHIGAN, 22 mi/35 km SSW of KALAMAZOO; 42°01′N 85°48′W. In farm and lake resort area. Manufacturing (logging, machining). Fish and Saddlebag lakes to NW. Incorporated 1879.

Marcellus, village (2006 population 1,783), ONONDAGA county, central NEW YORK, 10 mi/16 km WSW of SYRACUSE; 42°58′N 76°20′W. Feed milling and dairy products. Incorporated 1853.

Marcelova (mahr-TSE-lo-VAH), Slovak *Marcelová*, Hungarian *Marcelháza*, village, ZAPADOSLOVENSKY province, S SLOVAKIA, 8 mi/12.9 km ENE of KOMÁRNO; 47°47′N 18°18′E. Wheat, corn, sugar beets, and vegetables; vineyards and wine making. Food processing. Large Hungarian minority. Under Hungarian rule from 1938–1945.

March (MAHRKH), district, N SCHWYZ canton, central SWITZERLAND. Main town is LACHEN; population is German-speaking and Roman Catholic.

March (MAHCH), town (2001 population 19,042), N CAMBRIDGESHIRE, E ENGLAND, in the FENS, on NENE RIVER, and 13 mi/21 km NW of ELY; 52°33′N 00°05′E. Railroad junction. Former agricultural market, with agricultural-machinery works. Has 14th-century church of St. Wendreda.

March (MAHRKH), village, LOWER BAVARIA, GER-MANY, in BOHEMIAN FOREST, 3 mi/4.8 km SW of REGEN; 48°59′N 13°04′E. Quartz quarries.

March Air Force Base, California: see RIVERSIDE, city.

Marchairuz, Col du, pass (4,747 ft/1,447 m) in JURA mountains, VAUD canton, W SWITZERLAND, between LAKE GENEVA and Joux valley.

Marchamalo (mahr-chah-MAH-lo), town, GUADALA-JARA province, central SPAIN, 3.5 mi/5.6 km NW of GUADALAJARA, across the HENARES RIVER; 40°40′N 03°12′W. Cereals, olives, vegetables; livestock.

Marchand (mahr-SHAHND), community, SE MANI-TOBA, W central CANADA, 46 mi/73 km from WINNI-PEG, and in LA BROQUERIE rural municipality; 49°26′N 96°23′W.

Marchand (mahr-SHAHND), former village, S QUEBEC, E CANADA, on SAINT-MAURICE River, and 10 mi/16 km NW of TROIS-RIVIÈRES; 49°26′N 96°23′W. Mining. Merged into the city of RIVIÈRE-ROUGE in 2002.

Marchand, MOROCCO: see ROMMANI.

Marche (MAHRSH), region and former province of central FRANCE, on the NW margin of the MASSIF CENTRAL, now coextensive with CREUSE department, and parts of the HAUTE-VIENNE department; ⊙ GUÉRET. Marche is primarily an agricultural region specializing in sheep raising. The wool is manu-factured into carpets and tapestries at FELLETIN and AUBUSSON. The name of the region derived from its location as a N border fief (march) of the duchy of AQUITAINE. Passed (13th century) to the house of Lusignan but was seized (early 14th century) by Philip IV of France. Briefly united with the crown lands, it ultimately became an appanage of the house of Bourbon. It came definitively to France in 1531, fol-lowing the confiscation (1527) of the lands of Con-stable Charles de Bourbon by Francis I.

Marche (MAHR-kai) or **The Marches**, region (□ 3,742 sq mi/9,729.2 sq km), E central ITALY, extending from the E slopes of the APENNINES to the ADRIATIC Sea; ⊙ ANCONA 43°30′N 13°50′E. Divided into the prov-inces of ANCONA, ASCOLI PICENO, MACERATA, and PESARO E URBINO (named after their chief cities). The Marche is mostly hilly or mountainous, with many fortified hill towns and deep gorges, except for a narrow coastal strip, and is drained by the METAURO, POTENZA, TRONTO, and NERA rivers. Farming is the chief occupation; cereals, olives, grapes, vegetables, and tobacco are the main products, and livestock is raised. Industry has expanded in the 20th century with the construction of hydroelectric facilities. Manufacturing includes ships, textiles, chemicals, musical instruments, and pottery. Commercial and fishing ports are located at Ancona, PESARO, FANO, and SENIGALLIA. The beaches of Ancona and Pesaro attract summer tourism. The Umbri and the Picentes (Greek colonists for whom part of the region was called Picenum) lived in the region when it was col-onized (3rd century B.C.E.) by Rome. After the fall of Rome the area was invaded by the Goths. In the 6th century the N section, including four of the cities of the Pentapolis and adjoining territories, came under Byzantine rule; the S section became a part of the Lombard duchy of SPOLETO. In the 8th century the region passed, as part of the donations of Pepin the Short (754) and Charlemagne (774), under the nominal rule of the papacy, but later emperors granted fiefs in the area until the 13th century. The name *Marche* comes from *marca* [Italian= borderland] and originated around the 10th century, because the fiefs of Ancona, Fermo, and Camerino were established at the border of the Holy Roman Empire. Despite the strength of the popes and the emperors, who contested for control of the region, some cities established free communes or were gov-erned by noble families (including the Malatesta, the Varano, and the Montefeltro). From the 13th to the 16th centuries the popes gradually established their rule here and ended local autonomy. The region was occupied by the French from 1797 to 1815, when it was restored to the papacy. The Marche was united with the kingdom of Sardinia in 1860. There are universi-ties at Macerata and Urbino.

Marche-en-Famenne (MAHRSH–awn–fah-MEN), commune (□ 368 sq mi/956.8 sq km; 2006 population 17,023), ⊙ Marche-en-Famenne district, LUX-EMBOURG province, SE central BELGIUM, 20 mi/32 km E of DINANT; 50°12′N 05°20′E. Agricultural market. Major highway crossroad.

Marchegg (mahrkh-EK), town, E LOWER AUSTRIA, on Slovak border, on MARCH RIVER, and 28 mi/45 km ENE of VIENNA; 48°17′N 16°55′E. Railroad junction and railroad border station opposite SLOVAKIA. Gas

fields nearby. Founded 13th century, by Ottocar II of Bohemia, as a border fortress. Medieval fortifications preserved. Castle from the 14th and 15th centuries (reconstructed 17th–18th century) is today a hunting and African museum.

Marche, Monts de la (MAHRSH, MON duh lah), highlands of the W MASSIF CENTRAL, central FRANCE, extending c.50 mi/80 km. E-W across CREUSE and HAUTE-VIENNE departments. Average elevation 1,800 ft/549 m; highest point Puy de Sauvagnac (2,300 ft/701 m). Bounded by LIMOUSIN hill country (S), AU-VERGNE MOUNTAINS (SE), and the valleys of left tributaries of the LOIRE RIVER. Crossed by CREUSE and CHER rivers. Extensive grazing. Also known as Plateau de la Marche.

Marchena (mahr-CHAI-nah), city, SEVILLE province, SW SPAIN, near the GUADALQUIVIR RIVER, 31 mi/50 km E of SEVILLE. Railroad junction. Processing and trading center in agricultural region (cereals, olives, cotton, grapes, melons; livestock). Sawmilling, meat processing, olive-oil pressing, liquor distilling; clay and sand quarrying. Old historic town with remains of a wall; also has San Juan church, and palace be-longing to dukes of Arcos. Mineral springs nearby.

Marchena, Isla (mahr-CHAI-nah, EES-lah), island (□ 45 sq mi/117 sq km), N GALAPAGOS ISLANDS, ECUADOR, in the PACIFIC OCEAN, 100 mi/161 km NW of PUERTO BAQUERIZO; 00°20′N 90°30′W. Volcano here was active in 1991. Also known as Bindloe Island.

Marchfeld (MAHRKH-feld), plain, NE AUSTRIA, NE of VIENNA, between DANUBE and MORAVA (Ger. *March*) rivers, on SLOVAK border. One of the most important agricultural regions in Austria, producing grain, sugar beets, and vegetables. It also contains major oil and gas deposits. A strategic approach to Vienna, it was the site of several important battles. In 1260, Ottocar II of Bohemia defeated Bela IV of Hungary here, and in 1278, Ottocar was defeated and slain by the forces of Rudolf I of the house of Hapsburg. In 1809, Napoleon I was defeated here at Aspern by Archduke Charles but was victorious at WAGRAM.

Marchienne-au-Pont (MAHR-shyen–O–PAWN), town in commune of Charleroi, Charleroi district, HAINAUT province, S central BELGIUM, on SAMBRE RIVER, and 3 mi/4.8 km WSW of CHARLEROI; 50°24′N 04°23′E. Manufacturing (ceramics, furniture).

Marchiennes (mahr-SHYEN), town (□ 8 sq mi/20.8 sq km; 2004 population 4,660), NORD department, NORD-PAS-DE-CALAIS region, N FRANCE, inland port on the SCARPE RIVER, and 9 mi/14.5 km ENE of DOUAI; 50°24′N 03°17′E. Market for fresh garden produce. Founded in 7th century as site of a monas-tery which later became a Benedictine abbey. Also called Marchiennes-Ville.

Marchiennes-Ville, FRANCE: see MARCHIENNES.

Marchigüe (mahr-chee-GWAI), village, ⊙ Marchigüe comuna, CARDENAL CARO province, LIBERTADOR GENERAL BERNARDO O'HIGGINS region, central CHILE, on railroad, and 35 mi/56 km WNW of SAN FER-NANDO. In agricultural area (cereals, vegetables; live-stock). Sometimes spelled Marchihüe.

Marchin (mahr-SHA), commune (2006 population 5,153), Huy district, LIÈGE province, E central BEL-GIUM, 4 mi/6.4 km S of HUY; 50°28′N 05°14′E. Steel-rolling mills.

Mar Chiquita (mahr chee-KEE-tah), lagoon (□ 18 sq mi/46.8 sq km) in SE BUENOS AIRES province, AR-GENTINA, 25 mi/40 km NNE of MAR DEL PLATA, connected with ATLANTIC OCEAN; extends 15 mi/24 km NE-SW. Resort area.

Mar Chiquita (mahr chee-KEE-tah), salt lake (□ 10 sq mi/26 sq km) in N BUENOS AIRES province, ARGEN-TINA, on the upper Río Salado between ARENALES and JUNÍN; 10 mi/16 km long, 2 mi/3.2 km wide.

Mar Chiquita, salt lake (□ 580 sq mi/1,508 sq km) in NE CÓRDOBA province, ARGENTINA, in swampy region, 90 mi/145 km NE of CÓRDOBA; c.45 mi/72 km long, 15 mi/

24 km wide. Receives the RÍO PRIMERO (SW), RÍO SEGUNDO (SE), and RÍO DULCE (N); has no outlet. Adjoining it (N) are the PORONGOS salt lakes. Contains sodium, calcium, and magnesium salts. Resort at MIRAMAR, on S shore. Has group of islets, the largest EL MÉDANO.

Mar Chiquita, ARGENTINA: see CORONEL VIDAL.

March River, CZECH REPUBLIC: see MORAVA RIVER.

Marchtrenk (mahrkh-TRENK), township, E central UPPER AUSTRIA, near TRAUN RIVER, 4 mi/6.4 km NE of WELS; 48°12′N 14°07′E. Market center; manufacturing of transformers, heating plants, artificial wool, plastics, paper products; hydropower station.

Marcianise (mahr-chah-NEE-ze), town, CASERTA province, CAMPANIA, S ITALY, 3 mi/5 km SSW of CASERTA; 41°02′N 14°17′E. Highly diversified secondary industrial center; agricultural center (fruit, vegetables), wine.

Marcianopolis, Bulgaria: see DEVNYA.

Marcigny (mahr-see-NYEE), town (□ 3 sq km; 2004 population 1,933), SAÔNE-ET-LOIRE department, in BURGUNDY, E central FRANCE, near LOIRE RIVER, 16 mi/26 km N of ROANNE; 46°17′N 04°03′E. Cattle market. Has museum of old faïence in an old priory, and several 16th-century wooden houses.

Marcilla (mahr-THEE-lyah), town, NAVARRE province, N SPAIN, near ARAGON RIVER, 20 mi/32 km NNW of TUDELA; 42°20′N 01°44′W. Cereals; food processing (vegetable canning, sugar and flour mills, slaughterhouses), manufacturing of agricultural machinery and textiles. Medieval castle nearby.

Marcinelle (mahr-si-NEL), town in commune of Charleroi, Charleroi district, HAINAUT province, S central BELGIUM, just S of CHARLEROI, near SAMBRE RIVER; 50°24′N 04°26′E.

Marcionílio Souza (MAHR-see-o-NEEL-yo SO-sah), city (2007 population 10,690), E central BAHIA state, BRAZIL, on SALVADOR–BELO HORIZONTE railroad, 33 mi/53 km SW of IAÇU; 13°00′S 40°32′W.

Marck (MAHRK), town (□ 12 sq mi/31.2 sq km), PAS-DE-CALAIS department, NORD-PAS-DE-CALAIS region, N FRANCE; 50°57′N 01°57′E. SE suburb of CALAIS.

Marckolsheim (mahr-kol-ZEM), German, *Markolsheim* (MAHR-kols-heim), commune (□ 13 sq mi/33.8 sq km), BAS-RHIN department, in lowland of ALSACE, E FRANCE, on RHÔNE-RHINE CANAL, 8 mi/12.9 km SE of SÉLESTAT; 48°10′N 07°33′E. Quarrying. Major hydroelectric power plant on a canal parallel to the RHINE. There is a historical museum in one of the remaining strongholds of the MAGINOT LINE facing the Rhine.

Marco (mahr-ko), city (2007 population 23,262), N CEARÁ state, BRAZIL, 12 mi/19 km SW of ACARAÚ; 03°13′S 40°10′W.

Marco de Canaveses (MAHR-koo dai kah-nah-VAI-sesh), town, PÔRTO district, N PORTUGAL, on TÂMEGA RIVER, on railroad, and 23 mi/37 km E of OPORTO; 41°11′N 08°09′W. Watering place with mineral springs; paper milling.

Marcoing (mahr-KWAN), commune (□ 5 sq mi/13 sq km), NORD department, NORD-PAS-DE-CALAIS region, N FRANCE, on the Escaut (SCHELDT) River, on SAINT-QUENTIN CANAL, 4 mi/6.4 km SSW of CAMBRAI; 50°07′N 03°11′E. In sugar beet district.

Marco Island (MAHR-ko), city (2005 population 16,109), COLLIER county, SW FLORIDA, 15 mi/24 km SSE of NAPLES; 25°56′N 81°43′W. Center of resort complex that covers Marco Island.

Marcoma (mahr-KO-mah), canton, CHAYANTA province, POTOSÍ department, W central BOLIVIA, E of MARAGUA; 18°54′S 65°41′W. Elevation 13,635 ft/4,156 m. Agriculture (potatoes, yucca, bananas, barley); cattle. Iron deposits (though insignificant).

Marcona (mahr-KO-nah), town, ICA region, SW PERU, near the PACIFIC; 15°04′S 75°03′W. Ore deposits.

Marconi, Mount (mahr-KO-nee) (10,190 ft/3,106 m), SE BRITISH COLUMBIA, W CANADA, near ALBERTA border, in ROCKY MOUNTAINS, 60 mi/97 km N of FERNIE; 50°23′N 115°07′W.

Marco Polo Bridge, CHINA: see FENGTAI.

Marco Rondon (MAHR-ko RON-don), city, E RONDÔNIA state, BRAZIL, 108 mi/174 km SE of JI-PARANÁ; 12°15′S 60°50′W.

Marcos Juárez (MAHR-kos HWAH-rez), town (1991 population 22,487), ⊙ Marcos Juárez department, E CÓRDOBA province, ARGENTINA, 70 mi/113 km ESE of VILLA MARÍA; 32°42′S 62°06′W. On railroad. Agricultural center (wheat, soybeans, corn, flax; livestock) with tanneries, flour mills. Formerly Espinillos.

Marcos Parente (MAHR-kos PAH-ren-che), town (2007 population 4,176), SW PIAUÍ state, BRAZIL, 62 mi/100 km SW of FLORIANO; 07°07′S 43°57′W.

Marcos Paz (MAHR-kos PAHZ), town, ⊙ Marcos Paz district (□ 168 sq mi/436.8 sq km), NE BUENOS AIRES province, ARGENTINA, 27 mi/43 km WSW of BUENOS AIRES; 34°50′S 58°50′W. Railroad junction and agricultural center (grain, alfalfa; livestock); dairying.

Marcoule, FRANCE: see BAGNOLS-SUR-CÈZE.

Marcoussis (mahr-koo-SEE), town (□ 7 sq mi/18.2 sq km), ESSONNE department, ÎLE-DE-FRANCE region, N central FRANCE, on outer circumferential highway, 14 mi/23 km SSW of PARIS; 48°38′N 02°14′E. Electricity research center.

Marcovia (mahr-KO-vee-ah), town, CHOLUTECA department, S HONDURAS, on CHOLUTECA RIVER, and 7 mi/11.3 km W of CHOLUTECA; 13°17′N 87°19′W. Livestock; sawmilling, hardwood lumbering. Saltworks on nearby coast of GULF OF FONSECA.

Marcq-en-Baroeul (MAHRK–awn–bah-RU-yuh), town (□ 5 sq mi/13 sq km), NORD department, NORD-PAS-DE-CALAIS region, N FRANCE, 4 mi/6.4 km NNE of LILLE; 50°40′N 03°05′E. Textile-milling center in densely populated Lille-ROUBAIX-TOURCOING urban complex; also electronics and food-processing plants.

Marcus, town (2000 population 1,139), CHEROKEE county, NW IOWA, 14 mi/23 km WNW of CHEROKEE; 42°49′N 95°48′W. Feed. Incorporated 1892.

Marcus, village (2006 population 168), STEVENS county, NE WASHINGTON, 11 mi/18 km NW of COLVILLE, and on COLUMBIA River (FRANKLIN D. ROOSEVELT LAKE reservoir), opposite mouth of KETTLE River; 48°40′N 118°04′W. Terminus of railroad spur from KETTLE FALLS. Agriculture (wheat, alfalfa, barley; hogs). COULEE DAM National Recreation Area on both shores of reservoir; part of Colville National Forest to W. City moved (1941) from site c.1.5 mi/2.4 km S, now covered by reservoir.

Marcus Baker, Mount (13,176 ft/4,016 m), S ALASKA, in CHUGACH MOUNTAINS, 55 mi/89 km WNW of VALDEZ; 61°26′N 147°46′W.

Marcus Hook, borough (2006 population 2,258), DELAWARE county, SE PENNSYLVANIA, 15 mi/24 km SW of downtown PHILADELPHIA, and 2 mi/3.2 km SW of CHESTER, on DELAWARE RIVER (NEW JERSEY state line), at DELAWARE state line (SW); 39°48′N 75°25′W. Manufacturing (petroleum refining; vinyl products, chemicals, lumber, chemicals). Early 18th-century pirate rendezvous. Settled c.1640 by Swedes, laid out c.1701, incorporated 1893.

Marcus Island, Japanese *Minami-tori-shima* (mee-NAH-mee–TO-ree–SHEE-mah), volcanic island (□ 1 sq mi/2.6 sq km), W PACIFIC OCEAN, Ogasawara district, Tokyo prefecture, easternmost part of JAPAN, 806 mi/1,297 km E of BONIN ISLANDS; c.2 mi/3.2 km long; rises to 204 ft/62 m; 26°32′N 142°10′E. Phosphate deposits. Discovered 1896 by Japanese; annexed 1899 by Japan. In World War II, site of Japanese naval and air bases; after Japan's defeat, placed under U.S. military government.

Marcy, Mount (5,344 ft/1,629 m), ESSEX county, NE NEW YORK, in the ADIRONDACK MOUNTAINS, c.12 mi/19 km SSE of LAKE PLACID village; 44°07′N 73°56′W. Highest peak in the state. Lake TEAR OF THE CLOUDS, on its S slope, is the source of the main headstream

(Feldspar Creek–Opalescent River) of the HUDSON RIVER. Was first ascended in 1837.

Marda, Arab village, Nablus district, 9.3 mi/15 km SW of NABLUS, in the SAMARIAN Highlands, WEST BANK; 32°06′N 35°11′E. Agriculture (olives, fruit, cereals).

Mardakert (muhr-duh-KYERT), urban settlement and center of Mandakert region, N NAGORNO-KARABAKH autonomous oblast, AZERBAIJAN, 27 mi/43 km N of XANKANDI; 40°15′N 46°45′E. Wheat; butter and cheese processing; winery; small industries.

Mardakyany (muhr-dahk-YAH-nee), town, in Azizbekov district of Greater BAKY, AZERBAIJAN, on E APSHERON Peninsula, 18 mi/29 km ENE of Baky; 40°27′N 50°07′E. On electric railroad; seaside resort; vineyards. Oil fields offshore.

Mardan, district, NORTH-WEST FRONTIER PROVINCE, N PAKISTAN; ⊙ MARDAN. Bordered by the GRAND TRUNK ROAD, the Vale of PESHAWAR, the INDUS River, and the MALAKAND Pass (N). Sugarcane, tobacco, wheat, potatoes; agricultural processing. Orchard crops recently introduced. Area is gateway to SWAT, DIR, and CHITRAL districts. Military towns of RISALPUR and NOWSHERA to S. Home of such Afghan tribes as the Yusefzai.

Mardan (MUHR-dahn), city, ⊙ MARDAN district, N central NORTH-WEST FRONTIER PROVINCE, N PAKISTAN, on GRAND TRUNK ROAD (linked to districts further N) and on Kalagarmi River, 28 mi/45 km NE of PESHAWAR, on Kalagarmi River; 34°12′N 72°03′E. Market center (wheat, sugarcane); sugar milling, hand-loom weaving. An industrial center, manufacturing textiles, vegetable oil, and refined sugar. Sugar mill said to be one of ASIA's largest. Asokan rock edict 7 mi/11.3 km ENE, at SHAHBAZGARHI. Site of a military garrison and a fort built by the British in 1854. Sometimes called HOTI MARDAN (village of Hoti is just E; headgear manufacturing).

Mar da Palha, PORTUGAL: see LISBON BAY.

Marda Pass (MAR-dah), pass (c.6,500 ft/1,981 m), E central ETHIOPIA, on plateau, 10 mi/16 km W of JIJIGA and crossed by Jijiga-Harar road; 09°22′N 42°41′E.

Mardela Springs (mahr-DE-lah), town (2000 population 364), WICOMICO county, E MARYLAND, 11 mi/18 km NW of SALISBURY; 38°28′N 75°46′W. Originally called Barren Springs, the name was changed to the combination of the names of Maryland and DELAWARE in 1906. A natural spring (now polluted) was reputed to have health-giving qualities.

Mar del Plata (MAHR del PLAH-tah), city, E central ARGENTINA, on the ATLANTIC OCEAN; 38°00′S 57°33′W. One of the most popular seaside resorts in SOUTH AMERICA. Fishing and fish processing are also important industries. Founded in the 1850s. Seat of two universities.

Mar del Sur, ARGENTINA: see MIRAMAR (in BUENOS AIRES province).

Marden (MAH-duhn), village (2001 population 4,920), central KENT, SE ENGLAND, 7 mi/11.3 km S of MAIDSTONE; 51°10′N 00°30′E. The courthouse and church date from the 13th–14th century. In apple-growing area.

Mardin, town (2000 population 65,072), SE TURKEY, 50 mi/80 km SE of DIYARBAKIR, 14 mi/23 km from Syrian border; 37°19′N 40°43′E. Railroad terminus from KAHRAMANMARAŞ. Agricultural center; barley, wheat, onions, tobacco, mohair goats. On nearby hill are ruins of a once-strongly fortified castle. Ulu Mosque is Mardin's oldest, built in 1186, in the time of the Artukid ruler Kutbeddin Ilgaz.

Mare (MA-rai), village, WESTERN province, S central NEW GUINEA island, SW PAPUA NEW GUINEA, on coast of ARAFURA SEA near entrance to TORRES STRAIT, 100 mi/161 km W of DARU; Fish, trepang; crocodile skins; cattle.

Mare (MAH-rai), volcanic island (2 mi/3.2 km long), N Maluku province, INDONESIA, in MALUKU SEA, just W of HALMAHERA, 13 mi/21 km S of TERNATE; 00°34′N 127°24′E. Hilly, rising to 1,010 ft/308 m. Fishing.

Maré (mah-rai), raised coral island with volcanic fragments, southernmost of LOYALTY ISLANDS, NEW CALEDONIA, SW PACIFIC OCEAN, 35 mi/56 km S of LIFOU; c.22 mi/35 km long, 18 mi/29 km wide; 21°32′S 168°00′E. Rises to 300 ft/91 m. Copra, oranges.

Marea del Portillo (mah-RAI-ah dail por-TEE-yo), small town, GRANMA province, SE CUBA, near mouth of Silantros River, on CARIBBEAN coast; 19°55′N 77°11′W.

Mare aux Vacoas, MAURITIUS: see VACOAS.

Mareb, YEMEN: see MARIB.

Mareb River, Eritrea: see GASH RIVER.

Marecchia River (mah-REK-kyah), 40 mi/64 km long, N central ITALY; rises in ETRUSCAN APENNINES on MONTE FUMAIOLO, 9 mi/14 km SE of BAGNO DI RO-MAGNA; flows NE to the ADRIATIC Sea at RIMINI.

Marechal Cândido Rondon (MAH-re-shahl KAHN-zhee-do RON-don), city (2007 population 9,023), W PARANÁ state, BRAZIL, 34 mi/55 km SSE of GUAÍRA; 24°33′S 54°04′W. Timber.

Marechal Deodoro (MAH-re-shahl DAI-o-DO-ra), city, (2007 population 45,144), ALAGOAS state, NE BRAZIL, at S tip of MANGUABA LAGOON, 10 mi/16 km SW of MACEIÓ (canal connection); 09°13′S 35°52′W. In intensive sugar-growing district (sugar mills). Until 1939, called Alagoas. Former state capital.

Marechal Floriano (MAH-re-shahl FLO-ree-ah-no), city (2007 population 12,685), SE ESPÍRITO SANTO state, BRAZIL, on VITÓRIA–RIO DE JANEIRO railroad, 26 mi/42 km SW of Vitória; 20°25′S 40°41′W.

Marechal Floriano, Brazil: see PIRANHAS.

Marechal Taumaturgo, town, ACRE state, extreme W BRAZIL, 95 mi/153 km S of CRUZEIRO DO SUL, on upper Rio JURUÁ, near PERU border; 08°53′S 72°47′W. Highway BR 307 (under construction) from Cruzeiro do Sul passes through. Livestock raising and extractive activities. Airport.

Mare Chicose, village (2000 population 409), central GRAND PORT district, MAURITIUS, 21 mi/33.6 km SE of PORT LOUIS, and 4.4 mi/7.1 km E of ROSE BELLE. Sugarcane. Site of new landfill (102,473,498 cu ft/2,900,000 cu m capacity), operational April 1997.

Mare d'Albert, village (2000 population 4,424), GRAND PORT district, SE MAURITIUS, 5 mi/8 km WSW of MAHÉBOURG. Sugarcane.

Mareeba (muh-REE-buh), town, NE QUEENSLAND, NE AUSTRALIA, 22 mi/35 km WSW of CAIRNS; 17°00′S 145°26′E. Railroad junction. Major irrigation area; tobacco-growing center; sugar plantations; coffee; fruit; vegetables; cattle; mining. Granite gorge W of town. Mosque built by the town's Albanian community.

Maree, Loch (MER-ee), lake (13 mi/21 km long, 1 mi/1.6 km to 3 mi/4.8 km wide), HIGHLAND, NW Scotland; 57°40′N 05°30′W. Drains into the MINCH through the Ewe River and LOCH EWE. Isle Maree, near the N shore, has a primitive burial ground and ruins of 7th-century chapel.

Mareg, township, E central SOMALIA, on INDIAN Ocean, 175 mi/282 km NE of MOGADISHO. Fishing port. Has fort and mosque. Also spelled MEREGH.

Marehra, India: see MARAHRA.

Mare, Ilha da, islet in TODOS OS SANTOS BAY, E BAHIA state, BRAZIL, 15 mi/24 km N of SALVADOR.

Mare Ionium, Greece and Italy: see IONIAN SEA.

Mare Island, California: see VALLEJO.

Marek, Bulgaria: see DUPNITSA.

Mareka, LESOTHO: see MAKHEKA.

Mare La Chaux, village (2000 population 1,885), FLACQ district, MAURITIUS, 23 mi/36.8 km E of PORT LOUIS. Sugarcane, vegetables; livestock.

Maremma (mah-RAIM-mah), coastal area in TUSCANY, central ITALY, along the TYRRHENIAN Sea and extending E to the APENNINES. A flourishing region in Etruscan and early Roman times, it became marshy and was largely abandoned in the Middle Ages because of malaria. Reclamation was begun (19th cen-

tury) by the grand dukes of Tuscany and was continued in the 20th century by the Italian government. There are now wide fertile areas, rich borax mines, and good hunting grounds; cattle and a noted breed of horses are raised. Cities include PIOMBINO (a port) and Grosetto (an inland agricultural center).

Maréna, town, FIRST REGION/KAYES, MALI, 50 mi/80 km NE of KAYES; 14°38′N 11°55′W.

Maréna (mah-RAI-nah), village, THIRD REGION/SI-KASSO, MALI, 41 mi/68 km SE of SAN.

Marendego (ma-rain-DAI-go), village, LINDI region, E TANZANIA, 45 mi/72 km NNW of KILWA MASOKO, on Mohoto Bay, INDIAN OCEAN; 08°20′S 39°16′E. Fish; sisal, cashews, bananas, copra; goats, sheep.

Marengo (muh-REN-go), county (□ 982 sq mi/2,553.2 sq km; 2006 population 21,842), W Alabama, in the Black Belt; ⊙ Linden. Bounded W by Tombigbee River Corn, hay; cattle; lumber milling; textiles. Demopolis is in N. Formed 1818. Named in honor of Napoleon's victory over the Austrian army at Marengo in northern Italy on June 14, 1800.

Marengo (mo-REN-go), city (2000 population 6,355), MCHENRY county, N ILLINOIS, on KISHWAUKEE RIVER, and 11 mi/18 km SW of WOODSTOCK; 42°15′N 88°35′W. In dairy and farm area; makes fabricated metal products Settled 1835; incorporated as town in 1857, as city in 1893.

Marengo, city (2000 population 2,535), ⊙ IOWA county, E central IOWA, on IOWA RIVER, and 24 mi/39 km WSW of CEDAR RAPIDS; 41°47′N 92°04′W. Manufacturing (building materials; printing). AMANA COLONIES to E. Incorporated 1859.

Marengo, town (2000 population 829), CRAWFORD county, S INDIANA, on a tributary of BLUE RIVER, and 7 mi/11.3 km ENE of ENGLISH; 38°22′N 86°20′W. In agricultural area; limestone quarries; timber. Marengo Cave here is a tourist attraction.

Marengo (mah-RENG-go), village, PIEDMONT, NW ITALY, near ALESSANDRIA; 45°15′N 10°44′E. Site of a famous battle (June 14, 1800) between the French under Napoleon Bonaparte and the Austrians under Melas. Melas had almost won when Desaix arrived with fresh troops to bolster the French; Desaix lost his life, but the Austrians were completely defeated and retired to the MINCIO.

Marengo (muh-RENG-go), village (2006 population 312), MORROW county, central OHIO, 22 mi/35 km SE of MARION; 40°24′N 82°49′W.

Marenisco (mahr-en-IS-ko), village, GOGEBIC county, W UPPER PENINSULA, NW MICHIGAN, 24 mi/39 km ESE of IRONWOOD, on PRESQUE ISLE RIVER, in Ottawa National Forest; 46°22′N 89°41′W. In lumbering, and agricultural area. Manufacturing (wood products). Lake Gogebic State Park on W shore of large LAKE GOGEBIC, 8 mi/12.9 km NE.

Marennes (mah-REN), town (□ 7 sq mi/18.2 sq km), CHARENTE-MARITIME department, POITOU-CHAR-ENTES region, W France, near mouth of SEUDRE RIVER, on BAY of BISCAY, opposite S end of Île d'OLÉRON (bridge), 10 mi/16 km SW of ROCHEFORT; 45°50′N 01°06′W. Has France's largest oyster and mussel beds; saltworks. Its church has a tall Gothic tower. Involved (1568–1570) in Wars of Religion.

Mareotis, EGYPT: see MARYUT.

Maresfield (MERZ-feeld), village (2001 population 3,282) East SUSSEX, SE ENGLAND, 2 mi/3.2 km N of UCKFIELD; 50°59′N 00°05′E. Has 13th-century church.

Mareshah (mah-re-SHAH), biblical locality, ISRAEL, at W foot of JUDEAN HIGHLANDS, 20 mi/32 km WSW of JERUSALEM, just SSE of BEIT GUVRIN. Repeatedly mentioned in the Bible; scene of important excavations (Hellenistic finds). Also spelled Marissa or Marisa. Modern locality called Tell Sandahannah (originally Santa Ana after Jesus' grandmother).

Maresias (MAH-re-see-ahs), city, SE SÃO PAULO state, BRAZIL, 50 mi/80 km SE of SÃO JOSÉ DOS CAMPOS, on Atlantic Coast; 23°48′S 45°33′W.

Mare Tabac, village (2000 population 2,513), GRAND PORT district, MAURITIUS, 24 mi/38.4 km SE of PORT LOUIS, and 5 mi/8 km S of ROSE BELLE. Sugarcane, vegetables.

Ma Retraite (MAH RAI-trai-tuh), village, PARA-MARIBO district, N SURINAME, 4 mi/6.4 km NE of PARAMARIBO; 05°49′N 55°12′W. Coffee plantations.

Marettimo (mah-RET-tee-mo), ancient *Hiera*, island (□ 4 sq mi/10.4 sq km), one of EGADI Islands, in the MEDITERRANEAN Sea, off W SICILY, ITALY, 28 mi/45 km W of TRAPANI; 4.5 mi/7.2 km long, 2 mi/3 km wide; 37°58′N 12°04′E. Rises to 2,244 ft/684 m. Fisheries (blue fin tunny, coral). Chief port, Marettimo. Also spelled Marittimo.

Mareuil-sur-Lay-Dissais (mah-RU-yuh–syoor–lai–dee-SAI), commune (□ 9 sq mi/23.4 sq km; 2004 population 2,580), VENDÉE department, PAYS DE LA LOIRE region, W FRANCE, 14 mi/23 km SE of La ROCHE-SUR-YON; 46°32′N 01°14′W. Known for its vineyards.

Marevo (MAH-ree-vuh), village (2005 population 2,550), SW NOVGOROD oblast, NW European Russia, on crossroads, 55 mi/89 km SE of STARAYA RUSSA; 57°19′N 32°05′E. Elevation 351 ft/106 m. Lumbering.

Marfa (MAHR-fuh), town (2006 population 1,929), ⊙ PRESIDIO county, extreme W TEXAS, c.175 mi/282 km SE of EL PASO, just S of DAVIS MOUNTAINS; 30°18′N 104°01′W. Elevation 4,688 ft/1,429 m. Market, shipping center for ranching (cattle, horses), silver-mining region; manufacturing (solar pump jacks, feeds); tourist trade. Old Fort D. A. Russell was founded here in 1833. Hunting nearby. Marfa Mystery Lights, unexplained phenomenon first reported in 1883. Founded 1881, incorporated 1887.

Marfa Peninsula (mahr-fah), NW promontory of MALTA; c.3 mi/4.8 km long, 1 mi/1.6 km wide.

Marfin Brod (MAHR-feen BROT), urban settlement, SW MOSCOW oblast, central European Russia, on the MOSKVA River, on road, 3 mi/5 km NW, and under administrative jurisdiction, of MOZHAYSK; 55°32′N 35°58′E. Elevation 544 ft/165 m. Manufacturing (medical tools).

Marfino (MAHR-fee-nuh), village (2006 population 3,680), central MOSCOW oblast, central European Russia, on road and railroad, 11 mi/18 km WSW of MOSCOW, and 5 mi/8 km ENE of ODINTSOVO, to which it is administratively subordinate; 55°42′N 37°22′E. Elevation 554 ft/168 m. Holiday getaway spot for residents of Moscow and neighboring cities; hotels, sanatoria, summer homes.

Marfino (MAHR-fee-nuh), village, SE ASTRAKHAN oblast, S European Russia, port on the BUZAN' arm of the VOLGA RIVER delta mouth, on road, 6 mi/10 km SW of the RUSSIA-KAZAKHSTAN border, and 34 mi/55 km ENE of ASTRAKHAN; 46°24′N 48°43′E. Below sea level. Fisheries.

Marfrance (MAHR-frans), unincorporated town, GREENBRIER county, SE WEST VIRGINIA, 23 mi/37 km NNW of LEWISBURG; 38°03′N 80°41′W.

Margam (MAH-guhm), district (2001 population 2,389) of PORT TALBOT, Neath Port Talbot, S Wales; 51°33′N 03°44′W. Former steel works. Mountain of Mynydd Margam 1,130 ft/344 m is 2 mi/3.2 km NE.

Marganets, UKRAINE: see MARHANETS'.

Margao, town, ⊙ SOUTH GOA district, GOA state, W INDIA, on railroad, 17 mi/27 km SSE of PANAJI. Trade center (rice, copra, cashew nuts, mangoes, timber); sheep raising.

Margaree (mahr-guh-REE), village, NE NOVA SCOTIA, E CANADA, on CAPE BRETON ISLAND, on MARGAREE RIVER, and 15 mi/24 km NW of INVERNESS; 46°23′N 61°04′W. Elevation 196 ft/120 m. Salmon-fishing center. At mouth of Margaree River, on the GULF OF ST. LAWRENCE, 4 mi/6 km NW, is fishing port of Margaree Harbour.

Margaree River, (mahr-guh-REE) 10 mi/16 km long, NE NOVA SCOTIA, E CANADA, on NE CAPE BRETON ISLAND; rises in two branches: Southwest Margaree

River issues from LAKE AINSLIE, 8 mi/13 km ESE of INVERNESS; flows 15 mi/24 km N; Northeast Margaree River rises 20 mi/32 km SW of INGONISH; flows 40 mi/64 km S and E; the branches unite 4 mi/6 km S of MARGAREE, forming Margaree River, which flows NNW to the GULF OF ST. LAWRENCE at Margaree Harbour.

Margaret (MAHR-guh-ret), community, SW MANITOBA, W central CANADA, 6 mi/9 km W of DUNREA, in RIVERSIDE rural municipality; 49°24′N 99°51′W.

Margaret, village (2000 population 1,169), St. Clair co., N central Alabama, 12 mi/19 km SW of Ashville. Coal mines. Founded and named by Charles F. DeBardeleben. Named for his wife. Inc. in 1960.

Margareten (mahr-gah-RAI-ten), district (□ 1 sq mi/2.6 sq km) of VIENNA, AUSTRIA, 1.5 mi/2.4 km SW of city center.

Margaret River (MAHR-gret), village, SW WESTERN AUSTRALIA state, W AUSTRALIA, 140 mi/225 km SSW of PERTH, N of Cape LEEUWIN; 33°57′S 115°04′E. Dairy center (butter factory); vineyards; timber. MAMMOTH CAVES (limestone caves containing bones of prehistoric animals) nearby. Tourism. Maritime climate.

Margarettsville (MAHR-gruhts-vil), unincorporated village, NORTHAMPTON county, NE NORTH CAROLINA, 18 mi/29 km ENE of ROANOKE RAPIDS, at VIRGINIA state line, near MEHERRIN RIVER; 36°31′N 77°20′W. Agriculture (tobacco, grain; livestock).

Margaretville, summer-resort village (2006 population 641), DELAWARE county, S NEW YORK, in the CATSKILL MOUNTAINS, on East Branch of DELAWARE RIVER, and 37 mi/60 km NW of KINGSTON; 42°08′N 74°39′W.

Margarita (mahr-gah-REE-tah), town, ⊙ Margarita municipio, BOLÍVAR department, N COLOMBIA, on MARGARITA ISLAND, on the BRAZO SECO de MOMPÓS, and 108 mi/174 km SE of CARTAGENA; 09°09′N 74°17′W. Agriculture includes tobacco, sugarcane, plantains, corn.

Margarita, suburb, COLÓN province, PANAMA, 2.5 mi/4 km S of COLÓN; 09°20′N 79°54′W. Bananas, rubber, corn, rice; livestock.

Margarita Belén (mahr-gah-REE-tah be-LAIN), town, SE Chaco province, ARGENTINA, on railroad, and 12 mi/19 km N of RESISTENCIA; 27°16′S 58°58′W. Agriculture (cotton, corn, tobacco; citriculture); livestock raising; cotton ginning.

Margarita Island (mahr-gah-REE-tah), Spanish, *Isla de Margarita* (□ 444 sq mi/1,150 sq km), in the CARIBBEAN SEA off the coast of VENEZUELA; 11°01′N 63°53′W. With many smaller islands it constitutes the Venezuelan state of NUEVA ESPARTA.The economic center is PORLAMAR, where an important pearl-fishing industry exists. LA ASUNCIÓN, the state capital, exhibits distinguished architecture. Other island industries produce canned fish, salt, fishing boats, ceramics. Test holes drilled E of Margarita yielded gas reservoirs of high commercial interest. Margarita has become a popular tourist resort, especially for Venezuelans. Sighted by Columbus in 1498 and was used (1561) as a base of operations by the Spanish adventurer Lope de Aguirre. Because the people supported Simón Bolívar, Margarita and its neighboring islands were made a state after independence was won from SPAIN. Near Parque Nacional LAGUNA DE LA RESTINGA. International Aiport, Internacional del Caribe.

Margarita Island (mahr-gah-REE-tah), BOLÍVAR department, N COLOMBIA; formed by arms of middle MAGDALENA River, the BRAZO DE MOMPÓS (E) and the BRAZO DE LOBA (W); extends 60 mi/97 km NW from EL BANCO; c.20 mi/32 km wide; 09°05′N 74°30′W. Marshy lowlands with savannas and tropical forests. MOMPÓS is its largest town.

Margarita Island, MEXICO: see SANTA MARGARITA ISLAND.

Margaritas, Las, MEXICO: see LAS MARGARITAS.

Margariti (mahr-yah-REE-tee), village, THESPROTIA prefecture, S EPIRUS department, NW GREECE, 13 mi/21 km SE of IGOUMENITSA; 39°21′N 20°26′E. Olive oil; almonds; timber; livestock. Called Margalich under Turkish rule.

Margate (MAHR-gait), city (□ 8 sq mi/20.8 sq km; 2005 population 56,002), BROWARD county, SE FLORIDA, 5 mi/8 km W of POMPANO BEACH; 26°15′N 80°12′W. Commercial printing, light manufacturing.

Margate, town, SE KWAZULU-NATAL province, SOUTH AFRICA, on INDIAN OCEAN, 80 mi/129 km SW of DURBAN; 30°52′S 30°23′E. Popular seaside resort. Part of 15-mi/25-km section extending S of Margate coast.

Margate (MAH-gait), town (2001 population 58,465), in the Isle of THANET, NE KENT, SE ENGLAND, 16 mi/25.6 km NE of CANTERBURY; 51°23′N 01°24′E. Seaport with light industries and, since the 18th century, a popular resort, especially for Londoners. Of interest is the Church of St. John the Baptist (partly Norman). Cliftonville, Birchington, and Westgate-on-Sea are suburbs.

Margate (MAHR-gait), village, SE TASMANIA, AUSTRALIA, 12 mi/19 km SSW of HOBART, and on North West Bay of D'ENTRECASTEAUX CHANNEL; 43°02′S 147°16′E. Agriculture; fruit.

Margate City (MAHR-gait), resort and residential city (2006 population 8,601), ATLANTIC county, SE NEW JERSEY, on the ATLANTIC OCEAN, c.4 mi/6.4 km S of ATLANTIC CITY; 39°19′N 74°30′W. Known for its pleasant beaches and for its large, old homes, intermixed with expensive, renovated modern residences. Incorporated 1897.

Margaux (mahr-GO), commune (□ 3 sq mi/7.8 sq km), GIRONDE department, AQUITAINE region, SW FRANCE, in MÉDOC, near the GIRONDE ESTUARY, 15 mi/24 km NNW of BORDEAUX; 45°03′N 00°40′W. Produces fine red wines, including the noted Château-Margaux. Wine cellars can be viewed.

Margecany, SLOVAKIA: see KROMPACHY.

Marg, El, village, QALYUBIYA province, Lower EGYPT, 9 mi/14.5 km NE of CAIRO city center; 30°09′N 31°20′E. Cotton, flax, cereals, fruits. Site of some XVIII dynasty ruins. Also Al-Marj.

Margelan, UZBEKISTAN: see MARGHILON.

Margeride Mountains (mahr-zhuh-REED), granitic range in MASSIF CENTRAL, S central FRANCE, bounded by ALLIER (E), upper TRUYÈRE (W), and LOT (S) rivers; extends c.35 mi/56 km NNW-SSE along borders of HAUTE-LOIRE, CANTAL and LOZÈRE departments in SE AUVERGNE region; 44°50′N 03°25′E. Rises to 5,089 ft/1,551 m.

Marggrabowa, POLAND: see OLECKO.

Margherita (mahr-GAI-ri-tuh), town, DIBRUGARH district, NE ASSAM state, NE INDIA, on BURHI DIHING RIVER, and 50 mi/80 km ESE of DIBRUGARH; 27°17′N 95°41′E. Major Assam coal-mining center; trades in Burmese amber and rubber. Was capital of former Tirap frontier tract.

Margherita, SOMALIA: see JAMAME.

Margherita, UGANDA: see RUWENZORI, mountain range.

Margherita di Savoia (mahr-ge-REE-tah dee sah-VO-yah), town, FOGGIA province, APULIA, S ITALY, on the ADRIATIC Sea, near mouth of OFANTO River, 21 mi/34 km SE of MANFREDONIA; 41°22′N 16°09′E. Swimming resort. Has extensive saltworks, including drained Salpi lagoon; most important on Italian mainland.

Margherita, Lake, ETHIOPIA: see ABAYA, LAKE.

Margherita, Mount (MAHR-gah-REE-tah), highest summit (16,762 ft/5,109 m) of the RUWENZORI RANGE, in E central AFRICA on CONGO-UGANDA border, 30 mi/48 km ESE of BENI (CONGO); 00°22′N 29°51′E. Africa's third-highest peak (after KILIMANJARO and KENYA). Part of the RUWENZORI MOUNTAINS NATIONAL PARK, with guided climbing tours available. Was first ascended by the duke of Abruzzi and his expedition in 1906, and named for Queen Margherita of ITALY. Was thoroughly explored and mapped by a Belgian expedition in 1932.

Marghilon, city, FERGANA wiloyat, E UZBEKISTAN, N of FERGANA; 40°21′N 71°42′E. Center for textile and silk industries. Also MARGELAN.

Marghine, Catena del (mahr-GEE-ne, kah-TAI-nah del), mountain range, W central SARDINIA, ITALY; extends 15 mi/24 km SW from Catena del GOCEANO. Rises to 3,936 ft/1,200 m at Monte PALAI.

Marghita (mahr-GEE-tsah), Hungarian *Margitta*, town, BIHOR county, W ROMANIA, on Beretău River, on railroad, and 28 mi/45 km NE of ORADEA; 47°21′N 22°20′E. White-wine production; pottery making; manufacturing (construction materials). Under Hungarian rule, 1940–1945.

Margiana, TURKMENISTAN: see MERV.

Margibi (mar-GEE-bee), county (□ 1,260 sq mi/3,276 sq km; 1999 population 219,417), LIBERIA, on ATLANTIC OCEAN (SW), ⊙ KAKATA; 06°30′N 10°15′W. Borders MONTSERRADO (NW), BONG (N, E), and GRAND BASSA (SE) counties. A part of the coastal plain. Agriculture (oil palms). Rubber plantations inland. Small fishing industry along coast. Main centers include HARBEL. Margibi is divided into four districts: Gibi (NE), Kakata (N), Firestone (central), and Mambah-Kaba (S).

Margina (MAHR-jee-nah), Hungarian *Marzsina*, village, TIMIŞ county, W ROMANIA, on BEGA RIVER, on railroad, and 22 mi/35 km NE of LUGOJ; 45°51′N 22°16′E. Chemicals, charcoal manufacturing.

Marginea (MAHR-jee-na), village, SUCEAVA county, N ROMANIA, 6 mi/9.7 km SW of RĂDĂUŢI. Agriculture center with pottery manufacturing. Fortified 15th–16th-century monastery of SUCEVIŢA is 7 mi/11.3 km SW; it is famous for its 16th-century church covered with Byzantine frescoes, and for its valuable religious collections.

Margit Island, HUNGARY: see BUDAPEST.

Margit Sziget, HUNGARY: see BUDAPEST.

Margitta, ROMANIA: see MARGHITA.

Margny-lès-Compiègne (mahr-NYEE-lai-kon-PYEN), N suburb (□ 2 sq mi/5.2 sq km) of COMPIÈGNE, OISE department, PICARDIE region, N FRANCE, on the OISE RIVER; 49°26′N 02°49′E. The Forest of Compiègne lies E.

Margonin (mahr-GO-neen), town, Piła province, W POLAND, on small lake, 40 mi/64 km N of POZNAŃ; 52°58′N 17°06′E. Flour milling.

Margos (MAHR-gos), town, Huánuco province, HUÁNUCO region, central PERU, in Cordillera CENTRAL, 17 mi/27 km SW of HUÁNUCO; 10°04′S 76°26′W. Cereals, potatoes; livestock.

Margosatubig (MAHR-go-sah-TOO-big), town, ZAMBOANGA DEL SUR province, W MINDANAO, PHILIPPINES, 25 mi/40 km SW of PAGADIAN; 07°33′N 123°11′E. Corn, rice, coconuts. Sawmill.

Margraten (mahrkh-GRAH-tuhn), village, LIMBURG province, SE NETHERLANDS, 6 mi/9.7 km ESE of MAASTRICHT; 50°49′N 05°49′E. Belgian border 4 mi/6.4 km to S. Dairying; livestock; grain, vegetables, fruit.

Margrethe, Lake (mahr-GRETH), CRAWFORD county, N central MICHIGAN, 4 mi/6.4 km SW of GRAYLING; c.3 mi/4.8 km long, 1 mi/1.6 km wide; 44°39′N 84°47′W. Drained from W by a headstream of MANISTEE RIVER. Located within Camp Grayling National Guard Reservation.

Marguareis, Monte, highest peak (8,697 ft/2,651 m) in LIGURIAN ALPS, on French-Italian border, 6 mi/9.7 km E of TENDA PASS; 44°10′N 07°41′E. Until 1947, border passed 11 mi/18 km SW. Also known as Cima Marguareis.

Marguerite Bay (mah-guhr-REET), inlet of ANTARCTICA, on W coast of ANTARCTIC PENINSULA, in the South PACIFIC, between ADELAIDE and ALEXANDER islands, FALLIÉRES COAST, and WORDIE and GEORGE VI ice shelves; 68°30′S 68°30′W. Discovered 1909 by Jean B. Charcot, French explorer. Also known as Margaret Bay.

Cross-references are shown in SMALL CAPITALS. The pronunciation guide is shown on page xix. The sources of population figures are shown on page xvii.

Marguerittes (mahr-guh-REET), E suburb (□ 9 sq mi/ 23.4 sq km) of NÎMES, GARD department, LANGUEDOC-ROUSSILLON region, S FRANCE, 4 mi/6.4 km ENE of Nîmes; 43°51′N 04°27′E.

Marhanets' (MAHR-hah-nets), (Russian *Marganets*) [=manganese], city (2001 population 49,592), S DNIPROPETROVS'K oblast, UKRAINE, near N shore of KAKHOVKA RESERVOIR, 12 mi/19 km ENE of NIKOPOL'; 47°38′N 34°38′E. Elevation 249 ft/75 m. Major manganese-mining center in Nikopol' basin; mines; enrichment plants; mining machinery repair plant; railroad shops. Established at the end of the 19th century as a mining settlement, Horodyshche (Russian *Gorodishche*), later (around 1926), Komintern, and then amalgamated (1938) with several other settlements to form the current city.

Mari (mah-ree), city (2007 population 20,535), E PARAÍBA state, NE BRAZIL, on railroad, and 26 mi/42 km W of JOÃO PESSOA; 07°14′S 35°13′W. Cotton, sugar, fruit. Until 1944, called Aracá.

Mari (MAH-ree), ancient Semitic city of MESOPOTAMIA (modern SYRIA), on the right bank of the middle EUPHRATES RIVER, S of its junction with the HABOR (KHABUR), 10 mi/16 km from what is now IRAQI border, and 6 mi/9.7 km N of present-day Abu-KEMAL. The site at Tell el Hariri was discovered by chance in the early 1930s by Arabs digging graves and has subsequently been excavated by the French. The earliest evidence of habitation goes back to the Jemdet Nasr period in the 3rd millennium B.C.E. and Mari remained prosperous throughout the early dynastic period. The temple of Ishtar and other works of art show that Mari was at this time an artistic center with a highly developed style of its own. As the commercial and political focus of W ASIA c.1800 B.C.E., its power extended more than 300 mi/480 km from the frontier of BABYLON proper, up the EUPHRATES River to the border of Syria. The inhabitants were referred to as Amorites in the Old Testament and spoke a language related to the Hebrew of the patriarchs. The archives of the great King Zimri-lim, a contemporary of Hammurabi in the 18th century B.C.E., were discovered in 1937. They contain more than 20,000 clay documents, which have made it possible to fix the dates of events in Mesopotamia in the 2nd millennium B.C.E. Also found here is the great palace complex of Zimri-lim consisting of more than 200 rooms and covering 5 acres/2 hectares. Hammurabi conquered Mari c.1700 B.C.E.; and BABYLON then became the center of W Asia. Mari never regained its former status.

María (mah-REE-ah), town, ALMERÍA province, S SPAIN, 7 mi/11.3 km NW of VÉLEZ RUBIO. Chemical works (essential oils, resins). Lumbering; livestock raising; cereals.

Maria (mah-REE-uh), village (□ 37 sq mi/96.2 sq km), GASPÉSIE—ÎLES-DE-LA-MADELEINE region, E QUEBEC, E CANADA, S GASPÉ PENINSULA, on Cascapedia Bay of CHALEUR BAY, 20 mi/32 km ENE of DALHOUSIE; 48°10′N 65°59′W. Fishing port; lumbering. Resort.

Maria Alm am Steinernen Meer (mah-REE-ah AHLM ahm shtei-ner-nen MER), village, W central SALZBURG, W AUSTRIA, at SW foot of STEINERNES MEER, 12 mi/19 km NE of ZELL AM SEE; 47°24′N, 12°54′E. Elevation 2,731 ft/832 m. Center of winter and summer tourism; pilgrim church with extremely high spire.

Mariabé (mah-ree-ah-BAI), village and minor civil division, Pedasí district, LOS SANTOS province, S central PANAMA, in PACIFIC lowland, 6 mi/9.7 km NW of PEDASÍ; 07°35′N 80°04′W. Sugarcane; livestock.

Maria-Chapdelaine (mah-REE-ah–shah-duh-LEN), county (□ 14,796 sq mi/38,469.6 sq km; 2006 population 26,424), SAGUENAY—LAC-SAINT-JEAN region, S central QUEBEC, E CANADA; ☉ DOLBEAU-MISTASSINI; 48°59′N 72°17′W. Agriculture (potatoes); forestry; dairying; cornflowers. Composed of fourteen municipalities. Formed in 1983.

María Chiquita (mah-REE-ah chee-KEE-tah), village and minor civil division of Portobelo district, COLÓN province, central PANAMA, minor port of CARIBBEAN SEA, 11 mi/18 km NE of COLÓN; 09°27′N 79°45′W. Bananas, cacao, abacá; livestock.

María Cleófas Island, MEXICO: see TRES MARÍAS, LAS.

Maria Cristina Falls, LANAO DEL NORTE province, N MINDANAO, PHILIPPINES, near ILIGAN BAY, 5.6 mi/9 km SE of ILIGAN, on the AGUS RIVER. Harnessed for a hydroelectric project that is a main power source for W and NW Mindanao.

María de la Salud (mah-REE-ah dai lah sah-LOOD), town, MAJORCA island, BALEARIC ISLANDS, SPAIN, 23 mi/37 km ENE of PALMA; 39°40′N 03°05′E. Almonds, cereals, figs, grapes; livestock; sawmilling.

María Elena (mah-REE-ah ai-LAI-nah), town, ☉ María Elena comuna, TOCOPILLA province, ANTOFAGASTA region, N CHILE, on railroad, and 38 mi/61 km SE of TOCOPILLA. Large nitrate-processing center.

Maria Enzersdorf am Gebirge (mah-REE-ah ENtsers-dorf uhm ge-BIR-ge), township, E LOWER AUSTRIA, at E foot of WIENERWALD FOREST, and urban fringe of VIENNA, 10 mi/16 km SW of city center; 48°06′N 16°17′E. National sports and training center. Mission center Sant Gabriel has notable ethnographic collection. Manufacturing of machinery and pharmaceutical products. Monastery; castle Liechtenstein nearby; theater festival.

Mariager (mah-ree-YER), city and port, ÅRHUS county, E JUTLAND, DENMARK, on MARIAGER FJORD, and 34 mi/55 km NNW of ÅRHUS; 56°40′N 10°10′E. Dairying; hogs.

Mariager Fjord (mah-ree-YER), inlet of the KATTEGAT strait, E JUTLAND, DENMARK; c.25 mi/40 km long. MARIAGER on S shore, HOBRO at head. Inner part sometimes called Hobro Fjord.

María Grande (mah-REE-ah GRAHN-dai) or **Valle María**, town, W ENTRE RÍOS province, ARGENTINA, 18 mi/29 km S of PARANÁ. In agricultural area (grain; livestock, poultry).

María Grande, ARGENTINA: see VILLA MARÍA GRANDE.

Maria Helena (MAH-ree-ah E-le-nah), town (2007 population 6,012), NW PARANÁ state, BRAZIL, 8 mi/ 12.9 km NW of UMUARAMA. Coffee, cotton, rice, corn; livestock.

Mariahilf (mah-ree-ah-HILF), district (□ 1 sq mi/2.6 sq km) of VIENNA, AUSTRIA, just SW of city center; 46°55′N 12°17′E. Its main street (Mariahilferstabe) is the most popular shopping street in Vienna.

Mariahu (muhr-YAH-huh), town, JAUNPUR district, SE UTTAR PRADESH state, N central INDIA, 11 mi/18 km SSW of JAUNPUR; 25°37′N 82°37′E. Barley, rice, corn, wheat, sugarcane.

Maria Island (ma-ree-uh), uninhabited atoll of four islets, westernmost of AUSTRAL ISLANDS, FRENCH POLYNESIA, S PACIFIC; 21°45′S 154°30′W. Visited for copra, fish. Sometimes called Hull Island.

Maria Island (muh-REI-uh), AUSTRALIA, in TASMAN SEA, 3 mi/5 km off E coast of TASMANIA; 11 mi/18 km long, 6 mi/10 km wide; 42°38′S 148°05′E. Consists of two mountainous parts joined by narrow isthmus; limestone. Largest town, Maria Island, on NW coast.

Maria Islands, two small islands, off SSE SAINT LUCIA, CARIBBEAN SEA; 13°43′N 60°55′W. Nature preserve inhabited by rare species of snakes, lizards, birds.

Mariakani (mah-ree-AH-kah-nee), town (1999 population 10,987), Kilifi district, COAST province, KENYA, at road junction, 19 mi/30 km NW of MOMBASA; 03°52′S 39°29′E. Market and trading center; coconut and sisal in region.

Mariakerke (mah-REE-ah-KER-kuh), village, in commune of Ghent, Ghent district, EAST FLANDERS province, NW BELGIUM, on Bruges-Ghent Canal, and 2 mi/3.2 km NW of GHENT. Manufacturing.

María la Baja (mah-REE-ah lah BAH-hah), town, ☉ María la Baja municipio, BOLÍVAR department, N COLOMBIA, in CARIBBEAN LOWLANDS, 35 mi/56 km SSE of CARTAGENA; 09°59′N 75°17′W. Sugarcane center; plantains.

María Linda River, GUATEMALA: see MICHATOYA RIVER.

María Lionza, Monumento Natural (mah-REE-ah lee-ON-sah mo-noo-MEN-to nah-too-RAHL), national natural monument (□ 45 sq mi/117 sq km), SW YARACUY state, N VENEZUELA; 09°58′N 68°53′W. Created 1960.

Maria Luggau (mah-REE-ah lug-GOU), village, CARINTHIA, S AUSTRIA, in the LESACHTAL valley, 9 mi/14.5 km S of Lienz; 46°42′N, 12°44′E. Elevation 3,593 ft/ 1,095 m. Prominent pilgrim place with a catchment area including adjacent ITALY and SLOVENIA; Serviten monastery with a fine church, built 1520–1536, reconstructed in the baroque style.

María Madre Island, MEXICO: see TRES MARÍAS, LAS.

María Magdalena Island, MEXICO: see TRES MARÍAS, LAS.

Mariampol, LITHUANIA: see MARIJAMPOLĖ.

Mariana (mah-ree-ah-nah), city (2007 population 52,235), S central MINAS GERAIS state, BRAZIL, in Serra do ESPINHAÇO, on railroad, and 6 mi/9.7 km ENE of OURO PRÊTO, 45 mi/72 km SE of BELO HORIZONTE; 20°21′S 43°32′W. Historic city, first episcopal see; archiepiscopal see since 1905. Early 18th-century cathedral, ornate churches, and city hall. Formerly known for its important gold mines (still worked). Tungsten and kaolin deposits. Formerly spelled Marianna.

Mariana Islands, see: GUAM.

Mariana Lake (ma-ree-A-nuh), hamlet (2006 population 11), NE ALBERTA, W CANADA, on Highway 63, 59 mi/95 km S of FORT MCMURRAY, and included in WOOD BUFFALO; 55°57′N 112°01′W.

Marianao (mah-ree-ah-NOU), county and city,, CIUDAD DE LA HABANA province, W CUBA, a commercial and residential suburb of HAVANA in W metropolitan area; 23°05′N 82°25′W. Marianao encloses the N. American-built military base of Columbia, now called Ciudad Libertad. Chemicals, beer, and textiles are produced here. Has a fine beach. Founded in 1719 by Dominican and Augustinian monks, the city was destroyed by fire in 1726. Rebuilt in 1765 as Quemados de Marianao and grew with the sugar boom in the 19th century.

Marianas Trench (mah-ree-AH-nahs), **Marianas Trough**, or **Marianas Deep**, elongated depression on the floor of the PACIFIC OCEAN, c.210 mi/340 km SW of GUAM. Deepest known depression on the earth's surface. First founded in 1959 by SOVIET scientists; its bottom was reached in 1960 by two men in the U.S. Navy bathyscaphe *Trieste*. In March 1995 the remote-controlled Japanese deep-sea probe Kaiko touched bottom at 35,798 ft/10,911 m, c.3.3 ft/1 m short of the depth recorded by *Trieste*, but possibly marking the actual limit of the deep. The Japan Marine Science and Technology Centre, while not disputing the *Trieste* recording, suggested that technology for measuring the depth of the seabed had improved greatly between 1960 and 1995.

Marianna (MER-ee-A-nah), city (□ 6 sq mi/15.6 sq km; 2005 population 6,275), ☉ JACKSON county, NW FLORIDA, on CHIPOLA RIVER, and c.60 mi/ 97 km WNW of TALLAHASSEE; 30°46′N 85°14′W. Railroad junction. Lumber milling; manufacturing.

Marianna, town (2000 population 5,181), ☉ LEE county, E ARKANSAS, c 50 mi/80 km SW of MEMPHIS (TENNESSEE), and on L'ANGUILLE RIVER; 34°46′N 90°46′W. Agriculture (cotton, rice, soybeans); manufacturing (transportation equipment, fabricated metal products). The small Saint Francis National Forest is to SE. Incorporated 1877.

Marianna (MER-ee-A-nah), borough (2006 population 591), WASHINGTON county, SW PENNSYLVANIA,

13 mi/21 km SE of WASHINGTON on Tenmile Creek; 40°00′N 80°06′W. Agriculture (corn, hay; dairying). Incorporated 1901.

Marianna, Brazil: see MARIANA.

Marianne Island, E of RODRIGUEZ Island, dependency of MAURITIUS, 0.9 mi/1.4 km from Pointe Afine; 19°44′S 63°20′E. Small rocky island covering 4.9 acres/ 2 ha; c.16 ft/5 m high.

Mariannelund (mahr-EE-ahn-ne-LUND), village, JÖNKÖPING county, S SWEDEN, 30 mi/48 km E of NÄSSJÖ; 57°37′N 15°34′E. Manufacturing (metal goods, windows). Folk high-school.

Mariano Arista, Nancamilpa de, MEXICO: see NANA-CAMILPA.

Mariano Comense (mah-ree-AH-no ko-MEN-se), town, COMO province, LOMBARDY, N Italy, 9 mi/14 km SSE of COMO; 45°42′N 09°11′E. Furniture-manufacturing center; silk mills; aluminum industry; textile machinery, hardware.

Mariano Escobedo (mah-ree-AH-no es-ko-BAI-do), town, VERACRUZ, E MEXICO, in SIERRA MADRE ORIENTAL, 5 mi/8 km N of ORIZABA. Coffee, fruit.

Mariano I. Loza (mah-ree-AH-no ee LO-zah), town, S central CORRIENTES province, ARGENTINA, on railroad (Solari station), and 15 mi/24 km SSW of MERCEDES; 29°22′S 58°12′W. Livestock raising, farming; stone quarries. Formerly called Solari.

Mariano Moreno (mah-ree-AH-no mo-RAI-no), town, central NEUQUÉN province, ARGENTINA, 10 mi/ 16 km N of ZAPALA. Livestock-raising and oil-producing center; coal deposits and other minerals. Formerly Covunco Centro.

Marianopol', UKRAINE: see MARIUPOL'.

Marianópolis (MAH-ree-ah-NO-po-lees), town, W TOCANTINS state, BRAZIL, 194 mi/312 km SW of PALMAS; 09°50′S 49°50′W.

Mariano Roque Alonso (mah-ree-AH-no RO-kai ah-LON-so), town, (2002 population 65,229), Central department, S PARAGUAY, on Paraguay River, and 10 mi/ 16 km NE of ASUNCIÓN; 25°09′S 57°32′W. In agricultural area (sugarcane, fruit; livestock). Growing suburb of Asunción.

Marianske Lazne (MAH-ri-AHN-ske LAHZ-nye), Czech *Mariánské Lázně*, German *Marienbad*, town, ZAPADOCESKY province, W BOHEMIA, CZECH REPUBLIC, 33 mi/53 km NW of PLZEŇ; 49°58′N 12°42′E. Railroad junction. Food processing. World-famous spa, with many curative mineral springs and baths, situated on the grounds of a 12th century abbey. Site of numerous international congresses.

Mariapa, VENEZUELA: see MARIPA.

Maria Pereira, Brazil: see MOMBAÇA.

Mariapfarr (mah-ree-ah-PFAHR), village, SE SALZBURG, S central AUSTRIA, in the LUNGAU region, 27 mi/ 43 km NNE of SPITTAL AN DER DRAU; 47°09′N 13°45′E. Elevation 3,413 ft/1,040 m. Summer and winter tourism. Fine Gothic parish church, castle.

María Pinto (mah-REE-ah PEEN-to), village, ⊙ María Pinto comuna, MELIPILLA province, METROPOLITANA DE SANTIAGO region, central CHILE, 28 mi/45 km WSW of SANTIAGO; 33°32′S 71°08′W. In agricultural area (cereals, fruit, alfalfa; livestock).

Mariapu (mah-ree-AH-poo), canton, LARECAJA province, LA PAZ department, W BOLIVIA, 6 mi/10 km E of GUANAY, N of TIPUANI River; 15°47′S 68°40′W. Elevation 8,750 ft/2,667 m. Tungsten mining at Minas Ucumarini Mercedes San Antonio (Susana); clay, limestone, and gypsum deposits. Agriculture (potatoes, yucca, bananas, rye); cattle.

Máriaradna, ROMANIA: see RADNA.

Marias (muh-REI-uhs), river, c.210 mi/338 km long, N MONTANA; formed by joining of Cut Bank Creek and Two Medicine River on border between GLACIER and PONDERA counties; flows E, passes 8 mi/12.9 km S of SHELBY and through Lake ELWELL (Tiber Reservoir), then SE and receives TETON River just before entering MISSOURI River at Loma. Used for irrigation. Tiber

Dam (completed 1956), is part of the Missouri River Basin Project.

Maria Saal (mah-REE-ah SAHL), SLOVENIA *Gaspa Sveta*, township, CARINTHIA, S AUSTRIA, in the ZOLLFELD, near GLAN RIVER, 4 mi/6.4 km NNE of KLAGENFURT; 46°41′N 14°21′E. Fortified church founded in the middle of the 8th century as an ecclesiashical center of Carantania (Carinthia); until 945 was the see of a bishop. One of the most important ecclesiastical monuments in Austria with remarkable altars from different periods. Remains of Roman city of Virunum nearby.

Marías, Las, PUERTO RICO: see LAS MARÍAS.

Marias Pass (muh-REI-uhs) (5,280 ft/1,609 m), in LEWIS RANGE of CONTINENTAL DIVIDE, on border between GLACIER and FLATHEAD counties, NW MONTANA, at SUMMIT, on SE border of GLACIER NATIONAL PARK, and 50 mi/80 km SW of CUT BANK. Nearby are Lewis and Clark (NE) and Flathead (SW) national forests. Discovered 1889 by John F. Stevens, the pass is now crossed by U.S. Highway 2 and Burlington Northern railroad.

Maria Taferl (mah-REE-ah TAH-ferl), village, W LOWER AUSTRIA, on a steep slope near the DANUBE RIVER, 15 mi/24 km ENE of AMSTETTEN; 48°14′N 15°09′E. Well-known baroque pilgrimage church, receives visitors from throughout central Europe.

María Trinidad Sánchez, province (□ 508 sq mi/1,320.8 sq km; 2002 population 135,727), NE DOMINICAN REPUBLIC, on the ATLANTIC OCEAN coast; ⊙ NAGUA; 19°30′N 70°00′W. Agricultural (bananas) in S part.

Mari Autonomous Soviet Socialist Republic, RUSSIA: see MARI EL REPUBLIC.

Maria Van Dieman, Cape (mah-REI-a VAN DEE-man), extreme Far North district, NEW ZEALAND, 25 mi/40 km W of NORTH CAPE; 34°29′S 172°39′E. Automated lighthouse.

Mariaville (muh-REI-uh-vil), town, HANCOCK county, S MAINE, 12 mi/19 km N of ELLSWORTH; 44°45′N 68°23′W. In recreational area.

Maria Wörth (mah-REE-ah VUHRT), village, CARINTHIA province, S AUSTRIA, on the S shore of the Wörther See (largest lake in Carinthia), 7 mi/11.3 km W of KLAGENFURT; 46°37′N 14°4–10′E. Popular resort. Pilgrimage center, with two romanesque 12th century churches.

Mariazell (mah-ree-ah-TSEL), town, STYRIA province, central AUSTRIA; 47°46′N 15°19′E. Winter and summer resort with cable cars to the Bergalpe (elevation 3,859 ft/1,176 m). Chiefly noted as a place of pilgrimage, it is famous for its wood carving of the Virgin and Child (1370), a gift from King Ludwig of HUNGARY. Main place of pilgrimage in Austria, attracting thousands of pilgrims annually from throughout central EUROPE. Temporary grave site of Hungarian Cardinal J. Mindszenty. Several pilgrim paths with churches and chapels lead here.

Marib (MAH-rib), ancient city, E YEMEN, 75 mi/121 km E of SANA, on the edge of the desert; 15°25′N 45°21′E. Elev. 3,900 ft/1,190 m. It was one of the chief cities, perhaps even the capital, of ancient SHEBA. It was the site of a dam, built in the 6th century B.C.E., that was one of the great engineering feats of antiquity; ruins include walls of the dam and 2 Saberan temple sites. Trade center on frankincense route from Vivan to JERUSALEM. Declined with rise of Islam. The dam collapsed in the 6th century C.E., flooding the countryside. At present, Marib is a small outlying frontier township. Large oil fields were discovered in the vicinity in the 1980s. Production began in 1987 after a pipeline to the RED SEA port Selif was inaugurated. Also spelled Mareb.

Maribios, Cordillera de los (mah-REE-bee-os, kor-dee-YAI-rah dai los), range of volcanic peaks in W NICARAGUA; extends c.30 mi/48 km NW from LAKE MANAGUA, parallel to PACIFIC coast. Includes volcanoes of CHONCO, SAN CRISTÓBAL, CASITA, EL VIEJO,

SANTA CLARA, TELICA, ROTA, CERRO NEGRO, LAS PILAS, DEL HOYO, and ASOSOSCA. Sometimes spelled Marabios or Marrabios.

Maribo (MAH-ree-bo), city (2000 population 5,922), STORSTRØM county, SE DENMARK, on Lake Søndersø; 54°49′N 11°25′E. Commercial and industrial center with sugar refineries. The playwright Kaj Munk (1898–1944) was born here. Knuthenborg Safari Park nearby.

Maribondo Falls (MAH-ree-bon-do), S central BRAZIL, on the RIO GRANDE (border between SÃO PAULO and MINAS GERAIS states), 35 mi/56 km N of SÃO JOSÉ DO RIO PRÊTO (São Paulo); 20°18′S 49°10′W. Sometimes spelled Marimbondo.

Maribor (MAH-ree-bor), German *Marburg*, city (2002 population 92,284), in SLOVENIA, on the Drava River; 46°33′N 15°38′E. The second-largest city in SLOVENIA. Heavy industrial center with chemical, engineering, automotive, and electrical industries; manufacturing (Railroad cars, textiles; food and wood processing). International airport c.6 mi/9.7 km S. Known as early as the 12th century, it was an important city of STYRIA. Has a 12th-century Gothic cathedral, a 15th-century castle, and a fine Renaissance town hall.

Maribyrnong (MAR-i-buh-nahng), municipality (□ 12 sq mi/31.2 sq km), suburb 4 mi/6 km NW of MELBOURNE, VICTORIA, SE AUSTRALIA, bordering Melbourne Docklands area; 37°47′S 144°53′E. Immigration detention center here.

Maricá (MAH-ree-kah), city (2007 population 105,455), S RIO DE JANEIRO state, BRAZIL, on coastal lagoon of same name, 18 mi/29 km E of NITERÓI; 22°55′S 42°49′W. Fish processing; rice milling; sugar, brandy, vegetables; poultry. Feldspar and quartz deposits.

Maricaban Island (mah-ree-KAH-bahn) (□ 12 sq mi/ 31.2 sq km), BATANGAS province, PHILIPPINES, in VERDE ISLAND PASSAGE, off SW coast of LUZON, at entrance to BATANGAS BAY, 10 mi/16 km SW of BATANGAS; 8 mi/12.9 km long, 2 mi/3.2 km wide. Rises to 1,469 ft/448 m. Fishing; rice growing. Chief town is Papaya on S coast. Part of BAUAN municipality.

Maricao (mah-ree-KOU), town (2006 population 6,300), W PUERTO RICO, in W outliers of the Cordillera Central, 11 mi/18 km E of Mayagüez. Elevation 1,416 ft/ 432 m. Coffee-trading and producing center; citrus fruits, plantains; resort. Pharmaceuticals factory. Fish hatchery nearby. Adjoining S is extensive reforestation project. Maricao Forest Reserve nearby (S).

Marica River, SE Europe: see MARITSA RIVER.

Marichchukkaddi (MAH-rich-chuk-KUH-di), village, NORTHERN PROVINCE, NW SRI LANKA, near Gulf of MANNAR, 30 mi/48 km N of PUTTALAM; 08°35′N 79°56′E. Rice and coconut-palm plantations. Center for pearl fishing in banks off coast during early 20th century. Has country's lowest recorded rainfall.

Maricopa, county (□ 9,224 sq mi/23,982.4 sq km; 2006 population 3,768,123), central and central ARIZONA; ⊙ PHOENIX, the state capital; 33°20′N 112°29′W. MCDOWELL MOUNTAINS, part of MAZATZAL MOUNTAINS, and part of Tonto National Forest are in NE. Irrigated agricultural region extends along banks of SALT, GILA, SANTA CRUZ, VERDE, and AGUA FRIA (which forms Lake PLEASANT reservoir and part of N border) rivers. Long-staple cotton, citrus, vegetables, hay, alfalfa, wheat, barley; cattle, sheep, hogs; tourism. Manufacturing at Phoenix, GLENDALE, TEMPE, SCOTTSDALE, MESA, CHANDLER, other suburban cities. Sixteenth-largest county in U.S. Highly urbanized in NE around Phoenix; large retired population; Phoenix area has attracted large numbers of businesses and residents to its warm climate. Hummingbird Springs, BIG HORN MOUNTAINS, and parts of Eagletail Mountains and HARQUAHALA MOUNTAINS wilderness areas in NW; part of Tohono O'odham (Papago) Indian Reservation in S; Gila Bend Indian Reservation in SW; Salt River and Fort McDowell Reservations in NE. Luke Air Force Base W of Phoenix; part of large Barry

---

Cross-references are shown in SMALL CAPITALS. The pronunciation guide is shown on page xix. The sources of population figures are shown on page xvii.

M. Goldwater (formerly Luke) Air Force Range in SW. Gila River Indian Reservation to S. Nuclear power plants, Palo Verde One (initial criticality May 25, 1985), Palo Verde Two (initial criticality April 18, 1986), and Palo Verde Three (initial criticality October 25, 1987), are 36 mi/58 km W of Phoenix; use cooling water from a sewage treatment facility, and each has a maximum dependable capacity of 1221 Mwe. Formed 1871.

Maricopa, city (2000 population 1,111), KERN county, S central CALIFORNIA, 30 mi/48 km SW of BAKERSFIELD, in S part of San Joaquin Valley; 35°04′N 119°24′W. Oil wells. Cattle; dairying; fruit, grain. Los Padres National Forest to S; Buena Vista Lake irrigation reservoir to NE.

Maricopa Mountains, MARICOPA county, SW central ARIZONA, E of GILA River and GILA BEND; rise to c.3,000 ft/914 m. Javelina Mountain in S, 3,571 ft/1,088 m.

Marico River (MAH-ree-ko), SOUTH AFRICA and BOTSWANA; rises in three headstreams in NORTHWEST province in NE South Africa; flows N to Botswana border c.36 mi/58 km ENE of GABORONE (Botswana), turns NE forming c.40 mi/64 km section of South Africa–Botswana border; joins LIMPOPO (Crocodile) River. Source of primary water supply for the KGATLENG district (Botswana).

Maricourt (mah-ree-KOOR), village (□ 24 sq mi/62.4 sq km; 2006 population 521), ESTRIE region, S QUEBEC, E CANADA, 3 mi/5 km from VALCOURT; 45°30′N 72°15′W.

Maricunga, Salar de (mah-ree-KOON-gah, sah-LAHR dai), salt desert in S ATACAMA DESERT, ATACAMA region, N CHILE; extends 13 mi/21 km N-S (c.7 mi/11 km wide) along SW foot of CORDILLERA CLAUDIO GAY; 27°00′S. Elevation c.12,000 ft/3,658 m. Borax deposits.

Maridi, SUDAN: see MERIDI.

Marie Byrd Land, area of WEST ANTARCTICA, E of the ROSS ICE SHELF and the ROSS SEA and S of the PACIFIC OCEAN to E end of the AMUNDSEN SEA; 80°00′S 120°00′W. Discovered by Richard E. Byrd in 1929. Much of this region was explored during the second Byrd expedition (1933–1935) and the U.S. Antarctic Service Expedition (1939–1941). Geologically it is an island connected to the continent by the deep ice of the BYRD SUBGLACIAL BASIN. Also known as Byrd Land.

Marie Byrd Seamount, ANTARCTICA, elongated seamount on continental rise N of AMUNDSEN SEA, S PACIFIC OCEAN; 70°00′S 118°00′W.

Mariefred (mah-EE-e-FRED), town, SÖDERMANLAND county, E SWEDEN, on S shore of LAKE MÄLAREN, 30 mi/48 km W of STOCKHOLM; 59°16′N 17°13′E. Has seventeenth-century church. GRIPSHOLM Castle neaby. Incorporated 1605.

Marie-Galante (mah-REE–gah-LAHNT), island (□ 61 sq mi/158.6 sq km), POINTE-À-PITRE *arrondissement*, GUADELOUPE, French WEST INDIES; 15°56′N 61°16′W. Its highest peak, Morne Constant, rises to 670 ft/204 m. The island, 9 mi/15km wide, is commonly known as "la grande galette" [French=big pancake] due to its round shape. Sugarcane farming; rum distilling.

Marie-Galante: see GUADELOUPE.

Mariehamn, FINLAND: see MAARIANHAMINA.

Marieholm (mahr-EE-e-HOLM), village, SKÅNE county, S SWEDEN, 18 mi/29 km NNE of MALMÖ; 55°52′N 13°09′E.

Mariel (mah-ree-EL), town, LA HABANA province, W CUBA, on sheltered bay, 27 mi/43 km WSW of HAVANA; 23°00′N 82°46′W. In agricultural region (sugarcane, tobacco; cattle). Manufacturing of cement and cigars. Shark fishing. Nearby are Cuba's principal asphalt reserves and a geothermal power plant. There are also limestone and guano deposits; sulphurous springs. In the outskirts is the Cuban Naval Academy Site of 1980 Mariel boat lift, when more than 100,000

Cubans left for FLORIDA through its port. The Central San Ramón is 3 mi/5 km SW.

Marie Louise Island, in S AMIRANTES, outer group of the SEYCHELLES, 190 mi/306 km SW of MAHÈ ISLAND; 1 mi/1.6 km long, 0.5 mi/0.8 km wide; 06°11′S 53°09′E. Coral formation.

Mari El Republic (mah-REE–EL), constituent republic (□ 9,024 sq mi/23,462.4 sq km; 2006 population 716,330), E central European RUSSIA, in the middle VOLGA RIVER valley, 534 mi/862 km ENE of MOSCOW; ⊙ YOSHKAR-OLA. The region is a rolling plain, heavily forested (half of the land) with fir and pine and rich in mushrooms and berries; woods called "the lungs" of the Mid-Volga region and known for their pristine environmental quality. Drained by the ILET, Great KOKSHAGA, Yushut, and KUNDYSH rivers. Forest lakes include Yalchik, Kichiyer, and Karas. There is an extensive lumbering industry, and the republic produces paper and pulp and varied wood products. In the nonforested agricultural areas, grain and flax are grown, and there is dairy farming and livestock raising. The main industry, however, is machinery and machine tool manufacturing; also, food processing and consumer goods production (most of the country's commercial refrigeration equipment). Minerals include peat, sand, clay, building stone, and lime. Resorts with mineral springs. Local fauna include elk, wild boar, bear, hare, squirrel, marten, capercaillie, grouse. The population is mainly Russian (47.5%) and Mari (43.5%), with Tatar (5.9%), Chuvash, Udmurt, Mordovian, and Ukrainian minorities. In the 8th century, the Mari were under Khazar rule. Ruled by the Eastern Bulgars (9th–12th centuries), the Mari were then conquered (1236) by the Golden Horde. The Russians under Ivan IV assumed control in 1552. The autonomous region was organized in 1920 and the autonomous republic in 1936. A sovereign republic was declared on October 22, 1990, and was a signatory, under the name Republic of Mariy-El or Mariya-El, to the March 31, 1992, treaty that created the RUSSIAN FEDERATION. It has a 150-member parliament. Previously called Cheremiss, the Mari speak a Finno-Ugric language and are known for their wood and stone carving and embroidery. Also Mari-El.

Mariemont (muh-REE-mahnt), village (2006 population 3,056), HAMILTON county, extreme SW OHIO; suburb 8 mi/12.8 km E of CINCINNATI; 39°08′N 84°22′W. Bakery products, electrical apparatus, beverages. Laid out 1922, incorporated 1941.

Marienbad, CZECH REPUBLIC: see MARIANSKE LAZNE.

Marienberg (mah-REE-en-berg), town, SAXONY, E central GERMANY, in the ERZGEBIRGE, 17 mi/27 km SE of CHEMNITZ, near Czech border; 50°39′N 13°10′E. Textile milling (cotton, ribbon, lace), metal- and woodworking, button manufacturing. Climatic health resort. Has 16th-century church and town hall. Founded (1521) as silver-mining settlement, later became important staging point on Leipzig-Prague (Czech Republic) trade route. Chartered 1523.

Marienberg, Bad, GERMANY: see BAD MARIENBERG.

Marienborn (mah-REE-en-born), village, SAXONY-ANHALT, central GERMANY, 23 mi/37 km W of MAGDEBURG, 5 mi/8 km ESE of HELMSTEDT; 52°12′N 11°06′E. In lignite-mining region. Between 1945 and 1990 served as railroad and road traffic checkpoint between former East and West Germany.

Marienburg (MAH-ree-uhn-berg), town, COMMEWIJNE district, N SURINAME, on COMMEWIJNE RIVER, and 7 mi/11.3 km ENE of PARAMARIBO; 05°51′N 55°06′W. Sugar-milling center.

Marienburg, LATVIA: see ALŪKSNE.

Marienburg, POLAND: see MALBORK.

Marienburg, ROMANIA: see FELDIOARA.

Marienhausen, LATVIA: see VILAKA.

Marienheide (mah-REE-en-hei-de), town, North Rhine-Westphalia, W GERMANY, 6 mi/9.7 km N of

GUMMERSBACH; 51°05′N 07°32′E. Lead mining nearby. Castle here.

Marienmünster (mah-REE-en-myoon-ster), town, WESTPHALIA-LIPPE, North Rhine-Westphalia, W GERMANY, on W foot of the Weserbergland, 15 mi/24 km ESE of DETMOLD; 51°50′N 09°13′E. Health resort; manufacturing of textiles; woodworking; livestock. Has 12th-century church (rebuilt 1661 in baroque style). Benedictine monastery of Marienmünster was founded in 1128 (existed until 1803). Town was chartered in 1582; incorporated 1970 the towns Bredenborn and Vörden and eleven villages.

Mariensberg, village, EAST SEPIK province, N central NEW GUINEA island, NW PAPUA NEW GUINEA, on N bank of SEPIK River, 8 mi/13 km S of PACIFIC OCEAN coast, and 50 mi/80 km SE of WEWAK. Bananas, coconuts, copra, sago; fish.

Marienshöhe, SWITZERLAND: see SEELISBERG.

Marienstein, GERMANY: see WAAKIRCHEN.

Mariental (mah-REE-en-tahl), town, HARDAP REGION, S central NAMIBIA, 160 mi/257 km S of WINDHOEK; 24°38′S 17°58′E. Judicial and economic (distribution) center of region. In sheep-raising area; irrigation farming from Hardap Dam (largest dam in Namibia, capacity 405 million cu yd/310 million cu m). Airfield. Large Iceland-spar deposits nearby. Namibia's major ostrich-raising center.

Mariental, RUSSIA: see SOVETSKOYE, SARATOV oblast.

Marienville (MER-ee-ahn-vil), unincorporated town, FOREST county, NW PENNSYLVANIA, 28 mi/45 km NE of CLARION; 41°28′N 79°07′W. Allegheny National Forest immediately to N; Clear Creek State Forest to S.

Marienwerder, POLAND: see KWIDZYŃ.

Marie Reine (mah-REE REN), hamlet, central ALBERTA, W CANADA, 6 mi/10 km W of NAMPA, in NORTHERN SUNRISE county; 56°04′N 117°17′W.

Maries (MER-reez), county (□ 526 sq mi/1,367.6 sq km; 2006 population 9,099), central MISSOURI, in the OZARK MOUNTAINS; ⊙ VIENNA; 38°10′N 91°55′W. Drained by GASCONADE RIVER. Agriculture (wheat, corn) and livestock (cattle) region; timber, charcoal. Rolla Airport and weather station at Vichy in SE. Formed 1855.

Mariestad (mahr-EE-e-STAHD), town, ⊙ SKARABORG county, S SWEDEN, on LAKE VÄNERN; 58°42′N 13°50′E. Commercial and industrial center; manufacturing (paper, tissues, packaging, metal goods). Rebuilt after 1895 fire. Chartered 1583.

Marietta (mar-ee-ET-uh), city (2000 population 58,748), ⊙ COBB county, NW GEORGIA; 33°57′N 84°32′W. A growing and important suburb of Atlanta, its industry is largely involved in the production of aircraft. Other manufacturing includes building materials, plastics, food and beverages, consumer goods, textiles, chemicals, hardware and security equipment; poultry processing, marble and granite production; printing and publishing. Near KENNESAW MOUNTAIN, Marietta was the scene of a Union defeat in the Civil War; Kennesaw Mountain National Battlefield Park marks the site. Many Civil War dead are buried in the city's large national cemetery. Kennesaw House played a role in the Civil War. The Zion Baptist Church, built in 1888, is an important African-American institution. Seat of Kennesaw and Southern Polytechnic universities. Dobbins Air Force Base is nearby. Incorporated 1834.

Marietta (MER-ee-et-ah), city (□ 9 sq mi/23.4 sq km; 2006 population 14,189), ⊙ WASHINGTON county, SE OHIO, at the confluence of the MUSKINGUM and OHIO rivers, 90 mi/144 km SE of COLUMBUS; 39°25′N 81°26′W. Trading center for an agricultural and dairying area. Manufacturing. The first planned, permanent settlement in Ohio and the NORTHWEST TERRITORY. Founded in 1788 by the Ohio Company of Associates, among local Mound Builders' earthworks, Marietta grew as a shipbuilding and shipping center for a farm area. First houses were in stockaded enclosure called Campus Martius. Seat of Marietta College. Points of interest include the Ohio River Museum (established 1972); Mound Cemetery, where numerous Revolutionary officers are buried; and the Campus Martius Memorial State Museum. Incorporated 1801.

Marietta (mar-ee-ET-uh), town (2006 population 2,551), ⊙ LOVE county, S OKLAHOMA, 16 mi/26 km S of ARDMORE; 33°55′N 97°07′W. Elevation 843 ft/257 m. In diversified agricultural area (peanuts; sheep, hogs); manufacturing (machinery, apparel, foods); gas and oil. Lake Murray State Park and Lodge is 10 mi/16 km N; headwaters of Lake TEXOMA on Red River to E. Founded c.1887.

Marietta (mar-ee-ET-uh), village (2000 population 150), FULTON county, W central ILLINOIS, 15 mi/24 km NW of Lewiston; 40°30′N 90°23′W. In agricultural and bituminous-coal area.

Marietta, village (2000 population 174), LAC QUI PARLE county, SW MINNESOTA, near SOUTH DAKOTA state line, 11 mi/18 km W of MADISON; 45°00′N 96°25′W. Grain; livestock; dairying.

Marietta, village (2000 population 248), PRENTISS county, NE MISSISSIPPI, 20 mi/32 km NE of TUPELO, near East Fork TOMBIGBEE River; 34°30′N 88°28′W. John Bell Williams Wildlife Management Area to SE. NATCHEZ TRACE PARKWAY passes to SE. Manufacturing (furniture, apparel).

Marietta (mar-ee-E-tuh), village (□ 1 sq mi/2.6 sq km; 2006 population 170), ROBESON county, SE NORTH CAROLINA, 18 mi/29 km SSW of LUMBERTON, near SOUTH CAROLINA state line; 34°22′N 79°07′W. Retail trade; manufacturing (air filters); agriculture (grain, sunflowers, tobacco; livestock). Lumber River State Park to NE.

Marietta (mar-ee-ET-uh), unincorporated village, GREENVILLE county, NW SOUTH CAROLINA, and 13 mi/21 km NNW of GREENVILLE, on N SALUDA RIVER. Manufacturing includes fabric finishing, synthetic textiles, wood products. Agriculture includes timber; livestock, poultry; grain.

Marietta (mer-ee-E-tah), borough (2006 population 2,599), LANCASTER county, SE PENNSYLVANIA, 13 mi/21 km W of MARIETTA, on SUSQUEHANNA RIVER; 40°03′N 76°32′W. Agr. (grain, soybeans, potatoes; livestock, poultry; dairying); manufacturing (bldg. materials, fabricated metal products, pharmaceuticals). ELIZABETHTOWN-Marietta Airport to N. Marietta Ordnance Depot to W. Settled 1718, incorporated 1812.

Marieville (MAH-ree-vil, mah-ree-VEEL), town (□ 25 sq mi/65 sq km), ⊙ ROUVILLE county, MONTÉRÉGIE region, S QUEBEC, E CANADA, near RICHELIEU River, 20 mi/32 km ESE of MONTREAL; 45°26′N 73°10′W. Manufacturing, woodworking; dairying.

Marigaon (mah-ree-goun), district (□ 2,137 sq mi/5,556.2 sq km), ASSAM state, NE INDIA; ⊙ MARIGAON.

Marigaon (mah-ree-goun), town, district, ASSAM state, NE INDIA.

Marigat (mah-ree-GAHT), town, Baringo district, RIFT VALLEY province, KENYA, W of LAIKIPIA ESCARPMENT, on road, 53 mi/85 km NNW of NAKURU; 00°29′N 35°59′E. Market center.

Marigliano (mah-reel-YAH-no), town, NAPOLI province, CAMPANIA, S ITALY, 11 mi/18 km ENE of NAPLES; 40°56′N 14°27′E. In grape- and vegetable-growing region; pasta manufacturing, light manufacturing (machinery). Has mineral waters.

Marignane (mahr-ee-NYAHN), town (□ 8 sq mi/20.8 sq km), BOUCHES-DU-RHÔNE department, PROVENCE-ALPES-CÔTE D'AZUR region, SE FRANCE, near E shore of Étang de BERRE (a lagoon), 12 mi/19 km NW of MARSEILLE, on Marseille-Rhône Canal; 43°25′N 05°13′E. Site of Marseille-Provence international airport. Aeronautical workshops. Has an industrial park connected by railroad and expressway with Marseille.

Marignier (mah-ree-NYE), town (□ 7 sq mi/18.2 sq km), HAUTE-SAVOIE department, RHÔNE-ALPES region, SE FRANCE, on the GIFFRE RIVER near its influx into ARVE RIVER, and 5 mi/8 km ENE of BONNEVILLE, in FAUCIGNY valley; 46°06′N 06°31′E. Hydroelectric plant and electrochemical works.

Marigny (mah-ree-NYEE), commune (□ 4 sq mi/10.4 sq km; 2004 population 1,983), MANCHE department, BASSE-NORMANDIE region, NW FRANCE, 7 mi/11.3 km WSW of SAINT-LÔ; 49°06′N 01°14′W. Dairying. Here Americans broke through German defenses in Saint-Lô offensive (July 1944) of World War II. German military cemetery.

Marigot (mah-ree-GO), agricultural town, SUD-EST department, S HAITI, on the CARIBBEAN SEA, 18 mi/29 km E of JACMEL; 18°14′N 72°19′W. Coffee; fruit. Fishing port.

Marigot (mah-ree-GO), town, NE MARTINIQUE, French WEST INDIES, on the ATLANTIC OCEAN, 14 mi/23 km NE of FORT-DE-FRANCE; 14°49′N 61°02′W. Rum distilling.

Marigot (MA-ree-guht), village, NE DOMINICA, BRITISH WEST INDIES, 18 mi/29 km NNE of ROSEAU; 15°32′N 61°18′W. Coconuts; limes. At the Carib Reserve, 4 mi/6.4 km S, a few survivors of aboriginal inhabitants live. Nearby (W) is Melville Hall Airport.

Marigot (mah-ree-GO), village, SAINT MARTIN island, French WEST INDIES, on W coast of island, 160 mi/257 km NW of BASSE-TERRE (Guadeloupe); 18°04′N 63°05′W. Principal settlement of the French section of the island, with a good harbor. Produces some sugarcane, tropical fruit, cotton, and cattle for local use.

Marigot Bay (mahr-ee-GO), W coast of SAINT LUCIA, 4 mi/6.4 km S of CASTRIES; 13°58′N 61°01′W. One of the most beautiful and protected yachting anchorages in CARIBBEAN SEA. Hotel and yacht chartering facilities. Also known as Hurricane Hole.

Mariguana, BAHAMA Islands: see BAHAMAS

Marigüita, VENEZUELA: see MARIGÜITAR.

Marigüitar (mahr-ee-GWEE-tahr), town, ⊙ Bolívar municipio, SUCRE state, NE VENEZUELA, on Gulf of CARIACO, 18 mi/29 km E of CUMANÁ; 10°27′N 63°53′W. Coconuts, cacao, sugarcane, vegetables. Also known as Marigüita.

Marihovo (MAH-ree-ho-vo), region in MACEDONIA, extending c.25 mi/40 km SW-NE, between the CRNA REKA and Greek border. Chief village, ROZDEN. Also called Marikhovo, Mariovo, Marijovo, or Mariyovo.

Mari Indus, PAKISTAN: see KALABAGH.

Mariinsk (mah-ree-EENSK), city (2005 population 43,190), NE KEMEROVO oblast, S central SIBERIA, RUSSIA, on the KIYA river (OB' RIVER basin), on road and the TRANS-SIBERIAN RAILROAD, 227 mi/365 km NE of KEMEROVO; 56°13′N 87°45′E. Elevation 475 ft/144 m. Supply point for the upper KIYA River gold mines; woodworking, knitting; food industries (creamery, meat preserving, distillery). Founded in 1698 as a village of Kiyskoye; chartered in 1856; current name since 1857; railroad station since 1898.

Mariinskiy Posad (mah-ree-EEN-skeeye puh-SAHT), city (2005 population 10,180), N CHUVASH REPUBLIC, central European Russia, port on the Kuybyshev Reservoir on the VOLGA River, 30 mi/48 km E of CHEBOKSARY; 56°06′N 47°43′E. Elevation 219 ft/66 m. Terminus of a highway branch. Agricultural supplies, cable products, machinery, auto repair; distillery; logging, lumbering, timbering. Chartered in 1856.

Mariinskoye, KAZAKHSTAN: see MARYEVKA.

Marijampolė (mah-ri-yahm-PO-lai), Polish *Mariampol*, city (2001 population 48,675), S LITHUANIA, on the SHESHUPE RIVER, and 32 mi/51 km SW of KAUNAS, on the VIA BALTICA HIGHWAY from TALLINN (ESTONIA) to WARSAW (POLAND); 54°34′N 23°21′E. Road center. Major agricultural market; industries (sugar refining, flour milling, brewing, sawmilling); manufacturing (woolens, cotton goods, leather, automotive parts, furniture). Passed 1795 to PRUSSIA,

1815 to Russian Poland; in Suvalki government until 1920. Known during Soviet occupation as Kapsukas.

Marijovo, MACEDONIA: see MARIHOVO.

Marikhovo, MACEDONIA: see MARIHOVO.

Marikina (mah-ree-KEE-nah), town (2000 population 391,170), NATIONAL CAPITAL REGION, S LUZON, PHILIPPINES, 8 mi/12.9 km ENE of MANILA; 14°38′N 121°06′E. Agricultural center (rice, sugarcane, fruit); shoemaking. Part of the metropolitan Manila area.

Marikostinovo (mah-REEK-kos-to-no-vo), village, SOFIA oblast, PETRICH obshtina, BULGARIA; 41°25′N 23°26′E.

Marikuppam, India: see KOLAR GOLD FIELDS.

Marilac (MAH-ree-lahk), town (2007 population 4,278), E central MINAS GERAIS state, BRAZIL, 56 mi/90 km NNW of GOVERNADOR VALADARES; 18°28′S 42°15′W.

Marilândia (MAH-ree-LAHN-zhee-ah), town (2007 population 10,226), central ESPÍRITO SANTO state, BRAZIL, in RIO DOCE Valley, 16 mi/26 km NE of COLATINA; 19°28′S 40°30′W.

Marilândia do Sul (MAH-ree-lahn-zhee do SOOL), town (2007 population 8,932), N PARANÁ state, BRAZIL, 44 mi/71 km SE of MARINGÁ, on railroad; 23°45′S 51°19′W. Coffee, cotton, rice, corn; livestock. Also called Araruva.

Marilena (MAH-ree-LE-nah), town (2007 population 6,541), far NW PARANÁ state, BRAZIL, near junction of Parapanema and PARANÁ rivers; 22°43′S 53°03′W. Coffee, cotton, corn, rice; livestock.

Marília (MAHR-ree-lee-ah), city (2007 population 218,113), W central SÃO PAULO state, BRAZIL, on railroad, 55 mi/89 km W of BAURU; 22°13′S 49°56′W. Rich agricultural district along watershed between AGUAPEÍ and PEIXE rivers (leading cotton-growing center; also coffee, corn, rice, fruit; dairy products). Settled by pioneer farmers since 1920s; also trades in livestock (hogs and beef cattle). Cotton ginning, cottonseed-oil extracting, sawmilling, furniture manufacturing

Mariluz (MAH-ree-loos), city (2007 population 10,482), W PARANÁ state, BRAZIL, 55 mi/89 km W of CAMPO MOURÃO; 24°02′S 53°13′W. Coffee.

Marimba (mah-REEM-buh), town, MALANJE province, N ANGOLA, near Democratic Republic of the CONGO border, 100 mi/161 km NNE of MALANJE. Manioc, cotton.

Marimbondo (MAH-reem-BON-do), city, E central ALAGOAS state, BRAZIL, in Serra da Pedra Talhada; 09°37′S 36°16′W.

Marimbondo Falls, Brazil: see MARIBONDO FALLS.

Marin, county (□ 520 sq mi/1,352 sq km; 2006 population 248,742), W CALIFORNIA, ⊙ SAN RAFAEL; 38°02′N 122°45′W. Contains many residential suburbs of SAN FRANCISCO, linked by GOLDEN GATE BRIDGE to SAN FRANCISCO county (S) and by RICHMOND–SAN RAFAEL BRIDGE to CONTRA COSTA county (E). Wooded, hilly Marin Peninsula reaches S to the GOLDEN GATE strait, and between San Pablo and San Francisco bays (E) and Pacific Ocean (W); Pacific coast is indented by Bodega, Tomales, Drake's, and Bolinas bays. Petaluma River forms NE border. Dairying; poultry and livestock raising (cattle, sheep), farming (oats, nuts, fruit); fishing (oysters, clams, mussels); nurseries. Stone, sand, gravel, clay quarrying; mercury. Hamilton Air Force Base (now closed) in E. Mount Tamalpais State Park and MUIR WOODS NATIONAL MONUMENT are in S, W of Mill Valley. Includes Angel Island State Park, largest in San Francisco Bay; POINT REYES NATIONAL SEASHORE in SW; part of GOLDEN GATE NATIONAL RECREATION AREA in S. State prison at San Quentin. Formed 1850.

Marín, city, PONTEVEDRA province, NW SPAIN; Atlantic seaport on PONTEVEDRA bay, 4 mi/6.4 km SW of PONTEVEDRA. Exports fish, lumber, salt. Fishing; shipbuilding. Naval installations.

Marin (mah-RANG), town, SE MARTINIQUE, French WEST INDIES, minor port on bay, 16 mi/26 km SE of FORT-DE-FRANCE; 14°28′N 60°52′W. Trading and processing (alcohol), in agricultural region (cacao). Sometimes called Le Marin.

Marín, town, NUEVO LEÓN, N MEXICO, 22 mi/35 km NE of MONTERREY; 25°55′N 100°00′W. Cereals, cactus fibers; livestock.

Marín (mah-REEN), town, YARACUY state, N VENE- ZUELA, 5 mi/8 km NE of SAN FELIPE; 10°23′N 68°40′W. Cacao, corn, sugar, fruit; livestock.

Marina, city (2000 population 25,101), MONTEREY county, W CALIFORNIA, 8 mi/12.9 km W of SALINAS, on MONTEREY BAY; 36°41′N 121°48′W. Agriculture (artichokes, vegetables, fruit); dairying; cattle. Manufacturing (printing and publishing). Fort Ord military reservation to S.

Marina del Rey (mah-REE-nah del RAI), unincorporated town (2000 population 8,176), LOS ANGELES county, S CALIFORNIA; residential suburb 11 mi/18 km WSW of downtown LOS ANGELES (borders it on all sides except NE), 3 mi/4.8 km SE of SANTA MONICA, near Pacific Ocean; 33°59′N 118°27′W. Manufacturing (electronic equipment, machinery); service industry. Marina del Rey boat harbor extends inland here from coast. Hughes Airport here; LOS ANGELES INTERNA- TIONAL AIRPORT to S.

Marina di Besca, CROATIA: see BAŠKA.

Marina di Carrara (mah-REE-nah dee kah-RAH-rah), town, MASSA CARRARA province, TUSCANY, N ITALY, port on LIGURIAN Sea, 4 mi/6 km SW of CARRARA; 44°02′N 10°02′E. Swimming resort. Exports marble.

Marina di Pisa (mah-REE-nah dee PEE-sah), town, PISA province, TUSCANY, central ITALY, port on LI- GURIAN Sea, at mouth of the ARNO River, 7 mi/11 km SW of PISA; 43°40′N 10°16′E. Swimming resort. Sur- rounding dense pinewoods yield pignoli.

Marina di Ravenna (mah-REE-nah dee rah-VEN- nah), village, RAVENNA province, EMILIA-ROMAGNA, N central ITALY, on the ADRIATIC Sea, at mouth of LAMONE River, 6 mi/10 km NE of RAVENNA (con- nected by canal); 44°28′N 12°17′E. Swimming resort. Across the river is Porto Corsini, port of Ravenna.

Marina Falls, on Ipobe Creek (affluent of POTARO RIVER), central GUYANA, and 15 mi/24 km NW of KAIETEUR FALLS; 05°22′N 59°29′W. Drops c.500 ft/152 m in several leaps.

Marina Hemingway, recreational marine facility, at W edge of HAVANA, CIUDAD DE LA HABANA province, W CUBA; 23°05′N 82°31′W. An important sports-fishing and leisure-boating facility. Hosts annual spring fishing. Small craft can moor here for 72 hours without immigration visa or tourist card.

Marinaleda (mah-ree-nah-LAI-dah), town, SEVILLE province, SW SPAIN, 12 mi/19 km NE of OSUNA; 37°22′N 04°58′W. Cereals, vegetables, olives, acorns; livestock.

Marincho, Arroyo (mah-REEN-cho, ah-ROI-o), river, 37 mi/60 km long, FLORES department, SW URUGUAY; rises in the CUCHILLA GRANDE INFERIOR 7 mi/11.3 km W of TRINIDAD; flows N to YÍ River 6 mi/9.7 km above confluence with the RÍO NEGRO; 33°11′S 57°04′W.

Marinduque (mah-ree-DOO-ke), province (☐ 370 sq mi/962 sq km), in SOUTHERN TAGALOG region, the PHILIPPINES, between MINDORO and S LUZON; ⊙ BOAC; 13°25′N 121°57′E. Includes Marinduque Island and twenty-four other islands (seventeen of which are named). Population 16.3% urban, 83.7% rural; in 1991, all urban and 70% of rural settlements had electricity. In typhoon belt. Major industry is mining (gold, iron, and copper). Agriculture includes coconuts as cash crops and rice for food. Cattle; fishing; aquaculture. Airport at Boac; largest town and market center is SANTA CRUZ. Ports at Laylay and Balanacan. Marin- duque Island is volcanic, with coastal plain; several resorts.

Marine, village (2000 population 910), MADISON county, SW ILLINOIS, 10 mi/16 km E of EDWARDS- VILLE; 38°47′N 89°46′W. In agricultural area (wheat; dairy products; poultry; cattle.

Marine, MINNESOTA: see MARINE ON SAINT CROIX.

Marine City, town and port (2000 population 4,652), SAINT CLAIR county, E MICHIGAN, 18 mi/29 km S of PORT HURON, and on SAINT CLAIR RIVER at mouth of BELLE RIVER; 42°42′N 82°30′W. Manufacturing (transportation equipment; machining); grain eleva- tor. Salt mines. Ferry to Sombra (ONTARIO). In- corporated as a village 1865; named Marine City 1867; incorporated as city 1887.

Marinella (mah-ree-NEL-lah), village, TRAPANI prov- ince, W SICILY, ITALY, port on MEDITERRANEAN Sea, 5 mi/8 km SE of CAMPOBELLO DI MAZARA. Tunny fishing. Ruins of Selinunte (ancient *Selinus*), west- ernmost Greek settlement of island (founded 628 B.C.E.; destroyed by Carthaginians 409 B.C.E.) just W. Remains of eleven temples, several necropolises. Also called Marinella Selinunte.

Marine on Saint Croix (KROI), village, WASHINGTON county, E MINNESOTA, 23 mi/37 km N of downtown ST. PAUL, and 10 mi/16 km N of STILLWATER, on ST. CROIX River; 45°22′N 92°45′W. One of first European settlements in Minnesota. William O'Brian State Park to N; LOWER ST. CROIX NATIONAL SCENIC RIVERWAY on St. Croix River; numerous small lakes in area, es- pecially Big Marine Lake to NW. Also known as Marine.

Marin-Epagneir (MAH-ra–e-pah-NYAI), commune, NEUCHÂTEL canton, W SWITZERLAND, on LAKE NEU- CHÂTEL, and 4 mi/6.4 km NE of NEUCHÂTEL.

Marine Parade, town, SE Singapore island, SINGAPORE, residential suburb 4 mi/6.4 km ENE of downtown SINGAPORE, on Straits of SINGAPORE, on East Coast Parkway; 01°18′N 103°55′E. Changi International Air- port to NE.

Mariners' Harbor, neighborhood, in borough of STA- TEN ISLAND, N Staten Island on W side of Willow- brook Expressway, NEW YORK city, SE NEW YORK state; 40°38′N 74°09′W. Formerly a prosperous community whose economy was based on oyster fishing, little evidence now remains of the fine Classic Revival homes of Captain's Row. It became an in- dustrial, warehousing, and transshipment center owing to its location on S end of NEWARK BAY–KILL VAN KULL. In 1972, the first Mitchell-Lama housing subsidy project, known as North Shore Plaza, was planned for Staten Island in an attempt to revive the economically depressed North Shore communities. Goethals and Bayonne bridges to New Jersey are nearby.

Marines (mah-REEN), commune (☐ 3 sq mi/7.8 sq km), VAL-D'OISE department, ÎLE-DE-FRANCE region, N central FRANCE, 8 mi/12.9 km NW of PONTOISE; 49°09′N 01°59′E. Manufacturing of porcelain ware. Has 16th-century church.

Marines, Los (mah-REE-nes, los), village, HUELVA province, SW SPAIN, in the SIERRA MORENA, 3 mi/4.8 km WNW of ARACENA; 37°54′N 06°36′W. Chestnuts, olives, cork, fruit; timber; hogs, goats.

Marinette, county (☐ 1,550 sq mi/4,030 sq km; 2006 population 43,208), NE WISCONSIN, ⊙ MARINETTE; 45°20′N 88°00′W. Lumbering; wheat, corn, beans. Manufacturing at Marinette. Lake resorts. Bounded E by MENOMINEE RIVER (MICHIGAN state line and bor- der between Central and Eastern time zones), SE by GREEN BAY. Wooded region, drained by PESHTIGO RIVER. Thunder River State Fish Hatchery and Win- terset Ski Area in W; Mount LeBett Ski Area in SW; Nicolet National Forest immediately beyond W bor- der. Formed 1879.

Marinette (2006 population 11,009), ⊙ MARINETTE county, NE WISCONSIN, 43 mi/69 km NNE of GREEN BAY city, on GREEN BAY of LAKE MICHIGAN, at the mouth of the MENOMINEE RIVER; 45°05′N 87°37′W. Manufacturing (consumer goods, transportation equipment, wood products, paper products, tanks and vessels, chemicals). Railroad junction. A port of entry, it is the center of a tri-city area embracing

PESHTIGO and MENOMINEE (MICHIGAN). Fur trading began here c.1795 and gave way to lumbering, which flourished until the 1930s. The city was named for a Menominee Native American queen, who established a trading post on the river and built the first frame house here. Incorporated 1887.

Maringá (MAH-reen-GAH), city (2007 population 325,968), PARANÁ state, SE BRAZIL, 51 mi/82 km WSW of LONDRINA; 23°25′S 51°55′W. Agricultural center (coffee, maize, beans, rice, wheat, and sugarcane); main industry is coffee processing. Much of popula- tion of Japanese descent.

Maringa River (mah-RING-gah), c.325 mi/523 km long, W CONGO; rises 70 mi/113 km SE of DJOLU; flows W and NW, past BEFORI, MOMPONO, BEFALE, and WAKA, to join LOPORI RIVER at BASANKUSU, forming the LULONGA RIVER. Navigable for 250 mi/402 km below BEFORI.

Maringouin (MER-in-gwin), town (2000 population 1,262), IBERVILLE parish, SE central LOUISIANA, 19 mi/ 31 km W of BATON ROUGE, on Bayou Maringouin; 30°30′N 91°31′W. Manufacturing (lumber). Atch- afalaya National Wildlife Refuge to W.

Maringue District, MOZAMBIQUE: see SOFALA.

Maringues (mah-RANG), commune (☐ 8 sq mi/20.8 sq km), PUY-DE-DÔME department, in AUVERGNE, cen- tral FRANCE, in the LIMAGNE lowland, 10 mi/16 km ENE of RIOM; 45°55′N 03°20′E. Wheat, corn, and poultry market; flour milling. Has a long history as a tanning center; a revival is being fostered under a development plan for the regional park of LIVRADOIS- FOREZ (S).

Marinha Grande (mah-REEN-yah GRAHND), town, LEIRIA district, W central PORTUGAL, on railroad, and 7 mi/11.3 km W of LEIRIA; 39°45′N 08°56′W. Amidst pine forest. Former famed (since mid-18th century) glass- and crystal-manufacturing center; factory re- cently closed by government.

Marinilla (mah-ree-NEE-yah), town, ⊙ Marinilla mu- nicipio, ANTIOQUIA department, NW central CO- LOMBIA, in Cordillera CENTRAL, 17 mi/27 km ESE of MEDELLÍN; 06°10′N 75°20′W. Elevation 6,961 ft/2,122 m. Potatoes, coffee; livestock. Musical instruments. Old colonial town with typical Antioquia architecture.

Mar'inka, UKRAINE: see MAR'YINKA.

Marinkovitsa (MAH-reen-ko-VEE-tsah), peak (8,294 ft/2,528 m), in the RILA MOUNTAINS, SW BULGARIA; 42°08′N 23°32′E. Formerly called Marina Vala.

Mar'ino (MAHR-yee-nuh), village, central KURSK ob- last, SW European Russia, on road, 16 mi/26 km E of OBOYAN; 51°35′N 34°56′E. Elevation 574 ft/174 m. Sugar beets. Rzhava village lies 5 mi/8 km N; sugar refinery.

Marino, RUSSIA: see LENINSKIY, MARI EL Republic.

Marinópolis (MAH-ree-o-po-lees), town (2007 popu- lation 2,114), NW SÃO PAULO state, BRAZIL, 19 mi/31 km SW of JALES; 20°02′S 50°50′W. Coffee growing.

Marinuka, Lake, Wisconsin: see GALESVILLE.

Marinwood, unincorporated town, MARIN county, W CALIFORNIA; residential suburb 17 mi/27 km NNW of downtown SAN FRANCISCO, 2 mi/3.2 km W of San Pablo Bay, on Miller Creek. Lucas Valley to W. Ha- milton Air Force Base (now closed) to NE. Big Rock Ridge to N. Statistically reported as Lucas Valley– Marinwood.

Mario Dary Biotope (MAH-ree-o DAH-ree) (☐ 4 sq mi/10.4 sq km), BAJA VERAPAZ department, GUATE- MALA, 10 mi/16 km NW of SALAMÁ. The only preserve in Guatemala for the resplendent quetzal, a bird that is the country's endangered national symbol. Also preserves cloud forest environments. Open to the public.

Marion (MAR-ee-uhn), county (☐ 743 sq mi/1,931.8 sq km; 2006 population 30,165), NW ALABAMA; ⊙ Hamil- ton; 34°08′N 87°52′W. Agr. area bordering on Mississippi (W), drained by Buttahatchee River, crossed (N-S) by fall line. Corn, soybeans, wheat;

poultry. Coal mines; lumber; textiles. Formed 1818. Originally known as 'Muckle's Ridge' for Michael Muckle, an early settler, it was renamed for Francis Marion, a Revolutionary War general. Inc. in 1835.

Marion, county (□ 640 sq mi/1,664 sq km; 2006 population 16,931), N ARKANSAS, in the OZARK region; ⊙ YELLVILLE; 36°15′N 92°40′W. Bounded N by MISSOURI state line; intersected by WHITE RIVER (site of Bull Shoals Dam in NE), drained by BUFFALO RIVER and small Crooked Creek. Agriculture (cattle, hogs, turkeys). Some manufacturing at Yellville. Lead, zinc mines; timber. Bull Shoals State Park at dam, large BULL SHOALS LAKE dominates N quarter of county; part of BUFFALO NATIONAL RIVER in SE. Formed 1835.

Marion (MER-ee-uhn), county (□ 1,663 sq mi/4,323.8 sq km; 2006 population 316,183), N central FLORIDA; ⊙ OCALA; 29°12′N 82°03′W. Flatwoods area with scattered lakes, including Lakes Weir and Kerr; drained by OKLAWAHA RIVER. Ocala National Forest occupies E part. Agriculture (citrus fruit, vegetables, corn, peanuts); livestock raising (cattle, hogs); forestry (lumber, naval stores); and quarrying (limestone, phosphate). Formed 1844.

Marion, county (□ 368 sq mi/956.8 sq km; 2006 population 7,276), W GEORGIA; ⊙ BUENA VISTA; 32°21′N 84°32′W. Drained by KINCHAFOONEE RIVER Coastal plain agriculture (soybeans, wheat, corn, peanuts); cattle, poultry, hogs; timber. Formed 1827.

Marion, county (□ 575 sq mi/1,495 sq km; 2006 population 40,088), S central ILLINOIS; ⊙ SALEM; 38°39′N 88°56′W. Largest town is CENTRALIA in SW corner. Oil-producing, and gas area. Agriculture (wheat, soybeans, corn, sorghum; cattle, poultry; dairying). Some manufacturing (metal products, plastics products, machinery, transportation equipment, electronic equipment, food products, clothing). Drained by SKILLET FORK, CROOKED CREEK, and East Fork of KASKASKIA RIVER. Formed 1823. Stephen A. Forbes State Park in NE.

Marion, county (□ 403 sq mi/1,047.8 sq km; 2006 population 865,504), central INDIANA; ⊙ INDIANAPOLIS; 39°47′N 86°08′W. Transportation (railroad, highway), commercial, market, political, and manufacturing center at Indianapolis. Drained by West Fork of WHITE RIVER and small Eagle, Fall, and Buck creeks. Largely urbanized; some farming (corn, vegetables, soybeans); livestock raising (cattle, hogs); dairying. Eagle Creek (NW) and GEIST (NE) reservoirs. Indianapolis International Airport in W. Fort Benjamin Harrison at LAWRENCE in NE. On January 1, 1970, Indianapolis consolidated with county (except for municipalities of Lawrence, SPEEDWAY, SOUTHPORT, and BEECH GROVE). Indianapolis 500 Speedway (auto race track) at Speedway in W. Formed 1822.

Marion, county (□ 570 sq mi/1,482 sq km; 2006 population 32,987), S central IOWA; ⊙ KNOXVILLE; 41°19′N 93°05′W. Rolling prairie agricultural area (hogs, cattle, poultry, sheep; corn) drained by SKUNK and DES MOINES rivers and by WHITEBREAST CREEK. Many bituminous-coal mines, some limestone quarries. Sunset Ski Area in NE, W of PELLA; large Red Rock Reservoir (Des Moines River) NE of Knoxville where there is a Corps of Engineers recreational area and Elk Rock State Park on N shore. Widespread flooding along rivers in 1993. Formed 1845.

Marion, county (□ 953 sq mi/2,477.8 sq km; 2006 population 12,760), E central KANSAS; ⊙ MARION; 38°24′N 97°09′W. Gently rolling to hilly area, drained by COTTONWOOD RIVER. Wheat, sorghum, apples; livestock (cattle), poultry. Industrial machinery; food processing; marble and granite. Formed 1860.

Marion, county (□ 346 sq mi/899.6 sq km; 2006 population 18,979), central KENTUCKY; ⊙ LEBANON; 37°33′N 85°16′W. Drained by ROLLING FORK and BEECH FORK rivers. Rolling upland agricultural area, partly in SW part of BLUEGRASS REGION (burley tobacco, corn, hay, alfalfa, soybeans; wheat; hogs, cattle,

poultry; dairying); timber; stone quarries. Some manufacturing at Lebanon. Formed 1834.

Marion, county (□ 548 sq mi/1,424.8 sq km; 2006 population 25,730), S MISSISSIPPI; ⊙ COLUMBIA; 31°13′N 89°49′W. Partly bounded S by LOUISIANA state line. Drained by PEARL RIVER. Agriculture(cotton, corn; cattle; dairying); timber. Lake Bill Waller and Lake Columbia, state lakes in E. Formed 1811.

Marion, county (□ 440 sq mi/1,144 sq km; 2006 population 28,425), NE MISSOURI; ⊙ PALMYRA; 39°49′N 91°36′W. Bounded E by MISSISSIPPI RIVER, drained by NORTH and SOUTH Fabius rivers. Corn, wheat, soybeans; dairying; hogs, cattle. Manufacturing at HANNIBAL and Palmyra; limestone. Formed 1826.

Marion (MER-ee-uhn), county (□ 405 sq mi/1,053 sq km; 2006 population 65,583), central OHIO; ⊙ MARION; 40°35′N 83°10′W. Intersected by SCIOTO RIVER; also drained by OLENTANGY and LITTLE SCIOTO rivers and small Tymochtee Creek. Agriculture (livestock; dairy products; grain); manufacturing at Marion; limestone quarries, sand and gravel pits. Formed 1823.

Marion, county (□ 1,195 sq mi/3,107 sq km; 2006 population 311,304), NW OREGON; ⊙ SALEM; 44°54′N 122°34′W. Bounded S by North SANTIAM River, forms DETROIT LAKE Reservoir in SE (Detroit Lake State Park). Bounded N, in part, by Butte Creek. Manufacturing (food processing; electronic components, wood products). Agriculture (fruit, peas, onions, potatoes, mint, wheat, barley; poultry); wineries. Ankeny National Wildlife Refuge in SW; Champoeg and Willamette Mission state parks in NW; Silver Falls State Park in center; North Santiam State Park in S; part of MOUNT HOOD and Willamette national forests in E, including parts of MOUNT JEFFERSON and Bull of the Woods Wilderness Area. CASCADE RANGE in E. Formed 1843.

Marion, county (□ 494 sq mi/1,284.4 sq km; 2006 population 34,684), E SOUTH CAROLINA; ⊙ MARION; 34°05′N 79°26′W. Bounded GREAT PEE DEE (W) and LITTLE PEE DEE (E) rivers, which join at SE tip of county. Manufacturing of sand and clay. Agricultural area (tobacco, corn, oats, sorghum, hay; hogs, cattle). Formed 1798.

Marion, county (□ 507 sq mi/1,318.2 sq km; 2006 population 27,942), SE TENNESSEE; ⊙ JASPER; 35°08′N 85°37′W. Partly in the CUMBERLAND MOUNTAINS; bounded S by ALABAMA and GEORGIA; drained by TENNESSEE and SEQUATCHIE rivers and small Little Sequatchie River. Includes Hales Bar Reservoir. Manufacturing. Formed 1817.

Marion (MA-ree-uhn), county (□ 420 sq mi/1,092 sq km; 2006 population 10,970), E Texas; ⊙ JEFFERSON; 32°48′N 94°21′W. Bounded E by LOUISIANA state line, S in part by BIG CYPRESS CREEK and LITTLE CYPRESS BAYOU; drained by Big Cypress Creek, includes part of CADDO LAKE (hunting, fishing). Timber, oil, natural-gas wells; gravel; lignite. Agriculture (peaches, pecans, vegetables, blueberries; horticulture; hay); cattle, horses, hogs, ratites (ostriches, emus, rheas). Most of Lake O' the Pines reservoir in W. Formed 1860.

Marion, county (□ 312 sq mi/811.2 sq km; 2006 population 56,706), N WEST VIRGINIA; ⊙ FAIRMONT; 39°30′N 80°14′W. On ALLEGHENY PLATEAU; TYGART and WEST FORK rivers join at Fairmont in SE to form MONONGAHELA RIVER. Coal mining declined rapidly in the 1990s; gas and oil fields. Aluminum and iron industries. Manufacturing at Fairmont. Agriculture (corn, alfalfa, hay, apples); cattle, poultry, sheep. Pricketts Fort State Park in NE; part of Valley Falls State Park in SE. Formed 1842.

Marion (MAR-ee-uhn), city (2000 population 3,511), ⊙ Perry co., W central Alabama, 23 mi/37 km NW of Selma, near Cahaba River Lumber; clothing, cheese; poultry; catfish processing. Seat of Judson College and Marion Military Inst. Settled 1817. Talladega National Forest is E and N. U.S. fish hatchery nearby. Named for Revolutionary War general Francis Marion of SC.

Marion, city (2000 population 16,035), ⊙ WILLIAMSON county, S ILLINOIS; 37°43′N 88°56′W. Commercial and retail center of a farm and coal area and has a large soft drink-bottling plant. A maximum-security Federal prison is 8 mi/12.8 km S, known as "the new Alcatraz." Robert Ingersoll and John A. Logan lived here. Incorporated 1841.

Marion, city (2000 population 31,320), ⊙ GRANT county, E central INDIANA, on the MISSISSINEWA RIVER; 40°33′N 85°40′W. Trade, processing, and industrial center in a farm area. Manufacturing (transportation equipment, glassware, paper products, electronic equipment, corn products, dry ice, fabricated metal products, machinery, wood products; printing). Developed with the discovery of gas and oil in the late 1880s. Seat of Indiana Wesleyan University (formerly Marion College); Taylor University is in nearby UPLAND. Settled 1826, incorporated 1889.

Marion, city (2000 population 26,294), LINN county, E central IOWA, adjoining CEDAR RAPIDS; 42°01′N 91°35′W. A chiefly residential city, home construction and concrete manufacturing are its main industries. Manufacturing also includes furniture, machinery, feeds and fertilizers, plastic products, electrical products. Flour and dairy products are also processed here. Airfield to E; Squaw Creek Regional Park to S. Incorporated 1865.

Marion, city (2000 population 3,196), ⊙ CRITTENDEN county, W KENTUCKY, 34 mi/55 km ENE of PADUCAH, on Crooked Creek; 37°19′N 88°04′W. In oak-timber and agricultural (corn, burley tobacco, oats, wheat; dairying) area. Manufacturing (industrial ceramics, plastic products, building materials, electrical equipment); limestone quarrying and crushing, lumber milling. County museum, Clement Mineral Museum. Free ferry crosses OHIO RIVER to CAVE IN ROCK (Illinois), 10 mi/16 km to NNW. Amish community to N. Incorporated 1844.

Marion (MER-ee-uhn), city (2006 population 36,138), ⊙ MARION county, central OHIO, 42 mi/67 km N of COLUMBUS; 40°35′N 83°07′W. Was an industrial center noted for its production of major machinery and consumer goods until 1970s. Known as "The Popcorn Capital of the World." Limestone quarries in the area. A branch of Ohio State University is here. Home of President Warren G. Harding; his house is preserved as a museum, and his burial place is marked by a circular marble monument. Incorporated 1830.

Marion (MAR-ee-uhn), town (2000 population 8,901), ⊙ CRITTENDEN county, E ARKANSAS, 12 mi/19 km WNW of MEMPHIS (TENNESSEE); 35°12′N 90°12′W. In cotton- and rice-growing area. Manufacturing (building materials, wood products).

Marion, town (2000 population 2,110), ⊙ MARION county, E central KANSAS, on COTTONWOOD RIVER, and 45 mi/72 km NNE of WICHITA; 38°21′N 97°01′W. Railroad junction. Shipping center for grain and livestock area; food processing. County fair takes place here annually in October. Damaged by flood of July 1951. Marion Lake Reservoir to NW. County museum Settled 1860, laid out 1866, incorporated 1875.

Marion, town, PLYMOUTH county, SE MASSACHUSETTS, on W shore of BUZZARDS BAY, and 10 mi/16 km NE of NEW BEDFORD; 41°42′N 70°45′W. Resort. Formerly shipbuilding. Settled 1679, set off from ROCHESTER 1852.

Marion, unincorporated town (2000 population 1,305), LAUDERDALE county, E MISSISSIPPI, residential suburb 5 mi/8 km NE of MERIDIAN; 32°25′N 88°39′W. Cotton, corn; cattle, hogs; light manufacturing

Marion (MER-ee-uhn), town (□ 3 sq mi/7.8 sq km; 2000 population 4,943), ⊙ MCDOWELL county, W NORTH CAROLINA, 31 mi/50 km ENE of ASHEVILLE, in the BLUE RIDGE MOUNTAINS, near CATAWBA RIVER, forms lake; 35°40′N 82°00′W. JAMES reservoir to NE; Lake James State Park to NE. Railroad junction. Manufacturing (medical supplies, lumber and paper products, electronic and transportation equipment, machinery, tools, apparel, textiles, furniture, food processing); timber; stone quarrying; service industries; agriculture (corn; poultry). Linville Caverns are N; Pisgah National Forest and Lake Tahoma reservoir to NW. Incorporated 1844.

Marion, unincorporated town, FRANKLIN county, S PENNSYLVANIA, 5 mi/8 km SSW of CHAMBERSBURG; 40°38′N 77°16′W. Light manufacturing. Agriculture includes dairying; livestock, poultry; grain, potatoes, apples.

Marion, town (2006 population 6,959), ⊙ MARION county, E SOUTH CAROLINA, 20 mi/32 km E of FLORENCE; 34°10′N 79°24′W. Trade center in agricultural area (livestock; grain, tobacco, cotton, sorghum). Manufacturing (transportation equipment, textiles, apparel, millwork, ceramic tile, foods).

Marion, town (2006 population 836), TURNER county, SE SOUTH DAKOTA, 7 mi/11.3 km WNW of PARKER; 43°25′N 97°15′W. Hospital and clinic for persons suffering with bone disorders. Manufacturing (feeds, machinery).

Marion (MA-ree-uhn), town (2000 population 1,209), GUADALUPE county, S central TEXAS, 24 mi/39 km NE of SAN ANTONIO, and on GUADALUPE RIVER; 29°34′N 98°08′W. In agricultural area (peanuts, cotton; poultry). Manufacturing (machinery).

Marion (MA-ree-yuhn), town (2006 population 6,130), ⊙ SMYTH county, SW VIRGINIA, 38 mi/61 km NE of BRISTOL, near South Fork of HOLSTON RIVER, between WALKER MOUNTAIN (N) and IRON MOUNTAINS (S); 36°50′N 81°30′W. Manufacturing (machining, printing and publishing; beverages, wood products, furniture, transportation equipment, consumer goods, apparel). Agricultural area (tobacco, grain; livestock; dairying); limestone quarrying. Hungry Mother State Park to N; MOUNT ROGERS, highest point in Virginia (5,729 ft/1,746 m) 12 mi/19 km to S, within Mount Rogers National Recreation Area (part of Jefferson National Forest). Art Deco Mayan Revival Lincoln Theatre. Incorporated 1832.

Marion, town (2006 population 1,242), WAUPACA and SHAWANO counties, E central WISCONSIN, on small Pigeon River, and c.40 mi/64 km WNW of GREEN BAY; 44°40′N 88°53′W. In timber, dairy, and grain area. Dairy products; manufacturing (wood products, foods, chemicals, transportation equipment). Settled 1878; incorporated as village in 1898, as city in 1939.

Marion (MER-ee-uhn), village (2006 population 770), UNION parish, N LOUISIANA, 29 mi/47 km NNW of MONROE; 32°54′N 92°14′W. In agricultural area (cotton); lumbering; sawmill; pulpwood. Upper Ouachita National Wildlife Refuge to E, Union State Wildlife Area to W.

Marion, village (2000 population 836), OSCEOLA county, central MICHIGAN, 16 mi/26 km SE of CADILLAC; 44°06′N 85°09′W. In farm area. Manufacturing (light manufacturing).

Marion, hamlet, WAYNE county, W NEW YORK, 20 mi/32 km E of ROCHESTER; 43°09′N 77°12′W. Part of town of Marion. In fruit-growing region.

Marion, village (2006 population 130), LA MOURE county, SE central NORTH DAKOTA, 18 mi/29 km N of LA MOURE; 46°36′N 98°19′W. Founded in 1900 as Elmo and the name was changed in 1902 to Marion after Marion Mellon, daughter of a railroad executive. Incorporated in 1911.

Marion Bay (ME-ryun BAI), resort village, SOUTH AUSTRALIA state, S central AUSTRALIA, 199 mi/320 km W of ADELAIDE, on S tip of YORKE PENINSULA; 35°14′S 137°01′E. Near Innes National Park.

Marion Center, borough (2006 population 425), INDIANA county, W central PENNSYLVANIA, 11 mi/18 km NNE of INDIANA; 40°46′N 79°02′W. Agricultural area (grain, soybeans; dairying); light manufacturing; subsurface bituminous coal.

Marion Heights, borough (2006 population 693), NORTHUMBERLAND county, E central PENNSYLVANIA, 1 mi/1.6 km N of KULPMONT; 40°47′N 76°28′W. Agriculture (corn, hay; poultry); anthracite coal.

Marion Island, small subantarctic island in S INDIAN OCEAN, c.1,200 mi/1,931 km SE of CAPE TOWN, just SW of PRINCE EDWARD ISLAND, with which it forms the Prince Edward Islands; 13 mi/21 km long, 8 mi/12.9 km wide; 46°51′S 37°52′E. Rises to 4,200 ft/1,280 m. Formally annexed by SOUTH AFRICA in December 1947. Has meteorological station. First mapped in 1878 by HMS *Challenger*.

Marion, Lake, reservoir (□ 156 sq mi/405.6 sq km), on border between BERKELEY (N) and CLARENDON and CALHOUN (S) counties, E central SOUTH CAROLINA, on SANTEE RIVER, 25 mi/40 km NNW of CHARLESTON; 33°28′N 80°09′W. Largest lake in South Carolina Max. capacity 1,230,000 acre-ft. Connected to Lake MOULTRIE on SE. Formed by SANTEE DAM (50 ft/15 m high), built (1942) for power generation; owned by South Carolina Public Service Authority. Santee National Wildlife Refuge on N shore, Santee State Park on W shore.

Marion, Mount (ME-ree-uhn), (9,750 ft/2,972 m), SE BRITISH COLUMBIA, W CANADA, in SELKIRK MOUNTAINS, 55 mi/89 km N of NELSON; 50°17′N 117°13′W.

Marion Station, village, SOMERSET county, SE MARYLAND, on the EASTERN SHORE, near BIG ANNEMESSEX RIVER, 25 mi/40 km SSW of SALISBURY. Named for a daughter of the developer, John C. Horsey, it was the self-proclaimed strawberry capital of the world until the 1920s when growing centers shifted to the West Coast.

Marionville, city (2000 population 2,113), LAWRENCE county, SW MISSOURI, in the OZARK MOUNTAINS, 24 mi/39 km SW of SPRINGFIELD; 37°00′N 93°38′W. Apples, berries, peaches, vegetables; cattle; dairying. Manufacturing (wood products, apparel, magnetic coils). Laid out 1854.

Marionville (mah-ree-on-VEEL), unincorporated village (2006 population 900), SE ONTARIO, E central CANADA, 15 mi/24 km from OTTAWA, and included in RUSSELL township; 45°10′N 75°21′W. Primarily French-speaking population.

Mariópolis (MAH-ree-O-po-lees), town (2007 population 5,805), SW PARANÁ state, BRAZIL, on border with SANTA CATALINA state; 26°20′S 52°33′W.

Mariovo, MACEDONIA: see MARIHOVO.

Maripa (mah-ree-PAH), town, E FRENCH GUIANA, on OYAPOCK RIVER (BRAZIL border), and 45 mi/72 km SW of SAINT-GEORGES-DE-L'OYAPOCK. Customhouse.

Maripa (mah-REE-pah), town, ⊙ Sucre municipio, BOLÍVAR state, SE VENEZUELA, on CAURA RIVER, and 125 mi/201 km WSW of CIUDAD BOLÍVAR; 07°25′N 65°09′W. Rubber, balata gum. Also known as Mariapa.

Maripasoula (mah-ree-pah-soo-LAH), town, SW FRENCH GUIANA, on LAWA (MARONI) River (SURINAME border); 03°39′N 54°05′W. Gold placers in vicinity.

Maripí (mah-ree-PEE), town, ⊙ Maripí municipio, BOYACÁ department, central COLOMBIA, 45 mi/72 km W of TUNJA; 05°33′N 74°00′W. Coffee, sugarcane, corn; livestock.

Maripipi Island (mah-ree-PEE-pee) (□ 11 sq mi/28.6 sq km), BILIRAN province, PHILIPPINES, in SAMAR SEA, between Masbate and LEYTE islands, 5 mi/8 km NW of BILIRAN ISLAND; 4 mi/6.4 km in diameter; 11°47′N 124°19′E. Mountainous, rising to 3,020 ft/920 m. Extinct volcano. Fishing. Coral gardens. Famous for its ceramics made with local white clay.

Mariposa, county (□ 1,451 sq mi/3,772.6 sq km; 2006 population 18,401), central CALIFORNIA; ⊙ MARIPOSA; 37°35′N 119°55′W. On W slope of the SIERRA NEVADA, at S end of MOTHER LODE gold country; has peaks over 10,000 ft/3,050 m in NE. Part of YOSEMITE NATIONAL

PARK in NE. Also includes parts of Sierra and Stanislaus national forests NW-SE through center. Drained by MERCED RIVER (forms McClure Reservoir) in W; CHOWCHILLA RIVER forms part of S border. Includes Merced and Mariposa groves of the big trees *Sequoia gigantea*. Region famed for scenery and recreational resources (resorts; hunting, lake and stream fishing, camping, hiking, winter sports). A leading gold-mining county (quartz mines) of California. Timber (pine, fir, spruce). Livestock raising (some cattle, sheep, hogs, poultry); little farming. Sand and gravel pits; silver mining. Mariposa, Hornitos, Coulterville, and ruins of old gold camps are reminders of gold rush. Badger Pass Ski Area in Yosemite National Park. Formed 1850.

Mariposa, unincorporated town (2000 population 1,373), ⊙ MARIPOSA county, central CALIFORNIA, 32 mi/51 km NE of MERCED; 37°30′N 119°58′W. Cattle; manufacturing (electronic components). Gateway to YOSEMITE NATIONAL PARK and Sierra National Forest (NE). Airport nearby. An old gold-rush town; its courthouse (1854) is said to be oldest in state. Stanislaus National Forest to N; Lake McClure reservoir to NW.

Mariposa Grove, California: see YOSEMITE NATIONAL PARK.

Mariposa Monarca Ecological Reserve (mah-ree-PO-sah mo-NAHR-kah), a biological reserve in E MICHOACÁN, MEXICO, 3.7 mi/6 km S of the village of ANGANGUEO, near EL ROSARIO. This is the winter grounds for a migrant species of monarch butterfly (*Danaus plexipus*). Some 100 million butterflies make a c.5,000-mi/8,000-km trip from CANADA and the UNITED STATES to stay three months in this area. Logging and pollution threaten survival of these insects.

Mariposas (mah-ree-PO-sahs), canton, CARRASCO province, COCHABAMBA department, central BOLIVIA. Elevation 9,150 ft/2,789 m. Limestone deposits. Agriculture (potatoes, yucca, bananas, corn, rice, rye, sweet potatoes, soy, coffee); cattle for dairy products and meat.

Mariposas (mah-ree-PO-sahs), village, TALCA province, MAULE region, central CHILE, 20 mi/32 km ESE of TALCA. Railroad terminus in agricultural area (wheat, barley, wine; livestock).

Mariposa Vieja, Pampa la (mah-ree-PO-sah vee-AH-hai lah), plain, LAMBAYEQUE region, NW PERU; 06°19′S 79°54′W.

Mariquita (mah-ree-KEE-tah), town, ⊙ Mariquita municipio, TOLIMA department, W central COLOMBIA, in MAGDALENA valley, on railroad, and 50 mi/80 km NNE of IBAGUÉ; 05°12′N 74°54′W. Agricultural region (sugarcane, corn, coffee, sorghum). Gold mines nearby. Founded 1551.

Marisa, ISRAEL: see MARESHAH.

Mariscala (mah-rees-KAH-lah), town, LAVALLEJA department, SE URUGUAY, on highway, and 35 mi/56 km NE of MINAS; 34°03′S 54°47′W. Wheat, oats, corn; livestock. Also known as La Mariscala.

Mariscala de Juárez (mah-rees-KAH-lah dai HWAH-rez), town, in far NW OAXACA, MEXICO, 25 mi/40 km W of HUAJUAPAM DE LEÓN. Elevation 3,675 ft/1,120 m. Mountainous terrain on the Mixteco River. Temperate to hot climate. Agriculture (corn, beans, sugarcane, chilies), straw textiles. Cattle. Connected by unpaved road to Huajuapam de León.

Mariscal Braun (mah-rees-KAHL brahn), canton, Azurduy province, CHUQUISACA department, SE BOLIVIA, 12 mi/20 km SW of AZURDUY; 20°04′S 64°34′W. Elevation 8,163 ft/2,488 m. Gas and oil wells in area to E. Clay deposits. Agriculture (potatoes, yucca, bananas, corn, rye, sweet potatoes, peanuts); cattle and hog raising.

Mariscal Cáceres (mah-rees-KAHL KAH-sai-res), province (□ 8,186 sq mi/21,283.6 sq km), W central SAN MARTÍN region, N central PERU; ⊙ JUANJUÍ;

07°50′S 76°40′W. Bordering on LA LIBERTAD region. Site of PARQUE NACIONAL ABISEO.

Mariscal Estigarribia (mah-rees-KAHL es-tee-gah-ree-BEE-ah), town, ⊙ Boquerón department, N PARAGUAY, in the CHACO, 300 mi/483 km NW of Asunción; 22°01′S 60°38′W. Intersection of seven major roads. Until 1945, López de Filippis.

Mariscal Luzuriaga (mah-rees-KAHL loo-soo-ree-AH-gah), province, ANCASH region, W central PERU; ⊙ PISCOBAMBA; 08°50′S 77°51′W. In the Cordillera BLANCA of the ANDES Mountains, crossed by PARQUE NACIONAL HUASCARÁN. Cereals, potatoes; livestock.

Mariscal Nieto (mah-rees-KAHL nee-YE-to), province (□ 4,005 sq mi/10,413 sq km), central MOQUEGUA region, S PERU; ⊙ MOQUEGUA; 17°05′S 71°00′W. Between AREQUIPA (W) and PUNO (E) regions.

Mariscal Ramón Castilla (mah-rees-KAHL rah-MON kahs-TEE-yah), province, easternmost province of LORETO region, NE PERU, on the AMAZON River (COLOMBIA and BRAZIL borders); ⊙ CABALLOCOCHA.

Marishka (mah-REESH-kah), obshtina, PLOVDIV oblast, BULGARIA; 42°12′N 25°58′E.

Mariski Chal (mah-REES-kee CHAHL), peak (9,071 ft/ 2,765 m), in the RILA MOUNTAINS, SW BULGARIA; 42°10′N 23°36′E.

Marismas, Las (mah-REEZ-mahs, lahs), coastal plain in ANDALUSIA, S SPAIN, along GUADALQUIVIR RIVER estuary on the Atlantic Ocean, c.10 mi/16 km S of SEVILLE. Sparsely inhabited alluvial marshland, used as pasture; some cotton and rice grown.

Marissa, village (2000 population 2,141), SAINT CLAIR county, SW ILLINOIS, 24 mi/39 km SE of BELLEVILLE; 38°15′N 89°45′W. Stone and wood products; bituminous-coal mines; agriculture (corn, wheat; dairy products; poultry, livestock). Incorporated 1882.

Marissa, ISRAEL: see MARESHAH.

Maristova (MAH-ris-taw-vah), tourist station, SOGN OG FJORDANE county, W NORWAY, 34 mi/55 km ESE of SOGNDAL. Elev. 2,635 ft/803 m. A hospice has been here since fourteenth-century Borgund stave church (12th century), 10 mi/16 km WSW, is Norway's best-preserved wooden church.

Marith (mah-RET), village (2004 population 10,923), QABIS province, SE TUNISIA, between JABAL MATMATA (SW) and the GULF OF QABIS (NE), 23 mi/37 km S of QABIS; 33°38′N 10°18′E. Olive and palm trees; sheep, camels; administrative center. In World War II, it was N anchor point of a fortified defense line held by the Germans after their retreat from TRIPOLITANIA, until outflanked and broken by the British in March 1943.

Maritime, region (2005 population 2,196,857), extreme S TOGO ⊙ LOMÉ; 06°30′N 01°20′E. Bordered by PLATEAUX region (N), BENIN (E), BIGHT OF BENIN (S), and GHANA (W). MONO RIVER in E (forms BENIN border). LAKE TOGO in S. Togodo Forest Nature Reserve in NE. Railroad runs N from LOMÉ through PLATEAUX region to BLITTA in CENTRALE region; secondary railroad runs NW from LOMÉ to SW PLATEAUX region. Airport at LOMÉ. Maritime region is composed of six prefectures: AVÉ, GOLFE (sometimes divided into Golfe prefecture and Lomé commune), LACS, VO, YOTO, and ZIO. Also called Région Maritime.

Maritime Alps (MA-ri-teim ALPS), French *Alpes Maritimes* (ahlp mah-ree-teem), Italian *Alpi Marittime*, SE section of the FRENCH ALPS, in SE FRANCE, on the MEDITERRANEAN SEA (S) and the Italian border (E); 44°15′N 07°10′E. Rise to 10,817 ft/3,299.2 m. The Maritime Alps are bounded by the upper valley of the DURANCE RIVER. Their S outliers reach the Mediterranean in the form of coastal ranges such as the massifs of ESTÉREL and MAURES. The VAR and VERDON rivers drain the Maritime Alps S. The highest summits are now included in the MERCANTOUR NATIONAL PARK along Italian border. The population is sparse except for the mountain resorts. Sometimes called Provence Alps.

Maritime Provinces, E CANADA, term applied to NOVA SCOTIA, NEW BRUNSWICK, and PRINCE EDWARD ISLAND, which before the formation of the Canadian confederation 3973:51 (1867) were politically distinct from Canada proper.

Maritime Territory, Russian *Primorskiy Kray* (pree-MOR-skeeyee KREI), administrative division (□ 62,500 sq mi/162,500 sq km; 2006 population 2,006,040), RUSSIAN FAR EAST, extreme SE Siberian Russia, between CHINA (Northeast) in the W and the Sea of JAPAN (E); ⊙ VLADIVOSTOK. Most of the territory has a monsoon-type climate, influenced by a cold current in the Sea of Japan; winters are short and cold, summers are long, cool, and foggy, with plenty of precipitation. The territory's coastal mountain range contains coal, iron ore, lead, zinc, lignite, tin, silver, and prime timber. Fisheries (salmon and sardines) along the shore. An agricultural plain with wheat, sugar beets, millet and rice crops extends around Lake KHANKA along the Chinese border. The population (constituting 50% of the people of the Russian Far East) is predominantly Russian, Ukrainian, and Byelorussian, with small indigenous ethnic groups. The TRANS-SIBERIAN RAILROAD links Vladivostok with USSURIYSK, the territory's other major city. For history of the Maritime Territory, see Russian Far East.

Maritsa (mah-REET-sah), obshtina, PLOVDIV oblast, BULGARIA; 42°12′N 25°58′E. Also known as Marishka.

Maritsa, Bulgaria: see SIMEONOVGRAD.

Maritsa-Iztok (mah-REET-sah–EEZ-tok), coal mining and power complex, HASKOVO oblast, GULUBOVO obshtina, BULGARIA; 42°08′N 25°51′E.

Maritsa River (mah-REET-sah), c.300 mi/483 km long, in BULGARIA, NW TURKEY, and NE GREECE; rises in the RILA MOUNTAINS S of SOFIA, W BULGARIA; flows SE between the STARA PLANINA and RODOPI Mountains, past PLOVDIV, to EDIRNE, Turkey (head of navigation by small boats), where it turns S to enter the AEGEAN SEA near ENEZ (Turkey), 12 mi/19 km SE of Alexandroúpolis (Greece); 42°08′N 24°45′E. The TUNDZHA RIVER is its chief tributary. The lower course forms part of the borders between Bulgaria and Greece (WEST MACEDONIA and THRACE departments) and between Greece and Turkey. The upper Maritsa valley is a principal E-W route in Bulgaria. Used for power production and irrigation. For its last 20 mi/32 km, it widens into a delta, eventually forming a wetland some 7 mi/11 km wide and splitting into two mouths. The delta is crossed with dikes and paths supporting land where cotton and maize are grown. It is also a refuge for about 250 species of birds and other wildlife that compete with local fishermen for the fish here. Known as the Évros (also Hebros) River by the Greeks and the Meriç River by the Turks. Also spelled Marica River.

Mari-Turek (mah-REE too-RYEK), town (2006 population 5,875), E MARI EL REPUBLIC, E central European Russia, on road, 65 mi/105 km E of YOSHKAR-OLA; 56°47′N 49°37′E. Elevation 406 ft/123 m. In agricultural area; produce processing. Population mostly Udmurt.

Maritzburg, SOUTH AFRICA: see PIETERMARITZBURG.

Mariupol' (mah-ree-OO-pol), city (2001 population 492,176), S DONETS'K oblast, UKRAINE, on the Sea of AZOV, and at the mouth of the KAL'MIUS RIVER; 47°06′N 37°33′E. Elevation 101 ft/30 m. A seaport and railroad terminus, Mariupol' is also an iron and steel center with machine plants, chemical works, and shipyards. Coal, salt, and grain are the chief exports. Site of a 16th-century Cossack fort Kal'mius, renamed Pavlovsk by Russians (1775), then, with resettlement of the Crimean Greeks, renamed (1779) Marianopol' or Mariyupil' (Russian *Mariupol'*). Was called Zhdanov between 1948 and 1992.

Mariusa, Caño (mah-ree-OO-sah, KAHN-yo), c.100 mi/161 km long, E central arm of ORINOCO RIVER delta, DELTA AMACURO state, NE VENEZUELA; branches off from the Caño ARAGUAITO; flows NE and NW to the ATLANTIC OCEAN; 09°43′N 61°26′W.

Marivan (mah-ree-VAHN), town (2006 population 92,993), Kordestān province, W IRAN, 45 mi/72 km WNW of SANANDAJ, 5 mi/8 km from IRAQ border; 35°30′N 46°10′E. Grain; sheep raising; high-quality wool exported.

Mariveles (mah-ree-VE-les), town (2000 population 85,779), Bataan province, S LUZON, PHILIPPINES, on S BATAAN Peninsula, port on Mariveles Harbor (small inlet of SOUTH CHINA SEA), at N side of entrance to MANILA BAY; 14°29′N 120°29′E. Ships sugar, rice. Mariveles Export Processing Zone here, with about 600 companies (textiles, clocks, electrical appliances, automotive parts).

Mariveles Mountains (mah-ree-VE-les), S BATAAN Peninsula, S LUZON, PHILIPPINES, near MOUNT BATAAN, just N of MARIVELES; rise to 4,200 ft/1,280 m.

Mariyental, RUSSIA: see SOVETSKOYE, SARATOV oblast.

Mariyets (mah-REE-yets), town, E MARI EL REPUBLIC, E central European Russia, on the Shora River (VOLGA RIVER basin), on road, 55 mi/89 km NNE of KAZAN' (TATARSTAN Republic); 56°31′N 49°50′E. Elevation 587 ft/178 m. Glassworks (dating from 1842).

Mariyovo, MACEDONIA: see MARIHOVO.

Mariyupil', UKRAINE: see MARIUPOL'.

Marj, Al (MAHR-izh, el), ancient *Barca*, town, CYRENAICA region, N LIBYA, 55 mi/89 km ENE of BANGHAZI, on plateau. Road junction. Agricultural center (flour, olive oil, meat products). Resort (elevation c.930 ft/283 m). Colonized by Greeks mid-6th century B.C.E.; declined under Ptolemies. Port was Ptolemais. Italian agricultural settlements in vicinity founded late 1930s abandoned to Libyans. Scene of fighting (1941–1942) between Axis and British in World War II; bombed 1942. Restored Turkish fort (1842); mosque, hospital, power station. Was capital of former Al Fatih province. Formerly Barqa or Barqah (Arabic), Barce (Italian).

Marj, Al-, EGYPT: see MARG, EL.

Märjamaa (MAHR-yah-maw), town, W ESTONIA, on railroad, and 15 mi/24 km WSW of RAPLA; 58°54′N 24°25′E. Agricultural market; barley, potatoes. Also spelled MARYAMA.

Marjan, AFGHANISTAN: see WAZA KHWA.

Marjioun, LEBANON: see MARJ 'UYUN.

Marj 'Uyun (MAHRZH ei-YOON), town, S LEBANON, on a well-irrigated plain, 24 mi/39 km ENE of SAIDA; 33°21′N 35°35′E. Elevation 2,500 ft/762 m. Tobacco, cereals, oranges, olives. Former tourist center. Located 5 mi/8 km N of the Israeli border, it is the main urban and administrative center of the Israeli-controlled area ("security zone") of S Lebanon. Also spelled Merj 'Uyun and Marjioun.

Mark (MAHRK), former county of W GERMANY, now in North Rhine-Westphalia, astride middle RUHR RIVER. Passed 1614 (ratified 1666) to BRANDENBURG.

Mark, village (2000 population 491), PUTNAM county, N central ILLINOIS, 5 mi/8 km SSW of SPRING VALLEY; 41°15′N 89°15′W. In agricultural area (corn, wheat, barley; cattle; mushroom processing.

Marka, town, ⊙ SHABEELAHA HOOSE region, SE central SOMALIA, port on INDIAN OCEAN, 45 mi/72 km SW of MOGADISHO; 01°43′N 44°46′E. Banana exporting, oilseed pressing, manufacturing (textiles, packing boxes, boats). Fishing, food processing. Near unexploited salt reserves. Also spelled MARCA, formerly MERCA.

Markagunt Plateau, KANE county, SW UTAH, high tableland (rising to 11,307 ft/3,446 m in BRIAN HEAD PEAK) in IRON, GARFIELD, and Kane counties, E of CEDAR CITY. Bounded E by PAUNSAUGUNT PLATEAU, S by PINK CLIFFS. Covered by a unit of Dixie National

Forest; includes CEDAR BREAKS National Monument in SW.

Markakol (mahr-kah-KOL), lake (□ 165 sq mi/429 sq km), EAST KAZAKHSTAN region, KAZAKHSTAN, in ALTAI MOUNTAINS, near Chinese border, 85 mi/137 km NNE of ZAISAN. Elev. 5,710 ft/1,740 m. Abounds in fish. Outlet is Kaldzhir River (tributary of the Cherny-Irtysh River); site of gold placers. Also spelled Marka-Kul.

Markala, township, FOURTH REGION/SÉGOU, S MALI, on right bank of the NIGER RIVER opposite SANSANDING, and 25 mi/40 km NE of SÉGOU; 13°41′N 06°05′W. Peanuts, shea nuts, cotton, kapok; livestock. An irrigation dam here serves the SAHEL and MACINA irrigation canals. Sometimes called Markala-Kirango. High school.

Markam (MONG-KONG), Tibetan *Dzong*, town, E TIBET, SW CHINA, 125 mi/201 km SE of QAMDO, and 40 mi/64 km SW of BATANG (Paan); 29°41′N 98°33′E. Machinery; livestock. Sometimes spelled Markham. Also called Gartog.

Markandi, India: see GARHCHIROLI.

Markanum, India: see MARAKKANAM.

Markapur (MAHR-kah-poor), town, PRAKASAM district, ANDHRA PRADESH state, S INDIA, in foothills of EASTERN GHATS, on GUNDLAKAMMA RIVER, and 55 mi/89 km ENE of NANDYAL; 15°44′N 79°17′E. Cotton ginning, hand-loom weaving, slate and lead mining, livestock grazing; peanuts, turmeric. Timber, bamboo, fibers in nearby forests.

Markaryd (MAHR-kah-REED), town, KRONOBERG county, S SWEDEN, on LAGAN RIVER, 20 mi/32 km NNW of HÄSSLEHOLM; 56°27′N 13°36′E. Railroad junction; manufacturing (paper mill; heat pumps).

Markazī, province (□ 15,403 sq mi/40,047.8 sq km), central IRAN, occupying part of the interior plateau NE of the ZAGROS Mountains and SW of TEHRAN; ⊙ ARAK; 34°30′N 50°00′E. Pastureland and farming (wheat, barley, sugar beets) predominate in the W, with nomadic herding in the E. Manganese deposits in several areas. Arak is an important industrial city (oil refining; chemicals, textiles).

Mark Canal, NORTH BRABANT province, SW NETHERLANDS, 4 mi/6.4 km long E-W, from MARK RIVER, 3 mi/4.8 km N of BREDA, to WILHELMINA CANAL, 1 mi/1.6 km W of OOSTERHOUT.

Markdale (MAHRK-dail), former village (□ 2 sq mi/5.2 sq km; 2001 population 1,433), S ONTARIO, E central CANADA, on Rocky Saugeen River, and 22 mi/35 km SE of OWEN SOUND; 44°19′N 80°39′W. Dairying; woodworking, lumbering; manufacturing (frozen foods, footwear); in mixed farming area. Amalgamated into GREY HIGHLANDS township in 2001.

Markdorf (MAHRK-dorf), town, region of LAKE CONSTANCE-UPPER SWABIA, BADEN-WÜRTTEMBERG, GERMANY, 11 mi/18 km ESE of Überlingen; 47°43′N 09°24′E. Wine growing. Heavily damaged by fire 1842. Chartered in 13th century.

Marke (MAHR-kuh), village in commune of WEVELGEM, Kortrijk district, WEST FLANDERS province, W BELGIUM, on Leie River, and 2 mi/3.2 km SW of KORTRIJK; 50°48′N 03°13′E. Flax growing, linseed-oil pressing.

Marked Tree, town (2000 population 2,800), POINSETT county, NE ARKANSAS, on SAINT FRANCIS River, at mouth of LITTLE RIVER, and 26 mi/42 km SE of JONESBORO; 35°31′N 90°25′W. In agricultural area (cotton, corn). Manufacturing (industrial storage tanks, furniture, machinery, leather products). Saint Francis Sunken Lands Wildlife Management Area to N. Settled c.1870.

Markelo (MAHR-kuh-LAW), town, OVERIJSSEL province, E NETHERLANDS, 13 mi/21 km W of HENGELO; 52°14′N 06°30′E. Dairying; cattle, hogs; grain, vegetables. Formerly spelled Markeloo.

Marken (MAHR-kuhn), island (1 sq mi/2.6 sq km), NORTH HOLLAND province, W NETHERLANDS, in the

IJSSELMEER, 11 mi/18 km NE of AMSTERDAM, and 2 mi/3.2 km E of mainland, in the MARKERMEER; 2 mi/3.2 km long SW-NE; 52°27′N 05°06′E. Lighthouse at E end; village of Kerkbuurt at center. Linked to mainland since 1959 by 2-mi/3.2-km breakwater; another 2.5-mi/4-km breakwater extends N to entrance of Gouw Zee, bay W of island.

Markermeer (MAHR-kuhr-mer), lake, NW NETHERLANDS, part of former Zuyder Zee, between NORTH HOLLAND and FLEVOLAND provinces; enclosed by ENKHUIZEN-LELYSTAD dike, which separates it from IJSSELMEER. Land reclaimed from draining of lake is planned to become the Markerwaard, future polder of Flevoland province.

Markerville (MAHR-kuhr-vil), hamlet, S central ALBERTA, W CANADA, 19 mi/30 km from RED DEER, in RED DEER COUNTY; 52°07′N 114°10′W.

Markesan (MAHRK-i-san), town (2006 population 1,340), GREEN LAKE county, central WISCONSIN, on GRAND RIVER, and 28 mi/45 km WSW of FOND DU LAC; 43°42′N 88°59′W. In farming, dairying, and livestock-raising region. Manufacturing (canned vegetables, fabricated metal products). Railroad terminus.

Market Bosworth (MAH-kit BAHZ-wuhth), town (2001 population 1,906), W LEICESTERSHIRE, central ENGLAND, 12 mi/19 km W of LEICESTER; 52°37′N 01°24′W. Former agricultural market. To the S (2 mi/3.2 km) is BOSWORTH FIELD, scene of battle (1485) in which Richard III was killed and Henry Tudor became Henry VII.

Market Deeping (MAH-kit DEE-ping), town (2001 population 6,477), S LINCOLNSHIRE, E ENGLAND, on WELLAND RIVER, and 7 mi/11.3 km NNW of PETERBOROUGH; 52°40′N 00°18′W. Has 13th–14th-century church.

Market Drayton (MAH-kit DRAI-tuhn), town (2001 population 10,407), NE SHROPSHIRE, W ENGLAND, on Tern River, and 17 mi/27 km NE of SHREWSBURY; 52°54′N 02°28′W. Former agricultural market and dairying center. Has 12th-century church and grammar school (founded 1558) attended by Lord Clive.

Market Harborough (MAH-kit HAH-buh-ruh), town (2001 population 20,127), S LEICESTERSHIRE, central ENGLAND, on WELLAND RIVER, and 17 mi/27 km N of NORTHAMPTON; 52°28′N 00°55′W. Manufacturing of electrical equipment, machine tools, rubber goods, textiles, agricultural machinery. Has church begun in 14th century and 17th-century grammar school.

Markethill (MAHR-ket-hil), town (2001 population 1,290), central ARMAGH, S Northern Ireland, 6 mi/9.7 km SE of ARMAGH; 54°18′N 06°31′W. Market (cattle, potatoes). Site of 19th-century Gosford Castle.

Market Lavington (MAH-kit LAV-ing-tuhn), agricultural village (2001 population 2,840), central WILTSHIRE, S ENGLAND, 5 mi/8 km S of DEVIZES; 51°17′N 01°59′W. Has 14th-century church.

Market Rasen (MAH-kit RAI-zuhn), town (2001 population 4,675), N central LINCOLNSHIRE, E ENGLAND, 14 mi/23 km NE of LINCOLN; 53°23′N 00°20′W. Former agricultural market. Racecourse, golf course.

Market Warsop, ENGLAND: see WARSOP.

Market Weighton (MAH-kit WAI-tuhn), town (2001 population 5,212), East Riding of Yorkshire, NE ENGLAND, 17 mi/27 km ESE of YORK; 53°52′N 00°40′W. Has 12th–15th-century church. Agricultural village of Arras is 3 mi/4.8 km E.

Markgräfler Land (mahrk-GRAIF-ler LAHND), region in BADEN-WÜRTTEMBERG, GERMANY, S of FREIBURG, E of the RHINE RIVER, extending up to W slope of BLACK FOREST. Noted for its white wine. Chief town is MÜLLHEIM.

Markgröningen (mahrk-GRUH-ning-uhn), town, STUTTGART district, BADEN-WÜRTTEMBERG, GERMANY, 5 mi/8 km W of LUDWIGSBURG; 48°54′N 09°05′E. Wine. Has 16th-century church.

Markha (mahr-HAH), village (2006 population 11,520), central SAKHA REPUBLIC, central RUSSIAN FAR EAST, on the W bank of the LENA RIVER, on road, 5 mi/8 km N of YAKUTSK, of which it is a residential suburb; 62°06′N 129°43′E. Elevation 324 ft/98 m. The region's main airport is 1 mi/1.6 km to the E. Also known as Bol'shaya Markha.

Markham, city (2000 population 12,620), COOK county, NE ILLINOIS, suburb 19 mi/31 km SSW of downtown CHICAGO; 41°36′N 87°41′W. Manufacturing (building materials, industrial brushes, wood products). Native prairie preserved at Indian Boundaries Prairies. Incorporated 1925.

Markham (MAHR-kuhm), town (□ 82 sq mi/213.2 sq km; 2001 population 208,615), YORK region, S ONTARIO, E central CANADA, on ROUGE RIVER, and 15 mi/24 km NNE of TORONTO; 43°52′N 79°16′W. Suburban community. High technology industries; manufacturing (chemicals, medical supplies, transportation equipment, machinery, ordnance, computers, motor vehicles, frozen foods, paper goods, furniture, apparel, flour), book publishing, printing.

Markham (MAHRK-uhm), village, CREEK county, central OKLAHOMA, 10 mi/16 km NE of CUSHING.

Markham, CHINA: see MARKAM.

Markhamat, city, S ANDIJAN wiloyat, UZBEKISTAN, 12 mi/19 km SSE of ASSAKE; 40°30′N 72°19′E. In cotton- and silk-growing area.

Markham, Mount (MAHR-kuhm), a twin-peaked mountain, ANTARCTICA; 82°51′S 161°21′E. Highest peak in TRANSANTARCTIC MOUNTAINS at 14,270 ft/4,349 m. Discovered 1902 by Robert F. Scott, British explorer.

Markham River, 75 mi/121 km long, NE NEW GUINEA; rises circa 50 mi/80 km SSE of MADANG, in FINISTERRE RANGE, SE MADANG province, near source of RAMU RIVER; flows S then E to HUON GULF (SOLOMON SEA); just W of LAE, MOROBE province. Drains agricultural land. Rice, bananas (upper river); coconuts, sugarcane plantations (lower river).

Markham Sound (mahr-HAHM), strait of the ARCTIC OCEAN in central FRANZ JOSEF LAND, ARCHANGEL oblast, extreme N European Russia; separates S islands (Champ, Luigi) of Zichy Land (N) and MacClintock and Hall islands (S); 50 mi/80 km long, 10 mi/16 km wide; 80°33′N 55°42′E.

Markhlevsk, UKRAINE: see DOVBYSH.

Markiezaatsmeer (MAHR-kee-zahts-mer), lake, SW NETHERLANDS formed on E end of EASTERN SCHELDT River by the OESTER Dam (part of the Delta Project), between NORTH BRABANT province coast and Oester Dam.

Markinch (MAHR-kinch), town (2001 population 2,254), FIFE, E Scotland, on LEVEN RIVER, and 5 mi/8 km W of LEVEN; 56°12′N 03°08′W. Previously woolen and paper milling, textile bleaching. Lies on E side of GLENROTHES. Just NW is Balbirnie.

Märkisch Friedland, POLAND: see MIROSLAWIEC.

Markit (MEI-GEI-TEE), town and oasis, ⊙ MARKIT county, SW XINJIANG UYGUR AUTONOMOUS REGION, NW CHINA, 30 mi/48 km N of Shache (Yarkant), and on YARKANT River; 38°53′N 77°35′E. Livestock; textile manufacturing, food processing. Also appears as Maigaiti.

Markivka (MAHR-keef-kah), (Russian *Markovka*), town, NE LUHANS'K oblast, UKRAINE, 35 mi/56 km NE of STAROBIL'S'K; 49°32′N 39°34′E. Elevation 374 ft/113 m. Raion center. Cheese; feed mills. Vocational technical school; heritage museum. Established in 1690 as a Cossack settlement; town since 1960.

Markkleeberg (mahrk-KLEH-berg), town, SAXONY, E central GERMANY, 4 mi/6.4 km S of LEIPZIG city center; 51°16′N 12°20′E. Printing, woolen milling; manufacturing of electrical equipment. Just E is site of early Stone Age excavations.

Markland, village, SE NEWFOUNDLAND AND LABRADOR, E CANADA, in central part of AVALON PENINSULA, 40 mi/64 km WSW of St. JOHN's. Scene of agricultural resettlement scheme for unemployed that began in 1934.

Markle, town (2000 population 1,102), HUNTINGTON and WELLS counties, NE central INDIANA, on the WABASH RIVER, and 9 mi/14.5 km SE of HUNTINGTON; 40°50′N 85°20′W. Manufacturing (machinery, fabricated metal products).

Markleeville, unincorporated village, ⊙ ALPINE county, E CALIFORNIA, 18 mi/29 km SSE of SOUTH LAKE TAHOE, on East Fork of CARSON RIVER, in the SIERRA NEVADA, 32 mi/51 km S of CARSON CITY (Nevada). Mineral springs (Grover Hot Springs State Park) to NW. Hunting and fishing in region. Kirkwood Ski Area to W; Toiyabe National Forest to W, S, and E; Pacific Crest Trail to W.

Marklesburg (MAHR-kuhls-buhrg), borough (2006 population 211), HUNTINGDON county, S central PENNSYLVANIA, 11 mi/18 km SW of HUNTINGDON; 40°22′N 78°10′W. Agriculture (corn, hay; dairying). Post office is James Creek.

Markleville, town (2000 population 383), MADISON county, E central INDIANA, 9 mi/14.5 km SE of ANDERSON; 39°59′N 85°37′W. Agricultural area. Manufacturing (building materials, wood products). Laid out 1852.

Markleysburg (MAHR-klees-buhrg), borough (2006 population 272), FAYETTE county, SW PENNSYLVANIA, 18 mi/29 km SE of UNIONTOWN, 1 mi/1.6 km N of MARYLAND state line, and 2 mi/3.2 km NE of corner of WEST VIRGINIA state line; 39°44′N 79°27′W. Agriculture (dairying); manufacturing (concrete). In area are YOUGHIOGHENY RIVER LAKE (E) and Lake Courage (W) reservoirs.

Marklissa, POLAND: see LESNA.

Marklkofen (mahr-kuhl-KO-fuhn), village, LOWER BAVARIA, BAVARIA, SE GERMANY, on the VILS RIVER, 6 mi/9.7 km SSE of DINGOLFING; 48°33′N 12°33′E. Grain; textile industry.

Marklohe (MAHRK-lo-e), village, LOWER SAXONY, GERMANY, near the WESER RIVER, 3 mi/4.8 km NW of NIENBURG; 52°40′N 09°09′E. Forestry; food canning; glassworks.

Marknesse (MAHRK-nes-suh), village, NORTH-EAST POLDER, FLEVOLAND province, central NETHERLANDS, 4.5 mi/7.2 km E of EMMELOORD, on ZWOLSEVAART canal.

Markneukirchen (mahrk-NOI-kir-khuhn), town, SAXONY, E central GERMANY, in the ERZGEBIRGE, 15 mi/24 km SE of PLAUEN; 50°19′N 12°28′E. Musical-instrument-manufacturing center.

Markopoulo (mahr-KO-poo-lo), town, ATTICA prefecture, ATTICA department, E central GREECE, on railroad, and 13 mi/21 km SE of ATHENS. Wine center; wheat, olives; cattle and sheep raising.

Markopoulos (mahr-KO-poo-los), village, in SW KEFALLINÍA island, KEFALLINÍA prefecture, IONIAN ISLANDS department, off W coast of GREECE; 38°05′N 20°44′E. Every year on the Orthodox feast of the Dormition of the Virgin Mary (Aug. 15th), small harmless snakes with black cross markings on their heads crawl into the local church.

Markovka, UKRAINE: see MARKIVKA.

Markovo (MAHR-kuh-vuh), town (2005 population 1,240), W central IVANOVO oblast, central European Russia, on road and railroad, 18 mi/29 km W of IVANOVO; 57°01′N 40°30′E. Elevation 469 ft/142 m. Peat. Until 1940, called Markovo-Sbornoye.

Markovo (MAHR-kuh-vuh), town, S CHUKCHI AUTONOMOUS OKRUG, NE SIBERIA, NE RUSSIAN FAR EAST, near the ANADYR′ RIVER, at the terminus of a local highway, 215 mi/346 km SW of ANADYR′; 64°40′N 170°25′E. Air base; trading post; in reindeer-producing area.

Markovo (MAHR-ko-vo), village, PLOVDIV oblast, Rodopi obshtina, BULGARIA; 42°03′N 24°42′E.

Markovo (MAHR-kuh-vuh), industrial settlement (2005 population 6,285), central IRKUTSK oblast, E

central SIBERIA, RUSSIA, on the LENA RIVER, on road, 47 mi/76 km NE of UST′-KUT; 57°19′N 107°05′E. Elevation 1,532 ft/466 m. Natural gas extraction and processing.

Markovo-Sbornoye, RUSSIA: see MARKOVO, IVANOVO oblast.

Markranstädt (MAHRK-rahn-shtet), town, SAXONY, E central GERMANY, 6 mi/9.7 km W of LEIPZIG; 51°18′N 12°09′E. Machinery manufacturing, brewing, sugar refining.

Mark River (MAHR), 50 mi/80 km long, N BELGIUM and S NETHERLANDS; rises 5 mi/8 km N of TURNHOUT (Belgium); flows N, entering the Netherlands 6 mi/9.7 km S of BREDA, past Breda, then W to the VOLKERAK 4 mi/6.4 km SW of WILLEMSTAD; its lower course is called Dintel Mark River. Receives MARK CANAL 3 mi/4.8 km N of Breda. Navigable below Breda.

Marks (MAHRKS), city (2006 population 33,195), central SARATOV oblast, SE European Russia, grain port on the left bank of the VOLGA RIVER, on highway, 40 mi/64 km NE of SARATOV; 51°42′N 46°46′E. Agricultural and industrial center; manufacturing (diesel fuel apparatus; ferro-concrete), woodworking, food processing (dairy, brewery). Founded in 1765 under Catherine II by Dutch baron Beauregard. Called Baronsk or Yekaterinenshtadt (German *Katharinenstadt*) until 1918 and later (in German VOLGA ASSR), Marksshtadt (German *Marxstadt*) until 1941. Made city in 1918. Sometimes spelled Marx.

Marks, town (2000 population 1,551), ⊙ QUITMAN county, NW MISSISSIPPI, 18 mi/29 km ENE of CLARKSDALE, and on COLDWATER RIVER; 34°15′N 90°16′W. Trade center in rich agricultural area (soybeans, cotton, corn); manufacturing (apparel, electrical equipment; soybean processing). O'Keefe Wildlife Management Area to SE. Incorporated 1906.

Markstay (MAHRK-stai), former town, E central ONTARIO, E central CANADA, 46°29′N 80°32′W. Amalgamated into town of MARKSTAY-WARREN in 1999.

Markstay-Warren (MAHRK-stai–WAH-ren), town (□ 197 sq mi/512.2 sq km; 2001 population 2,627), central ONTARIO, E central CANADA; 46°29′N 80°29′W. Agriculture, forestry, tourism. Formed in 1999 from WARREN, MARKSTAY, AWREY, STREET, HAWLEY, LOUGHRIN, and HENRY; HAGAR and RATTER AND DUNNET are also included.

Marksville, city (2000 population 5,537), ⊙ AVOYELLES parish, E central LOUISIANA, 24 mi/39 km SE of ALEXANDRIA, near RED RIVER; 31°08′N 92°04′W. In agricultural area (cotton, corn, sugarcane, soybeans, sweet potatoes); food processing, manufacturing (apparel, foods); logging. Fort De Russey, site of Civil War fighting, nearby. Marksville State Commemorative Area (covers 43 acres/17 ha), with Native American mounds dating to 140 B.C.E., is here. Spring Bayou State Wildlife Area to E, with Lake Ophelia (NE) and Grand Cote (W) national wildlife refuges in area. Near giant casino on land owned by Tunica, Native Biloxi tribe in Avoyelles parish. Settled in late eighteenth century by Acadians.

Marktbreit (MAHRKT-breit), town, LOWER FRANCONIA, W BAVARIA, GERMANY, on the MAIN RIVER, and 13 mi/21 km SE of WÜRZBURG; 49°39′N 10°08′E. Manufacturing of machinery, brewing, tanning. Has Gothic church (15th–16th century), and castle (16th century). Portions of mid-16th-century wall still stand. Was first mentioned 1258; chartered 1557. Town since 1819.

Markt Erlbach (MAHRKT ERL-bahkh), village, MIDDLE FRANCONIA, W BAVARIA, GERMANY, 14 mi/23 km W of FÜRTH; 49°29′N 10°38′E. Agriculture (rye). Industry includes brewing; carp hatching.

Marktheidenfeld (mahrkt-HEI-duhn-felt), town, LOWER FRANCONIA, NW BAVARIA, GERMANY, on the MAIN RIVER, 15 mi/24 km WNW of WÜRZBURG; 49°51′N 09°36′E. Manufacturing of precision instruments, metalworking, brewing. Has early-17th-century church. Chartered 1397; town since 1949.

Markt Indersdorf (MAHRKT IN-ders-dorf), village, UPPER BAVARIA, BAVARIA, S GERMANY, on the Glonn River, 17 mi/27 km NNW of MUNICH; 48°21′N 11°22′E. Hops.

Marktleugast (mahrkt-LOI-gahst), village, UPPER FRANCONIA, NE BAVARIA, GERMANY, on SE slope of the FRANCONIAN FOREST, 10 mi/16 km NE of KULMBACH; 50°10′N 11°38′E. Forestry; tourism.

Marktleuthen (mahrkt-LOI-tuhn), town, UPPER FRANCONIA, NE BAVARIA, GERMANY, in the FICHTELGEBIRGE, on the EGER RIVER, and 7 mi/11.3 km WSW of SELB; 50°08′N 11°59′E. Glassworks; brewing, tanning, lumber milling. Granite quarries in area.

Marktoberdorf (mahrkt-O-buhr-dorf), town, SWABIA, SW BAVARIA, GERMANY, near the Werta River, 7 mi/11.3 km S of KAUFBEUREN; 47°47′N 10°37′E. Manufacturing of textiles, metalworking; jewelry. Has former 18th-century castle of prince-bishops of AUGSBURG and Gothic church. Chartered 1954.

Markt Piesting (mahrkt PEE-sting), township, SE LOWER AUSTRIA, E AUSTRIA, on PIESTING RIVER, 6 mi/9.7 km NW of WIENER NEUSTADT; 47°52′N, 16°08′E. Elevation 1,036 ft/316 m. Manufacturing of driving gears. Ruins of a castle (Starhemberg) nearby.

Marktredwitz (mahrkt-RED-vits), town, UPPER FRANCONIA, NE BAVARIA, GERMANY, in the FICHTELGEBIRGE, 23 mi/37 km ENE of BAYREUTH; 50°00′N 12°05′E. Manufacturing of glass, porcelain, textiles, machine tools; woodworking, brewing. Has 14th- and 16th-century churches.

Markt Rettenbach (MAHRKT RET-tuhn-bahkh), village, SWABIA, SW BAVARIA, GERMANY, 10 mi/16 km ESE of MEMMINGEN; 47°57′N 10°24′E. Grain; cattle. Sometimes called RETTENBACH.

Marktrodach (mahrkt-RO-dahkh), village, UPPER FRANCONIA, NE BAVARIA, GERMANY, on the RODACH RIVER, and in FRANCONIAN FOREST, 3 mi/4.8 km E of KRONACH; 50°14′N 11°24′E. Forestry; tourism.

Markt Sankt Florian (mahrkt sahnkt FLO-ree-ahn), township, E UPPER AUSTRIA, 7 mi/11.3 km SSE of LINZ; 48°12′N 14°23′E. Pilgrimage church. Augustine abbey, founded in 1071, reconstructed 1686–1708 in baroque style (main opus of C. A. Carlone). Grave of the composer Anton Bruckner, who played here as an organist. Large library and art gallery.

Markt Sankt Martin (mahrkt sahnkt MAHR-teen), township, central BURGENLAND, E AUSTRIA, 10 mi/16 km SW of SOPRON (HUNGARY); 47°34′N, 16°26′E. Elevation 930 ft/283 m. Manufacturing of knit goods. Impressive ruins of a castle (Landsee) nearby.

Markt-Schelken, ROMANIA: see ŞEICA MARE.

Marktschellenberg (mahrkt-SHEL-len-boorg) or **Schellenberg**, village, UPPER BAVARIA, GERMANY, on SE slope of the UNTERSBERG, 3 mi/4.8 km NW of HALLEIN, 5 mi/8 km NNE of BERCHTESGADEN, near Austrian border; 47°41′N 13°02′E. Lumber milling; resort (elevation 1,575 ft/480 m).

Markt Schwaben (MAHRKT SHVAH-ben), village, UPPER BAVARIA, GERMANY, 15 mi/24 km E of MUNICH; 48°11′N 11°51′E. Manufacturing includes brewing, woodworking.

Marktsteft (MAHRKTS-teft), town, LOWER FRANCONIA, W BAVARIA, GERMANY, on the MAIN RIVER, 11 mi/18 km SE of WÜRZBURG; 49°42′N 10°08′E. Brewing.

Mark Twain Lake, reservoir (□ 13 sq mi/33.8 sq km), RALLS county, NE MISSOURI, on SALT RIVER, 48 mi/77 km NE of COLUMBIA; 39°32′N 91°38′W. Maximum capacity 1,861,889 acre-ft. Formed by Clarence Cannon Dam, built (1984) by Army Corps of Engineers for flood control; also used for power generation and recreation. Mark Twain State Park and Mark Twain State Historical Site near center of reservoir.

Marl (MAHRL), city, North Rhine-Westphalia, W GERMANY, in the RUHR industrial district; 51°40′N 07°05′E. Industrial and coal-mining center; also supports a number of chemical factories. City first mentioned 9th century; chartered 1936.

Cross-references are shown in SMALL CAPITALS. The pronunciation guide is shown on page xix. The sources of population figures are shown on page xvii.

Marla (MAHR-luh), locality, SOUTH AUSTRALIA state, S central AUSTRALIA, 420 mi/676 km N of PORT AUGUSTA, and 99 mi/159 km S of NORTHERN TERRITORY border, on Stuart Highway; 27°19′S 133°39′E. Roadhouse, stopping point (opened 1982). Mintabie opal settlement 22 mi/35 km W.

Marlagne (mahr-LAHN-yuh), low plateau forming the border of the ARDENNES PLATEAU, on the border between HAINAUT and NAMUR provinces, BELGIUM, between the SAMBRE RIVER on the NW and FAMENNE depression on the SE.

Marland (MAHR-land), village (2006 population 277), NOBLE county, N OKLAHOMA, 10 mi/16 km SSW of PONCA CITY; 36°33′N 97°09′E. In oil- and natural-gas-producing area.

Marlboro, county (□ 485 sq mi/1,261 sq km; 2006 population 29,152), NE SOUTH CAROLINA, on NORTH CAROLINA (N and NE) state line; ⊙ BENNETTSVILLE; 34°36′N 79°40′W. Bounded SW by GREAT PEE DEE RIVER. Manufacturing of sand, gravel, clay. Mainly agriculture, including hogs; corn, rye, oats, tobacco, soybeans, hay, cotton, vegetables; some timber. Manufacturing at Bennettsville. Formed 1785.

Marlboro (MAWRL-buh-ruh), township, MONMOUTH county, E NEW JERSEY, 4 mi/6.4 km N of FREEHOLD; 40°20′N 74°15′W. In suburbanizing agricultural area. Nearby is site of former Marlboro State Hospital.

Marlboro (MAHRL-buh-ro), town, WINDHAM co., SE VERMONT, 8 mi/12.9 km W of BRATTLEBORO; 42°53′N 72°44′W. Home of Marlboro College. Likely named for Marlboro, Massachusetts.

Marlboro, village (2000 population 2,339), ULSTER county, SE NEW YORK, on W bank of the HUDSON RIVER, and 7 mi/11.3 km N of NEWBURGH; 41°36′N 73°58′W. Light manufacturing and agricultural products. Also spelled Marlborough.

Marlboro (MAHRL-buh-ro), hamlet, central ALBERTA, W CANADA, 13 mi/21 km W of EDSON, in YELLOWHEAD COUNTY; 53°33′N 116°45′W.

Marlborough, district (□ 4,824 sq mi/12,542.4 sq km), NE SOUTH ISLAND, NEW ZEALAND; ⊙ BLENHEIM; 41°30′S 173°30′E. PICTON is chief port. Largely mountainous, with sheep ranching; small but productive WAIRAU plain fronting on CLOUDY BAY. QUEEN CHARLOTTE and PELORUS sounds, in N, are inlets from COOK STRAIT. Agriculture (grain, fruits, grapes). Area generally corresponds to former Marlborough province and land district, now with unitary district council.

Marlborough, city (2000 population 36,255), MIDDLESEX county, E MASSACHUSETTS; 42°21′N 71°33′W. Shoe-manufacturing center for many years; manufacturing also includes plastic and paper products, consumer goods, machinery, computer equipment, processed foods, and chemicals. Growing site of suburban office activity. Skiing at Jericho Hill. Almost destroyed (1676) in King Philip's War. Settled on the site of a Native American village in 1657, incorporated as a city 1890.

Marlborough (MAHRL-buh-ro), township, central QUEENSLAND, NE AUSTRALIA, 53 mi/102 km N of ROCKHAMPTON; 22°49′S 149°53′E. In cattle-raising area.

Marlborough (MAWL-buh-ruh), town (2001 population 8,009), E WILTSHIRE, S ENGLAND, on KENNET RIVER, and 10 mi/16 km SSE of SWINDON; 51°25′N 01°43′W. Former agricultural market in dairying and sheep-raising region; bacon and ham curing, manufacturing of agricultural implements. Seat of Marlborough College, public school founded 1843 on site of historic castle. On school grounds is prehistoric mound, said to contain body of Merlin of Arthurian legend. Has two Norman churches and old St. Peter's church. Formerly important stage point on London-Bath road.

Marlborough (MAHRL-buh-ro), town, HARTFORD county, central CONNECTICUT, 15 mi/24 km SE of

HARTFORD; 41°38′N 72°27′W. In rural area that mainly is a residential community. Has 18th-century tavern.

Marlborough, town (2000 population 2,235), SAINT LOUIS county, E MISSOURI, 10 mi/16 km SW of downtown ST. LOUIS; 38°34′N 90°20′W. Residential and commercial suburb of St. Louis. Watson Road (old Route 66) runs length of town; known for its old motels, restaurants, and drive-in theatres.

Marlborough, town, CHESHIRE county, SW NEW HAMPSHIRE, 4 mi/6.4 km ESE of KEENE; 42°53′N 72°10′W. Drained by Minnewawa Brook. Agriculture (cattle, poultry; vegetables, nursery crops; dairying); manufacturing (plastics, consumer goods, other light manufacturing). Incorporated 1776.

Marlborough, region (□ 6,853 sq mi/17,817.8 sq km), NE SOUTH ISLAND, NEW ZEALAND, facing COOK STRAIT; BLENHEIM. As defined in 1989, corresponds closely with former Marlborough province and provincial district, minus the KAIKOURAS. The main geographical units include moist, rugged MARLBOROUGH SOUND, formed by marine invasion of a N tilting earth-block, creating intricate inlets now used for vacationing and, at PICTON, for inter-island ferry transport. Occasional dairying, extensive sheep grazing, and reversion to forest prevail. The core of settlement occurs in the down-faulted WAIRAU Valley, infilled with alluvial soils and gravels, productive with mixed crop and livestock farming in notably warm, dry climate. Blenheim is the urban regional center. The AWATERE river also occupies a faulted valley, and offers a parallel, but smaller, settlement area, with salt drying at Lake Grassmere. The remainder of the region consists of grass covered "High Country" with extensive sheep and beef-cattle grazing. Administratively, regional, urban, and district activities are focussed in a unitary authority.

Marlborough, for places in the United States: see also under "Marlboro".

Marlborough Sounds, off COOK STRAIT coast, between TASMAN and CLOUDY BAYS, SOUTH ISLAND, MARLBOROUGH district, NEW ZEALAND. Intricate shoreline and islands formed geologically by N tilting of crustal block, enabling marine penetration of river valleys, creating QUEEN CHARLOTTE and PELORUS sounds with coves and islands. Some incorporated in Marlborough Sounds Maritime Park.

Marle (MAHRL), commune (□ 5 sq mi/13 sq km), AISNE department, PICARDIE region, N FRANCE, on Serre River, and 14 mi/23 km NNE of LAON; 49°44′N 03°47′E. Sugar mill. Gothic church of 12th–13th century. Also called Marle-sur-Seine.

Marlenheim (mahr-luhn-EM), German (MAHR-luhn-heim), commune (□ 5 sq mi/13 sq km), BAS-RHIN department, in ALSACE, E FRANCE, at E foot of the VOSGES MOUNTAINS, 13 mi/21 km W of STRASBOURG; 48°37′N 07°30′E. Noted Alsatian wine center on the "Route des Vins" (the Wine Route).

Marles-les-Mines (MAHRL–lai–MEEN), town (□ 1 sq mi/2.6 sq km), PAS-DE-CALAIS department, NORD-PAS-DE-CALAIS region, N FRANCE, 6 mi/9.7 km WSW of BÉTHUNE; 50°30′N 02°31′E. In former coal-mining district.

Marle-sur-Seine, FRANCE: see MARLE.

Marlette (mahr-LET), town (2000 population 2,104), SANILAC county, E MICHIGAN, 21 mi/34 km N of IMLAY; 43°19′N 83°04′W. In farm area (grain, sugar beets, apples; livestock; dairy products). Manufacturing (building materials, hose clamps, transportation equipment); grain elevator. Incorporated 1881.

Marlin (MAHR-lin), town (2006 population 6,158), ⊙ FALLS county, E central TEXAS, near BRAZOS River, 23 mi/37 km SE of WACO; 31°18′N 96°53′W. Elevation 383 ft/117 m. Health resort (mineral springs); trade center for agricultural area (cotton, corn; cattle). Hot artesian wells discovered 1890s. Established 1830s.

Marlinton (MAHR-lin-tuhn), town (2006 population 1,235), ⊙ POCAHONTAS county, E WEST VIRGINIA, on

GREENBRIER RIVER, 50 mi/80 km S of ELKINS, within W edge of Monongahela National Forest; 38°13′N 80°05′W. Manufacturing (lumber, wood products). Agriculture (grain, apples); livestock; poultry; timber. Summer resort in hunting area. Greenbrier River State Trail passes through town; Watoga State Park to S; Seneca State Forest to NE; Edray State Fish Hatchery to N. Settled 1747 as Marlin's Bottom, first settlement W of ALLEGHENY MOUNTAINS.

Marlo (MAHR-lo), resort village, VICTORIA, SE AUSTRALIA, 246 mi/396 km E of MELBOURNE, 9 mi/15 km W of ORBOST, at mouth of SNOWY RIVER; 37°47′S 148°32′E. Holiday, fishing village.

Marlow (MAHR-lo), town (2001 population 17,522), S BUCKINGHAMSHIRE, central ENGLAND, on the THAMES (here crossed by 225-ft/69-m suspension bridge), and 4 mi/6.4 km S of HIGH WYCOMBE; 51°34′N 00°47′W. Paper industry. Has church built by Augustus Pugin. The poet Percy Shelley once lived here.

Marlow (MAHR-lo), town, CHESHIRE county, SW NEW HAMPSHIRE, 13 mi/21 km NNE of KEENE; 43°07′N 72°12′W. Drained by ASHUELOT RIVER. Agriculture (cattle, sheep, poultry; vegetables, apples; dairying; nursery crops). Manufacturing (electronic equipment).

Marlow, town (2006 population 4,566), STEPHENS county, S OKLAHOMA, 10 mi/16 km N of DUNCAN; 34°38′N 97°57′W. In agricultural area (corn, watermelons, peanuts; cattle); manufacturing (medical supplies, meat products). Settled 1892.

Marlton, village (2000 population 10,260), BURLINGTON county, W NEW JERSEY, 11 mi/18 km E of CAMDEN; 39°53′N 74°55′W. Has Baptist church (1805). Largely residential.

Marly (mahr-LEE), commune, FRIBOURG canton, W SWITZERLAND, on Ärgera River, just S of FRIBOURG.

Marly (mahr-LEE), town (□ 3 sq mi/7.8 sq km), E suburb of VALENCIENNES, NORD department, NORD-PAS-DE-CALAIS region, N FRANCE; 50°21′N 03°33′E. Metalworking, manufacturing of building materials, and textile printing.

Marly (mahr-LEE), town (□ 4 sq mi/10.4 sq km), MOSELLE department, LORRAINE region, NE FRANCE; 49°04′N 06°09′E. S suburb of METZ. Also called Marly-sur-Seille.

Marly-la-Ville (mahr-LEE-lah-VEEL), town (□ 3 sq mi/7.8 sq km), VAL-D'OISE department, ÎLE-DE-FRANCE region, N central FRANCE, 7 mi/11 km N of Charles de Gaule Airport; 49°05′N 02°30′E.

Marly-le-Roi (mahr-LEE-luh-RWAH), town (□ 2 sq mi/5.2 sq km), YVELINES department, ÎLE-DE-FRANCE region, N central FRANCE, on the left bank of the SEINE RIVER, 4 mi/6.4 km NNW of VERSAILLES, and 12 mi/19 km from Notre Dame Cathedral in PARIS; 48°52′N 02°05′E. Together with its neighbor, LOUVECIENNES, Marly is a growing residential suburb expanding beyond the old center (a favorite gathering place of artists and writers in 19th century) and the former country estate of Louis XIV. It owes its fame to the king's desire for a more rustic and informal retreat than could be found at Versailles. Accordingly, Mansart, the great architect of the time, proposed building a relatively modest royal castle to be surrounded by twelve pavilions each evoking a sign of the zodiac and focused upon the Sun King's residence. The project was built at great cost despite a shrinking national treasury, and Louis XVI occupied it in 1688. For much of the rest of his life, the king involved himself in planning many embellishments, particularly a cascade of waterfalls that would be fed by water pumped up from the Seine. The famous *machine de Marly*, a hydraulic engine to supply such water not only to Marly but also to the waterworks at Versailles, was constructed and remained in operation until 1804. The residence itself was destroyed after the French Revolution. The surrounding park, with intersecting allées and perspectives, was not completed until the early 19th century. Beyond (W) lies the Forest of Marly (□ c.8 sq mi/20.7 sq km), a wooded tableland, now surrounded by modern development.

Marly-sur-Seille, FRANCE: see MARLY.

Mar-Mac (MAHR–mak), unincorporated town (□ 4 sq mi/10.4 sq km; 2000 population 3,004), WAYNE county, E central NORTH CAROLINA, residential suburb 6 mi/9.7 km SSW of downtown GOLDSBORO, near NEUSE RIVER; 35°19′N 78°03′W. Retail trade; construction.

Marmaduke (MAHR-muh-dook), town (2000 population 1,158), GREENE county, NE ARKANSAS, 10 mi/16 km NE of PARAGOULD, near SAINT FRANCIS River; 36°11′N 90°23′W. In agricultural area. Manufacturing (plastic products).

Marmagao, town, SOUTH GOA district, GOA state, W India, seaport on ARABIAN SEA, 7 mi/11.3 km SSW of PANAJI; 15°24′N 73°48′E. Railroad terminus. Outlet for products of interior and of S MAHARASHTRA state; exports rice, cotton, betel and cashew nuts, fish, salt, copra, manganese, timber; fish curing, fruit canning, wood carving; coir work. Consists of two sections, harbor and industrial area and the modern commercial and residential development of Vasco da Gama (airport nearby). Also spelled Mormugao.

Marmande (mahr-MAWND), town (□ 23 sq mi/59.8 sq km), LOT-ET-GARONNE department, AQUITAINE region, SW FRANCE, on the GARONNE RIVER, and 36 mi/58 km SE of BORDEAUX; 44°30′N 00°10′E. Agricultural market and brandy-distilling center; canning (tomatoes, plums, peaches, and melons) and fruit preserving; woodworking, machine shops. Has 13th–16th-century church.

Marmara Island, ancient *Proconnesus* (□ 29 sq mi/75 sq km), in Sea of MARMARA, NW TURKEY, 24 mi/39 km S of TEKIRDAG; 12 mi/19 km long, 5 mi/8 km wide. Rises to 1,991 ft/607 m in Llyas Dag. Extensive marble quarries from which its modern name is derived; visitors may witness the quarrying process. Most of the white marble used in ISTANBUL in the Byzantium and Ottoman periods was shipped from Marmara Island. An open-air museum displays Roman and Byzantine artifacts. Also spelled Marmora.

Marmara, Lake (□ 17 sq mi/44 sq km), W TURKEY, 28 mi/45 km E of MANISA; 7 mi/11.3 km long, 5 mi/8 km wide.

Marmara, Sea of (□ c.4,430 sq mi/11,474 sq km), NW TURKEY, between EUROPE (N) and ASIA (S); c.175 mi/280 km long and 50 mi/80 km wide. Connected on the NE with the BLACK SEA through the BOSPORUS and on the SW with the AEGEAN SEA (part of the MEDITERRANEAN SEA) through the DARDANELLES. ISTANBUL is located at the entrance of the Bosporus into the Sea of Marmara. The sea has no strong currents and the tidal range is minimal. In ancient times the sea was known as Propontis. Its modern name is derived from the small island of MARMARA, or Marmora.

Marmaris, township, SW TURKEY, port on a sheltered bay, 28 mi/45 km N of RHODES, 26 mi/42 km SSW of MUGLA; 36°52′N 28°17′E. Chromium mines nearby. Legumes, onions.

Marmarth (MAHR-muhth), village (2006 population 126), SLOPE county, SW NORTH DAKOTA, 26 mi/42 km WNW of BOWMAN, near MONTANA state line, on LITTLE MISSOURI RIVER, in Little Missouri National Grassland; 46°17′N 103°55′W. Ships livestock; wheat, hay. Fort Dilts Historic Site to E. Founded in 1908 and incorporated in 1909. Named for (Mar)garet (Marth)a Fitch, granddaughter of railroad executive Albert J. Earling.

Marmato (mahr-MAH-to), town, ⊙ Marmato municipio, CALDAS department, W central COLOMBIA, in CAUCA valley, 27 mi/43 km NNW of MANIZALES; 05°28′N 75°35′W. Elevation 4,390 ft/1,338 m. Coffee growing; plantains, sugarcane. Gold placer mines.

Marmaton River (MAHR-muh-tuhn), 73 mi/117 km long, including longest fork, SE KANSAS and MIS-SOURI; rises in E ALLEN county (Kansas), NE of MORAN; flows S then E, past FORT SCOTT, to LITTLE OSAGE RIVER 7 mi/11.3 km SSE of RICH HILL (MISSOURI).

Marmaverken (MAHR-mah-VER-ken), village, GÄVLEBORG county, E SWEDEN, on LJUSNAN RIVER, 5 mi/8 km WSW of SÖDERHAMN; 61°17′N 16°54′E.

Marmelade (mahr-muh-LAHD), town, ARTIBONITE department, N HAITI, in MASSIF DU NORD, 20 mi/32 km SSW of CAP-HAÏTIEN; 19°31′N 72°21′W. Fruits; coffee growing.

Marmeleiro (MAHR-me-LAI-ro), city, SW PARANÁ state, BRAZIL, 8 mi/12.9 km SE of FRANCISCO BELTRÃO; 26°08′S 53°02′W.

Mar Menor (MAHR mai-NOR), saltwater lagoon (□ 63 sq mi/163.8 sq km) on coast of MURCIA province, SE SPAIN, NE of CARTAGENA; maximum length 14 mi/23 km, width 1 mi/1.6 km, depth 20 ft/6 m. Separated from the MEDITERRANEAN SEA by flat, sandy spit cut by few passages. Some rocky islets. Fishing and swimming.

Marmet (mahr-MET), town (2006 population 1,620), KANAWHA county, W central WEST VIRGINIA, on KANAWHA RIVER, 9 mi/14.5 km SSE of CHARLESTON; 38°15′N 81°34′W. Coal-mining region. Agriculture (corn, tobacco); cattle, poultry. Manufacturing (chemicals). Kanawha State Forest to W. Incorporated 1921.

Marmion Lake (MAHR-mee-uhn), W ONTARIO, E central CANADA, 30 mi/48 km W of Lac des Milles Lacs, 110 mi/177 km WNW of THUNDER BAY; 14 mi/23 km long, 6 mi/10 km wide; 48°54′N 91°30′W. Elevation 1,363 ft/415 m. Drained S by SEINE RIVER.

Marmolada, ITALY: see DOLOMITES.

Marmolejo (mahr-mo-LAI-ho), town, JAÉN province, S SPAIN, near GUADALQUIVIR RIVER, 7 mi/11.3 km WNW of ANDÚJAR; 38°03′N 04°10′W. Olive-oil-production center. Manufacturing of willow articles, nougat candy, liqueurs; honey and wax processing. Cereals, wine, vegetables; livestock. Has mineral springs.

Marmolejo, Cerro (mahr-mo-LAI-ho, SER-ro), Andean peak (20,000 ft/6,096 m) on ARGENTINA-CHILE border, 27 mi/43 km SSW of TUPUNGATO, 50 mi/80 km SE of SANTIAGO; 33°44′S 69°53′W.

Marmora (MAHR-muh-ruh), township (□ 206 sq mi/535.6 sq km; 2001 population 3,985), SE ONTARIO, E central CANADA, 25 mi/40 km from BELLEVILLE; 44°32′N 77°40′W.

Marmora (MAHR-muh-ruh), unincorporated village (□ 1 sq mi/2.6 sq km; 2001 population 1,589), SE ONTARIO, E central CANADA, 27 mi/43 km NW of BELLEVILLE, and included in MARMORA township; 44°29′N 77°41′W. In mining area (gold, iron, marble); stellite refining, lumber milling.

Marmora Island, TURKEY: see MARMARA ISLAND.

Marmot Peak (11,735 ft/3,577 m), in ROCKY MOUNTAINS, CHAFFEE and PARK counties, central COLORADO, 7 mi/11.3 km N of BUENA VISTA.

Marmoutier (mahr-moo-TYAI), commune (□ 4 sq mi/10.4 sq km), BAS-RHIN department, in ALSACE, E FRANCE, in the VOSGES MOUNTAINS, 4 mi/6.4 km S of SAVERNE; 48°41′N 07°23′E. Footwear manufacturing. Its Benedictine abbey (founded c.600) was abolished in French Revolution; only abbatial church (with 12th-century facade) remains and is one of finest edifaces in Alsace.

Mar Muerto (mahr moo-ER-to), lagoon of Gulf of TEHUANTEPEC, on coast of OAXACA and CHIAPAS, S MEXICO, 60 mi/97 km ESE of JUCHITÁN de Zaragoza; 45 mi/72 km long, 2 mi/3.2 km–7 mi/11.3 km wide; 16°15′N 94°15′W. Connected to MAR MUERTO INFERIOR by marshes and channels.

Mar Muerto Inferior (mahr moo-ER-to een-fai-REE-or), lagoon in OAXACA, S MEXICO, on Isthmus of TEHUANTEPEC, 18 mi/29 km SE of Juchitán de Zaragoza Bay; connected through narrow channels with MAR MUERTO SUPERIOR and Gulf of TEHUANTEPEC.

Mar Muerto Superior (mahr moo-ER-to soo-pai-REE-or), inlet of Gulf of TEHUANTEPEC, in OAXACA, S MEXICO, S of JUCHITÁN DE ZARAGOZA; 18 mi/29 km long, 5 mi/8 km–12 mi/19 km wide. Connected by narrow channel with MAR MUERTO INFERIOR.

Marnaz (mahr-NAHZ), town (□ 3 sq mi/7.8 sq km), HAUTE-SAVOIE department, RHÔNE-ALPES region, SE FRANCE, near ARVE RIVER, 3 mi/4.8 km W of CLUSES, and 20 mi/32 km SE of GENEVA, in FAUCIGNY valley of the Savoy Alps (ALPES FRANÇAISES); 46°04′N 06°32′E. Manufacturing of precision tools.

Marne (MAHRN), department (□ 3,151 sq mi/8,192.6 sq km), in old CHAMPAGNE province, N FRANCE; ⊙ CHÂLONS-EN-CHAMPAGNE; 48°55′N 04°10′E. Across it, E-W, are the ARGONNE hills, the humid Champagne (dairying), the Champagne badlands (*Champagne Pouilleuse*), and a wooded crest. Drained by the MARNE, the AISNE, and their tributaries. Here are the famous vineyards that supply the champagne industry of REIMS and ÉPERNAY and the extensive sheep ranges of the Champagne badlands that furnish raw material for the region's woolen manufacturing. Other agricultural products include wheat, oats, sugar beets. Lumbering in the Argonne forests and in the more recently planted pine stands of the Champagne badlands. Chief towns are Reims, Châlons-en-Champagne (commercial center), Épernay (champagne processing and storing), VITRY-LE-FRANÇOIS (transportation center). Most of department was battlefield in World War I. Forms part of the CHAMPAGNE-ARDENNE region.

Marne (MAHR-me), town, SCHLESWIG-HOLSTEIN, NW GERMANY, 20 mi/32 km W of ITZEHOE, in the S DITHMARSCHEN; 53°57′N 09°01′E. Food processing (sauerkraut, canned fish), brewing. Seed-selection station. Market center for surrounding reclaimed polder land (cattle; grain, vegetables). Chartered 1891.

Marne, town (2000 population 149), CASS county, SW IOWA, 6 mi/9.7 km NW of ATLANTIC; 41°27′N 95°06′W.

Marne-Aisne Canal, FRANCE: see AISNE-MARNE CANAL.

Marne, Haute-, FRANCE: see HAUTE-MARNE.

Marne-la-Vallée (MAHRN–lah–vah-LAI), town, SEINE-ET-MARNE department, ÎLE-DE-FRANCE region, N central FRANCE, on left bank of MARNE RIVER, and 20 mi/32 km E of central PARIS; 48°50′N 02°38′E. Site of Disneyland-Paris (formerly known as Euro-Disney) and of the Descartes science park. One of 5 new towns planned (since 1960s) to relieve congestion in the major suburbs and Paris itself. Reached via the Eastern expressway and the high-speed suburban railroad line (R.E.R.). Construction of residential developments and commercial and light industry facilities began here in 1972 with the support of the national government. Among the growth nodes are NOISY-LE-GRAND (college of electrical and electronics engineering); Val Maubuée (housing and sports complex, a chocolate factory, center of art and culture); Val de Bussy (planned office and research parks); and Val d'Europe where Disneyland-Paris and related accommodation facilities are located.

Marne-Rhine Canal (MAHRN–REIN), French *Canal de la Marne au Rhin*, (kah-nahl duh lah mahrn o rhan), 195 mi/314 km long, E FRANCE; 48°35′N 07°47′E. Connects the MARNE (at VITRY-LE-FRANÇOIS) with the RHINE RIVER (at STRASBOURG); follows ORNAIN RIVER upstream, past BAR-LE-DUC; crosses into MEUSE RIVER valley in tunnel (3 mi/4.8 km long); reaches the Meuse at Troussey (junction with Canal de l'EST); joins the MOSELLE RIVER at TOUL, paralleling it to FROUARD; then ascends the MEURTHE valley, past NANCY; traverses Lorraine plateau N of LUNÉVILLE; crosses the N VOSGES in SAVERNE GAP, entering Alsatian lowland E of SAVERNE. Built 1841–1852.

Marne River (MAHRN), river, 325 mi/523 km long, CHAMPAGNE-ARDENNE region, NE FRANCE; rises in the LANGRES plateau; flows in an arc generally NW to the SEINE RIVER just above PARIS; 48°55′N 04°11′E. It passes through CHAUMONT, SAINT-DIZIER (storage lake of DER nearby), VITRY-LE-FRANÇOIS, CHÂLONS-EN-CHAMPAGNE, and MEAUX. The MARNE-RHINE and MARNE-SAÔNE canals connect this river with the AISNE, MEUSE, MOSELLE, and the RHÔNE. In both world wars, the Marne region was the scene of fighting between the Allies and the Germans.

Marne-Saône Canal (mahrn–SON), c.130 mi/209 km long, E FRANCE, between SAÔNE and MARNE rivers; begins at Pontailler-sur-Saône; runs N across LANGRES PLATEAU (tunnel, 3.5 mi/5.6 km long), enters Marne River valley near Langres and follows it downstream past CHAUMONT and SAINT-DIZIER to VITRY-LE-FRANÇOIS. From there, it is continued by lateral canal to the Marne, thus forming a link between RHÔNE-SAÔNE River valley and PARIS.

Marneuli (mahr-ne-OO-lee), city (2002 population 20,065) and center of Marneuli region, S GEORGIA, on railroad, on Algeti River (KURA RIVER BASIN), and 15 mi/24 km S of TBILISI; 41°28′N 44°48′E. Food processing; building-material enterprises; haberdashery factory. Until 1947, Borchalo.

Marnhull (MAH-nuhl), village (2001 population 1,951), N DORSET, SW ENGLAND, 6 mi/9.7 km WSW of SHAFTESBURY; 50°58′N 02°18′W. Has 14th-century church.

Maro (mah-RO), town, MOYEN-CHARI administrative region, S CHAD, near border of CENTRAL AFRICAN REPUBLIC, 75 mi/121 km SE of SARH; 08°25′N 18°46′E.

Maroa (mah-RO-ah), city (2000 population 1,654), MACON county, central ILLINOIS, 12 mi/19 km N of DECATUR; 40°02′N 88°57′W. In agricultural area (corn, soybeans).

Maroa (mah-RO-ah), town, ⊙ Maroa municipio, AMAZONAS state, S VENEZUELA, landing on Río Guainía (see Río NEGRO) near COLOMBIA border, and 200 mi/322 km S of PUERTO AYACUCHO; 02°44′N 67°33′W. In tropical forest region (rubber, balata, vanilla).

Maroantsetra (mah-roo-ahn-TSAICH), town, TOAMASINA province, NE MADAGASCAR, small port in ANTONGIL BAY, on coast, 190 mi/306 km N of TOAMASINA; 15°25′S 49°45′E. Very rainy climate. Rice, coffee, cloves, vanilla. Hosp. Near Nosy Mangabe (island nature reserve in ANTONGIL BAY) and MASOALA NATIONAL PARK (established 1995).

Marochak, AFGHANISTAN: see MARUCHAK.

Maroim, Brazil: see MARUIM.

Marojejy Massif (mah-roo-ZAIDZ), mountain range, ANTSIRANANA province, NE MADAGASCAR, 30 mi/48 km WSW of SAMBAVA; 14°27′S 49°46′E. Steep forested range near ANDAPA, rising to 6996 ft/2133 m. Location of MAROJEJY NATURE RESERVE.

Marojejy Nature Reserve (mah-roo-ZAIDZ), ANTSIRANANA province, NE MADAGASCAR, 30 mi/48 km WSW of SAMBAVA; 14°26′S 49°45′E. Protects diverse lowland and montane rain forests, mountain heath, and lemur on steep MAROJEJY MASSIF (250 ft/76 m–7,000 ft/2,134 m).

Maroldsweisach (mah-rolds-VEI-sahkh), village, LOWER FRANCONIA, BAVARIA, GERMANY, 15 mi/24 km WSW of COBURG; 50°12′N 10°39′E. Manufacturing of synthetic fiber.

Marolles-en-Brie (mah-ROL–zahn–BREE), town (□ 1 sq mi/2.6 sq km), VAL-DE-MARNE department, ÎLE-DE-FRANCE region, N central FRANCE, 13 mi/21 km SE of PARIS; 48°44′N 02°33′E.

Marolles-en-Hurepoix (mah-ROL–zahn–oor-PWAH), town (□ 2 sq mi/5.2 sq km), ESSONNE department, ÎLE-DE-FRANCE region, N central FRANCE, 3 mi/5 km ESE of ARPAJON; 48°34′N 02°18′E.

Marolles-les-Braults (mah-ROL–lai–BRO), commune (□ 7 sq mi/18.2 sq km), SARTHE department, PAYS DE LA LOIRE region, W FRANCE, 17 mi/27 km SE of ALENÇON; 48°15′N 00°19′E. Rope making; cider mill.

Maromandia (mah-roo-mahn-DEE-ah), village, MAHAJANGA province, NW MADAGASCAR, near W coast, on highway, and 40 mi/64 km SSW of AMBANJA; 14°12′S 48°06′E. Rice market; fiber growing.

Maromitsa (mah-ruh-MEE-tsah), village (2005 population 1,815), NW KIROV oblast, E central European Russia, on road and railroad, 9 mi/14 km W of OPARINO; 59°52′N 48°01′E. Elevation 721 ft/219 m. In agricultural area (wheat, oats, flax, rye, vegetables; livestock); sawmilling.

Maromme (mah-ROM), town (□ 1 sq mi/2.6 sq km), SEINE-MARITIME department, HAUTE-NORMANDIE region, N FRANCE, 4 mi/6.4 km NW of ROUEN; 49°28′N 01°02′E. Metalworking, textile dyeing. Market for dairy products.

Marondera (mah-RO-en-dai-rah), town, ⊙ MASHONALAND EAST province, E ZIMBABWE, 40 mi/64 km SE of HARARE, on railroad; 18°11′S 31°32′E. Elevation 5,446 ft/1,660 m. Manufacturing (farm supplies). Dairying; cattle, sheep, goats; tobacco, peanuts, corn, citrus fruit. Site of Grasslands and Horticultural research centres and Kushinga-Phikelela Agricultural Institute. Formerly called Marandellas.

Marongora (mah-RO-en-go-rah), village, MASHONALAND WEST province, N ZIMBABWE, 105 mi/169 km NW of CHINHOYI, on ZAMBEZI escarpment, in Hurungwe Safari Area, at junction of road into MANA POOLS NATIONAL PARK; 16°14′S 29°10′E. Tourism.

Maroni River (MAH-ro-nee), Dutch *Marowijne* (MAH-RO-vay-nuh), c.450 mi/724 km long, in the Guianas, SOUTH AMERICA; rises in TUMUC-HUMAC MOUNTAINS of FRENCH GUIANA near BRAZIL border; flows N through tropical forest region for its greater part along French Guiana–SURINAME border, past ALBINA and SAINT-LAURENT-DU-MARONI, to the ATLANTIC OCEAN at GALIBI POINT; 05°30′N 54°00′W. Navigable for smaller vessels c.60 mi/97 km upstream. Interrupted by many waterfalls. Its upper course is called the ITANY (Litani); its mid-course, to the mouth of the TAPANAHONI, is called the LAWA or Aoua. Gold placers along the Aoua. Area between the Maroni and the Itany in French Guiana is claimed by Suriname. Sometimes also spelled Morouini.

Maroochydore (ma-ROOCH-i-dor), town, SE QUEENSLAND, AUSTRALIA, 70 mi/113 km N of BRISBANE, 10 mi/16 km E of NAMBOUR, on PACIFIC OCEAN, near Maroochy River estuary; 26°39′S 153°06′E. Popular beach resort; beaches, fishing, and boating. Sugarcane, tropical fruit, and citrus. Light industry. Maroochy Airport to N at Mudgimba.

Maroondah (muh-ROON-duh), city, E suburb of MELBOURNE, VICTORIA, SE AUSTRALIA. Established 1994.

Maroon Peak (14,156 ft/4,315 m), in ELK MOUNTAINS, PITKIN and GUNNISON counties, W central COLORADO, 12 mi/19 km SW of ASPEN. One of three peaks, collectively called the Maroon Bells, SW of Maroon Lake, among the most photographed areas of Colorado. In Maroon Bells Snowman Wilderness Area.

Maroon Town, town, SAINT JAMES parish, NW JAMAICA, in the COCKPIT COUNTRY, 14 mi/23 km SE of MONTEGO BAY; 18°21′N 77°49′W. Here the Maroon leader Cudjoe signed a peace treaty with the British in 1739 and was given 1,500 acres/607 ha, which they named "Trelawny Town." Later rebellious Maroons made their last stand (1795). Formerly Trelawny Town.

Maros (MAH-ross), town, ⊙ Maros district, Sulawesi Seletan province, INDONESIA, 18 mi/29 km NW of UJUNG PANDANG.

Marosberkes, ROMANIA: see BIRCHIŞ.

Maróshéviz, ROMANIA: see TOPLIŢA.

Marosillye, ROMANIA: see ILIA.

Marosludas, ROMANIA: see LUDUŞ.

Maros River, ROMANIA and HUNGARY: see MUREŞ RIVER.

Marostica (mah-RO-stee-kah), town, VICENZA province, VENETO, N ITALY, near BRENTA River, 15 mi/24 km NNE of VICENZA; 45°45′N 11°39′E. Machinery, fabricated metals, plastics, clothing.

Marosújvár, ROMANIA: see OCNA MUREŞ.

Marosvásárhely, ROMANIA: see TÎRGU MUREŞ.

Marotiri Isles or **Bass Isles**, small, uninhabited islets, FRENCH POLYNESIA, S PACIFIC, 50 mi/80 km SE of RAPA; 27°55′S 143°30′W. Governed with AUSTRAL ISLANDS. Also spelled Morotiri.

Maroua (MAH-roo-uh), town (2001 estimated population 271,700; 2004 estimated population 297,600), ⊙ DIAMARÉ department and Far-North province, CAMEROON, 500 mi/805 km NNE of Yaoundé, 120 mi/193 km SSW of N'Djamena (CHAD); 10°35′N 14°20′E. Elevation 2,395 ft/730 m. Administrative and tourist center. Livestock-raising center, communications point. Peanuts, cotton; food processing and textile industries. Has hospital, airport, agricultural school, and veterinary station.

Maroussi, Greece: see AMAROUSION.

Marovoay (mah-roo-VOO-ei), town, MAHAJANGA province, NW MADAGASCAR, on highway, at head of BETSIBOKA RIVER estuary, and 56 mi/94 km SSE of MAHAJANGA; 16°09′S 46°38′E. Fertile agricultural center specializing in quality rice, cotton, tobacco, peanuts; cattle. Rice processing.

Marowijne (MAH-RO-vai-nuh), district (1,786 sq mi/4,627 sq km; 2004 population 16,642), NE SURINAME, ⊙ ALBINA. Populated area with large-scale agriculture, especially rice. Bauxite mining.

Marowijne River, wide estuary of the MARONI RIVER, leading to ATLANTIC OCEAN, forming NE SURINAME border and NW FRENCH GUIANA border. Ferry connects ALBINA (Suriname) to Saint-Laurent (French Guiana).

Marpingen (MAHR-ping-uhn), town, SAARLAND, GERMANY, on small Alsbach River, 9 mi/14.5 km NW of NEUNKIRCHEN; 49°27′N 07°04′E. Metalworking.

Marple (MAH-puhl), town (2001 population 23,480), GREATER MANCHESTER, W ENGLAND, 4 mi/6.4 km E of STOCKPORT; 53°24′N 02°04′W.

Marquand (MAHR-kwahnd), town (2000 population 251), MADISON county, SE MISSOURI, in the SAINT FRANCOIS MOUNTAINS, on CASTOR River, and 11 mi/18 km SE of FREDERICKTOWN; 37°25′N 90°10′W. Cattle; timber. Surrounded by Mark Twain National Forest.

Marquard, town, ⊙ FREE STATE province, SOUTH AFRICA, 80 mi/129 km ENE of MANGAUNG (BLOEMFONTEIN), on the Laaispruit River (tributary of Vet River); 28°40′S 27°27′E. Elevation 4,977 ft/1,517 m. Railroad terminus. Grain; livestock. Airfield.

Marquesado (mahr-ke-SAH-do), town, ⊙ Rivadavia department, S SAN JUAN province, ARGENTINA, in SAN JUAN RIVER valley (irrigation area), 5 mi/8 km W of SAN JUAN; 31°31′S 68°37′W. Railhead in wine-grape- and fruit-growing area; wine making.

Marquesas Islands (mahr-KE-zuhs), volcanic group, South PACIFIC, a part of FRENCH POLYNESIA, c.740 mi/1,190 km NE of TAHITI; 09°00′S 139°30′W. There are twelve islands in the group, the largest being NUKU HIVA, site of TAIOHAE, the administrative center, and HIVA OA (second-largest), site of ATUONA, a former administrative center. The Marquesas, famous for their rugged beauty, are fertile and mountainous, rising directly from the sea, without barrier coral reefs, reaching 3,904 ft/1,190 m on Hiva Oa. There are breadfruit, pandanus, and coconut trees; wild cattle and hogs. The chief exports are copra, tobacco, and vanilla. Taiohae Bay, on Nuka Hiva, and the Bay of Traitors, on Hiva Oa, are the major harbors. The islands are divided into two groups. The S cluster (sometimes called the Mendaña Islands), including FATU HUKU, Hiva Oa, TAHUATA, MOHOTANI, and FATU HIVA, was discovered for Europeans in 1595 by the Spanish navigator Alvaro de Mendaña de Neira;

the N group (sometimes called the Washington Islands), including HATUTAA, EIAO, MOTU ITI, Nuku Hiva, Ua Huku, UA POU, and MOTU ONE, was discovered in 1791 by the American navigator Captain Joseph Ingraham. In 1813, Commodore David Porter claimed Nuku Hiva for the U.S., naming it Madison Island, but the U.S. Congress never ratified the claim. France took possession of the islands in 1842 and established a settlement on Nuku Hiva, which was abandoned in 1859. In 1870 the French administration over the Marquesas was reinstated. Of all the Polynesian peoples, the Marquesans suffered one of the greatest declines from the spread of European diseases; before the 1850s they may have approximated 20,000, about four times the present population. The islands are the setting for Herman Melville's novel *Typee*.

Marquês de Valença, BRAZIL: see VALENÇA.

Marqués, Villa de, MEXICO: see LA CAÑADA.

Marquetalia (mahr-kai-TAH-lee-ah), town, ⊙ Marquetalia municipio, CALDAS department, W central COLOMBIA, 33 mi/53 km ENE of MANIZALES; 05°17′N 75°03′W. Elevation 4,196 ft/1,278 m. Coffee, plantains, sugarcane; livestock.

Marquette (mahr-KET), county (□ 3,426 sq mi/8,907.6 sq km; 2006 population 64,675), NW UPPER PENINSULA, NW MICHIGAN, on LAKE SUPERIOR (N); 46°39′N 87°35′W; ⊙ MARQUETTE. Bordered S and SW by MENOMINEE, DICKINSON and IRON counties and Central time zone (county in Eastern time zone). Drained by DEAD and MICHIGAMME rivers, and by several branches of ESCANABA RIVER. Includes the MARQUETTE IRON RANGE and HURON MOUNTAINS. Some mining; lumbering; food processing. Manufacturing at Marquette. Cattle; forage, potatoes. Resorts (fishing, hunting, camping). A national experimental forest, fish hatchery and several lakes are in co. Small sub-unit of Ottawa National Forest on E border; Van Ripen State Park in W; Huron Mountains Wilderness in NW; Moose Range Cliffs Ridge Ski Area in NE. Nearby K.I. Soyer Air Force Base was closed in 1995 and there is an ongoing attempt to redevelop its facilities into commercial and industrial services. Organized 1851.

Marquette, county (□ 464 sq mi/1,206.4 sq km; 2006 population 154), S central WISCONSIN; ⊙ MONTELLO; 43°49′N 89°23′W. Agriculture (corn, soybeans; hogs, sheep); manufacturing (processing of dairy products); granite quarries. Drained by FOX RIVER and its tributaries; includes BUFFALO LAKE at center of county; several small lakes in N. Westfield State Fish Hatchery in W. Formed 1836.

Marquette (mahr-KET), city (2000 population 19,661), ⊙ MARQUETTE county, N MICHIGAN, UPPER PENINSULA, 140 mi/225 km W of SAULT SAINTE MARIE, on LAKE SUPERIOR; 46°32′N 87°24′W. Once an iron ore shipping port, it is now a shipping center for a lumber, cattle, and resort region. Railroad spur terminus (ship/RR transfer). Manufacturing includes chemicals, wood products, and mining machinery, dairy and bakery products; publishing. Ore Docks handle 7 million tons/6.4 million metric tons of iron ore annually. Seat of Northern Michigan University (has Olympic Training Center). Marquette County Airport to W. A branch of the state prison is also here. Maritime Museum. Cliffs Ridge Ski Area to S. Settled 1849, incorporated as a city 1871.

Marquette, town (2000 population 421), CLAYTON county, NE IOWA, on MISSISSIPPI RIVER (bridged here), opposite PRAIRIE DU CHIEN (WISCONSIN), and 13 mi/21 km NE of ELKADER; 43°02′N 91°10′W. Asphalt plant equipment; storage silos. Upper Mississippi National Wildlife Refuge; Effigy Mounds National Monument 3 mi/4.8 km N; Yellow River State Forest 10 mi/16 km N.

Marquette (mahr-KET), village (2000 population 542), MCPHERSON county, central KANSAS, on SMOKY HILL RIVER, and 23 mi/37 km SSW of SALINA; 38°32′N 97°49′W. Grain milling.

Marquette, village (2006 population 276), HAMILTON county, SE central NEBRASKA, 9 mi/14.5 km N of AURORA, near PLATTE RIVER; 41°00′N 98°00′W. Manufacturing (foods).

Marquette Heights, city (2000 population 2,794), TAZEWELL county, central ILLINOIS, residential suburb 5 mi/8 km S of downtown PEORIA, and S of CREVE COEUR, near ILLINOIS RIVER; 40°37′N 89°35′W.

Marquette Iron Range (mahr-KET), NW UPPER PENINSULA, NW MICHIGAN, low range in MARQUETTE county, lying generally W of MARQUETTE. Once a rich iron-mining region with mining center at Ishpeming and NEGAUNEE. National Mine still active with 7,000,000 tons/7,714,000 metric tons shipped out of Marquette annually. Also known as Marquette Range. Michigan Iron Industry Museum here.

Marquette Island, MICHIGAN: see LES CHENEAUX ISLANDS.

Marquette Range, MICHIGAN: see MARQUETTE IRON RANGE.

Marquez (MAHR-kez), village (2006 population 233), Leon county, E central TEXAS, 40 mi/64 km N of BRYAN, near NAVASOTA RIVER (forms LAKE LIMESTONE to NW); 31°14′N 96°15′W. Railroad junction to SW. Agricultural area.

Marquèze, FRANCE: see SABRES.

Marquez River (mahr-KES), 70 mi/113 km long, W BOLIVIA; rises on W slopes of Cordillera de los FRAILES, 20 mi/32 km E of Río Mulato, POTOSÍ department; flows WNW to POOPÓ Lake just E of PAMPA AULLAGAS; 19°10′S 67°03′W. Receives Mulato River.

Marquina (mahr-KEE-nah), town, VIZCAYA province, N SPAIN, 21 mi/34 km E of BILBAO. Manufacturing of arms, handballs; sawmilling. Mineral springs. Marble and stone quarries nearby.

Marquirivi (mahr-kee-REE-vee), canton, AROMA province, LA PAZ department, W BOLIVIA. Elevation 12,851 ft/3,917 m. Gas wells in area. Salt extraction; copper deposits; gypsum, clay, limestone deposits. Agriculture (potatoes, yucca, bananas, rye); cattle.

Marquirivi, canton, SAN PEDRO DE TOTORA province, ORURO department, W central BOLIVIA, 14 mi/22 km E of Cerro de Carangas, on the LA PAZ–ORURO road; 18°01′S 68°02′W. Elevation 12,795 ft/3,900 m. Gas wells in area. Copper deposits; clay, limestone, gypsum deposits. Agriculture (potatoes, yucca, bananas, rye); cattle.

Marquise (mahr-KEEZ), town (□ 5 sq mi/13 sq km), PAS-DE-CALAIS department, NORD-PAS-DE-CALAIS region, N FRANCE, 8 mi/12.9 km NNE of Boulogne; 50°49′N 01°42′E. Marble quarrying. Has 12th–16th-century church.

Marrabios, Cordillera de los, NICARAGUA: see MARIBIOS, CORDILLERA DE LOS.

Marracuene (mah-rah-KWAIN-ai), village, MAPUTO province, S MOZAMBIQUE, on MAPUTO BAY, at mouth of Incompati River, 18 mi/29 km N of MAPUTO (linked by railroad); 25°44′S 32°41′E. Manufacturing of cement, crating; peanuts, cotton, rice, corn. Formerly VILA LUIZA.

Marrakech (mah-rah-KESH), city (2004 population 823,154), ⊙ MARRAKECH-TENSIFT-AL HAOUZ administrative region, W central MOROCCO; 31°38′N 08°00′W. Renowned for leather goods and crafts; one of the principal commercial centers of Morocco and a major tourist center. Founded (1062) by the Almoravid leader Youssef ben Tachfine and was capital of Morocco until 1147, and again 1550–1660. Captured by the French in 1912. Beautifully situated near the snow-capped ATLAS MOUNTAINS, Marrakech has extensive gardens, a 16th-century palace, and a former palace that is now a museum of Moroccan art and crafts. Favorite winter vacation destination of Winston Churchill who made several paintings of the local landscape. The 253-ft/77-m minaret (completed 1195) of the Koutoubia mosque dominates the city. Seat of the Université Ben Youssef, a center of Islamic studies. Royal residence, and a public university here. International airport to SW.

Marrakech-Tensift-Al Haouz, administrative region (2004 population 3,102,652), W MOROCCO; ⊙ MARRAKECH. Bordered by Doukkala-Abda (NWN), Chaouia-Ouardigha (NNE), Tadla-Azilal (E), and Souss-Massa-Draâ (S) administrative regions and ATLANTIC OCEAN (W). Railroad travels NNE from Marrakech through CASABLANCA, RABAT, and KENITRA, then SE to MEKNES (joining railroad from TANGER) where it joins the Meknes-OUJDA railroad. Roads run through the region, connecting it to the rest of Morocco. International airport at Marrakech. Further divided into five secondary administrative divisions called prefectures and provinces: Al Haouz, Chichaoua, El Kelaâ des Sraghna, Essaouira, and Marrakech.

Marrak Point (MAH-rahk), Greenlandic *Marraq*, small peninsula, SW GREENLAND, on DAVIS STRAIT, 55 mi/89 km SSE of NUUK (Godthåb); 63°25′N 51°15′W. Airfield in World War II, called Bluie West 4.

Marra Mountain (mah-RAH), Arabic *Jabal Marra*, highest section of NILE–Lake CHAD watershed, in Central region, W SUDAN, SW of AL-FASHER; c.100 mi/161 km long, 20 mi/32 km wide. Rises to c.10,130 ft/3,088 m. Of volcanic origin.

Marraquene District, MAPUTO, MOZAMBIQUE.

Marrargiu, Cape (mah-RAHR-jyoo), point on W coast of SARDINIA, ITALY, NW of BOSA; 40°20′N 08°22′E. Fisheries (tunny, lobster, coral).

Marratxí (mah-rah-CHEE), town, MAJORCA island, BALEARIC ISLANDS, on railroad, and 5 mi/8 km NE of PALMA; 39°39′N 02°48′E. In agricultural region (cereals, almonds, grapes, raisins); wine making, oil milling.

Marrawah (MA-ruh-wah), village, NW TASMANIA, AUSTRALIA, 130 mi/209 km WNW of LAUNCESTON; 40°56′S 144°42′E. Tasmania's westernmost settlement. Agriculture center. Important Aboriginal sites in area.

Marrazes (mahr-RAH-zesh), village, LEIRIA district, W central PORTUGAL, just N of LEIRIA; 39°46′N 08°48′W.

Marree (muh-REE), settlement, E central SOUTH AUSTRALIA, on PORT PIRIE–ALICE SPRINGS railroad, 230 mi/370 km N of PORT PIRIE, in CENTRAL AUSTRALIA desert area; 29°39′S 138°04′E. Lake Eyre is N. Cattle.

Marrero (MER-ER-o), unincorporated city (2000 population 36,165), JEFFERSON parish, SE LOUISIANA, on the MISSISSIPPI RIVER, suburb opposite NEW ORLEANS, 3 mi/5 km SSW of downtown; 29°54′N 90°07′W. Manufacturing (machinery, building materials). Bayou Segnette State Park to SW.

Marri-Bugti Territory (MUHR-ee-BUG-tee), tribal area in SIBI district, SIBI division, BALUCHISTAN province, SW PAKISTAN, at junction of NE end of CENTRAL BRAHUI and S SULAIMAN ranges. Barren, hilly region, with scanty rainfall; inhabited by Baluch tribes (Marris, Bugtis).

Marrickville (MA-rik-vil), municipality, E NEW SOUTH WALES, SE AUSTRALIA, inner suburb 4 mi/6 km SW of SYDNEY; 33°55′S 151°10′E. In metropolitan area, near airport. Railway station. Manufacturing (pet food, footwear). Ethnically diverse.

Marromeu (mah-RO-mai-oo), village, SOFALA province, central MOZAMBIQUE, on right bank of lower ZAMBEZI River, and 130 mi/209 km NNE of BEIRA; 18°17′S 35°56′E. Terminus of railroad spur from CAIA. Sugar-growing and processing center amidst large plantations.

Marromeu District, MOZAMBIQUE: see SOFALA.

Marroquí, Point (mah-ro-KEE), southernmost cape of European mainland, in CÁDIZ province, S SPAIN, on STRAIT OF GIBRALTAR, just S of TARIFA; 36°00′N

05°36′W. Lighthouse. Generally considered dividing point between the MEDITERRANEAN SEA and the AT-LANTIC OCEAN. Sometimes written Marroqui. Also called Tarifa Point.

Marrowbone, unincorporated village, CUMBERLAND county, S KENTUCKY, 7 mi/11.3 km WNW of BURKESVILLE, on Marrowbone Creek. Burley tobacco, corn; cattle; diarying. Manufacturing (apparel).

Marruás, Brazil: see PÔRTO.

Marrupa (mah-ROO-puh), village, NIASSA province, N MOZAMBIQUE, on road, and 200 mi/322 km W of PEMBA; 13°11′S 37°29′E. Corn, sorghum.

Marrupa District, MOZAMBIQUE: see NIASSA.

Mars, borough (2006 population 1,693), BUTLER county, W PENNSYLVANIA, 16 mi/26 km N of PITTS-BURGH, on Breakneck Creek; 40°42′N 80°00′W. Agriculture (corn, hay, apples; livestock; dairying). Manufacturing (consumer goods, fabricated metal products, building materials, machinery; steel fabrication). Incorporated 1882.

Marsa (mahr-sah), town (2005 population 5,344), E MALTA, at head of GRAND HARBOUR, 2 mi/3.2 km SW of VALLETTA; 35°52′N 14°29′E. Ship repairing. Severely damaged during World War II, but its monuments suffered little.

Marsa, Al (mahr-SAH, el), town (2004 population 77,890), TUNIS province, N TUNISIA, on GULF OF TUNIS, 10 mi/16 km NE of TUNIS. Fashionable resort; beylical and archiepiscopal residences, summer palace of French resident general. Administrative center. A convention was held here in 1883 which implemented the treaty of Al Bardu (1881) establishing a French protectorate in TUNISIA.

Marsa ʾAlam (mahr-SAH ah-ah-LAM), port and township, RED SEA PROVINCE, E EGYPT, 85 mi/137 km SSE of KOSSEIR; 25°05′N 34°54′E. Gold mining. Previously known as Es-Sukkari.

Marsa Al Burayqah (MAHR-suh el buhr-AI-kuh), oldest and largest oil terminal, CYRENAICA region, NE LIBYA, E coast of Gulf of SURT, 125 mi/201 km WSW of BANGHAZI, and 50 mi/80 km WSW of AJDABIYAH. Natural-gas-liquefaction plant; petrochemicals (ammonia, ethanol). Pipeline to NASSIR (Zaltan) oil fields.

Marsa al Harīga (MAHR-sul el HAHR-ree-guh), oil terminal with refinery, CYRENAICA region, NE LIBYA, 1 mi/1.6 km E of TOBRUK, on MEDITERRANEAN coast. Natural-gas-liquefaction plant; petrochemical industries. Pipeline to SARIR field.

Mar Saba, ISRAEL: see KIDRON STREAM.

Marsabit (mahr-SAH-beet), town, EASTERN province, district administrative center, N KENYA, in volcanic Marsabit range (rising to 5,594 ft/1,705 m), on road, and 140 mi/225 km N of ISIOLO; 02°21′N 37°59′E. Nomadic livestock raising within Marsabit National Reserve. Airfield.

Marsa el Hamra, EGYPT: see ALAMEIN, EL.

Marsafa, village, QALYUBIYA province, Lower EGYPT, 5 mi/8 km SE of BENHA. Cotton, flax, cereals, fruits.

Marsal (mahr-SAHL), commune (□ 4 sq mi/10.4 sq km), MOSELLE department, in LORRAINE, NE FRANCE, on the SEILLE RIVER, 1 mi/8 km SE of CHÂTEAU-SALINS; 48°47′N 06°36′E. Its fortifications date back to Gallo-Roman times, when salt was mined here. LOR-RAINE NATURAL REGIONAL PARK is nearby.

Marsala (mahr-SAH-lah), ancient *Lilybaeum*, city (2001 population 77,784), W SICILY, ITALY, a port on the MEDITERRANEAN Sea, on Cape BOEO; 37°48′N 12°26′E. Noted for its sweet wine. Named Marsah al Allah [port of God] by the Arabs. In 1860, Garibaldi landed here at the start of his successful campaign to conquer the kingdom of the Two Sicilies.

Marsamuscetto Harbour (mahr-sah-moo-SHET-to), MEDITERRANEAN inlet (c.1.5 mi/2.4 km long) of E MALTA, on W shore of peninsula on which VALLETTA is built. Branches off into several creeks which enclose GZIRA island. Lined by naval installations and dock-yards.

Marsannay-la-Côte (mahr-sah-NAI–lah–KOT), town (□ 5 sq mi/13 sq km), CÔTE-D′OR department, in BURGUNDY, E central FRANCE, on E slopes of the CÔTE D′OR hills, 4 mi/6.4 km SSW of DIJON; 47°16′N 04°59′E. Burgundy, rosé, and red wines.

Marsa Scala (MAHR-sah SKAH-lah), fishing port (2005 population 9,346), SE MALTA, 3.5 mi/5.6 km SE of VALLETTA; 35°52′N 14°33′E.

Marsassoum (MAHR-sah-soom), village (2004 population 6,677), KOLDA administrative region, SW SE-NEGAL, landing on left bank of SOUNGROUGROU River, and 25 mi/40 km NE of ZIGUINCHOR; 12°50′N 16°00′W. Peanuts, timber.

Marsa Susa (MAHR-suh SOO-suh), township, CYRE-NAICA region, N LIBYA, port on MEDITERRANEAN SEA, on coastal road, 40 mi/64 km WNW of Darnah. Manufacturing (flour milling; domestic weaving). Port of nearby CYRENE in ancient times; flourished 4th century B.C.E. Greek and Roman ruins (walls, theater); remains of 5th-century. Christian basilica. Reestablished 1897 by Muslim refugees from CRETE. Formerly Apollonia.

Marsaxlokk (MAHR-sahsh-LAWK), town, SE MALTA, fishing port on Marsaxlokk Bay (c.2 mi/3.2 km long, 1 mi/1.6 km wide), 4.5 mi/7.2 km SSE of VALLETTA; 35°50′N 14°32′E. Has 18th-century fortifications, 17th-century palaces and churches. Port of Marssaxlokk is a free port, with no excise taxes on goods.

Marsberg (MAHRS-berg), town, WESTPHALIA-LIPPE, North Rhine-Westphalia, W GERMANY, in the SAUERLAND, on the DIEMEL RIVER, 18 mi/29 km SSE of PADERBORN; 51°28′N 08°51′E. Manufacturing (machinery, glass, textiles); metalworking. Has baroque church (originally built 13th century). Created by unification of OBERMARSBERG, NIEDERMARSBERG (chartered in 12th century) and fifteen smaller villages in 1975.

Marsciano (mahr-SHYAH-no), town, PERUGIA province, UMBRIA, central ITALY, 14 mi/23 km SSW of PERUGIA; 42°54′N 12°20′E. Foundry, cotton dyeworks; manufacturing of nails, agricultural tools, tower clocks.

Marsden (MAHZ-duhn), town (2001 population 3,575), WEST YORKSHIRE, N ENGLAND, on COLNE RIVER, and 7 mi/11.3 km WSW of HUDDERSFIELD; 53°36′N 01°55′W. Site of Huddersfield reservoirs.

Marsdiep (MAHRZ-deep), strait, NORTH HOLLAND province, NW NETHERLANDS, connects NORTH SEA (W) to the WADDENZEE (E); 2 mi/3.2 km wide, 3 mi/4.8 km long. Separates TEXEL island (N) from mainland. Port of DEN HELDER on S shore; car ferry to ′t-Horntje, Texel. W part also known as Helsdeur (Hell′s Door). Noorderhaaks, sand island (4 mi/6.4 km long E-W) lies outside of strait to W.

Marseillan (mahr-sai-YAWN), town (□ 9 sq mi/23.4 sq km), HÉRAULT department, LANGUEDOC-ROUSSILLON region, S FRANCE, small fishing port on SW shore of the Étang de THAU (a lagoon linked to the MEDI-TERRANEAN SEA), 9 mi/14.5 km SW of SÈTE; 43°21′N 03°32′E. Known for its white and rosé wines, and its liqueur distilleries (vermouth). Has marinas for pleasure craft. Marseillan-Plage has miles of sandy beaches on the Mediterranean shore.

Marseille or **Marseilles** (both: mahr-SAI), Greek *Massilia*, Latin *Massalia*, port city (□ 92 sq mi/239.2 sq km); ⊙ BOUCHES-DU-RHÔNE department and of PROVENCE-ALPES-CÔTE D′AZUR administrative region, S FRANCE, in PROVENCE, on the MEDITERRANEAN SEA, 400 mi/644 km SSE of PARIS (linked by high-speed railroad (TGV), expressways, and by air (Marseille-Provence airport at MARIGNANE); 43°18′N 05°24′E. Second-largest city of France and its chief Mediterranean port (oil imports from MIDDLE EAST, commerce with N Africa, other Mediterranean ports, and Latin America) on a bay surrounded by almost barren limestone hills—Chaîne de l′ESTAQUE (NW), Chaîne de l′ÉTOILE (N), Massif de Marseilleveyre (S; ending at

Cape CROISETTE which bounds Marseille′s S expansion). Marseille′s climate is atypical of the Mediterranean region; rainfall peaks during spring and fall rather than during its mild winters. It is also especially susceptible to the *Mistral*, a long-lasting, forceful wind funneled through the RHONE valley. In addition to its economic domination of W Provence, from the Rhône delta (W) to the naval base of TOULON (E), Marseille is the capital of a region encompassing six departments of SE France, with the entire reach of the French RIVIERA, including the large city of NICE, under its administrative oversight. The city′s major industrial focus is the Étang de BERRE, a large coastal lagoon with a navigable outlet to the sea at the Gulf of FOS which also receives the waters of the W distributary of the Rhône delta. Thus, the metropolitan area, with a population of 1,230,000, extends chiefly NW of Marseille proper, taking in the industrial district on the Étang de Berre and the Gulf of Fos, as well as the urbanizing belt to the N, which links the city with and includes AIX-EN-PROVENCE, the great cultural center of Provence, only 16 mi/26 km inland. Further, the port of Marseille has a navigation link with the lower Rhône River by way of the Marseille-Rhône canal, which crosses under the Estaque range in a long tunnel, then runs along the S shore of the Étang de Berre to MARTIGUES, where it runs NW, past the port of Fos to ARLES at the head of the Rhône delta. The port of Marseille extends from Vieux Port [the old port], which is the focal point of the city, some 5 mi/8 km NW to l′Etasque. Regular passenger service serves CORSICA, N Africa, and other Mediterranean destinations. Imports passing through the port include wine, olives, citrus fruit, spices, and tropical crops; exports include manufactured products, soap, flour, refined sugar, fruits and legumes, and many luxury items. Marseille′s chemical industry includes manufacturing of superphosphates, pharmaceuticals, and sulphur products. Petroleum refineries and tank farms, cement mills, petrochemical and metallurgical plants, lumber mills and storage depots, and related industries are located in the Gulf of Fos area next to the Marseille-Rhône canal and along the oil and gas pipelines running N and E from the petroleum terminals on deep water. There are more industrial complexes on the NE shore of the Étang de Berre, in the vicinity of the airport and along the railroad line and expressway leading NW to the Rhône valley. About half of France′s ship maintenance and repair facilities are located here. In addition to its administrative and commercial activities (international fair), Marseille is a religious (old bishopric) and educational center (the science and medicine faculties of the University of Aix-Marseille are here). The picturesque Vieux Port, with its fishing fleet and pleasure craft, is located at the foot of Marseille′s famed Canebière, its central thoroughfare and commercial hub. Much of the waterfront, damaged in World War II, has been rebuilt according to a modern plan. Here, too, is Le Corbusier′s famed *cité radieuse*, a residential complex contemplated to provide most of the services under one roof. Since 1977, Marseille has had a subway linking outlying quarters to the center. Though it is France′s oldest city, Marseille preserves relatively few relics of its past. Principal public buildings are the 19th-century cathedral of La Major in the old quarter, a museum of fine arts housed in a 19th-century mansion, and the 19th-century neo-Byzantine basilica of Notre-Dame-de-la-Garde, which rises to 500 ft/152 m atop an isolated hill just S of city center. The huge cable freight conveyor across the entrance to the old harbor, formerly a landmark of Marseille, was destroyed by the Germans along with the once bustling waterfront. Known for its teeming, exotic atmosphere, Marseille has a citizenry who, with their individualism and manner of speech, have contributed a distinctive flavor to French culture at large and to

Provence in particular. Settled by Phocaean Greeks from ASIA MINOR c.600 B.C.E. Later annexed by Rome (49 B.C.E.), it languished after the fall of ROME, regaining importance during the Crusades (11th–14th century) as a commercial center and an embarkation port for the Holy Land. Marseille and the rest of Provence came under French rule in 1481. In the 1700s commerce revived, mainly with the SW Asia and the BARBARY STATES of N Africa. Although the plague wiped out almost half its population in 1720, Marseille continued to enjoy prosperity, and in the 19th century the French conquest of ALGERIA and the opening of the SUEZ CANAL led to a tremendous expansion of the port and to the city's industrialization. A landmark of Marseille harbor is the Château d'IF, a fortress built on a small, rocky isle. Recent excavations have uncovered what are believed to be vestiges of the ramparts of old Massilia. The population of Marseille, always cosmopolitan, has been enlarged in recent times by immigrants from N and W Africa, many of whom ae Muslims who have endured racial hostility, and seek to live apart in their own districts, particularly N suburbs. Marseille-Provence-Marigane international airport 13 mi/5 km NW of city. The city's inhabitants are called *les Marseillais.*

Marseilles (mahr-SAILZ), city (2000 population 4,655), LA SALLE COUNTY, N ILLINOIS, on ILLINOIS RIVER (water power), and 8 mi/12.9 km E of OTTAWA; 41°19′N 88°41′W. Manufacturing (building materials, food products, chemicals). Marseilles Canal carries ILLINOIS WATERWAY shipping around rapids in Illinois River here. Illini State Park and Marseilles National Guard Training Area are nearby. Incorporated 1861.

Marseilles (MAHR-se-luhs), village (2006 population 119), WYANDOT COUNTY, N central OHIO, 12 mi/19 km ENE of KENTON; 40°42′N 83°23′W.

Marseilles, FRANCE: see MARSEILLE.

Marseilleveyre, Massif de, FRANCE: see MARSEILLE.

Marsella (mahr-SAI-yah), town, ⊙ Marsella municipio, RISARALDA department, W central COLOMBIA, in CAUCA valley, 7 mi/11 km NNW of PEREIRA, near the CAUCA River; 04°56′N 75°44′W. Agriculture includes coffee, sugarcane, plantains.

Mar, Serra do (MAHR, SE-rah do), great coastal escarpment of S and SE BRAZIL, skirting the shoreline for more than 800 mi/1,287 km from RIO GRANDE DO SUL (NE of PÔRTO ALEGRE) to the PARAÍBA delta in NE RIO DE JANEIRO, and traversing E part of SANTA CATARINA, PARANÁ, and SÃO PAULO states. It forms an effective barrier (average elevation 3,000 ft/914 m) between the coastal strip and the interior plateau, all major streams draining W toward the PARANÁ River from the gentle back slope of the escarpment. It rises to 7,365 ft/2,245 m (Pedra do Sino peak) in the Serra dos ORGÃOS (one of its sections). Other components include the Serra da ESTRÊLA, Serra do Cubatão, Serra PARANAPIACABA. As a result of heavy precipitation (reaching 150 in/381 cm per year), the front range is clad with dense, subtropical vegetation. With the exception of several reentrants which surround the valleys of short coastal streams (RIBEIRA DE IGUAPE, ITAJAÍ AÇU, JOINVILLE lowland) and a few deep inlets (GUANABARA BAY, PARANAGUÁ BAY), the escarpment closely parallels the coast and reappears offshore in the form of rocky islands (ILHA GRANDE, SÃO SEBASTIÃO, SANTA CATARINA). In NE, it effectively separates the Paraíba valley from the Atlantic Ocean. N of RIO, several well-known resorts (PETRÓPOLIS, TERESÓPOLIS, NOVA FRIBURGO) have been established here because of its cooler summer temperatures. Using the sheer drop from the crest of the escarpment, a large hydroelectric plant stands at CUBATÃO, near SANTOS. Historically, the Serra do Mar has been a major impediment to the development of Brazil's interior. Most inland cities were frontier towns until the railroad reached them in mid-19th century, especially Santos–São Paulo railroad.

Marsh, ENGLAND: see HUDDERSFIELD.

Marshall, county (□ 623 sq mi/1,619.8 sq km; 2006 population 87,185), NE Alabama; ⊙ Guntersville, 34°23′N 86°20′W. Bounded N by Paint Rock River Agr. area (poultry, cattle; corn, soybeans, hay); textiles. Wheeler and Guntersville reservoirs are on Tennessee River. Formed 1836. Named for John Marshall, fourth chief justice of the U.S.

Marshall, county (□ 398 sq mi/1,034.8 sq km; 2006 population 13,003), N central ILLINOIS; ⊙ LACON; 41°02′N 89°22′W. Agriculture (corn, wheat, soybeans, fruit; livestock). Some manufacturing (food products, books, chemicals, clothing). Drained by ILLINOIS RIVER and small Sandy Creek. Formed 1839. Includes Goose and Billsbach lakes. Southernmost county to span Illinois River.

Marshall, county (□ 449 sq mi/1,167.4 sq km; 2006 population 47,295), N INDIANA; ⊙ PLYMOUTH; 41°20′N 86°16′W. Agricultural area (grain, soybeans, oats, vegetables, fruit; dairy products; cattle, poultry, hogs), especially noted for vegetable and mint growing; processing of spearmint and peppermint oil. Several lakes, glacial in origin, concentrated in SW part of county, some with resorts; largest is Lake MAXINKUCKEE near CULVER. Drained by YELLOW and TIPPECANOE rivers. Formed 1835.

Marshall, county (□ 573 sq mi/1,489.8 sq km; 2006 population 39,555), central IOWA; ⊙ MARSHALLTOWN, 42°02′N 93°00′W. Prairie agricultural area (cattle, hogs, poultry; corn, oats) drained by IOWA and NORTH SKUNK rivers. Bituminous-coal deposits (W), limestone quarries (E). Manufacturing at Marshalltown. Formed 1846.

Marshall, county (□ 904 sq mi/2,350.4 sq km; 2006 population 10,349), NE KANSAS, on NEBRASKA (N) state line; ⊙ MARYSVILLE; 39°48′N 96°33′W. Gently rolling to hilly area; drained by BIG BLUE, Black Vermillion, and LITTLE BLUE rivers. Wheat, sorghum, strawberries, soybeans; hogs, cattle. Transportation equipment; gypsum products. N extremity of Tuttle Creek Reservoir in S. Formed 1855.

Marshall, county (□ 340 sq mi/884 sq km; 2006 population 31,278), W KENTUCKY; ⊙ BENTON; 36°52′N 88°21′W. Bounded N and E by TENNESSEE RIVER, forms KENTUCKY LAKE reservoir on E (Kentucky Dam); drained by East and West forks of CLARKS RIVER. Agricultural area (corn, dark and burley tobacco, alfalfa, hay, soybeans, wheat, corn; hogs, cattle; timber; clay pits. Manufacturing at Benton and CALVERT CITY. Includes Kentucky Dam State Resort Park in NE. Formed 1842.

Marshall, county (□ 1,812 sq mi/4,711.2 sq km; 2006 population 9,951), NW MINNESOTA, on RED RIVER (W; NORTH DAKOTA state line); ⊙ WARREN; 48°21′N 96°22′W. Drained by SNAKE, THIEF, TAMARAC, and MIDDLE rivers. Agricultural area (wheat, oats, flax, hay, alfalfa, sugar beets, beans, sunflowers, potatoes; poultry, cattle; sheep); timber. Thief Lake Wildlife Area in NE; Agassiz National Wildlife Refuge, surrounds MUD LAKE, in E center; Old Mill State Park in center. Formed 1879.

Marshall, county (□ 709 sq mi/1,843.4 sq km; 2006 population 35,853), N MISSISSIPPI, on TENNESSEE (N) state line, and on TALLAHATCHIE RIVER (S); ⊙ HOLLY SPRINGS; 34°46′N 89°30′W. Drained by COLDWATER RIVER. Hilly agricultural area (cotton, corn, hay, soybeans, wheat; cattle; dairying); clay products; processing of farm products. Includes part of Holly Springs National Forest in SE; Wall Doxey State Park in S center. Formed 1836.

Marshall, county (□ 426 sq mi/1,107.6 sq km; 2006 population 14,558), S OKLAHOMA; ⊙ MADILL; 34°01′N 96°46′W. Bounded E and S by Lake TEXOMA, formed by DENISON DAM in RED RIVER, and E by Washita Arm of Lake Texoma. Cattle-raising, recreation, and agricultural area (corn, pecans, barley); tourism. Includes Lake Texoma State Park in SE; part of Tishomingo National Wildlife Refuge in NE. Formed 1907.

Marshall, county (□ 885 sq mi/2,301 sq km; 2006 population 4,430), NE SOUTH DAKOTA, on NORTH DAKOTA (N) state line; ⊙ BRITTON; 45°46′N 97°36′W. Rich farming and cattle-raising region, with numerous lakes. Corn, wheat, flax, hay; dairy produce; poultry. Fort Sisseton, historic military outpost, and part of Lake Traverse (Sisseton Wahpeton) Indian Reservation in E. Sica Hollow and Roy Lake state parks and Clear Lake State Lakeside Use Area in E. Formed 1885.

Marshall, county (2006 population 28,884), central TENNESSEE; ⊙ LEWISBURG; 35°28′N 86°46′W. Drained by DUCK RIVER and its tributaries. Agriculture; timber; some manufacturing. Formed 1836.

Marshall, county (□ 312 sq mi/811.2 sq km; 2006 population 33,896), N WEST VIRGINIA, southernmost county of N PANHANDLE, on OHIO RIVER (W; OHIO state line) and on PENNSYLVANIA (E) state line; ⊙ MOUNDSVILLE; 39°52′N 80°40′W. Drained by Wheeling, Fish, and Grave creeks. Industrial area; manufacturing at MOUNDSVILLE, CAMERON, and BENWOOD is based on region's coal, natural gas, oil, glass-sand, clay, and timber. Some agriculture (honey, corn, potatoes, alfalfa, hay, nursery crops); cattle, hogs, sheep, poultry. Retail trade. Grave Creek Mound State Park in W; Burches Run Wildlife Management Area in NE. Formed 1835.

Marshall, city (2000 population 3,771), ⊙ CLARK county, E ILLINOIS, 16 mi/26 km WSW of TERRE HAUTE (INDIANA); 39°23′N 87°41′W. In agricultural area; oil wells. Incorporated 1853. On historic NATIONAL ROAD.

Marshall, city (2000 population 12,735), ⊙ LYON county, SW MINNESOTA, 87 mi/140 km WNW of MANKATO, on REDWOOD RIVER; 44°27′N 95°47′W. Elevation 1,174 ft/358 m. Trade and shipping center for agricultural area (grain; livestock, poultry; dairying); manufacturing (honey and beeswax, building materials, lumber, corn products, feed supplements, electronic products, foods and beverages; printing and publishing, custom welding). Municipal Airport to W. Camden State Park to SW. Settled 1871, plotted 1872, incorporated 1901.

Marshall, city (2000 population 12,433), ⊙ SALINE county, N central MISSOURI; 39°06′N 93°12′W. Railroad junction. Grain-, egg-, and meat-processing center of a large farm area (corn, wheat, oats, soybeans; hogs, poultry). Manufacturing (frozen foods, animal fats, gloves, egg products, building materials). Seat of Missouri Valley College. Nearby are Van Meter (NW) and Arrow Rock (SW) state parks. Incorporated 1839.

Marshall (MAHR-shuhl), city (2006 population 23,965), ⊙ HARRISON county, E TEXAS, 23 mi/37 km E of LONGVIEW, and 35 mi/56 km W of SHREVEPORT (LOUISIANA); 32°32′N 94°21′W. Elevation 375 ft/114 m. In a pine-covered hill and lake area. Oak-shaded streets and mansions recall the plantation past of the city, which has since declined economically. Railroad junction. Manufacturing (stoneware pottery, ceramic tiles, consumer goods, wood products, chemicals). Seat of East Texas Baptist College, Texas State Technical College and Wiley University. Lake O' the Pines reservoir to NW, CADDO LAKE reservoir and State Park to NE; Ginocchio National Historic District, three-block area, Starr Mansion State Historic Site. Incorporated 1844.

Marshall, town, MARGIBI county, S LIBERIA, port on ATLANTIC OCEAN, at mouth of FARMINGTON River, and 30 mi/48 km ESE of MONROVIA; 06°08′N 10°22′W. Liberia's chief rubber-shipping center. Sometimes called Junk; formerly called Little Bassa. International airport N at ROBERTSFIELD.

Marshall, town (2000 population 1,313), ⊙ SEARCY county, N ARKANSAS, 34 mi/55 km SE of HARRISON, in the OZARK Mountains; 35°54′N 92°38′W. Manufacturing (lumber, apparel). BUFFALO NATIONAL RIVER to N.

Cross-references are shown in SMALL CAPITALS. The pronunciation guide is shown on page xix. The sources of population figures are shown on page xvii.

Marshall, town (2000 population 360), PARKE county, W INDIANA, 6 mi/9.7 km NNE of ROCKVILLE; 39°51'N 87°11'W. Agricultural area; manufacturing (plastic products). Tourist area around TURKEY RUN STATE PARK to N; canoeing on SUGAR CREEK. Covered bridges in area. Laid out 1878.

Marshall, town (2000 population 7,459), ⊙ CALHOUN county, S MICHIGAN, 12 mi/19 km SE of BATTLE CREEK, and on KALAMAZOO RIVER; 42°15'N 57°84'W. In farm area (livestock, poultry; grain). Manufacturing (plastic products, transportation equipment, foods, fabricated metal products, chemicals); vehicle-component testing lab. Historical homes. Settled 1831; incorporated as village 1836, as city 1859.

Marshall (MAHR-shuhl), town (□ 3 sq mi/7.8 sq km; 2006 population 838), ⊙ MADISON county, W NORTH CAROLINA, 15 mi/24 km NNW of ASHEVILLE, on FRENCH BROAD RIVER; 35°47'N 82°40'W. Service industries; manufacturing (fabricated metal products, electrical products, apparel). In agricultural area (corn, tobacco; cattle). Pisgah National Forest to NW. Originally called Lapland. Incorporated 1863 and named for John Marshall (1755–1835), Chief Justice of the US Supreme Court.

Marshall, town (2006 population 3,583), DANE county, S WISCONSIN, on small Waterloo Creek, and 17 mi/27 km ENE of MADISON; 43°10'N 89°03'W. Farm area. Dairy products; feed mill; manufacturing (construction; pet foods).

Marshall, village (2000 population 349), W ALASKA, on YUKON River, and 75 mi/121 km N of BETHEL; 61°52'N 162°03'W. Placer gold mining. Scene of gold rush, 1913. Also called Fortuna Ledge.

Marshall or **New Marshall**, village (2006 population 278), LOGAN county, central OKLAHOMA, 22 mi/35 km NW of GUTHRIE; 36°09'N 97°37'W. In agricultural area.

Marshall (MAHR-shuhl), unincorporated village, FAUQUIER county, N VIRGINIA, 10 mi/16 km N of WARRENTON; 38°51'N 77°51'W. Manufacturing (clothing); in agricultural area (grain, soybeans, apples; livestock). Nearby is "Oak Hill" (1773), home of John Marshall.

Marshall (MAHR-shuhl), suburb 3 mi/5 km S of GEELONG, VICTORIA, SE AUSTRALIA; 38°12'S 144°22'E. Also called Marshalltown.

Marshall Archipelago, a group of large, ice-covered islands within the SULZBERGER ICE SHELF, on the SAUNDERS COAST of MARIE BYRD LAND, WEST ANTARCTICA; 77°00'S 148°30'W.

Marshall Hall, excursion resort, CHARLES county, S MARYLAND, on the POTOMAC RIVER, c.16 mi/26 km below WASHINGTON, D.C. An amusement park is located on an estate granted to William Marshall in 1690. The Marshall Hall House, containing period furniture, dates from the Colonial era when George Washington was a frequent visitor. Cruise ships from Washington, D.C. stop here daily most of the year and at MOUNT VERNON on the other side of the river. A jousting (Md. state sport) tournament is held here every August, as knights on horseback try to spear suspended rings.

Marshall Island, MAINE: see SWANS ISLAND.

Marshall Islands, officially the Republic of the Marshall Islands (□ 70 sq mi/182 sq km; 2007 population 61,815), central PACIFIC; ⊙ MAJURO; 09°00'N 168°00'W.

Geography
The Marshalls extend over 700 mi/1,127 km and include two major groups of atolls and coral islands: the RATAK (E) and RALIK (W) chains, with a total of thirty-one atolls and c.900 reefs. The major atolls are MAJURO, ARNO, AILINGLAPALAP, JALUIT (fine natural harbor; the archipelago's chief trade center), and KWAJALEIN (the largest atoll and site of a U.S. intercontinental ballistic missile test range).

Population
The population of the Marshalls is largely Micronesian.

Economy
The chief industries are fishing and tourism; coconut oil, copra, and fish are the major exports. A large portion of the Marshallese income derives from U.S. aid.

History
Some of the islands were visited by Spanish explorers in the early 16th century and were named after a British captain who visited in 1788. Much mapping was done on Russian expeditions under Adam Johann von Krusenstern (1803) and Otto von Kotzebue (1815 and 1823). Germany annexed the group in 1885 and tried with little success to establish a colony. The administrative affairs of the islands continued to be managed largely by private German and Australian interests. In 1914, JAPAN seized the Marshalls and in 1920 received a League of Nations mandate over them. In World War II the islands were taken by U.S. forces (1943–1944); they were included in the Trust Territory of the Pacific Islands in 1947. After the war both ENEWETAK and BIKINI atolls were used as U.S. nuclear weapons test sites. In 1983, the U.S. gave $183.7 million to the Marshalls for damages caused by the tests. The Marshalls became (1979) self-governing under U.S. military protection and achieved free association status in 1986. The first president, Amata Kabua, died in December 1996. Imata Kabua was elected to succeed him in January 1997. Kabua was succeeded in 2000 by Kessai H. Note, who began a second term in 2004. An amended compact of free association, extending the defense relationship with the U.S. and the lease on the U.S. base on Kwajalein, took effect in 2004.

Government
The Marshall Islands are governed under the constitution of 1979. The president, who is both head of state and head of government, is elected by the unicameral legislature (*Nitijela*) from among its members for a four-year term. The thirty-three legislators are popularly elected for four-year terms. The current head of state, President Kessai Hesa Note, was first elected in January 2000. Administratively, the country is divided into thirty-three municipalities.

Marshall Pass, Colorado: see SAWATCH MOUNTAINS.

Marshallton, unincorporated town, NEW CASTLE county, N DELAWARE, residential suburb 5 mi/8 km W of WILMINGTON, on White Clay Creek; 39°44'N 75°38'W. Elevation 75 ft/22 m. Delaware Park Horse Race Track to SW.

Marshallton, unincorporated town (2000 population 1,437), NORTHUMBERLAND county, E central PENNSYLVANIA, residential suburb 1 mi/1.6 km E of SHAMOKIN, near Shamokin Creek; 40°47'N 76°31'W.

Marshalltown, city (2000 population 26,009), ⊙ MARSHALL county, central IOWA, on the IOWA RIVER; 42°02'N 92°54'W. Railroad junction. Railroad and trade center of a rich grain and livestock area, as well as a busy manufacturing city. Manufacturing (plastic products, rubber products, paints, machinery, fabricated metal products, seeds, shipping containers, furnaces, canned goods; pork processing). The Iowa Veterans' Home and Marshalltown Community College are here; Union Grove State Park is to the NE. Incorporated 1863.

Marshalltown, AUSTRALIA: see MARSHALL.

Marshallville, town, MACON county, W central GEORGIA, 7 mi/11.3 km SSW of FORT VALLEY; 32°27'N 83°56'W. In a peach-growing area.

Marshallville, town, MACON county, GEORGIA, 13 mi/21 km W of PERRY; 32°28'N 83°56'W.

Marshallville (MAHR-shuhl-vil), village (2006 population 814), WAYNE county, N central OHIO, 13 mi/21 km NE of WOOSTER; 40°54'N 81°44'W. In agricultural area; meat products.

Marshes Siding, unincorporated town, MCCREARY county, S KENTUCKY, 2 mi/3.2 km N of WHITLEY CITY. Timber; tobacco; cattle. Area surrounded by Daniel Boone National Forest. BIG SOUTH FORK National River and Recreation Area to SW.

Marshfield, city (2000 population 5,720), ⊙ WEBSTER county, S central MISSOURI, in the OZARK MOUNTAINS, 22 mi/35 km ENE of SPRINGFIELD; 37°20'N 92°54'W. Cattle; dairying; hay, wheat, fruit. Manufacturing (apparel, machinery, caskets). Settled c.1830.

Marshfield, city (2006 population 19,136), WOOD and MARATHON counties, central WISCONSIN; 44°39'N 90°10'W. In a dairy area. Manufacturing (food and beverages, machinery, furniture, fabricated metal products, apparel, hydraulic cylinders, building materials). Has agricultural research station. Incorporated 1883.

Marshfield, town, PLYMOUTH county, SE MASSACHUSETTS, on the ATLANTIC coast, 9 mi. NNW of PLYMOUTH; 42°07'N 70°43'W. Sand and gravel are chief products, also electronic products. Resort. Has several colonial buildings. Daniel Webster lived and is buried here. Includes villages of Brant Park, Green Harbor, Humarock, Marshfield Hills, Ocean Bluff, Sea View. Settled 1632, incorporated 1640.

Marshfield, town (2006 population 270), including Marshfield village, WASHINGTON CO., central VERMONT, on Winooski River, 15 mi/24 km NE of MONTPELIER; 44°18'N 72°22'W. Includes part of Plainfield village. Named for Isaac Marsh who bought the land from local Indians in 1790.

Marshfield (MAHSH-feeld), village (2001 population 1,616), South Gloucestershire, SW ENGLAND, 12 mi/19 km E of BRISTOL; 51°28'N 02°19'W. Some engineering and manufacturing. Has 14th-century church, 17th-century almshouses, and 17th-century inn.

Marshfield or **Base Station**, COOS county, N central NEW HAMPSHIRE, valley station at W end of 3-mi/4.8-km-long Mount Washington cog railroad, 15 mi/24 km SSW of BERLIN; completed 1869 (first of its kind in the world).

Marshfield, Oregon: see COOS BAY.

Marshfield Hills, MASSACHUSETTS: see MARSHFIELD.

Marsh Harbour, town, N BAHAMAS, on E central shore of GREAT ABACO Island (5 mi/8 km W of HOPE TOWN; 26°33'N 77°04'W; 2000 population 4,700). Lumbering; fishing.

Mars Hill, agricultural town, AROOSTOOK county, E MAINE, 27 mi/43 km N of HOULTON, and on PRESQUE ISLE RIVER, near CANADIAN (NEW BRUNSWICK) border; 46°33'N 67°50'W. Ships potatoes. Takes name from Mars Hill (1,550 ft/472 m), 2 mi/3.2 km E. Includes Mars Hill village.

Mars Hill (MAHRS HIL), town (□ 1 sq mi/2.6 sq km; 2006 population 1,813), MADISON county, W NORTH CAROLINA, 16 mi/26 km N of ASHEVILLE; 35°49'N 82°32'W. Service industries; manufacturing (electrical products, apparel). In agricultural area (tobacco, corn; cattle). Seat of Mars Hill College. Parts of Pisgah National Forest to N and E.

Marsh Island, low marshy island, IBERIA parish, S LOUISIANA, 25 mi/40 km S of NEW IBERIA, between GULF OF MEXICO (S) and VERMILION, WEST COTE BLANCHE, and EAST COTE BLANCHE bays (NW, N, and E); 21 mi/34 km long, 2 mi/3 km–10 mi/16 km wide; 29°33'N 91°50'W. SOUTHWEST PASS separates island and mainland on W. Several small lakes, including Ferme, Oyster (S), and Sand (E). Entire island comprises Russell Sage National Wildlife Refuge.

Marsh Lake, reservoir, MINNESOTA RIVER, LAC QUI PARLE, SWIFT, and BIG STONE counties, W MINNESOTA, 13 mi/21 km ESE of ORTONVILLE; 7.5 mi/12.1 km long, 2 mi/3.2 km wide. Elevation 938 ft/286 m. Created by dam at SE end. In Lac qui Parle Wildlife Area. Nearby are LAC QUI PARLE (SE) and BIG STONE (NW) reservoirs.

Marsh Peak (12,240 ft/3,731 m), UINTA MOUNTAINS, NW UINTAH county, UTAH, in Ashley National Forest, 22 mi/35 km NNW of VERNAL. Highest point in E part of range.

Marshville (MAHRSH-vil), town (□ 2 sq mi/5.2 sq km; 2006 population 2,969), UNION county, S NORTH CAROLINA, 10 mi/16 km E of MONROE; 34°59′N 80°22′W. Manufacturing (wood products, textiles; turkey processing); service industries; agriculture (timber; cotton, grain; poultry, livestock; dairying).

Marshyhope Creek, c.40 mi/64 km long, W central DELAWARE and E MARYLAND; rises in swamps in SW central KENT county, 10 mi/16 km SW of DOVER; flows generally SSW, into Maryland, past FEDERALSBURG (head of navigation), and through DORCHESTER county, turns to SE to the NANTICOKE RIVER 2 mi/3.2 km below SW of Sharpstown. Formerly called NORTHWEST FORK OF NANTICOKE RIVER.

Marsico Nuovo (MAHR-see-ko NWO-vo), town, PO-TENZA province, BASILICATA, S ITALY, on upper AGRI River, and 16 mi/26 km SSW of POTENZA; 40°25′N 15°44′E. Wine, olive oil. Bishopric. Marsicovetere (1991 population 4,098) is 6 mi/10 km ESE.

Marsillargues (mahr-see-YAHRG), town (□ 16 sq mi/41.6 sq km), HÉRAULT department, LANGUEDOC-ROUSSILLON region, S FRANCE, near RHÔNE RIVER delta, and 15 mi/24 km ENE of MONTPELLIER; 43°40′N 04°11′E. Vineyards.

Marsing, town (2000 population 890), OWYHEE county, SW IDAHO, on SNAKE RIVER, and 30 mi/48 km W of BOISE, near Deer Flat Reservoir; 43°33′N 116°49′W. Railroad terminus and trading point in agricultural area (potatoes, grain; livestock). Served by Owyhee project. LAKE LOWELL RESERVOIR, surrounded by Deer Flat National Wildlife Refuge, to E; Jump Creek Canyon to SW, Squaw Creek Canyon to S.

Marsivan, TURKEY: see MERZIFON.

Marske-by-the-Sea (MAHSK–BI–THUH–SEE), town (2001 population 8,921), Redcar and Cleveland, NE ENGLAND, on NORTH SEA, 2 mi/3.2 km SE of REDCAR; 54°35′N 01°01′W. Museum.

Marsland, village, DAWES county, NW NEBRASKA, 30 mi/48 km SSW of CHADRON, and on NIOBRARA RIVER; 42°26′N 103°17′W. Box Butte Reservoir State Recreation Area to E.

Marsoui (mahr-SWEE), town (□ 71 sq mi/184.6 sq km; 2006 population 385), GASPÉSIE—ÎLES-DE-LA-MADE-LEINE region, SE QUEBEC, E CANADA, 6 mi/9 km from LA MARTRE; 49°13′N 66°04′W.

Marstal (MAHRS-tahl), town and port, FYN county, DENMARK, on E shore of ÆRO island; 54°51′N 10°30′E. Shipbuilding; fishing.

Marstal Bugt (MAHRS-tahl) or **Vejsnaes Bay** (VEIS-nes), bay, DENMARK, between S tips of ÆRO and LANGELAND islands.

Marston (MAHRS-tuhn), canton (□ 28 sq mi/72.8 sq km; 2006 population 630), ESTRIE region, S QUEBEC, E CANADA, 8 mi/13 km from LAC-MÉGANTIC; 45°30′N 71°00′W. Marina.

Marston, city (2000 population 610), NEW MADRID county, in the bootheel of extreme SE MISSOURI, near MISSISSIPPI RIVER, 7 mi/11.3 km SW of NEW MADRID; 36°31′N 89°36′W. Soybeans, rice, cotton; rice milling.

Marston Moor (MAH-stuhn), moorland area in NORTH YORKSHIRE, N ENGLAND, (2001 population 2,946), 7 mi/11.3 km W of YORK; 53°57′N 01°17′W. Site of Civil War victory of Parliamentarians over Royalists (1644). Just SE is agricultural village of Long Marston.

Marston Moretaine (MAH-zen MAW-tuhn), agricultural village (2001 population 3,684), W BED-FORDSHIRE, central ENGLAND, 6 mi/9.7 km SW of BEDFORD; 52°03′N 00°33′W. Its ancient church (re-built in 15th century) has separate Norman tower. Also spelled Marston Moreteyne.

Marstons Mills, MASSACHUSETTS: see BARNSTABLE.

Marstrand (MAHR-STRAHND), village (□ 3 sq mi/7.8 sq km), GÖTEBORG OCH BOHUS county, SW SWEDEN, on island (□ 3 sq mi/7.8 sq km) of same name in SKAGERRAK strait, 18 mi/29 km NW of GÖTEBORG; 57°53′N 11°35′E. Fishing port; marinas. Fashionable seaside resort. Founded (c.1225) by Norwegian king (c. 1225); passed to Sweden (1658), but lost its trade to Göteborg. Has fourteenth-century church; seven-teenth-century fortress.

Marsyangdi (muhr-SYAHNG-dee), river, central NEPAL; flows from the MANANG district to the TRISULI RIVER at 27°53′N 84°34′E.

Marsyata, RUSSIA: see MARSYATY.

Marsyatskoye, RUSSIA: see MARSYATY.

Marsyaty (mahr-SYAH-ti), town, N SVERDLOVSK ob-last, W Siberian Russia, on the SOS′VA RIVER (OB′ RIVER basin), on road and railroad, 30 mi/48 km N of SEROV; 60°02′N 60°28′E. Elevation 269 ft/81 m. Saw-milling; manganese mining, supplying Serov metal-lurgy. Also known as Marsyata or Marsyatskoye.

Mart (MARHT), town (2006 population 2,484), MCLENNAN county, E central TEXAS, 17 mi/27 km E of WACO; 31°32′N 96°49′W. Cotton, corn; dairying; cattle, hogs. Tradinghouse Creek Reservoir to W. Settled 1875, incorporated 1903.

Marta (MAHR-tah), town, VITERBO province, LATIUM, central ITALY, on Lake BOLSENA, near efflux of MARTA River, 12 mi/19 km NW of VITERBO; 42°32′N 11°55′E. Fishing; noted for fine wines. Has Farnese palace.

Martaban (MAHR-tah-BAHN), village, Poung town-ship, MON STATE, MYANMAR, on mouth of SALWEEN RIVER, opposite MAWLAMYINE (ferry service); head of railroad to BAGO. Trading center; pottery manufac-turing; railroad workshops. Founded 6th century C.E.; was capital of independent kingdom (1281). Invaded by British forces in First (1824) and Second (1852) Anglo-Burmese Wars.

Martaban, Gulf of (MAHR-tah-BAHN), arm of the ANDAMAN SEA, indenting S MYANMAR and receiving the waters of the SITTANG and THANLWIN rivers. The small port of MARTABAN, located at the mouth of the THANLWIN across the river from MAWLAMYINE, is famous for its glazed pottery.

Martadi (MAHR-tuh-dee), town, ⊙ BAJURA district, W NEPAL; 29°27′N 81°07′E.

Martan-Chu (mahr-TAHN–CHOO), village (2005 population 5,390), S central CHECHEN REPUBLIC, in the foothills of the NE CAUCASUS Mountains, on road, 16 mi/26 km SSW of GROZNYY, and 4 mi/6 km S of URUS-MARTAN, to which it is administratively sub-ordinate; 43°04′N 45°33′E. Elevation 1,102 ft/335 m. Renamed Grushevoye in 1944, when Chechen and Ingush residents were deported eastward for collab-oration with the German army during World War II; indigenous population allowed to return, and the old name reinstated, in the late 1950s. The village has changed hands repeatedly between the Russian army and Chechen separatist forces since 1994.

Martand, India: see ANANTNAG.

Martanis, TURKEY: see CATAK.

Martano (mahr-TAH-no), town, LECCE province, APULIA, S ITALY, 13 mi/21 km SSE of LECCE; 40°12′N 18°18′E. Wine, olive oil, cheese.

Martapura (MAHR-tah-POO-rah), town, ⊙ Banjar district, Kalimantan Selatan province, SE BORNEO, INDONESIA, on Riamkiwa River, 24 mi/39 km ESE of BANJARMASIN.

Marta River (MAHR-tah), 30 mi/48 km long, outlet of Lake BOLSENA, central ITALY; leaves lake near MARTA; flows S and SSW to TYRRHENIAN Sea 3 mi/5 km SW of TARQUINIA.

Martel (mahr-TEL), commune (□ 13 sq mi/33.8 sq km), LOT department, MIDI-PYRÉNÉES region, SW FRANCE, in the Causse de Martel (a limestone upland), near DORDOGNE RIVER, 16 mi/26 km SSE of BRIVE-LA-GAILLARDE; 44°56′N 01°35′E. Trade in nuts and can-ned (preserved) regional agricultural products. Has 12th–13th-century ramparts and the medieval fortress of the viscounts of Turenne, transformed into a Gothic palace with belfry and small corner towers. Named for Charles Martel, the defender of Chris-tendom in the battle of Poitiers (C.E. 732) against invading Muslims.

Martelange (mahr-tuh-LAWNZH), commune (2006 population 1,560), Arlon district, LUXEMBOURG province, SE BELGIUM, on SÛRE RIVER, and 11 mi/18 km NNW of ARLON, in the ARDENNES, on LUX-EMBOURG border; 49°50′N 05°44′E. Stone quarrying.

Martelange (mahr-tuh-LAWNZH), village, W Lux-embourg, in the ARDENNES, on SÛRE r., and 7 mi/11.3 km NW of REDANGE, on Belgian border, opposite Belgian town of MARTELANGE. Village of ROMBACH is contiguous.

Martell, unincorporated village, AMADOR county, cen-tral CALIFORNIA, 2 mi/3.2 km NW of JACKSON, near Jackson Creek, in 1849 California Gold Rush region. Logging; walnuts, grapes, grain; cattle.

Martelle, town (2000 population 280), JONES county, E IOWA, 14 mi/26 km E of CEDAR RAPIDS; 42°01′N 91°21′W. Soybean products.

Marten (MAHR-ten), village, RUSE oblast, SLIVO POLE obshtina, NE BULGARIA, on the DANUBE River, 8 mi/13 km NE of RUSE; 43°55′N 26°05′E. Sugar beets, sun-flowers; vineyards. Machine-building factory.

Marten (MAHR-ten), island, in the DANUBE River, NE BULGARIA; 43°58′N 26°08′E.

Martensdale, town (2000 population 467), WARREN county, S central IOWA, 6 mi/9.7 km W of INDIANOLA; 41°22′N 93°44′W.

Mártfü (MAHR-tah-foo), city, HUNGARY, 11 mi/18 km SSW of SZOLNOK; 47°01′N 20°17′E. Wheat, corn, barley, sunflowers, sugar beets; cattle, hogs, sheep, poultry. Hungary's largest vegetable oil and soap factory; other manufacturing (rubber wear, foot-wear).

Martha, village (2006 population 191), JACKSON county, SW OKLAHOMA, 7 mi/11.3 km NNW of ALTUS, near the SALT FORK OF RED RIVER; 34°43′N 99°23′W. In cotton and grain area.

Martha Brae River, c 20 mi/32 km long, TRELAWNY parish, N JAMAICA; rises in COCKPIT country; flows E and N to the CARIBBEAN SEA at FALMOUTH, for which it supplies water; 18°28′N 77°38′W. The village of Martha Brae is 2 mi/3.2 km S of Falmouth.

Marthasville, town (2000 population 837), WARREN county, E central MISSOURI, near MISSOURI River, 6 mi/9.7 km NW of WASHINGTON; 38°37′N 91°02′W. Grain; livestock. Electronic products, paper products. Original Daniel and Rebecca Boone graves 2 mi/3.2 km to E. Area of German settlement beginning 1830s.

Martha's Vineyard, island (□ c.100 sq mi/260 sq km; 1990 established population 8,850), DUKES county, SE MASSACHUSETTS, separated from the ELIZABETH IS-LANDS and CAPE COD by VINEYARD and NANTUCKET sounds. As a result of glaciation, the island has mo-rainal hills composed of boulders and clay deposits in the N, and low, sandy plains in the S. The English were the first Europeans to settle here (1642); they engaged in farming, brickmaking, salt production, and fishing. Martha's Vineyard became an important commercial center, with whaling and fishing as the main occupa-tions, in the eighteenth and early nineteenth centuries. In the late 1800s the island, with its harbors, beaches, and scenic attractions, developed into a summer re-sort. Ferry service from WOODS HOLE, FALMOUTH, and HYANNIS; Dukes County Airport. Divided into the towns of CHILMARK, EDGARTOWN (to which CHAP-PAQUIDDICK ISLAND is connected), GAY HEAD, OAK BLUFFS, TISBURY, and WEST TISBURY. Much of the island's interior is set aside as a state forest. Edgartown is the largest harbor and site of annual Edgartown Regatta. Gay Head glacial clay cliffs are most famous physical feature. Summer homes of many famous people include the late Jacqueline Kennedy Onassis. State beach and park.

Marthaville (MAHR-thuh-vil), unincorporated town, NATCHITOCHES parish, NW central LOUISIANA, 18 mi/29 km W of NATCHITOCHES; 31°44′N 93°23′W. In agricultural area; cotton gins, sawmills. In area are Rebel (NW) and Los Adaes (E) state commemorative areas.

Martí (mahr-TEE), town, CAMAGÜEY province, E CUBA, on Central Highway, and 33 mi/53 km ESE of CAMAGÜEY; 21°10′N 77°27′W. Railroad junction in sugarcane region.

Martí (mahr-TEE), town, MATANZAS province, W CUBA, on railroad, and 17 mi/27 km ESE of CÁRDENAS; 22°56′N 80°55′W. In agricultural region (sugarcane, oranges, sisal). Asphalt deposits and mineral springs in vicinity. Nearby is the sugar mill of Esteban Hernández (WNW). Formerly called Lacret.

Martiago (mahr-TYAH-go), village, SALAMANCA province, W SPAIN, 10 mi/16 km SSE of CIUDAD RODRIGO; 40°27′N 06°29′W. Cereals, wine; livestock.

Martignas-sur-Jalle (mahr-teen-YAH–syoor–ZHAHL), town (□ 10 sq mi/26 sq km) NW suburb of BORDEAUX, GIRONDE department, AQUITAINE region, SW FRANCE, at edge of Haut-Médoc wine-producing district; 44°50′N 00°46′W. Makes components for aeronautical industry.

Martigné-Ferchaud (mahr-teen-YAI–fer-SHO), commune (□ 28 sq mi/72.8 sq km), ILLE-ET-VILAINE department, in BRITTANY, W FRANCE, 8 mi/12.9 km NNE of CHÂTEAUBRIANT; 47°50′N 01°19′W. Dairying; distilling.

Martigny (mahr-teen-YEE), district, SW VALAIS canton, S SWITZERLAND. Main town is MARTIGNY; population is French-speaking and Roman Catholic.

Martigny (mahr-teen-YEE), Roman *Octodurum*, town (2000 population 14,361), VALAIS canton, SW SWITZERLAND, on DRANCE RIVER, 1 mi/1.6 km S of the RHÔNE RIVER, and 16 mi/26 km SW of SION; 46°06′N 07°05′E. Elevation 1,545 ft/471 m. Aluminum, calcium carbide; flour milling, printing. Hydroelectric plant at neighboring Martigny-Bourg. An old town, it has relics of a Roman amphitheater and a 13th-century castle tower. Martigny-Bourg and Martigny-Combe to S. Road leads S to GREAT ST. BERNARD PASS and ITALY; railroad to CHAMONIX (France).

Martigues (mahr-TEEG), city (□ 27 sq mi/70.2 sq km), BOUCHES-DU-RHÔNE department, PROVENCE-ALPES-CÔTE D'AZUR region, S FRANCE, port on Étang de BERRE at its outlet (the short Caronte canal) to the MEDITERRANEAN SEA, 18 mi/29 km WNW of MARSEILLE; 43°24′N 05°03′E. Forms part of a vast maritime industrial complex anchored by large oil refineries (at Lavéra and La Mède) and petrochemical facilities. The canal linking Marseille with the RHÔNE RIVER delta passes through the town. Martigues is divided into three parts (left bank of the canal; island between two branches of canal; right bank with beach on the Berre lagoon) linked by two revolving bridge crossings. At W end of Caronte Canal, on the open sea, are the large docks for oil tankers supplying the local refineries and pipelines leasing to the interior. When it was still a smaller community, Martigues attracted artists and writers by its bright hues and Provençal ambiance. It has 17th-century town hall and churches as well as a museum. A highway with an elevated viaduct over the canal now bypasses the town center.

Martil (MAHR-teel), town, Tetouan province, Tanger-Tétouan administrative region, NW MOROCCO, TETOUAN's port on the MEDITERRANEAN SEA, 6 mi/9.7 km NE of Tetouan; 35°37′N 05°16′W. Fish and palm fiber processing. Called Río Martín during Spanish Protectorate.

Martim Vaz, rocky islet in South Atlantic Ocean, c.750 mi/1,207 km E of VITÓRIA, ESPÍRITO SANTO state, BRAZIL; 28°51′N 20°31′S. TRINDADE island is c.35 mi/56 km W. Formerly spelled Martin Vaz.

Martin (MAHR-tin), county (□ 752 sq mi/1,955.2 sq km; 2006 population 139,393), E central FLORIDA, between LAKE OKEECHOBEE (W) and the ATLANTIC OCEAN (E), and partly sheltered E by JUPITER ISLAND (barrier beach); ⊙ STUART; 27°04′N 80°24′W. Lowland area, with swamps and many small lakes in W; crossed by ST. LUCIE CANAL. Produce and citrus-fruit region, with some cattle raising and fishing. Formed 1925.

Martin, county (□ 340 sq mi/884 sq km; 2006 population 10,340), SW INDIANA; ⊙ SHOALS; 38°43′N 86°48′W. Drained by LOST RIVER and East Fork of WHITE RIVER. Agriculture (corn, hay; cattle, hogs); gypsum, limestone; timber. Manufacturing at LOOGOOTEE, Shoals. Hoosier National Forest in SE quarter; Hindostan Falls State Fishing Area in SW. Martin State Forest in E. Crane Naval Weapons Support Center (including Greenwood Lake reservoir) in N quarter of county. Formed 1820.

Martin, county (□ 230 sq mi/598 sq km; 2006 population 12,093), E KENTUCKY, in the CUMBERLAND MOUNTAINS; ⊙ INEZ; 37°47′N 82°31′W. Bounded E by TUG FORK of BIG SANDY RIVER (WEST VIRGINIA state line); drained by several creeks. Mountain agricultural area (livestock; tobacco); bituminous-coal mines. Formed 1870.

Martin, county (□ 729 sq mi/1,895.4 sq km; 2006 population 20,768), S MINNESOTA, on IOWA (S) state line; ⊙ FAIRMONT; 43°40′N 94°33′W. Watered by numerous lakes, most in four distinct groupings, including MIDDLE CHAIN OF LAKES and East Chain of Lakes, also in S center (including Okamanpedan Lake on Iowa state line), and chain of lakes in NW. Agricultural area (corn, oats, soybeans, alfalfa; hogs, sheep, cattle). Formed 1857.

Martin (MAHR-tin), county (□ 461 sq mi/1,198.6 sq km; 2006 population 24,342), E North Carolina; ⊙ WILLIAMSTON; 35°50′N 77°05′W. Coastal plain; bounded N by ROANOKE RIVER. Manufacturing (fishing; tobacco and peanut processing); service industries; agricultural area (corn, wheat, soybeans, cotton, peanuts, tobacco; poultry, hogs); timber (pine, gum). Fort Branch Battlefield State Historical Site in NW. Formed 1774 from Halifax and Tyrrell counties. Named for Josiah Martin, (1737–1786) last royal governor of North Carolina.

Martin (MAHR-tuhn), county, (□ 915 sq mi/2,379 sq km; 2006 population 4,441), W TEXAS; ⊙ STANTON; 32°17′N 101°57′W. On S LLANO ESTACADO, with E-facing CAPROCK ESCARPMENT in NW; elevation 2,600 ft/790 m–3,000 ft/915 m. Cattle-ranching area; also hogs, sheep, goats; some agricultural (grains, sorghum). Some oil and gas production. Formed 1876.

Martin (mahr-TIN), Hungarian *Márton*, city (2000 population 60,133), STREDOSLOVENSKY province, NW SLOVAKIA; 49°04′N 18°56′E. Noted for largest Slovak machinery factory (tanks, construction vehicles) and woodworking (cellulose, bentwood furniture); brewery. Contains a railroad junction and military base. Former cultural center and cradle of Slovak independence movement; Matica slovenska cultural and educational organization founded here in 1863. Center of former TURIEC region. Has 13th-century church with 12th-century sacristy, Slovak National Museum (extensive ethnographic collections). Union of CZECH and Slovak lands was proclaimed here, 1918. Formerly Turciansky Sväty Martin Slovak *Turčiansky Svätý Martin*, Hungarian *Turócszentmárton*.

Martin, city (2006 population 1,035), ⊙ BENNETT county, S SOUTH DAKOTA, 90 mi/145 km SE of RAPID CITY; 43°10′N 101°43′W. Trading point for farm and livestock region; wheat, flax; honey; feeds. Lacreek Lake and National Wildlife Refuge to E. Pine Ridge Indian Reservation to N and W; Rosebud Indian Reservation to E (both in neighboring counties). Founded 1912, incorporated 1926.

Martin, city (2006 population 10,104), WEAKLEY county, NW TENNESSEE, 13 mi/21 km ESE of UNION CITY; 36°21′N 88°51′W. Agriculture; diversified manufacturing; recreation. University of Tennessee branch campus here. Founded 1873.

Martin, village (2000 population 311), STEPHENS county, NE GEORGIA, 9 mi/14.5 km SE of TOCCOA, near SOUTH CAROLINA state line; 34°29′N 83°11′W.

Martin, village (2000 population 633), FLOYD county, E KENTUCKY, in CUMBERLAND foothills, 15 mi/24 km WNW of PIKEVILLE, on Beaver Creek; 37°34′N 82°45′W. In bituminous-coal area; manufacturing (machinery, electrical products). One of two places so named in Kentucky.

Martin, village (2000 population 625), RED RIVER parish, NW LOUISIANA, 36 mi/58 km SW of SHREVEPORT, near BLACK LAKE BAYOU; 32°06′N 93°13′W. In agricultural area (cotton, soybeans, vegetables; cattle); timber.

Martin, village (2000 population 435), ALLEGAN county, SW MICHIGAN, 11 mi/18 km E of ALLEGAN; 42°32′N 85°38′W. Manufacturing (plastic products).

Martin, village (2006 population 80), SHERIDAN county, central NORTH DAKOTA, 10 mi/16 km NW of HARVEY; 47°49′N 100°06′W. Founded in 1896 as Casselman and in 1902 the name was changed to Martin, after William Leslie Martin, a railroad executive.

Martina Franca (mahr-TEE-nah FRAHNG-kah), town, Taranto province, APULIA, S ITALY, 17 mi/27 km NNE of TARANTO; 40°42′N 17°20′E. Railroad junction. Highly diversified secondary industrial center; winemaking center; olive oil, cheese; hosiery.

Martin Bay, N NEWFOUNDLAND AND LABRADOR, E CANADA, arm of LABRADOR SEA, 255 mi/410 km NNW of NAIN. Site of unmanned Nazi weather station located on SE lobe of Hutton Peninsula. On October 22, 1943, crew of U-Boat 537 landed and set up station. It remained operational for two weeks before signal was jammed.

Martin Bluff, unincorporated town, JACKSON county, SE MISSISSIPPI, residential suburb 10 mi/16 km NNW of PASCAGOULA, on PASCAGOULA RIVER; 30°27′N 88°38′W. Cotton, corn, fruit. Mississippi Sandhill Crane National Wildlife Refuge to SW; Ward Bayou Wildlife Management Area to N.

Martinborough, town, S Wairarapa district (□ 949 sq mi/2,467.4 sq km), WELLINGTON region, NORTH ISLAND, NEW ZEALAND, 35 mi/56 km E of WELLINGTON; 41°13′S 175°26′E. Level land, services intensive sheep-farming area, with some dairy plants.

Martin, Cape (mahr-TAN), headland on French RIVIERA, in township of ROQUEBRUNE-CAP-MARTIN, ALPES-MARITIMES department, MIDI-PYRÉNÉES region, SE FRANCE, jutting into the MEDITERRANEAN SEA between the resorts of MENTON and MONTE CARLO (Monaco). Linked by a scenic walkway along rocky coast to Monte Carlo (3 mi/4.8 km W).

Martin City, village, FLATHEAD county, NW MONTANA, 18 mi/29 km NE of KALISPELL, and 1 mi/1.6 km E of HUNGRY HORSE. Tourism. Timber. Area surrounded by Flathead National Forest; FLATHEAD RANGE to E, SWAN RANGE to SW, GLACIER NATIONAL PARK to NE; Hungry Horse Reservation to SE.

Martindale (MAHR-tuhn-dal), town (2006 population 1,096), CALDWELL county, S central TEXAS, 6 mi/9.7 km ESE of SAN MARCOS, on SAN MARCOS RIVER; 29°50′N 97°50′W. Oil and natural gas. Agricultural area (turkeys; eggs; cotton, sorghum, corn).

Martín de la Jara (mahr-TEEN dai lah HAH-rah), town, SEVILLE province, SW SPAIN, in the SIERRA DE YEGUAS, near MÁLAGA province border, 12 mi/19 km SE of OSUNA; 37°07′N 04°59′W. Cereals, olives; livestock.

Martín de Yeltes (mahr-TEEN dai YEL-tes), village, SALAMANCA province, W SPAIN, 18 mi/29 km NE of CIUDAD RODRIGO; 40°46′N 06°17′W. Cereals; livestock.

Martinengo (mahr-tee-NENG-go), town, BERGAMO province, LOMBARDY, N ITALY, near SERIO River,

10 mi/16 km SSE of BERGAMO; 45°34′N 09°46′E. Silk mill; machinery, clothing.

Martínez (mahr-TEE-nez), city, residential suburb, in Greater BUENOS AIRES, ARGENTINA, on the RÍO DE LA PLATA, and 12 mi/19 km NW of BUENOS AIRES; 34°29′S 58°30′W.

Martinez, city (2000 population 35,866), ⊙ CONTRA COSTA county, W CALIFORNIA; suburb 12 mi/19 km NNE of downtown OAKLAND, on CARQUINEZ STRAIT (Benicia-Martinez Bridge) between San Pablo and Suisun bays; 38°01′N 122°07′W. Railroad junction. In farm area. Its major industry is petroleum refining; also machinery, consumer goods, construction materials, steel. Seat of John F. Kennedy University. Home of the naturalist John Muir; JOHN MUIR NATIONAL HISTORIC SITE in S. Part of the Central Valley project is nearby. Incorporated 1884.

Martinez (mahr-teen-EZ), suburb (2000 population 27,749), on border between COLUMBIA and RICHMOND counties, GEORGIA; 33°30′N 82°04′W. Fast-growing suburb of AUGUSTA. Commercial center. Manufacturing (building materials, textiles); granite and marble processing, steel fabricating.

Martínez de la Torre (mahr-TEE-nes dai lah TO-rai), city and township, VERACRUZ, E MEXICO, in GULF lowland, 33 mi/53 km SE of PAPANTLA DE OLARTE; 20°05′N 97°02′W. Corn, sugarcane, coffee.

Martín García Island (mahr-TEEN gahr-SEE-ah), granite island, BUENOS AIRES province, ARGENTINA, in URUGUAY RIVER estuary, off coast of COLONIA department (SW URUGUAY), 30 mi/48 km NNE of BUENOS AIRES. Rises to 160 ft/49 m; covers 410 acres/166 ha.

Martinho Campos (MAHR-cheen-yo KAHM-pos), city (2007 population 12,165), central MINAS GERAIS state, BRAZIL, near Rio São Francisco, 17 mi/27 km SE of ABAETÉ; 19°19′S 45°25′W.

Martinique (mahr-tee-NEEK), French overseas department and administrative region (□ 425 sq mi/1,105 sq km), in the WINDWARD ISLANDS, WEST INDIES, coextensive with the island of Martinique; ⊙ FORT-DE-FRANCE; 14°40′N 61°00′W. Of volcanic origin, the island is rugged and mountainous and reaches its greatest height in PELÉE volcano. Most agriculture exists in the hot valleys and along the coastal strips; about 80% of this area is devoted to sugarcane, which was introduced from Brazil in 1654 and which provides Martinique's major export, rum. The island's industries consist mainly of sugar and rum production, pineapple canning, and petroleum refining. Tourism constitutes a major sector of the economy. Visited by Columbus, probably in 1502, the island was ignored by the Spanish; colonization began in 1635, when the French, who had promised the native Caribs the W half of the island, established a settlement. The French proceeded to eliminate the Caribs and later imported African slaves as sugar-plantation workers. In the 18th century, Martinique's sugar exports made it one of France's most valuable colonies; although slavery was abolished in 1848, sugar continued to hold a dominant position in the economy. A target of dispute during the Anglo-French worldwide colonial struggles, Martinique was finally confirmed as a French possession after the Napoleonic wars. Martinique supported the Vichy regime after France's collapse in World War II, but in 1943 a U.S. naval blockade forced the island to transfer its allegiance to the Free French. It became a department of France in 1946 and an administrative region in 1974. Although the island has recovered from the extensive damage caused by a hurricane in 1980, France has continued its attempts to improve the economic life of Martinique, which is plagued by overpopulation and a lack of development.

Martinique Passage, channel in the WINDWARD ISLANDS, WEST INDIES, between DOMINICA (N) and MARTINIQUE (S); c.25 mi/40 km wide.

Martin, Lake (□ 62 sq mi/161.2 sq km), TALLAPOOSA county, E ALABAMA, on TALLAPOOSA RIVER, 30 mi/48 km NE of MONTGOMERY; average width 5 mi/8 km; 32°40′N 85°54′W. Large and irregular-shaped. Formed by Martin Dam (168 ft/51 m high, 2,000 ft/610 m long; completed 1927), used for hydroelectric power and river control. Wind Creek State Park on W shore.

Martin Luther King Jr. Historic Site, ATLANTA, N GEORGIA; 33°45′N 84°22′W. The site is run by the National Park Service and is located on Auburn Avenue in downtown Atlanta. Dr. King was born, reared, and buried here. The MLK center contains Dr. King's gravesite, his church (the Ebenezer Baptist Church), Dr. King's boyhood home, Freedom Hall, the Martin Luther King, Jr. Center for Nonviolent Social Change, and a visitors center. Authorized 1980.

Martín Muñoz de las Posadas (mahr-TEEN moon-YOTH dai lahs po-SAH-dahs), town, SEGOVIA province, central SPAIN, 25 mi/40 km W of SEGOVIA; 40°59′N 04°36′W. Cereals, tubers, pepper, tomatoes, onions, vegetables; livestock. Lumbering; tile manufacturing.

Martinniemi (MAHR-tin-NEE-e-mee), village, in HAUKIPUDAS commune, OULUN province, W FINLAND, 15 mi/24 km NNW of OULU, on GULF OF BOTHNIA; 65°13′N 25°18′E. Elevation 33 ft/10 m. Railroad terminus; lumber and pulp milling.

Martinópole (mahr-chee-NO-po-le), town, (2007 population 10,301), N CEARÁ state, BRAZIL, on railroad, and 22 mi/35 km SSE of CAMOCIM; 03°15′S 40°45′W. Cotton.

Martinópolis (MAHR-chee-NO-po-lees), city (2007 population 23,981), W SÃO PAULO state, BRAZIL, on railroad, and 15 mi/24 km E of PRESIDENTE PRUDENTE; 22°11′S 51°12′W. In coffee zone; manufacturing (manioc flour, pottery, furniture, explosives).

Martinovo (mahr-TEE-no-vo), village, MONTANA oblast, CHIPROVTSI obshtina, BULGARIA; 43°24′N 22°51′E.

Martin Peninsula, mostly ice-covered peninsula jutting out 60 mi/100 km from the MARIE BYRD LAND coast of ANTARCTICA between the GETZ and DOTSON ice shelves, 20 mi/30 km wide; 74°25′S 114°10′W.

Martin Point, NE ALASKA, near CANADA (YUKON) border, cape on BEAUFORT SEA, 10 mi/16 km E of BARTER ISLAND; 70°08′N 143°12′W.

Martin River, 25 mi/40 km long, S ALASKA; rises in Martin River Glacier at 60°29′N 144°19′W; flows SW to Gulf of ALASKA E of CORDOVA.

Martins (MAHR-cheens), city (2007 population 8,063), W RIO GRANDE DO NORTE state, NE BRAZIL, in small hill range (elevation 2,500 ft/762 m), 75 mi/121 km SW of MOSSORÓ; 06°05′S 37°55′W. Health resort with mineral springs. Cotton, sugar, vegetables.

Martinsburg, city (2006 population 16,392), ⊙ BERKELEY county, NE WEST VIRGINIA, in the EASTERN PANHANDLE; 39°27′N 77°58′W. Railroad center in agricultural region (apples, peaches). Manufacturing (building materials, glassware, textiles, fiberglass, fabricated metal products, ceramics, wood products, modular homes, transportation equipment, ordnance; printing and publishing). Limestone is quarried nearby. During the Civil War, the city's strategic location on the railroad made it a frequent military objective. Belle Boyd, the Confederate spy, lived here and was imprisoned in the old courthouse. Sleepy Creek Wildlife Management Area to W. Settled 1732, incorporated as a city 1859.

Martinsburg, town (2000 population 126), KEOKUK county, SE IOWA, 11 mi/18 km S of SIGOURNEY; 41°10′N 92°15′W. Livestock; grain.

Martinsburg, town (2000 population 326), AUDRAIN county, NE central MISSOURI, 13 mi/21 km ESE of MEXICO; 39°06′N 91°39′W. Grain; livestock; lumber.

Martinsburg, village (2006 population 101), DIXON county, NE NEBRASKA, 7 mi/11.3 km WSW of PONCA; 42°30′N 96°49′W.

Martinsburg (MAHR-tinz-buhrg), village (2006 population 188), KNOX county, central OHIO, 11 mi/18 km SE of MOUNT VERNON; 40°16′N 82°21′W.

Martinsburg, borough (2006 population 2,151), BLAIR county, S central PENNSYLVANIA, 13 mi/21 km SSE of ALTOONA; 40°18′N 78°19′W. Railroad terminus. Agriculture (dairying; livestock; corn, hay, apples). Manufacturing (leather products, food products). Altoona Blair County Airport to S. Settled c.1793, laid out 1815, incorporated 1832.

Martins Creek, unincorporated town, Lower Mount Bethel township, NORTHAMPTON county, E PENNSYLVANIA, 6 mi/9.7 km N of EASTON, on Martins Creek (1 mi/1.6 km N of its mouth on DELAWARE RIVER); 40°47′N 75°11′W. Agriculture (corn, hay, potatoes, soybeans, apples; livestock; dairying). Manufacturing (flour, textiles). Franklin Hill Winery to N.

Martinsdale, village, MEAGHER county, central MONTANA, on SOUTH FORK 3 mi/4.8 km SW of confluence with NORTH FORK (forming MUSSELSHELL RIVER), and 30 mi/48 km ESE of WHITE SULPHUR SPRINGS. Sheep, cattle, poultry; hay. Mfg (pottery). Martinsdale Reservoir to E; parts of Lewis and Clark National Forest to N, S, and NW.

Martins Ferry (MAHR-tuhnz FER-ee), city (□ 2 sq mi/5.2 sq km; 2006 population 6,804), BELMONT county, E OHIO, on the OHIO RIVER, opposite WHEELING (WEST VIRGINIA); 40°06′N 80°43′W. Formerly an industrial coal-mining and steel-manufacturing city. The novelist William Dean Howells was born here. In Walnut Grove Cemetery are the graves of Elizabeth (Betty) and Ebenezer Zane. Settled 1780, incorporated as a city 1885.

Martin Siding (MAHR-tin SEI-deeng), unincorporated village, S ONTARIO, E central CANADA, in MUSKOKA lakes region, 35 mi/56 km E of PARRY SOUND, and included in town of HUNTSVILLE. Diatomite mining.

Martins Location, land grant, COOS county, N central NEW HAMPSHIRE, 9 mi/14.5 km S of BERLIN, in White Mountain National Forest. Drained by PEABODY RIVER.

Martinsville, unincorporated rural community, HARRISON county, NW MISSOURI, 8 mi/12.9 km NW of BETHANY.

Martinsville, city (2000 population 1,225), CLARK county, E ILLINOIS, 12 mi/19 km WSW of MARSHALL; 39°20′N 87°52′W. Agriculture (corn, apples); oil wells; foundry, oil-storage and oil-pumping station. Plotted 1833, incorporated 1875. On historic NATIONAL ROAD.

Martinsville, city (2000 population 11,698), ⊙ MORGAN county, central INDIANA, on West Fork of WHITE RIVER, and 28 mi/45 km SSW of INDIANAPOLIS; 39°25′N 86°25′W. In grain-growing area. Manufacturing (electronic products, wood products, furniture); timber; sand and gravel. Artesian springs. Settled 1822.

Martinsville (MAHR-tinz-vil), independent city (□ 11 sq mi/28.6 sq km; 2006 population 14,945), ⊙ surrounding HENRY county, S VIRGINIA, separate from surrounding Henry county, in BLUE RIDGE foothills near NORTH CAROLINA state line, 27 mi/43 km WNW of DANVILLE, on SMITH RIVER; 36°40′N 79°51′W. Railroad junction. Manufacturing (furniture, prefabricated homes, textiles, clothing, building materials, wood products, machinery, plastic products, chemicals, wood products, consumer goods, transportation equipment; tobacco processing). Patrick Henry Community College to N. Virginia Museum of Natural History. Philpott Reservoir and FAIRY STONE STATE PARK to NW; Martinsville Speedway to S. Founded 1793, incorporated as a city 1928.

Martinsville (MAHR-tinz-vil), village (2006 population 444), CLINTON county, SW OHIO, 33 mi/53 km ENE of CINCINNATI; 39°19′N 83°49′W.

Martinswand, AUSTRIA: see ZIRL.

Martinton, village (2000 population 375), IROQUOIS county, E ILLINOIS, 10 mi/16 km N of WATSEKA;

40°55′N 87°43′W. In agricultural area (corn, soybeans, sorghum; cattle, hogs).

Martintown (MAHR-tin-toun), unincorporated village, SE ONTARIO, E central CANADA, 9 mi/14 km NE of CORNWALL, on the Raisin River banks, and included in SOUTH GLENGARRY township; 45°09′N 74°42′W. Dairying. Hosts the longest canoe race in Ontario.

Martin Van Buren National Historic Site, home of the eighth U.S. president, COLUMBIA county, SE NEW YORK, 9 mi/14.5 km NE of HUDSON, and 2 mi/3.2 km S of KINDERHOOK, on KINDERHOOK CREEK; 42°22′N 73°42′W. Estate and thirty-six-room mansion. Legend has it that Van Buren first used the abbreviation *O.K.* ("Old Kinderhook") to indicate approval of paperwork. Authorized 1974.

Martinville (MAHR-tin-vil, mahr-tan-VEEL), village (□ 19 sq mi/49.4 sq km; 2006 population 480), ESTRIE region, S QUEBEC, CANADA, 11 mi/18 km from COOKSHIRE-EATON; 45°16′N 71°43′W.

Mártir de Cuilapan, MEXICO: see APANGO.

Mártires, ARGENTINA: see LAS PLUMAS.

Mártires de Tacubaya (MAHR-tee-res dai tah-koo-BAH-yah), town, in far NW OAXACA, MEXICO, on the border of the state of GUERRERO, on unpaved road 11 mi/18 km ENE of CUAJINICUILAPA (Guerrero). Hot climate. Agriculture (corn, coffee, sugarcane, beans, fruits), woods.

Martland, CANADA: see MONETVILLE.

Martock (MAH-tuhk), town (2001 population 5,454), S SOMERSET, SW ENGLAND, near PARRETT RIVER, 6 mi/9.7 km WNW of YEOVIL; 50°58′N 02°45′W. Former agricultural market; manufacturing (leatherworking machinery, gloves); dairying. Has 13th–15th-century church.

Marton, town, RANGITIKEI district (□ 1,729 sq mi/4,495.4 sq km), S NORTH ISLAND, NEW ZEALAND, 90 mi/145 km NNE of WELLINGTON; 40°05′S 175°23′E. Railroad junction; servicing center for intensive sheep and dairying area.

Martonvásár (MAHR-ton-vah-sahr), village, FEJER county, N central HUNGARY, 17 mi/27 km SW of BUDAPEST; 47°19′N 18°47′E. Wheat, corn; cattle, hogs.

Martorell (mahr-to-RAIL), city, BARCELONA province, NE SPAIN, on LLOBREGAT RIVER, and 14 mi/23 km NW of BARCELONA; 41°28′N 01°56′E. Potatoes, legumes, fruit; manufacturing of textiles, paper, machinery, metal furniture, chemicals, plastics, tires, ceramics. Motor-vehicle plant. Restored Roman bridge just S.

Martos (MAHR-tos), city, JAÉN province, S SPAIN, 10 mi/16 km WSW of JAÉN; 37°43′N 03°58′W. Olive-oil-production center. Cement works; manufacturing of soap, plaster, cotton textiles, pottery and tiles; flour milling. Cereals, vegetables; livestock in area. Mineral springs 3 mi/4.8 km SSW. Was Roman colony; fell to Moors (8th century), taken (1225) by Ferdinand III.

Martre, Lac la (MAHR-truh, lahk lah), lake (□ 840 sq mi/2,184 sq km), NORTHWEST TERRITORIES, N CANADA, 100 mi/161 km WNW of YELLOWKNIFE; 50 mi/80 km long, 12 mi/19 km–28 mi/45 km wide; 63°20′N 118°W. Drains SE into GREAT SLAVE LAKE

Martres-Tolosane (MAHR-truh–to-lo-ZAHN), commune (□ 9 sq mi/23.4 sq km), HAUTE-GARONNE department, MIDI-PYRÉNÉES region, S FRANCE, near the GARONNE RIVER, 16 mi/26 km ENE of SAINT-GAUDENS; 43°12′N 01°00′E. Pottery and porcelain manufacturing. Roman relics have been unearthed here.

Martubah (MAHR-tuh-bah), town, CYRENAICA region, NE LIBYA, near coastal road, 15 mi/24 km SE of Darnah, on NE fringe of Jebel Al AKHDAR plateau; 32°35′N 22°46′E. Agriculture (sheep, goats; grain, fruit, vegetables). World War II battleground (1942).

Martuk (mahr-TUHK), city, NW AKTÖBE region, KAZAKHSTAN, on TRANS-CASPIAN railroad, near ILEK RIVER, 40 mi/64 km NNW of AKTÖBE (Aktyubinsk); 50°45′N 56°30′E. Tertiary-level administrative center. In wheat and dairy-cattle area; metalworks.

Martuni (muhr-TOO-nyee), urban settlement and administrative center of Martuni region (2001 population 11,756), central ARMENIA, on S shore of LAKE SEVAN, 40 mi/64 km E of YEREVAN. Fisheries; wheat, potatoes; cannery; cheese processing, rug weaving, manufacturing of electric bulbs. Formerly Nizhni Karanlug.

Martuni (muhr-TOO-nyee), urban settlement and center of Martuni region, SE NAGORNO-KARABAKH autonomous oblast, AZERBAIJAN, 20 mi/32 km E of XANKANDI; 40°40′N 45°56′E. Sericulture; manufacturing (canning; cheese, woven rugs, electric bulbs).

Martvili (mahr-TVEE-lee), urban settlement and administrative center of Gegechkori region, W GEORGIA, on the Abasha River (RIONI River Basin), 20 mi/32 km NW of KUTAISI; 42°24′N 42°21′E. Food processing. Formerly Gegechkori.

Martvitsa, Bulgaria: see PODEM.

Martwa Wisła River, POLAND: see VISTULA River.

Martwick, unincorporated village, MUHLENBERG county, W KENTUCKY, 5 mi/8 km NE of CENTRAL CITY. In bituminous-coal-mining and agricultural area.

Martynovskoye, RUSSIA: see BOL'SHAYA MARTYNOVKA.

Martyush (MAHR-tyoosh), town (2006 population 4,090), S SVERDLOVSK oblast, W URALS, W Siberian Russia, near highway, 2 mi/3.2 km W of KAMENSK-URAL'SKIY, to which it is administratively subordinate; 56°24′N 61°53′E. Elevation 511 ft/155 m. Manufacturing (electric components for various industries; door locks and handles).

Maruchak (ma-ROO-chahk), village, BADGHIS province, NW AFGHANISTAN, 18 mi/29 km NW of BALA MURGHAB, on Murghab River, and TURKMENISTAN border; 35°48′N 63°09′E. Frontier post; transit trade. Also spelled Marochak or Meruchak.

Marudu Bay (mah-ROO-doo), inlet of SOUTH CHINA SEA, in extreme N BORNEO, MALAYSIA; 30 mi/48 km long, 15 mi/24 km wide. KUDAT is on NW shore; SAMPANMANGIO POINT is at W side of entrance.

Maruf (mah-ROOF), town, ZABUL province, SE AFGHANISTAN, 80 mi/129 km E of KANDAHAR, on ARGHASTAN RIVER, near BALUCHISTAN province (PAKISTAN) border; 31°34′N 67°03′E

Marugame (mah-ROO-gah-me), city, KAGAWA prefecture, N SHIKOKU, W JAPAN, on the INLAND SEA, 16 mi/25 km W of TAKAMATSU; 34°17′N 133°48′E. Nori; paper fans. The stone wall (197 ft/60 m high) of Marugame Castle is the highest such wall in Japan.

Marui (ma-RYOO-ee) or **Papui**, village, EAST SEPIK province, N central NEW GUINEA island, NW PAPUA NEW GUINEA, on N bank of SEPIK RIVER, 60 mi/97 km SW of WEWAK; 04°04′S 143°02′E. Marshy area. Road terminus. Sago, yams, taro, bananas.

Maruim (MAH-roo-eem), city (2007 population 15,150), E SERGIPE state, NE BRAZIL, on railroad, 12 mi/19 km NNW of ARACAJU; 10°45′S 37°05′W. Cotton shipping; corn. Formerly spelled Maroim.

Maruko (mah-roo-KO), town, Chisagata county, NAGANO prefecture, central HONSHU, central JAPAN, 22 mi/35 km S of NAGANO; 36°19′N 138°16′E. Medicinal plants; raw silk, silk yarn.

Marula, village, MATABELELAND NORTH province, SW ZIMBABWE, 40 mi/64 km SW of BULAWAYO, on railroad; 20°28′S 28°05′E. Elevation 4,765 ft/1,452 m. Cattle, sheep, goats; corn, soybeans, tobacco.

Marulan (muh-ROO-luhn), township, NEW SOUTH WALES, SE AUSTRALIA, 102 mi/164 km SW of SYDNEY; 34°43′S 150°00′E.

Marulanda (mah-roo-LAHN-dah), town, ⊙ Marulanda municipio, CALDAS department, W central COLOMBIA, 22 mi/35 km NE of MANIZALES; 05°17′N 75°15′W. Elevation 9,189 ft/2,800 m. Coffee, plantains, sugarcane; livestock.

Marull (mah-ROOL), village, NE CÓRDOBA province, ARGENTINA, 85 mi/137 km ENE of CÓRDOBA, near the MAR CHIQUITA; 30°58′S 62°50′W. Grain, flax, alfalfa; livestock.

Marum (MAH-rum), village, GRONINGEN province, N NETHERLANDS, 14 mi/23 km WSW of GRONINGEN; 53°08′N 06°16′E. Dairying; agriculture (vegetables, grain); cattle, sheep.

Marumba (ma-ROOM-bah), village, RUVUMA region, S TANZANIA, 90 mi/145 km SE of SONGEA; 11°18′S 34°53′E. Livestock; grain; timber.

Marumori (mah-roo-MO-ree), town, Nigu county, MIYAGI prefecture, N HONSHU, NE JAPAN, on ABUKUMA RIVER, 25 mi/40 km S of SENDAI; 37°54′N 140°56′E.

Marungu (ma-ROON-goo), village, KIGOMA region, NW TANZANIA, 110 mi/177 km ENE of KIGOMA; 03°44′S 30°54′E. Timber; tobacco, grain; sheep, goats.

Marungu Mountains (mah-RUNG-goo), highlands (6,000 ft/1,830 m–9,000 ft/2,740 m) in KATANGA province, SE CONGO, extending c.50 mi/80 km inland from SW shore of LAKE TANGANYIKA, just SE of MOBA. The region is especially suitable for livestock raising and temperate-climate agriculture.

Maruoka (mah-ROO-o-kah), town, Sakai county, FUKUI prefecture, central HONSHU, W central JAPAN, 6 mi/10 km N of FUKUI; 36°15′N 136°20′E. Textiles. Maruoka Castle is Japan's second-oldest castle.

Marupatti, India: see MORUPPATTI.

Maruseppu (mah-roo-SEP), town, Abashiri district, Hokkaido prefecture, N JAPAN, 118 mi/190 km N of SAPPORO; 44°00′N 143°20′E. Woodwork.

Marutea North (mah-roo-TAI-ah), atoll, TUAMOTU ARCHIPELAGO, FRENCH POLYNESIA, S PACIFIC; 17°00′S 143°11′W. Pearl-fishing lagoon. Formerly Furneaux Island.

Marutea South (mah-roo-TAI-ah), atoll, S TUAMOTU ARCHIPELAGO, FRENCH POLYNESIA, S PACIFIC; 21°30′S 135°40′W. Copra, pearls. Formerly Lord Hood Island.

Maruter (MAH-ruh-tuhr), town, WEST GODAVARI district, ANDHRA PRADESH state, S INDIA, in GODAVARI RIVER delta, 40 mi/64 km E of ELURU, on irrigation canal; 16°38′N 81°44′E. Rice milling; oilseeds, tobacco, sugarcane, coconuts.

Maruyama (mah-ROO-yah-mah), town, Awa county, CHIBA prefecture, E central HONSHU, E central JAPAN, 31 mi/50 km S of CHIBA; 35°01′N 139°57′E.

Marvão (mahr-VOU), town, PORTALEGRE district, central PORTUGAL, 8 mi/12.9 km NNE of PORTALEGRE, near Spanish border. Olives; vineyards. Old frontier fortress.

Marvão, Brazil: see CASTELO DO PIAUÍ.

Marvdasht (mahrv-DAHSHT), town, Fārs province, SW IRAN, 25 mi/40 km NE of SHIRAZ; 29°50′N 52°40′E.

Marvdasht (mahrv-DAHSHT), irrigated plain, Fārs province, S IRAN, 25 mi/40 km NE of SHIRAZ, along road to ESFAHAN. Sugar-beet cultivation; sugar refinery. Ruins of Persepolis are here. The city of Marv Dasht is the main center of the plain. Also spelled MERVDASHT.

Marvejols (mahrv-uh-ZHOL), town (□ 4 sq mi/10.4 sq km), LOZÈRE department, LANGUEDOC-ROUSSILLON region, S FRANCE, on SE slope of AUBRAC Mountains, in S MASSIF CENTRAL, 11 mi/18 km WNW of MENDE; 44°33′N 03°17′E. Wool-spinning center; also noted for its medical education institutions. Has gates of medieval fortifications dating from 14th century when Marvejols was declared a "royal city." Town was almost destroyed in 1586 during repression of Protestant communities.

Marvel, village, Bibb co., central Alabama, c.17 mi/27 km NE of Centreville. Named by Elizabeth Roden for her favorite poet, Andrew Marvell. Roden was the wife of Benjamin Franklin Roden, operator of the coal mines in the area.

Marvell, town (2000 population 1,395), PHILLIPS county, E ARKANSAS, 18 mi/29 km W of HELENA; 34°33′N 90°54′W. In agricultural area.

Marvin, village (2006 population 63), GRANT county, NE SOUTH DAKOTA, 14 mi/23 km W of MILBANK;

Area is shown by the symbol □, and capital city or county seat by ⊙.

45°15′N 96°54′W. Lake Traverse (Sisseton Wahpeton) Indian Reservation to W.

Marvine, Mount, highest peak (11,610 ft/3,539 m) in Fish Lake Mountains, S SEVIER county, S central UTAH, 25 mi/40 km ESE of RICHFIELD.

Marvin, Lake (MAHR-vuhn), HEMPHILL county, extreme N TEXAS, impounded by dam in a small N tributary of CANADIAN RIVER, 11 mi/18 km E of town of CANADIAN; c.1 mi/1.6 km long; 35°52′N 100°11′W. Fishing, hunting in area.

Mar Vista, SW residential section of LOS ANGELES, LOS ANGELES county, S CALIFORNIA, 10 mi/16 km W of downtown Los Angeles, 1 mi/1.6 km W of CULVER CITY.

Marwar, India: see JODHPUR.

Marwar Mundwa, India: see MUNDWA.

Marwayne (mahr-WAIN), village (2001 population 495), E ALBERTA, W CANADA, near SASKATCHEWAN border, near VERMILION RIVER, 24 mi/39 km NE of VERMILION, in VERMILION RIVER COUNTY NO. 24; 53°31′N 110°20′W. Dairying; grain; livestock. Incorporated 1952.

Marx, RUSSIA: see MARKS.

Marxstadt, RUSSIA: see MARKS.

Marxzell (MAHRKS-tsel), village, MIDDLE UPPER RHINE, BADEN-WÜRTTEMBERG, SW GERMANY, on the Alb River, and on NW slope of BLACK FOREST, 12 mi/19 km NE of BADEN-BADEN; 48°52′N 08°27′E. Forestry; wine making; distillery; agriculture (fruit).

Mary (mah-REE), weloyat (□ 34,700 sq mi/90,220 sq km), SE TURKMENISTAN; ⊙ MARY. Extends into the KARA KUM desert (N); borders on AFGHANISTAN (S); watered by MURGAB RIVER. Intensive cotton cultivation along Murgab River and in Murgab oasis; wheat; cattle, horses. Pistachio woods near Afghanistan border. Goat and karakul-sheep raising in desert. Cotton-ginning industry and cottonseed-oil extraction at Mary, BAIRAM-ALI, IOLOTAN. TRANS-CASPIAN RAILROAD passes through Murgab oasis. Population consists of Turkmen, Russians. Formed 1939.

Mary (mah-REE), city, ⊙ MARY weloyat, SE TURKMENISTAN, on the MURGAB RIVER delta; 37°36′N 61°50′E. Lying in a large oasis of the KARA KUM desert, it is the center of a rich cotton-growing area. Railroad junction. Extensive trade in cotton, wool, grain, and hides; also a major center of the natural-gas industry. Arose in 1884 as a Russian military-administrative center 20 mi/32 km from the site of ancient MERV and was itself called Merv until 1937.

Mary Alice, village, HARLAN county, SE KENTUCKY, 3 mi/4.8 km S of HARLAN, in the CUMBERLAND MOUNTAINS. Bituminous coal.

Maryama, ESTONIA: see MÄRJAMAA.

Mar'yanivka (mahr-YAH-neef-kah), (Russian *Mar'-yanovka*), town (2004 population 4,900), S VOLYN' oblast, UKRAINE, on left bank of the Lypa River (tributary of the STYR RIVER), on road and on railroad, 4 mi/7 km SW of HOROKHIV; 50°27′N 24°49′E. Elevation 662 ft/201 m. Railroad station for Horokhiv. Sugar refinery, vegetable cannery, cheese factory. Established around 1900, with present name since 1921; town since 1958.

Mar'yanivka (mahr-YAH-neef-kah), (Russian *Mar'-yanovka*), town (2004 population 4,600), W central ZHYTOMYR oblast, UKRAINE, in VOLHYNIAN POLISSYA, 36 mi/58 km W of ZHYTOMYR, and 9 mi/15 km ENE of BARANIVKA; 50°18′N 27°52′E. Elevation 790 ft/240 m. Glass factory based on locally abundant quartzite sands. Established at the end of the 16th century; town since 1977.

Mar'yanovka (mahr-YAH-nuhf-kah), village (2006 population 8,270), SW OMSK oblast, SW SIBERIA, RUSSIA, on road and the TRANS-SIBERIAN RAILROAD, 30 mi/48 km W of OMSK; 54°58′N 72°38′E. Elevation 337 ft/102 m. Grain processing.

Mar'yanovka, UKRAINE: see MAR'YANIVKA, Volyn' oblast; or MAR'YANIVKA, Zhytomyr oblast.

Mar'yanskaya (mahr-YAHN-skah-yah), village (2005 population 10,405), S central KRASNODAR TERRITORY, S European Russia, on the right bank of the KUBAN' River, on highway, 22 mi/35 km W of KRASNODAR; 45°06′N 38°29′E. In oil- and gas-producing region.

Maryborough (ME-ree-buh-ro), municipality, central VICTORIA, AUSTRALIA, 85 mi/137 km NW of MELBOURNE; 37°03′S 143°45′E. Railroad and commercial center for agricultural area (grain; sheep, cattle). Manufacturing (food processing, clothing); printing. Some gold mined in vicinity. Former gold-mining center. Several national parks nearby. Airstrip.

Maryborough (ME-ree-buh-ro), city and port (2001 population 21,191), SE QUEENSLAND, AUSTRALIA, on MARY RIVER, 23 mi/37 km from its mouth, and 135 mi/217 km NNW of BRISBANE; 25°32′S 152°42′E. Gateway to FRASER ISLAND, the world's largest sand island. Commercial center; shipyards, dairy plant, steel mill, sawmills, sugar mill, butter factory; wood products; vegetables, pineapples, and grain. Exports sugar, fruit, coal, and timber. Tourism. School of Arts (1861). Coal mines nearby.

Maryborough (ME-ree-buh-ro), former township (□ 89 sq mi/231.4 sq km; 2001 population 3,030), S ONTARIO, E central CANADA, from GUELPH; 43°45′N 80°44′W. Amalgamated into MAPLETON township in 1999.

Maryborough, IRELAND: see PORTLAOISE.

Marydale, town, NORTHERN CAPE province, SOUTH AFRICA, 42 mi/68 km NW of PRIESKA, on N10 highway and railroad link to UPINGTON, 10 mi/16 km SW of ORANGE RIVER. Center for farming community, producing crops under irrigation and sheep rearing. Nearby at Koegas is world's largest mine producing blue asbestos.

Marydel, town (2000 population 147), CAROLINE county, E MARYLAND, on DELAWARE state line, 12 mi/19 km WSW of DOVER (Del.); 39°07′N 75°45′W. Named for its position on the state line. Scene of a duel (Jan. 3, 1877) between James Gordon Bennett, owner of The New York *Herald*, and Frederick May, a well-known explorer, over May's broken engagement to Bennett's sister. Both men missed, and, subsequently, Bennett exiled himself to PARIS.

Mary Esther, city (□ 1 sq mi/2.6 sq km; 2005 population 4,086), OKALOOSA county, NW FLORIDA, on the GULF OF MEXICO, 32 mi/51 km E of PENSACOLA; 30°24′N 86°39′W.

Maryevka (mahr-YEF-kuh), village, SW NORTH KAZAKHSTAN region, KAZAKHSTAN, on ISHIM RIVER, 100 mi/161 km SW of PETROPAVLOVSK; 53°48′N 67°54′E. Wheat; cattle. Until 1939, Mariinskoye.

Maryfield (MEHR-ee-feeld), village (2006 population 347), SE SASKATCHEWAN, W CANADA, on MANITOBA border, 22 mi/35 km SSE of MOOSOMIN; 49°50′N 101°32′W. Railroad junction.

Mary Harmsworth, Cape, westernmost point of FRANZ JOSEF LAND, ARCHANGEL oblast, extreme N European Russia, in the ARCTIC OCEAN, on ALEXANDRA LAND; 80°37′N 42°10′E.

Maryhill (ME-ree-hil), unincorporated village, WATERLOO region, S ONTARIO, E central CANADA, 9 mi/14 km from KITCHENER, and included in the township of WOOLWICH; 43°32′N 80°23′W.

Maryhill, Scotland: see GLASGOW.

Mar'yina Gorka (MAH-ri-na GOR-kuh), town, ⊙ Pukhovski region, central MINSK oblast, BELARUS, near railroad (Pukhovichi station), 36 mi/58 km SE of MINSK. Manufacturing (paper, starch products). Pukhovichi village lies 5 mi/8 km NE.

Mar'yinka (MAHR-yeen-kah), (Russian *Mar'inka*), city, central DONETS'K oblast, UKRAINE, in the DONBAS, 15 mi/24 km W of DONETS'K (city center). Raion center. Tire vulcanizing, dairy and feed processing, grain milling, food-flavoring plant. Established in the 1840s, city since 1977.

Maryinsk Canal (mah-ree-YEENSK), 5 mi/8 km long, in N VOLOGDA oblast, NW European Russia, linking canalized VYTEGRA (NW) and KOVZHA (SE) rivers; 60°55′N 36°48′E. Built in 1808; reconstructed in the 1940s. Forms the watershed section of the Maryinsk canal system, which joins the VOLGA RIVER and the RYBINSK RESERVOIR (S) with the NEVA RIVER and SAINT PETERSBURG (NW). Entire waterway (N-S; repeatedly reconstructed and deepened) consists of Neva River, Ladoga Canals, Svir River, Onega Canal, Vytegra River, Mariinsk Canal, Kovzha River, Belozersk Canal, Sheksna River, and Rybinsk Reservoir. Lies on a historic Baltic Sea–Volga route, in use since the 9th century C.E. Formerly called Novo-Maryinsk Canal.

Maryinsk System, RUSSIA: see VOLGA-BALTIC WATERWAY.

Mary Island, SE ALASKA, in GRAVINA ISLANDS, in REVILLAGIGEDO CHANNEL, 25 mi/40 km SE of KETCHIKAN; 4 mi/6.4 km long, 2 mi/3.2 km wide; 55°05′N 131°12′W.

Mary Island, NEW YORK: see WELLESLEY ISLAND.

Mary Kathleen (ME-ree kath-LEEN), town site, NW QUEENSLAND, NE AUSTRALIA, 40 mi/64 km E of MOUNT ISA, on Barkly Highway; 20°44′S 139°30′E. Former uranium-mining town. Controversy led to mine closure (1982), all buildings removed by 1983, and site returned to its natural state. Burke and Wills Monument 20 mi/32 km E commemorates their ill-fated 1861 expedition to Gulf of CARPENTARIA. Mary Kathleen Memorial Park and Museum in CLONCURRY.

Maryknoll, locality, WESTCHESTER county, SE NEW YORK, near OSSINING. Catholic Foreign Mission Society of America trains missionaries (the Maryknoll Fathers) here. Seat of Maryknoll School of Theology, a graduate-level institution with distinctive degree programs in Justice and Peace, and in Mission Ministry. Founded 1911.

Mary, Lake, S DOUGLAS county, W MINNESOTA, 5 mi/8 km SW of ALEXANDRIA; 4 mi/6.4 km long, 1.5 mi/2.4 km wide; 45°49′N 95°28′W. Fishing resorts. Andrew Lake to E.

Maryland, state (□ 12,406 sq mi/32,132 sq km; 2000 population 5,296,486; 1995 estimated population 5,042,438), E UNITED STATES, in the Middle Atlantic region, one of the original 13 Colonies; ⊙ ANNAPOLIS; 38°59′N 76°34′W. BALTIMORE, with a large percentage of the state's population, is the dominant metropolis. George Washington is said to have bestowed the state's nickname of "Old Line State" in reference to its regular line troops who served in many Revolutionary War battles.

Geography

A seaboard state, E Maryland is divided by CHESAPEAKE BAY, which runs almost to the N border, separating the EASTERN SHORE from the main part of the state. Bounded N by PENNSYLVANIA (see MASON-DIXON LINE) and E by DELAWARE and the ATLANTIC OCEAN. For the most part, the erratic course of the POTOMAC River separates the main part of Maryland from VIRGINIA (to the S) and the long, narrow W handle from WEST VIRGINIA (to the S and W). The Potomac River lies entirely within Maryland, as the state line with Virginia on the S shore. The DISTRICT OF COLUMBIA cuts a rectangular indentation into the state just below the falls of the Potomac. The main part of the state is divided by the FALL LINE, which runs between the upper end of CHESAPEAKE BAY and WASHINGTON, D.C.; to the N and W is the rolling PIEDMONT, rising to the BLUE RIDGE and to the Pennsylvania hills. The heavily indented shores of Chesapeake Bay fringe the land with bays and estuaries, which helped in the development of a farm economy relying on water transport. In the mild winters and hot summers of the coastal plains typically Southern trees, such as the loblolly pine and the magnolia, flourish, while the cooler uplands have woods of black and white oak and beech. Maryland has nearly 3 million acres/1.2 million ha of forest land.

Economy

Chesapeake Bay dominates the E section of the state. Although the fishing industry is declining, the catch of fish and shellfish from Chesapeake Bay yields an annual income in the millions of dollars, and the state's annual catch of crabs is the largest in the nation. The coastal marshes abound in wild fowl. In the W part of Maryland are the mineral resources of building stone and coal, which have declined significantly. The iron mines, active in the 19th century, have declined along with other mining activity. Important industries include manufacturing of primary metals, food products, missiles, transportation equipment, clothing, and electrical machinery. Shipping (Baltimore is a major port), tourism (especially along Chesapeake Bay), and printing and publishing are also big industries. Services are the largest sector of the economy, ahead of finance, insurance, and real estate, all of which have surpassed government work and manufacturing. Although manufacturing well exceeds agriculture as a source of income, Maryland's, farms yield corn, hay, tobacco, soybeans, and other crops. Income from livestock, especially cattle and chickens, and livestock products, is almost twice that from crops; dairy and poultry farms thrive, and Maryland is famous for breeding horses.

Tourism

Maryland has become increasingly popular as a vacation area—OCEAN CITY is a popular seashore resort, and both sides of Chesapeake Bay are lined with beaches and small fishing towns. The Chesapeake Bay Bridge has brought the culture of the Eastern Shore, formerly quite distinctive, into a more homogeneous unity with that of the rest of the state; the area, however, is still noted for its unique rural beauty and architecture, strongly reminiscent of the English countryside left behind by early settlers. Annapolis, with its well-preserved Colonial architecture and 18th-century waterfront, is the site of the U.S. Naval Academy. Tourists are also attracted to the ANTIETAM NATIONAL BATTLEFIELD SITE and the national cemetery at SHARPSBURG; the FORT MCHENRY National Monument, near Baltimore's inner harbor; and the historic towns of FREDERICK and St. Marys City. Racing enthusiasts attend the annual Preakness and Pimlico Cup horse races at Baltimore. There are several military establishments, including FORT GEORGE G. MEADE and ANDREWS AIR FORCE BASE. The National Institute of Health in BETHESDA is a civilian government establishment. A National Agricultural Research Center (19 sq mi/49 sq km) is located at BELTSVILLE.

History to 1730

Giovanni da Verrazano, an Italian navigator in the service of France, probably visited (1524) the Chesapeake region, which was certainly later explored (1574) by Pedro Menéndez Marqués, governor of Spanish Florida. In 1603 the region was visited by an Englishman, Bartholomew Gilbert, and it was charted (1608) by Captain John Smith. In 1632, Charles I granted a charter to George Calvert, the first Baron Baltimore, yielding him feudal rights to the region between latitude 40°00′N and the Potomac River. Disagreement over the boundaries of the grant led to a long series of border disputes with Virginia. The territory was named "Maryland" in honor of Henrietta Maria, queen consort of Charles I. Before the great seal was affixed to the charter, George Calvert died, but his son Cecilius Calvert, the second Baron Baltimore, undertook development of the colony as a haven for his persecuted fellow Catholics and also as a source of income. In 1634 the ships *Ark* and *Dove* brought settlers (both Catholic and Protestant) to the Western Shore, and a settlement called SAINT MARY'S was set up. The Algonquian-speaking Native Americans

withdrew gradually and for the most part peacefully from the area during the colonial period, sparing Maryland the conflicts other colonies experienced. Religious conflict, however, was strong in ensuing years as the Puritans, growing more numerous in the colony and supported by Puritans in England, set out to destroy the religious freedom guaranteed with the founding of the colony.

History - 1730 to 1829

Tobacco became the basis of the economy by 1730. In 1767 the demarcation of the Mason-Dixon Line ended a long-standing border dispute with Pennsylvania. Economic and religious grievances led Maryland to support the growing colonial agitation against England. At the time of the American Revolution most Marylanders were stalwart patriots and vigorous opponents of the Britsh colonial policy. In 1776, Maryland adopted a declaration of rights and a state constitution and sent soldiers and supplies to aid the war for independence. At Annapolis, Congress ratified the Treaty of PARIS ending the Revolutionary War in 1783. In 1791 Maryland and Virginia contributed land and money for the new national capital. Industry, already growing in conjunction with renewed commerce, was furthered by the skills of German immigrants. The 18th century saw the emergence of commercially oriented Baltimore, which by 1800 had a population of over 30,000 and a flourishing coastal trade. The War of 1812 was marked by the British attack of 1814 on Baltimore and the defense of FORT MCHENRY, immortalized in Francis Scott Key's "The Star-Spangled Banner." After the war the state entered a period of great commercial and industrial expansion.

History - 1829 to 1865

This was accelerated by the building of the NATIONAL ROAD, which tapped the rich resources of the West; the opening of the Chesapeake and Delaware Canal (1829); and the opening (1830) of the Baltimore and Ohio Railroad, the first railroad in the U.S. open for public traffic. Southern ways and sympathies persisted, however, among the plantation owners, and as the rift between North and South widened, Maryland was torn by conflicting interests and the intense internal struggle of the true border state. In 1860 there were 87,000 slaves in Maryland, but industrialists and businessmen had special interests in adhering to the Union; and despite the urgings of Southern sympathizers, the state remained in the Union. At the beginning of the Civil War, President Lincoln suspended habeas corpus and sent troops to Maryland who imprisoned large numbers of secessionists. Nevertheless, Marylanders fought on both sides, and families were often split. General Lee's Army of Northern Virginia invaded Maryland in 1862 and was repulsed by Union forces at ANTIETAM. In 1863, Lee again invaded the North and marched across Maryland on the way to and from GETTYSBURG. Throughout the war, Maryland was the scene of many minor battles and skirmishes.

History - 1865 to 1980

With the end of the Civil War, industry was quickly revived and became a dominant force, economically and politically. New railroad lines traversed the state, making it more than ever a crossing point between North and South. Labor troubles hit Maryland with the Panic of 1873, and four years later railroad wage disputes resulted in large-scale rioting in CUMBERLAND and Baltimore. During the 20th century, however, Maryland became a leader in labor and other reform legislation. The great influx of population into the state during World War I was repeated and accelerated in World War II—war workers poured into Baltimore, where vital shipbuilding and aircraft plants were in operation, and military and other government employees moved into the area around Washington, D.C. Since World War II, public-works legislation, particularly that concerning roads and other traffic

arteries, has brought major changes. The opening of the CHESAPEAKE BAY BRIDGE in 1952 spurred significant industrial expansion on the Eastern Shore; a parallel bridge was opened in 1973. The PATAPSCO RIVER tunnel under Baltimore harbor was completed in 1957, and the Francis Scott Key Bridge (1977), crosses the Patapsco. Other projects include the Baltimore-Washington International Airport, formerly called Friendship International Airport (1950), S of Baltimore, and the Baltimore-Washinton Parkway (1954).

History - 1980 to Present

Maryland experienced tremendous suburban growth in the 1980s, especially the metropolitan Washington, D.C. area. This growth occured in spite of a decline in government jobs, as service sector employment rose dramatically. Suburban Baltimore grew as well while the city proper lost 6.4% of its population during the 1980s. Baltimore undertook major revitalization projects in the 1980s and the early 1990s, including the construction of Orioles Park at Camden Yards, the new home of the Baltimore Orioles baseball team. Maryland's medical, educational, and cultural institutions greatly benefited from philanthropic gifts in the late 19th century from Johns Hopkins, George Peabody, and Enoch Pratt. Institutions of higher learning in the state include Johns Hopkins University, at Baltimore; Saint John's College, at Annapolis; Towson State University and Goucher College at TOWSON; and the University of Maryland, at COLLEGE PARK and Baltimore.

Government

In 1968, Maryland Governor Spiro T. Agnew was elected Vice President, but had to resign the office because of a scandal. Maryland is governed under a constitution adopted in 1867. The general assembly consists of forty-seven senators and 141 delegates, all elected for four-year terms. The governor, also elected for a four-year term, may succeed himself once. The current governor is Martin O'Malley. The state elects two U.S. Senators and eight Representatives. It has ten electoral votes.

Maryland has twenty-three counties: ALLEGANY, ANNE ARUNDEL, BALTIMORE, CALVERT, CAROLINE, CARROLL, CECIL, CHARLES, DORCHESTER, FREDERICK, GARRETT, HARFORD, HOWARD, KENT, MONTGOMERY, PRINCE GEORGES, QUEEN ANNES, SAINT MARYS, SOMERSET, TALBOT, WASHINGTON, WICOMICO, and WORCESTER.

Maryland, county (□ 1,202 sq mi/3,125.2 sq km; 1999 population 71,977), SE LIBERIA, on ATLANTIC OCEAN coast; ⊙ HARPER. Bounded N by River Gee county, E by CAVALLY RIVER (CÔTE D'IVOIRE border), S by Atlantic Ocean, and W by GRAND KRU county. Agriculture (rubber, palm oil); cattle raising. Iron deposits. Main center is Harper. Maryland is divided into two districts: Barrobo (NW) and Pleebo/Sodeken (remainder of county).

Maryland, village, MASHONALAND WEST province, N ZIMBABWE, 40 mi/64 km WNW of HARARE; 17°38′S 30°29′E. Junction of railroad spur NE to KWEKWE; shipping point for chromite-mining district. Also called Maryland Junction.

Maryland Heights, town, Saint Louis county, E central MARYLAND, on Missouri River, W of Saint Louis and NE of Chesterfield; 38°43′N 90°28′W. Manufacturing (machinery, rubber products, electronic equipment, building materials, fertilizers, plastic products; printing, metal processing).

Maryland Junction, ZIMBABWE: see MARYLAND.

Maryland Park, village, PRINCE GEORGES county, central MARYLAND, E suburb of WASHINGTON, D.C.

Marylhurst, locality, CLACKAMAS county, NW OREGON, on WILLAMETTE RIVER, between WEST LINN and LAKE OSWEGO. Seat of Marylhurst College.

Mary McLeod Bethune Council House Historic Site, WASHINGTON, D.C. Home and political headquarters of the educator and activist. The carriage house contains the Bethune Archives. Authorized 1982.

Marymont (mah-REE-mont), residential district of WARSAW, Warszawa province, E central POLAND, on left bank of the VISTULA River, and 3 mi/4.8 km N of city center.

Maryport (MER-i-pawt), town (2001 population 11,090) and port, CUMBRIA, NW ENGLAND, on SOLWAY FIRTH, at mouth of ELLEN RIVER, 25 mi/40 km SW of CARLISLE; 54°42′N 03°30′W. Leather tanning, manufacturing (chemicals, shoes, electrical goods), beer brewing. Imports chemicals, petroleum products, cement. The town is of Roman origin and a number of Roman altars have been found here. Dearham is a suburb.

Mary River (ME-ree), 165 mi/266 km long long, SE QUEENSLAND, AUSTRALIA; rises in hills S of GYMPIE; flows N, past Gympie, and E at MARYBOROUGH, to HERVEY BAY; 25°26′S 152°55′E. Navigable 23 mi/37 km below MARYBOROUGH.

Mary's Harbour, settlement (2006 population 417), E NEWFOUNDLAND AND LABRADOR, E CANADA, on inlet of the ATLANTIC OCEAN, 10 mi/16 km WNW of BATTLE HARBOUR; 52°19′N 55°51′W. Fish plant.

Marys Igloo, Alaska: see IGLOO.

Marys River (ME-reez), c.45 mi/72 km long, NE NEVADA; rises in N ELKO county in Humboldt Natl Forest; flows S to HUMBOLDT River 16 mi/26 km W of WELLS. Also written Mary's River

Marystown, town (□ 24 sq mi/62.4 sq km; 2001 population 5,908), NEWFOUNDLAND AND LABRADOR, E CANADA, on E side of BURIN PENINSULA; 47°10′N 55°10′W. Fine natural harbor; shipbuilding, fish processing.

Marysvale, village (2006 population 342), PIUTE county, SW central UTAH, on SEVIER RIVER, and 23 mi/37 km SSW of RICHFIELD; 38°26′N 112°15′W. Elevation 5,866 ft/1,788 m. Trading point for mining and agricultural area (alfalfa, corn, potatoes; dairying; sheep, cattle). Gold, silver, potash, and alunite mines nearby. Railroad terminus. TUSHAR MOUNTAINS are just W. Marysvale Peak (10,943 ft/3,335 m) is 7 mi/11.3 km ENE, in SEVIER PLATEAU. Parts of Fishlake National Forest to E and West Piute Reservoir and State Park to S. Mount Belknap (12,137 ft/3,699 m) to W. Settled 1860s.

Marysville, city (2000 population 12,268), ⊙ YUBA county, N central CALIFORNIA, opposite YUBA CITY, on FEATHER RIVER, at mouth of YUBA RIVER, in Sacramento Valley, and 35 mi/56 km N of SACRAMENTO; 39°09′N 121°35′W. Railroad junction; trade and shipping center for fruit-growing region (peaches, prunes, kiwi fruit); walnuts, almonds, wheat, corn, rice. Manufacturing (meat processing; beverages, valves and fittings). Gold dredging on Yuba River. Hydraulic mining raised Yuba River bed above town, necessitating large levees (begun 1875). City was supply point in gold rush when it was head of Feather River navigation. Yuba College (two-year); Beale Air Force Base to E. Founded 1849, incorporated 1851.

Marysville (MER-eez-vil), city (□ 16 sq mi/41.6 sq km; 2006 population 17,621), ⊙ UNION county, central OHIO, 27 mi/43 km NW of COLUMBUS; 40°14′N 83°22′W. Former site of the Ohio State Reformatory for Women. Assembly plant for motor vehicles. Settled 1816.

Marysville, city (2006 population 31,938), SNOHOMISH county, NW WASHINGTON, 5 mi/8 km N of EVERETT, and on Possession Sound (arm of PUGET SOUND), at mouth of SNOHOMISH RIVER; 48°04′N 122°09′W. Timber; strawberries, vegetables; poultry, cattle; dairying; logging. Manufacturing (steel foundry, sawmill; machinery parts, log homes, abrasive products, fabricated metal products; printing and publishing, leather tanning). Railroad junction to N. Tulalip Indian Reservation to W (has casino); Wenberg State Park to NW. Incorporated 1890.

Marysville, town (2000 population 54), MARION county, S central IOWA, near CEDAR CREEK, 13 mi/21 km SSE of KNOXVILLE; 41°10′N 92°57′W. In agricultural area.

Marysville, town (2000 population 3,271), ⊙ MARSHALL county, NE KANSAS, on BIG BLUE RIVER, and 70 mi/113 km NW of TOPEKA; 39°51′N 96°38′W. Railroad division point in grain and livestock region; dairying, poultry packing. Manufacturing (honey, meats, machinery, paper products). Pony Express Station. Former ferry crossing (1849) on Oregon Trail. Incorporated 1861.

Marysville, town (2000 population 9,684), SAINT CLAIR county, E MICHIGAN, 5 mi/8 km SSW of PORT HURON, and on SAINT CLAIR RIVER; 42°54′N 82°28′W. Manufacturing (machinery, plastic products, fabricated metal products, transportation equipment, plastics and resins, rubber products); salt mining. Saint Clair County International Airport to W. Incorporated as village 1921; city 1924.

Marysville (ME-reez-vil), resort village, S central VICTORIA, SE AUSTRALIA, 45 mi/72 km NE of MELBOURNE, near HEALESVILLE, in GREAT DIVIDING RANGE; 37°31′S 145°45′E. Mountain resort. Home to 276-ft/84-m Steavenson Falls.

Marysville (ME-reez-vil), former village, SE BRITISH COLUMBIA, W CANADA, on SAINT MARY RIVER; 49°38′N 115°57′W. Timber; tourism. Amalgamated into KIMBERLEY in 1968.

Marysville, unincorporated village, FREMONT county, E IDAHO, 15 mi/24 km NE of SAINT ANTHONY, and 2 mi/3.2 km E of ASHTON, between FALLS and HENRYS FORK rivers; 44°04′N 111°52′W. Elevation 5,245 ft/1,599 m. Targhee National Forest to NE.

Marysville, village, LEWIS AND CLARK county, W central MONTANA, 15 mi/24 km NW of HELENA. Helena National Forest to SW; Great Divide Ski Area to W, just NE of CONTINENTAL DIVIDE. Nearby Drumlummon mine, once immensely rich in gold and silver.

Marysville, borough of FREDERICTON, S NEW BRUNSWICK, E CANADA, on NASHWAAK RIVER, 3 mi/5 km NE of FREDERICTON; 45°59′N 66°35′W. Lumber milling, former cotton milling.

Marysville, borough (2006 population 2,435), PERRY county, S central PENNSYLVANIA, 6 mi/9.7 km NW of HARRISBURG, on SUSQUEHANNA RIVER (Rockville Bridge, railroad), at mouth of Fishing Creek; 40°20′N 76°55′W. Agriculture area (corn, hay; poultry; dairying); manufacturing (food products). APPALACHIAN TRAIL passes to W and N. Laid out 1861, incorporated 1866.

Marysville Buttes, California: see SUTTER BUTTES.

Marytown, unincorporated village, MCDOWELL county, S WEST VIRGINIA, on TUG FORK RIVER, 6 mi/9.7 km NW of WELCH; 37°28′N 81°40′W.

Maryut (mahr-YOOT), salt lake (□ 95 sq mi/247 sq km), N EGYPT, in the NILE delta; 31°08′N 29°56′E. Separated from the MEDITERRANEAN SEA by the narrow isthmus on which ALEXANDRIA is situated. There are fisheries and saltworks on the lakeshore. Also called Lake Mareitis, Mareotis.

Maryville, city (2000 population 10,581), ⊙ NODAWAY county, NW MISSOURI, near ONE HUNDRED AND TWO RIVER, and 40 mi/64 km N of SAINT JOSEPH; 40°20′N 94°52′W. Agricultural service center. Corn, wheat; hogs. Manufacturing (milk products, electrical goods, transportation equipment, fabricated metal products, building materials; machining). Seat of Northwest Missouri State University. Benedictine convent nearby. Settled c.1845.

Maryville (MER-ee-vil), city (2006 population 26,433), ⊙ BLOUNT county, E TENNESSEE, 15 mi/24 km S of KNOXVILLE; 35°45′N 83°59′W. With its twin city, ALCOA, Maryville is an important center for the production of aluminum and aluminum products; other light manufacturing; agriculture. Seat of Maryville College. The Great Smoky Mountains National Park

and the Tuckaleechee Caverns are in the area. Settled around Fort Craig (built 1785), incorporated as a town 1830, as a city 1927.

Maryville, village (2000 population 4,651), MADISON county, SW ILLINOIS, suburb of SAINT LOUIS, 12 mi/19 km ENE of EAST SAINT LOUIS; 38°43′N 89°57′W. Agriculture (corn, wheat; cattle); manufacturing marble products.

Marzabotto (mahr-tsah-BOT-to), town, BOLOGNA province, EMILIA-ROMAGNA, N central ITALY, on RENO River, 13 mi/21 km SE of BOLOGNA; 44°20′N 11°12′E. Agricultural center (fruits and grains). Hemp and paper mills; machinery manufacturing. Has ruins (temples, tombs) of Etruscan town (6th century B.C.E.; excavated 1871); museum.

Marzagão (MAHR-sah-GOUN), town (2007 population 2,041), SE GOIÁS state, BRAZIL, 22 mi/35 km NW of CORUMBAÍBA; 17°55′S 48°38′W.

Marzo, 1 de, PARAGUAY: see PRIMERO DE MARZO.

Marzo, 28 de, ARGENTINA: see AÑATUYA.

Marzo, Cabo (MAHR-so, KAH-bo), headland on PACIFIC coast of CHOCÓ department, W COLOMBIA, 55 mi/89 km SW of RÍOSUCIO; 06°50′N 77°41′W. Also referred to as Punta Marzo.

Marzsina, ROMANIA: see MARGINA.

Marzuq (MAHR-zuhk), town, FAZZAN region, SW LIBYA; 25°55′N 13°55′E. With SABHA, among chief settlements of Fazzan region. Center of a group of oases. Developed around fort built c. 1310 (now in ruins); declined with advent of modern transportation. S terminal of modern road to MEDITERRANEAN SEA coast. Was capital of former Marzuq province. Sometimes spelled Murzuk or Mourzouk.

Marzuq Basin (MAHR-zuhk), sedimentary basin (□ 154,440 sq mi/401,544 sq km), SW LIBYA and N NIGER, S of Jabal As SAWDA and AL HARUJ AL ASWAD uplands. Rich oil fields, especially MARZUQ complex (under development since 1994).

Marzuq Field (MAHR-zuhk), rich oil field, FAZZAN region, SW LIBYA, 125 mi/201 km NW of GHAT. Development began 1994.

Masachapa (mah-sah-CHAH-pah), village, MANAGUA department, SW NICARAGUA, on the PACIFIC OCEAN, 30 mi/48 km SW of MANAGUA (linked by road); 11°47′N 86°31′W. Seaside resort. PUERTO SANDINO is just NW.

Masada (mah-SAH-dah), ancient mountaintop fortress in the JUDEAN DESERT, ISRAEL, the final outpost of the Zealot Jews in their rebellion against Roman authority (C.E. 66–73). Sits atop a mesa-shaped rock that towers some 1,475 ft/450 m above the W shore of the DEAD SEA. According to the ancient historian Josephus, Masada was first fortified sometime during the 1st or 2nd century B.C.E. Between 37 B.C.E. and 31 B.C.E. Herod the Great, king of JUDEA, further strengthened Masada, building two ornate palaces, a bathhouse, aqueducts, and surrounding siege walls. In C.E. 66, with the outbreak of the Jewish war against Rome, the Zealots, an extremist Jewish sect, seized the fortress in a surprise attack and massacred its Roman garrison. Masada remained under Zealot control until C.E. 73, when, after a siege of almost two years, the 15,000 soldiers of Rome's 10th legion finally subdued the 1,000 men, women, and children holding the fortress. In a final act of defiance, however, almost all of the Jewish defenders had killed themselves rather than be captured and enslaved by the Romans. Only two women and five children survived to tell of the Zealots' last action. Excavated (1963–1965) by Yigael Yadin and an international team of volunteer archaeologists, Masada is now a major tourist site and an Israeli historical shrine.

Mas'ade (mahs-AHD), Druze village, ISRAEL, in N GOLAN HEIGHTS. Previously served as an important road junction between KUNEITRA, BANIAS, and MAJDAL SHAMS. Birket Ram, a volcanic crater filled with fresh water, is nearby.

Más Afuera Island, CHILE: see JUAN FERNÁNDEZ.

Masagua (mah-SAH-gwah), town, ESCUINTLA department, S GUATEMALA, in Pacific piedmont, on GUACALATE RIVER, and 6 mi/9.7 km SSW of ES-CUINTLA, on Inter-coastal Highway and railroad; 14°12′N 90°51′W. Elevation 361 ft/110 m. Grain, fruit, sugarcane, cotton; livestock. Selected to be the site of a proposed new international airport that has been in planning stage since the late 1960s.

Masaguara (mah-sah-GWAH-rah), town, INTIBUCÁ department, SW HONDURAS, 12 mi/19 km ENE of LA ESPERANZA; 14°22′N 87°59′W. Coffee, tobacco, sugarcane; livestock.

Masahuat (mah-SAH-waht), municipality and town, N SANTA ANA department, EL SALVADOR, NNW of SANTA ANA city; 14°11′N 89°26′W.

Masai Steppe (mah-SAH-ee), plateau, NE central TANZANIA, mainly in MANYARA region, extending c. 175 mi/282 km S from KILIMANJARO and Mount MERU, in semiarid grassland; 125 mi/201 km wide. Elevation 3,937 ft/1,200 m. Livestock raising by nomadic Masai people. Simanjiro Plain in N, KITWE Plain in SE, Lossogoro Plateau in NE; TARANGIRE NATIONAL PARK in NW. Several wells, especially in S, provide water for people and livestock.

Masaka (mah-SAH-kuh), former administrative district (□ 4,097 sq mi/10,652.2 sq km), S UGANDA, along W shore of LAKE VICTORIA; capital was MASAKA; 00°15′S 31°40′E. As of Uganda's division into thirty-nine districts, was bordered by RAKAI (S and W), MBARARA, KABAROLE, and MUBENDE (NW), and MPIGI (N) districts and Lake Victoria (E, with KA-LANGALA district [SESE ISLANDS] not far from shore). Rich agricultural area (coffee, maize, bananas, beans); fishing industry at Lake Victoria. In 1997 E portion of district became current MASAKA district and W portion became SEMBABULE district.

Masaka, administrative district (2005 population 782,400), CENTRAL region, S UGANDA, on Lake VICTORIA (to E); ⊙ MASAKA; 00°30′S 31°45′E. As of Uganda's division into eighty districts, Masaka's border districts include SEMBABULE (NW), MPIGI (N), and KALANGALA (E, in Lake Victoria). Rich agricultural area (coffee, maize, bananas, beans); fishing industry at Lake Victoria. Marsh area in W. Main roads connect Masaka to KAMPALA city, SW Uganda, and TANZANIA. Created in 1997 from E portion of former MASAKA district (Sembabule district was created from W portion).

Masaka (mah-SAH-kuh), town (2002 population 67,768), ⊙ MASAKA district, CENTRAL region, S UGANDA, 75 mi/121 km SW of KAMPALA, on Trans-African Highway. Agricultrual trade center (cotton, coffee, bananas, corn, millet); coffee processing. An epicenter of the AIDS virus. Was capital of former SOUTH BUGANDA province.

Masakhane, SOUTH AFRICA: see JAMESTOWN.

Masaki (mah-SAH-kee), town, Iyo county, EHIME prefecture, NW SHIKOKU, W JAPAN, on IYO SEA, 5 mi/8 km SW of MATSUYAMA; 33°47′N 132°42′E. Marine food manufacturing.

Masakin (muh-sah-KEN), town (2004 population 55,721), SUSAH province, E TUNISIA, 7 mi/11.3 km SSW of SUSAH; 35°44′N 10°35′E. Railroad junction. Olive processing center; oil pressing, flour milling, artisan manufacturing (textiles, sandals, furniture). Horse raising.

Masalfasar (mah-sahl-fah-SAHR), town, VALENCIA province, E SPAIN, near the MEDITERRANEAN SEA, 7 mi/11.3 km NNE of VALENCIA; 39°34′N 00°19′W. Soap and glycerin manufacturing; rice, vegetables.

Masally (mah-SAH-lee), city and center of Masally region, SE AZERBAIJAN, on railroad, and 20 mi/32 km NNW of LÄNKÄRAN, in subtropical Lenkoran Lowland; 39°02′N 48°39′E. Manufacturing (tea, dairy products, building materials; canning, tobacco curing, processing of agricultural products).

Masambolahun, LIBERIA: see BOLAHUN.

Masampo, SOUTH KOREA: see MASAN.

Masan (MAH-SAHN), city (2005 population 428,079), SOUTH KYONGSANG province, SOUTH KOREA, on Masan Bay, 28 mi/45 km WNW of PUSAN; 35°11′N 128°35′E. Fishing and commercial transportation center. Port opened in 1899 to foreign trade; since 1970, has one of the largest free-trade zones. Its well-sheltered port is ice-free. Sake brewing, soy-sauce production, metalworking, cotton and wool weaving. Thermoelectric power station. Exports cotton, fish, salt. Formerly sometimes MASAMPO.

Masanasa (mah-sah-NAH-sah), outer S suburb of VALENCIA, VALENCIA province, E SPAIN; 39°25′N 00°24′W. In rich vegetable-farming area; brandy and liqueur manufacturing.

Masandam, Cape, northernmost point of OMAN PROMONTORY, on STRAIT OF HORMUZ, on small offshore Masandam Island; 26°23′N 56°31′E. The Oman Promontory is sometimes known as Masandam Peninsula. Sometimes spelled Cape Musandam.

Masandra, UKRAINE: see MASSANDRA.

Masang Gang (MAI-sahng GAHNG), mountain peak (23,619 ft/7,199 m), in W HIMALAYA, BHUTAN, near border of TIBET (CHINA), 19 mi/30 km N of GASA.

Masanjor (mah-SUHN-jor), or **Mesanjore**, village and tourist center, DUMKA district, NE Jharkhand state, E India, on Mor River, and 11 mi/18 km SE of DUMKA. Site of 125-ft/38-m-high dam and reservoir for Mor River irrigation project. Picnic spot; Tilpara Barrage (also called Canada Dam) on the Mayurakshi River is a popular scenic spot; Tatloie contains hot water springs.

Masanki (mah-SAHN-kee), village, SOUTHERN province, SW SIERRA LEONE, 32 mi/51 km SE of FREETOWN; 08°29′N 12°21′W. Oil-palm plantation; oil mill.

Masantol (mah-sahn-TOL), town, PAMPANGA province, LUZON, PHILIPPINES, 9 mi/14.5 km S of SAN FERNANDO, near PAMPANGA RIVER; 14°51′N 120°42′E. Agricultural center (sugarcane, rice).

Ma'sara, El (MAH-sah-ruh, el), town, GIZA province, Upper EGYPT, on E bank of the NILE River, 10 mi/16 km SSE of CAIRO city center, near site of ancient MEMPHIS. Has important quarries, which are thought to have furnished part of the material for the pyramids. Consists of town proper and river port.

Ma'sara, El, township, KAFR ESH SHEIKH province, Lower EGYPT, 13 mi/21 km WNW of SHIRBIN. Cotton.

Masardis (muh-SAHR-dis), town, AROOSTOOK county, NE MAINE, on the AROOSTOOK RIVER, and 21 mi/34 km SW of PRESQUE ISLE; 46°31′N 68°21′W.

Masaryk Peak, SLOVAKIA: see GERLACHOVSKY PEAK.

Masasi (mah-SAH-si), town, MTWARA region, SE TANZANIA, 80 mi/129 km SW of LINDI, W of MAKONDE PLATEAU; 10°45′S 38°38′E. Road junction, airstrip. Cashews, corn, bananas, beans, sweet potatoes; sheep, goats; timber.

Masatepe, town (2005 population 15,482), MASAYA department, SW NICARAGUA, 4 mi/6.4 km SW of MASAYA; 11°55′N 86°09′W. Health resort. Agriculture center; coffee and rice processing, tobacco.

Más a Tierra Island, CHILE: see JUAN FERNÁNDEZ.

Masaya (mah-SAH-yah), department (□ 230 sq mi/598 sq km), SW NICARAGUA; ⊙ MASAYA; 12°00′N 86°10′W. Smallest Nicaraguan department, located in plateau sloping to NE; contains volcanoes MASAYA and SANTIAGO. TIPITAPA RIVER forms N border. Agriculture (coffee and rice on slopes; tobacco, manioc, corn, beans, and sugarcane in lowlands; livestock (N; near TISMA). Industry (mainly Native American handicrafts) is concentrated at Masaya. Main centers are Masaya, MASATEPE. Cool summer climate makes department popular resort area.

Masaya (mah-SAH-yah), city (2005 population 92,598) and township, W NICARAGUA, ⊙ MASAYA department; 11°58′N 86°06′W. Connected by highway to GRANADA and MANAGUA, Masaya is a commercial and light manufacturing center in a rich agricultural district. It is noted for its handicrafts.

Masaya (mah-SAH-yah), twin volcanic peaks, Masaya and Santiago (1,804 ft/550 m), MASAYA department, SW NICARAGUA. Masaya is dormant but Santiago erupted in 1986, causing damage to crops and farmland.

Masaya Volcano National Park (mah-SAH-yah) (□ 21 sq mi/54.6 sq km), NICARAGUA's most important national park, includes Masaya and Santiago volcanoes.

Masbate, province (□ 1,563 sq mi/4,063.8 sq km), in BICOL region, the PHILIPPINES, comprised of Masbate, TICAO, and BURIAS islands; ⊙ MASBATE; 12°20′N 123°30′E. Population 25% urban, 75% rural; in 1991, 38% of urban settlements had electricity. Masbate Island, one of the VISAYAN ISLANDS, has a wet and a dry season; there are high, grassy plains in the SE and hills in the N and SW (maximum elevation 1,312 ft/400 m). Ticao and Burias are rugged and have year-round rain. All are in the typhoon belt. Major industry is mining (gold, silver, coal); iron, manganese, copper, and bauxite deposits are undeveloped. Masbate Island is the cattle-raising center of the Philippines; other agricultural industries are fish and fish processing and aquaculture. Corn and rice are grown for food; coconuts are cash crops. Ticao and Burias produce only copra. Masbate (1990 population 58,714) is the province's market center; there's also an airport and an inter-island ferry port. BALUD in SE Masbate Island is commercial fishing port. Immigration from other areas of the country here after World War II has produced a multilingual and multicultural community.

Masbate, town (2000 population 71,441), ⊙ MASBATE province, PHILIPPINES, on NE Masbate island, port on Masbate pass; 12°20′N 123°35′E. Exports copra, cattle.

Mascalucia (mahs-kah-LOO-chah), town, CATANIA province, E SICILY, ITALY, on N slope of Mount ETNA, 5 mi/8 km N of CATANIA; 37°34′N 15°03′E. In grape-growing region; manufacturing of fireworks, wine; stoneworks.

Mascara (mah-skah-RAH), wilaya, in W ALGERIA; ⊙ MASCARA; 35°25′N 00°09′E. Separates the N coastal wilaya of ORAN and MOSTAGANEM from TIARET and SAÏDA wilaya of the steppe region. The region includes vast self-managed farms with renowned vineyards; also, grains and olives.

Mascara (mah-skah-RAH), town, ⊙ MASCARA wilaya, W ALGERIA, on S slopes of Béni Chougrane Mountains, and 50 mi/80 km SE of ORAN; 35°20′N 00°09′E. The center of the vast, fertile Eghris plains. Trades in cereals, leather goods; manufacturing of shoes for most of the country. In the 18th century, under the Turks, was capital of the Western Beylik. Capital of Algeria's anti-French hero, Emir Abd El Kader, from 1832 to 1847. French settlers here produced much wine during the colonial period. Suffered an earthquake in 1994. The fertile plain of GHRISS is to the S.

Mascaraque (mahs-kah-RAH-kai), village, TOLEDO province, central SPAIN, on railroad, and 15 mi/24 km SE of TOLEDO; 39°43′N 03°49′W. Olives, cereals; livestock.

Mascardi, Lake (mahs-KAHR-dee) (□ 14 sq mi/36.4 sq km), SW RÍO NEGRO province, ARGENTINA, in Nahuel Huapí national park, 15 mi/24 km SW of SAN CARLOS DE BARILOCHE; c.10 mi/16 km long, c.2 mi/3.2 km wide. Elevation 2,618 ft/798 m. Surrounded by dense forests. Fishing.

Mascareen Peninsula (mas-kuh-REEN), SW NEW BRUNSWICK, E CANADA, on E side of PASSAMAQUODDY BAY, opposite ST. ANDREWS; 8 mi/13 km long, 5 mi/8 km wide; 45°03′N 65°49′W. Cod, scallops, lobster; aquaculture industry (salmon farms). Includes villages of Latete, Letang, Back Bay. Ferry to DEER ISLAND.

Mascareignes, Indian Ocean: see MASCARENE ISLANDS.

Mascarene Islands, in the INDIAN OCEAN, E of MADAGASCAR. They include MAURITIUS, RÉUNION, and RODRIGUES. Apparently known to the Arabs, they were rediscovered by the Portuguese at the beginning of the 16th century The islands are named after Pedro Mascarenhas, who visited them c.1512. Also spelled Mascareignes.

Mascat, OMAN: see MUSCAT.

Maschito (mahs-KEE-to), village, POTENZA province, BASILICATA, S ITALY, 11 mi/18 km SE of MELFI; 40°54′N 15°50′E. Wine, olive oil. Hot sulphur springs nearby. Albanian refugees settled here in 15th–16th century.

Mascoma Lake (mas-KO-muh), GRAFTON county, W NEW HAMPSHIRE, resort lake 5 mi/8 km E of LEBANON; 4.5 mi/7.2 km long. Mascoma River (c.30 mi/48 km long) rises in S central GRAFTON county c.16 mi/26 km NE of LEBANON, flows S then W through MASCOMA LAKE and LEBANON to CONNECTICUT RIVER at W. Lebanon. Shaker village and Shrine of Our Lady of LaSalette on SW shore.

Mascot, village, KNOX county, E TENNESSEE, 13 mi/21 km ENE of KNOXVILLE; 36°04′N 83°45′W. Manufacturing.

Mascot (MAS-kaht), suburb, E NEW SOUTH WALES, AUSTRALIA, 5 mi/8 km S of SYDNEY; 33°56′S 151°12′E. In metropolitan area; sugar mills; manufacturing (coffee, dairy products); aviation services industries. Kingsford Smith International Airport, Sydney's principal airport, on NW side of BOTANY BAY. Horse racing.

Mascota (mas-KO-tah), town and township, JALISCO, W MEXICO, 48 mi/77 km W of AMECA; 20°34′N 104°49′W. Agricultural center (grain, sugarcane, cotton, tobacco, fruit, rice).

Mascote (mahs-KO-te), city (2007 population 16,062), SE BAHIA, BRAZIL, on RIO PARDO, 62 mi/100 km S of ITABUNA; 15°35′S 39°19′W.

Mascotte (muh-SKAHT-tee), city (2005 population 4,647), Lake county, central FLORIDA, 16 mi/26 km S of LEESBURG; 28°34′N 81°53′W. In citrus-fruit region.

Mascouche (mas-KOOSH), city (□ 42 sq mi/109.2 sq km), LANAUDIÈRE region, S QUEBEC, E CANADA, 18 mi/29 km NNE of MONTREAL; 45°45′N 73°35′W. Agriculture. Part of the Metropolitan Community of Montreal (*Communauté Metropolitaine de Montréal*).

Mascoutah (mass-COO-tah), city (2000 population 5,659), SAINT CLAIR county, SW ILLINOIS, satellite community of SAINT LOUIS, 20 mi/32 km ESE of EAST SAINT LOUIS; 38°30′N 89°48′W. Bituminous coal mines; agriculture (wheat, soybeans, corn, apples, poultry, hogs; dairy products). Incorporated 1839.

Mas-d'Agenais, Le (MAH–dah-zhuh-NAI, luh), agricultural commune (□ 8 sq mi/20.8 sq km), LOT-ET-GARONNE department, AQUITAINE region, SW FRANCE, on the GARONNE RIVER and its lateral canal, and 7 mi/11.3 km SSE of MARMANDE. In rich agricultural valley. Has Romanesque church noted for its crucifixion painting by Rembrandt.

Mas-d'Azil, Le (MAH–dah-ZEEL, luh), commune (□ 15 sq mi/39 sq km), ARIÈGE department, MIDI-PYRÉNÉES region, S FRANCE, in foothills of the PYRENEES Mountains, on ARIZE RIVER, and 13 mi/21 km WSW of PAMIERS; 43°05′N 01°22′E. Wool spinning. A nearby tunnel-shaped grotto is traversed by the Arize River and by a road; contains prehistoric relics (primitive tools and weapons, mammoth bones).

Mas de Barberáns (MAHS dai bahr-vai-RAHNS), village, TARRAGONA province, NE SPAIN, 10 mi/16 km SW of TORTOSA; 40°44′N 00°22′E. Olive-oil processing; sheep raising.

Mas de las Matas (MAHS lahs MAH-tahs), town, TERUEL province, E SPAIN, 16 mi/26 km SSW of AL-CAÑIZ; 40°50′N 00°15′W. Olive-oil processing, brandy manufacturing; fruit, wine, alfalfa; livestock.

Masein (MAH-sain), village, KALEWA township, SA-GAING division, MYANMAR, on E bank of CHINDWIN RIVER, and 12 mi/19 km N of KALEWA.

Maselheim (MAH-sel-heim), village, Danube-Iller, BADEN-WÜRTTEMBERG, S GERMANY, 19 mi/30 km SSW of ULM; 48°08′N 09°53′E.

Masel'skaya (mah-SYEL-skah-yah), Finnish *Maaselkä*, village, S central Republic of KARELIA, NW European Russia, on road and the Murmansk railroad, 20 mi/32 km N of MEDVEZHYEGORSK; 63°09′N 34°18′E. Elevation 574 ft/174 m. Woodworking; granite quarry. Formerly called Masel'gskaya.

Maserada sul Piave (mah-ze-RAH-dah sool PYAH-ve), town, TREVISO province, VENETO, N ITALY, near PIAVE River, 7 mi/11 km NNE of TREVISO. Cotton and hemp mills.

Maseru (mah-SE-roo), district (□ 1,652 sq mi/4,295.2 sq km; 2001 population 477,599), W central and NW LESOTHO; ⊙ MASERU. Bounded W by CALEDON RIVER (SOUTH AFRICAN border), E by SENQUNYANE RIVER (forms Mohale Dam in NE); MALUTI MOUNTAINS in NE, THABA PUTSOA range in S, both ranges join in district. Sheep, goats, cattle, hogs, chickens; corn, sorghum, vegetables, peaches. Sandstone, clay quarrying.

Maseru (mah-SE-roo), city (2004 population 178,300), ⊙ Kingdom of LESOTHO and MASERU district, NW Lesotho, on the CALEDON RIVER (SOUTH AFRICAN border); 29°19′S 27°29′E. A trade and transportation hub, it is Lesotho's only sizable city. It lies on Lesotho's main road and is linked to South Africa's railroad network (1 mi/1.6 km railroad section connects with Maseru West Industrial Estate). Light manufacturing. Tourist center for Lesotho, wholesale distribution, vehicle distribution, and printing and publishing. Seat of Lesotho Agricultural College; National University of Lesotho is at ROMA, 17 mi/27 km to SE. Moshoeshoe I International Airport is 12 mi/19 km SSE on one of the few level areas near the city. Lesotho Government Archives are here, and the country's major hospital. Maseru was a small trading town when it was made the capital of the Basuto people by Moshoeshoe I, their paramount chief, in 1869. It was the capital of British Basutoland protectorate from 1869 to 1871 and from 1884 to 1966, when Lesotho achieved independence.

Masevaux (mah-zuh-VO), commune (□ 9 sq mi/23.4 sq km), HAUT-RHIN department, in ALSACE, E FRANCE, on S slopes of the VOSGES, 11 mi/18 km NNE of BELFORT; 47°47′N 07°00′E. Cotton milling. Has ruins of ancient abbey.

Mashaba, ZIMBABWE: see MASHAVA.

Mash'abbe Sade (mah-shah-BE sah-DE), kibbutz, S ISRAEL, 16 mi/25 km S of BEERSHEBA, in NEGEV; 31°00′N 34°47′E. Elevation 1,125 ft/342 m. Near the site of the Arab village Bir 'Asluj, which was evacuated during the 1948 war. It then became a base for the Egyptian army, which invaded Israel. Mixed farming and a metals factory. Two middle-Bronze Age sites were unearthed nearby, Ro'i 1 and 2, revealing one of the largest settlements in the Negev from this era. W of the kibbutz there are two nature reserves: Hulot and Narkissim Nahal Mass'ad (Mass'ad stream narcissi). Originally founded in 1947 as Kibbutz Mash'abbim, it took on its present name and location in 1949.

Mashad Ali, Iraq: see NAJAF.

Masham (MAS-uhm), town (2006 population 1,250), NORTH YORKSHIRE, N ENGLAND, on URE RIVER and 8 mi/12.9 km NW of RIPON; 54°13′N 01°39′W. Former agricultural market. Has 15th century church.

Mashan (MAH-SHAN), town, ⊙ Mashan county, W central GUANGXI ZHUANG AUTONOMOUS REGION, CHINA, 70 mi/113 km SW of LIUZHOU; 23°49′N 108°15′E. Rice, millet, beans, potatoes, sugarcane, medicinal herbs; electric power generation.

Mashapaug, CONNECTICUT: see UNION.

Masharij, Am (em muh-shuh-REEZ), village, SUBEIHI tribal area, S YEMEN, 26 mi/42 km WNW of LAHEJ. Also spelled Am Misharij and Am Musharij.

Mashava, town, MASVINGO province, SE central ZIMBABWE, 22 mi/35 km W of MASVINGO; 20°03′S 30°29′E. Asbestos-mining center. GATH MINE and Mushwe Reservoir to N. MUSHANDIKE RESERVOIR and Sanctuary to SE. Formerly spelled Mashaba.

Mashego (mah-SHAI-go), town, NIGER state, W central NIGERIA, on road, and 70 mi/113 km WNW of MINNA. Market town. Sorghum, millet, and maize.

Masherbrum (MUH-shuhr-broom), peak (25,660 ft/7,821 m) in KAILAS-KARAKORAM RANGE of KARAKORAM mountain system, in NORTHERN AREAS, NE PAKISTAN, in KASHMIR region, 45 mi/72 km NE of Sakardu; 35°39′N 76°18′E. English expedition in 1938 climbed to within 1,500 ft/457 m of the top.

Mashevka, UKRAINE: see MASHIVKA.

Mashewa (mah-SHAI-wah), village, TANGA region, NE TANZANIA, 40 mi/64 km NW of TANGA; 04°45′S 38°41′E. Timber; livestock; grain. USAMBARA Mountains to SW.

Mashgharah, LEBANON: see MESHGHARA.

Mashhad (mahsh-HAHD), city (2006 population 2,427,316), ⊙ Khorāsān province, NE IRAN; 29°52′N 52°48′E. It is an industrial and trade center and a transportation hub, and the second-largest city in IRAN. Manufacturing includes carpets, textiles, and processed foods. Connected by railroad to TEHRAN and the PERSIAN GULF. On ancient SILK ROAD to central ASIA. In 1996, a c.100-mi/160-km-long railroad NE to SARAKHS was completed, connecting with the TURKMENISTAN railroad system; the route is called the "silk road railroad" and carried mainly cotton from central Asia and consumer goods from the Persian Gulf. Mashhad is a religious center visited annually by large numbers of Muslim (Shiite) pilgrims. Formerly known as Sanabadh, it is the site of the beautiful shrine of the Imam Ali Riza, a Shiite holy person. Imam Riza died (C.E. 819) in the city after visiting the grave of Caliph Harun ar-Rashid, who had died here 10 years before; he was buried next to Harun, and the shrine was built over both graves. The city was attacked by the Oghuz Turks (12th century) and by the Mongols (13th century), but recovered by the 14th century, when it came to be known as Mashhad (in Arabic, means a place of martyrdom or shrine). It prospered under the Safavids, who were devout Shiite Muslims; Shah Abbas I embellished Mashhad with elaborate buildings. It reached its greatest glory in the 18th century, when Nadir Shah made Mashhad the capital of PERSIA. The city took on strategic importance in the late 19th century because of its proximity to the Russian and Afghan borders. The bombing of the sanctuary of the Imam Riza by the Russians in 1912 caused widespread resentment in the Shiite Muslim world. Near Mashhad are the remains of the former city of TUS, birthplace of the poet Firdausi and the philosopher al-Ghazali. Mashhad itself is the site of a university (founded 1947). The city is also known as Meshed.

Mashike (MAH-shee-ke), town, Rumoi district, W Hokkaido prefecture, N JAPAN, on SEA OF JAPAN, 56 mi/90 km N of SAPPORO; 43°51′N 141°31′E. Shrimp, octopus, herring; asparagus, fruit. Local sake. Uryu Marsh and Cape Ofuyu nearby.

Mashiki (mah-shee-KEE), town, Kamimashiki county, KUMAMOTO prefecture, W KYUSHU, SW JAPAN, 4.3 mi/7 km E of KUMAMOTO; 32°47′N 130°49′E.

Mashiko (mah-shee-KO), town, Haga county, TOCHIGI prefecture, central HONSHU, N central JAPAN, 12 mi/20 km N of UTSUNOMIYA; 36°27′N 140°05′E. Cameras.

Mashivka (MAH-shif-kah), (Russian *Mashevka*), town, E POLTAVA oblast, UKRAINE, 17 mi/27 km SE of POLTAVA; 49°26′N 34°52′E. Elevation 419 ft/127 m. Raion center; dairy; flour and feed mills. Heritage museum. Known since 1859, town since 1971.

Mashkil River (muhsh-KEEL), river, c.200 mi/322 km long, SE IRAN and W PAKISTAN; rises in several branches in easternmost arid ZAGROS Mountain ranges E of

IRANSHAHR; flows intermittently E and N to the HAMUN-I-MASHKEL (PAKISTAN). Receives Rakhman River.

Mashobra, India: see SHIMLA.

Mashonaland, region, NE ZIMBABWE. Chiefly of historical and ethnological interest as the region inhabited by the Mashona, a Bantu-speaking tribe dominated after 1837 by the warlike Matabele. Region was acquired 1889 by Cecil Rhodes for the British South Africa Company. Now part of MANICALAND, MASHONALAND CENTRAL, MASHONALAND EAST, and MASHONALAND WEST provinces.

Mashonaland Central, province (□ 10,945 sq mi/28,457 sq km; 2002 population 998,265), NE ZIMBABWE; ⊙ BINDURA. Major towns include Bindura and MOUNT DARWIN. Bounded N by MOZAMBIQUE, MVURWI RANGE on W; part of high veld (plateau) in S, ZAMBEZI ESCARPMENT crosses N part of province, with ZAMBEZI River plain to N of it. Drained by MAZOWE, RUYA, and Manyane rivers. Umfurudzi Safari Area in E, Dande Safari Area in NW corner. Gold, asbestos, nickel, and copper mining. Timber. Agriculture (tobacco, cotton, corn, wheat, peanuts, sorghum); cattle, sheep, goats. Formerly administered from HARARE.

Mashonaland East, province (□ 12,444 sq mi/32,354.4 sq km; 2002 population 1,125,355), NE ZIMBABWE; ⊙ MARONDERA. Other major towns include Chitungwiza and MARONDERA; bounded NE by of MOZAMBIQUE; part of high veld in S; Lakes MANYAME (Robertson) and CHIVERO (McIlwaine) reservoirs (both recreational parks) in W; Ewanrigg Botanical Gardens in W center. Province is source of several rivers, including SAVE, MAZOWE, and Nyangidzi. Timber. Gold mining. Agriculture (coffee, tea, tobacco, fruit, macadamia nuts, cotton, corn, wheat, soybeans, sorghum, peanuts, sunflowers, vegetables); dairying; cattle, sheep, goats, hogs, poultry. Former capital was HARARE.

Mashonaland West, province (□ 22,178 sq mi/57,662.8 sq km; 2002 population 1,222,583), N ZIMBABWE; ⊙ CHINHOYI. Other major towns include KARIBA, KADOMA, and CHEGUTU; bounded N and NW by ZAMBEZI River (ZAMBIA border), KARIBA DAM on Zambezi forms Lake KARIBA in NW; ZAMBEZI ESCARPMENT crosses province in N, part of high veld (plateau) in S, MVURWI RANGE on E boundary. Drained by MUPFURE, Manyane, and SANYATI rivers. MANA POOLS and MATUSADONA national parks and several safari areas in NW, Umfuli Recreation Area and Hartley Safari Area in W. Gold chromite, and copper mining; platinum mining at CHEGUTU (Hartley Mine). Agriculture (cotton, corn, wheat, soybeans, sorghum, peanuts, tobacco); cattle, sheep, goats; timber.

Mashpee (MASH-pee), resort town, BARNSTABLE county, SE MASSACHUSETTS, on W CAPE COD, 10 mi/16 km WSW of BARNSTABLE; 41°37′N 70°29′W. Includes villages of Seconsett, New Seabury, and Popponesset. Two state parks in vicinity. Site of Native American church and praying ground and lands of Wampanoag tribe. South Cape Beach State Park, Popponesset Beach.

Mashtagi (mush-TAH-gee) or **Mashtaya**, town in Mashtagi district of Greater BAKY, AZERBAIJAN, on N APSHERON Peninsula, 14 mi/23 km NE of BAKY. On electric railroad; manufacturing (canning, woodworking, dairy products); oil wells (developed in World War II), stone quarries.

Mashuk (mah-SHOOK), laccolithic mountain (3,258 ft/993 m) of the N CAUCASUS foothills, STAVROPOL TERRITORY, S European Russia, overlooking (NE) PYATIGORSK.

Mashur, Bandar, IRAN: see BANDAR MAHSHAHR.

Masicurí (mah-see-koo-REE), canton, VALLEGRANDE province, SANTA CRUZ department, E central BOLIVIA, 41 mi/70 km SW of VALLEGRANDE, on the Masicurí River; 18°59′S 63°46′W. Elevation 6,660 ft/2,030 m. Agriculture (potatoes, bananas, yucca, soy, rye, corn); cattle.

Masi Manimba (MAH-see mah-NEEM-bah), village, BANDUNDU province, SW CONGO, 65 mi/105 km WNW of KIKWIT; 04°46′S 17°55′E. Elev. 1,253 ft/381 m. Palm-oil milling, rice processing.

Masimpur (MUH-seem-poor), village, CACHAR district, S ASSAM state, NE INDIA, on BARAK (Surma) River, 4 mi/6.4 km NW of Silcha. Rice, tea, rape, and mustard. Oil wells nearby.

Masindi, administrative district, WESTERN region, W central UGANDA, on VICTORIA NILE RIVER to E and N; ⊙ MASINDI. As of Uganda's division into eighty districts, borders AMURU and OYAM (N, formed by Victoria Nile River), APAC (E, formed by Victoria Nile River), NAKASONGOLA and NAKASEKE (SE), KIBOGA (S), HOIMA (SW), and BULIISA (W) districts. Marsh area in E. KABALEGA NATIONAL PARK extends into N of district (from Amuru district), including KABALEGA FALLS on Victoria Nile River. Agriculture (including tobacco). Roads branch out from Masindi town to all parts of Uganda. Formed in 2006 from all but WNW portion of former MASINDI district (Buliisa district formed from WNW portion).

Masindi (mah-SIN-dee), former administrative district (□ 3,601 sq mi/9,362.6 sq km; 2005 population 538,400), WESTERN region, W central UGANDA, between LAKE ALBERT and LAKE KYOGA, on VICTORIA NILE RIVER (to E and N); capital was MASINDI; 01°50′N 31°50′E. As of Uganda's division into fifty-six districts, was bordered by NEBBI (NW tip, opposite Lake Albert), GULU (N, formed by Victoria Nile River), APAC (E, formed by Victoria Nile River), NAKASONGOLA (SE), LUWERO (SES), KIBOGA (S), and HOIMA (SW) districts and Lake Albert (W, with DEMOCRATIC REPUBLIC OF THE CONGO on opposite shore). Included forests with large chimpanzee population; marsh area in E. KABALEGA NATIONAL PARK extended into N of district (from Gulu district), including KABALEGA FALLS on Victoria Nile River. Tobacco and cassava were grown; timber industry was also important. In 2006 NWN portion of district was carved out to form BULIISA district and remainder of district was formed into current MASINDI district.

Masindi (mah-SIN-dee), town (2002 population 28,300), ⊙ MASINDI district, WESTERN region, W central UGANDA, 110 mi/177 km W of KAMPALA, and on road from LAKE KYOGA to LAKE ALBERT. Agricultural trade center (cotton, tobacco, coffee, bananas, corn); tobacco factory. Hospital. Was part of former WESTERN province.

Masindi Port (mah-SIN-dee), town, MASINDI district, WESTERN region, W central UGANDA, port on VICTORIA NILE RIVER at its issuance from LAKE KYOGA, 26 mi/42 km E of MASINDI; 01°41′N 32°04′E. Steamer-road transfer point; cotton, tobacco, coffee, bananas, corn. Has cotton gin, sisal factory. Was part of former WESTERN province.

Masingbi, town, SIERRA LEONE, 40 mi/64 km ESE of KABALA; 08°38′N 11°28′W. On road to towns in the NORTHERN province. Also spelled Masingbe.

Masinloc, town, ZAMBALES province, central LUZON, PHILIPPINES, port on SOUTH CHINA SEA, 45 mi/72 km W of TARLAC; 15°32′N 120°03′E. Mining and lumbering center. Nearby are chrome-ore mines and sawmills.

Masira Island, in ARABIAN SEA, off coast of OMAN, 140 mi/225 km SSW of RAS AL HADD; 40 mi/64 km long, 10 mi/16 km wide; 20°30′N 58°45′E. It is separated from mainland by Masira Channel (10 mi/16 km wide). Airfield on N tip. Masira Bay, or Bahr al Hadri, is an inlet of ARABIAN SEA, extending from Masira Island 100 mi/161 km SW to the RAS MADRAKA.

Masis, urban settlement and center of Masis region (2001 population 21,736), ARMENIA; 40°03′N 44°24′E.

Railroad station; cannery; manufacturing of corrugated cardboard, wood products.

Masisea (mah-see-SAI-ah), town, CORONEL PORTILLO province, UCAYALI region, E central PERU, landing on UCAYALI River, and 23 mi/37 km ESE of PUCALLPA; 08°36′S 74°19′W. In agricultural region (sugarcane, yucca, plantains, bananas).

Masisi (mah-SEE-see), village, NORD-KIVU province, E CONGO, 40 mi/64 km NNW of BUKAVU; 01°24′S 28°49′E. Elev. 5,419 ft/1,651 m. Trading center. Protestant mission. Tin and tantalite mining; coffee plantations in vicinity.

Masjed Soleyman (muhs-JEED sul-ai-MAHN), city (2006 population 108,682), Khuzestān province, SW IRAN, on the KARUN RIVER; 31°55′N 49°18′W. The site of the first discovery of petroleum in Iran (1908), it is now an oil-refining center.

Mas, Kali, INDONESIA: see MAS RIVER.

Maskall, village, BELIZE district, BELIZE, on Old Northern Highway 40 mi/64 km N of BELIZE CITY; 17°53′N 88°19′W. Local commercial and tourist center.

Maskanah, SYRIA: see MESKENE.

Maskat, OMAN: see MUSCAT.

Maskeliya (MUHS-KE-li-yuh), town, CENTRAL PROVINCE, SRI LANKA, on HATTON PLATEAU, 4.5 mi/7.2 km SSW of Hatton; 06°50′N 80°34′E. Tea processing; extensive tea plantations.

Maskell, village (2006 population 65), DIXON county, NE NEBRASKA, 16 mi/26 km NW of PONCA, near MISSOURI RIVER; 42°41′N 96°58′W.

Maski (MAHS-kee), town, RAICHUR district, KARNATAKA state, S INDIA, 50 mi/80 km WSW of RAICHUR; 15°58′N 76°40′E. Millet, rice, oilseeds; cotton ginning. Sometimes spelled Muski.

Maskinongé (mah-skee-non-ZHAI), county (□ 2,378 sq mi/6,182.8 sq km), MAURICIE region, S QUEBEC, E CANADA, extending NW from the SAINT LAWRENCE RIVER; ⊙ LOUISEVILLE; 46°23′N 73°05′W. Composed of seventeen municipalities. Regional county municipality (*municipalité régionale de comté*) created in 1982 from the former Maskinongé county.

Maskinongé (mah-skee-non-ZHAI), village (□ 29 sq mi/75.4 sq km), MAURICIE region, S QUEBEC, E CANADA, on MASKINONGE RIVER near its mouth on the SAINT LAWRENCE RIVER, and 24 mi/39 km WSW of TROIS-RIVIÈRES; 46°13′N 73°01′W. Dairying; pig raising. Restructured in 2001.

Maskinongé River (mah-skee-non-ZHAI), 35 mi/56 km long, S QUEBEC, E CANADA; issues from Lake Maskinongé (4 mi/6 km long), 25 mi/40 km NNW of SOREL; flows SE and then S, past MASKINONGÉ, to the SAINT LAWRENCE RIVER 10 mi/16 km NNE of SOREL-TRACY; 46°09′N 73°01′W. Above its mouth it has a falls of over 300 ft/91 m.

Mask, Lough, Gaelic *Loch Measca*, lake (10 mi/16 km long, 5 mi/8 km wide), S MAYO county and NW GALWAY county, W IRELAND, 3 mi/4.8 km N of N end of LOUGH CORRIB, with which it is connected by partly subterranean stream; 53°37′N 09°21′W. Contains several small islands and receives ROBE RIVER.

Maskoutains, Les (mahs-koo-TAN, lai), county (□ 506 sq mi/1,315.6 sq km; 2006 population 80,745), MONTÉRÉGIE region, S QUEBEC, E CANADA; ⊙ SAINTE-HYACINTHE; 45°37′N 72°57′W. Composed of seventeen municipalities. Formed in 1982.

Maslen Nos (MAS-len NOS), cape, BURGAS oblast, BULGARIA, on the BLACK SEA; 42°19′N 27°47′E.

Maslova Pristan' (MAHS-luh-vah PREES-tahn), village (2005 population 5,700), S central BELGOROD oblast, SW European Russia; on road and railroad; 12 mi/19 km S of BELGOROD; 50°28′N 36°45′E. Elevation 456 ft/138 m. Brewery.

Maslovka (MAHS-luhf-kah), village (2006 population 7,485), N VORONEZH oblast, S central European Russia, on the VORONEZH RIVER, on highway and railroad, 7 mi/11 km S of (and administratively subordinate to) VORONEZH; 51°33′N 39°14′E. Elevation 344 ft/104 m. Produce processing, woodworking. Holiday getaway spot for residents of neighboring cities; vacation homes.

Maslovo (MAHS-luh-vuh), town, N SVERDLOVSK oblast, W Siberian Russia, on the SOS′VA RIVER, on railroad, 40 mi/64 km S (and under jurisdiction) of IVDEL; 60°07′N 60°28′E. Elevation 344 ft/104 m. Limonite mining. Formerly known as Verkhne-Maslovo.

Maslovo (MAHS-luh-vuh), village, W YAROSLAVL oblast, central European Russia, near highway, 40 mi/64 km WSW of RYBINSK; 58°04′N 38°02′E. Elevation 488 ft/148 m. Flax.

Maslovskiy (MAHS-luhf-skeeyee), rural settlement (2006 population 3,420), N central VORONEZH oblast, S central European Russia, in the DON River basin, 11 mi/18 km SE of VORONEZH; 51°32′N 39°20′E. Elevation 439 ft/133 m. In agricultural area; produce processing.

Maslyanino (mahs-LYAH-nee-nuh), town (2006 population 13,335), SE NOVOSIBIRSK oblast, SW SIBERIA, RUSSIA, on the Berd River (tributary of the OB′ RIVER), on road junction, 62 mi/100 km N of BARNAUL, and 35 mi/56 km E of CHEREPANOVO; 54°20′N 84°13′E. Elevation 597 ft/181 m. Fruit and vegetable processing, cheese making, woodworking, flax processing.

Masna′a (muhs-NAH-uh), township, N OMAN, port on GULF OF OMAN, 60 mi/97 km WNW (deep inland) of MUSCAT. Trade center for Rustaq district.

Masna′a Aura, YEMEN: see AURA, MASNA′A.

Masnedø, DENMARK: see SMÅLANDSFARVAND.

Masnedsund, DENMARK: see SMÅLANDSFARVAND.

Masnières (mah-NYER), industrial town (□ 4 sq mi/10.4 sq km), NORD department, NORD-PAS-DE-CALAIS region, N FRANCE, on Escaut (SCHELDT) River, on SAINT-QUENTIN CANAL, and 4 mi/6.4 km S of CAMBRAI; 50°07′N 03°13′E.

Masnóu (mahs-NO), city, BARCELONA province, NE SPAIN, on the MEDITERRANEAN SEA, and 10 mi/16 km NE of BARCELONA; 41°29′N 02°19′E. Wine-producing center; manufacturing of cotton and silk fabrics. Trades in hazelnuts and fruit. Swimming resort.

Masny (mah-NEE), residential town (□ 1 sq mi/2.6 sq km), NORD department, NORD-PAS-DE-CALAIS region, N FRANCE, 6 mi/9.7 km ESE of DOUAI; 50°21′N 03°12′E. In former coal-mining district.

Maso (mah-SO), town (2000 population 2,780), BRONG-AHAFO REGION, GHANA, near Côte d'Ivoire border and source of BIA RIVER, 35 mi/56 km WSW of SUNYANI; 07°14′N 02°53′W. Timber, coffee, cocoa. Sometimes spelled Maaso.

Masoala National Park (mah-SWAHL) (□ 811 sq mi/2,108.6 sq km), TOAMASINA and ANTSIRANANA provinces, NE MADAGASCAR, 30 mi/48 km E of MAROANTSETRA; 15°38′S 50°09′E. Former nature reserve (1927–1964) reestablished as a national park in 1995 to protect largest unbroken block of lowland rain forest in the country.

Masoala Peninsula (mah-SWAHL), on NE coast of MADAGASCAR, in ANTSIRANANA and TOAMASINA provinces; 40 mi/64 km long, maximum 30 mi/48 km wide. Terminates in Cape Masoala (15°55′S 50°10′E). Bordered by ANTONGIL BAY (W). MASOALA NATIONAL PARK is located here. Exceptionally rainy climate.

Masohi (MAH-saw-hee), port, ⊙ Maluku Tengah district, Maluku province, E INDONESIA, on S coast of SERAM island; 03°17′S 129°02′E. Chief town and port of Seram Island.

Masoko (ma-SO-koh), village, MBEYA region, SW TANZANIA, 35 mi/56 km SE of MBEYA, SW of KIPENGERE RANGE; 09°19′S 33°47′E. Coffee, tea, grain; cattle, sheep, goats; timber.

Mason, county (□ 563 sq mi/1,463.8 sq km; 2006 population 15,503), central ILLINOIS; ⊙ HAVANA; 40°14′N 89°54′W. Bounded W by ILLINOIS RIVER and S by SANGAMON RIVER and SALT CREEK. Slough lakes along the ILLINOIS. Includes Chautauqua National Wildlife Refuge and Sand Ridge State Forest. Agriculture (corn, wheat, soybeans, vegetables; melons). Diversified manufacturing; river, railroad shipping. Formed 1841. Much agriculture benefits from outer-pivot irrigation.

Mason, county (□ 246 sq mi/639.6 sq km; 2006 population 17,271), NE KENTUCKY, on OHIO RIVER (N; OHIO state line); ⊙ MAYSVILLE; 38°36′N 83°49′W. Drained by North Fork of LICKING RIVER. Gently rolling upland agricultural area (burley tobacco, corn, wheat, soybeans, hay, alfalfa; cattle, poultry; dairying), in N part of BLUEGRASS REGION; limestone. Manufacturing at Maysville. Formed 1788.

Mason, county (□ 1,241 sq mi/3,226.6 sq km; 2006 population 29,045), W MICHIGAN, on LAKE MICHIGAN (W); 44°01′N 86°30′W; ⊙ LUDINGTON. Drained by PERE MARQUETTE, BIG SABLE and LITTLE MANISTEE rivers and short Lincoln River. Agriculture (cattle; apples, cherries, peaches, green beans; dairy products). Manufacturing at Ludington. Fisheries. Resorts. Ludington State Park, with its large sand dunes, is in W between Lake Michigan and large HAMLIN LAKE; Manistee National Forest in E and N margins of county. Organized 1855.

Mason (MAI-suhn), county (□ 932 sq mi/2,423.2 sq km; 2006 population 3,902), central TEXAS; ⊙ MASON; 30°43′N 99°13′W. On EDWARDS PLATEAU; elevation c.1,200 ft/366 m–2,300 ft/701 m. Drained by SAN SABA RIVER, LLANO RIVER and its tributaries. Ranching (beef cattle, sheep, goats); wool, mohair marketed; agriculture (peanuts, watermelons; hay). Topaz found in area. Hunting, fishing, scenery attract visitors. Formed 1858.

Mason, county (□ 1,051 sq mi/2,732.6 sq km; 2006 population 55,951), W WASHINGTON; ⊙ SHELTON; 47°21′N 123°11′W. Mountainous area indented in NE by HOOD CANAL. Timber; fish, clams, oysters; hay; dairying; poultry. Part of OLYMPIC NATIONAL FOREST in NW (includes Wonder Mountains and Mount Skokomish wilderness areas), Squaxin Island (E) and Skokomish (N) Indian reservations; Lake Cushman reservoir in N. Formed 1854.

Mason, county (□ 445 sq mi/1,157 sq km; 2006 population 25,756), W WEST VIRGINIA, on OHIO RIVER (N and W; OHIO state line); ⊙ POINT PLEASANT; 38°46′N 82°01′W. Drained by KANAWHA RIVER, which joins Ohio in NW. Bituminous-coal mines; some natural-gas wells. Agriculture (honey, corn, wheat, oats, tobacco, potatoes, alfalfa, hay, vegetables, nursery crops); cattle, hogs, sheep; dairying. Manufacturing at Point Pleasant and NEW HAVEN. Chief Cornstalk (S center) and Clifton F. McClintic (NW) wildlife management areas. Formed 1804.

Mason (MAI-suhn), city (□ 18 sq mi/46.8 sq km; 2006 population 29,491), WARREN county, SW OHIO, 21 mi/34 km NE of CINCINNATI; 39°22′N 84°18′W.

Mason, town (2000 population 396), EFFINGHAM county, SE central ILLINOIS, 12 mi/19 km SSW of EFFINGHAM; 38°57′N 88°37′W. Agriculture (cattle, hogs; wheat, soybeans, sorghum).

Mason, town (2000 population 6,714), ⊙ INGHAM county, S central MICHIGAN, 12 mi/19 km SSE of LANSING; 42°34′N 84°26′W. In farm area (dairying; livestock; beans, cabbage, vegetables, apples). Manufacturing (printing; fabricated metal products, consumer goods).

Mason, town, HILLSBOROUGH county, S NEW HAMPSHIRE, on MASSACHUSETTS (S) state line, 14 mi/23 km W of NASHUA; 42°45′N 71°45′W. Drained by Spaulding Brook. Agriculture (livestock, poultry; fruit, vegetables, corn; dairying; nursery crops, sugar maples); manufacturing (maple syrup).

Mason, town (2006 population 1,163), TIPTON county, W TENNESSEE, 34 mi/55 km NE of MEMPHIS; 35°25′N 89°32′W. In agricultural and manufacturing area.

Mason (MAI-suhn), town (2006 population 2,228), ⊙ MASON county, central TEXAS, on EDWARDS PLATEAU, 85 mi/137 km SE of SAN ANGELO, and on Comanche Creek in LLANO RIVER valley; 30°45′N 99°13′W. Elevation 1,550 ft/472 m. Shipping center for cattle, sheep, goats; peanuts, watermelons; wool, mohair. Manufacturing (beverages). Resort (hunting, fishing nearby). Site of historic Fort Mason, foundations of twenty-three buildings. Settled by Germans before Civil War. Incorporated after 1940.

Mason, town (2006 population 1,047), MASON county, W WEST VIRGINIA, near the OHIO RIVER (bridged), 14 mi/23 km NNE of POINT PLEASANT, and 2 mi/3.2 km SE of (opposite) POMEROY (OHIO); 39°01′N 82°01′W. Agriculture (grain); livestock; dairying. Coal-mining area. Clifton F. McClintic Wildlife Management Area to S.

Mason, village (2006 population 81), BAYFIELD county, N WISCONSIN, 13 mi/21 km SW of ASHLAND; 46°26′N 91°03′W. Lumber. Chequamegon National Forest to N, W and S.

Mason and Dixon's Line, U.S.: see MASON-DIXON LINE.

Masonboro, unincorporated town (□ 6 sq mi/15.6 sq km; 2000 population 11,812), NEW HANOVER county, SE NORTH CAROLINA, residential suburb 6 mi/9.7 km ESE of downtown WILMINGTON, on INTRACOASTAL WATERWAY, at Hewletts Creek estuary; 34°10′N 77°52′W. Masonboro Inlet to ATLANTIC OCEAN 2 mi/3.2 km to E. Service industries; manufacturing.

Mason City, city (2000 population 2,558), MASON county, central ILLINOIS, 28 mi/45 km N of SPRINGFIELD; 40°12′N 89°42′W. In agricultural (corn, wheat, soybeans) and clay area; manufacturing (edible oils). Incorporated 1869.

Mason City, city (2000 population 29,172), ⊙ CERRO GORDO county, N central IOWA; 43°08′N 93°12′W. Major railroad junction and trade and industrial center of a large agricultural area. The major industries are food processing, meatpacking, and the manufacturing of cement and fertilizers. Also manufacturing of soybean products, foods, feeds, electrical goods, fabricated metal products, paper products, building materials, machinery; printing, meat processing. Seat of North Iowa Area Community College (1918; oldest in the state). A large band festival is held here annually. McIntosh Woods and Clear Lake state parks to W. Incorporated 1874.

Mason City, village (2006 population 171), CUSTER county, central NEBRASKA, 20 mi/32 km SE of BROKEN BOW, and on MUD CREEK; 41°13′N 99°17′W. Grain; cattle.

Mason Dam, Oregon: see PHILLIPS RESERVOIR.

Mason-Dixon Line, state line between MARYLAND and both PENNSYLVANIA and DELAWARE, surveyed by the English astronomers Charles Mason and Jeremiah Dixon between 1763 and 1767. The ambiguous description of the borders in the Maryland and Pennsylvania charters led to a protracted disagreement between the proprietors of the two colonies; the dispute was submitted to the English court of chancery in 1735. A compromise between the Penn and Calvert families in 1760 resulted in the appointment of Mason and Dixon. By 1767 the surveyors had run their line 244 mi/393 km W from the Delaware border, every fifth milestone bearing the Penn and Calvert arms. The survey was completed to the W limit of Maryland in 1773; in 1779 the line was extended to mark the S border of Pennsylvania with VIRGINIA (the present-day WEST VIRGINIA state line). Before the Civil War the term "Mason-Dixon Line" popularly designated the boundary dividing the slave from the free states, and it is still used to distinguish the South from the North. Also known as Mason and Dixon's Line.

Ma, Song (MAH, SOUNG), river, more than 250 mi/402 km long, in N VIETNAM; rises in high plateau near DIEN BIEN PHU; flows SE, cutting wild, tortuous gorges through mountains and plateaus inhabited by the Black Thai, Meo, H'mong, Muong, and other minority peoples, to the THANH HOA plain and the Gulf of BAC BO near THANH HOA; 20°32′N 104°32′E. The ancient heart of DONG SON culture with its rich material of heritage of bronze drums, jewelry, diverse tools, and other domestic goods is situated in its lower reaches.

Mason Hall, village, central TOBAGO, TRINIDAD AND TOBAGO, 3 mi/4.8 km NW of SCARBOROUGH. Cacao-growing.

Masontown, town (2000 population 647), PRESTON county, N WEST VIRGINIA, 10 mi/16 km SE of MORGANTOWN. Agriculture (grain); livestock, poultry. Coal-mining area. Manufacturing (coal processing; building materials). Upper Decker Creek Wildlife Management Area to SW; Coopers Rock State Forest to N.

Masontown (MAI-suhn-toun), borough (2006 population 3,446), FAYETTE county, SW PENNSYLVANIA, 11 mi/18 km WSW of UNIONTOWN, near MONONGAHELA RIVER; 39°51′N 79°54′W. Agricultural area (corn, hay; dairying); manufacturing (apparel, crushed stone, machinery); bituminous coal, natural gas, lumber. Friendship Hill National Historic Site to S. Incorporated 1876.

Masonville, town, Delaware co., E Iowa, 7 mi/11.3 km W of Manchester; 42°28′N 91°35′W. Fertilizers. Limestone quarries nearby.

Masovia, Polish *Mazowsze*, historic region, almost coextensive with Warsaw province, central POLAND. At the death (1138) of Boleslaus III, Masovia became an independent duchy under the Piast dynasty. It became a suzerainty of Great Poland in 1351 and was finally united with it in 1526. Masovia passed to PRUSSIA during the 18th century partitions of Poland and was later a part of the Russian Empire. Reverted to Poland in 1918. Also spelled Mazovia.

Maspalomas (mahs-pah-LO-mahs), town, Grand Canary, CANARY ISLANDS, on S shore, 25 mi/40 km SSW of LAS PALMAS; 27°47′N 15°34′W. Resort with fine beach. Maspalomas Point is 2 mi/3.2 km S. Sometimes spelled Más Palomas.

Masparro River (mahs-PAHR-ro), c.100 mi/161 km long, BARINAS state, W VENEZUELA; rises in ANDEAN spur on TRUJILLO state border; flows SE through LLANOS, past LIBERTAD, to APURE RIVER 9.0 mi/14.5 km W of PUERTO DE NUTRIAS; 08°04′N 69°26′W.

Maspeth (MAS-puhth), a residential and industrial section of W QUEENS borough of NEW YORK CITY, SE NEW YORK, at head of NEWTOWN CREEK; 40°43′N 73°55′W. First European settlement in Queens, c.1642. Several cemeteries here. Mainly working-class residents.

Masqat, OMAN: see MUSCAT.

Masquefa (mahs-KAI-fah), town, BARCELONA province, NE SPAIN, 20 mi/32 km NW of BARCELONA; 41°30′N 01°49′E. Wine-producing center; wheat, olive oil.

Masr al-Qadimah, EGYPT: see FUSTAT, EL.

Mas River (MAHS) or *Kali Mas* (KAH-lee MAHS), 40 mi/64 km long, principal stream in BRANTAS RIVER delta, E JAVA, INDONESIA; branches from the Brantas. 8 mi/12.9 km W of MOJOKERTO; flows NE, past SURABAYA, to MADURA Strait 3 mi/4.8 N of Surabaya; 07°12′S 112°44′E.

Mass, village, ONTONAGON county, NW UPPER PENINSULA, NW MICHIGAN, 13 mi/21 km SE of ONTONAGON; 46°45′N 89°05′W. Also known as Mass City. Ottawa National Forest to S; Adventure Mountain Ski Area to N.

Massa (MAHS-ah), city (2001 population 66,769), ⊙ MASSA CARRARA province, TUSCANY, N central ITALY, near the LIGURIAN Sea. Marble is quarried; chemicals, metals, and machinery are produced; and food processed. From the 15th to the 19th centuries, Massa was the capital of the independent principality, later duchy, of Massa and Carrara, which was ruled by the Malaspina and the Cybo-Malaspina families. In 1829 the city passed through marriage to the house of

Austria-Este, dukes of Modena. It united with the kingdom of SARDINIA in 1859. The old town centers around the 15th-century Malaspina castle; in the new section are the Cybo-Malaspina Palace, a 15th-century cathedral, and a fine marble fountain.

Massabesic Lake (ma-suh-BEE-sik), HILLSBOROUGH and ROCKINGHAM counties, SE NEW HAMPSHIRE, 4 mi/6.4 km E of downtown MANCHESTER; 4 mi/6.4 km long; 3 mi/4.8 km wide; 42°59′N 71°21′W. Irregularly shaped. Drained from SW end by Cohas Brook (c.5 mi/8 km long) to MERRIMACK RIVER.

Massac (MAH-SAK), county (□ 242 sq mi/629.2 sq km; 2006 population 15,135), extreme S ILLINOIS, on OHIO River (S; KENTUCKY state line), across from PADUCAH (Kentucky); ⊙ METROPOLIS; 37°13′N 88°42′W. Bordered NW by CACHE RIVER. Agricultural area (wheat, sorghum, corn, soybeans; cattle), with some manufacturing (clothing, chemicals, cerment). Includes part of Shawnee National Forest along NE border; and FORT MASSAC STATE PARK and Mermet Lake Conservation Area. Lock and Dam Number 52 near BROOKPORT. Formed 1843. One of 17 Illinois counties to retain Southern-style commission form of government.

Massa Carrara (MAHS-sah kahr-RAH-rah), province (□ 446 sq mi/1,159.6 sq km), TUSCANY, central ITALY, on LIGURIAN Sea; ⊙ MASSA; 44°15′N 10°03′E. Mountainous terrain, including W APUANE Alps; watered by MAGRA River and its affluents. Center of Italian marble industry, with several hundred quarries in Apuane Alps. CARRARA and MASSA are chief producers. Agriculture and livestock raising in the LUNIGIANA and along coast. Called Apuania c.1938–1945.

Massac, Fort, ILLINOIS: see METROPOLIS.

Massachusetts (MAS-suh-CHOO-suhts), state, officially the Commonwealth of Massachusetts (□ 10,554 sq mi/27,335 sq km; 2000 population 6,349,097; 1995 estimated population 6,073,550), NE U.S., in NEW ENGLAND, on ATLANTIC OCEAN (E and SE),first of the thirteen Colonies; ⊙ BOSTON; 42°10′N 71°11′W. Massachusetts is known as the "Bay State" because of its proximity to several large bays, and the "Old Colony State" due to the original Plyouth colony.

Geography
Bordered NE by VERMONT, NW by NEW HAMPSHIRE, S by RHODE ISLAND (which also borders SE Massachusetts on the W) and CONNECTICUT, and W by NEW YORK. The E part, including the CAPE COD peninsula and the islands lying off it to the S—the ELIZABETH ISLANDS, MARTHA'S VINEYARD, and NANTUCKET—is a low coastal plain. In this area short, swift rivers such as the MERRIMACK have long supplied industry with power, and an indented coastline provides many good natural harbors, with Boston a major port. In the interior rise uplands separated by the rich CONNECTICUT River valley, and farther W lies the Berkshire valley, surrounded by the BERKSHIRE HILLS, part of the TACONIC Mountains. The W streams feed both the HUDSON and the HOUSATONIC rivers. The state has a mean elevation of c.500 ft/150 m, and Mount GREYLOCK in the Berkshires is the highest point (3,491 ft/1,064 m). The climate is variable with four seasons-cool winter, warm summer, and a spectacular autumn foliage season.

Economy
Traditionally an industrial state, and, with its predominantly urban population, one of the most densely settled in the nation. It has diverse manufacture, chiefly electrical and electronic equipment, computers, industrial equipment, plastic products., shoes and leather goods, clothing and textiles, paper and paper products, machinery, tools, and metal and rubber products. Shipping, printing, and publishing are important, and the jewelry industry dates from before the American Revolution. Leading agricultural products include cranberries, apples, vegetables, greenhouse and nursery items, and milk and other

dairy goods; poultry is also raised. The fishing fleets of GLOUCESTER and NEW BEDFORD still bring in a large and varied catch, and the coastal waters abound in shellfish. Lime, clay, sand, gravel, and stone are the chief mineral resources. High-technology research and development, finance, insurance, and trade industries have become the mainstays of the Massachusetts economy. Service sector industries, of which education and tourism are primary, made up over one-fifth of the state's gross state product (GSP) in 1986. Important cities include Boston (the largest), WORCESTER, SPRINGFIELD, LOWELL, NEW BEDFORD, CAMBRIDGE, BROCKTON, FALL RIVER, and QUINCY.

History: Settlement to 1676
The coast of what is now Massachusetts was probably skirted by Norsemen in the 11th century, and Europeans of various nationalities (but mostly English) sailed offshore in the late 16th and early 17th centuries. Settlement began when the Pilgrims arrived on the Mayflower and landed first at PROVINCETOWN (1620) then to a point which they named PLYMOUTH (for their port of embarkation in England). The Plymouth Colony took firm hold and eventually prospered. Other Englishmen soon established fishing and trading posts nearby, such as Naumkeag (SALEM), which in 1628 became the nucleus of a Puritan colony led by John Endecott of the New England Company (later, the Massachusetts Bay Company) and chartered by the private Council for New England.

In 1630, John Winthrop led the first large Puritan migration from ENGLAND (900 settlers on eleven ships). Boston supplanted Salem as capital of the colony, and Winthrop replaced Endecott as governor. After some initial adjustments to allow greater popular participation and the representation of outlying settlements in the General Court (consisting of a governor, deputy governor, assistants, and deputies), the "Bay Colony" continued to be governed as a private company for the next fifty years. It was also a thoroughgoing Puritan theocracy, in which clergymen such as John Cotton enjoyed great political influence. The status of freeman was restricted (until 1664) to church members, and the state was regarded as an agency of God's will on earth. Due to a steady stream of newcomers from England, the South Shore (i.e., area S of Boston), the North Shore, and the interior were soon dotted with firmly rooted communities.

The early Puritans were primarily an agricultural people, although a merchant class soon formed. Most of the inhabitants lived in villages, beyond which lay their privately owned fields. The typical village was composed of houses (also individually owned) grouped around the common—a plot of land held in common by the community. The dominant structure on the common was the meetinghouse, where the pastor, the most important figure in the community, held long Sabbath services. In the meetinghouse of the chief village of a town (in New England a town corresponds to what is usually called a township elsewhere in the U.S.) was also held the town meeting, traditionally regarded as a foundation of American democracy. In practice the town meeting served less to advance democracy than to enforce unanimity and conformity, and participation was as a rule restricted to male property holders and church members.

Because they valued the ability of everyone to study scripture and always insisted on a learned ministry, the Puritans also zealously promoted the development of educational facilities. The Boston Latin School was founded in 1635, one year before Harvard University was established, and in 1647 a law was passed requiring elementary schools in towns of 50 families. These were not free schools, but they were open to all and are considered the beginning of popular education in the U.S. Native American resentment of the Puritan presence resulted in the Pequot War of 1637, after which the four Puritan colonies

(Massachusetts Bay, Plymouth, Connecticut, and New Haven) formed the New England Confederation, the first voluntary union of American colonies. In 1675–1676, the confederation broke the power of the Native Americans of S New England in King Philip's War.

History: 1684 to 1761

In the course of the French and Indian Wars, however, frontier settlements such as Deerfield were devastated. The withdrawal of the charter of the Massachusetts Bay Colony (1684) occurred because the colony had consistently violated the terms of the charter and repeatedly evaded or ignored royal orders by operating an illegal mint, establishing religious rather than property qualifications for suffrage, and discriminating against Anglicans. In 1691, a new charter united Massachusetts Bay, Plymouth, and MAINE into the first royal colony of Massachusetts. This charter abolished church membership as a test for voting, although Congregationalism remained the established religion. Widespread anxiety over loss of the original charter contributed to the witchcraft panic that reached its climax in Salem in the summer of 1692. Nineteen persons were hanged and one crushed to death for refusing to confess to the practice of witchcraft. The Salem trials ended abruptly when colonial authorities, led by Cotton Mather, became alarmed at their excesses.

By the mid-eighteenth century the Massachusetts colony had come a long way from its humble agricultural beginnings. Fish and lumber were exported along with farm products in a lively trade carried by ships built here and manned by local seamen. That the menace of French Canada was removed by 1763 was due in no small measure to the unstinting efforts of the mother country, but the increasing British tendency to regulate colonial affairs, especially trade, without colonial advice, was most unwelcome. Because of the colony's extensive shipping interests, e.g., the traffic in molasses, rum, and slaves (the "triangular trade"), it sorely felt these restrictions.

History: 1761 to 1780

In 1761, James Otis opposed a Massachusetts superior court's issue of the writs of assistance (general search warrants to aid customs officers in enforcing collection of duties on imported sugar), arguing that this act violated the natural rights of Englishmen and was therefore void. He thus helped set the stage for the political controversy which, coupled with economic grievances, culminated in the American Revolution. The Stamp Act (1765) and the Townshend Acts (1767) preceded the Boston Massacre (1770), and the Tea Act (1773) brought on the Boston Tea Party. The rebellious colonials were punished for this with the Intolerable Acts (1774), which troops under General Thomas Gage were sent to enforce. Through Committees of Correspondence, Massachusetts and the other colonies had been sharing their grievances, and in 1774 they called the First Continental Congress at Philadelphia for united action. The mounting tension in Massachusetts exploded in April 1775, when General Gage decided to make a show of force. Warned by Paul Revere and William Dawes, the Massachusetts militia engaged the British force at Lexington and Concord. Patriot militia from other colonies hurried to Massachusetts, where, after the battle of BUNKER HILL (June 17, 1775), George Washington took command of the patriot forces. The British remained in Boston until March 17, 1776, when General William Howe evacuated the town.

History: 1780 to 1807

In 1780 a new constitution, drafted by a constitutional convention under the leadership of John Adams, was ratified by direct vote of the citizenry. Victorious in the Revolution, the colonies faced depressing economic conditions. Nowhere were those conditions worse than in W Massachusetts, where discontented Berkshire farmers erupted in Shays' Rebellion in 1786.

The uprising was promptly quelled, but it frightened conservatives into support of a new national constitution that would displace the weak government under the Articles of Confederation; this constitution was ratified by Massachusetts in 1788 (the sixth state to do so). Independence had closed the old trade routes within the British Empire, but newer ones were soon opened up, and trade with China became especially lucrative. Boston and lesser ports boomed, and the prosperous times were reflected politically in the commonwealth's unwavering adherence to the Federalist party, the party of the dominant commercial class. European wars at the beginning of the nineteenth century further stimulated the carrying trade until it led to interference with American shipping.

History: 1807 to 1820

To avoid war Congress resorted to Jefferson's Embargo Act of 1807, a severe blow to the economy of Massachusetts and the rest of the nation. War with Great Britain came anyway in 1812, and it was violently unpopular in New England. There was talk of secession at the abortive Hartford Convention of New England Federalists. As it transpired, however, the embargo and the War of 1812 had an unexpectedly favorable effect on the economy of Massachusetts. With English manufactured goods shut out, the U.S. had to begin manufacturing on its own, and the infant industries that sprang up after 1807 tended to concentrate in New England, and especially in Massachusetts. These industries, financed by money made in shipping and shielded from foreign competition by protective tariffs after 1816, grew rapidly, transforming the character of the commonwealth and its people. Labor was plentiful and often ruthlessly exploited. The power loom, perfected by Francis Cabot Lowell, as well as English techniques for textile manufacturing (based on plans smuggled out of England) made Massachusetts an early center of the American textile industry. The water power of the Merrimack River became the basis for the Lowell's cotton textile industry in the 1820s. Agriculture, on the other hand, went into a sharp decline because Massachusetts could not compete with the new agricultural states of the West, a region more readily accessible after the opening of the ERIE CANAL (1825).

History: 1820 to 1860

Farms were abandoned by the score; some farmers turned to work in the new factories, others moved to the West. In 1820, Maine was separated from Massachusetts and admitted to the Union as a separate state under the terms of the Missouri Compromise. In the same year the Massachusetts constitution was considerably liberalized by the adoption of amendments that abolished all property qualifications for voting, provided for the incorporation of cities, and removed religious tests for officeholders. (Massachusetts is the only one of the original thirteen states that is still governed under its original constitution, the one of 1780, although this was extensively amended by the constitutional convention of 1917–1919.) In the 1830s and 1840s the state became the center of religious and social reform movements, such as Unitarianism and transcendentalism. Of the transcendentalists, Ralph Waldo Emerson and Henry Thoreau were quick to perceive and decry the evils of industrialization, while Bronson Alcott, Margaret Fuller, Nathaniel Hawthorne, and Emerson had some association with BROOK FARM, an outgrowth of Utopian ideals.

Horace Mann set about establishing an enduring system of public education in the 1830s. The first normal school in the U.S. was established in 1839 and is now Farmington State College. During this period Massachusetts gave to the nation the architect Charles Bulfinch; such writers and poets as Richard Henry Dana, Emily Dickinson, Oliver Wendell Holmes, Henry Wadsworth Longfellow, James Russell Lowell,

and John Greenleaf Whittier; the historians George Bancroft, John Lothrop Motley, Francis Parkman, and William Hickling Prescott; and the scientist Louis Agassiz. In the 1830s reformers began to devote energy to the antislavery crusade. This was regarded with great displeasure by the mill tycoons, who feared that an offended South would cut off their cotton supply. The Whig party split on the slavery issue, and Massachusetts turned to the new Republican party and voted for John C. Frémont in 1856 and Abraham Lincoln in 1860.

History: 1860 to 1912

Massachusetts was the first state to answer Lincoln's call for troops after the firing on Fort Sumter. Massachusetts soldiers were the first to die for the Union cause when the Sixth Massachusetts Regiment was fired on by a secessionist mob in Baltimore. In the course of the war over 130,000 men from the state served in the Union forces, including an African-American regiment. After the war Massachusetts, with other Northern states, experienced rapid industrial expansion. Massachusetts capital financed many of the nation's new railroads, especially in the West. Although people continued to leave the state for the West, labor remained cheap and plentiful as European immigrants streamed into the state. The Irish, oppressed by both famine and the British, began arriving in droves even before the Civil War (beginning in the 1840s), and they continued to land in Boston for years to come. After them came French Canadians, arriving later in the 19th century, and followed in the early 20th century by Portuguese, Italians, Poles and other Slavs, Russian Jews, and Scandinavians. Also from the British Isles came Englishmen, Scots, and Welshmen.

Of all the immigrant groups, English-speaking and non-English-speaking, the Irish came to be the most influential, especially in politics. Their Roman Catholic religion and their Democratic political faith definitely set them apart from the old native Yankee stock. Practically all of the immigrants went to work in the factories. The halcyon days of shipping were over. The carrying trade had bounded back triumphantly after the War of 1812, but the supplanting of sail by steam, the growth of railroads and the destruction caused by Confederate cruisers in the Civil War helped reduce shipping to its present negligible state—a far cry from the colorful era of the clipper ships, which were perfected by Donald McKay of Boston. Whaling, once the glory of New Bedford and Nantucket, faded quickly with the introduction of petroleum.

History: 1912 to 1980

The rise of industrialism was accompanied by a growth of cities, although the small mill town, where the factory hands lived in company houses and traded in the company store, remained important. Labor unions struggled for recognition in a long, weary battle marked by strikes, sometimes violent, as was the Lawrence textile strike of 1912. World War I, which caused a vast increase in industrial production, improved the lot of workingmen, but not of Boston policemen, who staged and lost their famous strike in 1919. For his part in breaking the strike, Governor Calvin Coolidge won national fame and went on to become Vice President and then President, the third Massachusetts citizen (after John and John Quincy Adams) to hold the highest office in the land. The Sacco-Vanzetti Case, following the police strike, attracted international attention, as liberals raged over the seeming lack of regard for the spirit of the law in a state that had given the nation such an eminent jurist as Oliver Wendell Holmes (1841–1935). Labor unions finally came into their own in the 1930s under the New Deal. Industry spurted forward again during World War II, and in the postwar era the state has continued to develop. The decline of the textile industry has been offset by the growth of the electronics industry, attracted by the skilled labor in the Boston area.

History: 1980 to Present

This growth in the computer and electronics industries helped Massachusetts to gain economic prosperity throughout much of the 1980s and was a significant factor in the major population growth of suburban Boston. In 1989, the state economy was devastated by the effects of a nationwide recession and the burden of a huge state budget. Unemployment rose and the state's real estate market collapsed after years of growth.

Tourism and Recreation

As a recreation and vacation area, Massachusetts has great stretches of seashore in the E and many lakes and streams in the wooded Berkshire Hills in the W. There are numerous state parks, forests, and beaches, and Cape Cod is the site of a national seashore. PROVINCETOWN, on Cape Cod, and ROCKPORT, on Cape ANN, are artist colonies; MARBLEHEAD is a noted yachting center. The state is also famed for its historic points of interest, among them being those at STURBRIDGE, SALEM, CONCORD (Minute Man Historical Park), and LEXINGTON; and at six historic sites—Adams, Salem Maritime, Longfellow, John Fitzgerald Kennedy, Saugus Iron Works, and Dorchester Heights. Cultural attractions include the noted TANGLEWOOD music festival and the many educational facilities of the state.

Higher Learning

In the field of higher learning Massachusetts continues strong. Besides Harvard University and the Massachusetts Institute of Technology, at Cambridge, educational institutions include Radcliffe College, also at Cambridge; Amherst College, at Amherst; the University of Massachusetts, at Amherst and Boston; Boston College, at Newton; Boston University, Simmons College, and Northeastern Universtiy, at Boston; Brandeis University, at Waltham; Clark University, College of the Holy Cross, and Worcester Polytechnic Institute, at Worcester; Lowell University, at Lowell; Mount Holyoke College, at South Hadley; Smith College, at Northampton; Tufts University, at Medford; Wellesley Colege, at Wellesley; Wheaton College, at Norton; Williams College, at Williamstown; and several state colleges. The state is also renowned for its excellent private secondary schools, such as Phillips Academy.

Government

On the political scene, Massachusetts was home of the nation's 35th President, John F. Kennedy. The governor of Massachusetts is elected for a four-year term. The legislature (the General Court) has a senate of forty members and a house of representatives with one hundred and sixty members, all of whom serve two-year terms. Massachusetts sends ten Representatives and two Senators to the U.S. Congress, and has twelve electoral votes. The current governor is Mitt Romney.

Massachusetts has fourteen counties: BARNSTABLE, BERKSHIRE, BRISTOL, DUKES, ESSEX, FRANKLIN, HAMPDEN, HAMPSHIRE, MIDDLESEX, NANTUCKET, NORFOLK, PLYMOUTH, SUFFOLK, and WORCESTER.

Massachusetts Bay (MAS-suh-CHOO-suhts), inlet of the ATLANTIC OCEAN. The bay, with its arms (BOSTON, CAPE COD, and PLYMOUTH bays), extends 65 mi/105 km from Cape ANN on the N to CAPE COD on the S. Its coastline varies from the irregular, rocky shore of the N to the sandy beaches of the S. In the War of 1812, the battle between the *Chesapeake* and the *Shannon* took place here off Boston Harbor.

Massaciuccoli, Lago di (mahs-sah-CHOOK-ko-lee, LAH-go dee), shallow coastal lake (□ 2 sq mi/5.2 sq km) in TUSCANY, central ITALY, 8 mi/13 km W of LUCCA, between APUANE Alps and LIGURIAN Sea, into which it discharges; 43°50′N 10°20′E. Used for irrigation of land reclaimed in vicinity. Rice grown in remaining marshes.

Massacre Bay, NEW ZEALAND: see GOLDEN BAY.

Massacre River, c. 35 mi/56 km long, along DOMINICAN REPUBLIC–HAITI border; flows N, past DAJABÓN (Dominican Republic) and OUANAMINTHE (Haiti), to the ATLANTIC OCEAN at PEPILLO SALCEDO (Dominican Republic); 19°42′N 71°45′W.

Massada (mah-SAH-dah), kibbutz, Jordan Valley, NE ISRAEL, near Jordanian border, near S shore of SEA OF GALILEE, between the JORDAN (W) and the YARMUK (SE) rivers, 8 mi/12.9 km SSE of TIBERIAS; 32°41′N 35°36′E. Below sea level 708 ft/ 215 m. Manufacturing of mechanical equipment; mixed farming, dairying, banana growing; fish ponds. Founded 1937.

Massafra (mahs-SAH-frah), town, Taranto province, APULIA, S ITALY, 10 mi/16 km NW of TARANTO; 40°35′N 17°06′E. Agricultural center (figs, vegetables, tobacco); olive oil, wine. Has medieval castle. Numerous grottos. Nearby 16th-century sanctuary.

Massagam, town, West province, CAMEROON, 40 mi/64 km E of BAFOUSSAM; 05°25′N 11°01′E.

Massagno (mah-SAH-nyoh), commune, TICINO canton, S SWITZERLAND; N suburb of LUGANO; 46°01′N 08°57′E. Elevation 1,270 ft/387 m.

Massaguet (mah-sah-GAI), town, HADJER-LAMIS administrative region, W CHAD, 31 mi/50 km NE of N'DJAMENA, on road; 12°28′N 15°26′E.

Massakory (mah-sah-ko-REE), town, ⊙ HADJER-LAMIS administrative region, W CHAD, on the BAHR EL GHAZAL, in the drying marshes of Lake CHAD, 75 mi/121 km NNE of N'DJAMENA; 13°00′N 15°44′E. Agriculture (livestock; millet); manufacturing (butter).

Massalia, FRANCE: see MARSEILLE.

Massa Lombarda (MAHS-sah lom-BAHR-dah), town, RAVENNA province, EMILIA-ROMAGNA, N central ITALY, 4 mi/6 km WNW of LUGO; 44°27′N 11°49′E. Wine, marmalade; beet-sugar processing; paper, lumber.

Massamagrell (mah-sah-mah-GRAIL), town, VALENCIA province, E SPAIN, 8 mi/12.9 km NNE of VALENCIA. In rich vegetable-farming area; meat processing, flour and sawmilling; cereals, rice.

Massa Marittima (MAHS-sah mah-REET-tee-mah), town, GROSSETO province, TUSCANY, central ITALY, 23 mi/37 km NNW of GROSSETO; 43°03′N 10°53′E. Mining center (copper, iron, lead, borax); school of mining, mineralogical museum Bishopric. Has 13th-century Pisan cathedral, communal palace with altarpiece by Ambrogio Lorenzetti.

Massa Martana (MAHS-sah mahr-TAH-nah), village, PERUGIA province, UMBRIA, central ITALY, 6 mi/10 km E of TODI; 42°46′N 12°31′E.

Massambaba Beach, Brazil: see ARARUAMA LAGOON.

Massána, La, Catalan *La Maçana*, town, ANDORRA, on Valira del Nord River, 2 mi/3.2 km N of ANDORRA LA VELLA; 42°34′N 01°29′E. Tourism.

Massandra (mah-SAHN-drah) (Ukrainian *Masandra*), town, S Republic of CRIMEA, UKRAINE, just E of YALTA, on S coast of CRIMEA; 44°31′N 34°11′E. Elevation 702 ft/213 m. BLACK SEA swimming and climatic resort; wine production. Upper Massandra is a site of a former castle of Alexander III, now a sanatorium; has grotto formations (views). Lower Massandra has large park and tuberculosis institute with sanatorium.

Massangena, MOZAMBIQUE: see GAZA.

Massango (mah-SAHN-go), town, MALANJE province, ANGOLA, on road, and 100 mi/161 km N of MALANJE; 08°01′S 16°10′E. Market center.

Massantola, village, SECOND REGION/KOULIKORO, MALI, 16 mi/26 km SE of KOLOKANI.

Massanutten (mas-uh-NUT-uhn), unincorporated town, ROCKINGHAM county, NW VIRGINIA, 9 mi/15 km ESE of HARRISONBURG; 38°24′N 78°44′W. Tourism. Part of George Washington National Forest to N. Massanutten Resort is here.

Massanutten Mountain (mas-uh-NUT-uhn), N VIRGINIA, ridge in center of N SHENANDOAH VALLEY, parallels ALLEGHENY MOUNTAINS (W) and BLUE RIDGE (E); extends from point E of HARRISONBURG c.45 mi/72 km NNE to point W of FRONT ROYAL in section of George Washington National Forest; rises to c.3,000 ft/914 m; 38°23′N 78°46′W. North and South forks of SHENANDOAH RIVER meet at N end. Massanutten Resort and Caverns, at S end, Luray Caverns at E side. Observation tower near N end, E of WOODSTOCK. Masanutten Visitors' Center in center of ridge.

Massa, Oued (MAHS-suh, wahd), small stream, Souss-Massa-Draâ administrative region, MOROCCO, in arid coastal region between AGADIR and TIZNIT, 31 mi/50 km S of Agadir. Estuary reed beds and sandbanks, as well as adjacent dunes, beaches, and coastal forest, are site of Sous Massa National Park (□ 51 sq mi/132 sq km). Location on flyway means park is visited by migratory birds in the winter, including bald ibis, osprey, pochard, coot, and flamingo. Adjacent semiarid haitats support a diverse array of flora and fauna. Beach of Sidi Rbat with its marabout (Muslim hermit) tomb is within the park. Active irrigation scheme and small dam at Massa.

Massapê (mah-sah-PAI), city, (2007 population 33,305), NW CEARÁ state, BRAZIL, on CAMOCIM-CRATEÚS railroad, and 11 mi/18 km N of SOBRAL; 03°35′S 40°22′W. Cotton, sugar, carnauba wax.

Massapequa (ma-suh-PEE-kwuh), unincorporated city (□ 4 sq mi/10.4 sq km; 2000 population 22,652), NASSAU county, SE NEW YORK, on S shore of LONG ISLAND; 40°40′N 73°28′W. Chiefly residential, with waterfront properties to S. Comedian Jerry Seinfeld grew up here.

Massapequa Park (ma-suh-PEE-kwuh), residential village (□ 2 sq mi/5.2 sq km; 2006 population 17,115), NASSAU county, SE NEW YORK, on LONG ISLAND; 40°40′N 73°27′W. Bethpage State Park is 5 mi/8 km to N. Bordered by the Massapequa Preserve greenbelt to W. Incorporated 1931.

Massaranduba (MAH-sah-rahn-DOO-bah), city (2007 population 12,494), E central PARAÍBA state, BRAZIL, 10 mi/16 km ENE of CAMPINA GRANDE; 07°11′S 35°44′W. Also spelled Maçaranduba.

Massarossa (mahs-sah-ROS-sah), town, LUCCA province, TUSCANY, central ITALY, 5 mi/8 km E of VIAREGGIO. Shoe factory, woolen mill. Marble quarry nearby.

Massaua, Eritrea: see MASSAWA.

Massawa (mah-SAH-wah), former administrative division, central and SE ERITREA, on the RED SEA; capital was MASSAWA. Located in hot, arid coastal plain; E slope of central plateau (N); and DANAKIL desert in SE. Chief towns, GINDA, HIRGIGO (Arkiko), and ZULA. Traversed (E-W) by railroad and highway. Called Bassopiano Orientale [Ital.=eastern lowland] under Italian administration (until 1941).

Massawa (mah-SAH-wah) or **Mitsiwa'e** (mit-SOO-uh), city (2003 population 25,000), ⊙ SEMENAWI KAYIH BAHRI region, E central ERITREA, a port and the only deep-water harbor on the African side of the RED SEA; 15°35′N 39°25′E. The main port for Eritrea and N ETHIOPIA; linked by road with ASMARA; a narrow-gauge railroad line has been reconstructed. Major industries include meat processing, cement and salt production; gum arabic and sesame seeds are its major exports. Fishing is important to the city's economy. Once a thriving port with a population four times its present size.

The lengthy war against the Ethiopian government brought its commercial trade to a near halt and the retreating Ethiopian army looted the city in 1990. Long a commercial port, Massawa was part of the kingdom of Aksum (c.1st–8th century A.D.). Its initial port site was Basta Island—the heart of Old Massawa. In 1870, causeways were built to the mainland and Massawa, which quickly outstripped the old Ottoman port town of Argigio to its immediate N, which is now only a ruin. In 1577, it was captured by the Ottoman Turks, who, in 1868, transferred it to Egyptian control. In 1885, Massawa was taken by Italy, and it was capital of the Italian colony of Eritrea (1889–1900). Eritrean secessionists fought against the Ethiopian government here, and Massawa's port was especially important for receiving arms shipments.

The heart of the city is located on two islands (Twalet and Batsa) linked together and to the mainland by the causeways. Most public buildings, hotels, and shops are on these islands. Also spelled Massaua, Mitsiwa, or Massowa.

Massawippi, Lake (ma-suh-WI-pee), S QUEBEC, E CANADA, 10 mi/16 km SSW of SHERBROOKE; 8 mi/13 km long, 2 mi/3 km wide. Noted for scenic beauty. Drains N to SAINT FRANCIS RIVER.

Massbach (MUHS-bahkh), village, LOWER FRANCO-NIA, NW BAVARIA, GERMANY, 10 mi/16 km NNE of SCHWEINFURT; 50°11′N 10°17′E. Manufacturing (machinery).

Masse Island, MARQUESAS ISLANDS: see EIAO.

Massena, city (□ 4 sq mi/10.4 sq km; 2006 population 10,800), ST. LAWRENCE county, extreme N NEW YORK, on the SAINT LAWRENCE River; 44°55′N 74°53′W. Two locks and two dams of the SAINT LAWRENCE SEAWAY are nearby. In a summer resort area and has a state park. Metals manufacturing and services. An international bridge connects the city with CORNWALL (Ontario). Settled 1792, incorporated 1886.

Massena, town (2000 population 414), CASS county, SW IOWA, on WEST NODAWAY RIVER, and 16 mi/26 km SE of ATLANTIC; 41°15′N 94°46′W. In agricultural area.

Massénya (mah-sain-YAH), town, CHARI-BAGUIRMI administrative region, W CHAD, on arm of CHARI River, and 90 mi/145 km SE of N'DJAMENA; 11°24′N 16°10′E. Trade center (livestock; cotton, millet) on old caravan route to MAKKA (Mecca). Was capital of former BAGUIRMI state and famous slave market (notably for eunuchs). Reached by German explorers Heinrich Barth in 1855 and Gustav Nachtigal in 1872. Sometimes spelled Massénia.

Masset, village (□ 8 sq mi/20.8 sq km; 2001 population 926), W BRITISH COLUMBIA, W CANADA, on N GRAHAM ISLAND, on an inlet of DIXON ENTRANCE, 80 mi/129 km WSW of PRINCE RUPERT across HECATE STRAIT, in SKEENA-QUEEN CHARLOTTE regional district; 54°02′N 132°09′W. W terminus of YELLOWHEAD HIGHWAY (Route 16). Trade center, inhabited largely by Native Haida. Salmon fishing, once a mainstay of the economy, is now declining, with a ban on commercial chinook fishing in effect and fishermen selling their licenses back to the government in a plan to reduce fishing activity. Tourist center. Haida Village native cultural center, totem poles. Scheduled seaplane service; airport. Incorporated 1961. Originally named Graham City.

Masset Inlet, W BRITISH COLUMBIA, W CANADA, central GRAHAM ISLAND, 20 mi/32 km S of MASSET, connected with DIXON ENTRANCE by Masset Sound (25 mi/40 km long), in SKEENA-QUEEN CHARLOTTE regional district; 20 mi/32 km long, 2 mi/3 km–8 mi/13 km wide; 53°42′N 132°20′W. In lumbering and fishing area. PORT CLEMENTS is on E shore; Masset village is at mouth of Masset Sound.

Masseube (mah-SUB), commune (□ 8 sq mi/20.8 sq km; 2004 population 1,496), GERS department, MIDI-PYRÉNÉES region, SW FRANCE, on the GERS RIVER, and 10 mi/16 km SE of MIRANDE, in upper ARMAGNAC vineyard area; 43°26′N 00°35′E. Was a fortified medieval village known as a *bastide*.

Massey (MA-see), unincorporated town (□ 2 sq mi/5.2 sq km; 2001 population 1,035), SE central ONTARIO, E central CANADA, on SPANISH RIVER, 55 mi/89 km WSW of SUDBURY, and included in SABLES-SPANISH RIVERS township; 46°12′N 82°05′W. Nickel, copper, and gold mining; lumbering.

Massey, village, KENT county, E MARYLAND, 19 mi/31 km NW of DOVER, DELAWARE.

Massiac (mah-SYAHK), commune (□ 13 sq mi/33.8 sq km; 2004 population 1,838), resort in CANTAL department, AUVERGNE region, S central FRANCE, in MASSIF CENTRAL, 9 mi/14.5 km WSW of BRIOUDE; 45°15′N 03°13′E. Woodworking. The gorges of ALAGNON RIVER are downstream.

Massicault (mah-see-KO), village, BAJAH province, N TUNISIA, 17 mi/27 km WSW of TUNIS; 36°43′N 09°53′E. Agricultural settlement. Here during World War II, the German army made a last stand before TUNIS in May 1943.

Massico, Monte (MAHS-see-ko, MON-te), mountain (2,661 ft/811 m) on coast of GULF OF GAETA, central ITALY, 18 mi/29 km E of GAETA; 41°10′N 13°55′E. Pierced by railroad tunnel. The wine from its vineyards was famous in Roman times.

Massif Armoricain, FRANCE: see ARMORICAN MASSIF.

Massif Central (mah-SEEF sahn-TRAHL) [French= central highlands], vast highland in central and S FRANCE; rising to 6,184 ft/1,885 m in the PUY DE SANCY. The basic structure consists of old Paleozoic rocks strongly affected by the Tertiary folding and uplift of the ALPS which tilted the massif toward the NW. The S and E crystalline edges (Morvan, Charolais, MÂCONNAIS, beaujolais, Monts du LYONNAIS, vivarais, CÉVENNES, MONTAGNE NOIRE) were strongly uplifted; the center (AUVERGNE and Velay) were transformed by volcanic action and fault lines mark the edges of down-faulted foundered basins (Limagnes); the W (LIMOUSIN), more distant from the Alps, was less affected by Tertiary folding and subsequent vulcanism. The climate is rather severe, somewhat oceanic in the W, more continental on the E, and definitely MEDITERRANEAN on the S. Hard times in this difficult environment have led to much outmigration. Population is unevenly distributed throughout the massif, although it covers about one-seventh the surface of France; some areas are virtually uninhabited. Agriculture is still characterized by general farming and quasi-subsistence. Manufacturing is limited to the cities (CLERMONT-FERRAND, LIMOGES). Tourism; many health resorts, of which VICHY is the best known.

Massif de la Vanoise, FRANCE: see VANOISE, MASSIF DE LA.

Massif de Pelvoux, FRANCE: see ÉCRINS, MASSIF DES.

Massif des Écrins, FRANCE: see ÉCRINS, MASSIF DES.

Massilia, FRANCE: see MARSEILLE.

Massillon (MAS-i-lahn), city (□ 17 sq mi/44.2 sq km; 2006 population 32,315), STARK county, NE OHIO, on the TUSCARAWAS RIVER, 6 mi/9.6 km W of CANTON; 40°47′N 81°31′W. Manufacturing includes fabricated metal products, transportation equipment, medical supplies, food products, plastics. Jacob S. Coxey, the social reformer, lived here and was mayor in the early 1930s. State mental hospital and prison facility are nearby. Incorporated 1853.

Massinga (mah-SEEN-guh), village, INHAMBANE province, SE MOZAMBIQUE, on road, and 38 mi/61 km N of INHAMBANE; 23°20′S 35°22′E. Cotton, mafura, cashew nuts.

Massinga District, MOZAMBIQUE: see INHAMBANE.

Massinger District, MOZAMBIQUE: see INHAMBANE.

Massive, Mount, peak (14,421 ft/4,396 m), in W LAKE county, W central COLORADO, in the SAWATCH MOUNTAINS, at CONTINENTAL DIVIDE, 4 mi/6.4 km NNW of MOUNT ELBERT; in San Isabel National Forest. It is the second-highest peak in the U.S. ROCKY MOUNTAINS.

Masson-Angers (mah-SON-ahn-ZHE), former village, SW QUEBEC, E CANADA, on Lièvre River, just above its mouth on OTTAWA River, and 16 mi/26 km ENE of OTTAWA; 45°33′N 75°25′W. Lumbering; dairying; livestock raising. Amalgamated into the city of GATINEAU in 2002.

Masson Island (MA-suhn), off ANTARCTICA, in SHACKLETON ICE SHELF, off QUEEN MARY COAST; 20 mi/32 km long, 14 mi/23 km wide; 66°08′S 96°35′E. Elevation 1,525 ft/465 m. Discovered 1912 by Sir Douglas Mawson. Also known as Mission Island.

Massow, POLAND: see MASZEWO.

Massowa, Eritrea: see MASSAWA.

Massueville (mah-syoo-VEEL), town (2006 population 556), MONTÉRÉGIE region, S QUEBEC, E CANADA, 11 mi/17 km from SAINT-OURS; 45°55′N 72°56′W.

Massy or LENIN-DZHOL (le-nin-JOL), village, JALAL-ABAD region, KYRGYZSTAN, in FERGANA VALLEY, 21 mi/34 km WNW of JALAL-ABAD; 41°04′N 72°38′E. Cotton. Tertiary-level administrative center.

Massy (mah-SEE), residential SSW suburb (□ 3 sq mi/7.8 sq km) of PARIS, ESSONNE department, ÎLE-DE-FRANCE region, N central FRANCE, 10 mi/16 km from Notre Dame Cathedral, and just W of ORLY international airport; 48°44′N 02°17′E. Electronics industry. T.G.V. (high-speed rail) station near major highway interchanges.

Mastanli, Bulgaria: see MOMCHILGRAD.

Masterton, town, Masterton district (□ 888 sq mi/2,308.8 sq km; 2006 population 19,494), NEW ZEALAND, 55 mi/89 km NE of WELLINGTON; 40°57′S 175°39′E. Agricultural center; sheep and dairy area, with dairy plants, woolen mills, slaughterhouses.

Mastgat (MAHST-khaht), channel, 3 mi/4.8 km long, ZEELAND province, SW NETHERLANDS; extends NE-SW between DUIVELAND and SINT-Philipsland regions; joins the KRAMMER and GREVELINGENMEER channels in N, continues as KEETEN channel in S. Joined by Krabbe Creek estuary from E; Keeten joins with EASTERN SCHELDT estuary. Grevelingen Dam and Philips Dam, joined at mid-channel, block N entrance. Mastgat narrows to 0.5 mi/0.8 km.

Mastic (MAS-tik), residential village (□ 4 sq mi/10.4 sq km; 2000 population 15,436), SUFFOLK county, SE NEW YORK, on an inlet of Moriches Bay, on S shore of E LONG ISLAND, 2 mi/3.2 km W of CENTER MORICHES; 40°47′N 72°50′W.

Mastic Beach (MAS-tik), village (□ 5 sq mi/13 sq km; 2000 population 11,543), SUFFOLK county, SE NEW YORK, 10 mi/16 km E of PATCHOGUE; 40°45′N 72°50′W.

Mastuj (muhs-TUHJ), village, CHITRAL district, N NORTH-WEST FRONTIER PROVINCE, N PAKISTAN, on headstream of KUNAR RIVER, and 165 mi/266 km NNE of PESHAWAR; 36°17′N 72°31′E. Important distribution point serving N CHITRAL district and road route over SHANDUR PASS. Popular overnight stop for tourists on Chitral-Gilgit road.

Mastung (muhs-TUHNG), town, ⊙ Pischin district, BALUCHISTAN province, SW PAKISTAN, 28 mi/45 km SSW of QUETTA; 29°48′N 66°51′E. Was headquarters of former Sarawan division.

Masuda (mah-SOO-dah), city, SHIMANE prefecture, SW HONSHU, W JAPAN, on the SEA OF JAPAN, at the mouth of the Takutsu River, 87 mi/141 km S of MATSUE; 34°40′N 131°50′E. Beef cattle.

Masuda (mah-SOO-dah), town, Hiraka county, Akita prefecture, N HONSHU, NE JAPAN, 43 mi/70 km S of AKITA city; 39°11′N 140°32′E.

Masue, village, MATABELELAND NORTH province, W ZIMBABWE, 12 mi/19 km SSW of VICTORIA FALLS town, on railroad; 18°05′S 25°51′E. ZAMBEZI (to N) and KAZUMA PAN (to SW) national parks. Livestock. Also spelled Masuie.

Masueco (mah-SWAI-ko), village, SALAMANCA province, W SPAIN, near DUERO RIVER, 52 mi/84 km WNW of SALAMANCA; 41°12′N 06°35′W. Olive-oil processing; cereals, fruit, wine.

Masuho (mahs-HO), town, South Koma county, YAMANASHI prefecture, central HONSHU, central JAPAN, 9 mi/15 km S of KOFU; 35°33′N 138°27′E.

Masuleh (mah-soo-LAI), village, Gilān province, N IRAN, 35 mi/56 km WSW of RASHT, in ELBURZ range;

37°09′N 48°59′E. Sheep raising; wool weaving. Ironworking (nearby deposits).

Masuria, German *Masurenland*, Polish *Mazury*, region, N POLAND. It is a low-lying area covered by large lakes and forests and drained by many small rivers. The original population of the region was expelled by the Teutonic Knights and replaced (14th century) with Polish settlers. Later became part of EAST PRUSSIA and was largely Germanized by the early 20th century. After Masuria passed to Poland in 1945, most of the German-speaking population was expelled and replaced by Poles. The Masurian Lakes region, where over 2,700 lakes are located, was the scene of heavy fighting early in World War I. Two Russian armies, commanded by generals Samsonov and Rennenkampf, were defeated in the region—Samsonov by Hindenburg at Tannenburg (August 1914) and Rennenkampf by Mackensen in the lake country (September 1914). The Russians were also repulsed (February 1915) in Masuria in the so-called Winter Battle.

Masurian Canal (mah-SOO-ree-ahn), Polish *Kanał Mazurski* (kahn-ahl mah-zoo-skee), Russian *Mazurskiy Kanal*, 32 mi/51 km long, NE POLAND and KALININGRAD oblast, NW European RUSSIA; from LYNA RIVER at DRUZHBA (Kaliningrad oblast) extends SSE to Lake MAMRY (northernmost of Masurian Lakes), W of WĘGORZEWO. Difference in water level of 364 ft/111 m is overcome by ten locks. Formerly in German EAST PRUSSIA; divided between Russia and Poland after 1945.

Masury (MA-zhuh-ree), unincorporated village (□ 4 sq mi/10.4 sq km; 2000 population 2,618), TRUMBULL county, NE OHIO, 11 mi/18 km NE of YOUNGSTOWN, on PENNSYLVANIA state line; 41°12′N 80°32′W.

Masvingo (mah-SHVI-en-go), province (□ 21,840 sq mi/56,784 sq km; 2002 population 1,318,705), SE ZIMBABWE, on MOZAMBIQUE (SE) and SOUTH AFRICA (on S corner; LIMPOPO River, border); ⊙ MASVINGO. Other towns include CHIREDZI and MASHAVA. Bordered E by SAVE RIVER; SW by BUBI RIVER. Drained by TOKWE, RUNDE, and Mwenzi rivers. GREAT ZIMBABWE NATIONAL MONUMENT, Lake Mtirikwe (Lake Kyle) reservoir, in KYLE RECREATIONAL PARK, and Mushandike Sanctuary in N center, GONAREZHOU NATIONAL PARK in SE. Part of high veld in N, semiarid grassland in S. Gold, chromite, asbestos mining. Agriculture (wheat, corn, tobacco, cotton, peanuts, soybeans, sugarcane); cattle, sheep, goats. Formerly Victoria province.

Masvingo (mah-SHVI-en-go), town, ⊙ MASVINGO province, SE central ZIMBABWE, 80 mi/129 km SE of GWERU; 20°04′S 30°50′E. Elevation 3,571 ft/1,088 m. Airport to E. GREAT ZIMBABWE NATIONAL MONUMENT to SE; MUSHANDIKE RESERVOIR (recreational park) to SW, Lake Mtirikwe reservoir, in KYLE RECREATIONAL PARK, to SE. Makoholi Experiment Station, cattle and crop research, is here. Gold and asbsestos mining to W. Cattle, sheep, goats; dairying; tobacco, corn, peanuts, citrus fruit. Formerly called Fort Victoria.

Masyaf (mahs-YAHF), town, HAMA district, W SYRIA, 45 mi/72 km SE of LATTAKIA, 23 mi/37 km WSW of HAMA, on W slopes of Jebel Nasaria; 35°03′N 36°20′E. Cotton, tobacco, cereals; sericulture. Old fortress here, prominent in Crusades.

Maszewo (mah-SHE-vo), German *Massow*, town, Szczecin province, NW POLAND, 11 mi/18 km N of STARGARD SZCZECINSKI. Yeast manufacturing; chemical industry. Ruins of 13th century city walls. Formerly in German POMERANIA; passed to Poland in 1945.

Mat, village, MATHURA district, W UTTAR PRADESH state, N central INDIA, on the YAMUNA RIVER, 10 mi/16 km N of MATHURA. Gram, jowar, wheat, barley, cotton.

Mat, ALBANIA: see BURREL.

Mata (MAH-tah), town, VILLA CLARA province, central CUBA, on railroad, and 15 mi/24 km N of SANTA CLARA; 22°40′N 79°56′W. Sugar growing. Sugar mills Braulio Coroneaux (2 mi/3 km S) and El Vaquerito (2 mi/3 km NW).

Mata'aga, Mount, SAMOA: see MAUGA SILISILI.

Mataana, El, EGYPT: see ASFUN EL MATA'NA.

Matabeleland, region, W ZIMBABWE. Chiefly of historic and ethnographic interest as the region inhabited by the Matabele, a Bantu-speaking tribe of Zulu origin. Driven out of Natal (1823) and TRANSVAAL (1837), the Matabele occupied area N of LIMPOPO River and absorbed surrounding tribes (especially the Mashona). From them Cecil Rhodes obtained permission (1890) to exploit gold deposits, but in 1893, Lobengula, the Matabele chief, attacked the settlers. The revolt was suppressed in 1897, and the Matabele subsequently became herdsmen and farmers. Now part of MATABELELAND NORTH and MATABELELAND SOUTH provinces.

Matabeleland North, province (□ 28,967 sq mi/75,025 sq km; 1992 population 640,957; 2002 population 701,359), W ZIMBABWE; ⊙ LUPANE. Other major towns and cities include HWANGE and VICTORIA FALLS. Bounded by ZAMBEZI River (ZAMBIA border) on N, which forms Lake KARIBA reservoir in NE; by BOTSWANA on W and SW. Part of high veld in SE, semiarid grassland in W with numerous salt pans and water holes, gathering places for wildlife and livestock. HWANGE NATIONAL PARK in W, KAZUMA PAN, ZAMBEZI and Victoria Falls national parks in NW, Matetsi Safari Area in NW, CHIZARIRA NATIONAL PARK and Chete Safari Area in N. Tin, chromite, gold; coal mining at HWANGE. Agriculture (corn, soybeans, tobacco, cotton, sorghum); cattle, sheep, goats. Tourism. Former capital was BULAWAYO.

Matabeleland South, province (□ 20,916 sq mi/54,381.6 sq km; 2002 population 654,879), SW ZIMBABWE; ⊙ GWANDA. Major towns include Gwanda and ZVISHAVANE. Bounded SW by SHASHE RIVER (forms part of BOTSWANA border); SE by LIMPOPO River (SOUTH AFRICA border), E by BUBI RIVER. Drained by UMZINGWANI, THULI, and SHASHANI rivers. Part of high veld, including Matopos Hills in N, Thuli Safari Area in S. Gold; diamond mining at BEITBRIDGE (opened 1996). Agriculture (corn, wheat, cotton, soybeans, tobacco); cattle, sheep, goats. Formerly administered from BULAWAYO.

Matabhanga (mah-tah-BAHN-gah), town, KOCH BIHAR district, NE WEST BENGAL state, E INDIA, on the JALDHAKA RIVER, 15 mi/24 km W of KOCH BIHAR; 26°20′N 89°13′E. Rice, jute, tobacco, oilseeds, sugar. Also spelled Mathabhanga.

Mataca, Caño, VENEZUELA: see IMATACA, BRAZO.

Matacawa Levu, FIJI: see YASAWA GROUP.

Matachel River (mah-tah-CHEL), c.80 mi/129 km long, BADAJOZ province, W SPAIN; rises in outliers of the SIERRA MORENA; flows NW to the GUADIANA RIVER 6 mi/9.7 km SE of MÉRIDA.

Matachewan, township (□ 210 sq mi/546 sq km; 2001 population 308), E central ONTARIO, E central CANADA, 32 mi/52 km from KIRKLAND LAKE, and on MONTREAL RIVER shores; 47°56′N 80°38′W. Tourism, forestry; formerly mining. The Native American township name means "meeting of the waters."

Matachi (mah-TAH-chee), town, CHIHUAHUA, N MEXICO, on headstream of YAQUI RIVER, in SIERRA MADRE OCCIDENTAL, on railroad, and 105 mi/169 km W of CHIHUAHUA. Corn, wheat, beans, fruit, cattle. Formerly MATACHIC.

Matachic, MEXICO: see MATACHI.

Matacos, ARGENTINA: see LA FLORENCIA.

Mata de Alcántara (MAH-tah dai ahl-KAHN-tah-rah), village, CÁCERES province, W SPAIN, 29 mi/47 km NW of CÁCERES; 39°43′N 06°49′W. Cereals, wine, olive oil.

Mata de São João (mah-tah dee SOUN zho-OUN), city, (2007 population 37,175), E BAHIA state, BRAZIL, on railroad, and 35 mi/56 km NNE of SALVADOR; 12°32′S 38°15′W. Sugar, oranges, coconuts, manioc. Formerly spelled Matta de São João.

Matadi (mah-TAH-dee), city, ⊙ BAS-CONGO province, W CONGO, on the CONGO RIVER; 05°49′S 13°27′E. With one of the largest harbors in central Africa, MATADI is the main port of the country. Situated c.80 mi/129 km from the mouth of the CONGO, at the farthest point navigable by ocean-going vessels, the city is linked by railroad with KINSHASA. Chief exports are palm products, coffee, cotton, rubber, and bananas. Also spelled MATIDI.

Matador (MA-tuh-dor), town (2006 population 660), ⊙ MOTLEY county, NW TEXAS, just below CAPROCK ESCARPMENT of LLANO ESTACADO, 45 mi/72 km SW of CHILDRESS, near huge Matador Ranch; 34°01′N 100°49′W. Elevation 2,347 ft/715 m. In cattle-ranching region, also producing cotton, wheat, peanuts; oil and gas. Matador State Wildlife Management Area to E (Cottle county).

Matafao Peak (mah-tah-FOU), TUTUILA island, AMERICAN SAMOA, S PACIFIC. Highest peak (2,303 ft/ 702 m) on island.

Matagalpa (mah-tah-GAHL-pah), department (□ 3,380 sq mi/8,788 sq km), central NICARAGUA; ⊙ MATAGALPA; 12°55′N 85°40′W. Astride CORDILLERA DARIENSE; drained by navigable TUMA RIVER (N), a tributary of the RÍO GRANDE (S). Gold mining (SAN RAMÓN). An important coffee region, with chief plantations in central portion; corn, beans, and livestock raising (W); vegetables, potatoes, fruit, and sugarcane (center); lumbering and rubber (E). Major export, coffee. Main centers (Matagalpa, CIUDAD DARÍO, SÉBACO) served by INTER-AMERICAN HIGHWAY.

Matagalpa (mah-tah-GAHL-pah), city and township, (2005 population 80,228), ⊙ MATAGALPA department, W central NICARAGUA, 60 mi/97 km NNE of MANAGUA, on spur of INTER-AMERICAN HIGHWAY; 12°55′N 85°55′W. Agricultural and commercial center; coffee processing, flour milling, manufacturing of turpentine, soap, bricks, shoes, clothing. Has cathedral dating from colonial period; bishopric. Noted as one of Nicaragua's most appealing cities, it was badly damaged in the 1978–1979 civil war.

Matagami, village (□ 25 sq mi/65 sq km; 2006 population 1,886), NORD-DU-QUÉBEC region, W central QUEBEC, E CANADA, 110 mi/177 km NNE of VAL-D'OR, on S side of Lake Matagami; 49°45′N 77°38′W. Mining center; foresty. Road connection from S. Scheduled air service.

Matagorda (ma-tuh-GOR-duh), county (□ 1,612 sq mi/4,191.2 sq km; 2006 population 37,824), S TEXAS; ⊙ Bay City; 28°46′N 96°00′W. On MATAGORDA and East Matagorda bays, sheltered from GULF OF MEXICO by MATAGORDA PENINSULA, a sand barrier paralleling coast, and traversed by GULF INTRACOASTAL WATERWAY (also parallels coast 1 mi/1.6 km inland); drained by COLORADO RIVER. E border formed in part by Linville Bayou and Cedar Lake Creek. Railroad terminus. Cattle; agriculture (especially rice; also cotton, grain). Oil, natural gas, salt. Tourism; beaches, fishing attract visitors. Formed 1836. Nuclear power plants include South Texas One (initial criticality March 8, 1988) and South Texas Two (initial criticality March 12, 1989), both 12 mi/19 km SSW of BAY CITY; they use cooling water from the Colorado River, and each has a max. dependable capacity of 1251 MW. Big Boggy National Wildlife Refuge in SE; Matagorda Peninsula State Park in S.

Matagorda (ma-tuh-GOR-duh), unincorporated village, MATAGORDA county, S TEXAS, on INTRACOASTAL WATERWAY, at E end of MATAGORDA BAY and on SW East Matagorda Bay, at mouth of COLORADO RIVER, and 20 mi/32 km S of BAY CITY; 28°41′N 95°58′W. Fisheries; seafood, shell market. Matagorda Peninsula State Park 7 mi/11.3 km S on Gulf Coast; Big Boggy National Wildlife Refuge to NE. Christ Church here was built 1839. Settled 1825; served as port for Stephen F. Austin's colony.

Matagorda Bay (ma-tuh-GOR-duh), inlet of the GULF OF MEXICO, SE TEXAS, separated from Gulf by a long sand spit, MATAGORDA PENINSULA; 30 mi/48 km long and 3 mi/4.8 km–12 mi/19 km wide; 28°42′N 95°48′W. Receives the COLORADO RIVER at E end, is separated from peninsula with mainland, and is crossed by the INTRACOASTAL WATERWAY. MATAGORDA ISLAND is a sandbar farther S extending beyond tip of Matagorda Peninsula to entrance of SAN ANTONIO BAY. Lavaca Bay in NW arm of MATAGORDA BAY, East Matagorda Bay, inlet of Gulf of Mexico, extends 20 mi/32 km NE, from neck of land separating it from Matagorda Bay (SW); separated from gulf by E extension of Matagorda Peninsula. At E end of Matagorda Bay is the site of the village of MATAGORDA, which was settled in 1825 and served as a port for Stephen F. Austin's colony. Matagorda is known principally for fishing and oyster gathering. The area is often struck by hurricanes. Big Boggy National Wildlife Refuge on N shore. INTRACOASTAL WATERWAY is 1 mi/1.6 km inland from mainland shore.

Matagorda Island (ma-tuh-GOR-duh), CALHOUN county, S TEXAS, low sandy barrier island between SAN ANTONIO and ESPIRITU SANTO bays (NW; between island and mainland), and the GULF OF MEXICO (SE); 36 mi/58 km long, 1 mi/1.6 km–4 mi/6.4 km wide; 28°13′N 96°38′W. Separated by channel from ST. JOSEPH ISLAND (SW) and by Cavallo Pass from tip of MATAGORDA PENINSULA (NE), both continuations on the sand barrier that rings much of the Texas Gulf Coast. At E end, Matagorda Island State Park and Wildlife Area; former air force base used for bombing practice, now restored to natural status.

Matagorda Peninsula (ma-tuh-GOR-duh), MATAGORDA county, S TEXAS, sand barrier paralleling Texas Gulf Coast; c.50 mi/80 km. Connected to mainland by 7 mi/11.3 km neck of land, 25 mi/40 km S of BAY CITY. Extends NE c.20 mi/32 km, separating East Matagorda Bay from GULF OF MEXICO, and c.30 mi/48 km SW, separating MATAGORDA BAY from gulf. The SW extension is breached in three places, the work of numerous hurricanes and tropical storms. SW tip separated from MUSTANG ISLAND by Pass Cavallo. NE tip separated from mainland by small entrance to East Matagorda Bay. Matagorda Peninsula State Park at base of peninsula, S of village of MATAGORDA.

Mata Grande (MAH-tah GRAHN-zhee), city, (2007 population 24,478), W ALAGOAS state, NE BRAZIL, 80 mi/129 km WNW of PALMEIRA DOS ÍNDIOS; 09°10′S 37°45′W. Cotton, coffee, fruit. Until 1939, called Paulo Affonso. Also spelled Matta Grande.

Mataguá (mah-tah-GWAH), small town, VILLA CLARA province, central CUBA, on railroad line, and secondary highway, 11 mi/18 km S of SANTA CLARA; 22°14′N 80°00′W. Between small ranges of Sierra del Escambray (NE) and Sierra de Potrerillo (SW).

Matahambre, CUBA: see MINAS DE MATAHAMBRE.

Matai (mah-TEI), town, N TALDYKORGAN region, KAZAKHSTAN, on TURK-SIB RAILROAD, on the AKSU RIVER, 60 mi/97 km N of TALDYKORGAN. In desert area. Also spelled Motai.

Matai (mah-TEI), village, MINYA province, Upper EGYPT, on railroad, 9 mi/14.5 km NE of SAMALUT; 28°25′N 30°46′E. Cotton ginning, sugar milling; cotton, cereals, sugarcane. Also spelled Matay.

Matak, INDONESIA: see ANAMBAS ISLANDS.

Matala (mah-TAH-luh), town, HUILA province, ANGOLA, on railroad and road junction, 110 mi/177 km E of LUBANGO; 14°45′S 15°02′E. Market center. Electric power station.

Mata, La (MAH-tah, lah), town, TOLEDO province, central SPAIN, 22 mi/35 km W of TOLEDO. Cereals, chickpeas, olives, grapes; sheep. Olive-oil pressing, meatpacking; tile manufacturing.

Matala (MAH-tah-lah), village, IRÁKLION prefecture, S central coast of CRETE, GREECE, 8 mi/12.9 km S of Phaestos; 34°59′N 24°45′E. Also the name of a small bay with a semicircular beach, which is enclosed by two arms of hard-packed earth into which chambers have been cut probably beginning with the early Christian era. The chambers were used by transients over the centuries, most recently by hippies in the 1960s, but are now open only for viewing due to heavy tourist traffic.

Matala, village, CENTRAL province, S central ZAMBIA, 65 mi/105 km WNW of LUSAKA; 15°09′S 27°16′E. Gold-mining center. Cattle. Blue Lagoon National Park to SW.

Matalaque (mah-tah-LAH-kai), town, MOQUEGUA region, GENERAL SÁNCHEZ CERRO province, S PERU, at E foot of Nevado de PICHU PICHU, on TAMBO RIVER, and 50 mi/80 km NNE of MOQUEGUA; 16°29′S 70°49′W. Cereals; livestock.

Matalascañas, beach resort on the ATLANTIC OCEAN, SE HUELVA province, SW SPAIN, midway between HUELVA and CÁDIZ.

Matale, district (□ 754 sq mi/1,960.4 sq km; 2001 population 441,328), CENTRAL PROVINCE, SRI LANKA, in MATALE VALLEY, in N SRI LANKA HILL COUNTRY; ⊙ MATALE; 07°40′N 80°45′E.

Matale (MAH-thuh-LE), town (2001 population 36,451), ⊙ MATALE district, CENTRAL PROVINCE, SRI LANKA, in MATALE VALLEY, 12 mi/19 km N of KANDY; 07°28′N 80°37′E. Railroad spur terminus. Trade (tea, rubber, cacao, rice, coconuts, vegetables) and cattle center. Limestone quarries nearby. Buddhist rock temple of Aluwihare is 2 mi/3.2 km N; here (1st century A.C.E.) Buddhist scriptures (Vinayapitakas) were first transcribed. Matale is main town in Matale Valley, in N SRI LANKA HILL COUNTRY, W of KNUCKLES GROUP hills; c.20 mi/32 km long N-S, 5 mi/8 km wide; extensive tea, rubber, and cacao plantations; limestone quarries. Average rainfall 75 in/191 cm–100 in/254 cm.

Mataleng, township, SOUTH AFRICA: see BARKLY WEST.

Matale Valley, SRI LANKA: see MATALE.

Matallana de Torío (mah-tah-LYAH-nah dai TO-ryo), town, LEÓN province, NW SPAIN, 19 mi/31 km NNE of LEÓN; 42°52′N 05°31′W. In coal-mining area.

Matam, administrative region (□ 11,434 sq mi/29,728.4 sq km; 2004 population 461,836), NE SENEGAL; ⊙ MATAM town. Bordered by TAMBACOUNDA (S), LOUGA (W), and SAINT-LOUIS (NW) administrative regions, and MAURITANIA (NNE). Matam is the second largest administrative region by area (Tambacounda is the largest). Several roads branch out of and around Matam town. Airport at Matam town. Formed in 2002 out of Matam department of SE Saint-Louis region and the rural areas of Lougré Thioly and Vélingara of SE Louga region.

Matam (MAH-tahm), town, ⊙ MATAM administrative region, NE SENEGAL, landing on SENEGAL River (MAURITANIA border), and 215 mi/346 km ESE of SAINT-LOUIS. Produces millet, corn, cotton; livestock; fishing. Administrative center. Airfield.

Matama (mah-TAH-mah), town, West Kunisaki county, OITA prefecture, E KYUSHU, SW JAPAN, 25 mi/40 km N of OITA; 33°36′N 131°28′E.

Matamata (MA-tuh-MA-tuh), town (2001 population 6,078), Matamata-Piako district, WAIKATO region, Hawaki Plains, NORTH ISLAND, NEW ZEALAND, 90 mi/145 km SE of AUCKLAND; 38°05′S 175°50′E. Dairy and prime lamb farming.

Matamba (ma-TAM-bah), village, IRINGA region, SW TANZANIA, 32 mi/51 km E of MBEYA, in Poroto Mountains; 09°00′S 33°58′E. Subsistence crops; goats, sheep.

Mataméyé (MAH-tuh-mai-ai), town, ZINDER province, NIGER, 40 mi/64 km SW of ZINDER; 13°26′N 08°28′E. Food processing. Administrative center.

Matamoras (me-tuh-MO-ruhs), village (2006 population 916), WASHINGTON county, SE OHIO, on the OHIO RIVER (WEST VIRGINIA state line), 21 mi/34 km ENE of MARIETTA; 39°31′N 81°04′W. In agricultural area.

Matamoras (MA-tah-MOR-uhs), borough (2006 population 2,623), PIKE county, NE PENNSYLVANIA, 1 mi/1.6 km SW of PORT JERVIS (New York), on DELAWARE RIVER (bridged NE to NEW YORK, SE to NEW JERSEY); 41°22′N 74°42′W. Agriculture (dairying; cattle); manufacturing (medical supplies; commercial printing). Easternmost point in Pennsylvania Part of Delaware State Forest to W; High Point State Park (N.J.) to S.

Matamoros (mah-tah-MO-ros), city and township, COAHUILA, N MEXICO, in Laguna district, 12 mi/19 km E of TORREÓN; 25°33′N 103°15′W. Railroad junction. Agricultural center (cotton, corn, wheat, wine, vegetables).

Matamoros (mah-tah-MUH-ros), town, CHIHUAHUA, N MEXICO, 13 mi/21 km SSE of HIDALGO DEL PARRAL, on Mexico Highway 45; 26°48′N 105°35′W. Silver and lead mining. Also VILLA MATAMOROS. Formerly LAS CUEVAS or SAN ISIDRO DE LAS CUEVAS.

Matamoros, MEXICO: see IZÚCAR DE MATAMOROS.

Matamoros, MEXICO: see HEROICA MATAMOROS.

Matamur (mah-tah-MOOR), village, MADINIYINA province, SE TUNISIA, 4 mi/6.4 km W of MADINIYINA. Fortified village (qusur) built by Berbers to store grain; Arabs fortified the outer walls with residences in storage areas.

Mata'na, Asfun El, EGYPT: see ASFUN.

Matana, Lake (MAH-tah-nah), in NE Sulawesi Selatan province, INDONESIA, 30 mi/48 km S of KOLONODALE; 20 mi/32 km long, up to 5 mi/8 km wide; 02°28′S 121°20′E. Also spelled Matano.

Matandu River (ma-TAN-doo), 150 mi/241 km long, SE TANZANIA; rises c. 35 mi/56 km NW of LIWALE; flows generally ENE to INDIAN OCEAN 20 mi/32 km NNW of KILWA MASOKO. Upper course forms body of SELOUS GAME RESERVE.

Matane (muh-TAHN), county (□ 1,303 sq mi/3,387.8 sq km), BAS-SAINT-LAURENT region, E QUEBEC, E CANADA, on N shore of GASPÉ PENINSULA, on the SAINT LAWRENCE RIVER; ⊙ MATANE; 48°45′N 67°05′W. Deep-water port. Shrimp processing; pulp, paperboard mills. Composed of twelve municipalities. Regional county municipality (municipalité régionale de comté) created in 1982 from the former Matane county. Name is from the indigenous meaning "beaver breeding grounds."

Matane (muh-TAHN), town (□ 83 sq mi/215.8 sq km), ⊙ MATANE county, BAS-SAINT-LAURENT region, SE QUEBEC, E CANADA, on the SAINT LAWRENCE RIVER at the mouth of the Matane River, at the beginning of the GASPÉ PENINSULA; 48°51′N 67°32′W. Fishing (shrimp), lumbering, and pulpwood-shipping center; manufacturing (concrete, prepared meats, fresh and frozen fish); summer resort. Restructured in 2001.

Matanga Pwani (ma-tan-gah PWA-nee), town, PEMBA NORTH region, NE TANZANIA, in NW part of PEMBA island, 19 mi/31 km NNW of CHAKE CHAKE; 04°46′S 39°42′E. Cloves, copra; livestock; fish.

Matanilla Reef, northernmost part of the BAHAMAS, BRITISH WEST INDIES, 50 mi/80 km NNE of WEST END (GRAND BAHAMA Island); 27°25′N 78°42′W.

Matano, Lake, INDONESIA: see MATANA, LAKE.

Matanuska (ma-duh-NOO-skuh), village, S ALASKA, 30 mi/48 km NE of ANCHORAGE.

Matanuska River (ma-duh-NOO-skuh), 75 mi/121 km long, S ALASKA; rises in CHUGACH MOUNTAINS, near 61°47′N 147°40′W; flows SW past MOOSE CREEK, PALMER, and MATANUSKA, to KNIK ARM 30 mi/48 km NE of ANCHORAGE. Lower course flows through MATANUSKA VALLEY agricultural region. Source includes Matanuska glacier. Paralleled by GLENN HIGHWAY.

Matanuska-Sustina, borough (□ 24,694 sq mi/64,204.4 sq km; 2000 population 59,322), S ALASKA. Main town is PALMER. Bounded on S in part by COOK INLET. Part of ALASKA RANGE to NW and TALKEETNA

MOUNTAINS are in SE center. Part of DENALI NATIONAL PARK AND PRESERVE to NW, small part of LAKE CLARK National Park and Preserve in SW corner. MATANUSKA VALLEY, Alaska's main agricultural area, in the S. Agriculture (vegetables, potatoes, rutabagas, hay); dairying. Fishing.

Matanuska Valley (ma-duh-NOO-skuh), region (□ 1,000 sq mi/2,600 sq km) of S ALASKA, on lower MATANUSKA RIVER, NE of ANCHORAGE, extends c.40 mi/64 km ENE from head of KNIK ARM, between TALKEETNA (NNW) and CHUGACH (SSE) mountains; c.9,000 acres/3,642 ha are under cultivation, producing oats, barley, fodder grasses, vegetables, potatoes, berries, hogs; dairy products. Growing suburban region for Anchorage. Region served by GLENN HIGHWAY. Market center is PALMER. Climate is temperate, with adequate rainfall. University of Alaska has Matsu Community College. Transportation Museum, Palmer Fair, Tourism. Valley became site of Federal experiment in rural resettlement (May 1935), when 208 families from Middle Western drought areas were established here with aid of Federal loans. Matanuska Valley Farmers' Cooperating Association was established (1937) at Palmer. Control of project transferred (September 1938) to Department of the Interior.

Matanza, district, ARGENTINA: see SAN JUSTO (in BUENOS AIRES province).

Matanza (mah-TAHN-sah), town, ⊙ Matanza municipio, SANTANDER department, N central COLOMBIA, 10 mi/16 km NNE of BUCARAMANGA; 07°19′N 73°01′W. Elevation 5,889 ft/1,794 m. Coffee, corn; livestock.

Matanza de Acentejo, La (mah-TAHN-thah dai ahthen-TAI-ho, lah), town, SANTA CRUZ DE TENERIFE province, TENERIFE island, CANARY ISLANDS, SPAIN, 13 mi/21 km W of SANTA CRUZ DE TENERIFE; 28°26′N 16°27′W. Wine growing.

Matanzas (mah-TAHN-sahs), province (□ 30,225 sq mi/78,585 sq km; 2002 population 670,417), W central CUBA; ⊙ MATANZAS. Well endowed with natural resources. The N coast is lined with ports and bays and contains one of the world's finest beaches, at VARADERO ("Playa Azul"). In the N half of the province are plains of deep red-clay soil with good drainage; in the S are low-lying wetlands. The S coast includes Ciénaga de ZAPATA (a swamp with over 500,000 tons/454,000 metric tons of dry peat) and COCHINOS BAY, the site of the Bay of Pigs invasion (1961). Sugarcane (with 21 mills) and henequen are the major crops; subsistence agriculture is also practiced, and there is some cattle raising. The province produced 30% of Cuba's citrus crop in 1994. Has the French-designed Antonio Guiteras thermoelectric generating station, which can generate 330 MW, the fifth-most powerful in Cuba. Tourist industry; many mineral springs. Important cities include CÁRDENAS, COLÓN, JOVELLANOS, and JAGÜEY GRANDE.

Matanzas (mah-TAHN-sahs), city (2002 population 127,287), ⊙ MATANZAS province, W central CUBA; 23°04′N 81°33′W. A port with a large, deep harbor, it exports sugar (16% of Cuba's total 1994 sugar exports, second only to CIENFUEGOS), fruits, and sisal. Industries in the city include sugar refineries and textile mills. Petrochemical and petroleum-storage and -distribution facilities on port. The port has capacity to handle 6 ships simultaneously on 3,630-ft/1,106-m-long piers. On the highway between HAVANA and VARADERO Beach; popular stopover for vacationers, who explore the picturesque YUMURÍ River valley and the caves of BELLAMAR, famous for their calcite crystal formations. Founded in 1693, it was once a pirate haven but by the early 19th century had become Cuba's second city, mainly due to the growth of the sugar industry. As the industry moved E, the city's importance declined. Matanzas remains an important cultural center for the area.

Matanzas (mah-TAHN-zahs), officially San José de Matanzas, town, SAMANÁ province, N DOMINICAN REPUBLIC, on ESCOCESA BAY, 15 mi/24 km NW of SÁNCHEZ. In agricultural region (rice, cacao, corn, coconuts, coffee, fruit). Damaged by 1946 earthquake and tidal waves.

Matanzas River (mah-TAHN-zahs), c.25 mi/40 km long, NE BUENOS AIRES province, ARGENTINA; formed near CAÑUELAS; flows NE along SE border of the Federal district (BUENOS AIRES proper)—where it is also called the Riachuelo—to the RÍO DE LA PLATA 2 mi/3.2 km NE of AVELLANEDA at DOCK SUD. On its W bank is the oldest part of Buenos Aires and its port.

Matanzas River (muh-TAN-zuhs), narrow lagoon, c.17 mi/27 km long, ST. JOHNS county, NE FLORIDA; sheltered from the ATLANTIC OCEAN by ANASTASIA ISLAND; extends from ST. AUGUSTINE (N end; ocean outlet) to Matanzas Inlet, which connects its S end with the ocean. Followed by INTRACOASTAL WATERWAY.

Matão (mah-TOUN), city (2007 population 74,416), N central SÃO PAULO state, BRAZIL, on railroad, and 45 mi/72 km SW of RIBEIRÃO PRÊTO; 25°47′S 49°15′W. Manufacturing (agricultural equipment, macaroni, furniture); cotton, rice, coffee. Formerly spelled Mattão.

Matapa, MEXICO: see VILLA PESQUEIRA.

Mata Palacio (MAH-tah pah-LAH-see-o), town, HATO MAYOR province, E DOMINICAN REPUBLIC, 15 mi/24 km N of SAN PEDRO DE MACORÍS. In agricultural region (sugarcane, cacao, coffee, rice, fruit; cattle).

Matapalo, Cape (mah-tah-PAH-lo), SE extremity of OSA PENINSULA, on the PACIFIC OCEAN, S COSTA RICA, at entrance to the GOLFO DULCE.

Matapan, Cape (mah-tah-PAHN), LAKONIA prefecture, S PELOPONNESE department, extreme S mainland GREECE, at the tip of the central peninsula of the PELOPONNESUS, projecting into the IONIAN SEA, at S tip of TAIYETOS mountains. Known to the ancients as Taenarum. In World War II the British won an important naval battle (1941) over the Italians offshore here.

Matapeake, MARYLAND: see KENT ISLAND.

Matapedia (ma-tuh-PEE-dee-uh), French *Matapédia* (mah-tah-PAI-dee-ah), county (□ 2,075 sq mi/5,395 sq km; 2006 population 19,677), BAS-SAINT-LAURENT region, E QUEBEC, E CANADA, in central part of GASPÉ PENINSULA; ⊙ AMQUI; 48°25′N 67°25′W. Composed of twenty five municipalities. Formed in 1982.

Matapedia (ma-tuh-PEE-dee-ah), French, *Matapédia* (mah-tah-PAI-dee-uh), village (□ 27 sq mi/70.2 sq km), GASPÉSIE—ÎLES-DE-LA-MADELEINE region, E QUEBEC, E CANADA, SE GASPÉ PENINSULA, on RESTIGOUCHE River, at mouth of MATAPEDIA RIVER, 12 mi/19 km WSW of CAMPBELLTON; 47°58′N 66°56′W. Railroad junction. Lumbering, dairying.

Matapedia, Lake (ma-tuh-PEE-dee-uh), French, *Matapédia* (mah-tah-PAI-dee-uh), E QUEBEC, E CANADA, at the base of the GASPÉ PENINSULA, and S of MATANE; 14 mi/23 km long and 2 mi/3 km wide. Drained S by the MATAPEDIA RIVER, famous for salmon fishing. Well-known tourist center.

Matapedia River (ma-tuh-PEE-dee-ah), French, *Matapédia River* (mah-tah-PAI-dee-uh), 50 mi/80 km long, E QUEBEC, E CANADA, at base of GASPÉ PENINSULA; issues from Lake MATAPEDIA; flows SE, past AMQUI and CAUSAPSCAL, to RESTIGOUCHE River at MATAPEDIA; 47°58′N 66°56′W. Noted salmon stream.

Mataporquera (mah-tah-por-KAI-rah), town, CANTABRIA province, N SPAIN, 33 mi/53 km SSW of TORRELAVEGA; 42°53′N 04°10′W. Cement manufacturing; lumbering; livestock raising; cereals.

Matapozuelos (mah-tah-poth-WAI-los), town, VALLADOLID province, N central SPAIN, 17 mi/27 km SSW of VALLADOLID; 41°25′N 04°47′W. Flour mills; cereals, wine.

Mataquescuintla (mah-tah-ke-SKEEN-tlah), town, JALAPA department, E central GUATEMALA, in highlands, at W foot of Alzatate volcano, on secondary road, 23 mi/37 km WSW of JALAPA by road; 14°32′N 90°11′W. Elevation 5,413 ft/1,650 m. Agriculture (corn, beans, coffee; livestock). Until 1935, in SANTA ROSA department.

Mataquito River (mah-tah-KEE-to), c.65 mi/105 km long, central CHILE; formed by union of TENO and LONTUÉ rivers 5 mi/8 km WNW of CURICÓ; flows W, past HUALAÑE and LICANTÉN, to the PACIFIC OCEAN 25 mi/40 km NNE of CONSTITUCIÓN; forms border between CURICÓ and TALCA provinces. Length of Lontué-Mataquito, 140 mi/225 km.

Matar (muh-TAHR), town, KHEDA district, GUJARAT state, W central INDIA, 4 mi/6.4 km SSW of KHEDA; 22°00′N 72°59′E. Local market for rice, millet.

Matara, district (□ 490 sq mi/1,274 sq km; 2001 population 761,370), SOUTHERN PROVINCE, near southernmost point of SRI LANKA; ⊙ MATARA; 06°10′N 80°30′E.

Matara (MAH-uh-ruh) [Sinhalese=great ford], town (2001 population 76,254), ⊙ MATARA district, SOUTHERN PROVINCE, SRI LANKA, on SW coast, 24 mi/39 km ESE of GALLE; 05°56′N 80°32′E. Railroad terminus; road junction; coastal trade (tea, rubber, rice, citronella grass, coconuts, cinnamon) center. Important under Portuguese (17th century) and Dutch (built Star fort, 1765). British general and administrator Sir Henry Lawrence born here, 1806. District has iron-ore, sapphire, and beryl deposits (notably near Morawaka); major citronella-grass area in Sri Lanka. University of Ruhuna nearby.

Matará (mah-tah-RAH), village, central SANTIAGO DEL ESTERO province, ARGENTINA, on railroad, and 32 mi/51 km NW of AÑATUYA; 28°06′S 63°12′W. Agriculture; lumbering center; cotton; livestock; sawmills.

Mataraca (MAH-tah-rah-kah), town (2007 population 6,984), NE PARAÍBA state, BRAZIL, near Atlantic coast of PARÁ, 28 mi/45 km NNE of MAMANGUAPE; 06°36′S 35°03′W.

Mataram (MAH-tah-rahm), former Muslim sultanate in central JAVA, INDONESIA. Founded 16th century in the Solo (SURAKARTA) area, it flourished in the 17th century, controlling most of Java. It fell to Dutch in mid-18th century.

Mataram (MAH-tah-rahm), town (2000 population 317,374), ⊙ Nusa Tenggara Barat province, chief town of LOMBOK Island, INDONESIA, near W coast of island; 08°35′S 116°07′E. Trade center for agricultural area. Its port is nearby AMPENAN. Near Mataram are several palaces built by Balinese princes.

Mataranka (ma-tuh-RAN-kuh), settlement, N central NORTHERN TERRITORY, AUSTRALIA, 225 mi/362 km SE of DARWIN, on Darwin-BIRDUM railroad; 14°56′S 133°04′E. Sheep. Thermal pools at nearby Elsey National Park.

Matari, PAKISTAN: see MATIARI.

Mataripe (mah-tah-ree-pai), town, E BAHIA state, BRAZIL, on N shore of TODOS OS SANTOS Bay, c.25 mi/40 km NNW of SALVADOR, near SÃO FRANCISCO DO CONDE; 12°34′S 38°37′W. Petroleum refinery (completed 1950). Called Socorro until 1944.

Matariya, El (MAH-TAH-ree-yuh, el), town, NE EGYPT, 6 mi/9.7 km NE of CAIRO. On site of ancient HELIOPOLIS. Part of metropolitan Cairo.

Matariya, El, town, DAQAHLIYA province, Lower EGYPT, on a peninsula shore of Lake MANZALA, 20 mi/32 km SE of DUMYAT. Connected by canals with PORT SAID and Dumyat. Important fishing industry.

Mataró (mah-tah-RO), city (2001 population 106,358), BARCELONA province, NE SPAIN, in CATALONIA, and 15 mi/24 km NE of BARCELONA; 41°32′N 02°27′E. Mataró is a Mediterranean port and a manufacturing center, producing knitted goods, wine, and chemicals. The first railroad in Spain was built (1848) from

Barcelona to Mataró. The city's baroque church of Santa María is notable.

Mata Roma (MAH-tah RO-mah), city (2007 population 13,823), NE MARANHÃO state, BRAZIL, 17 mi/27 km W of BREJO; 03°36′S 43°10′W.

Matar Taris, EGYPT: see MATIR TARIS.

Mataruge (MAHT-ah-roo-gai), village, central SERBIA, on IBAR RIVER, on railroad, and 6 mi/9.7 km WSW of KRALJEVO; 43°41′N 20°35′E. MATARUSKA BANJA health resort is just NE.

Mataruska Banja, SERBIA: see MATARUGE.

Matas, Las, DOMINICAN REPUBLIC: see LAS MATAS.

Matathia (mah-TAH-dee-ah), village, CENTRAL province, S central KENYA, at E edge of GREAT RIFT VALLEY, on railroad, and 24 mi/39 km NW of NAIROBI; 00°38′S 38°59′E. Elevation 7,390 ft/2,252 m. Hardwood industry; sisal, wheat, coffee, corn.

Matatiele, town, in EASTERN CAPE province, SOUTH AFRICA, near LESOTHO border, at foot of DRAKENSBERG range, 110 mi/177 km SW of PIETERMARITZBURG (Msunduzi); 30°19′S 28°38′E. Elevation 5,510 ft/1,680 m. Railroad terminus in farming region; corn, dairying, and horsebreeding. Airfield. Established 1865 by Adam Kok.

Mataúna, BRAZIL: see PALMEIRAS DE GOIÁS.

Mataura (ma-TOU-ruh), town, GORE district, SOUTHLAND region, SOUTH ISLAND, NEW ZEALAND, 33 mi/53 km NE of INVERCARGILL, at margin of MATAURA RIVER plain with CATLINS HILLS; 46°11′S 168°52′E. Main railroad and road passage. Paper mill, dairy plants. Lignite mine.

Mataura River (ma-TOU-ruh), 149 mi/240 km long, S SOUTH ISLAND, NEW ZEALAND; rises in Eyre Mountains SW of LAKE WAKATIPU; flows SE, past GORE, MATAURA, and SOUTHLAND Plain to FOVEAUX STRAIT.

Mata-Utu, chief village (2003 population 1,191) and ⊙ WALLIS AND FUTUNA ISLANDS, in the S PACIFIC, on Île 'Uvea, in WALLIS ISLANDS; 13°22′S 176°12′W.

Matawan (MA-tuh-WAHN), borough (2006 population 8,781), MONMOUTH county, E NEW JERSEY, 11 mi/18 km N of FREEHOLD, in suburban region; 40°24′N 74°14′W. Manufacturing. Has 18th-century buildings. Called New Aberdeen before 1715, incorporated 1895.

Matawinie (mah-tah-WI-nee), county (□ 4,098 sq mi/10,654.8 sq km; 2006 population 44,112), LANAUDIÈRE region, S QUEBEC, E CANADA; ⊙ RAWDON; 46°16′N 73°47′W. Composed of twenty seven municipalities, including unorganized territory. Formed in 1982.

Matawin River (MA-tuh-win), 100 mi/161 km long, S QUEBEC, E CANADA; rises near MONT TREMBLANT; flows E, through Lake TORO, to SAINT MAURICE River; 25 mi/40 km from SHAWINIGAN; 46°54′N 72°56′W. Several falls. Sometimes spelled Mattawin.

Matay, EGYPT: see MATAI.

Matcha, village, SE LENINOBOD region, NW TAJIKISTAN, on ZERAVSHAN RIVER, and 45 mi/72 km SE of URA-TYUBE; 39°27′N 69°39′E. Wheat; sheep; gold placers. Formerly called Madrushkent.

Matchedash, former township, S ONTARIO, E central CANADA, 31 mi/50 km from BARRIE; 44°48′N 79°34′W. Amalgamated into SEVERN township in 1994.

Mateare (mah-te-AH-re), town, (2005 population 25,313), MANAGUA department, SW NICARAGUA, port on LAKE MANAGUA, 12 mi/19 km NW of MANAGUA; 12°14′N 86°26′W. Agriculture; fisheries.

Mateba (muh-TE-buh), village, BAS-CONGO province, W CONGO, on S shore of Mateba Island, in mouth of CONGO RIVER (ANGOLA border), and 15 mi/24 km WSW of BOMA; 05°54′S 12°50′E. Cattle-raising center.

Mateguá (mah-tai-GWAH), town and canton, ITÉNEZ province, BENI department, NE BOLIVIA, on GUAPORÉ River, and 95 mi/153 km ENE of MAGDALENA, on BRAZIL border; 13°01′S 62°48′W.

Matehuala (mah-te-WAH-law), city and township, ⊙ Matehuala municipio, SAN LUIS POTOSÍ, N central MEXICO, on interior plateau, 105 mi/169 km NNE of

SAN LUIS POTOSÍ, on Mexico Highway 57; 23°38′N 100°38′W. Elevation 5,299 ft/1,615 m. Railroad terminus. Gold and silver mining; tanning; maguey processing (textile fibers, liquor). Airfield.

Matelândia (MAH-te-LAHN-zhee-ah), city (2007 population 15,404), SW PARANÁ state, BRAZIL, 55 mi/89 km NE of FOZ DO IGUAÇU; 25°05′S 53°59′W.

Matelica (mah-TAI-lee-kah), town, MACERATA province, THE MARCHES, central ITALY, on ESINO River, and 7 mi/11 km SE of FABRIANO; 43°15′N 13°00′E. Tanneries, shoe factory, cement works; cutlery manufacturing. Bishopric.

Matelot (MA-tai-lot), village, N TRINIDAD, TRINIDAD AND TOBAGO, 18 mi/29 km ENE of PORT OF SPAIN. Swimming, fishing.

Matera (mah-TAI-rah), province (□ 1,329 sq mi/3,455.4 sq km), BASILICATA, S ITALY; ⊙ MATERA; 40°30′N 16°25′E. Borders on Gulf of TARANTO; traversed by the APENNINES, which descend gradually to coastal plain. Watered by CAVONE and lower courses of AGRI, BASENTO, BRADANO, and SINNI rivers. Agriculture (cereals, grapes, olives, fruit, cotton); livestock raising (sheep, goats). Forestry. Formed 1927 from POTENZA province, to which it transferred (c.1947) 135 sq mi/350 sq km in NW.

Matera (mah-TAI-rah), city, ⊙ MATERA province, in BASILICATA, S ITALY, in the APENNINES. Agricultural and industrial center with woolen textile mills and fabricated metals, machinery, wood products, ceramics and food-processing factories. A Romanesque cathedral and a castle (both 13th century) are in the city.

Matese, Lago di (mah-TAI-ze, LAH-go dee), narrow lake in the APENNINES, CAMPANIA, S ITALY, 24 mi/39 km N of CASERTA; c.3 mi/5 km long; 41°25′N 14°25′E. Furnishes power to hydroelectric plant at Piedimonte d'Alife.

Mátészalka (MAH-tai-sal-kah), Hungarian *Mátészalka*, city, ⊙ SZABOLCS-SZATMÁR county, NE HUNGARY, on KRASZNA river, and 28 mi/45 km E of NYÍREGYHÁZA; 47°57′N 22°20′E. Railroad junction; apples, alfalfa, grain, tobacco; hogs; manufacturing (knitted wear, measuring instruments, dairy products; milling). Technical college for agriculture.

Matetete, CONGO: see BAYENGA.

Matetsi, town, MATABELELAND NORTH province, W ZIMBABWE, 25 mi/40 km SSE of VICTORIA FALLS (town), on Matetse River, on railroad; 18°17′S 25°57′E. HWANGE NATIONAL PARK to SE; Matetsi Safari Area to S and W. Cattle, sheep, goats; corn.

Mateus Leme (MAH-tai-oos LE-me), city (2007 population 25,627), central MINAS GERAIS state, BRAZIL, 12 mi/19 km W of BETIM, on railroad; 20°00′S 44°30′W.

Matewan (MAIT-wahn), village (2006 population 497), MINGO county, SW WEST VIRGINIA, on TUG FORK RIVER, at KENTUCKY state line, 7 mi/11.3 km SE of WILLIAMSON; 37°37′N 82°10′W. Bituminous-coal region. Manufacturing (coal processing). One site of Hatfield-McCoy feud. Scene of bloody confrontation (May 1920) between mine owners and miners. Incorporated 1895.

Matfield Green, village (2000 population 60), CHASE county, E central KANSAS, on S fork of COTTONWOOD RIVER, and 14 mi/23 km S of COTTONWOOD FALLS; 38°09′N 96°33′W. Livestock; grain.

Matfors (MAHT-FORSH), town, VÄSTERNORRLAND county, NE SWEDEN, on LJUNGAN river (falls), 8 mi/12.9 km WSW of SUNDSVALL; 62°21′N 17°01′E.

Math (MUHT), village, Sindhudurg district, MAHARASHTRA state, W central INDIA, in WESTERN GHATS, 7 mi/11.3 km NNE of SAVANTVADI. Soap and sodium factories; rice, mangoes. Teak, blackwood in nearby forests.

Matha (mah-tah), commune (□ 7 sq mi/18.2 sq km), CHARENTE-MARITIME department, POITOU-CHARENTES region, W FRANCE, 12 mi/19 km N of COGNAC;

45°52′N 00°19′W. Brandy distilling. Has church with Romanesque sculptured façade.

Mathabhanga, INDIA: see MATABHANGA.

Mathay (mah-TAI), commune (□ 5 sq mi/13 sq km), DOUBS department, FRANCHE-COMTÉ region, E FRANCE, on DOUBS RIVER, in JURA Mountains, and 5 mi/8 km S of MONTBÉLIARD; 47°26′N 06°47′E. Metallurgy.

Matheniko Game Reserve, sanctuary, MOROTO district, NORTHERN region, NE UGANDA, 60 km/37 mi NNW of MOROTO. Antelope, birds. Was part of former KARAMOJA province.

Mather, unincorporated town, GREENE county, SW PENNSYLVANIA, 7 mi/11.3 km ENE of WAYNESBURG, near South Fork of Tenmile Creek; 39°56′N 80°04′W. Agriculture (dairying; corn).

Mather (MA-duhr), unincorporated village, S MANITOBA, W central CANADA, 60 mi/97 km SE of BRANDON, and in ROBLIN rural municipality; 49°05′N 99°11′W. Grain; livestock.

Mather Air Force Base, California: see SACRAMENTO, city.

Matheran (muh-TAI-ruhn), town, THANE district, MAHARASHTRA state, W central INDIA, 28 mi/45 km E of MUMBAI (Bombay); 18°59′N 73°16′E. Health resort (sanatorium) on scenic outlier (elevation c.2,500 ft/762 m) of WESTERN GHATS; annual rainfall c.250 in/635 cm.

Mather, Mount (12,123 ft/3,695 m), S central ALASKA, in ALASKA RANGE, in MOUNT MCKINLEY NATIONAL PARK, 140 mi/225 km N of ANCHORAGE; 63°11′N 150°26′W.

Matherville, village (2000 population 772), MERCER county, NW ILLINOIS, on EDWARDS RIVER, and 9 mi/14.5 km ENE of ALEDO; 41°15′N 90°36′W. In agricultural area.

Matheson Point, E extremity of KING WILLIAM ISLAND, NUNAVUT territory, N CANADA, on Rae Strait; 68°49′N 95°10′W.

Mathews (MA-thyooz), county (□ 251 sq mi/652.6 sq km; 2006 population 9,184), E VIRGINIA; ⊙ MATHEWS; 37°25′N 76°16′W. In Tidewater region; bounded E by CHESAPEAKE BAY, N by PIANKATANK RIVER estuary and S by MOBJACK BAY. Agriculture (corn, soybeans, vegetables, melons, bulbs; poultry); fish, crabs, oysters. Seasonal homes in area; waterfowl hunting. Formed 1791.

Mathews, unincorporated town (2000 population 2,003), LAFOURCHE parish, SE LOUISIANA, on BAYOU LAFOURCHE, 4 mi/6.4 km SE of RACELAND; 29°41′N 90°33′W. Fishing; sugarcane.

Mathews (MA-thyooz), unincorporated village, ⊙ MATHEWS county, E VIRGINIA, near CHESAPEAKE BAY, 28 mi/45 km ESE of WEST POINT; 37°26′N 76°19′W. Manufacturing (wood products); agriculture (vegetables, corn, soybeans; poultry; bulb growing).

Mathews Dam, California: see MATHEWS, LAKE.

Mathews, Lake, reservoir (c.5 mi/8 km long), RIVERSIDE county, S CALIFORNIA, at W end of COLORADO RIVER AQUEDUCT (enters E end of lake through 8-mi/12.9-km-long Valverde Tunnel), 9 mi/14.5 km S of RIVERSIDE. Impounded by Mathews Dam (210 ft/64 m high, 2,170 ft/661 m long; completed 1938). From lake, gravity carries water to cities of Los Angeles metropolitan district. SANTA ANA MOUNTAINS to SW. Formerly Cajalco Reservoir.

Mathia, Greece: see LYKODIMO.

Mathias Barbosa, Brazil: see MATIAS BARBOSA.

Mathinna (muh-THI-nuh), village, NE TASMANIA, SE AUSTRALIA, 38 mi/61 km E of LAUNCESTON, near SOUTH ESK RIVER; 41°29′S 147°53′E.

Mathis (MATH-is), town (2006 population 5,473), SAN PATRICIO county, S Texas, 34 mi/55 km NW of CORPUS CHRISTI, near NUECES RIVER; 28°05′N 97°49′W. Railroad, trade, shipping point in agriculture and oil-producing area; manufacturing (fertilizers). Just

W is Lake Corpus Christi; Lake Corpus Christi and Lipantitlan state parks to SW. Incorporated 1937.

Mathiston (MATH-is-tuhn), town (2000 population 720), WEBSTER and CHOCTAW counties, central MISSISSIPPI, 18 mi/29 km WNW of STARKVILLE, near BIG BLACK RIVER; 33°32′N 89°07′W. Manufacturing (lumber, apparel, machinery). Seat of Wood Jr. College. NATCHEZ TRACE PARKWAY passes to NW.

Mathoura (muh-THOU-ruh), town, NEW SOUTH WALES, SE AUSTRALIA, 21 mi/34 km S of DENILIQUIN, in RIVERINA region, and near VICTORIA border; 35°49′S 144°54′E. Gateway to red-gum (eucalyptus) forest; bird sanctuary. Timber.

Mathraki (mah-THRAH-kee), island (□ 2 sq mi/5.2 sq km), KÉRKIRA prefecture, IONIAN ISLANDS department, off W coast of GREECE, in IONIAN SEA, 5 mi/8 km WNW of KÉRKIRA island; 2 mi/3.2 km long, 1 mi/1.6 km wide; 39°48′N 19°29′E. Fisheries. Also called Samothrake or Samothraki; also spelled Mathrake.

Mathūiyya (mah-too-EE-yah), village and oasis (2004 population 9,946), QABIS province, E TUNISIA, 8 mi/12.9 km NNW of QABIS. Date palms; sheep, camels.

Mathura (muh-TOO-rah), district (□ 1,471 sq mi/3,824.6 sq km), W UTTAR PRADESH state, N central INDIA; ⊙ MATHURA. On GANGA-YAMUNA DOAB (E); irrigated by AGRA and UPPER GANGA canals. Agriculture includes gram, jowar, wheat, barley, cotton, mustard, pearl millet, sesame, sugarcane, corn. Main centers are Mathura, VRINDAVAN, Kosi, Chhata, and pilgrimage centers of MAHABAN, Gokul, Baldeo, and Govardhan. Numerous places here connected with Krishna legend. Formerly called Muttra.

Mathura (muh-TOO-rah), city (2001 population 323,315), ⊙ Mathura district, Uttar Pradesh state, N central India, on the Yamuna River; 27°30′N 77°41′E. An agr. market town best known as a Hindu pilgrimage site, the reputed birthplace of the god Krishna. The region, which may have been inhabited since the 7th century B.C.E., is rich in archaeological remains. Muslim rulers (16th–18th century) destroyed many Hindu temples here and erected mosques. Has relics from Hindu, Buddhist, and Muslim kingdoms. Previously known as Muttra.

Mathurai, India: see MADURAI.

Mathwar (muht-WAHR), village, Madhya Pradesh state, central India, 80 mi/129 km WSW of DHAR. Was capital of former princely state of Mathwar of Central India agency, along right bank of NARMADA RIVER; in 1948, merged with Madhya Bharat, later with Madhya Pradesh state.

Mati (MAH-tee), town, ⊙ DAVAO ORIENTAL province, SE MINDANAO, PHILIPPINES, at head of PUJADA BAY, 45 mi/72 km ESE of DAVAO; 06°55′N 126°09′E. Iron mines. Port and agricultural center (corn, durian, citrus fruit). Citrus processing.

Matiali (muht-YAH-lee), town, JALPAIGURI district, N WEST BENGAL state, E INDIA, 29 mi/47 km N of JALPAIGURI; 26°56′N 88°49′E. Railroad spur terminus. Tea processing; extensive tea gardens. Copper ore deposits nearby.

Matiari (muht-YAH-ree), town, HYDERABAD district, central SIND province, SE PAKISTAN, 13 mi/21 km NNE of HYDERABAD; 25°36′N 68°27′E. Also spelled MATARI.

Matias Barbosa (MAH-chee-ahs bahr-bo-sah), city (2007 population 13,315), S MINAS GERAIS state, BRAZIL, in the SERRA DA MANTIQUEIRA, on RIO PARAIBUNA, on railroad, and 7 mi/11.3 km S of JUIZ DE FORA; 21°49′S 43°22′W. Manufacturing (pharmaceuticals; dairying). Formerly spelled Mathias Barbosa.

Matias Cardoso (MAH-chee-ahs KAHR-so-so), town (2007 population 10,270), N central MINAS GERAIS state, BRAZIL, on Rio SÃO FRANCISCO, 22 mi/35 km NE of Missoes; 14°50′S 43°50′W.

Matías de Galvez, GUATEMALA: see SANTO TOMÁS DE CASTILLA.

Matias Olímpio (MAH-chee-ahs O-leem-pee-o), city (2007 population 10,468), NW PIAUÍ state, BRAZIL, near PARNAÍBA River; 03°43′S 42°33′W.

Matías Romero, city and township, ⊙ Matías Romero municipio, OAXACA, S MEXICO, in foothills of SIERRA MADRE DEL SUR, on railroad, and 22 mi/35 km NNE of IXTEPEC; 16°52′N 95°21′W. Elevation 659 ft/201 m. Railroad center. Processing, lumbering, and agricultural center (cereals, sugarcane, fruit; livestock). Manufacturing materials and building equipment.

Matibis, town, MASVINGO province, SE central ZIMBABWE, 40 mi/64 km SE of MWENEZI, near MWENEZI RIVER, SW of GONAREZHOU NATIONAL PARK; 21°41′S 31°19′E. Livestock; grain.

Matidi, CONGO: see MATADI.

Matifou, Cape (mah-tee-FOO), headland, ALGIERS wilaya, N central ALGERIA, on the MEDITERRANEAN SEA, bounding Algiers Bay on E, 9 mi/14.5 km ENE of ALGIERS; 36°47′N 03°14′E. Lighthouse. The sizeable town of BORDJ EL BAHRI village (formerly Cape Matifou) is nearby.

Matignon, Hôtel (mah-teen-YON, o-TEL), public mansion in PARIS, ÎLE-DE-FRANCE, FRANCE, in 7th district (arrondissement). Built 1721; serves as the Prime Minister's office.

Matiguás (mah-tee-GWAHS), town, MATAGALPA department, central NICARAGUA, 37 mi/60 km E of MATAGALPA, on S slopes of CORDILLERA DARIENSE; 12°50′N 85°28′W. Sugarcane, potatoes; livestock.

Matilla de los Caños del Río (mah-TEE-lyah dai los KAH-nyos del REE-o), town, SALAMANCA province, W SPAIN, 18 mi/29 km WSW of SALAMANCA; 40°49′N 05°56′W. Cereals, legumes; livestock.

Matimbuka (ma-teem-BOO-kah), village, RUVUMA region, S TANZANIA, 40 mi/64 km SSE of SONGEA, near Likonde River; 11°09′S 35°57′E. Timber; livestock; corn.

Matin, India: see KATGHORA.

Matina (mah-CHEE-nah), town (2007 population 11,860), SW BAHIA state, BRAZIL, 22 mi/35 km NNW of GUANAMBI; 13°55′S 42°50′W.

Matina (mah-TEE-nah), town, ⊙ Matina canton, LIMÓN province, E COSTA RICA, 19 mi/31 km WNW of LIMÓN, on Matina River (a 50-mi/80-km-long CARIBBEAN coastal stream); 10°00′N 83°15′W. Trading center: cacao, corn, bananas.

Matinecock (mah-TIN-uh-kok), affluent residential village (□ 2 sq mi/5.2 sq km; 2006 population 842), NASSAU county, SE NEW YORK, on NW LONG ISLAND, just E of GLEN COVE; 40°51′N 73°34′W. In marine recreational area. Oyster Bay National Wildlife Refuge 3.5 mi/5.6 km to E. Incorporated 1928.

Matinha (mah-cheen-yah), city (2007 population 20,089), N MARANHÃO state, BRAZIL, 45 mi/72 km SSW of SÃO LUÍS; 03°10′S 44°50′W. Rice, cotton, babassu nuts.

Matinicock Point, small peninsula extending N into LONG ISLAND SOUND, SE NEW YORK, just N of GLEN COVE, and marking E side of entrance to HEMPSTEAD HARBOR; 40°54′N 73°38′W.

Matinicus Isle (muh-TIN-ik-uhs), plantation, KNOX county, S MAINE, in the ATLANTIC OCEAN, c.20 mi/32 km SE of ROCKLAND; 43°53′N 68°54′W. Includes Matinicus (c.1 sq mi/2.6 sq km), Ragged (with Criehaven village), Seal, and Wooden Ball islands, and Matinicus Rock (lighthouse).

Matir (mah-TUHR), town (2004 population 31,345), BIZERTE province, N TUNISIA, 19 mi/31 km SW of BIZERTE; 37°03′N 09°40′E. Administrative and road and railroad center; heart of rich agricultural region (wheat, barley, fruits, wine, cattle). Flour milling, macaroni manufacturing. Ruled by Berber tribes during most of its history.

Matir Taris (MAH-tir TAH-ris), village, FAIYUM province, Upper EGYPT, 6 mi/9.7 km NE of FAIYUM; 29°22′N 30°54′E. Cotton, cereals, sugar, fruits. Also spelled Matar Taris.

Matla, India: see PORT CANNING.

Matlalcueyatl, MEXICO: see MALINCHE.

Matlawi (met-lah-WEE), village (2004 population 37,099), QAFSAH province, W central TUNISIA, 29 mi/47 km WSW of QAFSAH; 34°20′N 08°24′E. Administrative center; headquarters of phosphate mining region at junction of phosphate mining railroads (mine 2 mi/3.2 km N of station). The Compagnie des Phosphates de Gafsa (CPG), established in 1885 by the French, ships phosphates by railroad to fertilizer plants at QABIS. Philippe-Thomas, an adjoining "company" village, houses miners and other employees.

Matlock (MAT-lahk), township, VICTORIA, SE AUSTRALIA, 68 mi/110 km ENE of MELBOURNE, on GREAT DIVIDING RANGE; 37°36′S 146°13′E. Former gold-mining center. Matlomola, residential development, SOUTH AFRICA: see GRIQUATOWN.

Matlock (MAT-lahk), town (2001 population 10,688), ⊙ DERBYSHIRE, central ENGLAND, on DERWENT River, and 18 mi/29 km S of SHEFFIELD; 53°08′N 01°32′W. Popular resort, previously with mineral springs and baths. Light manufacturing; high technology. Has church with 15th-century tower. Nearby are lead mines and caves. The first cotton mill in Derby was installed here 1771 by Arkwright. Also known as The Matlocks.

Matlock, town (2000 population 83), SIOUX county, NW IOWA, 17 mi/27 km NNE of ORANGE CITY; 43°14′N 95°55′W. Livestock; grain.

Matmar, El (maht-MAHR, el), village, RELIZANE wilaya, NW ALGERIA, in irrigated MINA valley, 5 mi/8 km W of RELIZANE. Wine, olives, cereals. Formerly called Clinchant.

Matmatah (met-me-TAH), village (2004 population 2,116), QABIS province, SE TUNISIA, 27 mi/43 km SSW of QABIS; 33°33′N 09°58′E. Olives, barley, dates, and figs are grown on crudely irrigated terraces; camels, sheep. Tourism; administrative center. Troglodyte dwellings of Berber people, described by Herodotus in the 4th century B.C.E.

Matnog (maht-NOG), town, SORSOGON province, extreme SE LUZON, PHILIPPINES, on SAN BERNARDINO STRAIT, 27 mi/43 km SSE of SORSOGON; 12°36′N 124°02′E. Fishing; agriculture (abaca, coconuts, rice).

Matoaka (muh-TO-kuh), town (2006 population 304), MERCER county, S WEST VIRGINIA, 9 mi/14.5 km WNW of PRINCETON; 37°25′N 81°14′W. Coal-mining and lumbering area.

Matobo National Park, MATABELELAND SOUTH province, SW ZIMBABWE, 30 mi/48 km S of BULAWAYO, in Matopos Hills. Rock formations; bushmen cave paintings at Nswaguti, MAMBATA, and Pomongwe caves; burial place of Cecil J. Rhodes at MALINDIDZIMU HILL (also called World's View). Wildlife includes vervet monkey, baboon, white rhino, zebra, wildebeest, impala, and sable. Also called Matopos National Park; formerly Rhodes-Matopos National Park.

Matochkin Shar (MUH-tuhch-kyin SHUHR), strait of ARCTIC OCEAN, 60 mi/97 km long, joining BARENTS and KARA seas; separates N and S islands of NOVAYA ZEMLYA, ARCHANGEL oblast, extreme N European RUSSIA; 1 mi/1.6 km–2 mi/3.2 km wide; 73°15′N 55°00′E. Government observation stations at Cape Stolbovoi (W entrance), at Matochkin Shar (airfield), and at Cape Vykhodnoi (E entrance).

Matões (mah-TOINS), city (2007 population 28,208), E central MARANHÃO state, BRAZIL, 34 mi/55 km SW of TIMON, near PIAUÍ state border; 05°30′S 43°13′W.

Matões, Brazil: see PARNARAMA.

Matogoro Mountain (ma-to-GO-ro), RUVUMA region, S TANZANIA, 5 mi/8 km NW of SONGEA; 10°45′S 35°38′E. Elevation c. 4,921 ft/1,500 km.

Mato Grosso (MAH-to GRO-so) [Portuguese=thick forest], state (□ 562,220 sq mi/1,461,772 sq km; 2007 population 2,854,456), central and W BRAZIL; ⊙ CUIABÁ. Bounded W by BOLIVIA and RONDÔNIA state, S by Mato Grosso do Sul state, E by GOIAS and

TOCANTINS states, and N by AMAZONAS and PARÁ state; ARAGUAIA River forms E border with Goiás. The great wetland of the Mato Grosso PANTANAL lies in the SW along the border with Bolivia. Much of the state lies on the central Brazilian plateau, which extends S into PARAGUAY border (SERRA DE AMAMBAÍ), and W toward Bolivia (SERRA AZUL, SERRA DOS PARECIS) forming major drainage divide between the AMAZON (N) and PARAGUAY (S) basins. State is well drained by Amazon tributaries in N (JURUENA, ARINOS, SÃO MANUEL, XINGU, Araguaia rivers), and by the CUIABÁ River and other tributaries of the Paraguay. Situated in tropical savanna climate zone which merges with the Amazon rain forest in N. Average temperature, 77°F/25°C; rainfall (60 in/152 cm–70 in/178 cm; less in S) heaviest November–April. Livestock raising is important as well as recent agricultural developments on the Plateaus (Chapadas) (sugarcane, soybeans). Ships rubber, medicinal plants, and lumber from N regions. Large, unexploited mineral resources; gold and diamonds (found here in 17th century) still washed in Cuiabá, CÁCERES, DIAMANTINO, and GARÇAS River areas. First explored by prospectors from SÃO PAULO. Capital was moved (1820) from VILA BELA DA SANTISSIMA TRINIDADE to Cuiabá. Mato Grosso became a province of Brazilian Empire in 1822, and a state of federal republic in 1889. Invaded by the Paraguayan army in 1860s. N part of state prospered during early-20th-century rubber boom after construction of Madeira-Mamoré railroad. State was split in 1977 with extreme S portion (S of Rio Cuiabá and Rio Correntes) becoming the modern state of Mato Grosso do Sul. Also spelled Matto Grosso.

Matola (mah-TO-luh), village (2004 population 520,500), S MOZAMBIQUE, on Umbeluzi River, and 7 mi/11.3 km W of MAPUTO city; 25°58′S 32°28′E. Meatpacking, cement manufacturing.

Matomb (mah-TOOM), town, Central province, CAMEROON, 32 mi/51 km W of Yaoundé; 03°49′N 11°06′E.

Matope (mah-to-pe), village, Southern region, S MALAWI, on SHIRE RIVER, and 20 mi/32 km W of ZOMBA. Corn, rice. In 19th century, end of Shire River navigation from Lake Malawi (LAKE NYASA).

Matopo Hills, mountain range, MATABELELAND SOUTH province, SW ZIMBABWE, c.30 mi/48 km S of BULAWAYO, at S edge of the central plateau, or high veld; 50 mi/80 km long (E-W), 20 mi/32 km wide. Rising to 5,091 ft/1,552 m. At MALINDIDZIMU HILL (World's View), on N slopes, is tomb of Cecil J. Rhodes.

Matopos, town, MATABELELAND SOUTH province, SW ZIMBABWE, 20 mi/32 km SSW of Bulawayo, in Matopos Hills; 20°25′S 28°29′E. Matapos Reservoir is 4 mi/6.4 km to NE; MATOBO NATIONAL PARK to S. Tomb of Cecil J. Rhodes 5 mi/8 km to SSE. Agriculture (peanuts, corn, soybeans); cattle, sheep, goats. Tourism.

Matos Costa (MAH-tos KO-stah), town (2007 population 2,818), N central SANTA CATARINA state, BRAZIL, 16 mi/26 km S of PÔRTO UNIÃO; 26°27′S 51°09′W. Wheat, corn; livestock.

Matou (MAH-to), town, W central TAIWAN, 13 mi/21 km N of TAINAN; 23°11′N 120°14′E. Sugar-milling center; rice, corn, vegetables. Also spelled Matow.

Matouba (mah-too-BAH), thermal springs, S BASSE-TERRE island, GUADELOUPE, near SAINT-CLAUDE, 2 mi/3.2 km NNE of BASSE-TERRE.

Matoury (mah-too-REE), town (□ 17.8 sq mi/46.1 sq km; 1990 population 10,157), N FRENCH GUIANA, 5 mi/8 km S of CAYENNE; 04°51′N 52°20′W. Cacao, sugar, tropical fruit.

Mato Verde (MAH-to VER-zhe), city (2007 population 12,642), NE MINAS GERAIS state, BRAZIL, 20 mi/32 km S of MONTE AZUL; 15°30′S 42°57′W.

Matow, TAIWAN: see MATOU.

Matozinhos (maht-too-ZEEN-yoosh), city, PÔRTO district, N PORTUGAL, at mouth of small Leça River, on the ATLANTIC OCEAN, and 5 mi/8 km NW of OPORTO. Fishing port. With adjoining LEÇA DA PALMEIRA (NW) it encloses artificial harbor of LEIXÕES; fish preserving, sugar refining, rope manufacturing, other light industry.

Matpalli, India: see METPALLI.

Matrafured (MAHT-rah-foo-red), Hungarian *Mátrafüred*, village, HEVES county, N HUNGARY, in MÁTRA MOUNTAINS, 4 mi/6 km N of Gyöngyös. Health resort; mineral springs.

Matrah (MUHT-rah), town (2003 population 153,526), W of MUSCAT, OMAN; 22°47′N 57°26′E. Chief commercial center of the sultanate. In the past a starting point for caravans to the interior; has trade in dates, dried limes, fresh fruit, and salted fish. Shipbuilding and repair. Population is Arab, Indian, Baluch, and African.

Mátra Mountains (MAHT-rah), Hungarian *Mátra*, N HUNGARY, in S innermost range of the CARPATHIANS; extend 25 mi/40 km between ZAGYVA and TARNA rivers, rise to 3,330 ft/1,015 m in Mount KÉKES; 47°53′N 19°57′E. Forested slopes; lignite, limestone, basalt, trachyte deposits in S foothills. Wheat, sunflowers, peaches; hogs; vineyards. Gyöngyös at S foot.

Matrei am Brenner (MAH-trei ahm BREN-ner), township, TYROL, W AUSTRIA, in the WIPPTAL, on SILL RIVER, 10 mi/16 km S of INNSBRUCK; 47°08′N 11°27′E. Elevation 3,024 ft/922 m. Hydropower station (Brennerwerk); electrotechnical industry.

Matrei in Osttirol (MAH-trei in AWST-ti-rol), township, EAST TYROL, S AUSTRIA, near ISEL RIVER, 16 mi/26 km NW of LIENZ; 47°00′N 12°32′E. Tourist center. Formerly called Windisch-Matrei. National Park Hohe Tauern is nearby.

Mat River (MAHT), c.60 mi/97 km long, N central ALBANIA; rises SE of KLOS; flows NW and W, past BURREL, to DRIN GULF of the ADRIATIC SEA, 9 mi/14.5 km SSW of LEZHË. Receives FAN RIVER (right). Also spelled Mati River

Matriz de Camarajibe (MAH-trees dee KA-mah-rah-zhee-be), city, E ALAGOAS state, NE BRAZIL, 34 mi/55 km NE of MACEIÓ; 09°10′S 35°31′W. In sugar-growing district.

Matroosberg, mountain (7,376 ft/2,249 m), WESTERN CAPE province, SOUTH AFRICA, highest peak of HEX RIVER MOUNTAINS, 20 mi/32 km E of CERES; 33°23′S 19°38′E.

Matrouh, province (□ 81,897 sq mi/212,932.2 sq km; 2004 population 262,210), NW EGYPT, on the MEDITERRANEAN SEA (N), ⊙ MATRUH (or Mersa Matruh); 30°00′N 27°00′E. Borders ALEXANDRIA and BEHEIRA provinces (NE); Al GIZA province (SE); and El-Wadi province, El-Gadid province, and LIBYA (W). Largely part of the WESTERN DESERT, it occupies a plateau which compromises most of NW Egypt. The Qattarah Depression is near the center. Some oil and gas fields exist. Most of the population is concentrated near the coast or near a few small oases. SIWAH Oasis is the site of the Temple of Amon. Also spelled Matruh.

Matru (MAH-troo), town, SOUTHERN province, SW SIERRA LEONE, on JONG RIVER, on road, and 38 mi/61 km SW of BO; 07°35′N 12°13′W. Also spelled MATTRU and also known as Mattru Jong.

Matruh (maht-ROO), town, Matruh province, NW EGYPT, near the MEDITERRANEAN SEA. Built at the site of the Roman town of Paraetonium, it is located on the coast road and is on a railroad that links small ports. There are oil and natural-gas fields in Matruh's outlying region. During World War II several battles between Axis and British forces were fought nearby. Naval base. Major tourist resorts; 199-mi/320-km highway runs SW to SIWA Oasis in the Libyan (WESTERN) DESERT. Also spelled Mersa Matruh.

Matsalu Nature Reserve (MUHT-suh-loo) (□ 31 sq mi/80.6 sq km), on W coast of ESTONIA, 50 mi/80 km SE of TALLINN. Bogs, reed beds, and meadows; over 100,000 species of birds.

Matsapa (ma-TSAH-pa), town, MANZINI district, central SWAZILAND, 16 mi/26 km SE of MBABANE, and 4 mi/6.4 km W of MANZINI, near GREAT USUTU RIVER, on railroad; 26°32′S 31°17′E. University of Swaziland, at Kwalusini, 2 mi/3.2 km to N; Matsapa Airport, Swaziland's main international airport to W. Matsapa Industrial Estate, on railroad spur, is here. Manufacturing (refrigerators, textiles, furniture, office equipment, soft drink concentrates, footwear). Cattle, goats, sheep, hogs; corn, vegetables, citrus. Also spelled Matsapha.

Matsena (mah-CHAI-nah), town, BORNO state, extreme N NIGERIA, near NIGER border, 35 mi/56 km NW of NGURU; 13°08′N 10°03′E. Peanuts, cotton, millet; cattle, skins. Also Machena.

Matsesta (mah-TSE-stah), resort settlement (2004 population 9,830), S KRASNODAR TERRITORY, in the foothills of the NW CAUCASUS, SE European Russia, near the BLACK SEA coast, 3 mi/5 km SE of SOCHI; 43°34′N 39°48′E. Elevation 626 ft/190 m. Subtropical health resort amid orchards and vineyards. Consists of 2 sections: Novaya [=new] Matsesta, a beach resort on the coastal railroad, and Staraya [=old] Matsesta, on railroad spur 2 mi/3.2 km inland, with warm sulphur springs.

Matseyevo, UKRAINE: see LUKIV.

Matseyiv, UKRAINE: see LUKIV.

Matsiatra River, MADAGASCAR: see MANGOKY RIVER.

Matsieng (mah-TSEE-yeng), town, MASERU district, W LESOTHO, 20 mi/32 km SSE of MASERU and 7 mi/11 km E of MORIJA; 29°37′S 27°34′E. Moshoeshoe I International Airport 9 mi/14.5 km to N. Corn, sorghum, vegetables; cattle, sheep, goats, hogs. Seat of King of Lesotho, Moshoeshoe II.

Matsqui (MAT-skwee), former village, SW BRITISH COLUMBIA, W CANADA, near FRASER River, 16 mi/26 km WSW of CHILLIWACK; 49°07′N 122°16′W. Lumbering; fruit, hops, tobacco. Merged with town of ABBOTSFORD in 1995 to form new city of Abbotsford.

Matsu (MAHD-zoo), group of twenty-eight islands, in the EAST CHINA SEA, off FUJIAN province, CHINA, E of FUZHOU, and c.100 mi/161 km W of TAIWAN; 26°10′N 119°59′E. The major islands include Pei-kan and Nankan (Mazu). Fishing is the main economic activity. The islands remained a Chinese Nationalist-held outpost after the Communist takeover of the mainland in 1949. The Nationalists have built extensive military facilities and underground shelters. The People's Republic of China began bombarding the islands again in 1958, but the deployment of the U.S. Seventh Fleet prevented an escalation of the hostilities. The name Matsu was derived from Mother Ancestor, a legend about a girl who drowned in the ocean when searching for her father and later was seen as a goddess to bless the fishermen. Sometimes appears as Ma-tsu or Mazu.

Ma-tsu, TAIWAN: see MATSU.

Matsubara (mah-TSOO-bah-rah), city, OSAKA prefecture, S HONSHU, JAPAN, 6 mi/10 km S of OSAKA; 34°34′N 135°33′E. Industrial and residential suburb of Osaka known for the production of cultured and artificial pearls. Also ivory (for personal and business stamps) processing. Ancient tomb at Otsukayama (one of Japan's largest).

Matsubase (mah-TSOO-bah-SE), town, Shimomashiki county, KUMAMOTO prefecture, W KYUSHU, SW JAPAN, 9 mi/15 km S of KUMAMOTO; 32°38′N 130°40′E.

Matsubushi (mah-TSOO-boo-shee), town, North Katsushika county, SAITAMA prefecture, E central HONSHU, E central JAPAN, 9 mi/15 km N of URAWA; 35°55′N 139°49′E.

Matsuda (mah-TSOO-dah), town, Ashigarakami county, KANAGAWA prefecture, E central HONSHU, E central JAPAN, 31 mi/50 km S of YOKOHAMA; 35°20′N 139°08′E.

Matsudai (mah-TSOO-dah-ee), town, East Kubiki county, NIIGATA prefecture, central HONSHU, N central JAPAN, 59 mi/95 km S of NIIGATA; 37°07′N 138°36′E. Noodles; rice.

Matsudo (mah-TSOO-do), city (2005 population 472,579), CHIBA prefecture, E central HONSHU, E central JAPAN, 12 mi/20 km N of CHIBA; 35°47′N 139°54′E. Spring onions. Motors. Rice flour.

Matsue (MAHTS-e), city (2005 population 196,603), ⊙ SHIMANE prefecture, SW HONSHU, W JAPAN, a port on the SEA OF JAPAN; 35°27′N 133°03′E. Important distribution center and a popular tourist spot. Broccoli, pears; dumplings. Landmarks include 17th-century Matsue Castle and a museum containing a collection of the manuscripts and letters of Lafcadio Hearn, an Irish-American writer and educator who lived here in the latter part of the 19th century.

Matsuida (mah-TSOO-ee-dah), town, Usui county, GUMMA prefecture, central HONSHU, N central JAPAN, near Usui mountain pass, 19 mi/30 km W of MAEBASHI; 36°18′N 138°47′E. Known for *kamemeshi*, a portable three-tiered dish in which food is cooked.

Matsu Island, Taiwan: see MATSU.

Matsukari (mahts-KAH-ree), village, Shiribeshi district, Hokkaido prefecture, N JAPAN, 34 mi/55 km S of SAPPORO; 42°45′N 140°48′E.

Matsukawa (mahts-KAH-wah), town, Shimoina county, NAGANO prefecture, central HONSHU, central JAPAN, 74 mi/120 km S of NAGANO; 35°35′N 137°54′E. Pears.

Matsukawa (mahts-KAH-wah), village, N Azumi county, NAGANO prefecture, central HONSHU, central JAPAN, 25 mi/40 km S of NAGANO; 36°25′N 137°51′E. Eggplant; rice (for sake).

Matsumae (mah-TSOO-mah-e), town, Oshima county, extreme SW HOKKAIDO, N JAPAN, on TSUGARU STRAIT, 130 mi/210 km S of SAPPORO; 41°25′N 140°06′E. Seat of powerful Matsumae family in feudal times. Ruins of Fukuyama Castle. First named Matsumae, later, until the early 1940s, called Fukuyama. Sometimes called Matsumai.

Matsumoto (mah-TSOO-mo-to), city (2005 population 227,627), NAGANO prefecture, central HONSHU, central JAPAN, 31 mi/50 km S of NAGANO; 36°14′N 137°58′E. Vegetables, grapes. Manufacturing includes computer components, electronic equipment, guitars; also rustic traditional furniture. Shiroyama Park has remains of a 16th century Matsumoto Castle. Utsukushigahara Highlands nearby.

Matsumoto (mah-TSOO-mo-to), town, Hioki county, KAGOSHIMA prefecture, SW KYUSHU, SW JAPAN, 6 mi/10 km W of KAGOSHIMA; 31°36′N 130°26′E.

Matsuno (MAHTS-no), town, N Uwa county, EHIME prefecture, NW SHIKOKU, W JAPAN, 43 mi/70 km S of MATSUYAMA; 33°13′N 132°42′E. Pickles; eel.

Matsunoyama (mah-TSOO-no-YAH-mah), town, East Kubiki county, NIIGATA prefecture, central HONSHU, N central JAPAN, 59 mi/95 km S of NIIGATA; 37°05′N 138°36′E. Traditional dolls. Hot springs nearby.

Matsuo (mah-TSOO-o), town, Sanbu county, CHIBA prefecture, E central HONSHU, E central JAPAN, on BOSO PENINSULA, 12 mi/20 km E of CHIBA; 35°38′N 140°27′E.

Matsuo (mah-TSOO-o), village, Iwate county, IWATE prefecture, N HONSHU, NE JAPAN, 19 mi/30 km N of MORIOKA; 39°57′N 141°04′E.

Matsuoka (mah-TSOO-o-kah), town, Yoshida county, FUKUI prefecture, central HONSHU, W central JAPAN, 5.6 mi/9 km N of FUKUI; 36°05′N 136°18′E. Ayu; onions.

Matsusaka (mahts-SAH-kah), city (2005 population 168,973), MIE prefecture, S HONSHU, central JAPAN, port on W shore of ISE BAY, 9 mi/15 km S of TSU; 34°34′N 136°31′E. Glass; cotton; nori; Matsusaka beef cattle. Motoori Norinaga, scholar and writer, born here in 1730.

Matsushige (mah-TSOO-shee-ge), town, Itano county, TOKUSHIMA prefecture, SE SHIKOKU, W JAPAN, 5 mi/8 km N of TOKUSHIMA; 34°07′N 134°34′E. Lotus root, potatoes, pears. Wood products (Buddhist altars, furniture).

Matsushima (mah-TSOO-shee-mah), town, Amakusa county, on NE corner of AMAKUSA KAMI-SHIMA island, KUMAMOTO prefecture, W KYUSHU, SW JAPAN, 25 mi/40 km S of KUMAMOTO; 32°30′N 130°25′E. Prawns.

Matsushima, town, Miyagi county, MIYAGI prefecture, N HONSHU, NE JAPAN, on ISHINOMAKI BAY, 12 mi/20 km N of SENDAI; 34°34′N 136°31′E. Oysters, clams; nori. Confections. One of the three most famous scenic areas in Japan. Tourist center for the hundreds of scenic pine-covered islets in the bay. One island is the site of the noted Buddhist temple of Zuiganji (founded 828).

Matsuto (mahts-TO), city, ISHIKAWA prefecture, central HONSHU, central JAPAN, 5.6 mi/9 km S of KANAZAWA; 36°31′N 136°34′E. Traditional *taiko* drums, pottery. Sushi trays with tea servers.

Matsuura (mah-TSOO-oo-rah), city, NAGASAKI prefecture, NW KYUSHU, SW JAPAN, 43 mi/70 km N of NAGASAKI; 33°20′N 129°42′E. Yellowtail. Thermal power station.

Matsuwa-kaikyo, RUSSIA: see GOLOVNIN STRAIT.

Matsuwa-to, RUSSIA: see MATUA ISLAND.

Matsuyama (mah-TSOO-yah-mah), city (2005 population 514,937), ⊙ EHIME prefecture, NW SHIKOKU, W JAPAN, a port on the INLAND SEA; 33°50′N 132°46′E. Important agricultural distribution point and fishing port. Cotton textiles (decorated with splashed pattern), agricultural machinery, synthetic fibers, and bamboo work are produced, and mandarin oranges are cultivated. Matsuyama Castle (built 1603), one of the best preserved in Japan, stands in a magnificent park. Hot springs and scenic Okudogo are nearby.

Matsuyama (mah-TSOO-yah-mah), town, Soo county, KAGOSHIMA prefecture, SW KYUSHU, SW JAPAN, 28 mi/45 km E of KAGOSHIMA; 31°34′N 131°02′E.

Matsuyama, town, Shida county, MIYAGI prefecture, N HONSHU, NE JAPAN, 19 mi/30 km N of SENDAI; 38°30′N 141°03′E.

Matsuyama, town, Akumi county, YAMAGATA prefecture, N HONSHU, NE JAPAN, 50 mi/80 km N of YAMAGATA city; 38°51′N 139°57′E.

Matsuzaki (mah-TSOO-zah-kee), town, Kamo county, SHIZUOKA prefecture, central HONSHU, central JAPAN, on W IZU PENINSULA, on SURUGA BAY, 28 mi/45 km S of SHIZUOKA; 34°44′N 138°46′E. Cherry leaves; silk cocoons. Shiitake mushroom production originated here.

Mattabesset River (MA-te-BAI-sit), c.12 mi/19 km long, central CONNECTICUT; rises SE of NEW BRITAIN; flows generally SE to the CONNECTICUT River just above MIDDLETOWN.

Matta de São João, Brazil: see MATA DE SÃO JOÃO.

Mattagami (muh-TA-guh-mee), river, 275 mi/443 km long; rises in the lake district, E ONTARIO, E central CANADA, SW of TIMMINS; flows N to join the MISSINAIBI River, with which it forms the Moose River; 50°43′N 81°29′W.

Mattagami Lake (muh-TA-guh-mee), (□ 88 sq mi/228.8 sq km), W QUEBEC, E CANADA, 120 mi/193 km N of VAL-D'OR; 26 mi/42 km long, 10 mi/16 km wide. Elevation 765 ft/233 m. Drained NW by NOTTAWAY River.

Matta Grande, Brazil: see MATA GRANDE.

Mattakkuliya, town, WESTERN PROVINCE, SRI LANKA. Leather goods factory.

Mattakuliya (MUHT-tahk-KOO-li-yah), section of COLOMBO, WESTERN PROVINCE, SRI LANKA, 3.5 mi/5.6 km NE of city center. Also spelled MATTAKKULIYA.

Mattamuskeet, Lake (ma-tuh-muh-SKEET), E NORTH CAROLINA, in central HYDE county, N and NW of PAMLICO SOUND; c.15 mi/24 km long E-W, 6 mi/9.7 km wide; 35°29′N 76°10′W. State Highway 94 causeway crosses center of lake. Entirely within Mattamuskeet National Wildlife Refuge. INTRACOASTAL WATERWAY (ALLIGATOR-Pungo River Canal) passes to N.

Mattamuskeet National Wildlife Refuge, (□ 78 sq mi/202.8 sq km), HYDE county, E NORTH CAROLINA, 9 mi/14.5 km NE of SWAN QUARTER; 35°28′N 76°17′W. Open water, freshwater marsh, wetlands, croplands, forested uplands habitats. Significant wintering bird populations, fishery resources. Main feature is Lake MATTAMUSKEET (62.5 sq mi/161.9 sq km), North Carolina's largest lake. Established 1934.

Mattancheri, India: see COCHIN.

Mattão, Brazil: see MATÃO.

Mattapan, MASSACHUSETTS: see BOSTON.

Mattapoisett (ma-duh-POI-set), town, PLYMOUTH county, SE MASSACHUSETTS, on W shore of BUZZARDS BAY, 6 mi/9.7 km E of NEW BEDFORD; 41°40′N 70°49′W. Resort; manufacturing (boats, apparel); good harbor. Formerly had shipbuilding and whaling. Includes villages of Antasawamock Neck and East Mattapoisett. Settled 1750, incorporated 1857.

Mattaponi River (ma-tah-po-NEI), 120 mi/193 km long, E VIRGINIA; formed in NW CAROLINE county by joining of Matta and Poni rivers; flows SE, past BOWLING GREEN, joins PAMUNKEY RIVER at WEST POINT to form YORK RIVER estuary, arm of CHESAPEAKE BAY; 38°06′N 77°26′W. Navigable for c.40 mi/64 km above mouth; chief cargoes are wood products.

Mattawa (MAH-tuh-wo), town (□ 2 sq mi/5.2 sq km; 2001 population 2,270), NE ONTARIO, E central CANADA, on OTTAWA River at mouth of MATTAWA RIVER, and 36 mi/58 km E of NORTH BAY; 46°19′N 78°42′W. Lumbering, plywood manufacturing, mica mining.

Mattawa (MAT-uh-wah), town (2006 population 3,189), GRANT county, E central WASHINGTON, 29 mi/47 km ENE of YAKIMA, near COLUMBIA River (PRIEST RAPIDS LAKE reservoir); 46°45′N 119°54′W. Irrigated agricultural area (vegetables, fruit, sugar beets). Priest Rapids (to S) and WANAPUM (to N) dams. U.S. Military Reservation–U.S. Dept. of Energy Hanford Site to E and SE.

Mattawamkeag (mat-uh-WAHM-keg), town, PENOBSCOT county, E central MAINE, on the PENOBSCOT RIVER, at mouth of the MATTAWAMKEAG RIVER, and c.50 mi/80 km NNE of BANGOR; 45°32′N 68°18′W. Manufacturing (wood products). Hunting, fishing.

Mattawamkeag Lake, AROOSTOOK county, E central MAINE, 18 mi/29 km SW of HOULTON; 7 mi/11.3 km long. In lumbering, recreational area. A source of MATTAWAMKEAG RIVER.

Mattawamkeag River, c.70 mi/113 km long, E central MAINE; rises in two branches in S AROOSTOOK county; flows generally S and SW to the PENOBSCOT RIVER at MATTAWAMKEAG.

Mattawan (MAH-tuh-won), township (□ 77 sq mi/200.2 sq km; 2001 population 114), NE ONTARIO, E central CANADA, 30 mi/48 km from NORTH BAY; 46°21′N 78°49′W.

Mattawa River (MAH-tuh-wo), 45 mi/72 km long, central ONTARIO, E central CANADA; issues from TROUT LAKE (8 mi/13 km long), 3 mi/5 km E of NORTH BAY; flows E to OTTAWA River at MATTAWA; 46°19′N 78°42′W.

Mattawin, CANADA: see MATAWIN RIVER.

Mattawoman Creek (ma-tah-WO-muhn), c.35 mi/56 km long, S MARYLAND; rises just SE of Brandywine; flows generally W through swampland, forming part of border between CHARLES and PRINCE GEORGES counties, to the POTOMAC River c.4 mi/6.4 km below INDIAN HEAD.

Matteawan, NEW YORK: see BEACON.

Matterhorn (MA-tuhr-horn), French *Mont Cervin* (mon ser-VA), Italian *Monte Cervino* (MON-te ser-VEE-no), peak (14,692 ft/4,478 m) in the PENNINE ALPS, on the Swiss-Italian border, near ZERMATT; 45°59′N 07°39′E. Its distinctive pyramidal peak was formed by the enlargement of several cirques. Edward Whymper, the English mountaineer, scaled it first (1865). The nearby Matterjoch (MAH-tuhr-yokh) or

Area is shown by the symbol □, and capital city or county seat by ⊙.

Théodule (tai-o-DOOL) is a pass (10,800 ft/3,292 m) that links ITALY with SWITZERLAND.

Matterhorn, Klein, SWITZERLAND: see KLEIN MATTER-HORN.

Mattersburg (MAHT-ters-boorg), town, N BURGENLAND, E AUSTRIA, 9 mi/14.5 km SE of WIENER NEUSTADT; 47°44′N 16°24′E. Market center; important road junction. Strawberries raised in vicinity. Called Mattersdorf until 1924.

Mattertal, SWITZERLAND: see VISPA RIVER.

Matteseunk Lake (MA-tuh-suhnk), AROOSTOOK county, E MAINE, 17 mi/27 km ESE of MILLINOCKET; 3 mi/4.8 km long.

Matteson (MAT-ih-sun), village (2000 population 12,928), COOK county, NE ILLINOIS, S commuter suburb of CHICAGO; 41°30′N 87°44′W. Lincoln Mall is here.

Matthew Island, SW PACIFIC OCEAN, S of ANATOM island, VANUATU. Uninhabited outcrop claimed by Vanuatu and by FRANCE on behalf of NEW CALEDONIA.

Matthews (MA-thyooz), city (□ 14 sq mi/36.4 sq km; 2006 population 26,296), MECKLENBURG county, S NORTH CAROLINA, a suburb 10 mi/16 km SE of CHARLOTTE; 35°07′N 80°42′W. Service industries; manufacturing (machinery, plastic and paper products, lab equipment, building materials, safety valves, textiles, consumer goods; metal fabricating). Agriculture to E and S (grain; livestock). Originally called Stumptown, then Fullwood. Incorporated 1879.

Matthews, town, JEFFERSON county, E GEORGIA, 25 mi/40 km SW of AUGUSTA; 33°12′N 82°18′W.

Matthews, town (2000 population 595), GRANT county, E central INDIANA, 14 mi/23 km SSE of MARION. In agricultural area; electrical equipment.

Matthews, town (2000 population 605), NEW MADRID county, extreme SE MISSOURI, 12 mi/19 km NNW of NEW MADRID; 36°45′N 89°34′W. Soybeans, cotton, corn.

Matthews, mountain range (peaks at 7,792 ft/2,375 m), RIFT VALLEY province, KENYA, 81 mi/130 km N of Mount KENYA.

Matthews Peak (9,512 ft/2,899 m), APACHE county, NE ARIZONA, in CHUSKA MOUNTAINS, near NEW MEXICO state line, c.55 mi/89 km SW of FARMINGTON (New Mexico).

Matthews Ridge, village, BARIMA-WAINI district, GUYANA; 06°25′N 58°38′W. Near site of Jonestown mass suicide. Area has manganese deposits.

Matthew Town, minor port in S BAHAMAS, on W tip of GREAT INAGUA ISLAND, 350 mi/563 km SE of NASSAU, 55 mi/89 km NNE of Cape Maisí (E CUBA); 20°57′N 73°40′W. Salt (company headquarters for Bahamas).

Mattice-Val Côté (MA-tis–VAL ko-TAI) township (□ 160 sq mi/416 sq km; 2001 population 891), central ONTARIO, E central CANADA; 49°38′N 83°17′W.

Mattig (MAHT-tik), river, 20 mi/32 km long, in N SALZBURG and SW UPPER AUSTRIA, W central AUSTRIA, drains Grabensee, Mattsee and Obertrumer See in the FLACHGAU region; flows N past MATTIGHOFEN to INN RIVER, 5 mi/8 km NE of BRAUNAU.

Mattighofen (maht-tig-HAW-fen), town UPPER AUSTRIA, in the INNVIERTEL region, 13 mi/21 km SSE of BRAUNAU; 48°06′N 16°09′E. Market center; manufacturing of motorcycles, household utensils. Gas fields nearby. Well-preserved historic burgher houses.

Mattituck (MA-ti-tuk), resort village (□ 10 sq mi/26 sq km; 2000 population 4,198), SUFFOLK CO., SE NEW YORK, on NE LONG ISLAND, on inlet of LONG ISLAND SOUND, 8 mi/12.9 km NE of RIVERHEAD; 41°00′N 72°32′W. Marine industries, berry growing, vinyards. Popular for summer recreational activities.

Mattole River, c.50 mi/80 km long, NW CALIFORNIA; rises in NW MENDOCINO county, 3 mi/4.8 km inland from PACIFIC OCEAN, 10 mi/16 km NW of Leggett; flows NW to the Pacific c.35 mi/56 km SW of Eureka, N of Punta Gorda cape. King Mountain Range separates river from ocean.

Mattoon (mat-TOON), city (2000 population 18,291), COLES county, E central ILLINOIS, 11 mi/18 km W of CHARLESTON; 39°28′N 88°22′W. Processing, railroad, and industrial center for a farming region. Agriculture (corn, soybeans, sorghum). Manufacturing (paper products, electrical equipment, machinery, consumer goods, fire-prevention equipment, signs, building materials; vegetable processing). Nearby are many oil wells, a fish hatchery, and Lake Mattoon (6 mi/9.7 km SW). The farm and grave of Abraham Lincoln's father and stepmother are SE of the city. Lake Land College 5 mi/8 km S. Incorporated 1859.

Mattoon (mah-TOON), village (2006 population 433), SHAWANO county, E central WISCONSIN, 29 mi/47 km ENE of WAUSAU; 45°00′N 89°02′W. Lumbering; manufacturing (wood and wood-related products). Menominee Indian Reservation (all of MENOMINEE county) to E.

Mattru, SIERRA LEONE: see MATRU.

Mattsee (MAHT-sai), township, N SALZBURG, W central AUSTRIA, in the FLACHGAU region, on Lake Mattsee, and near Obertrumer See, 11 mi/18 km NNE of SALZBURG; 47°58′N 13°06′E. Elevation 1,533 ft/467 m. Summer tourism; mud baths. Manufacturing of shoes, boats, pipes. Abbey, founded in the 8th century with remarkable church; castle. Center of a lake district. Benedictine abbey Michaelbeuern (4 mi/6.4 km W), founded around 785, with a romanesque basilica and a library containing Walther-Bibel (second quarter of the 12th century).

Matty Island, NUNAVUT territory, N CANADA, in JAMES ROSS STRAIT, between BOOTHIA PENINSULA (NE) and KING WILLIAM ISLAND (SW); 20 mi/32 km long, 15 mi/24 km wide; 69°30′N 95°30′W.

Matua Island (mah-TOO-ah), Japanese *Matsuwa-to* (□ 20 sq mi/52 sq km), one of central main KURIL ISLANDS group, SAKHALIN oblast, extreme E SIBERIA, RUSSIAN FAR EAST; separated from RAIKOKE ISLAND (N) by GOLOVNIN STRAIT, from RASSHUA ISLAND (S) by NADEZHDA STRAIT; 7 mi/11.3 km long, 4 mi/6.4 km wide; 48°05′N 153°13′E. Sarychev Peak, Japanese *Fuyo-yama* is here. Settlement on the SE coast; seal reserve; fisheries; farming.

Matua Strait, INDONESIA: see OMBAI STRAIT.

Matucana (mah-too-KAH-nah), city, ⊙ HUAROCHIRÍ province, LIMA region, W central Peru, in CORDILLERA OCCIDENTAL of the ANDES Mountains, on RÍMAC RIVER, on LIMA–LA OROYA railroad and highway, and 45 mi/72 km ENE of Lima; 11°51′S 76°24′W. Elevation 7,792 ft/2,375 m. Alfalfa, vegetables, potatoes, cereals; livestock.

Matuku (mah-TOO-koo), crescent-shaped volcanic island (□ 11 sq mi/28.6 sq km), LAU group, FIJI, SW PACIFIC OCEAN; 4 mi/6.4 km long; 19°10′S 179°46′E. Copra.

Matun, AFGHANISTAN: see KHOST.

Matunuck (ma-TOO-nuhck), summer resort village, in SOUTH KINGSTOWN, WASHINGTON county, RHODE ISLAND, on coast of BLOCK ISLAND SOUND. Fishing; many public beaches.

Matupá (mah-too-PAH), city (2007 population 12,928), MATO GROSSO state, BRAZIL.

Maturá, Brazil: see AMATURÁ.

Matureivavao, TUAMOTU ARCHIPELAGO: see ACTAEON ISLANDS.

Maturín (mah-too-REEN), city, ⊙ MONAGAS state, NE VENEZUELA, on GUARAPICHE RIVER, and 85 mi/137 km SE of CUMANÁ (SUCRE state), 250 mi/402 km ESE of CARACAS; 09°45′N 63°11′W. Trading center in agricultural region (cacao, cotton, tobacco, cereals; livestock); manufacturing. Petroleum fields nearby (N and W). Founded 1710 by Capuchin missionaries. Birthplace of the brothers José Tadeo and José Gregorio Monagas, both of whom became presidents of Venezuela. Airport.

Matusadona National Park (□ 543 sq mi/1,406 sq km), MASHONALAND WEST province, N ZIMBABWE, 20 mi/

32 km SW of KARIBA, on S shore of Lake KARIBA reservoir (ZAMBEZI River), between SANYATI (E) and UME (W) rivers. Protects Matsuadona Range (highest point 3,949 ft/1,204 m), part of ZAMBEZI ESCARPMENT. Bush camping; park largely undeveloped to protect wildlife; access by water or rough road.

Matutuine District, MOZAMBIQUE: see MAPUTO.

Matveyev Island (maht-VYE-eef) (□ 4 sq mi/10.4 sq km), in SE BARENTS SEA, part of ARCHANGEL oblast, N European Russia, 45 mi/72 km WSW of KHABAROVO; 69°28′N 58°29′E.

Matveyevka (mah-TVYE-eef-kah), village (2006 population 3,270), NW ORENBURG oblast, in the SW foothills of the S URALS, SE European Russia, on road, 40 mi/64 km ESE of BUGURUSLAN; 52°09′N 56°11′E. Elevation 623 ft/189 m. In agricultural area (sunflowers, grains, vegetables).

Matveyevka, UKRAINE: see PAVLOHRAD.

Matveyevka, UKRAINE: see MATVIYIVKA, Mykolayiv oblast.

Matveyev Kurgan (maht-VYE-eef koor-GAHN), town (2006 population 14,340), SW ROSTOV oblast, S European Russia, on road and railroad, on the MIUS RIVER (AZOV Sea basin), 42 mi/68 km W of ROSTOV-NA-DONU, and 25 mi/40 km N of TAGANROG; 47°34′N 38°52′E. Elevation 177 ft/53 m. Building materials, asphalt; food industries.

Matviyivka (mah-TVEE-yif-kah) (Russian *Matveyevka*), town, S MYKOLAYIV oblast, UKRAINE, on the left bank of the Southern BUH RIVER, on road and on railroad spur, 6 mi/10 km NW of MYKOLAYIV city center and subordinated to its city council; 47°01′N 31°55′E. Silicate products manufacturing, fish processing. Known since 1778, town since 1987.

Matviyivka, UKRAINE: see PAVLOHRAD.

Matyra River (mah-TI-rah), approximately 89 mi/143 km long, in TAMBOV and LIPETSK oblasts, S central European Russia. Rises in W central Tambov oblast, approximately 18 mi/29 km WSW of TAMBOV, and flows generally WSW and W, with many twists and turns, past Yablonovets, into SE Lipetsk oblast. Past Annino and for most of the remaining 20 mi/32 km of its course, the river has been artificially expanded to create the Matyrskoye Reservoir. Flows into the VORONEZH RIVER less than 4 mi/6 km NE of LIPETSK. Used for timber shipping. Pollution levels are high because of an active industrial complex along most of its course.

Matyrskoye Reservoir (mah-TIRS-kuh-ye) Russian *Matyrskoye Vodokhranilishche*, artificial lake (surface □ approximately 73 sq mi/189 sq km), E LIPETSK oblast, S central European Russia; 52°35′N 39°46′E. Created by expanding the MATYRA River in its lower course, just before its confluence with the VORONEZH RIVER. Its water is used for steelmaking and other industries along its shores, and industrial pollution is an ongoing concern.

Matzen (MAH-tsen), township, E LOWER AUSTRIA, 4 mi/6.4 km N of Gänserndorf; 48°24′N 16°42′E. Center of gas and oil production in AUSTRIA and the largest oil field in central EUROPE. Castle from 12th century.

Matzuba, ISRAEL: see MATZUVA.

Matzuva (mah-TSOO-vah), kibbutz, N ISRAEL, 5 mi/8 km NE of NAHARIYA, in WESTERN GALILEE; 33°03′N 35°09′E. Farming (orchards, field crops, and poultry); textile and furniture manufacturing. Remains of Byzantine settlement and church. Founded 1940 alongside ancient town, but began to thrive only after 1948 War for Independence.

Matzuva (mah-TSOO-vah), kibbutz, UPPER GALILEE, NW ISRAEL, near MEDITERRANEAN SEA, 19 mi/31 km NNW of HAIFA; 33°03′N 35°09′E. Elevation 357 ft/108 m. Textile industry; dairying; fruit, olives, bananas, mushrooms; poultry. Also spelled Matzuba. Founded 1940.

Mau (MOU), district (□ 661 sq mi/1,718.6 sq km), E Uttar Pradesh state, N central India; ⊙ Mau.

Mau (MOU), town, tahsil headquarters, BANDA district, S UTTAR PRADESH state, N central INDIA, on the YAMUNA RIVER, 31 mi/50 km WSW of ALLAHABAD. Gram, jowar, wheat, oilseeds. Important glass sand deposits at Bargarh (10 mi/16 km SSE) and Panhai (17 mi/27 km SW) villages.

Mau, town, JHANSI district, S UTTAR PRADESH state, N central INDIA, on tributary of the DHASAN RIVER, 37 mi/60 km ESE of JHANSI. Road center; brass ware manufacturing; trades in jowar, oilseeds, wheat, gram, barley, rice. Jain temple. Large annual livestock fair. Projected irrigation dam nearby on tributary of the Dhasan. Also called Mau-Ranipur.

Mau (MOU), town (2001 population 212,657), ⊙ Mau district, E Uttar Pradesh state, N central India, on Tons River, and 24 mi/39 km SE of Azamgarh; 25°57′N 83°33′E. Railroad and road junction; hand-loom cotton-weaving center; silk weaving, rice and flour milling. Has 17th-cent. serai.

Maua (MAH-oo-ah), city, N PARANÁ state, BRAZIL, 55 mi/89 km SE of MARINGÁ, on railroad; 24°03′S 50°43′W. Coffee, cotton, rice, corn; livestock.

Mauá (MAH-oo-AH), city (2007 population 402,643), SE SÃO PAULO state, BRAZIL, outer SE suburb of SÃO PAULO, just SE of SANTO ANDRÉ, on São Paulo–SANTOS railroad; 23°40′S 46°27′W.

Maua (mah-OO-ah), town (1999 population 9,763), ⊙ Nyambrne district, EASTERN province, KENYA, on E side on NYAMBENI range, and 6 mi/10 km NW of Meru National Park; 00°12′N 37°57′E. Market and trading center.

Maua District, MOZAMBIQUE: see NIASSA.

Mau Aimma (MOU EIM-muh), town, ALLAHABAD district, SE UTTAR PRADESH state, N central INDIA, 18 mi/29 km NNE of ALLAHABAD city center; 25°42′N 81°55′E. Hand-loom cotton-weaving center; trades in gram, rice, barley, wheat, sugarcane, cotton.

Mauban (mah-OO-bahn), town, QUEZON province, S LUZON, PHILIPPINES, on LOPEZ BAY, 29 mi/47 km ENE of SAN PABLO; 14°19′N 121°42′E. Fishing and agricultural center (coconuts, rice).

Maubeuge (mo-BUZH), industrial town (□ 7 sq mi/ 18.2 sq km), NORD department, NORD-PAS-DE-CALAIS region, N FRANCE, on the canalized SAMBRE RIVER, near the Belgian border, and 47 mi/76 km ESE of LILLE; 50°17′N 03°58′E. The Belgian town of MONS is 9 mi/14.5 km N, within a continuous border-straddling urban belt. Industrial activities include metalworking (railroad and heating equipment, machine tools), glass and china manufacturing. Founded as an abbey in 7th century, Maubeuge became capital of the HAINAUT province, which lies partly in BELGIUM. Passed to France in 1678 and fortified by the famed military engineer Vauban. Its ancient Jesuit college is a baroque building of interest. In World War I Maubeuge was occupied for a long time by the Germans following a bitter siege. Easily overrun in June 1940, when Germans outflanked the NW extension of France's MAGINOT LINE with their tanks.

Maubin (mah-oo-BIN), former district, currently a township (□ 1,642 sq mi/4,269.2 sq km), Ayeyarwady div., Myanmar, in AYEYARWADY R. delta; ⊙ MAUBIN. Rice and fishing region. Largely swamps; once the most malarial dist. in Myanmar. Constituted 1903 from former Thongwa dist.

Maubin, town, ⊙ MAUBIN township, MYANMAR, in AYEYARWADY RIVER delta, 60 mi/97 km E of BASSEIN. River port, center of rice area, especially deep-water rice.

Maubisse (MOU-bees-sai), town, EAST TIMOR, in central TIMOR Island, 19 mi/31 km S of DILI; 08°50′S 125°36′E. Agricultural (coffee, wheat, fruit); sheep raising. Also spelled Maobisse.

Maubourguet (mo-boor-GAI), commune (□ 8 sq mi/ 20.8 sq km; 2004 population 2,478), HAUTES-PYR-ÉNÉES department, MIDI-PYRÉNÉES region, SW FRANCE, on the ADOUR RIVER, and 17 mi/27 km N of TARBES, in ARMAGNAC wine-producing area; 43°28′N 00°02′E. Has partly Romanesque church.

Mauchline (MAWK-lein), small town (2001 population 4,105), East Ayrshire, S Scotland, near Ayr River, and 11 mi/18 km ENE of AYR; 55°31′N 04°22′W. Previously manufacturing of wooden snuff boxes and other wooden articles. Burns lived here for some time and married Jean Armour here. Many local scenes (e.g., Ballochmyle) and personalities are celebrated in Burns' poems. Just NW is Mossgiel, where Burns wrote first volume of poetry.

Mauckport, town (2000 population 83), HARRISON county, S INDIANA, on OHIO River, and 13 mi/21 km S of CORYDON; 38°01′N 86°12′W. In agricultural area; hardwood veneer.

Maud, town (2006 population 1,155), POTTAWATOMIE and SEMINOLE counties, central OKLAHOMA, 16 mi/26 km SE of SHAWNEE, near WOLF CREEK; 35°07′N 96°46′W. In agricultural area (oats, wheat; livestock); oil wells. Incorporated 1929.

Maud (MAWD), town (2006 population 1,021), BOWIE county, NE TEXAS, 18 mi/29 km WSW of TEXARKANA, near SULPHUR RIVER (WRIGHT PATMAN LAKE to S); 33°19′N 94°20′W. Dairying; blueberries, vegetables; timber. Manufacturing (electronics).

Maudaha (MOU-duh-huh), town, HAMIRPUR district, S UTTAR PRADESH state, N central INDIA, 30 mi/48 km NNE of MAHOBA; 25°41′N 80°07′E. Gram, jowar, sesame, wheat, pearl millet. Has eighteenth-century Muslim tomb.

Maude (MAWD), settlement, NEW SOUTH WALES, SE AUSTRALIA, 485 mi/780 km SW of SYDNEY, 32 mi/52 km W of HAY, on MURRUMBIDGEE RIVER; 34°27′S 144°21′E.

Maudit, Mont (mo-DEE, MON), peak (14,649 ft/4,465 m) of MONT BLANC massif, on French-Italian border, just NE of Mont Blanc, at head of Brema glacier, which flows into ITALY.

Maudits, Monts, SPAIN: see MALADETTA MOUNTAINS.

Maud Rise or **Queen Maud Rise**, ANTARCTICA, rise on S edge of ATLANTIC-INDIAN BASIN, N of FIMBUL ICE SHELF; 66°00′S 03°00′E.

Maud Seamount or **Maud Bank**, ANTARCTICA, seamount on S edge of ATLANTIC-INDIAN BASIN; peak of MAUD RISE; 65°00′S 02°35′E.

Maud Subglacial Basin, a large subglacial basin S of the WOHLTHAT MOUNTAINS in the S part of QUEEN MAUD LAND, EAST ANTARCTICA; 81°00′S 15°00′E.

Mauer (MOU-er), village, LOWER NECKAR, BADEN-WÜRTTEMBERG, GERMANY, on the ELSENZ RIVER, and 6 mi/9.7 km SE of HEIDELBERG; 49°21′N 08°50′E. Brickwork. Jawbone of Heidelberg man found nearby in 1907.

Mauer, POLAND: see PILCHOWICE.

Mauerbach (MOU-er-bahkh), township, E LOWER AUSTRIA, in the WIENERWALD (forest), 8 mi/12.9 km ENE of VIENNA; 48°15′N 16°10′E. Abandoned 14th century Carthusian monastery.

Mauerkirchen (mou-er-KIR-khuhn), township, SW UPPER AUSTRIA, W central AUSTRIA, in the IN-NVIERTEL region, and on MATTIG RIVER, 6 mi/9.7 km SE of Braunau am Inn; 48°11′N, 13°08′E. Elevation 1,240 ft/378 m. Manufacturing of perambulators; dairy farming.

Mauer Lake, POLAND: see MAMRY, LAKE.

Maués (MOU-es), city, (2007 population 47,001), E AMAZONAS state, BRAZIL, head of navigation on the Maués-Guassú (right tributary of the AMAZON River), and 160 mi/257 km E of MANAUS; 03°25′S 57°36′W. Rubber, guaraná.

Mau Escarpment (MAH-oo), section of W rim of GREAT RIFT VALLEY in W KENYA, W and S of NAKURU. Elevation 10,000 ft/3,048 m.

Maug (MOUG), uninhabited group of volcanic islands, SAIPAN district, NORTHERN MARIANA ISLANDS, W PACIFIC, c.50 mi/80 km SSE of PAJAROS; 20°01′N 145°13′E. Comprises three islands (each c.1 mi/1.6 km long); remnants of partially submerged caldera surround deep harbor. The N island rises to 748 ft/228 m.

Maugansville (MO-gans-vil), village, WASHINGTON county, W MARYLAND, 4 mi/6.4 km NNW of HA-GERSTOWN. Named for Jonathan and Abraham Maughan in the 1880s, its development is recent. Washington County Regional Airport nearby.

Mauga Silisili (MOU-gah SEE-lee-SEE-lee), extinct volcanic crater (6,070 ft./1850 m), SAVAI'I, SAMOA; highest in Samoan group. Also known as Mount Mata'aga.

Mauguio (mo-GYO), town, HÉRAULT department, LANGUEDOC-ROUSSILLON region, S FRANCE, near a lagoon of same name off the Gulf of LION, 7 mi/11.3 km E of MONTPELLIER; 43°37′N 04°01′E. Fruit-juice processing; vineyards.

Maui (MOU-ee), county (□ 2,398 sq mi/6,211 sq km; 1990 population 100,374; 2000 population 128,094), HAWAII, includes KAHOOLAWE, LANAI, MAUI, MOLOKAI islands; ⊙ WAILUKU, on Maui; 20°52′N 156°37′W. Locally administered by 13 districts, 1 on Maui, 2 on Molokai, and 1 on Lanai. Largest city is KAHULUI, Maui. KALAWAO county, on Molokai, is officially a district of MAUI county and is part of Kaulapapa National Historical Park; Kahoolawe owned by U.S. military.

Maui (MOU-ee), offshore gas field (with some oil condensate) on continental shelf SSW of TARANAKI, NEW ZEALAND, now (1994) producing c. three-fourths of New Zealand's local supply, sent to mainland by undersea pipeline, then distributed to electric power stations at STRATFORD, NEW PLYMOUTH, and HUNTLY, and to synthetic gasoline plant at MOTUNUI. Pipelines lead N to AUCKLAND and WHANGAREI, S to WELLINGTON.

Maui (MOU-ee), island (□ 728 sq mi/1,892.8 sq km), MAUI county, HAWAII, second-largest island in the state of Hawaii, separated from the island of Hawaii to SE by the ALENUIHAHA CHANNEL, from MOLOKAI (NW) by the PAILOLO CHANNEL, from LANAI (NW) by AUAU CHANNEL, from KAHOOLAWE (SW) by Alakeiki Channel. Locally administered by 10 districts, Maui is made up of 2 mountain masses, the Haleakala in SE, in main section of island, and West Maui Mountains in NW; on West Maui Peninsula, connected by isthmus 7 mi/11.3 km wide. The highest point on the island is PUU ULAULA (Red Hill) on SW rim of Haleakala crater (10,023 ft/3,055 m) in HALEAKALA NATIONAL PARK. In West Maui, PUU KUKUI rises to 5,788 ft/1,764 m. The island's chief industries are tourism and the cultivation of sugarcane (especially W), cattle, pineapples, and timber (pine, eucalyptus). The principal ports are KAHULUI (largest city on island and in county) and LAHAINA. WAILUKU (1990 population 10,688) is the second-largest town and the capital of Maui county (1990 population 100,374), which includes the islands of Maui, Lanai, Kahoolawe, and Molokai. Wailuku/Kahului and Lahaina have become major tourist centers. Launiupoko (W) and Waianapanapa Cave (E) state parks; Poli Poli Springs State Recreational Area in S; Kaumahina and Puaa Kaa state waysides in NE; West Maui (NW), KOOLAU and MAKAWAO (NE), HANA (E), and Kula, Kahikinui, and Kipahulu (S) forest reserves.

Mauke (MOU-KAI), composite volcanic and low coral island (□ 7 sq mi/18.2 sq km), most easterly of COOK ISLANDS, S PACIFIC, 150 mi/241 km NE of RAROTONGA; 4 mi/6.4 km long; 2.5 mi/4 km wide. Main settlement is Areora. Fertile volcanic soil, karstic makatea, fringing reef; exports citrus fruits, copra.

Maulavi Bazar (mo-lo-vee), town (2001 population 40,107), SYLHET district, E EAST BENGAL, BANGLADESH, in SURMA VALLEY, on tributary of the KUSIYARA RIVER, and 29 mi/47 km SSW of SYLHET; 24°29′N 91°47′E. Road junction; trades in rice, tea, oilseeds. Manufacturing of iron implements 6 mi/9.7 km ENE, at Rajnagar. College. Also spelled Maulvi Bazar and Moulvi Bazar.

Area is shown by the symbol □, and capital city or county seat by ⊙.

Maulbronn (moul-BRUHN), town, N BLACK FOREST, BADEN-WÜRTTEMBERG, GERMANY, on the Saalbach River near its source, and 13 mi/21 km SE of BRUCHSAL; 49°00′N 08°50′E. Grain; cattle. Site of noted former Cistercian abbey (founded 1147), now Protestant theological seminary; famous students were Friedrich Hölderlin and novelist Hermann Hesse. Since 1993, World Culture Heritage site. Chartered 1886.

Maulburg (MOUL-boorg), village, HIGH RHINE, BADEN-WÜRTTEMBERG, SW GERMANY, on the WIESE RIVER, 5 mi/8 km ENE of LÖRRACH; 47°38′N 07°47′E. Forestry; manufacturing (textiles).

Maulden (MAWL-duhn), village (2001 population 2,900), BEDFORDSHIRE, central ENGLAND, 7 mi/11.3 km S of BEDFORD; 52°02′N 00°27′W. Church has 15th-century tower.

Mauldin (MAWL-duhn), city (2006 population 19,806), GREENVILLE county, NW SOUTH CAROLINA, suburb 6 mi/9.7 km SSE of GREENVILLE, near REEDY RIVER; 34°47′N 82°17′W. Manufacturing includes machinery, textile and textile products, plastic products, computer equipment, food products. Agriculture includes grain, tomatoes, soybeans, sorghum, peaches; livestock, poultry.

Maule (MOU-lai) or **Maule** (MOU-le), region (□ 2,172 sq mi/5,647.2 sq km; 2005 population 967,107), S central CHILE; ⊙ TALCA; 35°30′S 71°30′W. Extends from the ANDES Mountains (E) through the Central Valley to the PACIFIC OCEAN (W); the LIBERTADOR GENERAL BERNARDO O'HIGGINS region (N) and the BÍO-BÍO region (S). Principally an agricultural region, it produces grapes, cereals, vegetables, marigolds, sugar beets, fruit; and livestock. Food procesing; forestry; fishing; mining (especially quartz, talc, clay). Main cities are CURICÓ, TALCA, CAUQUENES, and LINARES, and is composed of provinces of the same name.

Maule (MOL), town (□ 6 sq mi/15.6 sq km), YVELINES department, ÎLE-DE-FRANCE region, N central FRANCE, on the Mauldre River, 5 mi/8 km S of SEINE RIVER; 48°55′N 01°51′E. Has church with c.11th-century crypt.

Maule (MOU-lai), village, ⊙ Maule comuna, TALCA province, MAULE region, central CHILE, on railroad, on MAULE RIVER, and on PAN-AMERICAN HIGHWAY, 7 mi/11 km S of TALCA; 35°32′S 71°42′W. Cereals, grapes; livestock.

Maule, Lake (MOU-lai), Andean lake (□ 17 sq mi/44.2 sq km), in SE TALCA province, MAULE region, central CHILE, 75 mi/121 km SE of TALCA; c.6 mi/10 km long, 2 mi/3 km–4 mi/6 km wide; 36°04′S 70°30′W. Elevation 7,200 ft/2,195 m. MAULE RIVER rises here.

Mauléon (mo-lai-ON), town, DEUX-SÈVRES department, POITOU-CHARENTES region, W FRANCE, on the SÈVRE NANTAISE RIVER, and 40 mi/64 km SE of NANTES; 46°56′N 00°45′W. Supplies equipment for motor carriers and buses.

Mauléon-Licharre (mo-lai-ON–lee-SHAHR), commune (□ 4 sq mi/10.4 sq km), PYRÉNÉES-ATLANTIQUES department, AQUITAINE region, SW FRANCE, in BASQUE country, 14 mi/23 km W of OLORON-SAINTE-MARIE; 43°14′N 00°53′W. Footwear manufacturing (sandals, slippers); woodworking.

Maule River (MOU-lai), c.175 mi/282 km long, central CHILE; rises in LAKE MAULE in the ANDES Mountains, near ARGENTINA border; flows WNW to the PACIFIC OCEAN at CONSTITUCIÓN. Receives the RÍO CLARO and LONCOMILLA RIVER. Used for irrigation and hydroelectric power.

Maullín (mou-YEEN), town, ⊙ Maullín comuna, LLANQUIHUE province, LOS LAGOS region, S central CHILE, minor port on bay of the PACIFIC OCEAN, at mouth of MAULLÍN RIVER, 20 mi/32 km NNE of ANCUD; 41°37′S 73°35′W. Agricultural center (cereals, vegetables; livestock); dairying; lumbering.

Maullín River (mou-YEEN), c.50 mi/80 km long, LLANQUIHUE province, LOS LAGOS region, S central CHILE; rises in LAKE LLANQUIHUE; flows SW, past LAS QUEMAS and MAULLÍN, to a wide estuary on CORONADOS GULF of the PACIFIC OCEAN. Navigable c.25 mi/40 km upstream.

Maulmain, MYANMAR: see MAWLAMYINE.

Maumbury Rings, ENGLAND: see DORCHESTER.

Maumee (mo-MEE), residential city (□ 10 sq mi/26 sq km; 2006 population 14,149), LUCAS county, NW OHIO, on the MAUMEE RIVER, 8 mi/12.8 km SW of TOLEDO; 41°34′N 83°39′W. Site of Fort Miami, a British post surrendered to the Americans during the War of 1812. Nearby is Fallen Timbers, the historical monument commemorating the battle fought in 1794. The Maumee River courses N and NE through Toledo, where it enters LAKE ERIE through Maumee Bay and is navigable. Automotive stamping plant. Incorporated 1838.

Maumee River, c.130 mi/209 km long, in INDIANA and OHIO; formed at FORT WAYNE (Indiana) by junction of SAINT JOSEPH and SAINT MARY'S rivers; flows NE, past DEFIANCE and TOLEDO (Ohio), to Maumee Bay, an arm of Lake ERIE just NE of Toledo. For several miles above its mouth, the Maumee serves as harbor of Toledo. Receives AUGLAIZE RIVER at Defiance.

Maumelle, Lake (maw-MEL), PULASKI county, central ARKANSAS, on Big Maumelle River, 15 mi/24 km WNW of LITTLE ROCK, and 2 mi/3.2 km W of the ARKANSAS RIVER; c.12 mi/19 km; 34°50′N 92°30′W. Maximum capacity 220,000 acre-ft. Formed by Lake Maumelle Dam (62 ft/19 m high), built (1957) by the City of Little Rock for its water supply.

Maumere, INDONESIA: see FLORES.

Maumusson, Pertuis de (mo-myoo-SON, per-TWEE duh), strait in ATLANTIC OCEAN off CHARENTE-MARITIME department, POITOU-CHARENTES region, W FRANCE, between S end of Île d'OLÉRON and mainland, connecting the BAY OF BISCAY with the Pertuis d'ANTIOCHE; 1 mi/1.6 km wide; used by shipping to estuary of SEUDRE RIVER (4 mi/6.4 km E); 45°48′N 01°14′W. Oyster beds.

Maun (MAH-oon), town (2004 population 47,000), NORTH-WEST DISTRICT, N BOTSWANA, on SE edge of OKAVANGO Basin, 220 mi/354 km SW of MARAMBA; 19°59′S 23°25′E. Elevation 3,157 ft/962 m. Former capital of NGAMILAND DISTRICT. Headquarters of Batawana tribe; tourist and education center. Airfield; hospital.

Maunabo (mou-NAH-bo), town (2006 population 12,679), SE PUERTO RICO, near coast, N of Maunabo River, and 14 mi/23 km E of GUAYAMA. Agriculture (plantains; livestock). Tourism at its beaches. Its port (beach/resort) is 1.5 mi/2.4 km SSE.

Mauna Kea (MOU-nah KAI-ah), dormant volcano (13,796 ft/4,205 m), in the N central part of the island of HAWAII, HAWAII county, HAWAII, 25 mi/40 km WNW of HILO. Highest point in Hawaii and Hawaiian Island chain; highest seamount in the world, rising c.32,000 ft/9,754 m from the PACIFIC OCEAN floor; a second peak, Puu Makanaka (11,633 ft/3,546 m), is 3 mi/4.8 km to NE. It has many cinder cones on its flanks and a great crater at the summit. Its fertile lower slopes are used for agriculture, especially the growing of coffee beans. The upper slopes are snow-covered in winter. Onizuka Astronomy Center S of summit. Observatory complex at summit. Pohakuloa Military Training Area to SW. Surrounded by Mauna Kea Forest Reserve; Mauna Kea Ice Age Natural Area Reserve on S slopes; Mauna Kea State Recreational Area to SW; large Parker Ranch to NW.

Mauna Kea Observatory, astronomical observatory at MAUNA KEA peak, HAWAII county, HAWAII, c.25 mi/40 km WNW of HILO; at elevation of over 13,796 ft/4,205 m. Observatory complex immediately N of peak. Operated by the Institute for Astronomy of the University of Hawaii. Instruments include the Canada-France-Hawaii 142-in/361-cm telescope, the U.K. Infrared 150-in/381-cm telescope, the 120-in/305-cm

Infrared Telescope Facility, an 88-in/224-cm reflecting telescope, and twin 24-in/61-cm reflectors as well as 2 telescopes used for observations in the submillimeter portion of the electromagnetic spectrum. The W. M. Keck 33-ft/10-m telescope, with its array of 36 segmented mirrors, began observations in 1991. A computer adjusts each small mirror many times per second so that a single image is formed of the object under study. New twin telescopes (Keck I 1993 and Keck III 1996) are the world's largest, with 4 times the power of the 200-in/508-cm telescope at Mount PALOMAR. Onizuka Astronomy Center to S of peak.

Maunaloa (MOU-nah-LO-ah), village (2000 population 230), W MOLOKAI, MAUI county, HAWAII, inland c.5 mi/8 km from both N and W coasts, 3 mi/4.8 km from S coast, and 13 mi/21 km WNW of KAUNAKAKAI; 21°08′N 157°12′W. Terminus of Manualoa Highway (State Highway 460). Molokai Ranch Wildlife Park to NW.

Mauna Loa (MOU-nah LO-ah), volcano (13,677 ft/4,169 m), S central HAWAII island, HAWAII county, HAWAII, 35 mi/56 km WSW of HILO, in W part of HAWAII VOLCANOES NATIONAL PARK. MOKUAWEOWEO, Mauna Loa's crater, lies at N end of the Southwest Rift Zone, which extends to S end of island. Mauna Loa has erupted 4 times (1942, 1949, 1975, and 1984) since its period of greatest activity in 1881. Mauna Loa (atmospheric) Observatory 5 mi/8 km to N. Kau (SE), Mauna Loa (N), Kapapala (SE) forest reserves.

Maunalua Bay (MOU-nah-LOO-ah), SE OAHU island, HONOLULU county, HAWAII, between DIAMOND HEAD (W) and KOKO HEAD (E), in city of HONOLULU, 6 mi/9.7 km E of downtown.

Maunawili (MOU-nah-WEE-lee), town (2000 population 4,869), SE OAHU island, HONOLULU county, HAWAII, 8 mi/12.9 km ENE of HONOLULU, and 2 mi/3.2 km SW of KAILUA and E coast, at intersection of Kalanianaole and Pali highways. Waimanalo Forest Reserve to SW.

Maungdaw (MOUNG-daw), township, RAKHINE STATE, MYANMAR, in the Arakan, on NAAF RIVER (BANGLADESH border), and 60 mi/97 km NW of SITTWE. Rice port. Head of roads to BUTHIDAUNG and to COX'S BAZAR (BANGLADESH) via BAWLI BAZAR.

Maunie (maw-NEE), village (2000 population 177), WHITE county, SE ILLINOIS, on WABASH RIVER, and 8 mi/12.9 km ESE of CARMI; 38°02′N 88°02′W. In agricultural area (corn, wheat, soybeans); oil and natural gas.

Maunoir, Lac (mon-WAHR, lahk), lake, NORTHWEST TERRITORIES, N CANADA, NW of GREAT BEAR LAKE; 30 mi/48 km long, 1 mi/2 km–15 mi/24 km wide; 67°30′N 125°W. Drains NE into ANDERSON RIVER. Sometimes called Manoir Lake.

Maupihaa (mou-PEE-hah), uninhabited atoll, W SOCIETY ISLANDS, French Polynesia, S PACIFIC; 16°55′S 153°55′W. Owned by French copra company. Also Mopihaa and Mopelia.

Maupin (MAW-pin), village (2006 population 408), WASCO county, N OREGON, 30 mi/48 km S of THE DALLES, on DESCHUTES RIVER, at mouth of Bakeoven Creek; 45°10′N 121°04′W. Timber; wheat, fruit; sheep, cattle. Fish hatchery to N. Warm Springs Indian Reservation to SW. MOUNT HOOD National Forest to W.

Maupiti (mou-PEE-tee), volcanic island, Leeward group, SOCIETY ISLANDS, S PACIFIC, c.25 mi/40 km W of BORA-BORA; 16°27′S 152°15′W. Circumference 6 mi/9.7 km; rises to 800 ft/244 m. Known for deposits of jet-black basaltic rock. Sometimes called Maurua.

Maur, commune, ZÜRICH canton, N SWITZERLAND, near W shore of LAKE GREIFENSEE, 6 mi/9.7 km SE of ZÜRICH; 47°22′N 08°40′E.

Mau-Ranipur, India: see MAU.

Maurawan (mou-RAH-wuhn), town, UNNAO district, central UTTAR PRADESH state, N central INDIA, 26 mi/

42 km ESE of UNNAO; 26°26′N 80°53′E. Wheat, barley, rice, gram, oilseeds.

Maure-de-Bretagne (MOR–duh–bruh-TAHN-nyuh), commune, ILLE-ET-VILAINE department, BRITTANY, W FRANCE, 19 mi/31 km SW of RENNES; 47°54′N 01°59′W. Dairying.

Mauren (MOU-ren), village (2006 population 3,718), N LIECHTENSTEIN, 4.5 mi/7.2 km NNE of VADUZ; 47°13′N 09°33′E. Dairy farming. Has prehistoric remains.

Maurepas (mor-uh-PAH), town (□ 3 sq mi/7.8 sq km), YVELINES department, ÎLE-DE-FRANCE region, N central FRANCE, 8 mi/13 km WSW of VERSAILLES; 48°45′N 01°55′E.

Maurepas, Lake (MAHR-i-paw), SE LOUISIANA, c.28 mi/45 km NW of NEW ORLEANS; c.13 mi/21 km long; 30°15′N, 90°29′W. Receives navigable TICKFAW RIVER from N, AMITE and Petite Amite rivers from W; connected to LAKE PONTCHARTRAIN (E) by the PASS MANCHAC waterway.

Maures, Massif des, FRANCE: see MAURES, MONTS DES.

Maures, Monts des (MAWR, MON dai) [French= mountains of the Moors], massif in VAR department, PROVENCE-ALPES-CÔTE D'AZUR region, SE FRANCE, extending c.30 mi/48 km along the MEDITERRANEAN coast of PROVENCE from vicinity of FRÉJUS–SAINT-RAPHAËL (NE) to HYÈRES (SW); c.15 mi/24 km wide; 43°16′N 06°23′E. Rises to 2,556 ft/779 m at Notre-Dame-des-Anges. Heavily forested (cork oaks, pines). Its coastline, rocky and heavily indented (Gulf of SAINT-TROPEZ, several headlands), forms W part of French RIVIERA; it is dotted with swimming resorts. A scenic road (the Corniche des Maures), hugs the steeply rising coast in mid-sect. Also known as Massif des Maures.

Mauretania, ancient district of AFRICA in Roman times. In a vague sense it meant only "the land of the Moors" and lay W of NUMIDIA, but more specifically it usually included most of present-day N MOROCCO and W ALGERIA. The district was not the same as modern MAURITANIA. It was a complex of native tribal units, but by the 2nd century B.C.E. when Jugurtha of Numidia was rebelling against ROME, Jugurtha's father-in-law, Bocchus, had most of Mauretania under his control. The Roman influence became paramount, and Augustus, having met opposition in restoring Juba II to the throne of Numidia, placed him instead (25 B.C.E.) as ruler of Mauretania. Revolts later occurred, and Mauretania was subdued (C.E. 41–A.D. 42); Emperor Claudius I made it into two provinces—Mauretania Caesariensis, with Caesarea (modern CHERCHELL, Algeria) as capital, and Mauretania Tingitana, with Tingis (modern TANGER, Morocco) as capital. Roman influence was never complete, and native chieftains remained powerful. With the onset of the barbarian invasions, Roman control weakened, and by the end of the 5th century A.D. it had disappeared.

Mauri (MOW-ree), canton, INGAVI province, LA PAZ department, W BOLIVIA, in the W part of the province, near the PERU border; 17°28′S 69°22′W. Elevation 12,641 ft/3,853 m. Gas wells in area; clay, limestone, gypsum deposits. Agriculture (potatoes, yucca, bananas, rye); cattle.

Mauriac (mor-YAHK), town (□ 10 sq mi/26 sq km; 2004 population 3,963), CANTAL department, AUVERGNE region, S central FRANCE, near upper DORDOGNE RIVER, and 20 mi/32 km NNW of AURILLAC; 45°15′N 02°25′E. Tourist center and livestock market; cheese making, woodworking. Has fine 12th-century Romanesque basilica of Notre-Dame-des-Miracles and a monastery founded atop Gallo-Roman ruins. Nearby are the volcanic peaks of Auvergne.

Maurice, town (2000 population 254), SIOUX county, NW IOWA, 6 mi/9.7 km W of ORANGE CITY; 42°58′N 96°10′W. In livestock and grain area.

Maurice (MAHR-ees), village (2000 population 642), VERMILION parish, S LOUISIANA, 12 mi/19 km NNW of ABBEVILLE, and 8 mi/13 km SW of LAFAYETTE, on Indian Bayou; 30°07′N 92°07′W. Rice; cattle. Gas field nearby.

Maurice, INDIAN OCEAN: see MAURITIUS.

Maurice River (maw-REES), township (over □ 94 sq mi/243 sq km; 1990 population 6,648), CUMBERLAND county, S NEW JERSEY, 6 mi/9.7 km SW of MILLVILLE; 39°17′N 74°56′W. Sparsely populated. Incorporated 1798.

Maurice River (maw-REES), c.50 mi/80 km long, S NEW JERSEY; rises near GLASSBORO; flows S, past MILLVILLE (dam here forms 3-mi/4.8-km-long UNION LAKE), to Maurice River Cove on DELAWARE BAY, 2 mi/3.2 km S of PORT NORRIS. Navigable to Millville; oystering, fishing docks at mouth.

Mauricetown (maw-REES-toun), village, CUMBERLAND county, S NEW JERSEY, on MAURICE RIVER, and 8 mi/12.9 km SSE of MILLVILLE; 39°17′N 74°59′W. Oystering.

Mauriceville (MAW-rees-vil), unincorporated town (2000 population 2,743), ORANGE county, SE TEXAS, 12 mi/19 km NW of ORANGE; 30°13′N 93°52′W. Railroad junction. Agriculture (cattle; rice; horticulture). Timber. Oil and natural gas. Manufacturing (machinery).

Mauricie (mo-ree-SEE), region (□ 13,688 sq mi/35,588.8 sq km; 2005 population 260,078), S central QUEBEC, E CANADA, halfway between MONTREAL and QUEBEC city; 47°33′N 73°25′W. Composed of forty nine municipalities, centered on TROIS-RIVIÈRES. High-technology, hydrogen, metal processing industries; manufacturing (paper products).

Maurienne (mawr-YEN), Alpine valley of ARC RIVER, in SAVOIE department, RHÔNE-ALPES region, SE FRANCE, deeply entrenched between Savoy (ALPES FRANÇAISES) (N) and DAUPHINÉ ALPS (S); 45°15′N 06°30′E. Extends over 70 mi/113 km in a great arc from Col de l'ISERAN (E) to junction with ISÈRE RIVER (NW). Followed by road and railroad to Italy via MONT CENIS pass and tunnel. Its hydroelectric plants power steel and aluminum works, electrochemical plants. Winter sports. Chief town is SAINT-JEAN-DE-MAURIENNE.

Maurilândia (MOU-ree-LAHN-zhee-ah), town (2007 population 10,769), S central GOIÁS state, BRAZIL, 37 mi/60 km N of QUIRINÓPOLIS; 18°03′S 50°30′W. Air strip.

Mauripur (MAW-ree-puhr), village, KARACHI administration area, SE PAKISTAN, on KARACHI harbor, 5 mi/8 km W of KARACHI city center; 24°52′N 66°55′E. Suburban settlement and airport. Extensive brine-salt deposits here form one of PAKISTAN's chief sources of supply. Also spelled Maurypur.

Mauritania, Islamic Republic of (MAW-ri-TAI-nee-uh), republic (□ 397,953 sq mi/1,030,698 sq km; 2004 estimated population 2,998,563; 2007 estimated population 3,270,065), NW AFRICA, on the ATLANTIC OCEAN (W); ⊙ NOUAKCHOTT (largest city).

Geography

Mauritania is bordered on the NW and N by the WESTERN SAHARA, NE by ALGERIA, E and SE by MALI, and SW by SENEGAL. Most of Mauritania is made up of low-lying desert, which comprises part of the SAHARA DESERT. Along the SÉNÉGAL RIVER (which forms the border with Senegal) in the SW is the semiarid SAHEL with some fertile alluvial soil. A wide sandstone plateau (rising to c.1,500 ft/460 m) runs through the center of the country from N to S. In the SE is the Hodh, a large basin in the desert. The republic is divided into one capital district (Nouakchott) and twelve administrative regions: Adrar, Assaba, BRAKNA, Dakhlet Nouâdhibou, Gorgol, GUIDIMAKHA, Hodh Ech Chargui, Hodh El Gharbi, Inchiri, TAGANT, Tiris Zemmour, and Trarza. The most important towns are Nouakchott, ATAR, and KAÉDI.

Population

The majority of the population is made up of nomadic and seminomadic persons of Berber, Arab, Tuareg, and Fulani descent. Those of Berber, Arab, and mixed Berber-Arab background are sometimes called Moors or Maures. The remainder of the population mostly belong to the Tukolor, Soninke, Bambara, and Wolof ethnic groups and live as sedentary agriculturalists near the Sénégal River. Since the droughts of the 1960s and 1970s, many nomads have been forced into the urban area of Nouakchott. Virtually all the inhabitants of the country are Muslim, and many belong to the Qadirriya brotherhood. The great majority of Mauritanians speak Arabic, which, along with French, is an official language. Indigenous languages are widely spoken.

Economy

Mauritania's economy is sharply divided between a traditional agricultural sector and a modern mining industry that was developed in the 1960s. The great majority of the country's workers are engaged either in raising crops or pasturing livestock and are largely unaffected by the mining industry. The principal agriculture products, produced chiefly near the Sénégal River and in scattered oases, are millet, pulses, dates, maize, groundnuts, gum arabic, rice, sorghum, cotton, yams, and wheat. In times of drought food-production levels can drop dangerously low. Large numbers of sheep, goats, cattle, and camels are raised. There is a small but growing fishing industry based in the Atlantic Ocean and on the Sénégal River. Since 1980, all foreign commercial fishing must be carried out jointly with Mauritania; this policy has increased export earnings. A large deposit of high-grade iron ore was discovered in N Mauritania in the late 1950s, and production for export began in 1963. Mining was controlled by the Iron Mining Company of Mauritania (MIFERMA), which was jointly owned by French and other European concerns. In 1974, the iron industry was nationalized. Foreign sales of iron ore account for a large percentage of the country's export earnings. The country has large copper ore reserves and, although difficult mining conditions resulted in the closure of a large mine in 1978, in the early 1980s the JORDAN-based Arab Mining Company of Inchiri (Société des Mine de l'Inchiri, or SAMINI) laid out plans to reopen the copper mines. However, by 1987 SAMINI had stopped its plans to operate the copper mine because of low commodity prices and the high costs of processing the low-grade ore with its high arsenic content. Salt is also produced, and there are untapped deposits of gypsum, titanium, and phosphates. The country's few manufactured goods are made up principally of basic consumer items such as processed food (especially fish), clothing, and matches. Mauritania's transportation system was expanded with the construction of the Trans-Mauritania (NOUAKCHOTT-NEMA) highway that connects the capital with the SE regions. The chief exports, in addition to iron ore, are copper ore, cattle, processed fish, and gum arabic; the leading imports are machinery, transportation equipment, chemicals, foodstuffs, and refined petroleum. The principal trade partners are JAPAN, FRANCE, SPAIN, and ITALY. In 1974, Mauritania became a charter member of the West African Economic Community.

History: to 1440

By the beginning of the first millennium C.E. Sanhaja Berbers had migrated into Mauritania, pushing the earlier African inhabitants (especially the Soninké) S toward the Sénégal River. The Hodh region, which became desert only in the 11th century, was the center of the ancient empire of GHANA (700–1200), whose capital, Kumbi-Saleh, located near the present-day border with Mali, has been unearthed by archaeologists. Until the 13th century, OUALATA, Awdaghost, and Kumbi-Saleh, all in SE Mauritania, were major

---

Area is shown by the symbol □, and capital city or county seat by ⊙.

centers along the trans-Saharan caravan routes linking MOROCCO with the region along the upper NIGER RIVER. In the 11th century the Almoravid movement was founded among the Muslim Berbers of Mauritania. In the 14th and 15th centuries, SE Mauritania was part of the empire of Mali, centered along the upper Niger River. By this time the Sahara had encroached on much of Mauritania, consequently limiting agriculture and reducing the populattion.

History: 1440 to 1960

In the 1440s, Portuguese navigators explored the Mauritanian coast and established a fishing base on Arguin Island, located near the present-day border with the Western Sahara. From the 17th century, Dutch, British, and French traders were active along the S Mauritanian coast; they were primarily interested in the gum arabic gathered near the Sénégal River. Under Louis Faidherbe, governor of Senegal (1854–1861; 1863–1865), France gained control of S Mauritania, but parts of the N were not pacified until the 1930s. The French ruled through existing political authorities and did little to develop the country's economy or to increase educational opportunities for the population. National political activity began only after World War II. In 1958, Mauritania became an autonomous republic within the French Community, and on November 28, 1960, it became fully independent. Its leader at independence was Makhtar Ould Daddah, who in 1961 formed the Party of the Mauritanian People (which in 1965 became the country's only legal party) and was the leading force in establishing a new constitution.

History: 1960 to 1976

The 1960s were marked by tensions between the darker-skinned Africans of the S and the lighter-skinned Arabs and Berbers of central and N Mauritania, some of whom sought to join Mauritania with Morocco. By the early 1970s the main conflicts in the country were over economic and ideological rather than ethnic matters, as dissident workers and students protested what they considered an unfair wage structure and an undue concentration of power in Ould Daddah's hands. The long-term drought in the semiarid Sahel region in the S, which began in the late 1960s, caused the death of about 80% of the country's livestock; in addition, the drought caused extremely poor harvests in the region bordering on the Sénégal River, thus necessitating sharply increased imports of foodstuffs. Ould Daddah was a strong advocate of independence for Spanish Sahara. In order to resolve the conflict over Spanish Sahara, Ould Daddah agreed to partition the colony between Morocco and Mauritania in 1975. This move left Mauritania (as well as Morocco) in conflict with the Polisario Front, a group of nationalist guerrillas fighting for the region's independence (which after the end of Spanish control in 1976 was known as Western Sahara).

History: 1976 to Present

Ould Daddah's regime was overthrown in 1978, and Lieutenant-Colonel Mustapha Ould Mohamad Salek assumed power, promising to end involvement in the war. Salek's proposed Arabization of the country's educational system made him many non-Arab enemies. Salek resigned and was succeeded by Lieutenant-Colonel Mohamed Mahmoud-Ould Louly in 1979. In that year, Mauritania renounced all claims to Western Sahara. In 1980, Ould Louly was overthrown and replaced by Prime Minister Lieutenant-Colonel Mohamed Khouna Ould Heydalla. In 1981, Mauritania severed diplomatic relations with Morocco after it appeared Morocco had engineered a coup attempt against Heydalla. Mauritania's formal recognition of Western Saharan independence in 1984 caused civil unrest. Also in 1984, Lieutenant-Colonel Maaouiya Ould Sidi Ahmed Taya overthrew Heydalla's regime and consolidated power through a reshuffling of the cabinet. Taya restored relations with

Morocco in 1985. In 1989, racial tensions reached new heights as 40,000 Senegalese workers were driven out of the country. Rioting resulted and Mauritania broke off diplomatic relations with Senegal. A new constitution approved in 1991 called for an elected president and National Assembly, and the government legalized political parties. Taya won election as president in 1992 and 1997, in balloting widely considered to be unfair, and survived an attempted coup in 2003. Taya was reelected in 2003, again amid charges of vote fraud. Taya was ousted by a coup in 2005, and Colonel Ely Ould Mohamed Vall, became head of a ruling military council. After a new constitution was approved in 2006, legislative elections were held at the end of the year, and Sidi Ould Cheikh Abdellahi was elected president in March 2007 after a run-off.

Government

Mauritania is governed under the constitution of 2006. The president, who is the head of state, is popularly elected for a five-year term and is eligible for a second term. The prime minister is appointed by the president. The bicameral legislature consists of the fifty-six-seat Senate, whose members are indirectly elected for six-year terms, and the ninety-five-seat National Assembly, whose members are popularly elected for five-year terms. The current head of state is President Sidi Ould Cheikh Abdellahi (since April 2007); the current head of government is Prime Minister Zeine Ould Zeidane (since April 2007).

Mauriti (mou-ree-chee), city, (2007 population 41,653), SE CEARÁ state, BRAZIL, near PARAÍBA state border, 50 mi/80 km ESE of CRATO; 07°21′S 38°42′W. Cattle; carnauba. Formerly spelled Maurity.

Mauritius (maw-RI-shyuhs), republic (□ 790 sq mi/ 2,054 sq km; 2007 population 1,250,882), in the SW INDIAN OCEAN, part of the MASCARENE Island group, c.500 mi/805 km E of MADAGASCAR; ⊙ PORT LOUIS. The islands of RODRIGUES and two groups of small islands, AGALEGA and CARGADOS CARAJOS, are dependencies of Mauritius. The country is divided into nine districts: BLACK RIVER, FLACQ, GRAND PORT, MOKA, PAMPLEMOUSSES, PORT LOUIS, PLAINES WILHEMS, RIVIÈRE DU REMPART, and SAVANNE.

Geography

Surrounded by coral reefs. A central plateau is ringed by mountains of volcanic origin, which rise to c.2,700 ft/823 m in the SW. The island has a tropical, rainy climate.

Population

Two thirds of the population are of Indian descent, and one-third are either mixed French and African or of French descent. There is also a small Chinese community. Over half the population is Hindu, about 30% are Christian, and the remainder are Muslim. Overpopulation (partially a result of the eradication of malaria) was considered a serious problem in the 1960s and 1970s but emigration and a dramatic drop in fertility have slowed the growth rate. French and English are the official languages and Bhojpuri and Creole languages are widely spoken.

Economy

Mauritius has had one of the world's fastest-growing economies since the early 1980s. Once totally dependent on sugar exports, Mauritius now counts textiles as its main export. The whole island has been declared an export-processing zone, and a combination of local investors and foreign investors have begun to diversify the industrial base. Other light manufacturing, financial services, and data processing are growing, and an emphasis on high-technology manufacturing and high-end services is emerging. A 1992 free port declaration has helped to offset disadvantages Mauritius inherited through the passage of GATT. Investors and trading partners increasingly come from Hong Kong, Singapore, Australia, South Africa, and India as well as the traditional colonial powers, the U.K. and France. Tourism has expanded rapidly during the 1980s and

1990s as well, with GRAND BAY in the N of the island an emerging resort area, as well as MAHEBOURG in the SE; TAMARIN, FLIC EN FLAC, and Wolman on the W coast, LE MORNE in the S, Belle Mare and TROU D'EAU DOUCE in the E, and the capital, Port Louis.

History: to 1835

Mauritius was probably visited by Arabs and Malays in the Middle Ages. Portugal sailors visited it in the 16th century. The island was occupied by the Dutch (1598–1710) and named after Prince Maurice of Nassau (originally Maurice Island, later Mauritius). The French settled here in 1722 and called it Île de France. It became an important way station on the route to India. The French introduced the cultivation of sugarcane and imported large numbers of African slaves to work the plantations. The British captured the island in 1810 and restored the Dutch name. After the abolition of slavery in 1835, indentured laborers were brought from India; their descendants constitute a majority of the population today.

History: 1835 to Present

Politics on Mauritius was long the preserve of the French and the Creoles, but the extension of the franchise under the 1947 constitution gave the Indians political power. Indian leaders in the 1950s and 1960s favored independence, while the French and Creoles wanted continuing association with Britain, fearing domination by the Hindu Indian majority. The 1967 election gave a majority in the assembly to Sir Seewoosagur Ramgoolam's Independence party. Independence was granted in 1968, and Ramgoolam became the first prime minister. Mauritius joined the Commonwealth of Nations and the UN. In the 1960s a multiracial militant left-wing party became active. In the 1970s and 1980s unstable coalitions of parties formed along ethnic and class lines. The economic crisis of the late 1970s and early 1980s, after Cyclone Claudette and a drop in world sugar prices, intensified internal disputes. In 1982, a socialist coalition assumed power and Ramgoolam's regime was removed from office. Sir Aneerood Jugnauth became the second prime minister and he ruled the country from 1982 to 1995. Although its politics continue to be volatile, Mauritius has enjoyed relative economic prosperity since the early 1980s, as a result of its free-trade economic programs, its export-industrialization program, and a tourism boom. Mauritius became a republic in 1992, and Cassam Uteem became the first president. In 1995, Navinchandra Ramgoolam (son of the first prime minister) swept into power as the new prime minister with a landslide election victory. Another landslide in 2000, however, restored Jugnauth's coalition to power. Jugnauth resigned in 2003, and Paul Bérenger, his coalition partner, succeeded him as prime minister. The 2005 elections brought the younger Ramgoolam's coalition back into power.

Government

Mauritius is governed under the constitution of 1968 as amended. The president, who is head of state, is elected by the National Assembly for a five-year term and is eligible for a second term. The government is headed by the prime minister, who is appointed by the president. The unicameral legislature consists of the seventy-seat National Assembly; sixty-two members are elected, and eight, representing ethnic minorities, are appointed by the election commission. All serve five-year terms. The current president is Sir Aneerood Jugnauth (since October 2003). Prime Minister Navinchandra Ramgoolam is the head of government (since July 2005).

Mauritius, Cape, RUSSIA: see ZHELANIYE, CAPE.

Maurity, Brazil: see MAURITI.

Mauritzstad, Brazil: see RECIFE.

Maú River (mah-OO), c.175 mi/282 km long, on BRAZIL-GUYANA border, NE South America; rises in the Sierra PACARAIMA; flows S to TACUTÚ RIVER near 03°30′N 59°45′W. Diamond deposits.

Mauron (mo-RON), commune (□ 26 sq mi/67.6 sq km), MORBIHAN department, W FRANCE, in BRITTANY, 26 mi/42 km W of RENNES; 48°05′N 02°18′W. Dairying. Forest of PAIMPONT nearby.

Maurosouli, Greece: see POLIKASTRO.

Maurs (MAWR), commune (□ 11 sq mi/28.6 sq km; 2004 population 2,265), CANTAL department, S central FRANCE, 19 mi/31 km SW of AURILLAC, in AUVERGNE; 44°43′N 02°12′E. Agriculture market (grain, potatoes, and cattle); tanning. Has Gothic church with fine 15th-century wooden statues surrounding the altar.

Maurua, SOCIETY ISLANDS: see MAUPITI.

Maurui (ma-oo-ROO-ee), village, TANGA region, NE TANZANIA, 47 mi/76 km W of TANGA, near PANGANI RIVER, on railroad; 05°07′S 38°22′E. Sisal, rice, grain; sheep, goats.

Maury (MOR-ree), county (□ 614 sq mi/1,596.4 sq km; 2006 population 78,309), central TENNESSEE; ⊙ COLUMBIA; 35°37′N 87°05′W. Drained by DUCK RIVER. Agriculture; leading producer of beef cattle in state; some manufacturing. Formed 1807.

Maury (MOR-ree), unincorporated village, GREENE county, E central NORTH CAROLINA, 15 mi/24 km N of KINSTON; 35°28′N 77°35′W. Manufacturing (clothing, wood and fabricated metal products). In agricultural area (tobacco, grain, cotton; poultry, livestock).

Maury Bay, an embayment of the VOYEYKOV ICE SHELF into the BANZARE COAST of WILKES LAND, EAST ANTARCTICA, just W of NORTHS HIGHLAND; 66°33′S 124°42′E.

Maury City, town (2006 population 706), CROCKETT county, W TENNESSEE, 26 mi/42 km NW of JACKSON; 35°49′N 89°14′W. In diversified farm area.

Maury River, c.90 mi/145 km long, NW VIRGINIA; rises in SW AUGUSTA county, 10 mi/16 km W of STAUNTON; flows SSW, receiving CALFPASTURE RIVER from NW near GOSHEN, turns SE, continuing past LEXINGTON and BUENA VISTA, then S to JAMES RIVER at GLASGOW; 37°56′N 79°27′W. Section below Goshen formerly called North River.

Maushij, village, W YEMEN, c.30 mi/48 km N of MOCHA; 13°43′N 43°17′E. Local port; fishing.

Mauston (MAWS-tun), town (2006 population 4,253), ⊙ JUNEAU county, central WISCONSIN, on LEMONWEIR RIVER, and 55 mi/89 km E of LA CROSSE; 43°47′N 90°04′W. In agricultural area. Agriculture (grain, hay, potatoes); manufacturing (fabricated metal products, machine tools, machinery, pork and dairy products, beverages, furniture, cigars). Castle Rock Dam and Lake to NE. Settled c.1840, incorporated 1883.

Mautern (MOU-tern), town, central LOWER AUSTRIA, on the DANUBE RIVER (bridged), and at E end of WACHAU valley, 2 mi/3.2 km SW of KREMS AN DER DONAU; 48°24′N 16°10′E. Roman camp Favianis, former monastery of Saint Severin, who died here in 482. Benedictine abbey, Göttweig is 3 mi/4.8 km SE, founded 1074 and reconstructed in the 18th century by Lukas von Hildebrant as one of the finest baroque buildings in AUSTRIA; on a hill and offers a magnificent view.

Mauterndorf (MOU-tern-dorf), township, SE SALZBURG, S central AUSTRIA, in the LUNGAU, 25 mi/40 km NNE of SPITTAL AN DER DRAU; 47°08′N, 13°41′E. Elevation 3,417 ft/1,042 m. Railroad terminus; road junction; winter tourism. Remarkable castle, built 1253, extended 14th–16th century; fine burgher houses. Samson (giant figure) processions.

Mautern in Steiermark (MOU-tern in SHTEI-uhr-mahrk), township, STYRIA, central AUSTRIA, in the KAMMERTAL valley, 12 mi/19 km W of LEOBEN; 47°24′N 14°50′E. Fine historical structures; former monastery.

Mauth, CZECH REPUBLIC: see MYTO.

Mauthausen (mout-HOU-sen), township, E UPPER AUSTRIA, on left bank of the DANUBE RIVER, 3 mi/4.8 km NE of ENNS; 48°15′N 14°31′E. Site of Hitler's main Austrian concentration camp during World War II. Local granite quarry ("Wienergraben;" now aban-

doned) incorporated into prisoners' work routine. Harshest category of concentration camp and highest death rate, drawing Jewish and non-Jewish victims from other camps. Now a historic monument and museum in honor of those imprisoned. Old commercial place of salt trade, plus remarkable architecture of the church and burgher houses.

Mauzé-sur-le-Mignon (mo-ZAI–syoor–luh–meen-YON), commune (□ 9 sq mi/23.4 sq km), DEUX-SÈVRES department, POITOU-CHARENTES region, W FRANCE, 13 mi/21 km SW of NIORT; 46°12′N 00°40′W. Dairying, peat extracting. Also called Mauzé. The regional park of the Marsh of Poitou, the valley of the Sèvre Niortaise and Vendée (*marais poitevin, val de sèvre et de la vendée*) lies just NW.

Mavago District, MOZAMBIQUE: see NIASSA.

Mavégani, Comoros: see BENARA.

Mavéguani, Comoros: see BENARA.

Maveiturai, SRI LANKA: see DELFT.

Mavelikara (MAH-vai-li-kuh-ruh), city, KOLLAM district, KERALA state, S INDIA, 25 mi/40 km N of KOLLAM; 09°16′N 76°33′E. Trades in coir rope and mats, rice, cassava, betel; cashew nut processing, pottery manufacturing.

Maverick (MAV-uhr-ik), county (□ 1,291 sq mi/3,356.6 sq mi; 2006 population 52,288), SW TEXAS, on RIO GRANDE (SW; Mexican border); ⊙ EAGLE PASS; 28°45′N 100°19′W. Bridged at Eagle Pass. Rich agriculture area (part of Winter Garden region), irrigated by the Rio Grande; grain sorghum, oats, wheat, pecans, vegetables; uplands are ranching region (cattle); oil and gas; sand and gravel. Formed 1856.

Mavesyn Ridware (MAI-vis-uhn RID-wer), village (2006 population 1,200), STAFFORDSHIRE, W ENGLAND, on Trent-Mersey Canal, 3 mi/4.8 km S of RUGELEY; 52°44′N 01°52′W. The Old Hall has 14th-century gatehouse.

Mavinga (mah-VEEN-guh), town, CUANDO CUBANGO province, SE ANGOLA, 190 mi/306 km SE of MENONGUE; 15°47′S 20°21′E.

Mavingoni, Comoros: see BENARA.

Mavis Bank, village, SAINT ANDREW parish, SE JAMAICA, in interior mountains, 11 mi/18 km NE of Hope Botanical Gardens; 18°01′N 76°40′W. Elevation c.2,000 ft/610 m. Coffee district. Noted scenery.

Maviya, YEMEN: see MAWIYA.

Mavli, India: see NATHDWARA.

Mavroneri River, Greece: see KIFISSOS RIVER.

Mavropotamos River, Greece: see ACHERON RIVER.

Mavrosouli, Greece: see POLIKASTRO.

Mavrovsko, Lake, reservoir, MACEDONIA, 12 mi/19 km SW of GOSTIVAR and 36 mi/58 km SW of SKOPJE. Hydroelectric power station. In national park (□ 282 sq mi/730 sq km).

Mavuradonha Mountains, MASHONALAND CENTRAL province, NE ZIMBABWE, 90 mi/145 km N of HARARE; c.40 mi/64 km long. Forms part of ZAMBEZI ESCARPMENT on N slope; highest point in range, Banirembize, in W, exceeds 5,000 ft/1,524 m.

Maw, former NW state (*myosaship*), now part of YENGAN township (□ 741 sq mi/1,926.6 sq km), S SHAN STATE, MYANMAR, on edge of SHAN PLATEAU; capital was MYOGYI. Largely hilly and barren; drained by ZAWGYI RIVER.

Mawa (MAH-wah), village, ORIENTALE province, N CONGO, on railroad, and 120 mi/193 km E of BUTA; 02°43′N 26°42′E. Elev. 2,142 ft/652 m. Cotton ginning; coffee plantations. Also cotton gins at MAWA GEITU, 12 mi/19 km NNE. Also known as MAWA GARE.

Mawa Gare, CONGO: see MAWA.

Mawa Geitu, CONGO: see MAWA.

Mawana (muh-WAH-nah), town, MEERUT district, NW UTTAR PRADESH state, N central INDIA, near distributory of UPPER GANGA CANAL, 15 mi/24 km ENE of MEERUT. Wheat, millet, sugarcane, oilseeds.

Mawchi (maw-CHEE), village, Hpasawng township, KAYAH STATE, MYANMAR, on Loikaw-Toungoo road,

and 741 mi/1,192 km E of TOUNGOO. Major tungsten- and tin-mining center.

Mawddach, Afon (MOU–[th]uhkh), river, 21 mi/34 km long, GWYNEDD, NW Wales; rises 10 mi/16 km NE of DOLGELLAU; flows S and W to CARDIGAN BAY of IRISH SEA at BARMOUTH. Receives WNION RIVER just NW of Dolgellau.

Mawei (MAH-WAI), town, E FUJIAN province, SE CHINA, port on MIN River, near its mouth on EAST CHINA SEA, 10 mi/16 km SE of FUZHOU.

Mawenzi, mountain peak (17,564 ft/5,354 m) on Mount KILIMANJARO, KILIMANJARO region, N TANZANIA; 03°06′S 37°27′E. One of two peaks on the mountain; see UHURU.

Mawiya (MAH-wee-yuh), township, SW YEMEN, on central plateau, 20 mi/32 km E of TAIZ; 13°35′N 44°21′E. Elev. 4,000 ft/1,219 m. Center of agricultural area (coffee, millet, citrus fruit). Sometimes spelled Mawya and Maviya.

Mawkmai (mawk-MAI), fomer S state (*sawbwaship*); currently township (□ 2,803 sq mi/7,287.8 sq km) of SHAN STATE, MYANMAR; capital was MAWKMAI, a village 55 mi/89 km SE of Taungyi. Astride SALWEEN RIVER, on THAILAND border. Central plain well irrigated (rice); hill ranges (teak); tobacco plantations.

Mawlaik (maw-LEIK), town and township, SAGAING division, MYANMAR, river port on W bank of CHINDWIN RIVER; 23°38′N 94°25′E. In mountain and jungle country. Location of minor oil fields.

Mawlamyine (maw-lahm-YEE-nai), city, ⊙ MON STATE, SE MYANMAR, near the mouth of the THANLWIN RIVER. MYANMAR's third-largest city. A river port and commercial center, it has teak mills and shipyards; rice, tea, teak wood, and rubber are exported. From 1826 to 1852, Mawlamyine was the chief town of British BURMA. A pagoda here is referred to in Rudyard Kipling's poem *Mandalay*. Formerly spelled Moulmein, Maulmain.

Mawli River (MAW-lee), 60 mi/97 km long, NORTHERN REGION, GHANA; rises S of Gambaga Scarpe, 35 mi/56 km NE of TAMALE; flows SSW to join Kalaraki River, and thence into the VOLGA River; 08°38′N 00°42′W.

Mawoga (ma-WO-gah), town, MBEYA region, SW TANZANIA, 45 mi/72 km NW of MBEYA; 08°53′N 33°06′E. Corn, wheat, pyrethrum; cattle, sheep, goats. Former gold-mining region.

Mawphlang, town, EAST KHASI HILLS district, MEGHALAYA state, NE INDIA, on SHILLONG PLATEAU, on tributary of the SURMA RIVER, and 12 mi/19 km SW of SHILLONG; 25°28′N 91°46′E. Rice, sesame. Coal deposits nearby. Also spelled Maoflang.

Mawson Bank, ANTARCTICA, submarine plateau in W ROSS SEA, NE of CRARY BANK; 73°30′N 174°E.

Mawson Coast, between William Scoresby Bay and Murray Monolith, in MAC ROBERTSON LAND, EAST ANTARCTICA; 67°40′S 63°30′E.

Mawson Glacier, outlet glacier in VICTORIA LAND, ANTARCTICA, flows from the polar plateau into the ROSS SEA at the N end of the SCOTT COAST, into Nordenskjöld Ice Tongue; 76°13′S 162°05′E.

Mawson Station (MAW-suhn), ANTARCTICA, Australian station on MAWSON COAST of MAC ROBERTSON LAND; 67°36′S 62°52′E.

Mawsynram (maw-sin-ruhm), town, EAST KHASI HILLS district, MEGHALAYA state, NE INDIA, in KHASI HILLS, 28 mi/45 km SSW of SHILLONG; 25°18′N 91°35′E. Rice, cotton. Coal deposits nearby. Also spelled Maolang.

Mawya, YEMEN: see MAWIYA.

Max, village (2006 population 260), MCLEAN county, central NORTH DAKOTA, 29 mi/47 km S of MINOT; 47°49′N 101°17′W. LAKE SAKAKAWEA to S. Founded in 1904 and incorporated in 1907. Named for Max Freiberg, the postmaster's son.

Maxbass, village (2006 population 85), BOTTINEAU county, N NORTH DAKOTA, 33 mi/53 km WSW of BOTTINEAU; 48°43′N 101°08′W. Founded and incorported in 1905 and named for Max Bass, a railroad commissioner of immigration.

Maxcanú (mahsh-kahn-DO), town, YUCATÁN, SE MEXICO, on railroad, and 36 mi/58 km SW of MÉRIDA, at a highway junction (Mexico Highways 180 and 184); 20°35′N 90°00′W. Henequen-growing center. Oxkintoc ruins (NE) and Calcehtoc grotto (E) nearby.

Maxéville (MAHK-sai-VEEL), NNW suburb (□ 2 sq mi/5.2 sq km) of NANCY, MEURTHE-ET-MOSELLE department, LORRAINE region, NE FRANCE, on MARNE-RHINE CANAL; 48°43′N 06°10′E. Blast furnace, brewery, and glassworks. Old iron mines and limestone quarries nearby.

Maxeys (MAKS-eez), town (2000 population 210), OGLETHORPE county, NE GEORGIA, 19 mi/31 km SE of ATHENS; 33°45′N 83°10′W.

Maxfield, town, PENOBSCOT county, S central MAINE, 35 mi/56 km N of BANGOR; 45°17′N 68°45′W. Lumbering.

Maxhütte (MAHKS-hyoot-te) or **Maxhütte-Haidhof**, town, UPPER PALATINATE, E central BAVARIA, GERMANY, 9 mi/14.5 km S of SCHWANDORF; 49°12′N 12°05′E. Ironworks.

Maximiliano de Almeida (MAHK-mee-lee-ah-no zhe AHL-mai-dah), town (2007 population 5,059), N RIO GRANDE DO SUL state, BRAZIL, 30 mi/48 km E of ERECHIM; 27°38′S 51°47′W. Wheat, corn, potatoes, manioc; livestock.

Máximo Gómez (MAHKS-ee-mo GO-mais), town, MATANZAS province, W CUBA, 14 mi/23 km SE of CÁRDENAS; 22°53′N 81°03′W. Railroad junction in agricultural region (sugarcane, fruit; sisal; cattle).

Máximo Paz (MAHKS-ee-mo PAHZ), town, S SANTA FE province, ARGENTINA, 40 mi/64 km SSW of ROSARIO. Agricultural area (wheat, flax, corn, potatoes, soybeans; livestock, poultry). Formerly called Paz.

Máximo River (MAHKS-ee-mo), c.35 mi/56 km long, CAMAGÜEY province, E CUBA; rises at S foot of the SIERRA DE CUBITAS; flows ENE to N coast 15 mi/24 km NW of NUEVITAS.

Maxinkuckee, Lake, SW MARSHALL county, N INDIANA, at CULVER; c.3 mi/4.8 km long. Indiana's second-largest natural lake. Glacially formed. Average depth is 88 ft/27 m. Resort area.

Max Meadows (MAKS ME-doz), unincorporated village, WYTHE county, SW VIRGINIA, 7 mi/11 km E of WYTHEVILLE; 36°58′N 80°56′W. Manufacturing (machining); in rich agricultural area (grain, apples, soybeans; livestock; dairying).

Max Patch Mountains, NORTH CAROLINA and TENNESSEE: see BALD MOUNTAINS.

Max Toledo (mahks to-LAI-do), canton, PACAJES province, LA PAZ department, W BOLIVIA; 17°14′S 68°40′W. Elevation 12,989 ft/3,959 m. Gas wells in area. Copper (unmined) deposits; clay, gypsum, and limestone deposits. Agriculture (potatoes, yucca, bananas, rye); cattle.

Maxton (MAKS-tuhn), town (□ 2 sq mi/5.2 sq km; 2006 population 2,664), ROBESON and SCOTLAND counties, S NORTH CAROLINA, 6 mi/9.7 km ESE of LAURINBURG, near Drowning Creek (becomes LUMBER RIVER downstream from here, to E); 34°44′N 79°20′W. Railroad terminus. Manufacturing (foods, furniture, wood products, textiles, apparel, paper); service industries; agriculture (grain, tobacco, soybeans, cotton; poultry, cattle, hogs). Incorporated 1874 as Shoe Hill; named changed to Maxton in 1887.

Maxville (MAKS-vil), unincorporated village (□ 1 sq mi/2.6 sq km; 2001 population 864), SE ONTARIO, E central CANADA, 19 mi/31 km N of CORNWALL, and included in NORTH GLENGARRY township; 45°17′N 74°51′W. Dairying, mixed farming.

Maxwell, town (2000 population 807), STORY county, central IOWA, 10 mi/16 km S of NEVADA; 41°53′N 93°24′W. Building materials.

Maxwell, unincorporated village, COLUSA county, N central CALIFORNIA, 10 mi/16 km NW of COLUSA, on Stone Corral Creek. Rice, wheat, sugar beets, walnuts,

almonds; cattle; rice milling. Colusa River to E. Nearby are Sacramento (N), Delevan (NE), and Colusa (SE) national wildlife refuges; Colusa–Sacramento River State Recreation Area to E.

Maxwell, village, HANCOCK county, central INDIANA, 4 mi/6.4 km N of GREENFIELD. In agricultural area; manufacturing (concrete products).

Maxwell, village (2006 population 326), LINCOLN county, SW central NEBRASKA, 13 mi/21 km ESE of NORTH PLATTE, near PLATTE RIVER; 41°04′N 100°31′W. Fort McPherson National Cemetery and State Wayside Area to S.

Maxwell, village (2006 population 256), COLFAX county, NE NEW MEXICO, on CANADIAN RIVER, N of mouth of VERMEJO RIVER, near SANGRE DE CRISTO MOUNTAINS, and 26 mi/42 km S of RATON; 36°32′N 104°32′W. Elevation 5,909 ft/1,801 m. Shipping point in irrigated region; sheep, cattle. Maxwell National Wildlife Refuge is here; Dorsey Mansion State Monument to E; Stubblefield Lake reservoir to NW.

Maxwell (MAKS-wel), unincorporated village, CALDWELL county, S central TEXAS, 8 mi/12.9 km E of SAN MARCOS; 29°52′N 97°47′W. Poultry; eggs; cotton, sorghum. Oil and natural gas. Manufacturing (fabricated metal products).

Maxwell Air Force Base, ALABAMA: see MONTGOMERY.

Maxwell's Hill, Malaysia: see TAIPING.

Maxwelton House (MAX-wel-tuhn HOUS), mansion in DUMFRIES AND GALLOWAY, S Scotland, on the Cairn River, and 2 mi/3.2 km ESE of MONIAIVE (Annie Laurie's birthplace); 55°11′N 03°53′W. She is said to be buried in nearby Glencairn Church. Just N is agricultural village of Kirkland.

May, village (2006 population 31), HARPER county, NW OKLAHOMA, 22 mi/35 km WNW of WOODWARD, and on NORTH CANADIAN (Beaver) River; 36°37′N 99°45′W. In grain and livestock area.

May (MAI), unincorporated village, BROWN county, central TEXAS, 19 mi/31 km N of BROWNWOOD; 31°58′N 98°55′W. In rich farm area.

Maya Biosphere Reserve (MAH-yah) (□ 5,000 sq mi/13,000 sq km), GUATEMALA. Created in 1989 and set aside as an extractive reserve to regulate sustainable development. As yet, there are no effective controls on overall use of this area, which includes the entire PETÉN department N of the 17th parallel. Five special protected areas have been set aside within or adjoining the Maya Biosphere Reserve: LAGUNA DEL TIGRE–RÍO ESCONDIDO and EL ZOTZSAN–MIGUEL LA PELOTADA biotopes, and TIKAL, Sierra Lacandón, and EL MIRADOR–Dos Lagunas–Río AZUL national parks.

Mayachnyy (mah-YAHCH-niyee), workers settlement (2004 population 1,540), SW BASHKORTOSTAN Republic, E European Russia, on railroad spur, 54 mi/87 km N of ORENBURG, and 6 mi/10 km SSW, and under administrative jurisdiction, of KUMERTAU; 52°41′N 55°44′E. Elevation 1,062 ft/323 m. Manufacturing of oil-refining machinery.

Mayadin, SYRIA: see MEYADIN.

Mayafarkin, TURKEY: see SILVAN.

Mayaguana: see BAHAMAS.

Mayaguana Passage, ATLANTIC OCEAN channel, S BAHAMAS, between Mayaguana island (E) and cays off E ACKLINS ISLAND (W); c.25 mi/40 km wide.

Mayagüez (mah-yah-GWAIZ), city (2006 population 94,478), W PUERTO RICO, facing MONA PASSAGE. Port of entry and shipping and manufacturing center. Tuna-processing plants, breweries. Other manufacturing (food processing; candy, chemical products, machinery, electronics, scientific instruments). Recent expansion of pharmaceutical industries and offshore assembly plants. Tourism. Long known for its embroidery. Communications and cultural center with a growing population. Has campus of University of Puerto Rico, a U.S. government agriculture research station, and two television stations. Mayagüez Zoo here. International airport. Founded c. 1760.

Mayahi (MAI-ah-hee), town, MARADI province, NIGER, 55 mi/89 km NE of MARADI; 13°58′N 07°40′E. Administrative center.

Mayajigua (mei-yah-HEE-gwah), town, SANCTI SPÍRITUS province, central CUBA, on railroad, and 33 mi/53 km SE of CAIBARIÉN; 22°15′N 78°59′W. Resort with mineral springs.

Mayak, RUSSIA: see DONSKOYE.

Mayaki, UKRAINE: see MAYAKY.

Mayakonda (MAH-yah-kuhn-dah), town, CHITRADURGA district, KARNATAKA state, S INDIA, 22 mi/35 km WNW of CHITRADURGA; 14°17′N 76°05′E. Cotton ginning, hand-loom weaving.

Mayaky (mah-yah-KI) (Russian *Mayaki*), village, NW ODESSA oblast, SW UKRAINE, 27 mi/43.5 km N of ODESSA; 47°26′N 29°33′E. Elevation 587 ft/178 m. The first station on the LOWER DNIESTER IRRIGATION SYSTEM.

Maya, La, CUBA: see LA MAYA.

Mayala (muh-YAH-luh), village, BANDUNDU province, SW CONGO, on right bank of KWANGO RIVER, and 185 mi/298 km SW of INONGO; 04°36′S 18°50′E. Elev. 1,456 ft/443 m. Hardwood lumbering.

Mayáls (mei-AHLS), town, LÉRIDA province, NE SPAIN, 18 mi/29 km SSW of LÉRIDA; 41°22′N 00°30′E. Olive-oil processing; sheep raising; cereals, almonds, honey.

Mayamay, IRAN: see MEYAMEY.

Maya-Maya (mah-YAH–mah-YAH), POOL region, SE Republic of the Congo, site of airport for BRAZZAVILLE to SE.

Mayamey, IRAN: see MEYAMEY.

Maya Mountains, range in STANN CREEK and TOLEDO districts, SW BELIZE, 50 mi/80 km long (SW-NE); average elevation 2,000 ft/610 m–3,000 ft/915 m; 16°40′N 88°50′W. Rise to 3,681 ft/1,122 m in Victoria Peak of the COCKSCOMB MOUNTAINS, 30 mi/48 km SW of DANGRIGA. Timber. Chiquibul National Park, Cockscomb Basin Wildlife Sanctuary; ruins.

Mayang (MAH-YAHNG), town, ⊙ Mayang county, W HUNAN province, CHINA, near GUIZHOU province border, 15 mi/24 km N of ZHIJIANG; 27°53′N 109°48′E. Tea, rice, tobacco, oilseeds; food processing, papermaking. Non-ferrous ore mining.

Mayanja (MEI-ahn-juh), river, c.40 mi/64 km long, UGANDA; rises in S central Uganda; flows NW to KAFU RIVER.

Mayapán (mah-yah-PAHN), town, YUCATÁN, SE MEXICO, 24 mi/39 km SSE of MÉRIDA; 20°28′N 89°11′W. In henequen-growing area. Was capital of a post-classic Maya state; ruins are extensive but not reconstructed.

Mayarí (mah-yah-REE), town (2002 population 29,027), HOLGUÍN province, E CUBA, on small MAYARÍ RIVER, near NIPE BAY (ATLANTIC OCEAN), and 40 mi/64 km ESE of HOLGUÍN; 23°40′N 75°44′W. Lumbering and agricultural center (sugarcane, tobacco, fruit), in important quarry and mining region (laterites; chromium and iron refined at FELTON, 8 mi/13 km NE). The sugar central Guatemala is 7 mi/11 km NNE.

Mayarí Arriba (mah-yah-REE ahr-REE-bah), town, SANTIAGO DE CUBA province, SE CUBA, in mountains, 31 mi/50 km NNE of SANTIAGO DE CUBA; 20°25′N 75°32′W. Food processing.

Mayarí River (mah-yah-REE), 66 mi/107 km long, HOLGUÍN province, E CUBA; rises NE of ALTO SONGO; flows N, past MAYARÍ, to NIPE BAY. Drains 475 sq mi/1,231 sq km. Navigable for small boats below Mayarí. Fourth-longest river on N coast.

Maya River (MAH-yah), 660 mi/1,062 km long, RUSSIAN FAR EAST; rises in the N DZHUGDZHUR RANGE, N KHABAROVSK TERRITORY; flows SSW, past NELKAN (head of navigation; 340 mi/547 km from mouth), and NW to the ALDAN RIVER at UST'-MAYA (SE Sakha Republic), 190 mi/306 km ESE of YAKUTSK. Ice-free, May through October.

Mayaro (MA-yah-ro), county (□ 146 sq mi/379.6 sq km), SE TRINIDAD, TRINIDAD AND TOBAGO, on the

ATLANTIC OCEAN; 10°10′N 61°05′W. Forms (together with SAINT DAVID, SAINT ANDREW, and NARIVA) the administrative district of Eastern Cos.

Mayaro (MA-yah-ro), village, E TRINIDAD, TRINIDAD AND TOBAGO, on N MAYARO BAY, 30 mi/48 km E of SAN FERNANDO. In coconut-growing region. Fine beaches along the coast.

Mayaro Bay (MA-yah-ro), along SE coast of TRINIDAD, TRINIDAD AND TOBAGO, ribbon of fine sand beaches, c.12 mi/19 km long (N-S); 10°15′N 60°58′W. Lined by coconut palms.

Maybee, town (2000 population 505), MONROE county, extreme SE MICHIGAN, 9 mi/14.5 km NW of MONROE; 42°00′N 83°31′W. In farm area. Limestone.

Maybeury (MAI-be-ree), unincorporated village, MCDOWELL county, S WEST VIRGINIA, 13 mi/21 km ESE of WELCH; 37°22′N 81°22′W.

Maybole (MAI-bol), town (2001 population 4,552), South Ayrshire, SW Scotland, 8 mi/12.9 km S of AYR; 55°20′N 04°42′W. Light manufacturing. Formerly in Strathclyde, abolished 1996.

Maybrook, village (□ 1 sq mi/2.6 sq km; 2006 population 4,044), ORANGE CO., SE NEW YORK, 10 mi/16 km WSW of NEWBURGH; 41°29′N 74°12′W. Transportation and distribution center. Incorporated 1926.

May, Cape, NEW JERSEY: see CAPE MAY.

Maych'ew, town (2007 population 35,980), TIGRAY state, N ETHIOPIA, N of LAKE ASHANGE, on Dessie-Mek'elē road, and 16 mi/26 km N of KOREM; 12°47′N 39°32′E. In cereal and coffee growing region; salt market. Italians won decisive battle (1936) between here and LAKE ASHANGE (S), during Italo-Ethiopian War. Also spelled Mai Ceu.

Maydan (mah-ee-DAHN) (Czech *Majdan*), town (2004 population 4,300), N central TRANSCARPATHIAN oblast, UKRAINE, in the GORGANY, in the upper reaches of the Rika River, on road to the Torun' Mountain Pass, 6 mi/10 km N of MIZHHIR'YA; 48°36′N 23°28′E. Elevation 1,971 ft/600 m. Forestry; sawmilling, plastics manufacturing. Known since the 16th century; town since 1976.

Maydan, Al-, IRAQ: see MAIDAN, AL.

Maydanovo (mei-DAH-nuh-vuh), urban settlement, N MOSCOW oblast, central European Russia, on road, less than 2 mi/3.2 km N of KLIN, to which it is administratively subordinate; 56°21′N 36°43′E. Elevation 538 ft/163 m. Home to research, design, and technological institute of liquid fertilizers.

Maydanpek, SERBIA: see MAJDANPEK.

Maydi, YEMEN: see MAIDI.

Maydos, TURKEY: see ECEABAT.

Mayen (MEI-en), town, RHINELAND-PALATINATE, W GERMANY, in the EIFEL, 16 mi/26 km W of COBLENZ; 50°20′N 07°14′E. Manufacturing (textiles, machinery). Has Romanesque-Gothic church, 13th-century castle. Of Celtic origin, it became a Roman road station. Passed to electors of TRIER in 13th century. Heavily damaged (eighty-seven percent) in World War II. Was first mentioned in 1041; chartered 1291.

Mayence, GERMANY: see MAINZ.

Mayenne (mah-YEN), department (□ 1,998 sq mi/5,194.8 sq km), in old MAINE province, W FRANCE; ⊙ LAVAL; 48°05′N 00°40′E. A low-lying agricultural region, Mayenne specializes in livestock raising (for meat and dairy products). Traditional textile, footwear, and printing industries have been joined by modern metalworking and electrical equipment plants. Laval, the main urban center, lies on the E-W expressway and T.G.V. (high-speed rail) line from PARIS to BRITTANY. Administratively included in the PAYS DE LA LOIRE (French=Loire country) region.

Mayenne (mah-YEN), town (□ 7 sq mi/18.2 sq km), MAYENNE department, PAYS DE LA LOIRE region, W FRANCE, on both banks of MAYENNE RIVER, and 17 mi/27 km NNE of LAVAL; 48°20′N 00°40′W. Agricultural trade center; makes cotton goods and home appliances. Printshops. Has 11th-century feudal castle and

12th-century early-Gothic church. Severely damaged in World War II.

Mayenne River (mah-YEN), 125 mi/201 km long, in MAYENNE and MAINE-ET-LOIRE departments, PAYS DE LA LOIRE region, W FRANCE; rises in hills (Les AVALOIRS) within NORMANDIE-MAINE NATURAL REGIONAL PARK, 3 mi/4.8 km E of PRÉ-EN-PAIL; flows W, then S, past MAYENNE (head of navigation), LAVAL, and CHÂTEAU-GONTIER, joining the SARTHE RIVER above ANGERS to form the MAINE RIVER. With numerous old-fashioned locks, the Mayenne lends itself to riverboat excursions.

Mayer (MAI-yuhr), village (2000 population 554), CARVER county, S central MINNESOTA, on South Fork CROW RIVER, and 32 mi/51 km W of MINNEAPOLIS; 44°52′N 93°53′W. Dairying; poultry; grain, soybeans, alfalfa.

Mayerling (MEI-er-ling), village, LOWER AUSTRIA province, E AUSTRIA, on the SCHWECHAT RIVER, in the WIENERWALD (Vienna Woods); 48°03′N 16°06′E. Site of the hunting lodge (now a convent) where Crown Prince Rudolf and Baroness Maria Vetsera died mysteriously in January 1889.

Mayersville (MEI-uhrs-vil), village (2000 population 795), ⊙ ISSAQUENA county, W MISSISSIPPI, on the MISSISSIPPI RIVER, and 34 mi/55 km S of GREENVILLE; 32°53′N 91°02′W. Agr. (cotton, grain; cattle). Indian Bayou Waterfall Area to E; Anderson-Tully and Shipland wildlife management areas to S.

Mayerthorpe (MAI-uhr-thorp), village (□ 2 sq mi/5.2 sq km; 2001 population 1,570), central ALBERTA, W CANADA, 70 mi/113 km WNW of EDMONTON, in LAC SAINTE ANNE COUNTY; 53°57′N 115°08′W. Lumbering, mixed farming, oil, and gas. Formed as a village in 1927; became a town in 1961.

Mayes (MAIZ), county (□ 683 sq mi/1,775.8 sq km; 2006 population 39,774), NE OKLAHOMA; ⊙ PRYOR; 36°17′N 95°14′W. Intersected by NEOSHO RIVER (impounded near center by PENSACOLA DAM), forming LAKE OF THE CHEROKEES; FORT GIBSON LAKE reservoir downstream (S). Cherokee Plains in W. Cattle; recreation; agriculture (corn, soybeans, sorghum, wheat, oats; dairying). Manufacturing (food products, apparel, chemicals, machinery; metal industry). Salina and Snowdale state parks, near center; SPAVINAW, CHEROKEE, and Little Blue–Disney state parks in NE. Formed 1907.

Mayesville, village (2006 population 1,025), SUMTER county, central SOUTH CAROLINA, 8 mi/12.9 km ENE of SUMTER, near BLACK RIVER; 33°59′N 80°12′W. Manufacturing of wood products; agriculture includes livestock; grain, tobacco, cotton, peanuts.

Mayet (mah-YAI), commune (□ 20 sq mi/52 sq km), SARTHE department, PAYS DE LA LOIRE region, W FRANCE, 17 mi/27 km S of LE MANS; 47°45′N 00°16′E. Dairy and flour products; mushrooms.

Mayet-de-Montagne, Le (mah-YAI-duh-mon-TAHN-yuh, luh), commune (□ 11 sq mi/28.6 sq km), ALLIER department, in AUVERGNE, central FRANCE, in N foothills of the MASSIF CENTRAL, 12 mi/19 km S of LAPALISSE; 46°04′N 03°40′E. Livestock market; woodworking handicrafts. Quarries for building materials and granite for tombstones.

Mayetta (mai-ET-uh), village (2000 population 312), JACKSON county, NE KANSAS, 8 mi/12.9 km S of HOLTON; 39°20′N 95°43′W. Trading point in livestock and grain area. Potawatomi Indian Reservation is W.

Mayevitsa, BOSNIA AND HERZEGOVINA: see MAJEVICA.

Mayfa'ah, Wadi, YEMEN: see MEIFA'A, WADI.

Mayfah, YEMEN: see MEIFA.

Mayfair (MAI-fer), fashionable residential district of WESTMINSTER, LONDON, SE ENGLAND, N of the THAMES RIVER, 1.5 mi/2.4 km WNW of CHARING CROSS, bounded by PICCADILLY (S), Bond Street (E), Oxford Street (N), and Park Lane and HYDE PARK (W); 51°30′N 00°08′W. Named for fair held here in May from 17th–18th century. District includes Grosvenor Square.

Mayfair, W suburb of JOHANNESBURG, GAUTENG province, SOUTH AFRICA, 2 mi/3.2 km W of downtown Johannesburg; 26°12′S 28°01′E. Elevation 5,707 ft/1,704 m. Mainly light-industrial wholesale commercial area.

Mayfield (MAI-feeld), town (2001 population 12,820), MIDLOTHIAN, E Scotland, on South Esk River, 8 mi/12.9 km SE of EDINBURGH; 55°52′N 03°02′W.

Mayfield, town (2000 population 10,349), ⊙ GRAVES county, W KENTUCKY, 23 mi/37 km S of PADUCAH; 36°44′N 88°39′W. Railroad terminus. In an area of farms and clay deposits, it is an agriculture trade center with a tobacco market. Manufacturing (tobacco processing; transportation equipment, apparel, railroad ties, industrial packaging, plastics products). Mayfield–Graves County Airport to E. Woolridge Monument, in Maplewood Cemetery, eighteen statues designed by Harry C. Woolridge from 1840. Western Kentucky Museum. Kaler Brothers (to NE) and Obion Creek (to W) wildlife management areas. Founded 1823.

Mayfield (MAI-feeld), village (2001 population 2,677), NE East SUSSEX, SE ENGLAND, 8 mi/12.9 km S of ROYAL TUNBRIDGE WELLS; 51°01′N 00°16′E. Former agricultural market. Has 15th-century church and modern convent incorporating parts of 14th-century palace of archbishops of Canterbury.

Mayfield, village (2000 population 113), SUMNER county, S KANSAS, 8 mi/12.9 km W of WELLINGTON; 37°15′N 97°32′W. In wheat area.

Mayfield, village (□ 1 sq mi/2.6 sq km; 1990 population 817; 2000 population 800), in Mayfield town (□ 1 sq mi/2.6 sq km; 2006 population 796), FULTON county, E central NEW YORK, on SACANDAGA Reservoir, 6 mi/9.7 km NE of GLOVERSVILLE; 43°06′N 74°16′W.

Mayfield (MAI-feeld), village (□ 4 sq mi/10.4 sq km; 2006 population 3,191), CUYAHOGA county, N OHIO, 13 mi/21 km E of downtown CLEVELAND, and on CHAGRIN RIVER, just S of MAYFIELD HEIGHTS; 41°33′N 81°26′W. Insurance is a major employer.

Mayfield, village (2006 population 424), SANPETE county, central UTAH, on Twelve Mile Creek, 10 mi/16 km SSW of MANTI; 39°07′N 111°42′W. Alfalfa, barley; dairying; cattle, sheep, poultry. Elevation 5,500 ft/1,676 m. Manti–La Sal National Forest to E. Settled 1876.

Mayfield, borough (2006 population 1,708), LACKAWANNA county, NE PENNSYLVANIA, 11 mi/18 km NE of SCRANTON; 41°32′N 75°31′W. Agriculture (dairying; cattle; corn); manufacturing (fabricated metal products, plastics products). Former anthracite-coal center. Archbald Pothole State Park to SW; Merli Sarnoski Park to N. Founded c.1840.

Mayfield, ENGLAND: see ASHBOURNE.

Mayfield Creek, c.70 mi/113 km long, SW KENTUCKY; rises in W CALLOWAY county; flows generally NNW, past MAYFIELD, then W to the MISSISSIPPI RIVER 1 mi/1.6 km S of WICKLIFFE. Receives West Fork MAYFIELD CREEK (c.25 mi/40 km long; rises SW of Mayfield, 6 mi/9.7 km ESE of Wickliffe).

Mayfield Heights (MAI-feeld HEITZ), city (□ 4 sq mi/10.4 sq km; 2006 population 18,110), CUYAHOGA county, NE OHIO; suburb c.13 mi/21 km ESE of CLEVELAND; 41°31′N 81°27′W. It is primarily residential.

Mayflower, town (2000 population 1,631), FAULKNER county, central ARKANSAS, 18 mi/29 km NNW of LITTLE ROCK, near ARKANSAS RIVER; 34°58′N 92°25′W. Manufacturing (plastic and wood products). Lake Conway reservoir to NE; Camp Robinson Wildlife Management Area to E; Camp Robinson National Guard Training Area to SE.

Mayflower Village, unincorporated town (2000 population 5,081), LOS ANGELES county, S CALIFORNIA; residential suburb 12 mi/19 km ENE of downtown LOS ANGELES, 2 mi/3.2 km SE of ARCADIA; 34°07′N 118°01′W. El Monte Airport to SW.

Mayhill, unincorporated village, OTERO county, S NEW MEXICO, in E foothills of SACRAMENTO MOUNTAINS,

30 mi/48 km E of ALAMOGORDO, in Lincoln National Forest. Elevation 6,580 ft/2,006 m. Timber; sheep; fruit. Cloudcraft Ski Area to W; Mescalero Apache Indian Reservation to N.

Mayiladutrai, city, Thanjavur district, TAMIL NADU state, S INDIA, on arm of KAVERI RIVER delta, 40 mi/64 km NE of THANJAVUR. Railroad junction. Rice milling, cotton and silk weaving, pith carving of temple models; brass and copper vessels. Hindu pilgrimage center (Shivaite temple). Also Mayuram.

May, Isle of (MAI), island (1 mi/1.6 km long) at mouth of Firth of Forth, FIFE, E Scotland, 6 mi/9.7 km SE of CRAIL; 56°11′N 02°33′W. Lighthouse. There are ruins of 13th-century priory of St. Adrian, killed here by Danes in 9th century. It was formerly a place of pilgrimage, known as the Holy Wells.

Maykain, KAZAKHSTAN: see MAIKAIN.

Maykop (mah-ee-KOP), city (2005 population 158,450), ⊙ ADYGEY REPUBLIC, N CAUCASUS, S European Russia, at the foot of the Greater Caucasus, on the BELAYA RIVER (tributary of the KUBAN' River), on railroad, 1,035 mi/1,666 km SSE of MOSCOW, and 47 mi/76 km E of KRASNODAR; 44°36′N 40°06′E. Elevation 748 ft/227 m. Highway hub; local transshipment center. Has machinery, lumber, metal, wood pulp and paper, and food-processing industries. Regional electric power station. Nearby are the important Maykop oil fields, discovered in 1900–1901 and linked by pipeline with the refineries at Krasnodar and the BLACK SEA port of TUAPSE. Heritage culture institute. Founded in 1858 as a Russian fortress. Made city in 1870. Occupied by the German forces in 1942, during World War II; retaken by the Soviets in 1943. The name of the city also appears as Maikop.

Maykopskoye (mei-KOPS-kuh-ye), settlement (2005 population 4,190), E KRASNODAR TERRITORY, S European Russia, on the KUBAN' River, on road and near railroad, 9 mi/14 km E of KROPOTKIN, and 4 mi/6 km NE of GUL′KEVICHI, to which it is administratively subordinate; 45°23′N 40°46′E. Elevation 311 ft/94 m. In agricultural area (wheat, flax, sunflowers, castor beans; livestock).

Maykor (mei-KOR), town (2005 population 3,195), SE KOMI-PERMYAK Autonomous Okrug, PERM oblast, E European Russia, on the Inva River (right tributary of the KAMA River), near highway, 20 mi/32 km NNW of CHËRMOZ; 59°05′N 55°52′E. Elevation 495 ft/150 m. Former center of metallurgy. Pig iron; peat digging and charcoal burning. Formerly called Maykorskiy Zavod, then Maykorskiy.

Maykor, RUSSIA: see MAIKOR.

Maykorskiy, RUSSIA: see MAYKOR.

Maykorskiy Zavod, RUSSIA: see MAYKOR.

Maylands (MAI-luhndz), town, SW WESTERN AUSTRALIA state, W AUSTRALIA, NE residential suburb of PERTH; 31°56′S 115°53′E.

Maylisay, KYRGYZSTAN: see MAYLUU-SUU.

Mayluu-Suu (mei-loo–SOO), city (1999 population 20,365), JALAL-ABAD region, KYRGYZSTAN, in foothills of FERGANA RANGE, on small Mayluu-Suu River (tributary of the KARA DARYA), and 38 mi/61 km NW of JALAL-ABAD; 41°03′N 72°26′E. Lamps; oil and gas resources; mining of vanadium and radioactive ores (radioactivity has become a serious problem). Also called Maili-Sai or Maylisay.

Mayma (mei-MAH), town (2005 population 15,430), NW ALTAI REPUBLIC, S central SIBERIA, RUSSIA, on the KATUN River, on the CHUYA road (Russian *Chuyskiy Trakt*), 6 mi/10 km N of GORNO-ALTAYSK; 52°03′N 85°55′E. Elevation 839 ft/255 m. Local tourist center (mostly whitewater rafting). Also known as Mayma-Chergachak.

Mayma-Chergachak, RUSSIA: see MAYMA.

Maymyo (mei-MYO), town and township, MANDALAY division, MYANMAR, 25 mi/40 km E of MANDALAY, and on railroad to LASHIO. Elevation c.3,500 ft/1,067 m. Center for trade with N SHAN STATE and CHINA'S YUNNAN province (tea, tobacco, cotton goods, iron, foodstuffs); sericulture station; botanical garden; forestry school; army cantonment. During British administration, it served as summer seat of government.

Mayna (MEI-nah), town (2006 population 7,885), N ULYANOVSK oblast, E central European Russia, on crossroads and railroad, 35 mi/56 km NW of ULYANOVSK; 54°07′N 47°37′E. Elevation 938 ft/285 m. Garment factory; cheese-making factory.

Mayna (MEI-nah), village (2005 population 5,260), SE KHAKASS REPUBLIC, S SIBERIA, RUSSIA, on the E bank of the YENISEY RIVER, on highway, 40 mi/64 km S of ABAKAN, and 4 mi/6 km SSE of SAYANOGORSK; 53°04′N 91°29′E. Elevation 1,519 ft/462 m. Sayano-Shushenskaya GES (hydroelectric power station) is nearby.

Mayna, RUSSIA: see MAINA.

Maynard, town (2000 population 500), FAYETTE county, NE IOWA, 14 mi/23 km SSW of WEST UNION; 42°46′N 91°52′W. Dairying; limestone quarry.

Maynard, town (2000 population 10,433), MIDDLESEX county, NE central MASSACHUSETTS, on ASSABET RIVER, and 21 mi/34 km WNW of BOSTON; 42°26′N 71°28′W. Computer company headquarters. Site of former wool textile mill. Settled 1638, incorporated 1871.

Maynard (MAI-nuhrd), village (2000 population 381), RANDOLPH county, NE ARKANSAS, 12 mi/19 km NNE of POCAHONTAS, near MISSOURI state line; 36°25′N 90°54′W. Lumber.

Maynard, village (2000 population 388), CHIPPEWA county, SW MINNESOTA, 13 mi/21 km E of MONTEVIDEO, on Hawk Creek; 44°53′N 95°28′W. Grain, soybeans, sugar beets; hogs, sheep; manufacturing (plastic products).

Maynard (MAI-nuhrd), hamlet, SE ONTARIO, E central CANADA, 3 mi/5 km from PRESCOTT, and included in AUGUSTA township; 44°43′N 75°34′W.

Maynardville, city (2006 population 1,920), ⊙ UNION county, NE TENNESSEE, 20 mi/32 km NNE of KNOXVILLE; 36°15′N 83°48′W. In fertile farm area. Manufacturing.

Maynas (MEI-nahs), province, N LORETO region, NE PERU, on ECUADOR and COLOMBIA borders; ⊙ IQUITOS; 03°00′S 73°40′W.

Mayne Island (□ 9 sq mi/23 sq km; 2006 estimated population 880), SW BRITISH COLUMBIA, W CANADA, GULF ISLANDS, in Strait of GEORGIA just off VANCOUVER ISLAND, between GALIANO (NW) and SATURNA (SE) islands, 30 mi/48 km S of VANCOUVER; 6 mi/10 km long, 1 mi/2 km–3 mi/5 km wide; 48°50′N 123°18′W. Lumbering, farming. Village is at NW end. Part of ISLANDS TRUST regional district.

Maynooth (muh-NOOTH), Gaelic *Maigh Nuad*, town (2006 population 10,715), KILDARE county, E IRELAND, 4 mi/6.4 km WNW of LEIXLIP; 53°23′N 06°36′W. Seat of St. Patrick's College (1795), the principal institution in Ireland for training Roman Catholic clergy, and the National University of Ireland (1997). A. W. Pugin designed some of the buildings. Near St. Patrick's College are the ruins of Maynooth Castle, also called Geraldine Castle, founded c.1176; besieged in the reign of Henry VIII and dismantled in the 17th century.

Mayo (MAI-YO), Gaelic *Maigh Eo*, county (□ 2,157 sq mi/5,608.2 sq km; 2006 population 123,839), NW IRELAND; ⊙ CASTLEBAR. Borders ATLANTIC OCEAN to N and W, SLIGO and ROSCOMMON counties to E, and GALWAY county to S. The W portion, including large ACHILL ISLAND, is mountainous; the E part is more level. There are numerous lakes (MASK, CARROWMORE, CONN, and CARRA), and the irregular coastline is deeply indented by bays (KILLALA, BROAD HAVEN, BLACKSOD, and CLEW). Oats and potatoes are grown; cattle and sheep are raised. Tourism is developing. The region was granted to the De Burghs after the Anglo-Norman invasion of Ireland, but the county was not brought fully under English control until the late 16th century. Other important towns are WESTPORT, BALLINA, BALLINROBE, and CLAREMORRIS.

Mayo, municipality and island, CAPE VERDE: see MAIO.

Mayo (MAI-o), former township (□ 78 sq mi/202.8 sq km; 2001 population 441), SE ONTARIO, E central CANADA; 45°05′N 77°36′W. Merged with CARLOW in 2000 to form CARLOW MAYO township.

Mayo (MAI-yo), town (2005 population 1,032), ⊙ LAFAYETTE county, N central FLORIDA, c.60 mi/97 km WNW of GAINESVILLE; 30°03′N 83°10′W. Lumbering; limestone quarrying.

Mayo (MAI-o), unincorporated town (2000 population 1,842), SPARTANBURG county, NW SOUTH CAROLINA, near PACOLET RIVER, 10 mi/16 km NNE of SPARTANBURG; 35°04′N 81°50′W. Agriculture includes dairying; livestock; grain, peaches, apples.

Mayo, village (2006 population 248), central YUKON, NW CANADA, on STEWART RIVER, and 110 mi/177 km ESE of DAWSON; 63°36′N 135°53′W. Mining and fur-trapping region, tourism. Government services, Royal Canadian Mounted Police, and Mayo Airport. Formerly called Mayo Landing.

Mayo (MAI-o), village (□ 28 sq mi/72.8 sq km; 2006 population 462), OUTAOUAIS region, SW QUEBEC, E CANADA, 7 mi/11 km from THURSO; 45°40′N 75°21′W.

Mayo (MAI-YO), Gaelic *Maigheo na Sacsan*, agricultural village (2006 population 426), S central MAYO county, NW IRELAND, 10 mi/16 km SE of CASTLEBAR; 53°45′N 09°07′W. Has a few remains of monastery founded in 7th century by St. Colman of Lindisfarne, after his defeat at the Synod of Whitby.

Mayo, 1 de, ARGENTINA: see PRIMERO DE MAYO.

Mayo, 25 de (MAH-o, vain-tee-SEENG-ko dai), town, FLORIDA department, S central URUGUAY, near the ARROYO SANTA LUCÍA CHICO, 13 mi/21 km SW of FLORIDA, on railroad line; 34°12′S 56°20′W. Grape growing; dairying; wheat, corn; cattle, sheep. Granite and limestone deposits nearby.

Mayo, 25 de, ARGENTINA: see MAQUINCHAO.

Mayo, 25 de, ARGENTINA: see VEINTICINCO DE MAYO.

Mayo, 25 de, ARGENTINA: see VILLA SANTA ROSA.

Mayo-Banyo (MAI-o–bahn-yoo), department (2001 population 134,902), ADAMAOUA province, CAMEROON; ⊙ BANYO.

Mayo Daga (MAH-yo DAH-gah), town, TARABA state, SE NIGERIA, near CAMEROON border, 135 mi/217 km S of JALINGO. Market town. Millet; livestock.

Mayodan (MAI-o-dan), town (□ 1 sq mi/2.6 sq km; 2006 population 2,610), ROCKINGHAM county, N NORTH CAROLINA, 30 mi/48.3 km N of GREENSBORO, 29 mi/47 km NE of WINSTON-SALEM, and 2 mi/3.2 km N of MADISON, near DAN RIVER, just S of VIRGINIA border; 36°24′N 79°58′W. Manufacturing (textiles, apparel); service industries; agriculture (tobacco, grain, soybeans; livestock; dairying). BELEWS LAKE reservoir to SW. Named after the converging MAYO and Dan rivers. Settled 1894; incorporated 1899.

Mayo-Danay (MAI-o–DAH-nai), department (2001 population 522,782), Far-North province, CAMEROON; ⊙ YAGOUA. Kalfou National Game Reserve W of YAGOUA.

Mayo-Darlé (mai-o–dahr-LAI), village, ADAMAWA province, W CAMEROON, 28 mi/45 km SW of BANYO, near Nigerian border; 06°31′N 11°33′E. Tin mining.

Mayo-Kani, department (2001 population 338,448), FAR NORTH province, CAMEROON, ⊙ KAÉLÉ.

Mayo-Kébbi (mah-yo-kai-BEE), former prefecture (2000 population 1,025,003), SW CHAD, on CAMEROON (W) border; capital was BONGOR. Was bordered N by CHARI-BAGUIRMI (border, CHARI River), E by TANDJILÉ, and SE by LOGONE OCCIDENTAL prefectures. Drained by LOGONE RIVER (part of Cameroon border). Major centers included Bongor and PALA. This was a prefecture prior to Chad's administrative division reorganization from fourteen prefectures to twenty-eight departments. Following a decree in 2002 that reorganized Chad's administrative divisions into eighteen regions, this area is composed of MAYO-KÉBBI EST (NE) and MAYO-KÉBBI OUEST (SW) regions.

Mayo-Kébbi Est, administrative region, SW CHAD; ⊙ BONGOR. Borders CHARI-BAGUIRMI (NNE, border partly formed by CHARI River), TANDJILE (ESE), and MAYO-KÉBBI OUEST (SSW) administrative regions and CAMEROON (W, partly formed by LOGONE RIVER). Drained by Logone River in S. Formed following a decree in October 2002 that reorganized Chad's administrative divisions from twenty-eight departments to eighteen regions. This area made up the NE portion of former MAYO-KÉBBI prefecture.

Mayo-Kébbi Ouest, administrative region, SW CHAD; ⊙ PALA. Borders MAYO-KÉBBI EST (NE), TANDJILE (E), and LOGONE OCCIDENTAL (SE) administrative regions and CAMEROON (S, W, and N). Formed following a decree in October 2002 that reorganized Chad's administrative divisions from twenty-eight departments to eighteen regions. This area made up the SW portion of former MAYO-KÉBBI prefecture.

Mayo-Kébbi River (mah-yo-kai-BEE), c.150 mi/241 km long, in W CHAD, and N CAMEROON; issues from FIANGA and TIKEM swamps near Fianga (Chad); flows WSW, forming rapids and lakes, to the BENOUÉ (Benue) River 12 mi/19 km ESE of GAROUA (Cameroon). Through this watercourse, LOGONE RIVER is occasionally in communication in flood time with the Benoué. Sometimes spelled Mayo-Kabi; its upper course is also called Mayo-Pé.

Mayoko, village, LÉKOUMOU region, W Congo Republic, near GABON border, 60 mi/97 km N of MOSSENDJO, on railroad line to Gabon. Airfield.

Mayo-Louti (MAI-o-loo-TEE), department (2001 population 334,312), North province, CAMEROON; ⊙ GUIDDER.

Mayo-Oulo (MAI-o-oo-lo), town, North province, CAMEROON, 48 mi/77 km NNE of GAROUA; 09°58′N 13°38′E.

Mayo-Pé River, CHAD: see MAYO-KÉBBI RIVER.

Mayor Drummond, town, N MENDOZA province, ARGENTINA, in MENDOZA RIVER valley, on railroad, and 10 mi/16 km S of MENDOZA, adjoining LUJÁN; 33°01′S 68°53′W. Wine-making center.

Mayo-Rey (MAI-o-rai), department (2001 population 242,441), North province, CAMEROON; ⊙ Tchólliré.

Mayorga (mei-OR-gah), town, VALLADOLID province, N central SPAIN, 24 mi/39 km NNW of MEDINA DE RÍOSECO; 42°10′N 05°16′W. Livestock raising; cereals, wine, fruit.

Mayorga, TONGA: see VAVA'U.

Mayori, Latvia: see JŪRMALA.

Mayor, Isla, SPAIN: see ISLA MAYOR.

Mayor Island, uninhabited volcanic island, NEW ZEALAND, offshore in BAY OF PLENTY region, NE NORTH ISLAND, c. 25 mi/40 km N of TAURANGA; 37°18′S 176°16′E. Deep-sea fishing; marine reserve. Obsidian. Sometimes called Tuhua [Maori=obsidian].

Mayo River (MAH-yo), c.100 mi/161 km long, in PATAGONIA region, CHUBUT province, ARGENTINA; rises in the ANDES MOUNTAINS at CHILE border; flows E to SENGUERR RIVER 35 mi/56 km WSW of SARMIENTO.

Mayo River (MAH-yo), c.220 mi/354 km long, NW MEXICO; rises in CHIHUAHUA NE of OCAMPO; flows SW into SONORA, past NAVOJOA, and ETCHOJOA, to a lagoon inlet of Gulf of CALIFORNIA, 5 mi/8 km SW of HUATABAMPO. Used for irrigation (Mocuzari Dam); along lower course; agriculture (beans, corn, rice).

Mayo River (MAH-yo), rises around N border of SAN MARTÍN region, N central Peru; meanders SE past MOYOBAMBA to the HUALLAGA RIVER; 06°36′W 76°18′W.

Mayo River (MAI-yo), c.15 mi/24 km long, VIRGINIA and NORTH CAROLINA; formed on state line by joining of North and South Mayo rivers 13 mi/21 km SW of MARTINSVILLE (Virginia); flows S into North Carolina

past MAYODAN to DAN RIVER just E of MADISON; 36°32′N 79°59′W. North Mayo River rises in E PATRICK county, flows SE c. 20 mi/32 km; South Mayo River rises in SE Patrick county, flows c.15 mi/24 km ESE.

Mayor Luis J. Fontana (mei-OR loo-EES zhee fon-TAH-nah), town, SE Chaco province, ARGENTINA, on railroad, and 5 mi/8 km NW of RESISTENCIA; 27°45′S 60°40′W. Agriculture (cotton, peanuts, corn, citrus fruit); quebracho processing.

Mayor Otaño (mei-or o-TAHN-yo), town, E PARAGUAY, Itápua department, 60 mi/97 km S of CIUDAD DEL ESTE; 26°18′S 54°42′W. Agricultural center (soybeans, corn) along the Paraná River

Mayo-Sava (MAI-o-sah-vah), department (2001 population 313,413), Far-North province, CAMEROON; ⊙ MORA.

Mayo-Tsanaga (MAI-o-SAH-nah-guh), department (2001 population 574,864), Far-North province, CAMEROON; ⊙ MOKOLO.

Mayotte (mah-YOT), collective territory (☐ 154 sq mi/400.4 sq km); ⊙ MAMOUDZOU; 12°50′S 45°10′E. Composed of MAYOTTE island, the southeasternmost of 4 main islands of the Comoros island group; PAMANDZI island, off NE coast of Mayotte; and MTSAMBORO island, off NW coast. Economy is mainly agricultural; main export is ylang-ylang oil for perfume; other agriculture includes coffee, rice, vanilla, coconuts, cassava; livestock; fish. Highest point is Mount BENARA (2,165 ft/660 m), in S center of Mayotte island. First came under French rule in 1843. Previously part of the Comoros Islands territory; when the Comoros Islands voted to become independent in 1974, the people of Mayotte elected to remain a protectorate of FRANCE; though the island is still claimed by the Comoros Republic, status as French territory was reaffirmed by referendum in 1976 and 1992.

Mayotte (mah-YOT), island (☐ 144 sq mi/374.4 sq km), MAYOTTE territory (FRANCE), southeasternmost of the COMOROS island group, in MOZAMBIQUE CHANNEL, INDIAN OCEAN, 280 mi/451 km WSW of N tip of MADAGASCAR; 25 mi/40 km long, 10 mi/16 km wide; 12°50′S 45°10′E. Rises to 2,165 ft/660 m at Mount BENARA in S center. Fish; livestock; ylang-ylang, vanilla, cassava, rice, and coffee. Volcanic in origin, it is the only island of the Comoros surrounded by coral reefs; irregular coastline indented in SW by Boeni Bay. MAMOUDZOU, capital of Mayotte territory, on NE coast; several towns and villages connected by coastal road. Also called Moharé and Maore.

Mayoumba, GABON: see MAYUMBA.

Maypearl (MAI-puhrl), town (2006 population 911), ELLIS county, N central TEXAS, 11 mi/18 km SW of WAXAHACHIE; 32°19′N 97°00′W. In agricultural (cotton; corn, wheat; dairying; cattle, horses) area; oil and gas.

May Pen, town, ⊙ CLARENDON parish, S JAMAICA, on MINHO RIVER, on Jamaica railroad, 30 mi/48 km W of KINGSTON; 17°58′N 77°14′W. Railroad junction, road and marketing center; citrus-processing plant, canneries, rope factory.

Mayport (MAI-port), village, DUVAL county, extreme NE FLORIDA, 14 mi/23 km E of JACKSONVILLE, on ST. JOHNS RIVER near its mouth on the ATLANTIC OCEAN. Fishing. Major service center for adjacent Mayport U.S. Naval Station.

Mayraira Point (mei-REI-rah), northernmost point of LUZON, PHILIPPINES, in ILOCOS NORTE province, near entrance to Babuyan Channel in SOUTH CHINA SEA; 18°40′N 120°50′E.

Mayrán, Laguna de (mah-ee-RAHN, lah-GOO-nah dai), depression in Laguna district of COAHUILA, N MEXICO, 45 mi/72 km E of TORREÓN; 25 mi/40 km long, c.15 mi/24 km wide. Contains water only during rainy period, when NAZAS River flows to it. Construction of dams has left the laguna dry, covered with clay. Desert climate. Also known as Desierto de Mayran.

Mayreau (MEI-ro) islet (☐ 2 sq mi/5.2 sq km), SAINT VINCENT AND THE GRENADINES, WEST INDIES, between CANOUAN and UNION ISLAND, 35 mi/56 km S of Saint Vincent; 12°39′N 61°23′W. Sometimes called Mayaro or Mayero.

Mayrhofen (mei-er-HO-fen), township, TYROL, W AUSTRIA, on ZILLER RIVER, and 22 mi/35 km ESE of INNSBRUCK, between ZILLERTAL and Tuxer Alps; 47°10′N 11°52′E. Railroad terminus; three large hydropower stations nearby (Mayrhofen, Häusling, Rosshag); prominent winter and summer tourist center; cable cars to Perka (elevation 6,386 ft/1,946 m), Filzenalm (elevation 5,959 ft/1,816 m), and Hoarbergalm (elevation 6,943 ft/2,116 m). International summer schools here.

Mayrtup (mah-yeer-TOOP), village (2005 population 10,790), E CHECHEN REPUBLIC, S European Russia, in the foothills of the NE CAUCASUS Mountains, on road, 10 mi/16 km S of GUDERMES; 43°12′N 46°08′E. Elevation 698 ft/212 m. Construction materials. Agriculture (grain, livestock). Formerly known as Sulebkent.

Mayrubah, LEBANON: see MEIRUBA.

Maysan, province, SE IRAQ, bordering IRAN (E); ⊙ 'AMARA; 32°00′N 47°00′E. The lower part of the TIGRIS RIVER runs N-S through the province. Agriculture (rice, corn, millet, sesame).

Maysí, Cape, CUBA: see MAISÍ, CAPE.

Mayskiy (MAH-yee-skeeyee), city (2005 population 27,695), NE KABARDINO-BALKAR REPUBLIC, N CAUCASUS, S European Russia, on railroad (Kotlyarevskaya station), on the TEREK RIVER, on road, 25 mi/40 km NE of NAL'CHIK; 43°38′N 44°04′E. Elevation 711 ft/216 m. Railroad junction (branch to Nal'chik). Manufacturing (electric vacuum machinery, X-ray apparatus). Made city in 1965.

Mayskiy (MEI-skeeyee), town (2005 population 7,570), W central BELGOROD oblast, SW European Russia; on road and railroad; less than 2 mi/3.2 km S of BELGOROD; 50°36′N 36°35′E. Elevation 439 ft/133 m. A residential suburb of Belgorod city.

Mayskiy (MAH-yee-skeeyee), town (2006 population 12,135), W central ROSTOV oblast, S European Russia, near highway and railroad, 6 mi/10 km SW of SHAKHTY; 48°19′N 41°03′E. Elevation 439 ft/133 m. Coal mining; agricultural production research and development.

Mayskiy (MAH-yee-skeeyee), town, E AMUR oblast, SE SIBERIA, RUSSIAN FAR EAST, near the Salemdzha River (tributary of the ZEYA River), on road, 160 mi/257 km N of BLAGOVESHCHENSK; 52°17′N 129°36′E. Elevation 869 ft/264 m. Gold mines.

Mayskiy (MEIS-keeyee), village (2006 population 5,180), W central PERM oblast, E European Russia, in the W outliers of the W central URALS, in the KAMA River basin, on road and railroad, 5 mi/8 km WNW, and under administrative jurisdiction, of KRASNOKAMSK; 58°06′N 55°34′E. Elevation 344 ft/104 m. In agricultural area; pig farming, livestock feed.

Mayskiy (MEIS-keeyee), settlement (2005 population 2,855), SE KHABAROVSK TERRITORY, RUSSIAN FAR EAST, on the TATAR STRAIT, on railroad spur, 5 mi/8 km NW, and under administrative jurisdiction, of SOVETSKAYA GAVAN'; 49°01′N 140°12′E. Sawmilling, timbering; gold mining. Russian air force base in the vicinity. Formerly called Desna.

Mayskiy, RUSSIA: see MAISKI.

Mayskoye (MAH-ee-skuh-ye), village, E central KALININGRAD oblast, NW European Russia, on road, 9 mi/14 km N of GUSEV; 54°43′N 22°14′E. Elevation 147 ft/44 m. Under German administration until 1945, in EAST PRUSSIA, where it was called Mallwischken and, later (1938–1945), Mallwen.

Mayskoye, RUSSIA: see NOVYYE ATAGI.

Mays Landing, resort village (2000 population 2,321), ⊙ ATLANTIC county, S NEW JERSEY, on Great Egg Harbor River, and 16 mi/26 km WNW of ATLANTIC

CITY; 39°27'N 74°43'W. Poultry; vegetables. Seat of Atlantic Cape Community College. Settled c.1710.

Mays Lick, unincorporated village, MASON county, NE KENTUCKY, 9 mi/14.5 km SSW of MAYSVILLE. Burley tobacco, grain; livestock; dairying. Manufacturing (feeds, fertilizer).

May-sur-Èvre, Le (MAI–syoor–EV-ruh, luh), commune (□ 12 sq mi/31.2 sq km), MAINE-ET-LOIRE department, PAYS DE LA LOIRE region, W FRANCE, 5 mi/8 km N of CHOLET, on Èvre River.

Maysville, city (2000 population 1,212), ⊙ DE KALB county, NW MISSOURI, 26 mi/42 km ENE of SAINT JOSEPH; 39°53'N 94°21'W. Corn, wheat, soybeans; cattle; light manufacturing. Settled 1845.

Maysville, town (2000 population 163), SCOTT county, E IOWA, 10 mi/16 km NW of DAVENPORT; 41°38'N 90°43'W. In agricultural area.

Maysville, town (2000 population 8,993), ⊙ MASON county, NE KENTUCKY, 50 mi/80 km SE of CINCINNATI (OHIO), on the OHIO RIVER (bridged here to ABERDEEN, Ohio), in N part of BLUEGRASS REGION; 38°37'N 83°46'W. Elevation 514 ft/157 m. Railroad junction; trade and industrial center, with air, railroad, and river connections. Manufacturing (machinery, fabricated metal products, textiles, apparel, transportation equipment, building materials; lime processing, paper converting). Agriculture (burley tobacco, soybeans, grain; livestock, poultry; dairying). Daniel Boone and his wife operated a tavern here (c.1786–1789). Ulysses S. Grant attended a local school. Singer Rosemary Clooney's Childhood Home. Site of Kenton's station, a stockaded trading post, is nearby. Fleming–Mason County Airport to S. Maysville Community College (University of Kentucky). Mason County Museum, Opera House (1851). Valley Pike Covered Bridge to W. Settled c.1782 by Simon Kenton and others as Limestone; established 1787 by VIRGINIA legislature; incorporated 1833.

Maysville (MAIZ-vil), town (2006 population 979), JONES county, E NORTH CAROLINA, 18 mi/29 km SW of NEW BERN, on White Oak River; 34°53'N 77°13'W. Manufacturing (crushed stone, apparel); agriculture (grain, beans, tobacco, peanuts, cotton; livestock). Croatan National Forest to E; Hoffman Forest to W.

Maysville, town (2006 population 1,305), GARVIN county, S central OKLAHOMA, 12 mi/19 km WNW of PAULS VALLEY; 34°49'N 97°24'W. In agricultural area; manufacturing (fabricated metal products, paper products); oil and gas.

Maysville, village (2000 population 1,247), BANKS and JACKSON counties, NE GEORGIA, 14 mi/23 km E of GAINESVILLE; 34°15'N 83°34'W. Textile manufacturing.

Maytown (MAI-TOUN), ghost town, N QUEENSLAND, NE AUSTRALIA, 45 mi/72 km S of LAURA; 16°03'S 144°17'E. Ruins of 1893 gold-mining operations, part of Palmer River Gold Rush.

Maytown, unincorporated town (2000 population 2,604), LANCASTER county, SE PENNSYLVANIA, near SUSQUEHANNA RIVER, 15 mi/24 km W of LANCASTER; 40°04'N 76°34'W. In agricultural area (grain, soybeans; livestock, poultry; dairying). ELIZABETHTOWN-MARIETTA Airport to N, Marietta Ordnance Depot to S.

Mayuge, administrative district (□ 1,804 sq mi/4,690.4 sq km), 2005 population 358,400), EASTERN region, SE UGANDA; ⊙ Mayuge; 00°20'N 33°30'E. Elevation 3,937 ft/1,200 m–4,921 ft/1,500 m. As of Uganda's division into eighty districts, borders MUKONO (W), JINJA (NW), IGANGA (N), and BUGIRI (E) districts and Lake Victoria (S, with TANZANIA on S shore). Includes flat ridges, hills, and tropical forests, as well as six islands. A majority of Mayuge's area is made up of Lake Victoria. Fishing is most important economic activity. Agricultural activity is difficult because of over cultivation and soil erosion along with a growing population. Over 10% of the district's land is forest reserve. Created in 2000 from SW portion of former IGANGA

district (Bugiri district was created from SE portion in 1997 and [now former] new IGANGA district from N portion in 2000).

Mayumba (mei-YOOM-buh), city, NYANGA province, SW GABON, at mouth of M'Banio lagoon (inlet of the ATLANTIC OCEAN), 50 mi/80 km SSW of TCHIBANGA; 03°27'S 10°41'E. Small fishing port and customs station. French traders here in 17th and 18th centuries. Deep-water port is under construction. Sometimes spelled Mayouumba.

Mayumbe (muh-YOOM-bai), region in W central AFRICA, extending roughly between the ATLANTIC COAST and the CRYSTAL MOUNTAINS, N of mouth of CONGO RIVER, and S of KOUILOU river. Its rich agricultural and forest resources (coffee, cacao, palm products, bananas, rubber, hardwoods) are tapped in CONGO by BOMA-TSHELA railroad, with port of BOMA as the outlet; some gold is also mined here. In CONGO REPUBLIC lumber, copper, and lead are exported by BRAZZAVILLE-POINTE-NOIRE railroad.

Mayu Peninsula (mah-YOO), on Arakan coast, Rakhine State, Myanmar, NW of Sittwe, bet. Naaf (Bangladesh border) and Mayu rivers. Scene of fighting in World War II. The Mayu Range, 70 mi/113 km long, forms the peninsular backbone and rises to over 2,000 ft/610 m. The Mayu R, 70 mi/113 km long, forms E side of the peninsula; flows SSE, past Buthidaung (head of navigation) and Rathedaung to Bay of Bengal, NW of Sittwe.

Mayuram, India: see MAYILADUTRAI.

Mayurbhanj (muh-YOOR-buhnj), district (□ 4,019 sq mi/10,449.4 sq km), NE ORISSA state, E central INDIA; ⊙ MAYURBHANJ. Bordered NE by WEST BENGAL state, N by BIHAR state. Low alluvial tract (E); large hilly section (center), with extensive sal and bamboo forests. Rice is chief crop; corn, oilseeds also grown; lac, honey, beeswax from forests. Valuable iron ore deposits worked (N); limestone, pottery clay. Formerly a princely state in BENGAL STATES of EASTERN STATES agency; acceded 1948 to India; incorporated 1949 into Orissa state as a district.

Mayurbhanj (muh-YOOR-buhnj), town, ⊙ MAYURBHANJ district, ORISSA state, E INDIA; 21°45'N 86°30'E.

Mayview, town, LAFAYETTE county, W central MISSOURI, 9 mi/14.5 km S of LEXINGTON; 39°02'N 93°49'W. Corn, soybeans; cattle.

Mayville, town (2006 population 1,923), TRAILL county, E NORTH DAKOTA, 49 mi/79 km NNW of FARGO, and on GOOSE RIVER. Terminus of one of two railroad spurs; joins PORTLAND spur 4 mi/6.4 km NNW. Considered a twin community with Portland. Trade center; grain; livestock. Manufacturing (printing, concrete, alfafa pellets). Seat of Mayville State University. Founded in 1877 and incorporated in 1883.

Mayville, town (2006 population 5,367), DODGE county, S central WISCONSIN, on ROCK RIVER, and 15 mi/24 km ENE of BEAVER DAM; 43°30'N 88°32'W. In farming, dairying, and resort area. Manufacturing (metal products, dairy products, canned foods, cheese, maple syrup, cable and hose reels, work platforms, tool and die, furniture, fabricated metal products, paper goods; sheet metal fabricating, printing). Railroad spur terminus. Horicon National Wildlife Refuge is to NW; Horicon Marsh Wildlife Area to SW. Settled c.1844, incorporated 1885.

Mayville, village (2000 population 1,055), TUSCOLA county, E MICHIGAN, suburb 3 mi/4.8 km ESE of SAGINAW; 43°20'N 83°20'W. In agricultural area; manufacturing (calcium chloride; machining).

Mayville, village (□ 1 sq mi/2.6 sq km; 2006 population 1,718), ⊙ CHAUTAUQUA county, extreme W NEW YORK, at NW end of CHAUTAUQUA LAKE, 17 mi/27 km NW of JAMESTOWN; 42°15'N 79°30'W. Tourism associated with the Chautauqua Institution in nearby CHAUTAUQUA. Incorporated 1830.

Maywood, city (2000 population 28,083), LOS ANGELES county, S CALIFORNIA; suburb 5 mi/8 km SE of

downtown LOS ANGELES; 34°00'N 118°11'W. Although chiefly residential, it has plants that make a variety of products, such as chemicals, signs, foods, flat glass; industry base has declined. Incorporated 1924.

Maywood, village (2000 population 26,987), COOK county, NE ILLINOIS, a suburb of CHICAGO, on the Des Plaines River; 41°52'N 87°50'W. Agricultural marketing and processing point and manufacturing center. Nearby are the Loyola University Medical Center and a Veterans Affairs hospital. Incorporated 1881.

Maywood, village (2006 population 285), FRONTIER county, S NEBRASKA, 35 mi/56 km SSE of NORTH PLATTE; 40°39'N 100°37'W. Grain; livestock, poultry.

Maywood, suburban village, ALBANY county, E NEW YORK, 8 mi/12.9 km NW of downtown ALBANY; 42°45'N 73°52'W.

Maywood, borough (2006 population 9,374), BERGEN county, NE NEW JERSEY, a residential suburb between HACKENSACK and PATERSON; 40°53'N 74°03'W. Chemicals. Incorporated 1894.

Maywood Park, town (2006 population 747), MULTNOMAH CO., NW OREGON, residential suburb 4 mi/6.4 km ENE of downtown PORTLAND, near Columbia River; 45°32'N 122°33'W.

Mayya (MEI-yah), village (2006 population 6,720), E central SAKHA REPUBLIC, central RUSSIAN FAR EAST, in the LENA RIVER valley, on road junction, 40 mi/64 km SE of YAKUTSK; 61°44'N 130°17'E. Elevation 521 ft/158 m. In agricultural area.

Mayya, RUSSIA: see MAIYA.

Maza (MAH-zah), town, W BUENOS AIRES province, ARGENTINA, near LA PAMPA province border, 40 mi/64 km NW of CARHUÉ; 36°50'S 63°19'W. Railroad junction. Wheat, alfalfa; sheep, cattle.

Maza (MAIZ-uh), village (2000 population 5), TOWNER, BENSON, and RAMSEY counties, N NORTH DAKOTA, 8 mi/12.9 km S of CANDO; 48°23'N 99°12'W. Wheat and livestock area. Lake Alice National Wildlife Refuge to SE (formerly Lac Aux Mortes); Lake Alice and Chain Lake to SE; Lake Irvine to S. Founded in 1893 and incorporated in 1922. Named for Maza Chante, a Sioux Indian chief. Post office closed 1962.

Mazabuka (mah-zah-BOO-kah), township, SOUTHERN province, S central ZAMBIA, 190 mi/306 km NE of MARAMBA, on railroad; 15°52'S 27°46'E. Agriculture (cotton, tobacco, corn, sugarcane, peanuts) and livestock center (poultry, cattle). Manufacturing (cotton and coffee processing). Site of central agricultural research station.

Mazagan, MOROCCO: see EL JADIDA.

Mazagão (mah-sah-GOUN), city (2007 population 13,863), S AMAPÁ state, N BRAZIL, on northernmost channel of the AMAZON River delta, 20 mi/32 km SW of MACAPÁ; 02°00'S 51°17'W. Rubber; cacao, Brazil nuts, sarsaparilla; timber. Formerly also called Mazaganópolis.

Mazagão Velho (mah-sah-GOUN vel-yo), town, S AMAPÁ state, N BRAZIL, near N channel of the AMAZON River delta, 30 mi/48 km SW of MACAPÁ; 00°03'S 51°20'W. Rubber, Brazil nuts; timber.

Mazaleón (mah-thah-lai-ON), town, TERUEL province, E SPAIN, 12 mi/19 km E of ALCAÑIZ; 41°03'N 00°07'E. Olive-oil processing; cereals, fruit.

Mazalgaon, India: see MANJLEGAON.

Mazama, Mount, Oregon: see CRATER LAKE NATIONAL PARK.

Mazamet (mah-zah-MAI), town (□ 27 sq mi/70.2 sq km), TARN department, MIDI-PYRÉNÉES region, S FRANCE, on N slope of the MONTAGNE NOIRE (French=Black Mountain), 10 mi/16 km SE of CASTRES; 43°30'N 02°24'E. France's leading wool-cleaning and hide-processing center since 18th century. Imports woolen skins from sheep-raising countries in Southern Hemisphere. Supplies French and foreign woolen and leather industries. Spinning and weaving mills in valley of Thoré River nearby.

Mazamitla (mah-sah-MEET-lah), town, JALISCO, central MEXICO, 37 mi/60 km E of SAYULA, on Mexico Highway 110; 19°55′N 103°02′W. Grain, beans; livestock.

Māzandarān (MA-zahn-dah-RAHN), province (□ 18,292 sq mi/47,559.2 sq km), N IRAN, on CASPIAN SEA (N); ⊙ SARI. Other cities include BABOL, AMUL, and GORGAN. Traversed by the ELBURZ mountains, which run parallel to the CASPIAN SEA and divide the province into many isolated valleys as well as separating it from Tehrān and Semnān provinces. Rice, grain, fruits, cotton, tea, tobacco, sugarcane, and silk are produced in the lowland strip along the Caspian shore. Oil wealth has stimulated industries in food processing; manufacturing of cement, textiles, cotton; and fishing. Natural-gas fields in NE near the border with TURKMENISTAN. Changed hands often early in its history and was incorporated 1596 into the Persian Empire by Shah Abbas I. Formerly known as Tabaristan.

Mazapa de Madero (mah-SAH-pah dai mah-DAI-ro), town, CHIAPAS, S MEXICO, near GUATEMALA border, on Mexico Highway 190, 4 mi/6.4 km E of MOTOZINTLA DE MENDOZA; 15°23′N 92°11′W. Sugarcane. Also called Mazapa.

Mazapil (mah-SAH-peel), city and township, ⊙ Mazapil municipio, ZACATECAS, N central MEXICO, on interior plateau, near COAHUILA state border, 65 mi/105 km SW of SALTILLO; 24°40′N 101°35′W. Elevation 7,677 ft/2,340 m. Mining center (gold, silver, lead, zinc, mercury).

Mazapiltepec de Juárez (mah-sah-PEEL-tai-pek dai HWAH-res), town, PUEBLA, central MEXICO, 35 mi/56 km ENE of PUEBLA. Elevation 7,940 ft/2,420 m. Cereals, maguey, fruit; livestock.

Mazara del Vallo (mah-TSAH-rah del VAHL-lo), town (2001 population 50,377), TRAPANI province, W SICILY, ITALY, port on MEDITERRANEAN coast, and 25 mi/40 km S of TRAPANI; 37°39′N 12°35′E. Pasta. Exports Marsala wine, vegetables, corn, olive oil. Fisheries (tunny, coral). Oil refinery and gypsum quarries nearby. Bishopric. Has 11th-century cathedral (rebuilt in 17th and 20th centuries; damaged in World War II) and ruined castle. Formerly Mazzara.

Mazarambroz (mah-thah-rahm-BROTH), town, TOLEDO province, central SPAIN, 11 mi/18 km S of TOLEDO; 39°42′N 04°02′W. Cereals, olives, olive oil, potatoes; sheep.

Mazar, El (ME-zer, el), township, S central JORDAN, 8 mi/12.9 km S of AL KARAK. Grain (wheat, barley), vegetables, fruit. Center of surrounding village communities.

Mazar, El (mah-ZAHR, el), village, Sinai province, NE EGYPT, near the coast, 65 mi/105 km ESE of PORT SAID, on the road connecting CAIRO with the GAZA STRIP and ISRAEL.

Mazar-i-Sharif (mah-ZAHR–ee–shah-REEF)(Persian, *noble tomb*), city, ⊙ BALKH province, N AFGHANISTAN, on the BALKH RIVER, 190 mi/306 km NW of KABUL, near the UZBEKISTAN border; 36°42′N 67°06′E. Connected by road and air to Kabul. Ancient central Asian city and trade center in a fertile region irrigated from the Balkh River. Agriculture (grains, fruit, cotton), Manufacturing (silk and cotton goods, bricks, ordnance; flour); cotton ginning; power plant. Radio transmitter. Old town contains large bazaars selling goods imported from former Soviet Central Asian Republics. In 1480 the tomb of Caliph Ali, son-in-law and cousin of Muhammad, was believed to be discovered; the city is held sacred by some Muslims and a noted mosque of Ali is here. Most of the inhabitants are Uzbeks. During the Afghanistan War (1979–1992), the city was an important link on the line of defenses guarding the strategic road between Kabul and Termez in Uzbekistan. Headquarters of the Northern Alliance opposition to Taliban rule, the city was the focus of fighting in the civil war and take by Taliban forces; it fell to Northern Alliance forces, backed by U.S. airstrikes, in November 2001. Seat of Balkh University

Mazar-i-Sharif, AFGHANISTAN: see BALKH.

Mazarrón (mah-thah-RON), town, MURCIA province, SE SPAIN, near the MEDITERRANEAN SEA, 20 mi/32 km W of CARTAGENA; 37°36′N 01°19′W. Agriculture is principal economic activity. Fishing. Tourism. Lead and iron mines no longer exploited.

Mazaruni (ma-zah-ROO-nee), river, c.350 mi/560 km long, NW GUYANA; rises in the GUIANA HIGHLANDS; flows generally E to the ESSEQUIBO River at BARTICA; 06°25′N 58°38′W. The river is the center of Guyana's diamond industry.

Mazatán (mah-sah-TAHN), town, CHIAPAS, S MEXICO, in PACIFIC lowland, 13 mi/21 km WSW of TAPACHULA. Coffee, sugarcane, cacao, mangoes, livestock.

Mazatán, town, SONORA, NW MEXICO, 55 mi/89 km E of HERMOSILLO; 29°00′N 110°10′W. Grain, beans; cattle.

Mazateca, Sierra, MEXICO: see OAXACA, SIERRA MADRE DE.

Mazatecochco (mah-sah-te-KOCH-ko), town, ⊙ Mazatecochco de José María Morelos, TLAXCALA, central MEXICO, at W foot of MALINCHE volcano, 10 mi/16 km N of PUEBLA; 19°11′N 98°10′W. Grain; livestock. Also known as JOSÉ MARÍA MORELOS.

Mazatecochco de José María Morelos, MEXICO: see MAZATECOCHCO.

Mazatenango (mah-sah-te-NAHN-go), city (2002 population 43,600), ⊙ SUCHITEPÉQUEZ department, SW GUATEMALA, in Pacific piedmont, on railroad, on Sis River (a PACIFIC OCEAN coastal stream), and 21 mi/34 km S of QUEZALTENANGO; 14°32′N 91°30′W. Elevation 1,217 ft/371 m. Commercial center in cotton, coffee, and sugarcane district; light manufacturing; livestock raising nearby.

Mazatepec (mah-SAH-tai-pek), city and township, MORELOS, central MEXICO, 16 mi/26 km SW of CUERNAVACA. Sugarcane, rice, fruit, vegetables.

Mazatlán (mah-saht-LAHN), city (2005 population 352,471) and township, SINALOA state, W MEXICO, on the PACIFIC coast; 23°11′N 106°25′W. One of the largest commercial and industrial centers of W Mexico, Mazatlán is one of Mexico's major Pacific seaports. It is on a railroad between the U.S. and MEXICO CITY, and its location makes it the country's primary ferry link to BAJA CALIFORNIA. Although the climate is hot, Mazatlán is a popular resort with a beautiful setting. Spanish colonial trade with the PHILIPPINES stimulated the development of the port. Buelna international airport to N.

Mazatlan de Flores, MEXICO: see MAZATLÁN VILLA DE FLORES.

Mazatlán Villa de Flores (mah-zaht-LAHN VEE-yah dai FLO-res), town, in N central OAXACA, MEXICO, in S part of Sierra MAZATECA, on the Chiquito River (a tributary of the SANTO DOMINGO RIVER), on unpaved road to HUAUTLA, and 12 mi/20 km SE of TEOTITLÁN DE FLORES MAGÓN; 18°02′N 96°54′W. Elevation 3,937 ft/1,200 m. In Teotitlán judicial district. Temperate climate. Mazatec-speaking area. Agriculture (corn, sugarcane, mangoes). Formerly known as San Cristóbal Mazatlán.

Mazatzal Mountains, MARICOPA, GILA and YAVAPAI counties, central ARIZONA, NE of PHOENIX, extend c.50 mi/80 km N-S, E of along VERDE River from EAST VERDE to SALT rivers, in Tonto National Forest. Mazatzal Peak (7,903 ft/2,409 m) is highest in the range; FOUR PEAKS (7,645 ft/2,330 m) and Mount ORD (7,128 ft/2,173 m) are other high points.

Mazé (mah-ZAI), village (□ 12 sq mi/31.2 sq km), MAINE-ET-LOIRE department, PAYS DE LA LOIRE region, W FRANCE, 13 mi/21 km E of ANGERS; 47°27′N 00°16′W. Fruit growing (chiefly melons); vineyards.

Maze (MAH-ze), village, Mashita county, GIFU prefecture, central HONSHU, JAPAN, 40 mi/65 km N of GIFU; 35°52′N 137°10′E.

Mažeikiai (MUH-zhei-kyei), Russian *Mozheiki* or *Mozheyki*, city (2001 population 42,675), NW LITHUANIA, 45 mi/72 km NW of SIAULIAI, near Latvian border; 56°19′N 22°20′E. Industrial center; manufacturing (metalware, woolens, furniture, starches); flour milling, poultry raising. Large oil refinery at terminus of Novopolotsk-Mazeikiai oil pipeline. Until 1920 Mazeikiai was in Russian Kovno government. Also spelled Mazheykyay.

Mazelspoort, locality, W FREE STATE province, SOUTH AFRICA, near MODDER RIVER, 15 mi/24 km ENE of MANGAUNG (Bloemfontein); 29°12′S 26°29′E. Site of Boyden Station of Harvard Observatory on small hill (4,575 ft/1,394 m) now called Harvard Koppie. Begun 1927, when it was transferred from Arequipa (Peru), observatory was completed 1929; 60-in/152-cm telescope completed 1933. Nearby are Bloemfontein waterworks and reservoir. Recreation and watersport area.

Mazenod (mah-ZE-nod), town, MASERU district, NW LESOTHO, 8 mi/12.9 km SSE of MASERU, on LITTLE CALEDON RIVER; 29°25′S 27°34′E. MALUTI MOUNTAINS to SE; Mashoeshoe I International Airport 5 mi/8 km to S. Mazenod Institute, Roman Catholic publishing house, is here. Near junction of main A and E road routes. Corn, sorghum, vegetables; cattle, sheep, goats, hogs.

Mazeppa (muh-ZE-puh), village (2000 population 778), WABASHA county, SE MINNESOTA, 18 mi/29 km NNW of ROCHESTER, on North Fork ZUMBRO RIVER; 44°16′N 92°32′W. Grain; livestock, poultry; dairying; manufacturing (fabricated metal products, hog troughs). Richard J. Dorer Memorial Hardwood State Forest to E.

Mazeras (mah-ZE-rahs), town, COAST province, SE KENYA, on railroad, and 14 mi/23 km WNW of MOMBASA; 03°58′S 39°33′E. Copra, sugarcane, fruits. Lead deposits.

Mazères (mah-ZER), commune (□ 17 sq mi/44.2 sq km), ARIÈGE department, MIDI-PYRÉNÉES region, SW FRANCE, on HERS RIVER, and 10 mi/16 km NNE of PAMIERS; 43°16′N 01°40′E. Sawmilling; poultry raising. Medieval aristocrat Gaston de Foix born here.

Mazgirt, village, E central TURKEY, near MUNZUR RIVER, 32 mi/51 km NE of ELAZIG; c. 39°01′N 39°36′E. Wheat. Sometimes spelled Mazkirt.

Mazia Pata, CONGO: see DIBAYA.

Mazidagi, Turkish=*Mazidaği*, village, SE TURKEY, 18 mi/29 km NW of MARDIN; 37°29′N 40°29′E. Grain, chickpeas, lentils; mohair goats. Formerly Samrah.

Mazinan, township, Khorāsān province, NE IRAN, just SW of DAVARZAN, and 40 mi/64 km W of SABZEVAR; 36°18′N 56°48′E. Cotton, grain.

Mazinde (mah-ZEEN-dai), village, TANGA region, NE TANZANIA, 60 mi/97 km WNW of TANGA, at foot of USAMBARA Mountains; 04°48′S 38°09′E. Sisal, rice, grain; sheep, goats; timber.

Mazingarbe (mah-zan-gahr-buh), town (□ 4 sq mi/10.4 sq km), PAS-DE-CALAIS department, NORD-PAS-DE-CALAIS region, N FRANCE, 5 mi/8 km SE of BÉTHUNE; 50°28′N 02°42′E. Chemical industry. In old coal-mining district.

Mazkeret Batya (mahz-KE-ret BAHT-yah), Jewish village, W ISRAEL, in coastal plain, 3 mi/4.8 km SSE of REHOVOT; 31°51′N 34°50′E. Elevation 193 ft/58 m. Mixed farming. Founded 1883 and supported by Baron Edmond de Rothschild (named for his mother). Formerly known as Eqron or Ekron. Just W, the township of KIRYAT EKRON was established 1948.

Mazkirt, TURKEY: see MAZGIRT.

Mazo (MAH-tho) or **El Pueblo** (el PWAI-blo), town, PALMA island, CANARY ISLANDS, SPAIN, 4 mi/6.4 km S of SANTA CRUZ DE LA PALMA. Cereals, sweet potatoes, fruit, tobacco, wine; livestock; timber. Fishing.

Mazoe, ZIMBABWE: see MAZOWE.

Mazoe River, ZIMBABWE: see MAZOWE RIVER.

Mazomanie (maiz-o-MAIN-ee), town (2006 population 1,588), DANE county, S WISCONSIN, on tributary of WISCONSIN RIVER, and 22 mi/35 km WNW of MADISON; 43°10′N 89°47′W. Lumber; shipping point and trade center for agricultural area. Railroad junction. Timberline Ski Area to W.

Mazombe (mah-ZOM-bai), village, IRINGA region, central TANZANIA, 20 mi/32 km ENE of IRINGA; 07°41′S 35°56′E. Tobacco, corn, wheat; cattle, sheep, goats.

Mazon (MAH-ZON), village (2000 population 904), GRUNDY county, NE ILLINOIS, 8 mi/12.9 km S of MORRIS; 41°14′N 88°25′W. In agricultural area. Near the Mazon River, a classic source of fossils from a middle Pennsylvanian formation.

Mazouna (mah-zoo-NAH), village, RELIZANE wilaya, N ALGERIA, in the coastal DAHRA range, 24 mi/39 km W of CHLEF. Handicraft weaving. Until 1701, seat of W Turkish beys.

Mazovetsk, POLAND: see WYSOKIE.

Mazovia, POLAND: see MASOVIA.

Mazowe, town, MASHONALAND EAST province, N ZIMBABWE, 23 mi/37 km NNW of HARARE, on MAZOWE RIVER; 17°30′S 30°58′E. Mazowe Reservoir to SE, irrigation for citrus district. Agriculture (citrus fruit, tobacco, corn); cattle, sheep, goats; dairying. Henderson Research Station (grasslands research) located here. Iron Mask Range to E. Formerly spelled Mazoe.

Mazowe River, c. 200 mi/322 km long, SE Africa; rises c. 10 mi/16 km N of HARARE, NE ZIMBABWE; flows ENE, past MAZOWE (town) and through Bindura-Shamva gold-mining district; forms Zimbabwe-MOZAMBIQUE border for c. 15 mi/24 km before entering TETE province (Mozambique), where it joins Luenha (RUENYA) River c.30 mi/48 km S of its mouth on ZAMBEZI River. Mazoe Dam on its upper course, 20 mi/32 km NNW of Harare, irrigates citrus-growing district. Formerly spelled Mazoe.

Mazowsze, POLAND: see MASOVIA.

Mazra (muhz-RAH), village, W central JORDAN, near E shore of DEAD SEA, at mouth of the Wadi Kerak (river), 13 mi/21 km NW of AL KARAK. Elevation 981 ft/299 m below sea level. Vegetables; salt and gypsum deposits. Has grown considerably since the establishment of the Jordanian potash works nearby. Also called El Mazra'.

Mazra'a (MAHZ-rah-AH), Arab village, ISRAEL, N of AKKO (Acre), and S of NAHARIYA in WESTERN GALILEE; 32°59′N 35°05′E. Population mostly Muslim, with some Christians. Founded in 18th century. Roman and medieval ruins and a section of the Akko-Kabri aqueduct built in the 18th century and renovated in 1814.

Mazra'a El Nubani, Arab village, Ramallah district, 10.6 mi/17 km NW of RAMALLAH, in the SAMARIAN Highlands, WEST BANK; 32°02′N 35°10′E. Agriculture (wheat, fruit, olives).

Mazra'a El Qabaliya, [Arab.=southern ranch], Arab village, Ramallah district, 5 mi/8 km NW of RAMALLAH, in the SAMARIAN Highlands, WEST BANK. Agriculture (fruit, wheat, barley).

Mazra'a El Sharqiya, large Arab village, Ramallah district, 8.1 mi/13 km NE of RAMALLAH, in the highest part of the Samarian Highlands (elevation 3,084 ft/940 m), WEST BANK. Agriculture (cereal, olives, grapes, and other fruits).

Mazsalaca (MUHZ-suh-lah-tsuh), German *salisburg*, city, N LATVIA, in VIDZEME, on the SALACA RIVER, and 26 mi/42 km NW of VALMIERA; 57°52′N 25°03′E. Flax milling and forest products. Also spelled MAZSALATSA.

Mazsalatsa, LATVIA: see MAZSALACA.

Mazuecos (mah-THWAI-kos), village, GUADALAJARA province, central SPAIN, near TAGUS RIVER, 37 mi/60 km ESE of MADRID; 40°15′N 03°00′W. Grain, grapes; livestock; olive-oil pressing.

Mazul'skiy (mah-ZOOL-skeeyee), town (2005 population 1,300), SW KRASNOYARSK TERRITORY, SE SIBERIA, RUSSIA, 6 mi/10 km SW, and under administrative jurisdiction, of ACHINSK; 56°12′N 90°20′E. Elevation 1,036

ft/315 m. Founded in 1933, when manganese mining began in the area.

Mazunga, town, MATABELELAND NORTH province, S ZIMBABWE, 33 mi/53 km NNW of BEITBRIDGE; 21°43′S 29°54′E. Cattle-raising center; sheep, goats.

Mazurski Kanal, POLAND and RUSSIA: see MASURIAN CANAL.

Mazury, POLAND: see MASURIA, OLSZTYN.

Mazy (mah-ZEE), village, Namur district, NAMUR province, S central BELGIUM, 10 mi/16 km WNW of NAMUR; 50°31′N 04°40′E. Marble and chalk quarrying.

Mazza, El (MAHZ-zah, el), town, DAMASCUS district, SW SYRIA, S of the Barada River, 3 mi/4.8 km W of Damascus; 33°30′N 36°15′E. Satellite town of Damascus.

Mazzarino (mah-tsah-REE-no), town, CALTANISSETTA province, S central SICILY, ITALY, 15 mi/24 km SE of CALTANISSETTA; 37°18′N 14°13′E. Food processing. Has ancient castle, palace.

Mba, FIJI: see BA.

Mbabane (uhm-bah-BAH-nai), city, HHOHHO district, administrative ⊙ SWAZILAND, NW Swaziland, in the Mdimba Mountains on the edge of the high veld; 26°19′S 31°08′E. It is primarily an administrative center and commercial hub for the surrounding agricultural region. Timber; corn, tea, vegetables, citrus; cattle, goats, sheep, hogs. Manufacturing (food processing; handicrafts, wood products). Thriving market for curios. Usutu Pulp Mill to S near Bhunya International Airport to SW. Swaziland National Archives; Swaziland Inst. of Management and Public Administration are here; seat of Swaziland College of Technology. Casino to SE at Ezulwini; tourism. First settled in 1888 and established in 1902 by the British.

Mbacké (uhm-BAH-kai), town (2004 population 52,898), DIOURBEL administrative ⊙ region, W SENEGAL, on Diourbel-Touba branch line of Dakar-Niger railroad, 22 mi/35 km ENE of DIOURBEL; 14°48′N 15°55′W. Peanut growing. Also spelled Mbaké or M'Backé.

Mbagne (uhm-BAH-nyuh), village (2000 population 10,383), BRAKNA administrative region, SW MAURITANIA, on the banks of the SÉNÉGAL RIVER, 171 mi/275 km ENE of SAINT-LOUIS (SENEGAL); 16°09′N 13°47′W. Livestock; millet, potatoes, sorghum.

Mbahiakro (uhm-bah-hee-AH-kro), town, N'zi-Comoé region, CÔTE D'IVOIRE, 37 mi/60 km NW of DAOUKRO; 07°27′N 04°20′W. Agriculture (yams, bananas, maize, peanuts, coffee).

M'Baïki (uhm-bei-KEE), town, ⊙ LOBAYE prefecture, S CENTRAL AFRICAN REPUBLIC, near LOBAYE River, 55 mi/89 km SW of BANGUI; 03°53′N 18°01′E. Market for rubber and palm products; sawmills; gold mining.

Mbakaou (uhm-BAH-ko-oo), reservoir, ADAMAOUA province, central CAMEROON, just E of TIBATI; 06°22′N 12°52′E.

M'Baké, SENEGAL: see MBACKÉ.

Mbala, town, NORTHERN province, NE ZAMBIA, 12 mi/19 km E of S end of LAKE TANGANYIKA; 08°50′S 31°22′E. Elevation 5,400 ft/1,646 m. Gold to S. Agriculture (coffee, tobacco); cattle. Manufacturing (food processing). Tourism; Moto Moto Museum. Airfield. A former British trade post, established 1889. Was capital of former Tanganyika province In hilly region (highest elevation, 6,784 ft/2,068 m, lies to SE); KALAMBO FALLS (725 ft/221 m) are 20 mi/32 km NNW, on TANZANIA border. Diesel power station; Lunzua hydroelectric station to W. NSUMBU NATIONAL PARK to W. Formerly called Abercorn.

Mbalabala (uhm-BAH-lah-BAH-lah), town, MATABELELAND SOUTH province, SW ZIMBABWE, 35 mi/56 km SE of BULAWAYO, on railroad; 20°27′S 29°03′E. Elevation 3,603 ft/1,098 m. Road junction. Filabusi goldmining area to E. Cattle, sheep, goats; tobacco, corn, sorghum, soybeans; dairying. Formerly called Balla Balla.

Mbalangeti River (uhm-bah-lahn-GAI-tee), c. 85 mi/137 km long, in SERENGETI PLAIN, N TANZANIA; rises

125 mi/201 km E of MWANZA in SERENGETI NATIONAL PARK; flows NW through length of park, enters E end of SPEKE GULF, Lake VICTORIA, 65 mi/105 km ENE of Mwanza.

Mbale, former administrative district (☐ 571 sq mi/1,484.6 sq km; 2005 population 783,600), EASTERN region, E UGANDA; capital was MBALE; 00°57′N 34°20′E. Elevation (in the plain areas) averaged 5,906 ft/1,800 m. As of Uganda's division into fifty-six districts, was bordered by TORORO (SSW), PALLISA (W), KUMI (NW), and SIRONKO (N) districts and KENYA (E). Original inhabitants were the Bamasaba people; the Adholas, Banyoli, Etesots, and Sabiny peoples also lived here. Part of Mount ELGON (14,176 ft/4,321 m) here, on Kenya border. Varied vegetation, from tropical and grassland savannah in plain to alpine vegetation in mountains. Primarily agricultural (included beans, carrots, coffee, cotton, maize, onions, plantains, potatoes, and sweet potatoes). Formed c.2000–2001 from S portion of former MBALE district (Sironko district was formed from N portion). In 2005 E and SE portions of district were carved out to form MANAFWA district; in 2006 NE portion was carved out to form BUDUDA district and W portion was formed into current MBALE district.

Mbale, administrative district, EASTERN region, E UGANDA; ⊙ MBALE. As of Uganda's division into eighty districts, Mbale's border districts include SIRONKO (N), BUDUDA (NE), MANAFWA (E), TORORO (S), and BUTALEJA, BUDAKA, and PALLISA (W). Includes W part of MOUNT ELGON (14,176 ft/4,321 m). Agricultural area (including cotton). Main road travels through Mbale town, connecting it to KAMPALA city to SW, SOROTI and LIRA towns to NW, Nakapiripirit town to NNE, and KENYA to SSE. Formed in 2005–2006 from W portion of former MBALE district created in 2000–2001 (Manafwa district formed from E and SE portions in 2005 and Bududa district from NE portion in 2006).

Mbale (uhm-BAH-lai), former administrative district (☐ 983 sq mi/2,555.8 sq km), E UGANDA, along KENYA border (to E); capital was MBALE; 01°10′N 34°15′E. As of Uganda's division into thirty-nine districts, was bordered by TORORO (SSW), PALLISA (W), KUMI (NW), MOROTO (N), and KAPCHORWA (NE) districts. Rich agricultural areas (coffee, cotton millet, bananas). MOUNT ELGON was in E. Mbale is Uganda's third-largest city. In c.2000–2001 S portion of district was formed into (now former) new MBALE district and N portion formed into SIRONKO district.

Mbale (uhm-BAH-lai), town (2002 population 71,130), ⊙ MBALE district, EASTERN region, UGANDA, 80 mi/129 km NE of Jinja, at W foot of MOUNT ELGON. Agricultural trade center (cotton, corn, bananas, millet; dairying). Railroad station is 2 mi/3.2 km W. Was capital of former EASTERN province.

M'Balmayo (uhm-BAHL-mai-o), town (2001 population 65,400), ⊙ NYONG-ET-SOO department, Central province, CAMEROON, on NYONG River, and 25 mi/40 km S of Yaoundé; 03°32′N 11°35′E. Transshipment point, railroad terminus, and trading center; brick manufacturing; sawmilling. Nyong River is navigable between here and Abong-M'Bang. Also spelled Mbalmayo.

Mbam, department, CENTRAL province, CAMEROON: see MBAM-ET-INOUBOU and MBAM-ET-KIM departments.

Mbamba Bay (uhm-BAHM-bah bai-ee), village, RUVUMA region, S TANZANIA, port on Lake NYASA, 75 mi/121 km SW of SONGEA, 17 mi/27 km N of MOZAMBIQUE border. Road terminus, airstrip. Tobacco, subsistence crops; livestock; timber.

Mbam-et-Inoubou, department (2001 population 153,020), CENTRAL province, CAMEROON.

Mbam-et-Kim, department (2001 population 64,540), CENTRAL province, CAMEROON.

M'Bamou Island, REPUBLIC OF THE CONGO: see MALEBO POOL.

M'Bam River (uhm-BAHM), in central CAMEROON, c.175 mi/282 km long; formed by headstreams 40 mi/64 km WSW of TIBATI; flows SW and S to SANAGA River 25 mi/40 km SE of BAFIA. Also spelled MBAM.

Mbandaka (mbuhn-DAH-kuh), city, ⊙ Équateur province, W CONGO, a port on the CONGO RIVER; 00°04′N 18°16′E. Elev. 1,007 ft/306 m. It is a commercial and transportation center and has tanning and fishing industries. Formerly called COQUILHATVILLE.

Mbandjock (uhm-BAN-jok), town (2001 population 17,500), Central province, CAMEROON, 45 mi/72 km NE of Yaoundé; 04°26′N 11°56′E. Sugarcane is a major crop produced in the area. Food processing is an important industry.

M'Banga (uhm-BANG-uh), town (2001 population 29,400), LITTORAL province, W CAMEROON, on railroad, 40 mi/64 km SW of N'Kongsamba and 30 mi/48 km NNW of DOUALA; 04°31′N 09°32′E. Banana-growing center; banana drying and processing for export. Also cacao and coffee. Also spelled MBANGA.

Mbanga, CAMEROON: see M'BANGA.

Mbangala (uhm-ban-GAH-lah), village, MOROGORO region, central TANZANIA, 85 mi/137 km SE of IRINGA; 08°37′S 36°41′E. Timber; cattle, sheep, goats; corn, wheat. Mount Chikweta to SW.

Mbanza Congo (uhm-BAHN-zuh KON-go), town, ⊙ ZAIRE province, NW ANGOLA, 70 mi/113 km SE of MATADI (Democratic Republic of the CONGO); 06°22′S 14°17′E. Cashews, corn, cotton, manioc. Airfield. Jesuit mission in 19th century. Was capital of native kingdom of Congo 16th–18th century. Formerly called São Salvador.

Mbanza-Ngungu (MBAHN-zuh—NGOON-goo), town, KINSHASA region, W DEMOCRATIC REPUBLIC OF THE CONGO, on railroad, and 130 mi/209 km ENE of BOMA; 05°15′S 14°52′E. Elev. 1,984 ft/604 m. Commercial center with large railroad workshops; sugar manufacturing for export. Also has tourist facilities and is known as a resort. Roman Catholic and Baptist missions; hospital; airport. Formerly known as THYSVILLE or THYSTAD.

Mbao (uhm-BOU), village, DAKAR administrative region, W SENEGAL, 9 mi/14.5 km NE of DAKAR, on the CAPE VERDE peninsula; 15°48′N 16°31′W. Oil refinery.

Mbaracayú, Cordillera de, PARAGUAY and BRAZIL: see MARACAJU, SERRA DE.

Mbarara, administrative district, WESTERN region, SW UGANDA; ⊙ MBARARA. As of Uganda's division into eighty districts, borders IBANDA (N), KIRUHURA (E), ISINGIRO (SE), NTUNGAMO (S), and BUSHENYI (W) districts. Some marsh area E of Mbarara town. The Iro and Hima people live here. Agriculture (including bananas and coffee); cattle-raising. Several roads branch out of Mbarara town, connecting it to surrounding Uganda. Formed in 2005 from W portion of former MBARARA district (Ibanda district formed from NW portion, Isingiro district from S portion, and Kiruhura district from N, E, and central portions).

Mbarara (uhm-bah-RAH-ruh), former administrative district (□ 2,836 sq mi/7,373.6 sq km; 2005 population 1,172,800), WESTERN region, SW UGANDA, along TANZANIA border (to S); capital was MBARARA; 00°25′S 30°45′E. As of Uganda's division into fifty-six districts, was bordered by NTUNGAMO (SW), BUSHENYI (W), KAMWENGE and KYENJOJO (N), SEMBABULE (NE), and RAKAI (E) districts. Some marsh area to E of Mbarara town. Primary inhabitants were the Banyankore people. Cattle-raising area (milk production was important); agriculture (large producer of matooke, also millet and sweet potatoes, cash crops included bananas, beans, coffee, maize, and potatoes); tin deposits. The cow was the district symbol. In 1993, part of district was taken to form Ntungamo district. In 2005 NW portion of district was carved out to form IBANDA district, S portion was carved out to form ISINGIRO district, N, E, and central portions were carved out to form KIRUHURA district, and W portion was formed into current MBARARA district.

Mbarara (uhm-bah-RAH-ruh), town (2002 population 69,363), ⊙ MBARARA district, principle town of WESTERN region, SW UGANDA, 80 mi/129 km WSW of MASAKA. Agricultural trade center; coffee, corn, millet, bananas; cattle, sheep, goats. Inhabited by Iro and Hima people. Local industries include plywood, soap, textiles. Wood carving, weaving, pottery making. Was capital of former SOUTHERN province.

Mbarika Mountains (uhm-ba-REE-kah), MOROGORO region, S central TANZANIA, 100 mi/161 km SE of IRINGA. Range of low hills, mostly 1,640 ft/500 m–3,281 ft/1,000 m high, traversed by Lowego River. Mount Chikweta (4,974 ft/1,516 m) is highest point.

Mbase Pan Sanctuary, MATABELELAND NORTH province, W central ZIMBABWE, 10 mi/16 km NW of NKAYI, on SHANGANI RIVER; 18°52′S 28°48′E. Wildlife refuge surrounding backwater lake.

Mbashe River, c.100 mi/161 km long, EASTERN CAPE province, SOUTH AFRICA; rises in foothills of S DRAKENSBERG near ELLIOT; flows in winding course SE to enter INDIAN OCEAN on Wild Coast, 58 mi/92 km SW of Port St. John's.

Mbate (uhm-BAH-tai), village, LINDI region, SE TANZANIA, 25 mi/40 km WNW of KILWA MASOKO, near MATANDU RIVER; 08°52′S 39°09′E. Cashews, bananas, sweet potatoes, corn; sheep; goats; timber.

Mbatiki, FIJI: see BATIKI.

Mbau, FIJI: see BAU.

Mbé (uhm-BAI), town, North province, CAMEROON, 36 mi/58 km N of Ngaoundéré; 07°52′N 13°37′E.

Mbemba (uhm-BAI-uhm-bah), village, LINDI region, S TANZANIA, 75 mi/121 km W of LINDI, on Mbwemburu River, E of Mihumo River mouth; 10°03′S 38°30′E. Cashews, bananas, sweet potatoes, corn; sheep, goats.

Mbembesi, town, MATABELELAND NORTH province, SW central ZIMBABWE, 24 mi/39 km ENE of BULAWAYO, on railroad; 19°59′S 28°55′E. Elevation 4,482 ft/1,366 m. Cattle, sheep, goats; tobacco, corn, peanuts, soybeans.

Mbemkuru River (uhm-baim-KOO-roo), c. 165 mi/266 km long, LINDI region, SE TANZANIA; formed by joining of several headstreams c. 110 mi/177 km WSW of LINDI; flows NE, enters INDIAN OCEAN 35 mi/56 km N of Lindi. Its main tributary, the Mihumo River, enters from the W 80 mi/129 km W of Lindi.

Mbengga, FIJI: see BEQA.

Mbengwi (uhm-BANG-wee), town, ⊙ MOMO department, North-West province, CAMEROON, 11 mi/18 km NW of BAMENDA; 06°01′N 10°01′E.

Mbere, department (2001 population 185,473), ADAMAOUA province, CAMEROON, ⊙ MEIGANGA.

Mberengwa (uhm-bai-rai-en-GWAH), town, MATABELELAND SOUTH province, S central ZIMBABWE, 13 mi/21 km SW of ZVISHAVANE, near NGEZI RIVER; 20°28′S 29°55′E. Gold mining and limestone quarrying. Cattle, sheep, goats; dairying; corn, wheat, tobacco, soybeans. Formerly called Belingwa.

M'Béré River (uhm-BAI-rai), W headstream of LOGONE River, 275 mi/443 km long, in E CAMEROON, SW CHAD, and Central Afr. Republic; rises 65 mi/105 km SE of N'Gaoundéré (Cameroon); flows NE, entering Chad, past Baïbokoum and MOUNDOU, to join Pendé River (E branch of Logone) 28 mi/45 km SSE of Laï. Also spelled Mbéré; also called Western LOGONE River.

Mbeya (uhm-BAI-yah), region (2006 population 2,346,000), SW TANZANIA, bounded S by ZAMBIA, MALAWI, and Lake NYASA (part of Malawi border); ⊙ MBEYA. Part of Lake Ruaha, in Uwanda Game Reserve, in W; part of RUAHA NATIONAL PARK in NE; MBEYA RANGE in S center. LUPA GOLDFIELDS (closed) in center. Coffee, tea, pyrethrum, corn, wheat; cattle, sheep, goats; fish; timber. Part of former SOUTHERN HIGHLANDS province.

Mbeya (uhm-BAI-yah), town (2002 population 230,318), ⊙ MBEYA region, SW TANZANIA, 250 mi/402

km SW of DODOMA, on Tanzam railroad; 08°55′S 33°29′E. Road junction; airstrip to W. MBEYA RANGE to N. Tea, coffee, pyrethrum; cattle, sheep, goats. Manufacturing (coffee and tea processing, cement). LUPA GOLDFIELDS (abandoned 1956) 20 mi/32 km N. MBOZI Meteorite Crater 30 mi/48 km WSW.

Mbeya Range or **Mbeya Mountain** (both: uhm-BAI-yah), small ridge, MBEYA region, SW TANZANIA, N of MBEYA; 30 mi/48 km long, 10 mi/16 km wide. Highest point in W part.

Mbigou (uhm-BEE-goo), town, NGOUNIÉ province, S GABON, 60 mi/97 km E of MOUILA; 01°50′S 11°54′E. Coffee. Famous Mbigou carving stone quarried nearby.

Mbilua, Solomon Islands: see VELLA LAVELLA.

Mbindera (uhm-been-DAI-rah), village, LINDI region, SE central TANZANIA, 155 mi/249 km WNW of LINDI; 09°36′S 37°24′E. SELOUS GAME RESERVE to W and N. Livestock; grain.

Mbinga (uhm-BEEN-gah), village, RUVUMA region, S TANZANIA, 45 mi/72 km WSW of SONGEA; 10°57′S 34°59′E. Timber; goats, sheep; grain.

Mbingu (uhm-BEEN-goo), village, MOROGORO region, S central TANZANIA, 100 mi/161 km SE of IRINGA; 08°53′S 36°41′E. Livestock; grain. MBARIKA MOUNTAINS to SW; SELOUS GAME RESERVE to S and E.

Mbingué (uhm-ben-GWAI), town, Savanes region, N CÔTE D'IVOIRE, 24 mi/38 km SW of NIÉLÉ; 10°00′N 05°54′W. Agriculture (corn, millet, peanuts, beans, cotton). Also spelled Mbengué.

Mbini (uhm-BEE-nee), town (2003 population 11,600), LITORAL province, continental EQUATORIAL GUINEA, on GULF OF GUINEA at mouth of the MBINI RIVER, 20 mi/32 km SSW of BATA; 01°35′N 09°38′E. Port exports lumber, coffee, palm oil. Sawmilling. Site of Forest Institute. Formerly called Benito or Río Benito.

Mbini River (uhm-BEE-nee), circa 200 mi/322 km long, continental EQUATORIAL GUINEA; rises as the Woleu in GABON SE of OYEM; flows generally W to the Gulf of GUINEA at MBINI TOWN. Interrupted by rapids at Mandoc and Ngom. Formerly called Benito or San Benito River.

Mbizi, village, MASVINGO province, SE central ZIMBABWE, 92 mi/148 km SSE of MASVINGO; 21°23′S 31°01′E. Junction of railroad spur to MKWASINE. Livestock; grain.

Mbocayaty (uhm-bo-kah-yah-TEE), town, Guairá department, S PARAGUAY, at road junction 4 mi/6 km NNE of Villarrica; 25°43′S 56°25′W. Agricultural center (sugarcane, tobacco, maté, fruit, cotton; livestock).

M'Bomou (uhm-bo-MOO), prefecture (□ 23,604 sq mi/61,370.4 sq km; 2003 population 164,008), SE CENTRAL AFRICAN REPUBLIC; ⊙ BANGASSOU. Bordered N by HAUTE-KOTTO prefecture, E by HAUTE-M'BOMOU prefecture, S by CONGO, and W by BASSE-KOTTO prefecture. Drained by CHINKO, M'bari, M'Bomou, OUARRA, and Vovodo rivers. Agriculture (coffee, cotton); uranium, copper, and tin deposits. Main centers are Bangassou, BAKOUMA, and RAFAÏ.

M'Bomu River, CONGO: see BOMU RIVER.

Mboro (uhm-BOOR-o), town (2004 population 12,076), THIÈS administrative region, W SENEGAL, 48 mi/77 km NE of DAKAR; 15°09′N 16°54′W. Important vegetable production in inter-dune depressions with favorable microclimate and groundwater conditions. Fishing, phosphate mining nearby.

M'Boud (uhm-BOOD), village (2000 population 8,899), Gorgol administrative region, S MAURITANIA, 255 mi/410 km E of SAINT-LOUIS (SENEGAL); 16°02′N 12°35′W. Gum, millet; livestock. Also spelled Mbout.

Mbouda (uhm-BOO-duh), city (2001 population 101,100), ⊙ BAMBOUTOS department, West province, CAMEROON, 17 mi/27 km NW of Bafoussam; 05°38′N 10°16′E.

Mboulou, town, KOUILOU region, SW Congo Republic, 35 mi/56 km NE of POINTE-NOIRE.

Area is shown by the symbol □, and capital city or county seat by ⊙.

Mbour (uhm-BOOR), town (2004 population 165,719), THIÈS administrative region, W SENEGAL, on the ATLANTIC OCEAN, 37 mi/60 km SE of DAKAR; 14°24′N 16°28′W. Trade and processing center for peanuts grown in the area; has an important fishing industry. Important tourist destination. Also spelled M'bour.

Mbout, MAURITANIA: see M'BOUD.

Mbozi (uhm-BO-zee), village, MBEYA region, SW TANZANIA, 35 mi/56 km WSW of MBEYA, on railroad; 09°03′S 33°00′E. Mbozi Meteorite Crater to E. Cattle, goats, sheep; corn, coffee, tea, pyrethrum.

Mbridge River (uhm-BRIDJ), 220 mi/354 km long, NW ANGOLA; rises in central plateau SW of MAQUELA DO ZOMBO; flows WSW to the ATLANTIC OCEAN just N of N'zeto. Not navigable.

Mbugwe (uhm-BOO-gwai), village, N central TANZANIA, 80 mi/129 km SW of ARUSHA, 5 mi/8 km E of MBULU; 03°39′S 35°32′E. Cattle, goats, sheep; wheat, corn.

Mbuji-Mayi (MBOO-yee—MAH-yee), city, ⊙ KASAI-ORIENTAL province, S central CONGO, on the SANKURU RIVER; 06°09′S 23°36′E. Elev. 1,804 ft/549 m. A commercial center in LUBA country, it handles most of the industrial diamonds produced in CONGO, and is known as the industrial diamond capital of the world. After CONGO attained independence (1960), the city's population grew rapidly with the immigration of LUBA people from other parts of the country (especially SHABA, where they were suffering from conflicts with other ethnic groups). From 1960 to 1962, it was capital of the secessionist Mining State of South Kasai. Many still live in a refugee camp created in 1992–1994. University. Formerly called BAKWANGA. Also spelled MBUY-MAYI.

Mbulamuti (uhm-boo-lah-MOO-tee), town, EASTERN region, SE central UGANDA, 30 mi/48 km NNW of Jinja. Elevation 3,490 ft/1,064 m. Railroad junction (branch line to NAMASAGALI). Cotton, tobacco, coffee, bananas, corn. Was part of former BUSOGA province.

Mbulu (uhm-BOO-loo), town, N TANZANIA, 90 mi/145 km WSW of ARUSHA, SW of Lake MANYARA; 03°52′N 35°46′E. Cattle, sheep, goats; corn, wheat, millet.

Mbuluzi River (uhm-boo-LOO-zee), c.60 mi/97 km long, SWAZILAND and MOZAMBIQUE; formed by joining of BLACK MBULUZI and WHITE MBULUZI rivers 47 mi/76 km ENE of MBABANE, in E Swaziland; flows E into Mozambique at GOBA, then NE past BOANE; enters MAPUTO BAY, INDIAN OCEAN, through Espurito Santo Estuary SW of Maputo. Called Umbuluzi in Mozambique.

Mbunga (uhm-BOON-gah), village, RUVUMA region, S TANZANIA, 40 mi/64 km NNE of SONGEA, near RUTUKIRA RIVER; 10°06′S 35°52′E. Livestock; subsistence crops.

Mburucuyá, town (1991 population 4,366), ⊙ Mburucuyá department, N CORRIENTES province, ARGENTINA, 55 mi/89 km SE of CORRIENTES; 28°03′S 58°14′W. Agricultural center (oranges, cotton, tobacco, sugarcane; livestock).

Mbusi River (uhm-BOO-see), c. 40 mi/64 km long, N central TANZANIA, E SHINYANGA region, in SERENGETI PLAIN; rises c. 80 mi/129 km ENE of SHINYANGA; flows S, enters SIBITI RIVER 70 mi/113 km ESE of Shinyanga.

Mbuyapey, town, Paraguarí department, S PARAGUAY, 85 mi/137 km SE of Asunción, 30 mi/48 km SW of VILLARRICA, on road junction; 26°13′S 56°45′W. Agricultural center (fruit; livestock); lumbering. Trade in hides.

Mbuy-Mayi, CONGO: see MBUJI-MAYI.

Mbuyuni (uhm-boo-YOO-nee), village, MOROGORO region, central TANZANIA, 85 mi/137 km SW of MOROGORO, on GREAT RUAHA RIVER (bridged); 07°29′S 36°32′E. Cattle, goats, sheep, poultry; timber. Rubeho Mountains to N.

Mbwawa (uhm-BWAH-wah), village, PWANI region, E TANZANIA, 30 mi/48 km WNW of DAR ES SALAAM,

near RUVU RIVER; 06°04′S 38°14′E. Cashews, corn, wheat, bananas, sisal; goats, sheep, poultry; timber.

McAdam, village (2001 population 1,513), SW NEW BRUNSWICK, CANADA, 40 mi/64 km SW of FREDERICTON, 5 mi/8 km ENE of VANCEBORO (MAINE); 45°35′N 67°20′W. Railroad junction; lumbering; hunting; tourist center.

McAdenville (mik-AD-uhn-vil), town (☐ 1 sq mi/2.6 sq km; 2006 population 641), GASTON county, S NORTH CAROLINA, suburb 15 mi/24 km W of CHARLOTTE, 7 mi/11.3 km E of GASTONIA, on South Fork of CATAWBA RIVER; 35°15′N 81°04′W. Manufacturing (textiles); service industries. Incorporated 1883.

McAdoo (mahk-A-doo), borough (2006 population 2,123), SCHUYLKILL county, E central PENNSYLVANIA, 5 mi/8 km S of HAZLETON; 40°53′N 75°59′W. Agriculture (corn, hay; poultry; dairying); manufacturing (cigars, building materials, motor vehicles); anthracite coal. Tuscarora State Park to S. Founded 1880, incorporated 1896.

McAlester (muh-KAL-uh-stuhr), city (2006 population 18,333), ⊙ PITTSBURG county, SE OKLAHOMA, 55 mi/89 km SSW of MUSKOGEE, 26 mi/42 km SSW of EUFAULA; 34°55′N 95°45′W. Elevation 740 ft/226 m. Railroad junction. A former coal-mining and farming community, McAlester has become a regional distribution center with a busy stockyard. Agriculture; manufacturing (apparel, motor vehicles, chemicals, ammunition, animal feeds, concrete), printing and publishing, food processing. McAlester Army Munitions Plant to SW, at Savanna. McAlester Municipal Airport to S. Incorporated 1899.

McAlisterville (muh-KA-luhs-tuhr-vil), unincorporated town, JUNIATA county, central PENNSYLVANIA, 8 mi/12.9 km ENE of MIFFLINTOWN on Little Lost Creek; 40°38′N 77°16′W. Manufacturing of lumber, wooden products, railroad ties; agriculture includes dairying; livestock; grain; timber.

McAllen (MIK-a-luhn), city (2006 population 126,411), HIDALGO county, extreme S TEXAS, 50 mi/80 km WNW of BROWNSVILLE, on the RIO GRANDE; 26°13′N 98°14′W. It is a railroad junction, a port of entry, and a packing and processing center for the citrus fruit, vegetables and other produce, and flowers grown in the lower Rio Grande valley. The growing city has oil refineries and manufacturing (medical supplies, apparel, marble products, printing, steel fabrication). Increasing numbers of multinational corporations and corporate offices are located in McAllen. It is also a winter resort. Its nickname, "the City of Palms," came from the fact that forty palm varieties flourish here. McAllen, connected by bridge with REYNOSA, MEXICO (8 mi/12.9 km SSW), saw a 50% increase in population from 1970 to 1990. Miller International Airport in S part of city. Bentson–Rio Grande Valley State Park is to W; Santa Ana National Wildlife Refuge to SE. Incorporated 1911.

McAlpin, locality, RALEIGH county, S WEST VIRGINIA, 8 mi/12.9 km SW of BECKLEY.

McAndrews (muh-KAND-rooz), unincorporated village, PIKE county, E KENTUCKY, in the CUMBERLAND MOUNTAINS, 7 mi/11.3 km S of WILLIAMSON (WEST VIRGINIA). Bituminous coal.

McArthur (mik-AHR-thur), village (☐ 1 sq km; 2006 population 2,050), ⊙ VINTON county, S OHIO, 27 mi/43 km ESE of CHILLICOTHE; 39°14′N 82°28′W. In livestock area; chemicals, lumber, clay products. Plotted 1815, incorporated 1851.

McArthur–Burney Falls State Park, California: see BURNEY.

McArthur Falls (muhk-AHR-thur), waterfalls in SE MANITOBA, W central CANADA, on WINNIPEG River, at N end of Lac du BONNET, 65 mi/105 km NE of WINNIPEG; 50°23′N 95°59′W. Hydroelectric power center.

McArthur, Mount (muhk-AHR-thur), peak (9,892 ft/3,015 m), SE BRITISH COLUMBIA, W CANADA, near ALBERTA border, in ROCKY MOUNTAINS, in YOHO

NATIONAL PARK, 50 mi/80 km NW of BANFF (Alberta); 51°32′N 116°36′W.

McArthur, Mount, peak (14,400 ft/4,389 m), SW YUKON, CANADA, near ALASKA border, in ST. ELIAS MOUNTAINS, 180 mi/290 km W of WHITEHORSE; 60°36′N 140°12′W.

McArthur, Port (muhk-AHR-thur), harbor, NE NORTHERN TERRITORY, AUSTRALIA, in Gulf of CARPENTARIA; formed by SIR EDWARD PELLEW ISLANDS (N) and mainland (W, S); 25 mi/40 km long, 10 mi/16 km wide; 15°47′S 136°41′E. Receives MCARTHUR RIVER.

McArthur River (muhk-AHR-thur), 125 mi/201 km long, NE NORTHERN TERRITORY, AUSTRALIA; rises in N hills of BARKLY TABLELAND; flows NE to Port MCARTHUR of Gulf of CARPENTARIA; 15°54′S 136°40′E. Navigable 40 mi/64 km by barges below BORROLOOLA.

McAuley (muh-KAW-lee), unincorporated village, SW MANITOBA, W central CANADA, just E of SASKATCHEWAN border, in ARCHIE rural municipality; 50°15′N 101°23′W. Agricultural area.

McBain, village (2000 population 584), MISSAUKEE county, N central MICHIGAN, 10 mi/16 km SE of CADILLAC; 44°11′N 85°12′W. Manufacturing (sawmill; feeds).

McBaine, town, BOONE county, central MISSOURI, on MISSOURI RIVER, 7 mi/11.3 km SW of COLUMBIA. Virtually destroyed by 1993 flood.

McBee, town (2006 population 714), CHESTERFIELD county, NE SOUTH CAROLINA, 38 mi/61 km NW of FLORENCE; 34°28′N 80°15′W. Manufacturing includes printing equipment, wire, apparel, water heaters, steel products. Agriculture includes peaches, watermelons; livestock.

McBride (muhk-BREID), village (☐ 2 sq mi/5.2 sq km; 2001 population 711), E BRITISH COLUMBIA, W CANADA, on FRASER River, 120 mi/193 km ESE of PRINCE GEORGE, and between the ROCKY and CARIBOO mountain ranges; 53°18′N 120°10′W. Lumbering; agriculture; tourism. Surveyed in 1912.

McBride, village (2000 population 232), MONTCALM county, central MICHIGAN, 4 mi/6.4 km NNE of STANTON; 43°21′N 85°02′W. In agricultural area.

McCall, town (2000 population 2,084), VALLEY county, W IDAHO, on North Fork of PAYETTE RIVER, at S end of PAYETTE LAKE, 27 mi/43 km N of CASCADE in recreation area; 44°55′N 116°07′W. Elevation 5,025 ft/1,532 m. Lumber milling, concrete manufacturing. Headquarters, Idaho National Forest. Nearby, deposits of thorium; summer and winter tourism. Upper Payette Lake to N, CASCADE RESERVOIR to S; Ponderosa State Park to E; parts of Payette National Forest to E and W; Brundage Mountain Ski Area to N; Payette Lake Ski Resort to W.

McCallsburg, town (2000 population 318), STORY county, central IOWA, 11 mi/18 km NNE of NEVADA; 42°10′N 93°23′W. Livestock; grain.

McCamey (MIK-a-mee), town (2006 population 1,635), UPTON county, W TEXAS, 60 mi/97 km S of Midland, near PECOS RIVER; 31°07′N 102°13′W. Elevation 2,441 ft/744 m. Distributing center for oil and gas and cattle- and sheep-ranching region. Agriculture (cotton, pecans). Oil refinery; manufacturing (petroleum production). Founded 1925 after oil discovery; incorporated 1926.

McCammon, village (2000 population 805), BANNOCK county, SE IDAHO, 20 mi/32 km SE of POCATELLO, on PORTNEUF RIVER; 42°39′N 112°11′W. Elevation 4,751 ft/1,448 m. Railroad junction. Wheat, barley, alfalfa, hay; sheep, cattle; grain milling; manufacturing (grain-milling equipment). Parts of Caribou National Forest to NE, NW, and SW.

McCandless, township, ALLEGHENY county, W PENNSYLVANIA, residential suburb 10 mi/16 km N of PITTSBURGH; 40°34′N 80°01′W. Includes the communities of Highland and Fox Ridge.

McCarthy, village, S ALASKA, 120 mi/193 km NE of CORDOVA, at foot of WRANGELL MOUNTAINS, 3 mi/4.8

---

km S of KENNICOTT. Airstrip. Formerly terminus of COPPER RIVER and Northwestern railroad, closed down 1938. Sometimes called Shushanna Junction.

McCaskill (muh-KAS-kil), village (2000 population 84), HEMPSTEAD county, SW ARKANSAS, 18 mi/29 km N of HOPE; 33°55′N 93°38′W.

McCauley Island (muh-KAW-lee) (□ 108 sq mi/280.8 sq km), W BRITISH COLUMBIA, W CANADA, in HECATE STRAIT W of PITT ISLAND; 53°40′N 130°15′W. Island is 18 mi/29 km long, 2 mi/3 km–12 mi/19 km wide.

McCausland, town (2000 population 299), SCOTT county, E IOWA, near WAPSIPINICON RIVER, 15 mi/24 km NNE of DAVENPORT; 41°44′N 90°27′W.

McCaysville (muh-KAIZ-vil), town (2000 population 1,071), FANNIN county, N GEORGIA, 35 mi/56 km ENE of DALTON, near TENNESSEE-NORTH CAROLINA state line, in copper-mining region; 34°59′N 84°22′W. Once made barren from sulfuric acid by-products from smelting. Vegetation is now rebounding. Nearby Burra Mine site and Ducktown Basin Museum are located across the state line in Tennessee.

McChord Air Force Base, WASHINGTON: see TACOMA.

McClain (muh-KLAIN), county (□ 580 sq mi/1,508 sq km; 2006 population 31,038), central OKLAHOMA; ⊙ PURCELL; 35°00′N 97°26′W. Bounded NE by CANADIAN River and drained by small creeks; WASHITA River beyond S boundary. Agriculture (cotton, alfalfa, hay, soybeans, vegetables; cattle; dairying). Natural-gas wells. NW corner is partially urbanized, opposite OKLAHOMA CITY and NORMAN. Formed 1907.

McCleary, town (2006 population 1,575), GRAYS HARBOR county, W WASHINGTON, 17 mi/27 km W of OLYMPIA, on WILDCAT CREEK; 47°04′N 123°16′W. Manufacturing (lumber, building materials). Historic McCleary Hotel.

McClellan Air Force Base, California: see SACRAMENTO, city.

McClellan Creek (MIK-klel-uhn), c.40 mi/64 km long, extreme N TEXAS; rises in CARSON county; flows E and NE to North Fork of RED RIVER 10 mi/16 km N of MCLEAN. In upper course, dam impounds Lake McClellan, in McClellan Creek National Grassland. Recreational area (fishing, bathing, camping) here.

McClelland, town (2000 population 129), POTTAWATTAMIE county, SW IOWA, 10 mi/16 km ENE of COUNCIL BLUFFS; 41°19′N 95°40′W.

McClellan, Fort, ALABAMA: see ANNISTON.

McClellanville (muh-KLE-len-vil), village (2006 population 471), CHARLESTON county, SE SOUTH CAROLINA, on the coast, on INTRACOASTAL WATERWAY, and 23 mi/37 km SSW of GEORGETOWN; 33°05′N 79°28′W. Fishing industry. Manufacturing includes seafood processing. Nearby are Harrietta House and Gardens.

McClintock Channel, arm of the ARCTIC OCEAN, NORTHWEST TERRITORIES, CANADA, between VICTORIA ISLAND and PRINCE OF WALES ISLAND; 72°00′N 103°00′W. Opens N on VISCOUNT MELVILLE SOUND; 170 mi/274 km long, 65 mi/105 km–130 mi/209 km wide.

McCloud, unincorporated town (2000 population 1,343), SISKIYOU county, N CALIFORNIA, near MCCLOUD RIVER, 17 mi/27 km SE of WEED, at S base of MOUNT SHASTA; 41°15′N 122°08′W. Cattle, sheep, lambs; potatoes, onions. Pacific Crest Trail passes to S; parts of Shasta National Forest to N and S.

McCloud River, c.50 mi/80 km long, N CALIFORNIA; rises in S SISKIYOU county, at SHASTA county line; flows first N, then SW, passes SE of McCloud to join SACRAMENTO RIVER and its branch, the PIT RIVER, at SHASTA LAKE reservoir; forms long arm of the lake, 19 mi/31 km S of Dunsmuir.

McClure (mi-KLUHR), village (2006 population 739), HENRY county, NW OHIO, 10 mi/16 km E of NAPOLEON, near MAUMEE RIVER; 41°22′N 83°57′W.

McClure (muh-KLUHR), borough (2006 population 945), SNYDER county, central PENNSYLVANIA, 15 mi/24 km NE of LEWISTOWN; 40°42′N 77°18′W. In agriculture area (corn, hay; livestock; dairying); manufacturing (apparel, steel products, lumber, concrete products). Snyder-Middleswarth State Park to NE; parts of Bald Eagle State Forest to E and N; BLUE MOUNTAIN ridge to SE.

McClure, Cape, N extremity of BANKS ISLAND, BAFFIN region, NORTHWEST TERRITORIES, CANADA, on MCCLURE STRAIT; 74°28′N 120°41′W.

McClure Strait, arm of BEAUFORT SEA of the ARCTIC OCEAN, BAFFIN region, NORTHWEST TERRITORIES, CANADA, extending W from VISCOUNT MELVILLE SOUND, between MELVILLE and EGLINTON islands (N) and BANKS ISLAND (S); 75°00′N 118°00′W. Channel is 170 mi/274 km long, 60 mi/97 km wide. In 1954, U.S. icebreakers cut through the strait for the first time, opening the last obstacle to the shortest water route across the Canadian arctic region.

McClusky (muh-KLUHS-kee), village (2006 population 337), ⊙ SHERIDAN county, central NORTH DAKOTA, 50 mi/80 km NNE of BISMARCK, on McClusky Irrigation Canal; 47°28′N 100°26′W. Founded in 1903 and incorporated in 1908. It became the county seat when it was formed in 1909. Named for William Henderson McCluskey, a pioneer settler.

McColl (muh-KAHL), town (2006 population 2,384), MARLBORO county, NE SOUTH CAROLINA, 8 mi/12.9 km ENE of BENNETTSVILLE, at NORTH CAROLINA state line; 34°40′N 79°32′W. Manufacturing includes apparel; agriculture includes cotton, grains, soybeans, vegetables; hogs.

McCollum Lake, village (2000 population 1,038), MCHENRY county, NE ILLINOIS, residential suburb 2 mi/3.2 km N of MCHENRY.

McComas (mik-O-muhs), unincorporated town, MERCER county, S WEST VIRGINIA, 7 mi/11.3 km W of PRINCETON. Coal region.

McComb (muh-KOM), city (2000 population 13,337), PIKE county, SW MISSISSIPPI, 72 mi/116 km S of JACKSON; 31°14′N 90°28′W. It is the trade and railroad center of an agricultural area (cotton, corn, soybeans; cattle); timber; manufacturing (wire products, printing and publishing, textile products, concrete, lumber; poultry processing). Percy Quin State Park to SW. Incorporated 1872.

McComb (mi-KOM), village (2006 population 1,656), HANCOCK county, NW OHIO, 9 mi/14 km WNW of FINDLAY; 41°06′N 83°47′W. Corn, wheat, oats; glass products.

McComb Mountain (4,425 ft/1,349 m), ESSEX county, NE NEW YORK, in the High Peak section of the ADIRONDACKS, 9 mi/14.5 km SE of Mount MARCY, c.20 mi/32 km SE of LAKE PLACID village; 44°03′N 73°47′W.

McConaughy, Lake, Nebraska: see C. W. MCCONAUGHY, LAKE.

McCone, county (□ 2,682 sq mi/6,946 sq km; 1990 population 2,276; 2000 population 1,977), NE MONTANA; ⊙ CIRCLE; 47°39′N 105°48′W. Agricultural region bounded N by MISSOURI River; drained by REDWATER RIVER. Wheat, barley, hay; sheep, cattle; oil. In W is part of Dry Arm (Big Dry Creek) of FORT PECK Reservoir. Reservoir is surrounded by Charles M. Russell National Wildlife Refuge. Formed 1919.

McConnell (muh-KAH-nuhl), community, SW MANITOBA, W central CANADA, 39 mi/62 km NW of BRANDON, in HAMIOTA rural municipality; 50°16′N 100°31′W.

McConnells, village (2006 population 324), YORK county, N SOUTH CAROLINA, 12 mi/19 km WSW of ROCK HILL; 34°52′N 81°13′W. Agriculture includes cotton, grain; livestock. Formerly called MCCONNELLSVILLE.

McConnellsburg (muh-KAH-nuhls-buhrg), borough (2006 population 1,040), ⊙ FULTON county, S PENNSYLVANIA, 22 mi/35 km W of CHAMBERSBURG; 39°55′N 78°00′W. Agricultural area (corn, hay; livestock; dairying); manufacturing (construction equipment, ma-chinery parts, toys, food processors); mountain resort. James Buchanan's Birthplace State Historical Park to SE. Cowans Gap State Park to NE; Meadow Grounds Lake reservoir to SW; Tuscarora Mountain ridge and part of Buchanan State Forest to E. Settled c.1730, laid out 1786, incorporated 1814.

McConnellsville, SOUTH CAROLINA: see MCCONNELLS.

McConnelsville (mi-KAHN-uhlz-vil), village (□ 2 sq mi/5.2 sq km; 2006 population 1,735), ⊙ MORGAN county, SE OHIO, on MUSKINGUM RIVER, 20 mi/32 km SSE of ZANESVILLE; 39°39′N 81°51′W. Gas and oil wells. Plotted 1817.

McCook, county (□ 577 sq mi/1,500.2 sq km; 2006 population 5,851), SE SOUTH DAKOTA; ⊙ SALEM; 43°40′N 97°21′W. Agricultural area drained by East and West forks of VERMILLION RIVER. Corn, soybeans; cattle, hogs. Lake Vermillion State Recreational Area in E. Formed 1873.

McCook, city (2006 population 7,542), ⊙ RED WILLOW county, S NEBRASKA, 65 mi/105 km S of NORTH PLATTE, and on REPUBLICAN RIVER, near KANSAS state line; 40°12′N 100°37′W. Trade center and railroad-division point in rich grain-raising region; livestock. Manufacturing (fertilizers, rubber products, concrete, printing, bakery products). Museum of the High Plains. Red Willow Reservoir (Hugh Butler Lake) and State Recreation Area to N (both in Frontier county). Founded as Fairview 1881; named McCook 1882; incorporated 1883.

McCook, village (2000 population 254), COOK county, NE ILLINOIS, industrial suburb 12 mi/19 km WSW of CHICAGO; 41°47′N 87°50′W. Large locomotive division plant here; manufacturing (consumer goods, boxes, oils, aluminum products, chemicals, crushed stone).

McCool (muh-KOOL), village (2000 population 182), ATTALA county, central MISSISSIPPI, 17 mi/27 km NE of KOSCIUSKO; 33°12′N 89°20′W. Agriculture(cotton, grain, soybeans; cattle).

McCool Junction, village (2006 population 416), YORK county, SE NEBRASKA, 9 mi/14.5 km S of YORK, and on West Fork of BIG BLUE River; 40°44′N 97°35′W.

McCordsville, town (2000 population 1,134), HANCOCK county, central INDIANA, 15 mi/24 km NE of INDIANAPOLIS; 39°55′N 85°56′W. Manufacturing (machinery, fabricated metal products). Corn, soybeans. GEIST RESERVOIR to NW. Laid out 1865.

McCormick, county (□ 393 sq mi/1,021.8 sq km; 2006 population 10,226), W SOUTH CAROLINA; ⊙ MCCORMICK; 33°53′N 82°17′W. Bounded SW by SAVANNAH River; includes part of Sumter National Forest. Agricultural area, includes chickens, cattle; hay; timber; textile milling. Several state parks in area. Formed 1916.

McCormick, town (□ 3 sq mi/7.8 sq km; 1990 population 1,659; 2000 population 1,489), ⊙ MCCORMICK county, W SOUTH CAROLINA, 21 mi/34 km SSW of GREENWOOD; 33°54′N 82°17′W. Manufacturing of lumber, apparel, textiles. Large retirement community here. Sumter National Forest nearby.

McCoy, Camp, Wisconsin: see SPARTA.

McCracken (muh-KRAK-uhn), county (□ 268 sq mi/696.8 sq km; 2006 population 64,950), W KENTUCKY; ⊙ PADUCAH; 37°03′N 88°43′W. Bounded N by OHIO RIVER (ILLINOIS state line), NE by the TENNESSEE RIVER (enters Ohio River); drained by CLARKS RIVER and its East and West forks and by MAYFIELD CREEK. Gently rolling agricultural area (dark and burley tobacco, corn, sorghum, hay, soybeans, wheat; hogs, cattle, poultry; dairying). Clay, sand and gravel, coals; timber. Manufacturing at Paducah. Barkley Regional Airport in W. West Kentucky Wildlife Management Area in NW; Lock and Dam Number 52 in NE, on Ohio River at Paducah. Formed 1824.

McCracken (muh-KRAK-uhn), village (2000 population 211), RUSH county, W central KANSAS, 13 mi/21 km WNW of LA CROSSE; 38°34′N 99°34′W. Wheat; cattle.

Area is shown by the symbol □, and capital city or county seat by ⊙.

McCreary (muh-KRIR-ee), county (□ 430 sq mi/1,118 sq km; 2006 population 17,354), S KENTUCKY; ⊙ WHITLEY CITY. In the CUMBERLAND MOUNTAINS. Bounded S by TENNESSEE, N and NE by CUMBERLAND RIVER, W in part by Little South Fork river; drained by SOUTH FORK of the Cumberland River; 36°44′N 84°28′W. Bituminous-coal-mining and timber region; oil wells, some farms (burley tobacco, hay; cattle). Some lumber milling. Daniel Boone National Forest covers nearly all of county. Includes a natural bridge, caves, and parts of Cumberland Falls State Park in NE; part of Big South Fork National River and Recreation Area in S, extends into Tennessee. Formed 1912.

McCreary (muh-KREE-ree), rural municipality (□ 202 sq mi/525.2 sq km; 2001 population 525), S MANITOBA, W central CANADA, 41 mi/66 km E of RIDING MOUNTAIN NATIONAL PARK; 50°45′N 99°20′W. Agriculture (grain; livestock). Main center is the village of McCreary. Incorporated 1909.

McCreary (muh-KREE-ree), village (□ 1 sq mi/2.6 sq km; 2001 population 522), S MANITOBA, W central CANADA, 34 mi/55 km E of RIDING MOUNTAIN NATIONAL PARK; 50°46′N 99°29′W. Agriculture (livestock); tourism. Center of MCCREARY rural municipality. Incorporated 1964.

McCrory (muh-KROR-ee), town (2000 population 1,850), WOODRUFF county, E central ARKANSAS, 28 mi/45 km NW of FORREST CITY, and on CACHE RIVER; 35°15′N 91°12′W. Agriculture (rice, soybeans). Manufacturing (paper products, plastics, fertilizer, feeds). Settled 1886.

McCrosson and Tovell, former township (□ 68 sq mi/176.8 sq km; 2001 population 154), NW ONTARIO, E central CANADA; 48°57′N 94°25′W. Amalgamated into LAKE OF THE WOODS township in 1998. Also written McCrosson-Tovell.

McCuddin Mountains, small range, MARIE BYRD LAND, WEST ANTARCTICA, located 40 mi/60 km E of the AMES RANGE; 75°47′S 128°42′W. Includes Mountains Flint and Petras.

McCulloch (MIK-uh-luh), county (□ 1,073 sq mi/2,789.8 sq km; 2006 population 8,016), central TEXAS; ⊙ BRADY; 31°11′N 99°20′W. Geographical center of state, on N EDWARDS PLATEAU, with Brady Mountains (c.2,000 ft/610 m) crossing E-W; bounded N by COLORADO RIVER and drained by SAN SABA RIVER and BRADY CREEK (forms Brady reservoir in W center). Diversified agricultural and livestock raising; oats, peanuts, cotton, wheat; hay; cattle, sheep, and goats (wool, mohair). Oil and gas; sand and gravel. Formed 1856.

McCune (muh-KYOON), village (2000 population 426), CRAWFORD county, extreme SE KANSAS, 18 mi/29 km WSW of PITTSBURG; 37°21′N 95°01′W. In diversified agricultural area. Coal mines nearby.

McCurtain, county (□ 1,901 sq mi/4,942.6 sq km; 2006 population 34,018), extreme SE OKLAHOMA; ⊙ IDABEL; 34°06′N 94°46′W. Bounded E by ARKANSAS state line, S by RED RIVER, here forming TEXAS state line; drained by LITTLE RIVER and MOUNTAIN FORK; part of OUACHITA MOUNTAINS in N. Lumbering; agriculture (corn, alfalfa, hay, soybeans; cattle, poultry); tourism, recreation. PINE CREEK Lake reservoir in W; part of Ouachita National Forest in SE (separate unit beyond N boundary). BROKEN BOW LAKE reservoir in NE center; Beavers Bend and Hochatown state parks at S end of lake at dam. McCurtain County Wilderness Area on N end of lake. Formed 1907.

McCurtain, village (2006 population 478), HASKELL county, E OKLAHOMA, 20 mi/32 km WNW of POTEAU; 35°08′N 94°58′W. In agricultural area. Sansbois Mountains to SW.

McDade (MIK-daid), unincorporated village, BASTROP county, S central TEXAS, 30 mi/48 km E of AUSTIN; 30°17′N 97°14′W.

McDermitt (muhk-DUHR-mit), unincorporated village (2000 population 269), HUMBOLDT county, N

NEVADA, 68 mi/109 km N of WINNEMUCCA, on OREGON state line; 41°58′N 117°35′W. Cattle, sheep. On U.S. Highway 95, one of most remote gateways to Nevada, small gambling resort developed to attract travelers entering and leaving state. Main section of Fort McDermitt Indian Reservation to E; SANTA ROSA RANGE, in section of Humboldt National Forest, to SE. Trout Creek Mountains (Oregon) to NW.

McDonald, county (□ 540 sq mi/1,404 sq km; 2006 population 22,949), extreme SW MISSOURI, in OZARKS, borders OKLAHOMA on W, ARKANSAS on S; ⊙ PINEVILLE; 36°37′N 94°20′W. Drained by ELK RIVER. Fruit (berries, grapes, tomatoes), vegetables, grain; dairying; turkeys, chickens, broilers. Manufacturing (food); lumber; tourism; major broiler (chicken) producer. Huckleberry Ridge State Forest at center; Bluff Dwellers Cave S of NOEL. Formed 1849.

McDonald, village (2000 population 159), RAWLINS county, NW KANSAS, 16 mi/26 km W of ATWOOD; 39°47′N 101°22′W. In grain region.

McDonald (mik-DAHN-uhld), village (2006 population 106), ROBESON county, SE NORTH CAROLINA, 10 mi/16 km WSW of LUMBERTON; 34°32′N 79°10′W. Manufacturing; agriculture (grain, tobacco; livestock).

McDonald (mik-DAHN-uhld), village (□ 2 sq mi/5.2 sq km; 2006 population 3,319), TRUMBULL county, NE OHIO, just NW of YOUNGSTOWN, on MAHONING RIVER; 41°10′N 80°43′W.

McDonald, borough (2006 population 2,152), WASHINGTON and ALLEGHENY counties, SW PENNSYLVANIA, suburb 14 mi/23 km WSW of PITTSBURGH, on Robinson Run; 40°23′N 80°13′W. Agriculture (corn, apples; dairying); diversified light manufacturing; bituminous coal, oil (gas and oil field to SE).

McDonald Islands, tiny subantarctic rocks in S INDIAN (Southern) Ocean, c.20 mi/32 km W of HEARD ISLAND and c.250 mi/402 km SSE of the KERGUELEN Islands; 53°00′S 72°25′E. Rises to c.755 ft/230 m. Includes the principal McDonald Island, as well as Flat Rock, Meyer Rock. Formally annexed by AUSTRALIA in December 1947. Together with Heard Island, they are the only volcanically active subantarctic islands in the world. Intact ecosystem undisturbed by humans. Declared a UNESCO World Heritage Area in 1997. Also called McDonald Rocks, Macdonald Group.

McDonald, Lake, MONTANA: see GLACIER NATIONAL PARK.

McDonald Observatory (MIK-dahn-uhld), astronomical observatory, on Mount Locke (6,791 ft/2,070 m), near FORT DAVIS, TEXAS; 30°40′N 104°01′W. Founded in 1932, sponsored by the University of Texas in cooperation with the University of Chicago at the bequest of amateur astronomer William J. McDonald. Its equipment includes 107-in/272-cm, 82-in/208-cm, 32-in/81-cm, and 30-in/76-cm reflecting telescopes. The 107-in/272-cm reflecting telescope, which began operation in 1968 as the third-largest telescope in the world, was built under contract with NASA; it is housed in a large dome.

McDonald Peak, MONTANA: see MISSION RANGE.

McDonalds Rocks, AUSTRALIA: see MCDONALD ISLANDS.

McDonough (muhk-DUHN-uh), county (□ 590 sq mi/1,534 sq km; 2006 population 31,823), W ILLINOIS; ⊙ MACOMB; 40°27′N 90°40′W. Agriculture (corn, sorghum, soybeans, hay; dairy). Bituminous-coal mining; clay pits. Macomb is trade and manufacturing center, with diversified products. Drained by LA MOINE RIVER and branches. Includes Western Illinois University and Argyle Lake State Park. Formed 1826.

McDonough (muhk-DUHN-uh), town (2000 population 8,493), ⊙ HENRY county, N central GEORGIA, 24 mi/39 km SSE of ATLANTA; 33°27′N 84°08′W. Romanesque Revival courthouse in center square. Emerging Atlanta suburb. Manufacturing includes

apparel, machinery parts, plastics, cleaners, animal feeds, computer equipment; pulpwood processing. Incorporated 1823.

McDougall (muhk-DOO-guhl), township (□ 102 sq mi/265.2 sq km; 2001 population 2,608), S ONTARIO, E central CANADA, 4 mi/6 km from PARRY SOUND; 45°24′N 80°00′W. Surveyed 1866, incorporated 1872.

McDowell (mik-DOU-uhl), county (□ 446 sq mi/1,159.6 sq km; 2006 population 43,414), W central NORTH CAROLINA; ⊙ MARION; 35°40′N 82°02′W. In the BLUE RIDGE MOUNTAINS; drained by CATAWBA RIVER (forms Lake JAMES reservoir in NE); N part in Pisgah National Forest. Manufacturing at Marion; service industries; farming (corn, apples, hay; poultry, cattle). Stone quarrying. Resort area. BLUE RIDGE PARKWAY follows NW boundary. Formed 1842 from Rutherford and Burke counties. Named for Major Joseph McDowell (1758–1996), officer in American Revolution and member of Congress.

McDowell, county (□ 535 sq mi/1,391 sq km; 2006 population 23,882), S WEST VIRGINIA; ⊙ WELCH; 37°27′N 81°39′W. On ALLEGHENY PLATEAU; bounded W, S, and SE by VIRGINIA; drained by TUG FORK and DRY FORK rivers (headstreams of the BIG SANDY RIVER). Extensive semi-bituminous-coal mining (Pocahontas coalfield); some natural gas. Timber. Some agriculture (apples); retail trade. Panther State Forest to W; Berwind Lake Wildlife Management Area in S; Anawalt Wildlife Management Area in E. Formed 1858.

McDowell (muhk-DOU-uhl), unincorporated village, FLOYD county, E KENTUCKY, in the CUMBERLAND MOUNTAINS, 12 mi/19 km W of PIKEVILLE, on Beaver Creek. Bituminous coal.

McDowell (muhk-DOU-uhl), unincorporated village, HIGHLAND county, NW VIRGINIA, in the ALLEGHENY MOUNTAINS, 8 mi/13 km SE of MONTEREY, on BULL-PASTURE RIVER; 38°20′N 79°29′W. Manufacturing (lumber); agriculture (alfalfa); timber.

McDowell Mountains, MARICOPA county, central ARIZONA; rise to 4,035 ft/1,230 m in McDowell Peak, c.20 mi/32 km NE of PHOENIX. McDowell Mountains Park in S; VERDE River to E.

McDuffie (muhk-DUHF-ee), county (□ 266 sq mi/691.6 sq km; 2006 population 21,917), E GEORGIA; ⊙ THOMSON; 33°29′N 82°29′W. Bounded N by LITTLE RIVER. Intersected by the fall line. Agriculture (cotton, vegetables, fruit); cattle; timber. Formed 1870.

McEwen (muhk-YOO-uhn), city (2006 population 1,683), HUMPHREYS county, central TENNESSEE, 45 mi/72 km W of NASHVILLE; 36°07′N 87°38′W. In timber region.

McEwensville (muhk-YOO-uhns-vil), borough (2006 population 299), NORTHUMBERLAND county, E central PENNSYLVANIA, 14 mi/23 km N of SUNBURY; 41°04′N 76°49′W. Agriculture (corn, hay; poultry; dairying).

McFadden, village, CARBON county, SE WYOMING, near ROCK CREEK, near foothills of MEDICINE BOW MOUNTAINS, and 36 mi/58 km NW of LARAMIE. Elevation c. 7,200 ft/2,195 m. Oil wells.

McFall, city (2000 population 135), GENTRY county, NW MISSOURI, near GRAND RIVER, 40 mi/64 km NE of SAINT JOSEPH; 40°06′N 94°13′W. Corn, soybeans; hogs, cattle.

McFarlan (mik-FAHR-luhn), village (2006 population 84), ANSON county, S NORTH CAROLINA, 12 mi/19 km SSE of WADESBORO, near Great Pee Dee River (PEE DEE RIVER), near SOUTH CAROLINA state line; 34°48′N 79°58′W. Service industries; agriculture (cotton, grain; livestock).

McFarland, city (2000 population 9,618), KERN county, S central CALIFORNIA, 22 mi/35 km NNW of BAKERSFIELD, near FRIANT-KERN CANAL (irrigation); 35°41′N 119°14′W. Railroad junction to S. Cotton, fruit, nuts, vegetables, grain; dairying; cattle.

McFarland, town (2006 population 7,504), DANE county, S WISCONSIN, suburb 7 mi/11.3 km SE of MADISON, near

---

Cross-references are shown in SMALL CAPITALS. The pronunciation guide is shown on page xix. The sources of population figures are shown on page xvii.

YAHARA RIVER, on E end of LAKE WAUBESA; 43°01′N 89°17′W. In dairy and lake-resort region. Manufacturing (telecomunications equipment, metal fabricating, paper products).

McFarland, village (2000 population 271), WABAUNSEE county, NE central KANSAS, 19 mi/31 km ESE of MANHATTAN and 3 mi/4.8 km NE of ALMA; 39°02′N 96°14′W. Railroad junction. In cattle and grain region.

McGarry (muhk-GA-ree), township (□ 33 sq mi/85.8 sq km; 2001 population 787), E central ONTARIO, E central CANADA, 21 mi/34 km from KIRKLAND LAKE; 48°08′N 79°34′W.

McGehee (muh-GEE), town (2000 population 4,570), DESHA county, SE ARKANSAS, 25 mi/40 km NW of GREENVILLE (MISSISSIPPI); 33°37′N 91°23′W. Railroad junction. In livestock-raising and agricultural area. Manufacturing (apparel, paper). Incorporated 1906.

McGill (muh-GIL), town (2000 population 1,054), WHITE PINE county, E NEVADA, between SCHELL CREEK RANGE (E), and EGAN RANGE (W), 12 mi/19 km NNE of ELY, near Duck Creek; 39°23′N 114°46′W. Elev. 6,210 ft/1,893 m. Cattle, sheep; gold, silver. Terminus of railroad spur from Ely. Pony Express station site 25 mi/40 km to N. Goshute Indian Reservation (UTAH-Nevada) 45 mi/72 km to NE. Part of Humboldt Natl Forest to E. North Schell Park (elev. 11,883 ft/3,620 m) to E.

McGillivray (muhk-GI-li-vrai), former township (□ 109 sq mi/283.4 sq km; 2001 population 1,789), S ONTARIO, E central CANADA; 43°12′N 81°34′W. Organized 1843. Amalgamated into NORTH MIDDLESEX township in 2001.

McGillivray Falls (muh-GI-liv-ree), ghost town, locality, SW BRITISH COLUMBIA, W CANADA, in COAST MOUNTAINS, on ANDERSON LAKE, 23 mi/37 km WSW of LILLOOET; 50°37′N 122°26′W. Former gold-mining area.

McGovern, unincorporated town (2000 population 2,538), WASHINGTON county, SW PENNSYLVANIA, residential suburb 2 mi/3.2 km SSW of CANONSBURG, on Chartiers Creek; 40°14′N 80°13′W. Agriculture includes dairying; apples.

McGrann, unincorporated village, Manor township, ARMSTRONG county, W PENNSYLVANIA, on ALLEGHENY RIVER, residential suburb 1 mi/1.6 km NNE of FORD CITY; 40°46′N 79°31′W.

McGrath, village (2000 population 401), SW central ALASKA, on upper KUSKOKWIM River (head of navigation), 220 mi/354 km NW of ANCHORAGE; 62°58′N 155°35′W. Has airfield. Established 1905.

McGrath, village (2000 population 65), AITKIN county, E MINNESOTA, on SNAKE river, 30 mi/48 km SE of AITKIN; 46°14′N 93°16′W. Solano State Forest to NE; MILLE LACS LAKE to W.

McGraw, village (2006 population 967), CORTLAND county, central NEW YORK, 4 mi/6.4 km E of CORTLAND; 42°35′N 76°05′W. Incorporated 1869.

McGregor, town (2000 population 871), CLAYTON county, NE IOWA, on MISSISSIPPI RIVER almost opposite mouth of WISCONSIN RIVER, 3 mi/4.8 km SW of PRAIRIE DU CHIEN (WIS.), in hilly "Little Switzerland" region; 43°01′N 91°10′W. Manufacturing (dairy products; consumer goods, beverages). Upper Mississippi National Wildlife Refuge; Effigy Mounds National Monument 5 mi/8 km N; McGregor Heights and Pikes Peak State Parks to S. Settled 1836, incorporated 1857.

McGregor (MIK-greg-uhr), town (2006 population 4,845), MCLENNAN county, E central TEXAS, 18 mi/29 km SW of WACO, near branch of BOSQUE RIVER; 31°26′N 97°22′W. Manufacturing (motors, furniture and bedding, office supplies, air conditioning equipment). Mother Neff State Park is SW, on BELTON LAKE reservoir. An ordnance plant was nearby in World War II. Established 1882.

McGregor, village (2000 population 404), AITKIN county, E central MINNESOTA, 20 mi/32 km ENE of AITKIN; 46°36′N 93°18′W. Manufacturing (motor vehicle parts, wood products, machinery). Sandy Lake Fur Post (1794) to N; Savanna State Forest to NE; Savanna Portage State Park to NE; Rice Lake National Wildlife Refuge to S; East Lake Indian Reservation to S; Sandy Lake Indian Reservation to NE; BIG SANDY LAKE reservoir to N. Numerous small natural lakes in area.

McGregor Bay (muhk-GRE-guhr BAI), unincorporated village, ONTARIO, E central CANADA, and included in the town of NORTHEASTERN MANITOULIN AND THE ISLANDS; 46°05′N 81°38′W.

McGregor, Lake (muhk-GRE-guhr), S ALBERTA, W CANADA, 60 mi/97 km SE of CALGARY; 50°25′N 112°52′W. Drains N into BOW RIVER; 23 mi/37 km long, 1 mi/2 km–3 mi/5 km wide; 50°25′N 112°52′W. Irrigation dams at N and S extremities.

McGrew, village (2006 population 101), SCOTTS BLUFF county, W NEBRASKA, 15 mi/24 km SE of SCOTTS-BLUFF, and on NORTH PLATTE RIVER; 41°45′N 103°25′W.

McGuffey (muh-GUHF-ee), village (2006 population 527), HARDIN county, W central OHIO, 9 mi/14 km WNW of KENTON; 40°42′N 83°47′W.

McGuire 1 and 2 Nuclear Power Plants, NORTH CAROLINA: see NORMAN, LAKE.

McGuire, Mount (10,082 ft/3,073 m), LEMHI county, E IDAHO, in YELLOWJACKET MOUNTAINS, 32 mi/51 km W of SALMON, in Frank Church–River of No Return Wilderness Area. S of confluence of SALMON and Middle Fork Salmon rivers.

Mchangani (mchan-GAH-nee), town, ZANZIBAR NORTH region, E TANZANIA, 13 mi/21 km NE of ZANZIBAR, in N center ZANZIBAR island; 06°02′S 39°18′E. Cloves, copra, bananas; livestock.

McHenry, county (□ 611 sq mi/1,588.6 sq km; 2006 population 312,373), NE ILLINOIS, on WISCONSIN state line (N); ⊙ WOODSTOCK; 42°19′N 88°27′W. Urban growth has extended into county from CHICAGO. Larger communities are Woodstock, MCHENRY, CRYSTAL LAKE, and ALGONQUIN. Dairying area; also livestock; corn, hay. Several lakes; fishing. Drained by FOX and KISHWAUKEE rivers. Formed 1836.

McHenry, county (□ 1,879 sq mi/4,885.4 sq km; 2006 population 5,429), N central NORTH DAKOTA; ⊙ TOWNER; 48°13′N 100°38′W. Agricultural area with extensive lignite deposits; drained by SOURIS River (Mouse River), Wintering River and Spring Coulee Creek. Cattle, poultry; dairy products; wheat, rye, barley, hay. North Lake and Buffalo Lodge Lake at center; George Lake, Smoky Lake, and other small lakes in E; part of J. Clark Salyer National Wildlife Refuge in N. Formed 1873 and government organized in 1884. Named for James McHenry, and early settler.

McHenry, city (2000 population 21,501), MCHENRY county, NE ILLINOIS, on FOX RIVER (bridged here), satellite community of CHICAGO, 22 mi/35 km W of WAUKEGAN; 42°20′N 88°17′W. Trade and processing center in dairying, remnant agriculture; manufacturing (plastics, rubber products, fabricated metal products, packaging, electrical motors); fishing. Settled 1836; incorporated as village in 1855, as city in 1923.

McHenry (muh-KEN-ree), village, OHIO county, W KENTUCKY, 3 mi/4.8 km WSW of BEAVER DAM; 37°22′N 86°55′W. Bituminous coal; agriculture (tobacco, grain; livestock; timber); manufacturing (copper products, furniture).

McHenry, village (2006 population 70), FOSTER county, E central NORTH DAKOTA, 27 mi/43 km ENE of CARRINGTON; 47°34′N 98°35′W. Founded in 1899 and incorporated in 1903. Named for E.N. McHenry, a railroad engineer.

McHenry, summer resort, GARRETT county, W MARYLAND, in the ALLEGHENIES on DEEP CREEK LAKE, 32 mi/51 km WSW of CUMBERLAND. Named for James McHenry, Washington's secretary of war for whom Fort McHenry was also named. McHenry purchased land here about 1805. State game refuge near. Garrett Community College and Garrett County Fairgrounds.

McHenry, Fort, MARYLAND: see FORT MCHENRY.

Mchinga (uhm-CHEEN-gah), town, LINDI region, SE TANZANIA, 15 mi/24 km N of LINDI, on INDIAN OCEAN; 09°49′S 39°42′E. Lindi Airport to S. Peanuts, cashews, bananas, manioc; goats, sheep.

Mchinji (uhm-chi-nji), administrative center and district (2007 population 440,162), Central region, MALAWI, near ZAMBIA border, on road to CHIPATA (20 mi/32 km NW), 60 mi/97 km WNW of LILONGWE; 13°48′S 32°54′E. Elevation 4,200 ft/1,280 m. Customs station; tobacco, cotton, corn. Formerly called Fort Manning.

McIlwaine, Lake, ZIMBABWE: see CHIVERO, LAKE.

McIntire, town (2000 population 173), MITCHELL county, N IOWA, near MINNESOTA state line, on WAPSIPINICON RIVER, 15 mi/24 km NE of OSAGE; 43°26′N 92°35′W. Limestone quarries, sand pits nearby.

McIntosh, county (□ 575 sq mi/1,495 sq km; 2006 population 11,248), SE GEORGIA, ⊙ DARIEN; 31°29′N 81°22′W. Bounded SE by the ATLANTIC OCEAN, SW by ALTAMAHA RIVER; includes SAPELO ISLAND. Agricultural, fishing, and sawmilling area; seafood canning at Darien. Formed 1793.

McIntosh, county (□ 991 sq mi/2,576.6 sq km; 2006 population 2,956), S NORTH DAKOTA, borders SOUTH DAKOTA on S; ⊙ ASHLEY; 46°07′N 99°26′W. Rich prairie land watered by South Branch of BEAVER CREEK. Farm machinery; dairy products, wheat, rye; cattle. Doyle Memorial State Park, on Green Lake is to the N; Lake Hoskins is to the S. Formed 1883 for the southern half of LOGAN County. Named for Edward H. McIntosh (1822–1901), a member of territorial council. Hoskins was the county seat from 1884–1888.

McIntosh, county (□ 715 sq mi/1,859 sq km; 2006 population 19,899), E OKLAHOMA; ⊙ EUFAULA; 35°22′N 95°40′W. Bounded S by CANADIAN River (forming EUFAULA LAKE here; Eufaula Dam in SE corner); intersected by NORTH CANADIAN and DEEP FORK of Canadian rivers (forming arms of Eufaula Lake). Agriculture (soybeans, peanuts; cattle; dairying). Some manufacturing; timber; tourism. Fountainhead State Park at center on Lake Eufaula. Formed 1907.

McIntosh, city (2006 population 213), ⊙ CORSON county, N SOUTH DAKOTA, 55 mi/89 km NW of MOBRIDGE; 45°55′N 101°20′W. Livestock; grain.

McIntosh, village (2000 population 638), POLK county, NW MINNESOTA, on Poplar River, 35 mi/56 km ESE of CROOKSTON; 47°38′N 95°53′W. Grain, sunflowers; livestock; dairying.

McIntosh, Fort, Texas: see LAREDO.

McIntyre (MAK-in-teir), former township, W central ONTARIO, E central CANADA; 48°27′N 89°21′W. Amalgamated into city of THUNDER BAY in 1970.

McIntyre, village, WILKINSON county, central GEORGIA, 26 mi/42 km E of MACON; 32°51′N 83°11′W. Manufacturing includes wood products; kaolin clay processing.

McIntyre Mountain (MAK-in-teir) (1,030 ft/314 m), SW CAPE BRETON ISLAND, E NOVA SCOTIA, CANADA, 12 mi/19 km N of PORT HAWKESBURY. Highest in CRAIGNISH HILLS.

McKay Creek (muh-KAI), 20 mi/32 km long, NE OREGON; rises at the line between UMATILLA and UNION counties; flows W and N through McKay Reservoir to UMATILLA RIVER, just W of PENDLETON. McKay Dam (180 ft/55 m high, 2,700 ft/823 m long; completed 1927) is 5 mi/8 km S of Pendleton. Used for irrigation.

McKay Lake (muh-KEI), central ONTARIO, E central CANADA, 28 mi/45 km E of GERALDTON, 10 mi/16 km E of LONG LAKE; 49°36′N 86°26′W. Lake is 12 mi/19 km long, 3 mi/5 km wide; elevation 1,052 ft/321 m. Drains S into Lake SUPERIOR through Pie River.

McKay, Mount (muh-KEI), (1,581 ft/482 m), W ONTARIO, E central CANADA, overlooking entrance of

THUNDER BAY, 3 mi/5 km S of THUNDER BAY; 48°20'N 89°17'W.

McKean, county (□ 984 sq mi/2,558.4 sq km; 2006 population 44,065), N PENNSYLVANIA; ⊙ SMETHPORT; 41°47'N 78°34'W. Plateau area. Bounded N by New York state line; drained by ALLEGHENY RIVER. Largest producer of Pennsylvania lubricating oils. Agriculture (corn, oats, hay, alfalfa; hogs, cattle; dairying); petroleum, natural gas. Manufacturing at BRADFORD. Kinzua Bridge State Park in S center; small part of Susquehannock State Forest in SE; part of Allegheny National Forest in W half, including Kinzua Bay and Sugar Bay, arms of ALLEGHENY RESERVOIR. Formed 1804.

McKean, borough (2006 population 366), ERIE county, NW PENNSYLVANIA, 9 mi/14.5 km SSW of ERIE, on Elk Creek; 42°00'N 80°08'W. Agriculture (dairying); manufacturing (wood products). Formerly called Middleboro.

McKean Island or **M'Kean Island**, uninhabited coral islet, PHOENIX ISLANDS, S PACIFIC, 75 mi/121 km SW of KANTON Island; 03°36'S 174°08'W. Discovered 1840 by Americans, included 1937 in British GILBERT and ELLICE ISLANDS Colony, and in independent KIRIBATI, in 1979, with U.S.-Kiribati treaty following. Covers an area of 142 acres/57 ha.

McKee (muh-KEE), town (2000 population 878), ⊙ JACKSON county, E KENTUCKY, in the CUMBERLAND MOUNTAINS, 40 mi/64 km SSE of WINCHESTER, in Cumberland National Forest, surrounded by Daniel Boone National Forest; 37°25'N 83°59'W. Agriculture (corn, hay, tobacco; cattle; dairying); coal mines; timber; manufacturing (consumer goods; limestone processing).

McKeesport, city (2006 population 22,408), ALLEGHENY county, SW PENNSYLVANIA, suburb 10 mi/16 km SE of PITTSBURGH, on MONONGAHELA RIVER, at mouth of YOUGHIOGHENY RIVER; 40°20'N 79°50'W. A steel and industrial city, McKeesport has undergone rapid decline and increased unemployment since the collapse of the U.S. steel industry in the 1980s. Present-day manufacturing includes electrical equipment, solvents, metal fabrication, coal tar products, meat processing, tools). Pennsylvania State University–McKeesport Campus is here. Settled 1755, incorporated as a city 1890.

McKees Rocks, borough (2006 population 6,109), ALLEGHENY county, SW PENNSYLVANIA, industrial suburb 4 mi/6.4 km WNW of PITTSBURGH, on the OHIO RIVER (bridged); 40°28'N 80°03'W. Manufacturing (fabricated steel products, lubricants, food products, furniture, metal fabrication). The regional coal and steel industries formerly based here have collapsed, thus leading to unemployment and industrial decline. Pittsburgh State Correctional Institute is here. Settled c.1764, incorporated 1892.

McKellar (muh-KE-luhr), township (□ 69 sq mi/179.4 sq km; 2001 population 933), S ONTARIO, E central CANADA, 13 mi/21 km from PARRY SOUND; 45°29'N 79°51'W.

McKelvey Valley, dry valley (no snow or ice) in the MCMURDO DRY VALLEYS, VICTORIA LAND, EAST ANTARCTICA; 77°26'S 161°33'E.

McKenney (muhk-KE-nee), town (2006 population 483), DINWIDDIE county, SE central VIRGINIA, 23 mi/37 km SW of PETERSBURG; 36°59'N 77°43'W. Manufacturing (consumer goods; tobacco processing) in agricultural area (tobacco, peanuts, grain, soybeans; livestock). Fort Pickett Military Reservation to W.

McKenzie, county (□ 2,735 sq mi/7,111 sq km; 2006 population 5,700), W NORTH DAKOTA, borders MONTANA on W, borders MISSOURI RIVER (LAKE SAKAKAWEA) on N; ⊙ WATFORD CITY; 47°43'N 103°23'W. Agricultural area watered by YELLOWSTONE RIVER in W, LITTLE MISSOURI RIVER in SE; rich in lignite, oil, and natural gas. Wheat, sugar beets; cattle. Irrigation projects in NW along Missouri River. Little Missouri National Grassland in S and E; N unit of THEODORE ROOSEVELT NATIONAL PARK in S; part of Fort Berthold Indian Reservation, including Four Bears State Recreational Area in E. The boundary between Mountain and Central time zones follows the Montana state line for a short distance S then crosses S part of county; N three-quarters of county is in Central time zone, S quarter in Mountain. Formed 1883 but then eliminated in 1891 due to lack of settlement and reorganized in 1905. Named for Alexander McKenzie (1851–1922), a political leader in the state. ALEXANDER was the county seat from 1905-1907, then Shafer from 1907–1941, and then Watford City was made the county seat in 1941.

McKenzie, city (2006 population 5,420), in CARROLL, WEAKLEY, and HENRY counties, NW TENNESSEE, 10 mi/16 km N of HUNTINGDON; 36°08'N 88°32'W. In timber area; manufacturing. Bethel College here. Incorporated 1868.

McKenzie (muh-KEN-zee), former township, S ONTARIO, E central CANADA, 26 mi/42 km from PARRY SOUND; 45°43'N 79°59'W. Amalgamated into WHITESTONE township in 2000.

McKenzie, town (2000 population 629), Butler co., S Alabama, 20 mi/32 km S of Greenville. Named for Bethune B. McKenzie, a Confederate veteran who became engineer-in-chief for the Louisville and Nashville RR.

McKenzie Island (muh-KEN-zee), community, NW ONTARIO, E central CANADA, included in town of RED LAKE; 51°04'N 93°49'W.

McKenzie River (muh-KEN-zee), 86 mi/138 km long, W OREGON; formed in CASCADE RANGE in NE LANE county, near the THREE SISTERS peaks; flows W to WILLAMETTE RIVER N of EUGENE. South Fork, c. 30 mi/48 km long, rises in W Lane county, flows NNW through COUGAR RESERVOIR, joins main stream c.25 mi/40 km W of source.

McKerrow (muh-KE-ro), community, S central ONTARIO, E central CANADA, 41 mi/65 km from KILLARNEY, and included BALDWIN township; 46°17'N 81°45'W.

McKillop (muh-KI-luhp), former township (□ 85 sq mi/221 sq km; 2001 population 1,266), S ONTARIO, E central CANADA; 43°36'N 81°18'W. Amalgamated into town of HURON EAST in 2001.

McKinlay (muh-KIN-lee), town, QUEENSLAND, NE AUSTRALIA, 65 mi/104 km SE of CLONCURRY; 21°16'S 141°18'E. Home to the Walkabout Creek Hotel.

McKinley, county (□ 5,456 sq mi/14,185.6 sq km; 2006 population 71,875), NW NEW MEXICO, ⊙ GALLUP; 35°34'N 108°15'W. Livestock-grazing area, watered by RIO PUERCO and ZUÑI RIVER; borders on ARIZONA (W); CONTINENTAL DIVIDE crosses county from NE corner to S center, at angle. Uranium, molybdenum, copper; hay, alfalfa, timber; cattle, sheep; wool, Native American artifacts, pottery. Coal mines near Gallup. Includes parts of ZUÑI MOUNTAINS (S), parts of Cibola National Forest (S and SE), and parts of Navajo (NW), Zuni (SW), and Ramah (SW) Indian reservations. BLUEWATER LAKE State Park S of boundary; Red Rock State Park in W center. Formed 1899.

McKinley, village (2000 population 80), ST. LOUIS county, NE MINNESOTA, 7 mi/11.3 km E of VIRGINIA, in MESABI IRON RANGE; 47°30'N 92°24'W. Iron mines in area.

McKinley, Mount or **Denali** (=the great one), mountain (20,320 ft/6,194 m), S central ALASKA, in the ALASKA RANGE; 63°04'N 151°00'W. Highest point in NORTH AMERICA. McKinley features two main peaks: the higher is SOUTH PEAK (20,320 ft/6,194 m) and NORTH PEAK (19,470 ft/5,934 m); the two together have sometimes been known as the Churchhill Peaks. Other notable peaks on the mountain are SOUTH BUTTRESS (15,885 ft/4,842 m), EAST BUTTRESS (14,730 ft/4,490 m), and BROWNE TOWER (14,530 ft/4,429m). Permanent snowfields cover more than half the mountain and feed numerous glaciers. Mount McKinley was first scaled successfully by the American explorer Hudson Stuck in 1913. It is included in DENALI NATIONAL PARK AND PRESERVE.

McKinley Park, village, central ALASKA, on Alaska railroad and PARKS HIGHWAY, 100 mi/161 km SW of FAIRBANKS. Tourist gateway to MOUNT MCKINLEY NATIONAL PARK; numerous hotels.

McKinleyville, unincorporated city (2000 population 13,599), HUMBOLDT county, NW CALIFORNIA, 10 mi/16 km N of EUREKA, near Pacific Ocean and mouth of Mud River; 40°57'N 124°07'W. Trinidad State Beach, Patrick's Point State Park, and REDWOOD NATIONAL PARK to N. Cattle, sheep; dairying; timber.

McKinney (MIK-i-nee), city (2006 population 107,530), ⊙ COLLIN county, N TEXAS, suburb 30 mi/48 km NNE of DALLAS, on East Fork of TRINITY RIVER; 33°12'N 96°39'W. It is a shipping point for cotton, cattle, and grains (wheat, sorghum) and has grown as industrial center, manufacturing (electronic equipment, leather products, marble items, consumer goods, food, copper wire). Located on the blackland prairie, it was one of the principal cotton cities before the Civil War. The restored 1836 home of Collin McKinney, for whom the city was named, is here. Collin County Community College. A wildlife sanctuary and a museum of natural science are just outside the city. Inc. 1849.

McKinney, unincorporated village, LINCOLN county, central KENTUCKY, 7 mi/11.3 km SSE of STANFORD. Historic railroad depot. Tobacco, grain; livestock. Manufacturing (concrete products).

McKinney, Lake, KEARNY county, SW KANSAS, on Mattox Draw (intermittent sidestream N of ARKANSAS RIVER), 18 mi/29 km W of GARDEN CITY; 5 mi/8 km long, 1 mi/1.6 km wide; 37°57'N 101°16'W. Used for irrigation.

McKinnon (muh-KI-nuhn), suburb 8 mi/13 km SE of MELBOURNE, VICTORIA, SE AUSTRALIA, between BENTLEIGH and ORMOND; 37°55'S 145°03'E.

McKittrick, town (2000 population 72), MONTGOMERY county, E central MISSOURI, on MISSOURI River, across river from HERMANN (bridge); 38°44'N 91°26'W. Feed mill. Flooded in 1993.

McKittrick, unincorporated village, KERN county, S central CALIFORNIA, at E base of TEMBLOR RANGE, 35 mi/56 km W of BAKERSFIELD. Oil and natural-gas field. Cattle; grain, cotton.

McLain, village (2000 population 603), GREENE county, SE MISSISSIPPI, 31 mi/50 km SE of HATTIESBURG, on LEAF RIVER; 31°06'N 88°49'W. Cotton, corn; cattle, poultry; timber. Manufacturing (furniture). De Soto National Forest to W and S.

McLaren Vale (muh-KLA-ruhn VAIL), town, SE SOUTH AUSTRALIA, 22 mi/35 km S of ADELAIDE; 35°14'S 138°32'E. Vineyards.

McLaughlin (muhk-LAWF-lin), town (2006 population 757), CORSON county, N SOUTH DAKOTA, 27 mi/43 km NW of MOBRIDGE, near Oak Creek, in Standing Rock Indian Reservation; 45°48'N 100°48'W. Trading point for ranching and farming area.

McLaughlin (muhk-LAHF-lin), hamlet, E central ALBERTA, W CANADA, 39 mi/62 km from VERMILION, in VERMILION RIVER COUNTY NO. 24, near SASKATCHEWAN border; 52°59'N 110°10'W.

McLean (muh-KLAIN), county (□ 1,186 sq mi/3,083.6 sq km; 2006 population 161,202), central ILLINOIS; ⊙ BLOOMINGTON; 40°29'N 88°50'W. Drained by SANGAMON and MACKINAW rivers and by KICKAPOO, SALT, and small Money and Sugar creeks. Includes Lake Bloomington (with residences, resort), Evergreen Lake, and Moraine View State Park. Agriculture (corn, wheat, soybeans; livestock; dairy products). Gravel and sand pits. Manufacturing, includes motor vehicle plant at NORMAL. Includes Illinois State University, Illinois Wesleyan University, Heartland College, and a branch of Lincoln College. Formed 1830.

McLean, county (□ 256 sq mi/665.6 sq km; 2006 population 9,844), W KENTUCKY; ⊙ CALHOUN; 37°31′N 87°15′W. Bounded W by GREEN and POND rivers; drained by Green River and Cypress Creek. Agricultural area (soybeans, corn, wheat, dark and burley tobacco, tomatoes, hay, alfalfa; hogs, cattle; timber); bituminous-coal mines. Some manufacturing at LIVERMORE and Calhoun. Formed 1854.

McLean, county (□ 2,065 sq mi/5,369 sq km; 2006 population 8,543), central NORTH DAKOTA; ⊙ WASHBURN; 47°36′N 101°19′W. Agricultural area in W and S by MISSOURI RIVER. Lignite mines; cattle; dairy products; wheat, barley, rye, flax. Audubon National Wildlife Refuge on Lake Audubon, E part of LAKE SAKAKAWEA, at county center; several small lakes in NE quarter; Fort Stevenson State Park on N shore of Lake SAKAKAWEA; Fort Mandan Historic Site in S; MCCLUSKY Canal crosses E part; part of Fort Berthold Indian Reservation in W part. The boundary between Mountain and Central time zones follows river upstream as far as LITTLE MISSOURI RIVER, including Lake Sakakawea above GARRISON DAM. Formed 1883 and government organized the same year. Named for John A. McLean (1849–1916), the first mayor of BISMARCK.

McLean (muhk-LEEN), unincorporated city (2000 population 38,929), FAIRFAX county, N VIRGINIA, suburb 8 mi/12.9 km WNW of WASHINGTON, D.C., near POTOMAC RIVER; 38°56′N 77°10′W. Manufacturing (foods, satellite components, printing and publishing, computer and telecommunications equipment). Headquarters of the CIA, Federal Highway Administration Research Station to NE in LANGLEY. GEORGE WASHINGTON MEMORIAL PARKWAY (National Park unit) to E, follows Potomac River, includes Turkey Run Park. TYSONS CORNER Center, one of the largest shopping centers in the U.S., is here.

McLean (muhk-LEEN), former township, S ONTARIO, E central CANADA, 9 mi/15 km from BRACEBRIDGE; 45°08′N 79°06′W. Amalgamated into LAKE OF BAYS township in 1971.

McLean (MIK-lain), town (2006 population 820), GRAY county, extreme N TEXAS, in the PANHANDLE, 75 mi/121 km E of AMARILLO; 35°13′N 100°35′W. Elevation 2,812 ft/857 m. In cattle, wheat, cotton region; gas and oil wells. Lake McClellan and MCCLELLAN CREEK National Grassland to W. Settled 1901, incorporated 1909.

McLean (mik-LAIN), village (2000 population 808), MCLEAN county, central ILLINOIS, 14 mi/23 km SW of BLOOMINGTON; 40°19′N 89°10′W. In rich agricultural area.

McLean, village (2006 population 36), PIERCE county, NE NEBRASKA, 13 mi/21 km N of PIERCE; 42°23′N 97°28′W.

McLean Canyon, CANADA: see CHURCHILL FALLS.

McLeansboro (muh-KLAINZ-bur-oh), city (2000 population 2,945), ⊙ HAMILTON county, SE ILLINOIS, 25 mi/40 km SE of MOUNT VERNON; 38°05′N 88°31′W. In agricultural area; corn, wheat; livestock. Incorporated 1840.

McLeansville (muh-KLEENZ-vil), unincorporated town (□ 6 sq mi/15.6 sq km; 2000 population 1,080), GUILFORD county, N central North Carolina, residential suburb 10 mi/16 km ENE of GREENSBORO; 36°06′N 79°39′W. Service industries in agricultural area (tobacco, grain, soybeans; poultry; cattle; dairying).

McLellan Reservoir, ARAPAHOE and DOUGLAS counties, N central COLORADO, on Dad Clark Gulch (sidestream of SOUTH PLATTE RIVER), 11 mi/18 km S of DENVER; 39°35′N 104°54′W. Maximum storage capacity of 8,952 acre-ft; 1 mi/1.6 km long. Formed by McLellan Dam (111 ft/34 m high), built (1969) by the City of Englewood for water supply.

McLemoresville (mak-luh-MORZ-vil), town (2006 population 301), CARROLL county, NW TENNESSEE, 8 mi/13 km W of HUNTINGDON; 35°59′N 88°34′W.

McLennan (MIK-le-nuhn), county (□ 1,060 sq mi/2,756 sq km; 2006 population 226,189), E central TEXAS; ⊙ WACO; 31°33′N 97°12′W. Commercial, distribution, manufacturing center for wide region. Drained by BRAZOS RIVER and BOSQUE RIVER and its North and Middle branches, and Aquilla and Tradinghouse creeks; includes Lake Waco. Rich agricultural area (oats, wheat, hay, cotton, corn, grain sorghum, pecans); extensive dairying; beef cattle, hogs. Limestone, clay, stone, sand and gravel; oil and gas. Fishing. Tradinghouse Creek Reservoir in E; Lake Waco (Bosque River) in center. Formed 1850.

McLennan (muhk-LE-nuhn), town (□ 2 sq mi/5.2 sq km; 2001 population 804), W ALBERTA, W CANADA, on Kimiwan Lake (7 mi/11 km long, 4 mi/6 km wide), 40 mi/64 km SSE of PEACE River, 40 mi/64 km NW of LESSER SLAVE LAKE; 55°42′N 116°54′W. Lumbering; wheat; mixed farming. Bird sanctuary. First established in 1914, the town was a railway center and on the original route to ALASKA. Incorporated as a village in 1944, and as a town in 1948.

McLeod (muh-KLOUD), county (□ 505 sq mi/1,313 sq km; 2006 population 37,279), S central MINNESOTA; ⊙ GLENCOE; 44°49′N 94°16′W. Watered by South Fork of CROW RIVER and BUFFALO CREEK. Agricultural area (corn, oats, wheat, hay, alfalfa, soybeans, peas; hogs, cattle, poultry; dairying). Many small lakes in county. Formed 1856.

McLeod, Fort (muh-KLOUD), trading post, E central BRITISH COLUMBIA, W CANADA, on McLeod Lake (14 mi/23 km long), 75 mi/121 km N of PRINCE GEORGE. Established 1805 by Simon Fraser for the North West Company; taken over 1821 by Hudson's Bay Company.

McLoud, town (2006 population 4,153), POTTAWATOMIE county, central OKLAHOMA, 12 mi/19 km NW of SHAWNEE, 21 mi/34 km E of OKLAHOMA CITY, on NORTH CANADIAN River; 35°24′N 97°05′W. In rich agricultural area (corn, wheat, peanuts; dairy products).

McLoughlin House National Historic Site (muh-GLAHK-luhn), in McLoughlin Park, OREGON CITY, NW OREGON; an affiliated area of the National Park Service. Home of fur trader Dr. John McLoughlin, the "Father of Oregon." Authorized 1941.

McLoughlin, Mount (muh-GLAHK-luhn) (9,495 ft/2,894 m), E JACKSON county, near KLAMATH county line, SW OREGON, in CASCADE RANGE, W of UPPER KLAMATH LAKE, c.30 mi/48 km NW of KLAMATH FALLS.

McLouth (muh-KLOUTH), village (2000 population 868), JEFFERSON county, NE KANSAS, 6 mi/9.7 km E of OSKALOOSA; 39°12′N 95°12′W. In livestock-raising, dairying, and general farming region. Leavenworth State Fishing Lake to SE.

McMahon Line, c.2,500 mi/4,023 km long, border between N INDIA and SW CHINA (TIBET), in dispute since early twentieth century. Named for Sir Henry McMahon, British delegate to and instigator of the Simla (now Shimla) Conference, a 1913–1914 meeting of Britain, China, and Tibet. Britain and Tibetan representatives accepted the border between what was then British India and an autonomous Tibet, but China simply initialed the agreement, keeping the issue open, as no Chinese government had recognized Tibetan sovereignty. When India became independent in 1947, it supported the frontier as it stood, while the Chinese believed the line to be a bullying tactic by British colonial power. Border violations by China started in 1954. Longju (in the E sector) was captured in 1958 and an Indian patrol was attacked near the Kongka pass (Ladakh region, W sector). Indian support for the Tibetan revolt (1959) and the asylum provided to the fleeing Dalai Lama further strained relations with China. A twenty-one-day war broke out in 1962, resulting in Chinese occupation of parts of Aksai Chin (JAMMU AND KASHMIR state; still occupied) and NE Arunachal Pradesh state (since returned to India). Relations between the two nations have improved in the 1990s and although India still claims occupied Aksai Chin, the issue is no longer prominent.

McMasterville (muhk-MAS-tuhr-vil), village (□ 1 sq mi/2.6 sq km; 2006 population 4,476), MONTÉRÉGIE region, S QUEBEC, E CANADA, 19 mi/30 km from MONTREAL; 45°32′N 73°16′W. Part of the Metropolitan Community of Montreal (*Communauté Metropolitaine de Montréal*).

McMechen (mik-MEK-uhn), town (2006 population 1,798), MARSHALL county, N WEST VIRGINIA, in NORTHERN PANHANDLE, on the OHIO RIVER, 8 mi/12.9 km S of WHEELING; 39°59′N 80°43′W. Industrial area; manufacturing (concrete, machining). Incorporated 1895.

McMicken Heights, village, KING county, W WASHINGTON, residential suburb 13 mi/21 km SSE of SEATTLE. Seattle-Tacoma International Airport (SEA-TAC) to W.

McMillan, village, LUCE county, NE UPPER PENINSULA, MICHIGAN, 9 mi/14.5 km W of NEWBERRY; 46°20′N 85°41′W. In lumbering and agricultural area.

McMillan, Lake, NEW MEXICO: see PECOS, river.

McMinn, county (2006 population 52,020), SE TENNESSEE; ⊙ ATHENS; 35°25′N 84°37′W. In GREAT APPALACHIAN VALLEY; Cherokee National Forest lies along SE border; bounded SW by HIWASSEE River. Agriculture and diversified manufacturing. Formed 1819.

McMinnville (muhk-MIN-vil), city (2006 population 30,410), ⊙ YAMHILL county, NW OREGON, 33 mi/53 km SW of PORTLAND, at confluence of North and South Yamhill rivers (from YAMHILL RIVER); 45°12′N 123°11′W. Railroad junction to NE. Trade and processing center in fertile WILLAMETTE Valley. Manufacturing (foods, textiles, building materials, printing and publishing, rubber products, concrete, steel milling, medical equipment). Agriculture (apples, cherries, plums, peaches, pears, grapes, nuts, berries, wheat, barley, oats; poultry); dairy products; wineries. Linfield College Siuslaw National Forest to W; Erratic Rock Wayside State Park to SW. Incorporated 1876.

McMinnville, city (2006 population 13,311), ⊙ WARREN county, central TENNESSEE, on branch of CANEY FORK, 38 mi/61 km SE of MURFREESBORO; 35°41′N 85°46′W. In timber and farm area; manufacturing; tree nurseries. Great Falls Dam is NE, CENTER HILL Reservoir N. Settled 1800; incorporated 1808.

McMullen (MIK-muhl-uhn), county (□ 1,142 sq mi/2,969.2 sq km; 2006 population 913), S TEXAS; ⊙ TILDEN; 28°21′N 98°34′W. Drained by FRIO and NUECES rivers. Cattle ranching; agriculture (grain sorghum, corn). Oil, natural-gas wells; sulphur. Part of CHOKE CANYON LAKE reservoir (Frio River) in NE, Choke Canyon State Park on S shore, on county line. Formed 1858.

McMunn (muhk-MUHN), hamlet, SE MANITOBA, W central CANADA, 67 mi/107 km from WINNIPEG, and in REYNOLDS rural municipality; 49°37′N 95°42′W.

McMurdo Dry Valleys, region on the S SCOTT COAST, in the TRANSANTARCTIC MOUNTAINS of EAST ANTARCTICA; 77°30′S 162°00′E. Encompasses the largest assemblage of ice-free features on the continent; 120 mi/194 km long, 50 mi/81 km wide.

McMurdo Ice Shelf, the part of the ROSS ICE SHELF that lies between MCMURDO SOUND and ROSS ISLAND to the N and MINNA BLUFF to the S, in WEST ANTARCTICA; 78°00′S 166°30′E.

McMurdo Sound, in the SW corner of the ROSS SEA, WEST ANTARCTICA, between ROSS ISLAND and VICTORIA LAND; 35 mi/55 km long and wide; 77°30′S 165°00′E.

McMurdo Station, on ROSS ISLAND, ANTARCTICA, SW corner of ROSS SEA; 77°51′S 166°40′E. In January 1956 the Americans established an Air Operating Facility of the U.S. Navy; it was upgraded to a Naval Air Facility in 1957 and became a permanent station in 1961, al-

Area is shown by the symbol □, and capital city or county seat by ⊙.

though not a scientific station. It is now the largest of all the stations in Antarctica.

McMurray, unincorporated town (2000 population 4,726), WASHINGTON county, SW PENNSYLVANIA, industrial suburb 12 mi/19 km SSW of PITTSBURGH on urban fringe; 40°16′N 80°05′W. Manufacturing includes printing and publishing, medical equipment, aerospace components, TV and microwave transmitters, wooden products.

McMurray, CANADA: see FORT MCMURRAY.

McMurrich/Monteith (muhk-MUH-rich, mahn-TEETH), township (□ 105 sq mi/273 sq km; 2001 population 766), S ONTARIO, E central CANADA, 27 mi/43 km from PARRY SOUND; 45°27′N 79°30′W.

McNab, village (2000 population 37), HEMPSTEAD county, SW ARKANSAS, 14 mi/23 km W of HOPE, near LITTLE RIVER; 33°39′N 93°49′W. Grassy Lake (backwater lake) to NW.

McNabb, village (2000 population 310), PUTNAM county, N central ILLINOIS, 12 mi/19 km SSW of PERU; 41°10′N 89°12′W. Corn, soybeans.

McNab-Braeside (muhk-NAB–BRAI-seid), township (□ 98 sq mi/254.8 sq km; 2001 population 6,843), SE ONTARIO, E central CANADA, 40 mi/65 km from PEMBROKE; 45°25′N 76°30′W. Also written McNab/Braeside.

McNair (MIK-ner), unincorporated village, HARRIS county, S TEXAS, suburb 20 mi/32 km E of HOUSTON, on Goose Creek; 29°48′N 95°01′W. Area is gradually being annexed by BAYTOWN, to S. Highlands Reservoir to N.

McNairy (muhk-NAI-ree), county (□ 569 sq mi/1,479.4 sq km; 2006 population 25,722), SW TENNESSEE; ⊙ SELMER; 35°11′N 88°34′W. Bounded S by MISSISSIPPI RIVER; drained by tributaries of South Fork of FORKED DEER, HATCHIE, and TENNESSEE rivers. Agriculture; manufacturing. Formed 1823.

McNary (mik-NER-ee), village, APACHE county, E ARIZONA, near WHITE RIVER, in NE part of Fort Apache Indian Reservation, 40 mi/64 km SW of SAINT JOHNS. Elevation 7,505 ft/2,288 m. Lumber milling; wood products. National forests nearby. WHITE MOUNTAINS to E. Apache-Sitgreaves National Forest to N; source of White River to SE; fish hatchery is here.

McNary, village (2000 population 211), RAPIDES parish, central LOUISIANA, 23 mi/37 km SSW of ALEXANDRIA; 30°60′N 92°35′W. In agricultural area. Cocodrie Lake to E. Kisatchie National Forest to NW.

McNary (MIK-ner-ee), village, HUDSPETH county, extreme W TEXAS, 50 mi/80 km SE of EL PASO, near the RIO GRANDE (Mexican border); 31°14′N 105°47′W. In irrigated farm area (cotton, vegetables, alfalfa; cattle, hogs). Ruins of old Fort Quitman are SE.

McNary Lock and Dam, Oregon: see WALLULA, LAKE.

McNeil, village (2000 population 662), COLUMBIA county, SW ARKANSAS, 6 mi/9.7 km NNE of MAGNOLIA; 33°21′N 93°12′W. Railroad junction. In agricultural area; manufacturing (aluminum for aircraft industry). Logoly State Park is here.

McNeil Island, PIERCE county, W WASHINGTON, in PUGET SOUND WSW of TACOMA and just W of STEILACOOM; c.3 mi/4.8 km long. Site of McNeil Island Federal Penitentiary. Ferry from Steilacoon via Anderson Island to S end. Carr Inlet (N), Pitt Passage and Kitsap Peninsula (NW), Balch Passage and Anderson Island (S).

McNeill, unincorporated village, PEARL RIVER county, SE MISSISSIPPI, 10 mi/16 km NNE of PICAYUNE. Agricultural area (cotton, corn, berries; cattle; dairying). Manufacturing (apparel).

McNutt Island (mik-NUHT), in the ATLANTIC OCEAN, at entrance to Shelburne Harbour, SW NOVA SCOTIA, CANADA, 7 mi/11 km S of SHELBURNE; 43°38′N 65°47′W. Island is 4 mi/6 km long, 2 mi/3 km wide. Home to a group of yellow Birch trees that are centuries old, one of which is 1500 years old.

McPhee, former village, MONTEZUMA county, SW COLORADO, on DOLORES RIVER, 11 mi/18 km NNE of CORTEZ. Elevation c.7,000 ft/2,134 m. MESA VERDE NATIONAL PARK nearby. Now called McPhee Reservoir.

McPhee, reservoir (□ 8 sq mi/20.8 sq km), MONTEZUMA county, extreme SW COLORADO, on DOLORES RIVER, 10 mi/16 km N of CORTEZ; 37°34′N 108°35′W. Maximum capacity 399,200 acre-ft. Formed by McPhee Dam (295 ft/90 m high), built (1984) by the Bureau of Reclamation for irrigation; also used for water supply, power generation, recreation, and as a fish and wildlife pond. Escalante Ruins Historical Site and Anasazi Heritage Center on dam.

McPherson (muhk-FUHR-suhn), county (□ 901 sq mi/2,342.6 sq km; 2006 population 29,380), central KANSAS; ⊙ MCPHERSON; 38°23′N 97°39′W. Rolling plain, drained (NW) by SMOKY HILL RIVER. Wheat, soybeans, alfalfa, hay; poultry, cattle, hogs, sheep; chemicals, petroleum refining, plastics products, mineral wool. Oil and gas fields. Formed 1870.

McPherson, county (□ 860 sq mi/2,236 sq km; 2006 population 497), W central NEBRASKA, in Sand Hills region; ⊙ TRYON; 41°34′N 101°03′W. Grazing area; cattle, hogs; corn. Central/Mountain time zone boundary follows N and W boundaries of county, and W part of S boundary. Brown Lake and White Water Lake in NW, Diamond Bar Lake in SW (all of them small natural lakes). Formed 1887.

McPherson (mik-FIR-suhn), county (□ 1,151 sq mi/2,992.6 sq km; 2006 population 2,565), N SOUTH DAKOTA, on NORTH DAKOTA state line; ⊙ LEOLA; 45°46′N 99°13′W. Agricultural (wheat, flax) and cattle-raising region. Drained by Spring Creek in NW and Foot Creek in SE. Formed 1873.

McPherson (muhk-FUHR-suhn), city (2000 population 13,770), ⊙ MCPHERSON county, central KANSAS, in a farm area on the old SANTA FE TRAIL; 38°22′N 97°39′W. Manufacturing (plastics products, railroad equipment, motor vehicles, petroleum refining, animal feeds, plating, pharmaceuticals). The city is named for Gen. James B. McPherson, the highest ranking Union general to die in the Civil War. McPherson College here. Incorporated 1874.

McPherson Range (muhk-FEER-suhn), E spur of GREAT DIVIDING RANGE, E AUSTRALIA; extends c.140 mi/225 km W and SW from Point Danger on PACIFIC coast to WALLANGARRA; forms part of boundary between QUEENSLAND and NEW SOUTH WALES. Highest point, Mount Barney (4,300 ft/1,311 m); 28°20′S 153°00′E. Sometimes spelled MacPherson.

McQueen, MONTANA: see MEADERVILLE.

McQueeney (MIK-kwee-nee), town (2000 population 2,527), GUADALUPE county, S central TEXAS, 5 mi/8 km WNW of SEGUIN, on Lake McQueeney reservoir (c.1.5 mi/2.4 km long) to N, on GUADALUPE RIVER; 29°36′N 98°02′W. Agriculturally diverse area (cattle; cotton, peanuts); greyhound breeding; manufacturing (building materials, pottery); clay; recreational area.

McRae, town (2000 population 2,682), ⊙ TELFAIR county, S central GEORGIA, 32 mi/51 km S of DUBLIN, adjacent to HELENA (W), and on LITTLE OCMULGEE RIVER; 32°04′N 82°54′W. Manufacturing includes lumber, beverages, clothing; pecan processing, seed packaging. Little Ocmulgee State Park nearby. Established by Scottish settlers in mid-nineteenth century. Incorporated 1874.

McRae, village (2000 population 661), WHITE county, central ARKANSAS, 37 mi/60 km NE of LITTLE ROCK; 35°06′N 91°49′W. In agricultural area.

McSherrystown, borough (2006 population 2,821), ADAMS county, S PENNSYLVANIA, suburb 2 mi/3.2 km W of HANOVER, near South Branch of Conewago Creek; 39°47′N 77°01′W. Agriculture (grain, soybeans, potatoes, apples; livestock; dairying); manufacturing (electronic equipment, cigars). Hanover Airport to S.

Conewago Chapel (1787) to NW. Long Arm Reservoir to SE. Incorporated 1882.

Mctavish (muhk-TA-vish), community, S MANITOBA, W central CANADA, 7 mi/12 km N of town of MORRIS, in MORRIS rural municipality; 49°27′N 97°22′W.

McVeigh (muhk-VAI), unincorporated village, PIKE county, E KENTUCKY, in the CUMBERLAND MOUNTAINS, 9 mi/14.5 km S of WILLIAMSON (WEST VIRGINIA). In bituminous-coal-mining area.

McVeytown, borough (2006 population 389), MIFFLIN county, central PENNSYLVANIA, 12 mi/19 km SW of LEWISTOWN, on JUNIATA RIVER; 40°30′N 77°44′W. Agriculture (dairying; livestock; corn, alfalfa); manufacturing (feeds). Tuscarora State Forest to SE; Rothrock State Forest to NW.

McVille (mak-VIL), village (2006 population 417), NELSON county, E central NORTH DAKOTA, 22 mi/35 km N of COOPERSTOWN, near SHEYENNE RIVER; 47°46′N 98°10′W. Founded in 1894 and incorporated in 1908. Name reflects the large number of families with "Mc" in their names.

McWatters (muhk-WAH-tuhrz), former village, W QUEBEC, E CANADA; 48°13′N 78°55′W. Gold mining. Amalgamated into ROUYN-NORANDA in 2002.

Mdaburo (uhm-dah-BOO-ro), village, SINGIDA region, central TANZANIA, 25 mi/40 km S of MANYONI, on KISIGO RIVER; 06°13′S 34°39′E. Sheep, goats; wheat, corn.

Mdandu (uhm-dahn-DOO), village, IRINGA region, SW TANZANIA, 80 mi/129 km ESE of MBEYA, and E of KIPENGERE RANGE; 09°10′S 34°40′E. Coffee, tea, pyrethrum, corn; cattle, goats, sheep.

Mdantsane, township, EASTERN CAPE province, S SOUTH AFRICA, 10 mi/16 km NW of EAST LONDON (of BUFFALO CITY municipality); 32°57′S 27°46′E.

Mdedelelo Wilderness Area, SOUTH AFRICA: see CATHKIN PEAK.

Mdina, town, W central MALTA, near RABAT. Ancient capital of Malta during times of Arabs and Normans. Walled city; reconstructed Roman villa (now museum). Called "silent city." Arab architecture style in much of town. Also 13th-century Christian cathedral.

M'diq (mah-DEEK), city, Tetouan province, Tanger-Tétouan administrative region, N MOROCCO, 9 mi/15 km N of TETOUAN; 35°41′N 05°19′W. Tourist (and reputed smuggling) center near the MEDITERRANEAN SEA coast.

Mead, unincorporated town, SPOKANE county, E WASHINGTON, suburb 8 mi/12.9 km NNE of SPOKANE, on Deadmans Creek. Agricultural area (vegetables, wheat; dairying; cattle, sheep, hogs). Manufacturing (food; iron foundry; aluminum production). Mead Airport to N.

Mead, village (2000 population 2,017), WELD county, N COLORADO, 8 mi/12.9 km NE of LONGMONT; 40°13′N 104°59′W. Elevation 5,140 ft/1,567 m. Manufacturing (chemicals, machinery). Barbour Ponds State Park to S.

Mead, village (2006 population 610), SAUNDERS county, E NEBRASKA, 7 mi/11.3 km E of WAHOO, near PLATTE RIVER; 41°13′N 96°29′W. U.S. and University of Nebraska agricultural experiment station. Packaging.

Meade, county (□ 979 sq mi/2,545.4 sq km; 2006 population 4,561), SW KANSAS; ⊙ MEADE; 37°14′N 100°21′W. Rolling prairie region, in High Plains region, bordered S by OKLAHOMA; drained by CROOKED CREEK. Wheat, corn, sorghum; cattle; volcanic ash deposits. Meade State Park in SW center. Formed 1885.

Meade, county (□ 324 sq mi/842.4 sq km; 2006 population 27,994), N KENTUCKY; ⊙ BRANDENBURG; 37°58′N 86°13′W. Bounded N and NW by OHIO RIVER (INDIANA state line); drained in E by Otter Creek. Rolling agricultural area (burley tobacco, hay, alfalfa, soybeans, wheat; hogs, cattle; timber); limestone quarries. Otter Creek Park in E; part of FORT KNOX Military Reservation and Gold Depository in E. Formed 1823.

Cross-references are shown in SMALL CAPITALS. The pronunciation guide is shown on page xix. The sources of population figures are shown on page xvii.

Meade, county (□ 3,482 sq mi/9,053.2 sq km; 2006 population 24,425), W central SOUTH DAKOTA; ⊙ STURGIS; 44°34′N 102°42′W. Ranching and agricultural area rich in mineral resources. Drained by BELLE FOURCHE RIVER and Elk, Sulphur, Beaver Dam, and Cherry creeks; and bounded E by CHEYENNE RIVER. Gold, manganese, lignite, bentonite, fuller's earth; marble; corn, soybeans, hay, sugar beets; cattle, hogs, sheep; dairying. Bear Butte State Park in W; small part of Black Hills National Forest in SW. Ellsworth Air Force Base on S boundary. Formed 1889.

Meade, town (2000 population 1,672), ⊙ MEADE county, SW KANSAS, on CROOKED CREEK, 35 mi/56 km SSW of DODGE CITY; 37°17′N 100°20′W. Shipping point for cattle and grain area. Dalton Gang Hideout and Museum here. Meade State Park to SW. Incorporated 1885.

Meade, Fort George G., MARYLAND: see FORT GEORGE G. MEADE.

Meade River, c.250 mi/402 km long, N ALASKA, S of POINT BARROW; rises near 69°15′N 158°30′W; flows N to ARCTIC OCEAN at 70°50′N 155°46′W. Atkasuk, 60 mi/97 km SSW of BARROW. Inuit village.

Meaderville (MEE-duhr-vil), NE suburb of BUTTE, SILVER BOW county, SW MONTANA. Copper mines with precipitating plant that recovered pure copper from water pumped out of mines; closed 1983. Originally called Gunderson.

Mead, Lake (□ 247 sq mi/640 sq km), reservoir, on the NEVADA-ARIZONA border, on COLORADO RIVER, c.25 mi/40 km SE of LAS VEGAS; 36°01′N 114°46′W. Shoreline is 550 mi/885 km long; reservoir is 115 mi/185 km long, 1 mi/1.6 km–8 mi/12.9 km wide; max. depth 589 ft/180 m. Formed by HOOVER DAM. Surrounded by LAKE MEAD NATIONAL RECREATION AREA, which also extends S, below dam, to take in Lake MOHAVE (Devil Dam). Has 30-mi/48-km-long N arm formed by VIRGIN RIVER. Grand Canyon National Park at headwaters of lake in Arizona Valley of Fire State Park (Nevada) on Virgin River arm.

Meadow (ME-do), village (2000 population 658), TERRY county, NW TEXAS, on the LLANO ESTACADO, 25 mi/40 km SW of LUBBOCK; 33°20′N 102°12′W. In agricultural area; oil and gas.

Meadow, village (2006 population 247), MILLARD county, W UTAH, 8 mi/12.9 km SW of FILLMORE; 38°53′N 112°24′W. Elevation 5,000 ft/1,524 m. Wheat, barley, alfalfa; cattle. Fishlake National Forest to E; Kanosh Indian Reservation to SE. Clear Lake Waterfowl Management Area to NW. BLACK ROCK DESERT to W.

Meadow Bridge, village (2006 population 307), FAYETTE county, S central WEST VIRGINIA, 20 mi/32 km SE of FAYETTEVILLE; 37°51′N 80°51′W. Coal-mining and agricultural region. NEW RIVER GORGE NATIONAL RIVER to W.

Meadowbrook, unincorporated village, MADISON county, SW ILLINOIS, residential suburb 21 mi/34 km NNE of SAINT LOUIS (MISSOURI), 2 mi/3.2 km E of Behalto, near Indian Creek; 38°53′N 90°00′W.

Meadow Grove, village (2006 population 298), MADISON county, NE central NEBRASKA, 15 mi/24 km W of NORFOLK, on ELKHORN RIVER; 42°01′N 97°44′W. Grain; livestock. Millstone State Wayside Area to E.

Meadow Heights (ME-do HEITS), locality, residential suburb 11 mi/18 km N of MELBOURNE, VICTORIA, SE AUSTRALIA. Significant Turkish, Vietnamese populations.

Meadow Lake (ME-do), town (2006 population 4,771), W SASKATCHEWAN, CANADA, on MEADOW LAKE (6 mi/10 km long, 3 mi/5 km wide), 100 mi/161 km N of NORTH BATTLEFORD; 54°08′N 108°26′W. Grain elevators; woodworking; lumbering; fur trapping; fishing.

Meadow Lands, unincorporated town, Canton township, WASHINGTON county, SW PENNSYLVANIA, sub- urb 3 mi/4.8 km NE of WASHINGTON and 3 mi/4.8 km SW of CANONSBURG, on Chartiers Creek; 40°13′N 80°13′W. Agriculture area (corn, hay; dairying); manufacturing (fabricated metal products, polyurethane and rubber products, electrical equipment, motor vehicle parts); coal. Also spelled Meadowlands.

Meadowlands, village (2000 population 111), ST. LOUIS county, NE MINNESOTA, on WHITEFACE RIVER, 35 mi/56 km NW of DULUTH; 47°04′N 92°43′W. Dairying; poultry; oats, alfalfa; manufacturing (furniture); Whiteface River State Forest to E.

Meadow Mountain, ridge (c.3,000 ft/914 m) of the ALLEGHENIES, NW MARYLAND; extends NE c.20 mi/32 km from DEEP CREEK LAKE to state line just SE of SALISBURY, PENNSYLVANIA. On it is part of Savage River State Forest.

Meadowood, unincorporated town (2000 population 2,912), BUTLER county, NW PENNSYLVANIA, residential suburb 1 mi/1.6 km SSE of BUTLER; 40°50′N 79°53′W.

Meadow River, 53 mi/85 km long, S central WEST VIRGINIA; rising on KEENEY KNOB in N SUMMERS county; flows generally NW, through Meadow River Wildlife Management Area, to GAULEY RIVER 14 mi/23 km E of GAULEY BRIDGE; lower 5 mi/8 km in GAULEY RIVER NATIONAL RECREATION AREA.

Meadows (ME-doz), community, S MANITOBA, W central CANADA, 22 mi/36 km from WINNIPEG, and in ROSSER rural municipality; 50°01′N 97°35′W.

Meadows (ME-doz), town (2006 population 629), FORT BEND county, SE TEXAS, residential suburb 15 mi/24 km WSW of HOUSTON, near Keegans Bayou; 29°38′N 95°35′W.

Meadows of Dan (ME-doz, DAN), unincorporated village, PATRICK county, SW VIRGINIA, 30 mi/48 km W of MARTINSVILLE, near source of DAN RIVER. Manufacturing (furniture, food, lumber, clothing); agriculture (apples, grain; cattle; dairying); timber. Pinnacles of Dan mountain (2,655 ft/809 m) to SW, Lovers Leap mountain (3,300 ft/1,006 m) to E. BLUE RIDGE PARKWAY passes to W.

Meadow Valley Wash, c.110 mi/177 km long, SE NEVADA; rises in mt. region of E LINCOLN county; flows S, past PANACA and CALIENTE, to MUDDY RIVER N of Lake MEAD.

Meadowview (ME-do-vyoo), unincorporated town, WASHINGTON county, SW VIRGINIA, 7 mi/11 km NE of ABINGDON; 36°45′N 81°51′W. Manufacturing (machining); agriculture (dairying; livestock; corn, alfalfa). WALKER MOUNTAIN ridge to W. EMORY 2 mi/3 km to NE.

Meadow Vista, unincorporated town (2000 population 3,096), PLACER county, E CALIFORNIA, 8 mi/12.9 km NNE of AUBURN; 39°00′N 121°02′W. Fruit, nuts, grain; cattle, sheep.

Meadville (MEED-vil), city (2000 population 457), LINN county, N central MISSOURI, near GRAND RIVER, 12 mi/19 km W of BROOKFIELD; 39°47′N 93°17′W. Corn, wheat, soybeans; cattle.

Meadville, city (2006 population 13,421), ⊙ CRAWFORD county, NW PENNSYLVANIA, 33 mi/53 km S of ERIE, on FRENCH CREEK, at mouth of Cussewago Creek; 41°38′N 80°09′W. Manufacturing (metal fabrication, furniture, plastic products, pet food, electrical components, food products, concrete, printing and publishing, glass). Agricultural area (corn, hay, potatoes; dairying). Oil deposits are located near the city. Seat of Allegheny College Port Meadville Airport to W. Woodcock Lake reservoir to NE, Tamarack Lake reservoir to SE, CONNEAUT LAKE (natural) to W. Settled 1788, incorporated 1866.

Meadville (MEED-vil), village (2000 population 519), ⊙ FRANKLIN county, SW MISSISSIPPI, 30 mi/48 km ESE of NATCHEZ; 31°28′N 90°53′W. Timber area surrounded by Homochitto National Forest.

Meaford (MEE-fuhrd), town (□ 3 sq mi/7.8 sq km; 2001 population 4,524), S ONTARIO, E central CANADA, on Nottawasaga Bay, inlet of GEORGIAN BAY, 18 mi/29 km E of OWEN SOUND; 44°36′N 80°35′W. Port; fruit processing; woodworking; shipbuilding; dairying; apple-growing, trout-fishing region.

Meagher (MAI-guhr), county (□ 2,394 sq mi/6,200 sq km; 1990 population 1,819; 2000 population 1,932), central MONTANA; ⊙ WHITE SULPHUR SPRINGS; 46°35′N 110°52′W. Mountain region drained by SMITH RIVER (source in S center), also Sixteenmile Creek (source in SE); and SOUTH FORK and NORTH FORK (sources in E), form MUSSELSHELL RIVER just E of E boundary. Barley, wheat, hay; cattle, sheep, hogs. Smith River State Park in NW; part of LITTLE BELT MOUNTAINS in NE; part of BIG BELT MOUNTAINS in W; part of Lewis and Clark National Forest in NE, center, and SE; part of Helena National Forest in W; small part of Gallatin National Forest in SW corner. Formed 1867. Present boundaries established 1911.

Meaghers Grant (MEE-guhrz), village, S NOVA SCOTIA, CANADA, on MUSQUODOBOIT RIVER, 26 mi/42 km NE of HALIFAX; 44°55′N 63°14′W. Settled 1692. Named for landowner, Martin Meagher.

Mealhada (mai-ahl-YAD-ah), town, AVEIRO district, N central PORTUGAL, on railroad, 12 mi/19 km N of COIMBRA. Vineyards, olive oil pressing, tar pitch manufacturing.

Meall Fuar-mhonaidh (MAIL FUHR–mon-neid), mountain (2,284 ft/696 m), HIGHLAND, N Scotland, on W shore of LOCH NESS; 57°15′N 04°33′W.

Mealy Mountains, range, E NEWFOUNDLAND AND LABRADOR, E CANADA; extending c.120 mi/193 km SW-NE along S shore of Lake MELVILLE; rises to 4,300 ft/1,311 m. There are several peaks over 3,000 ft/914 m high; numerous small lakes.

Meámbar (me-AHM-bar), town, COMAYAGUA department, W central HONDURAS, 27 mi/43 km NNW of COMAYAGUA; 14°47′N 87°46′W. Sugarcane, coffee, tobacco, cacao, rubber. CERRO AZUL MEÁMBAR NATIONAL PARK is nearby.

Meander Creek Reservoir (□ 3 sq mi/7.8 sq km), TRUMBULL county, NE OHIO, on Meander Creek, 5 mi/8 km WNW of YOUNGSTOWN; 41°09′N 80°48′W. Maximum capacity 62,000 acre-ft. Formed by Mineral Ridge Dam (60 ft/18 m high), built (1932) for water supply.

Meander River (mee-AN-duhr), unincorporated village, NW ALBERTA, W CANADA, on MACKENZIE HIGHWAY, 175 mi/282 km N of PEACE RIVER, on SE side of Hay River, on railroad to town of HAY RIVER, NORTHWEST TERRITORIES, in MACKENZIE NO. 23 specialized municipality; 59°02′N 117°42′W. Transportation service center to and from Northwest Territories. Timber, furs.

Meanguera (mai-ahn-GAI-rah), municipality and town, MORAZÁN department, EL SALVADOR, N of SAN FRANCISCO GOTERA in N central part of department; 13°51′N 88°09′W.

Meanguera del Golfo (mai-ahn-GAI-rah del GOL-fo), municipality and town (1993 population 1,225), LA UNIÓN department, EL SALVADOR, on Island of MEANGUERA in GULF OF FONSECA; 13°11′N 87°42′W. Victimized by pirates in colonial period repopulated in mid-1800s, now a fishing village with subsistence farming.

Meanguera Island (mai-ahn-GAI-rah), in GULF OF FONSECA, LA UNIÓN department, E EL SALVADOR, 11 mi/18 km SE of LA UNIÓN; 13°11′N 87°43′W. Island is 3 mi/4.8 km long, 2 mi/3.2 km wide; rises to 1,660 ft/506 m. Agriculture; fisheries. Town of MEANGUERA DEL GOLFO, on shore.

Meansville (MEENZ-vil), town (2000 population 192), PIKE county, W central GEORGIA, 16 mi/26 km S of GRIFFIN; 33°03′N 84°19′W.

Meares, Cape (MIR), promontory (700 ft/213 m) with lighthouse, TILLAMOOK county, NW OREGON, 5 mi/8 km NW of TILLAMOOK. Site of CAPE MEARES State Park.

Meares Island (MEERZ) (□ 27 sq mi/70.2 sq km), SW BRITISH COLUMBIA, W CANADA, in CLAYOQUOT SOUND off W VANCOUVER ISLAND, 45 mi/72 km W of PORT ALBERNI; 49°10′N 125°50′W. Island is 10 mi/16 km long, 2 mi/3 km–7 mi/11 km wide. Kakawis village (W); lumbering.

Meath (MEE-[th]), Gaelic *An Mhí*, county (□ 902 sq mi/2,335 sq km; 2002 population 134,005; 2006 population 162,831), NE IRELAND; ⊙ NAVAN; 53°40′N 06°40′W. Borders LOUTH, MONAGHAN, and CAVAN counties to N, WESTMEATH county to W, OFFALY county to SW, KILDARE county to S, DUBLIN county to SE, and IRISH SEA to E. The land is mostly level, being a part of the central plain of IRELAND, with extensive fertile areas near the BOYNE and BLACKWATER, the principal rivers. There is a sandy coastline of some 10 mi/16 km along the Irish Sea. Grain and potato cultivation and cattle raising support the bulk of the population. Manufacturing in the larger towns. The region is important in Irish history. TARA was long the seat of the ancient high kings of Ireland. Meath was considered a fifth province of Ireland for many centuries and was not finally organized as a county until the 17th century. Westmeath county separated from Meath county via a government act in 1543. Remains of archaeological interest have been found in the NEWGRANGE mounds. Other significant towns are KELLS and TRIM.

Méaulte, FRANCE: see ALBERT.

Meauwataka (myoo-WAH-tah-kuh), village, WEXFORD county, NW MICHIGAN, 10 mi/16 km NW of CADILLAC, near small Meauwataka Lake; 44°21′N 85°32′W. In Manistee National Forest.

Meaux (MO), city (□ 5 sq mi/13 sq km), SEINE-ET-MARNE department, ÎLE-DE-FRANCE region, N central FRANCE, in BRIE district, on MARNE RIVER and OURCQ CANAL, 25 mi/40 km ENE of PARIS; 48°55′N 03°00′E. It is an industrial center where metals, chemicals, and foodstuffs are produced. It is also a market for dairy products and cereals of the region. An episcopal see since the 4th century, Meaux has a cathedral (13th–16th century) that contains the tomb of Bossuet, the city's most influential bishop. In the massacre of Meaux (1358), thousands of peasants who had participated in a revolt against the authority of noblemen (known as the Jacquerie) were slain. Meaux has a 13th-century cathedral and vestiges of Roman and medieval ramparts. The city attracts tourists with its "sound and light" evening spectacle performed in the episcopal square. The château of Montceaux (partly in ruins) is 5 mi/8 km E of Meaux. It was known as the residence of French queens in 16th century. Meaux is a growth center benefiting from the E expansion of Parisian suburbs. Meaux-ESBLY international airport 3 mi/4.8 km SW of Meaux.

Mebane (MEB-in), town (□ 5 sq mi/13 sq km; 2006 population 9,285), ALAMANCE and ORANGE counties, N central NORTH CAROLINA, 9 mi/14.5 km E of BURLINGTON; 36°05′N 79°16′W. Service industries; manufacturing (consumer goods, furniture, textiles, apparel, heating and automotive components, lumber, tools; plastic, metal, and rubber products); agriculture (tobacco, grain, soybeans; cattle, poultry; dairying). Founded 1854.

Mecanhelas District, MOZAMBIQUE: see NIASSA.

Mecapaca (mai-kah-PAH-kah), town and canton, MURILLO province, LA PAZ department, W BOLIVIA, on LA PAZ RIVER, 15 mi/24 km SSE of LA PAZ; 16°41′S 68°02′W. Elevation 9,321 ft/2,841 m.

Mecatina, Cape (me-kuh-TEE-nuh), on the Gulf of SAINT LAWRENCE, E QUEBEC, E CANADA; 50°44′N 59°00′W. Air base; lighthouse. Also spelled Meccatina.

Mecatlán (me-kaht-LAHN), town, VERACRUZ, E MEXICO, in SIERRA MADRE ORIENTAL foothills, 28 mi/45 km SW of PAPANTLA DE OLARTE; 20°13′N 97°41′W. Corn, coffee, tobacco, sugarcane. In Totomac Indian area.

Mecayapan, town, VERACRUZ, SE MEXICO, 29 mi/47 km W of COATZACOALCOS. Tobacco, fruit.

Mecca, unincorporated town (2000 population 5,402), RIVERSIDE county, S CALIFORNIA, 32 mi/51 km SE of PALM SPRINGS, in COACHELLA VALLEY, and 4 mi/6.4 km NW of N end of SALTON SEA; 33°35′N 116°04′W. Dates, citrus. Painted Canyon, a spectacular many-colored gorge in Orocopia Mountains, to E. Salton Sea State Recreation Area to SE; JOSHUA TREE NATIONAL MONUMENT to NE; Coachella Canal to E; Torrez Martinez Indian Reservation to S and SW. Mecca is below sea level.

Mecca, town (2000 population 355), PARKE county, W INDIANA, 17 mi/27 km NNE of TERRE HAUTE, on RACCOON CREEK, near WABASH RIVER; 39°44′N 87°20′W. Covered bridges in area. Hogs, cattle; wheat, corn. Laid out 1890.

Mecca, SAUDI ARABIA: see MAKKA.

Meccatina, CANADA: see MECATINA, CAPE.

Mecerreyes (mai-ther-AI-es), village, BURGOS province, N SPAIN, 18 mi/29 km SSE of BURGOS; 42°06′N 03°34′W. Cereals, vegetables; sheep, hogs; charcoal manufacturing.

Mecham, Cape (MEE-chuhm), S extremity of PRINCE PATRICK ISLAND, NORTHWEST TERRITORIES, CANADA, on BEAUFORT SEA of the ARCTIC OCEAN, at entrance of MCCLURE STRAIT; 75°44′N 121°23′W.

Mechanic Falls (muh-KAN-ik), town, ANDROSCOGGIN county, SW MAINE, on the Little Androscoggin, and 8 mi/12.9 km W of AUBURN; 44°06′N 70°24′W. Set off from MINOT and POLAND 1893.

Mechanicsburg, village (2000 population 456), SANGAMON county, central ILLINOIS, 13 mi/21 km E of SPRINGFIELD; 39°48′N 89°24′W. In agricultural area.

Mechanicsburg (muh-KAN-iks-buhrg), village (□ 1 sq mi/2.6 sq km; 2006 population 1,719), CHAMPAIGN county, W central OHIO, 18 mi/29 km ENE of SPRINGFIELD; 40°04′N 83°33′W. In agricultural area. Tools, farm equipment.

Mechanicsburg (me-KA-niks-buhrg), borough (2006 population 8,802), CUMBERLAND county, S central PENNSYLVANIA, suburb 8 mi/12.9 km WSW of HARRISBURG; 40°12′N 77°00′W. In agricultural area (grain, soybeans, apples; livestock; dairying); manufacturing (printing and publishing, machine parts, plastic products, microwave equipment, motor vehicle parts, concrete, fabricated steel, food products). Camp Hill State Correctional Inst. to E; Navy Ships Control Center to NE. Williams Grove amusement park to S. APPALACHIAN TRAIL passes to W. Settled c.1790, incorporated 1828.

Mechanicsville (me-KA-niks-vil), unincorporated city (2000 population 30,464), HANOVER county, E central VIRGINIA, suburb 5 mi/8 km NE of RICHMOND, on CHICKAHOMINY RIVER; 37°37′N 77°21′W. Manufacturing (sporting equipment, machinery, dry ice; printing and publishing); agricultural area (dairying; cattle, poultry; grain, soybeans, peanuts, tobacco). Site of inconclusive Civil War battle of Mechanicsville (or Battle of Beaver Dam Creek), one of Seven Days Battles (June 26, 1862) nearby.

Mechanicsville, town (2000 population 1,173), CEDAR county, E IOWA, 11 mi/18 km NW of TIPTON; 41°53′N 91°15′W. Feed manufacturing; corn, soybeans; pigs.

Mechanicsville, unincorporated town (2000 population 3,099), Mahoning township, MONTOUR county, central PENNSYLVANIA, residential suburb 1 mi/1.6 km NE of DANVILLE; 40°58′N 76°35′W.

Mechanicsville, borough (2006 population 494), SCHUYLKILL county, E central PENNSYLVANIA, residential suburb 1 mi/1.6 km ENE of POTTSVILLE, on SCHUYLKILL RIVER; 40°41′N 76°10′W.

Mechanicsville, CONNECTICUT: see THOMPSON.

Mechanicville, city (2006 population 4,923), SARATOGA county, E NEW YORK, on the canalized HUDSON, and 18 mi/29 km N of ALBANY; 42°53′N 73°41′W. Former trade-crafts hub, paper-mill town, hydroelectric power center, and distribution point. Population dwindled in the later 20th century. Railroad shops; in dairying area. Settled before 1700; incorporated as village in 1859, as city in 1915.

Mechant, Lake (mi-SHANT), TERREBONNE parish, SE LOUISIANA, 24 mi/39 km SW of HOUMA, in marshy coastal region; 29°18′N 90°57′W. Lake is c.5 mi/8 km long, 3 mi/5 km wide; waterways connect it with Caillou Lake and GULF OF MEXICO (both S).

Mechara, ETHIOPIA: see MACHARA.

Mechelen (ME-guh-luhn), French *Malines* (mah-LEEN), commune (□ 197 sq mi/512.2 sq km; 2006 population 78,680), ⊙ Mechelen district, ANTWERPEN province, N central BELGIUM, on the DIJLE RIVER; 51°02′N 04°28′E. It is a commercial, industrial, and transportation center and was formerly a famous lace-making center. Manufacturing includes textiles, furniture, and beer. Founded in the early Middle Ages, Mechelen was a fief of the prince-bishops of LIÈGE until 1356, when it passed to the dukes of BURGUNDY. Mechelen became an archiepiscopal see in 1559. It retains many noteworthy medieval buildings, including the Gothic Cathedral of St. Rombaut (thirteenth century), which contains Anthony Van Dyck's painting, the *Crucifixion*, and has a 319-ft/97-m tower and a carillon; the churches of Notre Dame and of St. John, both of which have paintings by Peter Paul Rubens; the archiepiscopal palace (sixteenth century); and the city hall (fourteenth century; rebuilt eighteenth century). In English it is also known as Mechlin.

Mechelen-aan-de-Maas, BELGIUM: see MAASMECHELEN.

Méchemiré (mai-she-mee-RAI), town, KANEM administrative region, W CHAD, 40 mi/64 km SE of MAO; 13°49′N 15°45′E.

Mécheria (mai-sher-YAH), town, NAAMA wilaya, NW central ALGERIA, in the High Plateaus, on ORAN–BÉCHAR railroad, and 55 mi/89 km N of AÏN SEFRA; 33°31′N 00°20′W. Livestock market.

Mechernich (ME-kher-nikh), town, RHINELAND, North Rhine-Westphalia, W GERMANY, 7 mi/11.3 km SW of EUSKIRCHEN; 50°36′N 06°33′E. Machinery.

Mechetinskaya (mee-CHYE-teen-skah-yah), village (2006 population 6,515), S ROSTOV oblast, S European Russia, in the KAGAL′NIK RIVER basin, on road and near railroad, 50 mi/80 km SE of ROSTOV-NA-DONU; 46°46′N 40°27′E. Elevation 164 ft/49 m. In wheat, sunflower, and sheep-raising area; fisheries.

Mechi (MAI-chee), administrative zone (2001 population 1,307,669), W NEPAL. Includes districts of ILAM, JHAPA, PANCHTHAR, and TAPLEJUNG.

Mechi (MAI-chee), river, on border of E NEPAL and NE INDIA (WEST BENGAL state); 26°38′N 88°10′E.

Mechita (me-CHEE-tah), town, N central BUENOS AIRES province, ARGENTINA, 7 mi/11.3 km NE of BRAGADO; 35°04′S 60°24′W. Grain; cattle raising.

Mechi Vruh (ME-chee), peak (8,587 ft/2,617 m) in the SW RILA MOUNTAINS, W BULGARIA, 17 mi/27 km E of BLAGOEVGRAD; 42°04′N 23°27′E. Formerly called Aigidik or Aygidik.

Mechka (MECH-kah), village, LOVECH oblast, PLEVEN obshtina, BULGARIA; 43°31′N 24°49′E.

Mechka (MECH-kah), river, S BULGARIA, tributary of the MARITSA RIVER; 42°04′N 25°11′E.

Mechlin, BELGIUM: see MECHELEN.

Mechra bel Ksiri (ME-shruh bel KSEE-ri), town, Gharb-Chrarda-Beni Hssen administrative region, Sidi Kacem province, N MOROCCO, on Oued SEBOU river, on FES-TANGER railroad. In fertile GHARB lowland; agricultural trade (cereals; livestock).

Mechtal, POLAND: see MIECHOWICE.

Mechtras (mesh-TRAHS), village, TIZI OUZOU wilaya, N central ALGERIA, in Great KABYLIA, 12 mi/19 km S of TIZI OUZOU.

Mechuque Island, CHILE: see CHAUQUES ISLANDS.

Mecidiye, TURKEY: see ESKIPAZAR.

Mecin (ME-cheen), Czech *Mečín*, town, ZAPADOCESKY province, SW BOHEMIA, CZECH REPUBLIC, 8 mi/12.9 km NE of KLATOVY; 49°29′N 13°24′E. Fish ponds; agriculture (wheat, barley, potatoes, rape); cattle.

Mecina–Bombarón (mai-THEE-nah bom-bah-RON), town, GRANADA province, S SPAIN, on S slope of the SIERRA NEVADA, 22 mi/35 km S of GUADIX; 36°59′N 03°09′W. Cereals, olive oil, wine, fruit. Livestock raising; lumbering. Iron deposits.

Mecitozu, Turkish *Mecitözü*, village, N central TURKEY, 18 mi/29 km E of Çorum; 40°31′N 35°17′E. Grain; mohair goats.

Meckenbeuren (mek-kuhn-BOI-ruhn), town, Upper Swabia, BADEN–WÜRTTEMBERG, GERMANY, on the SCHUSSEN, 5 mi/8 km NE of FRIEDRICHSHAFEN; 47°42′N 09°34′E. Hops, fruit.

Meckenheim (MEK-ken-heim), town, RHINELAND, North Rhine-Westphalia, W GERMANY, 8 mi/12.9 km SSW of BONN; 50°38′N 07°02′E.

Meckesheim (MEK-kes-heim), village, LOWER NECKAR, BADEN–WÜRTTEMBERG, GERMANY, on the ELSENZ, and 8 mi/12.9 km SE of HEIDELBERG; 49°19′N 08°50′E. Fruit.

Mecklenburg (MEK-luhn-buhrg), county (□ 546 sq mi/1,419.6 sq km; 2006 population 827,445), S North Carolina; ⊙ CHARLOTTE; 35°15′N 80°49′W. In PIEDMONT region; bounded SW by SOUTH CAROLINA state line, W by CATAWBA RIVER (forms Lake WYLIE reservoir, on South Carolina state line, and Mountain Island and Lake NORMAN reservoirs). Service industries; manufacturing at Charlotte, MATTHEWS, DAVIDSON, PINEVILLE, and HUNTERSVILLE; agriculture (corn, hay, soybeans, wheat; cattle; timber (pine, oak). Most of county is highly urbanized, especially in S and center, centered on Charlotte, largest city in North Carolina. James K. Polk Memorial State Historical Site at Pineville in S. Formed 1762.

Mecklenburg (MEK-len-buhrg), county (□ 679 sq mi/1,765.4 sq km; 2006 population 32,381), S VIRGINIA; ⊙ BOYDTON; 36°40′N 78°22′W. Bounded S by NORTH CAROLINA state line, N by MEHERRIN RIVER; drained by ROANOK (STAUNTON) RIVER, joined by DAN RIVER on W boundary. Manufacturing at CLARKSVILLE, CHASE CITY, SOUTH HILL; agriculture (tobacco, wheat, barley, corn, soybeans, cotton, hay, alfalfa, peanuts; cattle, hogs, poultry; dairying); some timber. BUFFALO SPRINGS resort area. KERR RESERVOIR (Buggs Island Lake), LAKE GASTON reservoir, both on Roanoke River, in S. Occoneechee State Park in SW; Staunton River State Park on W boundary, both on Kerr Reservoir. Formed 1765.

Mecklenburg Bay (ME-kluhn-burg), BALTIC bight between FEHMARN Island (W) and DARSS peninsula (E), MECKLENBURG–WESTERN POMERANIA, N GERMANY; 50 mi/80 km wide. LÜBECK BAY forms SW arm, WISMAR BAY S arm.

Mecklenburg-Schwerin, GERMANY: see MECKLENBURG–WESTERN POMERANIA.

Mecklenburg-Strelitz, GERMANY: see MECKLENBURG–WESTERN POMERANIA.

Mecklenburg-Vorpommern, GERMANY: see MECKLENBURG– WESTERN POMERANIA.

Mecklenburg–Western Pomerania (MEK-klen-burg–pah-muh-RAI-nee-yah), German *Mechlenburg-Vorpommern*, state (□ 8,948 sq mi/23,264.8 sq km; 2006 population 1,694,600) NE GERMANY, bordering on the BALTIC SEA; ⊙ SCHWERIN. The region is comprised of the former East German districts of Schwerin, ROSTOCK, and NEUBRANDENBURG, which existed from 1952 until German reunification in 1990. As constituted in 1945 under Soviet military occupation, Mecklenburg consisted of the former states of Mecklenburg-Schwerin (□ 5,068 sq mi/13,126 sq km) and Mecklenburg-Strelitz (□ 1,131 sq mi/2,929 sq km), and of that part of the former Prussian province of POMERANIA situated W of the ODER River (but not including STETTIN). After reunification this region

was renamed Mecklenburg–Western Pomerania. Generally a low-lying, fertile agricultural area, with many lakes and forests. Until the end of World War II it was characterized by great estates and farms, but after 1945 the region was divided into innumerable small farms. Under the East German government these were soon collectivized as huge farms. They are currently in the process of reorganization and disintegration. The region is beautiful but economically disadvantaged, with poor transportation routes and little industry. It is one of the most sparsely populated German states; however, its many fishing towns and scenic coastline has long made it a favorite German tourist destination. The state government has been promoting it as a destination for health and spa tourism, as well has been encouraging sustainable technology development initiatives with subsidies. There are seven universities here, including two of Europe's oldest: Ernst Moritz Arndt University of GREIFSWALD (1456) and the University of Rostock (1419). On the Baltic coast are the cities of Rostock, WISMAR, and STRALSUND, long important as HANSEATIC ports, and the island of RÜGEN. (Rügen and Stralsund were formerly in Pomerania.) Occupied (6th century) by the Wends. Subdued (1160) by Henry the Lion. In 1621 the duchy divided into Mecklenburg-Schwerin and Mecklenburg-Güstrow, but during the Thirty Years War both dukes were deposed (1555) and the entire duchy was given to Wallenstein, the imperial general, who had conquered it. However, it was retaken by Gustavus II of SWEDEN and restored (1631) to its former rulers. The line of Mecklenburg-Güstrow died out in 1701, and the line of Mecklenburg-Strelitz took its place. At the Congress of Vienna both divisions of Mecklenburg were raised (1815) to grand duchies. They both joined the GERMAN CONFEDERATION, sided with PRUSSIA in the Austro-Prussian War of 1866, and joined the German Empire at its founding in 1871. The grand dukes were deposed in 1918. In 1934 the separate states of Mecklenburg-Schwerin and Mecklenburg-Strelitz were united.

Meckling, village, CLAY county, SE SOUTH DAKOTA, 8 mi/12.9 km NW of VERMILLION; 42°50′N 97°04′W. In farming region.

Meco (MAI-ko), town, MADRID province, central SPAIN, 20 mi/32 km ENE of MADRID; 40°33′N 03°20′W. Grain growing; livestock raising.

Meconta (me-KON-tuh), village, NAMPULA province, N MOZAMBIQUE, on railroad, 40 mi/64 km ENE of NAMPULA; 14°59′S 39°50′E. Cotton, sesame, castor beans.

Meconta District, MOZAMBIQUE: see NAMPULA.

Mecosta (me-KOS-tuh), county (□ 571 sq mi/1,484.6 sq km; 2006 population 42,252), central MICHIGAN; 43°38′N 85°19′W; ⊙ BIG RAPIDS. Drained by MUSKEGON, LITTLE MUSKEGON, CHIPPEWA, and PINE rivers. Agriculture (cattle, hogs, sheep; potatoes, apples, corn; dairy products). Manufacturing at Big Rapids. Resorts. Fish hatchery. Manistee National Forest is in SW. Organized 1859.

Mecosta (me-KOS-tuh), village (2000 population 440), MECOSTA county, central MICHIGAN, 14 mi/23 km ESE of BIG RAPIDS; 43°37′N 85°13′W. In lake-resort and farm area.

Mecoya (me-KO-yah), canton, ANICETO ARCE province, TARIJA department, S central BOLIVIA, 18 mi/30 km SW of PADCAYA and SAN FRANCISCO, on the border of ARGENTINA; 22°06′S 64°55′W. Elevation 6,549 ft/1,996 m. Limestone deposits. Agriculture (potatoes, yucca, bananas, corn); cattle.

Mecsek Mountains (MA-chek), BARANYA county, S HUNGARY; extend NE from Szentlőrinc; consist of two ranges, E range rising to 2,237 ft/682 m in Mount Zengővár, W range to 2,007 ft/612 m in Mount Mecsektető; 46°10′N 18°18′E. Heavily forested slopes; coal and uranium mined in valley between ranges; orchards, vineyards on S slopes. Uranium mine is

being sealed with much technical difficulty. PÉCS at foot of E range. Also called BARANYA MOUNTAINS.

Mecúfi (me-KOO-fee), village, CABO DELGADO province, N MOZAMBIQUE, on MOZAMBIQUE CHANNEL, 25 mi/40 km S of PEMBA; 13°17′S 40°33′E. Salt panning; copra, almonds.

Mecufi District, MOZAMBIQUE: see CABO DELGADO.

Mecula District, MOZAMBIQUE: see NIASSA.

Meda (MAI-dah), town, MILANO province, LOMBARDY, N ITALY, 14 mi/23 km N of MILAN. Railroad junction; furniture manufacturing center; clothing; sausage factories.

Meda (MAI-dah), town, GUARDA district, N central PORTUGAL, 30 mi/48 km N of GUARDA; 40°58′N 07°16′W. Alcohol distilling.

Medain, Al, IRAQ: see MAIDAN, AL.

Medak (MAI-duhk), district (□ 3,745 sq mi/9,737 sq km), ANDHRA PRADESH state, SE INDIA, on DECCAN PLATEAU; ⊙ SANGAREDDI. Mainly lowland, drained by MANJRA RIVER; NIZAM SAGAR (reservoir) forms NW corner of district. Largely sandy red soil; millet, oilseeds (chiefly peanuts, castor beans), rice, sugarcane. Oilseed and rice milling, cotton ginning. Main towns include Sangareddi (experimental farm), SIDDIPET. Part of HYDERABAD from beginning (early eighteenth century) of state's formation; added to Andhra Pradesh state after Independence.

Medak (MAI-duhk), town, MEDAK district, ANDHRA PRADESH state, SE INDIA, 45 mi/72 km NNW of HYDERABAD; 18°02′N 78°16′E. Rice, sugarcane.

Medan (MAI-dahn), city (2000 population 1,911,997), ⊙ Sumatra Utara province, NE SUMATRA, INDONESIA, on the DELI RIVER, c.15 mi/24 km from its mouth, where the city's port (BELAWAN), now included in the city limits, is situated; 03°35′N 98°40′E. The largest city in Sumatra and the fourth-largest in Indonesia, Medan is the marketing, commercial, and transportation center of a rich agricultural area containing large tobacco, rubber, and palm oil estates. Coffee and tea are also grown in the vicinity. Industries include the production of machinery and tile, and motor vehicle assembly. Gateway to the beautiful Lake TOBA region; a tourist center, with an international airport; attractions include the Great Mosque (the largest in Sumatra) and the Palace of the Sultan of Deli. Seat of the University of Sumatra Utara, the Islamic University of Sumatra Utara, and Nommensen University.

Medang, INDONESIA: see RANGSANG.

Médano, El (MAI-dah-no, el), town, TENERIFE island, CANARY ISLANDS, SPAIN, 33 mi/53 km SW of SANTA CRUZ DE TENERIFE. Cereals, tomatoes, potatoes; livestock.

Médano, El, ARGENTINA: see EL MÉDANO.

Médanos (MAI-dah-nos), town (1991 population 4,744), ⊙ Villarino district (□ 3,598 sq mi/9,354.8 sq km), SW BUENOS AIRES province, ARGENTINA, 23 mi/37 km WSW of BAHÍA BLANCA. In agricultural zone (wheat; sheep, cattle; vineyards). Salt mines of Salina Chica are nearby (W). Irrigation on the lower Río Colorado (S).

Médanos de Coro, Parque Nacional (MAI-dah-nos dai KO-ro PAHR-kai nah-see-o-NAHL), national park (□ 352 sq mi/915.2 sq km), FALCÓN state, N VENEZUELA 11°37′N 69°48′W. Coastal environment of sand dunes, salt marshes, and dry forest vegetation. Severe regional flooding in December 1998 formed temporary lakes among the dunes. Created 1974.

Médanos, Isthmus of (MAI-dah-nos), FALCÓN state, NW VENEZUELA, links PARAGUANÁ PENINSULA with mainland, just N of CORO, separates Golfete de CORO (W) from the CARIBBEAN SEA (E); 15 mi/24 km long, c.3 mi/4.8 km wide. Salt deserts on W coast.

Medaryville, town (2000 population 565), PULASKI county, NW INDIANA, on a tributary of BIG MONON

CREEK, 15 mi/24 km W of WINAMAC; 41°05′N 86°53′W. Agriculture; apparel. Medaryville Correctional Unit nearby.

Medchal (MAID-chuhl), village, RANGAREDDI district, ANDHRA PRADESH state, SE INDIA, 17 mi/27 km N of HYDERABAD. Rice, oilseeds.

Meddybemps (MED-ee-bemps), town, WASHINGTON county, E MAINE, 14 mi/23 km SW of CALAIS; 45°02′N 67°21′W. On Meddybemps Lake (6 mi/9.7 km long); resort area.

Mede (MAI-de), town, PAVIA province, LOMBARDY, N ITALY, 11 mi/18 km S of MORTARA; 45°06′N 08°44′E. Agricultural center, in rice-growing area; manufacturing of agricultural machinery, textiles.

Médéa (mai-dai-AH), wilaya, central ALGERIA; ⊙ MÉDÉA; 36°05′N 03°00′E. A rich farming region with vast orchards. Also a wine-producing area.

Médéa (mai-dai-AH), town, MÉDÉA wilaya, central ALGERIA, 40 mi/64 km SW of ALGIERS, and 19 mi/31 km SSW of BLIDA; 36°15′N 02°48′E. To the NW is the long ridge of Djebel Nador, whose S slopes fall gently to the town. The surrounding fertile countryside produces cereals, olives, fruit, vegetables, and vine grapes. The town has expanded rapidly since independence, with an enhanced administrative role and new industries since becoming a wilaya capital. Known as Lambdia in Roman times, it was a major medieval urban center. Was capital of the Titteri region under the Turks. After the French occupation in 1840, the city kept an important religious role.

Medeba (MAH-de-buh), ancient Moabite town, JORDAN, on the site of the modern MADABA, E of the DEAD SEA; 31°43′N 35°48′E. Changed hands between MOAB and ISRAEL several times. In early Christian times it was a bishop's see. One of the oldest maps of the HOLY LAND (probably 6th century) was found here in a mosaic in a Byzantine church. Medeba is mentioned several times in the Hebrew Bible.

Medebach (MAI-de-bahkh), town, Westfalia Lippe, North Rhine-Westphalia, W GERMANY, 15 mi/24 km SSE of BRILON; 51°12′N 08°43′E. Grain.

Medeiros Neto (me-DAI-ros NE-to), city (2007 population 21,714), SE BAHIA, BRAZIL, on Rio do Pequi, near MINAS GERAIS border; 17°23′S 40°14′W.

Medellín (mai-dai-YEEN), city, ⊙ ANTIOQUIA department, W central COLOMBIA; 06°15′N 75°35′W. Elevation c.5,000 ft/1,524 m. Medellín, the third largest city in the country (after BOGOTÁ and CALI), is a principal manufacturing center (textiles, steel, food products, motor vehicles, pharmaceuticals, woodwork, rubber, chemicals, and coffee). Coal, gold, and silver are mined in the surrounding region. Medellín has also acquired a reputation as an international distribution point for illicit cocaine, heightened by the past presence of Pablo Escobar, the infamous drug lord during the 1970s and 1980s. The drug trade has led to outbreaks of violence in the city. The city, which was founded in 1675, is located in a small intermontane valley. Until the development of transportation in the 19th century, it was practically isolated; it has since developed into a transportation hub. Medellín, rich in cultural institutions, has 3 universities, many museums, and several 17th-century churches. Airport.

Medellin (mai-de-LEEN), town, CEBU province, N Cebu island, PHILIPPINES, on Visayan Sea, near entrance to TAÑON STRAIT, 60 mi/97 km N of CEBU; 11°07′N 123°58′E. Agricultural center (corn, tobacco); sugar mill.

Medellín (mai-dhe-YEEN), town, BADAJOZ province, W SPAIN, on GUADIANA River, 21 mi/34 km E of MÉRIDA; 38°57′N 05°58′W. Olives, cereals, grapes. Hernán Cortés born here.

Medellín, MEXICO: see MEDELLÍN DE BRAVO.

Medellín de Bravo (mai-de-YEEN), town, VERACRUZ, E MEXICO, in GULF lowland, on railroad, 11 mi/18 km S of VERACRUZ; 19°04′N 96°09′W. Popular weekend resort. Site of ancient Indian town with prehistoric

Xicalango (Xicalanco) ruins in forest nearby. Cortés, who founded the new town, naming it after his native town in ESTREMADURA, SPAIN, resided here briefly (1526).

Medelpad (ME-del-PAHD), province (□ 3,031 sq mi/ 7,880.6 sq km), NE SWEDEN, on GULF OF BOTHNIA, in S part of VÄSTERNORRLAND county.

Medel, Piz, SWITZERLAND: see PIZ MEDEL.

Medemblik (MAI-duhm-blik), town, NORTH HOLLAND province, NW NETHERLANDS, on the IJSSELMEER, at SE edge of WIERINGERMEER polder, and 18 mi/29 km NE of ALKMAAR; 52°46′N 05°06′E. Passenger ferry to ENKHUIZEN; pumping station to S; terminus of Westfriesevaart canal. Sheep, cattle; potatoes, sugar beets, flowers, fruit; manufacturing (yachts). Has thirteenth-century Radboud Castle. Former capital of WEST FRIESLAND, succeeded c.1400 by HOORN.

Medenice, UKRAINE: see MEDENYCHI.

Medenichi, UKRAINE: see MEDENYCHI.

Médenine, province, TUNISIA: see MADANIYINA.

Medenitsa, UKRAINE: see MEDENYCHI.

Meden Rudnik (ME-den ROOD-neek), suburb of BURGAS, BURGAS oblast, SE BULGARIA; 42°27′N 27°24′E.

Medenshor, village, SW BADAKHSHAN AUTONOMOUS VILOYAT, SE TAJIKISTAN, in the PAMIR, 24 mi/39 km SE of KHORUGH; 37°20′N 71°45′E.

Medenychi (me-DE-ni-chee) (Russian *Medenichi*) (Polish *Medenice*), town (2004 population 9,000), S central L′VIV oblast, UKRAINE, 21 mi/19 km NE of DROHOBYCH; 49°26′N 23°45′E. Elevation 889 ft/270 m. Flour and feed mills; chemicals. Vocational technical school. Known since 1395, town since 1940. Called Medenytsya (Russian *Medenitsa*) from 1946 to 1989.

Medenytsya, UKRAINE: see MEDENYCHI.

Mederdra (me-DER-drah), village (2000 population 6,858), Trarza administrative region, SW MAURITANIA, 80 mi/129 km NE of SAINT-LOUIS (SENEGAL); 16°55′N 15°39′W. Gum arabic, millet; livestock. Salt deposits nearby.

Medes, W Asia: see MEDIA.

Medesano (me-de-ZAH-no), village, PARMA province, EMILIA-ROMAGNA, N central ITALY, near TARO River, 9 mi/14 km SW of PARMA; 44°45′N 10°08′E. Food processing; sausage. Thermal springs.

Medeshamstede, ENGLAND: see PETERBOROUGH.

Medfield, residential town, NORFOLK county, E MASSACHUSETTS, on CHARLES RIVER, 18 mi/29 km SW of BOSTON; 42°11′N 71°19′W. Manufacturing (laboratory equipment, scientific instruments). Includes Medfield village. Settled and incorporated 1650.

Medford, city (2000 population 55,765), MIDDLESEX county, E MASSACHUSETTS, residential and industrial suburb of BOSTON, on the MYSTIC RIVER; 42°25′N 71°07′W. Wax, paper, clothing, and furniture are among its products. A shipping and shipbuilding center from the seventeenth to the nineteenth centuries, Medford was also known for its rum. It is the seat of Tufts University. Several eighteenth-century buildings stand in the city. Includes village of West Medway. Settled 1630, incorporated as a city 1892.

Medford, city (2006 population 71,168), ⊙ JACKSON county, SW OREGON, 115 mi/185 km S of EUGENE, on Bear Creek; 42°20′N 122°50′W. Elevation 1,382 ft/421 m. Junction of logging railroad to WHITE CITY. Growing trade, shipping, and medical center in an agriculture area. Manufacturing (food processing; lumber, furniture, veneer, electrical equipment; boatbuilding). Tourism. Between 1836 and 1856, the area was the scene of a number of bloody conflicts between white settlers and Native Americans of ROGUE RIVER descent. Gold was discovered nearby in 1851. The gold-mining town of JACKSONVILLE has been restored. MEDFORD is the headquarters for CRATER LAKE NATIONAL PARK, 50 mi/ 80 km to NE. Tou Velle State Park to N; parts of Rogue River National Forest to S and E. Pear blossom festival. Incorporated 1884.

Medford, township, BURLINGTON county, W central NEW JERSEY, c.15 mi/24 km E of CAMDEN, near Rancocas River; 39°51′N 74°49′W. Friends' meetinghouse here, built 1814.

Medford, town (2006 population 1,042), ⊙ GRANT county, N OKLAHOMA, 24 mi/39 km W of BLACKWELL; 36°47′N 97°44′W. Elevation 1,094 ft/333 m. Railroad junction. In agricultural area (wheat, alfalfa, oats, cotton; cattle); manufacturing (concrete).

Medford, town (2006 population 4,140), ⊙ TAYLOR county, N central WISCONSIN, on BLACK RIVER, 37 mi/ 60 km NW of WAUSAU; 45°08′N 90°20′W. Lumbering, livestock-raising, and dairying area. Dairy products; manufacturing (food processing; crushed concrete, building materials). S unit of Chequamegon National Forest to NW. Nearby is a Mennonite colony. Incorporated 1889.

Medford, village (2000 population 984), STEELE county, SE MINNESOTA, on STRAIGHT RIVER, 8 mi/12.9 km S of FARIBAULT; 44°10′N 93°14′W. Dairying; poultry; grain, soybeans, beans; manufacturing (concrete blocks, exercise equipment). Incorporated 1936.

Medford, village (□ 10 sq mi/26 sq km; 2000 population 21,985), SUFFOLK county, SE NEW YORK, on central LONG ISLAND, 4 mi/6.4 km N of PATCHOGUE; 40°49′N 72°58′W. Diversified economy of manufacturing, commercial services, and agricultural products.

Medford Lakes, borough (2006 population 4,161), BURLINGTON county, W central NEW JERSEY, 9 mi/14.5 km S of MOUNT HOLLY; 39°51′N 74°48′W. Small lakes here.

Medfra, village, S central ALASKA, on upper KUSKOKWIM River, 30 mi/48 km ENE of MCGRATH; 63°06′N 154°43′W.

Medgidia (me-dzhee-DEE-yah), town, CONSTANȚA county, SE ROMANIA, in DOBRUJA, on the DANUBE–BLACK SEA CANAL, 18 mi/29 km WNW of CONSTANȚA; 44°15′N 28°17′E. Railroad junction and trading center; manufacturing of agricultural tools, ceramics, bricks, cheese, and flour. Kaolin and limestone quarrying. Has 19th-century mosque.

Medgyes, ROMANIA: see MEDIAȘ.

Medha (MAI-duh), village, SATARA district, MAHARASHTRA state, W central INDIA, 12 mi/19 km NW of SATARA, in WESTERN GHATS. Rice, millet.

Medi, YEMEN: see MAIDI.

Media (MEE-dee-ah), ancient country of W ASIA, occupying generally what is now W IRAN and S AZERBAIJAN, but whose actual boundaries cannot be defined. It extended from the CASPIAN SEA to the ZAGROS Mountains. The Medes were an Indo-European people who spoke an Iranian language closely akin to old Persian. Some scholars claim they were an Aryanized people from TURAN. Since there are no Median records, Assyrian and Greek sources must be relied upon for Median history. The Medes extended their rule over PERSIA during the reign of Sargon (died 705 B.C.E.) and, under Cyaxares, captured NINEVEH in 612 B.C.E. They were the first people subject to ASSYRIA to secure their freedom. The dynasty continued until the rule of Astyages, when it was overthrown (c.550 B.C.E.) by Cyrus the Great and united with the Persian Empire. In the second century B.C.E. Media became part of the Parthian kingdom and was later ruled by the Romans.

Media, village (2000 population 130), HENDERSON county, W ILLINOIS, 13 mi/21 km SE of OQUAWKA; 40°46′N 90°49′W. In agricultural area.

Media, borough (2006 population 5,456), ⊙ Delaware county, SE PENNSYLVANIA, residential suburb 12 mi/ 19 km W of PHILADELPHIA, on Ridley Creek; 39°55′N 75°23′W. Manufacturing (printing and publishing, diversified light manufacturing). Tyler Arboretum to W. Ridley Creek State Park to NW; Springton Reservoir to N. Seat of Penn State University–Delaware County Campus. Settled 1682, laid out c.1848, incorporated 1850.

Media Agua, ARGENTINA: see VILLA MEDIA AGUA.

Media Luna (MAI-dee-ah LOO-nah), site of sugar mill Juan Manuel Márquez, GRANMA province, SE CUBA, with nearby small port on GULF OF GUACANAYABO; 20°09′N 77°27′W.

Media Luna, MEXICO: see OXCHUC.

Media Luna, Cayo (MAI-dee-ah LOO-nah, KEI-yo), islet in the GULF OF GUACANAYABO, E CUBA, 55 mi/89 km S of CAMAGÜEY; 20°34′N 77°53′W. Fishing.

Mediano (mai-DHYAH-no), village, HUESCA province, NE SPAIN, on S slopes of the central PYRENEES, 33 mi/53 km ENE of HUESCA, near CINCA RIVER gorge. Irrigation reservoir, hydroelectric plant nearby.

Mediapolis, town (2000 population 1,644), DES MOINES county, SE IOWA, 14 mi/23 km N of BURLINGTON; 41°00′N 91°09′W.

Mediaş (ME-dyahsh), Hungarian *Medgyes*, city, SIBIU county, central ROMANIA, in TRANSYLVANIA, on railroad, 21 mi/34 km WSW of SIGHIŞOARA; 46°10′N 24°21′E. In a noted wine-growing region. Wine and cattle trade, glass manufacturing; natural gas production. Also produces bicycles, textiles (cotton, rayon, and wool), enamelware, hardware, leather goods, earthenware, and salami; processes furs. Has military aviation school with airfield and agriculture school. It was one of the first seven towns established in 12th century by German colonists, and it still preserves 14th-century fortress and town walls. Has 17th-century Lutheran church. National assembly of Saxons of Transylvania voted here (1919) for union with Romania.

Medical Lake, town (2006 population 4,403), SPOKANE county, E WASHINGTON, 15 mi/24 km SW of SPOKANE, and on Medical Lake; 47°34′N 117°42′W. Light manufacturing. Several lakes in area, largest is Silver Lake, to E. Four Lakes Battle Monument to SE. Fairchild Air Force Base to N. Incorporated 1889.

Medicina (me-dee-CHEE-nah), town, BOLOGNA province, EMILIA-ROMAGNA, N central ITALY, 15 mi/24 km E of BOLOGNA. Manufacturing (agricultural machinery; furniture).

Medicine Bow, village (2006 population 264), CARBON county, S WYOMING, on MEDICINE BOW RIVER, 15 mi/24 km N of N end of MEDICINE BOW MOUNTAINS, 50 mi/80 km NW of LARAMIE; 41°53′N 106°12′W. Elevation c. 6,563 ft/2,000 m. Supply point in oil and livestock area. Petrified forest in vicinity. Shirley Mountains to NW. Como Bluff, dinosaur fossil site, to E.

The area was first used by trappers and mountain men during the 1830's. In 1868, the Union Pacific Railroad was built through the area, and a pumping station was established on the river. A store and saloon were the beginning of the small village. By the following year, Medicine Bow had become a major supply point and in the 1870's, the federal government operated a military post in Medicine Bow to protect the railroad and freight wagons from attack. A post office was built and in 1876, the first elementary school was established. By the late 1870's and early 1880's, Medicine Bow had become the largest shipping point for range livestock on the Union Pacific line. Cattle were being brought for shipping from as far away as IDAHO and MONTANA. An average of 2,000 head a day were being shipped. By the turn of the century, Medicine Bow was also a major shipping point for wool, averaging 1,000 tons a year. In 1909, Medicine Bow was incorporated when the U.P. Railroad transferred ownership to the town. In late 1913, the transcontinental "Lincoln Highway" passed right through Medicine Bow. In the 1930's it was paved bringing tourism to the area. In later years, lumber, uranium, coal, oil, and natural gas were found in the area which added to the prosperity of the region.

Medicine Bow Mountains, outlying E range of the ROCKY MOUNTAINS, SE WYOMING and N COLORADO. It extends from village of ELK MOUNTAIN, Wyoming, WNW of LARAMIE, S c. 80 mi/129 km to CAMERON PASS, Colorado. Peaks include Medicine Bow Peak (12,013 ft/3,662 m) and Elk Mountain (11,156 ft/3,400 m). Wyoming part is in Medicine Bow National Forest (national forest includes part of LARAMIE MOUNTAINS), Colorado part is in Colorado State Forest.

Medicine Bow River, 195 mi/314 km long, S WYOMING; rises in N MEDICINE BOW MOUNTAINS in SE CARBON county; flows N past village of MEDICINE BOW, then W, joining North Platte River at SEMINOE RESERVOIR.

Medicine Camp de Masque, village (2000 population 6,637), MOKA and FLACQ districts, MAURITIUS, 20 mi/32 km E of PORT LOUIS. Sugarcane; livestock.

Medicine Creek, c.100 mi/161 km long, S IOWA and N MISSOURI; rises in S Iowa; flows S to GRAND RIVER 10 mi/16 km SE of CHILLICOTHE.

Medicine Creek, SW central NEBRASKA; rises in LINCOLN county; flows SE, joining REPUBLICAN RIVER at CAMBRIDGE. Medicine Creek Reservoir (Harry D. Strunk Lake) has state recreation area.

Medicine Hat (ME-di-sin HAT), city (□ 43 sq mi/111.8 sq km; 2005 population 56,048), SE ALBERTA, W CANADA, on the South SASKATCHEWAN River; 50°02′N 110°41′W. Center of a farming and ranching area. Natural-gas deposits. Light industries; glassblowing; rubber plants. First incorporated as a village in 1894; became a city in 1906.

Medicine Lake, village (2000 population 368), HENNEPIN county, E MINNESOTA, residential suburb 7 mi/11.3 km W of MINNEAPOLIS, on small peninsula on S shore of Medicine Lake (2 mi/3.2 km long, 1 mi/1.6 km wide), W of outflow of Bassett Creek; 45°00′N 93°25′W. Village is surrounded by city of PLYMOUTH.

Medicine Lake, village (2000 population 269), SHERIDAN county, NE MONTANA, on MEDICINE LAKE, near BIG MUDDY CREEK, 20 mi/32 km S of PLENTYWOOD; 48°30′N 104°30′W. Hogs, sheep; wheat, barley, oats, sugar beets, hay. Fort Peck Indian Reservation to W, Medicine Lake National Wildlife Refuge to SE, Homestead Lake unit of refuge to SW. Holds state's high-temp. record (117°F/47°C). Winter temperature commonly falls below −40°F/−40°C.

Medicine Lake, reservoir, SHERIDAN county, NE MONTANA, on Lake Creek, in Medicine Lake National Wildlife Refuge, 20 mi/32 km S of PLENTYWOOD; 9 mi/14.5 km long, 4 mi/6.4 km wide; 48°28′N 104°24′W. Drains W into BIG MUDDY CREEK.

Medicine Lodge, town (2000 population 2,193), ⊙ BARBER county, S KANSAS, on MEDICINE LODGE RIVER, 70 mi/113 km WSW of WICHITA; 37°17′N 98°34′W. Trade and refining point in wheat and livestock area; manufacturing Oil and gas wells, gypsum mines in vicinity. Quinquennial pageant (since 1927) commemorates signing of treaty (1867) nearby with Plains tribe. Carry Nation started antisaloon crusade here in 1899. Medicine Lodge Stockade and Museum here. Barber State Fishing Lake to N. Founded 1873, incorporated 1879.

Medicine Lodge River, 101 mi/163 km long, in KANSAS and OKLAHOMA; rises in KIOWA county in S KANSAS; flows SE past MEDICINE LODGE (KANSAS), into ALFALFA county in OKLAHOMA, to SALT FORK OF ARKANSAS RIVER.

Medicine Park, village (2006 population 366), COMANCHE county, SW OKLAHOMA, 11 mi/18 km NNW of LAWTON, and N of FORT SILL Military Reservation. Located at S end of Lake LAWTONKA, W end of dam. WICHITA MOUNTAINS National Wildlife Refuge to W. Recreation area.

Medimurje (MED-jee-muhr-ye), Hungarian *Muraköz*, German *Mittelmurgebiet* or *Murinsel*, agricultural region, N CROATIA; bounded by DRAVA RIVER (S), MURA RIVER and Hungarian border (NE). Densely populated area; manufacturing (textiles; printing), agriculture (turkey raising), crafts (basketry, pottery). Chief town, Čakovec. Under Hungarian rule 1720–1918, except for 1848–1861.

Médina (MAI-dee-nuh), district (2004 population 135,759), NW quarter of DAKAR, DAKAR administrative region, W SENEGAL. Densely populated residential neighborhood. Originally established as a temporary living area for African population whose homes in the European quarter (the Plateau) were razed during the 1914 plague epidemic. Has since absorbed repeated waves of rural-to-urban migrants.

Medina (muh-DEI-nuh), county (□ 424 sq mi/1,102.4 sq km; 2006 population 169,353), N OHIO; ⊙ MEDINA; 41°06′N 81°53′W. Drained by Rocky and Black rivers and small Chippewa Creek. Includes Chippewa Lake (resort village). In the Till Plains and Glaciated Plain physiographic regions. Agricultural area (poultry, sheep; corn, vegetables); manufacturing. Formed 1818.

Medina (MAH-dee-nah), county (□ 1,334 sq mi/3,468.4 sq km; 2006 population 43,913), SW TEXAS; ⊙ HONDO; 29°21′N 99°06′W. Crossed E-W by BALCONES ESCARPMENT, separating EDWARDS PLATEAU (in N) from plains of S. Drained by MEDINA RIVER, source of ATASCOSA RIVER in SE. Ranching (cattle, sheep, goats); agriculture (corn, grain, sorghum, peanuts, and cotton); irrigated vegetable farming. Some oil, natural gas; clay mining; sand and gravel. Part of Medina Lake (forms part of N boundary), used for irrigation and recreation, is in NE; FRIO RIVER, SW corner. Landmark Inn State Historic Site in E at CASTROVILLE. Formed 1848.

Medina (ME-dee-nah), city (2007 population 20,630), NE MINAS GERAIS, BRAZIL, 31 mi/50 km SW of PEDRA AZUL; 16°20′S 41°28′W.

Medina (muh-DEE-nuh), Arabic *Medinat an-Nabi* [Arabic=city of the Prophet] or *Madinat Rasul Allah* [Arab=city of the apostle of Allah], city (2004 population 918,889), HEJAZ region, W SAUDI ARABIA; ⊙ MEDINA province; 24°28′N 39°36′E. It is situated c.110 mi/177 km inland from the RED SEA in a well-watered oasis where fruit, dates, vegetables, and grain are raised. The city is an important communications, commerce, and services center. Before the flight (hegira) of Muhammad from MAKKA to the city in 622, Medina was called Yathrib. Muhammad quickly gained control of Medina, successfully defended it against attacks from Makka, and used it as the base for converting and conquering Arabia. Medina grew rapidly until 661, when the Umayyad dynasty transferred the capital of the caliphate to DAMASCUS. Thereafter Medina was reduced to the rank of a provincial town, ruled by governors appointed by the distant caliphs. Local warfare drained the city's prosperity. It came under the sway of the Ottoman Turks in 1517. The Wahhabis captured it in 1804, but it was retaken for the Turks by Muhammad Ali in 1812. In World War I, the forces of Husayn ibn Ali, who revolted against TURKEY, captured Medina. In 1924 it fell to Ibn Saud, Husayn's rival, after a fifteen-month siege. The city is surrounded by double walls flanked by bastions and pierced by nine gates. The chief building is the Prophet's Mosque, which contains the tombs of Muhammad, his daughter Fatima, and the caliphs Umar and Abu Bakr. The pilgrimage to Makka usually includes a side trip to Medina. Medina is the seat of Islamic University (established 1962). International airport to NE.

Medina (muh-DEI-nuh), city (□ 11 sq mi/28.6 sq km; 2006 population 26,350), ⊙ MEDINA county, N OHIO, 18 mi/29 km WNW of AKRON; 41°08′N 81°52′W. Paints, roofing, industrial products; aluminum and lumber processing, light industry. Town center was restored in early 20th century. Laid out 1818, incorporated as a city 1950.

Medina (me-DEE-nah), town, S TUCUMÁN province, ARGENTINA, 45 mi/72 km SSW of TUCUMÁN. Railroad terminus and agriculture center (sugar, rice; livestock); sugar refining. Population figure includes TRINIDAD.

Area is shown by the symbol □, and capital city or county seat by ⊙.

Medina (me-DEE-nah), town, ⊙ Medina municipio, CUNDINAMARCA department, central COLOMBIA, at E foot of Cordillera ORIENTAL, 50 mi/80 km E of BOGOTÁ; 04°30′N 73°21′W. Coffee, corn.

Medina (muh-DEI-nuh), town (2000 population 4,005), HENNEPIN county, E MINNESOTA, residential suburb 15 mi/24 km WNW of MINNEAPOLIS; 45°01′N 93°35′W. Includes community of Hamel in NE. Manufacturing at Hamel (motor vehicle parts, paper products, fabricated metal products, machining, building equipment). Several small lakes especially in W; Lake MINNETONKA to S; Lake Independence on W boundary. Morris T. Baker Park Reserve in SW.

Medina, town (2006 population 1,506), GIBSON county, NW TENNESSEE, 12 mi/19 km N of JACKSON; 35°48′N 88°46′W. In timber and farm area with some manufacturing.

Medina, unincorporated town (2006 population 3,081), KING county, W WASHINGTON, residential suburb 4 mi/6.4 km E of SEATTLE, 2 mi/3.2 km W of BELLEVUE, on E shore of LAKE WASHINGTON; 47°37′N 122°14′W. Evergreen Point Bridge to Seattle to NW.

Medina (muh-DEI-nuh), industrial village (□ 3 sq mi/ 7.8 sq km; 2006 population 6,191), ORLEANS county, W NEW YORK, on the BARGE CANAL and OAK ORCHARD CREEK, 30 mi/48 km NE of BUFFALO; 43°13′N 78°23′W. Various manufacturing (metal products, machinery, medical equipment), mining (sandstone), and agricultural products. Incorporated 1832.

Medina, village (2006 population 305), STUTSMAN county, central NORTH DAKOTA, 29 mi/47 km W of JAMESTOWN; 46°53′N 99°17′W. Chase Lake National Wildlife Refuge to NW. Founded in 1888 and incorporated in 1906.

Medina (mah-DEE-nah), unincorporated village, BANDERA county, SW TEXAS, on MEDINA RIVER, c.50 mi/80 km NW of SAN ANTONIO; 29°48′N 99°15′W. In livestock-ranching area (cattle, sheep, goats); guest ranches; manufacturing (leather products). Hill Country State Park to SE, Lost Maples State Park to W.

Médina, MALI: see MÉDINE.

Medina Azahara, historic ruins and archaeological site, CÓRDOBA province, S SPAIN, 3 mi/4.8 km W of CÓRDOBA. Built c.1010 by Abd al-Rahman III as Islamic cultural and trade center. Ruins discovered 1911, but only approximately 10% of the site has been excavated. Preservation of unexplored portions are threatened by recent construction activity. Also called Madinat al-Zahra.

Medinaceli (mai-dhee-nah-THAI-lee), town, SORIA province, N central SPAIN, 40 mi/64 km S of SORIA; 41°10′N 02°26′W. Grain growing; sheep raising; flour milling. Has saltworks and hydroelectric plant nearby. Ancient town dating back to Romans. Once a Moorish stronghold, it was later seat of the dukes of Medinaceli, who built a Renaissance palace. Also has Roman arch and ancient grain exchange.

Medina Dam, Texas: see MEDINA RIVER.

Medina de las Torres (mai-DEE-nah dai lahs TO-res), town, BADAJOZ province, W SPAIN, on railroad, 40 mi/64 km S of MÉRIDA; 38°20′N 06°24′W. Agricultural center (olives, cereals, grapes; livestock). Flour milling, olive-oil pressing; tile manufacturing.

Medina del Campo (mai-DEE-nah del KAHM-po), city, VALLADOLID province, central SPAIN, in CASTILE-LEÓN; 41°18′N 04°55′W. It is a communications center and agricultural market with food-processing industries. The city was almost completely destroyed by fire in the 16th century. Medina del Campo was the favorite residence of Queen Isabella I.

Medina de Pomar (mai-DEE-nah dai po-MAHR), town, BURGOS province, N SPAIN, 40 mi/64 km NNE of BURGOS; 42°56′N 03°29′W. On fertile plain (*La Losa*); cereals, produce; livestock; timber. Sawmilling, tanning, flour milling; manufacturing of chocolate and meat products. Ancient historic city of Roman

origin; has 14th-century castle, 13th-century church, convent, and palace of the dukes of Frías.

Medina de Ríoseco (mai-DEE-nah REE-o-SAI-ko), town, VALLADOLID province, N central SPAIN, in LEÓN, 24 mi/39 km NW of VALLADOLID; 41°53′N 05°02′W. Agricultural-trade center (cereals, vegetables, wine) and terminus of a branch of CANAL OF CASTILE. Cheese processing; livestock raising.

Medina Lake, Texas: see MEDINA RIVER.

Medina River (muh-DEE-nuh), 12 mi/19 km long, Isle of WIGHT, S ENGLAND; rises 6 mi/9.7 km S of NEWPORT; flows N, past Newport, to The SOLENT at COWES. Navigable below Newport.

Medina River (MAH-dee-nah), c.100 mi/161 km long, S central TEXAS; rises on EDWARDS PLATEAU NW of MEDINA; flows generally SE to SAN ANTONIO RIVER 14 mi/23 km S of SAN ANTONIO. Medina Dam (180 ft/ 55 m high, 1,580 ft/482 m long), built in 1913, 26 mi/42 km WNW of San Antonio, impounds Medina Lake (capacity 327,000 acre-ft), first large Texas irrigation reservoir. Fishing, recreation.

Medina–Sidonia (mai-DHEE-nah see-DHO-nyah), city, CÁDIZ province, SW SPAIN, in ANDALUSIA, picturesquely located on a plateau 21 mi/34 km E of CÁDIZ; 36°27′N 05°55′W. Cereals, fruit; livestock. Notable Gothic cathedral. Probably of Phoenician origin. Seat of former Medina-Sidonia duchy. Sulphur springs and caves with prehistoric paintings nearby.

Medinat as Sadat [Arabic=Sadat city], new town, N EGYPT, 55 mi/89 km N of CAIRO on the main ALEXANDRIA-Cairo road. The initial plan was to create a city for 1,000,000 inhabitants; mid-1990s population is estimated at 50,000–70,000. Small industrial center with over 30 factories.

Medinat Sitta Uktuber [Arabic=Sixth October city], new town, N EGYPT, near the GIZA pyramids, 20 mi/32 km from CAIRO, on main road to FAIYUM depression. The initial plan was to build an industrial town with a population in excess of 250,000. Most apartments in the town are not occupied.

Médine (MAI-deen), village, FIRST REGION/KAYES, SW MALI, on the SÉNÉGAL RIVER, and 6 mi/9.7 km SE of KAYES (linked by railroad). Old fort. Captured (c.1855) by Faidherbe, who made it his base for further conquests of the SUDAN. Sometimes Médina.

Medinet el Faiyum, EGYPT: see FAIYUM, EL.

Medinilla (mai-dhee-NEE-lyah), village, ÁVILA province, central SPAIN, 36 mi/58 km S of SALAMANCA; 40°26′N 05°37′W. Cereals, acorns; livestock; flour milling.

Medinilla (mai-deen-EE-yah), uninhabited volcanic island, S NORTHERN MARIANA ISLANDS, W PACIFIC, 180 mi/290 km NNE of GUAM; 1.5 mi/2.4 km long. Phosphates. Also called Farallon de Medinilla.

Medinipur, district (□ 5,437 sq mi/14,136.2 sq km), SW WEST BENGAL state, E INDIA; ⊙ MEDINIPUR. Bounded NW by BIHAR state, SW by ORISSA state, and S by BAY OF BENGAL; SE boundary formed by HUGLI RIVER, E boundary formed by RUPNARAYAN RIVER. Drained by SUBARNAREKHA and KASAI rivers. Lateritic soil W, alluvial soil E. Agriculture (rice, corn, pulses, wheat, mustard, jute, potatoes, peanuts, castor); sal, mahua, dhak in forested area (W); silk growing (E). Railroad workshops at KHARAGPUR; rice milling at Medinipur; regional cotton-cloth distributing center near PINGLA; cotton weaving, metalware manufacturing, general engineering works. Railroad Institute and Institute of Technology at Kharagpur, college at Contai. Former Buddhist temple now dedicated to Kali at Tamluk. Buddhist stronghold in fifth century B.C.E. Hijili was site of Job Charnock's victory (1687) over Mogul army; ceded to English in 1760. Formerly spelled Midnapore.

Medinipur, city, ⊙ Medinipur district, West Bengal state, E India, on the Kasai River; 22°26′N 87°20′E. Textile and chemical manufacturing. Linked to Calcutta by canal. Seat of a university.

Medio River (MAI-dee-o), c.70 mi/113 km long, on border of BUENOS AIRES and SANTA FE provinces, ARGENTINA; rises 12 mi/19 km NNE of COLÓN; flows NE along border to the PARANÁ RIVER 5 mi/8 km NW of SAN NICOLÁS.

Mediouna (mai-dee-OO-nuh), village, Grand Casablanca administrative region, NW MOROCCO, 12 mi/19 km SSE of CASABLANCA; 33°27′N 07°31′W. Farming, palm fiber processing.

Mediterranean Sea [Latin=in the midst of lands], the world's largest inland sea (□ 965,000 sq mi/2,509,000 sq km), surrounded by S Europe, W Asia, and N Africa. It is c.2,400 mi/3,862 km long, with a maximum width of c.1,000 mi/1,609 km; its greatest depth is c.14,450 ft/4,404 m, off CAPE MATAPAN, Greece. Connects with the ATLANTIC OCEAN through the STRAIT OF GIBRALTAR; with the BLACK SEA through the DARDANELLES, the SEA OF MARMARA, and the BOSPORUS; and with the RED SEA through the SUEZ CANAL. Its chief divisions are the TYRRHENIAN, ADRIATIC, IONIAN, and AEGEAN seas; its chief islands are SICILY, SARDINIA, CORSICA, CRETE, CYPRUS, MALTA, RHODES, the DODECANESE, the CYCLADES, the SPORADES, the BALEARIC islands, and the IONIAN islands. Divided by shallows (Adventure Bank) between Sicily and CAPE BON, Tunisia, into two main basins. It is of higher salinity than the Atlantic and has little variation in tides. The largest rivers that flow into it are the PO, RHÔNE, EBRO, and NILE. The shores are chiefly mountainous. Earthquakes and volcanic disturbances are frequent. The region around the sea has a warm, dry climate characterized by abundant sunshine. Strong local winds, such as the hot, dry sirocco from the S and the cold, dry mistral and bora from the N, blow across the sea. Fish (about 400 species), sponges, and corals are plentiful. In addition, oil and natural gas have been found in several sections. The overuse of the sea's natural and marine resources continues to be a problem.

Some of the world's oldest civilizations flourished around the Mediterranean. It was opened as a highway for commerce by merchants trading from PHOENICIA. CARTHAGE, GREECE, SICILY, and ROME were rivals for dominance of its shores and trade; under the Roman Empire it virtually became a lake and was called *Mare Nostrum* [Latin=our sea]. Later, the BYZANTINE EMPIRE and the Arabs dominated the Mediterranean. Between the 11th and 14th centuries, BARCELONA and Italian city trading states such as GENOA and VENICE dominated the region; they struggled with the Ottomans for naval supremacy, particularly in the E Mediterranean. Products of Asia passed to Europe over Mediterranean trade routes until the establishment of a route around the CAPE OF GOOD HOPE (late 15th century). With the opening of the Suez Canal (1869) the Mediterranean resumed its importance as a link on the route to the East. The development of the N regions of Africa and of oil fields in the Middle East has increased its trade. Its importance as a trade link and as a route for attacks on Europe resulted in European rivalry for control of its coasts and islands and led to campaigns in the region during both World Wars. Since World War II the Mediterranean region has been of strategic importance to U.S., Western European countries, and, until its dissolution, the USSR.

Medja, CONGO: see MEDJE.

Medjana (me-jah-NAH), village, BORDJ BOU ARRÉRIDJ wilaya, NE ALGERIA, on SE slope of BIBAN range, 6 mi/ 9.7 km NW of BORDJ BOU ARRÉRIDJ; 36°09′N 04°42′E. Cereals, olives.

Medje (ME-jai), village, ORIENTALE province, N CONGO, 170 mi/274 km ESE of BUTA; 02°25′N 27°18′E. Elev. 2,027 ft/617 m. Market and tourist center in Mangbettu territory; cotton ginning. Also spelled MEDJA.

Medjerda Mountains (me-jer-DAH), coastal range of the ATLAS MOUNTAINS, NE ALGERIA (SOUK AHRAS and

ANNABA wilaya) and NW TUNISIA, extending c.100 mi/161 km WSW-ENE from SOUK AHRAS to Béja (Tunisia) between the MEDITERRANEAN SEA (N) and the MEDJERDA valley (S). Average elevation 3,200 ft/ 975 m. Abundant rainfall. Cork-oak forests.

Medjerda River, TUNISIA: see MAJARDAH RIVER.

Medjimurje, CROATIA: see MEDIMURJE.

Medkovets (med-KO-vets), village, MONTANA oblast, Medkovets obshtina (1993 population 6,697), NW BULGARIA, 13 mi/21 km SSW of LOM; 43°38′N 23°11′E. Grain, produce; livestock. Iron-casting shops for the forklift plant in Lom. Sometimes spelled Metkovets.

Medlow Bath (MED-lo BATH), village, NEW SOUTH WALES, SE AUSTRALIA, 66 mi/107 km from SYDNEY, 4 mi/6 km W of KATOOMBA, and in BLUE MOUNTAINS; 33°40′S 150°17′E. Noted resort hotel.

Mednaya Shakhta (MYED-nah-yah SHAHKH-tah) [Russian=copper mine], town, W SVERDLOVSK Russia, W Siberian Russia, 9 mi/14 km E of KARPINSK; 59°44′N 60°15′E. Elevation 606 ft/184 m. A railroad station.

Mednogorsk (myed-nuh-GORSK), city (2006 population 30,455), E central ORENBURG oblast, SE European Russia, in the foothills of the S URALS, on the Blyava River (left tributary of the SAKMARA RIVER), on railroad (Mednyy station), 140 mi/225 km SE of ORENBURG, and 45 mi/72 km WNW of ORSK; 51°25′N 57°35′E. Elevation 1,125 ft/342 m. In Orsk-KHALILOVO industrial district. Center for copper and sulphur mining and processing; electric motors. Within city limits is Blyava (6 mi/10 km E of city center, on railroad; copper and pyrite mines). Founded in 1934, as town of Mednyy, to exploit the Blyavinskiy copper ore deposit; in 1939, became city and renamed Mednogorsk.

Mednogorskiy (myed-nuh-GOR-skeeyee), settlement (2005 population 4,335), W KARACHEVO-CHERKESS REPUBLIC, S European Russia, in the NW CAUCASUS Mountains, on the URUP RIVER, on road, 3 mi/5 km S of PREGRADNAYA; 43°54′N 41°11′E. Elevation 2,903 ft/884 m. Copper mining in the vicinity.

Mednoye (MYED-nuh-ye), village, S central TVER oblast, W European Russia, on the TVERTSA RIVER (VOLGA RIVER basin), on road junction, 16 mi/26 km WNW of TVER; 56°55′N 35°28′E. Elevation 465 ft/141 m. Poultry processing.

Mednyy, RUSSIA: see MEDNOGORSK.

Mednyy Island (MYED-niyee) [Russian=copper], second-largest of the KOMANDORSKI ISLANDS, in SW BERING SEA, KAMCHATKA oblast, extreme E SIBERIA, RUSSIAN FAR EAST, 29 mi/47 km E of BERING Island; 54°45′N 167°35′E. Island is 34 mi/55 km long, 4 mi/6.4 km wide. Chief village is Preobrazhenskoye. Copper deposits; fur-seal preserve.

Medobors'kyy Zakaznyk, UKRAINE: see TOVTRY.

Medobory Nature Preserve (me-do-BO-ree) (Ukrainian *Medobors'kyy Zakaznyk* or *Zapovidnyk Medobory*) (☐ 34 sq mi/88 sq km), TERNOPIL' oblast, UKRAINE, protecting a small segment of the TOVTRY, a ridge of Miocene coral reefs on the PODOLIAN UPLAND, 31 mi/50 km ESE of TERNOPIL'. Established as a geological reserve (zakaznyk) in 1982, it was upgraded to a stricter status of a preserve (zapovidnyk) in 1990. Forested and rocky hills, on right bank of ZBRUCH RIVER, rising from 150 ft/46 m to 200 ft/61 m above surrounding area, with rich flora, including 150 rare endemic and relict species. Provides soil, water, flora, and fauna conservation and is of aesthetic significance (picturesque natural landscape).

Médoc (mai-dok), region of SW FRANCE, occupying a peninsula extending 50 mi/80 km NNW of BORDEAUX between the BAY OF BISCAY and the GIRONDE ESTUARY, in GIRONDE department, AQUITAINE region; 45°10′N 00°46′W. It is covered with some of France's most famous vineyards, including CHÂTEAU-LAFITE, Château-Rothschild, CHÂTEAU-LATOUR, and CHÂTEAU-MARGAUX are among the renowned wines produced here. Chief towns are PAUILLAC (outport for Bordeaux), SAINT-ESTÈPHE, and SAINT-JULIEN-BEYCHE-VELLE. W part of Médoc merges into the sand and dune country of the LANDES along the ATLANTIC coast. Haut-Médoc is the name given to S part of Médoc nearest Bordeaux.

Medo Island, islet in SÃO MARCOS BAY, off MARANHÃO, NE BRAZIL, 7 mi/11.3 km W of SÃO LUÍS.

Medolla (me-DOL-lah), village, MODENA province, EMILIA-ROMAGNA, N central ITALY, 3 mi/5 km S of MIRANDOLA; 44°51′N 11°04′E. Food processing; textiles.

Medomak, MAINE: see BREMEN.

Medomak River (muh-DAHM-uhk), c.15 mi/24 km long, S MAINE; rises in KNOX county; flows S, widening below Waldeboro, to MUSCONGUS BAY.

Medomsley (MED-uhmz-lee), village (2006 population 2,100), N DURHAM, NE ENGLAND, 10 mi/16 km SW of NEWCASTLE UPON TYNE; 54°53′N 01°48′W. Former coal-mining community.

Medon (MEE-duhn), town (2006 population 192), MADISON county, W TENNESSEE, 11 mi/18 km S of JACKSON; 35°27′N 88°52′W.

Medora, town (2006 population 565), JACKSON county, S INDIANA, 8 mi/12.9 km SW of BROWNSTOWN; 38°49′N 86°10′W. In agricultural area.

Medora, village (2000 population 501), MACOUPIN county, SW ILLINOIS, 16 mi/26 km SW of CARLINVILLE; 39°10′N 90°08′W. In agricultural and bituminous-coal area.

Medora, village (2006 population 95), ⊙ BILLINGS county, W NORTH DAKOTA, on LITTLE MISSOURI RIVER, 36 mi/58 km W of DICKINSON; 46°54′N 103°31′W. Cattle raising; grain. Nearby is site of Chimney Butte Ranch, where Theodore Roosevelt engaged in livestock raising from 1883 to 1886; it is at entrance to S Unit of THEODORE ROOSEVELT NATIONAL PARK; in Little Missouri National Grassland; Sullys Creek State Primitive Park to S; Camel Hump Reservoir to W; Chateau de Mores Historic Site to SW. Founded 1883 and named for Medora von Hoffman.

Médouneu (MAI-doo-nuh), town, WOLEU-NTEM province, N GABON, 70 mi/113 km SSW of OYEM, on EQUATORIAL GUINEA border; 00°58′N 10°47′E. Cacao plantations.

Medpalli, India: see METPALLI.

Medrano (me-DRAH-no), town, N MENDOZA province, ARGENTINA, on railroad, on TUNUYÁN RIVER, 22 mi/35 km SE of MENDOZA; 33°11′S 68°37′W. Agricultural center (wine, corn, alfalfa, potatoes, fruit; livestock).

Medstead (MED-sted), village (2006 population 148), W SASKATCHEWAN, CANADA, 40 mi/64 km NNE of NORTH BATTLEFORD; 53°18′N 108°05′W. Wheat; mixed farming.

Medulin, town, W CROATIA, in ISTRIA, on ADRIATIC SEA, in Medulin Riviera, 6 mi/9.7 km SE of PULA. Resort with trailer park and nudist beaches. Ruins of prehistoric settlement and Roman villa nearby.

Medulla (muh-DULL-uh), unincorporated town (☐ 5 sq mi/13 sq km; 2000 population 6,637), POLK county, central FLORIDA, 7 mi/11.3 km S of LAKELAND; 27°57′N 81°59′W.

Medum (MAI-dum), town, BENI SUEF province, Upper EGYPT, on the NILE River, 20 mi/32 km ENE of FAIYUM. Site of pyramid of Snefru (c.2900 B.C.E.). Also spelled Maidum.

Meduncook River (muh-DUHN-kuk), inlet of MUSCONGUS BAY, 5.5 mi/8.9 km long, KNOX county, S MAINE, between FRIENDSHIP and CUSHING.

Meduxnekeag River (muh-DUHKS-nuh-keg), 35 mi/56 km long, in MAINE and NEW BRUNSWICK, CANADA; North and South branches rise in SE AROOSTOOK county, MAINE; flow c.20 mi/32 km to junction 8 mi/12.9 km NE of HOULTON (on South Branch), in New Brunswick, then c.15 mi/24 km SE to SAINT JOHN RIVER at WOODSTOCK, New Brunswick.

Medvedevo (meed-VYE-dee-vuh), town (2006 population 16,345), central MARI EL REPUBLIC, E central European Russia, on railroad, 2 mi/3.2 km W of YOSHKAR-OLA; 56°38′N 47°49′E. Elevation 380 ft/115 m. Peatworks in the vicinity. Has a heritage museum.

Medveditsa (meed-VYE-dee-tsah), river, approximately 430 mi/692 km long, SE European Russia; rises NW of VOL'SK (SARATOV oblast); flows roughly parallel to the VOLGA RIVER beyond PETROVSK, passes ATKARSK, ZHIRNOVSK (VOLGOGRAD oblast), and Mikhaylovka and empties into the DON River near SERAFIMOVICH.

Medveditskoye, RUSSIA: see LINEVO.

Medvedka, RUSSIA: see MEDVEDOK.

Medvednica (MED-ved-nee-tsah) or **Zagreb Mountain**, Croatian *Zagrebačka Gora*, central CROATIA, with central section culminating in the SLJEME (3,395 ft/1,035 m), 6 mi/9.7 km N of ZAGREB. Winter-sports center. Cement-rock deposits. W and central area protected as a nature park, encompassing eight forest reserves.

Medvedok (meed-VYE-duhk), town (2005 population 2,020), S KIROV oblast, E European Russia, on the VYATKA River, on highway, 13 mi/21 km SSE of NO-LINSK; 57°23′N 50°02′E. Elevation 252 ft/76 m. Ship repair yard; flour mill, sawmill. Previously known as Medvedka (meed-VYET-kah).

Medvedovskaya (meed-VYE-duhf-skah-yah), village (2005 population 17,020), central KRASNODAR TERRITORY, S European Russia, on the KIRPILI River, on road and railroad, 25 mi/40 km N of KRASNODAR; 45°27′N 39°01′E. In agricultural area (wheat, sunflowers, grapes, fruits; livestock); produce processing. Founded by the Kuban' Cossacks.

Medveja, resort village, W CROATIA, in ISTRIA, on ADRIATIC SEA, in OPATIJA Riviera, 3.1 mi/5 km S of Opatija.

Medvenka (MYED-veen-kah), town (2005 population 4,455), central KURSK oblast, SW European Russia, on road, 20 mi/32 km SSW of KURSK; 51°25′N 36°06′E. Elevation 764 ft/232 m. In agricultural area; food processing.

Medvezhya Gora, RUSSIA: see MEDVEZHYEGORSK.

Medvezhyegorsk (meed-vyezh-ye-GORSK), Finnish *Karhumäki*, city (2005 population 16,610), S central Republic of KARELIA, NW European Russia, on the Povenets Gulf of Lake ONEGA, on the Murmansk railroad, 110 mi/177 km N of PETROZAVODSK; 62°55′N 34°28′E. Railroad shops; woodworking (prefabricated houses, furniture), rosin extraction. Shipyards at Pindushi, 5 mi/8 km to the E. Formerly called Medvezhya Gora. Made city in 1938.

Medvezhyi Ostrova, RUSSIA: see BEAR ISLANDS.

Medvode (med-VO-de), town, central SLOVENIA, on the Sava River, at Sora River mouth, on railroad, 7 mi/ 11.3 km NNW of LJUBLJANA; 46°08′N 14°26′E. Summer resort; manufacturing (paper, cartons, chemicals) here and at nearby Vevče. Hydroelectric power plant on Sava.

Medwar, El, JORDAN: see MADWAR, EL.

Medway, town, PENOBSCOT county, E central MAINE, on the PENOBSCOT, c.60 mi/97 km NNE of BANGOR; 45°37′N 68°30′W. In lumbering area.

Medway, town, NORFOLK county, E MASSACHUSETTS, on CHARLES RIVER, 22 mi/35 km N of PROVIDENCE (RHODE ISLAND); 42°09′N 71°26′W. Treadmills, tools, plastics. Settled 1657, set off from MEDFIELD 1713. Includes West Medway and Medway villages.

Medway River, c.75 mi/121 km long, W NOVA SCOTIA, CANADA; rises ESE of ANNAPOLIS ROYAL; flows SE, through MALAGA and Ponbook lakes, to the Atlantic 8 mi/13 km NE of LIVERPOOL.

Medway River (MED-wai), 70 mi/113 km long, SURREY, West SUSSEX, East SUSSEX, and KENT, SE ENGLAND; rises in two headstreams (the N branch called Eden River) in SE Surrey and West Sussex; flows NE into Kent, past MAIDSTONE, ROCHESTER, CHA-

THAM, and GILLINGHAM, to THAMES estuary at SHEERNESS.

Medyka (me-DEEK-ah), village, Rzeszów province, SE POLAND, frontier station on Ukrainian border, 7 mi/11.3 km E of Przemyśl, 9 mi/14.5 km W of MOSTISKA.

Medyka, village, SW Przemyśl Wojewodztwo, POLAND, custom railroad station on Polish border with UKRAINE, 7 mi/11.3 km E of Przemyśl. Train-servicing facilities for switching undercarriages from narrower Polish tracks to wider Ukrainian ones.

Medyn' (mye-DIN), city (2005 population 7,745), N KALUGA oblast, central European Russia, on the Medyn'ka River (OKA River basin), 35 mi/56 km NNW of KALUGA; 54°58′N 35°51′E. Elevation 620 ft/188 m. Furniture factory; dairying, flax retting, woodworking. Chartered in 1389; attacked and seized by the Lithuanians in 1450; site of a defeat of the Polish-Tatar forces at the hands of the combined armies of the Russian princes in 1480, signifying the end of the Golden Horde's domination over the region.

Medzev (med-ZEF), Hungarian *Mecenzéf*, town, VYCHODOSLOVENSKY province, SE SLOVAKIA, on BODVA RIVER, 16 mi/26 km W of KOŠICE; 48°42′N 20°54′E. Contains a railroad terminus; tools manufacturing, woodworking. Has 15th-century church, traditional folk architecture. Under Hungarian rule from 1938–1945.

Medzhibozh, UKRAINE: see MEDZHYBIZH.

Medzhybizh (med-ZHI-beezh) (Russian *Medzhibozh*), town (2004 population 4,000), E KHMEL'NYTS'KYY oblast, UKRAINE, on the Southern BUH, 19 mi/31 km E of KHMEL'NYTS'KYY; 49°26′N 27°25′E. Elevation 935 ft/284 m. Fruit canning; fish plant; sewing. Historical ethnographical museum; castle (14th–16th, 19th centuries), palace (16th century), ruins of a Roman Catholic church (1632). Known since 1146, important center of Halych-Volyn' principality; Prince Danylo defeated a Tatar army nearby (1258); passed to Lithuania (1362), and later (1569) to Poland; site of major battles during the Cossack-Polish War (1648–1657); captured by the Turks and held (1666–1693), passed to Russia (1793); site of battles between the Ukrainian National Republic Army and the Red Army (1918). Jewish community since 1540; it was the seat of the founder of Hasidism 1740–1760; the town's population was mostly Jewish into the mid-20th century (4,614 in 1939); destroyed by the Nazis in 1941–1942. Also called Mezhybizh.

Medzilaborce (med-ZI-lah-BOR-tse), Hungarian *Mezőlaborc*, town, VYCHODOSLOVENSKY province, E SLOVAKIA, on LABOREC RIVER, at POLISH border, on railroad, 35 mi/56 km NE of PREŠOV; 49°16′N 21°55′E. Lumbering, woodworking, manufacturing (machinery). Site of heavy fighting between Austro-Hungarians and RUSSIANS in winter of 1915. DUKLA Pass (1,647 ft/502 m) is 13 mi/21 km NW, LUPKOW PASS (2,155 ft/657 m) 8 mi/12.9 km ESE.

Meeder (MAI-der), village, UPPER FRANCONIA, N BAVARIA, GERMANY, 5 mi/8 km NNW of COBURG; 50°19′N 10°54′E. Manufacturing (electronic equipment).

Meekathara (MEE-kuh-THAR-uh), town, W central WESTERN AUSTRALIA state, W AUSTRALIA, 280 mi/451 km NE of GERALDTON; 26°36′S 118°29′E. On Geraldton-WILUNA railroad. Mining center of MURCHISON GOLDFIELD; also sheep, cattle.

Meeker, county (☐ 645 sq mi/1,677 sq km; 2006 population 23,405), S central MINNESOTA; ⊙ LITCHFIELD; 45°07′N 94°31′W. Drained by North and South forks of CROW RIVER. Agricultural area (corn, oats, barley, wheat, hay, alfalfa, soybeans, beans, peas; sheep, hogs, cattle, poultry; dairying). Numerous small lakes in county; Washington Lake in SE, Lake KORONIS in N boundary in NW. Formed 1856.

Meeker, town (2000 population 2,242), ⊙ RIO BLANCO county, NW COLORADO, on White River, and 75 mi/121 km NNE of GRAND JUNCTION; 40°02′N 107°53′W. Elevation 6,249 ft/1,905 m. Resort and trading point in grain and livestock area; sheep, cattle; hay. Mining. Nearby is Meeker Monument, at scene of "Meeker Massacre" (1879), in which Utes killed a small group of whites including Nathan Meeker, Indian agent and co-founder of GREELEY. White River National Forest to E and SE; Lake Avery reservoir to E; Piceans State Wildlife Area to W. Incorporated 1885.

Meeker, town (2006 population 997), LINCOLN county, central OKLAHOMA, 10 mi/16 km N of SHAWNEE; 35°29′N 96°53′W. Trading point for agricultural area; N terminus of railroad spur from Shawnee. Manufacturing (machinery, consumer goods).

Meelpaeg Lake (MEEL-puh-eg) (☐ 37 sq mi/96.2 sq km), S NEWFOUNDLAND AND LABRADOR, CANADA, 40 mi/64 km SSE of BUCHANS; 15 mi/24 km long, 5 mi/8 km wide. Contains numerous islets. Drained by Grey River; connected SE with Lake EBBEGUNBAEG by 4-mi/6-km-long stream.

Meeniyan (MEEN-yuhn), town, VICTORIA, SE AUSTRALIA, 83 mi/150 km E of MELBOURNE, and in Strzelecki Ranges foothills; 38°35′S 46°01′E.

Meerane (me-RAH-ne), town, SAXONY, E central GERMANY, 9 mi/14.5 km N of ZWICKAU; 50°52′N 11°58′E. Textile center (cotton, wool, silk); metalworking. Textile industry introduced in 16th century. Was first mentioned 12th century; chartered 1374 and 1405.

Meerbeke (MER-bai-kuh), village in commune of NINOVE, Aalst district, EAST FLANDERS province, W central BELGIUM, just SSE of Ninove. Agriculture.

Meerbusch (MER-bush), city, RHINELAND, North Rhine-Westphalia, W GERMANY, on left bank of the RHINE, 5 mi/8 km W of DÜSSELDORF; 51°16′N 06°42′E. Manufacturing (steel, machinery, textiles); metalworking; agriculture (vegetables). Formed 1970 by conjunction of eight villages. Has two castles (Desch and Dyckhoff).

Meerhout (MER-hou-tuh), commune, Turnhout district, ANTWERPEN province, N BELGIUM, on GROTE NETE River, 15 mi/24 km SSE of TURNHOUT; 51°08′N 05°05′E. Agriculture.

Meerle (MER-luh), village, HOOGSTRATEN commune, Turnhout district, ANTWERPEN province, N BELGIUM, near NETHERLANDS border, 8 mi/12.9 km S of BREDA (Netherlands); 51°28′N 04°48′E. Agriculture.

Meersburg (MERS-burg), town, Upper Swabia, BADEN-WÜRTTEMBERG, GERMANY, on the Überlinger See (a branch of LAKE CONSTANCE), 7 mi/11.3 km SE of Überlingen; 47°42′N 09°17′E. Ferry station; wine growing. Has late-Gothic chapel; 16th-century granary, town hall, Old Castle (12th–16th century); 18th-century New Castle was residence of bishops of Constance from the 16th to 19th centuries. Poetess Droste-Hülshoff buried here. Was first mentioned 988; chartered 1299.

Meerssen (MAIR-suhn), town, LIMBURG province, SE NETHERLANDS, 4 mi/6.4 km NE of MAASTRICHT; 50°53′N 05°45′E. JULIANA CANAL passes to W; Vliek Castle to NE. Dairying; cattle, hogs, poultry; grapes, vegetables, sugar beets; light manufacturing. Has thirteenth-century church (Kloosterkerk) and residence (ninth century) of Frankish kings. The Treaty of Mersen between Charles the Bald and Louis the German, which divided the realm of Lothair, was signed here (870).

Meerut (MEE-ruht), district (☐ 1,510 sq mi/3,926 sq km), NW UTTAR PRADESH state, N central INDIA; ⊙ MEERUT. On GANGA-YAMUNA DOAB; irrigated by EAST YAMUNA and UPPER GANGA canals. Agriculture (wheat, gram, jowar, sugarcane, oilseeds, cotton, corn, rice, barley); a leading sugar-processing district. Main centers are Meerut, GHAZIABAD, SARDHANA.

Meerut (MEE-ruht), city (2001 population 1,161,716), ⊙ Meerut district, Uttar Pradesh state, N central India; 28°59′N 77°42′E. An agr. market, it processes flour, sugar, cotton, and vegetable oil. It is also an industrial center, with manufacturing (including textiles and leather) and smelting concerns. Meerut was conquered by Muslims in 1192, ravaged by Tamerlane in 1399, and became part of the Mogul empire. An important town of the Jat Bharatpur kingdom (mid-18th century), it subsequently fell to the British, who made it a major military cantonment. The 1st outbreak of the Indian Mutiny occurred in Meerut in May 1857, but the British held the city.

Meerzorg (MIR-sawrg), village, COMMEWIJNE district, N SURINAME, on right bank of SURINAME RIVER; 05°46′N 55°11′W. Opposite PARAMARIBO (connected to city by river ferry); agricultural center (rice, coffee, corn, tropical fruit).

Mées, Les (MAI, lai), commune (☐ 25 sq mi/65 sq km), ALPES-DE-HAUTE-PROVENCE department, PROVENCE-ALPES-CÔTE D'AZUR region, SE FRANCE, near DURANCE RIVER, 14 mi/23 km WSW of DIGNE, in the Provence Alps (MARITIME ALPS); 44°01′N 05°58′E. Olive and grape growing; fruit shipping. The jumbled rocks of Mées overhang the valley of Durance about 300 ft/90 m above the plain.

Mée-sur-Seine, Le (MAI–syoor–SEN, luh), SW suburb (☐ 2 sq mi/5.2 sq km) of MELUN, SEINE-ET-MARNE department, ÎLE-DE-FRANCE region, on the SEINE RIVER, and 7 mi/11.3 km NNW of FONTAINEBLEAU, adjacent to the Forest of Fontainebleau; 48°32′N 02°38′E.

Meeteetse (muh-TEET-see), village (2006 population 347), Park county, NW WYOMING, on GREYBULL RIVER, in SE foothills of ABSAROKA RANGE, and 27 mi/43 km SSE of CODY; 44°09′N 108°52′W. Elevation 5,797 ft/1,767 m. In sheep-raising region. Barley, sugar beets.

Me'etia, FRENCH POLYNESIA: see MEHETIA.

Meeuwen-Guitrode, commune, Maaseik district, LIMBURG province, NE BELGIUM, 10 mi/16 km N of GENK.

Mefou, department, CENTRAL province, CAMEROON: see MEFOU-ET-AFAMBA and MEFOU-ET-AKONO departments.

Mefou-et-Afamba, department (2001 population 89,805), CENTRAL province, CAMEROON.

Mefou-et-Akono, department (2001 population 57,051), CENTRAL province, CAMEROON.

Meftah (mef-TAH), town, BLIDA wilaya, N central ALGERIA. Major farming area with citrus and industrial crops. One of the country's largest cement works; other construction materials are manufactured here as well. Formerly called Rivet.

Mega (MAI-gah), town (2007 population 9,817), OROMIYA state, S ETHIOPIA, near KENYA border, on the main road from ADDIS ABABA to KENYA, 145 mi/230 km SE of ARBA MINCH; 04°02′N 38°18′E. In agricultural (corn, barley, potatoes, chickpeas), and livestock raising region; road junction; commercial center. Salt extracting nearby; old British consulate and fort.

Megahatenna, SRI LANKA: see MIGAHATENNA.

Megale Delos, Greece: see RINIA.

Megalo (mai-GAH-lo), village, SOUTHERN NATIONS state, S ETHIOPIA, between MENA and WABĒ GESTRO rivers, 20 mi/32 km S of GINIR; 06°51′N 40°47′E. Trade center (hides; cereals; wax, honey).

Megalopolis (me-yah-LO-po-lees), town, ARKADIA prefecture, central PELOPONNESE department, S mainland GREECE, on railroad, on branch of Alpheus River, and 15 mi/24 km SW of TRÍPOLIS; 37°24′N 22°08′E. Trades in tobacco, wheat, wine, potatoes. Has archeological museum. Ruins of ancient city (just N) include large theater, temples and statues. Ancient Megalopolis was founded c.370 B.C.E. by Epaminondas as a fortress against SPARTA and the center of the Arcadian League. Home of Philopoemen and Polybius. After repeated Spartan assaults, it finally fell to Cleomenus III (222 B.C.E.) and was razed. Modern town was formerly called Sinano.

Meganesi (me-yah-NEE-see), island (☐ 8 sq mi/20.8 sq km), LEFKAS prefecture, IONIAN ISLANDS department, off W coast of GREECE and off SE coast of LEFKAS island (separated by narrow Meganesi Strait); 38°38′N

20°45'E. Island is 4 mi/6.4 km long, 2 mi/3.2 km wide, with 5-mi/8-km-long narrow peninsula (SW). Produces olive oil, wine, wheat. Main town (on NE shore) is Vathy or Vathi, formerly called Taphion or Tafion. Also spelled Meganisi. Sometimes called Taphos or Tafos.

Megantic (muh-GAN-tik) or **Mégantic** (mai-gahn-TEEK), former county (□ 780 sq mi/2,028 sq km), S QUEBEC, E CANADA, on Lake SAINT FRANCIS; county seat was INVERNESS; 46°15'N 71°30'W.

Megantic, Lake (muh-GAN-tik), French, *Lac Mégantic* (mai-gahn-TEEK), S QUEBEC, E CANADA; extends S from LAC-MÉGANTIC, 50 mi/80 km E of SHERBROOKE; 45°31'N 70°52'W. Elevation 1,294 ft/394 m. Drained N by CHAUDIÈRE River into the SAINT LAWRENCE RIVER; 9 mi/14 km long, 2 mi/3 km wide.

Megantic Mountain (muh-GAN-tik) (3,625 ft/1,105 m), French, *Mont Mégantic* (mon mai-gahn-TEEK), S QUEBEC, E CANADA, 16 mi/26 km SW of LAC-MÉGANTIC, near U.S. (NEW HAMPSHIRE) border; 45°28'N 71°09'W.

Mégara (ME-yah-rah), town (2001 population 23,032), ATTICA prefecture, ATTICA department, E central GREECE, on N coast of the SARONIC GULF, W of ATHENS; 38°00'N 23°21'E. Wine, olive oil, and flour are produced. Site of the ancient town of Mégara, capital of Mégaris, a small district between the Gulf of CORINTH and the Saronic Gulf. The Dorians who succeeded the earliest known inhabitants made Mégara a wealthy city by means of maritime trade, and they founded many colonies, including, in the 7th century B.C.E., Chalcedon and Byzantium. After the Persian Wars the citizens summoned the aid of Athens against CORINTH (459 B.C.E.), but soon thereafter expelled the Athenians. The mathematician Euclid was probably born here.

Megargel (MUH-gahr-guhl), village (2006 population 258), ARCHER county, N TEXAS, 40 mi/64 km SW of WICHITA FALLS; 33°27'N 98°55'W. In farm, ranch area.

Megaspelaion, Greece: see KALÁVRITA.

Megen (MAI-khuhn), village, NORTH BRABANT province, E central NETHERLANDS, on MEUSE RIVER, 15 mi/24 km NE of 's-HERTOGENBOSCH, 5 mi/8 km N of OSS; 51°49'N 05°34'E. Dairying; cattle raising; agriculture (vegetables, grain).

Meget (mee-GYET), town (2005 population 8,300), S IRKUTSK oblast, E central SIBERIA, RUSSIA, on the W bank of the ANGARA RIVER, on road and the TRANS-SIBERIAN RAILROAD, 14 mi/23 km NW of IRKUTSK; 52°25'N 104°03'E. Elevation 1,358 ft/413 m. Lumbering, metalworking. Oil and natural gas pipelines in the vicinity.

Megève (me-zhe-vuh), town (□ 17 sq mi/44.2 sq km), HAUTE-SAVOIE department, RHÔNE-ALPES region, SE FRANCE, on upper ARLY RIVER, W of MONT BLANC range, 13 mi/21 km SW of CHAMONIX; 45°52'N 06°37'E. Elev. 3,650 ft/1,113 m. A leading winter-sport and summer resort of French Alps (ALPES FRANÇAISES) established after World War I and expanded since 1960s. Aerial tramways to Mont d'ARBOIS and Mont ROCHEBRUNE. World-famous ski schools and a wide range of accommodations are found here.

Meggen, commune, LUCERNE canton, central SWITZERLAND, on LAKE LUCERNE, and 4 mi/6.4 km E of LUCERNE; 47°03'N 08°23'E.

Meggett (MEG-git), town (2006 population 1,325), CHARLESTON county, SE SOUTH CAROLINA, 18 mi/29 km WSW of CHARLESTON; 32°42'N 80°15'W. On WADMALAW ISLAND. Manufacturing includes vegetable products.

Meghalaya, state (□ 8,660 sq mi/22,516 sq km), NE INDIA; ⊙ SHILLONG. Bordered S by BANGLADESH, N and E by ASSAM state; W by BRAHMAPUTRA. Meghalaya is in the Garo, Khasi, and Jaintia hills; elevation 4,000 ft/1,219 m–6,000 ft/1,829 m. The S sides of the hills are among the world's wettest places, with the highest recorded rainfall in the world at CHER-

RAPUNJI. Primarily agricultural, although the state overall has low productivity and so depends on imports from other states. Generally considered underdeveloped, although some natural resource extraction: minerals include coal, limestone, clay, and corundum. The state is heavily forested and has much natural, unspoiled scenic beauty; the government is trying to cultivate tourism industry, although has not been very successful due to outbreaks of violence, which the state suffers sporadically as indigenous hill peoples try to repel illegal immigrants from Bangladesh. The inhabitants are Khasi, Synteng, and Pner tribesmen, who speak a Mon-Khmer language. Christian missionaries have had considerable influence among Meghalaya's inhabitants; a great majority of the population is Christian. Meghalaya was formerly part of Assam state; it became a separate state in 1972. It is governed by a chief minister and cabinet responsible to an elected unicameral legislature and by a governor appointed by the president of India. The forests contain several rare animals, such as civets, hoolocks, Himalayan black bears, leapords, Asian golden cats, several species of poisonous snake, birds, and a large variety of butterflies.

Meghauli (ME-gou-lee), village, central NEPAL; 27°35'N 84°14'E. Elevation 600 ft/183 m. Airport here serves CHITWAN district, including ROYAL CHITWAN NATIONAL PARK.

Meghna (meg-nah), river, c.130 mi/210 km long, NE BANGLADESH; formed at the outlet of the SURMA VALLEY by the branches of the SURMA RIVER; flows S, receiving arms of the GANGA and BRAHMAPUTRA rivers, to the BAY OF BENGAL. The Meghna River is an important inland waterway, navigable throughout its length by river steamers, although hazardous due to its high speed. In the springtime, at high tide, tidal bores (c.20 ft/6.1 m high) rush upstream with great destructive force.

Megiddo (mi-gee-DO), ancient city, ISRAEL, by the KISHON RIVER on the SW edge of the plain of JEZREEL; 32°34'N 35°10'E. Elevation 426 ft/129 m. It was inhabited from the 4th millennium B.C.E. to c.450 B.C.E. Situated in a strategic position, controlling the Via Maris route that connected EGYPT with SYRIA and MESOPOTAMIA, it has been the scene of many battles throughout history, from Thutmose III (c.1468 B.C.E.) to General Edmund Allenby (later Viscount Allenby of Megiddo) in World War I. Excavations have unearthed over twenty strata of settlements. Found in the latest six strata, from the Canaanite period to c.500 B.C.E., were the Megiddo Ivories, one of the most important examples of Canaanite art, Solomon's chariot stables, and a sophisticated water system dating back to Ahab's time. Megiddo was one of King Solomon's three main regional military centers—along with HAZOR (N) and GEZER (S). Believed to be the biblical Armageddon. The plain is sometimes called the valley of Megiddon. Alongside ancient site is modern Kibbutz Megiddo (1994 population 348), with mixed farming, plastics factory, and jewelry production (some of which is Roman-inspired). Important tourist site now.

Megiddon, valley, ISRAEL: see MEGIDDO.

Megion (MYE-gee-uhn), city (2005 population 49,420), central KHANTY-MANSI AUTONOMOUS OKRUG, central SIBERIA, RUSSIA, on the OB' RIVER, near railroad (9 mi/15 km from Megion station), 16 mi/26 km WNW of NIZHNEVARTOVSK; 61°03'N 76°06'E. In oil and gas area. Oil field discovered in 1961. Made a city in 1980.

Megorskiy Pogost (mee-GOR-skeeyee puh-GOST), village, NW VOLOGDA oblast, N central European Russia, on the S coast of Lake ONEGA, on road, 20 mi/32 km SW of VYTEGRA; 60°52'N 36°01'E. Elevation 206 ft/62 m. Wind-based power station in the vicinity. Sometimes called Megra.

Megra, RUSSIA: see MEGORSKIY POGOST.

Megri (MYE-gree), town, S ARMENIA, on ARAS RIVER (IRAN border), on railroad, 50 mi/80 km ESE of NAKHICHEVAN; 38°54'N 46°14'E. In orchard district; dried fruit; winery; cannery; cheese dairy; silkworm breeding; crushed rock.

Megunticook Lake (me-GUHN-tuh-kuk), reservoir, KNOX and WALDO counties, S MAINE, source of Megunticook River, 3 mi/4.8 km NW of CAMDEN, in recreational area; 44°15'N 69°05'W. Reservoir is 3 mi/4.8 km long, 1.5 mi/2.4 km wide; river (c.4 mi/6.4 km long) flows SE to ATLANTIC OCEAN at Camden.

Megunticook, Mount, MAINE: see CAMDEN HILLS.

Meguro (ME-goo-ro), ward of TOKYO city, Tokyo prefecture, E central HONSHU, E central JAPAN, SW of central Tokyo. Bordered N by Shibuya ward, E by Shinagwa ward, SE by OTA ward, and W and SW by SETAGAYA ward.

Mehabad, IRAN: see MAHABAD.

Mehadia (me-HAH-dyah), ancient *Ad Mediam*, village, CARAŞ-SEVERIN county, SW ROMANIA, at W extremity of the TRANSYLVANIAN ALPS, on railroad, 37 mi/60 km SE of REŞITA; 44°54'N 22°22'E. Lignite mining. Has picturesque Roman and 16th-century remains.

Mehallet, in Egyptian names: see MAHALLAT.

Meham, India: see MAHAM.

Mehamn (MAI-hah-muhn), fishing village, FINNMARK county, N NORWAY, on BARENTS SEA of ARCTIC OCEAN, 80 mi/129 km NW of VARDØ, 5 mi/8 km SE of NORDKYN cape; 71°02'N 27°51'E.

Mehar (MAI-hahr), village, DADU district, W SIND province, SE PAKISTAN, 31 mi/50 km N of DADU; 27°11'N 67°49'E. Market center (rice, millet, wheat).

Meharry, Mount, highest peak (elevation 4,085 ft/1,245 m) of WESTERN AUSTRALIA state, W AUSTRALIA, on SE boundary of KARIJINI NATIONAL PARK; 22°59'S 118°35'E.

Mehdawal (MAI-dah-wuhl), town, BASTI district, NE UTTAR PRADESH state, N central INDIA, 27 mi/43 km NE of BASTI. Trades in rice, wheat, barley, oilseeds, sugarcane. Also spelled Mehndawal.

Mehdedya, MOROCCO: see MEHDIA.

Mehdia (med-YAH), village, Kénitra province, Gharb-Chrarda-Beni Hssen administrative region, W central MOROCCO, on the ATLANTIC OCEAN at mouth of the Oued SEBOU river, 18 mi/29 km NE of RABAT; 34°15'N 06°40'W. Railroad-spur terminus; bathing beach; fish-processing plant. Outport for KENITRA (10 mi/16 km upstream). An early Carthaginian settlement, it was coveted (16th century) by Portuguese and held (17th century) by Spaniards. In disuse as a port at time of French occupation (1911), it was abandoned in favor of Kenitra. There are two jetties (over 1 mi/1.6 km long) flanking river mouth to aid navigation. Lighthouse. Sometimes spelled Meheydia, Mehdiya, or Mehedya.

Mehdishar (me-DEE-shahr), village, Semnān province, N IRAN, in ELBURZ mountains, 12 mi/19 km N of SEMNAN. Grain, cotton; sheep raising. Formerly called SANG-e-Sar or Sange-e-Sar.

Mehdiya, MOROCCO: see MEHDIA.

Mehedinţi (me-he-DEENTS), county, SW ROMANIA, in WALACHIA, on border with SERBIA; ⊙ DROBETA-TURNU SEVERIN; 44°40'N 22°50'E. Terrain ranges from flat to hilly; DANUBE River along border. Contains IRON GATE.

Mehedinţi Mountains (me-he-DEENTS), range, SW extension of TRANSYLVANIAN ALPS, MEHEDINŢI county, SW ROMANIA, NE of ORŞOVA; 45°00'N 22°35'E.

Mehekar (MAI-kuhr), town, BULDANA district, MAHARASHTRA state, W central INDIA, in AJANTA HILLS, on PENGANGA RIVER, 37 mi/60 km SE of BULDANA; 20°09'N 76°34'E. Cotton ginning. Sodium carbonate and salt extracted from lake 13 mi/21 km S, near village of Lonar. Sometimes spelled Mehkar.

Mehelav, India: see MEHLAV.

Meherpur (me-her-poor), town (2001 population 34,624), KUSHTIA district, W EAST BENGAL, BANGLADESH, on distributary of JALANGI RIVER, 32 mi/51 km WSW of KUSHTIA; 23°47′N 88°38′E. Trades in rice, jute, linseed, sugarcane, wheat; bell-metal manufacturing. Until 1947, Meherpur was in the NADIA district of British Bengal province. Formerly called Mihrpur.

Meherrin River (muh-HE-ruhn), 126 mi/203 km long, in VIRGINIA and NORTH CAROLINA; formed by headstreams joining on the border of LUNENBURG and MECKLENBURG counties, S Virginia; flows ESE to N of SOUTH HILL, past EMPORIA, and SE into North Carolina, past MURFREESBORO (head of navigation), to CHOWAN RIVER 8 mi/13 km E of Murfreesboro; 36°49′N 78°16′W.

Meheso, ETHIOPIA: see MIESO.

Mehetia (mai-hai-TEE-ah), uninhabited volcanic island (□ 1 sq mi/2.6 sq km), Windward group, easternmost of SOCIETY ISLANDS, FRENCH POLYNESIA, S PACIFIC, 60 mi/97 km E of TAHITI; 17°52′S 148°03′W. Elevation c.1,410 ft/430 m. Also spelled Me'etia.

Meheydia, MOROCCO: see MEHDIA.

Mehidpur (MAI-hid-puhr) or **Mahidpur**, town (2001 population 28,080), UJJAIN district, Madhya Pradesh state, central India, on Sipra River, 22 mi/35 km NNW of UJJAIN; 23°49′N 75°40′E. Market center (millet, cotton, wheat, opium); cotton ginning, sugar milling, hand-loom weaving; place of pilgrimage. In nearby battle, British defeated Marathas in 1817. Mehidpur Road, railroad station, is 11 mi/18 km NW.

Mehkar, India: see MEHEKAR.

Mehlauken, RUSSIA: see ZALES'YE.

Mehlav (MAI-lahv), town, KHEDA district, GUJARAT state, W central INDIA, 14 mi/23 km SSE of KHEDA; 22°34′N 72°49′E. Local agriculture market (millet, tobacco, cotton). Sometimes spelled Mehelav or Mehelao.

Mehlis, GERMANY: see ZELLA-MEHLIS.

Mehmadabad (mai-mah-DAH-bahd), town, KHEDA district, GUJARAT state, W central INDIA, 6 mi/9.7 km NE of KHEDA. Rice, millet; cotton ginning, dairy farming. Sometimes spelled Mohammadabad.

MehmetÇik, CYPRUS: see GALATIA.

Mehndawal, India: see MEHDAWAL.

Mehomia, Bulgaria: see RAZLOG.

Mehoopany, township (2000 population 993), WYOMING county, NE PENNSYLVANIA, 6 mi/9.7 km W of TUNKHANNOCK; 41°33′N 76°03′W. Manufacturing includes paper products. Agriculture includes dairying; timber.

Mehran (mah-RAHN), town, Īlām province, W central IRAN, in the PUSHT KUH, on IRAQ border, 100 mi/161 km SSW of KERMANSHAH, and on road from DEZFUL to BAGHDAD; 33°07′N 46°09′E. Formerly called MANSURABAD.

Mehrauli, India: see MAHRAULI.

Mehrgarh, group of archaeological sites, SIBI district, BALUCHISTAN province SW PAKISTAN, at the head of the North Kachchi plain, at the foot of the BOLAN PASS, on the Kandahar (Afghanistan)–Quetta-Shikarpur route. Sites dating from the 8th–6th millenia B.C.E. reveal the earliest evidence of agriculture and pastoralism in S Asia; easternmost location of Neolithic strains of barley. Wheat also found; remains of wild and domestic ovicaprids. Other discoveries include mud bricks, necklaces, baskets, semiprecious stones, and the use of copper and ivory.

Mehrshahr (mahr-SHAHR), town, Tehrān province, N IRAN, 4 mi/7 km W of KARA; 35°49′N 50°54′E. Growing suburban area just WNW of Kara.

Meh Sot, THAILAND: see MAE SOT.

Mehtarlam (me-tar-LAHM), town, ⊙ LAGHMAN province, E AFGHANISTAN, 20 mi/32 km NW of JALALABAD, on Alingar River, N of its junction with the ALISHANG RIVER; 34°39′N 70°10′E. Some agriculture

(rice, wheat); livestock; timber industry. Population primarily consists of Nuristani and Ghilzais.

Mehuín (mai-WEEN), village, VALDIVIA province, LOS LAGOS region, S central CHILE, on PACIFIC coast, 26 mi/42 km N of VALDIVIA; 39°26′S 73°10′W. Beach area with several hotels.

Mehun-sur-Yèvre (mai-un–syur–YE-vruh), town (□ 9 sq mi/23.4 sq km), resort in CHER department, CENTRE administrative region, central FRANCE, on YÈVRE RIVER and BERRY CANAL, 9 mi/14.5 km NW of BOURGES; 47°09′N 02°13′E. Noted porcelain factories; manufacturing of plumbing equipment and optical glass. In its 14th-century castle Charles VII was crowned (1422); he also received (1429–1430) Joan of Arc and starved himself to death (1461) here.

Meia Meia (mai-yah MAI-yah), village, DODOMA region, central TANZANIA, 28 mi/45 km N of DODOMA; Lake Hombolo to SE; 05°48′S 35°48′E. Cattle, goats, sheep; corn, wheat.

Meichuan (MAI-CHWAHN), town, SE HUBEI province, CHINA, 17 mi/27 km SE of QICHUN; 30°09′N 115°35′E. Rice, cotton, peanuts.

Meidan Ekbes (MAI-dahn EK-bes), township, ALEPPO district, NW SYRIA, on Turkish border, on railroad, 45 mi/72 km NNW of ALEPPO; 36°49′N 36°40′E. Cereals, cotton.

Meiderich (MEI-de-rikh), industrial district of DUISBURG, W GERMANY, N of RUHR RIVER, 2 mi/3.2 km N of city center, on RHINE-HERNE CANAL (W), adjoining RUHRORT (E); 51°26′N 06°45′E. Steel milling. Has been district of DUISBURG since 1905.

Meifa (muh-EE-fuh), village, S YEMEN, on Gulf of ADEN, at mouth of the Wadi HAJR, 40 mi/64 km SW of MUKALLA. Center of agricultural area (grain, dates, citrus fruit). Also spelled Maifa and Mayfah.

Meifa'a, Wadi (mei-FAH-uh, WAH-dee), intermittent sporadic stream, c.100 mi/161 km long, S YEMEN; rises near YESHBUM; flows past HABBAN and AZZAN, to Gulf of ADEN 90 mi/145 km WSW of MUKALLA. Used for irrigation. Wadi Habban is one of its tributaries. Also spelled Wadi Maifa'a and Wadi Mayfa'ah.

Meifod (MEI-vuhd), village (2001 population 1,323), POWYS, E Wales, on VYRNWY RIVER, 6 mi/9.7 km NE of WELSHPOOL; 52°42′N 03°15′W. Church (1155) with grave slabs from 9th–10th century.

Meiganga (mai-GANG-uh), town (2001 population 71,000), ⊙ MBERE department, ADAMAOUA province, E CAMEROON, near Central Afr. Republic border, 80 mi/129 km SE of N'Gaoundéré; 06°31′N 14°18′E. Livestock raising, butter, and cheese manufacturing.

Meighen Island (MEE-uhn) (□ 360 sq mi/936 sq km), Sverdrup group of the QUEEN ELIZABETH ISLANDS, NUNAVUT territory, CANADA, in the ARCTIC OCEAN, separated from ELLEF RINGNES and AMUND RINGNES islands (S) by PEARY CHANNEL and from AXEL HEIBERG ISLAND (E) by SVERDRUP CHANNEL; 80°00′N 99°00′W. Island is 30 mi/48 km long, 8 mi/12.9 km–15 mi/24 km wide; central plateau rises to over 1,000 ft/305 m. Named 1921 by Stefansson after Arthur Meighen, Canadian prime minister.

Meigs (MEGZ), county (□ 434 sq mi/1,128.4 sq km; 2006 population 23,092), SE OHIO; ⊙ POMEROY, 39°06′N 82°01′W. Bounded SE by OHIO RIVER, here forming WEST VIRGINIA line; drained by small Shade River and Leading Creek. In the Unglaciated Plain physiographic region. Agriculture (poultry; grain); manufacturing (electronic equipment); coal mines, limestone quarries. Formed 1819.

Meigs, county (□ 213 sq mi/553.8 sq km; 2006 population 11,698), SE TENNESSEE; ⊙ DECATUR; 35°31′N 84°49′W. In GREAT APPALACHIAN VALLEY; bounded NW by the TENNESSEE RIVER; drained by HIWASSEE River. Includes parts of CHICKAMAUGA and WATTS BAR reservoirs. Agriculture; manufacturing. Formed 1836.

Meigs (MEGZ), town (2000 population 1,090), THOMAS and MITCHELL counties, S GEORGIA, 17 mi/27

km NNW of THOMASVILLE; 31°04′N 84°05′W. Manufacturing includes textiles, apparel, consumer goods.

Meigs, Fort, Ohio: see FORT MEIGS.

Meihekou (MAI-HUH-KO), city (□ 839 sq mi/2,173 sq km; 1994 estimated urban population 252,400; 1994 estimated total population 605,400), S central JILIN province, CHINA, located between the Songhua Plain and the Changbei Mountains; 42°26′N 125°42′E. A trading point formed at the junction of two railroads and four highways. Agriculture is the largest sector in the city's economy. Light and heavy industries (food processing; beverages, paper). Also called Hailong.

Meiho (MAI-HO), village, Gujo county, GIFU prefecture, central HONSHU, central JAPAN, 37 mi/60 km N of GIFU; 35°51′N 137°02′E.

Meihuashan (MAI-HWAH-SHAN), mountain (1,968 ft/600 m), W FUJIAN province, CHINA; part of the WUYI mountains. A world-class natural conservation region, the mountain has a variety of animal and plant species, including the endangered S China tiger.

Meijel (MEI-uhl), village, LIMBURG province, SE NETHERLANDS, 13 mi/21 km W of VENLO; 51°21′N 05°53′E. Dairying; livestock; vegetables, grain, sugar beets. Also spelled Meiel.

Meije, La (MEZH, lah), mountain of the Massif du Pelvoux, DAUPHINÉ ALPS, on border of ISÈRE and HAUTES-ALPES departments, SE FRANCE, 3 mi/4.8 km S of La Grave, overlooking deeply entrenched OISANS valley; 45°00′N 06°18′E. Of its 3 serrate peaks, the Grand Pic de la Meije (13,080 ft/3,983 m) is highest. It was first climbed in 1877; today it remains a challenge for alpinists. An aerial tramway reaches the glacier at Col des Ruillans (10,550 ft/3,216 m) from La Grave; splendid alpine views across peals of the ÉCRINS NATIONAL PARK.

Meikle Bin, Scotland: see LENNOX HILLS.

Meikle Says Law, Scotland: see LAMMERMUIR HILLS.

Meiktila (MAIK-ti-lah), township (□ 2,232 sq mi/5,803.2 sq km), MANDALAY division, MYANMAR; ⊙ MEIKTILA. Astride SAMON RIVER. Consists largely of NE plateau (elevation c.800 ft/245 m) of PEGU YOMA; in dry zone (annual rainfall 35 in/89 cm), mostly agriculture (rice, sesame, cotton, peas); catechu and teak forests along Shan Hills (E); small coal seams. Served by Yangon-Mandalay railroad and Thazi-Myingyan railroad. Population is nearly all Burmese. Formerly a district.

Meiktila, town, ⊙ MEIKTILA township, MANDALAY division, MYANMAR, on Thazi-Myingyan railroad, 80 mi/129 km SSW of MANDALAY. On NE plateau of the PEGU YOMA and on small Meiktila Lake (□ 4 sq mi/10.4 sq km), ancient Burmese irrigation reservoir. Major road and railroad hub of central MYANMAR; cotton-trading center; ginning industry. Army cantonment; airfield. In World War II, badly damaged (1945).

Meilen (MEI-luhn), district, S ZÜRICH canton, N SWITZERLAND; 47°17′N 08°38′E. Main town is KÜSNACHT; population is German-speaking and Protestant.

Meilen (MY-luhn), town (2000 population 11,480), ZÜRICH canton, N SWITZERLAND, on NE shore of LAKE ZÜRICH, 8 mi/12.9 km SSE of ZÜRICH; 47°17′N 08°38′E. Elevation 1,378 ft/420 m. Coffee processing; printing.

Meiling Mountains, CHINA: see DAYU MOUNTAINS.

Meiling Pass (MAI-LING), pass (c.1,300 ft/396 m), in DAYU MOUNTAINS, on GUANGDONG-JIANGXI border, S CHINA. On road between DAYU (N) and NANXIONG (S), it lies on one of chief Guangdong-Jiangxi routes.

Meinberg, Bad, GERMANY: see HORN–BAD MEINBERG.

Meine (MEI-ne), village, LOWER SAXONY, central GERMANY, near MIDLAND CANAL, 8 mi/12.9 km N of BRUNSWICK; 52°23′N 10°33′E. Agriculture (sugar beets, vegetables); sugar refinery.

Meiners Oaks, unincorporated town (2000 population 3,750), VENTURA county, S CALIFORNIA, 12 mi/19 km

N of VENTURA, near Ventura River, in Ojai Valley; 34°27′N 119°16′W. Strawberries, citrus, vegetables; flowers, nursery stock. Los Padres National Forest to N; Lake Casitas Reservoir to SW; Pine Mountain (7,510 ft/2,289 m) to NE.

Meinerzhagen (mei-nerts-HAH-gen), town, WESTPHALIA-LIPPE, North Rhine-Westphalia, W GERMANY, 7 mi/11.3 km S of LÜDENSCHEID; 51°07′N 07°38′E. Forestry.

Meinhard (MEIN-HAHRT), village, HESSE, central GERMANY, near the WERRA, 2 mi/3.2 km NNE of ESCHWEGE; 51°12′N 10°04′E. Textile manufacturing; fruit growing.

Meiningen (MEI-ning-uhn), city, THURINGIA, E central GERMANY, on the WERRA RIVER; 50°34′N 10°25′E. Manufacturing includes textiles, furniture, electronic equipment, and metal products; railroad repair. Meiningen was first mentioned in 982 and passed to the dukes of SAXONY in 1583. Was the capital of the duchy of Saxe-Meiningen from 1680 to 1918. Historically renowned for its theater and opera. The ducal palace in Meiningen dates from the 16th and 17th centuries.

Meire Grove, village (2000 population 149), STEARNS county, central MINNESOTA, 35 mi/56 km WNW of ST. CLOUD; 45°37′N 94°52′W. Grain; livestock; dairying.

Meiringen (MEI-ring-uhn), commune, BERN canton, S central SWITZERLAND, on AARE RIVER, 8 mi/12.9 km ESE of BRIENZ; 46°44′N 08°11′E. Elevation 1,952 ft/595 m. Year-round resort; chief village of the HASLITAL valley. Gorge of the Aare and REICHENBACH FALLS are nearby.

Meirings Poort, rocky defile (2,400 ft/732 m), WESTERN CAPE province, SOUTH AFRICA; crosses Great SWARTBERG range 25 mi/40 km NE of OUDTSHOORN; 33°25′S 22°32′E. Noted for scenic beauty and rock formations; c.5 mi/8 km long. Produced by superimposed drainage of Groot River (tributary of OLIFANTS RIVER), just N of DeRust, linking town to BEAUFORT WEST by road.

Mei River (MAI), 125 mi/201 km long, E GUANGDONG province, CHINA; rises in Jiuling Mountains on Guangdong-JIANGXI border; flows S and NE, past PINGYUAN, XINGNING, and MEIZHOU, to HAN RIVER SW of DABU. Navigable below Meizhou.

Meiron or **Meron** (both: me-RON), moshav, UPPER GALILEE, N ISRAEL, 3 mi/4.8 km W of ZEFAT, at foot of MOUNT MEIRON; 32°59′N 35°26′E. Elevation 2,099 ft/639 m. Modern settlement founded 1949. Mixed farming and light industry. Remains of ancient synagogue, reputedly dating from time of destruction of Second Temple, is nearby. Here is grave of the cabalist Simon ben Yohai, his son Elazar, and other rabbis; place of pilgrimage.

Meiron, Har or **Meron, Har** (both: me-RON, HAHR), highest peak (3,963 ft/1,208 m) of UPPER GALILEE, N ISRAEL, 5 mi/8 km WNW of ZEFAT; 33°00′N 35°24′E. Elevation 3,963 ft/1,208 m. Formerly known as Jebel Jarmaq or Jermaq.

Meir Shfeya (MER shi-fei-YAH) or **Meir Shefeya**, children's institution, NW ISRAEL, SW part of the MENASHE PLATEAU, 2 mi/3.2 km NE of Zikhron Ya'aqov; 32°35′N 34°57′E. Elevation 570 ft/173 m. Has children's village, agricultural school, and training farm. Founded 1892.

Meiruba (mai-ROOB-uh), village, central LEBANON, 18 mi/29 km ENE of BEIRUT. Elevation 3,900 ft/1,189 m. Summer resort; apples, tobacco, lemons. Also spelled Mayrubah.

Meisari, YEMEN: see DATHINA.

Meise (MEI-suh), commune (2006 population 18,600), Halle-Vilvoorde district, BRABANT province, central BELGIUM, 6 mi/10 km NNW of BRUSSELS; 50°56′N 04°20′E.

Meisenthal (mei-zuhn-tahl), commune (□ 2 sq mi/5.2 sq km), MOSELLE department, LORRAINE region, NE FRANCE, in the N VOSGES, 16 mi/26 km SE of SARRE-GUEMINES; 48°58′N 07°21′E. Former glassworks, now an exhibition center of glass and crystal making.

Meiser (MAI-ser) or **Khirbet a-Sheikh Meiser**, Arab village, ISRAEL, 8 mi/12.9 km ENE of HADERA; 32°26′N 35°02′E. Elevation 337 ft/102 m. Founded in late 19th century.

Meishan (MAI-SHAN), town, ⊙ Meishan county, central SICHUAN province, CHINA, 35 mi/56 km N of LESHAN, and on right bank of MIN River; 30°03′N 103°51′E. Rice, tobacco, oilseeds, sugarcane; textiles, transport equipment; food processing.

Meissen (MEIS-sen), city, SAXONY, E central GERMANY, on the ELBE RIVER; 51°10′N 13°47′E. A porcelain manufacturing center since 1710, Meissen is famous for its delicate figurines (often called "Dresden" china); the industry is supported by local deposits of kaolin and potter's earth. Famous chinaware factory is located 12 mi/19 km NW of DRESDEN on left bank of ELBE. Other manufacturing includes metal products, ceramics, and leather goods. Meissen was founded (929) by Henry of Saxony (later the German king Henry I), and it became in the 10th century the seat of the margraviate of Meissen, where the Wettin dynasty of Saxony originated. The diocese of Meissen was founded in 968, was suppressed in 1581, and was restored in 1921 with its see at BAUTZEN. The Albrechtsburg (15th century), a large castle, dominates the city; it housed (1710–1864) the royal porcelain factory. Among the other noteworthy buildings of Meissen are the cathedral and the Church of St. Afra (both 13th–15th century). Chartered in 13th century. Passed to Saxony in 1423. Sacked in Thirty Years War.

Meissner (MEIS-ner), village, N HESSE, central GERMANY, in Hoher Meissner mountains, 6.5 mi/10.5 km W of ESCHWEGE; 51°12′N 09°54′E. Forestry; tourism.

Meitan (MAI-TAN), town, ⊙ Meitan county, N GUIZHOU province, CHINA, 34 mi/55 km E of ZUNYI; 27°48′N 107°28′E. Grain, oilseeds; tobacco industry, beverages.

Meithalun, Arab village, WEST BANK, 8.7 mi/14 km N of NABLUS, on the W fringes of the Sanur Valley (Marj Sanur) in the SAMARIAN Highlands. Agriculture (olives, cereals, fruit, vegetables).

Meitingen (MEI-ting-uhn), town, SWABIA, W BAVARIA, GERMANY, on the LECH, 12 mi/19 km N of AUGSBURG; 48°33′N 10°50′E. Manufacturing of electrotechnical machinery and equipment.

Meiwa (MAI-WAH), town, Taki county, MIE prefecture, S HONSHU, central JAPAN, 16 mi/25 km S of TSU; 34°32′N 136°37′E. Ruins of Saigu.

Meiwa (MAI-WAH), village, Oura county, GUMMA prefecture, central HONSHU, N central JAPAN, 34 mi/55 km S of MAEBASHI; 36°12′N 139°32′E.

Meix-devant-Virton (MEKS–duh-vaw–vir-TAWN), commune (2006 population 2,678), Virton district, LUXEMBOURG province, SE BELGIUM, 5 mi/8 km NNW of VIRTON, in the ARDENNES; 49°36′N 05°29′E.

Mei Xian (MAI SI-AN), town, ⊙ Mei Xian county, SW SHAANXI province, CHINA, on WEI RIVER, 70 mi/113 km W of XI'AN, near LONGHAI RAILROAD; 34°17′N 107°45′E. Grain, building materials, chemicals, rubber products, beverages; graphite ore mining.

Mei Xian, CHINA: see MEIZHOU.

Meizhou (MAI-JO), city (□ 115 sq mi/298 sq km; 1994 estimated urban population 171,700; 1994 estimated total population 270,200), E GUANGDONG province, CHINA, on the MEI RIVER, 70 mi/113 km NNW of SHANTOU; 23°19′N 116°13′E. Light industry is the largest sector of the city's economy. Crop growing, animal husbandry, commercial agriculture, forestry, and fishing. Grain, oil crops, vegetables, fruits; hogs, poultry, eggs; manufacturing (tobacco, chemicals). Also called Mei Xian.

Meja, town, ALLAHABAD district, SE UTTAR PRADESH state, N central INDIA, 27 mi/43 km SE of ALLAHABAD; 25°25′N 74°33′E. Gram, rice, wheat, barley.

Mejicana, Cumbre de la (me-hee-KAH-nah, KOOM-bre dai lah), Andean mountain (20,500 ft/6,248 m) in SIERRA DE FAMATINA, N central LA RIOJA province, ARGENTINA, 23 mi/37 km NW of CHILECITO; 29°03′S 67°51′W. Copper mines.

Mejicanos (me-hee-KAH-nos), residential town and municipality, SAN SALVADOR department, S central EL SALVADOR, 2 mi/3.2 km N of SAN SALVADOR; part of San Salvador metropolitan area; 13°43′N 89°12′W.

Mejillones (mai-hee-YO-nais), town, ⊙ Mejillones comuna, ANTOFAGASTA region, N CHILE, PACIFIC port on a well-sheltered bay, 38 mi/61 km N of ANTOFAGASTA, at terminus of railroad line; 23°06′S 70°27′W. Until 1948 was major terminal for shipment of nitrates and Bolivian tin and other metals. Formerly manufactured railroad equipment, now fishing is major source of livelihood. Here in 1879, during the War of the Pacific, the capture of the Peruvian ironclad *Huáscar* gave Chile control of the sea; Mejillones was ceded 1882 to Chile.

Mejillones de Machaca (me-hee-YO-nais dai mah-CHAH-kah), canton, INGAVI province, LA PAZ department, W BOLIVIA, 28 mi/45 km SW of LA PAZ; 16°39′S 68°18′W. Elevation 12,641 ft/3,853 m. Gas wells in area; clay, limestone, gypsum deposits. Agriculture (potatoes, yucca, bananas, rye); cattle.

Mejit (ME-jeet), coral island, Ratak Chain, Kwajalein district, Marshall Islands, W central Pacific, 240 mi/386 km ENE of Kwajalein; 10°17′N 170°54′E. Island is c.5 mi/8 km long. Formerly called Miadi.

Mejorada (mai-ho-RAH-dhah), town, TOLEDO province, central SPAIN, 4 mi/6.4 km NW of TALAVERA DE LA REINA; 40°01′N 04°53′W. Cereals, grapes, olives; livestock.

Mejorada del Campo (mai-ho-RAH-dhah dhel KAHM-po), town, MADRID province, central SPAIN, near JARAMA RIVER, 10 mi/16 km E of MADRID; 40°24′N 03°29′W. Cereals, olives, grapes, vegetables; livestock.

Mékambo (MAI-kahm-bo), town, OGOOUÉ-IVINDO province, NE GABON, 80 mi/129 km NE of MAKOKOU; 00°59′N 13°55′E. Native rubber market. Iron deposits nearby.

Mekane Selam, town (2007 population 8,883), AMHARA state, central ETHIOPIA, 75 mi/121 km SW of DESSIE; 10°38′N 38°42′E. Has airfield.

Mekane Yesus (me-KAHN-nai YAI-soos), town (2007 population 16,773), AMHARA state, N ETHIOPIA, 15 mi/24 km SE of DEBRE TABOR, near MOUNT GUNA; 11°38′N 38°04′E. Has a Lutheran missionary center.

Mek'elē (MAH-kah-lai), town (2007 population 177,090), ⊙ TIGRAY state, N ETHIOPIA; 13°30′N 39°26′E. 60 mi/97 km SE of ADWA. Elevation is c.6,700 ft/2,042 m. Trade center (salt, cereals, honey, beeswax, cotton goods); has a busy market, a history museum, and an airport. Also spelled Makale.

Mekene (mee-KEE-ne), ancient city, in what is now central MESSENIA prefecture, SW PELOPONNESE department, extreme SW GREECE. Founded (c.369 B.C.E.) under Theban auspices to be capital and fort for the Messenians, whom the battle of Leuctra had just freed from the Spartans. The ruins, notably of the city walls dating from the 4th century B.C.E., are well preserved. Modern city of MESSENIA (Messíni) is in the area. Also spelled Messene.

Mékerra, Oued (mai-ke-RAH, WED), stream, c.150 mi/241 km long, in NW ALGERIA; rises in High Plateaux S of RAS EL MA (SIDI BEL ABBÈS wilaya); flows NNE, past SIDI BEL ABBÈS and SIG (below which it is called the Sig), to the coastal Sig lowland where its waters (dammed at CHEURFAS DAM 12 mi/19 km above Sig) are used for irrigation.

Mekhadir, YEMEN: see MAKHADAR.

Mékhé (MAI-kai), town (2004 population 15,636), THIÈS administrative region, W SENEGAL, 62 mi/100 km NE of DAKAR; 15°07′N 16°38′W. Peanut, millet production.

Mekhel'ta (mee-khyel-TAH), village (2005 population 2,970), W central DAGESTAN REPUBLIC, NE CAUCASUS, SE European Russia, on the S slope of the ANDI RANGE, on road, 30 mi/48 km W of BUYNAKSK; 42°48′N 46°29′E. Elevation 4,396 ft/1,339 m. Has a mosque.

Mekhliganj (MAI-kli-guhnj), town, KOCH BIHAR district, NE WEST BENGAL state, E INDIA, on the TISTA RIVER, 33 mi/53 km W of KOCH BIHAR; 26°21′N 88°55′E. Trades in rice, jute, tobacco, oilseeds, sugarcane. Also spelled Mekliganj.

Mekhnatabad, town, LENINOBOD viloyat, TAJIKISTAN, 37 mi/60 km W of KHUDJAND; 40°10′N 69°01′E. On S edge of Mirza Chol, Russian *Golodnaya Steppe*, a region of "Virgin Lands" cotton-sowing expansion initiated in the Khrushchev period. Cotton.

Mekhonskoye (mye-HON-skuh-ye), village, N KURGAN oblast, W SIBERIA, RUSSIA, on the ISET′ RIVER, just below the mouth of the MIASS River, on road, 35 mi/56 km ENE of SHADRINSK; 56°09′N 64°34′E. Elevation 219 ft/66 m. In agricultural area (wheat, oats, rye, barley; livestock).

Mekhtar (mahk-HAH-tuhr), village, LORALAI district, NE BALUCHISTAN, SW PAKISTAN, 45 mi/72 km ENE of LORALAI; 30°28′N 69°22′E.

Mekī, town (2007 population 38,342), OROMIYA state, S central ETHIOPIA; 08°09′N 38°49′E. Local center on roadside N of LAKE ZIWAY.

Mékinac (mai-kee-NAHK), county (□ 2,165 sq mi/5,629 sq km; 2006 population 12,884), MAURICIE region, S QUEBEC, E CANADA; ☉ SAINT-TITE; 46°49′N 72°31′W. Composed of fourteen municipalities. Formed in 1982.

Mekkaw, NIGERIA: see MEKO.

Mekliganj, India: see MEKHLIGANJ.

Meklong, THAILAND: see SAMUT SONGKHRAM, town.

Meklong River, THAILAND: see MAE KLONG RIVER.

Meknes (mek-NES), city, ☉ MEKNÈS-TAFILALET administrative region, N central MOROCCO; 33°54′N 05°33′W. One of Morocco's "imperial cities," it has a noted carpet-weaving industry. There are also woolen mills, cement and metal works, oil distilleries, vineyards, and food-processing plants. Meknes was (c.1672) capital of Morocco under Sultan Moulay Ismael, who undertook such palatial building operations that the city was called the VERSAILLES of Morocco. Much of his construction was destroyed in the LISBON earthquake of 1755. A European town is laid out beside the old one. It has a public university.

Meknès-Tafilalet, administrative region (2004 population 2,141,527), central MOROCCO; ☉ MEKNES. Bordered by Souss-Massa-Draâ (SWW), Tadla-Azilal (W), Chaouia-Ouardigha (W), Rabat-Salé-Zemmour-Zaër (NWN), Gharb-Chrarda-Beni Hssen (N), Fes-Boulemane (NEE), and Oriental (SE) administrative regions and ALGERIA (SES). Part of High Atlas (S central) and Middle Atlas (N central) mountains (both of the ATLAS MOUNTAINS) in region. Railroad runs NW out of Meknes, splitting N to TANGER and SW through KENITRA, RABAT, and CASABLANCA before terminating at MARRAKECH; railroad also runs E out of Meknes through FES and TAZA before reaching OUJDA. Main roads run through the region, especially in N near Meknes and S near ERRACHIDIA, connecting it to the rest of the country. Airport near Errachidia. Further divided into five secondary administrative divisions called prefectures and provinces: El Hajeb, Errachidia, Ifrane, Khénifra, and Meknès.

Meko (ME-ko), town, OGUN state, extreme SW NIGERIA, on BENIN border (customs deport), 40 mi/64 km NW of ABEOKUTA; 07°27′N 02°51′E. Cotton weaving, indigo dyeing; cacao, cotton. Also spelled Mekkaw.

Mekong (MAI-KUNG), Mandarin *Lancang jiang*, c.2,600 mi/4,184 km long, one of the great rivers of SE ASIA; rises in the TIBETAN PLATEAU as the Za Qu and flows generally S through YUNNAN province (CHINA) in deep gorges and over rapids. Leaving Yunnan, the Mekong forms the MYANMAR-LAOS border, then curves E and S through NW Laos before marking part of the Laos-THAILAND border. From SW Laos the river descends onto the Cambodian plain, where it receives water from TÔNLÉ SAP during the dry season by way of the Tônlé Sap River; during the rainy season, however, the floodwaters of the Mekong reverse the direction of the Tônlé Sap River and flow into Tônlé Sap, a lake that is a natural reservoir. The Mekong River finally flows into the SOUTH CHINA SEA through many distributaries in the vast Mekong delta (c.75,000 sq mi/194,250 sq km), which occupies SE CAMBODIA and S VIETNAM. The delta, crisscrossed by many channels and canals, is one of the greatest ricegrowing areas of Asia. It is a densely populated region; VINH LONG, CAN THO, and LONG XUYEN (all in Vietnam) are the chief towns here. HO CHI MINH CITY (Saigon) is located just E of the delta. The Mekong River is navigable for large vessels c.340 mi/547 km upstream; PHNOM PENH is a major port. N of the Cambodian border, the Mekong is navigable in short sections. At KHONE Falls, a series of rapids (6 mi/9.7 km long) in S Laos, the Mekong drops 72 ft/22 m. The falls are the site of a hydroelectric power station, part of the Mekong Scheme, a project undertaken by the UN in the early 1960s to develop the potentials of the lower Mekong basin. The project seeks to improve navigation, provide irrigation facilities, and produce hydroelectricity. The Mekong delta was the scene of heavy fighting in the Vietnam War.

Mekoryok, village, SW ALASKA, on N shore of NUNIVAK Island; 60°24′N 166°11′W.

Mekran, IRAN and PAKISTAN: see MAKRAN.

Mé, La (MAI, lah), village, S CÔTE D'IVOIRE, on small Mé River, near coastal lagoon, and 15 mi/24 km NE of ABIDJAN.

Melada, CROATIA: see MOLAT ISLAND.

Melaka (muh-LAH-kah) or **malacca**, state (□ 640 sq mi/1,664 sq km; 2000 population 635,791), MALAYSIA, S MALAY PENINSULA, on the Strait of MALACCA; ☉ Melaka. Formerly one of the Straits Settlements, it was constituted a state of Malaya in 1957. Nearly half the population are Malay; about two-fifths are Chinese.

Melaka (muh-LAH-kah) or **Malacca**, city (2000 population 149,518), ☉ Melaka state, MALAYSIA, S MALAY PENINSULA, on the Strait of MALACCA. Until the 17th century, Malacca was one of the leading commercial centers of East ASIA. It was founded c.1400 by a Malay prince who had been driven from SINGAPORE after a brief reign there. The city quickly gained wealth as a center of trade with CHINA, INDONESIA, INDIA, and the MIDDLE EAST. Its sultans, aided by the decline of the Madjapahit empire of Java and by the friendship of China, extended their power over the nearby coast of SUMATRA and over the Malay Peninsula as far N as KEDAH and PATTANI. More importantly, Gujarati traders introduced Islam to the Malay world through Malacca. In 1511, Malacca was captured by the Portuguese under Alfonso de Albuquerque. The sultan fled to PAHANG and then to JOHOR. In the mid-16th century St. Francis Xavier preached in Malacca. Portugal's control of the area was frequently contested by ACEH and Johor. In the early 17th century the Dutch entered the region, allied themselves with Johor, and captured Malacca in 1641 after a long siege. They utilized the city more as a fortress guarding the strait than as a trading port. The Dutch retained nominal control until 1824, although during the wars of the French Revolution and the Napoleonic period (1795–1818) the British occupied Malacca at the request of the Dutch government-in-exile. In 1824 the Dutch formally transferred Malacca to GREAT BRITAIN. The modern city, of slight economic importance, retains lasting traces of its past in its Port and Dutch buildings and Portuguese-Eurasian community. The majority of the city's inhabitants are Chinese, some of whom have acquired many Malay customs.

Melakou (mai-lah-KOO), village, TIARET wilaya, N ALGERIA, on S slope of the TELL ATLAS, 10 mi/16 km SSW of TIARET. Vineyards. Formerly called Palat.

Melancthon (muh-LANK-thahn), township (□ 121 sq mi/314.6 sq km; 2001 population 2,796), S ONTARIO, E central CANADA, 7 mi/10 km NW of SHELBURNE; 44°08′N 80°17′W. Agriculture (primarily potatoes). Contains the hamlets of RIVERVIEW, CORBETTON, and HORNING'S MILLS.

Melanesia, one of the three commonly recognized ethno-geographic subdivisions (along with POLYNESIA and MICRONESIA) of the Pacific Islands, SW PACIFIC OCEAN. Meaning "black islands," it originally denoted either the dark landscapes or the dark skins of most inhabitants. S of the equator and NE of AUSTRALIA, it includes NEW GUINEA and "Island Melanesia" extending E to FIJI. The term thus includes Irian Jaya (Indonesian New Guinea), PAPUA NEW GUINEA and its associated islands in the BISMARCK ARCHIPELAGO and Northern Solomons, the SOLOMON ISLANDS, VANUATU, NEW CALEDONIA, and Fiji. Within this region people belong to Melanesian political units. "Melanesian" has several connotations. Sometimes it refers to taller coastal peoples vis-à-vis stockier inland groups. Sometimes (and more technically) it means those who speak one of the Austronesian languages which extend from MADAGASCAR to Micronesia and Polynesia, in contrast to those who speak an unrelated "Papuan" tongue. Melanesian-Papuan affinities appear in adjacent Indonesian islands and Polynesian characteristics in some E Melanesian islands, including Fiji.

Melanes Valley (me-lah-NES), valley, in central NAXOS island, CYCLADES prefecture, SOUTH AEGEAN department, GREECE, 5 mi/8 km SE of Khora Naxos. Has several villages; also, orchards and olive trees. Noted for two large abandoned ancient (6th century B.C.E.) statues of youths (*Kouri*) made of marble from nearby quarry.

Melapalaiyam (mai-luh-PAH-luh-yuhm), town, NELLAI KATTABOMMAN district, TAMIL NADU state, S INDIA, 3 mi/4.8 km SE of TIRUNELVELI, across TAMBRAPARNI RIVER; 08°42′N 77°43′E. In cotton- and palmyra-growing area; towel- and carpet-weaving center. Also spelled Melapalayam. Formerly also spelled Mel Palaiyam.

Melba, village (2000 population 439), CANYON county, SW IDAHO, 25 mi/40 km SW of BOISE, near SNAKE RIVER; 43°22′N 116°32′W. Center of irrigated area (cattle, sheep; seed growing); manufacturing (feeds, birdseed; meat processing). Deer Flat National Wildlife Refuge at LAKE LOWELL, to NW.

Melbeta (mel-BAI-tuh), village (2006 population 139), SCOTTS BLUFF county, W NEBRASKA, 10 mi/16 km ESE of SCOTTSBLUFF, and on NORTH PLATTE RIVER; 41°46′N 103°31′W.

Melbourn (MEL-buhrn), village (2001 population 5,228), S CAMBRIDGESHIRE, E ENGLAND, 10 mi/16 km SSW of CAMBRIDGE; 53°08′N 01°32′W. Has Congregational chapel and 14th-century church. Site of Roman camp.

Melbourne (MEL-buhrn), city (2006 population 3,592,591); ☉ VICTORIA, SE AUSTRALIA, on PORT PHILLIP BAY, at the mouth of the YARRA RIVER; 37°50′S 145°00′E. Melbourne, Australia's second-largest city, is a railroad and air hub and financial and commercial center. Wool and raw and processed agricultural goods are exported. The city is heavily industrialized; industries include shipbuilding and the manufacturing of motor vehicles, farm machinery, textiles, and electrical goods. Settled in 1835, it was named (1837) for Lord Melbourne, the British prime minister. From 1901 to 1927 the city was the seat of the Australian federal government. Melbourne has major campuses of several universities, including University of Melbourne (1853), Monash University (1958), La Trobe University (1964), Deakin University,

Royal Melbourne Institute of Technology, Swinburne University, and Victoria University. The city also has the Australian Ballet School, the National Gallery, the Victorian Arts Centre, and Federation Square, a central public space comprising a city block and including a "Fractal Façade" building complex of arts venues, cultural and visitors' centers, and restaurants. Melbourne is the seat of Roman Catholic and Anglican archbishops. Melbourne is multicultural, including large Italian, Greek, Turkish, Yugoslav, and Vietnamese populations. The botanical gardens are a notable attraction. The Melbourne Cup Race is run annually at the Flemington Racecourse. Melbourne was the site of the 1956 Summer Olympic games. Included in the Melbourne urban agglomeration are many coastal resorts.

Melbourne (MEL-buhrn), city (□ 36 sq mi/93.6 sq km; 2005 population 76,646), BREVARD county, E central FLORIDA, on INDIAN RIVER (lagoon); 28°05′N 80°36′W. Tourist and aerospace center near the AT-LANTIC OCEAN. The leading industries are fruit processing and shipping, electronic equipment and boat manufacturing. Since the development of nearby CAPE CANAVERAL, the aerospace industry has bolstered Melbourne's economy and population. Florida Institute of Technology is in the city, and Patrick Air Force Base is nearby. Incorporated 1888.

Melbourne (MEL-buhrn), former township, SW ON-TARIO, E central CANADA, 19 mi/31 km from LONDON, and included in STRATHROY-CARADOC township; 42°49′N 81°33′W.

Melbourne (MEL-buhrn), town (2001 population 4,599), SE DERBYSHIRE, central ENGLAND, 7 mi/11.3 km SSE of DERBY; 52°49′N 01°25′W. Market gardening, silk milling. Has Norman church and Melbourne Hall, rebuilt in early 18th century. Thomas Cook, pioneer of organized travel, born here.

Melbourne, town (2000 population 1,673), ⊙ IZARD county, N ARKANSAS, 23 mi/37 km NW of BATES-VILLE; 36°03′N 91°54′W. Stock raising, agriculture; manufacturing (aircraft parts, hardwood flooring, apparel). Ozark National Forest to W.

Melbourne, town (2000 population 794), MARSHALL county, central IOWA, 12 mi/19 km SW of MAR-SHALLTOWN; 41°56′N 93°05′W.

Melbourne (MEL-buhrn), village (□ 66 sq mi/171.6 sq km), ESTRIE region, S QUEBEC, E CANADA, on SAINT-FRANÇOIS River, opposite RICHMOND; 45°35′N 72°10′W. Dairying.

Melbourne (MEL-buhrn), village (2000 population 457), CAMPBELL county, N KENTUCKY, 8 mi/12.9 km SE of CINCINNATI, OHIO, on the OHIO RIVER; 39°01′N 84°22′W. Tobacco, alfalfa, soybeans, corn; cattle. Manufacturing (building materials).

Melbourne Beach (MEL-buhrn), town (□ 1 sq mi/2.6 sq km; 2005 population 3,314), BREVARD county, E central FLORIDA, 3 mi/4.8 km E of MELBOURNE; 28°04′N 80°33′W. Light manufacturing.

Melbourne Docklands (MEL-buhrn DAHK-luhndz) or **Docklands**, district, MELBOURNE, VICTORIA, SE AUSTRALIA, to W of Melbourne's central business district, along YARRA RIVER. Officially administered urban redevelopment project and waterfront residential, commercial, retail, and leisure area. Completion expected 2015.

Melbourne Island, NUNAVUT territory, CANADA, in QUEEN MAUD GULF, just E of base of KENT PENINSULA, opposite SE VICTORIA ISLAND; 68°30′N 104°15′W. Island is 18 mi/29 km long, 10 mi/16 km wide. On N coast is Eskimo winter camp.

Melbourne Island, FRENCH POLYNESIA: see ACTAEON ISLANDS.

Melbu (MEL-boo), village, NORDLAND county, N NOR-WAY, on S shore of HADSELØY in the VESTERÅLEN group, 18 mi/29 km NNE of SVOLVÆR; 68°30′N 14°49′E. Fishing industries, fish oil; woolen milling; wood working; manufacturing (oilcloth; oleo margarine).

Melcher, town, MARION county, S central IOWA, near WHITEBREAST CREEK, 9 mi/14.5 km SW of KNOXVILLE; 33 mi/53 km SE of DES MOINES; 41°13′N 93°14′W. Merged in mid-1980s with DALLAS (1 mi/1.6 km N). Manufacturing of wood products; coal mining.

Melchor de Mencos (mel-CHOR dai MEN-kos), town, PETÉN department, GUATEMALA, on the BELIZE-Guatemala Highway at the border crossing; 17°04′N 89°10′W. Border post for crossing into and out of Belize on FLORES-Belize Highway.

Melchor Island, CHILE: see CHONOS ARCHIPELAGO.

Melchor Ocampo (mel-CHOR o-KAHM-po) or **Ocampo**, town, MÉXICO state, central MEXICO, 20 mi/32 km N of MEXICO CITY, and in the ZONA ME-TROPOLITANA DE LA CIUDAD DE MÉXICO; 18°00′N 102°13′W. Cereals; livestock.

Melchor Ocampo, town, NUEVO LEÓN, N MEXICO, 55 mi/89 km ENE of MONTERREY; 26°03′N 99°33′W. Corn, cactus fibers. Formerly called CHARCO REDONDO.

Melchor Ocampo, town, ⊙ Melchor Campo munici-pio, ZACATECAS, N central MEXICO, on COAHUILA border, 55 mi/89 km SW of SALTILLO. Elevation 7,415 ft/2,260 m. Railroad terminus; mining center (copper, lead, gold, silver). Formerly called San Pedro Ocampo.

Melcombe Regis, ENGLAND: see WEYMOUTH.

Meldal (MEL-dahl), village, SØR-TRØNDELAG county, central NORWAY, on ORKLA River and 34 mi/55 km SW of TRONDHEIM; 59°11′N 09°29′E. Agriculture. LØKKEN village, 5 mi/8 km N, a pyrite-mining center since 1652, is terminus of electric railroad to TROND-HEIMSFJORDEN; has copper smelter.

Meldorf (MEL-dorf), town, in SCHLESWIG-HOLSTEIN, NW GERMANY, on small Miele River near its mouth (harbor) on the NORTH SEA, 7 mi/11.3 km S of HEIDE, in the S DITHMARSCHEN; 54°05′N 10°50′E. Food pro-cessing (vegetables, flour); woodworking; tourism. Market center (cattle, grain). Has 13th-century church; also DITHMARSCHEN museum. Was main town of Dithmarschen (from 13th century to 1447). Home of poet Boie. Residence (1778–1815) of explorer Carsten Niebuhr; his son Barthold Georg spent his youth here.

Meleai, Greece: see MILEAI.

Meleb, hamlet, S MANITOBA, W central CANADA, in ARMSTRONG rural municipality; 50°43′N 97°13′W.

Mele, Cape (MAI-le), on LIGURIAN coast, NW ITALY, 8 mi/13 km NE of IMPERIA; 44°03′N 08°10′E. Light-house.

Meleda, CROATIA: see MLJET ISLAND.

Melegnano (me-len-YAH-no), town, MILANO prov-ince, LOMBARDY, N ITALY, on LAMBRO River, 10 mi/16 km SE of MILAN; 45°21′N 09°19′E. Agricultural center; cheese, sausage; manufacturing (machinery, metal products, silk, linen, rope, furniture). Scene in 1515 of victory of Francis I and Venetians over Swiss under Cardinal Schinner, and in 1859 of battle between French and Austrians. Formerly called Marignano.

Melekess, RUSSIA: see DIMITROVGRAD.

Melekhovo (MYE-lee-huh-vuh), town (2006 popula-tion 6,870), N VLADIMIR oblast, central European Russia, on a short right tributary of the KLYAZ'MA RIVER, on highway and railroad spur, 5 mi/8 km S, and under administrative jurisdiction, of KOVROV; 56°17′N 41°17′E. Elevation 357 ft/108 m. Quartz quar-rying, lumbering and woodworking.

Melena del Sur (mai-LAI-nah del soor), town, LA HABANA province, W CUBA, on railroad, 26 mi/42 km S of HAVANA; 22°47′N 82°12′W. Sugar-growing center, with the Gregorio Arlee Mañalich sugar mill 1.5 mi/2.4 km N. Limekiln.

Melenci (me-LEN-tsee), Hungarian *Melence*, village, VOJVODINA, N SERBIA, on railroad, 30 mi/48 km NE of NOVI SAD, in the BANAT region; 45°30′N 20°18′E. In-cludes BANJA RUSANDA, health resort, on small Ru-sanda Lake. Also spelled Melentsi.

Melendiz Dag (Turkish=*Melendiz Dağ*) peak (9,630 ft/2,935 m), central TURKEY, 10 mi/16 km NW of NIGDE.

Melenki (MYE-leen-kee), city (2006 population 15,760), SE VLADIMIR oblast, central European Russia, on the UNZHA RIVER (tributary of the OKA River), 90 mi/145 km SE of VLADIMIR; 55°20′N 41°37′E. Elevation 456 ft/138 m. Railroad and highway junction. Flax milling; food processing; sawmilling. Chartered in 1778.

Melentyevskoye (mee-LYEN-tyeef-skuh-ye), former town, W central CHELYABINSK oblast, on the NE slope of the S URALS, W Siberian Russia, on the MIASS River, on railroad, now a suburb of MIASS, less than 6 mi/10 km N of the city center; 55°04′N 60°06′E. Elevation 1,177 ft/358 m. Gold mining.

Melesse (me-les), town (□ 12 sq mi/31.2 sq km), ILLE-ET-VILAINE department, in BRITTANY, NW FRANCE, 8 mi/13 km N of RENNES; 48°13′N 01°42′W. Agriculture market.

Melet' (MYE-leet), village, SE KIROV oblast, E central European Russia, on the E bank of the VYATKA River, near highway and on short railroad spur, 10 mi/16 km NE, and under administrative jurisdiction, of MAL-MYZH; 56°39′N 50°49′E. Elevation 278 ft/84 m. In ag-ricultural area (grain, potatoes, flax, vegetables); logging, lumbering.

Melet, TURKEY: see MESUDIYE.

Meleuz (mye-lee-OOS), city (2005 population 66,200), SW BASHKORTOSTAN Republic, SE European Russia, on the Meleuz River at its confluence with the BELAYA RIVER, on railroad, 125 mi/201 km S of UFA, and 45 mi/72 km S of STERLITAMAK; 52°58′N 55°55′E. Elevation 580 ft/176 m. Highway junction; local transshipment center. Chemical industry (mineral fertilizers); food processing (milk, meat, sugar; brewery), wood in-dustries (building materials). Made city in 1958.

Melfa (MEL-fuh), town, ACCOMACK county, E VIRGI-NIA, 6 mi/9.7 km SSW of ACCOMAC, in EASTERN SHORE area, between ATLANTIC OCEAN (E) and CHE-SAPEAKE BAY (W); 37°38′N 75°44′W. Manufacturing (waste disposal systems, bronze sculptures); agricul-ture (vegetables, grain; livestock). County Airport to W; Accomack Vineyards to S. Eastern Shore Com-munity College.

Melfi (mel-FEE), town, GUÉRA administrative region, central CHAD, 150 mi/241 km W of AM-TIMAN; 11°04′N 17°56′E. Agriculture. Airport.

Melfi (MEL-fee), town, in BASILICATA, S ITALY; 41°00′N 15°39′E. It is an agricultural and tourist cen-ter noted for its wine; food processing. In 1041 it was made the first capital of the Norman country of Apulia. At Melfi, Emperor Frederick II promulgated (c.1231) his important code, the Constitutions of Melfi, or *Liber Augustalis*. In 1528 the town was sacked by the French under Lautrec, and it never recovered its position as a flourishing commercial center. Earthquakes have damaged the Norman castle (11th-13th century) and the cathedral (reconstructed 18th century), but the campanile (1153) still stands.

Melfort (MEL-fuhrt), city (2006 population 5,192), central SASKATCHEWAN, CANADA, on Melfort Creek, 55 mi/89 km ESE of PRINCE ALBERT; 52°52′N 104°36′W. Livestock-shipping and oil-distributing center; flour and lumber milling, dairying; cold stor-age plant.

Melfort, town, MASHONALAND EAST province, NE cen-tral ZIMBABWE, 20 mi/32 km SE of HARARE, on railroad; 18°00′S 31°19′E. Elevation 4,957 ft/1,511 m. Agriculture (tobacco, wheat, corn, citrus fruit, macadamia nuts); cattle, sheep, goats, hogs, poultry; dairying.

Melgaço (MEL-gah-SO), city (2007 population 17,824), N central PARÁ, BRAZIL, on Baía de Melgaço, across bay from PORTEL; 01°48′S 50°46′W.

Melgaço (mel-GAH-soo), northernmost town of PORTUGAL, VIANA DO CASTELO district, near MINHO RIVER (Spanish border), 42 mi/68 km NE of VIANA DO CASTELO; 42°07′N 08°16′W. Noted for its hams. Mineral springs. Founded 12th century as fortified frontier post by Alfonso I.

Area is shown by the symbol □, and capital city or county seat by ⊙.

Melgar (mel-GAHR), province (□ 1,709 sq mi/4,443.4 sq km), PUNO region, SE PERU, ☉ AYAVIRI; 14°45′S 70°45′W. W province of Puno region, bordering on CUZCO region.

Melgar (mel-GAHR), town, ☉ Melgar municipio, TOLIMA department, W central COLOMBIA, in MAGDALENA valley, 35 mi/56 km ESE of IBAGUÉ; 04°12′N 74°39′W. Coffee growing; sorghum, plantains, sugarcane.

Melgar de Fernamental (mel-GAHR dhai fer-nahmen-TAHL), town, BURGOS province, N SPAIN, 28 mi/45 km W of BURGOS; 42°24′N 04°15′W. Cereals, vegetables, grapes; sheep, cattle, hogs. Lumbering, flour milling; manufacturing of dairy and meat products, chocolate, tiles.

Melghir, Chott, ALGERIA: see MELRHIR, CHOTT.

Meliana (mai-LYAH-nah), N suburb, of VALENCIA, VALENCIA province, E SPAIN, near the MEDITERRANEAN SEA; 39°32′N 00°20′W. In rich farming area; tile and furniture manufacturing.

Mélida (MAI-lee-dah), town, NAVARRE province, N SPAIN, on ARAGON RIVER, and 21 mi/34 km N of TUDELA. Sugar beets, cereals, alfalfa; sheep.

Meligalas (me-lee-yah-LAHS), town, MESSENIA prefecture, SW PELOPONNESE department, extreme SW mainland GREECE, on railroad, 15 mi/24 km NW of KALAMATA; 37°13′N 21°58′E. Livestock raising (sheep, goats); olive oil, cotton. Pyrolusite deposits nearby.

Melika (mai-lee-KAH), town, GHARDAÏA wilaya, Saharan central ALGERIA, on the left bank of Oued M'zab, 1.2 mi/2 km SE of GHARDAÏA. Built on rock. Once the religious center of the M'ZAB, its main attractions now are the tombs of SIDI AÏSSA built in the local style.

Melilla (me-LEE-yuh), city (2001 population 66,411), Spanish possession on the MEDITERRANEAN SEA coast of MOROCCO, NW AFRICA; 33°23′N 07°08′W. Like CEUTA it is a free port. Principal industry is fishing. Ferries link it with three other Spanish ports, Ceuta, MÁLAGA, and ALMERÍA. SPAIN has held the city since 1496 despite many attacks by Moroccans, and has a military garrison here. Claims to be the point of origin for Franco's rebellion in 1936 which led to Spanish civil war. The rural district of KETAMA (Morocco) makes the same claim.

Melimoyu, Monte (mai-lee-MO-yoo, MON-tai), Andean peak (7,875 ft/2,400 m), AISÉN province, AISÉN DEL GENERAL CARLOS IBAÑEZ DEL CAMPO region, S CHILE, N of MAGDALENA ISLAND, 85 mi/137 km N of PUERTO AISÉN; 44°05′S 72°52′W.

Melincué (me-leen-KOO-e), town (1991 population 2,099), ☉ General López department (□ 4,415 sq mi/ 11,479 sq km), S SANTA FE province, ARGENTINA, near a lake, 70 mi/113 km SW of ROSARIO; 33°39′S 61°27′W. Railroad junction, fishing and agriculture center (corn, soybeans, wheat, flax; livestock). Lake resort nearby. Formerly called San Urbano.

Melineşti (me-lee-NESHT), village, DOLJ county, S ROMANIA, on JIU RIVER, 19 mi/31 km NW of CRAIOVA; 44°34′N 23°43′E. Orchards. Railroad junction.

Melinka (mai-LEEN-kah), town, ☉ Guaitecas comuna, AISÉN province, AISÉN DEL GENERAL CARLOS IBAÑEZ DEL CAMPO region, CHILE, 125 mi/201 km NW of PUERTO AISÉN; 43°53′S 73°44′W. Minor port on GUAITECAS ISLANDS.

Meliorativnoye, UKRAINE: see MELIORATYVNE.

Melioratorov (mye-lee-uh-RAH-tuh-ruhf), settlement (2006 population 5,295), SW TYUMEN oblast, SW Siberian Russia, on road and railroad, 7 mi/11 km NNE, and under administrative jurisdiction, of TYUMEN; 57°12′N 65°36′E. Elevation 180 ft/54 m. Forest restoration services.

Melioratyvne (me-lee-o-rah-TIV-ne) (Russian *Meliorativnoye*), town (2004 population 4,710), N central DNIPROPETROVS'K oblast, UKRAINE, on the left bank of the SAMARA RIVER, on highway and on railroad, 9 mi/15 km E of NOVOMOSKOVS'K; 48°36′N 35°22′E.

Elevation 223 ft/67 m. Mechanical repair depot, reinforced-concrete-fabrication plant. Established in 1969, in conjunction with the construction of the DNIEPER-DONBAS CANAL; town since 1975.

Meli-Park, amusement park, near Adinkerke-DE-PANNE, WEST FLANDERS province, W BELGIUM, near NORTH SEA and FRENCH border.

Melipeuco (mai-lee-pai-OO-ko), town, ☉ Melipeuco comuna, CAUTÍN province, ARAUCANIA region, S central CHILE, at head of mountain valley at edge of the ANDES, 50 mi/80 km ESE of TEMUCO; 38°51′S 71°42′W. Livestock.

Melipilla (mai-lee-PEE-yah), SW province, METROPOLITANA DE SANTIAGO region, central CHILE; ☉ MELIPILLA; 33°45′S 71°10′W. Intensive agriculture, industry, mining.

Melipilla (mai-lee-PEE-yah), town, ☉ Melipilla comuna (2002 population 53,522) and province, METROPOLITANA DE SANTIAGO region, central CHILE, on MAIPO RIVER, on railroad, 35 mi/56 km SW of SANTIAGO; 33°42′S 71°13′W. Fruit, alfalfa, cereals, vegetables; livestock.

Melissa (MEL-i-suh), village (2006 population 3,014), COLLIN county, N TEXAS, 37 mi/60 km NNE of DALLAS, near East Fork of TRINITY RIVER; 33°16′N 96°34′W. Agricultural area just beyond fringe of DALLAS–FORT WORTH urban area. Cotton, wheat, sorghum; cattle. Manufacturing (concrete, steel fabricating).

Melissani Lake (me-lee-sah-NEE), near E coast of KEFALLINÍA island, KEFALLINÍA prefecture, IONIAN ISLANDS department, off W coast of GREECE, 2 mi/3.2 km N of SAMI. The lake is enclosed in a cave; its water comes from an underground fault that extends a great distance from the W side of the island, where water enters at the *katavothres* [Greek=deep holes] outside Argostoli. This phenomenon eluded scientific explanation until the 20th century.

Melita (muh-LI-tuh), town (□ 1 sq mi/2.6 sq km; 2001 population 1,111), SW MANITOBA, W CANADA, on SOURIS River, 60 mi/97 km SW of BRANDON, in ARTHUR rural municipality; 49°16′N 100°59′W. Dairying; mixed farming; livestock raising; oil production; mining.

Melita, CROATIA: see MLJET ISLAND.

Melitene, TURKEY: see MALATYA.

Melito di Napoli (MAI-lee-to dee NAH-po-lee), town, NAPOLI province, CAMPANIA, S ITALY, 5 mi/8 km NNW of NAPLES; 40°55′N 14°14′E. Agricultural center (fruit, grains); sausage; glassworks.

Melito di Porto Salvo (MAI-lee-to dee POR-to SAHL-vo), town, Reggio di Calabria province, CALABRIA, S ITALY, port on IONIAN Sea, 15 mi/24 km SE of REGGIO DI CALABRIA; 37°55′N 15°47′E. Alcohol distilling. At tip of the "toe" of Italy; southernmost town on Italian mainland.

Melitopil', UKRAINE: see MELITOPOL'.

Melitopol' (me-lee-TO-pol) or **(Melitopil')**, city (2001 population 160,657), S ZAPORIZHZHYA oblast, UKRAINE, on the MOLOCHNA RIVER; 46°50′N 35°22′E. A raion and manufacturing center, it produces heavy machinery (motor vehicle engines, refrigerators, tractor parts) and has flour mills and food-processing plants. It houses an institute of agricultural mechanization, a pedagogical museum, and a regional museum. Founded in 1784, officially named Novooleksandrivka (Russian *Novo-Aleksandrovka*) in 1816, but was renamed in 1841. Jewish community since the 19th century, comprised the majority of the city's population in the first half of the 20th century (85,900 in 1939); partially evacuated with the city's industrial complex in 1941, the majority of those remaining liquidated by the Nazis. Fewer than 10,000 Jews in Melitopol' in 2005.

Melk, town, LOWER AUSTRIA province, N central AUSTRIA, on the DANUBE RIVER; 48°14′N 15°20′E. The Danube is bridged here; location of hydropower station. A noted tourist spot, it was one of the earliest

residences of the Austrian rulers. The large Benedictine abbey here, founded in 1089, has a library whose holdings include about 2,000 old manuscripts and 80,000 volumes. The abbey was completely rebuilt by architect J. Prandtauer in the 18th century and is a splendid example of the baroque style. Graves of St. Koloman and the early Babenberg dukes of Austria are here. The area is well known for its wines.

Melka Rafu, town (2007 population 10,292), OROMIYA state, E ETHIOPIA, 10 mi/16 km N of HARAR; 09°26′N 42°07′E.

Melkovo (myel-KO-vuh), village, SE TVER oblast, W central European Russia, on the VOLGA RIVER, on highway, 10 mi/16 km WSW, and under administrative jurisdiction, of KONAKOVO; 56°39′N 36°27′E. Elevation 423 ft/128 m. Fur-animals breeding farm.

Melksham (MELK-shuhm), town (2001 population 14,372), W WILTSHIRE, S ENGLAND, on the AVON, 10 mi/16 km E of BATH; 51°22′N 02°08′W. Former agricultural market. Dairying center. Rubber tire manufacturing, bacon and ham curing. Has 14th–15th-century church and ancient bridge.

Mella (MAI-yah), village, SANTIAGO DE CUBA province, E CUBA, at S foot of SIERRA DE NIPE, 28 mi/45 km N of SANTIAGO DE CUBA; 20°22′N 75°55′W. Sugar mills. Formerly called Miranda.

Mellah (MEL-lah), village, SUWEIDA district, S SYRIA, in the mountains, 20 mi/32 km SE of ES SUWEIDA; 36°31′N 37°57′E. Cereals. Also called MALLAH.

Mellah, El (mel-LAH, el), socialist village, SÉTIF wilaya, central ALGERIA, 6 mi/9.7 km S of El EULMA; 36°52′N 08°19′E. Now a local administrative center.

Mellah, Oued (mel-LAH, wahd), coastal stream, 60 mi/97 km long, NW MOROCCO; flowing NW across the country and entering the ATLANTIC OCEAN at MOHAMMEDIA. Dam c.18 mi/29 km above its mouth (built 1931; made taller in 1940) irrigates farms in CASABLANCA area.

Mellan Fryken, SWEDEN: see FRYKEN.

Mella River (MEL-lah), 60 mi/97 km long, LOMBARDY, N ITALY; rises in the ALPS 5 mi/8 km ENE of BOVEGNO, flows S, through the VAL TROMPIA, across Lombard plain, past MANERBIO, to OGLIO River 12 mi/19 km NE of CREMONA. Chief tributary, Garza River (left).

Mellawi, EGYPT: see MALLAWI.

Melle (ME-luh), commune (2006 population 10,605), Ghent district, EAST FLANDERS province, NW BELGIUM, on SCHELDT RIVER, 5 mi/8 km SE of GHENT; 51°00′N 03°48′E. Agriculture.

Melle (MEL), ancient *Metallum*, town (□ 3 sq mi/7.8 sq km), DEUX-SÈVRES department, POITOU-CHARENTES region, W FRANCE, 16 mi/26 km SE of NIORT, on left bank of Béronne River; 46°13′N 00°08′W. Chemical industry; alcohol distilling. Has three Romanesque churches. A Roman mint (which gave the town its name) has been excavated nearby. Silver-bearing lead mine was exploited here in late Middle Ages.

Melle (MEL-le), town, LOWER SAXONY, NW GERMANY, at S foot of WIEHEN MOUNTAINS, 14 mi/23 km NW of BIELEFELD; 52°12′N 08°20′E. Manufacturing of agricultural machinery and cleaning products.

Mellen, town (2006 population 802), ASHLAND county, N WISCONSIN, on branch of BAD RIVER, 20 mi/32 km SSE of ASHLAND, in wooded lake region, near GOGEBIC RANGE; 46°19′N 90°39′W. Commercial center for dairying area; wood-working; manufacturing (lumber, plywood, pulpwood). Railroad junction. Copper Falls State Park is to N; Chequamegon National Forest to W and S; small lakes to SE. Settled 1886, incorporated 1907.

Mellerud (MEL-le-ROOD), town, ÄLVSBORG county, SW SWEDEN, near SW shore of LAKE VÄNERN, 20 mi/ 32 km NNE of VÄNERSBORG; 58°42′N 12°28′E. Railroad junction; manufacturing (plastic products; mechanical equipment).

Mellette, county (□ 1,309 sq mi/3,403.4 sq km; 2006 population 2,099), S SOUTH DAKOTA; ☉ WHITE RIVER;

43°34′N 100°45′W. Farming and cattle-raising region bounded N by WHITE RIVER. Wheat; cattle. Drained by LITTLE WHITE RIVER and Oak and Black Pipe creeks. In Mountain time zone; the boundary between Mountain and Central time zones follows E county boundary to E part of N boundary to U.S. Highway 83. Formed 1909.

Mellette (MEL-let), village (2006 population 225), SPINK county, NE central SOUTH DAKOTA, 20 mi/32 km N of REDFIELD; 45°08′N 98°30′W.

Mellid (me-LYEED), town, LA CORUÑA province, NW SPAIN, 26 mi/42 km E of SANTIAGO DE COMPOSTELA; 42°55′N 08°00′W. Agricultural trade center (livestock; cereals, fruit); shoe manufacturing; tanning; flour milling. Summer resort. Also spelled Melide.

Mellieha (mel-YE-hah), village (2005 population 7,676), N MALTA, near Mallieha Bay, 9 mi/14.5 km NW of VALLETTA; 35°57′N 14°21′E. Marine salt raking; fishing. Nearby are many coastal fortifications, including 17th-century Red and White Towers.

Mellor (MEL-uh), village (2001 population 2,505), GREATER MANCHESTER, W ENGLAND, 3 mi/4.8 km NW of BLACKBURN; 53°23′N 02°01′W.

Mellor Glacier, outlet glacier, EAST ANTARCTICA, flowing from the S PRINCE CHARLES MOUNTAINS into LAMBERT GLACIER; 73°30′S 66°30′E.

Mellott, town (2000 population 207), FOUNTAIN county, W INDIANA, 13 mi/21 km E of COVINGTON; 40°10′N 87°09′W. Agriculture.

Mellrichstadt (MEL-rikh-shtaht), village, LOWER FRANCONIA, N BAVARIA, GERMANY, on small Streu River, 11 mi/18 km SSW of MEININGEN; 50°26′N 10°13′E. Metalworking; brewing; malting. Surrounded by medieval wall; has Gothic church.

Melluzi, LATVIA: see JŪRMALA.

Melmerby (MEL-muh-bee), village (2006 population 200), E CUMBRIA, NW ENGLAND, 8 mi/12.9 km NE of PENRITH; 54°44′N 02°35′W. Mount of Melmerby Fell (2,331 ft/710 m) is 2 mi/3.2 km ENE.

Melmoth, town, E KWAZULU-NATAL province, SOUTH AFRICA, 20 mi/32 km N of ESHOWE; 28°35′S 31°25′E. Elevation 3,280 ft/1,000 m. Wattle-bark, sugarcane, and timber industry. Site of tungsten, gold, tin deposits. Umfolozi game reserve is 20 mi/32 km ESE. Established in 1887.

Melnik (MEL-neek), city, SOFIA oblast, SANDANSKI obshtina, SW BULGARIA, on the W slope of the PIRIN MOUNTAINS, 12 mi/19 km NE of PETRICH; 41°32′N 23°24′E. Vineyards; winemaking. Smallest city in Bulgaria. Museum village. Has ruins of an ancient fortress. Once an important wine center; population emigrated after Balkan Wars and an influx of phylloxera.

Mělník (MNYEL-nyeek), Czech Mělník, town, STREDOCESKY province, N central BOHEMIA, CZECH REPUBLIC, on ELBE RIVER, opposite VLTAVA RIVER mouth, 18 mi/29 km N of PRAGUE; 50°21′N 14°29′E. Railroad junction; trading port. Manufacturing (food processing; consumer goods); shipbuilding yard; noted for wine (burgundy) and sugar production. Hothouse vegetable growing, large orchards and vineyards in vicinity. Has a 14th century town hall with valuable 16th century archives. Gothic castle with famous wine cellars. Regulating sluices and locks in harbor. Has gardening and viticulture school. During World War II, an oil pipeline was laid by Germans from BRATISLAVA to here.

Mel'nikovo (MYEL-nee-kuh-vuh), village (2006 population 9,760), SE TOMSK oblast, S central SIBERIA, RUSSIA, on the OB' RIVER, on road junction, 35 mi/56 km W of TOMSK; 56°33′N 84°05′E. Elevation 308 ft/93 m. Logging, lumbering; food processing (bakery).

Melnishka (MEL-neesh-kah), river, near MELNIK, SW BULGARIA, tributary of the STRUMA River; 41°32′N 23°24′E.

Mel'nitsa-Podolskaya, UKRAINE: see MEL'NYTSYA-PODIL'S'KA.

Mel'nytsya-Podil's'ka (MEL-ni-tsyah–po-DEEL-skah) (Russian *Melnitsa-Podolskaya*) (Polish *Mielnica*), town, SE TERNOPIL' oblast, UKRAINE, in the DNIESTER River valley, 14 mi/23 km SSE of BORSHCHIV; 48°37′N 26°10′E. Elevation 895 ft/272 m. Railroad terminus; cannery; folk musical instruments, and clothing. Museum of ethnography; tobacco and makhorka research station. Has a palace, church with medieval paintings. Known as Mel'nytsya or Mel'nytsya-nad-Dnistrom [Ukrainian=Mill over the Dniester] since beginning of the 17th century; renamed in 1940; town since 1960. Jewish community since the late 18th century, numbering over 1,400 in 1939; liquidated during World War II.

Melo (MAI-lo), city (2004 population 50,578), ⊙ CERRO LARGO department, NE URUGUAY, on the Arroyo Conventes (left affluent of TACUARÍ RIVER), on railroad and highway, 200 mi/322 km NE of MONTEVIDEO; 32°22′S 54°11′W. Railroad terminus and road junction; airport. Distributing center for surrounding region; wool, hides, agricultural products (wheat, corn, oats); cattle. Bishopric. Has college, industrial school. Founded 1795.

Melocheville (muh-losh-VEEL), former village, S QUEBEC, E CANADA, on Lake SAINT LOUIS, near NE end of BEAUHARNOIS CANAL, 25 mi/40 km SW of MONTREAL; 45°19′N 73°56′W. Quartz mining; dairying; resort. Formerly called Lac Saint Louis. Amalgamated into BEAUHARNOIS in 2002.

Melocotón (mai-lo-ko-TON), village, SANTIAGO province, METROPOLITANA DE SANTIAGO region, central CHILE, on railroad, on upper MAIPO RIVER, in the ANDES, 27 mi/43 km SE of SANTIAGO; 33°42′S 70°21′W.

Melokletskiy (mye-luh-KLYETS-keeyee), settlement, W VOLGOGRAD oblast, SE European Russia, on the right bank of the DON River, on road junction, 3 mi/5 km E of (and administratively subordinate to) KLETSKAYA; 49°18′N 43°07′E. Elevation 291 ft/88 m. Grain processing.

Melón, CHILE: see EL MELÓN.

Mélong (MAI-long), town (2001 population 29,100), LITTORAL province, CAMEROON, 37 mi/60 km SW of BAFOUSSAM; 05°08′N 09°59′E.

Meloria (me-LO-ree-ah), islet, ITALY, in LIGURIAN Sea, off coast of TUSCANY, 4 mi/6 km W of LEGHORN. Near here, in 1284, Genoese overwhelmed Pisan fleet.

Melovatka (mye-luh-VAHT-kah), village, E VORONEZH oblast, S central European Russia, near highway, 11 mi/18 km WNW of KALACH; 51°49′N 38°39′E. Elevation 633 ft/192 m. Wheat, sunflowers.

Melovoye, UKRAINE: see MILOVE.

Meløya (MAI-luh-yah), island (□ 8 sq mi/20.7 sq km) in NORTH SEA, NORDLAND county, N NORWAY, just offshore, 40 mi/64 km NW of Missouri. Fishing; agriculture; cattle raising. Village of MELØY is on S shore.

Melozitna River, 180 mi/290 km long, central ALASKA; rises NW of TANANA, near 66°01′N 152°45′W; flows SW to YUKON River opposite RUBY.

Mel Palaiyam, India: see MELAPALAIYAM.

Melpatti, India: see AMBUR.

Melrakkaslétta (MEL-rah-kahs-LYE-tah), peninsula, NE ICELAND; extends 25 mi/40 km N into GREENLAND between AXARFJÖRÐUR (W) and THISTILFJÖRÐUR (E); 66°20′N 16°10′W. N tip, RIFSTANGI Cape, is N extremity of Iceland, near ARCTIC CIRCLE. Raufarhöfn fishing port, on NE coast.

Melrhir, Chott (mel-GIR, SHOT), shallow saline lake, BISKRA wilaya, E ALGERIA, the westernmost of a series of chotts [Arab.=lakes, or salt flats] reaching into the SAHARA from the Gulf of Gabès (off TUNISIA). Its center is 50 mi/80 km SE of BISKRA. Surface is c.60 ft/18 m below sea level. Length (E-W), c. 80 mi/129 km, including lesser chotts near Tunisian border. The Chott MEROUANE is a SW inlet in El OUED wilaya. Its marshy W edge is paralleled by Biskra-TOUGGOURT

railroad. Receives intermittent waters of the Oued DJEDI and of streams rising in the AURÈS massif of the Saharan ATLAS. Also spelled Melghir.

Melrose, city (2000 population 27,134), MIDDLESEX county, E MASSACHUSETTS, suburb of BOSTON; 42°27′N 71°04′W. It is chiefly residential. The opera star Geraldine Farrar was born here. Settled c.1629, set off from MALDEN and incorporated 1850.

Melrose (mel-ROS), town (2001 population 1,656), Scottish Borders, S Scotland, on the TWEED RIVER, and 4 mi/6.4 km ESE of GALASHIELS; 55°35′N 02°43′W. It is the site of one of the finest ruins in Scotland—Melrose Abbey, owned by the nation and founded for Cistercians by David I in 1136. Sir Walter Scott's *Lay of the Last Minstrel* has descriptions of its beauty. Several times partly destroyed and rebuilt, the abbey contains the heart of Robert the Bruce. Formerly in Borders, abolished 1996.

Melrose, town (2000 population 130), MONROE county, S IOWA, on CEDAR CREEK, 14 mi/23 km WSW of ALBIA; 40°58′N 93°02′W. In bituminous-coal–mining and livestock area.

Melrose, town (2000 population 3,091), STEARNS county, central MINNESOTA, on SAUK River, 32 mi/51 km WNW of ST. CLOUD; 45°40′N 94°48′W. Trade and shipping point (grain; livestock; dairying); manufacturing (cheese, feeds, machine parts, furniture; food processing). Little and Big Birch lakes to N; Birch Lake State Forest to N. Settled 1857, incorporated as city 1898.

Melrose (MEL-roz), village, S SOUTH AUSTRALIA state, S central AUSTRALIA, 17 mi/27 km NNE of PORT PIRIE; 32°49′S 138°11′E. Wheat; wool; dairying.

Melrose, village (2000 population 1,876), MOKA district, MAURITIUS, 18 mi/28.8 km SE of PORT LOUIS. Sugarcane, vegetables; livestock.

Melrose, village, SILVER BOW county, SW MONTANA, on BIG HOLE RIVER at mouth of Camp Creek, 28 mi/45 km SSW of BUTTE. Trout-fishing center. Cattle, sheep.

Melrose, village (2006 population 722), CURRY county, E NEW MEXICO, 20 mi/32 km W of CLOVIS; 34°25′N 103°37′W. Cattle, sheep; alfalfa, pumpkins, vegetables; grain (especially wheat); dairying.

Melrose (MEL-ros), village (2006 population 307), PAULDING county, NW OHIO, 13 mi/21 km SSW of DEFIANCE, near AUGLAIZE River; 41°05′N 84°25′W.

Melrose, village (2006 population 501), JACKSON county, W central WISCONSIN, 24 mi/39 km NNE of LA CROSSE; 44°07′N 91°00′W. Creamery; feeds.

Melrose Park, village (2000 population 23,171), COOK county, NE ILLINOIS, industrial suburb W of CHICAGO; 41°53′N 87°51′W. It has large railroad yards and shops, steel mills, TV manufacturing, and factories that make a wide variety of products. Incorporated 1893.

Melrose Park, village (□ 4 sq mi/10.4 sq km; 2000 population 2,359), CAYUGA county, W central NEW YORK; 42°53′N 76°31′W. A suburb of AUBURN.

Melrose Park, unincorporated village, CHELTENHAM township, MONTGOMERY county, SE PENNSYLVANIA, suburb 8 mi/12.9 km N of PHILADELPHIA (at the city limits); 40°03′N 75°07′W. Light manufacturing.

Mels (MUHLS), town, ST. GALLEN canton, E SWITZERLAND, on SEEZ River, near SW corner of LIECHTENSTEIN; 47°03′N 09°25′E. Elevation 1,631 ft/497 m. Textiles, foodstuffs, chemicals. Capuchin monastery (17th century), baroque church (18th century).

Melsele (MEL-zai-luh), village, BEVEREN commune, Sint-Niklaas district, EAST FLANDERS province, N BELGIUM, 7 mi/11.3 km W of ANTWERP; 51°13′N 04°17′E. Agriculture.

Melsetter, ZIMBABWE: see CHIMANIMANI.

Melsiripura, town, NORTH WESTERN PROVINCE, SRI LANKA; 07°39′N 80°30′E. Trades in coconuts and rice. Dairy farming. Mining.

Melstone (MELS-tuhn), village (2000 population 136), MUSSELSHELL county, central MONTANA, on MUSSEL-

SHELL RIVER, 66 mi/106 km NNE of BILLINGS; 46°36′N 107°52′W. Sheep, cattle; wheat, hay. Oil fields in area.

Melsungen (MEL-sung-uhn), town, HESSE, W GERMANY, on the FULDA, 12 mi/19 km S of KASSEL; 51°07′N 09°33′E. Manufacturing of surgical instruments. Has 16th-century castle. Health resort.

Meltham (MEL-thuhm), town (2001 population 8,079), WEST YORKSHIRE, N ENGLAND, 5 mi/8 km SW of HUDDERSFIELD; 53°35′N 01°51′W. Previously woolen and cotton milling. Textiles, machinery and equipment. Located 1 mi/1.6 km SW is site of ancient fort, thought to date from Iron Age.

Melton (MEL-tuhn), village (2001 population 3,718), E SUFFOLK, E ENGLAND, on DEBEN RIVER, and just NE of WOODBRIDGE; 52°06′N 01°20′E. Former agricultural market.

Melton (MEL-tuhn), satellite community of MELBOURNE (2001 population 32,071), S central VICTORIA, AUSTRALIA, 26 mi/42 km W of city center, on Western Highway; 37°41′S 144°35′E. Pastoral area. Wheat, sheep; dairying.

Melton Mowbray (MEL-tuhn MO-bree), town (2001 population 25,554), NE LEICESTERSHIRE, central ENGLAND, on WREAKE RIVER, 14 mi/23 km NE of LEICESTER; 52°46′N 00°53′W. Produces leather and leather goods, Stilton cheese, pork pies. Has church begun in 13th century. Just W is Sysonby, with iron foundries.

Mel'tsany (myel-TSAH-ni), village, N central MORDVA REPUBLIC, central European Russia, on road junction, 25 mi/40 km NW of SARANSK; 54°28′N 44°43′E. Elevation 511 ft/155 m. In hemp-growing area; distilling.

Meluco District, MOZAMBIQUE: see CABO DELGADO.

Melukote, town, MANDYA district, KARNATAKA state, SW INDIA, 24 mi/39 km N of MYSORE, on isolated, 3,500-ft/1,067-m hill; 12°40′N 76°39′E. Pilgrimage center, especially sacred to adherents of Vishnu as the abode of Ramanuja, eleventh-century Vishnuite philosopher, during later years of his life. Priceless temple jewels are publicly displayed at large annual festival. Has college.

Melun (me-LUHN), town (□ 3 sq mi/7.8 sq km), ⊙ SEINE-ET-MARNE department, N central FRANCE, on the SEINE, 26 mi/42 km SSE of PARIS, at N edge of Forest of Fontainebleau; 48°30′N 02°45′E. An important industrial center where automobile parts, airplane engines, leather products, pharmaceuticals, and processed foods are produced. It is the home of the famous BRIE cheese. An ancient town of ÎLE-DE-FRANCE, Melun was founded as *Melodunum* on an island in the Seine; during Gallo-Roman times it expanded to both banks of the river. It was ravaged often by the Normans. Melun became an early residence of the Capetian kings. The town has a Romanesque church (12th century), a Gothic church (16th century), and vestiges of a Roman fortress and a Capetian castle. Nearby (4 mi/6.4 km NE) is the famous Château of VAUX-LE-VICOMTE, built for Nicholas Fouquet in 1615–1680. It lies amidst a huge display of formal gardens, water bodies, and statuary, aimed to outshine the splendor of VERSAILLES. Melun also has a school for police officers, a prison, and (nearby) an airfield for test flights.

Melun-Sénart, FRANCE: see SÉNART, FOREST OF.

Melur (MAI-loor), town, MADURAI district, TAMIL NADU state, S INDIA, 16 mi/26 km NE of MADURAI; 10°03′N 78°20′E. Road center in rich rice and sugarcane area irrigated by PERIYAR LAKE project.

Melut (me-LUHT), township, UPPER NILE state, S central SUDAN, on right bank of the WHITE NILE River, 75 mi/121 km NNE of MALAKAL; 10°26′N 32°12′E. Cotton, sugarcane, and durra; livestock. Processing of agricultural products, food production.

Melvern, city (2000 population 429), OSAGE county, E KANSAS, on MARAIS DES CYGNES River, 3 mi/4.8 km downstream (E) of Melvin Dam, 8 mi/12.9 km SSE of LYNDON; 38°30′N 95°38′W. Livestock; grain.

Melvern Lake, OSAGE county, E KANSAS, on MARAIS DES CYGNES River, 24 mi/39 km WSW of OTTAWA; 38°30′N 95°42′W. Maximum capacity 363,000 acre-ft; 13 mi/21 km long. Formed by Melvern Dam (93 ft/28 m high) built (1972) for flood control. Eisenhower State Park on N shore.

Melville (MEL-vuhl), city (2006 population 4,149), SE SASKATCHEWAN, CANADA, 25 mi/40 km SW of YORKTON; 50°56′N 102°48′W. Grain elevators, flour mills, dairying.

Melville, town (2000 population 1,376), SAINT LANDRY parish, S central LOUISIANA, 23 mi/37 km ENE of OPELOUSAS and on ATCHAFALAYA River (toll ferry); 30°42′N 91°45′W. In agricultural area (cattle, vegetables, cotton, sugarcane, rice), catfish, crawfish; manufacturing of food products, wood products. Heavily damaged by floods in 1927. Famous for Melville crevase (levee break). Settled c.1875, incorporated 1911.

Melville, village, SWEET GRASS county, S MONTANA, 19 mi/31 km N of BIG TIMBER on Sweet Grass Creek. Cattle, sheep, and horses and other rodeo stock. Dude ranches. CRAZY MOUNTAINS to W; Upper and Lower Glaston lakes to SE.

Melville, NW suburb of JOHANNESBURG, GAUTENG province, SOUTH AFRICA.

Melville Bay, Danish *Melville Bugt*, broad indentation of the W coast of GREENLAND, opening to the SW into BAFFIN BAY. The inland ice comes down to the coast, and glaciers discharge much ice into its waters.

Melville, Cape (MEL-vil), NE QUEENSLAND, AUSTRALIA, in CORAL SEA, near PRINCESS CHARLOTTE BAY; 14°10′S 144°31′E. Range of granite hills extends S.

Melville Island (MEL-vil), (□ 2,240 sq mi/5,824 sq km), NORTHERN TERRITORY, N AUSTRALIA, in the TIMOR SEA 16 mi/26 km off the coast; 11°40′S 131°00′E. It is 65 mi/105 km long and 45 mi/72 km wide and is separated from BATHURST ISLAND by Apsley Strait. Bathurst and Melville islands comprise the TIWI ISLANDS, home to the Tiwi Aboriginal people. The reservation consists largely of mangrove jungle with sandy soil.

Melville Island (□ c.16,400 sq mi/42,500 sq km), NORTHWEST TERRITORIES, CANADA, N of VICTORIA ISLAND; largest of the QUEEN ELIZABETH ISLANDS; 75°30′N 112°00′W. Generally hilly (rising to c.1,500 ft/460 m), it has several ice-covered areas in the interior. There are musk oxen on the island. Sir William Parry, the British explorer, visited Melville Island in 1819, and its S coast was explored (1851) by Sir Francis McClintock.

Melville Island, FRENCH POLYNESIA: see HIKUERU.

Melville, Lake (□ 1,133 sq mi/2,934 sq km), SE LABRADOR, CANADA; extending c.120 mi/190 km inland from Hamilton Inlet, an arm of the ATLANTIC OCEAN; 53°45′N 59°30′W. The saltwater lake receives the CHURCHILL RIVER in Goose Bay, its SW arm, and the NASKAUPI RIVER towns of GOOSE BAY and HAPPY VALLEY at SW end.

Melville Peninsula (□ 24,156 sq mi/62,564 sq km), BAFFIN region, NUNAVUT territory, CANADA, between the Gulf of BOTHNIA and FOXE BASIN, and separated from BAFFIN ISLAND to the N by the FURY AND HECLA STRAIT; it is joined to the mainland by the RAE ISTHMUS; c.68°00′N 84°00′W. Peninsula is c.250 mi/400 km long and 70 mi/113 km–135 mi/217 km wide. Numerous streams radiate from the peninsula's central hilly section, which rises to 1,850 ft/564 m. Hall Lake (□ c.200 sq mi/520 sq km) lies near the NE coast, and in the S portion of the peninsula are many connected lakes. The tundra-covered region is virtually uninhabited and is of little importance economically. Weather station at Mackar Inlet on W coast; trading post at REPULSE BAY and HALL BEACH on the S coast; air station near Hall Beach.

Melville Sound, CANADA: see VISCOUNT MELVILLE SOUND.

Melville Water, AUSTRALIA: see SWAN RIVER.

Melvin, town (2000 population 243), OSCEOLA county, NW IOWA, 11 mi/18 km SE of SIBLEY; 43°17′N 95°36′W. In livestock and grain area.

Melvin, village (2000 population 465), FORD county, E central ILLINOIS, 12 mi/19 km NNW of PAXTON; 40°34′N 88°15′W. Agriculture (grain; livestock); feed milling.

Melvin, village (2000 population 160), SANILAC county, E MICHIGAN, 15 mi/24 km NE of IMLAY CITY; 43°10′N 82°51′W. In farm area.

Melvin (MEL-vuhn), village (2006 population 149), MCCULLOCH county, central TEXAS, on BRADY CREEK, 15 mi/24 km WNW of BRADY; 31°11′N 99°34′W. In cotton, cattle region.

Melvina (mee-VEI-nuh), village (2006 population 88), MONROE county, W WISCONSIN, 23 mi/37 km E of LA CROSSE; 43°47′N 90°46′W.

Melvindale, city (2000 population 10,735), WAYNE county, SE MICHIGAN, SW residential suburb 6 mi/9.7 km SW of DETROIT; 42°16′N 83°10′W. Borders Detroit on E, RIVER ROUGE on N, North Branch of George River on S. Manufacturing (animal oils, meat processing, fabricated metal products, paper products, metal plating). Settled 1870, incorporated as city 1932.

Melvin Village, NEW HAMPSHIRE: see TUFTONBORO.

Melzo (MEL-tso), town, MILANO province, LOMBARDY, N ITALY, 12 mi/19 km E of MILAN; 45°30′N 09°25′E. Fabricated metal products, dairy and meat products, leather goods, textiles, paper, machinery; food processing.

Memaliaj (me-MAH-lee-ei), new town, S ALBANIA, 25 mi/40 km S of BERAT; 40°20′N 19°58′E. Among Albania's most important coal-mining centers.

Memanbetsu (me-MAHN-bets), town, Abashiri district, Hokkaido prefecture, N JAPAN, 152 mi/245 km E of SAPPORO; 43°54′N 144°10′E. Smelts. Protected wetlands nearby.

Memba (mem-buh), village, NAMPULA province, N MOZAMBIQUE, on MOZAMBIQUE CHANNEL, 80 mi/129 km S of PEMBA; 21°22′S 34°14′E. Sisal, cotton, peanuts.

Memba District, MOZAMBIQUE: see NAMPULA, province.

Membrilla (mem-BREE-lyah), town, CIUDAD REAL province, S central SPAIN, in CASTILE–LA MANCHA, 2 mi/3.2 km SE of MANZANARES; 38°58′N 03°21′W. Agricultural center (grapes, potatoes, saffron, vegetables, corn, cereals, sheep, goats). Alcohol and liquor distilling; flour milling; plaster manufacturing; dairying; lumbering. Has old, ornate sanctuary and ruins of a castle. Taken from Moors by Alfonso VIII.

Membrío (mem-BREE-o), village, CÁCERES province, W SPAIN, 37 mi/60 km WNW of CÁCERES; 39°32′N 07°03′W. Lumbering; sheep raising; cereals, olive oil.

Memdovas River (MEM–[th]o-vahs), 47 mi/76 km long, W GREECE; rises in SW KARDITSA prefecture, THESSALY department, in PINDOS Mountains, 15 mi/24 km SW of KARDITSA; flows SW through EVRITANIA prefecture, CENTRAL GREECE department, to AKHELÓOS River 15 mi/24 km WSW of KARPENESION. Hydroelectric plants. There is also a a lake in the area called Memdovas. Also spelled Megdhova River

Meme, department (2001 population 300,318), SOUTHWEST province, CAMEROON; ⊙ KUMBA.

Memel, BELARUS: see NEMAN.

Memele River (MAI-me-lai), Lithuanian *Nemunėlis*, right headstream of LIELUPE RIVER, 118 mi/190 km long, in LITHUANIA and LATVIA; rises N of ROKISKIS, Lithuania; flows generally WNW, partly along Latvian-Lithuanian border, joining MUSA RIVER at BAUSKA, Latvia, to form Lielupe River.

Memel Territory (ME-muhl), German *Memelland*, name applied to the district (□ 1,092 sq mi/2,839.2 sq km) of former EAST PRUSSIA situated on the E coast of the BALTIC SEA and the right (N) bank of the Nemunas River. In 1919 the Treaty of Versailles placed the district, containing the city and port of Memel, under League of Nations–sponsored French administration.

Cross-references are shown in SMALL CAPITALS. The pronunciation guide is shown on page xix. The sources of population figures are shown on page xvii.

Lithuanian troops occupied the area in 1923, forcing the French garrison to withdraw. The Allied council of ambassadors then drew up a new status for the territory, which became an autonomous region (1924) within LITHUANIA with its own legislature. The 1938 electoral victory of the National Socialists in the Memel Territory was followed in March 1939, by a German ultimatum demanding the district's return. Lithuania complied. In 1945 the area was taken by Soviet forces and was restored to Lithuania, by then a part of the USSR.

Memleben (MEM-le-ben), village, SAXONY-ANHALT, central GERMANY, on the UNSTRUT, 10 mi/16 km SW of QUERFURT; 51°16′N 11°29′E. Has remains of Benedictine monastery; traces of ancient imperial castle. Henry I and Otto I died here.

Memmelsdorf (MEM-mels-dorf), village, UPPER FRANCONIA, N BAVARIA, GERMANY, on W foot of the FRÄNKISCHE SCHWEIZ [Ger.=Franconian Switzerland], 4 mi/6.4 km NE of BAMBERG; 49°56′N 10°56′E. Manufacturing of electronic equipment.

Memmingen (MEM-ming-uhn), city, SWABIA, BAVARIA, S GERMANY, 19 mi/31 km NNW of KEMPTEN; 47°59′N 10°10′E. Manufacturing includes textiles, synthetics, machinery. Historically a Swabian town, Memmingen was first mentioned in 1128 and became a free imperial city in the mid-13th century. The Twelve Articles of the Peasantry (1525) were drawn up here during the Peasants War. Memmingen passed to BAVARIA in 1803. Parts of the city's 15th-century walls and gates remain. There are also two 15th-century Gothic churches, a 16th-century city hall, and the 16th-century Fugger House.

Memo Nani, CHINA: see GURLA MANDHATA.

Mempawah (mem-PAH-wah), town, ⊙ Pontianak district, Kalimantan Barat province, W BORNEO, INDONESIA, 48 mi/77 km NW of PONTIANAK.

Memphis, ancient city of EGYPT, the capital of the Old Kingdom (c.3100 B.C.E.–c.2258 B.C.E.), at the apex of the NILE delta, and 12 mi/18 km from the center of CAIRO. It was reputedly founded by Menes, the first king of united Egypt. Its god was Ptah. The temple of Ptah, the palace of Apries, and two huge statues of Ramses II are among the most important monuments found at the site. The necropolis of SAKKARA, near Memphis, was a favorite burial place for pharaohs of the Old Kingdom. A line of pyramids begins near the necropolis, extending for 20 mi/32 km to Al GIZA. Memphis remained important during the long dominance by Thebes and became the Persian satraps capital (525 B.C.E.). Deferring only to Alexandria under the Ptolemies and under Rome, it finally declined with the founding of nearby FUSTAT by the Arabs, and its ruins were largely removed for construction in the new city and, later, in Cairo.

Memphis, city (2000 population 2,061) ⊙ SCOTLAND county, NE MISSOURI, on NORTH FABIUS RIVER, 28 mi/45 km NE of KIRKSVILLE; 40°27′N 92°10′W. Livestock (sheep, cattle, hogs); soybeans and grain (corn); manufacturing (clothing, beverages). Settled 1838.

Memphis, city (2006 population 670,902), ⊙ SHELBY county, extreme SW TENNESSEE, on the Fourth, or Lower, Chickasaw Bluff above the MISSISSIPPI RIVER, at the mouth of the WOLF RIVER, 235 mi/378 km S of ST. LOUIS (MISSOURI); 35°08′N 90°04′W. A river port with excellent anchorages on the Wolf, Memphis is the largest city in the state, a port of entry, a railroad and air distribution center, and a leading hardwood lumber, cotton, and livestock market. Its wide variety of manufacturing includes textiles, consumer goods, paints, and automotive parts. A number of corporations have their national headquarters in the city, and its international airport handles the largest amount of domestic freight of any airport in the UNITED STATES. With the rise of gambling casinos in nearby MISSISSIPPI, casino management has become a major industry here. De Soto possibly crossed the Mississippi

near the site of Memphis, and La Salle's Fort Prudhomme may have been built here. The area was strategically important during the time of the British, French, and Spanish rivalries in the 18th century. A U.S. fort was erected in 1797. The city was established 1819 by Andrew Jackson, Marcus Winchester, and John Overton. In the Civil War it fell, on June 6, 1862, to a Union force led by the elder Charles Henry Davis. Severe yellow-fever epidemics occurred in the 1870s, and thousands died. So many people fled the city that its charter had to be surrendered (1879); it was not restored until 1891. Memphis was a key site during the nation's civil rights struggles in the 1960s; the Lorraine Motel, where Reverand Dr. Martin Luther King was shot in 1968, is now a national civil rights museum. The city is the seat of the University of Memphis, the University of Tennessee Medical Units, Rhodes College, Christian Brothers University, Lemoyne-Owen College, the Memphis College of Art, Southern College of Optometry, and Shelby State Community College. It has a museum of natural history, a planetarium, an opera company, a ballet company, a symphony orchestra, an art gallery, a notable park system, botanical gardens, a nature center, a zoo, an aquarium, a professional theater, a speedway, and a coliseum shaped like an Egyptian pyramid (tribute to the city's Egyptian heritage). It is the seat of a large medical center, St. Jude Children's Research Hospital, and a state mental hospital. The Mid-South Fairgrounds and a modern convention hall are here. An annual weeklong cotton carnival is held, and the Liberty Bowl postseason college football game is played here each year. A number of Victorian homes in the city have been restored. Graceland, former home of Elvis Presley, is one of the nation's largest tourist attractions. Beale Street, another popular site, was made famous by W. C. Handy, the blues composer. A number of military installations are in and near the city, including Memphis Naval Air Station at MILLINGTON. A trans-Mississippi bridge connects Memphis with ARKANSAS. Incorporated 1826.

Memphis (MEM-fis), unincorporated town (□ 3 sq mi/ 7.8 sq km; 2000 population 7,264), MANATEE county, W central FLORIDA, 2 mi/3.2 km N of BRADENTON; 27°32′N 82°33′W.

Memphis, town, MACOMB and SAINT CLAIR counties, SE MICHIGAN, 18 mi/29 km SW of PORT HURON and on BELLE RIVER; 42°53′N 82°46′W. Manufacturing (electronic equipment, tools, machining).

Memphis (MEM-fus), town (2006 population 2,376), ⊙ HALL county, NW TEXAS, 28 mi/45 km NW of CHILDRESS; 34°43′N 100°32′W. Elevation 2,067 ft/630 m. Trade, processing center for agriculture area (peanuts; cattle); manufacturing (fabricated metal products, peanut processing). Founded 1889, incorporated 1906.

Memphis, village, CLARK county, SE INDIANA, 5 mi/8 km NW of CHARLESTOWN. Agricultural area; manufacturing (beef and pork processing).

Memphis, village (2000 population 87), DE SOTO county, NW MISSISSIPPI, residential suburb 13 mi/21 km SSW of MEMPHIS (Tennessee); 34°55′N 90°08′W.

Memphis, village (2006 population 111), SAUNDERS county, E NEBRASKA, 12 mi/19 km SE of WAHOO, near PLATTE RIVER; 41°05′N 96°25′W. Farm trade center in fertile valley. Recreation grounds and artificial lake nearby. Memphis Lake State Recreation Area to W.

Memphrémagog (mem-frai-MAI-gog), county (□ 511 sq mi/1,328.6 sq km; 2006 population 45,018), ESTRIE region, S QUEBEC, E CANADA; ⊙ MAGOG; 45°16′N 72°05′W. Composed of seventeen municipalities. Formed in 1982.

Memphrémagog, Lake (mem-frai-MAI-gog), in S QUEBEC, E CANADA, and N VERMONT, mainly in Quebec; 45°08′N 72°16′W. NEWPORT, Vermont, and MAGOG, Quebec, are trade centers and resorts here. Lake is c.30 mi/48 km long, with a maximum width of

4 mi/6 km. Drains through Magog River and Lake Magog into SAINT-FRANÇOIS River, Quebec.

Memramcook (MEM-ruhm-kahk), village (2001 population 4,719), SE NEW BRUNSWICK, CANADA, on MEMRAMCOOK RIVER, and 13 mi/21 km ESE of MONCTON; 46°00′N 64°33′W. Lumbering; potatoes, grain. Nearby is St. Joseph, site of Acadian cultural center.

Memramcook River (MEM-ruhm-kahk), 25 mi/40 km long, SE NEW BRUNSWICK, CANADA; rises E of MONCTON; flows ESE and S to SHEPODY BAY at DORCHESTER.

Memuro (ME-moo-ro), town, Tokachi district, S central Hokkaido, N JAPAN, on TOKACHI RIVER, 87 mi/140 km E of SAPPORO; 42°54′N 143°03′E. Sugar beets, adzuki beans, potatoes.

Mena (ME-nah), city, central CHERNIHIV oblast, UKRAINE, 39 mi/63 km E of CHERNIHIV; 51°31′N 32°13′E. Elevation 393 ft/119 m. Food processing (flour, fruits and vegetables, poultry, cheese); tobacco and tobacco products. Known since 1066, as a trade center protected by a large fortress; company center in Chernihiv regiment (1648–1700s); city since 1966. Jewish community since the 19th century, numbering over 1,300 in 1939; destroyed during World War II — fewer than 100 Jews remaining in 2005.

Mena (MEEN-uh), town (2000 population 5,637), ⊙ POLK county, W ARKANSAS, in OUACHITA MOUNTAINS, c.55 mi/89 km S of FORT SMITH; 34°34′N 94°14′W. In farming area. Manufacturing (wood products, food processing, apparel, printing, automobile parts). Ouachita National Forest surrounds the town and vicinity except to SW; Queen Wilhelmina State Park to NW; Lake Wilhelmina reservoir to W. Founded 1896.

Mena, ETHIOPIA: see MENNA.

Menabe (mai-nah-BAI), region, TOLIARY province, W MADAGASCAR; 18°00′S–21°00′S. Includes MORONDAVA, BELO SUR TSIRIBIHINA, and coastal hinterlands up to Bemaraha plateau. Home of part of Sakalava tribe. Rice, sugarcane, beans, cassava, corn; cattle. Fishing.

Menadir, SAUDI ARABIA: see ABHA.

Menado, INDONESIA: see MANADO.

Menafra (mai-NAH-frah), town, RÍO NEGRO department, W central URUGUAY, in the CUCHILLA DE HAEDO, on highway and railroad, 60 mi/97 km NE of FRAY BENTOS; 32°34′S 57°29′W. Cattle, sheep.

Menaggio (me-NAHD-jo), resort village, COMO province, LOMBARDY, N ITALY, port on W shore of LAKE COMO, 16 mi/26 km NNE of COMO; 46°01′N 09°14′E. Paper and silk mills, foundry; wood-carving industry.

Menahamiya (mah-nah-KHEM-yah) or **Menahemia**, Jewish village, NE ISRAEL, Jordan Valley, near W bank of the JORDAN, 8 mi/12.9 km S of TIBERIAS; 32°40′N 35°33′E. Below sea level 675 ft/205 m. Mixed farming. Founded 1902.

Menahga (me-NAH-guh), town (2000 population 1,220), WADENA county, W central MINNESOTA, 22 mi/35 km N of WADENA, on Blueberry River; 46°45′N 95°05′W. Sheep, cattle, poultry; oats, barley, rye, beans; dairying; manufacturing (animal feeds, consumer goods, hardwood lumber, concrete blocks, wood products, rubber products). Hutersville State Forest to E; Smoky Hills State Forest to NW; several small natural lakes to E and NE.

Menai Bridge (ME-nei), Welsh *Porthaethwy*, town (2001 population 3,146), GWYNEDD, Wales, 4 mi/6.4 km SW of BANGOR; 53°13′N 04°09′W. On ANGLESEY end of road bridge. Butterflies, tourism. 14th-century church on adjoining Church Island.

Menai Strait (ME-nei), channel of the IRISH SEA, 14 mi/23 km long and from 200 yd/183 m to 2 mi/3.2 km wide, between the island of ANGLESEY and mainland GWYNEDD, NW Wales. Thomas Telford's suspension bridge (1826; rebuilt 1938–1941) carries the road from BANGOR on the mainland to Anglesey, over the strait, and Robert Stephenson's tubular railroad bridge

(1850) now carries a road. CAERNARVON is on the strait; LLANFAIRPWLLGWYNGYLL is nearby.

Ménaka (MAI-nah-kah), town, SEVENTH REGION/GAO, E MALI, SAHARAN outpost, 150 mi/250 km ESE of GAO; 15°55′N 02°24′E.

Menakha, YEMEN: see MANAKHA.

Menam, in Thai names: see MAE NAM.

Menan, village (2000 population 707), JEFFERSON county, SE IDAHO, 5 mi/8 km NW of RIGBY, inside bend of SNAKE RIVER; 43°43′N 112°00′W. Elevation 4,798 ft/1,462 m. Irrigated agricultural area (dairying; cattle, sheep; sugar beets, potatoes, fruit; wheat); manufacturing (log homes).

Menands (muh-NANZ), suburban village (□ 3 sq mi/ 7.8 sq km; 2006 population 3,797), ALBANY county, E NEW YORK, on the HUDSON RIVER, just N of ALBANY; 42°41′N 73°43′W. Regional wholesale produce center. Burial site of President Chester A. Arthur. Incorporated 1924.

Menara (mi-nah-RAH), kibbutz, UPPER GALILEE, NE ISRAEL, on the NAFTALI range, on Lebanese border, 16 mi/26 km N of ZEFAT; 32°46′N 35°33′E. Elevation 3,018 ft/920 m. Dairying; fruit growing; sheep raising; machine shop. Tourist guest house. Also spelled Manara. Formerly known as Ramim. Founded 1943.

Menard (muh-NAHRD), county (□ 315 sq mi/819 sq km; 2006 population 12,588), central ILLINOIS; ⊙ PE-TERSBURG; 40°01′N 89°47′W. Agriculture (corn, wheat, soybeans). Some manufacturing. Drained by SANGAMON RIVER and SALT CREEK (both partly forming N boundary of county). Includes NEW SALEM Historic Site, a reconstruction of the town in which Lincoln lived during 1831–1837. One of 17 Illinois counties to retain the Southern-style commission form of county government. Formed 1839.

Menard (ME-nahrd), county (□ 902 sq mi/2,345.2 sq km; 2006 population 2,210), W central TEXAS; ⊙ ME-NARD; 30°53′N 99°49′W. Elevation 1,800 ft/549 m– 2,450 ft/747 m. On EDWARDS PLATEAU; drained by SAN SABA RIVER. Ranching (sheep, goats, cattle); some irrigated agriculture (grains, pecans). Oil and natural gas. Scenery, hunting, fishing attract tourists. Formed 1858.

Menard (ME-nahrd), town (2006 population 1,545), ⊙ MENARD county, W central TEXAS, on EDWARDS PLATEAU, c.55 mi/89 km SE of SAN ANGELO; 30°55′N 99°46′W. Elevation 1,960 ft/597 m. Ranching region (cattle, sheep, goats); irrigated agriculture (pecans, grain); oil and gas; manufacturing (feeds; meat processing); resort. Ruins of Spanish mission and a presidio (restored), both established 1757, are nearby.

Menarguéns (mai-nahr-GENS), town, LÉRIDA province, NE SPAIN, near SEGRE RIVER, 10 mi/16 km NE of LÉRIDA; 41°44′N 00°45′E. Sugar milling; cereals, alfalfa.

Menasalbas (mai-nahs-AHL-vahs), town, TOLEDO province, central SPAIN, on N slopes of the MONTES DE TOLEDO, 20 mi/32 km SW of TOLEDO; 39°38′N 04°17′W. Agricultural center (cereals, legumes, olives, grapes; livestock). Wine making, oil milling.

Menasha (men-ASH-uh), city (2006 population 16,709), WINNEBAGO and CALUMET counties, E WIS-CONSIN, suburb 5 mi/8 km S of APPLETON, on the FOX RIVER; forms a continuous community with twin city of NEENAH; 44°12′N 88°26′W. Manufacturing (wires, printing and publishing, asphalt, paper products, inks, machinery); dairy farms; summer resort. Railroad junction. Menasha's large papermaking industry, which is served by water power, dates from the late 19th century. The region at the lake outlet was visited by Jean Nicolet (c.1634; the site is marked) and other French explorers and was described by Jonathan Carver in his *Travels* (1778). University of Wisconsin–Fox Valley Center. Settled 1840s, incorporated 1874.

Menasheh Plateau (mi-nah-SHE) or **Menasseh Plateau**, area of low hills, NW ISRAEL, S of the CARAMEL RANGE. Bounded by MEDITERRANEAN coastal plain

(W) and PLAIN OF JEZREEL (E). Rises to c.650 ft/198 m. Also known as Ramat Menasheh.

Menchum, department (2001 population 157,173), NORTH-WEST province, CAMEROON, ⊙ WUM.

Mencué (men-KOO-e), village, W río NEGRO province, ARGENTINA, 110 mi/177 km NE of SAN CARLOS DE BARILOCHE; 40°25′S 69°38′W. Sheep. Coal deposits nearby.

Menda (MEN-DAH), town, Kuma county, KUMAMOTO prefecture, E central KYUSHU, SW JAPAN, 37 mi/60 km S of KUMAMOTO; 32°14′N 130°54′E. Melon; distilled alcoholic drink (*shochu*); lumber processing and wood products. Nearby Maru Pond is known for its *ryukinka* buttercups (their S limit in Japan).

Mendali, IRAQ: see MANDALI.

Mendaña Islands, FRENCH POLYNESIA: see MARQUESAS ISLANDS.

Mendanau (MUHN-dah-nou), island, INDONESIA, in Gaspar Strait (between SOUTH CHINA and JAVA seas), just W of BELITUNG Island; 02°51′S 107°26′E. Island is 10 mi/16 km long, 7 mi/11.3 km wide.

Mendarda (main-dahr-dah), town, JUNAGADH district, GUJARAT state, W central INDIA, 20 mi/32 km WSW of JUNAGADH; 21°19′N 70°26′E. Cotton, millet.

Mendavia (men-DAH-vyah), town, NAVARRE province, N SPAIN, near the EBRO, 13 mi/21 km E of LOG-ROÑO; 42°27′N 02°12′W. Cereals, wine. Gypsum quarries nearby.

Mende (mahn-duh), town (□ 14 sq mi/36.4 sq km); ⊙ LOZÈRE department, LANGUEDOC-ROUSSILLON region, S FRANCE, on the LOT RIVER, 65 mi/105 km NNW of MONTPELLIER; 44°40′N 03°30′E. Mende is a tourist resort in the MASSIF CENTRAL; Mont Lozère (5,600 ft/1,707 m), lies c.10 mi/16 km SE. Principal economic activity is tanning and wool manufacturing. Has administrative and educational functions and is a regional retail center. Originally a small Gallo-Roman town that became an episcopal see in the 5th century, it was ruled by its bishops until 1306, when they had to cede a portion of it to Philip the Handsome. During the Wars of Religion (1562–1598) the city was repeatedly sacked. Points of interest include a 13th-century bridge over the Lot, a 14th–16th-century Gothic cathedral, an 18th-century town hall, and a warren of narrow streets and ancient houses. The CÉVENNES NATIONAL PARK extends SE from Mende.

Mendebo Mountains, range, OROMIYA region, S ETHIOPIA; 06°55′N 39°44′E. Highest mountain range in S ETHIOPIA and site of the BALE MOUNTAINS NATIONAL PARK. Highest peak, MOUNT BATU (14,130 ft/ 4306 m), rises W of the town of GOBA, which was capital of the former BALE province. The Mendebo Mountains are the source of many of the tributaries of the GENALE RIVER, which, after entering SOMALIA, is known as the JUBBA RIVER (Somalia's main river).

Mendefera (MAN-da-fa-ruh), town (2003 population 25,000), ⊙ DEBUB region, central ERITREA, on AS-MARA-ADUWA road, 32 mi/51 km SSW of ASMARA; 14°53′N 38°49′E. Elevation c.6,500 ft/1,981 m. Commercial center; flour mills, tannery. Lignite mine. Sometimes called Adi Ugri.

Mendeleyevo (myen-dee-LYE-ee-vuh), village (2006 population 4,260), central TYUMEN oblast, SW Siberian Russia, on the IRTYSH River, less than 3 mi/5 km SE of TOBOL'SK, to which it is administratively subordinate; 58°10′N 68°18′E. Elevation 144 ft/43 m. In petroleum-producing region; woodworking.

Mendeleyevo (meen-dye-LYE-ee-vuh), town (2006 population 7,945), central MOSCOW oblast, central European Russia, near highway, 15 mi/24 km SE, and under administrative jurisdiction, of SOLNECHNO-GORSK; 56°02′N 37°13′E. Elevation 698 ft/212 m. Measuring and testing devices (electronic and mechanical).

Mendeleyevsk (myen-dye-LYE-eefsk), city (2006 population 23,300), NE TATARSTAN Republic, E European Russia, 15 mi/24 km NE of YELABUGA; 55°53′N 52°19′E.

Elevation 298 ft/90 m. Manufacturing of explosives, fertilizers, silicates; woodworking, grain processing. Has a sanatorium based on mineral springs. Founded in 1868 in conjunction with a chemical factory. Named for D.I. Mendeleyev, who devised the periodic table of chemical elements and who worked in the chemical factory here. Oil fields nearby. Before 1964, called Bondyuzhskiy Zavod or simply Bondyuzhskiy. Made city in 1967.

Mendeleyevskiy (meen-dye-LYE-eef-skeeyee), town (2006 population 9,265), central TULA oblast, W central European Russia, on highway and railroad, 3 mi/5 km S of TULA, to which it is administratively subordinate; 54°09′N 37°35′E. Elevation 777 ft/236 m. Moscow Coal Basin underground station for gasification of coal; equipment for gas pipeline construction. Beekeeping.

Mendeli, IRAQ: see MANDALI.

Menden (MEN-den), town, WESTPHALIA-LIPPE, North Rhine-Westphalia, W GERMANY, 5 mi/8 km NE of ISERLOHN; 51°27′N 07°48′E. Manufacturing of metal products, electronic equipment; some agriculture and forestry. Has Gothic church. Chartered 1276. Destroyed and rebuilt in 1344.

Mendenhall (MEN-duhn-hawl), town (2000 population 2,555), ⊙ SIMPSON county, S central MISSISSIPPI, 29 mi/47 km SE of JACKSON, near STRONG RIVER; 31°57′N 89°52′W. Agriculture (corn, cotton; cattle, poultry); timber; manufacturing (steel fabrication, lumber, equipment parts). Simpson County Legion Lake (state lake) to SE.

Mendenhall Glacier, SE ALASKA, 25 mi/40 km NW of JUNEAU; 58°26′N 134°33′W. Glacier (17 mi/27 km long, 3 mi/4.8 km wide) is accessible by highway from Juneau.

Menderes, name of two rivers in TURKEY. The BÜYÜK (Great) Menderes is the ancient Maeander; the KÜÇÜK (Little) Menderes is the ancient Scamander.

Mendes (MEN-zhes), town (2007 population 17,242), W RIO DE JANEIRO state, BRAZIL, on railroad, 7 mi/11.3 km SE of BARRA DO PIRAÍ; 22°32′S 43°44′W. Paper milling, meatpacking.

Méndez, province, TARIJA department, S central BO-LIVIA; ⊙ SAN LORENZO; 21°10′S 64°55′W.

Méndez (MEN-des), town, MORONA-SANTIAGO province, SE central ECUADOR, on E slopes of the ANDES Mountains, 30 mi/48 km SSW of MACAS, on road from CUENCA; 02°41′S 78°19′W. Produces high-carat gold powder.

Méndez (MEN-des), town, TAMAULIPAS, NE MEXICO, 85 mi/137 km SW of MATAMOROS; 25°05′N 98°32′W. Cereals, sugarcane; livestock. Also Villa de Méndez.

Mendham (MEN-duhm), residential borough (2006 population 5,176), MORRIS county, N central NEW JERSEY, 6 mi/9.7 km WSW of MORRISTOWN; 40°46′N 74°35′W. Has pre-Revolutionary tavern. Settled before 1750, incorporated 1906.

Mendhar (main-DAH-ruh), town, PUNCH district, JAMMU AND KASHMIR state, extreme N INDIA, in W KASHMIR region, in W foothills of PIR PANJAL Range, 11 mi/18 km SSE of PUNCH; 33°37′N 74°08′E. Corn, wheat, rice, pulses.

Mendhi, PAKISTAN: see RONDU.

Mendida, town (2007 population 4,717), OROMIYA state, central ETHIOPIA, 15 mi/24 km SW of DEBRE BERHAN; 09°34′N 39°20′E. In livestock raising area.

Mendig (MEN-dig), town, RHINELAND-PALATINATE, W GERMANY, 15 mi/24 km W of COBLENZ; 50°23′N 07°17′E. Manufacturing (synthetic fiber, textiles) metalworking; brewery. Town first mentioned 1041; 1969 chartered, also incorporated Obermendig and Niedermendig.

Mendigorría (men-dee-go-REE-ah), town, NAVARRE province, N SPAIN, on ARGA river, 10 mi/16 km ESE of ESTELLA; 42°38′N 01°50′W. Olive-oil processing; wine, cereals.

Mendip Hills (MEN-dip), range, across N SOMERSET, SW ENGLAND, extending SE from the vicinity of Hutton to the Frome valley. Composed primarily of limestone, the hills (c.25 mi/40 km long) have numerous caves, such as WOOKEY HOLE and Cheddar Caves, some of which show signs of prehistoric occupation. In the hills are ruins of Roman lead mines, an amphitheater, and a Roman road. The gorges near CHEDDAR are particularly notable. Livestock raising and quarrying.

Mendis, town, ⊙ SOUTHERN HIGHLANDS province, E central NEW GUINEA island, S central PAPUA NEW GUINEA, 100 mi/161 km NW of KIKORI. Located on S side of Great Plateau. Airstrip; road access. Tea, coffee, sweet potatoes, vegetables, sugarcane.

Mendocino, county (□ 3,509 sq mi/9,123.4 sq km; 2006 population 88,109), NW CALIFORNIA, on the PACIFIC OCEAN; ⊙ UKIAH; 39°26′N 123°26′W. Mountain and valley region, traversed by several of the COAST RANGES; in E are summits over 6,000 ft/1,829 m. Drained by Eel, Russian, Big, Noyo, and Navarro rivers. Extensive timber, sawmilling; wineries, breweries. Cattle; farms in valleys produce fruit (apples, pears), grapes, hops, beans; dairying. Ocean fisheries (urchins, fish). Hot springs (resorts); trout and steelhead fishing, deer hunting. Part of Mendocino National Forest (NE). Round Valley Indian Reservation in N. Large stands of redwood near coast; pine, fir, and oak are inland. MacKerricher, Russian Gulch, and Van Damme state parks in W, Manchester State Beach in SW (all on Pacific Ocean); Hendy Woods State Park in S. Formed 1850.

Mendocino, unincorporated village, MENDOCINO county, NW CALIFORNIA, at mouth of BIG RIVER, 31 mi/50 km WNW of UKIAH. Redwood and pine timber; fish, urchins; tourism. Jackson State Forest to NE; Russian Gulch State Park to N, Van Damme State Park to S; Point Cabrillo to N.

Mendocino, Cape, promontory, HUMBOLDT county, NW CALIFORNIA, 27 mi/43 km S of EUREKA, on PACIFIC OCEAN; 40°26′N 124°24′W. Westernmost point of California. Rainbow Ridge to E.

Mendol (MUHN-dol), island, RIAU province, INDONESIA, in Strait of MALACCA, just off E coast of SUMATRA, 55 mi/89 km SW of SINGAPORE, opposite mouth of KAMPAR RIVER; 12 mi/19 km long, 10 mi/16 km wide; 00°36′N 103°13′E. Low, swampy. Sometimes called Pendjalai or Penjalai.

Mendon (MEN-duhn), town, WORCESTER county, S MASSACHUSETTS, 17 mi/27 km SE of WORCESTER; 42°05′N 71°33′W. Agriculture. Settled 1660, incorporated 1667.

Mendon, town (2000 population 208), CHARITON county, N central MISSOURI, 13 mi/21 km S of BROOKFIELD; 39°35′N 93°07′W. Corn, wheat, soybeans. Dairying; livestock raising. Swan Lake National Wildlife Refuge on N side.

Mendon, town, RUTLAND CO., W central VERMONT, just NE of RUTLAND, partly in Green Mountain National Forest; 43°37′N 72°52′W. 40% of its 25,000 acres is reserved for federal or state forestland. Chartered as Medway and changed to Mendon in 1827.

Mendon, village (2000 population 883), ADAMS county, W ILLINOIS, 11 mi/18 km NNE of QUINCY; 40°05′N 91°16′W. In agricultural area (corn, sorghum, soybeans; cattle).

Mendon, village (2000 population 917), SAINT JOSEPH county, SW MICHIGAN, 21 mi/34 km SSE of KALAMAZOO, on SAINT JOSEPH RIVER; 42°00′N 85°27′W. Rich farm area. Manufacturing (plastic products).

Mendon (MEN-duhn), village (2006 population 700), MERCER county, W OHIO, 9 mi/14 km N of CELINA, on ST. MARYS RIVER; 40°40′N 84°31′W. In agricultural area.

Mendon, village (2006 population 925), CACHE county, N UTAH, 8 mi/12.9 km W of LOGAN, near LITTLE BEAR RIVER; 41°42′N 111°58′W. Elevation 4,520 ft/1,378 m.

Wheat; dairying; cattle. Part of Wasatch National Forest, including Wellsville Wilderness Area, to W. Settled 1857.

Mendonça (MEN-don-sah), town (2007 population 3,980), NW SÃO PAULO state, BRAZIL, 28 mi/45 km SW of SÃO JOSÉ DO RIO PRETO; 21°12′S 49°34′W. Coffee growing.

Mendong Gompa (MEN-DUNG GUM-PAH), lamasery, S central TIBET, SW CHINA, in ALING KANGRI mountain range, 250 mi/402 km NW of XIGAZE; 31°09′N 85°14′E. Elevation c.15,700 ft/4,785 m.

Mendooran (men-DO-ruhn), town, NW NEW SOUTH WALES, SE AUSTRALIA, 246 mi/396 km NE of SYDNEY, 36 mi/58 km NE of DUBBO, on CASTLEREAGH RIVER; 31°50′S 149°08′E. Sheep, cattle grazing; grain.

Mendota, city (2000 population 7,890), FRESNO county, central CALIFORNIA, in San Joaquin Valley, 33 mi/53 km W of FRESNO; 36°46′N 120°23′W. Melons, cotton, sugar beets, figs, vegetables, almonds, grain; cattle; manufacturing (beet sugar). Terminus of DELTA-MENDOTA CANAL near here; California Aqueduct to SW. Incorporated 1942.

Mendota, city (2000 population 7,272), LA SALLE county, N ILLINOIS, 19 mi/31 km NW of OTTAWA; 41°32′N 89°07′W. Processing and shipping center in agricultural area; manufacturing (food processing, woodworking machinery, tools, concrete products); corn, wheat, soybeans; livestock. Incorporated 1859.

Mendota (men-DO-duh), village (2000 population 197), DAKOTA county, SE MINNESOTA, suburb 5 mi/8 km SW of ST. PAUL and 7 mi/11.3 km SE of MINNEAPOLIS, at confluence of MINNESOTA (W) and MISSISSIPPI (N) rivers; 44°53′N 93°09′W. Manufacturing (consumer goods). The first permanent white settlement in Minnesota. Served as meeting place for traders and trappers before 1819 and known as St. Peter's. Settled 1834, name changed 1837. Homes of Henry Hastings Sibley (Sibley House Museum), first governor of Minnesota, and of Jean Baptiste Fairbault, early trader and fur trapper, are here. Reconstructed in 1930s, buildings date back, respectively, to 1835 and 1837. Mendota Bridge (4,119 ft/1,255 m long, completed 1926) crosses Minnesota River here. Minneapolis–St. Paul International Airport to W. FORT SNELLING State Park to W and SW.

Mendota (men-DO-tuh), unincorporated village, WASHINGTON county, SW VIRGINIA, on North Fork of HOLSTON River, 12 mi/19 km W of ABINGDON; 36°42′N 82°18′W. Agriculture (dairying; livestock; corn, alfalfa).

Mendota Heights (men-DO-duh), town, DAKOTA county, SE MINNESOTA, residential suburb 5 mi/8 km SSW of ST. PAUL and 8 mi/12.9 km SE of MINNEAPOLIS, near confluence of MISSISSIPPI (NW) and MINNESOTA (W) rivers (both bridged here); 44°52′N 93°08′W. Railroad junction; manufacturing (rubber products, concrete, printing, medical equipment, food, pharmaceuticals, machinery). Small lakes in area; Rogers Lake in S. Part of FORT SNELLING State Park in W.

Mendota, Lake (men-DO-tuh), largest of the FOUR LAKES, DANE county, S WISCONSIN; c.6 mi/9.7 km long, c.4 mi/6.4 km wide. A resort lake stocked with fish, it is fed by YAHARA RIVER from N and drained by it to SE to lake MONONA. Downtown MADISON is on SE shore, city also bounds it on NE and SW; city of MIDDLETON on W end. University of Wisconsin on S shore; Governor Nelson State Park on NW shore.

Mendoza (men-DO-zah), province (□ 58,239 sq mi/151,421.4 sq km; 2001 population 1,579,651), W ARGENTINA; ⊙ MENDOZA. Bordered W by the ANDES MOUNTAINS along CHILE frontier; slopes gradually E to DESAGUADERO RIVER and the Río Salado. Watered by MENDOZA, ATUEL, and TUNUYÁN rivers. Its mountainous W border includes MT. ACONCAGUA, the highest peak in the Western Hemisphere. The TRANSANDINE RAILWAY (built 1887–1910) runs to

Chile beneath USPALLATA PASS. In its inhabited lower valleys it has temperate, dry, Mediterranean climate. Among its abundant mineral resources are coal, petroleum, lime, lead, zinc, sulphur, marble, onyx, serpentine, talc, and uranium deposits at SAN RAFAEL. Agricultural activity in irrigated river valleys, producing mostly wine; also fruit, potatoes, olives, grain, vegetables. Livestock raising (cattle, sheep, goats) of secondary importance. Fisheries in HUANACACHE lakes. Petroleum refineries in GODOY CRUZ and TUPUNGATO, cement works at CAPDEVILA and PANQUEUA. Fruit and wine industries concentrated in Mendoza and San Rafael. Has great number of hydroelectric power stations. Major resorts include EL SOSNEADO, LOS MOLLES, Potreillos, and spas of VILLAVICENCIO and CACHEUTA. Until 1776, the area was part of Chile.

Mendoza (men-DO-zah), canton, BELISARIO BOETO province, CHUQUISACA department, SE BOLIVIA, 12 mi/20 km NE of TOMINA and NW of VILLA SERRANO. Elevation 6,929 ft/2,112 m. Agriculture (potatoes, yucca, bananas, corn, barley, sweet potatoes, peanuts); cattle and hog raising.

Mendoza, canton, CHAPARÉ province, COCHABAMBA department, central BOLIVIA, 12 mi/20 km SW of SANTA ROSA; 16°57′S 65°52′W. Elevation 8,940 ft/2,725 m. Clay, limestone, and gypsum deposits. Agriculture (potatoes, yucca, bananas, corn, rice, rye, sweet potatoes, soy, tobacco, coffee, coca, tea, citrus fruits); beef and dairy cattle.

Mendoza (men-DO-zah), city, ⊙ Mendoza province, W Argentina; 32°53′S 68°49′W. With a backdrop of snowcapped mountains, Mendoza is surrounded by a fertile oasis, known as the "Garden of the Andes," irrigated by the Mendoza River. It is an agricultural market and the center of a rich wine-producing region, largely settled by Italian immigrants. Food processing and petrochemicals are also important to the city's economy. Oil and gas deposits nearby. Mendoza was founded in 1561 and belonged to Chile until the creation of the viceroyalty of Río de la Plata (1776). Destroyed by earthquake in 1861, the town was rebuilt and expanded rapidly after the completion of the railroad to Buenos Aires late in the 19th century. It is also the E terminus of the Transandine railroad. It was in Mendoza that San Martín began (1817) the final liberation of Chile from Spain. The city has three universities. Its landmarks include a Franciscan monastery where several Argentine national heroes are buried.

Mendoza (men-DO-sah), city, ⊙ RODRÍGUEZ DE MENDOZA province, AMAZONAS region, N PERU, in E ANDEAN foothills, 30 mi/48 km E of CHACHAPOYAS; 06°18′S 77°25′W. Sugarcane, fruit, cereals, coca, plantains, bananas. Variant names include: Mondoza; San Nicolas.

Mendoza (men-DO-sah), town, TRUJILLO state, W VENEZUELA, in ANDEAN spur, 10 mi/16 km SSW of VALERA; 09°12′N 70°40′W. Elevation 3,970 ft/1,210 m. Wheat, corn, potatoes, coffee, fruit.

Mendoza, MEXICO: see CIUDAD MENDOZA.

Mendoza River (men-DO-zah), c.200 mi/322 km long, N MENDOZA province, ARGENTINA; rises on the ACONCAGUA massif near CHILE border; flows generally E, past USPALLATA, POTRERILLOS, and LUJÁN, then N to the HUANACACHE lakes. Used for hydroelectric power (Uspallata, CACHEUTA) and irrigation (S of MENDOZA). Transandine railroad follows its valley.

Mendrisio (men-DREE-zee-o), commune, TICINO canton, S SWITZERLAND, 6 mi/9.7 km NW of COMO (Italy); 45°52′N 08°59′E. Elevation 1,161 ft/354 m. Metal products, apparel.

Mendrisio (men-DREE-zee-o), district, S TICINO canton, S SWITZERLAND; 45°52′N 08°59′E. Main town is CHIASSO; population is Italian-speaking and Roman Catholic.

Area is shown by the symbol □, and capital city or county seat by ⊙.

Mendu (MAIN-doo), town, ALIGARH district, W UTTAR PRADESH state, N central INDIA, 4 mi/6.4 km ENE of HATHRAS. Wheat, barley, pearl millet, gram, cotton, mustard. Also spelled Maindu.

Mendut (MUHN-doot) or **Candi Mendut** (CHAHN-dee MUHN-doot), historic site, Java Tengah province, INDONESIA, E of BOROBUDUR, 21 mi/33 km NW of YOGYAKARTA; 07°35'S 110°12'E. Temple, constructed about 850 C.E., originally stood 88 ft/27 m tall. Restored 1897–1904.

Menea, El, ALGERIA: see GOLÉA, EL.

Mene Grande (MAI-nai GRAHN-dai), town, ZULIA state, NW VENEZUELA, in MARACAIBO lowlands, 10 mi/16 km E of SAN LORENZO (connected by railroad and pipeline); 09°51'N 70°55'W. Major oil field, opened 1914.

Menemen, town (2000 population 46,079), W TURKEY, on IZMIR-MANISA railroad, near GEDIZ RIVER, 13 mi/21 km NNW of Izmir; 38°34'N 27°03'E. Valonia, raisins, wheat, barley, sugar beets.

Menemsha, MASSACHUSETTS: see CHILMARK.

Menen (MAI-nun), French *Menin*, commune (2006 population 32,447), Kortrijk district, WEST FLANDERS province, SW BELGIUM, on the Leie River, near the FRENCH border; 50°48'N 03°07'E. Manufacturing. Founded in 1578, Menen was strongly fortified in the seventeenth century.

Menéndez, Lake (me-NAIN-dez) (□ 25 sq mi/65 sq km), in ANDES MOUNTAINS, W CHUBUT province, ARGENTINA, N of the CORDÓN DE LAS PIRÁMIDES. Elevation 1,690 ft/515 m; c.12 mi/19 km long, 1 mi/1.6 km–3 mi/4.8 km wide; has several arms. LAKE FUTA-LAUFQUÉN (SE) is connected to it by a river.

Menengai (mai-nain-GAH-ee), village, RIFT VALLEY province, W KENYA, on railroad, 8 mi/12.9 km WNW of NAKURU; 00°12'S 05°36'E. Pyrethrum, sisal, coffee, wheat, corn. Active Menengai volcano (7,440 ft/2,268 m) rises just E.

Menera, Sierra (mai-NAI-rah, SYE-rah), hill range, part of IBERIAN MOUNTAINS, between TERUEL and GUADALAJARA provinces, E central SPAIN; 40°42'N 01°32'W. Rich iron mines at Ojos Negros, Almohaja, and Setiles.

Ménerbes (mai-NERB), commune (□ 11 sq mi/28.6 sq km), VAUCLUSE department, PROVENCE-ALPES-CÔTE D'AZUR region, S FRANCE, 21 mi/34 km ESE of AVIGNON, on a rocky height along N slope of LUBÉRON mountain, and at the edge of the LUBÉRON NATURAL REGIONAL PARK (established 1977); 43°50'N 05°13'E. Has a 13th-century fort, rebuilt in later century, which played a strategic role in the Wars of Religion (briefly occupied by the Calvinist Protestants in the 1570s). Subject of a pioneering 1950s sociological study of PROVENCE village life, *A Village in the Vaucluse*.

Meneses (mai-NAI-ses), town, SANCTI SPÍRITUS province, central CUBA, in low Sierra de Meneses, 23 mi/37 km SE of CAIBARIÉN; 22°15'N 79°16'W. Sugarcane, tobacco; cattle.

Menevia, Wales: see SAINT DAVID'S.

Ménez, FRANCE: see LANDES DU MÉNÉ.

Ménez-Hom (mai-ne–zom), summit (1,100 ft/335 m) of MONTAGNES NOIRES (French=black mountains), in BRITTANY, NW FRANCE, at base of the CROZON PENINSULA, near the ATLANTIC OCEAN; 48°14'N 04°12'E. Offers one of the finest views of Brittany's W coastline with its many indentations. The city of BREST is 19 mi/31 km NW.

Menfi (MEN-fee), town, AGRIGENTO province, W SICILY, 11 mi/18 km SE of CASTELVETRANO; 37°36'N 12°58'E. In olive-, almond-, and grape-growing region; cement.

Menfro, unincorporated community, PERRY county, E MISSOURI, near MISSISSIPPI RIVER, 9 mi/14.5 km ENE of PERRYVILLE.

Mengabril (meng-gah-VREEL), town, BADAJOZ province, W SPAIN, 22 mi/35 km E of MÉRIDA; 38°56'N 05°56'W. Cereals, olives, grapes.

Mengcheng (MENG-CHENG), town, ⊙ Mengcheng county, N ANHUI province, CHINA, 50 mi/80 km ENE of FUYANG and on GUO RIVER; 33°16'N 116°33'E. Grain, oilseeds, cotton; tobacco industry, food and beverages.

Mengen (MENG-guhn), town, Upper Swabia, BADEN-WÜRTTEMBERG, GERMANY, near the DANUBE, 6 mi/9.7 km SE of SIGMARINGEN; 48°03'N 09°20'E. Half-timbered houses from 16th–18th century here. Chartered 1276.

Mengene Dagi (Turkish=*Mengene Dağı*) mountain range, E TURKEY, running SW-NE 40 mi/64 km ESE of VAN, near Iranian border; rises to 11,844 ft/3,610 m.

Mengerskirchen (meng-guhrs-KIR-khuhn), village, W HESSE, central GERMANY, 15 mi/24 km W of WETZLAR, on W slope of the WESTERWALD; 50°34'N 08°10'E. Forestry. Has ruined castle.

Menggala (MUHNG-gah-lah), town, Lampung province, INDONESIA, 110 mi/177 km SSE of PALEMBANG; 04°32'S 105°17'E. Trade center for agriculture and forested area (rubber, timber; coffee, pepper; fibers).

Menghai (MENG-HEI), town, ⊙ Menghai county, extreme S YUNNAN province, CHINA, on route to THAILAND; 21°58'N 100°28'E. Elevation 4,003 ft/1,220 m. Tea, rice, timber, sugarcane; food and beverages. Also called Fohai.

Mengíbar (meng-GEE-vahr), town, JAÉN province, S SPAIN, near the GUADALQUIVIR RIVER, 13 mi/21 km SW of LINARES; 37°58'N 03°48'W. Olive- oil processing; lumbering. Agricultural trade (cereals, vegetables, sugar beets; livestock).

Mengjin (MENG-JIN), town, ⊙ Mengjin county, NW HENAN province, CHINA, 15 mi/24 km NE of LUOYANG and on HUANG HE (Yellow River); 34°50'N 112°26'E. Grain, tobacco, oilseeds.

Mengkofen (MENG-ko-fuhn), village, LOWER BAVARIA, BAVARIA, GERMANY, on small Aiterach River, 18 mi/29 km NE of LANDSHUT; 48°43'N 12°26'E. Agriculture (grain, sugar beets); manufacturing of electronic equipment.

Menglembu (MUHNG-lem-boo), town, central PERAK, MALAYSIA, 3 mi/5 km SW of IPOH, on slopes of KLEDANG; 04°34'N 101°03'E. A tin-mining center of Kinta Valley; important peanut-growing area.

Mengshan (MENG-SHAN), town, ⊙ Mengshan county, E GUANGXI ZHUANG AUTONOMOUS REGION, CHINA, 75 mi/121 km ESE of LIUZHOU; 24°12'N 110°31'E. Rice, sugarcane; food processing; logging; papermaking; chemicals, textiles.

Mengwang (MENG-WANG), town, SW YUNNAN province, CHINA, 60 mi/97 km SSW of SIMAO (Pu'er) and on right bank of MEKONG River; 22°17'N 100°33'E. Rice, millet, sweet potatoes, sugarcane, tea.

Meng Xian (MENG SIAN), town, ⊙ Meng Xian county, central W HENAN province, CHINA, 40 mi/64 km SW of JIAOZUO, on HUANG HE (Yellow River), opposite MENGJIN (road ferry); 34°54'N 112°47'E. Grain, oilseeds; papermaking, chemicals, engineering; food industry.

Mengyin (MENG-YIN), town, ⊙ Mengyin county, SE central SHANDONG province, CHINA, on road, 50 mi/80 km NNW of LINYI; 35°43'N 117°55'E. Grain, oilseeds, tobacco; textiles; engineering, food industry.

Mengzhe (MENG-JUH), town, SW YUNNAN province, CHINA, near MYANMAR border, 85 mi/137 km SW of SIMAO (Pu'er), in mountain region. Sugar refining; rice, millet, beans, sugarcane.

Mengzi (MENG-ZI), town, ⊙ Mengzi county, SE YUNNAN province, CHINA, c.12 mi/19 km E of GEJIU; 23°20'N 103°21'E. It is the commercial hub of a district where tin and antimony are mined.

Menheniot (muhn-HEN-i-uht), village (2001 population 1,892), SE CORNWALL, SW ENGLAND, 3 mi/4.8 km SE of LISKEARD; 50°26'N 04°42'W. Has 15th-century church.

Menidi, Greece: see AKHARNAI.

Menidion, Greece: see AKHARNAI.

Menifee (MEN-uh-fee), county (□ 206 sq mi/535.6 sq km; 2006 population 6,788), E central KENTUCKY; ⊙ FRENCHBURG; 37°57'N 83°35'W. Bounded NE by LICKING RIVER (CAVE RUN LAKE reservoir), S by RED RIVER; drained by several creeks. Rolling agricultural area (burley tobacco, hay; cattle); oil and gas wells, timber; some sawmills. Most of county (except E and W ends) in Daniel Boone National Forest. Part of Red River Geological Area, including Red River Gorge, in S. Formed 1869.

Menikion (me-NEE-kee-on), mountain massif in CENTRAL and EAST MACEDONIA AND THRACE departments, NE GREECE, between ANGITIS RIVER (Dráma lowland) and STRUMA (Strymon) River (Sérrai lowland), 11 mi/18 km NE of SÉRRAI; 41°11'N 23°48'E. Rises to 6,437 ft/1,962 m. Also called Karagioz Giol and Smiginitsa or Smiginova (Smiyinitsa or Smiyinova). Also spelled Menoikion.

Menin (me-NEEN), village, DAMASCUS district, SW SYRIA, 9 mi/14.5 km N of Damascus; 33°38'N 36°18'E. Elevation 4,000 ft/1,219 m. Summer resort; walnuts, orchards. Also spelled MANIN, MININ, MUNIN, or MNINE.

Menin, BELGIUM: see MENEN.

Menindee (muh-NIN-dee), village, W NEW SOUTH WALES, AUSTRALIA, on DARLING RIVER, and 65 mi/105 km ESE of BROKEN HILL; 32°24'S 142°26'E. Sheep center; citrus fruit orchards, vegetables.

Menindee, Lake (muh-NIN-dee) (□ 60 sq mi/156 sq km), W NEW SOUTH WALES, AUSTRALIA, 55 mi/89 km ESE of BROKEN HILL; 9 mi/14 km long, 8 mi/13 km wide; 32°21'S 142°20'E. Usually dry. Declared national park in 1967.

Meningie (muh-NIN-jee), village, SE SOUTH AUSTRALIA, S central AUSTRALIA, 70 mi/113 km SE of ADELAIDE, and on SE shore of Lake ALBERT; 35°42'S 139°20'E. Dried and citrus fruits; dairy products; fish.

Menlo, town (2000 population 365), GUTHRIE county, W central IOWA, near source of NORTH RIVER, 12 mi/19 km SSE of GUTHRIE CENTER; 41°31'N 94°24'W. In livestock and grain area.

Menlo (MEN-lo), village (2000 population 485), CHATTOOGA county, NW GEORGIA, 24 mi/39 km NW of ROME, near ALABAMA line and LOOKOUT MOUNTAIN; 34°29'N 85°29'W. Manufacturing of jewelry, apparel, lumber.

Menlo, village (2000 population 57), THOMAS county, NW KANSAS, near source of South Fork of SOLOMON RIVER, 16 mi/26 km E of COLBY; 39°21'N 100°43'W. In agricultural and cattle region.

Menlo Park, city (2000 population 30,785), SAN MATEO county, W CALIFORNIA; residential suburb 25 mi/40 km SSE of SAN FRANCISCO, 4 mi/6.4 km SSW of SAN FRANCISCO BAY; 37°29'N 122°08'W. Bounded by San Francisquito Creek in SE. Manufacturing (electronic and communications equipment, liquor, pharmaceuticals, medical equipment, wire and plastic products, aerospace parts, computer equipment). Menlo College and Stanford University are located 1 mi/1.6 km SE at STANFORD. SANTA CRUZ MOUNTAINS to SW; HETCH HETCHY AQUEDUCT passes to N; Stanford Linear Accelerator to S. Incorporated 1874.

Menlo Park, unincorporated residential community, MIDDLESEX county, central NEW JERSEY; 40°33'N 74°20'W. It is the site of Edison Memorial Tower and state park, where Thomas Edison kept his laboratories (1876–1887). The laboratories have since been transferred to the Edison Institute of Technology in Greenfield Village Museum at DEARBORN, MICHIGAN. Menlo Park developed into a suburban community after World War II.

Menna, town (2007 population 13,038), OROMIYA state, S ETHIOPIA, 45 mi/72 km SSW of GOBA, in S MENDEBO MOUNTAINS; 06°22'N 39°49'E. Has airfield. Also spelled Mena.

Cross-references are shown in SMALL CAPITALS. The pronunciation guide is shown on page xix. The sources of population figures are shown on page xvii.

Mennbij (mem-BEEJ), town, ALEPPO district, NW SYRIA, 50 mi/80 km NE of ALEPPO; 36°31′N 37°57′E. Important road junction. Pistachios, cereals.

Mennecy (me-nuh-SEE), town (□ 4 sq mi/10.4 sq km), ESSONNE department, ÎLE-DE-FRANCE region, N central FRANCE, near junction of Juine and ESSONNE rivers just SW of CORBEIL-ESSONNES; 48°34′N 02°26′E.

Menno, town (2006 population 682), HUTCHINSON county, SE SOUTH DAKOTA, 27 mi/43 km NNW of YANKTON; 43°14′N 97°34′W. Cooperative creamery and grain elevator.

Mennonite Colonies, Spanish *Colonia Mennonita*, settlements, Boquerón department, N PARAGUAY, scattered colonies in the CHACO, c.120 mi/193 km W of PUERTO CASADO (connected by railroad). Cotton growing, cattle raising; also lumbering, grain growing, poultry farming; flour mills, sawmills. Founded 1926 by Russo-German Mennonites from Canada on a Paraguayan concession (1921). Later enlarged by new settlers from CANADA, GERMANY, and RUSSIA. Altogether about thiry-two villages, two of which are Colonia Newland (22°38′S 60°07′W) and Colonia Mennonita (22°30′S 60°00′W); both of these villages sit at road junctions. Filidelfia is considered the urban center of the colonies.

Meno (MEEN-o), village (2006 population 191), MAJOR county, NW OKLAHOMA, 17 mi/27 km W of ENID; 36°23′N 98°10′W. Wheat, corn; cattle; oil and natural gas.

Menoikion, Greece: see MENIKION.

Menominee (me-NAH-me-nee), county (□ 1,337 sq mi/3,476.2 sq km; 2006 population 24,696), SW UPPER PENINSULA, N MICHIGAN; ⊙ MENOMINEE; 45°31′N 87°31′W. Bounded SE by GREEN BAY and SW by WISCONSIN; drained by MENOMINEE, BIG CEDAR, and LITTLE CEDAR rivers. Agriculture (cattle, poultry; forage, corn, oats; dairy products). Manufacturing at MENOMINEE. Fishing, lumbering. Resorts. Michigan Pottawatomi Indian Reservation in NE. J. W. Wills State Park in E on Green Bay. One of four Michigan counties in Central time zone; border between Central and Eastern time zones follows N and E borders. Organized 1863.

Menominee, county (□ 365 sq mi/949 sq km; 2006 population 4,597), E central WISCONSIN, ⊙ KESHENA; 45°01′N 88°42′W. Formerly part of SHAWANO county (to S). Drained by WOLF and Red rivers. Several small lakes, especially in SE corner. Barley, oats, alfalfa; cattle, hogs, sheep; dairying. Lumber.

Menominee (me-NAH-me-nee), city (2000 population 9,131), ⊙ MENOMINEE county, N MICHIGAN, W UPPER PENINSULA, on GREEN BAY of LAKE MICHIGAN at the mouth of the MENOMINEE RIVER; 45°07′N 87°37′W. A distribution center for upper Michigan and N WISCONSIN. Manufacturing (fabricated metal products, machinery, consumer goods, paper products, lumber and wood products). County Airport to NW. Of interest is the "mystery ship," raised (1969) from the bottom of Green Bay, where it sank in 1864. A bridge connects MENOMINEE with MARINETTE (Wisconsin). Incorporated 1883.

Menominee, village (2000 population 237), JO DAVIESS county, extreme NW ILLINOIS, on short Little Menominee River (bridged here), and 8 mi/12.9 km NW of GALENA; 42°28′N 90°32′W.

Menominee Iron Range (me-NAH-me-nee), mainly in IRON county, SW UPPER PENINSULA, N MICHIGAN, along Michigan-WISCONSIN state line NW of IRON MOUNTAIN (Michigan). Timber; tourism (sport fishing, hunting). Iron Mountain Mine still active. Also called Menominee Range.

Menominee River (me-NAH-me-nee), 118 mi/190 km long, N MICHIGAN; formed by the union of the BRULE and the MICHIGAMME rivers above IRON MOUNTAIN, W UPPER PENINSULA; flows SE into GREEN BAY at MENOMINEE; 45°57′N 88°11′W. It passes through a once plentiful iron-ore region and forms part of the WISCONSIN-Michigan state line for its entire length. Numerous small dams. Piers Gorge, near NORWAY (Michigan) named for piers built to slow river's flow.

Menominee River, Wisconsin: see MENOMONEE RIVER.

Menomonee Falls, city (2006 population 34,370), WAUKESHA county, SE WISCONSIN, on the MENOMINEE and FOX rivers, a suburb 17 mi/27 km NW of MILWAUKEE, and 8 mi/12.9 km NE of WAUKESHA; 43°08′N 88°07′W. Manufacturing (publishing and printing; wire, fabricated metal products, paper products, concrete products, plumbing fixtures, furniture, fiberglass products, machine tools, machinery, oil lamps and lamp oils, marble products, acoustical materials, medical supplies and equipment; steel and aluminum foundry). Founded 1843; settled originally by German immigrants. Incorporated 1892.

Menomonee River, c.25 mi/40 km long, SE WISCONSIN; rises at GERMANTOWN in SE WASHINGTON county; flows SE through MENOMONEE FALLS and WAUWA-TOSA to MILWAUKEE RIVER at its mouth on LAKE MICHIGAN at MILWAUKEE. Sometimes spelled Menominee. Menomonee Parkway follows mid-course.

Menomonie, city (2006 population 15,318), ⊙ DUNN county, W WISCONSIN, 20 mi/32 km WNW of EAU CLAIRE, on the RED CEDAR RIVER (forms Lake Menomonie); 44°53′N 91°54′W. Once a lumber town, it is a trade center in an area of poultry and dairy farms. Manufacturing (fabricated metal products, foods, machinery). N terminus of Red Cedar State Trail; Hoffman Hills State Recreation Area to NE. The University of Wisconsin-Stout campus is here. The ornate civic center building was erected (1890s) by a lumber baron. Plotted 1859, incorporated 1882.

Menongue (men-ON-gai), town, ⊙ CUANDO CUBANGO province, S central ANGOLA, 170 mi/274 km SSE of CUITO; 14°39′S 17°48′E. Trading center (corn, manioc, beans). Airport. Formerly Serpa Pinto.

Menorca, SPAIN: see MINORCA.

Ménoua (MAI-noo-ah), department (2001 population 372,244), West province, CAMEROON; ⊙ DSCHANG.

Mens (MAWNS), commune (□ 11 sq mi/28.6 sq km), ISÈRE department, RHÔNE-ALPES region, SE FRANCE, in DAUPHINÉ ALPS, 25 mi/40 km S of GRENOBLE; 44°49′N 05°45′E. Elevation 2,543 ft/775 m. Health resort.

Mensabé or **Puerto Mensabé** (PWER-to men-sah-BAI), village, LOS SANTOS province, S central PANAMA, minor port (6 mi/9.7 km ESE) of LAS TABLAS, on GULF OF PANAMA of the PACIFIC OCEAN. Sugarcane, coffee; livestock raising.

Mensdorf (MENS-dorf), hamlet, BETZDORF commune, E LUXEMBOURG, 8 mi/12.9 km ENE of LUXEMBOURG city; 49°39′N 06°18′E.

Menshah, El, EGYPT: see MANSHAH, EL.

Menshikov, Cape (MYEN-shi-kuhf), SE extremity of S island of NOVAYA ZEMLYA, ARCHANGEL oblast, extreme N European RUSSIA; 70°43′N 57°37′E. Also written Cape Men'shikov.

Menstrie (MEN-stree), small town, (2001 population 2,007), Clackmannanshire, central Scotland, near DEVON RIVER, 2 mi/3.2 km W of ALVA; 56°09′N 03°51′W. Previously woolen milling, furniture making. Large bonded whiskey house nearby. Formerly in Central region, abolished 1996.

Mentakab (men-TAH-kahb), town (2000 population 32,413), S central PAHANG, MALAYSIA, 6 mi/10 km WNW of TEMERLOH; 03°29′N 102°21′E. Junction of highway and E coast railroad; agricultural center.

Mentana (main-TAH-nah), town, in LATIUM, central ITALY; 42°02′N 12°38′E. On November 3, 1867, Garibaldi was defeated here by French and papal troops during his unsuccessful campaign to capture nearby ROME.

Mentasta Lake, village (2000 population 142), E ALASKA, 40 mi/64 km SW of TANACROSS, in MEN-TASTA Pass, on TOK CUT-OFF; 62°51′N 143°45′W. Sometimes called Mentasta.

Mentasta Mountains, E ALASKA, SE extension of ALASKA RANGE, between TANANA River (NE) and WRANGELL MOUNTAINS (SW); extend 50 mi/80 km NW from upper NABESNA RIVER. Rise to 8,300 ft/2,530 m (62°35′N 142°50′W). Continued SE by NUTZOTIN MOUNTAINS.

Mentawai Islands (MUHN-tah-wei), volcanic island group (□ 2,354 sq mi/6,097 sq km), INDONESIA, off W coast of SUMATRA, West Sumatra province, in INDIAN OCEAN; 02°11′S 99°40′E. Comprises approximately seventy islands. Largest island is SIBERUT; other major islands are North and South Pagai of PAGAI ISLANDS, and SIPURA. Group is generally hilly and fertile. Agriculture (sago, sugar, tobacco, coconuts); fishing. First visited by Europeans (Dutch explorers) c.1600; became Dutch possession in 1825. Commercial forestry threatens to destroy local environment and inhabitants. Also spelled Mentawei Islands.

Menteith, Lake of (MEN-teeth), lake (up to 1.5 mi/2.4 km wide), STIRLING, central Scotland, 5 mi/8 km SW of CALLANDER; 56°10′N 04°17′W. Mary, Queen of Scots, as a child, was hidden at Inchmahome priory on the largest of the lake's three islands. Formerly in Tayside, abolished 1996.

Mentese, TURKEY: see MUGLA.

Mentese Mountains (Turkish=*Menteçe*), SW TURKEY, westernmost part of the TAURUS range, extending over an area 100 mi/161 km by 60 mi/97 km, on AEGEAN (W) and MEDITERRANEAN (S) seas. Border BÜYÜK MENDERES river on N. Rise to 7,943 ft/2,421 m in BOR DAG. Rich deposits include emery, chromium, and lignite in W; emery, manganese, and silver in center; chromium, manganese, and asbestos in E.

Menthon-Saint-Bernard (mawn-TON–san–ber-NAHR), resort village (□ 2 sq mi/5.2 sq km), HAUTE-SAVOIE department, RHÔNE-ALPES region, SE FRANCE, on E shore of Lake of ANNECY, 5 mi/8 km SE of AN-NECY; 45°51′N 06°12′E. St. Bernard of Menthon born in nearby 13th–16th-century château.

Mentok, INDONESIA: see MUNTOK.

Menton (mawn-TON), town (□ 5 sq mi/13 sq km), ALPES-MARITIMES department, PROVENCE-ALPES-CÔTE D'AZUR region, SE FRANCE, near Italian border, on MEDITERRANEAN SEA, and 12 mi/19 km ENE of NICE; 43°47′N 07°30′E. A popular all-season resort of the French RIVIERA, it was part of the principality of MONACO until 1848, when it declared itself a free city under the protection of SARDINIA. It passed to France after a plebiscite in 1860. Has a 16th-century fort overlooking its harbor and a fine 17th-century baroque church. The town, surrounded by subtropical vegetation and mature lemon and olive groves, is noted for its music and art festivals. Separated from MONTE CARLO by Cape MARTIN headland. It is backed by an amphitheater of mountains which shelter it from winter's cold winds. Menton and adjacent (SW) swimming resort of ROQUEBRUNE-CAP-MARTIN form a single urban resort community. The Tropical Garden is a major visitor attraction.

Mentone (MEN-tuhn), resort town (2000 population 530), De Kalb co., NE Alabama, on Lookout Mt., 12 mi/19 km NE of Fort Payne, near Georgia state line. Founded by John Mason and named for Menton, a vacation spot in France then frequented by European aristocracy. Inc. in 1971.

Mentone, unincorporated town (2000 population 7,803), SAN BERNARDINO county, S CALIFORNIA; suburb 9 mi/14.5 km SE of SAN BERNARDINO, and 63 mi/101 km E of downtown LOS ANGELES, just E of Redlands. Poultry; dairying; citrus fruit, vegetables; nursery products. Agriculture is being displaced by urban development. Manufacturing (machinery, wood products). Redlands Municipal Airport to NW.

SAN BERNARDINO MOUNTAINS, in San Bernardino National Forest, to N.

Mentone, town (2000 population 898), KOSCIUSKO county, N INDIANA, 11 mi/18 km SW of WARSAW; 41°10′N 86°02′W. In agricultural area; eggs, chickens; lumber. Manufacturing (livestock feed mixing, poultry processing; mechanical springs, powder coating).

Mentone (MEN-ton), unincorporated village, ☉ LOVING county, W TEXAS, 19 mi/31 km NNW of PECOS, and on PECOS RIVER; 31°42′N 103°35′W. Elevation 2,683 ft/818 m. Oil and gas; some cattle. One of the smallest county seats in U.S., only locality within county.

Mentone (men-TON), residential suburb 22 km SE of MELBOURNE, VICTORIA, SE AUSTRALIA, on PORT PHILLIP BAY; 37°59′S 145°04′E. MOORABBIN Airport here.

Mentor (MEN-tuhr), city (□ 28 sq mi/72.8 sq km; 2006 population 51,593), Lake county, NE OHIO, on LAKE ERIE, 21 mi/33.6 km NE of CLEVELAND; 41°42′N 81°20′W. Manufacturing. James Garfield was living here when elected President; his home, "Lawnfield," is preserved. Founded 1799, incorporated 1855.

Mentor, village (2000 population 150), POLK county, NW MINNESOTA, 23 mi/37 km ESE of CROOKSTON; 47°42′N 96°08′W. Grain; dairying. Maple Lake to S.

Mentor-on-the-Lake (MEN-tuhr), city (2006 population 8,293), Lake county, NE OHIO, on LAKE ERIE, just NW of MENTOR, 21 mi/34 km NE of CLEVELAND; 41°43′N 81°22′W.

Mentougou (MUN-TO-GO), town, ☉ Mentougou county, CHINA, 18 mi/29 km W of BEIJING; 39°56′N 116°02′E. An administrative unit of Beijing municipality. Coal-mining center.

Mėntrida (MEN-tree-dah), town, TOLEDO province, central SPAIN, 29 mi/47 km WSW of MADRID; 40°14′N 04°11′W. Olive-oil pressing, wine making; vegetables; livestock.

Menucos, Los, ARGENTINA: see LOS MENUCOS.

Menucourt (mahn-yoo-KOOR), town (□ 1 sq mi/2.6 sq km), VAL-D'OISE department, ÎLE-DE-FRANCE region, N central FRANCE, near SEINE RIVER, and 6 km WSW of PONTOISE; 49°01′N 01°59′E.

Menuf, EGYPT: see MINUF.

Menufiya, EGYPT: see MINUFIYA.

Ménuires, Les, FRANCE: see SAINT-MARTIN-DE-BELLEVILLE.

Menuma (ME-noo-mah), town, Osato county, SAITAMA prefecture, E central HONSHU, E central JAPAN, 28 mi/45 km N of URAWA; 36°13′N 139°22′E. Spring onions, potatoes.

Menyamya, village, MOROBE province, E central PAPUA NEW GUINEA, E NEW GUINEA island, on Tauri River, in the CENTRAL HIGHLANDS, 70 mi/113 km WSW of LAE; 07°10′S 146°00′E. Coffee; timber. MENYAMYA and village of ASEKI, 30 mi/48 km SSE, are in the Anga country, named for group of ferocious warrior people which once occupied region. There are eerie burial caves with mummified bodies on platforms that overlook the valley. McAdam National Park to W.

Menyuan (MEN-yuh-WAHN), town, ☉ Menyuan county, NE QINGHAI province, CHINA, 50 mi/80 km NNW of XINING, and on Datong River; 37°27′N 101°49′E. Grain, oilseeds; livestock; food processing. Copper-ore and coal mining.

Menzaleh, Lake, EGYPT: see MANZALA, LAKE.

Menzelinsk (myen-zye-LEENSK), city (2006 population 16,910), E TATARSTAN Republic, E European Russia, on the Menzel River near its confluence with the IK RIVER, on highway junction, 180 mi/290 km E of KAZAN'; 55°43′N 53°08′E. Local transshipment point and center of agricultural area; woodworking, butter and cheese making, distillery, bakery. Founded in 1645. Made city in 1781.

Menzie (MEN-zee), community, SW MANITOBA, W central CANADA, 10 mi/16 km from STRATHCLAIR village, in STRATHCLAIR rural municipality; 50°31′N 100°29′W.

Menzies (MEN-zeez), village, S central WESTERN AUSTRALIA, on PERTH–LAVERTON railroad, 350 mi/563 km NE of Perth; 29°41′S 121°02′E. Former gold-mining center. Virtually uninhabited.

Menziken (MEN-tzee-ken), commune, AARGAU canton, N SWITZERLAND, 11 mi/18 km SE of AARAU; 47°15′N 08°11′E. Elevation 1,775 ft/541 m. Aluminum; metalware; printing. Tobacco, clothes.

Menzingen (MEN-tsing-en), residential commune, ZUG canton, N central SWITZERLAND, 4 mi/6.4 km E of ZUG; 47°11′N 08°35′E.

Menznau (MENTS-nah-ou), agricultural commune, LUCERNE canton, central SWITZERLAND, 12 mi/19 km WNW of LUCERNE; 47°05′N 08°02′E.

Meona (me-YOO-nah) or **Me'ona**, moshav, NW ISRAEL, 9 mi/15 km W of NAHARIYA; 33°01′N 35°16′E. Elevation 1,755 ft/534 m. Mixed farming; fruit; poultry. Established 1949.

Meopham (MEP-uhm), town (2001 population 8,628), NW KENT, SE ENGLAND, 7 mi/11.3 km W of ROCHESTER; 51°22′N 00°22′E. Former agricultural market. Has 15th-century church.

Meoqui (me-O-kee), city and township, CHIHUAHUA, N MEXICO, on affluent of CONCHOS RIVER, and 45 mi/72 km SE of CHIHUAHUA, on Mexico Highway 45; 28°18′N 105°30′W. Cotton center; cereals, beans, fruit, cattle.

Meota (mee-O-tuh), village (2006 population 297), W SASKATCHEWAN, W CANADA, on JACKFISH LAKE (10 mi/16 km long, 6 mi/10 km wide), 20 mi/32 km NNW of NORTH BATTLEFORD. Resort.

Meppel (ME-puhl), city, DRENTHE province, N central NETHERLANDS, at junction of the MEPPELERDIEP, Hoogeveense Vaart, and DRENTSE HOOFDVAART (Smildevaart) canals, 13 mi/21 km NNE of ZWOLLE; 52°42′N 06°12′E. Railroad junction. Dairying; cattle, sheep, poultry; eggs; grain, vegetables, sugar beets. Manufacturing (printing and publishing, food processing; tobacco products). Has fifteenth-century church, seventeenth-century weighhouse.

Meppelerdiep (ME-puh-luhr-DEEP), canal, DRENTHE and OVERIJSSEL provinces, N central NETHERLANDS; extends 6.5 mi/10.5 km NE-SW from the DRENTSE HOOFDVAART (Smildevaart) canal at MEPPEL to the Zwartewater channel at ZWARTSLUIS.

Meppen (MEP-pen), town, LOWER SAXONY, NW GERMANY, at junction of EMS RIVER, and DORTMUND-EMS CANAL (here joined by HASE RIVER), 12 mi/19 km N of LINGEN; 52°42′N 07°18′E. Manufacturing includes machinery, textiles; ironworking. In oil region. Has 15th-century town hall, 18th-century church. Chartered in 1360.

Mequinenza (mai-kee-NEN-thah), town, ZARAGOZA province, NE SPAIN, on the EBRO RIVER at influx of SEGRE RIVER, and 20 mi/32 km ENE of CASPE; 41°22′N 00°18′E. Olive oil and meat processing, flour milling. Agricultural trade (cereals, almonds; sheep). Coal and lignite mines nearby.

Mequon (ME-kawn), city (2006 population 23,600), OZAUKEE county, SE WISCONSIN, a suburb 10 mi/16 km N of downtown MILWAUKEE, on LAKE MICHIGAN and the MILWAUKEE RIVER; 43°14′N 87°59′W. Manufacturing (transportation equipment, wire forms, fabricated metal products, levels and carpentry tools, machinery, consumer goods, building materials, glass products). Roman Catholic training center, a Lutheran seminary, and an automotive museum are here. Established 1846, incorporated 1957.

Mer (MER), town (2004 population 5,830), LOIR-ET-CHER department, CENTRE administrative region, N central FRANCE, near the LOIRE RIVER, 11 mi/18 km NE of BLOIS; 47°42′N 01°30′E. Manufacturing of bedding materials; dairying; vineyards.

Mera (MAI-rah), village, PASTAZA province, E central ECUADOR, in the ANDES Mountains, on PASTAZA RIVER, and 38 mi/61 km ESE of AMBATO; 01°28′S 78°08′W. Petroleum wells and refinery nearby.

Merabello, Gulf of (me-rah-BE-lo), Greek *Kolpos Merabello*, inlet of AEGEAN SEA, on N coast of LASITHI prefecture, NE CRETE department, GREECE; 20 mi/32 km wide, 15 mi/24 km long. Named for village of Merabello (airport), N of port of AYIOS NIKOLAOS. Also called Gulf of Mirabella.

Merad (me-RAHD), village, TIPAZA wilaya, N central ALGERIA, at W edge of MITIDJA plain, 22 mi/35 km W of BLIDA. Vineyards; essential-oil processing. Formerly Meurad.

Merai (me-REI), village, EAST NEW BRITAIN province, E NEW BRITAIN island, E PAPUA NEW GUINEA, on SOLOMON SEA, 45 mi/72 km SSE of RABAUL; 04°49′S 152°21′E. Road access. Cocoa, copra, palm oil; tuna.

Merak (MUH-rahk), port in extreme NW of West Java province, INDONESIA, on SUNDA STRAIT, 60 mi/97 km W of JAKARTA; 06°01′S 106°26′E. Terminus of ferry to BAKRAUHENI (SE tip of Lampung province, SUMATRA) and PANJANG (S Lampung province, Sumatra); southernmost railroad terminus. Basalt quarries nearby.

Meråker (MER-aw-kuhr), village, NORD-TRØNDELAG county, central NORWAY, on STJØRDAL River, on railroad, and 40 mi/64 km E of TRONDHEIM; 63°26′N 11°45′E. Mining and industrial center. Winter sports area. Hydroelectric station at waterfall (S) provides power.

Merala, PAKISTAN: see MARALA.

Meramec River (mer-uh-MAK), 207 mi/333 km long, E MISSOURI; rises in the OZARK MOUNTAINS E of SALEM; meanders N, NE, and SE to the MISSISSIPPI RIVER 20 mi/32 km below ST. LOUIS. Fishing, boating, recreation. Receives the BOURBEUSE and the BIG rivers. Numerous caves and springs along it, as well as several state parks.

Merand, IRAN: see MARAND.

Merangi (mai-rahn-gee), town, VIZIANAGARAM district, ANDHRA PRADESH state, S INDIA, 9 mi/14.5 km ENE of PARVATIPURAM. Rice, oilseeds, sugarcane. Graphite deposits nearby. Also called Chinna-Merangi.

Merano (me-RAH-no), German *Meran* (me-RAHN), town, BOLZANO province, TRENTINO-ALTO ADIGE, N ITALY, near the ADIGE RIVER, 15 mi/24 km NW of BOLZANO; 46°40′N 11°09′E. Tourist and highly diversified small industrial center with sulphur refinery, foundry, fruit cannery, pottery works; furniture, jewelry, sealing wax, soap, insecticides, sausage, beer, wine. Noted for its mild climate. Has Gothic church (1367–1495), 15th-century castle, and museum

Merapi, Mount (MUH-rah-pee) [Indonesian= mountain of fire], active volcanic peak (9,550 ft/2,911 m), central JAVA, INDONESIA, 20 mi/32 km NNE of YOGYAKARTA; 07°12′S 108°27′E. Eruption in 1006 destroyed Hindu kingdom on island, and eruption in 1867 severely damaged Yogyakarta. Last eruption in 1994.

Merapi, Mount, INDONESIA: see MARAPI, MOUNT.

Mera River (me-RAH), 30 mi/48 km long, SWITZERLAND and ITALY; rises in RHAETIAN ALPS, 5 mi/8 km W of MALOJA PASS, Switzerland; flows W, through VAL BREGAGLIA, into Italy, past CHIAVENNA, and S to N end of LAKE COMO. Receives LIRO RIVER near Chiavenna. Forms small lake, Lago di Mezzola, N of Lake Como.

Merasheen Island (mee-ruh-SHEEN) (□ 46 sq mi/ 119.6 sq km), SE NEWFOUNDLAND AND LABRADOR, E CANADA, in PLACENTIA BAY, 70 mi/113 km W of St. JOHN'S; 21 mi/34 km long, 5 mi/8 km wide; 47°30′N 54°15′W. At S end is former fishing settlement of Merasheen, which was abandoned as part of government resettling program.

Merate (me-RAH-te), town, COMO province, LOMBARDY, N ITALY, near ADDA River, 11 mi/18 km S of LECCO; 45°42′N 09°25′E. Highly diversified secondary industrial center; silk mill. Astronomical observatory nearby.

Merauke (MUH-rou-kai), town, ⊙ Merauke district, IRIAN JAYA province, southeasternmost town in INDONESIA, on SE coast of NEW GUINEA island, port on ARAFURA SEA, at mouth of Merauke River (c.220 mi/ 354 km long), near border of PAPUA NEW GUINEA; 08°27′S 140°22′E. Exports copra. Has fort built in 1902. Mopah Airport is here.

Meraux (muh-RO), unincorporated city (2000 population 10,192), SAINT BERNARD parish, extreme SE LOUISIANA, suburb 8 mi/13 km E of downtown NEW ORLEANS, on the MISSISSIPPI RIVER, between CHALMETTE and VIOLET; 29°56′N 89°55′W. Manufacturing (gasoline, kerosene). Holds Louisiana Shrimp Festival.

Merawi, town (2007 population 16,784), AMHARA state, NW ETHIOPIA, on road, 20 mi/32 km SW of BAHIR DAR; 11°25′N 37°10′E. In agricultural area S of LAKE TANA.

Merbabu, Mount (MUHR-bah-boo), volcanic peak (10,308 ft/3,142 m), central JAVA, INDONESIA, 30 mi/48 km WNW of SURAKARTA; 07°27′S 110°26′E. Also spelled Mount Merbaboe.

Merbein (muhr-BEEN), town, NW VICTORIA, SE AUSTRALIA, on MURRAY RIVER, and 205 mi/330 km ENE of ADELAIDE, near MILDURA; 34°11′S 142°04′E. Fruit-growing center; wineries. Agricultural experiment station. Zoo just outside town.

Merbes-le-Château (MERB–luh–shah-TO), commune (2006 population 4,130), Thuin district, HAINAUT province, S BELGIUM, on SAMBRE RIVER, and 6 mi/9.7 km WSW of THUIN; 50°19′N 04°09′E.

Merbok River (MER-bok), 25 mi/40 km long, MALAYSIA; flows SW to Strait of MALACCA at Tanjong Dawai.

Merca, SOMALIA: see MARKA.

Mercadal (mer-kah-DAHL), town, MINORCA island, BALEARIC ISLANDS, SPAIN, 12 mi/19 km NW of MAHÓN. Grain growing; livestock raising; flour milling; hunting.

Mercaderes (mer-kah-DAI-res), town, ⊙ Mercaderes municipio, CAUCA department, SW COLOMBIA, 52 mi/ 84 km SW of POPAYÁN; 01°48′N 77°10′W. Elevation 3,031 ft/923 m. Sugarcane, coffee, corn; livestock.

Mercado, Cerro de, MEXICO: see VICTORIA DE DURANGO.

Mercan Daği (Turkish=*Mercan Daği*), peak (11,315 ft/ 3,449 m), E central TURKEY, 13 mi/21 km SSE of ERZINCAN, in Mercan Mountains, a range extending 40 mi/64 km S of the EUPHRATES RIVER. The source of two tributaries of the Euphrates. Town of PULUMUR on S slope.

Mercantour (mer-kawn-TOOR), Alpine mountain range, ALPES-MARITIMES department, PROVENCE-ALPES-CÔTE D'AZUR region, SE FRANCE, extending NW-SE along Italian border; 43°55′N 07°10′E. Highest point is 10,200 ft/3,109 m. BARCELONNETTE (in UBAYE valley) is N entry point. In MERCANTOUR NATIONAL PARK.

Mercantour National Park (mer-kawn-TOOR), French *Parc National du Mercantour* (pahrk nah-syonahl dyoo mer-kahn-toor) (□ 264 sq mi/686.4 sq km), ALPES-MARITIMES department, PROVENCE-ALPES-CÔTE D'AZUR region, SE FRANCE; 44°10′N 07°00′E. Includes Mercantour crystalline massif, rising to 10,312 ft/3,143 m. Embraces the Mercantour mountain range rich in scenery and diverse flora and fauna. Many trails have been created to provide access for hikers. The nature preserve of ARGENTERA is a smaller E extension on the the Italian side of range. Park headquarters is in NICE. Established 1979.

Mercara (mer-KAH-rah), city, ⊙ KODAGU district, KARNATAKA state, S INDIA, 65 mi/105 km W of MYSORE, on central plateau of state. Trade center for products of surrounding coffee, tea, rubber, cardamom, and sandalwood plantations. Has eighteenth-

century stone fort. K. M. Cariappa, first Indian commander in chief of Indian army, born here (1900).

Mercato San Severino (mer-KAH-to sahn se-ve-REE-no), town, Salerno province, CAMPANIA, S ITALY, 8 mi/ 13 km N of SALERNO; 40°47′N 14°46′E. Railroad junction. Cotton milling, tomato canning, wine making.

Merced (muhr-SED), county (□ 1,929 sq mi/5,015.4 sq km; 2006 population 245,658), central CALIFORNIA; ⊙ MERCED; 37°11′N 120°43′W. Extends across San Joaquin Valley from DIABLO RANGE (W and SW) to foothills of the SIERRA NEVADA (E and NE). Fertile agricultural area, irrigated by Merced, San Joaquin, and Chowchilla rivers. Grapes, alfalfa, grain, sweet potatoes, tomatoes, corn, cantaloupes, wheat, barley, oats, rice, beans, sugar beets; dairying; cattle, turkeys and poultry raising. Sand and gravel. Processing of farm products (fruit drying and canning, meat and poultry packing), lumber milling, cement manufacturing. Crossed in SW (SE to NW) by DELTA-MENDOTA CANAL and California Aqueduct. San Luis and Merced national wildlife refuges in center; George Hatfield and Fremont Ford state parks in NW; San Luis Reservoir and State Recreation Area in W; Ortigalita Peak (3,305 ft/1,007 m) in S corner (Diablo Range). Formed 1855.

Merced (muhr-SED), city (2000 population 63,893), ⊙ MERCED county, central CALIFORNIA, 50 mi/80 km NW of FRESNO; 37°18′N 120°29′W. Growing city and center for tourism and farm trade (cotton, fruit, and dairying; poultry; grain, alfalfa, almonds, sugar beets). Concentration of Hmong immigrants. Manufacturing (prefabricated wood buildings, fabricated metal products, machinery, transportation equipment, paper products). Seat of Merced College (two-year). Merced and San Luis national wildlife refuges to SW; Castle Air Force Base (closed 1995) to NW, YOSEMITE NATIONAL PARK c.50 mi/80 km NE. Professional basketball player Ray Allen born here. Incorporated 1889.

Mercedario, Cerro (mer-sai-DAH-ree-o, SER-ro), ANDEAN peak (21,885 ft/6,671 m), in SW SAN JUAN province, ARGENTINA, near CHILE border, 60 mi/97 km SW of TAMBERÍAS; 31°58′S 70°07′W.

Mercedes (MUHR-sai-deez), city (2006 population 14,734), HIDALGO county, extreme S TEXAS, 12 mi/19 km WSW of HARLINGEN, in the lower Rio Grande valley, 70 mi/113 km N of Rio Grande (MEXICO border); 26°08′N 97°55′W. Manufacturing (footwear, wood products, cheese, clay products, sheet metal, machine-shop products), meatpacking. Irrigated citrus, vegetable region. Pipeline (1,840 mi/2,961 km long) to NEW YORK area from oil field here. Founded 1907, incorporated 1909.

Mercedes (mer-SAI-dees), city (2004 population 42,032), ⊙ SORIANO department, SW URUGUAY, a port on the Río NEGRO; 33°16′S 58°01′W. An agriculture and livestock center, the city has a shipyard and several fine beaches and resorts. Tourism is an important industry. Founded in 1781. Has a famous cathedral.

Mercedes (mer-SAI-des), town (1991 population 45,895), ⊙ Mercedes district (□ 405 sq mi/1,053 sq km), N BUENOS AIRES province, ARGENTINA, on LUJÁN river, and 60 mi/97 km W of BUENOS AIRES. Railroad center. Has national college, Gothic church. Founded 1779.

Mercedes (mer-SAI-des), town (1991 population 24,975), ⊙ Mercedes department, central CORRIENTES province, ARGENTINA, on railroad, 160 mi/257 km SSE of CORRIENTES. Agricultural center (tobacco, olives, citrus fruit; livestock.

Mercedes (mer-SAI-des), town, OCOTEPEQUE department, SW HONDURAS, 5 mi/8 km E of border with EL SALVADOR, 15 mi/24 km SW of NUEVA OCOTEPEQUE; 14°17′N 88°58′W. Small farming (grain); livestock.

Mercedes de Oriente (mer-SAI-des dai o-ree-EN-tai), town, LA PAZ department, S central HONDURAS, 5 mi/8 km N of border with EL SALVADOR, 26 mi/42 km S of

LA PAZ; 13°54′N 87°43′W. Subsistence agriculture; grain, beans; livestock.

Mercedes la Ceiba (mer-SAI-des lah SAI-bah), municipality and town, in extreme N LA PAZ department, EL SALVADOR, NNW of ZACATECOLUCA; 13°38′N 88°54′W.

Mercedes Umaña (mer-SAI-des oo-MAHN-yah), municipality and town, USULUTÁN department, EL SALVADOR, on INTER-AMERICAN HIGHWAY, N of TECAPA Volcano; 13°34′N 88°30′W.

Merced Falls, village, MERCED county, central CALIFORNIA, on MERCED RIVER, near SIERRA NEVADA foothills, and 17 mi/27 km NE of MERCED. Cattle; alfalfa, grain, fruit, nuts. Large tailings left by gold-dredging operations. Lake McClure formed by New Exchequer Dam, to NE.

Merced, La, ARGENTINA: see LA MERCED.

Merced River, c.45 mi/72 km long, central CALIFORNIA; rises in Edna Lake in the SIERRA NEVADA and in SW YOSEMITE NATIONAL PARK, extreme NE MADERA county; flows N briefly, then W through small Merced Lake and dramatic Yosemite Valley, at center of park, then SW; receives South Fork from E 10 mi/16 km W of park boundary, continues WSW through Lake McClure, formed by New Exchequer Dam, past Livingston, to SAN JOAQUIN RIVER, c.25 mi/40 km WNW of Merced. South Fork rises 1 mi/1.6 km S of source of main stream, flows SW then W through S part of Yosemite National Park, then NW to Merced River.

Mercer, county (□ 568 sq mi/1,476.8 sq km; 2006 population 16,786), NW ILLINOIS; ⊙ ALEDO; 41°12′N 90°44′W. Bounded W by MISSISSIPPI RIVER; drained by EDWARDS RIVER and POPE CREEK. Agriculture (cattle, hogs; corn, soybeans, hay; dairy products). Some manufacturing. Formed 1825.

Mercer, county (□ 253 sq mi/657.8 sq km; 2006 population 21,818), central KENTUCKY; ⊙ HARRODSBURG; 37°47′N 84°52′W. Bounded NE by KENTUCKY RIVER, SE by DIX RIVER; forms HERRINGTON LAKE by Dix Dam; drained by SALT and Chaplin rivers. Rolling agricultural area in BLUEGRASS REGION (burley tobacco, soybeans, wheat, corn, hay, alfalfa; hogs, cattle, poultry; dairying); calcite mines, limestone quarries. Manufacturing at Harrodsburg. Includes Old Fort Harrod (E center) and High Bridge state parks. Formed 1785.

Mercer, county (□ 456 sq mi/1,185.6 sq km; 2006 population 3,584), N MISSOURI; ⊙ PRINCETON; 40°25′N 93°34′W. Drained by WELDON and THOMPSON rivers. Soybeans, corn; hogs (corporate hog farms), cattle. Manufacturing at Princeton. Lake Paho W of Princeton. Formed 1845.

Mercer, county (□ 228 sq mi/592.8 sq km; 2006 population 367,605), W NEW JERSEY, bounded W by the DELAWARE RIVER; ⊙ TRENTON; 40°16′N 74°42′W. Varied manufacturing and agriculture. Crossed by DELAWARE AND RARITAN CANAL; drained by MILLSTONE RIVER and CROSSWICKS CREEK. Formed 1837.

Mercer, county (□ 1,041 sq mi/2,706.6 sq km; 2006 population 8,234), central NORTH DAKOTA; ⊙ STANTON; 47°17′N 101°49′W. Agricultural area drained by KNIFE RIVER; bounded N and E by MISSOURI RIVER. Lignite mines. Wheat; cattle. GARRISON DAM forms LAKE SAKAKAWEA on N; Sakakawea State Park at Garrison Dam in NE; Knife River Indian Village Historic Site is in E; part of Fort Berthold Indian Reservation in NW. Border between Mountain and Central time zones follows Missouri River; county is in Mountain time zone. Formed in 1875 and government organized in 1884. Named for William Henry Harrison Mercer (1884–1901), an early rancher.

Mercer (MUHR-suhr), county (□ 454 sq mi/1,180.4 sq km; 2006 population 41,303), W OHIO, on INDIANA state line; ⊙ CELINA; 40°33′N 84°38′W. Drained by WABASH and ST. MARYS rivers; part of GRAND LAKE ST. MARYS reservoir is in E. Includes Fort Recovery State Park. In the Till Plains physiographic region. Agricultural area (poultry, sheep; corn, soybeans). Manufacturing (meat products, machinery; printing and publishing). Limestone quarries; timber. Formed 1824.

Mercer, county (□ 682 sq mi/1,773.2 sq km; 2006 population 118,551), NW PENNSYLVANIA, on OHIO (W) state line; ⊙ MERCER; 41°18′N 80°15′W. Drained by SHENANGO RIVER and Neshannock and Wolf creeks. Agriculture (corn, wheat, oats, barley, hay, alfalfa, potatoes; hogs, cattle; dairying). Bituminous coal, sandstone, limestone. Manufacturing at SHARON, FARRELL, GREENVILLE, Mercer, and GROVE CITY. LAKE WILHELM reservoir in NE; large SHENANGO RIVER LAKE reservoir, on SHENANGO RIVER and its tributary, PYMATUNING CREEK in W. Settled by veterans of Revolution. Formed 1800.

Mercer, county (□ 421 sq mi/1,094.6 sq km; 2006 population 61,278), S WEST VIRGINIA, on ALLEGHENY PLATEAU, and on VIRGINIA (S) border; ⊙ PRINCETON; 37°24′N 81°06′W. Drained by BLUESTONE RIVER. BLUEFIELD, semibituminous-coal-mining center in POCAHONTAS coalfield, is partly in Virginia. Coal, limestone deposits. Timber. Manufacturing at Bluefield and Princeton. Agriculture (corn, oats, tobacco, potatoes, alfalfa, hay, nursery crops); cattle, sheep. Includes Camp Creek State Forest in N, and Pinnacle Rock State Park in SW. Formed 1837.

Mercer (MUHR-suhr), town, SOMERSET county, central MAINE, on SANDY RIVER, and 12 mi/19 km SW of SKOWHEGAN; 44°40′N 69°54′W. Farming, lumbering.

Mercer, town (2000 population 342), MERCER county, N MISSOURI, near WELDON RIVER, 9 mi/14.5 km N of PRINCETON; 40°30′N 93°31′W.

Mercer, village (2006 population 78), MCLEAN county, central NORTH DAKOTA, 48 mi/77 km NE of BISMARCK; 47°29′N 100°42′W. Several small lakes to NW. Founded in 1905 and named for William Henry Harrison Mercer, early settler and rancher.

Mercer, village, IRON county, N WISCONSIN, 20 mi/32 km SSE of HURLEY. In wooded lake region; fishing. Manufacturing (forest products). Nearby is a fish hatchery. Large Turtle–Flambeau flowage reservoir to SW; Lac du Flambeau Indian Reservation and Northern Highland State Forest to SE.

Mercer, borough (2006 population 2,269), ⊙ MERCER county, W PENNSYLVANIA, 22 mi/35 km ENE of YOUNGSTOWN (Ohio), near Otter Creek; 41°13′N 80°14′W. Agriculture (potatoes, corn, hay; livestock; dairying). Manufacturing (fabricated metal products, transportation equipment, food products, machinery); bituminous coal. SHENANGO RIVER LAKE reservoir to NW; Lake Latonka reservoir (residential development) to NE. Settled 1795, laid out 1803, incorporated 1814.

Mercer Island, city (2006 population 23,463), KING county, W WASHINGTON, residential suburb 4 mi/6.4 km ESE of downtown SEATTLE, including all of MERCER ISLAND, in S end of LAKE WASHINGTON; 47°34′N 122°14′W. In MERCER ISLAND (Morrow) Bridge connects N end of island to both shores of lake. Manufacturing (machinery, wood products, furniture, fabricated metal products).

Mercer Island, WASHINGTON: see WASHINGTON LAKE.

Mercersburg, borough (2006 population 1,554), FRANKLIN county, S PENNSYLVANIA, 15 mi/24 km SW of CHAMBERSBURG; 39°49′N 77°54′W. Agricultural area (grain, potatoes, apples; poultry, livestock; dairying). Manufacturing (lumber, machinery, medical equipment, apparel). Buchanan's Birthplace Historic State Park to NW; Buchanan State Forest to NW; COVE MOUNTAIN ridge to W. Settled c.1729, laid out 1780, incorporated 1831.

Mercês (mer-SES), city (2007 population 10,928), S MINAS GERAIS state, BRAZIL, 40 mi/64 km N of JUIZ DE FORA; 21°12′S 43°45′W. Terminus of railroad from SANTOS DUMONT. Coffee, dairy products Mica mining.

Merchantville, residential borough (2006 population 3,806), CAMDEN county, SW NEW JERSEY, just E of CAMDEN; 39°57′N 75°02′W. Settled 1852, incorporated 1874.

Merchtem (MERKH-tem), commune (2006 population 14,938), Halle-Vilvoorde district, BRABANT province, central BELGIUM, 9 mi/14.5 km NW of BRUSSELS; 50°58′N 04°14′E. Market center for poultry region.

Merchweiler (MERKH-vei-luhr), town, SAARLAND, W GERMANY, 6 mi/9.7 km WSW of NEUNKIRCHEN; 49°20′N 07°05′E. Coal mining; iron smelting.

Mercia (MUH-see-uh), one of the former kingdoms of Anglo-Saxon ENGLAND, consisting generally of the region of the MIDLANDS. Settled by Angles c.500, probably first along the Trent valley. Its history emerges from obscurity with the reign of Penda, who extended his power over Wessex (645) and East ANGLIA (650) to gain overlordship of England S of the HUMBER RIVER. Penda's son, Wulfhere, reestablished a Greater Mercia that finally extended over all S England. This hegemony was strengthened by Offa (reigned 757–796), who controlled East Anglia, KENT, and SUSSEX. He had the great Offa's Dyke built to protect W Mercia from the Welsh. After his death, Mercian power gradually gave way before that of Wessex. In 874, Mercia weakly succumbed to the invading Danish army, and ultimately the E part became (886) a portion of the Danelaw, while the W part was controlled by Alfred of Wessex. Thereafter Mercia had no independent history, although it had one more distinguished ruler in Æthelflæd, Lady of the Mercians.

Mercier (mer-see-ER), canton, NICOLÁS SUÁREZ province, PANDO department, NW BOLIVIA, at the intersection of the Ina and Chipamanu rivers, 6 mi/10 km NW of PORVENIR, and 10 mi/15 km SE of COBIJA, on the paved Cobija-Porvenir road; 11°13′S 68°43′W. Elevation 715 ft/218 m. Agriculture (rice, rubber, yucca, bananas, cacao, coffee, tobacco, cotton, peanuts); cattle and horse raising.

Mercier (mer-SYAI), city (□ 18 sq mi/46.8 sq km; 2006 population 10,082), MONTÉRÉGIE region, S QUEBEC, E CANADA, 15 mi/24 km from MONTREAL; 45°19′N 73°45′W. Part of the Metropolitan Community of Montreal (*Communauté Métropolitaine de Montréal*).

Mercier–Hochelaga-Maisonneuve (mer-SYAI—hahshuh-LA-guh—me-zo-NUV), borough (French *arrondissement*) of MONTREAL, S QUEBEC, E CANADA, on E coast of MONTREAL ISLAND; 45°33′N 73°32′W.

Mercoal, unincorporated village, W ALBERTA, W CANADA, in ROCKY MOUNTAINS, near E side of JASPER NATIONAL PARK, on McLeod River, and 40 mi/64 km SW of EDSON, in YELLOWHEAD COUNTY; 53°10′N 117°06′W. Coal mining; timber.

Mercur, town, TOOELE county, NW UTAH, 20 mi/32 km WSW of LEHI, in OQUIRRH MOUNTAINS. Elevation 6,700 ft/2,042 m. Deposits of gold and silver. Barnick Mercur Gold Mine Historical Site.

Mercurea, ROMANIA: see MIERCUREA.

Mercurea-Ciuc, ROMANIA: see MIERCUREA-CIUC.

Mercurea-Niraj, ROMANIA: see MIERCUREA-NIRAJ.

Mercurey (mer-kyoo-REE), village, SAÔNE-ET-LOIRE department, in BURGUNDY, E central FRANCE, 8 mi/12.9 km NW of CHALON-SUR-SAÔNE. Noted Burgundy wines.

Mercury (MUHR-cyuh-ree), unincorporated village, MCCULLOCH county, near geographical center of TEXAS, 22 mi/35 km NE of BRADY.

Mercury Bay, irregular inlet facing PACIFIC OCEAN, NORTH ISLAND, NEW ZEALAND, on E coast of COROMANDEL peninsula, W of BAY OF PLENTY; 7 mi/11.3 km wide (across mouth, N–S), 11 mi/18 km long. Partly in Whanganui-A-Hei marine reserve.

Mercury Islands, E of COROMANDEL peninsula, NORTH ISLAND, NEW ZEALAND, offshore from MERCURY BAY; 36°35′S 175°55′W. Includes Great Mercury (8 mi/13 km long), Red Mercury, several islets. Part of mainland and islands (except Great Mercury) in Whanganui-A-Hei marine reserve.

Mercy, Cape, SE BAFFIN ISLAND, BAFFIN region, NUNAVUT territory, N CANADA, on DAVIS STRAIT, on N side of entrance of CUMBERLAND SOUND; 64°56′N 63°39′W.

Mer de Glace (mer duh GLAHS) [French=sea of ice], glacier (□ 16 sq mi/41.6 sq km), HAUTE-SAVOIE department, RHÔNE-ALPES region, E FRANCE, on the N slope of MONT BLANC, formed by the junction of three smaller glaciers; 3.5 mi/5.6 km long. Reaches to within 1,200 ft/366 m of CHAMONIX proper at its lowest point. There are deep crevasses and high seracs (ice needles).

Merdenik, TURKEY: see GOLE.

Merdrignac (mer-dreen-YAHK), commune (□ 22 sq mi/57.2 sq km), CÔTES-D'ARMOR department, BRITTANY, NW FRANCE, 25 mi/40 km SW of DINAN; 48°12′N 02°25′W. Mixed grains for poultry and pig raising.

Mere (MIR), village (2001 population 2,633), SW WILTSHIRE, S ENGLAND, 7 mi/11.3 km NW of SHAFTESBURY; 51°06′N 02°16′W. Former agricultural market in dairying region. Previously flour mills, limestone quarries. Has 13th-century church and 15th-century chantry.

Meredith (ME-ruh-dith), town, VICTORIA, SE AUSTRALIA; 37°51′S 144°05′E. Annual music festival.

Meredith, town, BELKNAP county, central NEW HAMPSHIRE, 8 mi/12.9 km N of LACONIA; 43°38′N 71°30′W. Bounded NE by LAKE WINNIPESAUKEE, which dominates E part of town, SE by WINNISQUAM LAKE. Agriculture (cattle, poultry; dairying; nursery crops; timber); manufacturing (consumer goods, electronic goods; machining). Resort, water sports. Site of Annalee Doll Museum. WAUKEWAN LAKE on NW border, Wickwas Lake in S. Incorporated 1768.

Meredith, Lake, reservoir (□ 10 sq mi/26 sq km), CROWLEY county, SE COLORADO, on Bob Creek, 14 mi/23 km NNW of LA JUNTA; 38°10′N 103°44′W. Maximum capacity 41,412 acre-ft. Fed by Colorado Canal. Formed by LAKE MEREDITH DAM (30 ft/9 m high), built (1900) for irrigation; also used for recreation and as a fish and wildlife pond.

Meredith Lake (ME-ruh-dith), reservoir (□ 48 sq mi/124.8 sq km), POTTER, MOORE, and HUTCHINSON counties, NW TEXAS, on CANADIAN RIVER, in Lake Meredith National Recreation Area, 37 mi/60 km NNE of AMARILLO; c.30 mi/48 km long; 35°43′N 101°34′W. Maximum capacity 2,434,220 acre-ft. Formed by Sanford Dam (228 ft/69 m high), built (1965) by the Bureau of Reclamation for water supply; also used for flood control and recreation. ALIBATES FLINT QUARRIES NATIONAL MONUMENT on SE shore.

Meredosia (mer-i-DO-shah), village (2000 population 1,041), MORGAN county, W central ILLINOIS, on ILLINOIS RIVER, and Meredosia Lake (c.5 mi/8 km long; a slough lake of Illinois River), 18 mi/29 km WNW of JACKSONVILLE; 39°49′N 90°33′W. In agricultural area (corn, wheat, soybeans, sorghum; cattle, hogs; dairying); manufacturing (chemicals).

Mère et l'Enfant, La (MER ai lahn-FAWN, lah) [French=mother and child], spur of the S TRUONG SON RANGE, in central VIETNAM, extending NE c. 20 mi/32 km to CAPE DAI LANH (or Mui Dao Lanh); 20°55′N 109°13′E. Highest peak (6,729 ft/2,051 m high) rises 33 mi/53 km NW of NHA TRANG.

Merefa (me-RE-fah), city, N central KHARKIV oblast, UKRAINE, 12 mi/19 km SW of KHARKIV; 49°49′N 36°03′E. Elevation 413 ft/125 m. Railroad junction. Manufacturing (bricks, reinforced concrete fabrications, glass, footwear); flour milling, yeast making; woodworking. Ukrainian Scientific Research Institute of Vegetable and Melon Cultivation. Known since the 17th century as a Cossack settlement; company center in Kharkiv regiment (17th–18th century); city since 1938.

Meregh, SOMALIA: see MAREG.

Merelbeke (MAI-ruhl-bai-kuh), commune (2006 population 22,386), Ghent district, EAST FLANDERS province, NW BELGIUM, S suburb of GHENT; 51°00′N 03°45′E.

Merend, IRAN: see MARAND.

Merendón, Cordillera del (mai-ren-DON, kor-dee-YAI-rah del), N spur of main Continental Divide in W HONDURAS; extends from San Jerónimo peak on Continental Divide 12 mi/19 km NE to ERAPUCA peak (7,898 ft/2408 m), here joining SIERRA DEL GALLINERO; 14°33′N 89°09′W. The name Merendón is sometimes applied to the entire great mountain chain of W Honduras, including Sierra del Gallinero, SIERRA DEL ESPÍRITU SANTO, SIERRA DE LA GRITA, and SIERRA DE OMOA.

Mere Point (MIR), SW MAINE, peninsula extending 4 mi/6.4 km into CASCO BAY, near BRUNSWICK. Site of summer colony.

Mere, The, ENGLAND: see ELLESMERE.

Merevari River, VENEZUELA: see CAURA RIVER.

Méréville (mai-rai-veel), agricultural commune (□ 10 sq mi/26 sq km), ESSONNE department, ÎLE-DE-FRANCE region, N central FRANCE, 9 mi/14.5 km SSW of ÉTAMPES, in the BEAUCE region; 48°18′N 02°05′E. Cereals; poultry. Its park, laid out in 18th century by the architect Belanger, resembles a British garden with bridges and statuary.

Merewether (ME-ree-we-thur), residential SW suburb of NEWCASTLE, NEW SOUTH WALES, SE AUSTRALIA, on E coast; 32°57′S 151°46′E. Swimming beach.

Mergentheim, Bad, GERMANY: see BAD MERGENTHEIM.

Merghi, MYANMAR: see MERGUI.

Mergozzo (mer-GO-tso), village, NOVARA province, PIEDMONT, N ITALY, on Lake Mergozzo (1.5 mi/2.4 km long; separated from Lago MAGGIORE by TOCE RIVER delta), 6 mi/10 km NW of PALLANZA; 45°58′N 08°26′E. Granite quarries at Monte Orfano, 1 mi/1.6 km S. Marble quarries nearby furnished the marble for MILAN's cathedral.

Mergui (mer-GWEE), formerly southernmost district, currently a township (□ 11,325 sq mi/29,445 sq km) of TENASSERIM division, Lower BURMA; capital of MERGUI. Narrow strip of land (260 mi/418 km long, 50 mi/80 km wide) in TENASSERIM RANGE between ANDAMAN SEA and THAILAND border, ending S at KOW SONG; drained by TENASSERIM RIVER, MERGUI ARCHIPELAGO is off irregular coast (mangrove swamps). Densely forested; tin, tungsten mining, especially in N; iron ore (unexploited); rubber plantations.

Mergui, town, ⊙ MERGUI township, TANINTHARYI division, MYANMAR, in TENASSERIM, port on ANDAMAN SEA, on island at mouth of TENASSERIM RIVER, and 120 mi/193 km S of DAWEI; sheltered by KING ISLAND. Trade (rice, salt fish, tin, tungsten) with MYANMAR and MALAYSIA; pearl fisheries; birds' nests. Airport. Ancient capital of Thai province; occupied by British East India Company after massacre of European settlers (1695). Captured 1824 in first Anglo-Burmese War by British forces from DAWEI. Formerly spelled Merghi.

Mergui Archipelago (mer-GWEE), island group in ANDAMAN SEA, off TENASSERIM coast, MYANMAR, between 09°00′N and 13°00′N. Summits of submerged mountain ridges; consists of c.900 islands ranging in size from rocks to KING ISLAND. Includes TAVOY, ELPHINSTONE, ROSS, SELLORE, BENTINCK, DOMEL, KISSERAING, SULLIVAN, and SAINT MATTHEW's islands. Mountainous and jungle-covered irregular coast, often set in mangrove swamps. Sparsely inhabited by Salons, related to the Malays. Produces birds' nests, bêche-de-mer, pearls; tin and tungsten mines; rubber plantations on some islands. Trade with mainland. Archipelago known for its beauty.

Merhavya (mer-khahv-YAH), kibbutz, N ISRAEL, 1 mi/1.6 km E of AFULA, in JEZREEL VALLEY; 32°36′N 35°19′E. Elevation 164 ft/49 m. Mixed farming. Plastic products; printing and publishing (archive of the Hashomer Hatza'ir movement). Founded in 1922, kibbutz in 1929. Stands on the site of the Crusader fortress La Feve, where a fierce Crusader-Muslim battle was fought in 1183 and where Napoleon's soldiers fought the Turks in 1799.

Méri (MAI-ree), town, Far-North province, CAMEROON, 18 mi/29 km NW of MAROUA; 10°46′N 14°07′E.

Meribah, village, SE SOUTH AUSTRALIA, 125 mi/201 km E of ADELAIDE, near VICTORIA border; 34°42′S 140°51′E. Wheat, wool.

Méribel-les-Allues (mai-ree-bel–laiz–AHL-loo), winter-sport resort, SAVOIE department, RHÔNE-ALPES region, SE FRANCE, in the VANOISE Massif of the Savoy Alps (ALPES FRANÇAISES), 5 mi/8 km SSE of MOÛTIERS; 45°23′N 06°35′E. Elevation of ski terrain 4,800 ft/1,463 m–9,000 ft/2,743 m. Forms part of a vast interconnected ski complex known as Les TROIS-VALLÉES (French=three valleys), which extends from COURCHEVEL (N) to VAL-THORENS (S). A planned community created in 1960s, Méribel's ski village has architectural unity and blends in with the terrain.

Meric, Turkish *Meriç*, village, European TURKEY, 30 mi/48 km S of EDIRNE; 41°12′N 26°24′E. Grain, rice. Formerly called Büyükdoganca and Kavakli.

Meriç, SE Europe: see MARITSA.

Merichleri (me-REECH-ler-ee), city, HASKOVO oblast, DIMITROVGRAD obshtina, S central BULGARIA, 10 mi/16 km ESE of CHIRPAN; 42°08′N 25°29′E. Spa (mineral springs). Vineyards; cotton, sesame.

Mérida (MAI-ree-dah), state (□ 4,360 sq mi/11,336 sq km; 2001 population 763,700), W VENEZUELA; ⊙ MÉRIDA; 08°30′N 71°10′W. Mountainous state traversed SW-NE by great ANDEAN spur Cordillera de MÉRIDA, rising in LA COLUMNA to highest elevation (16,411 ft/5,002 m) in Venezuela. A narrow neck of state borders on Lake MARACAIBO. Between the high ranges flows CHAMA RIVER, its course being followed by the transandine highway; in river's fertile upland valley are most of the important settlements. Mineral resources include petroleum (on Lake Maracaibo), mica (near TIMOTES and CHACHOPO), gold (MESA BOLÍVAR, ZEA), emeralds (SAN RAFAEL). Predominantly an agricultural region; produces corn, coffee, sugarcane, cacao, tobacco, cotton, yucca, bananas, plantains, fruits; wheat, barley, potatoes; cattle in higher altitudes. Known for dairy products Exports coffee, fiber bags, tobacco, butter, hides. Mérida, a university city, is its commercial and manufacturing center. Other main cities include EL VIGÍA, Timotes, Santo Domingo, BAILADORES, MUCUCHÍES, Jají. Includes the Parque Nacional Sierra La Culata and parts of Paramos Batallón and La Negra, El Tamá and Tapó-Caparo, and SIERRA NEVADA. Sierra de la Culata

Mérida (ME-ree-dah), city (2005 population 734,153) and township, ⊙ YUCATÁN state, SE MEXICO; 20°59′N 89°39′W. It is the chief commercial, communications, and cultural center of the YUCATÁN peninsula. Founded (1542) by Francisco de Montejo (the younger), on the site of a ruined Mayan city, Mérida has many fine examples of Spanish colonial architecture, notably the sixteenth-century cathedral. Rooftop windmills, characteristic of this region, are used to pump water from underground wells and streams. Commercial, administrative, agricultural, and tourist center. Once dependent upon the large crops of henequen from the surrounding region. Tourists visiting nearby Mayan ruins, notably CHICHÉN ITZÁ and UXMAL, contribute work to the local economy. International airport to SW.

Mérida, city (2001 population 50,271), BADAJOZ province, SW SPAIN, in EXTREMADURA, on the GUADIANA RIVER, and c.40 mi/64 km S of CÁCERES; 38°55′N 06°20′W. It is a railroad hub and agricultural center, producing textiles, leather, and cork. The colony Emerita Augusta, founded by the Romans in the 1st century B.C.E, Mérida became the capital of LUSITANIA. Its Roman remains, among the most important in Spain, include a magnificent bridge, triumphal arch, theater with marble columns, aqueduct, temple, imposing circus, and amphitheater. Later the chief city of Visigothic Lusitania. It fell (713) to the Moors, under whom it prospered. Conquered (1228) by Alfonso IX of León, it was given to the Knights of Santiago; thereafter quickly declined.

Mérida (MAI-ree-dah), city, ⊙ MÉRIDA state, W VENEZUELA; 08°35′N 71°08′W. The highest city in Venezuela, Mérida has fishing and a variety of light manufacturing, in addition to such popular tourist activities as skiing, mountain biking, and mountaineering in the Cordillera de MÉRIDA. Founded in 1558, it became a religious and educational center. Seat of the Universidad de los Andes (1785). Famous teleferico (cable car) to top of Pico Espejo. Venezuela's highest mountain, PICO BOLÍVAR (LA COLUMNA), is nearby. Airport.

Merida (me-REE-dah), town, LEYTE province, W LEYTE, PHILIPPINES, on ORMOC BAY, 8 mi/12.9 km SW of ORMOC; 10°59′N 124°30′E. Agricultural center (coconuts, rice).

Mérida, Cordillera de (MAI-ree-dah, kor-dee-YAI-rah dai) or **Mérida, Sierra Nevada de** (MAI-ree-dah, see-ER-rah nai-VAH-dah dai), mountain range, NW VENEZUELA, a spur of the ANDES, extending c.200 mi/322 km NE from the COLOMBIAN border to the Caribbean coastal range; 08°40′N 71°00′W. From 30 mi/48 km to 50 mi/80 km wide, it rises between the Orinoco llanos and MARACAIBO lowlands to perpetually snow-capped peaks. PICO BOLÍVAR (LA COLUMNA) is the highest point in Venezuela. Coffee. Also sometimes called Sierra de Mérida.

Meriden (ME-ri-duhn), city (2000 population 58,244), NEW HAVEN county, S central CONNECTICUT; 41°32′N 72°47′W. Known for its silver industry. Silverware and pewter were made here in the 18th century by Samuel Yale and later by the Rogers Brothers and a forerunner of the International Silver Company. Industry now diversified. Settled 1661, incorporated as a town 1806, as a city 1867, town and city consolidated 1922.

Meriden (ME-ri-duhn), town (2000 population 184), CHEROKEE county, NW IOWA, 5 mi/8 km WNW of CHEROKEE; 42°47′N 95°37′W. In agricultural area.

Meriden, town (2000 population 706), JEFFERSON county, NE KANSAS, on branch of DELAWARE RIVER, and 13 mi/21 km W of OSKALOOSA; 39°11′N 95°34′W. Grain growing; dairying and general agriculture.

Meriden, NEW HAMPSHIRE: see PLAINFIELD.

Meridi (me-REE-dee), agricultural village, W EQUATORIA state, S SUDAN, near CONGO border, on road, and 150 mi/241 km W of JUBA; 04°55′N 29°28′E. Livestock. Also spelled Maridi.

Meridian (muh-RID-ee-uhn), city (2000 population 34,919), ADA county, SW IDAHO, suburb 10 mi/16 km W of downtown BOISE; 43°37′N 116°24′W. In agricultural area (fruit, grain; cattle, sheep, poultry). Manufacturing (machinery, building materials, millwork). Served by BOISE IRRIGATION PROJECT. Founded 1891, incorporated 1902.

Meridian, city (2000 population 39,968), ⊙ LAUDERDALE county, E MISSISSIPPI, 85 mi/137 km E of JACKSON, near ALABAMA state line; 32°22′N 88°42′W. Drained in W by OKATIBBEE CREEK, forms Okatibbee Lake reservoir to NW. Railroad junction. Important railroad and highway point and the trade, shipping, and industrial center for a farm and timber area. Agriculture (cotton, corn; livestock); manufacturing (building materials, wire, food and beverages, transportation equipment, industrial scales, furniture, apparel, paper products, electronic equipment, fabricated metal products, and wood products; printing and publishing. In the Civil War, Meridian was the temporary capital of Mississippi (1863); it was destroyed by General Sherman in February 1864. Two junior colleges and a state mental hospital are here. Key Field airport in SW; Meridian Naval Air Station to NE; Temple Theatre (1923); "Merrehope," Greek Revival cottage (1858); Museum of Art; Jimmie Rodgers Museum, "Father of Country Music"; Grand Opera House (c.1890); Highland Park Carousel (1890s); Sam Dale State Historical Site to N; Lake Tom Bailey (state lake) to E. Settled 1831, incorporated 1860.

Meridian, unincorporated town (2000 population 3,794), BUTLER county, W PENNSYLVANIA, residential suburb 5 mi/8 km WSW of BUTLER; 40°51′N 79°57′W. Agriculture area (corn, hay, apples; livestock; dairying).

Meridian (ME-ri-dee-uhn), town (2006 population 1,508), ⊙ Bosque county, central TEXAS, on BOSQUE RIVER, and c.40 mi/64 km NW of WACO; 31°55′N 97°39′W. Elevation 791 ft/241 m. In diversified livestock (cattle) and agricultural (corn, wheat, pecans, peaches) area; manufacturing of tile. Meridian State Park to SW; LAKE WHITNEY reservoir to E. Settled 1854, incorporated 1886.

Meridian, unincorporated village, SUTTER county, N central CALIFORNIA, on SACRAMENTO RIVER, and 16 mi/26 km W of YUBA CITY. Agriculture (grain, vegetables, fruit, nuts, sugar beets); waterfowl hunting. Sutter Buttes to NE.

Meridian, village (2006 population 343), CAYUGA county, W central NEW YORK, 17 mi/27 km N of AUBURN; 43°09′N 76°32′W. In agricultural area.

Meridian, village (2006 population 59), LOGAN county, central OKLAHOMA, 10 mi/16 km ESE of GUTHRIE; 35°50′N 97°15′W. In agricultural area.

Meridian Dam, Oregon: see MIDDLE FORK.

Meridian Hills, town (2000 population 1,713), MARION county, central INDIANA, suburb 7 mi/11.3 km N of downtown INDIANAPOLIS, near WHITE RIVER; 39°53′N 86°10′W. Former municipality, merged with Indianapolis 1970.

Meridiano (ME-ree-zhee-ah-no), town (2007 population 3,857), NW SÃO PAULO state, BRAZIL, 6 mi/9.7 km S of FERNANDÓPOLIS, on railroad; 20°22′S 50°11′W. Coffee growing.

Meridian Township, suburb, INGHAM county, S central MICHIGAN, 6 mi/9.7 km E of LANSING, on RED CEDAR RIVER; 42°41′N 84°21′W. Includes former suburbs of HASLETT and Okemos. LAKE LANSING in NE corner of city.

Meridianville, village (2000 population 4,117), Madison co., N Alabama, 11 mi/18 km N of Huntsville; 34°52′N 86°34′W. Electronics manufacturing. Named for the Huntsville meridian, which marked the division between GA and MS territory in 1809.

Mériel (mai-ree-yel), commune (□ 2 sq mi/5.2 sq km), VAL-D'OISE department, ÎLE-DE-FRANCE region, N central FRANCE, 16 mi/25 km NW of PARIS; 49°05′N 02°12′E.

Mérignac (mai-ree-NYAHK), city (□ 18 sq mi/46.8 sq km), W industrial suburb of BORDEAUX, GIRONDE department, AQUITAINE region, SW FRANCE; 44°50′N 00°36′W. Produces diverse consumer goods; footwear manufacture; pharmaceuticals. Site of 13th-century dungeon. Bordeaux airport is 2 mi/3.2 km W.

Merigold, village (2000 population 664), BOLIVAR county, NW MISSISSIPPI, 6 mi/9.7 km N of CLEVELAND; 33°50′N 90°43′W. Agriculture (cotton, corn, soybeans; cattle).

Merigomish (me-ri-guh-MISH), village, E NOVA SCOTIA, E CANADA, on NORTHUMBERLAND STRAIT, 12 mi/19 km ENE of NEW GLASGOW; 45°37′N 62°25′W. Elevation 180 ft/54 m. Fishing. Name likely variation of the Micmac name, Malegomich, meaning "the merrymaking place."

Merigomish Island (me-ri-guh-MISH), NE NOVA SCOTIA, E CANADA, in NORTHUMBERLAND STRAIT, sheltering small Merigomish Bay, 12 mi/19 km E of PICTOU; 5 mi/8 km long, 2 mi/3 km wide; 45°40′N 62°25′W.

Merimbula (muh-RIM-byoo-luh), town, S NEW SOUTH WALES, SE AUSTRALIA, 293 mi/471 km S of SYDNEY, and on Sapphire Coast; 36°54′S 149°54′E. Resort; oysters, shrimp, fishing.

Mering (MAI-ring), town, SWABIA, S central BAVARIA, GERMANY, 8 mi/12.9 km SSE of AUGSBURG; 48°15′N 10°57′E. Manufacturing of machinery; agriculture (vegetables).

Merín, Laguna, URUGUAY and BRAZIL: see MIRIM LAKE.

Merino (muh-REE-no), village, SW VICTORIA, SE AUSTRALIA, 185 mi/298 km W of MELBOURNE, near CASTERTON; 37°43′S 141°33′E. Sheep, cattle.

Merino, village (2000 population 246), LOGAN county, NE COLORADO, on SOUTH PLATTE RIVER, and 14 mi/23 km SW of STERLING; 40°28′N 103°20′W. Elevation 4,035 ft/1,230 m. Shipping point in irrigated sugar-beet region. Manufacturing (machinery). Prewitt Reservoir to S, Summit Springs Battlefield to SE.

Merino Jarpa Island (mai-REE-no HAHR-pah), off W coast of AISÉN province, AÍSEN DEL GENERAL CARLOS IBAÑEZ DEL CAMPO region, S CHILE, at mouth of BAKER RIVER, SE of GULF OF PEÑAS; 33 mi/53 km long, 3 mi/5 km–12 mi/19 km wide; 47°55′S 74°15′W. Elevation c.3,300 ft/1,006 m.

Merinos (mai-REE-nos), town, RÍO NEGRO department, W central URUGUAY, in the CUCHILLA DE HAEDO, on road and railroad, and 70 mi/113 km E of PAYSANDÚ; 32°22′S 56°54′W. Wheat; cattle, sheep.

Merino Village, MASSACHUSETTS: see DUDLEY.

Merion (MER-ee-ahn) or **Merion Station**, unincorporated town, in Lower Merion township, MONTGOMERY county, SE PENNSYLVANIA, residential suburb 5 mi/8 km WNW of downtown PHILADELPHIA; 39°59′N 75°15′W. Manufacturing (commercial printing). St. Charles Borromeo Seminary and Eastern Baptist Theological Seminary to W. Barnes Museum here.

Merion Station, Pennsylvania: see MERION.

Merir, coral island, Republic of PALAU (Belau), W CAROLINE ISLANDS, W PACIFIC, c.29 mi/47 km SE of PULO ANNA; 1.3 mi/2.1 km long, 0.26 mi/0.41 km wide; 05°40′N 132°20′E. Rises to 50 ft/15 m.

Merivälja (MER-i-val-yuh), residential district, TALLINN, ESTONIA, NE of downtown, on the GULF OF TALLINN. Predominantly single-family homes.

Meriwether (MER-ee-weth-uhr), county (□ 505 sq mi/1,313 sq km; 2006 population 22,881), W GEORGIA; ⊙ GREENVILLE; 33°02′N 84°41′W. Bounded E by FLINT RIVER. Manufacturing includes apparel, textiles, wood products; printing and publishing; lumber. Piedmont peach-growing area; also produces pecans, melons, peppers; cattle, hogs. Formed 1827.

Meriz (mah-REEZ), town, YAZD province, SE central IRAN, 30 mi/48 km SSE of YAZD. Also spelled Mehriz.

Merizo (me-REE-so), town and municipality, S GUAM, on coast. Livestock; yams. Tourism.

Merj 'Uyun, LEBANON: see MARJ 'UYUN.

Merkanam, India: see MARAKKANAM.

Merkaz Shapira (mer-KAHZ shah-PEE-rah), village, ISRAEL, 8 mi/12.9 km SE of ASHDOD, on S coastal plain. Orthodox Jewish population. Serves as administrative, cultural, health, and educational center for surrounding rural areas.

Merke (mer-KYE), village, SW ZHAMBYL region, KAZAKHSTAN, on branch of TURK-SIB RAILROAD, 90 mi/145 km E of ZHAMBYL; 42°48′N 73°10′E. Tertiary-level administrative center. In irrigated agricultural area (wheat, sugar beets). Sugar refinery at adjoining OITAL.

Merkel (MUHR-kuhl), town (2006 population 2,593), TAYLOR county, W central TEXAS, 16 mi/26 km W of ABILENE; 32°28′N 100°00′W. In agricultural (cotton, wheat), cattle-ranching area. Manufacturing (apparel). Settled c.1875, incorporated 1906.

Merklin, CZECH REPUBLIC: see HROZNĚTÍN.

Merksem (MERK-sem), commune, Antwerp district, ANTWERPEN province, N BELGIUM, on ALBERT CANAL, and 2 mi/3.2 km NE of ANTWERP; 51°15′N 04°27′E. Glass products. Grain elevators; electric power station. Formerly spelled Merxem; annexed to Antwerp in 1983.

Merksplas (MERKS-plahs), commune (2006 population 8,225), Turnhout district, ANTWERPEN province, N BELGIUM, 5 mi/8 km NWN of TURNHOUT; 51°22′N 04°52′E. Agriculture; lumbering. Formerly spelled Merxplas.

Merkwiller-Péchelbronn (merk-vee-LER–pai-shuhl-BRON), German, *Merkweiler-Pechelbronn* (MERK-vei-luhr–PEK-uhl-bron), commune (□ 2 sq mi/5.2 sq km), BAS-RHIN department, in ALSACE, E FRANCE, 9 mi/14.5 km N of HAGUENAU; 48°56′N 07°50′E. Oil wells, inactive since 1970. Has small thermal establishment.

Merlebach, FRANCE: see FREYMING-MERLEBACH.

Merlera, Greece: see ERIKUSA.

Merlin, unincorporated town, JOSEPH county, SW OREGON, 6 mi/9.7 km NW of GRANTS PASS, on Jumpoff Joe Creek. Timber; livestock, poultry; dairying; pears, apples, plums; grain. Manufacturing (lumber, veneer, wood stoves). Siskiyou National Forest to W.

Merlo (MER-lo), town, NE SAN LUIS province, ARGENTINA, at W foot of Sierra de COMECHINGONES, 100 mi/161 km NE of SAN LUIS; 34°40′S 58°45′W. Agricultural center (soybeans, fruit; livestock). Mineral deposit nearby.

Merlo (MER-lo), suburb, BUENOS AIRES province, E ARGENTINA. An administrative, trade, and agricultural center of the Greater BUENOS AIRES area, it was founded in 1730 by Francisco de Merlo y Barbossa.

Mermentau (MUHR-muhn-to), village (2000 population 721), ACADIA parish, S LOUISIANA, 13 mi/21 km W of CROWLEY, and shallow-draft port on navigable MERMENTAU RIVER; 30°11′N 92°35′W. In rice-growing area; commercial fishing (crawfish, alligators), manufacturing (milled rice). Oil and natural gas field nearby.

Mermentau River (MUHR-muhn-to), c.71 mi/114 km long, S LOUISIANA; formed just above MERMENTAU by junction of NEZPIQUE and DES CANNES bayous; flows SW to GULF OF MEXICO, 48 mi/77 km E of SABINE PASS; 29°43′N, 93°06′W. Navigable. Widens to form Lake Arthur (c.7 mi/11 km long) at LAKE ARTHUR town, GRAND LAKE (c.10 mi/16 km long, 2 mi/3 km–10 mi/16 km wide) in central CAMERON parish, and small Upper Mud and Lower Mud lakes in marshy coastal area above mouth. Intersected by INTRACOASTAL WATERWAY through N part of Grand Lake, which in turn is joined by navigation canals to WHITE LAKE (SE) and thence to Freshwater Bayou Canal.

Merna, village (2006 population 378), CUSTER county, central NEBRASKA, 9 mi/14.5 km NW of BROKEN BOW; 41°28′N 99°45′W. Grain; livestock. Victoria Springs State Recreation Area to N.

Merom, town (2000 population 294), SULLIVAN county, SW INDIANA, on the WABASH RIVER, between Wabash River (W) and Turtle Creek Reservoir (E), 8 mi/12.9 km WSW of SULLIVAN; 39°04′N 87°34′W. Oil and natural-gas wells nearby. Grain; melon-growing area. Spectacular view of the Wabash River (W edge of town) at Meron Bluff. Laid out 1817.

Meron, ISRAEL: see MEIRON.

Merouana (me-rwah-NAH), village, BATNA wilaya, NE ALGERIA, 15 mi/24 km WNW of BATNA; 35°38′N 05°55′E. Wheat. Formerly Corneille.

Merouane, Chott, ALGERIA: see MELRHIR, CHOTT.

Merredin (ME-ri-duhn), town, SW central WESTERN AUSTRALIA state, W AUSTRALIA, 145 mi/233 km ENE of PERTH; 31°29′S 118°16′E. Railroad junction. Wheat, oats.

Merriam (MER-ee-uhm), city (2000 population 11,008), JOHNSON county, E KANSAS, suburb 7 mi/11.3 km SW of KANSAS CITY; 39°01′N 94°41′W. Manufacturing (lumber, concrete, clothing, electrical and electronic goods).

Merrick, county (□ 494 sq mi/1,284.4 sq km; 2006 population 7,954), E central NEBRASKA; ⊙ CENTRAL CITY; 41°10′N 98°01′W. Irrigated agricultural region. Bounded S by PLATTE RIVER; N border approaches LOUP RIVER. Leading well-irrigation county. Manufacturing of mobile homes, chemicals, and fertilizer. Cattle, hogs; dairying; corn. Mormon Trail State Wayside Area in E, near Clarks. Hord Lake State Recreation Area at Central City. Formed 1858.

Merrick, unincorporated city (□ 5 sq mi/13 sq km; 2000 population 22,764), NASSAU county, SE NEW YORK, on LONG ISLAND; 40°38′N 73°32′W. Although chiefly residential, it has some light manufacturing and a variety of commercial services.

Merrick (MER-ik), mountain (2,766 ft/843 m), DUMFRIES AND GALLOWAY, S Scotland, 13 mi/21 km N of NEWTON STEWART; 55°07′N 04°28′W.

Merrickville (ME-rik-vil), former village (□ 2 sq mi/5.2 sq km; 2001 population 968), SE ONTARIO, E central CANADA, on Rideau River and RIDEAU CANAL, and 35 mi/56 km SSW of OTTAWA; included in village of MERRICKVILLE-WOLFORD; 44°55′N 75°50′W. Light manufacturing; resort area.

Merrickville-Wolford (ME-rik-vil–WOL-fuhrd), village (□ 82 sq mi/213.2 sq km; 2001 population 2,812), SE ONTARIO, E central CANADA, 19 mi/31 km from BROCKVILLE; 44°51′N 75°49′W.

Merricourt, village, DICKEY county, SE NORTH DAKOTA, 18 mi/29 km NW of ELLENDALE; 46°12′N 98°45′W. Whitestone Battlefield Historic Site to SW.

Merrifield (ME-ri-feeld), unincorporated town, FAIRFAX county, NE VIRGINIA, residential suburb 10 mi/16 km W of WASHINGTON, D.C., 4 mi/6 km ENE of FAIRFAX; 38°52′N 77°14′W. Manufacturing (commercial printing). National Memorial Park Cemetery to E.

Merrifield, unincorporated village, CROW WING county, central MINNESOTA, 7 mi/11.3 km N of BRAINERD, at E end of NORTH LONG LAKE; 46°27′N 94°10′W. Manufacturing (fabricated metal products, consumer goods, electronic products); agriculture (cattle; dairying; oats); timber. Numerous small lakes in area. Crow Wing State Forest to NE.

Merrill, city (2006 population 9,897), ⊙ LINCOLN county, N central WISCONSIN, at confluence of WISCONSIN and small Prairie rivers, 16 mi/26 km N of WAUSAU; 45°10′N 89°42′W. In dairying and farming area; manufacturing (paper and paper goods, apparel, wood products, wire products, fabricated metal products, consumer goods, shoes, furniture, textiles, beverages). Nearby are the Grandfather Falls of the Wisconsin and Council Grounds State Park to NW. Settled c.1847, Merrill grew as lumbering town; incorporated 1883.

Merrill, town (2000 population 754), PLYMOUTH county, NW IOWA, on FLOYD RIVER, near confluence of West Branch Floyd River, and 6 mi/9.7 km SW of LE MARS; 42°43′N 96°15′W. Livestock, and grain area. Sand, gravel pits nearby.

Merrill, township, AROOSTOOK county, E MAINE, 20 mi/32 km WNW of HOULTON; 46°09′N 68°13′W. In lumbering area.

Merrill, town (2006 population 893), KLAMATH county, S OREGON, 15 mi/24 km SE of KLAMATH FALLS, on Lost River; 42°01′N 121°35′W. Elevation 4,064 ft/1,239 m. Barley, wheat, potatoes; sheep, cattle; dairy products. LAVA BEDS NATIONAL MONUMENT and TULE LAKE National Wildlife Refuge to S, in CALIFORNIA. LOWER KLAMATH National Wildlife Refuge (California and Oregon) to SW.

Merrill, village, SAGINAW county, E central MICHIGAN, 19 mi/31 km W of SAGINAW; 43°24′N 84°20′W. In agricultural area. Manufacturing (machinery, tool and machine parts).

Merrill, resort village, CLINTON county, extreme NE NEW YORK, on UPPER CHATEAUGAY LAKE, 25 mi/40 km WNW of PLATTSBURGH; 44°46′N 73°57′W.

Merrillan (MER-i-lan), village (2006 population 576), JACKSON county, W central WISCONSIN, 40 mi/64 km SE of EAU CLAIRE; 44°27′N 90°50′W. Railroad junction. In dairying region. Black River State Forest to SE; Bruce Mound Ski Area to E.

Merrill Peak, ARIZONA: see PINALENO MOUNTAINS.

Merrillville, town, THOMAS county, S GEORGIA, 10 mi/16 km NE of THOMASVILLE; 30°56′N 83°52′W.

Merrillville, town (2000 population 30,560), Lake county, NW INDIANA, suburb 12 mi/19 km S of GARY; 41°28′N 87°20′W. Drained by Deep River in SE and Turkey Creek in NW. Manufacturing (chemicals, aluminum and plastic products, transportation equipment, foods). Agriculture to S and E (dairying; vegetables). Settled 1847, incorporated 1970.

Merrimac (ME-ri-mak), rural town, including Merrimac port village, ESSEX county, NE MASSACHUSETTS, on MERRIMACK RIVER, near NEW HAMPSHIRE state line, and 13 mi/21 km NE of LAWRENCE; 42°51′N 71°01′W. Settled 1638, set off from AMESBURY 1876.

Merrimac (ME-ri-mak), unincorporated town, MONTGOMERY county, SW VIRGINIA, residential suburb 4 mi/6 km N of CHRISTIANSBURG, 3 mi/5 km S of BLACKSBURG; 37°11′N 80°25′W.

Merrimac (ME-ri-mak), village (2006 population 427), SAUK county, S central WISCONSIN, on WISCONSIN RIVER at W end of LAKE WISCONSIN, and 24 mi/39 km NNW of MADISON; 43°22′N 89°37′W. Manufacturing (printing and publishing). Railroad junction to W. Agricultural research station to SW. Ferry (free). Devil's Lake State Park to NW; Devil's Head Ski Area to N. Post office name formerly Merrimack.

Merrimack (ME-ri-mak), county (□ 956 sq mi/2,485.6 sq km; 2006 population 148,085), S central NEW HAMPSHIRE; ⊙ CONCORD; 43°17′N 40°71′W. Manufacturing at CONCORD, FRANKLIN, HOOKSETT, and Suncook; agriculture (nursery crops, vegetables, corn, apples, sugar maples, hay; cattle, poultry; dairying). Granite quarrying, mica mining, sand and gravel. Resorts on lakes. Hilly region, drained by MERRIMACK, CONTOOCOOK, SUNCOOK, SOUCOOK, BLACKWATER, and Lane rivers. Winslow and Rollins state parks in center. Wadleigh and part of Mount Sunapee state parks in W; part of Low State Forest in SW; part of Bear Brook State Park in SE; SUNAPEE LAKE on W border; New Hampshire State Forest Nursery in center. Formed 1823.

Merrimack (ME-ri-mak), town, HILLSBOROUGH county, S NEW HAMPSHIRE, 8 mi/12.9 km S of MANCHESTER, and 7 mi/11.3 km N of NASHUA; 42°51′N 71°31′W. Bounded E by MERRIMACK RIVER; drained by SOUHEGAN RIVER and Baboosic and Naticook brooks. Manufacturing (machinery, beverages, chemicals, fiberglass, electronic products, computers and computer products; consumer goods, plastic products, wood products); agriculture (fruit, vegetables, corn, nursery crops; poultry, livestock; dairying). Town includes village of Thorntons Ferry in S. Covered bridge in NW. Busch Clydesdale Hamlet in S. Naticook Pond in SW. Incorporated 1746.

Merrimack, Wisconsin: see MERRIMAC.

Merrimack River (ME-ri-mak), c.110 mi/177 km long, SE NEW HAMPSHIRE and NE MASSACHUSETTS; formed at FRANKLIN, S central NEW HAMPSHIRE, by the junction of the PEMIGEWASSET (rises in the WHITE MOUNTAINS) and Winnipesaukee (flows SW out of LAKE WINNIPESAUKEE and WINNISQUAM LAKE) rivers; flows S past CONCORD, MANCHESTER, MERRIMACK, and NASHUA (all in NEW HAMPSHIRE), and into NE MASSACHUSETTS, past DRACUT, LOWELL, HAVERHILL, and LAWRENCE, then turns ENE to the ATLANTIC OCEAN at NEWBURYPORT, widens into 3-mi/4.8-km-wide bay before entering ocean through narrow channel. With its numerous tributaries, the river drains most of S NEW HAMPSHIRE and NE MASSACHUSETTS. The river was a source of power for textile mills from c.1820 to 1930 with several large dams; this traditional industrial base has since moved to the Carolinas and other states. Receives CONTOOCOOK RIVER from SW above CONCORD, NASHUA RIVER from SW at NASHUA, CONCORD RIVER from S at LOWELL, and Shawsheen River from S at LAWRENCE.

Merriman, village (2006 population 114), CHERRY county, N NEBRASKA, 60 mi/97 km W of VALENTINE, and on branch of NIOBRARA RIVER, near SOUTH DAKOTA state line; 42°55′N 101°42′W.

Merriman Dam, NEW YORK: see RONDOUT CREEK.

Merrionette Park, village (2000 population 1,999), COOK county, NE ILLINOIS, SW suburb of CHICAGO; 41°40′N 87°42′W. Incorporated 1947.

Merriott (ME-ree-uht), village (2001 population 2,020), S SOMERSET, SW ENGLAND, 2 mi/3.2 km N of CREWKERNE; 50°54′N 02°48′W. Sailcloth manufacturing. Has 13th–15th-century church.

Merritt (ME-rit), city (□ 10 sq mi/26 sq km; 2001 population 7,088), S BRITISH COLUMBIA, W CANADA, on NICOLA RIVER at mouth of Coldwater River, 45 mi/72 km SSW of KAMLOOPS, and in THOMPSON-NICOLA regional district; 50°07′N 120°47′W. Elevation 2,030 ft/619 m. Lumbering and plywood; fox farming; cattle raising. Craigmont copper mine closed 1980s. Economy boosted by 1986 opening of new Coquihalla Tollway. Incorporated 1911.

Merritt Island, city (2000 population 36,090), BREVARD county, E central FLORIDA, 2 mi/3.2 km E of COCOA; 28°22′N 80°39′W. Manufacturing includes fabricated metal products, machining, and fiberglass products.

Merritt Island (□ 47 sq mi/122 sq km), BREVARD county, E central FLORIDA, separated from the mainland by INDIAN RIVER (a lagoon) and from the CANAVERAL peninsula (E) by BANANA RIVER (a lagoon); c.40 mi/64 km long and c.6 mi/9.7 km wide; 28°25′N 80°39′W. It produces citrus fruits and is noted for its birds and other wildlife. The Merritt Island National Wildlife Refuge and the John F. Kennedy Space Center, a division of NASA, are here. Tourism is important.

Merritton (ME-ri-tuhn), unincorporated town, S ONTARIO, E central CANADA, on WELLAND SHIP CANAL, and included in city of SAINT CATHARINES; 43°08′N 79°12′W. Steel, paper, and pulp milling.

Merritt Parkway, landscaped limited-access road, in CONNECTICUT, part of state's express highway system; extends from NEW YORK state line E of WHITE PLAINS, where it joins NEW YORK parkway system. Generally parallel to shore as far as NEW HAVEN, then to NE, through HARTFORD, by Wilbur Cross Parkway, where it merges with Interstate 84. First limited-access highway in UNITED STATES.

Merritt Reservoir (□ 5 sq mi/13 sq km), CHERRY county, N central NEBRASKA, on Snake River, 24 mi/39 km SW of VALENTINE; 42°38′N 100°52′W. Maximum capacity 86,134 acre-ft. Extends E-W. Formed by Merritt Dam (115 ft/35 m high), built (1964) by the Bureau of Reclamation for irrigation; also used for recreation. Samuel River McKelvie National Forest to N. Merritt Reservoir State Recreational Area on N shore.

Merriwa (ME-ree-wuh), town, E central NEW SOUTH WALES, SE AUSTRALIA, 95 mi/153 km NW of NEWCASTLE; 32°09′S 150°21′E. Terminus of railroad from MUSWELLBROOK. Sheep, cattle, horse studs; agriculture center (wheat, olive trees); coal mining.

Mer Rouge (MER ROOZH), village, MOREHOUSE parish, NE LOUISIANA, 8 mi/13 km E of BASTROP; 32°47′N 91°48′W. In agricultural area (cotton, rice, sweet potatoes, vegetables; cattle); timber. Handy Brake National Wildlife Refuge to N.

Merrygoen, village, E central NEW SOUTH WALES, SE AUSTRALIA, on CASTLEREAGH RIVER, 190 mi/306 km

NW of SYDNEY, and 6 mi/10 km E of Mendooran; 31°49′S 149°14′E. Sheep and agriculture center.

Merrymeeting Bay, SAGADAHOC county, SW MAINE, tidal bay 5 mi/8 km NE of BRUNSWICK. Formed by junction of ANDROSCOGGIN and KENNEBEC rivers below BOWDOINHAM, it extends 16 mi/26 km further S to the ATLANTIC OCEAN. The surrounding marshes are noted for duck hunting.

Merrymeeting Lake (ME-ree-mee-ting), STRAFFORD county, E central NEW HAMPSHIRE, in NEW DURHAM, 27 mi/43 km NE of CONCORD; 3 mi/4.8 km long. Drains through Merrymeeting River (c.10 mi/16 km long), which turns NW at New Durham Village, flows past ALTON to S end of LAKE WINNIPESAUKEE to W. Powder Mill Fish Hatchery located below lake's outlet.

Merrymount, MASSACHUSETTS: see QUINCY.

Merry Oaks (MER-ee OKS), unincorporated village, CHATHAM county, central NORTH CAROLINA, 22 mi/35 km SW of RALEIGH; 35°38′N 79°00′W. Brick manufacturing. Recreation area. B. EVERETT JORDAN LAKE (Jordan Lake) reservoir to N, Harris Lake reservoir to E.

Merryville (MER-ee-vil), town (2000 population 1,126), BEAUREGARD parish, W LOUISIANA, 17 mi/27 km WSW of DE RIDDER, near SABINE RIVER (TEXAS state line); 30°45′N 93°32′W. In agricultural area (soybeans, squash, blueberries, watermelons; cattle; dairying); wool market; logging. Oil field nearby. Boise-Vernon State Wildlife Area to N.

Mersa Fatuma (MUHR-suh FAHT-mah), village, SEMENAWI KAYIH BAHRI region, SE ERITREA, fishing port on RED SEA, 75 mi/121 km SE of MASSAWA; 14°53′N 40°19′E. Once exported potash mined at, and transferred from, DALLOL (ETHIOPIA) by narrow-gauge railroad and by road. Railroad and road were heavily damaged during war with Ethiopia.

Mersa Matruh, EGYPT: see MATRUH.

Mersch (MERSH), town, Mersch commune (2001 population 7,012), central LUXEMBOURG, on ALZETTE RIVER, and 10 mi/16 km N of LUXEMBOURG city; 49°45′N 06°06′E. Manufacturing Market center for agricultural region.

Mersea Island (MUH-zee), E ESSEX, E ENGLAND, between the BLACKWATER and COLNE estuaries, 8 mi/12.9 km SSE of COLCHESTER; 5 mi/8 km long, 2 mi/3.2 km wide; 51°48′N 00°55′E. Island connected to mainland by causeway. At SW end is town of West MERSEA. Oyster beds; sailing.

Mersea, West (MUH-zee), town (2001 population 6,925), on MERSEA ISLAND, E ESSEX, E ENGLAND, on BLACKWATER estuary, and 8 mi/12.9 km S of COLCHESTER; 51°46′N 00°54′E. Seaside resort and yachting center.

Merseburg (MER-se-boorg), city, SAXONY-ANHALT, E central GERMANY, on the SAALE RIVER, 16 mi/26 km W of LEIPZIG; 51°22′N 11°59′E. Industrial city and a lignite-mining center. Manufacturing includes chemicals, paper, steel, bricks, aluminum foil, and beer. A fortress in the 9th century, Merseburg was a favorite residence of Henry I (Henry the Fowler) and of Emperor Otto I. It served as a German outpost for subduing the Slavs and Poles. Episcopal see from 968 until its suppression (1545) during the Reformation. Bishopric passed to SAXONY in 1561. From 1656 to 1738 the city was the seat of the dukes of Saxe-Merseburg. In 1815 it passed to PRUSSIA. Heavily bombed during World War II (forty-five percent destoyed). Among its noted buildings are the cathedral (founded 1015, rebuilt in the 13th and 16th centuries) and the episcopal palace (15th century).

Mers El Hadjadj (MERS eh ah-JAHJ), village, ORAN wilaya, NW ALGERIA, on the Gulf of ARZEW (MEDITERRANEAN SEA), on railroad, and 20 mi/32 km SW of MOSTAGANEM. Swimming resort; vineyards. Formerly Port aux Poules.

Mers El Kebir (MERS el ke-BIR) [Arab.=great port], town and harbor, ORAN wilaya, W ALGERIA, on the Gulf of Oran; 35°48′N 00°43′W. Originally Portus Magnus, a Roman port, it was one of the naval arsenals of the Almohads in the 12th century. The Spanish held the town from 1505 to 1792. A naval base for centuries, it was once settled by Italian fishermen. Famous for an episode in World War II when, after the fall of FRANCE, the British navy destroyed the French naval fleet here (killing many) because the fleet refused to join the Allied side.

Merse, The (MERS), lowland district of Scottish Borders, S Scotland, between LAMMERMUIR HILLS and the TWEED RIVER, noted for its fertility; 55°40′N 02°15′W. Formerly in Borders, abolished 1996.

Mersey River (MUH-zee), c.70 mi/113 km long, NW ENGLAND; formed at STOCKPORT by the confluence of the Etherow and Goyt rivers (in GREATER MANCHESTER); flows E to the IRISH SEA near LIVERPOOL. The estuary of the Mersey, which is 16 mi/26 km long and c.2 mi/3.2 km wide, is navigable for oceangoing vessels. Its chief tributaries are the IRWELL and Bollin rivers. The Manchester Ship Canal uses the waters of the Mersey. Mersey Tunnel, or Queensway, a vehicular tunnel (opened 1934) with a length of 2.3 mi/3.7 km, is the longest subaqueous tunnel in the world; it connects Liverpool and BIRKENHEAD. Kingsway Tunnel (1.5 mi/2.4 km long; opened 1971) connects Liverpool and WALLASEY. The Mersey River is of great commercial importance to the cities served by it, especially Liverpool and MANCHESTER. Milling and oil refining are important industries along the river.

Mersey River (MUHR-zee), 60 mi/97 km long, N TASMANIA, AUSTRALIA; rises in small lakes N of Lake SAINT CLAIR; flows generally N, past LATROBE, to BASS STRAIT at DEVONPORT; 41°10′S 146°22′E. Orchards on banks of estuary.

Merseyside (MUH-zee-seid), metropolitan county (□ 253 sq mi/657.8 sq km; 2001 population 1,362,026), NW ENGLAND; ⊙ LIVERPOOL; 53°24′N 03°05′W. Lying on the banks of the MERSEY RIVER, the county consists mainly of marshes, cliffs, and sand dunes. The two banks of the river are connected by underground railroad lines, tunnels, and ferries. Liverpool is the principal city, a port which has lost its function as England's trade has become more European. Margarine and soap are produced in the region. Norsemen from IRELAND and the Isle of MAN invaded the area in the 10th century. Throughout the 17th century, Liverpool was heavily involved in the slave trade. By the 19th century, however, the city had become a more general trade center. Cotton was previously a major import. Other towns of the region include BOOTLE, ST. HELENS, BIRKENHEAD, and SOUTHPORT.

Mershon (MUHR-shuhn), town, PIERCE county, GEORGIA, 12 mi/19 km N of BLACKSHEAR; 31°28′N 82°15′W. Manufacturing of hydraulic equipment.

Mersin, city, ⊙ S TURKEY, on the MEDITERRANEAN SEA, 40 mi/65 km WSW of ADANA; 36°47′N 34°37′E. A railroad terminus and modern seaport, it exports cotton, petroleum products, chrome, copper, and agricultural produce. Textile production and oil refining. Excavations here in the 1930s showed that the site was occupied in early Neolithic times (c.3600 B.C.E.). Connected by ferry to CYPRUS. Site of one of Turkey's largest oil refineries. Formerly İçel (EE-chel).

Mersing (MER-sing), town (2000 population 20,094), NE JOHOR state, S MALAYSIA; 02°26′N 103°50′E. Fishing center, domestic tourism centered on off-shore islands.

Mersivan, TURKEY: see MERZIFON.

Mers-les-Bains (MER-lai–BAN), commune (□ 2 sq mi/5.2 sq km), SOMME department, PICARDIE region, N FRANCE, 20 mi/32 km WSW of ABBEVILLE, swimming resort N of mouth of BRESLE RIVER, on ENGLISH CHANNEL, opposite Le TRÉPORT; 50°04′N 01°23′E. Glassworks.

Merta (MER-tah), town, NAGAUR district, central RAJASTHAN state, NW INDIA, 65 mi/105 km ENE of JODHPUR; 26°39′N 74°02′E. Railroad spur terminus. Exports oilseeds, cotton, millet; cotton ginning, oilseed milling; handicrafts (palm fans, ivory goods, felts, pottery). Taken 1562 by Akbar. In battle nearby, Marathas defeated joint forces of Jodhpur and Jaipur here in 1790. Merta Road, railroad junction, is 9 mi/14.5 km NW.

Mertens (MUHR-tenz), village (2006 population 161), HILL county, N central TEXAS, 14 mi/23 km E of HILLSBORO; 32°03′N 96°53′W. In farm area (cattle; cotton).

Mertert (MER-tert), village, Mertert commune, E LUXEMBOURG, on MOSELLE RIVER, at mouth of SYRE RIVER, and 2 mi/3.2 km NE of GREVENMACHER, on German border; 49°42′N 06°29′E. Sand and gravel quarrying; fruit growing.

Merthyr Tydfil (MUHR-thir TID-fil), town and county (2001 population 55,981), SE Wales, on the TAFF RIVER, 21 mi/34 km NW of Cardiff; 51°45′N 03°23′W. Former iron, steel, and coal-mining town now houses engineering, electrical, and light industry. Was major center of industrial innovation. Formerly in MID GLAMORGAN, abolished 1996.

Merthyr Vale (MUHR-thir), village (2001 population 3,925), MERTHYR TYDFIL, SE Wales, 5 mi/8 km S of Merthyr Tydfil, on TAFF RIVER; 51°41′N 03°20′W. Formerly in MID GLAMORGAN, abolished 1996.

Mértola (MER-too-lah), ancient *Myrtilis Julia*, town, BEJA district, S PORTUGAL, head of navigation on the GUADIANA RIVER, and 28 mi/45 km SSE of BEJA; 37°38′N 07°40′W. Grain and livestock market; alcohol distilling. Overlooked by castle and keep (1292).

Merto Lemariam, town (2007 population 11,534), AMHARA region, N central ETHIOPIA, 50 mi/80 km NE of DEBRE MARKOS; 10°50′N 38°16′E. In livestock raising region. Also called Mertule Maryami.

Merton, town (2006 population 2,791), WAUKESHA county, SE WISCONSIN, on BARK RIVER, and 20 mi/32 km NW of MILWAUKEE; 43°08′N 88°18′W. In dairying region. Manufacturing (cleaning products). Lakes nearby.

Merton (MUH-tuhn), outer borough (□ 15 sq mi/39 sq km; 2001 population 187,908) of GREATER LONDON, SE ENGLAND; 51°25′N 00°10′W. Largely residential with some industry, including tanning and the manufacturing of silk and calico prints, varnish and paint, and toys. An annual fair dating from Elizabethan times is held within the borough at Mitcham, and cricket is played on the town common. Merton contains Wimbledon, England's tennis headquarters; the first Wimbledon Championship match took place in 1877. Cricket and golf matches are also played. George Eliot lived here. Has remains of a priory founded in 1115. Walter de Merton, Lord High Chancellor to Henry III and founder of Merton College (Oxford), and Thomas à Becket were educated at Merton Priory. Admiral Horatio Nelson and Lady Emma Hamilton lived together in Merton Park. The borough includes Mitcham and Morden.

Mertule Maryami, ETHIOPIA: see MERTO LEMARIAM.

Mertz Glacier, a broad, heavily crevassed glacier on the GEORGE V COAST of WILKES LAND, EAST ANTARCTICA; flows into MERTZ GLACIER TONGUE, its floating extension; 45 mi/70 km long and 25 mi/40 km wide; 67°30′S 144°45′E.

Mertz Glacier Tongue, on GEORGE V COAST of EAST ANTARCTICA, the seaward extension of MERTZ GLACIER; 67°10′S 145°30′E. Configuration and extent change through time.

Mertzig (MERT-sik), village and commune, central LUXEMBOURG, on WARK RIVER, and 5 mi/8 km WSW of ETTELBRUCK; 49°50′N 06°01′E. River port.

Mertz-Ninnis Valley (MUHRTS–NI-nis), **adélie depression** or **Mertz-Ninnis Trough**, ANTARCTICA, submarine valley in continental shelf off GEORGE V COAST, associated with MERTZ and NINNIS glaciers, S INDIAN OCEAN; 67°30′S 146°00′E.

Cross-references are shown in SMALL CAPITALS. The pronunciation guide is shown on page xix. The sources of population figures are shown on page xvii.

Mertzon (MUHRT-zuhn), town (2006 population 860), ⊙ IRION county, W TEXAS, 25 mi/40 km SW of SAN ANGELO, and on a tributary of CONCHO RIVER; 31°15′N 100°49′W. Elevation 2,250 ft/686 m. Retail, shipping point for sheep-ranching region; warehouses; manufacturing (natural gas processing, meat products).

Mertzwiller (merts-vee-LER), German, *Merzweiler* (MERTS-vei-luhr), commune (□ 3 sq mi/7.8 sq km), BAS-RHIN department, ALSACE, E FRANCE, at W edge of Forest of Haguenau, 6 mi/9.7 km NW of HAGUENAU; 48°52′N 07°41′E. Woodworking.

Méru (mai-ru), town (□ 8 sq mi/20.8 sq km), OISE department, PICARDIE region, N FRANCE, 14 mi/23 km S of BEAUVAIS; 49°14′N 02°08′E. Historic center of French mother-of-pearl and bone industry; now shifted to electronics and plastics manufacturing.

Meru (MAI-roo), town (1999 population 42,677), EASTERN province, district administrative center, S central KENYA, on NE slope of Mount KENYA, 110 mi/177 km NNE of NAIROBI. Elevation 5,800 ft/1,768 m. Coffee, wheat, corn, tea, fruits.

Meruchak, AFGHANISTAN: see MARUCHAK.

Meru, Mount (MAI-roo), extinct volcano, ARUSHA region, N TANZANIA, 8 mi/12.9 km NNE of ARUSHA, and 42 mi/68 km WSW of KILIMANJARO; 03°15′S 37°44′E. Elevation 14,979 ft/4,566 m. The central peak and Ngurdoto Crater, 12 mi/19 km to E, comprise ARUSHA NATIONAL PARK.

Meruoca (ME-roo-O-kah), city (2007 population 12,137), CEARÁ state, BRAZIL.

Merv (MERV), ancient *Margiana* or *Antiochia Margiana*, ancient city, in what is now MARY weloyat, SE TURKMENISTAN, in a large oasis of the KARA KUM desert, on the MURGAB RIVER. The city was founded in the third century B.C.E. on the site of an earlier settlement. Its periods of greatness were C.E. 651–C.E. 821, when it was the seat of the Arab rulers of KHORASAN and Transoxania and one of the main centers of Islamic learning. From 1118 to 1157, it was the capital of Seljuk Empire under the last sultan, Sandzhar. The Mongols destroyed the city early in the 13th century, but it was slowly rebuilt, to be destroyed again by the Bukharans in 1790. The Russians conquered the area in 1884. Several mausoleums, mosques, and castles of the 11th and 12th centuries are preserved and are among the best monuments of Muslim art in central ASIA. Present-day MARY, 20 mi/32 km away, was called Merv until 1937.

Mervdasht, IRAN: see MARVDASHT.

Merville (mer-VEEL), town (□ 10 sq mi/26 sq km), NORD department, NORD-PAS-DE-CALAIS region, N FRANCE, on the LYS River, and 8 mi/12.9 km N of BÉTHUNE; 50°39′N 02°39′E. Market in area of intensive agriculture (potatoes, peas); metalworking. Just W is Nieppe Forest, which figured in battle of the Lys (1918) during World War I.

Merville-Franceville-Plage (mer-VEEL–frawns-VEEL–PLAHZH), commune (□ 4 sq mi/10.4 sq km), CALVADOS department, BASSE-NORMANDIE region, NW FRANCE, on the ENGLISH CHANNEL just E of mouth of ORNE RIVER, 10 mi/16 km NE of CAEN; 49°16′N 00°12′W. Small swimming resort at Franceville-Plage (pine-planted dunes). British landing forces overcame strong German defenses here (June 1944) in Allied Normandy invasion.

Mervino (MYER-vee-nuh), former village, NW RYAZAN oblast, central European Russia, now a suburb of RYAZAN, 4 mi/6 km WNW of the city center; 54°38′N 39°40′E. Elevation 354 ft/107 m.

Merwede Canal (MER-vai-duh), UTRECHT and SOUTH HOLLAND provinces, THE NETHERLANDS, connects Utrecht to MERWEDE RIVER at GORINCHEM, 13 mi/21 km NE of DORDRECHT.

Merwede River (MER-vai-duh), 5.5 mi/8.9 km long, SOUTH HOLLAND and ZEELAND provinces, SW NETHERLANDS; formed as UPPER MERWEDE RIVER by junction of MEUSE and WAAL rivers at WOUDRICHEM; flows W, forking into LOWER MERWEDE RIVER (which flows 9 mi/14.5 km W to form OLD MAAS RIVER and NOORD RIVER) and NEW MERWEDE RIVER (which flows 12 mi/19 km SW to form the HOLLANDS DIEP channel). Navigable.

Merwin, town (2000 population 83), BATES county, W MISSOURI, 17 mi/27 km NW of BUTLER; 38°23′N 94°35′W.

Merwin, Lake, reservoir (□ 6 sq mi/15.6 sq km), CLARK and COWLITZ counties, SW WASHINGTON, on LEWIS RIVER, 23 mi/37 km SE of LONGVIEW; 46°00′N 122°29′W. Maximum capacity 422,000 acre-ft. Formed by Merwin Dam (313 ft/95 m high), built for power generation; also used for recreation and as a fish and wildlife pond.

Merxem, BELGIUM: see MERKSEM.

Merxplas, BELGIUM: see MERKSPLAS.

Méry-sur-Oise (mai-ree–syoor–WAHZ), town (□ 4 sq mi/10.4 sq km), VAL-D'OISE department, ÎLE-DE-FRANCE region, N central FRANCE, on OISE RIVER 5 mi/8 km NE of PONTOISE; 49°04′N 02°12′E. Has 16th–18th-century chateau.

Merzbacher, lake, in SARY-JAZ mountain range of TIANSHAN mountain system, KYRGYZSTAN; 42°06′N 79°50′E. Elevation 10,824 ft/3,299 m. Formed by ice jams from melting of N and S arms of ENGILCHEK glacier.

Merzenich (MER-tse-nikh), village, WESTPHALIA-LIPPE, North Rhine-Westphalia, W GERMANY, on small Elle-Bach River, 3 mi/4.8 km NNE of DÜREN; 50°50′N 06°31′E. In coal-mining region. Grain; food processing.

Merzhausen (merts-HOU-suhn), village, S Upper Rhine, BADEN-WÜRTTEMBERG, SW GERMANY, in the BREISGAU, 3 mi/4.8 km S of FREIBURG; 47°58′N 07°50′E. Forestry; wine; woodworking.

Merzifon, town (2000 population 45,613), N central TURKEY, 25 mi/40 km NW of AMASYA; 40°52′N 35°28′E. Textiles; lignite; wheat. Important road junction. Sometimes spelled Marsivan and Mersivan.

Merzig (MER-tsig), city, NW Saarland, on Saar River, 22 mi/35 km NW of Saarbrücken; 49°27′N 06°35′E. Manufacturing of ceramics, glass, soap, chemicals; metalworking, brewing. Has 12th–13th-cent. church and 12th-cent. townhall.

Mesa, county (□ 3,341 sq mi/8,686.6 sq km; 2006 population 134,189), W COLORADO, on UTAH (W) border; ⊙ GRAND JUNCTION; 39°01′N 108°28′W. N part is extensively irrigated farming area. Drained by Colorado and GUNNISON rivers. Fruit (peaches, grapes), beans, wheat, hay, oats, barley, corn, potatoes; cattle, sheep. Marble quarrying. Oil and natural gas. Includes COLORADO NATIONAL MONUMENT and GRAND VALLEY and large part of Grand Mesa National Forest in E; part of White River National Forest in NE; part of Uncompahgre National Forest in S; small part of Manti La Sal National Forest in SW corner; Highline State Park in NW; COLORADO RIVER and Island Acres state parks in N; Vega State Park in NE. Part of Grand Mesa Lakes on SE border. Numerous reservoirs (c.100) on either side of border between Mesa and DELTA counties. Powderhorn Ski Area in NE. Formed 1883.

Mesa, city (2000 population 396,375), MARICOPA county, S central ARIZONA, suburb 12 mi/19 km ESE of downtown PHOENIX, on SALT RIVER (intermittent), in the irrigated SALT RIVER VALLEY; 33°25′N 111°44′W. Manufacturing includes electronic products, fabricated metals, aircraft, and machine tools. One of the fastest-growing U.S. cities, marked by a population increase of 89% between 1980 and 1990. Tourism is important, and the citrus and farm products of the area are packed and processed here. The Mormons who founded the city in 1878 used old Native American irrigation canals for farming in the Salt River Valley. A Mormon temple, Mesa community college, training camp of Chicago Cubs baseball team, and the chief agricultural experimental farm of the University of Arizona are here. Usery Mountain Park in E; Mesa Municipal Airport (Falcon Field) in NE; Williams Air Force Base in SE; Arizona Boys Ranch to SE. Incorporated 1883.

Mesa (MAI-suh), village (2006 population 431), FRANKLIN county, SE WASHINGTON, 24 mi/39 km N of PASCO; 46°34′N 119°00′W. Agricultural area (wheat, alfalfa, vegetables; sheep, hogs). Mesa Lake reservoir to W.

Mesaba (muh-SAH-buh), locality, ST. LOUIS county, NE MINNESOTA, near E end of MESABI IRON RANGE, 19 mi/31 km ENE of VIRGINIA, and 5 mi/8 km ENE of AURORA; 47°34′N 92°07′W. Site of extensive open-pit iron mining operations of Erie Mining Company. Railroad delivers ore to TACONITE HARBOR, on Lake SUPERIOR, 57 mi/92 km to E. Iron reserves and mining activity have declined in late 1900s.

Mesabi Iron Range (muh-SAH-bee), range of low hills, in iron-mining district, ST. LOUIS and ITASCA counties, NE MINNESOTA. The ores were found in a belt c.110 mi/177 km long and 1 mi/1.6 km–3 mi/4.8 km wide, between BABBITT and GRAND RAPIDS, occurring in horizontal layers (up to 500 ft/152 m thick) near the surface and mined by the open pit method (MESABA open-pit mine, 5 mi/8 km ENE of AURORA). Reserves of high-grade hematite iron are now exhausted, and lower-grade taconite deposits are being worked. The taconite contains mostly chert and magnetite (an iron-bearing mineral) and must undergo a costly and complex beneficiation process before being shipped in the form of pellets containing c.60% iron. Mining centers include communities of Babbitt, Aurora, BIWABIK, GILBERT, VIRGINIA, MOUNTAIN IRON, BUHL, CHISHOLM, HIBBING, MARBLE, and COLERAINE. Most of the ore found is shipped by company railroads to DULUTH, TACONITE HARBOR, SILVER BAY, and TWO HARBORS. The Mesabi iron ore deposits were first discovered in 1887 by Leonidas Merritt and his brothers, who organized the Mountain Iron Company in 1890 to mine the ore; John D. Rockefeller gained control of the company in the Panic of 1893. Productivity has declined due to dwindling reserves and environmental restrictions. VERMILION IRON RANGE parallels Mesabi Range, c.10 mi/16 km to N.

Mesa Bolívar (MAI-sah bo-LEE-vahr), town, MÉRIDA state, W VENEZUELA, on slopes of ANDEAN spur, 32 mi/51 km WSW of MÉRIDA; 08°28′N 71°35′W. Elevation 3,402 ft/1,036 m. Sugarcane, grain, fruit. Gold mines nearby.

Mesanjore, India: see MASANJOR.

Mesão Frio (mai-ZOU FREE-oo), town, VILA REAL district, N PORTUGAL, on right bank of DUERO RIVER, and 12 mi/19 km SW of VILA REAL. Vineyards (port wine); olives, figs, almonds, oranges also grown in area.

Mesaoria (me-sou-REE-yah), region, NE central CYPRUS, broadly defined as lowland between TROODOS (S) and KYRENIA (N) mountains and from MORPHOU (W) to Famagusta (E) bays. Extends c. 55 mi/89 km W-E; includes city of NICOSIA; drained by Serachis (W) and PEDHIEOS (E) rivers. Often narrowly defined as the E part of the lowland. The area is largely irrigated and is the island's main agricultural region (olives, citrus, deciduous fruits, grapes, almonds, grain, vegetables, potatoes). Sometimes spelled MESSAORIA.

Mesara, Greece: see MESSARA.

Mesas, Las (MAI-sahs, lahs), town, CUENCA province, E central SPAIN, 12 mi/19 km S of BELMONTE; 39°24′N 02°45′W. Cereals, grapes; livestock.

Mesa Verde National Park (□ 81 sq mi/210 sq km), MONTEZUMA county, 10 mi/16 km SE of CORTEZ, SW COLORADO. It includes the most notable and best-preserved cliff dwellings and relics in the U.S., covering four archaeological periods. There are museums and a library. The mesa rises 1,800 ft/549 m–2,000 ft/ 610 m above the surrounding land. It is cut by many canyons where ancient Native Americans built pit and cliff dwellings C.E. 500–A.D. 1300. Bounded by Ute Mountain Indian Reservation on S and W. Authorized 1906.

Mescala, Río, MEXICO: see BALSAS, RÍO.

Mescalero (mes-kuh-LER-o), town (2000 population 1,233), OTERO county, S NEW MEXICO, in SACRAMENTO MOUNTAINS, 29 mi/47 km NNE of ALAMOGORDO, in Mescalero Apache Indian Reservation; 33°08′N 105°47′W. Elevation 6,600 ft/2,012 m. Timber; sheep, cattle. Manufacturing (ponderosa pine lumber). Reservation headquarters here; reservation has ski resort and gambling. In 1996, proposed as temporary nuclear-waste-storage site. Parts of Lincoln National Forest to N and S; Lake Mescalero reservoir to NE; SIERRA BLANCA (11,977 ft/3,651 m) to NNW.

Meschede (ME-she-de), town, WESTPHALIA-LIPPE, North Rhine-Westphalia, W GERMANY, in the RUHR industrial district, 10 mi/16 km ESE of ARNSBERG; 51°22′N 08°17′E. Metalworking; synthetic fabrics. Tourism. Henne dam and reservoir just SW.

Mescit Dag, (Turkish=*Mescit Dağ*) peak (10,680 ft/3,255 m), NE TURKEY, in CORUH MOUNTAINS (E part of the Pontus Mountains), 13 mi/21 km SE of ISPIR.

Mesegar (mai-sai-GAHR), village, TOLEDO province, central SPAIN, 25 mi/40 km W of TOLEDO; 39°56′N 04°30′W. Olives, fruit, grapes, esparto; livestock.

Mesembria, Bulgaria: see NESEBAR.

Mesemvriya, Bulgaria: see NESEBAR.

Mesen (MAI-suhn), French *Messines*, commune (2006 population 977), Ypres district, WEST FLANDERS province, W BELGIUM, 6 mi/9.7 km S of YPRES; 50°46′N 02°54′E.

Meseritz, POLAND: see MIEDZYRZECZ.

Meservey, town (2000 population 252), CERRO GORDO county, N IOWA, 22 mi/35 km SW of MASON CITY; 42°55′N 93°28′W.

Meseta, SPAIN: see IBERIAN PENINSULA.

Meshchera (meesh-CHYE-rah), extensive wooded and swampy watershed in MOSCOW, RYAZAN, and VLADIMIR oblasts, central European Russia, between KLYAZ'MA and OKA rivers; bounded by SHATURA (NW), RYAZAN (SW), and KASIMOV (E). Has numerous lakes. Quartzite and peat deposits; lumbering, peat cutting, and glassworking are chief industries. Also spelled Meshchora (meesh-CHO-rah).

Meshcherino (mee-SHCHYE-ree-nuh), urban settlement (2006 population 4,195), S central MOSCOW oblast, central European Russia, on the Severka River (tributary of the MOSKVA River), on road, 13 mi/21 km SW of VOSKRESENSK; 55°12′N 38°21′E. Elevation 511 ft/155 m. Road and bridge construction enterprise. Summer residences and offices of various Russian federal ministries and agencies.

Meshchovsk (myesh-CHOFSK), city (2005 population 4,335), central KALUGA oblast, central European Russia, 53 mi/85 km WSW of KALUGA; 54°19′N 35°17′E. Elevation 754 ft/229 m. Highway junction. Fruit canning, dairying; printing; starch factory. Known since the 13th century when in Chernigov principality. Made city in 1776.

Meshed, IRAN: see MASHHAD.

Meshed-i-Sar, IRAN: see BABOLSAR.

Meshghara (MUSH-guh-ruh), village, central LEBANON, 26 mi/42 km SSE of BEIRUT; 33°31′N 35°29′E. Tanning; grapes, fruit, vegetables. Also spelled Mashgharah.

Meshhed (MESH-hahd), Arab town, ISRAEL, 3.7 mi/5 km NE of NAZARETH, in LOWER GALILEE. Mixed farming. Thought to be the birthplace of the prophet Jonah. The local mosque is built on what Muslims regard as the burial ground of Jonah's family. Ruins on the town's tel indicate settlement here in the early and middle Bronze Age, the Iron age and Byzantine period.

Meshkinshahr (mahsh-KEEN-shahr), town (2006 population 63,655), Ardabīl province, NW IRAN, 32 mi/51 km E of AHAR, and 85 mi/137 km ENE of TABRIZ, at N foot of the Savalan; 38°23′N 47°41′E. Dried fruit; rugmaking. Sometimes called KHIAV or Khiov, or spelled MISHKINSHAHR.

Meshkovskaya (myesh-KOF-skah-yah), village, N ROSTOV oblast, S European Russia, on the Tikhaya River (small right affluent of the DON River), on road junction, 40 mi/64 km ENE of CHERTKOVO; 49°32′N 40°59′E. Elevation 249 ft/75 m. Flour mill.

Meshoppen, borough (2006 population 435), WYOMING county, NE PENNSYLVANIA, 25 mi/40 km NW of SCRANTON, on SUSQUEHANNA RIVER, at Meshoppen Creek; 41°36′N 76°02′W. Agriculture (corn, hay; dairying); light manufacturing; quarrying.

Meshyrich, UKRAINE: see MEZHYRICH.

Mesic (ME-sik), village (□ 1 sq mi/2.6 sq km; 2006 population 248), PAMLICO county, E NORTH CAROLINA, 26 mi/42 km ENE of NEW BERN, near BAY RIVER estuary; 35°12′N 76°39′W. Service industries; construction; agriculture (cotton, grain; hogs).

Mesick (ME-sik), village (2000 population 447), WEXFORD county, NW MICHIGAN, 19 mi/31 km NW of CADILLAC, and on MANISTEE RIVER, at end of Hodenpyl Dam Pond; 44°23′N 85°43′W. Manufacturing (tool and die; machining). Manistee National Forest to S. Annual mushroom festival.

Mesilla Park (ME-SEE-uh), unincorporated town, DOÑA ANA county, S NEW MEXICO, suburb 2 mi/3.2 km SSE of downtown LAS CRUCES, near RIO GRANDE. Manufacturing (ceramic tile, tortillas, jewelry).

Mesilot, ISRAEL: see MESSILOT.

Meskena, SYRIA: see MESKENE.

Meskene (mes-KEEN), village, ALEPPO district, N SYRIA, on the W bank of Lake ASAD, and 60 mi/97 km SE of Aleppo; 36°01′N 38°04′E. Just S are extensive ruins of ancient BARBALISSUS and medieval BALIS. The irrigated area in the vicinity was greatly extended (with Soviet assistance). Also spelled MESKENA or MASKANAH.

Mesker-Yurt (mees-KYER–YOORT), village (2005 population 11,160), central CHECHEN REPUBLIC, S European Russia, on crossroads and local railroad spur, 9 mi/14 km ESE of GROZNYY, and less than 4 mi/6 km S of ARGUN; 43°15′N 45°54′E. Elevation 452 ft/137 m. Agriculture (grain, vegetables, livestock). Formerly called Rubezhnoye (1944–1960).

Meskheti Mountains (mes-KHE-tee) or **Adzhar-Imenti**, mountain range in the Lesser CAUCASUS, extends 93 mi/150 km from the Adzhar coast of the BLACK SEA to the BORZHOMI GORGE of the KURA RIVER; 41°48′N 42°30′E. Rises to 9,351 ft/2,850 m at Mount Mepistskaro.

Meskiana (mes-kyah-NAH), village, OUM EL BOUAGHI wilaya, NE ALGERIA, on railroad, and 30 mi/48 km NW of TEBESSA; 35°34′N 07°34′E. Flour milling. Copper and lead mining.

Meslay-du-Maine (mai-LAI–dyoo–MEN), commune (□ 9 sq mi/23.4 sq km; 2004 population 2,670), MAYENNE department, PAYS DE LA LOIRE region, W FRANCE, 13 mi/21 km SE of LAVAL; 47°57′N 00°33′W. Cheese making, sawmilling.

Mesnil-Esnard, Le (mai-NEEL–ai-NAHR, luh), suburb (□ 2 sq mi/5.2 sq km) of ROUEN, SEINE-MARITIME department, HAUTE-NORMANDIE region, N central FRANCE; 49°25′N 01°09′E.

Mesnil-le-Roi, Le (mai-NEEL–luh–RWAH, luh), residential suburb (□ 1 sq mi/2.6 sq km) of PARIS, YVELINES department, ÎLE-DE-FRANCE region, N central FRANCE, on a bend of SEINE RIVER, and 3 mi/4.8 km NNE of SAINT-GERMAIN-EN-LAYE; 48°56′N 02°08′E.

Mesnil-sur-Oger, Le (mai-NEEL–syoor–o-ZHAI, luh), commune (□ 3 sq mi/7.8 sq km), MARNE department, CHAMPAGNE-ARDENNE region, N FRANCE, 8 mi/12.9 km SSE of ÉPERNAY; 48°56′N 04°02′E. Noted wine-growing center (Chardonnay wines). Museum of champagne production.

Mesocco (me-SOK-ko), town, GRISONS canton, SE SWITZERLAND, on MOËSA RIVER, in VALLE MESOLCINA, and 17 mi/27 km NE of BELLINZONA, at S end of climb up to SAN BERNARDINO PASS.

Mesolcina, Valle (me-sol-zhee-nah, VAHL-le), German *Misoxertal*, valley of MOËSA RIVER, in LEPONTINE ALPS, GRISONS canton, SE SWITZERLAND. Tourist area. Ruins of Castello Misox, just S of MESOCCO. Italian-speaking Roman Catholic population.

Mesolonghi, Greece: see MESOLONGI.

Mesolongi (MEE-so-lon-ghee), Greek *Messolongion*, town, ⊙ AKARNANIA prefecture, W central GREECE, on lagoon of Gulf of PATRAS, 20 mi/32 km NW of PATRAS, and 130 mi/209 km WNW of ATHENS; 38°21′N 21°17′E. Trades in fish, wine tobacco, livestock. Accessible only to shallow-draught vessels, it has a deep water port at KRIONERI (lined by railroad) and also has railroad connections with AGRINION and Neochorion. Seat of Greek metropolitan. A fishing village under Turkish rule, it became a Greek stronghold in the Greek War of Independence. It was successfully defended (1822–1823) against an initial Turkish siege. The English poet Lord Byron, who supported the cause of Greek independence, died of a fever here in 1824. Finally taken (1826) by the Turks after a one-year siege. Rebuilt after 1828, it developed into an important trade center. The seiges are commemorated by the Heroon (a mass grave); statue to Byron (1881). Also spelled Misolonghi, Missolonghi, or Mesologhi.

Mesones Hidalgo (mai-SO-nes hee-DAHL-go), town, in far W OAXACA, MEXICO, 87 mi/140 km W of OAXACA DE JUÁREZ. On the border of the state of GUERRERO. Mountain valley with temperate climate. Agriculture (corn, beans, coffee, fruits), woods, mezcal, cattle. Fabric. Formerly HIDALGO, also San José Mesones.

Mesopotamia, [Greek=between rivers], ancient country of ASIA, the region around the TIGRIS and EUPHRATES rivers, included in modern IRAQ. The region extends from the PERSIAN GULF N to the moutains of ARMENIA, and from the ZAGROS and Kurdish mountains on the E to the SYRIAN DESERT. From the mountainous N, Mesopotamia slopes down through grassy steppes to a central alluvial plain, which was once rendered exceedingly fertile by a network of canals. The S was long thought to be the cradle of civilization until earlier settlements (which probably date from about 5000 B.C.E.) were found in N Mesopotamia; Jarmo, the earliest of these, was superseded by a succession of cultures: Tell Hassuna, Samarra, and Tell Halaf. Tell Halaf, the most advanced of these early cultures, is famous for Halaf ware, the finest prehistoric pottery in Mesopotamia. It is found at such sites as NINEVEH and TEPE GAWRA. While these advances were being made in the N, civilization was just beginning in the S, particularly at ERIDU. The Al Ubaid culture that followed flourished in both N and S Mesopotamia. During the next period (called the proto-literate phase) the S was the important region, and the transformation of the village culture into an urban civilization took place. Erech (modern Warka), the foremost site at the beginning of this period, has yielded such monumental architecture as the temple of Inanna and the ziggurat of Anu. Also found at Erech were tablets including the earliest pictographic writing. The early dynastic phase that followed saw the development of city-states all over the MIDDLE EAST as far as N SYRIA, N Mesopotamia, and probably ELAM. The famous sites of this period are Tell Asmar, Kafaje, UR, KISH, MARI, Farah, and Telloh (LAGASH). The Sumerians, the inhabitants of these city-states of S Mesopotamia, were unified at NIPPUR, where they gathered together to worship Enlil, the wind god. The famous first dynasty of Ur came at the end of the early dynastic period. Sargon founded (c.2340 B.C.E.) the Akkadian dynasty, the first empire in Mesopotamia, whose example of empire building was later followed by the old Babylonian dynasty and late Assyrian Empire. Mesopotamia still had prestige at the time of Alexander the Great, but later it was generally a part of the Roman Empire. The Arabs took it from the Sassanid Empire, and it rose to great prominence after Baghdad was made capital of Abbasid caliphate (C.E. 762). This glory was destroyed when the Mongols under Hulagu Khan devastated the area in 1258, destroying the ancient irrigation system. In the centuries following, Mesopotamia never regained its former prominence. In World War I, however, it was an important battlefield. The kingdom of Iraq was formed in 1921 (Iraq became a republic in 1958) and rose to international importance due to its rich oil fields and archaeological finds of the distant past. Scene of fighting in the IRAN-Iraq War in the late 1980s, the PERSIAN GULF War of the early 1990s, and the war in Iraq that began March 29, 2003.

Mesopotamia (me-so-po-TAI-mee-ah), region (□ 74,000 sq mi/192,400 sq km) of NE ARGENTINA, between PARANÁ (W) and URUGUAY (E) rivers. Comprises ENTRE RÍOS, CORRIENTES, and MISIONES provinces.

Mesquite (MES-keet), city (2006 population 131,447), DALLAS county, N TEXAS, a suburb, 10 mi/16 km E of downtown DALLAS; 32°46′N 96°35′W. Elevation 491 ft/150 m. Bounded in SE by East Fork of TRINITY RIVER. Manufacturing industrial power supplies, also building materials, medical equipment, bank supplies, machinery, paper products. Headquarters of several major Texas corporations, such as the Texas Power and Light Company and Lonestar Gas Company. Agriculture to E and SE (dairying; cotton, peanuts). It has been one of the fastest-growing U.S. cities, marked by a population increase of more than 51% between 1980 and 1990; it is now surrounded on all but SE by other municipalities. Eastfield College (two-year) is in NW of city, and an annual rodeo is held here every October. LAKE RAY HUBBARD reservoir to NE; Devil's Bowl Speedway is here; Phil Lake Anderson Airport in E. Incorporated 1887.

Mesquite (mes-KEET), town (2006 population 14,799), CLARK county, SE NEVADA, 70 mi/113 km NE of LAS VEGAS, on ARIZONA state line, on VIRGIN RIVER; 36°31′N 114°06′W. One of several towns developed on Nevada state line to offer gambling to travelers. Desert Valley Museum. VIRGIN MOUNTAINS to SE; Virgin Mountains Natural Area to S; Virgin River State Recreation Area to SW.

Mesquite (muh-SKEET), unincorporated town, DONA ANA county, S NEW MEXICO, 12 mi/19 km SSE of LAS CRUCES, on RIO GRANDE. Cattle, sheep; dairying; chilies, jalapenos, vegetables, grain, alfalfa, nuts. Manufacturing (dairy products, fertilizer).

Messa (MES-uh), village, CYRENAICA region, NE LIBYA, on road, 15 mi/24 km WSW of CYRENE, on Jebel Al AKHDAR plateau. Agricultural settlement established 1933 as Luigi Razza, or Razza, by Italians, who left after World War II, replaced by Libyan population. Ruins (sepulchers) of ancient Messa.

Messac (me-SAHK), commune (□ 16 sq mi/41.6 sq km; 2004 population 2,477), ILLE-ET-VILAINE, BRITTANY, W FRANCE, on the VILAINE RIVER, and 19 mi/31 km S of RENNES; 47°49′N 01°48′W. Alcohol distilling. Also called Port de Messac.

Messalonskee Lake (mes-uh-LAHN-skee), reservoir (□ 6 sq mi/15.6 sq km), KENNEBEC county, central MAINE, on Messalonskee Stream, c.7 mi/11 km W of WATERVILLE; 10 mi/16 km long, c.1 mi/1.6 km wide; 44°32′N 69°44′W. Max. capacity 118,300 acre-ft. Formed by Snow Pond Dam (13 ft/4 m high), built (1992) for power generation. One of BELGRADE LAKES.

Messamena (mess-uh-MEE-nuh), village, East province, S central CAMEROON, 30 mi/48 km SW of Abong-M'Bang; 03°47′N 12°48′E. In coffee-growing area.

Messancy (meh-zahn-SEE), commune (2006 population 7,405), Arlon district, LUXEMBOURG province, SE BELGIUM, 6 mi/9.7 km S of ARLON; 49°36′N 05°49′E. Residential commune for nearby industrial ATHUS.

Messaoria, CYPRUS: see MESAORIA.

Messapion, Greece: see KTYPAS.

Messara (me-sah-RAH), agricultural lowland, IRÁKLION prefecture, S CRETE department, GREECE, on E coast of Gulf of MESSARA. Chief town is Timbaki. Also written Mesara.

Messara, Gulf of, inlet on S coast of central CRETE, GREECE; 15 mi/24 km wide, 5 mi/8 km long. Small port of Ayia Galini is on N shore. Also spelled Gulf of Mesara.

Messei-Saint-Gervais (me-se–san–zher-vai) or **Messei** (me-se), commune (□ 5 sq mi/13 sq km; 2004 population 1,912), ORNE department, BASSE-NORMANDIE region, NW FRANCE, 3 mi/4.8 km SSE of FLERS; 48°43′N 00°33′W. Metalworking.

Messejana, town, BEJA district, S PORTUGAL, 24 mi/39 km SW of BEJA. Grain; sheep.

Messel (MES-sel), village, N HESSE, central GERMANY, in the Main Plain, 6 mi/9.7 km NE of DARMSTADT; 49°57′N 08°44′E. Manufacturing of synthetic fiber. Fossils were found here at the Grube Messel, which was recognized as a World Culture Heritage site in 1995.

Messene, Greece: see MEKENE.

Messene, Gulf of, Greece: see MESSENIA, GULF OF.

Messenia (me-see-NEE-ah), prefecture, SW PELOPONNESE department, extreme SW corner of GREECE, on Gulf of MESSENIA (S) and IONIAN SEA and Gulf of KIPARISSIA (W); ⊙ MESSENIA; 37°15′N 21°50′E. Bordered N by WESTERN GREECE, NE by ARKADIA prefecture, and SE by LAKONIA prefecture. Includes Oinussai islands. Also name of ancient region in the same area. Excavation has revealed an important center of Mycenaean culture at PYLOS dating from the 13th century B.C.E. From the 8th century B.C.E. the Messenians were engaged in a series of revolts against expanding SPARTA. After the First Messenian War, the Spartans annexed (c.700 B.C.E.) the E part of Messenia. With the Second Messenian War the remaining inhabitants were reduced (7th century B.C.E.) to helots. The Third Messenian War (464 B.C.E.–459 B.C.E.) was a failure for Messenia but very costly to Sparta. The battle of LEUCTRA (371 B.C.E.) freed Messenia, and Messene was founded (c.369 B.C.E.) as capital. The region gave its name to MESSINA, SICILY (Italy), because of an influx of Messenian colonists (c.490 B.C.E.). Also spelled Messinia.

Messenia or **Messíni**, city, ⊙ MESSENIA prefecture, SW PELOPONNESUS, S GREECE, on PAMISOS River, and 5 mi/8 km W of KALAMATA. Trade center (rice; also cotton, olives, figs, wheat) with railroad connection to Kalamata. Named for ancient city of MEKENE (Messene), 10 mi/16 km NNW, on W slope of Mount Ithome, site of modern village of Mavromati. Ancient Mekene was founded 369 B.C.E. by Epaminondas as capital of Messinia. Ruins, partly unearthed, include theater, acropolis, and temple.

Messenia, Gulf of (me-see-NEE-ah), inlet of IONIAN SEA, off S PELOPONNESUS, extreme SW GREECE, between capes AKRITAS and MATAPAN; 30 mi/48 km wide, 35 mi/56 km long. KALAMATA is on N shore. Also called Gulf of Messene. Formerly known as Gulf of Koroni, or Kalamata.

Messenia Peninsula (me-see-NEE-ah), MESSENIA prefecture, SW PELOPONNESE department, extreme SW tip of GREECE, on IONIAN SEA, W of Gulf of MESSENIA; terminates S in Cape AKRITAS; 20 mi/32 km long, 15 mi/24 km wide. PYLOS and METHONI on W coast, KORONI on E coast.

Messias (me-SEE-ahs), city (2007 population 15,108), ALAGOAS state, BRAZIL, 16 mi/25 km N of MACEÍO, on BR 101 Highway to RECIFE; 09°23′S 35°49′W.

Messilot (mi-see-LOT) or **Mesilot**, kibbutz, NE ISRAEL, in Jordan valley, 2 mi/3.2 km W of BEIT SHEAN; 32°29′N 35°28′E. Below sea level 380 ft/115 m. Manufacturing of steel cables since 1961; mixed farming. Founded 1938.

Messina (mais-SEE-nah), city (2001 population 252,026), ⊙ Messina province, NE SICILY, ITALY, on the STRAIT OF MESSINA, opposite the Italian mainland; 38°11′N 15°34′E. Busy seaport and a commercial and light industrial center. Manufacturing includes processed food, chemicals, pharmaceuticals, clothing, fabricated metals, machinery, wood products, and construction materials. Founded (late 8th century B.C.E.) by Greek colonists and named Zancle, the city was captured (5th century B.C.E.) and renamed Messana. It was taken in 282 B.C.E. by mercenaries called Mamertines. The Romans answered an appeal for help from the Mamertines and intervened in Sicily, thus precipitating the first of the Punic Wars. Messina was subsequently allied with Rome, and it shared the history of the rest of SICILY. Developed a thriving silk industry (which declined in the 18th century). Suffered a severe plague in 1743 and major earthquakes in 1783 and 1908. The earthquake of December 28, 1908, destroyed 90% of Messina's buildings, including fine churches and palaces, and cost about 80,000 lives; afterward the city was completely rebuilt in conformity with standards for quake-resistant construction. In World War II, the Sicilian campaign ended with the fall of Messina to the Allies on August 17, 1943. Of interest in the city are the Norman-Romanesque cathedral (rebuilt after 1908) and the National Museum Seat of a university (1548).

Messina, town, SOUTH AFRICA: see MUSINA.

Messina, Strait of, channel, separating the Italian peninsula from SICILY and connecting the IONIAN and TYRRHENIAN seas; c.20 mi/32 km long and 2 mi/3 km–10 mi/16 km wide. REGGIO DI CALABRIA (SW ITALY, mainland) and MESSINA (on NE Sicily) are the main ports. A ferry crosses the dangerous waters of the strait from Messina to VILLA SAN GIOVANNI. There is much spearfishing. The currents, whirlpools, and winds of the strait, which still hamper navigation, gave rise in ancient times to many legends about its dangers to navigators.

Messines (me-SEEN), village (□ 42 sq mi/109.2 sq km; 2006 population 1,521), OUTAOUAIS region, SW QUEBEC, E CANADA, 11 mi/17 km from MANIWAKI; 46°14′N 76°01′W.

Messines, BELGIUM: see MESEN.

Messíni, Greece: see MESSENIA.

Messinia, Greece: see MESSENIA.

Messis, TURKEY: see MISIS.

Messkirch (MES-kirkh), town, Upper Swabia, BADEN-WÜRTTEMBERG, S GERMANY, 14 mi/23 km E of TUTTLINGEN; 47°59′N 09°07′E. Manufacturing of clothing, precision instruments; woodworking. Has 16th-century castle. Chartered 1250. Sometimes spelled MÖSSKIRCH.

Messolonghion, Greece: see MESOLONGI.

Messtetten (MES-shtet-tuhn), town, NECKAR-JURA, BADEN-WÜRTTEMBERG, S GERMANY, 24 mi/38 km SSW of TÜBINGEN; 48°11′N 08°58′E. Agriculture (fruit, hops); textile manufacturing; metalworking. Site of a folk art museum. Chartered 1978.

Mestanza (mes-TAHN-thah), town, CIUDAD REAL province, S central SPAIN, on N slopes of the SIERRA MORENA, 29 mi/47 km SSW of CIUDAD REAL; 38°35′N 04°04′W. Cereals, olives; livestock; timber; apiculture. Lead, silver, copper mining.

Area is shown by the symbol □, and capital city or county seat by ⊙.

Mesta River (ME-stah), Greek *Nestos*, Turkish *Kara Su*, Latin *Nestus*, 150 mi/241 km long, SW BULGARIA and NE GREECE; rises on the Kolarov Peak of the RILA MOUNTAINS; flows generally SE between the PIRIN and RODOPI Mountains into GREECE (EAST MACEDONIA AND THRACE department), through sparsely inhabited mountains and inaccessible gorges, then along the coastal plain above and below KHRISOUPOLIS, entering the Thracian Sea, part of the AEGEAN SEA, opposite island of Thásos; 41°33'N 23°50'E. Formed the border between Greek Macedonia and Bulgarian Thrace from 1913 to 1919, when Thrace became part of Greece.

Mestec Kralove (MNYES-tets KRAH-lo-VE), Czech *Městec Králové*, German *Königstadtl*, town, STREDOCESKY province, central BOHEMIA, CZECH REPUBLIC, on railroad, and 17 mi/27 km N of KUTNÁ HORA; 50°12'N 15°18'E. In sugar-beet district; manufacturing (glass cutting). Museum.

Mesters Vig (harbor) and **Mestersvig** (airfield), S side of KING OSCAR FJORD, E. GREENLAND; 72°15'N 23°54'E. Harbor and airfield for former lead-zinc deposit Blyklippen. Deposits were mined 1956–1962. Airstrip serves still as auxiliary airstrip for traffic in region.

Mestersvig, GREENLAND: see MESTERS VIG.

Mestia (mes-TEE-ah), village, NW GEORGIA, in SVANETIA, on S slope of the Greater Caucasus, near upper INGUR River, 55 mi/89 km N of KUTAISI; 43°02'N 42°43'E. Livestock raising; lumbering. Airfield.

Mesto Albrechtice (MNYE-sto AL-brekh-TYI-tse), Czech *Město Albrechtice*, German *Olbersdorf*, village, SEVEROMORAVSKY province, N SILESIA, CZECH REPUBLIC, on railroad, 7 mi/11.3 km NW of KRNOV; 50°10'N 17°34'E. Manufacturing (machinery); woodworking; agriculture (oats). Has at 18th century church.

Mesto Libava (MNYES-to LI-bah-VAH), Czech *Město Libavá*, German *Stadt Liebau*, village, SEVEROMORAVSKY province, N central MORAVIA, CZECH REPUBLIC, 15 mi/24 km NE of OLOMOUC; 49°43'N 17°31'E. In oat-growing area. Known as a town since 1301.

Mesto Touskov (MNYES-to TOUSH-kof), Czech *Město Touškov*, German *Stadt Tuschkau*, town, ZAPADOCESKY province, W BOHEMIA, CZECH REPUBLIC, on MZE RIVER, on railroad, and 6 mi/9.7 km WNW of PLZEŇ; 49°47'N 13°15'E. Manufacturing (machinery). Baroque church.

Mestre (ME-stre), town, VENEZIA province, VENETO, N ITALY, 5 mi/8 km NW of VENICE; 45°29'N 12°14'E. Industrial center developed about adjacent PORTO MARGHERA; processed foods; metallurgy. Has 12th-century bell tower.

Mesudiye, township, N TURKEY, on Melet River, 36 mi/58 km S of ORDU; 40°28'N 37°45'E. Grain. Formerly called Melet and Hamidiye.

Mesudiye, TURKEY: see ALUCRA.

Mesurado, LIBERIA: see MONTSERRADO.

Mesurado, Cape, LIBERIA: see MONTSERRADO, CAPE.

Mesyagutovo (mye-syah-GOO-tuh-vuh), village (2005 population 9,790), NE BASHKORTOSTAN Republic, E European Russia, on the AI RIVER (tributary of the UFA RIVER), on highway, 105 mi/169 km NE of UFA; 56°05'N 55°22'E. Elevation 341 ft/103 m. School for gifted children; sanatorium. Weather station.

Meta (MAI-tah), department (□ 32,903 sq mi/85,547.8 sq km), central COLOMBIA; ⊙ VILLAVICENCIO; 03°30'N 73°00'W. Extends from Cordillera ORIENTAL (E) to 71°05'W, between META RIVER (N) and GUAVIARE RIVER (S). Apart from Cordillera Oriental and its E spurs, it consists of LLANO grasslands and dense forests. It is a pioneer agricultural zone and has a hot, tropical climate. Mainly a cattle-grazing region, with coffee, rice, corn, and sugarcane. Forests yield timber. Villavicencio is an important communication and trading center serving the entire Colombian ORINOCO

river basin. Most people live in the capital or municipio centers.

Meta (MEE-tuh), town (2000 population 249), OSAGE county, central MISSOURI, near OSAGE RIVER, 20 mi/32 km SSW of LINN; 38°18'N 92°10'W. Pet food; charcoal products.

Metabetchouan (mee-tuh-bech-WAHN) or **Saint-Jérôme** (san-zhai-ROM), former town, S central QUEBEC, E CANADA, on SE shore of Lake SAINT JOHN, 13 mi/21 km SE of ALMA; 48°26'N 71°52'W. Dairying; lumbering; pig raising. Also spelled Métabetchouan (mai-tuh-bech-WAHN). Merged with LAC-À-LA-CROIX in 1998 to form Métabetchouan–Lac-à-la-Croix.

Métabetchouan–Lac-à-la-Croix (mai-tuh-bech-WAHN–lahk-ah-lah-KWAH), city (□ 72 sq mi/187.2 sq km; 2006 population 4,277), SAGUENAY–LAC-SAINT-JEAN region, central QUEBEC, E CANADA; 48°26'N 71°52'W. Formed in 1998 from METABETCHOUAN, LAC-À-LA-CROIX.

Metabetchouan River (mee-tuh-bech-WAHN) or **Métabetchouan River** (mai-tuh-bech-WAHN), 50 mi/80 km long, S central QUEBEC, E CANADA; rises in N part of LAURENTIDES Provincial Park; flows N to Lake SAINT JOHN at DESBIENS; 48°25'N 71°58'W.

Meta Gafersa, ETHIOPIA: see META GEFERSA.

Meta Gefersa (MAi-tah gai-FUHR-sah), village (2007 population 7,687), OROMIYA state, S ETHIOPIA, on regional road, near affluent of DAWA RIVER, 60 mi/97 km NE of MEGA; 04°45'N 38°49'E. Formerly called Arero or Araro. Also spelled Meta Gafersa.

Meta Incognita or **Kingait**, peninsula, SE BAFFIN ISLAND, baffin region, NUNAVUT territory, CANADA, extending SE into the ATLANTIC Ocean, between FROBISHER BAY (NE) and HUDSON STRAIT (SW); 61°52'N 65°55'W–63°30'N 70°W. Mountainous surface rises to c.2,500 ft/762 m in center; 170 mi/274 km long, 30 mi/48 km–80 mi/129 km wide. On NE coast, near mouth of Frobisher Bay, are Grinnell Ice Cap (c.3,000 ft/914 m high) and Southeast Ice Cap (c.2,800 ft/853 m high), both extending tongues to Frobisher Bay.

Metairie (MET-uh-ree), unincorporated city (2000 population 146,136), JEFFERSON parish, SE LOUISIANA, NW suburb, 7 mi/11 km WNW of NEW ORLEANS, on S shore of LAKE PONTCHARTRAIN; 30°00'N 90°11'W. Manufacturing (cultured marble, fabricated metal products, motor vehicle parts, concrete blocks, dairy products), printing and publishing. MISSISSIPPI RIVER to S. Named for the metairies, or little farms, that developed from the original plantations.

Metaliferi, Munţii, ROMANIA: see APUSENI MOUNTAINS.

Metaline (met-ah-LEEN), village (2006 population 172), PEND OREILLE county, NE WASHINGTON, on PEND OREILLE RIVER, 31 mi/50 km NE of COLVILLE, 10 mi/16 km S of CANADA (BRITISH COLUMBIA) border; 48°51'N 117°23'W. Lead, zinc, copper; timber. Box Canyon Dam to S, Boundary Dam to N. Crawford State Park to N, on Canadian boundary; parts of Colville National Forest to E and W.

Metaline Falls (met-ah-LEEN), village (2006 population 235), PEND OREILLE county, NE WASHINGTON, port of entry near BRITISH COLUMBIA line, 33 mi/53 km NE of COLVILLE, near PEND OREILLE RIVER 9 mi/14.5 km S of CANADA (British Columbia) border, port of entry (Nelway, British Columbia); 48°52'N 117°22'W. Railroad terminus. Timber; lead, zinc, copper. SULLIVAN LAKE reservoir to SE; parts of Colville National Forest to E and W, including Salmo-Priest Wilderness Area, to NE. Box Canyon Dam to S, Boundary Dam to N.

Metallostroy (me-tah-luh-STRO-yee), town (2005 population 24,945), central LENINGRAD oblast, NW European Russia, on the NEVA RIVER, on road and railroad, 12 mi/20 km SE of SAINT PETERSBURG, of which it is a suburb; 59°48'N 30°35'E. Manufacturing (electrical machinery, refrigerators).

Metamma, ETHIOPIA: see METEMA.

Metamora (met-a-MORE-ah), village (2000 population 2,700), WOODFORD county, central ILLINOIS, 12 mi/19 km ENE of PEORIA; 40°47'N 89°21'W. In agricultural area; canned foods. Former capital of Woodford county. Old courthouse is now state memorial to Lincoln, who often argued cases here. Incorporated 1845.

Metamora, village, FRANKLIN county, SE INDIANA, on the West Fork of WHITEWATER RIVER, and 7 mi/11.3 km WNW of BROOKVILLE. Located on the old Whitewater Canal, a section of which still goes through the village and is a tourist attraction. Laid out 1838.

Metamora (ME-ta-MOR-uh), village (2000 population 507), LAPEER county, E MICHIGAN, 7 mi/11.3 km S of LAPEER; 42°56'N 83°17'W. In farm area. Manufacturing (plastic molding, machining). Metamura-Hadley State Recreation Area to W.

Metamora (me-tuh-MO-ruh), village (2006 population 602), FULTON county, NW OHIO, 19 mi/31 km W of TOLEDO, at MICHIGAN line; 41°43'N 83°54'W. Agriculture (tomatoes, corn); poultry hatcheries.

Metán (me-TAHN), town (1991 population 23,067), ⊙ Metán department (□ 2,380 sq mi/6,188 sq km), S central SALTA province, ARGENTINA, 55 mi/89 km SE of SALTA; 25°30'S 64°40'W. Railroad junction (Metán station), agricultural center (corn, rice, alfalfa, flax, cotton; livestock), and lime works. Formerly called San José de Metán.

Metangula (me-tahn-GOO-luh), port, NIASSA province, NW MOZAMBIQUE, near MALAWI border, 45 mi/72 km NNW of LICHINGA; 12°40'S 34°51'E. Road terminus; only Portuguese port on Lake NYASA.

Metapa, NICARAGUA: see CIUDAD DARÍO.

Metapa de Domínguez (me-TAH-pah dai do-MEEN-ges), town, ⊙ Metapa municipio, CHIAPAS, S MEXICO, near GUATEMALA border, 6 mi/9.7 km SE of TAPACHULA; 14°50'N 92°11'W. Coffee, sugarcane, bee keeping.

Metapán (me-tah-PAHN), city and municipality, SANTA ANA department, NW EL SALVADOR, just N of small LAKE METAPÁN, 21 mi/34 km N of SANTA ANA, at S foot of the Sierra de Metapán; 14°20'N 89°27'W. Coffee, grain; livestock raising.

Meta Pond (MEE-tuh), lake, SE NEWFOUNDLAND AND LABRADOR, CANADA; 8 mi/13 km long, 2 mi/3 km wide; 48°03'N 54°53'W.

Metapontum (me-tuh-PAHN-tuhm), ancient city of Magna Graecia, on the Gulf of TARANTO, SE ITALY. Settled by Greeks (c.7th century B.C.E.), it flourished and gave refuge to Pythagoreans expelled from CROTONA. Pythagoras taught and died here. There are remains of a Doric temple, called Tavole Paladine, and other ruins.

Metar (me-TAHR) or **Meitar**, town, S ISRAEL, 9 mi/15 km NE of BEERSHEBA. Founded 1984; most residents commute to jobs in the surrounding industrial zones.

Metara (me-TER-uh), town, DEBUB region, central ERITREA, 56 mi/90 km SSE of ASMARA; 14°38'N 39°26'E. Site of important archaeological finds, some dated to 3rd century and thought to be Axumite.

Metarica District, MOZAMBIQUE: see NIASSA.

Meta River (MAI-tah), c.650 mi/1,046 km long, central and E COLOMBIA; its headstreams (GUATIQUÍA and GUAYURIBA) rise E and S of BOGOTÁ in Cordillera ORIENTAL; then the river flows NE and E through LLANO lowlands, forms part of VENEZUELA-Colombia border, and enters ORINOCO RIVER at PUERTO CARREÑO; 06°12'N 67°28'W. Navigable, but scarcely used.

Metasville (MET-uhs-vil), town, WILKES county, NE GEORGIA, 8 mi/12.9 km ENE of WASHINGTON; 33°46'N 82°36'W.

Metauro River (mai-TOU-ro), ancient *Metaurus*, river of The MARCHES, c.68 mi/110 km long, central ITALY;

rising in the ETRUSCAN APENNINES from a double source (the Meta and the Auro); flowing NE into the ADRIATIC Sea near FANO. On its banks the Romans defeated (207 B.C.E.) the Carthaginians under Hasdrubal in the Second Punic War.

Metcalf (MET-kaf), town, THOMAS county, S GEORGIA, 9 mi/14.5 km S of THOMASVILLE, near FLORIDA state line; 30°42′N 83°59′W.

Metcalf, village (2000 population 213), EDGAR county, E ILLINOIS, 14 mi/23 km NNW of PARIS; 39°47′N 87°48′W. In agricultural area.

Metcalfe (MET-kaf), county (□ 290 sq mi/754 sq km; 2006 population 10,334), S KENTUCKY; ⊙ EDMONTON; 36°59′N 85°37′W. Drained by LITTLE BARREN RIVER and Beaver and Marrowbone creeks. Rolling agricultural area (corn, wheat, burley tobacco, soybeans, hay, alfalfa; cattle, poultry; dairying; timber). Formed 1860.

Metcalfe (MET-kaf), town (2000 population 1,109), WASHINGTON county, W MISSISSIPPI, residential suburb, 4 mi/6.4 km NE of GREENVILLE; 33°27′N 91°00′W. Greenville Municipal Airport here. Stoneville National Wildlife Refuge to E.

Metchosin, district municipality (□ 27 sq mi/70.2 sq km; 2001 population 4,857), SW BRITISH COLUMBIA, W CANADA, in CAPITAL REGIONAL DISTRICT; 48°23′N 123°32′W.

Metedeconk River (muh-TEE-duh-KUHNGK), E NEW JERSEY; rises SW of FREEHOLD in North Branch (c.20 mi/32 km long) and South Branch (c.15 mi/24 km long), which flow SE to junction c.6 mi/9.7 km above mouth on BARNEGAT BAY; navigable below the junction.

Meteghan (meh-TAI-guhn), village, (pop. 2001, 9,067) W NOVA SCOTIA, CANADA, on the ATLANTIC OCEAN, at entrance of ST. MARY BAY, 25 mi/40 km N of YARMOUTH; 44°11′N 66°09′W. Elevation 78 ft/23 m. Fishing; wood-working. Settled in 1785.

Metehara, ETHIOPIA: see METEHARA DEBRE SELAM.

Metehara Debre Selam, town (2007 population 22,366), OROMIYA state, central ETHIOPIA, on highway, 75 mi/121 km E of ADDIS ABABA; 08°54′N 39°54′E. Near AWASH NATIONAL PARK; in sugar growing area. Sometimes called Metehara.

Metelen (ME-te-len), village, WESTPHALIA-LIPPE, North Rhine-Westphalia, W GERMANY, 5 mi/8 km W of STEINFURT; 52°09′N 07°13′E. Manufacturing (textiles); dairying.

Metema, town (2007 population 5,846), AMHARA state, NW ETHIOPIA, on SUDANESE border, opposite GALLABAT, near ATBARA RIVER; 12°58′N 36°12′E. Trade center (coffee, honey, butter; livestock). Once flourished as a slave market. In 1889, Emperor John of Ethiopia was killed and his forces defeated here, in a battle with Mahdist troops from the SUDAN. Also spelled Metamma. Sometimes called Metema Yohannes.

Metema Yohannes, ETHIOPIA: see METEMA.

Meteora (me-TAI-o-rah), group of monasteries, at KASTRAKI, TRIKKALA prefecture, W THESSALY department, N GREECE, on pillarlike rocks just N of KALAMBAKA, 14 mi/23 km NNW of TRIKKALA. Flourished 14th–16th century, when they received special privileges under Turkish rule. Wall paintings, writings are preserved. Of the original twenty-three monasteries, five remain inhabited.

Meteor Crater, large crater, SE COCONINO county, N central ARIZONA, 35 mi/56 km ESE of FLAGSTAFF. Created by meteorites; 4,150 ft/1,265 m in diameter, 570 ft/174 m deep, with rim rising 120 ft/37 m–160 ft/49 m above surrounding plain, about 1 mi/1.6 km wide. Fragments of iron, containing some nickel, have been found. NASA used the crater as training ground for Apollo astronauts before their missions to the moon. Museum of Astrogeology is at site. Sometimes called Diablo Crater.

Metepec (ME-te-pek), town, HIDALGO, central MEXICO, 27 mi/43 km ENE of PACHUCA DE SOTO. Corn, maguey, livestock, manufacturing of metal pipes.

Metepec, town, MEXICO state, central MEXICO, on railroad, 5 mi/8 km SE of TOLUCA DE LERDO, on Mexico Highway 55. Known for pottery and colorful market. Agricultural center (grain, fruit; livestock); dairying.

Methana (ME-thah-nah), town, ATTICA prefecture, ATTICA department, E central GREECE, in PELOPONNESUS, 30 mi/48 km SSW of PIRAEUS, across SARONIC GULF, on SE shore of Methana Peninsula, a headland of the ARGOLIS Peninsula; 37°35′N 23°23′E.

Methil (METH-uhl), port, FIFE, E Scotland, on N coast of Firth of Forth, 8 mi/12.9 km NE of KIRKCALDY; 56°12′N 03°01′W. Manufacturing of steel production platforms for NORTH SEA oil field.

Methlick (METH-lik), village, Aberdeenshire, NE Scotland, on YTHAN RIVER, and 7 mi/11.3 km NNE of OLDMELDRUM; 57°25′N 02°15′W. Just S is 18th-century Haddo House of marquis of Aberdeen, built by William Adam. Nearby are remains of ancient Gight Castle or House of Gight. Formerly in Grampian, abolished 1996.

Methoni (me-THO-nee), town, MESSENIA prefecture, SW PELOPONNESE department, extreme SW GREECE, port on IONIAN SEA, opposite Oinoussai Islands, 26 mi/42 km SW of KALAMATA; 36°49′N 21°42′E. Fisheries; livestock (goats, sheep). Site of ancient city, which became important Venetian port after 1206 and was then called Modon. Ruled by Turks after 1500, except for brief Venetian occupation (1699–1718). Formerly spelled Methone.

Methow River (MET-hou), c.80 mi/129 km long, OKANOGAN county, N WASHINGTON; rises in CASCADE RANGE N of LAKE CHELAN; flows N then generally SE through irrigated agricultural valley (apples), past WINTHROP and TWISP, to COLUMBIA River (Laje Pateros reservoir) at PATEROS.

Methuen (me-THOO-uhn), town, ESSEX county, NE MASSACHUSETTS, suburb of BOSTON; 42°44′N 71°11′W. Methuen is industrial, and among its products are food items, computer and microwave components, medical supplies, and textiles. The Tenney Estate was converted into Saint Basil's Seminary and Presentation of Mary Academy. Settled c.1642, set off from HAVERHILL 1725.

Methven, township, Canterbury Plains, SOUTH ISLAND, NEW ZEALAND, 50 mi/80 km WSW of CHRISTCHURCH; 43°38′S 171°39′E. Agriculture; linen mill. Also called Mount Hutt.

Methven (METH-vhn), village (2001 population 1,162), PERTH AND KINROSS, E Scotland, 6 mi/9.7 km W of PERTH; 56°25′N 03°37′W. Nearby is Glenalmond College, public school founded 1847. Formerly in Tayside, abolished 1996.

Methy Lake (MEE-thee), NW SASKATCHEWAN, CANADA, 20 mi/32 km NW of PETER POND LAKE; 21 mi/34 km long, 5 mi/8 km wide; 56°25′N 109°30′W. Elevation 1,460 ft/445 m. Main headstream of CHURCHILL RIVER rises here.

Methymna (ME-theem-nah), town, on N shore of LESBOS island, LESBOS prefecture, NORTH AEGEAN department, GREECE, 26 mi/42 km NW of MITILÍNI; 39°22′N 26°10′E. Trade in olive oil, wine, vegetables; fisheries. Its adjoining Aegean port is Molybdos or Molivdhos. Home of lyric poet Arion (7th century B.C.E.). Rivaled (5th–4th century B.C.E.) Mitilíni for the leadership of the island. Also spelled Mithimna.

Metica River, Colombia: see GUAYURIBA RIVER.

Meting (ME-ting), village, TATTA district, SW SIND province, SE PAKISTAN, on railroad line, and 70 mi/113 km ENE of KARACHI; 25°10′N 68°07′E.

Metinic Island (muh-TIN-ik), KNOX county, S MAINE, 6 mi/9.7 km SE of TENANTS HARBOR; 2 mi/3.2 km long, 0.25 mi/0.5 km wide.

Metiskow (muh-TIS-ko), hamlet, SE ALBERTA, W CANADA, 16 mi/26 km from PROVOST, in PROVOST NO. 52 municipal district; 52°24′N 110°38′W.

Metis Shoal (ME-tis), TONGA, S PACIFIC OCEAN, near VAVA'U group. Lies between the active volcanic cones of KAO and LATE. Eruptions in the shoal have provided reports of new islands in 1858, 1878, 1886, 1894, 1967, 1979, and 1995; these tend to disappear within a few years as they are eroded by the sea. An island formed in May 1979 was named LATEIKI (by the side of Late) and disappeared two years later. The island produced by an eruption in June 1995, which rose to a height of 164 ft/50 m and was considered to be a reappearance of LATEIKI. Submarine volcanoes are known locally as *fonwafo'ow* or "jack-in-the-box" islands.

Métis-sur-Mer (mai-TEE–syur–MER) or **Metis Beach** (ME-tis), village (□ 19 sq mi/49.4 sq km), BAS-SAINT-LAURENT region, SE QUEBEC, E CANADA, on the SAINT LAWRENCE RIVER, 25 mi/40 km WSW of MATANE; 48°40′N 67°59′W. Resort. Restructured in 2002.

Metkovets, Bulgaria: see MEDKOVETS.

Metković (MET-ko-veech), town, S CROATIA, Adriatic port (via Neretva Channel) on NERETVA RIVER, on railroad, and 22 mi/35 km S of MOSTAR, in DALMATIA, near BOSNIA-HERZEGOVINA border. Largely supplanted by new port of Ploče (formerly Kardeljevo). Roman ruins nearby.

Metlakahtla, Alaska, see: METLAKATLA.

Metlakatla (MET-luh-KAT-luh), village, SE ALASKA, on W shore of ANNETTE ISLAND, 17 mi/27 km S of KETCHIKAN. A model cooperative village of the Tsimshian Indians. Fishing, logging; cooperative cannery and sawmill. Established 1887 by Reverend William Duncan, missionary, and Indians emigrating from FORT SIMPSON, BRITISH COLUMBIA. Also spelled Metlakahtla.

Metlatónoc (met-lahn-TO-nok), town, GUERRERO, SW MEXICO, in isolated zone of SIERRA MADRE DEL SUR, near OAXACA border, 25 mi/40 km SE of TLAPA DE COMONFORT; 17°11′N 98°20′W. No paved roads. Cereals, fruit, stock.

Metlatoyuca (me-tlah-YOO-kah), town, ⊙ Francisco Z. Mena municipio, PUEBLA, central MEXICO, in foothills of SIERRA MADRE ORIENTAL, 33 mi/53 km SW of TÚXPAM DE RODRÍGUEZ CANO; 20°44′N 97°51′W. Sugarcane, coffee, tobacco, fruit. Pre-Columbian ruins nearby.

Metlika (met-LEE-kah), ancient *Metulum*, village, S SLOVENIA, on KUPA RIVER, on railroad, 13 mi/21 km SSE of Novo Mesto, on CROATIA border; 45°40′N 15°20′E. Metal and plastic products; trade center (first mentioned in 1300). Old castle.

Metlili (met-lee-LEE), Saharan village, GHARDAÏA wilaya, central ALGERIA, 15 mi/24 km S of GHARDAÏA. One of the M'ZAB oases.

Metnitz (MET-nits), township, CARINTHIA, S AUSTRIA, in GURKTAL ALPS, 16 mi/26 km NW of SANKT VEIT AN DER GLAN; 46°59′N 14°26′E. Timber processing. Church with remarkable frescoes from the 14th century; famous danse-macabre fresco from 1500; Romanesque pilgrim church Grades nearby.

Metohija (MET-o-ee-yah), fertile valley, between NORTH ALBANIAN ALPS and SAR Mountains, KOSOVO province, SW SERBIA, among forested mountains, on ALBANIA border. Largely agricultural (fruit, chestnuts, beans); wine growing; meadows. Drained by the WHITE DRIN River and its tributaries. Chief towns include PRIZREN, PEC, DJAKOVICA. Under Turkish rule until 1913. Sometimes spelled Metochia or Metokhiya.

Metolius (met-O-lee-uhs), village (2006 population 737), JEFFERSON county, N central OREGON, 5 mi/8 km

SW of MADRAS; 44°35′N 121°10′W. Grain, livestock; frozen potatoes. CROOKED RIVER National Grassland to SE. Warm Springs Indian Reservation to NW. Cove Palisades State Park, on LAKE HINOOK Reservoir, to SW.

Metolius River (met-O-lee-uhs), c. 60 mi/97 km long, SW JEFFERSON county, N central OREGON; rises in CASCADE RANGE in DESCHUTES National Forest; flows N, then SE, forming S boundary of Warm Springs Indian Reservation and long W arm of LAKE CHINOOK Reservoir. Metolius Bench, level plateau area, to N.

Metompkin Inlet, Virginia: see METOMPKIN ISLAND.

Metompkin Island, ACCOMACK county, E VIRGINIA, 5 mi/8 km E of ACCOMAC, separated from mainland by narrow channel (Wire Passage; to W); c.6 mi/10 km long. Metompkin Inlet, CEDAR ISLAND to S; Gargathy Inlet, ASSAWOMAN ISLAND to N.

Metpalli (met-pah-LEE), town, KARIMNAGAR district, ANDHRA PRADESH state, S central INDIA, 19 mi/31 km W of JAGTIAL. Rice. Sometimes spelled Matpalli, Medpalli.

MetroManila, PHILIPPINES: see NATIONAL CAPITAL REGION.

Metropolis (meh-TROP-o-lis), port city (2000 population 6,482), ⊙ MASSAC county, extreme S ILLINOIS, on OHIO River, and 28 mi/45 km ENE of CAIRO; 37°08′N 88°42′W. In agricultural and lumbering area; manufacturing (clothing); shipping center for corn, wheat, livestock. FORT MASSAC STATE PARK (1,470 acres/595 ha), on site of French fort (1757); Kincaid Indian Mounds nearby. Incorporated 1859. Annual celebration honoring Superman.

Metropolis (me-TROP-o-lis), from the Greek= "mother city," generic term for a large or important city; also the name of a fictional city employed notable by Austrian filmmaker Fritz Lang in his pioneering 1927 film of a dystopian future, *Metropolis*, and widely known as the home of the comics superhero Superman. See also GOTHAM.

Metropolis, TURKEY: see TORBALI.

Metropolitana de Santiago (mai-tro-po-lee-TAH-nah dai sahn-tee-AH-go), region (2005 population 6,538,896), CHILE, consisting of SANTIAGO city and its surroundings; ⊙ Santiago. Lies between the VALPAR-AISO region to the N and the LIBERTADOR GENERAL BERNARDO O'HIGGINS region to the S. Overwhelmingly dominated by the city of Santiago; also consists of the provinces of CHACABUCO, CORDILLERA, MELIPILLA, TALAGANTE, and MAIPO. Political and economic heart of the nation. Site of Parque Nacional El Morado.

Metropolitan Community of Montreal, French, *Communauté Metropolitaine de Montréal* (ko-MYOO-no-tai me-tro-po-lee-TEN duh mon-trai-AHL) (□ 1,482 sq mi/3,853.2 sq km; 2006 population 3,523,924), S QUEBEC, E CANADA. Formed in June 2000 as a collective body of eighty two municipalities responsible for economic planning and coordination of public services across the greater MONTREAL metropolitan area.

Metropolitan Community of Quebec, French, *Communauté Metropolitaine de Québec* (ko-MYOO-no-tai me-tro-po-lee-TEN duh kai-BEK) (2006 population 700,000), S QUEBEC province, E CANADA; 46°46′N 71°17′W. Formed in January 2002 as a collective body of twenty six municipalities responsible for economic, social, cultural, and environmental development in the greater Quebec metropolitan area.

Metsäkylä, RUSSIA: see MOLODĚZHNOYE.

Metsamor, locality, ARMENIA, 20 mi/32 km from YEREVAN, near border with TURKEY; 40°11′N 44°17′E. Farming area. Has two SOVIET-built nuclear reactors accounting for 30% of Armenia's electricity. Plants closed in 1989 after 1988 earthquake. Refitted with Russian assistance and reopened in 1995.

Metsovo (ME-tso-vo), town, IOÁNNINA prefecture, S EPIRUS department, NW GREECE, in the PINDOS Mountains, 19 mi/31 km ENE of IOÁNNINA, near source of ARAKHTHOS River; 39°46′N 21°11′E. Livestock raising; olive oil. Center of Walachian (Vlach) population of Greece.

Mettawa, village (2000 population 367), Lake county, NE ILLINOIS, residential suburb, 29 mi/47 km NNW of downtown CHICAGO, 5 mi/8 km W of LAKE FOREST, on Des Plaines River; 42°14′N 87°55′W.

Mettawee River (ME-tuh-wee), c.50 mi/80 km long, in VERMONT and NEW YORK; rises in TACONIC MOUNTAINS near DORSET, Vermont; flows NW, past PAWLET, Vermont, and GRANVILLE, New York, to Lake Champlain near WHITEHALL, New York.

Mettelhorn (ME-tel-horn), peak (11,175 ft/3,406 m) in PENNINE ALPS, VALAIS canton, S SWITZERLAND, 2 mi/3.2 km N of ZERMATT.

Metten (MET-tuhn), village, LOWER BAVARIA, GERMANY, at SW foot of the BAVARIAN FOREST, near the DANUBE, 3 mi/4.8 km, NW of DEGGENDORF; 48°52′N 12°54′E. Agriculture includes wheat, vegetables, and livestock.

Metter, town (2000 population 3,879), ⊙ CANDLER county, E central GEORGIA, 17 mi/27 km W of STATESBORO, near CANOOCHEE River; 32°24′N 82°04′W. Welcome center on I-16. William Bartram-inspired Charles C. Harrold Nature Preserve nearby operated by nature conservancy features sand hill formations. Manufacturing includes clothing, boats, metal fabrication, commercial printing; tobacco market; lumber; pecan and peanut processing.

Mettet (ME-TE), commune (2006 population 11,997), Namur district, NAMUR province, S central BELGIUM, 13 mi/21 km SW of NAMUR; 50°19′N 04°40′E.

Mettingen (MET-ting-uhn), town, WESTPHALIA-LIPPE, North Rhine-Westphalia, NW GERMANY, 11 mi/18 km NW of OSNABRÜCK; 52°19′N 07°46′E. Grain.

Mettlach (MET-lahkh), town, NW Saarland, on Saar River, and 19 mi/31 km S of Trier; 49°30′N 06°36′E. Glass manufacturing; ceramics. Former Benedictine abbey (rebuilt in mid-18th century) now houses ceramics works. Ruined castle Montclair (9th century); castle from 1878–1879.

Mettmach (MET-mahkh), township, W UPPER AUSTRIA, in the INNVIERTEL region, 8 mi/12.9 km SW of Ried, in HAUSRUCK MOUNTAINS; 48°10′N 13°21′E. Timber processing, dairy farming.

Mettmann (MET-mahn), town, RHINELAND, North Rhine-Westphalia, W GERMANY, 8 mi/12.9 km ENE of DÜSSELDORF; 51°15′N 06°59′E. Manufacturing of metal goods.

Mettupalaiyam (met-tuh-PAH-lei-yuhm), town, COIMBATORE district, TAMIL NADU state, S INDIA, on railroad spur (to Ootacamund), 21 mi/34 km N of COIMBATORE; 11°10′N 78°27′E. Road center; livestock grazing; tannery; shellac manufacturing. Also spelled Mettuppalaiyam and Mettupalayam.

Mettupalayam, India: see METTUPALAIYAM.

Mettuppalaiyam, India: see METTUPALAIYAM.

Mettur (met-TOOR), town, Salem district, TAMIL NADU state, S INDIA, on KAVERI River, and 26 mi/42 km WNW of Salem. Industrial center; textile and sugar mills, chemical plants (alkalies, soap, vegetable ghee, fertilizers), cement factory; fish processing. Kaveri-Mettur hydroelectric and irrigation system consists of Mettur Dam (5,300 ft/1,615 m long, 176 ft/54 m high; terminus of railroad spur from Salem), across the Kaveri just N of METTUR, and power plant (in operation since 1937); dam impounds STANLEY RESERVOIR. The Mettur system (linked at ERODE, PERIYAR district, with Pykara transmission network) powers industries at Mettur and Salem and furthers industrial development in S Tamil Nadu state. Supplies and controls irrigation works (GRAND ANICUT and VADAVAR canals) of the Kaveri delta and nondeltaic tracts of Thanjavur district. Sometimes spelled Metur.

Mettur Canal, India: see GRAND ANICUT.

Metu, town (2007 population 36,194), OROMIYA state, SW ETHIOPIA, 90 mi/145 km NW of JIMMA; 08°18′N 35°36′E. Road junction and market town. Was the capital of the former ILUBABOR province.

Metuchen (muh-TUH-chuhn), borough (2006 population 13,216), MIDDLESEX county, NE NEW JERSEY; 40°32′N 74°21′W. Chiefly residential; light manufacturing. In June 1777, a brief but bloody skirmish occurred here between British troops under General William Howe and a small American force led by William Alexander. Settled before 1700, incorporated 1900.

Metulla (mi-TOO-lah), Jewish township, UPPER GALILEE, NE ISRAEL; frontier point on Lebanese border, at S foot of the NAFTALI range, 22 mi/35 km NNE of ZEFAT; 33°17′N 35°34′E. Elevation 1,705 ft/520 m. Mixed farming, apple orchards. Summer resort, guest houses; large indoor sports center (ice rink). Trade gateway to S LEBANON. In World War II British base for operations (1941) against French Vichy government forces in SYRIA. Founded 1896.

Metung (MEE-tuhng), village, VICTORIA, SE AUSTRALIA, near LAKES NATIONAL PARK; 37°53′S 147°51′E. Boat charter companies. Gallery of wood carvings.

Metur, India: see METTUR.

Metz (METS), city (□ 16 sq mi/41.6 sq km); ⊙ MOSELLE department, NE FRANCE, on the canalized MOSELLE RIVER; 49°08′N 06°10′E. It is also the capital of the administrative region of LORRAINE. A gateway to the PARIS BASIN, along historic invasion routes from the E, Metz today lies at the intersection of major N-S and E-W auto routes, with excellent railroad connections to PARIS (W) and STRASBOURG (E). A cultural, commercial, and communications center of Lorraine, second only to NANCY. Metz is also an industrial city known for its motor vehicle plants and production of machinery, clothing, and food products. It is one of 8 cities targeted by the French government for special planning and redevelopment in recognition of the decline of the region's iron mines and steel industry. Of pre-Roman origin, the city was the capital of the Mediomatrici, a Gallic people. One of the most important cities of Roman GAUL, it was invaded and destroyed by the Vandals (406) and the Huns (451). Metz was an early episcopal see and became the capital of Austrasia (the E portion of the Merovingian Frankish empire in the 6th century). After the division of the Frankish empire (8th century), the bishops of Metz greatly increased their power, ruling a relatively vast area as a fief of the Holy Roman Empire. Metz became a cultural center in the Carolingian Renaissance (8th century) and later (10th century) a prosperous commercial city with an important Jewish community. It became a free imperial city in 12th century and was then one of the richest and most populous cities of the Empire. In 1552, Henry II annexed the three bishoprics of Lorraine (Metz, TOUL, and VERDUN). An important fortress and garrison town, Metz was besieged (1870) by the Germans in the Franco-Prussian War, and after a two-month siege, 180,000 French soldiers under Marshall Achille Bazaine capitulated. During the German annexation of NE Lorraine (1871–1918), Metz, largely French-speaking, remained a center of pro-French sentiment. Metz suffered in both World Wars, particularly under German occupation (1940–1944) when many French-speaking citizens were deported. There are many Gallo-Roman ruins in Metz, including an aqueduct, thermal baths, and part of an amphitheater. An archaeological museum occupies the buildings of a 17th-century convent. Much has also been preserved from the medieval period. The celebrated Gothic Cathedral of SAINT-ÉTIENNE was built between c.1221 and 1516 with stained glass windows. The Place Sainte-Louis is a square surrounded by medieval houses (13th–15th century). Metz has many other churches, including the basilica of Saint-Pierre-aux-Nonnains, mansions from the Middle Ages, and fine views of the river from the Esplanade. A university was established here in 1972, and a European institute for ecological studies is attached to it. Metz-Frescapy international airport 4 mi/6.4 km SW of city.

Metz, town (2000 population 67), VERNON county, W MISSOURI, on LITTLE OSAGE RIVER, and 12 mi/19 km NNW of NEVADA; 38°00′N 94°26′W. Sorghum, hay; cattle.

Metzeral (MET-uh-rahl), commune (□ 11 sq mi/28.6 sq km), HAUT-RHIN department, ALSACE, E FRANCE, near the crest of the VOSGES, 14 mi/23 km WSW of COLMAR; 48°01′N 07°04′E. Cheese manufacturing; granite quarrying. Lies within the Regional Park of the Vosges (BALLON DES VOSGES NATURAL REGIONAL PARK).

Metzingen (MET-tsing-uhn), town, NECKAR, BADEN-WÜRTTEMBERG, GERMANY, at NW foot of SWABIAN JURA, 4 mi/6.4 km NE of REUTLINGEN; 48°53′N 09°17′E. Manufacturing of textiles (cotton, cloth; weaving), machinery, metal goods, leather gloves. Vineyards, fruit. Has 15th-century-Gothic church. Was first mentioned 1100. Chartered 1831.

Metzquitlitlán (metz-kee-tee-TLAHN), town, ⊙ SAN AGUSTÍN METZQUITITLÁN municipio, HIDALGO, central MEXICO, 30 mi/48 km NNE of PACHUCA DE SOTO on Mexico Highway 105; 20°32′N 98°39′W. Corn, beans, maguey, fruit, livestock.

Metztitlán (metz-tee-TLAHN), town, HIDALGO, central MEXICO, on central plateau, near Lake Metztitlán, 33 mi/53 km N of PACHUCA DE SOTO; 20°36′N 98°45′W. Elevation 4,413 ft/1,345 m. Agricultural center (corn, beans, oranges, melons, tomatoes, livestock).

Meudon (muh-don), city (□ 4 sq mi/10.4 sq km), HAUTS-DE-SEINE department, ÎLE-DE-FRANCE region, N central FRANCE, a SW suburb of PARIS, 7 mi/11.3 km from Notre Dame Cathedral, on a height overlooking the left bank of the SEINE, just SE of SÈVRES; 48°49′N 02°14′E. It adjoins the Forest of Meudon (S and W; □ 4.5 sq mi/11.7 sq km), a favorite Parisian excursion center reached by suburban railroad (R.E.R.). Residential development in Meudon-la-Forêt. The town has a wind-tunnel facility for testing aeronautical equipment, and machine shops. The astrophysics department of the Paris Observatory (with mushroom-shaped tower) is located in the pavilion of an 18th-century château, which commands a magnificent view of PARIS. François Rabelais, Richard Wagner, and Auguste Rodin lived in Meudon. Rodin is buried in the garden of his villa, now a museum containing many of his sculptures.

Meugia Pass, Laos and Vietnam: see MUGIA PASS.

Meulaboh (me-yoo-LAH-bo), town, ⊙ West Aceh district, ACEH province, NW SUMATRA, INDONESIA, port on INDIAN OCEAN, 115 mi/185 km SE of BANDA ACEH; 04°08′N 96°08′E. Ships timber, resin, copra, pepper, gold. Airport. Also spelled Mulabo.

Meulan (mu-lahn), town (□ 1 sq mi/2.6 sq km), YVELINES department, ÎLE-DE-FRANCE region, N central FRANCE, on right bank of SEINE RIVER, opposite industrial Les Mureaux, 23 mi/37 km WNW of PARIS; 49°01′N 01°54′E. Commercial center. Old fortress town.

Meulebeke (MU-luh-bai-kuh), commune (2006 population 11,038), Tielt district, WEST FLANDERS province, W BELGIUM, 9 mi/14.5 km N of COURTRAI; 50°57′N 03°17′E. Cotton manufacturing; agricultural market.

Meulín Island (mai-oo-LEEN) (□ 4 sq mi/10.4 sq km), off E coast of CHILOÉ ISLAND, S CHILE, 25 mi/40 km ENE of CASTRO; 3 mi/5 km long, 1 mi/2 km–3 mi/5 km wide; 42°25′S 73°20′W. Livestock raising; lumbering, fishing.

Meung-sur-Loire (mung–syur–lwahr), town (□ 7 sq mi/18.2 sq km), LOIRET department, CENTRE administrative region, NW central FRANCE, on right bank of the LOIRE, and 11 mi/18 km SW of ORLÉANS; 47°50′N 01°42′E. Foundries, tanneries. Has 11th–13th-century church and a partly medieval castle of bishops of Orléans. In Hundred Years War, the British took the bridge across the Loire, but Joan of Arc recaptured it in 1429. We first meet d'Artagnan of *Three Musketeers* fame as he comes to this town from his native GASCONY.

Meurchin (muhr-shan), commune (□ 1 sq mi/2.6 sq km; 2004 population 3,633), PAS-DE-CALAIS department, NORD-PAS-DE-CALAIS region, N FRANCE; 50°30′N 02°53′E.

Meursault (muhr-so), commune (□ 6 sq mi/15.6 sq km), CÔTE-D'OR department, BURGUNDY, E central FRANCE, on SE slope of the Beaune hills, 4 mi/6.4 km SW of BEAUNE; 46°59′N 04°46′E. Noted white Burgundy wines.

Meurthe (MUHRT), former department of NE FRANCE, part of which was ceded to GERMANY in 1871, the remainder incorporated in MEURTHE-ET-MOSELLE department. Upon the return of LORRAINE to France in 1919, this territory was included in MOSELLE department.

Meurthe-et-Moselle (muhrt–ai–mo-zel), department (□ 2,024 sq mi/5,262.4 sq km), in former Lorraine province, NE FRANCE; ⊙ NANCY; 48°35′N 06°10′E. Administratively, department forms part of the modern region of LORRAINE. Abutting the VOSGES (SE), it is chiefly occupied by the Lorraine tableland and Côtes de MOSELLE, a cuesta that parallels the MOSELLE RIVER to LUXEMBOURG and the Belgian border. Agriculture is of secondary importance; some vineyards along the Moselle (PAGNY-SUR-MOSELLE). Historically, this was a leading iron-mining region, with mining areas around BRIEY, LONGWY, and NANCY, where a metallurgical industry (pig-iron and steel mills) developed in 19th-century. At one time, these mines supplied most of W Europe's steel industry (including the RUHR). For its own steel production, coking coal was brought in from N France and the Saar (SAARLAND). With declining metallurgical industry, the department has closed many iron mines and has diversified its economic base. Chemical plants (based on salt mines between Nancy and LUNÉVILLE), glass works (Baccarat, Cirey), breweries, and textile mills now provide increasing employment. Department is crossed S to W by Canal de l'Est, W to E by MARNE-RHINE CANAL. Chief cities: Nancy, Lunéville (porcelain), TOUL (rubber tires), Longwy (metallurgy). Department was formed 1871 from those parts of Meurthe and Moselle departments that remained with France after loss of Lorraine to Germany following Franco-Prussian War. LORRAINE NATURAL REGIONAL PARK (W of METZ and Nancy) lies partially within this department.

Meurthe River (MUHRT), c.105 mi/169 km long; rises in the VOSGES MOUNTAINS, LORRAINE region, NE FRANCE; flows NW past SAINT-DIÉ and LUNÉVILLE to join the MOSELLE RIVER just N of NANCY; 48°47′N 06°09′E. Its very irregular flow has necessitated an intricate system of controls. In its lower course it traverses the industrial and mining district SE of Nancy.

Meuse (MUZ), department (□ 2,400 sq mi/6,240 sq km), in LORRAINE, NE FRANCE bordering on BELGIUM; ⊙ BAR-LE-DUC; 49°00′N 05°30′E. VERDUN is the only other large town. The MEUSE RIVER valley is dotted with smaller communities. Metalworking is the principal industry; also a significant cheese producer. Agriculture is concentrated in the deeply incised Meuse River valley, where most of the people of the department live. Part of the ARGONNE forest (NW) and the partly forested and central and E plateau and hill country are devoted to livestock raising, some wheat growing, and forestry. During World War I, the department was the scene of bitter fighting, with U.S. forces participating in SAINT-MIHIEL and Meuse-Argonne 1918 offensive.

Meuselwitz (MOI-sel-vits), town, THURINGIA, central GERMANY, 20 mi/32 km S of LEIPZIG; 51°03′N 12°18′E. Lignite-mining center; manufacturing of machinery, chemicals, textiles, china.

Meuse River (MUZ), Dutch *Maas* (MAHS), c.560 mi/901 km long; this, an international waterway, rises in the LANGRES PLATEAU of NE FRANCE; flows N past NEUFCHÂTEAU, VERDUN (head of navigation), SEDAN, and CHARLEVILLE-MÉZIÈRES into S Belgium; and other Dutch ports by the intricate system of Dutch waterways; thus it is one of the chief water thoroughfares of W Europe. The Belgian section of the Meuse valley, especially around Namur and Liège, is an important industrial region. Though coal mining is in decline, a strategic line of defense, particularly in Belgium and France, the Meuse valley has been a battleground in many wars, and most of the cities along its course have been strongly fortified since the Middle Ages. Between SAINT-MIHIEL and Sedan, in particular, heavy fighting occurred in World War I.

Mevagissey (mev-uh-GIS-ee), town (2001 population 2,221), S central CORNWALL, SW ENGLAND, on the ENGLISH CHANNEL, 5 mi/8 km S of ST. AUSTELL; 50°17′N 04°47′W. Fishing port, resort; tourism. Has medieval church.

Mevasseret Yerushalayim, ISRAEL: see MEVASSERET ZION.

Mevasseret Zion (mi-vah-SE-ret tsee-YON), township, ISRAEL, 3.7 mi/5 km W of metropolitan JERUSALEM; 31°48′N 35°08′E. Elevation 2,047 ft/623 m. Its name comes from the merger of two nearby settlements: Ma'oz Zion (established 1951) and Mevasseret Yerushalayim (established 1956). Now a commuter suburb of Jerusalem.

Mewar, India: see UDAIPUR, RAJASTHAN state, city.

Mewar and Southern Rajputana States, India: see RAJPUTANA STATES.

Mewatha Beach (muh-WA-thuh), summer village (2001 population 101), E ALBERTA, W CANADA, 23 mi/37 km from ATHABASCA; 54°36′N 112°44′W. Incorporated 1978.

Mewe, POLAND: see GNIEW.

Mew Island (MYOO), islet in the IRISH SEA, at SE entrance to BELFAST LOUGH, DOWN, Northern Ireland, 4 mi/6.4 km N of DONAGHADEE; 54°41′N 05°32′W. Covers 32 acres/13 ha. Lighthouse (54°42′N 05°30′W).

Mex (MEKS), W suburb of ALEXANDRIA, EGYPT, on isthmus between Lake MARYUT and MEDITERRANEAN SEA. Nearby are limestone quarries used since ancient times. Also has saltworks, fish-breeding ponds, and pumping works which maintain the level of water in Lake Maryut.

Mexaranguape (MAH-shah-rahn-gwah-pai), city, E RIO GRANDE DO NORTE state, BRAZIL, on Atlantic Coast, and 25 mi/40 km N of NATAL; 05°28′S 35°33′W. Also spelled Maxaranguape.

Mexborough (MEKS-buh-ruh), town (2001 population 14,620), SOUTH YORKSHIRE, on DON RIVER, and 11 mi/18 km NE of SHEFFIELD; 53°29′N 01°16′W. Former agricultural market; manufacturing (furniture); industries (pottery). Previously coal mining and flour milling.

Mexcala River, MEXICO: see BALSAS, RÍO.

Mexcaltitán (meks-kahl-teet-LAHN), village, W central NAYARIT, MEXICO, 21 mi/34 km NW of TEPIC. Some people believe this tiny fishing village was a stop-off for the Mexico (Aztec) people during their search for a homeland. It can be reached by a paved road or by dugout canoe. This canoe trip is through a mangrove forest. Some of the houses are built on platforms because the island floods during the rainy season. The economy is based on shrimp in the lagoon.

Mexia (MEK-see-uh), town (2006 population 6,708), LIMESTONE county, E central TEXAS, 39 mi/63 km ENE of WACO; 31°40′N 96°28′W. Elevation 534 ft/163 m. Commercial, processing center for agriculture (cotton, grain; cattle), oil-producing area; manufacturing of clay products Old Fort Parker State Historic Site and Fort Parker State Park to SW; Confederate Reunion Grounds State Historic Site to N; Lake Mexia to W. Settled 1873; oil discovery (1920) led to boom.

Mexiana Island (c.600 mi/966 km), NE PARÁ, BRAZIL, in AMAZON delta, just N of MARAJÓ island, and E of CAVIANA island. Crossed by the equator, 120 mi/193 km NW of BELÉM.

Mexicali (meks-ee-KAHL-ee), city (2005 population 653,046) and township, ⊙ Mexicali municipio and of BAJA CALIFORNIA state, NW MEXICO, across the border from CALEXICO, CALIFORNIA; 32°39′N 115°30′E. Once noted chiefly as the center of a cotton- and cereal-raising area, it has experienced extensive construction of foreign-owned manufacturing plants called *maquiladoras*. A large and rapidly growing labor force and lower employment and production costs in Mexico have stimulated the development of *maquiladoras* in the border areas. Also center of irrigated agriculture, lower COLORADO RIVER.

Mexicaltzingo (meks-hee-kahlt-seen-go), town, ⊙ Mexicalcingo municipio, MEXICO state, central MEXICO, on railroad, and 8 mi/12.9 km SE of TOLUCA DE LERDO, on Mexico Highway 55; 19°21′N 99°07′W. Cereals; livestock; dairying.

Mexico (MEKS-i-ko), Spanish *México* or *Méjico* (both: ME-hee-ko), officially called the United States of Mexico, Spanish *Estados Unidos Mexicanos*, republic (☐ 753,665 sq mi/1,951,992 sq km; 2004 estimated population 104,959,594; 2007 estimated population 108,700,891), S NORTH AMERICA, bordering on the UNITED STATES in the N, on the GULF OF MEXICO and the CARIBBEAN SEA in the E, on BELIZE and GUATEMALA in the SE, and on the PACIFIC OCEAN in the S and W; ⊙ MEXICO city.

Geography
Most of Mexico is highland or mountainous; only about 20% of the land is arable. Lowland areas include most of the YUCATÁN peninsula and the Isthmus of TEHUANTEPEC in the SE, and low-lying strips of land along the Gulf of Mexico, the Pacific Ocean, and the GULF OF CALIFORNIA (known in Mexico as *Mar de Cortés*) separating the BAJA, or Lower, California peninsula from the rest of the country. The Mexican Plateau (c.700 mi/1,130 km long and c.4,000 ft/1,219 m–8,000 ft/2,438 m high) makes up the heart of Mexico. It is fringed by two mountain ranges, the SIERRA MADRE ORIENTAL (E) and the SIERRA MADRE OCCIDENTAL(W). Some of the country's major cities are located within drainage basins contained within the plateau. The LAGUNA DISTRICT, for example, was the scene of a major irrigation development and land colonization project (1936). In the N the plateau is mostly arid; some of the area has been irrigated and is used principally for raising livestock. The broad, shallow lakes of the ANÁHUAC region, famous for its rich cultural heritage, lie S of the deserts. S of the Anáhuac, which includes the Valley of MEXICO, is the TRANSVERSE VOLCANIC AXIS, a chain of volcanoes, including PICO DE ORIZABA (Citlaltepétl) (elevation 18,700 ft/5,700 m; the highest point in Mexico); POPOCATÉPETL; and IZTACCIHUATL. Further S are jumbled masses of mountains and the SIERRA MADRE DEL SUR. Mexico contains a few large rivers including the Río Bravo del Norte (RIO GRANDE), which forms the boundary with TEXAS, and its tributaries; and the Río GRIJALVA, Río Papaloapan, and Río ÚSUMACINTA, which flow into the Gulf of Mexico. The climate of the country varies with the elevation: *tierra caliente* (up to c.3,000 ft/1,220 m), *tierra templada* (c.3,000 ft/914 m–c.6,000ft/1,829 m), and *tierra fría* (above c.6,000 ft/1,829 m).

Population
Most of the citizens are of mixed Spanish and Native American descent; Spanish is the country's official language. Since 1920 the population of Mexico has had a very high rate of growth; from 1940 to 1990 the population grew from 19.6 million to 81.1 million. However, declining fertility rates indicate a slow down in the population growth. The vast majority of people are Roman Catholic, with Protestant minorities. The country has numerous universities notably in Mexico city, SALTILLO, GUADALAJARA, MONTERREY, and PUEBLA.

Economy: Agriculture
The Mexican government plays a major role in planning the economy and owns and operates some basic industries (including the petroleum industry) and means of transport. About 18% of the country's workers are engaged in farming, which is slowly becoming modernized. Because rainfall is inadequate outside the coastal regions, agriculture depends largely on extensive irrigation. Mexico produces a wide variety of agricultural products (basic grains, sugarcane, citrus fruits, cotton, coffee, and tomatoes); maguey is widely grown and is processed into the alcoholic beverages pulque and mescal. Livestock raising and fishing are also significant sources of economic activity.

Economy: Minerals and Oil
Mexico is among the world's leading producers of many minerals, including silver, fluorite, zinc, and mercury, and its petroleum reserves are one of its most valuable assets. In the late 1970s and early 1980s petroleum constituted about three-quarters of Mexico's exports. That figure fell drastically in the mid-1980s. While diversification of industry has helped to keep Mexico's trade economy from becoming dependent once more on a single export, the petroleum industry has also recovered substantially.

Economy: Industry
The country's principal industrial centers include Mexico city, Guadalajara, MONTERREY, CIUDAD JUÁREZ, TIJUANA, DURANGO, LEÓN, QUERÉTARO, and Puebla. There is also a petrochemical center at COATZACOALCOS-MINATITLÁN and an iron-steel complex at Lázaro Cárdenas on the Isthmus of TEHUANTEPEC. Leading manufacture includes iron and steel, motor vehicles, cement, refined petroleum and petrochemicals, processed food, electronic products, and textiles. Mexico is also known for its handicrafts, especially pottery, woven goods, and silverwork.

Economy: Tourism
Tourism is now Mexico's second-greatest source of income. Favorite tourist centers include ACAPULCO, CANCÚN, COZUMEL, PUERTO VALLARTA, MAZATLÁN, CABO SAN LUCAS, and TIJUANA, as well as Mexico city itself and some of the highland centers like Guadalajara and Puebla.

Economy: Trade
The country's chief ports are VERACRUZ, TAMPICO, COATZACOALCOS, MAZATLÁN, and ENSENADA. The leading imports are machinery, motor vehicles, electronic equipment, chemicals, consumer goods, and grain; the main exports are petroleum, cotton, sugar, coffee, tomatoes, shrimp, sulfur, and zinc. Until recently, the annual value of Mexico's imports was considerably higher than the value of its exports. Since the early 1980s, however, there has been considerable foreign investment in *maquiladoras*, which take advantage of a large, low-cost labor force to produce finished goods for export to the U.S. The *maquiladoras* have increased Mexico's export production considerably, as well as contributing to the diversification of the industrial sector. The principal trade partners are the U.S., the EU, and Japan. Mexico is a member of the UN, the OAS, the Latin American Integration Association, and the Latin American Economic System.

History: to 18th Century
A number of great civilizations flourished in Mexico; earliest of these were the Olmec civilization, reaching its high point between 800 and 400 B.C.E. The Maya civilization flourished between about C.E. 300 and 900, followed by the Toltec (900–1200) and the Aztec (1200–1519). Other notable civilizations of pre-Columbian Mexico are the Mixtec and the Zapotec. The first Europeans to visit Mexico were Francisco Fernández de Córdoba in 1517 and Juan de Grijalva in 1518. The conquest of Mexico was begun from Cuba in 1519 by Hernán Cortés, who conquered the Aztec capital, TENOCHTITLÁN, in 1521. The territory was constituted the viceroyalty of New Spain in 1535. The Spanish had difficulty establishing control, as is evidenced by such events as the Mixtón War (1541), but at last they managed to establish their power over the indigenous population. The society slowly developed three different status groupings—Spaniards, Native Americans, and mestizos (mixed Spanish and Native Americans). The growth of an underprivileged mestizo class and the antagonism between those born in Spain (*gachupines*) and those born in America (*criollos*, or creoles) created internal stress and conflict. During this period, the Spanish continued to conquer new territory. Most of present-day Mexico and the former Spanish holdings in the present-day U.S. were occupied early. NE Mexico and Texas began to be occupied by Europeans in large numbers in the middle and late 18th century.

History: 18th Century to 1836
Around this time, discontent with Spanish rule began to grow, sparking a revolution in the early 19th century. A priest named Miguel Hidalgo y Costilla began the rebellion by issuing the *Grito de Dolores* [Spanish=cry of Dolores] on September 16, 1810, a revolutionary tract calling for racial equality and the redistribution of land. Armies made up mostly of mestizos and Native Americans fought against the Spanish army. Although successful at first, the rebels were ultimately overmatched; by 1815, their armies had either been defeated in battle or driven into the wilds. A few years later another more peaceful rebellion was fostered by the royalist general Augustín de Iturbide among others. The resulting Plan of Iguala (February 1821) called for an independent monarchy, equality for Spaniards and creoles, and the maintenance of the privileged position of the church. Spain accepted Mexican independence that September, and an empire was established in 1822 with Iturbide at its head. In 1823, the republican leaders drove out Iturbide and established Mexico as a republic. Political unrest and frequent turnover of governments continued for the next several years.

History: 1836 to 1911
In 1836, Texas, calling itself a republic, withdrew from Mexico leading to the Mexican-American War (1846–1848). By the terms of the Treaty of GUADALUPE HIDALGO, the U.S. took over all Mexican lands N of the Rio Grande. After the war, the Mexican government remained unstable, and in 1855 it was again overthrown by a group of reform-minded men led by Benito Juárez. They drafted a liberal constitution (1857) that secularized church property and reduced the privileges of the army. Conservative opposition began the War of Reform (1858–1861); however, the liberals emerged victorious, and Juárez became president. At the invitation of the conservatives, Napoleon III of France intervened in 1864, unseating Juárez and setting up an empire. But the empire quickly collapsed (1867), and Juárez again assumed control of Mexico. Porfirio Díaz led a successful armed revolt in 1876 and, except for the period from 1880 to 1884, firmly held the reins of power as president until 1911.

History: 1911 to 1929
It was a period of considerable economic growth, but social inequality increased. Despite some liberal opposition, repressive and dictatorial regimes continued into 1914 when a successful revolution broke out under the leadership of Venustiano Carranza, Francisco "Pancho" Villa, and Emiliano Zapata. With the influence of U.S. military intervention, Carranza

became president (1914). Civil war broke out again later that year, but by the end of 1915 Carranza had reestablished control. A constitution ratified in 1917 established Mexico as a federal republic with a president and a bicameral legislature. In 1920 Carranza was deposed by General Álvaro Obregón, his former military chief, who was subsequently elected president. The Obregón administration (1920–1924) redistributed lands and undertook educational reforms. His successor, Plutarco Elías Calles, continued the agrarian and educational programs, but he became embroiled in serious disagreements with the U.S. over rights to petroleum and with the church over the separation of church and state. Despite his loss of the 1928 presidency, Calles remained a powerful political influence.

History: 1929 to 1942

He organized the National Revolutionary Party (1929; renamed the Institutional Revolutionary Party in 1946), the chief political party in Mexico for the rest of the century. Lázaro Cárdenas (1934) instituted reforms to improve the lot of the underprivileged. He redistributed land and supported the Mexican labor movement. Under his political term, railroads were nationalized; foreign holdings, particularly in petroleum fields, were expropriated with compensation; educational opportunities were increased and illiteracy reduced; medical facilities were extended; transport and communications were improved; and plans were drawn up for land reclamation and for hydroelectric and industrial projects. Relations with the U.S. improved in the 1940s.

History: 1942 to 1985

In World War II, Mexico declared war (1942) on the Axis powers; it made substantial contributions to the Allied cause and also received considerable U.S. economic aid. Since the end of World War II, Mexico has embarked upon an ambitious program of economic development (especially of its industrial plants) and implemented many of Cárdenas plans. Most of the benefits of this economic progress, however, have accrued to the middle and upper classes, and the relative welfare of poorer people has remained the same or deteriorated. The improvements made in Mexico's railroad network and the opening of the INTER-AMERICAN HIGHWAY aided the tourist trade and thus increased the commercial value of one of the country's greatest assets: the beauty of its land. Mexico remained on friendly terms with the U.S., ratifying treaties that settled long-standing border disputes in the EL PASO, Texas, region (1964, 1967) and called (1965) for the U.S. to maintain the freshwater content of the COLORADO RIVER, whose waters are used for irrigation in Mexico. Mexico maintained diplomatic relations with post-Castro Cuba, but it supported the U.S. during the Cuban missile crisis (1962). In the 1970s Mexico continued to expand its economy, borrowing significantly on the strength of its petroleum reserves, but when oil prices fell sharply in the early 1980s the country's ability to meet its international debt obligations was severely strained. In addition, population increases and inflation contributed to food shortages and unemployment. Private and foreign investment dropped sharply, and the population began to migrate from rural areas into the cities and into the U.S. The government responded with economic austerity policies, a renegotiation of Mexico's international debt, and a loosening of direct foreign investment regulations. The economic crisis, the austerity measures, and the added blow of a major earthquake in the city of Mexico in 1985 all contributed to a popular discontent with the PRI.

History: 1985 to Present

Salinas continued economic reform, encouraging foreign investment, privatizing many national industries, investigating corruption in public offices, and working toward increased trade with the U.S. A continued problem in Mexico's relations with the U.S., however, has been the flow of illegal immigrants and drugs across the border. By 1990, debt relief, diversification and privatization of the economy, and foreign investment began to show positive effects, and Mexico's economic growth rate returned to historic levels. In 1992 Mexico, the U.S., and Canada negotiated the North American Free Trade Agreement (NAFTA), which was designed to erase many trade barriers among the three governments and create a trading bloc of 370 million people; it was inaugurated in January 1994. Also in January 1994, a guerrilla uprising by the Zapatista National Liberation Army (EZLN) began in the SE state of CHIAPAS; the often-violent conflict continues, and another rebel group, the Popular Revolutionary Army (EPR), has begun to agitate against the government as well. In the presidential election of 1994, the PRI candidate, Luis Donaldo Colosio Murrieta, was assassinated in Tijuana; his successor, Ernesto Zedillo Ponce de León, was subsequently elected. In early 1995, soon after Zedillo took office, Mexico suffered a major economic crisis, with devaluation of the peso and near-catastrophic financial instability. An austerity plan and U.S. aid relieved the situation and saved the peso from complete collapse. In 2000 Mexicans elected Vincente Fox president. This was the first time since 1910 that an opposition party (National Action Party; PAN) defeated the party in power (Institutional Revolutionary Party; PRI). President Fox also was the first Mexican head of state elected in free and fair elections. PAN did not win control of the congress, however, and much legislation proposed by Fox was passed. In 2006 the PAN candidate, Felipe Calderón, narrowly edged Andrés Manuel Lopéz Obrador, the Democratic Revolutionary Party (PRD) candidate, who called the result a fraud (though EU election monitors said there were no irregularities) and demanded a full recount. The PRI placed third in both the presidential and congressional balloting. A partial recount of the vote was made, but it did not alter the outcome.

Government

Under the constitution of 1917 as amended, Mexico is a federal republic whose head of state and government is the president, directly elected to a nonrenewable six-year term and assisted by a cabinet. The bicameral National Congress is made up of the Senate, with 128 members serving six-year terms, and the Chamber of Deputies, with 500 members serving three-year terms. Ninety-six of the senators and 300 of the deputies are directly elected, while thirty-two of the senators and 200 of the deputies are chosen by a system of proportional representation. The current head of state is President Felipe Calderón (since December 2006).

Administratively, Mexico is divided into thirty-one states and the Federal District, which includes MEXICO city. The states are AGUASCALIENTES, BAJA CALIFORNIA, BAJA CALIFORNIA SUR, CAMPECHE, CHIAPAS, CHIHUAHUA, COAHUILA, COLIMA, DURANGO, GUANAJUATO, GUERRERO, HIDALGO, JALISCO, MEXICO, MICHOACÁN, MORELOS, NAYARIT, NUEVO LEÓN, OAXACA, PUEBLA, QUERÉTARO, QUINTANA ROO, SAN LUIS POTOSÍ, SINALOA, SONORA, TABASCO, TAMAULIPAS, TLAXCALA, VERACRUZ, YUCATÁN, and ZACATECAS.

Mexico (MEKS-i-ko), Spanish *México* or *Méjico* (both: ME-hee-ko), state (□ 8,286 sq mi/21,543.6 sq km), S central MEXICO; ⊙ TOLUCA DE LERDO; 18°22′N 98°35′W. The N section of the state, containing most of the Valley of MEXICO (part of the ANÁHUAC plateau), has broad, shallow lakes and is broken by low mountains. There are steeper mountains and valleys in the E, and the S and W areas are dominated by the rugged volcanic belt extending across the center of the country. On the state's SE border are the POPOCATÉPETL and IZTACCÍHUATL volcanoes. The principal river is the LERMA. Except on the S, the state encircles the Federal District, with the nation's capital, MEXICO CITY, and most of the E part lies within the Mexico City Metropolitan Zone. Suburbs of Mexico City which lie within Mexico state include CIUDAD NEZAHUALCÓYOTL, a huge (more than 2 million) working-class neighborhood that developed as a squatter settlement in former bed of Lake TEXCOCO, and the major industrial centers of NAUCALPAN, TLALNE-PANTLA, ECATEPEC, TULTITLÁN, and CUAUTITLÁN. The state is highly industrialized; a leading producer of automobiles, paper, chemicals, textiles, other light manufactures, iron, and steel. Mining (gold, silver, lead, zinc), and agriculture (maguey, beans, and cereals) are other economic activities. Mexico is one of the country's most densely populated states.

Mexico, city (2000 population 11,320), ⊙ AUDRAIN county, central MISSOURI, 28 mi/45 km NW of COLUMBIA; incorporated 1857; 39°10′N 91°52′W. Regional farm service area. Livestock markets. Saddle horses. Refractory clay deposits. Wheat, corn, soybeans; hogs, cattle; manufacturing (soybean processing, prefabricated homes, cloth bags, fire-clay and aluminum brick and insulation, trailers, plastic products, ceramic tile and bricks, copper wire, carpet tack). A saddle-horse museum, Missouri Military Academy, and a Missouri Veterans' Home are also in the city. John D. Moore born here.

Mexico, town, PAMPANGA province, central LUZON, PHILIPPINES, on railroad, and 4 mi/6.4 km NE of SAN FERNANDO; 15°07′N 120°41′E. Sugarcane, rice.

Mexico, residential town, including Mexico village, W MAINE, OXFORD county, on ANDROSCOGGIN River, opposite industrial RUMFORD; 44°33′N 70°32′W. Incorporated 1818.

Mexico, village (2000 population 984), MIAMI county, N central INDIANA, 5 mi/8 km NW of PERU, on EEL RIVER; 40°49′N 86°07′W. Soybeans, corn; cattle; dairying. Laid out 1834.

Mexico, village (□ 2 sq mi/5.2 sq km; 1990 population 1,555; 2000 population 1,572) in Mexico town (□ 2 sq mi/5.2 sq km; 2006 population 1,564), OSWEGO county, central NEW YORK, 13 mi/21 km E of OSWEGO city; 43°27′N 76°14′W. In rich agricultural area; manufacturing of food products. Incorporated 1851.

Mexico Beach, city (□ 1 sq mi/2.6 sq km; 2005 population 1,192), BAY county, NW FLORIDA, 21 mi/34 km SE of PANAMA CITY; 29°56′N 85°24′W.

Mexico City (MEKS-i-ko), Spanish *Ciudad de México* or *Méjico* (both: ME-hee-ko), city (2005 population 19,231,829), central MEXICO, ⊙ and largest city of Mexico, near the S end of the plateau of ANÁHUAC; 19°26′N 99°07′W. Elevation c.7,349 ft/2,240 m; the horizons of the city are almost obscured by mountain barriers, and the peaks of POPOCATÉPETL and IZTACCÍHUATL are not far off. The climate is cool and dry. Much of the surrounding valley is a lake basin with no outlet, and in the past during the rainy seasons, floods of runoff swelled the lakes. From the time when the Aztec capital of Tenochtitlán stood on an island in Lake TEXCOCO—now the heart of the metropolis—measures have been taken to protect the city and provide for expansion by draining Texcoco and the other lakes, CHALCO and XOCHIMILCO. In 1607–1608 an 8-mi/13-km-long drainage canal and tunnel were built to drain floodwaters from the basin N to the TULA RIVER. In 1900 a central canal was completed that reached to the headwaters of the PÁNUCO River. The Caracol [Span.=snail], a 12-mi/19-km spiral system fed by longitudinal canals begun in 1936, acts as an evaporating basin, from which valuable minerals are taken. Drainage and artesian wells have lowered the water table so that the alluvial soil, formerly saturated by subsoil water, can no longer sustain the heavier buildings of the city, which are sinking c.4 in/10.2 cm–12 in/30 cm a year. Some of Mexico's finest buildings have been damaged, among them the old cathedral (begun in 1553 near the site of

Area is shown by the symbol □, and capital city or county seat by ⊙.

an Aztec temple) and the Palace of Fine Arts. Modern office buildings have been shored up with pilings. In addition to the soft subsoil, the city is located in a region of high seismic activity; earthquakes in 1957 and 1985 caused substantial damage. Nevertheless, many monuments of Spanish colonial architecture remain. The cathedral and the National Palace are on the great central square, or Plaza de la Constitución, where the streets of the old town crisscross in a rough gridiron. From the Plaza the great avenues span out to the far sections of the capital. Many colonial churches are to be found. The Paseo de la Reforma cuts across the city to CHAPULTEPEC. Public buildings of the nineteenth century have a ponderous grandeur that shows French influence, but the newly built edifices are starkly modern. Some old buildings as well as the newer (e.g., the Palace of Fine Arts, the National Palace, and the National Preparatory School) have murals by the modern artists Diego Rivera, José Clemente Orozco, and David Alfaro Siqueiros. The National University of Mexico, founded in the sixteenth century, is housed in University City (opened 1952), built on El Pedregal, a lava outcrop in the outskirts, with famous mosaics designed by Juan O'Gorman. The city was the metropolis of Mexico even before New Spain was created. It is built on the ruins of the Aztec city of TENOCHTITLAN, which was begun in c.1325 and razed by Hernán Cortés in 1521. During the colonial period Mexico City served as the capital of New Spain and was for a time the cultural and social center of Spain's American empire. It was taken in 1847 by Winfield Scott's American army, after an inland march from Veracruz in the Mexican-American War. The French army captured Mexico City in 1863, and Emperor Maximilian, crowned in 1864, did much to beautify it before it was recaptured by Mexicans under Benito Juárez. In the years of revolution after 1910, it was a magnet for divergent insurrectionary forces. Perhaps the most spectacular incidents were the occupations (1914–1915) by Pancho Villa and Emiliano Zapata. Today Mexico City forms the core of the Federal District and is the commercial, industrial, financial, political, and cultural center of the nation. Among its diverse and important manufactures are chemicals, petroleum, food products, textiles, motor vehicles, machinery, pharmaceuticals, and consumer goods. Population has increased rapidly in a city that had already spread out in many residential sections called *colonias*. Overcrowding has become a major problem and traffic concentrations, combined with the atmospheric conditions of the city's surrounding valley, have resulted in heavy air pollution. The Metro, Mexico City's subway system, helps to reduce traffic congestion and pollution. The first of the nine lines was opened in 1969. Ixtapalapa and Delegations A. Madero are the largest suburbs of the Federal District; COYOACÁN is the oldest, with a palace built by Cortés. Among noted religious and recreational centers are GUADALUPE HIDALGO and XOCHIMILCO. AZCAPOTZALCO is the transport and industrial hub of the city. Its rich local color and extraordinary cultural attractions, make it a focal point for tourists, especially from the U.S. The Olympics were held in Mexico City in 1968. Benito Juárez International Airport to E.

Mexico, Gulf of, arm of the ATLANTIC OCEAN (□ 700,000 sq mi/1,820,000 sq km), SE NORTH AMERICA; 25°00′N 90°00′W. The Gulf stretches more than 1,100 mi/1,770 km from W to E and c.800 mi/1,287 km from N to S. It is bordered by the Gulf coast of the U.S. from FLORIDA to TEXAS, and the E coast of MEXICO from TAMAULIPAS to YUCATÁN. Near the entrance of the Gulf is the island of CUBA. On the N side of Cuba the Gulf is connected with the Atlantic Ocean by the STRAITS OF FLORIDA (from which the GULF STREAM ocean current originates); on the S side of Cuba it is connected with the CARIBBEAN SEA by

the YUCATÁN CHANNEL. The BAY OF CAMPECHE (Bahía de Campeche), Mexico, and APALACHEE BAY, Florida, are the Gulf's largest arms. Sigsbee Deep (12,714 ft/3,875 m), the deepest part of the Gulf, lies off the Mexican coast. The shoreline is generally low, sandy, and marshy, with many lagoons. Chief of the many rivers entering the Gulf are the MISSISSIPPI, ALABAMA, BRAZOS, and RIO GRANDE. The U.S. INTRACOASTAL WATERWAY follows the Gulf's coastline from S Florida to the Rio Grande. Oil deposits from the CONTINENTAL SHELF are tapped by offshore wells, especially along the coast of Texas and LOUISIANA. Most of the U.S. shrimp catch comes from the Gulf Coast; menhaden is another important catch. The chief ports along the Gulf of Mexico are at TAMPA and PENSACOLA, Florida; MOBILE, Alabama; NEW ORLEANS, Louisiana; GALVESTON and CORPUS CHRISTI, Texas; and TAMPICO and VERACRUZ, Mexico.

Mexico, Valley of, oval basin of interior drainage in Federal District and MEXICO state, central MEXICO; part of the large central plateau S of PÁNUCO River system; 50 mi/80 km by 40 mi/64 km. Once occupied by a system of connected shallow lakes (Chaldo, TEXCOCO, XALTOCAN, Xochimiyco Andzumpango). The basin has been drained by a complex of canals and tunnels to control flooding and facilitate MEXICO CITY urban expansion. Average elevation c.7,500 ft/2,286 m; temperate to subtropical climate. Site of Mexico City, it is one of Mexico's most densely populated areas, with important agricultural and industrial activities.

Meximieux (meks-ee-myu), town (□ 5 sq mi/13 sq km), AIN department, RHÔNE-ALPES region, E FRANCE, near the AIN RIVER, 21 mi/34 km NE of LYON; 45°54′N 05°12′E. Agricultural trade center for the DOMBES region. Fishing; livestock raising; cereal growing; horse breeding. The village of Pérouges (1993 estimated pop. 856; 1 mi/1.6 km W) has been preserved as a model of medieval architecture; it has served as a set for French filmmakers. The Place de la Halle (town square) is one of France's most picturesque sites.

Mexquitic de Carmona (mash-KEE-teek dai kahr-MO-nah), town, SAN LUIS POTOSÍ, N central MEXICO, 12 mi/19 km NW of SAN LUIS POTOSÍ, on Mexico Highway 49; 22°16′N 101°07′W. Corn, beans, maguey. Sometimes spelled Mezquitic.

Mexticacán (mesh-tee-kah-KAHN), town, JALISCO, central MEXICO, 13 mi/21 km SW of TEOCALTICHE; 21°13′N 102°43′W. Grain, vegetables; livestock.

Meyadin (me-yah-DEEN), township, DEIR EZ ZOR district, E SYRIA, on right bank of EUPHRATES RIVER, and 28 mi/45 km SE of DEIR EZ ZOR; 35°01′N 40°27′E. Also spelled MAYADIN.

Meyamey (me-yahm-AI), town, Semnān Prov., NE IRAN, 38 mi/61 km E of SHAHRUD, and on Mashhad-Tehran road. Wheat, cotton; rugmaking. Airfield. Sometimes spelled MAYAMEY and MAIAMAI.

Meyasir, YEMEN: see DATHINA.

Meybod (me-ye-BOD), town (2006 population 58,872), Yadz province, central IRAN, 40 mi/64 km NW of YAZD; 32°13′N 54°00′E. Cotton, pistachio nuts, grain, pomegranates. Handmade earthenware. Also called Maibud.

Meybod, IRAN: see MAIBUD.

Meycauayan (mai-kah-WAH-yahn), town (2000 population 163,037), BULACAN province, S central LUZON, PHILIPPINES, on railroad, and 9 mi/14.5 km N of MANILA; 14°44′N 121°00′E. Rice-growing center; cutlery manufacturing

Meyers Chuck, Alaska: see MYERS CHUCK.

Meyersdale, borough (2006 population 2,322), SOMERSET county, SW PENNSYLVANIA, 14 mi/23 km SSE of SOMERSET, on CASSELMAN RIVER, at mouth of Flaugherty Creek; 39°48′N 79°01′W. Railroad junction. Manufacturing (wooden products, fire tankers, apparel); surface coal. Agriculture (corn, oats, hay, soybeans, potatoes; dairying. Laid out 1844, incorporated 1871.

Meyers Lake (MEI-uhrz), village (2006 population 559), STARK county, E central OHIO; suburb 3 mi/4.8 km W of CANTON, on small Meyers Lake (former site of amusement park at streetcar-line terminus); 40°49′N 81°25′W.

Meyerton, town, GAUTENG, SOUTH AFRICA, 9 mi/14.5 km NE of VEREENIGING on main highway to GERMISTON; 27°33′S 28°01′E. Elevation 4,889 ft/1,490 m. Coal mining. Airfield.

Meylan (mai-lahn), town (□ 4 sq mi/10.4 sq km) of GRENOBLE, ISÈRE department, RHÔNE-ALPES region, SE FRANCE, on ISÈRE RIVER, opposite the campus of Grenoble University; 45°14′N 05°47′E.

Meymac (mai-mahk), commune (□ 33 sq mi/85.8 sq km), CORRÈZE department, LIMOUSIN region, S central FRANCE, at edge of Plateau of MILLEVACHES, 8 mi/12.9 km W of USSEL; 45°32′N 02°10′E. Agricultural trade (wine; cattle; fruits); woodworking center. Forestry management school. Has 12th-century abbatial church and many 15th–16th-century houses. A contemporary art exhibit by young artists, open every summer in a wing of the old cloister, attracts tourists.

Meymeh (me-YEMAI), village, Esfahān province, N IRAN, 40 mi/64 km SSW of Kashanm, and on TEHRAN-ISFAHAN road; 33°26′N 51°10′E. Center of Jowsheqan agricultural area; wheat, barley, nuts (walnuts and almonds), dairy products. Also has rug making, exports wool, and has marble quarries. Also spelled Meymeh.

Meymeh, IRAN: see MEIMEH.

Meyrin (meh-RINH), town (2000 population 19,548), GENEVA canton, SW SWITZERLAND, W suburb of GENEVA; 46°13′N 06°04′E.

Meyronne (MAI-ruhn), village (2006 population 35), SW SASKATCHEWAN, CANADA, on Pinto Creek and 40 mi/64 km W of ASSINIBOIA; 49°40′N 106°51′W. Mixed farming; livestock.

Meyrueis (mai-ru-AI), commune, resort in LOZÈRE department, LANGUEDOC-ROUSSILLON region, S FRANCE, between Causse Méjan (N) and Causse Noir (SW) limestone hills, on the JONTE RIVER, and 13 mi/21 km SW of FLORAC; 44°11′N 03°26′E. Agricultural trade; cheese making. Mont AIGOUAL, highest summit of the CÉVENNES MOUNTAINS, is 9 mi/14.5 km SE. Well-known caverns of Dargilan (2 mi/3.2 km NW) and Aven Armand (5 mi/8 km NW) attract tourists. Meyrueis is also a gateway to the CÉVENNES NATIONAL PARK.

Meythiet (me-tye), town NW suburb of ANNECY, HAUTE-SAVOIE department, RHÔNE-ALPES region, SE FRANCE. Airport for Annecy resort area.

Meyzieu (me-zyu), town (□ 8 sq mi/20.8 sq km), RHÔNE department, RHÔNE-ALPES region, SE FRANCE, 8 mi/12.9 km E of LYON, and part of its urban region; 45°46′N 05°00′E. Industrial park; plastics industry. Lyon-Satolas international airport is 3 mi/4.8 km SE. Also spelled Meyzieux.

Meyzieux, FRANCE: see MEYZIEU.

Mezaghrane (me-zah-GRAHN), village, MOSTAGANEM wilaya, NW ALGERIA, near the MEDITERRANEAN SEA, 3 mi/4.8 km SSW of MOSTAGANEM. Vineyards. Scene of fighting (1839–1840) between French garrison and Abd El Kader's raiders.

Mezaligon (me-zah-lee-GON), village, Ingapu township, AYEYARWADY division, MYANMAR, on railroad, and 20 mi/32 km NW of HENZADA.

Mezam, department (2001 population 465,644), NORTH-WEST province, CAMEROON; ⊙ BAMENDA.

Meža River, c.35 mi/56 km long, N SLOVENIA; rises on Austrian border, 5 mi/8 km ESE of EISENKAPPEL (AUSTRIA); flows ENE, past Črna, Mežica, and PREVALJE, to DRAVA RIVER at DRAVOGRAD. Lower course followed by Klagenfurt (Austria)–Maribor railroad. There are four hydroelectric plants on the river.

Mezcala River, MEXICO: see BALSAS, RÍO.

Mezdra (MEZ-drah), city, MONTANA oblast, MEZDRA obshtina (1993 population 28,555), NW BULGARIA, on the ISKUR River, 8 mi/13 km. ESE of VRATSA; 43°08′N 23°43′E. Railroad junction; cloth making, liquor distilling, machine building. Quarries.

Mèze (MEZ), town (□ 13 sq mi/33.8 sq km), HÉRAULT department, LANGUEDOC-ROUSSILLON region, S FRANCE, port on NW shore of the Étang de THAU (lagoon near MEDITERRANEAN coast), 5 mi/8 km WNW of SÈTE; 43°25′N 03°36′E. Wine center. Distilling. Oyster beds.

Mezek, Bulgaria: see SVILENGRAD.

Mezen' (mee-ZYEN), city (2005 population 3,600), NE ARCHANGEL oblast, N European Russia, on the MEZEN' River, 30 mi/48 km from the WHITE SEA, and 130 mi/209 km NE of ARCHANGEL; 65°51′N 44°14′E. River and maritime port. Founded in the 16th century. Chartered in 1779. Place of exile in Tsarist Russia.

Mezen' (mee-ZYEN), river, approximately 565 mi/909 km long; rises in the Timan Hills, KOMI REPUBLIC, NE European Russia; flows NW into the MEZEN' BAY (off ARCHANGEL oblast) of the WHITE SEA. Its lower course is navigable from May to November. Near its mouth is the city of MEZEN', a river port exporting lumber.

Mezen' Bay (mee-ZYEN), inlet of the WHITE SEA, N ARCHANGEL oblast, N European RUSSIA, W of KANIN Peninsula; 60 mi/97 km wide, 180 ft/55 m deep; 66°40′N 43°45′E. Receives MEZEN and KULOY rivers.

Mézenc, Mont (mai-ZAHNK, mon), volcanic summit (5,758 ft/1,755 m) of the Velay Mountains, S central FRANCE, in the S MASSIF CENTRAL, and 16 mi/26 km SE of LE PUY-EN-VELAY; 44°55′N 04°11′E. High-grade cattle raised on its slopes. A panoramic view form the summit can reach the ALPS far to the E on a clear day. This mountain has given its name to the volcanic Massif du Mézenc. It offers excellent cross-country ski terrain. Extensive lava flows from volcanoes (now extinct).

Mezere, TURKEY: see ELAZIG.

Mezhden (mezh-DEN), village, RUSE oblast, DULOVO obshtina, BULGARIA; 43°48′N 27°12′E. Railroad station.

Mezhdugor'ye, RUSSIA: see ALKHAN-KALA.

Mezhdurechensk (myezh-doo-RYE-cheensk), city (2005 population 100,715), E central KEMEROVO oblast, S central SIBERIA, RUSSIA, at the confluence of the USA RIVER and the TOM' River, on the SOUTH SIBERIAN RAILROAD, 200 mi/322 km SE of KEMEROVO, and 50 mi/80 km E of NOVOKUZNETSK; 53°41′N 88°03′E. Elevation 797 ft/242 m. Coal mining and processing in the Kuznetsk coal basin; reinforced concrete forms, repair of mining machinery and radio equipment; food processing (bakery). Founded in 1946 as Ol'zheras. Made a city in 1955.

Mezhdurechensk (myezh-doo-RYE-cheensk), settlement (2005 population 1,825), W KOMI REPUBLIC, NE European Russia, on local road and railroad, 11 mi/18 km S of USOGORSK; 63°14′N 48°33′E. Elevation 521 ft/158 m. Sawmilling, timbering, lumbering.

Mezhdurechensk (myezh-doo-RYE-cheensk) [Russian=between rivers], settlement (2006 population 3,445), W SAMARA oblast, E European Russia, in the narrow strip of land between the USA (N) and VOLGA (S) rivers just W of the SAMARA BEND, on highway and railroad, 23 mi/37 km ENE of SYZRAN', and 14 mi/23 km NE of OKTYABR'SK, to which it is administratively subordinate; 53°16′N 49°06′E. Elevation 334 ft/101 m. Railroad enterprises.

Mezhdurechenskiy (myezh-doo-RYE-cheen-skeeyee), town (2005 population 11,200), SW KHANTY-MANSI AUTONOMOUS OKRUG, W central SIBERIA, RUSSIA, near Lake Tuman, on road and railroad, 47 mi/76 km SE of URAY; 59°36′N 65°56′E. Elevation 160 ft/48 m. Sawmilling, lumbering.

Mezhdurechye, RUSSIA: see SHALI.

Mezhevaya, UKRAINE: see MEZHOVA.

Mezhgor'ye, UKRAINE: see MIZHHIR'YA.

Mezhireche, POLAND: see MIEDZYRZEC, town.

Mezhova (me-zho-VAH) (Russian Mezhevaya), town, E DNIPROPETROVS'K oblast, UKRAINE, 45 mi/72 km

ESE of PAVLOHRAD, and on railroad, 20 mi/32 km W of KRASNOARMIYS'K; 48°15′N 36°44′E. Elevation 583 ft/177 m. Food processing (flour, cheese). Vocational technical school; heritage museum. Established in 1884 as Mezhova railroad station, with several settlements around it; amalgamated in 1956 to form the town of Mezhova.

Mezhybizh, UKRAINE: see MEDZHYBIZH.

Mezhyrich (me-ZHI-reech) (Russian Meshyrich) (Polish Miédzyrzecz), village (2004 population 5,600), S RIVNE oblast, UKRAINE, at the confluence of Svyten'ka and Viliya rivers, left affluent of the HORYN' River, and 1.5 mi/4 km SW of OSTROH; 50°39′N 26°52′E. Elevation 639 ft/194 m. Fortified monastery (15th century) and church of the Holy Trinity, built by the Ostroz'kyy family; church and towers of monastery are preserved. Remnants of a fortified settlement from the time of Kievan Rus' are nearby. Known since 1396; town in 1605, with rights of Magdeburg law. Organized Jewish community since the 18th century, with a large brush factory, town became center for Hasidism. Nearly three-quarters of the population was Jewish before the community was eliminated by the Nazis in 1941.

Mezhyrichka, UKRAINE: see YEMIL'CHYNE.

Mezibori (ME-zi-BO-rzhee), Czech Meziboří, town, SEVEROCESKY province, NW BOHEMIA, CZECH REPUBLIC, in the ORE MOUNTAINS, and 8 mi/12.9 km NNW of MOST; 50°37′N 13°36′E. Settlement and recreation center near a heavily polluted industrial area. Until 1956, the town was called SENBACH, Czech Šenbach (SHEN-bahkh), German Schönbach.

Mežica, village, N SLOVENIA, on Meža River, 4 mi/6.4 km SW of PREVALJE, in the KARAWANKEN mountains, near Austrian border; 46°31′N 14°51′E. Open-pit mining of lead ore (galena and wulfenite), zinc, smithsonite, pyrites, and marcasite dates from 15th century; ore-dressing plant, smelter (pig lead). Power supplied by hydroelectric, diesel, and steam plants. Summer resort.

Mézidon-Canon (mai-zee-DON–kah-NON), town (□ 4 sq mi/10.4 sq km), CALVADOS department, BASSE-NORMANDIE region, NW FRANCE, on the DIVES RIVER, and 15 mi/24 km SE of CAEN. Railroad yards; biscuit manufacturing. Has 18th-century château with formal gardens and statuary.

Mézières, FRANCE: see CHARLEVILLE-MÉZIÈRES.

Mézières-en-Brenne (mai-ZYER–on–BREN), commune (□ 25 sq mi/65 sq km), INDRE department, CENTRE administrative region, central FRANCE, on small, poplar-lined CLAISE River, and 15 mi/24 km NNE of LE BLANC; 46°49′N 01°13′E. Center for sports fishermen; horse breeding. Has 14th-century church with Renaissance chapel added in 16th century. The BRENNE district, noted for its many lakes and ponds, is included since 1989 in LA BRENNE NATURAL REGIONAL PARK. Fish culture in several ponds.

Mezimesti (ME-zi-MNYES-tyee), Czech Meziměstí, German Halbstadt, town, VYCHODOCESKY province, NE BOHEMIA, CZECH REPUBLIC, 15 mi/24 km NNE of NÁCHOD, on Polish border opposite MIEROSZOW (POLAND); 50°37′N 16°15′E. Railroad junction. Manufacturing (textiles, machinery). Empire-era stone bridge.

Mézin (mai-ZAN), comune (□ 12 sq mi/31.2 sq km), LOT-ET-GARONNE department, AQUITAINE region, SW FRANCE, on the small Gélise River, and 7 mi/11.3 km SW of NÉRAC; 44°03′N 00°16′E. Distilling of ARMAGNAC brandy.

Mezinovskiy (mee-ZEE-nuhf-skeeyee), town (2006 population 2,100), SW VLADIMIR oblast, central European Russia, on road and railroad (Torfoprodukt station), 13 mi/21 km WNW of GUS'-KHRUSTAL'NYY; 55°30′N 40°21′E. Elevation 403 ft/122 m. Peat works.

Mezöbánd, ROMANIA: see BAND.

Mezöbereny (MA-zuh-be-re-nyuh), Hungarian Mezöberény, city, BÉKÉS county, SE HUNGARY, 10 mi/16

km N of BÉKÉSCSABA; 46°49′N 21°02′E. Wheat, corn, alfalfa; hogs; cattle. Manufacturing (bricks, textiles, footwear, metal structures, food industry machinery).

Mezöcsát (MA-zuh-chaht), Hungarian Mezöcsát, city, BORSOD-ABAÚJ-ZEMPLÉN county, NE HUNGARY, 20 mi/32 km SSE of MISKOLC; 47°49′N 20°55′E. Grain, potatoes; cattle, sheep; flour mills, vineyards.

Mezöhegyes (MA-zuh-he-dyesh), Hungarian Mezöhegyes, city, CSONGRÁD county, SE HUNGARY, 17 mi/27 km ENE of MAKÓ; 46°19′N 20°49′E. Railroad center on secondary lines; agricultural experiment station. Corn, wheat, sugar beets; cattle, hogs; manufacturing (hemp, agricultural tools, sugar refineries, textiles, aluminum articles).

Mezökeresztes (MA-zo-ke-res-tesh), Hungarian Mezökeresztes, village, BORSOD-ABAÚJ-ZEMPLÉN county, NE HUNGARY, 20 mi/32 km S of MISKOLC; 47°50′N 20°42′E. Grain, lentils, sunflowers; dairy farming; small flour mill. Turks defeated Hungarians here, 1596.

Mezökovácsháza (MA-zo-ko-vach-hah-zah), Hungarian Mezökovácsháza, city, CSONGRÁD county, SE HUNGARY, 28 mi/45 km E of HÓDMEZÖVÁSÁRHELY; 46°24′N 20°55′E. Market center; grain; hogs, cattle; manufacturing (agr. machinery components, flour milling).

Mezökövesd (MA-zo-kuh-vesh), Hungarian Mezökövesd, city, BORSOD-ABAÚJ-ZEMPLÉN county, NE HUNGARY, at S foot of Bükk Mountains, 22 mi/35 km SSW of MISKOLC; 47°49′N 20°35′E. Manufacturing (electrical and transportation equipment, furniture, apparel, baked goods, vegetable oil); tobacco warehouses, aluminum foundry, flour mills; exports embroidery. Has 14th century church.

Mezölaborc, SLOVAKIA: see MEDZILABORCE.

Mezötúr (MA-zo-toor), Hungarian Mezötúr, city (2001 population 19,329), SZOLNOK county, E central HUNGARY, on branch of BERETTYÓ RIVER, and 24 mi/39 km SE of SZOLNOK; 47°00′N 20°38′E. Railroad junction. Silage corn, grain, sugar beets; cattle, hogs; manufacturing (bricks, pottery, apparel, furniture, baked goods, pasta, tiles).

Mezquital, MEXICO: see SAN FRANCISCO DEL MEZQUITAL.

Mezquital del Oro (mes-KEE-tahl del O-ro), town, ⊙ Mezquital del Oro municipio, ZACATECAS, N central MEXICO, 35 mi/56 km S of TLALTENANGO DE SÁNCHEZ ROMÁN, an isolated community in the SIERRA MADRE OCCIDENTAL. Elevation 3,839 ft/1,170 m. Grain, fruit, vegetables; livestock.

Mezquital River, c.250 mi/402 km long, in W MEXICO; rises in SIERRA MADRE OCCIDENTAL 45 mi/72 km ENE of DURANGO; flows S, past NOMBRE DE DIOS and MEZQUITAL, into NAYARIT, then W, past TUXPAN, to coastal lagoons which drain into the PACIFIC OCEAN. Sometimes called Tuxpan River. Its lower course is also called San Pedro River.

Mezquital River, MEXICO: see SAN PEDRO RIVER.

Mezquitic (mes-KEE-tik), town, JALISCO, W MEXICO, on headstream of BOLAÑOS River, near ZACATECAS border, and 38 mi/61 km NW of COLOTLÁN; 22°22′N 103°42′W. An isolated community in the S SIERRA MADRE OCCIDENTAL. Cereals, beans, chili, alfalfa; livestock.

Mezquitic, MEXICO: see MEXQUITIC DE CARMONA.

Mezre, TURKEY: see ELAZIG.

Mezzo, CROATIA: see LOPUD.

Mezzocorona (med-zo-ko-RO-nah), town, TRENTO province, TRENTINO-ALTO ADIGE, N ITALY, on NOCE River, opposite Mezzolombardo, and 10 mi/16 km N of TRENT; 46°13′N 11°07′E. Wine making.

Mezzogiorno, large region, S ITALY. Comprises the S Italian regions of ABRUZZI, CAMPANIA, MOLISE, APULIA, BASILICATA, CALABRIA, and the islands of SICILY and SARDINIA. The term Mezzogiorno [Italian= midday] is a reference to the strength of the midday sun in S Italy. The Appenine mountain system is a pervasive feature throughout S Italy. Steep slopes and

---

Area is shown by the symbol □, and capital city or county seat by ⊙.

poor or eroded topsoil render about half of the land unarable; nevertheless, agriculture employs most of the workforce and is the mainstay of the generally underdeveloped economy. The chief crops are grains, fruits, olives, grapes, and vegetables. Industrialization is not as extensive as in the S, and as a result the per capita income and standard of living in S Italy is considerably lower. Two of the larger industrial centers are the port cities of BARI, with chemical and petrochemical plants, and NAPLES, with manufactures of textiles, iron, steel, machinery, and automobiles. Illiteracy in the Mezzogiorno is significantly higher than the national average. During most of the 12th century, S Italy was united under the rule of the Normans, who in 1198 were succeeded by the Hohenstaufen of GERMANY. The French Angevins ruled the region from 1266 to c.1442. During Angevin rule, the capital was moved from PALERMO to Naples, and feudalism was strengthened as the powers of the clergy and the nobility grew. Alfonso V of Aragon had conquered the KINGDOM OF NAPLES by 1442, beginning more than three centuries of Spanish rule. In the early 19th century the region was annexed to the French empire under Napoleon, and under the ten-year rule (known as the *Decennio*) of his brother-in-law, Lucien Murat, many reforms were made, including the abolishment of feudalism and the codification of law. Yet even after the emancipation (1860) of S Italy by Garibaldi's forces, feudal traditions persisted and peasants were still tied to large estates. The Mezzogiorno remained an underdeveloped area as the government in the last half of the 1800s and the first half of the 1900s focused on the prosperous N. Large-scale land reforms were not instituted until 1946. In 1950 the Cassa per il Mezzogiorno [Italian=Fund for the South] was set up by the Italian government to stimulate social and economic development in the Mezzogiorno.

Mezzola, Lago di, ITALY: see MERA RIVER.

Mfou (mFOO), town, Central province, CAMEROON, 13 mi/21 km SE of Yaoundé; 03°43′N 11°39′E.

Mfoundi (mFOON-dee), department (2001 population 1,248,235), Central province, CAMEROON; ⊙ Yaoundé.

Mfumbiro Range, CONGO: see VIRUNGA.

Mga (MGAH), town (2005 population 9,550), central LENINGRAD oblast, NW European Russia, on the Mga River (left affluent of the NEVA), 30 mi/48 km ESE of SAINT PETERSBURG; 59°45′N 31°04′E. Railroad and highway junction; railroad shops; reinforced concrete forms; lumber mills.

Mgachi (MGAH-chee), town (2006 population 1,600), NW SAKHALIN oblast, RUSSIAN FAR EAST, on the TATAR STRAIT, on coastal highway, 10 mi/16 km N of ALEKSANDROVSK-SAKHALINSKIY; 51°03′N 142°16′E. Elevation 167 ft/50 m. Coal mines.

Mgeni (m-GAI-nee), village, SINGIDA region, N central TANZANIA, 3 mi/4.8 km S of SINGIDA; 04°54′S 34°44′E. Livestock; grain.

Mgera (m-GAI-rah), village, TANGA region, NE TANZANIA, 105 mi/169 km WSW of TANGA; 05°39′S 37°34′E. Sheep, goats; grain.

Mgeta (m-GAI-tah), village, MOROGORO region, central TANZANIA, 35 mi/56 km SE of IRINGA; 08°18′S 36°03′E. Timber; corn; wheat; cattle, sheep, goats.

M'Ghaïr, El (uhm-gah-IR, el), village and Saharan oasis, El OUED wilaya, E ALGERIA, near W edge of the Chott MEROUANE, on BISKRA-TOUGGOURT railroad, and 65 mi/105 km SSE of Biskra. Date palms. Formerly M'Raïer.

Mglin (MGLEEN), city (2005 population 8,410), W BRYANSK oblast, central European Russia, on the Sudynka River, 104 mi/167 km W of BRYANSK, and 15 mi/24 km NNE of UNECHA; 53°03′N 32°50′E. Elevation 639 ft/194 m. Highway hub; local transshipment point. Center of agricultural area; starch, butter; apparel factories. Known from the 14th century. Made city in 1781.

M'Goun, Jbel (uhm-GOON, zhe-BEL) or **Ighil M'Goun** (EE-gil uhm-GOON), peak (13,353 ft/4,070 m) of the High ATLAS mountains, Souss-Massa-Draâ administrative region, S central MOROCCO, 50 mi/80 km NE of OUARZAZATE; 31°31′N 06°26′W. Seasonally snow-covered.

Mhangura (mah-eng-GOO-rah), town, MASHONALAND WEST province, N ZIMBABWE, 33 mi/53 km N of CHINHOYI, on Ridziwi River, W of HUNYANI RANGE; 16°52′S 30°09′E. Copper and gold mining. Livestock; grain, cotton, tobacco. Formerly called Mangula Mine.

Mhasva (MUHS-vah), village, JALGAON district, MAHARASHTRA state, W central INDIA, 2 mi/3.2 km E of PAROLA. Cotton and oilseeds; noted for its ghee. Mhasva Lake, 1 mi/1.6 km N, has dam which supplies two small irrigation canals.

Mhasvad (MUHS-vahd), town, SATARA district, MAHARASHTRA state, W central INDIA, 50 mi/80 km E of SATARA; 17°38′N 74°47′E. Market center for grain (millet, wheat), livestock, handicraft cloth fabrics. Large irrigation tank 3 mi/4.8 km SE. Sometimes spelled Mhaswad.

Mhlambanyatsi (mlah-ba-nee-YA-tsee), town, HHOHHO district, W SWAZILAND, 9 mi/14.5 km SSW of MBABANE, on Mhlambanyatsi River; 26°26′S 31°04′E. Name means "where the buffalo swims." Mlilwane Game Reserve to SE. Cattle, goats, sheep, hogs; corn, vegetables, citrus. Sawmills on the edge of pine plantations (216 acres/50,000 ha) originally planted by the Colonial Development Corp. in 1949.

Mhlinzi, residential section, SOUTH AFRICA: see MIDDELBURG.

Mhlume (mh-LOO-me), town, LUBOMBO district, NE SWAZILAND, 45 mi/72 km ENE of MBABANE, near BLACK MBULUZI RIVER, in Swaziland Irrigation Scheme; 26°02′S 31°49′E. Hlane Royal National Park and Mlawula Nat. Reserve to S. Sugarcane, pineapples, cotton, citrus, corn; cattle, goats, sheep. Diamond mining to S at Hlane. Landing Strip railroad 3 mi/5 km W.

Mhow (muh-HOU), town (2001 population 105,544), INDORE district, SW Madhya Pradesh state, central India, 12 mi/19 km SSW of INDORE, on S MALWA plateau; 22°33′N 75°46′E. Important military station; airport. Dairy farming. Founded by British in 1818; formerly in princely state of INDORE.

Mhunze (m-HOON-zai), village, SHINYANGA region, N central TANZANIA, 25 mi/40 km ENE of SHINYANGA; 03°39′S 33°49′E. Cotton, corn, wheat; cattle, sheep, goats.

Miacatlán (mee-an-kat-LAHN), town, MORELOS, central Mexico, 12 mi/19 km SW of CUERNAVACA; 18°46′N 99°22′W. Sugar, rice, coffee, wheat, fruit; livestock.

Miadi, Marshall Islands: see MEJIT.

Miagao (mee-yahg-OU), town, ILOILO province, S PANAY island, PHILIPPINES, 23 mi/37 km W of ILOILO, on PANAY GULF; 10°42′N 122°11′E. Agricultural center (rice, sugarcane, hemp). Noted church and fortress.

Miahuatlán (mee-ah-waht-LAHN), town, VERACRUZ, E MEXICO, in SIERRA MADRE ORIENTAL, 13 mi/21 km NNE of XALAPA ENRÍQUEZ. Corn, sugarcane, coffee, tobacco. Also called San José Miahuatlán.

Miahuatlán, MEXICO: see SAN JOSÉ MIAHUATLÁN.

Miahuatlán, MEXICO: see SANTIAGO MIAHUATLÁN.

Miahuatlán de Porfirio Díaz (mee-ah-wat-LAHN dai por-FEE-ree-o DEE-yahs), city, in central OAXACA, MEXICO; 16°20′N 96°35′W. Elevation 5,272 ft/1,607 m. On the bank of the Miahuatlán River, 61.1 mi/98.4 km S of OAXACA de JUÁREZ, on Mexico Highway 175. Hot climate. Major production of castor-oil plant along with corn and beans. Significant cattle-raising and livestock industry. Forestry with pine and oak trees, coffee, *aguardiente* (liquor). The major commercial center for coffee production in the whole region. Coffee is exported through PUERTO ANGEL or Oaxaca de Juárez. Formerly known as Miahuatlán San Andrés.

Miajadas (myah-HAH-dhahs), town, CÁCERES province, W SPAIN, 22 mi/35 km SSW of TRUJILLO; 39°09′N 05°54′W. Olive-oil and wine processing, tanning, flour milling. Agricultural trade (cereals, fruit, vegetables; livestock). Phosphate deposits nearby.

Miami (mei-AM-ee), county (□ 377 sq mi/980.2 sq km; 2006 population 35,552), N central INDIANA; ⊙ PERU; 40°46′N 86°03′W. Intersected by WABASH, MISSISSINEWA, and EEL rivers, and by DEER CREEK. Agriculture (grain, fruit; livestock, poultry; dairy products). Manufacturing at Peru. GRISSOM AIR BASE in SW. MISSISSINEWA Reservoir and Miami State Recreation Area in SE. Formed 1832.

Miami, county (□ 590 sq mi/1,534 sq km; 2006 population 30,900), E KANSAS; ⊙ PAOLA; 38°33′N 94°49′W. Rolling plain region, bordering E on MISSOURI; drained by MARAIS DES CYGNES River and POTTAWATOMIE CREEK. Agriculture (corn, oats, soybeans, sorghum; hogs, cattle); metal products, navigation equipment. Hillsdale Lake Reservoir in NW. Formed 1861.

Miami (mei-AM-ee), county (□ 407 sq mi/1,058.2 sq km; 2006 population 101,914), W OHIO; ⊙ TROY; 40°03′N 84°14′W. Intersected by GREAT MIAMI and STILLWATER rivers. In the Till Plains physiographic region. Agricultural area (livestock; corn, tobacco, wheat); manufacturing (furniture, rubber and plastic products) mainly at PIQUA, Troy, and TIPP CITY; sand and gravel pits, stone quarries; nurseries. Formed 1807.

Miami (mei-AM-ee), city (□ 54 sq mi/140.4 sq km; 2005 population 386,417), ⊙ MIAMI-DADE county, SE FLORIDA, on BISCAYNE BAY, at the mouth of the Miami River; 25°46′N 80°12′W. The second-largest city in the state (JACKSONVILLE is the largest), a port of entry, and the transportation and business hub of SE Florida, it is also a popular and famous resort of the E U.S. One of the country's most important financial centers, Miami boasts a strong international business community. Tourism remains a major industry, closely followed by manufacturing and commerce. The international airport has more direct connections to Latin and SOUTH AMERICA than any other U.S. city. The city is the principal port for cruise ships to the Caribbean. Miami is also the processing and shipping hub of a large agricultural region and an aircraft service center. Manufacturing includes clothing, transportation equipment, machinery, plastics, and electronics. Other industries are printing and publishing, fishing, and shellfishing.

The first settlement was made here in the 1870s near the site of Fort Dallas, built in 1836 during the Seminole War. In 1895, Henry M. Flagler took charge of the area; he made Miami a railroad terminus in 1896, built a major hotel, dredged the harbor, began to develop a recreational center, and promoted tourism. The city received its greatest impetus during the Florida land boom of the mid-1920s. Since 1959 the large influx of Cubans to the city has created "Little Havana," an ethnic community of Miami; by the late 1990s about 50% of the city's population was Hispanic, heavily of Cuban descent. Since 1980 Miami's position in SE Florida has eroded as a result of massive suburban growth, spurred by the increase of high-technology industries in the Miami area. Miami is the seat of a number of institutions of higher education, such as Barry University, St. Thomas University (formerly Biscayne College), Florida Memorial University, Florida International University, and Miami-Dade College, the largest two-year college in the country. The University of Miami is in nearby CORAL GABLES. A number of state parks, gardens, and major tourist attractions such as the Seaquarium are in the area. Site of American Airlines Arena, home of the Miami Heat professional basketball team. The region of Greater Miami encompasses all of metropolitan Miami-Dade county, including Miami, MIAMI BEACH, Coral Gables, HIALEAH, and many smaller communities. EVERGLADES national park to W. Incorporated 1896.

Miami, city (2000 population 160), SALINE county, central MISSOURI, on MISSOURI River, and 14 mi/23 km N of MARSHALL; 39°19′N 93°13′W. Wheat, corn; cattle. Highway bridge.

Miami, city, ⊙ OTTAWA CO., 70 mi/113 km NE of TULSA, and 20 mi/32 km WSW of JOPLIN, MISSOURI, extreme NE OKLAHOMA, in the foothills of the OZARKS; 36°53′N 94°52′W. Elevation 801 ft/244 m. On the headwaters of Grand Lake, which provides both electric power and recreation. It is a trade, shipping, and marketing center for a tristate livestock and dairy region where lead and zinc are mined. Manufacturing (apparel, fabricated metal products, electronic materials, leather products, motor coaches, fiberglass boats; vegetable processing, publishing and printing). Northeast Oklahoma A&M College here. Twin Bridges State Park to SE; Spring River State Park to NE.

Miami, town (2000 population 1,936), GILA county, E central ARIZONA, 68 mi/109 km E of PHOENIX, and 5 mi/8 km W of GLOBE; 33°23′N 110°52′W. Mines (copper, gold, silver, lead, molybdenum, perlite) nearby. Manufacturing machine parts; cattle, hay. PINAL MOUNTAINS to S; TONTO National Forest to S, W, and N; San Carlos Indian Reservation to E. Founded 1908.

Miami (mei-A-mee), unincorporated village, S MANITOBA, W central CANADA, 40 mi/64 km S of PORTAGE LA PRAIRIE, in PEMBINA MOUNTAINS, and in THOMPSON rural municipality; 49°22′N 98°15′W. Lumbering; grain elevators; clay and bentonite quarrying.

Miami (MEI-a-mee), village (2006 population 553), ⊙ ROBERTS county, extreme N TEXAS, in high plains of the PANHANDLE, 23 mi/37 km NE of PAMPA; 35°41′N 100°38′W. Elevation 2,744 ft/836 m. Cattle, hogs; wheat, milo, corn; gas and oil.

Miami, ZIMBABWE: see MWAMI.

Miami Beach (mei-AM-ee), city (□ 18 sq mi/46.8 sq km; 2005 population 87,925), MIAMI-DADE county, SE FLORIDA, on an island between BISCAYNE BAY and the ATLANTIC OCEAN, 5 mi/8 km E of MIAMI; 25°48′N 80°08′W. It is connected to Miami by four causeways. Miami Beach is a popular year-round resort, famous for its hotel strip, Art Deco district, and recreational facilities. The city's chief source of income derives from tourism. The glamorous hotel and vacation industry, however, declined in the 1970s and 1980s; a spurt in less-expensive development led to the influx of a younger population, and the larger and wealthier retired community primarily moved to other developing Florida (resort) cities. Efforts in the late 1980s and early 1990s have been made for the architectural revival of the Art Deco buildings in the S part of the city known as SOUTH BEACH, a rebuilt beach that had lost much of its sand to erosion. The large Convention Hall complex in Miami Beach has hosted several national political conventions, including the Democratic and Republican conventions in 1972. Incorporated 1915.

Miami-Dade (MEI-ya-MEE–DAID), county (□ 2,429 sq mi/6,315.4 sq km; 2006 population 2,402,208), SE FLORIDA, ⊙ Miami; 25°36′N 80°30′W. Lowland area bordered by FLORIDA KEYS enclosing Biscayne Bay (E) and part of Florida Bay (S). Coastal fringe is the heavily urbanized MIAMI metropolitan area. Interior lies in the EVERGLADES and includes part of Everglades National Park. Manufacturing includes food and wood products, construction materials, textiles. Miami is a business hub with offices of multi-national corporations and banks. Port of Miami is a major port of entry. S anchor of the urban area that lines the coast of SE Florida; has received infusion of immigrants since 1970s, mainly from Central and S. America. National Hurricane Center of the National Weather Service is located here in the unincorporated suburb of Sweetwater, c.7 mi/11.3 km W of central Miami.

Formed 1836. In late 1997 its name was changed from Dade county to Miami-Dade county.

Miami Lakes (mei-AM-ee), planned town (□ 6 sq mi/ 15.6 sq km; 2006 population 22,321), MIAMI-DADE county, SE FLORIDA, 10 mi/16 km NW of MIAMI; 25°54′N 80°18′W. Manufacturing includes apparel, medical equipment, and plastic products.

Miamisburg (mei-AM-eez-buhrg), city (□ 11 sq mi/28.6 sq km; 2006 population 19,878), MONTGOMERY county, SW OHIO, on the MIAMI RIVER, 10 mi/16 km SW of DAYTON; 39°38′N 84°16′W. Metal and paper products. Site of Mound Advanced Technology Center. Large Native American mound nearby. Laid out 1818; incorporated as a city 1932.

Miamisburg Mound (mei-AM-eez-buhrg), prehistoric conical mound (68 ft/21 m high), MONTGOMERY county, W OHIO, just SSW of MIAMISBURG; 39°37′N 84°16′W. Largest of its kind in the state. Picnic grounds here.

Miami Shores (mei-AM-ee), village (□ 4 sq mi/10.4 sq km; 2005 population 10,040), MIAMI-DADE county, SE FLORIDA; suburb 7 mi/11.3 km N of MIAMI, from which it was separated in 1932; 25°52′N 80°10′W. Barry University is here.

Miami Springs (mei-AM-ee), city (□ 3 sq mi/7.8 sq km; 2005 population 13,170), MIAMI DADE county, SE FLORIDA; residential suburb 5 mi/8 km NW of MIAMI; 25°49′N 80°17′W. Miami International Airport just S of the city. Incorporated 1926.

Mian Channu (MYAHN chuh-NOO), town, MULTAN district, S PUNJAB province, central PAKISTAN, 55 mi/ 89 km ENE of MULTAN; 30°27′N 72°22′E. Also spelled Mian Channun.

Mianchi (MI-AN-CHI), town, ⊙ Mianchi county, NW HENAN province, CHINA, 40 mi/64 km W of LUOYANG, and on LONGHAI RAILROAD, near HUANG HE (Yellow River); 34°46′N 111°46′E. Grain, tobacco; engineering, electronics, food processing, building materials, coal mining.

Miandoab, IRAN: see MIYANDUAB.

Miandrivazo (mee-ahn-dree-VAHZ), town, TOLIARY province, W MADAGASCAR, on a headstream of TSIRIBIHINA RIVER, 225 mi/362 km NNE of TOLIARY; 19°31′S 45°28′E. Agricultural and market center. Gold deposits in vicinity. Airfield.

Mianeh (mee-AH-NAI), town (2006 population 89,796), ĀZERBĀYJĀN-e SHARQI province, NW IRAN, on headstream of the SEFID Rud, and 95 mi/153 km SE of TABRIZ, on railroad; 37°24′N 47°43′E. Grain, cotton, grapes. Was railroad terminus, 1941–1951. Formerly called GARMRUD or Garmarud.

Miangas (MYANG-gahs) or **Palmas**, island, INDONESIA, off SE tip of MINDANAO, PHILIPPINES; 05°33′N 126°35′E.

Miani (mee-YAH-nee), town, HOSHIARPUR district, N PUNJAB state, N INDIA, near BEAS RIVER, 23 mi/37 km NW of HOSHIARPUR. Wheat, gram, cotton, rice.

Miani (mee-YAH-nee), town, SARGODHA district, central PUNJAB province, central PAKISTAN, 38 mi/61 km NNE of SARGODHA; 32°32′N 73°04′E.

Miani (mee-YAH-nee), village, JUNAGADH district, GUJARAT state, W central INDIA, on inlet of ARABIAN SEA, 19 mi/31 km NW of PORBANDAR. Small fishing port.

Miani, PAKISTAN: see SONMIANI.

Mianning (MI-AN-NING), town, ⊙ Mianning county, SW SICHUAN province, CHINA, 50 mi/80 km N of XICHANG, and on highway; 28°35′N 102°11′E. Grain; logging, iron ore mining.

Mianus River, c.25 mi/40 km long, NEW YORK and CONNECTICUT; rises NE of ARMONK, New York; flows NE then S, through Connecticut, to LONG ISLAND SOUND at Cos Cob Harbor in GREENWICH town.

Mianwali (MYAHN-wah-lee), district (□ 5,401 sq mi/ 14,042.6 sq km), W PUNJAB province, central PAKISTAN; ⊙ MIANWALI. Bounded W by INDUS River, NE by W end of SALT RANGE; lies mainly in SIND-SAGAR

DOAB; includes large part of THAL region. Wheat, millet grown along fertile riverbank; hand-loom weaving, cattle grazing. Rock salt, coal, aluminum, gypsum, and limestone on N hills. Hydroelectric project on INDUS River. Chief towns are MIANWALI, KALABAGH. Constituted a district in 1901.

Mianwali (MYAHN-wah-lee), town, ⊙ MIANWALI district, W PUNJAB province, central PAKISTAN, 175 mi/282 km WNW of LAHORE, on the INDUS River; 32°35′N 71°33′E. Market for a district that produces food grains, oilseed, hides, and wool; cement factory.

Mian Xian (MI-ANSI-AN), town, ⊙ Mian Xian county, SW SHAANXI province, CHINA, 30 mi/48 km WNW of Nanzheng; 33°10′N 106°40′E. Rice, oilseeds; building materials; food industry, iron smelting, engineering.

Mianyang (MI-AN-YANG), city (□ 606 sq mi/1,570 sq km; 1994 estimated urban population 336,000; 1994 estimated total population 956,500), N central SICHUAN province, CHINA, 30 mi/48 km NNW of SANTAI, and right bank of the FU RIVER; 21°25′N 104°45′E. Agriculture is the largest sector of the city's economy. Crop growing (grain, oil crops, cotton, vegetables, fruits) and livestock (hogs, eggs, poultry); manufacturing (food, textiles, machinery, electronics).

Mianyang, CHINA: see XIANTAO.

Mianzhu (MI-AN-JU), town, ⊙ Mianzhu county, central SICHUAN province, CHINA, 45 mi/72 km NNE of CHENGDU, near source of To River; 31°20′N 104°12′E. Rice, wheat, oilseeds, tobacco; food and beverages, chemicals, paper making, phosphorus ore mining, coal mining.

Miaodao Islands (MI-OU-DOU) or **miao islands**, in YELLOW Sea, NE SHANDONG province, CHINA, guarding entrance to BOHAI Bay, between Liaodong peninsula (N) and SHANDONG PENINSULA (S). Consists of fifteen islands with Changdao Island (4 mi/6.4 km N of PENGLAI) as the largest (7 mi/11.3 km long, 2 mi/3.2 km wide).

Miao Islands, CHINA: see MIAODAO ISLANDS.

Miaoli (MI-OU-LEE), city, NW TAIWAN, 20 mi/32 km SSW of HSINCHU, and on railroad; 24°34′N 120°49′E. Center of Miaoli county oil field, with chief production at Chukwangkeng, 7 mi/11.3 km SE. Sugar milling and chemical fertilizer production; noted for watermelons and persimmons.

Miarinarivo, town, ANTANANARIVO province, central MADAGASCAR, 40 mi/64 km W of ANTANANARIVO; 18°57′S 46°54′E. Elev. 4,395 ft/1,340 m. Market center. Fishery station for Lake ITASY. Beryl, quartz, aragonite, marble.

Miasa (mee-AH-sah), village, N Azumi county, NAGANO prefecture, central HONSHU, central JAPAN, 19 mi/30 km S of NAGANO; 36°34′N 137°53′E.

Miass (mee-AHS), city (2005 population 155,960), W CHELYABINSK oblast, SW Siberian Russia, in the NE foothills of the S URALS, on the MIASS River, on road and railroad, 60 mi/96 km WSW of CHELYABINSK; 55°04′N 60°05′E. Elevation 1,046 ft/318 m. Center of a major gold-mining district; automobile manufacturing, metalworking, firefighting equipment, automated industrial machinery; concrete; wood distilling, flour milling, sawmilling, furniture; food processing (bakery). Marble and talc quarries nearby. Founded in 1773 as a copper-smelting plant (closed in the early 19th century); developed as gold-mining and grain-trading town in the 19th century, and in the 20th century as an industrial center. During World War II, a truck factory evacuated here from MOSCOW. Became city in 1926. Il'men' reservoir is nearby.

Miass (mee-AHS), former town, W CHELYABINSK oblast, S URALS, RUSSIA, just N of MIASS city, into which it has been incorporated; 54°59′N 60°06′E. On railroad; gold mining.

Miass (mee-AHS), river, approximately 390 mi/630 km long, W Siberian Russia; rises in CHELYABINSK oblast in the E slopes of the S URAL Mountains; flows N and

NE past CHELYABINSK into KURGAN oblast, to the Iset' River, a tributary of the OB' RIVER.

Miasskoye (mee-AHS-skuh-ye), village (2004 population 1,030), E CHELYABINSK oblast, SW SIBERIA, RUSSIA, on the MIASS River, on crossroads and near railroad, 15 mi/24 km NE of KOPEYSK, and 13 mi/21 km ENE of CHELYABINSK; 55°17′N 61°54′E. Elevation 518 ft/157 m. Lignite deposits. Bakery.

Miasteczko (mee-STECH-ko) [Polish=small town], German *Georgenberg*, district of TARNOWSKIE GÓRY, Katowice province, S POLAND, 17 mi/27 km NNW of KATOWICE.

Miasteczko Krainskie (mee-STECH-ko KRAHN-skee), Polish *Miasteczko Krainśkie*, German *Friedheim*, town, Piła province, NW POLAND, on railroad, and 12 mi/19 km ESE of Piła (Schneidemühl), near NOTEĆ River; 53°06′N 17°01′E.

Miastko (mee-AHST-ko), German *Rummelsburg*, town in POMERANIA, Słupsk province, NW POLAND, 25 mi/40 km NNE of SZCZECINEK. Linen milling; dairying; manufacturing (distilling equipment).

Miatli (mee-ah-TLEE), village (2005 population 4,530), central DAGESTAN REPUBLIC, SE European Russia, in the SULAK RIVER valley at the edge of the NE foothills of the Greater CAUCASUS Mountains, near highway, 30 mi/48 km W of MAKHACHKALA, and 12 mi/19 km SE of KHASAVYURT; 43°05′N 46°49′E. Elevation 190 ft/57 m. Power plant. Agriculture (grain, fruits, grapes).

Miava, SLOVAKIA: see MYJAVA.

Mi, Ban, THAILAND: see BAN MI.

Mibenge (mee-BEN-gai), village, BANDUNDU province, W CONGO, 35 mi/56 km N of KIKWIT; 03°57′S 19°08′E. Elev. 1,502 ft/457 m. Palm oil production.

Mibu (MEE-boo), town, Shimotsuga county, TOCHIGI prefecture, central HONSHU, N central JAPAN, 15 mi/24 km S of UTSUNOMIYA; 36°25′N 139°48′E. Strawberries.

Mica Mountain, PIMA county, SE ARIZONA, highest peak (8,666 ft/2,641 m) in RINCON MOUNTAINS, c.25 mi/40 km E of TUCSON, in SAGUARO NATIONAL MONUMENT.

Micani (mee-KAH-nee), canton, CHARCAS province, POTOSÍ department, W central BOLIVIA, NW of SAN MARCOS; 18°23′S 65°49′W. Elevation 7,431 ft/2,265 m. Clay, limestone, and gypsum. Agriculture (potatoes, yucca, bananas, barley, oats); cattle.

Micanopy (mik-uh-NO-pee), town (□ 1 sq mi/2.6 sq km; 2005 population 652), ALACHUA county, N central FLORIDA, 10 mi/16 km S of GAINESVILLE; 29°30′N 82°16′W. Citrus fruit; colony for artists and writers. A fort was built here and saw action in the Second Seminole War.

Micaville (MEIK-uh-vil), unincorporated village, YANCEY county, W NORTH CAROLINA, 8 mi/12.9 km W of SPRUCE PINE, near South Toe River; 35°54′N 82°12′W. Mining; stone quarrying. Manufacturing (building stone, fabricated metal products, apparel). Parts of Pisgah National Forest to N and SW.

Micay, Colombia: see LÓPEZ.

Micaya (mee-KAH-yah), canton, AROMA province, LA PAZ department, W BOLIVIA, 16°57′S 68°13′W. Elevation 12,851 ft/3,917 m. Gas wells in area. Salt extraction; copper deposits; gypsum, clay, and limestone deposits. Agriculture (potatoes, yucca, bananas, rye); cattle.

Micco (MIK-o), unincorporated town (□ 9 sq mi/23.4 sq km; 2000 population 9,498), BREVARD county, E central FLORIDA, 15 mi/24 km SSE of MELBOURNE; 27°52′N 80°31′W.

Miccosukee, Lake (mi-kuh-SOO-kee), JEFFERSON county, NW FLORIDA, 20 mi/32 km ENE of TALLAHASSEE; 30°31′N 83°58′W. Triangular shaped; c.7 mi/11.3 km long, 1 mi/1.6 km–5 mi/8 km wide. Plant life surrounds most of this prairie lake.

Michael J. Kirwan Reservoir (□ 4 sq mi/10.4 sq km), PORTAGE county, NE OHIO, on West Branch of MAHONING RIVER, 23 mi/37 km ENE of AKRON; 41°10′N 81°05′W. Maximum capacity 124,000 acre-ft. Formed

by Michael J. Kirwan Dam (77 ft/23 m high), built (1966) by Army Corps of Engineers for flood control; also used for recreation and water supply. West Branch State Park on SE shore.

Michalany (mi-KHAH-lyah-NI), Slovak *Michal'any*, Hungarian *Mihályi*, village, VYCHODOSLOVENSKY province, SE SLOVAKIA, 10 mi/16 km SSE of TREBIŠOV; 48°31′N 21°38′E. Contains a railroad junction; customs station on Hungarian border. Under Hungarian rule from 1938–1945.

Michalkovice (MI-khahl-KO-vi-TSE), Czech *Michálkovice*, German *Michalkowitz*, NE suburb of OSTRAVA, SEVEROMORAVSKY province, NE SILESIA, CZECH REPUBLIC; 49°50′N 18°21′E. Coal mining and industrial community.

Michalkowitz, CZECH REPUBLIC: see MICHALKOVICE.

Michalovce (mi-KHAH-lawf-TSE), German *Gross-Michl*, Hungarian *Nagymihály*, city (2000 population 39,948), VYCHODOSLOVENSKY province, E SLOVAKIA, on LABOREC RIVER, on railroad; 48°45′N 21°56′E. Agricultural center (corn, wheat, sugar beets, fruit); fish. Manufacturing (machinery, ceramics, textiles; wood processing, food processing, brewing). Has 13th-century castle; museum of ZEMPLIN region. Neolithic-era archaeological site.

Michatoya River (mee-chah-TOI-yah), c.55 mi/89 km long, S GUATEMALA; rises in Lake AMATITLÁN at AMATITLÁN; flows generally S, between volcanoes AGUA and PACAYA, past PALÍN (falls and power station), to the PACIFIC OCEAN at IZTAPA; 14°06′N 90°39′W. The Jurín-Marinalá power plant, located between Palín and ESCUINTLA, was completed c.1972. Navigable for 15 mi/24 km below mouth of María Linda River (left). Also called María Linda (mah-REE-ah LEEN-dah) in lower course.

Michel (MI-chuhl), unincorporated village, SE BRITISH COLUMBIA, W CANADA, near ALBERTA border, in ROCKY MOUNTAINS, on Elk River, 18 mi/29 km NE of FERNIE, and included in SPARWOOD; 49°43′N 114°49′W. Elevation 3,861 ft/1,177 m. Open-pit coal mining; cattle; timber.

Michelago, settlement, NEW SOUTH WALES, SE AUSTRALIA, 34 mi/54 km S of CANBERRA; 35°43′S 149°10′E.

Michelau (MI-khe-lou), village, UPPER FRANCONIA, N BAVARIA, GERMANY, on the MAIN, and 2 mi/3.2 km NE of LICHTENFELS; 50°10′N 11°07′E. Forestry.

Michelbach, GERMANY: see WALD-MICHELBACH.

Micheldorf in Oberösterreich (MI-khel-dorf in o-buhr-OES-te-reikh), village, S UPPER AUSTRIA, on KREMS River, and 16 mi/26 km ESE of GMUNDEN; 47°53′N 14°08′E. Manufacturing includes equipment, plastics; timber processing; furniture, tools. Museum of scythes production. Castle Alt-Pernstein.

Michelena (mee-chai-LAI-nah), town, ⊙ Michelena municipio, TÁCHIRA state, W VENEZUELA, in ANDEAN spur, 13 mi/21 km N of SAN CRISTÓBAL; 07°57′N 72°14′W. Elevation 4,251 ft/1,295 m. Coffee; grain; livestock.

Michelson, Mount (8,855 ft/2,699 m), NE ALASKA, in ROMANZOF MOUNTAINS, NE BROOKS RANGE; 69°19′N 144°15′W.

Michelstadt (MI-khel-shtaht), town, S HESSE, GERMANY, on the MÜMLING, and 16 mi/26 km ENE of HEPPENHEIM; 49°41′N 09°01′E. Woodworking. Has 15th-century town hall and a 9th-century basilica. Health resort here. Was first mentioned 741.

Miches (MEE-ches), town, SEIBO province, E DOMINICAN REPUBLIC, on the coast, at entrance of SAMANÁ BAY, 17 mi/27 km N of SEIBO; 18°55′N 69°00′W. Agricultural products (cacao, coffee, coconuts, rice, corn, fruit). Until 1936, JOVERO or EL JOVERO.

Michiana (mish-ee-AN-uh), village (2000 population 200), BERRIEN county, extreme SW MICHIGAN, at INDIANA line, suburb 3 mi/4.8 km NE of MICHIGAN CITY, INDIANA; 41°45′N 86°48′W.

Michiana Shores, town (2000 population 330), LA PORTE county, NW INDIANA, on Lake MICHIGAN, 11

mi/18 km NNW of LA PORTE, adjacent to MICHIGAN state boundary; 41°46′N 86°49′W. Tourist area.

Michichi, hamlet, S ALBERTA, W CANADA, 13 mi/21 km from DRUMHELLER, in STARLAND COUNTY; 51°35′N 112°32′W.

Michielsgestel, Sint, NETHERLANDS: see SINTMICHIELSGESTEL.

Michigamme, Lake (mish-i-GAM-ee), MARQUETTE and BARAGA counties, NW UPPER PENINSULA, MICHIGAN, c.26 mi/42 km W of ISHPEMING; c.6 mi/9.7 km long, 1.5 mi/2.4 km wide; 46°30′N 88°06′W. Resort. Michigamme village is on NW shore. Source of MICHIGAMME RIVER; Michigamme Reservoir is further downstream on Michigamme River, in IRON county; 10 mi/16 km long and 2.5 mi/4 km wide, formed by Way Dam. Van Ripen State Park at E end.

Michigamme River (mish-i-GAM-ee), c.60 mi/97 km long, SW UPPER PENINSULA, MICHIGAN; rises in LAKE MICHIGAMME; flows SSW through Michigamme Reservoir (Way Dam) and Peavy Pond reservoir, both in IRON county, joining BRULE RIVER in SE Iron county to form MENOMINEE RIVER. Has two dams in lower course; 46°29′N 88°04′W.

Michigan, state (□ 96,810 sq mi/250,738 sq km; 1995 estimated population 9,549,353; 2000 population 9,938,444), N UNITED STATES, in the GREAT LAKES region, admitted to the Union in 1837 as the twenty-sixth state; ⊙ LANSING. DETROIT is the largest city. Other major cities are GRAND RAPIDS, WARREN, FLINT, and ANN ARBOR. Michigan is known as the "Great Lake State" because its shores touch four of the five Great Lakes.

Geography

The Lower Peninsula, shaped like a mitten, thrusts N from INDIANA and OHIO. On the E it is separated from ONTARIO (CANADA) by Lake ERIE and Lake HURON, and by the DETROIT River and the SAINT CLAIR RIVER, which together link the two Great Lakes; on the W it is separated from WISCONSIN by Lake MICHIGAN. Across Lake Michigan, NE of Wisconsin, the UPPER PENINSULA stretches E, separating Lake Michigan from Lake SUPERIOR, and itself separated from Ontario only by the narrow SAINT MARYS RIVER; 44°43′N 85°32′W.

The Upper Peninsula

The Upper Peninsula is separated from the Lower Peninsula by the Straits of MACKINAC; a bridge connecting the two peninsulas was opened in 1957 and has spurred the development of the Upper Peninsula. The E portion of the Upper Peninsula has swampy flats and limestone hills on the Lake Michigan shore, while sandstone ridges rise abruptly from the rough waters on Lake Superior; in the W the land rises to forested mountains, rich in copper and iron. The whole of the Upper Peninsula is N woods country, with what has been described as "ten months of winter and two months of poor sledding." The abundance of furred animals and the trees early attracted fur traders and lumbermen. The animals were trapped out, forests were aggressively logged; copper and iron ore mining declined. Hunting (deer, bears, small game; fish). Modern reforestation efforts have aided in second growth.

The Lower Peninsula

The Lower Peninsula has different topography, flora, and fauna. Its forests were also cut over in the lumber boom of the late 19th century, when busy sawmills made Michigan the temporary leader in lumber production. The soil of these cut-over lands, unlike the productive earth in other areas of the Lower Peninsula, proved generally unsuitable for agriculture, and reforestation has been undertaken. There is also mineral mining (gypsum, sandstone, limestone, salt, cement, sand, and gravel). Warm climate, due to the surrounding lakes, allows for a long growing season. However, they also contribute to the heavy snowfall, especially on the windward (E) side of Lake Michigan, and on KEWEENAW Peninsula on Lake Supreme.

Cross-references are shown in SMALL CAPITALS. The pronunciation guide is shown on page xix. The sources of population figures are shown on page xvii.

Economy

Agriculture is important to the state's economy, especially in the S counties and Michigan's noted fruit belt lining the shore of Lake Michigan. Crops account for almost half of farm income; corn is the chief crop, followed by hay, soybeans, and apples; noted for its cherries. Livestock raising and dairying are also important. Manufacturing accounts for 30% of the state's economic production, more than twice as much as any other sector. The manufacture of transportation equipment is by far the state's chief industry, and Detroit, DEARBORN, Flint, PONTIAC, and Lansing are historically centers for motor vehicle manufacturing. Michigan is most readily identified with the automobile industry, and its mass-production methods were the core of the early 20th-century industrial revolution. Other industries produce nonelectrical machinery, fabricated metal products, primary metals, chemicals, and food products. Industrial centers include SAGINAW, BAY CITY, MUSKEGON, and JACKSON. The chemical industry in MIDLAND is one of the nation's largest; KALAMAZOO is an important paper-manufacturing and pharmaceuticals center; Grand Rapids is noted for its furniture, and BATTLE CREEK for its breakfast cereals. Michigan's lack of manufacturing diversity makes it particularly susceptible to the fluctuations of the national economy. The state has attempted in recent years to stabilize its economy by attracting high-technology industry and developing the service sector. Although mining contributes less to income in the state than either agriculture or manufacturing, Michigan in 1989 was the nation's fourth leading state in nonfuel mineral production. The chief minerals produced are iron ore, cement, sand, and gravel. Michigan is a leading producer of peat, bromine, calcium-magnesium chloride, gypsum, and magnesium compounds. The Upper Peninsula is generally preCambrian, an extension of the CANADIAN SHIELD. Many minerals such as copper and iron ore have been mined, but the industry has diminished. Economies of depleted mining areas have been partially sustained by tourism and to lesser extent by forestry and fishing.

Tourism

Michigan's abundant natural beauty makes it a popular destination for tourists. Much of lower Michigan (Lower Peninsula) is composed of series of moraines (formed during glaciation) and dune ridges (formed since glaciation) causing the formation of numerous lakes, the irregular flow of rivers, and marshland. Large sand dunes on Lake Michigan E shore (Berrien, Mason, Leelanau and other counties), sand blocks out flowing rivers, forming small lakes behind dunes. The N Michigan wilds, numerous inland lakes, and some 3,000 mi/4,828 km of shoreline, combined with a pleasantly cool summer climate, have long attracted vacationers. In the winter Michigan's snow-covered hills bring skiers from all over the MIDWEST. Places of interest in the state include GREENFIELD VILLAGE, a re-creation of a 19th-century U.S. village, and the Henry Ford Museum, both at Dearborn; PICTURED ROCKS and SLEEPING BEAR DUNES national lake shores; and ISLE ROYALE NATIONAL PARK. Major islands include: Isle Royale (Lake Superior), BEAVER, MANITOU N and S, FOX ISLANDS (N Lake Michigan), BOIS BLANC, DRUMMOND, MACKINAC (Lake Huron), SUGAR and NEEBISH (Saint Marys River). The state normally ranks among the top ten states annually in tourism expenditures. Many of the small-lakes districts outside Detroit (Oakland county) and other parts of S Michigan, and the S part of Lake Michigan shore, once resorts, have given over to permanent and seasonal residences.

Academic Institutions

Educational institutions include the University of Michigan, at Ann Arbor; Michigan State University, at EAST LANSING; the University of Detroit Mercy and Wayne State University, at Detroit; Western Michigan University and Kalamazoo College, at Kalamazoo;

Eastern Michigan University, at YPSILANTI; Northern Michigan University, at MARQUETTE; and many other private and state colleges.

History: to 1763

The Ojibwa, the Ottawa, the Potawatomi, and other Algonquin-speaking Native American groups were living in Michigan when the French explorer Étienne Brulé landed at the narrows of SAULT SAINTE MARIE in 1618, probably the first European man to have reached present Michigan. Later, French explorers, traders, and missionaries came, including Jean Nicolet, who was searching for the NW Passage; Jacques Marquette, who founded a mission in the Mackinac region; and the empire builder, Robert Cavelier, sieur de La Salle, who came on the *Griffon*, the first ship to sail the Great Lakes. French posts were scattered along the lakes and the rivers, and MACKINAC ISLAND (in the Straits of Mackinac) became a center of the fur trade. Fort Pontchartrain, later Detroit, was founded in 1701 by Antoine de la Mothe Cadillac. The vast region was weakly held by France until lost to Great Britain in the last conflict (1754–1763) of the French and Indian Wars.

History: 1763 to 1812

The Native Americans of Michigan, who had lived in peace with the French, resented the coming of the British, who were the allies of the much-hated Iroquois tribes. Under Pontiac they revolted against the British occupation. The rebellion, which began in 1763, was short-lived, ending in 1766, and the Native Americans subsequently supported the British during the American Revolution. Native American resistance to U.S. control was effectively ended at the Battle of FALLEN TIMBERS in 1794 with the victory of General Anthony Wayne. Despite provisions of the Treaty of PARIS which ended the American Revolution (1783), the British held stubbornly to Detroit and Mackinac until 1796. After passage of the Northwest Ordinance in 1787, Michigan became part of the NORTHWEST TERRITORY. However, even after the Northwest Territory was broken up and Detroit was made (1805) capital of Michigan Territory, British agents still maintained great influence over the Native Americans, who fought on the British side in the War of 1812.

History: 1812 to 1836

In that war Mackinac and Detroit fell almost immediately to the British as a result of the ineffective control of U.S. General William Hull and his troops. Michigan remained in British hands through most of the war until General William Henry Harrison in the battle of THAMES and Oliver Hazard Perry in the battle of Lake Erie restored U.S. control. After peace came, pioneers moved into Michigan. The policy of pushing Native Americans W and opening the lands for settlement was largely due to the efforts of General Lewis Cass, who was governor of Michigan Territory (1813–1831) and later a U.S. senator. Steamboat navigation on the Great Lakes and sale of public lands in Detroit both began in 1818, and the ERIE CANAL was opened in 1825. Farmers came to the Michigan fields, and the first sawmills were built along the rivers. The move toward statehood was slowed by the desire of Ohio and INDIANA to absorb parts of present S Michigan and by the opposition of S states to the admission of another free state.

History: 1836 to 1865

The Michigan electorate organized a government without U.S. sanction and in 1836 operated as a state, although outside the Union. To resolve the boundary dispute Congress proposed that the Toledo strip be ceded to Ohio and Indiana with compensation to Michigan of land in the Upper Peninsula. Though the Michigan electorate rejected the offer, a group of Democratic leaders accepted it, and by their acceptance Michigan became a state in 1837. (The admission of ARKANSAS as a slaveholding state offset that of Michigan as a free state.) Detroit served as the capital until 1847, when it was replaced by Lansing. After statehood, Michigan promptly adopted a program of internal

improvement through the building of railroads, roads, and canals, including the SOO LOCKS Ship Canal at Sault Sainte Marie. At the same time lumbering was expanding, and the population grew as German, Irish, and Dutch immigrants arrived. In 1854 the Republican party was organized at Jackson (Michigan). During the Civil War, Michigan fought on the side of the Union, contributing 90,000 troops to the cause.

History: 1865 to 1903

After the war the state remained firmly Republican until 1882. Then Michigan farmers, moved by the same financial difficulties and outrage at high transportation and storage rates that aroused other Western farmers, supported movements advocating agrarian interests, such as the Granger movement and the Greenback party. The farmers joined with the growing numbers of workers in the mines and lumber camps to elect a Greenback-Democratic governor in 1882 and succeeded in getting legislation passed for agrarian improvement and public welfare. Reforms influenced by the labor movement were the creation of a state board of labor (1883), a law enforcing a ten-hour day (1885), and a moderate child-labor law (1887). The lumbering business, with its yield of wealth to the timber barons, declined to virtually nothing. Some of the loggers joined the ranks of industrial workers, which were further swelled by many Polish and Norwegian immigrants. With the invention of the automobile and the construction of automotive plants, industry in Michigan was radically altered.

History: 1903 to 1960

Henry Ford established the Ford Motor Company in 1903 and introduced conveyor-belt assembly lines in 1918. General Motors and the Chrysler Corporation were established shortly after Ford. Along with the development of mass-production methods came the growth of the labor movement. In the 1930s, when the automobile industry was well established in the state, labor unions struggled for recognition. The conflict between labor and the automotive industry, which continued into the 1940s, included sit-down strikes and was sometimes violent. Walter Reuther, a pioneer of the labor movement, was elected president of the United Auto Workers (UAW) in 1946. During World War II, Michigan produced large numbers of tanks, airplanes, and other war materiel. Industrial production again expanded after the Korean War broke out in 1950.

History: 1960 to Present

In the early 1960s, however, economic growth lagged and unemployment became a problem in the state. The SAINT LAWRENCE SEAWAY opened in 1959 and increased export trade by bringing many oceangoing vessels to the port of Detroit. Detroit was shaken by severe race riots in 1967 that left forty-three persons dead and many injured, in addition to $200 million in damage. In the wake of the rioting, programs were undertaken to improve housing facilities and job opportunities in the city, but these failed as the city suffered massive outmigration which has continued into the 21st century. While Detroit deteriorated, the suburbs experienced dramatic growth, spreading throughout SE Michigan. Resistance to busing was a major political issue in the state in the early 1970s. The state's dependence on the auto industry was exhibited during the recession of the early 1980s, when car sales slumped, many factories closed, and Michigan's unemployment rose. The federal government helped bail out the Chrysler Corporation in 1979, authorizing $1.5 billion in loan guarantees. After a brief period of recovery through limited diversification of the state economy, Michigan was again especially hard hit by national recession and continuing foreign competition in the early 1990s, as General Motors laid off 35,000 employees (a large percentage in Michigan) in 1991.

Government

Michigan's constitution, adopted in 1963, provides for an elected governor as the state's chief executive. The governor serves for a term of four years and may succeed himself in the office. The current governor is Jennifer M. Granholm. The state legislature is made up of a senate and house of representatives. The senate has thirty-eight members elected for terms of four years and the house of representatives has 110 members elected for two-year terms. Michigan sends sixteen representatives and two senators to the U.S. Congress and has eighteen electoral votes in presidential elections.

Michigan has eighty-three counties: ALCONA, ALGER, ALLEGAN, ALPENA, ANTRIM, ARENAC, BARAGA, BARRY, BAY, BENZIE, BERRIEN, BRANCH, CALHOUN, CASS, CHARLEVOIX, CHEBOYGAN, CHIPPEWA, CLARE, CLINTON, CRAWFORD, DELTA, DICKINSON, EATON, EMMET, GENESEE, GLADWIN, GOGEBIC, GRAND TRAVERSE, GRATIOT, HILLSDALE, HOUGHTON, HURON, INGHAM, IONIA, IOSCO, IRON, ISABELLA, JACKSON, KALAMAZOO, KALKASKA, KENT, KEWEENAW, LAKE, LAPEER, LEELANAU, LENAWEE, LIVINGSTON, LUCE, MACKINAC, MACOMB, MANISTEE, MARQUETTE, MASON, MECOSTA, MENOMINEE, MIDLAND, MISSAUKEE, MONROE, MONTCALM, MONTMORENCY, MUSKEGON, NEWAYGO, OAKLAND, OCEANA, OGEMAW, ONTONAGON, OSCEOLA, OSCODA, OTSEGO, OTTAWA, PRESQUE ISLE, ROSCOMMON, SAGINAW, SAINT CLAIR, SAINT JOSEPH, SANILAC, SCHOOLCRAFT, SHIAWASSEE, TUSCOLA, VAN BUREN, WASHTENAW, WAYNE, and WEXFORD.

Michigan, village, NELSON county, E central NORTH DAKOTA, 10 mi/16 km E of LAKOTA; 48°01′N 98°07′W. Livestock, poultry; dairy products; wheat. Also called Michigan City. Founded and incorporated in 1883.

Michigan Center, village (2000 population 4,641) JACKSON county, S MICHIGAN, 4 mi/6.4 km SE of JACKSON, on small Michigan Center Lake; 42°13′N 84°19′W.

Michigan City, city (2000 population 32,900), LA PORTE county, NW INDIANA, on Lake MICHIGAN; 41°43′N 86°53′W. An area with sand-dune beaches and a state park, Michigan City has industries that produce goods such as machinery, consumer articles, alumina powder, kitchen equipment, concrete anchoring systems, wire products, transportation equipment, magnetic steel laminates, chemicals, apparel, and cast-iron boilers. Indiana State Prison is located here, as is the Lakeside Correctional Center. Laid out 1832. Incorporated 1836.

Michigan, Lake (□ 22,178 sq mi/57,441 sq km), bordered by MICHIGAN, INDIANA, ILLINOIS, and WISCONSIN; third-largest of the GREAT LAKES and the only one entirely within the U.S.; 307 mi/494 km long and 30 mi/48 km–120 mi/193 km wide; its surface is 581 ft/177 m above sea level, and the lake is 923 ft/281 m deep. The Straits of MACKINAC, its only natural outlet, connects the lake with Lake HURON to the NE; the ILLINOIS WATERWAY links Lake Michigan with the MISSISSIPPI RIVER and the Gulf of MEXICO. The ST. LAWRENCE SEAWAY provides access to international trade on Lake Michigan; ore, coal, and limestone are the main items moved on the lake. The S part of the lake does not freeze over in the winter, but storms and ice halt interlake movement from December to April. Many islands are found in the N part of the lake; the N shorelines are indented, with GREEN BAY and GRAND TRAVERSE BAY the largest bays. The S part of Lake Michigan has a regular shoreline, necessitating the building of artificial harbors such as the CALUMET HARBOR in NE Illinois. The MUSKEGON, GRAND, KALAMAZOO, FOX, and MENOMINEE are the chief rivers flowing into Lake Michigan; the lake's current tends to clog the mouths of the rivers with sand. The CHICAGO RIVER formerly flowed into the lake, but its course was reversed in 1900. Sand dunes border the E and S shores of the lake; INDIANA DUNES NATIONAL LAKESHORE is here. The forested N region of Lake Michigan is generally sparsely populated. The S portion, located near the heart of the Midwest, is industrially important; the GARY-CHICAGO-MILWAUKEE urbanized area extends along the SW shore. MICHIGAN CITY, Gary, Chicago, RACINE, MILWAUKEE, and ESCANABA are the major lakeside cities. Such urban and industrial concentration has led to growing pollution problems associated with the lake's waters. Prevailing westerly winds tempered by the lake give the E shore a moderate climate, making it a rich fruit belt and popular resort area. Lake Michigan was discovered in 1634 by the French explorer Jean Nicolet and was later explored by the French traders Marquette and Joliet. French missionary and trade centers thrived there by the late 1600s. As part of the bitterly contested NORTHWEST TERRITORY, the area passed to England in 1763 and later to the U.S. in 1796. The area was isolated until the 1830s, when improvements in transportation brought settlers there.

Michigantown, town (2000 population 406), CLINTON county, central INDIANA, 8 mi/12.9 km ENE of FRANKFORT; 40°20′N 86°23′W. Agricultural area. Laid out 1830.

Michikamau Lake (mi-chi-KAH-mo), W NEWFOUNDLAND AND LABRADOR, E CANADA, near QUEBEC border; 65 mi/105 km long, 30 mi/48 km wide; 53°50′N 63°22′W. Elevation 1,650 ft/503 m. It is largest body of water in region of numerous small lakes; drained by CHURCHILL RIVER.

Michilimackinac, MICHIGAN: see MACKINAC, region.

Michinmáhuida, CHILE: see MINCHINMÁVIDA.

Michipicoten (mi-chi-pi-KO-tuhn), township (□ 161 sq mi/418.6 sq km; 2001 population 3,668), central ONTARIO, E central CANADA; 48°01′N 84°48′W.

Michipicoten Harbour (mi-chi-pi-KO-tuhn), ghost village, central ONTARIO, E central CANADA, on Michipicoten Bay of Lake SUPERIOR, at mouth of MICHIPICOTEN RIVER, 110 mi/177 km NNW of SAULT SAINTE MARIE; 47°57′N 84°54′W. Formerly an iron and pulp shipping port; gold, iron mining. Became a ghost town from the early 1950s.

Michipicoten Island (mi-chi-pi-KO-tuhn), central ONTARIO, E central CANADA, in NE part of Lake SUPERIOR, at entrance of Michipicoten Bay, 110 mi/177 km NW of SAULT SAINTE MARIE; 17 mi/27 km long, 3 mi/5 km–6 mi/10 km wide. Rises to 1,598 ft/487 m (W).

Michipicoten River (mi-chi-pi-KO-tuhn), 70 mi/113 km long, central ONTARIO, E central CANADA; issues from WABATONGUSHI LAKE, 110 mi/177 km N of SAULT SAINTE MARIE; flows generally SW, through DOG and Manitowick lakes, to Michipicoten Bay of Lake SUPERIOR, 3 mi/5 km SE of MICHIPICOTEN HARBOUR; 47°56′N 84°49′W.

Michliffen (MISH-lee-fin), extinct volcanic crater, MOROCCO, in central Middle ATLAS mountains, 9 mi/15 km S of IFRANE; 33°25′N 05°07′W. Ski slope.

Michmas, settlement, circa 8 mi/12.9 km NE of JERUSALEM, outer suburb of Jerusalem, near the Arab village of MUKHMAS, WEST BANK; 31°52′N 35°17′E. Believed to be near or on the site of the Hebrew Bible town of Michmash.

Michmash, ISRAEL: see BOZEZ.

Michoacán (mee-cho-an-KAHN) or **Michoacán de Ocampo**, state (□ 23,202 sq mi/60,325.2 sq km), S MEXICO; ⊙ MORELIA; 17°55′N 100°44′W. Dominated by the TRANSVERSE VOLCANIC AXIS of central Mexico, Michoacán extends from the PACIFIC OCEAN northeastward into the central plateau. The LERMA River and Lake CHAPALA form part of its N boundary with the state of JALISCO; the Río BALSAS constitutes the S border with GUERRERO. The climate and soil variations caused by topography and differences in elevation make Michoacán a diverse agricultural state, producing both temperate and tropical cereals, fruits, and vegetables. Michoacán's forests yield fine cabinet woods and dyewoods. Mining is a leading industry; gold and silver are most important, but iron, coal, and zinc are also major minerals. Industrial development is modest, including iron and steel production. Michoacán ships its products from the cities of MORELIA, URUAPAN, and LÁZARO CÁRDENAS. Federally sponsored irrigation and hydroelectric power projects, especially the PRESA INFIERNILLO dam on Río Balsas, have aimed at developing Michoacán's Pacific coastal region. Lake PÁTZCUARO (where UNESCO and the OAS have a training center for Latin American rural teachers) and the PARÍCUTIN volcano attract many tourists. Most of the state's inhabitants are native Tarascans. Michoacán played a leading role in Mexico's revolution against SPAIN and in subsequent struggles.

Michurin, Bulgaria: see TSAREVO.

Michurinsk (mee-CHOO-reensk), city (2006 population 92,650), NW TAMBOV oblast, S central European Russia, on crossroads and railroad junction, on the Lesnoy Voronezh River (DON River basin), 45 mi/72 km NE of TAMBOV; 52°54′N 40°30′E. Elevation 442 ft/134 m. Machinery and metal goods (locomotives, automobile parts, pumps), furniture; food and light industries. Founded in 1636. City status since 1779. Known as Kozlov until 1932, when renamed in honor of I. F. Michurin (1855–1935), a biologist who founded a plant-breeding institute here.

Mickle Fell (MIK-uhl FEL), mountain (2,591 ft/790 m) in the PENNINES, SW DURHAM, NE ENGLAND, 16 mi/26 km WNW of BARNARD CASTLE; 54°36′N 02°18′W.

Mickleham (MIK-uhl-uhm), agricultural village (2001 population 570), central SURREY, SE ENGLAND, on MOLE RIVER, at foot of BOX HILL, and 2 mi/3.2 km N of DORKING; 51°16′N 00°20′W. Has remains of priory founded 1228. Fanny Burney lived here for some time.

Micomeseng (mee-ko-mee-SING), town, KIE-NTEM province, continental EQUATORIAL GUINEA, near CAMEROON border, 70 mi/113 km NE of BATA; 02°08′N 10°38′E. Cacao, coffee. Former site of large leper hospital. Also spelled Micomiseng or Mikomesen.

Micomiseng, EQUATORIAL GUINEA: see MICOMESENG.

Miconje, town, CABINDA province, ANGOLA, on road 90 mi/145 km NE of CABINDA; 04°26′S 12°48′E. Market town.

Mico River (MEE-ko), S NICARAGUA; rises near LA LIBERTAD; flows c.100 mi/161 km E, past SAN PEDRO DE LÓVAGO, MUELLE DE LOS BUEYES, and EL RECREO, joining SIQUIA and Rama rivers at RAMA to form ESCONDIDO RIVER. Navigable for launches below El Recreo.

Mico, Sierra del (MEE-ko see-ER-rah del), range in IZABAL department, E GUATEMALA, E continuation of Sierra de las MINAS. Extends 50 mi/80 km NE to Bay of AMATIQUE, forming divide between Lake IZABAL (N) and MOTAGUA River (S); rises to 4,156 ft/1,267 m at Cerro San Gil. Coal deposits.

Micoud (mee-KOO), village (2001 population 2,619), SE SAINT LUCIA, minor port 14 mi/23 km SE of CASTRIES; 13°49′N 60°56′W. Agriculture (bananas, other tropical fruit); fishing.

Micro (MEI-kro), village (2006 population 513), JOHNSTON county, central NORTH CAROLINA, 10 mi/16 km NE of SMITHFIELD; 35°33′N 78°12′W. Manufacturing; agriculture (tobacco, cotton, grain; livestock, poultry).

Micronesia, one of the three main divisions of Oceania, in W PACIFIC OCEAN, mostly N of the equator. Micronesia as a geographical region includes the CAROLINE ISLANDS, MARSHALL ISLANDS, PALAU (Belau), MARIANA Islands, GILBERT ISLANDS, and NAURU. The inhabitants are genetically mainly of Southeast Asian Mongoloid, Micronesian, and Polynesian stock. They speak Austronesian languages.

Micronesia, Federated States of, a confederation of the CAROLINE ISLANDS (apart from PALAU) which forms a U.S.-linked free association state (c.271 sq mi/702 sq km; 2004 estimated population 108,155; 2007 estimated population 107,862), in the W PACIFIC OCEAN; ⊙ Kolonia (island of POHNPEI). It is comprised of four states: KOSRAE, POHNPEI, CHUUK, and YAP.

Population

The population is predominantly Micronesian.

Economy

The U.S. pledged to spend $1 billion in the islands in the 1990s, making financial assistance the primary source of income, but in 2002 the country's future revenues were reduced so the economic outlook appears fragile. Other mainstays of the economy are subsistence farming and fishing.

History

Germany purchased the islands from Spain in 1898. They were occupied (1914) by Japan, which received them (1920) as a League of Nations mandate. During World War II U.S. forces captured the islands and in 1947 they became part of the UN U.S. Trust Territory of the Pacific Islands. In 1979, as negotiations for termination of the trusteeship continued, they became self-governing as the Federated States of Micronesia, which in 1986 assumed free association status with the U.S. An amended compact of free association came into force in 2004.

Government

The islands are governed under the constitution of 1979. The president, who is both head of state and head of government, is elected by Congress for a four-year term. There are fourteen members of the unicameral Congress; four are popularly elected for four-year terms and ten for two-year terms. Defense is the responsibility of the U.S. The current head of state is President Emmanuel Mori (since May 2007).

Midagalola, ETHIOPIA: see MĪDEGA.

Midai (MEE-dei), island (4 mi/6.4 km long, 3 mi/4.8 km wide), RIAU province, INDONESIA, in SOUTH CHINA SEA, 145 mi/233 km WNW of Cape DATU (BORNEO); 03°01′N 107°48′E. Coconuts; trepang.

Midale (MEE-dahl), town (2006 population 462), SE SASKATCHEWAN, CANADA, 27 mi/43 km SE of WEYBURN; 49°24′N 103°24′W. Mixed farming.

Mid-Atlantic Ridge, submarine mountain range; c.300 mi/483 km–600 mi/966 km wide. Extends c.10,000 mi/ 16,093 km from ICELAND, c.67°N, to near the ANTARCTIC CIRCLE, c.55°S, and occupying the axis of the ATLANTIC OCEAN, approximately midway between the opposite shores of the continents. It is part of the mid-oceanic ridge system which forms a belt more than 37,000 mi/59,544 km long that circles the world. The highest peaks of the Mid-Atlantic Ridge reach more than 13,000 ft/3,962 m above the ocean floor. The ridge, which is a center of volcanic activity and earthquakes, has a 6-mi/9.7-km–19-mi/31-km-wide trough that is constantly widening and filling with molten rock from the earth's interior. The bilateral movement of oceanic crust away from its source along the mid-oceanic ridge is called sea-floor spreading and results in the continents moving away from each other. A recent discovery has been a site c.280 mi/450 km SW of the Azores, 36°N, described as a 1,075 sq ft/ 100 sq m area with more than 100 smoking vents at 7,700 ft/2,128 m, one of the largest volcanic vent fields in the Atlantic Ocean.

Mid-Canada Line, CANADA: see DEW LINE.

Middelberg Range, pass (5,184 ft/1,580 m), SW CEDARBERG RANGE above CITRUSDAL, SOUTH AFRICA; 32°37′S 19°09′E.

Middelburg (MI-duhl-berkh), city, ⊙ ZEELAND province, SW NETHERLANDS, on the former island of WALCHEREN, 58 mi/93 km SW of AMSTERDAM; 51°30′N 03°37′E. NORTH SEA 5 mi/8 km to W; Berkenbos Dunes, on North Sea, to NW. Agriculture (dairying; cattle, hogs; vegetables, seeds, grain, sugar beets); manufacturing (lamp parts, machinery, chimney systems); tourism. Chartered in 1217, Middelburg developed into an important medieval trade center. The last Spanish fortress in Zeeland, it was captured (1574) by the Beggars of the Sea. Although heavily damaged in World War II and flooded in 1953, Middelburg retains many beautiful old (or rebuilt) buildings, including a twelfth-century abbey and the sixteenth-century town hall.

Middelburg, town, EASTERN CAPE province, SOUTH AFRICA, on Klein Brak River, and 61 mi/98 km NNW of CRADOCK; 31°30′S 25°00′E. Elevation 4,592 ft/1,400 m. Railroad junction; junction of N9 and N10 highways to North. Agricultural center (wheat, feed crops; stock). Just NE is Grootfontein Agricultural College Town established in 1852 as midpoint between Cradock and COLESBURG, hence the name. Airfield. Two new residential areas Kwanonzame (S) and Midros (SE).

Middelburg, town, MPUMALANGA province, SOUTH AFRICA, on Little OLIFANTS RIVER, and 80 mi/129 km E of TSHWANE (formerly Pretoria); 25°46′S 29°28′E. Elevation 4,971 ft/1,515 m. Established in 1864, named as halfway point between Tshwane and LYDENBURG. In coal-mining, agricultural region (wheat, tobacco, potatoes). Grain elevator. Copper, iron, cobalt deposits nearby. On N11 highway 5 mi/8 km N of intersection with N4. Railroad and airfield. Site of technical college; large residential area to NW called Mhlinzi.

Middelfart (MI-dthuhl-fahrt), city (2000 population 13,010), FYN county, central DENMARK, on the LILLE BÆLT, which is spanned there by a road and rail bridge; 55°30′N 09°45′E. Middelfart has long been a port and fishing base.

Middelharnis (MI-duhl-hahr-nis), town, SOUTH HOLLAND province, SW NETHERLANDS, on GOEREE-OVERFLAKKEE island, and 18 mi/29 km SW of ROTTERDAM; 51°45′N 04°10′E. HARINGVLIET estuary 2 mi/ 3.2 km to NE. Agriculture (dairying; cattle, poultry; vegetables, sugar beets); manufacturing (liqueurs).

Middelstum (MI-duhl-stum), village, GRONINGEN province, NE NETHERLANDS, 10 mi/16 km NNE of GRONINGEN; 53°21′N 06°39′E. Agriculture (dairying; cattle, sheep; grain, vegetables; manufacturing (food processing).

Middenmeer (MI-duhn-mair), village, WIERINGERMEER polder, NORTH HOLLAND province, NW NETHERLANDS, 13 mi/21 km NE of ENKHUIZEN.

Middle, township, CAPE MAY county, S NEW JERSEY, 20 mi/32 km N of CAPE MAY; 39°05′N 74°50′W. Incorporated 1798.

Middle, in Russian names: see also SREDNE-, SREDNEYE, SREDNI, SREDNIYE, or SREDNYAYA.

Middle America, a term sometimes used to designate the area between the UNITED STATES border and COLOMBIA and VENEZUELA, marking transition from NORTH AMERICA to SOUTH AMERICA; it differs from CENTRAL AMERICA in that it includes MEXICO and the WEST INDIES.

Middle Andaman Island, India: see ANDAMAN AND NICOBAR ISLANDS.

Middle Atlas, MOROCCO: see ATLAS MOUNTAINS.

Middleback Range (MI-duhl-bak), hill range, S SOUTH AUSTRALIA, AUSTRALIA, extends 40 mi/64 km S from IRON KNOB, parallel with E coast of EYRE PENINSULA. Rises to 500 ft/152 m. Includes iron-producing hills of Iron Knob, Iron Monarch, Iron Prince, Iron Baron; sandstone.

Middle Bass Island, Ohio: see BASS ISLANDS.

Middle Beskyd, UKRAINE: see UPPER DNIESTER BESKYDS.

Middleboro, town, PLYMOUTH county, SE MASSACHUSETTS; 41°53′N 70°53′W. Cranberry-processing is a major industry in the town; other manufacturing (fire apparatus, chemicals, and shoes). The town was destroyed by Native Americans in King Philip's War but later rebuilt. Of interest are a Native American site believed to date from 2500 B.C.E.; restored Revolutionary industries, such as a slitting mill and an iron foundry. The Tom Thumb historical museum. The name is also spelled Middleborough. Incorporated 1669.

Middleboro, Pennsylvania: see MCKEAN.

Middlebourne (MID-uhl-buhrn), town (2006 population 847), ⊙ TYLER county, NW WEST VIRGINIA, 37 mi/60 km NE of PARKERSBURG, on MIDDLE ISLAND CREEK. Agriculture (corn); cattle; dairying. Gas, oil region. Manufacturing (secondary aluminum processing, wooden mouldings). Jug and Conaway Run Wildlife Management Area to E; Lewis Wetzel Wildlife Management Area to E.

Middle Brewster Island, MASSACHUSETTS: see BREWSTER ISLANDS.

Middleburg (MI-duhl-buhrg), town (2006 population 917), LOUDOUN county, N VIRGINIA, 17 mi/27 km N of WARRENTON, near LITTLE RIVER; 38°58′N 77°44′W. Manufacturing (lumber, wine, leather goods; printing and publishing); agriculture (dairying; livestock; grain, apples, grapes, soybeans). Center of country-estate area known for fox hunting, horse breeding.

Middleburg, village (□ 1 sq mi/2.6 sq km; 2000 population 1,398), SCHOHARIE county, E central NEW YORK, on SCHOHARIE CREEK, and 30 mi/48 km WSW of ALBANY; 42°36′N 74°19′W. Trade center for dairying and farming area. Settled 1712, incorporated 1881.

Middleburg (MI-duhl-burhg), village (2006 population 164), VANCE county, N NORTH CAROLINA, 6 mi/ 9.7 km NE of HENDERSON; 36°23′N 78°19′W. Manufacturing (lumber); agriculture (grain, tobacco, soybeans; livestock). Large S arm of Kerr reservoir to NW.

Middleburg, borough (2006 population 1,349), ⊙ SNYDER county, central PENNSYLVANIA, 15 mi/24 km WSW of SUNBURY, on Middle Creek; 40°47′N 77°02′W. Agricultural area (corn, hay, apples; dairying); manufacturing (concrete products, cutting tools, sportswear, food products). Parts of Bald Eagle State Forest to S and NW. Settled c.1760, laid out 1800, incorporated 1856.

Middleburg Heights (MID-uhl-buhrg HEITZ), city (□ 8 sq mi/20.8 sq km; 2006 population 15,237), CUYAHOGA county, NE OHIO; SW suburb of CLEVELAND; 41°22′N 81°49′W.

Middlebury, town, NEW HAVEN county, SW CONNECTICUT, just W of WATERBURY; 41°31′N 73°07′W. In summer resort area; manufacturing (watches, clocks). Corporate headquarters. QUASSAPAUG POND just W. Settled in early 18th century, incorporated 1807.

Middlebury, town (2000 population 2,956), ELKHART county, N INDIANA, on small Little Elkhart River, and 9 mi/14.5 km NE of GOSHEN; 41°40′N 85°43′W. In agricultural area; livestock. Manufacturing (plastics, recreational vehicles, modular and prefabricated homes, furniture, wood products, transportation equipment, window blinds, wire harnesses, store display fixtures, and agricultural machinery; food processing). Laid out 1835.

Middlebury, town, including Middlebury village, ⊙ ADDISON CO., W VERMONT, on Otter Creek, and 32 mi/51 km S of BURLINGTON; 44°00′N 73°07′W. Agriculture (poultry; dairy products, fruit); manufacturing (wood products, electronic equipment). Resort area; winter sports at Middlebury College Snow Bowl. Home of Middlebury College, and art museum (1829). Partly in Green Mountain National Forest. Chartered 1761; First settled 1773; permanently settled 1783. Name derives from its location midway between NEW HAVEN (north) and SALISBURY (south) which were chartered on the same day.

Middle Caicos or **Grand Caicos**, island (□ 48 sq mi/ 124.8 sq km), TURKS AND CAICOS ISLANDS, crown colony of GREAT BRITAIN, WEST INDIES, just W of EAST CAICOS; 25 mi/40 km long, 12 mi/19 km wide; 21°50′N 71°45′W. Largest and least developed island of the archipelago. Conch Bar caves, with underground salt lakes, are open to visitors. Small airfield.

Middle Chain of Lakes, nineteen small natural lakes and several lesser bodies of water running N-S in nearly straight line c.20 mi/32 km long, mainly in MARTIN county, S MINNESOTA, extends into EMMET county, N IOWA. Located in relatively lake-free agricultural prairie. From Iowa Lake on S, on Iowa-Minnesota state line, the lakes drain N through South Silver, North Silver, Wilmert, Mud, Aires, Amber, Hall, Budd, Sisseton, and George lakes (latter four in FAIRMONT city), connected by Center Creek; Buffalo, Canright, and Kiester lakes drain S to ELM CREEK; Charlotte, High, and Martin lakes drain N to Elm Creek; Murphy Lake drains S into Elm Creek; Perch Lake drains N, source of Perch Creek. Lake chain follows course of retreating Pleistocene glacier. East Chain of Lakes, 5 mi/8 km to E, c.15 mi/24 km long N-S, includes Goose and Burt Lakes, KOSSUTH county (Iowa), Sway Lake (on Iowa-Minnesota border), East Chain Lake (2 mi/3.2 km long), Sager, Rose, Little Hall and Imogene lakes (Martin county, Minnesota).

Middlechurch (MI-duhl-chuhrch), community, SE MANITOBA, W central CANADA, 8 mi/12 km from WINNIPEG, and in WEST ST. PAUL rural municipality; 49°58′N 97°04′W.

Middle Congo: see CONGO, PEOPLE'S REPUBLIC OF THE.

Middle Country, CHINA: see MIDDLE KINGDOM.

Middle Creek, KENTUCKY: see PRESTONSBURG.

Middle East, term applied to the countries of SW Asia and NE Afr. lying W of Afghanistan, Pakistan, and India. Thus defined it includes Cyprus, the Asian part of Turkey, Syria, Israel, Jordan, Iraq, Iran, Lebanon, the countries of the Arabian peninsula (Saudi Arabia, Yemen, Oman, United Arab Emirates, Qatar, Bahrain, Kuwait), Egypt, Sudan, and Libya. The term is sometimes used in a cultural sense to mean the group of lands in that part of the world predominantly Islamic in culture, thus including the remaining states of N Afr. as well as Afghanistan until World War II generally known as Near East. In modern times the designation Middle East was applied by W Europeans who viewed the area as midway between Europe and East Asia, which they called the Far East.

Middle Fabius River, MISSOURI: see FABIUS RIVER.

Middlefield, industrial town, MIDDLESEX county, S CONNECTICUT, between MIDDLETOWN and MERIDEN; 41°31′N 72°42′W. Manufacturing (ordnances, transportation equipment, tools, cement products, thermometers, fabricated metals, fixtures, consumer goods); agriculture (apples, peaches). Includes Rockfall village. State park here. Dinosaur tracks found here now in Peabody Museum of Yale University. Settled c.1700, set off from Middletown 1866.

Middlefield, town, HAMPSHIRE county, W MASSACHUSETTS, 14 mi/23 km ESE of PITTSFIELD; 42°21′N 73°01′W.

Middlefield (MID-uhl-feeld), village (□ 3 sq mi/7.8 sq km; 2006 population 2,414), GEAUGA county, NE OHIO, 30 mi/48 km E of CLEVELAND; 41°27′N 81°04′W. In agricultural area; rubber, plastic, and metal products; lumber, food products.

Middle Fork or **Middle Fork of Wilamette River**, river, c.115 mi/185 km long, W OREGON; formed by confluence of several small branches in CASCADE RANGE near Emigrant Pass (5,000 ft/1,524 m); flows NW through HILLS CREEK Reservoir, past OAKRIDGE, through LOOKOUT POINT Reservoir; joins COAST FORK 3 mi/4.8 km SE of EUGENE to form WILLAMETTE River. About 20 mi/32 km SE of Eugene is Lookout Point Dam (250 ft/76 m high, 3,106 m/947 m long; begun 1947), a unit of Willamette River flood-control plan; also called Meridian Dam.

Middle Franconia (frang-KO-nee-uh), German *Mittelfranken* (MIT-tuhl-frahnk-uhn), administrative division (Ger. *Regierungsbezirk*) (□ 2,798 sq mi/7,274.8 sq km) of W BAVARIA, GERMANY; ⊙ ANSBACH. Bounded S by SWABIA, SE by UPPER BAVARIA, E by UPPER PALATINATE, N by UPPER and LOWER FRANCONIA, W by BADEN-WÜRTTEMBERG. Hilly region in FRANCONIAN JURA; drained by ALTMÜHL, REDNITZ, and PEGNITZ rivers. Agriculture (wheat, barley; cattle, hogs); industries (machinery, vehicles, precision instruments) centered at NUREMBERG (Bavaria's second-largest city) and FÜRTH; and ERLANGEN, has electromedical industry. Population is predominantly Protestant. Part of old historic region of FRANCONIA.

Middle Gobi, MONGOLIA: see DUNDGOVD.

Middle Granville, village, WASHINGTON county, E NEW YORK, near VERMONT line, 20 mi/32 km ENE of GLENS FALLS; 43°26′N 73°17′W. In slate-quarrying area.

Middle Grove, unincorporated community, MONROE county, NE central MISSOURI, 15 mi/24 km SW of PARIS.

Middle Haddam, CONNECTICUT: see EAST HAMPTON.

Middleham (MID-uhl-uhm), agricultural village (2001 population 1,304), NORTH YORKSHIRE, N ENGLAND, on URE RIVER, and 9 mi/14.5 km SSW of RICHMOND; 54°17′N 01°49′W. Has castle (built in late 12th century) which belonged to Warwick the Kingmaker. Racing stables.

Middle Island, village, W St. Kitts, ST. KITTS AND NEVIS, WEST INDIES, 7 mi/11.3 km NW of BASSETERRE. Has St. Thomas Church and tomb of Sir Thomas Warner, founder of the British colony, who died in 1648.

Middle Island (MI-duhl), S ONTARIO, E central CANADA, tiny island in Lake ERIE just S of PELEE ISLAND, near U.S. (OHIO) boundary; southernmost point of Canada (41°41′N).

Middle Island Creek, c.85 mi/137 km long, NW WEST VIRGINIA; rises in SE DODDRIDGE county, 15 mi/24 km SW of CLARKSBURG; flows NW past WEST UNION and MIDDLEBOURNE, and SSW to OHIO RIVER, 1 mi/1.6 km NW of SAINT MARYS.

Middlekerke (MI-duhl-ker-kuh), commune, Ostend district, WEST FLANDERS province, W BELGIUM, 5 mi/8 km SW of OSTENDE, near NORTH SEA; 51°11′N 02°49′E. Just NW, on North Sea, is seaside resort of Middlekerke-Bad.

Middlekerke-Bad, BELGIUM: see MIDDLEKERKE.

Middle Kingdom, Mandarin *Zhongguo* (JUNG-GAW), Chinese name for CHINA. It dates from c.1000 B.C.E., when it designated the Zhou empire (1126–771 B.C.E.) situated on the NORTH CHINA PLAIN. The Zhou people, unaware of ancient civilizations in other parts of the world, believed their empire occupied the middle of the earth, surrounded by barbarians. Since 1949, when the Communists took power, the official name for China has been *Zhonghua Renmin Gongheguo* (middle glorious people's republican country) or, in English, the People's Republic of China.

Middle Loup River, Nebraska: see LOUP RIVER.

Middle Nodaway River, IOWA: see NODAWAY RIVER.

Middle Park (MI-duhl PAHRK), suburb of MELBOURNE, VICTORIA, SE AUSTRALIA, between ALBERT PARK and SAINT KILDA suburbs. National Estate conservation area.

Middle Pease River, Texas: see PEASE RIVER.

Middle Point, village (2006 population 578), VAN WERT county, W OHIO, 7 mi/11 km E of VAN WERT, on LITTLE AUGLAIZE RIVER; 40°51′N 84°26′W. In agricultural area.

Middleport, village (2006 population 1,816), NIAGARA county, W NEW YORK, on the BARGE CANAL, and 30 mi/48 km NNE of BUFFALO; 43°12′N 78°28′W. Light industry, tourism, commercial services. Grew after completion of ERIE CANAL (1825). Settled 1812, incorporated 1859.

Middleport (MID-uhl-port), village (□ 2 sq mi/5.2 sq km; 2006 population 2,508), MEIGS county, SE OHIO, on OHIO RIVER, 18 mi/29 km NNE of GALLIPOLIS; 38°59′N 82°03′W. Coal mines, gas and oil wells. Manufacturing.

Middleport, borough (2006 population 439), SCHUYLKILL county, E central PENNSYLVANIA, 8 mi/12.9 km ENE of Pottsville, on SCHUYLKILL RIVER; 40°43′N 76°05′W. Anthracite coal. Locust State Park and part of Weiser State Forest to N.

Middle Raccoon River, IOWA: see RACCOON RIVER.

Middle River, unincorporated town (2000 population 23,958), BALTIMORE county, N MARYLAND; 39°20′N 76°26′W. An industrial and growing residential suburb of BALTIMORE. Martin Airport, named for John Martin, founder of the Martin-Marieta Company. The river for which the town is named is an inlet between the GUNPOWDER and BACK RIVER estuaries.

Middle River, village (2000 population 319), MARSHALL county, NW MINNESOTA, on MIDDLE RIVER, and 22 mi/35 km N of THIEF RIVER; 48°26′N 96°09′W. Falls. Wheat, sugar beets, potatoes, beans, sunflowers; manufacturing (flour milling, transportation equipment). Thief Lake Wildlife Area to NE; large MUD LAKE, and Agassiz National Wildlife Refuge to SE.

Middle River, 105 mi/169 km long, S central IOWA; rises in SW GUTHRIE county; flows SE and ENE to DES MOINES RIVER 13 mi/21 km SE of DES MOINES.

Middle River, 70 mi/113 km long, MINNESOTA; rises in marshy area of E central MARSHALL county, NW Minnesota, just E of village of MIDDLE RIVER; 48°24′N 96°00′W; flows W past NEWFOLDEN and ARGYLE, to SNAKE RIVER 12 mi/19 km N of ALVARADO and 6 mi/9.7 km SE of its confluence with RED RIVER.

Middle River (MI-duhl), c.60 mi/97 km long, NW VIRGINIA; rises in S AUGUSTA county; flows NE (W and N of STAUNTON) to NORTH RIVER, 2 mi/3 km SW of PORT REPUBLIC; 38°02′N 79°17′W.

Middle Saranac Lake, New York: see SARANAC LAKES.

Middlesboro, city (2000 population 10,384), BELL county, S KENTUCKY, c.105 mi/169 km SE of LEXINGTON, on Yellow River, in the CUMBERLAND MOUNTAINS, 2 mi/3.2 km W of the point where Kentucky, TENNESSEE, and VIRGINIA meet; 36°36′N 83°43′W. Elevation 1,138 ft/347 m. It is a coal-mining center, with manufacturing (plastic pipe, elastic webbing, apparel; printing and publishing, meat processing, leather tanning, coal processing). Said to be the only U.S. city built within a meteor crater. Airport to NW. Southeast Community College—Middlesboro Campus. Coal House and Museum. Cumberland Gap Tunnel Project (completed 1996). Cumberland Gap National Historical Park to E; Kentucky Ridge State Forest to NW; PINE MOUNTAIN State Resort Park to N. Established 1889.

Middlesbrough (MID-uhlz-bruh), town and county (2001 population 134,855), NE ENGLAND, on S bank of the TEES, and 3 mi/4.8 km ENE of STOCKTON-ON-TEES; 54°34′N 01°14′W. Previously a major iron and steel center, with blast furnaces, steel mills, and chemical works. Town developed after opening of DARLINGTON-Stockton railroad and after discovery of iron in the CLEVELAND HILLS, nearby. Has the University of Teeside, metallurgical schools, libraries, and museum. Teeside Airport nearby. The port here previously handled coal. Large breakwaters protect the city. Just E is the suburb of Cargofleet. Suburb of Ormesby is 3 mi/4.8 km SE. Local authority of Middlesbrough created in 1996 when county of Cleveland was abolished and split up into Middlesbrough, Stockton-on-Tees, Hartlepool, and Redcar and Cleveland.

Middlesex (MI-duhl-seks), county (□ 1,281 sq mi/3,330.6 sq km; 2001 population 403,185), S ONTARIO, E central CANADA, on THAMES River; ⊙ LONDON; 43°00′N 81°30′W. (2001 population excludes incompletely enumerated Indian reserves or settlements.) Composed of LONDON, NORTH MIDDLESEX, SOUTHWEST MIDDLESEX, THAMES CENTRE, ADELAIDE METCALFE, LUCAN BIDDULPH, MIDDLESEX CENTRE, STRATHROY-CARADOC, and NEWBURY.

Middlesex, county, central JAMAICA, between CORNWALL (W) and SURREY counties (E); 17°33′W–18°28′N 76°54′W. Consists of SAINT CATHERINE, SAINT MARY, CLARENDON, SAINT ANN, and MANCHESTER parishes. Set up 1758; no longer has administrative functions.

Middlesex, county (□ 439 sq mi/1,141.4 sq km; 2006 population 163,774), S CONNECTICUT, on LONG IS-

LAND SOUND, bisected by CONNECTICUT River; ☉ MIDDLETOWN; 41°25′N 72°31′W. Agriculture (tobacco, potatoes, produce, fruit; dairy products, poultry); manufacturing (tools, hardware, electrical equipment, boats, textiles, metal products, consumer goods, piano parts, paper and fiber products, clothing, transportation equipment, agricultural machinery, chemicals, asbestos, cigars); fishing; sandstone and feldspar quarries. Resorts on shore. Includes POCOTOPAUG LAKE, several state parks and forests. Drained by Connecticut (E boundary), Hammonasset (W boundary), SALMON, and MATTABESSET rivers. Constituted 1785. Former site of Haddam Neck nuclear power plant, built in the mid-1960s on the banks of the Connecticut River, closed in 1996 for safety reasons. Site heavily contaminated.

Middlesex, county (□ 847 sq mi/2,202.2 sq km; 2006 population 1,467,016), NE MASSACHUSETTS, bordering N on NEW HAMPSHIRE; ☉ CAMBRIDGE and LOWELL; 42°29′N 71°23′W. Intersected by MERRIMACK and NASHUA rivers, and drained by CHARLES, CONCORD, SUDBURY, and ASSABET rivers, which furnish water power. Industrial towns include Lowell, Cambridge, SOMERVILLE, FRAMINGHAM, EVERETT, MALDEN, WALTHAM. Produces shoes, textiles, machinery and other metal products, electronic and computer components, watches, food and wood products, rubber goods, and agricultural produce. Formed 1643.

Middlesex, county (□ 322 sq mi/837.2 sq km; 2006 population 786,971), E NEW JERSEY, bounded E by RARITAN BAY and ARTHUR KILL; ☉ NEW BRUNSWICK; 40°26′N 74°24′W. Industrial, agricultural, rapidly suburbanizing residential area, with extensive clay deposits and allied industries. Oil refineries, ore smelters and refineries; shipyards, drydocks on Raritan Bay. Research and development, financial, healthcare, pharmaceuticals, and service industries. Drained by RARITAN River (navigable) and MILLSTONE and SOUTH rivers. Formed 1675.

Middlesex (MI-duhl-seks), county (□ 210 sq mi/546 sq km; 2006 population 10,615), E VIRGINIA; ☉ SALUDA; 37°36′N 76°30′W. In Tidewater region; bounded N by RAPPAHANNOCK RIVER, S by short Dragon Run and PIANKATANK RIVER, E by CHESAPEAKE BAY. Agriculture (hay, barley, wheat, soybeans, tobacco, corn, melons; cattle, hogs, poultry); timber (pine, oak); fish, oysters. Resort area. Formed 1673.

Middlesex, town (□ 1 sq mi/2.6 sq km; 2006 population 855), NASH county, NE central NORTH CAROLINA, 25 mi/40 km E of RALEIGH; 35°47′N 78°12′W. Manufacturing (transportation equipment, apparel, building materials); agriculture (tobacco, cotton, peanuts, grain; poultry, livestock).

Middlesex, town, WASHINGTON CO., central VERMONT, on Winooski River, just NW of MONTPELIER; 44°18′N 72°38′W.

Middlesex, hamlet, YATES county, W central NEW YORK, near CANANDAIGUA LAKE, 19 mi/31 km SW of GENEVA; 42°43′N 77°17′W. In grape-growing region. Incorporated 1796.

Middlesex, borough (2006 population 13,746), MIDDLESEX county, N central NEW JERSEY; 40°34′N 74°29′W. Diversified manufacturing. Incorporated 1913.

Middlesex Centre (MI-duhl-seks SEN-tuhr), township (□ 227 sq mi/590.2 sq km; 2001 population 14,242), SW ONTARIO, E central CANADA, 12 mi/19 km from LONDON; 43°01′N 81°27′W. Primarily rural. Composed of thirteen hamlets (Arva, Ballymote, Birr, Bryanston, Coldstream, Delaware, Denfield, Ilderton, Kilworth, Komoka, Lobo Village, Melrose, Poplar Hill). Formed in 1998 from DELAWARE, LOBO, and London townships.

Middle Teton, WYOMING: see GRAND TETON NATIONAL PARK.

Middleton, city (2006 population 16,595), DANE county, S WISCONSIN, on W end of LAKE MENDOTA, suburb, 7

mi/11.3 km W of MADISON; 43°06′N 89°30′W. In farming and dairying area; manufacturing (brewery, furniture, textiles, transportation equipment, medical instruments, household furnishings; food and beverage processing, genetic engineering). Morey Airport to NW. Governor Dodge State Park to NE. Incorporated 1905.

Middleton (MI-dul-tun), town, W NOVA SCOTIA, CANADA, on ANNAPOLIS RIVER, and 25 mi/40 km ENE of ANNAPOLIS ROYAL; 44°57′N 65°40′W. Elevation 62 ft/ 18 m.

Middleton (MID-uhl-tuhn), town (2001 population 45,688), GREATER MANCHESTER, NW ENGLAND, on the Irk River; 5 mi/8 km NNE of MANCHESTER; 53°33′N 02°12′W. Manufacturing includes silks, chemicals, plastics, and soap.

Middleton, town, ELBERT county, NE GEORGIA, 6 mi/ 9.7 km ESE of ELBERTON, near SAVANNAH River; 32°05′N 82°46′W.

Middleton, town (2000 population 2,978), CANYON county, SW IDAHO, 5 mi/8 km NE of CALDWELL, and on BOISE RIVER; 43°43′N 116°37′W. In fruit and grain area.

Middleton, town, ESSEX county, NE MASSACHUSETTS, on IPSWICH RIVER, and 8 mi/12.9 km NW of SALEM; 42°36′N 71°01′W. Resort; manufacturing (chemicals). State forest nearby. Settled 1659, organized 1728.

Middleton, town, STRAFFORD county, SE NEW HAMPSHIRE, 12 mi/19 km NNW of ROCHESTER; 43°28′N 71°04′W. Manufacturing (lumber); timber; agriculture (produce, vegetables; cattle; dairying). Sunrise Lake in S.

Middleton, town (2006 population 621), HARDEMAN county, SW TENNESSEE, 40 mi/64 km S of JACKSON; 35°03′N 88°53′W. Major retail and industrial center.

Middleton, VICTORIA, AUSTRALIA: see APOLLO BAY.

Middleton in Teesdale (MID-uhl-tuhn in TEEZ-dail), town (2001 population 1,456), W DURHAM, N ENGLAND, on TEES RIVER, and 9 mi/14.5 km NW of BARNARD CASTLE; 54°37′N 02°05′W. Previously flour milling.

Middleton Island, S ALASKA, in Gulf of ALASKA, 80 mi/ 129 km SSW of CORDOVA; 5 mi/8 km long; 59°26′N 146°20′W.

Middletown, industrial city (2000 population 43,167), MIDDLESEX county, central CONNECTICUT, on the W bank of the CONNECTICUT River; 41°32′N 72°39′W. Manufacturing (transportation equipment, marine hardware, rubber footwear, apparel, computer industry, and textiles). Shipping brought early prosperity to Middletown, and during colonial days it was the state's leading shipping, commercial, and cultural center. Settled 1650; incorporated 1784; town and city consolidated 1923. It is the seat of Wesleyan University and Middlesex Community Technical College. Also in the city are a state mental hospital, a state correctional school, and a state park. A bridge (1938) spans the Connecticut River to PORTLAND.

Middletown, industrial city (□ 4 sq mi/10.4 sq km; 2000 population 25,388), ORANGE county, SE NEW YORK, on the WALLKILL RIVER; 41°27′N 74°25′W. At the intersection of E-W Route I-84, which gives access to and from NEW ENGLAND, the mid-HUDSON RIVER valley and the NEW YORK STATE THRUWAY, and NYS Route 17 (Southern Tier Expressway), which funnels travel between NEW YORK city and the CATSKILLS–Upstate New York region. Middletown serves as the major regional hub for the region S and E of the Catskills, which is also an easy distance away for New York city residents. Summer homes on the numerous natural lakes and ponds and in the mountains and the large number and variety of recreational facilities and activities draw many downstate New Yorkers for extended visits. These people heavily influence the character of the city and its hinterland. Dairying and farming on the extensive mucklands S of the city are still viable industries. Diversified light manufacturing and commercial

services. Seat of Orange County Community College. Rockland Psychiatric Center and Middletown Residential Center (state youth correctional facility) are here. Settled 1756, incorporated as a city 1888.

Middletown (MID-uhl-toun), city (□ 26 sq mi/67.6 sq km; 2000 population 51,605), BUTLER county, SW OHIO, on the GREAT MIAMI RIVER, and 20 mi/32 km N of CINCINNATI; 39°30′N 84°22′W. Previously manufacturing of steel, aircraft parts, and paper products. Major manufacturing employment sector was steel. Miami University has a branch in the city. Incorporated 1866.

Middletown, town (2000 population 6,161), NEW CASTLE county, W DELAWARE, 25 mi/40 km SSW of WILMINGTON, near SILVER LAKE, and 3 mi/4.8 km E of MARYLAND state line; 39°27′N 75°42′W. Elevation 49 ft/14 m. Marketing and shipping center in agricultural area; manufacturing. Nearby is pre-Revolutionary St. Anne's Episcopal Church. Incorporated 1861.

Middletown, town (2000 population 2,488), HENRY county, E central INDIANA, on small Fall Creek, and 13 mi/21 km NNW of NEW CASTLE; 40°04′N 85°32′W. Livestock; grain, tomatoes. Manufacturing (bedding materials). Laid out 1829.

Middletown, town (2000 population 535), DES MOINES county, SE IOWA, 8 mi/12.9 km W of BURLINGTON; 40°49′N 91°15′W. In agricultural area.

Middletown, town (2000 population 5,744), JEFFERSON county, N KENTUCKY, residential, 15 mi/24 km E of downtown LOUISVILLE; 38°14′N 85°31′W. Manufacturing (sheet-metal fabricating).

Middletown, town (2000 population 2,668), FREDERICK county, W MARYLAND, in MIDDLETOWN VALLEY, and 8 mi/12.9 km WNW of FREDERICK; 39°26′N 77°32′W. Trade center in agricultural area (grain; dairy products); manufacturing (shoes). Martenbox Church, Revolutionary War cemetery.

Middletown, town (2000 population 199), MONTGOMERY county, E central MISSOURI, near West Fork of CUIVRE River, 11 mi/18 km NNE of MONTGOMERY CITY; 39°07′N 91°24′W. Agriculture.

Middletown, township, MONMOUTH county, E NEW JERSEY, 4 mi/6.4 km NW of RED BANK; 40°23′N 74°04′W. Suburban community with office and research facilities. Marlpit Hall (c.1684) now a museum. Adjoins Gateway National Recreation Center; first Baptist church in New Jersey (1668) built here. Settled 1665.

Middletown, unincorporated town (2000 population 7,378), NORTHAMPTON county, E PENNSYLVANIA, residential suburb, 2 mi/3.2 km ENE of BETHLEHEM on LEHIGH RIVER; 40°38′N 75°19′W.

Middletown, residential and resort town (2000 population 17,335), NEWPORT county, SE RHODE ISLAND, on RHODE ISLAND (Aquidneck) and NARRAGANSETT BAY; set off from NEWPORT; 41°31′N 71°17′W. Nursery farms. Incorporated 1743. Its name is derived from its location between Newport and Portsmouth. During the American Revolution, MIDDLETOWN was pillaged (1776) by the British.

Middletown (MID-uhl-toun), town (2000 population 1,015), FREDERICK county, N VIRGINIA, in SHENANDOAH valley, 12 mi/19 km SSW of WINCHESTER; 39°01′N 78°16′W. Agriculture (apples, grain; livestock; dairying). Site of Union victory in Civil War on CEDAR CREEK, small N tributary of North Fork of SHENANDOAH RIVER (October 19, 1864). Belle Grove (1794) National Historic Landmark. Lord Fairfax Community College.

Middletown, unincorporated village (2000 population 1,020), Lake county, NW CALIFORNIA, 22 mi/35 km NNE of SANTA ROSA, in a valley of the COAST RANGES, near PUTAH CREEK. Agriculture (grapes, pears, walnuts, oats; cattle); winery. Mineral springs (resorts) nearby.

Middletown, village (2000 population 434), LOGAN county, central ILLINOIS, 20 mi/32 km N of SPRING-

FIELD; 40°06′N 89°35′W. Agriculture (corn, soybeans; cattle, hogs).

Middletown, borough (2006 population 8,858), DAU-PHIN county, S PENNSYLVANIA, 8 mi/12.9 km SE of HARRISBURG, on SUSQUEHANNA RIVER, at mouth of SWATARA CREEK; 40°12′N 76°43′W. Agriculture (grain, soybeans, apples; poultry, livestock; dairying); manufacturing (electronic connectors, power supplies; diesel engines). Penn State Harrisburg, Capital College is here. Harrisburg International Airport on Susquehanna River to W. Olmsted Air Force Base. THREE MILE ISLAND Nuclear Power Plant 3 mi/4.8 km to S. Laid out 1755; incorporated 1828.

Middletown Springs, town, RUTLAND CO., W VER-MONT, on POULTNEY RIVER, and 11 mi/18 km SW of RUTLAND; 43°28′N 73°07′W. Originally called Middletown; changed in 1884 for the mineral springs along the river.

Middletown Valley, FREDERICK county, NW MARY-LAND, fertile agricultural valley, drained by CATOCTIN CREEK; extends N from the POTOMAC between prongs (CATOCTIN MOUNTAIN on E, SOUTH MOUNTAIN on W) of the BLUE RIDGE; c.15 mi/24 km long, 6 mi/9.7 km wide. It includes MIDDLETOWN, BRUNSWICK, Myersville, Berkittville, and Petersville.

Middle Tunguska River, RUSSIA: see STONY TUNGUSKA RIVER.

Middle Upper Rhine (REIN), German *Mittlerer Oberrhein*, region (□ 825 sq mi/2,145 sq km), in BADEN-WÜRTTEMBERG, SW GERMANY, on French border. Borders on FRANCE (SW), RHINELAND-PA-LATINATE (NW), NECKAR JURA (N), FRANCONIA (NE), NORTHERN BLACK FOREST (SE), and SOUTHERN UPPER RHINE (S). Chief town is KARLSRUHE.

Middle Valley, town, HAMILTON county, SE TENNES-SEE, 12 mi/19 km NNE of CHATTANOOGA; 35°11′N 85°11′W.

Middle Veld, SOUTH AFRICA: see VELD.

Middle Village, a residential section of QUEENS borough of NEW YORK city, SE NEW YORK; 40°43′N 73°54′W. Large Italian-American population along with other ethnic groups. St. John's Cemetery holds remains of mobsters Lucky Luciano and John Gotti. Diversified urban economy.

Middleville, town (2000 population 2,721), BARRY county, SW MICHIGAN, 20 mi/32 km SE of GRAND RAPIDS, and on THORNAPPLE RIVER; 42°42′N 85°28′W. In farm area. Manufacturing (clothing, machinery, transportation equipment, metal fabrication). Middleville Ski Area to N; Yankee Springs State Recreational Area to SW.

Middleville, village (2006 population 533), HERKIMER county, central NEW YORK, on W. CANADA CREEK, and 13 mi/21 km ENE of UTICA; 43°08′N 74°58′W. In dairying area. Incorporated 1890.

Middle West, U.S.: see MIDWEST.

Middle West Side, New York: see CLINTON.

Middlewich (MID-uhl-wich), town (2001 population 13,101), central CHESHIRE, W ENGLAND, on DANE RIVER, on Trent and Mersey Canal, and 7 mi/11.3 km N of CREWE; 53°11′N 02°26′W. Salt refining and processing, Cheshire cheese; light manufacturing. Two battles from the Civil War took place here in 1643.

Middle Yuba River, California: see YUBA RIVER.

Mīdega (mee-DAI-gah), village, E central ETHIOPIA, 28 mi/45 km S of HARAR; 08°52′N 42°10′E. In millet growing region. Formerly called Midagalola.

Midelt (MEE-delt), town, Khénifra province, Meknès-Tafilalet administrative region, central MOROCCO, on N slope of the Jbel AYACHI peak (High ATLAS mountains), in upper MOULOUYA river valley; 32°41′N 04°45′W. First stop on one of the main routes through the High Atlas mountains to the desert via ERRA-CHIDIA. Lead mines at Aouli-Miblden (14 mi/23 km NE) and Aït Labbès (SW) are no longer functioning.

Midfield, suburb (2000 population 5,626), Jefferson co., N central Alabama, just W of Birmingham; 33°27′N 87°55′W. Manufacturing of chain-link fencing and glazed blocks.

Mid Glamorgan (mid gla-MOR-guhn), Welsh *Morgannwg Gandl*, county (□ 393 sq mi/1,021.8 sq km), S Wales; 51°30′N 03°27′W. Comprises most of S Wales former coalfield. Redundant coal-mining settlements strung out along valleys lined with slag heaps, between gritstone uplands. PONTYPRIDD and MER-THYR TYDFIL are economic centers.

Midhnab (mid-NAHB), township and oasis (2004 population 24,039), Qassim province of NAJ′D, SAUDI ARABIA, 20 mi/32 km SE of ′UNAIZA; 25°52′N 44°14′E. Trading center; grain (wheat, sorghum), dates, vegetables, fruit; livestock raising.

Midhurst (MID-huhst), market town (2001 population 4,889), West SUSSEX, SE ENGLAND, on ROTHER RIVER, and 10 mi/16 km N of CHICHESTER; 50°59′N 00°44′W. Has grammar school and 15th-century church.

Midi, YEMEN: see MAIDI.

Midia, TURKEY: see MIDYE.

Midiah, TURKEY: see MIDYE.

Midi, Aiguille du, FRANCE: see AIGUILLE DU MIDI.

Midian, SAUDI ARABIA: see MADAN.

Midi, Canal du (mee-dee, kah-nahl dyoo), French waterway in HÉRAULT, AUDE, and HAUTE-GARONNE departments, S and SW FRANCE, extends 150 mi/241 km from the MEDITERRANEAN (SW of sète) to TOULOUSE, whence it is continued NW by the GAR-ONNE LATERAL CANAL to BORDEAUX, thus linking the ATLANTIC with the Mediterranean Sea. Canal crosses a drainage divide at the Col de NAUROUZE by means of numerous locks. Chief towns on it are BÉZIERS, CARCASSONNE, and CASTELNAUDARY. Although built (1666–1681) to carry ocean shipping between the Mediterranean and the ATLANTIC, thus eliminating voyage through Strait of GIBRALTAR, its size and depth limit it to barge traffic. Designated a UNESCO World Heritage site in 1996.

Midi de Bigorre, Pic du, FRANCE: see PIC DU MIDI DE BIGORRE.

Midi, Dent du, SWITZERLAND: see DENTS DU MIDI.

Midi, Pic du, FRANCE: see PIC DU MIDI DE BIGORRE.

Midi-Pyrénées (mee-dee–pee-rai-nai), administrative region (□ 45,349 sq mi/117,907.4 sq km) of S and SW FRANCE; ⊙ TOULOUSE; 43°30′N 01°20′E. Composed of the departments of ARIÈGE, AVEYRON, HAUTE-GAR-ONNE, GERS, LOT, HAUTES-PYRÉNÉES, TARN, and TARN-ET-GARONNE. Within this region, the midvalley of the GARONNE RIVER along with its tributaries (ARIÈGE and GERS, rising in the PYRENEES; TARN, AVEYRON, and LOT, rising in the S part of the MASSIF CENTRAL), constitutes a wide corridor of communication (road, railroad, and water) between the regions of AQUITAINE (W) and LANGUEDOC (E). TOULOUSE, at its geographical center, provides the chief market outlet for a diverse agricultural region and is the center of France's aircraft and aerospace industry. Tourism is important in the Pyrenees (summer and winter resorts) and in the cave region of the S Massif Central.

Midjur (MEE-joor), peak, on the Serbian border, W boundary of BULGARIA; 43°24′N 22°42′E. Also spelled Miozhur.

Midland, county (□ 527 sq mi/1,370.2 sq km; 2006 population 83,792), E central MICHIGAN; 43°38′N 84°23′W; ⊙ MIDLAND. Drained by TITTABAWASSEE (SANFORD LAKE reservoir in N), PINE, and CHIPPEWA rivers. Agriculture (cattle, hogs; corn, wheat, oats, soybean, beans, sugar beets). Oil wells, salt deposits, coal mines. Midland city is a chemical and metallurgical center. Organized 1855.

Midland (MID-land), county (□ 902 sq mi/2,345.2 sq km; 2006 population 124,380), W TEXAS; ⊙ MIDLAND; 31°53′N 102°01′W. On S LLANO ESTACADO. Elevation c.2,500 ft/762 m–3,000 ft/914 m. Drained by tributaries of COLORADO RIVER. Important oil fields (development begun 1950–1951); ranching (cattle, horses),

irrigated agriculture (cotton; alfalfa, pecans). Manufacturing at Midland city, headquarters for many oil companies. Formed 1885.

Midland, city (2000 population 41,685), ⊙ MIDLAND county, central MICHIGAN, 17 mi/27 km W of BAY CITY, in the Saginaw valley, at the confluence of the TIT-TABAWASSEE and CHIPPEWA rivers; 43°37′N 84°13′W. Elevation 629 ft/192 m. Midland owes its development after 1890 to the Dow Chemical Company; corporate headquarters here. Manufacturing includes silicone products, chemicals, magnesium, and plastics; other manufacturing (exhibits, metal cutting machinery; printing). Oil, coal, and salt are found in the area. Chippewa Nature Center; Dow Gardens, original gardens at home of Dr. Herbert H. Dow, founder of Dow Chemical Corporation, and Dow Gardens Library and Center for Arts are in Midland; Saginaw Valley State University at University Center, 12 mi/19 km E. Incorporated in 1887.

Midland (MID-land), city (2006 population 102,073), ⊙ MIDLAND county, W TEXAS, 18 mi/29 km NE of ODESSA, and 110 mi/177 km S of LUBBOCK, on the S edge of the LLANO ESTACADO; 32°01′N 102°05′W. Elevation 2,779 ft/847 m. Midland has prospered partly because of its cattle ranches, but the city's reputation for spectacular wealth and its great spurt in population after 1940 resulted from the drilling of oil. Midland sits in the heart of the Permian Basin "oil patch" and has thus attracted numerous oil-company offices to the city. Prefabricated metal buildings, oil field equipment, transportation equipment, paving materials, gas processing. The busy city continued to increase in terms of population from 1970 into the 1980s but growth slowed by the early 1990s. Midland College (two-year), a symphony orchestra, a planetarium, Permian Basin Petroleum Museum, and Hall of Fame are in the city. Incorporated 1906.

Midland (MID-luhnd), town (□ 11 sq mi/28.6 sq km; 2001 population 16,214), S ONTARIO, E central CA-NADA, on GEORGIAN BAY, NW of TORONTO; 44°45′N 79°54′W. Midland is a port; manufacturing (grain elevators, textiles, cameras, optical goods). The Martyrs' Shrine commemorating the deaths of five Jesuit priests who were among the eight North American martyrs canonized in 1930 is nearby, along with other remembrances of the early colonial period. Other attractions include the world's largest collection of freshwater islands, and Canada's largest collection of outdoor murals. First settled 1840s.

Midland, town (2000 population 473), ALLEGANY county, W MARYLAND, in the ALLEGHENIES, 11 mi/18 km WSW of CUMBERLAND; 39°35′N 78°57′W. Once called Koontz for an early settler, it has always been a mining town. The Thrasher Museum displays horse-drawn carriages and wagons in a former school building.

Midland (MID-luhnd), unincorporated town (2006 population 2,978), CABARRUS county, S central NORTH CAROLINA, 20 mi/32 km E of CHARLOTTE, on ROCKY RIVER; 35°13′N 80°30′W. Manufacturing (apparel, feeds, machining, wood products); agriculture (cotton, grain, soybeans; poultry; livestock; dairying). Incorporated 2000.

Midland, unincorporated town, PIERCE county, W WASHINGTON, residential suburb, 5 mi/8 km S of downtown TACOMA; 47°10′N 122°25′W. Pacific Lutheran University in SW. MCCHORD AIR FORCE BASE and Fort Lewis Military Reservation to SW.

Midland, village (2000 population 253), SEBASTIAN county, W ARKANSAS, 20 mi/32 km S of FORT SMITH, near OKLAHOMA line; 35°05′N 94°20′W. Manufacturing (blasting agents). Sugarloaf Lake to W.

Midland, unincorporated village, RIVERSIDE county, S CALIFORNIA, 20 mi/32 km NNW of BLYTHE, in N part of PALO VERDE VALLEY. Gypsum quarrying and processing. Big Maria Mountains to NE, Palen Mountains to W.

Midland, village (2006 population 267), CLINTON county, SW OHIO, 35 mi/56 km ENE of CINCINNATI; 39°18'N 83°55'W.

Midland, unincorporated village, Chartiers township, WASHINGTON county, SW PENNSYLVANIA, residential suburb, 2 mi/3.2 km W of CANONSBURG; 40°15'N 80°13'W.

Midland, village (2006 population 152), HAAKON county, central SOUTH DAKOTA, 25 mi/40 km NE of KADOKA, and on BAD RIVER; 44°04'N 101°09'W. Livestock; grain.

Midland (MID-luhnd), unincorporated village, FAUQUIER county, N VIRGINIA, 8 mi/13 km SSE of WARRENTON; 38°35'N 77°43'W. Manufacturing (food processing, machining, machinery, concrete products); agriculture (grain, apples, soybeans; livestock).

Midland, borough (2006 population 2,926), BEAVER county, W PENNSYLVANIA, 28 mi/45 km NW of PITTSBURGH, on OHIO RIVER; 40°38'N 80°27'W. Agriculture (dairying); manufacturing (metal recovery, stainless steel finishing, machinery, steel foundry). Nuclear Power Plant at Shipping port 2 mi/3.2 km to E. Settled c.1820.

Midland (MID-luhnd), suburb, SW WESTERN AUSTRALIA state, W AUSTRALIA, 8 mi/13 km ENE of PERTH; 31°54'S 116°00'E. Railroad junction; vineyards, fruit, wheat. Formerly Midland Junction.

Midland Canal, German *Mittelland Kanal*, artificial waterway system of GERMANY, extends E along the N German plain, from the DORTMUND-EMS CANAL to MAGDEBURG, on the ELBE RIVER; c.202 mi/325 km long. Created from two former canals, the Ems-Weser Canal and the Elbe-Weser Canal. An eastward extension of the Midland Canal passes through BERLIN and connects with the ODER RIVER. The system is made up of a series of canals that join parallel N-flowing rivers. The canal facilitates E-W transportation of raw materials and manufacturing goods, with coal and iron constituting a large portion of the barge traffic.

Midland City, town (2000 population 1,703), Dale co., SE ALABAMA, 8 mi/12.9 km NW of Dothan. Manufacturing structural steel. Originally called 'Kennedy's Crossroads,' it was renamed in 1890 due to its location between Pinckard and Grimes on the RR that became known as the Atlantic Coast Line.

Midland Junction, AUSTRALIA: see MIDLAND.

Midland Park, borough (2006 population 6,906), BERGEN county, NE NEW JERSEY, 5 mi/8 km N of PATERSON; 40°59'N 74°08'W. Manufacturing. Incorporated 1894. Largely residential.

Midlands, province (□ 18,983 sq mi/49,355.8 sq km; 2002 population 1,466,331), central and NW ZIMBABWE, ⊙ GWERU. Other major towns include KWEKWE, CHIVHU, and Shurugwi. Province is dominated by high veld in S and center; Mafugabusi Plateau in NW. Part of Chirisa Safari Area in NW, SEBAKWE RECREATIONAL PARK in SE. Mining (chromite, gold, copper, iron ore, coal); agriculture (cotton, corn, wheat, soybeans, sorghum, sunflowers, peanuts, tobacco; cattle, sheep, goats, hogs, poultry; dairying). Formerly called GWELO province.

Midlands (MID-luhndz), region of central ENGLAND, usually considered to include the counties of DERBYSHIRE, LEICESTERSHIRE, NORTHAMPTONSHIRE, NOTTINGHAMSHIRE, STAFFORDSHIRE, WARWICKSHIRE, WEST MIDLANDS, Herefordshire, and Worcestershire. The region is highly industrialized. See BLACK COUNTRY; POTTERIES, THE.

Midlands, SOUTH AFRICA: see EASTERN CAPE.

Midleton (MI-duhl-tuhn), Gaelic *Mainistir na Coran*, town (2006 population 3,934), SE CORK county, SW IRELAND, near NE end of CORK HARBOUR, 13 mi/21 km E of CORK; 51°55'N 08°10'W. Woolen milling, whisky distilling; agricultural market.

Midlothian (MID-lo-thee-uhn), county (□ 137 sq mi/356.2 sq km; 2001 population 80,941), E Scotland; ⊙ DALKEITH; 55°50'N 03°08'W. The principal rivers

are the North Esk and the South Esk, which converge to form the ESK RIVER at Dalkeith. Industries include clothing, glass, high technology; manufacturing (television sets). Former administrative district of LOTHIAN, abolished 1996.

Midlothian (mid-LO-thee-uhn), town (2006 population 14,452), ELLIS county, N TEXAS, 25 mi/40 km SSW of DALLAS; 32°29'N 97°00'W. Railroad junction in rich cotton, grain, cattle area; manufacturing (industrial gases, structural steel, portland cement); limestone. Cedar Hill State Park and Lake Joe Pool reservoir to NW. Settled 1880; incorporated 1898.

Midlothian (mid-LO-thi-an), village (2000 population 14,315), COOK county, NE ILLINOIS, suburb, 16 mi/26 km SW of downtown CHICAGO; 41°37'N 87°43'W. Manufacturing (sportswear, cleaning compounds, water treatment chemicals). Incorporated 1927.

Midlothian (mid-LO-thee-uhn), unincorporated village, CHESTERFIELD county, E central VIRGINIA, suburb, 11 mi/18 km WSW of RICHMOND; 37°30'N 77°38'W. Manufacturing (crushed granite, machinery, lumber, concrete, commercial printing, business forms); agriculture (tobacco, grain; livestock). Granite quarrying. Pocahontas State Forest and Park to SE, Swift Creek Reservoir to S. John Tyler Community College (Midlothian Campus).

Midmar Dam, SOUTH AFRICA: see HOWICK.

Midnapore, India: see MEDINIPUR.

Midongy-Atsimo (mee-DOONG-guh–ah-TSEEM), town, FIANARANTSOA province, SE MADAGASCAR, 150 mi/241 km S of FIANARANTSOA; 23°34'S 47°00'E. Livestock raising; rice, coffee, sweet potatoes. Hospital. Formerly Midongy-du-Sud.

Midori (MEE-do-ree), town, Takata county, HIROSHIMA prefecture, SW HONSHU, W JAPAN, 25 mi/40 km N of HIROSHIMA; 34°43'N 132°37'E.

Midori, town, Mihara district, S central AWAJI-SHIMA island, HYOGO prefecture, W central JAPAN, 32 mi/52 km S of KOBE; 34°19'N 134°49'E. Onions, Chinese cabbage, rice. Tile.

Midouze River (mee-dooz), 27 mi/43 km long, LANDES department, AQUITAINE region, SW FRANCE; formed at MONT-DE-MARSAN by confluence of two small streams, the DOUZE (right) and the Midou (left); flows SW to the ADOUR 4 mi/6.4 km below TARTAS; 43°50'N 00°51'W. Partially navigable; timber floating.

Midsayap (mee-sah-YAHP), town, NORTH COTABATO province, central MINDANAO, PHILIPPINES, 20 mi/32 km E of COTABATO; 07°07'N 124°30'E. At edge of a marshy area in lower PULANGI valley; rice, coconuts.

Midsk, UKRAINE: see MYDS'K.

Midsomer Norton, ENGLAND: see NORTON-RADSTOCK.

Midu (MEE-DU), town, ⊙ Midu county, NW central YUNNAN province, CHINA, 35 mi/56 km SE of DALI, near road to MYANMAR; 25°21'N 100°32'E. Rice, wheat, millet, beans, tobacco; engineering, food processing, coal mining.

Midūn (mee-DON), village (2004 population 30,481), MADINIYINA province, SE TUNISIA, on JARBAH ISLAND, 9 mi/14.5 km SE of HAWMAT AS-SUQ, in rich fruit growing area (orchards, olive trees, date palms).

Midvale, city (2006 population 27,249), SALT LAKE county, N UTAH, suburb 10 mi/16 km S of SALT LAKE CITY, and on JORDAN RIVER; 40°36'N 111°54'W. In mining area (lead, zinc, copper, gold, silver); has large smelter. Alfalfa, sugar beets; dairying; cattle, sheep; manufacturing (men's outerwear, computer platforms and software, medical devices, circuit boards, soap, catheters; food processing). Railroad junction. Elevation 4,390 ft/1,338 m. Settled 1859, known as Brigham Junction and East Jordan, named Midvale and incorporated 1909.

Midvale, village (2000 population 176), WASHINGTON county, W IDAHO, 18 mi/29 km NNE of WEISER, and on WEISER RIVER; 44°28'N 116°44'W. Center of agricultural area. Crane Creek Reservoir to SE, Mann Creek Reservoir to SW; Payette National Forest to NW.

Midvale, village in WANAQUE borough, PASSAIC county, NE NEW JERSEY, on WANAQUE RESERVOIR, and 11 mi/18 km NW of PATERSON; 41°03'N 74°17'W. Largely residential.

Midvale (MID-vail), village (2006 population 590), TUSCARAWAS county, E OHIO, 5 mi/8 km SE of NEW PHILADELPHIA, and at junction of TUSCARAWAS RIVER and STILLWATER CREEK; 40°26'N 81°22'W.

Midvale, Pennsylvania: see PLAINS.

Midville, village (2000 population 457), BURKE county, E GEORGIA, 23 mi/37 km SSW of WAYNESBORO, and on OGEECHEE RIVER; 32°49'N 82°14'W. Steel fabrication and light industry.

Midwar, El, JORDAN: see MADWAR, EL.

Midway, town (2000 population 457), Bullock co., SE Alabama, 12 mi/19 km SE of Union Springs. Named because of its location midway between Union Springs and Clayton in Barbour Co. Inc. in 1871.

Midway, town (2000 population 1,100), LIBERTY county, SE GEORGIA, 27 mi/43 km SW of SAVANNAH; 31°48'N 81°26'W. Manufacturing of trailer panels, paper products, and textiles.

Midway, town (2000 population 1,620), WOODFORD county, central KENTUCKY, near ELKHORN CREEK, 12 mi/19 km WNW of LEXINGTON, in BLUEGRASS REGION; 38°08'N 84°40'W. Agriculture (burley tobacco, grain; cattle, horses); manufacturing (corn meal, flour). Midway College. Weisenberger Mill (1862), on Elkhorn Creek to N. Established 1833.

Midway, unincorporated town, LA SALLE parish, LOUISIANA, 2 mi/3.2 km W of JENA; 31°43'N 92°09'W. In agricultural area (cotton, soybeans; cattle); timber.

Midway, unincorporated town (2000 population 2,323), Conewago township, ADAMS county, S PENNSYLVANIA, residential suburb, 1 mi/1.6 km NW of HANOVER; 39°47'N 77°00'W.

Midway, town (2006 population 3,117), WASATCH county, N central UTAH, 5 mi/8 km W of HEBER CITY, and on PROVO RIVER; 40°30'N 111°28'W. Dairying; limestone. Wasatch Mountain State Park to W; Uinta National Forest to W. Deer Creek Reservoir and Midway Fish Hatchery to S. Elevation 5,567 ft/1,697 m. Settled in 1866 by Swiss.

Midway (MID-wai), village (□ 5 sq mi/13 sq km; 2001 population 638), S BRITISH COLUMBIA, W CANADA, on U.S. (WASHINGTON) border, on KETTLE RIVER, at mouth of Boundary Creek, 14 mi/23 km W of GRAND FORKS, in KOOTENAY BOUNDARY regional district; 49°02'N 118°45'W. Elevation 2,000 ft/610 m. Fruit, vegetables; forestry; farming; tourism, Mount Baldy ski resort. Established 1893.

Midway, village (2006 population 263), MADISON county, central OHIO, 10 mi/16 km S of LONDON; 39°44'N 83°29'W. Also called Sedalia.

Midway (MID-wai), village (2006 population 300), MADISON county, E central TEXAS, 26 mi/42 km SW of CROCKETT; 31°01'N 95°45'W. Livestock area (cattle, horses, hogs); oil and gas.

Midway, borough (2006 population 940), WASHINGTON county, SW PENNSYLVANIA, 15 mi/24 km WSW of PITTSBURGH, on Robinson Run; 40°22'N 80°17'W. Agriculture (corn, hay; dairying); coal, natural gas. Hillman State Park to NW; Cherry Valley Reservoir to SW.

Midway, island group (□ 2 sq mi/5.2 sq km), central PACIFIC, c.1,150 mi/1,851 km NW of HONOLULU, comprising Sand and Eastern islands with the surrounding atoll. It is an inactive U.S. military base with no indigenous population. Discovered by Americans in 1859, Midway was annexed in 1867. A cable station opened in 1903. Midway became a commercial air station of Pan American Airways in 1935, and a U.S. naval base opened here in 1941. The Battle of Midway (June 3–6, 1942), one of the decisive Allied victories of World War II, occurred nearby. The battle, fought mostly with aircraft, resulted in the destruction of four Japanese aircraft carriers, crippling the Japanese navy. The islands are now administered by the U.S. Department of the Interior. Known for its population of Laysan albatrosses, or "gooney" birds.

Midway City, unincorporated village, ORANGE COUNTY, S CALIFORNIA; suburb 26 mi/42 km SSE of downtown LOS ANGELES, and 10 mi/16 km E of LONG BEACH. Pacific Ocean 4 mi/6.4 km SW.

Midway Park (MID-wai PAHRK), unincorporated village, ONSLOW COUNTY, E NORTH CAROLINA, residential suburb 5 mi/8 km SE of downtown JACKSONVILLE, on N boundary of CAMP LEJEUNE Marine base; 34°43′N 77°20′W.

Midway Village, Virginia: see STEELES TAVERN.

Midwest, village (2006 population 428), NATRONA county, central WYOMING, on SALT CREEK, and 40 mi/64 km N CASPER; 43°24′N 106°16′W. Elevation 4,820 ft/1,469 m. Oil wells. TEAPOT DOME to SSE.

Midwest, region of the United States centered on the W GREAT LAKES and the upper-middle Mississippi valley. It is a somewhat imprecise term that has been applied to the N section of the land between the APPALACHIANS and the ROCKY MOUNTAINS. More often it is restricted to the Old NW TERRITORY and the neighboring states to the S border of MISSOURI, E of the GREAT PLAINS. Also called Middle West, it thus includes OHIO, INDIANA, ILLINOIS, MICHIGAN, WISCONSIN, MINNESOTA, IOWA, MISSOURI, KANSAS, and NEBRASKA. The area has some of the richest farmland in the world and is known for its corn and cattle. The extended area also includes great wheat fields, particularly W of the MISSOURI River. The heavily industrialized parts of the Midwest known as the RUSTBELT declined in the 1970s and 1980s but started on the road to recovery in the 1990s, as industries modernized on the basis of high technology; financial and other services increased in importance. The chief cities are CHICAGO, DETROIT, SAINT LOUIS, MILWAUKEE, and MINNEAPOLIS-SAINT PAUL.

Midwest City, city (2006 population 55,161), OKLAHOMA county, central OKLAHOMA, a suburb, 9 mi/14.5 km E of downtown OKLAHOMA CITY; 35°27′N 97°22′W. Manufacturing (sporting goods, transportation equipment, sheet-metal products, apparel, silica sand, crushed granite products; publishing and printing). Rose State College (two-year). The developer and builder W. P. Atkinson planned the city as a model for an area of spacious parks and curved streets. Founded 1942 with the activation of adjoining Tinker Air Force Base (to S), a logistics center.

Midwolda (MIT-vawl-duh), village, GRONINGEN province, NE NETHERLANDS, 4 mi/6.4 km NNW of WINSCHOTEN; 53°12′N 07°01′E. Agriculture (grain, vegetables); dairying; livestock.

Midyan, SAUDI ARABIA: see MADIAN.

Midyat, town (2000 population 56,669), SE TURKEY, 36 mi/58 km ENE of MARDIN; 37°25′N 41°20′E. Wheat, barley, vetch, lentils, pistachios, grapes. Known for its silver objects (telkari).

Midye, ancient *Salmydessus*, village, European TURKEY, port on BLACK SEA, 60 mi/97 km WNW of ISTANBUL; 41°36′N 28°06′E. Grain; timber. Also spelled Midia and Midiah.

Midzhor (MEE-jhoor), Serbian *Midžor*, highest peak (7,113 ft/2,168 m) in W STARA PLANINA mountains, on BULGARIA-SERBIA border, in CHIPROVSKA MOUNTAINS, 18 mi/29 km N of PIROT (Serbia).

Mie (MEE-e), prefecture (□ 2,230 sq mi/5,776 sq km; 1990 population 1,792,542), S HONSHU, central JAPAN, on ISE BAY (E); ⊙ TSU. Bordered N by SHIGA prefecture, NE by GIFU and AICHI prefectures, SW by WAKAYAMA prefecture, and W by NARA prefecture. Other centers include ISE (a major Shintoist center), YOKKAICHI, MATSUSAKA, and KUWANA. Traditional industries here include textiles, ceramics, woodworking, and fishing, in addition to modern industries such as chemicals, petroleum, iron, and steel. Ise-shima National Park and Ise Shrine are here.

Mie (MEE-e), town, Ono county, OITA prefecture, E KYUSHU, SW JAPAN, 19 mi/30 km S of OITA; 32°58′N 131°35′E. Shiitake mushrooms.

Miechow (MEE-hoov), Polish *Miechów*, town, Kielce province, S POLAND, 20 mi/32 km N of CRACOW. Railroad junction; manufacturing of agricultural machinery, tanning, flour milling.

Miechowice (mee-ho-VEE-tse), German *Mechtal*, commune in UPPER SILESIA, after 1945 in Katowice province, S POLAND, 3 mi/4.8 km W of BYTOM (Beuthen). Mining (coal, zinc, lead). Until 1936, called Miechowitz.

Miedwie Lake (MEED-vee), German *Madü*, (□ 14 sq mi/36.4 sq km), in POMERANIA, after 1945 in NW POLAND, 5 mi/8 km W of STARGARD SZCZECINSKI; 9 mi/14.5 km long, 2 mi/3.2 km wide; 53°17′N 14°54′E. Drained NW by PLONA RIVER.

Miedzyborz (mee-ZEE-bosh), Polish *Miedzybórz*, German *Neumittelwalde* (noo-meet-vahl-de), town, Kalisz province, SW POLAND, in LOWER SILESIA region, 35 mi/56 km NE of WROCŁAW (Breslau). Agricultural market (grain, potatoes; livestock). Was (1919–1939) German frontier station on Poland border.

Miedzychod (mee-JE-hood), Polish *Miedzychód*, German *Birnbaum* (beern-bahm), town, Gorzów province, W POLAND, on WARTA RIVER, and 45 mi/72 km WNW of POZNAŃ. Railroad junction; manufacturing of machinery, bricks; canning, brewing.

Miedzylesie (mee-je-LE-see), Polish *Miedzylesie*, German *Mittelwalde*, town, Wałbrzych province, SW POLAND, in LOWER SILESIA region, at S foot of HABELSCHWERDT MOUNTAINS, on the GLATZER NEISSE river, and 20 mi/32 km S of KŁODZKO (Glatz). Frontier station on CZECH REPUBLIC border. Silk milling. Has 16th-century castle.

Miedzyrzec (mee-JE-zets), Polish *Migdzyrzec Podlaski*, Russian *Mezhireche* or *Mezhirech'e*, town (1993 estimated population 17,800), Biała Podlaska province, E POLAND, on KRZNA RIVER, on railroad, and 55 mi/89 km N of LUBLIN. Trade center (grain, hides); manufacturing of household products, agricultural implements, tanning, flour milling, distilling. Before World War II, population 75% Jewish, dating from 18th century; established a brush factory whose products were marketed throughout RUSSIA; also the community produced famous religious leaders.

Miedzyrzec, UKRAINE: see VELYKI MEZHYRICHI.

Miedzyrzecz (mee-JE-zets), Polish *Migdzyrzecz*, German *Meseritz*, town (1993 estimated population 20,500) in BRANDENBURG, Gorzów province, W POLAND, on OBRA RIVER, and 25 mi/40 km SE of GORZÓW WIELKOPOLSKI (Landsberg); 52°26′N 15°35′E. Grain, livestock, and lumber market; distilling; oil and flour mills. Has 13th century church, remains of medieval town walls. First mentioned as site of abbey in 1005; chartered 1485; destroyed 1476 by Hungarians. In World War II, c.40% destroyed.

Miédzyrzecz, UKRAINE: see MEZHYRICH.

Miedzyzdroje (mee-see-ZDRO-ee), Polish *Międzyzdroje*, German *Misdroy* (meez-dro-ye), town in POMERANIA, after 1945 in Szczecin province, NW POLAND, on N shore of WOLIN island, 10 mi/16 km E of SWINEMUNDE; 53°56′N 14°27′E. Seaside resort. Chartered after 1945.

Miejska Gorka (MEE-skah GOOR-kah), Polish *Miejska Górka* [town hill], German *Görchen*, town, Leszno province, W POLAND, 39 mi/63 km N of WROCŁAW (Breslau); 51°39′N 16°57′E. Railroad junction; manufacturing (bricks, cement, beet sugar). Has cloister.

Miélan (myai-lahn), commune (□ 8 sq mi/20.8 sq km), GERS department, MIDI-PYRÉNÉES region, SW FRANCE, 7 mi/11.3 km SW of MIRANDE, in the N outliers of the PYRENEES; 43°26′N 00°19′E. Hog market; vineyards in the upper ARMAGNAC (brandy-producing) region. A nearby reservoir, in a widening of the

Osse River valley, offers panoramic views of the crest of the Pyrenees.

Mielau, POLAND: see MLAWA.

Mielec (MEE-lets), town (2002 population 61,728), Rzeszów province, SE POLAND, on WISŁOKA River, on railroad, and 31 mi/50 km NW of RZESZÓW; 50°17′N 21°25′E. Airport. Manufacturing of ceramics, perfume; lumbering; flour milling; tannery; oil deposits nearby. Former aircraft manufacturing center.

Mielnica, UKRAINE: see MEL'NYTSYA-PODIL'S'KA.

Mi'elya (mi-EL-yah), Arab village, ISRAEL, 8.8 mi/14 km ENE of NAHARIYA; 33°01′N 35°15′E. Elevation 1,410 ft/429 m. Mixed farming, olive growing. Population is mostly Christian; the only Catholic settlement in Israel. Remains of a Crusader fort, Chateau de Roi, built in the 12th century. Originally owned by the Crusader king, the fortress was later acquired by several noble families who sold it to the Teutonic Knights. Mameluke Sultan Babers conquered it in 1265.

Miena (mei-EE-nuh), village, central TASMANIA, SE AUSTRALIA, 65 mi/105 km NW of HOBART, and on S shore of GREAT LAKE; 41°59′S 146°44′E. Fishing, bushwalking. Site of dam for WADDAMANA hydroelectric plant.

Mien Ba Chua Xu, temple, AN GIANG province, S VIETNAM; 16°49′N 106°39′E. Near CHAU DOC city, the Temple of Lady Chua Xu was built in the 1820s and was refurbished in 1972. It stands opposite SAM MOUNTAIN. Tourist area.

Miepol, AUSTRALIA: see MIEPOLL.

Miepoll, farming district, VICTORIA, SE AUSTRALIA, N of EUROA, between MURCHISON and VIOLET TOWN; 36°37′S 145°28′E. Also spelled Miepol.

Mier (MEE-er), city and township, TAMAULIPAS, N MEXICO, near RIO GRANDE, S of Falcon Dam, and 90 mi/145 km NE of MONTERREY on Mexican Highway 2; 26°28′N 99°10′W. Agricultural center (cotton, sugarcane, corn; livestock).

Miera River (MYAI-rah), 27 mi/43 km long, CANTABRIA province, N SPAIN; rises in CANTABRIAN MOUNTAINS; flows N to inlet of BAY OF BISCAY opposite SANTANDER.

Miercurea (MYER-koo-ra), Hungarian *Szerdahely*, village, SIBIU county, central ROMANIA, on railroad, and 15 mi/24 km NW of SIBIU; 45°53′N 23°48′E. Agricultural center. Also spelled Mercurea and sometimes called Miercurea-Sibiu or Miercurea-Sibiului.

Miercurea-Ciuc (MYER-koo-rah–CHYOOK), Hungarian *Csíkszereda*, city; ⊙ HARGHITA county, E central ROMANIA, in TRANSYLVANIA, in W foothills of the Moldavian CARPATHIANS, on OLT RIVER, on railroad, and 52 mi/84 km NNE of BRAȘOV; 46°21′N 25°48′E. Trading center for lumber; flour milling, sawmilling, clay extraction; manufacturing (vinegar). Has old citadel restored in 18th century. In HUNGARY, 1940–1945. Pilgrimage center of Șimuleu, with 13th-century statue of the Virgin, an old Franciscan monastery, and 16th-century printing works, is 2 mi/3 km NE. Also spelled Mercurea Ciuc.

Miercurea-Niraj (MYER-koo-rah-nee-RAHZH), Hungarian *Nyárádszereda*, village, MUREȘ county, central ROMANIA, 10 mi/16 km E of TÎRGU MURES; 46°32′N 24°48′E. Natural gas production. In HUNGARY, 1940–1945. Also spelled Mercurea-Niraj; sometimes called Miercurea-Nirajului.

Mieres (mee-AI-res), city, OVIEDO province, N SPAIN, in ASTURIAS, on the Lena River. It is an important mining center for coal, sulfur, and cinnabar. Has iron and steel plants.

Mieroszow (mee-RO-shoov), Polish *Mieroszów*, German *Friedland*, town in LOWER SILESIA, Wałbrzych province, SW POLAND, at N foot of the SUDETES, 9 mi/14.5 km SSW of WAŁBRZYCH (Waldnburg). Textiles, furniture manufacturing. Frontier station on CZECH border. After 1945, briefly called Frydland.

Mier y Noriega (mee-er ee no-ree-AI-gah), town, NUEVO LEÓN, N MEXICO, in SIERRA MADRE ORIENTAL,

Cross-references are shown in SMALL CAPITALS. The pronunciation guide is shown on page xix. The sources of population figures are shown on page xvii.

37 mi/60 km ESE of Matehuala (ZACATECAS); 23°25′N 100°07′W. Elevation 5,515 ft/1,681 m. Grain; livestock; lumbering.

Mies, CZECH REPUBLIC: see STRIBRO.

Miesbach (MEES-bahkh), town, UPPER BAVARIA, GERMANY, in Bavarian Alps, 13 mi/21 km WSW of ROSENHEIM; 47°47′N 11°49′E. Textile manufacturing; brewing. Summer resort (elevation 2,287 ft/697 m).

Mieso (mi-AI-so), town (2007 population 10,820), OROMIYA state, E central ETHIOPIA; 09°14′N 40°45′E. In the GREAT RIFT VALLEY, on Addis Ababa–Djibouti railroad, and 50 mi/80 km WSW of DIRE DAWA. Road junction. Formerly called Meheso.

Mies River, CZECH REPUBLIC: see MZE RIVER.

Miesville (MEES-vil), village (2000 population 135), DAKOTA county, SE MINNESOTA, 9 mi/14.5 km S of HASTINGS, in Richard J. Dorer Memorial Hardwood State Forest; 44°36′N 92°48′W. Agriculture (grain; livestock; dairying). MISSISSIPPI RIVER to NE.

Mietingen (MEE-ting-uhn), village, Danube-Iller, SE BADEN-WÜRTTEMBERG, S GERMANY, 15 mi/24 km SSW of ULM; 48°11′N 09°54′E. Forestry; metalworking.

Mieza (MYAI-thah), village, SALAMANCA province, W SPAIN, near DUERO RIVER, 57 mi/92 km WNW of SALAMANCA; 41°10′N 06°41′W. Olive-oil processing, flour milling; ships olives, wine, potatoes.

Mifflin, county (☐ 413 sq mi/1,073.8 sq km; 2006 population 46,057), central PENNSYLVANIA; ☉ LEWISTOWN; 40°36′N 77°37′W. Stone Mountain ridge in NW boundary, Long Mountain ridge on N boundary, BLUE MOUNTAIN ridge on SE boundary; bisected NE-SW by JACKS MOUNTAIN ridge (2 mi/3.2 km wide) in center of county; drained by JUNIATA RIVER (forms part of SW boundary); also by Kishacoquilla Creek. Agriculture region, (corn, wheat, oats, hay, alfalfa; sheep, hogs, cattle, poultry and eggs, dairying); limestone, sand. Manufacturing at Lewistown. Reeds Gap State Park in NE center; part of Rothrock State Forest in NW, part of Tuscarora State Forest is S, part of Bald Eagle State Forest in NE. Formed 1789.

Mifflin, village (2006 population 145), ASHLAND county, N central OHIO, 8 mi/13 km E of MANSFIELD; 40°46′N 82°22′W. Nearby on Black Fork of MOHICAN RIVER is Charles Mill Reservoir (capacity 88,000 acre-ft), built for flood control.

Mifflin, borough (2006 population 623), Juniata county, central PENNSYLVANIA, 9 mi/14.5 km ESE of LEWISTOWN, on JUNIATA RIVER opposite (W of) MIFFLINTOWN; 40°34′N 77°24′W. Agriculture (corn, hay, apples; livestock, poultry; dairying); timber; light manufacturing. Part of Tuscarora State Forest and BLUE MOUNTAIN ridge to NW.

Mifflinburg, borough (2006 population 3,568), UNION county, central PENNSYLVANIA, 9 mi/14.5 km WSW of LEWISBURG, near Buffalo Creek; 40°55′N 77°02′W. Agriculture (poultry; dairying); timber; manufacturing (yarn, apparel, mobile and modular homes, furniture). Bald Eagle State Forest to SW and NW. Laid out 1792; incorporated 1827.

Mifflintown, borough (2006 population 842), ☉ JUNIATA county, central PENNSYLVANIA, 10 mi/16 km ESE of LEWISTOWN, on JUNIATA RIVER, opposite (E) of MIFFLIN; 40°34′N 77°24′W. Agriculture (corn, hay, apples; poultry, livestock; dairying); manufacturing (wooden products, picnic tables, food processing, apparel; printing and publishing; limestone and shale quarries. Mifflintown Airport to N. Part of Tuscarora State Forest and BLUE MOUNTAIN ridge to NW. Laid out 1791, incorporated 1833.

Mifflinville, unincorporated town (2000 population 1,213), COLUMBIA county, E central PENNSYLVANIA, 3 mi/4.8 km SW of BERWICK on SUSQUEHANNA RIVER; 41°01′N 76°17′W. Manufacturing (food products, sprinkler systems, swimming pools); agriculture (dairying; livestock, poultry; corn, hay, potatoes, apples).

Mifi (MEE-fee), department (2001 population 290,758), WEST province, CAMEROON; ☉ BAFOUSSAM.

Mifune (mee-FOO-ne), town, Kamimashiki county, KUMAMOTO prefecture, W KYUSHU, SW JAPAN, 6 mi/10 km S of KUMAMOTO; 32°42′N 130°48′E. Tea.

Migahatenna (MEE-gah-hah-tan-nuh), village, WESTERN PROVINCE, SRI LANKA, 18 mi/29 km SE of KALUTARA. Graphite-mining center; vegetables, rice, rubber. Sometimes spelled MEGAHATENNA.

Migdal (meeg-DAHL), Jewish village, LOWER GALILEE, NE ISRAEL, near NW shore of SEA OF GALILEE, 4 mi/6.4 km NNW of TIBERIAS; 32°49′N 35°30′E. Elevation 643 ft/195 m below sea level. Modern village founded 1909 on site of village of same name that was known at time of Second Temple.

Migdal-el (meeg-DAHL–EL), ancient fortified town, NE LOWER GALILEE, N ISRAEL. Believed to be the ruins of Khirbet al Majdal, W of TIBERIAS. Most prolific in Second Temple and later as fishing center.

Migdal-gad (meeg-DAHL–GAHD), in the Hebrew Bible, town, S ISRAEL, c.12 mi/19 km SE of BEIT GUVRIN, the present Tell al Majadil.

Migennes (mee-zhen), market town (☐ 6 sq mi/15.6 sq km), YONNE department, in BURGUNDY, N central FRANCE, near junction of YONNE RIVER and BURGUNDY CANAL, 13 mi/21 km N of AUXERRE; 47°58′N 03°31′E. Transshipment of wood and building materials. Railroad station and yards at Laroche-Migennes (1 mi/1.6 km SW).

Migeregere (mee-gai-rai-GAI-rai), village, LINDI region, SE TANZANIA, 14 mi/23 km WNW of KILWA MASOKO, near INDIAN OCEAN; 08°52′S 39°20′E. Road junction. Cashews, sisal, bananas, sweet potatoes; goats, sheep.

Migir, TURKEY: see MIGIR TEPE.

Migir Tepe, peak (7,418 ft/2,261 m), in AMANOS MOUNTAINS, S TURKEY, 10 mi/16 km E of DORTYOL. Also called Migir.

Migiurtinia, SOMALIA: see MIJIRTEIN.

Migné-Auxances (mee-nyai–og-zahns), town (☐ 10 sq mi/26 sq km), N suburb of Poitiers, VIENNE department, POITOU-CHARENTES region, W central FRANCE; 46°38′N 00°19′E. Contains modern industrial park and expansion area. Nearby is the theme park and technology center, known as Le Futuroscope (see POITIERS).

Mignon (MIN-yon), village (2000 population 1,348), TALLADEGA county, E central ALABAMA, 20 mi/32 km SW of TALLADEGA.

Migori (mee-GO-ree), town (1999 population 31,644), ☉ Migori district, NYANZA province, KENYA, on road 31 mi/50 km SW of KISII; 01°03′S 34°28′E. Market and trading center; sugarcane, maize, finger millet. Gold mining.

Miguasha National Park, SE QUEBEC, E CANADA, on S coast of GASPÉ PENINSULA. With its preserved specimens of lobe-finned fishes, the site is considered the world's most outstanding illustration of the Devonian Period (370 million years ago) as the "Age of Fishes"; designated a UNESCO World Heritage site in 1999.

Miguel Alemán, MEXICO: see CIUDAD MIGUEL ALEMÁN.

Miguel Alves (MEE-gel AHL-ves), city (2007 population 31,985), N PIAUÍ, BRAZIL, landing on right bank of PARNAÍBA River (MARANHÃO border), and 60 mi/97 km N of TERESINA; 04°10′S 42°54′W. Ships babassu nuts, tobacco, manioc.

Miguel Auza (mee-GEL AH-oo-sah), town, ☉ Miguel Auza municipio, ZACATECAS, N central MEXICO, on interior plateau, near DURANGO border, 3 mi/5 km W of JUAN ALDAMA; 24°20′N 103°30′W. Elevation 6,683 ft/2,037 m. Silver mining; livestock raising. Also known as San Miguel de Mezquital.

Miguel Calmon (mee-gel kahl-MON), city, (2007 population 27,225), E central BAHIA, BRAZIL, on railroad, and 70 mi/113 km SSW of SENHOR DO BONFIM; 11°26′S 40°34′W. Cattle; sugar, tobacco, coffee; manganese mines.

Miguel de la Borda (MEE-gel dai lah BOR-dah), town, ☉ Donoso District, COLÓN province, central PANAMA, on CARIBBEAN SEA, 28 mi/45 km SW of COLÓN; 09°09′N 80°19′W. Road terminus; corn, coconuts; livestock. Also called Donoso.

Miguel Esteban (es-TAI-vahn), town, TOLEDO province, central SPAIN, in upper LA MANCHA, 45 mi/72 km SE of ARANJUEZ; 39°31′N 03°05′W. Agriculture (cereals, grapes, potatoes; sheep); industries (sawmilling, plaster).

Miguel Hidalgo (mee-GEL hee-DAHL-go), delegación, in the W part of Distrito Federal, MEXICO, 10 mi/16 km NW of TLALPAN. Created in 1970. Part of the MEXICO CITY metropolitan area.

Miguel Hidalgo y Costilla, MEXICO: see ACUAMANALA.

Miguel Leão (MEE-gel LAI-oun), town (2007 population 1,192), W PIAUÍ state, BRAZIL, 25 mi/41 km S of TERESINA; 05°42′S 42°49′W. Cotton, manioc.

Miguel Pereira (MEE-gel PE-rai-rah), city (2007 population 24,644), W central RIO DE JANEIRO state, BRAZIL, in the Serra do MAR (here called "Brazilian Switzerland"), 30 mi/48 km NW of RIO; 22°27′S 43°22′W. Summer resort.

Miguelturra (mee-gel-TOO-rah), town, CIUDAD REAL province, S central SPAIN, on railroad, 2 mi/3.2 km E of CIUDAD REAL; 38°58′N 03°53′W. Processing and agricultural center (potatoes, cereals, grapes, olives; livestock). Olive-oil pressing, alcohol and liquor distilling, tanning, textile milling, soapmaking. Resort with mineral springs.

Migues (MEE-ges) town, CANELONES department, S URUGUAY, 40 mi/64 km NE of MONTEVIDEO; 34°29′S 55°39′W. Grain, livestock. Railroad station is 3 mi/4.8 km S. Sometimes spelled Miguez.

Migulinskaya (mee-GOO-leen-skah-yah), village, N ROSTOV oblast, S European Russia, on the DON River, terminus of local highway branch, 10 mi/16 km SE of KAZANSKAYA; 49°41′N 41°15′E. Elevation 223 ft/67 m. Agricultural products. Population largely Cossack.

Migyaungye (MEE-joun-JEE), village, MAGWE township, MAGWE division, MYANMAR, on W bank of AYEYARWADY RIVER, and 20 mi/32 km SSE of MAGWE.

Mihăileni (mee-huh-ee-LEN), agricultural village, BOTOŞANI county, NE ROMANIA, on UKRAINE border, 14 mi/23 km W of DOROHOI.

Mihăileşti (mee-huh-ee-LESHT), town, GIURGIU county, SE central ROMANIA, 12 mi/19 km SW of BUCHAREST.

Mihailovgrad Dam (mee-HEI-lov-grahd), dam, NW BULGARIA; 43°24′N 23°12′E.

Mihailovo (mee-HEI-lo-vo). village, HASKOVO oblast, STARA ZAGORA obshtina, S central BULGARIA, 11 mi/18 km ENE of CHIRPAN; 42°15′N 25°32′E. Railroad junction; tobacco, cotton, vineyards. Sometimes spelled Mikhailovo or Mikhaylovo. Formerly known as Gokpala.

Mihailovo (mee-KHAI-lo-vo), village, MONTANA oblast, HAIREDIN obshtina, BULGARIA; 43°34′N 23°36′E.

Mihalic, TURKEY: see KARACABEY.

Mihaliccik, Turkish *Mihalicçik*, village, W central TURKEY, 50 mi/80 km E of ESKISEHIR; 39°52′N 31°30′E. Rich deposits of potter's clay; grain, onions, tobacco.

Mihalkovo (mee-hahl-KO-vo), village, PLOVDIV oblast, DEVIN obshtina, BULGARIA; 41°51′N 24°27′E. Resort, naturally carbonated mineral spa.

Mihaltsi (mee-HAHL-tsee), village, LOVECH oblast, PAVLIKENI obshtina, N BULGARIA, 4 mi/6 km SE of Pavlikeni; 43°11′N 25°22′E. Grain, vineyards, produce. Sometimes spelled Mikhaltsi.

Mihama (mee-HAH-mah), town, Chita county, AICHI prefecture, S central HONSHU, central JAPAN, 28 mi/45 km S of NAGOYA; 34°46′N 136°54′E. Nori.

Mihama, town, Mikata county, FUKUI prefecture, central HONSHU, W central JAPAN, 37 mi/60 km S of FUKUI; 35°36′N 135°56′E. Mikata Lakes nearby.

Mihama, town, S Muro county, MIE prefecture, S HONSHU, central JAPAN, 68 mi/110 km S of TSU; 33°48′N 136°03′E. Citrus fruits. Shichiri Mihama coast nearby.

Mihama, town, Hidaka county, WAKAYAMA prefecture, S HONSHU, W central JAPAN, 24 mi/38 km S of WAKAYAMA; 33°50′N 135°12′E. Crepe; whitebait. Protected pine forest nearby.

Mihara (mee-HAH-rah), city, HIROSHIMA prefecture, SW HONSHU, W JAPAN, on the INLAND SEA, 34 mi/55 km E of HIROSHIMA; 34°23′N 133°04′E. Machine tools. Edo-era castle town.

Mihara (mee-HAH-rah), town, Mihara district, S central AWAJI-SHIMA island, HYOGO prefecture, W central JAPAN, 35 mi/57 km S of KOBE; 34°17′N 134°46′E. Dairying; onions, lettuce. Nearby Onokoro islet has a shrine with a large torii.

Mihara, town, South Kawachi county, OSAKA prefecture, S HONSHU, W central JAPAN, 5.6 mi/9 km S of OSAKA; 34°32′N 135°33′E.

Mihara (MEE-hah-rah), village, Hata county, KOCHI prefecture, S SHIKOKU, W JAPAN, 59 mi/95 km S of KOCHI; 32°54′N 132°51′E. Inkstones.

Mihara, Mount (MEE-hah-rah), Japanese *Miharayama* (mee-HAH-rah-YAH-mah), active volcanic cone (2,477 ft/755 m) on central O-SHIMA, of island group Izushichito, Tokyo prefecture, SE JAPAN; surrounded by wasteland. Known for numerous suicides committed there. In 1934, two Japanese made a 1,250-ft/381-m descent into crater in steel cage.

Miharu (mee-HAH-roo), town, Tamura county, FUKUSHIMA prefecture, N central HONSHU, NE JAPAN, 22 mi/35 km S of FUKUSHIMA city; 37°26′N 140°29′E. Chess pieces, dolls.

Mihaylovgrad, Bulgaria: see MONTANA, city.

Mihijam, India: see CHITTARANJAN.

Mihintale (MI-HIN-tuh-LE), isolated peak (1,019 ft/311 m) in NORTH CENTRAL PROVINCE, SRI LANKA, 7 mi/11.3 km E of ANURADHAPURA; 08°21′N 80°30′E. A Buddhist pilgrimage center; cradle of Buddhism in Sri Lanka and traditional site where Buddhist missionary Mahinda converted Devanampiyatissa, first Buddhist king of Sri Lanka. Remains include hundreds of stone steps leading to a stupa built in 1st century C.E., Ambastala stupa (2nd century B.C.E.), and stone (Mahinda's) bed, Lion Bath, convocation hall, stone ponds.

Miho (MEE-ho), village, Inashiki county, IBARAKI prefecture, central HONSHU, E central JAPAN, 28 mi/45 km S of MITO; 36°00′N 140°18′E. Computer components.

Mihonoseki (mee-HO-no-SE-kee), town, Yatsuka county, SHIMANE prefecture, SW HONSHU, W JAPAN, on SEA OF JAPAN, 11 mi/17 km N of MATSUE, on narrow peninsula N of Naka-no-umi; 35°33′N 133°11′E.

Mihrpur, BANGLADESH: see MEHERPUR.

Mihuru (mee-HOO-roo), village, LINDI region, S TANZANIA, 90 mi/145 km ENE of SONGEA, in S end of SELOUS GAME RESERVE, near Mbarangandu River; 10°08′S 36°56′E. Livestock.

Miiraku (MEE-rah-koo), town, South Matsuura county, NAGASAKI prefecture, NW KYUSHU, SW JAPAN, 71 mi/115 km W of NAGASAKI; 32°44′N 128°41′E.

Mijares (mee-HAH-res), town, Ávila province, central SPAIN, in SIERRA DE GREDOS, 25 mi/40 km SSW of ÁVILA. Olives, vegetables; flour milling, olive-oil pressing. Hydroelectric plant.

Mijares River, 65 mi/105 km long, TERUEL and Castellón de la Plana provinces, E SPAIN; rises 18 mi/29 km W of TERUEL; flows generally ESE to the MEDITERRANEAN SEA 6 mi/9.7 km SSE of CASTELLÓN DE LA PLANA. Two dams and reservoirs feed several canals (Canal de Castellón) irrigating orange groves.

Mijas (MEE-hahs), city, MÁLAGA province, S SPAIN, at S slopes of the Sierra de Mijas, near the MEDITERRANEAN SEA, 15 mi/24 km SW of MÁLAGA; 36°36′N 04°38′W. In agricultural region (cereals, raisins, fruit); oil milling. Many foreigners live here. Tourism.

Mijdrecht (MEI-drekht), town, UTRECHT province, W central NETHERLANDS, 13 mi/21 km NW of UTRECHT; 52°12′N 04°52′E. Crooked Mijdrecht River to W.

Agriculture (flowers, fruit, vegetables); cattle, poultry; dairying; manufacturing (chemicals, machinery, hardware).

Mijirtein (MEEJ-ree-teen), Italian *migiurtinia*, region, NE SOMALIA. Occupies E AFRICA's "horn," bordered by Gulf of ADEN (N), INDIAN Ocean (E). Hot, arid region with high plateau (1,500–3,000 ft/457–914 m) in W and low plateau (600 ft/183 m) in E. Has narrow coastal plain lined with sand dunes. Capes Hafun and GUARDAFUI are easternmost points of Africa. Fishing (tunny, mother of pearl) and pastoralism (sheep, cattle, camels). Frankincense and gum arabic are gathered in N. Extensive saltworks in Hafun bay. Chief towns include BOSSASSO, ALULA, HURDIO, HAFUN, GARDO.

Mijoux (mee-zhoo), resort (□ 8 sq mi/20.8 sq km), AIN department, RHÔNE-ALPES region, E FRANCE, on Valserine River, in S JURA Mountains, near GEX, and 11 mi/18 km NW of GENEVA; 46°22′N 06°00′E. Ski terrain at elevation 3,000 ft/914 m–5,500 ft/1,676 m. Mijoux lies at E edge of the Regional Park of the High Jura (established 1986) and just W of the Col de la FAUCILLE (one of the chief E-W passes crossing the Jura), over which a main road leads to Switzerland.

Mikame (mee-KAH-me), town, W Uwa county, EHIME prefecture, W SHIKOKU, W JAPAN, on HOYO STRAIT, 40 mi/65 km S of MATSUYAMA; 33°22′N 132°25′E.

Mikamo (mee-KAH-mo), town, Miyoshi county, TOKUSHIMA prefecture, SE SHIKOKU, W JAPAN, 37 mi/60 km W of TOKUSHIMA; 34°02′N 133°56′E.

Mikamo (mee-KAH-mo), village, Maniwa county, OKAYAMA prefecture, SW HONSHU, W JAPAN, 40 mi/65 km N of OKAYAMA; 35°09′N 133°37′E.

Mikasa (mee-KAH-sah), city, W central Hokkaido prefecture, N JAPAN, 28 mi/45 km E of SAPPORO; 43°14′N 141°52′E.

Mikashevichi (mi-kah-SHE-vee-chee), Polish *Mikaszewicze*, town, BREST oblast, BELARUS, in PRIPET Marshes, 29 mi/47 km E of LUNINETS, on former USSR-POLAND border; 52°16′N 27°20′E. Railroad station; rock crushing and sorting plant.

Mikata (mee-KAH-tah), town, Mikata county, FUKUI prefecture, central HONSHU, W central JAPAN, 40 mi/65 km S of FUKUI; 35°33′N 135°54′E. Japanese plums. Early Jomon-era shell mound at nearby Torihama.

Mikata, town, Mikata district, HYOGO prefecture, S HONSHU, W central JAPAN, 63 mi/102 km N of KOBE; 35°27′N 134°32′E. *Tajima ushi* variety of beef cattle.

Mikawa, former province in central HONSHU, JAPAN; now part of AICHI prefecture.

Mikawa (MEE-kah-wah), town, Ishikawa county, ISHIKAWA prefecture, central HONSHU, central JAPAN, on SEA OF JAPAN, 12 mi/20 km S of KANAZAWA; 36°29′N 136°29′E. Pickled fish. Buddhist altars.

Mikawa, town, Tamana county, KUMAMOTO prefecture, W KYUSHU, SW JAPAN, 19 mi/30 km N of KUMAMOTO; 33°03′N 130°37′E.

Mikawa, town, East Tagawa county, YAMAGATA prefecture, N HONSHU, NE JAPAN, 50 mi/80 km N of YAMAGATA city; 38°47′N 139°51′E. Rice.

Mikawa, town, Kuga county, YAMAGUCHI prefecture, SW HONSHU, W JAPAN, 31 mi/50 km E of YAMAGUCHI; 34°14′N 131°59′E.

Mikawa (MEE-kah-wah), village, Kamiukena county, EHIME prefecture, NW SHIKOKU, W JAPAN, 19 mi/30 km S of MATSUYAMA; 33°36′N 132°58′E.

Mikawa (mee-KAH-wah), village, East Kanbara county, NIIGATA prefecture, central HONSHU, N central JAPAN, 22 mi/35 km S of NIIGATA; 37°42′N 139°23′E.

Mikazuki (mee-KAHZ-kee), town, Sayo district, HYOGO prefecture, S HONSHU, W central JAPAN, 46 mi/74 km N of KOBE; 34°58′N 134°26′E. Grapes.

Mikazuki, town, Ogi county, SAGA prefecture, N KYUSHU, SW JAPAN, 5 mi/8 km N of SAGA; 33°16′N 130°13′E.

Mikeno, Mount (mee-KE-no), second-highest peak (c.14,600 ft/4,450 m) of the VIRUNGA RANGE, E

CONGO, SE section of VIRUNGA NATIONAL PARK, near RWANDA-UGANDA border, 20 mi/32 km S of RUTSHURU, and 3 mi/4.8 km NW of MT. KARISIMBI; 01°27′S 29°26′E. Extinct volcano.

Mikese (mee-KAI-sai), village, MOROGORO region, E central TANZANIA, 17 mi/27 km E of MOROGORO, on railroad; 06°45′S 37°44′E. Road junction. Cotton, sugarcane, wheat, corn; cattle, sheep, goats. Mica mining.

Mikhailovgrad, Bulgaria: see MONTANA, city.

Mikhailovka (mee-khei-LOF-kuh), village, N PAVLODAR region, KAZAKHSTAN, 110 mi/177 km N of PAVLODAR; 53°52′N 76°31′E. Wheat.

Mikhailovka (mee-khei-LOF-kuh), suburb, SW ZHAMBYL region, KAZAKHSTAN, 8 mi/12.9 km NE of ZHAMBYL; 43°01′N 71°32′E. Tertiary-level (raion) administrative center. Cotton.

Mikhailovo, Bulgaria: see MIHAILOVO.

Mikhailovo Dam, Bulgaria: see MIHAILOVO.

Mikhalëvo (mee-hah-LYO-vuh), town (2005 population 15,280), central IVANOVO oblast, central European Russia, on road and railroad, 8 mi/13 km W of IVANOVO, to which it is administratively subordinate; 56°59′N 40°48′E. Elevation 465 ft/141 m. Industrial rubber; agricultural equipment.

Mikhali (mee-hah-LEE), village, E MOSCOW oblast, central European Russia, on road, 5 mi/8 km S of YEGOR'YEVSK, to which it is administratively subordinate; 55°18′N 39°05′E. Elevation 574 ft/174 m. In agricultural region; poultry farm.

Mikhalitch, TURKEY: see KARACABEY.

Mikhal'pol', UKRAINE: see MYKHAYLIVKA, Khmel'nyts'kyy oblast.

Mikha Tskhakaya (MEE-hah tsuh-khah-KAH-yah) or **Senaki**, city, W GEORGIA, in MINGRELIA, on railroad, and 32 mi/51 km W of KUTAISI, in COLCHIS LOWLAND. Junction for railroad spur to POTI. Wine making, carpet manufacturing; marlpits.

Mikhayl-, in Russian names: see MIKHAIL.

Mikhaylo-Kotsyubinskoye, UKRAINE: see MYKHAYLO-KOTSYUBYNS'KE.

Mikhaylo-Semenovskoye, RUSSIA: see LENINSKOYE, JEWISH AUTONOMOUS Oblast.

Mikhaylov (mee-HEI-luhf), city (2006 population 12,780), W RYAZAN oblast, central European Russia, on the PRONYA River (OKA River basin), on crossroads and railroad junction, 42 mi/68 km SW of RYAZAN; 57°49′N 36°29′E. Oil processing, bricks, textiles, food industries (bakery, dairy, meat processing); limestone quarries nearby. Founded (1137) as MOSCOW fortress against steppe nomads. Made city in 1778.

Mikhaylovka (mee-HEI-luhf-kah), city (2006 population 60,000), central VOLGOGRAD oblast, SE European Russia, on the MEDVEDITSA River (DON River basin), on highway junction and railroad (Sebryakovo station), 130 mi/210 km NW of VOLGOGRAD; 50°04′N 43°15′E. Elevation 236 ft/71 m. Center of agricultural area; agricultural machinery, asbestos, cement, silicate bricks, transportation equipment, cardboard; food industries (flour-milling center, canning, meat packing, dairying). Limestone quarrying. Became city in 1948.

Mikhaylovka (mee-KHEI-luhf-kuh), village, NE ISSYK-KOL region, KYRGYZSTAN, on ISSYK-KOL lake, and 10 mi/16 km NW of KARAKOL; 42°37′N 78°20′E. Health resort.

Mikhaylovka (mee-HEI-luhf-kah), village (2006 population 9,025), SW MARITIME TERRITORY, SE RUSSIAN FAR EAST, on road and the TRANS-SIBERIAN RAILROAD (Dubininskiy station), 8 mi/13 km N of USSURIYSK; 43°56′N 132°01′E. Elevation 108 ft/32 m. In agricultural area. Founded in 1894 with the construction of the railroad.

Mikhaylovka (mee-HEI-luhf-kah), village (2005 population 7,870), S IRKUTSK oblast, E central Siberian Russia, near the ANGARA RIVER, on road and TRANS-SIBERIAN RAILROAD, 25 mi/40 km NNW of KULTUK;

52°57′N 103°17′E. Elevation 1,840 ft/560 m. Manufacturing of fire-resistant powders; forestry services. A military airbase in the vicinity.

Mikhaylovka (mee-HEI-luhf-kah), village, E BURYAT REPUBLIC, S SIBERIA, RUSSIA, on highway branch, 22 mi/35 km E of KIZHINGA; 51°53′N 110°28′E. Elevation 2,372 ft/722 m. Agricultural cooperative.

Mikhaylovka (mee-HEI-luhf-kah), village, NW KURSK oblast, SW European Russia, on the Svapa River (tributary of the SEYM RIVER), on road junction and near railroad, 14 mi/23 km NE of DMITRIYEV-L'GOVSKIY; 52°14′N 35°22′E. Elevation 541 ft/164 m. Woodworking.

Mikhaylovka, RUSSIA: see KIMOVSK.

Mikhaylovka, RUSSIA: see ZHELEZNOGORSK, KURSK oblast.

Mikhaylovka, UKRAINE: see MYKHAYLIVKA.

Mikhaylovka, UKRAINE: see MYKHAYLIVKA, Luhans'k oblast.

Mikhaylovka, UKRAINE: see MYKHAYLIVKA, Zaporizhzhya oblast; MYKHAYLIVKA, Luhans'k oblast; or MYKHAYLIVKA, Khmel'nyts'kyy oblast.

Mikhaylovo, Bulgaria: see MIHAILOVO.

Mikhaylovsk (mee-HEI-luhfsk), city (2006 population 10,105), SW SVERDLOVSK oblast, central URALS, extreme E European Russia, on small lake near the UFA RIVER, on road junction and near railroad, 100 mi/163 km SW of YEKATERINBURG, and 75 mi/121 km SSE of NIZHNIYE SERGI, to which it is administratively subordinate; 56°26′N 59°06′E. Elevation 853 ft/259 m. Processing of nonferrous metals (aluminum foil and bimetals). Before 1941, produced structural iron and steel. Founded in 1806. Until 1942, known as Mikhaylovskiy Zavod. Made city in 1961.

Mikhaylovka (mee-HEI-luhf-skah-yah), village, NW VOLGOGRAD oblast, SE European Russia, on the KHOPER River, on road, 11 mi/18 km NNW of URYUPINSK; 50°57′N 41°54′E. Elevation 255 ft/77 m. Agricultural products.

Mikhaylovskaya (mee-HEI-luhf-skah-yah), village (2005 population 8,090), E KRASNODAR TERRITORY, S European Russia, on highway junction, 7 mi/11 km N of KURGANINSK; 44°59′N 40°36′E. Elevation 475 ft/144 m. In agricultural area; food processing.

Mikhaylovskiy Rudnik, RUSSIA: see ZHELEZNOGORSK.

Mikhaylovskoye (mee-HAH-yee-luhf-skuh-ye), town (2005 population 11,470), SW ALTAI TERRITORY, S central SIBERIA, RUSSIA, in the KULUNDA steppe, on road and railroad, 65 mi/105 km WNW of RUBTSOVSK; 51°49′N 79°43′E. Elevation 564 ft/171 m. Dried milk, reinforced concrete forms. Nearby is soda extraction from a lake at MALINOVOYE OZERO. Also known as Mikhaylovskiy or Mikhaylovka.

Mikhaylovskoye (mee-HEI-luhf-skuh-ye), village (2006 population 10,545), E NORTH OSSETIAN REPUBLIC, RUSSIA, in the N CAUCASUS Mountains, on railroad, 4 mi/6 km N of VLADIKAVKAZ; 43°06′N 44°38′E. Elevation 2,007 ft/611 m. Light manufacturing and food industries.

Mikhaylovskoye (mee-HEI-luhf-skuh-ye), village, E NIZHEGOROD oblast, central European Russia, on the left bank of the VOLGA RIVER, on highway, 9 mi/14 km N of (and administratively subordinate to) VOROTYNETS; 56°11′N 45°47′E. Elevation 196 ft/59 m. Sawmilling, lumbering.

Mikhaylovskoye, RUSSIA: see SHPAKOVSKOYE.

Mikhmoret (meekh-MO-ret), village, ISRAEL, 3.7 mi/5 km N of NETANYA, on MEDITERRANEAN coast; 32°24′N 34°52′E. At sea level. Founded 1945, as a fishing village. Now, a wealthy summer resort. Just S is the NAHAL ALEXANDER (Alexander stream) nature reserve. Remains of an ancient port were found nearby, indicating that the area was populated from late-Bronze Age to Roman-Byzantine times. Maritime school nearby.

Mikhnëvo (meekh-NYO-vuh), town (2006 population 10,715), S MOSCOW oblast, central European Russia, 22

mi/35 km NNW of KASHIRA, and 13 mi/21 km N of STUPINO, to which it is administratively subordinate; 55°07′N 37°57′E. Elevation 613 ft/186 m. Railroad junction; ventilating equipment, electrical goods, furniture, ceramics; food processing (bakery).

Miki (MEE-kee), city, HYOGO prefecture, S HONSHU, W central JAPAN, 13 mi/21 km N of KOBE; 34°47′N 134°59′E. Cutlery.

Miki (MEE-kee), town, Kita county, KAGAWA prefecture, NE SHIKOKU, W JAPAN, 6 mi/10 km S of TAKAMATSU; 34°15′N 134°08′E.

Mikindani (mee-keen-DAH-nee), town, MTWARA region, SE TANZANIA, port on Mikindani Bay of INDIAN OCEAN, 5 mi/8 km W of MTWARA; 10°17′S 40°05′E. Mtwara International Airport (E). Cashews, sisal, copra, peanuts, bananas; sheep, goats; fish. Manufacturing (peanut oil milling). Mikindani Slave Prison historical site here.

Mikir Hills (MEE-kir), isolated hills in BRAHMAPUTRA RIVER valley, central ASSAM state, NE INDIA, ENE of SHILLONG; c.65 mi/105 km long, 40 mi/64 km wide; rises to more than 4,470 ft/1,362 m. Coal, hematite, and limestone deposits. Tea gardens (mostly N). Inhabited chiefly by tribal Mikirs. Formerly an autonomous district of Assam state, now divided among GOLAGHAT, KARBI-ANGLONG, NAGAON, and SIBSAGAR districts.

Mikkabi (meek-KAH-bee), town, Inasa county, SHIZUOKA prefecture, central HONSHU, E central JAPAN, on NW cove of LAKE HAMANA, 50 mi/80 km S of SHIZUOKA; 34°47′N 137°33′E. Mandarin oranges. Remains of prehistoric Mikkabi Manitoba.

Mikkeli (MIK-ke-lee), Swedish *Sankt Michel*, city, ⊙ MIKKELIN province, SE central FINLAND; 61°41′N 27°15′E. Elevation 297 ft/90 m. In the SAIMAA lake region. Important lake port, commercial center, transportation hub, and resort area. Woodworking, textiles, and metalworking. It was chartered in 1838 and is the seat of a Lutheran bishopric. Headquarters of Marshall Mannerheim in Finnish War of Independence (1917–1919) and Winter War 1939–1940. Site of 4,000-year-old Ashtuvansalmi cliff paintings; museum; International Music Festival. Airport.

Mikkelin (MIK-ke-lin), Swedish *Sankt Michel*, province (□ 8,353 sq mi/21,717.8 sq km), SE FINLAND; ⊙ MIKKELI. Almost one-third of area consists of lakes of the SAIMAA and PÄIJÄNNE systems. Land is generally low and marshy. Fishing, agriculture, and livestock raising; industries include lumbering, timber processing, woodworking and metalworking, machinery manufacturing. Minerals worked include granite, limestone, and graphite. Main towns are Mikkeli, SAVONLINNA, and HEINOLA.

Miklavž, village, NE SLOVENIA, SE of MARIBOR, in Dravsko Polje (Drava plain) area; 46°30′N 15°43′E. Hydroelectric plant nearby.

Mikolajki (mee-ko-WAH-kee), Polish *Mikołajki*, German *Nikolaiken* (nee-ko-lahi-ken), town in EAST PRUSSIA, Suwałki province, NE POLAND, on Lake ŚNIARDWY, 50 mi/80 km E of OLSZTYN (Allenstein). Lake fisheries; limestone quarrying.

Mikolajow, UKRAINE: see MYKOLAYIV, L'viv oblast.

Mikolayevka, UKRAINE: see MYKOLAYIVKA, Odessa oblast.

Mikolongwe (mee-ko-LAWNG-gwai), town, Southern region, S MALAWI, 10 mi/16 km ESE of LIMBE; 15°53′S 35°11′E. Tea, tobacco, cotton, corn, rice.

Mikolow (mee-KO-woov), Polish *Mikołów*, German *Nikolai*, town (2002 population 38,096), Katowice province, S POLAND, 8 mi/12.9 km SW of KATOWICE; 50°10′N 18°54′E. Brickworks, chemical factories, iron foundries, coal mines; paper, machining. During World War II, called Nikolei (under German control).

Mikomesen, Equatorial Guinea: see MICOMESENG.

Míkonos, Greece: see MYKONOS.

Mikope (mee-KAW-pai), village, KASAI-OCCIDENTAL province, SW CONGO, 38 mi/61 km E of LUEBO; 04°55′S 20°46′E. Elev. 2,017 ft/614 m.

Mikoyanovka, RUSSIA: see OKTYABR'SKIY, BELGOROD oblast.

Mikoyanovsk, RUSSIA: see KHINGANSK.

Mikoyanovskiy, RUSSIA: see OKTYABR'SKIY, KAMCHATKA oblast.

Mikoyan-Shakhar, RUSSIA: see KARACHAYEVSK.

Mikra Delos, Greece: see DELOS.

Mikre (MEEK-re), village, LOVECH oblast, UGURCHIN obshtina, BULGARIA; 43°02′N 24°37′E.

Mikrí Prespa, Greece and Albania: see PRESPA, LAKE.

Mikropolis (mee-KRO-po-lees), town, DRÁMA prefecture, EAST MACEDONIA AND THRACE department, NE GREECE, 17 mi/27 km WNW of DRÁMA; 41°12′N 23°49′E. Tobacco, barley, cotton; wine. Formerly called Karlikova.

Miksova, SLOVAKIA: see STIAVNIK.

Mikstat (MEEK-staht), German *Mixstadt*, town, Kalisz province, W central POLAND, 17 mi/27 km SSW of KALISZ; 51°32′N 17°59′E. Trades in horses, cattle; flour milling.

Mikulasovice (MI-ku-LAH-sho-VI-tse), Czech *Mikulášovice*, German *Nixdorf*, town, SEVEROCESKY province, N BOHEMIA, CZECH REPUBLIC, 26 mi/42 km NNE of ÚSTÍ NAD LABEM, on railroad; 50°57′N 14°22′E. Manufacturing (cutlery, needles, buttons). Granite quarry nearby.

Mikulczyce (mee-KOOL-chee-tse), German *Klausberg*, commune in UPPER SILESIA, after 1945 in Katowice province, S POLAND, 3 mi/4.8 km N of ZABRZE (Hindenburg). Railroad junction; coal mining. Until 1935, called Mikultschütz.

Mikulínce, UKRAINE: see MYKULYNTSI.

Mikulintsy, UKRAINE: see MYKULYNTSI.

Mikulov (MI-ku-LOF), German *Nikolsburg*, town, JIHOMORAVSKY province, S MORAVIA, CZECH REPUBLIC, on railroad, and 18 mi/29 km S of BRNO, near Austrian border; 48°48′N 16°38′E. Manufacturing (machinery, leather, textiles); wine production. Has an old Jewish cemetery, and a picturesque castle. Extensive vineyards are in the vicinity. Site of the signing in 1621 of a treaty between Emperor Ferdinand II and Gabriel Bethlen, who renounced his kingship of HUNGARY. Armistice agreements which ended the Franco-Austrian War were signed here in 1805. Peace treaty between PRUSSIA and AUSTRIA was signed here in 1866. Paleolithic-era archaeological site.

Mikultschütz, POLAND: see MIKULCZYCE.

Mikumi National Park (mee-KOO-mee), MOROGORO region, central TANZANIA, 35 mi/56 km SW of MOROGORO, in S part of MKATA Plains, drained by Mkata River. Tourist lodges at Kikoboga.

Mikumo (MEE-koo-mo), town, Ichishi county, MIE prefecture, S HONSHU, central JAPAN, 6 mi/10 km S of TSU; 34°37′N 136°31′E.

Mikun' (MEE-koon), city (2005 population 11,445), W KOMI REPUBLIC, NE European Russia, on the VYCHEGDA River, on highway and near railroad, 60 mi/97 km S of SYKTYVKAR, and 2 mi/3.2 km SW of AYKINO; 62°12′N 49°56′E. Elevation 239 ft/72 m. Railroad enterprises, woodworking industries; greenhouse agriculture. Arose in 1937 as a railroad station. Made city in 1959.

Mikuni (mee-KOO-nee), town, Sakai county, FUKUI prefecture, central HONSHU, W central JAPAN, on SEA OF JAPAN, 12 mi/20 km N of FUKUI; 36°12′N 136°09′E. Scallions.

Mikurajima (mee-koo-RAH-jee-mah), village, on MIKURA-JIMA island, BONIN islands, Miyake district, Tokyo prefecture, SE JAPAN, 12 mi/20 km S of SHINJUKU; 33°53′N 139°35′E. Orchids.

Mikura-jima (mee-KOO-rah-JEE-mah), island (□ 8 sq mi/20.8 sq km) of island group IZU-SHICHITO, Miyake district, Tokyo prefecture, SE JAPAN, in PHILIPPINE SEA, 50 mi/80 km NNW of Hachijo-jima; roughly circular, 3 mi/4.8 km in diameter. Has central volcanic cone rising to 2,798 ft/853 m. Agriculture (rice, wheat, sweet potatoes); raw silk.

Mikve Israel (meek-VE yees-rah-EL) or **Mikveh Israel**, Israel's oldest agricultural school, W ISRAEL, in coastal plain, 3 mi/4.8 km SE of TEL AVIV; 32°01′N 34°46′E. Elevation 144 ft/43 m. Wine production. Has agricultural-research station, seed nurseries, botanical gardens. Sometimes spelled Miqve Yisrael. Founded 1870 as Jewish agricultural school.

Mila (mee-LAH), wilaya, E ALGERIA, N of CONSTANTINE; ⊙ MILA; 36°25′N 06°10′E. Mainly a grain-producing region with flour mills, it also has construction-material industries using local materials and clays. Created in 1984 with communes carved out of CONSTANTINE and JIJEL wilaya.

Mila (mee-LAH), ancient *Mileu*, town, ⊙ MILA wilaya, E ALGERIA, 20 mi/32 km WNW of CONSTANTINE; 36°18′N 06°16′E. In cereal-growing region. Prospered under Romans and Turks. Many figures of the Algerian nationalist movement, including the intellectual Moubarek El Mili and the guerilla Boussouf, born here.

Milaca (muh-LAH-kuh), town (2000 population 2,580), ⊙ MILLE LACS county, E MINNESOTA, 28 mi/45 km ENE of ST. CLOUD, on RUM RIVER; 45°45′N 93°39′W. Elevation 1,084 ft/330 m. Trading point in agricultural area (grain; livestock, poultry; dairying); manufacturing (concrete, coin and currency wrappers, medical instruments; printing and publishing). Part of Rum River State Forest to NE. Settled 1888; incorporated 1897.

Miladhunmadulu Atoll (mi-luhd-UM-muh-DUH-loo), N group of islands in INDIAN OCEAN, between 05°38′N 72°58′E and 06°29′N 73°30′E, just S of THILADHUNMATHI ATOLL. Coconuts.

Milagres (mee-lah-gres), city (2007 population 27,298), SE CEARÁ, BRAZIL, 40 mi/64 km ESE of CRATO; 07°21′S 38°54′W. Cheese manufacturing; cotton, tobacco, livestock.

Milagres (MEE-lah-gres), town (2007 population 11,658), E central BAHIA, BRAZIL, 52 mi/84 km W of SANTO ANTÔNIO DE JESUS; 12°52′S 39°51′W.

Milagro (mee-LAH-gro) or **El Milagro**, town, SE LA RIOJA province, ARGENTINA, 45 mi/72 km SW of SERREZUELA (CÓRDOBA province); 31°01′S 65°59′W. Railroad junction, livestock raising (goats, cattle), lumbering.

Milagro (mee-LAH-gro), town (2001 population 113,440), GUAYAS province, W central ECUADOR, in tropical plain, on GUAYAQUIL-ALAUSÍ railroad, and 22 mi/35 km ENE of Guayaquil; 02°07′S 79°36′W. Processing and trading center for fertile agricultural center (sugarcane, rice, cacao, tropical fruit); sugar refining, rice milling.

Milagro, town, NAVARRE province, N SPAIN, on ARAGON RIVER near its influx into the EBRO, and 15 mi/24 km NW of TUDELA; 42°15′N 01°46′W. Vegetable canning; cherries, sugar beets, alfalfa.

Milagros (mee-LAHG-ros), town, MASBATE province, central Masbate Island, PHILIPPINES, at head of ASID GULF, 13 mi/21 km SW of MASBATE; 12°13′N 123°30′E. Agricultural center (rice). Nearby are copper mines.

Milak (MEE-luhk), town, RAMPUR district, N central UTTAR PRADESH state, N central INDIA, 16 mi/26 km SSE of RAMPUR; 28°37′N 79°11′E. Corn, wheat, rice, gram, millet, sugarcane.

Milakokia Lake (mi-luh-KO-kee-uh), MACKINAC county, SE UPPER PENINSULA, N MICHIGAN, 22 mi/35 km NE of MANISTIQUE, and 6 mi/9.7 km N of LAKE MICHIGAN; c.3 mi/4.8 km long, 1.5 mi/2.4 km wide; 46°04′N 85°48′W.

Milam (MEI-lem), county (□ 1,021 sq mi/2,654.6 sq km; 2006 population 25,286), central TEXAS; ⊙ CAMERON; 30°47′N 96°58′W. Bounded NE by BRAZOS RIVER, drained by LITTLE RIVER and SAN GABRIEL RIVER, and E YEGUA CREEK. Diversified agriculture (especially

cotton; also corn, grain sorghum, wheat); livestock (cattle, poultry, hogs, some sheep). Large lignite deposits. Oil wells; large aluminum plant at Rockdale. Formed 1836.

Milam (MEI-luhm), village, ALMORA district, N Uttarakhand state, N central India, in E Kumaon HIMALAYAS, on tributary of Mahakali (SARDA) River, 65 mi/105 km NNE of ALMORA. Barley, buckwheat. Summer trade headquarters of Tibetan Bhotiyas.

Milan (mil-AHN), Italian *Milano*, Latin *Mediolanum*, city (2001 population 1,256,211), ⊙ LOMBARDY and of MILANO province, N ITALY, at the heart of the PO basin; 45°28′N 09°12′E. Because of its strategic position in the Lombard plain, at the intersection of several major transportation routes, it has been since the Middle Ages an international commercial, financial, fashion, convention, and industrial center. Today Milan is Italy's second-largest city after ROME and its economic heart. It has the highest per capita income in Italy. Manufactures include textiles, clothing, machinery, chemicals, electric appliances, printed materials, motor vehicles, airplanes, and rubber goods. The city has a large construction industry, and it is one of the most important silk markets in EUROPE. Two international airports: one located outside city limits at Malpensa; one at Linate airport fifteen minutes from city center. Probably of Celtic origin, Milan was conquered by Rome in 222 B.C.E. In later Roman times it was the capital (C.E. 305–402) of the Western Empire and the religious center of N Italy. In 313, Constantine I issued the Edict of Milan, which granted religious toleration. From 374 to 379 the city's bishop was Saint Ambrose, known for the liturgy he wrote and for his eloquence. Milan was severely damaged by the Huns (c.450) and again by the Goths (539) and was conquered by the Lombards in 569. In the 12th century it became a free commune and gradually gained supremacy over the cities of Lombardy. From the 11th to the 13th century Milan suffered from internal warfare between rich and poor, from the Guelph and Ghibelline strife, and from the enmity of rival cities, which assisted Emperor Frederick I in destroying it (1163). As a member of the Lombard League, Milan later contributed to the defeat of Frederick I at Legnano (1176). The city's independence was recognized in the Peace of Constance (1183). In the 13th century Milan lost its republican liberties; first the Torriani, then the Visconti (1277) became its lords. Galeazzo Visconti received (1395) the title of Duke of Milan from the emperor, and under him the duchy became one of the most important states in Italy. After the death of the last Visconti (1447), the Sforza became dukes of Milan. The city flourished until it became involved in the Italian Wars and passed under Spanish domination (1535). At the end of the War of the Spanish Succession, Austrian rule of Milan was established (1713–1796). Napoleon I made the city the capital of the CISALPINE Republic (1797) and of the kingdom of Italy (1805–1814). In 1815, Milan again came under AUSTRIA. It was a leading center throughout the Risorgimento; after five days of heroic fighting in 1848 the citizens of Milan succeeded in expelling the Austrians, who returned, however, a few months later. In 1859 the city was united with the kingdom of SARDINIA. Its industrial importance grew until it was incorporated (1861) into Italy. In World War II, Milan suffered widespread damage from Allied air raids; many significant buildings were damaged beyond repair. The most striking feature of the city is the large, white-marble cathedral (1386–1813), which shows traces of many styles (especially Gothic). It is elaborately ornamented with 135 pinnacles and more than 200 marble statues. A statue of the Madonna is on the highest pinnacle (354 ft/108 m). Other points of interest in Milan include Brera Palace and Picture Gallery (17th century), which includes major works by Mantegna and Bellini; the Castello Sforzesco (15th century, with 19th-century

additions), which houses a museum of art; the Church of Santa Maria delle Grazie (1465–1490), containing the famous fresco, *The Last Supper*, by Leonardo da Vinci; the Basilica of Sant' Ambrogio (founded in the 4th century, rebuilt in the 11th–12th century); the Ambrosian Library, which houses a rich collection of paintings; the Church of Sant' Eustorgio (9th century); the Leonardo da Vinci Museum of Science and Technology; and the gallery of modern art. Long a center of music, Milan has a conservatory and a famous opera house, Teatro alla Scala (opened in 1778). Home of Verdi. The city also has three universities and a polytechnic institute.

Milan (MEI-luhn), city (2000 population 1,958), ⊙ SULLIVAN county, N MISSOURI, 28 mi/45 km W of KIRKSVILLE; 40°12′N 93°07′W. Corn, soybeans; sheep, cattle, hogs. Corporate hog farming; poultry processing. Laid out 1845.

Milan, city (2006 population 7,885), GIBSON county, NW TENNESSEE, 11 mi/18 km ESE of TRENTON; 35°55′N 88°46′W. In agricultural region; manufacturing. Museum housing large collection of porcelain teapots here. Site of U.S. arsenal and wildlife management area.

Milán (mee-LAHN), town, ⊙Milán municipio, CAQUETA department, S COLOMBIA, 24 mi/39 km SSE of FLORENCIA, in the Oriente; 01°20′N 75°30′W. Elevation 1,512 ft/460 m. Livestock; sugarcane, corn.

Milan (MEI-luhn), town (2000 population 1,012), TELFAIR and DODGE counties, S central GEORGIA, 13 mi/21 km SSE of EASTMAN; 32°01′N 83°04′W.

Milan, town (2000 population 1,816), RIPLEY county, SE INDIANA, 7 mi/11.3 km NE of VERSAILLES; 39°08′N 85°08′W. In agricultural area. Laid out 1854.

Milan, town, WASHTENAW and MONROE counties, SE MICHIGAN, 14 mi/23 km SSE of ANN ARBOR, and on small Saline River; 42°04′N 83°40′W. Railroad junction. In farm area (beans, sugar beets). Manufacturing (auto bumpers, plastic components, coated seals, surface measuring equipment, wood floors and trusses, nonclay refractories, corrugated containers, lumber). Incorporated 1885.

Milan, town, COOS county, N NEW HAMPSHIRE, 7 mi/11.3 km N of BERLIN; 44°33′N 71°13′W. Drained by ANDROSCOGGIN and UPPER AMMONOOSUC rivers. Manufacturing (lumber); timber; agriculture (poultry, livestock; dairying). Includes village of West Milan in NW. Part of White Mountain National Forest in SW; Milan Hill State Forest in NW.

Milan (mee-LAHN), town (2006 population 2,504), CIBOLA county, W NEW MEXICO, suburb 2 mi/3.2 km NW of GRANTS, on SAN JOSE River; 35°11′N 107°53′W. Cattle, sheep, alfalfa, triticale. Manufacturing (machine tool accessories, bottled water, wood molding). Part of Cibola National Forest to E and W. EL MALPAIS National Monument to S.

Milan (mi-LAHN), village (□ 50 sq mi/130 sq km; 2006 population 322), ESTRIE region, S QUEBEC, E CANADA, 12 mi/20 km from LAC-MÉGANTIC; 45°36′N 71°08′W.

Milan (MEI-luhn), village (2000 population 5,348), ROCK ISLAND county, NW ILLINOIS, on ROCK RIVER (bridged), suburb, 4 mi/6.4 km S of downtown ROCK ISLAND-MOLINE; 41°26′N 90°33′W. In agricultural (corn, soybeans; cattle, hogs; dairying); bituminous-coal-mining area. Manufacturing (forklift components, printing, packaging and assembly, lighting fixtures). Incorporated 1865. Quad Cities Airport to E.

Milan (mil-AHN), village (2000 population 137), SUMNER county, S KANSAS, 15 mi/24 km W of WELLINGTON, near CHIKASKIA RIVER; 37°15′N 97°40′W. Wheat.

Milan (MEI-luhn), village (2000 population 326), CHIPPEWA county, SW MINNESOTA, 14 mi/23 km NW of MONTEVIDEO, near LAC QUI PARLE LAKE (reservoir on MINNESOTA RIVER); 45°06′N 95°54′W. Grain, soybeans, sugar beets; manufacturing (dolls). Las qui

Parle Wildlife Area to SW and S, on river; Lac qui Parle State Park to S.

Milan (mi-LAHN), village (□ 1 sq mi/2.6 sq km; 2006 population 1,338), ERIE county, N OHIO, 12 mi/19 km SSE of SANDUSKY, and on HURON RIVER; 41°17′N 82°36′W. Beer. Thomas A. Edison born here. Settled 1804 by Moravian missionaries.

Milang (muh-LANG), village, SE SOUTH AUSTRALIA state, S central AUSTRALIA, 40 mi/64 km SE of ADELAIDE, and on W shore of Lake ALEXANDRINA; 35°25′S 138°58′E. Railroad terminus; citrus and dried fruits. Resort.

Milange (mi-LAHN-gai), village, Zambézia province, central MOZAMBIQUE, on MALAWI border, on road to Malawi, and 140 mi/225 km NW of QUELIMANE; 16°05′S 35°47′E. Tea-growing center.

Milano (mee-LAH-no), province (□ 1,065 sq mi/2,769 sq km), LOMBARDY, N ITALY; ⊙ MILAN; 45°30′N 09°30′E. Comprises most of fertile PO plain lying between ADDA and Tieino rivers; watered by LAMBRO and OLONA rivers and by many irrigation canals. Most industrialized province of Italy, with textile, iron, metallurgical, printing and publishing, furniture, chemical industries, and oil refineries. Has many centers, including Milan, MONZA, SESTO SAN GIOVANNI, LODI, LEGNANO, RHO, and ABBIATEGRASSO. Agriculture (cereals, rice, raw silk, fruit, vegetables); large dairy industry. In 1927 area reduced to help form VARESE province.

Milano (mi-LAHN-o), village (2006 population 420), MILAM county, central TEXAS, 29 mi/47 km W of BRYAN; 30°42′N 96°51′W. Railroad junction in cattle, cotton, corn area.

Milanovac (mee-LAHN-o-vahts), town, central SERBIA, on railroad, and 23 mi/37 km W of KRAGUJEVAC; 44°11′N 21°36′E. Also spelled Gornji Milanovac or Gornyi Milanovats.

Milanovo (me-LAHN-o-vo), village, SOFIA oblast, SVOGE obshtina, BULGARIA; 43°08′N 23°24′E.

Milas, Turkish *Milâs*, ancient *Mylasa*, town (2000 population 38,061), province, TURKEY, 32 mi/51 km WNW of MUGLA; 37°19′N 27°48′E. Tobacco, fruit, olives, cereals. Rich emery deposits nearby, with chromium. In ancient times, a city of CARIA, noted for its buildings of white marble quarried nearby.

Milatyn Nowy, UKRAINE: see NOVYY MYLYATYN.

Milawa, township, NE VICTORIA, SE AUSTRALIA, 10 mi/16 km SE of WANGARATTA; 36°26′S 146°26′E. Vineyards; cheese.

Milazzo (mee-LAHT-tso), town, NE SICILY, ITALY, on a peninsula in the TYRRHENIAN Sea; 38°13′N 15°14′E. The town is a wine-trade and tuna-fishing center and is the gateway to the nearby LIPARI Islands. Garibaldi completed his conquest of Sicily by defeating (June 1860) the Bourbon troops there. Milazzo has an imposing 13th-century castle (now a prison) and a 16th-century cathedral. Oil refinery nearby. It is the ancient port of MYLAE. It was settled by colonists from Messina. Here in 260 B.C.E. the Romans in a newly built fleet were led to victory over the Carthaginians by the consul Caius Duilius in the First Punic War; it was Rome's first naval triumph. Mylae was (36 B.C.E.) the scene of a naval victory of Marcus Vipsanius Agrippa over Sextus Pompeius.

Milazzo, Cape, NE SICILY, N headland of promontory separating gulfs of MILAZZO and PATTI; 38°16′N 15°14′E.

Milazzo, Gulf of, inlet of TYRRHENIAN Sea in NE SICILY, between CAPE RASOCOLMO (E) and CAPE MILAZZO (W); 16 mi/26 km long, 5 mi/8 km wide; 38°15′N 15°20′E. Tunny fisheries. Chief port, MILAZZO.

Milbank, city (2006 population 3,302), ⊙ GRANT county, NE SOUTH DAKOTA, 30 mi/48 km NE of WATERTOWN, near MINNESOTA state line; 45°13′N 96°37′W. Granite quarries furnish material for gravestones and monuments; manufacturing (orthopedic braces, cheese, printing and publishing). Inkapa-Du-Ta Ski Area to NE.

Milborne Port (MIL-bawn PAWT), town (2001 population 2,644), SE SOMERSET, SW ENGLAND, 3 mi/4.8 km ENE of SHERBORNE; 50°58′N 02°28′W. Glove manufacturing; dairying. Has 15th-century church.

Milbridge, town, WASHINGTON county, E MAINE, 25 mi/40 km WSW of MACHIAS, at mouth of the NARRAGUAGUS, and on PLEASANT BAY; 44°28′N 67°51′W. Light manufacturing; fishing, lumbering area. Incorporated 1848. Sometimes spelled Millbridge.

Milburn, village (2006 population 305), JOHNSTON county, S OKLAHOMA, on BLUE RIVER, and 8 mi/12.9 km E of TISHOMINGO; 34°14′N 96°32′W. Tishomingo National Wildlife Refuge and Lake TEXOMA (WASHITA River) to SW.

Milcupaya (meel-koo-PEI-ah), town and canton, JOSÉ MARÍA LINARES province, POTOSÍ department, S central BOLIVIA, 9 mi/14.5 km NE of PUNA; 19°42′S 65°20′W.

Milden (MIL-duhn), village (2006 population 172), SW central SASKATCHEWAN, CANADA, 21 mi/34 km E of ROSETOWN; 51°29′N 107°31′W. Wheat.

Mildenhall (MIL-duhn-hawl), town (2001 population 11,284), NW SUFFOLK, E ENGLAND, on LARK RIVER, and 8 mi/12.9 km NE of NEWMARKET; 52°21′N 00°30′E. Previously flour mills. Site of Royal Air Force station. Has 15th-century market cross. A Roman station was here.

Mildmay (MEILD-mai), former village (□ 1 sq mi/2.6 sq km; 2001 population 1,150), SW ONTARIO, E central CANADA, 7 mi/11 km SSE of WALKERTON; 44°02′N 81°07′W. Dairying; lumbering. Amalgamated into SOUTH BRUCE municipality in 1999.

Mildred, village (2000 population 36), ALLEN county, SE KANSAS, 14 mi/23 km NE of IOLA; 38°01′N 95°10′W. Livestock, grain; dairying.

Mildura (mil-DYUH-ruh), municipality (2001 population 28,062), NW VICTORIA, AUSTRALIA, on MURRAY RIVER (Weirs nearby), and 210 mi/338 km ENE of ADELAIDE, on NEW SOUTH WALES border; 34°12′S 142°09′E. Commercial, irrigation center for sheep raising, agricultural area. Flour mill, fruit and vegetable canneries, brickyards; wool, butter, dried fruit, wheat; vineyards. Tourism; resort, annual festivals. Zoo nearby.

Mile (MEE-LUH), town, WASHINGTON county, ⊙ Mile county, E YUNNAN province, CHINA, 60 mi/97 km SE of KUNMING; 24°24′N 103°27′E. Elevation 4,593 ft/1,400 m. Rice, wheat, millet, sugarcane, tobacco; food processing, logging, electric power generation.

Mileai (me-LEE-ah), town, MAGNESIA prefecture, SE THESSALY department, N GREECE, at foot of the PELION, 11 mi/18 km ESE of VÓLOS (linked by railroad); 39°20′N 23°09′E. Tobacco; olive oil. Also spelled Meleai.

Mile High City, Colorado: see DENVER.

Miles (MEILZ), town, QUEENSLAND, NE AUSTRALIA, 211 mi/339 km W of BRISBANE; 26°40′S 150°11′E. Historical village and museum.

Miles, town (2000 population 462), JACKSON county, E IOWA, 18 mi/29 km E of MAQUOKETA; 42°02′N 90°19′W. Livestock, grain.

Miles (MEI-uhlz), town (2006 population 796), RUNNELS county, W central TEXAS, 17 mi/27 km NE of SAN ANGELO, near CONCHO RIVER; 31°36′N 100°10′W. Elevation 1,800 ft/549 m. In cattle, sheep, grain, cotton area. Old Opera House (1904).

Milesburg (MAI-uhls-buhrg), borough (2006 population 1,142), CENTRE county, central PENNSYLVANIA, 2 mi/3.2 km N of BELLEFONTE, on Bald Eagle Creek; 40°56′N 77°47′W. Agriculture (corn, hay; livestock; dairying); manufacturing (electronic equipment, plastic products). Bald Eagle State Park, including Howard State Nursery, to NE.

Miles City, town (2000 population 8,487), ⊙ CUSTER county, SE MONTANA, on YELLOWSTONE RIVER, at mouth of TONGUE RIVER, and 140 mi/225 km ENE of BILLINGS; 46°25′N 105°51′W. In irrigated area. Trade center, shipping point for wool, livestock. Railroad shops; gas wells. Agriculture (wheat, oats, corn, alfalfa, sugar beets; horses and rodeo stock, cattle, sheep, hogs); manufacturing (leather goods, flour, feeds, concrete products, printing, explosives, pond liners). Pirogue Island State Park in Yellowstone River to NE. Fort Keogh, 3 mi/4.8 km to SW, (served from 1877–1908) has been rebuilt and, with former military reservation, is now the Fort Keogh Agricultural Experiment Station, livestock research, includes large tract of range. Rodeo and fair; Custer County Art Center, Range Riders Museum. Incorporated 1887. Originally called Milestown.

Miles, Fort, DELAWARE: see CAPE HENLOPEN.

Miles Glacier, in CHUGACH Mountains, S ALASKA, N of MARTIN RIVER Glacier; 60°40′N 144°45′W. Flows into COPPER RIVER ENE of CORDOVA.

Milesovka, CZECH REPUBLIC: see BOHEMIAN CENTRAL HIGHLANDS.

Miles Platting, ENGLAND: see MANCHESTER.

Miles River, E Maryland, irregular estuary entering EASTERN BAY (arm of CHESAPEAKE BAY) in TALBOT county; c.20 mi/32 km long. Originally called Saint Michael's River, the saintly reference was deleted and the name was abbreviated.

Milestone (MEI-uhl-ston), town (2006 population 562), S SASKATCHEWAN, CANADA, 32 mi/51 km S of REGINA, in the rich prairie region; 49°59′N 104°31′W. Grain elevators, lumbering.

Mileto (mee-LAI-to), town, CATANZARO province, CALABRIA, S ITALY, 5 mi/8 km SSW of VIBO VALENTIA; 38°36′N 16°04′E. Agricultural trade center (cereals, olives, grapes, citrus fruit). Bishopric. Has cathedral (built 1928–1929), seminary, meteorological and seismological observatory. Severely damaged by earthquake in 1905. Roger II was born here.

Miletus (mei-LEE-tuhs), ancient seaport of W ASIA MINOR, in CARIA, on the mainland, near the mouth of the BÜYÜK MENDERES river, SE of the island of SÁMOS (now Milet), c. 20 mi/30 km S of the modern city of SÖKE; 37°30′N 27°18′E. It was occupied by Greeks in the settlement of the E Aegean (c.1000 B.C.E.) and became one of the principal cities of IONIA. From the 8th century B.C.E. it led in colonization, especially on the BLACK SEA. The Milesians were strong enough to resist the Lydian kings and were not molested by the Persians. In 499 B.C.E., however, they stirred up the revolt of Ionian Greeks against Persia; the Persians sacked the city (494 B.C.E.). Although less flourishing, Miletus remained an important seaport until the harbor silted up early in the Christian era. Miletus produced some of the earliest Greek philosophers, including Thales and Anaximander. The site was excavated by German archaeologists; the most visible ruins are a Greco-Roman theater and adjoining Byzantine castle.

Milevska Mountains (MEE-lev-ska), in KRAISHTE highland, on the Bulgarian-Yugoslav border; extend c.35 mi/56 km between the RUI MOUNTAINS (N) and OSOGOV MOUNTAINS (S); 42°34′N 22°26′E. Rise to 5,686 ft/1,733 m at Milevets peak, 27 mi/43 km W of RADOMIR, BULGARIA. Formerly spelled Milevo Mountains.

Milevsko (MI-lef-SKO), German *Mühlhausen*, town, JIHOCESKY province, S BOHEMIA, CZECH REPUBLIC, on railroad, and 14 mi/23 km WNW of TÁBOR; 49°27′N 14°22′E. In a rye and timber region; manufacturing (machinery, textiles); food processing. The town has old abbey with a Romanesque church.

Milford, residential city, NEW HAVEN county, SW CONNECTICUT, on LONG ISLAND SOUND; 41°13′N 73°03′W. Oysters and clams are gathered there for commercial use, and the city also has light manufacturing, such as the production of writing pens and electrical products. Major retail shopping area adjacent to Interstate 95 and along U.S. Route 1. Milford Academy is here. Settled 1639, incorporated as a city 1959.

Milford (MIL-fuhrd), city (□ 4 sq mi/10.4 sq km; 2006 population 6,317), on CLERMONT-HAMILTON county line, SW OHIO, 14 mi/23 km ENE of downtown CINCINNATI, on Little Miami River; 39°10′N 84°17′W. Makes burial vaults, hospital supplies, wood products, and consumer goods.

Milford, town (2000 population 6,732), KENT and SUSSEX counties, E DELAWARE, 18 mi/29 km SSE of DOVER, and at head of navigation on MISPILLION RIVER, which divides city into North and South Milford; 38°54′N 75°25′W. Trade and shipping center in vegetable and fruit-farming area; manufacturing. Has several 18th century buildings Milford Neck Wildlife Area to NE. Incorporated 1867.

Milford, town (2006 population 1,545), DECATUR county, SE central INDIANA, 7 mi/11.3 km W of GREENSBURG; 41°25′N 85°51′W. Laid out 1835.

Milford, town (2006 population 120), KOSCIUSKO county, N INDIANA, 12 mi/19 km N of WARSAW; 39°21′N 85°37′W. Poultry area. Manufacturing (motor vehicles, grain bins, livestock feeding and watering equipment, poultry processing, feed mixing, mobile restrooms). Laid out 1836.

Milford, town (2000 population 2,474), DICKINSON county, NW IOWA, 6 mi/9.7 km S of SPIRIT LAKE, near OKOBOJI lakes and Little Sioux River; 43°19′N 95°09′W. Agricultural trade center and summer resort; wood products. Founded 1869, incorporated 1892.

Milford, town, PENOBSCOT county, S central MAINE, on the PENOBSCOT, 11 mi/18 km NNE of BANGOR; 44°59′N 68°34′W. Light manufacturing; hunting, fishing.

Milford, industrial town, WORCESTER county, S MASSACHUSETTS, on the CHARLES RIVER; 42°10′N 71°31′W. Infarm area; pink granite has been quarried there since the mid-1800s. Manufacturing (glass containers, electronics, metal fabrication, precision and analytical instruments). Settled 1662 and set off from MENDON; incorporated 1780.

Milford, town (2000 population 6,272), OAKLAND county, SE MICHIGAN, 16 mi/26 km WSW of PONTIAC, and on HURON RIVER; 42°35′N 83°35′W. Manufacturing (transportation equipment, machinery). General Motors proving ground is nearby. Highland State Recreational Area to N; Proud Lake State Recreational Area to SE; Island Lake State Recreational Area and Kennington Metropark to SW; numerous lakes in area. Incorporated in 1869.

Milford, town (2006 population 2,049), SEWARD county, SE NEBRASKA, 10 mi/16 km S of SEWARD, and on BIG BLUE RIVER. Grain. Community college.

Milford, town, HILLSBOROUGH county, S NEW HAMPSHIRE, 10 mi/16 km WNW of NASHUA; 42°49′N 71°40′W. Drained by SOUHEGAN RIVER. Manufacturing (diamond tools, crystal materials, corrugated packaging, medical equipment, printing and publishing, electronics, plastics products; metal fabrication, commercial printing); agriculture (nursery crops, fruit, vegetables, corn; poultry, livestock; dairying); granite quarries. Milford State Fish Hatchery in NW. Set off 1794.

Milford, town (2006 population 1,441), BEAVER county, SW UTAH, on BEAVER RIVER, and 21 mi/34 km WNW of BEAVER; 38°23′N 113°00′W. Railroad and trade center for dairying and irrigated agricultural area (alfalfa, peas, potatoes, barley). Ships cattle. Lead, silver, and gold deposits, copper mines nearby. Town was once dependent on mining. Elevation 4,957 ft/ 1,511 m. Squaw Springs to W. Settled 1870, incorporated 1903.

Milford, village (2000 population 1,369), IROQUOIS county, E ILLINOIS, 11 mi/18 km S of WATSEKA, on Sugar Creek; 40°37′N 87°42′W. Trade and shipping center in agricultural area (corn, soybeans, sorghum; cattle, hogs; dairy products). Manufacturing (food processing, electrical motors). Settled c.1830; plotted 1836; incorporated 1874.

Milford, village (2000 population 502), GEARY county, NE central KANSAS, on REPUBLICAN RIVER, and 11 mi/ 18 km NNW of JUNCTION CITY; 39°10′N 96°54′W. Livestock; grain. On W shore of Milford Lake Reservoir and W of Fort Riley Military Reserve.

Milford, village (2006 population 482), OTSEGO county, central NEW YORK, on the SUSQUEHANNA River, and 11 mi/18 km NNE of ONEONTA; 42°35′N 74°57′W. In dairying area. Brewery.

Milford (MIL-fuhrd), village (2006 population 742), ELLIS county, N central TEXAS, c.45 mi/72 km SSW of DALLAS, and 13 mi/21 km NE of HILLSBORO; 32°07′N 96°57′W. In cotton, grain, cattle, dairying area.

Milford (MIL-fuhrd), unincorporated village, CAROLINE county, E VIRGINIA, 20 mi/32 km SSE of FREDERICKSBURG, on MATTAPONI RIVER; 38°01′N 77°22′W. Manufacturing (stone products, concrete, lumber, ladders); in agricultural area (grain, potatoes; cattle); timber.

Milford, borough (2006 population 1,219), HUNTERDON county, W NEW JERSEY, on DELAWARE RIVER, and 13 mi/21 km WNW of FLEMINGTON; 40°34′N 75°05′W. Manufacturing and agriculture.

Milford, borough (2006 population 1,221), ⊙ PIKE county, NE PENNSYLVANIA, 7 mi/11.3 km SW of PORT JERVIS, New York, on DELAWARE RIVER, at mouth of Sawkill Creek; 41°19′N 74°47′W. Agriculture (dairying; cattle); manufacturing (electronics, apparel, diversified light manufacturing); resort area. DELAWARE WATER GAP RECREATION AREA to S; part of Delaware State Forest to NW. Settled 1733.

Milford (MIL-fuhrd), locality (□ 1 sq mi/2.6 sq km; 2001 population 3,934), SURREY, SE ENGLAND, 6 mi/ 9.7 km SW of GUILDFORD; 51°10′N 00°40′W. Milford and Witley Commons include 0.6 sq mi/1.5 sq km of national trust land. Has Victorian church.

Milford Center (MIL-fuhrd SEN-tuhr), village (2006 population 690), UNION county, central OHIO, 5 mi/8 km SSW of MARYSVILLE, and on DARBY CREEK; 40°10′N 83°26′W. In agricultural area.

Milford Haven (MIL-fuhrd), town (2001 population 13,086), Pembrokeshire, SW Wales; 51°43′N 05°02′W. Seaport on the N side of the estuary called Milford Haven. The bay forms a natural harbor that can handle large oil tankers, making the town a key oil port and refining center. Oil and gas industry and pipeline. Other imports include cattle, food, and fertilizer.

Milford Haven (MIL-fuhrd), Welsh *Aberdaugleddau*, inlet of the ATLANTIC, Pembrokeshire, SW Wales; forms harbor, extends 12 mi/19 km E from St. Anne's Head; 1 mi/1.6 km–2 mi/3.2 km wide; 51°42′N 05°07′W. The short Eastern and Western Cleddau rivers enter it. Major berth for large oil tankers serving three huge refineries on N and S shores. Chief ports: MILFORD HAVEN, PEMBROKE DOCK, NEYLAND. In February 1996 the 147,000-ton/133,329-metric-ton tanker *Sea Empress* ran aground at the tip of Milford Haven's harbor, spewing more than 70,000 gal/264,971 liters of oil into the water. More than 120 mi/193 km of the Welsh coastline have been affected by the spill, and many fish and sea birds have been killed. Formerly in DYFED, abolished 1996.

Milford on Sea (MIL-fuhrd ON SEE), town (2001 population 4,527), SW HAMPSHIRE, S ENGLAND, on the CHANNEL, 15 mi/24 km SW of SOUTHAMPTON; 50°43′N 01°35′W. Seaside resort. Has Norman church. Located on The SOLENT (2 mi/3.2 km ESE) is 16th-century Hurst Castle, where Charles I was imprisoned in 1648.

Milford Sound, fjord facing TASMAN SEA, indenting SW SOUTH ISLAND, NEW ZEALAND. Part of FIORDLAND NATIONAL PARK. Mountains rise steeply from the shore; maximum elevation 9,042 ft/2,756 m. Well-known resort area.

Mīlgrāvis (MEEL-grah-vis), German *Mühlgraben*, outer port of Riga, LATVIA, on right bank of the DVINA

(DAUGAVA) River, 3 mi/5 km from Gulf of Riga, and 6 mi/10 km N of Riga city center; 57°02′N 24°06′E.

Milhã (meel-YAH), city (2007 population 14,082), CEARÁ, BRAZIL.

Milhaud (meel-o), town (□ 7 sq mi/18.2 sq km) SW suburb of NÎMES, GARD department, LANGUEDOC-ROUSSILLON region, S FRANCE, on main road to MONTPELLIER; 43°47′N 04°18′E. Vineyards. Nearby source of bottled spring water; shipping plant.

Mili (MEE-lee), southernmost atoll (□ 6 sq mi/15.6 sq km; 1999 population 1,032) of RATAK CHAIN, MAJURO district, MARSHALL ISLANDS, W central PACIFIC, 325 mi/523 km SE of KWAJALEIN; 06°10′N 171°55′E. Atoll is c.30 mi/48 km long, with 102 islets. Japanese air base in World War II. Formerly called Mulgrave Islands.

Milia, El (mee-LYAH, el), village, JIJEL wilaya, NE ALGERIA, on the Oued El Kebir, 56 mi/90 km NW of CONSTANTINE; 36°48′N 06°14′E. Cork stripping; lead and zinc mining.

Miliana (mee-lyah-NAH), town, AÏN DEFLA wilaya, NW ALGERIA, on S slope of the Djebel ZACCAR, overlooking the CHÉLIFF valley, 55 mi/89 km SW of ALGIERS; 36°20′N 02°15′E. Elev. 2,400 ft/732 m. Situated amidst fruit orchards and citrus groves, it is also noted for its table grapes and wines. Commercial importance lost to KHEMIS MILIANA (4 mi/6.4 km S), located in lowland along railroad. Iron mined on Djebel ZACCAR. Founded probably in 10th century on site of Roman *Zucchabar*. Occupied by French in 1840, and besieged by Abd El Kader until 1842. Town is surrounded by walls pierced by two gates. Magnificent panorama from the esplanade. Teachers college.

Milicz (MEE-leech), German *Militsch* (mee-leetch), town, Wrocław province, SW POLAND, on BARYCZ RIVER, 30 mi/48 km NNE of WROCŁAW (Breslau); 51°32′N 17°16′E. Linen milling. Has remains of old castle of prince-bishops of Breslau. Was first mentioned 1136; chartered c.1300. Considerably damaged in World War II. In LOWER SILESIA until 1945.

Milidia, TURKEY: see MALATYA.

Milig (mel-LEEG), village, MINUFIYA province, Lower EGYPT, on the BAHR SHIBIN, and 4 mi/6.4 km NE of SHIBIN EL KOM; 30°36′N 31°03′E. Cereals, cotton, flax.

Mililani Town (MEE-lee-LAH-nee), city (2000 population 28,608), central OAHU island, HONOLULU county, HAWAII, 12 mi/19 km NW of HONOLULU, 2 mi/ 3.2 km S of WAHIAWA, on Waikele Stream, and on Kamehameha Highway; 21°26′N 158°01′W. Wheeler Air Force Base and Schofield Barracks Military Reservation to N.

Militello in Val di Catania (mee-lee-TEL-lo een vahl dee kah-TAH-nyah) or **Militello** (mee-lee-TEL-lo), town, CATANIA province, E SICILY, S ITALY, 16 mi/26 km ENE of CALTAGIRONE; 37°16′N 14°48′E. In cereal-growing region; citrus fruit, olive oil. Rebuilt after earthquake of 1693. Has 16th-century churches.

Militsch, POLAND: see MILICZ.

Miljacka River (meel-YAHT-skah), c. 15 mi/24 km long, central BOSNIA, BOSNIA AND HERZEGOVINA; rises in two headstreams joining 4 mi/6.4 km SE of SARAJEVO; flows W through Sarajevo, BOSNA RIVER, 3 mi/ 4.8 km N of Ilidža. Also spelled Milyatska River.

Miljevina (meel-ye-VEE-nah), mining town, E central BOSNIA, BOSNIA AND HERZEGOVINA, near Foča. Coal mine.

Milk, river, 729 mi/1,173 km long, MONTANA and ALBERTA (CANADA); rising in the ROCKY MOUNTAINS, formed by joining of South and Middle Forks 21 mi/34 km N of BROWNING, GLACIER county, N W Montana. South Fork (c.30 mi/48 km long) and Middle Fork (c.20 mi/32 km long) rise in the Blackfeet Indian Reservation just E of GLACIER NATIONAL PARK; then the river flows ENE into Alberta, where it receives North Fork of the Milk River, then curves E past town of MILK RIVER and Writing-on-Stone Provincial Park, SE into Montana again, through FRESNO RESERVOIR, past HAVRE, then forms N boundary of Fort Belknap

Indian Reservation. From there, it then flows past MALTA and GLASGOW to the MISSOURI River, 10 mi/16 km downstream, (NE) of FORT PECK Dam, in SW corner of Fort Peck Indian Reservation. The Milk River reclamation project (established 1911) irrigates c.134,000 acres/54,230 ha. The largest of several dams is the Fresno Dam (completed 1939). Malta, CHINOOK, Glasgow, and HARLEM (Montana) are in the project area.

Milka (MEEL-kah), island in the DANUBE, NE BULGARIA; 43°43′N 25°12′E.

Milkovitsa (mil-KOV-eet-sah), village, LOVECH oblast, GULYANTSI obshtina, N BULGARIA, 4 mi/6 km SSW of SOMOVIT; 43°37′N 24°45′E. Grain, vegetables; livestock. Formerly called Gavren or Gaurene.

Mil'kovo (MEEL-kuh-vuh), village (2005 population 8,560), central KAMCHATKA oblast, RUSSIAN FAR EAST, on S central KAMCHATKA PENINSULA, on the KAMCHATKA RIVER (head of shallow-draught navigation), on road, 115 mi/185 km N of PETROPAVLOVSK-KAMCHATSKIY; 54°43′N 158°37′E. Elevation 511 ft/155 m. In agricultural area. Has a museum of aboriginal arts and crafts.

Milk River (MILK), town (□ 1 sq mi/2.6 sq km; 2001 population 879), S ALBERTA, W CANADA, near MONTANA border, on MILK River, and 50 mi/80 km SW of LETHBRIDGE, in WARNER COUNTY NO. 5; 49°08′N 112°05′W. Coal mining; mixed farming; ranching; cattle; flax, wheat, sugar beets. Incorporated as a village in 1916; became a town in 1956.

Milk River, c.20 mi/32 km long, CLARENDON parish, S JAMAICA; rises N of PORUS; flows S, past village of Milk River (spa), to the CARIBBEAN SEA; 17°55′N 77°20′W. Abounds in fish. Navigable for 2 mi/3.2 km upstream. Spa waters known as curative.

Millaa Millaa (MI-luh MI-luh), village, NE QUEENSLAND, NE AUSTRALIA, 40 mi/64 km S of CAIRNS, on S edge of ATHERTON TABLELAND; 17°31′S 145°37′E. Railroad terminus; sugar, dairy products. Waterfalls nearby.

Milladore (MIL-uh-dor), village (2006 population 257), WOOD county, central WISCONSIN, 14 mi/23 km N of WISCONSIN RAPIDS; 44°36′N 89°50′W. In dairy belt.

Millard, county (□ 6,828 sq mi/17,752.8 sq km; 2006 population 12,390), W UTAH; ⊙ FILLMORE; 39°02′N 113°05′W. Agricultural area bordering on NEVADA (W) and watered by SEVIER River, which flows into intermittent SEVIER LAKE at center of county. Irrigated lands around DELTA produce alfalfa, wheat, sugar beets, barley. Dairying; beryllium ore, mining, limestone, precious metals. Fort Deseret State Historical Park in NE corner. Kanosh Indian Reservation in SE. Desert Range Experimental Station in SW. Fishlake National Forest and PAVANT MOUNTAINS in E, semiarid region in W. County formed 1852.

Millard, suburb of OMAHA, DOUGLAS county, E NEBRASKA, 8 mi/12.9 km WSW of downtown. Part of city of Omaha.

Millares (mi-YAH-res), town and canton, CORNELIO SAAVEDRA province, POTOSÍ department, S central BOLIVIA, on Mataca River (branch of the PILCOMAYO River), 38 mi/61 km ENE of POTOSÍ; 19°25′S 65°12′W.

Millares (mee-LYAH-res), town, VALENCIA province, E SPAIN, 20 mi/32 km WNW of ALZIRA. Olive oil, wine; sheep.

Millarton (MI-luhr-tuhn), unincorporated village, SW ONTARIO, E central CANADA, included in KINCARDINE; 44°09′N 81°34′W.

Millarville (MI-luhr-vil), hamlet, SW ALBERTA, W CANADA, 15 mi/24 km W of OKOTOKS, in FOOTHILLS NO. 31 municipal district; 50°45′N 114°19′W.

Millas (mee-YAHS), commune (□ 7 sq mi/18.2 sq km), PYRÉNÉES-ORIENTALES department, LANGUEDOC-ROUSSILLON region, S FRANCE, on the TÊT, 10 mi/16 km W of PERPIGNAN; 42°42′N 02°42′E. Wine trade; alcohol distilling, olive oil processing, fruit and vegetable shipping.

Millau (mee-yo), town (□ 65 sq mi/169 sq km), AVEYRON department, MIDI-PYRÉNÉES region, S FRANCE, on the TARN RIVER, 46 mi/76 km ENE of ALBI; 44°05′N 02°55′E. The historic center of the French glove industry (since 12th century), Millau specializes in the preparation of lamb skins, dyeing, tanning, and leatherworking, not only for the glove trade but also for clothing, accessories, and home furnishings. In recent years it has diversified into printing, hosiery, and electronics.

Millau is also a tourist center for visitations of the Gorges du Tarn (upstream of Le Rozier) and the severe limestone plateaus of the CAUSSES, noted for their *karst* topography, marked by caves and sinkholes. At the nearby village of ROQUEFORT-SUR-SOULZON, the famous cheese is made from sheep's milk. Millau was a Huguenot stronghold in the 16th century. Points of interest include ruins of a Gallo-Roman place where pottery was made, a 16th–17th-century church, a belfry (partly from the 12th century), and a picturesque medieval town square. The town has become a staging point for hang-gliding enthusiasts. In December 2004, the Millau Bridge, the world's highest (taller than the EIFFEL TOWER) opened. It spans the Tarn Valley, and is now part of the highway espansion that goes from northern to southern France.

Millbank (MIL-bank), unincorporated village, S ONTARIO, E central CANADA, on tributary of NITH RIVER, 18 mi/29 km WNW of KITCHENER, and included in PERTH EAST; 43°34′N 80°50′W. Dairying; mixed farming.

Mill Basin, SE section of BROOKLYN borough of NEW YORK city, SE NEW YORK, bounded on E, S, and W by Mill Creek Basin (an inlet of JAMAICA BAY), and on N by Avenue U. Land sold to Dutch by Native Americans in 1664. Until early 20th century economy was largely based on abundant shellfish from its tidal marshlands, estuaries, and Jamaica Bay. In 1906, industrial development of the marshlands, channel dredging, and construction of wharves began. The area's seedy industrial character began to become more residential after World War II. Now this is one of Brooklyn's most exclusive sections, with large homes, circular streets, and private boat docks. Historically the population was largely Italian and Irish, but this has begun to change in recent years.

Millboro (MIL-buh-ro), unincorporated village, BATH county, NW VIRGINIA, in ALLEGHENY MOUNTAINS, 15 mi/24 km NNW of LEXINGTON, in George Washington National Forest; 37°58′N 79°36′W. Manufacturing (handicrafts, apparel); in agricultural area (cattle); timber. Resort area.

Millboro Springs (MIL-buh-ro SPREENGZ), unincorporated village, BATH county, NW VIRGINIA, 17 mi/27 km NNW of LEXINGTON, on COWPASTURE RIVER, in George Washington National Forest. Agriculture (cattle); timber. Resort area.

Millbourne, borough (2006 population 915), DELAWARE county, SE PENNSYLVANIA, residential suburb, 4 mi/6.4 km W of PHILADELPHIA, on Cobbs Creek; 39°57′N 75°15′W.

Millbrae, city (2000 population 20,718), SAN MATEO county, W CALIFORNIA; residential suburb 12 mi/19 km S of SAN FRANCISCO, on SAN FRANCISCO BAY; 37°36′N 122°24′W. Light manufacturing. San Francisco International Airport adjoins city to NE. San Francisco State Fish and Game Refuge, including SAN ANDREAS reservoir, to SW (on SAN ANDREAS FAULT). Incorporated 1948.

Millbridge, MAINE: see MILBRIDGE.

Millbrook, city (2000 population 10,386), Elmore co., E central Alabama, 8 mi/12.9 km N of Montgomery; 32°29′N 86°22′W. Manufacturing of aircraft engine parts and apparel. Inc. in 1971. Name is descriptive of the location by a small stream.

Millbrook (MIL-bruk), village (□ 1 sq mi/2.6 sq km; 2001 population 1,338), SW ONTARIO, E central CA-

NADA, 15 mi/24 km SW of PETERBOROUGH, and included in CAVAN-MILLBROOK-NORTH MONAGHAN township; 44°09′N 78°27′W. Lumbering; dairying; mixed farming.

Millbrook (MIL-bruhk), village (2001 population 2,033), SE CORNWALL, SW ENGLAND, on inlet of TAMAR RIVER, and 4 mi/6.4 km WSW of PLYMOUTH; 50°22′N 04°11′W.

Millbrook, residential and resort village (□ 1 sq mi/2.6 sq km; 2006 population 1,543), DUTCHESS county, SE NEW YORK, 12 mi/19 km NE of POUGHKEEPSIE; 41°46′N 73°41′W. In dairying and limited stock-raising area. Seat of Institute of Ecosystem Studies; Millbrook Preparatory School. Many estates and second homes for New Yorkers. Noted for polo playing. Incorporated 1896.

Millbrook, MASSACHUSETTS: see DUXBURY.

Millburn, residential township (2000 population 19,765), ESSEX county, NE NEW JERSEY, on RAHWAY and PASSAIC rivers, 7 mi/11.3 km W of NEWARK; 40°44′N 74°19′W. Includes SHORT HILLS, site of shopping mall. The Paper Mill Playhouse, the oldest continually running nonprofit playhouse, is here. Settled c.1725, incorporated 1857.

Millbury, town (□ 16 sq mi/41.6 sq km), WORCESTER county, S MASSACHUSETTS, on BLACKSTONE River, 6 mi/9.7 km SSE of WORCESTER; 42°11′N 71°47′W. Woolens, textile supplies, wire, tools, castings, and veterinary pharmaceuticals. Includes village of W. Millbury. In 1870, town had the world's largest felt mill. Settled 1716, incorporated 1813.

Millbury (MIL-buhr-ee), village (□ 1 sq mi/2.6 sq km; 2006 population 1,151), Wood county, NW OHIO, 9 mi/14 km SE of TOLEDO; 41°34′N 83°25′W.

Mill City, town (2006 population 1,625), LINN and MARION counties, W OREGON, on North SANTIAM RIVER, 30 mi/48 km ESE of SALEM; 44°45′N 122°28′W. Lumber. Grain, fruit; dairy products. North Santiam State Park to W; Willamette National Forest to E. Incorporated 1947.

Millcreek, township, ERIE county, NW PENNSYLVANIA, residential suburb, 4 mi/6.4 km SW of ERIE; 42°04′N 80°09′W. Includes the communities of Glenruadh, Westminster, Lakewood, Eaglehurst, Charter Oaks, Highland Park, CHESTNUT HILL, Kearsarge, and Bell Valley. Erie International Airport in NW.

Mill Creek, unincorporated town (2006 population 15,586), SNOHOMISH county, NW WASHINGTON, residential suburb 18 mi/29 km NNE of SEATTLE, and 10 mi/16 km S of EVERETT; 47°52′N 122°13′W. Agricultural area (dairying; poultry; berries, vegetables) in rapidly growing urban fringe.

Mill Creek, village (2000 population 78), UNION county, S ILLINOIS, 8 mi/12.9 km S of JONESBORO; 37°20′N 89°15′W.

Mill Creek, village (2006 population 332), JOHNSTON county, S OKLAHOMA, 10 mi/16 km SE of SULPHUR; 34°23′N 96°49′W. Sand quarrying.

Mill Creek, village (2006 population 651), RANDOLPH county, E central WEST VIRGINIA, on TYGART RIVER, 15 mi/24 km SSW of ELKINS; 38°43′N 79°58′W. Manufacturing (lumber). Monongahela National Forest to E, includes Valley Bend Wetland to N.

Mill Creek, borough (2006 population 332), HUNTINGDON county, central PENNSYLVANIA, 5 mi/8 km SE of HUNTINGDON, on JUNIATA RIVER, at mouth of Mill Creek; 40°26′N 77°55′W. Agriculture (corn, hay, alfalfa; livestock); manufacturing (wood products, stoves); glass sand. Swigart Museum, vintage autos. Part of Rothrock State Forest to E.

Mill Creek, c.50 mi/80 km long, N CALIFORNIA; rises in LASSEN VOLCANIC NATIONAL PARK in SE SHASTA county; flows SW into TEHAMA county, through Shasta National Forest to SACRAMENTO RIVER near Tehama, 10 mi/16 km SSE of Red Bluff (in orchard region).

Mill Creek, c.50 mi/80 km long, W central INDIANA; rises in W HENDRICKS county; flows SW and NW to EEL RIVER in SW PUTNAM county.

Milldale, CONNECTICUT: see SOUTHINGTON.

Millecoquins Lake (mil-i-KAH-kinz), MACKINAC county, SE UPPER PENINSULA, MICHIGAN, c.40 mi/64 km ENE of MANISTIQUE; 46°09′N 85°30′W. Lake is c.2.5 mi/4 km long, 1.5 mi/2.4 km wide; drained by Furlong Creek. Resort.

Milledgeville, city (2000 population 18,757), ⊙ BALDWIN county, central GEORGIA, on the OCONEE River, in a fertile agricultural area; 33°05′N 83°14′W. Laid out in 1803 as the site of the state capital, which was there from 1807 to 1868. Manufacturing includes clothing, carpets, aircraft parts, printing and publishing, lumber. Many antebellum homes survive in the Federal Greek Revival and Classical Revival styles. The old state capitol (1807) is now part of Georgia Military College, a prep school. The city is also the site of Georgia College and State University, a unit of the University System of Georgia; the university president maintains an office in the former governor's mansion. Central State Hospital, founded in 1837 as a mental institution, is now used as a state prison. Incorporated 1836.

Milledgeville, village (2000 population 1,016), CARROLL county, NW ILLINOIS, 14 mi/23 km NNW of STERLING, on ELKHORN CREEK; 41°57′N 89°46′W. Dairying; corn; manufacturing (fabricated aluminum).

Milledgeville (MIL-uhj-vil), village (2006 population 117), FAYETTE county, S central OHIO, 9 mi/14 km WNW of WASHINGTON COURT HOUSE; 39°35′N 83°35′W.

Mille Îles River or **Milles Îles River** (both: meel EEL), S QUEBEC, E CANADA, branch of OTTAWA River, flowing from Lake of the TWO MOUNTAINS NE along shore of Jesus Island to the SAINT LAWRENCE RIVER; 45°42′N 73°31′W. Also spelled Mille Isles.

Mille-Isles (MEEL–EEL), village (□ 23 sq mi/59.8 sq km; 2006 population 1,206), LAURENTIDES region, S QUEBEC, E CANADA, 10 mi/16 km from SAINTE-ADÈLE; 45°49′N 74°13′W.

Mille Lacs (MIL laks), county (□ 681 sq mi/1,770.6 sq km; 2006 population 26,169), E central MINNESOTA; ⊙ MILACA; 45°55′N 93°37′W. Drained by RUM RIVER. Resort area; agriculture (alfalfa, hay, corn, oats, barley, rye; hogs, cattle, poultry; dairying); timber; peat; sand and gravel. Part of Mille Lacs Wildlife Area here. S half of MILLE LACS LAKE in N; parts of Mille Lacs Indian Reservation in NW and NE corners; Father Hennepin State Park in NE; Mille Lacs Kathio State Park in NW, both on shore of Mille Lacs Lake; parts of Rum River State Forest in NW and E center. Formed 1857.

Mille Lacs, Lac des (MEEL LAHK, lahk dai), lake (□ 102 sq mi/265.2 sq km), NW ONTARIO, E central CANADA, 60 mi/97 km WNW of PORT ARTHUR; 48°50′N 90°30′W. Elevation 1,496 ft/456 m; 18 mi/29 km long, 12 mi/19 km wide. Drains SW into RAINY LAKE.

Mille Lacs Lake (□ 207 sq mi/536 sq km), AITKIN, MILLE LACS, and CROW WING counties, NE central MINNESOTA, 90 mi/145 km NNW of MINNEAPOLIS; 46°16′N 93°48′W. Elevation 1,251 ft/381 m. Minnesota's third-largest lake, after RED and LEECH lakes. Drains S through the RUM RIVER Marshy area; lake fed by small side streams; numerous small lakes to W and N. Sieur Duluth, a French explorer, visited (1679) the Ojibwas who lived on the lake. In 1680, Louis Hennepin, a French friar and explorer of NORTH AMERICA, and his companions were held captive near the lake by the Ojibwas for several weeks. The region is a center for tourists and sportsmen. Part of Mille Lacs Indian Reservation is on the SW shore. Wealthwood State Forest on N shore; Father Hennepin State Park on SE shore; Mille Lacs Kathio State Park on SW shore.

Millen (MIL-uhn), town (2000 population 3,492), ⊙ JENKINS county, E GEORGIA, on OGEECHEE RIVER, c.45 mi/72 km S of AUGUSTA; 32°49′N 81°56′W. Big buckhead church established in 1787 located nearby. Manufacturing includes motor vehicles, clothing,

fertilizer. Magnolia Spring State Park nearby. Settled early 1830s; incorporated 1881.

Miller, county (□ 637 sq mi/1,656.2 sq km; 2006 population 43,055), extreme SW ARKANSAS; ⊙ TEXARKANA; 33°18′N 93°52′W. Bounded W by TEXAS, S by LOUISIANA, E and N by RED RIVER; drained by SULPHUR RIVER. Agriculture (wheat, soybeans; cattle, hogs, chickens). Timber; oil and gas. Manufacturing at Texarkana. Sulphur River Wildlife Management Area in S. Formed 1874.

Miller, county (□ 284 sq mi/738.4 sq km; 2006 population 6,239), SW GEORGIA; ⊙ COLQUITT; 31°10′N 84°44′W. Coastal plain agriculture (peanuts, sugarcane, cotton, oats); cattle, hogs. Area drained by SPRING CREEK. Formed 1856.

Miller, county (□ 603 sq mi/1,567.8 sq km; 2006 population 24,989), central MISSOURI; ⊙ TUSCUMBIA; 38°13′N 92°25′W. In OZARK region; drained by OSAGE RIVER. Resort, recreational, and commercial development at Osage Beach and LAKE OZARK. Corn, wheat; cattle, poultry; timber; hydroelectricity; manufacturing at ELDON, IBERIA, and SAINT ELIZABETH. Bagnell Dam on Osage River forms LAKE OF THE OZARKS. Major recreation area about equidistant between ST. LOUIS and KANSAS CITY. Lake of the Ozarks State Park in SW (largest state park in Missouri). Formed 1837.

Miller, city (2006 population 1,365), ⊙ HAND county, central SOUTH DAKOTA, 70 mi/113 km ENE of PIERRE, and on Turtle Creek; 44°31′N 98°59′W. In agricultural region. Lake Louise State Recreation Area to NW. Settled 1882, incorporated as city 1910.

Miller, town (2000 population 754), LAWRENCE county, SW MISSOURI, 7 mi/11.3 km N of MOUNT VERNON; 37°13′N 93°50′W. Agriculture, flour mills. Corn, hay. Cattle; dairying.

Miller, village (2006 population 153), BUFFALO county, S central NEBRASKA, 23 mi/37 km NW of KEARNEY, and on WOOD RIVER; 40°55′N 99°23′W.

Miller City, village (2006 population 129), PUTNAM county, NW OHIO, 7 mi/11 km NW of OTTAWA; 41°06′N 84°08′W.

Miller Field, tract of open parkland on E side of STATEN ISLAND borough of NEW YORK city, SE NEW YORK, 2.5 mi/4 km NE of Great Kills Harbor; 40°35′N 74°05′W. Area is 203 acres/82-ha. Part of the 6-mi/9.7-km GATEWAY NATIONAL RECREATION AREA fronting on Lower NEW YORK BAY and the Atlantic.

Miller, Mount (11,000 ft/3,353 m), S ALASKA, in Robinson Mountains 30 mi/48 km NNE of Cape YAKATAGA; 60°28′N 142°14′W.

Mille Roches (meel ROSH), unincorporated village, SE ONTARIO, E central CANADA, on Cornwall Canal, 5 mi/8 km W of CORNWALL, and included in SOUTH STORMONT township. Dairying, mixed farming.

Millerovo (MEEL-lye-ruh-vuh), city (2006 population 38,550), NW ROSTOV oblast, S European Russia, on the Glubokaya River (tributary of the N DONETS River), 133 mi/214 km NNE of ROSTOV-NA-DONU; 48°55′N 40°23′E. Elevation 465 ft/141 m. Railroad and road junction (shops); agricultural center; agricultural machinery, metal, rubber, and food industries. Founded in 1786. Became city in 1926.

Miller Peak (9,466 ft/2,885 m), COCHISE county, SE ARIZONA, highest in HUACHUCA MOUNTAINS, near Mexican border, 11 mi/18 km S of FORT HUACHUCA.

Millersburg, town (2000 population 868), ELKHART county, N INDIANA, 9 mi/14.5 km SE of GOSHEN; 41°32′N 85°42′W. Manufacturing (motor vehicles). Laid out 1855.

Millersburg, town (2000 population 184), IOWA county, E central IOWA, 15 mi/24 km S of MARENGO; 41°34′N 92°09′W. In agricultural area.

Millersburg, town (2000 population 842), BOURBON county, N central KENTUCKY, 8 mi/12.9 km NE of PARIS, on Hinkston Creek, in BLUEGRASS REGION; 38°17′N 84°09′W. Manufacturing (machinery). Colville Covered Bridge to NW.

Millersburg, village (2000 population 263), PRESQUE ISLE county, NE MICHIGAN, 13 mi/21 km SW of ROGERS CITY, and on short Ocqueoc River; 45°19′N 84°03′W. In farm area. Lumber.

Millersburg (MIL-uhrz-buhrg), village (□ 2 sq mi/5.2 sq km; 2006 population 3,581), ⊙ HOLMES county, central OHIO, 32 mi/51 km SW of CANTON, on KILLBUCK CREEK; 40°33′N 81°55′W. In agricultural area; dairy products, rubber products, furniture. Coal mines. Settled 1816.

Millersburg, village (2006 population 661), LINN county, W OREGON, suburb, 2 mi/3.2 km NE of ALBANY, on WILLAMETTE RIVER; 44°40′N 123°04′W. Albany Municipal Airport to SE.

Millersburg, borough (2006 population 2,467), DAUPHIN county, central PENNSYLVANIA, 20 mi/32 km NNW of HARRISBURG, on SUSQUEHANNA RIVER (ferry), at mouth of Wiconisco Creek; 40°32′N 76°57′W. Agriculture (grain, soybeans; livestock; dairying); manufacturing (plastic products, tools, contract embroidery, textiles, consumer goods). Berry Mountain ridge to S, Mahantango Mountain ridge to W. Settled c.1790, laid out 1807, incorporated 1850.

Millers Creek (MIL-uhrz KREEK), unincorporated town (□ 4 sq mi/10.4 sq km; 2000 population 2,071), WILKES county, NW NORTH CAROLINA, 6 mi/9.7 km NW of WILKESBORO; 36°11′N 81°14′W. Rendezvous Mountain State Educational Forest to W; W. Kerr Scott reservoir (Yadkin River) to S. Manufacturing (lumber); timber; agriculture (tobacco, soybeans; poultry, cattle; dairying).

Millers Creek Reservoir (MIL-uhrz) (□ 5 sq mi/13 sq km), THROCKMORTON and BAYLOR counties, N central TEXAS, on Millers Creek, 55 mi/88 km SW of WICHITA FALLS; 33°25′N 99°22′W. Maximum capacity 131,000 acre-ft. Formed by Millers Creek Dam (75 ft/23 m high), built (1974) for water supply.

Miller's Dale (MIL-uhz DAIL), town (2006 population 200), NW DERBYSHIRE, central ENGLAND, on WYE RIVER, and 5 mi/8 km E of BUXTON; 53°15′N 01°47′W. Just W is village of Wormhill.

Millers Falls, village (2000 population 1,072) in ERVING and MONTAGUE towns, FRANKLIN county, NW MASSACHUSETTS, on MILLERS RIVER, and 5 mi/8 km E of GREENFIELD; 42°35′N 72°29′W. Tool manufacturing.

Millers Ferry Dam, ALABAMA: see WILLIAM "BILL" DANNELLY RESERVOIR.

Millers Mills, hamlet, HERKIMER county, central NEW YORK, 15 mi/24 km SSE of UTICA, on Little Unadilla Lake; 42°55′N 75°05′W. In winter, cutting of slabs of lake ice is a tradition and popular attraction here.

Millersport, village (□ 1 sq mi/2.6 sq km; 2006 population 961), FAIRFIELD county, central OHIO, 24 mi/39 km E of COLUMBUS, near BUCKEYE LAKE reservoir (resort); 39°54′N 82°32′W. In agricultural area.

Millers River, c.60 mi/97 km long, N MASSACHUSETTS; rises in N WORCESTER county; flows SW and W to the CONNECTICUT River c.5 mi/8 km E of GREENFIELD.

Millers Tavern (MI-luhrz TA-vuhrn), unincorporated village, ESSEX county, E VIRGINIA, 37 mi/60 km NE of RICHMOND; 37°49′N 76°56′W. Manufacturing (lumber, motor vehicle parts); agriculture (grain, soybeans; cattle); timber.

Millerstown, borough (2006 population 684), PERRY county, central PENNSYLVANIA, 25 mi/40 km NW of HARRISBURG, on JUNIATA RIVER; 40°32′N 77°09′W. Agriculture (corn, hay; dairying); manufacturing (lumber, wood products, food products). TUSCARORA MOUNTAIN to NW.

Millersville, city (2006 population 6,233), SUMNER county, N central TENNESSEE, 15 mi/24 km NNE of NASHVILLE; 36°22′N 86°42′W. Incorporated 1981.

Millersville, borough (2006 population 7,271), LANCASTER county, SE PENNSYLVANIA, 4 mi/6.4 km SW of LANCASTER, near CONESTOGA RIVER; 40°00′N 76°20′W. Agriculture (grain, soybeans, apples; livestock; dairying); light manufacturing. Seat of Mill-ersville University of Pennsylvania. Incorporated 1932.

Millerton, town (2000 population 48), WAYNE county, S IOWA, 6 mi/9.7 km N of CORYDON; 40°51′N 93°18′W. Livestock; grain.

Millerton, village (2006 population 922), DUTCHESS county, SE NEW YORK, near CONNECTICUT border, 28 mi/45 km NE of POUGHKEEPSIE; 41°57′N 73°30′W. In dairying area. Working-class population. Incorporated 1875.

Millerton, village (2006 population 356), MCCURTAIN county, SE OKLAHOMA, 12 mi/19 km WNW of IDABEL; 33°58′N 95°01′W. In agricultural area.

Millerton Lake, reservoir (15 mi/24 km long), central CALIFORNIA, on SAN JOAQUIN RIVER (border of FRESNO and MADERA counties), 17 mi/27 km NNE of FRESNO; 37°00′N 119°41′W. Elevation 561 ft/171 m. Extends NE. Formed by Friant Dam (3,430 ft/1,045 m long, 320 ft/98 m high; completed 1944), key irrigation and flood-control unit of CENTRAL VALLEY project. MADERA and FRIANT-KERN canals extend from dam to irrigate the valley farms. Dam destroyed fish-spawning run. Millerton Lake State Recreation Area surrounds reservoir.

Millertown, NEW ZEALAND: see WESTPORT.

Millerville, village (2000 population 115), DOUGLAS county, W MINNESOTA, 16 mi/26 km NW of ALEXANDRIA, near CHIPPEWA RIVER, in region of small natural lakes; 46°04′N 95°32′W. Grain; livestock; dairying. Inspiration Peak State Park is to N. Aaron and Moses lakes to W.

Milles Îles River, CANADA: see MILLE ÎLES RIVER.

Millesimo (meel-LAI-zee-mo), village, SAVONA province, LIGURIA, NW ITALY, on Bormida di Millesimo River, 14 mi/23 km WNW of SAVONA; 44°22′N 08°12′E. Limekilns. Noted for victory of Napoleon over Austrians in April 1796.

Millet (MI-lit), town (□ 2 sq mi/5.2 sq km; 2001 population 2,037), central ALBERTA, W CANADA, 30 mi/48 km S of EDMONTON, in WETASKIWIN COUNTY NO. 10; 53°05′N 113°28′W. Coal mining; oil and gas; barley, wheat; cattle. Established as a village in 1903; became a town in 1983.

Millevaches, Plateau of (meel-vah-shuh), tableland of the MASSIF CENTRAL and LIMOUSIN region, CORRÈZE and CREUSE departments, central FRANCE, forming part of France's central watershed, which divides the LOIRE from the GARONNE drainage basins; 45°45′N 02°11′E. Rises to c.3,205 ft/977 m at Mont de Bessou.

Millfield, village, ATHENS county, SE OHIO, 8 mi/13 km N of ATHENS; 39°26′N 82°06′W. In coal region. Mine disaster here (1930) killed over eighty men.

Mill Grove, unincorporated community, MERCER county, N MISSOURI, on WELDON RIVER, 7 mi/11.3 km S of PRINCETON.

Millgrove (MIL-grov), township, VICTORIA, SE AUSTRALIA, suburb 39 mi/63 km E of MELBOURNE; 37°46′S 145°39′E. Formerly timber, agricultural industries.

Mill Hall, borough (2006 population 1,480), CLINTON county, N central PENNSYLVANIA, 2 mi/3.2 km SW of LOCK HAVEN, near Bald Eagle Creek; 41°06′N 77°29′W. Drained by Fishing Creek. Agriculture (grain; livestock; dairying); manufacturing (consumer goods, cosmetics, lumber, food products). Laid out 1806, incorporated 1850.

Millhaven (MIL-hai-vuhn), unincorporated village, SE ONTARIO, E central CANADA, 12 mi/19 km WSW of KINGSTON, on NORTH CHANNEL, Lake ONTARIO, and included in LOYALIST township; 44°12′N 76°45′W. Ferry to AMHERST ISLAND. Mixed farming; dairying; apples.

Millheim, borough (2006 population 735), CENTRE county, central PENNSYLVANIA, 16 mi/26 km E of BELLEFONTE, on Elk Creek, in Penns Valley; 40°53′N 77°28′W. Agriculture (corn, wheat, hay, vegetables; livestock, dairying; timber); manufacturing (elec-

tronic equipment). Poe Valley State Park to S; Penn's Cave to W; part of Bald Eagle State Forest to SE.

Mill Hill, ENGLAND: see BARNET.

Millhousen, town (2000 population 136), DECATUR county, SE central INDIANA, 9 mi/14.5 km S of GREENSBURG; 39°13′N 85°26′W. In agricultural area. Near MUSCATATUCK RIVER. Settled 1838, plotted 1858.

Millicent (MI-li-sent), town, SE SOUTH AUSTRALIA, AUSTRALIA, 210 mi/338 km SSE of ADELAIDE, NW of MOUNT GAMBIER; 37°36′S 140°21′E. Railroad terminus. Sheep, grain, oil seed, barley; cattle, lambs, crayfish; wool, Monterey pine plantation. Paper mill nearby. National Trust museum; shell garden.

Milligan, village (2006 population 293), FILLMORE county, SE NEBRASKA, 10 mi/16 km E of GENEVA, and on branch of BIG BLUE RIVER; 40°30′N 97°23′W. Grain; feeds and alfalfa pellets.

Milligan College, town, CARTER county, NE TENNESSEE, 4 mi/6 km ESE of JOHNSON CITY. Milligan College (1881) here.

Milliken, town (2000 population 2,888), WELD county, N COLORADO, on BIG THOMPSON RIVER, at mouth of Little Thompson River, 10 mi/16 km SW of GREELEY; 40°19′N 104°50′W. Elevation 4,760 ft/1,451m. Sugar beets, beans, wheat, barley, oats, vegetables, fruits; cattle. Manufacturing.

Millikenpark (MIL-i-kuhn-pahrk), town, Renfrewshire, W Scotland, just N of JOHNSTONE; 55°49′N 04°32′W. Previously paper milling. Formerly in Strathclyde, abolished 1996.

Millington, city (2006 population 10,336), SHELBY county, SW TENNESSEE, 14 mi/23 km N of MEMPHIS; 35°20′N 89°52′W. In an agricultural region; manufacturing. The U.S. Naval Air Station provides a major source of employment. Incorporated 1903.

Millington, town (2000 population 416), KENT and QUEEN ANNES counties, E MARYLAND, 18 mi/29 km WNW of DOVER (DELAWARE); 39°16′N 75°50′W. Was first called Head of Chester for its location at the head of navigation on CHESTER RIVER; the name was changed in 1827 because of the numerous mills in the area. The Higman Mill, whose foundations are believed to date back to the 1760s, ground corn with water wheel and millstone until the early 1950s. Millington Wildlife Refuge is nearby.

Millington, town (2000 population 1,137), TUSCOLA county, E MICHIGAN, 19 mi/31 km NNE of FLINT; 43°16′N 83°31′W. In agricultural area (potatoes, beans, wheat, corn, soybeans, sugar beets; poultry, hogs); manufacturing (fiberglass products, steel fabrication). Murphy Lake to E.

Millington, village, MORRIS county, N central NEW JERSEY, on PASSAIC River, 8 mi/12.9 km SSW of MORRISTOWN; 40°40′N 74°31′W. Suburbanizing area.

Millington, CONNECTICUT: see EAST HADDAM.

Millinocket (mil-uh-NAHK-et), town, PENOBSCOT county, central MAINE, on West Branch of PENOBSCOT RIVER; 45°38′N 68°42′W. Developed around paper mills built here 1899–1900. Wood products are still important here. MILLINOCKET LAKE (c.6 mi/9.7 km wide) and Mount Katahdin are NW. Incorporated 1901.

Millinocket Lake (mil-uh-NAHK-et), PISCATAQUIS county, N central MAINE, 43 mi/69 km NNW of MILLINOCKET; 46°13′N 68°50′W. Lake is 3 mi/4.8 km long. Logging, recreation (winter sports).

Millinocket Lake (mil-uh-NAHK-et), reservoir, PENOBSCOT and PISCATAQUIS counties, N central MAINE, on Sandy Stream, small branch of PENOBSCOT RIVER, 5 mi/8 km NNW of MILLINOCKET; 45°43′N 68°43′W. Formed by dam; 6 mi/9.7 km long, 4 mi/6.4 km wide. Mount Katahdin, in Baxter State Park, to NW.

Millis, town, NORFOLK county, E MASSACHUSETTS, on CHARLES River, 20 mi/32 km SW of BOSTON; 42°10′N 71°22′W. Settled 1657, incorporated 1885.

Mill Island, off E end of QUEEN MARY COAST, ANTARCTICA; 65°30′S 100°40′E. Ice-domed island is 25 mi/

40 km long; 16 mi/26 km wide. Discovered 1936 by British expedition.

Mill Island, NUNAVUT territory, CANADA, in HUDSON STRAIT, at S end of FOXE CHANNEL; 20 mi/32 km long, 14 mi/23 km wide; 63°59′N 78°00′W.

Millmerran (mil-ME-ruhn), village, SE QUEENSLAND, AUSTRALIA, 110 mi/177 km WSW of BRISBANE; 27°52′S 151°16′E. Railroad terminus; mixed agriculture (sheep, cattle, grains, cotton, and wheat).

Mill Neck, residential village (□ 2 sq mi/5.2 sq km; 2006 population 856), NASSAU county, SE NEW YORK, on N shore of LONG ISLAND, on an inlet of OYSTER BAY Harbor, 2 mi/3.2 km NW of OYSTER BAY village; 40°52′N 73°33′W.

Millom (MIL-uhm), town (2001 population 1,430), CUMBRIA, NW ENGLAND, on DUDDON RIVER estuary, 7 mi/11.3 km NNW of BARROW-IN-FURNESS; 54°12′N 03°17′W. Previously limestone quarrying and wool weaving. Has remains of 14th-century castle.

Mill Park (MIL PAHRK), suburb 11 mi/18 km N of MELBOURNE, VICTORIA, SE AUSTRALIA, N of BUNDOORA and THOMASTOWN.

Millport (MIL-port), town (2001 population 1,253), North Ayrshire, SW Scotland, on S coast of GREAT CUMBRAE island, 4 mi/6.4 km SW of LARGS; 55°45′N 04°56′W. Fishing port, resort. Site of marine biological station.

Millport, town (2000 population 1,160), Lamar co., W Alabama, 14 mi/23 km S of Vernon. Moved three miles from its original site in 1882 when the Georgia Pacific RR was built. Named for the sawmill, planing mill, and gristmill once located in the town. Inc. in 1887.

Millport, village (2006 population 288), CHEMUNG county, S NEW YORK, 12 mi/19 km N of ELMIRA; 42°16′N 76°50′W. In agricultural area.

Mill River, c.17 mi/27 km, SW CONNECTICUT; rises W of MONROE; flows S to LONG ISLAND SOUND, forming harbor at FAIRFIELD. Dam on river forms Easton Reservoir. 4 mi/6.4 km NW of TRUMBULL.

Mill River, c.25 mi/40 km, W central MASSACHUSETTS; rises in ponds in N HAMPSHIRE county; flows SE to the CONNECTICUT at NORTHAMPTON.

Mill River, MASSACHUSETTS: see NEW MARLBORO.

Millry, town (2000 population 615), Washington co., SW Alabama, 11 mi/18 km N of Chatom. Lumber; clothing. Named for Mill Creek. Inc. in 1848.

Mills, county (□ 439 sq mi/1,141.4 sq km; 2006 population 15,595), SW IOWA, on NEBRASKA state line (W; formed here by MISSOURI RIVER); ⊙ GLENWOOD; 41°01′N 95°37′W. Prairie agricultural area (hogs, cattle, poultry; corn, oats) drained by WEST NISHNABOTNA RIVER and by KEG and SILVER creeks. Bituminous-coal deposits. Widespread river flooding in 1993. Formed 1851.

Mills (MILZ), county (□ 749 sq mi/1,947.4 sq km; 2006 population 5,184), central TEXAS; ⊙ GOLDTHWAITE; 31°30′N 98°35′W. Bounded SW by COLORADO RIVER; drained by PECAN BAYOU and other tributaries. Ranching area: sheep, goats, beef and dairy cattle; grains, pecans. Formed 1887.

Mills, town (2006 population 2,890), NATRONA county, central WYOMING, on North Platte River, suburb, 2 mi/3.2 km W of CASPER; 42°51′N 106°22′W. Manufacturing (motor vehicles; fabricated metal). Fort Casper and fairgrounds here. Site of Mormon Ferry built (1847) by Brigham Young.

Mills, unincorporated village, HARDING county, NE NEW MEXICO, 26 mi/42 km NW of MOSQUERO, in section of Kiowa National Grasslands. CANADIAN RIVER Canyon to W; Chicosa Lake State Park to SE.

Millsap (MIL-sap), village (2006 population 401), PARKER county, N central TEXAS, 38 mi/61 km W of FORT WORTH; 32°45′N 98°00′W. Agricultural area (cattle, horses; peanuts, pecans). Stone, clay. Manufacturing (crushed rock and asphalt, brick). Lake Mineral Wells State Park to N.

Area is shown by the symbol □, and capital city or county seat by ⊙.

Millsboro, town (2000 population 2,360), SUSSEX county DELAWARE, 8 mi/12.9 km SSE of GEORGETOWN, and on INDIAN RIVER; 38°35′N 75°17′W. Elevation 19 ft/5 m. Agriculture, manufacturing. CYPRESS SWAMP to S. Founded 1792. In early 19th century, over 15 grist mills and sawmills were in the area.

Millsfield, town, COOS county, N NEW HAMPSHIRE, 23 mi/37 km NNW of BERLIN. Drained by Clear Stream. Timber; livestock; dairying.

Mill Shoals, village (2000 population 235), WHITE county, SE ILLINOIS, near SKILLET FORK, 15 mi/24 km NW of CARMI; 38°15′N 88°20′W. In agricultural area.

Mill Springs, village, WAYNE county, SE KENTUCKY, 9 mi/14.5 km NNE of MONTICELLO, on the CUMBERLAND RIVER (forms LAKE CUMBERLAND reservoir). Mill Springs Park, site of the opening battle of the Kentucky-TENNESSEE campaign of the Civil War and the first important Union victory in the West (January 19, 1862), is here. Includes 1840 grist mill.

Millstadt (MIL-staht), village (2000 population 2,794), SAINT CLAIR county, SW ILLINOIS, 11 mi/18 km SSE of EAST SAINT LOUIS; 38°27′N 90°05′W. Flour milling, manufacturing (food processing systems); bituminous-coal mines; agriculture (wheat, apples; dairy products; hogs, poultry). Incorporated 1878.

Millstatt (MIL-shtaht), village, CARINTHIA, S AUSTRIA, on N shore of the Millstätter See (Millstatt Lake), 3 mi/4.8 km E of SPITTAL AN DER DRAU; resort; 46°48′N 13°34′E. Site of 12th century Benedictine abbey (abandoned 1469), and 12th century Romanesque Gothic church with sculptures.

Millstätter See (MIL-shtet-ter SAI) or **Lake Millstatt** (□ 5 sq mi/13 sq km), in CARINTHIA, S AUSTRIA, just E of SPITTAL AN DER DRAU; 7 mi/11.3 km long, 1 mi/1.6 km wide; 46°48′N 13°35′E. Elevation 1,792 ft/546 m; maximum depth 427 ft/130 m. Resorts of Millstatt and Seeboden on N shore, Dellach and Döbriach on SE shore. Center of summer tourism.

Millstone, village, NEW LONDON county, SE CONNECTICUT. Site of the Millstone Nuclear Power Station, completed in 1969, which serves WATERFORD as well as the larger NEW ENGLAND area.

Millstone, village, LETCHER county, SE KENTUCKY, 3 mi/4.8 km E of WHITESBURG, in the CUMBERLAND MOUNTAINS, on North Fork of KENTUCKY RIVER. Bituminous coal. Jefferson National Forest to SE.

Millstone, borough (2006 population 431), SOMERSET county, central NEW JERSEY, on MILLSTONE RIVER, and 4 mi/6.4 km S of SOMERVILLE; 40°30′N 74°35′W.

Millstone 1, 2, and 3 Nuclear Power Plants, CONNECTICUT: see NEW LONDON county.

Millstone River, c.40 mi/64 km long, central NEW JERSEY; rises SW of FREEHOLD; flows NW and N, past HIGHTSTOWN, PRINCETON (dam here forms Lake Carnegie), and MILLSTONE, to RARITAN River below SOMERVILLE.

Millstream-Chichester National Park (□ 771 sq mi/1,997 sq km), NW WESTERN AUSTRALIA state, W AUSTRALIA, 185 mi/298 km SW of PORT HEDLAND, in PILBARA district; 50 mi/80 km long, 30 mi/48 km wide; 21°25′S 117°20′E. Most of park is dominated by low, stony hills with hummock grass cover. River red gums, paperbarks, fan palms along the FORTESCUE RIVER, part of Millstream Aquifer. Also snappy-gums, cork bark hakeas, coolabahs, date and cotton palms. Red kangaroos, rock wallabies, dingoes, emus, kookaburras, treecreepers, warblers. Camping, picnicking, canoeing, walking tracks. Homeland of the Yinjibarndi Aboriginal people. Established 1964.

Millstreet, Gaelic *Sráid na Mhuilinn*, town (2006 population 1,401), W CORK county, SW IRELAND, on the BLACKWATER RIVER, and 11 mi/18 km SW of KANTURK; 52°04′N 09°04′W. Agricultural market. MacCarthy Drinshane Castle, with adjoining tower built 1436, is nearby.

Millthorpe (MIL-thorp), town, NEW SOUTH WALES, SE AUSTRALIA, 153 mi/246 km NW of SYDNEY, 14 mi/22 km S of ORANGE; 33°27′S 149°11′E. Formerly Spring Grove.

Milltown, former town, SW NEW BRUNSWICK, CANADA, on St. Croix River (international bridge), opposite MILLTOWN and CALAIS, MAINE, and 60 mi/97 km W of SAINT JOHN; 45°10′N 67°18′W. Amalgamated with ST. STEPHEN.

Milltown, town (2000 population 932), CRAWFORD and HARRISON counties, S INDIANA, on BLUE RIVER, and 25 mi/40 km WNW of NEW ALBANY; 38°20′N 86°16′W. Limestone quarrying and processing; poultry hatcheries. Laid out 1839.

Milltown, town (2000 population 888), POLK county, NW WISCONSIN, 39 mi/63 km NNE of HUDSON; 45°31′N 92°30′W. Dairying area; vegetable canning. Light manufacturing.

Milltown (MIL-toun), village, MISSOULA county, W MONTANA, 6 mi/9.7 km E of MISSOULA, on CLARK FORK River, at mouth of BLACKFOOT RIVER. Small Milltown Dam at confluence. Largest lumber mill in Montana. Cattle, horses; hay. Lolo National Forest to N and S. Originally called Riverside, then Finntown.

Milltown, borough (2000 population 7,000), MIDDLESEX county, E NEW JERSEY, 3 mi/4.8 km S of NEW BRUNSWICK; 40°27′N 74°25′W. Manufacturing and agriculture. Settled before 1800, incorporated 1889.

Milltown, IRELAND: see DUBLIN, city.

Milltown, MAINE: see CALAIS.

Milltown Malbay, IRELAND: see MILTOWN MALBAY.

Milluhuaya (mee-yoo-HWAH-yah), canton, NOR YUNGAS province, LA PAZ department, W BOLIVIA, S of CORIPATA. Elevation 5,676 ft/1,730 m. Clay, limestone, phosphate, and gypsum deposits. Agriculture (potatoes, yucca, bananas, rye); cattle.

Millvale, borough (2006 population 3,716), ALLEGHENY county, SW PENNSYLVANIA, residential suburb, 3 mi/4.8 km NNE of downtown PITTSBURGH; 40°28′N 79°58′W. Settled c.1844, incorporated 1868.

Mill Valley, city (2000 population 13,600), MARIN county, W CALIFORNIA, 11 mi/18 km NNW of SAN FRANCISCO; 37°55′N 122°33′W. Residential suburb, set in heavily timbered hills and valleys; redwood trees predominate. Manufacturing (printing and publishing). Golden Gate Baptist Theological Seminary is here. MOUNT TAMALPAIS (2,572 ft/784 m) to NW; Mount Tamalpais Game Refuge in Richardson Bay, to SE; Mount Tamalpais State Park and MUIR WOODS NATIONAL MONUMENT to W; part of GOLDEN GATE NATIONAL RECREATION AREA to SW; GOLDEN GATE BRIDGE to S. Incorporated 1900.

Mill Village, borough (2006 population 397), ERIE county, NW PENNSYLVANIA, 17 mi/27 km SSE of ERIE, on French Creek; 41°52′N 79°58′W. Agriculture (corn, hay, potatoes; dairying); manufacturing (plant food).

Millville, city (2006 population 28,194), CUMBERLAND county, S NEW JERSEY, on the MAURICE River, in an agricultural area that is suburbanizing; 39°23′N 75°02′W. Settled 1756, incorporated 1866. Light industry.

Millville, residential town, WORCESTER county, S MASSACHUSETTS, on BLACKSTONE RIVER, and 20 mi/32 km SE of WORCESTER, at RHODE ISLAND state line; 42°02′N 71°35′W. Settled 1662, set off from BLACKSTONE 1916.

Millville, town (2006 population 1,392), CACHE county, N UTAH, 3 mi/4.8 km S of LOGAN; 41°40′N 111°49′W. Fruit, vegetables, wheat, barley; dairying; cattle. Part of Wasatch National Forest to E. Elevation 4,542 ft/1,384 m. Established 1859.

Millville, village (2000 population 259), SUSSEX county, SE DELAWARE, 18 mi/29 km SE of GEORGETOWN; 38°32′N 75°06′W. Elevation 13 ft/3 m. Vegetables, fruit; livestock. Holts Landing State Park to N, on INDIAN RIVER BAY.

Millville, village, HENRY county, E INDIANA, 6 mi/9.7 km NE of NEW CASTLE. Agricultural area. Wilbur Wright Birthplace State Memorial nearby to N. Laid out 1854.

Millville, village (2000 population 186), WABASHA county, SE MINNESOTA, on ZUMBRO RIVER, 18 mi/29 km NNE of ROCHESTER, in Richard J. Dorer Memorial Hardwood State Forest; 44°15′N 92°17′W. Dairying.

Millville, village (2006 population 882), BUTLER county, extreme SW OHIO, 5 mi/8 km W of HAMILTON; 39°23′N 84°39′W.

Millville, borough (2006 population 954), COLUMBIA county, E central PENNSYLVANIA, 10 mi/16 km NNW of BLOOMSBURG, on Little Fishing Creek; 41°07′N 76°31′W. Agriculture (grain, soybeans, apples; livestock, poultry, dairying); manufacturing (crates, washing equipment, paper filing products). Covered bridges in area.

Millwood, unincorporated town (2000 population 885), SUMTER county, central SOUTH CAROLINA, residential suburb, 2 mi/3.2 km S of SUMTER, on Pocotaligo River; 33°54′N 80°23′W.

Millwood, town (2006 population 1,606), suburb, 7 mi/11.3 km E of downtown SPOKANE, SPOKANE county, E WASHINGTON; 47°42′N 117°17′W. Railroad center. Foits Field Municipal Airport to W.

Millwood Lake, reservoir (□ 45 sq mi/117 sq km), on LITTLE RIVER–HEMPSTEAD county border, Little River county, SW ARKANSAS, on LITTLE RIVER, 7 mi/11.3 km ENE of ASHDOWN; 33°42′N 93°58′W. Maximum capacity 1,854,930 acre-ft. Fed by SALINE RIVER. Formed by Little River Dam (88 ft/27 m high), built (1966) by Army Corps of Engineers for water storage, recreation, and as a fish and wildlife pond. Millwood State Park on S shore.

Milly-la-Forêt (mee-yee-lah-fo-rai), town (□ 13 sq mi/33.8 sq km), ESSONNE department, ÎLE-DE-FRANCE region, N central FRANCE, 11 mi/18 km W of FONTAINEBLEAU, resort at W edge of Forest of Fontainebleau; 48°24′N 02°28′E. It is a leading center for the cultivation of medicinal plants, notably aromatic herbs for culinary use. Has 15th-century church, wooden market hall, and old houses.

Milmarcos (meel-MAHR-kos), village, GUADALAJARA province, central SPAIN, 17 mi/27 km N of MOLINA; 41°05′N 01°52′W. Grain growing; sheep raising.

Milnathort (MILN-uh-thort), town (2001 population 1,738), PERTH AND KINROSS, E Scotland, at foot of OCHIL HILLS, and 2 mi/3.2 km N of KINROSS; 56°14′N 03°25′W. Previously bacon and ham curing. Nearby are ruins of Burleigh Castle.

Milne Bay, province (2000 population 210,412), SE PAPUA NEW GUINEA, SE tip of NEW GUINEA island, and D'ENTRECASTEAUX, Laughlin, SAMARAI, Conflict, TROBRIAND, Louisiande, and WOODLARK island groups; ⊙ ALOTAU. The capital was moved from Samarai to Alotau at W end of MILNE BAY, New Guinea, 1968. Bounded on W by NORTHERN and CENTRAL provinces, N by SOLOMON SEA, S by CORAL SEA. Includes E end of OWEN STANLEY RANGE. FERGUSSON ISLAND noted for hot springs, geysers, volcanoes. Coconuts, palm oil, bananas, yams; lobsters, fish, trepang, prawns. Boatbuilding at Samarai.

Milne Bay, bay at easternmost point of NEW GUINEA island, circa 225 mi/362 km SE of PORT MORESBY, MILNE BAY province, SE PAPUA NEW GUINEA; 15 mi/24 km wide, 30 mi/48 km long; 10°25′S 150°27′E. ALOTAU, the province capital, is at NW end. Site of Allied air base in World War II.

Milne Land, E GREENLAND. Greenlandic *Ilimananngip Nunaa*; 70°45′N 26°15′W. Largest island in the SCORESBY SOUND fjord complex (71 mi/115 km long, 28–12 mi/45–20 km wide); first described and named by W. Scoresby 1822.

Milner, village (2000 population 522), LAMAR county, central GEORGIA, 10 mi/16 km SSE of GRIFFIN; 33°07′N 84°12′W. Manufacturing of lumber and pork rinds.

Milner Pass (10,758 ft/3,279 m), GRAND and LARIMER counties, N central COLORADO, across CONTINENTAL

Cross-references are shown in SMALL CAPITALS. The pronunciation guide is shown on page xix. The sources of population figures are shown on page xvii.

DIVIDE, on Trail Ridge Road, 15 mi/24 km WNW of ESTES PARK. In ROCKY MOUNTAIN NATIONAL PARK.

Milngavie (MILN-gah-vee), town (2001 population 12,795), East Dunbartonshire, W Scotland, 7 mi/11.3 km NNW of GLASGOW; 55°56′N 04°19′W. Commuter town to Glasgow. Light engineering and packaging. Previously paper milling. Formerly in Strathclyde, abolished 1996.

Milnor (MIL-nuhr), village (2006 population 704), SARGENT county, SE NORTH DAKOTA, 17 mi/27 km SE of LISBON; 46°15′N 97°27′W. Founded in 1883 and incorporated 1884. It was the county seat from 1883 to 1886. Named for two railroad employees, William Milnor Roberts and William E. Milnor.

Milnrow (MILN-ro), town (2001 population 11,561), GREATER MANCHESTER, W ENGLAND, near WEST YORKSHIRE border, 2 mi/3.2 km ESE of ROCHDALE; 53°36′N 02°06′W. Engineering and dyeing; previously textiles and brick manufacturing.

Milnthorpe (MILN-thorp), village (2001 population 2,106), CUMBRIA, NW ENGLAND, 7 mi/11.3 km S of KENDAL; 54°14′N 02°46′W. Dairy farming.

Milo (MEI-lo), town (2000 population 839), WARREN county, S central IOWA, 8 mi/12.9 km SE of INDIANOLA; 41°17′N 93°26′W. In agricultural area.

Milo, town, PISCATAQUIS county, central MAINE, at confluence of the PISCATAQUIS and the Sebec, 12 mi/19 km ENE of DOVER-FOXCROFT; 45°15′N 68°58′W. Trade center, with manufacturing (lumber products). Center for SCHOODIC, Seboois, and SEBEC lakes region. Settled 1803, incorporated 1823.

Milo, town (2000 population 84), VERNON county, W MISSOURI, 7 mi/11.3 km SSE of NEVADA; 37°45′N 94°18′W.

Milo (MEI-lo), village (2001 population 115), S ALBERTA, W CANADA, near N end of Lake MCGREGOR, 60 mi/97 km N of LETHBRIDGE, and in VULCAN COUNTY; 50°34′N 112°53′W. Wheat, flax, cattle. Incorporated in 1931.

Milo (MEE-lo), village, IRINGA region, SW TANZANIA, 80 mi/129 km NW of SONGEA; LIVINGSTONE MOUNTAINS and Lake NYASA to W; 09°53′S 34°40′E. Livestock; grain.

Milo River, c.200 mi/322 km long, GUINEA; rises in S outliers of the FOUTA DJALLON mountains E of MACENTA; flows N, past KANKAN, to the NIGER RIVER 20 mi/32 km S of SIGUIRI. Partly navigable.

Mílos (MEE-los), mountainous island (□ 58 sq mi/150.8 sq km), CYCLADES prefecture, SOUTH AEGEAN department, SE GREECE, in the AEGEAN SEA; one of the CYCLADES; 36°41′N 24°25′E. Rises to 2,533 ft/772 m. Of volcanic origin. The main town is Mílos, formerly known as Plaka. Products include grain, cotton, fruits, and olive oil. Sulphur. Airport. Flourished as a center of early Aegean civilization because of its deposits of obsidian and its strategic location between the Greek mainland and CRETE. Lost importance when bronze replaced obsidian as a material for tools and weapons. Despite its neutrality in the Peloponnesian War, Mílos fell victim to ATHENS, which conquered it in 416 B.C.E. and then massacred the men, enslaved the rest of the population, and founded an Athenian colony. Much excavation has been done here. The most famous find is the Venus de Milo (now in the Louvre, Paris), discovered in 1820. Also Milo.

Miloševo (mee-LOSH-e-vo), village, VOJVODINA, NE SERBIA, 11 mi/18 km SW of KIKINDA, in the BANAT region. Formed (1947) by union of Dragutinovo and Beodra. Also spelled Miloshevo.

Miloslavskoye (mee-luh-SLAHF-skuh-ye), town (2006 population 4,655), W RYAZAN oblast, central European Russia, on highway junction and railroad, 18 mi/29 km SSW of SKOPIN; 53°34′N 39°26′E. Elevation 583 ft/177 m. In agricultural area; produce processing, winery.

Miloslaw (mee-WOS-lahv), Polish *Miłosław*, town, Poznań province, W central POLAND, on railroad, and 28 mi/45 km ESE of POZNAŃ; 52°12′N 17°30′E. Brewing, tanning.

Milosna, POLAND: see LUBOMIERZ.

Milot (mee-LOT), village, W central ALBANIA, on MAT RIVER, and 8 mi/12.9 km SSE of LEZHË, on Lezhë-Durrës road; 41°41′N 19°43′E. Also spelled Miloti.

Milot (mee-LO), village, NORD department, N HAITI, in foothills of the MASSIF DU NORD, 9 mi/14.5 km S of CAP-HAÏTIEN; 19°37′N 72°13′W. In agricultural region (sugarcane, cacao, citrus fruit, tobacco; cattle). Base for visitors to nearby Sans Souci palace and Citadelle La Ferrière, both built by Henri Christophe, king of Northern Haiti (1811–1820).

Milove (mee-lo-VE) (Russian *Melovoye*), town, NE LUHANS'K oblast, UKRAINE, just SW of CHERTKOVO (Russia), 55 mi/89 km E of STAROBIL'S'K; 49°22′N 40°08′E. Elevation 636 ft/193 m. Raion center, sunflower oil press. Established at the end of the 19th century; town since 1938.

Milovice (MI-lo-VI-tse), town, STREDOCESKY province, NE central BOHEMIA, CZECH REPUBLIC, 20 mi/32 km NW of KOLÍN; 50°14′N 14°54′E. Sugar beet farming. A military camp and shooting range for Soviet Army was located here. Environmental damage (e.g., oil in groundwater) under repair since soldiers left in 1990.

Milpa Alta (MEEL-pah AHL-tah), town and delegación, Federal Distrito, central MEXICO, 18 mi/29 km SSE of MEXICO CITY; 19°11′N 99°01′W. Part of the Mexico City metropolitan area. Agricultural center (cereals, fruit, vegetables; stock).

Milparinka (MIL-puh-RING-kuh), locality, NW NEW SOUTH WALES, AUSTRALIA, 155 mi/249 km N of BROKEN HILL; 29°44′S 141°53′E. Sheep. Heritage center and walking track, native-plants park.

Milpitas, city (2000 population 62,698), SANTA CLARA county, W CALIFORNIA; growing suburb 7 mi/11.3 km N of downtown SAN JOSE; 37°26′N 121°54′W. Agricultural area to W (vegetables, strawberries, cherries, mushrooms, grain); dairying; poultry; nursery products. High-tech manufacturing; other industries include food distributing, paint production, and an automobile-assembly plant. Branch of HETCH HETCHY AQUEDUCT runs through city; COYOTE CREEK drains city on W. Incorporated 1954.

Milroy, village, RUSH county, E central INDIANA, 7 mi/11.3 km S of RUSHVILLE. Manufacturing (paper products, canned tomatoes). Hogs; corn, wheat. Laid out 1830.

Milroy, village (2000 population 271), REDWOOD county, SW MINNESOTA, 23 mi/37 km WSW of REDWOOD FALLS; 44°25′N 95°32′W. Grain, soybeans; livestock; dairying; manufacturing (feeds).

Milroy, unincorporated village (2000 population 1,386), MIFFLIN county, central PENNSYLVANIA, 9 mi/14.5 km N of LEWISTOWN, on Laurel Creek; 40°42′N 77°35′W. Agriculture (corn, hay; poultry, livestock, dairying); manufacturing (crushed limestone, lumber, cabinets, audio speakers, pharmaceuticals). Laurel Creek Reservoir to NW; Reeds Gap State Park to E; Rothrock State Forest to W.

Milspe, GERMANY: see ENNEPETAL.

Milstead, village, ROCKDALE county, N central GEORGIA, just N of CONYERS; 33°41′N 83°59′W.

Miltenberg (MIL-tuhn-berg), town, LOWER FRANCONIA, W BAVARIA, GERMANY, on the MAIN and 19 mi/31 km SSE of ASCHAFFENBURG; 49°42′N 09°15′E. Manufacturing of precision instruments, wood, food products; brewing; and flour milling. Chartered in second half of 13th century. Has 14th-century church, renovated in 19th century; 15th-century gate tower and chapel; many 16th-century houses. Town museum contains prehistoric and Roman relics. The 13th-century castle Mildenburg, surrounded by old-German double wall, towers above town.

Milton (MIL-tuhn), city (□ 4 sq mi/10.4 sq km; 2005 population 8,131), ☉ SANTA ROSA county, NW FLORIDA, on BLACKWATER RIVER, and 18 mi/29 km NE of PENSACOLA; 30°37′N 87°02′W. Lumber milling. Site of U.S. Navy base. Founded c.1825, incorporated 1844.

Milton (MIL-tuhn), town, NEW SOUTH WALES, SE AUSTRALIA, 137 mi/220 km S of SYDNEY; 35°19′S 150°24′E. Timber; dairy products, honey; tourism.

Milton (MIL-tuhn), town, industrial suburb (□ 142 sq mi/369.2 sq km; 2001 population 31,471) of TORONTO, HALTON region, S ONTARIO, E central CANADA; 43°31′N 79°53′W. Manufacturing (trucks, cranes, vehicle parts, screws and bolts); warehousing, mixed farming; copper foundry.

Milton, township, CLUTHA district, SE SOUTH ISLAND, NEW ZEALAND, 37 mi/60 km SW of DUNEDIN; 46°07′S 169°58′E. Road-railroad connection; in sheep and mixed crop and livestock area of Tokomairiro Plain; dairy plants, woolen mills.

Milton, town (2000 population 1,657), SUSSEX county, SE DELAWARE, 10 mi/16 km W of LEWES, on BROADKILL RIVER; 38°46′N 75°18′W. Elevation 19 ft/5 m. In farm area; vegetables, fruit; poultry, livestock; dairying; manufacturing. Lake Cannon Museum is here; Prime Hook National Wildlife Refuge to NE. The town is known for Victorian architecture.

Milton, town (2000 population 611), WAYNE county, E INDIANA, on WHITEWATER RIVER, and 14 mi/23 km W of RICHMOND; 39°47′N 85°10′W.

Milton, town (2000 population 550), VAN BUREN county, SE IOWA, near MISSOURI state line, 10 mi/16 km WSW of KEOSAUQUA; 40°40′N 92°09′W. In livestock and grain area.

Milton, town (2000 population 26,062), NORFOLK county, E MASSACHUSETTS, a residential suburb, just S of BOSTON, on the NEPONSET RIVER; 42°14′N 71°05′W. Granite quarries are nearby. Milton is the seat of Curry College and several preparatory schools, including Milton Academy (1798). Meteorological observatory is on Blue Hill (tallest coastal feature in Massachusetts). Includes village of East Milton. Blue Hill Ski Area; Blue Hill Reservation Park. Settled 1636, set off from Dorchester and incorporated 1662.

Milton, town, STRAFFORD county, SE NEW HAMPSHIRE, on SALMON FALLS RIVER, 7 mi/11.3 km N of ROCHESTER (separated from Rochester in 1802); 43°27′N 71°00′W. Bounded E by Salmon Falls River (Maine state boundary), which flows through Milton and Northeast ponds in E center. Agriculture (nursery crops, corn, apples; cattle; dairying). New Hampshire Farm Museum. Includes village of Milton Mills (in N).

Milton, town, including Milton village, CHITTENDEN co., NW VERMONT, on LAMOILLE RIVER and Lake CHAMPLAIN, and 12 mi/19 km N of BURLINGTON; 44°38′N 73°09′W. Sandbar Refuge Management Area is here. Settled 1782, organized 1788.

Milton, town (2006 population 6,702), PIERCE county, W central WASHINGTON, suburb, 6 mi/9.7 km E of TACOMA; 47°15′N 122°19′W. Manufacturing (engines, pleating and stitching); logging. Evergreen Airport to N.

Milton, town (2006 population 2,356), CABELL county, W WEST VIRGINIA, on MUD RIVER, 16 mi/26 km E of HUNTINGTON; 38°25′N 08°82′W. Agriculture (corn, tobacco); cattle; poultry. Coal, gas, and oil region. Known as "Glass Town" for its glass manufacturing (hand-blown glassware, stained glass); also plastic dies, machining, back supports. Mountaineer Opry House. Mill Creek Wildlife Management Area to N. Incorporated 1876.

Milton, town (2006 population 5,720), ROCK county, S WISCONSIN, near LAKE KOSHKONONG (resort), 8 mi/12.9 km NE of JANESVILLE; 42°46′N 88°57′W. In farming and dairying area; manufacturing; railroad junction. Milton House State Historical Site. Incorporated 1904.

Milton, village, SW NOVA SCOTIA, CANADA, on Mersey River and 3 mi/5 km NW of LIVERPOOL; 44°2′N 64°45′W. Elevation 98 ft/29 m. Lumbering center.

Area is shown by the symbol □, and capital city or county seat by ☉.

Milton (MIL-tuhn), village (2001 population 604), HIGHLAND, N Scotland, in GLEN URQUHART, and 1 mi/1.6 km W of Drumnadrochit; 57°20′N 04°30′W. LOCH NESS is about 3 mi/4.8 km E.

Milton, village, PIKE COUNTY, W ILLINOIS, near ILLINOIS RIVER, 9 mi/14.5 km ESE of PITTSFIELD; 39°33′N 90°39′W. In agricultural area.

Milton, village (2000 population 525), TRIMBLE county, N KENTUCKY, on the OHIO RIVER (bridge to MADISON, INDIANA), and 40 mi/64 km NNE of LOUISVILLE; 38°42′N 85°22′W. Agriculture (tobacco, grain, soybeans; cattle; dairying); manufacturing (sand and gravel processing).

Milton, village (□ 1 sq mi/2.6 sq km; 2000 population 1,251), ULSTER county, SE NEW YORK, on W bank of the HUDSON RIVER, and 4 mi/6.4 km S of POUGHKEEPSIE; 43°01′N 73°50′W. In agricultural area.

Milton (MIL-tuhn), village (2006 population 125), CASWELL county, N NORTH CAROLINA, 15 mi/24 km NW of ROXBORO, and 13 mi/21 km ESE of DANVILLE, on DAN RIVER, at VIRGINIA state line; 36°32′N 79°12′W. Agriculture (tobacco, grain). Hyco reservoir (Hyco River) to SE.

Milton, village (2006 population 73), CAVALIER county, NE NORTH DAKOTA, 18 mi/29 km SE of LANGDON; 48°37′N 98°02′W. Founded in 1882 and incorporated in 1888.

Milton, borough (2006 population 6,406), NORTHUMBERLAND county, E central PENNSYLVANIA, 11 mi/18 km NNW of SUNBURY, on West Branch of SUSQUEHANNA RIVER; 41°00′N 50°76′W. Agriculture (corn, hay; livestock, poultry; dairying); diversified manufacturing; limestone. Milton Airport to S. Tiadaghton State Forest to W, Bald Eagle State Forest to NW; Milton State Park, on island in river, to W. Laid out 1792, incorporated 1817.

Miltona (mil-TO-nuh), village (2000 population 279), DOUGLAS county, W MINNESOTA, E of Lake MILTONA, 12 mi/19 km NNE of ALEXANDRIA; 46°02′N 95°17′W. Grain; poultry; dairying; manufacturing (concrete, golf course repair tools). Lake Carlos State Park to S; Irene Lake to NW.

Miltona, Lake (mil-TO-nuh) (□ 8 sq mi/20.8 sq km), DOUGLAS county, W MINNESOTA, 10 mi/16 km N of ALEXANDRIA; 5.5 mi/8.9 km long, 2 mi/3.2 km wide; 46°02′N 95°21′W. Elevation 1,365 ft/416 m. MILTONA village is E. Drains S into Lake IDA. Resorts.

Milton Center (MIL-tuhn SEN-tuhr), village (2006 population 194), Wood county, NW OHIO, 10 mi/16 km WSW of BOWLING GREEN; 41°18′N 83°49′W. In agricultural area.

Milton-Freewater, town (2006 population 6,402), UMATILLA county, NE OREGON, 10 mi/16 km S of WALLA WALLA, WASHINGTON, on WALLA WALLA RIVER, 23 mi/37 km NE of PENDLETON. Manufacturing (canned vegetables, frozen fruit). Wheat. Timber. Site of Frazier Farmstead Museum Umatilla National Forest, including Umatilla Wilderness Area, to SE and E. Incorporated 1886.

Milton Junction, village, ROCK county, S WISCONSIN, between LAKE KOSHKONONG and JANESVILLE, 1 mi/1.6 km W of MILTON. Incorporated 1949.

Milton Keynes (MIL-tuhn KEENZ), town and county (2001 population 207,057), S central ENGLAND; 52°02′N 00°45′W. Designated one of the New Towns in 1967 to alleviate overpopulation in London. It is the seat of the Open University. Includes BLETCHLEY, STONY STRATFORD, and WOLVERTON. Formerly in N BUCKINGHAMSHIRE.

Milton, Lake, reservoir (□ 3 sq mi/7.8 sq km), MAHONING county, NE OHIO, on MAHONING RIVER, 10 mi/16 km NW of ALLIANCE; 41°08′N 80°59′W. Maximum capacity 46,605 acre-ft. Formed by Lake Milton Dam (30 ft/9 m high), built (1916) for recreation; also used for water supply. Lake Milton State Park on W shore.

Milton Lake, Ohio: see MAHONING RIVER.

Milton Mills, NEW HAMPSHIRE: see MILTON.

Milton of Campsie (MIL-tuhn uhv CAMP-see), village (2001 population 3,950), East Dunbartonshire, W Scotland, 2 mi/3.2 km N of KIRKINTILLOCH; 55°57′N 04°10′W. Campsie Fells just N. Previously textile printing. Formerly in Strathclyde, abolished 1996.

Milton Regis, ENGLAND: see SITTINGBOURNE.

Miltonsburg (MIL-tuhnz-buhrg), village (2006 population 28), MONROE county, E OHIO, 6 mi/10 km NNW of WOODSFIELD; 39°50′N 81°10′W. In agricultural area. Limestone quarry.

Miltonvale, village (2000 population 523), CLOUD county, N central KANSAS, 19 mi/31 km SE of CONCORDIA; 39°21′N 97°24′W. In wheat region.

Miltown or **Milltown**, Gaelic *Baile an Mhuilinn*, town (2006 population 415), central KERRY county, SW IRELAND, on MAINE RIVER, and 8 mi/12.9 km S of TRALEE; 52°08′N 10°17′W. Agricultural market. Remains of ancient Kilcolman Abbey nearby.

Miltown Malbay or **Milltown Malbay**, Gaelic *Sráid na Cathrach*, town (2006 population 570), W CLARE county, W IRELAND, on MAL BAY, 17 mi/27 km W of ENNIS; 52°51′N 09°24′W. Agricultural market and seaside resort; racecourse.

Milumba (mee-LOOM-bah), village, RUKWA region, W TANZANIA, 67 mi/108 km NNW of SUMBAWANGA, in S part of Kitwei Plains Game Reserve, near Mfusi River; 06°47′S 31°03′E. Livestock; grain.

Miluo (MEE-LU-uh), city (□ 603 sq mi/1,562 sq km; 1994 estimated urban population 81,000; 1994 estimated total population 689,700), NE HUNAN province, CHINA, on BEIJING-GUANGZHOU railroad, and on the Miluo River, 60 mi/97 km NE of YIYANG; 29°00′N 112°59′E. Agriculture is the largest source of income for the city. Industries include food processing, textiles, chemicals, and iron and steel.

Milverton (MIL-vuh-tuhn), town (2001 population 1,485), W SOMERSET, SW ENGLAND, 6 mi/9.6 km W of TAUNTON; 51°01′N 03°15′W. Former agricultural market. Has 16th-century church.

Milverton (MIL-vuhr-tuhn), former village (□ 1 sq mi/2.6 sq km; 2001 population 1,707), SW ONTARIO, E central CANADA, 15 mi/24 km NNE of STRATFORD; 43°34′N 80°55′W. Light manufacturing, mixed farming, dairying; flour, apples; cattle; wood products. Amalgamated into PERTH EAST in 1998.

Milwaukee, county (□ 1,189 sq mi/3,091.4 sq km; 2006 population 915,097), SE WISCONSIN, ⊙ MILWAUKEE; 42°58′N 87°40′W. All of county is incorporated. The city of Milwaukee comprises 96 sq mi/249 sq km of the county, mainly in center and NW; remainder is comprised of WAUWATOSA, WEST ALLIS, FRANKLIN, OAK CREEK, and fourteen other cities and villages. Bounded E by LAKE MICHIGAN; drained by MILWAUKEE, MENOMONEE, and ROOT rivers. Highly industrialized area, centered at Milwaukee. Small areas of remnant agriculture, mainly in far S (wheat, soybeans). Formed 1834.

Milwaukee, city (2006 population 573,358), ⊙ MILWAUKEE county, SE WISCONSIN, at the point where the MILWAUKEE, MENOMONEE, and Kinnickinnic rivers enter LAKE MICHIGAN; 43°03′N 87°58′W. Borders Lake Michigan on E; city is hemmed in on all sides by neighboring municipalities, preventing annexation (other then by merges). A metropolitan area, it is the largest city in the state, and it is a port of entry, shipping heavy cargo from the entire Midwest to other lake ports and world ports via the ST. LAWRENCE SEAWAY. It is a producer of heavy machinery and electrical equipment and a principal manufacturer of diesel and gasoline engines, tractors, and beer; Milwaukee once dominated the country's beer-brewing industry. Various light and heavy manufacturing. In 1673, Father Jacques Marquette visited the site, which was then a Native American gathering and trading center. In 1795 the North West Company established a fur-trading post. Solomon Juneau, a fur trader, arrived

in 1818, and in 1838 several settlements merged to form Milwaukee village. It grew as a shipping center and became famous for its numerous industries, notably brewing and meat packing. German refugees arrived in large numbers after 1848, stimulating the city's political, economic, and social growth. The Knights of St. Crispin foreshadowed the city's growing labor movement after the Civil War. Victor Lake Berger, the Socialist leader, exerted a dominant influence there, and Daniel W. Hoan made Milwaukee known for efficient administration. Among the city's educational institutions are Marquette University, the University of Wisconsin (Milwaukee), Alverno College, Cardinal Stritch University (in FOX POINT), Wisconsin Lutheran College (in WAUWATOSA), Milwaukee Institute of Art and Design, the Wisconsin Conservatory of Music, the Milwaukee School of Engineering, and the Milwaukee Area Technical College. Local attractions include the breweries, with their guided tours; a public library and museum; an art center; a church built by Frank Lloyd Wright; a performing arts center; and the water tower. Havenswood Environmental Awareness Center. Among the numerous parks are Washington Park; Mitchell Park, with enclosed botanical gardens; Juneau Park; Estabrook Park, containing one of the city's oldest houses. Milwaukee has professional basketball and baseball teams. Economically, the city was hit hard in the 1979–1982 recession years; more than 60,000 jobs were lost in the industrial sector. From the late 1980s through the 2000s, Milwaukee regained some of its former vitality: in the 2000s, the city serves as headquarters for thirteen Fortune 500 companies; about 80% of the non-farm labor force is engaged in service-sector jobs, with manufacturing jobs comprising the rest. Incorporated 1846.

Milwaukee (mil-WAW-kee), unincorporated village, NORTHAMPTON county, NE NORTH CAROLINA, 6 mi/39 km E of ROANOKE RAPIDS; 36°24′N 77°13′W. Agriculture (peanuts, cotton, tobacco; livestock).

Milwaukee Deep, submarine trench, ocean depth at 28,232 ft/8,605 m, in the PUERTO RICO TRENCH, 19°35′N 66°30′W. Deepest point in the ATLANTIC OCEAN.

Milwaukee River, c.75 mi/121 km long, SE WISCONSIN; rises in lake region of FOND DU LAC county; flows S, past WEST BEND, turns E, then generally S, past GRAFTON, reaching LAKE MICHIGAN at MILWAUKEE.

Milwaukie, city (2006 population 20,988), CLACKAMAS and MULTNOMAH counties, NW OREGON, suburb, 5 mi/8 km SSE of downtown PORTLAND on the WILLAMETTE RIVER, at mouths of Johnson and Kellogg Creeks; 45°26′N 122°37′W. Railroad junction. Manufacturing (apparel, furniture products, fabricated metal). The city is a distribution center for farms and orchards of the Willamette Valley and has numerous warehouse facilities. Fruit trees brought here by covered wagon from IOWA (1848) inaugurated the state's important cherry-growing industry. Incorporated 1903.

Milyana River (mee-LYAH-nah), c.80 mi/129 km long, N central and N TUNISIA; rises 8 mi/12.9 km NE of SILYANAH; flows NE past AL FAHS to the GULF OF TUNIS, 6 mi/9.7 km ESE of TUNIS. Its waters are diverted for the irrigation of the Mornag lowland.

Milyatska River, BOSNIA AND HERZEGOVINA: see MILJACKA RIVER.

Milyutinskaya (mee-LYOO-teen-skah-yah), village (2006 population 2,740), E central ROSTOV oblast, S European Russia, in the KALITVA RIVER basin, on road junction, 35 mi/56 km NNW of MOROZOVSK; 48°38′N 41°40′E. Elevation 347 ft/105 m. Agricultural products. Established by the Don Cossacks.

Milyutinskaya, UZBEKISTAN: see GALLYAARAL.

Mim (MEEM), town, BRONG-AHAFO REGION, GHANA, 60 mi/97 km WNW of BIBIANI, on BIA RIVER, Ashanti Uplands; 06°54′N 02°34′W. Timber; cocoa, kola nuts.

Mima (MEE-mah), town, N Uwa county, EHIME prefecture, NW SHIKOKU, W JAPAN, 40 mi/65 km S of MATSUYAMA; 33°17′N 132°36′E.

Cross-references are shown in SMALL CAPITALS. The pronunciation guide is shown on page xix. The sources of population figures are shown on page xvii.

Mima, town, Mima county, TOKUSHIMA prefecture, SE SHIKOKU, W JAPAN, 28 mi/45 km W of TOKUSHIMA; 34°02′N 134°03′E.

Mimasaka, former province in SW HONSHU, JAPAN. Now part of OKAYAMA prefecture.

Mimasaka (mee-MAH-sah-kah), town, Aida county, OKAYAMA prefecture, SW HONSHU, W JAPAN, 28 mi/ 45 km N of OKAYAMA; 35°00′N 134°09′E.

Mimata (mee-MAH-TAH), town, N Morokata county, MIYAZAKI prefecture, SE KYUSHU, SW JAPAN, 22 mi/35 km S of MIYAZAKI; 31°43′N 131°07′E. Beef cattle.

Mimbres (MIM-bruhs), unincorporated village, GRANT county, SW NEW MEXICO, 18 mi/29 km ENE of SILVER CITY, on Mimbres River. Elevation 5,977 ft/1,822 m. Resort with hot mineral springs. Gila National Forest to W, N, and E; BLACK RANGE, BLACK PEAK, on CONTINENTAL DIVIDE, to NW; MIMBRES MOUNTAINS to E.

Mimbres Mountains (MIM-bruhs), SW NEW MEXICO, extending N from Cooks Range to BLACK RANGE, on GRANT and SIERRA county line; extending S into LUNA county. Chief peaks: Pine Flat Mountain (7,875 ft/ 2,400 m), Thompson Cone (7,932 ft/2,418 m), SEVEN BROTHERS MOUNTAIN (8,690 ft/2,649 m). Cooke's Peak (8,408 ft/2,563 m) in S. Range is largely within Gila National Forest.

Mimizan (mee-mee-zahn), town (□ 43 sq mi/111.8 sq km; 2004 population 6,605), LANDES department, AQUITAINE region, SW FRANCE, near the BAY OF BISCAY, 30 mi/48 km S of ARCACHON; 44°12′N 01°14′W. Paper milling, turpentine extracting from pine forests. Mimizan-les-Bains (3 mi/4.8 km W) is a small bathing resort on the ATLANTIC coast.

Mim Lake Punit, MEXICO, classical Maya site, partially excavated, unrestored; 15°02′N 88°53′W. Noted for royal tomb discovered in 1986 and for large stelae.

Mimms, Fort, ALABAMA: see FORT MIMMS.

Mimon (MI-mon-yuh), Czech *Mimoň*, German *Niemes*, town, SEVEROCESKY province, N BOHEMIA, CZECH REPUBLIC, on railroad, and 30 mi/48 km E of ÚSTÍ NAD LABEM; 50°39′N 14°44′E. Agriculture (oats); manufacturing (furniture, textiles). Horse races; castle.

Mimongo (MEE-mahn-go), town, NGOUNIÉ province, S GABON, 60 mi/97 km ENE of MOUILA; 01°38′S 11°35′E. In a gold-mining area.

Mimosa (mee-MO-sah), city, N RONDÔNIA state, BRAZIL, 76 mi/123 km SE of PÔRTO VELHO; 09°55′S 62°15′W.

Mimosa Park (mi-MO-suh), unincorporated town, SAINT CHARLES parish, 5 mi/8 km SE of HAHNVILLE; 29°54′N 90°21′W.

Mimoso do Sul (mee-mo-so do sool), city (2007 population 26,218), S ESPÍRITO SANTO, BRAZIL, on railroad, and 20 mi/ 32 km SW of CACHOEIRO DE ITAPEMIRIM; 21°09′S 41°25′W. Coffee and rice hulling; corn-meal manufacturing; leatherworking. Until 1944, called João Pessoa.

Mimot (MI-MUT), town, KOMPONG CHAM province, S CAMBODIA, 45 mi/72 km S of KOMPONG CHAM, near VIETNAM border; 11°49′N 106°11′E. Market center. Rubber plantations in area. Khmer with Vietnamese and Kuy minorities.

Mims (MIMZ), city (□ 25 sq mi/65 sq km; 2000 population 9,147), BREVARD county, E central FLORIDA, 6 mi/9.7 km N of TITUSVILLE; 28°40′N 80°50′W. Manufacturing includes liquefied gases, heat-sealing products, fruit packing, structural steel fabrication, and chemically-treated fittings.

Min (MIN), chief river, c.350 mi/563 km long, of FUJIAN province, SE CHINA; rises in WUYI mountains; flows SE to the SOUTH CHINA SEA near FUZHOU. It receives several tributaries near NANPING. From SHUIKOU to its delta is agricultural area mainly for rice, tea, oilseeds, and sugarcane. Fuzhou, a transshipment point, has a deep-water anchorage for oceangoing vessels.

Min, river, c.500 mi/805 km long, W SICHUAN province, central CHINA; rising in the MIN SHAN; flowing S

through the CHENGDU PLAIN to CHANG JIANG (Yangzi River) at YIBIN. The DADU RIVER, c.400 mi/644 km long, is its chief tributary. In the second century B.C.E., Li Bing, governor of Shu State, built the Dujiang Dam and diverted the Min's water into numerous channels that reunite downstream near PENGSHAN. The irrigation system is still used today to water the fertile Chengdu Plain.

Min, CHINA: see FUJIAN.

Mina (MEE-nah), town, NUEVO LEÓN, N MEXICO, in foothills of SIERRA MADRE ORIENTAL, near SALINAS RIVER, 27 mi/43 km NW of MONTERREY, on Mexico Highway 53; 26°00′N 100°33′W. Cereals, cactus fibers, livestock.

Mina (MEE-nuh), unincorporated village, MINERAL county, W NEVADA, 34 mi/55 km SE of HAWTHORNE. Elev. 4,540 ft/1,384 m. Cattle. EXCELSIOR MOUNTAINS to W; Pilot Peak (9,184 ft/2,799 m), in Monte Cristo Range, to E. Southern Pacific Spring to NE; Rhodes Salt Marsh to S.

Mina al Ahmadi (mee-NAH el AH-me-dee), town in SE KUWAIT, port on PERSIAN GULF, 22 mi/35 km SSE of Kuwait City. Main port, large loading terminal for oil and natural gas from BURGAN oil field and AHMADI tank farm (linked by pipelines); large refinery and petrochemical industry.

Minab (mee-NAHB), town (2006 population 56,009), Kormozgān province, SE IRAN, 50 mi/80 km E of BANDAR ABBAS, and on MINAB River (small coastal stream); 27°09′N 57°02′E. Center of rich agricultural area; date groves, fruit orchards (oranges, mangoes, bananas); tobacco, wheat, millet, vegetables. At mouth of nearby Minab River was ancient Hormoz, PERSIAN GULF port.

Minabe (mee-NAH-be), town, Hidaka county, WAKAYAMA prefecture, S HONSHU, W central JAPAN, on PHILIPPINE SEA, on S KII PENINSULA, 33 mi/54 km S of WAKAYAMA; 33°45′N 135°19′E.

Minabegawa (mee-NAH-be-GAH-wah), village, Hidaka county, WAKAYAMA prefecture, S HONSHU, W central JAPAN, 32 mi/52 km S of WAKAYAMA; 33°47′N 135°19′E.

Mina Clavero (MEE-nah klah-VAI-ro), town, W CÓRDOBA province, ARGENTINA, 55 mi/89 km SW of CÓRDOBA; 31°43′S 65°00′W. Popular mountain resort on small Mina Clavero River.

Minaçu (MEE-nah-soo), city (2007 population 31,051), N central GOIÁS, BRAZIL, 42 mi/68 km NE of Campinaçu; 13°30′S 48°20′W.

Mina de São Domingos (MEE-nah dai SOU doo-MEEN-goosh) or **Minas de São Domingos**, town, BEJA district, S PORTUGAL, near the CHANZA RIVER (Spanish border), 30 mi/48 km SE of BEJA; 37°40′N 07°30′W. Copper mining; connected by mining railroad with POMARÃO on the GUADIANA RIVER.

Minador do Negrão (MEEN-ah-dor do ne-GROUN), town (2007 population 5,145), N central ALAGOAS state, BRAZIL, near PERNAMBUCO border; 09°17′S 36°50′W.

Minahasa (MEE-nah-hah-hah), name sometimes given to NE part of N peninsula of SULAWESI, Sulawesi Utara province, INDONESIA; 01°00′N 124°35′E. So called because area is largely inhabited by the Minahassa, a Malayan group converted to Christianity in the nineteenth century. Main town is TONDANO. Formerly spelled Minihassa.

Minakami (mee-NAH-kah-mee), village, Tone county, GUMMA prefecture, central HONSHU, N central JAPAN, near Mount Tanigawadake, 34 mi/55 km N of MAEBASHI; 36°46′N 138°58′E. Joshin-Etsu Highlands National Park nearby; also hot springs.

Minakuchi (mee-NAHK-chee), town, Koka county, SHIGA prefecture, S HONSHU, central JAPAN, 19 mi/30 km E of OTSU; 34°57′N 136°10′E.

Mina La India, NICARAGUA: see LA CRUZ DE LA INDIA.

Minalin (mee-nah-LEEN), town, PAMPANGA province, central LUZON, PHILIPPINES, 4 mi/6.4 km SSW of SAN

FERNANDO; 14°58′N 120°42′E. Long known as "the egg basket of central Luzon," the town suffered the loss of most of its poultry farms in the eruption of MOUNT PINATUBO in June 1991 and the subsequent mud flows.

Minamata (mee-NAH-mah-tah), city, KUMAMOTO prefecture, SW JAPAN, on YATSUSHIRO BAY, 43 mi/70 km S of KUMAMOTO; 32°12′N 130°24′E. Railroad junction. Chemical fertilizer. Outbreak of mercury poisoning here in the 1960s.

Minami (MEE-nah-mee), village, Gujo county, GIFU prefecture, central HONSHU, central JAPAN, 22 mi/35 km N of GIFU; 35°39′N 136°57′E.

Minamiaiki (mee-NAH-mee-ah-EE-kee), village, S Saku county, NAGANO prefecture, central HONSHU, central JAPAN, 47 mi/75 km S of NAGANO; 36°02′N 138°32′E.

Minamiarima (mee-NAH-mee-AH-ree-mah), town, South Takaki county, NAGASAKI prefecture, W KYUSHU, SW JAPAN, on SE coast of SHIMABARA PENINSULA, on SHIMABARA BAY, 22 mi/35 km S of NAGASAKI; 32°37′N 130°15′E. Ruins of Hara Castle (destroyed in 1641 Shimabara Rebellion).

Minamiashigara (mee-NAH-mee-ah-shee-GAH-rah), city, KANAGAWA prefecture, E central HONSHU, E central JAPAN, near Mount Kintoki, 31 mi/50 km S of YOKOHAMA; 35°19′N 139°06′E. Film.

Minamichita (mee-NAH-mee-chee-TAH), town, Chita county, AICHI prefecture, S central HONSHU, central JAPAN, 31 mi/50 km S of NAGOYA; 34°42′N 136°55′E. Plastic products.

Minamidaito (mee-NAH-mee-DAH-ee-TO), village, on Minami-daito-shima island, Shimajiri county, Okinawa prefecture, SW JAPAN, 223 mi/360 km E of NAHA; 25°46′N 131°13′E.

Minamifurano (mee-NAH-mee-foo-RAH-no), town, Kamikawa district, Hokkaido prefecture, N JAPAN, 65 mi/105 km E of SAPPORO; 43°09′N 142°34′E. Carrots, potatoes. Trout and smelts are caught in nearby Kanayama Lake.

Minamiizu (mee-NAH-mee-EEZ), town, Kamo county, SHIZUOKA prefecture, central HONSHU, E central JAPAN, 34 mi/55 km S of SHIZUOKA; 34°38′N 138°51′E.

Minamikata (mee-NAH-mee-KAH-tah), town, Tome county, MIYAGI prefecture, N HONSHU, NE JAPAN, 31 mi/50 km N of SENDAI; 38°39′N 141°09′E.

Minamikawachi (mee-NAH-mee-KAH-wah-chee), town, Kawachi county, TOCHIGI prefecture, central HONSHU, N central JAPAN, 12 mi/20 km S of UTSUNOMIYA; 36°22′N 139°52′E. Dried gourd shavings. Pongee.

Minamikawara (mee-NAH-mee-KAH-wah-rah), village, North Saitama county, SAITAMA prefecture, E central HONSHU, E central JAPAN, 25 mi/40 km N of URAWA; 36°10′N 139°26′E.

Minamikayabe (mee-NAH-mee-kah-YAH-be), town, Oshima county, Hokkaido prefecture, N JAPAN, 84 mi/135 km S of SAPPORO; 41°54′N 140°58′E. Kombu.

Minamikushiyama (mee-NAH-mee-koo-SHEE-yah-mah), town, South Takaki county, NAGASAKI prefecture, NW KYUSHU, SW JAPAN, 16 mi/25 km W of NAGASAKI; 32°40′N 130°08′E.

Minamimaki (mee-NAH-mee-MAH-kee), village, S Saku county, NAGANO prefecture, central HONSHU, central JAPAN, 47 mi/75 km S of NAGANO; 36°01′N 138°29′E. Vegetables.

Minamiminowa (mee-NAH-mee-mee-NO-wah), village, Kamiina county, NAGANO prefecture, central HONSHU, central JAPAN, 53 mi/85 km S of NAGANO; 35°52′N 137°58′E.

Minaminasu (mee-NAH-mee-NAHS), town, Nasu county, TOCHIGI prefecture, central HONSHU, N central JAPAN, 12 mi/20 km N of UTSUNOMIYA; 36°39′N 140°05′E. Beef cattle, pigs.

Minami-nayoshi, RUSSIA: see SHEBUNINO.

Minamioguni (mee-NAH-mee-O-goo-nee), town, Aso county, KUMAMOTO prefecture, W KYUSHU, SW

JAPAN, 28 mi/45 km N of KUMAMOTO; 33°05′N 131°04′E.

Minamishinano (mee-NAH-mee-shee-NAH-no), village, Shimoina county, NAGANO prefecture, central HONSHU, central JAPAN, near Japanese South Alps, 93 mi/150 km S of NAGANO; 35°18′N 137°56′E. Tea, shiitake mushrooms, Japanese apricots; trout. *Konnyaku* (paste made from devil's tongue).

Minamitane (mee-NAH-mee-tah-NE), town, Kumage county, KAGOSHIMA prefecture, SW KYUSHU, SW JAPAN, 90 mi/145 km S of KAGOSHIMA; 30°24′N 130°54′E. Site of Tanegashima Space Center.

Minami-tori-shima, JAPAN: see MARCUS ISLAND.

Minami-uruppu-suido, RUSSIA: see URUP STRAIT.

Minamiyamashiro (mee-NAH-mee-yah-MAH-SHEE-ro), town, Soraku county, KYOTO prefecture, S HONSHU, W central JAPAN, 22 mi/35 km S of KYOTO; 34°46′N 135°59′E. Tea.

Minano (mee-NAH-no), town, Chichibu county, SAITAMA prefecture, E central HONSHU, central JAPAN, 34 mi/55 km N of URAWA; 36°04′N 139°06′E.

Mina, Oued (mee-NAH, WED), stream, c.150 mi/241 km long, N ALGERIA; rises TIARET wilaya, in the High Plateaus, 30 mi/48 km S of TIARET; flows NW across the TELL ATLAS, past RELIZANE, to the CHÉLIFF River (an important tributary), 25 mi/40 km E of MOSTAGANEM. Divides the OUARSENIS and the Béni Chougrane Mountains. The BAKHADA DAM (148 ft/45 m high; W of Tiaret) stores water for irrigation in Relizane area.

Mina Pirquitas (MEE-nah peer-KEE-tahs) or **Pirquitas**, village, W JUJUY province, ARGENTINA, 70 mi/113 km SW of LA QUIACA; 22°41′S 66°31′W.

Mina Ragra (MEE-nah RAH-grah), town, PASCO region, central PERU; 10°51′S 76°34′W. The Mina Ragra vanadium deposit discovered in 1905 became the first vanadium mine in the Western Hemisphere.

Minarets, The, jagged summits (c.12,000 ft/3,658 m) in SIERRA NEVADA, MADERA–MONO county line, E CALIFORNIA, in scenic region SE of YOSEMITE NATIONAL PARK.

Minas (MEE-nahs), city (2004 population 37,925), ⊙ LAVALLEJA department, SE URUGUAY, on railroad, and highway, and 60 mi/97 km NE of MONTEVIDEO; 34°23′S 55°14′W. Railroad terminus, airport. Granite crushing, quarrying; marble; wool, wheat, corn, oats; cattle and sheep raising. Lead deposits nearby. Has public library, theater. Founded 1783.

Minas (MEE-nahs), town, CAMAGÜEY province, E CUBA, on railroad, and 20 mi/32 km ENE of CAMAGÜEY; 21°29′N 77°37′W. Asphalt, iron, and copper mining.

Minas (MEE-nahs), town, RIAU province, E central SUMATRA, 20 mi/33 km N of PEKANBARU; 00°50′N 101°29′E. Site of first oil field in INDONESIA, discovered in late 1930s.

Minas, ARGENTINA: see ANDACOLLO.

Minas, URUGUAY: see LAVALLEJA.

Mina Salman, port, BAHRAIN, near islands of Bahrain (E) and MANAMAH (S), in PERSIAN GULF. Industrial town, oil-loading terminal and port.

Minas Basin (MEI-nuhs), central NOVA SCOTIA, CANADA, central part of deep inlet of the BAY OF FUNDY, with which it is connected (W) by MINAS CHANNEL; 24 mi/39 km long, up to 25 mi/40 km wide. Continued E by COBEQUID BAY. Narrows between CAPE SHARP and CAPE SPLIT separate it from Minas Channel. On N coast is PARRSBORO. Receives CORNWALLIS, AVON, and several smaller rivers.

Minas Channel, (MEI-nuhs) inlet, N central NOVA SCOTIA, CANADA, connects the Bay of FUNDY (W) with the MINAS BASIN (E); 24 mi/39 km long, 10 mi/16 km–14 mi/23 km wide. Its mouth is between Cape CHIGNECTO and Cape SHARP. N shore is indented by Advocate Bay (4 mi/6 km long, 8 mi/13 km wide) and Greville Bay (5 mi/8 km long, 16 mi/26 km wide). On S side, Cape SPLIT extends 6 mi/10 km into the channel.

On N shore is PORT GREVILLE. High tides of Bay of Fundy extend into Minas Channel.

Minas da Panasqueira (MEE-nahsh dah pah-nah-shkai-rah), village, CASTELO BRANCO district, central PORTUGAL, near ZÊZERE RIVER, 14 mi/23 km SW of COVILHÃ. Tungsten mining.

Minas de Corrales, URUGUAY: see CORRALES.

Minas del Tauler, SPAIN: see TAULER.

Minas de Matahambre (MEE-nahs dai mah-tah-AHM-brai), town, PINAR DEL RÍO province, W CUBA, 20 mi/32 km NW of PINAR DEL RÍO; 22°41′N 83°56′W. Copper-mining center. Also called Matahambre.

Minas de Oro (MEE-nahs dai O-ro) [Spanish=*gold mines*], town, COMAYAGUA department, W central HONDURAS, 40 mi/64 km NNW of TEGUCIGALPA; 14°48′N 87°21′W. Footwear manufacturing; tobacco, coffee. Airfield. Nearby gold deposits were formerly exploited.

Minas de Ríotinto, SPAIN: see RÍO TINTO.

Minas do Rio de Contas, Brazil: see RIO DE CONTAS.

Minase (mee-nah-SE), village, Ogachi county, Akita prefecture, N HONSHU, NE JAPAN, 50 mi/80 km S of AKITA city; 39°04′N 140°35′E.

Minas Gerais (mee-nahs zhe-reis) [Portuguese=*various mines*], state (□ 226,707 sq mi/589,438.2 sq km; 2007 population 19,216,816), E BRAZIL; ⊙ BELO HORIZONTE. Mining of iron ore and other minerals (largest mines in the country producing more than half of Brazil's mineral wealth); agriculture (coffee, soybeans) and livestock; steel and pulp mills; automobile factory. Gold was discovered at the end of 17th century, largely exhausted by the end of the 18th century. Large industrial base around Belo Horizonte.

Minas, Las, PANAMA: see LAS MINAS.

Minas, Las, PANAMA: see PUERTO PILÓN.

Minas Novas (mee-nahs no-vahs), city (2007 population 30,293), NE central MINAS GERAIS, BRAZIL, 80 mi/129 km WNW of TEÓFILO OTONI; 17°15′S 42°37′W. Elevation 3,000 ft/914 m. Semiprecious stones found here. Known since 1727, when diamonds were discovered in nearby streams.

Minas, Sierra de las (MEE-nahs, see-ER-rah dai lahs), range along ALTA VERAPAZ–ZACAPA department border, E central GUATEMALA; extends c.60 mi/97 km E-W between POLOCHIC RIVER (N) and MOTAGUA River (S); rises to c.10,000 ft/3,048 m; 15°30′N 88°55′W. Sierra del MICO adjoins (E).

Minatare (MI-nuh-tahr-ai), village (2006 population 777), SCOTTS BLUFF county, W NEBRASKA, 8 mi/12.9 km ESE of SCOTTSBLUFF, and on NORTH PLATTE RIVER; 41°48′N 103°30′W. Beet sugar, dairy and poultry produce, grain. Lake Minatare, artificial lake created (1915) for irrigation, nearby. State recreation area to N (part of North Platte National Wildlife Refuge). Manufacturing (sickle sharpeners).

Minatitlán (mee-nah-tee-TLAHN), city and township, VERACRUZ, SE MEXICO, port on navigable COATZACOALCOS River (20 mi/32 km from mouth), on Isthmus of TEHUANTEPEC, 135 mi/217 km SE of VERACRUZ; 17°59′N 94°32′W. Railroad terminus; along with COATZACOALCOS, the center of Mexico's oil and petrochemical industry (pipeline to SALINA CRUZ); lumber mills. Agricultural products: coffee, rice, corn, sugarcane, fruit; livestock.

Minatitlán (mee-nah-tee-TLAHN), town, COLIMA, W MEXICO, 24 mi/39 km WNW of COLIMA; 17°59′N 94°32′W. Rice, corn, beans, sugarcane, cotton, coffee, fruit; livestock; iron mining. Also known as EL MAMEY.

Minato (mee-NAH-to), ward, at the S edge of central TOKYO, Tokyo prefecture, E central HONSHU, E central JAPAN, on TOKYO BAY (SE). Bordered N by SHINJUKU and CHIYODA wards, NE by CHUO ward, S by SHINAGAWA ward, and W by Shibuya ward. Tokyo Tower is located here, as is the entertainment area of Roppongi.

Minaya (mee-NEI-ah), town, ALBACETE province, SE central SPAIN, 32 mi/51 km NW of ALBACETE; 39°17′N

02°19′W. Esparto-rope manufacturing, brandy distilling, sawmilling; melons, saffron, wine, cereals.

Minayevka (mee-NAH-eef-kah), village, E TOMSK oblast, W SIBERIA, RUSSIA, on the CHULYM RIVER near its confluence with the Chichka-Yul River (one of its right tributaries), on highway, 25 mi/40 km NNW of (and administratively subordinate to) ASINO; 57°26′N 85°50′E. Elevation 291 ft/88 m. Logging, lumbering, woodworking.

Minbu (MIN-boo), former district, now township, Magwe div, Myanmar; ⊙ MINBU. Bet. Arakan, Yoma, and Ayeyarwady rivers. In dry zone (annual rainfall 25 in/64 cm) irrigated by Salin, Man, and Mon rivers. Agr. (rice, millet, sesame, beans, tobacco); fisheries; teak forests, oil fields. Served by Ayeyarwady steamers.

Minbu, town, ⊙ MINBU township, MYANMAR, river port on AYEYARWADY RIVER (opposite MAGWE), and 100 mi/161 km NNW of PROME. Head of road, through AN PASS, over ARAKAN YOMA to An (KYAUKPYU district). Oil field, mud volcanoes nearby.

Minburn, town (2000 population 391), DALLAS county, central IOWA, 10 mi/16 km N of ADEL; 41°45′N 94°01′W. In agricultural area.

Minburn (MIN-buhrn), village (2001 population 88), E ALBERTA, W CANADA, 22 mi/35 km W of VERMILION, in MINBURN COUNTY NO. 27; 53°19′N 111°12′W. Mixed farming, wheat, flax; cattle. Established 1919.

Minburn County No. 27 (MIN-buhrn), municipality (□ 1,124 sq mi/2,922.4 sq km; 2001 population 3,436), E central ALBERTA, W CANADA; 53°24′N 111°38′W. Includes VEGREVILLE, MINBURN, INNISFREE, MANNVILLE, LAVOY, and RANFURLY. Formed as a municipal district in 1942.

Minbya (MIN-byah), village and township, RAKHINE STATE, MYANMAR, in the ARAKAN, on LEMRO RIVER, and 30 mi/48 km NE of SITTWE.

Minch (MINCH) or **North Minch**, strait (20 mi/32 km to 45 mi/72 km wide) separating LEWIS and HARRIS from mainland of Scotland (HIGHLAND). Little Minch (14 mi/23 km to 20 mi/32 km wide) to the SW, separates island of SKYE from the middle OUTER HEBRIDES islands.

Mincha (MEEN-chah), village, COQUIMBO region, N central CHILE, on CHOAPA river, and 15 mi/24 km WNW of ILLAPEL; 31°35′S 71°27′W. In agricultural area (grain, fruit; livestock).

Minchinabad (MIN-chin-ah-BAHD), town, BAHAWALNAGAR district, BAHAWALPUR division, PUNJAB province, central PAKISTAN, 125 mi/201 km NE of BAHAWALPUR; 30°10′N 73°34′E. Sometimes MANCHINABAD.

Minchinhampton (MIN-chin-HAMP-tuhn), village (2001 population 3,937), central GLOUCESTERSHIRE, W ENGLAND, 3 mi/4.8 km SE of STROUD, in the COTSWOLD HILLS; 51°42′N 02°10′W. Former agricultural market. Has 14th-century church.

Minchinmávida (meen-cheen-MAH-vee-dah) or **Michinmáhuida**, Andean volcanic peak (8,100 ft/2,469 m), CHILOÉ province, LOS LAGOS region, S CHILE; 42°50′S. A massif with several peaks and glaciers.

Minch, Little, Scotland: see THE MINCH.

Minchumina, Lake, Alaska: see LAKE MINCHUMINA.

Mincio (MEEN-cho), river, c.47 mi/76 km long, in LOMBARDY, N ITALY; flows generally S from the S end of LAKE GARDA through MANTUA (where it forms three lakes) to the PO RIVER. Above Lake Garda it is called the SARCA. The Sarca-Garda-Mincio line, which is 120 mi/193 km long, marks the natural border between Lombardy and VENETIA and has been of strategic importance, especially in the wars of the Risorgimento (1848–1849).

Minco (MINK-o), town (2006 population 1,788), GRADY county, central OKLAHOMA, 18 mi/29 km N of CHICKASHA; 35°19′N 97°57′W. In agricultural area (cotton, corn, wheat, peanuts; livestock); honey, lye soap.

Mindanao (MEEN-dah-nou), island (□ 36,537 sq mi/94,996.2 sq km), second-largest island of the PHI-

LIPPINES, NE of BORNEO; 08°00′N 125°00′E. Area also includes 1,018 offshore islands (513 of which are named). The terrain is generally mountainous and heavily forested, rising to 9,690 ft/2,954 m at MOUNT APO, an active volcano and the highest point in the Philippines. The island is indented by several deep bays and has a large W peninsula, the ZAMBOANGA or Sibuguey Peninsula. Its main rivers are the Mindanao (known as the PULANGI in its upper course), c.200 mi/320 km long and navigable by small steamers for c.40 mi/64 km; and the AGUSAN, c.240 mi/386 km long. The largest lake is LAKE LANAO, for centuries the habitat of Muslim Moros. Off the NE coast in the PHILIPPINE SEA is the MINDANAO TRENCH (c.35,000 ft/10,670 m deep), one of the greatest known ocean depths. Mindanao lies below the typhoon belt, and its climate is more favorable than that of LUZON to the N. Pineapples, mangoes, bananas, and other fruits are grown, as well as rice and corn. ZAMBOANGA and DAVAO are the principal cities; Davao is the most important port. There was considerable industrial growth during the 1960s. The extensive development of the water resources of the Lake Lanao–AGUS RIVER basin, including the harnessing of MARIA CRISTINA FALLS, has resulted in the establishment of heavy industrial plants, especially in the ILIGAN area. Wood products, agriculture, and seafood processing facilities are being developed in the S areas (at Panang, Davao, and GENERAL SANTOS); mining is important in the NE, as is logging in the center and E. Important coal-mining center. Fishing is a major activity in all coastal areas. About ⅓ of the island's population is Muslim. In the middle of the 14th century, Islam spread from Malaya and Borneo to the SULU ARCHIPELAGO, and from there to Mindanao. The arrival of the Spanish in the late 16th century united the various Muslim groups in a holy war against the conquerors that lasted some 300 years. The Moros likewise resisted U.S. domination; fighting between U.S. garrisons and Muslim groups occurred early in the 20th century. Although many of the Philippine Islands suffered extensive damage in World War II, Mindanao emerged relatively unscathed. As the chief frontier left in the difficult reconstruction years, it was the object of government colonization projects. During the 1960s, it experienced a phenomenal population increase and very rapid development. These changes brought serious problems. In many cases, the native Moros were pushed off their lands. A group called the Muslim Moro National Liberation Front began to engage in terrorist activities, and the Philippine army often cracked down on them violently. A truce was reached in 1976, but it was often violated by both sides. In response to the crisis, the Aquino government established the AUTONOMOUS REGION IN MUSLIM MINDANAO in 1990, consisting of four provinces with a Muslim majority. Sporadic violence continues to this day, but the Ramas government is considering creating a second autonomous region for the Moro. In 1971 anthropologists reported the discovery of the Tasaday, whom they portrayed as a Stone Age people inhabiting caves in Mindanao's rain forest and threatened by the encroachment of lumbering, mining, and ranching interests. By the mid-1980s, however, evidence emerged indicating that this phenomenon was actually a hoax.

Mindanao Deep, Pacific Ocean: see MINDANAO TRENCH.

Mindanao, Rio Grande de, PHILIPPINES: see PULANGI RIVER.

Mindanao Sea (MEEN-dah-nou), S PHILIPPINES, between MINDANAO (S), LEYTE, BOHOL, and Cebu (N), and NEGROS (W), opening E via SURIGAO STRAIT to PHILIPPINE SEA, W to SULU SEA, and N to Visayan Sea via TAÑON STRAIT, BOHOL STRAIT, and CANIGAO CHANNEL; c.170 mi/274 km E–W.

Mindanao Trench (meen-dah-NAH-o), submarine depression in PACIFIC OCEAN, off NE MINDANAO. One

of the deepest (32,995 ft/10,057 m) areas on the surface of the globe. The trench was sounded in 1912 to 32,112 ft/9,788 m by the ship *Planet* at 09°56′N 126°50′E. In 1927, the German cruiser *Emden* obtained a depth of 35,400 ft/10,790 m at 09°41′N 126°50′E, a value long recognized internationally as the Mindanao Deep or the Philippine Deep. However, subsequent soundings in the same area were shallower, and the *Emden* reading was regarded as erroneous. In 1945, the USS *Cape Johnson* sounded a depth of 34,440 ft/10,497 m at 10°27′N 126°39′E, but the Challenger Deep in the MARIANAS TRENCH (35,839 ft/10,924 m) is now regarded as the greatest ocean depth.

Minde (MEEN-dai), town, SANTARÉM district, central PORTUGAL, 19 mi/31 km N of SANTARÉM; 39°31′N 08°41′W. Manufacturing of woolens.

Mindelheim (MIN-del-heim), town, SWABIA, SW BAVARIA, GERMANY, on MINDEL RIVER, and 15 mi/24 km ENE of MEMMINGEN; 48°03′N 10°30′E. Textile manufacturing, jewelry, and metalworking; summer resort. Has 17th-century church.

Mindêlo (meen-DAI-loo), city (2000 population 62,970) and main port of Cape Verde Islands, ⊙ São Vicente municipality, NW shore of São Vicente island, c.500 mi/805 km WNW of Dakar (Senegal); 16°53′N 25°00′W. Important coaling station on Pôrto Grande bay, the archipelago's best harbor. Educational center. Coal mining; fishing. Submarine cable station. Radio transmitter. Often called Pôrto Grande for its harbor. Formerly spelled Mindello.

Mindel River (MIN-del), 47 mi/76 km long, BAVARIA, GERMANY; rises 4 mi/6.4 km WNW of KAUFBEUREN; flows N to the DANUBE, 4 mi/6.4 km S of GUNDEL-FINGEN.

Minden (MIN-den), city, North Rhine-Westphalia, NW GERMANY, a port on the WESER RIVER and the MIDLAND CANAL, 16 mi/26 km NNE of HERFORD; 52°18′N 08°54′E. It is an industrial center and railroad junction. Manufacturing includes ceramics, chemicals, metal- and woodworking. Minden was the see of a bishopric founded c.800 by Charlemagne. In the 13th century it joined the HANSEATIC LEAGUE. Minden and the secularized bishopric passed to BRANDENBURG in the Peace of Westphalia (1648). The city passed to PRUSSIA in 1719. Noteworthy buildings include the cathedral (10th–13th century) and the city hall (13th–17th century). Amusement park, "Potts Park Minden," in former mining area.

Minden (MIN-DIN), city (2000 population 13,027), ⊙ WEBSTER parish, NW LOUISIANA, 23 mi/37 km ENE of SHREVEPORT; 32°37′N 93°17′W. Railroad junction. Shipping center of an area rich in timber, oil, and natural gas. In agricultural area (cotton, sweet potatoes, vegetables, watermelons; cattle; dairying); diversified manufacturing. Oil and gas field nearby. German colony. Established as a socialist-utopian commune in 1835. Kisatchie National Forest to N. Incorporated 1850.

Minden, city (2006 population 2,877), ⊙ KEARNEY county, S NEBRASKA, 14 mi/23 km SSE of KEARNEY, and on branch of LITTLE BLUE RIVER; 40°30′N 98°57′W. Railroad junction. Grain; livestock; dairy and poultry produce. Manufacturing (fiberglass parts, plastic aircraft components). Harold Warp's Pioneer Village here. Fort Kearney State Historical Park to NW. Founded 1876.

Minden (MIN-duhn), town (2000 population 564), POTTAWATTAMIE county, SW IOWA, on KEG CREEK, and 20 mi/32 km NE of COUNCIL BLUFFS; 41°28′N 95°32′W.

Minden, unincorporated town (2000 population 2,836), ⊙ DOUGLAS county, W NEVADA, on East Carson River, and 15 mi/24 km S of CARSON CITY; 38°57′N 119°46′W. Alfalfa, potatoes, grain; cattle; poultry; manufacturing (lightning protection equipment); tourism. Toiyabe National Forest to W and SE. Jacks Valley Wildlife Management Area to N. Douglas county Airport to N. Est. 1905 as railroad town.

Minden, unincorporated town, FAYETTE county, S central WEST VIRGINIA, near NEW RIVER, 2 mi/3.2 km NE of OAK HILL. Coal-mining region.

Minden (MIN-duhn) or **Minden Hills**, village (□ 327 sq mi/850.2 sq km; 2001 population 5,312), ⊙ HALI-BURTON county, S ONTARIO, E central CANADA, between MINDEN LAKE (N) and GULLFOOT LAKE (S), 40 mi/64 km N of LINDSAY; 44°56′N 78°44′W. Dairying, mixed farming.

Minden City, village (2000 population 242), SANILAC county, E MICHIGAN, 17 mi/27 km NNE of SANDUSKY, near source of BLACK RIVER; 43°40′N 82°46′W. In farm area. Light manufacturing. Also called Minden.

Minden Lake (MIN-duhn), S ONTARIO, E central Canada, 3 mi/5 km N of MINDEN, 50 mi/80 km NNW of PETERBOROUGH; 3 mi/5 km long, 1 mi/2 km wide; 44°57′N 78°41′W. Drains S through Gull River into BALSAM LAKE and TRENT CANAL.

Mindenmines (MIN-din-meinz), city (2000 population 409), BARTON county, SW MISSOURI, 8 mi/12.9 km NE of PITTSBURG, KANSAS; 37°28′N 94°35′W. In a former strip coal-mining area; oil field. Wheat, corn, sorghum; cattle. Prairie State Park to N.

Mindif (min-DEEF), village, Far-North province, CAMEROON, 15 mi/24 km SE of MAROUA; 10°24′N 14°25′E. Cotton; livestock and cattle raising. Rock climbing at the famous La Dent de Mindif (Mindif's Tooth).

Mindigi (meen-DEE-gee), village, KATANGA province, SE CONGO, 50 mi/80 km W of LIKASI. Copper and cobalt mining. Also copper mining at nearby MIR-UNGWE, 8 mi/12.9 km SE, and at Tantara, 25 mi/40 km E.

Mindiptana (MEEN-dip-tah-nah), town, SE IRIAN JAYA, INDONESIA, near border of PAPUA NEW GUINEA, 228 mi/367 km N of MERAUKE; 05°45′S 140°22′E. Airport.

Mindon (MIN-don), village and township, MAGWE division, MYANMAR, 30 mi/48 km W of THAYETMYO, at foot of the ARAKAN YOMA.

Mindoro (mee-DO-ro), island (□ 3,759 sq mi/9,773.4 sq km), seventh-largest of the PHILIPPINES, SW of LUZON; 12°50′N 121°05′E. Its mountainous interior rises to c.8,500 ft/2,590 m at MOUNT HALCON. What little arable land there is on the island is devoted largely to subsistence farming. Coal is mined, and lumbering is an important industry. PUERTO GALERA is a well-known beach area.

Mindoro Occidental (mee-DO-ro ok-see-den-TAHL), province (□ 2,270 sq mi/5,902 sq km), in SOUTHERN TAGALOG region, W coast of MINDORO, PHILIPPINES; ⊙ MAMBURAO; 13°00′N 120°55′E. Population 32.5% urban, 67.5% rural; in 1991, 100% of urban and 58% of rural settlements had electricity. Forested mountains, with grasslands on W slopes, coastal plain. Agriculture (rice, corn, coconuts, sugarcane), livestock (cattle), fishing, and aquaculture. Salt a major product. Possible oil field has been discovered in S. Major port is SAN JOSE; airports at San Jose and MAMBURAO.

Mindoro Oriental (mee-DO-ro or-yen-TAHL), province (□ 1,685 sq mi/4,381 sq km), in SOUTHERN TA-GALOG region, E coast of MINDORO, PHILIPPINES; ⊙ CALAPAN; 13°00′N 121°05′E. Population 25.6% urban, 74.4% rural; in 1991, 100% of urban and 67% of rural settlements had electricity. Forested mountains, with coastal plain. Agriculture (rice, corn, coconuts, sugarcane), livestock (cattle and pigs), fishing and aquaculture for both sustenance and commerce (to markets in MANILA). Timber and wood products. Calapan is main port. Fishing and ferry port at PUERTO GALERA, now also a tourist resort.

Mindoro Strait (mee-DO-ro), PHILIPPINES, separates MINDORO and CALAMIAN ISLANDS, leads from SOUTH CHINA SEA to SULU SEA via the CUYO PASSES; c.50 mi/80 km wide. Apo Reef divides it into Apo West Pass and Apo East Pass.

Mindouli (meen-doo-LEE), village, POOL region, S Congo Republic, on railroad, 100 mi/161 km W of

BRAZZAVILLE; copper-mining center. Zinc, silver, lead, and manganese also mined here. Stock raising. Roman Catholic and Protestant missions. Mindouli copper deposits were worked by natives before European penetration of central Africa; deposits are nearly exhausted.

Mindszent (MEND-sant), village, CSONGRÁD county, S HUNGARY, on TISZA RIVER, and 10 mi/16 km NW of Hódmezővásárhely; 46°32′N 20°12′E. Corn, wheat, alfalfa, clover; cattle, sheep; manufacturing (pasta, flour milling, preservatives).

Minduri (MEEN-doo-ree), town (2007 population 3,602), S central MINAS GERAIS, BRAZIL, 14 mi/23 km W of São Vicente de Minas, on railroad and Rio Capivari; 21°40′S 44°35′W.

Mindyak (meen-DYAHK), town (2005 population 2,960), E BASHKORTOSTAN Republic, W Siberian Russia, on the E slope of the S URALS, 17 mi/27 km ENE of BELORETSK; 54°01′N 58°47′E. Elevation 1,702 ft/518 m. Mining and concentrating iron ore.

Mine (MEE-ne), city, YAMAGUCHI prefecture, SW HONSHU, W JAPAN, 16 mi/25 km W of YAMAGUCHI; 34°09′N 131°12′E. Marble, lime, raw materials for making cement.

Mine (MEE-ne), town, Kamiagata county, NAGASAKI prefecture, NW KYUSHU, SW JAPAN, 121 mi/ 195 km N of NAGASAKI; 34°27′N 129°19′E.

Mine, town, Miyaki co., SAGA prefecture, N Kyushu, SW Japan, 9 mi/15 km E of Saga; 33°17′N 130°26′E.

Mine Centre (MEIN SEN-tuhr), unincorporated village, W ONTARIO, E central CANADA, on Little Turtle Lake (7 mi/11 km long, 5 mi/8 km wide), 40 mi/64 km ENE of FORT FRANCES; 48°46′N 92°37′W. Gold, iron mining.

Minehama (mee-ne-HAH-mah), village, Yamamoto county, Akita prefecture, N HONSHU, NE JAPAN, 40 mi/65 km N of AKITA city; 40°17′N 140°02′E.

Minehead (MEIN-HED), town (2001 population 7,214), NW SOMERSET, SW ENGLAND, on BRISTOL CHANNEL, and 21 mi/34 km NW of TAUNTON; 51°12′N 03°29′W. Seaside resort; small fishing port. Has 14th–15th-century church.

Mine Head, cape, S WATERFORD county, S IRELAND, 7 mi/11.3 km SSE of DUNGARVAN. Lighthouse (52°00′N 07°35′W).

Mine Hill, township, MORRIS county, N central NEW JERSEY, 8 mi/12.9 km NW of MORRISTOWN; 40°52′N 74°35′W. Suburban area. Iron mine here opened 1858, reopened 1939, since closed.

Mineiros (mee-nai-ros), city (2007 population 45,169), SW GOIÁS, central BRAZIL, 110 mi/177 km W of RIO VERDE; 17°41′S 52°32′W. Sugar, tobacco, coffee.

Mineiros, Brazil: see MINEIROS DO TIETÊ.

Mineiros do Tietê (MEE-nai-ros do CHEE-e-tai), city (2007 population 11,760), central SÃO PAULO, BRAZIL, on railroad, and 10 mi/16 km SSE of JAÚ; 22°24′S 48°26′W. Produces beer, beverages, macaroni; coffee and rice processing, tanning. Until 1944, Mineiros.

Mine La Motte (MEIN luh MAHT), unincorporated community, historic lead-mining village, MADISON county, SE MISSOURI, in SAINT FRANCOIS MOUNTAINS 5 mi/8 km N of FREDERICKTOWN. First lead mine in present-day Missouri was opened here in c.1720 by Antoine de la Motte Cadillac.

Mineola (MIN-ee-o-luh), town (2006 population 5,091), Wood county, NE TEXAS, near SABINE RIVER, 24 mi/39 km NNW of TYLER; 32°40′N 95°29′W. Elevation 414 ft/126 m. Railroad junction; in agricultural and manufacturing (sleeping bags, feeds, food processing, wooden cabinets) area; timber. Settled 1872, incorporated 1873.

Mineola, suburb (□ 1 sq mi/2.6 sq km; 2006 population 18,808) of NEW YORK city, ⊙ NASSAU county, SE NEW YORK, on LONG ISLAND, 18 mi/29 km E of midtown MANHATTAN; 40°45′N 73°38′W. Chiefly residential, it is a commercial center, with some light industry. Site of early aviation experiments. Comedian Lenny Bruce

and business executive Louis Gerstner born here. Incorporated 1906.

Miner, county (□ 572 sq mi/1,487.2 sq km; 2006 population 2,553), E central SOUTH DAKOTA; ⊙ HOWARD; 44°01′N 97°36′W. Agricultural area watered by West Fork of VERMILLION RIVER and Rock and Redstone creeks. Corn, wheat, soybeans; dairy produce; cattle, hogs, poultry. Lake Carthage State Lakeside Use Area in N. Formed 1873.

Miner (MEI-nuhr), town, SCOTT county, SE MISSOURI, suburb, 2 mi/3.2 km E of SIKESTON; 36°53′N 89°31′W. Residential. Highway service center.

Minera (mi-NE-ruh), village (2001 population 1,608), WREXHAM, NE Wales, 4 mi/6.4 km WNW of Wrexham; 53°03′N 03°06′W.

Mineral, county (□ 877 sq mi/2,280.2 sq km; 2006 population 929), SW COLORADO; ⊙ CREEDE; 37°41′N 106°55′W. CONTINENTAL DIVIDE forms part of N boundary, winds SSW through HINSDALE county, turns SSE in SAN JUAN county, crossing back through Hinsdale county and across S part of Mineral county. Mining and livestock-grazing region, drained by headwaters of RIO GRANDE. Silver, lead. Includes ranges of ROCKY MOUNTAINS. Most of N three fourths of county in Rio Grande National Forest; most of S part, S of divide, in San Juan National Forest. Wheeler National Monument is in NE. Formed 1893.

Mineral, county (□ 1,223 sq mi/3,168 sq km; 1990 population 3,315; 2000 population 3,884), W MONTANA; ⊙ SUPERIOR; 47°09′N 114°59′W. Forested region bordering on IDAHO; drained by the CLARK FORK and Saint Regis rivers. Some sheep, cattle; hay; forest industries; mining. Lolo National Forest covers all of county, except margins of CLARK FORK. BITTERROOT RANGE in W. Formed 1914.

Mineral, county (□ 3,813 sq mi/9,913.8 sq km; 2006 population 4,868), W NEVADA; ⊙ HAWTHORNE; 38°32′N 118°25′W. Mountain region bordering on CALIFORNIA. Gold, silver; sand and gravel; cattle; recreation. Formed 1911. WALKER LAKE (Walker Lake State Recreational Area on W shore) and WASSUK RANGE in W. U.S. naval ammunition depot at Hawthorne, in W center; part of Walker River Indian Reservation in NW. EXCELSIOR MOUNTAINS in S. Gillis Range in N center. Part of Toiyabe and Inyo National Forests in SW. Formerly entirely within ESMERALDA county.

Mineral, county (□ 329 sq mi/855.4 sq km; 2006 population 26,928), WEST VIRGINIA, in EASTERN PANHANDLE; ⊙ KEYSER; 39°25′N 78°56′W. Bounded N and NW by North Branch of the POTOMAC RIVER (forms MARYLAND state line); Jennings Randolph Lake reservoir in W); drained by PATTERSON CREEK; traversed by KNOBLY and PATTERSON CREEK mountain ridges (on W boundary) and others. Coal mines; timber. Agriculture (corn, wheat, oats, barley, rye, tobacco, alfalfa, hay); cattle, hogs, sheep, poultry. Manufacturing at Keyser, especially glass. Nancy Hanks Memorial in S. Parts of Springfield Wildlife Management Area in E. Formed 1866.

Mineral (MI-nuh-ruhl), town (2006 population 467), LOUISA county, central VIRGINIA, 30 mi/48 km E of CHARLOTTESVILLE; 38°00′N 77°54′W. Manufacturing (clothing, wooden pallets, lumber); agriculture (tobacco, grain, soybeans; cattle).

Mineral, unincorporated village (2000 population 143), TEHAMA county, N CALIFORNIA, 35 mi/56 km ENE of RED BLUFF, near MILL CREEK. Headquarters for LASSEN VOLCANIC NATIONAL PARK (to NE).

Mineral, village (2000 population 272), BUREAU county, N ILLINOIS, 20 mi/32 km W of PRINCETON; 41°22′N 89°50′W. In agricultural area.

Mineral Bluff, village, FANNIN county, N GEORGIA, 4 mi/6.4 km NE of BLUE RIDGE; 34°55′N 84°17′W.

Mineral City, village (2006 population 850), TUSCARAWAS county, E OHIO, 9 mi/14 km NE of NEW PHILADELPHIA; 40°36′N 81°22′W. Coal-mining area.

Mineral de Angangueo (mee-NAI-ral de ahn-gahn-GWAI-o), town, ⊙ Angangueo municipio, MICHOACÁN, central MEXICO, on central plateau, 45 mi/72 km NW of TOLUCA DE LERDO, on railroad; 19°39′N 100°25′W. Elevation 8,622 ft/2,628 m. In the past, one of richest mining centers (silver, lead, copper, zinc).

Mineral de la Reforma, MEXICO: see PACHUQUILLA.

Mineral del Chico (mee-NE-rahl del CHEE-ko) or **El Chico**, town, HIDALGO, central MEXICO, 6 mi/9.7 km N of PACHUCA DE SOTO; 20°12′N 98°43′W. Elevation 7,713 ft/2,351 m. located within EL CHICO National Park. Former mining center. Resort in majestic mountain setting nearby.

Mineral del Monte (mee-ne-rahl del MON-te), city and township, ⊙ municipio Mineral del Monte, HIDALGO, central MEXICO, on central plateau, 4 mi/6.4 km E of PACHUCA DE SOTO; 20°08′N 98°40′W. Elevation 8,789 ft/2,679 m. Important silver- and gold-mining center worked by Welsh miners in 1800s, active; foundries; manufacturing of explosives. Sometimes REAL DEL MONTE.

Mineral Hills, village (2000 population 214), IRON county, SW UPPER PENINSULA, MICHIGAN, 2 mi/3.2 km NNW of IRON RIVER city; 46°06′N 88°38′W.

Mineral Hot Springs, village, SAGUACHE county, S central COLORADO, on SAN LUIS CREEK, in S foothills of SAWATCH MOUNTAINS, and 13 mi/21 km ENE of SAGUACHE. Elevation 7,747 ft/2,361 m. Sheep, cattle, feed. Resort. OURAY PEAK is 19 mi/31 km NW. Parts of Rio Grande National Forest to NW and NE.

Mineral Mountains, in BEAVER county, SW UTAH, extend 25 mi/40 km N from MINERSVILLE. Maximum elevation 7,583 ft/2,311 m.

Mineralni Bani (mi-ner-AHL-nee BAH-nee), village, HASKOVO oblast, Mineralni Bani obshtina (1993 population 7,452), BULGARIA; 41°55′N 25°21′E.

Mineral'nyye Vody (mee-nee-RAHL-ni-ye VO-di) [Russian=mineral waters], city (2006 population 76,520), S STAVROPOL TERRITORY, N CAUCASUS, S European Russia, in the KUMA RIVER valley, on the Rostov-Baku railroad, 105 mi/169 km SE of STAVROPOL, and 12 mi/19 km NNE of PYATIGORSK; 44°12′N 43°08′E. Elevation 1,020 ft/310 m. Important transport center; railroad junction for Pyatigorsk resort district (S) (freight yards, workshops); regional airport serving a group of resorts and mineral springs of the N Caucasus. Manufacturing of mineral water bottles; food industries. Made city in 1920.

Mineral Point, town (2000 population 363), WASHINGTON county, E central MISSOURI, in the OZARKS, 3 mi/4.8 km E of POTOSI; 37°57′N 90°43′W. Former lead-mining district; manufacturing (valves). Missouri State Correctional Center to NW.

Mineral Point, town (2006 population 2,569), IOWA county, S WISCONSIN, 7 mi/11.3 km SSW of DODGEVILLE; 42°51′N 90°10′W. In dairy and livestock area; dairy products. Light manufacturing. Winery. Has restored Cornish miners' houses dating from lead-mining activity of mid-19th century. Incorporated 1857.

Mineral Ridge, unincorporated village (□ 3 sq mi/7.8 sq km; 2000 population 3,900), TRUMBULL county, NE OHIO, 7 mi/11 km SE of WARREN; 41°08′N 80°46′W. Steel products, canned foods.

Mineral Springs, town (2000 population 1,264), HOWARD county, SW ARKANSAS, 6 mi/9.7 km SW of NASHVILLE; 33°52′N 93°55′W. Manufacturing (floor and roof trusses). Millwood Lake reservoir to NW.

Mineral Springs (MIN-uhr-uhl SPREENGZ), unincorporated village (□ 7 sq mi/18.2 sq km; 2006 population 2,493), UNION county, S NORTH CAROLINA, 7 mi/11.3 km SW of MONROE; 34°56′N 80°40′W. Construction; manufacturing (yarn, fertilizer); agriculture (cotton; livestock; poultry).

Mineral Wells (MIN-uhr-uhl), city (2006 population 17,065), PALO PINTO and PARKER counties, N TEXAS, 45 mi/72 km W of FORT WORTH; 32°49′N 98°04′W.

Elevation 925 ft/282 m. Manufacturing (aluminum products, bottled mineral water, clothing, building materials, gas processing, pharmaceuticals). The mineral water (Waters of Crazy Well were discovered 1885) made this hill city a popular health resort in the late 19th and early 20th century, and oil activity in the area also spurred the city's growth. Cattle, horses; peanuts, wheat; cedar timber. To the E is Lake Mineral Wells State Park, a reservoir in the TRINITY RIVER system. Incorporated 1882.

Mineral Wells or **Mineralwells**, unincorporated town, WOOD county, NW WEST VIRGINIA, 5 mi/8 km S of PARKERSBURG, near LITTLE KANAWHA RIVER; 39°10′N 81°30′W. Agriculture (grain, tobacco); livestock; poultry. Manufacturing (trailer and truck bodies, sand and gravel processing). West Virginia Motor Speedway is here. West Virginia Interstate Fair held in July.

Mineros (mee-NE-ros), canton, OBISPO SANTIESTEBAN province, SANTA CRUZ department, E central BOLIVIA, NW of GENERAL SAAVEDRA, 52 mi/83 km NE of SANTA CRUZ DE LA SIERRA; 17°10′S 63°14′W. Elevation 961 ft/ 293 m. Abundant gas and petroleum in area. Clay and limestone deposits. Agriculture (potatoes, yucca, bananas, corn, rice, cotton, peanuts, citrus fruits); cattle.

Minersville, village (2006 population 848), BEAVER county, SW UTAH, on BEAVER RIVER, and 17 mi/27 km WSW of Beaver, just SW of MINERAL MOUNTAINS; 38°12′N 112°55′W. Irrigated agricultural area (alfalfa, corn, fruit, potatoes. barley); cattle. Town was once dependent on mining. Elevation 5,625 ft/1,715 m. Minersville Reservoir and State Park to E. Settled 1858; first mine in Utah.

Minersville, borough (2006 population 4,298), SCHUYLKILL county, E central PENNSYLVANIA, 3 mi/ 4.8 km NNW of POTTSVILLE; 40°41′N 76°15′W. Agriculture (poultry, dairying); manufacturing (concrete, apparel); anthracite coal. Schuylkill County Airport 5 mi/8 km to W. Settled c.1793, incorporated 1831.

Minerva (muh-NUHRV-ah), village (□ 2 sq mi/5.2 sq km; 2006 population 3,963), on STARK-CARROLL county line, E OHIO, 15 mi/24 km ESE of CANTON; 40°43′N 81°06′W. In agricultural area. Previously brick and tile. Founded c.1835.

Minerva, hamlet, ESSEX county, NE NEW YORK, in the ADIRONDACK MOUNTAINS, 37 mi/60 km NNW of GLENS FALLS; 43°51′N 74°02′W. Hunting nearby.

Minerva Park (muh-NUHRV-ah), village (2006 population 1,263), FRANKLIN county, central OHIO; suburb 5 mi/8 km N of COLUMBUS; 40°04′N 82°57′W.

Minervois, le (mee-ner-VWAH, luh) in AUDE and HÉRAULT departments, Languedoc-Roussillon region, S FRANCE; 43°21′N 02°42′E. Vineyards.

Minetto, village (□ 3 sq mi/7.8 sq km; 1990 population 1,252; 2000 population 1,086) in Minetto town (□ 3 sq mi/7.8 sq km; 2000 population 1,086), OSWEGO county, N central NEW YORK, on OSWEGO RIVER, just S of OSWEGO; 43°23′N 76°28′W.

Mineville-Witherbee, census-designated place (2000 population 1), ESSEX county, NE NEW YORK, near Lake CHAMPLAIN, 18 mi/29 km NNW of TICONDEROGA; 44°05′N 73°31′W. It is composed of the two hamlets Mineville and Witherbee. In iron-mining area.

Mineyama (mee-NE-yah-mah), town, Naka county, KYOTO prefecture, S HONSHU, W central JAPAN, 56 mi/ 90 km N of KYOTO; 35°37′N 135°03′E. Machine metal industries. *Tango* crepe, which was invented by Kinuya Saheiji, who was born here.

Minfeng, CHINA: see MINGFENG.

Minga (MEEN-guh), village, KATANGA province, SE CONGO, close to ZAMBIAN border; 11°08′S 27°57′E. Elev. 3,845 ft/1,171 m.

Minga Guazu (MEEN-gah gwah-ZOO), town (2002 population 14,806), E PARAGUAY, Alto Parana department, 15 mi/24 km W of CIUDAD DEL ESTE; 25°26′S 54°45′W. Industrial park with assembly plants; also, cotton, soybeans, cooking oil.

Mingaladon (MING-gah-lah-DON), town, YANGON township, MYANMAR, 10 mi/16 km N of YANGON. Site of Yangon civil and military airport, army cantonment, and the Institute of Medicine.

Mingalay, Scotland: see MINGULAY.

Mingan (MIN-guhn), village, CÔTE-NORD region, E QUEBEC, E CANADA, on Jacques Cartier Strait of Gulf of SAINT LAWRENCE, and 100 mi/161 km E of SEPT-ÎLES, 400 mi/644 km NE of QUEBEC city; 50°18′N 64°02′W. Radio station, airfield. Coastal highway extended from SEPT-ÎLES to HAVRE–SAINT-PIERRE (via MINGAN) in late 1960s; ferry continues beyond to S end of NEWFOUNDLAND AND LABRADOR.

Mingan Archipelago National Park Reserve (□ 58 sq mi/150 sq km), French, *l'Archipel-de-Mingan, Réserve de parc national de* (lahr-kee-PEL-duh–MIN-guhn, rai-ZERV duh pahrk nah-syo-NAHL duh), c. fourty islands in Jacques Cartier Passage at mouth of SAINT LAWRENCE RIVER, E QUEBEC, E CANADA, near HAVRE–SAINT-PIERRE, 93 mi/150 km E of SEPT-ÎLES; 50°13′N 63°10′W. Established 1985; national park status pending settlement of Aboriginal land claims. Flowerpot islands of limestone, eroded at base by tides. Atlantic puffins, other seabirds; wildflowers. Camping, hiking.

Minganie (meen-gah-NEE), county (□ 49,611 sq mi/ 128,988.6 sq km; 2006 population 5,510), CÔTE-NORD region, E QUEBEC, E CANADA; ⊙ HAVRE-SAINT-PIERRE; 50°14′N 63°36′W. Composed of ten municipalities. Formed in 1982.

Mingan Islands (MIN-guhn), group of fifteen small islands and many islets, E QUEBEC, E CANADA, in the SAINT LAWRENCE RIVER, N of ANTICOSTI island; 50°13′N 63°50′W. They were visited (1535) by Jacques Cartier, the French explorer. In 1836 the islands were owned by the Hudson Bay Company.

Mingan Passage (MIN-guhn), E QUEBEC, E CANADA, channel (30 mi/48 km wide) of the SAINT LAWRENCE RIVER, between MINGAN, on Quebec mainland, and ANTICOSTI Island; 50°15′N 63°49′W. Near N coast are MINGAN ISLANDS.

Mingaora (min-GOU-rah), town, SWAT district, NORTH-WEST FRONTIER PROVINCE, N PAKISTAN, 3 mi/ 4.8 km N of SAIDU SHARIF, on SWAT RIVER, and 70 mi/ 113 km NE of PESHAWAR; 34°47′N 72°22′E. Market center for wheat, fruit, barley, sugarcane, rice, vegetables; wool. Timber harvesting and milling. Airport.

Mingbulak (meeng-boo-LAHK), village, ANDIJAN wiloyat, E UZBEKISTAN; 40°52′N 71°40′E.

Mingchien, town, W central TAIWAN, 20 mi/32 km S of TAICHUNG; agricultural products (sugarcane, rice, tea, fruit, bamboo articles; livestock).

Mingechaur (meen-gyi-CHOOR), city, central AZERBAIJAN, on KURA river (dammed), at E end of MINGECHAUR RESERVOIR, 15 mi/24 km NW of YEVLAKH, on railroad spur from 28 Aprelya; 40 mi/64 km long, 10 mi/16 km wide. Site of earth dam (250 ft/76 m high) and hydroelectric station supplying BAKY industrial district. Developed after 1945. Connected by an 11-mi/ 18-km railroad branch to the Mingechaur station (TBILISI-Baky line). Mingechaur arose in 1945 in connection with the construction of hydroelectric complex. Manufacturing (road-building machinery, cables, fiberglass, industrial rubber articles, electrical insulation, reinforced concrete structural components, panels for prefabricated housing, wood products); machinery repair shop, textile combine, meat-packaging plant. Archaeological and historical sites. The most important complex in Transeaucagia is located near Mingechaur region. It includes four settlements and three large burial grounds dating from the 3rd milennium B.C.E. to the 17th century C.E. Mingechaur's remains, covering a period of 4,000 years, constitute important sources for the study of the cultural-historical and socioeconomic development of Azerbaijan and its neighboring countries.

Mingechaur Reservoir (meen-gyi-CHOOR) (□ 234 sq mi/608.4 sq km), AZERBAIJAN, formed by dam of Mingechaur Hydroelectric Power Plant on KURA RIVER. Designed to eliminate flooding in lower course of Kura River. Upper Karabach Canal and Upper Shirvan Canal originate here. Mingechaur Hydroelectric Power Plant located near city of MINGECHAUR. Construction began 1945; put into operation 1954. Part of integrated Transcaucasian power grid.

Mingenew (MIN-uhn-yoo), town, WESTERN AUSTRALIA state, W AUSTRALIA, 238 mi/383 km N of PERTH; 29°15′S 115°27′E. Wheat, lupins; sheep. Known for its 250-million-year-old marine fossils.

Mingfeng (MING-FENG), town and oasis, ⊙ Mingfeng county, S XINJIANG UYGUR AUTONOMOUS REGION, CHINA, 160 mi/257 km E of Heitian, and on highway skirting S edge of TAKLIMAKAN Desert; 37°04′N 82°46′E. Sericulture. Livestock; food processing. Also called Niya. Also appears as Minfeng.

Mingin (MING-GIN), village, MINGIN township, SAGAING division, MYANMAR, on W bank of CHINDWIN RIVER, and 120 mi/193 km NW of MANDALAY. Timber center.

Mingjiang (MING-JI-ANG), town, SW GUANGXI ZHUANG AUTONOMOUS Region, CHINA, on railroad, and 70 mi/113 km SW of NANNING; 22°08′N 107°10′E. Rice, sugarcane.

Ming-Kush (meeng-KUHSH), town, NARYN region, KYRGYZSTAN, by Moldo-Too range, and on Kökömeren River; 41°30′N 74°20′E. Thermal power station; coal mining. Also spelled Mingkush.

Minglanilla (ming-glah-NEE-lyah), town, CUENCA province, E central SPAIN, on Madrid-Valencia highway, and 45 mi/72 km SE of CUENCA; 39°32′N 01°36′W. Road junction in agricultural region (saffron, grapes; sheep, goats; lumbering. Flour milling, vegetable canning, liquor distilling. Saltworks.

Minglun (MING-LUN), town, N GUANGXI ZHUANG AUTONOMOUS REGION, CHINA, 50 mi/80 km NNW of YISHAN, near GUIZHOU border; 25°13′N 108°23′E. Grain. Also known as Yibei.

Mingo (MING-o), county (□ 423 sq mi/1,099.8 sq km; 2006 population 27,100), SW WEST VIRGINIA, ⊙ WILLIAMSON; 37°43′N 82°08′W. Bounded SW by TUG FORK RIVER (KENTUCKY state line). Extensive bituminous-coal fields; natural gas and oil wells; timber. Some agriculture. Laurel Creek Wildlife Management Area in N; part of River. D. Bailey Lake (reservoir) Wildlife Management Area in SE. Formed 1895.

Mingo, town (2000 population 269), JASPER county, central IOWA, 14 mi/23 km WNW of NEWTON; 41°46′N 93°16′W. Livestock, grain.

Mingo Junction (MING-go JUHNK-shuhn), village (□ 3 sq mi/7.8 sq km; 2006 population 3,403), JEFFERSON county, E OHIO, 4 mi/6 km S of STEUBENVILLE, and on OHIO RIVER; 40°19′N 80°37′W. Truck and fruit farming. Coal mines nearby. Settled 1809, incorporated 1882.

Mingorria (ming-GOR-yah), town, ÁVILA province, central SPAIN, on ADAJA RIVER, and 6 mi/9.7 km N of ÁVILA; 40°45′N 04°40′W. Cereals, melons, grapes. Stone quarrying; flour milling, dairying, chocolate manufacturing.

Mingoyo (meen-GO-yoh), town, LINDI region, SE TANZANIA, 10 mi/16 km SW of LINDI, near LUKULEDI RIVER. Road junction. Cashews, peanuts, copra, sisal; sheep, goats. Salt deposits.

Mingoyo (meen-GO-yoh), village, LINDI region, SE TANZANIA, 10 mi/16 km SW of LINDI, near LUKULEDI RIVER; 10°06′S 39°37′E. Road junction. Cashews, peanuts, bananas, sweet potatoes; goats, sheep.

Mingrelia (min-GRE-lee-yuh), lowland region, W GEORGIA, bordering the BLACK SEA. Tea and grapes are the chief products POTI is the main port. The COLCHIS of the ancients, Mingrelia was a vassal principality (with ZUGDIDI as capital) under the OTTOMAN EMPIRE. It was annexed to RUSSIA in 1803. The Mingrelians (also called Megrelians) are closely related in culture and language to the Georgians.

Area is shown by the symbol □, and capital city or county seat by ⊙.

Mingrel'skaya (MEEN-greel-skah-yah), village (2005 population 5,550), S central KRASNODAR TERRITORY, S European Russia, on the Adagum River (KUBAN' River basin), on road, 42 mi/68 km ENE of ANAPA, and 35 mi/56 km W of KRASNODAR; 45°00'N 38°20'E. In oil- and gas-producing region.

Mingshan (MING-SHAN), town, ⊙ Mingshan county, central W SICHUAN province, CHINA, 15 mi/24 km NE of YA'AN; 30°08'N 103°10'E. Tea, millet, rice, wheat, oilseeds, sugarcane; food processing, paper making, chemicals, engineering.

Mingshui (MING-SHWAI), town, ⊙ Mingshui county, W central HEILONGJIANG province, CHINA, 100 mi/161 km E of QIQIHAR; 47°10'N 125°55'E. Grain, soybeans, jute, sugar beets, sugar refining.

Mingulay (MIN-guh-lai) or **Mingalay**, uninhabited island (□ 3 sq mi/7.8 sq km) in S part of OUTER HEBRIDES, Eilean Siar, NW Scotland, between PABBAY island (N) and BERNERAY island (S), and 10 mi/16 km SW of BARRA island; 56°49'N 07°38'W. Rises to 891 ft/ 272 m. Sea bird refuge.

Mingus (MIN-guhs), village (2006 population 258), PALO PINTO county, N central TEXAS, c.65 mi/105 km WSW of FORT WORTH; 32°32'N 98°25'W. In agricultural area (cattle; peanuts, wheat); oil and gas.

Mingus Mountain, peak (7,743 ft/2,360 m) in BLACK HILLS, central ARIZONA, 25 mi/40 km ENE of PRESCOTT, and 3 mi/4.8 km S of JEROME. Copper was mined nearby. In Prescott National Forest.

Mingxi (MING-SEE), town, ⊙ Mingxi county, W FUJIAN province, CHINA, 65 mi/105 km NE of CHANGTING; 26°24'N 117°12'E. Logging; timber processing, furniture, chemicals, paper making, engineering, iron smelting.

Minhang (MIN-HANG), town, ⊙ Minhang county, CHINA, 17 mi/27 km S of SHANGHAI, an administrative unit of SHANGHAI municipality. Engineering, chemicals, coking coal industry; rice, oilseeds, cotton.

Minhe (MIN-HUH), town, ⊙ Minhe county, NE QINGHAI province, CHINA, on GANSU border, on XINING RIVER at mouth of the DATONG, and 55 mi/89 km ESE of XINING; 36°20'N 102°50'E. Cattle raising; non-ferrous metal smelting.

Minh Hai, province (□ 2,968 sq mi/7,716.8 sq km), S VIETNAM, in MEKONG Delta, N border with KIEN GIANG and CAN THO provinces, NE border with SOC TRANG province, SE border on SOUTH CHINA SEA, W border on Gulf of THAILAND; ⊙ CA MAU; 09°10'N 105°10'E. Southernmost province of Vietnam. Sprawling marshes, seasonally inundated swamplands, and extensive mangrove forests. Fine alluvial soils sustain a thriving agricultural economy (wet rice cultivation, vegetables, pineapples, and other fruits). Diverse economy includes aquaculture (shrimp), riverine and maritime fisheries, fish drying and canning, food processing, light manufacturing, lumbering, charcoal making, tannin extraction, forest products (honey, wax, medicinals, thatch). Seriously damaged by aerial defoliation during Vietnam War, the ecologically complex mangroves are now being replanted. Kinh population with significant Khmer minority. Future plans call for the province to be divided into two provinces: BAC LIEU (⊙ BAC LIEU) and CA MAU (⊙ Ca Mau).

Minhla (min-LAH), town and township, BAGO division, MYANMAR, on Yangon-Prome railroad, and 65 mi/105 km SSE of PROME.

Minhla (min-LAH), village, ⊙ MINHLA township, MYANMAR, on W bank of AYEYARWADY RIVER (landing), and 15 mi/24 km SSE of MAGWE. Linked by pipeline with YENANMA oil field (SW). Site of old fort captured 1885 in Third Anglo-Burmese War.

Minho (MEE-nyoo), Spanish *Miño*, (MEE-nyo), river, c.210 mi/340 km long, SPAIN and PORTUGAL; rising in GALICIA, NW Spain; flowing generally SW to the ATLANTIC OCEAN c.10 mi/16 km N of VIANA DO CASTELO, Portugal. The Sil is its chief tributary. The lower

part of the Minho forms a section of the border between NW Spain and far northernmost Portugal. Hydroelectricity is produced near Orense, Spain.

Minho-Lima (MEEN-yoo), traditional region and historical province, NW PORTUGAL, between the MINHO and DUERO rivers; ⊙ BRAGA; 41°40'N 08°30'W. Region was settled by the Celts, who left many hill forts, and by the Romans. Geological faults traverse the area. The land supports intensive agricultural activity. Corn, wine grapes, and fruits are grown here. Lumbering, fishing, cattle raising, and textile manufacturing are important in some localities.

Minho River, c.40 mi/64 km long, W central and S JAMAICA; rises just E of SPALDINGS; flows SE and S through a fertile valley, past FRANKFIELD, Chapleton, MAY PEN, and ALLEY, to the coast; 18°05'N 77°14'W. Not navigable. Also known as Rio Minho.

Minhou (MIN-HAW), town, ⊙ Minhou county, E FUJIAN province, CHINA, port on island in MIN River near its mouth, 12 mi/19 km SE of FUZHOU; 26°07'N 119°27'E. Trans-shipment point.

Minhsiung (MIN-SI-UNG), town, W central TAIWAN, on railroad, and 5 mi/8 km N of CHIAI; 23°33'N 120°25'E. Manufacturing (brick, bamboo paper), pineapple; rice. Sometimes spelled Minhiung or Minsiung.

Minia, EGYPT: see MINYA.

Minicoy Island, southernmost of Lakshadweep (Laccadive) Islands, LAKSHADWEEP Union Territory, INDIA, in ARABIAN SEA; 08°15'N 73°05'E. Separated from other islands of the Lakshadweep proper (W) by NINE DEGREE CHANNEL and from Maldive Islands (S) by EIGHT DEGREE CHANNEL. Coconuts. Culturally akin to the Maldives, Minicoy was presented to a Moslem ruler of the Lakshadweep by a Maldive sultan in sixteenth century and has since shared the history of the Lakshadweep.

Minidoka, county (□ 763 sq mi/1,983.8 sq km; 2006 population 19,041), S IDAHO; ⊙ RUPERT. Bounded by SNAKE RIVER on S and SE. Irrigated farmlands receive water from LAKE WALCOTT, formed by Minidoka Dam (SE boundary of co.) on Snake River. Potatoes, sugar beets, dry beans, alfalfa; sheep, cattle; dairying; oats, barley, wheat; wholesale and retail trade. Part of Minidoka National Wildlife Refuge in SE corner, at dam; part of Lava Crater in N; SNAKE RIVER PLAIN. Formed 1913.

Minidoka, village (2000 population 129), MINIDOKA county, S IDAHO, 15 mi/24 km NE of RUPERT; 42°45'N 113°29'W. Elevation 4,286 ft/1,306 m. Railroad center in irrigated agricultural area. Minidoka Dam (LAKE WALCOTT RESERVOIR; SNAKE RIVER) is 6 mi/9.7 km S. Minidoka National Wildlife Refuge to SE on reservoir.

Minidoka Dam, Idaho: see WALCOTT, LAKE.

Minieh, EGYPT: see MINYA.

Minier (mei-NEER), village (2000 population 1,244), TAZEWELL county, central ILLINOIS, 17 mi/27 km W of BLOOMINGTON; 40°25'N 89°18'W. In agricultural area.

Miniet el Heit (men-YET el HAIT), village, FAIYUM province, Upper EGYPT, 3 mi/4.8 km WSW of ITSA. Cotton, cereals, sugarcane, fruits. Also called El Minya locally.

Minihassa, INDONESIA: see MINAHASA.

Minija River (MI-nee-yah), 132 mi/213 km long, in LITHUANIA; starting in the uplands S of TELSIAI; flows W, turns S near KRETINGA, and discharges into the Nemunas River delta. The fourth-longest river in Lithuania.

Minim (MEE-neem), village, Adamawa province, CAMEROON, 55 mi/89 km SW of Ngaoundéré; 06°58'N 12°53'E. BAUXITE production.

Minin, SYRIA: see MENIN.

Mininco (mee-NEEN-ko), village, MALLECO province, ARAUCANIA region, S central CHILE, on railroad, and 13 mi/21 km E of ANGOL; 37°47'S 72°28'W. In agricultural area (grain, apples; wine; cattle).

Miniota (mi-nee-O-tuh), rural municipality (□ 322 sq mi/837.2 sq km; 2001 population 969), SW MANITOBA, W central CANADA; 50°10'N 100°55'W. Agriculture (grain, oilseed, legumes; livestock). Main center is village of MINIOTA; other communities include ARROW RIVER, ISABELLA, CRANDALL, and BEULAH. Founded 1900.

Miniota (mi-nee-O-tuh), unincorporated village, SW MANITOBA, W central CANADA, on ASSINIBOINE RIVER, 50 mi/80 km WNW of BRANDON, and in MINIOTA rural municipality; 50°08'N 101°02'W. Dairying; grain, stock. Founded 1899.

Minipe (MI-NI-pe), town, CENTRAL PROVINCE, SRI LANKA, on the MAHAWELI GANGA River, and 22 mi/35 km ESE of KANDY; 07°13'N 80°59'E. Two anicuts across the Mahaweli divert irrigation water along canals on the E and W banks of the river. Extensive rice and vegetable cultivation.

Minish, IRELAND: see MWEENISH.

Minitonas (mi-ni-TO-nuhs), rural municipality (□ 463 sq mi/1,203.8 sq km; 2001 population 1,152), W MANITOBA, W central CANADA, N of DUCK MOUNTAIN PROVINCIAL PARK; 52°10'N 100°59'W. Logging; agriculture; tourist, service industries. Town of Minitonas is main center; other communities include RENWER, Sevick, BOWSMAN.

Minitonas (mi-ni-TO-nuhs), town (□ 1 sq mi/2.6 sq km; 2001 population 538), W MANITOBA, W central CANADA, 10 mi/16 km E of SWAN RIVER town, and in MINITONAS rural municipality; 52°05'N 101°02'W. Dubbed the "northern gateway" to DUCK MOUNTAIN PROVINCIAL PARK. Logging; agriculture; tourism. Incorporated 1948.

Minj, village, WESTERN HIGHLANDS province, E central NEW GUINEA island, N central PAPUA NEW GUINEA, 18 mi/29 km E of town of MOUNT HAGEN in Great Plateau area; 05°54'S 144°41'E. Bananas, coffee, tea, maize, vegetables; cattle.

Mink Creek (MEENK), community, SW MANITOBA, W central CANADA, 25 mi/40 km ESE of DUCK MOUNTAIN PROVINCIAL PARK, in ETHELBERT rural municipality; 51°25'N 100°29'W.

Mink Creek, unincorporated village, FRANKLIN county, SE IDAHO, 11 mi/18 km NE of PRESTON, near BEAR RIVER; 42°13'N 111°42'W. Poultry, dairy. Cache National Forest to E.

Min'kovtsy, UKRAINE: see MYN'KIVTSI.

Minlaton (MIN-luh-tuhn), village, S SOUTH AUSTRALIA, S central AUSTRALIA, on S central YORKE PENINSULA, 55 mi/89 km W of ADELAIDE, across Gulf SAINT VINCENT; 34°46'S 137°36'E. Barley, wheat; wool, grazing. Large eucalyptus common to area. Formerly called Gum Flat.

Minle (MIN-LUH), town, ⊙ Minle county, central GANSU province, CHINA, near QINGHAI border, 38 mi/61 km SE of ZHANGYE; 38°26'N 100°54'E. Elevation 7,874 ft/2,400 m. Grain, oilseeds; food processing.

Minmaya (meen-MAH-yah), village, E Tsugaru county, Aomori prefecture, extreme N HONSHU, N JAPAN, near Cape Tappi, 31 mi/50 km N of AOMORI; 41°11'N 140°26'E.

Minna (mee-NAH), town (1991 estiated population 120,000), ⊙ NIGER state, W central NIGERIA, 200 mi/322 km SSW of KANO; 09°37'N 06°32'E. Railroad junction (branch to BARO); gold-mining center; sheanut processing; cotton, ginger, cassava, durra, and yams. Has hospital.

Minna Bluff, peninsula, jutting out into the W ROSS ICE SHELF from the SE foot of MOUNT DISCOVERY, EAST ANTARCTICA; 25 mi/40 km long and 3 mi/5 km wide; 78°31'S 166°25'E. Elevation 3,300 ft/1,000 m.

Minneapolis, city (2000 population 382,618), ⊙ HENNEPIN county, E MINNESOTA, at the head of navigation on the MISSISSIPPI RIVER, at St. Anthony Falls; 44°57'N 93°16'W. Lock and Dam Number 1 at falls; river flows through center of city. The largest city in the state and a port of entry, it is a major industrial and railroad hub. Served by Twin Cities international airport in nearby RICHFIELD. With adjacent ST. PAUL to E, the two are known as the Twin Cities; similar in size, Minneapolis is the dominant twin, downtown St. Paul is 8 mi/12.9 km ESE of downtown Minneapolis. It is the processing, distributing, and trade center for a vast grain and cattle area. Minneapolis is also a banking and financial center with a significant high-technology industry that primarily developed in the 1980s. Diversified manufacturing, both heavy and light. Although the central city's population has declined since the 1970s, the outlying suburbs have grown significantly, including BLOOMINGTON (S), PLYMOUTH, and MINNETONKA (W). The falls were visited by Louis Hennepin in 1683; FORT SNELLING was established in 1819; and a sawmill was built at the falls in 1821. The village of ST. ANTHONY was settled c.1839 on the E side of the river near the falls. Minneapolis originated on the W side of the river c.1847 and included much of the reservation of Fort Snelling. It annexed St. Anthony in 1872. The city became the country's foremost lumber center, and after the plains were planted with wheat and the railroads were built, flour milling developed, with the 50 ft/15 m falls supplying power. The city was laid out with wide streets and has twenty-two lakes and 153 parks. In Minnehaha Park, in SE, is the Stevens House (1849), the first frame house in Minneapolis, and MINNEHAHA FALLS, on Minnehaha Creek. Fort Snelling State Park and National Cemetery, also to SE, several art galleries and museum (including the American Swedish Institutue), the Guthrie Theater, and the Minneapolis Grain Exchange. The Minnesota Symphony was founded here in 1903. The city is the seat of the University of Minnesota, Augsburg College, Minneapolis Community College, and National Education Center-Brown Institute Campus. Hennepin County Historical Society Museum, Minneapolis Institute of Arts, Minneapolis College of Art and Design, all three at Morrison Park in S. In the early 1960s, the main shopping avenue was converted into a ten-block mall lined with trees and flowers; a skyway system of sidewalks was provided for pedestrians; and a fifty-one-story skyscraper and other noteworthy buildings were erected. State Theatre; Orchestra Hall; Convention Center; Bell Museum of Natural History; Weissman Art Museum, and Mariucci Area at University of Minnesota, APPLE VALLEY, 15 mi/24 km to S; Como Park Zoo in St. Paul to E; Dunwoody Industrial Institute. Hubert Humphrey Metrodome (1982), stadium in downtown, replaced Metropolitan Stadium in Bloomington, to S, old stadium site is now Mall of America, third-largest shoping center in world, a major tourist destination. Incorporated 1856.

Minneapolis (min-ee-AP-uh-lis), town (2000 population 2,046), ⊙ OTTAWA county, N central KANSAS, on SOLOMON RIVER and 20 mi/32 km NNW of SALINA; 39°07′N 97°42′W. Railroad junction. Trade and shipping center for livestock, grain, and poultry region; grain storage. Manufacturing (motor homes). Ottawa State Fishing Lake to E, large, smooth sandstone formations. Laid out 1866, incorporated 1871.

Minnedosa (mi-ni-DO-suh), town (□ 6 sq mi/15.6 sq km; 2001 population 2,426), SW MANITOBA, W central CANADA, on MINNEDOSA RIVER, 30 mi/48 km N of BRANDON, and in MINTO rural municipality; 50°15′N 99°50′W. Lumbering, mixed farming; manufacturing (farm machinery, ethanol plants); resort. Incorporated 1883.

Minnedosa River (mi-ni-DO-suh), SW MANITOBA, W central CANADA; rises in RIDING MOUNTAIN NATIONAL PARK; flows c.150 mi/241 km in a winding course generally S, past MINNEDOSA, to ASSINIBOINE RIVER, 8 mi/13 km W of BRANDON.

Minnehaha [=laughing water], county (□ 813 sq mi/2,113.8 sq km; 2006 population 163,281), SE SOUTH DAKOTA, on MINNESOTA state line; ⊙ SIOUX FALLS; 43°40′N 96°47′W. Highly productive agricultural area drained by BIG SIOUX RIVER and Pinestone Creek. Manufacturing at Sioux Falls. Corn, soybeans, hay; cattle, hogs, sheep; dairy produce; honey. Urbanized in S central part, site of Sioux Falls and adjacent towns. Formed 1862.

Minnehaha Falls (min-nee-HAH-huh), HENNEPIN county, SE MINNESOTA, 53 ft/16 m high, in Minnehaha Creek, which flows from Lake MINNETONKA (□ 23 sq mi/60 sq km) SE to the MISSISSIPPI RIVER (at its outflow to Mississippi River is Lock and Dam Number 1 just N of confluence); 44°54′N 93°12′W. The surrounding area, including the gorge cut by the receding falls, is in Minnehaha Park, SE corner of city of MINNEAPOLIS. Most of the year only a thin trickle of water passes over the falls. The name *Minnehaha* (meaning "laughing water") is immortalized in Longfellow's *The Song of Hiawatha.*

Minneiska (mi-nee-IS-kuh), village (2000 population 116), WABASHA county, SE MINNESOTA, on MISSISSIPPI RIVER, and 15 mi/24 km NW of WINONA; 44°11′N 91°52′W. Grain; livestock, poultry; dairying. Lock and Dam Number 5 to SE; Whitewater Wildlife Area to SW.

Minneola (mi-nee-O-luh), city (□ 1 sq mi/2.6 sq km; 2005 population 8,665), Lake county, central FLORIDA, 23 mi/37 km W of ORLANDO, on small lake; 28°34′N 81°45′W. Ships citrus fruit. This retirement community features an increasing population.

Minneola (min-ee-O-luh), village (2000 population 717), CLARK county, SW KANSAS, 22 mi/35 km NW of ASHLAND; 37°26′N 100°00′W. Shipping point in wheat and livestock region. Oil and gas exploration and refining.

Minneota (mi-nee-YO-duh), town (2000 population 1,449), LYON county, SW MINNESOTA, on South Branch of YELLOW MEDICINE RIVER, 12 mi/19 km NW of MARSHALL; 44°33′N 95°58′W. Grain; livestock, poultry; dairying; manufacturing (fertilizers, transformers, trusses and rafters). Settled 1868, incorporated 1881.

Minneriya (MIN-NE-RI-yuh), town, NORTH CENTRAL PROVINCE, SRI LANKA, 40 mi/64 km SE of ANURADHAPURA; 08°02′N 80°54′E. Rice, vegetables. Irrigation reservoir built in 3rd century C.E. by King Mahasena, now restored.

Minnesota, state (□ 86,943 sq mi/226,051.8 sq km; 2006 population 5,167,101), N central UNITED STATES, in the GREAT LAKES region, admitted as the thirty-second state of the Union in 1858; ⊙ ST. PAUL; 46°29′N 94°04′W. St. Paul and its twin city MINNEAPOLIS are second-largest and largest cities, respectively. BLOOMINGTON (a suburb of Minneapolis) and DULUTH (in NE, on Lake SUPERIOR) are other major cities.

Geography

Except for ALASKA, Minnesota is the northernmost of all the states (reaching latitude 49°24′N). Minnesota is bounded on the N by CANADA (MANITOBA to NW, ONTARIO to N and NE), on the E by Lake Superior (forms boundary with MICHIGAN and part of WISCONSIN) and Wisconsin (the ST. CROIX and MISSISSIPPI rivers form most of border), on the S by IOWA, and on the W by SOUTH DAKOTA and NORTH DAKOTA. Mountainous in the NE along North Shore (Arrow Country, referring to a triangular shape of the NE corner); lowest point is Lake Superior (602 ft/183 m), shared by the counties of ST. LOUIS, Lake, and COOK. Highest point is Eagle Mountain (2,301 ft/701 m), in Cook county, only 13 mi/21 km from Lake Superior. The climate is humid continental. Winter locks the land in snow, and spring is brief; summers are warm. Prehistoric glaciers left marshes, boulder-strewn hills, numerous lakes, and rich, gray drift soil stretching from the N pine wilderness to the broad S prairies. The state is referred to as "Land of 10,000 Lakes."

Economy

In the E part of the state are mountains from which iron ore is decreasingly extracted. The VERMILION and CUYUNA ranges (discovered in 1884 and 1911) are virtually depleted, and the once rich MESABI iron range (1890) has seen major decline because of the depletion and environmental restrictions; all three mining districts are in the N and NE. As richer ores diminished, new methods were developed to use lower-grade ores such as taconite. In spite of the decline, Minnesota led the nation in iron ore production in 1988. Granite (from ST. CLOUD) and sand and gravel production are also among the largest in the country. S of the iron country, famous for its former boom towns, lie rolling hills. In the S and the W are prairies, the fertile farming country of Minnesota. Wheat, once paramount in the fields, has yielded its preeminence to corn, soybeans, and livestock. The state is a leader in the production of creamery butter, dry milk, cheese, and sweet corn. In the early 1950s manufacturing displaced agriculture as the major source of income in Minnesota. Major industries in the state include the manufacture of processed foods, electronic equipment, machinery, paper products, chemicals, and stone, clay, and glass products. Minnesota also pioneered the development of computers and other high-technology manufacturing. Printing and publishing are also important. Reforestation and the use of smaller trees for pulpwood have helped to keep timber as one of Minnesota's assets, even though the "big woods" of the early 19th century have been to a large extent recklessly felled. The state is roughly 30% forestland and has the Chippewa Natlional Forest in the N center; Superior National Forest in NE, including BOUNDARY WATERS Canoe Area; VOYAGEURS NATIONAL PARK on RAINY LAKE, on the Canada border. There are numerous state forests, especially in the N; and the Richard J. Dorer Memorial Hardwood State Forest in the SE. The days of logging in Minnesota, immortalized in the stories of the legendary Paul Bunyan and his prized possession, Babe the Blue Ox, were brief, but they helped build a number of large fortunes, such as that of Frederick Weyerhaeuser.

Water Resources

Another great resource of Minnesota is its water, which has been extensively developed near industrial centers. The state has more than 10,000 lakes, many of which create chains of lakes; numerous streams and rivers. The rivers feed three major river systems: the RED RIVER and its tributaries in the W run N to HUDSON BAY; the streams that run E into Lake Superior are part of the ST. LAWRENCE River System (ATLANTIC OCEAN); and the Mississippi flows S from its humble beginning in Lake ITASCA, gathering volume from the waters of the MINNESOTA and St. Croix rivers and others before leaving the state. Other rivers include BIG SIOUX RIVER and the MISSOURI River, both part of Mississippi River system. Locks and other improvements enable barge traffic to pass around the ST. ANTHONY FALLS to reach upstream beyond Minneapolis. Duluth, at the western tip of Lake Superior, has the largest inland harbor in the U.S.; W head of navigation of ST. LAWRENCE—GREAT LAKES Seaway System. With the completion of the SAINT LAWRENCE SEAWAY (1959) and a marine terminal, the city became a key port for overseas trade.

History to the Louisiana Purchase

Archaeological evidence indicates that Minnesota was inhabited long before the time of the Mound Builders. A skeleton ("Minnesota Man"), found in 1931 near Pelican Falls, is believed to date from the Pleistocene epoch, c.20,000 years ago. Much important archaeological information concerning the early inhabitants of NORTH AMERICA has been found in Minnesota. There are some experts who argue on the basis of the Kensington Rune Stone and other evidence that the first Europeans to reach Minnesota were the Norsemen; that French fur traders came in the mid-17th century is undeniably true. Other traders, explorers, and missionaries of NEW FRANCE also penetrated the country. Among these were Radisson and Groseilliers, Verendrye, the Sieur Duluth, and Father Hennepin and Michel Aco, who discovered the Falls of St. Anthony (the site of Minneapolis). At the time the French arrived, the dominant groups of Native Americans were the Ojibwa in the E and the Sioux in the W. Both were friendly to the French and contributed to the fur-trading empire of New France. Minnesota remained excellent country for fur trade throughout the British regime that followed the French and Indian Wars and continued so after the War of 1812, when the American Fur Company became dominant and the company's men helped to develop the area.

History: Louisiana Purchase to 1851
The E part of Minnesota had been included in the NORTHWEST TERRITORY and was governed under the Ordinance of 1787; the W part was joined to the U.S. by the Louisiana Purchase. Further exploration was pursued by Jonathan Carver (1766–1767), Zebulon M. Pike (1805–1806), Henry Schoolcraft (1820, 1829), and Stephen H. Long (1823). Only after the War of 1812, however, did settlement begin in earnest. In 1820, Fort St. Anthony (later FORT SNELLING) was founded as a guardian of the frontier. A gristmill established there in 1823 initiated the industrial development of Minneapolis. Treaties (1837, 1845, 1851, and 1855) with the Ojibwa and the Sioux, by which the U.S. government took over Native American lands, and the opening of a land office at St. Croix Falls in 1848 initiated a period of real expansion. In 1849 Minnesota became a territory. The Missouri and WHITE EARTH rivers were the W boundary. A land boom grew as towns were plotted, railroads chartered, and roads built. Attention was turned to education, and the University of Minnesota was started in 1851. The school, with its many associated campuses, exerts a great influence on the cultural life of the state.

History: 1851 to 1870
The building (1851–1853) of the SOO SHIP CANAL at SAULT SAINTE MARIE, Michigan, opened an E water route for lake shipping. The Panic of 1857 hit Minnesota particularly hard because of land speculation, but difficult times did not prevent the achievement of statehood in 1858, with St. Paul as the capital and Henry Hastings Sibley as the state's first governor. The population had swelled from 6,000 in 1850 to more than 150,000 in 1857; by 1870 there were nearly 440,000 inhabitants. Chiefly a land of small farmers (mainly of British, German, and Irish extraction), Minnesota supported the Union in the Civil War and supplied much wheat to the Northern armies. During the war years and afterward, the Sioux reacted to broken promises, fraudulent dealings, and the encroachment of settlers on their lands with violent resistance. A Sioux force under Little Crow was defeated by H. H. Sibley, virtually ending Native American resistance. Meanwhile, settlement boomed, aided by the Homestead Act of 1862. Later in the century came immigrants from SCANDINAVIA—Swedes, Norwegians, and Finns. Lumbering, which had begun in 1839 with a sawmill on the St. Croix, became paramount, and logging camps were established.

History: 1870 to 1915
Fortunes were made quickly in the 1870s and 1880s as the railroad pushed W. A boom in wheat made the Minnesota flour mills famous across the world and brought wealth to flour producers such as John S. Pillsbury. Farmers, however, suffered from such natural disasters as the blizzard of 1873 and insect plagues from 1874 to 1876. To these were added the miseries that accompanied the downward trend of the national economy, and Minnesota became a center of farmers' discontent, expressed in the Granger movement. The

opening of the iron mines gave new impetus to Minnesota's economy but also created discontent among the laborers. They joined forces with the farmers in the 1890s in the Populist party, one of several third-party movements that challenged the Republican party's traditional leadership in Minnesota. Ignatius Donnelly was one of the Populists' most powerful figures. Renewed agrarian discontent led to the founding of the Nonpartisan League in 1915.

History: 1915 to Present
Farmers and laborers joined forces again in 1920 in the Farmer-Labor party, which was dominant in the 1930s. The Republicans returned to power in 1939 with the election of Harold Stassen as governor. In 1944 the Farmer-Labor party and the Democrats merged. The most successful leader of the new party, the Democratic Farmer Labor party (DFL), has been Hubert H. Humphrey, who was elected to the U.S. Senate four times and was Vice President from 1965 to 1969. Orville Freeman, DFL governor from 1955 to 1961, was Secretary of Agriculture from 1961 to 1969. Walter F. Mondale, a Humphrey protégé, was a U.S. senator from 1964 to 1977. He was elected Vice President as Jimmy Carter's running mate in 1976 and ran for President in 1984, losing to incumbent Ronald Reagan. Since the 1950s the DFL and the Republicans have vied sharply in contests for state offices. In the 1970s the Republican party changed its name to the Independent Republican party. With the exception of 1952, 1956, and 1972, Minnesota has voted Democratic in every presidential election since 1932.

Cooperative
The state has been notable for experimentation in novel features of local government and has also been a leader in the use of cooperatives. This phenomenon is perhaps explained by the cooperative heritage present among its many people of Scandinavian descent. Credit unions, cooperative creameries, grain elevators, and purchasing associations were supported by legislation in 1919 that protected the institutions and instructed the state department of agriculture to encourage them. There are several thousand cooperative associations in Minnesota serving diversified needs. A nuclear power plant built by the Atomic Energy Commission is located at ELK RIVER, on the Mississippi River NW of Minneapolis. Since the mid-19th century the state has become progressively more urban. In 1970 the urban population was two-thirds of the total. Since 1970 dramatic suburban growth has taken place, especially in the Minneapolis–St. Paul metropolitan area. Minneapolis–St. Paul International Airport has become an important hub for the region. Nearby is the massive Mall of America (1992), in suburban Bloomington, one of the largest in the U.S. Many people come to Minnesota for treatment at the famous Mayo Clinic in ROCHESTER, and surgeons at the University of Minnesota have won recognition for their development of new heart-surgery techniques.

Places of Interest and Culture
The beauty of Minnesota's lakes and dense green forests has long attracted vacationers, and the abundant fish in the state's many rivers, lakes, and streams provide excellent fishing. Also of interest to tourists are the GRAND PORTAGE (in NE) and PIPESTONE (SW) national monuments, Itasca State Park, in NW (site of the headwaters of the Mississippi River), the Minnesota Museum of Mining (near CHISHOLM), in N, and the world's largest open-pit iron mine at HIBBING. The Minnesota Symphony Orchestra is nationally known, and a theater in Minneapolis houses the professional company of Tyrone Guthrie. Many Minnesotans are of Scandinavian descent; one local tradition is *lutefisk*, cod cured in lye, served during the holiday season at church dinners, attracting thousands of people. Minnesota has contributed important literary figures to the nation, including Sinclair Lewis, F. Scott Fitz-

gerald, and Old English Rølvaag. The economist Thorstein Veblen and Charles A. Lindbergh were also born in the state.

Government
The state is governed under the 1858 constitution. The legislature has sixty-seven senators elected for four-year terms and 134 representatives elected for two-year terms. The governor is elected for a four-year term and may succeed himself. The current governor is Tom Pawlenty. Minnesota sends two senators and eight representatives to Congress; it has ten electoral votes.

Minnesota has eighty-seven counties: AITKIN, ANOKA, BECKER, BELTRAMI, BENTON, BIG STONE, BLUE EARTH, BROWN, CARLTON, CARVER, CASS, CHIPPEWA, CHISAGO, CLAY, CLEARWATER, COOK, COTTONWOOD, CROW WING, DAKOTA, DODGE, DOUGLAS, FARIBAULT, FILLMORE, FREEBORN, GOODHUE, GRANT, HENNEPIN, HOUSTON, HUBBARD, ISANTI, ITASCA, JACKSON, KANABEC, KANDIYOHI, KITTSON, KOOCHICHING, LAC QUI PARLE, Lake, LAKE OF THE WOODS, LE SUEUR, LINCOLN, LYON, MCLEOD, MAHNOMEN, MARSHALL, MARTIN, MEEKER, MILLE LACS, MORRISON, MOWER, MURRAY, NICOLLET, NOBLES, NORMAN, OLMSTED, OTTER TAIL, PENNINGTON, PINE, PIPESTONE, POLK, POPE, RAMSEY, RED LAKE, REDWOOD, RENVILLE, RICE, ROCK, ROSEAU, ST. LOUIS, SCOTT, SHERBURNE, SIBLEY, STEARNS, STEELE, STEVENS, SWIFT, TODD, TRAVERSE, WABASHA, WADENA, WASECA, WASHINGTON, WATONWAN, WILKIN, WINONA, WRIGHT, and YELLOW MEDICINE.

Minnesota City, village (2000 population 235), WINONA county, SE MINNESOTA, 7 mi/11.3 km NW of WINONA, near MISSISSIPPI RIVER (Lock and Dam Number 5 to NW, Lock and Dam Number 5A to SE); 44°05′N 91°45′W. Railroad junction. Grain; livestock, poultry; dairying; light manufacturing. John A. Latsch State Park to NW; Richard J. Dorer Memorial Hardwood State Forest to SW.

Minnesota Glacier, in WEST ANTARCTICA; flows E through the ELLSWORTH MOUNTAINS, and separates the SENTINEL and HERITAGE Ranges into RUTFORD ICE STREAM, 40 mi/65 km long and 5 mi/8 km wide; 79°00′S 83°00′W.

Minnesota Lake, village (2000 population 681), FARIBAULT county, S MINNESOTA, 27 mi/43 km NW of ALBERT LEA, on E shore of Minnesota Lake (3 mi/4.8 km long, 2 mi/3.2 km wide); 43°50′N 93°49′W. Lake has two outflows (S and W) to MAPLE RIVER; small dams at both. Corn, oats, peas, soybeans; livestock, poultry.

Minnesota River, 332 mi/534 km long, S MINNESOTA; rising in BIG STONE LAKE (lake forms part of Minnesota-SOUTH DAKOTA state line; fed from NW by Little Minnesota River), exits lake through small dam at ORTONVILLE, Minnesota, and BIG STONE CITY, South Dakota, immediately entering Minnesota; flows SE through MARSH LAKE and LAC QUI PARLE LAKE reservoirs, past MONTEVIDEO, GRANITE FALLS, and NEW ULM, turns N at MANKATO and passes ST. PETER, and LE SUEUR, then flows NE past BELLE PLAINE, SHAKOPEE, and BLOOMINGTON; enters MISSISSIPPI RIVER 6 mi/9.7 km SW of downtown ST. PAUL and 6 mi/9.7 km SSE of downtown MINNEAPOLIS, in FORT SNELLING State Park; 45°18′N 96°27′W. Minneapolis–St. Paul International Airport is W of mouth. Minnesota River enters Mississippi River at its head of commercial navigation and is itself navigable only by small craft. Minnesota Valley State Trail and National Wildlife Refuge follow lower course of river. Earlier called the St. Peter or St. Pierre, it was an important route for explorers and fur traders. The river follows the valley of the prehistoric River Warren, the outlet of prehistoric Lake AGASSIZ.

Minnesott Beach (MIN-uh-saht), village (□ 1 sq mi/ 2.6 sq km; 2006 population 298), PAMLICO county, E NORTH CAROLINA, 10 mi/16 km NE of HAVELOCK, on NEUSE RIVER estuary (ferry); 34°58′N 76°49′W.

Cross-references are shown in SMALL CAPITALS. The pronunciation guide is shown on page xix. The sources of population figures are shown on page xvii.

Croatan National Forest to S; Cherry Point Marine Corps Air Station to SW. Service industries; manufacturing.

Minnetonka (mi-ne-TAWN-kuh), city (2000 population 51,301), HENNEPIN county, SE MINNESOTA, a suburb, 9 mi/14.5 km WSW of downtown MINNEAPOLIS, E of Lake MINNETONKA (Gray's Bay on W boundary), receives Minnehaha Creek; 44°55′N 93°27′W. Diversified manufacturing. Its population has increased significantly since 1970 due to the influx of former central-city Minneapolis residents to the outlying suburbs. Glen Lake Sanitorium in S. Incorporated 1956.

Minnetonka Beach (mi-ne-TAWN-kuh), village (2000 population 614), HENNEPIN county, E MINNESOTA, residential suburb, 16 mi/26 km W of downtown MINNEAPOLIS, on small peninsula in Lake MINNETONKA between Crystal Bay (N) and Lafayette Bay (S); 44°56′N 93°35′W.

Minnetonka, Lake (mi-ne-TAWN-kuh) (□ 23 sq mi/59.8 sq km), E MINNESOTA, largely in HENNEPIN county, extends S into CARVER county, 12 mi/19 km W of downtown MINNEAPOLIS; 10 mi/16 km long, maximum width 2.5 mi/4 km; 44°56′N 93°36′W. Has deeply indented shoreline (97 mi/156 km long), including peninsulas and land bridges, some manmade, and several small islands. Residential on all but SW shore, complex shoreline provides maximum access to residents. Drains through Minnehaha Creek (E) into MISSISSIPPI RIVER. Lake is divided into Upper (W) and Lower (E) lakes by isthmus crossed by narrow channels; lake is further divided into arms and bays, some of them lakes in their own right, especially in NW. Lake is on W urban fringe of Minneapolis–ST. PAUL (Twin Cities); municipalities on lake include ORONO and WAYZATA (N), SHOREWOOD and VICTORIA (S), MOUND and MINNETRISTA (W), SPRING PARK and TONKA BAY (center). Lake is celebrated in songs by Thurlow Lieurance ("By the Waters of Minnetonka") and Charles W. Cadman ("From the Land of the Sky-Blue Water"). Recreation area.

Minnetonka, Lake, Minnesota: see MINNEHAHA FALLS.

Minnetrista (mi-ne-TRIS-tuh), town (2000 population 4,358), HENNEPIN county, E MINNESOTA, residential suburb, 19 mi/31 km WSW of downtown MINNEAPOLIS, bounded on SE by Upper Lake and NE by Jennings Bay, both on Lake MINNETONKA; 44°56′N 93°42′W. Oak Lake in N. Carver Park Reserve to SE.

Minnewanka, Lake (mi-ni-WAHNG-kuh), SW ALBERTA, W CANADA, near BRITISH COLUMBIA border, in ROCKY MOUNTAINS, in BANFF NATIONAL PARK, 6 mi/10 km NE of BANFF, at foot of mountains AYLMER and GIROUARD; 12 mi/19 km long, 1 mi/1.6 km wide; 51°15′N 115°23′W. Elevation 4,769 ft/1,454 m.

Minnewaska, Lake (mi-nee-WAW-skuh) (□ 19 sq mi/49.4 sq km), POPE county, W MINNESOTA; 7.5 mi/12.1 km long, 2 mi/3.2 km wide; 45°36′N 95°27′W. Elevation 1,138 ft/347 m. Town of GLENWOOD at NE end; town of STARBUCK at SW end; resorts. Drains SW through Outlet Creek to Lake EMILY, 6 mi/9.7 km SW. Fed by short stream from W through Pelican Lake, enters at NE end.

Minnewaska, Lake, NEW YORK: see LAKE MINNEWASKA.

Minnewaukan (min-uh-WAW-kuhn), village (2006 population 297), ⊙ BENSON county, central NORTH DAKOTA, 18 mi/29 km WSW of the city of DEVILS LAKE, near W end of DEVILS LAKE; 48°04′N 99°15′W. Grain area. Devils Lake Sioux Indian Reservation to SE. Founded in 1884 and incorporated in 1897. Name is from the Indian name for Devils Lake, Mini Waukon Chante, which means water of bad spirits.

Minnewawa, Lake, AITKIN county, NE central MINNESOTA, just SE of Sandy Lake, 22 mi/35 km NE of AITKIN in Savanna State Forest; 5 mi/8 km long, maximum width 2 mi/3.2 km irregular shoreline;

46°42′N 93°16′W. Resorts. Sheshabee village on SE shore. Drains S and W to Sandy River through Minnewawa Creek. Separated from BIG SANDY LAKE to NW by 0.5 mi/0.8 km neck of land.

Minnipa (MIN-i-puh), village, S SOUTH AUSTRALIA state, S central AUSTRALIA, on W central EYRE PENINSULA, 135 mi/217 km NNW of PORT LINCOLN; 32°51′S 135°09′E. On Port Lincoln–PENONG railroad; wheat, wool. Granite hill formations nearby.

Minnitaki (mi-ni-TA-kee), unincorporated village, NW ONTARIO, E central CANADA, and included in MACHIN township, 49°48′N 93°05′W.

Mino (MEE-no), city, GIFU prefecture, central HONSHU, central JAPAN, 16 mi/25 km N of GIFU; 35°32′N 136°54′E.

Mino (mee-NO), town, Mitoyo county, KAGAWA prefecture, NE SHIKOKU, W JAPAN, 22 mi/35 km S of TAKAMATSU; 34°11′N 133°42′E.

Mino (MEE-no), town, Miyoshi county, TOKUSHIMA prefecture, N SHIKOKU, W JAPAN, on YOSHINO RIVER, 34 mi/55 km W of TOKUSHIMA; 34°02′N 133°58′E.

Minoa (min-O-ah), residential village (□ 1 sq mi/2.6 sq km; 2006 population 3,297), ONONDAGA county, central NEW YORK, 8 mi/12.9 km E of SYRACUSE; 43°04′N 76°00′W. In dairying area; former rail transfer site. Incorporated 1913.

Minobu (mee-NO-boo), town, South Koma county, YAMANASHI prefecture, central HONSHU, central JAPAN, near Mount Minobu, 22 mi/35 km S of KOFU; 35°22′N 138°26′E. Myohorengein Kuon Temple here is the main temple of the Buddhist Nichiren section.

Minocqua (min-AHK-waw), resort village, ONEIDA county, N WISCONSIN, 21 mi/34 km NW of RHINELANDER, in lake region. Lumbering. Light manufacturing. N terminus of Bear Skin State Trail. American Legion State Forest to E.

Minokamo (mee-NO-kah-mo), city, GIFU prefecture, central HONSHU, central JAPAN, 19 mi/30 km N of GIFU; 35°26′N 137°01′E. Video equipment. Persimmons.

Minong (MEI-nahng), village (2006 population 544), WASHBURN county, NW WISCONSIN, 43 mi/69 km SSE of SUPERIOR, in wooded region with numerous lakes; 46°06′N 91°49′W. Manufacturing (meat snacks, mailing machines).

Minonk (mi-NUNK), city (2000 population 2,168), WOODFORD county, central ILLINOIS, 18 mi/29 km NE of EUREKA; 40°53′N 89°02′W. Manufacturing (paper products). Agriculture (dairy products; livestock; grain). Incorporated 1867.

Minoo (mee-NO), city, OSAKA prefecture, S HONSHU, W central JAPAN, 9 mi/15 km N of OSAKA; 34°49′N 135°28′E.

Minooka (mi-NOO-kah), village (2000 population 3,971), GRUNDY county, NE ILLINOIS, 11 mi/18 km WSW of JOLIET; 41°27′N 88°15′W. In agricultural area.

Minooka (mi-NOO-kah), suburb, LACKAWANNA county, NE PENNSYLVANIA, residential section, 3 mi/4.8 km SW of downtown SCRANTON, on LACKAWANNA RIVER; 41°22′N 75°41′W. Part of city of Scranton.

Minor, suburb, JEFFERSON county, N central ALABAMA, just W of BIRMINGHAM; 33°31′N 86°57′W.

Minorca (mi-NOR-kuh), Spanish *Menorca*, island (□ 271 sq mi/704.6 sq km), Baleares province, SPAIN, in the W MEDITERRANEAN SEA, the second largest of the BALEARIC ISLANDS; 40°00′N 04°00′E. MAHÓN is the chief city and port. The terrain is mostly low but has a hilly center. Small farms produce cereals, almonds, and potatoes. Much of the agriculture is irrigated. Lobster fishing, livestock, and grain are sent to the mainland. Manufacturing of shoes and costume jewelry have become important sources of employment. Tourism is also important, based on the island's fine beaches. A great number of megalithic monuments have been found. Minorca shared the history of the other Balearic Islands until 1708, when it was occupied by the English during the War of the

Spanish Succession. England retained it until the Seven Years War, when it was seized by the French. The Treaty of Paris (1763) restored Minorca to Britain, but the French and Spanish again seized it (1782) in the American Revolution. In 1798, in the French Revolutionary Wars, England regained control; the Peace of Amiens (1802) awarded Minorca to Spain. The island still has a somewhat British flavor. The island is now easily accessible from BARCELONA and MAJORCA by air and ferry.

Minori (mee-NO-ree), town, Salerno province, CAMPANIA, S ITALY, port on GULF OF SALERNO, 7 mi/11 km WSW of SALERNO, on the Amalfi coast; 40°39′N 14°37′E. Paper milling. Bathing resort.

Minor Lane Heights, town (2000 population 1,435), JEFFERSON county, N KENTUCKY, residential suburb, 8 mi/12.9 km S of downtown LOUISVILLE; 38°07′N 85°43′W.

Minorskiy (mee-NOR-skeeyee), town, SE SAKHA REPUBLIC, RUSSIAN FAR EAST, on local road, 45 mi/72 km S of ALLAKH-YUN; 60°28′N 137°53′E. Elevation 2,677 ft/815 m. Gold mining. Also known as Minor.

Minot (MEI-naht), city (2006 population 34,745), seat of WARD county, NW NORTH DAKOTA, on the SOURIS RIVER; 48°13′N 101°17′W. Incorporated 1887. It is a commercial and transportation center for an extensive agricultural area. Railroad junction. There are lignite mines and oil basins in the region. Industries include building materials, petroleum compounds, farm machinery; dairy and meat products. Minot State University (1913) and a state agricultural experiment station are there. Minot Air Force Base is 10 mi/16 km N; Upper Souris National Wildlife Refuge to NW. Founded in 1887 and named for Henry Davis Minot (1859–1890), a director of the Great Northern Railroad. Replaced Burlington as a county seat in 1888.

Minot (MEI-naht), town, ANDROSCOGGIN county, SW MAINE, on the Little Androscoggin just W of AUBURN; 44°08′N 70°19′W. Light manufacturing.

Minot, MASSACHUSETTS: see SCITUATE.

Minots Ledge (MEI-nuhts), E MASSACHUSETTS, reef in MASSACHUSETTS BAY, c.2.5 mi/4 km off COHASSET; 42°16′N 70°46′W. The first lighthouse here (built 1850) destroyed by gale in 1851; present 114-ft/35-m structure built 1860.

Miño Volcano (MEE-nyo), Andean peak (18,440 ft/5,621 m), N CHILE, near BOLIVIA border; 20°11′S. At its NE foot rises LOA RIVER.

Minowa (mee-NO-wah), town, Kamiina county, NAGANO prefecture, central HONSHU, central JAPAN, 50 mi/80 km S of NAGANO; 35°54′N 137°59′E.

Minqin (MIN-CHIN), town, S Minqin county, central GANSU province, CHINA, at S edge of the GOBI Desert, 60 mi/97 km NNE of WUWEI, at the GREAT WALL; 38°42′N 103°11′E. Elevation 4,484 ft/1,367 m. Grain, sugar beets, oilseeds; food processing, engineering, textiles and clothing, coal mining.

Minqing (MIN-CHING), town, ⊙ Minqing county, E FUJIAN province, CHINA, 30 mi/48 km WNW of FUZHOU, and on MIN River; 26°13′N 118°51′E. Rice, sugarcane, oilseeds, tea; paper making, pharmaceuticals, logging, timber processing, kaolin quarrying.

Minquan (MIN-CHU-AN), town, ⊙ Minquan county, NE HENAN province, CHINA, 50 mi/80 km ESE of KAIFENG, and on LONGHAI railroad; 34°40′N 115°08′E. Grain, cotton, oilseeds; food and beverages.

Minsen, GERMANY: see HOOKSIEL.

Minshah, El, EGYPT: see MANSHAH, EL.

Min Shan (MIN-SHAN), outlier of the KUNLUN MOUNTAIN system, CHINA, on QINGHAI-GANSU-SICHUAN border, forming an extension of A'nyêmaqên (AMNE MACHIN) Mountains E of upper Huang Ho (YELLOW); 33°47′N 103°31′E. Elevation 8,200 ft/2,499 m. MIN River of W Sichuan rises on S slopes.

Minshat el Bakkari (mahn-SHAI-yet el bak-REE), village, GIZA province, Upper EGYPT, 7 mi/11.3 km WSW of CAIRO's center.

Area is shown by the symbol □, and capital city or county seat by ⊙.

Minshat Sabri (mahn-SHAI-yet sahb-REE), village, MINUFIYA province, Lower EGYPT, 7 mi/11.3 km E of SHIBIN EL KOM; 30°33'N 31°08'E. Cereals, cotton, flax.

Minshat Sultan (mahn-SHAI-yet sol-TAHN), village, MINUFIYA province, Lower EGYPT, on railroad, 5 mi/8 km N of MINUF; 30°32'N 30°55'E. Cereals, cotton, flax.

Minsiung, TAIWAN: see Minhsiung.

Minsk (MINSK), city and municipality (□ 61 sq mi/158.6 sq km; 2005 population 1,780,700), ⊙ BELARUS and MINSK oblast, on both banks of the SVISLOCH RIVER (a tributary of the BEREZINA RIVER); 53°51'N 27°30'E. Divided into seven districts, it is a railroad junction, with machine construction and metalworking the leading branch of industry. Light industry (fine fiber and worsted combines, leather, knitwear, garments); food processing (confectionary goods, tobacco, brewery, margarine plant, meat, milling, dairy); construction materials (plaster and porcelain, prefabricated metal equipment); chemical industry (medical and endocrinal products, varnishes and paints). More than half the output of Belarussian publishing industry comes from Minsk. A large thermal electric power plant connected to Belarussian power system is city's electric power base. Since 1960, gas has been supplied to Minsk from Dashawa (UKRAINE). Minsk automotive plant (MAZ) was major enterprise of SOVIET motor vehicle industry; producing vehicles of large load capacity. It is the headquarters of the COMMONWEALTH OF INDEPENDENT STATES. First mentioned in 1067, it was an outpost on the road from KIEV (Ukraine) to POLOTSK and was part of the Polotsk principality. It became the capital of the Minsk principality in 1101 and part of LITHUANIA in 1326. At the end of the 15th century it became a great craft and trade center. Joined to POLAND in 1569, it passed to RUSSIA in the second partition of Poland (1793). The city's industrial development began in the 1870s. It was one of the largest Jewish centers of Eastern EUROPE in the Middle Ages, and before World War II some 40% of the population was Jewish. From 1941 to 1943, Minsk was a concentration center for Jews prior to their extermination by the Nazis. Although the city was heavily damaged in the war, several monuments remain. These include a former 17th-century Bernardine convent and the 17th-century Ekaterin Cathedral (formerly called the Petropavlovsk church). Minsk is a major cultural, educational, and artistic center. It is the site of the Minsk Art Museum and thirteen higher educational institutions (the largest, Belarussian State University).

Minsk (MINSK), oblast (□ 8,500 sq mi/22,015 sq km; 2005 estimated population 1,474,100), central BELARUS; ⊙ MINSK. In LITHUANIAN-BELORUSSIAN UPLAND; drained by BEREZINA, upper PTICH, and SVISLOCH rivers. Highly industrialized (machine building and chemicals, food processing, building materials). Agriculture is intensive and diversified (potatoes, grain, flax). Energy is supplied by local fuel (peat) and by imports (coal, petroleum, natural gas). Formed 1938.

Minsk Mazowiecki (meensk mah-zo-VYETS-kee), Polish *Mińsk Mazowiecki*, Russian *Novo Minsk*, town (2002 population 36,341), Siedlce province, E central POLAND, 24 mi/39 km E of WARSAW; 52°11'N 21°34'E. Railroad junction; manufacturing (machinery, shingles, flour).

Minskoye (MEEN-skuh-ye), town, SW KOSTROMA oblast, central European Russia, on the E bank of the VOLGA RIVER, on road, 7 mi/11 km SE of (and administratively subordinate to) KOSTROMA; 57°42'N 41°05'E. Elevation 390 ft/118 m. Weekend and holiday getaway spot; summer homes, sanatorium.

Minster (MIN-stuh), town (2001 population 3,267), NE KENT, SE ENGLAND, on Isle of THANET, 5 mi/8 km W of RAMSGATE; 51°20'N 01°19'E. Has Norman church and some remains of 8th-century Saxon abbey.

Minster (MIN-stuhr), village (□ 2 sq mi/5.2 sq km; 2006 population 2,794), AUGLAIZE county, W OHIO, 10 mi/16 km S of ST. MARYS, near Lake Loramie; 40°23'N 84°22'W. Incorporated 1833.

Minster (MIN-stuh), suburb (2001 population 12,772), N KENT, SE ENGLAND, on Isle of SHEPPEY, on THAMES estuary, 3 mi/4.8 km ESE of SHEERNESS; 51°26'N 00°49'E.

Minsterley (MIN-stuh-lee), agricultural village (2001 population 1,597), W SHROPSHIRE, W ENGLAND, 9 mi/14.5 km SW of SHREWSBURY; 52°39'N 02°55'W. Dairying, milk canning. Has 17th-century church. In parish, 2 mi/3.2 km S, is village of Snailbeach.

Minster Lovell (MIN-stuh LUHV-uhl), village (2001 population 1,348), W OXFORDSHIRE, S central ENGLAND, 3 mi/4.8 km WNW of WITNEY; 51°47'N 01°32'W. Previously woolen milling. Has 15th-century church and remains of 12th-century monastery.

Minta (MEEN-tuh), town, Central province, CAMEROON, 102 mi/164 km NE of Yaoundé; 04°33'N 12°50'E.

Mintaro, town, SOUTH AUSTRALIA state, S central AUSTRALIA, 78 mi/126 km from ADELAIDE, in Clare Valley; 33°55'S 138°43'E. Slate quarries.

Minter, town, LAURENS county, central GEORGIA, 10 mi/16 km ESE of DUBLIN; 32°29'N 82°45'W.

Mint Hill (MINT HIL), city (□ 21 sq mi/54.6 sq km; 2006 population 18,663), MECKLENBURG county, S NORTH CAROLINA, residential suburb, 10 mi/16 km ESE of downtown CHARLOTTE; 35°10'N 80°39'W. Retail trade; service industries; agricultural area to E. Incorporated 1971.

Minthi Mountains, S ILIA prefecture, S WESTERN GREECE department, W central PELOPONNESUS, S mainland GREECE, between ANDRITSAINA and Gulf of KIPARISSIA. Rise to 4,000 ft/1,219 m. Also spelled Minthes.

Mintlaw (MINT-law), village (2001 population 2,647), Aberdeenshire, NE Scotland, 8 mi/12.9 km W of PETERHEAD; 57°31'N 02°00'W. Tourism, visitor center; Deer Abbey (7th century) 1 mi/1.6 km to W. Formerly in Grampian, abolished 1996.

Minto (MIN-to), rural municipality (□ 141 sq mi/366.6 sq km; 2001 population 684), SW MANITOBA, W central CANADA, N of BRANDON and S of RIDING MOUNTAIN NATIONAL PARK; 50°20'N 99°45'W. Agriculture; tourism. MINNEDOSA is main center; other communities include AMEER, BETHANY, LARGS, and CLANWILLIAM. Established 1903.

Minto (MIN-to), town (□ 116 sq mi/301.6 sq km; 2001 population 8,164), S ONTARIO, E central CANADA, 42 mi/68 km from GUELPH; 43°55'N 80°52'W. Agricultural area; also brickyards, lime kilns, dairy factories, sawmills. Composed of the communities of CLIFFORD, HARRISTON, and PALMERSTON.

Minto (MIN-to), unincorporated village, SW MANITOBA, W central CANADA, 30 mi/48 km S of BRANDON, in WHITEWATER rural municipality; 49°24'N 100°01'W. Grain, stock.

Minto, village (2001 population 2,776), central NEW BRUNSWICK, CANADA, near NW shore of GRAND LAKE, 60 mi/97 km N of SAINT JOHN; 46°05'N 66°05'W. Declining coal mining center. Coal was first shipped to NEW ENGLAND from here in 1643.

Minto (MIN-to), agricultural village, Scottish Borders, S Scotland, on TEVIOT River, 5 mi/8 km NE of HAWICK; 55°29'N 02°44'W. Just N are craggy Minto Hills (905 ft/276 m), site of remains of ancient Fatlips Castle. Also here is "Barnhill's Bed," described in Scott's *The Lay of the Last Minstrel.*

Minto, village (2000 population 258), central ALASKA, in Minto Flats, 40 mi/64 km W of FAIRBANKS, on Elliot Highway; 65°01'N 149°31'W. Gold mining.

Minto, village (2006 population 613), WALSH county, NE NORTH DAKOTA, 9 mi/14.5 km S of GRAFTON, and on FOREST RIVER; 48°17'N 97°22'W. Manufacturing (farm equipment). Lake Ardoch to SE. Founded 1880 and incorporated in 1883.

Minto Inlet, W VICTORIA ISLAND, NORTHWEST TERRITORIES, CANADA, arm of AMUNDSEN GULF, at S end of PRINCE OF WALES STRAIT; 75 mi/121 km long, 8 mi/13 km–25 mi/40 km wide; 71°15'N 117°W.

Minto Island, FRENCH POLYNESIA: see ACTAEON ISLANDS.

Minto, Lake (MIN-to) (□ 485 sq mi/1,261 sq km), N QUEBEC, E CANADA; 60 mi/97 km long, 15 mi/24 km wide; 57°25'N 74°30'W. Drained by LEAF RIVER.

Minto Mine (MIN-to), town, SW BRITISH COLUMBIA, W CANADA, in COAST MOUNTAINS, on Bridge River, 40 mi/64 km WNW of LILLOOET, in SQUAMISH-LILLOOET regional district; 50°54'N 122°46'W. Site has been flooded by Carpenter Lake Reservoir (1970s). Cattle, timber in area.

Mintraching (MIN-trah-khing), village, UPPER PALATINATE, BAVARIA, GERMANY, 7.5 mi/12 km SE of REGENSBURG; 48°57'N 12°13'E. Agriculture (sugar beets, grain).

Minturn, town (2000 population 1,068), EAGLE county, W central COLORADO, on EAGLE RIVER, just W of GORE RANGE, and 25 mi/40 km NNW of LEADVILLE; 39°34'N 106°25'W. Elevation 7,817 ft/2,383 m. Manufacturing (printing and publishing). Surrounded by White River National Forest. To NE is VAIL ski resort.

Minturn (MIN-tuhrn), village (2000 population 114), LAWRENCE county, NE ARKANSAS, 20 mi/32 km WNW of JONESBORO; 35°58'N 91°01'W.

Minturnae (min-TUHR-nee), ancient town of Latium, ITALY, 7 mi/11.3 km E of FORMIA. It was important because it controlled the bridge on the APPIAN WAY over the LIRI River. Founded by a people called the Aurunci or Ausones, it became a Roman colony (295 B.C.E.) and a flourishing commercial center. There are important ruins (including an aqueduct), two theaters, forums, and other buildings N of modern MINTURNO.

Minturno (meen-TOOR-no), town, LATINA province, LATIUM, S central ITALY, on hill overlooking GULF OF GAETA, 7 mi/11 km E of FORMIA; 41°15'N 13°45'E. Has castle and mid-12th-century church, both damaged in World War II. In the plain below are ruins (aqueduct, temples) of ancient *Minturnae* and a British military cemetery.

Minudasht (mee-NOO-dahsht), town, Māzandarān providence, in GORGAN, NE IRAN, 10 mi/16 km ESE of Gonbad-e-Qāvūs; 37°13'N 55°22'E.

Minuf (me-NOOF), town, MINUFIYA province, N EGYPT, between the RASHID and DUMYAT branches of the NILE River, 8 mi/12.9 km SW of SHIBIN EL KOM; 30°28'N 30°56'E. It is the trade center for an irrigated agricultural region that produces corn, grain, cotton, and dairy products. Also spelled Menuf.

Minufiya (ME-noo-FAI-yuh), province (□ 613 sq mi/1,593.8 sq km; 2004 population 3,171,058), Lower EGYPT, in NILE Delta; ⊙ SHIBIN EL KOM. Bounded N by GHARBIYA province, E and W by the DUMYAT and RASHID branches of the Nile, and S by the CAIRO province. Rich agricultural area: cotton, flax, cereals. Industries: cotton ginning, textile milling; manufacturing of belts, handkerchiefs, straw mats. Main urban centers: ASHMUN, TALA, Shibin el Kom, MINUF. Served by railroad from Cairo. Irrigated mainly by the canals Raiyah el MINUFIYA and BAHR SHIBIN. Sometimes spelled Menufiya.

Minufiya, Raiyah (me-noo-FAI-yuh, RAH-yah), navigable canal of the NILE RIVER Delta, Lower EGYPT, extends c.19 mi/31 km from Delta Barrage to the DUMYAT branch of the Nile 6 mi/9.7 km SW of BENHA. Irrigation.

Minusinsk (mee-noo-SEENSK), city (2005 population 72,855), SW KRASNOYARSK TERRITORY, SE SIBERIA, RUSSIA, on the YENISEY RIVER, 410 mi/660 km S of KRASNOYARSK; 53°42'N 91°41'E. Elevation 800 ft/243 m. River port, highway junction, and the center of the MINUSINSK agricultural and gold- and coal-mining basin. Electrical goods, food processing. Founded in 1739; made city in 1822.

Minusio (me-NOO-see-o), commune, TICINO canton, S SWITZERLAND, on LAGO MAGGIORE; 46°11'N 08°49'E.

Cross-references are shown in SMALL CAPITALS. The pronunciation guide is shown on page xix. The sources of population figures are shown on page xvii.

Elevation 794 ft/242 m. Residential suburb E of LO-CARNO; mineral spring.

Minute Man Historical Park (□ 1 sq mi/2.6 sq km), LEXINGTON and CONCORD E MASSACHUSETTS, 10 mi/16 km ENE of BOSTON. Authorized 1959. Scene of fighting on the opening day of the Revolutionary War; includes North Bridge, Minute Man statue, Battle Road, and the home of Nathaniel Hawthorne.

Minuwangoda (MI-nu-wahn-GO-duh), town, WES-TERN PROVINCE, SRI LANKA, 16 mi/26 km NNE of COLOMBO; 07°10′N 79°57′E. Road junction; trades in coconuts, rubber, rice.

Minvoul (MEEN-vool), town, WOLEU-NTEM province, N GABON, on NTEM RIVER, and 55 mi/89 km NE of OYEM, near CAMEROON border; 02°11′N 12°07′E. Coffee and cacao plantations are nearby.

Min Xian (MIN SI-AN), town, ⊙ Min Xian county, SE GANSU province, CHINA, on TAO RIVER, and 90 mi/145 km SW of TIANSHUI; 34°26′N 104°02′E. Grain, medicinal herbs, oilseeds.

Minya (MEN-yuh), province (□ 873 sq mi/2,269.8 sq km; 2004 population 3,960,656), N Upper EGYPT, in NILE valley; ⊙ MINYA. Bounded S by ASYUT province, E by ARABIAN DESERT, N by BENI SUEF province, W by LIBYAN DESERT. Cotton ginning, woolen and sugar milling; agriculture (cotton, cereals, sugar cane). Main urban centers, besides Minya, are BENI MAZAR, MAGHAGHA, El FASHN, ABU QURQAS, SAMA-LUT. Served by railroad along W bank of Nile River. Important archaeological finds at Beni Hassan and OXYRHYNCUS. Also spelled Minia.

Minya, EGYPT: see MINIET EL HEIT.

Minya, Al (MEN-yuh, el), city, ⊙ MINYA province, N central EGYPT, on the W bank of the NILE River; 28°06′N 30°45′E. About half the city's population is Coptic Christian. It is a tourist spot and an agricultural trade center. Products include ginned cotton, flour, and rugs. Al Minya has a university.

Minya Konka, CHINA: see GONGGA.

Minyar (meen-YAHR), city, W CHELYABINSK oblast, RUSSIA, on the NW slope of the S URALS, on the SIM RIVER (tributary of the BELAYA RIVER), on road and railroad, 230 mi/370 km W of CHELYABINSK, and 10 mi/16 km ENE of ASHA; 55°04′N 57°33′E. Elevation 557 ft/169 m. Mining of construction materials, hardware; former metallurgical center (pig and sheet iron), with metalworking and charcoal burning. Founded in 1784; became city in 1943. Until about 1928, called Minyarskiy Zavod.

Minyarskiy Zavod, RUSSIA: see MINYAR.

Minyet el Qamh (MEN-yet el KAHM), town, SHAR-QIYA province, Lower EGYPT, on the BAHR MUWEIS, on railroad, and 10 mi/16 km SW of ZAGAZIG. Cotton ginning.

Minyet Mahallet Damana (MEN-yet MAH-HAHL-let dah-MAH-nuh), village, DAQAHLIYA province, Lower EGYPT, on El BAHR ES SAGHIR (a delta canal) opposite MAHALLET DAMANA and 7 mi/11.3 km ENE of MAN-SURA; 31°05′N 31°29′E. Cotton, cereals.

Minyip (MIN-yip), town, W central VICTORIA, AUS-TRALIA, 160 mi/257 km WNW of MELBOURNE; 36°27′S 142°35′E. In wheat-raising area. Was setting for "Flying Doctors" television series.

Mio (MEE-o), village (2000 population 2,016), ⊙ OS-CODA county, NE central MICHIGAN, c.45 mi/72 km SW of ALPENA, and on AU SABLE RIVER; 44°39′N 84°08′W. Manufacturing (pipe bending, thermocouple alloys); resort. Surrounded by Huron National Forest on W, S, and E. Mio Mountain Ski Area to S.

Mionica (mee-ON-ee-tsah), village (2002 population 16,513), ⊙ Kolubara county, W SERBIA, 10 mi/16 km E of VALJEVO; 44°15′N 20°05′E. Also spelled Mionitsa.

Mions (mee-on), town (□ 4 sq mi/10.4 sq km) residential suburb of LYON, RHÔNE department, RHÔNE-ALPES region, E central FRANCE, 6 mi/9.7 km SE of city's center; 45°40′N 04°57′E. Industrial district of VÉNISSIEUX is just NW.

Miory (mi-O-ree), town, VITEBSK oblast, BELARUS, ⊙ MIORY region, 45 mi/72 km WNW of POLOTSK. In poorly drained area; railroad station; manufacturing (meat packing, flax mill, mixed feed plant).

Mios (mee-os), commune, GIRONDE department, AQUITAINE region, SW FRANCE, on LEYRE RIVER, and 10 mi/16 km E of ARCACHON; 44°36′N 00°56′W.

Miqdadiyah, IRAQ: see MUQDADIYAH.

Miquelon, French Island: see SAINT PIERRE AND MI-QUELON.

Miquihuana (mee-kee-WAH-nah), town, TAMAULI-PAS, NE MEXICO, in SIERRA MADRE ORIENTAL, 45 mi/72 km WSW of CIUDAD VICTORIA; 23°35′N 99°46′W. Cereals, livestock.

Miquon (ME-kwahn), unincorporated village, MON-TGOMERY county, SE PENNSYLVANIA, suburb, 9 mi/14.5 km NW of PHILADELPHIA, on SCHUYLKILL RIVER; 40°03′N 76°15′W. Manufacturing of fine paper.

Miqve Yisrael, ISRAEL: see MIKVE ISRAEL.

Mir, village, ASYUT province, central Upper EGYPT, 10 mi/16 km SW of DAIRUT; 27°27′N 30°44′E. Cereals, dates, sugarcane. About 3 mi/4.8 km SW are the ruins of the Necropolis of Gosu.

Mir (MIR), urban settlement, GRODNO oblast, BELARUS, 11 mi/18 km W of STOLBTSY. Manufacturing (parchment, pitch processing, lumbering, flour milling, dry milk). Has old Gothic palace, ruins of sixteenth-century castle. Noted horse-trading center until World War I.

Mira (MEE-rah), town, CARCHI province, ECUADOR, W of EL ANGEL; 00°33′N 78°02′W. Produces some of the highest quality woolens of highland Ecuador.

Mira (MEE-rah), town, VENEZIA province, VENETO, N ITALY, on Naviglio di Brenta, and 10 mi/16 km W of VENICE; 45°25′N 12°07′E. Highly diversified secondary industrial center includes manufacturing (fabricated metal products, machinery, chemicals, food).

Mira (MEE-rah), town, COIMBRA district, N central PORTUGAL, near the ATLANTIC OCEAN, 22 mi/35 km NW of COIMBRA. Agriculture and fishing. Has 17th-century church. Just W, along coast, are pine-covered dunes.

Mira (MEE-rah), town, CUENCA province, E central SPAIN, 45 mi/72 km SE of CUENCA. Cereals, grapes, saffron, fruit; sheep, goats; flour milling.

Mira Bay (MEI-ruh), inlet of the ATLANTIC, NE NOVA SCOTIA, CANADA, on NE coast of CAPE BRETON Island, 12 mi/19 km SE of SYDNEY. Leads inland into MIRA River.

Mirabel (mee-rah-BEL), county (□ 185 sq mi/481 sq km; 2006 population 31,832), LAURENTIDES region, S QUEBEC, E CANADA; ⊙ MIRABEL; 45°39′N 74°05′W. Composed of one municipality, Mirabel. Formed in 1985.

Mirabel (mee-rah-BEL), city (□ 185 sq mi/481 sq km; 2006 population 31,832), ⊙ MIRABEL county, LAUR-ENTIDES region, S QUEBEC, E CANADA, 23 mi/37 km from MONTREAL; 45°38′N 74°04′W. Agriculture. MIRABEL INTERNATIONAL AIRPORT. Established 1971. Part of the Metropolitan Community of Montreal (*Communauté Metropolitaine de Montréal*).

Mirabel (mee-rah-VEL), town, CÁCERES province, W SPAIN, 14 mi/23 km SW of PLASENCIA; 39°52′N 06°14′W. Olive-oil processing; fruit, wine, flax. Has some Roman remains and medieval castle.

Mirabela (MEE-rah-be-lah), city (2007 population 12,781), N central MINAS GERAIS, BRAZIL, 46 mi/74 km NNW of MONTES CLAROS; 16°20′S 44°10′W.

Mirabel International Airport (mee-rah-BEL), S central QUEBEC, E CANADA, 25 mi/40 km NW of downtown MONTREAL, SW of Laurentian Autoroute. Montreal's second international airport. Airport Code YMX.

Mirabella Eclano (mee-rah-BEL-lah ek-LAH-no), village, AVELLINO province, CAMPANIA, S ITALY, 14 mi/23 km NE of AVELLINO; 41°02′N 14°59′E. Agricultural

center (grains, olives, grapes, fruit; cattle); food processing.

Mirabella, Gulf of, Greece: see MERABELLO, GULF OF.

Mira Bhayandar (mee-RAH buh-YUHN-duhr), city, Thane district, Maharashtra state, W central India.

Miracatu (MEE-rah-KAH-too), city (2007 population 22,796), S SÃO PAULO, BRAZIL, on railroad, and 70 mi/113 km WSW of SANTOS; 24°17′S 47°28′W. Fruit and rice processing, sawmilling. Until 1944, Prainha.

Miracema (MEE-rah-se-mah), city (2007 population 26,241), NE RIO DE JANEIRO state, BRAZIL, near MINAS GERAIS border, 60 mi/97 km WNW of CAMPOS; 21°25′S 42°11′W. Textile milling, coffee and rice processing.

Miracema do Norte, city, TOCANTINS state, N central BRAZIL, on left bank of TOCANTINS River, and 80 mi/129 km N of PÔRTO NACIONAL. Until 1944, called Miracema, and, 1944–1948, Miracema do Norte.

Mira Daire (MEE-rah DEI-rai) or **Mira de Aire**, village, LEIRIA district, central PORTUGAL, 15 mi/24 km SSE of LEIRIA. Textiles.

Mirador (mee-rah-dor), city (2007 population 2,336), central MARANHÃO, BRAZIL, on upper ITAPECURU River, and 120 mi/193 km SW of CAXIAS; 06°15′S 44°25′W. Cotton; cattle. Roads to Caxias, LORETO, and NOVA IORQUE.

Mirador Nacional (mee-rah-DOR nah-syo-NAHL), mountain (1,644 ft/501 m), MALDONADO department, S URUGUAY, in the SIERRA DE LAS ÁNIMAS, 45 mi/72 km E of MONTEVIDEO; 34°44′S 55°20′W. Highest point in Uruguay.

Miradouro (MEE-rah-do-ro), town (2007 population 10,197), SE MINAS GERAIS, BRAZIL, 19 mi/31 km NE of MURIAE; 20°45′S 42°28′W.

Mira Estrêla (MEE-rah E-stre-lah), town, extreme NW São Paulo, 12 mi/19 km N of Fernandópolis, on AguaVermelha Reservoir; 19°59′S 50°10′W. Coffee growing.

Miraflores (mee-rah-FLO-res), canton, Litoral province, ORURO department, W central BOLIVIA. Elevation 12,451 ft/3,795 m. Copper, tin, and gypsum deposits in area. Agriculture (potatoes, yucca, bananas); cattle.

Miraflores (mee-rah-FLO-res), town, ⊙ Miraflores municipio, BOYACÁ department, central COLOMBIA, in valley of Cordillera ORIENTAL, along the Vaupés River (or UAUPÉS), 25 mi/40 km SE of TUNJA; 05°11′N 73°08′W. Elevation 4,698 ft/1,432 m. Once a major producer of rubber, it became a boomtown from coca production. However, the government's destruction of the region's coca acreage and cocaine laboratories in the mid-1990s caused the town's economy to collapse. Agriculture includes cacao, coffee, tobacco, rice; livestock.

Miraflores (mee-rah-FLO-res), town, AREQUIPA region, Arequipa province, S PERU; NE suburb of AREQUIPA; 16°23′S 71°31′W. Elevation 8,267 ft/2,519 m. Agricultural center in irrigation area (cereals, alfalfa, potatoes, vegetables).

Miraflores (mee-rah-FLO-res), town, HUAMALÍES province, HUÁNUCO region, central PERU, on E slopes of Cordillera BLANCA of the ANDES, near MARAÑÓN RIVER, 5 mi/8 km N of LLATA; 09°20′S 76°45′W. Cereals, vegetables, potatoes; livestock.

Miraflores (mee-rah-FLO-res), village, PANAMA province, CANAL AREA, on small artificial MIRAFLORES LAKE of the PANAMA CANAL, on transisthmian railroad, and 5 mi/8 km NW of PANAMA city.

Miraflores (mee-rah-FLO-res), S residential section of LIMA, Lima department, W central PERU; 12°07′S 77°02′W. Beach resort on the Pacific just N of BAR-RANCO.

Miraflores de la Sierra (dhai lah SYE-rah), town, MADRID province, central SPAIN; resort on E slopes of the SIERRA DE GUADARRAMA, 28 mi/45 km N of MA-DRID; 40°49′N 03°47′W. Livestock raising, dairying, apiculture. Hydroelectric plant.

Area is shown by the symbol □, and capital city or county seat by ⊙.

Miraflores Lake (mee-rah-FLO-res), tiny artificial lake (c.1 mi/1.6 km long; elevation 54 ft/16 m above sea level) in S PANAMA CANAL AREA, 5 mi/8 km NW of PANAMA city; 09°01′N 79°36′W. Used as part of the canal route, linking GAILLARD CUT (NW) with the PACIFIC section. PEDRO MIGUEL LOCKS at NW end raise (to 85 ft/26 m) and lower (to 54 ft/16 m) vessels in one step. The Miraflores Locks (SE) overcome in two sets of locks the level between Pacific Ocean and the lake.

Miraflores Locks, PANAMA: see MIRAFLORES LAKE.

Miragoâne (mee-rah-GWAHN), town, GRANDE-ANSE department, SW HAITI, minor port on N JACMEL PENINSULA, 50 mi/80 km W of PORT-AU-PRINCE; 18°27′N 73°06′W. Port ships coffee, fruit; logwood. Tobacco growing; sisal processing. Fishing port. Bauxite deposits in vicinity.

Miraí (mee-rah-ee), city (2007 population 13,000), SE MINAS GERAIS, BRAZIL, 15 mi/24 km NNE of CATA-GUASES; 21°14′S 42°43′W. Railroad terminus; coffee, tobacco. Formerly spelled Mirahy.

Miraíama (mee-REI-ah-mah), city, CEARÁ, BRAZIL.

Miraj (mee-RUHJ), city, Sangli district, Maharashtra state, W central India, 6 mi/9.7 km ESE of Sangli; 16°50′N 74°38′E. Railroad and road junction; trades in grain, cotton, sugarcane, oilseeds, cloth fabrics; cotton milling, hand-loom weaving, manufacturing of chemicals, consumer goods. Seat of large mission hospital and medical college. Was the capital of former Deccan state of Miraj Senior and headquarters of former Wadi Estate.

Miraj Junior, former princely state in DECCAN STATES, India. Incorporated 1949 into what are now KOLHA-PUR, SANGLI, SATARA, and SOLAPUR districts (MA-HARASHTRA state) and BELGAUM and DHARWAD districts (KARNATAKA state).

Miraj Senior, former princely state in DECCAN STATES, INDIA. Incorporated 1949 into what are now KOLHA-PUR, SANGLI, SATARA, and SOLAPUR districts (MA-HARASHTRA state) and BELGAUM and DHARWAD districts (KARNATAKA state).

Miraki, town, NE KASHKADARYO wiloyat, S UZBEKI-STAN, 15 mi/24 km E of KITAB on AKSU RIVER; 39°02′N 67°08′E.

Mira Loma, unincorporated city (2000 population 17,617), RIVERSIDE county, S CALIFORNIA, 10 mi/16 km W of RIVERSIDE; 33°59′N 117°31′W. Vineyards; manufacturing (printing and publishing, fabricated metal products, transportation equipment).

Miramar (MI-ruh-mahr), city (□ 31 sq mi/80.6 sq km; 2005 population 106,623), BROWARD county, SE FLORIDA, 15 mi/24 km N of MIAMI; 25°58′N 80°19′W. Residential community in the rapidly growing I-75 corridor. Incorporated 1955.

Miramar (mee-rah-MAHR), town (□ 463 sq mi/1,203.8 sq km), SE BUENOS AIRES province, ARGENTINA, 24 mi/39 km SW of MAR DEL PLATA; ⊙ General Alvarado district (□ 463 sq mi/1,199 sq km; 1991 population 30,043). Seaside resort and agricultural center (grain, flax; livestock); dairying. Beach resort Mar del Sur is 10 mi/16 km WSW.

Miramar, town, NE CÓRDOBA province, ARGENTINA, beach resort on S shore of the MAR CHIQUITA, 100 mi/161 km NE of CÓRDOBA. Hotels, sanitariums.

Miramar (mee-rah-MAHR), town, PUNTARENAS province, W COSTA RICA, 10 mi/16 km NW of PUN-TARENAS; ⊙ Montes de Oro canton. Commercial center. Former gold-mining district.

Miramar (mee-rah-MAHR), village and minor civil division of Santa Isabel District, COLÓN province, central PANAMA, on CARIBBEAN SEA, 1 mi/1.6 km E of PALENQUE. Bananas, cacao, coconuts, corn; livestock.

Miramar (mee-rah-MAHR), resort, TAMAULIPAS, NE MEXICO, on GULF at mouth of PÁNUCO River, 6 mi/9.7 km NE of TAMPICO.

Miramar Beach (MI-ruh-mahr), unincorporated town (□ 4 sq mi/10.4 sq km; 2000 population 2,435), WALTON county, NW FLORIDA, ON GULF OF MEXICO, and 8 mi/12.9 km ESE of DESTIN; 30°22′N 86°21′W.

Miramare (mee-rah-MAH-rai), seaside resort, NE ITALY, 4 mi/6 km NW of TRIESTE, on GULF OF TRIESTE of N ADRIATIC Sea. Site of park and former castle (built 1856) of Emperor Maximilian of MEXICO. Headquarters of U.S. troops in TRIESTE after World War II.

Miramas (mee-rah-mah), town (□ 9 sq mi/23.4 sq km), BOUCHES-DU-RHÔNE department, PROVENCE-ALPES-CÔTE D'AZUR region, SE FRANCE, near NW tip of Étang de berre, N of ISTRES, and 27 mi/43 km NW of MARSEILLE; 43°35′N 05°00′E. Railroad junction; chemicals. Istres-le-Tubé military airport and testing ground is just SW. Old quarter of Miramas has ruins of 13th-century castle.

Mirambo (mee-RAHM-bo), village, RUVUMA region, S TANZANIA, 45 mi/72 km SSW of SONGEA; 11°36′S 35°25′E. Tobacco, subsistence crops; goats, sheep; timber.

Miramichi (mi-ruh-muh-SHEE), city (□ 69 sq mi/179.4 sq km; 2001 population 18,505), NEW BRUNSWICK, CANADA, 75 mi/121 km NNW of MONCTON; 47°01′N 65°30′W. Formed in 1995 from the amalgamation of NEWCASTLE, CHATHAM, DOUGLASTOWN, Nelson, and LOGGIEVILLE and several other communities. For-estry, tourism, and cultural festivals.

Miramichi (mi-ruh-muh-SHEE), river system in N central NEW BRUNSWICK, CANADA, consisting of several streams rising in N central highlands of province and flowing E to estuarial section, which begins at NEWCASTLE and extends 15 mi/24 km ENE, past CHATHAM, to Miramichi Bay, inlet (20 mi/32 km long, 15 mi/24 km wide at mouth) of the Gulf of St. Lawr-ence. Main river of Miramichi system, the Southwest Miramichi, is 135 mi/217 km long. Miramichi Bay, visited by Cartier in 1534, contains several small is-lands; on shore are Acadian fishing settlements. Noted for salmon fishing. Ongoing struggle to keep salmon population stable; pressure from increased tourism and sport fishing. Logging, lumber mills.

Miramont-de-Guyenne (mee-rah-mon–duh–gee-yen), commune (□ 6 sq mi/15.6 sq km; 2004 population 3,263), LOT-ET-GARONNE department, AQUITAINE re-gion, SW FRANCE, 12 mi/19 km NE of MARMANDE; 44°36′N 00°22′E. Agriculture market (peaches, to-matoes); manufacturing of apparel.

Miram Shah (mi-RUHM SHAH), town, in NORTH WAZIRISTAN centrally administered tribal area, NORTH-WEST FRONTIER PROVINCE, W central PAKI-STAN, 105 mi/169 km SW of PESHAWAR; 33°01′N 70°04′E. Also spelled MIRANSHAH.

Miranda (mee-RAHN-dah), state (□ 3,070 sq mi/7,982 sq km; 2001 population 2,330,872), N VENEZUELA, on the CARIBBEAN; ⊙ LOS TEQUES; 10°15′N 66°25′W. Bounded N by DISTRITO CAPITAL; here along border, the CORDILLERA DE LA COSTA rises to its highest ele-vation. Predominantly mountainous, apart from al-luvial valley of lower TUY RIVER, it includes Lago de TACARIGUA along Caribbean coast. Climate is hot and tropical in E lowlands, with rains all year round; fertile higher sections are semitropical and dry, with rainy season June–October. An agricultural region, it pro-duces coffee, cacao, sugarcane, corn, rice, yucca, po-tatoes, bananas, plantains, coconuts. Some fishing and cattle-raising. Sugar milling and saw milling are its main industries. Asbestos, marble, copper, iron, gold, coal, and asphalt deposits. Other main cities include GUARENAS, OCUMARE DEL TUY, PETARE, San Antonio, San José, San Diego. Includes PARQUE NA-CIONAL LAGUNA DE TACARIGUA and parts of Guatopo, EL AVILA, and Macarao.

Miranda (mee-RAHN-dah), city (□ 120 sq mi/312 sq km; 2004 population 24,459), S MATO GROSSO DO SUL, BRAZIL, head of navigation on RIO MIRANDA at edge of RIO PARAGUAY flood plain, on SÃO PAULO-CORUMBÁ railroad, and 120 sq mi/311 sq km SE of Corumbá;

20°13′S 56°36′W. Ships dried meat and maté. Marble quarries. Captured by Paraguayans in 1865. In (for-mer) Ponta Porã territory, 1943–1946.

Miranda (mee-RAHN-dah), town, ⊙ Miranda muni-cipio, CAUCA department, SW COLOMBIA, on W slopes of Cordillera CENTRAL, 57 mi/92 km NNW of POPAYÁN; 03°15′N 76°13′W. Elevation 3,389 ft/1,032 m. Sugarcane, coffee, corn.

Miranda (mee-RAHN-dah), town, ⊙ Miranda muni-cipio, CARABOBO state, N VENEZUELA, 27 mi/43 km W of VALENCIA; 10°09′N 68°23′W. Elevation 2,857 ft/870 m. Agricultural center (coffee, sugarcane, corn, fruit; livestock).

Miranda de Arga (mee-RAHN-dah dhai AHR-gah), town, NAVARRE province, N SPAIN, on ARGA RIVER, and 15 mi/24 km SE of ESTELLA; 42°29′N 01°50′W. Wine, cereals, vegetables.

Miranda de Ebro (mee-RAHN-dah dhai AI-vro), city, BURGOS province, N SPAIN, in CASTILE-LEÓN region, on the EBRO (bridges), at ÁLAVA province border, and 45 mi/72 km NE of BURGOS; 42°41′N 02°57′W. Rail-road and road junction. Manufacturing of chemicals and textiles; metallurgy; food processing. The region produces potatoes, cereals, grapes, fruit; livestock. An ancient city, with remains of castle and walls.

Miranda del Castañar (mee-RAHN-dah dhel kahs-tahn-YAHR), town, SALAMANCA province, W SPAIN, 15 mi/24 km NW of BÉJAR; 40°29′N 06°00′W. Olive oil, fruit, wine.

Miranda do Corvo (mee-RAHN-dah doo KOR-voo), town, COIMBRA district, N central PORTUGAL, on railroad, and 9 mi/14.5 km SSE of COIMBRA; 40°06′N 08°20′W. Light manufacturing.

Miranda do Douro (mee-RAHN-dah doo DOR-oo), ancient *Sepontia*, city, BRAGANÇA district, northeast-ernmost PORTUGAL, above gorge of DUERO RIVER (Spanish border), and 34 mi/55 km SE of BRAGANÇA; 41°30′N 06°16′W. Accessible only by road. Has Re-naissance cathedral and ruins of 13th-century fort.

Miranda do Norte (MEE-rahn-dah do NOR-chee), city (2007 population 17,724), N central MARANHÃO state, BRAZIL, 74 mi/120 km S of SÃO LUÍS; 03°28′S 44°33′W.

Miranda, Lo, CHILE: see LO MIRANDA.

Mirande (mee-rahn-duh), commune (□ 9 sq mi/23.4 sq km), GERS department, MIDI-PYRÉNÉES region, SW FRANCE, on BAISE RIVER, and 13 mi/19 km SW of AUCH; 43°31′N 00°25′E. ARMAGNAC brandy distilling; market for poultry, hogs, and horses; tannery. Has 15th-century church and a fine art museum. Founded 1281 as a planned fortified community (*bastide*).

Mirandela (mir-ahn-DAI-lah), town, BRAGANÇA dis-trict, N PORTUGAL, on TUA RIVER, on railroad, and 30 mi/48 km SW of BRAGANÇA; 41°29′N 07°11′W. Pro-duces port wine; manufacturing of ceramics; tin- and ironworking, cork processing.

Mirandiba (MEE-rahn-zhee-bah), city (2007 popula-tion 13,513), W PERNAMBUCO state, BRAZIL, 34 mi/55 km SW of Serra Taihada, on railroad; 08°06′S 38°44′W.

Mirandilla (mee-rahn-DEE-lyah), town, BADAJOZ province, W SPAIN, 6 mi/9.7 km NNE of MÉRIDA; 39°00′N 06°17′W. Olives, cereals, livestock.

Mirando City (mi-RAN-do), unincorporated village, WEBB county, S TEXAS, 30 mi/48 km E of LAREDO; 27°26′N 99°00′W. In oil field; some manufacturing. Cattle.

Mirandola (mee-RAHN-do-lah), town, MODENA prov-ince, EMILIA-ROMAGNA, N central ITALY, 18 mi/29 km NNE of MODENA; 44°53′N 11°04′E. Railroad terminus; manufacturing (transportation equipment, fabricated metal products, textiles, plastics, apparel, food). Has cathedral, Jesuit church, old palace of dukes of Mir-andola.

Mirandópolis (MEE-rahn-DO-po-lees), city (2007 population 25,867), NW SÃO PAULO, BRAZIL, on rail-road, and 40 mi/64 km W of ARAÇATUBA; 21°09′S 51°06′W. Coffee, rice; forest products. Until 1944, Comandante Arbues.

Cross-references are shown in SMALL CAPITALS. The pronunciation guide is shown on page xix. The sources of population figures are shown on page xvii.

Mirangaba (MEE-rahn-GAH-bah), city (2007 population 17,598), N central BAHIA, BRAZIL, on Chapada da diamantina, 19 mi/30 km NNW of JACOBINA; 10°58′S 40°35′W.

Mirani (mi-RAN-ee), town, QUEENSLAND, NE AUSTRALIA, 23 mi/37 km W of MACKAY; 21°10′S 148°52′E. Sugar.

Mirano (mee-RAH-no), town, VENEZIA province, VENETO, N ITALY, 11 mi/18 km WNW of VENICE; 45°29′N 12°06′E. Characterized by numerous homes of nobility with extensive gardens.

Miranorte (MEE-rah-nor-che), city (2007 population 11,858), central TOCANTINS state, BRAZIL, 164 mi/264 km S of PALMAS, on Highway BR-153; 09°28′S 48°40′W.

Miranpur (MEE-rahn-poor), town, MUZAFFARNAGAR district, N UTTAR PRADESH state, N central INDIA, 19 mi/31 km SE of MUZAFFARNAGAR; 29°18′N 77°56′E. Wheat, gram, sugarcane, oilseeds; hand-loom woolen weaving.

Miranpur Katra (MEE-rahn-poor KAH-trah), town, SHAHJAHANPUR district, central UTTAR PRADESH state, N central INDIA, 19 mi/31 km NW of SHAHJAHANPUR; 28°02′N 79°39′E. Trades in wheat, rice, gram, oilseeds, sugarcane. Rohilla defeated (1774) nearby by combined forces of nawab of Oudh and British. Also called Katra.

Miran Sahib (MEE-rahn SAH-hib), village, JAMMU district, JAMMU AND KASHMIR state, N INDIA, in SW KASHMIR region, 7 mi/11.3 km SSW of JAMMU. Manufacturing of chemicals, apparel; indianite (source for rubies) works.

Miranshah, PAKISTAN: see MIRAM SHAH.

Mirante (MEE-rahn-CHEE), town (2007 population 9,218), S central BAHIA, BRAZIL, in Serra GERAL, 37 mi/60 km N of VITÓRIA DA CONQUISTA; 14°14′S 40°44′W.

Mirante do Paranapanema (MEE-rahn-che do PAH-rah-nah-pah-ne-mah), city (2007 population 16,679), W SÃO PAULO state, BRAZIL, 38 mi/61 km SW of PRESIDENTE PRUDENTE; 22°17′S 51°54′W.

Mira Por Vos [Spanish=*look out for yourself*], islets and reefs, S central BAHAMAS, 15 mi/24 km W of S tip of ACKLINS ISLAND, 270 mi/435 km SE of NASSAU; 22°10′N 74°32′W. Low, dangerous rocks.

Mira River (MEI-ruh), tidal inlet in E part of CAPE BRETON Island, NE NOVA SCOTIA, CANADA, extending 30 mi/48 km W and S from MIRA BAY; 1 mi/2 km–2 mi/3 km wide.

Mira River (MEE-ruh), c.150 mi/241 km long, in ECUADOR and COLOMBIA; rises in the ANDES SW of IBARRA (Ecuador). Called Chota River in its upper course and serves as border between CARCHI and IMBABURA provinces (Ecuador) during part of its course; flows NW to the PACIFIC in a large delta, its main arm reaching the ocean at MANGLES POINT (Colombia); 01°35′N 78°57′W.

Mira River (MEE-rah), 80 mi/129 km long, in BEJA district, S PORTUGAL; rises 4 mi/6.4 km SW of ALMODÓVAR; flows NW past ODEMIRA (head of navigation), to the ATLANTIC OCEAN at VILA NOVA DE MILFONTES.

Mira River, Colombia: see MIRA RIVER, Ecuador.

Mirasaka (mee-RAH-sah-kah), town, Futami county, HIROSHIMA prefecture, SW HONSHU, W JAPAN, 37 mi/60 km N of HIROSHIMA; 34°48′N 132°53′E.

Mirassol (MEE-rah-sol), city (2007 population 51,660), NW SÃO PAULO, BRAZIL, on railroad, and 7 mi/11.3 km W of SÃO JOSÉ DO RIO PRÊTO; 20°46′S 49°28′W. Coffee and cotton grown here; cotton ginning, coffee processing, livestock shipping. Formerly Mirasol.

Mirassolândia (MEE-rah-so-LAHN-zhee-ah), town (2007 population 4,107), N SÃO PAULO state, BRAZIL, 16 mi/26 km NW of SÃO JOSÉ DO RIO PRETO; 20°38′S 49°29′W. Coffee growing.

Miravalles (mee-rah-VAH-yes), town, VIZCAYA province, N SPAIN, 5 mi/8 km SSE of BILBAO. Metalworking. Cereals, jute; wine; livestock.

Miravalles (mee-rah-VAH-yais), active volcano (6,653 ft/2,028 m), in the CORDILLERA DE GUANACASTE, NW COSTA RICA, 16 mi/26 km NNE of BAGACES; 10°45′N 85°10′W. Sulphur springs.

Mira Vista, unincorporated town, VENTURA county, S CALIFORNIA, 10 mi/16 km N of VENTURA, on Ventura River (forms Lake Casitas reservoir to SW). Citrus, avocados, vegetables, flowers; nursery products. Los Padres National Forest to N; Emma Wood State Beach, on Pacific Ocean, to S.

Mir-Bashir, city, central AZERBAIJAN, on TERTER River, on railroad, 20 mi/32 km SSW of YEVLAKH, on border of NAGORNO-KARABAKH autonomous oblast; 40°20′N 47°00′E. Cotton district. Until 1949, called Terter. Butter, cheese; petroleum extracted nearby.

Mirboo North, settlement, VICTORIA, SE AUSTRALIA, 99 mi/160 km E of MELBOURNE, and in Strzelecki Ranges; 38°24′S 146°09′E. Dairying; potatoes, corn, onions.

Mirceşti (mir-CHESHT), village, IAŞI county, NE ROMANIA, on railroad, near SIRET RIVER, 11 mi/18 km NNW of ROMAN.

Mirditë (mir-DEET), tribal region of N ALBANIA, Mirditë district, c.25 mi/40 km ESE of SHKODËR. Highland area drained by FAN RIVER Pastoral, Roman Catholic population long known for independent spirit. Main villages are OROSH and Blinisht. Also spelled Mirdita.

Mirditë, ALBANIA: see RRËSHEN.

Mirdjaveh, IRAN: see MIRJAVEH.

Mirebalais (meer-bah-LAI), town, CENTRE department, S central HAITI, on ARTIBONITE River, and 25 mi/40 km NE of PORT-AU-PRINCE; 18°50′N 72°06′W. Agriculture (coffee, limes, sugarcane, sisal, cotton, rice); sugar processing.

Mirebeau (meer-uh-bo), commune (□ 5 sq km), VIENNE department, POITOU-CHARENTES region, W central FRANCE, 12 mi/19 km W of CHÂTELLERAULT; 46°47′N 00°11′E. Ruins of medieval stronghold. Also called Mirebeau-en-Poitou.

Mirebeau-en-Poitou, FRANCE: see MIREBEAU.

Mirecourt (meer-koor), town (□ 4 sq mi/10.4 km), VOSGES department, LORRAINE region, FRANCE, 17 mi/27 km NW of ÉPINAL; 48°18′N 06°08′E. Manufacturing of consumer goods, furniture; cotton milling. Also known for its handicrafts (lace, embroidery). Has 16th-century church and 17th-century arcaded market halls.

Mirepoix (meer-pwah), commune (□ 18 sq mi/46.8 sq km; 2004 population 3,060), ARIÈGE department, MIDI-PYRÉNÉES region, S FRANCE, on HERS RIVER, and 16 mi/26 km E of FOIX; 43°05′N 01°53′E. Agriculture market; woodworking. A 13th-century stronghold. Mirepoix has 13th–16th-century cathedral of Saint-Maurice. Was episcopal see (1317–1789). Its town square is surrounded by 13th–15th-century houses with store fronts.

Mirfield (MUH-feeld), town (2001 population 18,621), WEST YORKSHIRE, N ENGLAND, on CALDER RIVER, and 4 mi/6.4 km ENE of HUDDERSFIELD; 53°40′N 01°41′W. Railroad junction. Previously cotton milling.

Mirgorod, UKRAINE: see MYRHOROD.

Miri (ME-ree), town (2000 population 167,535), N SARAWAK, NW BORNEO, MALAYSIA, port on SOUTH CHINA SEA, 75 mi/121 km WSW of BRUNEI; 04°23′N 113°50′E. Outlet for oil refined at nearby Lutong. Has important oil fields. Severely damaged during World War II.

Miri, PAKISTAN: see KOH-I-SULTAN.

Mirialguda, India: see MIRYALGUDA.

Miribel (meer-ee-bel), town (□ 9 sq mi/23.4 sq km), AIN department, RHÔNE-ALPES region, E central FRANCE, 8 mi/12.9 km NE of LYON, on the Miribel Canal (an arm of RHÔNE RIVER); 45°50′N 04°57′E. Paper and container plant and chemical works.

Mirigama (MEE-ri-GUH-muh), village, WESTERN PROVINCE, SRI LANKA, 30 mi/48 km NE of COLOMBO; 07°14′N 80°07′E. Coconut processing; trades in co-

conut, rubber, rice. Buddhist rock temple (1st century C.E.) nearby.

Miri Hills (mee-REE), extension of E ASSAM HIMALAYAS in S central ARUNACHAL PRADESH state, NE INDIA, N of DAFLA HILLS. Inhabited by Miri tribe of Tibeto-Burman origin.

Mirik, Cape (mee-REEK) or **Cape Timiris**, headland on ATLANTIC OCEAN coast of MAURITANIA, 230 mi/370 km N of SAINT-LOUIS (SENEGAL); 19°22′N 16°30′W.

Mirim Doce (MEE-reen DO-see), town (2007 population 2,545), E SANTA CATARINA state, BRAZIL, 38 mi/62 km WSW of BLUMENAU; 27°12′S 50°05′W. Rice, manioc, corn; livestock.

Mirimire (mee-ree-MEE-ree), town, ⊙ San Francisco municipio, FALCÓN state, NW VENEZUELA, 38 mi/61 km NW of TUCACAS; 11°10′N 68°43′W. Corn, fruit.

Mirim Lake, Portuguese *Lagoa Mirim*, Spanish *Laguna Merín*, shallow tidewater lagoon (□ 1,145 sq mi/2,977 sq km), in extreme S BRAZIL and E URUGUAY, separated from the Atlantic by low, marshy bar (10 mi/16 km–35 mi/56 km wide; dotted with smaller lagoons; it discharges (at N end) into the Lagoa dos PATOS, through SÃO GONCALO Canal; 110 mi/177 km long, 25 mi/40 km wide. Navigable for small vessels. International boundary traverses lake's S half from mouth of JAGUARÃO River to the extreme S tip of lake W of CHUÍ. More than three-quarters of lake is in Brazil (RIO GRANDE DO SUL).

Miriñay River (mee-ree-NEI), flows c. 100 mi/161 km S, SE CORRIENTES province, ARGENTINA; rises in swamps S of Esteros del IBERÁ; flows to URUGUAY RIVER, 5 mi/8 km N of MONTE CASEROS, opposite mouth of QUARAÍ RIVER at BRAZIL-URUGUAY border.

Miringoni (mee-reen-GO-nee), town, MWALI island and district, SW COMOROS, 7 mi/11.3 km W of FOMBONI, at W end of island, on MOZAMBIQUE CHANNEL, INDIAN OCEAN; 12°17′S 43°39′E. Manufacturing (ylang-ylang oil). Fish; livestock; ylang-ylang, vanilla, bananas, coconuts.

Mirinzal (MEE-reen-sahl), city (2007 population 13,903), N MARANHÃO state, BRAZIL, 25 mi/41 km N of BEQUIMÃO, near Atlantic coast; 02°06′S 44°45′W.

Miritiba, Brazil: see HUMBERTO DE CAMPOS.

Miriti Paraná River (mee-REE-tee pah-rah-NAH), flows c.175 mi/282 km SE, in the AMAZONAS department, SE COLOMBIA, in densely forested lowlands; flows to the CAQUETÁ River; 01°11′S 70°02′W.

Mirjaveh (mir-JAHV-ai), village, Sīstān va Balūchestān providence, SE IRAN, 50 mi/80 km SE of ZAHEDAN, and on railroad to QUETTA, on PAKISTAN border, and on short intermittent MIRJAVEH River (frontier stream); 35°36′N 61°15′E. Customs station. Sometimes spelled Mirjawa.

Mirkovo (MEER-ko-vo), village (1993 population 1,925), SOFIA oblast, obshtina center, BULGARIA, E of SOFIA; 42°42′N 24°00′E.

Mirle (mir-LEE), town, MYSORE district, KARNATAKA state, SW INDIA, 28 mi/45 km NW of MYSORE; 12°32′N 76°18′E. Tobacco, rice, millet.

Mirna River (MIR-nah), Italian *Quieto*, c.30 mi/48 km long, W CROATIA, in the ISTRIA region; rises 6 mi/9.7 km SE of BUZET; flows W, past Buzet and MOTOVUN, to ADRIATIC SEA near NOVIGRAD.

Mirnaya Dolina, UKRAINE: see MYRNA DOLYNA.

Mirnoye, UKRAINE: see MYRNE, Donets'k oblast; MYRNE, Luhans'k oblast; MYRNE, Zaporizhzhya oblast; or MYRNE, Kherson oblast.

Mirnyy (MEER-niyee), city (2005 population 29,000), W central ARCHANGEL oblast, N European Russia, on road and railroad, 4 mi/6.4 km N of PLESETSK; 62°47′N 40°20′E. Elevation 370 ft/112 m. Along with Plesetsk, the site of the Soviet—and now Russian—missile technology research, development, and testing, both for military and space exploration. Established in 1957 as a closed city with a secret designation of Leningrad-300; opened and renamed in 1991.

Area is shown by the symbol □, and capital city or county seat by ⊙.

Mirnyy (MEER-niyee), city (2006 population 40,425), W SAKHA REPUBLIC, NW RUSSIAN FAR EAST, on the Irelyakh River (tributary of the VILYUY River), on highway junction, 600 mi/966 km W of YAKUTSK; 62°32′N 113°57′E. Elevation 1,092 ft/332 m. Founded in 1955, when diamonds were discovered (Kimberlite pipe "Mir"). Became a city in 1959 and is the center of Soviet diamond mining. Has an airport.

Mirnyy (MEER-niyee) [Russian=peaceful], town (2005 population 3,475), NW KIROV oblast, E central European Russia, on highway, 15 mi/24 km NNE of SVECHA; 58°29′N 47°39′E. Elevation 544 ft/165 m. Gas pipeline service station; peatworks.

Mirnyy (MEER-niyee), settlement, E NIZHEGOROD oblast, central European Russia, on the VETLUGA RIVER, on road and cargo railroad spur, 24 mi/39 km SE of KRASNYYE BAKI; 56°50′N 45°26′E. Elevation 298 ft/90 m. Logging, lumbering. Site of a penal labor colony. Formerly called Lesozavod.

Mirnyy (MEER-niyee), settlement (2006 population 7,375), central SAMARA oblast, E European Russia, in the KONDURCHA RIVER valley, near highway, 24 mi/39 km NNE of SAMARA, and 7 mi/11 km W of KRASNYY YAR, to which it is administratively subordinate; 53°33′N 50°13′E. Elevation 380 ft/115 m. Military base and airfield. The region's main airport is 3 mi/5 km to the SW.

Mirnyy (MEER-niyee), settlement, SW TVER oblast, W European Russia, on railroad, 8 mi/13 km W of OLENINO, to which it is administratively subordinate; 56°14′N 33°10′E. Elevation 725 ft/220 m. Logging, lumbering, woodworking.

Mirnyy (MEER-niyee), settlement (2006 population 3,165), E ULYANOVSK oblast, E central European Russia, on highway and near railroad, 14 mi/23 km E of ULYANOVSK, and less than 4 mi/6 km W of CHERDAKLY; 54°22′N 48°45′E. Elevation 282 ft/85 m. Region's major airport is 2 mi/3.2 km to the NNE.

Mirnyy, UKRAINE: see MYRNYY.

Mirnyy Station (MIR-nyee), ANTARCTICA, Russian station on coast of QUEEN MARY LAND; 66°33′S 93°01′E. Opened February 13, 1956.

Mirola (mee-ROH-lah), village, LINDI region, S TANZANIA, 50 mi/80 km SW of LIWALE, in S end of SELOUS GAME RESERVE; 10°07′S 35°21′E. Livestock.

Mironovka, UKRAINE: see MYRONIVKA.

Mironovskiy, UKRAINE: see MYRONIVS'KYY.

Miropol', UKRAINE: see MYROPIL'.

Miropol'ye, UKRAINE: see MYROPILLYA.

Miroschau, CZECH REPUBLIC: see MIROSOV.

Miroslav (MI-ro-SLAHF), German Misslitz, town, JIHOMORAVSKY province, S BOHEMIA, CZECH REPUBLIC, 13 mi/21 km NE of Znojmo. Agriculture (wheat, fruit, vegetables); food processing; vineyard. Located here are a 16th century castle, and an 18th century church.

Miroslawiec (mee-ro-SLAH-veets), Polish Mirosławiec, German Märkisch Friedland, town, in POMERANIA, PIŁA province, NW POLAND, 45 mi/72 km E of STARGARD SZCZECINSKI; 53°20′N 16°06′E. Grain, sugar beets, potatoes; livestock. Until 1938, in former Prussian province of Grenzmark Posen-Westpreussen.

Mirosov (MI-ro-SHOF), Czech Mirošov, German Miroschau, town, ZAPADOCESKY province, SW BOHEMIA, CZECH REPUBLIC, on railroad, and 13 mi/21 km ESE of PLZEŇ; 49°41′N 13°40′E. Summer resort. Woodworking. Has a 16th century castle.

Mirovice (MI-ro-VI-tse), German Mirowitz, town, JIHOCESKY province, S BOHEMIA, CZECH REPUBLIC, 12 mi/19 km SSE of PŘÍBRAM, on railroad; 49°39′N 13°05′E. Agriculture (rye, potatoes); poultry; food processing.

Mirovskoye, UKRAINE: see MYRIVS'KE.

Mirow (MEE-rou), town, Mecklenburg-Western Pomerania, N GERMANY, on Mirow Lake (8 mi/12.9 km long), 12 mi/19 km WSW of NEUSTRELITZ; 53°16′N 12°50′E. Manufacturing (furniture); agricultural market (grain, potatoes; livestock). Health resort. Has 18th-century former castle of grand dukes of Mecklenburg-Strelitz. Founded 1277 as monastery. Chartered 1919.

Mirowitz, CZECH REPUBLIC: see MIROVICE.

Mirpur (MIR-puhr), town, AZAD KASHMIR, NE PAKISTAN, in W KASHMIR region, 18 mi/29 km NNE of JHELUM; ⊙ MIRPUR district. Cotton weaving; trades in wheat, bajra, corn, pulses. Occupied in 1948 by PAKISTAN.

Mirpur Khas (MIR-puhr khahs), town, THAR PARKAR district, HYDERABAD division, SIND province, SE PAKISTAN, 45 mi/72 km WNW of UMARKOT, on the Let War canal; 25°32′N 69°00′E. Railroad junction. Cotton ginning, handicrafts, cloth weaving; agricultural farming; fruit research station. Ruins of Buddhist stupa and monastery. Founded in 1806.

Mirpur Sakro (MIR-puhr SUK-ro), town, TATTA district, SW PAKISTAN, 45 mi/72 km SE of KARACHI; 24°33′N 67°37′E. Rice, millet.

Mirrassol d'Oeste (MEE-rah-sol DO-es-che), city, SW Mato Grosso, 49 mi/79 km NW of Caceres; 15°35′S 58°10′W. Soybeans, sugarcane.

Mirria (MEER-ee-uh), town, ZINDER province, NIGER, 10 mi/16 km E of ZINDER; 13°43′N 09°07′E. Administrative center.

Mirror, village, S central Alberta, near Buffalo Lake, 32 mi/51 km ENE of RedDeer. Lumbering, dairying, mixed farming.

Mirror Lake, village, CARROLL county, E NEW HAMPSHIRE, in town of TUFTONBORO, 11 mi/18 km NE of LACONIA, and 5 mi/8 km SW of town center, between Mirror Lake (E) and N end of Winter Harbor of LAKE WINNIPESAUKEE. Resort area. Libby Museum to SE.

Mirror Lake (MI-ruhr), small alpine lake, SW ALBERTA, W CANADA, near BRITISH COLUMBIA border, in ROCKY MOUNTAINS, in BANFF NATIONAL PARK, 4 mi/6 km WSW of Lake LOUISE; 51°25′N 116°14′W. Drains E into BOW RIVER.

Mirror Lake, NEW JERSEY: see BROWNS MILLS.

Mirror Lake, NEW YORK: see LAKE PLACID.

Mirror Lake, WISCONSIN: see DELL CREEK.

Mirror Landing, ALBERTA, CANADA: see SMITH.

Mirs Bay, Chinese Taipang Wan, large bay, HONG KONG, CHINA, E of New Territories, 9 mi/15 km across from GUANGDONG province, China.

Mirsk (meersk), German Friedeberg, town, in LOWER SILESIA, Jelenia Góra province, SW POLAND, near CZECH border, at N foot of the ISERGEBIRGE, on KWISA RIVER, and 16 mi/26 km WNW of JELENIA GÓRA (Hirschberg); 50°58′N 15°23′E. Textiles. Town hall with 16th century tower. After 1945, briefly called Spokojna Gora, Polish Spokojna Góra.

Mirskoy (meers-KO-yee), settlement (2005 population 3,110), E KRASNODAR TERRITORY, S European Russia, on road and railroad, 12 mi/19 km NW of KROPOTKIN; 45°32′N 40°24′E. Elevation 308 ft/93 m. In agricultural area. Developed around a collective farm called Imeni Kaganovicha.

Mirtag, TURKEY: see MUTKI.

Mirungwe, CONGO: see MINDIGI.

Miryalguda (mir-YUHL-guh-duh), town, NALGONDA district, ANDHRA PRADESH state, SE INDIA, 19 mi/31 km SE of NALGONDA. Rice milling, cotton ginning, castor-oil extraction. Also spelled Mirialguda.

Miryang (MEEL-YAHNG), city (□ 296 sq mi/769.6 sq km), S Kyongsang province, S Korea, bordering on N Kyongsang province; 35°29′N 128°45′E. Small rivers run S off Taebaek Mountains to Nakdong River, forming a fertile plain. Agr. includes rice, barley, wheat, beans, vegetables, persimmon, pears. Kyongchon railroad connects to Kyongbu railroad; Kajisan Provincial Park; Buddhist temples.

Mirzaani (mir-zah-AH-nee), town, SE GEORGIA, 60 mi/97 km ESE of TBILISI, on SHIRAKI STEPPE. Railroad terminus; center of petroleum region; oil-cracking plant.

Mirzachul Steppe, UZBEKISTAN: see GULISTAN.

Mirzapur (mir-ZAH-poor), district (□ 1,746 sq mi/4,539.6 sq km), SE UTTAR PRADESH state, N central INDIA; ⊙ MIRZAPUR. Bounded N by the GANGA RIVER; drained by SON RIVER and RIHAND RIVER (hydroelectric project). VINDHYA RANGE in S (lac cultivation in sal jungle). Agriculture (rice, gram, barley, wheat, oilseeds, millet, sugarcane, corn). Sandstone quarries near CHUNAR. Main centers are Mirzapur, AHRAURA, Chunar.

Mirzapur (mir-ZAH-poor), city and joint municipality (total 1991 population 169,336) with Vindhyachal (former suburb), Uttar Pradesh state, N central India; ⊙ Mirzapur district; 25°09′N 82°35′E. Major road and railroad connections. Manufacturing of carpets and brassware. Many Hindu pilgrims visit the shrine of the goddess Vindhyeshwari.

Misahohé (mee-sah-HO-hai), village, PLATEAUX region, S TOGO, 4 mi/6.4 km E of KLOUTO, on GHANA border; 06°57′N 00°35′E. Cacao, palm oil and kernels, and cotton. Customhouse.

Misaka (mee-SAH-kah), town, E Yatsushiro county, YAMANASHI prefecture, central HONSHU, central JAPAN, 5.6 mi/9 km S of KOFU; 35°37′N 138°39′E. Peaches.

Misaki (mee-SAH-kee), town, Isumi county, CHIBA prefecture, E central HONSHU, E central JAPAN, 19 mi/30 km S of CHIBA; 35°17′N 140°23′E. Wetlands on nearby Taito coast.

Misaki, town, W Uwa county, EHIME prefecture, westernmost point of SHIKOKU, near the tip of Cape Sada, on Japan's narrowest peninsula, W JAPAN, 50 mi/80 km SW of MATSUYAMA; 33°23′N 132°07′E. Oranges; lobsters. Sea urchin canning. Seto-Naikai National Park nearby.

Misaki, town, Sennan county, OSAKA prefecture, S HONSHU, W central JAPAN, 21 mi/34 km S of OSAKA; 34°18′N 135°08′E.

Misakubo (mee-SAHK-bo), town, Iwata county, SHIZUOKA prefecture, central HONSHU, E central JAPAN, on TENRYU RIVER, 31 mi/50 km N of SHIZUOKA; 35°09′N 137°52′E. Cryptomeria.

Misamis Occidental (mee-SAH-mees ok-see-den-TAHL), province (□ 749 sq mi/1,947.4 sq km), in NORTHERN MINDANAO region, W MINDANAO, PHILIPPINES, bounded N by MINDANAO SEA, E by ILIGAN BAY; ⊙ OROQUIETA; 08°20′N 123°42′E. Population 32% urban, 68% rural. Mountainous terrain, rising to 7,965 ft/2,428 m. Has fertile coastal strip producing corn and coconuts. TANGUB and OZAMIS cities are in the province.

Misantla (mee-SAHNT-lah), city and township, VERACRUZ, E MEXICO, in SIERRA MADRE ORIENTAL foothills, 29 mi/47 km N of XALAPA ENRÍQUEZ; 19°56′N 96°51′W. Agricultural center (corn, sugarcane, coffee, tobacco). Ancient ruins nearby.

Misasa (mee-SAH-sah), town, Tohaku county, TOTTORI prefecture, S HONSHU, W JAPAN, near Mount Mitoku, 21 mi/34 km S of TOTTORI; 35°24′N 133°51′E. Health resort (radioactive hot springs); agriculture (fruits, pepper); uranium ore. Ancient temple and Ningyo mountain pass nearby.

Misato (mee-SAH-to), city, SAITAMA prefecture, E central HONSHU, E central JAPAN, 12 mi/20 km E of URAWA; 35°49′N 139°52′E.

Misato (mee-SAH-to), town, Gumma county, GUMMA prefecture, central HONSHU, N central JAPAN, 6 mi/10 km W of MAEBASHI; 36°23′N 138°57′E. Japanese plum processing, concrete block manufacturing.

Misato, town, Kodama county, SAITAMA prefecture, E central HONSHU, E central JAPAN, 34 mi/55 km N of URAWA; 36°10′N 139°11′E.

Misato, town, Kaiso county, WAKAYAMA prefecture, S HONSHU, W central JAPAN, 12 mi/19 km S of WAKAYAMA; 34°08′N 135°21′E.

Misato (mee-SAH-to), village, Age county, MIE prefecture, S HONSHU, central JAPAN, 6 mi/10 km W of TSU; 34°43′N 136°23′E.

Misato, village, S Azumi county, NAGANO prefecture, central HONSHU, central JAPAN, 31 mi/50 km S of NAGANO; 36°15′N 137°53′E. Apples.

Misato, village, Oe county, TOKUSHIMA prefecture, SE SHIKOKU, W JAPAN, 16 mi/25 km W of TOKUSHIMA; 34°01′N 134°15′E.

Misau (mee-SHOU), town, BAUCHI state, N NIGERIA, 35 mi/56 km SE of AZARE. Agricultural trade center; cotton, peanuts, millet, durra. Also spelled Missau.

Misawa (mee-SAH-wah), city, Aomori prefecture, N HONSHU, N JAPAN, 34 mi/55 km E of AOMORI; 40°40′N 141°22′E. Whitebait. Kosui Festival held here.

Misaz, TURKEY: see KOYULHISAR.

Mischabelhörner (MEE-shah-buhl-HER-nuhr) or **Mischabel**, French les Mischabels, group of peaks in PENNINE ALPS, in canton of VALAIS, S SWITZERLAND, NE of ZERMATT, 46°07′N 07°52′E. Highest peak is the Dom (14,911 ft/4,545 m). Other peaks include Täschhorn (14,734 ft/4,491 m; first ascended 1862), Nadelhorn (14,196 ft/4,327 m), Alphubel (13,799 ft/4,206 m), and Allalinhorn (13,212 ft/4,027 m), whose summit is reachable by funicular.

Miscouche (mis-KOOSH), village (2001 population 766), W PRINCE EDWARD ISLAND, CANADA, 5 mi/8 km WNW of SUMMERSIDE; 46°26′N 63°52′W. Mixed farming, dairying; potatoes.

Miscou Island (MI-skoo), in Gulf of St. Lawrence, NE NEW BRUNSWICK, CANADA, at entrance to CHALEUR BAY just N of Shippagan Island; 9 mi/14 km long, 5 mi/8 km wide; 47°55′N 64°30′W. Lobster, clams, oysters, cod fisheries. Lighthouse in operation since 1856. Beach areas.

Misdroy, POLAND: see MIEDZYZDROJE.

Miseno, Cape (mee-ZE-no), S ITALY, at the NW end of the Bay of NAPLES. Augustus founded (1st century B.C.E.) a naval station (Misenum) here, which was destroyed by the Arabs (9th century C.E.). Remaining are ruins of the imperial villa, baths, a theater, and a reservoir.

Misere (mee-ZER), village on central MAHÉ ISLAND, SEYCHELLES, on road, and 3.25 mi/5.3 km SSE of VICTORIA, in central range. Cinnamon, coconuts, other agriculture.

Misericórdia, Brazil: see ITAPORANGA.

Misery Island, MASSACHUSETTS: see GREAT MISERY ISLAND.

Misery, Mount (3,711 ft/1,131 m), NW St. Kitts, ST. KITTS AND NEVIS, LEEWARD ISLANDS, 7 mi/11.3 km NW of BASSETERRE.

Misery Point, N extremity of BELLE ISLE, NEWFOUNDLAND AND LABRADOR, CANADA, 40 mi/64 km NE of CAPE NORMAN; 52°01′N 55°17′W. Lighthouse.

Misgar (mis-GUHR), village, GILGIT district, NORTHERN AREAS, extreme NE PAKISTAN, in NW KASHMIR region, in NW KARAKORAM mountain system, on right tributary of HUNZA RIVER, and 65 mi/105 km NNE of GILGIT; 36°47′N 74°47′E. Elevation c.10,150 ft/3,094 m. On important trade route from GILGIT, via KILIK PASS, into CHINA. Formerly part of HUNZA state, GILGIT agency.

Mishal (mish-AL), village, S YEMEN, 22 mi/35 km NE of SHUQRA, and on road to LODAR; 13°39′N 45°47′E. Agricultural.

Mishan (MEE-SHAN), city (□ 2,982 sq mi/7,723 sq km; 1994 estimated urban population 146,600; estimated total population 427,000), E HEILONGJIANG province, CHINA, on the MULING RIVER, near N shore of the Xingkai Lake (KHANKA Lake), 140 mi/225 km NE of MUDANJIANG; 45°33′N 131°57′E. Agriculture and light industry are the main sources of income for the city. Grain, tobacco, sugar beets, oilseeds, milk, hogs, aquatic products; manufacturing (food, paper, chemicals).

Misharij, Am, YEMEN: see MASHARIJ, AM.

Mishawaka, city (2000 population 46,557), SAINT JOSEPH county, N INDIANA, on both banks of the SAINT JOSEPH RIVER, and adjacent to SOUTH BEND; 41°40′N 86°10′W. A growing industrial city, Mishawaka's industries are closely associated with those of South Bend. Manufacturing of military vehicles, aerospace parts, chemicals, plastics, transportation equipment, fabricated metal products. National Steel Company headquarters. The city is the seat of Bethel College. Settled c.1830, laid out 1833, incorporated 1899.

Mishelevka (mee-shi-LYEF-kah), town (2005 population 7,465), S IRKUTSK oblast, E central SIBERIA, RUSSIA, on the BELAYA River (tributary of the ANGARA RIVER), on highway, 20 mi/32 km S of CHEREMKHOVO; 52°51′N 103°10′E. Elevation 1,450 ft/441 m. Porcelain industry.

Misheronskiy (mee-shi-RON-skeeyee), town (2006 population 3,780), E MOSCOW oblast, central European Russia, on road and near railroad, 13 mi/21 km NNE of SHATURA; 55°43′N 39°44′E. Elevation 433 ft/131 m. Glassworks.

Mishicot (MISH-i-kaht), town (2006 population 1,401), MANITOWOC county, E WISCONSIN, 8 mi/12.9 km NW of TWO RIVERS; 44°13′N 87°38′W. Point Beach State Forest to SE. Dairying. Cherries, cranberries, grain. Light manufacturing.

Mishima (mee-SHEE-mah), city, SHIZUOKA prefecture, central HONSHU, E central JAPAN, 31 mi/50 km N of SHIZUOKA; 35°06′N 138°55′E. Hot-spring resort and transportation hub. Noted for its Mishima (Shinto) shrine and Rakujuen Park.

Mishima (mee-SHEE-mah), town, Onuma county, FUKUSHIMA prefecture, N central HONSHU, NE JAPAN, 56 mi/90 km S of FUKUSHIMA city; 37°28′N 139°38′E. Paulownia furniture.

Mishima, town, Santo county, NIIGATA prefecture, central HONSHU, N central JAPAN, 31 mi/50 km S of NIIGATA; 37°29′N 138°47′E.

Mi-shima (mee-SHEE-mah), island (□ 3 sq mi/7.8 sq km), YAMAGUCHI prefecture, W JAPAN, in SEA OF JAPAN, 27 mi/43 km NNW of HAGI, off SW HONSHU; 3 mi/4.8 km long, 1.5 mi/2.4 km wide. Hilly, fertile region. Extensive livestock raising; rice, raw silk.

Mishkar (mish-KAHR), oil and gas field, Tunisia, in Gulf of Qabis, c.75 mi/121 km SSE of SAFAQIS in Tunisian territorial waters.

Mishkino (MEESH-kee-nuh), town (2005 population 8,650), central KURGAN oblast, SW SIBERIA, RUSSIA, on road and the TRANS-SIBERIAN RAILROAD, 55 mi/89 km W of KURGAN; 55°20′N 63°55′E. Elevation 492 ft/149 m. Flour milling, dairying, lumbering.

Mishkino (MEESH-kee-nuh), village (2005 population 5,815), N BASHKORTOSTAN Republic, E European Russia, 55 mi/89 km N of UFA; 55°32′N 55°58′E. Elevation 357 ft/108 m. Flour milling, lumbering.

Mishkinshahr, IRAN: see MESHKINSHAHR.

Mishmar Ha'emeq (meesh-MAHR hah-NE-gev), kibbutz, NW ISRAEL, at the N foot of the Samarian Highlands, at W edge of PLAIN OF JEZREEL, 16 mi/26 km SSE of HAIFA; 32°36′N 35°08′E. Elevation 741 ft/225 m. Manufacturing of plastic products; mixed farming. Known for its educational institutions. Served as one of the main Palmah bases prior to 1948. Also Mishmar Haemek. Founded 1926.

Mishmar Hanegev (mish-MAR ha-NEH-gev), kibbutz, S ISRAEL, in N part of the NEGEV, 10 mi/16 km NNW of BEERSHEBA; 31°21′N 34°43′E. Elevation 741 ft/225 m. Light manufacturing; field crops, fodder, dairy products. Founded 1946.

Mishmar Hasharon (meesh-MAHR hah-shah-RON), kibbutz, W ISRAEL, in PLAIN OF SHARON, 4 mi/6.4 km NE of NETANYA; 32°21′N 34°54′E. Elevation 154 ft/46 m. Flower-growing center, including sunflowers; citrus, wheat, corn, cotton; fish ponds, turkey and chicken farms; large bakery serves the area. Founded 1933.

Mishmar Hashiv'a (meesh-MAHR hah-shee-VAH), moshav, central ISRAEL, 4 mi/6.4 km SE of TEL AVIV; 32°00′N 34°49′E. Elevation 114 ft/34 m. Founded 1949. Remains of a Turkish fort nearby.

Mishmar Hayarden (meesh-MAHR hah-yahr-DEN), moshav, Upper Jordan Valley, NE ISRAEL, 8 mi/12.9 km ENE of ZEFAT; 33°01′N 35°35′E. Elevation 305 ft/92 m. The present settlement is situated 2 mi/3.2 km W of the old site, which was totally destroyed by the Syrians in 1948. Mixed farming. Modern village founded 1890 on site of medieval Le Chasteilet, fortified 1178 by Knights Templars, captured 1179 by Saladin. Ruins. During Arab invasion settlement was captured (June 1948) by Syrians; became demilitarized zone under Israeli-Syrian armistice (July 1949). Jacob's Ford, just N, is historically important gateway.

Mishmarot (meesh-mah-ROT) or **Mishmaroth**, kibbutz, W ISRAEL, in PLAIN OF SHARON, 5 mi/8 km NE of HADERA; 32°29′N 34°59′E. Elevation 180 ft/54 m. Manufacturing of construction materials, consumer goods; citriculture, banana growing; mixed farming. Founded 1933.

Mishmi Hills (mee-SHUH-mee) or **Misimi**, hill range (rising to c.15,000 ft/4,572 m), in DIBANG VALLEY and LOHIT districts, ARUNACHAL PRADESH state, NE INDIA, N and E of SADIYA (ASSAM state); drained by DIBANG RIVER. Limestone deposits. Formerly tribal district of Assam's NE frontier tract. Inhabited by four Mishmi tribes.

Misho (MEE-shyo), town, S Uwa county, EHIME prefecture, SW SHIKOKU, W JAPAN, on HOYO STRAIT, 62 mi/100 km S of MATSUYAMA; 32°57′N 132°34′E. Oranges; pearls.

Mishongnovi, Hopi Indian pueblo, NE ARIZONA, in Hopi Indian Reservation c.55 mi/89 km N of WINSLOW. Elevation 6,230 ft/1,899 m. The Snake Dance is held here biennially.

Mishta (MESH-tuh), village, SOHAG province, central Upper EGYPT, on railroad, 7 mi/11.3 km NNW of TAHTA; 26°52′N 31°28′E. Cotton, cereals, dates, sugarcane.

Misida, MALTA: see MSIDA.

Misilmeri (MEE-zeel-MAI-ree), town, PALERMO province, NW SICILY, ITALY, 7 mi/11 km SE of PALERMO; 38°02′N 13°27′E. Cereals, citrus fruits; wine. Site of first botanical garden in Sicily.

Misima (mee-SEE-mah), volcanic island (□ 100 sq mi/260 sq km), main island of LOUISIADE ARCHIPELAGO, MILNE BAY province, PAPUA NEW GUINEA, SW PACIFIC OCEAN, 125 mi/201 km SE of NEW GUINEA; 10°41′S 152°42′E. Most important gold-bearing island of group, new major gold mine in operation 1990s; site of chief town, BWAGAOIA. Gold rush between World War I and World War II. In World War II, battle of CORAL SEA (1942) was fought nearby. Scheduled air service.

Misimi Hills, India: see MISHMI HILLS.

Misión, MEXICO: see LA MISIÓN.

Misiones (mee-see-O-nes), province (□ 11,514 sq mi/29,936.4 sq km; 2001 population 965,522), NE ARGENTINA, in MESOPOTAMIA, between the URUGUAY (SE), PARANÁ (W), and IGUASSÚ (N) rivers; ⊙ POSADAS. Subtropical, densely forested region, with low mountain ranges. Some agricultural (maté, tobacco, citrus fruit, rice, cotton, corn, tung trees, tea, sugarcane); small-scale livestock raising. Lumbering. Processing of maté, tobacco, and tung oil. Tourists are attracted to the great IGUASSÚ FALLS on the Brazilian line. Many of the farmers are from PARAGUAY, BRAZIL, and EUROPE. Jesuit missions founded here in 17th century were later dissolved. Site of several huge hydroelectric dams.

Misiones (mee-see O-nes), department (□ 3,690 sq mi/9,594 sq km; 2002 population 101,783), S PARAGUAY, between TEBICUARY RIVER (N) and Paraná River (S); ⊙ SAN JUAN BAUTISTA; 27°S 57°W. Forested lowlands (marshy), drained by numerous small rivers. Has subtropical, humid climate. Among its little exploited mineral resources are iron and copper ores, talc, ocher. Predominantly a cattle-raising area; also agr. (maté, oranges, rice, corn, sugarcane); timber. Processing at San Juan Bautista, SAN IGNACIO, SANTA ROSA, and Santiago. Many Jesuit missions were founded here during 17th century and early 18th century.

Misiones, Sierra de (mee-see-O-nes, see-ER-rah dai), low mountain range (c. 1,500 ft/457 m) in central MISIONES province, ARGENTINA, extends c. 110 mi/177 km SW from BERNARDO DE IRIGOYEN, on Brazilian border, forming watershed between the ALTO PARANÁ and URUGUAY rivers.

Misir (ME-suhr), village, KAFR ESH SHEIKH province, Lower EGYPT, 5 mi/8 km SE of KAFR ESH SHEIKH. Cotton.

Misis, town, S TURKEY, on CEYHAN RIVER, and 20 mi/32 km E of ADANA; 36°57′N 35°35′E. Ruins of ancient Mopsuestia, a free city of CILICIA under the Romans, are here. Home of the Misis Mosaic Museum. Formerly spelled Missis and Messis.

Misivri, Bulgaria: see NESEBUR.

Miskhor (mees-KOR), suburb of KOREYIZ, S Republic of CRIMEA, UKRAINE, on the BLACK SEA coast, at the foot of AY-PETRI, 8 mi/13 km SW of YALTA; 44°26′N 34°05′E. Elevation 301 ft/91 m. Has fourteen sanatoriums, dealing with respiratory problems, heart and circulatory diseases, and nervous disorders; terminus for cable car to Ay-Petri; site of the Swallows' Nest castle (built in 1912) on cape Ay-Todor, one of the most picturesque constructions in Crimea, formerly a villa of Baron Shteingel, a German industrialist, now containing a first-class Italian restaurant.

Miskin, Wales: see MOUNTAIN ASH.

Miskindzha (mees-keen-JAH), village (2005 population 3,700), S DAGESTAN REPUBLIC, E CAUCASUS, extreme SE European Russia, 8 mi/13 km NW of the RUSSIA-AZERBAIJAN border, on road, 93 mi/150 km S of MAKHACHKALA; 41°25′N 47°50′E. Elevation 3,520 ft/1,072 m. Has a mosque.

Miskito Cays (mees-KEE-to), islands, 25 mi/40 km E of Punta Gorda, NICARAGUA. A collection of small uninhabited limestone islands. Morrison Dennis Key is largest.

Miskolc (MISH-kolts), city (2001 population 184,125), NE HUNGARY, on the SAJÓ RIVER; ⊙ BORSOD-ABAÚJ-ZEMPLÉN county Hungary's third-largest city and a major industrial center, Miskolc has large iron and steel mills, lime and cement works, and a large food-processing plant. Manufacturing (machinery, cement and concrete, glass, food); large power plant. The region's numerous limestone caves are used as cellars by local winemakers. Miskolc also has an important trade in metal products and agricultural goods. As the country's largest heavy industrial center after BUDAPEST, Miskolc found the restructuring of its economic base in the 1990s particularly difficult. Many of its plants were supplying the Soviet market or making components for products destined for that market and couldn't satisfy Western consumers. Unemployment is one of the highest among major cities. The city is the seat of a Protestant bishopric. An old settlement, Miskolc was granted the status of a free city in the 15th century. Frequent invasions (by Mongols in the 13th century, Turks in the 16th and 17th centuries, and Habsburg imperial forces in the 17th and 18th centuries) marked the city's history. Industrialization began in the second half of the 19th century. Present-day landmarks include the Avas Reformed Church (15th century), the remains of a 13th century castle, and a museum containing Scythian art. The city also has a law school and a technical university.

Mislata (mees-LAH-tah), W suburb of VALENCIA, VALENCIA province, E SPAIN, on TURIA RIVER; 39°28′N 00°25′W. Tanning, flour milling, meat processing; light manufacturing. Cereals, vegetables.

Misolonghi, Greece: see MESOLONGI.

Misono (mee-SO-no), village, Watarai county, MIE prefecture, S HONSHU, central JAPAN, 19 mi/30 km S of TSU; 34°30′N 136°42′E.

Misool (MEE-sool), island, RAJA AMPAT ISLANDS, W IRIAN JAYA province, INDONESIA, in SERAM SEA, 40 mi/64 km SW of DOBERAI peninsula (NW NEW GUINEA); 01°55′S 130°00′E. Waigam is the principal settlement. Partly hilly, rises to c.3,250 ft/991 m in central area. Sago growing; trepang fishing. Became Dutch possession 1667.

Misore Islands, archipelago (□ 1,231 sq mi/3,188 sq km), INDONESIA, at entrance of CENDERAWASIH BAY, off NW NEW GUINEA; 00°20′–01°20′S 135°12′–136°50′E. Consists of BIAK, SUPIORI, NUMFOOR, and several small islands. Agriculture, fishing. First visited (1616) by Dutch navigator Schouten. Sometimes called Schouten Islands.

Mispillion River (mis-PIL-yun), 15 mi/24 km long, E DELAWARE; rises in streams W of MILFORD; flows E and NE past Milford (head of navigation) to DELAWARE BAY, 16 mi/26 km NW of CAPE HENLOPEN.

Misquah Hills (MIS-kwah), COOK county, extreme NE MINNESOTA, between CANADA border (N) and LAKE SUPERIOR (SE), in the Arrowhead region; 47°59′N 90°33′W. Small group of monadnock-type hills, rising to 2,230 ft/680 m. Eagle Mountain is highest point in Minnesota.

Misquamicut (mis-KWAW-mi-kut), summer resort village in WESTERLY town, WASHINGTON county, SW RHODE ISLAND, on BLOCK ISLAND SOUND, and 4 mi/6.4 km S of WESTERLY village. Renamed from Pleasant View 1928.

Misquihué (mees-kee-WAI), village, LLANQUIHUE province, LOS LAGOS region, S central CHILE, 28 mi/45 km WSW of PUERTO MONTT; 41°32′S 73°27′W. Agricultural center (wheat, potatoes; livestock; dairying); lumbering.

Misr: see EGYPT.

Misratah (mis-RUHT-uh), city, TRIPOLITANIA region, NW LIBYA, in oasis; 32°23′N 15°06′E. Seaport on MEDITERRANEAN SEA, near NW end of Gulf of SURT; exports fruit, grain; manufacturing (oil refining; pottery, textiles). Airport. Thermal power station. Known to Romans as Tubartis. Port built by Italians (1930s). Was capital of former Misratah province. Also known as Misurata.

Misrikh (MIS-rik), town, SITAPUR district, central UTTAR PRADESH state, N central INDIA, 13 mi/21 km SW of SITAPUR; 27°27′N 80°31′E. Wheat, rice, gram, barley.

Missafou, town, POOL region, S CONGO REPUBLIC, 45 mi/72 km W of BRAZZAVILLE.

Missão Catrimani (MEE-soun KAH-tree-mah-nee), city, W RORAIMA state, BRAZIL, 131 mi/211 km SW of BOA VISTA; 02°00′N 62°45′W.

Missão Velha (mee-SOUN vel-yah), city (2007 population 33,686), S CEARÁ, BRAZIL, on FORTALEZA-CRATO railroad, and 18 mi/29 km E of Crato; 07°15′S 39°13′W. Cotton, sugar; livestock.

Missau, NIGERIA: see MISAU.

Missaukee (mi-SAW-kee), county (□ 573 sq mi/1,489.8 sq km; 2006 population 15,197), N central MICHIGAN; ⊙ LAKE CITY; 44°20′N 85°05′W. Drained by MUSKEGON RIVER and its affluents. Cattle, hogs; corn, oats, forage, dairy products. Manufacturing at Lake City. Resorts. Cluster of about seven small lakes W of Lake City, in W, largest is LAKE MISSAUKEE. A state forest is in county. Organized 1871.

Missaukee, Lake, MICHIGAN: see LAKE CITY.

Misserghin (mee-ser-GEEN), village, ORAN wilaya, NW ALGERIA, at N edge of the ORAN SEBKHA, on railroad, and 8 mi/12.9 km SW of ORAN. Tanning; vegetable gardens, olive groves, and vineyards.

Missilya, large Arab village, WEST BANK, 5.3 mi/8.5 km S of JENIN, in Sanur Valley of the SAMARIAN HIGH-LANDS; 32°23′N 35°17′E. Olives, sesame, and cereals. Archaeological excavations found caves, gravestones, and water cisterns from the Mamluk (Turkish) period.

Missinaibi (mi-si-NAI-bee), river, c.265 mi/430 km long, central ONTARIO, E central CANADA; rising in Missinaibi Lake; flowing N and NE to the MATTAGAMI RIVER, SW of MOOSONEE, to form the MOOSE River; 50°44′N 81°28′W.

Missinipi River, CANADA: see CHURCHILL RIVER.

Mission (MI-shuhn), district municipality (□ 87 sq mi/226.2 sq km; 2001 population 31,272), SW BRITISH COLUMBIA, W CANADA, c.44 mi/70 km E of VANCOUVER, in FRASER VALLEY regional district; 49°14′N 122°20′W. Forestry. First Nations archeological site. Founded in 1892.

Mission, city (2000 population 9,727), JOHNSON county, E KANSAS, a suburb, 4 mi/6.4 km SSW of downtown KANSAS CITY, Kansas; 39°01′N 94°39′W. Manufacturing (printing and publishing, consumer goods, plastic products). Area referred to as Shawnee Mission. Incorporated after 1950.

Mission (MI-shuhn), city (2006 population 63,272), HIDALGO county, extreme S TEXAS, suburb, 5 mi/8 km W of McAllen, near RIO GRANDE (Mexican border); 26°12′N 98°19′W. Elevation 134 ft/41 m. It is a processing and canning center for citrus fruits (especially grapefruit) and vegetables grown in the irrigated lower Rio Grande valley, Anzalduas Dam (irrigation) to S. Manufacturing (consumer goods, concrete). Oil wells are also in Mission, which has been marked by a population growth since 1970. The city was founded on property that had belonged to the Oblate Fathers; their chapel still stands on the Rio Grande. Bentsen–Rio Grande State Park to SW. Incorporated 1910.

Mission, town (2006 population 971), TODD county, S SOUTH DAKOTA, on Antelope Creek, 75 mi/121 km S of PIERRE, and 33 mi/53 km NNW of VALENTINE, Nebraska; 43°18′N 100°39′W. Trading point near center of Rosebud Indian Reservation (comprises all of county); cattle feed, livestock.

Mission (MI-shuhn), unincorporated rural village, SW BRITISH COLUMBIA, W CANADA, on FRASER River, residential satellite community, 35 mi/56 km E of VANCOUVER; 49°08′N 122°18′W. Elevation 100 ft/30 m. Fruit and vegetable canning, fruit-jam making, lumbering, dairying, fishing. Established c.1860 as mission station. Seminary of Christ the King is here.

Mission, ZAMBIA: see CHISEKESI.

Missionary Ridge (c.1,000 ft/305 m), in TENNESSEE and GEORGIA, in GREAT APPALACHIAN VALLEY, E of CHATTANOOGA; 34°59′N 85°16′W. A Civil War battleground (1863) where Union forces won a costly victory; partly included in CHICKAMAUGA AND CHATTANOOGA NATIONAL MILITARY PARK.

Mission Bay, shallow lagoon (c.3 mi/4.8 km long and wide), SAN DIEGO county, S CALIFORNIA, at mouth of SAN DIEGO RIVER, 6 mi/9.7 km NW of downtown SAN DIEGO, within San Diego limits. Islands and small peninsula to S. Mission Bay Park (city park) between bay and city. SeaWorld theme park on S shore of bay.

Mission Beach (MI-shuhn BEECH), resort village, QUEENSLAND, NE AUSTRALIA, along CORAL SEA; 17°56′S 146°06′E. Rainforest, which reaches the shoreline here, is home to many animals, including the large, flightless cassowary and other birds, wallabies, and irridescent butterflies. Tourism; rainforest walks.

Mission Beach, suburban section of SAN DIEGO, SAN DIEGO county, S CALIFORNIA, 7 mi/11.3 km NW of downtown San Diego, on narrow strip of land between PACIFIC OCEAN (W) and MISSION BAY; mouth of bay and SAN DIEGO RIVER to S. Mission Bay Park to E; Sea World theme park to SE. Beach resort.

Mission Beach, California: see SAN DIEGO, city.

Cross-references are shown in SMALL CAPITALS. The pronunciation guide is shown on page xix. The sources of population figures are shown on page xvii.

Mission Hill, village (2006 population 172), YANKTON county, SE SOUTH DAKOTA, 7 mi/11.3 km ENE of YANKTON; 42°55′N 97°16′W. In farming region.

Mission Hills, city (2000 population 3,593), JOHNSON county, E KANSAS, a suburb, 5 mi/8 km S of KANSAS CITY, Kansas; 39°00′N 94°37′W. Borders MISSOURI on E.

Mission Hills, suburban section of LOS ANGELES, LOS ANGELES county, S CALIFORNIA, 18 mi/29 km NW of downtown Los Angeles, and 2 mi/3.2 km SW of San Fernando, in SAN FERNANDO VALLEY; 34°42′N 120°27′W. Area suffered heavy damage in Northridge Earthquake (January 17, 1994). Northridge area to W. Van Nuys Airport and Busch Gardens theme park to S. San Fernando Mission to N.

Mission Range, in ROCKY MOUNTAINS of NW MONTANA, Lake and MISSOULA counties, rises between FLATHEAD LAKE and SWAN RIVER, extends c.45 mi/72 km S toward Missoula. Numerous small lakes throughout center and S part of range; mountain divide forms E boundary of Flathead Indian Reservation. Mission Mountains Wilderness Area and Mission Mountains Tribal Wilderness Area at center of range. Highest point, McDonald Peak (9,820 ft/2,993 m).

Mission River (MI-shuhn), c.25 mi/40 km long, in REFUGIO county, S TEXAS; coastal stream formed by Blanco (c.50 mi/80 km) and Medio (c.65 mi/105 km) creeks; both rise in KARNES county, to NW, NW of REFUGIO; flow SE past Refugio, through oil fields, to Copano Bay.

Mission San Jose, unincorporated village, ALAMEDA county, W CALIFORNIA; suburb 14 mi/23 km N of SAN JOSE, 3 mi/4.8 km SE of FREMONT. Site of San Jose Mission (Mission San Jose de Guadalupe), built 1797.

Mission Viejo, city (2000 population 93,102), ORANGE county, S CALIFORNIA; residential suburb 43 mi/69 km SE of downtown LOS ANGELES, and 8 mi/12.9 km NE of LAGUNA BEACH and PACIFIC OCEAN; 33°37′N 117°39′W. Irrigated agriculture (citrus, nursery products, flowers; dairying; poultry). Diverse light manufacturing. Historic Mission Viejo to SE; Mission San Juan Capistrano, site of annual return of swallows March 19, to S. Santa Ana Mountains and Cleveland National Forest to NE. Seat of Saddleback College (two-year).

Mission Woods, village (2000 population 165), JOHNSON county, E KANSAS, a suburb, 4 mi/6.4 km S of downtown KANSAS CITY, Kansas; 39°01′N 94°36′W. Borders MISSOURI on E. Flour milling.

Missiquoi River (MIS-i-koi), c.100 mi/161 km long, in N VERMONT and S QUEBEC; rises near LOWELL, Vermont; flows N, past TROY, into Quebec, thence W and SW, reentering Vermont near Richford, thence generally W, through GREEN MOUNTAINS, to Lake CHAMPLAIN N of SWANTON.

Missir, EGYPT: see MISIR.

Missis, TURKEY: see MISIS.

Missisquoi (mi-SI-skwoi), former county (□ 375 sq mi/975 sq km), S QUEBEC, E CANADA, on U.S. (VERMONT) border; county seat was BEDFORD; 45°10′N 73°00′W.

Mississagi River (mi-si-SAH-gee), flows in a wide arc 170 mi/274 km W, S, and SE, in central ONTARIO, E central CANADA; issues from Mississagi Lake (6 mi/10 km long) at 47°09′N 82°32′W; flows through several small lakes, to Lake HURON 3 mi/5 km W of BLIND RIVER. Several rapids: Aubrey Falls (108 ft/33 m high), 50 mi/80 km NNW of Blind Riverr; GRAND FALLS (150 ft/46 m), 30 mi/48 km NW of Blind River; and Lake Falls (55 ft/17 m), 20 mi/32 km WNW of Blind River.

Mississauga (mi-si-SAW-guh), city (□ 111 sq mi/288.6 sq km; 2001 population 612,925), PEEL region, S ONTARIO, E central CANADA, 12 mi/20 km W of TORONTO, on Lake ONTARIO; 43°09′N 79°30′W. A residential suburb of Toronto and a growing transportation and industrial center, it is one of Canada's fastest-growing cities. The biomedical, financial services, and information and communications technologies sectors help sustain the local economy; manufacturing includes aircraft, motor vehicles, engines, chemicals, petroleum, steel and rubber products, cement, appliances, and printing and publishing. It has a port and is the site of LESTER B. PEARSON INTERNATIONAL AIRPORT. Originally an agricultural and then residential area, it had a population of 15,000 in 1945. Named after the Aboriginal people who lived in the area; the word describes the river mouth.

Mississinewa Lake, reservoir (c.12 mi/19 km long), MIAMI and WABASH counties, N central INDIANA, on MISSISSINEWA RIVER, 5 mi/8 km SE of PERU; 40°42′N 85°56′W. Maximum capacity of 368,400 acre-ft. Formed by Mississinewa Dam (122 ft/37 m high), built (1967) by the Army Corps of Engineers for flood control. State recreation areas along W shore.

Mississinewa River (MI-si-SI-nuh-wah), c.100 mi/161 km long, in OHIO and INDIANA; rises in DARKE county, W Ohio; flows W into Indiana, then generally NW, past MARION, to the WABASH RIVER at Peru; 40°45′N 86°01′W.

Mississippi, state (□ 48,434 sq mi/125,444 sq km; 2000 population 2,844,658; 1995 estimated population 2,697,243), S UNITED STATES, admitted as the twentieth state of the Union in 1817; ⊙ and largest city JACKSON; 32°45′N 89°31′W. Other important cities are BILOXI, GREENVILLE, HATTIESBURG, and MERIDIAN; also part of suburban MEMPHIS, TENNESSEE, extends into De Soto county in the NW. Mississippi is known as the "Magnolia State" because of the abundance of magnolia flowers and trees in the state. The magnolia is the official state flower and the official state tree.

Geography

Bounded W by the MISSISSIPPI RIVER (ARKANSAS and part of LOUISIANA state lines, actual state boundary follows the river channel that existed when the boundary was established; the present-day channel has left stranded numerous oxbow, or horseshoe, lakes on both sides of the river and with them pieces of opposite states); bounded S by Louisiana; bounded E by ALABAMA (including NE corner of TENNESSEE RIVER, which forms PICKWICK LAKE reservoir); bounded N by Tennessee; S extension of the Mississippi, in SE bounded by Gulf of MEXICO (MISSISSIPPI SOUND) and W by Louisiana (Pearl River). The generally hilly land reaches its highest point at Woodall Mountain (806 ft/246 m), in Tishomingo county, in the NE corner. The most distinctive region in the state's varied topography is the Delta, a flat alluvial plain between The Mississippi and the YAZOO rivers; the Yazoo follows the former channel of the Mississippi. A wide belt of longleaf yellow pine (the piny woods) covers most of S Mississippi to within a few mi/km of the coastal-plain grasslands. Important there are lumbering and allied industries. Most of the state's rivers belong to either the Mississippi or the Alabama river system, with the Pontotoc Ridge the divide and include the Pearl, PASCAGOULA, TALLAHATCHIE, and YALOBUSHA rivers. The climate of Mississippi is subtropical along the Gulf Coast to temperate in the northern part; the average annual rainfall is more than 50 in/127 cm.

Economy

Traditionally one of the more rural and economically depressed U.S. states. Since the 1960s, manufacturing has taken over as the leading industrial sector; many companies from the industrial NE states have relocated here to take advantage of cheaper labor. In 1990, though declining, Mississippi ranked third (after TEXAS and CALIFORNIA) in the nation in the production of cotton. Soil erosion, resulting from overcultivation of the crop, and the destruction caused by the boll weevil have led to the increased adoption of scientific farming techniques and to agricultural diversification. The most important crops are soybeans, rice, and hay. There has been a great rise in cattle, poultry, and hog raising and, especially, dairying. The state's most important and valuable mineral resources, petroleum and natural gas, have been developed only since the 1930s; especially important is the Tinsley Oil Field S of YAZOO CITY, in the W center. Sand and gravel and clays are also produced. Industry has grown rapidly since oil development began and has been helped by the Tennessee Valley Authority and the state's program to balance agriculture with industry. Under this program many communities have subsidized and attracted new industries; industrial products, including wood products, foods, and chemicals, have exceeded in value those of agriculture in recent years. The state has become a major producer of furniture, especially upholstered furniture, and also of apparel. On the Gulf coast there is a profitable fishing and seafood-processing industry, including shrimp and crabs. Catfish farming is a major industry in the Delta region in the W. Mississippi ranks among the country's highest gambling revenue states. Despite modernization efforts, however, the state's per capita income is the lowest in the nation. The TENNESSEE-TOMBIGBEE WATERWAY in the NE, extended from theTennessee River in the N, SSE along the TOMBIGBEE River into Alabama, allows and alternate route to the Mississippi River for barge traffic to the Gulf.

National Forests and Wildlife Refuges

National forests include two sections of HOLLY SPRINGS National Forest in the N, two sections of Tombigbee National Forest in the NE center, Bienville National Forest in the E center, two sections of De Soto National Forest in the SE, Delta National Forest in the W, and Homochitto National Forest in the SW. Also there are several national wildlife refuges, especially Mississippi Sandhill Crane National Wildlife Refuge in the SE; Panther Swamp National Wildlife Refuge in W, and Noxubee National Wildlife Refuge in NE center. The Choctaw Indian Reservation is in Neshoba county, and several smaller reservations are in the E center.

History: to 1795

Hernando De Soto's expedition undoubtedly passed (1540–1542) through the region, then inhabited by the Choctaw, Chickasaw, and Natchez, but the first permanent European settlement was not made until 1699, when Pierre le Moyne, sieur d'Iberville, established a French colony on BILOXI BAY. Settlement accelerated in 1718, when the colony came under the French Mississippi Company, headed by the speculator John Law. The region was part of Louisiana until 1763, when, by the Treaty of PARIS England received practically all the French territory E of the Mississippi River and also E FLORIDA and W Florida, which had belonged to Spain. English colonists, many of them retired soldiers, had made the NATCHEZ district a thriving agricultural community, producing tobacco and indigo, by the time Bernardo de Gálvez captured it for Spain in 1779. By the Treaty of Paris of 1783 at the end of the American Revolution, the U.S., with English approval, claimed as its S boundary in the W latitude 31°N (most of the present-day state of Mississippi was included in the area). Spain denied this claim, and the long, involved W Florida Controversy ensued.

History: 1795 to 1840

In the Pinckney Treaty (1795), Spain accepted latitude 31°N as the N boundary of its territory but did not evacuate Natchez until the arrival of U.S. troops in 1798. Congress immediately created the Mississippi Territory, with Natchez as the capital and William C. C. Claiborne as the governor. After Georgia's cession (1802) of its Western lands to the U.S. and the Louisiana Purchase (1803), a land boom swept Mississippi. The high price of cotton and the cheap, fertile land brought settlers thronging in, most of them via

---

Area is shown by the symbol □, and capital city or county seat by ⊙.

the NATCHEZ TRACE, from the Southern PIEDMONT region and even from NEW ENGLAND. A few attained great wealth, but most simply managed a living. In 1817, Mississippi became a state, with substantially its present-day boundaries; the E section of Mississippi was organized as Alabama. The aristocratic planter element of the Natchez region initially dominated Mississippi's government, as the state's first constitution (1817) showed. With the spread of Jacksonian democracy, however, the small farmer came into his own, and the new constitution adopted in 1832 was quite liberal for its time. Land hunger increased as more new settlers arrived, lured by the continuing cotton boom. By a series of treaties (1820, 1830, 1832), the Native Americans in the state were pushed westward across the Mississippi. Mississippians were among the leading Southern expansionists seeking new land for cotton and slavery.

History: 1840 to 1869

After 1840 slaves in the state outnumbered Whites. On January 9, 1861, Mississippi became the second state to secede from the Union. State pride was highly gratified by the choice of Jefferson Davis as president of the Confederacy. Civil War fighting did not reach Mississippi until April 1862, when Union forces were victorious at Corinth and IUKA. Grant's brilliant VICKSBURG campaign ended large-scale fighting in the state, but further destruction was caused by General W. T. Sherman in his march from Vicksburg to MERIDIAN. Moreover, cavalry of both the North and the South, particularly the Confederate forces of General N. B. Forrest, remained active. After the war, Mississippi abolished slavery but refused to ratify the Thirteenth and Fourteenth Amendments, and in March 1867, under the Congressional plan of Reconstruction, it was organized with Arkansas into a military district commanded by General E.O.C. Ord.

History: 1869 to 1904

After much agitation, a Republican-sponsored constitution guaranteeing basic rights to blacks was adopted in 1869. Mississippi was readmitted to the Union early in 1870 after ratifying the Fourteenth and Fifteenth Amendments and meeting other Congressional requirements. While Republicans were in power, the state government was composed of new immigrants from the North, blacks, and cooperative white Southerners. A. K. Davis became the state's first black lieutenant governor in 1874. The establishment of free public schools was a noteworthy aspect of Republican rule. As former Confederates were permitted to return to politics and blacks were increasingly intimidated, the Democrats regained strength. The Republicans were defeated in the bitter election of 1875. Lucius Q. C. Lamar figured largely in the Democratic triumph and was the state's most prominent national figure for many years. In Reconstruction days the Republicans could win only with solid black support. After Reconstruction, blacks were virtually disenfranchised. White supremacy was bolstered by the Constitution of 1890, later used as a model by other Southern states; under its terms a prospective voter could be required to read and interpret any of the constitution's provisions. Because at the turn of the century most Mississippi blacks could not read (neither could many whites, but the test was rarely applied to them) and because the county registrar could disqualify a prospective voter if that person disagreed with the interpretation of the constitution, blacks were legally disenfranchised. On the ruins of the shattered plantation economy rose the sharecropping system, and the merchant and the banker replaced the planter in having the largest financial interest in farming. Too often the system made the sharecroppers, white as well as black, little more than economic slaves.

History: 1904 to 1948

The landowners, however, maintained their hold on politics until 1904, when the small farmers, still the dominant voting group, elected James K. Vardaman governor. Nevertheless this agrarian revolt did not alter a deep-seated obscurantism that was reflected in the Jim Crow laws (1904) and in the ban on the teaching of evolution in the public schools (1926). Another reflection of the social structure of the state was prohibition, put into effect in 1908 and not repealed until 1959. Since the disastrous flood of 1927 the Federal government has taken over flood-control work—constructing levees, floodwalls, floodways, and reservoirs; stabilizing riverbanks; and improving channels. Navigation, too, has not been neglected; the INTRACOASTAL WATERWAY provides a protected channel along the entire Mississippi coastline and links the state's ports with all others along the Gulf coast and with all inland waterway systems emptying into the Gulf of Mexico. The Tennessee-Tombigbee Waterway, opened in 1985, connects the Tennessee River in NE Mississippi with the Tombigbee River in W Alabama. The state has made attempts to wipe out illiteracy, but it still has the highest illiteracy rate in the country. Mississippi is still plagued by racial problems, which have changed the state's alignment in national politics.

History: 1948 to 1963

In 1948, Mississippi abandoned the Democratic party because of the national Democratic party's stand on civil rights, and the state supported J. Strom Thurmond, the States' Rights party candidate, for President. The 1954 Supreme Court ruling against racial segregation in public schools occasioned massive resistance. Citizens Councils, composed of White men and dedicated to maintaining segregation, began to spring up throughout the state. In the 1960 presidential election Mississippians again rebelled against the Democratic national platform by giving victory at the polls to unpledged electors, who cast their electoral college votes not for John F. Kennedy but for Harry F. Byrd, the conservative senator from Virginia. In 1964 the conservative Republican Barry Goldwater carried the state; in 1968 Govermnor George Wallace of Alabama, who had become famous for opposing integration, won the state. In 1961 mass arrests and violence were touched off when "freedom riders," actively seeking to spur integration, made Mississippi a major target. However, there was not even token integration of public schools in Mississippi until 1962, when the state government under the leadership of Governor Ross River Barnett tried unsuccessfully to block the admission of James H. Meredith, a black, to the University of Mississippi at OXFORD. In the conflict the Federal and state governments clashed, and the U.S. Department of Justice took legal action against state officials, including Barnett; two persons were killed in riots, and Federal troops had to restore order. Racial antagonisms resulted in many more acts of violence.

History: 1963 to Present

Churches and black homes were bombed. Medgar Evers, an official of the National Association for the Advancement of Colored People, was killed in 1963; three civil rights workers (two White, one Black) were murdered the next year. After the Federal Voting Rights Act of 1965, many blacks succeeded in registering and voting. In 1967, for the first time since 1890, a Black was elected to the legislature, and blacks are now as well represented in Mississippi politics as any state, with a large degree of cross-racial voting. However, in 1992 the U.S. Supreme Court ordered the state college system to end its tradition of segregation. Mississippi's economic problems continued in the 1980s as the state was unable to shift emphasis from manufacturing to the service sector and was unable to avoid the national trend of industrial decline. In August 1969, Missississpi and Louisiana were devastated by Camille, one of the 20th-century's worst hurricanes. In April 1973, the Mississippi River rose to record levels in the state. The floodwaters from the

river and its tributaries covered about 9% of the state, including parts of Vicksburg and Natchez, causing millions of dollars of damage to property. S Mississippi (particularly the Gulf Coast region, Biloxi, and Gulfport) suffered widespread catastrophic damage from Hurricane Katrina in August 2005 with tens of thousands left homeless and hundreds killed.

Places of Interest

U.S. government installations include Meridian Naval Air Station in E, KEESLER AIR FORCE BASE and the Gulfport Naval Center in the SE, and NASA's Stennis Space Center, on the Pearl River, in the SE. Among the institutions of higher learning are the University of Mississippi at Oxford, Mississippi State University, near STARKVILLE, the University of Southern Mississippi, at Hattiesburg, Jackson State University, at Jackson, and Mississippi University for Women, at COLUMBUS. Historical sites in the state include Old Spanish Fort, the oldest house on the Mississippi River, near PASCAGOULA, and Vicksburg National Military Park, BRICES CROSS ROADS National Battlefield Site, and TUPELO National Battlefield Site. Mississippi, in the path of waterfowl migrations down the Mississippi valley, is noted for its duck and quail hunting. Along the Gulf Coast, a favorite fishing area, are several resort cities and part of Gulf Islands National Seashore. Long, narrow sandy islands lie generally 10 mi/16 km offshore in Gulf of Mexico, including SHIP, CAT, HORN, and PETIT BOIS islands, and Dog Keys; all except Ship Island are in the Gulf Islands National Seashore (Mississippi and Florida). The Intracoastal Waterway runs parallel to the coast in Mississippi Sound. In Natchez and Biloxi are many fine antebellum mansions. Natchez Trace Parkway (a unit of the National Park system) traverses the state from NE corner to Natchez in the SW and follows an old Native American road and former colonial trade route.

Government

Mississippi is governed under the 1890 constitution. The bicameral legislature consists of fifty-two senators and 122 representatives, all elected for four-year terms. The governor is also elected for a four-year term. The current governor is Haley Barbour. The state has two U.S. senators, five representatives, and seven electoral votes.

Mississippi has eighty-two counties: ADAMS, ALCORN, AMITE, ATTALA, BENTON, BOLIVAR, CALHOUN, CARROLL, CHICKASAW, CHOCTAW, CLAIBORNE, CLARKE, CLAY, COAHOMA, COPIAH, COVINGTON, DE SOTO, FORREST, FRANKLIN, GEORGE, GREENE, GRENADA, HANCOCK, HARRISON, HINDS, HOLMES, HUMPHREYS, ISSAQUENA, ITAWAMBA, JACKSON, JASPER, JEFFERSON, JEFFERSON DAVIS, JONES, KEMPER, LAFAYETTE, LAMAR, LAUDERDALE, LAWRENCE, LEAKE, LEE, LEFLORE, LINCOLN, LOWNDES, MADISON, MARION, MARSHALLM MONROE, MONTGOMERY, NESHOBA, NEWTON, NOXUBEE, OKTIBBEHA, PANOLA, PEARL RIVER, PERRY, PIKE, PONTOTOC, PRENTISS, QUITMAN, RANKIN, SCOTT, SHARKEY, SIMPSON, SMITH, STONE, SUNFLOWER, TALLAHATCHIE, TATE, TIPPAH, TISHOMINGO, TUNICA, UNION, WALTHALL, WARREN, WASHINGTON, WAYNE, WEBSTER, WILKINSON, WINSTON, YALOBUSHA, and YAZOO.

Mississippi, county (□ 919 sq mi/2,389.4 sq km; 2006 population 47,517), NE ARKANSAS; ⊙ BLYTHEVILLE and OSCEOLA; 35°45′N 90°02′W. Bounded N by MISSOURI line, E by MISSISSIPPI RIVER; drained by East Hand and West Hand Chutes of LITTLE RIVER. Agriculture (cattle, hogs; cotton, soybeans, sorghum, wheat, rice); timber. Industries at Blytheville and Osceola. Hampson Museum State Park at WILSON in E; BIG LAKE NATIONAL WILDLIFE REFUGE and Wildlife Management Area in N. Formed 1833.

Mississippi, county (□ 411 sq mi/1,068.6 sq km; 2006 population 13,770), extreme SE MISSOURI; on MISSISSIPPI RIVER (levees), with drainage channels; ⊙ CHARLESTON; 36°49′N 89°17′W. Agricultural region (corn, cotton, soybeans, wheat, potatoes, popcorn, melons; livestock); cotton processing; lumber; manufacturing at Charleston and EAST PRAIRIE. Big Oak Tree State Park in S; Towasahgy Archaeological Site at Dorena. Formed 1845.

Mississippi Mills (mi-si-SI-pee MILZ), town (□ 197 sq mi/512.2 sq km; 2001 population 11,647), SE ONTARIO, E central CANADA, 25 mi/40 km from PERTH; 45°15′N 76°17′W. Formed in 1998 from PAKENHAM, RAMSAY, and ALMONTE.

Mississippi Palisades State Park, ILLINOIS: see SAVANNA.

Mississippi River (mi-si-SI-pee), c.100 mi/160 km long, S ONTARIO, E central CANADA; rising E of the KAWARTHA LAKES; flowing NE through Mississippi Lake, then N to the OTTAWA River near ARNPRIOR; 45°26′N 76°17′W.

Mississippi River, principal river of the U.S. c.2,350 mi/3,780 km long, exceeded in length only by the MISSOURI River, chief of its numerous tributaries. The combined Missouri-Mississippi system (from the Missouri's headwaters in the ROCKY MOUNTAINS to the mouth of the Mississippi River) is c.3,740 mi/6,020 km long and ranks as the world's third-longest river system after the NILE and the AMAZON. With its tributaries, the Mississippi drains c.1,231,000 sq mi/3,188,290 sq km of the central U.S., including all or part of thirty-one states and c.13,000 sq mi/33,670 sq km of ALBERTA and SASKATCHEWAN in CANADA. The Mississippi River rises in small streams that feed Lake ITASCA (altitude 1,463 ft/446 m) in N MINNESOTA; flows generally S to enter the Gulf of MEXICO through a huge delta in SE LOUISIANA. A major economic waterway, the river is navigable from the sediment-free channel maintained through South Pass in the delta to the Falls of SAINT ANTHONY in MINNEAPOLIS, with canals circumventing the rapids near ROCK ISLAND, ILLINOIS, and KEOKUK, IOWA. For the low-water months of July, August, and September, there is a 45-ft/13.7-m channel navigable by oceangoing vessels from Head of the Passes to BATON ROUGE, LOUISIANA, and a 9-ft/2.7-m channel from Baton Rouge deep enough for barges and towboats to Minneapolis. The Mississippi connects with the INTRACOASTAL WATERWAY in the S and with the GREAT LAKES-St. Lawrence Seaway system in the N by way of the ILLINOIS WATERWAY. Along the river's upper course shipping is interrupted by ice from December to March; thick, hazardous fogs frequently settle on the cold waters of the unfrozen sections during warm spells from December to May. In its upper course the river is controlled by numerous dams and falls (c.700 ft/210 m) in the 513-mi/826-km stretch from Lake Itasca to Minneapolis and then falls (c.490 ft/150 m) in 856 mi/1,378 km from Minneapolis to CAIRO, Illinois The Mississippi River receives the Missouri River 17 mi/27 km N of ST. LOUIS and expands to a width of c.3,500 ft/1,070 m; it swells to c.4,500 ft/1,370 m at Cairo, where it receives the OHIO River. The lower Mississippi meanders in great loops across a broad alluvial plain (25 mi/40 km–125 mi/201 km wide) that stretches from CAPE GIRARDEAU, MISSOURI, to the delta region S of NATCHEZ, MISSISSIPPI. The plain is marked with oxbow lakes and marshes that are remnants of the river's former channels. Natural levees, built up from sediment carried and deposited in times of flood, border the river for much of its length; sediment has also been deposited on the river bed, so that in places the surface of the Mississippi is above that of the surrounding plain, as evidenced by the SAINT FRANCIS, BLACK, YAZOO, and TENSAS river basins. Breaks in the levees frequently flood the fertile bottomlands of these and other low-lying areas of the plain. After receiving the ARKANSAS and RED rivers, the Mississippi enters a birdsfoot-type delta, which was built outward by sediment carried by the main stream since C.E. c.1500. It then discharges into the

Gulf of Mexico through a number of distributaries, the most important being the ATCHAFALAYA RIVER and Bayou LAFOURCHE. The main stream continues SE through the Delta to enter the Gulf through several mouths, including Southeast Pass, South Pass, and Pass à Loutre. Indications that the Mississippi River might abandon this course and divert through the Atchafalaya River have led to the construction of a dam by the U.S. Army Corps of Engineers, known as the OLD RIVER CONTROL STRUCTURE, to prevent such an occurrence. Regarding the Delta, environmentalists and those in the seafood industry are concerned with the fact that it loses 25 sq mi/65 sq km–45 sq mi/117 sq km of marsh a year. The loss has been attributed to subsidence and a decrease in sediment largely due to dams, artificial channeling, and land conservation measures. Pollution and the cutting of new waterways for petroleum exploration and drilling have also taken their toll on the Delta. Louisiana has enacted environmental protection laws that are expected to slow, but not halt, the loss of the Delta marshes. Sluggish bayous and freshwater lakes dot the Delta region. The flow of the river is greatest in the spring, when heavy rainfall and melting snow on the tributaries (especially the Missouri and the Ohio) cause the main stream to rise and frequently overflow its banks and levees, inundating vast areas of the plain. Since the disastrous flood of 1927 the U.S. Congress has authorized the construction of dams on the upper Mississippi and its tributaries to regulate the flow; the building of c.1,600 mi/2,580 km of levees below Cape Girardeau to contain the swollen river; and the establishment of floodways to divert water at critical points, such as the Cairo–NEW MADRID, Atchafalaya, and MORGANZA floodways and the BONNET CARRE SPILLWAY at NEW ORLEANS, which diverts water into Lake PONTCHARTRAIN. Cutoffs have eliminated the dangerous winding channels, and an improved main channel has increased the river's flood-carrying capacity. Nonetheless, serious, record-breaking floods again occurred in the rainy spring of 1973, when the river crested at St. Louis at 43.3 ft/13.2 m, remained at flood stage for 77 days, and drove about 50,000 persons from their homes and again in 1993. In 1988 a severe drought brought water levels down to their lowest point in recorded history and halted most river traffic. The Spanish explorer Hernando De Soto is credited with the European discovery of the Mississippi River in 1541. The French explorers Jacques Marquette and Louis Joliet reached it through the WISCONSIN RIVER in 1673, and in 1682, La Salle traveled down the river to the Gulf of Mexico and claimed the entire territory for FRANCE. The French founded New Orleans in 1718 and effectively extended control over the upper river basin with settlements at CAHOKIA, KASKASKIA, PRAIRIE DU CHIEN, and St. Louis. France ceded the river to SPAIN in 1763 but regained it in 1800; the U.S. acquired the Mississippi River as part of the Louisiana Purchase in 1803. A major artery for Native Americans and the fur-trading French, the river became in the nineteenth century the principal outlet for the newly settled areas of mid-America; exports were floated downstream with the current, and imports were poled or dragged upstream on rafts and keelboats. The first steamboat plied the river in 1811, and successors became increasingly luxurious as river trade increased in profitability and importance. Traffic from the north ceased after the outbreak of the Civil War. During the Civil War the Mississippi was an invasion route for Union armies and the scene of many important battles. Especially decisive were the capture of New Orleans (1862) by Admiral David Farragut, the Union naval commander, and the victory of Union forces under General Grant at VICKSBURG in 1863. River traffic resumed after the end of the war; it is colorfully described in Mark Twain's *Life on the Mississippi* (1883). However, much of the trade was

lost to the railroad in the mid-1800s. With modern improvements in the river channels, traffic has increased, especially since the mid-1950s, with principal freight items being petroleum products, chemicals, sand, gravel, and limestone. Cotton and rice are important crops in the lower Mississippi valley; sugarcane is raised in the Delta. The Mississippi is abundant in freshwater fish; shrimp are taken from the briny Delta waters. The Delta also yields sulfur, oil, and gas. A 220-acre/89-ha model of the Mississippi River basin is located at CLINTON, Mississippi, which has been used by the U.S. Corps of Engineers to simulate various conditions in the basin.

Mississippi River Gulf Outlet, canal, ORLEANS and SAINT BERNARD parishes, SE LOUISIANA, SE extension of GULF INTRACOASTAL WATERWAY; 29°50′N 89°39′W. Begins in N, 8 mi/13 km ENE of downtown NEW ORLEANS, flows SE c.30 mi/48 km into shallow Breton Sound. Roughly parallels SW end of LAKE BORGNE, coming within 1 mi/2 km of its shore. Provides direct access between offshore oil and natural-gas fields and processing facilities in New Orleans and BATON ROUGE. It also reserves the MISSISSIPPI RIVER natural channel for deepwater shipping from Port of New Orleans.

Mississippi Sound, arm (c.100 mi/161 km long and 7 mi/11 km–15 mi/24 km wide) of the GULF OF MEXICO, extending from LAKE BORGNE in LOUISIANA, on the W, to MOBILE BAY in ALABAMA, on the E; 30°18′N 88°55′W. It is part of the INTRACOASTAL WATERWAY and is separated from the Gulf by a series of narrow islands and sandbars. Main islands are Cat, Ship, Horn, PETIT BOIS (Mississippi), and DAUPHIN (Alabama). GULFPORT, BILOXI, and PASCAGOULA (all in Mississippi), are on the Mississippi Sound. Ship, Horn, and Petit Bois, along with a small mainland unit, comprise Gulf Islands National Seashore.

Mississippi State, village, OKTIBBEHA county, E MISSISSIPPI; suburb 1 mi/1.6 km SE of STARKVILLE. Seat of Mississippi State University. Formerly called State College.

Misslitz, CZECH REPUBLIC: see MIROSLAV.

Missolonghi, Greece: see MESOLONGI.

Misson Hills, unincorporated town (2000 population 3,142), SANTA BARBARA county, SW CALIFORNIA, 3 mi/4.8 km NNE of LOMPOC, near SANTA YNEZ RIVER, in Purisima Hills. Fruit, avocados, grain, cattle. La Purísima Mission State Historic Park to SE.

Missoula (muh-SUL-uh), county (□ 2,618 sq mi/6,781 sq km; 1990 population 78,687; 2000 population 95,802), W MONTANA, borders IDAHO in SW; ⊙ MISSOULA; 47°02′N 113°56′W. Irrigated agricultural region (wheat, barley, hay; cattle; dairying; poultry); timber; mining. Drained by the CLARK FORK, BITTERROOT, SWAN, CLEARWATER, and BLACKFOOT rivers. Frenchtown Pond and Council Group state parks in W; Placid Lake and Salmon Lake state parks in NE; part of Flathead National Forest in far N; part of MISSION Mountains Wilderness and Flathead Indian Reservation in N; small part of Bitterroot National Forest and SELWAY-BITTERROOT Wilderness Area on S boundary; parts of Lolo National Forest throughout county, especially W and N. Formed 1865.

Missoula (muh-SUL-uh), city (2000 population 57,053), W MONTANA, 89 mi/143 km W of HELENA, on the CLARK FORK river, 5 mi/8 km E of mouth of BITTERROOT RIVER; ⊙ MISSOULA county; 46°52′N 114°01′W. Elevation 3,210 ft/978 m. Third-largest city in Montana. In the midst of well-watered valleys, large forests, and dairy and cattle area. Missoula is a commercial center with a busy lumber and paper industry; manufacturing (printing and publishing, food, chemicals, construction materials, furniture, lumber, fabricated metal products). The "Salish Council" of 1855 opened the area to white settlement. Hell Gate town was founded nearby in 1860 and moved to the Missoula site six years later. The coming of the railroad (1883, 1908) stimulated Missoula's growth. In the city are the Missoula campus of the University of Montana; Historical Museum at Fort Missoula to SW, Missoula Museum of the Arts; and a regional headquarters of the U.S. Forest Service. Smokejumper Visitor Center located at Missoula County International Airport NW of city. Rocky Mountain Elk Foundation Visitor Center. Montana Snowbowl Ski Area to N; parts of Lolo National Forest to N, W, and SE. Incorporated 1889.

Missour (MEE-soor), village, Boulemane province, Fes-Boulemane administrative region, E central MOROCCO, in upper MOULOUYA river valley, between the Middle ATLAS (N) and the High Atlas (S) mountains, 50 mi/80 km NE of MIDELT, on road; 33°03′N 03°59′W. Sheep, esparto.

Missouri, state (□ 69,709 sq mi/180,546 sq km; 2000 population 5,595,211; 1995 estimated population 5,323,523), central UNITED STATES, admitted as the 24th state of the Union in 1821; ⊙ JEFFERSON CITY; 38°13′N 92°25′W. Largest cities are KANSAS CITY, SAINT LOUIS, SPRINGFIELD, and INDEPENDENCE. Missouri has been called the "Show Me State," which conotates a certain self-deprecating stubbornness and devotion to simple common sense.

Geography

Missouri is bounded on the N by IOWA; on the W by NEBRASKA, KANSAS, and OKLAHOMA; on the S by ARKANSAS; and on the E, where the MISSISSIPPI RIVER forms the border, by ILLINOIS, KENTUCKY, and TENNESSEE. The state lies N of latitude 36°30′N except for a small area (called the "bootheel") in the extreme SE that protrudes into Arkansas. The center of population of the U.S., for both 1980 and 1990, was located in the N OZARKS of Missouri. Two great rivers, the Mississippi and the MISSOURI, have had a great influence on the development of Missouri. Missouri is a diverse state, topographically, culturally, and economically, often considered a microcosm of the nation, possessing characteristics of the U.S.'s four quadrants. The Mississippi River tied the region to the South, particularly to NEW ORLEANS. The Missouri crosses the state from W to E and enters the Mississippi near Saint Louis. The portion of the Missouri Valley between Saint Louis and Kansas City was the greatest avenue of pioneer advance westward across the continent. The region N of the Missouri River is largely prairie land, where, not unlike the Iowa plains to the N, corn and livestock are raised. Most of the region S of the Missouri is covered by foothills and by the dissected plateau of the Ozark highlands, a unique region of hill scenery originally populated by a relatively isolated, self-reliant people. The rough, heavily forested eastern section of the Ozarks extends into the less hilly farming region in the W and encompasses the irregular, twisting Lake of the OZARKS to the NW. In SW Missouri is a long, narrow area of flat land, the Osage Plains, which are part of the GREAT PLAINS, where livestock and forage crops are raised. In the SE, S of CAPE GIRARDEAU, are the cotton, rice, and soybean fields of the Mississippi flood plain, an area that was once swampy but was converted to agriculture after the establishment of a drainage system in the early part of the 20th century. The state's rivers have periodically flooded. Record flood levels were attained at many places in floods of 1993. Missouri's share of the flood of 1993 was first brought about by heavy rains in IOWA and MINNESOTA in spring, which swelled the Mississippi and Missouri rivers downstream in Missouri, then exacerbated by locally heavy rains, which also caused flash flooding in smaller streams. Ironically SE Missouri was being affected by the severe drought that gripped the SE U.S. that same year.

Economy

Missouri has extensive bituminous coal deposits in the W and N central sections. Fire, or refractory, clays occur in central and NE Missouri, and barite in E

central Missouri. The Ozarks is a great metalliferous region, including lead, zinc, silver, manganese, copper, and iron. Much of the state is underlain by limestone and dolomite. Lead, cement, and stone are the chief minerals. Missouri is, by far, the leading state in the production of lead. Missouri's economy, however, rests chiefly on industry. The manufacture of aerospace and transportation (cars, vans, trucks, railroad) vehicles and equipment is the major industry in the state; food and chemicals are next in commercial importance, followed by printing and publishing. Machinery, fabricated metal products, and electrical equipment are also produced. Saint Louis is an important center for the manufacture of planes, cars, metals, and chemicals. In Kansas City, long a leading market and agricultural business center for livestock and wheat, the manufacture of vending machines and of cars and trucks are leading industries. Missouri remains important agriculturally, and farming contributes substantially to the state's income. The most valuable farm products are cattle, hogs, soybeans, corn, and dairy items. After soybeans, the chief crops are corn, hay, and wheat. Missouri also has important wine producing areas, especially along Missouri River between Saint Louis and Jefferson City. The development of resorts and recreation facilities in the Ozarks has encouraged tourism and retirement communities and added to the state's income. Services and wholesale and retail trade closely follow manufacturing in economic importance.

History: to 1800

Missouri's recorded history begins in the latter half of the 17th century when the French explorers Jacques Marquette and Louis Joliet descended the Mississippi River, followed by Robert Cavelier, sieur de La Salle, who claimed the whole area drained by the Mississippi River for France and called the territory Louisiana. When the French explorers came the area was inhabited by Native Americans of the Osage and the Missouri groups and by the end of the 17th century, French trade with the Native Americans flourished. In the early 18th century the French worked the area's lead mines and made numerous trips through Missouri in search of furs. Trade down the Mississippi prompted the settlement of SAINTE GENEVIÈVE before 1750 and the founding of Saint Louis in 1764 by Pierre Laclede and René Auguste Chouteau, who were both in the fur-trading business. Although not involved in the last conflict (1754–1763) of the French and Indian Wars, Missouri was affected by the French defeat when in 1762, France secretly ceded the territory W of the Mississippi to Spain.

History: 1800 to 1821

In 1800 the Louisiana Territory (including the Missouri area) was retroceded to France, but in 1803 it passed to the U.S. as part of the Louisiana Purchase. French influence remained dominant, even though by this time Americans had filtered into the territory. At the time of the Lewis and Clark expedition (1803–1806), Saint Louis was already known as the gateway to the Far West. The U.S. Territory of Missouri was set up in 1812, and settlement proceeded rapidly after the War of 1812. The coming of the steamboat increased traffic and trade on the Mississippi. Planters from KENTUCKY and VIRGINIA brought slaves into the territory in which the French had been using Black slaves since the 1720s. The question of admitting the Missouri Territory as a state became a burning national issue because it involved the question of extending slavery into the territories west of the Mississippi River. The dispute was resolved by the Missouri Compromise, which admitted (1821) Missouri to the Union as a slave state but excluded slavery from other lands of the Louisiana Purchase N of latitude 36°30′. Slaveholding interests became politically powerful in the new state.

History: 1821 to 1854

In 1822, W. H. Ashley (who later made a fortune in fur trading) led an expedition of the adventurous trappers, who became known as mountain men, up the Missouri River to explore the West for furs. From Missouri traders established a thriving commerce over the SANTA FE TRAIL with the inhabitants of NEW MEXICO, and pioneers followed the OREGON TRAIL to settle the NW. FRANKLIN, Westport, Independence, and SAINT JOSEPH became famous as the points of origin of these expeditions. Settlement of Missouri itself quickened, spreading in the 1820s over the river valleys into central Missouri and by the 1830s into W Missouri. The final boundaries of the state were formed after Native Americans gave up their claim to the Platte country in 1836; this strip of land in the NW corner of Missouri was added to the state. Mormon immigrants came to settle Missouri in the 1830s, but their opposition to slavery, their friendliness with Native Americans, and their growing numbers made them unwelcome, and they were driven by force from the state in 1839. German immigrants, however, were cordially received during the 1840s and 1850s, settling principally in the counties in the Saint Louis area.

History: 1854 to 1904

In 1854 the problem of slavery was made acute with the passage of the Kansas-Nebraska Act, leaving the question of slavery in the Kansas and Nebraska territories to the settlers themselves. The proslavery forces in Missouri became very active in trying to win Kansas for the slave cause and contributed to the violence and disorder that tore the territory apart in the years just prior to the Civil War. Nevertheless Missouri also had leaders opposed to slavery, including one of its senators, Thomas Hart Benton. During the Civil War most Missourians remained loyal to the Federal government. A state convention, which met in March 1861, voted against secession, and in 1862, the convention set up a provisional government because the pro-Southern governor had set up a separate state government in SW Missouri. Guerrilla activities persisted during this period, and the lawlessness bred by civil warfare persisted in Missouri after the war in the activities of outlaws such as Jesse James. A new Missouri rose out of the war. The semi-Southern atmosphere, along with the river life and steamboating, began to decline, but the flavor of the period was preserved in the works of one of Missouri's most celebrated sons, Mark Twain (Samuel Lake Clemens). The coming of the railroad brought the eventual decay of many of Missouri's river towns and tied the state more closely to the East and North.

History: 1904 to Present

Urbanization and industrialization progressed, and the Louisiana Purchase Exposition, held at Saint Louis in 1904, dramatically revealed Missouri's economic growth. Saint Louis was the nation's fourth-largest city. Although during World War I general prosperity prevailed in the state, the Depression years of the 1930s sent farm values crashing down, and many banks, especially in rural areas, failed. Prosperity returned during World War II, when both Saint Louis and Kansas City served as vital midcontinental transportation centers. After the war Missouri's industrialization increased enormously. During this period, Missouri became the second-largest (behind MICHIGAN) producer of automobiles in the nation. Although most industry remains centered around the major urban centers of Kansas City and Saint Louis, the smaller cities and towns have had success in attracting light and heavy industry. The central cities of the two metropolitan areas have experienced dramatic outmigration, often to nearby suburbs. The population of Saint Louis declined 53% from 1950 to 1990. Since the brief period of radical Republican rule from 1864 to 1870, Missouri has been permanently wedded to neither major party. While tending toward the Republicans in the days of Theodore Roosevelt, it turned solidly Democratic for Franklin D. Roosevelt and helped to elect Missourian Harry S. Truman to the presidency in 1948. Political machines in the large cities have attracted national attention, notably the machine of Thomas J. Pendergast (1872–1945) in Kansas City. Missouri has contributed to the U.S. such outstanding statesmen as Champ Clark, James Reed, and W. Stuart Symington. Thomas Hart Benton, a descendant of the Missouri senator of the same name, was one of the country's important artists. Places of cultural and historic interest in Missouri include the JEFFERSON NATIONAL EXPANSION MEMORIAL, a national historic site, in Saint Louis; GEORGE WASHINGTON CARVER National Monument, in DIAMOND; WILSON'S CREEK NATIONAL BATTLEFIELD, near Springfield; the William Rockhill Nelson Gallery of Art, in Kansas City; the Harry S. Truman Memorial Library, in Independence; the Museum of the American Indian, in Saint Joseph and the State Capitol in Jefferson City. Missouri's schools were desegregated following the Supreme Court decision in 1954. Institutes of higher learning include the University of Missouri, with four campuses, including the main one at COLUMBIA; Saint Louis University, Washington University, and Webster College, at Saint Louis; Rockhurst College, at Kansas City; and Westminster College (where Winston Churchill made his famous "Iron Curtain" speech), at Fulton.

Government

In 1945, Missouri adopted a new state constitution that remains in effect. As an independent city, Saint Louis is prohibited by the Missouri Constitution to annex into adjacent Saint Louis county, thereby unable to balance its inner city decline with suburban growth. The governor of the state is elected for a term of four years. The current governor is Matt Blunt. The general assembly, or legislature, has a senate with thirty-four members elected for four years and a house of representatives with 163 members elected for two years. The state also elects nine representatives and two senators to the U.S. Congress and has eleven electoral votes in presidential elections.

Missouri has one independent city and 114 counties: ADAIR, ANDREW, ATCHISON, AUDRAIN, BARRY, BARTON, BATES, BENTON, BOLLINGER, BOONE, BUCHANAN, BUTLER, CALDWELL, CALLAWAY, CAMDEN, CAPE GIRARDEAU, CARROLL, CARTER, CASS, CEDAR, CHARITON, CHRISTIAN, CLARK, CLAY, CLINTON, COLE, COOPER, CRAWFORD, DADE, DALLAS, DAVIESS, DE KALB, DENT, DOUGLAS, DUNKLIN, FRANKLIN, GASCONADE, GENTRY, GREENE, GRUNDY, HARRISON, HENRY, HICKORY, HOLT, HOWARD, HOWELL, IRON, JACKSON, JASPER, JEFFERSON, JOHNSON, KNOX, LACLEDE, LAFAYETTE, LAWRENCE, LEWIS, LINCOLN, LINN, LIVINGSTON, MCDONALD, MACON, MADISON, MARIES, MARION, MERCER, MILLER, MISSISSIPPI, MONITEAU, MONROE, MONTGOMERY, MORGAN, NEW MADRID, NEWTON, NODAWAY, OREGON, OSAGE, OZARK, PEMISCOT, PERRY, PETTIS, PHELPS, PIKE, PLATTE, POLK, PULASKI, PUTNAM, RALLS, RANDOLPH, RAY, REYNOLDS, RIPLEY, SAINT CHARLES, SAINT CLAIR, SAINT FRANCOIS, SAINT LOUIS, SAINT LOUIS (independent city), SAINTE GENEVIEVE, SALINE, SCHUYLER, SCOTLAND, SCOTT, SHANNON, SHELBY, STODDARD, STONE, SULLIVAN, TANEY, TEXAS, VERNON, WARREN, WASHINGTON, WAYNE, WEBSTER, WORTH, and WRIGHT.

Missouri, river, c.2,565 mi/4,130 km long (including its Jefferson-Beaverhead-Red Rock headstream), the longest river of the U.S. and the principal tributary of the MISSISSIPPI RIVER. The length of the combined Missouri-Mississippi system from the headwaters of the Missouri to the mouth of the Mississippi is c.3,740 mi/6,020 km, making it the world's third-longest river after the NILE River and the AMAZON River. The Missouri River drains an area of c.580,000 sq mi/1,502,200 sq km, including 2,550 sq mi/6,600 sq km in

CANADA. The principal headwaters of the Missouri are the JEFFERSON, MADISON, and GALLATIN rivers, which rise high in the ROCKY MOUNTAINS, SW MONTANA, and join to form the Missouri near THREE FORKS, Montana. The Missouri's upper course flows N through scenic mt. terrain including Gate of the Mountains, a deep gorge. At GREAT FALLS, Montana, the river enters a 10-mi/16-km stretch of cataracts that prevented navigation to the upper river and effectively established FORT BENTON, Montana, as the head of navigation for 19th-cent. riverboats. Below Fort Benton, the Missouri follows a meandering course E and then SE across the GREAT PLAINS of W-central U.S., crossing Montana, NORTH DAKOTA, and SOUTH DAKOTA and forming part of the boundaries of NEBRASKA, KANSAS, and IOWA before crossing MISSOURI and entering the Mississippi River 17 mi/27 km N of SAINT LOUIS. Nicknamed "Big Muddy" for its heavy load of silt, the brown waters of the Missouri do not readily mix with the gray waters of the Mississippi until c.100 mi/160 km downstream. The YELLOWSTONE, PLATTE, KANSAS, and OSAGE rivers are the Missouri's chief tributaries. Above SIOUX CITY, Iowa, the Missouri's fluctuating flow is regulated by seven major dams (Gavins Point, FORT RANDALL, Big Bend, Oahe, GARRISON, Fort Peck, and CANYON FERRY) and more than eighty other dams on tributary streams. These dams, with their reservoirs, are part of the coordinated, basinwide Missouri River basin project (authorized by the U.S. Congress in 1944), which provides for flood control, navigation, hydroelectric power, irrigation water, and recreational facilities. The dams serve to impound for later use the spring rains and snow melt that swell the volume of the river in March and April and also the second flood stage that frequently occurs in June as the snow melts in the more remote mountain regions. Because the dams have no locks, Sioux City is the head of navigation for the 9-ft/2.7-m channel maintained over the 760-mi/1,223-km stretch downstream to the Mississippi. Tugboats pushing strings of barges move freight along this route. From December to March, navigation is interrupted by ice and low water levels (resulting from upstream freezing); summer water levels, which frequently fall so low as to cause riverboats to go aground, are now maintained at safe levels by the release of water from Gavins Point Dam. Silt, fertilizers, and pesticides, which are contained in the runoff from agricultural lands, and urban areas pollute the river at selected times of the year. The Missouri River was an important artery of commerce for Native American villages of the Plains culture long before the French explorers Jacques Marquette and Louis Joliet passed the mouth of the river in 1683 and the Canadian explorer Vérendrye visited the upper reaches of the river in 1738. David Thompson, a Canadian fur trader, explored part of the river in 1797. Meriwether Lewis and William Clark followed the Missouri on their journey (1803–1806) to the Pacific Ocean and described it at length. The first steamboat ascended the river in 1819 and hundreds more later navigated the uncertain waters to Fort Benton. Mormons bound for UTAH and pioneers bound for OREGON and CALIFORNIA followed the Missouri valley and that of the Platte overland to the West. River traffic declined with the loss of freight to the railroad after the Civil War, but it has been revitalized in the 20th century, in the section below Sioux City, through the navigational improvements and flood control efforts of the Missouri River basin project. The Missouri River is the water supply for several million persons. Occasional high floods cause considerable damage. The Great Flood of 1993 on the river below Omaha, which set record crests and record discharges, and another flood in 1995, have prompted reevaluation of river management, goals, and strategies.

Missouri City, city (2000 population 295), CLAY county, W MISSOURI, on MISSOURI River, and 20 mi/32 km NE of KANSAS CITY; 39°14′N 94°17′W. Concrete and crushed limestone.

Missouri City (mi-ZUH-ree), city (2006 population 73,679), FORT BEND and HARRIS counties, SE TEXAS, suburb, 14 mi/23 km SW of downtown HOUSTON; 29°34′N 95°32′W. Near BRAZOS RIVER. Drained by Oyster Creek. Agricultural area (rice, cotton, nursery crops, vegetables). Oil and natural gas. Some manufacturing Blue Ridge State Prison Farm to E.

Missouri Mountain (14,067 ft/4,288 m), CHAFFEE county, central COLORADO, in COLLEGIATE range of ROCKY MOUNTAINS, in San Isabel National Forest, E of CONTINENTAL DIVIDE, 15 mi/24 km WNW of BUENA VISTA.

Missouri National Wild and Scenic River, free-flowing portion (59 mi/95 km long) of MISSOURI RIVER with islands, bars, and chutes. From Gavins Point Dam, near YANKTON, SOUTH DAKOTA, downstream to PONCA, NEBRASKA. Authorized 1978. Limited recreational facilities.

Missouri Valley, city (2000 population 2,992), HARRISON county, W IOWA, near BOYER RIVER, 7 mi/11.3 km SW of LOGAN; 41°33′N 95°54′W. Manufacturing (feed, beverages, concrete blocks). Settled 1854, incorporated 1871.

Mistassibi River (mis-tuh-SI-bee), 200 mi/322 km long, central QUEBEC, E CANADA; rises E of Lake MISTASSINI; flows S to MISTASSINI river at DOLBEAU-MISTASSINI, near Lake SAINT JOHN; 48°53′N 72°13′W.

Mistassini (mis-tuh-SEE-nee), former village, S central QUEBEC, on MISTASSIBI RIVER, near its mouth on MISTASSINI RIVER; 48°54′N 72°12′W. Agriculture (dairying; pigs); blueberry processing. Amalgamated into DOLBEAU-MISTASSINI in 1997.

Mistassini, Lake (mis-tuh-SEE-nee) (□ 840 sq mi/2,184 sq km), S QUEBEC, E CANADA, NW of Lake SAINT JOHN; 51°00′N 73°37′W. In sparsely settled country, it drains W to JAMES BAY by way of the RUPERT RIVER (380 mi/612 km long).

Mistassini, Réserve de (mis-tuh-SEE-nee, rai-ZERV duh), Cree village, W QUEBEC, E CANADA, SW of Lake MISTASSINI, within BAIE-JAMES municipality; 51°30′N 73°15′W. Site discovered in 1672.

Mistassini River (mis-tuh-SEE-nee), 200 mi/322 km long, central QUEBEC; rises E of Lake MISTASSINI; flows S, past DOLBEAU-MISTASSINI, where it receives MISTASSIBI RIVER, to Lake SAINT JOHN; 48°42′N 72°19′W. On upper course are numerous rapids.

Mistek, CZECH REPUBLIC: see FRÝDEK-MISTEK.

Mistelbach (MIS-el-bahkh) or **Mistelbach an der Zaya**, town, NE LOWER AUSTRIA, in the WEINVIERTEL, 27 mi/43 km NNE of VIENNA; market center for corn and wine region in the Weinviertel; 48°34′N 16°34′E. Asparn an der Zaya, with former moated castle, a place of prehistoric findings, 4 mi/6.4 km WNW.

Misterbianco (MEE-ster-BYAHNG-ko), town, CATANIA province, E SICILY, ITALY, 4 mi/6 km W of CATANIA; 37°31′N 15°00′E. In grape- and orange-growing region. Manufacturing (nonmetallics, fabricated metals, machinery, wine, soap); food processing. Largely destroyed by eruption of Mount ETNA in 1669.

Misterton (MIS-tuh-tuhn), village (2001 population 2,318), N NOTTINGHAMSHIRE, central ENGLAND, 5 mi/8 km NW of GAINSBOROUGH; 53°27′N 00°51′W. Chemical and tileworks. Has 13th-century church, rebuilt in 19th century.

Misti, El, PERU: see EL MISTI.

Mistky (meest-KI) (Russian *Mostki*), village, NW LUHANS'K oblast, UKRAINE, on the Borova River, and road, 18 mi/29 km WNW of STAROBIL'S'K; 49°20′N 38°29′E. Elevation 311 ft/94 m. Wheat, sunflowers, sugar beets.

Mistley (MIST-lee), village (2001 population 2,474), NE ESSEX, SE ENGLAND, on STOUR RIVER estuary, and 9 mi/14.5 km ENE of COLCHESTER; 51°56′N 01°05′E.

Area is shown by the symbol □, and capital city or county seat by ⊙.

Mist Mountain (MIST) (10,303 ft/3,140 m), SW ALBERTA, W CANADA, near BRITISH COLUMBIA border, in Misty Range of ROCKY MOUNTAINS, 50 mi/80 km SW of CALGARY; 50°33′N 114°55′W.

Mistras (mees-TRAHS), medieval fortress in LAKONIA prefecture, S PELOPONNESE department, extreme SE mainland GREECE, on Mistra Hill, a spur of TAIYETOS Mountains, 3 mi/4.8 km W of SPARTA; 37°04′N 22°22′E. Founded 1248–1249 by French crusaders. Site of extensive Greco-Byzantine ruins (palaces, tower, monasteries). Modern village of Mistras is just ESE. Also Misitra.

Mistrató (mees-trah-TO), town, ⊙ Mistrató municipio, RISARALDA department, W central COLOMBIA, 34 mi/55 km NNW of PEREIRA; 05°18′N 75°53′W. Coffee, sugarcane; livestock.

Mistretta (mees-TRET-tah), ancient *Amestratus*, town, Messina province, N SICILY, ITALY, in NEBRODI Mountains, 13 mi/21 km N of NICOSIA; 37°56′N 14°22′E. In livestock-raising region; olive oil, wine. Has government mule-breeding station. Baroque churches. Annual festival of the Madonna held here in September.

Misugi (mee-SOO-gee), village, Ichishi county, MIE prefecture, S HONSHU, central JAPAN, 16 mi/25 km S of TSU; 34°33′N 136°15′E.

Misumba (mee-SOOM-buh), village, KASAI-OCCIDENTAL province, SW CONGO, 80 mi/129 km NNE of LUEBO; 04°34′S 20°29′E. Elev. 1,604 ft/488 m.

Misumi (MEE-soo-mee), town, Uto county, KUMAMOTO prefecture, W KYUSHU, SW JAPAN, port on SHIMABARA BAY, 19 mi/30 km S of KUMAMOTO, on W tip of small peninsula, opposite OYANO-SHIMA; 32°36′N 130°28′E. Commercial center in agricultural area; mandarin oranges.

Misumi (mee-SOO-mee), town, Naka county, SHIMANE prefecture, SW HONSHU, W JAPAN, 78 mi/126 km S of MATSUE; 34°46′N 131°58′E. Traditional paper making.

Misumi (MEES-mee), town, Otsu county, YAMAGUCHI prefecture, SW HONSHU, W JAPAN, 19 mi/30 km N of YAMAGUCHI; 34°21′N 131°15′E.

Misungwi, TANZANIA: see MANTARE.

Misurata, LIBYA: see MISRATAH.

Mit Abu Ghalib (MEET ah-BOO GAH-leb), village, KAFR ESH SHEIKH province, Lower EGYPT, on DUMYAT branch of the NILE, 11 mi/18 km NE of SHIRBIN; 31°17′N 31°40′E. Cotton.

Mitad del Mundo (mee-TAHD del MOON-do), monument marking the Equatorial Line, 14 mi/23 km N of QUITO, ECUADOR, surrounded by a park and leisure area, including an ethnographic museum. The Pululagua crater (extinct) and the Inca ruins of Rumicucho are nearby.

Mitagawa (mee-tah-GAH-wah), town, Kanzaki county, SAGA prefecture, N KYUSHU, SW JAPAN, 6 mi/10 km N of SAGA; 33°19′N 130°24′E. Yoshinogari Iseki (historical site) is here.

Mitaka (mee-TAH-kah), city (2005 population 177,016), Tokyo prefecture, E central HONSHU, E central JAPAN, immediately W of TOKYO city, and 22 mi/35 km W of SHINJUKU; 35°40′N 139°33′E. Communication equipment.

Mitake (mee-TAH-ke), town, Kani county, GIFU prefecture, central HONSHU, central JAPAN, 25 mi/40 km E of GIFU; 35°25′N 137°08′E.

Mitake (mee-TAH-ke), village, Kiso county, NAGANO prefecture, central HONSHU, central JAPAN, 65 mi/105 km S of NAGANO; 35°50′N 137°37′E.

Mitama (mee-TAH-mah), town, W Yatsushiro county, YAMANASHI prefecture, central HONSHU, central JAPAN, 6 mi/10 km S of KOFU; 35°30′N 138°35′E.

Mitan (mee-TAHN), village, S SAMARKAND wiloyat, UZBEKISTAN, on the Ak Darya River (S arm of ZERAVSHAN RIVER), and 33 mi/53 km NW of SAMARKAND; 40°00′N 66°35′E. Cotton; metalworks.

Mitanni (mi-TAN-ee), ancient kingdom est. in the 2nd millennium B.C.E. in NW MESOPOTAMIA. It was founded by Aryans but was later made up predominantly of Hurrians. Washshukanni was its capital. Mitanni controlled ASSYRIA for a period and was engaged in military efforts to hold back Egyptian forces intent on conquering Syria. In c.1450 B.C.E. the army of Thutmose III of EGYPT successfully advanced as far as the EUPHRATES RIVER; the king of Mitanni surrendered, sending tribute to Egypt, which halted its invasion. Friendly relations later developed between the two powers as evidenced by correspondence between King Tushratta of Mitanni and Amenhotep III of Egypt. In the 14th century B.C.E., Mitanni became involved in struggles with the Hittites and c.1335 fell to the Hittites as well as to resurgent Assyrian forces.

Mita Point (MEE-tah), cape on the PACIFIC OCEAN, at NE entrance of BANDERAS BAY, NAYARIT, W MEXICO, 65 mi/105 km SW of TEPIC; 20°46′N 105°33′W. Lighthouse.

Mitare (mee-TAH-rai), town, FALCÓN state, NW VENEZUELA, in Gulf of VENEZUELA lowlands, 25 mi/40 km WSW of CORO; 11°21′N 70°01′W. Saltworks.

Mitau, LATVIA: see JELGAVA.

Mit Badr Halawa (MEET BA-duhr hah-LAH-wuh), village, GHARBIYA province, Lower EGYPT, 8 mi/12.9 km S of SAMANNUD; 30°51′N 31°14′E. Cotton.

Mit Bashshar (MEET bahsh-SHAHR), village, SHARQIYA province, Lower EGYPT, 7 mi/11.3 km SW of ZAGAZIG; 30°31′N 31°24′E. Cotton.

Mit Bera (MEET BEE-ruh), village, MINUFIYA province, Lower EGYPT, on railroad, just N of BENHA. Cotton ginning; cereals; cotton, flax. Also spelled Mit Bira.

Mitcham (MICH-uhm), town, residential suburb S of ADELAIDE, SE SOUTH AUSTRALIA state, S central AUSTRALIA; 34°59′S 138°36′E. In metropolitan area; wineries. Orchards to S.

Mitcham (MICH-uhm), E central suburb of MELBOURNE, S VICTORIA, SE AUSTRALIA; 37°49′S 145°12′E. In fruit-growing area.

Mitcham, ENGLAND: see MERTON.

Mitchel Air Force Base, former U.S. installation at Mitchel Field, NASSAU county, SE NEW YORK, just E of GARDEN CITY; 40°43′N 73°36′W. Established 1918. Air Defense Command headquarters here until 1951. Adjacent to ROOSEVELT FIELD civilian air facility, Mitchel Field was largest military airfield on the East Coast before it closed in 1961. Now part of UNIONDALE, New York. Cradle of Aviation Museum nearby.

Mitchell (MI-chuhl), unincorporated town (□ 2 sq mi/ 5.2 sq km; 2001 population 4,022), S ONTARIO, E central CANADA, on THAMES River, 12 mi/19 km NW of STRATFORD, and included in township of WEST PERTH; 43°28′N 81°01′W. Food processing.

Mitchell, county (□ 514 sq mi/1,336.4 sq km; 2006 population 23,852), SW GEORGIA; ⊙ CAMILLA; 31°13′N 84°11′W. Bounded NW by FLINT RIVER. Coastal plain agriculture (corn, pecans, soybeans, peanuts; cotton, tobacco; cattle, hogs, poultry); in sawmilling area. Manufacturing at CAMILLA and PELHAM. Formed 1857.

Mitchell, county (□ 469 sq mi/1,219.4 sq km; 2006 population 10,856), N IOWA, on MINNESOTA line; ⊙ OSAGE; 43°21′N 92°47′W. Prairie agricultural region (dairying; cattle, hogs; corn, hay) drained by WAPSI-PINICON, CEDAR, and LITTLE CEDAR rivers. Has many limestone quarries, sand, clay and gravel pits. Pioneer State Park in SE. General flooding in 1993. Formed 1851.

Mitchell, county (□ 718 sq mi/1,866.8 sq km; 2006 population 6,299), N KANSAS; ⊙ BELOIT; 39°23′N 98°12′W. Smoky Hills region, drained by SOLOMON RIVER. Agriculture (cattle, sheep, hogs; wheat, soybeans). Manufacturing (farm machinery). Waconda Lake Reservoir and Glen Elder State Park (at dam) in W. Formed 1870.

Mitchell (MI-chuhl), county (□ 222 sq mi/577.2 sq km; 2006 population 15,681), W NORTH CAROLINA; ⊙ BA-KERSVILLE; 36°00′N 82°09′W. Bounded N by TENNESSEE state line, W by NOLICHUCKY RIVER, SW by Toe River; UNAKA MOUNTAINS in N, the BLUE RIDGE MOUNTAINS in S. Largely in Pisgah National Forest in N half and SE corner. Manufacturing; service industries; agricultural area (tobacco, apples, hay; cattle). Mining (mica, feldspar, quartz, kaolin). Resort area. APPALACHIAN TRAIL (Appalachian National Scenic Trail) follows state line in N. BLUE RIDGE PARKWAY follows SE county line. Formed 1861 from Yancey, Watauga, Caldwell, Burke, and McDowell counties. Named for Elisha Mitchell (1793–1857), professor at the University of North Carolina.

Mitchell (MI-chuhl), county (□ 915 sq mi/2,379 sq km; 2006 population 9,327), W TEXAS; ⊙ COLORADO CITY; 32°17′N 100°55′W. Elevation 1,900 ft/579 m–2,600 ft/ 792 m. Rolling prairies, drained by COLORADO RIVER and BEALS CREEK. Ranching (cattle, sheep), agricultural region (cotton, grains). Oil and natural gas wells. Chamion Creek Reservoir and Lake Colorado City (State Park) at center of county. Formed 1876.

Mitchell, city (2000 population 4,567), LAWRENCE county, S INDIANA, 9 mi/14.5 km S of BEDFORD; 38°44′N 86°29′W. Agriculture (fruit, grain). Manufacturing (cement, transportation equipment, machinery, crushed stone, lime); limestone quarrying. Spring Mill State Park (recreation), with restored pioneer village, is nearby to E. Karst topography. Settled 1813, laid out 1853.

Mitchell, city (2006 population 14,857), ⊙ DAVISON county, SE central SOUTH DAKOTA, c.70 mi/110 km WNW of SIOUX FALLS; 43°43′N 98°01′W. Trade, distribution, and shipping center for dairy and livestock area. Manufacturing (printing, trailers, transportation equipment, machinery, computer equipment); food processing. LAKE MITCHELL reservoir to N; railroad junction. Dakota Wesleyan University; Vocational Technical Institute; Corn Palace. Opera house and music hall. Friends of the Middle Border Museum; Oscar Howe Art Center. Incorporated 1881.

Mitchell (MI-chuhl), town, S central QUEENSLAND, NE AUSTRALIA, 110 mi/177 km E of CHARLEVILLE, and on Maranoa River; 26°30′S 147°56′E. Sheep-raising center; wheat. Mineral springs spa.

Mitchell, town (2000 population 155), MITCHELL county, N IOWA, 4 mi/6.4 km NW of OSAGE; 43°19′N 92°52′W. Livestock; grain.

Mitchell, town (2006 population 1,781), SCOTTS BLUFF county, W NEBRASKA, 8 mi/12.9 km NW of SCOTTS-BLUFF, and on NORTH PLATTE RIVER; 41°56′N 103°48′W. In irrigated agricultural region; beet sugar, honey, dairy products, potatoes.

Mitchell, village (2000 population 173), GLASCOCK county, E GEORGIA, c.45 mi/72 km WSW of AUGUSTA, near OGEECHEE RIVER; 33°13′N 82°42′W. Lumber.

Mitchell, village (2006 population 149), WHEELER county, N central OREGON, 30 mi/48 km S of Fossil on Bridge Creek; 44°34′N 120°09′W. Ochoco National Forest and Bridge Creek Wilderness Area to S. JOHN DAY FOSSIL BEDS NATIONAL MONUMENT (Painted Hills Unit) to NW.

Mitchell and Alice Rivers National Parks (MI-chuhl, A-lis) (□ 143 sq mi/371.8 sq km), N QUEENSLAND, NE AUSTRALIA, 420 mi/676 km NW of CAIRNS on W side of CAPE YORK PENINSULA; 15°30′S 142°05′E. 30 mi/56 km inland from Gulf of CARPENTARIA, at confluence of MITCHELL and Alice rivers. Nearly inaccessible wilderness area. Mixture of open forest and grassland; rainforest along rivers. No facilities. Established 1977.

Mitchell Heights, village (2006 population 280), LOGAN county, SW WEST VIRGINIA, 5 mi/8 km N of LOGAN, on GUYANDOTTE RIVER. Residential. Chief Logan State Park to W.

Mitchell Island, TUVALU: see NUKULAELAE.

Mitchell Lake (MI-chuhl), E BRITISH COLUMBIA, W CANADA, in CARIBOO MOUNTAINS, 75 mi/121 km E of

QUESNEL; 10 mi/16 km long, 2 mi/3 km wide; 52°53′N 120°36′W. Elevation 3,170 ft/966 m. Drains SW into QUESNEL LAKE.

Mitchell Lake (□ 9 sq mi/23.3 sq km), on border of Chilton and Coosa counties, central Alabama, in Coosa River, 30 mi/48 km NNW of Montgomery; c.14 mi/23 km long; 32°48′N 86°26′W. Extends NNW to Lay Dam; has 4-mi/6.4-km E arm. Formed by privately built (1923) Mitchell Dam (106 ft/32 m high, 1,264 ft/385 m long). Both the lake and dam were named for James Mitchell, a well-known promoter of water power.

Mitchell, Lake, WEXFORD county, NW MICHIGAN, 3 mi/4.8 km W of CADILLAC; c.3 mi/4.8 km long, 3 mi/4.8 km wide; 44°15′N 85°29′W. In resort area. Joined to LAKE CADILLAC (E) by short stream.

Mitchell, Lake, reservoir (3 mi/4.8 km long, 1 mi/1.6 km wide), DAVISON county, SE central SOUTH DAKOTA, on Firesteel Creek, 2 mi/3.2 km W of MITCHELL; 43°39′N 98°00′W. Formed by dam. Recreational area.

Mitchell, Mount (MI-chuhl), peak (6,684 ft/2,037 m), YANCEY county, W NORTH CAROLINA, in the BLACK MOUNTAINS of the APPALACHIAN system; 35°45′N 82°15′W. Highest mountain in North America E of ROCKY MOUNTAINS. In Mount Mitchell State Park, surrounded by Pisgah National Forest. Road access to summit from SW; restaurant and observation tower. BLUE RIDGE PARKWAY passes to S and SE.

Mitchell Peak (7,951 ft/2,423 m), GREENLEE county, E ARIZONA, in BLUE RANGE, 10 mi/16 km N of MORENCI. Apache-Sitgreaves National Forest.

Mitchell River (MI-chuhl), 350 mi/563 km long, N QUEENSLAND, AUSTRALIA; rises in GREAT DIVIDING RANGE near RUMULA; flows generally WNW to Gulf of CARPENTARIA 165 mi/266 km N of NORMANTON; 15°12′S 141°35′E. PALMER and Lynd rivers, main tributaries.

Mitchell River (MI-chuhl), 60 mi/97 km long, SE VICTORIA, AUSTRALIA; formed by two headstreams rising in AUSTRALIAN ALPS S of Mount HOTHAM; flows S and ESE, past BAIRNSDALE, to Lake KING on SE coast; 37°53′S 147°41′E.

Mitchell's Plain, town, WESTERN CAPE province, SOUTH AFRICA. 34°2′60S 18°37′0E. Elevation 55 ft/16 m.

Mitchellville, town (2000 population 1,715), POLK county, central IOWA, near SKUNK RIVER, 14 mi/23 km ENE of DES MOINES; 41°39′N 93°21′W. In agricultural and coal-mining area. Seat of Iowa Correctional Institute for Women.

Mitchellville, town (2006 population 203), SUMNER county, N TENNESSEE, near KENTUCKY line, 22 mi/35 km NE of SPRINGFIELD; 36°38′N 86°32′W.

Mitchelstown (MI-chuhlz-toun), Gaelic *Baile Mhistéala*, town (2006 population 3,365), NE CORK county, SW IRELAND, 15 mi/24 km SSW of TIPPERARY, at foot of GALTY MOUNTAINS; 52°16′N 08°16′W. Agricultural market in dairying region, with dairy-processing plants. It is a 19th-century planned town. Noted limestone caves nearby.

Miteja (mee-TAI-jah), village, LINDI region, E TANZANIA, 30 mi/48 km NNW of KILWA MASOKO, near INDIAN OCEAN; 08°35′S 39°13′E. Cashews, bananas, sisal; goats, sheep.

Mit el 'Amil, village, DAQAHLIYA province, NE Lower EGYPT, 10 mi/16 km S of MANSURA; 30°54′N 31°21′E. Cotton; cereals.

Mit el Ghuraqa (MEET el GOO-rah-kuh), village, KAFR ESH SHEIKH province, Lower EGYPT, on DUMYAT branch of the NILE, on railroad, and 2 mi/3.2 km WSW of TALKHA; 31°02′N 31°20′E. Cotton.

Mit el Nasara (MEET en NAH-sah-ruh), village, DAQAHLIYA province, NE Lower EGYPT, 13 mi/21 km ENE of MANSURA. Cotton; cereals.

Mitémélé (MEE-tai-mai-lai), river, 100 mi/161 km long, CENTRO SUR and LITORAL provinces, S Río Muni,

EQUATORIAL GUINEA; enters RÍO MUNI. Also called Temboni River and Utamboni River.

Mit Ghamr (MEET GA-muhr), town, DAQAHLIYA province, Lower EGYPT, on DUMYAT branch of the NILE River, on railroad, and 15 mi/24 km NW of ZAGAZIG; 31°08′N 31°38′E. Cotton ginning, cottonseed-oil extraction.

Mithankot (MIT-uhn-kot), town, DERA GHAZI KHAN district, SW PUNJAB province, central PAKISTAN, near INDUS River, 75 mi/121 km SSW of DERA GHAZI KHAN; 28°57′N 70°22′E.

Mithapur, India: see OKHA.

Mitha Tiwana (MIT-uh ti-WAH-nah), town, SARGODHA district, W central PUNJAB province, central PAKISTAN, 34 mi/55 km WNW of SARGODHA; 32°15′N 72°07′E. Railroad station. Cotton, oilseeds; wheat.

Mithimna, Greece: see METHYMNA.

Mitiaro (mee-tee-AH-ro), composite karstic coral island with small pockets of volcanic rock (□ 4 sq mi/10.4 sq km), S COOK ISLANDS, S PACIFIC, 142 mi/229 km NE of RAROTONGA; 4 mi/6.4 km long, 1 mi/1.6 km wide. Exports copra, dried bananas, eels.

Mitidja (mee-tee-JAH), fertile alluvial plain, N central ALGERIA, enclosed between the northernmost range of TELL ATLAS and the Sahel (a hilly coastal strip) S of ALGIERS; 65 mi/105 km long, 20 mi/32 km wide. Well-endowed with water from the various oueds and wells. Intensive farming, including livestock rearing to increase milk production. This region has the longest colonial history in the country, with many European settler villages on its rich soils. During the War of Independence (1954–1962), many rural settlements were created, causing a decline in local livestock rearing. After independence, following the departure of French settlers, the government established large self-managed farms run by peasants formerly employed on settler lands. Socialist villages with modern amenities were also set up for the benefit of the peasants. Following difficulties with FRANCE over the export of Algerian wine there, 82 sq mi/212 sq km of Mitidja vineyards were destroyed and the land converted to other crops, such as vegetables grown in plastic greenhouses. The Mitidja has suffered major losses of its best lands in recent decades to the expanding cities of Algiers and BLIDA, new roads, and the construction of new industrial plants. The region's agriculture has also lost large supplies of water, which is now devoted to urban and industrial needs.

Mitidja Atlas (mee-tee-JAH), coastal range of the TELL ATLAS, in N central ALGERIA, overlooking the MITIDJA lowland, S of ALGIERS. Rises to 5,344 ft/1,629 m in ABD EL KADER peak. City of BLIDA at NW foot. N slopes covered with vineyards and citrus groves. Also called Atlas of Blida.

Mitilíni (mee-tee-LEE-nee), city (2001 population 27,247), on LESBOS island, ⊙ LESBOS prefecture, NORTH AEGEAN department, E GREECE, in AEGEAN SEA; 39°06′N 26°33′E. Port. Airport. Roman remains. Also Mytilene, Mitilini.

Mitilini, Greece: see LESBOS, island.

Mitino (MEE-tee-nuh), village, central KIROV oblast, near the VYATKA River, on road and near railroad, 15 mi/24 km WSW of (and administratively subordinate to) SLOBODSKOY, and 7 mi/11 km NE of KIROV; 58°40′N 49°50′E. Elevation 626 ft/190 m. Weekend and holiday retreat for residents of neighboring cities; has a sanatorium.

Mitis, La (mee-TEE, lah), county (□ 893 sq mi/2,321.8 sq km; 2006 population 19,452), BAS-SAINT-LAURENT region, E QUEBEC, E CANADA; ⊙ MONT-JOLI; 48°32′N 68°05′W. Composed of eighteen municipalities. Formed in 1982.

Mitiyagoda (MEE-ti-yah-GO-DUH), village, SOUTHERN PROVINCE, SRI LANKA, 4 mi/6.4 km SE of AMBALANGODA; 06°11′N 80°06′E. Major moonstone-

mining center of Sri Lanka. Agriculture (vegetables, cinnamon, coconuts, rice; rubber).

Mit Khaqan (MEET kah-KAHN), village, MINUFIYA province, Lower EGYPT, 5 mi/8 km N of SHIBIN EL KOM; 30°34′N 31°02′E. Cereals; cotton, flax.

Mit Kinana (MEET ke-NAH-nah), village, QALYUBIYA province, S Lower EGYPT, 20 mi/32 km N of CAIRO; 30°23′N 31°16′E. Cotton, flax; cereals, fruits.

Mitkof Island, SE ALASKA, in ALEXANDER ARCHIPELAGO, between KUPREANOF ISLAND (W) and mainland (E), 10 mi/16 km NW of Wrangell; 24 mi/39 km long, 7 mi/11.3 km–17 mi/27 km wide; 56°40′N 132°47′W. Rises to 3,960 ft/1,207 m (SE). PETERSBURG town, N.

Mit'kovo (MEET-kuh-vuh), settlement, S TVER oblast, W European Russia, on the VOLGA RIVER, on highway, 8 mi/13 km NW of (and administratively subordinate to) RZHEV; 56°18′N 34°08′E. Elevation 620 ft/188 m. Gravel and stone quarrying.

Mitla (MEE-tlah) [Nahuatl=abode of the dead], archaeological site, religious center of the Zapotec at SAN PABLO VILLA DE MITLA, near OAXACA, SW MEXICO; 16°56′N 96°19′W. Probably built in the thirteenth century, the buildings, unlike the pyramidal structures of most MIDDLE AMERICA architecture, are low, horizontal masses enclosing the plazas. Mitla is thought to represent the highest expression of Zapotec architectural talent, although some decorative elements have been attributed to the Mixtec, who conquered Mitla as well as MONTE ALBÁN.

Mitla, EGYPT: see SINAI.

Mit Mihsin (MEET ME-sen), village, DAQAHLIYA province, Lower EGYPT, 2 mi/3.2 km NE of MIT GHAMR; 30°44′N 31°16′E. Cotton; cereals. Also spelled Mit Mohsin.

Mito (mee-TO), city (2005 population 262,603), ⊙ IBARAKI prefecture, central HONSHU, E central JAPAN, on the Naka River; 36°21′N 140°28′E. Chiefly a communications center. Natto [fermented soybeans]. From 1606, MITO was the seat of a branch of the Tokugawa family. The city's Tokiwa Park, with its Kairakuen Garden, is one of the greatest landscape gardens of Japan.

Mito (MEE-to), town, Hoi county, AICHI prefecture, S central HONSHU, central JAPAN, 34 mi/55 km S of NAGOYA; 34°48′N 137°19′E.

Mito (mee-TO), town, Mino county, SHIMANE prefecture, SW HONSHU, W JAPAN, 82 mi/132 km S of MATSUE; 34°39′N 131°59′E.

Mito, town, Mine county, YAMAGUCHI prefecture, SW HONSHU, W JAPAN, 9 mi/15 km N of YAMAGUCHI; 34°13′N 131°20′E. Marble products. Nearby Akiyoshi Dai is the largest karst area in Japan.

Mitoginskiy (mee-TO-geen-skeeye) town, W KAMCHATKA oblast, RUSSIAN FAR EAST, on SW KAMCHATKA PENINSULA, on the Sea of OKHOTSK, 15 mi/24 km N of UST'-BOL'SHERETSK; 53°11′N 156°06′E. Fish-processing plant.

Mitomi (MEE-to-mee), village, E Yamanashi county, YAMANASHI prefecture, central HONSHU, central JAPAN, 16 mi/25 km N of KOFU; 35°47′N 138°44′E. Beans, fruit. Karisaka mountain pass is nearby.

Mitomoni (mee-to-MO-nee), village, RUVUMA region, S TANZANIA, 60 mi/97 km SSW of SONGEA, on Ruvuma River, near MOZAMBIQUE border; 11°33′S 35°25′E. Livestock; timber.

Mitontic (mee-to-TEEK), town, CHIAPAS, S MEXICO, in Sierra de HUEYTEPEC, 10 mi/16 km NNE of SAN CRISTÓBAL DE LAS CASAS. Elevation 6,135 ft/1,870 m. Agriculture (wheat, fruit). A Tzotzil Maya community. Also known as SAN MIGUEL.

Mitoya (MEE-to-yah), town, Ishi county, SHIMANE prefecture, SW HONSHU, W JAPAN, 28 mi/45 km S of MATSUE; 35°17′N 132°52′E.

Mitre, department, SANTIAGO DEL ESTERO province, ARGENTINA.

Mit Riheina (MEET rah-HEE-nah), village, GIZA province, Upper EGYPT, 20 mi/32 km S of CAIRO. Corn, cotton. On part of site of ancient MEMPHIS, it has two colossal statues of Ramses II.

Mitrofanovka (mee-truh-FAH-nuhf-kah), village (2006 population 5,620), S VORONEZH oblast, S central European Russia, on road junction and railroad, 17 mi/27 km SSE of ROSSOSH; 49°58′N 39°41′E. Elevation 583 ft/177 m. Flour mill.

Mitrovica, SERBIA: see KOSOVSKA MITROVICA.

Mitrovica, SERBIA see SREMSKA MITROVICA.

Mit Salsil (MEET sahl-SEEL), village, DAQAHLIYA province, Lower EGYPT, on El BAHR ES SAGHIR (a delta canal), and 8 mi/12.9 km WNW of MANZALA; 31°10′N 31°48′E. Cotton; cereals.

Mitsamiouli (mee-tsah-mee-OO-lee), town, NJAZIDJA island and district, NW COMOROS, 24 mi/39 km N of MORONI, near N end of island, on MOZAMBIQUE CHANNEL, INDIAN OCEAN; 11°23′S 43°18′E. Agriculture (ylang-ylang, vanilla, bananas, coconuts; livestock); fish. Remains of 18th-century wall built as defense against pirates.

Mitsikeli (mee-tsee-KE-lee), mountain outlier of central PINDOS system, S EPIRUS department, NW GREECE, N of Lake IOÁNNINA; 15 mi/24 km long; rises to 5,936 ft/1,809 m 4 mi/6.4 km NE of IOÁNNINA; 39°45′N 20°50′E.

Mitsinjo (mee-TSEENZ), town, MAHAJANGA province, NW MADAGASCAR, on MAHAVAVY RIVER, near W coast N of Lake KINKONY, 35 mi/56 km WSW of MAHAJANGA; 16°00′S 45°52′E. Cattle market.

Mitsiwa, Eritrea: see MASSAWA, city.

Mitsiwa'e, Eritrea: see MASSAWA, city.

Mitsu (MEETS), town, Ibo district, HYOGO prefecture, S HONSHU, W central JAPAN, 34 mi/55 km W of KOBE; 34°46′N 134°33′E.

Mitsu, town, Mitsu county, OKAYAMA prefecture, SW HONSHU, W JAPAN, 9 mi/15 km N of OKAYAMA; 34°47′N 133°56′E. Agriculture (fruit, shiitake mushrooms, potato, chestnuts; orchids); fish. The Buddhist Myokaku Temple here is the main temple of the Fuju-Fuse sect of the Nichiren school.

Mitsue (mee-TSOO-e), village, Uda district, NARA prefecture, S HONSHU, W central JAPAN, 24 mi/39 km S of NARA; 34°29′N 136°10′E.

Mitsugi (MEETS-gee), town, Mitsugi county, HIROSHIMA prefecture, SW HONSHU, W JAPAN, 40 mi/65 km E of HIROSHIMA; 34°30′N 133°08′E.

Mitsuhashi (mee-TSOO-hah-shee), town, Yamato county, FUKUOKA prefecture, N KYUSHU, SW JAPAN, 31 mi/50 km S of FUKUOKA; 33°09′N 130°26′E.

Mitsuishi (mee-TSOO-ee-shee), town, Hidaka district, Hokkaido prefecture, N JAPAN, 84 mi/135 km S of SAPPORO; 42°14′N 142°33′E. Kombu; horses; lumber.

Mitsukaido (meets-KAH-ee-do), city, IBARAKI prefecture, central HONSHU, E central JAPAN, 37 mi/60 km S of MITO; 36°01′N 139°59′E.

Mitsuke (meets-KE), city, NIIGATA prefecture, central HONSHU, N central JAPAN, 28 mi/45 km S of NIIGATA; 37°31′N 138°54′E. Sometimes spelled Mituke.

Mitsuse (MEETS-se), village, Kanzaki county, SAGA prefecture, N KYUSHU, SW JAPAN, 12 mi/20 km N of SAGA; 33°25′N 130°16′E.

Mitsushima (mee-TSOO-shee-mah), town, Shimoagata county, NAGASAKI prefecture, NW KYUSHU, SW JAPAN, 112 mi/180 km N of NAGASAKI; 34°15′N 129°18′E. Clams; cuttlefish drying. Known for *tsukudani*, a dish in which food is boiled in soy sauce.

Mittaghorn, SWITZERLAND: see PETERSGRAT.

Mittagong (MI-tuh-gahng), town, E NEW SOUTH WALES, AUSTRALIA, 60 mi/97 km SW of SYDNEY; 34°27′S 150°27′E. Coal, trachyte marble. Agriculture (dairying; fruits, vegetables); cattle, sheep, poultry; steel; sawmilling. Formerly called New Sheffield.

Mittagskogel (MIT-tahgs-kaw-gel), Slovenian *Kepa* (KE-pah), peak (c. 6,532 ft/1,991 m), in the KAR-

AWANKEN mountain range, on AUSTRO-SLOVENIAN border, 9 mi/14.5 km SE of VILLACH, Austria; 46°30′N 13°57′E.

Mitta Mitta River (MI-tuh MI-tuh), 125 mi/201 km long, NE VICTORIA, SE AUSTRALIA; rises in AUSTRALIAN ALPS SSW of OMEO; flows N and NNW, past TALLANDOON, to HUME RESERVOIR near TALLANGATTA; 36°12′S 147°11′E.

Mittelberg (MIT-tel-berg), village, VORARLBERG, W AUSTRIA, in the Kleinwalsertal, 20 mi/32 km ESE of DORNBIRN; 47°20′N 10°09′E. Seasonal tourist center that uses German currency and is only accessible from GERMANY. Cable railroad to the Walmendinger Horn at elevation 6,075 ft/1,852 m.

Mittelfranken, GERMANY: see MIDDLE FRANCONIA.

Mittelland (MEE-tel-lahnd), district, central APPENZELL Ausser Rhoden half-canton, SWITZERLAND. Main town is TEUFEN; population is German-speaking and Protestant.

Mittelland (MEE-tel-lahnd), rolling plateau, SWITZERLAND, between the ALPS and the JURA mountains, extending from LAKE GENEVA to CONSTANCE LAKE. Elevation 1,300 ft/396 m–2,600 ft/792 m. Contains most of the large towns and cities (BASEL excepted), most of the population, and most of the agriculture and industrial production.

Mittelland Kana, GERMANY: see MIDLAND CANAL.

Mittelmurgebiet, CROATIA: see MEDIMURJE.

Mittelwalde, POLAND: see MIEDZYLESIE.

Mittenaar (MIT-tuhn-ahr), village, W HESSE, central GERMANY, 15 mi/24 km WSW of MARBURG; 50°42′N 08°28′E. Agriculture (grain); forestry. Manufacturing (electronic equipment).

Mittenwald (MIT-tuhn-vahlt), village, UPPER BAVARIA, GERMANY, between the Karwendelgebirge and the WETTERSTEINGEBIRGE, on the ISAR, and 8 mi/12.9 km ESE of GARMISCH-PARTENKIRCHEN; 47°26′N 11°14′E. Frontier station (Austrian border) on Innsbruck-Munich railroad. Violins, zithers (home industry). Summer and winter health resort (elevation 3,051 ft/930 m). Cable cars to KARWENDEL PEAK (7,825 ft/2,385 m). Has baroque church (1738–1740). Was first mentioned 1080. Chartered 1361; was important trade center in 15th and 16th centuries.

Mitterbach (MIT-ter-bahkh), village, S LOWER AUSTRIA, 30 mi/48 km SSW of SANKT POLTEN, at styria province border, near MARIAZELL; 47°49′N 15°18′E. Hydroelectric station is just S. Summer resort and winter sports; Erlauf Lake and Ötscher mountain with gorge and caves nearby.

Mitterburg, CROATIA: see PAZIN.

Mittersill (MIT-ter-sil), township, SW SALZBURG, W central AUSTRIA, in the PINZGAU, on the SALZACH river, and 12 mi/19 km SSE of Kitzbühel; 47°17′N 12°29′E. Market center; important road junction. Manufacturing (skis, flags); tungsten mining in vicinity (Felbertal). Seasonal tourism. Old restored castle and Hohe Tauern National Park are nearby.

Mitterteich (MIT-ter-teikh), town, UPPER PALATINATE, NE BAVARIA, GERMANY, on SE slope of the FICHTELGEBIRGE, 4.5 mi/7.2 km SW of WALDSASSEN; 49°57′N 12°14′E. Manufacturing (optical glass, porcelain, precision instruments). Chartered 1501, town since 1932.

Mittweida (mit-VEI-dah), town, SAXONY, E central GERMANY, on Zschopau River, and 11 mi/18 km NNE of CHEMNITZ; 50°59′N 12°59′E. Manufacturing (machinery, precision mechanics, glass, apparel); metalworking. Has 15th-century church. Chartered 1286.

Mitú (mee-TOO), town, ⊙VAUPÉS department, SE COLOMBIA, on Vaupés River (or UAUPÉS), in region of tropical forests, 14 mi/23 km W of BRAZIL border, 375 mi/603 km SE of BOGOTÁ; 01°07′N 70°02′W. Forest products. Airport.

Mitumbiri (mee-too-BEE-ree), village, CENTRAL province, S central KENYA, on railroad, and 30 mi/48 km

NE of NAIROBI; 00°59′S 37°09′E. Coffee, wheat, corn; wattle growing.

Mitunguu (mee-TOON-goo), town, EASTERN province, S central KENYA, on road, and 15 mi/24 km SSE of MERU, E of Mount KENYA; 00°08′S 37°49′E. Coffee, wheat, corn. Airfield.

Mitwaba (meet-WAH-buh), village, KATANGA province, SE CONGO, 170 mi/274 km NNE of LIKASI; 08°38′S 27°20′E. Elev. 5,088 ft/1,550 m. Tin-mining and trading center; tin concentrating. Has hydroelectric plant, airfield.

Mityana, administrative district, CENTRAL region, S central UGANDA. As of Uganda's division into eighty districts, borders KIBOGA (N), NAKASEKE (NE), WAKISO (E), MPIGI (S), and MUBENDE (W) districts. Agricultural (including bananas, coffee, cotton, and tea) and livestock area. Some fishing in Lake Wamala in SW of district. Railroad between KASESE town and KAMPALA city (and continuing E then SE to MOMBASA [KENYA]) travels across district. Main road between Kyenjojo town and Kampala city also runs W-E across district. Formed in 2005 from E portion of former MUBENDE district (current Mubende district formed from W and central portions).

Mityana (mit-YAH-nuh), town (2002 population 34,116), MITYANA district, CENTRAL region, S UGANDA, 38 mi/61 km WNW of KAMPALA, on swampy Lake Wamala. Coffee and tea center; cotton, bananas, corn; dairy products. Was part of former Buganda province.

Mitzic (MIN-zeek), town, WOLEU-NTEM province, N GABON, 60 mi/97 km S of OYEM; 00°47′N 11°31′E. Cacao, coffee, rubber plantations.

Mitzpe Adi (ah-DEE), community, ISRAEL, 10 mi/16 km ESE of HAIFA, in LOWER GALILEE. Established 1980 as agricultural community, but now predominantly a residential suburb for workers employed in the industries located around HAIFA BAY.

Miura (mee-OO-rah), city, KANAGAWA prefecture, E central HONSHU, E central JAPAN, at S tip of MIURA Peninsula, 22 mi/35 km N of YOKOHAMA; 35°08′N 139°37′E. Agriculture (watermelon, cabbage, daikon); pickled fish. The Miura coast is nearby.

Miura Peninsula (mee-OO-rah), Japanese *Miura-hanto* (mee-OO-rah-HAHN-to), KANAGAWA prefecture, E central HONSHU, E central JAPAN, E of SAGAMI Bay, W of TOKYO BAY and URAGA STRAIT; 14 mi/23 km long, 2 mi/3.2 km–5 mi/8 km wide. YOKOSUKA (naval base) is on E coast.

Miusinsk, UKRAINE: see MIUSYNS'K.

Mius River (mee-OOS), 160 mi/258 km long, E UKRAINE and ROSTOV oblast, RUSSIA; rises in the DONETS RIDGE 7 mi/11 km ESE of DEBAL'TSEVE; flows SSE, past MIUSYNS'K near KRASNY LUCH in Ukraine, and then S past Matveyev-Kurgan and Pokrovskoye, to Mius Liman (inlet of Sea of AZOV), NW of TAGANROG in Russia. A fortified Russian border (1695–1711); during World War II, a Soviet-German battle line (1941–1942, 1943).

Miusyns'k (mee-oo-SINSK) (Russian *Miusinsk*), city, SW LUHANS'K oblast, UKRAINE, in the DONBAS, on the MIUS RIVER, 3 mi/5 km S of, and subordinated to the city council of, KRASNYY LUCH; 48°04′N 38°54′E. Elevation 337 ft/102 m. Large coal-fed power station, power equipment repair depot. Energy technical school. Established in 1923 as Shtergres (Ukrainian *Shterhres*); city since 1965 and renamed. Also spelled, in Ukrainian, Miusyns'ke.

Miusyns'ke, UKRAINE: see MIUSYNS'K.

Mivtahim (meev-tah-KHEEM) or **Mivtachim**, moshav, SW ISRAEL, in the NEGEV, 20 mi/32 km S of GAZA; 31°14′N 34°23′E. Elevation 439 ft/133 m. Mixed farming. Founded 1947. Abandoned during the 1948 war, reestablished 1990.

Miwa (MEE-wah), town, Ama county, AICHI prefecture, S HONSHU, central JAPAN, 6 mi/10 km W of NAGOYA; 35°11′N 136°47′E.

Miwa, town, Asakura county, FUKUOKA prefecture, N KYUSHU, SW JAPAN, 19 mi/30 km S of FUKUOKA; 33°25′N 130°38′E.

Miwa, town, Futami county, HIROSHIMA prefecture, SW HONSHU, W JAPAN, 50 mi/80 km N of HIROSHIMA; 34°39′N 132°50′E.

Miwa, town, Amata county, KYOTO prefecture, S HONSHU, W central JAPAN, 34 mi/55 km N of KYOTO; 35°12′N 135°14′E.

Miwa, town, Kuga county, YAMAGUCHI prefecture, SW HONSHU, W JAPAN, 37 mi/60 km E of YAMAGUCHI; 34°13′N 132°04′E. Chestnuts.

Miwa (MEE-wah), village, Naka county, IBARAKI prefecture, central HONSHU, E central JAPAN, 22 mi/35 km N of MITO; 36°39′N 140°18′E.

Miwani (mee-WAH-nee), village, NYANZA province, W KENYA, on railroad, and 15 mi/24 km E of KISUMU; 00°04′S 34°59′E. Sugar mill.

Mi-Wuk Village (ME–wook), unincorporated town (2000 population 1,485), TUOLUMNE county, central CALIFORNIA, 12 mi/19 km NE of SONORA, in SIERRA NEVADA; 38°04′N 120°11′W. Timber, cattle, hay. Area surrounded by Stanislaus National Forest.

Mixco (MEESH-ko), town, GUATEMALA department, S central GUATEMALA, on INTER-AMERICAN HIGHWAY, and 4 mi/6.4 km W of GUATEMALA city; 14°38′N 90°36′W. Elevation 5,705 ft/1,739 m. Market center; pottery making; truck (fruit, vegetables). Now a part of Guatemala city metropolitan area with a total population of over 200,000 in the municipio.

Mixcoac (meesh-ko-AHK), SW section of MEXICO CITY, central MEXICO, and part of BENÍTO JUÁREZ Delegacion; 19°23′N 99°12′W. Manufacturing suburb (cement plant, processing industries); nursery gardens. ALVARO OBREGÓN adjoins (S).

Mixco Viejo (MEESH-ko vee-AI-ho), archaeological site, 22 mi/35 km NNW of GUATEMALA city, GUATEMALA; 14°52′N 90°40′W. Postclassic site; was the principal center of the Pokomam people at the time of conquest. Reconstructed in 1960s.

Mi Xian (MEE-SIAN), town, N Mi Xian county, N HENAN province, CHINA, 30 mi/48 km SW of ZHENGZHOU; 34°31′N 113°22′E. Grain, oilseeds, tobacco; building materials, paper making, coal mining.

Mixistlán de la Reforma (meesh-ko-AHK de lah re-FOR-mah), town, in E central OAXACA, MEXICO, 48 km ENE of TLACOLULA DE MATAMOROS. Elevation 5,906 ft/1,800 m. Mountainous region with temperate climate. Agriculture (cereals and fruits), and woods. A Mixe-speaking town. Formerly known as SANTA MARÍA MIXISTLÁN.

Mixnitz (MIKS-nits), village, central STYRIA, E central AUSTRIA, on MUR RIVER, and 20 mi/32 km N of GRAZ; 47°20′N 15°22′E. Hydroelectric station; summer resort. Gorge and cave nearby.

Mixquiahuala (meesh-kee-ah-WAH-lah), town, ⊙ Mixquiahuala de Juárez municipio, HIDALGO, central MEXICO, on TULA RIVER, and 32 mi/51 km WNW of PACHUCA DE SOTO, on railroad; 20°11′N 99°10′W. Elevation 6,549 ft/1,996 m. Grain, beans, potatoes, fruit, livestock.

Mixquic (meesh-KEEK), town, Federal Distrito, TLÁHUAC delegación, central MEXICO, 18 mi/29 km SE of MEXICO CITY, and part of the Mexico City metropolitan area. Cereals, fruit, vegetables; livestock.

Mixstadt, POLAND: see MIKSTAT.

Mixtla, MEXICO: see SAN FRANCISCO MIXTLA.

Mixtla de Altamirano (MEESH-tlah dai ahl-mee-RAH-no), town, VERACRUZ, E MEXICO, in Sierra ZONGOLICA, 21 mi/34 km SSW of CÓRDOBA. An isolated town in mountainous area. Coffee, sugarcane, fruit.

Mixtlán (meesh-TLAHN), town, JALISCO, W MEXICO, 23 mi/37 km WSW of AMECA; 20°26′N 104°25′W. Isolated town in Sierra Verde. Corn, chickpeas, beans, sugarcane.

Miya (MEE-yah), village, Ono county, GIFU prefecture, central HONSHU, central JAPAN, 56 mi/90 km N of GIFU; 36°04′N 137°14′E. Traditional hinoki hats.

Miyada (mee-YAH-dah), village, Kamiina county, NAGANO prefecture, central HONSHU, central JAPAN, 62 mi/100 km S of NAGANO; 35°45′N 137°56′E.

Miyagawa (mee-YAH-gah-wah), village, Yoshiki county, GIFU prefecture, central HONSHU, central JAPAN, 68 mi/110 km N of GIFU; 36°19′N 137°08′E. *Nijimasu* -trout farming (boiled trout is a local specialty); ginger.

Miyagawa, village, Taki county, MIE prefecture, S HONSHU, central JAPAN, 25 mi/40 km S of TSU; 34°21′N 136°20′E. Yoshino-Kumano National Park and Osugidani Gorge are nearby.

Miyagi (mee-YAH-gee), prefecture (□ 2,808 sq mi/7,273 sq km; 1990 population 2,248,521), N HONSHU, NE JAPAN, on PACIFIC OCEAN (E); ⊙ SENDAI. Bordered N by IWATE prefecture, NW by Akita prefecture, S by FUKUSHIMA prefecture, and W by YAMAGATA prefecture. A mountainous prefecture, it is known for the more than two hundred pine-covered islands in Matsushima Bay. Yields farm products, fish, lumber, raw silk, electrical machinery, and transport equipment. ISHINOMAKI and SHIOGAMA are important ports.

Miyagi (MEE-yah-gee), village, Seta county, GUMMA prefecture, central HONSHU, N central JAPAN, 9 mi/15 km N of MAEBASHI; 36°26′N 139°10′E. Livestock breeding.

Miyahara (mee-YAH-hah-rah), town, Yatsushiro county, KUMAMOTO prefecture, W KYUSHU, SW JAPAN, 16 mi/25 km S of KUMAMOTO; 32°33′N 130°41′E. Sometimes called Miyanoharu.

Miyajima (mee-YAH-jee-mah), town, Saeki county, HIROSHIMA prefecture, SW HONSHU, W JAPAN, 9 mi/15 km S of HIROSHIMA; 34°17′N 132°19′E. Confections. Itsukushima (Shinto) shrine is located here.

Miya-jima, JAPAN: see ITSUKU-SHIMA.

Miyake (mee-YAH-ke), town, Shiki district, NARA prefecture, S HONSHU, W central JAPAN, 9 mi/14 km S of NARA; 34°34′N 135°46′E.

Miyake (mee-YAH-ke), village, Miyake district, Tokyo prefecture, E central HONSHU, E central JAPAN, 12 mi/20 km W of SHINJUKU; 34°04′N 139°33′E. Hydrangeas, dropworts; potatoes.

Miyako (mee-YAH-ko), city, IWATE prefecture, N HONSHU, NE JAPAN, on the HEI RIVER, and Miyako Bay, 43 mi/70 km E of MORIOKA; 39°38′N 141°57′E. Important fishing port; seaweed (kombu, *wakame*).

Miyako-gunto (mee-YAH-ko–GUN-to), island subgroup (□ 96 sq mi/249.6 sq km) of SAKISHIMA Islands, in the S Ryukyus, Miyako county, Okinawa prefecture, extreme SW JAPAN. Includes MIYAKO-JIMA, IRABU-SHIMA, and Tarama-shima. islands Also called Miyako-retto.

Miyakoji (mee-YAH-ko-jee), village, Tamura county, FUKUSHIMA prefecture, N central HONSHU, NE JAPAN, 28 mi/45 km S of FUKUSHIMA city; 37°25′N 140°47′E. Shiitake mushrooms.

Miyako-jima (mee-YAH-ko–JEE-mah), volcanic island (□ 70 sq mi/182 sq km) of SAKISHIMA Islands, in the RYUKYU ISLANDS, Okinawa prefecture, extreme SW JAPAN, between EAST CHINA (W) and PHILIPPINE (E) seas, 60 mi/97 km ENE of Ishigaki-shima; 13 mi/21 km long, 12 mi/19 km wide; roughly triangular. Hilly, fertile; surrounded by coral reef. Produces sugarcane, sweet potatoes, soybeans, some rice. Formerly called Taipinsan. Chief town is HIRARA.

Miyakonojo (mee-YAH-ko-NO-jo), city (2005 population 170,955), MIYAZAKI prefecture, S KYUSHU, SW JAPAN, 25 mi/40 km S of MIYAZAKI; 31°42′N 131°03′E. Important railroad junction and commercial center. Agriculture (pigs, beef cattle). Castle town of the Shimazu family in the 11th century.

Miyakubo (mee-YAH-koo-bo), town, Ochi county, EHIME prefecture, NW SHIKOKU, W JAPAN, 28 mi/45 km N of MATSUYAMA; 34°10′N 133°04′E.

Miyama (mee-YAH-mah), town, Asuwa county, FUKUI prefecture, central HONSHU, W central JAPAN, 9 mi/15 km S of FUKUI; 35°59′N 136°21′E.

Miyama (mee-YAH-mah), town, Yamagata county, GIFU prefecture, central HONSHU, central JAPAN, 12 mi/20 km N of GIFU; 35°35′N 136°44′E.

Miyama (mee-YAH-mah), town, N Kuwata county, KYOTO prefecture, S HONSHU, W central JAPAN, 22 mi/35 km N of KYOTO; 35°16′N 135°33′E.

Miyama, town, N Muro county, MIE prefecture, S HONSHU, central JAPAN, 47 mi/75 km S of TSU; 34°06′N 136°14′E. Fish (yellowtail, sea bream).

Miyama (mee-YAH-mah), village, Hidaka county, WAKAYAMA prefecture, S HONSHU, W central JAPAN, 22 mi/35 km S of WAKAYAMA; 33°58′N 135°22′E.

Miyamori (mee-YAH-mo-ree), village, Kamihei county, IWATE prefecture, N HONSHU, NE JAPAN, 28 mi/45 km S of MORIOKA; 39°20′N 141°21′E. Wasabi.

Miyanduab (mee-YAHN-doo-ahb), town (2006 population 114,153), Āžerbāyjān-e GHARBI province, NW IRAN, 80 mi/129 km S of TABRIZ, and on ZARINEH RIVER, SE of Lake URMIA. Elev. 4,200 ft/1,280 m. Agriculture (grain, fruit, sugar beets; sheep); beet-sugar refinery. Also spelled MIANDOAB.

Miyan Kaleh Peninsula (mee-YAHN kah-LAI), narrow sandspit, on SE CASPIAN SEA, Māzandarān province, NE IRAN; 35 mi/56 km long, 2 mi/3.2 km–3 mi/4.8 km wide, nearly closing off GORGAN LAGOON. ASHURADEH ISLANDS are off E tip.

Miyanojo (mee-yah-NO-JO), town, Satsuma county, KAGOSHIMA prefecture, SW KYUSHU, SW JAPAN, near Mount Shibi, 22 mi/35 km N of KAGOSHIMA; 31°54′N 130°27′E. Bamboo work, distilled spirits (*shochu*). Hot springs.

Miyashiro (mee-YAH-shee-ro), town, S Saitama county, SAITAMA prefecture, E central HONSHU, E central JAPAN, 12 mi/20 km N of URAWA; 36°01′N 139°43′E.

Miyata (mee-YAH-TAH), town, Kurate county, FUKUOKA prefecture, N KYUSHU, SW JAPAN, 19 mi/30 km N of FUKUOKA; 33°43′N 130°40′E. Computer components.

Miyazaki (mee-YAH-zah-kee), prefecture (□ 2,998 sq mi/7,765 sq km; 1990 population 1,184,047), E KYUSHU, SW JAPAN, on HYUGA SEA (N arm of PHILIPPINE SEA; E) and ARIAKE BAY (S); ⊙ MIYAZAKI. Bordered NW by Kumamoto prefecture, NE by Oita prefecture, and SW by Kagoshima prefecture. Mountainous terrain; rises to 5,650 ft/1,722 m at ICHIBUSA-YAMA. N drained by GOKASE RIVER. Primarily agricultural (rice, wheat, soybeans, sweet potatoes, millet). Extensive forested area, with pine and Japan cedar trees. Iino, in interior, is horse-breeding center. Numerous small fisheries along coast. Manufacturing (fertilizer, medicines, lumber, raw silk, charcoal). Many small ports export produce and lumber. Chief centers include Miyazaki (E), NOBEOKA (NE), MIYAKONOJO (S). Major honeymoon spot. Contains the crater-filled Kirishima-Yaku National Park.

Miyazaki (mee-YAH-zah-kee), city (2005 population 366,897), ⊙ MIYAZAKI prefecture, SE KYUSHU, SW JAPAN, on the HYUGA SEA and Oyodo River; 31°54′N 131°25′E. Peppers, pumpkin. Popular tourist and resort center and the seat of the great Shinto shrine, Miyazaki-jingu (with an archaeological museum, including a haniwa gallery), dedicated to Jimmu, first emperor of Japan. The famous (Shigaya) resort is here; includes indoor artificial sea and surfing. Ao islet (semitropical plants) and Hajo rock (both in Nichinan Coast quasi-national park). *Yakko so* plant originated here.

Miyazaki (mee-YAH-zah-kee), town, Kami county, MIYAGI prefecture, N HONSHU, NE JAPAN, 25 mi/40 km N of SENDAI; 38°36′N 140°45′E. Lumber.

Miyazaki (mee-YAH-zah-kee), village, Nyu county, FUKUI prefecture, central HONSHU, W central JAPAN, 12 mi/20 km S of FUKUI; 35°56′N 136°04′E. Echizen ceramic art village.

Miyazu (mee-YAHZ), city, KYOTO prefecture, S HONSHU, W central JAPAN, on Miyazu Bay, 47 mi/75 km N of KYOTO; 35°31'N 135°11'E. Fishing port; sardine oil processing. Also *mochi* rice cake. Nearby is AMA-NO-HASHIDATE [=heaven's bridge], a long promontory covered with pine trees whose fantastic shapes are reflected in the waters of the bay. This was the site, according to legend, where Izanagi and Izanami stood while they created the islands of Japan. Wakasa Bay quasi-national park is nearby.

Miyoshi (mee-YO-shee), city, HIROSHIMA prefecture, SW HONSHU, W JAPAN, 37 mi/60 km N of HIROSHIMA; 34°48'N 132°51'E. Beef cattle; grapes; tea; fish (ayu). Traditional dolls.

Miyoshi (mee-YO-shee), town, West Kamo county, AICHI prefecture, S central HONSHU, central JAPAN, 9 mi/15 km S of NAGOYA; 35°05'N 137°04'E. Motor vehicles.

Miyoshi, town, Iruma county, SAITAMA prefecture, E central HONSHU, E central JAPAN, 6 mi/10 km S of URAWA; 35°49'N 139°31'E.

Miyoshi, town, Miyoshi county, TOKUSHIMA prefecture, SE SHIKOKU, W JAPAN, 40 mi/65 km W of TOKUSHIMA; 34°02'N 133°52'E. Persimmons; bamboo work.

Miyoshi (mee-YO-shee), village, Awa county, CHIBA prefecture, E central HONSHU, E central JAPAN, 31 mi/50 km S of CHIBA; 35°01'N 139°53'E.

Miyota (mee-YO-tah), town, S Saku county, NAGANO prefecture, central HONSHU, central JAPAN, 28 mi/45 km S of NAGANO; 36°19'N 138°30'E. Bearings. Lettuce.

Miyun (MI-YUN), town, ⊙ Miyun county, CHINA, on BAI RIVER, 40 mi/64 km NE of BEIJING, and on railroad; 40°22'N 116°49'E. An administrative unit of Beijing. Grain, oilseeds, fruits; food processing; textiles; engineering.

Mizan Teferī, town, SOUTHERN NATIONS state, W central ETHIOPIA, 105 mi/169 km SW of JIMMA; 06°59'N 35°35'E. Airfield.

Mizata (mee-ZAH-tah), village, LA LIBERTAD department, SW EL SALVADOR, 24 mi/39 km WSW of NUEVA SAN SALVADOR. Minor Pacific post; coastal trade.

Mizdah (MIZ-duh), village, TRIPOLITANIA region, NW LIBYA, on road, 50 mi/80 km S of GHARYAN, in oasis; 31°26'N 12°59'E. Manufacturing (carpets, tents, bags); agriculture (fruit, grain, spices). Ruins of forts and walls in region.

Mize (MEIZ), village (2000 population 285), SMITH county, S central MISSISSIPPI, 27 mi/43 km WNW of LAUREL, on Oakahay Creek; 31°52'N 89°32'W. Agriculture (corn, cotton; poultry, cattle); manufacturing (electrical components, lumber, polishing machinery).

Mizen Head (MI-zuhn), Gaelic *Carn Uí Néid*, Atlantic cape, SW CORK county, at SW extremity of IRELAND, 25 mi/40 km WSW of SKIBBEREEN; 51°27'N 09°49'W.

Mizen Head (MI-zuhn), Gaelic *Ard an Fhéaraigh*, promontory on the IRISH SEA, SE WICKLOW county, E IRELAND, 9 mi/14.5 km N of WICKLOW; 52°41'N 06°03'W.

Mizhhir'ya (meezh-HIR-yah) [Ukrainian=among the mountains] (Russian *Mezhgor'ye*) (Czech *Volové*) (Hungarian *Ökörmező*), town, N TRANSCARPATHIAN oblast, UKRAINE, 37 mi/60 km ENE of MUKACHEVE; 48°31'N 23°30'E. Elevation 1,640 ft/499 m. Raion center. Forest products complex, electronics component manufacturing. Tourist center in Carpathian mountains. Founded in the 13th century as Volove (Russian *Volovo*); town since 1947; renamed in 1953.

Mizhi (MEE-JI), town, ⊙ Mizhi county, NE SHAANXI province, CHINA, 40 mi/64 km SSE of YULIN; 37°45'N 110°11'E. In mountain region. Grain; food processing, chemicals, coal mining.

Mizil (mee-ZEEL), town, PRAHOVA county, SE central ROMANIA, on railroad, and 22 mi/35 km ENE of PLOIEŞTI; 45°01'N 26°27'E. Metallurgy; furniture manufacturing, food processing.

Mizobe (mee-ZO-be), town, Aira county, KAGOSHIMA prefecture, SW KYUSHU, SW JAPAN, 19 mi/30 km N of KAGOSHIMA; 31°49'N 130°40'E.

Mizoch (MEE-zoch), (Polish *Mizocz*), town (2004 population 9,500), S RIVNE oblast, UKRAINE, 15 mi/24 km SSW of RIVNE, on railroad spur; 50°24'N 26°09'E. Elevation 839 ft/255 m. Agricultural processing (sugar beets, cereals; dairy); electric fittings; sawmilling. Forest preserve, visitors' center. Known since 1322, town since 1940. Jewish community since the 18th century, numbering 3,000 in 1939; destroyed by the Nazis in 1942 (the mass grave on the outskirts of town also contains the remains of the Jews from the neighboring towns of Bilashev and Pivni)—fewer than 100 Jews remaining in 2005.

Mizocz, UKRAINE: see MIZOCH.

Mizokuchi (mee-ZOK-chee), town, Hino county, TOTTORI prefecture, S HONSHU, W JAPAN, 47 mi/76 km S of TOTTORI; 35°20'N 133°26'E.

Mizoram (mee-ZO-ruhm), state (□ 8,139 sq mi/21,161.4 sq km; 2001 population 888,573), NE INDIA; ⊙ AIZAWL. Bordered E by MYANMAR, W by BANGLADESH and TRIPURA state, and E by ASSAM and MANIPUR states. The state contains jungles, forests, and the Mizo Hills. Shifting agriculture still practiced in some places. Mizoram became a union territory in 1972 and twenty third state in 1987. Formerly part of Assam state. The Mizos and Lushai tribal population is closely related to the Chins of Myanmar. More than 80% of the population is Christian.

Mizots'kyy Ridge, UKRAINE: see VOLHYNIAN UPLAND.

Mizpah (MIZ-puh), village (2000 population 78), KOOCHICHING county, N MINNESOTA, 60 mi/97 km SW of INTERNATIONAL FALLS; 47°55'N 94°13'W. In forest area. Timber, alfalfa; cattle; manufacturing (valve lifters). Pine Island State Forest to N and E.

Mizpah, village, ATLANTIC county, S NEW JERSEY, 5 mi/8 km NW of MAYS LANDING; 39°29'N 74°50'W. In agricultural area. Founded by Jewish immigrants in the late 19th century.

Mizpe Ramon (meets-PE rah-MON) or **Mitzpeh Ramon**, Jewish township, S ISRAEL, S of BEERSHEBA in NEGEV highlands; 30°36'N 34°48'E. Elevation 2,844 ft/866 m. Overlooks the Ramon erosional crater [=Makhtesh Ramon]. Geological museum. Israel's main astronomic-research station with large telescope nearby. Prehistoric sites with tools from the Chalcolithic era and the early Stone Age nearby. Remains of a Nabatean settlement on the ridge above the crater. Founded 1954.

Mizque (MEE-skai), province (□ 1,054 sq mi/2,740.4 sq km), COCHABAMBA department, central BOLIVIA; ⊙ MIZQUE; 17°54'S 66°21'W. Elevation 6,463 ft/1,970 m. Limestone and gypsum deposits. Potatoes, yucca, bananas, corn, oats, rye, sweet potatoes, soy, coffee; cattle; dairying.

Mizque (MEE-skai), city and canton, ⊙ MIZQUE province, COCHABAMBA department, central BOLIVIA, on S outliers of Cordillera de COCHABAMBA, Valles Altos region of SE Cochabamba, on MIZQUE River, and 70 mi/113 km SE of COCHABAMBA; 17°56'S 65°19'W. Elevation 6,693 ft/2,040 m. On highway from Cochabamba to SUCRE, with a branch to SANTA CRUZ; on Cochabamba-AIQUILE railroad. Wheat, corn, potatoes, oats, coffee; livestock.

Mizque River, 130 mi/209 km long, in COCHABAMBA department, central BOLIVIA; rises in S outliers of Cordillera de COCHABAMBA near Villa Viscarra; flows E and SE past MIZQUE to Río GRANDE 18 mi/29 km SW of VALLEGRANDE; 18°39'S 64°20'W. Forms border between Cochabamba and SANTA CRUZ departments in lower course.

Mizra (MEEZ-rah), kibbutz, N ISRAEL, 2.5 mi/4 km N of AFULA, 4 mi/6.4 km S of NAZARETH, JEZREEL VALLEY; 32°39'N 35°17'E. Elevation 278 ft/84 m. Named for ancient Mizra, which was located in this area. Mixed farming; hydraulic tools; meat and pro-

cessed meat products. It was the first Jewish organization in Israel to mass produce non-kosher meat products. School serves region. Founded 1923.

Mizuho (MEEZ-ho), town, Funai county, KYOTO prefecture, S HONSHU, W central JAPAN, 25 mi/40 km N of KYOTO; 35°10'N 135°22'E.

Mizuho (mee-ZOO-ho), town, South Takaki county, NAGASAKI prefecture, NW KYUSHU, SW JAPAN, 25 mi/40 km E of NAGASAKI; 32°51'N 130°14'E.

Mizuho (MEEZ-ho), town, Ochi county, SHIMANE prefecture, SW HONSHU, W JAPAN, 51 mi/83 km S of MATSUE; 34°51'N 132°32'E.

Mizuho (mee-ZOO-ho), town, West Tama county, Tokyo prefecture, E central HONSHU, E central JAPAN, just W of HACHIOJI, and 19 mi/30 km W of SHINJUKU; 35°46'N 139°12'E. Pongee; tea.

Mizuho Plateau, ice plateau, E of the QUEEN FABIOLA MOUNTAINS, and S of the SHIRASE GLACIER, in QUEEN MAUD LAND, ANTARCTICA.

Mizukami (mee-zoo-KAH-mee), village, Kuma county, KUMAMOTO prefecture, W KYUSHU, SW JAPAN, 37 mi/60 km S of KUMAMOTO; 32°18'N 131°00'E.

Mizuma (MEE-zoo-mah), town, Mizuma county, FUKUOKA prefecture, N KYUSHU, SW JAPAN, 25 mi/40 km S of FUKUOKA; 33°15'N 130°28'E. Garden plants.

Mizumaki (mee-ZOO-mah-kee), town, Onga county, FUKUOKA prefecture, N KYUSHU, SW JAPAN, 25 mi/40 km N of FUKUOKA; 33°51'N 130°41'E.

Mizunami (mee-ZOO-nah-mee), city, GIFU prefecture, central HONSHU, central JAPAN, 31 mi/50 km E of GIFU; 35°21'N 137°15'E. Poultry. Uranium deposits.

Mizur (mee-ZOOR), town (2006 population 2,955), S NORTH OSSETIAN REPUBLIC, in the N central CAUCASUS Mountains, RUSSIA, on road in the Alatyr canyon, on the upper ARDON RIVER, 30 mi/48 km WSW of VLADIKAVKAZ; 42°51'N 44°05'E. Elevation 3,507 ft/1,068 m. Concentrating mill for the Sadon lead and zinc mine (linked by cableway).

Mizusawa (mee-ZOO-sah-wah), city, IWATE prefecture, N HONSHU, NE JAPAN, on KITAKAMI RIVER, and 37 mi/60 km S of MORIOKA; 39°08'N 141°08'E. *Manbu*-style ironware. Shoho Temple here is one of the three main temples of the Buddhist Soto section. Ruins of Isawa Castle.

Mizzen Topsail, mountain (1,761 ft/537 m), W NEWFOUNDLAND AND LABRADOR, CANADA, 25 mi/40 km E of NE end of GRAND LAKE; 49°05'N 56°37'W.

Mjällom (MYEL-loom), fishing village, VÄSTERNORRLAND county, NE SWEDEN, on small inlet of GULF OF BOTHNIA, 25 mi/40 km SSW of ÖRNSKÖLDSVIK; 62°59'N 18°26'E. Food processing.

Mjanji (um-JAHN-jee), village, BUGIRI district, EASTERN region, SE UGANDA, minor port on LAKE VICTORIA, 31 mi/50 km SSW of TORORO; 00°15'N 33°58'E. Potash deposits. Was part of former EASTERN province.

Mjölby (MYUHL-BEE), town, ÖSTERGÖTLAND county, S SWEDEN, on Svartån River, 18 mi/29 km WSW of LINKÖPING; 58°20'N 15°8'E. Railroad center on main lines from MALMÖ to STOCKHOLM and N Sweden. Metalworking; motor vehicles, windows; food processing. Known since thirteenth century, prominent since railroad (1870s). Incorporated 1920.

Mjøndalen (MYUHN-dah-luhn), village, BUSKERUD county, SE NORWAY, on DRAMMENSELVA River (falls), on railroad, and 8 mi/12.9 km W of DRAMMEN; 59°45'N 10°01'E. Paper, cellulose, and textile mills; sawmilling. Hydroelectric plant.

Mjörn (MYUHRN), lake (8 mi/12.9 km long, 1 mi/1.6 km–3 mi/4.8 km wide), expansion of SÄVEÅN RIVER, SW SWEDEN, 14 mi/23 km NE of GÖTEBORG. ALINGSÅS at NE end.

Mjøsa (MYUH-sah), largest lake of NORWAY (□ 142 sq mi/368 sq km; 1,453 ft/443 m deep), on the OPPLAND-HEDMARK county border, SE Norway. It is fed by the LÅGEN River and is drained by the Vorma River into the GLOMMA River. The lake is the center of a fertile agricultural region; grains. HAMAR, GJØVIK, and LIL-LEHAMMER are the principal cities on the lake.

Mjumbe Salim's (um-joom-bai SAH-lee-um), village, RUVUMA region, S TANZANIA, 55 mi/89 km E of SONGEA; 10°20′S 36°31′E. Timber; livestock; subsistence crops.

Mkangira (um-kan-GEE-rah), village, LINDI region, S central TANZANIA, 140 mi/225 km S of MOROGORO, on Luwego River, in SELOUS GAME RESERVE; 08°56′S 37°24′E. Livestock.

Mkasu (um-KAH-soo), village, MOROGORO region, S central TANZANIA, 90 mi/145 km SSE of IRINGA, and near Ruhudgi River; 09°06′S 35°54′E. Timber; corn, wheat; sheep, goats.

Mkata (um-KAH-tah), village, MOROGORO region, E central TANZANIA, 20 mi/32 km W of MOROGORO, on Mkata River, in Makata Plain; 06°42′S 37°20′E. Cotton, sugarcane, corn, wheat; cattle, sheep.

Mkhaya Game Reserve (um-KHA-yah) (□ 24 sq mi/62.4 sq km), LUBOMBO district, E central SWAZILAND, c.25 mi/40 km ESE of MANZINI, and 5 mi/8 km NE of Sipofani; 26°35′S 31°42′E. Elephants (reintroduced to Swaziland in 1986 after fifty years), leopards, hyena, white and black rhinoceros, birds (including purple-crested lourie, the national bird); park covers 15,314 acres/6,198 ha. A breeding program for indigenous Nguni cattle began here in 1976.

Mkhondvo, river, South Africa: see ASSEGAAI RIVER

Mkoani (um-ko-AH-nee), town, PEMBA SOUTH region, NE TANZANIA, on W coast of PEMBA island, Tanzania, 11 mi/18 km SW of CHAKE CHAKE; 05°23′S 39°16′E. Road terminus. Cloves, copra, bananas; livestock; fish.

Mkokotoni (um-ko-ko-TO-nee), town, ⊙ ZANZIBAR NORTH region, E TANZANIA, on W coast of ZANZIBAR island, 20 mi/32 km NNE of ZANZIBAR city; 05°53′S 39°16′E. Copra, cloves; fish. TUMBATU ISLAND to NW.

Mkoma (um-KO-mah), village, MTWARA region, SE TANZANIA, 65 mi/105 km SW of MTWARA, on MAKONDE PLATEAU; 10°51′S 39°13′E. Cashews, bananas, sweet potatoes; goats, sheep.

Mkomazi (um-ko-MAH-zee), village, TANGA region, NE TANZANIA, 70 mi/113 km NW of TANGA, near PANGANI RIVER; 04°40′S 38°05′E. Road junction. Timber; sisal, rice; livestock. PARE MOUNTAINS to NW, USAMBARA Mountains to E.

Mkomazi, river, SOUTH AFRICA: see DONNYBROOKE.

Mkondo River, SOUTH AFRICA and SWAZILAND: see ASSEGAAI RIVER.

Mkulwe (um-KOO-lwai), village, MBEYA region, SW TANZANIA, 60 mi/97 km SE of SUMBAWANGA, on MOMBA RIVER, opposite Sangama; 08°35′S 32°17′E. Sheep, goats; corn, wheat; timber.

Mkumbara (um-KOO-MBA-rah), village, TANGA region, NE TANZANIA, 35 mi/56 km NW of KOROGWE, on railroad; 04°45′S 38°11′E. Connected by cable car with Mount Lukome (6,694 ft/2,040 m) 5 mi/8 km to N. Timber; rice, sisal; livestock. Tourism.

Mkuranga (um-koo-RAN-gah), village, PWANI region, E TANZANIA, 15 mi/24 km S of DAR ES SALAAM; 07°05′S 39°12′E. Cashews, bananas, corn; goats, sheep, poultry.

Mkushi, township, CENTRAL province, central ZAMBIA, 65 mi/105 km ENE of KABWE, on Mkushi River; 13°37′S 29°24′E. Tobacco, wheat, corn; cattle. Beryl mining. Lunsemfwa (Mita Hills) Dam and Reservoir to W; Muchinga Escarpment to SE. Also called Old Mkushi.

Mkuyuni (um-koo-YOO-nee), village, MOROGORO region, E central TANZANIA, 12 mi/19 km SE of MOROGORO; 06°59′S 37°48′E. Cotton, sugarcane, corn, wheat; cattle, goats, sheep. ULUGURU MOUNTAINS to W.

Mkwaja (um-KWAH-jah), village, TANGA region, NE TANZANIA, 50 mi/80 km SSW of TANGA on ZANZIBAR CHANNEL; 05°46′S 38°50′E. Fish; sisal, corn, bananas, copra; goats, sheep; timber. SADANI Game Reserve to S.

Mkwasine (em-gwah-SEE-nai), village, MASVINGO province, SE ZIMBABWE, 90 mi/145 km SE of MASVINGO, on Mkwasine River; 20°48′S 31°58′E. Terminus

of railroad spur from MBIZI. Sugarcane, cotton, peanuts, corn, wheat; cattle, sheep, goats.

Mkwaya (um-KWAH-yah), village, LINDI region, SE TANZANIA, 12 mi SW of LINDI, on LUKULEDI RIVER; 10°03′S 39°38′E. Was railroad-river transfer point during failed groundnut scheme of late 1940s and early 1950s. Cashews, bananas, peanuts, copra; sheep, goats.

Mkwera (um-KWAI-rah), village, RUVUMA region, S TANZANIA, 12 mi/19 km ESE of SONGEA; 10°48′S 35°49′E. Timber; tobacco, subsistence crops; livestock.

Mladá Boleslav (MLAH-dah BO-les-LAHF), German *Jungbunzlau*, city (2001 population 44,255), STREDO-CESKY province, CZECH REPUBLIC, N central BOHEMIA, on JIZERA RIVER; 50°25′N 14°54′E. Railroad junction. Industrial center; manufacturing (motor vehicles [Škoda]); food processing; wood. Founded in the 10th century, it became a center of the Bohemian Brethren (15th–16th centuries). Has a 16th century Renaissance town hall and museum.

Mlada Vozice (MLAH-dah VO-zhi-TSE), Czech *Mladá Vožice*, German *Jungwoschitz*, town, JIHOCESKY province, S BOHEMIA, CZECH REPUBLIC, 10 mi/16 km NE of TÁBOR; 49°32′N 14°49′E. Agriculture (oats, barley, potatoes); manufacturing (buttons, food processing, metal products). Has an 18th century baroque castle.

Mlade Buky (MLAH-de BU-ki), Czech *Mladé Buky*, German *Jungbuch*, village, VYCHODOCESKY province, NE BOHEMIA, CZECH REPUBLIC, 5 mi/8 km NW of TRUTNOV; 50°36′N 15°52′E. Manufacturing (textiles); paper mill; agriculture (barley and flax); tourist resort.

Mladenovac (mlah-DEN-o-vahts), town (2002 population 52,490), N central SERBIA, 27 mi/43 km SSE of BELGRADE, in the Sumadija region; 44°26′N 20°42′E. Railroad junction. Manufacturing (jute bags). Mineral waters. Also spelled Mladenovats.

Mlala Hills (MLA-lah), hill range, rising to c. 5,577 ft/1,700 m, RUKWA region, W TANZANIA, 75 mi/121 km N of SUMBAWANGA. KATAVI PLAINS to W, Lake RUKWA to S.

Mlali (MLA-lee), village, DODOMA region, E central TANZANIA, 65 mi/105 km E of DODOMA; 06°19′S 36°44′E. Cattle, sheep, goats; wheat, corn. Mount Mumwere (7,425 ft/2,263 m) to E.

Mlangali (mlan-GAH-leeh), village, IRINGA region, SW TANZANIA, 95 mi/153 km NW of SONGEA, in LIVING-STONE MOUNTAINS; 09°45′S 34°29′E. Subsistence crops; livestock. Iron deposits. Lake NYASA to SW.

Mlangeni (mlah-ngeh-ni), village, Central region, MALAWI, on MOZAMBIQUE border, on road, and 25 mi/40 km SE of DEDZA; 15°56′S 34°46′E. Tobacco, wheat, corn, peanuts.

Mlava, POLAND: see MLAWA.

Mlava River (MLAH-vah), c.60 mi/97 km long, in E SERBIA; rises in the CRNI VRH near ZAGUBICA; flows NNW past PETROVAC to an arm of the DANUBE River N of KOSTOLAC.

Mlawa (MLAH-vah), Polish *Mława*, Russian *Mlava*, town (2002 population 29,422), Ciechanów province, NE central POLAND, 65 mi/105 km NNW of WARSAW; 53°07′N 20°23′E. Railroad junction. Cement, thread, candy; flour milling, tanning. During World War II, in EAST PRUSSIA, and called Mielau.

Mlawula Nature Reserve (mlah-WOO-lah) (□ 46 sq mi/119.6 sq km), LUBOMBO district, NE SWAZILAND, c.15 mi/24 km SE of MHLUME; 26°15′S 32°02′E. Bounded E by MOZAMBIQUE, N byMBULUZI RIVER. It incorporates Ndzindza Nature Reserve and covers c.29,500 acres/12,000 ha. Crocodiles, hyenas, antelope, sambala monkeys, wildebeest. railroad cuts through center of park. White-water rafting on the Usutu in the Bulungu Gorge is a tourist attraction.

Mlazovice, CZECH REPUBLIC: see LAZNE BELOHRAD.

Mligasi River (mlee-GAH-see), c. 90 mi/145 km long, in E TANZANIA; rises c.100 mi/161 km WSW of TANGA;

flows SE then E, enters ZANZIBAR CHANNEL, INDIAN OCEAN, 68 mi/109 km SSW of TANGA in SADANI Game Reserve.

Mlilwane Wildlife Sanctuary (mlee-WAH-ne) (□ 17 sq mi/44.2 sq km), HHOHHO district, W central SWAZILAND, c.5 mi/8 km S of MBABANE; 26°30′S 31°09′E. Hippopotamus, crocodiles, zebra, antelope, birds, giraffe, leopards. First wildlife preserve in the country (1964); covers 10,992 acres/4,448 ha. National Environmental Education Center visited by 30,000 students each year.

Mlini, resort village, S CROATIA, on ADRIATIC coast of DALMATIA, on Dubrovnik Riviera, S of DUBROVNIK.

Mlinov, UKRAINE: see MLYNIV.

Mljet Island (MLYET), Italian *Meleda*, ancient *Melita*, (□ 38 sq mi/98.8 sq km) Dalmatian island in ADRIATIC SEA, S CROATIA, 30 mi/48 km W of DUBROVNIK; 24 mi/39 km long E-W; rises to 1,686 ft/514 m. Has three freshwater lakes and two grottoes; emits sulphur fumes. Was Roman place of exile. Has Roman palace, Benedictine cloister (now a hotel). Chief village, Babino Polje. Mljet Channel (Croatian *Mljetski Kanal*), separates island from the Pelješac peninsula (N). Mljet National Park located here.

Mljet National Park (□ 12 sq mi/31 sq km), on W part of Mljet Island, in ADRIATIC SEA, DALMATIA, S CROATIA. Includes two lakes linked by canal, Veliko jezero [=big l.] and Malo jezero [=small l.]. Canal also connects Veliko jezero with Adriatic. Benedictine monastery (12th century) on small island in Veliko jezero has been a hotel since 1961. Lush Mediterranean flora.

Mlola (MLO-lah), village, TANGA region, NE TANZANIA, 45 mi/72 km NW of TANGA, in USAMBARA Mountains; 04°39′S 38°25′E. Timber; sheep, goats; corn.

Mlowe (mlo-we), port on W shore of Lake Malawi (LAKE NYASA), Northern region, MALAWI, 25 mi/40 km N of RUARWE, and 62 mi/100 km N of NKHATA BAY; 10°45′S 34°13′E.

Mlumati River, SOUTH AFRICA and Swaziland: see LO-MATI RIVER

Mlynany, SLOVAKIA: see ZLATE MORAVCE.

Mlyniv (MLI-neef) (Russian *Mlinov*) (Polish *Mlynów*), town, SW RIVNE oblast, UKRAINE, on IKVA RIVER, and 7 mi/11 km NNW of DUBNO; 50°30′N 25°36′E. Elevation 616 ft/187 m. Food processing (grain, dairy, vegetables), furniture manufacturing. Heritage museum. Known since the beginning of the 16th century; town since 1959.

Mlynów, UKRAINE: see MLYNIV.

Mmabatho, town, NORTH-WEST province, N SOUTH AFRICA, c.10 mi/16 km S of BOTSWANA border, 5 mi/8 km N of MAFIKENG, its twin town; 25°51′S 25°38′E. On railroad and highway. Regional airport. Was capital of North-West province before it was switched to Mafikeng. Was capital of BOPHUTHATSWANA.

Mnero (MNAI-roh), village, LINDI region, SE TANZANIA, 75 mi/121 km WSW of LINDI; 10°15′S 38°36′E. Cashews, subsistence crops; livestock.

Mnichovo Hradiste (MNYI-kho-VO HRAH-dyish-TYE), Czech *Mnichovo Hradiště*, German *München-grätz*, town, STREDOCESKY province, N BOHEMIA, CZECH REPUBLIC, on left bank of JIZERA RIVER, on railroad, and 17 mi/27 km SSW of Liberec; 50°32′N 14°59′E. Manufacturing (machinery, textiles). Noted for Prussian victory in 1866 over the Austrians. Wallenstein is buried here in a castle chapel. On right bank, across Jizera River, is the village of KLASTER HRADISTE, which is home to a brewery (established in 1852).

Mnine, Syria: see MENIN.

Mnischek unter Brdy, CZECH REPUBLIC: see MNISEK POD BRDY.

Mnisek nad Hnilcom (mnyee-SHEK NAHD hnyil-TSOM), Slovak *Mníšek nad Hnilcom*, Hungarian *Szepesremete*, village, VYCHODOSLOVENSKY province,

E central Slovakia, on railroad, and 16 mi/26 km NE of ROZNAVA; 48°48'N 20°48'E. Former iron mining.

Mnisek pod Brdy (MNYEE-shek POD BUHR-di), Czech *Mníšek pod Brdy*, German *Mnischek unter*, town, STREDOCESKY province, central BOHEMIA, CZECH REPUBLIC, on railroad, and 16 mi/26 km SSW of PRAGUE; 49°52'N 14°16'E. Metallurgy of aluminum. Has a 17th century baroque castle, and an 18th century baroque church.

Mnogovershinnyy (mno-guh-veer-SHIN-niyee) [Russian=of many summits], settlement (2005 population 2,690), E KHABAROVSK TERRITORY, RUSSIAN FAR EAST, 57 mi/92 km NW, and administrative jurisdiction of NIKOLAYEVSK-NA-AMURE (connected by local highway); 53°56'N 139°55'E. Elevation 1,282 ft/ 390 m.

Mnyusi (MNYOO-see), village, TANGA region, NE TANZANIA, 35 mi/56 km WSW of TANGA, near PANGANI RIVER, on railroad; 05°15'S 38°34'E. Sisal, rice; livestock.

Mo (MAW) or **Mo i Rana**, town, NORDLAND county, N central NORWAY, at head of RANA Fjord, at mouth of RANA River, on railroad, and 70 mi/113 km S of BODØ; 63°16'N 09°01'E. Steel-milling center in mining (zinc, copper, pyrite, lead) region. Lumbering; furniture manufacturing.

Moa (MO-ah), deep-water port city, on NE coast of HOLGUÍN province, E CUBA; 20°40'N 74°56'W. Ships nickel and sugar; nickel smelter and quarries. Airport (runway 6,000 ft/1,829 m).

Moa (MO-ah), town, TANGA region, NE TANZANIA, 20 mi/32 km N of TANGA, on Moa Bay, INDIAN OCEAN; 04°45'S 39°13'E. Fish; sisal, rice, corn; sheep, goats; timber.

Moa (MO-wah), largest island (☐ 169 sq mi/439.4 sq km) of the LETI ISLANDS, S MALUKU province, INDONESIA, in BANDA SEA, 40 mi/64 km ENE of E tip of TIMOR; 25 mi/40 km long, 7 mi/11.3 km wide; 08°20'S 127° 57'E. Fishing; coconuts. Dutch fort built here 1734.

Moab (MO-ab), town (2006 population 4,875), ⊙ GRAND county, E UTAH, on COLORADO RIVER (crossed here by bridge), and 100 mi/161 km SE of PRICE; 38°34'N 109°32'W. Elevation 4,025 ft/1,227 m. Tourist point; trade center for cattle, sheep; irrigated agricultural area. Oil and natural-gas processing; salt, potassium mining. Vanadium and uranium mines nearby. LA SAL MOUNTAINS are E, in section of La Sal National Forest (headquarters at Moab); ARCHES NATIONAL PARK is N. CANYONLANDS NATIONAL PARK and Dead Horse Point State Park to SW. Settled 1855; grew as ranching point after 1876.

Moab (MO-ab), ancient nation located in the uplands E of the DEAD SEA, now part of JORDAN. High plateau descending in the W by a steep escarpment to the shores of the Dead Sea, and sloping gently E toward the SYRIAN DESERT. The area is unprotected from the E; hence its history is a chain of raids by the Bedouin. The Moabites were close kin to the Hebrews, and the language of the Moabites is practically the same as biblical Hebrew. The relations of Moab with Judah and ISRAEL are continually mentioned in the Bible. As a political entity, Moab came to an end after the invasion (c.733 B.C.E.) of Tiglath-pileser III. Its people were later absorbed by the Nabataeans. The Moabite religion was much like that of CANAAN. Archaeological exploration in Moab has shown that settlements first occurred in the 13th century B.C.E.

Moabi (mo-ah-BEE), town, NYANGA province, SW GABON, 35 mi/56 km S of MOUILA; 02°22'S 10°58'E.

Moabit (mo-ah-BIT), residential district, N central BERLIN, GERMANY. Site of large prison.

Moa, Cayo Grande de, CUBA: see GRANDE DE MOA, CAYO.

Moa, Cuchillas de (MO-ah, koo-CHEE-yahs dai), small range, HOLGUÍN province, NE CUBA, NW of BARACOA, extends c.20 mi/32 km NW along ATLANTIC

coast; rises to 1,175 m at Pico de Toldo. Yields timber; has iron deposits. Has 300 ft/91 m cascade on small Moa River.

Moa Island (MO-uh) (☐ 35 sq mi/91 sq km), in TORRES STRAIT, 32 mi/51 km N of CAPE YORK PENINSULA, N QUEENSLAND, NE AUSTRALIA, just E of BADU ISLAND (formerly Mulgrave Island); 10°11'S 142°16'E. Circular, 28 mi/45 km in circumference; rises to 1,310 ft/399 m. Fertile, wooded. Pearl shell, trepang. Aboriginal reserve. Native population mainly Malays and Melanesians. Formerly called Banks Island.

Moak Lake (MOK), community, central MANITOBA, W central CANADA, 16 mi/26 km from THOMPSON, in MYSTERY LAKE local government district; 55°55'N 97°35'W.

Moala (mo-AH-lah), volcanic island (☐ 24 sq mi/62.4 sq km), LAU group, FIJI, SW PACIFIC OCEAN; 7 mi/11.3 km long; 18°36'S 179°53'E. Rises to 1,535 ft/468 m. Copra.

Moala Group (mo-AH-lah), island group, FIJI, SW PACIFIC OCEAN, with three main high volcanic islands: MOALA, TOTOYA, and MATUKU.

Moalboal (mo-AHL-bo-AHL), town, CEBU province, S Cebu island, PHILIPPINES, on TAÑON STRAIT, 40 mi/ 64 km SW of CEBU; 09°55'N 123°21'E. Agricultural center (corn, coconuts). Was severely damaged in 1984 typhoon.

Moama (mo-A-muh), municipality, S NEW SOUTH WALES, AUSTRALIA, 120 mi/193 km N of MELBOURNE, on MURRAY RIVER, on VICTORIA border; 36°07'S 144°47'E. Sheep, agriculture center.

Moamba (mo-AHM-buh), village, MAPUTO province, S MOZAMBIQUE, 35 mi/56 km NW of MAPUTO; 25°36'S 32°15'E. Railroad junction. Cattle-raising center; beans, corn.

Moamba District, one of seven districts of MAPUTO province, MOZAMBIQUE.

Moanda (MO-ahn-duh), town, HAUT-OGOOUÉ province, SE GABON, 25 mi/40 km NW of FRANCEVILLE; 01°32'S 13°13'E. Important manganese production center.

Moanda (MWAHN-dah), village, BAS-CONGO province, W CONGO, on ATLANTIC coast, 50 mi/80 km W of BOMA; 05°56'S 12°20'E. Tourist center with beach and baths. Palm-oil milling, cattle raising. Lighthouse. Airport. Roman Catholic mission. Also spelled MUANDA.

Moapa (MO-puh) or **Moapa Valley**, unincorporated village, CLARK county, SE NEVADA, 44 mi/71 km NE of LAS VEGAS; 36°34'N 114°28'W. Small farming community at entrance to Moapa Indian Reservation, to SW. Junction of railroad spur to OVERTON. Cattle; poultry; vegetables; manufacturing (concrete). Valley of Fire State Park and Lake Mead National Recreation Area to SE. Warm Springs Resort, desert tourist oasis, is here.

Moa River (MO-ah), 125 mi/201 km long; rises in S GUINEA, N of KAILAHUN (SIERRA LEONE); flows into Sierra Leone SW past DARU to the ATLANTIC OCEAN at SULIMA; 07°40'N 15°11'W. Also called Galhina River or Gallina River (for Gallina tribe settled along lower course).

Moate (MOT), Gaelic *An Móta*, village (2006 population 1,888), SW WESTMEATH county, S IRELAND, 10 mi/16 km ESE of ATHLONE; 53°33'N 07°43'W. Agricultural market. Ancient rath (earthwork) nearby.

Moatize (mo-ah-TEE-zai), village, NW MOZAMBIQUE, 12 mi/19 km ENE of TETE; 16°10'S 33°46'E. Coal mines (linked by spur railroad with BENGA on the ZAMBEZI).

Moatize District, one of twelve districts of TETE province, MOZAMBIQUE.

Moba (MO-buh), village, KATANGA province, SE CONGO, on W shore of LAKE TANGANYIKA, 85 mi/137 km SSE of KALEMIE; 07°03'S 29°47'E. Elev. 2,506 ft/ 763 m. Terminus of navigation on the TANGANYIKA; customs station. Formerly known as BAUDHUINVILLE or BAUDHOUINVILLE.

Mobara (mo-BAH-rah), city, CHIBA prefecture, E central HONSHU, E central JAPAN, on E central BOSO PENINSULA, 12 mi/20 km S of CHIBA; 35°25'N 140°17'E. Appliances. Peanuts.

Mobarakeh (muh-bah-RAHK-ai), town (2006 population 62,728), Esfahán province, W central IRAN, 27 mi/43 km SSW of ESFAHAN, and on Zaindeh River Rice, cotton.

Mobaye (mo-bah-YAI), town, ⊙ BASSE-KOTTO prefecture, S CENTRAL AFRICAN REPUBLIC, on UBANGI RIVER opposite MOBAYI-MBONGO (CONGO), and 100 mi/161 km SSE of BAMBARI; 04°19'N 21°11'E. Cotton ginning; fishing; agriculture (coffee) and timber plantations (rubber, elaeis-palm). River port and customs station.

Mobayi-Mbongo (mo-BAH-yee—MBON-go), village, Équateur province, NW CONGO, on UBANGI RIVER (CENTRAL AFRICAN REPUBLIC border), opposite MOBAYE, 145 mi/233 km NNW of LISALA; 04°18'N 21°11'E. Elev. 1,404 ft/427 m. Customs station and trading center; cotton ginning. Has churches and schools. UBANGI RIVER is navigable between here and YAKOMA. Formerly called BANZYVILLE or BANZYSTAD. Also known as YASANYAMA.

Mobeetie (MO-bee-tee), village (2006 population 101), WHEELER county, extreme N TEXAS, in the PANHANDLE, 30 mi/48 km E of PAMPA; 35°31'N 100°26'W. In farming and cattle region; oil. Nearby is Old Mobeetie, site of old Fort Elliot. Historic jail is now a museum.

Mobeka, CONGO: see GUMBA.

Mobendi, CONGO: see BRABANTA.

Moberly, city (2000 population 11,945), RANDOLPH county, N central MISSOURI; 39°25'N 92°26'W. Manufacturing (instant ice packs, automobile components, gym equipment). Limestone quarries. Former coal-mining region. Former important railroad center. Moberly Area Junior College. State Correctional Center (penitentiary) to S. Incorporated 1868.

Mobile (MO-beel), county (☐ 1,644 sq mi/4,274.4 sq km; 2006 population 404,157), extreme SW ALABAMA; ⊙ MOBILE. Coastal plain, bounded S by MISSISSIPPI SOUND, E by MOBILE BAY and Mobile River, W by MISSISSIPPI. Corn, soybeans, pecans, berries, subtropical fruits; seafood; paper mills; crude-oil and natural-gas production. City of Mobile (manufacturing center) is Alabama's only seaport. Formed 1812.

Mobile (MO-beel), city (2000 population 198,915), ⊙ Mobile co., SW Alabama, at the head of Mobile Bay, at the mouth of the Mobile River. It is one of the country's major ports, the only seaport in Alabama, and the 2nd-largest city in the state. Oil refineries; paper, textiles, aluminum, and chemicals; iron smelting. After the Tennessee-Tombigbee waterway was completed in 1984, connecting N Mississippi's Tennessee River with the Tombigbee River in W Alabama and providing access to the Gulf of Mexico, Mobile enjoyed a boom in business growth and redevelopment. Mobile was the capital of French Louisiana (1710–1719). The British held it (1763–1780), before Bernardo de Gálvez took it for Spain. Mobile was seized by the U.S. in 1813. During the Civil War, ships from Mobile evaded the Federal blockade until Admiral Farragut's victory at Mobile Bay (1864); Gen. E. River S. Canby captured the city in April, 1865. Mobile has many beautiful antebellum homes and magnificent gardens. Also noteworthy are a Roman Catholic cathedral, the city hall (1858), and Marine Hosp. (1842). Of historical interest are forts Morgan and Gaines at the entrance to Mobile Bay. Mobile is the seat of Spring Hill College, Universityof Mobile, the University of South Alabama, and Bishop State Community College Brookley, a coast guard station, and a coast guard aviation training center are here. The USS *Alabama* Battleship Memorial Park, USS-*Drum* submarine, and numerous aircraft are here. The colorful annual Mardi Gras was begun in the early 1700s; the Azalea Trail. Festival dates from 1929. The Bankhead Tunnel lies under the Mobile River. Inc. 1814.

Mobile Bay (MO-beel), arm of the GULF OF MEXICO, SW ALABAMA, extending c.35 mi/56 km from the Gulf to the mouth of the Mobile River; 8 mi/12.9 km–18 mi/29 km wide. A ship channel connects Mobile Bay with the Gulf. The INTRACOASTAL WATERWAY passes through the S part of the bay. MOBILE is on the NW shore. Admiral David Farragut, a Civil War naval hero, won the celebrated battle of Mobile Bay on August 5, 1864.

Mobilong, AUSTRALIA: see MURRAY BRIDGE.

Mobjack Bay (MAHB-jak), arm of CHESAPEAKE BAY, GLOUCESTER and MATHEWS counties, E VIRGINIA, 25 mi/40 km N of NEWPORT NEWS; c.5 mi/8 km wide at SE entrance, up to 10 mi/16 km long. Has several inlets.

Mobridge, city (2006 population 3,231), WALWORTH county, N central SOUTH DAKOTA, on MISSOURI RIVER, and 80 mi/129 km N of PIERRE; 45°32′N 100°26′W. Trade and distribution point for agricultural region; lignite coal; cement blocks, concrete; beverages; livestock, dairy produce. Lutheran academy is here. Sports fishing center. Mountain-Central time zone boundary runs through center of river. Large Standing Rock Indian Reservoir to W (including Sitting Bull's Grave and Sacajawea Monument across river); large Cheyenne River Indian Reservation to SW. Founded 1906, incorporated 1908.

Mobutu Sese Seko, Lake, CONGO: see ALBERT, LAKE.

Moca (MO-kah), city (2002 population 59,172), ⊙ ESPAILLAT province, N DOMINICAN REPUBLIC, on S slope of Cordillera SEPTENTRIONAL, on highway, and 13 mi/21 km ESE of SANTIAGO; 19°25′N 70°30′W. Coffee-growing center; also cacao and tobacco. Founded 1780.

Moca (MO-kah), town (2006 population 43,664), NW PUERTO RICO, 3 mi/4.8 km SE of AGUADILLA. Livestock; light manufacturing; rope hammock-making artisan shops. Famous for mundillo (lace-making).

Mocaboc Point (mo-KAH-bok), easternmost point of NEGROS island, PHILIPPINES, in TAÑON STRAIT near its entrance; 10°50′N 123°33′E. Shelters small Escalante Bay.

Mocajuba (mo-kah-zhoo-bah), city (2007 population 23,184), E PARÁ, BRAZIL, on right bank of TOCANTINS River, and 100 mi/161 km SW of BELÉM; 02°45′S 49°28′W. Cacao, rubber, medicinal plants.

Mocal River (mo-KAHL), c.50 mi/80 km long, in W HONDURAS; rises in Sierra de Celaque 12 mi/19.2 km NE of SAN MARCOS (OCOTEPEQUE department); flows SSE through LEMPIRA department to RÍO LEMPA 7 mi/11.2 km SW of CANDELARIA.

Moçambique (mo-zam-BEEK) or **Mozambique**, city, NAMPULA province, NE MOZAMBIQUE, a seaport on a small coral island in the MOZAMBIQUE CHANNEL (an arm of the INDIAN OCEAN); 15°02′N 40°44′E. It is c.3 mi/4.8 km from the mainland town of LUMBO, a terminus of a railroad into the interior. The city is a trade center; exports include cashew nuts and timber. It was occupied by the Portuguese (1505) and was the capital of the Portuguese holdings in Mozambique until 1907. Still standing are three old forts and the governor's palace, which attract numerous tourists.

Mocambo (MO-kahm-bo), city, N PIAUÍ state, BRAZIL, 75 mi/121 km NW of TERESINA; 04°05′S 42°08′W.

Mocambo, MEXICO: see BOCA DEL RÍO.

Mocanaqua (MO-kah-NAH-kwah), unincorporated town, Conyngham township, LUZERNE county, E central PENNSYLVANIA, 15 mi/24 km SW of WILKES-BARRE, on SUSQUEHANNA RIVER, at mouth of Turtle Creek; 41°08′N 76°08′W. Light manufacturing

Mocanguê Pequeno Island (MO-kahm-goo-ai), in GUANABARA BAY, RIO DE JANEIRO state, SE BRAZIL, 3.5 mi/5.6 km NE of RIO DE JANEIRO, near harbor of NITERÓI; 22°52′S 43°08′W. Dry docks.

Mo Cay (MO KAI), village, BEN TRE province, S VIETNAM, in MEKONG delta, 8 mi/12.9 km SSW of BEN TRE; 10°08′N 106°20′E. Market hub; rice milling; rice-growing center. Formerly Mocay.

Moccasin (MAH-kuh-suhn), unincorporated village, MOHAVE county, NW ARIZONA, 18 mi/29 km SW of KANAB, UTAH, headquarters of Kaibab Indian Reservation. PIPE SPRING NATIONAL MONUMENT is to S, within reservation.

Moccasin, village, JUDITH BASIN county, central MONTANA, 22 mi/35 km WSW of LEWISTOWN. Railroad junction. Wheat, barley, hay; livestock. Agricultural research center to W. Ackley Lake State Park to S.

Moccasin Creek, NORTH CAROLINA: see CONTENTNEA CREEK.

Mocejón (mo-thai-HON), town, TOLEDO province, central SPAIN, on canal of the TAGUS, and 8 mi/12.9 km NE of TOLEDO; 39°55′N 03°54′W. Potatoes, sugar beets, cereals, melons; sheep; manufacturing (furniture).

Mocenok (mo-CHE-nok), Slovak *Močenok*, Hungarian *Mocsonok*, village, ZAPADOSLOVENSKY province, SW SLOVAKIA, 9 mi/14.5 km SW of NITRA; 48°14′N 17°56′E. Sugar beets, barley, wheat, corn. Food processing. Has a castle with a park. Called Sladeckovce, Slovak *Sládečkovce* (1951–1990).

Mocha (MO-kah), town, S YEMEN, a port on the RED SEA, 40 mi/64 km N of BAB EL MANDEB; 13°19′N 43°14′E. It was noted for the export of the coffee to which it gave its name but declined as a trading port in the late 19th century with the rise of HODEIDA and ADEN. Also spelled Mokha.

Mocha Island (MO-chah), 20 mi/32 km off coast of ARAUCO province, BÍO-BÍO region, S central CHILE, 50 mi/80 km SSW of LEBU, in the PACIFIC; 8 mi/13 km long, c.3 mi/4.8 km wide; 38°22′S 73°56′W. Rises to 1,768 ft/539 m.

Mochalishche (muh-CHAH-lee-shchye), village (2006 population 2,215), S central MARI EL REPUBLIC, E central European Russia, on the Yushut River (right tributary of the ILET' RIVER), on railroad, 28 mi/45 km SE of YOSHKAR-OLA; 56°20′N 48°21′E. Elevation 328 ft/99 m. Agricultural products; lumber.

Moche (MO-chai), town (2005 population 24,853), TRUJILLO province, LA LIBERTAD region, NW PERU, on coastal plain, on railroad, and 4 mi/6 km SSE of TRUJILLO, on PAN-AMERICAN HIGHWAY; 08°11′S 79°02′W. Corn, sugarcane, fruit, cotton, rice. Near Temple of the Sun (largest pre-Columbian building in SOUTH AMERICA) and the Moon, from the Moche culture.

Moche River (MO-chai), LA LIBERTAD region, NW PERU; 08°09′S 79°02′W.

Mochigase (mo-CHEE-gah-se), town, Yazu county, TOTTORI prefecture, S HONSHU, W JAPAN, 12mi/19 km S of TOTTORI; 35°20′N 134°12′E.

Mochima, Parque Nacional (mo-CHEE-mah PAHR-kai nah-see-o-NAHL), national park (□ 365 sq mi/949 sq km), SUCRE and ANZOÁTEGUI states, NE VENEZUELA; 10°17′N 64°30′W. Includes coastal and island zone between CUMANÁ and PUERTO LA CRUZ. Created 1973.

Mochis, MEXICO: see LOS MOCHIS.

Mochishche (MO-chee-shchye), village (2006 population 3,300), NE NOVOSIBIRSK oblast, SW Siberian Russia, on road and railroad, 7 mi/11 km NE of NOVOSIBIRSK; 55°09′N 83°07′E. Elevation 626 ft/190 m. In agricultural area (livestock). Has an airfield.

Mochitlán (mo-chee-TLAHN), town, GUERRERO, SW MEXICO, in SIERRA MADRE DEL SUR, 9 mi/14.5 km ESE of Chilpancingo de Dravo; 17°30′N 99°18′W. Cereals, sugarcane, fruit, forest products (resin, vanilla).

Mochito, El, HONDURAS: see EL MOCHITO.

Mochizuki (mo-CHEEZ-kee), town, S Saku county, NAGANO prefecture, central HONSHU, central JAPAN, 28 mi/45 km S of NAGANO; 36°15′N 138°21′E. Highland vegetables, medicinal plants. Edo-period streetscape. Hot-spring and skiing area.

Mochovce, SLOVAKIA: see TLMACE.

Mochrum (MO-kruhm) or **Kirk of Mochrum**, village, DUMFRIES AND GALLOWAY, S Scotland, 8 mi/12.9 km SW of WIGTOWN; 54°47′N 04°34′W. Nearby is mansion of marquess of Bute, with two 15th-century towers.

Mo Chu (MO CHOO), river, BHUTAN, tributary of PUNATSANG; 26°23′N 89°48′E.

Mochudi (MO-choo-dee), town (2001 population 36,962), ⊙ KGATLENG district, SE BOTSWANA, near SOUTH AFRICA border, 26 mi/42 km NE of GABORONE; 24°20′S 26°10′E. Elevation 3,320 ft/1,012 m. Agricultural center. Headquarters of Bakgatla tribe. Hospital. Museum located in oldest school building in country.

Mochuly (muh-CHOO-li), village, S central SMOLENSK oblast, W European Russia, near highway and railroad, 6 mi/10 km SE of (and administratively subordinate to) POCHINOK; 54°19′N 32°33′E. Elevation 659 ft/200 m. Flax processing. Also spelled Machuly.

Mochumí (mo-choo-MEE), town, LAMBAYEQUE province, LAMBAYEQUE region, NW PERU, on coastal plain, on PAN-AMERICAN HIGHWAY, and 10 mi/16 km NNE of LAMBAYEQUE; 06°33′S 79°52′W. In irrigated La Leche River valley; rice, corn, cotton; livestock. Variant spelling: Muchumi.

Mochuritsa (mo-CHOO-reet-sah), river, BURGAS oblast, E central BULGARIA; 43°35′N 26°50′E. Tributary of the TUNDZHA RIVER.

Mocímboa da Praia (mo-seem-BO-uh dah PREI-uh), village, CABO DELGADO province, northernmost MOZAMBIQUE, on MOZAMBIQUE CHANNEL of the INDIAN OCEAN, 135 mi/217 km N of PEMBA, and 60 mi/97 km S of Tanzanian border; 11°20′S 40°21′E. Ships copra, sisal, castor beans. Airfield.

Mocímboa da Praia, one of 16 districts of CABO DELGADO province, MOZAMBIQUE.

Mociu (MO-chyoo), Hungarian *Mocs*, village, CLUJ county, W central ROMANIA, 20 mi/32 km E of CLUJ-NAPOCA; 46°48′N 24°02′E. Flour milling.

Mockau (MOK-kou), industrial suburb N of LEIPZIG, SAXONY, E central GERMANY; 51°23′N 12°55′E.

Möckeln (MUHK-keln), lake, S SWEDEN, extends N from ÄLMHULT; 9 mi/14.5 km long, 1 mi/1.6 km–4 mi/6.4 km wide. Drained W by HELGE Å RIVER.

Möckeln, lake expansion of LETÄLVEN RIVER, S central SWEDEN, extends SW from KARLSKOGA and BOFORS; 6 mi/9.7 km long, 1 mi/1.6 km–2 mi/3.2 km wide.

Möckern (MUK-kern), industrial NW suburb of LEIPZIG, SAXONY, E central GERMANY; 51°23′N 12°50′E. Scene (October 1813) of major engagement during battle of LEIPZIG.

Mock Horn Island (MAHK HORN), marshy island, NORTHAMPTON county, E VIRGINIA, off ATLANTIC OCEAN coast, 6 mi/10 km E of CAPE CHARLES town; 8 mi/13 km long, 2 mi/3.2 km wide; 37°14′N 75°53′W. Separated from DELMARVA (EASTERN SHORE) peninsula by Mockhorn (NW) and Magothy (SW) bays; separated from WRECK and SHIP SHOAL islands to E by South Bay. Mockhorn Island Wildlife Management Area covers island.

Mockingbird Valley, town (2000 population 190), JEFFERSON county, N KENTUCKY, residential suburb, 4 mi/6.4 km NE of LOUISVILLE, and near OHIO RIVER; 38°16′N 85°40′W.

Möckmühl (MUK-myool), town, FRANCONIA, BADEN-WÜRTTEMBERG, GERMANY, on the JAGST RIVER, and 14 mi/23 km NNE of HEILBRONN; 49°20′N 09°22′E. Paper milling.

Mocksville (MAHKS-vil), town (□ 6 sq mi/15.6 sq km; 2006 population 4,525), ⊙ DAVIE county, central NORTH CAROLINA, 22 mi/35 km NW of WINSTON-SALEM; 35°53′N 80°33′W. Light manufacturing; service industries; agriculture (tobacco, soybeans, grain; chickens, livestock; dairying). Joppa Cemetery, burial site of Squire and Sarah Boone, parents of pioneer Daniel Boone. Settled before 1750, incorporated 1839.

Area is shown by the symbol □, and capital city or county seat by ⊙.

Moclín (mo-KLEEN), town, GRANADA province, S SPAIN, on small Moclín River, and 15 mi/24 km NW of GRANADA; 37°20′N 03°47′W. In fertile agricultural region (olives, cereals, tubers, sugar beets, beans, potatoes, livestock); olive-oil pressing, flour milling. Old town was once a Moorish stronghold.

Moclinejo (mo-klee-NAI-ho), town, MÁLAGA province, S SPAIN, 9 mi/14.5 km ENE of MÁLAGA; 36°46′N 04°15′W. Exports raisins, lemons, olives. Goat raising.

Moclips (MO-klips), unincorporated town, GRAYS HARBOR county, W WASHINGTON, 23 mi/37 km NW of HOQUIAM, on PACIFIC, at mouth of Moclips River railroad junction. Logging. Copalis National Wildlife Refuge, including coastal rocks, offshore. Quinault Indian Reservation to N.

Mocoa (mo-KO-ah), town, ⊙PUTUMAYO department, SW COLOMBIA, on affluent of CAQUETÁ River, in ANDEAN foothills, and 45 mi/72 km E of PASTO (NARIÑO department), 300 mi/483 km SW of BOGOTÁ; 01°08′N 76°38′W. Trading post in tropical forest and livestock region; some sugarcane. Airport.

Mococa (MO-ko-kah), city (2007 population 66,102), E SÃO PAULO, BRAZIL, near MINAS GERAIS border, 55 mi/89 km ESE of RIBEIRÃO PRÊTO; 21°28′S 47°01′W. Dairying center; coffee processing.

Mocochá (mo-ko-CHAH), town, YUCATÁN, SE MEXICO, 13 mi/21 km NE of MÉRIDA, on Mexico Highway 281. Henequen.

Mocomoco (mo-ko-MO-ko), town and canton, CAMACHO province, LA PAZ department, W BOLIVIA, 21 mi/34 km ENE of PUERTO ACOSTA, in the ALTIPLANO; 15°22′S 68°59′W. Elevation 10,193 ft/3,107 m.

Mocorito (mo-ko-REE-to), city and township, SINALOA, NW MEXICO, on small Evora River, and 55 mi/89 km NW of CULIACÁN (Rosales); 25°30′N 107°53′W. Agricultural center (corn, sugarcane, tomatoes, chickpeas, fruit).

Moçoró, Brazil: see MOSSORÓ.

Mocs, ROMANIA: see MOCIU.

Moctezuma (mok-te-SOO-mah), city and township, SAN LUIS POTOSÍ, N central MEXICO, on interior plateau, 43 mi/69 km N of SAN LUIS POTOSÍ; 22°46′N 101°06′W. Grain, beans, cotton, maguey. Thermal springs nearby.

Moctezuma (mok-te-SOO-mah), town, SONORA, NW MEXICO, on MOCTEZUMA RIVER, and 90 mi/145 km NE of HERMOSILLO; 29°50′N 109°40′W. Copper-mining, wheat-growing center.

Moctezuma River (mok-te-SOO-mah), c.175 mi/282 km long, in N central and NE MEXICO; rises in SIERRA MADRE ORIENTAL SW of San Juan del Río; flows NE along QUERÉTARO-HIDALGO border, past TAMAZUNCHALE and TANQUIÁN in fertile La HUASTECA plains (SAN LUIS POTOSÍ), to join SANTA MARÍA (or Tamuin) River in forming PÁNUCO River on VERACRUZ border, 50 mi/80 km SW of TAMPICO. Used for irrigation.

Moctezuma River, c.110 mi/177 km long, in SONORA, NW MEXICO; rises in W outliers of SIERRA MADRE OCCIDENTAL; flows S past CUMPAS and MOCTEZUMA to YAQUI RIVER at PLUTARCO ELIAS CALLES Reservoir; 21°59′N 98°34′W.

Mocuba (mo-KOO-buh), village, ZAMBÉZIA province, central MOZAMBIQUE, 70 mi/113 km N of QUELIMANE (linked by railroad); 16°51′S 36°56′E. Agr. center; ships sisal, cotton.

Mocuba District, one of 15 districts in Zambézia province, MOZAMBIQUE.

Mocupe (mo-KOO-pai), town, LAMBAYEQUE province, LAMBAYEQUE region, NW PERU, on coastal plain, on irrigated SAÑA RIVER, and 20 mi/32 km SE of CHICLAYO; 06°59′S 79°37′W. Rice, corn, cotton, sugarcane.

Modale, town (2000 population 303), HARRISON county, W IOWA, near mouth of SOLDIER RIVER on MISSOURI RIVER, 11 mi/18 km W of LOGAN; 41°37′N 96°00′W. In agricultural area.

Modaliarpet (mo-dahl-YAHR-pet), French Modéliarpeth, town, PUDUCHERRY UNION TERRITORY, S INDIA; industrial suburb of PUDUCHERRY, 1.5 mi/2.4 km SW of city center. Cotton, rice, and oilseed milling, manufacturing of copra; dyeworks; pottery. Also spelled Mudaliarpet, Mudaliyarpettai.

Modane (mo-dahn), town (□ 27 sq mi/70.2 sq km), SAVOIE department, RHÔNE-ALPES region, SE FRANCE, in Alpine MAURIENNE valley, on the ARC RIVER, and 17 mi/27 km ESE of SAINT-JEAN-DE-MAURIENNE; 45°12′N 06°40′E. Elevation 3,468 ft/1,057 m. International railroad station at N entrance of MONT CENIS (or Fréjus) tunnel (built 1857–1872) linking France and Italy. Road tunnel, built 1972–1980, also cuts through frontier range from Modane to the Doria valley in Italy. A large military wind-tunnel test installation for planes, missiles, and space vehicles is close by to the E, near AVRIEUX. The VANOISE NATIONAL PARK extends N and NE of Mondane.

Modasa (mo-dah-sah), town, SABAR KANTHA district, GUJARAT state, W central INDIA, 23 mi/37 km ESE of HIMATNAGAR; 23°28′N 73°18′E. Trade center for cotton, millet, wheat, peanuts; calico-cloth dyeing and printing, oilseed pressing. Gujarat fortress in fifteenth century.

Modave (mo-DAHV), commune (2006 population 3,743), Huy district, LIÈGE province, E BELGIUM, 6 mi/10 km SSW of HUY; 50°27′N 05°18′E.

Modbury (MAHD-buh-ree), village (2001 population 1,454), S DEVON, SW ENGLAND, 11 mi/18 km E of PLYMOUTH; 50°21′N 03°53′W. Former agricultural market. Has 13th-century church.

Modderfontein, town, GAUTENG, SOUTH AFRICA, on WITWATERSRAND, 8 mi/12.9 km NNE of JOHANNESBURG, near KEMPTON PARK; 26°05′S 28°06′E. Elevation 5,543 ft/1,690 m. Chemical manufacturing (explosives, nitrates, ammonia, insecticides).

Modder River, Afrikaans Modderrivier, village, NORTHERN CAPE province, SOUTH AFRICA, on FREE STATE border, on MODDER RIVER, near confluence of RIET RIVER, 22 mi/35 km SSW of KIMBERLEY; 29°02′S 24°05′E. Elevation 3,739 ft/1,140 m. Resort for water sports. Scene (November 28, 1899) of battle in South African War called Battle of Modderrivier. Two residential areas have developed to SW on Modder River: Morswedimosa and Ritchie.

Modder River, c.225 mi/362 km long, FREE STATE and NORTHERN CAPE provinces, SOUTH AFRICA; rises NW of WEPENER near DEWETSDORP; flows in a wide arc NW and W, past BOTSHABELO glen and SW past PAARDEBERG, to its confluence with the RIET RIVER at town of MODDER RIVER, 22 mi/35 km SSW of KIMBERLEY, from where the combined rivers enter the VAAL near DOUGLAS. Along its course are sites of numerous battles between British and Boers during South African War (1899–1902).

Modéliarpeth, India: see MODALIARPET.

Modena (MO-de-nah), province (□ 1,038 sq mi/2,698.8 sq km), EMILIA-ROMAGNA, N central ITALY; ⊙ MODENA; 44°30′N 10°54′E. Extends from ETRUSCAN APENNINES N to the PO, with plain occupying about 50% of area. Drained by SECCHIA and PANARO rivers. Agriculture (cereals, fodder, grapes, hemp, sugar beets, cherries, peaches, strawberries, tomatoes); livestock raising (cattle, pigs, sheep). Hydroelectric plant at Farneta. Manufacturing at Modena, MIRANDOLA, and SASSUOLO. Ferrari, Maserati automobile factories; sausage making, ceramic works, food processing, manufacturing of irrigated pumps, tractors, animal feed. Extensive wine making; known for Lambrusco wine.

Modena (MO-de-nah), city (2001 population 175,502), ⊙ MODENA province, EMILIA-ROMAGNA, N central ITALY, on the PANARO River; 44°40′N 10°55′E. An agricultural, commercial, and highly diversified, important industrial center. Manufactures include motor vehicles, cast-iron, machine tools, chemicals, and leather; food processing. An Etruscan settlement, the city was the site of a Roman colony called Mutina, founded in the early 2nd century B.C.E. and located on the AEMILIAN WAY. Modena became a free commune in the 12th century and in 1288 permanently passed to the Este family of FERRARA. The duchy of Modena, established in 1452, became the seat of the Este family after it lost Ferrara (1598). From the fall of Napoleon I in 1814 until 1859 the house of Austria-Este ruled harshly. Among the city's notable structures are the cathedral (12th century), which has a massive white marble campanile (289 ft/88 m high) called the Ghirlandina; the Palazzo dei Musei (1753–1767), which contains several art collections and the Este library; and the ducal palace (17th century). The nearby Nonantola abbey (founded 752) was a center of learning in the Middle Ages. Modena has a university. Natural-gas fields nearby.

Modena (mo-DEE-nuh), borough (2006 population 602), CHESTER county, SE PENNSYLVANIA, 1 mi/1.6 km S of COATESVILLE, on West Branch of BRANDYWINE CREEK; 39°57′N 75°47′W. Agriculture (grain, apples; livestock; dairying); manufacturing (metal bins, paper). Formerly Paperville.

Modena (mo-DEE-nuh), unincorporated rural community, MERCER county, N MISSOURI, between THOMPSON and WELDON rivers, 8 mi/12.9 km SW of PRINCETON.

Moder River (MO-duhr), c.35 mi/56 km long, BAS-RHIN department, in ALSACE, E FRANCE; rises in the N VOSGES, 2 mi/3.2 km N of La PETITE-PIERRE; flows generally ESE, past INGWILLER, HAGUENAU, and BISCHWILLER, to the RHINE below DRUSENHEIM. Receives the ZORN near its mouth.

Modesto, city (2000 population 188,856), ⊙ STANISLAUS county, central CALIFORNIA, 50 mi/80 km ENE of SAN JOSE, on TUOLUMNE RIVER (S) and Modesto Main Canal (N), near N part of San Joaquin Valley; 37°40′N 121°00′W. Elevation 27 ft/8 m. Center of a farming and fruit-growing area; food processing, tomato canning, wineries, fruit orchards, diversified manufacturing. Agriculture (dairying; cattle, poultry; nuts, vegetables, beans, pumpkins, melons, rice). Fishing (codfish). Railroad junction. Riverbank Army Ammunition Plant 7 mi/11.2 km to NE, on Stanislaus River; HETCH HETCHY AQUEDUCT to N. Modesto's population increased 54% between 1980–1990, and is still climbing steadily. Modesto Junior College (two-year) here. Incorporated 1884.

Modesto, village (2000 population 252), MACOUPIN county, SW central ILLINOIS, 16 mi/26 km NNW of CARLINVILLE; 39°28′N 89°58′W. In agricultural (cattle, hogs; corn, wheat, sorghum, soybeans) and bituminous-coal area. OTTER LAKE 5 mi/8 km SE.

Modesto Omiste (mo-DES-to o-MEES-tai), province (□ 873 sq mi/2,269.8 sq km), POTOSÍ department, W central BOLIVIA; ⊙ VILLAZÓN; 22°06′S 65°34′W. Elevation 11,296 ft/3,443 m. Agriculture (potatoes, yucca, bananas); cattle.

Modesto Reservoir, California: see TUOLUMNE RIVER.

Modica (MO-dee-kah), city (2001 population 52,639), SE SICILY, ITALY; 36°52′N 14°46′E. It is the center of an agricultural region where livestock is raised. Known in ancient times as Motyca, it was a feudal county in the 12th century and enjoyed a high degree of independence from the 14th to 18th century. Nearby are the Cava d'Ispica (a series of limestone grottoes containing cave dwellings) and prehistoric and early Christian tombs.

Modigliana (mo-deel-YAH-nah), town, FORLÎ province, EMILIA-ROMAGNA, N central ITALY, 10 mi/16 km SSW of FAENZA; 44°09′N 11°04′E. Bishopric.

Modimolle, town, LIMPOPO province, SOUTH AFRICA, on upper MOGALAKWENA River, and 75 mi/121 km N of TSHWANE, in the foothills of the Waterberge range; 24°43′S 28°24′E. Elevation 3,952 ft/1,205 m. Railroad junction; airfield. Agriculture center (wheat, tobacco, peanuts). Formerly called Nylstroom.

Modinagar (mo-dee-nuh-guhr), city, Ghaziabad district, Uttar Pradesh state, N central India.

Modjokerto, INDONESIA: see MOJOKERTO.

Modlimb, town, SOLAPUR district, MAHARASHTRA state, W central INDIA, 37 mi/60 km NW of SOLAPUR. Local trade center (millet, wheat, cotton). Also spelled Modnimb.

Modlin (mod-LEEN), part of the city of NOWY DWÓR MAZOWIECKI, Warszawa province, E central POLAND, on the VISTULA, at Narew River mouth, and 20 mi/32 km NW of WARSAW. In Russian Poland (1815–1919), called NOVOGEORGIEVSK, it was site of strong fortress built (1807–1812) by Napoleon; captured (1915, 1939) by Germans after heavy attacks.

Mödling (MUD-ling), town, E LOWER AUSTRIA, at E foot of WIENERWALD (forest), at entrance of picturesque Brühl valley, in the suburban zone of VIENNA, 9 mi/14.5 km SSW of city center; 48°05′N 16°23′E. Market and educational center with a technical high school. Vineyards. Has 15th century Gothic church.

Modnimb, India: see MODLIMB.

Modoc, county (□ 3,944 sq mi/10,254.4 sq km; 2006 population 9,597), NE CALIFORNIA; ⊙ ALTURAS; 41°36′N 120°43′W. Bounded N by OREGON, E by NEVADA. On high, semiarid volcanic plateau (lowest elevation in county is 4,000 ft/1,219 m), with extensive lava beds; rises to Eagle Peak (9,892 ft/3,015 m) in WARNER MOUNTAINS (E). Cattle, sheep; timber; potatoes, onions, horseradish, wheat, barley, oats, sugar beets; pumice, sand and gravel. Drained by PIT RIVER. Includes CLEAR LAKE RESERVOIR (for irrigation) and Clear Lake National Wildlife Refuge in NW; part of Goose Lake on N boundary, extends into Oregon; SURPRISE VALLEY (in E), which contains intermittently dry upper, middle, and lower Alkali lakes; Cow Head Lake (dry) in NE; Fort Bidwell Indian Reservation in NE; part of Klamath irrigation project (N); Modoc National Forest covers large part of county, especially in center and NW, and Warner Mountains in E; small part of Shasta National Forest in SW corner; Big Sage Reservoir in center; part of Tule Lake National Wildlife Refuge and part of LAVA BEDS NATIONAL MONUMENT on W boundary; parts of XL Ranch Indian Reservation in E center and at S end of Goose Lake. Good waterfowl and deer hunting, fishing. Formed 1874.

Modoc (MO-dahk), town, EMANUEL county, E central GEORGIA, 5 mi/8 km N of SWAINSBORO; 32°39′N 82°18′W.

Modoc, town (2000 population 225), RANDOLPH county, E INDIANA, 11 mi/18 km SW of WINCHESTER; 40°02′N 85°08′W.

Modohn, LATVIA: see MADONA.

Modon, Greece: see METHONI.

Modra (mod-RAH), Hungarian *Modor*, town, ZAPADOSLOVENSKY province, W SLOVAKIA, 15 mi/24 km NE of BRATISLAVA; 48°20′N 17°19′E. Noted for its ceramics and wine making (vineyards); also for food processing and building-material manufacturing. Has wine school. Also has Gothic church and museum of L'udovít Štúr.

Modrača Lake (mo-DRAH-chah), Serbo-Croatian *Modraško jezero*, artificial lake, NE BOSNIA, BOSNIA AND HERZEGOVINA, on Spreča River, near TUZLA.

Modran, CZECH REPUBLIC: see MODRANY.

Modrany (MOD-rzhah-NI), Czech *Modřany*, German *Modran*, S district of PRAGUE, PRAGUE-CITY province, central BOHEMIA, CZECH REPUBLIC, on right bank of VLTAVA RIVER, 5 mi/8 km from city center; 50°01′N 14°25′E. Manufacturing (machinery, furniture, pharmaceuticals, confections).

Modriča (mo-DREE-chah), town, N BOSNIA, BOSNIA AND HERZEGOVINA, on BOSNA RIVER, on railroad, and 18 mi/29 km NE of DOBOJ; 44°57′N 18°18′E. Also spelled Modricha.

Modrice (MOD-rzhi-TSE), Czech *Modřice*, German *Mödritz*, village, JIHOMORAVSKY province, S MOR-

AVIA, CZECH REPUBLIC, on SVRATKA RIVER, on railroad, and 5 mi/8 km S of BRNO; 49°08′N 16°37′E. Foundry; food processing (beverages, meat, fruit, vegetables); agriculture (noted for vegetable and fruit growing).

Mödritz, CZECH REPUBLIC: see MODRICE.

Modry Kamen (mod-REE kah-MEN-yuh), Slovak *Modrý Kameň*, Hungarian *Kékkö*, town, STREDOSLOVENSKY province, S SLOVAKIA, 35 mi/56 km SSE of BANSKÁ BYSTRICA; 48°15′N 19°20′E. Sugar beets, wheat; vineyards. Has 18th-century Baroque castle.

Modugno (mo-DOO-nyo), town, BARI province, APULIA, S ITALY, 5 mi/8 km SW of BARI; 41°05′N 16°47′E. Machinery, transport equipment, plastics; food processing; wine, olive oil.

Moe (MO-ee), town (2001 population 15,512), S VICTORIA, AUSTRALIA, 75 mi/121 km ESE of MELBOURNE; 38°10′S 146°16′E. Forested region. Sawmill; dairying; lignite mining. Former gold-mining area. Gippsland Heritage Park.

Moearaenim, INDONESIA: see MUARAENIM.

Moei River, MYANMAR: see THAUNGYIN RIVER.

Moëlan-sur-Mer (muh-e-lahn–syoor–mer), town (□ 18 sq mi/46.8 sq km), summer resort in FINISTÈRE department, in BRITTANY, NW FRANCE, on an inlet of the BAY OF BISCAY, 13 mi/21 km NW of LORIENT; 47°49′N 03°38′W.

Moel Famau, Wales: see CLWYDIAN HILLS.

Moel Sych, Wales: see BERWYN MOUNTAINS.

Moelv (MAW-elv), village, HEDMARK county, SE NORWAY, on E shore of Lake MJØSA, on railroad, and 13 mi/21 km SE of LILLEHAMMER; 60°56′N 10°42′E. Lumber mills; prefabricated houses.

Moelwyn Mawr (MOIL-win MAWR), mountain (2,527 ft/770 m), GWYNEDD, NW Wales, 3 mi/4.8 km W of BLAENAU-FFESTINIOG; 52°59′N 04°00′W.

Moema (MO-e-mah), town (2007 population 6,746), W central MINAS GERAIS state, Brazil, 16 mi/26 km NE of LAGOA DA PRATA; 19°43′S 45°30′W.

Moen, island, State of CHUUK, CAROLINE, Federated States of MICRONESIA, W PACIFIC. Largest urban area in the Federated States of MICRONESIA.

Möen, DENMARK: see MØN.

Moena (mo-AI-nah), village, TRENTO province, TRENTINO-ALTO ADIGE, N ITALY, on AVISIO River, and 17 mi/27 km SE of BOLZANO; 46°22′N 11°39′E. Manufacturing (furniture, packing boxes). Downhill-ski resort, known for annual "Marcialonga" ski race held on the last Sunday in January.

Moena, INDONESIA: see MUNA.

Moenge (mo-EN-gai), village, Équateur province, NW CONGO, along Zaire River SE of BUMBA; 02°02′N 22°56′E. Elev. 1,394 ft/424 m. Subsistence and cash crops.

Moengo (MOONG-goo), town, MAROWIJNE district, NE SURINAME, on upper COTTICA RIVER, and 55 mi/89 km ESE of PARAMARIBO; 05°35′N 54°26′W. Center of bauxite mining, with large crushing and drying plant and docking facilities for ocean vessels. Most of the ore is shipped to U.S.

Moenkopi, unincorporated town (2000 population 901), COCONINO county, N central ARIZONA, 64 mi/103 km NNE of FLAGSTAFF, and 2 mi/3.2 km S of TUBA CITY, in W part of Navajo Indian Reservation, on Moenkopi Wash and Moenkopi Plateau. Crafts. Dinosaur tracks site to W.

Moerbeke (MUR-bai-kuh), commune (2006 population 5,907), Ghent district, EAST FLANDERS province, N BELGIUM, 13 mi/21 km NE of GHENT, near NETHERLANDS border; 51°10′N 03°56′E. Beet-sugar refining. Has remains of ancient Benedictine monastery.

Moerbeke, CONGO: see KWILU-NGONGO.

Moerdijk (MOOR-deik), village, NORTH BRABANT province, SW NETHERLANDS, on the HOLLANDS DIEP channel, and 11 mi/18 km NNW of BREDA; 51°42′N 04°38′E. At S end of MOERDIJK BRIDGES (railroad, highway). Dairying; livestock; grain, vegetables, sugar

beets; manufacturing (ship containers, cat litter). To the W of the village, a port and industrial zone (including a thermal power plant and energy-recovery waste-incineration plant) was developed in the 1970s.

Moerdijk Bridges (MOOR-deik), SW NETHERLANDS, span the HOLLANDS DIEP river channel between MOERDIJK (S) and WILLEMSDORP (N); 51°44′N 04°39′E. Railroad bridge (originally about 4,500 ft/1,372 m long; completed 1871; after reconstruction in 1946 about 3,300 ft/1,006 m long) and road bridge (about 3,300 ft/1,006 m; completed 1937) are on main routes between AMSTERDAM and ROTTERDAM (N) and ANTWERP and BRUSSELS, BELGIUM (S). Both bridges destroyed (1944) in World War II by retreating Germans; rebuilt 1946.

Moeris, EGYPT: see BIRKET QARUN.

Moero, Lake, CONGO and ZAMBIA: see MWERU, LAKE.

Moers (muhrs), city, North Rhine-Westphalia, W GERMANY, in the RUHR industrial district, 6 mi/9.7 km WNW of DUISBURG; 51°27′N 06°40′E. Metalworking, construction, coal-mining center. Has 14th–15th-century castle. Was first mentioned in the 9th century; chartered 1300; passed to PRUSSIA 1702. Formerly also spelled MÖRS.

Moerzeke (MUR-zai-kuh), village in Dendermonde district, EAST FLANDERS province, N BELGIUM, near SCHELDT RIVER, and 3 mi/4.8 km NE of DENDERMONDE; 51°04′N 04°09′E. Tobacco, early potatoes.

Moësa (mo-E-sah), district, SW GRISONS canton, SWITZERLAND. Population is Italian speaking and Roman Catholic.

Moesala, INDONESIA: see SIBOLGA.

Moësa River (mo-E-sah), 27 mi/43 km long, in SE SWITZERLAND; rises near SAN BERNARDINO PASS; flows SSW through VALLE MESOLCINA to TICINO RIVER just N of BELLINZONA, and drains 184 sq mi/477 sq km.

Moesia (MOI-see-yah), ancient region of SE EUROPE, S of the lower DANUBE River; 43°25′N 24°35′E. Inhabited by Thracians, it was captured by the Romans in 29 B.C. It was later organized as a Roman province, comprising roughly what is now SERBIA (Upper Moesia) and BULGARIA (Lower Moesia). Under the empire, Roman colonies flourished in the Danube valley.

Moeskroen, BELGIUM: see MOUSCRON.

Moetis, Mount, Indonesia: see MUTIS, MOUNT.

Moffat, county (□ 4,751 sq mi/12,352.6 sq km; 2006 population 13,680), extreme NW COLORADO; ⊙ CRAIG; 40°36′N 108°12′W. Livestock-grazing area; borders on UTAH (W) and WYOMING (N); drained by YAMPA, LITTLE SNAKE, and Green rivers. ELKHEAD RESERVOIR on E boundary; Elkhead Creek forms part of E county line. Cattle; wheat, oats, barley. Part of DINOSAUR NATIONAL MONUMENT in W (extends into Utah); parts of Routt National Forest (in NE); parts of White River National Forest (SE); Browns Park National Wildlife Reserve in NW. Formed 1911.

Moffat (MUH-faht), small town (2001 population 2,135), DUMFRIES AND GALLOWAY, S Scotland, on ANNAN RIVER, at foot of MOFFAT HILLS, and 20 mi/32 km NNE of DUMFRIES; 55°19′N 03°27′W. Spa resort, with medicinal springs. Nearby Dumcrieff mansion was residence of John McAdam. Also nearby is Burns's Cottage and Craigieburn House, birthplace of Jean Lorimer, Burns's "Chloris."

Moffat, village (2000 population 114), SAGUACHE county, S COLORADO, on SAN LUIS CREEK; SAGUACHE CREEK to W, and 14 mi/23 km ESE of SAGUACHE; 38°00′N 105°54′W. Elevation 7,561 ft/2,305 m. Shipping point in livestock region. CRESTONE PEAK, 18 mi/29 km E; GREAT SAND DUNES NATIONAL MONUMENT to SE; SANGRE DE CRISTO MOUNTAINS to E; parts of Rio Grande National Forest to NE, NW, and W.

Moffat Hills (MUH-faht HILZ), mountain range, Scottish Borders and DUMFRIES AND GALLOWAY, S Scotland, extending 25 mi/40 km NE-SW between

MOFFAT and PEEBLES; 10 mi/16 km wide. Chief peaks: Hart Fell (2,651 ft/808 m), 5 mi/8 km NNE of Moffat; Broad Law (2,754 ft/839 m), 12 mi/19 km NNE of Moffat; Dollar Law (2,680 ft/817 m), 9 mi/14.5 km SSW of Peebles; and Dun Law (2,650 ft/808 m), 10 mi/16 km SSW of Peebles. On E slope is ST. MARY'S LOCH. CLYDE, TWEED, YARROW, and ANNAN rivers rise in Moffat Hills.

Moffat Tunnel, railroad tunnel, N central COLORADO, on the CONTINENTAL DIVIDE, NW of DENVER; 24 ft/7 m high, 18 ft/5 m wide, and 6.2 mi/10 km long. One of the country's longest railroad tunnels, it was built between 1922 and 1927. At an elevation of 9,094 ft/2,772 m, it passes through JAMES PEAK. An adjacent bore carries water to Denver. Extends from WINTER PARK, GRAND county, on W, to East Portal, Gilpinco, on E, S of Rollins Pass.

Moffet (mo-FAI), village (□ 167 sq mi/434.2 sq km; 2006 population 218), ABITIBI-TÉMISCAMINGUE region, S QUEBEC, E CANADA, 10 mi/16 km from LA-FORCE; 47°33′N 78°57′W.

Moffet Inlet, Anglican mission station, N BAFFIN ISLAND, BAFFIN region, NUNAVUT territory, CANADA, on E side of ADMIRALTY INLET, 65 mi SSE of ARCTIC BAY trading post; 77°11′N 84°28′W.

Moffett, village (2006 population 178), SEQUOYAH county, E OKLAHOMA, on ARKANSAS RIVER (bridged), suburb, 3 mi/4.8 km WNW of Fort Smith, Ark; 35°23′N 94°27′W. Residential, commercial, and agricultural area.

Moffett Air Force Base, California: see SUNNYVALE.

Moga (mo-gah), city, Faridkot district, W Punjab state, N India, 35 mi/56 km ESE of Faridkot; 30°48′N 75°10′E. Agr. market center (gram, wheat, cotton, oilseeds); hand-loom weaving, palm-mat making. Has college Annual festival fair.

Mogadisho (moo-GAH-dee-SHO), Italian *Mogadiscio*, city (2005 population 1,320,000), ☉ SOMALIA, on the INDIAN Ocean. The country's largest city, a port, and a commercial and financial center. Mogadisho has little industry except for food and beverage processing and cotton ginning. Uranium ore has been discovered nearby. The city is linked by road with KENYA and ETHIOPIA and has an international airport. Important trade center for the E coast of AFRICA. Heavily affected by the nation's devastating drought and famine in the early 1990s. Rebel forces entered the city in 1990 and UN forces in 1993 during Somalia's long civil war. UN and U.S. task force left in 1995 after unsuccessful attempts to maintain stability. Historic buildings include Mosque of Fakr ad-Din (1269) and Garesa Palace (currently museum and library). Somalia National University headquarters here. Vast soviet-built airbase of Bale Dogle is 55 mi/89 km to W. Also spelled Mugdisho and Mogadishu.

Mogador, MOROCCO: see ESSAOUIRA.

Mogadore (MAH-guh-dor), village (□ 2 sq mi/5.2 sq km; 2006 population 3,946), SUMMIT county, NE OHIO, just E of AKRON; 41°03′N 81°24′W. Tools, clay products, composites, plastic products.

Mogadouro (mo-gah-DOR-oo), town, BRAGANÇA district, N PORTUGAL, in hill region, 32 mi/51 km S of BRAGANÇA. Agricultural trade (vegetables, white wine); woodcrafts, blankets; lumbering, leatherworking.

Mogalakwena River, 250 mi/402 km long, LIMPOPO province, REPUBLIC OF SOUTH AFRICA; rises W of NYLSTROOM where its called Nyl River; flows generally N past Nylstroom through Nyluley Nature Reserve, W past POTGIETERSRUS through Glen Alpine Dam, to LIMPOPO RIVER 70 mi/113 km W of BEITBRIDGE on BOTSWANA border. Formerly spelled Magalakwin River.

Mogale, CONGO: see MOGALO.

Mogale City, municipality (2001 population 289,720), GAUTENG province, NE SOUTH AFRICA. Created in 2001 by merging the city of KRUGERSDORP with smaller surrounding areas including Kagiso, Muldersdrift, Munsieville, Tarlton, Hekpoort, MAGALIESBURG, and Azaadville, as well as the industrial areas including Chamdor, Factoria, and Boltonia.

Mogalo (mo-GAH-lo), village, Équateur province, NW CONGO, 45 mi/72 km W of GEMENA, on left bank of LUA RIVER; 03°10′N 19°04′E. Elev. 1,292 ft/393 m. Also known as MOGALE.

Mogalturru, India: see NARSAPUR.

Mogami (MO-gah-mee), town, Mogami county, YAMAGATA prefecture, N HONSHU, NE JAPAN, 37 mi/60 km N of YAMAGATA city; 38°45′N 140°13′E.

Mogami River (MO-gah-mee), Japanese *Mogami-gawa* (mo-GAH-mee-GAh-wah), 134 mi/216 km long, YAMAGATA prefecture, N HONSHU, NE JAPAN; formed by union of two headstreams S of NAGAI; flows generally NNE, past Nagai, Yachi, and OISHIDA, and NW to SEA OF JAPAN at SAKATA. Drains large rice-growing area.

Mogán (mo-GAHN), town, Grand Canary island, CANARY ISLANDS, 24 mi/39 km SW of LAS PALMAS; 27°53′N 15°43′W. Resort with nearby beach and landing. Region produces grapes, cereals, bananas, tomatoes, tobacco. Fishing.

Mogan Shan (MO-GAN), mountain (2,500 ft/762 m) in N ZHEJIANG province, CHINA, 30 mi/48 km NW of HANGZHOU; 30°35′N 119°51′E. Noted summer resort.

Mogarraz (mo-gah-RAHTH), village, SALAMANCA province, W SPAIN, 13 mi/21 km NW of BÉJAR; 40°29′N 06°03′W. Olive-oil processing; lumbering; wine, chestnuts.

Mogaung (mo-GOUNG), village and township, KACHIN STATE, MYANMAR, on railroad, on Mogaung River (W affluent of the AYEYARWADY RIVER), and 30 mi/48 km W of MYITKYINA. Shipping point for jade mines of LONKIN and amber mines of HUKAWNG VALLEY; carving handicrafts. Was capital of a petty SHAN STATE (13th century), rivaling MOHNYIN; passed to BURMA (18th century) under Alaungpaya. Scene of heavy fighting (1944) during World War II.

Mogeiro (mo-GAI-ro), city (2007 population 12,310), SE PARAÍBA, BRAZIL, 11 mi/18 km NW of Itabiana; 07°18′S 35°29′W. Also spelled Mojeiro.

Mogelnitsa, POLAND: see MOGIELNICA.

Mogelsberg (MO-guhls-berg), commune, ST. GALLEN canton, NE SWITZERLAND, 12 mi/19 km WSW of ST. GALLEN; 47°22′N 09°07′E. Embroideries, metal products.

Mogente (mo-HEN-tai), town, VALENCIA province, E SPAIN, 15 mi/24 km SW of JÁTIVA; 38°52′N 00°45′W. Olive-oil and meat processing; toy and plaster manufacturing; lumbering; wine, cereals, fruit. Summer resort.

Mögglingen (MUG-ling-guhn), village, E WÜRTTEMBERG, BADEN-WÜRTTEMBERG, S GERMANY, 5 mi/8 km E of SCHWÄBISCH GMÜND; 48°50′N 10°11′E. Forestry.

Moghalpura (muh-GUHL-puhr-ah), industrial E suburb of LAHORE, LAHORE district, E PUNJAB province, central PAKISTAN; 31°34′N 74°23′E. Also spelled MUGHALPUR.

Moghal-Sarai, India: see MUGHAL SARAI.

Mogi das Cruzes (MO-zhee dahs KROO-ses), city (2007 population 362,991), SÃO PAULO state, SE BRAZIL, on the TIETÊ RIVER; 23°31′S 46°11′W. Industrial center and agricultural distribution point for São Paulo and RIO DE JANEIRO states. Manufacturing includes paper, textiles, ceramics, and chemical products. Established in the early 17th century. Also spelled Moji das Cruzes.

Mogielnica (mo-geel-NEE-tsah), Russian *Mogelnitse* or *Mogel'nitse*, town, Warszawa province, E central POLAND, 40 mi/64 km SSW of WARSAW. Tanning, flour milling.

Mogi-Guaçu (MO-zhee–GWAH-soo), city (2007 population 131,879), E SÃO PAULO state, BRAZIL, 4 mi/6.4 km N of MOJI-MIRIM, and on MOGI-GUAÇU railroad; 22°22′S 46°57′W. Dairying, pottery manufacturing, sawmilling; agriculture (coffee, sugar) and livestock (cattle). Formerly spelled Mogy-Guassú; also spelled Moji-Guaçu.

Mogi-Guaçu River (MO-zhee GWAH-soo), 220 mi/354 km long, NE SÃO PAULO state, BRAZIL; rises in MINAS GERAIS near OURO FINO; flows NW, past MOGI-GUAÇU and Pôrto Ferreira (head of navigation), to the RIO PARDO 30 mi/48 km NW of RIBEIRÃO PRÊTO.

Mogilev (mo-gee-LYOF), oblast (□ 8,000 sq mi/20,720 sq km; 2005 estimated population 1,146,800), E BELARUS; ☉ MOGILEV. In DNIEPER LOWLAND; drained by DNIEPER and SOZH rivers. Forested and agricultural region. Manufacturing (building materials, machinery, metalworking). Peat is the main local fuel. Formed 1938.

Mogilev (mo-gee-LYOF), city, ☉ MOGILEV region, in E BELARUS, on the DNEPR River, 53°54′N 30°20′E. It is an important rail and highway junction, and a river port. It is an industrial center with metalworking, machine building, and chemical industries (synthetic fibers); food processing (meat combines, confectionary plant, canning, creameries, dairies); light industry (garment, knitwear, ribbon, footwear); building materials (silicate goods, reinforced concrete products, wood products). There are two heat and electric power plants here. Arising in the thirteenth-century on the territory of SMOLENSK principality, the city grew around a castle dating from 1267 and became a noted commercial center from the fourteenth-century. Mogilev was part of the grand duchy of LITHUANIA (united with POLAND in 1569), was later held by SWEDEN, and passed to RUSSIA during the first partition of POLAND (1772). It was occupied and heavily damaged by the Germans during World War II. A tower built by the Tatars and several old churches survive. Several higher education institutions are here. Major Jewish center from the sixteenth-century; Jews were half of population in nineteenth-century and one one-third until WW I.

Mogilev-Podol'skiy, UKRAINE: see MOHYLIV-PODIL'S'KYY.

Mogilno (mo-GEEL-no), town, Bydgoszcz province, central POLAND, 16 mi/26 km SW of INOWROCŁAW. Railroad junction; machine manufacturing, distilling, flour milling; salt mine nearby.

Mogincual (mo-geen-kwahl), village, NAMPULA province, E MOZAMBIQUE, on MOZAMBIQUE CHANNEL, 40 mi/64 km SSW of Mozambique city; 15°34′S 40°24′E. Cotton, sisal.

Moglia (MO-lyah), village, MANTOVA province, LOMBARDY, N ITALY, near SECCHIA River, 17 mi/27 km SSE of MANTUA. Alcohol distillery; textiles, clothing, wood products.

Mogliano Veneto (mo-LYAH-no VAI-ne-to), town, TREVISO province, VENETO, N ITALY, 7 mi/11 km SE of TREVISO; 45°33′N 12°14′E. In peach-growing region; manufacturing (shoes, agricultural machinery, bicycles, wax, silk textiles). Agricultural center noted for its aviculture.

Möglingen (MUG-ling-guhn), town, STUTTGART district, BADEN-WÜRTTEMBERG, SW GERMANY, 9 mi/14.5 km NNW of STUTTGART; 48°54′N 09°06′E. Grain; wine; manufacturing (electronic equipment).

Mogo, town, NEW SOUTH WALES, SE AUSTRALIA, 181 mi/291 km S of SYDNEY, 6 mi/10 km S of BATEMAN'S BAY. Tourism; former mining center. Zoo just S.

Mogocha (muh-GO-chah), city (2005 population 12,450), NE CHITA oblast, S SIBERIA, RUSSIA, on the Mogocha River at its confluence with the Amazar River (AMUR River basin), on the TRANS-SIBERIAN RAILROAD, 440 mi/708 km NE of CHITA, and 170 mi/274 km W of SKOVORODINO; 53°44′N 119°46′E. Elevation 2,053 ft/625 m. Terminus of a local highway. In gold-mining area; railroad shops; forestry. Arose in 1910 with the construction of the railroad. Made city in 1950.

Mogochin (muh-GO-cheen), town (2006 population 3,195), SE central TOMSK oblast, W SIBERIA, RUSSIA, on the OB' RIVER near the mouth of the CHULYM RIVER, on road, 45 mi/72 km SE of KOLPASHEVO; 57°42'N 83°34'E. Elevation 187 ft/56 m. Sawmilling.

Mogoituy, RUSSIA: see MOGOYTUY.

Mogok (MO-gok), township and village, MANDALAY division, MYANMAR, on the SHAN PLATEAU. It is the centuries-old center of the Myanmarese ruby trade.

Mogollon (muh-guh-YON), unincorporated village, CATRON county, SW NEW MEXICO, near SAN FRANCISCO RIVER, W of MOGOLLON MOUNTAINS, 52 mi/84 km NNW of SILVER CITY. Elevation 6,620 ft/2,018 m. Silver mines in vicinity. Snow Lake to E; the Catwalk to SW.

Mogollon Mountains (muh-guh-YON), GRANT county, SW NEW MEXICO and GILA county, ARIZONA, just E of SAN FRANCISCO RIVER, in Gila National Forest, near Arizona state line. Prominent points: Granite Peak (8,731 ft/2,661 m), Mogollon Baldy (10,778 ft/3,285 m), WHITEWATER BALDY (10,892 ft/3,320 m). Silver mining.

Mogollon Plateau (muh-guh-YON) or **Mogollon Mesa,** tableland, E central ARIZONA, part of the COLORADO PLATEAU. Elevation from 7,000 ft/2,134 m to 8,000 ft/2,438 m. It is covered by pine forests, parts of which are included in Coconino, Tonto, and Apache-Sitgreaves national forests. Its S edge is a rugged S-facing escarpment called the Mogollon Rim, which forms COCONINO-GILA county line in NW, and N boundary of Fort Apache Indian Reservation in E. The plateau is not directly connected with the MOGOLLON MOUNTAINS in W NEW MEXICO.

Mogol-Tau, mountain range, a section of TIANSHAN mountain system, on UZBEKISTAN-TAJIKISTAN border, at W end of FERGANA VALLEY; 25 mi/40 km long (NE-SW). Rises to c.5,000 ft/1,524 m. Rich iron (magnetite) deposits. The SYR DARYA River flows around S end of range; forms BEGOVAT rapids.

Mogonawri, GHANA: see MOGONORI.

Mogonori (mo-go-NOOR-ee), village, UPPER EAST REGION, GHANA; 11°07'N 00°17'W. Located 5 mi/8 km NNW of BAWKU, on BURKINA FASO border; border and cattle-quarantine station. Sometimes spelled MOGONAWRI.

Mogoşoaia (mo-go-SHWAH-yah), village, ILFOV AGRICULTURAL SECTOR, S ROMANIA, on COLENTINA RIVER, and 8 mi/12.9 km NW of BUCHAREST; 44°32'N 26°00'E. Poultry research farm. Has 18th-century palace (now a children's home), built by Constantine Brancovan.

Mogotes (mo-GO-tais), town, ⊙ Mogotes municipio, SANTANDER department, N central COLOMBIA, in Cordillera ORIENTAL, 40 mi/64 km SE of BUCARAMANGA; 06°29'N 72°59'W. Elevation 5,728 ft/1,746 m. Cassava, coffee, corn; livestock. Consumer goods; food processing.

Mogovolas District, MOZAMBIQUE: see NAMPULA, province.

Mogoytuy (muh-guhyee-TOO-yee), town (2005 population 8,370), NE AGIN-BURYAT AUTONOMOUS OKRUG, S central CHITA oblast, S SIBERIA, RUSSIA, 75 mi/121 km SE of CHITA; 50°45'N 114°07'E. Elevation 2,719 ft/828 m. In agricultural area. Also spelled Mogoituy.

Mograt Island (MO-graht), long, narrow island in NILE River, River Nile state, SUDAN, between 4th and 5th cataracts, opposite ABU HAMED; 17 mi/27 km long, 2 mi/3.2 km wide; 19°30'N 33°15'E.

Mogre, SUDAN: see ANDHERI.

Moguer (mo-GER), town, HUELVA province, SW SPAIN, in ANDALUSIA, on left bank of the RÍO TINTO estuary, and 6 mi/9.7 km E of HUELVA; 37°16'N 06°50'W. Viticultural center. Alcohol distilling; cereals; livestock. An old city noted for its fine buildings, such as church with Mozarabic tower and the Santa Clara convent in whose archives are documents pertaining to discovery of America.

Mogul Empire (MOO-guhl), N and central India, 1526–1857. The dynasty was founded by Babur, a Turkish chieftain who had his base in Afghanistan. Babur's invasion of India culminated in the battle of PANIPAT (1526) and the occupation of DELHI and AGRA. Babur was succeeded by his son, Humayan, who soon lost the empire to the Afghan Sher Khan. Akbar, the son of Humayan and the greatest of the Mogul emperors, reestablished Mogul power in India. At the time of Akbar's death (1605), the empire occupied a vast territory from Afghanistan E to what is now ORISSA STATE and S to the DECCAN PLATEAU. Mogul expansion continued under Akbar's son Jahangir and under his grandson Shah Jahan, who built many architectural marvels at Delhi and at Agra (including the TAJ MAHAL). Aurangzeb expanded Mogul territory to its greatest extent, but at the same time the empire suffered the blows of major revolts. The most serious of these was the Maratha uprising. Weakened by the Maratha wars, dynastic struggles, and invasions by Persian and Afghan rulers, the empire came to an effective end as the British established control of India in the late eighteenth and early nineteenth century. However, the British maintained puppet emperors until 1857. Many features of the Mogul administrative system were adopted by Great Britain in ruling India, but the most lasting achievements of the Moguls were in art and architecture.

Mogy das Cruzes, Brazil: see MOGI DAS CRUZES.

Mogy-Guassú, Brazil: see MOGI-GUAÇU.

Mogy-Mirim, Brazil: see MOJI-MIRIM.

Mogzon (muhg-ZON), town (2005 population 4,200), SW CHITA oblast, S SIBERIA, RUSSIA, on the KHILOK RIVER (tributary of the SELENGA River), on road and the TRANS-SIBERIAN RAILROAD, 65 mi/105 km WSW of CHITA; 51°45'N 111°58'E. Elevation 3,008 ft/916 m. Lumbering.

Mohács (MO-hahch), town (2001 population 19,223), S HUNGARY, on the DANUBE River; 45°59'N 18°42'E. An important river port and railroad terminus. Has steel foundry; manufacturing (furniture, brick, tile, apparel, silk, lumber). Mohács is best known for the crushing defeat (Aug. 29, 1526) here of Louis II of Hungary and BOHEMIA by Suleyman I of the Ottoman Empire. Hungary was ill-prepared for the attack, and when Louis hastily tried to unite Hungary and Christendom behind him, only the pope sent help. With a poorly equipped and badly organized army of 28,000, Louis joined battle with a Turkish army of 200,000. The king and almost 25,000 of his army were killed in the battle; the rest were taken captive and massacred. The defeat brought with it more than 150 years of Ottoman domination in Hungary. At Mohács are monuments to the slain, regarded ever since as martyrs to Christianity and to Hungarian independence. Mohács also was the scene (1687) of a Turkish defeat by Charles V of LORRAINE, which two years later ended Ottoman rule in all of Hungary, except the BÁNÁT.

Mohács Island (MO-hahch), S HUNGARY and NW SERBIA, in the DANUBE; c.35 mi/56 km long, greatest width 11 mi/18 km. Swampy. Wheat, corn; hogs; vineyards.

Mohales Hoek (mo-HA-les HO-ek) [Lesotho=Mohale's corner], district (□ 1,363 sq mi/3,543.8 sq km; 2001 population 206,842), S central LESOTHO, bounded W by SOUTH AFRICA (border formed by ORANGE and Kornetspruit rivers), S by Orange (Senqu) River; ⊙ MOHALES HOEK. Drained in E by Maletsunyane and Senqunyane rivers. THABA PUTSUA range crosses district center N-S; CENTRAL RANGE in E. Sheep, goats, horses, cattle; corn, beans, peas.

Mohales Hoek (mo-HA-les HO-ek), town, ⊙ MOHALES HOEK district, SW LESOTHO, 57 mi/92 km S of MASERU, on main N-S road, 4 mi/6.4 km W of SOUTH AFRICAN border; 30°09'S 27°29'E. Terminal point of S-bound main road. Aerodome. Sheep, goats, horses; vegetables, corn. Cannibal Caves to NW. Land granted to British administration by chief Mohale in 1884. Also spelled Mohaleshoek or Mohale's Hoek.

Mohall (MO-hawl), town (2006 population 756), ⊙ RENVILLE county, N NORTH DAKOTA, 38 mi/61 km N of MINOT; 48°46'N 101°30'W. Founded in 1902 and named for the town's founder Martin O. Hall (1853–1925). It became the county seat 1920 when the county was formed.

Mohamadabad-Bidar, INDIA: see BIDAR.

Mohamdi, India: see MUHAMDI.

Mohammadabad, India: see MEHMADABAD.

Mohammadbad, India: see MUHAMMADABAD.

Mohammadia (mo-ahm-mai-DYAH), town, MASCARA wilaya, NW ALGERIA, in the TELL, 24 mi/39 km S of MOSTAGANEM; 35°35'N 00°05'E. Railroad hub (junction of E-W line and of spur inland to BÉCHAR) with railroad workshops; center of irrigated HABRA lowland (dams on HAMMAM and FERGOUG rivers), growing citrus fruit, cotton, vegetables. Agricultural experiment station nearby. Has experienced recent population growth and increased employment, especially in electrical engineering plants. Formerly Perrégaux.

Mohammadpur, India: see ROORKEE.

Mohammareh, IRAN: see KHORRAMSHAHR.

Mohammedabad, IRAN: see DARREGAZ.

Mohammedia (mo-ahm-mai-DYAH), city, Mohammedia prefecture, GRAND CASABLANCA administrative region, W MOROCCO, on the ATLANTIC OCEAN, 14 mi/23 km NE of CASABLANCA; 33°41'N 07°23'W. Morocco's chief petroleum-importing, refining, and storage center, second-largest port, and a rapidly growing industrial city. A corsair refuge in 17th–18th centuries, it was but a hamlet in 1912. The modern port (completed 1934), protected by two converging breakwaters, accommodates ships drawing up to 20 ft/6 m. Located on trunk railroad, Mohammedia is rapidly becoming Casablanca's outer manufacturing suburb. Chief industries are fish and vegetable (peas, beans, spinach) canning, fruit preserving, meat packing, cork processing (including manufacturing of bottle corks), oil refining, and chemical and brick manufacturing. There is a large modern cotton mill. Important tourism center with major beach, yacht club, water sports, golf course, and casino. Branch of Casablanca's university here. Saltworks nearby. Formerly called Fédala.

Mohanga, CONGO: see LUTUNGURU.

Mohanganj, BANGLADESH: see SHAMGANJ.

Mohanlalganj (mo-huhn-LAHL-guhnj), town, LUCKNOW district, central UTTAR PRADESH state, N central INDIA, 12 mi/19 km SSE of LUCKNOW; 26°41'N 80°58'E. Road center; trades in wheat, rice, gram, millet.

Mohanpur (mo-huhn-poor), town, ETAH district, W UTTAR PRADESH state, N central INDIA, 20 mi/32 km ESE of KASGANJ. Wheat, pearl millet, barley, corn, jowar, oilseeds.

Mohapa, India: see MOHPA.

Moharé, Comoros: see MAYOTTE (island).

Moharraq, BAHRAIN: see MUHARRAQ.

Mohattanagar (mo-TAH-NUH-guhr), village, NAWABSHAH district, central SIND province, SE PAKISTAN, 17 mi/27 km E of NAWABSHAH; 26°13'N 68°41'E. Railroad station. Formerly called PRITAMABAD.

Mohave, county (□ 13,470 sq mi/35,022 sq km; 2006 population 193,035), NW ARIZONA; ⊙ KINGMAN; 35°42'N 113°45'W. Mining (lead, silver, zinc, gold, copper); agriculture (cattle; alfalfa, wheat, barley, sorghum, lettuce, honeydews). HUALAPAI, MOHAVE, and BLACK mountains in SW. Drained by BIG SANDY RIVER in SE. Parts of Grand Canyon National Park in N center and NE; parts of LAKE MEAD NATIONAL RECREATION AREA in NW and N center. COLORADO RIVER bounds county in large part on W, foms lakes HAVASU, MOHAVE, and MEAD and state boundary of NEVADA (part) and CALIFORNIA; entire W state line forms Pacific-Mountain time zone boundary (Arizona is in Mountain); Colorado River crosses county in N and forms part of E boundary; bounded by Utah on N, with Kaibab Creek forming NE boundary; BILL WILLIAMS and SANTA MARIA rivers form S boundary. Part of large Hualapai Indian Reservation in NE center, small sections of reservation NE and SW of Kingman; part of Kaibab Indian Reservation in far NE; part of Fort Mojave Indian Reservation in SW corner, on Colorado River, next to California and Nevada. Mount Nuff, Warm Springs, and Wabayuma Peak wilderness areas in SW corner, two sections of Havasu National Wildlife Refuge in SW; Paints and Beaver Dam wilderness areas in NW corner; Hualapai Mountain Park in center, SE of Kingman; Lake Havasu State Park in SW. HOOVER DAM forms Lake Mead on Nevada border. Fifth-largest county in U.S. in land area. Formed 1864.

Mohave, California: see MOJAVE.

Mohave, Lake, ARIZONA: see DAVIS DAM.

Mohave Mountains, MOHAVE county, W ARIZONA, E side of COLORADO RIVER, just N of LAKE HAVASU CITY, Arizona, and E of NEEDLES, CALIFORNIA. Rise to 5,102 ft/1,555 m (Grossman Peak) in Arizona, to 3,688 ft/1,124 m in California.

Mohawk, village, KEWEENAW county, NW UPPER PENINSULA, MICHIGAN, 17 mi/27 km NE of HOUGHTON; 47°18′N 88°42′W. Railroad terminus. Manufacturing (cedar furniture); acorn nuts.

Mohawk, village (2006 population 2,539), HERKIMER county, central NEW YORK, on MOHAWK RIVER and the BARGE CANAL, and 13 mi/21 km SE of UTICA; 43°00′N 75°00′W. Light manufacturing. Settled 1826, incorporated 1844.

Mohawk, village, Coshocton co., central Ohio, 12 mi/19 km WNW of Coshocton.

Mohawk Dam; Mohawk Reservoir, Ohio: see WALHONDING RIVER.

Mohawk, Lake, reservoir, SPARTA, SUSSEX county, NW NEW JERSEY, on WALLKILL RIVER, 7 mi/11.3 km ESE of NEWTON; c.2.5 mi/4 km long; 41°01′N 74°38′W. Recreational area.

Mohawk River (MO-hawk), 10 mi/16 km long, COOS county, N NEW HAMPSHIRE; rises in Mud Pond, 3 mi/4.8 km N of DIXVILLE NOTCH pass; flows SW and W to the CONNECTICUT RIVER at Colebrook.

Mohawk River, largest tributary of the HUDSON RIVER, c.140 mi/225 km long, central and E NEW YORK; rises in ONEIDA county; flows S and SE, past ROME, UTICA, AMSTERDAM, and SCHENECTADY, to the Hudson River at COHOES (falls here). Drains 3,412 mi/5,491 km. From Rome to its mouth, river is either paralleled by or part of NEW YORK STATE BARGE CANAL joining GREAT LAKES and East Coast ports via Hudson River. The beautiful and fertile Mohawk valley, as the E-W passage between the ADIRONDACK MOUNTAINS (N) and the ALLEGHENY PLATEAU, or Upland (S), was scene of many battles in the French and Indian War and in the American Revolution, and was an important route for westbound pioneers; the old ERIE CANAL followed the river in much the same manner as the Barge Canal.

Mohawk Trail (MO-hawk), motor highway extending c.30 mi/50 km across N MASSACHUSETTS from GREENFIELD to NORTH ADAMS. Follows a trail blazed originally by the Mohawks. Traversing the scenic HOOSAC Mountains and BERKSHIRE HILLS, the route is popular with tourists. A number of state forests are along the route.

Mohawk Trail, old road in central NEW YORK state following the MOHAWK RIVER in part; c.100 mi/160 km long. It was the major route through the APPALACHIANS by which thousands of settlers emigrated from the E seaboard to the MIDWEST. It traverses territory once occupied by the Iroquois Confederacy, and thus is also known as the Iroquois Trail. In the Colonial period it was a series of turnpikes that began at SCHENECTADY and extended to ROME, with lesser trails stretching W. The ERIE CANAL rendered the road less important, and when the railroads were built, its value was further diminished. The identically named

MOHAWK TRAIL in Massachusetts and the Iroquois Trail were only loosely connected.

Mohawk Valley, California: see BLAIRSDEN.

Mohe (MO-HUH), town, ⊙ Mohe county, northernmost HEILONGJIANG province, Northeast, CHINA, 320 mi/515 km NW of HEIHE, and on AMUR River (RUSSIAN border); 53°01′N 122°29′E. Logging. Northernmost custom city of China.

Moheda (MOO-HED-ah), town, KRONOBERG county, S SWEDEN, on MÖRRUMSÅN RIVER, 12 mi/19 km NW of VÄXJÖ; 57°00′N 14°34′E.

Mohedas (mo-AI-dhahs), town, CÁCERES province, W SPAIN, 18 mi/29 km NNW of PLASENCIA; 40°16′N 06°12′W. Olive oil, wine, sheep.

Mohedas de la Jara (mo-AI-dhahs dhai lah HAH-rah), town, TOLEDO province, central SPAIN, 30 mi/48 km SW of TALAVERA DE LA REINA; 39°36′N 05°08′W. Olives, cereals, fruit; livestock. Olive-oil pressing.

Mohegan (mo-hee-GIN), community, NEW LONDON county, CONNECTICUT. Attained Native American reservation status in 1995. Casino gambling complex under development in early 1996.

Mohegan, village, BURRILLVILLE town, PROVIDENCE county, N RHODE ISLAND; 41°59′N 71°37′W.

Mohegan Lake, village, WESTCHESTER county, SE NEW YORK, on Lake Mohegan (c.1 mi/1.6 km long), 5 mi/8 km NE of PEEKSKILL; 41°19′N 73°52′W.

Moheli, COMOROS: see MWALI.

Mohéli Island, COMOROS: see MWALI.

Mohelnice (MO-hel-NYI-tse), German *Müglitz*, town, SEVEROMORAVSKY province, NW central MORAVIA, CZECH REPUBLIC, on railroad, and 19 mi/31 km NW of OLOMOUC; 49°47′N 16°55′E. Agriculture (sugar beets, oats); manufacturing (electronical goods, machinery), wood processing. Has a Gothic church, and museum. Paleolithic-era archaeological site.

Mohelno (MO-hel-NO), village, JIHOMORAVSKY province, S MORAVIA, CZECH REPUBLIC, 14 mi/23 km ESE of TŘEBÍČ; 49°07′N 16°11′E. Has a 16th century Gothic church. A Neolithic barrow was found here. Military airport. Nuclear power plant (4 reactors of PWR type; total output 1,632 mw) at DUKOVANY (DU-ko-VAH-ni), 2 mi/3.2 km S.

Mohenjo-Daro (MON-jo-DAH-ro) [Sindhi=mound of the dead], a famous site of Indus valley civilization, LARKANA district, NW SIND province, SE PAKISTAN, near right bank of the INDUS River, 17 mi/27 km S of Larkana; 27°19′N 68°07′E. Since its antiquity became clear in 1922, excavations have uncovered remains of 6 or 7 successive cities, generally considered to have existed between 4000 and 1500 B.C.E., in Chalcolithic Era. Ruins include brick dwellings, baths, drains, and streets; artifacts are engraved seals, implements, weapons, jewelry, sculpture, and pottery. Inhabitants apparently enjoyed a sophisticated civilization. In NW sector is Buddhist stupa of a 2nd-cent. C.E. Kushan king. Nearby is archaeological camp and museum with numerous antiquities.

Mohespur, town, JESSORE district, W EAST BENGAL, BANGLADESH, 8 mi/12.9 km NNW of JESSORE; 23°25′N 88°51′E. Trades in rice, jute, sugarcane.

Mohican, Cape (mo-HEE-kin), W ALASKA, W extremity of NUNIVAK Island, on BERING SEA; 60°12′N 167°24′W.

Mohican River (MO-i-kahn), c.40 mi/64 km long, central OHIO; formed by junction of forks in region E of MANSFIELD; flows S, joining KOKOSING RIVER to form WALHONDING RIVER 16 mi/26 km NW of COSHOCTON; 40°21′N 82°09′W. Among its headstreams and tributaries (some with flood-control works) are Black, Lake, Clear, and Jerome forks.

Mohicanville Dam (MO-i-kahn-vil), ASHLAND county, N central OHIO, on Lake Fork of MOHICAN RIVER, 13 mi/21 km WSW of WOOSTER; 40°43′N 82°08′W. Reservoir is dry except during flood periods; maximum potential capacity 102,000 acre-ft; extends NNE into WAYNE county.

Mohinora, Cerro (mo-ee-NO-rah), peak (10,663 ft/3,250 m) in SIERRA MADRE OCCIDENTAL, CHIHUAHUA, N MEXICO, 80 mi/129 km NNE of CULIACÁN ROSALES; near 00°26′N 107°04′W. Sometimes known as Muinora.

Mohinora River, MEXICO: see SINALOA RIVER.

Möhlin, commune, AARGAU canton, N SWITZERLAND, just S of the RHINE RIVER (German border), 12 mi/19 km E of BASEL; 47°33′N 07°51′E. Shoes. Riburg-Schwörstadt hydroelectric plant is N.

Möhne River (MUN-nuh), 34 mi/55 km long, W GERMANY; rises 2 mi/3.2 km NW of BRILON; flows W to the RUHRAT NEHEIM-HÜSTEN. Dammed at Günne (4,767 million cu ft/135 million cu m)

Möhnesee (MUN-ne-sai), village, WESTPHALIA-LIPPE, North Rhine-Westphalia, GERMANY, on the MÖHNE LAKE, on N slope of the SAUERLAND, 18 mi/29 km E of UNNA; 51°31′N 08°06′E. Formed 1969 by union of fifteen villages. Agriculture (fruit, grain, sugar beets); forestry.

Mohnton (MON-tuhn), borough (2006 population 3,093), BERKS county, SE central PENNSYLVANIA, suburb, 5 mi/8 km SW of READING; 40°17′N 75°59′W. Manufacturing (apparel, communications systems, food products, commercial printing, machinery). Maple Grove Raceway (drag racing). Founded 1850, incorporated 1907.

Mohnyin (mawn-YIN), village and township, SW KACHIN STATE, MYANMAR, on railroad, 75 mi/121 km SW of MYITKYINA. Former capital of a petty Shan kingdom rivaling MOGAUNG.

Moho (MO-ho), province, PUNO region, SE PERU; ⊙ MOHO.

Moho (MO-ho), village, MOHO province, PUNO region, SE PERU; minor port on NE shore of Lake TITICACA, 19 mi/31 km ESE of HUANCANÉ; 15°21′S 69°30′W. Elevation 12,595 ft/3,839 m. Fibers, cereals; livestock.

Mohol, town, SOLAPUR district, MAHARASHTRA state, W central INDIA, 20 mi/32 km NW of SOLAPUR; 17°49′N 75°40′E. Millet, cotton; handicraft cloth weaving.

Moholm (MOO-HOLM), village, SKARABORG county, S SWEDEN, 10 mi/16 km NE of MARIESTAD; 58°36′N 14°03′E. Railroad junction.

Mohonk, Lake (MO-hawnk), small lake in scenic resort area of the SHAWANGUNK range, ULSTER county, SE NEW YORK, 4 mi/6.4 km WNW of NEW PALTZ; 41°46′N 74°09′W. Mohonk Lake village is here. Mohonk Mountain House, a private resort on the 12-sq-mi/31-sq-km public Mohonk Preserve, is a major vacation destination.

Moho River (MO-ho), 24 mi/39 km long, in GUATEMALA and BELIZE; rises in MAYA MOUNTAINS 30 mi/48 km W of PUNTA GORDA; flows E to Bay of AMATIQUE of CARIBBEAN SEA, 5 mi/8 km SSW of Punta Gorda. Navigable for small craft; timber floating.

Mohoro (mo-HO-ro), town, PWANI region, E TANZANIA, 90 mi/145 km S of DAR ES SALAAM, 8 mi/12.9 km S of RUFIJI RIVER (delta on INDIAN OCEAN to NE); 08°10′S 39°06′E. Sisal, copra, rice; livestock.

Mohotani (mo-ho-TAH-nee), uninhabited volcanic island, MARQUESAS ISLANDS, FRENCH POLYNESIA, S PACIFIC, 11 mi/18 km SE of HIVA OA; 5 mi/8 km long, 2 mi/3.2 km wide; 09°59′S 138°49′W. Rises to 1,706 ft/520 m. Sometimes spelled Motane.

Mohpa (mo-pah), town, NAGPUR district, MAHARASHTRA state, central INDIA, 20 mi/32 km NW of NAGPUR; 21°19′N 78°49′E. Cotton ginning; millet, wheat, oilseeds. Sometimes spelled Mohapa.

Mohpani, India: see GADARWARA.

Möhrendorf (MUHR-ren-dorf), village, MIDDLE FRANCONIA, BAVARIA, GERMANY, on the MAIN-DANUBE CANAL, 5 mi/8 km N of ERLANGEN; 49°40′N 11°00′E. Forestry; food processing; paper.

Mohrin, POLAND: see MORYN.

Möhringen (MUHR-ring-guhn), S suburb of STUTTGART, BADEN-WÜRTTEMBERG, SW GERMANY; 48°44′N 09°06′E. Incorporated 1942 into STUTTGART.

Cross-references are shown in SMALL CAPITALS. The pronunciation guide is shown on page xix. The sources of population figures are shown on page xvii.

Mohrungen, POLAND: see MORAG.

Mohyliv-Podil's'kyy (mo-hi-LEEF–po-DEEL-skee) (Russian *Mogilev-Podol'skiy*), city (2004 population 38,100), SW VINNYTSYA oblast, UKRAINE, on the DNIESTER River, less than 2 mi/3.2 km NE of the border with MOLDOVA; 48°27′N 27°48′E. Elevation 469 ft/142 m. Raion center; industrial center with machine building (tools, gas equipment), food processing (flour, dairy, cannery); clothing manufacturing. Technical schools, medical school; regional museum, St. Nicholas Cathedral (1757), St. George's Church (1809–1819). Founded in 1595 by S. Potocki on site of Ivankivtsi village, the town was named after his father-in-law, Prince B. Mohyla (Movila) of Moldavia. The town became an important trading center on the road from Ukraine to Moldavia, and was periodically ruled by the Ukrainian Cossacks, Poles, and Turks. From the 17th to 19th century, the town was known under various names, such as Mohyliv, Mohyliv-na-Dnistri, and Mohyliv-Dnistrovs'kyy. Home of the Orthodox brotherhood press (established in 1616), which printed books in Ukrainian, Russian, Greek, and Moldavian. Held by the Turks (1672–1699); passed to Russia (1795); site of battles (1918–1919), including the defeat of the Red Army by the Ukrainian National Republic Army (1919). By the 20th century, half of the population was Jewish. No mass killings of the Jews took place in the city during World War II, but the majority of the community was destroyed through starvation, exposure, and disease living in the ghetto conditions from 1941 to 1944.

Moidart (MOI-dahrt), barren upland portion of HIGHLAND, N Scotland, extending N-S between the Sound of Arisaig (N) and LOCH SHIEL (S), rising to 2,852 ft/869 m on DRUIM FIACLACH, 10 mi/16 km SE of Arisaig. At SW extremity of region is Loch Moidart, sea inlet (6 mi/9.7 km long) containing island of EILEAN SHONA.

Moiese (mo-EEZ), village, Lake county, W MONTANA, 40 mi/64 km NW of MISSOULA, on Mission Creek in Flathead Indian Reservation. Headquarters for National Bison Range to S.

Moikovats, MONTENEGRO: see MOJKOVAC.

Moimenta da Beira (moi-MEN-tah dah BAI-rah), town, VISEU district, N central PORTUGAL, 13 mi/21 km SE of LAMEGO; 40°59′N 07°37′W. Rye, potatoes, wheat; oil; wine.

Moín (mo-EEN), port, COSTA RICA, 1 mi/1.6 km N of LIMÓN. Formerly a small fishing village; site of modern port with container loading facilities which has taken over much of the cargo that was handled by Limón. Costa Rica's only oil refinery is here.

Moinabad (MO-ee-nah-BAHD), town, BIDAR district, KARNATAKA state, SW INDIA, 25 mi/40 km SW of BIDAR; 17°42′N 77°13′E. Millet, cotton, rice, tobacco. Formerly called Chitgopa.

Moindou (mwan-DOO), coastal village, W NEW CALEDONIA, 65 mi/105 km NW of NOUMÉA, adjacent to PANIÉ range; 21°42′S 165°41′E. Coffee; livestock; former coal mining.

Moines, Île aux (MWAHN, EEL o), island in Gulf of MORBIHAN (BAY OF BISCAY), MORBIHAN department, off S BRITTANY (*Bretagne*), W FRANCE, 5 mi/8 km SW of VANNES; 4 mi/6.4 km long, 2 mi/3.2 km wide; 48°53′N 03°29′W. Oyster beds. A quiet resort reached by motor launch from Port Blanc. Megalithic monuments.

Moineşti (mo-NESHT), town, BACĂU county, E central ROMANIA, in E foothills of the Moldavian CARPATHIANS, 20 mi/32 km SW of BACĂU; 46°28′N 26°29′E. Railroad terminus; oil and natural-gas center; oil refining, woodworking, tanning, manufacturing of candles, chemicals. Also a health resort.

Mointy (moin-TUH), town, E ZHEZKAZGAN region, KAZAKHSTAN, junction on Trans-Kazakhstan railroad, 70 mi/113 km NW of BALKASH (linked by railroad

branch); 47°10′N 73°18′E. Railroad shops. Iron, lead-zinc, and arsenic deposits.

Moira (MOI-ruh), Gaelic *Mag Roth*, town (2001 population 3,669), NW DOWN, SE Northern Ireland, near LAGAN RIVER, 4 mi/6.4 km ENE of LURGAN; 54°29′N 06°14′W. Former linen and market plantation settlement. Here in 637 Domhnall, King of Tara, defeated the men of Ulster and Argyll in the most celebrated battle of Gaelic legend.

Mo i Rana, NORWAY: see MO.

Moirans (mwah-rahn), town (□ 7 sq mi/18.2 sq km), ISÈRE department, RHÔNE-ALPES region, SE FRANCE, near the ISÈRE, 12 mi/19 km NW of GRENOBLE; 45°20′N 05°34′E. Railroad junction; paper manufacturing.

Moirans-en-Montagne (mwah-rahn–zahn–mon-tahn-yah), commune (□ 10 sq mi/26 sq km; 2004 population 7,810), JURA department, FRANCHE-COMTÉ region, E FRANCE, near AIN RIVER (here forming long, narrow lake behind Vouglans Dam), 7 mi/11.3 km WNW of SAINT-CLAUDE; 46°26′N 05°44′E. Toy manufacturing center; woodworking, plastics. Village lies at W edge of the Regional Park of the High Jura.

Moira River (MOI-ruh), SE ONTARIO, E central CANADA; rises SW of SHARBOT LAKE; flows 60 mi/97 km SW to the Bay of QUINTE at BELLEVILLE; 44°09′N 77°23′W. Course is rapid; supplies water power.

Mõisaküla (MUH-ee-sah-kuh-luh) or **Myzakyula**, German *Moisaküll*, city, SW ESTONIA, 33 mi/53 km SE of PÄRNU, on Latvian border; 58°05′N 25°11′E. Flax-growing center; lumbering; railroad junction (repair shops). Also spelled Myyzakyula.

Moi's Bridge, town, RIFT VALLEY province, W KENYA, on railroad, 12 mi/19 km SSE of KITALE; 00°52′N 35°08′E. Agriculture (maize, beans) and livestock; maize storage silos. Market and trade center.

Moisei (moi-SAI), Hungarian *Majszin*, village, MARAMUREŞ county, NW ROMANIA, on W slopes of the CARPATHIANS, on railroad, and 35 mi/56 km SE of SIGHETU MARMAŢIEI; 47°39′N 24°33′E. Pilgrimage center with 18th-century monastery. In HUNGARY, 1940–1945.

Moisés S. Bertoni, **Moisés Bertoni** or **Doctor Moisés Bertoni** (mo-ee-SES ber-TO-nee), town, Caazapá department, S PARAGUAY, 110 mi/177 km SE of Asunción, S of Caazapá, and c.20 mi/32 km N of YEGROS; 26°21′S 56°26′W. Access to major roads and railroad. Lumber, fruit, livestock.

Moisés Ville (mo-ee-SES VEE-ye), town, central SANTA FE province, ARGENTINA, 31 mi/50 km SSW of SAN CRISTÓBAL; 30°43′S 61°29′W. Agricultural center (alfalfa, corn, wheat; livestock; dairying). Named after Moses Montefiore, French Jewish philanthropist who supported establishment of agricultural settlement in 1899 by E European Jews fleeing pogroms. Jewish population was once 90% of total. Although only a small Jewish community of about 300 remains, the entire town still celebrates Jewish holidays.

Moisie (mwah-ZEE), former town, E QUEBEC, E CANADA, on the SAINT LAWRENCE RIVER, near mouth of MOISIE River, 10 mi/16 km E of SEPT-ÎLES; 50°11′N 66°06′W. Hudson's Bay Company trading post. Amalgamated into SEPT-ÎLES in 2003.

Moisie (mwah-ZEE), river, 210 mi/338 km long; rises in E QUEBEC, E CANADA, near the NEWFOUNDLAND and LABRADOR border; flows S to the SAINT LAWRENCE RIVER; 50°11′N 66°05′W. The Hudson's Bay Company has an important trading post near the river's mouth.

Moissac (mwah-sahk), town (□ 33 sq mi/85.8 sq km), TARN-ET-GARONNE department, MIDI-PYRÉNÉES region, SW FRANCE, on the GARONNE LATERAL CANAL, on TARN RIVER near its influx into the GARONNE, and 13 mi/21 km W of MONTAUBAN; 44°07′N 01°05′E. A market center for fresh fruits and vegetables, it is noted for vineyards producing white/golden grapes of high quality. Apricots, cherries, peaches, and asparagus are also shipped, especially to PARIS. Built around

an 11th-century abbey, Moissac suffered repeatedly during Wars of Religion. The church of SAINT-PIERRE, part of the now demolished abbey, has a fine 12th-century portal. The older cloister has a remarkable inner garden with an arcade supported by simple columns in marble of various hues. It is a major tourist attraction.

Moïssala (mo-ees-sah-LAH), town, MANDOUL administrative region, S CHAD, on the BAHR SARA river (tributary of CHARI River), and 75 mi/121 km SW of SARH; 08°21′N 17°46′E. Cotton ginning. Until 1946, in Ubangi-Chari colony.

Moissy-Cramayel (mwah-see–krah-mah-yel), town (□ 5 sq mi/13 sq km) outer SE suburb of PARIS, in SEINE-ET-MARNE department, ÎLE-DE-FRANCE region, near outer circumferential highway of Greater Paris; 48°37′N 02°36′E. It forms part of the "new town" of Sénart. Has sugar mill and an industrial park along railroad line to MELUN.

Moita (MOI-tah), town, SETÚBAL district, S central PORTUGAL, near S bank of TAGUS RIVER estuary, on railroad, and 10 mi/16 km SE of LISBON. Fruit- and vegetable-growing center; resin processing, pottery and cheese manufacturing.

Moita Bonita (MO-ee-tah BO-nee-tah), city (2007 population 10,910), central SERGIPE state, BRAZIL, 31 mi/50 km NW of ARACAJU; 10°35′S 37°22′W. Manioc.

Moitaco (moi-TAH-ko), town, BOLÍVAR state, SE VENEZUELA, landing on ORINOCO RIVER, and 55 mi/89 km WSW of CIUDAD BOLÍVAR; 08°03′N 64°21′W. Livestock.

Mojácar (mo-HAH-kahr), town, ALMERÍA province, S SPAIN, 18 mi/29 km SSE of HUÉRCAL-OVERA; 37°08′N 01°51′W. Almonds, cereals, potatoes; sheep raising. Tourist resort, visited for its medieval Moorish quality; has ruined castle.

Mojada, Sierra, low NW spur of SIERRA MADRE ORIENTAL, on COAHUILA-CHIHUAHUA border, N MEXICO, W of ESMERALDA; c.30 mi/48 km long NW-SE; average elevation c.5,000 ft/1,524 m. Rich in minerals (silver, gold, lead, copper, zinc).

Mojados (mo-HAH-dhos), town, VALLADOLID province, N central SPAIN, 16 mi/26 km SSE of VALLADOLID; 41°26′N 04°39′W. Cereals, pine nuts, wine.

Mojave, unincorporated village (2000 population 3,836), KERN county, S central CALIFORNIA, in MOJAVE DESERT, c.47 mi/76 km ESE of BAKERSFIELD; 35°03′N 118°11′W. Supply point for mines (tungsten, silver, borax, gold); cattle and wheat ranches. Manufacturing (cement, aircraft parts). Railroad junction. Los Angeles Aqueduct and Pacific Crest National Scenic Trail to W; EDWARDS AIR FORCE BASE and Rogers Lake (dry) to SE; Red Rock Canyon State Recreation Area to N.

Mojave Dam, California: see SILVERWOOD LAKE.

Mojave Desert or **Mohave** (both: mo-hah-vai) (□ 15,000 sq mi/39,000 sq km), region of low, barren mountains and flat valleys, 2,000 ft/610 m–5,000 ft/1,525 m high, S CALIFORNIA; part of the GREAT BASIN of the U.S. (all rivers leading into basin have no outlet to sea). Bordered on the N and W by the SIERRA NEVADA and the TEHACHAPI; on SW by SAN GABRIEL and SAN BERNARDINO mountains; it merges with the COLORADO DESERT in SE; N part includes DEATH VALLEY (282 ft/86 m below sea level), lowest point in Western Hemisphere. Beneath the Mojave lies the Fenner Basin, a large aquifer 3,506 ft/1,069 m below the surface. Once a part of an ancient interior sea, the desert was formed by volcanic action (lava surfaces with cinder cones are present) and by material deposited by the COLORADO RIVER. The temperature is uniformly warm to very hot, with cold nights throughout the year, and with a wide variation from day to night. Strong, dry winds blow in the afternoon and evening. Located in the rain shadow of the COAST RANGES, the Mojave receives an average annual rainfall of 5 in/12.7 cm, mostly in winter. Juniper and

---

Area is shown by the symbol □, and capital city or county seat by ⊙.

Joshua trees are found on the higher, outer mountain slopes; desert-type vegetation, and numerous intermittent lakes and streams are present in the valleys. The MOJAVE RIVER enters from SW and is the largest stream. Minerals found in the desert include borax and other salines; gold, silver, and iron. The desert is crossed by two railroad lines and two highways. Military installations were established in the Mojave during World War II; modern istallations include China Lake Naval Air Weapons Station, Fort Irwin Military Reservation, and Twentynine Palms Marine Corps Base. A solar-power plant has been built using molten salt to store energy from sunlight. Ward Valley, 20 mi/ 32 km from the Colorado River and Arizona border, contains a 1.5-sq mi/4-sq km tract that is scheduled to become a low-level nuclear waste dump. Death Valley National Monument in N; JOSHUA TREE NATIONAL MONUMENT in S.

Mojave River or **Mohave** (MO-hah-vai), c.100 mi/161 km long; rises in the SAN BERNARDINO MOUNTAINS, SAN BERNARDINO county, S CALIFORNIA, at LAKE ARROWHEAD reservoir (as Deep Creek for first c.20 mi/32 km); flows N past Victorville, then NE past Barstow to disappear in the MOJAVE DESERT, SW of Soda Lake (dry). Due to the porous soil and rapid evaporation, it is intermittent for much of its course except during the short wet season.

Mojiang (MO-JI-ANG), town, ⊙ Mojiang county, S YUNNAN province, CHINA, on road, and 45 mi/72 km NE of PU'ER; 23°25′N 101°44′E. Elevation 6,332 ft/1,930 m. Timber, rice, millet, beans; chemicals, plastics, food industry; logging, nonferrous ore mining.

Moji-Mirim (MO-zhee–MEE-reen), city (2007 population 85,390), E SÃO PAULO, BRAZIL, 33 mi/53 km N of CAMPINAS; 22°26′S 46°57′W. Railroad junction; cotton ginning, meat packing, rice processing; brewing, brandy distilling. Manufacturing (flour products); agriculture (coffee, sugar). Sericulture. Airfield. Established in 18th century. Formerly spelled Mogy-Mirim.

Mojkovac (MOI-ko-vahts), village (2003 population 4,120), E MONTENEGRO, on TARA RIVER, on BIJELO POLJE–KOLASIN road, and 10 mi/16 km NNE of KOLASIN; 42°57′N 19°34′E. Also spelled Moikovats or Moykovats.

Mojo (MO-ho), canton, MODESTO OMISTE province, POTOSÍ department, W central BOLIVIA, 2 mi/3 km S of MORAYA, and 18 mi/30 km N of VILLAZÓN, on the UYUNI-VILLAZÓN railroad and highway; 21°54′S 65°33′W. Elevation 11,296 ft/3,443 m. Agriculture (potatoes, yucca, bananas); cattle.

Mojo (MO-jo), town (2007 population 41,191), OROMIYA state, central ETHIOPIA; 08°36′N 39°07′E. On Addis Ababa-Djibouti railroad, and 40 mi/64 km SE of ADDIS ABABA at road junction; in cereal growing and cattle raising region.

Mojo, INDONESIA: see MOYO.

Mojocoya (mo-ho-KOI-ah), town and canton, ZUDAÑEZ province, CHUQUISACA department, S central BOLIVIA, 28 mi/45 km NNE of ZUDAÑEZ; on banks of San Lucas River; 18°45′S 64°37′W. Wheat.

Mojokerto (maw-jaw-KUHR-taw), municipality (2000 population 109,073), ⊙ Mojokerto district, Java Timur province, INDONESIA, on BRANTAS RIVER, and 27 mi/43 km SW of SURABAYA; 07°28′S 112°26′E. Trade center in agricultural area (sugar, rice, peanuts, corn); textile mills, railroad shops. Extensive irrigation works nearby. Also spelled Modjokerto.

Mojones, Cerro (mo-HO-nes, SER-ro), Andean volcano (19,650 ft/5,989 m) in N CATAMARCA province, ARGENTINA, 30 mi/48 km N of ANTOFAGASTA.

Mojos (MO-hos), canton, FRANZ TAMAYO province, LA PAZ department, W BOLIVIA, 19 mi/30 km NW of APOLO, and 12 mi/20 km W of PATA; 14°34′S 68°53′W. Elevation 4711 ft/1,436 m. Clay, limestone, and gypsum deposits. Agriculture (potatoes, yucca, bananas, rye); cattle.

Mojotoro (mo-ho-TO-ro), canton, OROPEZA province, CHUQUISACA department, SE BOLIVIA, 6 mi/10 km S of CHUQUI-CHUQUI, on the Chico River (hot springs); 19°10′S 65°16′W. Elevation 8,212 ft/2,503 m. Iron (Mina Okekhasa and Mina Virgén de Copacabana), and manganese (Mina Lourdes) mining; limestone and gypsum deposits. Agriculture (potatoes, yucca, bananas, corn, barley, oats, rye, peanuts); cattle.

Mojstrana (mo-ye-STRAH-nah), village, NW SLOVENIA, on the SAVA DOLINKA RIVER, on railroad, and 39 mi/63 km NW of LJUBLJANA, between JULIAN ALPS and the Karawanken mountains; 46°27′N 13°56′E. Climatic resort; sports center; cement works. Includes hamlet of Dovje.

Moju (mo-zhoo), city (2007 population 64,385), E PARÁ, BRAZIL, on RIO MOJU (navigable), and 40 mi/64 km SSW of BELÉM; 01°55′S 48°46′W. Ships rubber, lumber. Road to ABAETETUBA. Formerly spelled Mojú.

Moka, district (□ 89 sq mi/231.4 sq km; 2004 population 78,600), central MAURITIUS; ⊙ MOKA; 20°15′S 57°35′E. Bounded on E and NE by FLACQ district, on N by PAMPLEMOUSSES district, on NW by Port Louis district, on SW by PLAINES WILHEMS district, and on S by GRAND PORT district. Sugarcane, vegetables, flowers, fruits; livestock; light manufacturing (sugar refining; apparel). Mahatma Gandhi Institute; Moka Eye Hospital.

Moka (MO-ka), village (2000 population 8,286); ⊙ MOKA district, NW MAURITIUS, 4 mi/6.4 km S of PORT LOUIS; 20°14′S 57°30′E. Dairying; sugarcane; glass manufacturing. Site of Mahatma Gandhi Institute.

Mokabe Kasari, CONGO: see MOKABE-KASARI.

Mokabe-Kasiri (mo-KAH-bai–kuh-SEE-ree), village, KATANGA province, SE CONGO, close to MANIKA PLATEAU (UPEMBA NATIONAL PARK); 09°58′S 26°16′E. Elev. 3,720 ft/1,133 m. Also known as MOKABE KASARI or MUKABE KASARI.

Mokama (mo-kah-mah), **Mukama** or **Mokameh**, town (2001 population 56,400), PATNA district, W central BIHAR state, E INDIA, in GANGA PLAIN, on GANGA RIVER (railroad ferry), and 51 mi/82 km ESE of PATNA; 25°24′N 85°55′E. Trade center (rice, gram, wheat, barley, oilseeds, corn, sugarcane, millet).

Mokambo (mo-KAHM-bo), village, KATANGA province, SE CONGO, on ANGOLA border, on railroad, and 85 mi/137 km SE of Lubambashi; 12°25′S 28°21′E. Elev. 4,324 ft/1,317 m. Customs station.

Mokane (mo-KAIN), town (2000 population 188), CALLAWAY county, central MISSOURI, on MISSOURI River, and 12 mi/19 km S of FULTON; 38°40′N 91°52′W. Grain, livestock.

Mokapu Point (MO-KAH-poo), E OAHU Island, HONOLULU county, HAWAII, 11 mi/18 km NE of HONOLULU; 21°27′N 157°43′W. NE corner of Mokapu Peninsula; Kaneohe Marine Corps Air Station occupies most of peninsula, which is bounded by KANEOHE BAY (W), by KAILUA BAY (SSE). Moku Manu Island 1 mi/1.6 km NNE of point.

Moka Range, mountain range, N central MAURITIUS; rises to 2,700 ft/823 m in the Pieter Both, 4 mi/6.4 km SE of PORT LOUIS.

Mokau River (muh-KOW), W NORTH ISLAND, NEW ZEALAND; rises in KING COUNTRY ranges, S of TE KUITI; flows 98 mi/158 km SW to NORTH TARANAKI BIGHT. Small Mokau coal mines.

Mokelumne Hill, unincorporated village (2000 population 774), CALAVERAS county, central CALIFORNIA, near MOKELUMNE RIVER, and c.40 mi/64 km NE of STOCKTON, in 1849 California Gold Rush region. Cattle; walnuts olives, honey. During gold rush, it was county seat for a time and an important freighting point for the mines. Pardee Reservoir to W.

Mokelumne Peak (9,332 ft/2,844 m), AMADOR county, E CALIFORNIA, in the SIERRA NEVADA, 27 mi/43 km S of LAKE TAHOE. In Eldorado National Forest.

Mokelumne River, 80 mi/129 km long, E central CALIFORNIA; rises in the SIERRA NEVADA in S central AL-

PINE county. Formed by joining of North and South forks 9 mi/14.5 km E of Jackson. North Fork rises in Sierra Nevada, 30 mi/48 km S of SOUTH LAKE TAHOE; flows c.60 mi/97 km WSW through Salt Springs Reservoir, to South Fork, which rises in Sierra Nevada 40 mi/64 km E of Jackson; flows c.30 mi/48 km SW, then W. The combined Mokelumne River flows SW, past village of Mokelumne Hill through Pardee and Camanche reservations and past Lodi, then NW and SW to SAN JOAQUIN RIVER 20 mi/32 km NW of Stockton (San Joaquin River forms shipping channel to SUISUN BAY at this point). Near Mokelumne Hill, Pardee Dam (358 ft/109 m high, 1,337 ft/408 m long; completed 1929) impounds PARDEE RESERVOIR, which supplies water to cities on E shore of SAN FRANCISCO BAY. On a headstream (North Fork), on Calaveras-Amador county line, is Salt Springs Dam (328 ft/100 m high, 1,260 ft/384 m long; completed 1931; for power). Mokelumne Aqueduct diverts water from CAMANCHE RESERVOIR to Stockton and N San Francisco Bay area.

Mokena (mo-KEE-nah), village (2000 population 14,583), WILL county, NE ILLINOIS, 10 mi/16 km E of JOLIET; 41°31′N 87°52′W. In agricultural area (corn, soybeans; dairying). Manufacturing (water pollution control equipment; galvanized ducts and fittings; food processing equipment, plastic injection molds).

Mokha, YEMEN: see MOCHA.

Mokhotlong (mo-KHOT-lawng) [Lesotho=place of the bald-headed ibis], district (□ 1,573 sq mi/4,089.8 sq km; 2001 population 89,705), E and NE LESOTHO, ⊙ Mokhotlong, bounded E and NE by SOUTH AFRICA; border formed in DRAKENSBERG RANGE. Drained by Mokhotlong River and ORANGE (Senqu) River (source near NE boundary, forms Mashai Dam in SW corner). Horses, sheep, goats, cattle; peas, beans, corn. Diamond reserve in N, mined until 1982.

Mokhotlong (mo-KHOT-lawng), town, ⊙ MOKHOTLONG district, NE LESOTHO, 95 mi/153 km E of MASERU, in E foothills of DRAKENSBERG, near Mokhotlong River; 29°17′S 29°05′E. Airstrip to N. Goats, sheep, horses; peas, beans. A 35 mi/56 km 4x4 track connects the town to Sani Pass to Kwagulu-Natal; closest town to Thalon-Ntlenyana (highest peak in SOUTH AFRICA).

Mokhovaya (muh-huh-VAH-yah), town, SE KAMCHATKA oblast, RUSSIAN FAR EAST, on the N shore of the Avachinsk Bay, 3 mi/5 km NW of PETROPAVLOVSK-KAMCHATSKIY, to which it is administratively subordinate; 53°04′N 158°37′E. Elevation 505 ft/153 m. Fish combine.

Mokhovo (MO-huh-vuh), industrial settlement, W central KEMEROVO oblast, S central SIBERIA, RUSSIA, near cargo railroad spur, 10 mi/16 km N, and under administrative jurisdiction, of BELOVO; 54°35′N 86°22′E. Elevation 698 ft/212 m. Open-pit coal mining. Also called Mokhova (same pronounciation).

Mokhovoye (MO-huh-vuh-ye), village, central ORËL oblast, central European Russia, on road junction and near railroad, 19 mi/31 km E of ORËL; 52°57′N 36°34′E. Elevation 823 ft/250 m. Agricultural products.

Mokhovoye, RUSSIA: see PARFËNOVO.

Mokino (MO-kee-nuh), village, S KIROV oblast, E central European Russia, 11 mi/18 km SE of SOVETSK, to which it is administratively subordinate; 57°26′N 49°11′E. Elevation 554 ft/168 m. In agricultural area (wheat, flax, oats, vegetables, potatoes; livestock). Logging and lumbering.

Mokokchung, district (□ 624 sq mi/1,622.4 sq km), Nagaland state, NE India; ⊙ Mokokchung.

Mokokchung, town, ⊙ MOKOKCHUNG district, NAGALAND state, NE INDIA, 57 mi/92 km NNE of KOHIMA; 26°20′N 94°32′E. Rice, cotton, oranges, potatoes.

Mokolo (MO-ko-lo), town (2001 population 29,100), ⊙ MAYO-TSANAGA department, Far-North province, CAMEROON, near Nigerian border, 40 mi/64 km WNW of MAROUA; 10°48′N 13°50′E. Market town and administrative center; livestock raising; peanuts, millet; tourism. Has hospital, Protestant mission.

Mokopane, town, LIMPOPO province, SOUTH AFRICA, near MOGALAKWENA River, at foot of STRYDPOORT Mountains, 35 mi/56 km SW of POLOKWANE (Pietersburg), on Nyl River, and on N1 highway to ZIMBABWE; 24°11'S 29°01'E. Elevation 3,729 ft/1,135 m. Tin mining; agriculture center (oranges, peaches, groundnuts; cattle). ZEBEDIELA (citrus estate) to ESE. Founded 1855. Also known as Potgietersrus until early 2000s; both or either name officially used.

Mokotow (mo-KO-toov), Polish *Mokotów*, residential suburb of WARSAW, Warszawa province, E central POLAND, 2 mi/3.2 km S of city center. Site of airfield; before World War I, Russian military training ground.

Mokpalin (MOK-pah-lin), village, in former THATON district, MYANMAR, on left bank of SITTOUNG RIVER estuary, on Bago-Martaban railroad, and 60 mi/97 km NE of YANGON.

Mokpo (MOK-PO), city (2005 population 244,888), SOUTH CHOLLA province, SW SOUTH KOREA, port on YELLOW SEA, 190 mi/306 km SSW of SEOUL; 34°47'N 126°23'E. Commercial center for area producing rice and cotton. Rice refineries, cottonseed-oil and cotton-ginning factories, canneries, sake breweries, fish-processing plants; ceramics, glass industries. Exports rice, cotton, marine products, hides. Port was opened 1897 to foreign trade. Ferry links Mopko to CHEJU ISLAND. The Koha Island on the SW of the city associated with Admiral Yi Sun-shin. The Yongsang reservoir was built on the Yongsan River in 1981. Makpo National University (1979).

Mokra, MACEDONIA: see JAKUPICA.

Mokra-Horakov (MOK-rah–HO-rah-KOF), Czech *Mokrá-Horákov*, village, JIHOMORAVSKY province, S MORAVIA, CZECH REPUBLIC, 6 mi/9.7 km ENE of BRNO. Large cement- and lime-works; limestone quarry. OCHOZSKA CAVE (Czech *Ochozská*) is just N.

Mokra Kalyhirka (MO-krah kah-LI-heer-kah) (Russian *Mokraya Kaligorka*), town (2004 population 4,800), S central CHERKASY oblast, UKRAINE, on the DNIEPER UPLAND, on road, 13 mi/21 km SE of KATERYNOPIL; 48°51'N 31°13'E. Elevation 636 ft/193 m. Food processing, including dairy products; vocational school, park. Known since the beginning of the 18th century; town since 1965.

Mokran, IRAN: see MAKRAN.

Mokra Planina (MO-krah PLAH-mee-nah), NE spur of NORTH ALBANIAN ALPS, SW SERBIA and SE MONTENEGRO, just N of ALBANIA border, between upper IBAR RIVER (N) and the METOHIJA valley (S). Highest peak, Zljeb or Zhlyeb, (Serbian *Žljeb*) is 7,813 ft/2,381 m high and on the Serbia-Montenegro border, 6 mi/9.7 km NNW of PEC, Serbia. LIM RIVER flows at W foot; 42°43'N 20°00'E.

Mokraya Kaligorka, UKRAINE: see MOKRA KALYHIRKA.

Mokresh (mo-KRESH), village, MONTANA oblast, VULCHITRUN obshtina, BULGARIA; 43°44'N 23°24'E.

Mokrin (MOK-rin), village, VOJVODINA, NE SERBIA, near ROMANIA border, 8 mi/12.9 km NNW of KIKINDA, in the BANAT region; 45°56'N 20°24'E.

Mokrishte (mo-KREESH-te), village, PLOVDIV oblast, PAZARDZHIK obshtina, BULGARIA; 42°11'N 24°15'E.

Mokrous (muh-kruh-OOS), town (2006 population 6,725), central SARATOV oblast, SE European Russia, on road and railroad, 32 mi/51 km WSW of YERSHOV; 51°14'N 47°31'E. Elevation 360 ft/109 m. Food products, asphalt.

Mokrousovo (muh-kruh-OO-suh-vuh), village (2005 population 4,815), E KURGAN oblast, SW SIBERIA, Russia, on a right tributary of the TOBOL River, on road, 37 mi/60 km NNE of LEBYAZHYE; 55°48'N 66°46'E. Elevation 416 ft/126 m. Dairy plant, flour mill.

Mokrye Gory, GEORGIA: see DZHAVAKHETI Mountains.

Moksha (muhk-SHAH) river, approximately 375 mi/603 km long; rises NW of PENZA, central PENZA oblast,

---

S central European Russia; flows generally NW through central and NW MORDVA REPUBLIC, into the OKA River in NE RYAZAN oblast. Its lower course is navigable.

Mokshan (muhk-SHAHN), town (2005 population 11,600), central PENZA oblast, S central European Russia, on the MOKSHA River (OKA River basin), on road, 27 mi/43 km NW of PENZA; 53°26'N 44°36'E. Elevation 728 ft/221 m. Road center; hemp processing; dairying. Made city in 1780; reduced to status of a village in 1925; made town in 1960.

Mokuaweoweo (MO-koo-AH-WAI-o-WAI-o), crater and caldera of MAUNA LOA volcano, S central HAWAII island, HAWAII county, HAWAII, at W end of HAWAII VOLCANOES NATIONAL PARK. Second-largest active crater (□ 3.7 sq mi/9.6 sq km, length 3.7 mi/6 km, width 1.7 mi/2.7 km) in world; KILAUEA, also on Hawaii island, is larger. Lava flow of 1880–1881 was 50 mi/80 km long.

Mokuleia (MO-koo-LAI-ee-ah), town (2000 population 1,839), NW OAHU island, HONOLULU county, HAWAII, 26 mi/42 km NW of HONOLULU, on N coast at mouth of Makaleha Stream; 21°34'N 158°10'W. Kukui nuts; cattle; fish. Dillingham Air Force Base (inactive) to W, Kaena Military Reservation lies farther W at KAENA POINT. Mokuleia Beach Park to W; WAIANAE Mountains to S.

Mokwa (MO-kwah), town, NIGER state, W NIGERIA, on railroad, and 110 mi/177 km WSW of MINNA; 09°17'N 05°03'E. Market center. Millet, maize, rice, yams, and sorghum.

Mol (MAWL), commune (2006 population 32,907), Tournhout district, ANTWERPEN province, N BELGIUM, 12 mi/19 km SSE of TURNHOUT, near the DESSEL-KWAADMECHELEN canal; founded in the ninth century; 51°11'N 05°06'E. It is a manufacturing city and has Euratom's nuclear research center. Provincial Parks "ZILVERMEER" and "Zilverstrand" nearby.

Mol (MAWL), Hungarian *Mohol*, village, VOJVODINA, N SERBIA, on TISZA RIVER, and 12 mi/19 km S of SENTA, in the BACKA region; 45°45'N 20°07'E.

Mola, Cabo de la (KAH-vo dhai lah MO-lah), headland, SW MAJORCA island, BALEARIC ISLANDS, 15 mi/24 km W of PALMA; 39°32'N 02°43'E.

Molagavita (mo-lah-gah-VEE-tah), town, ⊙ Molagavita municipio, SANTANDER department, N central COLOMBIA, 30 mi/48 km SE of BUCARAMANGA; 06°41'N 72°50'W. Coffee, corn, cassava; livestock.

Molai (mo-LAH-ee), town, LAKONIA prefecture, SE PELOPONNESE department, extreme SE mainland GREECE, 30 mi/48 km SE of SPARTA; 36°48'N 22°51'E. Livestock. Also Molaoi.

Molakalmuru (mol-KAHL-muh-roo), town, CHITRADURGA district, KARNATAKA state, SW INDIA, 40 mi/64 km NW of CHITRADURGA. Hand-loom silk weaving, handicraft glass bangles. Asokan edicts of third century B.C. carved on rocks, 7 mi/11.3 km NNE, near village of Siddapura. Also spelled Molkalmuru.

Molale, town (2007 population 5,823), AMHARA state, central ETHIOPIA; 30 mi/48 km NNE of DEBRE BERHAN; 10°06'N 39°42'E.

Molalla (mo-LAL-uh), town (2006 population 7,012), CLACKAMAS county, NW OREGON, 14 mi/23 km S of OREGON CITY, near Molalla River; 45°08'N 122°34'W. Railroad terminus. Timber. Berries, apples, potatoes; poultry, hogs, sheep, cattle; nurseries. Table Rock Wilderness Area to SE; MOUNT HOOD National Forest to E.

Molango (mo-LAHN-go), town, ⊙ Molango de Escamilla municipio, HIDALGO, central MEXICO, 45 mi/72 km N of PACHUCA DE SOTO, on Mexico Highway 105; 20°48'N 98°44'W. Corn, wheat, beans, fruit, livestock.

Molaoi, Greece: see MOLAI.

Molara Island (mo-LAH-rah), (□ 2 sq mi/5.2 sq km), off NE SARDINIA, in TYRRHENIAN Sea, in SASSARI province; rises to 518 ft/158 m; 40°52'N 09°43'E.

Molar, El (mo-LAHR, el), town, MADRID province, central SPAIN, 22 mi/35 km N of MADRID. Spa. Grain growing and wine producing.

Molares, Los (los mo-LAH-res), town, SEVILLE province, SW SPAIN, 21 mi/34 km SE of SEVILLE; 37°09'N 05°43'W. Olive industry. Ancient castle in vicinity.

Molat Island (MO-laht), Italian *Melada*, Dalmatian island in ADRIATIC SEA, S CROATIA, 18 mi/29 km NW of ZADAR; 6 mi/9.7 km long. Molat village on SE shore.

Molay-Littry, Le (MO-lai—lee-tree, luh), commune (□ 10 sq mi/26 sq km; 2004 population 55), CALVADOS department, BASSE-NORMANDIE region, NW FRANCE, 9 mi/14.5 km WSW of BAYEUX; dairying (cheese and casein manufacturing). Coal mine, worked since 1743, is now inactive. There is a small museum tracing the mine's history.

Molbergen (MAWL-ber-gen), village, LOWER SAXONY, NW GERMANY, in the Oldenburger Münsterland, 6 mi/9.7 km W of CLOPPENBURG; 52°51'N 07°56'E. Manufacturing (machinery).

Molcaxac (mol-KAH-hak), town, PUEBLA, central MEXICO, near ATOYAC RIVER, 29 mi/47 km SE of PUEBLA. Grain, maguey.

Molchanovo (muhl-CHAH-nuh-vuh), village (2006 population 6,115), S central TOMSK oblast, W SIBERIA, Russia, on the OB' RIVER, above the CHULYM RIVER mouth, on road junction, 90 mi/145 km NW of TOMSK; 57°35'N 83°46'E. Elevation 285 ft/86 m. In flax-growing area; produce processing, lumbering.

Mold (MUHLD), Welsh *Yr Wyddgrug*, town (2001 population 9,568), ⊙ Flintshire, NE Wales, on ALYN RIVER, and 6 mi/9.7 km S of FLINT; 53°10'N 03°08'W. Market town; light industry. Has 15th-century church and remains of ancient castle.

Moldau River, CZECH REPUBLIC: see VLTAVA RIVER.

Moldauthein, CZECH REPUBLIC: see TYN NAD VLTAVOU.

Moldava nad Bodvou (mol-DAH-vah NAHD bod-VOU), Hungarian *Szepsi*, town, VYCHODOSLOVENSKY province, SE SLOVAKIA, on BODVA RIVER, and 14 mi/23 km SW of KOŠICE; 48°37'N 21°00'E. Contains railroad junction. Electrical goods manufacturing, food and wood processing, wine making, brick kiln. Fruit; vineyard. Gothic church. Under Hungarian rule between 1938–1945.

Moldavian Autonomous Soviet Socialist Republic, former republic, existed 1924–1940 as a part of Ukrainian Soviet Socialist Republic, within the USSR; capital was TIRASPOL. It included TRANSNISTRIA (Pridnestrovie) plus a few localities subsequently transferred to UKRAINE. The bulk of this entity's area was incorporated in the Moldavian Soviet Socialist Republic (1940–1991) after the "liberation" of BESSARABIA in 1940.

Molde (MAWL-luh), city (2007 population 24,254), ⊙ MØRE OG ROMSDAL county, W NORWAY, on MOLDEFJORDEN (an arm of ROMSDALFJORDEN). Commanding a panoramic view of the snow-capped ROMSDAL Mountains, it is a favorite tourist center and fishing port. Textiles and furniture are manufactured.

Moldefjorden (MAWL-luh-fyawr-uhn), inlet of NORTH SEA, in MØRE OG ROMSDAL county, W NORWAY, extending c.10 mi/16 km E from MOLDE, and branching into ROMSDALFJORDEN (E), FANNEFJORDEN (NE; 16 mi/26 km long), and TRESFJORDEN (S; 8 mi/12.9 km long). OTTERØY guards its mouth.

Moldgreen, ENGLAND: see HUDDERSFIELD.

Moldova (mol-DO-vah), republic (□ c.13,000 sq mi/33,670 sq km; 2004 estimated population 4,446,455; 2007 estimated population 4,320,490), E EUROPE; ⊙ CHISINAU.

Geography

Moldova is landlocked. The PRUT RIVER separates it from ROMANIA in the W. In the N and E, the DNIESTER (Nistru) River forms its approximate boundary with UKRAINE, on which it also borders in the S. Mostly a hilly plain, Moldova occupies all but the

---

southernmost and northernmost sections of former BESSARABIA. Its proximity to the BLACK SEA gave it one of the mildest climates in the former USSR.

Population

The majority of the population is Moldovan/Romanian (64.5%), but there are large Ukrainian (13.8%) and Russian (13%) populations, and smaller Gagauz (3.5%), and Bulgarian (2%). Only a remnant of the former substantial Jewish minority is left. The Moldovan language is virtually indistinguishable from Romanian, and the two groups are ethnically identical.

Economy

The fertile soil supports wheat, corn, barley, tobacco, sugar beets, soybeans, and sunflowers, as well as extensive fruit orchards, vineyards, and walnut groves. Horticulture is important for the production of such essences as rose oil and lavender. Beef and dairy cattle are raised, and beekeeping and silk breeding are widespread. Food processing is the main industry; others include metalworking, engineering, and the manufacture of electrical equipment.

History

For history prior to independence, see MOLDOVA (province). The Moldavian Soviet Socialist Republic was declared an independent republic in August 1991. A guerrilla war broke out in the predominantly Russian-speaking TRANSNISTRIA (Pridniestrovyie) region, which declared its own independence from Moldova. Russian forces have remained on Moldovan territory east of the Dniester helping the Ukrainians and Russians there. There was also a separatist movement in the area populated by the Turkic Gagauz ethnic group in SE Moldova; the Gagauz area has been granted special status. Some intellectuals and politicians have advocated that Moldova unite with Romania, an unlikely development at present. Mircea Snegur was president from 1991 to 1997, when he was defeated by Petru Lucinschi. Vladimir Voronin, a Communist (the party now is pro-Western and favors freer markets), was elected president in 2001 and reelected in 2005. Moldova is a member of the COMMONWEALTH OF INDEPENDENT STATES (CIS).

Government

Moldova is governed under the constitution of 1994. The president, who is the head of state, is elected by the legislature for a four-year term and is eligible for a second term. The prime minister, who is the head of government, is appointed by the president, as is the cabinet. Members of the 101-seat Parliament are elected by popular vote to serve four-year terms. The current head of state is President Vladimir Voronin (since April 2001). Prime Minister Vasile Tarlev has been the head of government since April 2001. Administratively, Moldova is divided into thirty-two raions (districts or counties), three municipalities, and two territorial units, one of which (Gagauzia) is autonomous.

Moldova (mol-DO-vah), region and historic province (□ c.14,700 sq mi/38,100 sq km), extending from the CARPATHIANS in ROMANIA E to the DNIESTER (Nistru) River in independent MOLDOVA. The region borders on UKRAINE in the NE and on WALACHIA in the S. In Romania it comprises roughly the modern administrative divisions of BACAU, GALATI, and IASI. SUCEAVA and IASI, its historic capitals, and GALATI, its port on the DANUBE RIVER, are the chief cities (all in Romania). Moldova, a fertile plain drained by the SIRETUL RIVER, is the granary of Romania. Besides farming there is livestock raising; orchards and vineyards dot the countryside. Lumbering and petroleum extraction are the main industries. It was part of the Roman province of DACIA and has retained its Latin speech despite the centuries of invasion and foreign rule. It has been a frequent battleground because of its position as a historic passageway between Asia and S Europe. Greek, Slavic, Turkish, Jewish, and other elements have influenced its culture. Moldova was part

of the Kievan state from the 9th to the 11th centuries. In the 13th century the Cumans, who then held Moldova, were expelled by the Mongols. When the Mongols withdrew, Moldova became (early 14th century) a principality under native rulers. It then included BUKOVINA and BESSARABIA. Like its sister principality, Walachia, it was torn by strife among the boyars—the great landowners and officeholders—and among rival claimants to the throne. The rural population was reduced to misery and virtual slavery (which lasted well into the 19th century) by the princes, whose absolute rule was marked by cruelty. Moldova reached its height under Stephen the Great (1457–1504), who routed the Turks in 1475, but became tributary to the sultans in 1504. The Jewish population, dating back to 15th century, played a key role in developing the region's cities and towns beginning in the 17th century; by 1900s, Jews were half the population of Iasi, one-third of Suceavea, and more than one-fifth of Galati; many were killed by the Nazis in World War II.

Although it was frequently occupied by foreign powers in the continuous wars among the OTTOMAN EMPIRE, AUSTRIA, TRANSYLVANIA, POLAND, and RUSSIA, Moldova remained a highly fortified border region under the Ottoman Empire. S Bessarabia early passed under the rule of the khans of CRIMEA. Early in the 18th century the Turks ended the rule by native princes—who had sided with the enemy as often as with Turkey—and appointed governors (hospodars), mostly Greek Phanariots. The Greeks surpassed their predecessors in avarice, while the nobility fell into total decay and corruption. Their rule was ended (1822) after the Greek insurrection instigated by Alexander Ypsilanti, and native hospodars were appointed. Meanwhile, Bukovina had been taken (1775) by Austria and E. Moldavia and Bessarabia by Russia (1791–1812). After the Russo-Turkish War of 1828–1829, Moldova and Walachia were made virtual protectorates of Russia, although they continued to pay tribute to the sultan. A Romanian national uprising (1848–1849) was suppressed by Russian intervention. In the Crimean War, Moldova was again occupied by Russia, but in 1856 the two Danubian principalities, Walachia and Moldova, were guaranteed independence under the nominal suzerainty of Turkey. With the accession (1859) of Alexander John Cuza as prince of both Moldova and Walachia, the history of modern Romania began. In 1878, S Bessarabia was ceded to Russia following the Russo-Turkish War. Following World War I, Bessarabia, along with Bukovina, was reincorporated into Romania, which had seized Bessarabia in 1918. The USSR, refusing to sanction the Romanian takeover, established the MOLDAVIAN AUTONOMOUS SOVIET SOCIALIST REPUBLIC (ASSR) in Ukraine, with TIRASPOL as the capital. The USSR forced Romania in 1940 to cede Bessarabia and N Bukovina. The predominantly Ukranian districts in the S and around KHOTIN in the N were incorporated into Ukraine, as were parts of the Moldavian ASSR; the rest was merged with what remained of the Moldavian ASSR and made a constituent republic. Taken by Romania in 1941, the area was reconquered by the USSR in 1944. The Moldavian SSR adopted a measure in June 1990 calling for greater sovereignty within the USSR. Moldova declared its independence in August 1991. See also Moldova, republic.

Moldova Nouă (mol-DO-vah NO-wuh), Hungarian *Újmoldova*, town, CARAŞ-SEVERIN county, SW ROMANIA, 20 mi/32 km S of ORAVIŢA; 44°44′N 21°41′E. Center for wine production; copper pyrite extraction; flour milling, lumbering. Dates from Roman times.

Moldova River (mol-DO-vah), German *Moldau*, E ROMANIA in Moldavian Carpathians, 45 mi/72 km WSW of RĂDĂUŢI; c.110 mi/177 km long; flows E past CÎMPULUNG MOLDOVENESC and SE past ROMAN to SIRET RIVER, 4 mi/6.4 km SW.

Moldova-Veche (mol-DO-vah–VE-ke), Hungarian *Ómoldova*, village, CARAŞ-SEVERIN county, SW ROMANIA, on the DANUBE (Yugoslav border), and 22 mi/35 km S of ORAVIŢA; 44°43′N 21°37′E. White-wine production, sericulture. Built on ruins of ancient Roman city.

Moldoveanu (mol-do-VA-noo), peak, central ROMANIA, 38 mi/61 km W of BRAŞOV in the TRANSYLVANIAN ALPS; highest point in country (8,343 ft/2,543 m).

Moldoviţa (mol-do-VEE-tsah), village, SUCEAVA county, N ROMANIA, in the Moldavian Carpathians, 9 mi/14.5 km N of CÎMPULUNG MOLDOVENESC; 47°41′N 25°32′E. Lumbering center. Has 16th-century monastery with notable Byzantine frescoes, valuable religious collections.

Mole Creek (MOL), village, N central TASMANIA, SE AUSTRALIA, 40 mi/64 km W of LAUNCESTON; 41°33′S 146°24′E. Dairying and agriculture center; honey. Limestone caves.

Moledet (mo-LE-det), moshav, N ISRAEL, 7 mi/11.3 km NW of BEIT SHEAN in LOWER GALILEE; 32°35′N 35°26′E. Elevation 98 ft/29 m. Mixed farming (beef, dairy, sheep, poultry, and field crops); factory manufacturing; hoists. Founded 1937 as a kibbutz; it changed its status in 1944.

Molegbe (mo-LEG-bai), village, Équateur province, NW CONGO, 20 mi/32 km W of MOBAYI-MBONGO; 04°14′N 20°53′E.

Molegbwe (mo-LEG-bwai), village, Équateur province, NW CONGO, 15 mi/24 km W of MOBAYI-MBONGO; 04°14′N 20°53′E. In cotton area; has Capuchin and Franciscan missions and mission schools, hospital. Seat of vicar apostolic of CONGO-UBANGI district.

Moleke (mo-LE-kai), village, BANDUNDU province, W CONGO, along FIMI RIVER N of BANDUNDU; 02°59′S 17°15′E. Elev. 915 ft/278 m.

Molena (mo-LEE-nuh), village (2000 population 475), PIKE county, W central GEORGIA, 13 mi/21 km NW of THOMASTON, near FLINT RIVER; 33°01′N 84°30′W. Fabric dyeing.

Molenbeek-Saint-Jean (MO-luhn-baik–sen–ZHAW), Flemish *Sint-Jans-Molenbeek*, suburban commune (2006 population 80,576), in Capital district of BRUSSELS, BRABANT province, central BELGIUM; 50°51′N 04°19′E. Manufacturing.

Molepolole (MO-lee-po-LO-lai), town (2004 population 58,600), ☉ KWENENG District, SE BOTSWANA, 31 mi/50 km NE of GABORONE. Road junction at edge of KALAHARI DESERT. Headquarters of Bakwena tribe; hospital, main educational center of Kweneng district. College of Education is here.

Mole River (MOL), ENGLAND; rises in West SUSSEX 3 mi/4.8 km S of HORLEY; flows 30 mi/48 km NW into SURREY, past DORKING, LEATHERHEAD, COBHAM, and ESHER, to the THAMES at East MOLESEY.

Môle-Saint-Nicolas (mol–sang–nee-ko-LAH), coastal town, NORD-OUEST department, NW HAITI, port on inlet of WINDWARD PASSAGE, 115 mi/185 km NW of PORT-AU-PRINCE, near NW extremity of HISPANIOLA island (site of Columbus's landing on December 6, 1492); 19°48′N 73°23′W. Banana growing; beekeeping. Fishing port.

Molesey, East and West (MOL-zee), residential location (2001 population 18,565), NE SURREY, SE ENGLAND, 3 mi/4.8 km WSW of KINGSTON UPON THAMES; 51°24′N 00°21′W. East Molesey, on the THAMES, at mouth of MOLE RIVER. Has electric-cable and pharmaceutical works. Hurst Park race course is here. Incorporated 1933 in ESHER.

Moléson (mo-LAI-son), Alpine peak (6,568 ft/2,002 m), in BERNESE ALPS, FRIBOURG canton, W SWITZERLAND, 5 mi/8 km SSW of BULLE; 46°33′N 07°01′E. Reached by chairlift; views of Bernese Alps, winter ski runs.

Molesworth (MOLZ-wuhrth), town, VICTORIA, SE AUSTRALIA, 53 mi/85 km NE of MELBOURNE; 37°10′S 145°32′E. Agriculture.

Cross-references are shown in SMALL CAPITALS. The pronunciation guide is shown on page xix. The sources of population figures are shown on page xvii.

Molfetta (mol-FET-tah), city (2001 population 62,546), in APULIA, S ITALY, on the ADRIATIC Sea; 41°12′N 16°36′E. It is a fishing port and light industrial center. Manufactures include cement, boats, wood products, clothing, and food products. An Apulian-Romanesque cathedral (12th–13th century) is here.

Molfsee (MUHLF-sai), village, SCHLESWIG-HOLSTEIN, N GERMANY, on the EIDER, 4 mi/6.4 km SSW of KIEL; 54°16′N 10°04′E. Ship construction; various agriculture.

Molha (MOL-yah), city, S TOCANTINS state, BRAZIL, 203 mi/327 km S of PALMAS on TOCANTINS River; 10°05′S 48°30′W.

Mølholm (MUL-holm), town, VEJLE county, E JUTLAND, DENMARK, 1 mi/1.6 km S of VEJLE; 56°51′S 10°01′W.

Moliagul, township, VICTORIA, SE AUSTRALIA, c.124 mi/200 km NW of MELBOURNE; 36°45′S 143°40′E. A large gold nugget called "Welcome Stranger" was discovered near here, 1869.

Molina (mo-LEE-nah), city, ⊙ Molina comuna (2002 population 27,203), CURICÓ province, MAULE region, central CHILE, 31 mi/50 km NE of TALCA. Agricultural center (grapes, cereals). Lumbering. Known for its wines and waterfalls. Founded 1834.

Molina (mo-LEE-nah), town, GUADALAJARA province, central SPAIN, in CASTILE-LA MANCHA, at foot of a ridge, on affluent of the TAGUS, and 50 mi/80 km NW of TERUEL. Elevation c.3,450 ft/1,052 m. Historic city surrounded by pine forests. It is considered one of the coldest cities in all of Spain. Region produces cereals, hemp, fruit, livestock. Flour milling; manufacturing of woolen goods. Besides remains of an old castle, it has a superb alcazar, Aragon tower, and pantheon of Molina family. Religious college. Formerly Molina de Aragón.

Molina de Aragón, SPAIN: see MOLINA.

Molina de Segura (mo-LEE-nah dhai sai-GOO-rah), city, MURCIA province, SE SPAIN, on SEGURA RIVER, and 6 mi/9.7 km NW of MURCIA; 38°03′N 01°12′W. Fruit, vegetables. Iron and saltpeter mines. Mineral springs. Limestone and gypsum quarries nearby.

Molina, Parameras de (mo-LEE-nah pah-rah-MAI-rahs dhai), high tableland of the CORDILLERA IBÉRICA, GUADALAJARA province, central SPAIN, on E edge of central plateau (Meseta), just E of MOLINA. Extends c.30 mi/48 km WNW from TERUEL province border; rises to 4,980 ft/1,518 m.

Molina Pass (mo-LEE-nah) (12,500 ft/3,810 m), in the ANDES, on ARGENTINA-CHILE border, 20 mi/32 km SW of MAIPO VOLCANO, on road between SAN RAFAEL (Argentina) and RANCAGUA(Chile); 34°24′S 70°03′W.

Moline (mo-LEEN), city (2000 population 43,768), ROCK ISLAND county, NW ILLINOIS, on the MISSISSIPPI RIVER (N), ROCK RIVER (S); 41°29′N 90°29′W. In a coal area. It is a transportation and industrial center. Manufacturing (dairy products, concrete, metal fabricating, printing and publishing; elevators and elevator door openers; industrial equipment; nonferrous die castings, presses and punches, small airplanes, ice cream). Has been a major producer of farm machinery since the industrialist John Deere moved there in 1847. A military arsenal is nearby. Moline, with BETTENDORF, IOWA; ROCK ISLAND, Illinois; and DAVENPORT, Iowa, is part of an economic unit called the Quad Cities. Site of Black Hawk College Quad Cities Campus. Incorporated 1855.

Moline, village (2000 population 457), ELK county, SE KANSAS, 7 mi/11.3 km S of HOWARD; 37°21′N 96°17′W. Shipping point in cattle and grain region; dairying. Crushed stone.

Moline Acres, town (2000 population 2,662), SAINT LOUIS county, E MISSOURI; residential suburb, 11 mi/18 km N of downtown ST. LOUIS; 38°45′N 90°14′W.

Molinella (mo-lee-NEL-lah), town, BOLOGNA province, EMILIA-ROMAGNA, N central ITALY, near RENO River, 15 mi/24 km S of FERRARA; 44°37′N 11°40′E.

Secondary industrial center including beet-sugar refinery. Manufacturing (machinery, fabricated metals, food products, clothing).

Molinero (mo-lee-NE-ro), canton, MIZQUE province, extreme S COCHABAMBA department, central BOLIVIA, 31 mi/50 km S of MIZQUE; 18°15′S 65°24′W. Elevation 6,463 ft/1,970 m. Clay, limestone, and gypsum deposits. Agriculture (potatoes, yucca, bananas, corn, oats, rye, sweet potatoes, soy, coffee); cattle for meat and dairy products.

Molines-en-Queyras (mo-leen–zahn–kai-rah), village (□ 20 sq mi/52 sq km), HAUTES-ALPES department, PROVENCE-ALPES-CÔTE D'AZUR region, SE FRANCE, 13 mi/21 km SE of BRIANÇON, in the QUEYRAS district of the French Alps (alpes françaises); 44°44′N 06°51′E. Elevation 5,700 ft/1,737 m. Ancient mountain village with large barns on top of old stone houses; summer resort for alpinists and nature lovers; situated in the Queyras regional park, established 1977.

Molini, Cape (mo-LEE-nee), point on E coast of SICILY, S ITALY, at N end of GULF OF CATANIA; 37°34′N 15°11′E.

Molino (mo-LEE-no), unincorporated town (□ 6 sq mi/15.6 sq km; 2000 population 1,312), ESCAMBIA county, extreme NW FLORIDA, 22 mi/35 km NNW of PENSACOLA; 30°43′N 87°19′W. Agricultural-produce shipping point. Previously brickworks and clay pits.

Molino de las Flores National Park (mo-LEE-no dai lahs FLOR-es), a national park, in TEXCOCO DE MORA, in E MÉXICO, MEXICO. Park is partly restored seventeenth-century hacienda and nearby pre-Hispanic ruins, the baths of King Netzahualcóyotl.

Molino del Rey (mo-LEE-no del rai), group of massive stone buildings, SW section of MEXICO CITY, MEXICO, just W of CHAPULTEPEC. Scene (September 8, 1847) of battle in Mexican War in which U.S. forces were victorious.

Molino, El, MEXICO: see VISTA HERMOSA DE NEGRETE.

Molinopampa (mo-lee-no-PAHM-pah), town, CHACHAPOYAS province, AMAZONAS region, N PERU, in E ANDEAN foothills, 16 mi/26 km E of CHACHAPOYAS; 06°11′S 77°37′W. Cereals, sugarcane.

Molinos (mo-LEE-nos), town (1991 population 505), ⊙ Molinos department (□ 1,060 sq mi/2,756 sq km), S SALTA province, ARGENTINA, in CALCHAQUÍ valley, 70 mi/113 km SW of SALTA, in livestock region; 25°25′S 66°19′W.

Molinos, Los (mo-LEE-nos, los), town, MADRID province, central SPAIN; summer resort in the SIERRA DE GUADARRAMA, 30 mi/48 km NW of MADRID. Cereals, potatoes, livestock. Stone quarries.

Molíns de Rey (mo-LEENS dhai RAI), city, BARCELONA province, NE SPAIN, in CATALONIA, on LLOBREGAT RIVER, and 9 mi/14.5 km WNW of BARCELONA; 41°25′N 02°01′E. In wine-making area; grapes, fruit, vegetables. River spanned by fifteen-arch bridge.

Moliro (mo-LEE-ro), village, KATANGA province, SE CONGO, on SW shore of LAKE TANGANYIKA, on ANGOLA border, 185 mi/298 km SE of KALEMIE; 08°13′S 30°34′E. Elev. 2,509 ft/764 m. Terminus of lake navigation; customs station.

Molise (mo-LEE-zai), region (□ 1,714 sq mi/4,456.4 sq km), S central ITALY, bordering on the ADRIATIC Sea in the E; 41°40′N 14°30′E. CAMPOBASSO is ⊙ of the region, which is divided into the provinces of CAMPOBASSO and Isèrnia. Mostly mountainous, Molise is crossed by the APENNINES; there is a narrow coastal strip. The main occupation in the generally poor region is farming; cereals, pigs, and sheep are raised. Molise's few industries include the processing of food and the manufacture of clothing. Molise was conquered by the Romans in the 4th century B.C.E. After the fall of Rome, it came under the Lombard duchy of Benevento (6th–11th centuries). From the 12th century, it shared the history of ABRUZZI.

Molitg-les-Bains (mo-lich–lai–ban), spa, town (□ 5 sq mi/13 sq km), PYRÉNÉES-ORIENTALES department, LANGUEDOC-ROUSSILLON region, S FRANCE, 4 mi/6.4 km NW of PRADES, in the CONFLENT district of the Pyrenean foothills; 42°39′N 02°23′E. Specializes in treating dermatological ailments in its thermal establishment.

Molkalmuru, India: see MOLAKALMURU.

Mölkau (MUL-kou), village, SAXONY, E central GERMANY, 5 mi/8 km E of LEIPZIG; 51°19′N 12°27′E. Agriculture (grain).

Molkom (MOOL-kom), village, VÄRMLAND county, W SWEDEN, between two small lakes, 17 mi/27 km NNE of KARLSTAD; 59°36′N 13°43′E.

Molla-Kara, TURKMENISTAN: see JEBEL.

Möllbrücke (mul-BRYOO-kuh), township, CARINTHIA, S AUSTRIA, in the LURNFELD, near confluence of Möll River into DRAU RIVER, 7 mi/11.3 km NW of SPITTAL AN DER DRAU; 46°50′N 13°22′E. Elevation 1,701 ft/518 m. Road junction. Summer tourism.

Mölle (MUHL-le), fishing village, SKÅNE county, SW SWEDEN, on W shore of KULLEN peninsula, on KATTEGATT strait, 18 mi/29 km NNW of HELSINGBORG; 56°17′N 12°30′E. Seaside resort.

Mollebamba (mo-ye-BAHM-bah), canton, GUALBERTO VILLARROEL province, LA PAZ department, W BOLIVIA, NW of PAPEL PAMPA; 17°44′S 67°50′W. Elevation 12,723 ft/3,878 m. Gas wells in area. Clay, limestone, and gypsum deposits. Agriculture (potatoes, yucca, rye, bananas); cattle.

Molle Island, AUSTRALIA: see SOUTH MOLLE ISLAND.

Mollendo (mo-YEN-do), town (2005 population 22,650), ⊙ ISLAY province, AREQUIPA region, S PERU; port on the PACIFIC OCEAN; 17°02′S 72°01′W. On railroad to AREQUIPA and JULIACA (PUNO region). Mollendo exports wool and has industries producing cement, textiles, canned fish, and cheese. It is also a popular beach resort.

Mollepata (mo-yai-PAH-tah), town, ANTA province, CUSCO region, S central PERU, 40 mi/64 km W of CUSCO; 13°31′S 72°32′W. Cereals, sugarcane, potatoes.

Möller Ice Stream, ice stream, WEST ANTARCTICA; flows into the RONNE ICE SHELF W of FOUNDATION ICE STREAM; 82°20′S 63°30′W.

Moller, Port (MO-luhr), bay (20 mi/32 km long, 10 mi/16 km wide at mouth), SW ALASKA, on ALASKA PENINSULA, on BRISTOL BAY; 56°N 160°26′W. PORT MOLLER village is on N shore.

Mollerusa (mo-lyai-ROO-sah), town, LÉRIDA province, NE SPAIN, 14 mi/23 km E of LÉRIDA; 41°38′N 00°54′E. Chief center of irrigated Urgel plain (cereals, wine, almonds, cherries); olive-oil processing. Manufacturing (soda water, soap).

Molles, URUGUAY: see CARLOS REYLES.

Mollet (mo-LYET), satellite city, BARCELONA province, NE SPAIN, 12 mi/19 km NNE of BARCELONA; 41°33′N 02°13′E. Cotton; silk milling and dyeing; brandy manufacturing. Wine, livestock, cereals, potatoes in area.

Mollevilque (mo-ye-VEEL-kai), canton, GENERAL B. BILBAO province, POTOSÍ department, W central BOLIVIA, NW of ARAMPAMPA; 17°54′S 66°06′W. Elevation 9,974 ft/ 3,040 m. Agriculture (potatoes, yucca, bananas, barley); cattle.

Mollina (mo-LYEE-nah), town, MÁLAGA province, S SPAIN, at S foot of the SIERRA DE YEGUAS, 30 mi/48 km NNW of MÁLAGA; 37°08′N 04°40′W. Agricultural center (olives, cereals, livestock). Olive-oil pressing.

Mollis (MAW-lis), commune, GLARUS canton, E central SWITZERLAND, on LINTH RIVER, opposite NÄFELS; 47°05′N 09°05′E. Cotton textiles, knit goods.

Molln (MAWLN), township, SE UPPER AUSTRIA, 13 mi/21 km SSW of STEYR, near the STEYR RIVER; 47°53′N 14°15′E. Manufacturing (ski and mountain boots; furniture); hydropower station.

Mölln (MULN), town, in SCHLESWIG-HOLSTEIN, NW GERMANY, harbor on the E bank of ELBE-LÜBECK

CANAL, and 16 mi/26 km S of LÜBECK, between two small lakes; 53°37′N 10°42′E. Manufacturing of textiles, furniture, mattresses; food processing. Wood and grain trade on canal. Has 13th-century church, 14th-century town hall. Town first mentioned in 1188. Chartered c.1202. According to legend, Till Eulenspiegel died here in 1350.

Mollösund (MOL-luh-SUND), fishing village, GÖTEBORG OCH BOHUS county, SW SWEDEN, on SW coast of ORUST island, on SKAGERRAK, 14 mi/23 km S of LYSEKIL; 58°04′N 11°28′E. Seaside resort.

Möll River (MUL), 50 mi/80 km long, in CARINTHIA, S AUSTRIA; rises in PASTERZE glacier; flows S and E, past HEILIGENBLUT and OBERVELLACH, to the DRAU RIVER 6 mi/9.7 km W of Spittal. Hydroelectric station near Kolbitz. High alpine valley, where a combination of alpine agriculture and tourism prevails.

Mölltorp (MUHL-TORP), village, SKARABORG county, S SWEDEN, at S end of Lake Bottensjön (5 mi/8 km long, 2 mi/3.2 km wide), on GÖTA Canal route, 20 mi/32 km W of MOTALA; 58°30′N 14°25′E. Medieval church.

Mollwitz, Polish *Malujowice*, village in LOWER SILESIA, after 1945 in OPOLE province, SW POLAND, 5 mi/8 km W of BRZEG (Brieg). In War of the Austrian Succession, Austrians were defeated here (April 1741) by Prussians under Frederick the Great. The battle was important in military history as it demonstrated the superiority of modern infantry over cavalry.

Mollymook, town, NEW SOUTH WALES, SE AUSTRALIA, just N of ULLADULLA, 140 mi/225 km S of SYDNEY. Tourism; resort.

Mölndal (MUHLN-DAHL), city, GÖTEBORG OCH BOHUS county, SW SWEDEN, industrial suburb, 5 mi/8 km SSE of GÖTEBORG; 57°39′N 12°02′E.

Mölnlycke (MUHLN-LIK-ke), town, GÖTEBORG OCH BOHUS county, W SWEDEN, 6 mi/9.7 km SE of GÖTEBORG; 57°39′N 12°07′E. Manufacturing (fluff products).

Molo (MO-lo), town (1999 population 20,944), RIFT VALLEY province, W KENYA, in picturesque highlands forming W rim of GREAT RIFT VALLEY, on railroad, and 20 mi/32 km W of NAKURU; 00°15′N 35°45′E. Elevation 8,064 ft/2,458 m. Agriculture (coffee, tea, wheat, corn, pyrethrum, potatoes). Manufacturing (dairying, saw milling).

Molo (MO-lo), town, EASTERN region, SE UGANDA, near railroad, 10 mi/16 km N of TORORO; 00°49′N 34°10′E. Cotton, corn, millet, sweet potatoes. Busumbu apatite is shipped by railroad via Magodes station, 1 mi/1.6 km W. Was part of former EASTERN province.

Moloacán (mo-lo-ah-KAHN), town, VERACRUZ, SE MEXICO, on Isthmus of TEHUANTEPEC, 13 mi/21 km E of MINATITLÁN. Fruit. Petroleum production.

Molochans'k (mo-lo-CHAHNSK) (Russian *Molochansk*), city (2004 population 10,400), central ZAPORIZHZHYA oblast, UKRAINE, on MOLOCHNA RIVER, and 6 mi/10 km SW of TOKMAK, and subordinated to its city council; 47°12′N 35°35′E. Elevation 170 ft/51 m. Railroad shops; metalworks; milk canning, juice packing, asphalt and cement works; furniture manufacturing. Established in 1803 as Halbstadt by German Mennonite colonists; renamed Molochans'k in 1915; city since 1938.

Molochansk, UKRAINE: see MOLOCHANS'K.

Moloch, Mount (MO-lahk) (10,195 ft/3,107 m), SE BRITISH COLUMBIA, W CANADA, in SELKIRK MOUNTAINS, near GLACIER NATIONAL PARK, 16 mi/26 km NNW of REVELSTOKE; 51°20′N 117°56′W.

Molochna Lagoon (mo-LOCH-nah) (□ 65 sq mi/169 sq km) (Ukrainian *Molochnyy Lyman*) (Russian *Molochnyy Liman*), a lagoon on N coast of Sea of AZOV, SW ZAPORIZHZHYA oblast, SE UKRAINE, 10 mi/16 km–20 mi/32 km S of MELITOPOL'; 22 mi/35 km long, maximum width (at S end) 6 mi/10 km, depth 2 ft/0.5 m–10 ft/3 m; receives MOLOCHNA RIVER from N; sep-

arated (except for a narrow passage) from the Sea of Azov by a sandy-shell bar; freezes over on the surface in winter. Bottom muds have medicinal properties. Marine and freshwater fishes; abundant shoreline aquatic vegetation.

Molochna River (mo-LOCH-nah) (Russian *Molochnaya*), ZAPORIZHZHYA oblast, UKRAINE; rises in the AZOV UPLAND 12 mi/19 km NE of CHERNIHIVKA; flows W past Chernihivka, TOKMAK and MOLOCHANS'K, and S past MELITOPOL', to MOLOCHNA LAGOON, NW inlet of the Sea of AZOV; approximately 120 mi/197 km long. Also called Tokmak, especially near and above the city of Tokmak.

Molochnaya River, UKRAINE: see MOLOCHNA RIVER.

Molochnoye (muh-LOCH-nuh-ye), town (2006 population 8,370), S VOLOGDA oblast, N central European Russia, on the Vologda River (SUKHONA RIVER basin), on railroad, 10 mi/16 km NW of VOLOGDA, to which it is administratively subordinate. Milk processing.

Molochnyy (muh-LOCH-niyee), town (2006 population 5,205), N MURMANSK oblast, NW European Russia, near the mouth of the TULOMA RIVER, on railroad and near highway, 6 mi/10 km S of MURMANSK, and 2 mi/3.2 km S of KOLA, to which it is administratively subordinate; 68°51′N 33°00′E. Elevation 167 ft/50 m.

Molochnyy Liman, UKRAINE: see MOLOCHNA LAGOON.

Molochnyy Lyman, UKRAINE: see MOLOCHNA LAGOON.

Molodechno (mo-lo-DECH-nah), Polish *Molodeczno*, city, W MINSK oblast, BELARUS, ⊙ MOLODECHNO region, on USHA RIVER (tributary of Viliia), and 40 mi/64 km NW of MINSK; 54°16′N 26°50′E. Railroad junction; important industrial center (metal structural components, semiconductor valves, machine tools, metal articles, reinforced concrete goods); food processing (meat packing, fruit canning, food combine, milk plant); light industry (clothing and footwear, furniture); enterprises for railroad transport maintenance. Has old palace.

Molodezhnaya Station (mo-lo-dezh-NAH-yuh), ANTARCTICA, Russian station in W ENDERBY LAND, on the PRINCE OLAV COAST; 67°40′S 45°51′E. It is Russia's biggest Antarctic Station; main study is meteorology, also rocketry and geophysics. Opened January 14, 1963.

Molodezhnoye, UKRAINE: see MOLODIZHNE, Kirovohrad oblast; or MOLODIZHNE, Republic of Crimea.

Molodëzhnoye (muh-luh-DYOZH-nuh-ye), town (2005 population 1,395), W LENINGRAD oblast, NW European Russia, on the N shore of the Gulf of FINLAND, on road and railroad, 26 mi/42 km WNW, and administratively a suburb, of SAINT PETERSBURG; 60°12′N 29°32′E. Popular holiday and vacation retreat; has a sanatorium. Until 1945, in FINLAND and called Metsäkylä.

Molodizhne (mo-lo-DEEZH-ne) (Russian *Molodezhnoye*), town, central Republic of CRIMEA, UKRAINE, on road and railroad, on the right bank of the SALHYR RIVER, 16 km NW of SIMFEROPOL' city center; 45°00′N 34°05′E. Elevation 839 ft/255 m. Experimental farm of the Crimean Agricultural Institute. Established in 1929; town since 1972.

Molodizhne (mo-lo-DEEZH-ne) (Russian *Molodezhnoye*), town, S central KIROVOHRAD oblast, UKRAINE, on railroad spur near headwaters of the Berezivka River, left tributary of the INHUL River, 29 mi/47 km SE of KIROVOHRAD, and 6 mi/9 km NW of DOLYNS'KA; 48°10′N 32°39′E. Elevation 593 ft/180 m. Sugar refinery. Established in 1960; town since 1967.

Molodogvardeysk, UKRAINE: see MOLODOHVARDIYS'K.

Molodohvardiys'k (mo-lo-do-hvahr-DEESK) (Russian *Molodogvardeysk*), city (2004 population 35,500), SE LUHANS'K oblast, UKRAINE, 5 mi/8 km N of, and subordinated to the city council of, KRASNODON; 48°21′N 39°43′E. Elevation 646 ft/196 m. City has four bituminous-coal mines, coal-enrichment plant; building-materials plant; professional-technical schools. Established in 1954; city since 1961.

Molodoy Tud (muh-luh-DO-yee TOOT), village, SW TVER oblast, W European Russia, on local road junction, 28 mi/45 km WNW of RZHEV; 56°25′N 33°36′E. Elevation 685 ft/208 m. Flax processing.

Mologa (muh-LO-gah), former city, YAROSLAVL oblast, central European Russia, on the VOLGA RIVER, at the mouth of the MOLOGA RIVER. Known since the 15th century. Trading center from the 15th to the 20th century. Population evacuated (1940) and city flooded by filling of the RYBINSK RESERVOIR.

Mologa River (muh-LO-gah), approximately 200 mi/322 km long, W European Russia; rises in TVER oblast in the hills W of BEZHETSK; flows E and N, past BEZHETSK, W and N, past PESTOVO (NOVGOROD oblast; head of navigation), and generally E, past USTYUZHNA (VOLOGDA oblast), to the RYBINSK RESERVOIR N of VESYEGONSK, at the administrative border of Tver and Vologda oblasts. Receives Chagodoshcha River (left), a section of the Tikhvin canal system.

Molo, Gulf of (MO-lo), inlet of IONIAN SEA in E ITHÁKI island, IONIAN ISLANDS department, off W coast of GREECE, penetrates deeply into island, almost dividing it into two parts; 3 mi/4.8 km long, 2 mi/3.2 km wide. Fisheries. Town of Itháki on SE shore.

Molokai (MO-lo-KAH-ee), island (□ 261 sq mi/678.6 sq km), MAUI county, HAWAII, between OAHU (W, separated by KAIWI CHANNEL), LANAI (S, separated by KALOHI Channel), and MAUI (separated by PAILOLO CHANNEL). Molokai is generally mountainous, with Mount KAMAKOU (4,970 ft/1,515 m) the highest peak in E end. On the N coast is the KALAUPAPA peninsula, separated by a rocky mountain wall from the rest of the island and accessible only over a 2,000 ft/610 m pass, on which the Kalaupapa leper colony existed (1886–1969) and is now included in KALAUPAPA NATIONAL HISTORIC PARK; much of E part of island is in Molokai Forest Reserve. Molokai has many cattle ranches. Most pineapple plantations have reverted to hayfields and other crops. The chief town and port is KAUNAKAKAI, on central S shore, from which the island's products are shipped to HONOLULU for export. Palaau State Park in N center, just SW of Kalaupapa; Molokai Ranch Wildlife Park at W end.

Molokini (MO-lo-KEE-nee), island, MAUI county, HAWAII, between MAUI and KAHOOLAWE islands, in Alalakeiki Channel. Barren, rocky; lighthouse.

Molokovo (muh-luh-KO-vuh), village (2006 population 2,465), NE TVER oblast, W central European Russia, near the MOLOGA RIVER, on highway, 28 mi/45 km N of BEZHETSK. Elevation 600 ft/182 m. In agricultural area (flax).

Molokovo (muh-luh-KO-vuh), village, central MOSCOW oblast, central European Russia, on the MOSKVA River, 5 mi/8 km NE of (and administratively subordinate to) GORKI-LENINSKIYE; 55°33′N 37°52′E. Elevation 518 ft/157 m. Cardboard products. In agricultural area; livestock breeding.

Moloma River (muh-luh-MAH), approximately 138 mi/222 km long, NW and W KIROV oblast, E central European Russia. Begins on the administrative border between VOLOGDA and Kirov oblasts and meanders generally S and SSE, gathering tributaries along the way, to feed into the VYATKA River 7 mi/11 km ENE of KOTEL'NICH. Free of ice between late April and early November, at which time it is used for timber floating. Non-commercial fishing along both banks.

Molong (MO-lahng), municipality, E central NEW SOUTH WALES, SE AUSTRALIA, 145 mi/233 km WNW of SYDNEY; 33°06′S 148°52′E. Railroad junction. Beef cattle, hogs, lambs; wool; fruit, wheat, fodder crops (oats, hay); vineyards.

Molopo River, c.600 mi/966 km long, SOUTH AFRICA and BOTSWANA; rises W of Mafikeng-Mmabatho (in what was BOPHUTHATSWANA) at W end of WITWATERSRAND; flows in a winding course generally W and

WNW, then forms border between NORTHERN CAPE province and Botswana, where it is joined by NOSSOB RIVER (N) and KURUMAN RIVER (SE) before entering ORANGE RIVER 70 mi/113 km W of UPINGTON, below AUGHRABIES FALLS. A largely intermittent stream that flows only in rainy periods.

Molotov, RUSSIA: see PERM, city.

Molotovabad, KYRGYZSTAN: see UCH-KORGON.

Molotovo (mo-lo-TO-vo), town, S GEORGIA, on KHRAM RIVER and 35 mi/56 km WSW of TBILISI, just S of TSALKA. Site of Khram hydroelectric station (*Khramges*).

Molotovo, RUSSIA: see OKTYABR'SKOYE, LIPETSK oblast.

Molotovo, RUSSIA: see PERM, city.

Molotovo, UZBEKISTAN: see UCHKUPRIK.

Molotovsk, city, NW ARCHANGEL oblast, RUSSIA, lumber port on DVINA Bay, 20 mi/32 km W of ARCHANGEL (connected by railroad spur); shipbuilding center; pulp mill. Became city in 1938; formerly called Sudostroi.

Molotovsk, RUSSIA: see SEVERODVINSK.

Moloundou (MO-loon-doo), village, East province, CAMEROON, on N'Goko River, along Cameroon-Congo border, and 180 mi/290 km SSE of BATOURI; 02°03'N 15°13'E. Center of trade; head of navigation on the N'Goko.

Mols (mols), peninsula, E JUTLAND, DENMARK, extending SW from DJURSLAND peninsula, between Kalø Bay (W) and Ebeltoft Bay (E); c.8 mi/12.9 km long, 4 mi/6.4 km wide. Mols (or Agri) Hills (55 ft/17 m) in center. HELGENÆS peninsula (S extension; 5 mi/8 km long) forms BEGTRUP VIG between Mols peninsula and S Mols.

Molsheim (mawls-heim), town (□ 4 sq mi/10.4 sq km), BAS-RHIN department, E FRANCE, in ALSACE, near E foot of the VOSGES, on the small BRUCHE, and 12 mi/19 km WSW of STRASBOURG; 48°33'N 07°30'E. Noted for its Riesling wine. Manufacturing of airplane equipment. Has remains of medieval fortifications and a fine 16th-century guild hall.

Molson, village, OKANOGAN county, N WASHINGTON, 1 mi/1.6 km S of CANADA (BRITISH COLUMBIA) border, 11 mi/18 km ENE of OROVILLE. Old Mission, Molson Museum and Ghost Town.

Molson (MOL-suhn), hamlet, SE MANITOBA, W central CANADA, 39 mi/62 km E of WINNIPEG, in REYNOLDS rural municipality; 50°01'N 96°18'W.

Molson Lake (MOL-suhn), central MANITOBA, W central CANADA, 50 mi/80 km NE of Lake WINNIPEG; 27 mi/43 km long, 12 mi/19 km wide; 54°12'N 96°45'W. Drains N into HAYES River.

Moltay Lake (muhl-TEI) (□ 1 sq mi/2.6 sq km), in central SVERDLOVSK oblast, W Siberian Russia, 16 mi/26 km NNE of REZH, near the REZH RIVER. Health resort (mud baths) developed in the late 1940s.

Molteno, town, EASTERN CAPE province, SOUTH AFRICA, in STORMBERG range, 45 mi/72 km NW of QUEENS-TOWN on the Stormberg river; 31°25'S 26°22'E. Elevation 5,740 ft/1,750 m. Railroad junction; agricultural center (stock, grain); sawmilling, coal mining (formerly major industry, now worked out). Airfield. Recent residential suburb development called Nomonde (NW).

Möltenort, GERMANY: see HEIKENDORF.

Moluccas, INDONESIA: see MALUKU.

Molunkus Lake (mo-LUHNK-uhs), c.35 mi/56 km long, AROOSTOOK county, E central MAINE, 20 mi/32 km E of MILLINOCKET; 3 mi/4.8 km long. Drains into Molunkus Stream, which rises to NW, flows SE to MATTAWAMKEAG RIVER.

Molvitino, RUSSIA: see SUSANINO.

Molvízar (mol-VEE-thahr), town, GRANADA province, S SPAIN, 6 mi/9.7 km NW of MOTRIL; 36°47'N 03°37'W. Olive oil, wine, raisins, almonds.

Molvotitsy (muhl-VO-tee-tsi), village, S NOVGOROD oblast, NW European Russia, on road, 55 mi/89 km SE of STARAYA RUSSA; 57°24'N 32°20'E. Elevation 242 ft/73 m. Flax processing.

Molybdos, Greece: see METHYMNA.

Molyneux River, NEW ZEALAND: see CLUTHA RIVER.

Moma (MO-muh), town, NAMPULA province, MO-ZAMBIQUE; 16°44'S 39°14'E. It is important mainly as a harbor for the export of tropical produce.

Moma (MO-muh), village, Équateur province, central CONGO, on TSHUAPA RIVER, and 230 mi/370 km SE of BOENDE; 01°36'S 23°57'E. Elev. 1,459 ft/444 m. Terminus of steam navigation and trading post in copal-gathering region.

Moma, mountain range, RUSSIA: see RUSSIAN FAR EAST.

Moma, RUSSIA: see KHONUU.

Moma District, MOZAMBIQUE: see NAMPULA, province.

Momaligi, town, 12 mi/19 km N of BONTHE, SOUTHERN province, SIERRA LEONE; 07°37'N 12°21'W. Road terminus; rice.

Momauguin, CONNECTICUT: see EAST HAVEN.

Momauk (mo-MOUK), village and township, BHAMO district, KACHIN STATE, MYANMAR, E of BHAMO, bordering on CHINA.

Momax (mo-MAKS), town, ZACATECAS, N central MEXICO, 12 mi/19 km NNW of TLALTENANGO DE SÁNCHEZ ROMÁN, on Tlaltenango River. Grain, chickpeas, tobacco; alfalfa; livestock.

Mombaça (mom-bah-sah), city (2007 population 44,242), central CEARÁ, BRAZIL, 22 mi/35 km SW of SENADOR POMPEU; 05°45'S 39°36'W. Cattle; cotton, sugar. Until 1944, called Maria Pereira.

Mombach (MUHM-berg), NW suburb of MAINZ, RHINELAND-PALATINATE, W GERMANY, on left bank of the RHINE; 50°01'N 08°12'E.

Mombacho (mom-BAH-cho), extinct volcano (4,472 ft/1345 m), SW NICARAGUA, 6 mi/9.7 km S of GRAN-ADA, near NW shore of LAKE NICARAGUA. Coffee is grown on its slopes.

Momba River (MOM-bah), length c. 110 mi/177 km, SW TANZANIA; formed by joining of several head-streams c. 50 mi/80 km SW of MBEYA, in MBEYA region; flows c. 90 mi/145 km NW and N, enters Lake RUKWA 85 mi/137 km NW of Mbeya.

Mombasa (mom-BAH-sah), city (2004 population 777,100), ⊙ COAST province, SE KENYA, mostly on Mombasa island in the INDIAN OCEAN and partly on the mainland (with which it is connected by a causeway); 03°59'S 39°40'E. It is Kenya's chief port and an important commercial and industrial center. Manufacturing (processed food, cement, and glass); oil refining, tourism. From the 8th to the 16th century Mombasa was a center of the Arab trade in ivory and slaves. The city was visited (1498) by Vasco da Gama on his first voyage to INDIA. Mombasa was burned three times by the Portuguese who then controlled the city until 1698, when it was regained by the Arabs. The Portuguese briefly regained control in 1729. It came under ZANZIBAR in the mid-19th century and passed to GREAT BRITAIN in 1887. Mombasa was the capital of the British East Africa Protectorate from 1887 to 1907. Of note are the remains of Fort Jesus, built by the Portuguese in 1593–1594. The city's extensive beaches and resorts attract thousands of tourists annually. International airport.

Mombeltrán (mom-bel-TRAHN), town, ÁVILA province, central SPAIN, in the SIERRA DE GREDOS, 31 mi/50 km SW of ÁVILA; 40°16'N 05°00'W. In fertile region (grapes, cereals; livestock). Flour mill; wine making. Hydroelectric plant. Castle of dukes of Albuquerque nearby.

Mombin-Crochu (mong-BANG–kro-SHYOO), town, NORD-EST department, HAITI, 29 mi/47 km SE of CAP-HAÏTIEN; 18°18'N 73°13'W. Coffee, citrus fruit growing.

Mombo (MOM-bo), town, TANGA region, NE TANZA-NIA, 55 mi/89 km WNW of TANGA, on railroad, at SW foot of USAMBARA Mountains; 05°15'S 38°17'E. Road junction. Sisal, rice, grain; livestock; timber.

Momboyo River (mom-BO-yo), 315 mi/507 km long, in central and W CONGO; rises as the LUILAKA 50 mi/80 km SW of LOMELA; flows c.270 mi/435 km NW and NNW, past IKALI and MONKOTO, to a point 8 mi/12.9

km upstream from Waka-sur-Momboyo, where it becomes the MOMBOYO and flows NW to join BUSIRA RIVER at INGENDE, forming the RUKI; 01°25'S 17°43'E. MOMBOYO-LUILAKA is navigable for 170 mi/274 km below IKALI.

Mömbris (MUM-bris), town, LOWER FRANCONIA, NW BAVARIA, GERMANY, on W slope of the SPESSART, 6 mi/9.6 km N of ASCHAFFENBURG; 50°04'N 09°10'E. Fruit; forestry. Tourism.

Momchilgrad (mom-CHEEL-grahd), city, HASKOVO oblast, Momchilgrad obshtina (1993 population 20,826), S BULGARIA, in the E RODOPI Mountains, near right branch of the ARDA RIVER, 8 mi/13 km S of KURDZHALI; 41°32'N 25°24'E. Agricultural center (cotton), wool; tobacco processing, machine building, manufacturing agricultural tools, aluminum goods. Once linked with Komotine (GREECE) via the Makaz Pass road. Until 1934, called Mustanli. Sometimes spelled Mastanli. Railroad spur terminus Podkova is 11 mi/18 km to the S.

Momchilovtsi (mom-CHEEL-ov-tsee), village, PLOV-DIV oblast, SMOLYAN obshtina, BULGARIA. Museum and vacation village; 41°40'N 24°46'E.

Momeik, MYANMAR: see MOGMIT.

Momence (mo-MENS), city (2000 population 3,171), KANKAKEE county, NE ILLINOIS, on KANKAKEE RIVER (bridged here) and 10 mi/16 km E of KANKAKEE; 41°09'N 87°39'W. Railroad junction. In agricultural area (corn, soybeans; dairying); manufacturing (food products). Plotted 1844, incorporated 1874.

Momignies (MO-meen-YEE), commune (2006 population 5,137), Thuin district, HAINAUT province, S BELGIUM, 29 mi/47 km SSW of CHARLEROI, near FRENCH border; 50°02'N 04°10'E. Glass industry (since sixteenth century); agriculture.

Momil (mo-MEEL), town, ⊙ Momil municipio, CÓR-DOBA department, N COLOMBIA, in CARIBBEAN LOW-LANDS, on the CIÉNAGA GRANDE, 36 mi/58 km NNE of MONTERÍA; 09°14'N 75°41'W. Rice, corn, sugarcane, fruit; livestock.

Mominabad (MO-mee-nah-bahd), town, BID district, MAHARASHTRA state, W central INDIA, 45 mi/72 km ESE of BID. Trades in agriculture products (cotton, millet, wheat, oilseeds). Site of former cavalry cantonment.

Momina Banya (MO-mee-na BAHN-yah), former village, SOFIA district, W central BULGARIA, now part of KOSTENETS; 42°21'N 23°52'E.

Momina Banya, Bulgaria: see HISARYA.

Momin Brod (MO-meen BROD), former village, BULGARIA, merged with LOM; 43°48'N 23°14'E.

Momin Dvor (MO-meen DVOR), peak (8,940 ft/ 2,725 m) in the PIRIN MOUNTAINS, SW BULGARIA; 41°41'N 23°35'E.

Momino (MO-mee-no), plateau, DANUBE Plain, NE BULGARIA; 43°07'N 27°45'E.

Momino, Bulgaria: see AVREN.

Mömlingen (MUM-ling-guhn), village, LOWER FRAN-CONIA, NW BAVARIA, GERMANY, on the MÜMLING RIVER, 9 mi/14.5 km S of ASCHAFFENBURG; 49°52'N 09°06'E. Forestry; tourism.

Mömling River, GERMANY: see MÜMLING RIVER.

Momo, department (2001 population 213,402), North-West province, CAMEROON, ⊙ MBENGWI.

Momoishi (mo-MO-ee-shee), town, Kamikita county, Aomori prefecture, N HONSHU, N JAPAN, 40 mi/65 km S of AOMORI; 40°35'N 141°25'E.

Momostenango (mo-mos-te-NAHN-go), town, TO-TONICAPÁN department, W central GUATEMALA, on headstream of Río Negro (CHIXOY RIVER), and 11 mi/18 km N of TOTONICAPÁN; 14°30'N 89°52'W. Elevation 7,333 ft/2,235 m. Wool-weaving and market center; corn, wheat, beans; sheep raising; textile (blanket) making. Quiché-speaking population.

Momotombito (mo-mo-TOM-bee-to), volcano (2,549 ft/389 m) and island in LAKE MANAGUA, W NICAR-AGUA, 19 mi/31 km NW of MANAGUA.

Momotombo (mo-mo-TOM-bo), volcano (4,128 ft/ 1280 m), W NICARAGUA, on NW shore of LAKE MANAGUA, 23 mi/37 km E of LEÓN. Its 1609 eruption destroyed original city of León, situated at its W foot, on lake shore, at site of modern village and minor port of Momotombo. Other eruptions occurred in 1764, 1849, 1885, and 1905.

Momoyama (mo-MO-yah-mah), town, Naga county, WAKAYAMA prefecture, S HONSHU, W central JAPAN, 12 mi/19 km E of WAKAYAMA; 34°14′N 135°21′E. Peaches, garden plants; horticulture.

Mompog Pass (mom-POG), channel in PHILIPPINES, connecting TAYABAS BAY of S LUZON with SIBUYAN SEA, between MARINDUQUE island (W) and BONDOC PENINSULA (E); c.40 mi/64 km long, 11 mi/18 km wide.

Mompono (mom-PO-no), village, Équateur province, W CONGO, on MARINGA RIVER, and 230 mi/370 km E of MBANDAKA; 00°04′N 21°48′E. Elev. 1,354 ft/412 m. Palm products. Has Roman Catholic and Baptist missions.

Mompós (mom-POS), town, ⊙ Mompós municipio, BOLÍVAR department, N COLOMBIA, on MARGARITA ISLAND, on the right arm of MAGDALENA River (BRAZO DE MOMPÓS), and 96 mi/154 km SE of CARTAGENA; 09°14′N 74°25′W. River port. Cattle-raising; plantains, yucca, corn. Market for forest products. Although having lost most of its commercial importance, it retains a rich colonial character. Furniture manufacturing, gold filigree work. Site of impressive religious ceremonies during Holy Week. During War of Independence the town resisted a siege by royalist forces. Formerly sometimes spelled Mompox.

Mompox, Colombia: see MOMPÓS.

Mon, district (□ 690 sq mi/1,794 sq km), Nagaland state, NE India; ⊙ Mon.

Mon, town, ⊙ Mon district, Nagaland state, NE India; 26°45′N 95°06′E.

Mon (MOON), village, Älvsborg county, SW SWEDEN, 20 mi/32 km E of STRÖMSTAD, on Norwegian border, opposite KORNSJØ, NORWAY. Frontier station.

Møn (MUH-uhn), island (□ 84 sq mi/218.4 sq km), SE DENMARK, in the BALTIC SEA, S of SJÆLLAND and NE of FALSTER; 55°00′N 12°20′E. Stege is the main town. Møn is largely agricultural; sugar beets are the main crop, and cattle are also raised. At the island's E point are the Møns Klint, scenic white chalk cliffs that rise to 420 ft/128 m. This area is also covered with beech forests. Also written Möen.

Mon, MYANMAR: see MON RIVER.

Mona, village and institution, SAINT ANDREW parish, SE JAMAICA, 7 mi/11.3 km from KINGSTON, an amalgam of Mona Heights, University Heights, the College Commons, and Mona Reservoir lands; 18°00′N 76°44′W. Bounded SW by Long Mountain and Beverley Hills, N by Hope Botanical Gardens and the University of Technology and separated from the HOPE RIVER by Papine Village. Home of University of the West Indies and University College Hospital. The university was established in 1948 as the first degree-granting institution in the WEST INDIES and today has campuses in PORT OF SPAIN, TRINIDAD, and Cave Hill, BARBADOS. In 1962 the college received its charter to be an independent university. Mona Heights, W of the university, is a private housing development.

Mona, village (2006 population 1,198), JUAB county, central UTAH, 8 mi/12.9 km N of NEPHI, on Current Creek, at S end of Mona Reservoir; 39°49′N 111°51′W. Alfalfa, barley, wheat; cattle. Mona Reservoir (5 mi/8 km long, 1 mi/1.6 km wide) is just N, formed by dam on small tributary of UTAH LAKE. Mount NEBO (11,928 ft/3,636 m) is E, in WASATCH RANGE. In Mount Nebo Wilderness Area of Uinta National Forest. Elevation 4,025 ft/1,227 m. Settled 1851.

Mona (MO-nuh), Roman name for ANGLESEY, NW Wales. It was also sometimes used to designate the ISLE OF MAN.

Mona, PUERTO RICO: see MONA ISLAND.

Monaca (MAH-nah-kah), borough (2006 population 5,886), BEAVER county, W PENNSYLVANIA, on OHIO RIVER, opposite mouth of BEAVER RIVER, 22 mi/35 km NW of PITTSBURGH, and 1 mi/1.6 km SE of BEAVER; 40°40′N 80°16′W. Railroad junction. Agriculture (corn, hay; livestock, dairying); manufacturing (metal fabrication and products, glassware, printing). Community College of Beaver County here; Penn State University Beaver Campus to W. Settled 1813, incorporated 1839.

Monachil (mo-nah-CHEEL), town, GRANADA province, S SPAIN, 5 mi/8 km SE of GRANADA; 37°08′N 03°32′W. Olive-oil processing. Hydroelectric power plant nearby.

Monach Isles, Scotland: see HEISKER.

Monaco (MAH-nah-ko), independent principality (□ 0.7 sq mi/1.9 sq km; 2004 estimated population 32,270; 2007 estimated population 32,671), between the MARITIME ALPS and the MEDITERRANEAN SEA; an enclave within ALPES-MARITIMES department, SE FRANCE.

Geography

This famous resort, stretching for about 3 mi/5 km along the Côte d'Azur (French RIVIERA), consists of the reclaimed land of FONTVIEILLE and Monaco, and the beaches of MONTE CARLO and Larvotto (NE). The pleasant Mediterranean climate (74°F/23°C in August; annual precipitation 31 in/787 mm) and its magnificent situation have helped to make Monaco one of the best-known resorts in the world.

Population

Although French is the official language and is spoken by almost everyone, it is the mother tongue of only about 40% of the population; 15% speak Italian as their first language, 15% Monégasque (a Romance dialect closely related to Provençal and the dialects of Liguria), and 30% a scattering of other languages; 90% are Roman Catholics.

Economy/Tourism

Monaco today consists of four quarters (French= quartiers). From SW to NE they are: Fontvieille, Monaco-Ville, La Condamine, and Monte Carlo. Fontvieille, a relatively new district partly constituted of reclaimed land, is an area of light industry (manufacturing pharmaceuticals, processed foods, clothing, and precision instruments). Monaco-Ville, the old town atop a rocky promontory (Rocher de Monaco) which separates the harbor of Fontvieille from the Harbor of Monaco (Port de Monaco) includes the 16th-century Renaissance palace of the prince of Monaco, the 19th-century cathedral, and the world-renowned Oceanographic Museum (founded in 1910). La Condamine, the central business district of Monaco, abuts the W edge of the Port of Monaco, a natural harbor greatly modified into a square shape protected by breakwaters. N of the port lies MONTE CARLO, the site of the famous casino (where Monegasque citizens are forbidden to gamble), an opera house, numerous luxury hotels, and the International Hydrographic Bureau. Monaco, once a winter resort only, presently attracts many summer vacationers to its beaches, marinas, well-known road races (Monte Carlo rally and Grand Prix de Monaco), and other international sporting events. But although Monaco's major source of revenue is tourism, taxes levied on French citizens and on foreign corporations represent an appreciable share of the government's revenues. Monaco's television station, like that of Luxembourg, is aimed at the French market and has an audience out of proportion to its national population.

History: to 1865

The presence of Paleolithic humans in Monaco is attested by exhibits in the Museum of Prehistoric Anthropology; the headland was known to Phoenician, Greek, and Carthaginian traders (the name itself is probably of Phoenician origin); and its was eventually absorbed in to the Roman Empire and Chris-

tianized in the 1st century C.E. Briefly under Saracen domination in the 8th century, it was ruled, beginning in the 13th century by the Grimaldi family of GENOA whose line died out in 1731, but whose name was adopted by the French family who succeeded by marriage. The principality, variously under French, Spanish, and Italian protection in the 17th and 18th centuries, became dependent on France when Sardinia ceded the county of NICE to France in 1860. Monaco gave up half its territory to France in 1861; it was the transfer (of the communes of MENTON and ROQUEBRUNE which lie between present-day Monaco and ITALY) that made the principality an enclave entirely within French territory.

History: 1865 to Present

In return, France recognized Monaco's independence, and in 1865 a customs union was established between the two countries. Until Monaco's first constitution in 1911, the prince was an absolute ruler. The treaty of 1918 made the succession to the throne subject to French approval. Prince Rainier III succeeded to the throne in 1949. Monaco's policy of not levying income taxes on its residents or on international businesses with headquarters in the principality led to serious disagreement with the French government in the 1960s since four-fifths of Monaco's population were French citizens. In 1962 a compromise was reached by which French citizens of less than five years residence were taxed at French rates, and taxes were imposed on companies that did more than a one-quarter of their business outside the principality. Monaco also came into conflict with Aristotle Onassis, the Greek shipping magnate, who owned a majority interest in most Monegasque enterprises; the issue was settled in 1967 when Monaco bought Onassis out. Rainier III died in 2005; he was succeeded by his son, Albert II.

Government

Monaco is governed under the constitution of 1962. The heredity monarch is the head of state. The minister of state, selected by the monarch from three candidates nominated by France, is the head of government. The unicameral legislature is the National Council, which is elected by universal suffrage every five years. The monarch may initiate legislation, but all laws must be approved by the National Council. The current head of state is Prince Albert II (since April 2005). The current head of government is Minister of State Jean-Paul Proust (June 2005).

Monadhliath Mountains (MAH-nahd-lee-ahth), range, HIGHLAND, N Scotland, between the SPEY RIVER (E) and LOCH NESS (W). Carn Bán (3,087 ft/941 m) is the highest point.

Monadnock Mountain (muh-NAD-nawk), peak (3,165 ft/965 m), CHESHIRE county, SW NEW HAMPSHIRE, 9 mi/14.5 km SE of KEENE, in Monadnock State Park. It is a popular hiking and cross-country skiing area. Geomorphic term monadnock referring to a hill or mountain isolated from larger highland mass by erosion is derived from this mountain, which is a classical example of an erosional remnant above the base level of a peneplane surface.

Monadnock Mountain (muhn-AD-nawk), isolated peak (3,140 ft/957 m), NE VERMONT, near the CONNECTICUT River, in LEMINGTON town, opposite COLEBROOK, NEW HAMPSHIRE.

Monagas (mo-NAH-gahs), state (□ 11,160 sq mi/29,016 sq km; 2001 population 712,626), NE VENEZUELA; ⊙ MATURÍN; 09°20′N 63°00′W. Bounded by ORINOCO RIVER (SE), the Caño MÁNAMO (E), SAN JUAN RIVER and Gulf of PARIA (NE). Apart from CORDILLERA DE LA COSTA (N), it consists of LLANOS and low tablelands, with marshes in Orinoco River delta area. Climate is tropical, with rains in some parts all year round. Mineral resources include petroleum (wells at QUIRIQUIRE, JUSEPÍN, SANTA BÁRBARA, and TEMBLADOR fields), coal (Santa Bárbara), zinc and cadmium (CHAGUARAMAL), asphalt, sulphur, marble, and salt

---

Cross-references are shown in SMALL CAPITALS. The pronunciation guide is shown on page xix. The sources of population figures are shown on page xvii.

deposits. Predominantly a cattle-raising region. N uplands grow coffee, tobacco, sugarcane, cacao, cotton, yucca, vegetables, corn. Its vast forests abound in a variety of hardwood and palm trees. Maturín is its trading center. CARIPITO has oil-refining plant. Main cities include CARIPE, CAICARA DE MATURÍN, San Antonio de Maturín, BARRANCAS. Includes Parque Nacional EL GUÁCHARO.

Monaghan (MAH-nuh-huhn), Gaelic *Muineachán*, county (□ 500 sq mi/1,300 sq km; 2006 population 55,997), NE IRELAND; ⊙ MONAGHAN. Bordered on the N by NORTHERN IRELAND, on the SE by LOUTH county, and on the S by CAVAN county. The NW portion of the county is a part of the fertile central plain of Ireland; to the S and E are hilly sections. It is primarily an agricultural county. The raising of beef and dairy cattle is the main enterprise. Potatoes, oats, and turnips are the chief crops; pigs, sheep, and poultry, as well as cattle, are raised in large numbers. Other industries are bacon curing and the manufacturing of furniture and footwear. Significant towns are CARRICKMACROSS, CASTLEBLAYNEY, and CLONES.

Monaghan (MAH-nuh-huhn), Gaelic *Muineachán*, town (2006 population 6,221), ⊙ MONAGHAN county, NE IRELAND, 12 mi/19.2 km NE of CLONES; 54°15′N 06°58′W. It is a farm market with some manufacturing. It houses the cathedral of the Roman Catholic diocese of Clogher.

Monagrillo (mo-nah-GREE-yo), village and minor civil division of Chitré District (2000 population 8,418), HERRERA province, S central PANAMA, in PACIFIC lowland, adjoins CHITRÉ and part of that city's urban area.

Monahans (MAH-nuh-hanz), town (2006 population 6,392), ⊙ WARD county, extreme W TEXAS, in the Pecos valley, 35 mi/56 km WSW of ODESSA; 31°38′N 103°03′W. Elevation 2,613 ft/796 m. Trade, shipping, processing center for oil fields; cattle ranches nearby; manufacturing (gas processing, light manufacturing). Established 1881; incorporated 1928.

Mona Island (MO-nah), Spanish *Isla de Mona* (□ c. 20 sq mi/52 sq km), part of the Commonwealth of PUERTO RICO, c. 50 mi/80 km W of MAYAGUEZ. Uninhabited island. It was discovered by Columbus in 1493, and in 1508 Ponce de León stopped here. In 1511 the island was ceded to Columbus's younger brother Bartolomé, but it soon became a haven for pirates and corsairs. Here is a nature reserve (bird sanctuary), turtle nesting sites, high cliff walls (up to 200 ft/61 m), and caves. Lighthouse on N side. Administered by the Department of Natural and Environmental Resources of the Commonwealth of Puerto Rico.

Monango (mo-NANG-o), village (2006 population 26), DICKEY county, SE NORTH DAKOTA, 12 mi/19 km NNW of ELLENDALE; 46°10′N 98°35′W. Whitestone Battlefield Historic Site to W. Founded in 1886.

Mona Passage (MO-nah), strait, between PUERTO RICO and the DOMINICAN REPUBLIC; c. 80 mi/129 km wide. Connecting the N ATLANTIC OCEAN with the CARIBBEAN SEA, it is a favored shipping lane.

Monapo District, MOZAMBIQUE: see NAMPULA, province.

Monarch, village, CHAFFEE county, central COLORADO, on branch of ARKANSAS RIVER, in SAWATCH MOUNTAINS, and 16 mi/26 km W of SALIDA. Elevation c.10,000 ft/3,048 m. Limestone quarries. MONARCH Ski Resort here. In San Isabel National Forest; CONTINENTAL DIVIDE to W. Gunnison National Forest beyond DIVIDE. MONARCH PASS to SW, Mount Shavano 7 mi/11.3 km NNE.

Monarch, SOUTH CAROLINA: see MONARCH MILLS.

Monarch Mills, village, UNION county, N SOUTH CAROLINA, suburb, 2 mi/3.2 km E of UNION; 34°43′N 81°34′W. Agricultural area of poultry, livestock, dairying, grain, peaches, apples. Sumter National Forest to S.

Monarch Mountain (MAH-nahrk) (11,590 ft/3,533 m), W BRITISH COLUMBIA, W CANADA, in COAST MOUNTAINS, 200 mi/322 km NW of VANCOUVER; 51°56′N 125° 56′W. Highest point in CENTRAL COAST regional district.

Monarch Pass (11,312 ft/3,448 m), central COLORADO, in SAWATCH MOUNTAINS, between CHAFFEE and GUNNISON counties. Crossed by U.S. Highway 50. View from highest point on pass includes twelve peaks exceeding 14,000 ft/4,267 m to N. Gunnison National Forest (S); San Isabel National Forest (N).

Monarch, The (MAH-nahrk), mountain (9,528 ft/2,904 m), SE BRITISH COLUMBIA, W CANADA, near ALBERTA border, in ROCKY MOUNTAINS, on SE edge of KOOTENAY NATIONAL PARK, 16 mi/26 km SW of BANFF (Alberta); 51°03′N 115°51′W.

Monashee Mountains (muh-NA-shee), range of ROCKY MOUNTAINS, SE BRITISH COLUMBIA, W CANADA, W of SELKIRK MOUNTAINS, extending c.200 mi/ 322 km N from WASHINGTON state line between COLUMBIA River and ARROW LAKES (E) and upper North Thompson River, SHUSWAP LAKE, and OKANAGAN LAKE (W); 51°00′N 119°00′W. Peaks include HALLAM PEAK (10,560 ft/3,219 m) and Cranberry Mountain (9,470 ft/2,886 m). Between REVELSTOKE and SHUSWAP LAKE, the range is crossed by Canadian Pacific railroad. In S part is important mining (gold, silver, copper, lead, zinc) region.

Monasterboice (mah-nuhs-tuhr-BOIS), Gaelic *Mainistear Bhuíthin*, village (2006 population 1,164), LOUTH county, NE IRELAND; 53°46′N 06°26′W. It is one of the oldest monastic sites in Ireland, established near the end of the 5th century. There are ruins of a round tower and two churches. Celebrated for its 10th-century High Crosses.

Monasterevan (mah-nuhs-tuhr-RE-vuhn) or **Monasterevin**, Gaelic *Mainistir Eimhín*, town (2006 population 3,649), W KILDARE county, E IRELAND, on BARROW RIVER, on branch of GRAND CANAL, and 7 mi/11.3 km W of KILDARE; 53°08′N 07°03′W. Agricultural market. Nearby Moore Abbey is on site of 12th-century monastery.

Monasterio, SPAIN: see MONESTERIO.

Monasterzyska, UKRAINE: see MONASTYRYS'KA.

Monastier-sur-Gazeille, Le (mo-nah-stee-ye–syur– GAH-zai-yuh, luh) or **Monastier, Le**, commune (□ 15 sq mi/39 sq km), HAUTE-LOIRE department, AUVERGNE region, S central FRANCE, 9 mi/14.5 km SE of Le PUY-EN-VELAY; 44°56′N 04°00′E. Cattle market. Has Romanesque and Gothic Benedictine abbey dating from 7th century, built with volcanic stone of various colors. The village was visited by River Lake Stevenson in 1878 during his travels on foot in the CÉVENNES MOUNTAINS.

Monastir, MACEDONIA: see BITOLA.

Monastir, TUNISIA: see MUNASTIR.

Monastir Gap, MACEDONIA: see PELAGONIJA.

Monastyrishche (muh-nahs-TI-ree-shchye), village (2006 population 4,160), SW MARITIME TERRITORY, SE RUSSIAN FAR EAST, on road and railroad, 10 mi/ 16 km S, and under administrative jurisdiction, of CHERNIGOVKA; 44°12′N 132°29′E. Elevation 436 ft/132 m. In agricultural area (grains, soybeans; livestock).

Monastyrishche, UKRAINE: see MONASTYRYSHCHE.

Monastyriska, UKRAINE: see MONASTYRYS'KA.

Monastyrshchina (muh-nahs-TIR-shchee-nah), town (2006 population 4,460), SW SMOLENSK oblast, W European Russia, on the Vikhra River (tributary of the SOZH RIVER, DNIEPER River basin), on road junction, 32 mi/51 km SSW of SMOLENSK; 54°21′N 31°50′E. Elevation 623 ft/189 m. Flax processing, distilling, bakery.

Monastyrskoye, RUSSIA: see TURUKHANSK.

Monastyryshche (mo-nah-sti-RI-shche) (Russian *Monastyrishche*), city, W CHERKASY oblast, UKRAINE, 25 mi/40 km NW of UMAN'; 48°59′N 29°49′E. Elevation 797 ft/242 m. Raion center; boiler works, phar-

maceutical, asphalt making, food processing, feed concentrate; professional-technical school; heritage museum. Known since the late 16th century. Site of the defeat of the Poles by the Ukrainian Cossacks (1653); city since 1985.

Monastyrys'ka (mo-nah-sti-RIS-kah) (Russian *Monastyriska*) (Polish *Monasterzyska*), city (2004 population 7,300), SW TERNOPIL' oblast, UKRAINE, 10 mi/16 km W of BUCHACH; 49°05′N 25°10′E. Elevation 1,079 ft/328 m. Raion center; tobacco-growing center; cigar manufacturing, sewing, grain milling, brickworking. Has ruins of an old palace. Ruins of a Kievan Rus' fortified settlement found nearby. Known since 1454. Important 16th-century trade center; declined after destruction by the Turks (1672). Passed from Poland to Austria (1772); part of West Ukrainian National Republic (1918); reverted to Poland (1919); part of Ukrainian SSR since 1939; independent Ukraine since 1991.

Monatélé (mawn-uh-TAI-lai), town, ⊙ LEKIE department, Central province, CAMEROON, 34 mi/55 km NW of Yaoundé; 04°15′N 11°14′E.

Mona Vale (MO-nuh VAIL), town, E NEW SOUTH WALES, SE AUSTRALIA, residential suburb, 15 mi/24 km NNE of SYDNEY; 33°41′S 151°19′E.

Monaville (MO-nuh-vil), unincorporated village, LOGAN county, SW WEST VIRGINIA, 2 mi/3.2 km S of LOGAN.

Monbazillac (mon-bah-zee-yahk), commune (□ 17 sq mi/44.2 sq km), DORDOGNE department, AQUITAINE region, SW FRANCE, 4 mi/6.4 km S of BERGERAC; 44°47′N 00°30′E. Produces famed white dessert wine. Its château, built c.1550 in Renaissance style, is owned by the Monbazillac wine cooperative; it overlooks the Dordogne valley.

Monbetsu (MON-bets), city, N Hokkaido prefecture, N JAPAN, on Sea of OKHOTSK, 133 mi/215 km N of SAPPORO; 44°21′N 143°21′E. Dairying.

Monbetsu (MON-bets), town, Saru county, Hidaka district, Hokkaido prefecture, N JAPAN, 56 mi/90 km ESE of SAPPORO; 42°28′N 142°04′E. Lumber; horses; smelts.

Moncada (mong-KAH-dhah), city, VALENCIA province, E SPAIN, 6 mi/9.7 km NNW of VALENCIA. In rich produce-farming area; manufacturing of silk textiles, burlap, candy; olive-oil processing.

Moncada (mong-KAH-dah), town, TARLAC province, central LUZON, PHILIPPINES, on railroad, and 17 mi/27 km N of TARLAC; 15°45′N 120°33′E. Agricultural center (coconuts, rice, sugarcane).

Moncada y Reixach (mong-KAH-dah ee rai-SHAHK), satellite city, BARCELONA province, NE SPAIN, on BESÓS RIVER, and 7 mi/11.3 km NNE of BARCELONA. Iron smelting, meat processing; manufacturing of cement, beer, soap, gloves; cotton spinning. Stone quarries nearby.

Moncagua (mon-KAH-gwah), municipality and town, SAN MIGUEL department, EL SALVADOR, WNW of SAN MIGUEL city.

Mon Cai, VIETNAM: see HAI NINH.

Moncalieri (mong-kah-LYAI-ree), town (2001 population 53,350), TORINO province, PIEDMONT, NW ITALY, on PO RIVER, and 4 mi/6 km S of TURIN; 45°00′N 07°41′E. Railroad junction; industrial and commercial center; foundries, canneries; machinery, fabricated metals, textiles, paper products, plastics. Has royal palace (built 1789; now a military academy) and meteorological observatory.

Moncalvo (mong-KAHL-vo), village, ASTI province, PIEDMONT, NW ITALY, 11 mi/18 km NNE of ASTI; 45°03′N 08°16′E. Important wine center.

Monção (mon-SOUN), city (2007 population 27,586), N central MARANHÃO, BRAZIL, on RIO PINDARÉ, and 90 mi/145 km SW of SÃO LUÍS; 03°29′S 45°15′W. Rice, cotton; carnauba wax.

Monção (mon-SOU) or **Monsão**, town, VIANA DO CASTELO district, northernmost PORTUGAL, on left

---

Area is shown by the symbol □, and capital city or county seat by ⊙.

bank of MINHO RIVER (SPAIN border), 35 mi/56 km N of BRAGA. Portuguese railroad terminus; agricultural trade; vineyards. Cold sulphur springs. Resisted Spanish siege in 1658.

Moncayo, Sierra del (mong-KEI-o, SYE-rah dhel), range of the CORDILLERA IBÉRICA, N SPAIN, on Aragon-Castile border, 35 mi/56 km W of ZARAGOZA; 41°46′N 01°50′W. Rises to c.7,590 ft/2,313 m. Forms watershed between Ebro and Duero basins.

Monceau-sur-Sambre (maw-SO–sur–SAWM-bruh), town in commune of Charleroi, Charleroi district, HAINAUT province, S central BELGIUM, on SAMBRE RIVER, and 3 mi/4.8 km W of CHARLEROI; 50°25′N 04°22′E. Metal industry; electric-power station.

Mönch (MUNKH) [German=monk], Alpine peak (13,448 ft/4,099 m), in BERNESE ALPS, S central SWITZERLAND, on border between cantons of BERN and VALAIS, 2 mi/3.2 km NE of the JUNGFRAU; 46°34′N 07°59′E.

Monchegorsk (muhn-chi-GORSK), city (2006 population 49,075), central MURMANSK oblast, NW European Russia, on the KOLA PENINSULA, on Lake IMANDRA, on spur of the Murmansk railroad, 90 mi/145 km S of MURMANSK; 67°56′N 32°52′E. Elevation 419 ft/127 m. Nickel and copper mining and smelting. Mines at Malaya Sopcha, just SW. Arose in 1937 with the development of mining. Made city in 1937.

Mönchengladbach (mun-khen-GLAHD-bahkh), city, North Rhine-Westphalia, W GERMANY; 51°11′N 06°28′E. Major center of German cotton-textile industry; manufacturing (textile machinery, iron, and chemicals). Road, railroad, and air terminus, it is the N central European headquarters of the NATO. Developed around a Benedictine abbey (founded c.972), which was rebuilt several times between the 14th and the 18th century and which now serves as the city hall. Formerly twin city with RHEYDT (first mentioned 1180); Rheydt became district of Mönchengladbach in 1975. Sometimes called MÜNCHEN-GLADBACH.

Mönchhof (MUNKH-hof), village, BURGENLAND, E AUSTRIA, in the SEEWINKEL, 21 mi/34 km NE of SOPRON (HUNGARY), across LAKE NEUSIEDL; 47°53′N 16°56′E. Vineyards. Kneipp spa. Modern Cistercian abbey.

Monchique (mon-SHEEK), town, FARO district, S PORTUGAL, in the SERRA DE MONCHIQUE, 13 mi/21 km N of PORTIMÃO; 37°19′N 08°33′W. Fruits, potatoes, vegetables; cork. Spa of CALDAS DE MONCHIQUE is 3 mi/4.8 km S.

Monchique, Serra de (mon-SHEEK, SER-rah dah), hill range in FARO district, S PORTUGAL, near SW extremity of IBERIAN PENINSULA; 37°19′N 08°36′W. Rises to 2,960 ft/902 m at LA FOIA. Pine, oak, chestnut trees.

Monción (mon-see-ON), town, MONTE CRISTI province, NW DOMINICAN REPUBLIC, in N outliers of the Cordillera CENTRAL, 32 mi/51 km W of SANTIAGO; 19°28′N 71°10′W. Agricultural region (tobacco, coffee, cacao; beeswax; hides). Sometimes Benito MONCIÓN; formerly Guaraguanó.

Moncks Corner, town (2006 population 6,572), ⊙ BERKELEY county, SE SOUTH CAROLINA 28 mi/45 km N of CHARLESTON, and on COOPER RIVER; 33°12′N 80°00′W. PINOPOLIS DAM and Hydroelectric plant of SANTEE-COOPER power and navigation development nearby. Manufacturing includes plastic tubing, timing belts, noise control equipment, cargo shipping liners. Agriculture includes cotton, tobacco, corn, soybeans; hogs, cattle, poultry.

Monclova, city and township, ⊙ Monclova municipio, COAHUILA state, MEXICO; 26°54′N 101°25′W. Elevation 1,923 ft/586 m. Situated within outlier ranges of the SIERRA MADRE ORIENTAL on Mexico Highways 30, 53, and 57, it is a regional commercial and industrial center. Monclova's chief industry is the production of iron and steel. It is the third-largest city in Coahuila state.

Monções (MON-ses), town (2007 population 2,054), NW SÃO PAULO state, BRAZIL, 33 mi/53 km NE of ARAÇATUBA; 20°52′S 50°05′W. Coffee growing.

Moncófar (mong-KO-fahr), town, Castellón de la Plana province, E SPAIN, 14 mi/23 km SSW of CASTELLÓN DE LA PLANA; 39°48′N 00°09′W. Ships oranges and onions.

Moncontour (mon-kon-toor), commune, CÔTES-D'ARMOR department, W FRANCE, in BRITTANY, 12 mi/19 km SSE of SAINT-BRIEUC; dairying; stone quarries near by. Retains some of its 11th-century fortifications and a 12th-century dungeon. Has 16th-century church with remarkable stained-glass windows. Also called Moncontour-de-Bretagne.

Moncoutant (mon-koo-tahn), commune (□ 9 sq mi/23.4 sq km; 2004 population 3,019), DEUX-SÈVRES department, POITOU-CHARENTES region, W FRANCE, near SÈVRE NANTAISE river, 17 mi/27 km WNW of PARTHENAY, in the Gâtine Heights; 46°43′N 00°35′W. Manufacturing of hosiery.

Moncreiffe Hill (MAHN-kreef HIL) (725 ft/221 m), PERTH AND KINROSS, E Scotland, 3 mi/4.8 km SE of PERTH; 56°21′N 03°23′W. Formerly in Tayside, abolished 1996.

Moncton (MUHNGK-tuhn), city (2001 population 61,046), SE NEW BRUNSWICK, CANADA, on the PETITCODIAC RIVER; 46°06′N 64°47′W. It is an air and railroad transportation center. Service and distribution center for region and province; IT firms and customer call centers. It was called The Bend until 1833, when it was renamed in honor of the British general Robert Monckton. Magnetic Hill, an optical illusion, and the Tidal Bore, a high tide occurring twice daily, are features of the city. The Université de Moncton (1963) is here.

Moncure (MAHN-kyor), unincorporated village, CHATHAM county, central NORTH CAROLINA, 28 mi/45 km SW of RALEIGH, and 1 mi/1.6 km W of HAYWOOD, near confluence of HAW and DEEP rivers, which form CAPE FEAR RIVER; 35°37′N 79°04′W. Timber. Manufacturing (industrial yarns and resins, fiberboard, crushed stone).

Monda (MON-dah), town, MÁLAGA province, S SPAIN, in outliers of the CORDILLERA PENIBÉTICA, 24 mi/39 km WSW of MÁLAGA; 36°38′N 04°50′W. Olives, olive oil, grapes, almonds.

Mondamin, town (2000 population 423), HARRISON county, W IOWA, near SOLDIER RIVER, 14 mi/23 km NW of LOGAN; 41°42′N 96°01′W.

Mondego, Cape (mon-DAI-goo), rocky headland on the ATLANTIC OCEAN, in COIMBRA district, N central PORTUGAL, 4 mi/6.4 km NW of FIGUEIRA DA FOZ (at mouth of MONDEGO RIVER); 40°11′N 08°54′W. Lighthouse. Buarcos hills (just E) rise to 700 ft/213 m.

Mondego River (mon-DAI-goo), 137 mi/220 km long, N central PORTUGAL; rises in SW GUARDA district in the SERRA DA ESTRÊLA 10 mi/16 km N of COVILHÃ; flows generally SW, through fertile Coimbra plain, to the ATLANTIC OCEAN at Figueira da Foz (COIMBRA district). Navigable seasonally for barges to influx of DÃO RIVER (above COIMBRA). Largest river flowing entirely in Portugal. Forms border of VISEU and GUARDA districts and Viseu and Coimbra districts.

Mondéjar (mon-DAI-hahr), town, GUADALAJARA province, central SPAIN, 30 mi/48 km E of MADRID; 40°20′N 03°07′W. Agricultural center (cereals, grapes, olives, honey, livestock). Tanning, soap manufacturing.

Mondercange (MON-der-kawzh), commune (2001 population 6,089), 3 mi/4.8 km NE of ESCH-SUR-ALZETTE, LUXEMBOURG; 49°32′N 05°59′E. In the plains region between LUXEMBOURG city and ESCH-sur-Alzette. Historic church.

Mondeville (mon-duh-veel), town (□ 3 sq mi/7.8 sq km), E industrial suburb of CAEN, CALVADOS department, BASSE-NORMANDIE region, NW FRANCE; 49°10′N 00°19′W. Manufacturing of steel sheeting for local auto industry.

Mondim de Basto (mon-DEENG dai BAHSH-too), town, VILA REAL district, N PORTUGAL, on TÂMEGA RIVER, and 13 mi/21 km NW of VILA REAL; 41°25′N 07°58′W. Livestock, agriculture.

Mondo (mon-DO), village, KANEM administrative region, W CHAD, 135 mi/217 km NNE of N'DJAMENA; 13°47′N 15°32′E. Formerly part of LAC prefecture.

Mondo (MON-do), village, DODOMA region, N central TANZANIA, 85 mi/137 km N of DODOMA; 04°59′S 35°55′E. Livestock, grain.

Mondolfo (mon-DOL-fo), town, PESARO E URBINO, The MARCHES, central ITALY, 7 mi/11 km SSE of FANO; 43°45′N 13°06′E. Manufacturing clothing, wood products.

Mondolkiri (MON-DUL-KEE-REE), province (□ 5,516 sq mi/14,341.6 sq km; 2007 population 44,913), E CAMBODIA, borders KRATIE (W), STUNG TRENG (NW), and RATANAKIRI (N) provinces, and VIETNAM (E and S); ⊙ SENMONOROM; 12°30′N 107°00′E. In forested region (which has been having problems with deforestation recently). Shifting cultivation, agroforestry, forest products. Khmer population with Mmong, Pmong, and other minorities. Also spelled Mondol Kiri and Mondul Kiri.

Mondoñedo (mon-do-NYAI-dho), town, LUGO province, NW SPAIN, in GALICIA, 30 mi/48 km NNE of LUGO; 43°26′N 07°22′W. Agricultural trade center (livestock; potatoes; wine; lumber); brewery. Mineral springs. Bishopric with 13th-century Gothic cathedral. Graphite quarries nearby.

Mondorf-les-Bains (mong-DORF-lai-BANG), town, Mondorf-les-Bains commune, SE LUXEMBOURG, 10 mi/16 km SE of LUXEMBOURG city, on French border; 49°30′N 06°17′E. Manufacturing; tourism (resort with mineral springs).

Mondoubleau (mon-doo-blo), commune (□ 1 sq mi/2.6 sq km), LOIR-ET-CHER department, CENTRE administrative region, N central FRANCE, 15 mi/24 km NW of VENDÔME; 47°59′N 00°54′E. Livestock market; basket making. Has the ruins of 12th-century keep.

Mondovì (mon-do-VEE), town, CUNEO province, PIEDMONT, NW ITALY, on Ellero River, and 14 mi/23 km E of CUNEO; 44°23′N 07°49′E. Railroad junction. Comprised of lower town, Breo (industrial section); upper town, Piazza; and, just NE, Carassone (ceramics). Has food, clothing, paper, wood, chemical, non-metallics, fabricated metals, iron and steel industries. Bishopric with cathedral (1763). Santuario di Vicoforte, large pilgrimage church built 1596–1731, is 3 mi/5 km SE. Napoleon defeated Austro-Sardinian forces at Mondovì in 1796. In World War II, bombed (1942–1943).

Mondovi (mahn-DO-vee), town (2006 population 2,638), BUFFALO county, W WISCONSIN, on BUFFALO RIVER (hydroelectric plant), and 19 mi/31 km SSW of EAU CLAIRE; 44°34′N 91°40′W. Dairy products, poultry hatcheries; manufacturing (lumber, cabinets). W terminus of Buffalo River State Trail. Settled 1855, incorporated 1889.

Mondoza, PERU: see MENDOZA.

Mondragón (mon-drah-GON), city, GUIPÚZCOA province, N SPAIN, in the Basque Provinces, 20 mi/32 km WSW of TOLOSA; 43°04′N 02°29′W. Metalworking (home appliances, metal furniture); flour milling. Agricultural trade (cereals, chestnuts, hazelnuts; livestock). Iron deposits nearby. Mineral springs 3 mi/4.8 km WSW at Santa Agueda.

Mondragone (mon-drah-GO-ne), town, CASERTA province, CAMPANIA, S ITALY, between MONTE MASSICO and GULF OF GAETA, 17 mi/27 km W of CAPUA; 41°07′N 13°53′E. Textiles, canned foods, wine; marble works. Hot mineral baths nearby. Remains of Roman APPIAN WAY nearby.

Cross-references are shown in SMALL CAPITALS. The pronunciation guide is shown on page xix. The sources of population figures are shown on page xvii.

Mondrain Island, AUSTRALIA: see RECHERCHE ARCHIPELAGO.

Mond River (MUHND), coastal stream, 300 mi/483 km long, S IRAN; rises in several branches in the ZAGROS ranges W of SHIRAZ; flows SE, S, and W, to PERSIAN Gulf 60 mi/97 km SSE of Bushuhr.

Mondsee (MOND-sai), lake (□ 5 sq mi/13 sq km), SW UPPER AUSTRIA and NE SALZBURG, in the SALZKAMMERGUT, 15 mi/24 km E of SALZBURG; c. 4 mi/6.4 km long, 1.7 mi/2.7 km wide; 47°49′N 13°23′E. Maximum depth 207 ft/63 m; elevation 1,466 ft/447 m. A short stream connects it (N) with small ZELLER SEE or Irrsee and E with Attersee or KAMMERSEE. Remains of neolithic lake dwellings.

Monduli (mon-DOO-lee), village, ARUSHA region, N TANZANIA, 16 mi/26 km WNW of ARUSHA, Mount Monduli (8,727 ft/2,660 m) to NE; 03°19′S 36°26′E. Coffee, corn, wheat, vegetables; cattle, sheep, goats.

Mondul Kiri, province, CAMBODIA: see MONDOLKIRI.

Mondy (MON-di), settlement (2004 population 420), SW BURYAT REPUBLIC, S SIBERIA, RUSSIA, on the IRKUT RIVER, on the Irkutsk-Uliastay highway, 150 mi/241 km WSW of IRKUTSK; 51°40′N 100°59′E. Elevation 4,291 ft/1,307 m. Transit point in the Eastern Sayan Mountains, near (approximately 4 mi/6.4 km N of) the Mongolian border. Asbestos deposits.

Moneague, village, SAINT ANN parish, central JAMAICA, resort at N foot of MOUNT DIABLO, 30 mi/48 km NW of KINGSTON; 18°17′N 77°06′W.

Moneasa, ROMANIA: see SEBIŞ.

Monédières (mo-nai-dyer), range in the MASSIF CENTRAL, CORRÈZE department, LIMOUSIN region, S central FRANCE, extends c.20 mi/32 km N-S, between upper valleys of the CORRÈZE and the VÉZÈRE rivers, rising to more than 3,000 ft/914 m; 45°30′N 01°52′E. Blueberries shipped from here.

Monee (mo-KNEE), village (2000 population 2,924), WILL county, NE ILLINOIS, 19 mi/31 km ESE of JOLIET; 41°25′N 87°45′W. In agricultural area.

Monein (mo-nen), town (□ 31 sq mi/80.6 sq km), PYRÉNÉES-ATLANTIQUES department, AQUITAINE region, SW FRANCE, 11 mi/18 km W of PAU; 43°20′N 00°35′W. Vineyards and orchards. Natural gas is recovered from local well field.

Monemvasía (mo-nem-vah-SEE-ah), village, LAKONIA prefecture, SE PELOPONNESE department, extreme SE mainland GREECE, on the Mirtoan Sea, on a rocky island joined to the mainland by a mole; 36°41′N 23°03′E. In the Middle Ages it was a fortress and an important commercial port, occupied in turn by the Normans, Byzantines, and Venetians; exporting Malvasian or malmsey wine, a type now made in many places. Well-preserved medieval city with a wealth of Byzantine churches. Seat of the first Greek national assembly in 1829. Formerly Malvasia.

Moneragala, district (□ 2,127 sq mi/5,530.2 sq km; 2001 population 397,375), WESTERN PROVINCE, SRI LANKA; ⊙ MONERAGALA; 06°40′N 81°20′E.

Moneragala, town, ⊙ MONERAGALA district, WESTERN PROVINCE, SRI LANKA, 30 mi/48 km ESE of BADULLA; 06°52′N 81°21′E. Trades in rice and vegetables.

Moneron Island (muh-nye-RON), Japanese *Kaiba-to*, in the Sea of JAPAN, 30 mi/48 km off SW SAKHALIN Island, SAKHALIN oblast, extreme E SIBERIA, RUSSIAN FAR EAST; 4 mi/6.4 km long, 2 mi/3.2 km wide; 46°15′N 141°15′E. Under Japanese rule, 1905–1945.

Monessen (mo-NE-suhn), city (2006 population 8,219), WESTMORELAND county, SW PENNSYLVANIA, 20 mi/32 km S of PITTSBURGH, on the MONONGAHELA RIVER (bridged); 40°08′N 79°52′W. Manufacturing (wire products, printing and publishing, fencing). Monessen's steel mill was closed as a result of the industry's decline. Founded 1898, incorporated 1921.

Monesterio (mo-ne-STAI-ryo) or **Monasterio**, town, BADAJOZ province, W SPAIN, in SIERRA MORENA, 32 mi/51 km SE of JEREZ DE LOS CABALLEROS; 38°05′N 06°16′W. Stock raising (sheep, goats), lumbering, and

agricultural center (grain, olives, grapes; apiculture). Manufacturing of plaster; meat products.

Monestier-de-Clermont (mo-nest-yai–duh–kler-mon), commune (□ 1 sq mi/2.6 sq km), resort in ISÈRE department, RHÔNE-ALPES region, SE FRANCE, in DAUPHINÉ ALPS, 19 mi/31 km SSW of GRENOBLE; 44°54′N 05°38′E. Elevation 2,730 ft/832 m. Water-based sports on Monteynard Lake (formed by dam on DRAC RIVER, just E). Nearby Mont Aiguille (6,900 ft/2,103 m) is a peak climbed by alpinists since 1500.

Monestier-les-Bains, Le, FRANCE: see MONÉTIER-LES-BAINS, LE.

Moneta, town, O'BRIEN county, NW IOWA, 13 mi/21 km WE of PRIMGHAR; 43°07′N 95°23′W. In agricultural area.

Moneta (muh-NE-tuh), unincorporated village, BEDFORD county, SW central VIRGINIA, 18 mi/29 km ESE of ROANOKE; 37°10′N 79°37′W. Manufacturing (lumber, concrete, meat processing, printing and publishing); agriculture (dairying; livestock; grain, tobacco). SMITH MOUNTAIN LAKE reservoir and State Park to S.

Moneta, village, FREMONT county, central WYOMING, on Poison Creek, on branch of Big Horn River, and 75 mi/121 km WNW of CASPER. Elevation 5,428 ft/1,654 m. Castle Gardens, area of picturesque sandstone formations, and Gas Hills Uranium Mining District to S.

Monéteau (mo-nai-to), town (□ 7 sq mi/18.2 sq km), YONNE department, in BURGUNDY, central FRANCE, N suburb of AUXERRE; 47°51′N 03°35′E.

Monêtier-les-Bains, Le (mo-ne-tye–lai–ban, luh), town (□ 37 sq mi/96.2 sq km), Alpine spa in HAUTES-ALPES department, PROVENCE-ALPES-CÔTE D'AZUR region, SE FRANCE, on the GUISANE RIVER, and 9 mi/14.5 km NW of BRIANÇON, at NE foot of Massif du Pelvoux (Massif des ÉCRINS); 44°58′N 06°30′E. Elevation 3,602 ft/1,098 m. Mineral springs. Winter sports at SERRE-CHEVALIER ski terrain; 15th-century church. Also spelled Le Monestier-les-Bains.

Monetnyy (muh-NYET-niyee), town (2006 population 5,510), S SVERDLOVSK oblast, E URALS, W Siberian Russia, on road and railroad (Monetnaya station), 18 mi/29 km NE (under jurisdiction of) BERËZOVSKY; 57°03′N 60°53′E. Elevation 892 ft/271 m. Peat digging; tungsten mining, tractor repair shop. Developed in the 1930s.

Monett (mo-NET), city (2000 population 7,396), BARRY and LAWRENCE counties, SW MISSOURI, in the OZARKS, 35 mi/56 km ESE of JOPLIN; 36°55′N 93°55′W. Cattle and dairying. Ships fruit; light manufacturing. Surveyed 1887.

Monetta (muh-NET-ah), village (2006 population 220), AIKEN and SALUDA counties, SW SOUTH CAROLINA, 21 mi/34 km NNE of AIKEN; 33°51′N 81°36′W. Manufacturing includes fiberglass truck bodies. Agriculture includes livestock, poultry; grain, cotton, peanuts.

Monette (mo-NET), town (2000 population 1,179), CRAIGHEAD county, NE ARKANSAS, 21 mi/34 km E of JONESBORO; 35°53′N 90°20′W. In agricultural area; manufacturing (plastic molding, shoes). Saint Francis Sunken Lands Wildlife Management Area to W. Incorporated 1900.

Monetville (mo-NE-vil), settlement, SE central ONTARIO, E central CANADA, and included in town of FRENCH RIVER; 46°09′N 80°21′W. First settled 1895. Formerly called Martland.

Moneva, Embalse de, reservoir, ZARAGOZA province, NE SPAIN, on Aguasvivas River, 1.2 mi/1.9 km N of Moneva, 30 mi/48 km S of ZARAGOZA; 41°10′N 00°50′W.

Money Island, CHINA: see CRESCENT GROUP.

Moneymore (muh-nee-MOR), Gaelic *Muine Mór*, town (2001 population 1,371), SE LONDONDERRY, Northern Ireland, 30 mi/48 km S of COLERAINE; 54°42′N 06°41′W. Plantation market and linen settlement founded by the London Drapers' Company.

Monfalcone (mon-fahl-KO-nai), city, in FRIULI-VENEZIA GIULIA, extreme NE ITALY, near the ADRIATIC Sea; 45°48′N 13°32′E. Modern industrial center; manufacturing (ships, airplanes, textiles, chemicals, food products, refined oil).

Monflanquin (mon-flahn-kan), commune (□ 24 sq mi/62.4 sq km; 2004 population 2,352), LOT-ET-GARONNE department, AQUITAINE region, SW FRANCE, 9 mi/14.5 km NNE of VILLENEUVE-SUR-LOT; 44°32′N 00°46′E. Felt manufacturing; plums. Established as planned fortified community (*bastide*) in 13th century.

Monforte (mom-FOR-tai), city, LUGO province, NW SPAIN, in GALICIA, 22 mi/35 km NE of ORENSE. Railroad junction. Agricultural trade center (cereals, vegetables, potatoes; wine); meat processing, tanning, sawmilling, manufacturing of dairy products. Stock raising and lumbering in area. Iron mines nearby. Dominated by hill with remains of medieval castle. Has former Benedictine monastery (now a hospital) and Jesuit college and church.

Monforte (mon-FORT), town, PORTALEGRE district, central PORTUGAL, 16 mi/26 km S of PORTALEGRE. Cheese manufacturing.

Monforte da Beira (mon-FORT dah BAI-rah), town, CASTELO BRANCO district, central PORTUGAL, 12 mi/19 km SE of CASTELO BRANCO; 39°44′N 07°18′E. Pottery manufacturing; grain, olives; livestock. Oak woods.

Monforte del Cid (mom-FOR-tai dhel THEED), town, ALICANTE province, E SPAIN, 8 mi/12.9 km NNW of ELCHE; 38°23′N 00°43′W. Wine-production center; alcohol, brandy, and liqueur distilling. Olive oil, cereals.

Monfurado, Serra de (mon-foo-RAH-doo, SER-rah dah), hills in ÉVORA district, S PORTUGAL, SW of ÉVORA, rising to 1,400 ft/427 m; 38°36′N 08°06′W. Iron deposits.

Monga (MAWNG-gah), village, ORIENTALE province, N CONGO, on Bili River, a headstream of UBANGI RIVER, and 155 mi/249 km NW of BUTA; 04°12′N 22°49′E. Elev. 1,587 ft/483 m. Customs station near CENTRAL AFRICAN REPUBLIC border; cotton ginning, palm-oil milling. Has Roman Catholic and Baptist missions.

Monga (MON-gah), village, LINDI region, SE central TANZANIA, 110 mi/177 km W of KILWA MASOKO, near MATANDU RIVER, in SELOUS GAME RESERVE; 09°06′S 37°51′E.

Mongala River (mawng-GAH-lah), c.205 mi/330 km long, NW CONGO; formed by three headstreams at BUSINGA; flows S and SW, past LIKIMI and BINGA, to CONGO RIVER at GUMBA-MOBEKA; 01°53′N 19°46′E. Navigable downstream from BUSINGA. Was first explored in 1886.

Mongalla (mahng-GAHL-lah), township, central EQUATORIA state, S SUDAN, on right bank of the BAHR AL-GABAL (WHITE NILE River), on road, and 25 mi/40 km NNE of JUBA; 05°12′N 31°46′E. Cotton center. Former capital of Mongalla province (later called EQUATORIA province). Township and surroundings inhabited by Bari tribe.

Mongar (MAWN-gahr), village, E central BHUTAN, 124 mi/200 km E of JAKAR, by road; 27°15′N 91°14′E. Site of future hydroelectric project on the Kuru Chu.

Mongbwalu (mawng-BWAH-loo), village, ORIENTALE province, NE CONGO, 38 mi/61 km NNE of IRUMU; 01°57′N 30°02′E. Elev. 4,511 ft/1,374 m. Gold mining and trading center; gold processing. Hospital.

Mongeri (mawng-GE-ree), town, SOUTHERN province, central SIERRA LEONE, on TEYE RIVER (headstream of JONG RIVER), and 25 mi/40 km N of BO. Palm oil and kernels.

Monghopung, MYANMAR: see TAHKILEK.

Monghsu (MONG-SHOO), township (□ 470 sq mi/1,222 sq km), SHAN STATE, MYANMAR, on the NAM PANG; ⊙ MONGHSU. The capital of the township is a village 80 mi/129 km SE of LASHIO, borders on THANLWIN RIVER to E. Formerly the NE state (myo-saship).

Area is shown by the symbol □, and capital city or county seat by ⊙.

Mongicual District, MOZAMBIQUE: see NAMPULA, province.

Mongie, La (mon-zhee, lah), wintersport resort, HAUTES-PYRÉNÉES department, MIDI-PYRÉNÉES region, SW FRANCE, in the PYRENEES (elevation 5,906 ft/ 1,800 m), 11 mi/18 km S of BAGNÈRES-DE-BIGORRE, near the source of ADOUR RIVER. It lies just E of TOURMALET PASS (6,939 ft/2,115 m), and SE of the isolated PIC DU MIDI DE BIGORRE Mountain (9,400 ft/ 2,865 m). Ski terrain extends to BARÈGES (6 mi/9.7 km W). Aerial tramway to Le Taoulet (7,680 ft/2,341 m).

Monginevro, FRANCE: see MONTGENÈVRE PASS.

Monginevro Pass, FRANCE and ITALY: see MONTGENÈVRE PASS.

Mongkol Borey (MUNG-KUL BO-REI), town, W CAMBODIA, 5 mi/8 km SE of SISOPHON. Transportation hub on Mongkol Borey River and PHNOM PENH-BANGKOK railroad (formerly terminus, extension built 1942); rice-growing and trading center; orchards. In THAILAND, 1941–1946.

Mongkung (MONG-KOONG), township (□ 1,593 sq mi/4,141.8 sq km), S SHAN STATE, MYANMAR; ⊙ MONGKUNG. The capital of the township is a village 65 mi/105 km NNE of TAUNGGYI. Forests and rice; pottery. Formerly the N state (sawbwaship).

Mongla, town (2001 population 56,746), KHULNA district, SW EAST BENGAL, BANGLADESH, on the PUSUR RIVER, and 24 mi/39 km S of KHULNA; 22°35′N 89°35′E. Anchorage port; loading (raw jute) and unloading (fodgrains, coal, salt) is overside into barges and lighters.

Mongmit (MONG-MIT), Burmese *Momeik*, small township, SHAN STATE, MYANMAR; ⊙ MONGMIT. Astride SHWELI RIVER and on edge of SHAN PLATEAU, it consists (W) of flat, jungle-covered country. Rice, tea; timber. Small coal deposits. Served by Mogok-Bhamo road. Just N of ruby-mine area. Formerly the NW state (sawbwaship).

Mongmit (MONG-MIT), village, ⊙ MONGMIT township, SHAN STATE, MYANMAR, 85 mi/137 km NNE of MANDALAY. Head of road (N) to BHAMO and trade center. Founded 1279; destroyed by Kachins 1858.

Mongnai (MONG-NEI), township, SHAN STATE, MYANMAR; ⊙ Mongna. The capital of the township is a village 60 mi/97 km ESE of TAUNGGYI. Rice, tobacco; paper manufacturing. Formerly the E central state (sawbwaship).

Mongnawng (MONG-NAWNG), former NE state (myosaship) (□ 1,646 sq mi/4,279.6 sq km), now incorporated within KEHSI-MANSAM township, SHAN STATE, MYANMAR; capital was MONGNAWNG, a village 90 mi/145 km NE of TAUNGGYI.

Mongo (mon-GO), town, ⊙ GUÉRA administrative region, central CHAD, 75 mi/121 km SSE of ATI; 12°11′N 18°42′E. Road junction, airport. Was capital of former GUÉRA prefecture.

Mongobele (mon-go-BE-lai), village, BANDUNDU province, W CONGO, on left bank of FIMI RIVER, 20 mi/32 km NE of BANDUNDU; 02°47′S 17°52′E. Elev. 994 ft/302 m.

Mongol Altayn Nuuru, MONGOLIA: see MONGOLIAN ALTAY.

Mongolia, republic (□ 604,000 sq mi/1,566,500 sq km; 2004 estimated population 2,751,314; 2007 estimated population 2,951,786), NE central ASIA; ⊙ ULAAN-BAATAR. The republic encompasses more than half the area that is traditionally called MONGOLIA. The other part of Mongolia, called INNER MONGOLIA, which adjoins the State of MONGOLIA on the S and E, lies within the political unit of CHINA.

Geography

It is bordered by China on the W, S, and E, and by RUSSIA (East SIBERIA) on the N. As a sparsely settled area bordered by populous and powerful neighbors, Mongolia was long under foreign influence (Chinese, 1691–1911; Russian, 1921–1992). From 1924 to 1992 it was called the People's Republic of Mongolia and was

in the political and economic sphere of the former Soviet Union. Mongolia has high mountains, high plateaus, and large intermontane basins in the N and W, with wide lower plateaus and upland plains in the S and E. The low population density reflects the geographical position of the land-locked country in the continental interior of Asia at relatively high latitudes (42°N 88°E–52°N 120°E) and average elevation of 5,200 ft/1,580 m. The climate is therefore cold and very dry.

Geography: Mountain Ranges

Only 1% of the land is under cultivation, a result of Mongolia's lack of rainfall. Most of the country consists of steppes, semi–deserts, or mountain slopes, all of which are utilized for grazing large numbers of livestock. For each person, including men, women, and children, there are in a typical year six sheep, two goats, one head of cattle, one horse, and one-third of a camel. There are five main mountain areas in the country. High mountain ranges lie in W and SW Mongolia; the ALTAY extend NW-SE within Mongolia for about 1,000 mi/1,600 km. The MONGOLIAN ALTAY (Mongol Altayn Nuruu) (88°–95°E) in W Mongolia reaches 14,350 ft/4,374 m in the peak of KUYTEN-UUL, the highest in the country, in the great mountain knot where the boundaries of Mongolia, China, Russia and KAZAKHSTAN converge. Its E extension, the GOBI ALTAY (Govd Altayn Nuruu) (95°–106°E) in SW Mongolia, reaches 12,987 ft/3,957 m. The HANGAYN NURUU (Khangai Mountains) form a large high mountainous area in W central Mongolia and reach 13,192 ft/4,021 m. In N Mongolia, mountain masses without a common name adjoin the E SAYAN MOUNTAINS of Russian East Siberia; these mountains have very steep slopes and lie around lake HÖVSGÖL NUUR. The HENTIYN NURUU (Kentei Mountains) in N central Mongolia are much lower but on their W edges lie the most densely settled areas of Mongolia, including Ulaanbaatar. Amid the high mountains of the W and S lie some enormous depressions; the Basin of the Great Lakes in the NW with numerous large fresh- or saltwater lakes, and the Vale of Lakes in the SW with several salt lakes or playas. S of the GOBI ALTAY mountains lies the GOBI DESERT, one of the largest and driest deserts in the world. In the S and E the land forms plains that slope gently eastward toward N China, but at no point fall below 1,830 ft/560 m. The country suffers frequent and severe earthquakes, particularly along fault lines that lie parallel to the mountain ranges.

Geography: Rivers/Lakes

Mongolia lies in three hydrographic basins: the Arctic, the Pacific, and the Central Asian areas of interior drainage. In the N central part of the country the SELENGA River and its tributaries such as the ORHON GOL, drain northward into Lake BAYKAL and from there into the ARCTIC OCEAN. In the NE the ONON RIVER and KERULEN drain eastward to the AMUR River and then to the PACIFIC OCEAN. But a much larger area in the W, SW, and SE parts of the country lies in basins of interior drainage, where more widely spaced rivers, often intermittent in their lower courses, drain into salt lakes, marshes, or playas (or simply disappear into desert gravels or sandy expanses). But even here there are some long rivers, notably the DZAVHAN GOL, 502 mi/808 km long, in the W. In its basin lie the HOVD LAKES, the freshwater HAR US NUUR and HAR NUUR, and the salt lake HYARGAS NUUR. The two other notable lakes are the freshwater, very deep tectonic HÖVSGÖL NUUR, 780 ft/238 m deep, amid mountains in the extreme N part of the country, and the highly mineralized UVS NUUR in the W, the largest lake in the country, with a variable area of 1,290 sq mi/3,350 sq km and a fluctuating depth. Only the Selenga River is navigable and used for shipping in the N central part of the country. The rivers and lakes freeze for many months during the year.

Climate

The cool, dry, markedly continental climate is characterized by very cold and long winters, warm but short summers, low and irregular precipitation, high annual and diurnal range of temperature, and an exceptional number of clear, sunny days. The frost-free season varies from its absence in the mountains to c.150 days in the Gobi desert in the S. In winter, the ground freezes to a great depth everywhere. The higher elevations in the N part of the country are covered by permafrost or permanently frozen ground. The amount of precipitation is generally low but there is a very sharp gradient with drops in rainfall from N to S and from higher elevations to lower. The N central part of Mongolia has precipitation of about 16 in/40 cm, which declines dramatically to virtually nonexistent in the Gobi desert of the S and more gradually to about 8 in/20 cm in the E part of the country. In the mountainous areas of the W, the amount of precipitation depends on elevations, attaining a maximum of about 16 in/40 cm in parts of the Altay mountain and falling to 4 in/10 cm or less in the intermontane basins. On average, there are more than 250 clear days each year and only 15–25 days with rain. The highly variable rainfall occurs mainly during summer thunderstorms. Ulaanbaatar has a mean January temperature of −17°F/−27°C, a mean July temperature of 64°F/ 18°C, and a mean annual rainfall of 10 in/25 cm. Because of the dry climate, the predominant vegetation is grassland, or steppe. In the N parts, forests of larch, cedar, fir, and pine occupy higher elevations, especially on cooler N–facing slopes. Most of the lower–lying S and E parts of the country are semidesert and desert.

Population

The population is more than half urban (57% in 1989) and very low in density; half the urban population is concentrated in Ulaanbaatar. Main urban centers are Ulaanbaatar, Darhan (site of heavy industry), and Erdenet (a mining center); each hold the status equivalent to that of an *aymag* (province). The eighteen towns that are centers of *aymags* represent regional centers; they have population between 10,000 and 30,000, but CHOYBALSAN, the regional center of E Mongolia, is somewhat larger. The small towns or villages that serve as centers of the 333 local administrative units called *sums* are much smaller, generally with populations between 1,000 and 2,000. The main international transport line is the Trans–Mongolian Railroads, running through Mongolia into Russia in the N (1950) and China in the S (1955). Ulaanbaatar also has an international airport. A network of about 50,000 mi/80,000 km of roads serves all major parts of the country, but because only about 10% of the roads are surfaced, most places lie on unimproved dirt roads. The population is predominantly Khalka Mongol (79%). The largest minority, Turkic Kazakhs (6%) is located in the extreme W part of the country near Kazakhstan, to which some emigration took place in the 1990s. Twenty Mongol groups, other than the Khalka, constitute most of the remaining 15% of the popultion. Although the density of population is low, the rate of natural increase is very high. Khalka Mongolian, the official language, was written in the old Uigur Turkic script until 1946, when it was replaced by the Cyrillic alphabet. After 1992, both the old Mongol writing and the Cyrillic alphabets were taught in the schools. The dominant religion is Lamaist Buddhism. The country's major institution of higher learning is the Mongolian State University at Ulaanbaatar. The Academy of Sciences of Mongolia, also in UlaanBaatar has research institutes.

Economy

The main economic activity of the country is livestock herding, utilizing the predominant grassland and the sparse vegetation of desert margins; sheep, horses, and cattle are the most important. They are raised in the more moist N central part of the country. Goats and camels are raised in the deserts of the S and SW. Some yaks are raised in the cold mountains of the W. The grassland vegetation and lightness of the snow cover permits year-round grazing. Because of the variability of the rainfall and thus of the forage, the numbers fluctuate. The very limited area of cropland is devoted mainly to forage crops, wheat, barley, and oats. Vegetables and potatoes are also grown for local consumption. Because only a small fraction of the land is farmed, many wild animals, more than a hundred species of mammals, have survived. Hunting is practiced both in the forested mountains and on the steppes. Sable, squirrel, fox, wolf, lynx, snow leopard, and marmots are hunted or trapped for sport or furs. The forests of the mountains in N central Mongolia are the source of timber. Mineral resources have been developed particularly with aid from the former Soviet Union. The largest mining enterprise is at ERDENET, 150 mi/240 km NW of Ulaanbaatar, for the mining and dressing of copper and molybdenum ore for export to Russia. The mining and processing of fluorspar is centered at Bor Under, 175 mi/280 km SE of Ulaanbaatar. Coal is mined from several deposits for use within Mongolia; the largest is SHARIN GOL, 100 mi/160 km N of Ulaanbaatar. All are connected by branch railroad lines to the Trans-Mongolian Railroad. Industry is based mainly on livestock resources: wool and woolen textiles, blankets, hides, leather and leather goods, clothing, meat and dairy goods. A wide range of consumers' goods is produced for domestic consumption. The main industrial center is Ulaanbaatar. The new industrial city of DARHAN specializes in heavy industry with cement and steel, utilizing nearby coal deposits. The main exports are based on livestock (wool, cashmere, and hides) or mines (copper, molybdenum, and fluorspar).

History: to 1945

For the early history of the region, see the separate article on Mongolia. In 1911, a group of Mongol princes ousted the Manchu governor from China and proclaimed an autonomous Mongolia with the Living Buddha of Urga (now Ulaanbaatar) as ruler. Red Army troops from the Russian and Mongolian units under Mongolian Communist leaders Sühbaatar and Khorloin Choybalsan proclaimed an independent state in 1921, which remained a monarchy until the Living Buddha died in 1924. The establishment in November 1924 of the Mongolian People's Republic was followed by campaigns against groups owning land and livestock and by persecution of the Lama priests. In the Lama Rebellion of 1932, priests led thousands of people with seven million hed of livestock across the border to Inner Mongolia. A new constitution, adopted in 1940, consolidated the power of the Communist regime.

History: 1945 to Present

In 1945, a plebiscite was held under a Sino–Soviet agreement, and the republic voted overwhelmingly for continued independence from China. The border between Mongolia and China was fixed by treaties of 1962 and 1988. With the waning power of the Soviet Union, a series of demonstrations in the late 1980s called for freedom and human rights. In 1990 a reformist Communist government, headed by Punsalmaagiyn Ochirbat, came to power. Increased political, economic, and religious freedom resulted, and many abandoned monasteries were reoccupied by Lamaist monks. In 1991 privatization of state enterprises was begun. Mongolia adopted a new democratic constitution in 1992, and in 1993 Ochirbat, running as a non-Communist, won Mongolia's first free presidential election. The Mongolian People's Revolutionary party (MPRP; former Communist) majority in parliament, however, continued until the 1996 elections. Following a downturn in the economy, Natsagiyn Bagabandi, the MPRP candidate, won a decisive victory against Ochirbat in 1997, and the MPRP won most of the

parliamentary seats in 2000. Bagabandi was reelected in 2001. Parliamentary electoral losses forced the MPRP into a unity government with the opposition in 2004. The MPRP retained control of the presidency in 2005 when Nambaryn Enkhbayar was elected to the post. In 2006 the MPRP withdrew from the coalition and formed a new government without the opposition.

Government

Mongolia is governed under the constitution of 1992. The president, who is head of state, is popularly elected for a four-year term and is eligible for a second term. The government is headed by the prime minister. The unicameral legislature consists of the 76-seat State Great Hural, whose members are popularly elected for four-year terms. Following legislative elections, the leader of the majority party or majority coalition is usually elected prime minister by the legislature. Administratively, the country is divided into twenty-one provinces and the capital district. The current head of state is President Nambaryn Enkhbayar (since June 2005). The current head of government is Prime Minister Miegombyn Enkhbold (since January 2006).

Mongolia, ASIAN region (□ 900,000 sq mi/2,340,000 sq km), bordered roughly by XINJIANG UYGUR AUTONOMOUS REGION, CHINA, on the W; the Manchurian provinces of China on the E; SIBERIA on the N; and the GREAT WALL of China on the S; 38′–52′N 88′–126′E. It now comprises the independent republic of MONGOLIA (OUTER MONGOLIA) on the N and the INNER MONGOLIA AUTONOMOUS REGION of China on the S and E. MONGOLIA is chiefly a region of desert and of steppe plateau from c.3,000 ft/900 m–5,000 ft/1,500 m high. Winters are cold and dry, and summers are warm and brief. The GOBI DESERT, which is entirely wasteland, is in the central section. To the W are the ALTAI mountains, which rise to 15,266 ft/4,653 m (14,351 ft/4,374 m inside Mongolia in the MONGOLIAN ALTAY). Rivers include a section of the HUANG HE (Yellow River) in the S and the SELENGA, ORHON, and KERULEN in the N. Rainfall averages much less than 15 in/40 cm a year, but irrigation has made some cultivation possible along streams or on the fringes; wheat and oats are the chief crops. Mongolia has traditionally been a land of pastoral nomadism; livestock raising and the processing of animal products are the main industries. Wool, hides, meat, cloth, and leather goods are exported. Coal, iron ore, copper, molybdenum, fluorspan, gold, and oil are mineral resources. Mongolia is crossed N-S by a railroad linking BEIJING with RUSSIA. The region has an adequate system of roadways, although most roads are unpaved. Camels and yaks are used often in desert and mountain areas, respectively. Great hordes of horsemen have repeatedly swept S from Mongolia into N China, establishing vast, although generally short-lived, empires. In the 1st century C.E., Mongolia was inhabited by various Turkic tribes who dwelt mainly along the upper course of the Orhon River. It was also the home of the Hsiung-nu (the Huns) who ravaged (1st–5th century) N China. The Uigur Turks founded their first empire (744–856) with its capital near KARAKORUM. The Khitan, who founded the Liao dynasty (947–1125) in N China, were from Mongolia. Many smaller territorial states followed until (c.1205) Jenghiz Khan conquered all Mongolia, united its tribes, and from his capital at Karakorum led the Mongols in creating one of the greatest empires of all time. His successors established the Golden Horde in SE Russia and founded the Hulagid dynasty of PERSIA and the Yuan dynasty (1260–1368) of China. After the decline of the Mongol empire, Mongolia intruded less in world affairs. China, which earlier had gained control of Inner Mongolia, subjugated Outer Mongolia in the late 17th century but in the succeeding years struggled with Russia for control. Outer Mongolia finally broke away in 1921 to form the

Mongolian People's Republic, later the independent state, of Mongolia. Inner Mongolia remained under Chinese control, although the JAPANESE conquered REHE (1933), which they included in MANCHUKUO, and Chahar and SUIYUAN (1937), which they formed the Mengjiang (Mongol Border Land). These areas were returned to China after World War II. In 1949 the Chinese Communists joined most of Inner Mongolia to N Rehe province and W HEILONGJIANG province to form the Inner Mongolian Autonomous Region. In 1945 Tannu Tuva, long recognized as part of Mongolia, was incorporated into the former Soviet Union and later became the TUVA REPUBLIC within the Russian Federation. On July 7, 1996, after 76 years of rule, the Communist regime finally toppled.

Mongolian Altay or **Mongolian Altai** (both: al-TEI), Mongolian *Mongol Altayn Nuruu*, in W MONGOLIA, form an E extension of the ALTAI mountain range of RUSSIA stretching as a great elongated series of mountain ranges from the point where the borders of Mongolia, Russia, CHINA, and KAZAKHSTAN converge at the W tip of Mongolia. The mountains run SE for about 1,000 mi/1,600 km; about 400 mi/650 m as the Mongolian Altay proper and another 600 mi/1,000 km as the GOBI ALTAY (Govd Altayn Nuruu); the NW part lies along the Mongolia-China (XINJIANG UYGUR AUTONOMOUS REGION) border but the SE part is largely within Mongolia. The Mongolian Altay join the SAILYUGEM mountain range (Siylügemiyn Nuruu), which extends NE along the boundary between Russia (ALTAI Republic) and Mongolia. The S Altai extends W into the Altai Republic and along the Russian border with Kazakhstan. The Mongolian Altay proper lie between 49°–45′N and 88°–95′E and consist of parallel ridges separated by long valleys. The mountains are made up of high plateaus surmounted by peaks. The highest point lies in the TAVAN BOGD UUL mountain knot, where the peak Nayramadlin Orgil (Kuyten-Uul) reaches 14,350 ft/4,374 m; other high peaks include Türgen Uul (13,009 ft/3,965 m) and HARHIRA UUL to the NE, TSAST UUL (13,757 ft/4,193 m) and TSAMBAGARAV UUL (13,665 ft/4,165 m) to the E, and Mönh Hayrhan Uul (Munkhe Khayrkhan mountain) (13,793 ft/4,204 m) and SUTAY UUL (13,430 ft/4,090 m) to the SE. The peaks are spread over an area 300 mi/500 km long NW-SE and 150 mi/240 km wide SW-NE. Their high latitude and high elevation result in low temperatures that support the formation of 36 glaciers covering an area of 62 sq mi/160 sq km. The Potanini Mösön Gol is the largest, 12 mi/19 km long, up to 2 mi/5 km wide, covering 20 sq mi/50 sq km. The modest amount of precipitation, however, acts as a limiting factor on the development of glaciers. Vegetation varies by elevation, exposure, and amount of moisture. The peaks are mostly bare rocks. High plateaus have alpine meadows. The more moist slopes have forests of fir and larch in the middle reaches and steppes in the lower reaches. The drier slopes are mainly steppes and semideserts. The intermontane basins and valleys are also semideserts.

Mongolküre, CHINA: see ZHAOSU.

Mongomo (mahn-GO-mo), town, ⊙ WELE-NZAS province, E EQUATORIAL GUINEA, on GABON border; 01°39′N 11°20′E.

Mongono (mahn-GOO-noo), town, BORNO state, extreme NE NIGERIA, near Lake CHAD, 65 mi/105 km NNW of MAIDUGURI. Road center; cattle raising; gum arabic, cassava, millet, durra. Also Mongonu.

Mongoumba (mong-goom-BAH), village, LOBAYE prefecture, SW CENTRAL AFRICAN REPUBLIC, on UBANGI RIVER, at mouth of LOBAYE RIVER, opposite LIBENGE (CONGO), and 55 mi/89 km S of BANGUI; 03°40′N 18°35′E. Agriculture (coffee); manufacturing (sawmilling); exports (hardwoods, coffee, palm products). Customs station and small river port.

Mongpai (MONG-PEI), formerly southernmost state (sawbwaship) (□ 730 sq mi/1,898 sq km) of SHAN

STATE, MYANMAR, now part of Pekon township. Borders both KAYIN STATE and KAYAH STATE; ⊙ Mongpai, a village on the NAM PILU and 10 mi/16 km NW of LOIKAW.

Mongpan, township (□ 2,988 sq mi/7,768.8 sq km), SHAN STATE, MYANMAR, 90 mi/145 km ESE of TAUNGGYI, bordering the W bank of the THANLWIN RIVER. Former SE state (sawbwaship); ⊙ was Mongpan.

Mongpawn, village (□ 502 sq mi/1,305.2 sq km), SHAN STATE, MYANMAR, LOILEM township, on Thazi-Kengtung road, and 25 mi/40 km E of TAUNGGYI, on the NAM PAWN. A former central state (sawbwaship).

Mongu (MO-eng-goo), township, ⊙ WESTERN province, W ZAMBIA, near ZAMBEZI RIVER, 250 mi/402 km NW of MARAMBA; 15°13′S 23°08′E. Cattle; timber; agriculture (corn, cashews). Manufacturing (cashew processing). Airfield. Linwa Plain National Park to NW. Village of Lealui, native capital of Barotseland, is 8 mi/12.9 km W, on the Zambezi River.

Mongua (mon-GWAH), town, ⊙ Mongua municipio, BOYACÁ department, central COLOMBIA, in the Cordillera ORIENTAL, 40 mi/64 km NE of TUNJA; 05°45′N 72°48′W. Elevation 9,967 ft/3,037 m. Coffee, corn, sugarcane; livestock.

Monguel (mawn-GEL), village (2000 population 4,895), Gorgol administrative region, S MAURITANIA, 217 mi/349 km NE of SAINT-LOUIS (SENEGAL), and 31 mi/50 km NE of KAÉDI; 16°24′N 13°11′W. Livestock, millet.

Monguelfo (mon-GWEL-fo), German *Welsberg*, village, BOLZANO province, TRENTINO-ALTO ADIGE, N ITALY, on RIENZA River, and 8 mi/13 km ESE of BRUNICO; 46°45′N 12°06′E. Summer resort (elevation 3,365 ft/1,026 m). Has 12th–century castle.

Monguí (mon-GEE), town, ⊙ Monguí municipio, BOYACÁ department, central COLOMBIA, in the ANDES, 40 mi/64 km NE of TUNJA 05°43′N 72°49′W. Elevation 10,662 ft/3,249 m. Noted monastic religious center.

Mongyai (MONG-yei), village and township, SHAN STATE, MYANMAR, on Lashio-Loilem road, and 40 mi/64 km SSE of LASHIO, near head of the NAM PANG. Road junction. Formerly known as South Hsenwi; was capital of SOUTH HSENWI STATE.

Mongyang (MONG-yang), village and township, SHAN STATE, MYANMAR, 40 mi/64 km N of KENGTUNG, on route to CHINA (YUNNAN province). Borders CHINA.

Monhegan (mahn-HEE-guhn), island (□ 2 sq mi/5.2 sq km), LINCOLN county, c.10 mi/16 km off the coast of S MAINE. It is a summer resort favored by artists for its scenery. In the War of 1812 the USS *Enterprise* defeated the HMS *Boxer* SE of the island. Settled c.1622.

Monheim (MAWN-heim), town, SWABIA, W BAVARIA, GERMANY, 9 mi/14.5 km NNE of DONAUWÖRTH; 48°51′N 10°51′E. Brewing, lumber. Chartered before 1350.

Monheim am Rhein (MAWN-heim ahm REIN), town, Rhineland, North Rhine-Westphalia, on right bank of the Rhine, 11 mi/18 km NNW of Cologne; 51°06′N 06°54′E. Chemical and pharmaceutical industry; paper manufacturing; brewery.Town 1st mentioned 1150, chartered 1960, incorporated Baumberg (1951) and Hitdorf (1961). Has ruins of 15th-cent. fortifications.

Mönh Hayrhan Uul, MONGOLIA: see Mongolian ALTAI.

Monh Saridag, MONGOLIA: see MUNKU-SARDYK.

Moniaive (MAH-nee-aiv), agricultural village, DUMFRIES AND GALLOWAY, S Scotland, in the Cairn valley, and 15 mi/24 km NW of DUMFRIES; 55°11′N 03°55′W. Nearby is MAXWELTON HOUSE, birthplace of Annie Laurie.

Monianga (mon-YAHNG-gah), village, Équateur province, NW CONGO, on left bank of GIRI RIVER, and 175 mi/282 km W of LISALA; 02°05′N 19°10′E. Terminus of navigation. Also spelled MANIANGA.

Mönichkirchen (mun-ikh-KIR-khuhn), township, SE LOWER AUSTRIA, E AUSTRIA, on E slope of the WECHSEL mountain, and on Styrian border, 14 mi/23 km S of NEUNKIRCHEN; 47°31′N, 16°02′E. Elevation 2,911 ft/887 m. Biseasonal tourism.

Monida (muh-NEI-duh), village, BEAVERHEAD county, extreme SW MONTANA, 50 mi/80 km SSE of DILLON, at SE extremity of BITTERROOT RANGE, on IDAHO state line, just N of Monida Pass (6,870 ft/2,094 m), crosses the CONTINENTAL DIVIDE into Idaho. Targhee National Forest (Idaho) to S; Beaverhead National Forest to SW; LIMA RESERVOIR to N (RED ROCK RIVER); Red Rock Lakes National Wildlife Refuge to E; important nesting site of Trumpeter swan, Sandhill crane, Barrow's Goldeneye, and Great Blue heron.

Monikie (MAH-ni-kee), agricultural village, ANGUS, NE Scotland, 9 mi/14.5 km NE of DUNDEE; 56°32′N 02°49′W. Site of reservoirs supplying Dundee. Nearby are ruins of ancient Affleck Castle, a well-preserved 15th-century towerhouse. Formerly in Tayside, abolished 1996.

Monimbó Urban District (mo-neem-BO), area of MASAYA, NICARAGUA, S of city center. Noted for production of Indian crafts.

Monimpébougou (mon-EEM-pai-boo-goo), village, FOURTH REGION/SÉGOU, MALI, 68 mi/109 km NE of SÉGOU; 14°09′N 05°31′W.

Monino (MO-nee-nuh), town (2006 population 19,805), E central MOSCOW oblast, central European Russia, on railroad, 30 mi/49 km E of MOSCOW, 9 mi/14 km W of NOGINSK, and 7 mi/11 km SE of SHCHËLKOVO, to which it is administratively subordinate; 55°50′N 38°11′E. Elevation 508 ft/154 m. Railroad terminus; sanatorium, rest home.

Moniquirá (mo-nee-kee-RAH), town, ⊙ Moniquirá municipio, BOYACÁ department, central COLOMBIA, in Cordillera ORIENTAL, 29 mi/47 km NW of TUNJA; 05°38′N 73°33′W. Elevation 5,767 ft/1,758 m. Agriculture includes coffee, sugarcane, corn, potatoes.

Monistrol or **Monistrol de Montserrat** (mo-nees-TROL dhai mon-se-RAHT), town, BARCELONA province, NE SPAIN, on LLOBREGAT RIVER, and 7 mi/11.3 km S of MANRESA; 41°37′N 01°51′E. Cotton milling, olive-oil processing. Starting point for ascension of the Montserrat.

Monistrol-sur-Loire (mo-nee-strol–syur–lwahr), town (□ 18 sq mi/46.8 sq km), HAUTE-LOIRE department, AUVERGNE region, S central FRANCE, near gorges of upper LOIRE, 14 mi/23 km SW of SAINT-ÉTIENNE; 45°17′N 04°10′E. Metalworks (bicycle parts). Paper mill nearby. Has 15th–17th-century castle of the bishops of Le Puy.

Moniteau (mah-ni-TAW), county (□ 418 sq mi/1,086.8 sq km; 2006 population 15,092), central MISSOURI; ⊙ CALIFORNIA; 38°37′N 92°34′W. Bounded NE by MISSOURI River. Agriculture (wheat, corn, soybeans); cattle, poultry, limestone; manufacturing at California and TIPTON. Formed 1845.

Monitor Range, central NEVADA, largely in NYE county, extends N into LANDER county, E of Toquema Range. Lies in Toiyabe National Forest. ANTELOPE PEAK (10,220 ft/3,115 m) and SUMMIT MOUNTAIN. (10,461 ft/3,189 m) are in N.

Monjas (MON-hahs), town, JALAPA department, E central GUATEMALA, in highlands, 12 mi/19 km SE of JALAPA; 14°30′N 89°52′W. Elevation 3,153 ft/961 m. Corn, wheat, beans; livestock.

Monjas (MON-hahs), town, in S central OAXACA, MEXICO, 50 mi/80 km S of OAXACA DE JUÁREZ, on Mexico Highway 175. Elevation 4,921 ft/1,500 m. A mountainous region with a temperate climate. Agriculture on the Atoyac River.

Monkey Bay, village, MANGOCHE district, Southern region, MALAWI, port on Monkey Bay, inlet of Lake Malawi (LAKE NYASA), 33 mi/53 km NW of Mangoche town; 14°05′S 34°55′E. Tourist resort (hotel); roadship transfer point; airport.

Monkey Bay Wildlife Sanctuary, BELIZE district, BELIZE, 31 mi/50 km W of BELIZE CITY. Privately owned sanctuary and government reserve. Dedicated to preserving and reintroducing native wildlife by restoring their habitat.

Monkey Mia (MUHN-kee MEI-uh), hamlet, WESTERN AUSTRALIA state, W AUSTRALIA, 16 mi/26 km N of DENHAM, in SHARK BAY area; 25°48′S 113°43′E. Wild dolphins.

Monkey Point, W TRINIDAD, TRINIDAD AND TOBAGO, on the Gulf of Paria, just W of village of CALIFORNIA, 18 mi/29 km S of PORT OF SPAIN; 10°25′N 61°30′W.

Monkey Point, NICARAGUA: see MONO POINT.

Monkeys Eyebrow, unincorporated village, BALLARD county, W KENTUCKY, 20 mi/32 km WNW of PADUCAH, near OHIO RIVER. Lock and Dam Number 53 to W, on Ohio River. Ballard County Wildlife Management Area to SW.

Monkland (MUHNK-luhnd), unincorporated village, SE ONTARIO, E central CANADA, 14 mi/22 km from CORNWALL, and included in NORTH STORMONT township; 45°11′N 74°52′W.

Monkoto (mawng-KO-to), village, Équateur province, NW CONGO, on Luilaba River, and 100 mi/161 km S of BOENDE; 01°38′S 20°39′E. Elev. 1,332 ft/405 m. Agricultural and trading center with rubber and coffee plantations; also palm products. SALONGA NATIONAL PARK nearby.

Monkstown, IRELAND: see PASSAGE WEST.

Monkton, town, ADDISON CO., W VERMONT, 17 mi/27 km S of BURLINGTON; 44°13′N 73°07′W. Named for General Robert Monckton (1726–1782).

Monkton, Scotland: see PRESTWICK.

Monkwearmouth, ENGLAND: see SUNDERLAND.

Monmouth (MUHN-muth), county (□ 665 sq mi/1,729 sq km; 2006 population 635,285), E NEW JERSEY, bounded E by the ATLANTIC OCEAN, N by RARITAN and SANDY HOOK bays; ⊙ FREEHOLD; 40°17′N 74°09′W. Many coastal resorts, including ASBURY PARK, MANASQUAN, and LONG BRANCH. Agricultural area; some manufacturing. Drained by METEDECONK, MANASQUAN, and SHARK rivers; NAVESINK RIVER and SHREWSBURY RIVER estuaries, NAVESINK HIGHLANDS, and SANDY HOOK are in NE. Formed 1675, Largely residential and seasonal resort communities along the coast.

Monmouth, city (2000 population 9,841), ⊙ WARREN county, W ILLINOIS; 40°54′N 90°38′W. Located in a farm area, it is a trade center with a packing plant. Manufacturing (pottery, farm tools, feed). Monmouth College is in the city. Wyatt Earp was born here. Incorporated 1852.

Monmouth (MAHN-muth), former township (□ 87 sq mi/226.2 sq km; 2001 population 944), S ONTARIO, E central CANADA, 23 mi/36 km from MINDEN; 44°57′N 78°15′W. Amalgamated into HIGHLANDS EAST township in 2001.

Monmouth (MAHN-muth), town (2000 population 180), JACKSON county, E IOWA, 11 mi/18 km W of MAQUOKETA; 42°04′N 90°52′W. Livestock, grain.

Monmouth, town, KENNEBEC county, S MAINE, 15 mi/24 km SW of AUGUSTA, near LAKE COBBOSSEECONTEE; 44°14′N 70°00′W. Orchard center, with University of Maine agricultural experiment station. Seat of Monmouth Academy. Settled 1775, incorporated 1792.

Monmouth, town (2006 population 9,476), POLK county, NW OREGON, 10 mi/18 km SW of SALEM near WILLAMETTE river; 44°51′N 123°13′W. Agriculture (fruit, hops, corn, brans; poultry, hogs, sheep, cattle); dairy products; wineries. Site of Paul Jensen Arctic Museum and Western Oregon State College Baskett Slough National Wildlife Refuge to N; Helmick State Park to S; McDonald State Forest to S. Plotted in 1855.

Monmouth (MUHN-muth), town (2001 population 8,877), ⊙ MONMOUTHSHIRE, SE Wales, at the junction of the MONNOW and WYE rivers; 51°49′N 02°43′W. Popular tourist center with cattle and produce markets. Food processing, paper manufacturing. On site of Roman settlement; medieval castle and church. Formerly in GWENT, abolished 1996.

Cross-references are shown in SMALL CAPITALS. The pronunciation guide is shown on page xix. The sources of population figures are shown on page xvii.

Monmouth Beach (MUHN-muhth), resort borough (2006 population 3,574), MONMOUTH county, E NEW JERSEY, between the coast and SHREWSBURY RIVER inlet, 2 mi/3.2 km N of LONG BRANCH, and 16 mi/26 km ENE of FREEHOLD; 40°20′N 73°59′W.

Monmouth, Fort, NEW JERSEY: see RED BANK.

Monmouth, Mount (MAHN-muhth) (10,470 ft/3,191 m), SW BRITISH COLUMBIA, W CANADA, in COAST MOUNTAINS, 120 mi/193 km NNW of VANCOUVER; 50°59′N 123°47′W.

Monmouthshire, county (□ 329 sq mi/855.4 sq km; 2001 population 84,885), SE Wales, borders England to the E; SEVERN RIVER to the S; counties of Newport, TORFAEN, Blaenau Gwent to the W, and POWYS to the N; ⊙ CWMBRAN; 51°50′N 02°45′W. Includes part of BRECON BEACONS NATIONAL PARK, USK, and ABERGAVENNY. Conquered by the Romans c.75 C.E. who built a fortress at Caerleon and town at Caerwent. Conquered by Normans who built Chepstow Castle c.1066. Henry VIII took control of the area for England in 1536. Many ruins, including the Cistercian abbey at Tintern made famous in the Wordsworth poem *Tintern Abbey*. Agriculture, dairying, printing, fishing on the USK RIVER; commuters to NEWPORT. The Severn cable-stayed bridge is the longest bridge of its kind Great Britain (3 mi/5 km).

Monnaie (mo-nai), commune (□ 15 sq mi/39 sq km), INDRE-ET-LOIRE department, CENTRE administrative region, W central FRANCE, 8 mi/12.9 km NNE of TOURS, on highway to ORLÉANS; 47°30′N 00°47′E. Vegetables for PARIS markets.

Monnerie-le-Montel, La (mon-ne-ree–luh–mon-tel, lah), commune (□ 1 sq mi/2.6 sq km), PUY-DE-DÔME department, AUVERGNE region, central FRANCE, 3 mi/4.8 km NE of THIERS. Handicraft cutlery manufacturing.

Monnetier-Mornex (mon-uh-tyai–mor-nai), township (□ 4 sq mi/10.4 sq km), HAUTE-SAVOIE department, RHÔNE-ALPES region, SE FRANCE, on N slope of Mont SALÈVE, near Swiss border, 5 mi/8 km SE of GENEVA; 46°10′N 06°12′E. Monnetier (2,300 ft/701 m.) is a small resort.

Monnikendam (MAW-nee-kuhn-dahm), town, NORTH HOLLAND province, W NETHERLANDS, on the Grous Meer (manmade bay of Markemeer, S section of IJSSELMEER), and 8 mi/12.9 km NE of AMSTERDAM; 52°27′N 05°02′E. Fishing; dairying; cattle, sheep, poultry; vegetables, flowers, fruit; manufacturing (yachts, food processing).

Monnow River (MUHN-o), Welsh *Mynwÿ*, on WALES-ENGLAND border; rises in BLACK MOUNTAINS; flows about 25 mi/40 km SE through Herefordshire, then into the WYE RIVER at MONMOUTH, GWENT.

Mono, department (□ 539 sq mi/1,401.4 sq km; 2002 population 360,037), SW corner of BENIN; ⊙ LOKOSSA; 06°35′N 01°50′E. Bordered on N by COUFFO department, E by ATLANTIQUE department, S by BIGHT OF BENIN, W by TOGO. Includes the towns of ATHIÉMÉ, COMÉ, GRAND-POPO, and Lokossa and the village of AGOUÉ. The MONO RIVER flows N to S in the W of the department. In 1999 the N portion of Mono was separated and established as Couffo department.

Mono, county (□ 3,046 sq mi/7,919.6 sq km; 2006 population 12,754), E CALIFORNIA; ⊙ BRIDGEPORT; 37°55′N 118°52′W. Rugged SIERRA NEVADA country; crest of range (with peaks over 13,000 ft/3,962 m) forms W and is crossed by scenic TIOGA PASS (E entrance to YOSEMITE NATIONAL PARK). Bounded E by NEVADA state line; drained by Owens, East Walker, and West Walker rivers; hydroelectric plants. Sweetwater Mountains in NE, White Mountains in SE. Much of county is in Toiyabe (N) and Inyo (S and SE) national forests. Bodie State Historical Park in center. Recreational region (hiking, camping, hunting, fishing, winter sports); includes saline MONO LAKE, in center; Mammoth Lakes and June Lake in SW; Mammoth Mountain Ski Area, small lakes (fishing), part of JOHN MUIR TRAIL, and

scenic wilderness preserves on W boundary. Stock raising (cattle, sheep), mining (pumice, gold, lead, silver, andalusite). Timber stands (chiefly pine) remain largely intact. Formed 1861.

Mono, town (□ 107 sq mi/278.2 sq km; 2001 population 6,922), S ONTARIO, E central CANADA; 44°01′N 80°03′W. Four rivers flow through the town: the CREDIT, Humber, GRAND, and NOTTAWASAGA.

Mono (mo-NO), town, Mono county, MIYAGI prefecture, N HONSHU, NE JAPAN, 28 mi/45 km N of SENDAI; 38°33′N 141°14′E. N limit of Japanese tea cultivation.

Mono, SOLOMON ISLANDS: see TREASURY ISLANDS.

Monobe (mo-NO-BE), village, Kami county, KOCHI prefecture, S SHIKOKU, W JAPAN, 25 mi/40 km N of KOCHI; 33°41′N 133°52′E. Citrons.

Monocacy (mah-no-kah-see), river, c.60 mi/97 km long; formed at PENNSYLVANIA-MARYLAND state line 8 mi/12.9 km S of GETTYSBURG by joining of Marsh and Rock Creeks (39°43′N 77°13′W); flows S across Maryland, passes E of FREDERICK, to join the POTOMAC RIVER 13 mi/21 km S of Frederick. On its banks, just E of Frederick, the Civil War battle of Monocacy was fought on July 9, 1864. Although the Union forces under General Lew Wallace were defeated, they delayed the Confederate forces under Gen. J. A. Early long enough to give General Ulysses Grant time to dispatch troops to defend WASHINGTON, D.C. and drive Early back into VIRGINIA.

Monocacy National Battlefield (c.2 sq mi/5.2 sq km), FREDERICK county, W MARYLAND, 3 mi/4.8 km SSE of FREDERICK. Authorized 1976. On July 7, 1864, General Lew Wallace, who later wrote *Ben Hur*, took a position at the railroad junction on the Moncacy River with 2,650 Union soldiers facing 17,500 Confederate troops under General Jubal Early. On July 8, Wallace was reinforced by an an additional 3,500 making the odds three, rather than seven, to one. When the two forces clashed the next day, the Union forces were decisively defeated, but the Confederates lost their best opportunity to seize the capital. Both sides lost more than 1,300 men each.

Mono Craters, group of about twenty geologically recent volcanic cones (maximum elevation c.9,000 ft/2,743 m), MONO county, E CALIFORNIA, just S of MONO LAKE.

Monok (MO-nok), village, BORSOD-ABAÚJ-ZEMPLÉN county, NE HUNGARY, 17 mi/27 km ENE of MISKOLC; 48°13′N 21°09′E. Apples, alfalfa, sugar beets, grain. Louis Kossuth was born here.

Monoklissia (mo-no-klee-SYAH), village, SÉRRAI prefecture, CENTRAL MACEDONIA department, NE GREECE, 10 mi/16 km SE of SÉRRAI; 41°04′N 23°24′E. Known for *yinekokratia* [Greek=rule of the women], when each January 8, women take over the streets and taverns and men perform domestic chores. This custom was brought here in 1922 by refugees from E Thrace; some claim that its origins are in ancient Dionysian rites.

Mono Lake, saline lake (□ 87 sq mi/226.2 sq km), MONO county, E CALIFORNIA, c.50 mi/80 km NW of BISHOP, just E of SIERRA NEVADA crest. Elevation 6,407 ft/1,953 m. It has no outlet, and its waters, which contain many natural impurities, support brine shrimp. Supports various migratory birdlife, principally California gull population. Has two small volcanic islands; larger is Paoha Island. Fed mainly by small mountain streams of Sierra Nevada.

Monolith, village, KERN county, S central CALIFORNIA, in TEHACHAPI MOUNTAINS, just E of TEHACHAPI.

Monomonac, Lake (muh-NAW-muh-nak), on NEW HAMPSHIRE-MASSACHUSETTS state line, N of RINDGE, New Hampshire, 17 mi/27 km NW of FITCHBURG, Massachusetts; 2.5 mi/4 km long. Drains SW to MILLERS RIVER.

Monomoy Island, MASSACHUSETTS: see CHATHAM.

Monon, town (2000 population 1,733), WHITE county, NW central INDIANA, on small Little Monon Creek,

and 10 mi/16 km NW of MONTICELLO; 40°52′N 86°53′W. Railroad junction. Agricultural area (corn, oats, soybeans); diversified manufacturing; stone quarrying. Laid out 1853, incorporated 1879.

Monona, county (□ 698 sq mi/1,814.8 sq km; 2006 population 9,343), W IOWA, on NEBRASKA state line (W; formed here by MISSOURI RIVER); ⊙ ONAWA; 42°03′N 95°57′W. Prairie agricultural area (corn; hogs, cattle, poultry) drained by Little Sioux, MAPLE, and SOLDIER rivers; BLUE LAKE, Oxbow Lake of Missouri River in W. Bituminous-coal deposits (E, S), sand and gravel pits. Lewis and Clark State Park on N end of Blue Lake, near Missouri River in W; Preparation Canyon State Park in S. Formed 1851; lost territory in 1943 to Burt county, Nebraska.

Monona (muh-NO-nuh), city (2006 population 7,938), DANE county, S WISCONSIN, on S shore of LAKE MONONA, a suburb, 2 mi/3.2 km SE of downtown MADISON; 43°02′N 89°19′W. In dairy region; manufacturing (wood products). YAHARA RIVER flows through city connecting Lake Monona with LAKE WAUBESA in S. Incorporated 1938.

Monona, town (2000 population 1,550), CLAYTON county, NE IOWA, 14 mi/23 km N of ELKADER; 43°02′N 91°23′W. In grain, dairy, and timber area; manufacturing (farm equipment, electrical wiring harnesses). Airfield here. Incorporated 1897.

Monona Lake (muh-NO-nuh), one of the FOUR LAKES, DANE county, S WISCONSIN; downtown MADISON is on NW shore (between Lake Monona and LAKE MENDOTA); roughly triangular, c.4 mi/6.4 km long, c.3 mi/4.8 km wide. City of MONONA is on S shore. City of Madison borders on all other sides. YAHARA RIVER drains into it from Lake Mendota, to NW, and out of it to LAKE WAUBESA, to SE.

Monongah (muh-NAHN-guh), town (2006 population 914), MARION county, N WEST VIRGINIA, on the WEST FORK RIVER, 5 mi/8 km WSW of FAIRMONT; 39°27′N 80°13′W. Bituminous-coal-mining region. Agriculture (corn, apples); livestock; poultry. In 1907, a mine disaster here killed 361 men. Founded c.1768.

Monongahela (mah-nahn-gah-HEE-lah), city (2006 population 4,502), WASHINGTON county, SW PENNSYLVANIA, 17 mi/27 km S of PITTSBURGH, on MONONGAHELA RIVER, at mouth of Pigeon Creek; 40°12′N 79°55′W. Agriculture (grain, soybeans; livestock; dairying); manufacturing (business forms, sheet metal fabrication, concrete, specialty chemicals, scrap metal processing, sledge hammers); bituminous coal, gas. A center of Whisky Rebellion, 1794. Settled 1770, incorporated as borough 1833, as city 1873.

Monongahela (mah-nahn-gah-HEE-lah), river, 128 mi/206 km long, N WEST VIRGINIA; formed at FAIRMONT by the joining of the WEST FORK and TYGART rivers; flows NE past MORGANTOWN and STAR CITY and enters PENNSYLVANIA; receives CHEAT RIVER from SE and turns N, flows past FRIENDSHIP HILL NATIONAL HISTORIC SITE, past FREDERICKTOWN, CALIFORNIA, MONESSEN, MONONGAHELA, and CLAIRTON; receives YOUGHIOGHENY RIVER from SE at MCKEESPORT, turns NW at DUQUESNE, joins ALLEGHENY RIVER (enters from NE) in downtown PITTSBURGH to form OHIO RIVER. The channelized river is navigable for most of its length. Iron, steel, and coal are the chief products moved on the river. The Monongahela River was the first river in the U.S. to be improved for navigation.

Monongalia (muh-nahn-GAIL-yuh), county (□ 366 sq mi/951.6 sq km; 2006 population 84,752), N WEST VIRGINIA; ⊙ MORGANTOWN; 39°37′N 80°02′W. On ALLEGHENY PLATEAU; bounded N by PENNSYLVANIA; drained by MONONGAHELA and CHEAT rivers. Coal mining; gas and oil fields; limestone quarries, sand pits. Manufacturing (lumber, glass, metal products, chemicals) at Morgantown. Agriculture (honey, corn, oats, alfalfa, hay, strawberries, nursery crops); cattle; poultry; sheep. Part of CHESTNUT RIDGE Park and part of Coopers Rock State Forest in NE. Formed 1776.

Mono Pass (c.10,650 ft/3,246 m), TUOLUMNE-MONO county line, E CALIFORNIA, in the SIERRA NEVADA, c.10 mi/16 km SW of MONO LAKE. Foot trail passes through. On E boundary of YOSEMITE NATIONAL PARK.

Mono Point (MO-no), E headland of NICARAGUA, on CARIBBEAN SEA, 30 mi/48 km S of BLUEFIELDS. Also known as Mico Point. Sometimes called Monkey Point, Spanish *Punta del Mono*.

Monopoli (mo-NO-po-lee), town, BARI province, APULIA, S ITALY, port on the ADRIATIC, 26 mi/42 km ESE of BARI; 40°57'N 17°18'E. Industrial and commercial center; textile and flour mills; metal fabrication, nonmetallics; food cannery; manufacturing of pasta, soap. Exports olive oil, wine, cherries. Bishopric. Has 12th-century cathedral, 16th-century castle.

Monor (MO-nor), village (2001 population 20,560), PEST county, N central HUNGARY, 20 mi/32 km SE of BUDAPEST; 47°21'N 19°27'E. Corn, vegetables; hogs; manufacturing (baked goods, machinery for food industry, aluminum, flour milling); granaries.

Monori (mo-NO-ree), town, E Ibaraki county, IBARAKI prefecture, central HONSHU, E central JAPAN, 12 mi/20 km S of MITO; 36°13'N 140°21'E. Chestnuts.

Mono River (MO-no), c.250 mi/402 km long, TOGO and BENIN; rises near BENIN border NE of SOKODÉ; flows S past TCHAMBA, through central TOGO to form part of TOGO-BENIN border before entering Bight of BENIN coast in Gulf of GUINEA, just W of GRAND-POPO in Benin's coastal panhandle. Its mouth is connected with Lake TOGO (coastal lagoon) through a channel. Lower course forms TOGO-BENIN border. A dam at NANGBETO provides hydroelectric power to TOGO and BENIN. Navigable for small vessels near mouth.

Monos Island (MO-nos) (□ 2 sq mi/5.2 sq km), off NW TRINIDAD, TRINIDAD AND TOBAGO, in the DRAGON'S MOUTH, 11 mi/18 km W of PORT OF SPAIN; 10°41'N 61°41'W. Elev. 942 ft/287 m. Bathing and fishing resort. Its central section was leased to U.S. in 1941. None a forest reserve.

Monóvar (mo-NO-vahr), town, ALICANTE province, E SPAIN, in VALENCIA, 20 mi/32 km WNW of ALICANTE; 38°26'N 00°50'W. Wine-production center. Brandy distilling, oil processing; manufacturing of footwear. Mineral springs. Marble and stone quarries. Saltwater lagoon 4 mi/6.4 km NNW. Dominated by hill crowned by old castle. Also spelled Monóver.

Mono Vista, unincorporated town (2000 population 3,072), TUOLUMNE county, E central CALIFORNIA, 6 mi/9.7 km ENE of SONORA, in W foothills of SIERRA NEVADA; 38°01'N 120°16'W. Timber. Stanislaus National Forest to E; Tuolumne Indian Rancheria (Indian reservation) to SE.

Monowi (MI-no-wee), village, BOYD county, N NEBRASKA, 27 mi/43 km ESE of BUTTE, and on PONCA CREEK, near MISSOURI RIVER; 42°49'N 98°19'W.

Monpazier (mon-pah-zyai), commune, DORDOGNE department, AQUITAINE region, SW FRANCE, 23 mi/37 km SE of BERGERAC; 44°41'N 00°54'E. Woodworking. Preserves gates and ramparts of its *bastide* (fortified medieval village) built 1284 by Edward I of England. The central square is lined with arcaded buildings.

Monreal del Campo (mon-rai-AHL dhel KAHM-po), town, TERUEL province, E SPAIN, on the JILOCA RIVER, and 33 mi/53 km NNW of TERUEL; 40°47'N 01°21'W. Agricultural trade center (sugar beets, wine, cereals, sheep); chocolate and brandy manufacturing, sawmilling.

Monreale (mon-rai-AH-lai), town, NW SICILY, ITALY, near PALERMO; 38°05'N 13°17'E. An agricultural market and tourist center, it commands a magnificent view of the fertile Conca d'Oro plain. A famous cathedral, one of the masterpieces of Norman-Sicilian architecture, was begun there (1174) by William II of Sicily. The cathedral has fine copper doors by Bo-

nanno Pisano; its interior is decorated with exceptional Byzantine mosaics. Nearby is a lovely cloister with about 200 twin columns and an Arabian fountain formerly used as a lavabo.

Mon River, Burmese *Mon Chaung*, 150 mi/241 km long, CHIN STATE and MAGWE division, MYANMAR; rises in S CHIN HILLS; flows SE and E, past SIDOKTAYA and PWINBYU, to the AYEYARWADY RIVER 12 mi/19 km N of MINBU. Used for irrigation in lower course; the Mon canals (on left and right banks) taking off at headworks 28 mi/45 km WNW of MINBU.

Monroe (muhn-RO), county (□ 1,034 sq mi/2,688.4 sq km; 2006 population 23,342), SW Alabama; ⊙ Monroeville. Coastal plain, bounded SW by Alabama River, S by the Little River Cotton, peanuts, corn, soybeans; timber; crude-oil and natural-gas production. Formed 1815. Named for James Monroe, then secretary of state under James Madison and later president of the U.S.

Monroe, county (□ 621 sq mi/1,614.6 sq km; 2006 population 9,095), E central ARKANSAS; ⊙ CLARENDON; 34°40'N 91°12'W. Drained by WHITE (forms SW boundary) and CACHE rivers. Agriculture (cotton, rice, wheat, soybeans; hogs); timber. Manufacturing at BRINKLEY and Clarendon. Commercial fishing. Louisiana Purchase State Historical Monument on E boundary; part of large White River National Wildlife Refuge at S end; Dagmar Wildlife Management Area in N. Formed 1829.

Monroe (MUHN-ro), county (□ 3,737 sq mi/9,716.2 sq km; 2006 population 74,737), SW FLORIDA and FLORIDA KEYS, at tip of peninsula; ⊙ MARATHON; 25°07'N 81°09'W. Consists of sparsely populated EVERGLADES area on the Florida peninsula (including CAPE SABLE, WHITEWATER BAY, and part of Everglades National Park) and all of the Florida Keys, enclosing FLORIDA BAY. Has large Seminole Indian Reservation in N. Fishing, dairying, and poultry raising; citrus-fruit growing (especially limes) on Florida Keys. Major tourist industry (Keys). Formed 1823.

Monroe, county (□ 398 sq mi/1,034.8 sq km; 2006 population 24,443), central GEORGIA; ⊙ FORSYTH; 33°01'N 83°55'W. Bounded E by OCMULGEE River Piedmont agriculture (corn, wheat, vegetables, pecans, fruit); cattle, poultry; in timber area. Textile manufacturing at Forsyth. Formed 1821.

Monroe, county (□ 397 sq mi/1,032.2 sq km; 2006 population 31,876), SW ILLINOIS; ⊙ WATERLOO; 38°16'N 90°10'W. Bounded W by MISSISSIPPI RIVER, E by KASKASKIA RIVER. In N, is part of SAINT LOUIS metropolitan area. Agriculture (wheat, hay, barley, sorghum; poultry); dairy products); limestone quarries; caves. Formed 1816. Dominant town is COLUMBIA in N. A large portion of W part of county damaged by floods, 1993. VALMEYER relocated itself to bluff site as result. One of 17 Illinois counties to retain Southern-style commission form of county government.

Monroe, county (□ 411 sq mi/1,068.6 sq km; 2006 population 122,613), S central INDIANA; ⊙ BLOOMINGTON; 39°10'N 86°31'W. Drained by WHITE RIVER, SALT CREEK, and small Beanblossom and Clear creeks. Agriculture (corn; cattle; dairying); limestone quarrying; clay, timber. Manufacturing at Bloomington. Hoosier National Forest in SE corner (including Hardin Ridge Recreation Area and MONROE Reservoir); Paynetown and Fairfax State Recreation Areas (both on Monroe Reservoir) in S. Part of Morgan-Monroe State Forest in NE. Karst topography in S. Formed 1818.

Monroe, county (□ 434 sq mi/1,128.4 sq km; 2006 population 7,725), S IOWA; ⊙ ALBIA; 41°01'N 92°52'W. Prairie agricultural (hogs, cattle, poultry, sheep; corn, oats, hay) and coal-mining area. Unit of Stephens State Forest in NW. Formed 1843.

Monroe, county (□ 332 sq mi/863.2 sq km; 2006 population 11,771), S KENTUCKY; ⊙ TOMPKINSVILLE; 36°42'N 85°43'W. Bounded S by TENNESSEE; drained

by CUMBERLAND and BARREN rivers and Skaggs creek. Hilly agricultural area (corn, wheat, hay, alfalfa, burley tobacco, soybeans; cattle, poultry; dairying; timber); limestone quarries. Includes Old Mulkey Meeting House State Historic Site, SW of Tompkinsville, in S center of county. Formed 1820.

Monroe, county (□ 680 sq mi/1,768 sq km; 2006 population 155,035), SE corner of MICHIGAN; 41°55'N 83°30'W; ⊙ MONROE. Bounded S by OHIO state line, E by LAKE ERIE, NE by HURON RIVER; drained by RIVER RAISIN. Cattle, hogs, sheep, poultry, dairying; agriculture (corn, wheat, soybeans, sugar beets, apples); nurseries. Manufacturing at Monroe and DUNDEE. Limestone quarrying. Formed 1817. Fermi 2 nuclear-power plant, initial criticality June 21, 1985; is 25 mi/40 km NE of TOLEDO, OHIO, uses cooling water from LAKE ERIE, and has a maximum dependable capacity of 1060 MWe.

Monroe, county (□ 772 sq mi/2,007.2 sq km; 2006 population 37,572), E MISSISSIPPI; ⊙ ABERDEEN; 33°53'N 88°29'W. Bordered E by ALABAMA state line. Drained by Buttahatchie River and East Fork of the TOMBIGBEE River, which forms Aberdeen Lake reservoir in center. TENNESSEE-TOMBIGBEE Waterway runs parallel to the river. Agriculture (cotton, corn, soybeans, wheat; cattle; dairying); timber. Formed 1821.

Monroe, county (□ 669 sq mi/1,739.4 sq km; 2006 population 9,396), NE central MISSOURI; ⊙ PARIS; 39°30'N 92°00'W. Drained by SALT RIVER. Agriculture (corn, soybeans, wheat; cattle, sheep and saddle horses; lumber; manufacturing at MONROE CITY. Union Covered Bridge SW of Paris. MARK TWAIN LAKE in E; Mark Twain State Park and Birthplace at FLORIDA. Formed 1831.

Monroe, county (□ 1,364 sq mi/3,546.4 sq km; 2006 population 730,807), W NEW YORK; ⊙ ROCHESTER; 43°17'N 77°41'W. Bounded N by Lake ONTARIO (resorts); crossed by the BARGE CANAL; drained by GENESEE River and HONEOYE and other creeks. Extensive manufacturing, especially in Rochester, EAST ROCHESTER, FAIRPORT, and WEBSTER. Dairy products, horticultural and vegetables, grain. Named for James Monroe, who was U.S. President at time county was created (1821). George Eastman brought fame to area with invention of instant photography in 1878. With the coming of the ERIE CANAL in 1823, Rochester became known as the Flour City for its milling of the W New York wheat region (nation's leading wheat-growing region in the mid-1800s). The growth of Rochester as a science and manufacturing center never overshadowed the agricultural wealth of the county. Even today, despite suburbanization and regional economic change, the agricultural landscape remains vital. Several large optics/imaging corporations (Eastman Kodak, Bauch & Lomb, Xerox) continue to operate in the county, although in some cases (e.g., Kodak) their work forces have been downsized.

Monroe (MAHN-ro), county (□ 455 sq mi/1,183 sq km; 2006 population 14,606), E OHIO; ⊙ WOODSFIELD, 39°44'N 81°04'W. Bounded SE by OHIO RIVER, here forming WEST VIRGINIA state line; also drained by small Sunfish Creek and by LITTLE MUSKINGUM RIVER. In the Unglaciated Plain physiographic region. Agriculture (cattle; dairy products; corn); manufacturing (primarily metal industries); coal mines, limestone quarries. Formed 1813.

Monroe, county (□ 616 sq mi/1,601.6 sq km; 2006 population 165,685), E PENNSYLVANIA; ⊙ STROUDSBURG; 41°03'N 75°20'W. Bounded E by DELAWARE RIVER (N.J. state line). POCONO MOUNTAINS plateau is in W; series of long high ridges in S, separated by narrow valleys. Agriculture (corn, wheat, oats, hay, alfalfa; cattle; dairying); resort area. Manufacturing at STROUDSBURG and EAST STROUDSBURG. Parts of DELAWARE WATER GAP RECREATION AREA in SE and NE; Delaware National Scenic River on E boundary;

part of Delaware State Forest in NE; Gouldsboro and Tobyhanna state parks in N; Big Pocono State Park in center; APPALACHIAN TRAIL follows S boundary on KITTATINNY and BLUE MOUNTAIN ridges. The first settlers (1725) worked copper mines. Formed 1836.

Monroe, county (□ 665 sq mi/1,729 sq km; 2006 population 44,163), SE TENNESSEE; ⊙ MADISONVILLE; 35°27′N 84°15′W. Bounded SE and E by NORTH CAROLINA, NE by LITTLE TENNESSEE RIVER; drained by its tributaries; UNICOI MOUNTAINS lie along S border. Includes Tellico Wildlife Management Area and part of Cherokee National Forest. Lumbering; agriculture; natural resources; recreation. Some gold was mined here in the 19th century. Formed 1819.

Monroe, county (□ 474 sq mi/1,232.4 sq km; 2006 population 13,510), SE WEST VIRGINIA; ⊙ UNION; 37°33′N 80°32′W. Bounded SE and SW by VIRGINIA; drained by Indian and Potts creeks. Mountainous region, with PETERS MOUNTAIN along Virginia state line and summits of the ALLEGHENY MOUNTAINS (including BICKETT KNOB) to W. Agriculture (corn, oats, barley, rye, alfalfa, hay); cattle; dairying; hogs; poultry; sheep. Limestone and iron-ore deposits; some natural gas. Manufacturing (aircraft parts). Moncove State Park and Wildlife Management Area in NE; part of Jefferson National Forest in SE; APPALACHIAN TRAIL follows part of Virginia and West Virginia state line in S. Formed 1799.

Monroe, county (□ 908 sq mi/2,360.8 sq km; 2006 population 43,028), W central WISCONSIN; ⊙ SPARTA; 43°57′N 90°37′W. Dairying and farming area (tobacco, cranberries, corn, soybeans, alfalfa, hay; cattle); timber. Processing of dairy products, lumber. Drained by LEMONWEIR, Black (bounds NW corner), LA CROSSE, and KICKAPOO rivers. Part of Central Wisconsin Conservation Area in NE corner. Fort McCoy Military Reserve in N center. Mill Bluff State Park on E boundary. Formed 1854.

Monroe (muhn-RO), city, ⊙ OUACHITA parish, SE LOUISIANA, 75 mi/121 km NNE of ALEXANDRIA, on the OUACHITA RIVER; 32°31′N 92°05′W. The center of the great Monroe Natural Gas Field (discovered 1916). Manufacturing (apparel, food and beverages, construction materials), printing and publishing. The first settlers founded (c.1785) Fort Miró. The community was renamed in 1819 after the *James Monroe*, the first steamship to come up the Ouachita. Northeast Louisiana University is in the city. Of interest are Masur Museum of Art, Louisiana Purchase Garden and Zoo, and antebellum houses in area. Site of the Louisiana Folklife Festival. Russell Sage State Wildlife Area to E, D'Arbonne National Wildlife Refuge to NW, Cheniere Brake State Fish Preserve and lake to SW. Founded c.1785, incorporated as a city 1900.

Monroe, city (2000 population 22,076), ⊙ MONROE county, SE MICHIGAN, 19 mi/31 km NE of TOLEDO, OHIO, and 38 mi/61 km SW of DETROIT, on LAKE ERIE; 41°54′N 83°23′W. Paper products, heating equipment, plastic tubing, flour, and auto parts are made. The city has large nurseries and is the shipping point for a farm region. MONROE was the scene of the RIVER RAISIN massacre during the War of 1812 and the center of the "Toledo War." George A. Custer lived here, and the local museum has a large collection of Custer memorabilia. A community college is in the city, General Custer Historic Site, and Sterling State Park to NE. Settled 1778, incorporated 1837.

Monroe (muhn-RO), city (□ 24 sq mi/62.4 sq km; 2006 population 30,871), ⊙ UNION county, S NORTH CAROLINA, 24 mi/39 km SE of CHARLOTTE, in the PIEDMONT region; 34°59′N 80°32′W. Railroad junction. Manufacturing (metal fabricating and casting, textiles and apparel, plastic and stone products, pharmaceuticals, printing and publishing, industrial machinery, lighting fixtures, aviation and electronic equipment; food processing; service industries. Has

diverse agriculture (cotton, soybeans, grain; poultry; livestock). Wingate College (4-year) to E at WINGATE. Source of LYNCHES RIVER to S. Settled 1751; incorporated 1844.

Monroe (MAHN-ro), city (□ 15 sq mi/39 sq km; 2006 population 11,226), BUTLER county, extreme SW OHIO, 11 mi/18 km ENE of HAMILTON; 39°26′N 84°20′W.

Monroe, city (2006 population 10,599), ⊙ GREEN county, S WISCONSIN, 35 mi/56 km SSW of MADISON; 42°36′N 89°38′W. Dairying region. One of state's leading cheese-producing centers; manufacturing (food and beverages, electronic equipment, wood products). Inhabitants mostly of Swiss descent; annual cheese fair is held. Brownstone-Cadiz State Recreation Area (former Cadiz Springs State Park) to W. Incorporated as village c.1859, as city in 1882.

Monroe (muhn-RO), town, FAIRFIELD county, SW CONNECTICUT, on the HOUSATONIC River, and 10 mi/16 km N of BRIDGEPORT; 41°20′N 73°13′W. Chiefly residential; some light industry. Settled c.1775, incorporated 1823.

Monroe, town (2000 population 11,407), ⊙ WALTON county, N central GEORGIA, 37 mi/60 km E of ATLANTA; 33°47′N 83°43′W. Manufacturing of apparel, printing and publishing, steel molds, plastics, roofing materials; machining. Has some fine old houses in classic-revival style. Birthplace of eight former governors. The Georgia Trust for Historic Preservation maintains the 1887 McDaniel-Tichenor house. Incorporated 1821.

Monroe, town (2000 population 734), ADAMS county, E INDIANA, 6 mi/9.7 km S of DECATUR; 40°44′N 84°56′W. Manufacturing (truck trailers). There is also a Monroe in TIPPECANOE county, 13 mi/21 km SE of LAFAYETTE.

Monroe, town (2000 population 1,808), JASPER county, central IOWA, 13 mi/21 km SSW of NEWTON; 41°31′N 93°05′W. In agricultural area. Manufacturing (seeds, plastic containers, feeds). Laid out 1851.

Monroe, town, WALDO county, S MAINE, 11 mi/18 km N of BELFAST; 44°36′N 69°02′W.

Monroe, rural town, FRANKLIN county, NW MASSACHUSETTS, on DEERFIELD River (power dam) and 9 mi/14.5 km E of NORTH ADAMS; 42°43′N 72°59′W. Includes Monroe Bridge village. Former paper mill town. Several state forests in vicinity.

Monroe, town, GRAFTON county, NW NEW HAMPSHIRE, 13 mi/21 km WSW of LITTLETON; 44°17′N 72°00′W. Bounded W and NW by CONNECTICUT RIVER (Vermont state line). Manufacturing (egg and poultry processing); agriculture (nursery crops, vegetables; cattle; poultry; dairying). Commerford Dam, hydroelectric plant. Gardner Mountain (2,330 ft/710 m) in NE corner.

Monroe, township, MIDDLESEX county, central NEW JERSEY, 12 mi/19 km S of NEW BRUNSWICK; 40°19′N 74°25′W. Incorporated 1838.

Monroe, town, OVERTON county, N TENNESSEE, 5 mi/8 km NE of LIVINGSTON; 36°26′N 85°14′W.

Monroe, town (2000 population 1,845), SEVIER county, SW central UTAH, in Sevier River Valley, 10 mi/16 km S of RICHFIELD; 38°37′N 112°07′W. Alfalfa, barley; dairying; cattle. Hot springs nearby. Units of Fishlake National Forest to E and NW. Elevation 5,395 ft/1,644 m. SEVIER PLATEAU is E. Settled 1863.

Monroe, town (2006 population 16,152), SNOHOMISH county, NW WASHINGTON, suburb, 15 mi/24 km SE of EVERETT, and 24 mi/39 km NE of downtown SEATTLE, on SKYKOMISH RIVER, in W foothills of CASCADE RANGE; 47°52′N 121°59′W. Railroad junction. Vegetables, berries; dairying; poultry; salmon. Granite quarrying; manufacturing (aircraft parts, business forms, salmon processing, wood furniture, boat building, feeds and fertilizer); food processing. Mount Baker-Snoqualmie National Forest to E; Lake Roesiger to N. Evergreen State Fair.

Monroe, village (2006 population 310), PLATTE county, E central NEBRASKA, 12 mi/19 km W of COLUMBUS, and on LOUP RIVER; 41°28′N 97°35′W.

Monroe, summer residential and recreational village (□ 3 sq mi/7.8 sq km; 2006 population 8,141), ORANGE county, SE NEW YORK, 15 mi/24 km SW of NEWBURGH; 41°19′N 74°11′W. Manufacturing of steel products. Small lakes nearby. Incorporated 1894.

Monroe, village (2006 population 578), BENTON county, W OREGON, 20 mi/32 km NNW of EUGENE; 44°19′N 123°17′W. Farm trade center. Berries, grapes; wineries. Timber. Washburne Wayside State Park to SE; William Lake Finley National Wildlife Refuge to N.

Monroe, village (2006 population 157), TURNER county, SE SOUTH DAKOTA, 7 mi/11.3 km NNW of PARKER; 43°29′N 97°13′W.

Monroe (muhn-RO), unincorporated village, AMHERST county, central VIRGINIA, 6 mi/10 km NNE of LYNCHBURG; 37°30′N 79°07′W. Manufacturing (wood-burning stoves, lumber, steel fabrication); agriculture (corn, apples; cattle); timber.

Monroe, PENNSYLVANIA: see MONROETON.

Monroe Bridge, MASSACHUSETTS: see MONROE.

Monroe City (muhn-RO), city (2000 population 2,588), MONROE and MARION counties, NE central MISSOURI, near SALT RIVER, 20 mi/32 km W of HANNIBAL; 39°38′N 91°43′W. Railroad junction. Poultry-shipping center; corn, soybeans; hogs; manufacturing (aluminum and magnesium die castings); lumber. Mark Twain born in nearby FLORIDA; MARK TWAIN LAKE to S. Incorporated 1869.

Monroe City, town (2000 population 548), KNOX county, SW INDIANA, 10 mi/16 km SE of VINCENNES; 38°37′N 87°21′W. In agricultural and bituminous-coal area.

Monroe, Fort, VIRGINIA: see FORT MONROE.

Monroe, Lake (MUHN-ro), a shallow widening (c.5 mi/8 km long, 3 mi/4.8 km wide) of ST. JOHNS RIVER, E central FLORIDA, on SEMINOLE-VOLUSIA county line; 28°48′N 81°15′W. SANFORD is on S shore.

Monroe Lake, reservoir, MONROE and BROWN counties, S central INDIANA, on SALT CREEK (formed here by North and South forks), 12 mi/19 km N of BLOOMINGTON; 15 mi/24 km long; 39°00′N 86°31′W. Has two arms. Formed by Monroe Dam, built by Army Corps of Engineers for water supply. On shores are state recreation areas and Whitewater State Park (NE).

Monroe, Mount, NEW HAMPSHIRE: see PRESIDENTIAL RANGE.

Monroe Peak (11,227 ft/3,422 m), in SEVIER PLATEAU, S SEVIER county, SW central UTAH, 7 mi/11.3 km SSE of MONROE. In Fishlake National Forest.

Monroeton (MUHN-ro-tuhn), borough, BRADFORD county, NE PENNSYLVANIA, 4 mi/6.4 km SSW of TOWANDA, on Towanda Creek; 41°42′N 76°28′W. Manufacturing includes dowels, commercial printing, feeds. Agriculture includes dairying; livestock; corn, hay. Towanda Airport to SE. Formerly called Monroe.

Monroeville (muhn-RO-vil), town (2000 population 6,862), ⊙ Monroe co., SW ALABAMA, 75 mi/121 km NE of Mobile. Trade center in agr. area (cotton, corn); clothing manufacturing, lumber, paper, concrete. Factory outlet mall located here. Tourism. Alabama Southern Community College here. Setting for Harper Lee's novel *To Kill a Mockingbird*. Settled c.1815.

Monroeville, town (2000 population 1,236), ALLEN county, NE INDIANA, 14 mi/23 km ESE of FORT WAYNE, near OHIO line; 40°59′N 84°52′W. In agricultural area (corn, soybeans). Manufacturing (fertilizer blending). Settled 1841, incorporated 1865.

Monroeville (MAHN-ro-vil), village (□ 1 sq mi/2.6 sq km; 2006 population 1,381), HURON county, N OHIO, 14 mi/23 km S of SANDUSKY, and on West Branch of HURON RIVER; 41°14′N 82°42′W.

Monroeville (MUHN-ro-vil), borough (2000 population 29,349), ALLEGHENY county, SW PENNSYLVANIA, suburb, 12 mi/19 km E of downtown PITTSBURGH;

40°25′N 79°45′W. Bounded by Turtle Creek on S. Primarily residential, spurred by the suburban growth in 1980s and 1990s from the Pittsburgh area. Monroeville has chemical and nuclear research centers. Community College of Allegheny County–Boyce Campus. Pittsburgh Monroeville Airport in N. Koppers Research Center in NE. Boyce Regional Park and Ski Area to N. Settled 1810, incorporated 1952.

Monrovia (muhn-RO-vee-uh), city (2003 population 550,200), MONTSERRADO county, ⊙ Republic of LIBERIA, in NW, a port on the ATLANTIC OCEAN at the mouth of the SAINT PAUL RIVER; 06°18′N 10°47′W. Monrovia is Liberia's largest city and its administrative, commercial, communications, and financial center. The city's economy revolves around its harbor, which was substantially improved by U.S. forces under Lend-Lease during World War II. In 1948 the first port capable of handling oceangoing vessels was opened; there are now several ports, including a free port. The main exports are rubber and iron ore. The city also has extensive storage and ship-repair facilities. Manufacturing includes cement, refined petroleum, food products, bricks and tiles, furniture, and pharmaceuticals. Roads, railroad, and ROBERTSFIELD International Airport (E) connect Monrovia with Liberia's interior. Monrovia was founded in 1822 by the American Colonization Society as a haven for freed slaves from the U.S. and the BRITISH WEST INDIES and was named for James Monroe, then President of the U.S. The University of Liberia (1862) and Cuttington College and Divinity School (1889; Episcopal) are in Monrovia.

Monrovia, city (2000 population 36,929), LOS ANGELES county, S CALIFORNIA; suburb 16 mi/26 km NE of downtown LOS ANGELES, in the foothills of the SAN GABRIEL MOUNTAINS; 34°10′N 118°00′W. Diversified manufacturing. Bounded N and NE by Angeles National Forest, including San Gabriel Wilderness to N, MOUNT WILSON OBSERVATORY to NW; San Gabriel and Morris reservoirs to NE, on San Gabriel River. Incorporated 1886.

Monrovia, town, MORGAN county, central INDIANA, 5 mi/8 km SW of MOORESVILLE. Hogs; soybeans, corn, apples. Laid out 1834.

Monroy (mon-ROI), town, CÁCERES province, W SPAIN, 14 mi/23 km NE of CÁCERES; 39°38′N 06°12′W. Meat processing, flour milling; cereals, olive oil, sheep. Has 13th-century castle.

Mons (MAWNS), Flemish *Bergen*, commune (□ 225 sq mi/585 sq km; 2006 population 90,984), ⊙ Mons district and HAINAUT province, SW BELGIUM, near the FRENCH border; 50°27′N 03°56′E. At the junction of CANAL DU CENTRE and the CONDÉ-MONS CANAL, Mons is the processing and shipping center of the BORINAGE and Centre districts and is also a manufacturing center. Closing of coal mines has caused severe economic hardship. Known since the seventh century, Mons became (1295) the seat of the counts of Hainaut. Of note in Mons are the Gothic Church of St. Waltrude (fifteenth–sixteenth century), the city hall (fifteenth century), and many beautiful houses of the sixteenth–eighteenth century. Mons contains several educational institutions including the Polytechnic Faculty, the Academy of Beaux Arts, the Royal Conservatory of Music, and the Higher Institute of Architecture.

Monsalvat, SPAIN: see MONTSERRAT.

Monsanto, Brazil: see MONTE SANTO DE MINAS.

Monsanto, ILLINOIS: see SAUGET.

Monsâo, Portugal: see MONÇÃO.

Monsarás, Ponta de, headland of ESPÍRITO SANTO, E BRAZIL, on the ATLANTIC, at mouth of the RIO DOCE, and 60 mi/97 km NE of VITÓRIA; 19°35′S 38°47′W.

Monsaraz (mon-sah-RASH), town, ÉVORA district, S central PORTUGAL, near the GUADIANA RIVER and Spanish border, 31 mi/50 km ESE of ÉVORA; 38°26′N 07°23′W. Walled castle town in area settled successively by Megalithic peoples, Iron Age cultures, Romans, and Moors. Ruins; tourism.

Monschau (MAWN-shou), town, RHINELAND, North Rhine-Westphalia, W GERMANY, on the RUR, and 14 mi/23 km SSE of AACHEN, near Belgian border; 50°34′N 06°45′E. Textiles. Until 1918, called MONTJOIE.

Monsefú (mon-sai-FOO), city (2005 population 22,018), CHICLAYO province, LAMBAYEQUE region, NW PERU, on coastal plain, on Reque River, on railroad, and 7 mi/11 km S of CHICLAYO; 06°52′S 79°52′W. Rice milling, manufacturing; weaving of native textiles; rice, sugarcane, fruit.

Monselice (mon-SAI-lee-che), town, PADOVA province, VENETO, N ITALY, at SE foot of EUGANEAN HILLS, 13 mi/21 km SSW of PADOVA; 45°14′N 11°45′E. Jute mills; manufacturing (nonmetallics, fabricated metals, plastics, clothing, shoes, agricultural tools, food processing, marmalade, candy, liquor). Has cathedral (1256; restored 1931) and ruins of castle built by Frederick II. Trachyte quarries nearby.

Mons-en-Barœul (mon–zahn–bah-ruh-yuhl), town (□ 1 sq mi/2.6 sq km), outer NE suburb of LILLE, NORD department, NORD-PAS-DE-CALAIS region, N FRANCE, on road to ROUBAIX; 50°38′N 03°07′E. Textile industry; printing, brewing, food products.

Monsenhor Hipólito (MON-sen-yor EE-po-lee-to), town (2007 population 7,206), E PIAUÍ state, BRAZIL, 28 mi/45 km E of PICOS; 06°59′S 41°07′W.

Monsenhor Tabosa (MON-sen-yor TAH-bo-sah), city (2007 population 16,557), E central CEARÁ, BRAZIL, 26 mi/42 km NNE of TAMBORIL; 04°58′S 40°08′W.

Monseñor Nouel (mon-sen-YOR no-WEL), province (□ 388 sq mi/1,008.8 sq km; 2002 population 167,618), central DOMINICAN REPUBLIC, located on N flank of Cordillera CENTRAL; ⊙ BONAO; 18°55′N 70°25′W. Agricultural (sugarcane, sugarcane refining, coffee, cocoa, fruit.) Formerly part of LA VEGA province.

Monserrat (mon-se-RAHT), town, VALENCIA province, E SPAIN, 15 mi/24 km SW of VALENCIA; 39°22′N 00°36′W. Olive-oil processing, plaster manufacturing; wine; cereals, oranges, grapes. Gypsum quarries nearby.

Monserrat, SPAIN: see MONTSERRAT.

Monsey, residential suburb (□ 2 sq mi/5.2 sq km; 2000 population 14,504) of SUFFERN, ROCKLAND county, SE NEW YORK, 5 mi/8 km E of city center; 41°07′N 74°04′W. Large population of Orthodox Jews.

Møns Klint, DENMARK: see MØN.

Monson (MAHN-suhn), town, PISCATAQUIS county, central MAINE, on small Hebron Pond, and 15 mi/24 km NW of DOVER-FOXCROFT; 45°17′N 69°30′W. Known for its slate quarries.

Monson, town, including Monson village, HAMPDEN county, S MASSACHUSETTS, 13 mi/21 km E of SPRINGFIELD; 42°05′N 72°19′W. Woollens granite; dairying; poultry, vegetables. Settled 1715, incorporated 1760.

Mon State, MYANMAR, on ANDAMAN SEA, one of seven states in MYANMAR; ⊙ MAWLAMYINE. Contains ten townships.

Monster, town, SOUTH HOLLAND province, W Netherlands, 7 mi/11.3 km SW of The HAGUE, in WESTLAND region; 52°01′N 04°30′E. NORTH SEA 0.5 mi/0.8 km to NW; entrance to NEW WATERWAY on North Sea 4 mi/6.4 km to SW. Dairying; cattle, poultry; flowers, potted plants, tropical plants, vegetables, fruit; manufacturing (food processing).

Mönsterås (MUHN-ster-OS), town, KALMAR county, SE SWEDEN, on small bay of KALMAR SOUND of BALTIC SEA, 15 mi/24 km S of OSKARSHAMN; 57°03′N 16°26′E. Port; manufacturing (pulp mill; metal goods, ceramics).

Monsummano Terme (mon-soom-MAH-no TERme), town, PISTOIA province, TUSCANY, central ITALY, 7 mi/11 km SW of PISTOIA. Manufacturing (shoes, fertilizer, brooms, clothing, paper, plastics); food processing. Nearby is Giusti grotto (natural vapor baths), named after G. Giusti who was born here.

Montabaur (MUHN-tah-bou-er), town, RHINELAND-PALATINATE, W GERMANY, 11 mi/18 km ENE of KO-BLENZ; 50°26′N 07°50′E. Main town of WESTERWALD district. Iron foundries. Health resort. Has late-Gothic church, and former castle of electors of TRIER.

Montaberner (mon-tah-ver-VER), town, VALENCIA province, E SPAIN, 8 mi/12.9 km S of JÁTIVA; 38°53′N 00°30′W. Alcohol, vermouth manufacturing; cereals, wine, olive oil.

Montadas (MON-tah-dahs), town (2007 population 4,558), E central PARAÍBA, BRAZIL, 7 mi/11.3 km SW of ESPERANÇA; 07°05′S 35°57′W.

Montafon (mawn-tah-FAWN), deep upper valley of ILL RIVER, in vorarlberg, W AUSTRIA; 15 mi/24 km long; 47°02′N 9°57′E. Dairy farming. Tourist trade, with centers at SCHRUNS and GASCHURN. Hydroelectric works at Lünersee and Rodund Kops and Vermunt.

Montagnac (mon-tah-nyahk), commune (□ 15 sq mi/39 sq km), HÉRAULT department, LANGUEDOC-ROUSSILLON region, S FRANCE, near the HÉRAULT, 12 mi/19 km NW of SÈTE. Vineyards. Has Gothic church.

Montagnana (mon-tah-NYAH-nah), town, PADOVA province, VENETO, N ITALY, 23 mi/37 km SW of PADUA; 43°40′N 11°06′E. Manufacturing (textile machinery, textiles, wood products, fabricated metal products, beet sugar, sausage, pasta). Has Gothic cathedral (1431–1502) and picturesque medieval walls (13th–14th century) with twenty-four towers.

Montagne Blanche, village (2000 population 8,116), MOKA and FLACQ districts, MAURITIUS, 19 mi/31 km E of PORT LOUIS. Sugarcane.

Montagne de Reims Natural Regional Park (monTAH-nyuh duh ram) (□ 193 sq mi/501.8 sq km), MARNE department, CHAMPAGNE-ARDENNE region, N FRANCE. Vineyards on N slopes of Montagne de REIMS.

Montagne du Droit (mohn-TAH-nyuh du DRWAH), range in the Bernese JURA, NW SWITZERLAND, extending generally NE from LA CHAUX-DE-FONDS to NW of BIEL, between FRANCHES MONTAGNES and Vallon de St. IMIER (SUZE RIVER). Mont Soleil (German *Sonnenberg*) (4,236 ft/1,291 m), is highest point.

Montagne, La (mon-tah-nyuh), town (□ 1 sq mi/2.6 sq km), LOIRE-ATLANTIQUE department, PAYS DE LA LOIRE region, W FRANCE, on right bank of LOIRE RIVER, and 6 mi/9.7 km W of NANTES. Armaments manufacturing.

Montagne Noire (mon-tah-nyuh nwahr), extreme S range of MASSIF CENTRAL, S FRANCE, along border between TARN and AUDE departments, mostly in HAUT LANGUEDOC NATURAL REGIONAL PARK. Rises to 3,973 ft/1,211 m in Pic de Nore. Densely forested. Textile manufacturing district of MAZAMET on N slope.

Montagne Noire, FRANCE: see MONTAGNES NOIRE.

Montagnes Noires (mon-tahn-yuh nwahr), low granitic hills on border of range in FINISTÈRE and MORBIHAN departments, in BRITTANY, W FRANCE, a part of the eroded ARMORICAN MASSIF. Rises to 1,070 ft/326 m. Also spelled Montagne Noire.

Montagne Tremblante Park (mon-TAH-nyuh trahnBLAHNT), provincial park in the LAURENTIANS, S QUEBEC, E CANADA, c.70 mi/113 km NW of MONTREAL. Mount TREMBLANT (Trembling Mountain) rises to 3,150 ft/960 m. Resort area, notably for skiing.

Montagu, town, WESTERN CAPE province, SOUTH AFRICA, in the LANGEBERG range, 40 mi/64 km E of WORCESTER; 33°46′S 20°06′E. Elevation 1,574 ft/580 m. Health resort with hot radioactive mineral springs (temp. 96°F/35.5°C); viticulture, fruit drying. Popular resort town. Bushman paintings in nearby caves. Montagu gardens contain fine examples of many indigenous "fynbos" species of the Western Cape. Near Kogmans Kloof Pass through Langeberg range from ROBERTSON. Airfield. Established in 1851; named for John Montagu, colonial secretary.

Montagu (MAHN-tuh-gyoo), village, NW TASMANIA, SE AUSTRALIA, 120 mi/193 km WNW of LAUNCESTON; 40°46′S 144°58′E. Cheese.

Cross-references are shown in SMALL CAPITALS. The pronunciation guide is shown on page xix. The sources of population figures are shown on page xvii.

Montague (MAHN-tuh-gyoo), county (□ 938 sq mi/ 2,438.8 sq km; 2006 population 19,810), N TEXAS; ⊙ MONTAGUE; 33°40′N 97°43′W. Bounded N by RED RIVER (here the OKLAHOMA state line); source of Elm Fork of the TRINITY RIVER, and Clear, DENTON, and Big Sandy creeks. Diversified agriculture (peanuts, cotton, corn, grains, fruit, truck); large poultry, dairy industries; cattle ranching. Oil, natural gas fields; timber. Manufacturing, processing at BOWIE, NO- CONA. Lake Nocona reservoir in N (on tributary of Red), Lake Carter reservoir in SW (Big Sandy Creek). Formed 1857.

Montague, city (2000 population 1,456), SISKIYOU county, N CALIFORNIA, at W base of CASCADE RANGE, 6 mi/9.7 km E of YREKA, and on SHASTA RIVER; 41°44′N 122°32′W. Dairying; cattle, sheep, lambs; grain. Lake Shastina reservoir to SE; part of Klamath National Forest to NW.

Montague (MAHN-tuh-gyoo), township (□ 107 sq mi/ 278.2 sq km; 2001 population 3,671), SE ONTARIO, E central CANADA, on Rideau River to S, and 15 mi/24 km from PERTH; 44°57′N 75°57′W.

Montague (MAHN-tuh-goo), town (2001 population 1,945), E. PRINCE EDWARD ISLAND, CANADA, on Montague River, near its mouth on CARDIGAN BAY, 24 mi/39 km ESE of CHARLOTTETOWN; 46°10′N 62°39′W. Agricultural market in dairying, cattle-raising, pota- to-growing region.

Montague, town, FRANKLIN county, NW MASSACHU- SETTS, on CONNECTICUT River just SE of GREENFIELD; 42°33′N 72°31′W. Machinery, dies, fishing tackle. Large hydroelectric plant. Villages include Montague City, Turners Falls (1990 population 4,731) (site of first dam across the Connecticut), Millers Falls, and Lake Pleasant. Settled 1715, set off from SUNDERLAND 1754.

Montague (MON-tawg), town (2000 population 2,407), MUSKEGON county, SW MICHIGAN, 14 mi/23 km NNW of MUSKEGON, at head of White Lake, near LAKE MICHIGAN, opposite WHITEHALL; 43°24′N 86°21′W. Shipping point for agriculture (fruits and vegetables; poultry) and dairying area; manufacturing (food products); manufacturing (horticultural labels, pharmaceuticals, cast aluminum products, machin- ing). Resort; winter ice fishing. Duck Lake State Park to S. Incorporated as village 1883, as city 1935.

Montague (MAHN-tuh-gyoo), unincorporated village, ⊙ MONTAGUE county, N TEXAS, 47 mi/76 km ESE of WICHITA FALLS; 33°39′N 97°43′W. Elevation 1,075 ft/ 328 m. In agricultural area (dairying; cattle; water- melons, cantaloupes; peanuts; wheat); oil and gas; stone.

Montague Island (MAHN-tuh-gyoo) (□ 0.3 sq mi/0.8 sq km), in TASMAN SEA, 4 mi/6 km off SE coast of NEW SOUTH WALES, SE AUSTRALIA; 2 mi/3 km long, 1 mi/ 2 km wide; 36°15′S 150°13′E. Rises to 250 ft/76 m. Lighthouse. Largely composed of granite. Seals, birdlife, Little Penguins colony. Also spelled Montagu Island; variant name: Barunguba.

Montague Island (□ 18 sq mi/47 sq km), LOWER CA- LIFORNIA, NW MEXICO, at mouth of COLORADO RIVER, at head of Gulf of CALIFORNIA, 75 mi/121 km SE of MEXICALI; 6 mi/9.7 km long, 3 mi/4.8 km wide. Flat, alluvial, uninhabited.

Montague Island, S ALASKA, on W side of entrance of PRINCE WILLIAM SOUND, 60 mi/97 km E of SEWARD; 50 mi/80 km long, 5 mi/8 km–12 mi/19 km wide; 60°05′N 147°23′W.

Montague Sound (MAHN-tuh-gyoo), inlet of TIMOR SEA, NE WESTERN AUSTRALIA, between Cape VOL- TAIRE (E) and BIGGE ISLAND (W); 20 mi/32 km long, 30 mi/48 km wide; 14°28′S 125°20′E.

Montague Strait, S ALASKA, entrance from Gulf of ALASKA (S) to PRINCE WILLIAM SOUND (N), between MONTAGUE ISLAND (E) and LATOUCHE ISLAND (W); 60°08′N 147°38′W.

Montagu Island, AUSTRALIA: see MONTAGUE ISLAND.

Montagu Pass (2,348 ft/716 m), WESTERN CAPE prov- ince, SOUTH AFRICA, crosses OUTENIQUA MOUNTAINS 5 mi/8 km N of GEORGE; 33°52′S 22°26′E. Railroad from George climbs c.1,600 ft/488 m over distance of 17 mi/27 km. No longer pass for road traffic over Outeniqua Mountains, superceded by newer, wider Outeniqua Pass.

Montaigu (mon-te-gyoo), town (□ 1 sq mi/2.6 sq km), VENDÉE department, PAYS DE LA LOIRE, W FRANCE, 20 mi/32 km SSE of NANTES; 46°58′N 01°18′W. Road center; diverse manufacturing (pottery, wooden fur- niture). Horse breeding. Museum.

Montaigu, BELGIUM: see SCHERPENHEUVEL-ZICHEM.

Montalbán or **Montalbán de Córdoba** (mon-tahl- VAHN dhai KOR-dho-vah), town, CÓRDOBA prov- ince, S SPAIN, 22 mi/35 km S of CÓRDOBA. Agricultural trade center; olive-oil processing; cereals, vegetables, melons.

Montalbán, town, TERUEL province, E SPAIN, 38 mi/61 km NNE of TERUEL. Produces wine; cereals, almonds, saffron, hemp. Summer resort. Notable church. Lig- nite, alum, coal, and marble mines in vicinity.

Montalbán (mon-tahl-BAHN), town, ⊙ Montalbán municipio, CARABOBO state, N VENEZUELA, 23 mi/37 km W of VALENCIA; 10°12′N 68°19′W. Elevation 2,431 ft/740 m. Agricultural center (sugarcane, coffee, to- bacco, corn, fruit).

Montalbanejo (mon-tahl-vah-NAI-ho), village, CUENCA province, E central SPAIN, 30 mi/48 km SW of CUENCA; 39°44′N 02°30′W. Cereals, saffron, grapes; sheep, goats; apiculture. Lumbering (pine).

Montalbano Ionico (YO-nee-ko), town, MATERA province, BASILICATA, S ITALY, near AGRI River, 7 mi/ 11 km S of PISTICCI; 40°17′N 16°34′E. In agricultural region (cereals, vegetables, citrus fruit). Known for liquorice production.

Montalbo (mon-TAHL-vo), town, CUENCA province, E central SPAIN, 32 mi/51 km WSW of CUENCA; 39°52′N 02°41′W. Cereals, grapes, potatoes, vegetables, olives; sheep.

Montalcino (mon-tahl-CHEE-no), town, SIENA prov- ince, TUSCANY, central ITALY, 20 mi/32 km SSE of SIENA; 43°03′N 11°29′E. Manufacturing center for clothing, shoes, pasta. Famous for wine production, particularly "Brunello," considered Italy's finest by experts. Known also for traditional white and tur- quoise pottery, honey. Bishopric. Has cathedral, pal- ace (13th–14th century) with picture gallery.

Montalegre (mon-tah-LAI-grai), town, VILA REAL district, northernmost PORTUGAL, on S slope of Serra do Larouco, 19 mi/31 km WNW of CHAVES, near Spanish border; 41°49′N 07°48′W. Olives; vineyards; livestock.

Montalieu-Vercieu (mon-tah-lyu–ver-syu), commune (□ 3 sq mi/7.8 sq km), ISÈRE department, RHÔNE- ALPES region, SE FRANCE, near left bank of the RHÔNE, 17 mi/27 km N of La TOUR-DU-PIN; 45°49′N 05°24′E. Cement works.

Montalivet (mon-tah-lee-vai), small beach resort, GIRONDE department, AQUITAINE region, SW FRANCE, on BAY OF BISCAY, 12 mi/19 km S of Le VERDON- SUR-MER (which lies at N point of the sandy LANDES coastline); 45°23′N 01°09′W. Also known as Montalivet-les-Bains.

Montalivet-les-Bains, FRANCE: see MONTALIVET.

Mont Alto (MAHNT AL-to), borough (2006 popula- tion 1,817), FRANKLIN county, S PENNSYLVANIA, 6 mi/ 9.7 km N of WAYNESBORO, on West Branch Antietam Creek; 39°50′N 77°32′W. Agricultural area (grain, apples; poultry, livestock; dairying). Mont Alto Ar- boretum. Mont Alto State Park to E; APPALACHIAN TRAIL passes to E.

Montalto (mon-TAHL-to), highest peak (6,417 ft/1,956 m) in CALABRIA, S ITALY, in the ASPROMONTE, 15 mi/ 24 km ENE of REGGIO DI; 38°10′N 15°55′E.

Montalvânia (mon-tahl-VAH-nee-ah), city (2007 population 15,944), N central MINAS GERAIS, BRAZIL,

on Rio Cocha, 40 mi/64 km NW of MANGA; 14°29′S 44°18′W.

Montalvão (mon-tah-VOU), agriculture village, POR- TALEGRE district, central PORTUGAL, near Spanish border, 22 mi/35 km NNW of PORTALEGRE; 39°36′N 07°32′W.

Montamarta (mon-tah-MAHR-tah), town, ZAMORA province, NW SPAIN, 10 mi/16 km NNW of ZAMORA; 41°39′N 05°48′W. Livestock raising, lumbering; cere- als, wine.

Montaña (mon-TAHN-yah), region, NE and E PERU; 10°00′S 73°00′W. Originally a name for the E forested slopes of the ANDES, it now also includes the tropical lowlands of the Peruvian AMAZON basin.

Montana, state (□ 147,046 sq mi/382,319.6 sq km; 2006 population 944,632), NW UNITED STATES, in the ROCKY MOUNTAIN region; ⊙ HELENA; 47°15′N 109°13′W. Admitted as the forty-first state of the Union in 1889; it is the fourth-largest state in U.S. Helena, BILLINGS, and GREAT FALLS are the largest cities; other places of importance include MISSOULA and BUTTE. Montana is known as the "Treasure State," which refers to the importance of mining in the state.

Geography

The state lies on the N border of the contiguous U.S., S of the Canadian provinces of BRITISH COLUMBIA, ALBERTA, and SASKATCHEWAN. It is bounded by NORTH DAKOTA and SOUTH DAKOTA on the E, by WYOMING and IDAHO on the S, and by Idaho on the W. Montana is thinly populated and has many remote areas. Life in the state's W mountain area differs greatly from that on its E plains. In the E half of the state are broad rolling plains, punctuated by buttes and outlier ranges of Rocky Mountains drained by the MISSOURI River, which originates in SW Montana, and by its tributaries, the MILK, the MARIAS, the SUN, and especially the YELLOWSTONE. Much of Montana's W boundary is marked by the crest of the lofty BIT- TERROOT RANGE, part of the Rocky Mountains, which dominate the W section of the state and along which, in SW Montana, runs the CONTINENTAL DIVIDE (forms the Idaho and Montana boundary in the S and SW, and continues N through W Montana to GLA- CIER NATIONAL PARK). Montana's very name is de- rived from the Spanish word *montaña*, meaning mountain country.

Economy

High granite peaks, green forests, blue lakes, and such natural wonders as those of Glacier National Park in the NW and part of YELLOWSTONE NATIONAL PARK in the SW have helped make tourism the state's second- ranking industry. The mountains, moreover, offer more than massive beauty, for in and around the mountainous W region are the large mineral deposits for which Montana is famous—copper, silver, gold, platinum, zinc, lead, and manganese. The E part of the state is noted for its petroleum and natural gas, and there are also vast coal deposits. In addition, Montana mines vermiculite, chromite, cadmium, talc, molyb- denum, and tungsten. Manufacturing includes forest products (timber is important in the W half of the state), processed food, refined petroleum, and coal products. In E Montana the high grass of the GREAT PLAINS once nourished herds of buffalo and later sustained the cattle and sheep of huge ranches; much of the high grass is gone, but the cattle and sheep remain, grazing mainly on short grass. Despite the dangers of drought and the severe years that drove many farmers out of the state—turning farming communities into ghost towns—agriculture, both rain-fed and irrigated, provides an important share of Montana's income. Agriculture, coal, and petroleum contribute to the growth of Billings, Montana's fastest-growing city. Cattle are the most valuable farm item; sheep, horses, and hogs are regionally important. The principal crops raised are wheat, hay, barley, potatoes, and sugar beets.

Important for hydroelectric development and for irrigation.

Tourism

Places of interest, besides Glacier National Park, include Little Bighorn (formerly Custer) Battlefield National Monument, BIG HOLE NATIONAL BATTLEFIELD, GRANT-KOHRS RANCH National Historic Site, and part of FORT UNION TRADING POST NATIONAL HISTORIC SITE, on the NORTH DAKOTA state line. Also the National Bison Range, near RAVALLI, where herds of bison may still be seen. Strips of Yellowstone National Park, including the N and W entrances, are also in Montana. BIGHORN CANYON NATIONAL RECREATION AREA on the Wyoming boundary in the S. The Upper Missouri (N center) and Flathead (NW) are national wild and scenic rivers. National forests include: Kootenai, Kaniksu, Lolo, and Flathead in the NW; Beaverhead, Bitterroot, and Deerlodge in the SW; Helena and Lewis and Clark in the W center; Custer National Forest in the SE. The many kinds of fish found in the rushing mountain streams and innumerable lakes bring fishing enthusiasts to the state, and the abundant wildlife—wapiti, elk, deer, antelope, moose, bear, and waterfowl—attract hunters. The state's outstanding recreational areas also include facilities for skiing, hiking, boating, and swimming.

Indian Presence

Montana's rivers were once avenues of travel for the Native Americans known to have inhabited the region at the time Europeans first explored it. Ethnic groups included the Blackfeet, the Sioux, the Shoshoni, the Atsina (Gros Ventre), the Kootenai, the Cheyenne, the Salish, Crow, the Pend D'Oreille, Assiniboine, the Ojibway, the Cree, and others. Major Indian Reservations: Blackfeet Indian Reservation and Flathead Indian Reservation in the NW, Rocky Boy's Indian Reservation and Fort Belknap Indian Reservation in the N center, Fort Peck Indian Reservation in the NE, Crow and Northern Cheyenne Indian reservations in the S.

History to 1846

Early explorers of the country also traveled along the rivers. Exploration of the region began in earnest after most of Montana had passed to the U.S. under the Louisiana Purchase (1803). The Lewis and Clark expedition traveled W across Montana in 1805, and François Antoine Larocque, along with his North West Company of Canada, explored the YELLOWSTONE RIVER after 1805. The first trading post in Montana was established at the mouth of the BIGHORN in 1807 by a trading expedition under Manuel Lisa that came up the MISSOURI from ST. LOUIS. For some years both Canadian and American fur traders continued to open the territory. David Thompson of the North West Company built several trading posts in NW Montana between 1807 and 1812, and beaver in the mountain streams and lakes attracted adventurous trappers, the so-called "mountain men." The American Fur Company, with its posts on the Missouri and the Yellowstone, dominated the later years of the region's fur trade, which diminished in the 1840s.

History - 1846 to 1876 (The Battle of the Little Bighorn)

The U.S. claim to NW Montana, the area between the Rockies and the N Idaho border, was legalized in the Oregon Treaty of 1846 with the British. Montana was then still a wilderness of forest and grass, with a few trading posts and some missions. Montana's first period of growth was the rapid, boisterous, and unstable expansion brought on by a gold rush. The discovery of gold, made initially in 1852, brought many people to mushrooming mining camps such as those at BANNACK (1862) and VIRGINIA CITY (1864). Crude shanty towns were built, complete with saloons and dance halls—ephemeral settlements as colorful as the earlier gold-rush camps in California and per-

haps even more lawless. Previously part of, successively, the territories of Oregon, Washington, Nebraska, Dakota, and Idaho, Montana itself became a territory in 1864. The territory was still a rough frontier, however, and the first governor, Sidney Edgerton, was driven out of the region, and later Thomas Francis Meagher, appointed temporary governor, died mysteriously. After the Civil War, the grasslands attracted ranchers. Although cattle had been introduced into Montana by missionaries, traders, and the Métis before the gold rushes, large-scale ranching dates from the acquisition of cattle by Richard Grant in the Beaverhead Basin in 1850. Yet it was not until after wars with the Sioux that ranching was safe.

History - 1876 to 1920

The Sioux did not tamely submit to having their lands taken from them; in 1876 at the battle of the LITTLE BIGHORN, they defeated Gendfal George A. Custer and his force in one of the greatest of Native American victories. The Sioux were eventually subdued, and the gallant attempt of Chief Joseph of the Nez Percé to lead his people into Canada to escape pursuing U.S. troops had its pitiful end in the Bears Paws Mountains in Montana. Great ranches spread out across the plains, and cow towns that were to grow into cities, such as Billings, sprang up as the railroads were built in the West (c.1880–c.1910). Achievement of statehood in 1889 and the building of the railroad put an end to the era of the open range. Mining continued to dominate Montana: the discovery of silver at Butte (1875) had been followed (c.1880) by discovery of copper at that same "richest hill on earth." Montana's fate was subsequently linked to copper, and the Anaconda Copper Mining Company came to play a major role in Montana life. The titans of the mines, Marcus Daly and William A. Clark, fought bitterly not only for ownership of the mineral deposits but for political control, and their rivalry was physically fought out by the miners. Fritz Augustus Heinze also entered the scramble for copper claims, challenging the claims of the Amalgamated Copper Company. Amalgamated ruled triumphant, however, exercising control over state affairs. Struggles between the company and the workingmen led to strikes, disorder, and bloodshed, but they also resulted in the enactment of some early measures for social security. This was an important achievement, for over the years the livelihood of the residents of the mining towns has been dependent on the market price of copper. Despite fluctuating metals prices, the mines have contributed a large amount to the state's wealth.

History - 1920 to 1950

After the coming of the railroads, farmers came by the trainload to develop the lands of E Montana. They planted their fields in the second decade of the 20th century; the initial yield of wheat was great, but it did not last long. The calamitous drought of 1919 and the consequent dust storms seared the fields, and in the 1920s the farms began to disappear as rapidly as they had been established. When the great national depression began in 1929, Montana was already accustomed to depression. In subsequent years vigorous measures were taken to aid agriculture in the state, and by the late 1940s, Federal dam and irrigation projects—on the Missouri, the Yellowstone, the MARIAS, the SUN, and elsewhere—opened many acres to cultivation. Some of the vast grazing lands were brought under planned use, and the development of hydroelectric power continued. Major multipurpose dams in Montana producing power include Fort Peck Dam (Missouri River, NE central Montana), Hungry Horse Dam (South Fork FLATHEAD River, NW), and Canyon Ferry (Missouri River, W center). Large FORT PECK LAKE reservoir is surrounded by Charles M. Russell National Wildlife Refuge. FLATHEAD LAKE, in NW, is Montana's largest natural lake.

History - 1950 to Present

The demand for copper in World War II and the E Montana oil boom of the early 1950s stimulated Montana's economy. The state, however, still faces problems regarding high transportation costs, lack of manpower, and the necessary regulation of resources. There has been a beneficial if slow trend toward a more diversified economy, with manufacture growing in importance in relation to farming and mining. The latter sector of the economy has declined in importance, while tourism has been growing. Development of vast recreational facilities along the GALLATIN River stirred protests from conservationists in the early 1970s. In the 1970s the exploitation of coal increased dramatically. The Anaconda Mining Company closed its largest copper smelter in the 1970s and closed down operations in Butte altogether in 1983, primarily the results of strict environmental regulations. Much of Montana is still largely undeveloped and unpopulated. There are also many reclamation projects. The state's major institutions of higher learning are included in the University of Montana and Montana State University systems.

Government

In 1973, Montana implemented a new constitution, which replaced the one adopted in 1889. The governor of the state is elected for a term of four years and may be reelected. The current governor is Brian Schweitzer. The Legislative Assembly is made up of a senate with fifty members and a house of representatives with 100 members. State senators are elected for terms of four years and representatives for terms of two years. Montana is represented in the U.S. Congress by one representative and two senators, and the state has three electoral votes in presidential campaigns.

Montana has fifty-six counties: BEAVERHEAD, BIG HORN, BLAINE, BROADWATER, CARBON, CARTER, CASCADE, CHOUTEAU, CUSTER, DANIELS, DAWSON, DEER LODGE, FALLON, FERGUS, FLATHEAD, GALLATIN, GARFIELD, GLACIER, GOLDEN VALLEY, GRANITE, HILL, JEFFERSON, JUDITH BASIN, LAKE, LEWIS AND CLARK, LIBERTY, LINCOLN, MCCONE, MADISON, MEAGHER, MINERAL, MISSOULA, MUSSELSHELL, PARK, PETROLEUM, PHILLIPS, PONDERA, POWDER RIVER, POWELL, PRAIRIE, RAVALLI, RICHLAND, ROOSEVELT, ROSEBUD, SANDERS, SHERIDAN, SILVER BOW, STILLWATER, SWEET GRASS, TETON, TOOLE, TREASURE, VALLEY, WHEATLAND, WIBAUX, and YELLOWSTONE.

Montana (mohn-TAH-nah), commune, VALAIS canton, S SWITZERLAND, 2 mi/3.2 km W of SIERRE; 46°18′N 07°28′E. Elevation 4,905 ft/1,495 m. Includes Montana Village, Vermala, and CRANS. Resort with sanatoria.

Montana (mon-TAHN-ah), oblast (□ 4,093 sq mi/ 10,641.8 sq km), □ c.4,093 sq mi/10,601 sq km or 9.5% of BULGARIA; 43°25′N 23°15′E. In NW Bulgaria, bordering SERBIA to the W, the DANUBE and ROMANIA to the N, LOVECH oblast to the E, STARA PLANINA (SOFIA oblast) to the S. It is the smallest oblast. Undulated Danubian plain includes 60% of the region. LOM and OGOSTA rivers intersect the area. Has 24 cities. Provides 7.6% of nation's industrial and 9.2% of agricultural production. Iron ore in Chiprovska Stara Planina, copper in the VRACHANSKA MOUNTAINS, marble, gypsum, limestone, clays for construction industries. Machine construction and chemical industry. Manufacturing forklifts, chemicals-nitrogen fertilizers, cellulose. Canning, flour milling, wine making, dairy and meat. Agriculture (grain, livestock, corn, hemp, wheat, sunflowers, tomatoes, wine grapes). Interesting caves: Magurata, Ledenika, and rock formations; old fortress near VIDIN. Ferry boat at Vidin across the Danube.

Montana (mon-TAHN-ah), city (2001 population 49,368), MONTANA oblast, Montana obshtina, NW BULGARIA, on the OGOSTA RIVER, 22 mi/35 km NW of VRATSA; 43°25′N 23°15′E. Agricultural and cattle center; manufacturing (construction machines, electroacoustics, foodstuffs, textiles). Has ruins of a Roman town of Montanensia. Formerly called Golyama Kutlovitsa, later Ferdinand (1891–1945), later Mikhailovgrad or Mikhaylovgrad or Mihailovgrad (1945–1990).

Montaña Blanca (mon-TAH-nyah BLAHNG-kah), village and mountain, LANZAROTE island, CANARY ISLANDS, 3 mi/4.8 km NW of ARRECIFE; 28°59′N 13°38′W. Fruit growing and wine producing.

Montaña Clara Island (mon-TAH-nyah KLAH-rah), tiny islet, N CANARY ISLANDS, SPAIN, just N of GRACIOSA ISLAND, 150 mi/241 km NE of LAS PALMAS; 29°18′N 13°32′W. A pinnacle rising to c.790 ft/241 m.

Montaña de Fuego or **Montañas del Fuego** (mon-TAH-nyah dhel FWAI-go), semiactive volcanoes in SW LANZAROTE island, CANARY ISLANDS, SPAIN, 12 mi/19 km W of ARRECIFE. Rise above 1,600 ft/488 m. Visited by tourists.

Montaña de Yoro National Park (mon-TAHN-yah de YO-ro), park (□ 60 sq mi/156 sq km), YORO department, N central HONDURAS, 19 mi/30 km S of YORO; 15°03′N 87°04′W. Elevation 7,800 ft/2,378 m. One of Honduras's largest surviving cloud forests.

Montañas del Fuego, SPAIN: see MONTAÑA DE FUEGO.

Montañas La Sierra (mon-TAHN-yahs lah see-EH-rah), part of continental divide in SW HONDURAS, extending c.20 mi/32 km SW-NE between MARCALA and OPATORO; 14°04′N 87°54′W. Rises to over 7,380 ft/2,250 m. Forms watershed between upper ULÚA (NW) and Goascorán (SE) rivers.

Montánchez (mon-AHN-cheth), town, CÁCERES province, W SPAIN, 22 mi/35 km SE of CÁCERES; 39°13′N 06°09′W. Agricultural center noted for its cured hams and sausage; olive oil and cheese processing. Wine, cereals; livestock; cork in area. Has ruined medieval castle.

Montanha (MON-tahn-yah), city (2007 population 17,983), N central ESPÍRITO SANTO, BRAZIL; 18°08′S 40°26′W.

Montara, unincorporated town (2000 population 2,950), SAN MATEO county, W CALIFORNIA; suburb 17 mi/27 km SSW of downtown SAN FRANCISCO, 1 mi/1.6 km N of MOSS BEACH, on PACIFIC OCEAN; 37°33′N 122°30′W. Artichokes, brussel sprouts, flowers, grains. Nearby Montara Point has lighthouse and radio-compass station. Montara Mountain (extension of Santa Cruz Mountains) and San Francisco State Fish and Game Refuge to E; Montara State Beach is here; Grey Whale Cove State Beach to N.

Montargil (mon-tahr-JEEL), reservoir and lake, SW PORTALEGRE district, E central PORTUGAL, on Sor River, SW of PONTE DE SOR, and E of Montargil mountain; 39°08′N 08°07′W. Montargil is on W shore.

Montargis (mon-tahr-zhee), town (□ 1 sq mi/2.6 sq km), LOIRET department, CENTRE administrative region, N central FRANCE, 39 mi/63 km ENE of ORLÉANS near Montargis Forest, and on LOING RIVER and BRIARE CANAL near its junction with ORLÉANS CANAL; 48°00′N 02°44′E. Industrial center and agriculture market (poultry, dairy produce); manufacturing (machinery, furniture, shoes, and rubber goods), tanneries. Montargis was the capital of old GÂTINAIS district (known for hunting and fishing), and a royal residence in 14th–15th century. Has preserved its medieval aspect, including small canals penetrating town center (many bridges). Girodet art museum is surrounded by a fine garden. Montargis figured prominently in the Hundred Years War. Mirabeau born at nearby castle of Bignon.

Montasio, Jôf del (mon-TAH-zyo. jof del), second-highest peak (9,035 ft/2,754 m) in JULIAN ALPS, NE ITALY, 9 mi/14 km SW of TARVISIO.

Montastruc-la-Conseillère (mon-tahs-truk–lah–konse-yer), agricultural commune (□ 5 sq mi/13 sq km), HAUTE-GARONNE department, MIDI-PYRÉNÉES region, S FRANCE, 11 mi/18 km NE of TOULOUSE; 43°43′N 01°36′E.

Montataire (mon-tah-tair), town (□ 4 sq mi/10.4 sq km), OISE department, PICARDIE region, N FRANCE, on the THÉRAIN and 8 mi/12.9 km NW of SENLIS; 49°16′N 02°26′E. Metalworking center; production of chemicals. Has medieval church.

Montauban (mon-to-bahn) or **Montauban-de-Bretagne** (mon-to-bahn–duh–bre-tahn-yuh), commune (□ 16 sq mi/41.6 sq km), ILLE-ET-VILAINE department, BRITTANY, W FRANCE, 18 mi/29 km WNW of RENNES; 48°12′N 02°03′W. Dairying and cheese manufacturing.

Montauban (mon-to-bahn), city (□ 52 sq mi/135.2 sq km), ⊙ TARN-ET-GARONNE department, MIDI-PYRÉNÉES region, S FRANCE, on the TARN RIVER, and 27 mi/43 km N of TOULOUSE; 44°01′N 01°20′E. A commercial and industrial center where aeronautical and electrical equipment is made and regional foods are sold and processed; brickyards. Founded in 1144, Montauban was a stronghold (a *bastide*) of the Albigenses in 13th century and of the Huguenots in 16th century. It enjoyed prosperity until Louis XIV's religious persecutions (17th century). Points of interest include a 14th-century brick bridge over the Tarn, a cathedral (17th–18th century) containing a celebrated painting by Jean Ingres, born 1780 in Montauban, and the well-known Ingres museum. The central square (Place Nationale), with brick facades and arcades is site of a daily market. The city has several military instruction centers.

Montauban (mon-to-BAHN), unincorporated village, S central QUEBEC, E CANADA, on BATISCAN RIVER, and 30 mi/48 km from SHAWINIGAN, and included in NOTRE-DAME-DE-MONTAUBAN; 46°52′N 72°18′W. Dairying; cattle, pig, poultry raising.

Montauban-les-Mines (mon-to-BAHN–lai MEEN), unincorporated village, S central QUEBEC, E CANADA, 27 mi/43 km from SHAWINIGAN, and included in NOTRE-DAME-DE-MONTAUBAN; 46°49′N 72°20′W. Mining.

Montauk, resort village (□ 19 sq mi/49.4 sq km; 2000 population 3,851), SUFFOLK county, SE NEW YORK, on E LONG ISLAND near tip of its S peninsula, 14 mi/23 km ENE of EAST HAMPTON; 41°02′N 71°57′W. E terminus of S Shore line of Long Island railroad. Commercial- and sport-fishing center, with world records for marlin, shark, and tuna. Just E is Lake Montauk (c.2 mi/3.2 km long), sheltered inlet of LONG ISLAND SOUND; yacht harbor. MONTAUK POINT State Park is 5 mi/8 km E; Hither Hills State Park (camping) is 4 mi/6.4 km WSW. Name derived from Montauk word for "hilly land." Founded on land bought from Montauks in 1686 by settlers from nearby East Hampton to raise cattle. Site of oldest cattle ranch in U.S. Owners of estates have included artist Andy Warhol and musician Billy Joel. Surfing beaches.

Montauk Point, E extremity of the S peninsula of LONG ISLAND, the "South Fork," SE NEW YORK; 41°04′N 71°52′W. Approximately 115 mi/185 km E of MANHATTAN, it is the easternmost point of land of the state. In 1792 President George Washington signed order for construction of the Montauk Point Lighthouse. Built by John McComb Jr., a famous early American naval architect, on spot where Royal Navy had kept signal bonfires for its ships during the American Revolution. Lighthouse completed in 1796 at cost of $22,300. The area is included in Montauk Point State Park. Coast Guard station.

Mont-aux-Sources, mountain (10,822 ft/3,299 m), W KWAZULU-NATAL province, REPUBLIC OF SOUTH AFRICA, on LESOTHO and FREE STATE province borders, 50 mi/80 km SE of BETHLEHEM; 28°46′S 28°53′E. For a time thought to be highest peak of DRAKENSBERG RANGE and of South Africa, but in 1951 THABANA-NTLENYANA (11,421 ft/3,481 m), to S in Lesotho, was found to be higher.

Mont aux Sources National Park (mont-O), MOKHOTLONG district, NE LESOTHO, in N part of DRAKENSBERG, 95 mi/153 km ENE of Maseru, bounded on

N and E by SOUTH AFRICA, 35 mi/56 km SSW of HARRISMITH, opposite ROYAL NATAL NATIONAL PARK; 28°48′S 28°56′E. Source of Khubelu River; Mont-aux-Sources (10,768 ft/3,282 m) in N.

Mön Tawang, India: see TAWANG.

Montbard (mon-bahr), town (□ 17 sq mi/44.2 sq km; 2004 population 5,815), CÔTE-D'OR department, in BURGUNDY, E central FRANCE, on small Brenne River and BURGUNDY CANAL, and 40 mi/64 km NW of DIJON; 47°35′N 04°30′E. Steel pipe factory, tile works. Buffon (born here) acquired in 1740 the 14th-century ruined castle of the dukes of Burgundy; it is now an archaeology museum, which also honors the work of Buffon. The 12th-century Cistercian abbey of FONTENAY is 2 mi/3.2 km NE.

Montbazon (mon-bah-zon), commune (□ 2 sq mi/5.2 sq km; 2004 population 3,713), INDRE-ET-LOIRE department, CENTRE administrative region, W central FRANCE, on INDRE RIVER, and 7 mi/11.3 km SSE of TOURS; 47°17′N 00°43′E. Flour milling. Has 11th-century keep built by counts of ANJOU.

Mont Beauviay, FRANCE: see MORVAN.

Montbeillard (mon-be-YAHR), former place, SW QUEBEC, E CANADA; 48°02′N 79°15′W. Amalgamated into ROUYN-NORANDA in 2002.

Montbéliard (mon-bail-yahr), industrial town (□ 5 sq mi/13 sq km), DOUBS department, FRANCHE-COMTÉ region, E FRANCE, on the RHÔNE-RHINE CANAL, and 10 mi/16 km SSW of BELFORT; 47°20′N 06°40′E. Automobiles are the town's primary product. With its surrounding countryside it constituted a county (after the 12th century) of the Holy Roman Empire. The county passed (1397) to the counts (later dukes) of WÜRTTEMBERG, who held it, with interruptions, until its capture by French Revolutionary troops in 1793. The town was a Huguenot refuge during the Reformation. It was formally ceded to France by the Treaty of LUNÉVILLE (1801). The 15th-century castle of the Montbéliard counts was rebuilt in the 18th century. Peugeot auto museum is located in industrial suburb of SOCHAUX (auto assembly plant). There are technical institutes and new industrial parks in the urban complex whose growing population exceeds 115,000.

Mont-Bellevue (MON–bel-VYOO), borough (French *arrondissement*) of SHERBROOKE, S QUEBEC, E CANADA.

Mont Belvieu (BEL-vyoo), town (2006 population 2,603), CHAMBERS county, SE TEXAS, suburb, 28 mi/45 km E of HOUSTON, on Cedar Bayou, 6 mi/9.7 km inland from Trinity Bay; 29°51′N 94°52′W. Urban growth area. In oil-producing and agricultural area (rice, soybeans; cattle); natural-gas and gas-liquids processing.

Mont Blanc (mon blahnk), Italian *Monte Bianco*, mountain, highest Alpine massif, on the French-Italian and French-Swiss border, culminating in Mont Blanc (15,771 ft/4,807 m), highest peak in FRANCE and in EUROPE W of the CAUCASUS; 45°55′N 06°55′E. The Massif extends c.30 mi/48 km N from vicinity of LITTLE SAINT BERNARD PASS to a point overlooking the great right-angle bend of the RHÔNE at MARTIGNY (Switzerland). The SE (Italian) face is a sheer wall; on the NW slopes are numerous glaciers, the largest of which (the *mer de glace*) flows into the valley of Chamonix (Haute-Savoie department), a famous French resort region and starting point for mountain climbers. There are many accommodations along the base of Mont Blanc for both summer and winter tourists and athletes. ARGENTIÈRE, SAINT-GERVAIS-LES-BAINS, and MEGÈVE are the main resorts on the French side of Mont Blanc, in addition to Chamonix. The first successful ascent of Mont Blanc was made in 1786. Today, it is climbed with relative ease, with guides, from the mountain terminus of AIGUILLE DU MIDI aerial tramway (12,572 ft/3,832 m). From that point (a lookout created by excavating a chamber within the highest peak of the Aiguille [French=needle] du Midi), the finest views of the Alpine landscape can be found. Other summits of the Mont Blanc Massif include, from N to S, AIGUILLE D'ARGENTIÈRE (12,795 ft/3,900 m), AIGUILLES ROUGES (9,728 ft/2,965 m), AIGUILLE VERTE (13,524 ft/4,122 m), GRANDES JORASSES (13,806 ft/4,208 m), AIGUILLE DU GÉANT (13,166 ft/4,013 m), Aiguille du Midi (12,605 ft/3,842 m), Mont Blanc (15,771 ft/4,807 m), Mont MAUDIT (14,649 ft/4,465 m), Aiguille des Glaciers (12,520 ft/3,816 m). In 1965 a highway tunnel (7.2 mi/11.6 km long) under the Mont Blanc Massif (directly under Aiguille du Midi) from CHAMONIX to COURMAYEUR in Italy, was opened to traffic. It provides a short, year-round route between PARIS and ROME, as well as a link between GENEVA and N ITALY. A road circuit of Mont Blanc, via GREAT and Little Saint Bernard passes, is a favorite tourist route (200 mi/322 km).

Mont Blanc de Seilon (mohn BLAHN duh sai-YON), peak (12,697 ft/3,870 m) in PENNINE ALPS, S SWITZERLAND, 17 mi/27 km S of SION.

Montblanch (mont-BLAHNCH), town, TARRAGONA province, NE SPAIN, in CATALONIA, 19 mi/31 km NNW of TARRAGONA; 41°22′N 01°10′E. Manufacturing of cement, alcohol, liqueurs, chocolate; olive-oil processing. Agricultural trade (wine, cereals, fruit, filberts; lumber). Has old walls, gates, and towers, and 14th-century church. Cistercian abbey of POBLET is 4 mi/6.4 km W.

Montbrió de Tarragona (mont-BREE-o dhai tah-rah-GO-nah), town, TARRAGONA province, NE SPAIN, 6 mi/9.7 km WSW of REUS; 41°07′N 01°00′E. Manufacturing of cotton textiles, needles; olive oil and wine processing. Wheat, almonds, filberts in area.

Montbrison (mon-bree-son), town (□ 6 sq mi/15.6 sq km), LOIRE department, RHÔNE-ALPES region, E central FRANCE, 8 mi/12.9 km NW of SAINT-ÉTIENNE; 45°40′N 04°05′E. Road and market center (bakery products, wines of FOREZ); manufacturing of precision tools (especially drills) and toys (chiefly dolls). Has 13th–16th-century church of Notre-Dame-d'Espérance, the medieval Diana chapter house, which now contains a lapidary museum, and old houses surrounded by a belt of boulevards. After 1441, Montbrison was the seat of the counts of Forez, and for a time (1801–1856) ⊙ Loire department. Town is built in circular fashion surrounding a volcanic mound.

Montbron (mon-bron), commune (□ 16 sq mi/41.6 sq km), CHARENTE department, POITOU-CHARENTES region, W FRANCE, on the TARDOIRE river, and 17 mi/27 km E of ANGOULÊME; 45°40′N 00°30′E. Textiles. Has 12th-century Romanesque church.

Mont Brûlé, peak (11,762 ft/3575 m), in PENNINE ALPS, VALAIS canton, SW SWITZERLAND, on Italian border, 6 mi/10 km W of MATTERHORN.

Mont-Brun (mon—BRUN), former municipality, SW QUEBEC, E CANADA; 48°22′N 78°43′W. Amalgamated into ROUYN-NORANDA in 2002.

Mont Buxton, district, NE MAHÉ ISLAND, SEYCHELLES; 04°37′S 55°27′E. Borders LA RIVIÈRE ANGLAISE (N and E), SAINT LOUIS (S), and BEAU VALLON (W and N) districts. Formed c.1979.

Montcalm (mon-KAHLM), rural municipality (□ 181 sq mi/470.6 sq km; 2001 population 1,400), S MANITOBA, W central CANADA, 51 mi/81 km S of WINNIPEG; 49°09′N 97°20′W. Agriculture (grain, oilseed, legumes; livestock). Composed of the communities of Saint Joseph, SAINT JEAN BAPTISTE, and LETELLIER.

Montcalm (mahnt-KAHLM, French, mon-KAHLM), county (□ 3,894 sq mi/10,124.4 sq km), LANAUDIÈRE region, SW QUEBEC, E CANADA, N of the SAINT LAWRENCE RIVER, on GATINEAU river; ⊙ SAINTE-JULIENNE; 47°28′N 76°00′W. Composed of eleven municipalities. Formed in 1982.

Montcalm, county (□ 721 sq mi/1,874.6 sq km; 2006 population 63,977), central MICHIGAN; ⊙ STANTON; 43°18′N 85°09′W. Drained by FLAT and PINE rivers,

FISH CREEK, and short Tamarack River. Agriculture (potatoes, apples, corn, wheat, beans; cattle, hogs, sheep; dairy products). Manufacturing at GREENVILLE; lake resorts. Has state game area. Wintersköl Ski Area in NW; small part of Manistee National Forest in NW corner. Organized 1850.

Montcalm (mahnt-KAHM), unincorporated town (2000 population 885), MERCER county, S WEST VIRGINIA, 6 mi/9.7 km N of BLUEFIELD, near BLUESTONE RIVER; 37°21′N 81°15′W. Railroad junction. Agriculture (grain, tobacco); livestock; bituminous coal. Pinnacle Rock State Park to SW.

Montcalm (mon-KAHLM), village (□ 46 sq mi/119.6 sq km; 2006 population 575), LAURENTIDES region, S QUEBEC, E CANADA, 12 mi/19 km from SAINTE-AGATHE-DES-MONTS; 45°58′N 74°30′W.

Mont-Carmel (mon-kahr-MEL), village (□ 168 sq mi/436.8 sq km; 2006 population 1,222), BAS-SAINT-LAURENT region, S QUEBEC, E CANADA, 9 mi/15 km from KAMOURASKA; 47°26′N 69°52′W.

Montceau-les-Mines (mon-SO-lai-MEEN), town (□ 6 sq mi/15.6 sq km), SAÔNE-ET-LOIRE department, in BURGUNDY, E central FRANCE, on BOURBINCE RIVER and Canal du CENTRE, 9 mi/14.5 km SSW of Le CREUSOT; 46°40′N 04°22′E. Coal mining has declined but other industries are in place making boilers, rubber tires, faucets, pipes, construction materials, plastics, and hosiery. Industrial development also occurs in suburban BLANZY.

Montcenis (mon-suh-nee), commune (□ 4 sq mi/10.4 sq km; 2004 population 2,221), SW suburb of Le CREUSOT, SAÔNE-ET-LOIRE department, in BURGUNDY, E central FRANCE; 46°47′N 04°23′E.

Mont Cenis, Col du (mon se-NEE, kol dyoo), Italian *Monte Cenisio*, Alpine pass (6,831 ft/2,082 m), in SAVOIE department, RHÔNE-ALPES region, SE FRANCE, 5 mi/8 km NW of Italian border, and 2 mi/3.2 km SSE of LANSLEBOURG-MONT-CENIS; 45°15′N 06°54′E. Road between Lanslebourg and Susa (Italy), across the MAURIENNE valley (France) with DORA RIPARIA RIVER valley (Italy). Lake Mont Cenis (2 mi/3.2 km SE of pass) was first converted (1901–1921) into a reservoir behind a dam (later enlarged) and hydroelectric station producing power shared by France and Italy. Historic Mont Cenis hospice is near the lake. Known through history as a major invasion route into France, Napoleon ordered the first vehicular road to be built (1803–1811) over the pass. In 1872 the FRÉJUS railroad tunnel was completed c.12 mi/19 km SW of the pass, with its French terminus near MODANE; the tunnel is c.8.5 mi/13.7 km long. A double-lane highway tunnel in the same vicinity was completed in 1980, providing faster access to Italy than the older road over the pass, which may be closed in winter. At an elevation of 3,940 ft/1,201 m—4,260 ft/1,298 m, the new tunnel is 8 mi/12.9 km long under the Fréjus range; it connects Modane French with BARDONECCHIA (Italy).

Montcerf (mon-SERF), former village, SW QUEBEC, E CANADA; 46°32′N 76°03′W. Amalgamated into MONTCERF-LYTTON in 2001.

Montcerf-Lytton (mon-SERF–lee-TON), village (□ 138 sq mi/358.8 sq km; 2006 population 712), OUTAOUAIS region, SW QUEBEC, E CANADA; 46°32′N 76°03′W. Formed in 2001 from MONTCERF and LYTTON.

Mont Cervin, SWITZERLAND: see MATTERHORN.

Montchanin (mon-shah-NAN), town (□ 3 sq mi/7.8 sq km) S suburb of Le CREUSOT, SAÔNE-ET-LOIRE department, in BURGUNDY, E central FRANCE, on the Canal du CENTRE; 46°45′N 04°27′E. Railroad yards; forges, manufacturing of refractories. Formerly Mont Chanin-les-Mines.

Mont Chanin-les-Mines, FRANCE: see MONTCHANIN.

Montclair, city (2000 population 33,049), SAN BERNARDINO county, SE CALIFORNIA; suburb 32 mi/51 km E of downtown LOS ANGELES, and 3 mi/4.8 km W of ONTARIO; 34°04′N 117°42′W. Light manufacturing.

Former citrus fruit-area; replaced by urbanization in 1990s. San Gabriel Mountains and San Bernardino National Forest to N; Angeles National Forest to NW; Cable Airport to N; Mount Baldy Ski Area to N. Incorporated 1956.

Montclair (mahnt-KLER), unincorporated city, PRINCE WILLIAM county, NE VIRGINIA, residential suburb 25 mi/40 km SSW of WASHINGTON, D.C., near POTOMAC RIVER. PRINCE WILLIAM FOREST PARK (national park unit) to W, Leesylvania State Park to E; 38°36′N 77°20′W.

Montclair, town (2000 population 38,977), ESSEX county, NE NEW JERSEY; 40°49′N 74°12′W. A suburb of NEWARK and NEW YORK city, 6 mi/9.7 km NNW of Newark, on a slope of the WATCHUNG MOUNTAINS. Although chiefly residential, it has some manufacturing. The art museum contains several paintings by George Inness, who lived here. Montclair State University is in UPPER MONTCLAIR (notable for its many large mansions). Settled c.1666 as part of Newark, set off from Newark 1812, set off from BLOOMFIELD and incorporated 1868.

Mont Clare (MAHNT KLER), unincorporated town, UPPER PROVIDENCE township, MONTGOMERY county, SE PENNSYLVANIA, on SCHUYLKILL RIVER, residential suburb, opposite (1 mi/1.6 km N of) PHOENIXVILLE; 40°08′N 75°30′W.

Mont-Dauphin (mon–do-fan), fortified village, HAUTES-ALPES department, PROVENCE-ALPES-CÔTE D'AZUR region, SE FRANCE, in DAUPHINÉ ALPS at influx of GUIL RIVER into the DURANCE, adjacent to community of GUILLESTRE, and 16 mi/26 km S of BRIANÇON; 44°40′N 06°37′E. Built and fortified by Vauban in 1693 at the order of Louis XIV to protect the Durance valley from invasion. Has arsenal (built 18th century) with an exhibit of Vauban's architectural work in the ALPS.

Mont-de-Lans (mon-duh–lahn), commune (□ 12 sq mi/31.2 sq km), ISÈRE department, RHÔNE-ALPES region, SE FRANCE, near upper ROMANCHE river, 5 mi/8 km ESE of Le BOURG-D'OISANS; 45°02′N 06°08′E. Elevation 4,200 ft/1,280 m. Popular Alpine winter sport resort in the Massif du Pelvoux (Massif des ÉCRINS). Cable cars to glacier of Mont-de-Lans (10,499 ft/3,200 m–11,155 ft/3,400 m) from base station at Les DEUX-ALPES. CHAMBON DAM (with 2 mi/3.2 km-long flood control reservoir) is 1 mi/1.6 km N.

Mont-de-l'Enclus (maw–duh–law-KLOO), commune, Tournai district, HAINAUT province, SW BELGIUM, 5 mi/8 km W of RONSE.

Mont-de-Marsan (mon–duh–mahr-sahn), town (□ 14 sq mi/36.4 sq km), ⊙ LANDES department, AQUITAINE region, SW FRANCE, at confluence of Midou and DOUZE rivers, and 65 mi/105 km S of BORDEAUX; 43°54′N 00°30′W. Administrative and commercial center where important fairs are held. Sawmills process the timber of the LANDES district. Market for poultry and goose-liver pâté. Hippodrome and Basque fronton. A military air base for test flights is nearby. Sculpture museum in a medieval tower.

Mont des Cats (mon dai kah), one of the few hills dominating the Flanders plain, NORD department, NORD-PAS-DE-CALAIS region, N FRANCE, overlooking a vast World War I battlefield, which lies about midway between the cities of LILLE and DUNKERQUE, near Belgian border; 50°47′N 02°40′E. Atop hill (518 ft/158 m) is a Trappist monastery, founded in 1826.

Montdidier (mon-dee-dyai), town (□ 4 sq mi/10.4 sq km), SOMME department, PICARDIE region, N FRANCE, 21 mi/34 km SE of AMIENS; 49°45′N 02°35′E. Road and market center (cattle, poultry); tanning, luggage and footwear manufacturing. Has two 15th–16th-century churches.

Mont Dol (mon dol), granite rock in ILLE-ET-VILAINE department, BRITTANY, W FRANCE, 1 mi/1.6 km N of DOL-DE-BRETAGNE, and 13 mi/21 km SE of SAINT-MALO; 48°34′N 01°46′E. Situated in a reclaimed coastal marsh (Marais de DOL). Originally a Druid center, Mont Dol later gave asylum to St. Malo, St. Sampson, and other apostles of Brittany. It rises to 210 ft/64 m and is topped by a modern chapel. Remains of prehistoric animals and fish have been excavated from this rock. Commune of Mont-Dol with 12th–14th-century church is on S slope.

Mont Dolent, SWITZERLAND: see DOLENT, MONT.

Mont-Dore (mon–dor), commune (□ 13 sq mi/33.8 sq km), PUY-DE-DÔME department, AUVERGNE region, central FRANCE, in the AUVERGNE MOUNTAINS, on DORDOGNE RIVER, and 19 mi/31 km SW of CLERMONT-FERRAND; 45°34′N 02°49′E. Elevation 3,445 ft/1,050 m. Noted thermal resort frequented since Roman times for respiratory diseases. Dominated by the volcanic PUY DE SANCY (cable car), Capucin peak (funicular railroad), and several other volcanic cones, it has become an important winter-sports center. Also called Mont-Dore-les-Bains.

Mont-Dore, Le (mon–DOR, luh), now suburbanized settlement, NEW CALEDONIA, on SW coast, 8 mi/12.9 km E of NOUMÉA; 22°17′S 166°35′E. Agriculture, market gardening.

Mont-Dore-les-Bains, FRANCE: see MONT-DORE.

Mont-Dore, Massif du (mon–dor, mah-seef dyoo), mountain mass of volcanic cones and peaks of the AUVERGNE, central FRANCE. PUY DE SANCY (6,184 ft/1,885 m) is highest peak. Tourism and livestock raising. Town of MONT-DORE lies amidst these mountains.

Monte (MON-tai) or **San Miguel del Monte**, town (1991 population 9,905), ⊙ Monte district (□ 671 sq mi/1,744.6 sq km), E central BUENOS AIRES province, ARGENTINA, 65 mi/105 km SSW of BUENOS AIRES; 35°30′S 58°45′W. Agricultural center (grain; livestock).

Monte (MON-tai), township, MADEIRA, PORTUGAL, 2 mi/3.2 km N of FUNCHAL, overlooking Funchal Bay. Elevation 1,965 ft/599 m. Resort. Church, founded 1470, contains tomb of Emperor Karl of Austria.

Monte (MON-tai), fishing village, AVEIRO district, N central PORTUGAL, on Aveiro lagoon, 7 mi/11.3 km N of AVEIRO. Saltworks.

Monteagle (mawnt-EE-guhl), town (2006 population 1,215), GRUNDY and MARION counties, S central TENNESSEE, 32 mi/51 km NW of CHATTANOOGA, on Monteagle Mountain in the CUMBERLAND MOUNTAINS; 35°15′N 85°50′W. Elevation c.1,900 ft/579 m. Summer resort; vineyards.

Monteagudo, town and canton, ⊙ HERNANDO SILES province, CHUQUISACA department, S BOLIVIA, 100 mi/161 km ESE of SUCRE; 19°49′S 63°59′W. Elevation 3,733 ft/1,138 m. On Sucre-CAMIRI road. On Camiri-Tintín oil and gas pipeline; gas pipelines, Taquipirenda-Camiri; MONTEAGUDO-Sucre. Corn, potatoes, tobacco, bananas. Formerly called Sauces.

Monteagudo, town, NAVARRE province, N SPAIN, 8 mi/12.9 km SW of TUDELA. Olive-oil processing; wine, hemp, cereals, fruit.

Monteagudo (mon-tai-ah-GOO-do), village, ⊙ Guaraní department, SE MISIONES province, ARGENTINA, on URUGUAY RIVER, opposite ALTO URUGUAI (BRAZIL), and 110 mi/177 km E of POSADAS; 27°31′S 65°17′W. Tung plantation center.

Monteagudo de las Vicarías (mon-tai-ah-GOO-do dhai lahs vee-kah-REE-ahs), village, SORIA province, N central SPAIN, near ZARAGOZA province border, on railroad, and 32 mi/51 km SE of SORIA; 41°22′N 02°10′W. Cereals. Historic palace. Irrigation reservoir nearby.

Monte Aguila (MON-te ah-GEE-lah), village, CONCEPCIÓN province, BÍO-BÍO region, S central CHILE, 40 mi/64 km ESE of CONCEPCÍON; 37°04′S 72°27′W. Railroad junction; wheat, corn, wine, vegetables; livestock; flour milling.

Monte Albán (MOH-te ahl-BAHN), ancient city, c.7 mi/11.3 km from OAXACA, SW MEXICO; ⊙ Zapotec; 17°02′N 96°46′W. Monte Albán was built on an artificially leveled, rocky promontory above the Valley of OAXACA. Located around an enormous plaza about 1,000 ft/305 m long and 650 ft/198 m wide are long, low buildings set off by sunken courts and stairways. The tombs, particularly Tomb 7, have yielded great archaeological treasure—jewelry of gold, copper, jade, rock crystal, obsidian, and turquoise mosaic, and bone and wood carving showing elaborate religious symbolism. Excavation was begun (1931) by the Mexican archaeologist Alfonso Caso. The Zapotec apparently had an advanced culture here c.200 B.C.E. and already were using the bar and dot system of numerals used by the Maya. The final epoch (c.1300–1521), terminated by the Spanish conquest, covers the ascendancy of the Mixtec, when the Zapotec were driven from Monte Albán and Mitla. Tomb 7 belongs to the final period. Cultural links with the Olmec and the Toltec have been found.

Monte Alegre (mon-chee al-le-grai), city (2007 population 62,073), W central PARÁ state, BRAZIL, on height near left bank of the AMAZON, 55 mi/89 km NE of SANTARÉM; 02°00′S 54°03′W. Ships rubber, fish, alcohol; grain and sugarcane growing, cattle raising. Jasper deposits nearby.

Monte Alegre (MON-che AH-le-grai), city, E central PARANÁ state, BRAZIL, 15 mi/24 km NNW of TIBAGI, 110 mi/177 km NW of CURITIBA; 24°17′S 50°25′W. Large woodpulp and newsprint plant built in 1940s, using Paraná pine. Hydroelectric power generated at Mauá Falls on TIBAGI RIVER (24 mi/39 km NW).

Monte Alegre, city (2007 population 20,675), E RIO GRANDE DO NORTE state, BRAZIL, 22 mi/35 km SW of NATAL; 06°04′S 35°20′W. Corn, cotton, aloe; livestock.

Monte Alegre, Brazil: see TIMBAÚBAS.

Montealegre del Castillo (mon-tai-ah-LE-grai dhel kahs-TEE-lyo), town, ALBACETE province, SE central SPAIN, 13 mi/21 km WSW of ALMANSA; 38°47′N 01°19′W. Flour milling, plaster manufacturing; livestock raising; cereals, wine. Fine hunting grounds in vicinity.

Monte Alegre do Minas (MON-chee AH-le-grai do MEE-nahs), city, W central MINAS GERAIS state, BRAZIL, in TRIÂNGULO MINEIRO, 16 mi/26 km SW of TUPACIGUARA; 18°48′S 48°47′W.

Monte Alegre do Piauí (MON-che AH-le-gre do PEE-ah-oo-ee), town (2007 population 10,336), extreme SW PIAUÍ state, BRAZIL; 09°46′S 45°18′W.

Monte Alegre do Sergipe (MON-che AH-le-gre do SER-zhee-pe), city, N SERGIPE state, BRAZIL, 73 mi/117 km NW of ARACAJU; 10°02′S 36°33′W. Corn, manioc; sheep.

Monte Alegre do Sul (MON-che AH-le-gre do sool), city (2007 population 6,954), E SÃO PAULO state, BRAZIL, 28 mi/45 km NE of CAMPINAS; 22°40′S 46°41′W.

Monte Alto (MON-che AHL-to), city (2007 population 44,085), N central SÃO PAULO state, BRAZIL, 45 mi/72 km W of RIBEIRÃO PRÊTO; 21°17′S 48°29′W. Manufacturing of flour products; coffee, rice, and cotton processing.

Monte Alto (MAHN-tai AHL-to), unincorporated town, HIDALGO county, S TEXAS, 23 mi/37 km NE of MCALLEN; 26°22′N 97°58′W. Located in irrigated Rio Grande Valley agricultural area (citrus, vegetables, cotton). Manufacturing (frozen foods).

Monte Aprazível (MON-che AH-prah-see-vel), city (2007 population 19,706), NW SÃO PAULO state, BRAZIL, 21 mi/34 km W of SÃO JOSÉ DO RIO PRÊTO; 20°45′S 49°42′W. Pottery manufacturing; processing of brown sugar, corn meal, rice, and coffee; distilling.

Montearagón (MON-tai-ah-rah-GON), village, TOLEDO province, central SPAIN, near the TAGUS, on railroad, and 11 mi/18 km E of TALAVERA DE LA REINA; 39°58′N 04°38′W. Cereals; livestock; apiculture.

Monte Azul (mon-chee ah-sool), city (2007 population 22,645), N MINAS GERAIS state, BRAZIL, 110 mi/177 km NE of MONTES CLAROS; 15°14′S 42°51′W. On railroad from SALVADOR to RIO DE JANEIRO; cotton growing. Until 1939, called Tremedal.

Monte Azul, Brazil: see MONTE AZUL PAULISTA.

Monte Azul do Turvo, Brazil: see MONTE AZUL PAULISTA.

Monte Azul Paulista (MON-che AH-sool POU-lee-stah), city (2007 population 19,187), N SÃO PAULO state, BRAZIL, 25 mi/40 km S of BARRETOS; 20°55′S 48°38′W. Coffee, rice, tobacco. Until 1944, Monte Azul; and Monte Azul do Turvo, 1944–1948.

Montebello, city (2000 population 62,150), LOS ANGELES county, S CALIFORNIA; residential and industrial suburb 7 mi/11.3 km ESE of LOS ANGELES; 34°01′N 118°07′W. Diversified manufacturing. A wide variety of products are manufactured here, developed significantly during the 1970s along with the Southern California area. The city is also known for its oil wells. Bounded on E by branch of LOS ANGELES RIVER, bounded W by Los Angeles. Incorporated 1920.

Montebello (mon-te-BAI-yo), town, ⊙ Montebello municipio, ANTIOQUIA department, NW central COLOMBIA, 20 mi/32 km ESE of MEDELLÍN; 07°05′N 75°34′W. Agriculture includes coffee, plantains, sugarcane; livestock.

Montebello (mahn-ti-BEL-o), village (□ 3 sq mi/7.8 sq km), OUTAOUAIS region, SW QUEBEC, E CANADA, on the OTTAWA River NE of OTTAWA; 45°39′N 74°56′W. Summer resort in a lumbering and farming area.

Montebello, village (□ 4 sq mi/10.4 sq km; 2006 population 3,690), ROCKLAND county, SE NEW YORK, 1 mi/1.6 km NNE of SUFFERN, 9 mi/14.5 km NW of NYACK; 41°07′N 74°07′W.

Monte Bello Islands (MON-te BE-lo), group on North West Shelf of INDIAN OCEAN, 50 mi/80 km off NW coast of WESTERN AUSTRALIA, 10 mi/16 km N of BARROW ISLAND; 14 mi/23 km long, 8 mi/13 km wide; 20°28′S 115°32′E. Oil and natural gas. Site of British nuclear experiments, 1952.

Montebello Vicentino (MON-te-BEL-lo vee-chen-TEE-no), village, VICENZA province, VENETO, N ITALY, near CHIAMPO River, 10 mi/16 km SW of VICENZA; 45°27′N 11°23′E. Manufacturing (machinery, wine, leather, clothing).

Montebelluna (MON-te-bel-LOO-nah), town, TREVISO province, VENETO, N ITALY, at SW foot of the MONTELLO, 12 mi/19 km NW of TREVISO; 45°46′N 12°02′E. Railroad junction; cotton and hemp mills; manufacturing of agricultural tools, metal furniture, fabricated metals, plastics, machinery, clothing, leather belts, hosiery.

Monte Bianco, FRANCE: see MONT BLANC.

Montebourg (mon-tuh-boor), commune (□ 2 sq mi/5.2 sq km), MANCHE department, BASSE-NORMANDIE region, NW FRANCE, near E coast of COTENTIN PENINSULA, 15 mi/24 km SE of CHERBOURG; 49°29′N 01°23′W. Livestock market; dairying. Here Germans sought to stop American advance on Cherbourg during NORMANDY campaign (June 1944) of World War II. UTAH BEACH is a few miles SE.

Monte Brè (MOHN-teh BREH), peak (3,035 ft/925 m), in LEPONTINE ALPS, TICINO canton, S SWITZERLAND; panoramic view of Lepontine Alps and Lake Lugano reached by cog railroad from LUGANO.

Monte Buey (MON-tai BWAI), town, E CÓRDOBA province, ARGENTINA, 28 mi/45 km SW of MARCOS JUÁREZ; 32°55′S 62°27′W. Wheat, corn, soybeans, cattle.

Monte Carlo (MON-tai KAHR-lo) or **Puerto Monte Carlo**, town, ⊙ San Pedro department, central MISIONES province, ARGENTINA, river port on PARANÁ River (PARAGUAY border), and 85 mi/137 km NE of POSADAS; 26°34′S 54°47′W. Agricultural center (corn, tobacco, maté, tung); sawmills.

Monte Carlo (mahn-TE-kahr-LO), tourist quarter of the principality of MONACO, SE FRANCE, lying along the MEDITERRANEAN SEA on the *Côte d'Azur* (FRENCH RIVIERA); 43°44′N 07°25′E. World-famous gambling casino in a resort setting (subtropical vegetation, luxurious villas and hotels). In 1863, Prince Charles III established the *Société des Bains de Mer* [French= seaside resort corporation] now known as SBM, whch then consisted of a few hotels, a theater, and a casino (first opened in 1861). In 1866 Charles created the eponymous district of Monte-Carlo [Italian=mount of Charles]. In 1878, Charles Garnier, architect of the Garnier Opera House in Paris, designed the present Casino and its included Opera House (*Salle Garnier*). Francois Blanc, director of the casino in Bad Homburg (Germany), directed the development of the resort during the *Belle Époque*. In 1954 the gambling concession came under the control of Aristotle Onassis, the Greek shipping magnate; it was taken over by the government in 1967, but contributes less than 5% of the annual state budget. Since World War II, high-rise hotels and apartment houses have replaced many of the spectacular villas characteristic of the *Belle Époque*. Numerous festivals continue to be held here. Famous events that bear the name of the district include the Monte Carlo car rally, the Monaco Grand Prix, an international tennis championship, the Monte Carlo Open golf tournament, and performances by the opera and Monte Carlo symphony orchestra.

Monte Carmelo (mon-chee kahr-me-lo), city (2007 population 44,428), W MINAS GERAIS state, BRAZIL, in the TRIÂNGULO MINEIRO, on railroad, and 50 mi/80 km ENE of UBERLÂNDIA; 18°40′S 47°27′W. Cattle, soybeans. Formerly spelled Monte Carmello.

Monte Carmelo (MON-tai kahr-MAI-lo), town, ⊙ Monte Carmelo municipio, TRUJILLO state, W VENEZUELA, in ANDEAN spur, 16 mi/26 km SW of VALERA; 09°11′N 70°49′W. Elevation 3,451 ft/1,051 m. Coffee, sugarcane, corn.

Monte Caseros (MON-tai kah-SAI-ros), town (1991 population 19,649), ⊙ Monte Caseros department, SE CORRIENTES province, ARGENTINA, port on URUGUAY RIVER, opposite BELLA UNIÓN (URUGUAY), and 80 mi/ 129 km SSE of MERCEDES; 30°15′S 57°39′W.

Montecassiano (MON-te-kahs-SYAH-no), town, MACERATA province, The MARCHES, central ITALY, 5 mi/8 km N of MACERATA; 43°21′N 13°26′E. Manufacturing clothing.

Monte Cassino (MON-tai kahs-SEE-no), monastery, in LATIUM, central ITALY, E of the RAPIDO River. Situated on a hill (1,674 ft/510 m) overlooking CASSINO, it was founded c.529 by Saint Benedict of Nursia, whose rule became that of all Benedictine houses in the world. Monte Cassino was throughout the centuries one of the great centers of Christian learning and piety; its influence on European civilization is immeasurable. Its greatest abbot after Saint Benedict was Desiderius (later Pope Victor III) in the 11th century. The buildings of the abbey were destroyed four times: by the Lombards (c.581); by the Arabs (883); by an earthquake (1349); and, after their restoration in the 17th century, by a concentrated Allied aerial bombardment in 1944. The German garrison, who had used the abbey as a fortress, survived the bombing in previously dug caves, but the buildings were flattened and most of their art treasures destroyed. The abbey's monks were able to save a considerable portion of the library's collection of invaluable manuscripts. The monastery was rebuilt again after World War II.

Monte Castelo (MON-che KAH-ste-lo), town (2007 population 8,113), NE SANTA CATARINA state, BRAZIL, 36 mi/58 km SW of MAFRA; 26°30′S 50°11′W. Rice, manioc; fruit; livestock.

Montecatini Terme (TER-me), town, PISTOIA province, TUSCANY, central ITALY, at foot of ETRUSCAN APENNINES, 8 mi/13 km WSW of PISTOIA; 43°53′N 10°46′E. Health resort, noted for thermal saline springs; manufacturing (cotton textiles, cork products, hats, gloves, pasta). Nearby copper mine now closed.

Montecchio Emilia (mon-TEK-kyo e-MEE-lyah), town, REGGIO NELL'EMILIA province, EMILIA-ROMAGNA, N central ITALY, near ENZA River, 9 mi/14 km W of REGGIO NELL'EMILIA; 44°42′N 10°27′E. Railroad terminus; fabricated metals, machinery, food products.

Montecchio Maggiore (mahd-JO-re), town, VICENZA province, VENETO, N ITALY, 7 mi/11 km SW of VICENZA; 45°30′N 11°24′E. Manufacturing (electric motorcars, celluloid, chemicals, plastics, textiles, clothing, leather); food processing (canned fruit, marmalade, liquor). Has two castles.

Monte Ceneri (MOHN-teh tcheh-NEH-ree), railroad and highway pass (1,818 ft/554 m), TICINO canton, S central SWITZERLAND, between BELLINZONA and LUGANO; transmitter for national Italian-language broadcasts above the pass.

Monte Cervino, SWITZERLAND: see MATTERHORN.

Montech (mon-TESH), commune (□ 19 sq mi/49.4 sq km), TARN-ET-GARONNE department, MIDI-PYRÉNÉES region, SW FRANCE, on GARONNE LATERAL CANAL, 7 mi/11.3 km WSW of MONTAUBAN; 43°58′N 01°14′E. Paper mill. Since 1974 an inclined boat-lift has replaced the locks on this canal, which enjoys heavy commercial traffic.

Montechiarugolo (MON-te-kyah-ROO-go-lo), town, PARMA province, EMILIA-ROMAGNA, N central ITALY, on ENZA River, and 9 mi/14 km SE of PARMA; 44°42′N 10°25′E. Small diversified manufacturing center. Has fine castle.

Montecillos, Cordillera de (mon-te-SEE-yos, kor-dee-YE-rah de), N spur on Continental Divide, W central HONDURAS, extending c.20 mi/32 km NW-SE between JESÚS DE OTORO and LA PAZ; 14°49′N 88°03′W. Forms watershed between Río Grande de Otoro section of ULÚA RIVER (W) and COMAYAGUA RIVER (E). Elevation 7,874 ft/2,400 m.

Montecito, unincorporated town, SANTA BARBARA county, SW CALIFORNIA; residential suburb near Pacific Ocean, 3 mi/4.8 km E of SANTA BARBARA.

Monte Comán (MON-tai ko-MAHN), town, central MENDOZA province, ARGENTINA, on DIAMANTE River (irrigation area), and 25 mi/40 km E of SAN RAFAEL; 34°36′S 67°54′W. Railroad junction and agricultural center (wine, alfalfa, corn, fruit, tomatoes; apiculture).

Monte Cristi (MON-te KREE-stee) or **Montecristi**, province (□ 1,150 sq mi/2,990 sq km; 2002 population 111,014), NW DOMINICAN REPUBLIC, on the coast, bordering W on HAITI; ⊙ MONTE CRISTI; 19°40′N 71°25′W. Watered by the YAQUE DEL NORTE, the irrigated valley of which here forms W section of fertile CIBAO region; adjoined by Cordillera SEPTENTRIONAL (N) and Cordillera CENTRAL (S). Lumbering; livestock raising; agriculture (rice, bananas, cocoa, corn; beeswax, honey; dairy products; hides; hardwood.) Main center is Monte Cristi city, with nearby harbor. Port of PEPILLO SALCEDO, built 1945, ships bananas. Province was set up 1879; in 1938 DAJABÓN province was separated from it.

Monte Cristi (MON-te KREE-stee), officially San Fernando de Montecristi, city, ⊙ MONTE CRISTI province, NW DOMINICAN REPUBLIC, near coast and near mouth of the YAQUE DEL NORTE, 65 mi/105 km WNW of SANTIAGO, 150 mi/241 km NW of SANTO DOMINGO; 19°51′N 71°40′W. Trading and agricultural center (rice, cotton, coffee, bananas; goats) in irrigated W section of the fertile CIBAO region. Exports hides and skins through its fine harbor (1 mi/1.6 km away). Founded 1533 by Spanish peasants on a site explored and named by Columbus in 1493. Leveled 1606 by the Spanish because of illegal trade with pirates; rebuilt c. 1750. Also spelled Montecristi. Sometimes called San Fernando de Monte Cristi.

Montecristi (mon-te-KREES-tee), town (2001 population 14,636), MANABÍ province, W ECUADOR, on MANTA-PORTOVIEJO railroad, and on Manta-GUAYAQUIL highway, 15 mi/24 km W of Portoviejo; 01°03′S 80°40′W. Famous for manufacturing of Panama hats and wickerwork; copra-trading center. President Eloy Alfaro born here.

Monte Cristo (MON-te KREES-to), village, W central AMAZONAS state, BRAZIL, 56 mi/90 km SW of CARAUARI, on Rio JURUÁ; 05°14′S 67°17′W.

Monte Cristo (MAHN-tee KRIS-to), unpopulated, rocky island (□ 6 sq mi/15.6 sq km), belonging to ITALY, in the TYRRHENIAN Sea between CORSICA and the Italian coast; 42°20′N 10°19′E. It owes its fame to the novel by Alexandre Dumas, *The Count of Monte Cristo*.

Montecristo, Cerro (mon-te-KREES-to SE-ro), peak (7,920 ft/2,414 m) in Sierra de Metapán, on GUATEMALA–EL SALVADOR–HONDURAS border, 10 mi/16 km W of NUEVA OCOTEPEQUE (Honduras); 14°25′N 89°21′W.

Montecristo National Nature Reserve, EL SALVADOR, N CHALATENANGO and Santa Rosa, extends into HONDURAS and GUATEMALA. Salvadoran portion of tri-national park. In El Salvador this park protects the country's last remaining cloud forest.

Montecristo-Trifinio National Park, HONDURAS: see EL TRIFINIO INTERNATIONAL PARK.

Monte do Carmo (MON-che do KAHR-mo), town (2007 population 6,387), S TOCANTINS state, BRAZIL, 242 mi/389 km S of PALMAS; 10°55′S 48°01′W.

Monte, El, CHILE: see EL MONTE.

Monte Escobedo (MON-te es-ko-BE-do), town, ZACATECAS, N central MEXICO, 45 mi/72 km SW of JÉREZ DE GARCÍA SALINAS; 22°20′N 103°30′W. Grain, alfalfa, sugarcane, livestock.

Monte Estoril (MONT esh-to-REEL), village, LISBOA district, W central PORTUGAL, on the ATLANTIC OCEAN, 14 mi/23 km W of LISBON (electric railroad); 38°42′N 09°24′W. Fashionable year-round resort between ESTORIL (just E) and CASCAIS (SW).

Montefalcione (MON-te-fahl-CHO-ne), village, AVELLINO province, CAMPANIA, S ITALY, 5 mi/8 km NE of AVELLINO; 40°58′N 14°53′E. Wine.

Montefalco (MON-te-FAHL-ko), town, PERUGIA province, UMBRIA, central ITALY, 12 mi/19 km NNW of SPOLETO; 42°54′N 12°39′E. In former church of San Francesco (mid-14th century; now a museum) are frescoes by B. Gozzoli and Perugino.

Montefiascone (MON-te-fyah-SKO-ne), town, VITERBO province, LATIUM, central ITALY, near Lake BOLSENA, 9 mi/14 km NNW of VITERBO; 42°32′N 12°01′E. Wine, olive oil. Bishopric. Has cathedral designed by Sanmicheli and ruins of castle. Nearby is 11th-century church of San Flaviano (remodeled 1262).

Monteforte d'Alpone (MON-te-FOR-te dahl-PO-ne), town, VERONA province, VENETO, N ITALY, 14 mi/23 km E of VERONA; 45°25′N 11°17′E. Secondary diversified manufacturing center. Wine, silk textiles, tomato paste.

Montefrío (mon-tai-FREE-o), town, GRANADA province, S SPAIN, 14 mi/23 km NE of Loja; 37°19′N 04°01′W. Olive-oil processing; wheat; livestock.

Monte Generoso (MOHN-teh djeh-neh-ROH-soh), peak (5,581 ft/1,701 m), in LEPONTINE ALPS, TICINO canton, S SWITZERLAND, S and E of Lake LUGANO on Italian border.

Montegiorgio (MON-te-JOR-jo), town, ASCOLI PICENO province, The MARCHES, central ITALY, 9 mi/14 km WSW of FERMO; 43°08′N 13°32′E. Manufacturing fabricated metals, clothing.

Montegnée (maw-ten-YAI), town in commune of SAINT-NICOLAS, Liège district, LIÈGE province, E BELGIUM, 3 mi/4.8 km W of LIÈGE; 50°39′N 05°31′E.

Montego Bay, city, ⊙ SAINT JAMES parish, NW JAMAICA; 18°28′N 77°55′W. One of the most popular tourist resorts in the CARIBBEAN Sea with highly developed tourism facilities, Montego Bay is also a port and commercial center. Active trade in sugar, bananas, coffee, rum. Various light manufacturing industries. International airport.

Monte Gordo (mon-tai GOR-doo), village, FARO district, S PORTUGAL, 2 mi/3.2 km WSW of VILA REAL DE SANTO ANTÒNIO, on the ATLANTIC OCEAN (S coast). Seaside resort; fisheries.

Montegranaro (MON-te-grah-NAH-ro), town, ASCOLI PICENO province, The MARCHES, central ITALY, 7 mi/11 km NW of FERMO; 43°14′N 13°38′E. Shoe manufacturing.

Monte Grande (MON-tai GRAHN-de), town, ☉ Esteban Echeverría district (□ 151 sq mi/392.6 sq km), NE BUENOS AIRES province, ARGENTINA, 16 mi/26 km SSW of BUENOS AIRES. Cattle-raising and meat-packing center; tree nurseries.

Montegut (MAHN-ti-guht), town (2000 population 1,803), TERREBONNE parish, LOUISIANA, 13 mi/21 km SE of HOUMA; 29°28′N 90°33′W. Fishing (shrimp, crabs, finfish).

Monte Hermoso (MON-tai er-MO-so), town, S BUENOS AIRES province, ARGENTINA, beach resort on ATLANTIC coast, 28 mi/45 km S of CORONEL DORREGO; 38°55′S 61°33′W.

Montehermoso (MON-tai-er-MO-so), village, CÁCERES province, W SPAIN, near ALAGÓN RIVER, 15 mi/24 km WNW of PLASENCIA; 40°05′N 06°21′W. Agricultural trade center (cereals, olive oil, pepper, tomatoes). Lead mines nearby.

Monte Horebe (MON-che O-re-be), town (2007 population 4,345), W PARAÍBA state, BRAZIL, on CEARÁ, border, 11 mi/18 km NW of BONITO DE SANTA FÉ; 07°13′S 35°34′W.

Monteiro (Mon-TAI-ro), city (2007 population 29,967), central PARAÍBA, NE BRAZIL, on BORBOREMA PLATEAU, 60 mi/97 km S of Patos; 07°49′S 37°10′W. Sugar, rice, fruit. Formerly called Alagoa do Monteiro.

Monteirópolis (MON-tai-RO-po-lees), town (2007 population 7,090), W central ALAGOAS state, BRAZIL, 11 mi/18 km NW of BATALHA; 09°38′S 37°15′W.

Monteith (mahn-TEETH), village, SE SOUTH AUSTRALIA, 50 mi/80 km ESE of ADELAIDE. Citrus fruit, dairy products.

Montejaque (mon-tai-HAH-kai), town, MÁLAGA province, S SPAIN, on W slope of the SIERRA DE RONDA, 5 mi/8 km W of RONDA; 36°43′N 05°16′W. Wheat, barley, chickpeas, wine, fruit, livestock; manufacturing of liquor, meat products. Cueva del Gato (cave) is 2 mi/3.2 km E.

Montejícar (mon-tai-HEE-kahr), town, GRANADA province, S SPAIN, 28 mi/45 km NNE of GRANADA; 37°34′N 03°30′W. Brandy distilling; livestock raising; lumbering; cereals; wine.

Montejunto, Serra de (MON-tai-JOON-too, SER-rah dai), mountain range of LISBOA district, W central PORTUGAL, extending c.20 mi/32 km NE from TÔRRES VEDRAS; 39°11′N 09°03′W. Rises to 2,178 ft/664 m.

Monte, Laguna del (MON-tai, lah-GOO-nah del), salt lake (□ 60 sq mi/156 sq km) in W BUENOS AIRES province, ARGENTINA, 20 mi/32 km NE of CARHUÉ; 10 mi/16 km long, 7 mi/11.3 km wide. GUAMINÍ is on S shore. Tourist resort.

Monte Leone, SWITZERLAND: see LEPONTINE ALPS.

Montelíbano (mon-te-LEE-bah-no), town, ☉ Montelíbano municipio, CÓRDOBA department, N COLOMBIA, on the SAN JORGE RIVER, 60 mi/97 km SE of MONTERÍA; 08°05′N 75°29′W. Cassava, rice; livestock.

Montélimar (mon-tai-lee-mahr), town (□ 18 sq mi/46.8 sq km), DRÔME department, RHÔNE-ALPES region, S FRANCE, on the small Roubion River near its influx into the RHÔNE, and 27 mi/43 km SSW of VALENCE; 44°34′N 04°45′E. Noted nougat-manufacturing center based on locally grown almonds. Trade in fruits, wines, candy. Manufacturing of cement and other construction materials nearby. Major hydroelectric installations in this stretch of the Rhône valley. The 12th-century fortress was expanded (14th century) into a castle under the rule of the popes of AVIGNON. There is an arcaded market square of typical Provence style.

Montelimar (mon-te-lee-MAHR), village, MANAGUA department, SW NICARAGUA, near PACIFIC coast, 4 mi/6.4 km SW of SAN RAFAEL DEL SUR; 11°49′N 86°31′W. Former estate of President Anastasio Somoza. Was nationalized after the 1978–1979 civil war, now developed as a tourist center.

Monte Lirio (mon-tai LEE-ree-o), village, COLÓN province, CANAL AREA, PANAMA, on JUAN GALLEGOS ISLAND of GATÚN LAKE, on transisthmian railroad, and 9 mi/14.5 km SSE of COLÓN; 09°14′N 79°51′W. Bananas; livestock.

Montellano (mon-te-LYAH-no), town, SEVILLE province, SW SPAIN, in W spur of the CORDILLERA PENIBÉTICA, 35 mi/56 km SW of SEVILLE. Processing and trading center for agricultural region (cereals, olives, legumes, livestock). Vegetable oil and liquor distilling; tanning.

Montello, town (2006 population 1,472), ☉ MARQUETTE county, central WISCONSIN, at E end of BUFFALO LAKE, at confluence of FOX RIVER and small Montello River (E end of Montrello Lake), 45 mi/72 km W of FOND DU LAC; 43°47′N 89°19′W. Resort; livestock, grain; granite quarries; manufacturing (lumber mill, wood products). Settled 1849, incorporated 1938.

Montello (mon-TEL-lo), ridge (□ 25 sq mi/65 sq km) in VENETO, N ITALY, 10 mi/16 km NE of TREVISO, W of PIAVE River; extends 15 mi/24 km E-W, 4 mi/6 km wide; rises to 1,207 ft/368 m. Heavy fighting here (June 1918) in World War I.

Montelupo Fiorentino (MON-te-LOO-po fyo-ren-TEE-no), town, FIRENZE province, TUSCANY, central ITALY, on ARNO River, at mouth of PESA River, and 12 mi/19 km W of FLORENCE; 43°44′N 11°01′E. Manufacturing non-metallics, chemicals, glass, clothing.

Monte Maíz (MON-tai mah-EEZ), town, SE CÓRDOBA province, ARGENTINA, 70 mi/113 km SE of VILLA MARÍA; 33°12′S 62°36′W. Wheat, corn, soybeans, alfalfa; hogs, cattle.

Montemar, CHILE: see VIÑA DEL MAR.

Montemarciano (MON-te-mahr-CHAH-no), town, ANCONA province, The MARCHES, central ITALY, 10 mi/16 km WNW of ANCONA; 43°38′N 13°19′E. Highly diversified secondary manufacturing center.

Montemayor (mon-tai-mei-OR), town, CÓRDOBA province, S SPAIN, 17 mi/27 km SSE of CÓRDOBA. Olive-oil processing, plaster manufacturing. Cereals, vegetables; livestock.

Montemayor de Pililla (mon-tai-mei-OR dhai pee-LEE-lyah), town, VALLADOLID province, N central SPAIN, 17 mi/27 km SE of VALLADOLID; 41°30′N 04°27′W. Lumbering; sheep raising; sugar beets, cereals, wine, chicory.

Montemolín (mon-tai-mo-LEEN), town, BADAJOZ province, SW SPAIN, 8 mi/12.9 km SE of FUENTE DE CANTOS; 38°09′N 06°12′W. Cereals, grapes, olives; livestock.

Montemorelos (mon-te-mo RE-los), city and township, NUEVO LEÓN, N MEXICO, in E foothills of SIERRA MADRE ORIENTAL, on railroad, and 45 mi/72 km SE of MONTERREY on INTER-AMERICAN HIGHWAY (Mexico Highway 85); 25°12′N 99°50′W. Agricultural center (oranges, sugarcane, pecans, cactus fibers, livestock). Known for yearly fiesta in July.

Montemor-o-Novo (MON-tai-mor–oo–NO-voo), town, ÉVORA district, S central PORTUGAL, on railroad spur, and 18 mi/29 km WNW of ÉVORA; 38°39′N 08°13′W. Cork-processing center. Has a Moorish castle.

Montemor-o-Velho (MON-tai-mor–oo–VAIL-yoo), town, COIMBRA district, N central PORTUGAL, on lower MONDEGO RIVER, and 14 mi/23 km W of COIMBRA; 40°10′N 08°41′W. In poorly drained rice-growing region. Has 11th-century feudal castle.

Montemuro, Serra de (MON-tai-MOO-roo, SER-rah dai), mountain range in VISEU district, N central PORTUGAL, 10 mi/16 km SW of LAMEGO; 40°58′N 08°01′W. Rises to 4,535 ft/1,382 m.

Montendre (mon-TAHN-druh), commune (□ 10 sq mi/26 sq km), CHARENTE-MARITIME department, POITOU-CHARENTES region, W FRANCE, 28 mi/45 km N of BORDEAUX; 45°18′N 00°25′W. Woodworking, plastics. Founded as a Roman stronghold, it has a 12th–15th-century square tower at village entrance.

Montenegro, Serbian *Crna Gora* [=black mountain], republic, (5,332 sq mi/13,810 sq km; 2004 estimated population 630,548; 2007 estimated population 684,736); ☉ PODGORICA (formerly Titograd). In early 1990, secessionist movements in four of the six Yugoslavian republics left only SERBIA and Montenegro in Yugoslavia. Their status as Yugoslavia remained uncertain, however, as the UN declared in September 1992 that the newly re-formed Yugoslav federation could not claim the international rights and duties of the former Yugoslavia. In 2003, Yugoslavia was formally abolished in favor of a loose federation of Serbia and Montenegro. In 2006, Montenegro voted to leave the federation as a completely independent republic. In addition to the capital, other principal cities are CETINJE, NIKŠIĆ, the town of BAR, Serbia-Montenegro's largest port, and KOTOR, the only naval base remaining since the breakup of the Yugoslavian federation.

Geography

Situated at the S end of the DINARIC ALPS, Montenegro is almost entirely mountainous. It consists of two regions: the barren karst of Montenegro proper (W) is separated by the ZETA RIVER and its plain from the higher BRDA region (E), which has forests and pastures.

Economy

Sheep and goat raising are important occupations. Only about 6% of the area is cultivated, and agriculture, mainly in the ZETA valley and near SKADARSKO Lake (Lake SHKODËR), which forms part of the Albanian border, is poorly developed. Industry is also relatively underdeveloped. Montenegro has significant deposits of iron, bauxite, and petroleum.

Population

The Montenegrin people are mostly Serbs, but they are recognized as a separate ethnic nationality. They belong mostly to the Orthodox faith.

History to 1696

From the 14th to the 19th century their principal activity was fighting the Turks, who never entirely conquered their mountain stronghold. The region constituting present Montenegro was, in the 14th century, the virtually independent principality of Zeta in the Serb empire. After Serbia was defeated by the Turks in the battle of KOSOVO (1389), Montenegro continued to resist and became a refuge for Serb nobles who fled Turkish rule. The sultans did not recognize Montenegrin independence, but they never succeeded in making Montenegro a tributary (although they thrice destroyed Cetinje). However, the princes of Montenegro ruled only a small part of the present republic, the rest being governed by TURKEY after 1499 and by VENICE, which held Kotor. From 1515 until 1851 the rule of Montenegro was vested in the prince-bishops (*vladikas*) of Cetinje, who were assisted by civil governors. Social organization, geared almost exclusively to the needs of war, was largely military and patriarchal.

History: 1696 to 1910

With Danilo I, who ruled from 1696 to 1735, the episcopal succession was made hereditary in the Niegosh family; the office ordinarily passed from uncle to nephew because the bishops could not marry. Danilo I also inaugurated (1715) the traditional alliance of Montenegro with RUSSIA; the emperors of Russia were henceforth considered as at least the spiritual suzerains of the *vladikas*. Peter I, who reigned from 1782 to 1830, defied both France and Austria when the Treaty of Campo Formio (1797) transferred the Venetian possession of Kotor to Austria, but he failed to obtain the coveted port. However, in 1799,

Sultan Selim III recognized the independence of Montenegro. Peter I instituted internal reforms and sought to end the blood feuds and lawlessness that had become a traditional way of life. He was canonized as a saint after his death. Peter II (reigned 1830–1851), a gifted poet, continued his predecessor's work of reform and fostered a revival of learning and culture; aside from occasional border warfare, he lived in relative peace with his neighbors, Turkey and Austria. Danilo II, who succeeded him, secularized his principality in 1852 and transferred his ecclesiastic functions to an archbishop. Under Nicholas I (reigned 1860–1918) Montenegro was formally recognized as an independent state at the Congress of BERLIN (1878), which increased its territory and gave it a narrow outlet on the ADRIATIC SEA.

History: 1910 to Present

In 1910, Nicholas proclaimed himself king. He fought Turkey in the Balkan Wars and took Shkodër in 1913, but was forced by the pressure of the European powers to evacuate the city. Montenegro did, however, receive part of the territory claimed by newly independent ALBANIA, and when World War I broke out (1914), the Montenegrins invaded Albania. Montenegro declared war on Austria in August 1914, but late in 1915 it was overrun by Austro-German forces. In November 1918, a national assembly declared Nicholas deposed and effected the union of Montenegro with Serbia. In 1946, Montenegro became one of the six republics of Yugoslavia, and its territory was enlarged with the addition of part of the Dalmatian coast. As Yugoslavia began to disintegrate in the early 1990s, Montenegro was the only republic other than Serbia in which the electorate voted to keep the former Communists in power. Montenegro sided with Serbia during the breakup of Yugoslavia, but its support lessened during the subsequent fighting and the KOSOVO war. Montenegrin leaders sought greater autonomy within Yugoslavia in the late 1990s, leading to strained relations with Serbia. Milo Djukanović, a moderate Montenegrin nationalist who became president in 1997 (and resigned to become prime minister in 2002) actively worked for Montenegro's independence. Tensions eased after Vojislav Koštunica became (October 2000) Yugoslav president, but many Montenegrins continued to favor greater autonomy or independence. A March 2002 Serbian-Montenegrin accord led to the reconstitution of Yugoslavia as Serbia and Montenegro in February 2003, giving the constituent republics greatly increased autonomy, but the new looser union was short-lived. In May 2006 Montenegrins voted for independence in a referendum, and the following month Montenegro declared itself independent. In 2006 Željko Šturanović succeeded Djukanović as prime minister; Filip Vujanović has been president since 2003.

Government

Montenegro is governed under the constitution of 1992. The president, who is the head of state, is popularly elected for a five-year term and is eligible for a second term. The government is headed by the prime minister, who is proposed by the president with the approval of the legislature. Members of the unicameral legislature, the 81-seat Assembly, are popularly elected to serve four-year terms. The current head of state is Filip Vujanović (since May 2003). The current head of government is Željko Šturanović (since November 2006). Administratively, Montenegro is divided into 21 municipalities.

Montenegro (MON-te-ne-gro), city (2007 population 56,790), E RIO GRANDE DO SUL, BRAZIL, on CAÍ river, and 28 mi/45 km NW of PÔRTO ALEGRE; 29°42′S 51°28′W. Railroad junction (spur to CAXIAS DO SUL); modern industrial center (meat packing and by-products processing); wine growing. Settled by Italian immigrants in mid-19th century. Until 1930s called São João de Montenegro.

Montenegro (mon-te-NAI-gro), town, ⊙ Montenegro municipio, QUINDÍO department, W central COLOMBIA, on W slope of Cordillera CENTRAL, on railroad, and 6.0 mi/9.7 km WNW of ARMENIA; 04°34′N 75°46′W. Elevation 4,239 ft/1,292 m. Coffee-growing center; also sugarcane, cereals, fruit; livestock.

Monte Negro, MEXICO: see SANTIAGO JOCOTEPEC.

Montenero (MON-te-NAI-ro), peak (6,171 ft/1,881 m) in La SILA mountains, S ITALY, 19 mi/31 km ESE of COSENZA.

Monte Nievas (MON-tai nee-AI-vahs), village, NE LA PAMPA province, ARGENTINA, on railroad, 28 mi/45 km SW of GENERAL PICO; 35°52′S 64°09′W. Grain; livestock.

Monte Pascoal (MON-chee PAH-sko-AHL), village, SE BAHIA state, BRAZIL, near Atlantic coast. In Parque Nacional de MONTE PASCOAL; 16°55′S 39°23′W.

Monte Patria (MON-tai PAH-tree-ah), town, ⊙ Monte Patria comuna, LIMARÍ province, COQUIMBO region, N central CHILE, on railroad, and Embalse La Paloma, 18 mi/29 km ESE of OVALLE; 30°41′S 70°56′W. Cereals, fruit; livestock.

Monte Plata (MON-te PLAH-tah), province (□ 841 sq mi/2,186.6 sq km; 2002 population 180,376), S central DOMINICAN REPUBLIC; ⊙ MONTE PLATA; 18°50′N 69°50′W. Includes S flank of Cordillera CENTRAL and coastal plain. Agricultural (rice, cacao, coffee, bananas.) Formerly part of SAN CRISTÓBAL province, also called TRUJILLO.

Monte Plata (MON-te PLAH-tah), town (2002 population 14,850), ⊙ MONTE PLATA province, S DOMINICAN REPUBLIC, at S foot of Cordillera CENTRAL, 23 mi/37 km NNE of SANTO DOMINGO; 18°45′N 69°51′W. Agricultural (rice, cacao, coffee.)

Monteponi (mon-te-PO-nee), village, CAGLIARI province, SW SARDINIA, ITALY, 2 mi/3 km SW of IGLESIAS.

Montepuez (mon-TAI-poo-ez), village, CABO DELGADO province, N MOZAMBIQUE, on road, and 95 mi/153 km W of PÔRTO AMÉLIA; 12°33′S 40°16′E. Cotton, kapok.

Montepuez District, MOZAMBIQUE: see CABO DELGADO, province.

Montepulciano (MON-te-pool-CHAH-no), town, SIENA province, TUSCANY, central ITALY, 28 mi/45 km SE of SIENA, W of Lake of Montepulciano (1.5 mi/2.4 km long; connected with the CHIANA); 43°05′N 11°47′E. Highly diversified secondary manufacturing center, including wine making, woodworking. Bishopric. Rich in Renaissance buildings; has several palaces and churches, including some designed by the elder A. da Sangallo. Cardinal Roberto Bellarmine was born here.

Monte Quemado (MON-tai kai-MAH-do), town (1991 population 7,835), ⊙ Copo department (□ 5,160 sq mi/13,416 sq km), N SANTIAGO DEL ESTERO province, ARGENTINA, on railroad, and 85 mi/137 km N of TINTINA, near Chaco border. Stock-raising center. Formerly called Kilómetro 1243.

Montereau-fault-Yonne, FRANCE: see MONTEREAU-FAUT-YONNE.

Montereau-faut-Yonne (mon-tuh-ro-fo-yon), town (□ 3 sq mi/7.8 sq km), SEINE-ET-MARNE department, ÎLE-DE-FRANCE region, N central FRANCE, inland port on the SEINE at influx of YONNE river, and 17 mi/27 km SE of MELUN; 48°23′N 02°57′E. Industrial center specializing in electronics, metal and refractory products (agricultural equipment, cable, bricks, tiles, and pottery). Sugar milling and food processing. In 19th century, the town became known for its artistic ceramics. Between two 18th-century bridges spanning the Seine and the Yonne is the equestrian statue of Napoleon I, commemorating his nearby victory (1814) over Austrian and Prussian forces prior to his capitulation in PARIS. Also spelled Montereau-fault-Yonne.

Montérégie (mon-tai-rai-ZHEE), region (□ 4,290 sq mi/11,154 sq km; 2005 population 1,371,731), S QUEBEC, E CANADA, in SE SAINT LAWRENCE plain; 45°23′N 73°06′W. Composed of 180 municipalities, centered on LONGUEUIL. Noted biotechnology center.

Monterey (mahn-tuhr-RAI), county (□ 3,322 sq mi/8,637.2 sq km; 2006 population 410,206), W CALIFORNIA, on PACIFIC OCEAN and MONTEREY BAY; ⊙ SALINAS; 36°14′N 121°19′W. Bounded N by Pajaro River valley; Salinas River valley, in its center, is flanked E by Gabilan and Diablo ranges, W by SANTA LUCIA RANGE (rising here to 5,862 ft/1,787 m at Junipero Serra Peak). MONTEREY PENINSULA (site of historic towns of MONTEREY and CARMEL) in NW is famed scenic resort region. County includes BIG SUR (redwoods); scenic Big Sur coastal drive (State Highway 1) S from Monterey; Monterey State Beach, Asilomar State Beach and Point Lobos State Reserve in NW; Fremont Peak State Park and Moss Landing State Beach in N; Andrew Molera, Pfeiffer Big Sur and Julia Pfeiffer Burns state parks in SW; old Soledad (center), Carmel, and San Antonio de Padua (SW) missions; U.S. Fort Ord Military Reservation (NW), Hunter Liggett Military Reservation (S); part of Los Padres National Forest (SW); and part of Pinnacles National Monument on NE boundary. Agriculture (grain, tomatoes, asparagus, peppers, olives, celery, spinach, mushrooms, artichokes, carrots, broccoli, lettuce, strawberries, grapes); wineries; cattle; dairying; asbestos mining at King City; sand and gravel. Formed 1850.

Monterey (mahn-tuhr-RAI), city (2000 population 29,674), MONTEREY county, W CALIFORNIA, 12 mi/19 km W of SALINAS; port on MONTEREY BAY, at base of MONTEREY PENINSULA to W; 36°36′N 121°53′W. Manufacturing (printing and publishing, calculating machines, computers, electrical aparatus, paperboard boxes). It is a popular resort, the home of many artists and writers, and one of the oldest cities in California. Juan Cabrillo visited the bay in 1542; Sebastián Vizcaíno entered and named it in 1602. An expedition under Gaspar de Portolá arrived and established a presidio in 1770. Junípero Serra remained to found a Franciscan mission. Monterey was the capital of ALTA CALIFORNIA during many of the years between 1775 and 1846. A U.S. naval force under Commodore John D. Sloat took the city in 1846, and the state constitutional convention met there in 1849. An early whaling and fishing center, Monterey's 20th-century economy is based on tourism (Fisherman's Wharf, restaurants, shops, seafood markets) and the revenues and employment derived from nearby military installations, such as Fort Ord to E. California's first theater (1844) and first brick building (1847) still stand, and it was in Monterey that California's first newspaper was established in 1846. Numerous museums are in the city, as is the Presidio of Monterey (1770) and the Monterey Bay Aquarium. Monterey Peninsula College (two-year), the Monterey Institute of International Studies, the Naval Postgraduate School, and California State University, Monterey Bay (1995). Carmel Bay to SW; Monterey Pennisula Airport to SE; Monterey State Beach. Founded 1770; incorporated 1850.

Monterey, town (2000 population 231), PULASKI county, NW INDIANA, on TIPPECANOE RIVER, and 9 mi/14.5 km NE of WINAMAC; 41°10′N 86°29′W. In agricultural area. Tippecanoe River State Park to W. Laid out 1849.

Monterey, town, BERKSHIRE county, SW MASSACHUSETTS, in the BERKSHIRES, 19 mi/31 km S of PITTSFIELD; 42°11′N 73°14′W. Beartown State Forest.

Monterey (MON-tuh-rai), town (2006 population 2,838), PUTNAM county, central TENNESSEE, 14 mi/23 km E of COOKEVILLE; 36°09′N 85°16′W. In timber, coal, and farm area; light manufacturing. Small Monterey Lake (resort) is nearby.

Monterey (mahnt-uh-RAI), town (2006 population 150), ⊙ HIGHLAND county, NW VIRGINIA, in ALLEGHENY MOUNTAINS, 32 mi/51 km NW of STAUNTON; 38°24′N 79°34′W. Manufacturing (printing and publishing, fish processing); agriculture (alfalfa); timber; freshwater fish.

Monterey (mahn-tuh-RAI), village (2000 population 167), OWEN county, N KENTUCKY, on KENTUCKY RIVER, and 16 mi/26 km N of FRANKFORT; 38°25′N 84°52′W. Agriculture (tobacco, corn; dairying); recreation (fishing, camping). Larkspur Press, small book printer using handset type.

Monterey, MINNESOTA: see TRIMONT.

Monterey Bay, crescent-shaped inlet (26 mi/42 km long) of the PACIFIC OCEAN, MONTEREY and SANTA CRUZ counties, W CALIFORNIA, c.65 mi/105 km S of SAN FRANCISCO. Formed by break in COAST RANGES. Santa Cruz Mountains (with Santa Cruz at their base) rise steeply to NE, Santa Lucia Range (behind Monterey, on Monterey Peninsula) to S. PAJARO RIVER enters from E; fertile Salinas River valley extends SE from bay. Fisheries; beaches. Sighted (1542) by Cabrillo; entered (1602) and named by Vizcaíno.

Monterey Park, city (2000 population 60,051), LOS ANGELES county, S CALIFORNIA; residential suburb 6 mi/9.6 km E of LOS ANGELES; 34°03′N 118°08′W. Wholesale, retail, and financial-service center. Some light industry and manufacturing. East Los Angeles College (two-year) here. California State University–Los Angeles to W, in Los Angeles. Was the first city in continental U.S. to have Asian majority and suburban Chinatown. Incorporated 1916.

Monterey Peninsula, rugged, almost square peninsula jutting 4 mi/6.4 km NW into the PACIFIC between MONTEREY BAY (N) and CARMEL BAY (S), W CALIFORNIA. Communities of MONTEREY, PACIFIC GROVE, DEL MONTE FOREST, Del Ray Oaks, PEBBLE BEACH, CARMEL, Asilomar, other resorts. Seventeen-Mile Drive is a scenic toll road around shore. Point Piños (lighthouse) is N tip; Cypress Point is its W extremity; Pescadero Point in S. Pebble Beach, site of world golf classics and other popular golf courses; Asilomar and Monterey state beaches in N.

Montería (mon-te-REE-ah), city, ⊙ CÓRDOBA department, N COLOMBIA, inland port on SINÚ RIVER, in savannas, and 120 mi/193 km SSW of CARTAGENA; 08°45′N 75°53′W. Center for communication, trading, lumbering, and livestock-raising; forest products. Airport. Gold placer mines and natural gas deposits nearby. Sometimes called San Jerónimo de Buenavista.

Monteriggioni (MON-te-reed-JO-nee), village, SIENA province, TUSCANY, central ITALY, 7 mi/11 km NW of SIENA; 43°23′N 11°13′E. Manufacturing machinery, non-metallics; wine making, alcohol distilling. Has 13th-century walls.

Monte Rio, unincorporated town (2000 population 1,104), SONOMA county, W CALIFORNIA, on RUSSIAN RIVER, and 15 mi/24 km W of SANTA ROSA; 38°28′N 123°01′W. Grapes, apples, grain; dairying; sheep. ARMSTRONG REDWOODS STATE PARK and Austin Creek State Recreation Area to N.

Montero (mon-TAI-ro), town (2001 population 78,294) and canton, ⊙ OBISPO SANTIESTEBAN province, SANTA CRUZ department, central BOLIVIA, 30 mi/48 km N of SANTA CRUZ; 17°21′S 63°17′W. Elevation 961 ft/293 m. Agriculture (sugarcane, rice, corn, potatoes, peanuts, coffee, coca). Formerly San Ramón de Víbora.

Monteroni di Lecce (mon-TAI-ro dee LET-che), town, LECCE province, APULIA, S ITALY, 4 mi/6 km WSW of LECCE; 40°19′N 18°06′E. Olive oil, wine. Has 16th-century ducal palace.

Monteros (mon-TAI-ros), town (1991 population 19,816), ⊙ Monteros department, S central TUCUMÁN province, ARGENTINA, on railroad, and 30 mi/48 km SW of TUCUMÁN; 27°10′S 65°30′W. Agricultural and lumbering center; sugar refineries, flour mills, sawmills; tobacco, sugar, citrus fruit, potatoes. Tobacco research station.

Monte Rosa (MON-teh ROH-sa), massif in the PENNINE ALPS, on the Swiss-Italian border, 9 mi/14.5 km SE of ZERMATT, in the canton of VALAIS. Its highest peak, the DUFOURSPITZE (15,203 ft/4,634 m), is the highest point in SWITZERLAND. Other peaks include Lyksamm (14,698 ft/4,480 m) on the W. The Swiss side is the head of the GORNER GLACIER. A nuclear research laboratory has been established on the Italian side of Monte Rosa.

Monte Rosa (MON-te RO-zah), French *Mont Rose*, highest mountain group of the PENNINE ALPS, on Italian-Swiss border, 28 mi/45 km SSW of BRIG. Has ten summits, of which the DUFOURSPITZE (15,203 ft/4,634 m) is highest. Several glaciers (the GORNER is the largest) on N slope converge upon head of Matter Visp River valley near ZERMATT.

Monterosso al Mare (MON-te-ROS-so ahl MAH-re), village, La SPEZIA province, LIGURIA, N ITALY, port on GULF OF GENOA, 9 mi/14 km NW of SPEZIA; 44°09′N 09°39′E. Noted for white wine. Bathing resort.

Monterotondo, town, ROMA province, LATIUM, central ITALY, near the TIBER, 11 mi/18 km NW of TIVOLI. Bricks, tiles, pottery, wine. Here Garibaldi defeated papal forces in 1867.

Monterrey (mon-te-RAI), city (2005 population 1,133,070) and township, ⊙ NUEVO LEÓN state, NE MEXICO, c.150 mi/241 km S of LAREDO (TEXAS), at the mouth of a valley surrounded by mountains on three sides; 25°40′N 100°20′W. The third-largest city of Mexico, Monterrey is the railroad and highway hub of NE Mexico. It is also Mexico's second-most important industrial center. The site of the nation's largest iron and steel foundries and a major producer of cement. Monterrey's modern industrial complex also includes a wide range of light manufacturing (including glass and beverages). The city has experienced further growth with the construction of maquiladoras, foreign-owned plants that use low-wage labor for goods exported to the U.S. Natural gas, coal and petroleum from the neighboring states of COAHUILA and TAMAULIPAS are also major sources of industrial activity. Its moderate, dry climate, cool mountains, and hot springs make the Monterrey area a popular resort. Monterrey is the home of the University of Nuevo León and Monterrey Technical Institute, one of Mexico's most prestigious institutions of higher education. The city was founded in 1579.

Monterrey (mon-ter-RAI), town, ⊙ Monterrey municipio, CASANARE department, S COLOMBIA, 42 mi/68 km SW of YOPAL on the E slopes of the Cordillera ORIENTAL; 04°52′N 72°53′W. Sugarcane, coffee, plantains; livestock.

Monterrico-Hawaii Biotope (mon-te-REE-ko), sanctuary (□ 24 sq mi/62.4 sq km), SANTA ROSA department, GUATEMALA, on PACIFIC OCEAN coast 22 mi/35 km E of SAN JOSE. Protects mangrove forests and coastal wetlands, and includes a portion of the CHIQUIMULILLA CANAL. One of a number of protected areas created in 1989–1990.

Monterrubio de la Serena (mon-tai-ROO-vyo dhai lah sai-RAI-nah), town, BADAJOZ province, W SPAIN, on LA SERENA plain, 9 mi/14.5 km SE of CASTUERA; 38°35′N 05°26′W. Spa; agricultural center (olives, cereals, grapes; stock).

Monterubbiano (MON-te-roob-BYAH-no), village, ASCOLI PICENO province, The MARCHES, central ITALY, 5 mi/8 km S of FERMO; 43°05′N 13°43′E. Manufacturing shoes and wood, food processing.

Montes (MON-tes), town, CANELONES department, S URUGUAY, on railroad, and 45 mi/72 km ENE of MONTEVIDEO; 34°30′S 55°35′W. Grain, beetroots; livestock.

Montesa (mon-TAI-sah), town, VALENCIA province, E SPAIN, 9 mi/14.5 km SW of JÁTIVA; 38°57′N 00°39′W. Olive oil, wine, oranges. Dominated by hill topped by ruins of fortress that belonged (14th century) to military order of Montesa.

Montes Altos (MON-chees AHL-tos), town (2007 population 8,787), W central MARANHÃO state, BRA-ZIL, near indigenous area and RIO TOCANTINS; 05°46′S 47°07′W.

Monte San Juan (MON-te sahn hwahn), municipality and town CUSCATLÁN department, EL SALVADOR, NNW of Cojuyepeque; 13°46′N 88°57′W.

Montesano (mahn-tuh-SAI-no), town (2006 population 3,525), ⊙ GRAYS HARBOR county, W WASHINGTON, 10 mi/16 km E of ABERDEEN, and on CHEHALIS RIVER, near mouth of Wyandotte River; 47°01′N 123°59′W. Railroad junction. Peas, hay; lumber, dairy products; manufacturing (lumber, pumps). Lake Sylvia State Park 1 mi/1.6 km to N. Incorporated 1883.

Montesano sulla Marcellana (MON-te-SAH-no SOOL-lah mahr-chel-LAH-nah), town, Salerno province, CAMPANIA, S ITALY, 10 mi/16 km SSE of SALA CONSILINA; 40°16′N 15°42′E. Agricultural center. Thermal springs; resort.

Monte San Savino (MON-te sahn sah-VEE-no), town, AREZZO province, TUSCANY, central ITALY, 12 mi/19 km SW of AREZZO. Small diversified industrial center, including broom manufacturing. Has municipal palace designed by the elder A. da Sangallo and Renaissance loggia. A Sansovino was born here.

Monte Sant'Angelo (MON-te sahn-TAHN-je-lo), town, FOGGIA province, APULIA, S ITALY, on S slope of GARGANO promontory, 27 mi/43 km NE of FOGGIA; 41°42′N 15°57′E. Agricultural center (cereals, wine, olive oil; livestock); chemicals. Has ancient pilgrimage church of San Michele.

Monte Santo (mon-chee shan-to), city (2007 population 52,115), NE BAHIA state, BRAZIL, 60 mi/97 km E of SENHOR DO BONFIM; 10°27′S 39°19′W. Tobacco, beans; cattle.

Monte Santo (MON-che SAHN-to), town, S TOCANTINS state, BRAZIL, 200 mi/322 km SSW of PALMAS; 09°54′S 49°03′W.

Monte Santo, Greece: see AKTI.

Monte Santo, Cape (MON-te SAHN-to), point on E coast of SARDINIA, ITALY, at S end of GULF OF OROSEI; 40°06′N 09°43′E. Lobster fisheries. Sometimes called Monte Santu.

Monte Santo de Minas (MON-chee sahn-to dee MEE-nahs), city (2007 population 20,084), SW MINAS GERAIS state, BRAZIL, near SÃO PAULO border, on railroad, and 50 mi/80 km E of RIBEIRÃO PRÊTO; 21°14′S 46°57′W. Coffee-growing center. Called Monsanto, 1944–1948.

Montescaglioso (MON-te-skah-LYO-zo), town, MATERA province, BASILICATA, S ITALY, 8 mi/13 km SSE of MATERA; 40°33′N 16°40′E. Agricultural center; olive oil, wine, cheese, fruit, wool. Has ancient Benedictine monastery (now a public building).

Montes Claros (mon-chees klah-ros), city (2007 population 352,384), MINAS GERAIS state, E central BRAZIL; 16°42′S 43°50′W. Cattle breeding; agr; manufacturing (cloth, furniture). Commercial airport, good road, railroad facilities.

Montesclaros (mon-tes-KLAH-ros), village, TOLEDO province, central SPAIN, 12 mi/19 km NW of TALAVERA DE LA REINA; 40°06′N 04°56′W. Cereals, acorns, forage; livestock. Marble and lime quarries; limekilns.

Montes Claros (mon-tesh KLAR-oos), locality, ÉVORA district, S central PORTUGAL, 9 mi/14.5 km SE of ESTREMOZ. Column commemorates final Portuguese victory over Spaniards (1665).

Montes de Oca (MON-tes dai O-kah), town, S SANTA FE province, ARGENTINA, 70 mi/113 km WNW of ROSARIO. Agricultural center (wheat, flax, soybeans, corn; livestock); dairying.

Montes de Oca, San Pedro de, COSTA RICA: see SAN PEDRO, town.

Montes de Oro, COSTA RICA: see MIRAMAR.

Montes de Toledo, SPAIN: see TOLEDO, MONTES DE.

Monte Sereno, city (2000 population 3,483), SANTA CLARA county, W CALIFORNIA; residential suburb 7 mi/11.3 km SW of downtown SAN JOSE; 37°14′N

122°00′W. Santa Cruz Mountains to SW; Los Gatos Creek to E.

Montesilvano (MON-te-seel-VAH-no), town, PESCARA province, ABRUZZI, S central ITALY, near the ADRIATIC, 4 mi/6 km WNW of PESCARA; 42°29′N 14°08′E. Fabricated metal products, wood products, clothing; food processing.

Montespertoli (MON-te-SPER-to-lee), town, FIRENZE province, TUSCANY, central ITALY, 9 mi/14 km SE of EMPOLI; 43°38′N 11°04′E. Highly diversified secondary industrial center.

Montesquieu-Volvestre (mon-tes-kyu–vol-vestruh), commune (□ 23 sq mi/59.8 sq km), HAUTE-GARONNE department, MIDI-PYRÉNÉES region, S FRANCE, on the ARIZE, and 28 mi/45 km S of TOULOUSE; 43°13′N 01°14′E. Horse breeding, agricultural trade. Has fortified old church.

Montesson (mon-te-son), town (□ 2 sq mi/5.2 sq km), NW outer suburb of PARIS, YVELINES department, ÎLE-DE-FRANCE region, N central FRANCE, inside great bend of SEINE RIVER, and 10 mi/16 km from Notre Dame Cathedral; 48°55′N 02°09′E. Lies near new highway linking SAINT-GERMAIN-EN-LAYE (3 mi/4.8 km W) directly with Paris in La Défense district.

Monte Tamaro (MON-teh tah-MAH-ro), peak (6,437 ft/1,962 m) in the LEPONTINE ALPS, TICINO canton, S SWITZERLAND, 8 mi/12.9 km NW of LUGANO, near Italian border.

Montets, Col des (mon-te, kol dai), pass (1,461 ft/445 m), in HAUTE-SAVOIE department, RHÔNE-ALPES region, in the Savoy Alps (*alpes françaises*) of SE FRANCE, which link the Chamonix valley (SW) with the RHÔNE valley at MARTIGNY (Switzerland, NE); 46°00′N 06°56′E. It is crossed by Chamonix-Martigny highway and offers excellent ski terrain between it and resort of ARGENTIÈRE (1.5 mi/2.4 km S). The Swiss border is 4 mi/6.4 km N.

Monteux (mon-tu), town (□ 15 sq mi/39 sq km; 2004 population 10,597), VAUCLUSE department, PROVENCE-ALPES-CÔTE D'AZUR region, SE FRANCE, 11 mi/18 km NE of AVIGNON, and 3 mi/4.8 km SW of CARPENTRAS; 44°02′N 05°00′E. Fruit and vegetable canning; manufacturing of pyrotechnic devices. Has 14th-century walls.

Montevallo (mahn-tuh-VAH-lo), town (2000 population 4,825), Shelby co., central Alabama, 17 mi/27 km WSW of Columbiana; 33°06′N 86°51′W. Cotton area; lumber milling and wood products. University of Montevallo here. Talladega National Forest is S. Settled c.1815, incorporated 1848. Originally named 'Wilson's Hill' after Jesse Wilson, one of Andrew Jackson's soldiers in the Creek Indian War and an early settler in the area, it was renamed using the combination of Italian 'monte' (mountain) and 'valle' (valley).

Montevallo (mahn-tee-VAL-o), unincorporated town, VERNON county, W MISSOURI, 16 mi/26 km SE of NEVADA. Cattle, hay, sorghum, corn. Destroyed during the Civil War. Rebuilt 1 mi/1.6 km from original site.

Montevarchi (MON-te-VAHR-kee), town, AREZZO province, TUSCANY, central ITALY, near the ARNO, 16 mi/26 km WNW of AREZZO; 43°31′N 11°34′E. Manufacturing (machinery, shoes, consumer goods, textiles).

Monteverde (mon-tai-VEHR-dai), village, CORDILLERA DE TILARÁN, COSTA RICA, 20 mi/32 km W of JUNTAS; 10°06′N 83°26′W. Dairying community established 1950s by Quaker farmers from U.S. Produces cheese and other dairy products.

Monteverde Biologica Reserve (mon-tai-VEHR-dai bee-oh-LO-hee-ka), CORDILLERA DE TILARÁN, COSTA RICA. Privately managed biological reserve (□ 26,000 acres/10,522 ha) preserves cloud forest with unique fauna. Popular tourist destination.

Monteverdi Peninsula, between the BACH ICE SHELF and the GEORGE VI SOUND, forms the S part of ALEXANDER ISLAND, WEST ANTARCTICA; 72°30′S 72°00′W. Ice-covered.

Montevideo (mahn-tuh-vi-DAI-o), department (□ 205 sq mi/533 sq km; 2004 population 1,325,968), S URUGUAY, on N bank of the Río de la PLATA; ⊙ MONTEVIDEO; 34°50′S 56°10′W. The smallest department of the country with the largest population, almost coextensive with Montevideo city. Its indented coastline, bounded by SANTA LUCÍA R. mouth (W), has the well-protected bay of Montevideo port and a string of fine beaches (RAMÍREZ, POCITOS, Carrasco), which owe their fame also to the temperate climate of the region. The department is, despite its size, important for cattle and sheep raising; agricultural crops include grain, flax, all kinds of vegetables and grapes. Montevideo city, with its large meat-packing and other basic processing industries (steel, chemicals, textiles), is the undisputed center for all commercial, industrial, cultural, and political activities of the country.

Montevideo (mon-tai-vee-DAI-o), city (2004 population 1,269,552), S URUGUAY, ⊙ Uruguay, on the Río de La PLATA; 34°53′S 56°11′W. It is the largest city of Uruguay and one of the major ports of S. America and the governmental, financial, and commercial center of Uruguay. Much of the S Atlantic fishing fleet is based in Montevideo, and Uruguay's exports—frozen and canned meats and fish, wool, and grains—pass through the port. Industries include textiles, dairy items, wines, and meat packing; oil refineries, metals fabrication, and railroad factories. Tourism is also important. Montevideo's origins lay in the colonial rivalry of the Spanish and Portuguese. The Portuguese constructed (1717) a fort on top of the hill that overlooks the harbor. Captured by the Spanish in 1724, the fort became the nucleus of the settlement founded in 1724 by the governor of BUENOS AIRES. Montevideo became the capital of Uruguay in 1828. It suffered during Uruguay's 19th-century civil wars and was besieged from 1843 to 1851. Today Montevideo is spacious, modern, and attractive, with broad, tree-lined boulevards, numerous beautiful parks, and fine buildings and residences. Notable among the parks is the Prado, which, with its lovely botanical gardens containing many thousands of plant species, is a popular promenade; among the impressive buildings are the cabildo, the legislative palace, the government palace, and the cathedral. Montevideo is the seat of Uruguay's two universities. There are fine beaches and luxurious hotels along the PLATA estuary E to PUNTA DEL ESTE on the ATLANTIC OCEAN. Many old buildings and winding streets can be found in the old heart of the city (Ciudad Vieja).

Montevideo (MAWN-te-vi-dee-yo), town (2000 population 5,346), ⊙ CHIPPEWA county, SW MINNESOTA, 35 mi/56 km SW of WILLMAR on MINNESOTA RIVER at mouth of CHIPPEWA RIVER; 44°57′N 95°43′W. Elevation 955 ft/291 m. Agricultural trade center; grain, soybeans, sugar beets; livestock; dairying; manufacturing (tarpaulins, fertilizers, computer sub-assemblies, mobile systems, miniature aircraft motors). To W is Camp Release State Memorial with granite monument commemorating release (1862) of 269 white captives of Sioux Indians. Lac qui Parle Wildlife Area and Lac qui Parle State Park, both of Minnesota River, to NW. Plotted 1870, incorporated as village 1879, as town 1908.

Monte Vista, town (2000 population 4,529), RIO GRANDE county, S COLORADO, on RIO GRANDE, just E of SAN JUAN MOUNTAINS, in SAN LUIS VALLEY, and 17 mi/27 km NW of ALAMOSA; 37°34′N 106°08′W. Elevation 7,663 ft/2,336 m. Shipping point in potato area; manufacturing (starches, light manufacturing). Headquarters of Rio Grande National Forest to W. Dude ranches, gold and silver mines in vicinity. Nearby are state soldiers' and sailors' home, Rio Grande National Forest, and Picture Rocks, with pictographs.

Monte Vista National Wildlife Refuge. Junction of railroad spur to center to SE. Incorporated 1886.

Montezinho Nature Park (mon-tai-ZEEN-yoo), natural area, BRAGANÇA, in NE corner of PORTUGAL, 6 mi/10 km S of BRAGANÇA; 41°53′N 06°45′W. Microclimates prove a range of environments, including heather-covered hills, forests, and grassy plains. Hiking trails. Birds; boars.

Montezuma, county (□ 2,040 sq mi/5,304 sq km; 2006 population 25,217), SW COLORADO; ⊙ CORTEZ; 37°20′N 108°35′W. Bordering on NEW MEXICO (S) and UTAH (W); SW corner of county forms Four Corners Area (Utah, ARIZONA, New Mexico, Colorado meet), bounded E by LA PLATA MOUNTAINS; drained by DOLORES RIVER. SAN JUAN RIVER crosses SW corner, Dolores River forms MCPHEE RESERVOIR in N. Sheep, wheat, hay, beans, oats, barley, cattle, peaches; aspen timber. MESA VERDE NATIONAL PARK in SE; YUCCA HOUSE NATIONAL MONUMENT in W; part of Ute Mountain Indian Reservation in S; part of San Juan National Forest in NE; Mancos State Park in E; unit of Havenwood National Monument in W (other parts in Utah). Formed 1889.

Montezuma (mahn-tuh-ZOO-muh), a railroad community, SAN MIGUEL county, NEW MEXICO, 6 mi/9.7 km N of LAS VEGAS. Site of an historic seventy-five room hotel built to receive guests from the SANTA FE railroad. In 1982 Armand Hammer bought the building and donated it to the United World College as the site for one of its campuses.

Montezuma (MON-te-SOO-muh), town (2007 population 7,287), N central MINAS GERAIS state, BRAZIL, near border with BAHIA, 31 mi/50 km NE of MATO VERDE; 15°12′S 42°38′W.

Montezuma (mahn-tuh-ZOO-muh), town (2000 population 3,999), MACON county, central GEORGIA, c.45 mi/72 km SW of Macon, and on FLINT RIVER opposite OGLETHORPE; 32°18′N 84°02′W. Manufacturing includes plastic pipe, motor vehicle seats, clothing, frozen fruits and vegetables. Mennonite farm community to the E. Incorporated 1854.

Montezuma, town (2000 population 1,179), PARKE county, W INDIANA, on WABASH RIVER, and 7 mi/11.3 km NW of ROCKVILLE; 39°47′N 87°22′W. In agricultural and bituminous-coal area; manufacturing (lumber); fisheries; gravel and clay pits; mineral springs. Settled c.1821, plotted 1849, incorporated 1851.

Montezuma, town (2000 population 1,440), ⊙ POWESHIEK county, central IOWA, 21 mi/34 km NNE of OSKALOOSA, near source of South Fork of ENGLISH RIVER; 41°34′N 92°31′W. Manufacturing (wood products, chemicals, epoxy products, polishing compounds, automotive parts, balsa wood and model airplane kits, printing, tankage). Fun Valley Ski Area to W. Incorporated 1868.

Montezuma (mahn-tuh-ZOO-muh), village (2000 population 42), SUMMIT county, central COLORADO, on branch of BLUE RIVER, in FRONT RANGE, and 5 mi/8 km SW of GRAYS PEAK, 12 mi/19 km NE of BRECKENRIDGE; 39°34′N 105°52′W. Elevation c.10,280 ft/3,133 m. CONTINENTAL DIVIDE 1 mi/1.6 km to E. Skiing. Arapaho Basin Ski Area to N. In Arapaho National Forest; Pike National Forest E of DIVIDE.

Montezuma, village (2000 population 966), GRAY county, SW KANSAS, 14 mi/23 km SSW of CIMARRON; 37°36′N 100°26′W. Wheat-shipping point.

Montezuma (mahn-te-ZOOM-ah), village (2006 population 189), MERCER county, W OHIO, 4 mi/6 km SSE of CELINA, on GRAND LAKE ST. MARYS reservoir; 40°29′N 84°32′W.

Montezuma Castle National Monument, YAVAPAI county, on Wet Beaver Creek, central ARIZONA. Consists of 858 acres/347 ha. Montezuma Castle, built by a farming culture, c.1250, is a five-story, twenty-room apartment house perched high in the cavity of a cliff. It was named by early settlers who believed it had been built by the Singua. The monument has two sections; main section 2 mi/3.2 km NNE of CAMP VERDE has cliff dwellings, Montezuma Well, which has a continuous spring, 1,500,000 gals/5,677,950 liters per day, is 9 mi/14.5 km NE of Camp Verde. Both units are bounded by Presott National Forest. Authorized 1906.

Mont-Fat (maw-FAHT), theme park, NAMUR province, S BELGIUM, 15 mi/24 km S of NAMUR.

Montfaucon, FRANCE: see MONTFAUCON-D'ARGONNE.

Montfaucon-d'Argonne (mon-fo-kon–dahr-gon), commune (□ 9 sq mi/23.4 sq km), MEUSE department, in LORRAINE, NE FRANCE, 13 mi/21 km NW of VERDUN; 49°16′N 05°08′E. Atop Montfaucon hill, amidst ruins of old Montfaucon village (destroyed 1918), rises Meuse-Argonne American Memorial (180 ft/55 m high), largest U.S. World War I memorial in Europe. Formerly named Montfaucon.

Montfaucon-en-Velay (mon-fo-kon–ahn–vuh-lai), commune (□ 1 sq mi/2.6 sq km; 2004 population 1,154), HAUTE-LOIRE department, AUVERGNE region, S central FRANCE, in Monts du VIVARAIS, 16 mi/26 km SSW of SAINT-ÉTIENNE; 45°11′N 04°19′E. Lumber market. Has chapel with religious paintings by a Flemish artist (1592).

Montfermeil (mon-fer-me), town (□ 2 sq mi/5.2 sq km), SEINE-SAINT-DENIS department, ÎLE-DE-FRANCE region, N central FRANCE, an outer ENE residential suburb of PARIS, 11 mi/18 km from Notre Dame Cathedral; 48°54′N 02°34′E. Metalworks.

Montferrand, FRANCE: see CLERMONT-FERRAND.

Montferrat (mahnt-fuh-RAT), Italian *Monferrato*, historic region of PIEDMONT, NW ITALY, S of the PO RIVER, now mostly in ALESSANDRIA province. It is largely hilly and produces wine, fruit, and cereals. In the late 10th century Montferrat was created as a marquisate held by the Aleramo family, and its rulers played an important role in the Crusades. Casale became the capital of the marquisate in 1435. Emperor Charles V gave (1536) Montferrat to the Gonzaga family of MANTUA, despite the claims of the house of Savoy. After Francesco Gonzaga's death in 1612, Savoy renewed its claims on Montferrat and invaded (1613) the region. SPAIN and FRANCE intervened. The Treaty of Cherasco (1631) assigned parts of Montferrat to the house of Savoy, and the rest (including Casale) followed the fortunes of the duchy of Mantua and passed to the Nevers (French) branch of the Gonzaga family. All of Montferrat was recognized by the Peace of Utrecht (1713) as belonging to the house of Savoy.

Mont Fleuri, district, NE MAHÉ ISLAND, SEYCHELLES, on INDIAN OCEAN (to E); 04°38′S 55°27′E. Borders PLAISANCE (S) and BEL AIR (W and N) districts. Formed c.1979.

Montfoort (MAWN-fawrt), town, UTRECHT province, W central NETHERLANDS, on HOLLANDSCHE IJSSEL River, and 8 mi/12.9 km WSW of UTRECHT; 52°03′N 04°57′E. Dairying; cattle, poultry, hogs; vegetables, sugar beets; manufacturing (food processing, remote control units). Has old city walls.

Montfort (mon-for), town (□ 5 sq mi/13 sq km), ILLE-ET-VILAINE department, in BRITTANY, W FRANCE, 13 mi/21 km WNW of RENNES; 48°08′N 01°57′W. Market center; meat packing, dairying, horse raising. Also known as Montfort-sur-Meu.

Montfort, village (2006 population 647), GRANT county, SW WISCONSIN, 16 mi/26 km N of PLATTEVILLE; 42°58′N 90°25′W. In agricultural area (dairying; poultry; grain farms); cheese.

Montfort (mon-FOR) [French=strong mountain], crusaders fort (12th century) built predominately in Gothic style, NW ISRAEL, 20 mi/32 km NE of HAIFA; 33°03′N 35°14′E. Overlooking large part of W GALILEE. Seigneurial castle built in early 12th century by Templar knights. Destroyed 1187 by Saladin. Retaken by Crusaders in 1192. Sold to Crusader Knights of the German Teutonic Order in 1220 and renamed Starkenburg (also means strong mountain). Captured and

destroyed (1271) by Mamluks. Ruins partly reconstructed. Tourist attraction.

Montfort-l'Amaury (mon-for–lah-mo-ree), commune (□ 2 sq mi/5.2 sq km), YVELINES department, ÎLE-DE-FRANCE region, N central FRANCE, 10 mi/16 km N of RAMBOUILLET, resort at N edge of Forest of Rambouillet; 48°47′N 01°49′E. Has late 15th-century church with 16th-century stained-glass windows. Composer Maurice Ravel lived here (museum).

Montfort-le-Gesnois (mon-for–luh–zhe-nwah), agricultural commune (□ 7 sq mi/18.2 sq km), SARTHE department, PAYS DE LA LOIRE region, W FRANCE, near the HUISNE, 10 mi/16 km ENE of Le MANS; 48°03′N 00°25′E. Horse breeding. Has castle rebuilt 1820 in Italian style. Formerly called Montfort-le-Rotrou.

Montfort-le-Rotrou, FRANCE: see MONTFORT-LE-GESNOIS.

Montfort-sur-Meu, FRANCE: see MONTFORT.

Montfrin (mon-fran), commune (□ 5 sq mi/13 sq km), GARD department, LANGUEDOC-ROUSSILLON region, S FRANCE, on the GARD, near its influx into the RHÔNE, and 11 mi/18 km ENE of NÎMES; 43°52′N 04°35′E. Vineyards, fruit, and olives.

Montgenèvre Pass (mon-zhe-ne-vruh), Italian *Monginevro*, in ALPS, in HAUTES-ALPES department, PROVENCE-ALPES-CÔTE D'AZUR region, at French-Italian border, 5 mi/8 km ENE of BRIANÇON (FRANCE); 44°56′N 06°43′E. Elevation 6,070 ft/1,850 m. Connects valleys of DORA RIPARIA (Italy) and DURANCE (France) rivers. Crossed by Cesana-Briançon road built (1802–1807) by Napoleon. Hospice founded 14th century. One of easiest Alpine passes open year-round, it was frequently used by invading armies. Also spelled Mont-Genèvre. Commune of Montgenèvre just W, in Hautes-Alpes department, is a resort with excellent ski facilities that link French and Italian terrain with interconnected ski-lift system.

Montgeron (mon-zher-on), town (□ 4 sq mi/10.4 sq km), ESSONNE department, ÎLE-DE-FRANCE region, N central FRANCE, near SEINE RIVER just S of VILLE-NEUVE-SAINT-GEORGES, and 11 mi/18 km SSE of PARIS; 48°42′N 02°27′E. Railroad junction with industrial area.

Montgiscard (mon-zhee-skahr), agricultural commune (□ 5 sq mi/13 sq km), HAUTE-GARONNE department, MIDI-PYRÉNÉES region, S FRANCE, on the CANAL DU MIDI, and 12 mi/19 km SSE of TOULOUSE; 43°27′N 01°35′E.

Montgomery, county (□ 800 sq mi/2,080 sq km; 2006 population 223,571), E central ALABAMA; ⊙ MONTGOMERY; 32°14′N 86°14′W. In the BLACK BELT. Bounded NW by ALABAMA RIVER, N by TALLAPOOSA RIVER. Cotton, livestock, grain. Manufacturing at Montgomery. Formed 1816.

Montgomery, county (2006 population 9,272), W ARKANSAS; ⊙ MOUNT IDA; 34°32′N 93°39′W. Drained by OUACHITA, CADDO, and LITTLE MISSOURI rivers; in OUACHITA MOUNTAINS region. Agriculture (cattle, hogs, chickens); dairy products; cotton ginning, sawmilling, food processing, stone quarrying. All of county except a narrow margin in SE is within Ouachita National Forest, (eight forest campgrounds), W half of large Lake OUACHITA reservoir in E; Little Missouri Falls in SW. Formed 1842, organized 1844.

Montgomery, county (□ 247 sq mi/642.2 sq km; 2006 population 9,067), E central GEORGIA; ⊙ MOUNT VERNON; 32°10′N 82°32′W. Bounded W by OCONEE River, S by ALTAMAHA RIVER. Manufacturing (apparel, textiles, beer and distilled beverages); agriculture (cotton, corn, tobacco, peanuts); cattle, hogs; in timber area. Formed 1793.

Montgomery, county (□ 709 sq mi/1,843.4 sq km; 2006 population 30,367), S central ILLINOIS; ⊙ HILLSBORO; 39°14′N 89°29′W. Agriculture (corn, wheat, sorghum, soybeans, cattle; dairy products). Some manufacturing (food products, paper boxes, glass jars, metal products, industrial machinery, transportation equipment).

Drained by SHOAL and MACOUPIN creeks. Has several recreational reservoirs, including Lake Lou Yaeger, Lake Glenn Shoals, Coffeen Lake. Largest town is LITCHFIELD in W. Formed 1821.

Montgomery, county (□ 505 sq mi/1,313 sq km; 2006 population 38,173), W central INDIANA; ⊙ CRAWFORDSVILLE; 40°02′N 86°53′W. Drained by SUGAR and RACCOON creeks. Agriculture (soybeans, corn, wheat, cattle, hogs, sheep, poultry; dairy products). Clay pits; timber. Commerce and manufacturing at Crawfordsville. Shades State Park in SW; canoeing on Sugar Creek. Formed 1822.

Montgomery, county (□ 424 sq mi/1,102.4 sq km; 2006 population 11,365), SW IOWA; ⊙ RED OAK, 41°02′N 95°10′W. Prairie agricultural area (hogs, cattle; corn, wheat, oats) drained by EAST NISHNABOTNA, WEST NODAWAY, and TARKIO rivers and by WALNUT CREEK; bituminous-coal deposits. Formed 1851.

Montgomery, county (□ 651 sq mi/1,692.6 sq km; 2006 population 34,692), SE KANSAS; ⊙ INDEPENDENCE; 37°12′N 95°44′W. Level to hilly region, bordering S on OKLAHOMA; drained by VERDIGRIS and ELK rivers. Livestock; grain; dairying; wheat, sorghum, soybeans, hay, strawberries; petroleum refining, plastics products, industrial machinery. Numerous oil and gas fields. Formed 1869.

Montgomery, county (□ 198 sq mi/514.8 sq km; 2006 population 24,887), E central KENTUCKY; ⊙ MOUNT STERLING; 38°03′N 83°55′W. Drained by Hinkston and Slate creeks. Rolling upland agricultural area, mainly in NE of BLUEGRASS REGION (burley tobacco, corn, soybeans, hay, alfalfa; cattle, poultry; dairying). Manufacturing at Mount Sterling. Formed 1796.

Montgomery, county (□ 506 sq mi/1,315.6 sq km; 2006 population 932,131), central MARYLAND; ⊙ ROCKVILLE; 39°08′N 77°12′W. Bounded NE by PATUXENT River, S by DISTRICT OF COLUMBIA, W and SW by POTOMAC River (forms VIRGINIA state line here); drained by ROCK CREEK. Rolling piedmont area containing many residential suburbs (including CHEVY CHASE, BETHESDA, TAKOMA PARK) of WASHINGTON, D.C. Agricultural hinterland produces dairy products, truck, apples, wheat, hay, cattle, poultry; some manufacturing (especially scientific instruments, wood products). Formed in 1776 and named for General Richard Montgomery, killed leading the American attack on QUEBEC in 1775, the county ceded 60 sq mi/155 sq km to the Federal government in 1790 to form the District of Columbia. One of the most affluent communities in the country, the county operates 168 parks with a total area of 19,000 acres/7,689 ha. Government installations, especially the Department of Health and Human Services and the Food and Drug Administration in Rockville, Bethesda, and Silver springs employ thousands. The Takoma Park Campus of Montgomery College was founded in 1945 as the first junior college in the nation. Includes some of National Capital Parks (W), among them Great Falls Park. Josiah Henson, the inspiration for Harriet Beecher Stowe's *Uncle Tom's Cabin*, grew up as a slave in Montgomery county His mother had been sold to a Montgomery county planter and his father to a Southern plantation owner. Henson fled to Canada in 1830, and wrote a book, *Truth Stranger than Fiction*, describing his life. An oak log cabin restored near Maryland 355 and Old Georgetown Road may have been the home in which Henson, who became a Methodist preacher and Abolitionist speaker, grew up. His descendant, Matthew Henson, accompanied Admiral Robert E. Peary to the North Pole in 1902.

Montgomery, county (□ 407 sq mi/1,058.2 sq km; 2006 population 11,754), central MISSISSIPPI; ⊙ WINONA; 33°30′N 89°36′W. Drained by BIG BLACK RIVER. Agriculture (cotton, corn, soybeans; cattle); timber. Manufacturing at Winona. Formed 1871.

Montgomery, county (□ 533 sq mi/1,385.8 sq km; 2006 population 12,170), E central MISSOURI; ⊙ MON-

TGOMERY CITY; 38°58'N 91°29'W. On MISSOURI River and drained by the Loutre. Agriculture (corn, wheat, soybeans; cattle, sheep, hogs; manufacturing at Montgomery City and JONESBURG; limestone, fire-clay pits. Graham Cave State Park in W. Flooding in 1993 destroyed RHINELAND and MCKITTRICK in S. Formed 1818.

Montgomery, county (□ 410 sq mi/1,066 sq km; 2006 population 49,112), E central NEW YORK; ⊙ FONDA; 42°53'N 74°25'W. Lies in fertile MOHAWK RIVER valley; traversed by the NEW YORK STATE BARGE CANAL; also drained by SCHOHARIE CREEK. Dairying region, with light manufacturing centers in AMSTERDAM, CANAJOHARIE, FORT PLAIN, and SAINT JOHNSVILLE. Formed 1772.

Montgomery (mahnt-GUHM-uhr-ee), county (□ 501 sq mi/1,302.6 sq km; 2006 population 27,638), central North Carolina; ⊙ TROY; 35°19'N 79°54'W. Bounded W by Yadkin (PEE DEE) River (joined by ROCKY RIVER, from W, to form Pee Dee River in SW corner) which forms BADIN and TILLERY reservoirs; UWHARRIE and Little rivers. Bounded SE by Drowning Creek (LUMBER RIVER). In forested PIEDMONT region; agricultural area (wheat, soybeans, hay, peaches, cotton, tobacco; poultry, cattle, hogs). Some manufacturing at Troy and BISCOE; service industries. Within the Piedmont Triad region. Part of Uwharrie National Forest covers most of county, in S center, W, and N. Town Creek Indian Mound State Historical Site in S (16th-century reconstruction). North Carolina Peach Festival held here annually. Formed 1778 from Anson County. Named for General Richard Montgomery (1736–1775), a leader in the Revolutionary War.

Montgomery (MAHNT-guhm-uhr-ee), county (□ 465 sq mi/1,209 sq km; 2006 population 542,237), SW OHIO; ⊙ DAYTON; 39°45'N 84°18'W. Intersected by GREAT MIAMI, STILLWATER, and MAD rivers, and by small Bear, Wolf, and Twin creeks. MIAMISBURG MOUND and ENGLEWOOD DAM are here. In the Till Plains physiographic region. Agricultural area (hogs, corn, soybeans, tobacco); extensive manufacturing, especially at Dayton. Sand and gravel pits; cement plants. Formed 1803. Population has declined steadily since late 20th century.

Montgomery, county (□ 487 sq mi/1,266.2 sq km; 2006 population 775,688), SE PENNSYLVANIA; ⊙ NORRISTOWN; 40°12'N 75°22'W. Bounded SE by PHILADELPHIA; drained by SCHUYLKILL RIVER (forms part of SW boundary, in W) and PERKIOMEN CREEK (forms Green Lane Reservoir in N) and Pennypack Creek. County is highly urbanized in SE and center rural in NW. Agriculture (corn, wheat, barley, hay, alfalfa, soybeans; hogs, cattle, dairying); sand and gravel, limestone. Manufacturing at LANSDALE, Norristown, HATFIELD, TELFORD, CONSHOHOCKEN, and COLLEGEVILLE. Part of VALLEY FORGE NATIONAL HISTORICAL PARK in SW; Evansburg State Park in center, Fort Washington State Park in S center. Nuclear power plants: Limerick 1 (initial criticality December 22, 1984) and Limerick 2 (initial criticality August 12, 1989) are 21 mi/34 km NW of Philadelphia. Use cooling water from the Schuylkill River, and each has a maximum dependable capacity of 1055 MWe. First settled by Swedes and Welsh; later by Germans. Formed 1784.

Montgomery, county (2006 population 147,114), N TENNESSEE; ⊙ CLARKSVILLE; 36°30'N 87°23'W. Bounded N by KENTUCKY; drained by CUMBERLAND and RED rivers. Agriculture; manufacturing; trade and retail center. Clarksville one of fastest growing cities in S. Has part of FORT CAMPBELL Army base. Formed 1796.

Montgomery (mahnt-GUHM-uh-ree), county (□ 1,076 sq mi/2,797.6 sq km; 2006 population 398,290), SE TEXAS; ⊙ CONROE; 30°18'N 95°30'W. Drained by West Fork (forms LAKE CONROE in NW) of SAN JACINTO RIVER and Peach Creek (forms NE boundary).

Bounded S by Spring Creek. Oil, natural gas production and processing; sand and gravel are important industries. Also cattle, horses, ratites (ostriches, emus, rheas); fruit, some corn, cotton; nursery and greenhouse products Hunting, fishing. Fringe of HOUSTON urbanized area and extension of Houston city limits are in S and SE edge of county Includes parts of Sam Houston National Forest in NW and NE. Formed 1837.

Montgomery (muhnt-GUHM-ree), county (□ 389 sq mi/1,011.4 sq km; 2006 population 84,541), SW VIRGINIA; ⊙ CHRISTIANSBURG, 37°10'N 80°23'W. Independent city of RADFORD, in W, separate from county. Mainly in GREAT APPALACHIAN VALLEY, traversed by ridges; ALLEGHENY MOUNTAINS are NW, BLUE RIDGE in SE; bounded W by NEW RIVER, SW and S by LITTLE RIVER, ROANOKE RIVER, and CRAIG CREEK. Manufacturing at Christiansburg, BLACKSBURG; agriculture (corn, alfalfa, hay, apples; cattle, sheep; dairying); some bituminous-coal mining; stone quarrying. Includes part of Jefferson National Forest. in NW. APPALACHIAN TRAIL passes through N corner. Formed 1776.

Montgomery, city (2000 population 201,568), ⊙ Alabama and Montgomery co., E central Alabama Near the head of navigation on the Alabama River just below the confluence of the Coosa and Tallapoosa rivers, and in the rich Black Belt. It is an important market center for lumber and agr. goods, especially livestock and dairy products. Manufacturing includes commercial fertilizer, furniture, air conditioning and heating units, automotive wiring, food items, and paper. Montgomery became the capital of Alabama in 1847 and boomed as a river port and cotton market. The city has been called the "Cradle of the Confederacy." In the capitol building (erected 1857) the convention met (Feb. 1861) that formed the Confederate States of America. Jefferson Davis was inaugurated president on the capitol steps, and the city served as the Confederate capital until the seat was moved to Richmond in May 1861. The city was occupied by Federal troops in the spring of 1865. During the civil rights movement in the 1950s and 1960s, Montgomery was marked by Afr.-Amer. demonstrations, led by Martin Luther King, Jr., who was a minister in Montgomery in the mid-1950s. In December 1955, Afr.-Americans organized a nonviolent boycott of the segregated public bus system; by the following year a desegregation edict regarding public transportation was issued. Racial unrest ensued here in the 1960s. The city is the seat of Alabama State University, Auburn University at Montgomery, Huntingdon College, Troy State University, Faulkner University, 2 technical schools, and Southern Christian University. Maxwell Air Force Base, home of Air University, the center of professional military education for the U.S. Air Force, adjoins the city on the NW. In addition to the historic state capitol, points of interest in Montgomery include the "1st White House of the Confederacy" (built c.1825), preserved as a Confederate museum; a planetarium; a museum of fine arts; the state archives and history museum; a zoo; the Alabama Shakespeare festival and many antebellum homes and buildings. Inc. 1819. Named for Richard Montgomery, a major general killed in the Battle of Quebec in December of 1775.

Montgomery (MAHNT-guhm-uh-ree), city (□ 5 sq mi/13 sq km; 2006 population 9,856), HAMILTON county, extreme SW OHIO; suburb 12 mi/19.2 km NE of CINCINNATI; 39°14'N 84°20'W.

Montgomery, town (2000 population 787), GRANT parish, central LOUISIANA, 36 mi/58 km NW of ALEXANDRIA, and on RED RIVER; 31°40'N 92°54'W. In agricultural area (cotton, vegetables, soybeans, cattle); lumber; manufacturing (septic tanks). Kisatchie National Forest to N.

Montgomery, town, HAMPDEN county, SW MASSACHUSETTS, 14 mi/23 km NW of SPRINGFIELD; 42°13'N 72°49'W.

Montgomery, town (2000 population 2,794), LE SUEUR county, S MINNESOTA, 40 mi/64 km SSW of MINNEAPOLIS; 44°26'N 93°34'W. Agricultural trade center (grain, peas, soybeans; livestock, poultry; dairying); manufacturing (canned peas and corn, fireplaces, printing and publishing, hardware). Numerous small lakes in area. Platted 1877, incorporated 1902.

Montgomery, township, SOMERSET county, N NEW JERSEY, 10 mi/16 km N of PRINCETON; 40°25'N 74°40'W. Incorporated 1798.

Montgomery, town, FRANKLIN CO., NW VERMONT, on small Trout River, and 23 mi/37 km E of St. Albans, 44°51'N 72°36'W. Has seven covered bridges. Chartered 1789, settled 1793. Named for General Richard Montgomery (1738–1775).

Montgomery, town (2006 population 1,926), KANAWHA and FAYETTE counties, W central WEST VIRGINIA, on KANAWHA RIVER (bridged), 22 mi/35 km SE of CHARLESTON; 38°10'N 81°19'W. Bituminous-coal mines; gas and oil wells. Manufacturing (coal processing, concrete). Agriculture (grain); livestock. West Virginia University Institute of Technology. Incorporated 1890.

Montgomery (MAHNT-guhm-uh-ree), Welsh Trefaldwyn, village, POWYS, E Wales, 7 mi/11.3 km S of WELSHPOOL; 52°35'N 03°10'W. Sheep and cattle market. Nearby OFFA'S DYKE is very well preserved.

Montgomery, village (2000 population 5,471), Kane co., NE Illinois, on Fox River, just SW of Aurora, residential satellite community, c.35 mi/56 km WSW of Chicago; 41°43'N 88°20'W. In agr. area (dairy products; corn, soybeans); manufacturing

Montgomery, village (2000 population 368), DAVIESS county, SW INDIANA, 6 mi/9.7 km E of WASHINGTON; 38°40'N 87°03'W. Glendale State Fish and Wildlife Area to S. Manufacturing (post buildings, wood trusses). Turkeys, cattle; grain. Bituminous-coal area, surface mines. Laid out 1865.

Montgomery, village (2000 population 386), HILLSDALE county, S MICHIGAN, 13 mi/21 km SW of HILLSDALE, near INDIANA state line (NE corner of Indiana), OHIO state line 5 mi/8 km SE; 41°46'N 84°48'W. In farm area; egg processing; manufacturing (auto parts).

Montgomery, village (□ 1 sq mi/2.6 sq km; 2000 population 3,636), ORANGE county, SE NEW YORK, on small WALLKILL RIVER, and 11 mi/18 km W of NEWBURGH; 41°31'N 74°14'W. Light manufacturing; custom motorcycles.

Montgomery (mahnt-GUHM-uh-ree), village (2006 population 580), MONTGOMERY county, SE TEXAS, 15 mi/24 km WNW of CONROE; 30°23'N 95°42'W. Timber and agricultural area. Light manufacturing. LAKE CONROE reservoir to NE. Sam Houston National Forest to N.

Montgomery, borough (2006 population 1,600), LYCOMING county, N central PENNSYLVANIA, 10 mi/16 km ESE of WILLIAMSPORT, on West Branch of SUSQUEHANNA RIVER (bridged); 41°10'N 76°52'W. Agricultural area (grain, potatoes, soybeans; livestock, dairying); manufacturing (paper processing equipment, lumber, truck bodies, commercial printing, wooden furniture, aluminum and vinyl blinds, apparel, plastic products). Allenwood Federal Prison Camp to SW; Muncy State Correctional Institution to NE. Tiadaghton State Forest to N. Settled 1778, incorporated 1887.

Montgomery City, city (2000 population 2,442), ⊙ MONTGOMERY county, E central MISSOURI, 25 mi/40 km ENE of FULTON; 38°58'N 91°30'W. Corn, wheat, soybeans; cattle; manufacturing (commercial printing, toll booths, feeds, apparel). Laid out 1853.

Montgomery Pass (mahnt-GUHM-ree) (7,132 ft/2,174 m), SW NEVADA, in N WHITE MOUNTAINS, N of BOUNDARY PEAK, 60 mi/97 km W of TONOPAH, 8 mi/

12.9 km NE of CALIFORNIA state line. U.S. Highway 6 passage. In Inyo National Forest.

Montgomery Peak, California: see WHITE MOUNTAINS.

Montgomeryville, unincorporated town (2000 population 12,031), MONTGOMERY county, SE PENNSYLVANIA, suburb 20 mi/32 km N of PHILADELPHIA; 40°15′N 75°14′W. Manufacturing includes fabricated metal products, infrared detectors, cement paste, commercial printing, caulking compounds, plastic products, clothing, fencing, microwave components.

Montguyon (mon-gee-yon), commune (□ 7 sq mi/18.2 sq km), CHARENTE-MARITIME department, POITOU-CHARENTES region, W FRANCE, 17 mi/27 km S of BARBEZIEUX-SAINT-HILAIRE; 45°13′N 00°12′W. Lime-kilns; dairying, distilling. Has ruins of 12th–15th-century castle.

Monthermé (mon-ter-mai), commune (□ 12 sq mi/31.2 sq km), ARDENNES department, CHAMPAGNE-ARDENNE region, N FRANCE, in the Ardennes Mountains, 9 mi/14.5 km N of CHARLEVILLE-MÉZIÈRES, on a bend of the entrenched MEUSE near mouth of the Semoy (SEMOIS) River; 49°53′N 04°44′E. Center of tourism in Meuse River valley.

Monthey (mohn-TEY), district, W VALAIS canton, SWITZERLAND; 46°15′N 06°57′E. Main town is MONTHEY; population is French-speaking and Roman Catholic.

Monthey (mohn-TEY), town (2000 population 13,933), VALAIS canton, SW SWITZERLAND, near the RHÔNE, 10 mi/16 km SSE of LAKE GENEVA; 46°15′N 06°57′E. Elevation 1,391 ft/424 m. Chemicals, metal and cement products, tobacco. Has 14th-century castle.

Monticello, city (□ 3 sq mi/7.8 sq km; 2006 population 2,546), ⊙ JEFFERSON county, N FLORIDA, 27 mi/43 km ENE of TALLAHASSEE, near GEORGIA state line and LAKE MICCOSUKEE; 30°32′N 83°52′W. Farm trade center, plywood manufacturing. Settled in early 19th century.

Monticello, city (2000 population 5,138), ⊙ PIATT county, central ILLINOIS, on SANGAMON RIVER (bridged here) and 24 mi/39 km ENE of DECATUR; 40°01′N 88°24′W. In rich agricultural area; manufacturing (drugs, food products); corn, soybeans, wheat, livestock, dairy products. Incorporated 1841.

Monticello, city (2000 population 5,723), ⊙ WHITE county, NW central INDIANA, on TIPPECANOE RIVER, between SHAFER and FREEMAN lakes, and 21 mi/34 km W of LOGANSPORT; 40°45′N 86°46′W. In agricultural area (corn, oats, soybeans; hogs); Light manufacturing. Resort area. Settled 1831, laid out 1834, incorporated 1853.

Monticello, city (2000 population 3,607), ⊙ JONES county, E IOWA, on MAQUOKETA RIVER, and 10 mi/16 km NE of ANAMOSA; 42°14′N 91°11′W. Manufacturing (farm equipment, feed, concrete products). Limestone quarries nearby. Settled 1836, incorporated 1867.

Monticello (mahn-tuh-SEL-o), town (2000 population 9,146), ⊙ DREW county, SE ARKANSAS, c.45 mi/72 km SSE of PINE BLUFF; 33°37′N 91°47′W. Industrial center for agricultural area; lumber milling, manufacturing (printing, tables, elastomeric liquid containers, recreational boats, polyester and nylon textile yarns, trash bags, lumber). University of Arkansas at Monticello campus SW of town.

Monticello, town (2000 population 2,428), ⊙ JASPER county, central GEORGIA, 32 mi/51 km N of MACON; 33°19′N 83°41′W. Manufacturing of lumber, industrial axles, plywood; feldspar mining and processing. Incorporated 1810. Oconee National Forest in area.

Monticello, town (2000 population 5,981), ⊙ WAYNE county, S KENTUCKY, in CUMBERLAND foothills, 23 mi/37 km SW of SOMERSET; 36°50′N 84°50′W. In area of agriculture (corn, burley tobacco, wheat; livestock; timber). Manufacturing (poultry processing, steel pipes, sleepwear, shirts and blouses, houseboats, kitchen cabinets, blue jeans). Wayne County Airport

to NW. Camp Earl Wallace Conservation Education Center to SW. Large LAKE CUMBERLAND (CUMBERLAND RIVER) to NW. Settled before 1800.

Monticello, town, AROOSTOOK county, E MAINE, on MEDUXNEKEAG RIVER, and 13 mi/21 km N of HOULTON; 46°19′N 67°50′W. Potatoes shipped. Settled 1830, incorporated 1846.

Monticello (mawn-ti-SE-lo), town (2000 population 7,868), WRIGHT county, E MINNESOTA, on MISSISSIPPI RIVER, and 35 mi/56 km NW of MINNEAPOLIS; 45°17′N 93°47′W. Grain, soybeans; livestock; dairying; manufacturing (molded products, hand tools, cabinets, fabicated structural metals). Several small lakes in area. Lake Maria State Park to W; Sherburn National Wildlife Refuge and Sand Dunes State Forest to NE. Nuclear Power Plant. Incorporated 1856.

Monticello (mahnt-uh-SEL-o), town (2000 population 1,726), ⊙ LAWRENCE county, S central MISSISSIPPI, 20 mi/32 km E of BROOKHAVEN, and on PEARL RIVER; 31°33′N 90°06′W. In agricultural (cotton, corn; cattle, hogs, poultry) and timber area; manufacturing (linerboard, sportswear, industrial screens, lumber). railroad junction to N. Lake Mary Crawford State Lake to W; Armstrong-Lee House (1856); Fox House (1848). Founded 1798.

Monticello (mahn-teh-SEL-o), town (2000 population 126), ⊙ LEWIS county, NE MISSOURI, on NORTH FABIUS RIVER, near the MISSISSIPPI, and 20 mi/32 km NW of QUINCY, ILLINOIS; 40°07′N 91°42′W. Corn, soybeans; cattle, hogs.

Monticello (2006 population 1,922), ⊙ SAN JUAN county, SE UTAH, 50 mi/80 km SSE of MOAB; 37°52′N 109°20′W. Flour-milling point in wheat and cattle area; mining of uranium and vanadium ores. ABAJO MOUNTAINS are just W, in section of La Sal National Forest. Elevation 7,066 ft/2,154 m. CANYONLANDS NATIONAL PARK and Newspaper Rock State Park to NW. Settled 1888.

Monticello, town (2006 population 1,132), GREEN county, S WISCONSIN, 10 mi/16 km N of MONROE; 42°45′N 89°35′W. In agricultural area; manufacturing cooperative (feed, cheese, wire garment hangers). On Sugar River State Trail.

Monticello, village (□ 3 sq mi/7.8 sq km; 2006 population 6,635), ⊙ SULLIVAN county, SE NEW YORK, 20 mi/32 km NW of MIDDLETOWN; 41°38′N 74°41′W. In timber, dairying, and recreational lake area; light manufacturing and quarry products. Incorporated 1830.

Monticello (mahn-tuh-SE-lo) [Italian=little mountain], estate, ALBEMARLE county, central VIRGINIA, 4 mi/6 km SE of CHARLOTTESVILLE, on RIVANNA RIVER; home of Thomas Jefferson for 56 years; 38°00′N 78°27′W. Covers 640 acres/259 ha. The mansion, which Jefferson designed, was begun in 1770 on property inherited from his father. The building materials—stone, brick, lumber, and nails—were prepared on the estate, and most of the construction work was carried out by Jefferson's artisan slaves. By 1772, when Jefferson took his bride there to live, part of the house was ready for occupancy; for many years afterward, he added to the building. The house is one of the earliest examples of the American classic revival style. Not long after Jefferson's death, his daughter, unable to maintain the property, sold it, retaining only the family burial plot in which Jefferson is interred. Monticello was later bought by Uriah P. Levy, a naval officer, who bequeathed it to "the people of the United States," but his heirs successfully contested the will. By 1879, Jefferson P. Levy was in full ownership, but he sold Monticello in 1923 to the Thomas Jefferson Memorial Foundation. Dedicated as a national shrine in 1926, and extensively renovated during the next 30 years, the estate was opened to the public in 1954. Designated a World Heritage Site in 1987.

Monticello Conte Otto (mon-tee-CHEL-lo KON-te OT-to), village, VICENZA province, VENETO, N ITALY,

4 mi/6 km NNE of VICENZA; 45°35′N 11°35′E. Highly diversified secondary industrial center.

Monticello Nuclear Power Plant, MINNESOTA: see WRIGHT, county.

Montichiari (MON-tee-KYAH-ree), town, BRESCIA province, LOMBARDY, N ITALY, on CHIESE River, and 12 mi/19 km SE of BRESCIA; 45°25′N 10°23′E. Agricultural and livestock market; manufacturing (machinery, agricultural tools, hardware, fabricated metals, textiles, clothing); food processing.

Montiel (mon-TYEL), town, CIUDAD REAL province, S central SPAIN, 26 mi/42 km E of VALDEPEÑAS; 38°42′N 02°52′W. Cereals, grapes, olives, carob beans; livestock. Wine making. Ruined castle in which Peter the Cruel was held prisoner by his half brother towers over the town. The surrounding region, a plain of LA MANCHA stretching into ALBACETE province, is called Campo de Montiel.

Montiel, Selvas de, hilly forests in N ENTRE RÍOS province, ARGENTINA, extend c.60 mi/97 km SSW from CORRIENTES province border to area S of VILLA FEDERAL. Subtropical woods.

Montiér-en-Der (mon-tyai-ahn-der), commune (□ 10 sq mi/26 sq km), HAUTE-MARNE department, CHAMPAGNE-ARDENNE region, NE FRANCE, 14 mi/23 km SW of SAINT-DIZIER; 48°29′N 04°46′E. Agricultural machinery manufacturing. Stud farm. It has a 10th-12th-century abbatial church. Also spelled Montier-en-Der.

Montier-en-Der, FRANCE: see MONTIÉR-EN-DER.

Montieri (mon-TYAI-ree), village, GROSSETO province, TUSCANY, central ITALY, 9 mi/14 km NE of MASSA MARITTIMA; 43°08′N 11°01′E. In mining region (copper, iron, lead, "Soffioni," pyrite).

Montignac (mon-tee-nyahk), commune (□ 14 sq mi/36.4 sq km), DORDOGNE department, AQUITAINE region, SW FRANCE, on the VÉZÈRE and 12 mi/19 km N of SARLAT-LA-CANÉDA; 45°04′N 01°10′E. Woodworking center (gun stocks, furniture); tanning. The LASCAUX CAVE, with its 600 walls of famed 17,000-year-old paintings of pre-historic animals, was discovered 1 mi/1.6 km S of Montignac in 1940. It was opened to the public in 1948, but had to be closed in 1980 because over one million visitors in 15 years threatened the integrity of the paintings. In 1983, Lascaux II, a true reproduction of the grotto, was opened above ground nearby and has become an international tourist attraction. Also called Montignac-sur-Vézère.

Montignac-sur-Vézère, FRANCE: see MONTIGNAC.

Montignies-le-Tilleul (maw-teen-YEE–le–tee-YUH), commune, Charleroi district, HAINAUT province, S central BELGIUM, 4 mi/6.4 km WSW of CHARLEROI, near l'Eau d'Heure River; 50°23′N 04°22′E. Metallurgical industry.

Montignies-sur-Sambre (maw-teen-YEE–sur–SAWM-bruh), town in commune of Charleroi, Charleroi district, HAINAUT province, S central BELGIUM, on SAMBRE RIVER, and 2 mi/3.2 km ESE of CHARLEROI; 50°24′N 04°30′E. Metallurgy.

Montigny-en-Gohelle (mon-tee-nyee–ahn-go-el), residential town (□ 1 sq mi/2.6 sq km), PAS-DE-CALAIS department, NORD-PAS-DE-CALAIS region, N FRANCE, 4 mi/6.4 km E of LENS, in former coal-mining district; 50°26′N 02°56′E.

Montigny-en-Ostrevent (mon-tee-nyee–ahn-o-struh-vahn), town (□ 2 sq mi/5.2 sq km) E suburb of DOUAI, NORD department, NORD-PAS-DE-CALAIS region, N FRANCE; 50°22′N 03°11′E. In industrial metalworking district.

Montigny-le-Bretonneux (mon-tee-nyee–luh-bre-to-nu), town (□ 4 sq mi/10.4 sq km) E suburb of SAINT-QUENTIN-EN-YVELINES (a new town), YVELINES department, ÎLE-DE-FRANCE region, N central FRANCE, 7 mi/11.3 km SW of VERSAILLES; 48°46′N 02°02′E. Electronics and food processing industry.

Montigny-lès-Cormeilles (mon-tee-nyee–lai–kor-mai), town (□ 1 sq mi/2.6 sq km), NW suburb of

PARIS, VAL-D'OISE department, ÎLE-DE-FRANCE region, N central FRANCE, 11 mi/18 km NW of Notre Dame Cathedral; 48°59′N 02°11′E. Located on main NW motor route to NORMANDY and on right bank of SEINE RIVER, opposite the Forest of Saint-Germain.

Montigny-lès-Metz (mon-tee-nyee-lai-me), town (□ 2 sq mi/5.2 sq km), SSW residential suburb of Metz, MOSELLE department, LORRAINE region, NE FRANCE, on Moselle's right bank; 49°06′N 06°09′E. Railroad yards. Airport just S.

Montijo (mon-TEE-ho), town, ⊙ Montijo district, VERAGUAS province, W central PANAMA, in PACIFIC lowland, near MONTIJO GULF, 8 mi/12.9 km S of SANTIAGO; 07°59′N 81°03′W. Sawmilling. Mercury deposits. Former gold-mining center.

Montijo (mon-TEE-joo), town, SETÚBAL district, S central PORTUGAL, near LISBON BAY, on railroad spur, 10 mi/16 km E of LISBON; 38°42′N 08°58′W. Cork; other manufacturing: pottery, chemical fertilizer, hardware; fruit preserving, distilling.

Montijo (mon-TEE-ho), town, BADAJOZ province, W SPAIN, on railroad, and 19 mi/31 km E of BADAJOZ; 38°55′N 06°37′W. Processing and agricultural center (olives, olive oil, grapes, liquor, vegetables; livestock). Has palace of counts of Montijo.

Montijo Gulf, inlet of the PACIFIC in W central PANAMA, between Las Palmas and AZUERO peninsulas; 15 mi/24 km wide, 20 mi/32 km long. Barred by CÉBACO ISLAND, off its mouth, it receives SAN PABLO RIVER.

Montilla (mon-TEE-lyah), city, CÓRDOBA province, S SPAIN, in ANDALUSIA; 37°35′N 04°38′W. It is the center of an agricultural district famous for wines which resemble sherry.

Montillana (mon-tee-LYAH-nah), town, GRANADA province, S SPAIN, 28 mi/45 km NNW of GRANADA; 37°30′N 03°40′W. Cereals, legumes.

Montividiu (MON-chee-vee-zhe-oo), town (2007 population 9,243), SW GOIÁS, BRAZIL, on Rio Verdao Alegre, 50 mi/80 km NE of JATAÍ; 17°30′S 51°13′W.

Montivilliers (mon-tee-veel-yai), town (□ 7 sq mi/18.2 sq km) NE residential suburb of Le HAVRE, SEINE-MARITIME department, HAUTE-NORMANDIE region, N FRANCE; 49°33′N 00°12′E. Textiles. Has 11th–16th-century church.

Montjean-sur-Loire (mon-zhon-syoor-lwahr), commune (□ 7 sq mi/18.2 sq km), MAINE-ET-LOIRE department, PAYS DE LA LOIRE region, W FRANCE, on left bank of LOIRE RIVER, and 15 mi/24 km WSW of ANGERS; 47°23′N 00°52′W. Vineyards.

Montjoie, GERMANY: see MONSCHAU.

Mont-Joli (mon-zho-LEE), town (□ 9 sq mi/23.4 sq km), ⊙ LA MITIS county, BAS-SAINT-LAURENT region, E QUEBEC, E CANADA, near the SAINT LAWRENCE RIVER, 18 mi/29 km NE of RIMOUSKI; 48°35′N 68°11′W. On railroad spur to MATANE; railroad junction, highway junction. Railroad workshops, metal foundries, lumber mill; hydroelectric station. Dairying, pig-raising region. Airfield. Merged with SAINT-JEAN-BAPTISTE in 2001.

Montjoly (mon-zho-lee), locality, N FRENCH GUIANA; 04°55′N 52°16′W. A 1933 farm colony and settlement of RÉMIRE, 5 mi/8 km E of CAYENNE.

Montjuich (mont-jweek), isolated hill (575 ft/175 m), BARCELONA province, NE SPAIN, on MEDITERRANEAN SEA between S district of BARCELONA city and Llobregat plain. Crowned by citadel (18th century).

Mont-Laurier (mon-LO-ree-ai), town (□ 228 sq mi/592.8 sq km), ⊙ ANTOINE-LABELLE county, LAURENTIDES region, SW QUEBEC, E CANADA, on the LIÈVRE River, N of OTTAWA; 46°33′N 75°30′W. Located in the LAURENTIAN MOUNTAINS, it is a winter resort in a lumbering and potato-growing region and has a hydroelectric-power station. Was seat of historic LABELLE county. Restructured in 2003.

Mont-Lebel (mon-luh-BEL), former village, S QUEBEC, E CANADA; 48°20′N 68°24′W. Amalgamated into RIMOUSKI in 2002.

Montlhéry (mon-lai-ree), town (□ 1 sq mi/2.6 sq km), ESSONNE department, ÎLE-DE-FRANCE, N central FRANCE, 16 mi/26 km SSW of PARIS; 48°38′N 02°16′E. Market center. Auto racetrack. Has massive 13th-century keep of former feudal castle once occupied by Edward III of England; from its top (500 ft/152 m), the higher buildings of Paris can be seen.

Mont-Louis (mon-loo-ee), commune, PYRÉNÉES-ORIENTALES department, LANGUEDOC-ROUSSILLON region, S FRANCE, in E PYRENEES, road junction 2 mi/3.2 km ENE of Col de la PERCHE (a pass, elevation 5,200 ft/1,585 m), and 17 mi/27 km WSW of PRADES; 42°31′N 02°07′E. Elevation 5,280 ft/1,609 m. Winter sports. Has massive 17th-century ramparts built by Vauban against possible invasion from Spain (border is 10 mi/16 km SW). First solar oven using parabolic mirrors was built here.

Montlouis-sur-Loire (mon-loo-ee-syur–lwahr), town (□ 9 sq mi/23.4 sq km), INDRE-ET-LOIRE department, CENTRE administrative region, W central FRANCE, on LOIRE RIVER, and 6 mi/9.7 km E of TOURS; 47°23′N 00°50′E. Noted for its white Pinot wines.

Montluçon (mon-loo-son), city (□ 8 sq mi/20.8 sq km), ALLIER department, AUVERGNE region, central FRANCE, on the CHER RIVER, and 45 mi/72 km NNW of CLERMONT-FERRAND; 46°25′N 02°40′E. The iron and steel industry developed here in 19th century due to nearby coal mines and iron-ore deposits, linked by the BERRY CANAL. Following the decline of that industry, Montluçon succeeded in diversifying its economic base, which now includes electrical and metalworking facilities, tire manufacturing, chemical, furniture, and clothing factories. The industrial area spreads northward along Cher River. Points of interest in the old town include the Romanesque Church of Saint-Pierre, the Flamboyant Gothic church of Notre Dame, and many houses dating from the 15th and 16th century; the castle (15th–16th century) of the dukes of Bourbon now houses a crockery museum.

Montluel (mon-loo-el), town (□ 15 sq mi/39 sq km; 2004 population 6,505), AIN department, RHÔNE-ALPES region, E FRANCE, 13 mi/21 km NE of LYON; 45°51′N 05°03′E. Military camps (E).

Montmagny (mon-mahn-YEE), county (□ 661 sq mi/1,718.6 sq km), CHAUDIÈRE-APPALACHES region, SE QUEBEC, E CANADA, on the SAINT LAWRENCE RIVER and U.S. (MAINE) border; ⊙ MONTMAGNY; 46°50′N 70°20′W. Composed of fourteen municipalities. Formed in 1982.

Montmagny (mon-mahn-YEE), town (□ 49 sq mi/127.4 sq km), ⊙ MONTMAGNY county, CHAUDIÈRE-APPALACHES region, SE QUEBEC, E CANADA, on the SAINT LAWRENCE RIVER; 46°58′N 70°33′W. Manufacturing includes textiles, furniture, and household appliances.

Montmagny (mon-mah-nyee), town (□ 1 sq mi/2.6 sq km), VAL-D'OISE department, ÎLE-DE-FRANCE region, N central FRANCE, an outer N suburb of PARIS, N of SAINT-DENIS, and 8 mi/12.9 km from Notre Dame Cathedral; 48°58′N 02°21′E. Chocolate factory.

Montmagny Island, CANADA: see LAVAL.

Montmartre (MAHNT-mahr-truh), village (2006 population 413), SE SASKATCHEWAN, near small Chapleau Lakes, 23 mi/37 km SE of INDIAN HEAD; 50°13′N 103°27′W. Mixed farming.

Montmartre (mon-mahr-truh) [French=hill of the martyrs], a N quarter of PARIS, Ville de Paris department, ÎLE-DE-FRANCE, FRANCE, occupying a height (Butte de Montmartre) that rises 330 ft/101 m from right bank of SEINE RIVER; 48°53′N 02°20′E. Named after the presumed martyrdom of St. Denis, first bishop of Paris. It is the highest point of Paris, topped by the famous 19th-century basilica of SACRÉ-COEUR, with a Byzantine-style dome. Parts of the ancient quarter have long been a favorite residence of the bohemian world. Until the 20th century Montmartre retained a rural look, providing material for Van Gogh, Pissarro, Utrillo, and other impressionist

artists. Montmartre is also famed for its night life; among its many nightclubs is the Moulin Rouge. The cemetery of Montmartre contains the tombs of Stendhal, Renan, Heine, Berlioz, and Alfred de Vigny. The church of Saint-Pierre (founded 1134) is also here. Montmartre was annexed to Paris in 1860. The hill, a natural fortress, played a military role during the Paris Commune (1871).

Montmaurin (mon-mo-ran), commune (□ 3 sq mi/7.8 sq km), HAUTE-GARONNE department, MIDI-PYRÉNÉES region, SW FRANCE, on the LANNEMEZAN PLATEAU N of the PYRENEES, and 8 mi/12.9 km NNW of SAINT-GAUDENS; 43°13′N 00°38′E. Here are remains of a Gallo-Roman "villa," in fact a marble palace with 200 rooms. Part of an ancient human skull, one of the oldest found in France, was discovered in a nearby cavern.

Montmédy (mon-mai-dee), commune (□ 9 sq mi/23.4 sq km), MEUSE department, LORRAINE region, NE FRANCE, on the CHIERS, and 25 mi/40 km N of VERDUN; 49°31′N 05°21′E. Agriculture market. Upper town (atop rock 330 ft/101 m high) was fortified by Vauban after its capture (1656) for Louis XIV by Turenne. Fell to Germans in 1870 and again in 1914. AVIOTH, a tiny village with noted 14th–15th-century Flamboyant Gothic basilica, is 4 mi/6.4 km N, near Belgian border. It has been a pilgrimage place since 12th century.

Montmélian (mon-mai-lyahn), town (□ 2 sq mi/5.2 sq km), SAVOIE department, RHÔNE-ALPES region, SE FRANCE, on the ISÈRE, and 8 mi/12.9 km SE of CHAMBÉRY, at N end of GRÉSIVAUDAN valley (red wines); 45°30′N 06°04′E. Railroad junction; manufacturing of electrical equipment. Was stronghold of dukes of Savoy.

Montmerle-sur-Saône (mon-mer-luh–syoor–so-nuh), commune (□ 1 sq mi/2.6 sq km), AIN department, RHÔNE-ALPES region, E FRANCE, on the SAÔNE, and 7 mi/11.3 km N of VILLEFRANCHE-SUR-SAÔNE; 46°05′N 04°45′E. Shipping of Beaujolais wines.

Montmirail (mon-mee-re), commune (□ 18 sq mi/46.8 sq km; 2004 population 3,742), MARNE department, CHAMPAGNE-ARDENNE region, N FRANCE, on the PETIT MORIN, and 14 mi/23 km SSE of CHÂTEAU-THIERRY; 48°52′N 03°34′E. Champagne wines. Here Napoleon defeated Prussians in 1814. First battle of the Marne (1914) was also fought in area. Has a 17th-century castle in Louis XIII style.

Montmoreau-Saint-Cybard (mon-mo-ro–san–see-bahr), commune (□ 4 sq mi/10.4 sq km), CHARENTE department, POITOU-CHARENTES region, W FRANCE, 17 mi/27 km S of ANGOULÊME; 45°24′N 00°08′E. Cattle market. Romanesque church (12th century); châteaux (15th century).

Montmorenci, village, TIPPECANOE county, W central INDIANA, 7 mi/11.3 km NW of LAFAYETTE. Manufacturing. Corn, soybeans, cattle, hogs. Laid out 1838.

Montmorenci (mahnt-muh-REN-see), unincorporated village, AIKEN county, SW SOUTH CAROLINA, 5 mi/8 km ESE of AIKEN. Agriculture includes livestock; grain, cotton, peanuts, poultry.

Montmorency, county (□ 562 sq mi/1,461.2 sq km; 2006 population 10,478), N MICHIGAN; 45°01′N 84°07′W; ⊙ ATLANTA. Drained by THUNDER BAY, RAINY, and BLACK rivers. Dairying, corn; cattle; industrial machinery; also a recreation area, with state forest, game refuge, small lakes. Clear Lake State Park in N, Sheridan Valley Ski Area in SW; large FLETCHER POND on E boundary. Organized 1881.

Montmorency (mahnt-muh-REHN-see, French, mon-mo-rahn-SEE), unincorporated town, S QUEBEC, E CANADA, at the confluence of the SAINT LAWRENCE and MONTMORENCY rivers; now part of the municipality of BEAUPORT; 46°53′N 71°09′W. It is a suburb of QUEBEC city and the site of the scenic Montmorency Falls.

Montmorency (mon-mo-rahn-see), town (□ 11 sq mi/28.6 sq km), VAL-D'OISE department, ÎLE-DE-FRANCE region, N central FRANCE, a N suburb of PARIS, 9 mi/

Cross-references are shown in SMALL CAPITALS. The pronunciation guide is shown on page xix. The sources of population figures are shown on page xvii.

14.5 km from Notre Dame Cathedral; 48°59′N 02°19′E. Diverse manufacturing. Market for fresh vegetables, fruit. Rousseau lived here (1756–1762). Forest of Montmorency (7,000 acres/2,833 ha), extending 7 mi/11.3 km NW toward OISE RIVER, is a favorite Parisian excursion place.

Montmorency (mahnt-muh-REN-see), locality, suburb 11 mi/18 km NE of MELBOURNE, VICTORIA, SE AUSTRALIA, near ELTHAM; 37°43′S 145°07′E.

Montmorency (mon-mo-rahn-SEE), river, c.60 mi/100 km long, E CANADA; rising in the LAURENTIAN MOUNTAINS, S QUEBEC; flowing generally S to the SAINT LAWRENCE RIVER; 46°53′N 71°09′W. Near its mouth are Montmorency Falls (275 ft/84 m high), providing hydroelectric power.

Montmorency No. 1 and **Montmorency No. 2** (mon-mo-rahn-SEE), former county (□ 2,198 sq mi/5,714.8 sq km), S QUEBEC, E CANADA, on the SAINT LAWRENCE RIVER. Montmorency No. 1 (47°30′N 71°15′W) consisted of mainland, from Saint Lawrence NW into LAURENTIDES provincial park. Montmorency No. 2 (46°55′N 70°58′W) consisted of Île d'ORLÉANS in the Saint Lawrence; its county seat was SAINTE-FAMILLE.

Montmorillon (mon-mor-ree-yon), town (□ 22 sq mi/57.2 sq km), VIENNE department, POITOU-CHARENTES region, W central FRANCE, on GARTEMPE RIVER, and 27 mi/43 km ESE of POITIERS; 46°25′N 00°40′E. Agriculture trade center specializing in production of macaroons. Has a curious two-storied, 12th-century sepulchral chapel called the Octagon, and a 12th–14th-century church with mural paintings.

Montmorot (mon-mo-ro), commune (□ 4 sq mi/10.4 sq km), W suburb of LONS-LE-SAUNIER, JURA department, FRANCHE-COMTÉ region, E FRANCE. Paper milling. Extensive nearby salt mines are no longer worked.

Mont Mosan Recreation Park (MAW mo-ZAHN), near HUY, LIÈGE province, E BELGIUM, 17 mi/27 km SW of LIÈGE.

Monto (MAHN-to), town, SE QUEENSLAND, NE AUSTRALIA, 80 mi/129 km W of BUNDABERG on Burnett Highway; 24°52′S 151°07′E. Former gold-mining district (1875). Fossicking (gold hunting). Dairying; beef cattle; citrus, cotton. Irrigated from underground water sources.

Montoir-de-Bretagne (mon-twahr–duh–bre-tah-nyuh), industrial town (□ 14 sq mi/36.4 sq km), LOIRE-ATLANTIQUE department, PAYS DE LA LOIRE region, W FRANCE, 5 mi/8 km NNE of SAINT-NAZAIRE; 47°20′N 02°09′W. Terminal for natural gas and methane transport. Saint-Nazaire airport. The regional park of BRIÈRE, a vast marshland, lies just N.

Montoire-sur-le-Loir (mon-twahr–syur–luh–lwahr), town (□ 8 sq mi/20.8 sq km; 2004 population 4,186), LOIR-ET-CHER department, CENTRE administrative region, N central FRANCE, on LOIR RIVER, and 10 mi/16 km WSW of Vendôme; 47°45′N 00°52′E. Horse and cattle market. Has 11th-century chapel of SAINT-GILLES and ruins of a feudal castle.

Montois-la-Montagne (mon-twah–lah–mon-tahn-yuh), commune (□ 2 sq mi/5.2 sq km), MOSELLE department, LORRAINE region, NE FRANCE, near the ORNE, 4 mi/6.4 km SE of BRIEY; 49°13′N 06°01′E. In old iron-mining district.

Montona, CROATIA: see MOTOVUN.

Montone River (mon-TO-ne), 53 mi/85 km long, N central ITALY; rises in ETRUSCAN APENNINES 10 mi/16 km SSE of Marradi; flows NNE, past Rocca San Casciano and FORLÌ, joining RONCO River near RAVENNA to form the FIUMI UNITI, which flows 6 mi/10 km E to the ADRIATIC. Used for irrigation. Chief tributary, Rabbi River.

Mont-Organisé (mong–or-gah-nee-ZAI), agricultural town, NORD-EST department, NE HAITI, near DOMINICAN REPUBLIC border, 17 mi/27 km SSE of FORT LIBERTÉ; 19°24′N 71°47′W. Coffee growing and processing; citrus fruit; gold deposits.

Montoro (mon-TO-ro), town, CÓRDOBA province, S SPAIN, in ANDALUSIA, on the GUADALQUIVIR RIVER, and 24 mi/39 km ENE of CÓRDOBA; 38°01′N 04°23′W. Olive-oil center. Wine making. Trades in cereals, vegetables, olives, livestock. Has 16th-century bridge. Was important fortress under Moors. Copper, iron, and lead deposits nearby.

Montour (mahn-TOOR), county (□ 132 sq mi/343.2 sq km; 2006 population 17,934), central PENNSYLVANIA; ⊙ DANVILLE; 41°01′N 76°39′W. Drained by SUSQUEHANNA RIVER (forms part of S boundary) and Chillisquaque Creek. Agriculture (corn, wheat, oats, barley, hay, alfalfa, soybeans; sheep, hogs, cattle, poultry; dairying); limestone. Manufacturing at Danville. Smallest county in terms of land area in Pennsylvania. Lake Chillisquaque reservoir in Montour Preserve in N. Formed 1850.

Montour, town (2000 population 285), TAMA county, central IOWA, on IOWA RIVER and 7 mi/11.3 km E of TOLEDO; 41°58′N 92°43′W. Limestone quarries nearby.

Montour Falls, village (□ 3 sq mi/7.8 sq km; 2006 population 1,784), SCHUYLER county, W central NEW YORK, in FINGER LAKES region, just SE of WATKINS GLEN, and on small Catherine Creek; 42°21′N 76°50′W. Summer resort. Chequaga (or Shequaga) Falls (156 ft/48 m) here attract tourists, especially when snowmelt in the spring enhances the spectacle. Incorporated 1890.

Montoursville (mahn-TOORS-vil), borough (2006 population 4,647), LYCOMING county, N central PENNSYLVANIA, suburb 4 mi/6.4 km E of WILLIAMSPORT, on West Branch of SUSQUEHANNA RIVER, at mouth of LOYALSOCK CREEK; 41°15′N 76°55′W. Manufacturing (paper, fabricated metal products, concrete, printing, furniture). Williamsport-Lycoming County Airport in S. Parts of Tiadaghton State Forest to N and S. Settled 1807, laid out 1820, incorporated 1850.

Montoz (mo-TO), mountain range in the JURA, BERN canton, NW SWITZERLAND, between Vallée de Tavannes and AARE valley, NNE of BIEL; 47°13′N 07°17′E. Highest point reaches 4,357 ft/1,328 m.

Montparnasse (mon-pahr-nahs), quarter of PARIS, Ville de Paris department, ÎLE-DE-FRANCE, FRANCE, on left bank of the SEINE RIVER, centering on the intersection of the Boulevard de Montparnasse and the Boulevard Raspail, in the 14th *Arrondissement*. Its famous cafés have long been centers of the Parisian artistic and intellectual world. It is an active area of commerce and business. The quarter contains the Pasteur Institute, the Gallo-Roman catacombs, and the Montparnasse cemetery, with the tombs of Saint-Saëns, Houdon, Baudelaire, Poincaré, André Citroën, Jean-Paul Sartre, César Franck, Guy de Maupassant, and Leconte de Lisle, among others.

Mont-Pèlerin (mo–PE-le-rai), resort, VAUD canton, W SWITZERLAND, 2 mi/3.2 km NNE of VEVEY. Elevation c.2,600 ft/792 m. Reached by road or by cog railroad, which continues to top of Mont Pèlerin (3,543 ft/1,080 m); views of FRENCH ALPS.

Montpelier, city (2000 population 1,929), BLACKFORD county, E INDIANA, on SALAMONIE RIVER, and 8 mi/12.9 km NE of HARTFORD CITY. In agricultural area (livestock; dairy products; soybeans, corn, grain); manufacturing (grip nuts, corrugated containers, glass products, gloves); old natural gas and oil wells; stone quarries. Settled 1836, laid out 1837, incorporated 1937.

Montpelier (mahnt-PEEL-yuhr), city (2006 population 7,954), ⊙ WASHINGTON CO. and VERMONT (since 1805), central Vermont, at the junction of the WINOOSKI and North Branch rivers; 44°16′N 72°34′W. The economy is dominated by state government and insurance industries. It is also a trading center in a lumber, granite, and winter-resort area. Granite is processed here; manufacturing includes printing, plastics, textiles, and sawmill machinery. Montpelier shares an airport with

BARRE to SE. Vermont College of Norwich University, a community college, the state historical society, and a state school for children with special needs are here. Of interest are the state capitol and an art gallery for wood sculpture. Surrounded by mountains, the city has an excellent view of Mt. Mansfield, the highest point in the state. Admiral George Dewey b. here. Airport. Inc. 1855.

Montpelier, town, SAINT JAMES parish, NW JAMAICA, on Jamaica railroad, and 8 mi/12.9 km S of MONTEGO BAY; 18°23′N 77°56′W. In fertile agricultural region, growing principally bananas.

Montpelier, town (2000 population 2,785), BEAR LAKE county, extreme SE IDAHO, near BEAR RIVER, 70 mi/113 km SE of POCATELLO; 42°19′N 111°18′W. Elevation c.5,943 ft/1,811 m. Railroad division point, trade and shipping center for agricultural and grazing area; dairying; manufacturing (dairy feeds, light manufacturing). Phosphate deposits nearby. Founded 1864 by Mormons. Early names: Clover Creek, Belmont. Renamed (1865) by Brigham Young in honor of Vermont capital. Has Mormon tabernacle. Caribou National Forest to NE, Cache National Forest to W.

Montpelier (mahnt-PEEL-yuhr), village (2000 population 214), SAINT HELENA parish, SE LOUISIANA, 33 mi/53 km ENE of BATON ROUGE, on TICKFAW RIVER; 30°41′N 90°39′W. Agricultural and timber area. Manufacturing (feeds, dog food, sausage).

Montpelier, village (2006 population 92), STUTSMAN county, SE central NORTH DAKOTA, 16 mi/26 km SSE of JAMESTOWN, and on JAMES RIVER; 46°42′N 98°35′W. Founded in 1885 and named for the capital of VERMONT, the home state of two early settlers.

Montpelier (MAHNT-peel-yuhr), village (□ 3 sq mi/7.8 sq km; 2006 population 4,111), WILLIAMS county, extreme NW OHIO, on ST. JOSEPH RIVER, and 8 mi/13 km NNW of BRYAN; 41°35′N 84°36′W. Wood fixtures, metal stampings, truck bodies. Settled 1855, incorporated 1875.

Montpelier (mahnt-PEEL-yuhr), unincorporated village, HANOVER county, E VIRGINIA, 25 mi/40 km NNW of RICHMOND; 37°49′N 77°41′W. Agriculture (dairying; livestock; grain).

Montpellier (mon-pel-yai), city (□ 21 sq mi/54.6 sq km); ⊙ HÉRAULT department, S FRANCE, near the MEDITERRANEAN, and 80 mi/129 km WNW of MARSEILLE; 43°40′N 03°50′E. It is the commercial and cultural center of the LANGUEDOC region, which extends along the coast form the RHÔNE delta almost to the Spanish border. The wine business dominates the region's economy, primarily for French domestic consumption. While not primarily an industrial center, Montpellier has developed (especially since World War II) a diverse mix, including food processing, textiles, printing, electronics, and chemicals (based on salt-extraction nearby). It is a major center of the computer industry. The old city has narrow, winding streets and many fine 17th–18th-century public and private buildings. It is now linked to the commercial center by the post-modern Antigone district of daring architectural design. Of principal interest are the Fabre museum (fine arts) and an archeological museum; the botanical garden founded 1593 by Henry IV; the restored 14th-century cathedral (a former abbey), and the Promenade du Peyrou (with triumphal arch and fountain) overlooking the city. The well-known university was founded in 1289; Rabelais was its most famous student. The faculty of medicine, in existence since the Middle Ages, occupies the old episcopal palace. Other educational institutions include the school of military administration (1948), an infantry school (1967), and an agriculture college. The city hosts a dance festival, international fair, and a European medical congress. Dating from the 8th century, Montpellier was the seat of a fiefdom under the counts of Toulouse; it passed (13th century) to the kings of Majorca, who sold it (1349) to Philip VI of France. A Huguenot center, it was besieged and captured by Louis XIII in 1622. It was later the seat of the provincial estates (a form of parliament) of Languedoc. In recent times, Montpellier has grown with the influx of North Africans following the end of French colonial domination in 1960s. Its Mediterranean climate and ambitious development of the coast have made tourism an important contributor to the economy. With government regionalization in the 1980s, Montpellier was designated as the capital of the LANGUEDOC-ROUSSILLON administrative region, which embraces 5 French departments from the PYRENEES to the Rhône valley. Montpellier-Fréjorgues international airport 5 mi/8 km SE of the city.

Montpellier (mon-pel-YAI), village (□ 96 sq mi/249.6 sq km; 2006 population 792), OUTAOUAIS region, SW QUEBEC, E CANADA, 6 mi/10 km from CHÉNÉVILLE; 45°51′N 75°10′W.

Montpezat-de-Quercy (mon-pe-zah–duh–ker-see), commune (□ 16 sq mi/41.6 sq km), TARN-ET-GARONNE department, MIDI-PYRÉNÉES region, SW FRANCE, 14 mi/23 km S of CAHORS; 44°15′N 01°28′E. Has a 14th-century collegiate church noted for its huge 16th-century tapestries (from N France).

Montpon-Ménestérol (mon-pon–mai-ne-stai-rol), town (□ 18 sq mi/46.8 sq km), DORDOGNE department, AQUITAINE region, SW FRANCE, in PÉRIGORD region, on the Isle, and 19 mi/31 km NW of BERGERAC; 45°00′N 00°10′E. Center of tobacco cultivation; paper milling (for tobacco industry), canning, cheese making. Formerly known as Monpont-sur-l'Isle. Ancient Chartreuse de Vauclaire nearby.

Montrachet, FRANCE: see PULIGNY-MONTRACHET.

Mont Raimeux (mon rai-MUH), ridge, in JURA mountains, W SWITZERLAND, between BERN and JURA cantons. Highest point (4,272 ft/1302 m) is NE of MOUTIER.

Montréal (mon-trai-ahl), commune, AUDE department, LANGUEDOC-ROUSSILLON region, S FRANCE, 11 mi/18 km W of CARCASSONNE; 43°12′N 02°09′E. Vineyards. Has 14th-century church with fine interior features.

Montreal (mahn-tree-OL), city, S QUEBEC, E CANADA, on MONTREAL ISLAND, surrounded by SAINT LAWRENCE RIVER and Rivière des PRAIRIES; 45°31′N 73°34′W. Montreal is the second-largest metropolitan area in Canada, after TORONTO, and is a cultural, tourist, commercial, financial, and industrial center. It is the second-largest French-speaking city in the world, though most of its inhabitants also speak English. Montreal lies at the foot of MOUNT ROYAL—the source of its name—and has an excellent harbor on the SAINT LAWRENCE SEAWAY, which connects the city to the great industrial centers of the GREAT LAKES. Canada's most important port, Montreal is a transshipment point for oil, grain, sugar, machinery, and manufactured goods. It is also an important railroad hub, and it has two international airports, DORVAL and Mirabel. Montreal's underground railroad system, the Métro, was inaugurated in 1966. Manufacturing includes steel, electronic equipment, refined petroleum, transportation materials and equipment, raw textiles, clothing, food and beverages, printed materials, and tobacco. Once Canada's pre-eminent city, Montreal has been eclipsed by Toronto as the country's economic center. Tensions over Quebec's insistence on enforcing its francophone culture have caused an outmigration of English-speaking people to ONTARIO and to the growing W provinces. Despite these changes, Montreal remains one of North America's great cosmopolitan cities, and had a burst of prosperity in the 1980s due to its growth as a financial service center. A stockaded Native-American village, Hochelaga, was found on the site (1535) by Cartier, and the island was visited in 1603 by Champlain, but it was not settled by the French until 1642, when a band of priests, nuns, and settlers under Paul de Chomedey, Sieur de Mai-

sonneuve, founded the Ville Marie de Montréal. The settlement grew to become an important center of the fur trade and the starting point for the W expeditions of Jolliet, Marquette, La Salle, Vérendrye, and Duluth. It was fortified in 1725 and remained in French possession until 1760, when Vaudreuil de Cavagnal surrendered it to British forces under Amherst. Americans under Richard Montgomery occupied it briefly (1775–1776) during the American Revolution. The city's growth was aided by the opening in 1825 of the LACHINE CANAL, making possible water communications with the Great Lakes. From 1844 to 1849, Montreal was the capital of United Canada. The Canadian Pacific railroad established its headquarters here in the 1880s. The area of Old Montreal has undergone extensive restoration and few buildings from the French period are extant. Among the city's notable buildings are the Gothic Church of Notre Dame (c.1820), St. Sulpice Seminary (1685), the Château de Ramezay (1705), and the Place Ville Marie (1962). Montreal is the seat of McGill University, the University of Montreal, the University of Quebec at Montreal, and Concordia University. Expo '67, the international exposition of 1967, was held in the city. Montreal hosted the 1976 Summer Olympics, which created a great financial burden. Its professional hockey team, the Montreal Canadiens, has won 23 Stanley Cups, making it one of the athletic world's most enduring dynasties. In late 1960s, Montreal experienced a high-rise building boom; a system of underground promenades of shops and restaurants was created, linking office buildings (Place Ville Marie and Place Bonaventure), the Queen Elizabeth Hotel, train station, subway, arts center, and other facilities. As part of a recent national and provincial trend toward municipal reorganization, the city of Montreal has restructured, and as of January 1, 2006, is composed of 19 boroughs (*arrondissements*): Ahuntsic-Cartierville, ANJOU, Côte-des-Neiges–Notre-Dame-de-Grâce, LACHINE, LASALLE, Le Plateau-Mont-Royal, Le Sud-Ouest, L'Île-Bizard–Sainte-Geneviève, Mercier–Hochelaga-Maisonneuve, MONTRÉAL-NORD, OUTREMONT, Pierrefonds-Roxboro, Rivière-des-Prairies–Pointe-aux-Trembles, Rosemont–La Petite-Patrie, SAINT-LAURENT, SAINT-LÉONARD, VERDUN, Ville-Marie, Villeray–Saint-Michel–Parc-Extension. The following 15 municipalities voted to regain their independence after having been part of Montreal from 2002–2005: BAIE-D'URFÉ, BEACONSFIELD, Côte-Saint-Luc, Dollard-des-Ormeaux, DORVAL, HAMPSTEAD, Kirkland, L'Île-Dorval, Montréal-Est, MONTRÉAL-OUEST, MONT-ROYAL, POINTE CLAIRE, SAINTE ANNE DE BELLEVUE, SENNEVILLE, and WESTMOUNT. A new agglomeration council oversees Montreal along with the 15 "reconstituted cities." Further, Montreal is now just one of 82 municipalities comprising the Metropolitan Community of Montreal (*Communauté Métropolitaine de Montréal*), formed in June 2000, with a population of 3.4 million and covering 1,474 sq mi/3,818 sq km.

Montreal (mahn-tree-AWL), town (2006 population 771), IRON county, N WISCONSIN, in GOGEBIC RANGE, 2 mi/3.2 km W of HURLEY, on Gile Flowage near Gile Falls; 46°25′N 90°14′W. Iron mining; forestry. Incorporated 1924.

Montréal (mon-trai-AHL), administrative region (□ 192 sq mi/499.2 sq km; 2005 population 1,873,813), S QUEBEC, E CANADA; 45°30′N 73°41′W. Composed of sixteen municipalities, centered on the city of MONTREAL. It is the most heavily populated of Quebec's regions.

Mont Réal, CANADA: see MOUNT ROYAL.

Montréal-Est (mon-trai-AHL–EST), city (□ 5 sq mi/13 sq km; 2006 population 3,796), Montréal administrative region, S QUEBEC, E CANADA, on NE MONTREAL ISLAND; 45°37′N 73°30′W. Part of the Metropolitan Community of Montreal (*Communauté Metropolitaine de Montréal*).

Montreal Island, NUNAVUT territory, Canada, in CHANTREY INLET, on E side of ADELAIDE PENINSULA; 13 mi/21 km long, 2 mi/3 km–4 mi/6 km wide; 67°52′N 96°20′W.

Montreal Island (mahn-tree-OL), French, *Île de Montréal* (EEl duh mon-trai-AHL), S QUEBEC, E Canada, bounded by Lake SAINT LOUIS (S), the SAINT LAWRENCE RIVER (E), and River des PRAIRIES (NW), branch of OTTAWA River; 30 mi/48 km long, up to 10 mi/16 km wide. Island is composed of MONTREAL and 15 independent municipalities.

Montreal Lake (muhn-tree-AWL) (□ 162 sq mi/421.2 sq km), central SASKATCHEWAN, Canada, 60 mi/97 km N of PRINCE ALBERT; 32 mi/51 km long, 7 mi/11 km wide; 54°03′N 105°48′W. Drains N into Churchill River through LAC LA RONGE.

Montréal-Nord (mon-TRAI-ahl–NORD), former city, borough (French *arrondissement*) of MONTREAL, S QUEBEC, E Canada, on N coast of MONTREAL ISLAND, on SE bank of PRAIRIES River; 45°36′N 73°37′W. Bounded on S by SAINT-LEONARD.

Montréal-Ouest (mon-trai-AHL–WEST), city, Montréal administrative region, S QUEBEC, E Canada, 5 mi/8 km SSW of downtown MONTREAL, on S central MONTREAL ISLAND; 45°27′N 73°39′W. LACHINE CANAL on SE. Part of the Metropolitan Community of Montreal (*Communauté Metropolitaine de Montréal*).

Montreal River (mahn-tree-AHL), 90 mi/145 km long, central ONTARIO, E central Canada; flows S and WSW to Lake SUPERIOR 50 mi/80 km NNW of SAULT SAINTE MARIE; 47°14′N 84°38′W. Montreal Falls (150 ft/46 m) is 10 mi/16 km above its mouth; hydroelectric power.

Montreal River, c.40 mi/64 km long, in WISCONSIN and MICHIGAN; rises in IRON county, N Wisconsin, in small Pine Lake; flows N and NW, past IRONWOOD, Michigan, and HURLEY, Wisconsin, to LAKE SUPERIOR; forms part of Michigan-Wisconsin state line for most of its length.

Montreat (MAHN-treet), village (□ 2 sq mi/5.2 sq km; 2006 population 698), BUNCOMBE county, W NORTH CAROLINA, 14 mi/23 km E of ASHEVILLE, in the BLUE RIDGE MOUNTAINS; 35°38′N 82°17′W. Pisgah National Forest to NE. Service industries; retail trade; agriculture (corn, tobacco; cattle). Montreat College (4-year). Founded as a spiritual retreat. Montreat Conference Center, affiliated with the Presbyterian Church (USA). Incorporated 1897.

Montredon-Labessonnié (mon-truh-don–lah-be-so-nee-ai), commune (□ 42 sq mi/109.2 sq km), TARN department, MIDI-PYRÉNÉES region, S FRANCE, 9 mi/14.5 km NNE of CASTRES; 43°44′N 02°21′E. Livestock. Ruins of an old fort. Nearby is the regional park of the Upper Languedoc (HAUT LANGUEDOC NATURAL REGIONAL PARK), which contains the gorge of the AGOUT RIVER and a granite tableland with huge quarries worked for tombstones and monuments.

Montréjeau (mon-trai-zho), commune (□ 3 sq mi/7.8 sq km), HAUTE-GARONNE department, MIDI-PYRÉNÉES region, S FRANCE, on the GARONNE, near mouth of the NESTE, and 8 mi/12.9 km W of SAINT-GAUDENS; 43°05′N 00°35′E. Agriculture market for fruit and wool trade. Was founded (13th century) as a fortified village (a *bastide*).

Montreuil (mon-troi), commune (□ 1 sq mi/2.6 sq km), market town in PAS-DE-CALAIS department, NORD-PAS-DE-CALAIS region, N FRANCE, on left bank of CANCHE RIVER, and 19 mi/31 km SSE of BOULOGNE-SUR-MER; 50°28′N 01°46′E. It once stood by the sea and preserves a 16th–17th-century citadel and 13th–17th-century ramparts. It was an important center (1993 estimated pop. c.40,000) in Middle Ages. Neuville-sous-Montreuil (on opposite bank of Canche River) has a Carthusian monastery. The nearby Channel coast is a resort area. Also known as Montreuil-sur-Mer.

Montreuil (mon-troi), town (□ 3 sq mi/7.8 sq km), SEINE-SAINT-DENIS department, ÎLE-DE-FRANCE region, N central FRANCE, an E suburb of PARIS, 4 mi/6.4 km from Notre Dame Cathderal and just N of VINCENNES; 50°30′N 01°55′E. Long famous for its peaches and pears, Montreuil has a variety of light industries, including tanneries, food processing, and container manufacturing. It was founded before 1000. In the center of town is the Gothic church of Saints Peter and Paul (12th century), where Charles V was baptized. A nearby public park has a museum dedicated to the Socialist workers' movement. Also known as Montreuil-sous-Bois.

Montreuil-Bellay (mon-troi–be-lai), town (□ 18 sq mi/46.8 sq km), MAINE-ET-LOIRE department, PAYS DE LA LOIRE region, W FRANCE, on the THOUET, and 9 mi/14.5 km SSW of SAUMUR; 47°08′N 00°09′W. Wine-shipping center; distilling. Its imposing 13th–15th-century feudal castle and 15th-century town walls give the town a medieval aspect. Fine gardens along the river below the castle.

Montreuil-Juigné (mon-troi–zhwee-nyai), NW suburb (□ 5 sq mi/13 sq km), of ANGERS, MAINE-ET-LOIRE department, PAYS DE LA LOIRE region, W FRANCE, on MAYENNE RIVER; 47°32′N 00°36′W. Metalworking.

Montreuil-sous-Bois, FRANCE: see MONTREUIL.

Montreuil-sur-Mer, FRANCE: see MONTREUIL.

Montreux (mo-TRUH), town (2000 population 22,454), on the NE shore of LAKE GENEVA, VAUD canton, W SWITZERLAND; 46°26′N 06°55′E. Elevation 1,306 ft/398 m. Montreux was once a pre-eminent meeting place of the rich and famous as attested by its grand hotels. It is still a leading resort on the VAUD Riviera.

Montrevel-en-Bresse (mon-tre-vel–ahn-bre-suh), commune (□ 4 sq mi/10.4 sq km), AIN department, RHÔNE-ALPES region, E central FRANCE, 10 mi/16 km NNW of BOURG-EN-BRESSE; 46°20′N 05°07′E. Poultry and dairy market.

Montrichard (mon-tree-shahr), commune (□ 5 sq mi/13 sq km), LOIR-ET-CHER department, CENTRE administrative region, N central FRANCE, on the CHER, and 18 mi/29 km SSW of BLOIS; 47°21′N 01°11′E. Well known for its white wines. Has remains (keep) of a medieval castle, with view of Cher River valley. Château of CHENONCEAUX is 5 mi/8 km downstream.

Montroig (mont-ROIG), town, TARRAGONA province, NE SPAIN, 10 mi/16 km SW of REUS. Alcohol manufacturing, olive-oil processing; sheep raising.

Mont-Rond, FRANCE: see FAUCILLE, COL DE LA.

Montrond-les-Bains (mon-tron–lai-ban), commune (□ 3 sq mi/7.8 sq km; 2004 population 421), resort in LOIRE department, RHÔNE-ALPES region, E central FRANCE, in the Forez basin, on the upper LOIRE, 8 mi/12.9 km ENE of MONTBRISON, and 30 mi/48 km W of LYON; 45°39′N 04°14′E. Health spa (for diabetics) with thermal springs.

Montrose, county (□ 2,242 sq mi/5,829.2 sq km; 2006 population 38,559), W COLORADO, ⊙ MONTROSE; 38°24′N 108°16′W. Agricultural region bordering UTAH; drained by DOLORES, SAN MIGUEL, and UNCOMPAHGRE rivers. Fruit, beans, hay, livestock. Mining of uranium, radium, coal, silver, copper. Includes parts of Uncompahgre, Gunnison, and La Sal national forests and, in NE, BLACK CANYON OF THE GUNNISON NATIONAL MONUMENT. Crystal Reservoir and part of Morrow Point Reservoir National Recreation Area (part of Curecanti National Recreation Area) in E; part of Gunnison National Forest in far E; part of Uncompahgre National Forest in center, part of Manti La Sal National Forest on W boundary (with Utah). Uncompahgre Plateau bisects county NW-SE; regular highway connections via neighboring counties. Formed 1883.

Montrose (mahn-TROZ), city (2000 population 417), HENRY county, W central MISSOURI, 14 mi/23 km SW of CLINTON; 38°15′N 93°58′W. Corn, wheat, soybeans; cattle; manufacturing (lead castings, fishing lures). Montrose Conservation Area to N. Strip-coal mining.

Coal-fired electric power generating plant (KANSAS CITY).

Montrose (MAHN-tros), town (2001 population 10,845), ANGUS, NE Scotland, on the NORTH SEA at mouth of South Esk River, 26 mi/42 km NE of DUNDEE; 56°42′N 02°28′W. Open to water on three sides, it is a spacious resort town, with agricultural equipment, pharmaceuticals, boat yards, and a fishing industry. Previously brick making, distilling, flax and jute mills, and fruit canneries. Montrose was the scene of John de Balliol's surrender of the Scottish throne to Edward I of England in 1296. Formerly in Tayside, abolished 1996.

Montrose, unincorporated town, LOS ANGELES county, S CALIFORNIA; suburb 11 mi/18 km N of downtown LOS ANGELES, just N of Glendale, and in foothills of SAN GABRIEL MOUNTAINS. Manufacturing (electronic coils, plastics products).

Montrose, town (2000 population 12,344), ⊙ MONTROSE county, SW COLORADO, on UNCOMPAHGRE River, and 55 mi/89 km SE of GRAND JUNCTION; 38°28′N 107°52′W. Elevation 5,794 ft/1,766 m. Railroad terminus spur from DELTA. Trade center in irrigated fruit, potato, sugar-beet region; meat and dairy products, flour; manufacturing (signs, lumber, hydroelectric testing equipment, boxed chocolates). Carnotite deposits nearby are source of radium and uranium. BLACK CANYON OF THE GUNNISON NATIONAL MONUMENT to NE. Nearby Gunnison Tunnel conducts water from GUNNISON River to Uncompahgre valley. Ute Indian Museum to S, Currant National Recreation Area to E. Founded and incorporated 1882.

Montrose, town (2000 population 957), LEE county, extreme SE IOWA, on MISSISSIPPI RIVER and 10 mi/16 km SE of FORT MADISON, in livestock area; 40°31′N 91°25′W. One of the first permanent white settlements in Iowa was made here in 1799, when Louis Tesson, a French Canadian, established a trading post. Town was laid out in 1837.

Montrose (mahn-TROS), town (2000 population 1,619), GENESEE county, SE central MICHIGAN, suburb 15 mi/24 km NW of FLINT; 43°10′N 83°53′W. In farm area; lumber.

Montrose (MAWN-troz), town (2000 population 1,143), WRIGHT county, S central MINNESOTA, 8 mi/12.9 km S of BUFFALO; 45°04′N 93°54′W. Poultry; dairying; manufacturing (speaker systems, color dispensers). Small natural lakes in area.

Montrose (MAHN-tros), unincorporated town, HENRICO county, E central VIRGINIA, residential suburb 4 mi/6 km ESE of downtown RICHMOND; 37°31′N 77°22′W. Richmond National Cemetery to W, Richmond International Airport to E.

Montrose (MAHN-tros), village (□ 1 sq mi/2.6 sq km; 2001 population 1,067), S BRITISH COLUMBIA, W Canada, 3 mi/5 km SE of TRAIL, in Kootenay Boundary regional district; 49°05′N 117°35′W. Elevation 1,944 ft/580 m. Residential settlement near smelter town. Incorporated 1956.

Montrose (MAHN-troz), village (2000 population 526), ASHLEY county, SE ARKANSAS, 12 mi/19 km W of LAKE VILLAGE; 33°17′N 91°30′W. Manufacturing (barbecue sauce).

Montrose (mahn-TROZ), village (2000 population 154), LAURENS county, central GEORGIA, 15 mi/24 km W of DUBLIN; 32°34′N 83°09′W.

Montrose, village (2000 population 257), EFFINGHAM county, SE central ILLINOIS, 8 mi/12.9 km ENE of EFFINGHAM; 39°10′N 88°22′W. In agricultural area.

Montrose (mahnt-ROZ), village (2000 population 127), JASPER county, E central MISSISSIPPI, 35 mi/56 km WSW of MERIDIAN, in Bienville National Forest; 32°07′N 89°14′W. Agricultural and timber area.

Montrose, village (2006 population 464), MCCOOK county, SE SOUTH DAKOTA, 10 mi/16 km E of SALEM; 43°42′N 97°10′W. Lake Vermillion State Recreation Area to S.

Montrose (MAHN-troz), unincorporated village (2006 population 157), RANDOLPH county, E WEST VIRGINIA, 11 mi/18 km N of ELKINS; 39°04′N 79°48′W. Monongahela National Forest to E.

Montrose (MAHN-tros), borough (2006 population 1,581), ⊙ SUSQUEHANNA county, NE PENNSYLVANIA, 31 mi/50 km NNW of SCRANTON; 41°49′N 75°52′W. Agriculture (corn, hay; livestock; dairying); manufacturing (wooden products, cabinet parts, machinery, printing and publishing, cement reinforcing materials, textile products); mountain resort area with numerous small lakes. Salt Springs State Park to N. Settled 1799, incorporated 1824.

Montrose (MAHNT-roz), suburb 20 mi/32 km E of MELBOURNE, VICTORIA, SE AUSTRALIA, at NW foothills of DANDENONG RANGES; 37°49′S 145°21′E.

Montrose-Ghent (MAHN-troz–GENT), unincorporated village (□ 10 sq mi/26 sq km; 2000 population 5,261), SUMMIT county, NE OHIO, just NW of AKRON; 41°10′N 81°38′W. Large-scale retailing.

Montross (mahn-TRAHS), town (2006 population 300), ⊙ WESTMORELAND county, E VIRGINIA, 38 mi/61 km ESE of FREDERICKSBURG; 38°05′N 76°49′W. Manufacturing (printing and publishing, seafood processing, computer cable assemblies, lumber); agriculture (grain, soybeans; cattle). GEORGE WASHINGTON BIRTHPLACE National Monument to NW; "STRATFORD HALL," birthplace of Robert E. Lee, Westmoreland State Park to N.

Montrouge (mon-troozh), industrial suburb, just S of PARIS, HAUTS-DE-SEINE department, ÎLE-DE-FRANCE region, N central FRANCE, 3 mi/4.8 km from Notre Dame Cathedral; 48°49′N 02°19′E. Electronics is now the leading industrial activity, followed by aeronautics, surgical instruments, textiles, and chemicals.

Montroy (mont-ROI), town, VALENCIA province, E SPAIN, 16 mi/26 km SW of VALENCIA; 39°20′N 00°37′W. Wine, cereals, olive oil.

Mont-Royal (MON–rwah-YAHL) or **Mount Royal** (MOUNT ROI-yuhl), city (□ 3 sq mi/7.8 sq km), Montréal administrative region, S QUEBEC, E Canada, 3 mi/5 km W of downtown MONTREAL, on central MONTREAL ISLAND; 45°31′N 73°39′W. Bounded N and S by Montreal, E by OUTREMONT, W by SAINT-LAURENT. Part of the Metropolitan Community of Montreal (*Communauté Metropolitaine de Montréal*).

Monts (mon), town (□ 9 sq mi/23.4 sq km; 2004 population 6,953), INDRE-ET-LOIRE department, CENTRE administrative region, NW central FRANCE, 13 mi/21 km SSW of TOURS; 47°17′N 00°37′E. Pharmaceutical industry.

Mont-Saint-Aignan (mon–san-ten-YAHN), town (□ 3 sq mi/7.8 sq km), NW residential suburb of ROUEN, SEINE-MARITIME department, HAUTE-NORMANDIE region, N FRANCE; 49°28′N 01°05′E. Site of University of Rouen campus.

Mont-Saint-Amand, BELGIUM: see SINT-AMANDSBERG.

Mont-Saint-Éloi (mon–san-tai-lwah), village (□ 6 sq mi/15.6 sq km) and ridge (400 ft/122 m), PAS-DE-CALAIS department, NORD-PAS-DE-CALAIS region, N FRANCE, 6 mi/9.7 km NW of ARRAS; 50°21′N 02°42′E. Figured prominently in Lens-Arras battlefield of World War I. Atop ridge are remains of Augustinian abbey founded by St. Eloi in 7th century.

Mont-Saint-Grégoire (mon–san-grai-GWAHR), village (□ 31 sq mi/80.6 sq km; 2006 population 3,103), Montérégie region, S QUEBEC, E Canada, 5 mi/8 km from SAINT-JEAN-SUR-RICHELIEU; 45°20′N 73°10′W.

Mont-Saint-Guibert (maw–sen-yee-BER), commune (2006 population 6,442), Nivelles district, BRABANT province, central BELGIUM, 6 mi/9.7 km S of WAVRE; 50°38′N 04°37′E. Paper manufacturing.

Mont-Saint-Hilaire (MON–SAN–tee-LER) or **Saint-Hilaire**, city (□ 17 sq mi/44.2 sq km; 2006 population 15,338), Montérégie region, S QUEBEC, E Canada, on RICHELIEU RIVER, 16 mi/26 km E of MONTREAL; 45°34′N 73°12′W. Food processing; in agricultural region. Part of the Metropolitan Community of Montreal (*Communauté Metropolitaine de Montréal*).

Mont-Saint-Martin (mon–san–mahr-tan), town (□ 3 sq mi/7.8 sq km), N suburb of LONGWY, MEURTHE-ET-MOSELLE department, LORRAINE region, NE FRANCE, near Belgian border; 49°32′N 05°47′E. Steel milling in iron-mining region.

Mont-Saint-Michel (mon–san–mee-SHEL), village (□ 53 sq mi/137.8 sq km; 2006 population 635), Laurentides region, S QUEBEC, E Canada, 4 mi/6 km from Lac-Saint-Paul; 46°47′N 75°20′W.

Mont-Saint-Michel, Bay of, FRANCE: see SAINT-MICHEL, BAY OF.

Mont-Saint-Michel, Le (mon–san–mee-shel, luh), commune (□ 1 sq mi/2.6 sq km), rocky islet in Bay of Mont-Saint-Michel of ENGLISH CHANNEL, 1 mi/1.6 km off coast of NW FRANCE, in MANCHE department, BASSE-NORMANDIE region, 8 mi/12.9 km SW of AVRANCHES and E of SAINT-MALO; 48°38′N 01°30′W. It is accessible from PONTORSON (6 mi/9.7 km S) by causeway 1 mi/1.6 km long; parking area. A cone-shaped tombola (256 ft/78 m), girt at its base by medieval walls and towers, above which rise, in three levels, the clustered buildings of the village, crowned by the graceful Benedictine abbey church. Six abbatial buildings facing the sea form the unit called La Merveille (French=the marvel), built in 13th century, including almonry and cellar (first story); refectory and Hall of Knights where order of St. Michael was founded in 1469 (second story); and dormitory and cloister (third story), the whole representing a marvel of Gothic architecture. The celebrated abbey was founded in 708 by Aubert, bishop of Avranches. Frequently assaulted by the English in the Hundred Years War, but never captured. Surviving World War II undamaged (despite proximity of Allies' Normandy landing), it remains one of France's leading cultural icons and tourist attractions. Major engineering project is underway to reverse the silting of the inner bay and maintain Mont-Saint-Michel an island except at low tide. Mont-Saint-Michel and its bay were designated a UNESCO World Heritage site in 1979.

Mont-Saint-Pierre (mon–san–PYER), town (□ 23 sq mi/59.8 sq km; 2006 population 225), Gaspésie—Îles-de-la-Madeleine region, SE QUEBEC, E Canada, 6 mi/10 km from Rivière-à-Claude; 49°13′N 65°49′W.

Montsalvy (mon-sahl-vee), commune (□ 7 sq mi/18.2 sq km), CANTAL department, AUVERGNE region, S central FRANCE, in AUVERGNE MOUNTAINS, 15 mi/24 km SSE of AURILLAC; 44°42′N 02°30′E. Abbatial church (12th century) with Romanesque nave. A volcanic summit (panoramic view) is 1 mi/1.6 km E.

Monts d'Arrée, FRANCE: see ARMORICAN MASSIF.

Monts d'Aubrac, FRANCE: see AUBRAC, MONTS D'.

Monts d'Auvergne, FRANCE: see AUVERGNE MOUNTAINS.

Montsec (mon-sek), village (□ 2 sq mi/5.2 sq km), MEUSE department, LORRAINE, NE FRANCE, 8 mi/12.9 km E of SAINT-MIHIEL; 48°53′N 05°43′E. Atop Montsec hill (900 ft/274 m) is Saint-Mihiel American Memorial of World War I.

Montsech (mont-SECH), foothills of the PYRENEES, in LÉRIDA province, NE SPAIN, between crest of the central Pyrenees and Urgel plain, c.16 mi/26 km N of BALAGUER; rise to 5,500 ft/1,676 m. Gorge in foothills traversed by NOGUERA PALLARESA RIVER.

Montségur (mon-sai-guhr), commune (□ 14 sq mi/36.4 sq km), ARIÈGE department, MIDI-PYRÉNÉES region, SW FRANCE, in foothills of the PYRENEES, 13 mi/22 km SE of FOIX; 42°51′N 01°49′E. The ruins of Montségur castle, the last stronghold of the Albigensian revolt, stand on a rocky outcrop. After surrendering in 1244, its defenders were massacred in the terminal horror of the Albigensian Crusade. There-

after the LANGUEDOC fell into a severe cultural and economic decline.

Montserrado (mahnt-suhr-AH-do), county (□ 1,058 sq mi/2,750.8 sq km; 1999 population 843,783), W LIBERIA, on ATLANTIC OCEAN coast; ⊙ BENSONVILLE. Bounded E by MARGIBI county, NE by BONG county, and NW by BOMI county. Rubber (extensive plantations on right bank of FARMINGTON RIVER), palm oil and kernels, coffee, citrus fruit; fisheries. Main centers (served by roads): MONROVIA and CAREYSBURG. Formerly called Mesurado. Montserrado is divided into four districts: Todee (N), Careysburg (central), Saint Paul River (SW), and Greater Monrovia (SE).

Montserrado, Cape (mahnt-suhr-AH-do), headland of W LIBERIA, on ATLANTIC OCEAN; 06°19′N 10°49′W. Rises to 290 ft/88 m. Site of MONROVIA. Formerly called Cape Mesurado.

Montserrat (mawn-suh-RAHT), island and British dependency in LESSER ANTILLES (□ 38 sq mi/98 sq km; 1994 est. population 13,000; 2001 population 4,482; 2002 est. population 8,995), WEST INDIES, one of the LEEWARD ISLANDS; ⊙ BRADES ESTATE (interim), was PLYMOUTH (up to 1997). It is a rugged, scenic island of volcanic origin; the eruption of the SOUFRIÈRE Hills volcano in 1995 and subsequent eruptions have devastated the island and caused many inhabitants to be evacuated. Plymouth was the only outlet for the cotton and other agricultural products of the island. Tourism had become the economic mainstay, accounting for 25% of the island's GNP but diminished after the 1995 eruption. Mainly known for producing calypso music records (there's a recording studio on the island). Highest point was CHANCE'S PEAK in the SOUFRIÈRE Hills (elevation 3,002 ft/915 m); the Soufrière Hills' volcano has at times measured higher since its first eruption in 1995, but continued eruptions and collapse of the volcanic dome mean the elevation of the volcano changes. Columbus sighted the island in 1493 and named it Santa María de Montserrat after a monastery near Barcelona. Was first settled in 1632 by Irish Catholics from St. Kitts, possibly seeking religious freedom. It developed a prosperous sugar industry, which suffered from the abolition of slavery in 1834 and was replaced by raising limes. After changing hands several times between France and Britain, it was ceded to Great Britain in 1783. The island was a member of the former Leeward Islands colony and of the Federation of the West Indies. In 1966, Montserrat rejected self-government. The island suffered extensive damage from a hurricane in 1989 and again when the Soufrière Hills volcano erupted in July 1995 after 400 years of dormancy. A subsequent eruption in 1997 buried the S half of the island under lava and ash, and most of the population had to be evacuated. Some two-thirds of the inhabitants fled the island but some had begun to return by 2003, though the S half of the island has been designated an exclusion zone where entry is illegal. The economy of the island is slowly recovering from the devastation of the eruption, but is still reliant on foreign aid.

Montserrat or Monserrat (both: mon-se-RAHT), mountain (4,054 ft/1,236 m), NE SPAIN, rising abruptly from a plain in CATALONIA, NW of BARCELONA; 41°36′N 01°49′E. On a narrow terrace, more than halfway up its precipitous cliffs, is a celebrated Benedictine monastery, one of the greatest religious shrines of Spain. Only ruins are left of the old monastery (11th century). The present monastery was built in the 18th century and restored after being destroyed by French troops in 1812. It has a valuable painting collection, library, and museum. The Renaissance church (16th century; largely restored in the 19th and 20th centuries) contains the black wooden image of the Virgin, which, according to tradition, was carved by St. Luke, brought to Spain by St. Peter, and hidden in a cave near Montserrat during the Moorish occupation. In the Middle Ages the mountain, also called

Monsalvat, was thought to have been the site of the castle of the Holy Grail. At Montserrat, St. Ignatius of Loyola devoted himself to his religious vocation just before the founding of the Society of Jesus. Montserrat has acquired symbolic status as the center of Catalan nationalism.

Montserrat Hills (mon-sai-RAHT), low range in W central TRINIDAD, TRINIDAD AND TOBAGO, 20 mi/32 km SE of PORT OF SPAIN. Rises to 918 ft/280 m.

Montsinéry (mon-san-ai-ree), town, N FRENCH GUIANA; 04°54′N 52°30′W. Located on small Montsinéry River near the ATLANTIC, and 13 mi/21 km W of CAYENNE.

Mont Soleil, SWITZERLAND: see MONTAGNE DU DROIT.

Montsoult (mon-SOO), commune (□ 1 sq mi/2.6 sq km; 2004 population 3,525), VAL-D'OISE department, ÎLE-DE-FRANCE region, N central FRANCE, 14 mi/23 km N of PARIS; 49°04′N 02°19′E.

Monts, Pointe des (MON, pwant dai), cape on the Gulf of SAINT LAWRENCE, E QUEBEC, E Canada, on N side of mouth of the SAINT LAWRENCE RIVER, opposite GASPÉ PENINSULA, 36 mi/58 km ENE of BAIE COMEAU; 49°20′N 67°22′W; lighthouse.

Mont-sur-Lausanne, Le, commune, VAUD canton, SW SWITZERLAND, a N suburb of LAUSANNE. Elevation 2,303 ft/702 m.

Mont-sur-Marchienne (maw–sen–mahr-SHYEN), town in commune of Charleroi, Charleroi district, HAINAUT province, S central BELGIUM, 2 mi/3.2 km SW of CHARLEROI; 50°23′N 04°24′E.

Montsûrs (mon-syur), commune (□ 4 sq mi/10.4 sq km), MAYENNE department, PAYS DE LA LOIRE region, W FRANCE, 11 mi/18 km ENE of LAVAL; 48°08′N 00°33′W. Cheese making, flour milling.

Mont Tendre (mon TAHN-druh), peak (5,509 ft/1679 m), W SWITZERLAND; highest peak in Swiss JURA, between Vallée de Joux and LAKE GENEVA.

Mont-Tremblant (MON–trahn-BLAHN), village (□ 91 sq mi/236.6 sq km), Laurentides region, SW QUEBEC, E Canada, in the LAURENTIANS, at foot of Mont TREMBLANT, on small Lake Mercier, 20 mi/32 km NW of SAINTE-AGATHE-DES-MONTS; 46°12′N 74°38′W. Dairying; skiing center. Resort. Restructured in 2000.

Montuiri (mon-TWEE-ree), town, MAJORCA island, BALEARIC ISLANDS, on railroad, and 18 mi/29 km E of PALMA; 39°34′N 02°59′E. Cereals, vegetables, almonds, figs, apricots; livestock; timber; meat products.

Montuoso Forest Reserve, small inland forest area on AZUERO PENINSULA, VERAGUAS province, PANAMA, E of MONTIJO GULF.

Monturque (mon-TOOR-kai), town, CÓRDOBA province, S SPAIN, 7 mi/11.3 km NW of LUCENA; 37°27′N 04°34′W. Olive-oil processing, soap and plaster manufacturing; cereals.

Montvale, borough (2006 population 7,308), BERGEN county, NE NEW JERSEY, 10 mi/16 km N of HACKENSACK; 41°02′N 74°02′W. Diverse commercial establishments. Incorporated 1894. In suburban area.

Mont Vélan (mon VAI-lahn), peak (12,241 ft/3,731 m), in PENNINE ALPS, VALAIS canton, SW SWITZERLAND, 4 mi/7 km ENE of the GREAT SAINT BERNARD PASS, on Italian border.

Montverde (mahn-tuh-VER-dee), town (□ 2 sq mi/5.2 sq km; 2005 population 956), Lake county, central FLORIDA, near LAKE APOPKA, 19 mi/31 km W of ORLANDO; 28°36′N 81°40′W. Grapes.

Montville (mon-vel), town (□ 4 sq mi/10.4 sq km; 2004 population 4,558), SEINE-MARITIME department, HAUTE-NORMANDIE region, N FRANCE, 7 mi/11.3 km N of ROUEN; 48°15′N 02°14′E. Manufacturing of electrical apparatus. Formerly spelled Monville.

Montville, town, NEW LONDON county, SE CONNECTICUT, 7 mi/11.3 km NNW of NEW LONDON; 41°27′N 72°09′W. Paper products, sheet metal, computer boards, tachometers, aluminum doors and windows, and boxes are made. High security state prison; former nuclear submarine building site. Nearby are the

Tantaquidgeon Native American Museum and a state park. Founded 1670, incorporated 1786.

Montville, township, MORRIS county, N NEW JERSEY, 10 mi/16 km W of PATERSON; 40°54'N 74°21'W. Largely residential.

Montville, MASSACHUSETTS: see SANDISFIELD.

Monument, town (2000 population 1,971), EL PASO county, central COLORADO, on MONUMENT CREEK, in SE foothills of FRONT RANGE, 15 mi/24 km N of COLORADO SPRINGS, and 45 mi/72 km S of DENVER; 39°04'N 104°51'W. Elevation 6,961 ft/2,122 m. Manufacturing (cable products, light manufacturing). Pike National Forest to W. U.S. Air Force Academy to S. National Carvers Museum to S.

Monument, unincorporated village, LEA county, SE NEW MEXICO, 13 mi/21 km WSW of HOBBS. Diversified irrigated agricultural area on LLANO ESTACADO. Cattle, sheep, cotton, grain, alfalfa, dairying. Manufacturing (oil and natural gas production).

Monument, village (2006 population 132), GRANT county, NE central OREGON, 27 mi/43 km NW of CANYON CITY, on North Fork of JOHN DAY RIVER; 44°49'N 119°25'W. Part of Umatilla National Forest to N.

Monument Beach, MASSACHUSETTS: see BOURNE.

Monument Creek, 34 mi/55 km long, central COLORADO; rises in FRONT RANGE in NW EL PASO county; flows generally S, past PALMER LAKE and MONUMENT towns, through E side of U.S. Air Force Academy, to FOUNTAIN CREEK at COLORADO SPRINGS.

Monument Mountain (1,710 ft/521 m), peak of the BERKSHIRES, SW MASSACHUSETTS, in Monument Mountain State Reservation (260 acres/105 ha), 4 mi/6.4 km N of GREAT BARRINGTON.

Monveda (mon-VE-duh), village, Équateur province, NW CONGO, along DUA RIVER, 20 mi/32 km N of LISALA; 02°57'S 21°27'E. Elev. 1,354 ft/412 m. Subsistence farming. Also known as MONYEDA.

Monville, FRANCE: see MONTVILLE.

Monyeda, CONGO: see MONVEDA.

Môn, Ynys, Wales: see ANGLESEY.

Monywa (mon-YWAH), town and township, SAGAING division, MYANMAR, on E bank of CHINDWIN RIVER, 60 mi/97 km W of MANDALAY, on railroad to YE-U. Timber-clearing station; copper deposits nearby.

Monza (MON-tsah), city (2001 population 120,204), MILANO province, LOMBARDY, N ITALY; 45°35'N 09°16'E. Manufacturing of this highly diversified major industrial center include felt hats, carpets, clothing, textiles, glass, chemicals, plastics, wood products, paper, and machinery; food processing. The history of Monza is closely related to that of MILAN. The cathedral, founded (6th century) by the Lombard queen Theodolinda, contains the iron crown of Lombardy, which was made, according to tradition, from a nail of Christ's cross and which was used to crown Charlemagne, Charles V, Napoleon I, and other emperors as kings of Lombardy or of Italy. An expiatory chapel was built (1910) at the place where King Humbert I was assassinated in 1900. Monza has the Autodromo, a major automobile racetrack (rebuilt 1955).

Monze, township, SOUTHERN province, S ZAMBIA, 35 mi/56 km SSW of MAZABUKA; 16°17'S 27°29'E. On railroad and at junction of road to NAMWALA (to NW). Agricultural area (tobacco, soybeans, corn).

Monze, Cape (MON-zai), headland on ARABIAN SEA, KARACHI administration area, SE PAKISTAN, 22 mi/35 km W of KARACHI; 24°49'N 66°40'E. Lighthouse. Mouth of HAB RIVER is just N. Also called RAS MUARI.

Monzen (MON-zen), town, Fugeshi county, ISHIKAWA prefecture, central HONSHU, central JAPAN, on NW NOTO Peninsula, 50 mi/80 km N of KANAZAWA; 37°17'N 136°46'E.

Monzón (mon-SON), town, HUAMALÍES province, HUÁNUCO region, central PERU, on E slopes of Cordillera CENTRAL, on Monzón River (left affluent of the Huallaga), and 33 mi/53 km NE of LLATA, at edge of tropical forests; 09°10'S 76°23'W. Rubber, lumber, rice, cacao, coca.

Monzón (mon-SON), town, HUESCA province, NE SPAIN, near CINCA RIVER, 34 mi/55 km SE of HUESCA; 41°55'N 00°12'E. Agricultural center (olive oil, wine, sugar beets, livestock, cereals); sugar and flour mills; manufacturing of soap, chocolate. Has Gothic church. On hill above town is 10th–11th-century fortified castle that belonged to the Knights Templars.

Monzón River (mon-SON), affluent, HUÁNUCO region, central PERU; rises in the ANDES; flows E past MONZÓN to the HUALLAGA RIVER; 09°11'S 75°58'W.

Monzur, TURKEY: see MUNZUR RIVER.

Mooar (MOO-are), village, LEE county, extreme SE IOWA, 15 mi/24 km SW of FORT MADISON. Explosives plant.

Moodiesburn (MOO-dees-buhrn), town (2001 population 6,614), North Lanarkshire, central Scotland, 4 mi/6.4 km NW of COATBRIDGE; 55°54'N 04°06'W. Formerly in Strathclyde, abolished 1996.

Moodus (MOO-duhs), village (2000 population 1,263), EAST HADDAM town, MIDDLESEX county, S CONNECTICUT, near the CONNECTICUT River; 41°30'N 72°27'W. "Moodus noises," subterranean rumblings about which Indians had legends, believed to be caused by minor earthquakes beneath hill here. Summer camps and resorts. Former poultry raising.

Moody, county (□ 521 sq mi/1,354.6 sq km; 2006 population 6,644), E SOUTH DAKOTA on MINNESOTA state line; ⊙ FLANDREAU; 44°01'N 96°40'W. Rich farming and livestock-raising region drained by BIG SIOUX RIVER and Pipestone Creek. Corn, soybeans; hogs, cattle, sheep; dairying. Flandreau Indian Reservation N of Flandreau. Formed 1873.

Moody, town (2000 population 8,053), St. Clair co., N central Alabama, c.15 mi/24 km E of Birmingham; 33°35'N 86°29'W. Named for Epps Moody from NC, a settler in the 1820's. Also known as 'Moody's' and 'Moody's Crossroads.' Inc. in 1962.

Moody (MOO-dee), town (2006 population 1,395), MCLENNAN county, E central TEXAS, 21 mi/34 km SSW of WACO; 31°18'N 97°21'W. Elevation 783 ft/239 m. In farm area (cattle; grains; dairying); manufacturing (pecan and honey processing). Mother Neff State Park on BELTON LAKE reservoir to W. Incorporated 1901.

Moody Air Force Base, U.S. military base, LOWNDES county, S GEORGIA, N of VALDOSTA. Established at the beginning of World War II. Largest employer in county.

Mooers, village (2000 population 440), CLINTON county, extreme NE NEW YORK, on GREAT CHAZY RIVER, and 20 mi/32 km NNW of PLATTSBURGH; 44°58'N 73°35'W. Port of entry, at QUEBEC border 3.3 mi/5.3 km NNW.

Mooi River, Afrikaans *Mooirivier*, town, KWAZULU-NATAL, SOUTH AFRICA, in DRAKENSBERG RANGE, on Mooi River, and 35 mi/56 km NW of MSUNDUZI (Pietermaritzburg); 29°13'S 30°00'E. Elevation 4,556 ft/1,389 m. Established in 1921. Resort; dairying, bacon curing. In livestock-raising region. Educational and conference center. On N3 highway between DURBAN and JOHANNESBURG.

Mooka (MO-o-kah), city, TOCHIGI prefecture, central HONSHU, N central JAPAN, 9 mi/15 km S of UTSUNOMIYA; 36°26'N 140°00'E. Aluminum products; peanuts, strawberries.

Mook en Middelaar (MAWK uhn MI-duh-lahr), village, LIMBURG province, E NETHERLANDS, on MEUSE RIVER (car ferry), 7 mi/11.3 km S of NIJMEGEN; 51°45'N 05°53'E. SE entrance of MAAS-WAAL Canal 1 mi/1.6 km to NW; German border 3 mi/4.8 km to E. Dairying; livestock; grain, vegetables. Battle here (1574), in which Henry and Louis of Nassau, brothers of William the Silent, were defeated and killed by Spaniards.

Mookgophong, town, LIMPOPO province, SOUTH AFRICA, on Naboomspruit River (tributary of Nyl River), and 60 mi/97 km SW of POLOKWANE (Pietersburg); 24°31'S 28°43'E. Elevation 3,608 ft/1,100 m. Established 1923 as a center for platinum production. RR junction; agriculture center (wheat, tobacco, peanuts). On N1 highway; railroad to N; airfield. Formerly called Naboomspruit; name changed in 2002.

Moolap (MOO-lap), locality, residential and industrial suburb 3 mi/5 km ES of GEELONG, VICTORIA, SE AUSTRALIA; 38°11'S 144°26'E. Aluminum smelter nearby.

Mooltan, PAKISTAN: see MULTAN.

Moomaw, Lake (□ 4 sq mi/10.4 sq km), ALLEGHANY and BATH counties, W VIRGINIA, on JACKSON RIVER, in Monongahela National Forest, 10 mi/16 km N of COVINGTON; 37°58'N 79°57'W. Maximum capacity 421,500 acre-ft. Formed by Gathright Dam (also known as Moomaw Dam; 257 ft/78 m high), built (1978) by Army Corps of Engineers for flood control; also used for recreation.

Moon, ESTONIA: see MUHU.

Moon, PENNSYLVANIA: see CARNOT-MOON.

Moonachie (moo-NAH-KEE), borough (2006 population 2,797), BERGEN county, NE NEW JERSEY, 3 mi/4.8 km S of HACKENSACK; 40°50'N 74°03'W. Incorporated 1910. Largely residential; some industrialized areas.

Moonah (MOO-nah), town, SE TASMANIA, SE AUSTRALIA, N residential and commercial N suburb of HOBART; 42°51'S 147°18'E. Abattoirs, electrolytic zinc works, oil-storage facilities.

Moonbeam (MOON-beem), township (□ 91 sq mi/236.6 sq km; 2001 population 1,201), central ONTARIO, E central Canada; 49°22'N 82°10'W.

Moondyne Cave, AUSTRALIA: see AUGUSTA.

Moonie (MOO-nee), locality, oil field, QUEENSLAND, NE AUSTRALIA, 201 mi/323 km W of BRISBANE; 27°42'S 150°20'E. Surrounding area is known as Surat-Bowen Basin.

Moon Island, in Boston Harbor, E MASSACHUSETTS, 5 mi/8 km S of BOSTON, E of QUINCY. Connects to Quincy by causeway. Former site of sewage treatment plant. Later Boston Police Department outdoor firearms range and bomb disposal, also Boston Fire Department training facility. Connected to LONG ISLAND by Long Island Bridge. Restricted use by the city of Boston.

Moon Lake, MISSISSIPPI: see LULA.

Moon Lake Dam, UTAH: see LAKE FORK.

Moon, Mountains of the, UGANDA: see RUWENZORI, mountain range.

Moonridge, unincorporated town, SAN BERNARDINO county, S CALIFORNIA, c.25 mi/40 km ENE of SAN BERNARDINO, and on S shore of BIG BEAR LAKE reservoir. Recreation area. Tourism. Area surrounded by San Bernardino National Forest.

Moon Run, unincorporated town, Robinsoe township, ALLEGHENY county, W PENNSYLVANIA, suburb 5 mi/8 km W of downtown PITTSBURGH; 40°27'N 80°06'W. In bituminous-coal and agricultural area.

Moonta (MOON-tuh), town, S SOUTH AUSTRALIA state, S central AUSTRALIA, on W YORKE PENINSULA, and 65 mi/105 km SSW of PORT PIRIE, near Port HUGHES inlet of SPENCER GULF; 34°04'S 137°35'E. Railroad terminus; wheat, barley, sheep, and wool. Former copper-mining center.

Moora, town, WESTERN AUSTRALIA state, W AUSTRALIA, 117 mi/189 km N of PERTH; 30°40'S 116°01'E. Sheep, cattle, wheat; light industry; beekeeping. Wildflowers.

Moorabbin (moo-RA-bin), residential and commercial suburb, S VICTORIA, SE AUSTRALIA, 10 mi/16 km SSE of MELBOURNE; 37°56'S 145°02'E. In metropolitan area. Manufacturing. Moorabbin's N section united with CAULFIELD in 1994 to form GLEN EIRA CITY.

Moorcroft, town (2006 population 854), CROOK county, NE WYOMING, on BELLE FOURCHE RIVER, and 30 mi/48 km WSW of SUNDANCE; 44°15'N 104°57'W. Elevation c. 4,206 ft/1,282 m. Trade and shipping center in livestock, timber region; oil refinery. Oil wells nearby.

Area is shown by the symbol □, and capital city or county seat by ⊙.

KEYHOLE RESERVOIR (Belle Fourche River) and Keyhole State Park to NE.

Moordrecht (MAWR-drekht), town, SOUTH HOLLAND province, W NETHERLANDS, on HOLLANDSCHE IJSSEL River (car ferry), and 3 mi/4.8 km SW of GOUDA; 51°59′N 04°40′E. Dairying; cattle, sheep, poultry; vegetables, flowers, fruit; manufacturing (food processing).

Moore (MOR), county (□ 705 sq mi/1,833 sq km; 2006 population 83,162), central NORTH CAROLINA; ⊙ CARTHAGE; 35°18′N 79°28′W. Forested sand hills in PIEDMONT region; drained by DEEP RIVER, bounded SE by Little River, SW by Drowning Creek (LUMBER RIVER). Service industries; manufacturing (timber, sawmilling, textiles); agriculture (tobacco, peaches, corn, wheat, soybeans, hay; cattle, hogs). Bounded in SE by FORT BRAGG MILITARY RESERVATION. House in the Horseshoe State Historic Site in NE. Weymouth Wood State Park (Sandhills Nature Preserve) in SE. Formed 1784 from Cumberland County. Named for Alfred Moore (1755–1810), soldier in the American Revolutionary War and later an associate justice of the United State Supreme Court.

Moore, county (□ 122 sq mi/317.2 sq km; 2006 population 6,070), S TENNESSEE; ⊙ LYNCHBURG; 35°17′N 86°22′W. Bounded SE by ELK RIVER. Agriculture; timber; whiskey distilleries, the most famous is Jack Daniel Distillery. Formed 1871.

Moore (MOR), county (□ 909 sq mi/2,363.4 sq km; 2006 population 20,591), extreme N TEXAS; ⊙ DUMAS; 35°50′N 101°53′W. Elevation 3,000 ft/914 m–4,000 ft/1,219 m. In high plains of the PANHANDLE; drained in SE by CANADIAN RIVER. One of richest areas in the huge Panhandle natural gas and oil field; helium production; wheat, corn, sorghum, pinto beans farming, cattle ranching. Part of MEREDITH LAKE reservoir (Canadian River) and Meredith Lake National Recreation Area in SE corner. Formed 1876.

Moore, city (2006 population 49,277), CLEVELAND county, central OKLAHOMA, suburb 9 mi/14.5 km S of downtown OKLAHOMA CITY; 35°19′N 97°28′W. Manufacturing (lightning and surge protection equipment, packaging for food condiments, bumpers and grill guards, printing). Incorporated 1887.

Moore, village (2000 population 196), BUTTE county, SE central IDAHO, 7 mi/11.3 km NNW of ARCO, on BIG LOST RIVER; 43°44′N 113°22′W. Parts of Challis National Forest to E and N.

Moore, village (2000 population 186), FERGUS county, central MONTANA, 13 mi/21 km WSW of LEWISTOWN; 46°58′N 109°42′W. Livestock; wheat, barley, hay. Ackley Lake State Park to W. Part of Lewis and Clark National Forest to SE.

Moore, unincorporated village, SPARTANBURG county, NW SOUTH CAROLINA, 8 mi/12.9 km SSW of SPARTANBURG, near TYGER RIVER. Manufacturing includes silicon wafers, contract embroidery, recycled cotton waste.

Moore (MOR), unincorporated village, FRIO county, SW TEXAS, c.40 mi/64 km SW of SAN ANTONIO; 29°03′N 99°00′W. Winter vegetables, melons, corn, peanuts; hogs, cattle; natural gas wells.

Moorea (mo-RAI-ah), volcanic island (□ 50 sq mi/130 sq km), South PACIFIC, second-largest of the Windward group of the SOCIETY ISLANDS, FRENCH POLYNESIA; 17°32′S 149°50′W. The island is mountainous, with Mount Tohivea (3,960 ft/1,207 m) the highest peak. On the N coast are Cook Bay and Papetoai Bay. Afareaitu, the chief town, is on the E coast. Tourism is significant. Copra, coffee, and pineapples are the main products. Agricultural training school at Opunohu Bay. Formerly known as Eimeo.

Moore Embayment, an embayment of the ROSS ICE SHELF, WEST ANTARCTICA, into the HILLARY COAST of EAST ANTARCTICA; 78°45′S 165°00′E.

Moorefield, town (2006 population 2,426), ⊙ HARDY county, NE WEST VIRGINIA, 27 mi/43 km S of KEYSER,

in EASTERN PANHANDLE, on South Branch of the POTOMAC RIVER, at mouth of South Fork, South Branch Potomac River. In hunting, fishing area. Agriculture (grain, soybeans); livestock; poultry. Manufacturing (kitchen cabinets and counter tops, limestone processing, marble products, poultry processing, poultry feed). LOST RIVER State Park to SE. Settled 1777.

Moorefield (MOOR-feeld), unincorporated village, S ONTARIO, E central Canada, on Conestoga River, and 25 mi/40 km NNW of KITCHENER, and included in Mapleton township; 43°45′N 80°44′W. Dairying, mixed farming.

Moorefield, village, SWITZERLAND county, SE INDIANA, 6 mi/9.7 km NW of VEVAY. In agricultural area.

Moorefield, village (2006 population 46), FRONTIER county, S NEBRASKA, 5 mi/8 km NE of CURTIS; 40°41′N 100°24′W.

Moore Haven, city (□ 1 sq mi/2.6 sq km; 2005 population 1,751), ⊙ GLADES county, S central FLORIDA, c.50 mi/80 km ENE of FORT MYERS, on W shore of LAKE OKEECHOBEE near entrance (lock here) to CALOOSAHATCHEE RIVER; 26°49′N 81°05′W. Small-scale farming, fishing.

Moore, Lake (MOR) (□ 449 sq mi/1,167.4 sq km), W central WESTERN AUSTRALIA state, W AUSTRALIA, 150 mi/241 km NE of PERTH; 60 mi/97 km long; 29°50′S 117°35′E. Usually dry.

Mooreland, town (2000 population 393), HENRY county, E INDIANA, 8 mi/12.9 km NE of NEW CASTLE; 40°00′N 85°15′W. In agricultural area.

Mooreland, town (2006 population 1,228), WOODWARD county, NW OKLAHOMA, 10 mi/16 km E of WOODWARD; 36°26′N 99°12′W. In grain, oil and natural gas area; also alfalfa, livestock; manufacturing (farm supplies, feeds). Boiling Springs State Park to W.

Moore Park (MOR), community, S MANITOBA, W central Canada, 17 mi/27 km NNE of BRANDON, and in Odanah rural municipality; 50°03′N 99°46′W.

Moore Reservoir, c.10 mi/16 km long, NW NEW HAMPSHIRE and NE VERMONT, on CONNECTICUT RIVER, 4 mi/6.4 km WNW of LITTLETON, New Hampshire; 44°20′N 71°51′W. Maximum capacity 115,000 acre-ft. Formed by Moore Dam (178 ft/54 m), built (1956) by the Connecticut River Power Company for power generation.

Mooresboro, village (□ 1 sq mi/2.6 sq km; 2006 population 330), CLEVELAND county, SW NORTH CAROLINA, 8 mi/12.9 km W of SHELBY; 35°17′N 81°42′W. Manufacturing (rugs, seat covers); service industries; agriculture (cotton, grain; livestock).

Moores Creek National Battlefield (MORZ KREEK), PENDER county, SE NORTH CAROLINA, 20 mi/32 km NW of WILMINGTON, near BLACK RIVER; 34°27′N 78°06′W. Park consists of 87 acres/35 ha. The patriot victory over the Loyalists at Moores Creek Bridge on February 27, 1776, prevented the intended British invasion of North Carolina and spurred revolutionary sentiment in the South; the battle is often called the Lexington and Concord of the South. Established 1926.

Moores Hill, town (2000 population 635), DEARBORN county, SE INDIANA, 12 mi/19 km W of LAWRENCEBURG; 39°07′N 85°05′W. In agricultural area. Laid out 1838.

Moore's Mill, NE suburb of HUNTSVILLE, MADISON county, N ALABAMA; 34°51′N 86°31′W.

Moores Mills, village, SW New Brunswick, Canada, 7 mi/11 km N of St. Stephen; 45°17′N 67°16′W.

Moorestown (MAWRZ-toun), township, BURLINGTON county, SW NEW JERSEY, an industrial suburb of the CAMDEN, New Jersey–Philadelphia area, 9 mi/14.5 km E of Camden; 39°58′N 74°56′W. Electronic equipment, metal products, and chemicals are the principal manufacturing. Of interest are several 18th-century houses. Settled 1682 by Quakers, incorporated 1922.

Mooresville, town (2000 population 33), Limestone co., N Alabama, 5 mi/8 km E of Decatur, across Tennessee River. Named for William Moore, who settled in the area in 1808. Inc. in 1818.

Mooresville, town (2000 population 9,273), MORGAN county, central INDIANA, on WHITELICK CREEK, and 16 mi/26 km SW of downtown INDIANAPOLIS; 39°37′N 86°22′W. Agricultural area (grain, fruit; dairy products); varied manufacturing. Laid out 1824.

Mooresville, town (2000 population 89), LIVINGSTON county, N central MISSOURI, near GRAND RIVER, 9 mi/14.5 km WSW of CHILLICOTHE; 39°45′N 93°43′W.

Mooresville, town (□ 14 sq mi/36.4 sq km; 2006 population 20,944), IREDELL county, W central NORTH CAROLINA, 14 mi/23 km SSE of STATESVILLE, near Lake NORMAN; 35°34′N 80°48′W. Ironworks. Varied manufacturing; service industries. Lake Norman reservoir (CATAWBA RIVER) to W; Duke Power State Park, on Lake Norman, named for Duke Power Company, which built lake to NW. Bahari Racing to W. Town known as Race City USA. Founded 1868; incorporated 1873.

Mooresville, NORTH CAROLINA: see WARSAW.

Mooreton (MOR-tuhn), village (2006 population 192), RICHLAND county, SE NORTH DAKOTA, 13 mi/21 km W of WAHPETON, on Antelope Creek; 46°16′N 96°52′W. Founded in 1884 and incorporated in 1911.

Moore Town, town, PORTLAND parish, E JAMAICA, on the RIO GRANDE, 7 mi/11.3 km SSE of PORT ANTONIO; 18°04′N 76°25′W. Site of ancient Maroon settlement.

Moorfoot Hills (MOR-fuht HILZ), mountain range, Scottish Borders, S Scotland, extends 12 mi/19 km NE from PEEBLES; 55°44′N 03°05′W. Highest point is Blackhope Scar (2,136 ft/651 m), 7 mi/11.3 km NE of Peebles.

Moorhead (MOOR-hed), city (2000 population 32,177), ⊙ CLAY county, NW MINNESOTA, on the RED RIVER; 46°51′N 96°45′W. Elevation 910 ft/277 m. A sister city of FARGO, NORTH DAKOTA, 1 mi/1.6 km E of Fargo. Railroad junction. It is a shipping and processing center for an agricultural area; cattle, poultry, sheep, hogs; dairying. Manufacturing (sugar and molasses, barley malt, soft drinks, printing and publishing, dairy bottles, fiberglass tanks). Seat of Moorhead State University and Concordia College. The Plains Art Museum; Archie's West Limited Art Gallery; Rourke Art Gallery are here. Buffalo River State Park to E. Incorporated 1881.

Moorhead, town (2000 population 232), MONONA county, W IOWA, on SOLDIER RIVER, and 14 mi/23 km SE of ONAWA; 41°55′N 95°50′W. In livestock and grain area. State park nearby.

Moorhead, town (2000 population 2,573), SUNFLOWER county, W MISSISSIPPI, 20 mi/32 km W of GREENWOOD; 33°27′N 90°30′W. In rich soybean and cotton-growing area; also corn, rice; catfish; manufacturing (canned vegetables, catfish feed). Mississippi Delta Community College. Incorporated 1899.

Mooringsport (MOR-eeng-sport), town (2000 population 833), CADDO parish, extreme NW LOUISIANA, on S shore of CADDO LAKE, 17 mi/27 km NW of SHREVEPORT; 32°41′N 93°58′W. In oil-producing and agricultural area; timber, clay; manufacturing (bricks). Soda Lake State Wildlife Area to SE.

Moorland, town (2000 population 197), WEBSTER county, central IOWA, 7 mi/11.3 km SW of FORT DODGE; 42°26′N 94°17′W. In agricultural area.

Moorland, village (2000 population 464), JEFFERSON county, N KENTUCKY, residential suburb 10 mi/16 km E of downtown LOUISVILLE; 38°16′N 85°34′W.

Moormerland (MAWR-mer-lahnd), commune, LOWER SAXONY, NW GERMANY, in EAST FRIESLAND, on right bank of the EMS, and on DORTMUND-EMS CANAL, 4 mi/6.4 km NE of LEER; 53°17′N 07°30′E. Dairying; shipyard (barges).

Mooroopna, AUSTRALIA: see SHEPPARTON.

Cross-references are shown in SMALL CAPITALS. The pronunciation guide is shown on page xix. The sources of population figures are shown on page xvii.

Moorpark, city (2000 population 31,415), VENTURA county, S CALIFORNIA, 18 mi/29 km ENE of OXNARD, and 42 mi/68 km NW of LOS ANGELES; 34°17′N 118°53′W. Heavy and high-technology manufacturing; citrus orchards, vegetables, strawberries; flowers, nursery products; oil fields nearby. Moorpark College (two-year). Santa Susana Mountains to N.

Moorreesburg, town, Swartland district, WESTERN CAPE province, SOUTH AFRICA, 20 mi/32 km N of MALMESBURY; 33°08′S 18°39′E. Elevation 1,180 ft/360 m. Wheat-growing center; grain elevator; on N highway to SPRINGBOK.

Moorrege (MAWR-re-ge), village, SCHLESWIG-HOLSTEIN, NW GERMANY, 5 mi/8 km S of ELMSHORN, on the Pinnau; 53°40′N 09°41′E. Fruit; food processing.

Moorsel (MOR-suhl), commune, EAST FLANDERS province, N central BELGIUM, 3 mi/4.8 km ENE of AALST; 50°57′N 04°06′E. Hops, flowers.

Moorsele (MOR-sai-luh), village, WEST FLANDERS province, W BELGIUM, 5 mi/8 km W of KORTRIJK; 50°50′N 03°09′E. Cotton industry; market for flax and tobacco.

Moorslede (MORS-lai-duh), commune (2006 population 10,672), Roeselare district, WEST FLANDERS province, W BELGIUM, 5 mi/8 km SW of ROESELARE; 50°53′N 03°04′E.

Moosburg an der Isar (MAWS-burg ahn der EE-sahr), town, UPPER BAVARIA, GERMANY, on the ISAR, 10 mi/16 km NE of FREISING; 48°28′N 11°55′E. Manufacturing of chemicals, machinery, electronics. Has late-13th- and mid-15th-century churches. Chartered 1331.

Moose, village, TETON county, NW WYOMING, near SNAKE RIVER, and 12 mi/19 km N of JACKSON. Service center located at center of GRAND TETON NATIONAL PARK. TETON RANGE to W.

Moose (MOOS), river, c.50 mi/80 km long, Canada; formed in central ONTARIO, by the MATTAGAMI and MISSINAIBI rivers; flows NE to its confluence with the ABITIBI RIVER and into SW JAMES BAY near MOOSONEE; 51°21′N 80°23′W.

Moose Creek (MOOS KREEK), unincorporated village, SE ONTARIO, E central Canada, 20 mi/32 km NW of CORNWALL, and included in North Stormont township; 45°15′N 74°58′W. Dairying, mixed farming. Bilingual (English, French) community. Settled by people of Scottish origin.

Moose Creek, village, S ALASKA, in MATANUSKA VALLEY, on MATANUSKA RIVER, and 40 mi/64 km NE of ANCHORAGE. Fishing; tourism.

Moose Factory (MOOS), trading post, NE ONTARIO, E central Canada, near the mouth of the MOOSE RIVER on JAMES BAY; 51°15′N 80°36′W. A fort was built here by Charles Bayly, governor of the Hudson's Bay Company, in the early 1670s. In the struggle between the English and French in Canada, the fort changed hands several times and shortly after 1696 was destroyed. In 1730 the company built a post close to the ruins of the original fort. This post has been in continuous operation since.

Moosehead Lake, reservoir (□ 120 sq mi/312 sq km), on PISCATAQUIS and SOMERSET county border, W central MAINE, on KENNEBEC RIVER, 2 mi/3.2 km S of ROCKWOOD; c.30 mi/48 km long, max. 10 mi/16 km wide; 49°34′N 69°42′W. Elevation 1,029 ft/314 m. Max. capacity 714,980 acre-ft. Extensions include LILY and Spencer bays (E) and North Bay (N). MOOSE RIVER enters from W 2 mi/3.2 km WNW of ROCKWOOD. Formed by Western Outlet Dam (gravity dam; 7 ft/2.1 m high), built (1960) by Kennebec Logging Company for logging activities. New outlet at Moosehead, 6 mi/9.7 km S of Rockwood. Has an irregular shoreline and numerous islands, including Sugar (SE), Deer (S), and Famr (NW) islands. Lily Bay State Park on SE shore. Mount Kineo (1,789 ft/545 m high) is located on a peninsula that extends into the lake.

Mooseheart, ILLINOIS: see BATAVIA.

Moose Island, WASHINGTON county, E MAINE, at mouth of Indian River, just NW of JONESPORT; 2 mi/3.2 km long, 0.75 mi/1.2 km wide.

Moose Jaw (MOOS JAW), city (2006 population 32,132), S central SASKATCHEWAN, CANADA; 50°24′N 105°33′W. Railroad and distribution center. Oil refineries; meat-packing and dairy-processing plants, flour; lumber, woolen mills; stockyards. Canada's largest jet-training base.

Moose Lake, town (2000 population 2,239), CARLTON county, E MINNESOTA, 38 mi/61 km SW of DULUTH; 46°27′N 92°46′W. Oats, alfalfa; timber; poultry; dairying; manufacturing (feeds, fishing tackles). Moose Lake State Park to SE; Sand and Island lakes to S. Founded as lumber town before 1875, rebuilt after destruction by forest fire, 1918.

Moose Lake (MOOS), E BRITISH COLUMBIA, W Canada, near ALBERTA border, in ROCKY MOUNTAINS, in MOUNT ROBSON PROVINCIAL PARK, 30 mi/48 km W of JASPER (Alberta); 8 mi/13 km long, 2 mi/3 km wide. Elevation 3,386 ft/1,032 m. Drained NW by FRASER River.

Moose Lake (MOOS) (□ 525 sq mi/1,365 sq km), W MANITOBA, W central Canada, 32 mi/51 km E of THE PAS; 43 mi/69 km long, 30 mi/48 km wide; 53°55′N 99°45′W. Drained S into LAKE WINNIPEG by SASKATCHEWAN River, through CEDAR LAKE.

Mooselookmeguntic Lake (moos-luk-mi-GUHN-tik), reservoir (□ 25 sq mi/65 sq km), OXFORD county, W MAINE, on Rapid River, adjacent to RICHARDSON LAKE, 26 mi/42 km NNW of RUMFORD; 44°53′N 70°52′W. Max. capacity 192,039 acre-ft. Formed by Upper Dam.

Moose Mountain (MOOS), range, SE SASKATCHEWAN, Canada; extends 30 mi/48 km E-W near MANITOBA border; rises to 2,725 ft/831 m, 50 mi/80 km NNE of ESTEVAN; 49°45′N 102°37′W. Moose Mountain Provincial Park (□ 152 sq mi/394 sq km), a region of woods, lakes, resorts.

Moose Pass, village, S ALASKA, on E KENAI PENINSULA, 25 mi/40 km N of SEWARD. On SEWARD HIGHWAY.

Moose Pond, reservoir, at DENMARK, OXFORD and CUMBERLAND counties, SW MAINE, on small stream; c.8 mi/12.9 km long; 44°02′N 70°47′W. Drains S into the SACO RIVER.

Moose River, township, SOMERSET county, W MAINE, just N of JACKMAN in wilderness area; 45°42′N 70°13′W. Lumbering, recreation.

Moose River, 62 mi/100 km long, in W MAINE; rises in N FRANKLIN county; flows generally E to MOOSEHEAD LAKE, near ROCKWOOD.

Moose River, c.30 mi/48 km long, in N central NEW YORK; rises in the W ADIRONDACKS in North, Middle, and South branches, which join near FULTON. CHAIN OF LAKES (drained by Middle Branch); flows SW and W to Black River at LYONS FALLS.

Moose River, c.30 mi/48 km long, NE VERMONT; rises near EAST HAVEN; flows S and W, past Concord, to the PASSUMPSIC river at St. Johnsbury.

Moosic (MOO-sik), borough (2006 population 5,765), LACKAWANNA county, NE PENNSYLVANIA, suburb 5 mi/8 km SW of SCRANTON, on LACKAWANNA river; 41°21′N 75°42′W. Manufacturing (explosives, cardboard cartons, contract sewing, food products, textile printing). Wilkes-Barre Scanton International Airport to S. Lackawanna County Stadium to E. Montage Ski Area and Waterslide to E; Moosic Mountains to SE.

Moosilauke, Mount (MOO-si-lawk), peak (4,810 ft/1,466 m), GRAFTON county, W NEW HAMPSHIRE, in WHITE MOUNTAINS, 7 mi/11.3 km W of North Woodstock, SE of KINSMAN NOTCH pass, in White Mountain National Forest.

Moosinning (MAW-sin-ning), village, UPPER BAVARIA, S GERMANY, on the Middle Isar Canal, and in the Erdinger Moos, 15 mi/24 km NE of MUNICH; 48°16′N 11°50′E. Agriculture (vegetables); canning.

Moosomin (MOO-so-min), town (2006 population 2,257), SE SASKATCHEWAN, CANADA, near MANITOBA border, 80 mi/129 km WNW of BRANDON; 50°09′N 101°40′W. Grain elevators; dairying, livestock raising.

Moosonee (MOO-suh-nee), village (□ 206 sq mi/535.6 sq km; 2001 population 936), NE ONTARIO, E central Canada, on the Moose River near JAMES BAY; 51°16′N 80°39′W. It is the N terminus of the Ontario Northland railroad and the province's only saltwater port. Popular tourist center. Meteorological station.

Moosthenning (MAWS-ten-ning), village, LOWER BAVARIA, SE GERMANY, 15 mi/24 km SSW of STRAUBING; 48°41′N 12°29′E. Forestry.

Moosup (MOO-suhp), industrial village (2000 population 3,237), PLAINFIELD town, WINDHAM county, E CONNECTICUT, on small MOOSUP RIVER (water power), and 16 mi/26 km NE of NORWICH; 41°43′N 71°52′W. Textiles, thread, metal and wood products, oil burners.

Moosup River (MOOS-up), c.25 mi/40 km long, RHODE ISLAND and CONNECTICUT; rises in W Rhode Island; flows S and W, past Sterling and MOOSUP, Connecticut (water power), to QUINEBAUG RIVER near WAUREGAN village.

Mootwingee National Park, AUSTRALIA: see MUTAWINTJI NATIONAL PARK.

Mopa (mo-PAH), town, KOGI state, W central NIGERIA, 25 mi/40 km NNW of KABBA; 08°06′N 05°54′E. Tin-mining center; shea-nut processing; cotton, cassava, durra, yams.

Mopangu (mo-PAHN-goo), town, KASAI-OCCIDENTAL province, SW CONGO; 05°14′S 21°22′E. Elev. 1,387 ft/422 m. Also known as MUPONGU BWANA.

Mopán River, GUATEMALA: see BELIZE RIVER.

Mopeia (mo-PAI-uh), village, Zambézia province, central MOZAMBIQUE, near left bank of ZAMBEZI River, 80 mi/129 km WSW of QUELIMANE; 17°59′S 35°43′E. Sugar, corn, cotton.

Mopeia District, MOZAMBIQUE: see ZAMBÉZIA.

Mopelia, FRENCH POLYNESIA: see MAUPIHAA.

Mopihaa, FRENCH POLYNESIA: see MAUPIHAA.

Mopsuestia, ASIA MINOR: see MISIS.

Mopti, city, ⊙ FIFTH REGION/MOPTI, central MALI, a port at the confluence of the NIGER and BANI rivers; 14°30′N 04°12′W. The city is built on three islets linked by dikes; one dike, 8 mi/12.9 km long, connects the islets with the river bank. Mopti is the market center for a region where rice, millet, peanuts, and cassava are produced and cattle are raised. Camel caravans bring salt to market from the SAHARA DESERT. Livestock; fishing. High school; port; radio station; tourism.

Mopti, administrative region, MALI: see FIFTH REGION.

Moqor, AFGHANISTAN: see MUKUR.

Moquegua (mo-KAI-gwah), region (□ 6,075 sq mi/15,795 sq km), S PERU, bordering W on the PACIFIC; ⊙ MOQUEGUA; 16°50′S 70°55′W. Crossed by ridges of CORDILLERA OCCIDENTAL, it includes several snow-capped volcanic peaks. Watered by TAMBO and Moquegua rivers. Climate is dry, semitropical on coast, cooler in uplands. Among its mineral resources are borax, salt, coal, sulphur. Predominantly agricultural; irrigated fertile W section grows grapes, fruit, and olives on large scale; also cotton, sugarcane, figs, corn; wheat, barley, potatoes and livestock in uplands. Moquegua city and Pacific port of ILO are processing centers (wine, liquor, olive oil, canned fruit, flour). CARUMAS and OMATE have thermal springs. Main cities are Omate, Ilo, Moquegua.

Moquegua (mo-KAI-gwah), city (2005 population 52,979), ⊙ MARISCAL NIETO province and MOQUEGUA region, S PERU, in oasis on MOQUEGUA RIVER, at W slopes of CORDILLERA OCCIDENTAL, on PAN-AMERICAN HIGHWAY, and 37 mi/60 km NE of its Pacific port ILO, 550 mi/885 km SE of LIMA; 17°12′S 71°02′W. Elevation 4,715 ft/1,437 m. Processing and agricultural center (olives, grapes, cotton, fruit); cotton ginning, wine making, liquor distilling, olive-oil processing, flour milling, bottling; manufacturing. Extraction of carbon, copper, zinc, and silver. Exports wine, olive oil, cotton, copper. Has an excellent, though dry, mild climate. Cathedral with a roof made of sugarcane stalks and mud. Airport.

Moquegua River (mo-KAI-gwah), MOQUEGUA region, S PERU; rises in the ANDES; flows SW past the capital MOQUEGUA, enters Osmore valley where it is called Osmore River, then flows to the PACIFIC OCEAN at the port of ILO; 17°30'S 71°09'W.

Mór, Hungarian *Mór*, city, FEJÉR county, central HUNGARY, at SW foot of VÉRTES MOUNTAINS, 15 mi/24 km NW of SZÉKESFEHÉRVÁR; 47°23'N 18°12'E. Wheat, corn; hogs, cattle; manufacturing (flour milling; farinaceous foods); manufacturing of textiles nearby.

Mora (MOR-uh), county (□ 1,933 sq mi/5,025.8 sq km; 2006 population 5,151), NE NEW MEXICO; ⊙ MORA; 36°01'N 104°56'W. Livestock and agricultural region, watered by MORA RIVER; bounded E by CANADIAN RIVER. Fruit, hay, alfalfa, wheat, nuts; cattle, some sheep. Part of SANGRE DE CRISTO MOUNTAINS in W (crest forms most of W boundary); part of Santa Fe National Forest in SW. Charette Lakes in N; part of Kiowa National Grasslands in E; Morphy Lake State Park in SW; Coyote Creek State Park in NW; FORT UNION NATIONAL MONUMENT in S center. Formed 1860.

Mora (MOR-uh), city (2001 population 24,200), ⊙ MAYO-SAVA department, Far-North province, N CAMEROON, 30 mi/48 km NNW of MAROUA; 11°03'N 14°10'E. Well-known market town. Cattle; peanuts, millet. Customs station near Nigerian border.

Mora (MOR-ah), town, ÉVORA district, S central PORTUGAL, 30 mi/48 km NNW of ÉVORA. Cheese and pottery manufacturing.

Mora (MO-rah), town, TOLEDO province, central SPAIN, in CASTILE-LA MANCHA, 18 mi/29 km SE of TOLEDO; 39°41'N 03°46'W. Agricultural center (olives, grapes, cereals, legumes). Among its numerous industries are wine making, alcohol distilling, and olive-oil pressing. Also manufacturing of marble, metal transformers, ceramics. Once famous for its swords. A ruined castle to E; airfield to S.

Mora (MOO-rah) or **Morastrand**, town, KOPPARBERG county, central SWEDEN, at N end of LAKE SILJAN, at mouth of Österdalälven River, 45 mi/72 km NW of FALUN; 61°01'N 14°33'E. Railroad junction; manufacturing (metalworking; electrical fittings, carpentries). Airport. Tourist resort. Gustavus Vasa took refuge (c.1520) from Danes here, commemorated by *Vasaloppet*, 53-mi/85-km cross-country ski race. Has thirteenth-century church with seventeenth-century tower.

Mora (MOR-ruh), town (2000 population 3,193), ⊙ KANABEC county, E MINNESOTA, c.60 mi/97 km N of MINNEAPOLIS on SNAKE RIVER, N of mouth of Ann River; 45°52'N 93°16'W. Elevation 1,010 ft/308 m. Trading point in oats, barley, potatoes; livestock, poultry; dairying; manufacturing (yachts, plastic molding, printing and publishing, tool and die, dairy machinery). Museum of Izaak Walton League is here. Ann Lake Wildlife Area to NW; KNIFE LAKE to N; Fish Lake to S. Plotted 1881.

Mora (MOR-uh), unincorporated village, ⊙ MORA county, N NEW MEXICO, on MORA RIVER, in SANGRE DE CRISTO MOUNTAINS, and 29 mi/47 km N of LAS VEGAS. Elevation 7,179 ft/2,188 m. In irrigated fruit region; farming; cattle, some sheep; grain, alfalfa, fruits, nuts; resort. Santa Fe National Forest to W; Carson National Forest to NW; dude ranch nearby. Morphy Lake State Park to SW; Coyote Creek State Park to N; FORT UNION NATIONAL MONUMENT to E.

Mora, COSTA RICA: see CIUDAD COLÓN.

Mora, India: see URAN.

Moraca River (MOR-ah-tsah), Serbian *Morača* (MOR-a-chah), c.60 mi/97 km long, MONTENEGRO; rises in headstreams on STOZAC mountain, joining 11 mi/18 km SE of SAVNIK; flows S, past PODGORICA, to Lake SCUTARI 3 mi/4.8 km W of PLAVNICA. Receives ZETA RIVER. Also spelled Moracha River.

Morada, unincorporated town (2000 population 3,726), SAN JOAQUIN county, central CALIFORNIA, 4 mi/6.4 km NE of STOCKTON, on Mosher Creek; 38°02'N 121°14'W. Mokelumne Aqueduct passes to S. Fruit, nuts, olives, grain, vegetables; dairying; cattle. San Joaquin County Historic Museum to N.

Moradabad (mo-RAH-dah-bahd), district (□ 2,304 sq mi/5,990.4 sq km), ROHILKHAND division, N central UTTAR PRADESH state, N central INDIA; ⊙ MORADABAD. On W GANGA PLAIN; bounded W by the GANGA RIVER; drained by the RAMGANGA RIVER. Agriculture (wheat, rice, pearl millet, mustard, sugarcane, barley, gram, corn, cotton, jowar); cotton weaving, sugar milling. Main centers are Moradabad, Sambhal, Amroha, and Chandausi.

Moradabad (mo-RAH-dah-bahd), city (2001 population 641,583), ⊙ Moradabad district, Uttar Pradesh state, N central India; 28°50'N 78°47'E. Important railroad junction and an agr. market center. Manufacturing includes metalworking, electroplating, printing, sugar milling, and cotton weaving. Site of the Jama Masjid [Arab. and Persian=great mosque], built in 1631. Also Muradabad.

Morada Nova (mo-rah-dah no-vah), city (2007 population 61,908), E CEARÁ, BRAZIL, on left tributary of RIO JAGUARIBE, and 60 mi/97 km SW of ARACATI; 05°11'S 38°25'W. Dairying; ships carnauba wax, cotton, sugar. Airfield.

Morada Nova de Monas (mo-rah-dah NO-vah zhe MEE-nahs), town, W central MINAS GERAIS, BRAZIL, 28 mi/45 km NE of PAINEIRAS, on TRÉS MARIAS Reservoir; 18°30'S 45°27'W.

Mora de Ebro (MO-rah dhai AI-vro), town, TARRAGONA province, NE SPAIN, on right bank of the EBRO, and 11 mi/18 km ENE of GANDESA; 41°05'N 00°38'E. Agricultural trade center (wine, cereals, almonds, fruit); cement, soap manufacturing; olive-oil processing and shipping.

Mora de Rubielos (MO-rah dhai roo-VYAI-los), town, TERUEL province, E SPAIN, 29 mi/47 km ESE of TERUEL; 40°15'N 00°45'W. Cereals, hemp, livestock; wine making; pottery.

Morag (MO-rong), Polish *Morąg*, German *Mohrungen*, town, OLSZTYN province, NE POLAND, in lake region, 30 mi/48 km SE of ELBLĄG (Elbing); 53°55'N 19°56'E. Railroad junction; grain and cattle market. Herder born here. In EAST PRUSSIA until 1945.

Morag, a former Israeli settlement in the GAZA STRIP (under Palestinian Authority), 1.8 mi/3 km S of KHAN YUNIS. Founded in 1972. Destroyed in 2005 with Israel's unilateral withdrawal from Gaza Strip.

Moraga, city (2000 population 16,290), CONTRA COSTA county, W CALIFORNIA; residential suburb 7 mi/11.3 km E of downtown OAKLAND, in BERKELEY HILLS; 37°51'N 122°07'W. Light manufacturing. St. Mary's College of California to NE. Berkeley Hills to W; Upper San Leandro Reservoir to S; Redwood Regional Park to SW; Las Trampas Regional Park to E.

Morai (mo-REI), town, KWARA state, W NIGERIA, near BENIN border, 80 mi/129 km NW of ILORIN. Market center. Guinea corn, yams, groundnuts.

Moraine (muh-RAIN), city (□ 9 sq mi/23.4 sq km; 2006 population 6,595), MONTGOMERY county, SW OHIO, just S of DAYTON on U.S. Interstate 75; 39°42'N 84°13'W. Manufacturing (truck components, powertrain systems).

Moraine Lake (mo-RAIN), SW ALBERTA, W Canada; 51°19'N 116°11'W. Small glacial lake formed by moraine nestled in ROCKY MOUNTAINS, BANFF NATIONAL PARK, near BRITISH COLUMBIA border, 7 mi/11 km S of LAKE LOUISE. Flows E into BOW RIVER. About 0.5 mi/0.8 km in diameter. Popular tourist attraction.

Morakhi River, INDIA: see MOR RIVER.

Mora la Nueva (MO-rah lah NWAI-vah), town, TARRAGONA province, NE SPAIN, on left bank of the EBRO, and 12 mi/19 km ENE of GANDESA; 41°06'N 00°39'E. Olive-oil and wine processing; agricultural trade (cereals, almonds, fruit).

Moral de Calatrava (mo-RAHL dhai kah-lah-TRAH-vah), town, CIUDAD REAL province, S central SPAIN, in CASTILE-LA MANCHA, on railroad, and 22 mi/35 km SE of CIUDAD REAL; 38°50'N 03°35'W. Agricultural center on La Mancha plain (olives, cereals, grapes; livestock). Alcohol and liquor distilling, olive-oil pressing.

Moraleda Channel (mo-rah-LAI-dah CHAN-nel), strait of the PACIFIC, on coast of AISÉN province, AÍSEN DEL GENERAL CARLOS IBAÑEZ DEL CAMPO region, S CHILE, separating CHONOS ARCHIPELAGO from mainland and MAGDALENA ISLAND; c.80 mi/129 km long; between 44° and 45° S.

Moraleja (mo-rah-LAI-hah), town, CÁCERES province, W SPAIN, 45 mi/72 km NNW of CÁCERES; 40°04'N 06°39'W. Meat processing, olive pressing, carbon-disulphide and soap manufacturing. Fruit, honey; livestock; cork.

Moraleja del Vino (mo-rah-LAI-hah dhel VEE-no), town, ZAMORA province, NW SPAIN, 6 mi/9.7 km SE of ZAMORA; 41°28'N 05°39'W. Brandy distilling; wine, cereals; livestock.

Morales (mo-RAH-lais), town, ⊙ Morales municipio, BOLIVAR department, N COLOMBIA, on the BRAZO MORALES of the MAGDALENA River, 170 mi/274 km SE of CARTAGENA; 08°18'N 73°52'W. Plantains, cassava, cotton, corn; livestock. Sometimes called Corregimiento Morales.

Morales (mo-RAH-lais), town, ⊙ Morales municipio, CAUCA department, SW COLOMBIA, 18 mi/29 km N of POPAYÁN; 02°45'N 76°38'W. Elevation 5,951 ft/1,813 m. Agriculture includes coffee, corn, sugarcane; livestock.

Morales (mo-RAH-les), town, IZABAL department, E GUATEMALA, on MOTAGUA River, on railroad and Intercoastal Highway, and 22 mi/35 km SW of PUERTO BARRIOS; 15°29'N 88°49'W. Adjacent community of Bananera was former United Fruit Company Headquarters. Formerly an important railroad stop and banana shipping center. The surrounding area remains a prominent banana-growing area.

Morales del Vino (mo-RAH-les dhel VEE-no), village, ZAMORA province, NW SPAIN, 4 mi/6.4 km S of ZAMORA; 41°27'N 05°44'W. Cereals, wine, fruit.

Morales de Rey (mo-RAH-les dhai RAI), town, ZAMORA province, NW SPAIN, 8 mi/12.9 km NW of BENAVENTE. Cereals, flax, sugar beets.

Morales de Toro (mo-RAH-les dhai TO-ro), town, ZAMORA province, NW SPAIN, 5 mi/8 km E of TORO; 41°32'N 05°18'W. Brandy distilling; wine, cereals; cattle.

Morales Island (mo-RAH-lais), 40 mi/64 km long, up to 11 mi/18 km wide, BOLÍVAR department, N COLOMBIA, formed by arms of lower MAGDALENA River, 28 mi/45 km W of OCAÑA; 08°20'N 73°50'W.

Moralzarzal (mo-RAHL-thahr-THAHL), town, MADRID province, central SPAIN, 24 mi/39 km NW of MADRID; 40°41'N 03°58'W. Grain; livestock; apiculture. Mineral springs.

Moram (MO-ruhm), town, OSMANABAD district, MAHARASHTRA state, W central INDIA, 38 mi/61 km SE of OSMANABAD; 17°48'N 76°28'E. Millet, wheat, cotton.

Mora Manas River, India: see MANAS RIVER.

Moramanga (moo-rah-MAHNG-guh), town, TOAMASINA province, E central MADAGASCAR, 100 mi/161 km SW of TOAMASINA; 18°53'S 48°14'E. Market and agricultural center, railroad junction for Lake ALAOTRA. Charcoal and lumber production. Forest plantations; coffee, rice.

Cross-references are shown in SMALL CAPITALS. The pronunciation guide is shown on page xix. The sources of population figures are shown on page xvii.

Moran (mor-AN), village (2000 population 562), ALLEN county, SE KANSAS, 13 mi/21 km E of IOLA; 37°55'N 95°10'W. Livestock; grain; dairying. Manufacturing (hand tools, limestone).

Moran (MOR-an), village (2006 population 224), SHACKELFORD county, N TEXAS, 33 mi/53 km E of ABILENE; 32°32'N 99°10'W. In wheat, cattle, oil area.

Moranbah (MO-ruhn-buh), town, E central QUEENSLAND, NE AUSTRALIA, 110 mi/177 km WSW of MACKAY, 8 mi/13 km N of Peak Downs Highway; 21°58'S 148°03'E. Service center for GOONYELLA and PEAK DOWNS coal mines, BOWEN COAL BASIN. Cattle, sheep.

Morangis (mo-rahn-zhee), town (□ 1 sq mi/2.6 sq km), industrial S suburb of PARIS, ESSONNE department, ÎLE-DE-FRANCE region, N central FRANCE, directly S of ORLY airport, 10 mi/16 km from Notre Dame Cathedral; 48°42'N 02°20'E.

Moranhat, India: see NAZIRA.

Moran Junction (mor-AN), village, TETON county, NW WYOMING in GRAND TETON NATIONAL PARK, on SNAKE RIVER, at mouth of Blackrock Creek, and 28 mi/45 km NNE of JACKSON. Elevation 6,742 ft/2,055 m. Service center at E entrance to park. Jackson Lake Dam, unit in Minidoka reclamation project, to W, raises level of natural lake. TETON RANGE to W; Jackson Hole Biological Research Station also to W. Bridger-Teton National Forest to NE and SE.

Mora-Noret (MOO-rah–NOO-net), village, KOPPARBERG county, central SWEDEN, on N shore of Lake SILJAN, just E of MORA, 45 mi/72 km NW of FALUN. Tourist resort area.

Morant Bay, town, ⊙ SAINT THOMAS parish, SE JAMAICA, port with open roadstead, at mouth of small Morant River, 25 mi/40 km ESE of KINGSTON (linked by highway); 17°53'N 76°25'W. Ships cacao, coffee, pimento, ginger, coconuts, copra, honey, rum. Sea resort. Scene of 1865 rebellion.

Morant Cays, group of three CARIBBEAN islets, dependency of JAMAICA, at S entrance of JAMAICA CHANNEL, 45 mi/72 km SE of MORANT BAY (SE Jamaica). Northeast Cay is at 17°25'N 75°58'W, Southwest Cay at 17°23'N 75°58'W. The uninhabited islands are of little economic importance, though sea-bird eggs and guano are collected. They were occupied by the British in 1862 and annexed to Jamaica in 1882. Sometimes called, together with PEDRO CAYS (120 mi/193 km WSW), Guano Islands.

Morant Point, cape at E extremity of JAMAICA, on JAMAICA CHANNEL, and 40 mi/64 km E of KINGSTON; 17°55'N 76°10'W. Lighthouse. Another headland, SOUTH EAST POINT, is c. 1 mi/1.6 km S.

Mora Passage, channel, c.7 mi/11.3 km long, BARIMA-WAINI district, NW GUYANA; 08°21'N 59°46'W. Links BARIMA RIVER just above MORAWHANNA with WAINI RIVER mouth on the ATLANTIC OCEAN.

Morar (mo-RAHR), town (2001 population 38,881), GWALIOR district, Madhya Pradesh state, central INDIA, 4 mi/6.4 km NE of LASHKAR; E of GWALIOR; 26°14'N 78°14'E. Trades in millet, wheat, gram, leather goods; tanning, tent manufacturing. Founded 1844; scene of major uprising in Indian Mutiny of 1857.

Morar (MOR-ahr), village, HIGHLAND, N Scotland, 3 mi/4.8 km S of MALLAIG; 56°57'N 05°50'W. At W end of Loch Morar (12 mi/19 km long, up to 2 mi/3.2 km wide; 1,017 ft/310 m deep), one of the deepest known depressions of the European plateau. The Morar district extends 19 mi/31 km between LOCH NEVIS (N) and the Sound of Arisaig (S).

Mora River (MOR-uh), 75 mi/121 km long, N NEW MEXICO; rises 7 mi/11.3 km NW of MORA in SANGRE DE CRISTO MOUNTAINS; flows SE, past Mora and WATROUS, and E to CANADIAN RIVER, 45 mi/72 km ENE of LAS VEGAS, New Mexico. Receives Sapello Creek from W below Mora.

Mora River, CZECH REPUBLIC: see MORAVICE RIVER.

Morastrand, SWEDEN: see MORA.

Morat, SWITZERLAND: see MURTEN.

Morata de Jalón (mo-RAH-tah dhai hah-LON), town, ZARAGOZA province, NE SPAIN, on JALÓN RIVER, and 12 mi/19 km NE of CALATAYUD; 41°28'N 01°28'W. In wine-producing area; manufacturing (cement, alcohol, tartaric acid); olive oil, fruit, sugar beets.

Morata de Jiloca (mo-RAH-tah dhai hee-LO-kah), village, ZARAGOZA province, NE SPAIN, on JILOCA RIVER, and 8 mi/12.9 km SSE of CALATAYUD; 41°15'N 01°35'W. Wine, sugar beets.

Morata de Tajuña (tah-HOO-nyah), town, MADRID province, central SPAIN, on TAJUÑA RIVER (irrigation), on railroad, and 18 mi/29 km SE of MADRID; 40°14'N 03°26'W. Agricultural center (wine, cattle, sheep). Olive-oil pressing, wine making, alcohol distilling, vegetable canning; stone and lime quarrying.

Moratalla (mo-rah-TAH-lyah), town, MURCIA province, SE SPAIN, 7 mi/11.3 km NNW of CARAVACA DE LA CRUZ. Agricultural trade center (cereals, rice, wine, fruit). Olive-oil processing, flour milling, brandy distilling, fruit-conserve manufacturing; lumbering; livestock raising.

Morat Island (mo-RAHT), ISLAS DE LA BAHÍA department, N HONDURAS, in CARIBBEAN SEA, between SANTA ELENA (W) and BARBARETA (E) islands; 0.5 mi/0.8 km long; 16°24'N 86°11'W. Within BARBARETA MARINE NATIONAL PARK.

Morat, Lake (mo-RAH), German Murtensee, French Lac de Morat (□ 9 sq mi/23 sq km), W SWITZERLAND, bordering on cantons of FRIBOURG and VAUD; 5.5 mi/8.9 km long, maximum depth 151 ft/46 m. Elevation 1,407 ft/429 m. BROYE RIVER enters SW, leaves NW, flowing 4 mi/6.4 km W to LAKE NEUCHÂTEL. Main town on lake, MURTEN (French Morat).

Morattico (mour-A-ti-ko), unincorporated village, LANCASTER county, E VIRGINIA, 17 mi/27 km SE of TAPPAHANNOCK, on RAPPAHANNOCK RIVER estuary, at mouth of Lancaster Creek; 37°47'N 76°37'W. Manufacturing (crab-meat processing); fish, oysters, crabs.

Moratuwa (MO-rah-TU-wah), town (2001 population 177,563), WESTERN PROVINCE, SRI LANKA, on W coast, 10 mi/16 km S of COLOMBO city center; 06°47'N 79°52'E. Fishing; furniture manufacturing; trades in coconuts, rice, vegetables. Industries include pharmaceuticals, textiles, and rubber goods in N area (called LUNAWA). Seat of University of Moratuwa (engineering, architecture).

Moraújo (MO-rah-EW-zho), town (2007 population 8,254), CEARÁ, BRAZIL.

Morava, Serbian Moravicki Okrug, district (□ 1,164 sq mi/3,026.4 sq km; 2002 population 224,772), ⊙ CACAK, W central SERBIA; 43°50'N 20°05'E. Includes municipalities (opštinas) of Cacak, Gornkji Milanovac, IVANJICA, and Lucani. Manufacturing (metal processing, chemicals, paper); agriculture.

Morava (mo-RAH-vah), village, LOVECH oblast, SVISHTOV obshtina, BULGARIA.

Morava, CZECH REPUBLIC: see MORAVIA.

Morava River (MO-rah-VAH), German March, c.240 mi/390 km long, CZECH REPUBLIC; rising in the SUDETES, N CZECH REPUBLIC; flowing generally S past OLOMOUC into the DANUBE RIVER, W of BRATISLAVA, SLOVAKIA. Navigable in its lower course, which also forms part of the Slovak border with the Czech Republic and AUSTRIA. The Morava Valley is very fertile, producing sugar beets, grains, grapes, and tobacco. It is an important N-S thoroughfare.

Morava River (MOR-ah-vah), Serbian Morava or Velika Morava, 134 mi/216 km long, SERBIA, main river of Serbia, formed by junction of the SOUTHERN MORAVA RIVER and the WESTERN MORAVA RIVER (which receives the IBAR RIVER) near STALAC; flows N, past CUPRIJA, through the wide, fertile, and densely populated POMORAVLJE valley, to the DANUBE River 10 mi/16 km ENE of SMEDEREVO, forming delta mouth. Receives RESAVA RIVER (right). Includes in its basin nearly all Serbia, uniting mountain valleys into a natural region of communications. Most of its course is followed by Belgrade-Thessaloniki railroad. Also called Great Morava River.

Moraveh Tappeh (muh-RAH-vai tahp-PAI), town, Mäzandarän province, NE IRAN, on ATREK RIVER, and 60 mi/97 km NE of Gonbad-e Qābūs, near TURKMENISTAN border; 37°53'N 55°57'E.

Moravia (mo-RAI-vee-yah), Czech Morava (MO-rah-VAH), German Mähren, historic region (□ 8,632 sq mi/22,443.2 sq km), CZECH REPUBLIC, central part of former CZECHOSLOVAKIA. Bordered W by BOHEMIA, E by SLOVAKIA (across the LITTLE and WHITE CARPATHIAN MOUNTAINS), and N by CZECH SILESIA (across the SUDETES MOUNTAINS, which include the MORAVIAN GATE, a historically strategic N-S route). Central Moravia is a valley, opening in the S toward AUSTRIA; it is drained by the MORAVA RIVER and its tributaries. Moravia is a fertile agricultural area that encompasses the HANÁ REGION. It has important iron and steel industries as well as diverse light industries. Diverse mineral resources, such as lignite, coal, oil, iron, copper, silver, and lead, spurred industrialization in the 20th century. Major cities include: BRNO, the former Moravian capital and a leading machinery and textile center; ZLÍN, famous for its shoe industry; OSTRAVA, a coal-mining center with a large iron and steel industry; and OLOMOUC. With Bohemia and Czech Silesia, Moravia makes up the Czech Republic, or the Czech Lands, a portion of former Czechoslovakia traditionally occupied by the Czechs, a branch of the Western Slavs, who displaced Germanic tribes that occupied the region from the 1st to the 5th century C.E. Before then, Moravia had been inhabited by the Celtic Boii and Cotini. Subjugated by the Avars, the Czechs freed themselves under the leadership of Samo (627–c.660), who established the first state of the Western Slavs. The state disintegrated after his death, but by the 9th century the Moravians, again united, formed a great empire, which included what are now Bohemia, SILESIA, Slovakia, S POLAND, and N HUNGARY. In 863 C.E., the missionaries Cyril and Methodius were sent to Moravia on the appeal of Duke Rastislav; as a result, the Moravians accepted Christianity, placing themselves under the Roman Catholic Church. The Moravian empire reached its height under Svatopluk (d. 894), but after his death it broke apart and in the early 10th century, it fell to the Magyars. When Emperor Otto I defeated the Magyars in 955, Moravia became a march of the Holy Roman Empire. From the early 11th century, it was in effect a crownland of the kingdom of BOHEMIA, with which it came under AUSTRIAN rule in 1526. However, Moravia retained its separate diet and was at times separated from the Bohemian crown (e.g., at periods during the Hussite Wars of the 15th century and from 1608 to 1611, when Bohemia was ruled by Emperor Rudolf II and Moravia by his brother Matthias). Moravia, generally more tolerant of Hapsburg authority than Bohemia, suffered less in the religious and civil strife of the 16th century and even experienced a flowering of Protestantism during a period of religious toleration. In 1618, however, the Czechs of Bohemia revolted and were crushed at the battle of the White Mountain by the Hapsburgs, who thereafter took reprisals against the Moravian Czechs as well. Moravia's diet was reduced to total ineffectiveness. The Moravian towns underwent thorough Germanization beginning in the 13th century. Under Hapsburg rule nearly the entire upper and middle classes were German; cities such as the predominantly German-speaking Brno, were surrounded by a countryside of Czech-speaking people. In 1849, following an abortive revolution during which the Czechs of Bohemia and Moravia demanded unification of their historic lands and creation of a common diet, Moravia was made an Austrian crown land. Hapsburg rule was finally

Area is shown by the symbol □, and capital city or county seat by ⊙.

overthrown in 1918, and Moravia was incorporated into CZECHOSLOVAKIA. In 1927, Moravia, with CZECH SILESIA, was constituted into the province of Moravia and Silesia. The German element, however, continued to play an important part in Moravian life. The Munich Pact of 1938 resulted in the annexation by GERMANY of Czech Silesia, of NW and S Moravia, and of N and W Bohemia (the SUDETENLAND). In 1939 Moravia and Bohemia became a German "protectorate." After World War II the pre-1938 boundaries were restored, and the larger part of the German-speaking population was expelled. In 1949 the province of Moravia and Silesia was replaced by four administrative regions, and in 1960, in a new administrative reorganization, Moravia was divided into JIHOMORAVSKY and SEVEROMORAVSKY provinces. On January 1, 1969, the Moravian region, along with Bohemia and Czech Silesia, was incorporated into the Czech Socialist Republic, renamed the Czech Republic in 1990, which in turn became the independent Czech Republic when the Czechoslovakian Federation was dismantled on January 1, 1993.

Moravia, town (2000 population 713), APPANOOSE county, S IOWA, 11 mi/18 km NNE of CENTERVILLE; 40°53′N 92°49′W. Manufacturing of cement blocks. Limestone quarry nearby.

Moravia, village (□ 1 sq mi/2.6 sq km; 2006 population 1,319), CAYUGA county, W central NEW YORK, near S end of OWASCO LAKE, 26 mi/42 km WSW of SYRACUSE; 42°42′N 76°25′W. Agricultural produce; dairying. Fillmore Glen State Park and the birthplace of President Millard Fillmore are nearby. Industrialist John D. Rockefeller lived here as a boy. Incorporated 1837.

Moravian Falls (mor-AIV-ee-uhn), unincorporated town (□ 5 sq mi/13 sq km; 2000 population 1,440), WILKES county, NW NORTH CAROLINA, 5 mi/8 km S of NORTH WILKESBORO; 36°06′N 81°10′W. Manufacturing (beehives); retail trade; agriculture (tobacco, soybeans; poultry; dairying; honey). WILKESBORO RESERVOIR to W.

Moravian Gate (mo-RAI-vee-yuhn), Czech *Moravská brána* (MO-rahf-SKAH BRAH-nah), wide pass (c.900 ft/274 m) between E end of the SUDETES and W end of the Carpathian Mountains, SEVEROMORAVSKY province, N MORAVIA, CZECH REPUBLIC. Natural communications channel and important central European trade route since pre-Roman times. Drained by upper ODER (N) and BECVA (S) rivers.

Moravian Karst (mo-RAI-vee-yuhn KAHRST), Czech *Moravský kras* (MO-rahf-SKEE KRAHS), picturesque region of limestone formations, JIHOMORAVSKY province, W central MORAVIA, CZECH REPUBLIC, just E of BLANSKO. Noted for stalactite caves, precipitous cliffs, underground streams, and prehistoric sites; famous MACOCHA chasm is 453 ft/138 m deep. Main tourist centers are OSTROV U MACOCHY (E) and Sloup (N) of Blansko.

Moravian-Silesian Beskids, CZECH REPUBLIC: see LYSA HORA.

Moravian Slovakia (mo-RAI-vee-yuhn slo-VAH-kee-yah), Czech *Moravské Slovácko* (MO-rahf-ske SLO-vahts-KO), German *Mährische Slowakei*, region of SE MORAVIA, CZECH, extending roughly from BŘECLAV NNE to NAPAJEDLA. Noted for colorful national costumes, song and dance festivals (especially in STRAZNICE), and folk fairs. Trade and cultural centers are UHERSKÉ HRADIŠTĚ and UHERSKY BROD.

Moravia, San Vicente de, COSTA RICA: see SAN VICENTE.

Moravica River, SERBIA: see SOUTHERN MORAVA RIVER and WESTERN MORAVA RIVER.

Moravice River (MO-rah-VI-tse), German *Mora*, 63 mi/101 km long, SEVEROMORAVSKY province, N MORAVIA and central SILESIA, CZECH REPUBLIC; rises in the JESENIKS on SE slope of PRADED Mountain; flows SE, past RYMAROV, and NE to OPAVA RIVER just E of OPAVA city. SLEZSKA HARTA DAM, Czech *Slezská Harta*, SSE of BRUNTAL, forms reservoir with maximum capacity 2,155 acres/872 ha.

Moraviţa (mo-rah-VEE-tsah), village, TIMIŞ county, SW ROMANIA, in TRANSYLVANIA, 30 mi/48 km E of REŞIŢA; 45°16′N 21°16′E. Border crossing with SERBIA.

Moravitsa (mo-RAH-veet-sah), village, MONTANA oblast, MEZDRA obshtina, BULGARIA; 43°09′N 23°39′E.

Moravská brána, CZECH REPUBLIC: see MORAVIAN GATE.

Moravska Trebova (MO-rahf-SKAH TRZHE-bo-VAH), Czech *Moravská Třebová*, German *Mährisch-Trübau*, town, VYCHODOCESKY province, W MORAVIA, CZECH REPUBLIC, on railroad, and 28 mi/45 km NW of OLOMOUC; 49°45′N 16°40′E. Textile manufacturing, glass- and woodworking. Clay quarrying nearby. Has a picturesque castle and a military police school.

Moravske Budejovice (MO-rahf-SKE BU-dye-YO-vi-TSE), Czech *Moravské Budějovice*, German *Mährisch-Budwitz*, town, JIHOMORAVSKY province, S MORAVIA, CZECH REPUBLIC, 25 mi/40 km SSE of JIHLAVA; 49°03′N 15°48′E. Railroad junction. Agriculture (barley, oats); poultry; manufacturing (machinery, wood and food processing). Has a 16th century church and castle.

Moravské Slovácko, CZECH REPUBLIC: see MORAVIAN SLOVAKIA.

Moravsky Beroun (MO-rahf-SKEE BE-roun), Czech *Moravský Beroun*, German *Bärn*, town, SEVEROMORAVSKY province, N MORAVIA, CZECH REPUBLIC, on railroad, and 16 mi/26 km NNE of OLOMOUC; 49°48′N 17°27′E. Manufacturing (plastics, textiles).

Moravský kras, CZECH REPUBLIC: see MORAVIAN KARST.

Moravsky Krumlov (MO-rahf-SKEE KRUM-lof), Czech *Moravský Krumlov*, German *Mährisch-Kromau*, town, JIHOMORAVSKY province, S MORAVIA, CZECH REPUBLIC, on railroad, and 17 mi/27 km SW of BRNO; 49°03′N 16°19′E. Agriculture (barley, oats); manufacturing (glass, furniture, food processing). Built as a fortress, it still retains its old castle (now a museum of prehistoric artifacts) and remnants of fortifications.

Moravsky Pisek (MO-rahf-SKEE PEE-sek), Czech *Moravský Písek*, German *Mährisch Pisek*, village, JIHOMORAVSKY province, S MORAVIA, CZECH REPUBLIC, on railroad, and 8 mi/12.9 km SW of UHERSKÉ HRADIŠTĚ; 48°59′N 17°20′E. Agriculture (wheat, vegetables); manufacturing (electrical and building materials); woodworks.

Morawaka (MO-rah-wah-kuh), town, SOUTHERN PROVINCE, SRI LANKA, 24 mi/39 km NE of GALLE; 06°15′N 80°29′E. Precious and semiprecious stone-mining center (including alexandrite, aquamarine, ruby, sapphire); beryl deposits.

Morawhanna (mor-rah-WAH-nah), village, BARIMA-WAINI district, NW GUYANA; 08°16′N 59°45′W. Located 153 mi/246 km NW of GEORGETOWN (has ferry port). Coffee, citrus fruits.

Moray (MOR-ai), county (□ 864 sq mi/2,246.4 sq km; 2001 population 86,940), NE Scotland; ☉ ELGIN; 57°30′N 03°10′W. Borders HIGHLAND to W, MORAY FIRTH and NORTH SEA to N, and Aberdeenshire to E. Main industries are agriculture, fishing, food processing, paper, whiskey, and oil. Formerly part of Grampian, abolished 1996.

Moraya (mo-RAH-yah), canton, MODESTO OMISTE province, POTOSÍ department, W central BOLIVIA, 25 mi/40 km N of VILLAZÓN on the UYUNI-Villazón railroad and highway; 21°45′S 65°32′W. Elevation 11,296 ft/3,443 m. Agriculture (potatoes, yucca, bananas); cattle.

Moray Firth (MOR-ai FUHRTH), inlet of the NORTH SEA, HIGHLAND, on NE coast of Scotland; 57°45′N 03°30′W. It is usually considered to cover the inlet between KINNAIRDS HEAD and Duncansby Head, a distance of 78 mi/126 km. In its more restricted sense the firth covers the inlet between LOSSIEMOUTH (S) and TARBAT NESS (N), a distance of 21 mi/34 km. At its head the firth is continued W by BEAULY FIRTH. Moray Firth is noted for its fish; important riparian fishing ports are BANFF, BUCKIE, CROMARTY, FINDHORN, HELMSDALE, and Lybster. At head of firth is INVERNESS. Main inlets of firth are DORNOCH and CROMARTY firths.

Morazán (mo-rah-ZAHN), department (□ 909 sq mi/2,363.4 sq km), E EL SALVADOR, on HONDURAS border; ☉ SAN FRANCISCO GOTERA. Mainly mountainous; drained by Torola River (N; left affluent of the LEMPA) and Río Grande de San Miguel (S). Its henequen industry (rope, hammocks) is important. Agriculture (grain, fruit, sugarcane). Site of heavy fighting during 1978–1990 civil war. Main centers: SAN FRANCISCO GOTERA, GUATAJIAGUA, CACAOPERA. Formed 1875.

Morazán (mo-rah-SAHN), town, EL PROGRESO department, E central GUATEMALA, on short branch of upper MOTAGUA River, and 7 mi/11.3 km NW of EL PROGRESO; 14°56′N 90°09′W. Elevation 1,148 ft/350 m. Corn, wheat, sugarcane; livestock.

Morazán (mo-rah-SAHN), town, YORO department, NW HONDURAS, on paved road, 31 mi/50 km WNW of YORO; 15°17′N 87°34′W. Elevation 1,318 ft/400 m. Airport. Small farming, grain, beans; livestock; tropical fruit.

Morazán, NICARAGUA: see PUERTO MORAZÁN.

Morbach (MOR-bahkh), town, RHINELAND-PALATINATE, W GERMANY, in the HUNSRÜCK, 22 mi/35 km ENE of TRIER; 49°49′N 07°11′E. Health resort. Wood- and metalworking. Formed by the union of nineteen small villages.

Morbegno (mor-BAI-nyo), town, SONDRIO province, LOMBARDY, N ITALY, in the VALTELLINA, 15 mi/24 km WSW of SONDRIO; 46°08′N 09°34′E. Resort; base for Alpine excursions. Small diversified industrial center includes food cannery, cheese factory, lumber and silk mills; hydroelectric plant.

Morbi (mo-bee), city, RAJKOT district, GUJARAT state, W central INDIA, on KATHIAWAR peninsula, 35 mi/56 km N of RAJKOT; 22°49′N 70°50′E. Trade center (cotton, millet, salt, ghee, wool); cotton ginning and milling, manufacturing of pottery, matches, shoes, paint, bone fertilizer; glass- and metalworks. Airport. Technical institute. Was the capital of the former princely state of Morvi of WESTERN INDIA STATES agency; state merged 1948 with SAURASHTRA and later with Gujarat state.

Morbihan (mor-bee-yahn) [Breton=little sea], department (□ 2,634 sq mi/6,848.4 sq km), in BRITTANY, W FRANCE; ☉ VANNES; 47°55′N 02°55′W. Extends along BAY OF BISCAY, with much indented coastline (Gulf of MORBIHAN, QUIBERON PENINSULA) and several offshore islands (BELLE-ÎLE, GROIX, HOUAT, HOËDIC). Drained by short streams (BLAVET, SCORFF, OUST), which rise in ARMORICAN MASSIF (N) and flow generally S. Chief crops: rye, buckwheat, oats, barley, artichokes, green peas; apple orchards, vineyards (on Rhuis Peninsula); livestock raising. Extensive fisheries and oyster beds (in coastal inlets). Chief industries are fish canning, tanning, metalworking (HENNEBONT, PLOËRMEL, Vannes), and handicrafts (lace, clothing, furniture, and ceramics). Tourist trade centers around megalithic monuments at CARNAC and LOCMARIAQUER, shrine of Sainte-Anne-d'Auray, and many quaint Breton villages. Chief towns: Vannes, LORIENT (seaport, submarine base), PONTIVY. Morbihan is included in the administrative region of Bretagne (Brittany).

Morbihan, Gulf of (mor-bee-yahn), tidewater basin (□ 40 sq mi/104 sq km) in MORBIHAN department, BRITTANY, W FRANCE, linked with Bay of BISCAY by channel (1 mi/1.6 km wide), and enclosed by Rhuis Peninsula. Contains numerous sandy islands, including Île aux MOINES. Port of VANNES is on N shore. The gulf has given its name to the department. Boat trips around the gulf are a popular tourist attraction.

Mörbisch am See (MUHR-bish ahm SAI), village, N BURGENLAND, E AUSTRIA, near W shore of LAKE NEUSIEDL, 9 mi/14.5 km SE of EISENSTADT; 47°45′N 16°40′E. Vineyards; summer tourism. Mineral spring. Large Protestant community.

Mörbylånga (MUHR-BEE-LONG-ah), village, KALMAR county, SE SWEDEN, SW ÖLAND island, on KALMAR SOUND of BALTIC SEA, 9 mi/14.5 km S of KALMAR; 56°31′N 16°24′E. Manufacturing (foods, cement, rubber).

Morcellement St. André, village (2000 population 5,616), central PAMPLEMOUSSES district, MAURITIUS, 10 mi/16 km N of PORT LOUIS. Sugarcane.

Morcenx (mor-SAHNS), town (□ 24 sq mi/62.4 sq km), LANDES department, AQUITAINE region, SW FRANCE, 23 mi/37 km WNW of MONT-DE-MARSAN; 44°02′N 00°55′W. Railroad yards; lumber distribution center; manufacturing of resinous products.

Morchenstern, CZECH REPUBLIC: see SMRZOVKA.

Morcles, Grande Dent de, SWITZERLAND: see DENT DE MORCLES.

Morcone (mor-KO-ne), town, BENEVENTO province, CAMPANIA, S ITALY, 16 mi/26 km NNW of BENEVENTO; 41°20′N 14°40′E. Agricultural center; food processing; textiles (traditional methods with wool), shoes. Bathing resort.

Mordab Lagoon, IRAN: see MURDAB LAGOON.

Mor Dag, peak (12,500 ft/3,810 m), SE TURKEY, in HAKKARI MOUNTAINS, 13 mi/21 km N of YUKSEKOVA, near IRAN border.

Mordelles (mor-del), town (□ 11 sq mi/28.6 sq km), ILLE-ET-VILAINE department, in BRITTANY, W FRANCE, on the Meu, and 9 mi/14.5 km WSW of RENNES; 48°04′N 01°51′W. Market center (apples, pears, and vegetables).

Morden (MOR-duhn), town (□ 5 sq mi/13 sq km; 2001 population 6,142), S MANITOBA, W central Canada, 65 mi/103 km SW of WINNIPEG, and surrounded by Stanley rural municipality; 49°11′N 98°06′W. In an agricultural region. Manufacturing includes farm machinery and food- and fiber-processing plants. Government experimental farm in the town. Incorporated 1882.

Morden, ENGLAND: see MERTON.

Mordialloc (MOR-dee-A-luhk), suburb, S VICTORIA, AUSTRALIA, 15 mi/24 km SSE of MELBOURNE and on E shore of PORT PHILLIP BAY; 38°00′S 145°05′E. Beaches, creeks. In metropolitan area; residential; light industry; gardening.

Mordino (MOR-dee-nuh), village, S KOMI REPUBLIC, NE European Russia, near the Lokchim River (tributary of the VYCHEGDA River), on road and railroad spur, 39 mi/63 km S of STOROZHEVSK; 61°21′N 51°54′E. Elevation 500 ft/152 m. Logging, lumbering.

Mordov-, in Russian names: see also MORDV-.

Mordovian Preserve (muhr-DO-vee-yahn) (□ 124 sq mi/322.4 sq km), wildlife refuge, in the S part of the Oka-Klyaz′ma forest area, MORDVA REPUBLIC, European Russia. Pine forests and meadows. Local fauna include fox, bear, mink, marten, badger, lynx, otter, elk, and beaver. Established in 1935.

Mordovian Republic, RUSSIA: see MORDVA REPUBLIC.

Mordovo (muhr-DO-vuh), town (2006 population 7,045), SW TAMBOV oblast, S central European Russia, on the BITYUG RIVER (tributary of the DON River), on road junction and railroad (Oborona station), 52 mi/84 km NNW of TAMBOV; 52°05′N 40°47′E. Elevation 498 ft/151 m. Food industries.

Mordovshchikovo (muhr-duhf-shchee-KO-vuh), former town, SW NIZHEGOROD oblast, central European Russia, near the OKA River, 7 mi/11 km E of MUROM; 55°32′N 42°12′E. Elevation 278 ft/84 m. Now a suburb of NAVASHINO city, less than 1 mi/1.6 km N of the city center.

Mordovshchikovo, RUSSIA: see NAVASHINO.

Mordoy (muhr-DO-yee), town, SW CHITA oblast, S Siberian Russia, less than 20 mi/32 km N of the Mongolian border, on one of the left tributaries of the ONON RIVER, near highway, 8 mi/13 km NE, and under administrative jurisdiction, of KYRA; 49°42′N 112°02′E. Elevation 3,146 ft/958 m. Meat processing.

Mordva Republic (muhr-DVAH) or **Mordovia** (muhr-DO-vee-yah), constituent republic (□ 10,100 sq mi/26,260 sq km; 2006 population 862,030), E central European Russia; ⊙ SARANSK. Once a forested steppe, it consists of the VOLGA HILLS (E) and the Oka-Don lowland (W). Borders RYAZAN oblast (W), PENZA oblast (S), ULYANOVSK oblast (SE), CHUVASH REPUBLIC (NE), and NIZHEGOROD oblast (N), with all of which it is connected by a highly developed communication infrastructure of roads and railroads. Agricultural processing and manufacturing of machinery, electricity, construction materials, furniture, paper, and wood chemicals are the major industries. Beekeeping is a long-established economic activity. Cattle and sheep are raised, and grain, hemp, potatoes, and flax are grown. Major cities include Saransk and ARDATOV. The population is composed of Russians (61%), Mordvinians (32%), Tatars (5%), and smaller minorities like Ukrainians, Byelorussians, Chuvashi, Udmurti, and Armenians. The Mordvinians, an ancient Finno-Ugric ethnic group, were first mentioned by the Gothic historian Jordanes in the 6th century C.E. They were land tillers and herders, with close ties to the Slavs. In the mid-13th century, they came under the political control of the Golden Horde and, when it disintegrated, passed to the Kazan khanate. Russia annexed the territory of the Mordvinians in 1552. The Mordvinian Autonomous SSR, also known as the Mordovian Autonomous SSR, was formed in 1934. It was a signatory, under the name Mordva Republic, to the March 31, 1992, treaty that created the Russian Federation. The Mordvinians (Russian *Mordva*) speak a Finno-Ugric language and are Orthodox Christians. Also known as Mordovian Republic.

Mordves (muhr-DVYES), village, NE TULA oblast, central European Russia, on road and railroad, 18 mi/29 km S of KASHIRA (MOSCOW oblast), and 14 mi/23 km N of VENËV, to which it is administratively subordinate; 54°34′N 38°12′E. Elevation 636 ft/193 m. Distilling.

Mordy (MOR-dee), town, Siedlce province, E POLAND, on railroad, and 11 mi/18 km ENE of SIEDLCE; 52°13′N 22°31′E. Brewing, flour milling, brick manufacturing.

Møre (MUH-ruh), mountainous coastal region, MØRE OG ROMSDAL county, W NORWAY, cut by numerous fjords. Its N part between TROLLHEIMEN mountains (E) and NORTH SEA (W), includes islands of SMØLA, TUSTNA, ERTVÅGØY, FREI, and AVERØY, and is separated by the Romsdal from SW part of region, which lies between the DOVREFJELL and North Sea near ÅLESUND. Off the coast are the NORDØYANE and SØRØYANE islands.

More (MO-rai), canton, MAMORÉ province, BENI department, NE BOLIVIA; 14°14′S 64°50′W. Elevation 459 ft/140 m. Agriculture (bananas, coffee, tobacco, cotton, cacao, peanuts, yucca); cattle and horse raising.

Morea, Greece: see PELOPONNESOS.

Moreau River (MOR-o), NW SOUTH DAKOTA; North and South forks rise in SE HARDING county and converge in PERKINS county; flows E through Cheyenne River Indian Reservation to MISSOURI RIVER S of MOBRIDGE; 45°08′N 102°49′W.

Moreauville (MOR-o-vil), village (2000 population 922), AVOYELLES parish, E central LOUISIANA, 33 mi/53 km SE of ALEXANDRIA; 31°02′N 91°59′W. In cotton-growing area; manufacturing (fabricated structural metal).

Morecambe (MOR-kuhm), hamlet, E central ALBERTA, W Canada, 12 mi/19 km E of TWO HILLS, in Two Hills County No. 21; 53°40′N 111°28′W.

Morecambe and Heysham (MAW-kuhm and HAI-shuhm), town (2001 population 48,227), LANCASHIRE, N ENGLAND, on MORECAMBE BAY; 54°04′N 02°52′W. Morecambe is a resort famous for its shrimp. Heysham is a port with service to BELFAST. Nearby is an unusual single-celled chapel dating from before the Norman conquest. Site of the Heysham I (two units, 1,150 MW capacity) and Heysham II (two units, 1,250 MW capacity) nuclear power plants, which went online in 1983 and 1988, respectively.

Morecambe Bay (MAW-kuhm), shallow inlet of the IRISH SEA, NW ENGLAND, separating FURNESS peninsula from LANCASHIRE; 16 mi/26 km long and 10 mi/16 km wide. The bay receives the KENT and LUNE rivers. Abounds in shrimp.

Moreda (mo-RAI-dhah), town, GRANADA province, S SPAIN, 15 mi/24 km NW of GUADIX. Rail junction. Cereals, olive oil.

Moreda, town, OVIEDO province, NW SPAIN, in Aller valley, 15 mi/24 km SE of OVIEDO. Steel mill; tin-plate manufacturing; bituminous-coal mines.

Moree (mo-REE), municipality, N NEW SOUTH WALES, SE AUSTRALIA, on GWYDIR river, and 265 mi/426 km NNW of NEWCASTLE; 29°28′S 149°51′E. Railroad junction; wool; wheat, cotton, oil seeds, pecans, olives; farm machinery distribution. Hot springs.

Morehead, town (2000 population 5,914), ⊙ ROWAN county, NE KENTUCKY, 58 mi/93 km ENE of LEXINGTON, surrounded by Daniel Boone National Forest; 38°11′N 83°23′W. In timber, clay, burley tobacco, corn, and cattle area; manufacturing (apparel, lumber, wood products, metal production, furniture, printing and publishing). Seat of Morehead State University, including Kentucky Folk center. CAVE RUN LAKE reservoir to S, including Minor Clark State Fish Hatchery.

Morehead City (MOR-hed SIT-ee), town (□ 5 sq mi/13 sq km; 2006 population 9,293), CARTERET county, E NORTH CAROLINA, 32 mi/51 km SSE of NEW BERN, on W side of mouth of Beaufort Harbor (bridged to BEAUFORT 3 mi/4.8 km E; receives Newport River from NW) and N shore of BOGUE SOUND (bridged to BOGUE ISLAND), and just W of Beaufort; 34°43′N 76°43′W. Ocean port (with shipping terminal built 1935–1937); resort, fishing center. Service industries; manufacturing (seafood processing, diversified light manufacturing). Carteret Community College. Fort Macon State Park to SE, on Bogue Island, one of 2 North Carolina state parks (other in WILMINGTON). Croatan National Forest to NW. Cape LOOKOUT 14 mi/23 km to SE. Founded 1857; incorporated 1860.

Moreh, Hill of (mo-RE), Hebrew *Givath Hamoreh* (1,690 ft/515 m), N ISRAEL, at NE edge of PLAIN OF JEZREEL, 3 mi/4.8 km NE of AFULA. In biblical history Gideon camped here before defeating the Midianites. Modern Upper Afula is built on the SW slopes of this hill.

Morehouse, city (2000 population 1,015), NEW MADRID county, extreme SE MISSOURI, on LITTLE River, and 6 mi/9.7 km WSW of SIKESTON; 36°51′N 89°41′W. Cotton, rice, soybeans; woodworking plant. Settled 1880.

Morehouse, parish (□ 804 sq mi/2,090.4 sq km; 2006 population 29,761), NE LOUISIANA; ⊙ BASTROP; 32°46′N 91°55′W. Bounded E by BOEUF river, W by OUACHITA RIVER, N by ARKANSAS state line; intersected by BAYOU BARTHOLOMEW and Bayou Bonne Idee; lies mostly in MISSISSIPPI RIVER delta land. Agricultural (cotton, corn, wheat, sorghum, home gardens, rice, soybeans, sweet potatoes, vegetables, cattle, horses), catfish. Logging, some manufacturing, including processing of agricultural products, timber, paper products, apparel. Includes Chemin-a-Haut State Park in N (recreation), Georgia Pacific State Wildlife Area in NW, small part of Upper Ouachita National Wildlife Refuge in NW corner, Handy Brake National Wildlife Refuge in N center,

Coulee State Game Refuge in S, and part of Russell Sage State Wildlife Refuge in far S. Abraham Morehouse was one of its early settlers. Formed 1844.

Moreira Sales (MO-rai-rah SAH-les), city (2007 population 12,946), W PARANÁ state, BRAZIL, 42 mi/68 km W of CAMPO MOURÃO; 24°01'S 53°03'W. Coffee.

Moreland (MOR-luhnd), city (□ 20 sq mi/52 sq km), S VICTORIA, SE AUSTRALIA; 37°45'S 144°58'E. Consists of the suburbs of BRUNSWICK, Brunswick East, Brunswick West, COBURG, Coburg North, Fawkner, Glenroy, Gowanbrae, Hadfield, Merlynston, Newlands, Oak Park, Pascoe Vale, Pascoe Vale South, and Westbreen. City created in 1994 from the amalgamation of the suburbs Brunswick and Coburg and the S part of Broadmeadows. Manufacturing; many small businesses.

Moreland, town (2000 population 393), COWETA county, W GEORGIA, 6 mi/9.7 km S of Newman; 33°17'N 84°46'W.

Moreland, unincorporated village, BINGHAM county, SE IDAHO, 4 mi/6.4 km W of BLACKFOOT; 43°13'N 112°26'W. Irrigated agricultural area near SNAKE RIVER. Sheep, cattle, wheat, oats, barley, potatoes, sugar beets. Junction of railroad spur to ABERDEEN.

Morelganj, BANGLADESH: see MORRELGANJ.

Morelia (mo-RAI-lee-ah), city (2005 population 608,049) and township, ⊙ MICHOACÁN state, W MEXICO; 19°40'N 101°11'W. It is the commercial and processing center of an irrigated agricultural and cattle-raising area. Founded as Valladolid in 1541 by Antonio de Mendoza, Morelia is built on a rocky hill and is surrounded by a fertile valley at the W edge of the central plateau. High peaks border the valley on three sides. The climate is warm and healthful. The city is supplied with water by an aqueduct dating from the colonial period. The most imposing Spanish structure is the cathedral, begun in 1640; colonial architecture, some modern buildings, and shaded plazas give the city a pleasant atmosphere. The Colegio de San Nicolás, founded (1540) in PÁTZCUARO and transferred in 1580 to Morelia, is the oldest institution of higher learning in Mexico. Morelia was the birthplace of Agustín de Iturbide and of the patriot José María Morelos y Pavón, for whom it was renamed in 1828.

Morelia (mo-RAI-lee-ah), town, ⊙ Morelia municipio, CAQUETA department, S COLOMBIA, in the Oriente, 6.0 mi/9.7 km S of FLORENCIA; 01°29'N 75°43'W. Elevation 1,788 ft/544 m. Sugarcane, corn; livestock.

Morell (mo-RAIL), town, TARRAGONA province, NE SPAIN, 6 mi/9.7 km NNW of TARRAGONA; 41°12'N 01°13'E. Woolen textiles; agricultural trade (olive oil, wine, carob beans, hazelnuts, almonds).

Morell (mo-REHL), village (2001 population 332), NE Prince Edward Island, Canada, on St. Peters Bay, 23 mi/37 km ENE of CHARLOTTETOWN; 46°25'N 62°42'W. Lobster, fisheries.

Morella (mo-RAI-lyah), town, Castellón de la Plana province, E SPAIN, 32 mi/51 km WNW of VINAROZ; 40°37'N 00°06'W. Chief town of MAESTRAZGO district. Manufacturing of wool textiles; lumbering; cereals, legumes, livestock; honey, wax. Coal mines. Dominated by medieval castle; has Gothic church (14th century), tower of Saloquia, and ancient walls. Founded by Romans; was fortress of kingdom of Valencia in Middle Ages; belonged to military order of Montesa; was Carlist stronghold (1838–1840) until it fell to General Espartero.

Morelos (mo-RAI-los), state (□ 1,917 sq mi/4,984.2 sq km), S MEXICO; ⊙ CUERNAVACA. Morelos is separated from the Federal District and from MEXICO state by the TRANSVERSE VOLCANIC AXIS crossing central CIUDAD DE MEXICO. Morelos itself is mountainous, with many broad, semiarid valleys in the S. The climate is cold in the mountains and hot in the valleys. Chiefly agricultural, the state grows sugarcane, rice, cereals, tropical fruits, and vegetables. Industry is

progressing; automobile manufacturing is significant, and mining is being developed. The principal towns are Cuernavaca and CUAUTLA, which is famous for its defense (1812) by José María Morelos y Pavón in the war against SPAIN. The state, created in 1869, was named in his honor. It is one of Mexico's most densely populated states.

Morelos (mo-RAI-los), town, CHIHUAHUA, N MEXICO, in valley of SIERRA MADRE OCCIDENTAL, 120 mi/193 km W of HIDALGO DEL PARRAL; 26°42'N 107°40'W. Extremely isolated, no road access. Corn; cattle; timber. Sometimes REAL MORELOS.

Morelos, town, COAHUILA, N MEXICO, on railroad and 27 mi/43 km SW of PIEDRAS NEGRAS (TEXAS border); 28°28'N 100°52'W. Cattle raising; wheat, bran, istle fibers, candelilla wax.

Morelos, town, ZACATECAS, N central MEXICO, 8 mi/12.9 km N of ZACATECAS, at junction of Mexico Highways 45, 49, and 54; 22°51'N 102°32'W. Elevation 7,621 ft/2,323 m. Active silver mining; agriculture (cereals, maguey; livestock).

Morelos, MEXICO: see MORELOS CAÑADA.

Morelos, MEXICO: see VILLA MORELOS.

Morelos Cañada, town, ⊙ Cañada Morelos municipio, PUEBLA, central MEXICO, on railroad, and 18 mi/29 km S of SERDÁN; 18°44'N 97°25'W. Elevation 7,667 ft/2,337 m. Wheat, corn, vegetables. Sometimes Morelos.

Morelos, Ciudad, MEXICO: see CUAUTLA.

Morelos, Villa, MEXICO: see VILLA MORELOS.

Moremi Game Reserve (MO-rai-MEE), park (□ 1,888 sq mi/4,908.8 sq km), central NORTH-WEST DISTRICT, N central BOTSWANA, on E part of OKAVANGO DELTA. Swampland rich in animal, bird, and plant life. Major tourist attraction. Nearest town is MAUN (SE).

Morena (MO-rai-nah), district (□ 1,920 sq mi/4,992 sq km; 2001 population 1,279,090), Madhya Pradesh state, central India; ⊙ MORENA. About half the district's land is devoted to agriculture; wheat, famous for mustard production. Some large-scale industry: tire manufacturing, dairy processing, steel production. National Chambal Sanctuary, set up to preserve river aquaculture, has rare Ganges River Dolphin; Crocodile Centre at Deori. Sheopur district carved out of SW part (area of district prior to bifurcation was 4,476 sq mi/11,594 sq km). Formerly called Tonwarghar.

Morena (mo-RAI-nah) or **Pech Morena**, town (2001 population 150,890), ⊙ MORENA district, N central MADHYA PRADESH state, central INDIA, 23 mi/37 km NNW of LASHKAR; 26°29'N 78°01'E. Agriculture market (millet, gram, wheat, barley); oilseed milling, hand-loom cotton weaving.

Morena Dam, California: see COTTONWOOD CREEK.

Morena, Sierra, SPAIN: see SIERRA MORENA.

Morenci, unincorporated town, GREENLEE county, SE ARIZONA, at S tip of BLUE RANGE, 110 mi/177 km NE of TUCSON. Elevation 4,710 ft/1,436 m. Cattle; grain, alfalfa. Built on steep hillside. Rich copper mines here, discovered 1872. San Carlos Indian Reservation to NW; Gila Box Riparian National Conservation Area to S; Apache-Sitgreaves National Forest to N.

Morenci (muhr-EN-see), town (2000 population 2,398), LENAWEE county, SE MICHIGAN, on TIFFIN RIVER, and 15 mi/24 km SW of ADRIAN, near OHIO state line; 41°43'N 84°13'W. In diversified farm area (corn, grain, apples, sugar beets; livestock; dairy). Manufacturing (broaching tools, metal fabrication). Incorporated as city 1934.

Moreni (mo-REN), town, DÎMBOVIȚA county, S central ROMANIA, on railroad, in WALACHIA, on IALOMITA RIVER, and 17 mi/27 km WNW of PLOIEȘTI. Petroleum center; oil production and refining; machine shops. Lignite mines and extensive vineyards in vicinity.

Moreno (mo-RAI-no), city, BUENOS AIRES province, E ARGENTINA, residential and district administrative center in the Greater BUENOS AIRES area; 34°35'S 58°50'W. The district was the scene of several major

battles during the Argentine War of Independence and the mid-19th-century Unitarian-Federalist conflict.

Moreno (MO-re-no), city (2007 population 52,780), E PERNAMBUCO, NE BRAZIL, on railroad, and 17 mi/27 km WSW of RECIFE; 08°07'S 35°06'W. Agriculture (sugar, corn, potatoes, fruit, manioc); manufacturing (sugar milling, coconut processing).

Moreno, ARGENTINA: see TINTINA.

Moreno, Colombia: see PAZ DE ARIPORO.

Moreno Valley, city (□ 49 sq mi/127.4 sq km; 2000 population 142,381), RIVERSIDE county, S CALIFORNIA; suburb 12 mi/19 km ESE of RIVERSIDE, in Moreno Valley; 33°56'N 117°13'W. Manufacturing (office chairs, printing and publishing). As of 1990, Moreno Valley was California's fastest growing city, with a population increase of more than 300% between 1980 and 1990; statistics include former wine area of Morena, Sunnymead and Edgemont. Among its developing industries are high-technology electronics, steel and other metals, home-construction and improvement enterprises, engineering, and retailing. March Air Force Base (in SW), the oldest base in the W U.S., is headquarters for the Strategic Air Command's 15th Air Force; it employs a large proportion of the city's workers. Moreno Valley is the seat branch of Chapman University and a growing number of cultural institutions. Lake Perris reservoir and State Recreation Area to S; Riverside Auto Race Track in W; COLORADO RIVER AQUEDUCT runs E-W to S. Incorporated 1984.

Møre og Romsdal (MUH-ruh aw RAWMS-dahl), county (□ 5,832 sq mi/15,104 sq km; 2007 estimated population 245,407), W NORWAY, on the ATLANTIC OCEAN (W); ⊙ MOLDE. Scenic mountainous region, with deep valleys, including the ROMSDALEN valley, numerous fjords, and many islands. Fishing, tourism, and farming are important, and there are aluminum, textile, shoe, and furniture industries.

Morera, La (lah mo-RAI-rah), town, BADAJOZ province, W SPAIN, 29 mi/47 km SE of BADAJOZ. Cereals, olives, grapes; livestock.

Morés (mo-RAIS), town, ZARAGOZA province, NE SPAIN, on JALÓN RIVER, and 10 mi/16 km NE of CALATAYUD; 41°28'N 01°34'W. Cereals, wine, olive oil, fruit.

Moresby, PAPUA NEW GUINEA: see PORT MORESBY.

Moresby Island (MORZ-bee) (□ 1,060 sq mi/2,756 sq km), W BRITISH COLUMBIA, W Canada, QUEEN CHARLOTTE ISLANDS, in the PACIFIC, separated from mainland by HECATE STRAIT, and just S of GRAHAM ISLAND, from which it is separated by SKIDEGATE INLET; 85 mi/137 km long, 4 mi/6 km–34 mi/55 km wide; 52°25'N 131°30'W. Queen Charlotte Mountains here rise to 3,810 ft/1,161 m. Lumbering; fishing; cattle. Chief villages are SANDSPIT and Aliford Bay, both on NE coast. Inhabitants are mostly Haida Indians.

Moresnet (maw-rez-NE), district (□ 2 sq mi/5.2 sq km), LIÈGE province, E BELGIUM, near the GERMAN border; 50°43'N 05°59'E. It was formerly a lead- and zinc-mining center. Under joint PRUSSIAN and DUTCH (after 1830, Belgian) suzerainty from 1816, it was awarded (1919) to Belgium under the Treaty of VERSAILLES.

Morestel (mor-est-tel), commune (□ 3 sq mi/7.8 sq km), ISÈRE department, RHÔNE-ALPES region, SE FRANCE, 8 mi/12.9 km N of La TOUR-DU-PIN; 45°40'N 05°28'E. Aluminum products. The hilly setting has attracted painters since mid-19th century.

Moreton Bay (MOR-tuhn), inlet of the PACIFIC OCEAN, QUEENSLAND, NE AUSTRALIA, nearly enclosed by MORETON and STRADBROKE islands; 65 mi/105 km long and 20 mi/32 km wide; 27°15'S 153°15'E. Receiving the BRISBANE RIVER, the bay is the entrance to the port of BRISBANE. There are many resort towns on its W shore.

Moreton, Cape, AUSTRALIA: see MORETON ISLAND.

Cross-references are shown in SMALL CAPITALS. The pronunciation guide is shown on page xix. The sources of population figures are shown on page xvii.

Moretonhampstead (maw-tuhn-HAMP-sted), town (2001 population 1,536), central DEVON, SW ENGLAND, 12 mi/19 km WSW of EXETER; 50°39′N 03°45′W. Tourist resort. Has 17th-century almshouses, 13th-century inn, and 15th-century church.

Moreton-in-Marsh (MAW-tuhn–in–MAHSH), town (2001 population 3,198), NE Gloucestershire, SW ENGLAND, 13 mi/21 km ESE of EVESHAM; 51°59′N 01°42′W. Located at a junction of the FOSSE WAY.

Moreton Island (MOR-tuhn) (□ 71 sq mi/184.6 sq km), in PACIFIC OCEAN 15 mi/24 km off SE coast of QUEENSLAND, NE AUSTRALIA. Forms E shore of MORETON BAY; 25 mi/40 km long (N-S), 5 mi/8 km wide (E-W); rises to 910 ft/277 m; sandy. Cape Moreton, its N point (27°02′S 153°28′E), forms E side of entrance to Moreton Bay; lighthouse. Fishing. Of environmental interest due to its relatively untouched state and steep sand dunes, perched lakes, wetlands, and other features.

Moretown, town, WASHINGTON CO., central VERMONT, on MAD RIVER, 11 mi/18 km W of MONTPELIER; 44°15′N 72°43′W.

Moret-sur-Loing (mo-rai–syur–lwahng), town (□ 1 sq mi/2.6 sq km), SEINE-ET-MARNE department, ÎLE-DE-FRANCE region, N central FRANCE, on the LOING (canalized) near its mouth into the SEINE, and 6 mi/9.7 km SE of FONTAINEBLEAU; 48°22′N 02°49′E. It is known for its barley sugar. Precision metalworks; manufacturing (stretch materials, ceramics, photo equipment). Has 13th–15th-century church, old houses, and 14th-century fortified gates and bridge across the Loing.

Moreuil (mo-roi), town (□ 9 sq mi/23.4 sq km), SOMME department, PICARDIE region, N FRANCE, on the AVRE, and 12 mi/19 km SE of AMIENS; 49°46′N 02°29′E. Manufacturing (hosiery, wallpaper).

Morey, Lake, in town of FAIRLEE, ORANGE CO., E VERMONT, 26 mi/42 km SE of BARRE, and 1 mi/1.6 km W of Connecticut River; c.2 mi/3.2 km long; 43°55′N 72°10′W. Resort.

Morey-Saint-Denis (mo-rai–san–duh-nee), village (□ 3 sq mi/7.8 sq km), CÔTE-D'OR department, BURGUNDY region, E central FRANCE, on E slope of the CÔTE D'OR, 9 mi/14.5 km SSW of DIJON; 47°12′N 04°58′E. Noted Burgundy wines (CHAMBOLLE-MUSIGNY, GEVREY-CHAMBERTIN vineyards).

Morez (mo-rez), town (□ 3 sq mi/7.8 sq km), JURA department, FRANCHE-COMTÉ region, E FRANCE, near Swiss border, in gorge of BIENNE RIVER, and 12 mi/19 km NE of SAINT-CLAUDE, in the E JURA Mountains; 46°31′N 06°02′E. Elevation 2,303 ft/702 m. Leading center of optical industry (dating from end of 18th century). Also produces clocks, enamelware, and plastic frames. Morbier cheese is made in area. Town has a museum of the history of eyeglass manufacturing.

Morfa Dyffryn (MUHR-vuh DUH-frin), GWYNEDD, NW Wales, extends 2 mi/3.2 km along coast N of BARMOUTH; 52°54′N 04°07′W. Area of marshland on CARDIGAN BAY. National nature reserve known for dunes, flora.

Morfa Harlech, Wales: see HARLECH.

Mörfelden-Walldorf (MUHR-fel-den–VUHL-dorf), town, S HESSE, W GERMANY, 8 mi/12.9 km NNW of DARMSTADT; 49°59′N 08°34′E. Metalworking.

Morgab River, AFGHANISTAN and TURKMENISTAN: see MORGHAB RIVER.

Morgan, county (□ 599 sq mi/1,557.4 sq km; 2006 population 115,237), N ALABAMA; ⊙ DECATUR; 34°27′N 86°51′W. Agricultural area drained in N by WHEELER LAKE (in TENNESSEE RIVER). Cotton, corn, soybeans; poultry, livestock; textiles. Deposits of coal, sandstone, fuller's earth, asphalt. Formed 1818.

Morgan, county (□ 1,293 sq mi/3,361.8 sq km; 2006 population 28,109), NE COLORADO; ⊙ FORT MORGAN; 40°16′N 103°49′W. Irrigated agricultural region, drained by SOUTH PLATTE RIVER. Cattle, wheat, hay, sunflowers, brans, barley, corn, sugar beets. Empire Reservoir on W boundary; JACKSON RESERVOIR and Jackson State Park in NW; Bijou Reservoir in W. Formed 1889.

Morgan, county (□ 355 sq mi/923 sq km; 2006 population 17,908), N central GEORGIA; ⊙ MADISON; 33°35′N 83°29′W. Bounded NE by APALACHEE RIVER, drained by LITTLE RIVER Piedmont. Agriculture (cotton, corn, wheat, peaches; cattle, hogs, poultry); in lumbering area. Formed 1807.

Morgan, county (□ 572 sq mi/1,487.2 sq km; 2006 population 35,666), W central ILLINOIS; ⊙ JACKSONVILLE; 39°43′N 90°13′W. Bounded W by ILLINOIS RIVER; drained by APPLE, Sandy, Mauvaise Terre, and Indian creeks. Agriculture (corn, wheat, sorghum, soybeans, cattle, hogs; dairying). Food processing; manufacturing of bookbinding, paper products, chemicals; coal. Includes part of Lake Meredosia. Formed 1823. One of 17 Illinois counties to retain Southern-style commission form of county government. Includes Illinois College and MacMurray College.

Morgan, county (□ 409 sq mi/1,063.4 sq km; 2006 population 70,290), central INDIANA; ⊙ MARTINSVILLE; 39°29′N 86°27′W. Agricultural area (hogs, wheat, corn, fruit; poultry). Manufacturing at Martinsville and MOORESVILLE. Clay deposits; timber; artesian springs. Part of Morgan-Monroe State Forest on S boundary; Cikana State Fish Hatchery E of Martinsville. Drained by West Fork of WHITE RIVER, WHITELICK CREEK, and small Indian Creek. Formed 1821.

Morgan, county (□ 383 sq mi/995.8 sq km; 2006 population 14,306), E KENTUCKY; ⊙ WEST LIBERTY; 37°55′N 83°16′W. Drained by LICKING RIVER (upper reach of CAVE RUN LAKE reservoir in NW) and several creeks. Hilly agricultural area in CUMBERLAND foothills (corn, burley tobacco, sorghum, hay, alfalfa; cattle); bituminous-coal mines. Includes part of Daniel Boone National Forest in NW; part of PAINTSVILLE Lake reservoir in SE. Formed 1822.

Morgan, unincorporated rural community, LACLEDE county, S central MISSOURI, in the OZARKS, near Osage Fork of GASCONADE RIVER, 12 mi/19 km S of LEBANON.

Morgan, county (□ 596 sq mi/1,549.6 sq km; 2006 population 20,716), central MISSOURI; ⊙ VERSAILLES; 38°25′N 92°52′W. In the OZARKS, on LAKE OF THE OZARKS; drained N by LAMINE RIVER; drained by OSAGE RIVER. Commercial, recreational, and residential development along lake at Laurie, and Gravois Mills. Wheat, corn; cattle, poultry; manufacturing at Versailles; former barite mines; timber; tourist region in S. Large Amish community N of Versailles. Formed 1833.

Morgan (MOR-guhn), county (□ 418 sq mi/1,086.8 sq km; 2006 population 14,821), SE OHIO; ⊙ MCCONNELSVILLE; 39°37′N 81°50′W. Intersected by MUSKINGUM RIVER and small Meigs and Wolf creeks. Agricultural area (livestock; dairy products; corn, wheat, cabbages); manufacturing at McConnelsville; limestone quarries, coal mines. Formed 1817.

Morgan, county (□ 539 sq mi/1,401.4 sq km; 2006 population 20,108), NE central TENNESSEE; ⊙ WARTBURG; 36°08′N 84°39′W. On CUMBERLAND PLATEAU. Light manufacturing; traditional extractive industries; recreation. Formed 1817. Part of Obed Wild and Scenic River system is here.

Morgan, county (□ 610 sq mi/1,586 sq km; 2006 population 8,134), N UTAH; ⊙ MORGAN; 41°06′N 111°38′W. Irrigated agricultural area watered by WEBER RIVER. Alfalfa, barley, sugar beets, vegetables; dairying; cattle. WASATCH RANGE throughout. Wasatch National Forest in NE, N, and W. East Canyon Reservoir and State Park in S. Lost Creek Reservoir and State Park in NE. Formed 1862.

Morgan, county (□ 223 sq mi/579.8 sq km; 2006 population 28,654), NE WEST VIRGINIA, in EASTERN PANHANDLE; ⊙ BERKELEY SPRINGS; 39°33′N 78°15′W. Bounded N and W by POTOMAC RIVER (MARYLAND state line), S, in part, by VIRGINIA; drained by CACAPON RIVER. Agriculture (honey, corn, wheat, oats, barley, alfalfa, hay, apples); cattle; poultry. Manufacturing (furniture). Glass-sand pits; timber. Cacapon State Park is here, includes CACAPON MOUNTAIN; Berkeley Springs State Park in N; part of Sleepy Creek Wildlife Management Area in E. Formed 1820.

Morgan, city (2000 population 1,464), ⊙ CALHOUN county, SW GEORGIA, 23 mi/37 km W of ALBANY, and on ICHAWAYNOCHAWAY CREEK; 31°32′N 84°36′W. Declining agricultural service center symbolized by abandoned cotton gin.

Morgan, town (2000 population 903), REDWOOD county, SW MINNESOTA, 24 mi/39 km WNW of NEW ULM; 44°25′N 94°55′W. Grain, soybeans, sugar beets, alfalfa; livestock, poultry; dairying; manufacturing (electrical components, feeds, fertilizer). Lower Sioux Indian Reservation to NW; on MINNESOTA RIVER.

Morgan, unincorporated town, South Fayette township, ALLEGHENY county, W PENNSYLVANIA, suburb 10 mi/16 km SW of downtown PITTSBURGH, on Millers Run creek; 40°21′N 80°08′W. Manufacturing (novelty items).

Morgan or **Morgan City**, town (2006 population 3,101), ⊙ MORGAN county, N UTAH, on WEBER RIVER at mouth of East Canyon Creek, in WASATCH RANGE, and 22 mi/35 km NNE of SALT LAKE CITY; 41°02′N 111°40′W. Trade center for agricultural region; sugar beets, alfalfa, vegetables, barley; dairying; manufacturing (cement, luggage, small arms). Elevation 5,068 ft/1,545 m. Lost Creek Reservoir and State Park to NE; East Canyon Reservoir and State Park to S. Wasatch National Forest to N, W, and S. Settled 1852 by Mormons. Surrounding area served by irrigation works on Weber River.

Morgan, town, ORLEANS CO., N VERMONT, on SEYMOUR LAKE and 9 mi/14.5 km E of NEWPORT; 44°53′N 71°58′W. Resorts. Chartered as Caldersburgh and name changed to Morgan in 1801.

Morgan (MOR-guhn), village, port, SE SOUTH AUSTRALIA, S central AUSTRALIA, 85 mi/137 km NE of ADELAIDE, and on MURRAY RIVER; 34°02′S 139°39′E. Railroad terminus; fruit; livestock.

Morgan, village, PHILLIPS county, N MONTANA, port of entry at CANADA (SASKATCHEWAN) border, c.44 mi/71 km N of MALTA. Mixed livestock and cash-crop farming.

Morgan (MOR-guhn), village (2006 population 519), BOSQUE county, central TEXAS, 7 mi/11.3 km N of MERIDIAN; 32°01′N 97°36′W. In farm area. LAKE WHITNEY reservoir to E.

Morgan Acres, unincorporated town, SPOKANE county, E WASHINGTON, suburb 5 mi/8 km NE of downtown SPOKANE. Railroad center.

Morgan City, city (2000 population 12,703), SAINT MARY parish, S LOUISIANA, 48 mi/77 km S of BATON ROUGE; 29°41′N 91°12′W. Elevation 8 ft/3 m. Fishing port on the ATCHAFALAYA RIVER (connected to the INTRACOASTAL WATERWAY). Headquarters for offshore petroleum drilling; oil and gas wells. Shipyards; large shrimp and crawfish fleet, an oyster industry, and alligator farms. Manufacturing (barges, propeller shafts, crew boats, yachts; printing and publishing. Holds Louisiana Shrimp and Petroleum Festival. LAKE PALOURDE to NE. Incorporated 1860.

Morgan City, village (2000 population 305), LEFLORE county, W central MISSISSIPPI, 14 mi/23 km SW of GREENWOOD, near YAZOO RIVER; 33°22′N 90°20′W. Cotton, grain; cattle.

Morganfield, town (2000 population 3,494), ⊙ UNION county, W KENTUCKY, 21 mi/34 km WSW of HENDERSON; 37°40′N 87°54′W. Railroad terminus. Bituminous-coal mining and agricultural (corn, wheat, hay, alfalfa; livestock) area; manufacturing (wire products, automobile parts and accessories, paper cones, meat processing, coal processing). Morganfield

Airport to E. U. S. Camp Breckinridge to E (inactive) was active in World War II. Higgins-Henry Wildlife Management Area to SE.

Morgan, Fort, historic site, BALDWIN COUNTY, SW ALABAMA, 21 mi/34 km W of GULF SHORES at the entrance to MOBILE BAY. Crucial to Confederate defense of MOBILE, Fort Morgan surrendered to the North in August 1864 after the Battle of Mobile Bay.

Morgan Hill, city (2000 population 33,556), SANTA CLARA county, W CALIFORNIA, 20 mi/32 km SE of SAN JOSE, in Santa Clara valley; 37°08′N 121°39′W. Varied manufacturing; mushrooms (abundant mines), fruit, vegetables, grapes, cherries; diversified agriculture; cattle, poultry; dairying, eggs; nursery products; timber. Anderson Reservoir to NE, Coyote Reservoir to SE (both on Coyote Creek); Henry W. Coe State Park to NE; Santa Cruz Mountains to SW. Incorporated 1906.

Morgan, Mount (13,748 ft/4,190 m), INYO county, E CALIFORNIA, in the SIERRA NEVADA, 19 mi/31 km W of BISHOP, in Inyo National Forest.

Morgannwg Gandl, Wales: see MID GLAMORGAN.

Morgans Point (MOR-guhnz), village, HARRIS county, S TEXAS, on GALVESTON BAY, and 20 mi/32 km ESE of downtown HOUSTON, on W side of mouth of SAN JACINTO RIVER–HOUSTON SHIP CHANNEL; 29°40′N 94°59′W.

Morgan's Point Resort (MOR-guhnz), town, BELL county, central TEXAS, residential and recreational suburb 7 mi/11.3 km WNW of TEMPLE, on E shore of BELTON LAKE; 31°09′N 97°27′W.

Morganton (MOR-guhn-tuhn), city (□ 18 sq mi/46.8 sq km; 2006 population 17,224), ⊙ BURKE county, W NORTH CAROLINA, 63 mi/101 km NW of CHARLOTTE, on the CATAWBA RIVER at upper reach of Lake Rhodhiss reservoir; Lake JAMES reservoir upstream (to W), in the foothills of the BLUE RIDGE MOUNTAINS; 35°44′N 81°42′W. Manufacturing (textiles, apparel, chemicals, furniture, foam rubber, auto and tool parts, metal and fiberglass fabricating, electronics, printing and publishing); service industries. South Mountain State Park to S; Tuttle Educational State Forest to NE. Founded 1784, incorporated 1885.

Morganton, village, FANNIN county, N GEORGIA, 4 mi/6.4 km E of BLUE RIDGE, and on BLUE RIDGE LAKE; 34°52′N 84°14′W. Manufacturing of lumber and clothing.

Morgantown, city (2000 population 26,809), ⊙ MONONGALIA county, N WEST VIRGINIA, near the PENNSYLVANIA state line, on the MONONGAHELA RIVER; 39°38′N 79°57′W. Shipping point for coal and limestone region. Manufacturing (glass, chemicals, printing and publishing, office furniture, mining equipment, organic phosphate, crushed limestone, lumber, pharmaceuticals, airport equipment, machining). Agriculture (strawberries, grain); livestock; poultry. Morgantown Airport to E. West Virginia University (two main campuses). Kennedy Youth Center, National Training Center for Boys, to S. Chestnut Ridge Park and Coopers Rock State Forest to NE. Fort Morgan built here (1772), and the first settlers arrived the same year. Iron, discovered in 1789, was the principal industry until the Civil War. Incorporated 1785.

Morgantown, town (2006 population 966), MORGAN county, central INDIANA, near small Indian Creek, 8 mi/12.9 km SE of MARTINSVILLE; 39°22′N 86°16′W. In agricultural area. Laid out 1831.

Morgantown, town (2000 population 2,544), ⊙ BUTLER county, W central KENTUCKY, on GREEN RIVER, and 20 mi/32 km NW of BOWLING GREEN; 37°13′N 86°42′W. Agriculture (dark and burley tobacco, grain; livestock; catfish); coal mining; manufacturing (apparel, rubber products, lumber, plastic molding, leather products). Read's Ferry crosses Green River to NW. Established 1813.

Morgantown, unincorporated town, BERKS county, SE central PENNSYLVANIA, 12 mi/19 km S of READING;

40°09′N 76°53′W. Manufacturing includes smelting, testing instruments, truck and van bodies, electronic assemblies, gantry cranes, crushed stone, log home kits; limestone. Agriculture includes dairying; livestock; grain, soybeans, potatoes.

Morganville, village (2000 population 198), CLAY county, N KANSAS, on REPUBLICAN RIVER, and 8 mi/12.9 km NNW of CLAY CENTER; 39°28′N 97°12′W. Grain; livestock. Manufacturing (farm machinery).

Morganza (mor-GAN-zuh), village (2000 population 659), POINTE COUPEE parish, SE central LOUISIANA, on the MISSISSIPPI RIVER, 32 mi/51 km NW of BATON ROUGE; 30°45′N 91°35′W. Agriculture (cotton, rice, vegetables, sugarcane, cattle), crawfish. Morganza Spillway (Floodway) 25 mi/40 km NW of Baton Rouge, diverts Mississippi River floodwaters into ATCHAFALAYA BASIN.

Morganza Spillway, LOUISIANA: see MORGANZA.

Morgårdshammar (MOOR-GORDS-hahm-mahr), village, KOPPARBERG county, central SWEDEN, on small lake, 6 mi/9.7 km E of LUDVIKA; 60°08′N 15°20′E.

Morgarten (MOR-gahr-tuhn), saddle on the ROSSBERG mountain range (c.3,300 ft/1,006 m), N central SWITZERLAND, on the border of SCHWYZ and ZUG cantons; 47°06′N 08°37′E. There, on November 15, 1315, a small Swiss force decisively defeated the Austrians, thus paving the way for Swiss independence. A monument commemorates the battle.

Morgat (mor-gah), small Breton seaside resort of FINISTÈRE department, BRITTANY, W FRANCE, on S shore of CROZON PENINSULA, 11 mi/18 km S of BREST; 48°14′N 04°30′W. Sardine fishing in DOUARNENEZ Bay. Harbor accommodates pleasure crafts.

Morgaushi (muhr-GAH-oo-shi), village (2005 population 2,915), NW CHUVASH REPUBLIC, central European Russia, on highway branch, 22 mi/35 km SW of CHEBOKSARY; 55°58′N 46°46′E. Elevation 485 ft/147 m. Brick factory; dairy plant.

Morgenroth, POLAND: see NOWY BYTOM.

Morges (mohr-JUH), district, S central VAUD canton, SWITZERLAND, on LAKE GENEVA; 46°31′N 06°30′E. Main town is MORGES; population is French-speaking and Protestant.

Morges (mohr-JUH), town (2000 population 14,154), VAUD canton, W SWITZERLAND, on LAKE GENEVA, 7 mi/11.3 km W of LAUSANNE; 46°31′N 06°30′E. Elevation 1,240 ft/378 m. Metalworking, fats, biscuits. Medieval castle with museum, 17th-century town hall, art museum. Small harbor.

Morghab River (moor-GAHB), 530 mi/853 km long, AFGHANISTAN and TURKMENISTAN; rises in the Paropamisus range, NE Afghanistan; flows NW into Turkmenistan at 35°50′N 63°07′E; runs through the KARA KUM desert, past Tashkepristroi to the Merv oasis; then disappears into desert. Forms part of the Afghanistan-Turkmenistan border. With the KUSHK RIVER, its main tributary, it is an important source of water; irrigation dams are at Tashkepristroi and Iolotan. Also spelled Morgab.

Morgins, Pas de (mor-ZHAN, PAH duh), pass between VALAIS and FRANCE (4,491 ft/1,369 m), SW SWITZERLAND, 4 mi/6.4 km W of MONTHEY. Morgins village, a health resort, is nearby on Swiss side (elevation 4,281 ft/1,305 m).

Morgongåva (MOR-gon-GO-vah), village, VÄSTMANLAND county, central SWEDEN, 13 mi/21 km E of SALA; 59°57′N 16°58′E.

Morhange (mo-rahnzh), town (□ 5 sq mi/13 sq km), MOSELLE department, LORRAINE region, NE FRANCE, 25 mi/40 km SE of METZ; 48°55′N 06°38′E. Manufacturing of rubber products. Here French were defeated by Germans in one of earliest battles (August 1914) of World War I.

Mori (MO-RI), town and oasis, ⊙ Mori county, E XINJIANG UYGUR AUTONOMOUS REGION, CHINA, 125 mi/201 km E of URUMQI, and on highway N of the

Bogdo Ola; 37°17′N 79°42′E. Oilseeds, grain, livestock; food processing, building materials. Also appears as Mulei.

Mori (MO-ree), town, Oshima county, Hokkaido prefecture, N JAPAN, 78 mi/125 km S of SAPPORO, on SW shore of UCHIURA Bay; 42°06′N 140°34′E. Scallops. Hot springs nearby.

Mori (mo-REE), town, Syuchi county, SHIZUOKA prefecture, central HONSHU, E central JAPAN, 28 mi/45 km S of SHIZUOKA; 34°49′N 137°55′E. Lettuce, melons, persimmons; tea; pottery, tile. There are three shrines here associated with *bugaku*, traditional court music and dance.

Mori (MO-ree), village, TRENTO province, TRENTINO-ALTO ADIGE, N ITALY, in VAL LAGARINA, 4 mi/6 km SW of ROVERETO; 45°51′N 10°59′E. Agriculture (tobacco, vineyards, asparagus); active electrochemical industry. Rebuilt since World War I. Marble caves.

Mori, GHANA: see MORIE.

Moriah (MO-rei-ah), village, central TOBAGO, TRINIDAD AND TOBAGO, 4.5 mi/7.2 km N of SCARBOROUGH. Cacao.

Moriah, village, ESSEX county, NE NEW YORK, near Lake CHAMPLAIN, 14 mi/23 km NNW of TICONDEROGA; 44°02′N 73°31′W. In former iron-mining area.

Moriah, Mount, NEVADA: see SNAKE RANGE.

Moriah, Mount, NEW HAMPSHIRE: see CARTER-MORIAH RANGE.

Moriarty (mor-ee-AHR-i-tee), town (2006 population 1,807), TORRANCE county, central NEW MEXICO, 37 mi/60 km E of ALBUQUERQUE; 35°00′N 106°02′W. Elevation 6,217 ft/1,895 m. Cattle, corn, pumpkins, grain, alfalfa; manufacturing (metal stampings). MANZANO MOUNTAINS to SW; SANDIA MOUNTAINS to NW; El Cuervo Butte (6,947 ft/2,117 m) to NE.

Morice Lake (□ 40 sq mi/104 sq km), W central BRITISH COLUMBIA, W Canada, in COAST MOUNTAINS, 50 mi/80 km SSW of SMITHERS, near TWEEDSMUIR PARK; 27 mi/43 km long, 1 mi/2 km–5 mi/8 km wide; 53°59′N 127°37′W. Drains N into BULKLEY River.

Morichal Largo River (mo-ree-CHAHL LAHR-go), c.150 mi/241 km long, MONAGAS state, NE VENEZUELA; rises in low tableland NNW of CIUDAD BOLÍVAR; flows NE to the Caño MÁNAMO (arm of ORINOCO RIVER delta) 38 mi/61 km SSW of PEDERNALES; 09°27′N 62°25′W.

Moriches, village, SUFFOLK county, SE NEW YORK, on SE LONG ISLAND, on Moriches Bay, 9 mi/14.5 km E of PATCHOGUE, in resort; 40°48′N 72°49′W. Farm produce and poultry.

Morie (mo-REE), town, CENTRAL REGION, GHANA; 05°08′N 01°12′W. On Gulf of GUINEA, 5 mi/8 km ENE of CAPE COAST; fishing; cassava, corn. First Dutch station in Ghana, founded 1598; fort built by Dutch in 1612 and named Fort Nassau, passed to British in 1872. Sometimes spelled MOUREE or MORI.

Morienval (mo-ree-ahn-vahl), commune (□ 9 sq mi/23.4 sq km), OISE department, PICARDIE region, N central FRANCE, 9 mi/14.5 km SE of COMPIÈGNE; 49°18′N 02°56′E. Has interesting 11th–12th-century church, first built as part of a 7th-century abbey.

Morières-lès-Avignon (mo-ryer-lez-ah-vee-nyon), town (□ 4 sq mi/10.4 sq km), E suburb of AVIGNON, VAUCLUSE department, PROVENCE-ALPES-CÔTE D'AZUR region, S FRANCE, on LYON-MARSEILLE expressway; 43°56′N 04°54′E. Market for early fruits and vegetables; wine grape growing.

Moriguchi (mo-REE-goo-chee), city (2005 population 147,465), OSAKA prefecture, S HONSHU, W central JAPAN, 4.3 mi/7 km N of OSAKA; 34°44′N 135°33′E.

Morija (moh-REE-jah), town, MASERU district, W LESOTHO, 20 mi/32 km S of MASERU; 29°37′S 27°31′E. Moshoeshoe I International Airport 10 mi/16 km NNE. Morija Museum Archives and Paris Evangelical mission are here. The earliest mission station bound by Paris Evangelical Mission (Protestant) in 1833. Moshoeshoe's son, Letsie, educated here. Manufacturing (handicrafts). Corn, sorghum, vegetables; cattle, sheep, goats, hogs.

Moriles (mo-REE-les), town, CÓRDOBA province, S SPAIN, 7 mi/11.3 km WNW of LUCENA. Wine, olive oil, cereals; livestock.

Morin (mo-ran), rivers of the PARIS BASIN, N central FRANCE, tributaries of the MARNE RIVER E of PARIS. They are the GRAND MORIN (70 mi/113 km long) and the PETIT MORIN (55 mi/89 km). Both rivers traverse the fertile BRIE region.

Morinda (MO-rin-dah), town, RUPNAGAR district, E PUNJAB state, N INDIA, 33 mi/53 km NW of AMBALA. Wheat, gram, cotton.

Moringen (MAW-ring-uhn), town, LOWER SAXONY, W GERMANY, 6 mi/9.7 km W of NORTHEIM; 51°43′N 09°53′E. Manufacturing of chemicals; grain.

Morin-Heights, village (□ 21 sq mi/54.6 sq km; 2006 population 3,023), Laurentides region, S QUEBEC, E Canada, 4 mi/6 km from Saint-Sauveur; 45°54′N 74°15′W.

Morinj, Bay of, MONTENEGRO: see KOTOR, GULF OF.

Morinville (MO-ruhn-vil), town (□ 4 sq mi/10.4 sq km; 2001 population 6,540), central ALBERTA, W Canada, 18 mi/29 km NNW of EDMONTON, in Sturgeon County; 53°48′N 113°39′W. Railroad junction; mixed farming, dairying. Established as a village in 1901; became a town in 1911.

Morioka (mo-ree-O-kah), city (2005 population 300,746), ⊙ IWATE prefecture, N HONSHU, NE JAPAN, on the KITAKAMI RIVER; 39°41′N 141°09′E. Terminus of Tohoku bullet train line. Industrial and commercial center. Seat of Iwate University. Ruins of 12th century Morioka Castle and Tokugawa-era tollgate at Kuriyagawa.

Mori River (MO-ree), c. 35 mi/56 km long, MARA region, N TANZANIA; rises c. 45 mi/72 km ENE of MUSOMA, near KENYA border; flows W, enters Mori Bay, Lake VICTORIA, 15 mi/24 km NE of Musoma.

Moris (MO-rees), town, ⊙ Moris municipio, CHIHUAHUA, N MEXICO, 165 mi/266 km ESE of HERMOSILLO; 28°08′N 108°35′W. Elevation 2,507 ft/764 m. Silver, gold, lead mining.

Morisset (MAH-ri-sit), town, E NEW SOUTH WALES, SE AUSTRALIA, 18 mi/29 km SW of NEWCASTLE, on SW shore of Lake MACQUARIE; 33°06′S 151°30′E. State forest tracts W of town. Coal-mining center; tourism.

Morita (MO-ree-tah), village, W Tsugaru county, Aomori prefecture, N HONSHU, N JAPAN, 22 mi/35 km E of AOMORI; 40°46′N 140°21′E.

Moritzberg (MAW-rits-berg), W suburb of HILDESHEIM, LOWER SAXONY, NW GERMANY.

Moritzburg (MAW-rits-burg), village, SAXONY, E central GERMANY, 8 mi/12.9 km NNW of DRESDEN; 51°09′N 13°40′E. Agriculture. Site of 16th-century Moritzburg palace, former royal hunting lodge, and since 1947 baroque museum, containing salvaged remains of Dresden art collections. Home of painter Käthe Kollwitz (1867–1945).

Moriya (MO-ree-yah), town, N Soma county, IBARAKI prefecture, central HONSHU, E central JAPAN, 40 mi/65 km S of MITO; 35°56′N 140°00′E.

Moriyama, city, SHIGA prefecture, S HONSHU, central JAPAN, 9 mi/15 km N of OTSU; 35°03′N 135°59′E. Melons; flowers. Known for 262-ft/80-m-high fountain.

Moriyama (mo-REE-yah-mah), town, North Takaki county, NAGASAKI prefecture, NW KYUSHU, SW JAPAN, 16 mi/25 km E of NAGASAKI; 32°49′N 130°07′E.

Moriyoshi (mo-REE-yo-shee), town, N Akita county, Akita prefecture, N HONSHU, NE JAPAN, 31 mi/50 km N of AKITA city; 40°06′N 140°23′E.

Morki (muhr-KEE), town (2006 population 9,530), SE MARI EL REPUBLIC, E central European Russia, near the ILET′ RIVER, on highway, 45 mi/72 km ESE of YOSHKAR-OLA; 56°26′N 49°00′E. Elevation 495 ft/150 m. Sawmilling, logging, timbering.

Morkovice (MOR-ko-VI-tse), German *morkowitz*, village, JIHOMORAVSKY province, central MORAVIA,

CZECH REPUBLIC, 9 mi/14.5 km SW of Kroměříž; 49°15′N 17°13′E. Railroad terminus. Manufacturing (basketwork, brushes, brooms). Castle.

Morkowitz, CZECH REPUBLIC: see MORKOVICE.

Morkvashi, RUSSIA: see ZHIGULËVSK.

Morlaàs (mor-lahs), village (□ 5 sq mi/13 sq km), PYRÉNÉES-ATLANTIQUES department, AQUITAINE region, SW FRANCE, 6 mi/9.7 km NE of PAU; 43°21′N 00°16′W. Livestock and poultry market. Has church (founded 1089) with highly detailed Romanesque portal. Village was the capital of BÉARN district in early Middle Ages.

Morlaix (mor-lai), town (□ 9 sq mi/23.4 sq km), FINISTÈRE department, BRITTANY, W FRANCE, seaport on inlet of ENGLISH CHANNEL, 33 mi/53 km ENE of BREST; 48°35′N 04°00′W. Produces cigars and chewing tobacco; manufacturing (carved woodwork, electrical appliances), flour milling, and dairy products. Market for leather, honey, and horses. Center of tourist trade and recreational boating. Has 15th- and 16th-century wooden houses along pedestrian mall. Railroad viaduct spans Morlaix River valley just above port.

Morland, village (2000 population 164), GRAHAM county, NW KANSAS, on South Fork of SOLOMON RIVER, and 13 mi/21 km W of HILL CITY; 39°21′N 100°04′W. Trade and shipping center for grain region.

Morlanwelz (MAWR-law-VELZ), commune (2006 population 18,631), Thuin district, HAINAUT province, S central BELGIUM, 9 mi/14.5 km WNW of CHARLEROI. Has ruins of thirteenth-century abbey.

Morla Vicuña, CHILE: see QUIDICO.

Mörlenbach (MUHR-luhn-bahkh), village, S HESSE, central GERMANY, in the ODENWALD, 6 mi/9.7 km SE of HEPPENHEIM; 49°36′N 08°50′E. Fruit; forestry; manufacturing (electronic equipment).

Morley (MOR-lee), township (□ 71 sq mi/184.6 sq km; 2001 population 447), NW ONTARIO, E central Canada; 48°42′N 94°11′W.

Morley (MAW-lee), town (2001 population 54,051), WEST YORKSHIRE, N ENGLAND, 4 mi/6.4 km SSW of LEEDS; 53°44′N 01°35′W. Woolen textiles; light and heavy engineering. Electrical and auto parts assembly. H. H. Asquith born here.

Morley, town (2000 population 88), JONES county, E IOWA, 12 mi/19 km S of ANAMOSA; 42°00′N 91°15′W. Livestock, grain.

Morley, town (2000 population 792), SCOTT county, SE MISSOURI, in MISSISSIPPI alluvial plain, 6 mi/9.7 km SSW of BENTON; 37°02′N 89°36′W. Manufacturing: motor vehicle parts.

Morley, village (2000 population 495), MECOSTA county, central MICHIGAN, 15 mi/24 km S of BIG RAPIDS, and on LITTLE MUSKEGON RIVER; 43°29′N 85°27′W. Machining. Newaygo State Park, on HARDY DAM POND and Manistee National Forest to W.

Mörlunda (MUHR-LUND-ah), village, KALMAR county, SE SWEDEN, near EMÅN RIVER, 20 mi/32 km W of OSKARSHAMN; 57°19′N 15°52′E.

Mormaço (MOR-mah-ko), city, N RIO GRANDE DO SUL state, BRAZIL, 33 mi/53 km SW of PASSO FUNDO; 28°38′S 52°31′W. Wheat, potatoes, corn; livestock.

Mormant (mor-mahn), commune (□ 6 sq mi/15.6 sq km), SEINE-ET-MARNE department, ÎLE-DE-FRANCE region, N central FRANCE, 12 mi/19 km NE of MELUN; 48°36′N 02°54′E. Small-scale manufacturing.

Mormon Flat Dam, ARIZONA: see CANYON LAKE.

Mormon Lake (□ 12 sq mi/31 sq km), COCONINO county, central ARIZONA, 20 mi/32 km SE of FLAGSTAFF; 4 mi/6.4 km long, 3 mi/4.8 km wide. Summer resorts. Mount Mormon (8,440 ft/2,573 m) is near W shore.

Mormon Pioneer Affiliated Area, historic trail running from ILLINOIS to UTAH. Follows the route of Brigham Young; authorized 1978. The general route is from NAUVOO, Illinois, to SALT LAKE CITY, Utah. This 1,300 mile long trail passes through Illinois, IOWA, NEBRASKA, WYOMING and Utah. From 1846–1869,

about 70,0000 Mormons traveled along an integral part of the road west, the Mormon Pioneer Trail. It started in Nauvoo, Illinois, traveled across Iowa, connected with the Great Platte River Road at the MISSOURI River, and ended near the GREAT SALT LAKE. In order to maintain their religious and cultural identity, it was necessary for the Mormons to find an isolated place where they could permanently settle and practice their religion in peace. Departing in February 1846, thousands of Mormons crossed into Iowa seeking refuge from religious persecution. They spent the next winter in the COUNCIL BLUFFS, Iowa and OMAHA, Nebraska area. Early in 1847, Brigham Young led an advanced party west, generally paralleling the OREGON TRAIL, to FORT BRIDGER, Wyoming, where they turned southwest and eventually came to the Great Salt Lake. Knowing that others would follow, pioneers improved the trail and built support facilities. Some of the ferries they built helped finance the movement. They planted crops and recorded pertinent information such as the topography, latitude, longitude, distances, flora, and fauna. Six thousand people died on the trail. The greatest threats to life were illness and accidents. They suffered from poor nutrition and exposure to the elements. Handcarts were used from 1856–1860. Nearly 3,000 migrants used this method of transportation. They could make 25–30 miles a day (a wagon would travel 10–15 miles a day). There were ten handcart companies.

Mormorque (mor-MOR-kai), canton, NOR CHICHAS province, POTOSÍ department, W central BOLIVIA; 20°47′S 65°26′W. Elevation 8,596 ft/2,620 m. Antimony mining at Mina Churquini. Agriculture (potatoes, yucca, bananas, rye); cattle.

Mormugao, India: see MARMAGAO.

Mornant (mor-nahn), commune (□ 6 sq mi/15.6 sq km), RHÔNE department, RHÔNE-ALPES region, E central FRANCE, in the Monts du LYONNAIS, 12 mi/19 km SW of LYON; 45°37′N 04°39′E. Woodworking. Remains of a Roman aqueduct.

Morne-à-l'Eau (mawr-nah-LO), town, W GRANDE-TERRE island, GUADELOUPE, French WEST INDIES, on the GRAND CUL DE SAC, 7 mi/11.3 km N of POINTE-À-PITRE; 16°20′N 61°31′W. Rum distilling, sugar milling.

Morne Diablotin, DOMINICA: see DIABLOTIN, MORNE.

Morne Fortune (morn for-too-NAY), hill, SAINT LUCIA, 1 mi/1.6 km E of CASTRIES; 14°00′N 61°00′W. Spectacular views overlooking Castries harbor. Site of governor general's mansion. Ruins of fortifications.

Morne Garu, central mountain range, Saint Vincent, SAINT VINCENT AND THE GRENADINES, West INDIES; 13°17′N 61°10′W. SOUFRIÈRE (4,049 ft/1,234 m) is an active volcano and major mountain on N side of this range.

Morne-Rouge (mawrn–ROOZH), town, N MARTINIQUE, French WEST INDIES, at S foot of Mont PELÉE, 12 mi/19 km NNW of FORT-DE-FRANCE; 15°53′N 61°17′W. In sugar-growing region; rum distilling. Destroyed (1902) by eruption of Pelée.

Morne Seychellois, SEYCHELLES; see SEYCHELLOIS, MORNE.

Morne Seychellois National Park, NW MAHÉ ISLAND, SEYCHELLES, 2 mi/3.2 km SW of VICTORIA; 04°38′N 55°26′E. National park covering a largely mountainous area of NW Mahé Island. Morne Seychellois and several lesser peaks are here.

Morningdale, MASSACHUSETTS: see BOYLSTON.

Morningside, town (2000 population 1,295), PRINCE GEORGES county, central MARYLAND, suburb SE of WASHINGTON, D.C.; 38°50′N 76°53′W. Formed in 1949, many residents work in nearby Andrews Air Force Base or government agencies such as the U.S. Census in SUITLAND.

Morningside, N residential suburb of DURBAN, E KWAZULU-NATAL, SOUTH AFRICA; 29°49′S 31°01′E. Now part of Greater Durban metropolitan area. Overlooking mouth of MGENI RIVER.

Area is shown by the symbol □, and capital city or county seat by ⊙.

Morningside (MOR-neeng-seid), hamlet, S central ALBERTA, W Canada, 23 mi/37 km from RED DEER, in Lacombe Couty; 52°35′N 113°38′W.

Morningside, Scotland: see EDINBURGH.

Morningside Heights, section of the W side of MANHATTAN borough of NEW YORK CITY, SE NEW YORK, lying between Riverside Park along the HUDSON RIVER (W) and Morningside Park (E), N of 110th St.; 40°49′N 73°57′W. Manhattanville section adjoins on N. Site of Columbia University, Teachers College (Columbia University), Cathedral of St. John the Divine, Riverside Church, Union Theological Seminary, Jewish Theological Seminary of America. Numerous famous educators, writers, artists, and politicians have lived here.

Morning Sun, town (2000 population 872), LOUISA county, SE IOWA, 7 mi/11.3 km SSW of WAPELLO, and 25 mi/40 km SSW of MUSCATINE; 41°05′N 91°15′W. Rendering works; limestone quarries nearby.

Mornington (MOR-neeng-tuhn), resort town, suburb, S VICTORIA, AUSTRALIA, on SE shore of PORT PHILLIP BAY, and 30 mi/48 km S of MELBOURNE; 38°13′S 145°02′E.

Mornington (MOR-neeng-tuhn), former township (□ 79 sq mi/205.4 sq km; 2001 population 3,406), SW ONTARIO, E central Canada; 43°36′N 80°52′W. Amalgamated into Perth East in 1998.

Mornington Island (MOR-neeng-tuhn), northernmost and largest of WELLESLEY ISLANDS, AUSTRALIA, in Gulf of CARPENTARIA, 15 mi/24 km off NW coast of QUEENSLAND; 40 mi/64 km long, 15 mi/24 km wide; 16°33′S 139°24′E. Rises to 300 ft/91 m. Rocky, wooded.

Mornington Island (MOR-neeng-tuhn), off coast of S CHILE, on TRINIDAD GULF just W of WELLINGTON ISLAND; 49°45′S 75°25′W; 28 mi/45 km long, c.8 mi/13 km wide.

Mornos River (MOR-nos), 38 mi/61 km long, in PHOCIS prefecture, W CENTRAL GREECE department, GREECE; rises in OITI massif, near border of FTHIOTIDA prefecture; flows SW to Gulf of CORINTH just SE of NAVPAKTOS. Forms border of WESTERN GREECE and CENTRAL GREECE departments in lower course. Hydroelectric plants.

Moro (MO-ro), village, NAWABSHAH district, central SIND province, SE PAKISTAN, 38 mi/61 km NW of NAWABSHAH; 26°40′N 68°00′E. Wheat, millet, rice.

Moro (MOR-o), village (2000 population 241), LEE county, E ARKANSAS, 19 mi/31 km SW of FORREST CITY; 34°47′N 90°59′W.

Moro (MOR-o), village (2006 population 291), ⊙ SHERMAN county, N OREGON, 23 mi/37 km ESE of THE DALLES; 45°29′N 120°43′W. Elevation 1,807 ft/551 m. Wheat, barley, oats; poultry, cattle. DESCHUTES RIVER State Recreation Area to W.

Moro (MOR-o), plantation, AROOSTOOK county, E MAINE, on Rockabema Lake, and 25 mi/40 km W of HOULTON; 46°10′N 68°22′W. In hunting, fishing area.

Moro Bayou (MOR-o), c.65 mi/105 km long, S ARKANSAS; rises in DALLAS county NW of FORDYCE; flows S past Fordyce to OUACHITA RIVER at Moro Bay State Park, 19 mi/31 km NE of EL DORADO.

Morobe (mo-RO-bai), province (2000 population 539,404), E central PAPUA NEW GUINEA, NE NEW GUINEA island; ⊙ LAE, Papua New Guinea's second-largest city. Bounded on S by NORTHERN, CENTRAL and GULF provinces, W by EASTERN HIGHLANDS province, N by MADANG province and BISMARCK SEA, E by SOLOMON SEA. Includes UMBOI, Tolokiwa, and Sukar islands, separated from HUON PENINSULA by VITIAZ STRAIT and from NEW BRITAIN island to E by DAMPIER STRAIT. Drained by MARKHAM and Watut rivers; Markham Valley is one of Papua New Guinea's primary agricultural areas. OWEN STANLEY RANGE in S. Coconuts, rice, sugarcane, coffee; copra; fish, prawns, lobster; cattle; timber; tourism. Papua New Guinea University of Technology ("Unitech") at Lae.

Morobe (mo-RO-bai), town, MOROBE province, E NEW GUINEA island, E central PAPUA NEW GUINEA, 80 mi/129 km SE of LAE; 07°45′S 147°37′E. Located on small peninsula in SOLOMON SEA. Airstrip, boat access. Coconuts, palm oil; mackerel, prawns, tuna; cattle.

Morocco or *El Maghreb el Aqsa* [Arabic=the extreme west of the Arab world], kingdom (□ 171,834 sq mi/ 445,050 sq km, with WESTERN SAHARA, 274,543 sq mi/ 711,066 sq km; 2004 estimated population 32,209,101; 2007 estimated population 33,757,175), NW AFRICA, on MEDITERRANEAN SEA and ATLANTIC OCEAN; ⊙ RABAT; 34°02′N 06°50′W.

Geography

Morocco is bordered, now that it has annexed Western Sahara, by MAURITANIA on the S and by ALGERIA on the S and E. IFNI, formerly a Spanish-held enclave on the Atlantic coast, was ceded to Morocco in 1969. Two cities in the N, CEUTA and MELILLA, and several small islands off the Mediterranean coast remain part of metropolitan SPAIN. Morocco claims and administers Western Sahara although its sovereignty remains unresolved. The population of Morocco is concentrated in the coastal regions, and mountains, where rainfall is most plentiful. Physiographically, Morocco falls into four main natural regions: the RIF, the ATLAS, the Atlantic coastal plains and plateaus, and the SAHARA. The Rif area is geologically a continuation of Spain's Cordillera PENIBETICA; it is separated from the Atlas ranges to the S by the Taza corridor (the chief route from Algeria to W Morocco). Following a decentralization/regionalization law passed in March 1997, Morocco is divided into fifteen primary administrative divisions called regions: Grand Casablanca, Chaouia-Ouardigha, Doukkala-Abda, Fes-Boulemane, Gharb-Chrarda-Beni Hssen, Guelmim-Es Smara, Laayoune-Boujdour-Sakia El Hamra, Marrakech-Tensift-Al Haouz, Meknès-Tafilalet, Oriental, Rabat-Salé-Zemmour-Zaër, Souss-Massa-Draâ, Tadla-Azilal, Tanger-Tétouan, and Taza-Al Hoceima-Taounate. Morocco also claims a sixteenth administrative region in the unresolved area of Western Sahara: Oued Ed-Dahab-Lagouira. Morocco is further divided into sixty-one secondary administrative divisions called prefectures and provinces. Prior to the 1997 reorganization Morocco was divided into thirty-nine primary administrative divisions called provinces.

Population

Morocco's population is predominantly Berber ethnically, and today is completely Islamized and dominantly Arabized. A formerly large Jewish population has been reduced to a remnant by migration to Israel, Europe, and Canada. The population has increased rapidly since independence, but the rate is beginning to slow. Islam is the state religion and Arabic is the official language, but French, Berber, and Spanish are also spoken. There are universities at Rabat, FES, MARRAKECH, MEKNES, KENITRA, EL JADIDA, AGADIR, OUJDA, Casablanca, TETOUAN, and IFRANE.

Economy

In parts of the Rif Mountains in the NE some 40 in/102 cm of rain fall each year, and wheat and other cereals can be raised without irrigation. On the Atlantic coast, where there are extensive plains, wheat, barley, olives, citrus fruits, grapes, almonds, and vegetables are grown. Most of Morocco's arable land, a majority of its population, and most of its major cities are found in this zone. Wine and vegetables are grown near the coastal cities, especially in the Dukkala S of El Jadida and in the coastal districts N and S of Casablanca.Cattle raising occurs in these areas, most often in small dairy operations producing for local urban markets. This coastal area is dissected by many short streams that originate in the well-watered Atlas Mountains. Of these the OUM ER RBIA is the longest, where citrus orchards in the TADLA area are irrigated,

but the Oued SEBOU crossing the GHARB plain waters the richer agricultural land. Other coastal streams are the Oued LOUKOS, Oued Beht (a Sebou tributary), BOU REGREG, Oued TENSIFT, and Oued Sous. These important permanently flowing rivers are controlled by a system of dams, allowing the irrigation of farmlands as well as providing flood protection and hydroelectric power. The Oued MOULOUYA drains semi-arid E Morocco to the Mediterranean. Morocco N of the Atlas has a Mediterranean subtropical climate modified by Atlantic influences. The coastal cities have an equitable but humid climate, while interior cities are extremely hot in summer. Cork oaks, dwarf palms, and jujube trees are native to the Atlantic coastal region; thuya, evergreen oaks, and juniper trees predominate in areas of Mediterrean climate. Forests yield cork (only the Iberian Peninsula and Algeria have larger stands), cabinetwood, and building materials. Part of the maritime population fishes for its livelihood. TANTAN, AGADIR, SAFI, ESSAOUIRA, AL HOCEIMA, El Jadida, and LARACHE are among the important fishing harbors. Casablanca is by far the largest port for the extensive overseas trade and, together with nearby MOHAMMEDIA, is the country's most important industrial center; modern factories producing building materials, superphosphates, textiles, and footwear, as well as an oil refinery and most of the country's food-processing plants are located here. The main exports are phosphates and phosphate products, clothing, shellfish, citrus fruits and vegetables. The main imports are petroleum, chemicals, machines, and plastics. France is the leading trade partner. Central Morocco consists largely of the Atlas Mountains. Forming three SW-NE trending ranges (Middle, High, and Anti-Atlas), these mountains extend from the Atlantic S of Agadir to the Algerian border. The High Atlas rises to 13,655 ft/4165 m at Jbel TOUBKAL, the highest peak in N Africa. Precipitation occurs in winter and amounts can vary significantly from year to year. Totals range from c.35 in/89 cm in the W Rif and Middle Atlas (where some sites exceed 59 in/150 cm) to less than the 10 in/25 cm in the dry lee of the Atlas (E and SE) and in areas near the coast S of Agadir. Argan trees, unique in Morocco, grow in the SW plains (especially in the SOUS). Heavy snowfall in the higher elevations of the Middle and High Atlas is common; the area supports large stands of oak and cedar, while esparto grass is widespread in the steppe region of E Morocco, N of the Sahara. Sheep and goats are the primary animals raised in the mountainous uplands and drier interior. S Morocco lies in the Sahara Desert. All streams that rise on the SE slopes of the Atlas (except the Oued DRA) disappear in the Sahara after watering a string of oases of which the TAFILALT is the largest. These date palm oases, as well as coastal fishing towns, are the main foci of settlement. Incorporation of Western Sahara into Morocco has been followed by a great deal of investment in infrastructure in the major towns of the S and a large influx of people from the N. Morocco is a country with considerable mineral deposits; phosphates are the most important (Morocco has three-quarters of the world's reserves), but iron, silver, zinc, copper, lead, manganese, cobalt, barytine, gold, and the only sizable coal deposits in all N. Afr. are also found. However, only phosphate deposits are sufficiently substantial to have an impact on world trade; phosphate primarily exported through the port of JORF-LASFAR S of El Jadida. Cash remittances from Moroccans working in Europe, primarily France, are the chief source of foreign exchange; tourism and phosphates (and phosphate products) are also important. Revenue from sales, primarily to Europe, of illegally cultivated cannabis is of growing importance to the economy of N Morocco. The coastal areas and the mineral-producing interior are linked by a good-quality road and railroad network; a limited access highway connecting Rabat and Casablanca was completed in the 1980s, and has been extended N toward TANGER. Port facilities are being further developed, which is essential to economic development since there are no natural harbors large enough to handle modern vessels.

History: to 1554

Berbers inhabited Morocco at the end of the 2nd millennium B.C. In Roman times, Morocco was roughly coextensive with the province of Mauretania Tingitania. The Arabs first swept into Morocco c.685, bringing with them Islam. Christianity was all but extirpated, but the Jewish colonies by and large retained their religion. Morocco became an independent state in 788 under the royal line founded by Idris I. After 900 the country again broke into small tribal states. The Almoravids overran (c.1062) Morocco and established a kingdom stretching from Spain to Senegal. The Almohads, who succeeded (c.1174) the Almoravids, ruled both Morocco and Spain, but the Merinid dynasty (1259–1550), after some triumphs, was limited to Morocco. In the decades following the conquest of Granada in 1492, Spain and Portugal made a number of attacks on the Moroccan coast. Beginning with the capture of Ceuta in 1415, Portugal took all the chief ports except Melilla and Larache, both of which fell to Spain.

History: 1554 to 1905

The Christian threat stimulated the growth of resistance under religious leaders, one of whom established (1554) the Saadian, or first Sherifian, dynasty. At the battle of ALCAZARQUIVIR (1578) the Saadian king decisively defeated Portugal. The present ruling dynasty, the Alawite, or second Sherifian, dynasty, came to power in 1660 and recaptured many European-held strongholds. In the 19th century the strategic importance and economic potential of Morocco excited the interest of the European powers. Spain invaded in 1860. In 1880 the major European nations and the U.S. decided at the Madrid Conference to preserve the territorial integrity of Morocco and to maintain equal trade opportunities for all. Political and commercial rivalries soon disrupted this cordial arrangement and brought on several international crises. France sought to gain Spanish and British support against the opposition of Germany. Thus, in 1904, France concluded a secret treaty with Spain to partition Morocco and secretly agreed with Great Britain (the Entente Cordiale) not to oppose British aims in Egypt in exchange for a free hand in Morocco.

History: 1905 to 1911

In 1905, after France had asked the sultan of Morocco for a protectorate, Germany moved quickly: Emperor William II visited Tanger and declared support for Morocco's integrity. At German insistence the ALGECIRAS Conference (January–March 1906) was called to consider the Moroccan question. The principles of the Madrid Conference were readopted and German investments were assured protection, but French and Spanish interests were given marked recognition by the decision to allow France to patrol the border with Algeria and to allow France and Spain to police Morocco. Under the claim of effecting pacification, the French steadily annexed territory. In 1908 friction arose at Casablanca, under French occupation, when the German consul gave refuge to deserters from the French Foreign Legion. This dispute was settled by the HAGUE Tribunal. Shortly afterward in a coup d'état Abd al-Aziz IV was unseated and his brother, Abd al-Hafid, was installed on the throne. He had difficulty maintaining order and received help from France and Spain, especially in a revolt that broke out in 1911.

History: 1911 to World War II

In this situation the appearance of the German warship *Panther* at Agadir on July 1, 1911, was interpreted by the French as a threat of war and hastened a final adjustment of imperial rivalries. On November 4, 1911, Germany agreed to a French protectorate in Morocco in exchange for the cession of French ter-

ritory in equatorial Africa. Finally, at Fes (March 30, 1912), the sultan agreed to a French protectorate, and on November 27 a Franco-Spanish agreement divided Morocco into four administrative zones—French Morocco, 90% of the country, a protectorate with Rabat as the capital; a Spanish protectorate, which included Spanish Morocco, with its capital at Tetouan; a Southern Protectorate of Morocco, administered as part of the Spanish Sahara; and the international zone of Tanger. The French protectorate was placed under the rule of General Lyautey, who remained in office until 1925. A strong threat to European rule was posed (1921–1926) by the revolt (the Rif War) of Adb-el-Krim. In 1934 a group of young Moroccans presented a plan for reform, marking the beginning of the nationalist movement. In 1937 the French crushed a nationalist revolt. Francisco Franco's successful revolt against the republican government of Spain began in Spanish Morocco in 1936.

History: World War II to 1961

During World War II, French Morocco remained officially loyal to the Vichy government after the fall of France in 1940. On November 8, 1942, Allied forces landed at all the major cities of Morocco and Algeria; on November 11, all resistance ended. In January 1943, Allied leaders met at Casablanca. An independence party, the Istiglal, was formed during the war. After the war the nationalist movement gained strength and received the active support of the sultan, Sidi Mohammad, who demanded a unitary state and the departure of the French and Spanish. Vast numbers of Jews migrated to the newly formed state of Israel in the early 1950s, although a small number remain. Faced with growing nationalist agitation, the French outlawed (1952) the Istiglal and, in August 1953, deposed and exiled Sidi Mohammad. These measures proved ineffective, and under the pressure of rebellion in Algeria and disorders in Morocco, the French were compelled (1955) to restore Sidi Mohammad. In March 1956, France relinquished its rights in Morocco; in April the Spanish surrendered their protectorate; in October Tanger was given to Morocco by international agreement. Spain ceded the Southern Protectorate (Ifni) in 1958. The sultan became (1957) King Mohammad V (Sidi Mohammad) and soon embarked on a foreign policy of "positive neutrality," which included support for the Muslim rebels in Algeria.

History: 1961 to 1976

After the king's death (February 1961), his son Hassan II ascended the throne. In December 1962, a constitution was promulgated that provided Morocco with a bicameral parliament. Border hostilities with Algeria in 1963 cost both sides many lives; Emperor Haile Selassie of Ethiopia aided the cease-fire negotiations. Final agreement on the border was reached in 1970. In June 1965, following a political crisis that threatened to undermine the monarchy, King Hassan declared a state of emergency and took over both executive and legislative powers. The country returned to a modified form of parliamentary democracy in 1970, with a revised constitution that strengthened the king's authority. Opposition groups, later called the National Front, rejected the constitution and boycotted elections for the national legislature held in August. An abortive coup by military leaders took place on July 10, 1971. Hassan announced another constitution in February 1972. The new constitution, which lessened the king's powers, was approved by referendum despite a boycott by the National Front. In August, an assassination attempt took place when the airplane carrying King Hassan was strafed on its way back from France. King Hassan continued to rule and maintained relative order. He achieved national consensus when, in 1974, Morocco pressed its claim to sovereignty over Spanish Morocco, and in November

1975, Hassan lead the "Green March" of 300,000 settlers to the disputed region.

History: 1976 to Present

In 1976, Spain released control of the Spanish Sahara, which became known as Western Sahara, and negotiations between Morocco and MAURITANIA that year resulted in the division of Western Sahara between them. However, the Polisario Front, a group of Western Saharan guerrillas with Algerian and Libyan backing, contested Morocco's claim. In 1979, Morocco took over Mauritania's portion of the Western Sahara. The Polisario's armed struggle, or "war of the sand," for self-determination continued throughout the 1980s and Morocco lost face with some countries for its allegedly imperialistic policies. The war slowed down when Morocco completed a 435-mi/700-km-long defensive wall around the disputed territory that halted Polisario attacks and access to the sea. Morocco has invested heavily to develop the territory, primarily in the city of LAAYOUNE. Normalization of relations with Algeria, in 1988, also lead to Algeria's cutting off support for the rebels. In 1991, the UN brokered a cease-fire and agreed to a referendum by the Saharans to decide between independence or union with Morocco. Since then, however, there has been no agreement on a formula to decide who is eligible to vote, and the referendum has not been held. Morocco is viewed by the U.S. and European countries as a moderating force in the region and an important power in the quest for Arab-Israeli peace. Morocco is a member of the UN, the League of Arab States, the Organization of African Unity, and the Union of the Arab Maghreb, and has had important trade and cooperation agreements with the EU since 1969. King Hassan died in 1999 and was succeeded by his son Mohamed VI.

Government

A constitutional monarchy, Morocco is governed under the constitution of 1972 as amended. The king, who is the head of state, holds effective power and appoints the prime minister, who is the head of government. The bicameral Parliament consists of the 270-seat Chamber of Counselors, whose members are elected by indirect vote for nine-year terms, and the 325-seat Chamber of Representatives, whose members are elected by popular vote for five-year terms. The current head of state is Mohamed VI (since July 1999). The current head of government is Prime Minister Abbas El Fassi (since September 2007).

Morocco, town (2000 population 1,127), NEWTON county, NW INDIANA, near ILLINOIS state line, 12 mi/19 km N of KENTLAND; 40°57′N 87°27′W. In agricultural area; corn, soybeans, cattle. Manufacturing (insert moldings). Settled 1833, laid out 1851, incorporated 1890.

Morocelí (mo-ro-sai-LEE), town, EL PARAÍSO department, SE HONDURAS, on paved highway, 19 mi/31 km NW of DANLÍ; 14°07′N 86°52′W. Small farming, corn, beans; livestock.

Morocelí Valley (mo-ro-sai-LEE), EL PARAÍSO and FRANCISCO MORAZÁN departments, S HONDURAS, between MOROCELÍ and VILLA DE SAN FRANCISCO; 14°07′N 86°52′W. Drained by the CHOLUTECA RIVER. Grain; livestock.

Morochata (mo-ro-CHAH-tah), town and canton, AYOPAYA province, COCHABAMBA department, W central BOLIVIA, 23 mi/37 km WNW of COCHABAMBA, in Cordillera de COCHABAMBA, at foot of TUNARI (mountain peak), on road; 17°17′S 66°29′W. Elevation 10,151 ft/3,094 m. Barley, potatoes.

Morochne (mo-ROCH-ne) (Russian *Morochno*) (Polish *Moroczno*), village (2004 population 2,400), NW RIVNE oblast, UKRAINE, in Prypyet (Ukrainian *Prypyat'*), 44 mi/71 km NW of SARNY; 51°50′N 25°54′E. Elevation 423 ft/128 m. Potatoes, flax; lumbering. Natural gas pipeline nearby.

Morochno, UKRAINE: see MOROCHNE.

Moroch' River, 93 mi/150 km long, MINSK oblast, BELARUS, tributary of Sluch' River; rises from KOPYL Ridge. Basin has an area of 888 sq mi/2,300 sq km. Used to float timber.

Morococala (mo-ro-ko-KAH-lah), town and canton, PANTALEÓN DALENCE province, ORURO department, W BOLIVIA, at S foot of Morococala peak (17,060 ft/5,200 m; highest point of Cordillera de AZANAQUES), 25 mi/40 km ESE of ORURO; 18°10′S 66°44′W. Elevation 14,760 ft/4,499 m. Tin-mining center (Minas HUANUNI, Santa Fe, Japo, and Morococala).

Morococha (mo-ro-KO-chah), town, YAULI province, JUNÍN region, central PERU, in CORDILLERA OCCIDENTAL, on LIMA-LA OROYA railroad, and 16 mi/26 km WSW of La Oroya; 11°37′S 76°09′W. Elevation 15,000 ft/4,572 m. Copper-, zinc- and silver-mining center, shipping ore to La Oroya smelter on railroad route. Potatoes, cereals; livestock.

Moroczno, UKRAINE: see MOROCHNE.

Morodomi (mo-RO-do-mee), town, Saga county, SAGA prefecture, N KYUSHU, SW JAPAN, 3.1 mi/5 km S of SAGA; 33°13′N 130°21′E.

Moroeni (mo-ro-EN), village, DÎMBOVIŢA county, S central ROMANIA, 20 mi/32 km N of TÎRGOVIŞTE, and on IALOMIŢA RIVER; 45°13′N 25°26′E. Hydroelectric station.

Morogoro (mo-ro-GO-ro), region (2006 population 1,929,000), E central and S central TANZANIA, ⊙ MOROGORO. GREAT RUAHA RIVER crosses center of region, Makata Plain and Nguru Mountains in N; ULUGURU MOUNTAINS in NE; part of MBARIKA MOUNTAINS in W; part of Rubeho Mountains in SW. MIKUMI NATIONAL PARK in N; part of SELOUS GAME RESERVE in center and S. Sugarcane, grain, sisal, tobacco, pyrethrum; sheep, goats; timber. Mica mining in NE and NW.

Morogoro (mo-ro-GO-ro), town (2002 population 206,868), ⊙ MOROGORO region, E central TANZANIA, 17 mi/27 km W of DAR ES SALAAM, near source of NGERENGERE River, on railroad; ULUGURU MOUNTAINS to S; 06°49′S 37°38′E. Airstrip. Cotton, sugarcane, sisal, wheat, corn; cattle, sheep, goats; timber. Mica mining to E. Manufacturing (cotton finishing, women's outerwear). Seat of Sokoine University of Agriculture.

Moro Gulf (MO-ro), large inlet of CELEBES SEA, PHILIPPINES, in S coast of W MINDANAO; SIBUGUEY BAY opens W along ZAMBOANGA peninsula. SULU ARCHIPELAGO stretches to SW.

Moroleón (mo-ro-le-ON), city and township, ⊙ Moroleón municipio, GUANAJUATO, central MEXICO, on central plateau, 38 mi/64 km N of Mexico Highway 430; 20°07′N 101°11′W. Elevation 5,814 ft/1,772 m. Agricultural center (grain, sugarcane, fruit, livestock); manufacturing (shoes, scarves, shawls).

Morolica (mo-ro-LEE-kah), town, CHOLUTECA department, S HONDURAS, on CHOLUTECA RIVER, and 26 mi/42 km NE of CHOLUTECA; 13°34′N 86°54′W. Ceramics (bricks, tiles, pottery); dairying.

Morombe (moo-room-BAI), town, TOLIARY province, W MADAGASCAR, on MOZAMBIQUE CHANNEL, 110 mi/177 km NNW of TOLIARY; 21°44′S 43°22′E. Cabotage port. Cotton, beans, rice, corn, cassava; cattle.

Moromoro (mo-ro-MO-ro), town and canton, VALLEGRANDE province, SANTA CRUZ department, central BOLIVIA, near MIZQUE RIVER, 15 mi/24 km NW of VALLEGRANDE; 18°23′S 64°18′W. Corn, potatoes, barley.

Morón (mo-RON), city, BUENOS AIRES province, E ARGENTINA, district administrative center in the Greater BUENOS AIRES area; 34°40′S 58°40′W. Settled in the early 16th century, Morón became an outpost on the route between Buenos Aires and CHILE and PERU. It has meat-packing and food-processing industries.

Morón (mor-ON), city (2002 population 53,551), CIEGO DE ÁVILA province, E CUBA, near N coast (LECHE LAGOON), 65 mi/105 km NW of CAMAGÜEY; 22°09′N 78°39′W. Railroad junction, trading and processing

center in rich agricultural region (sugarcane, tobacco, cacao, coffee, fruit; cattle). Lumbering; manufacturing of meat products. Airfield. Has several sugar mills in outskirts, including Patria o Muerte (SE), MÁXIMO GÓMEZ (SW), and Enrique Varona (NW). Town core possesses housing stock built by American sugar interests in early decades of 20th century.

Moron (mo-RONG), town, GRANDE-ANSE department, SW HAITI, on W JACMEL PENINSULA, 12 mi/19 km SW of JÉRÉMIE; 18°34′N 74°15′W. Cacao, coffee.

Mörön (MUH-ruhn), town (2000 population 28,903), ⊙ HÖVSGÖL province, NW MONGOLIA, at crossing of DELGER MÖRÖN (river) by Trans-Mongolian Highway, 330 mi/530 km WNW of ULAANBAATAR; 49°38′N 100°10′E. Site of former monastery. Food and wood industries; coal mining nearby. Also spelled Muren.

Morón (mo-RON), town, ⊙ Juan José Mora municipio, CARABOBO state, N VENEZUELA, near CARIBBEAN coast, 14 mi/23 km W of PUERTO CABELLO; 10°29′N 68°12′W. Coconuts, divi-divi, sugarcane, bananas, plantains.

Moron (moh-ROHN), ridge, in the JURA, BERN canton, NW SWITZERLAND, 8 mi/12.9 km N of BIEL. Reaches 4,383 ft/1,336 m.

Mörön, MONGOLIA: see DELGER MÖRÖN.

Morona River (mo-RO-nah), c.260 mi/418 km long, ECUADOR and PERU; rises in E Andean foothills of Ecuador NE of MACAS; flows S to MARAÑÓN RIVER 25 mi/40 km WNW of BARRANCA. Navigable for almost its whole course in Peru.

Morona-Santiago (mo-RO-nah–sahn-tee-AH-go), province (□ c.9,900 sq mi/25,641 sq km; 2001 population 115,412), SE ECUADOR in ORIENTE region, SW of PASTAZA province and E of CHIMBORAZO, CAÑAR, and AZUAY provinces; ⊙ MACAS. Its W border is on the E slopes of the ANDES (highest point is the very active SANGAY volcano, 17,460 ft/5,322 m). The UPANO RIVER basin is bordered on the E by the Cordillera de CUTUCÚ. Terrain descends E to the Amazonian plain. Climate is mostly warm and humid leading to forest vegetation. All its rivers flow toward the AMAZON. Economic activity is mainly agricultural (corn, yucca, bananas, some rice, vanilla, and tropical fruits; some cattle); tropical woods; placer gold is extracted in its many rivers. Many aboriginal groups. Frontier zone in area of disputed political boundary with PERU. Until 1953 this province and ZAMORA-CHINCHIPE province used to be Santiago-Zamora province.

Morondava (moo-roon-DAHV), town, TOLIARY province, W MADAGASCAR, on MOZAMBIQUE CHANNEL at mouth of Morondava River, 215 mi/346 km N of TOLIARY; 20°18′S 44°17′E. Seaport and commercial center. Exports fish, lima beans, lentils, corn, rice. Rice milling, sawmilling; rum distillery, sugar factory. Agricultural college (1991). Airport. Tourism (beach hotels, baobabs). Heart of Sakulava tribal territory; unusual cemeteries.

Morón de Almazán (mo-RON dhai ahl-mah-THAHN), village, SORIA province, N central SPAIN, on railroad, and 24 mi/39 km S of SORIA; 41°25′N 02°25′W. Cereals, chick peas, livestock, wool, timber.

Morón de la Frontera (mo-RON dhai lah fron-TAIrah), city, SEVILLE province, S SPAIN, in ANDALUSIA, in NW outliers of the CORDILLERA PENIBÉTICA, near GUADAIRA RIVER, 35 mi/56 km ESE of SEVILLE (linked by railroad); 37°08′N 05°27′W. Trading center for rich agricultural region (olives, cereals, grapes, legumes, livestock). Important livestock fair. Its leading industries are the manufacturing of cement and ceramics based on large lime and gypsum quarries. The ancient city has ruins of a large Moorish castle destroyed (1811) by the French.

Morondo (mo-RON-do), town, Worodougou region, NW CÔTE D'IVOIRE, 44 mi/70 km SW of BOUNDIALI; 08°57′N 06°47′W. Agriculture (manioc, rice, corn, beans, peanuts, tobacco, cotton).

Morong (MO-rong), town, Bataan province, S LUZON, PHILIPPINES, on SOUTH CHINA SEA, on W BATAAN Peninsula, 50 mi/80 km W of MANILA; 14°42′N 120°20′E. Sugarcane, rice. Site of a Refugee Processing Center that received hundreds of Vietnamese "boat people."

Morong, town, RIZAL province, S LUZON, PHILIPPINES, on LAGUNA DE BAY, 18 mi/29 km ESE of MANILA; 14°32′N 121°13′E. Rice, sugar, fruit.

Morongo Valley, unincorporated town (2000 population 1,929), SAN BERNARDINO county, S CALIFORNIA, 40 mi/64 km ESE of SAN BERNARDINO, in E part of SAN BERNARDINO MOUNTAINS; 34°04′N 116°36′W. San Bernardino National Forest to NW; Morongo Indian Reservation; JOSHUA TREE NATIONAL MONUMENT to E. Pacific Coast National Scenic Trail to W. Desert resort area. Cattle.

Moroni (mo-RO-nee), city (2003 population 41,557), ⊙ Njazidja island/administrative district and Comoros Republic, on SW coast of Njazidja island, on Mozambique Channel, Indian Ocean; 11°43′S 43°14′E. Moroni is the main port and administrative center of the islands Mt. Karthala, active volcano and highest point in Comoros, to SE. Manufacturing includes beverages, printing and publishing, soap, plastic products, lumber, ylang-ylang oil. Harbor has been a trading center for over 1,000 years. Iconi Airport, National Museum Palais du Peuple cultural center. Hahaya International Airport 10 mi/16 km to N.

Moroni (muh-RO-nei), town (2006 population 1,273), SANPETE county, central UTAH, 17 mi/27 km SE of NEPHI, near SAN PITCH RIVER; 39°31′N 111°34′W. Elevation c.5,600 ft/1,707 m. Poultry, cattle, sheep; barley, wheat, alfalfa. Manufacturing (turkey feed, turkey processing). Uinta National Forest to W. Terminus of railroad spur from EPHRAIM.

Moros (MO-ros) village, ZARAGOZA province, NE SPAIN, 11 mi/18 km NW of CALATAYUD; 41°24′N 01°49′W. Cereals, wine, fruit; livestock (horses, sheep).

Morosaglia (mo-ro-SAH-glyah), commune (□ 9 sq mi/23.4 sq km), N central CORSICA, HAUTE-CORSE department, FRANCE, 12 mi/19 km NNE of CORTE; 42°26′N 09°19′E. Paoli (the "father of Corsica") born and buried here.

Morotai (mo-RO-tei), island (□ c.695 sq mi/1,800 sq km), E INDONESIA, N Maluku province, N of HALMAHERA Islands; 02°20′N 128°25′E. Heavily wooded, it produces timber and resin.

Morotiri, FRENCH POLYNESIA: see MAROTIRI ISLES.

Moroto, administrative district (2005 population 194,300), NORTHERN region, E UGANDA; ⊙ MOROTO; 02°30′N 34°15′E. As of Uganda's division into eighty districts, borders NAKAPIRIPIRIT (S), KATAKWI (SW), AMURIA, LIRA, and ABIM (W), KOTIDO (NW), and KAABONG (N) districts and KENYA (E). The traditionally nomadic Karamajong people live here. District composed primarily of plains. OKOK River flows S in W part of district and OKERE River flows S through N central and W parts of district. Mount MOROTO in E (10,118 ft/3,084 m). Cattle raising and agriculture. Towns include Moroto and Lothaa. Formed in 2001 from N portion of former MOROTO district (Nakapiripirit district formed from S portion).

Moroto (mo-RO-to), former administrative district (□ 5,449 sq mi/14,167.4 sq km), E UGANDA, along KENYA border (E); capital was MOROTO; 02°20′N 34°30′E. As of Uganda's division into thirty-nine districts, was bordered by KAPCHORWA and MBALE (S), KUMI (SW), SOROTI (W), LIRA (NW tip), and KOTIDO (N) districts. Largely rural area known for cattle raising; sorghum was grown. In 2001 N portion of district was formed into current MOROTO district and S portion formed into NAKAPIRIPIRIT district.

Moroto (mo-RO-to), town (2002 population 7,380), ⊙ MOROTO district, NORTHERN region, E UGANDA, 120 mi/193 km ENE of LIRA, near KENYA border. Agricultural trade center (millet, beans, corn; cattle). Inhabited by Karamajong people. MOUNT MOROTO (10,118 ft/3,084 m) just E. Limestone and mica deposits nearby. Local crafts include: pottery, weaving, woodworking. Was capital of former KARAMOJA province.

Moroto, Mount (mo-RO-to), mountain (10,118 ft/3,084 m), MOROTO district, NORTHERN REGION, E UGANDA, N of MOUNT ELGON; 02°35′N 34°46′E. Was in former KARAMOJA province.

Morotsuka (MO-ROTS-kah), village, E Usuki county, MIYAZAKI prefecture, SE KYUSHU, SW JAPAN, 43 mi/70 km N of MIYAZAKI; 32°30′N 131°19′E. Shiitake mushrooms.

Morouni, FRENCH GUIANA: see MARONI RIVER.

Morovis (mo-RO-vis), town (2006 population 32,379), N central PUERTO RICO, 21 mi/34 km WSW of SAN JUAN. Industrial and commercial area. Agriculture (some coffee, plantains); dairying (milk production); manufacturing (clothing, plastic products).

Moroyama (mo-RO-yah-mah), town, Iruma county, SAITAMA prefecture, E central HONSHU, E central JAPAN, 19 mi/30 km W of URAWA; 35°36′N 139°18′E.

Morozova, Imeni (muh-RO-zuh-vah, EE-mye-nee), town (2005 population 10,610), central LENINGRAD oblast, NW European Russia, at the mouth of the NEVA river, on road and railroad spur, 11 mi/18 km ESE of (and administratively subordinate to) VSEVOLOZHSK, and 8 mi/13 km N of KIROVSK; 59°59′N 31°02′E. Chemical products (glues, adhesive, laquers, plastics); food processing.

Morozovsk (muh-RO-zuhfsk), city (2006 population 29,320), E ROSTOV oblast, S European Russia, on the BYSTRAYA River (tributary of the KALITVA river), 165 mi/266 km NE of ROSTOV-NA-DONU, and 75 mi/121 km E of KAMENSK-SHAKHTINSKIY; 48°21′N 41°49′E. Elevation 173 ft/52 m. Railroad and highway junction; center of agricultural area; agricultural machinery, food industries. Became city in 1941.

Morpará (mor-pah-RAH), town (2007 population 8,728), W central BAHIA, BRAZIL, on Rio SÃO FRANCISCO, 56 mi/90 km SW of XIQUE-XIQUE; 11°34′S 43°16′W.

Morpeth (MOR-puhth), town and river port, E NEW SOUTH WALES, SE AUSTRALIA, on HUNTER river, and 13 mi/21 km NNW of NEWCASTLE; 32°44′S 151°38′E. Coal-mining center. Tourism.

Morpeth (MAW-puhth), town (2001 population 13,833), ⊙ NORTHUMBERLAND, NE ENGLAND, on Wansbeck River, and 14 mi/23 km N of NEWCASTLE UPON TYNE; 55°10′N 01°41′W. Previously woolen mills, iron foundries. Of its ancient castle, only a 14th-century gatehouse remains. Has 14th-century church; town hall (1714) built by Vanbrugh. Just to ENE is town of Pegswood.

Morphou (MOR-foo), town, LEFKOSIA district, NW CYPRUS, on Serrakhis River near its mouth on MORPHOU BAY, 20 mi/32 km W of NICOSIA; 35°12′N 32°59′E. Occupied by Turks; Attila Line passes to S; EVRYKHOU VALLEY mining district farther S. Manufacturing (food processing); agricultural products include grain, vegetables, potatoes, citrus, olives; sheep, goats, poultry. Turkish name Guzelyurt.

Morphou Bay (MOR-foo), NW CYPRUS, arm of MEDITERRANEAN SEA; 15 mi/24 km–20 mi/32 km long, 7 mi/11.3 km wide. City of MORPHOU 4 mi/6.4 km inland to E. Turkish name Guzelyurt Bay.

Morral (MAH-ruhl), village (☐ 3 sq mi/7.8 sq km; 2006 population 379), MARION county, central OHIO, 8 mi/13 km NW of MARION; 40°41′N 83°13′W. Metal and food products.

Morrelganj (mor-rel-gawnj), village (2001 population 21,718), KHULNA district, SW EAST BENGAL, BANGLADESH, in the SUNDARBANS, on distributary of lower MADHUMATI (BALESWAR) River and 31 mi/50 km SE of KHULNA; 22°31′N 89°16′E. Trades in rice, jute, oil-

seeds, sugarcane. Also spelled Morelganj and Morrellganj.

Morrellganj, BANGLADESH: see MORRELGANJ.

Morretes (MO-re-ches), city (2007 population 16,198), SE PARANÁ, BRAZIL, 27 mi/43 km E of CURITIBA; 25°28′S 48°49′W. Railroad junction (spur to Paranguá). Paper manufacturing, distilling; rice, bananas. Iron deposits.

Morrice (MOR-is), village (2000 population 882), SHIAWASSEE county, S central MICHIGAN, 21 mi/34 km NE of LANSING, and 2 mi/3.2 km E of PERRY; 42°50′N 84°10′W. In agricultural area.

Morrill, county (☐ 1,429 sq mi/3,715.4 sq km; 2006 population 5,171), W NEBRASKA; ⊙ BRIDGEPORT. Located in Platte River valley. Irrigated farm area drained by NORTH PLATTE RIVER and its branches. Sugar beets, beans, cattle, hogs, corn, wheat. Several small natural lakes in NE corner, including McCarthy, Wildhorse, Rush, Storm, and Goose lakes. CHIMNEY ROCK NATIONAL HISTORIC SITE in W; Bridgeport State Recreation Area at center. Formed 1909.

Morrill (MAHR-uhl), town, WALDO county, S MAINE, just W of BELFAST; 44°25′N 69°10′W. In agricultural, recreational area.

Morrill, town (2006 population 938), SCOTTS BLUFF county, W NEBRASKA, 15 mi/24 km WNW of SCOTTSBLUFF, and on NORTH PLATTE river; 41°57′N 103°55′W. Dairy products; livestock; grain, sugar beets, potatoes. Manufacturing (feeds, dry beans and peas, rubber decoys).

Morrill (mor-EL), village (2000 population 277), BROWN county, NE KANSAS, 10 mi/16 km NW of HIAWATHA, near NEBRASKA state line; 39°55′N 95°41′W. In grain and livestock region (corn belt).

Morrilton (MOR-uhl-tuhn), town (2000 population 6,550), ⊙ CONWAY county, central ARKANSAS, 40 mi/64 km NW of LITTLE ROCK, and on ARKANSAS river (here bridged; Arthur V. Ormond Lock and Dam to SW); 35°09′N 92°44′W. Stock raising, sawmilling, dairying, manufacturing (light machinery, food products, textiles). PETIT JEAN STATE PARK to W; Brewer Lake reservoir to NE, on Petit Jean Mountain. Founded in 1870s.

Morrin (MAH-rin), village (2001 population 252), S ALBERTA, W Canada, 14 mi/23 km N of DRUMHELLER, in Starland County; 51°40′N 112°46′W. Wheat. Incorporated 1920.

Morrinhos (MOR-reen-yos), city (2007 population 21,264), NW CEARÁ, BRAZIL, 19 mi/31 km S of SÃO BENEDITO; 04°17′S 40°52′W.

Morrinhos (mo-REEN-yos), city (2007 population 38,991), S GOIÁS, central BRAZIL, 75 mi/121 km S of GOIÂNIA; 17°35′S 49°10′W. Cattle, tobacco, rice, coffee. Rutile deposits.

Morrinsville, town (2001 population 6,165), Matamata-Piako district, WAIKATO region, NORTH ISLAND, NEW ZEALAND, 70 mi/113 km SSE of AUCKLAND, on lowland linking Hamilton Basin with HAURAKI PLAINS; 37°39′S 175°32′E. Services intensive dairy and prime lamb-farming area.

Morris (MAH-ris), rural municipality (☐ 402 sq mi/1,045.2 sq km; 2001 population 2,723), S MANITOBA, W central Canada, 36 mi/58 km SW of WINNIPEG; 49°24′N 97°27′W. Agriculture (grain; fruits and vegetables; poultry; livestock); manufacturing (steel). Composed of the communities of Rosenort, Lowe Farm, Mctavish, Silver Plains, Kane, Aubigny, Sperling, Riverside, and Sewell. Incorporated 1880.

Morris, county (☐ 702 sq mi/1,825.2 sq km; 2006 population 6,046), E central KANSAS; ⊙ COUNCIL GROVE; 38°41′N 96°38′W. Located in FLINT HILLS region, watered by Neosho River. Hogs, cattle; wheat, barley, hay, soybeans; industrial machinery. Council Grove Lake reservoir in E. Formed 1859.

Morris, county (☐ 481 sq mi/1,250.6 sq km; 2006 population 493,160), N NEW JERSEY, bounded SE and E by

PASSAIC River; ⊙ MORRISTOWN; 40°52′N 74°32′W. Hilly estate and resort area, with many lakes and mountain ridges. Research, pharmaceuticals, and light manufacturing. Drained by PEQUANNOCK, ROCKAWAY, WHIPPANY, and MUSCONETCONG rivers, and branches of RARITAN River; includes MORRISTOWN NATIONAL HISTORICAL PARK and part of Lake HOPATCONG. E largely suburbanized. In W half still agricultural, but that is quickly becoming suburbanized. Formed 1739.

Morris (MOR-is), county (☐ 258 sq mi/670.8 sq km; 2006 population 13,002), NE TEXAS; ⊙ DAINGERFIELD; 33°07′N 94°43′W. Bounded N by SULPHUR RIVER, S by BIG CYPRESS CREEK and Ellison Creek (Ellison Creek Reservoir in S). Diversified agriculture, peanuts, watermelons; cattle, poultry. Iron-ore deposits. Timber (mainly pine). Daingerfield State Park in S. Formed 1875.

Morris, city (2000 population 11,928), ⊙ GRUNDY county, NE ILLINOIS, on ILLINOIS RIVER, and 21 mi/34 km SW of JOLIET; 41°22′N 88°25′W. Shipping and industrial center in agricultural, clay area; manufacturing (paper products, chemicals, plastics products). Plotted 1842, incorporated 1853. Nearby is Gebhard Woods State Park (30 acres/12 ha) along old Illinois and Michigan Canal Parkway.

Morris (MAH-ris), town (☐ 2 sq mi/5.2 sq km; 2001 population 1,673), S MANITOBA, W central Canada, on RED RIVER, at mouth of small Morris River, and 40 mi/64 km SSW of WINNIPEG; 49°21′N 97°22′W. Grain elevators; livestock raising, dairying; oil distributing point. Incorporated 1883.

Morris (MAH-ris), former township (☐ 89 sq mi/231.4 sq km; 2001 population 1,734), S ONTARIO, E central Canada; 43°46′N 81°18′W. Merged with Turnberry township in 2001 to form Morris-Turnberry.

Morris, resort town, LITCHFIELD county, W CONNECTICUT, on BANTAM LAKE, the state's largest natural lake, and 10 mi/16 km SW of TORRINGTON; 41°41′N 73°12′W. Includes LAKESIDE village resort, part of state park, game sanctuary, agriculture and light industry.

Morris, town (2000 population 5,068), ⊙ STEVENS county, W MINNESOTA, near POMME DE TERRE RIVER, and 50 mi/80 km NW of WILLMAR; 45°35′N 95°54′W. Elevation 1,136 ft/346 m. Railroad junction. Wheat, corn, oats, barley, soybeans, alfalfa, sunflowers; hogs, cattle, sheep; manufacturing (truck trailers, bulk ethanol, printing and publishing, belt conveyors). Plotted 1869, incorporated as village 1878, as city 1903.

Morris, township, MORRIS county, N NEW JERSEY; 40°47′N 74°29′W. Includes the capital of county, MORRISTOWN. Incorporated 1798. Pharmeceuticals.

Morris, town (2006 population 1,319), OKMULGEE county, E central OKLAHOMA, 6 mi/9.7 km E of OKMULGEE; 35°36′N 95°51′W. In oil-producing and coal-mining area; manufacturing (lead fishing weights).

Morris, village (2006 population 557), OTSEGO county, central NEW YORK, 13 mi/21 km NW of ONEONTA; 42°32′N 75°15′W. In dairying area; manufacturing of veterinary medicines. Gilbert Lake State Park is c.5 mi/8 km NE. Otsego county fairground here.

Morrisania, residential section of S BRONX borough of NEW YORK city, SE NEW YORK; 40°50′N 73°54′W. An economically distressed area that has undergone some improvement in recent years.

Morrisburg (MAH-ris-buhrg), unincorporated village (2001 population 2,568), SE ONTARIO, E central Canada, on the SAINT LAWRENCE RIVER, and included in South Dundas; 44°54′N 75°11′W. Manufacturing includes concrete products, brooms and brushes, and surgical and dental supplies. Just E of the village is the Upper Canada Village, a model of a typical 19th-century community.

Morris Canal, NEW JERSEY, abandoned canal (c.100 mi/161 km long) joining DELAWARE RIVER at PHILLIPSBURG and NEWARK BAY at NEWARK. Chartered 1824 as outlet for PENNSYLVANIA coal regions; opened 1831; extended to JERSEY CITY, 1836; declined after peak traffic in 1860s; finally abandoned 1923. Bed at Newark utilized for subway (1935).

Morrisdale, unincorporated village, CLEARFIELD county, central PENNSYLVANIA, 12 mi/19 km SE of CLEARFIELD; 40°56′N 78°13′W. Surface bituminous coal.

Morris Dam, California: see SAN GABRIEL RIVER.

Morris Island, CHARLESTON county, S SOUTH CAROLINA, one of the SEA ISLANDS at S side of entrance to CHARLESTON HARBOR; c.3.5 mi/5.6 km long; 32°42′N 79°53′W. Charleston Lighthouse on S tip. JAMES ISLAND to W, FOLLY ISLAND to S, ATLANTIC OCEAN to E.

Morris Jesup, Cape (JES-uhp), Danish *Kap Morris Jesup,* northernmost land point in the world, N GREENLAND. At lat. 83°39′N, it is 440 mi/708 km from the NORTH POLE. U.S. explorer Robert Peary reached the cape in 1892.

Morrison, county (□ 1,153 sq mi/2,997.8 sq km; 2006 population 32,919), central MINNESOTA; ⊙ LITTLE FALLS; 46°00′N 94°16′W. Agricultural area drained by MISSISSIPPI RIVER. Bounded in NW by CROW WING RIVER (Mississippi River forms parts of N and S county boundary). Alfalfa, hay, potatoes, corn, oats, barley, rye, sunflowers, beans; timber; sheep, hogs, cattle, poultry; dairying; deposits of marl and peat. Numerous small lakes in NW, including Lake ALEXANDER. Camp Ripley Military Reservation in NW. Charles A. Lindbergh State Park in W center. County formed 1855.

Morrison, city (2000 population 4,447), ⊙ WHITESIDE county, NW ILLINOIS, on ROCK CREEK (bridged here), and 12 mi/19 km E of CLINTON (IOWA); 41°48′N 89°58′W. In farming and dairying area; manufacturing (dairy products). Rockwood State Park nearby. Founded 1855, incorporated 1867.

Morrison, town, E central CÓRDOBA province, ARGENTINA, 28 mi/45 km SE of VILLA MARIA; 32°36′S 62°50′W. Wheat, soybeans, corn, alfalfa, livestock.

Morrison, town (2000 population 97), GRUNDY county, central IOWA, 5 mi/8 km E of GRUNDY CENTER; 42°20′N 92°40′W. In agricultural area.

Morrison, town (2000 population 123), GASCONADE county, E central MISSOURI, on MISSOURI River, 7 mi/11.3 km W of HERMANN; 38°40′N 91°37′W. Corn, soybeans, cattle. Damaged by flood of 1993.

Morrison, town (2006 population 705), WARREN county, central TENNESSEE, 10 mi/16 km SW of MCMINNVILLE; 35°36′N 85°55′W.

Morrison, village (2000 population 430), JEFFERSON county, N central COLORADO, on Bear Creek, suburb 10 mi/16 km SW of DENVER; 39°38′N 105°10′W. Elevation c.5,800 ft/1,768 m. Light manufacturing. Resort; entrance to Denver mountain park system. Red Rocks Natural Amphitheater, used for concerts, to NW. FRONT RANGE of ROCKY MOUNTAINS to W; Bear Creek Lake reservoir to E. Formerly Mount Morrison.

Morrison, village (2006 population 622), NOBLE county, N OKLAHOMA, 12 mi/19 km NNE of STILLWATER, on BLACK BEAR CREEK; 36°17′N 97°00′W. In agricultural area (grain, livestock).

Morrison City, village, SULLIVAN county, NE TENNESSEE, 4 mi/6 km N of KINGSPORT, near Tennessee-Virginia state line; 36°35′N 82°35′W.

Morrisonville, village (2000 population 1,068), CHRISTIAN county, central ILLINOIS, 14 mi/23 km SW of TAYLORVILLE; 39°25′N 89°27′W. Agriculture (grain; livestock); soybean products. Incorporated 1872.

Morrisonville, village (□ 2 sq mi/5.2 sq km; 2000 population 1,702), CLINTON county, extreme NE NEW YORK, on Saranac River, and 5 mi/8 km W of PLATTSBURGH; 44°41′N 73°32′W.

Morris Plains, residential borough (2006 population 5,601), MORRIS county, N NEW JERSEY, just N of MORRISTOWN; 40°50′N 74°28′W. Light manufacturing. State mental hospital (1871) nearby. Incorporated 1926.

Morris Run, unincorporated village, Hamilton township, TIOGA county, N PENNSYLVANIA, 2 mi/3.2 km E of BLOSSBURG, on Morris Run creek; 41°40′N 77°01′W. Manufacturing (wooden products).

Morristown, city (2000 population 24,965), ⊙ HAMBLEN county, NE TENNESSEE, 40 mi/64 km ENE of KNOXVILLE; 36°12′N 83°18′W. Diverse industry and retail; 2 community colleges (Walters State and Knoxville College at Morristown). CHEROKEE Reservoir (HOLSTON River) is N; state fish hatchery nearby. Settled 1783, incorporated 1855.

Morristown, town (2006 population 1,257), SHELBY county, central INDIANA, on BIG BLUE RIVER, and 11 mi/18 km NNE of SHELBYVILLE; 39°40′N 85°42′W. In agricultural area; manufacturing (motor vehicle leaf springs, seals, steering components, hamburger buns, plastic moldings, windows and doors). Laid out 1828.

Morristown, town (2000 population 981), RICE county, S MINNESOTA, 10 mi/16 km WSW of FARIBAULT, on CANNON RIVER; 44°13′N 93°26′W. Grain, soybeans; livestock; dairying. Small lakes in area. Sakatah Lake State Park to W, Cannon Lake to NE, both on Cannon River.

Morristown, town (2000 population 18,544), ⊙ MORRIS county, N NEW JERSEY, on the WHIPPANY RIVER; 40°47′N 74°28′W. Although chiefly residential, it has electronics and light industry; telecommunication research and development center. Morristown also has become a burgeoning center of corporate activity. It was a principal area of Revolutionary maneuvers, particularly in the winters of 1777 and 1779–1780, when the Continental army encamped here. Benedict Arnold was court-martialed in the town. S. F. B. Morse and Alfred Vail perfected (c.1837) the telegraph here. Of interest are the Schuyler-Hamilton House (1760), where Alexander Hamilton courted (1779–1780) Elizabeth Schuyler (it has become headquarters for the Daughters of the American Revolution); and the courthouse (1826). Other notable residents of Morristown were the cartoonist Thomas Nast, the writer Bret Harte, and the humorist Frank River Stockton. MORRISTOWN NATIONAL HISTORICAL PARK includes the Ford Mansion, which was Washington's headquarters in 1779–1780; a historical museum at the rear of the Ford Mansion; and the reconstructed sites of encampment of the Continental army at Fort Nonsense and at Jockey Hollow. The engine of the *Savannah*, the first steamship to cross the ATLANTIC OCEAN, was made near Morristown. Settled c.1710, incorporated 1865.

Morristown, town, LAMOILLE CO., N central VERMONT, on LAMOILLE RIVER, and just S of HYDE PARK, region; 44°32′N 72°38′W. Includes Morrisville village. Chartered 1781. Economic and social center of the county.

Morristown, village (□ 1 sq mi/2.6 sq km; 2000 population 456), ST. LAWRENCE county, N NEW YORK, on the SAINT LAWRENCE River, opposite BROCKVILLE, ONTARIO (ferry), and 12 mi/19 km SW of OGDENSBURG; 44°34′N 75°39′W. Fishing center; port of entry.

Morristown (MOR-uhs-toun), village (2000 population 299), BELMONT county, E OHIO, 8 mi/13 km W of ST. CLAIRSVILLE; 40°04′N 81°04′W. In coal-mining area.

Morristown, village (2000 population 82), CORSON county, N SOUTH DAKOTA, 18 mi/29 km W of MCINTOSH, on NORTH DAKOTA state line; 45°56′N 101°43′W. Supply point for ranching region; in Standing Rock Indian Reservation.

Morristown National Historical Park (□ 2 sq mi/5.2 sq km), N NEW JERSEY; 40°46′N 74°31′W. Authorized 1933. Site of military encampments during the Revolution; George Washington's headquarters, 1779–1780.

Morris-Turnberry (MAH-ris–TUHRN-be-ree), township (□ 146 sq mi/379.6 sq km; 2001 population 3,499), S ONTARIO, E central Canada, 23 mi/36 km from GODERICH; 43°49′N 81°16′W. Formed in 2001 from the former townships of Morris and Turnberry.

Morrisville, town (2000 population 344), POLK county, central MISSOURI, near LITTLE SAC RIVER, 8 mi/12.9 km S of BOLIVAR; 37°28′N 93°25′W.

Morrisville, town (□ 6 sq mi/15.6 sq km; 2006 population 12,513), DURHAM and WAKE counties, central NORTH CAROLINA, suburb 9 mi/14.5 km W of downtown RALEIGH, and 11 mi/18 km SSE of downtown DURHAM; 35°49′N 78°49′W. Service industries; manufacturing (concrete and glass blocks, pharmaceuticals, trade show exhibits, machine parts, carbide cutting tools, laminated wood products). RESEARCH TRIANGLE PARK to NW, Raleigh-Durham International Airport to NE. Lake Crabtree reservoir to NE.

Morrisville, unincorporated town (2000 population 1,443), GREENE county, SW PENNSYLVANIA, residential suburb 1 mi/1.6 km E of WAYNESBURG; 39°53′N 80°10′W. Agriculture includes dairying, hay. Greene County Airport to E.

Morrisville, village (□ 1 sq mi/2.6 sq km; 2006 population 2,096), MADISON county, central NEW YORK, 28 mi/45 km ESE of SYRACUSE, in dairying area; 42°53′N 75°38′W. State University of New York at Morrisville is here.

Morrisville, village (2006 population 2,053) in MORRISTOWN town, LAMOILLE CO., N central VERMONT; 44°33′N 72°35′W. Lumber, dairy products Winter sports.

Morrisville, borough (2006 population 9,746), BUCKS county, SE PENNSYLVANIA, suburb 1 mi/1.6 km SW of downtown TRENTON, NEW JERSEY, and 27 mi/43 km NE of downtown PHILADELPHIA, on the DELAWARE RIVER; 40°12′N 74°46′W. Manufacturing (water pumps, clutches, coatings, concrete products, rubber products, coated coils, foam fabrication, wire screens, printing and publishing, gases and chemicals). Steel fabricating to S at FAIRLESS HILLS. George Washington had his headquarters here December 8–14, 1776. Pennsbury Manor State Park, with reconstructed Pennsbury, which was William Penn's home, to S. Settled c.1624 by the Dutch West India Company, incorporated 1804.

Morrito (mor-REE-to) or **El Morrito,** town, RÍO SAN JUAN department, S NICARAGUA, port on E shore of LAKE NICARAGUA, 40 mi/64 km SE of JUIGALPA; 11°37′N 85°05′W. Lumbering; livestock.

Mor River or **Morakhi,** river, c.100 mi/161 km long, in JHARKHAND and WEST BENGAL states, E INDIA; rises in NE Chota Nagpur Plateau foothills, in headstreams joining 5 mi/8 km NNW of DUMKA (Jharkhand state); flows SE, past Dumka, Masanjor, and Sainthiya (West Bengal state), and E to Dwarka River (tributary of the Bhagirathi River) 7 mi/11.3 km E of KANDI. Irrigation project consists of dam at MASANJOR (impounding reservoir between Masanjor and Dumka) and barrage near SIURI to supply canal system of c.630 mi/1,014 km. Project mainly benefits Birbhum and Murshidabad districts (West Bengal).

Morroa (mor-RO-ah), town, ⊙ Morroa municipio, SUCRE department, N central COLOMBIA, 4.0 mi/6.4 km NNE of, and suburb of, SINCELEJO; 09°20′N 75°18′W. Corn, vegetables; livestock.

Morro Agudo (MO-ro AH-goo-do), city (2007 population 25,400) N SÃO PAULO, BRAZIL, 35 mi/56 km NNW of RIBEIRÃO PRÊTO; 20°44′S 48°04′W. Pottery manufacturing; rice processing; coffee, corn, rice.

Morro Bay, city (2000 population 10,350), SAN LUIS OBISPO county, SW CALIFORNIA, 12 mi/19 km NW of SAN LUIS OBISPO, on Morro Bay (an inlet of ESTERO BAY, Pacific Ocean); 35°22′N 120°52′W. A 4-mi/6.4-km long spit separates Morro Bay from the ocean, opening on the N. Guarded by Morro Rock (576 ft/176 m). Manufacturing (medical equipment); cattle; grain, flowers, nursery stock, apples, avocados, vegetables, strawberries. Section of Los Padres National Forest to NE; Montana de Oro State Park and Atascadero State Beach to S. Morro Bay State Park is here.

Morro Castle (MOR-ro), fort, at the entrances to HAVANA harbor, CIUDAD DE LA HABANA province, W CUBA; 23°09′N 82°19′W. Erected by the Spanish under the direction of architect-engineer Batista Antonelli in

1589 to protect the city from buccaneers and was also used as a prison. It was captured by the British in 1762.

Morro Castle (MOR-ro), fort, at entrance to SANTIAGO DE CUBA harbor, SANTIAGO DE CUBA province, SE CUBA; 21°33′N 75°17′W. Built shortly after the Morro Castle of HAVANA. It was taken by the American forces in the Spanish-American War (1898).

Morro Castle (MOR-ro) or **Castillo del Morro**, fort at harbor of SAN JUAN, PUERTO RICO.

Morro Channel (MOR-ro), Spanish *Canal de Morro*, off coast of GUAYAS province, SW ECUADOR, W of PUNÁ Island, linking Gulf of GUAYAQUIL with GUAYAS RIVER estuary; c.25 mi/40 km long.

Morro Chico (MOR-ro CHEE-ko), village, MAGALLANES province, MAGALLANES Y LA ANTARTICA CHILENA region, S CHILE, on Patagonian mainland, 85 mi/137 km NNW of PUNTA ARENAS. Resort.

Morrocoy, Parque Nacional (mor-RO-koi,PAHR-kai nah-see-o-NAHL), national park (□ 127 sq mi/330.2 sq km), FALCÓN state, N VENEZUELA; 10°48′N 68°13′W. Beach and island environment, protecting bird life. Created 1974.

Morro da Fumaça (MOR-ro dah FOO-mah-sah), city (2007 population 15,426), SE SANTA CATARINA state, BRAZIL, 13 mi/21 km E of CRICIÚMA, on railroad; 28°40′S 49°12′W. Rice, corn, manioc.

Morro da Mina, Brazil: see CONSELHEIRO LAFAIETE.

Morro do Chapéu (mo-ro do shah-PE-oo), city (2007 population 33,543), central BAHIA, BRAZIL, on the CHAPADA DIAMANTINA, 50 mi/80 km SW of JACOBINA; 11°31′S 41°12′W. Diamond mining; nitrate deposits. Has noteworthy church. Formerly spelled Morro do Chapéo.

Morro do Urucum, Brazil: see CORUMBÁ.

Morro, El, ARGENTINA: see EL MORRO.

Morro Grande, Brazil: see BARÃO DE COCAIS.

Mórrope (MOR-ro-pai), town, LAMBAYEQUE region, NW PERU, on coastal plain, on S edge of MÓRROPE DESERT, on Mórrope River, and 13 mi/21 km NW of LAMBAYEQUE, on PAN-AMERICAN HIGHWAY; 06°32′S 80°01′W. Corn, alfalfa, cotton, rice.

Mórrope Desert (MOR-ro-pai), Spanish *Pampa Mórrope*, LAMBAYEQUE region, NW PERU, on coastal plain, just NW of LA LECHE RIVER and the town of MÓRROPE; 40 mi/64 km long, 35 mi/56 km wide; 06°38′S 80°00′W. It is continued by Pampa del Salitre (NW) and Pampa la Mariposa Vieja (NE). Salt mining. Archaeological museum.

Mórrope River (MOR-ro-pai), LAMBAYEQUE region, NW PERU; 06°28′S 79°55′W.

Morro Plancho, Alto (MOR-ro PLAHN-cho, AHL-to), mountain, ANTIOQUIA department, W central COLOMBIA; 05°33′N 75°43′W.

Morro Point (MO-ro), headland on coast of CAMPECHE, SE MEXICO, on W YUCATÁN peninsula, 15 mi/24 km SW of CAMPECHE; 19°41′N 90°42′W. Occasionally Morros Point.

Morropón (mor-ro-PON), province (□ 1,315 sq mi/3,419 sq km), PIURA region, NW PERU; ⊙ CHULUCANAS; 05°15′S 80°00′W.

Morropón (mor-ro-PON), town, PIURA region, NW PERU, in irrigated PIURA RIVER valley, 15 mi/24 km ESE of CHULUCANAS; 05°11′S 79°58′W. Trade center in rice region; livestock.

Morros (mor-os), city (2007 population 17,099), N MARANHÃO, BRAZIL, head of navigation on Monim River, and 38 mi/61 km SE of SÃO LUÍS; 06°35′S 44°10′W. Cotton, sugar, rice.

Morros da Mariana (MOHR-ros dah MAH-ree-ah-nah), city, extreme N PIAUÍ state, BRAZIL, on Big Santa Isabel Island in Atlantic Ocean, 6 mi/9.7 km NW of PARNAÍBA; 02°52′S 41°49′W.

Morrosquillo, Golfo de (mor-ros-KEE-yo, GOL-fo dai), inlet of CARIBBEAN SEA in BOLÍVAR department, N COLOMBIA; 20 mi/32 km long from SAN BERNARDO POINT to mouth of SINÚ RIVER, c.8 mi/12.9 km wide; 09°35′N 75°40′W.

Morro Velho, Brazil: see NOVA LIMA.

Morrow (MOR-o), county (□ 404 sq mi/1,050.4 sq km; 2006 population 34,529), central OHIO; ⊙ MOUNT GILEAD; 40°32′N 82°47′W. Drained by KOKOSING RIVER and small Whetstone and Big Walnut creeks. In the Till Plains physiographic region. Agricultural area (hogs, cattle; corn); manufacturing mainly at Mount Gilead and CARDINGTON (machinery, paper products, electric lighting). Formed 1848.

Morrow (MAHR-o), county (□ 2,048 sq mi/5,324.8 sq km; 2006 population 11,753), N OREGON; ⊙ HEPPNER; 45°25′N 119°34′W. Bounded N by Columbia River (forms WASHINGTON state line). Drained by WILLOW CREEK. Agriculture (alfalfa, grapes, corn, wheat, oats, barley, onions, potatoes; sheep, cattle). Umatilla Ordnance Depot on NE boundary. Part of Umatilla National Forest in SE; parts of Umatilla National Wildlife Refuge in N. Formed 1885.

Morrow (MAHR-o), town (2000 population 4,882), CLAYTON county, NW central GEORGIA, 12 mi/19 km S of ATLANTA; 33°35′N 84°20′W. Emerging suburb of Atlanta. The Atlanta Beach Recreation Complex gained fame as the site of beach volleyball competition during the 1996 summer Olymic Games. Reynolds Memorial Nature Preserve and museum here. Manufacturing of cleaning chemicals, plastics, hydraulic cylinders; steel fabrication.

Morrow (MOR-o), village (□ 2 sq mi/5.2 sq km; 2006 population 1,534), WARREN county, SW OHIO, 7 mi/11 km SE of LEBANON, and on Little Miami River; 39°20′N 84°07′W.

Morrow Point Reservoir, MONTROSE and Garrison counties, W COLORADO, on GUNNISON RIVER, in CURECANTI NATIONAL RECREATION AREA, 18 mi/29 km S of MONTROSE; 11 mi/18 km long; 38°27′N 107°33′W. Extends E to Blue Mesa Dam. Maximum capacity of 121,500 acre-ft. Formed by Morrow Point Dam (400 ft/122 m high), built (1968) by the Bureau of Reclamation for power generation and flood control.

Morrowville, village (2000 population 168), WASHINGTON county, N KANSAS, 7 mi/11.3 km WNW of WASHINGTON; 39°50′N 97°10′W. Grain; livestock. Washington State Fishing Lake to N.

Mörrum (MUHR-ROOM), town, BLEKINGE county, S SWEDEN, on MÖRRUMSÅN RIVER, 5 mi/8 km W of KARLSHAMN; 56°12′N 14°45′E. Salmon fishing.

Morrumbala (mor-ROOM-bah-luh), village, Zambézia province, central MOZAMBIQUE, 95 mi/153 km WNW of QUELIMANE; 17°20′S 35°35′E. Cotton, rice.

Morrumbene (mor-ROOM-bai-nai), village, INHAMBANE province, SE MOZAMBIQUE, small port on MOZAMBIQUE CHANNEL, 15 mi/24 km N of INHAMBANE; 23°39′S 35°20′E. Copra; fishing.

Morrumbene District, MOZAMBIQUE: see INHAMBANE province.

Mörrumsån (MUHR-ROOMS-ON), river, 80 mi/129 km long, S SWEDEN; rises E of VÄRNAMO; flows S, past ALVESTA and RYD, to BALTIC SEA 4 mi/6.4 km W of KARLSHAMN. Salmon fishing.

Mors (mors), island (□ 54 sq mi/140.4 sq km), NW DENMARK, in the LIMFJORD; 56°50′N 08°45′E. NYKØBING is the chief city. The island has considerable fertile soil, and offshore there are oyster fisheries.

Mörs, GERMANY: see MOERS.

Morsang-sur-Orge (mor-san-syoor-orzh), town (□ 1 sq mi/2.6 sq km), residential S suburb of PARIS, ESSONNE department, ÎLE-DE-FRANCE region, N central FRANCE, 13 mi/21 km S of Notre Dame Cathedral; 48°40′N 02°21′E. It lies on the S-bound expressway to LYON and PROVENCE.

Morsbach (MORS-bahkh), town, RHINELAND, North Rhine-Westphalia, W GERMANY, 13 mi/21 km W of SIEGEN; 50°53′N 07°44′E. Forestry.

Morsbronn-les-Bains (mors-bron–lai–ban), spa (□ 2 sq mi/5.2 sq km), BAS-RHIN department, ALSACE, E FRANCE, 6 mi/9.7 km NNW of HAGUENAU; 48°54′N 07°44′E. Mineral springs for treatment of rheumatic diseases.

Morschach (mor-SHAH), village, SCHWYZ canton, central SWITZERLAND, near LAKE LUCERNE, 3 mi/4.8 km SW of SCHWYZ; 46°59′N 08°38′E. Elevation 2,116 ft/645 m. Health resort.

Morschen (MOR-shen), village, HESSE, central GERMANY, on right bank of the FULDA, 10 mi/16 km ENE of HOMBERG (Efze); 51°04′N 09°37′E. Forestry; manufacturing (chemicals).

Morschwiller-le-Bas (morsh-vee-LER–luh–BAH), commune (□ 3 sq mi/7.8 sq km), outer WSW suburb of MULHOUSE, HAUT-RHIN department, ALSACE, E FRANCE; 47°45′N 07°16′E. Chemical works based on local potash deposits.

Morse (MORS), town (2006 population 236), S SASKATCHEWAN, CANADA, 35 mi/56 km ENE of SWIFT CURRENT; 50°24′N 107°00′W. Grain elevators, lumbering.

Morse, town (2000 population 759), ACADIA parish, S LOUISIANA, 10 mi/16 km SW of CROWLEY; 30°07′N 92°30′W. In agricultural (rice) area; manufacturing (mops).

Morse Bluff, village (2006 population 131), SAUNDERS county, E NEBRASKA, 15 mi/24 km WSW of FREMONT, and on PLATTE RIVER; 41°25′N 96°46′W.

Morse Reservoir, water supply reservoir, HAMILTON county, central INDIANA, 22 mi/35 km NNE of downtown INDIANAPOLIS. Located on CICERO CREEK and built by the Indianapolis Water Company.

Morshansk (muhr-SHAHNSK), city (2006 population 42,930), NE TAMBOV oblast, S central European Russia, on the TSNA RIVER (landing), on highway junction and railroad, 50 mi/80 km NNE of TAMBOV; 53°27′N 41°48′E. Elevation 393 ft/119 m. Center of agricultural area; machinery and metal industries (chemical machinery, motor repair), woolen milling (coarse cloths), clothing, tobacco processing, woodworking, food processing (creamery, brewery, bakery and confectionery). City chartered in 1779.

Morshin, UKRAINE: see MORSHYN.

Morshyn (MOR-shin) (Russian *Morshin*) (Polish *Morszyn*), town (2004 population 8,200), S L′VIV oblast, UKRAINE, 7 mi/11 km S of STRYY, in coniferous woodland; 49°09′N 23°52′E. Elevation 1135 ft/345 m. Subordinated to the Stryy city council. Health resort with mineral springs. Potassium deposits nearby. Known since 1482; health resort by end of the 19th century; town since 1948.

Morsi (MOR-see), town, AMRAVATI district, MAHARASHTRA state, central INDIA, 32 mi/51 km NNE of AMRAVATI; 21°20′N 78°00′E. Trades in cotton, millet, wheat, oilseeds.

Mörsil (MUHR-sheel), village, JÄMTLAND county, NW SWEDEN, on Järpströmmen River (falls), tributary of INDALSÄLVEN RIVER, 30 mi/48 km WNW of ÖSTERSUND; 63°19′N 13°40′E.

Morson (MOR-suhn), former township (□ 39 sq mi/101.4 sq km; 2001 population 161), NW ONTARIO, E central Canada, 54 mi/87 km from FORT FRANCES; 49°06′N 94°19′W. Amalgamated into Lake of the Woods in 1998.

Morszyn, UKRAINE: see MORSHYN.

Mortagne-au-Perche (mor-tah-nyuh–o–persh), town (□ 3 sq mi/7.8 sq km), ORNE department, BASSE-NORMANDIE region, NW FRANCE, in the Perche hills, 22 mi/35 km ENE of ALENÇON; 48°30′N 00°40′E. Market for Percheron horses bred in area. Sausage making. Has restored 15th–16th-century Flamboyant Gothic church with 18th-century woodwork.

Mortagne River (mor-TAHN-yuh), c.30 mi/48 km long, VOSGES and MEURTHE-ET-MOSELLE departments, LORRAINE region, NE FRANCE; rises near

SAINT-DIÉ in the VOSGES MOUNTAINS; flows NW, past RAMBERVILLERS and Gerbéviller, to the MEURTHE 4 mi/6.4 km below LUNÉVILLE; 48°33′N 06°27′E.

Mortagne-sur-Sèvre (mor-tahn-yuh–syoor–se-vruh), town (□ 8 sq mi/20.8 sq km), VENDÉE department, PAYS DE LA LOIRE region, W FRANCE, on SÈVRE NANTAISE RIVER, and 5 mi/8 km SSW of CHOLET; 47°00′N 00°57′W. Tannery. Manufacturing of textiles and footwear.

Mortágua (mor-TAH-gwah), town, VISEU district, N central PORTUGAL, on railroad, and 18 mi/29 km NE of COIMBRA; 40°24′N 08°14′W. Wool milling, pottery manufacturing.

Mortain (mor-tan), commune (□ 2.9 sq mi/7.5 sq km; 1993 estimated pop. 2,612), MANCHE department, BASSE-NORMANDIE region, NW FRANCE, in NORMANDY HILLS, 19 mi/31 km E of AVRANCHES; 48°39′N 00°56′W. The 11th–13th-century Abbaye Blanche (now housing seminary) is 1 mi/1.6 km N near a waterfall on the Cance River. Site of crucial battle in Normandy campaign of World War II in which Germans failed to cut off Allied southerly advance (August 1944).

Mortara (mor-TAH-rah), town, PAVIA province, LOMBARDY, N ITALY, 21 mi/34 km WNW of PAVIA; 45°15′N 08°44′E. Railroad junction; agricultural center. Chief town of the Lomellina; highly diversified secondary industrial center including dairy products; manufacturing (rice mill machinery, textiles, hats, buttons). Has church built 1375–1380. Nearby is 5th-century abbey (rebuilt 8th and 16th centuries).

Morteau (mor-to), town (□ 5 sq mi/13 sq km; 2004 population 6,339), DOUBS department, FRANCHE-COMTÉ region, E FRANCE, on the DOUBS, and 15 mi/24 km NE of PONTARLIER, near Swiss border; resort in the E JURA MOUNTAINS; 47°03′N 06°35′E. Town has tradition of watchmaking, now expanded to manufacturing of cutlery, bicycle parts, and other metal products. The falls of the Doubs (*Saut du Doubs*) are 6 mi/9.7 km ENE. Local sausage production is a gastronomic specialty.

Mortehoe (MAWT-ho), village (2001 population 2,126), N DEVON, SW ENGLAND, on BRISTOL CHANNEL, and 5 mi/8 km WSW of ILFRACOMBE; 51°11′N 04°12′W. Has 15th-century church. Morte Point, an 800-ft/244-m-high promontory, is nearby.

Morter, CROATIA: see MURTER ISLAND.

Morteratsch, Piz, SWITZERLAND: see BERNINA ALPS.

Morteros (mor-TAI-ros), town, NE CÓRDOBA province, ARGENTINA, 50 mi/80 km N of SAN FRANCISCO; 30°42′S 62°00′W. Agricultural and industrial center; manufacturing of agricultural machinery, shoes, dairy products, food processing, soybeans, alfalfa, grain; livestock.

Mort-Homme, Le (mor–uhm, luh), a height (□ 2 sq mi/5.2 sq km) overlooking left bank of MEUSE RIVER, MEUSE department, LORRAINE region, NE FRANCE, 7 mi/11.3 km NW of VERDUN. Scene of violent combat (March 1916) between French and German forces in World War I. Monument to fallen soldiers bears inscription, "They did not pass." Also called Cumières-le-Mort-Homme.

Mortimer (MOR-ti-muhr), town, CALDWELL county, W central NORTH CAROLINA, 14 mi/23 km WNW of LENOIR; 35°59′N 81°45′W.

Mortimer's Cross (MAW-ti-muhz KRAHS), battlefield, NW Herefordshire, W ENGLAND, on LUGG RIVER, near LEOMINSTER; 52°16′N 02°51′W. Scene of a decisive Yorkist victory over Lancastrians (1461) in the Wars of the Roses.

Mortka (MORT-kah), village (2005 population 3,725), S KHANTY-MANSI AUTONOMOUS OKRUG, W central SIBERIA, Russia, on road and railroad, 28 mi/45 km S of Lugovoy; 59°20′N 66°01′E. Elevation 246 ft/74 m. Sawmilling, woodworking.

Mortlach (mort-LAKH), town (2006 population 254), S SASKATCHEWAN, CANADA, 23 mi/37 km W of MOOSE JAW; 50°27′N 106°04′W. Wheat.

Mortlake (MORT-laik), town, S VICTORIA, SE AUSTRALIA, 115 mi/185 km WSW of MELBOURNE; 38°06′S 142°52′E. Railroad terminus; sheep. Deposits of olivine, an olive-green mineral.

Mortlake, ENGLAND: see RICHMOND UPON THAMES.

Mortlock Island, PAPUA NEW GUINEA: see TAKUU.

Mortlock Islands, MICRONESIA: see NOMOI ISLANDS.

Morton (MOR-tuhn), rural municipality (□ 421 sq mi/1,094.6 sq km; 2001 population 760), SW MANITOBA, W central Canada, between SOURIS River and TURTLE MOUNTAIN, on U.S. (NORTH DAKOTA) border; 49°10′N 100°05′W. Mixed farming. Includes communities of Whitewater, BOISSEVAIN.

Morton, county (□ 729 sq mi/1,895.4 sq km; 2006 population 3,138), extreme SW KANSAS; ⊙ ELKHART; 37°12′N 101°48′W. Rolling plain, bordered W by COLORADO and S by OKLAHOMA; drained by CIMARRON RIVER. Wheat and grain sorghums; cattle, livestock. Gas field in E. Includes Cimarron National Grassland at center of county. (Mountain/Central time zone boundary). Formed 1886.

Morton, county (□ 1,920 sq mi/4,992 sq km; 2006 population 25,754), central NORTH DAKOTA; ⊙ MANDAN; 46°43′N 101°16′W. Agricultural area drained by BIG MUDDY CREEK and HEART RIVER; bounded by MISSOURI RIVER (LAKE OAHE Reservoir S of Mandan) on E (forms the border between the Mountain and Central time zones, except for NE corner of county, including Mandan, which is in Central). Manufacturing food products, petroleum refining; diversified farming, cattle; hogs; dairying; wheat, barley, hay. Fort Lincoln State Park in E; Fort Rice Historic Site in SE; Sweet Briar Reservoir in N. Formed 1873 and government organized in 1878. Named for Oliver Hazard Perry Throck Morton (1823–1877), a govenor and senator of Indiana. Lincoln was the county seat from 1878–1879.

Morton, town (2000 population 3,482), SCOTT county, central MISSISSIPPI, 32 mi/51 km E of JACKSON, in Bienville National Forest; 32°21′N 89°39′W. Manufacturing (paper mill equipment, wooden pallets and boxes, poultry processing). Roosevelt State Park to SW.

Morton (MOR-tuhn), town (2006 population 1,914), ⊙ COCHRAN county, NW TEXAS, on the LLANO ESTACADO, c.55 mi/89 km W of LUBBOCK; 33°43′N 102°45′W. Elevation 3,758 ft/1,145 m. Shipping, storage, trade center for agriculture and cattle-ranching area (cotton, wheat, sorghum); manufacturing (meat packing); oil and gas. Muleshoe National Wildlife Refuge to N. Incorporated 1934.

Morton, town (2006 population 1,095), LEWIS county, SW WASHINGTON, 34 mi/55 km E of CHEHALIS; 46°34′N 122°17′W. Agriculture, dairying; manufacturing (lumber, wood products). MOUNT SAINT HELENS NATIONAL VOLCANIC MONUMENT to SE; parts of Gifford Pinchot National Forest to N, NE, and SE. Riffe Lake reservoir (MOSSYROCK DAM) to S.

Morton (MAW-tuhn), village (2001 population 1,907), E DERBYSHIRE, central ENGLAND, 7 mi/11.3 km S of CHESTERFIELD; 53°08′N 01°23′W.

Morton, village (2000 population 15,198), TAZEWELL county, central ILLINOIS; 40°36′N 89°28′W. In a grain-farming and livestock area. Manufacturing (food processing, tractor parts, washing machines, pottery). Incorporated 1877.

Morton, village (2000 population 442), RENVILLE county, SW MINNESOTA, on MINNESOTA RIVER, and 7 mi/11.3 km E of REDWOOD FALLS; 44°32′N 94°58′W. Grain, soybeans, peas, sugar beets; livestock, poultry; dairying; manufacturing (rubber products). Lower Sioux Indian Reservation to S; Birch Coulee Battlefield Historic Site, set aside in commemoration of battle between Sioux Indians and U.S. Cavalry in 1862.

Morton, borough (2006 population 2,657), DELAWARE county, SE PENNSYLVANIA, suburb 9 mi/14.5 km WSW of downtown PHILADELPHIA; 39°54′N 75°19′W.

Manufacturing (railroad, wheels and axles, plastics products). Incorporated 1898.

Morton and Hanthorpe (MAW-tuhn and HAN-thorp), village (2001 population 2,352), LINCOLNSHIRE, E ENGLAND, 3 mi/4.8 km N of BOURNE; 52°48′N 00°23′W. Farming. Originally called just Morton, the village changed its name in 2004 to avoid confusion with another town called Morton, also in Lincolnshire.

Morton, Cape, GREENLAND: see WASHINGTON LAND.

Morton, East, ENGLAND: see BINGLEY.

Morton Grove, village (2000 population 22,451), COOK county, N of CHICAGO, NE ILLINOIS; 42°02′N 87°47′W. It has research laboratories and manufacturing plants (pumps, electrical equipment, cosmetics). On North Branch of CHICAGO RIVER. Incorporated 1895.

Morton National Park (MOR-tuhn) (□ 627 sq mi/1,630.2 sq km), E NEW SOUTH WALES, SE AUSTRALIA, 93 mi/150 km SW of SYDNEY, 10 mi/16 km S of MOSS VALE; 45 mi/72 km long, 30 mi/48 km wide; 34°55′S 150°15′E. Colorful sandstone cliffs and gorges; waterfalls; patches of sub-tropical rainforest. Platypus, fish. Visitor center, kiosk. Camping, picnicking, and hiking. Established 1938.

Mortons Gap, town (2000 population 952), HOPKINS county, W KENTUCKY, 7 mi/11.3 km S of MADISONVILLE; 37°14′N 87°28′W. Railroad junction. In coal-mining, agricultural (burley tobacco, grain; livestock; timber) area; manufacturing (barrel staves). White City Wildlife Management Area to NE.

Mortsel (MAWRT-suhl), commune (2006 population 24,386), Antwerp district, ANTWERPEN province, N BELGIUM, 3 mi/4.8 km SE of ANTWERP; 51°10′N 04°28′E. Manufacturing of photography products.

Mortugaba (MOR-too-GAH-bah), city (2007 population 13,628), S central BAHIA, BRAZIL, near MINAS GERAIS, BRAZIL, border; 14°59′S 42°39′W.

Morty (muhr-TI), village, N TATARSTAN Republic, E European Russia, near local higway, 13 mi/21 km W, and under administrative jurisdiction of YELABUGA; 55°49′N 51°44′E. Elevation 551 ft/167 m. Agricultural products.

Moruga (mo-ROO-ga), village, S TRINIDAD, TRINIDAD AND TOBAGO, 18 mi/29 km SE of SAN FERNANDO; 10°08′N 61°19′W. In cacao- and coconut-growing region; beach.

Morumbateman, AUSTRALIA: see MURRUMBATEMAN.

Morungaba (MO-roon-gah-bah), town (2007 population 12,007), S SÃO PAULO state, BRAZIL, 17 mi/27 km E of CAMPINAS; 22°52′S 46°48′W.

Morungole, Mount (MOR-oon-GO-lai), mountain (9,020 ft/2,749 m), KAABONG district, NORTHERN region, NE UGANDA, E of KIDEPO NATIONAL PARK; 03°48′N 34°03′E. Was in former KARAMOJA province.

Morunyaneng (moh-roon-YAH-neng), village, Mafeteng district, W LESOTHO, 35 mi/56 km SSE of MASERU, and 18 mi/29 km E of MAFETENG, near MAKHALENG RIVER; 29°50′S 27°32′E. Malealea Lodge to E. Corn, sorghum, vegetables; livestock. Tourism. Matelile park just to N (7,856 ft/2,395 m).

Moruppatti (MO-ruh-puht-tee), town, TIRUCHIRAPPALLI district, TAMIL NADU state, S INDIA, 30 mi/48 km NW of TIRUCHIRAPPALLI. Cotton weaving; sheep grazing. Magnetite deposits in PACHAIMALAI HILLS (N). Sometimes spelled Marupatti, Morupatti.

Moruroa, FRENCH POLYNESIA: see MURUROA.

Moruya (muh-ROO-yuh), town, SE NEW SOUTH WALES, SE AUSTRALIA, 70 mi/113 km SE of CANBERRA; 35°55′S 150°05′E. Dairying and agriculture center; blue-granite quarry; timber; oyster farming. Once the gateway to goldfields at ARALUEN and BRAIDWOOD.

Morvan (mor-vahn), mountainous region of E central FRANCE, forming northernmost spur of the MASSIF CENTRAL, mostly in NIÈVRE department; 47°10′N 04°00′E. This heavily forested upland with infertile soil rises to 2,959 ft/902 m at BOIS-DU-ROI. Cattle raising and lumbering are the main activities. The region is sparsely settled. AUTUN (in SAÔNE-ET-LOIRE department) is nearest larger urban center. The YONNE and CURE rivers flow N from the Morvan to the SEINE. The regional park of MORVAN (676 sq mi/1,751 sq km) was established in 1970 to stimulate tourism.

Morvan Natural Regional Park (mor-vahn), French *Parc Naturel Régional du Morvan* (pahrk nah-tyu-rel rai-zhyo-nahl dyoo mor-vahn) (□ 676 sq mi/1,757.6 sq km), in the NIÈVRE and SAÔNE-ET-LOIRE departments, BURGUNDY region, central FRANCE, northernmost spur of the MASSIF CENTRAL; 47°10′N 04°05′E. Within the park is Mont Beuvray (2,694 ft/821 m), Oppidum (pallisaded fort) and capital of the Eduii who called on Caesar for help when threatened by the Helvetii and thus stared the Gallic Wars.

Morvant (MOR-vah), village (2000 population 15,366), NW TRINIDAD, TRINIDAD AND TOBAGO, 1.5 mi/2.4 km E of PORT OF SPAIN. In coconut and citrus fruit region; saw milling. Has new housing development and residential suburbs.

Morven, village (2000 population 634), BROOKS county, S GEORGIA, 16 mi/26 km WNW of VALDOSTA; 30°56′N 83°30′W.

Morven, village (□ 1 sq mi/2.6 sq km; 2006 population 552), ANSON county, S NORTH CAROLINA, 8 mi/12.9 km SSE of WADESBORO; 34°51′N 80°00′W. Manufacturing (home furnishings); service industries; agriculture (cotton, grain; livestock).

Morven (MOR-vuhn), mountain (2,862 ft/872 m) of the GRAMPIANS, Aberdeenshire, NE Scotland, 5 mi/8 km N of BALLATER; 57°07′N 03°02′W. Formerly in Grampian, abolished 1996.

Morven (MOR-vuhn), mountain (2,313 ft/705 m), HIGHLAND, N Scotland, 9 mi/14.5 km NNW of HELMSDALE; 58°13′N 03°42′W. It is celebrated in poems by Ossian.

Morvern (MOR-vuhrn), mountainous district and peninsula (c.25 mi/40 km long, 14 mi/23 km wide), SW HIGHLAND, N Scotland, bounded by Sound of Mull (SW), LOCH SUNART (N), and LOCH LINNHE (SE). Site of Glensanda quarry, Europe's largest coastal granite quarry.

Morwell (MOR-wel), town (2001 population 13,823), S VICTORIA, SE AUSTRALIA, 85 mi/137 km ESE of MELBOURNE; 38°14′S 146°24′E. Railroad junction; coal-mining center (lignite); thermal power stations; coal and power museum, visitor center. Public rose gardens.

Moryakovskiy Zaton (muh-ryah-KOF-skeeyee zah-TON), town (2006 population 4,750), SE TOMSK oblast, W SIBERIA, RUSSIA, on the TOM′ River, on road, 18 mi/29 km NNW of TOMSK; 56°41′N 84°39′E. Elevation 298 ft/90 m. Ship repair yards and base of a river fleet; glass factory.

Moryn (MO-reen), Polish *Moryń*, German *Mohrin*, town in BRANDENBURG, after 1945 in Szczecin province, NW POLAND, 20 mi/32 km NNW of KOSTRZYŃ (Küstrin); 52°51′N 14°23′E. Grain, potatoes, vegetables; livestock. Ruins of old town walls. Tourist resort.

Morzhovoi (mor-SHO-vee), village, SW ALASKA, on SW ALASKA PENINSULA, on narrow strait between the PACIFIC OCEAN and BERING SEA, opposite UNIMAK Island; 54°55′N 163°18′W. Native population works in fishing industry during summer.

Morzine (mor-zeen), commune (□ 17 sq mi/44.2 sq km), HAUTE-SAVOIE department, RHÔNE-ALPES region, SE FRANCE, on Dranse River, 17 mi/27 km SE of THONON-LES-BAINS, in the Chablais region of the lower Savoy Alps (*alpes françaises*); 46°11′N 06°43′E. Summer and winter resort with fine excursion trails and good ski terrain for beginners at elevation of 2,000 ft/610 m–4,000 ft/1,219 m (aerial tramways). The modern high-rise village of AVORIAZ (5 mi/8 km E; steep, curvy road) lies at 6,000 ft/1,829 m. It has miles of ski trails connected with other stations in the hill area S of Lake GENEVA.

Mosabani, India: see MUSHABANI.

Mosal'sk (muh-SAHLSK), city (2005 population 4,300), W KALUGA oblast, central European Russia, on crossroads, 58 mi/93 km W of KALUGA; 54°29′N 34°59′E. Elevation 688 ft/209 m. Agricultural industries (drying of vegetables, dairy products, flax processing, bakery). Chartered in 1231. Became city in 1776.

Mosbach (MAWS-bahkh), town, LOWER NECKAR, BADEN-WÜRTTEMBERG, GERMANY, 21 mi/34 km ESE of HEIDELBERG; 49°21′N 09°05′E. Manufacturing of small locomotives. Has 16th-century town hall; old frame houses. Was first mentioned in 8th century; chartered in 1291.

Mosborough (MOS-buh-ruh), suburb (2001 population 34,711) of SHEFFIELD, SOUTH YORKSHIRE, N ENGLAND, 7 mi/11.2 km S of ROTHERHAM; 53°19′N 01°22′W. Residential. Former coal-mining site.

Mosby, town (2000 population 242), CLAY county, W MISSOURI, 8 mi/12.9 km NNE of LIBERTY; 39°19′N 94°17′W.

Mosby (MAWS-bu), village, VEST-AGDER county, part of KRISTIANSAND city, S NORWAY, on the OTRA. On railroad.

Mosca (MOS-kah), town, HUÁNUCO region, central PERU, in Cordillera CENTRAL, 22 mi/35 km SW of AMBO; 10°15′S 76°19′W. Elevation 9,842 ft/2,999 m. Potatoes, cereals; livestock.

Mosca, hamlet, ALAMOSA county, S COLORADO, 12 mi/19 km N of ALAMOSA. Elevation 7,550 ft/2,301 m. In agricultural region of the SAN LUIS VALLEY. GREAT SAND DUNES NATIONAL MONUMENT and San Luis Lakes State Park are E.

Mosca Pass (9,713 ft/2,961 m), S COLORADO, in SANGRE DE CRISTO MOUNTAINS, between HUERFANO and ALAMOSA counties. Crossed by trail, county road reaches Mosca Pass radio tower from E.

Mosca Peak, NEW MEXICO: see MANZANO MOUNTAINS.

Moscari (mo-SKAH-ree), town and canton, CHARCAS province, POTOSÍ department, W central BOLIVIA, 7 mi/11.3 km SSW of San Pedro; 18°19′S 66°02′W. Oca.

Moscavide (MOOSH-kah-veed), suburb of LISBON, LISBOA district, W PORTUGAL, 5 mi/8 km NNE of city center; 38°46′N 09°06′W. On W bank of TAGUS RIVER.

Moščenička Draga, resort village, W CROATIA, in ISTRIA, on ADRIATIC SEA, in Opatija Riviera, 4.3 mi/6.9 km S of OPATIJA.

Moschin, POLAND: see MOSINA.

Moscice (mos-TSEE-tse), Polish *Mościce*, village, Kraków province, SE POLAND, just W of TARNÓW; 50°01′N 20°56′E. Manufacturing of nitrogen fertilizers.

Mosciska, UKRAINE: see MOSTYSKA.

Moscow, state existing in W central RUSSIA from the late 14th to mid-16th centuries, with the city of MOSCOW as its nucleus. Its formation and eventual ascendancy over other Russian principalities and over the Tatars of the GOLDEN HORDE came about gradually and resulted particularly from its central location, its importance as a trade artery, its dynastic continuity, its circumspect loyalty to Tatar overlords, and its prestige as a religious center. After the decline of KIEV in the mid-12th century, Russian territory broke up into a number of separate political units, among which the principality of Vladimir-Suzdal was the most important. The rulers of Vladimir were the only Russian princes who bore the title "grand duke" and were regarded as suzerains of the other princes.

According to tradition, Moscow was founded on a strategic site on the MOSKVA River as a military outpost of Vladimir-Suzdal; by the mid-12th century, when its existence is first mentioned in Russian chronicles, it had become a walled town. The first known prince of Moscow was Daniel (d. in 1303), son of Grand Duke Alexander Nevskiy. Daniel received Moscow as a separate appanage. His son, Yuriy (reigned 1303–1325), launched the struggle for Moscow's predominance in Russia, competing for leadership with the prince of

TVER for both the title of grand duke and the allegiance of the less powerful Russian princes. Yuriy was temporarily appointed grand duke of Vladimir by the khan of the Empire of the Golden Horde. His younger brother, Ivan I (Ivan Kalita; reigned 1328–1341), was not only granted the title of grand duke (1328) but was given the right to collect Tatar tributes from neighboring principalities. Moreover, during Ivan's reign, Moscow became the seat of the Russian Orthodox Church. The adjacent areas were subdued or acquired, and Moscow's importance continued to increase, particularly under Ivan I's grandson, Dmitriy Donskoy (reigned 1359–1389), who was probably the first to bear the title "grand duke of Moscow." Dmitriy's successors, above all Ivan III (reigned 1462–1505) and Vasily III (reigned 1505–1533), laid the basis of Muscovite absolutism, built the Great Russian state, and threw off the Tatar yoke. By the mid-16th century, therefore, the unification of the Great Russian lands had been completed under the princely dynasty. The Muscovite rulers now bore the title "grand duke of Moscow and of all Russia," and the history of the grand duchy of Moscow became that of Russia.

Moscow, Russian *Moskva* (muhs-KVAH), city (2005 population 10,472,630), ☉ RUSSIAN FEDERATION and of MOSCOW oblast, central European Russia, on the MOSKVA River near its junction with the MOSCOW Canal; 55°45′N 37°37′E. Elevation 410 ft/124 m. Moscow is Russia's largest city and the leading political, economic, cultural, educational, and scientific center. The hub of the Russian railroad network, it is also an inland port and has several civilian and military airports. Moscow's major industries include machine building, metalworking, machine tools, precision instruments, automobiles, trucks, aircraft, chemicals, oil refining, wood and paper products, textiles, clothing, footwear, filmmaking, and publishing. Although archaeological evidence indicates that the site has been occupied since Neolithic times, the village of Moscow was first mentioned in the Russian chronicles in 1147. Around 1271, Moscow became the seat of the grand dukes of Suzdal-Vladimir, who later assumed the title of grand dukes of Moscow. The first stone walls of the Kremlin were built in 1367. Moscow, or Muscovy, achieved dominance over the Russian lands by virtue of its strategic location at the crossroads of medieval trade routes, its leadership in the struggle against, and final defeat of, the Tatars, and its gathering of neighboring principalities under Muscovite suzerainty. By the 15th century, Moscow had become the capital of the Russian state, and in 1547, Grand Duke Ivan IV became the first to assume the title of tsar. Moscow was also the seat of the Metropolitan (later Patriarch) of the Russian Orthodox Church from the early 14th century. It has been an important commercial center since the Middle Ages and the center of many crafts. Burned by the Tatars in 1381 and again in 1572, the city was taken by the Poles during the Time of Troubles. The Russian capital was transferred from Moscow to SAINT PETERSBURG in 1712; but Moscow's cultural and social life continued uninterrupted, and the city never ceased to be the religious center of Russia. Built largely of wood until the 19th century, Moscow suffered from numerous fires, the most notable of which occurred in the wake of Napoleon I's occupation in 1812. Nearly the entire city, except for the great stone churches and palaces, burned down. Rebuilt, Moscow developed from the 1830s as a major textile center. In 1918, the Soviet government transferred the capital back to Moscow and fostered spectacular economic growth in the city, whose population doubled between 1926 and 1939 and again between 1939 and 1992. During World War II, Moscow was the goal of a 2-pronged German offensive, but suffered virtually no war damage. (Kuybyshev, now Samara, served as the temporary wartime capital during the German advance toward Moscow.)Moscow is governed by a city council and a mayor and is divided into boroughs. The 5 major sections of Moscow form concentric circles, of which the innermost is the Kremlin, a walled city in itself. Its walls represent the city limits as of the late 15th century. Adjoining the Kremlin in the E is the huge Red Square, originally a marketplace and a meeting spot for popular assemblies; it is still used as a parade ground and for demonstrations. On the W side of the Red Square and along the Kremlin wall are the Lenin Mausoleum and the tombs of other Soviet political figures; on the N side is the historical museum; and at the S end stands the imposing cathedral of Basil the Beatified. It was built in the 16th century to commemorate the conquest of Kazan. One of the most exuberant examples of Russian architecture, the cathedral has numerous cupolas, each a different color, grouped around a central dome. To the E of the Red Square extends the old district of Kitaigorod [Russian=China city], once the merchant's quarter, later the banking section, and now an administrative hub with various government offices and ministries. Gorky Street (during the Soviet period called Tver Street or, in Russian, *Tverskaya ulitsa*), a main thoroughfare, extends N from the Kremlin and is lined with modern buildings, including the headquarters of the council of ministers; it is connected with the St. Petersburg highway, which passes the huge Dynamo stadium and the central airport. Near the beginning of Tver Street is Theater Square, containing the Bolshoy and Maly theaters. Encircling the Kremlin and Kitaigorod are the Bely Gorod [Russian=white city], traditionally the most elegant part of Moscow and now a commercial and cultural area; the Zemlyanoy Gorod [Russian=earth city], named for the earthen and wooden ramparts that once surrounded it; and the inner suburbs. A notable feature of Moscow are the concentric rings of wide boulevards and railroad lines on the sites of old walls and ramparts.Except for its historical core, Moscow was transformed into a sprawling but well-planned modern city under the Soviet government. Among its many cultural and scientific institutions are the University of Moscow (founded in 1755), the Russian Academy of Sciences (founded in 1725 in St. Petersburg and transferred to Moscow in 1934), a nuclear research reactor, a conservatory (1866), the Tretyakov art gallery (opened in the 1880s), the Museum of Oriental Cultures, the State Historical Museum, the Russian State Library (during the Soviet period called the Lenin Library), and the Agricultural Exhibition. Theaters include the Moscow Art Theater, the Bolshoy (opera and ballet), and the Maly Theater (drama). Moscow is also the see of a patriarch, head of the Russian Orthodox Church. A new Palace of Congresses was built (1961) inside the Kremlin walls for meetings of the Supreme Soviet. Moscow's numerous large parks and recreation areas include Gorky Central Park, the forested Izmailovo and Sokolniki parks, and Ostankino Park, with its botanical gardens. The ornate Moscow subway system opened in 1935.

Moscow, city (2000 population 21,291), ⊙ LATAH county, NW IDAHO, 23 mi/37 km N of LEWISTON, at the WASHINGTON state line, and 10 mi/16 km E of PULLMAN, Washington, near South Fork of PALOUSE RIVER; 46°44′N 117°00′W. Elevation 2,564 ft/782 m. It is a trade center for a lumber and farm area where wheat, peas, lentils, and dairy items are produced. Manufacturing (semiconductors, printing and publishing, erosion control blankets, concrete, wooden cabinets). Originally part of the Nez Perce Reservation, it was first settled by whites in 1871. The University of Idaho is there, as well as a historical museum and a U.S. government forest sciences laboratory. Nez Perce Indian Reservation to SE. Incorporated 1887.

Moscow (MAW-skou), town, SOMERSET county, W central MAINE, on the KENNEBEC and 22 mi/35 km NNW of SKOWHEGAN; 45°07′N 69°52′W.

Moscow, town (2006 population 556), FAYETTE county, SW TENNESSEE, on WOLF RIVER, and 34 mi/55 km E of MEMPHIS; 35°04′N 89°24′W.

Moscow, village, RUSH county, E central INDIANA, on FLATROCK RIVER, and 10 mi/16 km SW of RUSHVILLE. Agricultural area. Settled 1822, laid out 1832.

Moscow (MAHS-kou), village (2000 population 247), STEVENS county, SW KANSAS, 12 mi/19 km NE of HUGOTON; 37°19′N 101°12′W. In agricultural area.

Moscow (MAHS-ko), village (2006 population 241), CLERMONT county, SW OHIO, 15 mi/24 km S of BATAVIA, and on OHIO RIVER (here forming KENTUCKY state line); 35°52′N 84°13′W. Grant Memorial Bridge is nearby.

Moscow (MAHS-kou), borough (2006 population 1,944), LACKAWANNA county, NE PENNSYLVANIA, 9 mi/14.5 km SE of SCRANTON, on Roaring Brook, forms Elmhurst Reservoir to N; 41°20′N 75°31′W. Agricultural area (corn, hay; cattle, dairying); timber; manufacturing (lumber, metal fabrication). Lackawanna State Forest to SW; Moosic Mountains to NW. Settled 1830.

Moscow (MAWS-kuh-oo), Russian *Moskovskaya* (muhs-KOF-skah-yah), oblast (□ 17,783 sq mi/46,235.8 sq km; 2006 population 6,547,865), in the central East European Plain, between the VOLGA and OKA rivers, European Russia; ⊙ MOSCOW. Generally level; low hills alternating with flat depressions. Forests cover 40% of the oblast; reforestation work is in progress and industrial timber cutting severely restricted. The hilly SMOLENSK-MOSCOW UPLAND (N; up to 1,017 ft/310 m in the KLIN-DMITROV RIDGE) stretches from SW to NE and contains Trostenskoe, Nerskoe, and Krugloe lakes. Further N is the swampy Upper Volga Lowland, a flat low plain with ridges and dunes and including the Shosha and Dubna depressions. Meshchera Lowland, a swampy, mostly flat outwash plain is in the SE; contains Chernoye, Velikoye, Sviatoye, and Dubovoye lakes. In the S is the CENTRAL RUSSIAN UPLAND, dissected by river valleys and ravines and gullies. Continental climate, with cold winters and warm summers; precipitation reaches 26 in/65 cm. The growing season lasts 4–5 months. A dense river network, all within the Volga River basin, including the KLYAZMA and DUBNA (N) and the Oka, PROTVA, Nara, Lopasnia, TSNA, and Osetr (S) rivers. The MOSKVA River (Mozhaysk, the largest; Ruza; Ozery; Istra; Klyazma; and Ucha reservoirs; canal) flows through Greater Moscow. The Klyazma River and its tributaries originate here. The MOSCOW CANAL passes through the N. Largest railroad network in Russia. Regular navigation on Moscow Canal and Moskva and Oka rivers. Main gas pipeline. Local fauna include elk, marten, polecat, badger, fox, boar. An old industrial region, with manufacturing dating back to the 18th cent. (textiles). Today, industry is highly developed, with electric power, chemicals, oil refining, machines; metallurgy and metalworking; food processing. Intensive agriculture oriented toward local urban markets, including dairy and livestock. Population almost entirely Russian. Formed in January 1929.

Moscow Basin, lignite basin, TULA oblast, central European Russia, S of MOSCOW; approximately 200 mi/322 km long E to W and 50 mi/80 km wide N to S. TULA is the chief city of the region. Low-grade bituminous and lignite coals, suitable for the power plants of the Moscow industrial region, are mined here.

Moscow Canal (MAWS-kuh-oo), Russian *Kanal Imeni Moskvy*, 80 mi/129 km long; waterway in MOSCOW oblast, European Russia, linking VOLGA and MOSKVA rivers; leaves the Volga River at IVANKOVO, at the E end of the VOLGA RESERVOIR; extends generally S, past DMITROV and YAKHROMA, then climbs by means of locks onto the KLIN-Dmitrov Ridge, here forming a series of reservoirs (Iksha, Pestovo, Pyalovo, Ucha, Klyaz'ma), using courses of UCHA and KLYAZ'MA

rivers. From the Khimki Reservoir (N port of MOSCOW, in NW city limits; 5 mi/8 km long, 1 mi/1.6 km wide), the canal descends 100 ft/30 m, via two locks, to Moskva River just W of Moscow. Completed in 1937; called Moscow-Volga Canal (Russian *Kanal Moskva-Volga*) until 1947.

Moscow Mills (mahs-ko-MILZ), town (2000 population 1,742), LINCOLN county, E MISSOURI, on CUIVRE RIVER, and 4 mi/6.4 km SE of TROY; 38°57′N 90°55′W. Agriculture. Residential growth from ST. LOUIS metropolitan area.

Moscow Sea, RUSSIA: see VOLGA RESERVOIR.

Moscow-Volga Canal, RUSSIA: see MOSCOW CANAL.

Mosel, FRANCE and GERMANY: see MOSELLE.

Moseley, unincorporated community, POWHATAN county, central VIRGINIA; 37°28′N 77°46′W.

Moseley and Kings Heath (MOZ-lee and KINGZ HEETH), S industrial suburbs (2001 population 24,273) of BIRMINGHAM, West Midlands, central ENGLAND; 52°27′N 01°51′W.

Moselle (mo-zel), department (□ 2,400 sq mi/6,240 sq km), in LORRAINE, NE FRANCE; ⊙ METZ; 49°00′N 07°00′E. Borders on the German Saar (SAARLAND) (NE) and on Luxembourg (N). Bounded E by the N VOSGES MOUNTAINS. Drained by the MOSELLE and the SAAR rivers. Central section is occupied by Lorraine plateau, of limited fertility, and in areas dotted with shallow ponds (drained periodically for cultivation). Livestock raising. Chief crops: barley, oats, rye, potatoes, hops, and tobacco. Vineyards along left bank of the Moselle. Mineral deposits are still of some importance; iron has been mined for decades in THIONVILLE basin (at HAYANGE, KNUTANGE) and in ORNE RIVER valley; coal mines (S extension of Saar basin) are in decline in Forbach-Saint-Avold district; salt mines in Château-Salins-Dieuze area. Metallurgy, the department's chief industry, but now in rapid decline is found in the Metz-Thionville stretch of the Moselle valley; it is being replaced by modern rolling, finishing, and steel product manufacturing facilities. Glassworks in the N VOSGES, near SARREBOURG. Other industries: tanning and shoe manufacturing (Metz), porcelain (SARREGUEMINES), woodworking, brewing, fruit and vegetable preserving. Chief cities are Metz and Thionville, with numerous smaller agglomerations in the old mining districts. Department is the part of Lorraine that was twice annexed to Germany (1871–1918; 1940–1944). Has small number of people who speak a Germanic dialect. High unemployment has followed decline of primary industries. A large nuclear power plant operates at CATTENOM, near the Moselle NE of Thionville. The department forms part of the administrative region of Lorraine.

Moselle (mo-ZEL), unincorporated rural community, FRANKLIN county, E central MISSOURI, on MERAMEC RIVER, and 8 mi/12.9 km SE of UNION.

Moselle, Côtes de (mo-zel, kot duh), cuesta in NE FRANCE, extending from the E bank of MEUSE RIVER above NEUFCHÂTEAU (VOSGES department) NNE to NANCY, then N, principally along left bank of the MOSELLE, into Luxembourg; 49°30′N 05°45′E. These heights have been of great strategic importance as an obstacle to invasions from E. Major strongholds and (later) transportation and industrial centers (Nancy, METZ, THIONVILLE) have arisen where gaps provided access to PARIS BASIN. Some wine is grown near TOUL and PAGNY-SUR-MOSELLE. Iron mines near Nancy, BRIEY, and Thionville are in severe decline.

Moselle River (mo-zel), German *Mosel*, 320 mi/515 km long, in NE FRANCE and W GERMANY; rises in the central VOSGES MOUNTAINS, NE France; winds generally N across LORRAINE, past ÉPINAL, TOUL, and METZ; 50°10′N 07°15′E. Leaving France, it forms part of the border between Luxembourg and Germany, then enters Germany, passes TRIER, and cuts between the EIFEL and the HUNSRÜCK ranges to reach the RHINE RIVER at KOBLENZ. The Moselle receives the

MEURTHE above NANCY and the SAAR RIVER near Trier. The French section of the Moselle valley is partly occupied by old mining communities and a declining steel industry. The German section is dotted with numerous old castles and is covered with celebrated vineyards. The Moselle Canal, built in 1964, made the river navigable for 1,500-ton/1,361-metricton barges from below Nancy to Koblenz. It is overseen by representatives of France, Luxembourg, and Germany.

Moserboden, AUSTRIA: see KAPRUN.

Moses Lake, city (2006 population 17,272), GRANT county, central WASHINGTON, on E shore of MOSES LAKE, 68 mi/109 km NE of YAKIMA; 47°08′N 119°17′W. A distributing and shipping point for the COLUMBIA BASIN PROJECT, its chief products are sugar beets, potatoes, and milk. Heavy manufacturing. Grant County Airport to N; Municipal Airport to E. Columbia National Wildlife Refuge to S; Moses Lake State Park to W. Central part of city is on peninsula between Parker Horn and Pelican Horn, horn-shaped arms of lake. Large Potholes Reservoir to S. Adam East Museum; Big Bend Community College here. Settled 1897, incorporated 1938.

Moses Lake, reservoir, GRANT county, E WASHINGTON, on Crab Creek, 2 mi/3 km SW of MOSES LAKE; 47°05′N 119°20′W. Maximum capacity 50,000 acre-ft. Formed by Moses Lake South Dam (20 ft/6 m high), built (1962) by the Bureau of Reclamation for irrigation. Moses Lake State Park on S bank.

Moses Point, locality, W ALASKA, on SE SEWARD PENINSULA, on NORTON SOUND, at mouth of NORTON BAY, 100 mi/161 km E of NOME; 64°45′N 161°46′W. Air field. In the Elim Indian Reservation.

Mosetenes, Serranía de (se-rah-NEE-ah dai mo-sai-TAI-nes), N outlier of Cordillera de COCHABAMBA, in COCHABAMBA department, W central BOLIVIA; extends c.120 mi/193 km NW from area N of SACABA to area W of Huachi, parallel to Cordillera REAL. Rises to 8,200 ft/2,499 m at 16°40′S 66°03′W. Forms watershed between BENI and MAMORÉ river basins.

Moseushi (mo-SE-oo-shee), town, Sorachi district, Hokkaido prefecture, N JAPAN, 53 mi/85 km N of SAPPORO; 43°41′N 141°58′E.

Mosgiel (MOZ-geel), SW suburb within DUNEDIN city, SE SOUTH ISLAND, NEW ZEALAND, at NE margin of agriculturally productive TAIERI PLAINS. Woolen mills, small coal mines. Taieri airport nearby. Incorporated into Dunedin City in 1989.

Moshassuck River (mo-SHASS-uck), c.10 mi/16 km long, N RHODE ISLAND; rises in LINCOLN; flows generally SSE, past Saylesville and Central Falls, through PROVIDENCE, joining WOONASQUATUCKET RIVER just before entering PROVIDENCE RIVER.

Moshenskoye (muh-shin-SKO-ye), village (2006 population 2,675), E NOVGOROD oblast, NW European Russia, on the Uver' River (tributary of the MSTA River), on road junction, 25 mi/40 km ENE of BOROVICHI; 58°31′N 34°35′E. Elevation 501 ft/152 m. Food processing. Formerly called Nikolo-Moshenskoye.

Moshi (MO-shee), city, ⊙ KILIMANJARO region, N TANZANIA, 165 mi/266 km NW of TANGA, near S base of Mount KILIMANJARO, near KENYA, on railroad; 03°21′S 37°21′E. Airstrip to S, Kilimanjaro International Airport 16 mi/26 km SW to SE. Coffee, tea, sugarcane, wheat, corn, sisal, vegetables; cattle, sheep, goats. Manufacturing (coffee and tea processing, food processing, machinery). ARUSHA NATIONAL PARK to W. Nyumba ya Mungu reservoir (PANGANI RIVER) to S. Seat of Cooperative College and College of African Wildlife Management. The original town, now called Old Moshi and located nearby, was a capital of the 19th-century Chagga kingdom and became (late 19th century) an administrative center under the Germans. In the 20th century the British moved Moshi to its present site.

Moshkovo (muh-SHKO-vuh), town (2006 population 10,500), E NOVOSIBIRSK oblast, SW SIBERIA, RUSSIA,

on road and the TRANS-SIBERIAN RAILROAD, 30 mi/48 km NE of NOVOSIBIRSK; 55°18′N 83°36′E. Elevation 685 ft/208 m. In agricultural area; food processing. Railroad station since 1898; town since 1961.

Moshra Al-Rig, village, Warrap state, S SUDAN, on an island in Lake AMBADI, at head of navigation on the BAHR AL-GHAZAL, and 100 mi/161 km NE of WAU. Steamer landing.

Mosier (MO-zhuhr), village (2006 population 410), WASCO county, N OREGON, 12 mi/19 km NW of THE DALLES, and on Columbia River (Bonneville Reservoir) at mouth of Mosier Creek; 45°40′N 121°23′W. Agriculture (apples, cherries, grapes). Koberg Beach Wayside State Park to W; Memaloose and Mayer State Parks to E; MOUNT HOOD National Forest to S.

Mosigkau (MAW-sik-kou), SW suburb of DESSAU, SAXONY-ANHALT, central GERMANY; 51°49′N 12°12′E. Has former palace (1752) with noted art collection.

Mosina (mo-SEE-nah), German Moschin, town, Poznań province, W POLAND, on WARTA River, on railroad, and 11 mi/18 km S of POZNAŃ. Tanning, sawmilling. LUDWIKOWO, German Ludwigsberg, 2 mi/ 3.2 km W, is railroad spur terminus, health resort, near WIELKOPOLSKI NATIONAL PARK.

Mosinee, town (2006 population 4,078), MARATHON county, central WISCONSIN, on Mosinee Flowage of WISCONSIN RIVER, and 12 mi/19 km SSW of WAUSAU, at head of Du Bay Reservoir; 44°47′N 89°40′W. In lumbering and dairying area; paper and sawmilling, cheese making. Manufacturing (trailer components, papers, nylon slings). BIG EAU PLEINE RESERVOIR to SW. Incorporated 1931. Central Wisconsin Regional Airport to SE.

Mosi-oa-Toenja, ZIMBABWE and ZAMBIA: see VICTORIA FALLS.

Mosi-oa-Tunya National Park (□ 25 sq mi/65 sq km), SOUTHERN province, S ZAMBIA, on ZAMBEZI RIVER. On ZIMBABWE border, immediately W of MARAMBA (Livingstone). Park name means "smoke that thunders," refers to VICTORIA FALLS (343 ft/105 m), on Zambezi River, main feature of park. Wildlife (giraffes, buffalo, zebras, wildebeest, sable, baboons, vervet monkeys). Lookout Tree, baobab tree with observation platform; Knife Edge footbridge connects island with mainland. Park has Field Museum; Zoological Park upstream (to W). Zimbabwe's Victoria Falls National Park adjoins park to S.

Mosjøen (MAW-shuh-uhn), town, NORDLAND county, N central NORWAY, on VEFSNFJORD (inlet of NORTH SEA), at mouth of VEFSNA River, on railroad, and 100 mi/161 km NNE of NAMSOS; 65°50′N 13°12′E. Portuguese. Manufacturing (aluminum, textiles); lumber milling, Cold-storage plant. Steatite (used for Trondheim cathedral) quarried nearby. Incorporated 1875.

Moskalënki (muhs-kah-LYON-kee), town (2006 population 9,715), SW OMSK oblast, SW SIBERIA, RUSSIA, on road junction and the TRANS-SIBERIAN RAILROAD, 55 mi/89 km W of OMSK; 54°57′N 71°56′E. Elevation 364 ft/110 m. Dairy products, grain elevator. Formerly called Ol'gino.

Moskal'vo (muh-skahl-VO), village, NW SAKHALIN oblast, RUSSIAN FAR EAST, major oil port on a sheltered inlet of Sakhalin Gulf, terminus of highway branch and railroad spur, 18 mi/29 km W of OKHA; 53°34′N 142°30′E. Has pipeline connection with petroleum wells at Okha and Dagi on Sakhalin, and refinery at KOMSOMOL'SK-NA-AMURE on the mainland. Special oil-loading installations built here (1933–1934).

Moskenesøya (MAWSK-uh-nais-uh-u), island (□ 72 sq mi/187.2 sq km) in NORTH SEA, NORDLAND county, N NORWAY, one of the LOFOTEN Islands, 30 mi/48 km WSW of SVOLVÆR; 22 mi/35 km long, 6 mi/9.7 km wide. Rises to 3,392 ft/1,034 m. Has fishing villages.

Moskenstraumen (MAWSK-uhn-stroum-uh) or **Malstrøm**, tidewater whirlpool in the LOFOTEN Islands, NW Norway. Formed when a strong tidal

current flows through an irregular channel S of MOSKENESØYA island, it is c.2.5 mi/4 km wide and may at its center reach a speed of 10 ft/3 m per second. It is a danger to small ships and has been described with great imagination by Edgar Allan Poe and Jules Verne. The term malstrøm is applied to any whirlpool.

Moskovsk (mos-KOVSK), town, NE LEBAP region, E TURKMENISTAN, near the AMU DARYA River, 16 mi/26 km NW of CHARJEW; 39°13′N 63°19′E. Cotton. Formerly called Buyun-Uzun.

Moskovsk, town, LEBAP weloyat, TURKMENISTAN, alongside W bank of AMU DARYA, c.16 mi/25 km NW of CHARJEW; 39°13′N 63°19′E.

Moskovskiy, town, SE KHATLON viloyat, TAJIKISTAN, near PANJ RIVER (AFGHANISTAN border), 20 mi/32 km SSW of KULOB; 37°37′N 69°42′E. Center of agricultural area (cotton, jute) placed under cultivation in 1940s; tertiary level administrative center. Formerly called Chubek. Also spelled Moskovskii.

Moskovskiy (muhs-KOFS-keeyee), urban settlement (2006 population 3,445), SW TYUMEN oblast, SW Siberian Russia, on highway and near the TRANS-SIBERIAN RAILROAD, 4 mi/6 km SW, and under administrative jurisdiction, of TYUMEN; 57°06′N 65°25′E. Elevation 324 ft/98 m. Research and development center for agricultural industry; institute for re-qualification of agricultural professionals, pedigree livestock-breeding farm.

Moskovskiy (muhs-KOFS-keeyee), settlement (2006 population 15,395), central MOSCOW oblast, central European Russia, on highway, 4 mi/6 km SSE of ODINTSOVO; 55°36′N 37°21′E. Elevation 610 ft/185 m. In agricultural area (wheat, oats, flax, potatoes, vegetables).

Moskovskoye More, RUSSIA: see VOLGA RESERVOIR.

Moskushamn (MAWS-koos-HAHM-uhn), coalmining settlement, central West Spitsbergen, SPITSBERGEN group, NORWAY, on E shore of Advent Bay (small S arm of ISFJORDEN), 3 mi/4.8 km NE of Longyear City; 78°15′N 15°45′E. Formerly called Hiorthamn.

Moskva (MOSK-vah), village, central LEBAP weloyat, E TURKMENISTAN, on island in the AMU DARYA (river), 33 mi/53 km SE of CHARJEW; 38°30′N 64°30′E. Cotton.

Moskva (muhs-KVAH), river, approximately 310 mi/ 500 km long, in MOSCOW oblast, central European Russia; rising in the hills W of MOSCOW, near the border of SMOLENSK and Moscow oblasts; meandering generally E past MOZHAYSK and Moscow to join the OKA River near KOLOMNA. It is connected with the upper VOLGA RIVER at DUBNA by the MOSCOW-VOLGA CANAL (80 mi/130 km long).

Moskva, RUSSIA: see MOSCOW, city.

Moslavina (MOS-lah-vee-nah), region, central CROATIA, between Česma and ILOVA rivers, N of Lonjsko Polje; includes hilly area of Moslavačka gora and contact zone with alluvial plain around Sava River. Pop. concentrated in the contact zone along Zagreb–Slavonski Brod railroad. Oil and gas fields (Stružec, Goljlo) provide basis for petrochemical industry in KUTINA (chief town); other manufacturing includes lumber; furniture, metals, foods; wineries.

Mosman (MAHS-muhn), residential suburb, E NEW SOUTH WALES, SE AUSTRALIA, 4 mi/6 km NNE of SYDNEY across PORT JACKSON; 33°49′S 151°14′E. In metropolitan area. Ferry. Residential, with commercial and retail area.

Mosnang (mos-NAH), agricultural commune, ST. GALLEN canton, NE SWITZERLAND, 17 mi/27 km WSW of ST. GALLEN; 47°22′N 09°02′E. Elevation 2,382 ft/726 m.

Mosnov, CZECH REPUBLIC: see BRUSPERK.

Mosolovo (muh-suh-LO-vuh), village, central RYAZAN oblast, central European Russia, on short right tributary of the OKA River, on railroad and terminus of local highway branch, 40 mi/64 km SE of RYAZAN; 54°17′N 40°32′E. Elevation 456 ft/138 m. Woodworking.

Moson (MO-shon), former county of NE HUNGARY. Major portion ceded (1920) to CZECHOSLOVAKIA; now forms part of Győr-SOPRON county.

Mosonmagyaróvár (MO-shon-mah-dyahr-o-vahr), Hungarian *Mosonmagyaróvár*, city (2001 population 30,432), Győr-SOPRON county, NW HUNGARY, on arm of the DANUBE River (here bridged), at mouth of LAJTA RIVER, and 21 mi/34 km NW of Győr; 47°52′N 17°17′E. Near Hungary's W border and VIENNA; preserves many historical relics from the 16th–18th centuries (castles, homes of the aristocracy, churches) and some Roman, Hungarian, and Avar periods. Industrial, agriculture, market center. Manufacturing (agriculture machinery and tools, munitions, knitted wear, meat processing, farinaceous food products and baked goods, preservatives); aluminum plants, flour mills. The large aluminum industry is undergoing severe contraction with the end of the Hungarian-Soviet aluminum agreement. Summer resort; agricultural experiment station and technical college Grain, sugar beets; honey; cattle, hogs, ducks, geese in area. Formed 1939 by union of MOSON, German *Wieselburg*, and MAGYARÓVÁR, German *Ungarisch-Altenburg*.

Mosonszentjános (MO-shont-sant-yah-nosh), Hungarian *Mosonszentjános*, village, Győr-SOPRON coounty, NW HUNGARY, 24 mi/39 km WNW of Győr. Vegetable canneries; manufacturing of coffee substitute; distilleries, flour mills, brickworks.

Mosonszolnok (MO-shon-sol-nok), village, Győr-SO-PRON county, NW HUNGARY, 24 mi/39 km NW of Győr; 47°51′N 17°11′E. Grain, sugar beets; cattle, hogs. Bronze Age artifacts found nearby.

Mosor (MO-sor), mountain in DINARIC Alps, S CROATIA, near ADRIATIC SEA, in DALMATIA. Highest point is Veliki Kabal (4,393 ft/1,339 m), 10 mi/16 km E of SPLIT.

Mosovce (mo-SHAWF-tse), Slovak *Mošovce*, Hungarian *Mosóc*, village, STREDOSLOVENSKY province, W central SLOVAKIA, 23 mi/37 km SSE of Žilina; 48°54′N 18°54′E. Fur coats. Poet Ján Kollár (1793–1852) born here.

Mospino, UKRAINE: see MOSPYNE.

Mospyne (MOS-pi-ne) (Russian *Mospino*), city, central DONETS'K oblast, UKRAINE, in the DONBAS, 11 mi/18 km SE of MAKIYIVKA; 47°53′N 38°04′E. Elevation 462 m. Coal-mining center (mine enrichment plant, repair shop); flour, dairy; broiler factories. Established in 1800 as Makhorivka (Russian *Makhorovka*), name changed to Mospyne with city status in 1938.

Mosqueiro (MOS-kai-ro), city, E SERGIPE state, BRAZIL, 10 mi/16 km S of ARACAJU on Atlantic coast; 11°07′S 37°09′W.

Mosqueiro (mos-kai-ro), town, E PARÁ, BRAZIL, on island in RIO PARÁ (AMAZON delta), 20 mi/32 km N of BELÉM (steamer connection); 01°00′S 48°32′W. Resort.

Mosquera (mos-KAI-rah), town, ⊙ Mosquera municipio, CUNDINAMARCA department, central CO-LOMBIA, on railroad and highway, and 13 mi/21 km NW of BOGOTÁ; 04°42′N 74°13′W. Elevation 8,986 ft/2,739 m. Wheat, potatoes, fruit; livestock.

Mosquera (mos-KAI-rah), town, ⊙ Mosquera municipio, NARIÑO department, SW COLOMBIA, minor port on the PACIFIC OCEAN, 130 mi/209 km NW of PASTO; 02°32′N 78°24′W. Agriculture includes rice, plantains, and sugarcane.

Mosquero (mos-KER-o), village (2006 population 90), HARDING–SAN MIGUEL county line, ⊙ HARDING county, NE NEW MEXICO, 45 mi/72 km NNW of TU-CUMCARI; 35°46′N 103°57′W. Cattle, alfalfa, wheat, oats, barley, millet. Chicosa Lake State Park and section of Kiowa National Grasslands to NW.

Mosquitia, Central America: see MOSQUITO COAST.

Mosquito (mos-KEE-to), village on NW VIEQUES Island, E PUERTO RICO, 5 mi/8 km WSW of ISABEL SEGUNDA. Fishing.

Mosquito Coast or **Mosquitia**, region, E coast of NI-CARAGUA and HONDURAS. The name is derived from the Mosquito or Miskitto, the indigenous inhabitants. Never exactly delimited, the region is a belt c.40 mi/64 km wide extending from the SAN JUAN RIVER N into NE HONDURAS. It is sultry and swampy, rising to low hills in the W. Banana cultivation was a major economic activity in the 1920s and 1930s; today the main source of income is lobstering (centered at ROATAN, in the BAY ISLANDS). In the early colonial period English and Dutch buccaneers preyed on Spanish shipping from coastal bases, and English loggers exploited the forest products. England established a protective kingdom at BLUEFIELDS in 1678. Slaves from Jamaica were brought in to increase the labor supply in the mid-eighteenth century to work the sugarcane and tobacco plantations. Their main colonies were at Bluefields and Greytown (SAN JUAN DEL NORTE), both English-speaking. In the early twentieth century, a greater number of West Indian black laborers were imported for the banana plantations. In 1848 the British claimed and took San Juan del Norte to offset U.S. interest in a transisthmian route to CALIFORNIA. Nicaragua protested the seizure. The Clayton-Bulwer Treaty (1850) between the U.S. and Great Britain checked British expansion, but relinquishment of the coast was delayed until a separate treaty was concluded with Nicaragua (1860), which established the autonomy of the so-called Mosquito Kingdom. In 1894, José Santos Zelaya ended the anomalous position of the territory by forcibly incorporating it into Nicaragua. The N part, however, was awarded to Honduras in 1960 by the International Court of Justice, thus ending a long-standing dispute between the two countries. Miskito Indians refused to accept the Sandinista government that was established after 1979, and the Mosquito Coast became a center of anti-Sandinista activity. The region, which lies entirely within the department of ZELAYA, was given partial autonomy in 1985.

Mosquito Creek, c.60 mi/97 km long, SW IOWA; rises in SHELBY county; flows SW, to MISSOURI RIVER 5 mi/8 km S of COUNCIL BLUFFS.

Mosquito Creek (muh-SKEE-to), c.30 mi/48 km long, in NE OHIO; rises in ASHTABULA county; flows S, past CORTLAND, to MAHONING RIVER at NILES; 41°10′N 80°45′W. A flood-control dam (5,650 ft/1,722 m long, 47 ft/14 m high) impounds MOSQUITO CREEK LAKE reservoir c.9 mi/14 km above mouth.

Mosquito Creek Lake (muh-SKEE-to), reservoir (□ 12 sq mi/31.2 sq km), TRUMBULL county, NE OHIO, on MOSQUITO CREEK, 7 mi/11 km NE of WARREN; 41°24′N 80°46′W. Maximum capacity 180,000 acre-ft. Formed by Mosquito Creek Dam (43 ft/13 m high), built (1944) by Army Corps of Engineers for flood control; also used for recreation and water supply.

Mosquito Lagoon, in VOLUSIA and BREVARD counties, E central FLORIDA, separated from the ATLANTIC OCEAN by narrow barrier beach; 28°50′N 80°48′W. It is c.17 mi/27 km long and 2 mi/3.2 km wide. Contains many small islands. Followed by INTRACOASTAL WA-TERWAY.

Mosquito Pass (13,186 ft/4,019 m), PARK-LAKE counties, central COLORADO, in PARK RANGE, just N of MOUNT SHERMAN. County road passes through it.

Mosquitos, URUGUAY: see SOCA.

Mosquitos Gulf, Spanish *Golfo de los Mosquitos*, bight of CARIBBEAN SEA in W central PANAMA, E of VA-LIENTE PENINSULA; 80 mi/129 km wide, 10 mi/16 km long. Contains ESCUDO DE VERAGUAS island.

Moss (MAWS), city (2007 population 28,633), ØSTFOLD county, SE NORWAY, a port on the OSLOFJORD; 59°26′N 10°40′E. It is a commercial, industrial, and tourist center, with textile factories, glasswork; metalworks, cellulose and breweries. On August 14, 1814, the convention establishing the personal union of Sweden and Norway was signed there.

Mossaka (mo-sah-KAH), town, CUVETTE region, E central Congo Republic, on CONGO RIVER (CONGO border) at confluence of LIKOUALA RIVER and SAN-GHA RIVER, and 80 mi/129 km SE of OWANDO; 01°12′S 16°47′E. Ships hardwoods; palm-oil milling, elaeis-palm plantations. Customs station.

Mossamedes, ANGOLA: see NAMIBE, town.

Mossbank, town (2006 population 330), S SASKATCH-EWAN, CANADA, near OLD WIVES LAKE, 35 mi/56 km SW of MOOSE JAW; 49°56′N 105°58′W. Former coal mines. Grain elevators.

Mossbank (MAWS-bank), village, on NE coast of Mainland island, SHETLAND ISLANDS, extreme N Scotland, 21 mi/34 km N of LERWICK; 60°27′N 01°12′W. Fishing.

Moss Beach, unincorporated town (2000 population 1,953), SAN MATEO county, W CALIFORNIA; residential suburb c.18 mi/29 km SSW of downtown SAN FRAN-CISCO; 37°31′N 122°31′W. Fishing; ornamentals, flowers, artichokes, brussels sprouts, grain; marine gardens offshore. Half Moon Bay Airport to SE. Montana Mountain (extension of Santa Cruz Mountains) and San Francisco State Fish and Game Refuge to NE.

Mossblown (MAWS-blon), village (2001 population 2,038), South Ayrshire, SW Scotland, 3 mi/4.8 km E of PRESTWICK; 55°29′N 04°34′W. Farming and agriculture. Formerly in Strathclyde, abolished 1996.

Mossel Bay, Afrikaans *Mosselbaai*, town, WESTERN CAPE province, SOUTH AFRICA, seaport on Mossel Bay (20 mi/32 km wide) of the INDIAN OCEAN, 210 mi/338 km W of NELSON MANDELA METROPOLE (PORT ELI-ZABETH), 25 mi/40 km SW of GEORGE; 34°11′S 22°09′E. Serves agricultural and wool-producing region; fishing center, with important oyster and mussel beds; seaside resort on N2 highway. Airport. Yellow ochre and quartzite mined nearby; platinum, iridium deposits discovered (1934) in vicinity but no longer worked. Site of offshore natural gas reserves now being tapped by sea-based rigs. Considerable growth since mid-1980s. Cape St. Blaize (W); lighthouse; just N, in Mossel Bay, is Seal Island, with large seal colony. Bay was visited (1487) by Bartholomew Diaz and (1497) by Vasco da Gama; wrecked Portuguese crew found refuge here 1501. Used as postal clearing house by passing ships—Post Office Tree is a national monument. Formerly called Aliwal South.

Mossendjo (mo-sen-JO), town (2007 population 18,624), NIARI region, W Congo Republic, 130 mi/209 km NNE of POINTE-NOIRE; 02°57′S 12°43′E. Trading center; rice growing. Roman Catholic and Protestant missions.

Mosses, Col des (MOSS, col deh), pass (4,741 ft/1,445 m) in the W BERNESE ALPS, VAUD canton, W SWIT-ZERLAND. Road leads S from CHÂTEAU-D'OEX to AIGLE.

Mossey River (MAH-see), rural municipality (□ 434 sq mi/1,128.4 sq km; 2001 population 686), S MANITOBA, W central Canada, bounded on N by Lake WINNIPE-GOSIS; 51°40′N 100°00′W. Agriculture (wheat, barley, oats, canola; livestock); related farm businesses. Includes the communities of Fork River, Oak Brae, Volga, and WINNIPEGOSIS village. Incorporated 1906.

Mossi (MO-see), a densely populated region of BURKINA FASO, located on the CENTRAL PLATEAU. Named for Mossi people. Principal settlement is OUAGADOUGOU.

Mössingen (MUS-sing-guhn), town, NECKAR-ALB, BADEN-WÜRTTEMBERG, GERMANY, on NW slope of SWABIAN JURA, 7.5 mi/12.1 km S of TÜBINGEN; 48°25′N 09°03′E. Metalworking; fruit.

Mösskirch, GERMANY: see MESSKIRCH.

Moss Landing, unincorporated village (2000 population 300), MONTEREY county, W CALIFORNIA, on MONTEREY BAY, 15 mi/24 km NNE of MONTEREY. Fishing docks; artichokes, vegetables, fruit; dairying; cattle; oil-loading pipeline; power plant.

Mossleigh (MAHS-lee), hamlet, S ALBERTA, W Ca-nada, 22 mi/36 km N of VULCAN, in Vulcan County; 50°43′N 113°20′W.

Mossley (MOS-lee), town (2001 population 9,856), GREATER MANCHESTER, W ENGLAND, on TAME RIVER, near WEST YORKSHIRE border, and 9 mi/14.5 km ENE of MANCHESTER; 53°31′N 02°01′W. Textiles, light engineering.

Mossman (MAHS-muhn), town, NE QUEENSLAND, NE AUSTRALIA, 35 mi/56 km NNW of CAIRNS; 16°28′S 145°22′E. Sugar-producing center. Mossman Gorge W of town.

Mossø (MOSS-uh), largest lake in JUTLAND, DENMARK, 15 mi/24 km SW of ÅRHUS; 6 mi/9.7 km long, 1.5 mi/2.4 km wide. Surrounded by farm land; forests on SW.

Mossoró (MO-so-ro), city (2007 population 234,392), RIO GRANDE DO NORTE, NE BRAZIL, head of navigation on APODI (or Mossoró) River, and 150 mi/241 km WNW of NATAL; 05°11′S 37°20′W. Railroad to Natal. Saltworking and shipping center; important livestock, cotton, and carnauba market. Gypsum and marble quarries. Manganese deposits nearby. Airport. Also spelled Moçoró.

Mossoró River, Brazil: see APODI RIVER.

Moss Point, city (2000 population 15,851), JACKSON county, extreme SE MISSISSIPPI, 4 mi/6.4 km N of PASCAGOULA, on E channel of PASCAGOULA RIVER at mouth of Escatawba River; 30°25′N 88°31′W. Manufacturing (rubber and latex, gears and pumps, shipbuilding, metal fabrication, fish oil). Regional airport to N; Escatawba River forms Robertson and Beardsley lakes in W part of city; Grand Bay National Wildlife Refuge to SE; Mississippi Sandhill Crane National Wildlife Refuge to W.

Moss Side, ENGLAND: see MANCHESTER.

Moss Town, town, central BAHAMAS, on central GREAT EXUMA ISLAND, 5 mi/8 km WNW of GEORGE-TOWN; 23°30′N 75°51′W. Livestock raising (sheep, goats, hogs).

Mossuril (mo-SOO-reel), town, NAMPULA province, NE MOZAMBIQUE, small port on MOZAMBIQUE CHANNEL, 7 mi/11.3 km NW of Mozambique city; 14°58′S 40°39′E.

Mossuril District, MOZAMBIQUE: see NAMPULA.

Mossurize District, MOZAMBIQUE: see MANICA.

Moss Vale (MAHS VAIL), town, E NEW SOUTH WALES, SE AUSTRALIA, 65 mi/105 km SW of SYDNEY; 34°33′S 150°23′E. Dairy, agriculture center (meat; fruit; horses).

Mossyrock, village (2006 population 509), LEWIS county, SW WASHINGTON, on COWLITZ RIVER, and 25 mi/40 km SE of CHEHALIS; 46°32′N 122°29′W. Dairying; timber. Parts of Gifford Pinchot National Forest to N and SE; Ike Kinswa State Park, on Mayfield Lake reservoir, to NW. MOSSYROCK DAM, forms Riffe Lake, to E.

Mossyrock Dam, WASHINGTON: see RIFLE LAKE.

Most (MOST), German *Brüx*, city (2001 population 68,263), SEVEROCESKY province, CZECH REPUBLIC, NW BOHEMIA, near the German border; 50°32′N 13°39′E. Railroad junction and industrial city in a lignite-mining area. Manufacturing (chemicals, steel, and ceramics). City dates at least to the 11th century. It has several medieval churches and an old town hall, which was relocated in the 1960s due to an expansion of the coalfields.

Mosta (MAH-stah), town (2005 population 18,735), central MALTA, 5 mi/8 km W of VALLETTA; 35°55′N 14°26′E. Olive oil, vegetables, honey; sheep, goats. Roman remains. Mosta Church has the third largest dome in Europe. In April 1942, a German bomb crashed through the dome into the crowded church below. It failed to explode, an event considered a miracle by many Maltese. Airport to S.

Mosta (MOS-tah), town (2005 population 615), SE IVANOVO oblast, central European Russia, on road and railroad, 6 mi/10 km SE of YUZHA; 56°32′N 42°10′E. Elevation 374 ft/113 m. Lumbering.

Mostafa ben Brahim (mo-stah-FAH en brah-HEEM), village, SIDI BEL ABBÈS wilaya, W ALGERIA. One of many agricultural reform villages created in the 1970s to give peasants access to modern housing and amenities. Named after an illustrious poet from the region. Carpet sales and exhibitions.

Mostaganem, wilaya, in coastal W ALGERIA, E of ORAN wilaya; ⊙ MOSTAGANEM; 36°00′N 00°20′E. Plays an important agricultural and industrial role in its region. Outside of Mostaganem city, small towns and hamlets predominate. Lost 22 of its hinterland communes in 1984 to newly created RELIZANE wilaya.

Mostaganem (mo-stah-gah-NEM), city, ⊙ MOSTAGA-NEM department, NW ALGERIA, a port on the MEDITERRANEAN SEA, on the E side of the Gulf of Azrew; 35°54′N 00°05′E. A sugar refinery and a complex for paper pulp. The country's leading institute for the training of farm and forestry engineers is located here. Made up of an old city and a new one, which has expanded since the city became a wilaya capital. It has a rich history from Roman through medieval times, when it was an imposing Almoravid city. Occupied by SPAIN in the 16th century, passed to the Turks, and then to FRANCE in 1833. The port was used mostly for fishing but also for trade. The European population left after independence.

Mostar (MOS-tahr), city, in BOSNIA AND HERZEGOVINA, on the NERETVA RIVER; 43°20′N 17°48′E. The chief city of HERZEGOVINA, it produces tobacco, wine, and aluminum products. Bauxite mines and a hydroelectric plant are nearby. Known in 1442, it became (16th century) the chief Turkish administrative and commercial center in Herzegovina. It passed to AUSTRIA in 1878 and to the former YUGOSLAVIA in 1918. The city had a 16th-century Turkish stone bridge and numerous Turkish mosques and old houses. In 1992, most of these structures (including the bridge) were destroyed as Serb and Croat forces battled Muslim forces for control of the city before the Croats and Muslims established a coalition of their own. Having succeeded in expelling most of the Serb population, the coalition splintered as Croats and Muslims squared off in some of the bloodiest fighting of the war. In 1997 the city was divided between the Croats (W) and Muslims (E).

Mostardas (MO-stahr-dahs), town (2007 population 11,904), E RIO GRANDE DO SUL state, BRAZIL, on barrier island between Lagoa dos PATOS and Atlantic Ocean; 31°06′S 51°57′W. Onions.

Mostar Lake (MOS-tahr), region, in W HERZEGOVINA, BOSNIA AND HERZEGOVINA, c. 5 mi/8 km W of MOSTAR; 8 mi/12.9 km long, 2 mi/3.2 km wide. Flooded between November and June. Also called Mostar Plain.

Mostecka Basin, CZECH REPUBLIC: see SEVEROCESKY.

Mosteiros, municipality (2005 population 9,706), NE FOGO island, CAPE VERDE, in SOTAVENTO ISLANDS group, on ATLANTIC OCEAN (to N and E); ⊙ Mosteiros; 15°00′N 24°21′W. Bordered to S and W by São Filipe municipality.

Mosterhamn (MAWST-uhr-HAHM-uhn), village, HORDALAND county, SW NORWAY, port at SE tip of BØMLO, 20 mi/32 km N of HAUGESUND; 59°42′N 05°23′E. Limestone quarries. Olav I Tryggvason landed here (995) from DUBLIN to claim crown of NORWAY.

Mostiska, UKRAINE: see MOSTYSKA.

Mostki, UKRAINE: see MISTKY.

Móstoles (MO-sto-les), city (2001 population 196,524), MADRID province, central SPAIN, on railroad, and 11 mi/18 km SW of MADRID; 40°19′N 03°51′W. Now an important suburb of the capital. Meatpacking, vinegar distilling. Celebrated for its declaration of war on Napoleon, made by its mayor Andrés Torrejón.

Moston, ENGLAND: see MANCHESTER.

Mostove (mos-to-VE) (Russian *Mostovoye*), village, W MYKOLAYIV oblast, UKRAINE, on the Chychyklia River, right tributary of the Southern BUH RIVER, on road 56 mi/90 km NW of MYKOLAYIV, and 20 mi/32 km WSW of VOZNESENS'K; 47°25′N 31°00′E. Livestock, dairy.

Mostovoye, RUSSIA: see GERMENCHUK.

Mostovoye, RUSSIA: see MOSTOVSKOY.

Mostovoye, UKRAINE: see MOSTOVE.

Mostovskoy (muhs-tuhf-SKO-yee), town (2005 population 25,310), E KRASNODAR TERRITORY, S European Russia, in the foothills of the NW Greater CAUCASUS Mountains, on the LABA RIVER (tributary of the KUBAN' River), on road and railroad spur, 17 mi/27 km S of LABINSK; 44°24′N 40°47′E. Elevation 1,282 ft/390 m. Lumbering, furniture factory. Formerly a village called Mostovoye or Mostovskoye.

Mostovskoye (muh-STOFS-kuh-ye), village, NE KURGAN oblast, SW SIBERIA, RUSSIA, on road, 35 mi/56 km NNW of LEBYAZHYE; 55°44′N 66°08′E. Elevation 462 ft/140 m. Dairy farming.

Mostovskoye, RUSSIA: see MOSTOVSKOY.

Mosty (mo-STEE), town, E GRODNO oblast, BELARUS, ⊙ Mosty region, on NEMAN River, and 20 mi/32 km NNE of volkovysk, 53°25′N 24°32′E. Railroad junction; food industries.

Mosty bei Jablunka, CZECH REPUBLIC: see MOSTY U JABLUNKOVA.

Mostyn (MUHS-tin), village (2001 population 2,012), Flintshire, NE Wales, on the DEE estuary, and 6 mi/9.7 km ESE of PRESTATYN; 53°18′N 03°16′W. Formerly in CLWYD, abolished 1996.

Mostyska (mos-TIS-kah) (Russian *Mostiska*) (Polish *Mościska*), city (2004 population 7,420), W L'VIV oblast, UKRAINE, on VYSHNYA River, and 40 mi/64 km W of L'VIV; 49°48′N 23°09′E. Elevation 695 ft/211 m. Raion center; cement and brick manufacturing, electronic components; dairy. Known since 1244 as part of Halych principality; passed to Hungary (1372), to Poland (1387); city (Magdeburg Law) in 1404; passed to Austria (1772); part of West Ukrainian National Republic (1918); reverted to Poland (1919); incorporated into Ukrainian SSR (1939); independent Ukraine since 1991. About half of the population was Jewish from the late 19th century, until the community was destroyed in World War II.

Mosty Wielkie, UKRAINE: see VELYKI MOSTY.

Mosul (mo-SOOL), Arabic *al Mawsil*, city, ⊙ NINEVEH province, N IRAQ, on the TIGRIS RIVER, S of the ruins of NINEVEH; 36°20′N 43°08′E. It is the largest city in N Iraq and the third-largest city in the country. Trade in agricultural goods and exploitation of oil in the nearby oil fields are the two main occupations of the inhabitants. Mosul has an oil refinery; its productivity in the 1980s was hindered by the IRAN-IRAQ War. Other industries include textile, leather, food, building materials. An important road and railroad center. While most of the urban population is Arab, the surrounding region is inhabited largely by Kurds. MOSUL was the chief city of N MESOPOTAMIA from the 8th to 13th century, when it was devastated by the Mongols. The city remained poor and shabby through its occupation by the Persians (1508) and the Turks (1534–1918). Under the British occupation and mandate (1918–1932) it regained its stature as the chief city of the region. Its possession by Iraq was disputed by TURKEY (1923–1925) but was confirmed by the League of Nations (1926). The city is the seat of Mosul University and a center of Nestorian Christianity.

Møsvatn (MUHS-VAH-tuhn), lake (□ 23 sq mi/59.8 sq km), TELEMARK county, S NORWAY, W of RJUKAN; c.25 mi/40 km long, 1 mi/1.6 km–3 mi/4.8 km wide, with several branches; 148 ft/45 m deep. Fisheries. Outlet is MÅNE River Dam built 1904–1906, with later improvements.

Moswansicut River (mos-WAN-si-cut), N central RHODE ISLAND, short stream entering N arm of SCITUATE RESERVOIR. Formerly joined PONAGANSET RIVER to form North Branch of PAWTUXET RIVER in area now flooded by reservoir.

Mota (MO-tah), town (2007 population 32,974), AM-HARA state, NW ETHIOPIA, on N slope of the CHOKE MOUNTAINS, c.10 mi/16 km S of the BLUE NILE RIVER, 55 mi/89 km SSW of DEBRE TABOR; 11°05′N 37°52′E. Trade center; airfield.

Mota, RUSSIA: see KHONUU.

Mota, VANUATU: see BANKS ISLANDS.

Mota del Cuervo (MO-tah dhel KWER-vo), town, CUENCA province, E central SPAIN, in CASTILE-LA MANCHA, 55 mi/89 km SW of CUENCA; 39°30′N 02°52′W. Agricultural center surrounded by fine pastures and vineyards. Olive-oil pressing, pottery manufacturing.

Mota del Marqués (mahr-KAIS), village, VALLADOLID province, N central SPAIN, 24 mi/39 km W of VAL-LADOLID; 41°38′N 05°10′W. Tanning, wine making; cereals, grapes; livestock. Has medieval mansion.

Motagua (mo-TAH-gwah), river, c.250 mi/402 km long; rising in S central GUATEMALA; flowing NE to the Gulf of HONDURAS; 15°44′N 88°14′W. The longest river within Guatemala, it waters a valley where tomatoes, rice, bananas, and other tropical crops are raised. The Intercoastal Highway and trans-Guatemala railroad follow the valley.

Motai, KAZAKHSTAN: see MATAI.

Motajica (mo-tah-EET-sah), mountain (2,139 ft/652 m) in DINARIC ALPS, N BOSNIA, BOSNIA AND HERZEGO-VINA, along right bank of Sava River, 20 mi/32 km E of BOSANSKA Gradiška; 45°04′N 17°40′E. Also spelled Motayitsa.

Motala (MOO-TAH-lah), town, ÖSTERGÖTLAND county, S SWEDEN, on LAKE VÄTTERN and GÖTA Canal; 58°32′N 15°02′E. Important port and industrial center; manufacturing (weapons, appliances). Site of Sweden's first radio transmitter. Radio museum; Göta Canal museum.

Motala ström (MOO-TAH-lah STRUHM), river, 60 mi/97 km long, SE SWEDEN; issues from LAKE VÄT-TERN at MOTALA; flows E, through small Lake Boren, to LAKE ROXEN, then flows N to Lake Glan and E to NORRKÖPING, where it turns N and flows to Bråviken, 30-mi/48-km-long inlet of BALTIC SEA, 3 mi/4.8 km N of NORRKÖPING. Between lakes Vättern and Roxen course parallels Göta Canal.

Mota Lava, VANUATU: see BANKS ISLANDS.

Motane, FRENCH POLYNESIA: see MOHOTANI.

Motatán (mo-tah-TAHN), town, ☉ Motatán munici-pio, TRUJILLO state, W VENEZUELA, on MOTATÁN RIVER, and 6.0 mi/9.7 km N of VALERA; 09°23′N 70°35′W. Elevation 1,046 ft/318 m. In agricultural re-gion (sugarcane, corn, tobacco, coffee).

Motatán River (mo-tah-TAHN), c.100 mi/161 km long, W VENEZUELA; rises in ANDEAN spur at N foot of MUCUCHÍES PASS (MÉRIDA state); flows N and NW through TRUJILLO state, past VALERA and MOTATÁN, to Lake MARACAIBO 2.5 mi/4.0 km SE of SAN LORENZO (ZULIA state); 09°28′N 70°37′W.

Motavka, RUSSIA: see BOR.

Motayitsa, BOSNIA AND HERZEGOVINA: see MOTAJICA.

Motegi (mo-TE-gee), town, Haga county, TOCHIGI prefecture, central HONSHU, N central JAPAN, 12 mi/25 km E of UTSUNOMIYA; 36°31′N 140°11′E. *Konnyaku* (paste made from devil's tongue).

Motembo, Minas de (mo-TAIM-bo, MEE-nahs dai), village, MATANZAS province, W CUBA, 30 mi/42 km ESE of CÁRDENAS; 22°50′N 80°41′W. Petroleum wells.

Motenge-Boma (mo-TEN-gai—BAW-muh), town, Équateur province, NW CONGO, 90 mi/145 km W of GEMENA; 03°15′N 18°39′E. Elev. 1,062 ft/323 m.

Moth (MOT), town, JHANSI district, S UTTAR PRADESH state, N central INDIA, 30 mi/48 km NE of JHANSI; 25°43′N 78°57′E. Jowar, oilseeds, wheat, gram.

Mothe-Achard, La (mot-uh-ah-shahr, lah), commune (□ 3 sq mi/7.8 sq km), VENDÉE department, PAYS DE LA LOIRE region, W FRANCE, 11 mi/18 km WSW of La ROCHE-SUR-YON; 46°37′N 01°39′W. Livestock raising in field and hedgerow country (BOCAGE VENDÉEN).

Mother Lode, belt of gold-bearing quartz veins, MAR-IPOSA, CALAVERAS, AMADOR, EL DORADO, and PLACER counties, central CALIFORNIA, along the W foothills of the SIERRA NEVADA. The term is sometimes limited to a strip c.70 mi/113 km long and 1 mi/1.6 km–6.5 mi/10.5 km wide, running NW from Mariposa. It is used to mean the gold-bearing area E of the Sacramento and San Joaquin rivers and W of the Sierra Nevada. The discovery of alluvial gold at Sutter's Mill on the South Fork of the AMERICAN RIVER in 1848 led to the gold rush centered on this territory in 1849. Those traveling here to seek gold were nicknamed "The Fortyniners" and constructed the lawless boom towns of ANGELS CAMP, VOLCANO, and SONORA. Mark Twain and Bret Harte helped make the Mother Lode famous. Highway 49, named for the first year of the Gold Rush, runs through most of the historic towns. The area around Sonora and Tuolumne counties is now comprised of middle-income suburbs.

Motherwell (MUH-thuhr-wel), town (2001 population 30,311), ☉ North Lanarkshire, central Scotland, 11 mi/17.6 km SE of GLASGOW; 55°47′N 04°00′W. Formerly a center of heavy industry (coal and steel), Motherwell now houses light engineering. Extends to village of Cambusnethan. Formerly in Strathclyde, abolished 1996.

Mothe-Saint-Héray, La (mo-tuh–san–tai-rai, lah), commune (□ 5 sq mi/13 sq km), DEUX-SÈVRES de-partment, POITOU-CHARENTES region, W FRANCE, on the SÈVRE NIORTAISE, 17 mi/27 km E of NIORT. Dai-rying. Megalithic monuments and 17th-century castle nearby.

Moti (MO-tee), volcanic island, N Maluku province, INDONESIA, in MALUKU SEA, just W of HALMAHERA, 20 mi/32 km S of TERNATE; 4 mi/6.4 km wide; 00°28′N 127°24′E. Mountainous, rising to 3,117 ft/950 m. Also spelled Motir.

Môtiers (mo-TYAI), commune, NEUCHÂTEL canton, W SWITZERLAND, 16 mi/26 km WSW of NEUCHÂTEL, and on AREUSE RIVER, in Val de TRAVERS; 46°55′N 06°36′E. Watches.

Motihari (mo-tee-HAH-ree), town (2001 population 101,506), ☉ EAST champaran district, NW BIHAR state, E INDIA, on GANGA plain, 46 mi/74 km NW of MUZAFFARPUR; 26°39′N 84°55′E. Road center; oilseed and sugar milling, cotton weaving. Kesaria stupa here is biggest Buddhist stupa in the world; Gandhi Memorial commemorates where Gandhi began his campaign of nonviolent resistance to British rule in 1917. George Orwell born here in 1903.

Motilla del Palancar (mo-TEE-lyah dhel pah-lahng-KAHR), town, CUENCA province, E central SPAIN, 37 mi/60 km SSE of CUENCA; 39°34′N 01°53′W. Agri-cultural center (grapes, saffron, cereals, fruit, sheep); sawmilling.

Motilones, Cerro, Colombia, Venezuela: see MOTI-LONES, SERRANÍA DE LOS.

Motilones, Serranía de los (mo-tee-LO-nais, ser-rah-NEE-ah dai los), ANDEAN range on COLOMBIA-VE-NEZUELA border, rising to 12,300 ft/3,749 m. N part of Cordillera ORIENTAL, it forms together with Serranía de VALLEDUPAR and Montes de OCA, further N, the Sierra de PERIJÁ; c.80 mi/129 km long; 10°00′N 73°00′W. Variant names: Sierra de los Motilones; Sierra de Motilones; Cerro Motilones.

Motilones, Sierra de, Colombia, Venezuela: see MO-TILONES, SERRANÍA DE LOS.

Motilones, Sierra de los, Colombia, Venezuela: see MOTILONES, SERRANÍA DE LOS.

Motir, INDONESIA: see MOTI.

Motiti Island (mo-TEE-tee), volcanic island, NEW ZEALAND, in BAY OF PLENTY, N NORTH ISLAND, 11 mi/18 km ENE of TAURANGA; 3.7 mi/6 km long, 1.2 mi/2 km wide. Some farming; deep-sea fishing; Tuatara preserve.

Motlatse River Canyon Nature Reserve (Pulane=a river that is always full), wildlife reserve (□ 96 sq mi/249.6 sq km), MPUMALANGA, SOUTH AFRICA, 38 mi/60 km NE of LYDENBURG, in the DRAKENSBERG moun-tain range on edge of escarpment. Provincial nature reserve containing the world's third-largest canyon. Mainly popular for breathtaking scenery but has prolific plant, bird, and animal life, including all South African primate species. Famous for its natural potholes, some up to 20 ft/6 m deep, and its views of the Lowveld. Formerly called Blyde River Canyon Nature Reserve (Afrikaans=river of joy). Renamed in 2005 as part of a movement to revive tribal place names in South Africa.

Motley (MAHT-lee), county (□ 989 sq mi/2,571.4 sq km; 2006 population 1,276), NW TEXAS; ☉ MATADOR; 34°04′N 100°47′W. In broken plains just below CA-PROCK escarpment of LLANO ESTACADO; drained by North, South, and Middle Pease rivers. Chiefly cattle-ranching region, producing also cotton, peanuts, grain sorghum. Oil, gas fields; sand and gravel; tim-ber. Formed 1876.

Motley (MAWT-lee), village (2000 population 585), MORRISON county, central MINNESOTA, 22 mi/35 km W of BRAINERD on CROW WING RIVER, at mouth of LONG PRAIRIE RIVER; 46°20′N 94°38′W. Manufactur-ing (feeds, smoked fish, imitation seafood). Small lakes in area; Shamineau Lake to SE.

Moto (MO-to), village, ORIENTALE province, NE CONGO, 16 mi/26 km WSW of WATSA; 02°27′N 26°25′E. Elev. 2,027 ft/617 m. Gold-mining center; rice processing.

Motobu (mo-TO-boo), town, on W peninsula of OKI-NAWA island, in RYUKYU ISLANDS, Kunigami county, Okinawa prefecture, SW JAPAN, near Yaedake mountain, 34 mi/55 km N of NAHA; 26°39′N 127°53′E. Pineapples. The country's earliest cherry blossoms appear here. Location of Marine Expo Memorial Park, the only semi-tropical park in Japan. Shio River (saltwater spring) is nearby.

Motodomari, RUSSIA: see VOSTOCHNYY, SAKHALIN ob-last.

Motokolea (mo-to-ko-LAI-ah), village, NORD-KIVU province, E CONGO, 100 mi/161 km N of BUKAVU; 00°22′N 28°46′E. Elev. 2,969 ft/904 m. Gold-mining center. Also gold mining at MANGUREDJIPA, 15 mi/24 km NW.

Motol (MO-tol), town, BREST oblast, BELARUS, port on YASELDA RIVER (head of navigation), and 25 mi/40 km NW of PINSK, 52°02′N 29°10′E. Manufacturing (tan-ning, flour milling).

Motomiya (mo-TO-mee-yah), town, Adachi county, FUKUSHIMA prefecture, N central HONSHU, NE JAPAN, on ABUKUMA RIVER, 16 mi/25 km S of FUKUSHIMA city; 37°30′N 140°23′E. Electronic equipment.

Motono (MO-to-no), village, Inba county, CHIBA pre-fecture, E central JAPAN, 9 mi/15 km N of CHIBA; 35°48′N 140°12′E.

Motor City, MICHIGAN: see DETROIT.

Motosu (mo-TOS), town, Motosu county, GIFU pre-fecture, central HONSHU, central JAPAN, 6 mi/10 km N of GIFU; 35°29′N 136°42′E. Onions. Tile, cryptomeria products, cement, water valves.

Mototomari, RUSSIA: see VOSTOCHNYY, SAKHALIN ob-last.

Motovilikha (muh-tuh-VEE-lee-hah), former city, PERM oblast, W URALS, E European Russia. Arose in 1736 at a copper smelter (closed in 1863); developed in the late 19th century with two gun factories; incor-porated into PERM city in 1938.

Motovka Gulf (MO-tuhf-kah), Russian *Motovskiy Zaliv*, inlet of the BARENTS SEA between RYBACHI and KOLA peninsulas, MURMANSK oblast, extreme NW European RUSSIA; 35 mi/56 km long, 4 mi/6 km–8 mi/13 km wide; 69°35′N 32°30′E. Fisheries. Receives Ti-tovka and Zapadnaya Litsa rivers.

Motovun (MO-to-voon), Italian *Montona*, village, W CROATIA, on MIRNA RIVER, and 8 mi/12.9 km N of PAZIN, in the ISTRIA region. Has cathedral (1614) with

Romanesque campanile. Motovunska Šuma (forest) special reservation.

Motown, MICHIGAN: see DETROIT.

Motoyama (mo-TO-yah-mah), town, Nagaoka county, KOCHI prefecture, central SHIKOKU, W JAPAN, on YOSHINO River, 12 mi/20 km N of KOCHI; 33°45′N 133°35′E.

Motoyoshi (mo-TO-yo-shee), town, Motoyoshi county, MIYAGI prefecture, N HONSHU, NE JAPAN, 50 mi/80 km N of SENDAI; 38°47′N 141°30′E. Seaweed (*wakame*).

Motozinhos (MO-to-zheen-yos), city, central MINAS GERAIS, BRAZIL, 5 mi/8 km N of PEDRO LEOPOLDO, on railroad; 19°30′S 44°14′W.

Motozintla de Mendoza (mo-to-SEEN-tlah dai men-DO-sah), town, ☉ Motozintla municipio, CHIAPAS, S MEXICO, in SIERRA MADRE, near GUATEMALA border, 33 mi/53 km N of TAPACHULA on Mexico Highway 190; 15°21′N 92°14′W. Sugarcane, coffee, fruit.

Motrico (mo-TREE-ko), town, GUIPÚZCOA province, N SPAIN; fishing port on BAY OF BISCAY, 21 mi/34 km W of SAN SEBASTIÁN; 43°18′N 02°23′W. Fish processing, boatbuilding; makes fine jewelry. Limestone and gypsum quarries in vicinity.

Motril (mo-TREEL), city, GRANADA province, S SPAIN, in ANDALUSIA, near the MEDITERRANEAN SEA, 30 mi/48 km SSE of GRANADA; 36°45′N 03°31′W. Terminus of aerial tramway from DÚRCAL. Picturesquely situated in amphitheater facing the sea and backed by mountains. Fishing port. Wine making. Irrigated surrounding area yields sugarcane, beets, grapes, almonds, olives, potatoes. Small port, El Verdadero, is on the Mediterranean 1.5 mi/2.4 km S. Tourism. Maritime customs.

Motru (MO-troo), town, GORJ county, ROMANIA, 20 mi/32 km NE of DROBETA-TURNU SEVERIN; 44°48′N 22°58′E. Center of important lignite-mining district.

Motza (MO-tsah) or **Motsa**, suburb, ISRAEL, in JUDEAN HIGHLANDS, 4 mi/6.4 km W of JERUSALEM; 31°46′N 35°08′E. Elevation 1,732 ft/527 m. Brick, tile, and wine manufacturing; some farming. MOTZA ILLIT is situated just above. Also called Kalaniya, both names dating back to Talmudic times.

Motsa, ISRAEL: see MOTZA.

Mott, town (2006 population 725), ☉ HETTINGER county, SW NORTH DAKOTA, 42 mi/68 km SE of DICKINSON, and on CANNONBALL RIVER; 46°22′N 12°19′W. Flour, dairy products, wheat. Founded in 1904 and incorporated in 1910.

Motta di Livenza (MOT-lah dee lee-VEN-tsah), town, TREVISO province, VENETO, N ITALY, on LIVENZA River, and 20 mi/32 km ENE of Treviso; 45°46′N 12°36′E. Railroad junction; manufacturing (soap, lye, silk textiles, clothing, wood products). Has cathedral, completed 1650.

Mottarone, Monte (mot-tah-RO-ne, MON-te), mountain (4,890 ft/1,490 m), ITALY, between lakes MAGGIORE and ORTA, 4 mi/6 km W of STRESA; 45°53′N 08°27′E. Winter sports. Summit (ascended by light railroad from Stresa) commands superb view of the ALPS, from TENDA PASS (SW) to the ORTLES (NE). AGOGNA River rises here.

Motte, Grande, FRANCE: see GRANDE-MOTTE, LA.

Motte, Lac La (MOT, lahk lah), lake (8 mi/13 km long, 6 mi/10 km wide), W QUEBEC, E Canada, 16 mi/26 km NW of VAL-D'OR, in gold-mining region; 49°24′N 78°03′W. Drained N by HARRICANAW RIVER.

Motte-Servolex, La (mot-uh-ser-vo-leks, lah), town (☐ 11 sq mi/28.6 sq km), NW suburb of CHAMBÉRY, SAVOIE department, RHÔNE-ALPES region, SE FRANCE. Cheese manufacturing; retailing.

Mott Haven, industrial section of SW BRONX borough of NEW YORK city, SE NEW YORK, along HARLEM RIVER (bridges to MANHATTAN); 40°50′N 73°56′W. Manufacturing of machinery, metal products; railroad yards. Site of Yankee Stadium and Bronx County Building. Lower-income population, mainly African-American and Hispanic, with growing number of middle-class families moving into refurbished areas.

Mottingham, ENGLAND: see BROMLEY.

Mottola (MOT-to-lah), town, Taranto province, APULIA, S ITALY, 15 mi/24 km NW of TARANTO; 40°38′N 17°02′E. In agricultural region (cereals, grapes, olives, almonds). Has 15th-century cathedral.

Motueka (muh-too-EE-ka), township (2001 population 6,891), TASMAN district, N SOUTH ISLAND, NEW ZEALAND, on W shore of TASMAN BAY, and 34 mi/55 km NW of NELSON. Processing and servicing center for intensive fruit-, hops-, and tobacco-growing area; research centers.

Motu Iti (MO-too EE-tee), Rock islet, MARQUESAS ISLANDS, FRENCH POLYNESIA, S PACIFIC, c.25 mi/40 km NE of NUKU HIVA; 08°41′S 140°36′W. Rises to c.720 ft/220 m. Also Hatu Iti.

Motu Iti, SOCIETY ISLANDS: see TUPAI.

Motul de Carrillo Puerto (mo-TOOL dai kah-REE-yo poo-ER-to), town, ☉ Motul municipio, YUCATÁN, SE MEXICO, on railroad, 23 mi/37 km ENE of MÉRIDA; 21°09′N 89°19′W. Henequen, corn, tropical fruit, livestock. Archaeological remains nearby.

Motunui (mo-too-NYOO-ee), site of Motunui synthetic petroleum (gasoline) plant, NEW ZEALAND, c. 4.3 mi/7 km E of WAITARA, TARANAKI region. Drawing mainly on MAUI off-shore natural gas (and some condensate), the terminal produces petroleum products for direct use or further processing.

Motu One (MO-too O-nai), uninhabited atoll, Leeward group, SOCIETY ISLANDS, FRENCH POLYNESIA, S PACIFIC, 150 mi/241 km W of BORA-BORA; 15°48′S 154°33′W. Also known as Bellingshausen Island.

Motupe (mo-TOO-pai), town, LAMBAYEQUE region, NW PERU, in W foothills of CORDILLERA OCCIDENTAL, on PAN-AMERICAN HIGHWAY, and 40 mi/64 km NNE of LAMBAYEQUE, on Motupe River (affluent of La Leche); 06°09′S 79°44′W. Corn, fruit, cotton, rice, tobacco, alfalfa; livestock.

Motutapu Island (MO-too-TAH-poo), off N NORTH ISLAND, NEW ZEALAND, in HAURAKI GULF, 6 mi/10 km E of entrance to WAITEMATA HARBOUR of AUCKLAND; 3.1 mi/5 km long, 1.9 mi/3 km wide; 36°45′S 174°55′E. Causeway to RANGITOTO ISLAND vacation resort.

Motya, SICILY, ITALY: see STAGNONE ISLANDS.

Motygino (muh-TI-gee-nuh), village (2005 population 6,590), central KRASNOYARSK TERRITORY, SE SIBERIA, RUSSIA, at a confluence of the ANGARA RIVER and one of its short right tributaries, 84 mi/135 km E of LESOSIBIRSK; 58°11′N 94°45′E. Elevation 633 ft/192 m. Logging and lumbering. Site of exile for political opponents of the tsarist regime, and later part of the prison camp system under the Soviet rule. Has a heritage museum.

Motza Illit (mo-TZA ee-LEET), suburb of JERUSALEM, E ISRAEL, in the JUDEAN HIGHLANDS; 31°48′N 35°08′E. Elevation 2,047 ft/623 m. Founded initially in the 1930s (although a recuperative center had been operating alongside since 1923). Developed into a modern suburb in the 1960s. Vespasian settled Roman legions here and called it Colonia Amasa. It is mentioned in the Book of Joshua. Founded 1894 as an agricultural settlement; destroyed (1929) in Arab riots, then rebuilt. Key Israeli position (1948) during battle for Jerusalem road.

Mouanko (moo-AHN-ko), town, LITTORAL province, CAMEROON, 28 mi/45 km S of DOUALA; 03°39′N 09°48′E.

Moubray Bay, an indentation of the BORCHGREVINCK COAST of VICTORIA LAND, EAST ANTARCTICA, N of CAPE HALLETT; 72°11′S 170°15′E.

Moucha Islands (MOO-shah), coral group in GULF OF TADJOURA, DJIBOUTI, 9 mi/14.5 km NNE of DJIBOUTI; 11°43′N 43°13′E. Weekend resort; fishing; bathing beach. Also called Musha Islands.

Mouchoir Bank, shoal with reefs, in the WEST INDIES, 60 mi/97 km N of HISPANIOLA, separated from TURK ISLANDS (W) by Mouchoir Passage (c. 60 mi/97 km long). Silver Bank is 25 mi/40 km E, beyond Silver Bank Passage.

Moudhros, Greece: see MOUDROS.

Mouding (MO-DING), town, ☉ Mouding county, N central YUNNAN province, CHINA, 18 mi/29 km N of CHUXIONG; 25°20′N 101°35′E. Elevation 6,299 ft/1,920 m. Rice, wheat, millet, sugarcane, tobacco; copper-ore mining.

Moudjéria (moo-JE-ryah), village (2000 population 1,974), TAGANT administrative division, S central MAURITANIA, oasis in low Tagant massif of the SAHARA DESERT, 300 mi/483 km ENE of SAINT-LOUIS (SENEGAL); 17°55′N 12°20′W. Dates, millet; livestock.

Moudon (moo-DOHN), district, NE VAUD canton, SWITZERLAND. Population is French-speaking and Protestant.

Moudon (moo-DOHN), German *Milden*, town, VAUD canton, W SWITZERLAND, on BROYE RIVER, and 12 mi/19 km NE of LAUSANNE; 46°40′N 06°48′E. Elevation 1,683 ft/513 m. Market center for agricultural region. An old town (ancient *Minnodunum* or *Minidunum*), it has a Gothic church (13th century) and two castles (16th and 17th century), and museum.

Moudros (MOO–[th]ros), town, on LÍMNOS island, LESBOS prefecture, NORTH AEGEAN department, GREECE, port on E shore of Gulf of Moudros; 39°52′N 25°16′E. Trade in olive oil, wine; fisheries. Allied base (1915) in World War I Gallipoli (GALIBOLU) campaign. The Allied-Turkish armistice was signed here, 1918. Formerly spelled Mudros. Also Moudhros.

Mougins (moo-zhan), town (☐ 9 sq mi/23.4 sq km), resort in ALPES-MARITIMES department, PROVENCE-ALPES-CÔTE D'AZUR region, SE FRANCE, on the French RIVIERA, and 5 mi/8 km N of CANNES; 43°36′N 06°59′E. Fine views of the MEDITERRANEAN coastline. Picasso lived and worked here, 1961–1973, amidst flowers and subtropical vegetation.

Mouguerre (moo-ger), commune (☐ 8 sq mi/20.8 sq km; 2004 population 4,280), SE suburb of BAYONNE, PYRÉNÉES-ATLANTIQUES department, AQUITAINE region, SW FRANCE; 43°28′N 01°25′W. Saltworks, chemical plant. Predominantly Basque population.

Mouhoun, province (☐ 2,571 sq mi/6,684.6 sq km; 2005 population 299,947), BOUCLE DU MOUHOUN region, NW BURKINA FASO; ☉ DÉDOUGOU; 12°15′N 03°25′W. Borders KOSSI and NAYALA (N), SANGUIÉ (E), BALÉ and TUY (S), HOUET (SW), and BANWA (W) provinces. W border drained by MOUHOUN RIVER. Agriculture (cotton, groundnuts, shea nuts, millet) and livestock (cattle); industry (processing of shea-nut butter). Zinc and gold deposits in area. Thermal center. Main center is DÉDOUGOU. A portion of this province was excised in 1997 when fifteen additional provinces were formed.

Mouhoun River, c.500 mi/805 km long, in BURKINA FASO and GHANA, W AFRICA; rises in BURKINA FASO near BOBO-DIOULASSO; flows S and E into GHANA, past LAWRA and BAMBOI, joining the Nakanbe River 38 mi/61 km NW of Yeji to form VOLTA RIVER. Forms part of boundary between CÔTE D'IVOIRE and Ghana. Drains W section of N and upper Ghana; is dammed at BUI gorge, a hydroelectric plant. Formerly called Black Volta River and Volta Noire.

Mouila (moo-EE-luh), city, ☉ NGOUNIÉ province, S GABON, on NGOUNIÉ RIVER, and 190 mi/306 km SSE of LIBREVILLE; 01°52′S 11°01′E. Trading center; wood and timber products. Airfield.

Mouilleron-en-Pareds (moo-yu-ron–ahn—pah-red), agricultural commune (☐ 7 sq mi/18.2 sq km), VENDÉE department, PAYS DE LA LOIRE region, W FRANCE, 27 mi/43 km E of LA ROCHE-SUR-YON; 46°40′N 00°51′W. Georges Clemenceau born here; also General de Lattre de Tassigny, liberator of Alsace and signer (for France) of German surrender (May 1945) in World War II.

Moulamein (MOO-luh-min), township, NEW SOUTH WALES, SE AUSTRALIA, 523 mi/842 km SW of SYDNEY,

Area is shown by the symbol ☐, and capital city or county seat by ☉.

in RIVERINA region; 35°03′S 144°05′E. Livestock grazing; wheat, rice.

Moulara, Greece: see AMOULIANI.

Moulay Idriss (moo-lay id-REES), town, Meknès prefecture, MEKNÈS-TAFILALET administrative region, N central MOROCCO, picturesquely situated in a ravine of the ZERHOUN massif, 11 mi/18 km N of MEKNES; 34°03′N 05°31′W. Holy Moslem city containing the shrine of Moulay Idriss I (founder of first Arab dynasty in Morocco; father of Idriss II, the founder of FES). Most important Muslim pilgrimage site in Morocco. Ruins of Roman VOLUBILIS are 2 mi/3.2 km NW.

Moulay Yacoub (MOO-lay ya-KOOB), spa, Moulay Yacoub prefecture, Fes-Boulemane administrative region, N central MOROCCO, 10 mi/16 km WNW of FES; 34°05′N 05°10′W. Hot sulphur springs; thermal spa. Also spelled Moulay Yakoub.

Mould Bay, U.S.-Canada Arctic weather station, SE PRINCE PATRICK ISLAND, NORTHWEST TERRITORIES; 76°05′N 119°45′W.

Moule (MOOL), town, E GRANDE-TERRE island, GUADELOUPE, French WEST INDIES, minor port 14 mi/23 km ENE of POINTE-À-PITRE; 16°20′N 61°21′W. Trading (rum, fruit, lumber); distilling; fishing. Archaeological museum and hotels. Sometimes Le Moule.

Moule à Chique, Cape (MOOL ah SHEEK), headland on S SAINT LUCIA, 20 mi/32 km S of CASTRIES; 13°43′N 60°57′W. Forms a narrow neck of land running into the sea for 3 mi/4.8 km. Lighthouse.

Mouléngui Binza (moo-LAIN-gee BIN-zah), town, NYANGA province, SW GABON, near CONGO border; 03°29′S 11°45′E.

Moulin (MOO-lin), agricultural village, PERTH AND KINROSS, E Scotland, on TUMMEL RIVER just NW of PITLOCHRY; 56°43′N 03°44′W. Just SE are ruins of Castle Dhu (Gaelic, *Caisteal Dubh*), former stronghold of the Campbells. Formerly in Tayside, abolished 1996.

Moulins (moo-lan), town; ⊙ ALLIER department, AUVERGNE region, central FRANCE, on the ALLIER RIVER, 90 mi/145 km NW of LYON, and 165 mi/266 km SSE of PARIS; 46°30′N 03°20′E. Moulins has a diversified economy based on the manufacturing of footwear, machine tools, and food products. Its commercial role is also significant. Formerly the capital of the duchy of Bourbonnais, Moulins has remarkable artistic and historic treasures. The Gothic cathedral (an episcopal see) contains a late 15th-century triptych, considered one of the finest French paintings of the period. It also has an historic belfry. The tomb of Henri de Montmorency, designed by François Anguier, is at the former convent (now a school) of the Order of Visitation. Other historic buildings are the ruined castle of the dukes of Bourbon and a Renaissance pavilion now occupied by the museum of art and archaeology. Although the dukes resided at Moulins from the mid-14th century, it did not become the capital of the duchy until the late 15th century. The duchy was confiscated by the French crown in 1527. In 1566, Charles IX held a great assembly at Moulins at which important administrative and legal reforms were adopted. Moulins is surrounded by at least 3 forests within 4 mi/6.4 km of the town center.

Moulins, Les (moo-LAN, lai), county (□ 102 sq mi/265.2 sq km; 2006 population 121,066), Lanaudière region, S QUEBEC, E Canada; ⊙ Terrebonne; 45°45′N 73°36′W. Composed of 2 municipalities. Formed in 1982.

Moulins, Les, FRANCE: see OMAHA BEACH.

Moulins-lès-Metz (moo-lan-lai-mes), town (□ 3 sq mi/7.8 sq km), MOSELLE department, LORRAINE region, NE FRANCE, S suburb of METZ; 49°06′N 06°06′E.

Moulis-en-Médoc (moo-lee-ahn-mai-dok), commune (□ 7 sq mi/18.2 sq km), GIRONDE department, AQUITAINE region, SW FRANCE, 18 mi/29 km NW of BORDEAUX; 45°04′N 00°46′W. Noted red wines of MÉDOC region (Grand-Poujeaux vineyards).

Moulmein, MYANMAR: see MAWLAMYINE.

Moulmeingyun (MOOL-main-juhn), town and township, AYEYARWADY division, MYANMAR, on right bank of AYEYARWADY RIVER (ferry), and 45 mi/72 km SE of PATHEIN.

Moulouya, Oued (wahd mel-WEE-yuh), river, 320 mi/515 km long, in NE MOROCCO; rises in the High ATLAS mountains SW of MIDELT; flows NNE, through a semi-arid valley past GUERCIF (where it is crossed by FES-OUJDA railroad), to the MEDITERRANEAN SEA 35 mi/56 km SE of MELILLA. Morroco's longest river. There are two important dams and hydroelectric plants; the Mohammed V (completed 1967, N of Taourit) and Mechri Homadi (completed 1955), the control point for an irrigation system that waters the TRIFFA and BOUAREG plains. Its lower course formed border between French and Spanish Protectorates. Very irregular volume; not navigable.

Moulton (MUL-tuhn), town (2000 population 3,260), ⊙ Lawrence co., NW Alabama, 20 mi/32 km WSW of Decatur. Clothing manufacturing, lumber milling, limestone quarrying. Named for Michael Moulton, an officer killed during the Creek Indian War of 1813–1814. Inc. in 1819 and chosen as county seat in 1820.

Moulton, town (2000 population 658), APPANOOSE county, S IOWA, near MISSOURI state line and source of NORTH FABIUS RIVER, 11 mi/18 km ESE of CENTERVILLE; 40°41′N 92°40′W. Settled 1867.

Moulton (MUHL-tuhn), town (2006 population 971), LAVACA county, S TEXAS, 20 mi/32 km N of YOAKUM, near source of LAVACA RIVER; 29°34′N 97°09′W. In cattle area; some oil and gas; manufacturing (oil field equipment).

Moultonborough (MOL-tuhn-buh-ro), town, CARROLL county, E central NEW HAMPSHIRE, 15 mi/24 km NNE of LACONIA; 43°43′N 71°22′W. Bounded SW in part by LAKE WINNIPESAUKEE, SE by Moultonborough Bay, in NW corner by SQUAM LAKE. Manufacturing (machine parts, book publishing); agriculture (nursery crops, apples, vegetables; cattle, poultry; dairying). Castle in the Clouds in E; Red Hill (2,029 ft/618 m) in NW; Lake Kanasatka in E; Snow Peak (2,975 ft/907 m), part of Ossipee Mountains, in E; Ossipee Mountain Ski Area in E center. Includes Long Island (in S), bridged to mainland.

Moultrie (MOL-tree), county (□ 344 sq mi/894.4 sq km; 2006 population 14,383), central ILLINOIS; ⊙ SULLIVAN; 39°38′N 88°37′W. Agriculture (corn, wheat, soybeans; livestock). Manufacturing (candy, wood products, farm machinery). Drained by KASKASKIA RIVER. Lake SHELBYVILLE in S. Formed 1843.

Moultrie (MOL-tree), city (2000 population 14,387), ⊙ COLQUITT county, SW GEORGIA, on the OCHLOCKONEE RIVER; 31°10′N 83°46′W. Manufacturing includes furniture, printing and publishing, fertilizer, lighting fixtures, cotton processing, concrete, animal feeds, aircraft, meat processing, paper products, agr. equipment, steel and aluminum products. The town grew as a lumbering and naval stores center, but the timber has since been depleted. Now center of commercial produce production. Airport. The Colquitt County Farmer's Market is here. Incorporated 1890.

Moultrie, Fort, SOUTH CAROLINA: see FORT MOULTRIE.

Moultrie, Lake, reservoir (□ 94 sq mi/244.4 sq km), BERKELEY county, SE SOUTH CAROLINA, on West Branch COOPER RIVER, 25 mi/40 km N of CHARLESTON; 33°16′N 79°58′W. Max. capacity of 1,110,000 acre-ft. Connected to Lake MARION at NW. Formed by PINOPOLIS DAM (75 ft/23 m high), built (1942) for power generation; owned by South Carolina Public Service Authority. Francis Marion National Forest to E; Old Santee Canal State Park just S.

Moultrie, Lake, SOUTH CAROLINA: see SANTEE RIVER.

Moultrieville, SOUTH CAROLINA: see SULLIVANS ISLAND.

Moulvi Bazar, town, BANGLADESH: see MAULAVI BAZAR.

Mounana (moo-NAH-nah), town, HAUT-OGOOUÉ province, SE GABON, 32 mi/51 km NW of FRANCEVILLE; 01°22′S 13°10′E. Major uranium-mining and production center.

Mound, town (2000 population 9,435), HENNEPIN county, E MINNESOTA, residential suburb 20 mi/32 km W of downtown MINNEAPOLIS, bounded by extension of Lake MINNETONKA: Upper Lake and Cooks Bay in S, West Arm and Jennings Bay in NE; Langdon Lake in W; 44°55′N 93°39′W. Diverse light manufacturing. Dutch Lake on N boundary. Settled 1854, incorporated 1912.

Mound, village (2000 population 12), MADISON parish, NE LOUISIANA, near the MISSISSIPPI RIVER, 8 mi/13 km W of VICKSBURG, MISSISSIPPI; 32°20′N 91°01′W. Derives name for an Indian mound at the site. Agriculture.

Mound Bayou, town (2000 population 2,102), BOLIVAR county, NW MISSISSIPPI, 25 mi/40 km SSW of CLARKSDALE; 33°52′N 90°43′W. In rich agricultural area (cotton, corn, rice, soybeans; cattle).

Mound City, city (2000 population 692), ⊙ PULASKI county, extreme S ILLINOIS, on OHIO River, and 5 mi/8 km N of CAIRO; 37°05′N 89°09′W. Manufacturing of mineral products; agriculture (grains, vegetables; livestock). Important Union naval base in Civil War. National cemetery is nearby. City severely damaged in 1937 flood. Incorporated 1857.

Mound City, city (2000 population 1,193), HOLT county, NW MISSOURI, 32 mi/51 km NW of SAINT JOSEPH; 40°08′N 95°13′W. Fruit, grain; hogs, cattle. Big Lake State Park and Squaw Creek National Wildlife Refuge to SW. Laid out 1857.

Mound City, town (2000 population 821), ⊙ LINN county, E KANSAS, 22 mi/35 km NNW of FORT SCOTT; 38°08′N 94°49′W. In livestock and fruit region. Annual county fair is held here. Big Hill Lake Reservoir to N.

Mound City, town (2006 population 71), ⊙ CAMPBELL county, N SOUTH DAKOTA, 20 mi/32 km ENE of MOBRIDGE; 45°43′N 100°04′W.

Moundhouse, locality, CARSON CITY, independent city (formerly ORMSBY county), W NEVADA, 5 mi/8 km NE of Carson City proper, on CARSON RIVER. Agricultural area (dairying; poultry; vegetables, hay); manufacturing (truck bodies, fishing rods, aerospace fasteners; printed circuit boards).

Moundou (moon-DOO), city, ⊙ LOGONE OCCIDENTAL administrative region, SW CHAD, on M'BÉRÉ River, and 260 mi/418 km SSE of N'DJAMENA; 08°34′N 16°05′E. Cotton center; fishing. Until 1946, in Ubangi-Chari colony. Was the capital of former Logone Occidental prefecture.

Moundridge, town (2000 population 1,593), MCPHERSON county, central KANSAS, 14 mi/23 km SSE of MCPHERSON; 38°12′N 97°31′W. Flour, feed. Manufacturing (steel-processing equipment, lawn and garden equipment, fertilizers, farm machinery, meat packaging). Oil and gas wells nearby.

Mounds, city (2000 population 1,117), PULASKI county, extreme S ILLINOIS, 3 mi/4.8 km NNW of MOUND CITY; 37°06′N 89°12′W. In agricultural area (corn). Incorporated 1904.

Mounds, town (2006 population 1,274), CREEK county, central OKLAHOMA, 20 mi/32 km SW of TULSA; 35°52′N 96°03′W. In oil-producing and agricultural area.

Mound Station, village (2000 population 127), BROWN county, W ILLINOIS, 6 mi/9.7 km W of MOUNT STERLING; 40°00′N 90°52′W. In agricultural area. Also called TIMEWELL.

Mounds View, city (2000 population 12,738), RAMSEY county, E MINNESOTA, suburb 9 mi/14.5 km NNE of downtown MINNEAPOLIS and 12 mi/19 km NW of downtown ST. PAUL, E of MISSISSIPPI river, drained by Rice Creek; 45°06′N 93°12′W. Manufacturing (tools, skates, gaskets, data communications equipment, business checks). Spring Lake on W boundary.

Moundsville, city (2006 population 9,455), seat of MARSHALL county, N WEST VIRGINIA, 120 mi/193 km NNE of CHARLESTON, in the NORTHERN PANHANDLE, on the OHIO RIVER (bridged); 39°55′N 80°44′W. Coal was once the chief industry, and some is still mined. Manufacturing (asphalt, roofing materials, petroleum calcined coke, hygiene products). Agriculture (corn, potatoes); livestock; poultry. Moundsville State Penitentiary. Palace of Gold, temple of New Vrindaban Spiritual Community. Grave Creek Mound State Park (Native American burial site) to S. Burches Run Wildlife Management Area to NE. Settled 1771, incorporated 1865.

Mound Valley, village (2000 population 418), LABETTE county, SE KANSAS, 11 mi/18 km SW of PARSONS; 37°12′N 95°24′W. In grain and diversified-farming area. Big Hill Lake reservoir to N.

Moundville, town (2000 population 1,809), Hale co., W Alabama, on Black Warrior River, and 20 mi/32 km N of Greensboro; 33°00′N 37°87′W. Trade and shipping point in cotton, corn, and vegetable area; cotton ginning, lumber milling; feed. Nearby state monument includes group of large Native American mounds and museum (dedicated 1939). Originally named 'Carthage' for the ancient city in north Africa, it was renamed for the many Native American mounds nearby. Inc. in 1908.

Moundville, town (2000 population 103), VERNON county, W MISSOURI, 7 mi/11.3 km SW of NEVADA; 37°46′N 94°27′W.

Moung (MONG), town, BATTAMBANG province, W CAMBODIA, on PHNOM PENH–BANGKOK railroad, and 25 mi/40 km SE of BATTAMBANG. Rice-growing area; fisheries. Also called Moung Roessei.

Moungo or **Mungo**, department (2001 population 452,722), LITTORAL province; CAMEROON, ⊙ Nkongsamba.

Moungo River, CAMEROON: see MUNGO RIVER.

Mount, town, CARROLL and FREDERICK counties, N MARYLAND, 30 mi/48 km W of BALTIMORE. In agricultural area.

Mount, for names beginning thus, and not found here: see under name following.

Mount Abu, India: see ABU.

Mount Aetna, town (2000 population 838), WASHINGTON county, W MARYLAND; 39°37′N 77°40′W. A recently discovered iron furnace apparently lit up the hillside village; it most likely provided the inspiration for the name when it was founded in 1761. The largest limestone caverns in Maryland are also here. The numerous formations have been given names such as the Great Wall of China and Jefferson Davis, because of fancied resemblances. They were discovered by Boy Scouts in 1928, but closed to the public in 1932.

Mountain (MOUN-ten), rural municipality (□ 1,008 sq mi/2,620.8 sq km; 2001 population 1,589), W MANITOBA, W central Canada; 52°13′N 100°44′W. Divided into N and S regions. Mountain N is bounded in part by W shore of SWAN LAKE; Mountain S is located along W shores of Lake WINNIPEGOSIS. The N includes the communities of Mafeking, Novra, Birch River, and Bellsite.

Mountain, village (2006 population 125), PEMBINA county, NE NORTH DAKOTA, 13 mi/21 km SW of CAVALIER; 48°40′N 97°51′W. Icelandic State Park to NE. Founded in 1873 by immigrants from ICELAND and incorporated in 1940. Named for nearby Pembina Mountains.

Mountainair, town (2006 population 1,077), TORRANCE county, central NEW MEXICO, 65 mi/105 km NW of SOCORRO, between MANZANO RANGE (NW) and Chupadero Mesa (S); 34°31′N 106°14′W. Elevation 6,499 ft/1,981 m. Agricultural shipping point for cattle, corn, beans, pumpkins, grains, alfalfa. Manufacturing (hydro pressure filters, traffic control devices). Abo Pueblo Ruins to W, Quarai Pueblo Ruins to NW,

Gran Quivira Pueblo Ruins (SOCORRO county) to SE; all three are units comprising Salinas Pueblo Missions National Monument. Manzano Peak (10,098 ft/3,078 m) to W; MANZANO MOUNTAIN State Park to NW; parts of Cibola National Forest to W and E.

Mountain Ash (ASH), Welsh *Aberpennar*, town, Rhondda Cynon Taff, S Wales, in Cynon Valley extending S through Miskin to Penrhiwceibr, 4 mi/6.4 km SE of ABERDARE; 51°41′N 03°23′W. Former mining community. Formerly in Mid Glamorgan, abolished 1996.

Mountain Brook, suburb (2000 population 20,604), JEFFERSON county, N central ALABAMA. A residential suburb E of BIRMINGHAM. Incorporated 1942.

Mountainburg, village (2000 population 682), CRAWFORD county, NW ARKANSAS, 21 mi/34 km NE of FORT SMITH on Frog Bayou, at S edge of the OZARKS; 35°37′N 94°10′W. Manufacturing (printed circuit boards). Tourism and resort area. Lake Fort Smith State Park and Ozark National Forest to NE.

Mountain City, town (2000 population 829), RABUN county, extreme NE GEORGIA, 3 mi/4.8 km N of CLAYTON, in the BLUE RIDGE; 34°55′N 83°23′W. Manufacturing of apparel and wood products Black Rock Mountain Park has the highest elevation among nearby state parks.

Mountain City, town (2006 population 2,402), extreme NE TENNESSEE, 33 mi/53 km ENE of JOHNSON CITY, between STONE MOUNTAINS (E) and IRON MOUNTAINS (W); ⊙ JOHNSON county; 36°28′N 81°48′W. Agriculture in summer-resort region; lumbering; manufacturing.

Mountain City, unincorporated village, ELKO county, NE NEVADA, on EAST FORK OWYHEE RIVER, and 70 mi/113 km N of ELKO in Humboldt National Forest, beyond SE corner of Duck Valley Indian Reservation. Elev. c.5,620 ft/1,713 m. Trading point. RIO TINTO (just SW) is former company-owned, copper-mining town, 10 mi/16 km S of IDAHO state line.

Mountain Creek Lake (KREEK), Dallas county, N TEXAS, reservoir impounded 10 mi/16 km WSW of DALLAS by dam in small Mountain Creek (also forms Lake Joe Pool to S); a S tributary of West Fork of TRINITY River; 32°43′N 96°56′W. Dallas Naval Air Station on N shore; Dallas Baptist University on SE shore.

Mountain Dale, resort village, SULLIVAN county, SE NEW YORK, in the CATSKILLS, 8 mi/12.9 km ENE of MONTICELLO; 41°42′N 74°33′W.

Mountain Fork, OKLAHOMA and ARKANSAS: see LITTLE RIVER.

Mountain Grove, city (2000 population 4,574), WRIGHT county, S central MISSOURI, in the OZARKS, 55 mi/89 km E of SPRINGFIELD; 37°07′N 92°15′W. Agriculture, livestock, timber area; trade center; manufacturing (metal stamping, feeds and fertilizers, exercise equipment, caps); lumber; limestone quarries. State fruit experiment station nearby. Settled 1851, incorporated 1882.

Mountain Home, town (2000 population 11,012), ⊙ BAXTER county, N ARKANSAS, c.40 mi/64 km ENE of HARRISON, in the OZARKS, between WHITE RIVER (BULL SHOALS Dam and Reservoir) to W and the North Fork (NORFORK Reservoir) to E; 36°20′N 92°22′W. Manufacturing (printing, motor vehicle equipment, rubber, health care products, consumer goods). Bull Shoals State Park to W. Tourism and recreation, growing retirement population.

Mountain Home, town (2000 population 11,143), ⊙ ELMORE county, SW IDAHO, near SNAKE RIVER, 40 mi/64 km SE of BOISE; elevation 3,145 ft/959 m; 43°08′N 115°42′W. Wool-shipping point in irrigated agricultural area (sheep, cattle, fruit, potatoes, sugar beets; hay); manufacturing (potato skins). Mountain Home U.S. Air Force Base to SW. Plans for further land reclamation in vicinity have been made. Zinc, silver, gold mines to N. Boise National Forest to N;

Long Tom Reservoir to N; Bruneau Dunes State Park to S, at Bruneau. Incorporated 1896.

Mountain Home (MOUNT-uhn HOM), unincorporated town (□ 2 sq mi/5.2 sq km; 2000 population 2,169), HENDERSON county, SW NORTH CAROLINA, 16 mi/26 km S of ASHEVILLE, and 4 mi/6.4 km N of HENDERSONVILLE; 35°22′N 82°30′W. Pisgah National Forest to W. Manufacturing (metal, paper, and ceramic products); agricultural area (hay, corn; dairying, cattle).

Mountainhome, unincorporated town (2000 population 1,169), MONROE county, E PENNSYLVANIA, 13 mi/21 km NNW of STROUDSBURG, in POCONO MOUNTAINS; 41°10′N 75°15′W. Manufacturing includes food products and fabricated metal products. Tourism is important to the economy. Delaware State Forest to E. Ski area here.

Mountain Iron, town (2000 population 2,999), ST. LOUIS county, NE MINNESOTA, in MESABI IRON RANGE 4 mi/6.4 km W of VIRGINIA; 47°31′N 92°37′W. Manufacturing (crushers and shovels, taconite, pallets, steel fabrication). Superior National Forest to N. Grew with development of iron deposits. West Two Rivers Reservoir to S. Settled 1890, incorporated 1892.

Mountain Island Lake, NORTH CAROLINA: see CATAWBA RIVER.

Mountain-Karabakh, AZERBAIJAN: see NAGORNO-KARABAKH AUTONOMOUS REGION.

Mountain Lake, town (2000 population 2,082), COTTONWOOD county, SW MINNESOTA, 10 mi/16 km ENE of WINDOM, near WATONWAN RIVER, forms Mountain Lake to NW; 43°56′N 94°55′W. Grain, soybeans; livestock; manufacturing (slurry pumps, oak furniture, light manufacturing). Laid out 1872, settled by Mennonites.

Mountain Lake, in chain of lakes on CANADA-U.S. border, in COOK county, NE MINNESOTA, and Thunder Bay district, W ONTARIO, 25 mi/40 km NNE of GRAND MARAIS, Minnesota; 7 mi/11.3 km long, 1 mi/1.6 km wide; 48°06′N 90°13′W. Minnesota part in BOUNDARY WATERS Canoe Area of Superior National Forest. Fed From Watap Lake by small stream from W; drains E through small stream to Moose Lake and PIGEON River; all on U.S.-Canada border.

Mountain Lake, reservoir, GREENVILLE county, NW SOUTH CAROLINA, in the BLUE RIDGE MOUNTAINS, on South SALUDA RIVER, and 19 mi/31 km N of GREENVILLE. Resort area.

Mountain Lake Park, town (2000 population 2,248), GARRETT county, W MARYLAND, in the ALLEGHENIES just SE of OAKLAND; 39°24′N 79°23′W. Vegetable cannery. Founded in 1881, the village was originally an extensive complex of hotels, sporting facilities, and a tabernacle called the "Mountain Chautauqua." Thousands gathered here to listen to William Jennings Bryan, perennial Presidential candidate famed for his "Cross of Gold" speech; Samuel Gompers, president of the American Federation of Labor; and President William Howard Taft—when magicians, jugglers, and bell ringers were not performing.

Mountain Lakes, residential borough (2006 population 4,343), MORRIS county, N NEW JERSEY, 6 mi/9.7 km NNE of MORRISTOWN; 40°53′N 74°26′W. Has several artificial lakes. Settled 1915 as real estate development, incorporated 1924.

Mountain Meadow, located in Dixie National Forest, small valley in extreme SW UTAH, N central WASHINGTON county, near headwaters of SANTA CLARA RIVER where in 1857 a party of some 140 emigrants bound for CALIFORNIA were massacred by Indians led by fanatical Mormons settlers after a three-day battle; the incident exacerbated tensions between Mormons and non-Mormons.

Mountain Mesa, unincorporated town (2000 population 716), KERN county, S central CALIFORNIA, residential community 4 mi/6.4 km ENE of LAKE ISABELLA town, on S shore of Lake Isabella reservoir

(KERN RIVER), in SIERRA NEVADA; 35°38′N 118°25′W. Cattle, hay, timber. Area surrounded by parts of Sequoia National Forest.

Mountain Park, city, FULTON county, NW central GEORGIA, 12 mi/19 km NE of MARIETTA; 34°05′N 84°25′W.

Mountain Park (MOUN-ten), unincorporated village, W ALBERTA, W Canada, in ROCKY MOUNTAINS, near JASPER NATIONAL PARK, 35 mi/56 km E of JASPER, and in Yellowhead County; 52°56′N 117°17′W. Elevation 5,815 ft/1,772 m. Coal mining; timber; tourism. Near source of McLeod River.

Mountain Park, village (2006 population 380), KIOWA county, SW OKLAHOMA, 25 mi/40 km SSE of HOBART; 34°42′N 98°57′W. In agricultural area. Great Plains State Park nearby, on Tom Steed Lake reservoir, to N.

Mountain Pine, town (2000 population 772), GARLAND county, central ARKANSAS, 8 mi/12.9 km NW of HOT SPRINGS, and on OUACHITA RIVER, at E end of BLAKELY MOUNTAIN DAM; 34°34′N 93°10′W. Manufacturing (lumber, wooden furniture). Tourism and recreation. Large Lake OUACHITA reservoir to NW; Lake Ouachita State Park to N; Ouachita National Forest to W.

Mountain Pine Ridge Forest Reserve (□ 227 sq mi/ 590 sq km), CAYO district, BELIZE, 20 mi/32 km E of BENQUE VIEJO DEL CARMEN; 17°03′N 88°51′W. The term "ridge" refers to pine forests that develop on poor soils, not to landforms. The main feature of this managed forest is Hidden Valley Falls (also known as Thousand Foot Falls), the tallest waterfall in CENTRAL AMERICA (elev. c.1,200 ft/365 m).

Mountain Point, village, SE ALASKA, on S shore of REVILLAGIGEDO ISLAND, SE of KETCHIKAN; 55°18′N 131°32′W.

Mountain Province (□ 810 sq mi/2,106 sq km), in CORDILLERA ADMINISTRATIVE REGION, N LUZON, PHILIPPINES, KALINGA-APAYAO province to N, IFUGAO province to S; ⊙ BONTOC; 17°05′N 121°10′E. Population 8.5% urban, 91.5% rural; in 1991, 80% of urban and 63% of rural settlements had electricity. Mountainous, with three river valleys. Indigenous Bontoc Igorot people practice double crop irrigated and terraced rice cultivation; also river fishing. Settled by the Spanish in the mid-19th century. The province remains accessible only by a few roads of poor quality. An area of heavy conflict between the government and the Marxist New People's Army.

Mountain Rest, village, OCONEE county, NW SOUTH CAROLINA, in the BLUE RIDGE, 9 mi/14.5 km NW of WALHALLA; near GEORGIA state line in Sumter National Forest. OCONEE STATE PARK to E, fish hatchery to N.

Mountainside, residential borough (2006 population 6,644), UNION county, NE NEW JERSEY, 11 mi/18 km SW of NEWARK; 40°40′N 74°21′W. Park and bird sanctuary here. Incorporated 1895.

Mountains of the Moon, UGANDA: see RUWENZORI, mountain range.

Mountain State: see WEST VIRGINIA.

Mountain Top, unincorporated town, Fairview township, LUZERNE county, NE PENNSYLVANIA; industrial suburb 5 mi/8 km S of WILKES-BARRE; 41°10′N 75°52′W. Elevation 1,680 ft/512 m. Diversified manufacturing. Crystal Lake reservoir to E. Founded 1788.

Mountain View, city (2000 population 70,708), SANTA CLARA county, W CALIFORNIA, suburb 8 mi/12.9 km WNW of downtown SAN JOSE, and 31 mi/50 km SSE of SAN FRANCISCO, on SAN FRANCISCO BAY; 37°24′N 122°05′W. Located in an area that grew rapidly in 1960s–1970s, Mountain View is a major part of the Silicon Valley industrial complex. Varied high-technology manufacturing. It also has research organizations and diverse manufacturing industries. Fruits and vegetables are shipped from the city. Saint Patrick's College is there. Moffet Naval Air Station adjoins the city to NE. Drained by Stevens Creek; Stevens Creek Reservoir to S; Santa Clara Mountains

to SW; branch of HETCH HETCHY AQUEDUCT runs through city. Incorporated 1902.

Mountain View, town (2000 population 2,876), ⊙ STONE county, N ARKANSAS, 27 mi/43 km WNW of BATESVILLE, in the OZARKS; 35°52′N 92°06′W. Agriculture; varied manufacturing. Part of Ozark National Forest to N; Blanchard Springs Caverns and Recreation Area to NW; Ozark Folk Center State Historical Park is here.

Mountain View, town (2000 population 2,799), SE HAWAII island, HAWAII county, HAWAII, 12 mi/19 km S of HILO inland, 12 mi/19 km from NE coast, 15 mi/24 km from SE coast, on Volcano Road (Hawaii Belt Road); 19°31′N 155°09′W. HAWAII VOLCANOES NATIONAL PARK 12 mi/19 km SW, active craters and lava flows to sea 12 mi/19 km S. Waiakea Forest Reserve to N; Puna Forest Reserve and Wao Kele O Puna Natural Area Reserve to SE.

Mountain View, town (2000 population 2,430), HOWELL county, S MISSOURI, in the OZARKS, near ELEVEN POINT RIVER, 20 mi/32 km NNE of WEST PLAINS; 36°59′N 91°42′W. Grain, livestock, lumber products.

Mountain View (MOUNT-uhn VYOO), unincorporated town (□ 4 sq mi/10.4 sq km; 2000 population 3,768), CATAWBA county, W central NORTH CAROLINA, 5 mi/8 km SSW of HICKORY, between HENRYS FORK and Jacobs Fork rivers; 35°40′N 81°22′W. Manufacturing; service industries; agricultural area (grain, soybeans; poultry, livestock, dairying).

Mountain View, town (2006 population 823), KIOWA county, SW OKLAHOMA, 20 mi/32 km ENE of HOBART, and on WASHITA River, N of Wichita Mountains; 35°06′N 98°45′W. In wheat, livestock, cotton area.

Mountain View, town (2006 population 1,180), UINTA county, SW WYOMING, 5 mi/8 km SSW of LYMAN, on Williams Creek; 41°16′N 110°20′W. Elevation 6,795 ft/ 2,071 m. Cattle, sheep; alfalfa, hay; timber. Wasatch National Forest to S.

Mountain View (MOUN-ten VYOO), unincorporated village, S ALBERTA, W Canada, near U.S. (MONTANA) border, 50 mi/80 km SW of LETHBRIDGE, in Cardston County; 49°08′N 113°36′W. Gateway to WATERTON LAKES NATIONAL PARK. Cattle, sheep.

Mountain View, village (2000 population 569), JEFFERSON county, N central COLORADO, residential suburb 5 mi/8 km NW of downtown DENVER; 39°46′N 105°03′W. Elevation 5,385 ft/1,641 m. Bounded E by Denver, N by LAKESIDE, W and S by WHEAT RIDGE.

Mountain View, residential village, PASSAIC county, NE NEW JERSEY, on POMPTON RIVER, and 5 mi/8 km W of PATERSON.

Mountain View Acres, unincorporated town (2000 population 2,521), SAN BERNARDINO county, S CALIFORNIA, residential suburb 4 mi/6.4 km SW of VICTORVILLE, and 5 mi/8 km NNW of HESPERIA; 34°30′N 117°21′W. Agriculture (citrus, pears, apples, nursery products, grain); dairying; cattle. Roy Rogers/Dale Evans Museum to NE; GEORGE AIR FORCE BASE to N. Irrigated agricultural area at SW edge of MOJAVE DESERT.

Mountain View County (MOUN-ten VYOO), municipality (□ 1,469 sq mi/3,819.4 sq km; 2001 population 12,134), S ALBERTA, W Canada, 44 mi/71 km from CALGARY; 51°40′N 114°14′W. Agriculture (wheat, canola, oats, barley, seed and specialty crops, hay); livestock; oil and gas; mining; forestry; manufacturing. Includes CARSTAIRS, Sundre, DIDSBURY, Cremona, OLDS; and the hamlets of Eagle Hill, Water Valley, and Westward Ho. Formed 1912.

Mountain Village, village, W ALASKA, on YUKON River, and 110 mi/177 km NW of BETHEL; 62°05′N 163°43′W.

Mountainville, village, HUNTERDON county, W NEW JERSEY, 13 mi/21 km N of FLEMINGTON, in agricultural area.

Mountainville, village, ORANGE county, SE NEW YORK, 12 mi/19 km SSW of NEWBURGH, in HUDSON high-

lands; 41°24′N 74°05′W. Site of Storm King Art Center (see also STORM KING).

Mountain Zebra National Park (□ 25.2 sq mi/65.3 sq km; 16,144 acres/6,536 ha), EASTERN CAPE province, SOUTH AFRICA, 17 mi/27 km WNW of CRADOCK. Reserve established 1938. Popular tourist destination. This tiny park houses one of the rarest large mammals in the world, the subspecies of zebra *Equus zebra zebra*. This subspecies grew from near extinction in 1937 to a current herd of 200, but is still on endangered list. Other species include eland, springbok, black wildebeest, red hartebees, blesbok, a variety of smaller mammals and over 200 bird species.

Mount Airy, town (2000 population 6,425), CARROLL and FREDERICK counties, central MARYLAND, 6 mi/9.7 km SW of WESTMINSTER; 39°22′N 77°10′W. Situated on Parr's Ridge with a water tower visible for miles around, Mount Airy was settled by Henry Bussard in 1814, and annexed neighboring Ridgeville in 1966. A series of inclined planes had to be built by the BALTIMORE AND OHIO RAILROAD to allow its early engines to negotiate Parr's Ridge. A tunnel was built through the ridge in 1890. The once well-known Mary Garrett Childrens' Hospital located here has been converted into apartments. Mostly in Carroll county, the town splits into Frederick county.

Mount Airy (MOUNT ER-ee), town (□ 8 sq mi/20.8 sq km; 2006 population 8,457), SURRY county, NW NORTH CAROLINA, 34 mi/55 km NW of WINSTON-SALEM, in BLUE RIDGE foothills, near VIRGINIA state line; 36°30′N 80°36′W. Railroad terminus. Granite quarrying. Manufacturing (stone, granite, and plastic products; apparel, lumber, store displays, utility buildings, furniture, bucket trucks); service industries. Pilot Mountain State Park to SE. Incorporated 1885.

Mount Airy, village (2000 population 604), HABERSHAM county, NE GEORGIA, 11 mi/18 km WSW of TOCCOA; 34°31′N 83°30′W.

Mount Albert (MOUNT AL-buhrt), unincorporated village, York region, S ONTARIO, E central Canada, 32 mi/51 km N of TORONTO; 44°08′N 79°19′W. Now part of city of East Gwillimbury. Dairying, mixed farming.

Mount Albert, residential suburb within AUCKLAND City, N NORTH ISLAND, NEW ZEALAND; at base of extinct volcanic cone, Mount Albert (440 ft/134 m). Incorporated into Auckland City in 1989.

Mount Allen, unincorporated town, CUMBERLAND county, S PENNSYLVANIA, residential suburb 7 mi/11.3 km SW of HARRISBURG; 40°10′N 76°58′W.

Mount Angel, town (2006 population 3,389), MARION county, NW OREGON, 13 mi/21 km NE of SALEM near Pudding River, in WILLAMETTE RIVER Valley; 45°04′N 122°47′W. Trade center for fruit, grain, and dairy products. Manufacturing (bakery products, soft drinks, vitamins). Agriculture (vegetables, mint, berries, tulips, iris; poultry, cattle). Mount Angel Seminary (Roman Catholic) to SE. Incorporated 1905.

Mount Apo (AH-po), active volcano, NORTH COTABATO province, S MINDANAO island, the PHILIPPINES. It is the highest peak of the islands (elevation 9,690 ft/2,953 m). Mount Apo has a snow-capped appearance but is actually covered with white sulfur. Mount Apo National Park (□ 281 sq mi/728 sq km; established 1936) is here. Operational area for New People's Army.

Mount Arayat (ah-RAH-yaht), cone-shaped extinct volcano, PAMPANGA province, central LUZON, PHILIPPINES, near ARAYAT, 45 mi/72 km NNW of MANILA; 15°02′N 120°44′E. Elevation 3,378 ft/1,030 m.

Mount Arlington, borough (2006 population 5,708), MORRIS county, N NEW JERSEY, on Lake HOPATCONG, and 12 mi/19 km NW of MORRISTOWN; 40°55′N 74°38′W.

Mount Athos, Greece: see AKTI.

Mount Auburn, town (2000 population 75), SHELBY county, E INDIANA, 11 mi/18 km SSW of SHELBYVILLE; 39°49′N 85°11′W. In agricultural area. Laid out 1837.

Mount Auburn, town (2000 population 160), BENTON county, E central IOWA, near CEDAR RIVER, 9 mi/14.5 km NNW of VINTON; 42°15′N 92°05′W. In agricultural area.

Mount Auburn, village (2000 population 515), CHRISTIAN county, central ILLINOIS, near SANGAMON RIVER, 21 mi/34 km E of SPRINGFIELD; 39°46′N 89°15′W. Grain; livestock.

Mount Auburn, MASSACHUSETTS: see CAMBRIDGE.

Mount Augustus (MOUNT uh-GUS-tuhs), monolith, WESTERN AUSTRALIA state, W AUSTRALIA, 199 mi/320 km E of CARNARVON. Elevation 3,625 ft/1,105 m; its central ridge is c.5 mi/8 km long. Slopes have significant vegetation. Estimated to be 1000 million years old.

Mount Ayr, town (2000 population 147), NEWTON county, NW INDIANA, 15 mi/24 km NNE of KENTLAND; 40°57′N 87°18′W. In agricultural area.

Mount Ayr, town (2000 population 1,822), ⊙ RINGGOLD county, S IOWA, 24 mi/39 km SSE of CRESTON; 40°42′N 94°14′W. In livestock and grain area. Manufacturing (stadium bleachers, automotive wire harnesses, motor-home leveling systems). Airfield here; fish hatchery to N. Founded c.1855, incorporated 1875.

Mount Babuyan Claro, active volcano (3,871 ft/1,180 m), CAGAYAN province, BABUYAN Island, PHILIPPINES; 19°31′N 121°56′E. Last known eruption 1913.

Mount Banahao (bah-NAH-hou), active volcano (7,142 ft/2,177 m), S LUZON, PHILIPPINES, SE of MANILA, on border between LAGUNA and QUEZON provinces; 14°04′N 121°29′E. Eruptions occurred in 1730, 1743, and 1780, and mud flows occurred in 1730, 1843, and 1909. Fumeroles. Also spelled Banahaw.

Mount Barker (MOUNT BAHR-kuhr), town, SE SOUTH AUSTRALIA state, S central AUSTRALIA, 18 mi/29 km ESE of ADELAIDE; 35°06′S 138°52′E. Poultry, dairy products, salmon plant; metal trades; perfumery.

Mount Barker (MOUNT BAHR-kuhr), town, SW WESTERN AUSTRALIA state, W AUSTRALIA, 210 mi/338 km SSE of PERTH, at base of Mount Barker (829 ft/253 m); 34°36′S 117°37′E. Sheep, cattle; vineyards, apples. Tourism. Spongolite, a stone made mostly from fossilized sponges, in a quarry E of town.

Mount Bataan (bah-tah-AHN) (4,700 ft/1,433 m), highest peak on BATAAN Peninsula, Bataan province, S LUZON, PHILIPPINES, 6 mi/9.7 km NNW of MARIVELES; 14°31′N 120°28′E.

Mount Baw Baw, ski resort, VICTORIA, SE AUSTRALIA, 95 mi/153 km from MELBOURNE; 37°50′S 146°17′E. Walking tracks during summer. Grasslands, heathlands, and snow gums (eucalyptus) in area.

Mount Beacon, NEW YORK: see BEACON.

Mount Beauty (MOUNT BYOO-tee), town, VICTORIA, SE AUSTRALIA, 208 mi/335 km NE of MELBOURNE, at foot of Mount BOGONG, in Alpine National Park; 36°45′S 147°10′E. Ski center.

Mount Berlin, volcano (11,483 ft/3,500 m) in MARIE BYRD LAND, WEST ANTARCTICA, at the W end of the FLOOD RANGE, overlooking the HOBBS COAST; 76°03′S 135°52′W.

Mount Biliran, active volcano (3,048 ft/929 m), BILIRAN province, BILIRAN ISLAND, PHILIPPINES; 11°40′N 124°24′E. Last eruption 1939.

Mount Bischoff, AUSTRALIA: see WARATAH.

Mount Blanchard (BLAN-chuhrd), village (2006 population 447), HANCOCK county, NW OHIO, 10 mi/16 km SSE of FINDLAY, and on BLANCHARD RIVER; 40°54′N 83°33′W.

Mount Borradaile, N central NORTHERN TERRITORY, N central AUSTRALIA, 140 mi/225 km ENE of DARWIN; 12°05′S 132°51′E. Located in Arnhem Land Aboriginal Reserve. Aboriginal rock art site. Remote location; access limited to licensed tour operators and individuals holding permits from the Northern Aboriginal Land Council.

Mount Braddock, unincorporated village, Menallen township, FAYETTE county, SW PENNSYLVANIA, 6 mi/9.7 km NE of UNIONTOWN; 39°56′N 79°38′W. Connellsville Airport to NW.

Mount Bruce, in S central NORTH ISLAND, NEW ZEALAND, 16 mi/26 km N of MASTERTON. A reserve where rare native birds are held and bred in limited freedom, to save them from extinction and enable research. Species include kakapo, takahe, black stilt.

Mount Brydges (MOUNT BRI-juhz), unincorporated village, S ONTARIO, E central Canada, 15 mi/24 km WSW of LONDON, and included in Strathroy-Caradoc township; 42°54′N 81°29′W. Dairying, farming.

Mount Buddajo, active volcano (2,661 ft/811 m), SULU province, JOLO Island, PHILIPPINES; 05°55′N 121°10′E. Erupted in 1641, 1897. Also known as Mount Jolo; is sometimes spelled Bud Dajo.

Mount Bulusan (boo-LOO-sahn), active volcano (5,115 ft/1,559 m), SORSOGON province, extreme SE LUZON, PHILIPPINES, just W of BULUSAN, 15 mi/24 km SSE of SORSOGON; 12°46′N 124°03′E. A crater lake is near the summit. At least thirteen eruptions have occurred between 1852 and 1995; most have been mild.

Mount Cabalasan (kah-bah-LAH-sahn), highest peak (3,324 ft/1,013 m) of Cebu island, CEBU province, PHILIPPINES, in central part of island, 12 mi/19 km NW of CEBU; 10°26′N 123°43′E.

Mount Cagua (KAHG-wah), active volcano (3,800 ft/1,158 m), CAGAYAN province, NE LUZON, PHILIPPINES, in N part of the SIERRA MADRE, 32 mi/51 km ESE of APARRI; 18°13′N 122°07′E. Last eruption was in 1860.

Mount Calayo, PHILIPPINES: see MOUNT MUSUAN.

Mount Calm (KAHM), village (2006 population 345), HILL county, central TEXAS, 20 mi/32 km NE of WACO, near source of NAVASOTA RIVER; 31°45′N 96°52′W. In cotton area; dairying; horses.

Mount Calvary, village (2006 population 939), FOND DU LAC county, E WISCONSIN, near SHEBOYGAN RIVER, 10 mi/16 km ENE of FOND DU LAC, in dairying region; 43°49′N 88°15′W. Kettle Moraine State Forest to SE; Sheboygan Marsh County Park to NE.

Mount Cameroon, active volcano (13,354 ft/4,070 m), in the CAMEROON Highlands, SOUTH-WEST province, Cameroon. Highest point in W Africa. The W side of the mountain receives an average annual rainfall of more than 400 in/1,016 cm and is covered with tropical rain forest. Cocoa, banana, rubber, and tea plantations are found on the lower slopes. The volcano last erupted in 1959; minor eruption in 1982. Called FAKO by local residents.

Mount Camiguin de Babuyanes (kah-mee-GEEN de bah-BOO-yahns), active volcano, CAGAYAN province, CAMIGUIN ISLAND, BABUYAN ISLANDS, PHILIPPINES; 18°50′N 121°51′E. Elevation 2,336 ft/712 m; last eruption 1857.

Mount Canlaon (kahn-lah-ON), active volcano, N central NEGROS island, PHILIPPINES, in NEGROS ORIENTAL province; 10°24′N 123°07′E. Elevation 8,088 ft/2,465 m. Just W of Quezon; highest peak of island. There have been at least fifteen eruptions between 1866 and 1996; most have been mild. Also called Malaspina.

Mount, Cape, GRAND CAPE MOUNT county, headland of SW LIBERIA, on ATLANTIC OCEAN; 06°45′N 11°23′W. Rises to 1,068 ft/326 m. Site of ROBERTSPORT.

Mount Carbine, settlement, NE QUEENSLAND, NE AUSTRALIA, 50 mi/80 km NW of CAIRNS; 16°32′S 145°08′E. Mining district. Tungsten.

Mount Carbon, borough (2006 population 83), SCHUYLKILL county, E central PENNSYLVANIA, residential suburb, 1 mi/1.6 km SE of POTTSVILLE, on SCHUYLKILL RIVER; 40°40′N 76°11′W.

Mount Carmel, city (2000 population 7,982), ⊙ WABASH county, SE ILLINOIS, on the WABASH (bridged here), and 24 mi/39 km SSW of VINCENNES, INDIANA; 38°25′N 87°46′W. Center of rich agricultural area (corn, wheat, soybeans); manufacturing of electronic equipment, metal products; railroad shops. Oil obtained nearby. Musselshell industry flourished here after 1900. Indian mounds nearby have yielded artifacts. Incorporated 1825. Wabash Valley College is nearby.

Mount Carmel, town (2000 population 106), FRANKLIN county, SE INDIANA, 7 mi/11.3 km ESE of BROOKVILLE; 39°25′N 84°53′W. In agricultrual area.

Mount Carmel (KAHR-mel), town (2006 population 5,356), HAWKINS county, NE TENNESSEE, 6 mi/10 km W of KINGSPORT, on HOLSTON River, near Holston Army Ammunition Plant; 36°33′N 82°39′W. Incorporated 1961.

Mount Carmel (KAHR-mel), unincorporated village (□ 2 sq mi/5.2 sq km; 2000 population 4,308), CLERMONT county, extreme SW OHIO; far E suburb of CINCINNATI; 39°05′N 84°17′W.

Mount Carmel (KAHR-muhl), village (2000 population 237), MCCORMICK county, W SOUTH CAROLINA, 14 mi/23 km NW of MCCORMICK, and on W edge of Sumter National Forest; 34°00′N 82°30′W.

Mount Carmel, unincorporated village, KANE county, SW UTAH, on EAST FORK VIRGIN RIVER, and 15 mi/24 km NNW of KANAB. Elevation 5,314 ft/1,620 m. Coral Pink Sand Dunes State Park to S. ZION NATIONAL PARK to W; Dixie National Forest to N. Highway extends 25 mi/40 km W to Zion Canyon, in Zion National Park. MOUNT CARMEL JUNCTION 2 mi/3.2 km SW.

Mount Carmel, borough (2006 population 5,970), NORTHUMBERLAND county, E central PENNSYLVANIA, 20 mi/32 km ESE of SUNBURY, near North Branch Shamokin Creek (source to NE); 40°47′N 76°24′W. Agricultural area (grain, apples; poultry, livestock, dairying); manufacturing (aluminum doors and windows, corrugated boxes, plastic products); anthracite coal. Laid out 1835, incorporated 1864.

Mount Carmel, CONNECTICUT: see HAMDEN.

Mount Carmel Junction, unincorporated village, KANE county, S UTAH, 14 mi/23 km NNW of KANAB, on EAST FORK VIRGIN RIVER. Elevation 5,191 ft/1,582 m. Cattle. East gateway to ZION NATIONAL PARK. Coral Pink Cliffs State Park to S.

Mount Carroll, city (2000 population 1,832), ⊙ CARROLL county, NW ILLINOIS, 22 mi/35 km SW of FREEPORT; 42°06′N 89°58′W. In rich agricultural area (dairy products, grain; livestock, poultry). Founded 1843, incorporated 1867.

Mount Cauitan (kah-WEE-tahn), MOUNTAIN PROVINCE, N LUZON, PHILIPPINES, 45 mi/72 km SE of VIGAN; 17°15′N 120°59′E. Elevation 8,427 ft/2,569 m.

Mount Chase, plantation, PENOBSCOT county, N central MAINE, 31 mi/50 km NNE of MILLINOCKET; 46°04′N 68°30′W. In hunting, fishing area.

Mount Clare, unincorporated town, HARRISON county, N WEST VIRGINIA, 5 mi/8 km S of CLARKSBURG. Agriculture (corn); cattle; poultry.

Mount Clemens, city (2000 population 17,312), ⊙ MACOMB county, suburb 22 mi/35 km NE of downtown DETROIT, SE MICHIGAN, on the CLINTON RIVER; 42°36′N 82°52′W. The city is known for its mineral waters. It has a large floral industry and manufacture of tool and die, enamels, powder coatings, tool holders and publishing. National Guard Base to E. Settled c.1798, incorporated as a city 1879.

Mount Cobb, unincorporated town (2000 population 2,140), LACKAWANNA county, NE PENNSYLVANIA, 9 mi/14.5 km E of SCRANTON; 41°25′N 75°30′W. Agriculture includes dairying, livestock; corn, hay. Several reservoirs in area.

Mount Communism, TAJIKISTAN: see PEAK KOMMUNIZMA.

Mount Coolon (KOO-lahn), town, E QUEENSLAND, NE AUSTRALIA, 150 mi/241 km SSE of TOWNSVILLE; 21°23′S 147°20′E. Cattle. Former gold mines.

Mount Cory (KOR-ee), village (2006 population 195), HANCOCK county, NW OHIO, 11 mi/18 km SW of FINDLAY; 40°56′N 83°49′W. In agricultural area.

Mount Crawford (MOUNT KRAW-fuhrd), village (2006 population 285), ROCKINGHAM county, NW

VIRGINIA, in SHENANDOAH VALLEY, 6 mi/10 km SSW of HARRISONBURG, on NORTH RIVER; 38°21'N 78°56'W. Manufacturing (fertilizer, poultry packaging, dairy products); agriculture (poultry, livestock; dairying; grain, apples, peaches). Blue Ridge Community College is 6 mi/10 km to S.

Mount Croghan (KRO-guhn), village (2006 population 153), CHESTERFIELD county, N SOUTH CAROLINA, 20 mi/32 km WNW of CHERAW, near NORTH CAROLINA state line; 34°46'N 80°13'W. Manufacturing includes textile parts, knit goods; agriculture includes livestock, poultry, grain, peaches.

Mount Darwin, town, MASHONALAND CENTRAL province, NE ZIMBABWE, 80 mi/129 km NNE of HARARE, on Mufuri River, S of its confluence with RUYA RIVER; 16°48'S 31°36'E. MOUNT DARWIN (4,951 ft/1,509 m) to S. Cattle sheep, goats; corn.

Mount Desert (DEZ-zert), township, HANCOCK county, S MAINE, on MOUNT DESERT ISLAND; 44°19'N 68°19'W. Includes villages of Seal Harbor, Northeast Harbor, and Asticou. Settled 1762; formerly called Somesville; incorporated 1789.

Mount Desert Island (□ c.100 sq mi/259 sq km), largest island off the coast of MAINE, separated from the mainland by FRENCHMAN BAY, Mount Desert Narrows, and Western Bay. The island's rugged topography is a result of glacial action. It is almost equally divided into E and W halves by SOMES SOUND, a fjord. A chain of rounded granite peaks dominates the island, culminating in Cadillac Mountain (1,530 ft/467 m high). The peaks were named *Monts Deserts*, meaning "wilderness mountains," by the French explorer Samuel de Champlain, who landed on the island in 1604. The first French Jesuit mission and colony in America was established here in 1613; the first permanent English settlement began in 1762. The island developed as a fishing and lumbering center, and by the end of the nineteenth century it had become a famous resort area. A forest fire in 1947 damaged much of the E of the island BAR HARBOR, MOUNT DESERT, TREMONT, and SOUTHWEST HARBOR are the main towns. The major part of the island is in ACADIA NATIONAL PARK.

Mount Didicas, active volcano, CAGAYAN province, Didicas Island, PHILIPPINES; 19°04'N 122°12'E. Elevation 801 ft/244 m; six violent eruptions between 1773 and 1978.

Mount Discovery, extinct volcano in S VICTORIA LAND, EAST ANTARCTICA, E of KOETTLITZ GLACIER overlooking the NW portion of the ROSS ICE SHELF near MCMURDO SOUND; 78°23'S 165°00'E. Elevation 8,793 ft/ 2,680 m.

Mount Dora (DAW-ruh), city (□ 6 sq mi/15.6 sq km; 2005 population 11,474), Lake county, central FLORIDA, 23 mi/37 km NW of ORLANDO, on LAKE DORA; 28°48'N 81°38'W. Yachting resort; citrus-fruit shipping center. Previously packing houses, cannery, and box and fertilizer factories. Settled 1874, incorporated 1910.

Mount Dora (DO-ruh), village, UNION county, NE NEW MEXICO, 18 mi/29 km WNW of CLAYTON, near Cieneguilla del Burro or Seneca Creek. Elevation 5,686 ft/ 1,733 m. Livestock; alfalfa, and grain.

Mount Eagle, Gaelic *Sliabh an Iolair*, mountain (1,696 ft/517 m), W KERRY county, SW IRELAND, 2 mi/3.2 km NE of SLEA HEAD cape; 52°07'N 10°25'W.

Mount Eaton (EE-tuhn), village (2006 population 243), WAYNE county, N central OHIO, 14 mi/23 km ESE of WOOSTER; 40°42'N 81°42'W. In agricultural area.

Mount Eba, settlement, central SOUTH AUSTRALIA state, S central AUSTRALIA, 100 mi/161 km NW of WOOMERA-MARALINGA, and 200 mi/322 km NW of PORT AUGUSTA; 39°11'S 135°41'E. In arid district. Site of some installations of former Woomera rocket-launching range.

Mount Eden, suburb of HAYWARD, ALAMEDA county, W CALIFORNIA, 15 mi/24 km SSE of downtown OAK-LAND, E of SAN FRANCISCO BAY. At E end of San Mateo–Hayward Bridge (toll).

Mount Eden, residential suburb within AUCKLAND City, N NORTH ISLAND, New Zealand. At base of extinct volcanic cone, Mount Eden (643 ft/196 m). Incorporated into Auckland City in 1989.

Mount Enterprise (EN-tuhr-preiz), village (2006 population 536), RUSK county, E TEXAS, 16 mi/26 km SSE of HENDERSON; 31°55'N 94°40'W. In pine timber; cattle; vegetable-growing area.

Mount Ephraim (EE-free-uhm), residential borough (2006 population 4,437), CAMDEN county, SW NEW JERSEY, 5 mi/8 km S of CAMDEN; 39°52'N 75°05'W. Settled before 1800, incorporated 1926.

Mount Erie, village (2000 population 105), WAYNE county, SE ILLINOIS, 12 mi/19 km NE of FAIRFIELD; 38°30'N 88°13'W. In agricultural area.

Mount Etna, town (2000 population 110), HUNTINGTON county, NE central INDIANA, on SALAMONIE RIVER, and 10 mi/16 km SSW of HUNTINGTON; 40°44'N 85°34'W. In agricultural area.

Mount Evelyn (MOUNT E-vuh-lin), suburb 23 mi/37 km E of MELBOURNE, VICTORIA, SE AUSTRALIA, E of LILYDALE; 37°47'S 145°23'E.

Mount Forest (MOUNT FAH-ruhst), former town (□ 3 sq mi/7.8 sq km; 2001 population 4,584), S ONTARIO, E central Canada, on branch of SAUGEEN RIVER, and 40 mi/64 km NNW of KITCHENER; 43°58'N 80°44'W. Light manufacturing; dairying, woodworking. Originally Maitland Hills; name changed to Mount Forest in 1853. Amalgamated into Wellington North township in 1999.

Mount Gambier (MOUNT GAM-beer), town (2001 population 22,751), extreme SE SOUTH AUSTRALIA state, 235 mi/378 km SE of ADELAIDE, near VICTORIA line; 37°50'S 140°46'E. Railroad junction; sheep and agriculture center; tourism. Nearby is Mount Gambier, extinct volcano with collapsed craters (highest, 650 ft/198 m), one of which, called the Blue Lake, changes color seasonally from grey to bright blue. Limestone quarries nearby.

Mount Garnet (MOUNT GAHR-nuht), settlement, NE QUEENSLAND, NE AUSTRALIA, 70 mi/113 km SW of CAIRNS on S edge of ATHERTON TABLELAND; 17°41'S 145°07'E. Copper-mining center; also zinc, lead, silver, garnet, tin, and limestone. Timber.

Mount Gay or **Gay**, unincorporated town, LOGAN county, SW WEST VIRGINIA, suburb 2 mi/3.2 km W of LOGAN. Manufacturing (electrical boxes, machining). Agriculture (tobacco). Chief Logan State Park to N.

Mount Gerdine (guhr-DEEN), mountain (11,258 ft/ 3,834 m), S ALASKA, in Tordrillo Mountains, 90 mi/145 km NW of ANCHORAGE; 61°23'N 152°29'W. Named for topographic engineer in late nineteenth century.

Mount Gilead (GIL-ee-uhd), town (□ 3 sq mi/7.8 sq km; 2006 population 1,406), MONTGOMERY county, S central NORTH CAROLINA, 14 mi/23 km SE of ALBEMARLE; 35°13'N 80°00'W. Manufacturing (barcode equipment, boots, wood chips, socks and hosiery, fabricated metal products, laminate flooring); service industries. Agricultural area (cotton, tobacco, grain; livestock). Sawmilling, timber (pine). Lake TILLERY reservoir to W, on Yadkin (PEE DEE) River. Town Creek Indian Mound State Historical Site to SE. Originally called Providence. Incorporated 1898.

Mount Gilead, village (□ 3 sq mi/7.8 sq km; 2006 population 3,525), ☉ MORROW county, central OHIO, 16 mi/26 km ESE of MARION; 40°33'N 82°50'W. Manufacturing (hydraulic presses, electrical apparatus, chemicals); seed growing. Site of annual gourd festival. Founded c.1824; renamed Mount Gilead in 1832.

Mount Gretna, borough (2006 population 235), LEBANON county, SE central PENNSYLVANIA, 6 mi/9.7 km SSW of LEBANON, suburb of Mount Gretna Heights 0.5 mi/0.8 km to E; 40°15'N 76°28'W. Agriculture (grain; poultry, livestock, dairying). Conewago Lake reservoir, on Conewago Creek, in town.

Mount Guitinguitin, PHILIPPINES: see SIBUYAN ISLAND.

Mount Hagen, town (2000 population 27,782), ☉ WESTERN HIGHLANDS province, NE NEW GUINEA island, N central PAPUA NEW GUINEA, 200 mi/322 km NW of LAE; 05°52'N 144°13'E. Commercial and services center for Great Plateau region. Road access; Kagamuga International Airport. Coffee, tea, bananas, sweet potatoes, taro, maize, vegetables; timber; manufacturing (soft drinks, timber mills, sheet metal); tourism. Supply center to oil and gas fields to S and gold-mining areas to NW. City is commonly called Hagen.

Mount Halcon (hahl-KON), highest peak (8,484 ft/ 2,586 m) of MINDORO island, MINDORO ORIENTAL province, PHILIPPINES, in N part of island, 16 mi/26 km SW of CALAPAN; 13°16'N 121°00'E.

Mount Hamilton Range, California: see DIABLO RANGE.

Mount Hampton, an extinct volcano with an ice-filled crater that is the northernmost peak of the Executive Range of MARIE BYRD LAND, WEST ANTARCTICA; 76°29'S 125°48'W. Elevation 10,909 ft/3,325 m.

Mount Healthy (HEL-thee), city (□ 1 sq mi/2.6 sq km; 2006 population 6,461), HAMILTON county, extreme SW OHIO; N suburb of CINCINNATI; 39°14'N 84°32'W. Previously tools, brick. Founded 1817.

Mount Herbert, highest point (3,018 ft/920 m) on BANKS PENINSULA, Banks Peninsula district, NEW ZEALAND, S of Lyttelton Harbour. Formerly known as Herbert Peak.

Mount Hermon, resort village, SANTA CRUZ county, W CALIFORNIA, in SANTA CRUZ MOUNTAINS, 5 mi/8 km N of SANTA CRUZ. Redwood groves and a Narrow Gauge railroad nearby.

Mount Hermon, MASSACHUSETTS: see NORTHFIELD.

Mount Herzl (HUHR-tsuhl) or **Har Herzl** (Hebrew), JERUSALEM, E ISRAEL. Rises to 2,723 ft/830 m. Named after Theodor (Benjamin Zeev) Herzl, founder of modern Zionism. Israel's main military cemetery and the burial place of presidents and prime ministers lies on its S slope. On the summit is the central memorial and Herzl's grave.

Mount Hibok-Hibok, active volcano (4,265 ft/1,300 m), Camiguin province, CAMIGUIN Island, PHILIPPINES; 09°12'N 120°40'E. It has six eruptions between 1827 and 1953. The 1948–1953 eruption took more than 1,000 lives and led to mass outmigration from the island.

Mount Holly (MOUNT HAH-lee), city (□ 8 sq mi/20.8 sq km; 2006 population 9,804), GASTON county, S NORTH CAROLINA, suburb 12 mi/19 km WNW of CHARLOTTE, and 10 mi/16 km ENE of GASTONIA, on CATAWBA RIVER (Lake WYLIE reservoir) dam here forms Mountain Island Lake; 35°18'N 81°01'W. Railroad junction. Manufacturing (textiles, industrial chemicals, trucks, ironworks, printing and publishing); service industries; agriculture (cotton, grain; livestock, poultry, dairying). Hydroelectric plant nearby. Incorporated 1879.

Mount Holly, township, ☉ BURLINGTON county, W NEW JERSEY, on RANCOCAS CREEK, and 18 mi/29 km E of CAMDEN; 40°00'N 74°47'W. Manufacturing; trade center for agricultural region. Settled by Friends c.1680; occupied by British in Revolution. Friends' meetinghouse (1775), courthouse (1796), John Woolman memorial building (1771), and other 18th-century buildings survive.

Mount Holly, town, RUTLAND CO., S central VERMONT, 13 mi/21 km SE of RUTLAND; 43°26'N 72°48'W. Partly in Green Mountain National Forest. Population peaked around 1600 residents in 1870.

Mount Holly, unincorporated village, BERKELEY county, SE SOUTH CAROLINA, suburb 16 mi/26 km NNW of CHARLESTON. Manufacturing includes primary aluminum production, aluminum foil, scrap metal. Charleston Naval Base to SE; CYPRESS GARDENS to E.

Mount Holly Springs, borough (2006 population 1,912), CUMBERLAND county, S PENNSYLVANIA, 6 mi/ 9.7 km S of CARLISLE, near Yellow Breeches Creek;

Cross-references are shown in SMALL CAPITALS. The pronunciation guide is shown on page xix. The sources of population figures are shown on page xvii.

40°06′N 77°10′W. Manufacturing (filter paper, oscillators and quartz crystals, capacitor paper, candy). Kings Gap Environmental Training Center to W. APPALACHIAN TRAIL passes to S; Pine Grove State Park Furnace to SW. Laid out 1815.

Mount Hope, town (2000 population 830), SEDGWICK county, S central KANSAS, 22 mi/35 km NW of WICHITA; 37°52′N 97°39′W. In wheat, livestock, and poultry area. Oil wells nearby.

Mount Hope, town (2006 population 1,398), FAYETTE county, S central WEST VIRGINIA, 8 mi/12.9 km N of BECKLEY; 37°53′N 81°10′W. Railroad junction. Bituminous-coal fields. Manufacturing (lumber, rebuilt mining machinery, urethane castings). Agriculture (grain); cattle. NEW RIVER GORGE NATIONAL RIVER to E; Plum Orchard Wildlife Management Area to NW. Incorporated 1897.

Mount Hope (MOUNT HOP), village, central NEW SOUTH WALES, AUSTRALIA, 320 mi/515 km WNW of SYDNEY; 34°07′S 135°23′E.

Mount Hope, village, MORRIS county, N central NEW JERSEY, 9 mi/14.5 km NNW of Morrisville, in lake and hill region; 40°55′N 74°32′W. Once an important iron-mining area.

Mount Hope, village (2006 population 172), GRANT county, extreme SW WISCONSIN, 16 mi/26 km ESE of PRAIRIE DU CHIEN; 42°58′N 90°51′W. Livestock and dairy region. Eagle Valley Nature Preserve to E.

Mount Hope, industrial suburb, COLÓN province, on transisthmian railroad, and 1.5 mi/2.4 km S of COLÓN (PANAMA), adjoining (S) CRISTOBAL, on mouth of old French canal; 09°20′N 79°54′W. Has railroad yards, repair wharves, and drydocks; cold-storage plants.

Mount Hope Bay, RHODE ISLAND and MASSACHUSETTS, NE arm of NARRAGANSETT BAY, just W of FALL RIVER, Massachusetts; c.6 mi/9.7 km long, 3 mi/4.8 km wide. Opens S into Sakonnet River, SW into NARRAGANSETT BAY at N end of Rhode Island, where it is crossed by Mt. Hope Bridge (1929), one of largest in NEW ENGLAND.

Mount Hopkins Observatory, N SANTA CRUZ county, S ARIZONA, astronomical observatory located 35 mi/56 km S of TUCSON, W of Mount WRIGHTSON, in section of Coronado National Forest, at an elevation of 8,500 ft/2,591 m. It is operated jointly by the Smithsonian Astrophysical Observatory and the University of Arizona. The principal instrument is the multiple-mirror telescope (MMT), scheduled for conversion to a 6-m single mirror telescope. The MMT consists of six identical 72-in/183-cm reflecting telescopes mounted in a hexagonal array on a common mounting and feeding their images to a single focus. A 30-in/76-cm reflector in the center of the mounting serves as a guide telescope. The combined light-gathering power of the MMT is equal to that of a conventional 176-in/447-cm reflector. It is designed for observations in both the optical and the infrared part of the spectrum. Also at Mount Hopkins are a 60-in/152-cm and a 394-in/10-m dish with 248 small mirrors used for gamma-ray astronomy observations. Gamma rays come from particular objects such as supernovae remnants and black holes.

Mount Horeb (HO-reb), town (2006 population 6,573), DANE county, S WISCONSIN, 19 mi/31 km WSW of MADISON; 43°00′N 89°43′W. In dairying and farming area; processes dairy products, feed; manufacturing (plastic injection molding). On Military Ridge State Trail. Settled 1860 by Norwegians and Swiss; incorporated 1899.

Mount Ida, town (2000 population 981), ⊙ MONTGOMERY county, W ARKANSAS, 32 mi/51 km W of HOT SPRINGS, in Ouachita National Forest; 34°32′N 93°37′W. Manufacturing (machining, ladies' shoes). Large Lake OUACHITA National Forest to NW.

Mount Iraya, active volcano, Batanes province, BATAN ISLAND, PHILIPPINES; 20°27′N 120°01′E. Elevation 3,307 ft/1,008 m; seismic activity 1970.

Mount Iriga, active volcano (3,750 ft/1,143 m), CAMARINES SUR province, PHILIPPINES, 22 mi/35 km SE of NAGA; 13°27′N 123°27′E.

Mount Irvine Bay, SW TOBAGO, TRINIDAD AND TOBAGO; popular bathing resort, 7 mi/11.3 km W of SCARBOROUGH. Also known as Little Courland Bay.

Mount Isa (EI-zuh), town (2001 population 20,525), NW QUEENSLAND, NE AUSTRALIA, 65 mi/105 km W of CLONCURRY, and on LEICHHARDT RIVER; 20°44′S 139°30′E. Railroad terminus; cattle. Silver, copper, lead, and zinc mines are nearby; world's largest single mine for silver and lead. Aboriginal art sites.

Mount Isarog (ee-sah-ROG), extinct volcano (6,482 ft/1,976 m), CAMARINES SUR province, SE LUZON, PHILIPPINES, 10 mi/16 km E of NAGA; 13°39′N 123°23′E. Numerous streams rise here. Sulphur mining.

Mount Jackson (MOUNT JAK-suhn), town (2006 population 1,782), SHENANDOAH county, NW VIRGINIA, 24 mi/39 km NNE of HARRISONBURG, on North Fork of SHENANDOAH RIVER; 38°45′N 78°38′W. Manufacturing (fruit processing, concrete products, furniture, apparel, fertilizer); agriculture (apples, grain, soybeans; poultry, livestock; dairying). Shenandoah Caverns to SW are a tourist attraction.

Mount Jewett, borough (2006 population 1,021), MCKEAN county, N PENNSYLVANIA, 16 mi/26 km S of BRADFORD, near Kinzua Creek; 41°43′N 78°38′W. Railroad junction. Agriculture (hay; livestock, dairying); timber; manufacturing (particleboard, electronic components). Bradford Regional Airport to N. Kinzua Bridge State Park (bridge originally built of iron 1882, rebuilt of steel 1900; 301 ft/92 m high, 2,110 ft/643 m long; on abandoned railroad line); Allegheny National Forest to W. Settled c.1838, incorporated 1893.

Mount Jolo, PHILIPPINES: see MOUNT BUDDAJO.

Mount Joy, borough (2006 population 7,056), LANCASTER county, SE PENNSYLVANIA, 12 mi/19 km NW of LANCASTER, on Little Chickies Creek; 40°06′N 76°30′W. Agriculture (grain, potatoes, soybeans; livestock, poultry and eggs, dairying); manufacturing (telecommunications components, commercial printing, apparel, printing and publishing, machinery, flour, aluminum castings, food products). Covered bridges in area. Settled 1768, laid out 1812, incorporated 1851.

Mount Juliet, city (2006 population 19,369), WILSON county, central TENNESSEE, 15 mi/24 km E of NASHVILLE, 36°12′N 86°31′W. In farming area; some industry; near OLD HICKORY Reservoir (CUMBERLAND RIVER). Incorporated 1972.

Mount Kalatungan, active volcano (9,265 ft/2,824 m), BUKIDNON province, PHILIPPINES, 37 mi/60 km S of CAGAYAN DE ORO; 07°57′N 124°48′E.

Mount Kenya National Park (□ 277 sq mi/717 sq km), CENTRAL province, KENYA, NE of NYERI town; 00°10′N 37°20′E. Designated a national park in 1946.

Mount Kilimanjaro National Park (ki-li-mahn-JAH-ro), KILIMANJARO region, N TANZANIA, 10 mi/16 km N of MOSHI, near KENYA border. Protects Mount KILIMANJARO, highest mountain in AFRICA (19,340 ft/5,895 m), generally above the 6,562 ft/2,000 m level. Kilimanjaro (UHURU Peak) is a snowcapped, extinct volcano with remnant activity (03°05′S 37°21′E).

Mount Kisco, residential village in Mount Kisco town (□ 3 sq mi/7.8 sq km; 2006 population 10,441), WESTCHESTER county, SE NEW YORK, 13 mi/21 km N of WHITE PLAINS; 41°12′N 73°43′W. Light industry, commercial services, retail. An affluent community, it is the birthplace or home of a number of prominent Americans, including business leaders Carl Icahn, Michael Eisner, and Arthur Ochs Sulzberger, Jr. Incorporated 1874.

Mountlake Terrace, city (2006 population 20,225), SNOHOMISH and KING counties, NW WASHINGTON, a residential suburb 13 mi/21 km N of downtown SEATTLE; 47°48′N 122°19′W. Manufacturing (refrigeration systems, printing and publishing, switchgear,

netting, communications equipment). PUGET SOUND to W, N end of LAKE WASHINGTON to SE. Ballinger Lake (in Snohomish county) is in city. Incorporated in 1954.

Mount Laurel, township, BURLINGTON county, W central NEW JERSEY, 8 mi/12.9 km S of WILLINGBORO; 39°57′N 74°54′W. Site of housing dispute resolved by New Jersey Supreme Court in 1975 and clarified in Mount Laurel II (1985), which validated low- and moderate-income housing. Residential.

Mount Lavinia, town and residential suburb of COLOMBO, WESTERN PROVINCE, SRI LANKA, on W coast, 7 mi/11.3 km S of Colombo city center; 06°50′N 79°51′E. Tourist center. Municipal Council (1995 est. population 192,000) includes DEHIWALA. RATMALANA, 2 mi/3.2 km S, has regional airport and manufacturing.

Mount Lebanon, municipality (2000 population 33,017), ALLEGHENY county, SW PENNSYLVANIA; residential suburb 5 mi/8 km SSW of downtown PITTSBURGH; 40°22′N 80°02′W. Former township.

Mount Lebanon, town (2000 population 73), BIENVILLE parish, LOUISIANA, 9 mi/14.5 km SW of ARCADIA; 32°30′N 93°03′W. In timber area, oil and gas nearby.

Mount Lehman (LEE-muhn), unincorporated village, SW BRITISH COLUMBIA, W Canada, 4 mi/6 km WSW of MISSION across FRASER River, and included in ABBOTSFORD; 49°07′N 122°23′W. Dairying; livestock, vegetables, fruit.

Mount Leonard, town (2000 population 123), SALINE county, central MISSOURI, near MISSOURI River, 10 mi/16 km W of MARSHALL; 39°07′N 93°23′W.

Mount Leuser National Park (LAI-suhr) (□ 3,653 mi/9,461 sq km), South Aceh district, North Sumatra province, INDONESIA, c.62 mi/100 km W of MEDAN; 03°30′N 97°30′E. Established 1980, it is one of the largest protected forests in SE ASIA. Park boundary encompasses Mount LEUSER (elevation 11,092 ft/3,381 m). Fauna include tiger, elephant, Sumatran rhinoceros, and sun bear. Ketambe Research Station located in park; Bohorok Orangutan Rehabilitation Center adjoins park. Also called Gunung Leuser National Park.

Mount Lofty (MOUNT LAHF-tee), village, SE SOUTH AUSTRALIA state, S central AUSTRALIA, 9 mi/14 km SE of ADELAIDE, in MOUNT LOFTY RANGES; 35°00′S 138°42′E. Dairying center.

Mount Lofty Ranges (MOUNT LAHF-tee), SE SOUTH AUSTRALIA state, S central AUSTRALIA, extend 200 mi/322 km S from PETERBOROUGH to Cape JERVIS; rise to 3,063 ft/934 m (Mount Bryan); 35°00′S 138°50′E. Fertile valleys (olive plantations, vineyards). Phosphate rock, marble, barite. Source of TORRENS RIVER. Mount Lofty (2,384 ft/727 m) is near ADELAIDE. The area of the Mount Lofty Ranges closest to Adelaide is known as ADELAIDE HILLS.

Mount Lyell, TASMANIA: see QUEENSTOWN.

Mount Macedon, AUSTRALIA: see MACEDON.

Mount Magnet (MOUNT MAG-nuht), town, W central WESTERN AUSTRALIA, 200 mi/322 km ENE of GERALDTON; 28°04′S 117°49′E. On Geraldton–WILUNA railroad. Sheep station; mining center in MURCHISON GOLDFIELD.

Mount Makaturing, active volcano (5,495 ft/1,675 m), LANAO DEL SUR province, PHILIPPINES, 65 mi/105 km SSE of CAGAYAN DE ORO; 07°38′N 124°19′E.

Mount Makiling (mah-KEE-leeng) (3,750 ft/1,143 m), LAGUNA province, S LUZON, PHILIPPINES, 10 mi/16 km WNW of SAN PABLO, near LAGUNA DE BAY, in national forest; 14°08′N 121°11′E. Mineral pigments are mined. Also called Mount Maquiling.

Mount Mantalingajan (mahn-tah-leeng-AH-hahn), peak (6,839 ft/2,085 m), S PALAWAN, PHILIPPINES, NW of BROOKE'S POINT.

Mount Maquiling, PHILIPPINES: see MOUNT MAKILING.

Mount Margaret (MOUNT MAHR-gruht), village, S central WESTERN AUSTRALIA, 440 mi/708 km NE of

PERTH. On Perth–LAVERTON railroad. Mining center for Mount Margaret Goldfield (□ 42,154 sq mi/109,179 sq km).

Mount Marion, village, ULSTER county, SE NEW YORK, near ESOPUS CREEK and the HUDSON RIVER, 3 mi/4.8 km SSW of SAUGERTIES. In resort and agricultural area; 42°02′N 74°00′W.

Mount Matutum (7,523 ft/2,293 m), SOUTH COTABATO province, PHILIPPINES, 19 mi/30 km NNE of GENERAL SANTOS; 06°22′N 125°06′E. Possible eruption in 1911 (may be dormant).

Mount Mayon (mah-YON), active volcano (8,077 ft/2,462 m), ALBAY province, SE LUZON, the PHILIPPINES; 13°15′N 123°41′E. It is considered one of the world's most perfect cones. The town of Cagsawa was buried in an eruption in 1814. There have been forty-four eruptions between 1616 and 1993, with the most recent major occurrences in 1984 and 1993. In the 1993 eruption, sixty-eight people died and 60,000 were evacuated. Eruptions are frequently explosive.

Mount McClintock, the highest mountain (11,450 ft/3,490 m) in the BRITANNIA RANGE in the TRANSANTARCTIC MOUNTAINS of EAST ANTARCTICA; 80°13′S 157°26′E.

Mount McKinley National Park, Alaska: see DENALI NATIONAL PARK AND PRESERVE.

Mount Melbourne, a dormant volcano (8,957 ft/2,730 m) at the N end of the BORCHGREVINCK COAST in VICTORIA LAND, EAST ANTARCTICA; 74°21′S 164°42′E.

Mount Melleray (ME-luh-ree), locality in W WATERFORD county, S IRELAND, at foot of KNOCKMEALDOWN MOUNTAINS, 5 mi/8 km NE of LISMORE. Site of famous Trappist monastery, founded 1830, after explusion of foreign Trappists from France.

Mountmellick (mount-ME-lik), Gaelic *Móinteach Mílic*, town (2006 population 3,869), N LAOIGHIS county, E central IRELAND, on branch of the GRAND CANAL, and 6 mi/9.7 km NNW of PORTLAOISE; 53°07′N 07°19′W. Agricultural market; formerly important manufacturing town. Originally Quaker linen-making settlement.

Mount Menzies, in the PRINCE CHARLES MOUNTAINS, EAST ANTARCTICA, on the S side of Fisher Glacier; 73°30′S 61°50′E. Elevation 11,007 ft/3,355 m.

Mount Minto, lofty mountain in the central part of the ADMIRALTY MOUNTAINS of VICTORIA LAND, EAST ANTARCTICA; 71°47′S 168°45′E. Elevation 13,665 ft/4,165 m.

Mount Molloy (MOUNT muh-LOI), town, QUEENSLAND, NE AUSTRALIA, 66 mi/106 km from CAIRNS; 16°37′S 145°23′E. Cattle grazing; former copper-mining center.

Mount Morgan (MOUNT MOR-guhn), town, E QUEENSLAND, NE AUSTRALIA, 22 mi/35 km SSW of ROCKHAMPTON; 23°39′S 150°23′E. Gold, silver, and copper mines until late 1980s.

Mount Moriah (mo-REI-yuh), town (2000 population 143), HARRISON county, N MISSOURI, 13 mi/21 km NE of BETHANY; 40°19′N 93°47′W.

Mount Morris, city (2000 population 3,194), GENESEE county, SE central MICHIGAN, suburb 7 mi/11.3 km N of FLINT; 43°07′N 83°42′W. In farm area (grain, potatoes; dairy products); manufacturing (packaging, machining). Settled 1842; incorporated as village 1867, as city 1930.

Mount Morris, unincorporated town, GREENE county, SW PENNSYLVANIA, 13 mi/21 km SSE of WAYNESBURG, and 8 mi/12.9 km NNW of MORGANTOWN, WEST VIRGINIA, near West Virginia state line, on Dunkard Creek; 39°43′N 80°04′W. Manufacturing (lumber); subsurface coal mining; agriculture (hay; livestock; dairying).

Mount Morris, village (2000 population 3,013), OGLE county, N ILLINOIS, 6 mi/9.7 km WNW of OREGON; 42°02′N 89°25′W. In rich agricultural area; large printing plant. Settled 1838, incorporated 1857.

Mount Morris, village (□ 2 sq mi/5.2 sq km; 2006 population 2,963), LIVINGSTON county, W central

NEW YORK, on GENESEE River, and 34 mi/55 km SSW of ROCHESTER; 42°43′N 77°52′W. N entrance to Letchworth State Park is here. Mount Morris Dam (550 ft/168 m long, 216 ft/66 m high; for flood control) is on the Genesee here.

Mount Morrison, Colorado: see MORRISON.

Mount Morrison, TAIWAN: see YU SHAN.

Mount Mulligan (MOUNT MUH-li-guhn), locality, QUEENSLAND, NE AUSTRALIA, 31 mi/50 km N of Dimbulah; 16°51′S 144°52′E. Former gold-mining center. Virtually uninhabited.

Mount Murchison, behind the BORCHGREVINCK COAST of VICTORIA LAND, EAST ANTARCTICA; 73°25′S 166°18′E. Elevation 11,483 ft/3,500 m.

Mount Murphy, a massive, snow-covered volcano overlooking the WALGREEN COAST of MARIE BYRD LAND, WEST ANTARCTICA, just S of the BEAR PENINSULA; 75°20′S 110°44′W. Elevation 8,875 ft/2,705 m.

Mount Musuan, active volcano, BUKIDNON province, PHILIPPINES, 40 mi/65 km SSE of CAGAYAN DE ORO; 07°52′N 125°04′E. Elevation 2,199 ft/646 m. Last eruption c.1886, seismic activity 1976. Also known as Mount Calayo.

Mount Natazhat (NAH-zuhr-raht), mountain, S ALASKA, in SAINT ELIAS MOUNTAINS, 150 mi/241 km NNW of YAKUTAT; 61°31′N 141°09′W.

Mount Nemrut, TURKEY: see NEMRUT DAG.

Mount of Olives, ISRAEL: see OLIVES, MOUNT OF.

Mount Olive, city (2000 population 2,150), MACOUPIN county, SW central ILLINOIS, 18 mi/29 km SSE of CARLINVILLE; 39°04′N 89°43′W. In agricultural and bituminous-coal-mining area. Incorporated 1917. Has graves of "Mother" Jones and "General" Alexander Bradley, pioneers in miners' union movement.

Mount Olive, town (2000 population 893), COVINGTON county, S central MISSISSIPPI, 36 mi/58 km NNW of HATTIESBURG, on Okatoma Creek; 31°45′N 89°39′W. Railroad junction to N. Agriculture (cotton, corn; poultry, cattle); manufacturing (sportswear). Lake ROSS BARNETT State Lake to NE.

Mount Olive, township, MORRIS county, N NEW JERSEY, 10 mi/16 km W of MORRISTOWN; 40°52′N 74°44′W. Incorporated 1871.

Mount Olive (MOUNT AH-liv), town (□ 2 sq mi/5.2 sq km; 2006 population 4,410), WAYNE and DUPLIN counties, E central NORTH CAROLINA, 13 mi/21 km SSW of GOLDSBORO; 35°12′N 78°04′W. Manufacturing (furniture, food processing, electrical equipment, printing and publishing, plastic products); agriculture (grain, tobacco, cotton, cucumbers, peppers; poultry, livestock). Founded 1839–1840.

Mount Oliver, borough (2006 population 3,715), ALLEGHENY county, SW PENNSYLVANIA; residential suburb 2 mi/3.2 km S of downtown PITTSBURGH, near MONONGAHELA RIVER; 40°24′N 79°59′W. Bounded on all sides by Pittsburgh. Incorporated 1892.

Mount Olivet (AHL-uh-vet), village (2000 population 289), ⊙ ROBERTSON county, N KENTUCKY, 17 mi/27 km WSW of MAYSVILLE; 38°31′N 84°02′W. In BLUEGRASS agricultural region (dairying; poultry, cattle; corn, burley tobacco); manufacturing (hand-made soap). Johnson Creek Covered Bridge to S; Blue Licks Battlefield State Park to S.

Mount Orab (O-ruhb), village (□ 6 sq mi/15.6 sq km; 2006 population 2,815), BROWN county, SW OHIO, 11 mi/18 km N of GEORGETOWN; 39°01′N 83°55′W. In tobacco and grain area.

Mount Palung National Park (pah-LUNG) (□ 116 sq mi/301.6 sq km), Kalimantan Barat province, INDONESIA, 19 mi/30 km E of KETAPANG; 01°52′S 110°17′E. Park includes wide variety of habitat, from coast to mountain forest.

Mount Pearl, city (□ 6 sq mi/15.6 sq km; 2001 population 26,964), SE NEWFOUNDLAND AND LABRADOR, Canada, suburb 4 mi/6 km SW of St. JOHN'S, on both sides of Waterford River. Oil and gas, technology.

Mount Penn, borough (2006 population 3,002), BERKS county, SE central PENNSYLVANIA; residential suburb 2 mi/3.2 km ESE of READING, near SCHUYLKILL RIVER; 40°19′N 75°53′W. Mount Penn, with observation tower and the Pagoda, to NW. Laid out 1884, incorporated 1902.

Mount Perry (MOUNT PE-ree), village, SE QUEENSLAND, NE AUSTRALIA, 180 mi/290 km NNW of BRISBANE, 65 mi/104 km SW of BUNDABERG; 25°12′S 151°39′E. Railroad terminus. The rock Boolboonda Tunnel is near town. Mount Perry mountain rises to 2,461 ft/750 m.

Mount Pinatubo, active volcano, ZAMBALES province, PHILIPPINES, 56 mi/90 km NE of MANILA; 15°07′N 120°20′E. Elevation 5,840 ft/1,780 m. Dormant from 1380, the volcano had a massive eruption in June 1991, with ashfalls spreading as far as CAMBODIA (1,118 mi/1,800 km W) and stratospheric ash that affected global weather for two years. More than 1,000 people died either in the initial eruption or subsequent related events (such as typhoon-generated mud flows); and as many as 250,000 people were displaced. The towns of BACOLOR, MINALIN, and Santo Thomas de Zambales were completely buried by volcanic sludge (lahar) of sand, pebbles, and boulder, which hardens under monsoonal rains, and c.154 sq mi/399 sq km of agricultural and residential land was destroyed in Zambales, TARLAC, and PAMPANGA provinces. The cities of SAN FERNANDO and ANGELES (site of the now-abandoned Clark Air Force Base, the major U.S. Air Force facility in the W Pacific) are currently facing a threat from mud flows that are expected to continue until at least c.2005.

Mount Pleasant, city (2000 population 8,751), ⊙ HENRY county, SE IOWA, 25 mi/40 km WNW of BURLINGTON; 40°57′N 91°32′W. Manufacturing (consumer goods, food processing, printing, fabricated metal products, construction material). Limestone quarries nearby. Has Iowa Wesleyan College; Mount Pleasant Correctional Center; and Midwest Old Settlers and Threshers Heritage Museum. Oakland Mills State Park to SW; Grode State Park to SE. Founded 1839, incorporated 1842.

Mount Pleasant, city (2000 population 25,946), ⊙ ISABELLA county, central MICHIGAN, on the CHIPPEWA RIVER; 43°36′N 84°46′W. The city grew after oil was found nearby in 1928. Oil wells and refineries are there. Manufacturing (motor assemblies, machinery, foil stamping). Municipal airport to E. Mount Pleasant is the seat of Central Michigan University. Isabella Indian Reservation surrounds and includes city. Settled before 1860, incorporated as a city 1889.

Mount Pleasant, city (2006 population 59,113), CHARLESTON county, SE SOUTH CAROLINA, residential suburb 4 mi/6.4 km E of downtown CHARLESTON, across CHARLESTON HARBOR (here bridged); 32°49′N 79°51′W. Port facilities, boat repairs. Manufacturing includes marine equipment, plastic products, signs, and seafood processing. Upper middle-class suburb and one of the state's fastest-growing areas. A 300-year-old tradition of weaving baskets from local marsh grass and sweet grass by the local African-American community is continued today.

Mount Pleasant, city (2006 population 4,448), MAURY county, central TENNESSEE, 11 mi/18 km SW of COLUMBIA; 35°32′N 87°12′W. Diverse mix of domestic and foreign manufacturing. Meriwether Lewis National Monument is 14 mi/23 km W.

Mount Pleasant (PLE-zuhnt), city (2006 population 15,202), ⊙ TITUS county, NE TEXAS, c.50 mi/80 km SE of PARIS; 33°10′N 94°58′W. Elevation 416 ft/127 m. Trade, railroad junction, shipping center for agriculture (poultry; dairying; cattle; watermelons, corn), oil, timber region; manufacturing (food processing, trailers, concrete, insulated wire, doors, printing). Northwest Texas Community College, at nearby CHAPPELL HILL (serves CAMP, MORRIS and Titus

---

Cross-references are shown in SMALL CAPITALS. The pronunciation guide is shown on page xix. The sources of population figures are shown on page xvii.

counties). Formerly a lumbering center; settled before mid-19th century, incorporated 1900.

Mount Pleasant (MOUNT PLEZ-uhnt), town (□ 1 sq mi/2.6 sq km; 2006 population 1,532), CABARRUS county, S central NORTH CAROLINA, 8 mi/12.9 km E of CONCORD; 35°23′N 80°26′W. Service industries; manufacturing (hosiery, marble bathtubs and vanities, lumber, yarn); agricultural area (grain, soybeans; poultry, livestock, dairying).

Mount Pleasant, town (2006 population 2,698), SAN-PETE county, central UTAH, 22 mi/35 km NNE of MANTI, near SAN PITCH RIVER, in irrigated SANPETE VALLEY; 39°32′N 111°27′W. Shipping point for cattle, sheep, hogs, poultry and agricultural area (alfalfa, wheat, barley, oats; dairying); manufacturing (flour, cheese). Elevation c.5,924 ft/1,806 m. WASATCH RANGE in Manti–La Sal National Forest is E. Old Pioneer Museum Settled 1852 by Mormons, incorporated 1868.

Mount Pleasant (MOUNT PLE-zuhnt), village, SE SOUTH AUSTRALIA, 28 mi/45 km ENE of ADELAIDE, on N edge of ADELAIDE HILLS; 34°47′S 139°01′E. Heritage center.

Mount Pleasant, unincorporated village, NEW CASTLE county, W DELAWARE, 17 mi/27 km SSW of WIL-MINGTON; 39°31′N 75°43′W. Elevation 49 ft/14 m.

Mount Pleasant (PLEZ-uhnt), village (2006 population 510), JEFFERSON county, E OHIO, 17 mi/27 km SSW of STEUBENVILLE; 40°10′N 80°48′W.

Mount Pleasant, borough (2006 population 4,485), WESTMORELAND county, SW PENNSYLVANIA, 30 mi/48 km SE of PITTSBURGH; 40°08′N 79°32′W. Bituminous coal; manufacturing (food processing, technical equipment, fabricated metal products, consumer goods); timber. Agricultural area (corn, hay, apples, dairying). CHESTNUT RIDGE to SE; Bridgeport Reservoir (Jacobs Creek) to SE; Mount Pleasant–Scottdale Airport to S. Laid out c.1897, incorporated 1828.

Mount Pleasant, residential section in NW WA-SHINGTON, D.C., E of ROCK CREEK and N of Irving Street; 38°56′N 77°02′W. Early suburb of Washington, largely developed in the 1870s. Formerly the site of a Civil War hospital. The community today has a large Hispanic and first-generation immigrant population.

Mount Pleasant, AUSTRALIA: see FOREST HILL.

Mount Plymouth (PLI-muhth), unincorporated town (□ 3 sq mi/7.8 sq km; 2000 population 2,814), Lake county, central FLORIDA, 20 mi/32 km NW of OR-LANDO; 28°48′N 81°31′W. Golfing resort.

Mount Pocono, borough (2006 population 3,001), MONROE county, E PENNSYLVANIA, 12 mi/19 km NW of STROUDSBURG, in POCONO MOUNTAINS; 41°07′N 75°21′W. Elevation 1,658 ft/505 m. Tourism, large resort community. Light manufacturing. Numerous residential developments to N and W. Pocono Mountains Municipal Airport to N; Pennsylvania Dutch Farm to E; Pocono International Raceway to SW; Pocono Manor Ski Area to S; Mount Airy Ski Area to SE; Tobyhanna State Park to NW; Tobyhanna Depot to NW.

Mount Prospect, village (2000 population 56,265), COOK county, NE ILLINOIS, 42°04′N 87°56′W. Incorporated 1917. It is a large and growing residential suburb NW of CHICAGO.

Mount Pulaski (poo-LAS-kei), city (2000 population 1,701), LOGAN county, central ILLINOIS, 25 mi/40 km NE of SPRINGFIELD; 40°00′N 89°16′W. In agricultural area. Was ⊙ Logan county, 1847–1853; incorporated 1893. The old courthouse (1847) was made a state monument in 1936.

Mount Pulog (POO-log), peak (9,606 ft/2,928 m), on border between BENGUET and NUEVA VIZCAYA provinces, NW LUZON, the PHILIPPINES, in the CORDIL-LERA CENTRAL. It is the second-highest point in the Philippines. The mountain is sacred to highland peoples.

Mount Ragang (rah-GAHNG), active volcano, on the border between LANAO DEL SUR and NORTH COTA-BATO provinces, W central MINDANAO, PHILIPPINES, SE of LAKE LANAO; 07°43′N 124°32′E. Elevation 9,236 ft/2,815 m. There have been ten eruptions between 1756 and 1916, which have often been explosive.

Mountrail, county (□ 1,819 sq mi/4,729.4 sq km; 2006 population 6,442), NW central NORTH DAKOTA; ⊙ STANLEY; 48°12′N 102°22′W. Rich agricultural area borders MISSOURI RIVER (LAKE SAKAKAWEA) on SW, Van Hook Arm of Lake Sakakawea on S and drained by WHITE EARTH and LITTLE KNIFE rivers and Shell Creek. Lignite mines; cattle, dairy produce, wheat, rye. Powers, Cottonwood, White and Lower Lostwood lakes in N, Shell Lake in E; part of Fort Berthold Indian Reservation in S; Shell Lake National Wildlife Refuge in SE. Formed in 1873 as Mountraille County but eliminated by the state legislature in 1891. Named for Jopseh Mountraille. Re-established as Mountrail County in 1909 and government formed the same year.

Mount Rainier, city (2000 population 8,498), PRINCE GEORGES county, central MARYLAND, suburb NE of WASHINGTON, D.C.; 38°56′N 76°58′W. Land in the area once belonged to Army officers from SEATTLE, hence the name. Established 1902.

Mount Rainier National Park (rai-NIR) (□ 368 sq mi/956.8 sq km), PIERCE and LEWIS counties, W central Washington, in the CASCADE RANGE. The area is dominated by MOUNT RAINIER (14,411 ft/4,392 m), a dormant, glaciated volcanic peak. The mountain is snow-crowned and has 26 glaciers; its heavily forested lower slopes and alpine meadows are popular with hikers. Bounded W, N, and E by Snoqualmie National Forest, S by Gifford Pinchot National Forest. Numerous waterfalls; road access in E and S, NW corner. Established 1899.

Mountrath (mount-RAHTH), Gaelic *Maighean Rha*, town (2006 population 1,885), central LAOIGHIS county, E central IRELAND, 8 mi/12.9 km WSW of PORTLAOISE; 53°00′N 07°28′W. Furniture manufacturing. Formerly an important trade center.

Mount Remarkable National Park (MOUNT ree-MAHR-kuh-buhl), (□ 33 sq mi/85.8 sq km), E central SOUTH AUSTRALIA state, S central AUSTRALIA, 160 mi/257 km N of ADELAIDE, 30 mi/48 km SE of PORT AU-GUSTA, 8 mi/13 km E of N end of SPENCER GULF; 11 mi/18 km long, 6 mi/10 km wide. Includes large oval pound area, a concave natural amphitheater feature peculiar to region, perimeter cliffs face outward. Mount Remarkable (3,265 ft/995 m) situated in small detached sub-unit 3 mi/5 km E. Red kangaroos, euros; emus, kookaburras, cockatoos; lizards, snakes. Blue gums, sugar gums, river red gums, cypress pine; wildflowers. Camping, picnicking, hiking. Established 1972.

Mount Repose (ruh-POZ), unincorporated village (2000 population 4,102), CLERMONT county, extreme SW OHIO; suburb 16 mi/26 km NE of downtown CINCINNATI; 39°11′N 84°13′W. Adjacent to MULBERRY.

Mount Revelstoke National Park (REV-uhk-stok) (□ 100 sq mi/260 sq km), SE BRITISH COLUMBIA, W Canada, in the SELKIRK MOUNTAINS, just E of the COLUMBIA River valley; 51°06′N 118°04′W. Established 1914. Situated on a high plateau, rising to c.7,000 ft/2,134 m at Mount Revelstoke, the park has several small lakes and glaciers. A popular resort area, noted especially for winter sports.

Mount Robson Provincial Park (RAHB-suhn) (□ 803 sq mi/2,087.8 sq km), E BRITISH COLUMBIA, W Canada, in the ROCKY MOUNTAINS W of JASPER, AL-BERTA; 52°58′N 118°50′W. Established 1913. High peaks, glaciers, lakes, waterfalls, and headwaters of FRASER River. Mount Robson (12,972 ft/3,954 m), in the park, is the highest peak in the Canadian Rocky Mountains.

Mount Royal (MOUNT ROI-uhl) or **Mont Réal** (mon rai-AHL) (900 ft/274 m), S QUEBEC, E Canada, in N part of MONTREAL city. On its upper slopes is Mount Royal Park (□ 1 sq mi/2 sq km); on S slope is McGill University, on N slope Montreal University. Residential district.

Mount Rushmore National Memorial (□ 2 sq mi/5.2 sq km), PENNINGTON and CUSTER counties, SW SOUTH DAKOTA, in the BLACK HILLS at KEYSTONE, 17 mi/27 km SW of RAPID CITY, N of CUSTER STATE PARK; 43°52′N 103°27′W. Carved on the face of Mount Rushmore and visible for 60 mi/100 km are the enormous busts of four U.S. Presidents—Washington, Jefferson, Theodore Roosevelt, and Lincoln. The sculpture, nearly completed when the sculptor, Gutzon Borglum, died (1941), was finished later that year by his son Lincoln. It took fourteen years to complete the figures. Established 1925, dedicated 1927. Rushmore Cave, nearby, is still largely unexplored. Mountain and cave named after Charles E. Rushmore, a New York attorney who visited the Black Hills in the 1880s.

Mount Rushmore State: see SOUTH DAKOTA.

Mount Saint George, village, E TOBAGO, TRINIDAD AND TOBAGO, 4 mi/6.4 km ENE of SCARBOROUGH, opens onto Hillsborough Bay. Coconuts. Formerly called Georgetown, former capital of TOBAGO.

Mount Saint Helens, Washington: see SAINT HELENS, MOUNT.

Mount Saint Helens National Volcanic Monument (□ 172 sq mi/445 sq km); in SKAMANIA, COWLITZ, and LEWIS counties, SW WASHINGTON, 55 mi/89 km NNE of PORTLAND, OREGON, and 120 mi/193 km S of SEATTLE, Washington. Established 1982 to protect the area around MOUNT SAINT HELENS volcano and to commemorate the event of May 18, 1980, when Mount Saint Helens erupted, one of the most significant geologic events of recorded North Amererican history. SPIRIT LAKE in NE; Ape Cave in S.

Mount Savage, village, ALLEGANY county, W MARY-LAND, in the ALLEGHENIES near PENNSYLVANIA state line, 7 mi/11.3 km WNW of CUMBERLAND. Bituminous-coal and clay mining; firebrick plant, railroad shops. A blast furnace was built here in 1839 by English entrepreneurs, after coal and iron ore both were found nearby, making the village an early industrial center. They sold out to American investors in the Mount Savage Iron Company, which made the first iron rails in America. The iron works failed when the ore proved inferior, but a firebrick plant, headed by President Franklin Delano Roosevelt's father, James, in the 1880s, is still operating. BIG SAVAGE MOUNTAINS is just nearby.

Mount's Bay (MOUNTZ BAI), inlet of the ENGLISH CHANNEL, SW CORNWALL, SW ENGLAND, between LAND'S END and LIZARD Point; 21 mi/34 km long, 10 mi/16 km wide. PENZANCE and MARAZION are on it. Pilchard fisheries. SAINT MICHAEL'S MOUNT island is in the inlet.

Mount Selinda, village, MANICALAND province, E ZIMBABWE, 125 mi/201 km ESE of MASVINGO, at MOZAMBIQUE border, opposite ESPUNGABERA, Mozambique (border crossing closed due to rebel activity); 20°27′S 32°43′E. Chirinda Forest to S; Mount Selinda (3,599 ft/1,097 m) to SW. Coffee, tea, citrus fruit, macadamia nuts; livestock; timber.

Mount Shasta, city (2000 population 3,621), SISKIYOU county, N CALIFORNIA, tourist center at SW foot of MOUNT SHASTA; 41°19′N 122°19′W. Cattle, sheep, lambs; grain, potatoes, onions; manufacturing (printing and publishing); timber, fish hatchery nearby. Railroad junction. Settled in 1850s, incorporated 1905. Known as Sisson until 1925. CASTLE CRAGS State Park to S; near SACRAMENTO RIVER, its source is c.10 mi/16 km to SW; parts of Shasta National Forest to NE and SW. Sisson Museum.

Mount Smith, active volcano, CAGAYAN province, BA-BUYAN ISLAND, PHILIPPINES; 19°31′N 121°54′E. Elevation 2,257 ft/688 m; eight eruptions between 1652 and 1924.

Mount Snow, ski area and year-round resort, WIND-HAM CO., S VERMONT, in Dover town, and 15 mi/24 km ENE of BENNINGTON.

Area is shown by the symbol □, and capital city or county seat by ⊙.

Mount Solon, village, AUGUSTA county, NW Virginia; 38°20'N 79°05'W. Natural Chimneys Regional Park, with natural limestone formations resembling chimneys, is here.

Mount Somers, township, ASHBURTON district, SOUTH ISLAND, NEW ZEALAND, where Canterbury Plains meet foothills of SOUTHERN ALPS, near Mount Somers (elevation 5,535 ft/1,687 m), on S Branch of Upper Ashburton River; 43°43'S 171°24'E. Irrigation area. Rangitata Diversion Race. Mixed crop and livestock farming.

Mountsorrel (mount-SAH-ruhl), village (2001 population 6,662), N LEICESTERSHIRE, central ENGLAND, on SOAR RIVER, and 7 mi/11.3 km N of LEICESTER; 52°43'N 01°09'W. Granite quarries (the largest in England) nearby.

Mount Sterling, city (2000 population 2,070), ⊙ BROWN county, W ILLINOIS, 33 mi/53 km E of QUINCY; 39°58'N 90°45'W. In agricultural area (corn, soybeans, cattle, hogs). Settled 1830, incorporated 1837. Western Illinois Correctional Institution near here.

Mount Sterling, town (2000 population 40), VAN BUREN county, SE IOWA, near MISSOURI state line, on FOX RIVER, and 7 mi/11.3 km S of KEOSAUQUA; 40°37'N 91°56'W. Livestock, grain.

Mount Sterling, town (2000 population 5,876), ⊙ MONTGOMERY county, E central KENTUCKY, 31 mi/50 km E of LEXINGTON; 38°03'N 83°57'W. Elevation 940 ft/287 m. Trade center in NE part of BLUEGRASS REGION. Agricultural (dairying; livestock, poultry; burley tobacco, corn, wheat) area; varied manufacturing. Mount Sterling-Montgomery County Airport to W. Gaitskill Mounds, burial mounds of Adena Indians (c.800 B.C.–700 A.D.) to N. Ascension Church (1878). Platted 1793. Captured and sacked (1863) by Confederate General John Hunt Morgan in Civil War.

Mount Sterling (STUHR-ling), village (□ 1 sq mi/2.6 sq km; 2006 population 1,835), MADISON county, central OHIO, 15 mi/24 km SE of LONDON, and on DEER CREEK; 39°43'N 83°16'W. In agricultural area. Founded 1828.

Mount Sterling, village (2006 population 198), CRAWFORD county, SW WISCONSIN, 21 mi/34 km NNE of PRAIRIE DU CHIEN; 43°18'N 90°55'W. Hog-raising and dairying area. Goat cheese and milk. Quarry here.

Mount Stewart, village (2001 population 312), N Prince Edward Island, Canada, on HILLSBOROUGH RIVER, and 16 mi/26 km NE of CHARLOTTETOWN; 46°22'N 62°52'W. Mixed farming, dairying; potatoes.

Mount Stromlo Observatory, astronomical observatory located on Mount Stromlo, near CANBERRA, SE AUSTRALIA. Since 1957 it has been operated by the Australian National University. The observatory was destroyed by wildfire in 2003, but the university will rebuild. The principal instrument was a 74-in/188-cm reflecting telescope, controlled by computer and having a battery of auxiliary equipment. Other instruments included 50-in/127-cm and 30-in/76-cm reflectors, and a 26-in/66-cm refractor. Its programs include fundamental research, such as investigations of quasars, as well as teaching; the observatory's contribution to public education increased dramatically while Bart J. Bok was its director (1957–1966).

Mount Summit, town (2000 population 313), HENRY county, E INDIANA, 6 mi/9.7 km N of NEW CASTLE; 40°00'N 85°23'W. In agricultural area. Manufacturing (food processing). Laid out 1854.

Mount Surprise (MOUNT suhr-PREIZ), mining district, NE QUEENSLAND, NE AUSTRALIA, 130 mi/209 km SW of CAIRNS on Gulf Developmental Road; 18°09'S 144°19'E. Nickel, sapphires, diatonite, tin; cattle.

Mount Taal, PHILIPPINES: see VOLCANO ISLAND.

Mount Tabor, town, RUTLAND CO., S central VERMONT, in hunting, fishing area of Green Mountain National Forest, 16 mi/26 km S of RUTLAND; 43°21'N 72°55'W. Former charcoal center. Originally named Harwich.

Named changed in 1803 in honor of Gideon Tabor (d. 1824), veteran of American Revolutionary War.

Mount Takahe, an isolated, extinct volcano (11,352 ft/ 3,460 m), in MARIE BYRD LAND, WEST ANTARCTICA, S of MOUNT MURPHY and the KOHLER RANGE; 76°16'S 112°14'W.

Mount Terror, an extinct volcano and the second-highest on ROSS ISLAND, WEST ANTARCTICA, of which it forms the E portion, 20 mi/30 km E of MOUNT EREBUS; 77°29'S 168°32'E. Elevation 10,597 ft/3,230 m.

Mount Tom, MASSACHUSETTS: see EASTHAMPTON.

Mount Tremper, resort village, ULSTER county, SE NEW YORK, in the CATSKILL MOUNTAINS, on ESOPUS CREEK, and 16 mi/26 km WNW of KINGSTON; 42°03'N 74°17'W.

Mount Tyree, a high mountain peak (16,289 ft/4,965 m) NW of VINSON MASSIF, in the main ridge of the SENTINEL RANGE, ELLSWORTH MOUNTAINS, WEST ANTARCTICA; 78°24'S 85°55'W.

Mount Union, town (2000 population 132), HENRY county, SE IOWA, 10 mi/16 km NE of MOUNT PLEASANT; 41°03'N 91°23'W. In livestock area.

Mount Union, borough (2000 population 2,374), HUNTINGDON county, S central PENNSYLVANIA, 29 mi/47 km ESE of ALTOONA, on JUNIATA RIVER; 40°22'N 77°52'W. Manufacturing (screw-machine products, clothing, fiberglass, storage tanks); timber; agriculture (alfalfa, grain; poultry, dairying). Mount Union Airport to S; Tuscarora State Forest to E; part of Rothrock State Forest to N and NW. Laid out 1849, incorporated 1867.

Mount Utsayantha, New York: see STAMFORD.

Mount Van Hoevenberg, NEW YORK: see MOUNT VAN HOEVENBERG STATE RECREATION AREA.

Mount Van Hoevenberg State Recreation Area, ESSEX county, N NEW YORK, 5.5 mi/8.9 km SE of LAKE PLACID, on NE slopes and to NE of Mount Van Hoevenberg (elevation 2,940 ft/896 m); 44°13'N 73°56'W. Winter sports recreation destination. The first bobsled run in N. America was built here for the 1932 Winter Olympics. Site of the bobsled, luge, cross-country, and biathlon events of the 1980 Winter Olympic Games held at Lake Placid.

Mount Vernon, city (2000 population 16,269), ⊙ JEFFERSON county, SE ILLINOIS; 38°19'N 88°54'W. Settled 1819, incorporated 1837. It is a trade, railroad, and industrial center in a farm and coal region. Manufacturing (tools, tires, transformers, coal-mining equipment, neon signs, rebuilt locomotives); diversified agriculture. State game farm nearby. Rend Lake College 10 mi/16 km S.

Mount Vernon, city (2000 population 7,478), ⊙ POSEY county, extreme SW INDIANA, 18 mi/29 km W of EVANSVILLE, and on OHIO River near influx of the WABASH RIVER; 37°56'N 87°54'W. Trade center for agricultural area; oil refining; manufacturing of petroleum and coal products, flour grinding, asphalt roofing shingles, plastics, gases, feed mixing. Ohio River port. Settled 1816, incorporated 1865.

Mount Vernon, city (2000 population 4,017), ⊙ LAWRENCE county, SW MISSOURI, 30 mi/48 km W of SPRINGFIELD; 37°06'N 93°49'W. Wheat, corn, produce; dairying; cattle; manufacturing (woodburning stoves, apparel, aluminum cans, motor vehicle equipment). Missouri Rehabilitation Center and Missouri Veterans' Home. Laid out 1845.

Mount Vernon, city (2006 population 68,395), WESTCHESTER county, SE NEW YORK, between the BRONX and HUTCHINSON rivers and adjacent to the BRONX (borough); 40°55'N 73°50'W. Settled 1664, incorporated 1892. Primarily a residential suburb of NEW YORK city, with some light manufacturing and commercial services. John Peter Zenger was arrested there for libel in 1733. The city itself was not founded until 1851, when a cooperative group, the Industrial Home Association, bought the land and built a planned community. St. Paul's Church (c.1761), a national historic

site, is here. The city has a large African-American population and a sizeable Caribbean population. It has undergone significant economic development in recent years, with an influx of businesses and retail outlets and improvements to residential areas. Writer E.B. White was born here, as were Hollywood stars Dick Clark and Denzel Washington and musician P. Diddy (Sean Combs).

Mount Vernon (VUHR-nuhn), city (□ 8 sq mi/20.8 sq km; 2006 population 15,908), ⊙ KNOX county, central OHIO, on the KOKOSING RIVER, 40 mi/64 km NE of COLUMBUS; 40°23'N 82°28'W. Livestock and dairy farms; manufacturing. Knox County Historical Society here. Laid out 1805, incorporated as a city 1880.

Mount Vernon (MOUNT VUHR-nuhn), unincorporated city (2000 population 28,582), FAIRFAX county, NE VIRGINIA, residential suburb 7 mi/11 km S of ALEXANDRIA and 14 mi/23 km SSW of WASHINGTON, D.C., on POTOMAC RIVER; 38°42'N 77°05'W. MOUNT VERNON historical site is here, as well as the Woodlawn Plantation and George Washington Grist Mill Historic State Park to the W. Tourism.

Mount Vernon, city (2006 population 29,984), ⊙ SKAGIT county, NW WASHINGTON, 25 mi/40 km SSE of BELLINGHAM and on SKAGIT RIVER, near Skagit Bay of PUGET SOUND (to SW); 48°25'N 122°19'W. Berries, vegetables, tulips; dairying; poultry; manufacturing (fabricated metal products, fresh and frozen seafoods, farm machinery, service industry machinery, signs, printing and publishing, poultry processing); stone quarrying. Swinomish Indian Reservation to W; Bay View State Park to NW. Skagit Valley Community College is here. Settled c.1877.

Mount Vernon, town (2000 population 2,082), ⊙ MONTGOMERY county, E central GEORGIA, 9 mi/14.5 km W of VIDALIA, near OCONEE River; 32°11'N 82°35'W. Manufacturing includes apparel, textiles, lumber.

Mount Vernon, town (2000 population 3,390), LINN county, E IOWA, 13 mi/21 km ESE of CEDAR RAPIDS; 41°55'N 91°25'W. Manufacturing (feed). Cornell College (Methodist; 1853) is here. State park nearby. Incorporated 1869.

Mount Vernon, town (2000 population 2,592), ⊙ ROCKCASTLE county, central KENTUCKY, 32 mi/51 km SE of DANVILLE, on old Wilderness Road; 37°21'N 84°20'W. Railroad junction to E. In coal-mining, agricultural (corn, burley tobacco; livestock; dairying) area; coal mines; manufacturing (crushed limestone, plastic packaging materials, uniform pants). Daniel Boone National Forest to E (including Great Saltpeter Caves). Hunting in vicinity. RENFRO VALLEY Entertainment Center to N; Lake Linville reservoir to N. Settled 1810; incorporated 1818.

Mount Vernon, town, KENNEBEC county, S MAINE, 17 mi/27 km NW of AUGUSTA; 44°28'N 69°57'W. In lake district; agriculture, resorts, lumbering.

Mount Vernon, town, HILLSBOROUGH county, S NEW HAMPSHIRE, 13 mi/21 km SW of MANCHESTER; 42°53'N 71°40'W. Drained by Beaver Brook and Ceasans Brook. Manufacturing (canvas goods); agriculture (fruit, vegetables; poultry, livestock; dairying).

Mount Vernon (VUHR-nuhn), town (2006 population 2,633), ⊙ FRANKLIN county, NE TEXAS, 70 mi/113 km WSW of TEXARKANA, and near WHITE OAK CREEK; 33°10'N 95°13'W. Elevation 476 ft/145 m. In agricultural, oil, timber area; varied manufacturing (furniture, wiring). Lakes Bob Sandlin, CYPRESS SPRINGS and Monticello to SE, Lake Bob Sandlin State Park to SE. Incorporated 1910.

Mount Vernon, village (2000 population 844), Mobile co., SW Alabama, near Mobile River, 27 mi/43 km N of Mobile. Lumber, flooring. Named for the U.S. army located here in 1811. Inc. in 1832.

Mount Vernon, village (2006 population 524), GRANT county, NE central OREGON, on JOHN DAY RIVER, at mouth of Beech Creek, 6 mi/9.7 km W of JOHN DAY;

44°25′N 119°06′W. Elevation 2,871 ft/875 m. Trading point for agriculture and livestock. STRAWBERRY MOUNTAINS to SE. Clyde Holliday State Park to E; parts of Malheur National Forest to N and S.

Mount Vernon, village (2006 population 468), DAVISON county, SE central SOUTH DAKOTA, 10 mi/16 km W of MITCHELL; 43°42′N 98°15′W.

Mount Vernon (MOUNT VUHR-nuhn), historic site, FAIRFAX county, NE VIRGINIA, overlooking POTOMAC RIVER, 7 mi/11 km S of ALEXANDRIA, 14 mi/23 km SSW of downtown WASHINGTON, D.C.; home of George Washington from 1747 until his death in 1799. The land was patented in 1674, and the house was built in 1743 by Lawrence Washington, George Washington's half brother. Mount Vernon was named for Admiral Edward Vernon, Lawrence's commander in the British navy. George Washington inherited it in 1754 and made additions that were not completed until after the Revolution. The mansion is a wooden structure of Georgian design, two and a half stories high, with a broad, columned portico; wide lawns, fine gardens, and subsidiary buildings surround it. The mansion has been restored, after Washington's detailed notes, with much of the original furniture, family relics, and duplicate pieces of the period. Purchased in 1860 by the Mount Vernon Ladies' Association (organized 1856), its permanent custodian. Tombs (built 1831–1837) of George and Martha Washington and other family members.

Mount Victoria (MOUNT vik-TO-ryuh), township, NEW SOUTH WALES, SE AUSTRALIA, 75 mi/120 km from SYDNEY, in BLUE MOUNTAINS; 33°35′S 150°15′E.

Mount Victory (VIK-tuhr-ee), village (2006 population 602), HARDIN county, W central OHIO, 9 mi/14 km SSE of KENTON; 40°32′N 83°31′W. In agricultural area.

Mountville, town, TROUP county, W GEORGIA, 8 mi/12.9 km E of LA GRANGE, in agricultural area; 33°02′N 84°52′W.

Mountville, unincorporated village, LAURENS county, NW SOUTH CAROLINA, 9 mi/14.5 km SSE of LAURENS.

Mountville, borough (2006 population 2,799), LANCASTER county, SE PENNSYLVANIA, 6 mi/9.7 km W of LANCASTER; 40°02′N 76°25′W. Manufacturing (tool and die, plastic products); agriculture (grain, apples, soybeans; poultry, livestock, dairying). Laid out 1814.

Mount Vsevidof, snow-covered volcano, SW ALASKA, on W UMNAK Island, ALEUTIAN ISLANDS, 15 mi/24 km NNE of NIKOLSKI; 60°40′N 143°00′W.

Mount Waddington (WAH-deeng-tuhn), regional district (□ 7,810 sq mi/20,306 sq km; 2001 population 13,111), SW BRITISH COLUMBIA, W Canada, N VANCOUVER ISLAND; 50°45′N 127°00′W. Logging; cellulose mill; commercial fishing (salmon). Consists of small, unincorporated settlements and the municipalities of ALERT BAY, PORT ALICE, PORT HARDY, Port McNeill, Winter Harbour/Holberg, Hyde Creek, Coal Harbour, MALCOLM ISLAND, QUATSINO, and Woss. Incorporated 1966.

Mount Washington, town (2000 population 8,485), BULLITT county, NW KENTUCKY, 15 mi/24 km SE of LOUISVILLE; 38°02′N 85°32′W. Agricultural area (burley tobacco, grain, livestock, poultry, dairying). Manufacturing (concrete, materials handling equipment, executive coaches).

Mount Washington, resort town, BERKSHIRE county, SW MASSACHUSETTS, in the BERKSHIRES, near NEW YORK state line, 25 mi/40 km SSW of PITTSFIELD; 42°06′N 73°28′W. Includes Union Church village and Mt. Everett. Bash Bish Falls State Park nearby; state forest.

Mount Washington, MARYLAND: see BALTIMORE.

Mount Waverley (MOUNT WAI-vuhr-lee), suburb 9 mi/15 km SE of MELBOURNE, VICTORIA, SE AUSTRALIA; 37°53′S 145°08′E. Railway station.

Mount Wilson (MOUNT WIL-suhn), village, NEW SOUTH WALES, SE AUSTRALIA, 78 mi/126 km W of SYDNEY, in BLUE MOUNTAINS; 33°30′S 150°23′E.

Mount Wilson Observatory, LOS ANGELES county, S CALIFORNIA, at MOUNT WILSON, SAN GABRIEL MOUNTAINS, 16 mi/26 km NE of downtown LOS ANGELES, 7 mi/11.3 km NE of PASADENA, and in Angeles National Forest. George E. Hale founded Mount Wilson Observatory in 1904. Its equipment includes 100-in/254-cm and 60-in/152-cm reflecting telescopes and two solar tower telescopes 150 ft/46 m and 60 ft/18 m in length. Principal research programs that have been conducted at the observatory include studies of the structure and dimensions of the universe and the physical nature, chemical composition, and evolution of celestial bodies. An ongoing program on the 60-in/152-cm telescope is a long-term study of singly ionized calcium lines to monitor sunspot cycles on nearby solar-type stars. Formerly part of the Hale Observatories jointly administered by the California Institute of Technology and the Carnegie Institution. The Carnegie Institution closed the 100-in/254-cm telescope in 1985, and transferred the observatory's management to the newly formed Mount Wilson Institute in 1989.

Mount Wolf, borough (2006 population 1,344), YORK county, S PENNSYLVANIA, 7 mi/11.3 km NNE of YORK, and 2 mi/3.2 km SW of SUSQUEHANNA RIVER; 40°03′N 76°42′W. Manufacturing (food products, corrugated containers, steel-wire and fiberglass-mesh products); agriculture (grain, soybeans, apples; poultry, livestock, dairying).

Mountzinos (moon-tzee-NOS), mountain massif in EAST MACEDONIA AND THRACE department, NE GREECE, W of MESTA (Nestos) River, 15 mi/24 km NNE of KAVÁLLA. Also called Tsali.

Mount Zion, city (2000 population 1,275), CARROLL county, W GEORGIA, 7 mi/11.3 km NW of CARROLLTON, near ALABAMA state line; 33°38′N 85°11′W.

Mount Zion, village (2000 population 4,845), MACON county, central ILLINOIS, 6 mi/9.7 km SE of DECATUR; 39°46′N 88°52′W. In agricultural area. SPITLER Woods State Natural Area nearby.

Mouping, CHINA: see MOPING.

Moura (MOU-ruh), town, SE QUEENSLAND, NE AUSTRALIA, 100 mi/161 km SW of ROCKHAMPTON on Dawson Highway; 24°34′S 150°01′E. Open-cut coal mines. Grain, cotton, cattle.

Moura (MO-rah), town, N central AMAZONAS, BRAZIL, steamer landing on right bank of the RIO NEGRO below influx of the Rio Branco, and 160 mi/257 km NW of MANAUS; 01°30′S 61°35′W. Rubber. Airport.

Moura (MO-rah), town, BEJA district, S PORTUGAL, 24 mi/39 km ENE of BEJA. Railroad spur terminus; produces olive oil and cheese; trade in grain, sheep, figs. Has a Manueline church and a Moorish castle (rebuilt 1920).

Mourão (mor-OU), town, ÉVORA district, S central PORTUGAL, between the GUADIANA RIVER and Spanish border, 35 mi/56 km ESE of ÉVORA. Grain, cork; sheep.

Mourdiah (mor-JAH), village, SECOND REGION/KOULIKORO, W MALI, on SAHARAN desert road and 130 mi/209 km N of BAMAKO; 14°28′N 07°28′W. Peanuts, kapok, gum arabic; livestock.

Mouree, GHANA: see MORIE.

Mourenx (moo-rahn), town (□ 2 sq mi/5.2 sq km), PYRÉNÉES-ATLANTIQUES department, AQUITAINE region, SW FRANCE, 14 mi/23 km NW of PAU, in the AQUITAINE BASIN; 43°23′N 00°36′W. Founded in late 1950s, Mourenx provides quarters (high-rise apartment buildings) for employees of industrial plants established between here and LACQ (4 mi/6.4 km N), where a natural-gas field was discovered after World War II. The facilities include sulphur retrieval and storage units, an aluminum reduction plant, and a thermal power plant fired with natural gas.

Mouriès (moo-ryai), commune (□ 14 sq mi/36.4 sq km), BOUCHES-DU-RHÔNE department, PROVENCE-ALPES-CÔTE D'AZUR region, SE FRANCE, at S foot of the ALPILLES (a limestone range), 12 mi/19 km E of ARLES; 43°41′N 04°52′E. Irrigation agriculture; market gardens.

Mouriki (moo-REE-kee), mountain (5,587 ft/1,703 m), in WEST MACEDONIA department, N central GREECE, 12 mi/19 km ESE of KASTORÍA; 38°03′N 21°55′E. Formerly called Vlatse (or Vlatsi) for former name of Vlaste village at SE foot.

Mourik, Jbel (moor-EEK, zhe-BEL), mountain peak (10,607 ft/3,233 m), Tadla-Azilal administrative region, MOROCCO, in E High ATLAS mountains, 12 mi/20 km W of Plateau of the Lakes, and 31 mi/50 km SE of BENI MELLAL; 32°07′N 05°54′W.

Mourilyan (muh-RIL-yuhn), town, NE QUEENSLAND, NE AUSTRALIA, 55 mi/89 km SSE of CAIRNS, 4 mi/6 km from INNISFAIL, on coast; 17°34′S 146°03′E. Etty Bay to NE. Sugar port, sugar industry museum.

Mourmelon-le-Grand (moor-muh-lon–luh–grahn), town (□ 9 sq mi/23.4 sq km), MARNE department, CHAMPAGNE-ARDENNE region, N FRANCE, 13 mi/21 km N of CHÂLONS-EN-CHAMPAGNE; 04°22′N 04°19′E. Large military camp nearby.

Mourne Mountains (MOORN), Gaelic *Beanna Boirche*, range in S DOWN, SE Northern Ireland, extending 15 mi/24 km NE-SW between CARLINGFORD LOUGH and DUNDRUM BAY of the IRISH SEA, rising to 2,796 ft/852 m in Slieve Donard, 2 mi/3.2 km SW of NEWCASTLE; 54°10′N 06°05′W. BELFAST receives its main water supply from here.

Mouroux (MOO-roo), commune (□ 6 sq mi/15.6 sq km), SEINE-ET-MARNE department, ÎLE-DE-FRANCE region, N central FRANCE; 48°49′N 03°02′E.

Mourzouk, LIBYA: see MARZUQ.

Mousam Lake (MOOS-uhm), SW MAINE, W YORK county; c.6.5 mi/10.5 km long. Drains SE into MOUSAM RIVER.

Mousam River (MOOS-uhm), 23 mi/37 km long, SW MAINE; rises in MOUSAM LAKE (5 mi/8 km long), W YORK county; flows SE to the ATLANTIC 4 mi/6.4 km below KENNEBUNK, where it furnishes water power.

Mouscron (moo-KRAWN), Flemish *Moeskroen*, commune (□ 39 sq mi/101.4 sq km; 2006 population 53,023), ⊙ of Mouscron district, HAINAUT province, SW BELGIUM, on FRENCH border, opposite TOURCOING; 50°44′N 03°13′E. This commune was part of WEST FLANDERS province until linguistic census of 1960 resulted in a transfer to Hainaut province.

Mousehole (MOUZ-uhl), village (2001 population 830), SW CORNWALL, SW ENGLAND, on MOUNT'S BAY of the CHANNEL, and 3 mi/4.8 km SSW of PENZANCE; 50°05′N 05°33′W. Pilchard-fishing center and seaside resort. Offshore is St. Clement's Isle, with ruins of ancient chapel.

Moussaya (MOO-sah-yah), town, Kindia administrative region SW GUINEA, in Guinée-Maritime geographic region, 60 mi/97 km S of KINDIA; 09°59′N 13°49′W. Pineapples, citrus fruits, bananas.

Moussoro (moo-so-RO), town, KANEM administrative region, W CHAD, 140 mi/225 km NE of N'DJAMENA. Trading center; livestock; millet. Airfield.

Moustafouli, Greece: see PANAITOLION.

Moustier, Le, FRANCE: see EYZIES-DE-TAYAC, LES.

Moustiers-Sainte-Marie (moo-tyai–san-tuh–mah-ree), commune (□ 34 sq mi/88.4 sq km), resort in ALPES-DE-HAUTE-PROVENCE department, PROVENCE-ALPES-CÔTE D'AZUR region, SE FRANCE, in MARITIME ALPS, 17 mi/27 km S of DIGNE-LES-BAINS; 43°50′N 06°13′E. It is situated in a narrow ravine. Noted for its decorative faïence ware, whose origin dates from 1679. The chapel of Notre-Dame-de-Beauvoir is a pilgrimage center; it stands on a limestone ledge overlooking the village. The spectacular canyon of the VERDON RIVER is 3 mi/4.8 km S, where it opens onto St. Croix Lake.

Moutfort (moot-FOR), hamlet, CONTERN commune, SE LUXEMBOURG, 6 mi/9.7 km ESE of LUXEMBOURG city; 49°35′N 06°16′E.

Mouthe (moo-tuh), commune (□ 15 sq mi/39 sq km), DOUBS department, FRANCHE-COMTÉ region, E FRANCE, in the JURA Mountains, 15 mi/24 km SSW of PONTARLIER; 46°44′N 06°14′E. Gruyère cheese. Here is the source of DOUBS RIVER. Winter sports.

Mouthiers-sur-Boëme (moo-tyai–syoor–bo-em), commune (□ 13 sq mi/33.8 sq km), CHARENTE department, POITOU-CHARENTES region, W FRANCE, 7 mi/11.3 km S of ANGOULÊME; 45°33′N 00°07′E. Paper milling.

Moutier (moo-TYAI), district, NW BERN canton, SWITZERLAND; 47°17′N 07°22′E. Main town is MOUTIER; population is French-speaking and Protestant.

Moutier (moo-TYAI), German *Münster*, town, BERN canton, NW SWITZERLAND, on BIRS RIVER, and 6 mi/9.7 km NNW of GRENCHEN, to which it is connected by the GRENCHENBERG railroad tunnel; 47°17′N 07°22′E. Elevation 1,755 ft/535 m. Watches, metal products, glassware.

Moûtiers (moo-TYAI), town (□ 1 sq mi/2.6 sq km), SAVOIE department, RHÔNE-ALPES region, SE FRANCE, in TARENTAISE valley of the Savoy Alps (ALPES FRANÇAISES), 15 mi/24 km SSE of ALBERTVILLE; 45°29′N 06°32′E. Tourist center and old capital of former Tarentaise district. Electrometallurgy in ISÈRE RIVER valley. Moûtiers is a former episcopal see with 15th-century cathedral. The spas of SALINS-LES-BAINS (1 mi/1.6 km S) and BRIDES-LES-BAINS (3 mi/4.8 km S) have curative saltwater springs.

Mouvaux (moo-vo), town (□ 1 sq mi/2.6 sq km), W residential suburb of ROUBAIX, NORD department, NORD-PAS-DE-CALAIS region, N FRANCE; 50°42′N 03°08′E.

Mouy (moo-ee), town (□ 3 sq mi/7.8 sq km), OISE department, PICARDIE region, N FRANCE, on THÉRAIN RIVER, and 13 mi/21 km SE of BEAUVAIS; 49°19′N 02°19′E. Construction equipment.

Mouyondzi (moo-yon-DZEE), town, BOUENZA region, S Congo Republic, 90 mi/145 km WNW of BRAZZAVILLE; 03°59′S 13°54′E. Palm products. Roman Catholic and Protestant missions. Also spelled Muyondzi and Mouyoundzi.

Mouyoundzi, Congo Republic: see MOUYONDZI.

Mouzaïa (moo-zah-ee-AH), village, BLIDA wilaya, N central ALGERIA, in the MITIDJA plain, 8 mi/12.9 km W of BLIDA. Distilling of essential oils (geraniums); olive-oil pressing; mineral water bottling industry. Formerly Mouzaïaville.

Mouzaïa-les-Mines, ALGERIA: see DRAA ESMAR.

Mouzarak (moo-zah-RAH), village, KANEM administrative region, W CHAD, 85 mi/137 km NNE of N'DJAMENA.

Mouzon (moo-zon), commune (□ 13 sq mi/33.8 sq km; 2004 population 2,554), ARDENNES department, CHAMPAGNE-ARDENNE region, N FRANCE, on island formed by MEUSE RIVER, 9 mi/14.5 km SE of SEDAN; 49°36′N 05°05′E. Manufacturing of felt floor covering. Has restored former abbatial church (13th century) often visited by archbishops of REIMS.

Moville (mo-VIL), Gaelic *Bun an Phobail*, town (2006 population 2,174), NE DONEGAL county, N IRELAND, on W shore of LOUGH FOYLE, 18 mi/29 km NE of LONDONDERRY; 55°11′N 07°03′W. Resort, small port; furniture manufacturing.

Moville, town (2000 population 1,583), WOODBURY county, W IOWA, on West Fork Little Sioux River, and 17 mi/27 km E of SIOUX CITY; 42°29′N 93°04′W. In agricultural area.

Mowai, Al, SAUDI ARABIA: see MUWAIH.

Mowar (MO-wahr), town, NAGPUR district, MAHARASHTRA state, central INDIA, 45 mi/72 km NW of NAGPUR; 21°28′N 78°26′E. Cotton, millet, wheat, oilseeds; mango groves.

Mowbray, community, S MANITOBA, W central Canada, 17 mi/27 km S of MANITOU, in Pembina rural municipality, and on U.S. (NORTH DAKOTA) border; 49°00′N 98°29′W.

Moweaqua (mo-WEE-kwah), village (2000 population 1,923), SHELBY county, central ILLINOIS, 14 mi/23 km S of DECATUR; 39°37′N 89°01′W. Agriculture (corn, wheat, soybeans; dairy products; livestock). Incorporated 1877.

Mower (MOU-wuhr), county (□ 711 sq mi/1,848.6 sq km; 2006 population 38,666), SE MINNESOTA; ⊙ AUSTIN; 43°40′N 92°45′W. Agricultural area bordering IOWA on S, drained by CEDAR RIVER and headwaters of ROOT and LITTLE CEDAR rivers. Hay, soybeans, corn, oats, peas, alfalfa; sheep, hogs, cattle, poultry; dairying; meat processing at Austin; limestone. Lake Louise State Park in SE corner. Formed 1855.

Mowrystown (MOU-reez-TOUN), village (2000 population 373), HIGHLAND county, SW OHIO, 13 mi/21 km SSW of HILLSBORO; 39°02′N 83°45′W.

Moxee (MAHK-see), town (2006 population 1,836), YAKIMA county, S WASHINGTON, 8 mi/12.9 km SE of YAKIMA, near Roza Canal; 46°34′N 120°24′W. Railroad terminus. Fruits, vegetables, hops; dairying; manufacturing (fertilizers, millwork, furniture, mattresses). Yakima Ridge to N; Yakima Indian Reservation to SW. U.S. Military Reservation Yakima Training Center and Firing Range to NE.

Moxico (mo-SHEE-ko), province (□ 86,087 sq mi/223,826.2 sq km), E ANGOLA; ⊙ Luena. Bordered N by Rio Cassai and Lunda Sul, E by ZAMBIA, S by CUANDO CUBANGO and Rio Cuando, W by BIÉ province Drained by LUNGUE-BUNGO, Luena, Rio Cuando, ZAMBEZI, Rio Cuito. Includes Cameia National Park. Agriculture includes rice, manioc, corn, peanuts, fishing, poultry, wood. Minerals include coal, copper, manganese, iron. Main centers are Luena, LUCUSSE, CAZOMBO, CHIUME, CALUNDAU, LUATAMBA, KAVUNGO, Moxico.

Moxie Mountain (MAKS-ee) (2,925 ft/892 m), SOMERSET county, W central MAINE, 30 mi/48 km NNW of SKOWHEGAN.

Moxie Pond (MAKS-ee), reservoir, SOMERSET county, W central MAINE, on branch of KENNEBEC RIVER, 40 mi/64 km N of SKOWHEGAN; 7 mi/11.3 km long; 45°20′N 69°51′W. Formed by dam S of Lake Moxie village. Recreation area.

Moxos, province, BENI department, NE BOLIVIA; ⊙ SAN IGNACIO DE MOXOS; 15°10′S 65°30′W.

Moxotó (MO-sho-to), city, S central PERNAMBUCO state, BRAZIL, 70 mi/113 km NW of GARANHUNS; 08°44′S 37°31′W.

Moxotó River (MO-sho-to), c.120 mi/193 km long, NE BRAZIL; left tributary of SÃO FRANCISCO River, which it enters just above PAULO AFONSO Falls, after forming PERNAMBUCO-ALAGOAS border. Intermittent-flowing stream.

Moy (MOI), agricultural village, HIGHLAND, N Scotland, 9 mi/14.5 km SE of INVERNESS; 57°23′N 04°03′W. On Loch Moy, a small lake (1.5 mi/2.4 km long, 0.5 mi/0.8 km wide). Nearby Moy Hall is seat of the MacKintosh.

Moya (MOI-ah), town, Nzwani island and district, SE Comoros Republic, 12 mi/19 km SSE of Mutsamudu, on S coast of island, on Mozambique Channel, Indian Ocean; 12°19′S 44°27′E. Fish; livestock; ylang-ylang, vanilla, coconuts, bananas.

Moyá (moi-AH), town, BARCELONA province, NE SPAIN, 15 mi/24 km NE of MANRESA. Textiles (cotton, wool, hemp).

Moya (MOI-ah), town, Grand Canary island, CANARY ISLANDS, SPAIN, 10 mi/16 km W of LAS PALMAS. Bananas, cereals, corn, potatoes, livestock. Lumbering. Medicinal springs. Prehistoric caves nearby.

Moyahua, MEXICO: see MOYAHUA DE ESTRADA.

Moyahua de Estrada (mo-YAH-wah dai es-TRAH-dah), town, ZACATECAS, N central MEXICO, on JUCHIPILA RIVER, and 35 mi/56 km SSE of TLALTENANGO DE SÁNCHEZ ROMÁN on Mexico Highway 54; 21°18′N 103°09′W. Grain, sugarcane, fruit, vegetables, livestock.

Moyale (moi-YAH-lai), town (2007 population 26,232), OROMIYA state, S ETHIOPIA; 03°32′N 38°03′E. On border opposite MOYALE (KENYA), 60 mi/97 km SE of MEGA, on plateau. Caravan center and customs station.

Moyale (mo-YAH-lai), village, EASTERN province, N KENYA, on border opposite MOYALE (ETHIOPIA), on road, and 245 mi/394 km NNE of ISIOLO; 03°30′N 39°08′E. Customs station. Livestock raising. Airfield.

Moyamba (mo-YAHM-bah), town, ⊙ Moyamba district, SOUTHERN province, SW SIERRA LEONE, 60 mi/97 km ESE of FREETOWN; 08°10′N 12°26′W. Has United Methodist and United Brethren churches, hospital, all girls' boarding school.

Moyar River (mo-YAHR), c.90 mi/145 km long, in NILGIRI and COIMBATORE districts, TAMIL NADU state, S INDIA; rises in NILGIRI HILLS on MUKURTI peak, near KERALA–Tamil Nadu state border; flows N past PYKARA (falls and hydroelectric works 8 mi/12.9 km NNW), and E (forming part of KARNATAKA–Tamil Nadu state border) to BHAVANI RIVER, 10 mi/16 km WSW of SATYAMANGALAM (160-ft/49-m dam near confluence; built 1949). Construction of regulating dam and headworks 8 mi/12.9 km below falls (10 mi/16 km NE of Pykara) begun 1946 to develop power from tailwater of Pykara power station. Called Pykara River above Pykara Falls.

Moyen Atlas or **Middle Atlas**, mountains, MOROCCO: see ATLAS MOUNTAINS.

Moyen-Cavally, region (□ 5,460 sq mi/14,196 sq km; 2002 population 443,200), W CÔTE D'IVOIRE; ⊙ GUIGLO; 06°25′N 07°30′W. Bordered N by Dix-Huit Montagnes region (part of W portion of border formed by CAVALLY RIVER), NE by Haut-Sassandra region, ESE by Bas-Sassandra region, SWW by LIBERIA. Part of Lake Buyo in E, where Moyen-Cavally meets Bas-Sassandra and Haut-Sassandra. NZO RIVER in N and Cavally River in W. Small part of Mount Peko National park in NE (extending from Dix-Huit Montagnes) and part of TAÏ WORLD BIOSPHERE RESERVE NATIONAL PARK in SE (extending into Bas-Sassandra). Towns include DUÉKOUÉ, GUIGLO, and Taï. Regional airport near Guiglo.

Moyen-Chari (mwah-YANG–shah-REE), administrative region, S CHAD; ⊙ SARH. Borders MANDOUL (SWW), TANDJILE (W), CHARI-BAGUIRMI (NW), GUERA (N), and SALAMAT (E) administrative regions and CENTRAL AFRICAN REPUBLIC (S). Drained by CHARI and BAHR SALAMAT rivers. Lake Iro in NE. Major centers include Sarh and KYABÉ. Formed following a decree in October 2002 that reorganized Chad's administrative divisions from twenty-eight departments to eighteen regions. This area made up the central and E portions of former MOYEN-CHARI prefecture.

Moyen-Chari (mwah-YANG–shah-REE), former prefecture (2000 population 950,019), S CHAD; capital was SARH. Was bordered N by GUÉRA prefecture, E by SALAMAT prefecture, S by CENTRAL AFRICAN REPUBLIC, and W by CHARI-BAGUIRMI, TANDJILÉ, and LOGONE ORIENTAL prefectures. Drained by CHARI, OUHAM, and BAHR SALAMAT rivers. Major centers included Sarh, KOUMRA, KYABÉ, and MOÏSSALA. This was a prefecture prior to Chad's administrative division reorganization from fourteen prefectures to twenty-eight departments. Following a decree in 2002 that reorganized Chad's administrative divisions into eighteen regions, this area is composed of Moyen-Chari (E and central) and MANDOUL (W) regions.

Moyen-Comoé, region (□ 2,660 sq mi/6,916 sq km; 2002 population 488,200), E CÔTE D'IVOIRE; ⊙ ABENGOUROU; 06°30′N 03°25′W. Bordered N by Zanzan region, E by GHANA, S by Sud-Comoé region, and W by KOMOÉ RIVER (NNW by N'zi-Comoé region, WSW by Agnéby region, SW tip by Lagunes region). Towns include ABENGOUROU and AGNIBILÉKROU.

Moyeni, LESOTHO: see QUTHING.

Moyen-Mono, prefecture (2005 population 75,382), PLATEAUX region, S TOGO, ⊙ TOHOUN; 07°12′N 01°33′E.

Moyenmoutier (mwah-yahn-moo-tyai), commune (□ 13 sq mi/33.8 sq km; 2004 population 3,338), VOSGES department, LORRAINE region, E FRANCE, near the MEURTHE, 7 mi/11.3 km N of SAINT-DIÉ, in the NW VOSGES; 48°23′N 06°55′E. Textile weaving and bleaching. Has 18th-century abbatial church with fine sculpted woodwork.

Moyenne-Guinée, geographic region, W central GUINEA; ⊙ LABÉ. Bordered N by SENEGAL, NE tip by MALI, E by Haute-Guinée geographic region (Faranah administrative region), S by SIERRA LEONE, SWW by Guinée-Maritime geographic region (N by Boké and central and S by Kindia administrative regions), and NW by GUINEA-BISSAU. Mountainous area. FOUTA DJALLON region is here. Includes the administrative regions of Labé in the NE, Mamou in the S, and N half of Boké (Gaoual and Koundara prefectures) in the NW.

Moyenne Island (mwah-YEN), one of the SEYCHELLES, in the MAHÉ group, off NE coast of MAHÉ ISLAND 4 mi/6.4 km E of VICTORIA; 0.25 mi/0.4 km long, 0.25 mi/0.4 km wide; 04°37′S 55°31′E. Separated from SAINT ANNE ISLAND (N) by Saint Anne Channel; tourism.

Moyen Ogooué, province (□ 7,156 sq mi/18,605.6 sq km; 2002 population 54,600), W GABON; ⊙ LAMBARÉNÉ (largest city); 00°18′S 10°42′E. Bounded on S by NGOUNIÉ province, E by OGOOUÉ-IVINDO province, N by ESTUAIRE and WOLEU-NTEM provinces, W by OGOOUÉ-MARITIME province. A forested area where sawmills and the lumber industry is important.

Moyen-Ouest (mwah-YA–WEST) [French=middle west], region, ANTANANARIVO province, W central MADAGASCAR; 18°00′S 21°00′E–45°30′S 47°00′E. The W, lower (elevation 2,500 ft/762 m–4,000 ft/1,219 m), drier portion of the central highlands. Vast grasslands; country's prime cattle-raising zone. Main centers and cattle markets are TSIROANOMANDIDY and MANDOTO. Zone of agricultural settlement, some of which is government-planned (SAKAY). Historical no-man's-land between Merina and Sakalava tribes.

Moyeuvre-Grande (mwah-yu-vruh–grahnd), town (□ 3 sq mi/7.8 sq km), MOSELLE department, LORRAINE region, NE FRANCE, on the ORNE, and 11 mi/18 km NNW of METZ; 49°15′N 06°02′E. Former active iron-mining center; metalworks.

Moyhu (moi-YOO), township, NE VICTORIA, SE AUSTRALIA, 112 mi/180 km from MELBOURNE, 17 mi/27 km S of WANGARATTA, in King River valley; 36°36′S 146°24′E.

Moyie (MOI-ee), unincorporated village, SE BRITISH COLUMBIA, W Canada, in PURCELL MOUNTAINS, at S end of Moyie Lake (8 mi/13 km long), on Moyie River, 16 mi/26 km S of CRANBROOK, and included in East Kootenay regional district; 49°17′N 115°50′W. Timber; tourism.

Moyie Springs, village (2000 population 656), BOUNDARY county, N IDAHO, 8 mi/12.9 km E of BONNERS FERRY, on Moyie River, near its confluence with Kootanai River, near MONTANA state line; 48°43′N 116°11′W. Timber; manufacturing (heating logs, lumber). Moyie Falls to NE; Kaniksu National Forest to N and S.

Moyingyi Reservoir, MYANMAR: see PYINBONGYI.

Moykovats, MONTENEGRO: see MOJKOVAC.

Moyo, administrative district (□ 795 sq mi/2,067 sq km; 2005 population 248,200), NORTHERN region, NW UGANDA, on ALBERT NILE RIVER; ⊙ MOYO; 03°15′N 31°45′E. As of Uganda's division into eighty districts, borders ADJUMANI (E and S, formed by Albert Nile River) and YUMBE (S and W) districts and SUDAN (N). Albert Nile River drains area. Majority of population is agricultural (beans, cassava, maize, sweet potatoes, rice, sorghum, and millet, some cotton). There is also live-

stock raising (including poulty, cattle, goats, and pigs) and some fishing in Albert Nile River. Formed in 1997 from NWN portion of former MOYO district (Adjumani district was created from S and E portions).

Moyo (MOI-o), former administrative district (□ 1,933 sq mi/5,025.8 sq km), NW UGANDA, along SUDAN border (to N); capital was MOYO; 03°20′N 31°45′E. As of Uganda's division into thirty-nine districts, was bordered by GULU (E and S) and ARUA (W) districts. Largely rural area drained by ALBERT NILE RIVER. Millet, cassava was grown. In 1997 NWN portion of district became current MOYO district and S and E portions became ADJUMANI district.

Moyo (MOI-o), town (2002 population 12,074), ⊙ MOYO district, NORTHERN region, NW UGANDA, near SUDAN border, 25 mi/40 km W of NIMULE. Cotton, peanuts, sesame. Was part of former NILE province.

Moyo (MO-yo), island, Nusa Tenggara Barat province, INDONESIA, just off N coast of SUMBAWA, at entrance of Saleh Bay (inlet of FLORES SEA); 20 mi/32 km long, 9 mi/14.5 km wide; 08°15′S 117°34′E. Also spelled Mojo.

Moyobamba (mo-yo-BAHM-bah), province (□ 2,093 sq mi/5,441.8 sq km), SAN MARTÍN region, N central PERU; ⊙ MOYOBAMBA; 05°45′S 77°15′W. Northwesternmost province of San Martín region, bordering on AMAZONAS and LORETO regions.

Moyobamba (mo-yo-BAHM-bah), city (2005 population 38,530), ⊙ MOYOBAMBA province and SAN MARTÍN region, N central PERU, near MAYO RIVER (AMAZON basin), in E outliers of the ANDES, on road 60 mi/97 km NW of TARAPOTO; 06°03′S 76°56′W. Elevation 2,800 ft/853 m. Has a humid tropical climate. It is situated in a fertile agricultural area (cotton, sugarcane, tobacco, rice, cacao, coffee, grapes). Produces cocoa; excellent wines; alcohol and liquor distilling; cotton; manufacturing. Hot springs, gold placers, and petroleum seepages in vicinity. Though dating back to an ancient Indian settlement under Inca influence, it is the second town in Peru to be founded (1539) E of the Andes by the Spanish. Airport.

Moyogalpa (mo-yo-GAHL-pah), town, RIVAS department, SW NICARAGUA, on W OMETEPE ISLAND, and 6 mi/9.7 km SW of Alta Gracia, at W foot of volcano CONCEPCIÓN; 11°32′N 85°42′W. Boatbuilding; coffee, tobacco, cotton, corn. Tourism. Radio station.

Moyowasi River (mo-yo-WAH-see), c. 150 mi/241 km long, NW TANZANIA; rises c. 125 mi/201 km NE of KIGOMA in N KIGOMA region; flows NE, then S; receives NIKONGA and Kigosi rivers before joining GOMBE RIVER 125 mi/201 km E of Kigoma to form Malagalasi River.

Moy River, Gaelic *An Mhuaidh*, 40 mi/64 km long, NW IRELAND; rises in SLIEVE GAMPH mountains, SLIGO county; flows SW into MAYO county, then N, past Foxford and Ballina, to KILLALA BAY. Navigable below Ballina. Salmon fisheries.

Møysalen (MUH-oo-sah-luhn), highest peak (4,153 ft/1,266 m) on HINNØYA of the VESTERÅLEN group, NORDLAND county, N NORWAY, 30 mi/48 km NE of SVOLVÆR; 68°32′N 15°24′E. Permanently snow-clad.

Moyston, town, SW VICTORIA, SE AUSTRALIA, 13 mi/21 km SE of POMONAL, at foot of GRAMPIANS NATIONAL PARK; 37°18′S 142°46′E.

Moyu (MO-YOO-I) [=black jade], town, ⊙ Moyu county, SW XINJIANG UYGUR AUTONOMOUS REGION, CHINA, 20 mi/32 km NW of HOTAN, and on Karakax River (tributary of HOTAN RIVER); 37°17′N 79°43′E. Jade-mining center. Grain, cotton, livestock; textiles, food processing. Also known as Karakax.

Moyuta (mo-YOO-tah), town, JUTIAPA department, SE GUATEMALA, in Pacific piedmont, at E foot of volcano Moyuta (5,525 ft/1,684 m), 20 mi/32 km SW of JUTIAPA; 14°02′N 90°05′W. Elevation 4,209 ft/1,283 m. Corn, beans, coffee, sugarcane; livestock.

Mozac (mo-zahk), commune (□ 1 sq mi/2.6 sq km; 2004 population 3,529), suburb of PUY-DE-DÔME de-

partment, central FRANCE, in AUVERGNE region, 1.5 mi/2.4 km W of RIOM; 45°53′N 03°06′E. Stone quarries in volcanic lava field. Has 12th-century church of St. Peter, with remarkable enameled shrine.

Mozambique (mo-zam-BEEK), country (□ 302,328 sq mi/783,030 sq km; 2004 estimated population 18,811,731; 2007 estimated population 20,905,585), SE AFRICA, bordering on the INDIAN OCEAN in the E, on SOUTH AFRICA and SWAZILAND in the S, on ZIMBABWE, ZAMBIA, and MALAWI in the W, and on TANZANIA in the N; ⊙ MAPUTO. Other cities include BEIRA, INHAMBANE, MOÇAMBIQUE, NAMPULA, PEMBA, QUELIMANE, TETE, ANTÓNIO ENES, and XAIXAI. There are ten provinces with 129 administrative districts.

Geography

The MOZAMBIQUE CHANNEL separates the country from the island of MADAGASCAR. Mozambique's c.1,600 mi/2,575 km coastline is interrupted by the mouths of numerous rivers, notably the ROVUMA (which forms part of the boundary with Tanzania), LÚRIO, Incomati, LUGENDA, ZAMBEZI (which is navigable for c.290 mi/465 km within the territory), Revùe, SAVE (Sabi), and LIMPOPO. South of the Zambezi estuary the coastal belt is very narrow, and in the far N the coastline is made up of rocky cliffs. Along the N coast are numerous islets and lagoons; in the far S MAPUTO BAY. The N and central interior is mountainous; Monte Binga (7,992 ft/2,436 m), the country's loftiest point, is situated at the Zimbabwean border W of Beira. About one-third of Lake NYASA falls within Mozambique's boundaries; Lake CHIRWA (Lago Chirua) is at the border with Malawi. Much of the country is covered with savanna; there are also extensive hardwood forests, and palms grow widely along the coast and near rivers.

Population

Almost 95% of the population speak a Bantu language. The principal ethnic groups are, in the N, the Yao, Makonde, and Makua; in the center, the Thonga, Chewa, Nyanja, and Sena; and in the S, the Shona and Tonga. Small numbers of Swahili live along the coast. In addition, before the agreement (1974) to grant independence was announced, there were about 170,000 Europeans (largely Portuguese), most of whom lived in urban areas; approximately 40,000 *mestiços* (persons of mixed African and European descent); and a small number of Asians (persons of Indian and Pakistani background) and Chinese. By 1978, there were fewer than 15,000 Portuguese. Most of the inhabitants of Mozambique follow traditional religious beliefs; in addition, there are large numbers of Christians (mostly Roman Catholic) and Muslims (most of whom live in the N). Portuguese is the official language and African languages are widely spoken.

Economy

One of the world's poorest nations, Mozambique is an overwhelmingly agricultural country, with the majority of its workers engaged in subsistence cultivation. Also, many work in South African mines. The principal cash crops include cashews, sugar, copra, and tea, which are grown on government farms. Peasants grow cotton on private plots. Agricultural production has declined steadily since independence, when farms were nationalized. Cattle, sheep, and goats are raised but their numbers are kept low by the tsetse fly. There are small forestry and fishing industries, including shrimp. The territory's mineral wealth has not been determined fully, and mining is a minor factor in Mozambique's economy. Mozambique's rudimentary industrial sector is devoted largely to the processing of raw materials. In addition, refined petroleum, construction materials (particularly cement), steel, chemical fertilizer, clothing, and footwear are produced. Mozambique privatized several industries that were nationalized after 1975. Electricity from the giant CABORA BASSA hydroelectric project (located on the Zambezi near Tete) is exported to South Africa. A smaller hydroelectric plant is situated at CHICAMBA REAL (near Beira) on the Revùe River. The annual cost of Mozambique's imports is usually much higher than its earnings from foreign sales. The principal imports are machinery, foodstuffs, motor vehicles, crude petroleum and petroleum products, textiles, and metals; the chief exports are shrimp, cotton, sugar, cashew nuts, tea, copra, fruit, and timber. South Africa, BELGIUM, and ITALY are Mozambique's chief trading partners. Mozambique also derives considerable revenue from handling some of the foreign trade of nearby countries; goods are shipped on railroad lines that terminate at the ports of Maputo, Beira, NACALA, and LUMBO (near Moçambique). However, much of Mozambique's transportation system was destroyed by fighting in the nation's post-independence (ongoing) civil war.

History to 1509

Bantu-speakers began to migrate into the region of Mozambique in the middle of the first millennium C.E. From 1000, Arab and Swahili traders settled along parts of the coast, notably at Sofala (near modern Beira), at Cuama (near the Zambezi estuary), and on the site of present-day Inhambane. The traders had contact with the interior, and Sofala was particularly noted as a gold- and ivory-exporting center closely linked with—and at times controlled by—Kilwa (on the coast of modern Tanzania). In 1498, Vasco da Gama, a Portuguese navigator en route around Africa to INDIA, visited Quelimane and Moçambique. Between 1500 and 1502 Pedro Álvares Cabral and Sancho de Tovar, also Portuguese explorers, visited Sofala and Maputo Bay (formerly Delagoa Bay). In 1505, the Portuguese under Francisco de Almeida occupied Moçambique, and Pedro de Anaia established a Portuguese settlement at Sofala. The Portuguese also set up trading stations N of Cabo Delgado (near the mouth of the ROVUMA), but their main influence (especially after 1600) in E Africa was in the Mozambique region.

History: 1509 to 1574

Between 1509 and 1512 António Fernandes traveled inland and visited the Mwanamutapa kingdom, which controlled the region between the Zambezi and SAVE rivers and was the source of much of the gold exported at Sofala. Soon after, Swahili traders resident in Mwanamutapa began to redirect the kingdom's gold trade away from Portuguese-controlled Sofala and toward more N ports. Thus, Portugal became interested in directly controlling the interior. In 1531, posts were established inland at Sena and Tete on the Zambezi, and in 1544 a station was founded at Quelimane. In 1560 and 1561 Gonçalo da Silveira, a Portuguese Jesuit missionary, visited Mwanamutapa, where he quickly made converts, including King Nogomo Mupunzagato. However, the Swahili traders who lived there, fearing for their commercial position, persuaded Nogomo to have Silveira murdered. Between 1569 and 1572 an army of about 1,000 Portuguese under Francisco Barreto attempted to gain control of the interior, but Barreto and most of the soldiers died of disease at Sena.

History: 1574 to 1820

In 1574, an army of four hundred men under Vasco Fernandes Homem marched into the interior from Sofala, but most of the men were killed in fighting with Africans. In the late 16th and early 17th centuries the official Portuguese presence in the interior was limited to small trading colonies along the Zambezi. At the same time Portuguese adventurers began to establish control over large estates (called *prazos*), which resembled feudal kingdoms. They were ruled absolutely and often ruthlessly by their owners (called *prazeros*); Africans were forced to work on plantations, and considerable slave-raiding was undertaken (especially after 1650). Some of the *prazeros* maintained private armies, and they were generally independent of the Portuguese crown to which they were theoretically subordinate. From about 1628 the Portuguese gained increasing influence in Mwanamutapa, and they became intimately involved in the civil wars that led to the demise of that kingdom by the end of the 17th century. Mozambique was ruled as part of GOA in India until 1752, when it was given its own administration headed by a captain-general. Although the Portuguese helped introduce several American crops (notably maize and cashew nuts) that became staples of Mozambique's agriculture, the impact of their presence on African society was mainly destructive. From the mid-18th to the mid-19th century large numbers of Africans were exported as slaves, largely to the MASCARENE ISLANDS and to BRAZIL.

History: 1820 to 1895

In the 1820s and 1830s groups of Nguni-speaking people from S Africa invaded Mozambique; most of the Nguni continued northward into present-day Malawi and Tanzania, but one group, the Shangana, remained in S Mozambique, where they held effective control until the late 19th century. From the mid-19th century to the late 1880s the *mestiço* Joaquim José da Cruz and his son António Nicente controlled trade along the lower Zambezi. Thus, when the scramble for African territory among the European powers began in the 1880s, the Portuguese government had only an insecure hold on Mozambique. Nevertheless, Portugal tried to increase its nominal holdings, partly in an attempt to connect by land its territory in Mozambique and in ANGOLA (in SW Africa). Portuguese claims in present-day Zimbabwe and Malawi were strongly opposed by the British, who in 1890 delivered an ultimatum to Portugal demanding that it withdraw from these regions. Portugal complied, and in 1891 a treaty establishing the boundaries between British and Portuguese holdings in SE Africa was negotiated. Beginning in the 1890s and ending only around 1920, the Portuguese established their authority in Mozambique by force of arms against determined African resistance.

History: 1895 to 1951

Between 1895 and 1897 the Shangana were defeated; between 1897 and 1900 the Nyanja were conquered; in 1912 the Yao were pacified; and in 1917 control was established in extreme S Mozambique. In the 1890s several private companies were founded to develop and administer most of Mozambique. In 1910 the status of the territory was changed from province to colony. After the 1926 revolution in Portugal, the Portuguese government took a more direct interest in Mozambique. The companies lost the right to administer their regions, and at the same time the government furthered economic development by building railroads and by systematically forcing Africans to work on European-owned land. Portuguese colonial policy was based on the egalitarian theory of "assimilation": if an African became assimilated to Portuguese culture (i.e., if he was fluent in Portuguese, was Christian, and had a "good character"), he was to be given the same legal status as a Portuguese citizen. In practice, however, very few Africans qualified for citizenship (partly because there were inadequate educational opportunities), and they were directed to work for Europeans or to grow export crops.

History: 1951 to 1969

In 1951 the status of Mozambique was changed to "overseas province" in a move designed to indicate to world opinion that the territory would have increased autonomy; in a similar move in 1972, Mozambique was declared to be a "self-governing state." In both instances, however, Portugal maintained firm control over the territory. Between 1961 and 1963 several laws (one of which abolished forced labor) were passed to improve the living conditions of Africans. At the same time, many African nations were becoming independent, and nationalist sentiment was growing in Mozambique. In 1962 several nationalist groups were united to form the Mozambique Liberation Front (Frelimo), headed by Eduardo Mondlane. The Portuguese adamantly refused to give the territory independence, and in 1964 Frelimo initiated guerrilla warfare in N Mozambique. In 1969, Mondlane was assassinated in DAR ES SALAAM; he was succeeded by Uria Simango (1969) and then by Samora Moisès Machel (1970). By the early 1970s, Frelimo (which had a force of about 7,000 guerrillas) controlled much of central and N Mozambique and was engaged in often fierce fighting with the Portuguese (who maintained an army of about 60,000 in the territory). Frelimo's efficacy was hurt somewhat by internal dissension and by the defection of some of its leaders to the Portuguese side.

History: 1969 to 1979

Frelimo received aid from several foreign sources, including the government of Sweden (beginning in 1969). Roman Catholic missionaries accused the Portuguese of massacring about 400 inhabitants of the village of WIRIYAMU (near Tete) in December 1972; Portugal denied the charge. It was reported that beginning in mid-1973 Portugal had resettled about one million Africans in fortified villages to insulate them from Frelimo activities. In 1974 the government of Portugal was overthrown by the military. The new regime (which favored self-determination for all of Portugal's colonies) made an effort to resolve the conflict in Mozambique by implementing a number of reforms, by releasing political prisoners, by calling for a cease-fire, and by entering (June) into negotiations with Frelimo. The talks resulted in a mutual cease-fire (July 29) and an agreement (September 7) for Mozambique to become independent in June 1975. In reaction to the agreement, a group of white rebels attempted to seize control of the Mozambique government but were quickly subdued by Portuguese and Frelimo troops. As black rule of Mozambique became a reality (with Machel as president), the Portuguese settlers left in droves and they took their valuable skills, which had an adverse effect on the economy. Frelimo established a single-party Marxist state, nationalized all industry, and abolished private land ownership. Frelimo also instituted health and education reforms.

History: 1979 to 1986

Mozambique became a base for the nationalist rebels of the Zimbabwe African National Union (ZANU), a move that angered Rhodesia and South Africa. In 1979, Rhodesia invaded Mozambique, destroying communications facilities, agricultural centers, and transportation lines; many civilians were killed in the attacks. After Zimbabwe (formerly Rhodesia) obtained majority rule in 1980, the Mozambique National Resistance Movement (MNR or Renamo), a powerful dissident group financed in part by South Africa, waged guerrilla warfare against Frelimo. In addition to the chaos created by economic and political conditions, Mozambique was foundering under the weight of a large and inefficient bureacracy. In the 1980s, Machel cut the size of the government and began to privatize industry. In 1984 Mozambique signed a nonaggression pact (the Incomati accord) with South Africa; the terms of the pact prohibited South African support of Renamo and Mozambican support of the African National Congress (ANC). Mozambique accused South Africa of violating the accord and fighting continued between the government and Renamo throughout the 1980s.

History: 1986 to Present

In 1986 Machel was killed in a plane crash and succeeded by Joaquim Chissano. In 1992, Mozambique suffered from one of the worst droughts of the century and suffered from widespread famine. Renamo rebels, who controlled most of the nation's rural areas, blocked famine relief efforts and international peacekeeping forces were used to keep the Beira Railroad open. Civil war and starvation killed tens of thousands and over a million refugees have fled the country. The civil war ended with a peaceful accord between Frelimo and Renamo negotiated by the UN leading to multi-party elections. In free elections in 1994, Chissano won the presidency, and Frelimo won control of parliament. The government privatized a number of state-owned companies in the 1990s and appeared to be making progress in cutting inflation, stabilizing the currency, and stimulating economic growth. Repeat victories by Chissano and Frelimo in the 1999 elections were denounced as fraudulent by Afonso Dhlakama of Renamo; foreign observers were denied access to the vote-counting process. In February–March 2000, an estimated one million people in S Mozambique were affected by severe flooding due to a cyclone (hurricane). Frelimo and its presidential candidate, Armando Guebuza, again defeated Renamo and Dhlakama in the 2004 elections. Renamo again accused the government of fraud; foreign observers said that what fraud had occurred had not affected the outcome.

Government

Mozambique is governed under the constitution of 1990. The president, who is head of state, is elected by popular vote for a five-year term and is eligible for a second term. The government is headed by a prime minister, who is appointed by the president. The unicameral legislature consists of the 250-seat Assembly of the Republic, whose members are popularly elected for five-year terms. The current head of state is President Armando Guebuza (since February 2005). Prime Minister Luisa Diogo has been head of government since February 2004.

Mozambique, MOZAMBIQUE: see MOÇAMBIQUE, city.

Mozambique Channel (mo-zam-BEEK), strait in INDIAN OCEAN between MADAGASCAR and SE African mainland (MOZAMBIQUE); over 1,000 mi/1,609 km long, c. 600 mi/966 km wide in widest part (20°S lat.) Important shipping lane for E African navigation. Chief ports are Mozambique, BEIRA, and MAPUTO on mainland, MAJUNGA and TULÉAR on Madagascar's W coast.

Mozambique Current (mo-zam-BEEK), warm current flowing SW from INDIAN OCEAN, through MOZAMBIQUE CHANNEL, along coast of SE and S AFRICA; off Cape AGULHAS (S tip of Africa) it is deflected SE. S part also called AGULHAS CURRENT.

Mozarlândia (MO-sahr-LAHN-zhee-AH), city (2007 population 13,186), NW GOIÁS, BRAZIL, 31 mi/50 km NE of ARUANÁ; 14°50′S 50°36′W.

Mozdok (muhz-DOK), city (2006 population 44,105), N NORTH OSSETIAN REPUBLIC, SE European Russia, on railroad and highway junctions, on the left bank of the TEREK RIVER, 57 mi/92 km N of VLADIKAVKAZ; 43°45′N 44°39′E. Elevation 419 ft/127 m. In agricultural area; food industries (distilling, brewing, wine making, meat, milk). Pumping station on oil pipelines from MALGOBEK and GROZNYY. Founded in 1763 as a stronghold in the center of the fortified Caucasus line. City transferred from STAVROPOL Territory to North Ossetian Republic in 1944.

Mozema (mo-ZAI-mah), village, KOHIMA district, NAGALAND state, NE INDIA, in Naga Hills, 6 mi/9.7 km W of KOHIMA. Rice, cotton, oranges. Former stronghold of Naga tribes.

Mozhary (muh-ZHAH-ri), village, S RYAZAN oblast, central European Russia, on road junction, 40 mi/64 km ENE of RYAZHSK; 53°53′N 41°02′E. Elevation 377 ft/114 m. In agricultural area (wheat, hemp).

Mozhaysk (muh-ZHEISK), city (2006 population 31,595), SW MOSCOW oblast, central European Russia, on the MOSKVA River, on highway junction and railroad, 70 mi/113 km WSW of MOSCOW; 55°30′N 36°01′E. Elevation 744 ft/226 m. Pipe fitting, motor vehicle parts, medical instruments, woodworking, printing, clothing, food industries (dairy, meat packing). First mentioned in 1231, joined with Moscow principality in 1303, it became an important fortress and commercial center. Nearby is the site of the Battle of Borodino, September 7, 1812, between Napoleon's Grande Armeé and Russians under Gen. Mikhail Kutuzov. The city marked the furthest advance by the Germans, toward Moscow from the W, in World War II (October 15, 1941).

Mozhga (muhzh-GAH), city (2006 population 47,330), SW UDMURT REPUBLIC, E European Russia, on railroad (Syuginskaya station), 60 mi/97 km SW of IZHEVSK; 56°22′N 52°15′E. Elevation 456 ft/138 m. Tanning extracts, construction machinery parts, sawmilling, furniture, glassworking, flax processing, food processing (meat, dairy; bakery). Arose in 1835 as Syuginskiy, around a glassworks factory; renamed Krasnyy around 1920. In 1926, absorbed adjacent Syuginskiy Zavod, was granted city status and renamed Mozhga.

Mozirje (mo-ZIR-ye), village, N SLOVENIA, onsavinja RIVER, and 16 mi/26 km WNW of CELJE; 46°20′N 14°57′E. Local trade center; apparel industry. Summer resort. Has old monastery. Formerly called Prihova.

Mozoncillo (mo-thon-THEE-lyo), town, SEGOVIA province, central SPAIN, 12 mi/19 km NNW of SEGOVIA; 41°09′N 04°11′W. Cereals, grapes, chicory; flour milling; lumbering. Naval stores.

Mozonte (mo-ZON-te), town, NUEVA SEGOVIA department, NW NICARAGUA, 2 mi/3.2 km ENE of OCOTAL; 13°39′N 86°27′W. Coffee, sugarcane, livestock.

Mozyr' (MO-zir), city, GOMEL oblast, BELARUS, ☉ Mozyr' region, river port at Pkhov, on PRIPET River, and 145 mi/233 km SE of MINSK; 52°03′N 29°15′E. Railroad station. Irrigation and drainage machinery and cables produced; truck repair, foundry; other manufacturing (machinery, clothing, art items; food processing); oil refinery. Teachers' college.

Mpakani (uhm-pa-KAH-nee), village, PWANI region, E TANZANIA, 25 mi/40 km SSE of DAR ES SALAAM, on INDIAN OCEAN, near Ras Pembamnasi; 07°10′S 39°29′E. Fish; sisal, cashews, bananas, copra; sheep, goats.

MPal (uhm-PAHL), village, SAINT-LOUIS administrative region, NW SENEGAL, on Saint-Louis railroad, and 16 mi/26 km ESE of SAINT-LOUIS; 15°26′N 16°13′W. Peanut-growing; livestock raising. Sometimes spelled Pal or M'Pal.

Mpala (uhm-PAH-lah), village, KATANGA province, SE CONGO, on W shore of LAKE TANGANYIKA, 55 mi/89 km SSE of KALEMIE; 06°45′S 29°31′E. Elev. 2,542 ft/774 m. Cattle; vegetables. Roman Catholic mission with school of marine trades. Founded 1882. Also spelled PALA.

Mpanda (uhm-PAN-dah), village, RUKWA region, W TANZANIA, 112 mi/180 km NNW of SUMBAWANGA, terminus of railroad spur form KALIUA, to NE. Road junction; airstrip. KATAVI PLAINS NATIONAL PARK to S. Railroad line constructed to serve lead-mining operations; lead discovered in 1946, later abandoned. Gold still mined.

Mpanganya (uhm-pan-GAH-nyah), village, PWANI region, E TANZANIA, 75 mi/121 km SSW of DAR ES SALAAM, near RUFIJI RIVER; 09°54′S 38°43′E. Rice, bananas; sheep, goats.

Mpese (MPES-sai), village, BAS-CONGO province, W CONGO, 60 mi/97 km SSE of KINSHASA; 05°14′S 15°33′E. Elev. 1,889 ft/575 m.

Mpessoba (uhm-PES-o-bah), village, THIRD REGION/SIKASSO, S MALI, 150 mi/241 km E of BAMAKO; 12°40′N 05°43′W.

Mphoengs (uhm-PO-engs), town, MATABELELAND NORTH province, SW ZIMBABWE, 88 mi/142 km SW of BULAWAYO, near Ramaquabine River (BOTSWANA border); 21°11′S 27°53′E. Road terminus. Cattle, sheep, goats; corn. Also spelled M'phoengs.

Mpigi, administrative district (2005 population 429,500), CENTRAL region, S UGANDA; ☉ MPIGI; 00°05′N 32°00′E. As of Uganda's division into eighty districts, borders MASAKA (S), SEMBABULE (SWW), MUBENDE (NWN), MITYANA (N), and WAKISO (E) districts and Lake VICTORIA (SE). Agricultural area (including bananas, coffee, cotton, millet, and sugarcane). Mpigi town on road between KAMPALA city and MASAKA town. Formed in 2000 from W portion of former MPIGI district (Wakiso district formed from E portion).

Mpigi (uhm-PEE-gee), former administrative district (□ 2,402 sq mi/6,245.2 sq km), S UGANDA, along N shore of LAKE VICTORIA (to S), W of KAMPALA; capital was MPIGI; 00°10′N 32°15′E. As of Uganda's division into thirty-nine districts, was bordered by MASAKA (W), MUBENDE and LUWERO (N), and KAMPALA and MUKONO (E) districts. Agricultural area (tea, coffee, bananas). E section included ENTEBBE and suburbs of Kampala. In 2000 W portion of district was formed into current MPIGI district and E portion formed into WAKISO district.

Mpigi (uhm-PEE-gee), town (2002 population 10,272), ☉ MPIGI district, CENTRAL region, S UGANDA, 15 mi/24 km NW of ENTEBBE. Road junction; cotton, coffee, sugarcane, bananas, millet. Was part of former CENTRAL province.

Mpika (uhm-PEE-kah), township, NORTHERN province, NE central ZAMBIA, in MUCHINGA MOUNTAINS, 120 mi/193 km SSE of KASAMA. Road junction; on Tazara railroad. Corn, cattle. Isangano National Park to NW; NORTH LUANGWA NATIONAL PARK to E; SOUTH LUANGWA NATIONAL PARK to SE.

Mpitimbi (uhm-pee-TEEM-bee), village, RUVUMA region, S TANZANIA, 13 mi/21 km SSW of SONGEA; 10°50′S 35°31′E. Timber; tobacco, subsistence crops; livestock.

M'poko River (uhm-po-KO), 160 mi/257 km long, SW CENTRAL AFRICAN REPUBLIC; rises in OUHAM prefecture 30 mi/48 km S of BOSSANGOA; flows generally SSW through OMBELLA-M'POKO prefecture, to UBANGI RIVER 10 mi/16 km SW of BANGUI.

Mporokoso (uhm-po-lo-KO-so), township, NORTHERN province, N ZAMBIA, 100 mi/161 km NW of KASAMA; 09°23′S 30°08′E. Road junction. Agriculture (corn, coffee). NSUMBU NATIONAL to N; MWERU WANTIPA NATIONAL PARK to NW.

M'Pouia, Congo Republic: see MPOUYA.

Mpouya (uhm-poo-YAH), village, PLATEAUX region, E central Congo Republic, steamboat landing on CONGO RIVER, and 125 mi/201 km NE of BRAZZAVILLE. Has trypanosomiasis and leprosy treating station. Also spelled M'Pouia.

Mpraeso (uhm-PRAI-so), town, local council headquarters, EASTERN REGION, GHANA, near NKAWKAW railroad station, 45 mi/72 km NW of KOFORIDUA, on main road between ACCRA and KUMASI; 06°35′N 00°44′E. Bauxite mining. Mines on nearby Mount EJUANEMA were developed during early 1940s.

Mpui (uhm-POO-ee), village, RUKWA region, W TANZANIA, 35 mi/56 km SSE of SUMBAWANGA; 08°23′S 31°50′E. Road junction. Sheep, goats; wheat, corn.

Mpulungu (uhm-poo-LOO-eng-goo), township, NORTHERN province, NE ZAMBIA, port at S tip of LAKE TANGANYIKA, 20 mi/32 km WNW of ABERCORN; 08°46′S 31°08′E. Zambia's only port. Resort area; fishing base. Agriculture (coffee). Airfield and air service at Kasaba Bay Lodge to NW. KALAMBO FALLS to NE, on TANZANIA border; NSUMBU NATIONAL PARK to W.

Mpumalanga, province (□ 30,259 sq mi/78,673.4 sq km; 2007 population 3,536,300), NE SOUTH AFRICA; ☉ NELSPRUIT; 26°00′S 30°00′E. Former E TRANSVAAL, includes S Lowveld and S part of KRUGER NATIONAL park, bordered to E by SWAZILAND and MOZAMBIQUE, to S by KWAZULU-NATAL and FREE STATE provinces, to W by GAUTENG province, and to N by LIMPOPO province. Edge of escarpment and subtropical Lowveld area; most extensively watered and forest region of the republic. Rich in minerals. Its area is 6.4% of the republic's total land area.

Mpumalanga, town, KWAZULU-NATAL, SOUTH AFRICA, 25 mi/40 km W of DURBAN near Cato Ridge, and 19 mi/30 km SE of MSUNDUZI (Pietermaritzburg). Large dormitory town for workers employed in DURBAN and Pietermaritzburg, on main railroad link between the 2 cities in Mlazi River valley. Rich agricultural area producing poultry and dairy products. Known as "Little Beirut" during late 1980s because of fighting between ANC and Inkatha supports.

Mpurukasese (uhm-poo-roo-KAH-sai-sai), village, RUVUMA region, S TANZANIA, 60 mi/97 km NE of SONGEA; 10°14′S 36°28′E. Timber; grain; livestock.

Mputi, river, SOUTH AFRICA: see IDUTYWA.

Mpwapwa (uhm-PWAH-pwah), town, DODOMA region, E central TANZANIA, 50 mi/80 km ESE of DODOMA, near Mkondoa River; 06°24′S 36°40′E. Road junction. Peanuts, wheat, corn; cattle, sheep, goats. Limestone and phosphate deposits.

Mracaj, BOSNIA AND HERZEGOVINA: see GORNJI VAKUF.

Mragowo (mron-GO-vo), Polish *Mrągowo,* German *Sensburg,* town (2002 population 22,074), in EAST PRUSSIA, after 1945 in OLSZTYN province, NE POLAND, in Masurian Lakes region, 35 mi/56 km ENE of OLSZTYN (Allenstein); 53°52′N 21°18′E. Sawmilling. Founded 1348 by Teutonic Knights. In World War II, c.85% destroyed.

Mraijat, LEBANON: see MUREIJAT.

Mrakotin, CZECH REPUBLIC: see TELC.

Mrakovo (MRAH-kuh-vuh), village (2005 population 8,400), SW BASHKORTOSTAN Republic, on the W slopes of the S URALS, E European Russia, on the GREATER IK RIVER, 60 mi/97 km SSE of STERLITAMAK; 53°47′N 56°11′E. Elevation 360 ft/109 m. Flour mill. Also known as Pervoye Mrakovo [Russian=first Mrakovo], with another village named Mrakovo located further to the SE.

Mramoren (MRAH-mo-ren), village, MONTANA oblast, VRATSA obshtina, BULGARIA; 43°18′N 23°41′E.

Mrcajevci (MUHRTS-ei-yev-tsee), village, W central SERBIA, 9 mi/14.5 km ESE of CACAK; 43°52′N 20°31′E. Also spelled Mrchayevtsi.

Mrewa, ZIMBABWE: see MUREWA.

Mrežnica River (MREZH-nee-tsah), c.40 mi/64 km long, central CROATIA; rises 5 mi/8 km WSW of Slunj, in the KORDUN region; flows N, past DUGA RESA, to KORANA RIVER 2 mi/3.2 km S of KARLOVAC. Known for its scenic beauty, it is an important recreational area, especially for ZAGREB and Karlovac residents; hydroelectric plant.

Mrijo (m-REE-jo), village, DODOMA region, N central TANZANIA, 80 mi/129 km NNE of DODOMA, in W edge of MASAI STEPPE; 05°09′S 36°16′E. Livestock; wheat, corn.

Mrirasandu, UGANDA: see MWIRASANDU.

M'rirt (mah-RIRT), city, Khénifra province, Meknès-Tafilalet administrative region, MOROCCO, 16 mi/25 km NNE of KHENIFRA. Administrative and agricultural marketing center.

Mrkonjić Grad (muhr-ko-NEECH GRAHD), town, central BOSNIA, BOSNIA AND HERZEGOVINA, on Crna Reka River (tributary of VRBAS RIVER), and 20 mi/32 km S of Banja Luka; 44°24′N 17°05′E. Copper mining. Until 1930s, called Varcar Vakuf. Also spelled Mrkonyich Grad.

Mrocza (MROCH-nah), German *Mrotschen,* town, Bydgoszcz province, NW POLAND, 18 mi/29 km WNW of BYDGOSZCZ; 53°15′N 17°37′E. Flour and sawmilling.

Mruvila (uhm-ROO-vee-lah), village, KIGOMA region, W TANZANIA, 25 mi/40 km E of KIGOMA; 04°54′S 29°57′E. Timber; subsistence crops; livestock.

Msagali (uhm-sah-GAH-lee), village, DODOMA region, E central TANZANIA, 40 mi/64 km ESE of DODOMA, near Mkondoa River, on railroad; 07°11′S 36°22′E. Cattle, sheep, goats; wheat, corn, peanuts.

Msalato (uhm-sah-LAH-to), village, SINGIDA region, central TANZANIA, 35 mi/56 km WSW of MANYONI; 05°56′S 34°20′E. Timber; sheep, goats; corn, wheat.

Msanga (uhm-SAHN-gah), village, DODOMA region, central TANZANIA, 20 mi/32 km ENE of DODOMA; 06°04′S 36°02′E. Cattle, sheep, goats; peanuts, corn, wheat.

Msangasi River (uhm-san-GAH-see), c. 185 mi/298 km long, TANGA region, E TANZANIA; rises c. 110 mi/177 km WSW of TANGA; flows E, enters ZANZIBAR CHANNEL, INDIAN OCEAN, 40 mi/64 km SSW of Tanga.

Msasani (uhm-sah-SAH-nee), town, DAR ES SALAAM, suburb 5 mi/8 km N of DAR ES SALAAM, E TANZANIA, on ZANZIBAR CHANNEL, INDIAN OCEAN; 06°46′S 39°16′E. Government meat-packing plant to NW.

Mscheno, CZECH REPUBLIC: see MSENO.

Msene-lazne (MSHE-ne–LAHZ-nye), Czech *Mšené-lázně,* German *Bad Mscheno,* village, SEVEROCESKY province, N central BOHEMIA, CZECH REPUBLIC, on railroad, and 23 mi/37 km NNW of PRAGUE; 50°22′N 14°07′E. Health resort with peat baths. Founded in the 18th century.

Mseno (MSHE-no), Czech *Mšeno,* German *Mscheno,* town, STREDOCESKY province, N central BOHEMIA, CZECH REPUBLIC, on railroad, and 9 mi/14.5 km NE of MELNIK; 50°27′N 14°38′E. Manufacturing of baby carriages; malthouse. Summer resort, museum.

Mshchonov, POLAND: see MSZCZONOW.

Msida (mi-SEE-dah), town (2005 population 7,629), E central MALTA, at head of MEDITERRANEAN inlet, 1.5 mi/2.4 km W of VALLETTA; 35°53′N 14°28′E. Boat repairing, fishing. Its monuments escaped severe damage during World War II. Sometimes spelled Misida.

M'Sila (muh-see-LAH), wilaya, on the High Plains of interior central ALGERIA; ⊙ M'SILA; 35°20′N 04°20′E. An important crossroads among the central, W, E, and S parts of the country. Rich cereal-producing region; livestock rearing. Large-scale irrigation comes from the Oued K'sob Dam, 10 mi/16 km N, at S edge of HODNA MOUNTAINS. Here was launched the third phase of Algeria's agricultural revolution in the 1970s, which aimed at limiting herd ownership and giving herdsmen access to modern amenities.

M'Sila (muh-see-LAH), town, ⊙ M'Sila wilaya, central ALGERIA, in the HODNA depression, SE of ALGIERS, and 60 mi/97 km SW of SETIF; 35°40′N 04°31′E. Important wilaya administrative center and major crossroads connecting Bousâada, BATNA, and BORDJ BOU ARRÉRIDJ. Horse and sheep market; leather working. Founded by Fatimids; frequently visited by medieval historian Ibn Khaldoun. First head of state, Ahmed Ben Bella, placed under house arrest here after 1965 coup; town also survived an earthquake the same year. Many Roman remains in the vicinity.

Msindaji (uhm-seen-DAH-jee), village, PWANI region, E TANZANIA, 45 mi/72 km N of DAR ES SALAAM, on MAFIA CHANNEL, INDIAN OCEAN; 07°37′S 38°16′E. Fish; livestock; cashews, sisal, bananas, copra.

Msonedi, town, MASHONALAND CENTRAL province, NE ZIMBABWE, 48 mi/77 km WNW of HARARE; 17°08′S 30°53′E. Mvurwi range to W. Cattle, sheep, goats; tobacco, corn, soybeans. Also spelled Msonneddi.

Msoro, village, EASTERN province, E ZAMBIA, 55 mi/89 km W of CHIPATA; 13°36′S 31°54′E. Agriculture (tobacco, corn). Mission. SOUTH LUANGWA NATIONAL PARK to NW.

Msta (MSTAH), river, approximately 280 mi/451 km long, in TVER oblast, W European Russia; rising N of VYSHNIY VOLOCHEK; flowing generally NW into Lake IL'MEN near NOVGOROD (NOVGOROD oblast). Navigable in its lower course, it is included in the VYSHNEVOLOTSK canal system.

Mstera (MSTYE-rah), town (2006 population 5,430), NE VLADIMIR oblast, central European Russia, near the KLYAZ'MA RIVER, on road, 13 mi/21 km NW of VYAZNIKI; 56°22′N 41°55′E. Noted handicraft center developed out of religious icon painting (gold-leaf

working, miniature painting on lacquerware, stitching, jewelry making). Religious art museum.

Mstikhino (MSTEE-hee-nuh), settlement, central KALUGA oblast, central European Russia, on the OKA River, on road and near railroad, 6 mi/10 km W of KALUGA; 54°33′N 36°06′E. Elevation 518 ft/157 m. A weekend retreat for residents of Kaluga and neighboring towns; has a resort. Gas pipeline in the vicinity.

Mstislavl' (msti-SLAH-vuhl), city, E MOGILEV oblast, BELARUS, ⊙ Mstislavl' region, on the Vikhra River (DNIEPER basin), and 60 mi/97 km ENE of MOGILEV, 54°02′N 31°44′E. Manufacturing (food processing, flax processing). The oldest Russian city of Smolensk Land, mentioned in the chronicle in 1156. In 1180 it was center of principality. Noteworthy architectural monuments include Jesuit monastery with a Roman Catholic church and a collegium (17th and 18th centuries), a carmelite church with mural paintings (1654), the Troitskaia Church (19th century), Alexander Nevsky Church (19th century).

Msuna, village, MATABELELAND NORTH province, NW ZIMBABWE, 35 mi/56 km NE of HWANGE, on ZAMBEZI River (ZAMBIA border), upstream (SW of) headwaters of Lake KARIBA reservoir, at mouth of GWAYI RIVER; 18°00′S 26°53′E. Road terminus. Located in Msuna Safari Area. Livestock; fish.

Msunduzi, municipality (2001 population 553,223), KWAZULU/NATAL province, SOUTH AFRICA. Forged from PIETERMARITZBURG and surrounding areas as part of a movement to revive native place-names; often used colloquially for Pietermaritzburg.

Mszczonow (msh-CHO-noov), Polish *Mszczonów,* Russian *Mshchonov,* town, Skierniewice province, E central POLAND, 28 mi/45 km SW of WARSAW; 51°59′N 20°31′E. Manufacturing of matches, brewing, flour milling; grain trade.

Mtaka taka (uhm-tah-kah tak-kah), village, SALIMA district, Central region, MALAWI, on railroad, and 23 mi/37 km NE of DEDZA; 14°14′S 34°32′E. Tobacco, corn.

Mtakuja (uhm-tah-KOO-jah), town, RUKWA region, W TANZANIA, 75 mi/121 km NW of SUMBAWANGA, on Lake TANGANYIKA; 05°20′S 30°36′E. Lake port. Fish; livestock; grain.

Mtambo River (uhm-TAM-bo), c. 75 mi/121 km long, RUKWA region, W TANZANIA; rises c. 100 mi/161 km N of SUMBAWANGA; flows NNW, enters UGALLA RIVER 120 mi/193 km SE of KIGOMA.

Mtandawala (uhm-tan-da-WAH-lah), village, LINDI region, SE TANZANIA, 30 mi/48 km WSW of KILWA MASOKO, near Mavaji River; 09°09′S 39°08′E. Cashews, bananas, copra, manioc; goats, sheep.

Mtandi (uhm-TAN-dee), village, LINDI region, SE TANZANIA, 40 mi/64 km NNW of LINDI, near INDIAN OCEAN; 09°26′S 39°33′E. Cashews, peanuts, bananas, copra, manioc; goats, sheep.

Mtandura (uhm-tan-DOO-rah), village, LINDI region, SE TANZANIA, 10 mi/16 km S of KILWA MASOKO, on Sangurungu Haven (bay), INDIAN OCEAN; 09°07′S 39°33′E. Fish; cashews, copra, bananas, manioc; livestock.

Mtarazi Falls National Park, MANICALAND province, E ZIMBABWE, 35 mi/56 km NNE of MUTARE, in S part of NYANGA MOUNTAINS, near MOZAMBIQUE; 18°19′S 32°48′E. Mtararzi Falls (2,500 ft/762 m) on Honde River, near its source.

Mtera (uhm-TAI-rah), village, IRINGA region, central TANZANIA, 60 mi/97 km SSE of DODOMA, on GREAT RUAHA RIVER; 07°06′S 35°57′E. MTERA RESERVOIR to W. Cattle, sheep, goats; corn, wheat.

Mtera Reservoir (uhm-TAI-rah), DODOMA and IRINGA regions, central TANZANIA, 55 mi/89 km S of DODOMA; formed by Mtera Dam (07°06′S 35°57′E) on GREAT RUAHA RIVER, which enters lake from SW and drains to E; KISIGO RIVER, from W, and Bubu River, from N, form arms of lake.

Mtito Andei (uhm-TEE-to ahn-DAI-ee), town (1999 population 4,304), EASTERN province, S central

KENYA, on railroad, and 140 mi/225 km SE of NAIROBI; 02°41′S 38°10′E. Magnesite mining; sapphire and co-rundum deposits. Trading and transport center.

Mtoko, ZIMBABWE: see MUTOKO.

Mtondo (uhm-TON-do), village, LINDI region, SE central TANZANIA, 100 mi/161 km WNW of LINDI, near Nganga River; 09°36′S 38°19′E. Livestock; sub-sistence crops.

Mto Wa Mbu (uhm-to WAHM-boo), village, ARUSHA region, N central TANZANIA, 55 mi/89 km W of AR-USHA, at N end of Lake MANYARA; 03°23′S 35°53′E. LAKE MANYARA NATIONAL PARK to SW, Olkerii Escarpment to NE, NGORONGORO CRATER to NW. Road junction, airstrip. Tourism. Livestock, grain.

Mtsamboro, town (□ 5 sq mi/13 sq km), MAYOTTE island, MAYOTTE territory (France), Comoros Islands, MOZAMBIQUE CHANNEL, INDIAN OCEAN, 12 mi/19 km WNW of MAMOUDZOU, at NW end of island; 12°41′S 45°03′E. Fish; livestock; ylang-ylang, vanilla, coconuts. Mtzamboro island 4 mi/6.4 km to NW.

Mtsensk (MTSENSK), city (2006 population 47,550), N OREL oblast, SW European Russia, on the Zusha River (tributary of the OKA River), on highway and railroad, 30 mi/48 km NE of ORËL; 53°16′N 36°33′E. Elevation 551 ft/167 m. Aluminum casting, secondary non-ferrous metals, machinery. Chartered in 1147. Made city in 1778.

Mtskheta (muhts-KHE-tah), town (2002 population 7,718), W central GEORGIA, on the KURA River, and the GEORGIAN MILITARY ROAD; 41°50′N 44°41′E. It was the capital of ancient IBERIA until the 6th century C.E., when the capital was moved to TBILISI; Mtskheta remained the religious center of the country. The Sveti-Tskhoveli cathedral (11th century; destroyed by Timur; rebuilt 15th century) contains the burial vaults of Georgian rulers. The Samtavro cathedral (11th century) was restored in 1903. In the hills near the town are ruins of the Dzhvari temple of the late 6th or early 7th century. Hydroelectric power plant nearby.

Mtua (uhm-TOO-ah), village, LINDI region, SE TAN-ZANIA, 20 mi/32 km SW of LINDI, on LUKULEDI RIVER; 10°13′S 39°27′E. Cashews, peanuts, grain, bananas, sweet potatoes; sheep, goats.

Mtubatuba, town, N KWAZULU-NATAL, SOUTH AFRICA, 25 mi/40 km NN of RICHARD′S BAY, and 15 mi/24 km inland of ST. LUCIA on the MFOLOZI RIVER trading center in rich sugar-producing area; plantations of Wattles and Eucalyptus provide for a thriving lumber industry. Railroad link and access point to Umfolozi and St. Lucia nature reserves.

Mount Vale, mountain, SOUTH AFRICA: see KEISKAMMA RIVER.

Mtwara (uhm-TWAH-rah), region (2006 population 1,220,000), SE TANZANIA, ⊙ MTWARA, bounded S by RUVUMA River (MOZAMBIQUE border); E by INDIAN OCEAN; MAKONDE PLATEAU in center. Fish; timber; grain, cashews, copra, bananas. Part of former Mtwara province.

Mtwara (uhm-TWAH-rah), town, ⊙ MTWARA region, SE TANZANIA, 40 mi/64 km SE of LINDI, on MIKINDANI Bay, INDIAN OCEAN; 10°12′S 40°11′E. MOZAMBIQUE border 15 mi/24 km to SE. Seaport; international air-port; road junction. Timber; fish; cashews, peanuts, bananas, copra; livestock.

Mu, MYANMAR: see MU RIVER.

Mu: see ATLANTIS.

Muai, Mae, MYANMAR and THAILAND: see THAUNGYIN RIVER.

Mu'amaltayn, LEBANON: see MA'AMELTIN.

Muan (MOO-AHN), county (□ 166 sq mi/431.6 sq km), W central SOUTH CHOLLA province, SOUTH KOREA, adjacent to HAMPYONG on N, YONG-AM on E, MOKPO on S, and SHINAN on W. Low mountains in SW; Muan, Mangun, and Haeje peninsulas. Agriculture (rice, sweet potatoes, onions, tobacco, rape); seaweed and shell cultivation. Honam railroad and freeway

between KWANGJU and Mokpo pass. Known for coastal scenery.

Muaná (moo-ah-NAH), city (2007 population 28,813), E PARÁ, BRAZIL, near S shore of MARAJÓ island in AMAZON delta, 45 mi/72 km W of BELÉM; 01°25′S 49°10′W. Cattle and horse raising.

Muanda, CONGO: see MOANDA.

Muang [Thai=town], for names in THAILAND begin-ning thus: see under following part of the name.

Muang Samakhi Xai, Laos: see ATTAPU, town.

Muang Sen, VIETNAM: see KY SON.

Muanza District, MOZAMBIQUE: see SOFALA.

Muar (MOO-ahr), city (2000 population 102,273), NW Johore, SW MALAYSIA, port on Strait of MALACCA at mouth of MUAR RIVER, and 90 mi/145 km NW of Johore Bharu; 02°02′N 102°34′E. Rubber, coconuts, oil-palm, bananas; fisheries. Airfield; traditional Ma-laysian cultural center. Formerly Bandar Maharani.

Muara (moo-WAH-rah), town, BRUNEI MUARA dis-trict, NE BRUNEI DARUSSALAM, NW BORNEO, 12 mi/19 km NE of BANDAR SERI BEGAWAN, on Brunei Bay, near entrance to SOUTH CHINA SEA. Agriculture (rice, cassava, vegetables), livestock, chickens; fishing. Bru-nei′s primary international port with oil and natural-gas storage and procesing facilities. Coal was mined here 1888–1927. Completely destroyed during World War II. Site of Mergang Beach.

Muaraenim (mwah-rah-AI-nim), town, ⊙ Liot district, Sumatra Selatan province, SW SUMATRA, INDONESIA, on Lematang River (tributary of MUSI RIVER), and 80 mi/129 km SW of PALEMBANG; on spur of Palembang–BANDAR LAMPUNG railroad; 03°39′S 103°48′E. Trade center for oil-producing and coal-mining region. Also spelled Moeraenim.

Muarasiberut (mwah-rah-see-buh-ROOT), town, Sumatra Barat province, INDONESIA, on SE coast of SIBERUT island; 01°36′S 99°11′E. Also spelled Moer-asiberuet.

Muarasikabaluan (mwah-rah-see-kah-bah-LWAHN), town, Sumatra Barat province, MENTAWAI ISLANDS, INDONESIA, on NE coast of SIBERUT island; 01°07′S 98°59′E. Also spelled Muarasikabaluan.

Muarateweh (MWAH-rah-TAI-WAI), town, ⊙ Barito Utara district, Kalimantan Tengah province, central BORNEO, INDONESIA, on Barito River, 180 mi/290 km WNW of BALIKPAPAN. Extensive coal deposits nearby. Airport.

Muar River (MOO-ahr), 140 mi/225 km long, S MA-LAYSIA; rises in S outliers of central Malayan range E of SEREMBAN; flows E, past KUALA PILAH, and in an arc S through NW JOHOR rubber-growing district to Strait of MALACCA at MUAR; 02°05′N 102°35′E. Na-vigable in lower course.

Mu'awiye (muh-AH-wi-ye), Arab village, ISRAEL, 12.5 mi/20 km NE of HADERA, and SW of MEGIDDO; 32°31′N 35°06′E. Elevation 1,082 ft/329 m. Named for Sheikh Mu'awiye, whose burial site lies here.

Mubarakeh, IRAN: see MOBARAKEH.

Mubarakpur (moo-BAH-ruhk-POOR), town, AZAM-GARH district, E UTTAR PRADESH state, N central INDIA, 7 mi/11.3 km ENE of AZAMGARH. Hand-loom cotton-weaving center; silk weaving, sugar milling.

Mubarakpur (muh-BAH-rak-puhr), town, BAHA-WALNAGAR district, BAHAWALPUR division, PUNJAB province, central PAKISTAN, 20 mi/32 km SW of BA-HAWALPUR; 29°15′N 71°22′E.

Mubarraz, SAUDI ARABIA: see HOFUF.

Mubayira (moo-bah-YEE-rah), town, MASHONALAND WEST province, central ZIMBABWE, 45 mi/72 km SW of HARARE, on Nyangweni River, branch of MUPFURE RIVER; 18°19′S 30°35′E. Cattle, sheep, goats, hogs; to-bacco, coffee, tea, cotton, corn, wheat, soybeans, sorghum.

Mubende, administrative district, CENTRAL region, S central UGANDA; ⊙ MUBENDE. As of Uganda's division into eighty districts, borders KIBOGA (N), MITYANA (E), MPIGI and SEMBABULE (S), KYENJOJO (W), and

KIBAALE (NW) districts. Agricultural area (including bananas and coffee). Some fishing in Lake Wamala in SE of district. Railroad between KASESE town and KAMPALA city (and continuing E then SE to MOMBASA [KENYA]) travels through S part of district. Main road between Kyenjojo town and Kampala city runs W-E through Mubende town. Formed in 2005 from W and central portions of former MUBENDE district (Mi-tyana district formed from E portion).

Mubende (MOO-ben-dai), former administrative dis-trict (□ 2,392 sq mi/6,219.2 sq km; 2005 population 763,000), CENTRAL region, S UGANDA, midway be-tween KAMPALA and FORT PORTAL; capital was MU-BENDE; 00°30′N 31°40′E. As of Uganda's division into fifty-six districts, was bordered by KIBAALE (NW), KIBOGA (N), LUWERO (NE), WAKISO (E), MPIGI (S), SEMBABULE (SW), and KYENJOJO (W) districts. Pri-mary inhabitants were the Baganda people. Primarily agricultural area (included large coffee and tea plan-tations, also bananas, beans, cassava, cotton, ground-nuts, maize, passion fruit, pineapples, potatoes, sweet potatoes, tobacco, and tomatoes). Livestock also im-portant (cattle, goats, poultry, rabbits, and sheep). Some fishing in Lake Wamala in E of district. MI-TYANA was largest town. In 2005 E portion of district was carved out to form MITYANA district and remain-der of district was formed into current MUBENDE district.

Mubende (MOO-ben-dai), town (2002 population 15,996), ⊙ MUBENDE district, CENTRAL region, W central UGANDA, 85 mi/137 km WNW of KAMPALA. Agricultural trade center (cotton, coffee, bananas, corn). Tungsten, beryl, tantalite deposits. Was part of former Buganda province.

Mubi (MOOM-bee), town, ADAMAWA state, N NI-GERIA, 90 mi/145 km NE of YOLA; 10°16′N 13°16′E. Peanuts, pepper, hemp, rice, cotton; cattle, skins.

Mucajaí (MOO-kah-zhah-ee), city (2007 population 12,546), central RORAIMA state, BRAZIL, 25 mi/40 km S of BOA VISTA; 02°25′N 60°52′W.

Mucambo (MOO-kahm-bo), city (2007 population 14,235), NW CEARÁ, BRAZIL, 13 mi/21 km E of Ibia-pana; 03°55′S 40°48′W.

Much (MOOKH), town, RHINELAND, North Rhine-Westphalia, W GERMANY, 20 mi/32 km E of COLOGNE; 50°55′N 07°25′E.

Muchachos, Picos de los (PEE-kos dhai los moo-CHAH-chos), small volcanic range in N PALMA is-land, CANARY ISLANDS, rising to 7,730 ft/2,356 m.

Muchalat Inlet (moo-CHA-lit), bay, SW BRITISH CO-LUMBIA, W Canada, in W VANCOUVER ISLAND, W arm (24 mi/39 km long) of NOOTKA SOUND, 50 mi/80 km W of COURTENAY; 49°39′N 126°14′W.

Muchamiel (moo-chah-MYEL), town, ALICANTE prov-ince, E SPAIN, near the MEDITERRANEAN SEA, 6 mi/9.7 km NE of ALICANTE; 38°25′N 00°26′W. In irrigated area yielding produce, olive oil, wine, cereals.

Mücheln (MYOO-kheln), town, SAXONY-ANHALT, central GERMANY, 10 mi/16 km WSW of MERSEBURG; 51°19′N 11°48′E. Lignite mined; chemicals. Has 16th-century town hall.

Much Hadham (MUHCH HAD-huhm), village (2001 population 2,784), E HERTFORDSHIRE, E ENGLAND, 4 mi/6.4 km WSW of BISHOP′S STORTFORD; 51°51′N 00°04′E. Has 13th-century church.

Muchinga Mountains (moo-CHI-eng-gah), NORTH-ERN and CENTRAL provinces, NE ZAMBIA, between LUANGWA and LUAPULA river watersheds. Section of great African escarpment along a spur of GREAT RIFT VALLEY, extending 300 mi/483 km NNE-SSW, from ISOKA to SERENJE. Elevation over 6,000 ft/1,829 m.

Muchkapskiy (mooch-KAHP-skeeyee), town (2006 population 7,740), SE TAMBOV oblast, S central Eu-ropean Russia, on the VORONA RIVER (DON River basin), on crossroads and railroad, 70 mi/113 km SE of TAMBOV; 51°50′N 42°28′E. Elevation 406 ft/123 m. Flour milling, bakery, poultry and meat processing.

Area is shown by the symbol □, and capital city or county seat by ⊙.

Muchuan (MU-CHWAH), town, ⊙ Muchuan county, SW SICHUAN province, CHINA, 33 mi/53 km NW of YIBIN, in mountain area; 28°56′N 103°58′E. Grain, medicinal herbs, sugarcane.

Muchumí, PERU: see MOCHUMÍ.

Mucientes (moo-THYEN-tes), town, VALLADOLID province, N central SPAIN, 7 mi/11.3 km NNW of VALLADOLID; 41°45′N 04°45′W. Cheese processing, plaster manufacturing.

Muck (MUK), island (2.5 mi/4 km long, 1.5 mi/2.4 km wide), INNER HEBRIDES, W HIGHLAND, N Scotland, 3 mi/4.8 km SW of EIGG; 56°50′N 06°14′W. Rises to 451 ft/137 m.

Muckalee Creek (MUHK-uh-lee), c.65 mi/105 km long, SW central GEORGIA; rises SE of BUENA VISTA; flows SSE, past AMERICUS, to KINCHAFOONEE RIVER just N of ALBANY; 32°18′N 84°31′W.

Mücke (MYOOK-ke), town, HESSE, central GERMANY, on N slope of the Vogelsberg, 17 mi/27 km ENE of GIESSEN; 50°38′N 09°01′E. Forestry; metalworking.

Muckish (muh-KISH), mountain (2,197 ft/670 m) in the DERRYVEAGH MOUNTAINS, N DONEGAL county, N IRELAND, 6 mi/9.7 km SSW of DUNFANAGHY.

Muckle Flugga (MUH-kuhl FLUH-gah), islet, northernmost of the SHETLAND ISLANDS, extreme N Scotland, just N of UNST island; 60°51′N 00°53′W. Site of lighthouse. Island is most northerly habitation of Great Britain.

Muckle Roe (MUH-kuhl RO), island (circular, 3 mi/4.8 km diameter; rises to 557 ft/170 m), of the SHETLAND ISLANDS, extreme N Scotland, in ST. MAGNUS BAY, just off NW coast of Mainland island, and 16 mi/26 km NW of LERWICK; 60°22′N 01°27′W.

Muckno, Lough (MUHK-no, LAHK), lake (□ 1 sq mi/ 2.6 sq km), E MONAGHAN county, NE IRELAND, extending S from CASTLEBLAYNEY; 54°07′N 06°42′W. Fed and drained by FANE RIVER.

Muckross, IRELAND: see KILLARNEY, LAKES OF.

Muconda (moo-KON-duh), town, LUNDA SUL province, NE ANGOLA, on LUEMBE RIVER, and 90 mi/145 km SE of Saurime. Rice, manioc. Formerly Nova Chaves (NO-vah SHAH-vez).

Mucuchíes (moo-koo-CHEE-ais), town, ⊙ Rangel municipio, MÉRIDA state, W VENEZUELA, on upper CHAMA RIVER, at W foot of Pico MUCUÑUQUE, on transandine highway, and 19 mi/31 km NE of MÉRIDA; 08°50′N 70°58′W. Elevation 9,767 ft/2,977 m. Wheat, fruits, vegetables.

Mucuchíes Pass (moo-koo-CHEE-ais), MÉRIDA state, W VENEZUELA, on transandine highway, N of Pico MUCUÑUQUE, and 27 mi/43 km NE of MÉRIDA. Elevation c.13,500 ft/4,115 m. Often snowbound during wet season.

Mucugê (moo-koo-zhuh), city (2007 population 14,108), central BAHIA, BRAZIL, in the Serra do SINCORÁ, on upper PARAGUAÇU River, and 14 mi/23 km S of ANDARAÍ; 12°59′S 41°29′W. Diamond-mining center.

Muçum (MOO-koon), town (2007 population 4,538), E RIO GRANDE DO SUL state, BRAZIL, 72 mi/116 km NW of PÔRTO ALEGRE, on railroad; 29°10′S 51°53′W. Grapes, wheat, corn, potatoes; livestock.

Mucuñuque, Pico (moo-koon-YOO-kai, PEE-ko), peak (15,328 ft/4,672 m) in ANDEAN spur, MÉRIDA state, W VENEZUELA, 28 mi/45 km NE of MÉRIDA; 08°46′N 70°49′W. Highest elevation of Sierra de SANTO DOMINGO.

Mucur, village, central TURKEY, 13 mi/21 km ESE of KIRSEHIR; 39°05′N 34°25′E. Wheat, linseed, mohair goats. Noted for its handwoven prayer rugs.

Mucurapo (myoo-ko-RAH-po), NW suburb of PORT OF SPAIN, NW TRINIDAD, TRINIDAD AND TOBAGO.

Mucuri (MOO-koo-ree), city (2007 population 33,106), SE BAHIA, BRAZIL, on RIO MUCURI, 19 mi/31 km SW of NOVA VIÇOSA; 18°09′S 39°32′W.

Mucurici (MOO-koo-ree-SEE), town (2007 population 5,837), N central ESPÍRITO SANTO, BRAZIL, on Braço

Norte do Rio, 12 mi/19 km WNW of MONTANHA; 18°08′S 40°30′W.

Mudanjiang (MU-DAN-JIANG), city (□ 522 sq mi/ 1,351 sq km; 1994 estimated urban population 602,700; 1994 estimated total population 736,200), SE HEILONGJIANG province, CHINA; 44°35′N 129°36′E. It is a railroad junction and a lumbering center in a rich timber region. Major manufacturing includes rubber products, especially tires, in addition to aluminum and various construction materials. Also spelled Mutanjiang or Mutankiang.

Mudan River (MU-DAN), Manchu *Hurka*, 415 mi/668 km long, E HEILONGJIANG province and NE JILIN province, NORTHEAST, CHINA; rises in highlands 30 mi/48 km SW of DUNHUA (Jilin province); flows NNE, past Dunhua, through the Lake JINGPO (Heilongjiang and Jilin border; hydroelectric station), past NING'AN and MUDANJIANG, to SONGHUA River at YILAN (Sansing). Unnavigable because of swift current and rapids; frozen November–April.

Mudanya, town (2000 population 20,682), NW TURKEY, port on S shore of Gulf of Gemlik of Sea of MARMARA, 16 mi/26 km NW of BURSA; 40°23′N 28°53′E. Railroad terminus; copper, olives, Merinos.

Muda River (MOO-dah), 100 mi/161 km long, KEDAH, NW MALAYSIA; rises in Kalakhiri Mountains on THAILAND border; flows S and W, past Kuala Ketil, to Strait of MALACCA at KUALA MUDA, forming Kedah-PROVINCE WELLESLEY border in lower course; 05°33′N 100°22′E.

Mudau (MOO-dou), village, LOWER NECKAR, BADEN-WÜRTTEMBERG, central GERMANY, in the ODENWALD, 25 mi/40 km ENE of HEIDELBERG; 49°32′N 09°13′E. Fruit; forestry; metalworking.

Mudawara (mu-DO-wuh-ruh), village, S JORDAN, 60 mi/97 km ESE of AQABA. Last Jordan station on disused part of HEJAZ RAILROAD. Airfield. Important station on pilgrimage road to MAKKA. Also known as Qalat el Mudawara.

Mudaybi, OMAN: see MUDHAIBI.

Mud Creek, 77 mi/124 km long, central NEBRASKA; rises in CUSTER county near BROKEN BOW; flows SE to SOUTH LOUP RIVER near RAVENNA.

Mud Creek, c.65 mi/105 km long, S OKLAHOMA; rises SE of DUNCAN in STEPHENS county; flows SE, through JEFFERSON county, to RED RIVER in LOVE county, 23 mi/37 km SW of ARDMORE.

Muddebihal (muhd-DAI-bi-hahl), town, BIJAPUR district, KARNATAKA state, SW INDIA, 45 mi/72 km SE of BIJAPUR; 16°20′N 76°08′E. Cotton, millet, peanuts, wheat. Limestone deposits nearby.

Muddusjärvi (MOOT-toos-YHR-vee), lake, LAPIN province, N FINLAND, W of Lake INARI, 4 mi/6.4 km NW of INARI village; 10 mi/16 km long, 1 mi/1.6 km–6 mi/9.7 km wide; 69°00′N 27°00′E. Helsinki University arctic experimental station.

Muddy Boggy Creek, c.110 mi/177 km long, SE OKLAHOMA; rises E of ADA; flows SE, past ATOKA, to junction with Clear Boggy River in CHOCTAW county, c.18 mi/29 km WNW of HUGO.

Muddy Creek, AUSTRALIA: see TOORA.

Muddy River, c.60 mi/97 km long, SE NEVADA; rises in SHEEP RANGE; flows SE to N arm (VIRGIN RIVER) of Lake MEAD just SE of OVERTON. Sometimes called Muddy Creek. MEADOW VALLEY WASH is tributary.

Muddy River, c.70 mi/113 km long, S central UTAH; rises at SANPETE-SEVIER county line, in WASATCH PLATEAU; flows SE past EMERY, joining FREMONT RIVER in WAYNE county, N of HENRY MOUNTAINS, to form DIRTY DEVIL RIVER. Its main tributary, Muddy Creek, rises in SE SEVIER county, flows c.35 mi/56 km SSE, joining Muddy River 8 mi/12.9 km SSE of Emery.

Muden (Aller) (MOO-den (AHL-ler)), village, LOWER SAXONY, GERMANY, 13 mi/21 km ESE of CELLE, on the ALLER; 52°32′N 10°22′E. Manufacturing (chemical products, electronic equipment).

Mudersbach (MOO-ders-bahkh), village, RHINELAND-PALATINATE, W GERMANY, on the SIEG, 4 mi/6.4 km SW of SIEGEN; 50°50′N 07°57′E.

Mudgal (muhd-gahl), town, RAICHUR district, KARNATAKA state, SW INDIA, 60 mi/97 km WSW of RAICHUR; 16°01′N 76°26′E. Millet, cotton, oilseeds.

Mudgee (MU-jee), municipality, E central NEW SOUTH WALES, SE AUSTRALIA, 130 mi/209 km NW of SYDNEY; 32°26′S 149°35′E. Livestock, abattoir; wool, honey, cereals, lucerne, vegetables; wineries. Coal mine; former gold-mining center.

Mudhaibi (mud-HAI-bee), township, interior of OMAN, 30 mi/48 km W of IBRA, at foot of EASTERN HAJAR hills; 22°47′N 57°26′E. Date groves. Sometimes spelled Mudaybi.

Mudhol (MUHD-ol), town, BIJAPUR district, KARNATAKA state, SW INDIA, on GHATPRABHA RIVER, and 45 mi/72 km SW of BIJAPUR; 16°21′N 75°17′E. Market center for cotton, millet, peanuts, wheat; cotton ginning, hand-loom weaving, oilseed pressing. Was the capital of the former princely state of Mudhol in DECCAN STATES, Bombay; incorporated (1949) into Bijapur district.

Mudhol (MUHD-ol), town, NANDED district, MAHARASHTRA state, W central INDIA, 30 mi/48 km ESE of NANDED; 18°58′N 77°55′E. Extensive cotton ginning here and in nearby villages; millet, wheat, rice. Also spelled Mudhole.

Mudia (moo-DEE-yuh), broad plain in DATHINA tribal district of S YEMEN, at S foot of the Kaur al Audhilla, inhabited by Hasani (Hasanah) and Meisari (Meyasir) tribes. Also known as 'Amudiyah.

Mudigere (MUH-di-ge-re), town, CHIKMAGALUR district, KARNATAKA state, SW INDIA, 15 mi/24 km SW of CHIKMAGALUR; 13°08′N 75°38′E. Rice milling, beekeeping. Coffee, tea, and cardamom estates in nearby hills.

Mudkhed (muhd-KAID), town, NANDED district, MAHARASHTRA state, W central INDIA, 12 mi/19 km E of NANDED; 19°10′N 77°31′E. Cotton ginning; millet, wheat.

Mudki (MUHD-kee), village, FIROZPUR district, W PUNJAB state, N INDIA, 21 mi/34 km SE of FIROZPUR. British defeated (1845) Sikh forces nearby, in First Sikh War.

Mud Lake, village (2000 population 270), JEFFERSON county, E IDAHO, 32 mi/51 km NW of IDAHO FALLS; 43°51′N 112°29′W. Wheat, alfalfa, sheep, cattle. Manufacturing at Terreton, 2 mi/3.2 km E (feeds, conveyor chain). Mud Lake reservoir to NE; Camas National Wildlife Refuge to NE; large tract of Idaho National Laboratory to W.

Mud Lake, TRAVERSE county, W MINNESOTA and ROBERTS county, NE SOUTH DAKOTA, on BOIS DE SIOUX RIVER, 1 mi/1.6 km NE of outflow of Lake TRAVERSE reservoir, also on Minnesota-South Dakota state line; 5 mi/8 km long, 2 mi/3.2 km wide at S end. Marshy area.

Mud Lake, NW MINNESOTA, E central MARSHALL county, 17 mi/27 km NE of THIEF RIVER FALLS, in Agassiz National Wildlife Refuge; 6 mi/9.7 km wide, 11 mi/18 km long. Irregular shape. Fed by Mud River from E, drains into THIEF RIVER, which passes immediately to W. Marshy area; waterfowl.

Mud Lake, NW NEVADA, intermittent body of water in WASHOE county, NNE of PYRAMID LAKE, S of GERLACH; c.20 mi/32 km long, 6 mi/9.7 km wide. Sometimes known as Mud Flat. PAH-RUM PEAK to SW.

Mud Lake, in JEFFERSON county, N NEW YORK, 6 mi/9.7 km SE of ALEXANDRIA BAY; c.2.5 mi/4 km long, 5 mi/.8 km wide; 44°17′N 75°49′W. Resort; fishing.

Mud Mountain Dam, WASHINGTON: see WHITE RIVER.

Mudo Gashe (moo-DO gah-shai), village, NORTHEASTERN province, E central KENYA, on road, and 120 mi/193 km ENE of ISIOLO, near Lorian swamps; 00°43′N 39°10′E. Livestock raising. Airfield.

Cross-references are shown in SMALL CAPITALS. The pronunciation guide is shown on page xix. The sources of population figures are shown on page xvii.

Mudon (MOO-don), village and township, MON STATE, MYANMAR, on Mawlamyine-Ye railroad, and 15 mi/24 km SSE of MAWLAMYINE. Canal to THANLWIN RIVER mouth.

Mud River, c.70 mi/113 km long, S KENTUCKY; rises in LOGAN county E of RUSSELLVILLE; flows generally N to GREEN RIVER at ROCHESTER.

Mud River, 72 mi/116 km long, W WEST VIRGINIA; rises in W BOONE county W of MADISON; flows NNW and W past MILTON to GUYANDOTTE RIVER at BARBOURSVILLE, near OHIO RIVER.

Mudros, Greece: see MOUDROS.

Mudugh (moo-DOOG), region (□ 27,000 sq mi/70,200 sq km), central SOMALIA, bordering on INDIAN Ocean (E), ETHIOPIA(W); ⊙ GALKAYO; 07°00′N 48°00′E. Hot, arid plain with sand dunes along coast. Rises to 600 ft/183 m (N). Pastoralism (cattle, sheep, camels). Chief centers: OBBIA, HARARDERA, EL BUR.

Mudug Plateau, Somalia: see SOMALIA.

Mudukulattur (MUHD-uh-kuh-luht-TOOR), town, RAMANATHAPURAM district, TAMIL NADU state, S INDIA, 15 mi/24 km SSW of Paramagudi; 09°21′N 78°31′E. In cotton, palmyra, grain area.

Mudumalai (muh-duh-muh-lei), tourist center and wildlife sanctuary, NILGIRI district, TAMIL NADU state, S INDIA, in NILGIRI HILLS.

Mudungunj, BANGLADESH: see MADANGANJ.

Mudurnu, village, NW TURKEY, 28 mi/45 km. SW of BOLU; 40°27′N 31°12′E. Grain, flax.

Mudwara, India: see KATNI.

Mud'yuga (MOOD-yoo-gah), town (2006 population 925), NW ARCHANGEL oblast, N European Russia, on a tributary of the ONEGA River, near railroad, 35 mi/56 km E of ONEGA; 63°44′N 39°15′E. Railroad shops; sawmilling. Developed in the early 1940s.

Mudzi Maria, CONGO: see MANGO.

Mudzi River, c. 80 mi/129 km long, ZIMBABWE and MOZAMBIQUE; rises c. 10 mi/16 km E of MTOKO, NE Zimbabwe; flows ENE past NYAMAPANDA, enters TETE province, W Mozambique, and joins Luenha River 15 mi/24 km S of Changara, Mozambique.

Muecati (MWAI-kah-tee), village, NAMPULA, N MOZAMBIQUE, 33 mi/53 km NE of NAMPULA; 14°53′S 39°37′E. Cotton.

Muecati District, MOZAMBIQUE: see NAMPULA.

Mueda (MWAI-duh), village, CABO DELGADO, N MOZAMBIQUE, near TANZANIA border, 120 mi/193 km NW of PEMBA; 11°39′S 39°33′E. Cotton, peanuts. Handicraft industry. Site of an uprising and subsequent repression by colonialist forces in early 1960s.

Mueda District, MOZAMBIQUE: see CABO DELGADO.

Mueka, CONGO: see MWEKA.

Muel (mwel), town, ZARAGOZA province, NE SPAIN, 18 mi/29 km SW of ZARAGOZA; 41°28′N 01°05′W. Wine, cereals, sugar beets. In nearby monastery are paintings by Goya.

Muelle de los Bueyes (MWAI-yai dai los BWAI-yais), town, SOUTH ATLANTIC COAST AUTONOMOUS REGION, ZELAYA department, E NICARAGUA, on MICO RIVER, and 19 mi/31 km WSW of RAMA; 12°04′N 84°32′W. English widely spoken here. On road; sugarcane, livestock; lumbering.

Muelle de San Carlos (moo-AI-yai dai sahn KAHR-los), village, ALAJUELA province, N COSTA RICA, on SAN CARLOS RIVER, and 14 mi/23 km NNW of CIUDAD QUESADA. Livestock raising, lumbering.

Mueller Glacier, in MOUNT COOK National Park, SOUTHERN ALPS, NEW ZEALAND; rises at parting of Mount Cook and Ben Ohau ranges, bends 8 mi/13 km NE then S to E-flowing TASMAN River.

Muembe District, MOZAMBIQUE: see NIASSA.

Muenster (MUHN-stuhr), town (2006 population 1,683), COOKE county, N TEXAS, 14 mi/23 km W of GAINESVILLE, Elm Fork of TRINITY RIVER; 33°38′N 97°22′W. Elevation 970 ft/296 m. Manufacturing (cheese and dairy products, meat packing).

Muermos, Los, CHILE: see LOS MUERMOS.

Muerto, Cerro El (MWER-to, SER-ro el), Andean mountain (21,450 ft/6,538 m) on ARGENTINA-CHILE border, near the Nevados TRES CRUCES, 30 mi/48 km WSW of Cerro INCAHUASI; 27°04′S 68°29′W. Elevation sometimes cited at 21,286 ft/6488 m.

Muerto, Isla de, ECUADOR: see SANTA CLARA, ISLA.

Muerto, Mar, MEXICO: see MAR MUERTO.

Muertos, PUERTO RICO: see CAJA DE MUERTOS Island.

Muflahi, YEMEN: see MAFLAHI.

Mufulira (moo-foo-LEE-lah), city (2000 population 137,272), COPPERBELT province, N central ZAMBIA, on CONGO border; 12°33′S 28°14′E. Railroad junction connecting systems of both countries. City is a copper-mining center, located in the COPPERBELT REGION. Agriculture (peanuts, tobacco, cotton, grain); fish; cattle. Manufacturing (frozen fish, explosives, scrap metal).

Mufu Mountains (MU-FU), JIANGXI-HUBEI border, CHINA, 125 mi/201 km WNW of NANCHANG, in northeasternmost HUNAN province; 29°14′N 114°20′E. Sometimes called Jiugong Mountains after a lower peak, 90 mi/145 km NW of Nanchang.

Mugan (moo-GAHN), urban settlement, Ali-Bayramli region, AZERBAIJAN; 40°05′N 48°48′E.

Mugango (moo-GAN-go), village, MARA region, NW TANZANIA, 13 mi/21 km SSW of MUSOMA, on Lake VICTORIA; 01°41′S 33°42′E. Fish; cattle; cotton, corn, wheat.

Mugardos (moo-GAHR-dhos), town, LA CORUÑA province, NW SPAIN, in GALICIA, on S shore of Ferrol Bay, opposite (1 mi/1.6 km SW of) EL FERROL; 43°28′N 08°15′W. Fishing, boatbuilding, lumbering, agriculture. Granite and graphite quarries nearby.

Muge (MOO-zhi), town, SANTARÉM district, central PORTUGAL, near left bank of the TAGUS RIVER, 10 mi/16 km S of SANTARÉM; 39°06′N 08°43′W. Agriculture.

Mugegawa (moo-GE-gah-wah), town, Mugi county, GIFU prefecture, central HONSHU, central JAPAN, 12 mi/20 km N of GIFU; 35°31′N 136°50′E.

Mügeln (MYOO-geln), town, SAXONY, E central GERMANY, 12 mi/19 km SW of RIESA; 51°14′N 13°03′E. Manufacturing (chemicals, ceramics).

Mugera (moo-GE-rah), village, GITEGA province, central BURUNDI, near RUVUBU River, 7 mi/11.3 km N of Gitega; 03°19′S 29°58′E. Coffee. Has Roman Catholic mission with small seminary, convent for native nuns, and schools.

Müggel Lake (MYOOG-gel), German *Müggelsee* (MYOOG-gel-sai), lake (□ 3 sq mi/7.8 sq km), BRANDENBURG, E GERMANY, 10 mi/16 km ESE of BERLIN city center; 3 mi/4.8 km long, 1 mi/1.6 km–2 mi/3.2 km wide; maximum depth 26 ft/8 m, average depth 16 ft/5 m; 52°26′N 13°37′E. Traversed by the SPREE RIVER. Popular excursion resort.

Muggensturm (MUG-gen-shturm), village, MIDDLE UPPER RHINE, BADEN-WÜRTTEMBERG, GERMANY, 3 mi/4.8 km ENE of RASTATT; 48°52′N 08°17′E. Food processing, fruit.

Muggia (MOOJ-jah), Slovenian *Milje* (MEEL-lyai), town, Trieste province, ITALY, bordering SLOVENIA, on S shore of Muggia Bay (ADRIATIC inlet of the GULF OF TRIESTE) opposite TRIESTE; 45°35′N 13°45′E. Shipyards. Has 15th-century church.

Muggia, ETHIOPIA: see MUJA.

Muggiò (mood-JO), town, MILANO province, LOMBARDY, N ITALY, 2 mi/3 km W of MONZA; 45°35′N 09°13′E. Highly diversified secondary industrial center, including manufacturing furniture, machinery, hardware, fabricated metals, shoes.

Mughaiyir (mu-GAI-yer), village, N JORDAN, 5 mi/8 km NE of IRBID; 31°24′N 35°46′E. Grain, olives.

Mughalbhin, PAKISTAN: see JATI.

Mughalpur, PAKISTAN: see MOGHALPURA.

Mughal Sarai (MOO-guhl suh-REI), town, VARANASI district, SE UTTAR PRADESH state, N central INDIA, 7 mi/11.3 km ESE of VARANASI city center; 25°18′N 83°07′E. Railroad junction (workshops); rice, barley, gram, wheat, sugarcane. Also spelled Moghal-Sarai.

Mughar (muh-GAHR), township, N ISRAEL, 9.5 mi/15.3 km NW of TIBERIAS, in E LOWER GALILEE; 32°53′N 35°24′E. Elevation 554 ft/168 m. Most of its inhabitants are Druze, about one-third are Christian, and a small Muslim population. Surrounded by large area of olive groves. Hometown of Azzam Azzam.

Mugi (MOO-gee), town, Mugi county, GIFU prefecture, central HONSHU, central JAPAN, 22 mi/35 km N of GIFU; 35°33′N 137°00′E.

Mugi, town, Kaifu county, TOKUSHIMA prefecture, E SHIKOKU, W JAPAN, on PHILIPPINE SEA, 28 mi/45 km S of TOKUSHIMA; 33°39′N 134°25′E.

Mugi (moo-GEE), village (2005 population 3,460), S central DAGESTAN REPUBLIC, NE CAUCASUS, SE European Russia, on road, 41 mi/66 km S of MAKHACHKALA; 42°18′N 47°25′E. Elevation 5,186 ft/1,580 m. Indigenous handicrafts.

Mugía (moo-HEE-ah), town, LA CORUÑA province, NW SPAIN, near the ATLANTIC, 37 mi/60 km NW of SANTIAGO DE COMPOSTELA; 43°06′N 09°13′W. Fishing, lumbering, stock raising.

Mugia Pass (moo-gee-AH PAS) (elevation 1,371 ft/418 m), in TRUONG SON RANGE, running from central LAOS into VIETNAM, 55 mi/89 km WNW of Dong Hoi. Used by Tanap-Thakhek road. Also spelled MEUGIA.

Múgica, MEXICO: see NUEVA ITALIA DE RUIZ.

Mugla, Turkish *Muğla*, town (2000 population 43,845), SW TURKEY, 100 mi/161 km SSE of IZMIR, 12 mi/20 km N of the Gulf of GÖKOVA; 37°13′N 28°22′E. Irregular coastline, with gulfs and bays backed by mountains, among which small, cultivated basins are found. Millet, wheat, tobacco, olives, onions, citrus fruits, cotton; emery, chromium. Formerly, under the name of Mentese (Menteshe) it was the capital of an independent duchy whose pirates dominated the AEGEAN in the 13th century.

Mugla (muh-GLAH), village, PLOVDIV oblast, SMOLYAN obshtina, BULGARIA; 41°37′N 24°30′E.

Mugling, village, central NEPAL, on the PRITHVI RAJMARG; 27°50′N 84°34′E. Important road junction; the primary route from KATHMANDU to INDIA turns S here, following the NARAYANI RIVER to meet the MAHENDRA RAJMARG at BHARATPUR.

Müglitz, CZECH REPUBLIC: see MOHELNICE.

Muglizh (moog-LEEZH), village (1993 population 3,741), HASKOVO oblast, Muglizh obshtina, central BULGARIA, in the KAZANLUK BASIN, 8 mi/13 km E of KAZANLUK; 42°36′N 25°33′E. Grain, tobacco, vineyards, horticulture (roses, mint); chestnut and mulberry groves. Hydroelectricity. Labor force for Kazanluk. Also spelled Maglizh.

Mugnano di Napoli, town, NAPOLI province, CAMPANIA, S ITALY, 4 mi/6 km S of AVERSA; 40°55′N 14°12′E. In agricultural region (grapes, fruit, vegetables); manufacturing of shoes.

Mugodzhar Hills, range, E KAZAKHSTAN, S spur of the Ural Mountains, forming the divide between the CASPIAN and ARAL basins; c.275 mi/440 km long. Highest point c.2,150 ft/660 m. Coal, copper, nickel, and chrome deposits.

Mugombazi (moo-gom-BAH-zee), village, KIGOMA region, W TANZANIA, 70 mi/113 km SSE of KIGOMA; Lake TANGANYIKA to W; 05°53′S 30°08′E. Timber; livestock; grain.

Mugreyevskiy (moo-GRYE-eef-skeeye), town (2005 population 905), SE IVANOVO oblast, central European Russia, on the Lukh River (OKA River basin), on railroad spur, 8 mi/13 km ENE of YUZHA; 56°36′N 42°21′E. Elevation 324 ft/98 m. Peat working.

Mugtaa, El (MUG-tah, ahl), village, TRIPOLITANIA region, N central LIBYA, on CYRENAICA border, 150 mi/241 km ESE of SURT, near Gulf of SURT.

Mugu (MOO-goo), district, NW NEPAL, in KARNALI zone; ⊙ GAMGADI.

Mugu, Point, VENTURA county, S CALIFORNIA, low foreland just SE of PORT HUENEME, and 7 mi/11.3 km SSE of OXNARD. POINT MUGU Naval Air Station. Point Mugu State Park is here.

Mugur-Aksy (moo-GOOR–ahk-SI), village (2006 population 4,000), SW TUVA REPUBLIC, S SIBERIA, RUSSIA, near the border with MONGOLIA, on road, 195 mi/314 km SW of KYZYL; 50°21′N 90°30′E. Elevation 6,673 ft/2,033 m. In agricultural area (feed corn, hardy grain cultures; livestock).

Muguru (MOO-goo-roo), town, MYSORE district, KARNATAKA state, SW INDIA, 23 mi/37 km SE of MYSORE; 12°08′N 76°57′E. In silk-growing area; hand-loom silk and cotton weaving; sugarcane. Also spelled Mugur.

Muhamdi (moo-HUHM-dee), town, KHERI district, N UTTAR PRADESH state, N central INDIA, 35 mi/56 km W of LAKHIMPUR; 27°57′N 80°13′E. Sugar milling; trades in rice, wheat, gram, corn, oilseeds. Has seventeenth-century Muslim fort ruins. Sometimes spelled Mohamdi.

Muhammadabad (muh-HUHM-mah-dah-BAHD), town, AZAMGARH district, E UTTAR PRADESH state, N central INDIA, on TONS RIVER, and 13 mi/21 km E of AZAMGARH. Hand-loom cotton weaving, sugar milling; rice, barley, wheat.

Muhammadabad, town, GHAZIPUR district, E UTTAR PRADESH state, N central INDIA, 11 mi/18 km E of GHAZIPUR. Trades in rice, barley, gram, oilseeds. Also spelled Mohammadbad.

Muhammadgarh (muh-HUHM-muhd-gurh), former princely state of CENTRAL INDIA AGENCY. In 1948, merged with Madhya Bharat and later with Madhya Pradesh state.

Muhammad Ghul, SUDAN: see MUHAMMAD GOL.

Muhammad Gol (muh-HAHM-med KOL), village, Red Sea state, NE SUDAN, and 90 mi/145 km N of PORT SUDAN; 20°54′N 37°05′E. Minor port on RED SEA; exports gold from JABEIT mines. Sometimes spelled MUHAMMAD GHUL or MUHAMMAD QOL.

Muhammadpur, India: see ROORKEE.

Muhammad Qol, SUDAN: see MUHAMMAD GOL.

Muhammed, Ras (moo-HAHM-mahd, RAHS), southernmost tip of SINAI Peninsula, NE EGYPT, on RED SEA; 27°45′N 34°15′E. Tourism: Ras Muhammed Underwater National Park.

Muhammidiyah, Al (maw-HAHM-mi-DEE-yah, el), village, ZAGHWAN province, N TUNISIA, 8 mi/12.9 km S of TUNIS; 36°41′N 10°10′E. Wine-grape growing. Has extensive remains of an abandoned 19th-century beylical palace. Also spelled La Mohamédia.

Muharraq (moo-HUHR-ruhk), island (□ 6 sq mi/15.6 sq km; 2000 population 91,307), BAHRAIN archipelago, BAHRAIN, in PERSIAN GULF; separated by shallow 1.5-mi/2.4-km-wide channel from NE shore of main Bahrain island; 4 mi/6.4 km, 1 mi/1.6 km wide; 26°13′N 50°37′E. Site of main international airport and seaplane base of Bahrain. On SW point is Muharraq town, formerly chief residence of the sheik, linked by road causeway with MANAMAH. Fishing is still an important occupation. There are also some modern industries. Sometimes spelled Moharraq.

Muhazi, Lake (moo-HAH-zhee), lake, RWANDA, E of KIGALI; 01°51′S 30°23′E. At junction of former BYUMBA, KIBUNGO, and KIGALI provinces.

Muheza (moo-HAI-zah), town, TANGA region, NE TANZANIA, 25 mi/40 km WSW of TANGA, near railroad; USAMBARA Mountains to NW; 05°15′S 38°47′E. Highway junction. Sisal, rice, grain; livestock.

Mühlacker (MYOOL-ahk-kuhr), town, N BLACK FOREST, BADEN-WÜRTTEMBERG, GERMANY, on the ENZ RIVER, and 7 mi/11.3 km NE of PFORZHEIM; 48°57′N 08°50′E. Foundries; manufacturing of machine tools, tools, jewelry; metalworking. Has ruined castle.

Mühlbach, ROMANIA: see SEBEŞ.

Mühlburg (MYOOL-boorg), W suburb of KARLSRUHE, MIDDLE UPPER RHINE, BADEN-WÜRTTEMBERG, SW GERMANY; 49°01′N 08°21′E.

Mühldorf am Inn (MYOOL-dorf ahm IN), city, UPPER BAVARIA, GERMANY, on the INN RIVER, and 27 mi/43 km SE of LANDSHUT; 48°14′N 12°31′E. Hydroelectric plant; manufacturing of basic chemicals and textiles, brewing, woodworking, lumber milling. Was important river crossing in Roman times. Chartered c.954. Between here and AMPFING, Emperor Louis the Bavarian defeated his rival, Frederick of AUSTRIA (1322), in one of the most important battles of the Middle Ages. Gothic parish church (1432–1443) here.

Mühleberg (MYOO-le-berg), commune, BERN canton, W SWITZERLAND, 9 mi/14.5 km W of BERN; 46°57′N 07°15′E. Hydroelectric plant on nearby WOHLENSEE. Farming. Nuclear reactor.

Mühlenbach, Luxembourg: see LUXEMBOURG.

Muhlenberg (MYOO-luhn-buhrg), county (□ 479 sq mi/1,245.4 sq km; 2006 population 31,561), W KENTUCKY; ⊙ GREENVILLE; 37°12′N 87°09′W. Bounded NE by GREEN RIVER, E by MUD RIVER, W by POND RIVER; drained by Pond, Cypress, and Rocky creeks. Important bituminous-coal mining area; agriculture (soybeans, burley and dark tobacco, wheat, hay, alfalfa, corn; hogs, cattle; timber); limestone. Manufacturing at Greenville and CENTRAL CITY. Lake Malone reservoir (Rocky Creek) and State Park on S boundary. Formed 1798.

Mühlgraben, LATVIA: see MĪLGRĀVIS.

Mühlhausen (myool-HOU-suhn), city, THURINGIA, central GERMANY, on the UNSTRUT, 30 mi/48 km NW of ERFURT; 51°12′N 10°28′E. In barite-mining region; metalworking, textile manufacturing, dyeing, woodworking. Also manufacturing of electrical equipment, shoes, sewing machines. Has 13th- and 14th-centuy churches; town hall (1605); many 16th–18th-century houses. Fortified (remains of walls) by Henry I in 10th century. City first mentioned 967. Was a center of Teutonic Order in 13th century Created free imperial city in 1256; became member of HANSEATIC LEAGUE (1418). Anabaptist center after Reformation. During Peasants' War, it was headquarters of Thomas Müntzer (executed here in 1525); subsequently it was deprived of privileges as free city. Passed to PRUSSIA in 1815. Also called Thomas-Müntzer-Stadt (since 1974).

Mühlhausen (myool-HOU-suhn), village, LOWER NECKAR, BADEN-WÜRTTEMBERG, GERMANY, 11 mi/18 km S of HEIDELBERG; 49°15′N 08°43′E. Agriculture (sugar beets); wine making.

Mühlhausen, village, UPPER PALATINATE, central BAVARIA, GERMANY, near MAIN-DANUBE CANAL, 7 mi/11.3 km S of NEUMARKT; 49°10′N 11°26′E. Metal- and woodworking.

Mühlhausen, CZECH REPUBLIC: see MILEVSKO.

Mühlhausen, CZECH REPUBLIC: see NELAHOZEVES.

Mühlheim am Main (MYOOL-heim ahm MEIN), town, S HESSE, W GERMANY, on left bank of the MAIN RIVER, and 3 mi/4.8 km ENE of OFFENBACH; 50°08′N 08°51′E. Metal- and leatherworking, manufacturing of electrical equipment.

Mühlhofen, GERMANY: see UHLDINGEN-MÜHLHOFEN.

Mühlig-Hofmann Mountains, a major group of mountains in NEW SCHWABENLAND, EAST ANTARCTICA, between the GJELSVIK and ORVIN mountains; 65 mi/105 km long; 72°00′S 05°20′E.

Mühl River, AUSTRIA and GERMANY: see GROSSE MÜHL.

Mühltal (MYOOL-tahl), town, S HESSE, central GERMANY, 5 mi/8 km SE of DARMSTADT; 49°48′N 08°43′E. Chemical industry.

Mühlviertel (MYOOL-fir-tel) or **Mühlkreis**, region of UPPER AUSTRIA, N AUSTRIA; 48°09′–46′N, 13°44′–14°59′E. Bounded NW by BAVARIA (GERMANY), N by the CZECH REPUBLIC, E by LOWER AUSTRIA. Highland 1,500 ft/457 m–2,500 ft/762 m, dissected by tributaries of the Danube (e.g., Große Mühl, Kleine Mühl, Aist, Naarn). Agriculture (dairy farming) dominated the local economy; commuting to the central regions of Upper Austria is significant,

mainly to LINZ. FREISTADT and Rohrbach in Upper Austria are the urban centers. A main S-N traffic route (road, railroad) from LINZ via Kirschbaum Pass (2,286 ft/697 m) to Ceské Budejovice (Czech Republic). It follows the track of the first horse-drawn railroad in the Austrian Empire, opened 1827–1832.

Muhoroni (moo-ho-RO-nee), village (1999 population 13,664), NYANZA province, W KENYA, on railroad, and 30 mi/48 km ESE of KISUMU; 00°09′S 35°13′E. Hardwood and rubber center; coffee, corn. Limestone quarry and kilns nearby.

Muhu (MOO-hoo), German *Moon*, 3rd-largest island (□ 79 sq mi/205.4 sq km) of ESTONIA, in BALTIC SEA, between SAAREMAA island and mainland; 13 mi/21 km long, 10 mi/16 km wide; 58°34′N 23°15′E. Consists of Silurian limestone with thin top layer of marine sediments; agriculture, fishing. Connected by 3 mi/4.8 km-causeway with Saaremaa Island (SW) and the mainland through a ferry station on SE coast at Kivastu. Also called MUHUMAA.

Muhukuru (moo-hoo-KOO-roo), village, RUVUMA region, S TANZANIA, 20 mi/32 km SSW of SONGEA; 11°07′S 33°27′E. Timber; tobacco, subsistence crops; livestock.

Muhulu, CONGO: see MUTIKO.

Muhumaa, ESTONIA: see MUHU.

Muhumbika (moo-hoom-BEE-kah), village, LINDI region, SE TANZANIA, 20 mi/32 km SW of LINDI, on LUKULEDI RIVER; 10°08′S 39°38′E. Cashews, peanuts, bananas, manioc; livestock.

Muhu Sound (MOO-hoo), Estonian *Muhu Väin* or *Väinameri*, arm of BALTIC SEA, between ESTONIAN mainland (E) and MUHU and HIIUMAA islands (W); 4 mi/6.4 km–15 mi/24 km wide.

Muhutwe (moo-HOO-twai), village, KAGERA region, NW TANZANIA, 15 mi/24 km S of BUKOBA, on Bumbire Channel, Lake VICTORIA; 01°35′S 31°44′E. Road junction. Coffee, subsistence crops; livestock.

Muhuvesi River (moo-hoo-VAI-see), c. 115 mi/185 km long, RUVUMA region, S TANZANIA; rises c. 80 mi/129 km ENE of SONGEA; flows SE, enters Ruvuma River, on MOZAMBIQUE border, 150 mi/241 km SW of MTWARA, Tanzania.

Muiden (MOI-duhn), town, NORTH HOLLAND province, W central NETHERLANDS, on the IJMEER, S extension of MARKERMEER, 8 mi/12.9 km ESE of AMSTERDAM; 52°20′N 05°04′E. Mouth of VECHT RIVER on IJmeer 1 mi/1.6 km to N. Dairying; cattle, poultry; vegetables, fruit, nursery stock. Has thirteenth-century castle, Muiderslot, and Muiderslot Museum.

Muidumbe District, MOZAMBIQUE: see CABO DELGADO.

Muika (MWEE-kah), town, South Uonuma county, NIIGATA prefecture, central HONSHU, N central JAPAN, 59 mi/95 km S of NIIGATA; 37°03′N 138°52′E.

Muikaichi (moo-ee-KAH-ee-chee), town, Kanoashi county, SHIMANE prefecture, SW HONSHU, W JAPAN, 100 mi/161 km S of MATSUE; 34°21′N 131°56′E.

Muilrea (mweel-RAI) or **Mweelrea**, mountain range, SW MAYO county, NW IRELAND, extending 8 mi/12.9 km NW-SE along KILLARY HARBOUR. Rises to 2,688 ft/819 m.

Muinak (moo-ee-NAHK), town, N KARAKALPAK, W UZBEKISTAN, on former shoreline of ARAL SEA, 95 mi/153 km NNW of NUKUS; 43°44′N 59°00′E. Formerly a port and fish-processing center. The port vanished with the retreat of the Aral Sea and, in the ensuing ecological disaster, this city has lost the fishing-related basis of much of its economy. Also spelled MUYNAK.

Mui Ne (MWEE NAI), village, BINH THUAN province, S central VIETNAM, on SOUTH CHINA SEA coast, 11 mi/18 km E of PHAN THIET; 10°56′N 108°17′E. Fishing and fish curing. Formerly Muine.

Muine Bheag (mwi-nuh VYUHG), town (2006 population 2,735), W CARLOW county, SE IRELAND, on BARROW RIVER, and 8 mi/12.9 km ENE of KILKENNY; 52°42′N 06°57′W. Agricultural market, with sandstone and granite quarries. Formerly Bagenalstown, laid out in 18th century.

Muir (MYUR), community, S MANITOBA, W central Canada, 28 mi/44 km WNW of PORTAGE LA PRAIRIE, and in Westbourne rural municipality; 50°06′N 98°52′W.

Muir (MEER), unincorporated town, SCHUYLKILL county, E central PENNSYLVANIA, 2 mi/3.2 km E of TOWER CITY; 40°35′N 76°31′W. Agriculture includes dairying, livestock, poultry; corn, hay. Former anthracite-coal region.

Muir (MYOOR), village (2000 population 634), IONIA county, S central MICHIGAN, 7 mi/11.3 km E of IONIA; and on GRAND RIVER at mouth of MAPLE RIVER; 43°00′N 84°56′W. In farm area.

Muir Glacier, Alaska: see GLACIER BAY NATIONAL PARK AND PRESERVE.

Muirhead (MIR-heed), village (2001 population 1,389), North Lanarkshire, central Scotland, 4 mi/6.4 km N of COATBRIDGE; 55°56′N 04°01′W. Formerly in Strathclyde, abolished 1996.

Muirkirk (MIR-kuhrk), town (2001 population 1,630), East Ayrshire, S Scotland, on Ayr River, and 9 mi/14.5 km NE of CUMNOCK; 55°31′N 04°03′W. Previously ironworks. Village of Kaimes, just S. Formerly in Strathclyde, abolished 1996.

Muirkirk (MYOOR-kuhrk), village, PRINCE GEORGES county, central MARYLAND, 18 mi/29 km NE of WASHINGTON, D.C. William and Elias Ellicott built an iron furnace in 1847 that was the precursor of the present Mineral Pigments Corporation. Muirkirk iron, used in cannons, gun carriages and railroad wheels, was noted for its tensile strength. In 1924, the works were converted to the production of dry pigments from ores. The nearby Ammendale Normal Institute (R.C.), established in 1880, is the provincial house and novitiate of the Christian Brothers for the Diocese of Baltimore.

Muir, Mount (14,015 ft/4,272 m), on INYO-TULARE county line, E central CALIFORNIA, in the SIERRA NEVADA, just S of MOUNT WHITNEY, and on E boundary of SEQUOIA NATIONAL PARK.

Muir of Ord (MIR uhv ORD), village (2001 population 1,812), HIGHLAND, N Scotland, 6 mi/9.7 km S of DINGWALL, near head of BEAULY FIRTH; 57°31′N 04°28′W. Agricultural market, distillery.

Muir Woods National Monument, grove of virgin redwood trees (□ 1 sq mi/2.6 sq km), MARIN county, W CALIFORNIA, 12 mi/19 km NW of downtown SAN FRANCISCO. Bounded N, W, and S by Mount Tamalpais State Park; PACIFIC OCEAN 2 mi/3.2 km to SW. Named for natural conservationist John Muir, cofounder of the Sierra Club. Authorized 1908.

Muizen (MOI-zuhn), village in commune of MECHELEN, Mechelen district, ANTWERPEN province, N central BELGIUM, on DIJLE RIVER, and 2 mi/3.2 km SE of Mechelen; 51°01′N 04°31′E. Metal industry; vegetable market. Wildlife Park "Planckendael" nearby.

Muizenberg, residential town, WESTERN CAPE province, SOUTH AFRICA, on FALSE BAY, 12 mi/19 km S of CAPE TOWN; 34°07′S 18°29′E. Popular seaside resort. Scene (1795) of defeat of Dutch forces resisting British landing. Cecil Rhodes died here (1902); his cottage is now national historic memorial. Considered part of greater Cape Town metropolitan area.

Muja (MOO-zha), Italian *Muggia*, town, AMHARA state, NE ETHIOPIA, near source of TEKEZĒ RIVER, 27 mi/43 km E of LALIBELA; 12°03′N 39°29′E. Trade center (cattle, horses, salt, cereals).

Mujere, town, MATABELELAND NORTH province, NW ZIMBABWE, 60 mi/97 km NE of BINGA on Lake KARIBA shore; 17°06′S 27°56′E. Tourism, hunting, sport fishing.

Mujeres Island, MEXICO: see ISLA MUJERES.

Mujib, Wadi el (mu-ZHEEB, WAH-dee el), river, 35 mi/56 km long, W central JORDAN; formed 4 mi/6.4 km E of AL KARAK by union of 2 branches; flows N and W to DEAD SEA 21 mi/34 km NNW of Al Karak. Bitumen deposits near mouth. Biblically it is the ancient ARNON RIVER.

Muju (MOO-JOO), county (□ 244 sq mi/634.4 sq km), NE NORTH CHOLLA province, SOUTH KOREA, on border with SOUTH CHUNGCHONG, NORTH CHUNGCHONG, SOUTH KYONGSANG, and NORTH KYONGSANG provinces. The name Muju came from the merger of Mupung and Chugye in early Yi Dynasty. Surrounded by Sobaek mountains (elevation over 3,280 ft/1,000 m). Some agriculture (ginseng, garlic, tobacco, mushroom, pepper, medical herbs); sericulture; animal husbandry. Manners, customs, and dialects of this region reflect its location on border area of five provinces. Tokyu Mountain National Park.

Mukabe Kasari, CONGO: see MOKABE-KASIRI.

Mukacheve (moo-KAH-che-ve) (Russian *Mukachevo*) (Czech *Mukačevo*) (Hungarian *Munkács*), city (2001 population 82,346), central TRANSCARPATHIAN oblast, UKRAINE, 21 mi/34 km E of UZHHOROD; 48°27′N 22°43′E. Elevation 403 ft/122 m. It is a raion center, a railroad terminus, and highway junction; has food, tobacco, beer, wine, furniture, textile, and timber industries. From the 9th to the 11th century, Mukacheve was part of the Kievan state. Taken by the Hungarians in 1018, it became a dominion center of the Hungarian kings. It later (15th century) developed as a prominent trade and craft center, and served as seat of a Uniate (Greek Catholic, now commonly known as Ukrainian Catholic) eparchy until 1780. Part of the Transylvanian duchy from the 16th century, Mukacheve then came under Austrian control and was made a key fortress of the Austro-Hungarian empire. Mukacheve passed to Czechoslovakia in 1919, was under German-Hungarian occupation from 1938 to 1944, and was ceded to the Ukrainian SSR in 1945 (Ukraine since 1991). The city's architectural landmarks include a castle and a monastery (both 14th century) and a wooden church built in the Ukrainian architectural style (18th century). A center of Jewish learning, from the mid-19th century, approximately half of the population was Jewish during the first half of the 20th century until the killings and deportations between 1941 and 1944—fewer than 700 Jews remaining in the city in 2005. Also called, in Ukrainian, Mukachiv.

Mukacheve Pass, UKRAINE: see VERETS'KYY PASS.

Mukachevo, UKRAINE: see MUKACHEVE.

Mukachiv, UKRAINE: see MUKACHEVE.

Mukah, town, SARAWAK state, MALAYSIA, in W BORNEO, on SOUTH CHINA SEA, 45 mi/72 km NNE of SIBU; 02°54′N 112°06′E. Agriculture (sago, pineapples); livestock raising; fishing.

Mukaihara (moo-KAH-ee-hah-rah), town, Takata county, HIROSHIMA prefecture, SW HONSHU, W JAPAN, 22 mi/35 km N of HIROSHIMA; 34°36′N 132°43′E. Archaeological site (oldest tiles in Japan found here).

Mukaishima (moo-KAH-ee-shee-mah), town, on MUKAI-SHIMA island, Mitsugi county, HIROSHIMA prefecture, W JAPAN, 40 mi/65 km E of HIROSHIMA; 34°23′N 133°12′E. Onions, tomatoes; orchids, bamboo.

Mukai-shima (moo-KAH-ee-shee-mah), island (□ 9 sq mi/23.4 sq km), Mitsugi county, HIROSHIMA prefecture, W JAPAN, nearly connected with city of ONOMICHI on SW HONSHU; 4 mi/6.4 km long, 3 mi/4.8 km wide. Mountainous; fertile (fruit, rice, raw silk). Manufacturing (sailcloth, canned food, tatami).

Mukalla (MOO-KA-luh), town, S YEMEN, a port on the Gulf of ADEN. On ancient frankincense route. It was the capital of the former sultanate of Qaiti. Fish products, tobacco, and coffee are exported. Wooden dhow manufacturing Trading center for a large part of SE Yemen. Modern oil port. Also spelled Makalla.

Mukama, India: see MOKAMA.

Mukandwara Pass (muh-kuhnd-WAH-rah), in range of low hills, SE RAJASTHAN state, NW INDIA, 27 mi/43 km SSE of KOTA. Used by road and railroad. Scene of many battles in Rajput history; on route of British retreat before Jaswant Rao Holkar in 1804.

Muka Turi, town (2007 population 5,682), OROMIYA state, central ETHIOPIA, 27 mi/43 km N of ADDIS ABABA; 09°25′N 38°47′E. Road junction.

Mukawa (moo-KAH-wah), town, Iburi district, Hokkaido prefecture, N JAPAN, 43 mi/70 km S of SAPPORO; 42°34′N 141°55′E. Smelts.

Mukawa (moo-KAH-wah), village, North Koma county, YAMANASHI prefecture, central HONSHU, central JAPAN, 12 mi/20 km N of KOFU; 35°46′N 138°23′E.

Mukdahan (MUK-DAH-HAHN), province, Northeastern region, THAILAND; ⊙ MUKDAHAN; 16°35′N 104°30′E. Rice, tapioca, sugarcane. Formed in 1982 from NAKHON PHANOM and UBON RATCHATHANI provinces. The MEKONG RIVER forms a natural border between MUKDAHAN and LAOS. Mukdahan town is opposite the LAO city of Suwannakhet. Phu Pha Thep National Park, a forest with a variety of wildlife, is located 10 mi/16 km S of Mukdahan town.

Mukdahan (MUK-DAH-HAHN), village and district center, NAKHON PHANOM province, E THAILAND, on right bank of MEKONG RIVER (LAOS border), opposite SAVANNAKHET, 60 mi/97 km S of NAKHON PHANOM 398 mi/641 km ENE of BANGKOK; 16°32′N 104°43′E. Airport. Sometimes spelled Mukdahar or Mukdaharn.

Mukden (MOOK-den), town and canton, NICOLÁS SUÁREZ province, PANDO department, NW BOLIVIA, 23 mi/37 km WSW of COBIJA; 11°11′S 69°02′W. Rice.

Mukden, CHINA: see SHENYANG.

Mukha, YEMEN: see MOCHA.

Mukhaiba Foga (mu-KAI-buh FO-guh), village, N JORDAN, near the S bank of the YARMUK RIVER, 6 mi/9.7 km from where the Yarmuk River meets the JORDAN River. Site of planned dam (never built). Olives, fruit.

Mukhaiba Tehta (mu-KAI-buh TAH-tuh), village, N JORDAN, 1 mi/1.6 km S of the YARMUK River. Olives, fruit.

Mukhairas, YEMEN: see MUKHEIRAS.

Mukhanovo (moo-HAH-nuh-vuh), town (2006 population 650), NE MOSCOW oblast, central European Russia, in the E central KLIN-DMITROV RIDGE, on the administrative border with VLADIMIR oblast, near highway, 20 mi/32 km N of (and administratively subordinate to) SERGIYEV POSAD; 56°30′N 38°19′E. Elevation 715 ft/217 m.

Mukhavets River or **Belyi Mukhovets** (moo-hah-VETS), Polish *Muchawiec*, 75 mi/121 km long, in PRIPET Marshes, SW BREST oblast, BELARUS; rises N of PRUZHANY; flows S, past PRUZHANY, and WSW, past KOBRIN, to BUG RIVER at BREST; basin has an area of 2,452 sq mi/6,350 sq km. Connected with Pripiat' River by DNIEPER Bug Canal. Navigable for c.50 mi/80 km in lower course, up to DNIEPER-BUG CANAL; forms W part of DNIEPER-BUG waterway. Non-overflow dam at its mouth.

Mukhayras, YEMEN: see MUKHEIRAS.

Mukheiras (muh-KAI-rahs), township, S YEMEN, on plateau (elev. 7,000 ft/2,134 m) at N foot of the Kaur al Audhilla, 45 mi/72 km N of SHUQRA and 100 mi/161 km NE of ADEN, near former boundary between NORTH YEMEN and SOUTH YEMEN, 5 mi/8 km SE of Beida, the SE terminal of the N Yemen road system; center of major agricultural district, supplying Aden market with fresh fruit and vegetables. Also spelled Mukheras, Mukhairas, or Mukhayras.

Mukhen (MOO-heen), settlement (2005 population 4,570), S KHABAROVSK TERRITORY, RUSSIAN FAR EAST, on road, 46 mi/74 km ESE of KHABAROVSK; 48°01′N 136°06′E. Elevation 360 ft/109 m. Woodworking. Has a heritage museum.

Mukher (moo-KER), town, NANDED district, MAHARASHTRA state, W central INDIA, 31 mi/50 km S of

---

NANDED; 18°42′N 77°22′E. Millet, cotton, wheat. Also spelled Mukhed.

Mukheras, YEMEN: see MUKHEIRAS.

Mukhino (MOO-hee-nuh), village, E KIROV oblast, E European Russia, on road, 17 mi/27 km SSW of ZUYEVKA; 58°10′N 51°02′E. Elevation 452 ft/137 m. In agricultural area (grain, flax). Also known as Mukhinskaya.

Mukhinskaya, RUSSIA: see MUKHINO.

Mukhmas, Arab village, Jerusalem district, 6.3 mi/8 km NE of JERUSALEM, on the W fringes of the Judean Wilderness, WEST BANK; 31°52′N 35°17′E. Believed to be on the site of the biblical Makhmash. Cereal.

Mukhorshibir' (moo-huhr-shi-BEER), village (2005 population 5,340), S central BURYAT REPUBLIC, S SIBERIA, RUSSIA, on road, 55 mi/89 km S of ULAN-UDE; 51°03′N 107°50′E. Elevation 2,496 ft/760 m. In agricultural and livestock-raising area; dairy processing, logging and lumbering.

Mukhtara, El (muk-TAH-ruh, el), village, central LEBANON, 16 mi/26 km SSE of BEIRUT. Elevation 2,800 ft/853 m. Sericulture, cereals, oranges.

Mukhtolovo (mookh-TO-luh-vuh), town (2006 population 5,585), SW NIZHEGOROD oblast, central European Russia, on road and railroad, 21 mi/34 km W of ARZAMAS, and 14 mi/23 km N of ARDATOV, to which it is administratively subordinate; 55°28′N 43°11′E. Elevation 570 ft/173 m. Sawmilling, wood chemicals, specialized clothing. Has a sanatorium.

Mukhtuya, RUSSIA: see LENSK.

Mukilteo (muh-kuhl-TEE-o), town (2006 population 20,308), SNOHOMISH county, W WASHINGTON, on Possession Sound passage between PUGET SOUND (S) and Possession Sound, extension of Puget (N), suburb 5 mi/8 km SW of EVERETT, and 23 mi/37 km N of downtown SEATTLE; 47°56′N 122°19′W. Ferry to CLINTON, on WHIDBEY ISLAND (W). Varied manufacturing. Mukilteo State Park is here.

Mukinbudin (muhk-uhn-BOO-duhn), town, WESTERN AUSTRALIA state, W AUSTRALIA, c.186 mi/300 km NE of PERTH; 30°52′S 118°08′E. Grain; sheep. Wildflower displays.

Muko (moo-KO), city, KYOTO prefecture, S HONSHU, W central JAPAN, 6 mi/10 km S of KYOTO; 34°56′N 135°42′E. Residential suburb of Kyoto. Bamboo shoots.

Muko-jima, JAPAN: see PARRY ISLANDS.

Mukono, administrative district (□ 4,542 sq mi/11,764 sq km; 2002 population 795,393; 2005 estimated population 867,300), CENTRAL region, S UGANDA; ⊙ MUKONO; 00°15′N 32°55′E. As of Uganda's division into eighty districts, borders WAKISO and KAMPALA (W), LUWERO (NW), KAYUNGA (N), JINJA (NE), and MAYUGE (E) districts and Lake VICTORIA (S). Agriculture (cash crops include tea and sugarcane, also vanilla). Livestock (including poultry, cattle, goats, and pigs). Industrialization is growing (including textiles and tea and sugarcane manufacturing). Fishing in Lake Victoria. Mabira forest; gorilla reserve on Koome Islands. Towns include Mukono and NJERU. Railroad between KASESE town (W Uganda) and MOMBASA (SE KENYA) runs W-E through district. Highway runs through Mukono town, connecting district to KAMPALA city (and W Uganda) and Jinja city (and E Uganda), as well as Kenya. Formed in 2000 from S portion of former MUKONO district (Kayunga district formed from N portion).

Mukono (moo-KO-no), former administrative district (□ 5,499 sq mi/14,297.4 sq km), S UGANDA, along N shore of LAKE VICTORIA, E of KAMPALA; capital was MUKONO; 00°20′N 33°00′E. As of Uganda's division into thirty-nine districts, was bordered by MPIGI, KAMPALA, and LUWERO (W), LIRA (N), and KAMULI, JINJA, and IGANGA (E) districts. Rich agricultural area (cocoa, coffee, tea, bananas, sugar). Largest towns were NJERU, LUGAZI. In 2000 S portion of district was formed into current MUKONO district and N portion formed into KAYUNGA district.

Mukono (moo-KO-no), town (2002 population 46,506), ⊙ MUKONO district, CENTRAL region, S UGANDA, 12 mi/19 km ENE of KAMPALA, near railroad. Cotton, coffee, sugar, bananas, corn, millet; livestock. Was part of former NORTH BUGANDA province.

Mukran, IRAN: see MAKRAN.

Mukry (moo-REE), town, SE LEBAP weloyat, extreme SE TURKMENISTAN, port on the E bank of AMU DARYA, on railroad, and 25 mi/40 km SE of KERKI; 37°36′N 65°44′E. Cotton. MUKRY station, 12 mi/19 km SE, is junction of railroad spur to GAURDAK (NNE).

Muksu, river, 54 mi/87 km long, N TAJIKISTAN; rises in the FEDCHENKO GLACIER; flows W, through goldmining area, joining the KYZYL-SU 28 mi/45 km ENE of Khait to form Surkhob River. Receives the SAUKSAI.

Muktagacha (mook-tah-gah-chah), town (2001 population 37,762), MYMENSINGH district, central EAST BENGAL, BANGLADESH, 9 mi/14.5 km W of MYMENSINGH; 24°46′N 90°14′E. Rice, jute, oilseeds.

Mukteswar, India: see NAINI TAL.

Muktinath (MOOK-tee-naht), Tibetan *Chumik Gyatsa*, village, N NEPAL, 40 mi/64 km N of POKHARA; 28°49′N 83°53′E. Elevation 12,467 ft/3,800 m. One of most sacred of Nepalese shrines; Hindu and Buddhist pilgrimage center; has 108 sacred springs.

Muktsar (MUHK-tsuhr), town, FARIDKOT district, W PUNJAB state, N INDIA, 50 mi/80 km SW of FARIDKOT; 30°29′N 74°31′E. Trades in grain, cotton, oilseeds, cloth fabrics; cotton ginning, hand-loom weaving, ice manufacturing, oilseed milling. Annual Sikh festival.

Mukumari (moo-koo-MAH-ree), village, KASAI-ORIENTAL province, central CONGO, 160 mi/257 km NNW of LUSAMBO; 02°50′S 23°07′E. Elev. 1,998 ft/608 m. Agricultural research station and rubber plantations.

Mukumbura (moo-koo-em-BOO-rah), village, MASHONALAND CENTRAL province, NE ZIMBABWE, 120 mi/193 km NNE of HARARE, on Mkumbura River (MOZAMBIQUE border); 16°12′S 31°39′E. Border crossing. Livestock; grain.

Mukur (MOO-koor), town, GHAZNI province, E AFGHANISTAN, 65 mi/105 km SW of GHAZNI, on highway to KANDAHAR; 32°52′N 67°47′E. Irrigated agriculture. Sometimes spelled Moqor or Muqur.

Mukurti (moo-KOOR-tee), peak (8,380 ft/2,554 m) in NILGIRI HILLS, TAMIL NADU state, S INDIA, 13 mi/21 km WSW of OOTY, or UDAGAMANDALAM; gives rise (E) to MOYAR RIVER; 11°22′N 76°31′E. The name Mukurti also applies to a dam of PYKARA hydroelectric system. Also spelled Murkurti.

Mukwar, SUDAN: see SENNAR.

Mukwonago (muhk-WAHN-uh-go), town (2006 population 6,815), WAUKESHA county, SE WISCONSIN, on small Phantom Lake, 24 mi/39 km SW of MILWAUKEE; 42°52′N 88°19′W. In dairying, livestock-raising, and farming region; manufacturing (bottled water, pump components). Rail junction. Kettle Moraine State Forest (S unit) to W.

Mula (MOO-lah), town, MURCIA province, SE SPAIN, 21 mi/34 km W of MURCIA; 38°03′N 01°30′W. Agricultural center in fertile garden region producing citrus and other fruit, olives, wine, cereals, rice. Flour milling, olive-oil processing, fruit-conserve manufacturing, brandy distilling; lumbering, livestock raising. Has ruins of ancient castle; two Renaissance churches. Mineral springs 3 mi/4.8 km W. Irrigation reservoir nearby.

Mulabo, INDONESIA: see MEULABOH.

Mulag, India: see MULUG.

Mulainagiri, India: see BABA BUDAN RANGE.

Mulakatholhu Atoll, small central group of MALDIVES, in INDIAN OCEAN; 02°44′N 73°16′E–03°11′N 73°40′E. Also called Mulaku Atoll.

Mulaku Atoll, MALDIVES: see MULAKATHOLHU ATOLL.

Mulaly (moo-lah-LUH), town, TALDYKORGAN region, KAZAKHSTAN, on railroad, 25 mi/40 km N of TALDYKORGAN; 45°25′N 78°16′E. Also spelled Molaly.

Mula-Mutha River (MOO-tah–MOO-lah), c.80 mi/129 km long, MAHARASHTRA state, W central INDIA; rises in WESTERN GHATS in two headstreams joining at PUNE; flows E to BHIMA RIVER 17 mi/27 km NW of DAUND. Dam on N headstream (Mula River) supplies Bhira hydroelectric plant. Dam on S headstream (Mutha River) supplies canal irrigation system, with headworks 7 mi/11.3 km SW of Pune; left system, with headworks 7 mi/11.3 km SW of Pune; left (N) canal extends 18 mi/29 km NE, right (S) canal 70 mi/113 km E to point 8 mi/12.9 km WNW of Daund.

Mulan (MU-LAN), town, ⊙ Mulan county, central S HEILONGJIANG province, NORTHEAST, CHINA, on left bank of SONGHUA River, and 75 mi/121 km E of HARBIN; 45°56′N 128°04′E. Grain, tobacco, sugar beets.

Mulanay (moo-LAH-nei), town, QUEZON province, S LUZON, PHILIPPINES, on W BONDOC PENINSULA, on MOMPOG PASS, 60 mi/97 km ESE of LUCENA; 13°32′N 122°26′E. Fishing.

Mulanje (muh-lahn-jeh), administrative center and district (2007 population 557,138), Southern region, MALAWI, on road to QUELIMANE (MOZAMBIQUE), and 40 mi/64 km SE of BLANTYRE; 16°05′S 35°29′E. Tea-growing center; also tung, cotton, tobacco, corn, rice.

Mulanje Mountains (muh-lahn-jeh), syenite outcrop, Southern region, SE MALAWI, NE of MULANJE; rise abruptly from surrounding plateau; 12 mi/19 km across. Highest point (in center) is SAPITWA peak on Mount Mulanje (9,849 ft/3,002 m).

Mula Pass (MOO-lah) (elevation c.6,000 ft/1,829 m) at S end of CENTRAL BRAHUI RANGE, central BALUCHISTAN province, SW PAKISTAN, SE of KALAT; 28°20′N 67°30′E. Lies on trade route, formerly of some importance, between highlands of KALAT and plains (E). Name also applied to course of MULA RIVER which rises c.10 mi/16 km SE of KALAT and flows c.180 mi/290 km SSE and NE into KACHHI plain near GANDAVA.

Mularas (maw-LAH-rahs), mining settlement, QAFSAH province, W TUNISIA, on railroad 32 mi/51 km W of QAFSAH. Important phosphate mines and processing plant. Also called Moularés.

Mula River, India: see MULA-MUTHA RIVER.

Mulatas Islands, PANAMA: see SAN BLAS ISLANDS.

Mulatière, La (mu-lah-tyer, lah), town, S inner suburb of LYON, RHÔNE department, RHÔNE-ALPES region, E central FRANCE, on right bank of RHÔNE RIVER, at mouth of the SAÔNE. Produces electrical cables. Railroad yards along river are opposite artificial harbor of the city of Lyon.

Mulbagal (mool-BUH-guhl), town, KOLAR district, KARNATAKA state, SE INDIA, 17 mi/27 km E of KOLAR; 13°10′N 78°24′E. Tobacco curing, sheep grazing.

Mulbekh (muhl-BAIK), town, KARGIL district, JAMMU AND KASHMIR state, extreme N INDIA, in central KASHMIR Range, on right tributary of SURU River 19 mi/31 km SE of KARGIL; 34°23′N 76°22′E. Has Buddhist monastery, Dard castle ruins, rock inscriptions, and huge (c.eighth century A.D.) Buddhist rock sculpture.

Mulberry (muhl-BER-ee), city (□ 3 sq mi/7.8 sq km; 2005 population 3,233), POLK county, central FLORIDA, 10 mi/16 km S of LAKELAND; 27°53′N 81°58′W. Processes phosphate from nearby mines; also manufacturing of fertilizer, steel. Known as the "Phosphate Capital of the World."

Mulberry, town (2000 population 1,627), CRAWFORD county, NW ARKANSAS, 22 mi/35 km ENE of FORT SMITH, and on MULBERRY RIVER; 35°30′N 94°04′W. N of its confluence with ARKANSAS RIVER. Manufacturing (farm gates and corral panels); trade center for agricultural area. Ozark National Forest to N.

Mulberry, town (2000 population 1,387), CLINTON county, central INDIANA, 10 mi/16 km NW of FRANKFORT; 40°21′N 86°40′W. In agricultural area (corn, soybeans, hogs). Laid out 1858.

Mulberry (MUHL-ber-ee), unincorporated town (□ 5 sq mi/13 sq km; 2000 population 2,269), WILKES county, NW NORTH CAROLINA, 6 mi/9.7 km N of WILKESBORO; 36°13′N 81°10′W. Manufacturing; agricultural area (tobacco, soybeans; poultry, dairying); timber.

Mulberry, unincorporated town (2000 population 841), SUMTER county, central SOUTH CAROLINA, residential suburb 3 mi/4.8 km E of SUMTER; 33°57′N 80°19′W.

Mulberry, village (2000 population 577), CRAWFORD county, SE KANSAS, at MISSOURI state line, 11 mi/18 km NNE of PITTSBURG; 37°33′N 94°37′W. In livestock, poultry, and grain area. Coal mines nearby.

Mulberry (MUHL-ber-ee), unincorporated village (□ 2 sq mi/5.2 sq km; 2000 population 3,139), CLERMONT county, extreme SW OHIO; suburb 15 mi/24 km NE of downtown CINCINNATI; 39°11′N 84°14′W. Adjacent to MOUNT REPOSE (unincorporated village) and just off U.S. I-275.

Mulberry Fork, river, c.100 mi/161 km long, in N ALABAMA; rises in NE CULLMAN county; flows SW, past CORDOVA, to join LOCUST FORK river c.20 mi/32 km W of BIRMINGHAM, forming BLACK WARRIOR RIVER.

Mulberry Gap (MUHL-be-ree GAP) (3,100 ft/945 m), on WILKES and ALLEGHANY county line, NW NORTH CAROLINA, pass through the BLUE RIDGE, 15 mi/24 km N of WILKESBORO.

Mulberry Grove, village (2000 population 671), BOND county, S central ILLINOIS, 8 mi/12.9 km ENE of GREENVILLE; 38°55′N 89°16′W. In agricultural area (corn, wheat; livestock; dairy products); ships sand.

Mulberry River, c.35 mi/56 km long, in central ALABAMA; rises in CHILTON county, W of CLANTON; flows S to ALABAMA RIVER 9 mi/14.5 km E of SELMA.

Mulberry River, c.70 mi/113 km long, in NW ARKANSAS; rises NE in JOHNSON county, in the OZARKS; flows SW to ARKANSAS RIVER just S of MULBERRY. Most of upper course is in Ozark National Forest. Famous for canoeing.

Mulchatna River, Alaska: see NUSHAGAK RIVER.

Mulchén (mool-CHAIN), town, ⊙ Mulchén comuna (2002 population 21,819), BÍO-BÍO province, BÍO-BÍO region, S central CHILE, in S part of the central valley, 19 mi/31 km SSE of LOS ÁNGELES; 37°43′S 72°14′W. Railroad terminus; wheat and cattle center; food processing, lumbering.

Mul'da (mool-DAH), settlement (2005 population 171), NE KOMI REPUBLIC, NE European Russia, in the USA RIVER basin, on road, 4 mi/6 km W of VORKUTA; 67°30′N 63°51′E. Elevation 524 ft/159 m. Coal mines in the vicinity.

Mulde River (MUL-de), 77 mi/124 km long, central GERMANY; formed 2 mi/3.2 km N of COLDITZ by the FREIBERGER MULDE and the ZWICKAUER MULDE; flows NNW, past GRIMMA and WURZEN, to the ELBE at DESSAU. Not navigable.

Muldoon (MUHL-doon), unincorporated village, FAYETTE county, S central TEXAS, near COLORADO RIVER, 14 mi/23 km SE of SMITHVILLE; 29°48′N 97°04′W. Agricultural area.

Muldraugh (MUHL-duhr), town (2000 population 1,298), MEADE county, N KENTUCKY, 37 mi/60 km SSW of LOUISVILLE, near OHIO RIVER, within FORT KNOX Military Reserve, U.S. Gold Bullion Depository to S. Otter Creek Park to W; 37°56′N 85°59′W.

Muldrow (MUHL-dro), town (2006 population 3,202), SEQUOYAH county, E OKLAHOMA, 5 mi/8 km ESE of SALLISAW; 35°23′N 94°35′W. In agricultural area; manufacturing (furniture).

Muleba (moo-LAI-bah), village, KAGERA region, NW TANZANIA, 37 mi/60 km S of BUKOBA, on Lake VICTORIA; 01°53′S 31°40′E. Road junction. Livestock; coffee, subsistence crops; fish.

Mulegé, municipio, MEXICO: see SANTA ROSALÍA.

Mulegé (moo-le-HE), village, BAJA CALIFORNIA SUR, NW MEXICO, near Gulf of CALIFORNIA, 35 mi/56 km S of SANTA ROSALÍA on Mexico Highway 1; 26°54′N 112°00′W. Manganese-mining and agricultural center (sugarcane, coconuts, dates, figs, grapes). Mission Santa Rosade Mulegé (1705–1828) is nearby.

Mulei, CHINA: see MORI.

Mulembera (moo-laim-BAI-rah), village, KIGOMA region, NW TANZANIA, 65 mi/105 km NE of KIGOMA, on MAKERE River; 04°19′S 30°21′E. Timber; tobacco, grain; sheep, goats.

Mule Mountains, COCHISE county, SE ARIZONA, W of BISBEE, near Mexican border; rise to 7,500 ft/2,286 m in Mount BALLARD. Copper, gold, and silver are mined near Bisbee and WARREN, in S part of range.

Muleshoe (MYOOL-shoo), town (2006 population 4,486), ⊙ BAILEY county, NW TEXAS, on the LLANO ESTACADO, c.65 mi/105 km NW of LUBBOCK, on Blackwater Draw Creek; 34°13′N 102°43′W. In cattle-ranching and irrigated agricultural region; agribusiness; dairying; cotton, wheat, corn, sorghum; vegetables, potatoes; food processing (corn flour, seed packaging, tortillas, potato packaging). Muleshoe National Wildlife Refuge to S; Coyote Lake to SW. Settled 1913, incorporated 1926.

Muley-Hacén, SPAIN: see MULHACÉN.

Mulfingen (MUL-fing-uhn), village, FRANCONIA, BADEN-WÜRTTEMBERG, GERMANY, on the JAGST, 7 mi/ 11.3 km ENE of KÜNZELSAU; 49°20′N 09°48′E. Forestry.

Mulga, town (2000 population 973), Jefferson co., N central Alabama, just W of Birmingham. Name may be derived from the Creek word 'omalga,' meaning 'all.' Inc. in 1947.

Mulgirigala, village, SOUTHERN PROVINCE, SRI LANKA, 13 mi/21 km N of TANGALLA. Trades in rice, coconuts, and vegetables. On top of isolated rock (692 ft/211 m high) are cave temples, some dating to 2nd century C.E., a stupa, and a Buddhist monastery. Also called Mulkirigala.

Mulgrave (MUHL-graiv), town (2001 population 904), E NOVA SCOTIA, CANADA, on W shore of STRAIT OF CANSO, 30 mi/48 km E of ANTIGONISH; 45°37′N 61°24′W. Elevation 82 ft/24 m. Fishing; port. Named for the Earl of Mulgrave. Former and merged community names include: McNair's Cove, Port Mulgrave, Wylde's Cove.

Mulgrave-et-Derry (MUHL-graiv–ai–de-REE), village (□ 115 sq mi/299 sq km; 2006 population 254), Outaouais region, SW QUEBEC, E Canada, 9 mi/15 km from Saint-Sixte; 45°47′N 75°22′W.

Mulgrave Island, AUSTRALIA: see BADU ISLAND.

Mulgrave Islands, MARSHALL ISLANDS: see MILI.

Mulhacén (moo-lah-THAIN) or **Muley-Hacén** (moo-LAI–ah-THAIN), mountain peak (11,411 ft/3,478 m), GRANADA province, S SPAIN. Highest point of the SIERRA NEVADA and of Spain.

Mulhall (MUHL-hawl), village (2006 population 265), LOGAN county, central OKLAHOMA, 13 mi/21 km N of GUTHRIE; 36°03′N 97°24′W. In farming area.

Mülhausen, FRANCE: see MULHOUSE.

Mülheim an der Ruhr (MYOOL-heim an der ROOR), city, North Rhine-Westphalia, W GERMANY, 6 mi/9.6 km WSW of ESSEN, on the RUHR RIVER; 51°26′N 06°53′E. Industrial center and port of the RUHR district; road and railroad traffic hub. Once produced chiefly coal and steel, but in the mid-20th century its products were diversified to include machinery, electrical goods, leather products, and iron. At the city's noted institute for coal research, the Fischer-Tropsch process for coal liquefaction and the Ziegler process for the production of polyethylene plastics were discovered. Chartered in 1808.

Mülheim-Kärlich (MYOOL-heim–KER-likh), town, RHINELAND-PALATINATE, W GERMANY, near the RHINE RIVER, 4 mi/6.4 km WNW of KOBLENZ; 50°24′N 07°28′E. Agriculture (fruit); limestone quarries; nuclear power station. Mülheim and Kärlich (both first mentioned 1277) united in 1969.

Mulhouse (mul-OOZ), German *Mülhausen* (MUL-hou-zuhn), city (□ 9 sq mi/23.4 sq km), HAUT-RHIN department, E FRANCE, in ALSACE, on the ILL RIVER and RHÔNE-RHINE CANAL, 245 mi/394 km ESE of PARIS, and 18 mi/29 km NW of BASEL (Switzerland); 47°40′N 07°25′E. A leading manufacturing center, Mulhouse has a diversified textile industry, chemical plants, (converting locally mined potash into fertilizer), explosives, machine shops (auto parts, textile machinery, steel pipes), a paper mill (nearby), printing plants, breweries, and a variety of food-processing facilities. Nearby are the only important potash mines in W Europe. The city has a university established in 1975 and shares an international airport at St. Louis (12 mi/21 km SE) with Basel. Mulhouse became a free imperial city in the 13th century. In 1515 it became an allied member (but not a canton) of the Swiss Confederation, and in 1586 it became a neutral republic. In 1798, Mulhouse voted to unite with France. The city, like the rest of Alsace, was annexed by Germany (1871–1918, and again 1940–1944). Mulhouse has a 16th-century Renaissance town hall and narrow, winding streets and old houses. Its suburban residential workers' districts, laid out in an octagonal grid pattern, were considered a model of industry and labor cooperation in the late 19th century. In its many musuems, Mulhouse features a collection of classic and antique automobiles, a historic railroad exhibit, displays of electric energy, textile and paper prints, and a fine zoological and botanical garden. The metropolitan area's population exceeds 220,000; suburbs extend N to the potash mines, and SW along the Rhône-Rhine Canal. N-S and E-W expressways intersect at Mulhouse, providing easy access to Germany's BLACK FOREST, N Alsace, and to Switzerland.

Muli (MUHL-ee), town, SURENDRANAGAR district, GUJARAT state, W central INDIA, 15 mi/24 km WSW of WADHWAN; 22°39′N 71°28′E. Cotton, millet; handloom weaving. Was capital of the former EASTERN KATHIAWAR state of Muli of WESTERN INDIA STATES agency; state merged (1948) with SAURASHTRA, later with Gujarat state.

Muling (MU-LING), town, ⊙ Muling county, SE HEILONGJIANG province, NORTHEAST, CHINA, 32 mi/51 km E of MUDANJIANG, and on railroad; 44°56′N 130°32′E. Grain, tobacco, medicinal herbs.

Muling River (MU-LING), over 250 mi/402 km long, SE HEILONGJIANG province, NORTHEAST, CHINA; rises in Laoye Mountains on JILIN border, flows N and NE, through coal-mining region, past MULING, JIXI, and MISHAN, to USSURI River (RUSSIA border) S of HULIN. Sometimes called Muren River.

Mulino (MOO-lee-noh), town (2006 population 11,450), W NIZHEGOROD oblast, central European Russia, on road and near railroad, 36 mi/58 km W of NIZHNIY NOVGOROD, and 10 mi/16 km WNW of VOLODARSK, to which it is administratively subordinate; 56°17′N 42°56′E. Elevation 475 ft/144 m. Peat works; food enterprises.

Mulino (MOO-lee-nuh), village, NE KIROV oblast, E European Russia, on the VYATKA River, 3 mi/5 km E, and under administrative jurisdiction, of NAGORSK; 59°19′N 50°54′E. Elevation 419 ft/127 m. Sawmilling, lumbering, timbering. Also called Mulinskoye (MOO-leen-skuh-ye).

Mulki (MUHL-kee), town, Dakshin Kannad district, KARNATAKA state, SW INDIA, on MALABAR COAST of ARABIAN SEA, 16 mi/26 km N of MANGALORE; 13°06′N 74°48′E. Fish curing; coconuts, mangoes.

Mull (MUHL), island (□ 351 sq mi/912.6 sq km), Argyll and Bute, W Scotland, largest island of the INNER HEBRIDES, separated from the mainland by the Sound of Mull and the Firth of Lorn; 56°27′N 06°00′W. The land is mountainous, rising from the deeply indented coastline to 3,169 ft/966 m at BEN MORE. Farming, fishing, forestry, quarrying. TOBERMORY, a summer resort, is the chief town. A Spanish treasure galleon sank in its bay in 1588. Several medieval castles still stand. Isle of IONA to SW. Formerly in Strathclyde, abolished 1996.

Mullaghmore (muh-luhk-MOR), Gaelic *Mullach Mór*, promontory on DONEGAL BAY, NE SLIGO county, NW IRELAND, 14 mi/23 km N of SLIGO; 54°28′N 10°28′W. Fishing village of Mullaghmore on E side of promontory.

Mullaitivu, district (□ 467 sq mi/1,214.2 sq km), NORTHERN PROVINCE, SRI LANKA; ⊙ MULLAITIVU; 09°15′N 80°33′E.

Mullaitivu (MU-LAI-TI-vu), town, ⊙ MULLAITIVU district, NORTHERN PROVINCE, SRI LANKA, on E coast, 60 mi/97 km SE of JAFFNA; 09°16′N 80°49′E. Fishing port; trades in vegetables; rice and tobacco.

Mullan, town (2000 population 840), SHOSHONE county, N IDAHO, 6 mi/9.7 km E of WALLACE in COEUR D'ALENE MOUNTAINS, near MONTANA state line; 47°28′N 115°48′W. Elevation 3,277 ft/999 m. Terminus of railroad spur from Wallace. Lead, silver, zinc mines. Founded 1884 at time of lead-silver strike. Lookout Pass (4,725 ft/1,440 m) on state line and Lookout Pass Ski Area (Montana) to E. Incorporated 1904.

Mullen, village (2006 population 502), ⊙ HOOKER county, central NEBRASKA, 60 mi/97 km NNW of NORTH PLATTE, near MIDDLE LOUP RIVER; 42°02′N 101°02′W. In heart of Sand Hills region. Beef cattle, livestock.

Müllendorf (MYOOL-len-dorf), village, N BURGEN-LAND, E Austria, 3 mi/4.8 km W of EISENSTADT; 47°51′N 16°27′E. Chalk quarries and chalk manufacturing.

Müllendorf, LUXEMBOURG: see STEINSEL.

Mullens, town (2006 population 1,630), WYOMING county, S WEST VIRGINIA, on GUYANDOTTE RIVER, 23 mi/37 km NNW of BLUEFIELD; 37°34′N 81°23′W. Bituminous- and semibituminous-coal region. Agriculture (corn, potatoes); cattle. Timber. Twin Falls State Park to NW.

Mulleriyawa, town, WESTERN PROVINCE, SRI LANKA; 06°56′N 79°56′E. Market town and residential suburb of COLOMBO. Buddhist temple nearby.

Mullet Bay, SINT MAARTEN, NETHERLANDS ANTILLES, beach area on SW shore, 6 mi/9.7 km E of PHILIPS-BURG. Location of large resort and casino complex, near Princess Juliana International Airport.

Mullet Peninsula (MUH-let) or **The Mullet**, Gaelic *An Muirthead*, NW MAYO county, NW IRELAND, extending 17 mi/27 km S between BLACKSOD BAY (E) and the ATLANTIC. Connected with mainland by narrow isthmus at BELMULLET. BROAD HAVEN inlet N.

Mullett Lake, CHEBOYGAN county, N MICHIGAN, 5 mi/8 km S of CHEBOYGAN; c.11 mi/18 km long, 3 mi/4.8 km wide; 45°33′N 84°31′W. MULLETT LAKE, resort village, is on NW shore. Village of Aloha and Aloha State Park on E shore. Villages of Mullet Lake and TOPINABEE on W shore. Source of CHEBOYGAN RIVER.

Mullewa (MU-luh-wuh), town, W WESTERN AUSTRALIA, 60 mi/97 km ENE of GERALDTON; 28°32′S 115°31′E. Railroad junction; wheat, livestock. Many wall murals of town's history. Wildflowers.

Müllheim (MYOOL-heim), town, S UPPER RHINE, BADEN-WÜRTTEMBERG, GERMANY, at W foot of BLACK FOREST, 16 mi/26 km SW of FREIBURG; 47°49′N 07°37′E. Railroad junction; wine-trade center of MARKGRÄFLER. Land wine region; woodworking; tourism.

Mullhyttan (MUL-HIT-tahn), village, ÖREBRO county, S central SWEDEN, on Svartån River, 19 mi/31 km WSW of ÖREBRO; 59°09′N 14°42′E.

Mulliangiri, India: see BABA BUDAN RANGE.

Mullica (MUH-li-kuh), township, ATLANTIC county, S NEW JERSEY, along MULLICA RIVER, SW of BATSTO; 39°36′N 74°40′W. Incorporated 1838.

Mullica Hill (MUH-li-kuh), village (2000 population 1,658), GLOUCESTER county, SW NEW JERSEY, on RACCOON CREEK, and 15 mi/24 km SSW of CAMDEN, in rich fruit and produce region; 39°44′N 75°13′W. Residential development starting in area.

Mullica River (MUH-li-kuh), c.55 mi/89 km long, SE NEW JERSEY; rises near BERLIN; flows generally SE, forming part of BURLINGTON-ATLANTIC county line, to GREAT BAY SW of TUCKERTON. Receives BATSTO, WADING, and BASS rivers. Navigable for c.23 mi/37 km above mouth.

Mulliken, village (2000 population 557), EATON county, S central MICHIGAN, 17 mi/27 km W of LANSING; 42°45′N 84°54′W. In farm area; manufacturing (metal fabrication).

Mullin (MUHL-uhn), village (2006 population 175), MILLS county, central TEXAS, 21 mi/34 km SE of BROWNWOOD; 31°33′N 98°39′W. In grain, livestock area.

Mullingar (MUH-lin-GAHR), Gaelic *An Muileann gCearr*, town (2006 population 18,054), ⊙ WESTMEATH county, central IRELAND, on BROSNA RIVER and ROYAL CANAL, 50 mi/80 km WNW of DUBLIN; 53°32′N 07°21′W. Agricultural market in dairying, cattle-raising, potato-growing region; furniture manufacturing. Has important annual horse fair. Seat of Roman Catholic bishop of Meath and was former site of 13th-century Augustinian and Dominican abbeys. Loughs Owel and Ennell, frequented by anglers, nearby.

Mullins, town (2006 population 4,844), MARION county, E SOUTH CAROLINA, 28 mi/45 km E of FLOR-ENCE, near LITTLE PEE DEE RIVER; 34°12′N 79°15′W. Manufacturing includes clothing, meat processing, military food rations, light bulbs, lumber. Important tobacco market; grain, sorghum, cotton, timber. Little Pee Dee State Park to N.

Mullinville, village (2000 population 279), KIOWA county, S KANSAS, 10 mi/16 km W of GREENSBURG; 37°35′N 99°28′W. Grain, livestock.

Mull of Galloway (MUHL uhv GA-lo-wai), headland (239 ft/73 m), DUMFRIES AND GALLOWAY, SW Scotland, the southernmost extremity of Rinns of Galloway and of Scotland; 54°38′N 04°51′W. Lighthouse.

Mull of Kintyre (MUHL uhv KIN-tei-uhr), headland at SW extremity of KINTYRE peninsula, Argyll and Bute, W Scotland, on the NORTH CHANNEL, and 11 mi/18 km SW of CAMPBELTOWN; 55°17′N 05°47′W. Lighthouse. The cape is nearest British point to Ireland, 13 mi/21 km SW. Formerly in Strathclyde, abolished 1996.

Mull of Oa (MUHL uhv O-ah), headland at S tip of ISLAY, INNER HEBRIDES, Argyll and Bute, W Scotland; 55°35′N 06°19′W. Formerly in STRATHCLYDE, abolished 1996.

Mullovka (mool-LOF-kah), town (2006 population 6,365), NE ULYANOVSK oblast, E central European Russia, on highway, 10 mi/16 km W, and under administrative jurisdiction, of DIMITROVGRAD; 54°12′N 49°24′E. Elevation 236 ft/71 m. Woolen milling.

Mullsjö (MUL-SHUH), town, SKARABORG county, S SWEDEN, on Stråken Lake (10-mi/16-km-long expansion of upper TIDAN River), 14 mi/23 km NW of JÖNKÖPING; 57°55′N 13°53′E.

Mullumbimby (MU-luhm-BIM-bee), municipality, NE NEW SOUTH WALES, SE AUSTRALIA, 80 mi/129 km SSE of BRISBANE, and adjacent Brunswick River; 28°33′S 153°30′E. Banana plantations.

Mulmur (muhl-MYUR), township (□ 111 sq km; 2001 population 3,099), S ONTARIO, E central Canada, 19 mi/30 km from ORANGEVILLE; 44°11′N 80°06′W. Agriculture. Incorporated 1851.

Mulobezi, village, WESTERN province, SW ZAMBIA, 85 mi/137 km NW of MARAMBA, on Machile River; 16°47′S 25°10′E. Terminus of railroad from Maramba. Cattle; major teakwood area; diamond deposits to W. KAFUE NATIONAL PARK to NE.

Mulock Glacier, outlet glacier in EAST ANTARCTICA flowing from the polar plateau into the NW corner of the ROSS ICE SHELF; 79°00′S 160°00′E.

Mulroy Bay (mul-ROI), narrow inlet (12 mi/19 km long) of the ATLANTIC, N DONEGAL county, N IRE-LAND, between LOUGH SWILLY and SHEEP HAVEN; 55°09′N 07°41′W.

Mulsanne (mool-sahn), town (□ 5 sq mi/13 sq km; 2004 population 4,847), SARTHE department, PAYS DE LA LOIRE region, W central FRANCE; 47°54′N 00°15′E. Agricultural market town 8 mi/13 km SSE of Le MANS.

Multai (mool-TEI), town (2001 population 21,428), BETUL district, W Madhya Pradesh state, central India, 25 mi/40 km ESE of BETUL; 21°46′N 78°15′E. Trades in wheat, millet, oilseeds. Lac cultivation in nearby forested hills (teak, sal, myrobalan). Source of TAPI RIVER (just W) is place of Hindu pilgrimage.

Multan (muhl-TAHN), district (□ 5,653 sq mi/14,697.8 sq km), S PUNJAB province, E central PAKISTAN, in BARI DOAB; ⊙ MULTAN. Bounded S by SUTLEJ River, W by CHENAB River, N by RAVI River A hot, dry tract, largely of alluvial soil and widely irrigated (SIDHNAI CANAL; N). Formerly also spelled MOOLTAN.

Multan (muhl-TAHN), city, ⊙ MULTAN district, S PUNJAB province, E central PAKISTAN, near the CHE-NAB River; 30°11′N 71°29′E. Important road and railroad junction, an agricultural center, and a market for textiles, leather goods, and other products. Industries include metalworking, flour, sugar, and oil milling, and the manufacturing of textiles, fertilizer, soap, chemicals, glass, and engineered goods. Also known for its handicrafts, especially pottery and enamel work. Seat of a government college. One of the INDIAN SUBCONTINENT's oldest cities, Multan derives its name from an idol in the temple of the sun god, a shrine of the pre-Muslim period. Conquered (c.326 B.C.E.) by Alexander the Great, visited (C.E. 641) by the Chinese Buddhist scholar Hsüan-tsang, taken (8th century) by the Arabs, and captured by Muslim Turkish conqueror Mahmud of Ghazni in 1005 and by Tamerlane in 1398. In the 16th and 17th centuries, Multan enjoyed peace under the early Mogul emperors. In 1818 the city was seized by Ranjit Singh, leader of the Sikhs. The British held it from 1848 until PAKISTAN achieved independence in 1947. Landmarks include an old fort containing the monumental 14th-century tombs of two Muslim saints; one of which is an award-winning architectural conservation project.

Multnomah (muhlt-NO-muh), county (□ 465 sq mi/1,209 sq km; 2006 population 681,454), NW OREGON; ⊙ PORTLAND; 45°32′N 122°24′W. Bounded N by Columbia River, forms WASHINGTON state line. Drained by WILLAMETTE RIVER. Manufacturing and shipping center at Portland. Urbanized in W; agricultural areas in center and extreme NW. Agriculture (wheat, oats, barley, potatoes, corn, berries, apples, cherries, pears, plums, peaches, hops); nurseries, wineries. Part of MOUNT HOOD National Forest, including Columbia Wilderness Area, in E. County has 12 state parks, all E of Portland, along the Columbia River, most notably Bridal Veil Falls, Benson and John B. Yeon State Parks. Formed 1854.

Multnomah Falls (muhlt-NO-muh), waterfall (c. 850 ft/259 m high, including upper and lower falls), NW OREGON, in Multnomah Creek, rising on Larch Mountain (4,100 ft/1,250 m) and plunging into Columbia River gorge 9 mi/14.5 km WSW of BONNE-VILLE, near Benson State Park. Site of historic Multnomah Falls Lodge.

Muluá, GUATEMALA: see SANTA CRUZ MULUÁ.

Mulug (moo-LUHG), town, WARANGAL district, AN-DHRA PRADESH state, SE INDIA, 27 mi/43 km NE of WARANGAL; 18°11′N 79°57′E. Rice, oilseeds. Noted thirteenth-century temples 5 mi/8 km N, at Palampet. Also spelled Mulag.

Mulu, Gunung (MOO-loo, GOO-nong), highest peak (7,798 ft/2,377 m) of SARAWAK, N central BORNEO, MALAYSIA, 65 mi/105 km ESE of MIRI; 04°04′N 114°56′E.

Mulund (MOO-luhnd), town, MUMBAI (Bombay) sub-urban district, MAHARASHTRA state, W central INDIA, on SALSETTE ISLAND, 2 mi/3.2 km SW of Thane. Match manufacturing, salt drying; cement and metalworks. Suburb of Mumbai.

Mulungu (MOO-loon-goo), city (2007 population 9,317), E PARAÍBA, NE BRAZIL, 37 mi/60 km WNW of JOÃO PESSOA; 07°00′S 35°29′W. Junction (spur to ALAGOA GRANDE) on NATAL–João Pessoa railroad. Called Camarazal, 1944–1948.

Mulungu, city, W RIO GRANDE DO NORTE state, BRAZIL, 25 mi/40 km SW of MOSSORÓ; 05°29′S 37°42′W.

Mulungu, CONGO: see TSHIBINDA.

Mulungu do Morro (MOO-loon-goo do MOR-ro), city (2007 population 14,178), central BAHIA, BRAZIL, 56 mi/91 km SSE of IRECÊ; 11°58′S 41°36′W.

Mulungushi Dam, ZAMBIA: see KABWE.

Mulungu-Tshibinda, CONGO: see TSHIBINDA.

Mulungwishi (moo-lung-WEE-shee), town, KATANGA province, SE CONGO, 15 mi/24 km NW of LIKASI; 10°47′S 26°38′E. Elev. 4,484 ft/1,366 m.

Mulvane (muhl-VAIN), town (2000 population 5,155), on SEDGWICK-SUMNER county line, S KANSAS, on ARKANSAS RIVER, and 16 mi/26 km SSE of WICHITA; 37°28′N 97°14′W. Railroad junction. In diversified-farming area. Sand and gravel plant here. Laid out 1879, incorporated 1883.

Muma (MOO-muh), village, ORIENTALE province, NE CONGO, SW of BONDO; 03°24′N 23°15′E. Elev. 1,469 ft/447 m. Oil palms.

Mumbai, formerly **Bombay**, city and district (□ 27 sq mi/70.2 sq km; 2001 population 16,434,386), ⊙ MAHARASHTRA state, W INDIA, on Bombay and Salsette Islands; 18°58′N 72°50′E. Bombay Island was created in the 19th century by reclamation projects that combined seven basaltic islets and is a peninsula of the larger Salsette Island to the N. Salsette Island itself is connected to the mainland by causeways and railroad embankments. Mumbai has the only natural deepwater harbor in W India. It is a transportation hub and industrial center. Manufacturing includes cotton textiles, motor vehicles, machinery, clothing, chemicals, pharmaceutical products, electronic goods and refined petroleum; shipbuilding, fish processing. Home to India's largest banks, including the headquarters of the Reserve Bank of India and the State Bank of India and several financial houses. Also the center of India's domestic film and entertainment industry. Although it contains some of the world's largest slums (about half the city's population live in slum districts), Mumbai is the nation's financial center and is also a city of great wealth; most of India's tax revenues come from here. There is an extensive system of hydroelectric stations, and nearby at TROMBAY is a nuclear reactor. Seat of Bombay University (founded 1857), with its new campus near the domestic airport, and Shrimati Nathibai Thakersey (SNDT) University for Women. Tata Institute of Fundamental Research, Bhaba Atomic Research Centre, IIT Bombay (one of seven campuses of the prestigious Indian Institute of Technology), and several medical, engineering, and technical institutions are located here. National Centre of Performing Arts and Jahangir Art Gallery are some of Mumbai's cultural attractions. Victorian-style India Gate near the waterfront commemorates a 1911 visit by King George V. A modern international airport and a chain of luxury hotels add to the tourism facilities. Mumbai has many large suburbs, including Powaii, ANDHERI, THANE, and ULHASNAGAR, each with a population of over 100,000. An efficient suburban railroad system enables approximately six million suburban commuters to get to the city center for work every day. Mumbai's Central Business District (CBD) is thus thinned out every evening while office workers travel long distances to their living quarters in the suburbs. Bombay harbor was India's chief passenger port before the advent of airlines. Now Mumbai's airport is the country's busiest. Mumbai has the largest community of Parsis, descendants of Persian Zoroastrians, who have contributed significantly to the development of

Mumbai and also to the country as a whole. Historical remains exist from the period (320–184 B.C.E.) when much of the area belonged to the Buddhist Maurya empire. Buddhism was supplanted (c.5th century A.D.) by Hinduism, which remains the major religion. In 1534, the area was ceded by the Sultan of Gujarat to Portugal, then the leading power operating on the W coast of the Indian peninsula. After passing to Great Britain in 1661, Bombay was the W Indian headquarters (1668–1858) of the East India Company. By the early 19th century the British had formed the Bombay Presidency. During the American Civil War, the port expanded to meet the world demand for cotton and became a leading cotton-spinning and weaving center. In 1937, Bombay became a province in the new administration system but retained its presidency status. After India gained independence in 1947, all former native states within the provincial boundary joined Bombay. In 1956, Bombay was reorganized as a large, trilingual state and absorbed parts of Hyderabad and Madhya Pradesh and the princely states of Kutch and Saurashtra. In 1960, however, Bombay state was divided again into the new unilingual states of Gujarat and Maharashtra. On Salsette Island are Buddhist caves. The nearby small island of ELEPHANTA is noted for its hewn-stone temples. Although in 1995 the city's name was formally changed from Bombay to Mumbai, long-time residents of the city as well as most of its better-known institutions continue to use "Bombay."

Mumbai Suburban (mum-BAI), district (□ 206 sq mi/535.6 sq km; 2001 population 8,588,000), Maharashtra state, W central India, on Salsette Island just N of MUMBAI (Bombay); ⊙ BANDRA. Bounded W by Arabian Sea, E by Thane Creek and Bombay Harbour. Agriculture (rice, coconuts); fishing along coasts. Suburban towns of Bandra, Kurla, Ghatkopar, and Andheri have local industries. Santa Cruz Domestic Airport and Chhatrapati Shivaji International Airport are located here, in the Santa Cruz neighborhood. IIT Bombay, one of seven campuses of the prestigious Indian Institute of Technology, is located here in the suburb of Powai. Major film studios scattered throughout the district, especially in Andheri and Powai.

Mumbles Head (MUHM-buhlz), promontory on SWANSEA BAY, SWANSEA, S Wales, 5 mi/8 km SSW of Swansea; 51°34′N 03°58′W. Lighthouse. Formerly in WEST GLAMORGAN.

Mumbles, The, Wales: see SWANSEA.

Mumbwa (moo-em-bwah), township, CENTRAL PROVINCE, S central ZAMBIA, 90 mi/145 km WNW of LUSAKA; 14°59′S 27°04′E. Agriculture (peanuts, cotton, tobacco); cattle. Mining (phosphate, lead/zinc). Mumbwa Caves noted for artifacts and fossils. KAFUE NATIONAL PARK to W; Blue Lagoon National Park to S.

Mumena (moo-MEN-nuh), village, KATANGA province, SE CONGO, near ZAMBIA border, 60 mi/97 km W of LUBUMBASHI; 11°46′S 26°31′E. Elev. 4,793 ft/1,460 m. The Lufra and Luluaba rivers originate here.

Mumford, town, CENTRAL REGION, GHANA; 05°16′N 00°45′W. On Gulf of GUINEA, 10 mi/16 km WSW of WINNEBA; fishing. Also called Dwomba.

Mumford, village, MONROE county, W NEW YORK, 17 mi/27 km SSW of ROCHESTER, in agricultural area; 42°59′N 77°52′W.

Mumias (moo-MEE-ahs), town (1999 population 36,158), WESTERN province, W KENYA on NZOIA RIVER, on road, and 38 mi/61 km NNW of KISUMU; 00°20′N 34°29′E. Cotton, peanuts, sesame, corn. Planned sugarcane development and refinery.

Mümling River (MYOOM-ling), 30 mi/48 km long, W GERMANY; rises in the ODENWALD 7 mi/11.3 km N of EBERBACH; flows generally NNE to the MAIN RIVER, 1 mi/1.6 km S of OBERNBURG. Sometimes called MÖMLING RIVER.

Mümliswil-Ramiswil (MYOOM-lees-veel–RAH-mees-veel), commune of two villages, SOLOTHURN canton, N SWITZERLAND, 2 mi/3.2 km N and NNW of

BALSTHAL, on road from Balsthal to BASEL; 47°21′N 07°42′E. Metalworking.

Mummery, Mount (MUH-muhr-ee) (10,918 ft/3,328 m), SE BRITISH COLUMBIA, W Canada, near ALBERTA border, in ROCKY MOUNTAINS, in HAMBER PROVINCIAL PARK, 65 mi/105 km NW of BANFF (Alberta); 51°40′N 116°51′W.

Mummidivaram (moom-MEE-dee-VUH-ruhm), town, EAST GODAVARI district, ANDHRA PRADESH state, SE INDIA, in GODAVARI RIVER delta, 23 mi/37 km SSW of KAKINADA; 16°39′N 82°07′E. Rice milling; sugarcane, tobacco, coconuts.

Mummy Mountain, Colorado: see MUMMY RANGE.

Mummy Range, spur of FRONT RANGE in NE corner of ROCKY MOUNTAIN NATIONAL PARK, LARIMER county, N COLORADO. Prominent peaks include Mount Dunraven (12,571 ft/3,832 m), Mount Chiquita (c.13,069 ft/3,983 m), Mount Chapin (12,454 ft/3,796 m), Mummy Mountain (13,425 ft/4,092 m), Mount Fairchild (13,502 ft/4,115 m), Ypsilon Mt. (13,514 ft/4,119 m), Hague's Peak (13,562 ft/4,134 m). Rowe Glacier is in NE tip of range.

Mumra (MOOM-rah), town, SW ASTRAKHAN oblast, S European Russia, on the VOLGA RIVER delta mouth, on the Staraya Volga arm, 40 mi/64 km SSW of ASTRAKHAN; 45°46′N 47°39′E. Below sea level. Fish processing; ship repair.

Mumtrak (MUHM-trak), village, SW ALASKA, near N shore of GOODNEWS BAY, 13 mi/21 km NE of PLATINUM; 59°08′N 161°31′W. In mining area.

Muna (MOO-nah), town, YUCATÁN, S MEXICO, on railroad, and 35 mi/56 km SSW of MÉRIDA; 20°29′N 89°41′W. Agricultural center (henequen, sugarcane, corn, fruit, timber).

Muna (MOO-nah), island (□ 659 sq mi/1,713.4 sq km), SE Sulawesi Tanggara province, INDONESIA, between MALUKU and FLORES seas, just off SE extremity of SULAWESI and just W of BUTON; between 04°36′S–05°27′S and 122°16′E–122°46′E. Island is 60 mi/97 km long, 30 mi/48 km wide; generally level. Principal town and port is RAHA. Chief products are teak, rice, sago, trepang. Also spelled Moena.

Munakata (moo-NAH-kah-tah), city, FUKUOKA prefecture, N KYUSHU, SW JAPAN, 19 mi/30 km N of FUKUOKA; 33°48′N 130°32′E.

Munamägi (MOO-nuh-mag-ee), highest point (1,040 ft/317 m) in ESTONIA and in the BALTIC STATES, 9 mi/14 km SSE of VÕRU.

Muñano (moon-YAH-no) or **Kilómetro 1308**, village, W SALTA province, ARGENTINA, 12 mi/19 km SE of SAN ANTONIO DE LOS COBRES; 24°15′S 66°12′W. Elevation 12,950 ft/3,947 m. Railroad station.

Munastir (maw-nah-STIR), province (□ 393 sq mi/1,021.8 sq km; 2006 population 475,200), N central TUNISIA; ⊙ MUNASTIR; 35°37′N 10°45′E. Borders MEDITERRANEAN SEA (to E). Also Monastir and Al Munastīr.

Munastir (maw-nah-STIR), ancient *Ruspina*, town (2004 population 71,546), ⊙ MUNASTIR province, E TUNISIA, port on the headland in the GULF OF HAMMAMAT (central MEDITERRANEAN SEA), 16 mi/26 km ESE of SUSAH; 35°47′N 10°50′E. Olive oil pressing, distilling, soap manufacturing, tuna canning; handicraft textile industry. Saltworks; tourism; administrative center. Sakanas international airport. A Phoenician and Roman settlement. Its old cloister (founded C.E. 180) now forms the core of the impressive *qasbah*, which was built in 796 to protect the coast. Birthplace of Habib Abu Ruqaybah (Bourguiba). Formerly spelled Monastir.

Münchberg (MYOONKH-berg), town, UPPER FRANCONIA, NE BAVARIA, GERMANY, at N foot of the FICHTELGEBIRGE, 10 mi/16 km SW of HOF; 50°11′N 11°47′E. Textiles; weaving. Granite quarries in area.

Müncheberg (MYOON-khe-berg), town, BRANDENBURG, E GERMANY, 20 mi/32 km NW of FRANKFURT/

Oder, 30 mi/48 km E of BERLIN; 52°31′N 14°08′E. In lignite-mining region; market gardening. Has early-Gothic church, ancient town walls. Site of plant-research institute. Was first mentioned 1232.

München, GERMANY: see MUNICH.

Münchenbernsdorf (myoon-khen-BERNS-dorf), town, THURINGIA, central GERMANY, 8 mi/12.9 km SW of GERA; 50°50′N 11°59′E. Woolen and rayon milling, carpet manufacturing. First mentioned 1332. Chartered 1923.

Münchenbuchsee (MYOON-khen-bookh-zai), residential commune, BERN canton, NW SWITZERLAND, 5 mi/8 km N of BERN; 47°01′N 07°27′E. Former commandery of Knights of Malta. Has early-Gothic church.

München-Gladbach, GERMANY: see MÖNCHENGLAD-BACH.

Münchengrätz, CZECH REPUBLIC: see MNICHOVO HRA-DISTE.

Münchenstein (MYOON-khen-shtein), town (2000 population 11,702), Basel-Land half-canton, N SWIT-ZERLAND, on BIRS RIVER, and 3 mi/4.8 km SSE of BASEL; 47°31′N 07°37′E. Elevation 981 ft/299 m. Aluminum, chemicals, foodstuffs.

Münchhausen (myoonkh-HOU-suhn), village, HESSE, central GERMANY, 11 mi/18 km N of MARBURG; 50°58′N 08°43′E. Forestry.

Munchon (MOON-CHUHN), county, SOUTH HAM-GYONG province, NORTH KOREA, 12 mi/19 km NW of WONSAN. Coal mining.

Münchwilen (MYOONKH-vee-len), commune, THURGAU canton, N SWITZERLAND, 2 mi/3.2 km WNW of WIL; 47°27′N 09°00′E. Elevation c.1,640 ft/ 500 m.

Münchwilen (MYOONKH-vee-len), district, S THUR-GAU canton, SWITZERLAND. Main town is SIRNACH; population is largely German-speaking and mainly Protestant, but communes of Bichelse and Fischingen are Roman Catholic.

Muncie, city, ⊙ DELAWARE county, E INDIANA, on the WHITE RIVER; 40°12′N 85°23′W. It is a trade, processing, and manufacturing center. The city is in a fertile agricultural area with dairying, soybean, fruit, corn, oats, and vegetables. Manufacturing (machinery, electronic equipment, plastics, glass, motor vehicle equipment, metal fabrication, consumer goods). Area first settled by peoples of Native American Delaware culture; town was named for one of their tribes. In 1818 the land passed by treaty to the U.S. government. Industrialization came after the discovery (1886) of natural gas in the county. Muncie is the seat of Ball State University. Settled 1818, laid out 1827, incorporated 1854.

Muncie, village (2000 population 155), VERMILION county, E ILLINOIS, 10 mi/16 km W of DANVILLE; 40°07′N 87°50′W. In agricultural and bituminous-coal area.

Muncy (MUHN-see), borough (2006 population 2,506), LYCOMING county, N central PENNSYLVANIA, 14 mi/23 km E of WILLIAMSPORT, on West Branch of SUSQUEHANNA RIVER (bridged), at mouth of MUNCY CREEK; 41°12′N 76°47′W. Manufacturing (grain-processing equipment, aluminum and plastic products, printing, food products, modular-wood buildings, conveyor systems, pneumatic conveyors). Agricultural area (potatoes, soybeans, grain; livestock; dairying). Tiadaghton State Forest to W; Muncy State Correctional Institution to W. Laid out 1797, incorporated 1826.

Muncy Creek (MUHN-see), c.40 mi/64 km long, E central PENNSYLVANIA; rises in SE SULLIVAN county; flows SW past HUGHESVILLE to West Branch of SUS-QUEHANNA RIVER 1 mi/1.6 km NW of MUNCY; 41°12′N 76°48′W.

Mund, IRAN: see MOND RIVER.

Munda (MOON-dah), ancient town in ANDALUSIA, S SPAIN, known for Caesar's victory (45 B.C.E.) over Pompeian forces. Its location is disputed; believed to have been either in vicinity of RONDA (Málaga province) or MONTILLA (Córdoba province).

Munda, SOLOMON ISLANDS: see NEW GEORGIA.

Mundaca (moon-DAH-kah), fishing village, VIZCAYA province, N SPAIN, on BAY OF BISCAY, and 14 mi/23 km NE of BILBAO. Boatbuilding. Vegetables, *chacolí* wine, livestock in area.

Mundakayam (moon-dah-KAH-yuhm), town, KOT-TAYAM district, KERALA state, S INDIA, 25 mi/40 km ESE of KOTTAYAM, in foothills of WESTERN GHATS; 09°33′N 76°53′E. Rubber and tea processing, saw-milling.

Mundalla, AUSTRALIA: see MUNDULLA.

Mundame, village, SW CAMEROON, on MUNGO River, and 8 mi/12.9 km SE of KUMBA. Cacao, bananas.

Mundare, town (□ 1 sq mi/2.6 sq km; 2001 population 653), central ALBERTA, W Canada, 14 mi/23 km NW of VEGREVILLE, in Lamont County; 53°36′N 112°20′W. Mixed farming, dairying. Established as a village in 1907; became a town in 1951.

Mundaring (mun-DAH-ring), residential town, SW WESTERN AUSTRALIA, suburb 19 mi/31 km E of PERTH, and in DARLING RANGE; 31°54′S 116°10′E. Nearby Mundaring Weir (built 1903) on Helena River (tributary of SWAN RIVER) supplies water for gold fields 350 mi/563 km E.

Mundar-Yurt, RUSSIA: see ZNAMENSKOYE, CHECHEN REPUBLIC.

Mundawa, INDIA: see MUNDWA.

Munday (MUHN-dai), town (2006 population 1,318), KNOX county, N TEXAS, 72 mi/116 km WSW of WI-CHITA FALLS; 33°27′N 99°37′W. In cattle-ranching and irrigated agricultural area (cotton, sorghum, wheat; vegetables); manufacturing (portable buildings). Incorporated 1906.

Munday, Mount (11,000 ft/3,353 m), W BRITISH CO-LUMBIA, W Canada, in COAST MOUNTAINS, 170 mi/274 km NW of VANCOUVER, just SE of Mount WAD-DINGTON; 51°21′N 125°14′W.

Mundelein, village (2000 population 30,935), Lake county, NW of CHICAGO, NE ILLINOIS; 42°16′N 88°00′W. Founded 1835 as Mechanics Grove, incorporated 1909. The name was changed in 1926 to honor George Cardinal Mundelein. Its light manufacturing produces a variety of household goods. Saint Mary of the Lake Seminary is in Mundelein.

Mundemba (moon-DEM-buh), town, ⊙ NDIAN department, South-West province, CAMEROON, near NIGERIA border, 45 mi/72 km WNW of KUMBA; 04°57′N 08°53′E. Palm oil. Formerly called NDIAN.

Munden, village (2000 population 122), REPUBLIC county, N KANSAS, 8 mi/12.9 km NE of BELLEVILLE; 39°54′N 97°32′W. In corn and wheat region.

Münden, GERMANY: see HANNOVERISCH-MÜNDEN.

Münder am Deister, Bad, GERMANY: see BAD MÜNDER AM DEISTER.

Munderfing (MUN-der-fing), village W UPPER AUS-TRIA, 15 mi/24 km SSE of BRAUNAU, at SW foot HAUSRUCK MOUNTAINS in the INNVIERTEL region; 48°04′N 13°11′E. Gas fields.

Munderkingen (MUN-der-king-uhn), town, Danube-Iller, BADEN-WÜRTTEMBERG, GERMANY, on the DA-NUBE, and 5 mi/8 km SW of EHINGEN; 48°14′N 09°39′E. Was first mentioned 792. Late-Gothic chapel here.

Munderwa, INDIA: see BASTI.

Mundesley (MUHNZ-lee), village (2001 population 2,695), NE NORFOLK, E ENGLAND, on NORTH SEA, 7 mi/11.3 km SE of CROMER; 52°52′N 01°25′E. Seaside resort and small fishing port.

Mundgod (MOOND-god), town, UTTAR KANNAD district, KARNATAKA state, SW INDIA, 60 mi/97 km ENE of KARWAR; 14°58′N 75°02′E. Markets rice, millet.

Mundia (MOON-dee-yuh), town, BUDAUN district, N central UTTAR PRADESH state, N central INDIA, 4 mi/ 6.4 km NW of BISAULI. Wheat, pearl millet, mustard, barley, gram, jowar.

Mundolsheim (MOON-dols-heym), N suburb (□ 2 sq mi/5.2 sq km) of STRASBOURG, BAS-RHIN department, E FRANCE, near MARNE-RHINE CANAL, and major highway interchange; 48°39′N 07°42′E.

Mundo Novo (MOON-do NO-vo), city (2007 population 24,021), E central BAHIA, BRAZIL, on railroad, and 100 mi/161 km SSW of SENHOR DO BONFIM; 11°49′S 40°29′W. Coffee, tobacco, copaiba oil; amethysts found here.

Mundo Novo, city (2004 population 14,524), S Mato Grosso do Sul state, BRAZIL, near RIO PARANÁ and river ports; 23°57′S 54°18′W. Airport.

Mundo Novo, Brazil: see URUPÊS.

Mundo Nuevo (MOON-do NWAI-vo), town, Anzoá-tegui state, NE VENEZUELA, 45 mi/72 km ESE of BARCELONA. Cotton, coffee, sugarcane.

Mundo River (MOON-do), 75 mi/121 km long, ALBA-CETE province, SE central SPAIN; rises 8 mi/12.9 km NW of YESTE; flows E and S to the SEGURA RIVER 6 mi/9.7 km NNE of CALASPARRA. Used for irrigation and power.

Mundra (MOON-drah), town, KACHCHH district, GU-JARAT state, W central INDIA, near GULF OF KACHCHH, 29 mi/47 km S of BHUJ; 22°51′N 69°44′E. Exports wheat, barley, cloth fabrics; cotton ginning. Port facilities 3 mi/4.8 km S.

Mundrabilla, locality, WESTERN AUSTRALIA state, W AUSTRALIA; 31°51′S 127°51′E. Stopping point on Eyre Highway; roadhouse and other facilities.

Mundubbera (muhn-DUHB-uh-ruh), town, QUEENS-LAND, NE AUSTRALIA, 242 mi/390 km NW of BRIS-BANE; 25°35′S 151°18′E. Citrus fruits.

Mundulla (mun-DU-luh), village, SE SOUTH AUS-TRALIA, 155 mi/249 km SE of ADELAIDE, near BOR-DERTOWN; 36°22′S 140°41′E. Wheat. Also spelled Mundalla.

Mundwa (MOON-dwah), town, NAGAUR district, RA-JASTHAN state, NW INDIA, 75 mi/121 km NE of JODHPUR; 27°04′N 73°49′E. Local market. Sometimes called Mundawa or Marwar Mundwa.

Mundybash (moon-di-BAHSH), town (2005 population 5,820), S KEMEROVO oblast, S central SIBERIA, RUSSIA, in the GORNAYA SHORIYA, on road and railroad, 40 mi/64 km S of NOVOKUZNETSK; 53°14′N 87°19′E. Elevation 1,056 ft/321 m. Iron-ore concentrating plant, serving Temir-Tau, Telbes, and Odra-bash mines.

Munébrega (moo-NAI-vrai-gah), town, ZARAGOZA province, NE SPAIN, 8 mi/12.9 km SSW of CALA-TAYUD; 41°15′N 01°42′W. Wine, cereals, livestock.

Muñecas, province, LA PAZ department, W BOLIVIA, ⊙ CHUMA; 15°20′S 68°40′W.

Muñecas (moo-YAI-kahs), town, central TUCUMÁN province, ARGENTINA, on railroad, 3 mi/4.8 km NW of TUCUMÁN; 26°47′S 65°15′W. In agricultural area (corn, alfalfa, sugarcane). Hydroelectric station nearby.

Munera (moo-NAI-rah), town, ALBACETE province, SE central SPAIN, 18 mi/29 km SSE of VILLARROBLEDO. Agricultural center; olive-oil and cheese processing; wine, saffron, farm produce; livestock.

Munford, town (2006 population 6,062), TIPTON county, W TENNESSEE, 11 mi/18 km SSW of COVING-TON; 35°27′N 89°49′W. In agricultural area; timber.

Munford, village, Talladega co., E Alabama, 11 mi/18 km NE of Talladega. Cheaha Mt. (2,407 ft/734 m), the highest point in Alabama, is 10 mi/16 km E. Civil War battle here (1865). Named for Mrs. Munford, operator of a boardinghouse for workmen who were building the Alabama and Tennessee Rivers RR.

Munfordville, town (2000 population 1,563), ⊙ HART county, central KENTUCKY, 35 mi/56 km ENE of BOWLING GREEN, on GREEN RIVER, in limestone cave region; 37°16′N 85°54′W. Agriculture (poultry; dairying; burley tobacco, corn; timber); manufacturing (lumber, mattress pads). Indian relics found in vicinity. MAMMOTH ONYX CAVE, containing onyx formations is to S. In Civil War, Confederate General

Braxton Bragg captured Union fort and garrison here in September 1863. Hart County Historical Museum. MAMMOTH CAVE NATIONAL PARK to SW. Established 1801.

Mungana (mun-GAH-nuh), village, NE QUEENSLAND, NE AUSTRALIA, 90 mi/145 km WSW of CAIRNS; 17°07′S 144°24′E. Railroad connection to Cairns. Cattle; national parks; tourism.

Mungaoli (muhn-GOU-lee), town (2001 population 19,536), GUNA district, N central MADHYA PRADESH state, central INDIA, 50 mi/80 km ESE of GUNA; 24°25′N 78°06′E. Markets wheat, millet, gram.

Mungbere (mung-BE-rai), village, ORIENTALE province, NE CONGO, 120 mi/193 km NW of IRUMU; 02°38′S 28°30′E. Elev. 2,893 ft/881 m. Railroad terminus and trading post in cotton and coffee area.

Mungeli (moon-GAI-lee), town (2001 population 27,387), BILASPUR district, NW CHHATTISGARH state, central INDIA, 30 mi/48 km W of BILASPUR; 22°04′N 81°41′E. Agriculture market (rice, wheat, oilseeds, corn).

Munger (moon-GER), district (□ 545 sq mi/1,417 sq km; 2001 population 1,135,499), NE central BIHAR state, E INDIA; ⊙ MUNGER. On GANGA PLAIN; foothills of CHOTA NAGPUR PLATEAU in S; NW corner bounded by GANGA River; drained by tributaries of the Ganga. Mainly alluvial soil; rice, corn, wheat, gram, barley, oilseeds; mango, palmyra, and bamboo in forested hills. Mica (near JAMUI), slate (near JAMALPUR) quarries; lime extracting. Railroad workshops at JAMALPUR; firearms and cigarette manufacturing at Munger. Extensively damaged by earthquake of 1934. Lakhisarai district formed out of W Munger in 1994. Formerly spelled Monghyr.

Munger (moon-GER), city (2001 population 187,311), ⊙ MUNGER district, SE BIHAR state, E INDIA, on the GANGES RIVER; 25°23′N 86°28′E. Grain market, with important road, railroad, and ferry connections. Has one of India's largest cigarette factories and a firearms industry that dates back to the eighteenth century. Famous for its goldsmiths and silversmiths. Founded (according to tradition) during the Gupta dynasty (c.320–545). The Muslim leader Mir Kasim Ali used Munger as a base during his war against the British in 1764. Formerly spelled Monghyr.

Mu Ngiki, SOLOMON ISLANDS: see BELLONA.

Mungindi (MUHN-gin-dei), village, extending from N NEW SOUTH WALES to QUEENSLAND, AUSTRALIA, 320 mi/515 km NW of NEWCASTLE, and traversed by the MACINTYRE RIVER; 28°58′S 148°59′E. Railroad terminus; service center for sheep, beef cattle, cotton, and wheat.

Mungo Lake National Park (MUHN-go) (□ 108 sq mi/280.8 sq km), SW NEW SOUTH WALES, SE AUSTRALIA, 68 mi/109 km NE of MILDURA, VICTORIA. Part of UNESCO WILLANDRA Lakes World Heritage Area. Ancient lake system that existed some 15,000 years ago at end of last ice age. Evidence of Aboriginal occupation at least 40,000 years ago in vicinity of sandstone bluffs referred to as the Walls of China. Sand, sparse vegetation, and dunes and ridges form the landscape. Former sheep station. Casuarina trees near lake margins, other natural vegetation recovering from grazing. Kangaroos, wallabies; emus, cockatoos, mulga parrots, ducks, pelicans, lizards. Established 1979.

Mungo River, French Moungo (MUN-go), c.100 mi/161 km long, CAMEROON; rises in South-West province in the Rumpi Hills c. 24 mi/39 km NNW of KUMBA; flows S to Gulf of GUINEA 7 mi/11.3 km E of TIKO, forming the CAMEROON River with WOURI River Navigable in lower course for shallow-draught boats for 60 mi/97 km. Separated former West Cameroon and East Cameroon.

Mungra-Badshahpur (MOON-grah–BAHD-shah-poor), town, JAUNPUR district, SE UTTAR PRADESH state, N central INDIA, 25 mi/40 km NE of ALLAHABAD; 25°40′N 82°11′E. Trades in sugarcane, cotton, barley, rice, corn. Also called Badshahpur.

Munguía (moong-GEE-ah), town, VIZCAYA province, N SPAIN, on railroad, and 7 mi/11.3 km NNE of BILBAO; 43°21′N 02°50′W. Corn, wheat, hogs, cattle. Lumbering; potteries.

Munhall, borough (2006 population 11,358), ALLEGHENY county, SW PENNSYLVANIA; residential suburb 6 mi/9.7 km ESE of downtown PITTSBURGH, on the MONONGAHELA RIVER (bridged); 40°23′N 79°54′W. The once-large steel and iron works have declined significantly, as has the national steel industry. Part of large U.S. Steel complex is here. Manufacturing (office furniture, insulation tape). Munhall was a site of the Homestead Strike in 1892. Sandcastle water theme park is here. Incorporated 1901.

Munhango (moon-WAHN-go), village, BIÉ province, central ANGOLA, on Benguela railroad, and 125 mi/201 km ENE of CUITO. Road junction.

Munhoz de Melo (MOON-yos zhe ME-lo), town (2007 population 3,554), NW PARANÁ state, BRAZIL, 16 mi/26 km NE of MARINGÁ; 23°03′S 51°46′W. Coffee, cotton, corn, rice; livestock.

Munich (MYOO-nik), German *München* (MOOIN-khen), city (2005 population 1,259,677), ⊙ BAVARIA, S GERMANY, on the ISAR RIVER near the Bavarian Alps; 48°08′N 11°33′E. Financial, commercial, industrial, transportation, communications, and cultural center. Its industries produce precision and optical instruments, electrical appliances, clothing, chemicals, motor vehicles, and beer. Munich is also a major center for film production and book publishing, and is home to one of Europe's largest wholesale produce markets. The city is a major tourist and convention center; a new airport handling both domestic and international flights was opened in 1992. Munich (ancient *Monacium*), a monastic village under the Carolingians, was chartered 1158. Became (1255) residence of Wittelsbach family, and since then has always been the capital of Bavaria. Suffered disastrous fire in 1327. It was occupied 1632 by Gustavus Adolphus, and in 1705 and 1742 by Austrian forces. Developed into a center of German culture in 19th century, tripling its population between 1870 and 1900. In 20th century it became the birthplace and headquarters of National Socialist movement, and was scene (1938) of Munich Pact. Munich was badly damaged during World War II (c.40%), but after 1945 it was extensively rebuilt and many modern buildings were constructed. Among the city's chief attractions are the Frauenkirche (Church of Our Lady), a twin-towered cathedral built from 1468 to 1488; the Renaissance-style St. Michael's Church (1583–1597); the Theatinerkirche (17th–18th century), a baroque church; Nymphenburg castle (1663–1728), with a porcelain factory (founded 1747) and the nearby Amalienburg (1734–1739), a small rococo hunting chateau; the new city hall (1867–1908); Propyläen (1846–1862), a monumental neoclassic gate; and the large English Garden (laid out 1789). The city also has several leading museums, including the Old Pinakothek (built 1826–1836), which houses a distinguished collection of paintings; the Bavarian National Museum (built 1894–1899); the Schackgalerie; the Glyptothek (built 1816–1830); and the Deutsche Museum, which has wide-ranging exhibits on science, technology, and industry. The seat of an archbishop, Munich has a famous university (founded 1472 at INGOLSTADT; transferred in 1802 to LANDSHUT and in 1826 to Munich) in addition to a technical university, a conservatory of music, an opera, numerous theaters, and many publishing houses. Other educational institutions include academies of art, music, military studies, philosophy, film, and television. Munich is also known for its lively Fasching (Shrove Tuesday) and Oktoberfest (October festival, held in September) celebrations. The 1972 Olympic summer games were held (August–September) at Munich (except for the sailing and yachting events, held at Kiel).

Munich (MYOO-nik), village (2006 population 222), CAVALIER county, N NORTH DAKOTA, 23 mi/37 km SW of LANGDON; 48°40′N 98°49′W. Founded in 1904 and incorporated in 1910. Named for MUNICH, GERMANY.

Municipality of Murrysville, city (2006 population 19,472), WESTMORELAND county, SW PENNSYLVANIA; suburb 15 mi/24 km E of PITTSBURGH; 40°26′N 79°39′W. Manufacturing includes aluminum dies, construction equipment, metal stampings, gas detectors, patient-ventilation products, vinyl windows, roof and floor trusses. Agriculture includes dairying. Usually just called Murrysville.

Muniesa (moo-NYAI-sah), town, TERUEL province, E SPAIN, 35 mi/56 km W of ALCAÑIZ; 41°02′N 00°48′W. Brandy manufacturing; produces wine, cereals.

Munilla (moo-NEE-lyah), village, LA RIOJA province, N SPAIN, 20 mi/32 km SE of LOGROÑO; 42°11′N 02°18′W. Woolen mills; manufacturing of shoes and chocolate; cereals, sheep.

Munin, Syria: see MENIN.

Muni, Río (MOO-nee, ree-o), estuarine inlet of the Gulf of Guinea, forming border between EQUATORIAL GUINEA (N) and GABON (S); c.15 mi/24 km long, up to 4 mi/6.4 km wide. Navigable. Fed by many coastal streams, of which the Utamboni (MITÉMÉLÉ) is longest. COGO (Río Muni) and COCOBEACH (Gabon) are small ports on it. In CORISCO BAY, at its mouth, are Equatorial Guinea islands of CORISCO and ELOBEY. The name Río Muni is commonly applied to continental Equatorial Guinea.

Munising (MUHN-is-ing), town (2000 population 2,539), ⊙ ALGER county, N UPPER PENINSULA, MICHIGAN 36 mi/58 km ESE of MARQUETTE on small Munising Bay of LAKE SUPERIOR, and facing Grand Island Resort; 46°25′N 86°38′W. Lumbering and agriculture in area. Manufacturing (paper and wood products, including log homes); fisheries. Headquarters for Pictured Rocks National Lakeshore (to NE); Hiawatha National Forest to W, S, and E; Wagner Falls State Park to S. Incorporated as village 1897, as city 1916.

Muniz Ferreira (MOO-nees FAI-rai-rah), town (2007 population 6,990), E central BAHIA, BRAZIL, 9 mi/14 km W of NAZARÉ; 13°02′S 39°05′W.

Muniz Freire (moo-nees FRAI-ruh), city (2007 population 18,207), S ESPÍRITO SANTO, BRAZIL, 35 mi/56 km NW of CACHOEIRO DE ITAPEMIRIM; 20°26′S 41°22′W. Coffee, bananas, rice.

Munk'acs, UKRAINE: see MUKACHEVE.

Munka-Ljungby (MUNK-ah–YUNG-BEE), town, SKÅNE county, SW SWEDEN, 4 mi/6.4 km E of ÄNGELHOLM; 56°16′N 12°58′E. Has thirteenth-century church.

Munkedal (MUNK-e-DAHL), town, GÖTEBORG OCH BOHUS county, SW SWEDEN, 13 mi/21 km NW of UDDEVALLA; 58°28′N 11°41′E. Railroad junction; manufacturing (paper mill).

Munkfors (MUNK-FORSH), town, VÄRMLAND county, W SWEDEN, on Klarälven River, 20 mi/32 km WNW of FILIPSTAD; 59°50′N 13°33′E. Manufacturing (ironworks). Tourist resort.

Munkhafad al-Qattarah, EGYPT: see QATTARA DEPRESSION.

Munkhe Kharyrkhan, MONGOLIA: see MONGOLIAN ALTAY.

Munkholmen (MOONK-hawl-mun), tiny island in TRONDHEIM harbor, SØR-TRØNDELAG county, central NORWAY, 1 mi/1.6 km from the mainland. Has ruins of Nidarholm Benedictine abbey (12th century), fortress (1660). Used for political prisoners in eighteenth century.

Munksund (MUNK-SUND), village, NORRBOTTEN county, N SWEDEN, on PITE ÄLVEN River, just above its mouth on GULF OF BOTHNIA, 2 mi/3.2 km SE of PITEÅ, on island of Pitholmen; 65°18′N 21°30′E.

Munku-Sardyk (mun-KOO–suhr-DIK), Mongolian *Monh Saridzg*, mountain (11,453 ft/3,491 m), on border of MONGOLIA and BURYAT REPUBLIC of the RUSSIAN

FEDERATION. Highest peak in EASTERN SAYAN MOUNTAINS, N of HÖVSGÖL NUUR (Lake Khubsugul); 51°45′N 100°20′E. Forests occupy river valleys up to about 6,500 ft/2,000 m, above which are alpine meadows, mountain tundra, and rock streams. Small glaciers. The mountain has six peaks.

Munkyong (MOON-GYOUNG), city (□ 334 sq mi/868.4 sq km; 2005 population 70,926), NW NORTH KYONGSANG province, SOUTH KOREA, bordering CHUNGCHONG province Sobaek Mountains in N and W; small rivers run off mountains, forming a narrow field used for agriculture (rice, barley, tobacco, mushrooms, silk) in S. Mining of coal, graphite, limestone, and iron. Chinaware. Munkyongsaejae and Ihwaryong passes open traffic routes to other regions.

Munnar (MUH-nuhr), hill station, Idduki district, KERALA state, S INDIA, on right tributary of PERIYAR River, 60 mi/97 km NE of KOTTAYAM, in WESTERN GHATS. Tea-plantation area. Hydroelectric power plant (opened 1940) utilizes nearby falls; auxiliary power provided since 1946 by link with hydroelectric works at Papanasam village (TAMIL NADU state). Transmission network serves industrial centers. Wildlife sanctuary. Also Pallivasal.

Munn Bay, Canada: see CORAL HARBOUR.

Munn, Cape, N extremity of SOUTHAMPTON ISLAND, KIVALLIQ region, NUNAVUT territory, Canada, on ROES WELCOME SOUND; 65°55′N 85°28′W.

Münnerstadt (MYOON-ner-shtaht), town, LOWER FRANCONIA, N BAVARIA, GERMANY, on S slope of the RHÖN, 6 mi/9.7 km S of BAD NEUSTADT AN DER SAALE; 50°15′N 10°12′E. Manufacturing (glass, textiles, precision instruments, guns). Has churches (13th- and 17th-centuries), monastery, castle (16th–17th-centuries). Town first mentioned 770, chartered 1335.

Munnsville, village (2006 population 423), MADISON county, central NEW YORK, on Oneida Creek, and 19 mi/31 km SW of UTICA; 42°58′N 75°35′W. Quarry.

Muñoz (moon-YOS), town, ⊙ Munoz de Domingo Arenas Municipio, N TLAXCALA, MEXICO, 7 mi/11 km NW of APIZACO. Relatively flat region in the Pie Grande Valley with water from the Zahuapan River. Temperate climate. Agriculture (corn, beans, potatoes, pulque, fruits, medicinal plants). Population is mostly indigenous.

Muñoz (moon-YOS), town (2000 population 65,586), NUEVA ECIJA province, central LUZON, PHILIPPINES, on railroad, and 16 mi/26 km NNW of CABANATUAN; 15°44′N 120°53′E. Rice-growing center.

Muñoz de Domingo Arenas, MEXICO: see MUÑOZ.

Muñoz Gamero Peninsula (moon-YOS gah-MAI-ro), mountainous area of Chilean Patagonia, SW of PUERTO NATALES, bordering on STRAIT OF MAGELLAN (SW). ADELAIDE ISLANDS are just off W coast. Rugged in outline, with many smaller peninsulas, it is 80 mi/129 km long (N-S) and c. 30 mi/48 km wide. The uninhabited, snow-capped area rises to 5,740 ft/1,750 m in Mount Burney (NW).

Mun River (MOON), 300 mi/483 km long, E THAILAND; rises in SW extremity of SAN KAMPHAENG RANGE at elevation of c.1,500 ft/457 m; flows E through KORAT PLATEAU, past NAKHON RATCHASIMA and Ubon, to MEKONG RIVER NW of PAKSE. Main tributary, CHI RIVER. Navigable below Tha Chang, 10 mi/16 km E of NAKHON RATCHASIMA; used for irrigation.

Munro (moon-RO), town in Greater BUENOS AIRES, ARGENTINA, adjoining FLORIDA, 10 mi/16 km NW of Buenos Aires; 34°32′S 58°33′W.

Munroe Falls (MUHN-ro FALZ), city (□ 3 sq mi/7.8 sq km; 2006 population 5,260), SUMMIT county, NE OHIO, 6 mi/10 km NE of AKRON, and on CUYAHOGA RIVER; 41°08′N 81°26′W. Paper products.

Munsey Park, residential village (2006 population 2,568), NASSAU county, SE NEW YORK, on NW LONG ISLAND, between MANHASSET (W) and ROSLYN on N shore; 40°47′N 73°40′W. Incorporated 1930.

Munshar (mawn-SHAHR), village, BAJAH province, N TUNISIA, 7 mi/11.3 km E of BAJAH, in wheat growing area.

Munshiganj (moon-shee-gawnj), ancient *Idrakpur*, town (2001 population 53,202), DHAKA district, E central EAST BENGAL, BANGLADESH, on the DHALESWARI RIVER, and 13 mi/21 km SSE of DHAKA; 23°46′N 88°50′E. Trades in rice, jute, bananas, oilseeds, fish; ice manufacturing. Large annual fair nearby.

Münsing (MYOON-zing), village, UPPER BAVARIA, BAVARIA, S GERMANY, 19 mi/30 km SSW of MUNICH, near E tip of the STARNBERGER LAKE; 47°54′N 11°21′E. Tourism.

Münsingen (MYOON-zing-uhn), commune (2000 population 10,937), BERN canton, W central SWITZERLAND, 7 mi/11.3 km SE of BERN, near AARE RIVER; 46°53′N 07°34′E. Elevation 1,775 ft/541 m. Printing; woolen textiles, tobacco.

Münsingen (MYOON-zing-uhn), town, NECKAR-JURA, BADEN-WÜRTTEMBERG, GERMANY, in SWABIAN JURA Mountains, 7 mi/11.3 km SE of Bad Urach; 48°25′N 09°30′E. Cement works. Castle here.

Munson (MUHN-suhn), agricultural village (□ 1 sq mi/2.6 sq km; 2001 population 222), S central ALBERTA, W Canada, near RED DEER River, 7 mi/11 km N of DRUMHELLER, and in Starland County; 51°34′N 112°45′W. Established 1911.

Munson Corners, village, CORTLAND county, central NEW YORK, 1 mi/1.6 km SW of CORTLAND; 42°35′N 76°12′W.

Munster (MUHN-stuhr), Gaelic *Cúige Mumhan*, province (□ 9,501 sq mi/24,702.6 sq km; 2006 population 1,173,340), SW IRELAND. The largest of the Irish provinces, it comprises the counties of CLARE, CORK, KERRY, LIMERICK, TIPPERARY, and WATERFORD. One of the ancient kingdoms of Ireland, its control passed to the well-known families of the Fitzgeralds (earls of Desmond) and the Butlers (earls of Ormonde) after the Anglo-Norman invasion of Ireland.

Münster (MYOON-stuhr), city, WESTPHALIA-LIPPE, North Rhine-Westphalia, W GERMANY, 29 mi/47 km SW of OSNABRÜCK, a port and industrial center on the DORTMUND-EMS CANAL; 51°58′N 07°37′E. Manufacturing includes heavy machinery and textiles. An airport here services OSNABRÜCK and Münster. Also a trade center for the Westphalian cattle market. Founded (c.800) as a Carolingian episcopal see. Its bishops ruled a large part of WESTPHALIA as princes of the Holy Roman Empire from the 12th century until 1803, when the bishopric was secularized. From the 14th century the city was a prominent member of the HANSEATIC LEAGUE, trading especially with ENGLAND and RUSSIA. In 1534–1535 it was the scene of the Anabaptist experimental government under John of Leiden. In 1648 the Treaty of Münster was signed here. Passed to PRUSSIA in 1816 and became Westphalian provincial capital. It was severely damaged in World War II but was rebuilt after 1945. Still retains some of its medieval character. Its historical buildings include the cathedral (13th century), the Lambertikirche (14th–15th century), the Liebfrauenkirche (14th century), and several other churches, in addition to a baroque palace (1767–1773), a Gothic city hall (14th century), and several gabled houses. A park rings the city on the site of the old city walls. Seat of a university; contains the Westphalian state museum. Also known as Münster in Westfalen.

Munster (moon-STER), town (□ 3 sq mi/7.8 sq km; 2004 population 5,108), HAUT-RHIN department, ALSACE, NE FRANCE, on FECHT river; 48°03′N 07°09′E. Produces cheese and textile goods. VOSGES DU NORD NATURAL REGIONAL PARK nearby.

Münster (MYOON-stuhr), town, S HESSE, W GERMANY, on the GERSPRENZ RIVER, 9 mi/14.5 km ENE of DARMSTADT; 49°56′N 08°52′E. Grain.

Munster (MUN-ster), town, LOWER SAXONY, NW GERMANY, 11 mi/18 km E of SOLTAU; 52°59′N 10°06′E. Garrison town; tank manufacturing.

Munster, town (2000 population 21,511), Lake county, NW INDIANA, 13 mi/21 km NW of CROWN POINT on the ILLINOIS state line; 41°33′N 87°30′W. Light manufacturing (soft drinks, printing, aluminum alloys). It is a primarily residential suburb in the industrialized HAMMOND–EAST CHICAGO area. Settled 1855.

Münster, SWITZERLAND: see MOUTIER.

Münster am Stein, Bad, GERMANY: see BAD MÜNSTER AM STEIN.

Münsterberg, POLAND: see ZIEBICE.

Münstereifel, Bad, GERMANY: see BAD MÜNSTEREIFEL.

Münster in Westfalen, GERMANY: see MÜNSTER.

Münstertal (MYOON-stuhr-tahl), Romansch *Val Müstair*, district, E GRISONS canton, SWITZERLAND. Population is Romansch-speaking and Protestant, except commune of MÜSTAIR, which is German-speaking and Roman Catholic.

Münstertal (MYOON-stuhr-tahl), village, S Upper Rhine, Baden-Württemberg, in the Black Forest, 7 mi/11.3 km SSE of Freiburg; 47°53′N 07°52′E. Forestry; wine. Also known as Münstertal-Schwarzwald.

Münstertal (MYOON-stuhr-tahl), Romansh *Val Müstair*, valley, GRISONS canton, easternmost SWITZERLAND, SE of OFEN PASS. Watered by Rombach River. Münster and Ste. Maria im Münstertal are main towns.

Munsungan Lake (muhn-SUHN-guhn), PISCATAQUIS county, N central MAINE, 50 mi/80 km NNW of MILLINOCKET; 7.5 mi/12.1 km long, 2 mi/3.2 km wide; 46°22′N 68°55′W. Drained ESE by Munsungan Stream into Aroostook River Wilderness recreational area. Chase Lake immediately upstream (to NW).

Muntele Mic, ROMANIA: see TARCU MOUNTAINS.

Muntengene (mun-TAIN-gai-nai), city, South-West province, CAMEROON, 26 mi/42 km WNW of DOUALA; 04°05′N 09°20′E. Important trade center.

Muntenia (moon-TEN-yah) or **Greater Walachia**, ROMANIA, the E part (□ 20,270 sq mi/52,702 sq km) of WALACHIA.

Muntervary, IRELAND: see SHEEP HEAD.

Muntinglupa (moon-teeng-LOO-pah), town (2000 population 379,310), NATIONAL CAPITAL REGION, S LUZON, PHILIPPINES, on LAGUNA DE BAY, on railroad, and 15 mi/24 km SSE of MANILA; 14°23′N 121°03′E. Agricultural center (rice, sugarcane, fruit). Part of metropolitan Manila area. Site of large prison. Also spelled Muntinlupa.

Muntok (MOON-tak), chief port of BANGKA Island and district, Sumatra Selatan province, INDONESIA, on Bangka Strait, 65 mi/105 km W of PANGKALPINANG; 02°04′S 105°11′E. Tin refinery and exports tin. Has airfield. Also spelled Mentok.

Muntu (MOON-too), village, BANDUNDU province, W CONGO, on left bank of LUKENIE RIVER; 02°50′S 18°28′E. Elev. 1,089 ft/331 m.

Munuscong Lake (mi-NE-skahng), in S central ONTARIO, E central Canada, and E UPPER PENINSULA, MICHIGAN, a widening of SAINT MARYS RIVER, 23 mi/37 km SSE of SAULT SAINTE MARIE and W of SAINT JOSEPH ISLAND; c.15 mi/24 km long, 5 mi/8 km wide; 46°10′N 84°03′W. Michigan-Ontario line passes through lake. Receives Munuscong River (c.15 mi/24 km long) from SW.

Munyati River, c. 225 mi/362 km long, central ZIMBABWE; rises c. 60 mi/97 km S of HARARE; flows WNW and N, past UMNIATI town; joins MUPFURE RIVER 55 mi/89 km W of CHINHOYI to form SANYATI RIVER. Formerly Umniati River.

Münzenberg (MYOON-tsen-berg), town, HESSE, central GERMANY, 10 mi/16 km SSE of GIESSEN; 50°28′N 08°47′E. Agriculture (grain, vegetables, sugar beets). Has ruined castle Wetterauer Tintenfass (12th century).

Münzkirchen (myoonts-KIR-khuhn), township, NW UPPER AUSTRIA, in the Sauwald and INNVIERTEL, 7 mi/11.3 km ENE of Schärding; 48°29′N 13°34′E.

Munzur Dagi, (Turkish=*Munzur Dagi*) mountain range, E central TURKEY, extending 55 mi/89 km S and E of the EUPHRATES; rises to 10,460 ft/3,188 m in Munzur Dag, 13 mi/21 km E of KEMALIYE. Sometimes spelled Monzur.

Munzur River, 90 mi/145 km long, E central TURKEY; rises in MUNZUR DAGI 19 mi/31 km E of KEMALIYE; flows E and SE, past OVACIK, to Keban Lake (artificial), which covers the lower part of the MURAT valley. Sometimes spelled Monzur.

Muong Hiem (MOO-uhng HEE-uhm), town, HUA PHAN province, NE LAOS, SW of SAM NEUA; 20°06′N 103°22′E. Border town.

Muong Hou, LAOS: see PHONG SALI, town.

Muong Hung (MU-uhng HOONG), town, SON LA province, N VIETNAM, on the Song Ma River, 26 mi/42 km S of SON LA; 22°47′N 104°08′E. Shifting cultivation and forestry. Bloch Thai and other minority peoples. Formerly Muonghung.

Muong Kham (MOO-uhng KHAM), town, XIENG KHUANG province, central LAOS, 21 mi/33 km NE of Phonsavan; 19°40′N 103°35′E. Dist. capital.

Muong Kheung, Laos: see MUONG KHUN.

Muong Khoua (MOO-uhng koo-AH), town, PHONG SALI province, N LAOS, S of Nam Ou River; 21°05′N 102°31′E. Dist. capital.

Muong Khun (MOO-uhng KOON), town, XIENG KHUANG province, N central LAOS, SE of Phonsavan town; 19°21′N 103°22′E. Dist. center; former province and capital of Xieng Kuang; destroyed by U.S. during Laos air campaign. Also called Muong Kheung.

Muong Khuong (MU-uhng KU-uhng), town, LAO CAI province, N VIETNAM, near CHINA border, 21 mi/34 km NNE of LAO CAI; 22°46′N 104°07′E. Market center and transportation hub. Shifting cultivation, forest products, opium. H'mong, Dao, and other minorities. Formerly Muongkhong.

Muong May (MOO-uhng MAI), town, ATTAPU province, S LAOS, on the Se KONG, and 65 mi/105 km ESE of PAKSE; 14°49′N 106°56′E. Administrative and market center. Shifting cultivation, agroforestry. Lao, Halong, and other minority peoples.

Muong Ngoi (MOO-uhng NGOI), town, LUANG PHABANG province, N LAOS, E of Nam Ou River; 20°43′N 102°41′E. Dist. capital.

Muong Ou Nua (MOO-uhng OO NU-ah), town, PHONG SALI province, N LAOS, on the Nam Nou, and 40 mi/64 km NNW of Phong Sali; 22°18′N 101°48′E. Also called Ou Neua.

Muong Phalane, LAOS: see PHALANE.

Muong Phin (MOO-uhng PIN), town, district ⊙, SAVANNAKHET province, S LAOS, E of SAVANNAKHET town; 16°32′N 106°02′E. Transportation hub. Also called Muong Phine.

Muong Sai (MOO-uhng SEI), town, ⊙ UDOMSAI province, N LAOS; 20°42′N 101°59′E. Transportation, administrative, and market center. Shifting cultivation, agroforestry, forest products. Lao, Meo, Khmer, and other minorities.

Muong Sing (MOO-uhng SEENG), town, BOKEO province, NW LAOS, 260 mi/418 km NNW of VIENTIANE, near junction of LAOS, CHINA, and BURMA borders; 21°11′N 101°09′E. Elevation 2,608 ft/795 m. Shifting cultivation, poultry, cattle trading. Agriculture: castor beans, cotton, opium, sugar, rubber; lac. Lolo, H'mong, Tibeto-Burman, and other minority peoples.

Muong Song Khone, LAOS: see SONG KHONE.

Muong Soui (MOO-uhng SOI), town, Xieng Khwang province, central LAOS, on Tranninh Plateau, near the PLAIN OF JARS, 50 mi/80 km SE of LUANG PHABANG, on highway to VINH; 19°33′N 102°52′E. Shifting cultivation, agroforestry, forest products. Important monastery. H'mong, Thai, Phuteng, and other minorities. Also called Ban Khai.

Muong Xen, VIETNAM: see KY SON.

Muonio (MOO-o-nee-o), village, LAPIN province, NW FINLAND, on the Muoniojoki (river forming part of Swedish border), and 110 mi/177 km NW of ROVANIEMI; 67°57′N 23°42′E. Elevation 825 ft/250 m. Lumbering.

Muonio älven, SWEDEN and FINLAND: see TORNEÄLVEN.

Muotathal (moo-O-tah-THAL), commune, SCHWYZ canton, central SWITZERLAND, on Muota River, and 6 mi/9.7 km ESE of SCHWYZ; 46°58′N 08°46′E. Elevation 2,001 ft/610 m. Silk textiles, farming. Old nunnery, 18th-century church. Hölloch grottoes are E of town.

Mupa (moo-puh), town, CUNENE province, ANGOLA, S of MUPA NATIONAL PARK, on road and 65 mi/105 km N of ONDJIVA; 16°07′S 15°45′E. Market center, resort town.

Mupa National Park (moo-puh), park (□ 900 sq mi/2,340 sq km) CUNENE province, SE ANGOLA, 80 mi/207 km N of ONDJIVA. Traversed by Cuvelai River. Became a national park in 1964 and is important for its birdlife.

Mupfure Recreational Park (moo-PFOO-rai), MASHONALAND WEST province, N central ZIMBABWE, on MUPFURE RIVER, 30 mi/48 km NNW of KADOMA; 17°57′S 29°54′E. Formerly Umfuli Recreational Park.

Mupfure River (moo-PFOO-rai), 150 mi/241 km long, N central ZIMBABWE; rises c. 40 mi/64 km SSE of HARARE; flows WNW, past BEATRICE, through Hippo Pools; N of Chegutu; joins MUNYATI RIVER 55 mi/89 km W of CHINHOYI to form SANYATI RIVER. Formerly Umfuli River.

Muping (MU-PING), town, ⊙ Muping county, NE SHANDONG province, CHINA, on road, and 15 mi/24 km SE of YANTAI, near YELLOW SEA; 37°23′N 121°35′E. Fruit, grain; fisheries; transport equipment. Also appears as Mouping.

Muqaiyir, Tall al, IRAQ: see UR.

Muqdadiyah (muhk-duh-DEE-yuh), town, DIYALA province E IRAQ, near the E bank of the DIYALA RIVER, c.25 mi/40 km from IRANIAN border, and 55 mi/89 km NNE of BAGHDAD. Also known as Miqdadiyah.

Muqeibla (muh-KEB-lah) or **el-Mukeibla**, Arab village, N ISRAEL, S of AFULA in S JEZREEL VALLEY; 32°31′N 35°17′E. Elevation 351 ft/106 m. Mixed farming. Some ruins and a well from Roman and Byzantine periods. Founded in the 19th century and named for a sacred Muslim grave located here.

Muquém do São Francisco (moo-KAIN do SOUN FRAHN-sees-ko), town, BAHIA, BRAZIL.

Muqui (moo-kee), city (2007 population 13,931), S ESPÍRITO SANTO, BRAZIL, on railroad, and 7 mi/11.3 km SW of CACHOEIRO DE ITAPEMIRIM, in coffee-growing area; 20°54′S 41°22′W. Sawmilling. Bauxite deposits nearby. Until 1944, called São João do Muqui (formerly spelled Muquy).

Muquiyauyo (moo-kee-YOU-yo), city, JAUJA province, JUNÍN region, central PERU, on MANTARO RIVER, and 5 mi/8 km SE of JAUJA; 11°50′S 75°20′W. Hydroelectric plant; potatoes, cereals.

Muqur, AFGHANISTAN: see MUKUR.

Mur (MUR), township, NW YEMEN, on TIHAMA coastal plain, 20 mi/32 km ESE of LUHAIYA. Sheepskin tanning; manufacturing (shoes, other leather products, food processing).

Mur (MOOR), Slovenian, Croatian, and Hungarian *Mura* (MOO-ru), river, 300 mi/483 km long, in AUSTRIA, SLOVENIA, CROATIA, and HUNGARY; rises at Murtörl of Radstädter Tauern, SE of SALZBURG, W central Austria; flows E and NE to STYRIA past MURAU, JUDENBURG to BRUCK AN DER MUR; turns SSE, passing GRAZ; turns E near SPIELFELD, forming the Austro-Slovenian border for 20 mi/32 km, turning SE into Slovenia 4 mi/6.4 km SE of BAD RADKERSBURG (Austria) and GORNJA RADGONA (Slovenia), where it is called the MURA River; flows SE across NE Slovenia, forming the Slovenian-Croatian border, and flowing past Mursko Središče (Croatia) to form brief portion of the Croatian-Hungarian border; joins the DRAVA RIVER, 4 mi/6.4 km S of Murakeresztúr (Hungary). Major left tributaries are the Liesing, Mürz, and LEDAVA rivers; major right tributaries are the Kainach, Sulm, and Scavnica rivers. Used for hydroelectric power generation mainly in its middle course in Austria, between Bruck an der Mur and SPIELFELD.

Murabek (MOOR-ah-bek), city, KASHKADARYO wiloyat, S UZBEKISTAN.

Muradiye, village, SE TURKEY, 8 mi/12.9 km from NE shore of LAKE VAN, 40 mi/64 km NNE of VAN; 39°00′N 43°44′E. Wheat, naphtha. Formerly called Bargiri.

Muradnagar (muh-RAHD-nuh-guhr), town, MEERUT district, NW UTTAR PRADESH state, N central INDIA, on UPPER GANGA CANAL, 18 mi/29 km SW of MEERUT; 28°47′N 77°30′E. Wheat, gram, jowar, sugarcane, oilseeds.

Murafa River (moo-RAH-fah), river (length 100 mi/162 km), in SW UKRAINE, E tributary of the DNIESTER River. Part of the natural border between Ukraine and Romania. Its W shore was the sight of major Polish-Turkish battles in early 17th century and Polish-Cossack battles in mid-17th century.

Murafa Tovtry, UKRAINE: see TOVTRY.

Murafs'ki Tovtry, UKRAINE: see TOVTRY.

Murakami (moo-RAH-kah-mee), city, NIIGATA prefecture, central HONSHU, N JAPAN, 31 mi/50 km N of NIIGATA; 38°13′N 139°28′E. Manufacturing of lacquerware. Salmon layered in sake is a local specialty.

Murakeresztúr (MU-rah-ke-res-tur), Hungarian *Murakeresztúr*, village, ZALA county, W HUNGARY, on MURA RIVER, and 8 mi/13 km SW of NAGYKANIZSA, on railroad crossing to CROATIA; 46°22′N 16°52′E. Railroad center for agriculture (wheat, rye, corn); dairy area.

Muraköz, CROATIA: see MEDIMURJE.

Muralto, commune, TICINO canton, S SWITZERLAND, on LAGO MAGGIORE. Residential suburb E of LOCARNO. Jewelry. Romanesque church.

Muramatsu (moo-RAH-mahts), town, Nakakanbara county, NIIGATA prefecture, central HONSHU, N central JAPAN, 16 mi/25 km S of NIIGATA; 37°41′N 139°10′E.

Muramvia (moo-rahm-VYAH), village, BURUNDI, 50 mi/80 km SW of GITEGA; 02°56′S 30°07′E. Local trading center; one-time residence of *mwami* (king) of Urundi; dairying. Cinchona plantations nearby.

Muramvya, province (□ 269 sq mi/696 sq km; 1999 estimated population 481,848 [prior to the formation of MWARO province from S portion of Muramvya province]), W BURUNDI; ⊙ MURAMVYA; 03°15′S 29°40′E. Borders KAYANZA (N), GITEGA (E), Mwaro (S), BUJUMBURA RURALE (W), and BUBANZA (NW) provinces. Mubarazi River flows through province.

Muramvya (moo-rahm-VYAH), town, ⊙ MURAMVYA province, W BURUNDI, 20 mi/32 km NEE of BUJUMBURA; 03°15′N 29°36′E. Established 1923.

Muran, SLOVAKIA: see TISOVEC.

Murang'a (moo-RAHN-gah), town (1999 population 11,021), CENTRAL province, S central KENYA, on railroad, and 45 mi/72 km NNE of NAIROBI; 00°43′S 37°10′E. Elevation 4,050 ft/1,234 m. Agricultural center; coffee, sisal, wheat, corn; dairying. Nearby is a hydroelectric plant serving Nairobi.

Murano (moo-RAH-no), suburb of VENICE, VENETO region, NE ITALY, on five small islands in the Lagoon of Venice; 45°27′N 12°21′E. From the late 13th century it was the center of the Venetian glass industry, which reached a peak in the 16th century and was revived in the 19th century by Antonio Salviati. Today mirrors and optical instruments are also produced. With its old houses, canals, and bridges, Murano has the same quaint charm as Venice. Of note are a Venetian-Byzantine basilica (7th–12th century) and a museum of old and new Venetian glass.

Muraoka (moo-RAH-o-kah), town, Mikata district, HYOGO prefecture, S HONSHU, W central JAPAN, 62 mi/100 km N of KOBE; 35°28′N 134°35′E. *Tajima ushi*, a variety of beef cattle.

Murarai, INDIA: see NALHATI.

Mura River, central EUROPE: see MUR.

Murashi (moo-rah-SHI), city (2005 population 7,040), N KIROV oblast, E European Russia, on the Kirov-Syktyvkar highway and railroad, 70 mi/113 km NNW of KIROV; 59°24′N 48°58′E. Elevation 757 ft/230 m. Highway junction; mechanical repair shops and railroad depots. Gas pipeline service station; logging, lumbering, timbering; scrap metal processing; veterinary clinic; seed inspection. Arose in 1895 with the construction of the railroad. Became city in 1944.

Murat (myu-rah), commune (□ 2 sq mi/5.2 sq km), CANTAL department, AUVERGNE region, S central FRANCE, on NE slope of Massif du CANTAL, on ALAGNON RIVER, and 12 mi/19 km WNW of SAINT-FLOUR; 45°06′N 02°51′E. An industrial center with agriculture-processing plants and a nearby chemical facility that treats locally quarried diatomite. Murat also has an active handicraft industry (woodworking, stone-cutting). It serves as an entry point to the regional park of the Auvergne Volcanoes (*volcans d'auvergne*).

Murat, IRAQ: see EUPHRATES RIVER.

Murata (moo-RAH-tah), town, Shibata county, MIYAGI prefecture, N HONSHU, NE JAPAN, 12 mi/20 km S of SENDAI; 38°06′N 140°43′E. Racecourse.

Murat Dag, peak (11,545 ft/3,519 m), E TURKEY, 27 mi/43 km SE of AGRI.

Murat Dagi, peak (7,585 ft/2,312 m), W TURKEY, 19 mi/31 km SE of GEDIZ.

Murato (mu-RAH-to), commune (□ 7 sq mi/18.2 sq km), N CORSICA, FRANCE, 10 mi/16 km SW of BASTIA, in HAUTE-CORSE department; 42°35′N 09°19′E. Vineyards. Has church of San Michele built in 1280.

Murat River, 380 mi/612 km long, E central TURKEY, a principal headstream of the EUPHRATES RIVER; rises c.40 mi/64 km SW of MT. ARARAT, c.15 mi/24 km S of DIYADIN; flows W, past AGRI, TUTAK, GENC, and PALU, to join the W headstream of the Euphrates River at Keban Lake, formed by the Keban Dam near the confluence of the Murat and Euphrates rivers, 5 mi/8 km NE of KEBAN, WNW of ELAZIG. From the source of the Murat River to the SYRIA border is 685 mi/1,102 km. Receives Hinis and MUNZUR rivers (right). Sometimes spelled Murad.

Murau (MOO-rou), town, STYRIA, central AUSTRIA, on MUR RIVER, and 23 mi/37 km W of JUDENBURG; summer and winter resort (2,530 ft/771 m); 47°07′N 14°10′E. Old town, founded 1298; some medieval fortifications still intact; castle, old burgher houses, Gothic church with frescoes and baroque altar.

Muravlyanka (moo-rahv-LYAHN-kah), village, SE RYAZAN oblast, central European Russia, on road junction, 45 mi/72 km E of RYAZHSK, and 8 mi/13 km SE of SARAI, to which it is administratively subordinate; 53°39′N 41°11′E. Elevation 531 ft/161 m. Agricultural products.

Muravyev-Amurskiy Peninsula (moo-rah-VYEF–ah-MOOR-skee), S MARITIME TERRITORY, RUSSIAN FAR EAST, extreme SE Siberian Russia, in VLADIVOSTOK city limits, extending 25 mi/40 km SSW into PETER THE GREAT BAY, separating its two inlets, AMUR (W) and USSURI (E) bays; 43°10′N 132°00′E. Across the EASTERN BOSPHORUS (at S end) lie RUSSIAN, POPOVA, and Reineke islands. Amur Bay shore is lined with summer resorts. Vladivostok proper is on the SW tip of the peninsula, on the GOLDEN HORN BAY. Named for the 19th century governor-general of Eastern Siberia.

Murayama (moo-RAH-yah-mah), city, YAMAGATA prefecture, N HONSHU, NE JAPAN, 16 mi/25 km N of YAMAGATA city; 38°28′N 140°23′E. Cherries, watermelon; pond lilies, safflowers. MOGAMI RIVER rapids nearby.

Murayjat, LEBANON: see MUREIJAT.

Murbach, FRANCE: see GUEBWILLER.

Murbad (MOOR-bahd), town, THANE district, MAHARASHTRA state, W central INDIA, 28 mi/45 km E of Thane; 19°15′N 73°24′E. Local market for rice, timber.

Murbat, OMAN: see MARBAT.

Murça (MOOR-sah), town, VILA REAL district, N PORTUGAL, 18 mi/29 km NE of VILA REAL. Vineyards; olives, figs, almonds, oranges.

Murchante (moor-CHAHN-tai), town, NAVARRE province, N SPAIN, 4 mi/6.4 km SW of TUDELA; 42°02′N 01°39′W. Olive oil, wine, sugar beets.

Murcheh Khvort (moor-CHE khe-VUHRT), township, Esfahān province, W central IRAN, 32 mi/51 km NNW of ESFAHAN, and on Esfahan-Tehran road; 33°05′N 51°28′E. Cotton, barley, tobacco, madder root, melons. Nadir Shah defeated the Afghans here in 1729.

Murchevo (mur-CHAI-vo), village, BOICHINOVTSI obshtina, MONTANA oblast, BULGARIA; 43°31′N 21°23′E.

Murchinson (MUHR-chin-suhn), village, HENDERSON county, E TEXAS, 26 mi/42 km WSW of TYLER; 32°16′N 95°45′W. Oil and natural gas. Agricultural area. Manufacturing (gas processing). Lake Athens reservoir to S.

Murchison, township, TASMAN district, SOUTH ISLAND, NEW ZEALAND, on upper BULLER RIVER, 81 mi/130 km SW from NELSON city, between Nelson and WESTPORT; 41°48′S 172°20′E. Sawmills; a little livestock. Epicenter of massive (1929) earthquake.

Murchison (MUHR-chi-suhn), village, N central VICTORIA, AUSTRALIA, on GOULBURN RIVER, and 85 mi/137 km N of MELBOURNE; 36°38′S 145°14′E. Livestock; oats, barley.

Murchison, Cape, NE ELLESMERE ISLAND, BAFFIN region, NUNAVUT territory, Canada, on HALL BASIN, at S end of ROBESON CHANNEL, near N entrance of LADY FRANKLIN BAY; 81°42′N 64°05′W.

Murchison Falls, UGANDA: see KABALEGA FALLS and KABALEGA NATIONAL PARK.

Murchison Glacier, in MOUNT COOK National Park, SOUTHERN ALPS, NEW ZEALAND; rises in Classen Saddle on E side of Alps; flows between MALTE BRUN and Liebig ranges, 11 mi/17 km S to MURCHISON RIVER.

Murchison Goldfield (MUHR-chi-suhn) (□ 21,000 sq mi/54,600 sq km), W central WESTERN AUSTRALIA state, W AUSTRALIA. Mining center is MEEKATHARA. Gold discovered here 1891; area placed that year under government control and leased to mining interests.

Murchison Gorge, AUSTRALIA: see KALBARRI NATIONAL PARK.

Murchison Mount (MUHR-chi-suhn) (10,659 ft/3,249 m), SW ALBERTA, W Canada, near BRITISH COLUMBIA border, in ROCKY MOUNTAINS, in BANFF NATIONAL PARK, 70 mi/113 km NW of BANFF; 51°55′N 116°38′W.

Murchison Promontory, N extremity of BOOTHIA PENINSULA, NUNAVUT territory, Canada, on BELLOT STRAIT; 71°58′N 94°28′W. Rises steeply to c.2,500 ft/762 m.

Murchison Rapids, MALAWI: see SHIRE RIVER.

Murchison River (MUHR-chi-suhn), 440 mi/708 km long, rising in SE ROBINSON RANGES, W central WESTERN AUSTRALIA; flows generally SW to INDIAN OCEAN 80 mi/129 km N of GERALDTON. Intermittent. Roderick and Sanford rivers, main tributaries.

Murcia (MOOR-thyah), autonomous region, province, and former Moorish kingdom (□ 4,370 sq mi/11,362 sq km; 2001 population 1,197,646), SE SPAIN, on the MEDITERRANEAN SEA. It became an autonomous region in 1982. The area has a generally rugged terrain, except along its coastal plain, and it is one of the hottest and driest regions of Europe, resembling N Africa in climate and vegetation. However, an irrigation system (dating from Moorish times) and several fertile valleys (especially that of the SEGURA RIVER) permit the growing of large crops of citrus and other fruits, vegetables, almonds, olives, grains, and grapes. Hemp, esparto, and minerals (lead, silver, zinc) are exported. Sericulture was long a traditional occupation. There is some small-scale industry, including a petrochemical center, and coastal tourism is important. The region was settled by the Carthaginians, who founded there (3rd century B.C.E.) the port of Cartago Nova (modern CARTAGENA). It was taken (8th century A.D.) by the Moors and emerged as an independent kingdom after the fall (11th century) of the caliphate of Córdoba. Later occupied by the Almoravids and Almohads, the kingdom of Murcia also included parts of the modern provinces of ALICANTE and ALMERÍA. In 1243 it became a vassal state of Castile, which annexed it outright in 1266.

Murcia, city (2001 population 370,745, ⊙ MURCIA province, SE SPAIN, on the SEGURA RIVER, 42 mi/67 km SW of ALICANTE; 37°59′N 01°07′W. The city lies in one of the finest irrigated garden regions in Spain. The silk industry, a traditional occupation for many years, has declined. There are food-processing, tanning, textile, and other light industries. Lead, silver, sulfur, and iron are mined nearby, and aluminum is produced. Murcia rose to prominence under the Moors, when it was for a time the capital of the independent kingdom of Murcia. The Gothic cathedral (14th–15th century) and the episcopal palace are landmarks. Murcia is the see of a bishop and has a university (founded 1915).

Murcia (MOOR-shah), town, NEGROS OCCIDENTAL province, W NEGROS island, PHILIPPINES, 8 mi/12.9 km SE of BACOLOD; 10°35′N 123°06′E. Agricultural center (rice, sugarcane).

Murdab Lagoon (moor-DAHB), inlet of CASPIAN SEA, GILAN province, N IRAN, NW of RASHT; 25 mi/40 km long, 5 mi/8 km wide. Closed off by sandspit cut by shallow access channel. Ports of BANDAR-E ANZALI and GHAZIAN are opposite each other at entrance.

Mûr-de-Bretagne (MYOOR–duh–bruh-TAHN-yuh), commune (□ 11 sq mi/28.6 sq km), CÔTES-D'ARMOR department, NW FRANCE, 24 mi/39 km SSW of SAINT-BRIEUC, in central BRITTANY. Livestock market; slate quarries nearby. Guerlédan dam, hydroelectric plant and recreational reservoir, just SW, on BLAVET RIVER. The BREST-NANTES CANAL is no longer in operation W of the dam, but is limited use to S.

Murderkill River, 19 mi/31 km long, central DELAWARE; rises in W KENT county; flows E through Killen's Pond State Park and NE, past FREDERICA (head of navigation), to DELAWARE BAY at Bowers.

Murdo, city (2006 population 553), ⊙ JONES county, S central SOUTH DAKOTA, 40 mi/64 km SSW of PIERRE; 43°53′N 100°42′W. Mountain/Central time zone boundary to E.

Murdochville (MUHR-dahk-vil, muhr-dok-VEEL), city (□ 25 sq mi/65 sq km; 2006 population 825), Gaspésie—Îles-de-la-Madeleine region, SE QUEBEC, E Canada; 48°58′N 65°30′W.

Murdock, village (2000 population 303), SWIFT county, SW MINNESOTA, 19 mi/31 km WNW of WILLMAR; 45°13′N 95°23′W. Alfalfa, beans, sugar beets, grain; livestock, poultry; dairying.

Murdock, village (2006 population 272), CASS county, SE NEBRASKA, 23 mi/37 km ENE of LINCOLN, near PLATTE RIVER; 40°55′N 96°16′W.

Mure (MOO-re), town, Kita county, KAGAWA prefecture, NE SHIKOKU, W JAPAN, 5 mi/8 km W of TAKAMATSU; 34°20′N 134°08′E. Granite and granite processing.

Mure (MOO-re), village, Kamiminochi county, NAGANO prefecture, central HONSHU, central JAPAN, 6 mi/10 km N of NAGANO; 36°45′N 138°14′E. Rice; apples; edible wild plants.

Mureaux, Les (myu-RO, lai), industrial town (□ 4 sq mi/10.4 sq km), YVELINES department, ÎLE-DE-FRANCE region, N central FRANCE, on left bank of SEINE RIVER, and 22 mi/35 km WNW of PARIS; 48°59′N 01°55′E. Site of aerospace and aircraft industry.

Cross-references are shown in SMALL CAPITALS. The pronunciation guide is shown on page xix. The sources of population figures are shown on page xvii.

Mureck (MOOR-ek), town, STYRIA, SE AUSTRIA, on MUR RIVER, on Slovenian border, and 11 mi/18 km NNE of MARIBOR, SLOVENIA; 46°42′N 15°46′E. Town hall dating to 1669 with clock tower. Castle Mureck, Slovenian *Cmurek*, is on the Slovenian side of the border. Border station. Vineyards nearby.

Mure d'Isère, La, FRANCE: see MURE, LA.

Mureijat (me-rai-ZHAT), village, central LEBANON, 18 mi/29 km SE of BEIRUT; 33°50′N 35°32′E. Elevation 3,800 ft/1,158 m. Grapes, tobacco, cereals, fruit. Summer resort. Also spelled Murayjat and Mraijat.

Mureji (moo-rai-JEE), town, NIGER state, W central NIGERIA, on NIGER RIVER opposite PATEGI, at mouth of the KADUNA RIVER, and 35 mi/56 km W of KATCHA. River landing; shea-nut processing, twine, and sackmaking.

Mure, La (myoor, lah), town (□ 3 sq mi/7.8 sq km), ISÈRE department, RHÔNE-ALPES region, SE FRANCE, near the DRAC, 20 mi/32 km S of GRENOBLE, in the DAUPHINÉ ALPS; 44°53′N 05°47′E. Center of small anthracite-coal basin still being worked. Mines (3 mi/4.8 km–5 mi/8 km NW) are reached by narrow gauge railroad line through many tunnels and over several viaducts. It offers spectacular views of deep Drac River valley with hydroelectric impoundments. Railroad is now a tourist attraction. Also known as La Mure-d'Isère.

Muren, MONGOLIA: see DELGER MÖRÖN.

Muren, MONGOLIA: see MÖRÖN, town.

Muren River, CHINA: see MULING RIVER.

Mureş (MOO-resh), county, central ROMANIA, in TRANSYLVANIA; ☉ TÎRGU MUREŞ; 46°39′N 25°33′E. Terrain ranges from hilly to mountainous; MUREŞ RIVER runs through county. Industry, agriculture.

Mureş River (MOO-resh), Hungarian *Maros*, c.470 mi/756 km long, N central ROMANIA; rising in the CARPATHIAN Mountains; flows generally W, past DEVA and ARAD, into S HUNGARY, where it joins the TISZA RIVER at SZEGED. It is navigable for small craft below Deva. Also called Mureşul River.

Mureşul River, ROMANIA: see MUREŞ RIVER.

Muret (myu-rai), town (□ 22 sq mi/57.2 sq km), HAUTE-GARONNE department, MIDI-PYRÉNÉES region, SW FRANCE, on GARONNE RIVER, and 12 mi/19 km SSW of TOULOUSE; 43°20′N 01°15′E. Agriculture market and manufacturing center (foundry products, surgical instruments, and building materials). In 1213, Simon de Montfort, leader of the Albigensian Crusade, defeated the nobles of S France and Aragon at Muret, thus ending their independence. Muret has a 12th-century church and several 15th- and 16th-century houses. A modern quarter has been developed in the river plain.

Murewa, town, MASHONALAND EAST province, NE ZIMBABWE, 55 mi/89 km ENE of HARARE; 17°39′S 31°47′E. Highway junction. Cattle, sheep, goats; dairying; tobacco, corn. Formerly Mrewa.

Murfreesboro (MUHR-freez-BUHR-o), city (2006 population 92,559), ☉ RUTHERFORD county, central TENNESSEE, on STONES RIVER; 35°50′N 86°25′W. Manufacturing; diverse industry; residential. Murfreesboro, capital of Tennessee from 1819 to 1826, was the site of the Civil War battle of Murfreesboro (or Stones River) December 31, 1862–January 2, 1863. STONES RIVER BATTLEFIELD commemorates the battle, and Civil War dead are buried in Stones River National Cemetery. Oakland Mansion is another historic attraction. Middle Tennessee State University is here. Incorporated 1817.

Murfreesboro (MUHR-freez-BUHR-o), town (2000 population 1,764), ☉ PIKE county, SW ARKANSAS, c.45 mi/72 km SW of HOT SPRINGS, near LITTLE MISSOURI RIVER; 34°03′N 93°41′W. Narrows Dam to N; forms Lake Greerson. In agricultural area; mercury mines. Crater of Diamonds State Park is on the site of only diamond reserve in U.S. (N).

Murfreesboro, town (□ 2 sq mi/5.2 sq km; 2006 population 2,295), HERTFORD county, NE NORTH CAROLINA, on MEHERRIN RIVER, and 31 mi/50 km E of ROANOKE RAPIDS; 36°26′N 77°05′W. Service industries; manufacturing (socks, clothing, baskets and crates, steel fabricating); agriculture (tobacco, peanuts, soybeans, corn, cotton; poultry, hogs).

Murg (MURG), village, HIGH RHINE, BADEN-WÜRTTEMBERG, SW GERMANY, on left bank of the RHINE, on the Swiss border, 10 mi/16 km WSW of Waldshut-Tiengen; 47°33′N 08°02′E. Manufacturing (textiles).

Murgab (moor-GAHB), town, central MARY weloyat, SE TURKMENISTAN, on railroad (near Semenik station), and 10 mi/16 km SE of MARY; 37°30′N 61°58′E. Tertiary-level administrative center. Metalworks; cotton.

Murgab (moor-GAHB), village, SE BADAKHSHAN AUTONOMOUS VILOYAT, TAJIKISTAN, in the PAMIR, on Osh-Khorugh highway, on MURGAB RIVER, and 140 mi/225 km ENE of KHORUGH, in grazing area; 38°10′N 73°59′E. Elev. 11,940 ft/3,639 m. Meteorological station; tertiary level administrative center. Until c.1929 called PAMIRSKI POST. Also spelled Murghab, Murghob.

Murgab River, c.60 mi/97 km long, TAJIKISTAN; formed by junction of the Ak-Su and Ak-Baital headstreams at Murgab; flows W to SAREZ LAKE. Joins with Kokyibel River to form BARTANG RIVER joining the Paris River at RUSHAN.

Murgash Mountains (mur-GAHSH), part of the W STARA PLANINA Mountains, N central BULGARIA, extending c.20 mi/32 km between the ISKUR River gorge (W) and BOTEVGRAD PASS (E); 42°51′N 23°41′E. Rise to 5,534 ft/1,687 m at the Murgash peak, 7 mi/11 km SW of BOTEVGRAD. Anthracite mining at REBROVO.

Murgaşi (moor-GAHSH), village, DOLJ county, S ROMANIA, 15 mi/24 km NNE of CRAIOVA.

Murgeni (moor-JEN), village, VASLUI county, E ROMANIA, on railroad, and 18 mi/29 km ESE of BÎRLAD; 46°12′N 28°01′E. Flour milling; manufacturing of edible oils.

Murgenthal (MOOR-guhn-tahl), commune, AARGAU canton, N SWITZERLAND, on AAR RIVER, and 8 mi/12.9 km SSW of OLTEN; 47°16′N 07°49′E. Metal products, knit goods, textiles; woodworking.

Murghab, AFGHANISTAN: see MORGHAB.

Murgon (MUHR-guhn), town, SE QUEENSLAND, NE AUSTRALIA, 105 mi/169 km NW of BRISBANE, c.28 mi/45 km N of KINGAROY; 26°14′S 151°57′E. Railroad junction in agricultural area (corn, sugar, and bananas); beef cattle. Dairy industry museum. Emu farm.

Murg River (MURG), 60 mi/97 km long, S GERMANY; rises in the BLACK FOREST 4 mi/6.4 km ESE of Ottenhöfen; flows generally N, past RASTATT, to the RHINE 3 mi/4.8 km N of Plittersdorf. Large hydroelectric station at FORBACH; dam (Murg-Schwarembach; 505 million cu ft/ 14.3 million cu m).

Murguba (moor-GOO-bah), town, BORNO state, extreme NE NIGERIA, on railroad, 40 mi/64 km WSW of MAIDUGURI. Market center. Groundnuts, maize; livestock.

Murgud (MOOR-guhd), town, KOLHAPUR district, MAHARASHTRA state, W central INDIA, 22 mi/35 km S of KOLHAPUR; 16°24′N 74°12′E. Agricultural market (tobacco, sugarcane, chili).

Muri (MYOO-ree), commune, AARGAU canton, N SWITZERLAND, on Bünz River, and 11 mi/18 km SW of ZÜRICH. Elevation 1,503 ft/458 m. Metal products, chemicals. Remains of abbey (founded 1027), abbey church (13th century); remodeled 17th century).

Muri (MYOO-ree), district, SE AARGAU canton, SWITZERLAND. Main town is MURI; population is German-speaking and Roman Catholic.

Muri (MOO-ree), town, TARABA state, E NIGERIA, near BENUE River, 100 mi/161 km W of YOLA; 09°11′N 10°53′E. Agricultural trade; cassava, durra, yams; cattle. Formerly called Hamarua.

Muriaé (moo-ree-ah-E), city (2007 population 95,449), SE MINAS GERAIS, BRAZIL, terminus of railroad spur from PATROCÍNIO DO MURIAÉ, and 75 mi/121 km NW of CAMPOS (RIO DE JANEIRO), and on Rio de Janeiro-BAHIA highway; 21°11′S 42°29′W. Agricultural trade center (sugar, coffee, cereals, dairy products) with processing plants and distilleries. White-marble quarries. Formerly spelled Muriahé.

Murias de Paredes (MOO-ryahs dhai pah-RAI-dhes), town, LEÓN province, N central SPAIN, on ÓRBIGO RIVER, and 35 mi/56 km NW of LEÓN; 42°51′N 06°11′W. Livestock, lumber, cereals, flax. Antimony mining nearby.

Muribeca (MOO-ree-be-kah), town (2007 population 7,196), NE SERGIPE, NE BRAZIL, on railroad, 17 mi/27 km S of PROPRIÁ; 10°26′S 36°59′W.

Muri bei Bern (MYOO-ree bai BAIRN), town (2000 population 12,571), BERN canton, W central SWITZERLAND, SE suburb of BERN.

Murici (MOO-roo-see), city (2007 population 25,968), E ALAGOAS, NE BRAZIL, on railroad, and 27 mi/43 km NW of MACEIÓ, in sugar- and cotton-growing region; 09°17′S 35°57′W. Sugar mills. Formerly spelled Muricy.

Muricilândia (MOO-ree-see-LAHN-zhee-ah), town (2007 population 2,850), N TOCANTINS state, BRAZIL, 31 mi/50 km W of PALMAS; 07°05′S 48°57′W.

Muriedas (moo-RYAI-dhahs), town, CANTABRIA province N SPAIN, 3 mi/4.8 km SW of SANTANDER; 43°26′N 03°51′W. Manufacturing.

Muriege (moo-REE-ai-gai), town, LUNDA SUL province, N ANGOLA, on major road junction, and 60 mi/97 km ESE of SAURIMO; 09°58′S 21°13′E. Market center.

Murillo, province, LA PAZ department, W BOLIVIA; ☉ PALCA; 16°20′S 68°00′W.

Murillo (moo-REE-yo), town, ☉ Murillo municipio, TOLIMA department, W central COLOMBIA, 27 mi/43 km N of IBAGUÉ in the Cordillera CENTRAL; 04°52′N 75°10′W. Elevation 9,419 ft/2,870 m. Coffee, corn, livestock.

Murillo (moo-RI-lo), unincorporated village, W ONTARIO, E central Canada, 12 mi/19 km W of THUNDER BAY, and included in township of Oliver Paipoonge; 48°24′N 89°30′W. Dairying; grain.

Murillo de Río Leza (moo-REE-lyo dhai REE-o LAI-thah), town, LA RIOJA province, N SPAIN, 8 mi/12.9 km SE of LOGROÑO; 42°24′N 02°19′W. Olive-oil processing, distilling (alcohol, brandy); wine, fruit, livestock, lumber.

Murillo el Fruto (moo-REE-lyo el FROO-to), town, NAVARRE province, N SPAIN, on ARAGON RIVER, and 16 mi/26 km SE of TAFALLA; 42°23′N 01°28′W. Olive-oil processing; sugar beets, wine, cereals, sheep.

Murilo (moo-REE-lo), atoll, HALL ISLANDS, State of CHUUK, E CAROLINE ISLANDS, MICRONESIA, W PACIFIC, 5 mi/8 km ENE of NOMWIN; c.20 mi/32 km long, 10 mi/16 km wide; 08°41′N 152°15′E. Consists of 11 low islets.

Murindó (moo-reen-DO), town, ☉ Murindó municipio, ANTIOQUIA department, NW central COLOMBIA, on Murindó River, 90 mi/145 km WNW of MEDELLÍN; 06°59′N 76°45′W. Sugarcane, rice, plantains; livestock.

Murino (MOO-ree-nuh), town (2005 population 4,970), W central LENINGRAD oblast, NW European Russia, on railroad, 15 mi/24 km NNE of SAINT PETERSBURG; 60°02′N 30°27′E. Manufacturing (tools, household appliances).

Murinsel, CROATIA: see MEDIMURJE.

Muritiba (moo-ree-chee-bah), city (2007 population 27,213), E BAHIA, BRAZIL, on right bank of RIO PARAGUAÇU, 2 mi/3.2 km above CACHOEIRA, in tobacco-growing region; 12°39′S 39°01′W. Cigar manufacturing. Formerly spelled Murityba.

Müritz, Bad, GERMANY: see GRAAL-MÜRITZ.

Müritz Lake (MYOO-rits) (□ 45 sq mi/117 sq km), largest lake entirely in GERMANY, 13 mi/21 km W of NEUSTRELITZ, Mecklenburg; 19 mi/30 km long, 1 mi/1.6 km–8.7 mi/14 km wide; maximum depth 108 ft/33 m, average depth 21 ft/6 m; 53°26′N 12°41′E. Elevation 207 ft/63 m. WAREN at N tip.

Mu River (MOO), 300 mi/480 km long, MYANMAR; rises in N SAGAING division; flows S past PINLEBU, KYUNHLA, and YE-U to the AYEYARWADY RIVER just E of MYINMU. Extensively used for irrigation; the Shwebo and Ye-u irrigation canals branch off at KABO headworks. The old left-bank Mu canal (dating from 18th century) is abandoned.

Murkurti, INDIA: see MUKURTI.

Murlungu (MOOR-loon-GOO), town, N central CEARÁ, BRAZIL, in Serra de Baturité, 14 mi/23 km NW of BATURITÉ; 04°20′S 39°00′W.

Murman Coast (MOOR-mahn), Russian *Murmanskiy Bereg*, ice-free N shore of the KOLA PENINSULA, MURMANSK oblast, extreme NW European Russia, on the BARENTS SEA; approximately 200 mi/322 km long. Deeply indented (W) by fjord-like inlets; KOLA GULF (site of MURMANSK), Ura, Titovka, and Zapadnaya Litsa bays. Sometimes called Norman Coast, of which Murman is a corrupt form.

Murmansk (MOOR-mahnsk), oblast (□ 55,287 sq mi/143,746.2 sq km; 2006 population 825,535), on the KOLA PENINSULA, an extension of the Scandinavian Peninsula, and mainland (W of the Kola River), in extreme NW corner of European Russia, between the WHITE (S) and BARENTS (N) seas; ⊙ MURMANSK. Bordered W by NORWAY and FINLAND, S by Republic of KARELIA; tundra (N), swampy forests (S). Gulf Stream action tempers climate along the ice-free Barents Sea coast, Russia's only unrestricted access to the ATLANTIC OCEAN. Severely cold climate inland. Although suffering from mismanagement and closures following the breakup of the Soviet Union, mining remains the most important activity, especially in the KHIBINY MOUNTAINS, at KIROVSK (apatite and nephelite), APATITY (apatite), and MONCHEGORSK (nickel and copper); also at NIKEL′ (nickel; far NW, near the Norwegian border). Coastal fishing by Russians (trawling along N coast; canneries at MURMANSK, PORT VLADIMIR, and TERIBERKA), lumbering (sawmills and woodworking plants at Murmansk, KANDALAKSHA, and LESOZAVODSKIY) and metallurgy (Monchegorsk, nickel smelters at Nikel′, aluminum and superphosphate works at Kandalaksha). Shipbuilding and submarine pens (Murmansk), net making. Agriculture (vegetable and dairy farming) along Murmansk railroad. Much of Russia's navy (the Northern Fleet) and part of its commercial fleet is based on the Kola Peninsula, including nuclear submarines and icebreakers. Air pollution at Nikel′. Power supplied by hydroelectric stations along TULOMA and NIVA rivers; nuclear power station. Indigenous peoples (both raise reindeer) include Lapps in interior and Saami, spread throughout N Scandinavia. Originally part of Archangel government; formed separate government (1921–1927), and later an okrug within Leningrad oblast. Constituted as oblast in 1938; acquired former Finnish territories off the W coast of the Rybachiy Peninsula in 1940 and Pechenga (Petsamo) district in 1944. Main entrepot for lend-lease materials during World War II, and terminus of the N Atlantic "lifeline" to the USSR.

Murmansk (MOOR-mahnsk), city (2006 population 313,700), ⊙ MURMANSK oblast, NW European Russia, on the KOLA GULF of the BARENTS SEA, 1,220 mi/1,963 km N of MOSCOW; 68°58′N 33°05′E. Elevation 164 ft/49 m. The terminus of the NORTHEAST PASSAGE, and the beginning point for the NORTHERN SEA ROUTE, it is a leading Russian freight port and a base for fishing fleets. It is the world's largest city N of the ARCTIC CIRCLE. It is a major naval base, and the main home port of the Russian nuclear submarine fleet. The port

at Murmansk is ice-free. The city is also a railroad terminus and is linked by railroad with Moscow and SAINT PETERSBURG. Murmansk has fish canneries, shipyards, textile factories, breweries, and sawmills. Lumber, fish, and apatite are exported, and machinery and coal are imported. Oil and gas exploration in the vicinity is underway. Murmansk was a small village before World War I, named Romanov-na-Murmane, until 1917. The port and its railroad line inland from Petrograd (now Saint Petersburg) were built in 1915–1916, when the Central Powers cut off the Russian Baltic and BLACK SEA supply routes. Allied forces occupied the Murmansk area from 1918 to 1920, during the Russian civil war. A major World War II supply base and port for Anglo-American convoys. During the 1970s and 1980s, the adjacent sea of Murmansk was the dump site of the exhausted cores of Soviet nuclear reactors. The city has a polar research institute.

Murmansk-130, RUSSIA: see GADZHIYEVO.

Murmansk-140, RUSSIA: see OSTROVNOY.

Murmansk-150, RUSSIA: see ZAOZERSK.

Murmansk-60, RUSSIA: see SNEZHNOGORSK, MURMANSK oblast.

Murmashi (moor-mah-SHI), town (2006 population 15,115), NW MURMANSK oblast, NW European Russia, on the TULOMA River, on road and railroad, 12 mi/19 km SW of MURMANSK; 68°49′N 32°50′E. Elevation 187 ft/56 m. Site of the Tuloma hydroelectric station. Has a sanatorium. The region's main international airport is 2 mi/3.2 km to the SSW.

Murmino (MOOR-mee-nuh), town (2006 population 3,480), W central RYAZAN oblast, central European Russia, on the left bank of the OKA River, on road junction, 13 mi/21 km E of RYAZAN; 54°36′N 40°03′E. Elevation 380 ft/115 m. Woolen milling.

Murmuntani (moor-moon-TAH-nee), canton, LOAYZA province, LA PAZ department, W BOLIVIA. Elevation 8,333 ft/2,540 m. Some gas wells in area. Lead-bearing lode; tin mining at Mina Viloco; clay, limestone, and gypsum deposits. Agriculture (potatoes, yucca, bananas, rye); cattle.

Murnau am Staffelsee (MUR-nou ahm SHTAHF-fel-saih), town, UPPER BAVARIA, GERMANY, at N foot of the Bavarian Alps, at SE tip of the STAFFELSEE, 11 mi/18 km S of WEILHEIM; 47°40′N 11°11′E. Textile manufacturing; woodworking, lumber and paper milling. Summer health resort and winter-sports center (elevation 2,257 ft/688 m). Completely destroyed during the Thirty Years War. Has early-18th-century church. Was first mentioned 1180. Chartered 1322.

Muro (MOO-ro), town, MAJORCA island, BALEARIC ISLANDS, 24 mi/39 km ENE of PALMA. Wine-growing and livestock-raising center. Rice processing; lime quarrying.

Muro, Capo di (moo-ro, KAH-po dee), headland of SW Corsica, on the MEDITERRANEAN, 13 mi/21 km SSW of AJACCIO, between Gulf of AJACCIO (N) and Gulf of VALINCO (S); 41°45′N 08°39′E.

Muro de Alcoy (MOO-ro dhai ahl-KOI), town, ALICANTE province, E SPAIN, 6 mi/9.7 km NNE of ALCOY; 38°46′N 00°26′W. Cement and soap manufacturing, cottonseed and olive-oil processing; honey, wine, cereals, flax, vegetables.

Murol (myu-rol), commune (□ 5 sq mi/13 sq km), resort in PUY-DE-DÔME department, central FRANCE, 7 mi/11.3 km E of MONT-DORE; 45°35′N 02°57′E. It is surrounded by the extinct volcanoes (a regional park) of the AUVERGNE region; 2,700 ft/830 m above sea level, near Lake Chambon. Its 13th-century castle, surrounded by ramparts, has been restored to its medieval appearance; there are organized evening spectacles for tourists.

Muro Lucano (MOO-ro loo-KAH-no), town, POTENZA province, BASILICATA, S ITALY, 19 mi/31 km WNW of POTENZA; 40°45′N 15°29′E. In agricultural region (cereals, fruit). Bishopric. Has castle partly destroyed

(1694, 1783) by earthquakes. Hydroelectric plant nearby.

Murom (MOO-ruhm), city (2006 population 126,975), SE VLADIMIR oblast, central European Russia, on the OKA River, on the administrative border with NIZHEGOROD oblast, on highway junction, 85 mi/137 km SE of VLADIMIR; 55°34′N 42°02′E. Elevation 400 ft/121 m. It is a port and a railroad junction, with railroad repair shops, manufacturing of machinery (diesel engines, refrigerators, radios), woodworking, and cotton and linen textile industries. It was first mentioned in the chronicles in 862 and became the capital of the Murom principality in the 12th century. In medieval times, Murom was an important trade center on the Oka-Volga water route. The city has a cathedral and some monasteries dating from the 16th and 17th centuries.

Muromtsevo (MOO-ruhm-tsi-vuh), town (2006 population 11,105), E OMSK oblast, SW SIBERIA, RUSSIA, on the TARA RIVER (tributary of the IRTYSH River), on road, 80 mi/129 km NNW of TATARSK (NOVOSIBIRSK oblast), and 27 mi/43 km NE of BOL′SHERECHYE; 56°22′N 75°15′E. Elevation 239 ft/72 m. Furniture; creamery.

Murone (moo-RO-ne), village, E Iwai county, IWATE prefecture, N HONSHU, NE JAPAN, 56 mi/90 km S of MORIOKA; 38°56′N 141°27′E.

Murongo (moo-RON-go), village, KAGERA region, NW TANZANIA, 80 mi/129 km WNW of BUKOBA on KAGERA RIVER (UGANDA border), opposite KIKAGATI ferry, and 13 mi/21 km E of RWANDA border; 01°05′S 30°38′E. Rumanyika Game Reserve to SE, Ibanda Game Reserve to SW. Tin-mining center.

Muroran (moo-RO-rahn), city, SW Hokkaido prefecture, N JAPAN, on UCHIURA BAY, 53 mi/85 km S of SAPPORO; 42°18′N 140°58′E. Port on Cape Etomo and major industrial center with steel works and an oil refinery; sword and ceramics manufacturing. Fishing; seaweed cultivation and processing. Hot-spring resorts are nearby.

Muros (MOO-ros), town, LA CORUÑA province, NW SPAIN; fishing port on inlet of the ATLANTIC, 27 mi/43 km WSW of SANTIAGO DE COMPOSTELA; 42°47′N 09°02′W. Fish processing, boatbuilding; lumber; livestock.

Muros de Nalón (MOO-ros dhai nah-LON), town, OVIEDO province, NW SPAIN, near NALÓN RIVER estuary, 22 mi/35 km W of GIJÓN; 43°32′N 06°06′W. Flour milling, meat processing. Agricultural trade (corn, potatoes, cattle; lumber). Iron and copper mines in vicinity.

Muroto (moo-RO-TO), city, KOCHI prefecture, S SHIKOKU, W JAPAN, on TOSA BAY, near MUROTO POINT, 40 mi/65 km S of KOCHI; 33°17′N 134°09′E. Fishing port (tuna, bonito); loquats, chestnuts; tea; charcoal. Muroto Anan Coast quasi-national park is nearby. Sometimes called Murotsu.

Muroto Point (moo-RO-TO), Japanese *Muroto-mizaki*, cape in KOCHI prefecture, S SHIKOKU, W JAPAN, at E side of TOSA BAY; 33°14′N 134°11′E. Lighthouse.

Murou (moo-RO-oo), village, Uda district, NARA prefecture, N central JAPAN, 14 mi/23 km S of NARA; 34°33′N 136°01′E. Murou Temple here is the smallest outdoor five-storied pagoda in Japan.

Murovani Kurylivtsi (moo-RO-vah-nee koo-RI-leeft-tsee) (Russian *Murovanyye Kurilovtsy*), town, W VINNYTSYA oblast, UKRAINE, 22 mi/35 km NW of MOHYLIV-PODIL′S′KYY; 48°44′N 27°31′E. Elevation 643 ft/195 m. Raion center; food processing (flour, fruit canning, dairy), mineral water, sewing. Known since 1493; town in 1775 and urban settlement (town) in 1956.

Murovanyye Kurilovtsy, UKRAINE: see MUROVANI KURYLIVTSI.

Murov Vruh (MOOR-ov VRUHK), peak (8,005 ft/2,440 m) in the PIRIN MOUNTAINS, SW BULGARIA.

Murowana Goslina (moo-ro-VAH-nah go-SLEE-nah), Polish *Murowana Goślina*, German *Murowana Goslin*, town, Poznań province, W POLAND, 12 mi/19 km NNE of POZNAŃ; 52°34′N 17°01′E. Flour milling, sawmilling.

Murphy (MUHR-fee), town (□ 2 sq mi/5.2 sq km; 2006 population 1,574), ⊙ CHEROKEE county, extreme W NORTH CAROLINA, on HIWASSEE RIVER, and 90 mi/145 km WSW of ASHEVILLE, in Nantahala National Forest, near GEORGIA state line; 35°05′N 84°01′W. Railroad terminus. Retail trade; manufacturing (motors, wooden cabinets and pallets, apparel, thread, veneers). Timber; agriculture (tobacco, corn, apples; cattle). Hiwassee and APPALACHIA reservoirs are near. Founded c.1830.

Murphy (MUHR-fee), town (2006 population 12,789), COLLIN county, N TEXAS, residential suburb 17 mi/27 km NE of downtown Dallas, in urban growth area; 33°00′N 96°36′W.

Murphy, unincorporated village, ⊙ OWYHEE county, SW IDAHO, 25 mi/40 km S of NAMPA; 43°13′N 116°33′W. One of the smallest county seats in U.S.; county museum here.

Murphy Islands, group of six small islands, ST. LAWRENCE county, N NEW YORK, in the SAINT LAWRENCE River, 11 mi/18 km SW of MASSENA; 44°53′N 75°09′W.

Murphys, unincorporated town (2000 population 2,061), CALAVERAS county, central CALIFORNIA, 45 mi/72 km ENE of STOCKTON, on Angels Creek; 38°09′N 120°28′W. Walnuts, olives, grapes, oats, cattle; timber. Manufacturing (printing and publishing). El Dorado National Forest and CALAVERAS BIG TREES STATE PARK to NE.

Murphysboro, city (2000 population 13,295), ⊙ JACKSON county, S ILLINOIS, on the BIG MUDDY RIVER, 7 mi/11.3 km WNW of CARBONDALE; 37°46′N 89°20′W. It is a trade and distributing center for a fertile farm area. Shoes, feed, and fertilizer are made there. A memorial to John A. Logan is in the city. Nearby are a state park and a national forest. Incorporated 1867.

Murr (MUR), village, STUTTGART district, BADEN-WÜRTTEMBERG, GERMANY, on the MURR, 12 mi/19 km N of STUTTGART; 48°57′N 09°13′E. Grain; winery.

Murra (MOOR-rah), town, NUEVA SEGOVIA department, NW NICARAGUA, 36 mi/58 km ENE of OCOTAL; 13°46′N 86°01′W. Sometimes called San Juan de Murra.

Murray (MUHR-ee), county (□ 347 sq mi/902.2 sq km; 2006 population 41,398), NW GEORGIA; ⊙ CHATSWORTH; 34°47′N 84°45′W. Bounded N by TENNESSEE state line, W by CONASAUGA RIVER. Manufacturing of textiles; agricultural area for corn, hay, soybeans, fruit; cattle, hogs, poultry; sawmilling; talc mining. Part of CHATTAHOOCHEE National Forest (E). Formed 1832.

Murray, county (□ 719 sq mi/1,869.4 sq km; 2006 population 8,778), SW MINNESOTA; ⊙ SLAYTON; 44°01′N 95°45′W. Agricultural area drained by DES MOINES RIVER, which forms Lake SHETEK (state park) in N. Corn, oats, alfalfa, soybeans; hogs, sheep, cattle, poultry; dairying. Several small lakes in county. Formed 1857.

Murray, county (□ 424 sq mi/1,102.4 sq km; 2006 population 12,945), S OKLAHOMA; ⊙ SULPHUR; 34°29′N 97°04′W. Intersected by WASHITA River (forms border on N and S). Includes part of ARBUCKLE MOUNTAINS (to W, including Turner and Price falls), and Chickasaw National Recreation Area and Arbuckle Reservoir in E. Livestock raising, agriculture (corn, wheat, poultry, fruit), dairying; manufacturing (oil and gas field machinery). Mining (limestone, sand). Recreation. Sulphur is resort, with mineral springs. Formed 1907.

Murray (MUHR-ee), city (2000 population 14,950), ⊙ CALLOWAY county, W KENTUCKY, 35 mi/56 km SSE of PADUCAH on East Fork Clarke River; 36°36′N 88°19′W. Elevation 515 ft/157 m. Varied light Manufacturing, agriculture (dark and burley tobacco, grain; livestock, poultry; dairying); and tourism are important; recreation and retirement area. Murray State University, including Weather-West Museum (his-

tory of W Kentucky) is here. Kenlake State Resort Park to NE, KENTUCKY LAKE reservoir (Kenlake; TENNESSEE RIVER) to E. Established 1822; incorporated 1844.

Murray, city (2006 population 44,844), SALT LAKE county, N central UTAH, suburb 7 mi/11.3 km S of downtown SALT LAKE CITY on JORDAN RIVER, at mouths of Big Cottonwood and Little Cottonwood creeks; 40°38′N 111°53′W. Elevation 4,300 ft/1,311 m. Retail center; dairying; cattle, sheep; manufacturing (fish food, cabinets). WASATCH RANGE and National Forest to E, including Mount Olympus Wildlife Area. The county fairgrounds are in Murray. Settled 1890s, incorporated 1903.

Murray (MUH-ree), former township (□ 82 sq mi/213.2 sq km; 2001 population 8,157), SE ONTARIO, E central Canada; 44°08′N 77°41′W. Amalgamated into Quinte West in 1998.

Murray, town (2000 population 766), CLARKE county, S IOWA, 9 mi/14.5 km W of OSCEOLA; 41°02′N 93°57′W. In livestock and grain area; manufacturing of automotive wiring harnesses.

Murray (MUHR-ee), village (2006 population 525), CASS county, SE NEBRASKA, 7 mi/11.3 km S of PLATTSMOUTH, near MISSOURI RIVER; 40°55′N 95°55′W.

Murray Bay, Canada: see LA MALBAIE.

Murray Bridge (MUH-ree BRIJ), town and river port (2001 population 12,831), SE SOUTH AUSTRALIA state, S central AUSTRALIA, on MURRAY RIVER, and 40 mi/64 km ESE of Adelaide; 35°07′S 139°16′E. Railroad junction; irrigated agriculture and dairying center; citrus and dried fruits, vegetables (snow peas, tomatoes), chickens, pigs; building stone. Town laid out 1883. Formerly called Mobilong.

Murray Canal (MUHR-ee), SE ONTARIO, E central Canada, connects Bay of QUINTE with Lake ONTARIO, cutting across narrow isthmus that connects peninsula of PRINCE EDWARD county with mainland; 44°02′N 77°37′W. Completed 1889.

Murray City, village (2006 population 465), HOCKING county, S central OHIO, 13 mi/21 km E of LOGAN; 39°30′N 82°10′W. In coal-mining area.

Murray Harbour, village (2001 population 357), SE Prince Edward Island, Canada, on Murray Harbour of the Gulf of St. Lawrence, at mouth of MURRAY RIVER, 33 mi/53 km ESE of CHARLOTTETOWN; 46°00′N 62°32′W. Lobster, oyster, fisheries.

Murray Hill, business and residential district of S central MANHATTAN borough of NEW YORK city, SE NEW YORK, on the EAST SIDE S of 42nd Street; 40°44′N 73°58′W.

Murray Hill, town, UNION county, NEW JERSEY, 10 mi/16 km from ELIZABETH; 40°41′N 74°24′W. Major center for Lucent Technologies, Inc.

Murray Isle, one of the THOUSAND ISLANDS, JEFFERSON county, N NEW YORK, in the SAINT LAWRENCE county, between WELLESLEY and GRINDSTONE islands, 7 mi/11.3 km SW of ALEXANDRIA BAY; 44°17′N 76°03′W. Island is 1.1 mi/1.8 km long, elongated NE-SW; highest elevation c.345 ft/105 m. Popular for summer-resort homes. Post office in CLAYTON.

Murray, Lake, reservoir and resort lake, LOVE and CARTER counties, S OKLAHOMA, on small tributary of RED RIVER, 8 mi/12.9 km SSE of ARDMORE; c.15 mi/24 km long; 34°01′N 97°01′W. Lake Murray State Park on W shore.

Murray, Lake, reservoir, LEXINGTON, SALUDA, NEWBERRY counties, central SOUTH CAROLINA, on SALUDA RIVER, 12 mi/19 km WNW of COLUMBIA; 30 mi/48 km long; 34°02′N 81°12′W. Formed by SALUDA Dam (formerly DREHER SHOALS DAM), built for power generation. Dreher Island State Park along W part of lake.

Murray Lock and Dam, PULASKI county, central ARKANSAS. Dam is 68 ft/21 m high and is on the ARKANSAS RIVER. It was built by the Army Corps of Engineers in 1969 for navigational purposes. The

reservoir it creates has a maximum capacity of 108,500 acre-ft.

Murray River, village (2001 population 435), SE Prince Edward Island, Canada, on small Murray River, and 30 mi/48 km ESE of CHARLOTTETOWN; 46°00′N 62°38′W. Mixed farming, dairying; potatoes.

Murray River (MUH-ree), principal river of AUSTRALIA, 1,609 mi/2,589 km long; rising in the AUSTRALIAN ALPS, SE NEW SOUTH WALES, SE Australia; flowing W to form the New South Wales–VICTORIA border, then flows SW across SOUTH AUSTRALIA state through Lake ALEXANDRINA, a lagoon, into the INDIAN OCEAN; 35°22′S 139°22′E. It receives its main tributary, the DARLING RIVER, at WENTWORTH and secondary tributary, the MURRUMBIDGEE, adjacent to Boundary Bend. Navigable below ECHUCA, 1,100 mi/1,770 km by paddlewheel steamers; series of twelve weirs and locks aid navigation to mouth, control snow meltwater and for irrigation. Irrigation schemes in Murray–Darling Basin have led to vast production of vineyards, citrus groves, vegetables, olives, rice, cotton, and livestock. Irrigation has led to salting of soil by raising ground water table, which dissolves natural salts and brings the salts to the surface. Hydroelectric station at HUME Weir and Reservoir.

Murray River, Canada: see MALBAIE RIVER.

Murraysburg, town, WESTERN CAPE province, SOUTH AFRICA, on buffalo RIVER, and 50 mi/80 km WNW of GRAAFF-REINET, on the GREAT KAROO, close to border intersection of Northern, Western, and Eastern Cape; 31°58′S 23°45′E. Elevation 4,100 ft/1,250 m. Established in 1859; named for Andrew Murray, famous Dutch reformed minister from Scotland. Wool-producing center; livestock, feed crops; horse-breeding center.

Murrayville (MUH-ree-vil), town, NW VICTORIA, SE AUSTRALIA, 15 mi/24 km E of SOUTH AUSTRALIA border; 35°16′S 141°11′E. Has underground basin used by farmers of potatoes and other crops.

Murrayville, village (2000 population 644), MORGAN county, W central ILLINOIS, 10 mi/16 km S of JACKSONVILLE; 39°34′N 90°15′W. In agricultural area (corn, wheat, soybeans, cattle, hogs).

Murree (mu-REE), town, RAWALPINDI district, N PUNJAB province, central PAKISTAN, on W spur of PUNJAB Himalayas, 24 mi/39 km NE of ISLAMABAD; 33°54′N 73°24′E. Popular hill resort and military garrison; bakery; sericulture. Nearby are several forest-clad peaks (over 8,000 ft/2,438 m), vegetable fields, and fruit orchards. At UPPER TOPA (E) is forest college, opened in 1948. Nearby are many resorts including Ayubia, a winter ski resort.

Murrels Inlet (MUH-ruhls), unincorporated town GEORGETOWN county, E SOUTH CAROLINA, 18 mi/29 km NE of GEORGETOWN; 33°32′N 79°02′W. Fishing, boatbuilding. Brookgreen Gardens to W. Manufacturing includes fish processing, printing, concrete.

Mürren (MYOO-ruhn), hamlet, in LAUTERBRUNNEN commune, BERN canton, S central SWITZERLAND, in the BERNESE ALPS, 9 mi/14.5 km S of INTERLAKEN; 46°34′N 07°53′E. Health and sports center with a splendid view of the JUNGFRAU and neighboring peaks; funicular to top of SCHILTHORN. Highest village in Bern canton that has year-round residents (elevation 5,397 ft/1,645 m). Reached by cog railroad from Lauterbrunnen. Also noted for its annual international ballooning week.

Murrhardt (MUR-hahrt), town, STUTTGART district, BADEN-WÜRTTEMBERG, GERMANY, 7 mi/11.3 km ENE of BACKNANG; 48°59′N 09°35′E. Manufacturing of tanning materials; metal- and woodworking, tanning. Has Romanesque–late-Gothic church. Was Roman castrum. In 1973 more than fifty graves of abbots and noblemen were discovered under the 15th-century town church.

Murrieta, unincorporated village (□ 2 sq mi/5.2 sq km; 2000 population 44,282), RIVERSIDE county, S CALI-

Area is shown by the symbol □, and capital city or county seat by ⊙.

FORNIA, 30 mi/48 km S of RIVERSIDE on Murrieta Creek; 33°34′N 117°13′W. Resort, with thermal springs. Manufacturing (optical instruments, screw machine parts, biological products, medical instruments, signs). This part of SW Riverside county has grown rapidly from 1980 to mid-1990s and is projected to continue to grow in 21st century. Murrieta and Temecula hot springs to E; SANTA ANA MOUNTAINS to SW; parts of Cleveland National Forest to W and SE.

Murrieta Springs, unincorporated town, RIVERSIDE county, S CALIFORNIA, 30 mi/48 km S of RIVERSIDE, 2 mi/3.2 km E of MURRIETA; 33°34′N 117°09′W. Murrieta and TEMECULA hot springs are here. Established resort area now at fringe between LOS ANGELES and SAN DIEGO.

Murringo (muh-RIN-go), village, NEW SOUTH WALES, SE AUSTRALIA, 16 mi/25 km E of YOUNG; 34°18′S 48°31′E.

Murro di Porco, Cape (MOOR-ro dee POR-ko), point on SE coast of SICILY, ITALY, 4 mi/6 km SE of SYRACUSE; 37°N 15°20′E.

Murrumbala District, MOZAMBIQUE: see ZAMBÉZIA.

Murrumbateman (MUH-ruhm-BAIT-muhn), village, NEW SOUTH WALES, SE AUSTRALIA, 25 mi/40 km NW of CANBERRA, 11 mi/17 km S of YASS; 34°58′S 149°02′E. Vineyards; wool. The Ngunnawal Aboriginal people inhabited the area before European settlement, 1820s. Variant spelling: Morumbateman.

Murrumbidgee River (muh-ruhm-BIJ-ee), c.1,050 mi/1,690 km long, AUSTRALIA; rising in the AUSTRALIAN ALPS, SE NEW SOUTH WALES; flowing generally W to the MURRAY RIVER on the VICTORIA border; 34°43′S 143°12′E. Used extensively for irrigation, the river receives water from the SNOWY MOUNTAINS Hydroelectric Scheme. Fruits, rice, and vegetables are grown, notably in Murrumbidgee Irrigation Area. BURRINJUCK Dam and hydroelectric station are on the river.

Murrumpula District, MOZAMBIQUE: see NAMPULA.

Murrurundi (MUHR-uh-RUN-dee), municipality, E NEW SOUTH WALES, SE AUSTRALIA, 90 mi/145 km NW of NEWCASTLE, and at foot of LIVERPOOL RANGE; 37°22′S 145°32′E. Agriculture (wheat; sheep).

Murrysville, PENNSYLVANIA: see MUNICIPALITY OF MURRYSVILLE.

Mursa, CROATIA: see OSIJEK.

Mursan (moor-SAHN), town, ALIGARH district, W UTTAR PRADESH state, N central INDIA, 7 mi/11.3 km W of HATHRAS; 27°35′N 77°56′E. Wheat, barley, pearl millet, gram, corn. Formerly principal Jat estate in Uttar Pradesh. Cutlery manufacturing at NAYA or Nayaganj Hathras, 8 mi/12.9 km W.

Murshidabad (moor-SHEE-dah-bahd), district (□ 2,056 sq mi/5,345.6 sq km), central WEST BENGAL state, E INDIA; ⊙ BAHARAMPUR. Bounded NE by the PADMA (Ganga) River and BANGLADESH; drained by BHAGIRATHI and JALANGI Rivers; MOR RIVER irrigation project in W. Undulating plain interspersed with marshes. Area E of the Bhagirathi, in GANGA DELTA, is major agricultural region of district (rice, gram, oilseeds, jute, barley, mangoes, wheat); extensive mulberry cultivation (silk growing) in W area. Main industrial towns, Baharampur and BELDANGA (silk weaving, rice, sugar, and oilseed milling, match manufacturing). Silk-weaving center at MURSHIDABAD; cotton weaving (JANGIPUR, DHULIAN), metalware (KANDI), shellac (NIMTITA). Bengal Silk Technological Institute, Krishnanath College at Baharampur. Part of Gaur kingdom (1197). COSSIMBAZAR was major English trade center of BENGAL in 17th century. Invaded in 18th century by Marathas; in 1751, nawab of Bengal ceded ORISSA to the Marathas. Baharampur was scene of first overt act of Indian Mutiny in 1857.

Murshidabad (moor-SHEE-dah-bahd), town, MURSHIDABAD district, E WEST BENGAL state, E INDIA, on the BHAGIRATHI RIVER, 5 mi/8 km N of BAHARAMPUR; 24°11′N 88°16′E. Silk-weaving center; trades in rice, gram, oilseeds, jute, wheat, barley. Has a magnificent

palace of titular nawab of Bengal; burial place of old nawabs of Bengal. Reputedly founded by Akbar. In 1704 became the capital of BENGAL following its removal from DACCA; continued as the capital of Bengal under British until 1790. The defeat of the last reigning nawab, Siraj-ud-Daula, by Robert Clive at PLASSEY (1757) marked the beginning of British rule in India. Murshidabad once famous for carved ivory and embroideries. Formerly called Maksudabad, Makhsusabad, or Muxadabad; sometimes Lalbagh.

Murska Sobota (MOOR-skah SO-bo-tah), town (2002 population 12,393), NE SLOVENIA, on LEDAVA RIVER, on railroad, and 25 mi/40 km ENE of MARIBOR; 46°39′N 16°09′E. Chief center of the PREKMURJE; trade center for grain-growing region; poultry raising, meatpacking.

Murtaugh, village (2000 population 139), TWIN FALLS county, S IDAHO, 15 mi/24 km ESE of TWIN FALLS, and on SNAKE RIVER; 42°29′N 114°10′W. Elevation 4,082 ft/1,244 m. Agriculture (potatoes); manufacturing (potato starch). Commercial fish hatcheries. Milner Dam to E; Murtaugh Lake to S. Incorporated 1937.

Murtazapur (moor-tuh-ZAH-poor), town, AKOLA district, MAHARASHTRA state, W central INDIA, 25 mi/40 km E of AKOLA. In major cotton-growing area; railroad junction; cotton ginning, oilseed milling. Also spelled Murtizapur.

Murten (MOOR-tuhn), French *Morat*, commune, FRIBOURG canton, W SWITZERLAND, on LAKE MORAT; 46°56′N 07°07′E. Elevation 1,503 ft/458 m. It is best known as the scene of the defeat (1476) of Charles the Bold of Burgundy by the Swiss. Founded by the dukes of ZÄHRINGEN in the 12th century, Murten has preserved much of its historic architecture. It has a 13th-century castle, town walls (14th–15th century), and a 15th-century French Gothic church.

Murtensee, SWITZERLAND: see MORAT, LAKE.

Murter, town, S CROATIA, in DALMATIA, on N coast of MURTER ISLAND, 3 mi/4.8 km NW of TISNO. Sea resort and marina; tourism.

Murter Island (MOOR-ter), Italian *Morter*, ancient *Colentum*, Dalmatian island in ADRIATIC SEA, S CROATIA, 13 mi/21 km WNW of SIBENIK; max. 7 mi/11.3 km long, max. 2 mi/3.2 km wide. Bridged to mainland. Chief villages are TISNO on E coast; MURTER on N coast.

Murtizapur, INDIA: see MURTAZAPUR.

Murtle Lake (MUHR-tuhl), E BRITISH COLUMBIA, W Canada, in CARIBOO MOUNTAINS, in WELLS GRAY PROVINCIAL PARK, 100 mi/161 km NNE of KAMLOOPS; 17 mi/27 km long, 1 mi/2 km–6 mi/10 km wide; 52°08′N 119°38′W. Elevation 3,650 ft/1,113 m. Drains W into North Thompson River.

Murtoa (muhr-TO-uh), town, W central VICTORIA, AUSTRALIA, 160 mi/257 km WNW of MELBOURNE; 36°36′S 142°29′E. Railroad junction; wheat-growing center; flour mill. Bird museum housed in old water tower.

Murton (MUH-tuhn), town (2001 population 7,243), DURHAM, NE ENGLAND, 2 mi/3.2 km to NORTH SEA coast, 7 mi/11.3 km NE of DURHAM; 54°49′N 01°24′W. Previously coal mining.

Murtosa (moor-TO-sah), town, AVEIRO district, N central PORTUGAL, on Aveiro lagoon, 6 mi/9.7 km N of AVEIRO; 40°44′N 08°38′W. Sardine fisheries, saltworks.

Murtvitsa, Bulgaria: see PODEM.

Murud (MUHR-uhd), town, RAIGAD district, MAHARASHTRA state, W central INDIA, on ARABIAN SEA, at mouth of wide creek, 45 mi/72 km SSE of MUMBAI (Bombay). Fish-supplying center (mackerel, pomfrets); rice, coconuts; coir rope and furniture manufacturing; bakery. Teak forests in N. Was the capital of former Deccan state of JANJIRA.

Muruhuta (moo-roo-HOO-tah), canton, LOAYZA province, LA PAZ department, W BOLIVIA. Elevation 8,333 ft/2,540 m. Some gas wells in area. Lead-bearing

lode; tin mining at Mina Viloco; clay, limestone, and gypsum deposits. Agriculture (potatoes, yucca, bananas, rye); cattle.

Muruntau (MOOR-oon-taw), town, NAWOIY wiloyat, central UZBEKISTAN; 41°30′N 64°37′E.

Muruntinga do Sul (MOO-roon-cheen-gah do sool), town, NW SÃO PAULO state, BRAZIL, 8 mi/12.9 km SE of ANDRADINA, on railroad; 20°59′S 51°18′W. Coffee growing.

Mururata (moo-roo-RAH-tah), canton, NOR YUNGAS province, LA PAZ department, W BOLIVIA, 13 mi/20 km NW of COROICO; 16°07′S 67°45′W. Elevation 5,705 ft/1,739 m. Clay, limestone, gypsum, phosphate deposits. Agriculture (potatoes, yucca, bananas, rye); cattle.

Mururata (moo-roo-RAH-tah), peak (19,255 ft/5,869 m) in Cordillera de LA PAZ, SUD YUNGAS province, LA PAZ department, W BOLIVIA, 19 mi/31 km ESE of LA PAZ; 16°30′S 67°50′W. Tin-tungsten deposits on E slopes.

Mururoa (moo-roo-RO-ah), atoll, S FRENCH POLYNESIA; 17 mi/27 km long, 18 mi/29 km wide; 21°52′S 138°55′W. Has an entrance for ships to enter central lagoon. For thirty years from 1966, Mururoa was at various periods the main site of French nuclear testing, first in the atmosphere and then in shafts sunk deep into the atoll. It has occasionally been subject to anti-nuclear protests. Quay, airstrip. Also called Moruroa.

Murutungura (moo-roo-toon-GOO-roo), village, MWANZA region, NW TANZANIA, 35 mi/56 km NNE of MWANZA, on E shore of UKEREWE ISLAND, Lake VICTORIA; 02°01′S 33°05′E. Fish; livestock; cotton, grain.

Murviedro, SPAIN: see SAGUNTO.

Murviel-lès-Béziers (myur-vyel–lai–bai-zye), commune (□ 12 sq mi/31.2 sq km), HÉRAULT department, LANGUEDOC-ROUSSILLON administrative region, S FRANCE, 8 mi/12.9 km NW of BÉZIERS; 43°26′N 03°08′E. Vineyards in LANGUEDOC region.

Murwar, INDIA: see KATNI.

Mürwik (MYOOR-vik), N suburb of FLENSBURG, SCHLESWIG-HOLSTEIN, N GERMANY.

Murwillumbah (muhr-WI-luhm-buh), municipality, NE NEW SOUTH WALES, SE AUSTRALIA, near QUEENSLAND border, on TWEED RIVER, and 65 mi/105 km SSE of BRISBANE; 28°19′S 153°24′E. Railroad terminus; dairying center; banana plantations; sugar cane; tourism. World Heritage Rainforest Centre.

Murygino (moo-RI-gee-nuh), town (2005 population 7,915), N central KIROV oblast, E central European Russia, on the VYATKA River, on railroad spur, 15 mi/24 km NW of KIROV; 58°44′N 49°27′E. Elevation 439 ft/133 m. Paper milling; garment factory; bakery. Has a sanatorium.

Mürz River (MYOORTS), 67 mi/108 km long, E central AUSTRIA; rises in the SCHNEEALPE massif in S LOWER AUSTRIA; flows SE to Mürzzuschlag, thence SW, past KAPFENBERG, to the MUR river at BRUCK AN DER MUR. From Mürzzuschlag to Bruck an der Mur, the Mürz forms a wide valley with old industries. It is part of an important traffic route between VIENNA and VENICE (ITALY).

Murzuk, LIBYA: see MARZUQ.

Mürzzuschlag (myoorts-TSOO-shlahg), town, NE STYRIA, E central AUSTRIA, on Mürz River, and 24 mi/39 km NE of BRUCK AN DER MUR; 47°36′N 15°40′E. Railroad junction; market center; iron and steelworks.

Muş, city (2000 population 67,962), E TURKEY; 38°45′N 41°30′E. It is in a valley with many vineyards. Founded c.400 B.C.E., it was an important town of ARMENIA. Called Tarun by the Arabs, Muş was captured by the Seljuk Turks, the Mongols, and Tamerlane before being annexed by the OTTOMAN EMPIRE in 1515.

Mus, INDIA: see CAR NICOBAR ISLAND.

Musa (MOO-suh), village, Équateur province, NW CONGO, 45 mi/72 km SW of GEMENA; 02°40′N 19°18′E. Elev. 1,505 ft/458 m.

Musa Daği, mountain peak (4,445 ft/1,355 m), S TUR-KEY; rising from the MEDITERRANEAN SEA, W of AN-TAKYA. The resistance of the Armenians against the Turks at Musa Daği in World War I is the subject of Franz Werfel's novel *The Forty Days of Musa Dagh*.

Musafirkhana (moo-SAH-fir-KAH-nah), town, SUL-TANPUR district, E central UTTAR PRADESH state, N central INDIA, 18 mi/29 km WNW of SULTANPUR; 26°22′N 81°48′E. Rice, wheat, gram, barley. Also written Musafir Khana.

Musa, Gebel (MOO-sah, ge-BEL) [Arabic=mount of Moses], peak (c.7,400 ft), S SINAI Peninsula, EGYPT, 2 mi/3.2 km N of Gebel KATHERINA. Sometimes iden-tified with Mt. SINAI. On its N slope is the famous monastery of St. Catherine, where the Codex Sinai-ticus was found.

Musaia (moo-SAH-yah), village, NORTHERN province, N SIERRA LEONE, near GUINEA border, 70 mi/113 km NNE of MAKENI; 09°46′N 11°37′W. Livestock station.

Musaiyib, Al (moo-SAH-yib, ahl), town, BABYLON province, central IRAQ, on the E bank of the EU-PHRATES, on railroad, and 20 mi/32 km NNW of HILLA, 5 mi/8 km N of the point where the Euphrates divides into the SHATT HILLA and SHATT HINDIYA; 32°46′N 44°17′E. Dates, wheat, corn, vegetables. Sometimes spelled Al-Musayah.

Musa, Jbel (MOO-suh, zhe-BEL), mountain (2,790 ft/ 850 m), Tanger-Tétouan administrative region, N MOROCCO, on the Strait of GIBRALTAR overlooking CEUTA (just E; Spanish posession). Northernmost extremity of the RIF mountains, opposite SPAIN's Cordillera PENIBÉTICA (of which it is the geological extension). Often identified as one of antiquity's PILLARS OF HERCULES (the other being GIBRALTAR). Also called Jabal Sidi Musa.

Musa Khel Bazar (MOO-sah KHAIL), town, LORALAI district, NE BALUCHISTAN province, SW PAKISTAN, 80 mi/129 km NE of LORALAI; 30°52′N 69°49′E. Wheat. Olives grown in SULAIMAN Range (SE). Sometimes called Musakhel.

Musakin (moo-sah-KEEN), town, SUSAH province, E TUNISIA, near the coast of the MEDITERRANEAN SEA, on railroad, and 7 mi/11.3 km SE of SUSAH; olive oil pressing, citrus growing; artisan industry (textiles, esparto products); administrative center. Saltworks nearby. Known until the 19th century for its Byzan-tine-style jewelry.

Musala (moo-sah-LAH), highest peak (9,596 ft/2,925 m) of BULGARIA, in the E RILA, W Bulgaria, 9 mi/15 km S of SAMOKOV; 42°11′N 23°35′E. Meteorological observatory. Called Musala or Mus-Allah until 1949, later Stalin Peak (until 1960).

Musala, INDONESIA: see SIBOLGA.

Mus-Allah, Bulgaria: see MUSALA.

Musandam (muh-SUHN-dem), governate (2003 pop-ulation 23,378), exclave of OMAN, on STRAIT OF HOR-MUZ, at tip of OMAN PROMONTORY, separating the PERSIAN GULF and the GULF OF OMAN. Its main town is KHASAB. Musandam is a mountainous region with a rocky, deeply indented coast, terminating in CAPE MUSANDAM and separated from the rest of Oman sultanate by an emirate of the UNITED ARAB EMI-RATES. It is a strategic area; 90% of the PERSIAN GULF oil passes through the Straits of Hormuz. The Co-mazari tribe live in a secluded area at the tip of Mu-sandam. Fishing; date cultivation. Sometimes called Ras el Jebel.

Musandam, Cape, OMAN: see MASANDAM, CAPE.

Musa Qala (moo-SAH KAH-lah), town, HELMAND province, S central AFGHANISTAN, 40 mi/64 km NNE of GIRISHK, on right tributary of HELMAND River, in outliers of the Hindu Kush; 32°22′N 64°46′E.

Mūsa River (MOO-shah), left headstream of LIELUPE River, 112 mi/180 km long, in LITHUANIA and LATVIA; rises W of Ioniskis (Lithuania); flows E and NNW, joining Nemunėlis (Latvian *Memele*) River at BAUSKA (Latvia), to form Lielupe River. Also spelled Musha.

Musashi, former province in central HONSHU, JAPAN; now SAITAMA prefecture and part of KANAGAWA prefecture.

Musashi (moo-SAH-shee), town, East Kunisaki county, OITA prefecture, NE KYUSHU, SW JAPAN, on E KUNISAKI PENINSULA, on IYO SEA, 19 mi/30 km N of OITA; 33°29′N 131°43′E.

Musashimurayama (moo-SAH-shee-moo-rah-YAH-mah), city, Tokyo prefecture, E central HONSHU, E central JAPAN, 9 mi/15 km S of SHINJUKU; 35°45′N 139°23′E. Motor vehicles.

Musashino (moo-SAH-shee-no), city, Tokyo prefec-ture, E central HONSHU, E central JAPAN, on the Su-mida River, 22 mi/35 km W of SHINJUKU; 35°42′N 139°34′E. Suburb of TOKYO. Medical products.

Musayab, Al-, IRAQ: see MUSAIYIB, AL.

Musaymir, YEMEN: see MUSEMIR.

Muscat (MUHS-ket), city (2003 population 26,668); ⊙ OMAN, SE ARABIA, on the GULF OF OMAN; 23°37′N 58°36′E. Flanked by rugged mountains, Muscat is the heart of an urban agglomeration of nearly a half million inhabitants; has a fine harbor. Dates, dried fish, mother-of-pearl, and frankincense are exported, although much of Muscat's trade had been taken over by neighboring MATRAH, which has better land communications. Muscat has expanded and mod-ernized its port since country began exportation and gained much of its trade back from Matrah. Highway to international airport in SIB (NW). Was seized by the Portuguese Afonso de Albuquerque in 1508 and kept by PORTUGAL until 1648. Persian princes held it until 1741, when it became the capital of the Oman sultanate. Sometimes spelled Masqat or Maskat.

Muscat, governate (2003 population 632,073), N OMAN, on the GULF OF OMAN, S part of the BATINA plain. Main city is Muscat. The political, economic, and administrative heart of the Omani sultanate. Contains six sub-provinces. Most densely populated governate, with high non-Omani population. Rapid growth.

Muscat and Oman, sultanate: see OMAN.

Muscatatuck National Wildlife Refuge (□ c.12 sq mi/ 31 sq km), in E JACKSON county and W JENNINGS county, SE INDIANA. Diversified habitat of farmland, timber, lakes, marshes, and ponds. Ducks, geese, wild turkeys, deer. Only national wildlife refuge in Indiana.

Muscatatuck River, 45 mi/72 km long, S INDIANA; formed in W JEFFERSON county by junction of Big Graham Creek and Big Chicken Run; flows SW and W to East Fork of WHITE RIVER 3 mi/4.8 km S of MEDORA.

Muscatine, county (□ 449 sq mi/1,167.4 sq km; 2006 population 42,883), SE IOWA, bounded SE by MIS-SISSIPPI RIVER (forms ILLINOIS state line here); ⊙ MUSCATINE; 41°29′N 91°06′W. Prairie agricultural area (sheep, hogs, cattle, poultry; corn, soybeans) drained by CEDAR RIVER. Limestone; sand and gravel pits; bituminous-coal deposits. Manufacturing and Lock and Dam No. 16 at Muscatine. Wildcat Den State Park in SE. Formed 1836.

Muscatine, city (2000 population 22,697), ⊙ MUSCA-TINE county, SE IOWA, on the MISSISSIPPI RIVER; 41°25′N 91°04′W. An early center of river traffic and lumbering, Muscatine is the shipping and processing center of a rich agricultural area. Grains; diversified light manufacturing. Muscatine Community College, an airfield, and Lock and Dam No. 16 are there; Wildcat Den State Park to E. Serious flooding oc-curred along Mississippi River in 1993. Incorporated 1851.

Muscle Shoals, town (2000 population 11,924), Col-bert co., NW Alabama, on the Tennessee River S of Florence. Center of experimental development of phosphate and nitrate fertilizers and animal foods. Various products are made in the chemical works here; manufacturing (truck trailers, nuts, screws, and bolts). Inc. 1923. Originally laid out by backers of Henry Ford, who had planned to build an automobile

plant here, the town was named for the shoals now inundated by the impounded water of the Tennessee River. Cherokee Indians had called the shoals 'mussel place,' and the U.S. Board on Geographic Names chose 'muscle' as the official spelling of the word.

Muscoda (muhs-KO-duh), town (2006 population 1,355), GRANT county, SW WISCONSIN, on WISCONSIN River, and 35 mi/56 km WNW of PRAIRIE DU CHIEN; 43°11′N 90°25′W. In livestock and dairy region; manufacturing (lumber, cheese, whey protein).

Muscogee (muhs-KO-gee), county (□ 221 sq mi/574.6 sq km; 2006 population 188,660), W GEORGIA; ⊙ CO-LUMBUS; 32°31′N 84°52′W. Bounded W by ALABAMA state line; drained by CHATTAHOOCHEE RIVER. Inter-sected by the fall line. Manufacturing at Columbus; dairying, cattle, vegetable farming. FORT BENNING is S. Formed 1826.

Musconetcong (MUHS-kuh-NET-kuhng), river, c.44 mi/71 km long, NW NEW JERSEY; rises NE of Lake HOPATCONG, in SE SUSSEX county; flows generally SW forming part of the Sussex-Morris and Warren-Morris and Warren-HUNTERDON county lines, joining the DELAWARE RIVER at Riegezsuille.

Musconetcong, Lake (MUHS-kuh-NET-kuhng), res-ervoir, SUSSEX and MORRIS counties, NW NEW JERSEY, on MUSCONETCONG River, 12 mi/19 km NW of MOR-RISTOWN; 40°54′N 74°41′W. Formed SW of Lake HOPATCONG.

Musconetcong Mountain (MUHS-kuh-NET-kuhng) (c.800 ft/244 m–900 ft/274 m), ridge of Appalachians; extends NE from DELAWARE RIVER near RIEGELS-VILLE, paralleling on SE the fertile valley of Musco-netcong River, which drains Lake Musconetcong and LAKE HOPATCONG in SE Sussex county. The river flows c.45 mi/72 km SW to the Delaware at Riegels-ville.

Muscongus Bay (muhs-KAHNG-uhs), S MAINE, be-tween Pemaquid Point and PORT CLYDE; c.15 mi/24 km across, 8 mi/12.9 km long.

Muscongus Island, MAINE: see LOUDS ISLAND.

Muscotah (muhs-KO-tuh), village (2000 population 200), ATCHISON county, NE KANSAS, on DELAWARE RIVER, and 21 mi/34 km W of ATCHISON; 39°32′N 95°31′W. Diversified farming.

Muscovy, RUSSIA: see MOSCOW, former state.

Muscoy, unincorporated town (2000 population 8,919), SAN BERNARDINO county, S CALIFORNIA, residential suburb 54 mi/87 km ENE of downtown LOS ANGELES, and 4 mi/6.4 km NW of SAN BERNARDINO, near LYTLE CREEK, in S foothills of SAN BERNARDINO MOUNTAINS (San Bernardino National Forest); 34°09′N 117°21′W. Glen Helen Regional Park to NW. California State University (San Bernardino) to NW.

Muse, unincorporated town, WASHINGTON county, SW PENNSYLVANIA, 14 mi/23 km SW of PITTSBURGH, and 2 mi/3.2 km N of CANONSBURG; 40°17′N 80°12′W. Manufacturing (printing).

Muse (MOO-SAI), village and township, SHAN STATE, MYANMAR, on SHWELI RIVER (on border of CHINA), 45 mi/72 km N of HSENWI, and 9 mi/14.5 km from BURMA ROAD.

Museimir (moo-sai-MIR), township, S YEMEN, on road, and 55 mi/89 km NW of ADEN, near former border between NORTH YEMEN and SOUTH YEMEN, on the Wadi TIBAN. Former caravan center; agricultural. Airfield. Was center of the former HAUSHABI tribal area. Also spelled Musemir and Musaymir.

Muselievo (moo-se-LEE-vo), village, LOVECH oblast, NIKOPOL obshtina, BULGARIA; 43°38′N 24°51′E.

Müsellim Remma, arm of AEGEAN SEA between LESBOS island (Greece) and TURKEY; 10 mi/16 km long, 6 mi/ 9.7 km wide. Mythemna on SW shore. Connects E with Gulf of EDREMIT, SE with MYTILENE CHANNEL.

Musemir, YEMEN: see MUSEIMIR.

Musengezi River, c. 150 mi/241 km long, ZIMBABWE and MOZAMBIQUE; rises in N central Zimbabwe c. 50 mi/80 km NNW of HARARE, in MVURWI RANGE; flows

NNE entering W Mozambique; joins ZAMBEZI River in Lago Cabora Bassa reservoir 10 mi/16 km SW of Carinde.

Museros (moo-SAI-ros), town, VALENCIA province, E SPAIN, 8 mi/12.9 km N of VALENCIA; 39°34′N 00°21′W. In produce-farming area; vegetable canning.

Musgrave Ranges (MUZ-graiv), NW SOUTH AUSTRALIA state, S central AUSTRALIA, extend 50 mi/80 km E-W, parallel with NORTHERN TERRITORY border; 26°10′S 131°50′E. Rise to 4,970 ft/1,515 m (Mount Woodroffe; highest peak in state). Granite. Aboriginal reservation in W part.

Mush, TURKEY: see MUS.

Musha (MOO-shah), village, ASYUT province, central Upper EGYPT, 7 mi/11.3 km NW of ABU TIG; 27°07′N 31°14′E. Cereals, dates, sugarcane.

Mushabani (moo-shah-bah-nee) or **Mosabani** or **Musabani**, town (2001 population 33,892), EAST SINGHBHUM district, SE Jharkhand state, E INDIA, near SUBARNAREKHA RIVER, 25 mi/40 km SE of JAMSHEDPUR; 22°31′N 86°27′E. Copper-mining center; major copper-smelting plant. Cyanite mining 5 mi/8 km NNE, near Ghatsila.

Mushandike Reservoir Recreational Park (moo-shah-n-DEE-kai) (☐ 51 sq mi/132.6 sq km), MASVINGO province, S central ZIMBABWE, formed on TOKWE RIVER by Mushandike Dam 15 mi/24 km WSW of MASVINGO. Surrounded by Mushandike Sanctuary. Formerly Mushandike National Park. Wildlife; recreation.

Musharij, Am, YEMEN: see MASHARIJ, AM.

Mushenge (moo-SHEN-gai), town, KASAI-OCCIDENTAL province, SW CONGO, 70 mi/113 km S of DEKESE; 04°32′S 21°21′E. Elev. 1,505 ft/458 m. Oil palms, rice.

Musherfe (mu-SHER-fe), Arab village, N ISRAEL, SW of AFULA; 32°33′N 35°09′E. Elevation 715 ft/215 m. Founded in early 20th century by former residents of UMM AL-FAHM.

Mushie (MOOSH-yai), village, BANDUNDU province, W CONGO, on right bank of the KWA, at mouth of FIMI RIVER, and 130 mi/209 km SW of INONGO; 03°01′S 16°54′E. Elev. 593 ft/180 m. Steamboat landing and trading center in sesame-growing region; rubber plantations in vicinity. Roman Catholic mission.

Mushin (moo-SHEEN), city, SW NIGERIA, an industrial and residential suburb of LAGOS; 06°31′N 03°21′E. Manufacturing includes textiles, furniture, printed materials, metal products, plastics, milk products, and shoes. Motor vehicles are assembled. Mushin grew mainly after World War II, with the industrialization of the Lagos area.

Mushiru-Kaikyo, RUSSIA: see KRUZENSHTERN STRAIT.

Mushiru-retsugan, RUSSIA: see LOVUSHKI ISLANDS.

Mushtuhur (MUSH-tuh-huhr), village, QALYUBIYA province, Lower EGYPT, 20 mi/32 km N of CAIRO. Cotton, flax, cereals, fruits.

Mushu or **Muschu**, island, EAST SEPIK province, NW PAPUA NEW GUINEA, 15 mi/24 km NW of WEWAK, in PACIFIC OCEAN, 3 mi/4.8 km N of NEW GUINEA coast; circa 8 mi/12.9 km long. Volcanic origin; Coconuts, palm oil; fish. KAIRIRU to N.

Musi (MOO-see), river, c.325 mi/523 km long, Sumatra Selatan province, INDONESIA; rising in BARISAN MOUNTAINS, 02°20′S 104°56′E; flows SE to PALEMBANG (head of ocean-going navigation), then NE through the swampy coastal plain to the BANGKA Strait. The Rawas, Ogan, and Komering rivers, are its chief tributaries. Rubber and a variety of tropical crops are raised in the valley.

Music Mountain, peak (6,761 ft/2,061 m), MOHAVE county, NW ARIZONA; rises from tableland (elevation c.6,000 ft/1,829 m) at S end of GRAND WASH CLIFFS, 30 mi/48 km NE of KINGMAN. In Music Mountains, small range at SW edge of Hualapai Indian Reservation.

Musidora, hamlet, E central ALBERTA, W Canada, 7 mi/11 km from TWO HILLS, in Two Hills County No. 21; 53°42′N 111°35′W.

Musikot (MOO-see-kot), town, ⊙ RUKUM district, S NEPAL, 35 mi/56 km NNW of BUTWAL; 28°37′N 82°15′E.

Musina [Venda=the spoiler], town, LIMPOPO PROVINCE, SOUTH AFRICA, near ZIMBABWE border, 115 mi/185 km NNE of POLOKWANE, northernmost town in the republic; 22°21′S 30°03′E. Elevation 2,214 ft/675 m. Copper-mining and -refining center. LIMPOPO RIVER is crossed, 10 mi/16 km NNW, by OTTO BEIT BRIDGE to Zimbabwe border station of BEITBRIDGE. Name's origin lies in the spoiling of iron ore by colocated copper deposits. Located in Baobab Tree Reserve. Formerly Messina.

Musiri (MOO-see-ree), town, TIRUCHCHIRAPPALLI district, Tamil Nadu state, S INDIA, on KAVERI River opposite KULITTALAI (ferry), 20 mi/32 km WNW of Tiruchchirappalli; 10°56′N 78°27′E. Rice, plantains, coconut palms; cotton textiles, wicker coracles.

Musi River (MOO-see), 180 mi/290 km long, in ANDHRA PRADESH state, SE INDIA; rising in isolated S hills of DECCAN PLATEAU, near VIKARABAD; flows E past VIKARABAD and HYDERABAD and S to KRISHNA RIVER 37 mi/60 km SSE of NALGONDA. Numerous irrigation canals.

Muskau, Bad, GERMANY: see BAD MUSKAU.

Muskeget Channel (muh-SKEE-get), SE MASSACHUSETTS, separates NANTUCKET and its adjacent islands from MARTHA'S VINEYARD, opening N on NANTUCKET SOUND; c.7 mi/11.3 km wide. Muskeget Island (c.1.5 mi/2.4 km long) is just off NW tip of TUCKERNUCK ISLAND.

Muskego (muhs-KEE-go) or **Muskego Center**, city (2006 population 22,760), WAUKESHA county, SE WISCONSIN, a suburb 14 mi/23 km SW of downtown MILWAUKEE, on Big Muskego, LITTLE MUSKEGO, and Denmoor lakes in city; 42°53′N 88°07′W. Varied light manufacturing.

Muskegon (mus-KEE-guhn), county (☐ 1,459 sq mi/3,793.4 sq km; 2006 population 175,231), SW MICHIGAN; 43°17′N 86°27′W; ⊙ MUSKEGON. Bounded W by LAKE MICHIGAN; drained by MUSKEGON and WHITE rivers, and by small Crockery Creek. Agriculture (cattle, hogs, apples, cherries, vegetables, corn, wheat, oats, grain; dairy products). Manufacturing at Muskegon. Fisheries. Resorts. Muskegon County Airport at Morton Shores, S of Muskegon. P.F. Hoffmaster, Muskegon, and Duck Lake state parks, in W, on or near Lake Michigan; part of Manistee National Forest is in N. Formed and organized 1859.

Muskegon (mus-KEE-guhn), city (2000 population 40,105), ⊙ MUSKEGON county, W MICHIGAN, 33 mi/53 km WNW of GRAND RAPIDS, on LAKE MICHIGAN; 43°13′N 86°15′W. Elevation 620 ft/189 m. A port of entry, the city is a car-ferry terminus and a ship/RR transfer point for a farm, fruit, and industrial region. Manufacturing (motor vehicle equipment and parts, foundry products, chemicals, paper products, sporting goods equipment, ink pigments, gasoline pumps, and heavy machinery). A fur-trading post was established here c.1810. The first sawmill was built in 1837, and the lumber industry thrived until 1890, when the city was swept by fire. Muskegon Community College is here. Manistee National Forest to NE; P.F. Hoffmaster State Park to S at Norton Shores, Muskegon State Park to W, both on Lake Michigan. Incorporated as a city 1869.

Muskegon Heights (mus-KEE-guhn), city (2000 population 12,049), MUSKEGON county, W MICHIGAN, a suburb 2 mi/3.2 km S of MUSKEGON; 43°12′N 86°14′W. Light manufacturing (foundry items, restaurant furniture, tanks and hoppers, lubricants). Mona Lake on S border. Incorporated 1903.

Muskegon River (mus-KEE-guhn), 227 mi/365 km long, N central MICHIGAN; rising in HOUGHTON LAKE; flowing SW past EVART and BIG RAPIDS, through ROGERS, HARDY and CROTON DAM ponds, then past NEWAYGO to LAKE MICHIGAN at MUSKEGON; 44°23′N 84°47′W. At its mouth the river widens into Muskegon Lake, forming a harbor c.2.5 mi/4 km wide and c.5.5 mi/8.9 km long.

Musketov, glacier (☐ 28 sq mi/72.8 sq km), in Sary-Jaz Range, KYRGYZSTAN; c.13 mi/21 km long. Elevation 11,283 ft/3,439 m. Source of SARY-JAZ RIVER.

Muski, INDIA: see MASKI.

Muskingum (mus-KING-guhm), county (☐ 667 sq mi/1,734.2 sq km; 2006 population 86,125), central OHIO; ⊙ ZANESVILLE; 39°57′N 81°57′W. Intersected by MUSKINGUM and LICKING rivers and small Salt and Jonathan creeks. In the Unglaciated Plains physiographic region. Agriculture (hogs, cattle; dairy products; corn, fruit); manufacturing mainly at Zanesville and ROSEVILLE (lumber and wood products; stone, clay, and glass products; pottery products; engines and other electrical equipment). Coal mines, limestone quarries; sand, gravel, and clay pits. There are a number of state parks at the various dams on the Muskingum River. Formed 1804.

Muskingum River (muhs-KING-guhm), 111 mi/179 km long, formed in NE OHIO, at COSHOCTON, by the union of the WALHONDING and TUSCARAWAS rivers and flowing S through ZANESVILLE, then SE to the OHIO RIVER at MARIETTA; 39°24′N 81°27′W. The Muskingum River system has extensive flood-control projects. The canalized lower river is navigable. The upper river was a link between the OHIO AND ERIE CANAL.

Muskogee (muh-SKO-gee), county (☐ 839 sq mi/2,181.4 sq km; 2006 population 71,018), E OKLAHOMA; ⊙ MUSKOGEE; 35°36′N 95°22′W. Bounded N in part by ARKANSAS RIVER, S by CANADIAN River. Agriculture (corn, hay, soybeans; cattle; dairying). Varied manufacturing especially food products. Manufacturing at Muskogee. Oil and natural-gas fields. Fort Gibson National Cemetery and Stockade in NE; Fort Gruber National Guard Training Facility in E. Greenleaf State Park in E; part of Sequoyah National Wildlife Refuge in far SE. Formed 1907.

Muskogee (muh-SKO-gee), city (2006 population 40,004), ⊙ MUSKOGEE county, 38 mi/61 km SE of TULSA, E OKLAHOMA, on ARKANSAS RIVER, opposite confluences of VERDIGRIS and NEOSHO (Grand) rivers, near the junction of the Arkansas, Verdigris, and Grand rivers; 35°45′N 95°20′W. It is an important transportation, trade, and industrial center in the agricultural Arkansas valley, with a modern port (port of Muskogee, NE part of city; opened 1971) on the McClellan–Kerr Arkansas River Navigation System. Manufacturing (machinery, rubber products, food products, consumer goods). Bacone College (two-year) here. Fairgrounds. Chouteau Dam (on Verdigris River) to N; Sequoyah and Bay state parks on FORT GIBSON reservoir to NE. Fort Gruber National Guard Training Facility to SE. Hatbox Airport in W. Five Civilized Tribes Museum and Fort Gibson (1824, restored) to E, with its national cemetery. Incorporated 1898.

Muskoka (muh-SKO-kuh), district (☐ 1,502 sq mi/3,905.2 sq km; 2001 population 53,106), S ONTARIO, E central Canada, on GEORGIAN BAY of Lake HURON; ⊙ BRACEBRIDGE; 45°00′N 79°20′W. Established 1971.

Muskoka, Lake (muh-SKO-kuh), S ONTARIO, E central Canada, 30 mi/48 km SE of PARRY SOUND; 15 mi/24 km long, 5 mi/8 km wide; 45°03′N 79°28′W. Drained W by MUSKOKA RIVER into GEORGIAN BAY. It is one of the Muskoka lakes (which also include ROSSEAU Lake, Lake JOSEPH, and Lake of BAYS) in a popular resort region of forests, ponds, and rivers, drained by Muskoka River into Georgian Bay (W).

Muskoka Lakes (muh-SKO-kuh), township (☐ 302 sq mi/785.2 sq km; 2001 population 6,042), S ONTARIO, E central Canada, at S tip of CANADIAN SHIELD, and 20 mi/32 km from BRACEBRIDGE; 45°06′N 79°35′W. Formed in January 1971 from Watt, Cardwell, Medora and Wood, WINDERMERE, PORT CARLING, BALA, and portions of Monck Township.

Muskoka River (muh-SKO-kuh), 120 mi/193 km long, S ONTARIO, E central Canada; rises in ALGONQUIN PROVINCIAL PARK in 2 branches that unite at BRACEBRIDGE (waterfalls nearby); flows W through Lake MUSKOKA to GEORGIAN BAY 25 mi/40 km S of PARRY SOUND; 45°01′N 79°23′W. Drains Muskoka lake region.

Muskwa (MUH-skwuh), unincorporated village, NE BRITISH COLUMBIA, W Canada, on MUSKWA RIVER at mouth of Prophet River, on ALASKA HIGHWAY, and 5 mi/8 km W of FORT NELSON, included in Fort Nelson Indian reserve; 58°45′N 122°41′W. Lumbering.

Muskwa River (MUH-skwuh), 160 mi/257 km long, NE BRITISH COLUMBIA, W Canada; rises in STIKINE MOUNTAINS near 57°45′N 124°50′W. Flows generally NE to confluence with SIKANNI CHIEF RIVER at FORT NELSON, forming FORT NELSON RIVER.

Muslimbagh, town, ZHOB district, NE BALUCHISTAN province, SW PAKISTAN, on ZHOB RIVER, and 60 mi/97 km NE of QUETTA; 30°49′N 67°45′E. Formerly HINDUBAGH.

Muslyumovo (moo-SLYOO-muh-vuh), village (2006 population 7,415), E TATARSTAN Republic, E European Russia, on the IK RIVER (tributary of the KAMA River), 28 mi/45 km SSE of MENZELINSK; 55°18′N 53°11′E. Elevation 265 ft/80 m. In agricultural area; flour mill, dry and skim milk plant.

Musmar (muhs-MAHR), village, RED SEA state, NE SUDAN, on railroad and road, 120 mi/193 km ENE of ATBARAH; 18°13′N 35°38′E.

Musmus (MUHS-muhs), Arab village, ISRAEL, SW of AFULA in Irron Valley; 32°32′N 35°09′E. Elevation 1,263 ft/384 m.

Musoma (moo-SO-mah), town, MARA region, NW TANZANIA, 95 mi/153 km NE of MWANZA, on SE shore of Lake VICTORIA, on Mara Bay, mouth of MARA RIVER 10 mi/16 km to E; 01°30′S 33°48′E. Small lake port. SERENGETI NATIONAL PARK to SE. TARIME Goldfield 40 mi/64 km to NE. Cotton, corn, millet; livestock; fish. Airport.

Musomishta (moo-SO-mee-tsah), village, SOFIA oblast, GOTSE DELCHEV obshtina, BULGARIA; 41°33′N 23°45′E.

Musone River (moo-ZO-ne), 45 mi/72 km long, The MARCHES, central ITALY; rises in the APENNINES 4 mi/6 km E of MATELICA; flows ENE to the ADRIATIC 3 mi/5 km NE of LORETO.

Musonoi (moo-so-NOI), village, KATANGA province, SE CONGO, on railroad, and 3 mi/4.8 km W of KOLWEZI; 10°41′S 25°26′E. Elev. 4,724 ft/1,439 m. Copper-mining center; also palladium, platinum, and cobalt mining; cattle raising.

Musquacook Lakes (muhs-KWAH-kuk), PISCATAQUIS and AROOSTOOK counties, N MAINE, 13-mi/21-km chain of five lakes (First to Fifth Musquacook lakes), c.58 mi/93 km W of PRESQUE ISLE; drain N, through Musquacook Stream, into ALLAGASH RIVER.

Musquodoboit Harbour (muh-sko-DO-bit), village, S NOVA SCOTIA, CANADA, at head of an inlet of the Atlantic, at mouth of MUSQUODOBOIT RIVER, 25 mi/40 km ENE of HALIFAX; 44°46′N 63°8′W. Elevation 52 ft/15 m. Salmon and trout fishing. Located. Name comes from Mi'kmaq, Mooskudoboogwek, meaning "suddenly widening out after a narrow entrance at the mouth". English settlement began in 1754 and grew with the arrival of Loyalists and American Revolutionary War veterans.

Musquodoboit River (muh-sko-DO-bit), 60 mi/97 km long, central NOVA SCOTIA, CANADA; rises in COBEQUID MOUNTAINS NW of SHEET HARBOUR; flows SW and S to the Atlantic, 22 mi/35 km E of HALIFAX.

Mussa Ali (MOO-sah-ah-lee), volcanic mountain (elevation 6,768 ft/2,063 m), in DEBUBAWI KAYIH BAHRI region, S ERITREA, at S extremity of DANAKIL mountains, near junction of Eritrea, ETHIOPIA, and

DJIBOUTI borders, 55 mi/89 km SW of ASEB; 12°29′N 42°24′E. Also known as Mussa-Ali or Mussaali.

Mussau Islands (moo-SOU), NEW IRELAND province, BISMARCK ARCHIPELAGO, PAPUA NEW GUINEA, SW PACIFIC OCEAN, 100 mi/161 km NW of NEW IRELAND; 01°24′S 149°38′E. There are 2 volcanic islands, MUSSAU (□ c.160 sq mi/414 sq km) and EMIRAU (□ 20 sq mi/52 sq km); several coral islets. Coconut plantations; canoe building, handicraft.

Musselburgh (MUS-uhl-BUHRG), town (2001 population 22,112), East Lothian, E Scotland, on Firth of Forth at mouth of ESK RIVER, 6 mi/9.7 km E of EDINBURGH; 55°56′N 03°03′W. Seaside resort with golf links and race course; fishing, vegetables. Previously net making, paper milling. Has Loretto School (public school on site of 16th-century shrine) and Pinkie House, noted Jacobean mansion. Nearby are fishing villages of Fisherrow (W) and Duddington. Just S is Inveresk, site of the battle of Pinkie (1547).

Musselshell (MUH-suhl-SHEL), county (□ 1,870 sq mi/4,843 sq km; 1990 population 4,106; 2000 population 4,497), central MONTANA; ⊙ ROUNDUP; 46°29′N 108°24′W. Agricultural region drained by MUSSELSHELL RIVER. Wheat, oats, hay, some corn, dairying, cattle, sheep, hogs. Oil, gas, and coal. Formed 1911.

Musselshell (MUH-suhl-SHEL), village, MUSSELSHELL county, central MONTANA, 58 mi/93 km NNE of BILLINGS, on MUSSELSHELL RIVER, at mouth of Hawk Creek. Grain, livestock. Bull Mountains to S.

Musselshell River (MUH-suhl-SHEL), 292 mi/470 km long, central MONTANA; rises in several branches in the CRAZY, Castle, and LITTLE BELT mountains, formed by joining of NORTH and SOUTH Forks in W WHEATLAND county, just E of MEAGHER county line; flows E past HARLOWTON, NE past ROUNDUP, and N to FORT PECK LAKE. Not navigable.

Mussidan (mu-see-dahn), town (□ 1 sq mi/2.6 sq km), DORDOGNE department, AQUITAINE region, SW FRANCE, on the ISLE RIVER, and 18 mi/29 km SW of PÉRIGUEUX; 45°02′N 00°22′E. Diverse small industries, especially poultry and fruit preserving and packing. Tobacco grown here. A Huguenot stronghold, it was besieged in 1569 during the Wars of Religion.

Mussomeli (MOOS-so-MAI-lee), town, CALTANISSETTA province, central SICILY, ITALY, 19 mi/31 km WNW of CALTANISSETTA; 37°35′N 13°45′E. Agricultural and zootechnical center. Restored 14th-century castle nearby.

Musson (muh-SAW), commune (2006 population 4,190), Virton district, LUXEMBOURG province, SE BELGIUM, in the ARDENNES, 10 mi/16 km SSW of ARLON; 49°34′N 05°42′E. Blast furnaces.

Mussoorie (MUHS-oo-ree), town (2001 population 26,069), DEHRADUN district, NW Uttarakhand state, N central INDIA, in W Kumaon Himalaya foothills, 9 mi/14.5 km N of DEHRADUN; 30°27′N 78°05′E. Noted hill resort; breweries. Seat of National Academy of Public Administration, training center for members of Indian Administration Service.

Mustachal (moo-sta-CHAL), peak in the RILA MOUNTAINS, SW BULGARIA; 42°10′N 23°37′E. Elevation 8,638 ft/2,633 m.

Mustadfors (MOO-STAHD-FORSH), village, Älvsborg county, SW SWEDEN, on Dalsland Canal (locks), 16 mi/26 km SW of ÅMÅL. Manufacturing (horseshoe nails).

Mustafa Kemalpasa, Turkish *Mustafa Kemal Paşa*, town (2000 population 46,731), NW TURKEY, on KEMALPASA RIVER, and 35 mi/56 km WSW of BURSA; 40°03′N 28°25′E. Grain; lignite deposits. Formerly Kirmasti.

Mustafa Kemalpasa, TURKEY: see KEMALPASA RIVER.

Mustafa Pasha, Bulgaria: see SVILENGRAD.

Mustahil (moo-STAH-hil), village (2007 population 8,122), SOMALI state, SE ETHIOPIA; 05°15′N 44°44′E. On the E bank of the WABĒ SHEBELĒ RIVER, on road, and 45 mi/72 km SE of K'ELAFO, in camel and sheep raising region.

Müstair (myoo-STER), commune, GRISONS canton, SE SWITZERLAND, on the Italian border, in the VAL MÜSTAIR, the valley of the Rom River (Italian *Ram*; affluent of the ADIGE RIVER); 46°38′N 10°26′E. Elevation 4,091 ft/1,247 m.

Müstair, Val, SWITZERLAND: see MÜNSTERTAL.

Mustamäe (MOOST-uh-ma-ai), residential district, ESTONIA, SE of downtown TALLINN. High-rise apartment complex built largely in the late 1960s and early 1970s.

Mustang (MOOS-tahng), district, N central NEPAL, in DHAULAGIRI zone; ⊙ JOMSOM.

Mustang, city (2006 population 16,443), CANADIAN county, central OKLAHOMA, suburb 13 mi/21 km WSW of downtown OKLAHOMA CITY, near Canadian River Manufacturing (electronics).

Mustang Creek, New Mexico and Texas: see RITA BLANCA CREEK.

Mustang Island (MUHS-tang) NUECES county, S TEXAS, barrier island, across entrance to CORPUS CHRISTI BAY, 17 mi/27 km E of CORPUS CHRISTI; 25 mi/40 km long, 1 mi/1.6 km–4 mi/6.4 km wide; 27°44′N 97°07′W. PORT ARANSAS is on ARANSAS PASS, shipping channel to Corpus Christi dug through N end of island, at NE end; narrow channel that separated island from PADRE ISLAND (S) is now filled. Causeway from Corpus Christi connects Padre and Mustang islands with mainland; ferry to Harbor Island from N end connects with causeway to town of ARANSAS PASS. Mustang Island State Park straddles S part.

Mustang Mountain, SW NEVADA, in N WHITE MOUNTAINS, near CALIFORNIA state line, 60 mi/97 km WNW of GOLDFIELD. Elev. 10,316 ft/3,144 m.

Mustasaari, FINLAND: see KORSHOLM.

Mustayevo (moo-STAH-ee-vuh), village, SW ORENBURG oblast, SE European Russia, on a right tributary of the URAL River, on local road junction, 19 mi/31 km N of ILEK; 51°48′N 53°24′E. Elevation 311 ft/94 m. In agricultural area (wheat; livestock). Until 1940, called Mustayevka.

Mustèr, SWITZERLAND: see DISENTIS.

Musters, Lake, large freshwater lake (□ 168 sq mi/436.8 sq km) in Patagonian highlands, CHUBUT province, ARGENTINA, just NW of SARMIENTO, 6 mi/9.7 km W of Lake COLHUÉ HUAPÍ (linked by short stream); 26 mi/42 km long, 3 mi/4.8 km–8 mi/12.9 km wide. Elevation 890 ft/271 m. Receives SENGUERR.

Mustiala (MOOS-tee-AH-lah), village, HÄMEEN province, SW FINLAND, in lake region, 5 mi/8 km E of FORSSA; 60°49′N 23°46′E. Elevation 330 ft/100 m. Agricultural school.

Mustinka River (muhs-TEEN-kuh), 80 mi/129 km long, W MINNESOTA; rises SE of FERGUS FALLS, SW OTTER TAIL county; flows S through Lightning Lake, and to W of ELBOW LAKE town, then W past NORCROSS, receiving West Branch Mustinka River (c.30 mi/48 km long) from S, then SW past WHEATON to NE end of TRAVERSE LAKE, in BOIS DE SIOUX RIVER, on SOUTH DAKOTA state line, 8 mi/12.9 km SW of Wheaton; 46°12′N 96°06′W.

Mustique (mus-TEEK), island, SAINT VINCENT AND THE GRENADINES, WEST INDIES, SE of BEQUIA, 18 mi/29 km S of Saint Vincent; 12°53′N 61°11′W. Privately managed tourist island under government control. Sea-island cotton, coconut, and citrus trees. Britannia Bay is small port and center.

Mustla (MOOST-luh), German *Tarvast*, city, S ESTONIA, near W shore of Lake VÕRTSJÄRV, 13 mi/21 km SE of VILJANDI; 58°14′N 25°52′E. Agricultural market.

Mustoh (MUHS-to), village, East Khasi Hills district, MEGHALAYA state, NE INDIA, in KHASI HILLS, 29 mi/47 km SW of SHILLONG. Rice, cotton. Coal deposits nearby.

Mustvee (MOOST-vai), city, E ESTONIA, port on NW shore of LAKE PEIPSI, 40 mi/64 km SE of RAKVERE; 58°50′N 26°56′E. Flax area; sawmilling.

Area is shown by the symbol □, and capital city or county seat by ⊙.

Musún (moo-SOON), highest peak (5,600 ft./1,450 m) of CORDILLERA DARIENSE, central NICARAGUA, 30 mi/48 km ENE of MATIGUÁS; 12°59′N 85°15′W.

Muswellbrook (MUS-wuhl-bruk), municipality, E NEW SOUTH WALES, SE AUSTRALIA, on HUNTER RIVER, and 65 mi/105 km NW of NEWCASTLE; 32°16′S 150°54′E. Railroad junction; coal-mining center, power stations; sheep, cattle, horses; vineyards; dairying.

Muswell Hill, ENGLAND: see HARINGEY.

Muszyna (moo-SHEE-nah), town, Nowy Sącz province, S POLAND, in the CARPATHIANS, on POPRAD RIVER, and 5 mi/8 km SSW of KRYNICA, near SLOVAKIA border; 49°21′N 20°55′E. Railroad junction; health resort. Large linden forest nearby. Tourist resort.

Mut, ancient *Claudiopolis*, township, S TURKEY, near Goksu, 65 mi/105 km W of MERSIN; 36°38′N 33°27′E. Wheat, barley, sesame, beans, onions. Some ancient ruins. Named for the Emperor Claudius.

Mut (MOOT), village, NEW VALLEY province, S central EGYPT, in DAKHLA oasis, 16 mi/26 km SSE of El QASR; 25°29′N 28°59′E. Dates, oranges, wheat, barley. Airfield.

Mutambara (moo-tah-em-BAH-rah), town, MANICALAND province, E ZIMBABWE, 40 mi/64 km S of MUTARE, in CHIMANIMANI MOUNTAINS, on Umvumvumvu River, near MOZAMBIQUE border; 19°33′S 32°40′E. Cattle, sheep, goats; corn, wheat, tobacco, tea, coffee.

Mutanda (moo-TAHN-duh), lake (□ 9 sq mi/23.4 sq km), KISORO district, WESTERN region, SW UGANDA, 4 mi/6.4 km N of KISORO; 01°12′S 29°40′E. Elevation 5,852 ft/1,784 m; maximum depth 164 ft/50 m. Formed by volcanic activity. Lowest lake in the Bufumbiro bay. Was in former SOUTHERN province.

Mutankiang, CHINA: see MUDANJIANG.

Mutarara (moo-tuh-RAH-ruh), village, TETE province, central MOZAMBIQUE, on the ZAMBEZI (left bank), and 170 mi/274 km N of BEIRA; 17°27′S 35°04′E. Between here and SENA (on right bank), the Trans-ZAMBEZIA railroad (serving MALAWI) crosses ZAMBEZI River on a steel bridge (12,064 ft/3,677 m long; opened 1935). Airfield.

Mutarara District, MOZAMBIQUE: see TETE.

Mutare (moo-TAH-rai), city (2002 population 170,106), ⊙ MANICALAND province, E ZIMBABWE, 130 mi/209 km SE of HARARE, near the MOZAMBIQUE border; 18°59′S 32°39′E. Commercial center for rich agricultural (coffee, tea, citrus fruit, macadamia nuts, corn, tobacco, peanuts) and gold-mining region. Timber; cattle, sheep, goats; dairying. Manufacturing (wood preserving, forest products, textiles, automobile assembly, petroleum refining). On railroad to port of BEIRA (Mozambique). Founded in 1890, Mutare grew with the construction, completed in 1899, of the railroad. VUMBA MOUNTAINS and Vumba Botanical Garden (494 acres/200 ha) to SE; Cecil Kop Nature Reserve at E edge of city. Mutare Museum, Utopia House Museum. It is the seat of Africa University. Formerly Umtali.

Mutas (MOO-tahs), village, S central BAHIA, BRAZIL, 16 mi/26 km SW of GUANAMBÍ; 14°19′S 42°59′W.

Mutatá (moo-tah-TAH), town, ⊙ Mutatá municipio, ANTIOQUIA department, NW central COLOMBIA, on Río SUCIO, 90 mi/145 km NW of MEDELLÍN; 07°14′N 76°25′W. Sugarcane, rice, plantains; livestock.

Mutawintji National Park (□ 266 sq mi/691.6 sq km), NEW SOUTH WALES, SE AUSTRALIA, 800 mi/1,287 km WNW of SYDNEY, 80 mi/129 km NE of BROKEN HILL; 30 mi/48 km long, 20 mi/32 km wide; 31°07′S 142°23′E. Semi-arid sandstone range; waterholes in narrow gorges lined with red river gums. Aboriginal painting galleries and engravings. Evidence of early European exploration. Hiking, picnicking. No water or services. In 1998, the park was transferred to Aboriginal ownership. Formerly spelled Mootwingee.

Mutenice (MU-tye-NYI-tse), Czech *Mutěnice*, German *Mutienitz*, village, JIHOMORAVSKY province, S MOR-

AVIA, CZECH REPUBLIC, 5 mi/8 km NW of HODONÍN. Railroad junction. Noted vineyard, wine making and research (Grapevine Institute). Has an 18th century baroque church.

Mutepatepa (moo-tai-pah-TAI-pah), village, MASHONALAND CENTRAL province, NE ZIMBABWE, 25 mi/40 km N of BINDURA; 16°57′S 31°15′E. Cattle, sheep, goats; tobacco, peanuts, cotton. Also spelled Mtepatepa.

Muthanna, Arabic *al Muthanna*, province, S IRAQ, bordering on SAUDI ARABIA to S, NAJAF province to W, BASRA province to E, and QADISSIYA and THI-GAR provinces to N; ⊙ Samawa. Agriculture (rice, dates); manufacturing (building materials). SAMAWA is major trading center.

Muthill (MUT-hil), agricultural village (2001 population 675), Perth and Kinross, E SCOTLAND, near EARN RIVER, and 3 mi/4.8 km S of CRIEFF; 56°20′N 03°51′W. Has 15th-century church. Nearby are remains of Drummond Castle (1491) with famous gardens. Formerly in Tayside, abolished 1996.

Muthupet, INDIA: see MUTTUPET.

Muthuramalinga Thevar, Pasumpan, INDIA: see PASUMPAN MUTHURAMALINGA THEVAR.

Muti'a, El, village, ASYUT province, central Upper EGYPT, on W bank of the NILE River, 6 mi/9.7 km N of ABU TIG; 27°08′N 31°18′E. Cereals, dates, sugarcane.

Mutienitz, CZECH REPUBLIC: see MUTENICE.

Mutiko (moo-TEE-ko), village, NORD-KIVU province, E CONGO, 42 mi/68 km NW of BUKAVU; 01°38′S 28°13′E. Elev. 3,684 ft/1,122 m. Tin-mining center; tantalite and wolfram mining in vicinity. Tin mines of MUHULU 30 mi/48 km NW.

Mutirikwi River (moo-tee-REE-kwee), c. 145 mi/233 km long, SE central ZIMBABWE; rises c.40 mi/64 km NNE of MASVINGO; flows S through Lake Mtirikwe (Lake KYLE) reservoir, then SSE through Lake Bangala reservoir, entering RUNDE RIVER 85 mi/137 km SSE of Masvingo. Also spelled Mtirikwe and Mtilikwe.

Mutiscua (moo-TEES-koo-ah), town, ⊙ Mutiscua municipio, NORTE DE SANTANDER department, N COLOMBIA, 42 mi/68 km WSW of CÚCUTA, in the ANDES; 07°18′N 72°45′W. Elevation 9,189 ft/2,800 m. Coffee, plantains; livestock.

Mutis, Mount (MOO-tees), highest peak (7,888 ft/2,404 m) of Indonesian TIMOR, in W part of island, 80 mi/129 km NE of KUPANG; 09°34′S 124°14′E. Also spelled MOUNT MUTUS and MOUNT MOETIS.

Mutivir (moo-tee-VIR), river, N central BULGARIA; 42°28′N 23°57′E. Tributary to the TOPOLNITSA RIVER.

Mutki, village, SE TURKEY, 10 mi/16 km WNW of BITLIS; 38°25′N 41°54′E. Millet. Also called Hur and Mirtag.

Mutlangen (MOOT-lang-uhn), village, E WÜRTTEMBERG, BADEN-WÜRTTEMBERG, GERMANY, on the N border of SCHWÄBISCH GMÜND; 48°49′N 09°48′E. Forestry.

Mutnitsa (MOOR-nee-tsah), river, PLOVDIV oblast, BULGARIA; 41°28′N 23°47′E. Tributary of the CHEPINSKA RIVER (itself a tributary of the MARITSA RIVER).

Mutoko, town, MASHONALAND EAST province, NE ZIMBABWE, 85 mi/137 km ENE of HARARE; 17°24′S 32°14′E. Road junction. Cattle, sheep, goats. Formerly spelled Mtoko.

Mutomo (moo-TO-mo), town, Kitui district, EASTERN province, KENYA, on road, and 37 mi/60 km SSW of KITUI; 01°50′S 38°13′E. Market and trading center.

Mutorashanga (moo-too-rah-SHAH-eng-gah), town, MASHONALAND WEST province, N central ZIMBABWE, 50 mi/80 km NW of HARARE, on W side of NYANGA Range; 17°09′S 30°40′E. Mutorashanga Pass to E. Coffee, tea, corn, wheat, tobacco; cattle, sheep, goats, hogs. Also spelled Mtoroshanga.

Mutoshi (moo-TO-shee), village, KATANGA province, SE CONGO, near railroad, 5 mi/8 km NE of KOLWEZI; 10°38′S 25°33′E. Elev. 4,645 ft/1,415 m. Copper- and

gold-mining center; palladium-platinum metals also occur here. Also known as RUWE.

Mutsalaul (moo-tsahl-ah-OOL), village (2005 population 6,370), W central DAGESTAN REPUBLIC, SE European Russia, on road and near railroad, 57 mi/92 km NW of MAKHACHKALA, and 7 mi/11 km ENE of KHASAVYURT, to which it is administratively subordinate; 43°16′N 46°44′E. Elevation 127 ft/38 m. Agriculture (gardening, fruit orchards, vineyards).

Mutsamudu (moo-tsah-MOO-doo), city (2002 population 21,558), ⊙ Nzwani island and administrative district, Comoros Republic, c.85 mi/137 km ESE of Moroni, on NW coast of island; 12°10′S 44°24′E. Manufacturing (ylang-ylang oil, lumber). Fish; livestock; ylang-ylang, vanilla, coconuts, bananas, cassava; timber. Citadel of Abdullah I (1860) is here. Ouani Airport 4 mi/6.4 km to NNE; Mt. Ntingui to SE. Also spelled Moutsamoudou.

Mutsing, village, MOROBE province, NE NEW GUINEA island, N central PAPUA NEW GUINEA, 60 mi/97 km NW of LAE. Located in MARKHAM RIVER valley, S of FINISTERRE RANGE, on main road from Lae to highlands. Copra, maize, rice, sugarcane; cattle; timber. Formerly called KAIAPIT (KEI-a-pit).

Mutsu, former province in N HONSHU, JAPAN; now Aomori prefecture.

Mutsu (MOOTS), city, Aomori prefecture, N HONSHU, N JAPAN, near Mount Osore, 40 mi/65 km N of AOMORI; 41°17′N 141°12′E. Osore-zan Reitai Festival held here.

Mutsu Bay (MOOTS), Japanese *Mutsu-wan* (moo-TSOO-wahn), inlet of TSUGARU STRAIT, Aomori prefecture, N HONSHU, N JAPAN; 25 mi/40 km long. Merges W with AOMORI BAY. Famous for scallops.

Mutsumi (moo-TSOO-mee), village, Abu county, YAMAGUCHI prefecture, SW HONSHU, W JAPAN, 19 mi/30 km N of YAMAGUCHI; 34°26′N 131°34′E.

Mutsuzawa (moo-TSOO-zah-wah), town, Chose county, CHIBA prefecture, E central HONSHU, E central JAPAN, 16 mi/25 km S of CHIBA; 35°21′N 140°19′E.

Muttaburra, township, QUEENSLAND, NE AUSTRALIA, 52 mi/84 km N of ARAMAC; 22°33′S 144°31′E.

Muttenz (MOO-tuhnz), town (2000 population 16,654), Basel-Land half-canton, N SWITZERLAND, 3 mi/4.8 km SSE of BASEL; 47°31′N 07°39′E. Elevation 922 ft/281 m. Chemicals. Medieval church.

Mutterstadt (MUT-ter-shtaht), town, RHINELANDPALATINATE, W GERMANY, 5 mi/8 km SW of LUDWIGSHAFEN; 49°27′N 08°18′E. Wine; grain, tobacco, sugar beets.

Muttler (MOO-luhr), peak (10,807 ft/3,294 m) in RHAETIAN ALPS, E SWITZERLAND, 4 mi/6.4 km N of Vnà, overlooking lower ENGADINE, and 5 mi/8 km NW of Swiss-Austrian-Italian border; 46°54′N 10°23′E.

Mutton Bird Islands, group of islands off SW coast of STEWART ISLAND and also in FOVEAUX STRAIT off NE Stewart Island, NEW ZEALAND. Many scattered islets where Maori families have right to take "titi" or young of sooty shearwater. Sometimes called TITI ISLANDS.

Mutton Island, Gaelic *Oiléan Caorach*, islet in MAL BAY, W CLARE county, W IRELAND, 3 mi/4.8 km SW of SPANISH POINT; 52°48′N 09°31′W. According to legend, island was severed from mainland c.800 by a storm. Ruins of oratory of St. Senan.

Muttontown, affluent residential village (□ 6 sq mi/15.6 sq km; 2000 population 3,412), NASSAU county, SE NEW YORK, on NW LONG ISLAND, just SE of GLEN COVE; 40°49′N 73°32′W. Former sheep-raising area. Incorporated 1931.

Muttra, INDIA: see MATHURA.

Muttupet (MUHT-uh-pait), town, Thanjavur district, TAMIL NADU state, S INDIA, port (6 mi/9.7 km inland; served by distributary of VENNAR RIVER) of PALK STRAIT, 35 mi/56 km SE of THANJAVUR, 8 mi/12.9 km NE of Adirampatnam; 10°24′N 79°29′E. Exports rice. Also spelled Muthupet, Mutupet.

Mutual (MYOO-choo-uhl), village (2006 population 130), CHAMPAIGN county, W central OHIO, 6 mi/10 km ESE of URBANA; 40°05′N 83°38′W.

Mutual, village (2006 population 78), WOODWARD county, NW OKLAHOMA, 19 mi/31 km SE of WOODWARD; 36°13′N 99°10′W. In livestock, grain, and oil and natural-gas area.

Mutubis (mo-TO-bahs), township, KAFR ESH SHEIKH province, Lower EGYPT, on RASHID branch of the NILE, on railroad, 7 mi/11.3 km NNW of FUWA; 31°18′N 30°31′E. Cotton.

Mutuípe (moo-too-EE-pai), city (2007 population 20,871), E central BAHIA, BRAZIL, 25 mi/41 km E of SANTA INÉS; 13°12′S 39°30′W.

Mutum (moo-toon), city (2007 population 26,331), E MINAS GERAIS, BRAZIL, near ESPÍRITO SANTO border, 30 mi/48 km SW of AIMORÉS; 19°45′S 41°29′W. Tobacco, coffee, cattle. Until 1939, São Manuel do Mutum.

Mutum or **Mutúm** (both: moo-TOON), military post, CHIQUITOS province, SANTA CRUZ department, E BOLIVIA, near BRAZIL border, 15 mi/24 km SSW of PUERTO SUÁREZ; 19°10′S 57°54′W. Great iron and manganese deposits.

Mutum Paraná (MOO-toon pah-rah-NAH), city, W RONDÔNIA state, BRAZIL, 108 mi/174 km SW of PÔRTO VELHO; 09°38′S 64°57′W.

Mutupet, INDIA: see MUTTUPET.

Mutus, Mount, INDONESIA: see MUTIS, MOUNT.

Mutwal (MUT-twol), section of COLOMBO, WESTERN PROVINCE, SRI LANKA, 2.5 mi/4 km NE of city center; 06°56′N 79°51′E. Fishing center.

Mutwanga (moot-WAHNG-gah), village, NORD-KIVU province, E CONGO, at W foot of the RUWENZORI, near UGANDA border, 25 mi/40 km ESE of BENI; 00°20′N 29°45′E. Elev. 4,393 ft/1,338 m. Tourist center in SEMLIKI section of VIRUNGA NATIONAL PARK and point of departure for ascents of the RUWENZORI.

Mutzig (moo-tsik), town (□ 3 sq mi/7.8 sq km), BASRHIN department, E FRANCE, in ALSACE, on the BRUCHE, and 2 mi/3.2 km W of MOLSHEIM; 48°32′N 07°28′E. Small-scale manufacturing (machinery, hardware). Brewery. Vineyards (Riesling wine) extend from here to Molsheim. Military camp nearby.

Muuga (MOO-guh), port, ESTONIA, 9.3 mi/15 km NE of TALLINN, on MUUGA Bay. Largest port in Estonia, built by Soviets to receive grain shipments from the West. Handles wheat and container shipments.

Muujarvi, RUSSIA: see MUYEZERSKIY.

Muvattupula (MOO-vah-too-poo-lah), town, ERNAKULAM district, KERALA state, S INDIA, 27 mi/43 km N of KOTTAYAM; 09°58′N 76°35′E. Trades in rice, cassava, coir rope, and mats. Also spelled Muvatupuzha.

Muveran, Grand (myoo-vuh-RAHN, GRAHN), peak (10,010 ft/3,051 m) in BERNESE ALPS, on border of VAUD and VALAIS cantons, SW SWITZERLAND, 11 mi/18 km W of SION.

Muwaih (moo-WAIH), village, W NAJ'D, MAKKA province, SAUDI ARABIA, 150 mi/241 km NE of Makka, and on road to RIYADH. Also spelled Al Mowai or Muwayh.

Muwailih (moo-WAI-li), village, N HEJAZ, Tabuk province, SAUDI ARABIA, minor port on MADIAN coast of RED SEA, 100 mi/161 km SSE of AQABA, on the road between Aqaba and JIDDA; 27°40′N 35°30′E. Bedouin trade center. Also spelled Muwaila.

Muwale (moo-WAH-lai), village, SINGIDA region, central TANZANIA, 70 mi/113 km SW of MANYONI; 06°22′S 33°45′E. Timber; cattle, sheep, goats; wheat, corn.

Muxadabad, INDIA: see MURSHIDABAD.

Muxima (moo-SHEE-muh), village, BENGO province, NW ANGOLA, on left bank of Cuanza River, and 70 mi/113 km SE of LUANDA. Oil palms, cotton, manioc, fishing.

Muxupip (moo-shoo-PEEP), town, YUCATÁN, SE MEXICO, 20 mi/32 km E of MÉRIDA; 21°02′N 82°20′W. Henequen, corn.

Muyezerskiy (moo-YE-zeer-skeeyee), Finnish *Muujarvi*, town (2005 population 3,865), central Republic of KARELIA, NW European Russia, on railroad, 57 mi/92 km SE of KOSTOMUKSHA; 63°56′N 31°39′E. Elevation 574 ft/174 m. Ceded by FINLAND to the USSR in 1940, following the Winter War, and called Muyozero; current name since 1948.

Muyinga, province (□ 709 sq mi/1,843.4 sq km; 1999 population 485,347), NE BURUNDI, on RWANDA (to NE) and TANZANIA (to E) borders; ⊙ MUYINGA, 02°50′S 30°20′E. Borders CANKUZO (SE), KARUZI (S), NGOZI (W), and KIRUNDO (N) provinces.

Muyinga (moo-YEEN-gah), town (2004 population 45,300), ⊙ MUYINGA province, NE BURUNDI, near TANZANIA border, 50 mi/81 km NE of GITEGA; 02°51′S 30°19′E. Customs station, local trading center; cattle raising. Has veterinary laboratory. Established 1928.

Muyinga (moo-YEEN-gah), village, BURUNDI, 130 mi/209 km E of BUJUMBURA. Administrative center.

Muy, Le (mwee, luh), commune (□ 25 sq mi/65 sq km), VAR department, SE FRANCE, in PROVENCE, on the ARGENS, and 7 mi/11.3 km SE of DRAGUIGNAN; 43°28′N 06°31′E. Market center; fruit shipping.

Muy Muy (MOO-ee MOO-ee), town, MATAGALPA department, central NICARAGUA, 30 mi/48 km ESE of MATAGALPA; 12°46′N 85°38′W. Coffee, livestock.

Muynak, UZBEKISTAN: see MUINAK.

Muyondzi, REPUBLIC OF THE CONGO: see MOUYONDZI.

Muyuka (moo-YOO-kuh), town (2001 population 19,200), South-West province, CAMEROON, 25 mi/40 km NW of DOUALA; 04°17′N 09°25′E.

Muyumba (moo-YOOM-bah), village, KATANGA province, SE CONGO, on LUALABA RIVER, and 175 mi/282 km SW of KALEMIE; 07°15′S 26°59′E. Elev. 1,814 ft/552 m. Trans-shipment point (steamer-railroad) for MANONO tin-mining area. Site of Protestant mission.

Muyuni (moo-YOO-nee), village, ZANZIBAR SOUTH region, E TANZANIA, 23 mi/37 km SE of ZANZIBAR city, ZANZIBAR island; 06°24′S 39°27′E. Cloves, copra; fish.

Muyun Kum (moo-YUN KUM), sandy desert, ZHAMBYL region, KAZAKHSTAN, extending S of CHU RIVER, and E of the KARATAU MOUNTAINS. Average elevation 1,000 ft/305 m. Sheep, camels.

Muyurina (moo-yoo-REE-nah), canton, VALLEGRANDE province, SANTA CRUZ department, E central BOLIVIA, S of TRIGAL; 18°21′S 64°08′W. Elevation 6,660 ft/2,030 m. Agriculture (corn, potatoes, yucca, soy, rye); cattle.

Muzaffarabad (mu-ZAH-fah-rah-BAHD), district (□ 2,408 sq mi/6,260.8 sq km), AZAD KASHMIR, NE PAKISTAN, in W PUNJAB Himalayas; ⊙ MUZAFFARABAD. Bounded W by NORTH-WEST FRONTIER PROVINCE; drained by JHELUM and KISHANGANGA rivers. Agriculture. Prevailing languages are Kashmiri and Panjabi.

Muzaffarabad (mu-ZAH-fah-rah-BAHD), town, ⊙ MUZAFFARABAD district and AZAD KASHMIR, NE PAKISTAN, at the confluence of the JHELUM and Neelam rivers; 34°22′N 73°28′E. Chief city of AZAD KASHMIR. Trading center.

Muzaffargarh (mu-ZAH-fahr-guhr), district (□ 5,605 sq mi/14,573 sq km), SW PUNJAB province, central PAKISTAN; ⊙ MUZAFFARGARH. In S SIND-SAGAR DOAB, between INDUS (W), CHENAB (E), and PANJNAD (S) rivers. S section is alluvial tract subject to periodic floods; N section consists mostly of low sand hills (S THAL region).

Muzaffargarh (mu-ZAH-fahr-guhr), town, ⊙ MUZAFFARGARH district, SW PUNJAB province, central PAKISTAN, near CHENAB River, 15 mi/24 km SW of MULTAN; 30°04′N 71°12′E.

Muzaffarnagar (moo-ZUHF-fuh-nuh-guhr), district (□ 1,547 sq mi/4,022.2 sq km), N UTTAR PRADESH state, N central INDIA; ⊙ MUZAFFARNAGAR. On Ganga-Yamuna Doab; irrigated by East YAMUNA and UPPER GANGA canals; jungle in NW. Agricultural (wheat, gram, sugarcane, oilseeds, corn, rice, barley, cotton); a leading sugar-processing district of India.

Main towns: Muzaffarnagar, KAIRANA, KANDHLA, SHAMLI.

Muzaffarnagar (moo-ZUHF-fuh-nuh-guhr), town (2001 population 331,668), ⊙ MUZAFFARNAGAR district, UTTAR PRADESH state, N central INDIA; 29°28′N 77°41′E. In a wheat- and sugarcane-growing area, with road and railroad connections to DELHI.

Muzaffarpur (moo-ZUHF-fuhr-poor), district (□ 1,225 sq mi/3,185 sq km; 2001 population 3,743,836), N BIHAR state, E INDIA; ⊙ MUZAFFARPUR. On GANGA PLAIN; bounded N by NEPAL, W by GANDAK RIVER, S by GANGES RIVER; drained by its tributaries. Famed for litchi fruit production: over half of India's litchi comes from this region. Alluvial soil; rice, wheat, barley, corn, tobacco, sugarcane, cotton, oilseeds. Rice and sugar milling, livestock raising; hides. Main trade centers: Muzaffarpur, HAJIPUR. Buddhist pilgrimage center near LALGANJ.

Muzaffarpur (moo-ZUHF-fuhr-poor), city (2001 population 305,465), ⊙ MUZAFFARPUR district, N central BIHAR state, E INDIA, on GANGA PLAIN, on Burhi Ghandak River, 39 mi/63 km NNE of PATNA; 26°07′N 85°24′E. Railroad and road junction; river trade; rice, wheat, barley, corn, tobacco, sugarcane, oilseeds; rice and sugar milling, cutlery manufacturing. Has college and university.

Muzambinho (moo-sahm-been-yo), city (2007 population 19,925), SW MINAS GERAIS, BRAZIL, near SÃO PAULO border, on railroad, and 14 mi/23 km ESE of GUAXUPÉ; 15°46′S 29°21′W. Elevation 3,500 ft/1,067 m. Coffee growing, cattle raising.

Muzhi (MOO-zhi), village (2006 population 3,215), W YAMALO-NENETS AUTONOMOUS OKRUG, TYUMEN oblast, NW SIBERIA, RUSSIA, on the Malaya Ob' River (left arm of the OB' RIVER), 90 mi/145 km SSW of SALEKHARD; 65°22′N 64°40′E. Reindeer raising.

Muzhichi, RUSSIA: see GALASHKI.

Muzillac (myu-zee-yahk), village (□ 15 sq mi/39 sq km; 2004 population 4,218), MORBIHAN department, W FRANCE, in S BRITTANY, near mouth of VILAINE RIVER, 14 mi/23 km ESE of VANNES; 47°33′N 02°29′W. Fisheries, woodworking.

Muzo (MOO-so), town, ⊙ Muzo municipio, BOYACÁ department, central COLOMBIA, in W foothills of Cordillera ORIENTAL, 50 mi/80 km W of TUNJA; 05°32′N 74°06′W. Elevation 4,068 ft/1,240 m. Famed for emerald mines. Coffee, potatoes, corn; livestock.

Muzoka (moo-ZO-kah), township, SOUTHERN province, S ZAMBIA, 23 mi/37 km NE of CHOMA; 16°42′S 27°20′E. On railroad. Agriculture (tobacco, corn, peanuts, soybeans); cattle. MAAMBA coal-mining district to SE, near LAKE KARIBA.

Muzon, Cape, SE ALASKA, S tip of DALL ISLAND, ALEXANDER ARCHIPELAGO, in DIXON ENTRANCE, 60 mi/97 km SW of KETCHIKAN; 54°40′N 132°41′W.

Múzquiz (MOOS-kees), city and township, COAHUILA, N MEXICO, 25 mi/40 km W of SABINAS, on Mexico Highway 53; 27°52′N 101°30′W. Elevation 1,654 ft/504 m. Railroad terminus. Mexico's major coal-mining center; also silver, gold, lead, zinc. Agricultural center (grain, cattle). Formerly CIUDAD MELCHOR MÚZQUIZ.

Muztag (MU-ZI-TAH-GUH), peak (23,890 ft/7,282 m) in the Karatax range of the KUNLUN MOUNTAINS, S XINJIANG UYGUR AUTONOMOUS REGION, CHINA, near KASHMIR border; 36°36′N 87°40′E. Formerly called K5, it was regarded as highest peak of Kunlun mountains until discovery of the ULUGH.

Muztagata Range (MUHZ-TAH-GUH-TAH), massif in westernmost XINJIANG UYGUR AUTONOMOUS REGION, CHINA, extends 200 mi/322 km NNW-SSE parallel to E edge of the PAMIR; rises to 25,146 ft/7,665 m in the KUNGUR, 70 mi/113 km SW of KASHI. The peak Muztagata (24,388 ft/7,433 m) is 100 mi/161 km SW of Kashi. Sometimes called Kashi Range and Bolor Tagh.

Muztagh-Karakoram, PAKISTAN: see KARAKORAM.

Muztor, KYRGYZSTAN: see TOKTOGUL.

Mvila, department (2001 population 163,826), South province, CAMEROON.

Mvomero (uhm-vo-MAI-ro), village, MOROGORO region, E central TANZANIA, 35 mi/56 km NNW of MOROGORO; 06°23′S 37°25′E. NGURU Mountains to N. Road junction. Cotton, sugarcane, sisal, grain; livestock.

Mvoti River, SOUTH AFRICA: see GREYTOWN.

Mvouti (uhm-voo-TEE), town, KOUILOU region, SW Congo Republic, on railroad, and 60 mi/97 km NE of POINTE-NOIRE; 04°14′S 12°29′E. Hardwood lumbering center.

Mvuha (uhm-VOO-hah), village, MOROGORO region, E central TANZANIA, 35 mi/56 km SE of MOROGORO; ULUGURU MOUNTAINS to NW; 07°16′S 37°54′E. Grain, cotton, sisal; livestock.

Mvuma, town, MIDLANDS province, central ZIMBABWE, 50 mi/80 km ENE of GWERU, on railroad; 19°17′S 30°32′E. Elevation 4,540 ft/1,384 m. Agriculture area (dairying; cattle, sheep, goats; citrus fruit, tobacco, wheat, corn). Copper mining through 1920s, site of Falcon Mine. Formerly Umvuma.

Mvurwi, town, MASHONALAND CENTRAL province, NW central ZIMBABWE, 55 mi/89 km NNW of HARARE, E of MVURWI RANGE; 17°01′S 30°53′E. Chromite mining to W. Manufacturing (farm supplies). Agriculture (corn, wheat, cotton, tobacco, sorghum, soybeans); cattle, sheep, goats, hogs.

Mvurwi Range, N ZIMBABWE, extending c. 100 mi/161 km N from MANYAME RIVER, 35 mi/56 km NW of HARARE. Forms N section of the Great Dyke, rises to 5,731 ft/1,747 m; forms part of boundary between MASHONALAND CENTRAL and Mashonaland Western provinces. Major chromite-mining district, with chief centers at DARWENDALE, MUTORASHANGA, and KILDONAN; town of MVURWI to NE. Formerly Umvukwe Range.

Mwadingusha (mwah-ding-GOO-shah), village, KATANGA province, SE CONGO, on LUFIRA RIVER, and 35 mi/56 km NE of LIKASI; 10°45′S 27°15′E. Elev. 3,802 ft/1,158 m. Site of leading hydro-electric power station in CONGO, with reservoir (□ c.25 sq mi/65 sq km) and dam. MWADINGUSHA plant is the oldest in the country and supplies most of energy needs of copper industry of SHABA region. It is also known as CORNET FALLS, French *Chutes-Cornet*. Additional hydroelectric installations are located at KONI, 5 mi/8 km N.

Mwadui (mwah-DOO-ee), village, SHINYANGA region, NW central TANZANIA, 15 mi/24 km NE of SHINYANGA, on railroad spur; 03°26′S 32°29′E. Noted Williamson diamond mine discovered here, 1940; largest pipe diamond mine in Africa.

Mwaleshi, river, ZAMBIA. Originates in the Muchinga Escarpment, flows down in a series of rapids and small waterfalls, ending with Chomba Waterfall, into the Luangwa Valley, where it bisects the NORTH LUANGWA NATIONAL PARK; ends when it flows into the LUANGWA RIVER.

Mwali (muh-WAI-lee), island and administrative district (□ 112 sq mi/291.2 sq km; 2003 population 35,751), one of four main islands of Comoros island group and one of three main islands comprising the Comoros Republic, in Mozambique Channel, Indian Ocean, c.340 mi/547 km WNW of Madagascar; ⊙ Fomboni; 18 mi/29 km long, 8 mi/12.9 km wide; 12°20′S 43°40′E. Njazidja (Grande Comore) island to NW; Nzwani (Anjouan) island to E. It rises to 2,592 ft/790 m at Mt. Koukoule, at center. Ruled by the Persians until 1830, when the Malagasy assumed control. In 1886 it became French protectorate; became part of Comoros Republic in 1974. Mwali Airport on N coast, the Marine Reserve off S coast. Called Moheli under French rule. A secessionist movement on the island seeks a return to French rule, arguing that the economic favoritism by the government directs resources to Njazidja.

Mwambo (MWAM-bo), village, MTWARA region, SE TANZANIA, 17 mi/27 km SSE of MTWARA, on RUVUMA River (MOZAMBIQUE border), 5 mi/8 km W of Ruvuma Bay, INDIAN OCEAN; 10°31′S 40°22′E. Port of entry. Fish; livestock; cashews, bananas, grain; timber.

Mwami, village, MASHONALAND WEST province, N ZIMBABWE, 55 mi/89 km NNW of CHINHOYI; 16°46′S 29°46′E. Mica mining. Cattle, sheep, goats; cotton, soybeans, tobacco, corn. Formerly spelled Miami.

Mwanza (MWAHN-zah), administrative center and district (2007 population 179,019), Southern region, MALAWI, near MOZAMBIQUE border, on road, and 35 mi/56 km WNW of BLANTYRE; 15°40′S 34°34′E. Police and customs station; cotton, tobacco.

Mwanza (MWAHN-zah), city (2002 population 378,327), ⊙ MWANZA region, NW TANZANIA, 510 mi/821 km NW of DAR ES SALAAM, main port in S part of Lake VICTORIA; 02°33′S 32°55′E. Terminus of railroad spur from TABORA. Shipping lines connect with ports in Tanzania, KENYA, and UGANDA; railroad ferry links Mwanza to KISUMU, Kenya and JINJA, Uganda. International airport. Cotton, sugarcane, wheat, corn; cattle, sheep, goats; fish. Manufacturing (food processing, peanut oil milling). Seat of National Institute for Medical Research (Mwanza Centre). Saanane Game Reserve to S. Complex Casino on lake front.

Mwanza (MWAHN-zah), village, KATANGA province, SE CONGO, near LUALABA RIVER, 125 mi/201 km NE of KAMINA; 07°51′S 26°41′E. Elev. 2,942 ft/896 m. Protestant mission.

Mwanza (MWAHN-zah), region (2006 population 3,169,000), NW TANZANIA, ⊙ MWANZA, including S part of Lake VICTORIA (RUBONDO NATIONAL PARK) and MAISOME islands in W; UKEREWE and UKARA islands in N; EMIN PASHA GULF in W; SPEKE GULF in E; Mwanza Gulf extends S. Sugarcane, cotton, corn, wheat, millet; cattle, sheep, goats; timber; fish. Part of former Lake province.

Mwarazi River, c. 80 mi/129 km long, NE ZIMBABWE; rises c.30 mi/48 km ESE of MARONDERA; flows NE, joins NYANGADZI RIVER 30 mi/48 km ESE of MTOKO to form RUENYA (Luenha River).

Mwaro, province (□ 324 sq mi/842.4 sq km), W BURUNDI; ⊙ MWARO; 03°30′S 29°40′E. Borders MURAMVYA (N), GITEGA (E, partly formed by Luvironza River), BURURI (S), and BUJUMBURA RURALE (W) provinces. Formed c.1999 from S portion of Muramvya province.

Mwaru (MWAH-roo), village, SINGIDA region, N central TANZANIA, 40 mi/64 km W of SINGIDA, near Mwaru River; 04°53′S 34°08′E. Timber; livestock; corn, wheat.

Mwatate (mwah-TAI-tai), village, COAST province, SE KENYA, on railroad, and 18 mi/29 km WSW of VOI; 03°30′S 23°38′E. Sisal, coffee, tea, corn. Also spelled Mwatate.

Mwaya (MWAH-yah), village, MBEYA region, SW TANZANIA, small port at N end Lake NYASA, 55 mi/89 km SSE of MBEYA; 09°33′S 33°57′E. Tea, coffee, rice; livestock. Coal deposits to N.

Mweelrea, IRELAND: see MUILREA.

Mweenish (MWEE-nish) or **Minish**, island (□ 1 sq mi/2.6 sq km) in GALWAY BAY, at entrance to KILKIERAN BAY, SW GALWAY county, W IRELAND, 14 mi/23 km SW of CLIFDEN; 53°18′N 09°51′W. Connected by road bridge.

Mweiga (mwei-gah), town, Nyeri district, CENTRAL province, KENYA, 68 mi/110 km N of NAIROBI; 00°32′S 37°21′E. Market center.

Mweka (MWE-kah), village, KASAI-OCCIDENTAL province, central CONGO, on railroad, and 40 mi/64 km NNE of LUEBO; 04°51′S 21°34′E. Elev. 1,532 ft/466 m. Trading center for Bakuba tribe; also cotton center. Rubber plantations nearby. Also spelled MUEKA.

Mwene-Ditu (MWE-nai–DEE-too), village, KASAI-ORIENTAL province, central CONGO, on railroad, and 90 mi/145 km SW of KABINDA; 07°00′S 23°27′E. Elev. 2,988 ft/910 m. Sawmilling, cotton ginning.

Mwenemutapa (em-wai-NAI-moo-TAH-pah), former state, SE Afr. The Mwenemutapa empire, headed by a ruler of the same name, was founded c.1420 among the Karanga people (a subgroup of the Bantu-speaking Shona) and was centered at Great Zimbabwe in present-day SE ZIMBABWE. The empire was ruled in pyramidal fashion, with the Mwenemutapa appointing regionally based vassals. In about 1490 the empire split into 2 parts—Changamire in the S (including Great Zimbabwe) and Mwenemutapa in the N. The latter stretched from the Indian Ocean in the E to present-day central ZAMBIA in the W and from central Zimbabwe in the S to the ZAMBEZI R. in the N. An important source of gold and ivory, the area attracted Swahili traders from the E coast of Afr. (in modern TANZANIA). Beginning in the early 16th cent., Port. traders and soldiers from MOZAMBIQUE established contact with the empire, and by the mid-17th cent. the Portuguese controlled Mwenemutapa, which continued to exist in nominal form until the late 19th cent. During this time, however, the social structure of the empire was severely dislocated by the ravages of slave traders.

Mwenezi, town, MASVINGO province, SE ZIMBABWE, 90 mi/145 km S of MASVINGO, on MWENEZI RIVER; 21°25′S 30°44′E. Cattle, sheep, goats; corn. Formerly called Nuanetsi.

Mwenezi River, c. 260 mi/418 km long, ZIMBABWE and MOZAMBIQUE; rises in MATABELELAND SOUTH province, S central Zimbabwe, c. 55 mi/89 km ESE of BULAWAYO; flows past MWENEZI town, entering Mozambique 13 mi/21 km E of Pafari (Mozambique) and joins LIMPOPO River 25 mi/40 km ESE of Pafari. Formerly called Nuanetsi River.

Mwenga (MWENG-gah), village, SUD-KIVU province, E CONGO, 45 mi/72 km SW of BUKAVU; 03°02′S 28°26′E. Elev. 5,393 ft/1,643 m. In tin-mining area.

Mwera (MWAI-rah), town, ZANZIBAR SOUTH region, central ZANZIBAR island, E TANZANIA, 7 mi/11.3 km E of ZANZIBAR city; 06°07′S 39°19′E. Road junction. DUNGA ruins nearby. Cloves, copra.

Mwera (MWAI-rah), village, TANGA region, NE TANZANIA, 35 mi/56 km SSW of TANGA, on INDIAN OCEAN; mouth of MSANGASI RIVER 3 mi/4.8 km to S; 05°35′S 38°55′E. Fish; livestock; grain, sisal, bananas, copra.

Mwerihari River, c. 90 mi/145 km long, E central ZIMBABWE; rises c. 15 mi/24 km ENE of CHIVHU, in E Midlands Southern province; flows ESE, entering SAVE RIVER 45 mi/72 km WSW of MUTARE.

Mweru, Lake (MWE-roo), central AFRICA, on the CONGO-ZAMBIA border; 70 mi/113 km long and 30 mi/48 km wide; 09°00′S 28°45′E. Elevation 3,000 ft/914 m. It is drained to the N by the LUVUA RIVER. The lake has large fisheries. Also known as LAKE MOERO.

Mweru-Luapula, ZAMBIA: see MANSA.

Mweru Wantipa National Park (MWE-roo wa-en-TEE-pah) (□ 1,209 sq mi/3,143.4 sq km), NORTHERN and LUAPULA provinces, N ZAMBIA, 135 mi/217 km NW of KASAMA. Surrounds LAKE MWERU and adjacent marsh; hilly in W; Lake Mweru, on CONGO border, to W. Abundance of wildlife. Difficult access.

Mwilambongo (MWEE-lahm-BON-go), town, BANDUNDU province, W CONGO, 50 mi/80 km W of KIKWIT; 04°56′S 19°48′E. Elev. 1,446 ft/440 m.

Mwimbi (MWEEM-bee), village, RUKWA region, W TANZANIA, 45 mi/72 km S of SUMBAWANGA, near ZAMBIA border; 08°31′S 31°38′E. Livestock; subsistence crops.

Mwingi (MWEEN-gee), town (1999 population 10,138), Kitui district, EASTERN province, KENYA, on road, and 87 mi/140 km ENE of NAIROBI; 00°56′S 38°05′E. Market and trading center.

Mwinilunga (MWEE-nee-LOO-eng-gah), township, NORTH-WESTERN province, NW ZAMBIA, 150 mi/241 km NW of KASEMPA on West Lunga River; 11°44′S 24°26′E. Agriculture (beeswax; corn); hardwood. Diesel power station. West Lunga National Park to S. Transferred 1946 from KAONDE-LUNDA province.

Cross-references are shown in SMALL CAPITALS. The pronunciation guide is shown on page xix. The sources of population figures are shown on page xvii.

Mwirasandu (MWEE-ruh-sahn-doo), village, WESTERN region, SW UGANDA, 30 mi/48 km SE of MBARARA. Tin-mining center. Was part of SOUTHERN province. Sometimes spelled Mrirasandu.

Mwitikira (mwee-tee-KEE-rah), village, DODOMA region, central TANZANIA, 22 mi/35 km S of DODOMA; 06°34′S 35°41′E. Cattle, sheep, goats; corn, wheat.

Mwokil (MO-keel), atoll, State of POHNPEI, E CAROLINE ISLANDS, Federated States of MICRONESIA, W PACIFIC, 88 mi/142 km E of POHNPEI; 2 mi/3.2 km long, 1 mi/1.6 km wide. Consists of 3 wooded islets. Formerly Mokil, and Duperry Island.

Mwomboshi (mwom-BO-shee), township, CENTRAL province, S central ZAMBIA, on railroad 16 mi/26 km NNE of LUSAKA. Agriculture (tobacco, corn, cotton, soybeans); cattle.

Myadel (MYAH-del), Polish *Miadziol* or *Miadziol Nowy*, urban settlement, N MINSK oblast, BELARUS, ⊙ MYADEL region, on small lake just E of Lake NAROCH, 27 mi/43 km N of VILEIKA. Fisheries, lumbering.

Myaing (MYAING), village and township, MAGWE division, MYANMAR, 25 mi/40 km NW of PAKOKKU. Road hub.

Myakka River (mei-AK-kuh), c.50 mi/80 km long, W central FLORIDA; rises in E MANATEE county; flows SW, through swamps of SARASOTA county, then SE, into CHARLOTTE HARBOR c.7 mi/11.3 km W of PUNTA GORDA. Last 10 mi/16 km an estuary.

Myaksa (MYAH-ksah), village, S VOLOGDA oblast, N central European Russia, on the NE shore of the RYBINSK RESERVOIR (landing), on road junction, 18 mi/29 km SSE of CHEREPOVETS; 58°54′N 38°12′E. Elevation 505 ft/153 m. Flax processing, dairy products.

Myanaung (MYAH-noun), town and township, AYEYARWADY division, MYANMAR, on right bank of AYEYARWADY RIVER, and 40 mi/64 km S of PROME, on Henzada-Kyangin railroad. Former capital of the HENZADA district.

Myanmar (MYAHN-MAHR), republic (□ 261,789 sq mi/ 678,034 sq km; 2004 estimated population 42,720,196; 2007 estimated population 47,373,958), SE ASIA; ⊙ YANGON (Rangoon). Naypyidaw, a new town under construction in central Myanmar, began functioning as the administrative capital in 2005. The country's name was changed from the Union of Burma to the Union of Myanmar in 1989.

Geography and Population
Myanmar is bounded on the W by BANGLADESH, INDIA, and the Bay of BENGAL; on the N and NE by CHINA; on the E by LAOS and THAILAND; and on the S by the ANDAMAN SEA. The most densely populated part of the country is the valley of the AYEYARWADY River, which, with its vast delta, is one of the main rice-growing regions of the world. MANDALAY, the country's second largest city, is on the Ayeyarwady in central Myanmar. The Ayeyarwady basin is inhabited by the Burmans proper, a Mongoloid race who came S from Tibet by the 9th century. The valley is surrounded by a chain of mountains that stem from the E HIMALAYAS and spread out roughly in the shape of a giant horseshoe; the ranges and river valleys of the CHINDWINN (a tributary of the Ayeyarwady) and of the SITTANG and the THANLWIN (both to the E of the Ayeyarwady) run from N to S. Between the Bay of Bengal and the hills of the ARAKAN YOMA is the ARAKAN, a narrow coastal plain with the port of SITTWE. In the mountains of N Myanmar (rising to more than 19,000 ft/5,791 m) and along the India-Myanmar frontier live various Mongoloid peoples; the most important are the Kachins (in the KACHIN STATE in the N) and the Chins (in the CHIN Special State in the W). These peoples practice shifting cultivation (*taungya*) and cut teak in the forests. In E Myanmar on the Shan Plateau is the SHAN STATE, home of the Shans, a Tai race closely related to the Siamese. S of the Shan State are the mountainous KAYAH STATE and the KAYIN STATE; the Karens, who inhabit this region, are of Tai-Chin. origin, and many are Christians. S of the Kayin State is the TENASSERIM region, a long, narrow strip of coast extending to the Isthmus of KRA. At its N end is the port of MAWLAMYINE, Myanmar's fourth-largest city. Most of Myanmar has a tropical, monsoon climate; however, N of the Pegu Hills around Mandalay is the so-called dry zone with a rainfall of 20 in/51 cm to 40 in/102 cm. On the Shan Plateau temperatures are moderate.

Religion, Language, and Education
Theravada Buddhism is the religion of about 85% of the population Burmese (the tongue of the Burmans) is the official language, but the Shans, Kachins, and Karens speak their own languages; in all, over 100 languages are spoken in Myanmar. There are colleges and universities in Yangon, Mandalay, and MAWLAMYINE.

Economy
Myanmar suffered extensive damage in World War II, when it was called Burma, and some sectors of its economy have not yet fully recovered. Most of the population work in agriculture and forestry, and rice accounts for about half of the agricultural output. (Until 1964, Burma was the world's largest rice exporter.) Other important crops are maize, groundnuts, and pulses. It is also a producer of illegal opium in a N region bordering Laos and Thailand, known as the "Golden Triangle" and its laboratories produce two-thirds of the world's heroin. Myanmar's forests, which are government-owned, are the source of teak and other hardwoods. The country is rich in minerals. Petroleum is found E of the Ayeyarwady in the dry zone. Tin and tungsten are mined in E Myanmar; the MAWCHI mines in Kayah State are also rich in tungsten. In the Shan State, NW of LASHIO, are the BAWDWIN mines, the source of lead, silver, and zinc. Coal and iron deposits have also been found in Myanmar. Gems (notably rubies and sapphires) are found near MOGOK. Since the 13th century, Myanmar has exported to China jade from the HUKAWNG VALLEY in the N. Aside from food-processing establishments, there are few manufacturing industries in Myanmar. The country's chief trade partners are Southeast Asia, the EU, India, and China. Rice and teak are the leading exports, and machinery, transportation equipment, and textiles are the chief imports.

History: to 1900
Myanmar's early history is mainly the story of the struggle of the Burmans against the Mons, or Talaings (of Mon-Khmer origin, now assimilated). In 1044, King Anawratha established Burman supremacy over the Ayeyarwady delta and over THATON, capital of the Mon kingdom. Anawratha adopted Hinayana Buddhism from the Mons. His capital, PAGAN, "the city of a thousand temples," was the seat of his dynasty until it was conquered by Kublai Khan in 1287. Then Shan princes predominated in upper Burma, and the Mons revived in the S. In the 16th century the Burman Toungoo dynasty unified the country and initiated the permanent subjugation of the Shans to the Burmans. In the 18th century the Mons of the Ayeyarwady delta overran the dry zone. In 1758, Alaungapaya rallied the Burmans, crushed the Mons, and established his capital at Rangoon. He extended Burman influence to areas in present-day India (ASSAM and MANIPUR) and Thailand. Burma was ruled by his successors (the Konbaung dynasty) when friction with the British over border areas in India led to war in 1824. The Treaty of YANDABO (1826) forced Burma to cede to British India the Arakan and Tenasserim coasts. In a second war (1852) the British occupied the Ayeyarwady delta. Fear of growing French strength in the region, in addition to economic considerations, caused the British to instigate the Third Anglo-Burman War (1885) to gain complete control of Burma. The Burman king was captured, and the remainder of the country was annexed to India. Under British rule rice cultivation in the delta was expanded, an extensive railroad network was built, and the natural resources of Burma were developed. Exploitation of the rich oil deposits of Yenangyaung in central Burma was begun in 1871; the export of metals also became important.

History: 1900 to 1950
Until the 20th century, however, Burma was allowed no self-government. In 1923 a system of "dyarchy," already in effect in the rest of British India, was introduced, whereby a partially elected legislature was established and some ministers were made responsible to it. In 1935 the British gave Burma a new constitution (effective 1937), which separated the country from British India and provided for a fully elected assembly and a responsible cabinet. During World War II, Burma was invaded and quickly occupied by the Japanese, who set up a nominally independent Burman regime under Dr. Ba Maw. Disillusioned members of the Burmese Independent Army (which the Japanese had formed secretly before the war to assist in expelling the British) under Aung San formed an anti-Japanese resistance movement, the Anti-Fascist People's Freedom League (AFPFL). Allied forces drove the Japanese out of Burma in April 1945. In 1947 the British and Aung San reached agreement on full independence for Burma. Most of the non-Burman peoples supported the agreement, although the acquiescence of many proved short-lived. Despite the assassination of Aung San in July 1947, the agreement went into effect on January 4, 1948. Burma became an independent republic outside the British Commonwealth of Nations. The new constitution provided for a bicameral legislature with a responsible prime minister and cabinet. Non-Burman areas were organized as the Shan, Kachin, Kawthule, and Kayah states and the Chin Special Division; each possessed a degree of autonomy. The government, controlled by the socialist AFPFL, was soon faced with armed risings of Communist rebels and of Karen tribesmen, who wanted a separate Karen nation. International tension grew over the presence in Burma of Chinese Nationalist troops who had been forced across the border by the Chinese Communists in 1950 and who were making forays into China.

History: 1950 to 1990
Burma took the matter to the UN, which in 1953 ordered the Nationalists to leave Burma. In foreign affairs Burma has followed a generally neutralist course. It refused to join the Southeast Asia Treaty Organization and was one of the first countries to recognize the Communist government in China. In the elections of 1951–1952 the AFPFL triumphed. In 1958 the AFPFL split into two factions; with a breakdown of order threatening, Premier U Nu invited General Ne Win, head of the army, to take over the government. After the 1960 elections, which were won by U Nu's faction, civilian government was restored. In March 1962, Ne Win staged a military coup, discarded the constitution, and established a Revolutionary Council, made up of military leaders who ruled by decree. While the federal structure was retained, a hierarchy of workers' and peasants' councils was created. A new party, the Burma Socialist Program party, was made the only legal political organization. The Revolutionary Council fully nationalized the industrial and commercial sectors of the economy and imposed a policy of international isolation. Discussions were entered into with the minority peoples in 1963, but no agreement was reached. Insurgency became a major problem of the Ne Win regime. Pro-Chinese Communist rebels—the "White Flag" Communists—were active in the N part of the country, where, from 1967 on, they received aid from Communist China; the Chinese established links with the Shan and Kachin insurgents as well. The deposed U Nu, who managed to leave Burma in 1969, also used minority rebels to

organize an anti-Ne Win movement, the National Liberation Council, among the Shans, Karens, and others in the E. By the early 1970s the various insurgent groups controlled about one-third of Burma. A new constitution, providing for a unicameral legislature and one legal political party, took effect in March 1974. At that time the Revolutionary Council was disbanded and Ne Win was installed as president. Economic strife and ethnic tensions throughout the 1970s and 1980s led to anti-government demonstrations in 1988, which caused Ne Win to resign from office. The series of governments that followed failed to restore order, and the military seized control as the State Law and Order Restoration Council (SLORC), and the protests were brutally suppressed. In June 1989 the military government officially changed the name of the country to the Union of Myanmar to better reflect the nation's ethnic diversity.

History: 1990 to 1994

Elections were held in May 1990 and the National League for Democracy (NCD) won a large majority of seats in the assembly. However, the military government declared the election results invalid, and arrested many leaders and members of the NCD. Aung San Suu Kyi, the leader of the NCD, was placed under house arrest. She won the Nobel Peace Prize in 1991. In 1992, under new SLORC head General Than Shwe, many political prisoners were released, most martial law decrees were lifted (the ban on assembly by more than five persons still stands), and plans to draft a new constitution were announced. But there was little evidence that the army was prepared to return the government to civilian control, and Aung San Suu Kyi remained under house arrest. In December 1992, a UN General Assembly committee unanimously condemned the Myanmar military regime for its refusal to surrender power to a democratically elected parliament. The committee also called for the unconditional release of Aung San Suu Kyi. She was released on July 10, 1995, and continued to work for social and political change. Non-violent protest and the holding of meetings in defiance of the government decree are the tactics used. By the end of 1995 the UN had condemned extensive human rights offenses by the government, and calls for a boycott had grown. While the international situation remained tense, relations improved with ASEAN and particularly with neighboring India and Thailand over the issues of the hot pursuit of rebels, refugee movement, and sporadic violent border clashes. In 1992, India took steps to improve relations in regards to refugee movement and actions, and Thailand agreed to delineate its border with Myanmar in 1993, as well as constructing a "friendship bridge" in 1994.

History: 1994 to Present

Myanmar was invited to attend the ASEAN Foreign Ministers Annual Meeting in July of 1994, marking the first time the new government has been invited. Positive economic changes in mid-1990s include allowing the private sector to engage in most economic activities; legalizing border trade and allowing it to be conducted at market exchange rates; signing border trade agreements with India, Thailand, China, and Bangladesh; privatizing some state-owned enterprises; allowing farmers to sell around 80% of their rice crop on the free market and permitting most agricultural crops other than rice to be exported by the private sector; and easing procedures for issuing business and tourist visas. Also in 1996, the Myanmar government forces negotiated the surrender of the ailing drug lord Khun Sa and his Mong Tai Army. It was widely thought by his followers that, ailing from high blood pressure, diabetes, and a heart attack, he had sold them out when the government occupied his headquarters in July of 1996 without resistance. In 1997 the SLORC changed its name to the State Peace and Development Council (SPDC). Myanmar was ac-

cepted as a member of ASEAN in 1999. Although the junta has signed cease-fires with most of the insurgent ethnic minorities, the Shan revoked their agreement in 2005. Human rights groups have continued to report numerous abuses, including the use of members of ethnic minority groups in forced labor. In November 2005 the government announced that the capital would be moved to near PYINMANA from Yangon and that it had begun relocating ministries there; the new capital was named Naypyidaw in 2006. Beginning in August 2007 there were increasingly large, peaceful antigovernment demonstrations led by Buddhist monks in Yangon and elsewhere; the military brutally crushed the demonstrations in late September. The UN Security Council adopted a resolution deploring the crackdown.

Government

Myanmar's 1974 constitution has been suspended since 1988, when a military junta assumed power. The process of drafting a new constitution began in 1994, was suspended in 1996, and resumed in 2004; however, it has not included participation of the democratic opposition. The chairman of the State Peace and Development Council serves as head of state. The government is headed by the prime minister. The unicameral legislature consists of the 485-seat People's Assembly, whose members are popularly elected for four-year terms; however, the military junta has never allowed the legislature to convene. The current head of state (since 1992) is Chairman of the State Peace and Development Council Senior General Than Shwe. Administratively, the country is divided into seven divisions and seven states.

Myatlevo (MYAHT-lee-vuh), town (2005 population 1,960), N KALUGA oblast, central European Russia, on road and railroad, 35 mi/56 km NW of KALUGA; 54°53′N 35°40′E. Elevation 646 ft/196 m. Transportation-service establishments.

Myaundzha (myah-oon-JAH), town (2006 population 1,925), NW MAGADAN oblast, E RUSSIAN FAR EAST, in the E central CHERSKIY RANGE, in the KOLYMA River basin, on highway, 31 mi/50 km WNW of SUSUMAN; 63°03′N 147°11′E. Elevation 2,467 ft/751 m. In gold-, iron-, and coal-mining region. Formerly known as Bilikan, established around a forced-labor camp under Stalin's Gulag prison system.

Myaungbwe (MYOUNG-BWAI), village, Mrauk-oo township, RAKHINE STATE, MYANMAR, in the ARAKAN, on LEMRO RIVER (head of navigation), and 35 mi/56 km NE of SITTWE.

Myaungmya (MYOUNG-MYAH), former district (□ 2,835 sq mi/7,371 sq km), AYEYARWADY division, MYANMAR; ⊙ MYAUNGMYA. Between BASSEIN RIVER (W) and AYEYARWADY RIVER (E), embracing part of AYEYARWADY RIVER delta on ANDAMAN SEA. Important rice production.

Myaungmya (MYOUNG-MYAH), town, ⊙ MYAUNGMYA township, AYEYARWADY division, MYANMAR, in AYEYARWADY RIVER delta, 20 mi/32 km SE of BASSEIN. Rice-cultivation center, both shallow and deep water. Steamer landing.

Myawaddy (MYAH-wah-DEE), village and township, KAYIN STATE, MYANMAR, in TENASSERIM, on THAUNGYIN RIVER (THAILAND border) opposite MAE SOT, 60 mi/97 km ENE of MAWLAMYINE. Here Japanese forces invaded BURMA from THAILAND in World War II. Also spelled Myawadi.

Mycale (MI-kuh-lee), promontory, W ASIA MINOR, opposite SAMOS island. The center of the Ionian League was there, in the temple of Poseidon. In 479 B.C.E. the Greeks destroyed the Persian fleet at Mycale. This ended the Persian Wars for European Greece and began the rapid liberation of the Greeks of Asia Minor. Mycale, in modern TURKEY, is called Samsun Daği. It was also known as Mount Lydia.

Mycenae (mee-KEE-ne), ancient city of Argolis, S mainland GREECE, in what is now ARGOLIS prefecture,

NE PELOPONNESE department, 7 mi/11.3 km N of ARGOS. Natural rock citadel on N edge of Argive Plain. In historical times it had little importance and was usually dependent on Argos. Its significance is in its remote past as a center of Mycenaean civilization. The famous Lion Gate, which led into the city, and the Treasury of Atreus, the largest of the beehive tombs outside the walls of the city, are the most notable of its ancient remains.

Mydrecht, NETHERLANDS: see MIJDRECHT.

Myds'k (MIDSK) (Russian *Midsk*), town, N central RIVNE oblast, UKRAINE, 28 mi/45 km E of RIVNE and 8 mi/13 km NE of STEPAN'; 51°05′N 26°09′E. Elevation 547 ft/166 m. Copper mining.

Myebon (MYAI-BON), village and township, RAKHINE STATE, MYANMAR, on Arakan coast, at mouth of LEMRO RIVER, and 30 mi/48 km E of SITTWE.

Myers Chuck, village, SE ALASKA, on W side of CLEVELAND PENINSULA, on CLARENCE STRAIT 37 mi/60 km NW of KETCHIKAN. Trading center; fishing. Also spelled Meyers Chuck.

Myerstown (MEI-uhrz-toun), borough (2006 population 3,107), LEBANON county, SE central PENNSYLVANIA, 8 mi/12.9 km NE of LEBANON, on Tulpehocken Creek; 40°22′N 76°18′W. Varied heavy manufacturing. Agriculture (grain, soybeans, apples; poultry, livestock, dairying). Lebanon Valley Airport to W. Laid out 1768, incorporated c.1910.

Myersville, town, SAINT ELIZABETH parish, SW JAMAICA, 26 mi/42 km W of MAY PEN; 17°58′N 77°38′W. Corn, spices; livestock.

Myersville, town (2000 population 1,382), FREDERICK county, W MARYLAND, on CATOCTIN CREEK, between CATOCTIN and SOUTH Mountains, and 11 mi/18 km N of FREDERICK; 39°31′N 77°34′W. Founded by James Stottlemeyer in 1742, it has escaped development despite being between Routes 40 and 70.

Myingyan (MYIN-JAHN), former district (□ 2,707 sq mi/7,038.2 sq km), MANDALAY division, MYANMAR; ⊙ MYINGYAN. On left bank of AYEYARWADY RIVER and N plateau of PEGU YOMA; MOUNT POPA is the most prominent feature. In dry zone (annual rainfall 26 in/66 cm) and poorly irrigated; produces rice, sesame, cotton, beans, peanuts. Oil fields at SINGU; manufacturing lacquerware.

Myingyan (MYIN-JAHN), town, ⊙ MYINGYAN township, MANDALAY division, MYANMAR, on left bank of AYEYARWADY RIVER, and 55 mi/89 km SW of MANDALAY. River port, head of railroad to THAZI and MANDALAY. Cotton-trading center; cotton-spinning mill.

Myinmu (MYIN-moo), town and township, SAGAING division, MYANMAR, landing on right bank of AYEYARWADY RIVER, and 35 mi/56 km W of MANDALAY. On railroad to YE-U.

Myitche (MYIT-chai), village, PAKOKKU township, MAGWE division, MYANMAR, on right bank of AYEYARWADY RIVER, and 16 mi/26 km SW of PAKOKKU, opposite Pagay. Also spelled Myitchay.

Myitkyina (myit-CHEE-NAH), former N district (□ 29,723 sq mi/77,279.8 sq km) of KACHIN STATE, MYANMAR, on the INDIA (ASSAM) and CHINA (YUNNAN) borders; ⊙ MYITKYINA. Drained by MALI and NMAI headstreams of the AYEYARWADY RIVER and upper course of CHINDWIN RIVER (HUKAWNG VALLEY), it is bounded by a horseshoe of high ranges (to nearly 20,000 ft/6,090 m), and includes the KUMON RANGE and THE TRIANGLE. Agriculture (rice, tobacco, vegetables, sugarcane); jade mining (LONKIN, TAWMAW), amber (on S edge of HUKAWNG VALLEY); sugar mill at Sawmaw. Served by Mandalay-Myitkyina railroad, LEDO (Stilwell) Road, and by Ayeyarwady steamers during low water season.

Myitkyina (myit-CHEE-NAH), township and city, N MYANMAR; ⊙ KACHIN STATE, on the AYEYARWADY RIVER; 25°23′N 97°24′E. A leading town of N MYANMAR, it is a trade center (including teak and jade), the extreme N terminus of a railroad line from YANGON, and formerly an important town on the LEDO ROAD. In World War II, its capture (August 1944) by Allied troops after a siege of seventy-eight days marked an important stage in the liberation of BURMA from the Japanese.

Myitmaka River (myit-mah-KAH), over 100 mi/160 km long, MYANMAR; rises in marshy lake, 15 mi/24 km SSE of PROME; flows parallel to AYEYARWADY RIVER, past TANTABIN and INSEIN, to YANGON, where it joins the PEGU RIVER to form the RANGOON RIVER. Called HLAING RIVER in lower course.

Myitnge (MYIT-uhng-AI), town, MANDALAY division, MYANMAR, on MYITNGE RIVER (railroad and road bridges), and 8 mi/12.9 km S of MANDALAY, on Yangon-Mandalay railroad. Railroad workshops.

Myitnge River (MYIT-uhng-AI), over 250 mi/400 km long, MYANMAR; rises in SHAN PLATEAU E of LASHIO; flows SW, past NAMTU, HSIPAW, and MYITNGE, to the AYEYARWADY RIVER opposite SAGAING. Crossed by GOKTEIK VIADUCT in its middle course. Called Nam Tu in upper course.

Myittha (myit-THAH), village and township, MANDALAY division, MYANMAR, on railroad, and 38 mi/61 km S of MANDALAY. Considerable production of irrigated rice.

Myittha River (myit-THAH), over 150 mi/240 km long, in MYANMAR; rises in S CHIN HILLS W of TILIN; flows N, past GANGAW and KALEMYO, to CHINDWIN RIVER at KALEWA. Receives MANIPUR RIVER (left). Lower valley is sometimes considered part of KABAW VALLEY.

Myjava (mi-YAH-vah), Hungarian *Miava*, town, ZAPADOSLOVENSKY province, W SLOVAKIA, in SE foothills of the WHITE CARPATHIAN MOUNTAINS, on railroad, and 23 mi/37 km SW of TRENČÍN; 48°45′N 17°34′E. Refractory-clay deposits in vicinity; manufacturing (machinery; housing frames; food processing; brick kiln. Traditional folk architecture.

Mykhaylivka (mi-HEIL-leef-kah) (Russian *Mikhaylovka*), town, SW LUHANS'K oblast, UKRAINE, in the DONBAS, 2 mi/3 km ENE of ALCHEVS'K. Quarry; poultry. Established in the 17th century, town since 1938.

Mykhaylivka (mi-HEIL-leef-kah) (Russian *Mikhaylovka*), town, W central ZAPORIZHZHYA oblast, UKRAINE, 35 mi/56 km S of ZAPORIZHZHYA; 47°16′N 35°13′E. Elevation 288 ft/87 m. Raion center; feed mill; butter. Vocational technical school; heritage museum. Established at the beginning of the 19th century, town since 1965.

Mykhaylivka (mi-HEIL-leef-kah) (Russian *Mikhaylovka*), town, S LUHANS'K oblast, UKRAINE, in the DONBAS, 33 mi/53 km SSE of LUHANS'K, and 6 mi/9.7 km E of ROVEN'KY and subordinated to the Roven'ky city council. Coal mine, footwear factory, asphalt-cement plant. Established in 1910, town since 1954.

Mykhaylivka (mi-HEI-leef-kah) (Russian *Mikhailovka*), village (2004 population 6,000), E central KHMEL'NYTS'KYY oblast, UKRAINE, 13 mi/21 km SE of KHMEL'NYTS'KYY; 50°23′N 26°24′E. Elevation 761 ft/231 m. Sugar beets, fruit. Until 1946, known as Mykhal'pil' (Russian *Mikhal'pol'*).

Mykhaylivs'ka Tsilyna, UKRAINE: see UKRAINIAN STEPPE NATURE PRESERVE.

Mykhaylivs'kyy, Khutir, UKRAINE: see DRUZHBA, Sumy oblast.

Mykhayliv Virgin Land, UKRAINE: see UKRAINIAN STEPPE NATURE PRESERVE.

Mykhaylo-Kotsyubyns'ke (mi-KEI-lo–ko-tsyoo-BIN-ske) (Russian *Mikhaylo-Kotsyubinskoye*), town (2004 population 2,820), W CHERNIHIV oblast, UKRAINE, 8 mi/12.9 km WSW of CHERNIHIV; 51°27′N 31°05′E. Elevation 436 ft/132 m. Feed mill; dry goods; forestry. Known since 1667 as Kozel until around 1937, town since 1960.

Mykines, Danish *Myggenæs*, island (□ 4 sq mi/10.4 sq km) of the W FAEROE ISLANDS, DENMARK, separated from W VÁGAR by Mykinesfjørður. On W is a bridge to MYKINES (□ c.0.25 sq mi/0.65 sq km), a small island. Terrain mountainous and rocky, less than 1% cultivated; highest point is 1,837 ft/560 m. Fishing, sheep raising.

Mykolayiv (mi-ko-LAH-yeef) (Russian *Nikolayev*), city (2001 population 514,136), ⊙ MYKOLAYIV oblast, S UKRAINE, at the confluence of the Southern BUH and INHUL rivers, and on the bank of the Buh estuary; 46°58′N 32°00′E. Elevation 121 ft/36 m. A major seaport and railroad junction, Mykolayiv exports grain, iron, and manganese. It has shipyards, machinery plants, an aluminum plant, and a cotton textile factory. Founded in 1784 as a fortress near the site of the ancient Greek colony of Olbia, the city was named Mykolayiv (Russian *Nikolayev*) in 1788 when it became a shipbuilding center; in 1850, 1870–1900, it served as command post of the Black Sea fleet and became the third-largest port in the Russian Empire; oblast capital since 1937. Scientific and cultural institutions include hydrometeorological, shipbuilding, agricultural, educational, and cultural institutes, three theaters and two museums.

Mykolayiv (mi-ko-LAH-yeef) (Russian *Nikolayev*) (Polish *Mikolajów*), city (2004 population 13,500), central L'VIV oblast, UKRAINE, 21 mi/34 km S of L'VIV; 49°31′N 23°59′E. Elevation 1,187 ft/361 m. Cement works, ceramics (pottery, tiles, bricks), dry goods, sewing, flour mill. Has an old town hall and churches. Founded in 1552; passed from Poland to Austria (1772); training center for the Ukrainian Sich Riflemen (1917); part of West Ukrainian National Republic (1918–1919); reverted to Poland (1919); ceded to USSR in 1945; part of independent Ukraine since 1991.

Mykolayiv (mi-ko-LAH-yif) (Russian *Nikolayev*), village, central KHMEL'NYTS'KYY oblast, UKRAINE, on the Buzhok River, and 13 mi/21 km NW of KHMEL'NYTS'KYY. Sugar beets, wheat.

Mykolayivka (mi-ko-LAH-yif-kah) (Russian *Nikolayevka*), town (2004 population 2,700), W central Republic of CRIMEA, UKRAINE, on KALAMIT BAY of the BLACK SEA, connected by highway 25 mi/40 km W of SIMFEROPOL'; 44°58′N 33°37′E. Seaside resort with pansionates (vacation homes), youth camps; winery, cannery. Established in 1857; town since 1988.

Mykolayivka (mi-ko-LAH-yif-kah) (Russian *Nikolayevka*), town, central DNIPROPETROVS'K oblast, UKRAINE, on left bank of DNIEPER (Ukrainian *Dnipro*) River, and 10 mi/16 km W of DNIPROPETROVS'K; 48°23′N 34°40′E. Elevation 344 ft/104 m. Established in the late 18th century; town since 1938.

Mykolayivka (mi-ko-LEI-yeev-kah) (Russian *Nikolayevka*), town (2004 population 3,350), NE ODESSA oblast, UKRAINE on the Chychykliya River, 22 mi/35 km NNW of BEREZIVKA; 47°43′N 29°27′E. Elevation 672 ft/204 m. Raion center; animal feed and flour milling, cheese making, food processing. Established at the end of the 18th century; town since 1965. Previously called Mykolayivka Druha (Russian *Nikolayevka Vtoraya*).

Mykolayivka (mi-ko-LEI-yeev-kah) (Russian *Nikolayevka*), town, N DONETS'K oblast, UKRAINE, in the DONBAS, 7 mi/11 km E of SLOV'YANSK and subordinated to its city council; 48°47′N 37°13′E. Elevation 590 ft/179 m. Site of the Slov'yansk state regional thermal-electric power station. Electric station building technical school. Settlement established in the early 18th century; town since 1956.

Mykolayivka (mi-ko-LEI-yeev-kah) (Russian *Nikolayevka*), village (2004 population 2,200), W LUHANS'K oblast, UKRAINE, in the DONBAS, 11 mi/18 km SSW of LYSYCHANS'K; 49°46′N 38°57′E. Elevation 255 ft/77 m. Formerly a mining town serving nearby coal mines; deteriorated to a village after the mines were exhausted.

Mykolayivka, UKRAINE: see NOVOVORONTSOVKA.

Mykolayivka Druha, UKRAINE: see MYKOLAYIVKA, Odessa oblast.

Mykolayivka-Vyrivs'ka, UKRAINE: see ZHOVTNEVE.

Mykolayiv oblast (mi-ko-LAH-yif) (Ukrainian *Mykolayivs'ka*) (Russian *Nikolayevskaya*), oblast (□ 9,497 sq mi/24,692.2 sq km; 2001 population 1,264,743), S UKRAINE; ⊙ MYKOLAYIV. In BLACK SEA LOWLAND; bounded S by BLACK SEA; drained by lower Southern BUH and INHUL rivers; steppe region. Population mostly Ukrainian (75.6%), with large Russian (19.4%), as well as Moldavian (1.3%), Belorussian (1.1%), Jewish (0.9%), and Bulgarian (0.5%) minorities. Wheat, barley, and corn are chief grain crops; sunflowers, sugar beets, castor beans (N); truck produce near Mykolayiv; dairy farming. Mykolayiv is a major port and industrial center. Machine building (PERVO-MAYS'K), light industries (VOZNESENS'K), and fish processing (OCHAKIV). Flour milling, dairy, and meat processing in major cities; butter and cheese making, sugar refining in rural areas. Formed in 1937; has (1993) nine cities, twenty towns, nineteen rural raions.

Mykolayivs'ka oblast, UKRAINE: see MYKOLAYIV oblast.

Mykonos (MEE-ko-nos), mountainous island (□ 35 sq mi/91 sq km), CYCLADES prefecture, SOUTH AEGEAN department, SE GREECE, in the AEGEAN SEA; 37°27′N 25°23′E. One of the CYCLADES. Tourist resort; fisheries. Many churches. Airport. Also Míkonos.

Mykulyntsi (mi-KOO-lin-tsee) (Russian *Mikulintsy*) (Polish *Mikulińce*), town (2004 population 6,800), S central TERNOPIL' oblast, UKRAINE, on the SERET RIVER, and 11 mi/18 km S of TERNOPIL'; 49°24′N 25°36′E. Elevation 941 ft/286 m. Agricultural trading center; food processing (pork, cheese, flour, canning, honey), brewing, brick manufacturing, lumbering, furniture manufacturing. Has ruins of a 16th-century castle, 18th–19th-century palace. Known since 1096 as Mykulyn, first in Kievan Rus', then Halych (12th century) and Halych-Volyn' (13th century) principalities; passed to Poland (1387); renamed Mykulyntsi (1389); fortified (1550); town status (1595); destroyed by the Turks (1672); passed to Austria (1772); part of West Ukrainian National Republic (1918); reverted to Poland (1919); incorporated in Ukrainian SSR (1939); in independent Ukraine since 1991. Jewish population since the 18th century; one of the few in Eastern Europe to suffer from intra-community violence, during the discord between the Hasidim and Haskala followers in the mid-19th century; numbering 1,900 in 1939, destroyed during World War II.

Mykytivka (mi-KI-teef-kah) (Russian *Nikitovka*), NNW suburb of HORLIVKA, central DONETS'K oblast, UKRAINE, in the DONBAS, 4 mi/6.4 km NNW of city center; 48°22′N 38°03′E. Elevation 807 ft/245 m. Railroad center; limestone (dolomite) quarries, chemical works; mercury refinery (mines 3 mi/5 km W).

Mykytyne, UKRAINE: see NIKOPOL'.

Mylae, ITALY: see MILAZZO.

My Lai, VIETNAM: see son MY.

Mylasa, TURKEY: see MILAS.

Mylliem (MEIL-lee-uhm), town, East Khasi Hills district, MEGHALAYA state, NE INDIA, on SHILLONG PLATEAU, 5 mi/8 km SSW of SHILLONG; 25°30′N 91°50′E. Rice, sesame, cotton.

Myllykoski (MUL-lu-KOS-kee), village, KYMEN province, SE FINLAND, on the KYMIJOKI (river), and 6 mi/9.7 km SSE of KOUVOLA; 60°47′N 26°48′E. Elevation 149 ft/45 m. Pulp and paper mills, hydroelectric station.

Mylo (MEI-lo), village (2006 population 19), ROLETTE county, N NORTH DAKOTA, 16 mi/26 km S of ROLLA; 48°38′N 99°37′W. Group of small lakes to SW. Founded in 1905 and incorporated in 1907. Named for the county's first Catholic priest, Father John E. Malo.

Mylor (MEI-luhr), village, SOUTH AUSTRALIA state, S central AUSTRALIA, 17 mi/27 km from ADELAIDE; 35°03′S 138°45′E.

Mylor (MEI-luh), village (2001 population 2,533), CORNWALL, SW ENGLAND, on Mylor Creek (an inlet of CARRICK ROADS), and 2 mi/3.2 km NNE of FAL-

MOUTH; 50°10′N 05°03′W. Mylor Pool (N) is anchorage for small vessels. Formerly site of royal dockyard. Has 15th-century church.

Mymensingh (MEI-mon-sing), district, E central EAST BENGAL, BANGLADESH; ⊙ MYMENSINGH. Bounded N by ASSAM and MEGHALAYA states (INDIA), SE by MEGHNA RIVER, W by JAMUNA RIVER (main course of the BRAHMAPUTRA RIVER); drained by the JAMUNA, old BRAHMAPUTRA, and MEGHNA rivers. Alluvial soil; agriculture (rice, jute, oilseeds, sugarcane, tobacco, cotton); timber (red cotton trees, bamboo, sal in MADHUPUR JUNGLE). Jute pressing at MYMENSINGH and SARISHABARI, rice and oilseed milling at GOURIPUR, sugar milling at Kishorganj; cotton weaving (BAJITPUR), metalware manufacturing (ISLAMPUR). College at MYMENSINGH. Part of 14th-century independent kingdom of Bengal under Muslim ruler; passed in 1765 to the British. Present district formed in 1787; part of former British Bengal province until it was transferred in 1947 to East Pakistan. Also spelled Maimansingh.

Mymensingh (MEI-mon-sing), town (□ 22 sq mi/57.2 sq km; 2001 population 227,201), MYMENSINGH district, N central EAST BENGAL, BANGLADESH, on an old channel of the BRAHMAPUTRA RIVER; 24°47′N 90°24′W. Trading center for rice, jute, sugarcane, oilseeds, tobacco, mustard, and pulses. Once noted for the manufacturing of glass bangles; now has jute-pressing and electrical-supply industries. In the town are Ananda Mohan College (an affiliate of Dhaka University), an agricultural university, a veterinary training institute, and the Institute of Radiation Genetics and Plant Breeding.

Mynaral (mee-nah-RAHL), town, ZHAMBYL region, KAZAKHSTAN, on SW shore of LAKE BALKASH; 45°24′N 73°40′E. Fishing; fish processing.

Myndd-Bwlch-Groes (MUHN-[th]–BUHLKH–GROS), mountain (1,449 ft/442 m), POWYS, E Wales, 5 mi/8 km E of LLANDOVERY. BLACK MOUNTAINS adjoin to S.

Myn'kivtsi (MIN-kif-tsee) (Russian *Min'kovtsy*), village (2004 population 6,300), SE KHMEL'NYTS'KYY oblast, UKRAINE, on the Ushytsya River, left tributary of the DNIESTER River, and 25 mi/40 km ENE of KAMYANETS'-PODIL'S'KYY; 48°51′N 27°06′E. Elevation 646 ft/196 m. Clothing industry.

Mynydd Eppynt (MUH-ni-[th] E-pint), mountain range, POWYS, E Wales, extends 15 mi/24 km E-W. Highest point is Drum ddu (1,554 ft/474 m), 11 mi/18 km NNW of BRECON.

Mynydd Margam, Wales: see MARGAM.

Mynydd Moel, Wales: see CADER IDRIS.

Mynydd Preseli (MUH-ni-[th] pre-SE-lee) or **Prescelly Mountains**, mountain range, Pembrokeshire, SW Wales, extends 5 mi/8 km E to W; 51°56′N 04°46′W. Highest point is Foel Cwmcerwyn (1,760 ft/ 536 m), 12 mi/16 km SW of CARDIGAN. Formerly in DYFED, abolished 1996.

Myogi (MYO-gee), town, Kanra county, GUMMA prefecture, central HONSHU, N central JAPAN, 19 mi/30 km S of MAEBASHI; 36°16′N 138°49′E. Mount Myogi is in nearby Myogi Arafune Saku Highland quasi-national park.

Myogyi (myo-JEE), village, in the township of YENGAN, SHAN STATE, MYANMAR, on ZAWGYI RIVER, and 40 mi/ 64 km SSE of MANDALAY. Was capital of former MAW STATE.

Myohaung (MYO-houng) or **Mrauk-oo**, village and township, RAKHINE STATE, MYANMAR, in the ARAKAN, between LEMRO and KALADAN rivers, 40 mi/64 km NE of SITTWE. Former capital (15th–18th centuries) of old Arakan kingdom, and later of Burmese Arakan province, until supplanted (1826) because of its unhealthy climate by Akyab (now Sittwe). Ruins of fortifications and palace.

Myohyang, Mount (MYO-HAHNG), collective name for six mountain peaks in NORTH PYONGAN province,

NORTH KOREA, c.50 mi/80 km NE of SINUIJU. Highest peak is Mount Piro (6,263 ft/1,909 m).

Myoko (MYO-ko), village, Nakakubiki county, NIIGATA prefecture, central HONSHU, N central JAPAN, 78 mi/125 km S of NIIGATA; 36°55′N 138°13′E. Hot springs and skiing area (Japan's oldest) nearby. Site of Sekiyama railroad station, the country's oldest switchback-railroad station.

Myokokogen (MYO-KO-KO-gen), town, Nakakubiki county, NIIGATA prefecture, central HONSHU, N central JAPAN, 84 mi/135 km S of NIIGATA; 36°52′N 138°12′E. Myoko Highlands (skiing area) nearby.

Myongchon (MYONG-CHUHN), county, NORTH HAMGYONG province, NORTH KOREA, 43 mi/69 km SSW of CHONGJIN, in coal-mining area.

Myongju (MYOUNG-JOO), county (□ 365 sq mi/949 sq km), E central KANGWON province, SOUTH KOREA, on E slope of TAEBAEK Mountains, facing East Sea. Some agriculture; fishery (pollack, squid, pike, anchovies). CHUMUNJIN Beach; lime grottos; Taekwanryong pass.

Myothit (MYO–[th]IT), village, MOMAUK township, KACHIN STATE, MYANMAR, on Taping River, on road to MYITKYINA, and 15 mi/24 km NE of BHAMO.

Myothit (MYO–[th]IT), village and township, MAGWE division, MYANMAR, on Pyinmana-Kyaukpadaung railroad, and 34 mi/55 km E of MAGWE. Important agricultural area producing cattle, sesame, and millet.

Mypolonga, village, SE SOUTH AUSTRALIA state, S central AUSTRALIA, 45 mi/72 km ESE of ADELAIDE, and on MURRAY RIVER, near MURRAY BRIDGE; 35°03′S 139°21′E. Dairy products; livestock; fruit.

Myra, ancient city and seaport of LYCIA, on the MEDITERRANEAN SEA, SW ASIA MINOR, now SW TURKEY near KAS; 36°17′N 29°58′E. The Acts of the Apostles reports that the city was visited by Paul. According to tradition, it was the see of Saint Nicholas. Ruins of a theater are on the acropolis, and the necropolis has many grand tombs. Modern name is Demre.

Mýrasýsla (MEE-rahs-EES-lah), county, SW ICELAND, on E shore of FAXAFLOI ⊙ BORGARNES. HVITA River forms S border. Has rocky coastline; marshy highlands in interior (the Fiskivötn), an area of small lakes glacial in origin from which flow several rivers to W and N. Sheep, cattle, horses; fishing.

Mýrdalsjökull (MIR-dahls-yuh-kuh-tuhl), extensive glacier, S ICELAND, near the coast; 30 mi/48 km long (E-W), 10 mi/16 km–20 mi/32 km wide; 63°39′N 19°10′W. W extension, called Eyjafjallajokull, rises to 5,466 ft/1,666 m at 63°37′N 19°36′W. At S edge of glacier is Katla, active volcano (2,382 ft/726 m) whose eruption in 1918 caused great floods.

Myrhorod (MIR-ho-rod) (Russian *Mirgorod*), city (2004 population 46,600), central POLTAVA oblast, UKRAINE, on KHOROL River, and 50 mi/80 km WNW of POLTAVA; 49°58′N 33°36′E. Elevation 347 ft/105 m. Raion center; food processing (flour, cheese, fruit canning, mineral water); manufacturing (broiler, building materials); brickworks. Ceramics technical school and factory. Peat bogs nearby. Health resort (salt and carbonic springs) with seven sanatoria. Established in the mid-16th century, town status (1575); sometime center of the Myrhorod regiment (1575–1638, 1648–1700s); immortalized by Nikolai Gogol by naming his second volume of Ukrainian stories *Mirgorod*. Jewish community since the 18th century, reduced by the 1905 pogroms and 1918–1919 civil war pogroms; numbering 2,000 in 1939; destroyed by the Nazis in November 1941—fewer than 100 Jews remaining in 2005.

Myrina, Greece: see KÁSTRON.

Myrivs'ke (MI-reev-ske) (Russian *Mirovskoye*), town (2004 population 5,150), W DNIPROPETROVS'K oblast, UKRAINE, on road and railroad, 12 mi/20 km N of KRYVYY RIH city center and subordinated to its city council; 48°05′N 33°24′E. Elevation 469 ft/142 m. Re-

inforced concrete, housing panel construction. Established as town in 1958.

Myrna Dolyna (MIR-nah do-LI-nah) [Ukrainian= peaceful valley] (Russian *Mirnaya Dolina*), town, W central LUHANS'K oblast, UKRAINE, in a shallow valley of high right bank of the DONETS River, on road, and near highway, 6 mi/10 km S of LYSYCHANS'K city center; 48°48′N 38°28′E. Elevation 580 ft/176 m. State farm. Established in 1773, town since 1964.

Myrnam (MUHR-nuhm), village (□ 1 sq mi/2.6 sq km; 2001 population 322), E central ALBERTA, W Canada, near North SASKATCHEWAN River, 26 mi/42 km NW of VERMILION, in Two Hills County No. 21; 53°40′N 111°14′W. Mixed farming, grain, livestock. Fort De L'Isle Historical Site nearby. Ukranian heritage. Incorporated 1930.

Myrne (MIR-ne) (Russian *Mirnoye*), town, SW KHERSON oblast, UKRAINE, on road, on railroad, 14 mi/22.5 km N of PRYMORS'KYY; 46°16′N 33°27′E. Grain processing. Railroad cargo depot.

Myrne (MIR-ne) [Ukrainian=peaceful] (Russian *Mirnoye*), town (2004 population 3,020), SE DONETS'K oblast, UKRAINE, on road 16 mi/26 km SSE of VOLNOVAKHA; 47°25′N 37°44′E. Elevation 406 ft/123 m. Quarry, 2 asphalt plants. Established in 1951 as Karans'kyy Kam'yanyy Kar'yer [Ukrainian=Karans'kyy Rock Quarry] (Russian *Karanskiy Kamennyy Kar'er*), re-named in 1958; town since 1967.

Myrne (MIR-ne) [Ukrainian=peaceful] (Russian *Mirnoye*), town, SE LUHANS'K oblast, UKRAINE, in the DONBAS, on road, and near railroad 4 mi/7 km W of KRASNODON; 48°18′N 39°41′E. Elevation 429 ft/130 m. Poultry state farm. Established in 1929, town since 1938.

Myrne (MIR-ne) [Ukrainian=peaceful] (Russian *Mirnoye*), town, S ZAPORIZHZHYA oblast, UKRAINE, on right bank of MOLOCHNA RIVER, on road, and near highway 7 mi/12 km NNE of MELITOPOL'. Lumber depot. Archeological point of interest, "Kam'yana Mohyla," nearby. Established in 1951; town since 1987.

Myrnyy (MIR-nee) (Russian *Mirnyy*) [Ukrainian and Russian=peaceful], town, W Republic of CRIMEA, UKRAINE, between S tip of DONUZLAV LAGOON and BLACK SEA coast, and on highway, 17 mi/28 km WNW of, and subordinated to, YEVPATORIYA; 45°19′N 33°01′E. Fishing; cargo trucks repair shops. Town since 1977.

Myronivka (mi-RO-nif-kah) (Russian *Mironovka*), town, S KIEV oblast, UKRAINE, on the Rosava River, tributary of the ROS' RIVER, and 40 mi/64 km ESE of BILA TSERKVA; 49°39′N 30°59′E. Elevation 383 ft/116 m. Raion center; railroad junction; food processing (sugar, flour, powdered skim milk); manufacturing (asphalt, cement); brickworks, auto repair. Home of Myronivka Institute of Wheat Selection and Seed Cultivation; museum. Established in the first half of the 17th century, city since 1968.

Myronivs'kyy (mi-RO-nif-skyee) (Russian *Mironovskiy*), town, E DONETS'K oblast, UKRAINE, on the LUHAN' RIVER, right tributary of the DONETS, near highway, and on railroad spur 10 mi/16 km NNW of and subordinated to DEBAL'TSEVE; 48°27′N 38°16′E. Elevation 580 ft/176 m. Regional thermal electric station, reinforced concrete building materials; vocational school. Myronivs'kyy Reservoir of Luhan' River outlines the W side of town. Established in 1950, when the station was built; town since 1953.

Myropil' (mi-RO-peel) (Russian *Miropol'*), town (2004 population 4,200), SW ZHYTOMYR oblast, UKRAINE, on SLUCH RIVER, on highway, and railroad 43 mi/70 km WSW of ZHYTOMYR and 11 mi/18 km WSW of DZERZHYNS'K; 50°07′N 27°42′E. Elevation 744 ft/226 m. Paper mill, silicate and asphalt manufacturing, toy factory; forestry. Established in first half of 16th century; town since 1957.

Myropillya (mi-ro-PEEL-lyah) (Russian *Miropol'ye*), village, E SUMY oblast, UKRAINE, on the PSEL RIVER, and 20 mi/32 km ENE of SUMY; 51°01′N 35°16′E. Elevation 590 ft/179 m. Livestock, dairy.

Myrtilis Julia, PORTUGAL: see MÉRTOLA.

Myrtle (MUHR-tuhl), village (2000 population 63), FREEBORN county, S MINNESOTA, near IOWA state line, 12 mi/19 km SE of ALBERT LEA; 43°33′N 93°09′W. Dairying.

Myrtle, village (2000 population 407), UNION county, N MISSISSIPPI, 8 mi/12.9 km NW of NEW ALBANY; 34°33′N 89°07′W. Agriculture (cotton, corn; cattle; dairying); timber; manufacturing (upholstered furniture). Holly Springs National Forest to W and N.

Myrtle Beach, city (2006 population 28,597), HORRY county, E SOUTH CAROLINA, on the coast, 14 mi/23 km SE of CONWAY; 33°42′N 78°53′W. Year-round beach resort; largest seashore resort in state. Socastee yacht basin, on INTRACOASTAL WATERWAY, is nearby. Myrtle Beach Air Force Base. Manufacturing includes ceramic capacitors, concrete, wire harnesses, industrial component parts, printing and publishing, signs, bricks. Incorporated 1938.

Myrtle Creek (MUHR-tuhl), town (2006 population 3,539), DOUGLAS county, SW OREGON, 13 mi/21 km S of ROSEBURG, on South UMPQUA RIVER, at mouth of Myrtle Creek; 43°01′N 123°16′W. Timber.

Myrtleford (MUHR-tuhl-fuhrd), town, NE central VICTORIA, SE AUSTRALIA, on OVENS RIVER, and 130 mi/209 km NE of MELBOURNE; 36°33′S 146°44′E. In tobacco, hops, vegetable (asparagus), mint, nuts (walnuts, chestnuts), and livestock area; dairy plant; former gold-mining center. Agricultural experiment station.

Myrtle Grove (MUHR-tuhl GROV), unincorporated town (□ 7 sq mi/18.2 sq km), NEW HANOVER county, SE North Carolina, residential suburb 7 mi/11.3 km SSE of WILMINGTON, near INTRACOASTAL WATERWAY 1 mi/1.6 km to E, ATLANTIC OCEAN 2 mi/3.2 km to E; 34°07′N 77°52′W. Service industries; manufacturing.

Myrtle Grove (MER-tuhl), suburb (□ 7 sq mi/18.2 sq km; 2000 population 17,211) of PENSACOLA, ESCAMBIA county, extreme NW FLORIDA, 6 mi/9.7 km W of city center; 30°25′N 87°17′W.

Myrtle Point (MUHR-tuhl), town (2006 population 2,501), COOS county, SW OREGON, South Fork of COQUILLE RIVER (forms main stream) at confluence of North Fork, 23 mi/37 km S of COOS BAY; 43°03′N 124°07′W. PACIFIC OCEAN 15 mi/24 km to W. Trade center for dairy products and livestock; timber; fish hatchery to NE. Maria C. Jackson State Park to NE; Hoffman Memorial Wayside and Coquille and Myrtle Grove State Parks to S; Siskiyou National Forest to S. Settled 1858, incorporated 1903.

Myrtletowne, unincorporated town, HUMBOLDT county, NW CALIFORNIA, residential suburb 4 mi/6.4 km E of EUREKA, near Humboldt Bay; 40°48′N 124°09′W. Dairying, cattle, sheep; timber.

Myrtoan Sea, Greece: see AEGEAN SEA.

Mysanov Chal (myah-SAHN-ov CHAHL), peak (7,697 ft/2,346 m) in the RILA MOUNTAINS, SW BULGARIA.

Mysen (MU-suhn), village, ØSTFOLD county, SE NORWAY, on railroad, and 20 mi/32 km NNE of SARPSBORG; 59°33′N 11°20′E. Agriculture, foodstuff industries; lumber. Administration center for the inner region of the county. Site of domestic-industry schools, music college, museum, and geriatric hospital. The racetrack here hosts the annual Momarkedet-festival.

Myshega (MI-shi-gah), former town, NW TULA oblast, W central European Russia, on the OKA River, 2 mi/3.2 km W of ALEKSIN (across the river), into which it has been incorporated; 54°31′N 37°02′E. Elevation 446 ft/135 m. Iron foundry, metalware.

Myshkin (MISH-keen), city (2006 population 6,055), W YAROSLAVL oblast, central European Russia, on the VOLGA RIVER (landing), on highway junction, 78 mi/126 km NW of YAROSLAVL, and 22 mi/35 km SSW of RYBINSK; 57°47′N 38°27′E. Elevation 354 ft/107 m.

Brickworks, agricultural machinery. Nearby is compressor station for the natural-gas pipeline, Northern Light. Made city in 1777; reduced to status of a village in 1917; made town in 1943; made city again in 1991. As a village, was also called Myshkino.

Mysia (MI-shee-uh), ancient region, NW ASIA MINOR. It was N of LYDIA and its coast faced LESBOS. Mysia was not a political unit, and it passed successively to LYDIA, PERSIA, Macedon, SYRIA, PERGAMUM, and ROME. It is an area entirely within present-day TURKEY.

Myskhako (mis-hah-KO), village (2005 population 5,770), SW KRASNODAR TERRITORY, S European Russia, on the NE coast of the BLACK SEA, 27 mi/43 km SE of ANAPA, and 5 mi/8 km S of NOVOROSSIYSK; 44°39′N 37°46′E. Winemaking.

Myski (mis-KEE), city (2005 population 43,970), S central KEMEROVO oblast, S central SIBERIA, RUSSIA, in KUZNETSK BASIN, at a confluence of the Mrassu and Tom′ rivers, on railroad, 217 mi/349 km SE of KEMEROVO, and 28 mi/45 km E of NOVOKUZNETSK; 53°42′N 87°48′E. Elevation 935 ft/284 m. Mining and processing of coal; sawmilling; building materials; food industries (poultry, meat processing). Tom′-Usinskaya power station nearby. Founded in 1826. Made city in 1956.

Myslenice (mees-lee-NEE-tse), Polish *Myślenice*, town, Kraków province, S POLAND, on RABA RIVER, and 15 mi/24 km S of KRAKÓW; 49°50′N 19°56′E. Manufacturing of hats; brewing, tanning, flour milling; stone quarrying. Hydroelectric plant. Winter resort; ruins of 14th century castle; museum.

Mysliborz (mees-LEE-boz), Polish *Myślibórz*, German *Soldin*, town, in BRANDENBURG, Gorzów province, NW POLAND, on small lake, 25 mi/40 km NNE of KOSTRZYŃ (Küstrin). Dairy center; sawmilling. Has 13th century basilica of former Dominican monastery (founded 1275) and remains of medieval town gates.

Myslowice (mees-lo-VEE-tse), Polish *Mysłowice*, German *Myslowitz*, city (2002 population 75,712), Katowice province, S POLAND, on the CZARNA PRZEMSZA, near its confluence with the Biała Przemsza, and 5 mi/8 km E of KATOWICE; 50°14′N 19°09′E. In dense railroad network. Metal industry, brickworks, coal mines, chemical factory. In GERMANY (on Russian Poland–AUSTRIA border) until World War I.

Mys Mart′yan zapovednik, UKRAINE: see CAPE MARTYAN NATURE RESERVE.

Mys Mart′yan zapovidnyk, Crimea: see CAPE MARTYAN NATURE RESERVE.

My Son (MEE SON), ruins, QUANG NAM-DA province, central VIETNAM; 15°38′N 108°08′E. This ancient political capital of CHAMPA is regarded by archaeologists to be on par with SOUTHEAST ASIA's other ancient cities, such as ANGKOR (CAMBODIA) and PAGAN (MYANMAR). Structures date from the 4th to the 12th centuries. Originally comprised of 68 different structures, only twenty remain, and these are undergoing restoration. Tourist area.

Mysore (mei-SOR), district (□ 4,615 sq mi/11,999 sq km), KARNATAKA state, SE INDIA; ☉ MYSORE.

Mysore (mei-SOR), city (2001 population 799,228), ☉ MYSORE district, KARNATAKA state, S INDIA. Industrial, commercial, and educational center; 12°18′N 76°39′E. Manufacturing includes silk saris, textiles, leather, chemical goods, and cigarettes. Magnificent palace of maharajas. Earlier capital of the wealthy Muslim state of Mysore (from sixteenth century) and later became the capital of a Hindu state.

Mysore, INDIA: see KARNATAKA.

Mysovka (MI-suhf-kah), village, N KALININGRAD oblast, RUSSIA, on the E shore of the Courland Lagoon, in the NEMAN River delta, on road, 26 mi/42 km WNW of SOVETSK (linked by narrow-gauge railroad); 55°11′N 21°16′E. Seaside resort amid pinewoods. Until 1945, in EAST PRUSSIA and called Karkel′n.

Mysovsk, RUSSIA: see BABUSHKIN, BURYAT Republic.

Mys Shmidta (MIS SHMEET-tah) [Russian=Cape Shmidt], settlement (2005 population 640), on the coast of the CHUKCHI SEA, N CHUKCHI AUTONOMOUS OKRUG, N RUSSIAN FAR EAST, N of the ARCTIC CIRCLE, 264 mi/425 km NW of PROVIDENIYA; 68°56′N 179°30′W. Government arctic station; trading post, airfield.

Mystery Lake (MIS-tuhr-ee), local government district (□ 1,338 sq mi/3,478.8 sq km; 2001 population 79), N central MANITOBA, W central Canada; 55°34′N 97°59′W. Nickel mining and smelting, logging, light manufacturing, transportation industry. At its center is city of THOMPSON; other communities include Sipiwesk, Parlee, Moak Lake, and Johnson. Incorporated 1956.

Mystic, city (2000 population 588), APPANOOSE county, S IOWA, 4 mi/6.4 km W of CENTERVILLE; 40°46′N 92°56′W. Incorporated 1899.

Mystic, town, IRWIN county, S central GEORGIA, 9 mi/14.5 km SW of FITZGERALD; 31°37′N 83°20′W.

Mystic, village (2000 population 4,001), at the mouth of the MYSTIC RIVER, SE CONNECTICUT; 41°21′N 71°57′W. Mystic is a postal section of both GROTON and STONINGTON since the village of Mystic is split between the two towns. It is a major tourist area, known for the Mystic Marinelife Aquarium, from which Robert Ballard and the Institute for Underseas Exploration operate. Mystic Seaport Museum chronicles the village's historical importance in colonial Connecticut's fishing and whaling industries and offers river and sound cruises aboard historic ships. Olde Misticke retail center on outskirts of village. Tourist traffic on the narrow main street in village center often comes to a patient standstill in the summertime as the drawbridge over the Mystic River opens and closes, allowing sailboats to go in and out of the Mystic harbor. Major sidewalk art festival annually in August.

Mystic (MIS-tik), river, c.7 mi/11 km long; rising in MYSTIC LAKES, E MASSACHUSETTS; and flowing SE, past MEDFORD, into Boston Harbor at CHARLESTOWN. Medford was one of the important early settlements on its banks. Former inner section of Boston Harbor and spanned by Mystic River Bridge between BOSTON and CHELSEA.

Mystic Lakes (MIS-tik), reservoir consisting of two lakes, on MEDFORD-ARLINGTON city border, and in WINCHESTER town, MIDDLESEX county, E MASSACHUSETTS, on MYSTIC RIVER, 6 mi/9.7 km NNW of downtown BOSTON; c.2 mi/3.2 km long; 42°26′N 71°09′W. Includes Upper Mystic and Lower Mystic ponds. Residential area.

Mystic River, c.10 mi/16 km long; rising in SE CONNECTICUT; flowing S past OLD MYSTIC to LONG ISLAND SOUND, where it divides the village of MYSTIC into two municipal districts, one in town of GROTON (W), another in STONINGTON (E).

Mytho (MEE-TO), city, ☉ TIEN GIANG province, S VIETNAM, in MEKONG Delta, on left bank of Mekong delta arm and 35 mi/56 km SW of HO CHI MINH CITY (linked by railroad); 10°21′N 106°21′E. Situated in fertile and well-irrigated area. Rice milling and trading center, transportation hub, administrative and educational center; coconut (oil extraction; soap manufacturing) and food processing; light manufacturing; fruit, rice, sugarcane. Former Khmer district; colonized late-17th century by Vietnamese.

Mytholmroyd (MI-[th]uhm-roid), town (1991 population 4,019; 2006 estimated population 4,200), WEST YORKSHIRE, central ENGLAND, on CALDER RIVER, on Rochdale Canal, and 5 mi/8 km W of HALIFAX; 53°43′N 01°59′W.

Mytikas, Greece: see OLYMPOS.

Mytilene, Greece: see LESBOS.

Mytilene Channel (mit-il-EE-nee), arm of AEGEAN SEA, between LESBOS island (Turkish=*Midilli*) and TURKEY; 30 mi/48 km long, 10 mi/16 km wide. Its ports are Mytilene (on Lesbos) and Ayvlik and DIKILI (Turkey).

Mytilini (mee-tee-LEE-nee), town, on SÁMOS island, SÁMOS prefecture, NORTH AEGEAN department, GREECE, 4 mi/6.4 km WSW of SÁMOS town; 37°44′N 26°54′E. Olive oil, tobacco, wine. Formerly Mytilenoi or Mitilinoi.

Mytishchi (mi-TEE-shchee), city (2006 population 160,805), central MOSCOW oblast, central European Russia, on highway, 12 mi/19 km NNE of MOSCOW; 55°54′N 37°44′E. Elevation 508 ft/154 m. Railroad junction, including the TRANS-SIBERIAN RAILROAD; car building center (tramway and subway cars), power-generating equipment, industrial cables, bread-baking equipment, woodworking tools and equipment, synthetic fiber, haberdashery, mineral fertilizers; food industries. Has a number of technology institutes and vocational schools. Founded in 1460; major textile milling and brickmaking center by the mid-19th century. Became city in 1925. Architectural landmarks include the churches of Annunciation (built in 1677) and Vladimir (1713).

Myto (MEE-to), Czech *Mýto*, German *Mauth*, town, ZAPADOCESKY province, SW central BOHEMIA, CZECH REPUBLIC, on railroad, and 15 mi/24 km ENE of PLZEŇ; 49°48′N 13°44′E. Agriculture (wheat, barley, potatoes); livestock (cattle). Has Gothic (14th century) and baroque churches.

Myton, village (2006 population 567), DUCHESNE county NE UTAH, 18 mi/29 km E of DUCHESNE, and on DUCHESNE RIVER; 40°11′N 110°03′W. Wheat, barley, alfalfa; cattle, sheep; uintaite deposits, oil and natural gas. Elevation 5,084 ft/1,550 m. At S edge of Uintah and Ouray Indian Reservation. Originally an Indian trading post. Established 1905.

Myts'ko, UKRAINE: see RADOMYSHL'.

Myuregi, RUSSIA: see MYUREGO.

Myurego (myoo-RYE-guh), village (2005 population 4,655), E DAGESTAN REPUBLIC, SE European Russia, in the foothills of the E Greater CAUCASUS Mountains, 33 mi/53 km SSE of MAKHACHKALA; 42°24′N 47°41′E. Elevation 1,522 ft/463 m. Agriculture (hardy grain, vegetables; livestock). Also known as Myuregi (myoo-RYE-gee).

Mývatn (MEE-vah-tuhn), lake (□ 10 sq mi/26 sq km), NE ICELAND, 30 mi/48 km E of AKUREYRI; 60 mi/97 km long, 1 mi/2 km–4 mi/6 km wide; maximum depth 23 ft/7 m; 65°36′N 17°00′W. Drains N into GREENLAND SEA. Depression in volcanic basalt rock, surrounded by small craters and lava formations.

Myvatn Og Laxa, nature reserve (□ 1,699 sq mi/4,400 sq km), SUDUR-THINGEYJARSYSLA county, NE ICELAND, 35 mi/56 km SSE of HUSAVIK. Conservation area since late 1700s; attempt to build power plant averted leading to declaration as reserve, 1974. Encompasses a very active volcanic area about 60 mi/97 km long and 35 mi/56 km wide. Includes Lake MYVATN, E part of the Odadahraun lava desert, and the ASKJA volcano, which erupted in 1961. The table mt., Herdubreid, was formed by subglacial eruptions during Ice Age. DETTIFOSS waterfall on NE edge of reserve. NW extension follows Laxa river on NE edge of reserve. NW extension follows Laxa River 25 mi/40 km to Skjalfandi Bay. Huge swarms of midges (gnats) provide sustenance for population of ducks, grebes, and other fowl.

Myzakyula, ESTONIA: see MÕISAKÜLA.

Myzeqe (myoo-ze-CHAI), fertile agriculture plain of S central ALBANIA, traversed by Seman River. Produces mainly grain and cotton (irrigated). Chief towns are LUSHNJE (N) and FIER (S). Important agricultural district in Roman times. Drained and irrigated under Communist regime. Also spelled Myzeqeja.

M'zab (muh-ZAHB), region in N SAHARA, GHARDAÏA wilaya, central ALGERIA, centered around GHARDAÏA. Stony, barren valley. The M'Zab River runs mostly underground, watering the region. Inhabitants are called Mozabites and are renowned for their austerity, strict discipline, and great trading skills. Mozabites are also found in other parts of Algeria, where they run successful shops and businesses. Each M'zab community forms a pyramid, with houses rising in terraces and clustered on hills. Traditional construction uses only local stone, mud, and palm trunks. Local architecture is renowned for having inspired the French architect Le Corbusier. Mozabite towns include BÉNI ISGUEN, Bou Noura, El ATTEUF, GHARDAÏA, and MELIKA. Settled by the Muslim Kharidjite sect in the 11th century, the area was occupied by the French in 1853.

Mze River (MZHE), Czech *Mže*, German *Mies*, c.45 mi/72 km long, ZAPADOCESKY province, W BOHEMIA, CZECH REPUBLIC; rises in BOHEMIAN FOREST 13 mi/21 km SW of MARIANSKE LAZNE; flows generally W, past TACHOV and STRIBRO, joining RADBUZA RIVER at PLZEŇ to form BEROUNKA RIVER. HRACHOLUSKY DAM, just E of STRIBRO, impounds reservoir with a maximum capacity of 1,161 acres/470 ha.

Mzi, Djebel, ALGERIA: see KSOUR MOUNTAINS.

Mziha (uhm-ZEE-hah), village, TANGA region, NE central TANZANIA, 105 mi/169 km SW of TANGA, on Lukigura River; 05°53′S 37°45′E. Subsistence crops; livestock.

Mzimba (uhm-zim-bah), administrative center and district (□ 10,670 sq mi/27,742 sq km; 2007 population 614,453), Northern region, MALAWI, 145 mi/233 km N of LILONGWE; 11°55′S 33°39′E. Corn, cassava. Mica deposits.

Mzuzu (uhm-zoo-zoo), city (2007 population 150,065), ⊙ Northern region, MALAWI, 38 mi/61 km NE of MZIMBA administrative center; 11°31′S 34°00′E. Tung oil processing; airport.

Mzymta River (MZIM-tah), 51 mi/82 km long, S KRASNODAR TERRITORY, S European Russia; rises on the S slope of the W Greater CAUCASUS Mountains at 9,780 ft/2,981 m; flows W and SW, past KRASNAYA POLYANA, to the BLACK SEA at ADLER.

N

Naab River (NAHB), 103 mi/166 km long, BAVARIA, GERMANY; rises as the WALDNAAB, 2 mi/3.2 km S of BÄRNAU; flows generally S, past WEIDEN and SCHWANDORF, to the DANUBE RIVER, 3 mi/4.8 km W of REGENSBURG. Receives the Fichtelnaab, the Schwattach, and the VILS rivers (right).

Naafkopf (NAHF-kopf), alpine peak (7,833 ft/2,387 m) on AUSTRIAN-SWISS border, in the RÄTIKON, 7 mi/11.3 km SE of VADUZ; 47°04′N 09°36′E.

Naaf River (NAHF), tidal inlet of Bay of BENGAL, 30 mi/ 48 km long, on MYANMAR-BANGLADESH border; extends from Taungbro past TEKNAF and MAUNGDAW to the sea. Served by steamer route. Crossed by Muslim refugees fleeing Myanmar during the late 1980s and early 1990s. Also spelled Naf R.

Naaldwijk (NAHLT-veik), town, SOUTH HOLLAND province, W NETHERLANDS, 7 mi/11.3 km SW of The HAGUE; 51°59′N 04°13′E. NORTH SEA 3 mi/4.8 km to NW; NEW WATERWAY shipping channel 3 mi/4.8 km to SW. Cattle; flowers, nursery stock, vegetables, fruit; manufacturing (dye materials, cable).

Naalehu (NAH-AH-LAI-hoo), town (2000 population 919), S HAWAII island, HAWAII county, HAWAII, 1 mi/ 1.6 km from coast, 53 mi/85 km SSW of HILO; 19°04′N 155°34′W. Whittington (to E) and Punaluu Black Sand (to NE), beach parks; Kau Forest Reserve to NW.

Naama (nah-ah-MAH), wilaya, W ALGERIA, bordering MOROCCO; ⊙ NAAMA; 33°15′N 00°45′E. Major activities include sheep rearing, grain farming, wool manufacturing, and esparto grass production; also trade and handicrafts. Gateway to the SAHARA Desert, it has important links with Morocco. Major towns include AÏN SEFRA and Mechria. Recently created from communes formerly in SW SAÏDA wilaya.

Naama, town, ⊙ NAAMA wilaya, SW ALGERIA, 40 mi/65 km N of AÏN SEFRA, near the Moroccan border; 33°16′N 00°19′E. In 1984 it was made capital of this vast, sparsely populated wilaya, and has adopted many new administrative functions. It occupies a strategically favorable position in the SW part of the country. On the junction between towns of AÏN SEFRA and MECHERIA.

Naan (NAH-ahn), kibbutz, central ISRAEL, in coastal plain, 3 mi/4.8 km ESE of REHOVOT; 31°52′N 34°51′E. Elevation 265 ft/80 m. Irrigation technology, metalworking; mixed farming. Founded 1930. Also spelled Na'an.

Naantali (NAHN-TAH-lee), Swedish *Nådendal*, town, TURUN JA PORIN province, SW FINLAND, on inlet of GULF OF BOTHNIA, 8 mi/12.9 km W of TURKU; 60°27′N 22°02′E. Railroad and ferry terminus; major oil refinery, plywood mills, sugar refinery; seaside resort. Founded 1445, it was formerly site of convent. Has become part of the Turku conurbation.

Naarden (NAHR-duhn), town, NORTH HOLLAND province, W central NETHERLANDS, on the GOOIMEER channel (arm of Markemeer), 2 mi/3.2 km N of BUSSUM, and 12 mi/19 km ESE of AMSTERDAM; 52°18′N 05°00′E. Hollands Bridge to FLEVOLAND polder, 2 mi/ 3.2 km to NW. Dairying; cattle, poultry; nursery stock, vegetables, fruit. Manufacturing (food processing; signs). Has seventeenth-century town hall. Museum.

Naas (NAIS), Gaelic *An Nás*, town (2006 population 20,044), ⊙ KILDARE county, E central IRELAND, on branch of GRAND CANAL, and 19 mi/31 km WSW of DUBLIN; 53°13′N 06°40′W. Previously cotton milling, shoe manufacturing. It was capital of kings of LEINSTER in ancient times. Site of castle, monastic estab-

lishments, and ancient ruins. Punchestown racecourse 3 mi/4.8 km SE.

Naauport, SOUTH AFRICA: see NOUPOORT.

Naba (nah-BAH), village, INDAW township, SAGAING division, MYANMAR, 110 mi/177 km SW of MYITKYINA. At junction of Mandalay-Myitkyina railroad and branch line to KATHA.

Nababeep, town, NORTHERN CAPE province, SOUTH AFRICA, 15 mi/24 km NW of SPRINGBOK, in NAMAQUALAND; 29°34′S 17°46′E. Elevation 2657 ft/809 m. Major copper-mining and -smelting center for the region, having replaced O'KIEP.

Nabadwip (nuh-bahd-WEEP), city, NADIA district, E WEST BENGAL state, E INDIA, at confluence of BHAGIRATHI and JALANGI Rivers, 8 mi/12.9 km W of KRISHNAGAR. Metalware and pottery manufacturing, sari weaving; trades in rice, jute, linseeds, sugarcane. Noted Sanskrit schools. Pilgrimage center. Birthplace (1485) of Chaitanya, famous Vaishnava saint. Was capital of Sen kingdom following its transfer (12th century) from Gaur. Known as the "Benares (Varanasi) of Bengal." Formerly called Nadia.

Naband Bay (nah-BAHND), inlet of PERSIAN GULF in S IRAN, 155 mi/249 km SE of Bushuhr, sheltered S by Cape NABAND; 27°23′N 52°35′E. A good natural harbor; Bandar-e Assaluyeh, a new port, has been developed here.

Nabari (NAH-bah-ree), city, MIE prefecture, S HONSHU, central JAPAN, 40 mi/64 km S of TSU; 34°37′N 136°06′E. *Matsutake* mushrooms. Paint. Nearby attractions include Kochi Gorge and Akame waterfalls.

Nabaroh (na-BAH-ro), village, KAFR ESH SHEIKH province, Lower EGYPT, on BAHR SHIBIN, and 5 mi/8 km NW of TALKHA; 31°06′N 31°18′E. Cotton.

Nabataea (nuh-buh-TEE-ye), ancient kingdom of Arabia, S of EDOM, in present-day JORDAN, extending W into the NEGEV (S ISRAEL). Flourished from the 4th century B.C.E. to C.E. 106, when it was conquered by ROME. The history of Nabataea consists mainly of the struggle to control the trade routes between ASIA and the MEDITERRANEAN SEA. In the Negev, its rulers established a number of fortified cities, the ruins of which can be seen to the present day. PETRA, the capital, is noted for its unique rock-cut monuments, tombs, and temples.

Nabatiye, En (nuhb-uh-TEE-ye, en), township, ⊙ En Nabatiye district, S LEBANON, 14 mi/23 km SE of SAIDA. Tobacco, cereals, fruit. Commercial center for the region. De facto capital for the Shiites. Also known as Nabatiye Takhta (Lower Nabatiye). Nearby is a large village called Nabatiye Fauqa (Upper Nabatiye).

Nabburg (NAHB-boorg), town, UPPER PALATINATE, E BAVARIA, GERMANY, on the NAAB RIVER, 9 mi/14.5 km NNE of SCHWANDORF; 49°27′N 12°10′E. Wood- and metalworking, food canning, lumber milling. Surrounded by 15th-century walls; has Romanesque and Gothic churches. Was first mentioned 929. Chartered 1296.

Naberera (nah-bai-RAI-rah), village, MANYARA region, N central TANZANIA, 55 mi/89 km SSE of ARUSHA, in N center of MASAI STEPPE, E of Laviera Well; 04°14′S 36°55′E. Cattle, sheep, goats; corn, wheat.

Naberezhnoye (NAH-be-ryezh-nuh-ye), village, SW LIPETSK oblast, S central European Russia, on the OLYM River (right tributary of the SOSNA RIVER), on railroad, 9 mi/14 km ESE of (and administratively subordinate to) VOLOVO; 51°57′N 38°07′E. Elevation 629 ft/191 m. In agricultural area; grain storage. Formerly called Knyazhnoye (KNYAZH-nuh-ye).

Naberezhnoye, RUSSIA: see KULARY.

Naberezhnyye Chelny (NAH-bee-ryezh-ni-ye cheel-NI), city (2006 population 521,800), N central TATARSTAN Republic, E European Russia, on the KAMA River, on road and railroad, 140 mi/225 km E of KAZAN; 55°45′N 52°25′E. Elevation 433 ft/131 m. Site of the Kamaz assembly plant for heavy trucks, among the world's largest such plants. Production expanded

recently to medium trucks and passenger cars. Power is supplied by the lower Kama hydroelectric plant. Other industries include metalworking, glassworking, manufacturing (composite materials, medical equipment, fire-resistant materials, plastics, furniture, cement and mortar, bricks, textiles, garments) and food processing (dairy, confectionery). Has a number of institutes and technical academies. Once a small town named Chelny, it became a city and was renamed in 1930. From 1982 to 1988, called Brezhnev.

Nabesna (nah-BES-nuh), village, E ALASKA, near CANADIAN (YUKON) border, on NABESNA RIVER, and 70 mi/113 km S of TANACROSS, on N slope of WRANGELL MOUNTAINS. Airfield. Another Nabesna is c.50 mi/80 km downstream on Nabesna River opposite NORTHWAY.

Nabesna Glacier (nah-BES-nuh), E ALASKA, extends N from WRANGELL MOUNTAINS, 20 mi/32 km long, 3 mi/ 4.8 km wide; near 62°05′N 142°55′W. Flows into NABESNA RIVER.

Nabesna River (nah-BES-nuh), 70 mi/113 km long, E ALASKA; rises in NABESNA GLACIER, in the WRANGELL MOUNTAINS, near 62°14′N 142°55′W; flows NNE, past NABESNA, to TANANA River at 63°02′N 141°53′W.

Nabesna Village (nah-BES-nuh), Native American village, E ALASKA, near CANADIAN (YUKON) border, on NABESNA RIVER, and 50 mi/80 km SE of TANACROSS, near ALASKA HIGHWAY.

Nabeul, TUNISIA: see NABUL.

Nabha (NAH-bah), town, PATIALA district, PUNJAB state, N INDIA, 14 mi/23 km WNW of PATIALA; 30°22′N 76°09′E. Trades in millet, wheat, gram, cloth fabrics, sugar; cotton ginning, wool carding, handloom weaving, embroidering. Has college. Was capital of former princely state of Nabha of PUNJAB STATES; state formed 1763 by Sikhs during breakup of MOGUL EMPIRE; in 1948 merged with PATIALA AND EAST PUNJAB STATES UNION, later with Punjab state.

Nabiac (NA-bee-ak), town, NEW SOUTH WALES, SE AUSTRALIA, 179 mi/288 km N of SYDNEY; 32°06′S 152°23′E.

Nabiagali (nah-bee-AHG-ah-lee), village, W AZERBAIJAN, 6 mi/10 km N of GYANDZHA. In wine-growing and cotton district; grape processing.

Nabire, town, ⊙ Paniai district, IRIAN JAYA province, on NEW GUINEA island, INDONESIA, S side of CENDERAWASIH BAY, 252 mi/406 km SW of Manokwari. Airport.

Nabisar Road (nuh-BEE-suhr), village, THAR PARKAR district, S SIND province, SE PAKISTAN, on railroad, and 26 mi/42 km SSW of UMARKOT; 25°02′N 69°31′E.

Nabk, En, SYRIA: see NEBK, EN.

Nablus (NAH-blus), Heb. *Shechem* (she-KHEM), city, in the Samarian highlands, WEST BANK; 32°13′N 35°15′E. Chief city of SAMARIA, situated in a narrow valley between Mounts EBAL (N) and GERIZIM (S). The modern city has spread well up the slopes of both mountains. It is the market center for a region where wheat and olives are grown and sheep and goats are grazed. Manufacturing includes soap made from olive oil and colorful shepherds' coats. Pop. mostly Palestinian Arab Muslim. Linked by highway with JERUSALEM. Nablus, an ancient Canaanite town, has remains dating from circa 2000 B.C.E., about the time when the city was held by EGYPT. The Samaritans made it their capital and built a temple on Mount GERIZIM to rival that of Jerusalem. Nablus still has a small community of Samaritans. Destroyed (129 B.C.E.) by John Hyrcanus I. Under Hadrian it was rebuilt on an adjacent site and named Neapolis, from which the present name derives. Nearby are the reputed sites of the tomb of Joseph and the well of Jacob. Much of the city was destroyed by an earthquake in 1927. It suffered a similar fate several times in its history. The city came under Israeli occupation following the Arab-Israeli War of 1967. The city's refugee camps have exacerbated tensions between residents and Israeli troops. Nablus has long been a center of Arab nationalism. During the Intifada beginning 1987, it was the scene of ongoing violent clashes between Arabs and Jews. Handed over to the Palestinian Authority in 1995 as part of the Israel-Palestinian agreement. Al-Fariah UN Relief Workers Administration (UNRWA) Palestinian refugee camp (1995 population 4,400) is located 1.9 mi/3 km to the S; NABLUS (1995 population 5,600), Balata (1995 population 16,600), and Askar UNRWA Palestinian refugee camps are on the outskirts of the city.

Nabón (nah-BON), village, AZUAY province, S ECUADOR, on PAN-AMERICAN HIGHWAY, and 32 mi/51 km S of CUENCA; 03°20′S 79°04′W. Elev. 8,865 ft/2,702 m. Cereals, potatoes; livestock.

Naboomspruit, town, SOUTH AFRICA: see MOOKGOPHONG.

Nabotas, PHILIPPINES: see NAVOTAS.

Nabowla, village, TASMANIA, SE AUSTRALIA, 9 mi/15 km W of SCOTTSDALE; 41°10′S 147°22′E. Bridestowe Lavender Farm, established 1921; produces lavender flowers, oil.

Nabua (nan-BOO-wah), town, CAMARINES SUR province, SE LUZON, PHILIPPINES, 19 mi/31 km SE of NAGA; 13°23′N 123°20′E. Agricultural center (rice, corn, abaca); gypsum.

Nabul (nah-BUHL), province (□ 1,076 sq mi/2,797.6 sq km; 2006 population 714,300), NE TUNISIA, on MEDITERRANEAN SEA (E); ⊙ NABUL; 36°40′N 10°40′E. Also Nabeul and Nābul.

Nabul (nah-BUHL), ancient *Neapolis*, town (2004 population 56,387), ⊙ NABUL province, NE TUNISIA, on GULF OF HAMMAMAT, and 6.2 mi/10 km N of AL HAMMAMAT; 36°27′N 10°44′E. Largest town on CAPE BON peninsula. Center of well-known Andalusian handicraft pottery industry (established in the 16th century); perfume factories. Distilling, macaroni manufacturing, stone cutting; tourism. Chalk quarries nearby. Extensive citrus plantations and perfume-flower gardens in area. Phoenician Neapolis was destroyed by Romans in 146 B.C.E. and then rebuilt as a Roman colony; ruins of the settlement are at the edge of town. Also spelled Nabeul.

Nabūr (nuh-BUHR), ancient *Castellum*, village (2004 population 3,435), AL KAF province, NW TUNISIA, 7 mi/11.3 km N of Al Kaf; 36°17′N 08°46′E. Administrative center. Railroad terminus. Important iron mines; zinc and lead deposits.

Nabuyonge Island (nah-boo-YON-gai), KAGERA region, NW TANZANIA, in Lake VICTORIA, 53 mi/85 km E of BUKOBA; 2 mi/3.2 km long, 1 mi/1.6 km wide; 01°29′S 32°45′E.

Nacabí, PERU: see MACABÍ ISLAND.

Nacajuca (nah-kah-HOO-kah), city and township, TABASCO, SE MEXICO, on arm of GRIJALVA River, and 13 mi/21 km NNW of VILLAHERMOSA; 18°12′N 93°01′W. In an important oil-producing region. Corn, rice, beans, tobacco; fruit; livestock.

Nacala (nah-KAH-luh), village, NAMPULA province, NE MOZAMBIQUE, on MOZAMBIQUE CHANNEL of INDIAN OCEAN, 35 mi/56 km N of Mozambique city; 15°42′S 38°17′E. Railroad-spur terminus; port; ships sisal, copra.

Nacaome (nah-kah-O-mai), city (2001 population 13,931), ⊙ VALLE department, S HONDURAS, on INTER-AMERICAN HIGHWAY, 45 mi/72 km SSW of TEGUCIGALPA; 13°32′N 87°29′W. Commercial center; beverages, bricks; tanning. Has colonial church (rebuilt 1867). Dates from 16th century; became city in 1845.

Nacaome River (nah-kah-O-mai), c.50 mi/80 km long, S HONDURAS; rises in SIERRA DE LEPATERIQUE; flows S, past REITOCA, PESPIRE, and NACAOME, to GULF OF FONSECA; 13°32′N 87°26′W. Called Río Grande de Reitoca in upper course, Pespire River in middle course. Navigable below Nacaome in rainy summer season.

Nacascolo, NICARAGUA: see PUERTO MORAZÁN.

Nacebe (nah-SAI-bai), town and canton, ABUNÁ province, PANDO department, N BOLIVIA, on ORTON River, and 25 mi/40 km S of SANTA ROSA; 10°57′S 67°25′W.

Nacfa, Eritrea: see NAKFA.

Nachalovo (nah-CHAH-luh-vuh), village (2005 population 4,800), SE ASTRAKHAN oblast, S European Russia, in the VOLGA RIVER delta, on road, 8 mi/13 km E of ASTRAKHAN; 46°20′N 48°11′E. Below sea level. Technical control office for the GROZNYY oil pipeline.

Na Cham, VIETNAM: see VAN LANG.

Naches (NA-cheez), village (2006 population 691), YAKIMA county, S WASHINGTON, 12 mi/19 km NW of YAKIMA, and on NACHES RIVER; 46°44′N 120°42′W. Railroad terminus. Fruit; dairying; manufacturing (wood products, lumber); logging. Snoqualmie National Forest to W.

Naches Pass (NA-cheez) (4,923 ft/1,501 m), on border between PIERCE and YAKIMA counties, central WASHINGTON, gateway through CASCADE RANGE used by pioneers, c.50 mi/80 km E of TACOMA, NE of MOUNT RAINIER NATIONAL PARK.

Naches River (NA-cheez), c.75 mi/121 km long, central WASHINGTON; rises in CASCADE RANGE near NACHES PASS as Little Naches River; flows c.20 mi/32 km SE, joins American River (from SW) to form Naches River, 20 mi/32 km SW of CLE ELUM, flows another 55 mi/89 km SE past NACHES to YAKIMA RIVER between YAKIMA and SELAH.

Nachikatsuura (NAH-chee-kah-TSOO-rah), town, East Muro county, WAKAYAMA prefecture, S HONSHU, W central JAPAN, on SE KII PENINSULA, port on KUMANO SEA, 61 mi/99 km S of WAKAYAMA; 33°37′N 135°56′E. Nachi-no waterfall (430 ft/131 m high) on small Nachi River is highest in Japan. Site of 6th-century Buddhist temple. Hot springs. Sometimes called Kii-katsuura.

Nachingwea (nah-cheen-GWAI-ah), village, LINDI region, SE TANZANIA, 70 mi/113 km WSW of LINDI; 10°23′S 38°45′E. Former terminus of railroad from ground-nut growing scheme at MKWAYA. Airport. Peanuts, corn, cashews; sheep, goats.

Náchod (NAH-khot), city (2001 population 21,400), VYCHODOCESKY province, NE BOHEMIA, CZECH REPUBLIC, in NNW foothills of the Eagle Mountains, on railroad, near Polish border; 50°25′N 16°10′E. Manufacturing (textiles, tires); brewery (established in 1872). Has an historic castle with paintings and Gobelin tapestries, and a 14th-century cathedral. Prussians defeated Austrians here in 1866. Health resort of BELOVES, Czech *Béloves* (BYE-lo-VES), with peat baths and acidulous springs is just ENE.

Nachrodt-Wiblingwerde (NAHKH-rot–VIB-LING-ver-de), village, WESTPHALIA-LIPPE, North Rhine-Westphalia, W GERMANY, on the LENNE RIVER, and 4 mi/6.4 km SW of ISERLOHN; 51°20′N 07°38′E. Steel and metalworking.

Nachtigal, Cape (NAH-tee-gal), on Gulf of Guinea coast of CAMEROON, South-West province, 6 mi/9.7 km SE of LIMBE; 03°57′N 09°15′E.

Nachvak Fiord (NACH-vak), inlet of the ATLANTIC OCEAN, NE LABRADOR-UNGAVA, E CANADA, at foot of CIRQUE MOUNTAIN; 30 mi/48 km long, 3 mi/5 km wide; 59°03′N 63°40′W.

Nacimiento (nah-see-mee-EN-to), town, ⊙ Nacimiento comuna (2002 population 20,884), BÍO-BÍO province, BÍO-BÍO region, S central CHILE, in S part of the central valley, on BÍO-BÍO RIVER where it meets the VERGARA RIVER, 18 mi/29 km WSW of LOS ÁNGELES; 37°30′S 72°40′W. Agricultural center (cereals, vegetables, grapes); lumbering, wood processing. Founded as a fort in 1603.

Nacimiento (nah-thee-MYEN-to), village, ALMERÍA province, S SPAIN, 21 mi/34 km NNW of ALMERÍA; 37°06′N 02°38′W. Olive-oil processing. Ships grapes.

Nacimiento, Cerro del (nah-see-mee-AIN-to, SER-ro del), Andean peak (21,300 ft/6,492 m), W CATAMARCA

province, ARGENTINA, near CHILE border, 23 mi/37 km SW of Cerro INCAHUASI; 27°16′S 68°32′W.

Nacimiento Mountains (nah-see-MYEN-to), range in SANDOVAL and RIO ARRIBA counties, NW NEW MEXICO, just E of RIO PUERCO; extend c.30 mi/48 km S from CUBA. Prominent points include Pajarito Peak (9,040 ft/2,755 m), San Miguel Mountain (9,473 ft/2,887 m), Nacimiento Peak (9,801 ft/2,987 m). Range is partly in Santa Fe National Forest.

Nacimiento Reservoir (□ 8 sq mi/20.7 sq km), SAN LUIS OBISPO county, SW CALIFORNIA, on Nacimiento River, in SANTA LUCIA RANGE, 12 mi/19 km NW of PASO ROBLES; 35°44′N 120°53′W. Maximum capacity 470,000 acre-ft. Formed by Nacimiento Dam (215 ft/66 m high), built (1957) for power generation; also used for irrigation, water supply, recreation, and flood control. Hunter Liggett Military Reservation nearby.

Naciria (nah-see-RYAH), village, BOUMERDES wilaya, N central ALGERIA, in Great KABYLIA, on railroad, and 12 mi/19 km W of TIZI OUZOU. Tobacco; vineyards. Formerly Haussonvillers.

Nacka (NAHK-kah), town, STOCKHOLM county SWEDEN, on BALTIC SEA, suburb of STOCKHOLM (5 mi/8 km E); 59°19′N 18°7′E. Manufacturing (compressors, steam turbines, chemicals, processed food). Radio and television transmitters.

Nackara, village, E SOUTH AUSTRALIA state, S central AUSTRALIA, 75 mi/121 km ENE of PORT PIRIE; 32°48′S 139°14′E. On Port Pirie–BROKEN HILL railroad. Wool, some wheat.

Nackenheim (NAHK-ken-heim), village, RHENISH HESSE, RHINELAND-PALATINATE, W GERMANY, on left bank of the RHINE RIVER, and 7 mi/11.3 km SE of MAINZ; 49°55′N 08°21′E. Wine.

Nacmine (NAK-mein), unincorporated village, S ALBERTA, W Canada, on RED DEER River, included in DRUMHELLER; 51°28′N 112°47′W. Coal mining. Cattle; wheat.

Naco (NAH-ko), town, SONORA, NW MEXICO, in NW spurs of SIERRA MADRE OCCIDENTAL, on U.S. border adjoining NACO (ARIZONA), on railroad, and 60 mi/97 km E of NOGALES; 31°20′N 109°53′W. In rich copper belt; cattle.

Naco (nah-ko), unincorporated village, COCHISE county, SE ARIZONA, on Mexican border, opposite and adjacent to NACO (SONORA, MEXICO), 24 mi/39 km W of DOUGLAS. Elevation 4,680 ft/1,426 m. Agriculture (cattle sheep). U.S. customs and Mexican consulate are here. MULE MOUNTAINS are N.

Nacogdoches (NAH-kah-do-ches), county (□ 981 sq mi/2,550.6 sq km; 2006 population 61,079), E TEXAS, ⊙ NACOGDOCHES; 31°36′N 94°36′W. Rolling, wooded area; chief industries are timber; dairying, poultry raising; vegetables. Bounded W and S by ANGELINA RIVER, E by ATTOYAC BAYOU. Oil found here 1866; natural gas; clay, some manufacturing Part of Angelina National Forest in SE; Lake Nacogdoches in W. Formed 1836.

Nacogdoches (NAH-kah-do-ches), city (2006 population 31,135), ⊙ NACOGDOCHES county, E TEXAS, 60 mi/97 km S of LONGVIEW; 31°36′N 94°39′W. In a pine and hardwood forest area. Highly industrialized city including lumbering, livestock and poultry raising and processing, manufacturing (feed, wood products, motor homes, electronic products, furniture). Tourism is also important; to the SE is the large SAM RAYBURN RESERVOIR and Angelina National Forest; Lake Nacogdoches is to W. Area explored by La Salle (1687), and a Spanish mission was founded near in 1716. The state's first oil wells were drilled nearby in 1859. Seat of Stephen F. Austin State University, Spanish presidio built in 1779 on campus.

Nácori Chico (NAH-ko-ree CHEE-ko), town, SONORA, NW MEXICO, on affluent of YAQUI river, in W outliers of SIERRA MADRE OCCIDENTAL, and 135 mi/217 km ENE of HERMOSILLO; 29°38′N 108°49′W. Elevation 3,550 ft/1,082 m. Livestock raising and wheat growing.

Area is shown by the symbol □, and capital city or county seat by ⊙.

Nacozari de García (nah-ko-SAH-ree dai gahr-SEE-ah), town, ⊙ Nacozari de García muncipio, SONORA, NW MEXICO, in broad valley of W outliers of SIERRA MADRE OCCIDENTAL, on railroad, and 70 mi/113 km S of DOUGLAS (ARIZONA); 30°22′N 109°41′W. Elevation 3,412 ft/1,040 m. Cattle-raising and copper-mining center; silver, gold, lead, zinc mines.

Nacunday (nah-koon-DEI), town, Alto Paraná department, SE PARAGUAY, on upper Paraná River, and 45 mi/72 km S of HERNANDARIAS; 26°01′S 54°46′W. Maté; lumber.

Nada (NAH-DAH), town, NW HAINAN province, SE CHINA, 20 mi/32 km SE of DAN XIAN. Tropical crops, oilseeds; fisheries; animal husbandry.

Nada, ETHIOPIA: see OMO NADA.

Nadachi (NAH-dah-chee), town, West Kubiki county, NIIGATA prefecture, in central HONSHU, N central JAPAN, 71 mi/115 km S of NIIGATA; 37°09′N 138°05′E. Moxa.

Nadadores (nah-dah-DO-RES), town, COAHUILA, N MEXICO, 15 mi/24 km NW of MONCLOVA, on railroad, on Mexico Highway 30; 27°02′N 101°38′W. Cereals, wine, livestock.

Nadasaki (nah-DAH-sah-kee), town, Kojima county, OKAYAMA prefecture, SW HONSHU, W JAPAN, 9 mi/15 km S of OKAYAMA; 34°32′N 133°52′E. Eggplants, lotus root.

Nádasfö, SLOVAKIA: see ROHOZNIK.

Nadbai (NAHD-bei), town, tahsil headquarters, BHARATPUR district, E RAJASTHAN state, NW INDIA, 17 mi/27 km W of BHARATPUR; 27°14′N 77°12′E. Local market for millet, gram, oilseeds.

Naddnepryanskoye, UKRAINE: see NADDNIPRYANS'KE.

Naddnipryans'ke (nahd-dnee-PRYAHN-ske) [Ukrainian=overlooking the Dnieper] (Russian *Naddnepryanskoye*), town, W KHERSON oblast, UKRAINE, on the right bank of the INHULETS' River, near its confluence with the DNIEPER River, on highway and near railroad, 6 mi/10 km NE of and subordinated to KHERSON; 46°44′N 32°42′E. Elevation 147 ft/44 m. Research institute for irrigational agriculture, with test plots and park with many kinds of trees and shrubs. Established in 1966; town since 1979.

Nadelhorn, SWITZERLAND: see MISCHABELHÖRNER.

Nådendal, FINLAND: see NAANTALI.

Naden Harbour (NAI-duhn), inlet of Virago Sound, DIXON ENTRANCE, in N GRAHAM ISLAND, W BRITISH COLUMBIA, W Canada, 17 mi/27 km W of MASSET; 8 mi/13 km long, 1 mi/2 km–3 mi/5 km wide; 54°00′N 132°35′W. Salmon fishing.

Nadezhda (nah-DEZH-dah), former village, now housing quarter in SOFIA, SOFIA oblast, W BULGARIA; 42°43′N 23°16′E.

Nadezhda Strait (nah-DYEZH-dah), Japanese *Rashowa-kaikyo*, in central main KURIL ISLANDS group, SAKHALIN oblast, extreme E RUSSIAN FAR EAST, between MATUA (N) and RASSHUA (S) islands; 17 mi/27 km wide; 47°56′N 153°13′E.

Nadezhinsk, RUSSIA: see SEROV.

Nadezhinskiy Zavod, RUSSIA: see SEROV.

Nadi (NAHN-dee), town, W Viti Levu, FIJI, SW PACIFIC OCEAN, 70 mi/113 km WNW of SUVA. The third largest town of Fiji, it is a focus for tourism. Sugarcane. International airport. Pronounced and often spelled as Nandi.

Nadia (NUHD-yah), district (□ 1,516 sq mi/3,941.6 sq km), E WEST BENGAL state, E INDIA, on BANGLADESH (E) border; ⊙ KRISHNAGAR. In GANGA DELTA; bounded W by BHAGIRATHI RIVER; drained by the JALANGI RIVER. Alluvial plain; bamboo, moringa, and areca palm groves; rice, jute, linseeds, sugarcane, wheat, tobacco, chili, turmeric. Dispersed swamps, largely responsible for district's high malarial mortality. Sugarcane processing is important industry (main center, Krishnagar); hand-loom cotton weaving (SANTIPUR); clay-figure manufacturing (Krishnagar). Cultural and educational center. Noted Sanskrit schools at NABADWIP. Part of Sen kingdom, defeated

(13th century) by a Delhi sultan. Robert Clive's decisive victory over nawab of Bengal (1757) at PLASSEY resulted in English acquisition of BENGAL. Original district reduced 1947, when E area was separated to form new district of KUSHTIA, EAST BENGAL (now in Bangladesh), following creation of PAKISTAN.

Nadia, INDIA: see NABADWIP.

Nadiad (NUHD-ee-ahd), city, KHEDA district, Gujarat state, W central INDIA; 22°42′N 72°52′E. Center of modern agricultural production and an important road and railroad hub.

Nădlac (NUHD-lahk), Hungarian *Nagylak*, town, ARAD county, W ROMANIA, on MUREŞ RIVER, and 27 mi/43 km W of ARAD; 46°10′N 20°45′E. Border-crossing point with HUNGARY. Agriculture center; hemp processing; manufacturing furniture, food processing.

Nador (NAH-door), city (1994 population112,450), ⊙ Nador province, Oriental administrative region, E MOROCCO, port on a lagoon of the MEDITERRANEAN SEA, on mining railroad, and 8 mi/12.9 km S of MELILLA; 35°10′N 02°55′W. Cement, sugar refinery. Trade in livestock, barley, olives, fruits. Taouima airfield just S. Beni Bouyafrour and Ouiksane iron mines just S. Formerly called Villa Nador.

Nădrag (nuh-DRAHG), Hungarian *Nadrág*, village, TIMIŞ county, W ROMANIA, 14 mi/23 km ESE of LUGOJ; 45°39′N 22°11′E. Iron foundries and rolling mills, producing sheet iron, steel tubes, saws, and nickel-ware.

Nadrechye, RUSSIA: see CHIRI-YURT.

Nádszeg, SLOVAKIA: see TRSTICE.

Nadterechnaya, RUSSIA: see NADTERECHNOYE.

Nadterechnoye (naht-TYE-reech-nuh-ye) or **Nadterechnaya** (naht-TYE-reech-nah-yah), settlement (2005 population 7,885), NW CHECHEN REPUBLIC, S European Russia, on the right bank of the TEREK River, opposite Naurskaya (2 mi/3.2 km to the N), on road, 28 mi/45 km NNW of GROZNYY; 43°37′N 45°20′E. Elevation 265 ft/80 m. Until 1944, called Nizhniy Naur.

Nădudvar (NAHD-ud-vahr), city, HAJDU-BIHAR county, E Hungary, on Kösely River, and 23 mi/37 km WSW of DEBRECEN; 47°25′N 21°10′E. Wheat, corn, sugar beets; hogs; cattle. Agricultural experiment station.

Nadur (nah-DOOR), town, SE GOZO, MALTESE ISLANDS, 3 mi/4.8 km E of VICTORIA; 36°02′N 14°17′E.

Năduşita (nuh-doo-SHEE-tah), village, N MOLDOVA, 19 mi/31 km SW of SOROCA. Wheat, sunflowers. Served by Drochia railroad station, 4 mi/6.4 km W. Formerly spelled Nadushita.

Naduvattam (nah-duh-VUHT-tuhm), town, NILGIRI district, TAMIL NADU state, S INDIA, 12 mi/19 km WNW of Ooty, or UDAGAMANDALAM; 11°29′N 76°34′E. In tea- and cinchona-estate area; manufacturing of quinine. Also spelled Naduvatam.

Nadvirna (nahd-VIR-nah) (Russian *Nadvornaya*) (Polish *Nadworna*), city, central IVANO-FRANKIVS'K oblast, UKRAINE, on the Bystrytsya Nadvirnyans'ka River, and 22 mi/35 km SSW of IVANO-FRANKIVS'K; 48°38′N 24°34′E. Elevation 1,404 ft/427 m. Raion center. In petroleum district; manufacturing center; petroleum refining, lumbering, building materials manufacturing, flour milling; dairy. Has ruins of an old castle. Known since 1595; passed from Poland to Austria (1772); site of Russo-German battles in 1915; part of West Ukrainian National Republic (1918–1919); reverted to Poland (1919); incorporated into Ukrainian SSR in 1939 and granted city status. Jewish population since the 17th century; received full citizenship rights under the Austro-Hungarian rule in 1867; accounted for two-thirds of the total population in 1880, many of them engaged in farming. Emigration in the 1890–1900s and pogroms of 1905 and 1918–1921 reduced the community to 2,040 members in 1939; eliminated by the Nazis during World War II—fewer than 100 Jews remaining in 2005.

Nadvoitsy (nah-DVO-ee-tsi), town (2005 population 10,685), E central Republic of KARELIA, NW European Russia, port on the WHITE SEA-BALTIC CANAL, at the N end of VYGOZERO (lake), on railroad, 45 mi/72 km SSW of BELOMORSK; 63°52′N 34°19′E. Elevation 236 ft/71 m. Aluminum works (built after World War II) based on alumina from LENINGRAD oblast.

Nadvornaya, UKRAINE: see NADVIRNA.

Nadworna, UKRAINE: see NADVIRNA.

Nadym (nah-DIM), city (2006 population 46,420), central YAMALO-NENETS AUTONOMOUS OKRUG, TYUMEN oblast, NW SIBERIA, RUSSIA, port on the lower NADYM RIVER (flows into the OB' BAY of the KARA SEA), on road junction and railroad spur, 760 mi/1,223 km N of TYUMEN, and 350 mi/563 km E of SALEKHARD; 65°32′N 72°31′E. On a system of natural gas pipelines to Urals and central European Russia. Port facilities; manufacturing (pipes, building panels). Made city in 1972.

Nadym River (nah-DIM), 155 mi/249 km long, YAMALO-NENETS AUTONOMOUS OKRUG, TYUMEN oblast, NW Siberian Russia; rises in Num-to (lake) at 63°45′N 72°00′E; flows N to the Ob' Bay, forming the Nadym Bay (approximately 90 mi/145 km long E-W) W of NYDA.

Naejang Mountain National Park, Korean *Naejangsan Kungnip Kongwon,* located between CHANGSONG-GUN, SOUTH CHOLLA province and Chungup city, NORTH CHOLLA province, SW SOUTH KOREA, 22 mi/35 km N of KWANGJU. Best known for its autumn colors. Steep rocky peaks (Shinson-bong at 2,503 ft/763 m), strange rock formations, Kumson and Todok waterfalls, Kumson and Pulchul valleys, and virgin forest of maple and nutmeg trees. Has two Buddhist temples (Naejang-sa in the center and Paegyang-sa in Changsong-gun). The nutmeg forest around Paegyang-sa Temple is protected as a natural monument (No. 153). Lodges, campgrounds, picnic area, hiking trails, observatory, cable car. Established 1971.

Naenwa (NEIN-wah), town, tahsil headquarters, BUNDI district, E RAJASTHAN state, NW INDIA, 26 mi/42 km NNE of BUNDI. Millet, oilseeds. Also spelled Nainwa.

Nærøyfjorden (NAR-uh-oo-FYAWR-uhn), inlet, S arm of AURLANDSFJORDEN, SOGN OG FJORDANE county, W NORWAY; c.12 mi/19 km long, less than 1 mi/1.6 km wide. Entrance cuts through 5,600-ft/1,707-m-high mountains. On its W shore lies village of Bakka, 24 mi/39 km SSW of SOGNDAL; at head of fjord, 2 mi/3.2 km SSW, lies GUDVANGEN village, over which 6,000-ft/1,829-m-high mountains tower vertically. Nearby is scenic Kilofoss, a waterfall (500 ft/152 m).

Nærum (NAI-room), town, Copenhagen county, SJÆLLAND, DENMARK, 9 mi/14.5 km N of COPENHAGEN; 55°49′N 12°33′E.

Næsby (NES-buh), town, FYN county, DENMARK, on FYN island, suburb, 2 mi/3.2 km NNW of ODENSE. Agriculture (fruit, sugar beets; hogs).

Næstved (NEST-vedth), city (2000 population 39,408), STORSTRØM county, SE DENMARK; 55°15′N 11°40′E. Seaport, linked (since 1938) with the Karrebæk Fjord (an arm of the STORE BÆLT) by a 5-mi/8.1-km canal. Also an industrial center and a railroad junction.

Nafada (nah-FAH-dah), town, BAUCHI state, NE NIGERIA, on GONGOLA River, and 50 mi/80 km NNE of GOMBE. Cassava, millet, durra.

Nafadié (na-fa-JEE-ai), village, FIRST REGION/KAYES, SW MALI, on DAKAR (SENEGAL)-NIGER railroad, and 55 mi/89 km WNW of BAMAKO. Peanuts, shea nuts; livestock.

Näfels (NAI-fuhls), commune, GLARUS canton, E central SWITZERLAND; 47°06′N 09°04′E. Elevation 1,444 ft/440 m. Has a magnificent baroque palace. Site of one of the many battles in which the infantry of Swiss Confederates defeated the Hapsburg cavalry (1388).

Cross-references are shown in SMALL CAPITALS. The pronunciation guide is shown on page xix. The sources of population figures are shown on page xvii.

Nafplion, Greece: see NÁVPLION.

Naf River, S ASIA: see NAAF RIVER.

Nafta (nef-TAH), town and oasis (2004 population 20,308), TAWZAR province, SW TUNISIA, in the BILAD AL JARID, at NW edge of SHATT AL JARID, 14 mi/23 km SW of TAWZAR; 33°53'N 07°53'E. Administrative center; date growing (over 400,000 date palm trees, 70,000 of which produce top-quality dates). Artisan industries; wool and camel hair trade. Trans-Saharan caravan terminus. The second most important Islamic center in TUNISIA; Sufisim introduced to country here. There are twenty-four mosques, tombs of 100 Marabuts; pilgrimage center.

Naftalan (nuf-tah-LAHN), town, central AZERBAIJAN, 27 mi/43 km ESE of GYANDZHA; 40°30'N 46°49'E. In oil-bearing district. Health resort with petroleum baths.

Naftali, Hills of (NUHF-tuh-lee), small narrow range of hills, the easternmost part of the Upper Galilee Highlands, extending c.15 mi/24 km N-S along border between NE ISRAEL and SW LEBANON, W of headwaters of the JORDAN RIVER; rise to c.2,900 ft/884 m. The N part of this range is called the Ramin Mountains, a name that in Israel is often applied to the entire range.

Naft-e-Shāh (nahft–ei–SHAH), oil town, Kermānshahān province, W IRAN, 95 mi/153 km WSW of KERMANSHAH, on IRAQ border. Oil field, adjoining Iraqi. Naft Khaneh field (W), is linked by pipeline with KERMANSHAH refinery. Also called Naftshahr.

Naftia, Lake, ITALY: see PALAGONIA.

Naft Khaneh (NUFT KAH-ne), oil-mining town, DIYALA province, E IRAQ, at IRAN border opposite NAFT-E-SHAH (Iran), 75 mi/121 km NE of BAGHDAD; 34°02'N 45°28'E. Connected by pipeline to refinery in Baghdad.

Naft Safid (nahft SAH-feed), oil town, Khuzestān province, SW IRAN, 45 mi/72 km NE of AHVAZ, and 105 mi/169 km NE of ABADAN (linked by pipeline); 31°37'N 49°17'E. Oil field opened 1945.

Nafud (NUH-fid), desert area in the N part of the ARABIAN PENNINSULA, SAUDI ARABIA, occupying a great oval depression; 180 mi/290 km long and 140 mi/225 km wide. This area of red sand is surrounded by sandstone outcrops that have eroded into peculiar shapes. The Nafud is noted for its sudden violent winds and dust storms, which have formed lines of crescent-shaped dunes, rising up to 600 ft/183 m. Rainfall occurs on few days in winter, the average annual quantity is 2 in/5.1 cm–4 in/10.2 cm. In some lowland basins, especially those near the HEJAZ Mountains (to the W), there are oases where dates, vegetables, barley, and fruits are raised. The Nafud is connected to the RUB' AL KHALI, the great desert of S Arabia, by the DAHNA, a corridor of gravel plains and sand dunes, 800 mi/1,287 km long and 15 mi/24 km–50 mi/80 km wide. Generally known as the Great Nafud (or Nefud); also known as Northern Nafud or Nufud and An Nafud.

Nafzawiyah (nef-zaw-WEE-yah), group of oases, QABILI province, S TUNISIA, near E edge of the SHATT AL JARID. Chief center is QABILI. Date palms.

Naga, city (□ 30 sq mi/78 sq km; 2000 population 137,810), ⊙ CAMARINES SUR province, SE LUZON, PHILIPPINES, on BICOL RIVER, on railroad, and 50 mi/80 km NW of LEGASPI; 13°37'N 123°11'E. Trade center for agricultural area (rice, abaca, corn). Formerly Nueva Caceres.

Naga (NAH-gah), town, Naga county, WAKAYAMA prefecture, S HONSHU, W central JAPAN, 17 mi/27 km E of WAKAYAMA; 34°16'N 135°26'E.

Naga, town, central Cebu island, CEBU province, PHILIPPINES, on BOHOL STRAIT, on railroad, and 12 mi/19 km SW of CEBU; 10°15'N 123°43'E. Agricultural center (corn, coconuts). Coal and limestone mined for ce-

ment factory. Multiuse economic zone (residential, commercial, industrial) under development.

Naga, town, ZAMBOANGA DEL SUR province, W MINDANAO, PHILIPPINES, small port on small Maligay Bay, 40 mi/64 km SW of PAGADIAN; 07°50'N 122°45'E. Sawmill; rice, corn, coconuts.

Nagaga (nah-GAH-gah), village, MTWARA region, SE TANZANIA, 80 mi/129 km SW of MTWARA, on MAKONDE PLATEAU; 10°56'S 39°08'E. Timber; cashews, manioc, sweet potatoes; sheep, goats.

Nagahama (nah-GAH-hah-mah), city, SHIGA prefecture, S HONSHU, central JAPAN, on NE shore of LAKE BIWA, 37 mi/60 km N of OTSU; 35°22'N 136°16'E.

Nagahama (nah-GAH-hah-mah), town, Kita county, EHIME prefecture, W SHIKOKU, W JAPAN, port on IYO SEA, 22 mi/35 km S of MATSUYAMA; 33°36'N 132°28'E. Mandarin oranges; fish. Japan wax, traditional footwear. Sometimes called Iyo-nagahama.

Nagahama, RUSSIA: see OZËRSKIY.

Nagai (nah-GAH-ee), city, YAMAGATA prefecture, N HONSHU, NE JAPAN, on MOGAMI RIVER, and 19 mi/30 km S of YAMAGATA; 38°06'N 140°02'E. Furs. Pongee, housewares and appliances, pottery. Ayame Park known for its irises.

Nagai Island (nuh-GEI), SHUMAGIN ISLANDS, SW ALASKA, off SW ALASKA PENINSULA; 32 mi/51 km long, 1 mi/1.6 km–11 mi/18 km wide; 55°06'N 160°00'W.

Nagaizumi (nah-GAH-ee-zoo-mee), town, Sunto county, SHIZUOKA prefecture, central HONSHU, E central JAPAN, 31 mi/50 km N of SHIZUOKA; 35°08'N 138°54'E.

Nagakute (nah-GAHK-te), town, Aichi county, AICHI prefecture, S central HONSHU, central JAPAN, 9 mi/15 km E of NAGOYA; 35°10'N 137°03'E.

Nagalama (nah-gah-LAH-muh), town, MUKONO district, CENTRAL region, S UGANDA, 19 mi/31 km NNE of KAMPALA; 00°30'N 32°45'E. Cotton, coffee, sugarcane; cattle, sheep, goats. Was part of former Buganda province.

Nagaland (nah-GUH-land), state (□ 6,401 sq mi/16,642.6 sq km; 2001 population 1,990,036), NE INDIA, on MYANMAR (E) border; ⊙ KOHIMA. Formerly called the Naga Hills–Tuensang area in ASSAM state, it gained full state status in 1961. It is a wild, forested, and undeveloped region bounded S by MANIPUR, NW by Assam, and N by ARUNACHAL PRADESH states. The region is inhabited by Nagas, a Tibeto-Burman tribe, who once practiced head-hunting. Naga and English are official languages. Intense missionary activity has resulted in a state with the largest Christian population, percentage-wise, in India. An independence movement was terminated by Shillong Peace Agreement of 1975, though an insurgency started up again in 1995. The state is governed by a chief minister and cabinet responsible to a bicameral legislature with one elected house and by a governor appointed by the president of India.

Nagamangala (nah-guh-MUHN-guh-luh), town, MANDYA district, KARNATAKA state, S INDIA, 21 mi/34 km NNW of MANDYA; 12°49'N 76°46'E. Training center for hand-loom weaving; handicraft brass ware; millet, rice.

Nagambie (nuh-GAM-bee), village, central VICTORIA, SE AUSTRALIA, on GOULBURN RIVER, and 70 mi/113 km N of MELBOURNE; 36°47'S 147°10'E. Livestock; cereals; vineyards; tourism.

Nagano (NAH-gah-no), prefecture [Japanese *ken*] (□ 5,261 sq mi/13,626 sq km; 1990 population 2,156,656), central HONSHU, central JAPAN; ⊙ NAGANO. Bordered N by TOYAMA and NIIGATA, E by GUMMA, S by YAMANASHI, SHIZUOKA, and AICHI, and W by GIFU prefectures. Landlocked and extremely mountainous, with an average elevation of over 2,600 ft/792 m. Mountain range along W border, rising to 10,527 ft/3,209 m at MOUNT HOTAKA; drained by TENRYU RIVER. Hot springs in LAKE SUWA area.

Known for its traditional raw-silk industry; more recently it has emerged as a major inland high-technology center. Extensive rice growing, lumbering. Manufacturing (textiles, sake), woodworking. Main centers are Nagano, MATSUMOTO, UEDA, Iida, OKAYA, SUWA. Site of 1998 Winter Olympics, southernmost site ever chosen.

Nagano (NAH-gah-no), city (2005 population 378,512), ⊙ NAGANO prefecture, central HONSHU, central JAPAN, on the TENRYU RIVER; 36°38'N 138°11'E. Asparagus; peaches; apples. Produces electronic equipment. Also a religious center, the site of Zenkoji, a 7th-century Buddhist temple that houses statues sent from the king of Korea in 552. Myoko Highlands nearby. Site of 1998 Winter Olympics.

Naganohara (nah-GAH-no-HAH-rah), town, Agatsuma county, GUMMA prefecture, central HONSHU, N central JAPAN, 31 mi/50 km N of MAEBASHI; 36°32'N 138°38'E. Onioshidashi (lava flow of Mount ASAMA). Hot springs.

Naganuma (nah-gah-NOO-mah), town, Iwase county, FUKUSHIMA prefecture, N central HONSHU, NE JAPAN, 34 mi/55 km S of FUKUSHIMA city; 37°17'N 140°12'E. Cucumbers.

Naganuma (nah-GAH-noo-mah), town, Sorachi district, Hokkaido prefecture, N JAPAN, 19 mi/30 km E of SAPPORO; 43°00'N 141°41'E.

Nagao (nah-gah-O), town, Okawa county, KAGAWA prefecture, NE SHIKOKU, W JAPAN, 9 mi/15 km S of TAKAMATSU; 34°15'N 134°10'E. Bamboo work. Frozen foods.

Nagaoka (nah-GAH-o-kah), city (2005 population 283,224), NIIGATA prefecture, central HONSHU, N central JAPAN, 34 mi/55 km S of NIIGATA; 37°26'N 138°50'E. Produces Buddhist altars and traditional confections. Transportation hub and a center for the surrounding skiing industry. Nagaoka Festival is held here. Nagaoka New Town is nearby.

Nagaokakyo (nah-gah-O-kah-KYO), city, KYOTO prefecture, S HONSHU, W central JAPAN, 9 mi/15 km S of KYOTO; 34°55'N 135°41'E. Televisions. Bamboo shoots.

Nagaon (nah-GOUN), district, central ASSAM state, NE INDIA; ⊙ NAGAON. Mainly in left-bank BRAHMAPUTRA RIVER valley; alluvial soil; agriculture (rice, jute, rape and mustard, tea, sugarcane); silk and lac growing in dispersed forest areas; tea processing. Original district reduced in 1950, when E area with tribal Mikir majority was separated to form MIKIR HILLS district (now KARBI-ANGLONG district). Formerly spelled Nowgong.

Nagaon (nah-GOUN), town, ⊙ NAGAON district, central ASSAM state, NE INDIA, in BRAHMAPUTRA RIVER valley, on KALANG RIVER, 60 mi/97 km ENE of GUWAHATI; 18°36'N 72°55'E. Trades in rice, jute, rape and mustard, tea, lac; tea processing. Railroad junction 3 mi/4.8 km SW, at SENCHOA, with spur to Mairabari, 17 mi/27 km WNW. Formerly spelled Nowgong.

Nagappatinam (nah-gah-PUHT-ee-nuhm), city, ⊙ NAGAPPATINAM Quaid-E-Milleth district, TAMIL NADU state, S INDIA, port (roadstead with wharves) on COROMANDEL COAST of BAY OF BENGAL, in KAVERI River delta, 160 mi/257 km S of CHENNAI (MADRAS); 10°46'N 79°50'E. Exports peanuts, cotton, and silk goods, livestock, tobacco; timber landing and storing on extensive foreshore. Metallurgical center (electric furnaces; steel-rolling mills); industries powered by Mettur and Pykara hydroelectric systems; steel trunk manufacturing, textile dyeing and printing; salt factory. Known to Ptolemy as Nigamos. Twin port of NAGORE is 4 mi/6.4 km N; railroad terminus. Port of VELANGANNI is 6 mi/9.7 km S; seat of Roman Catholic missions. One of earliest Portuguese settlements in India (founded 1612); captured by Dutch in 1660, by English in 1781. Formerly spelled Negapattinam, Negapatam.

Area is shown by the symbol □, and capital city or county seat by ⊙.

Nagappatinam Quaid-E-Milleth (nah-gah-PUHT-ee-nuhm), district (□ 1,776 sq mi/4,617.6 sq km), TAMIL NADU state, S INDIA; ⊙ NAGAPPATINAM.

Nagar (nuh-GUHR), former feudatory state in GILGIT Agency, NW KASHMIR, NE PAKISTAN, in NW KARAKORAM mountain system. History parallels that of HUNZA. Mother tongue was Burushaski.

Nagar (nuh-GUHR), town, KULLU district, HIMACHAL PRADESH state, N INDIA, 50 mi/80 km ESE of Dharmsala, in Kullu valley. Trades in fruit (apples, pears) and tea. Trout hatchery in BEAS RIVER, just W. Site of Urusvati Himalayan Research Institute of Roerich Museum. Sometimes spelled Naggar.

Nagar (nuh-GUHR), village, SHIMOGA district, KARNATAKA state, S INDIA, 7 mi/11.3 km S of HOSANAGARA. As capital of a seventeenth-century kingdom, known as Bednur or Bidaruhalli; in 1763, Hyder Ali made it an important arsenal and renamed it Haidarnagar (or Hydernagar). Called Nagar since 1789. Declined after 1893, when HOSANAGARA replaced it as a subdivisional administrative headquarters of district.

Nagar (nuh-GUHR), village and tehsil, HUNZA division, NORTHERN AREAS, extreme NE PAKISTAN, NW KASHMIR region, on left bank of HUNZA RIVER, and 35 mi/56 km NE of GILGIT. Fort. Was capital of former NAGAR state, GILGIT agency. Also spelled NAGIR.

Nagar, INDIA: see RAJNAGAR.

Nagara (nah-GAH-rah), town, Chose county, CHIBA prefecture, E central HONSHU, E central JAPAN, 9 mi/15 km S of CHIBA; 35°25′N 140°13′E.

Nagara River, JAPAN: see KISO RIVER.

Nagar Devla (nuh-GUHR DAIV-lah), town, JALGAON district, MAHARASHTRA state, W central INDIA, 14 mi/23 km NE of CHALISGAON. Cotton, millet, peanuts. Sometimes written Nagardevla.

Nagareyama (nah-GAH-re-YAH-mah), city (2005 population 152,641), CHIBA prefecture, E central HONSHU, E central JAPAN, 16 mi/25 km N of CHIBA; 35°51′N 139°54′E. Sweet-sake brewing. Tone Canal (5.3 mi/8.5 km long) is nearby.

Nagar Haveli, INDIA: see DADRA AND NAGAR-HAVELI.

Nagarjunakonda, INDIA: see GURUZALA.

Nagarkarnul (nuh-GUHR-kuhr-nool), town, tahsil headquarters, MAHBUBNAGAR district, ANDHRA PRADESH state, S INDIA, 28 mi/45 km SE of MAHBUBNAGAR. Millet, rice, oilseeds. Sometimes spelled Nagerkurnool.

Nagarote (nah-gah-RO-te), town (2005 population 19,614), LEÓN department, W NICARAGUA, near LAKE MANAGUA, on railroad, and 21 mi/34 km WNW of MANAGUA; 12°16′N 86°34′W. Agriculture (corn, sesame, rice, beans); livestock.

Nagar Parkar (nuh-GUHR PAHR-kahr), town, THAR PARKAR district, SE SIND province, SE PAKISTAN, 95 mi/153 km SE of UMARKOT, in small hilly area extending into NE RANN OF CUTCH; 24°22′N 70°45′E.

Nagartse, CHINA: see NAGARZE.

Nagarze (nah-GAHR-ze), town, ⊙ Nagarze county, S TIBET, SW CHINA, near YAMZHO YUMCO Lake, on main INDIA-LHASA trade route, and 65 mi/105 km SW of Lhasa. Noted lamasery of Samding, 4 mi/6.4 km E, headed by abbess was, briefly, residence of 13th Dalai Lama after his return in 1912 from India; 28°58′N 90°24′E. Also spelled as Nagartse or Nangkartse, or Langkazi.

Nagasaka (nah-GAH-sah-kah), town, North Koma county, YAMANASHI prefecture, central HONSHU, central JAPAN, 16 mi/25 km N of KOFU; 35°49′N 138°22′E.

Nagasaki (nah-gah-SAH-kee), prefecture [Japanese ken] (□ 1,574 sq mi/4,077 sq km; 1990 population 1,563,015), NW KYUSHU, SW JAPAN; chief port and ⊙ NAGASAKI. Bounded NW by KOREA STRAIT, W by EAST CHINA SEA, S by AMAKUSA SEA, E by the ARIAKE and SHIMABARA bays. SAGA prefecture to E. Includes TSUSHIMA, IKI-SHIMA, O-SHIMA (two islands with the latter name), IKITSUKI-SHIMA, HIRADO-SHIMA, GOTO-RETTO, and numerous scattered islets. Comprises part of hilly HIZEN PENINSULA, which rises to 4,460 ft/1,359 m at Mount Unzen in hot-springs distrtict on SHIMABARA PENINSULA. Numerous streams drain many fertile valleys. Deeply indented coast provides numerous natural harbors. Coalfields near SASEBO and on TAKA-SHIMA. Primarily agricultural (rice, sweet potatoes, grain, soybeans). Widespread production of raw silk. Whaling and fishing on Goto-retto, Tsushima, and other islands. Heavy industry concentrated at Nagasaki; fine porcelain ware produced on Hirado-shima. Sasebo (W) is important naval base. Atomic bomb dropped on Nagasaki city during World War II. Important cities are Nagasaki and HIRADO.

Nagasaki (nah-gah-SAH-kee), city (2005 population 455,206), ⊙ NAGASAKI prefecture, W KYUSHU, SW JAPAN, on Nagasaki Bay; 32°44′N 129°52′E. One of Japan's leading ports. Shipbuilding is the chief industry and machinery plants are also important. Loquats and tortoise shells are other local products Nagasaki's port, the first to receive Western trade, was known to Portuguese and Spanish traders before it was opened to the Dutch in 1567. After the Portuguese and Spanish merchants were forced to leave Japan in 1637, the Dutch traders were restricted (1641–1858) to DE-JIMA, an island in the harbor. Gradually reopened to general foreign trade during the 1850s. Long a center of Christianity, the city had until 1945 Japan's largest Roman Catholic cathedral. During World War II, on August 9, 1945, Nagasaki became the target of the second atomic bomb ever detonated on a populated area; about 75,000 people were killed or wounded, and more than half of the city was devastated. Among Nagasaki's landmarks is Glover Mansion, scene of Puccini's opera *Madama Butterfly*. Peace Park is another attraction.

Nagashima (nah-GAH-shee-mah), town, on NAGA-SHIMA island, Izumi count., KAGOSHIMA prefecture, SW JAPAN, 47 mi/75 km N of KAGOSHIMA; 32°09′N 130°07′E. Sweet potatoes; *shochu* (distilled alcoholic drink).

Nagashima, town, Kuwana county, MIE prefecture, S HONSHU, central JAPAN, port on KUMANO SEA, 28 mi/45 N of TSU; 35°05′N 136°42′E. Hot springs nearby.

Naga-shima (nah-GAH–shee-mah), island (□ 35 sq mi/91 sq km) of AMAKUSA, in EAST CHINA SEA, Izumi county, KAGOSHIMA prefecture, SW JAPAN, off W coast of KYUSHU, just SE of AMAKUSA SHIMO-JIMA; 9.5 mi/15.3 km long, 5.5 mi/8.9 km wide. Irregular coastline; mountainous, fertile. Chief products are grain and fish.

Naga-shima, island (□ 11 sq mi/28.6 sq km), Kumage county, YAMAGUCHI prefecture, W JAPAN, in IYO SEA, nearly connected with peninsula S of YANAI (on SW HONSHU); 5 mi/8 km long, 3 mi/4.8 km wide. Mountainous, fertile. Produces sweet potatoes, plums, oranges. Manufacturing (shell buttons, insect powder, soy sauce) at KAMINOSEKI on NE coast. Nearby islets include YASHIMA (SSE), 2.5 mi/4 km long, 0.5 mi/0.8 km wide; Iwai-shima (W), 2 mi/3.2 km long, 1.5 mi/2.4 km wide.

Nagasu (nah-GAHS), town, Tamana county, KUMAMOTO prefecture, W KYUSHU, SW JAPAN, on ARIAKE BAY, 19 mi/30 km N of KUMAMOTO; 32°55′N 130°27′E. Shipbuilding. Pickles. Goldfish and marine-product processing.

Nagate Point (NAH-gah-te), Japanese *Nagate-misaki* (NAH-gah-te-mee-SAH-kee), ISHIKAWA prefecture, central HONSHU, central JAPAN, at NE tip of NOTO Peninsula, at NW entrance to TOYAMA BAY; forms easternmost point of peninsula; 37°27′N 137°22′E. Suzu Point is 4.5 mi/7.2 km NNW, Rokko Point is 6 mi/9.7 km NNW.

Nagato (nah-GAH-to), city, YAMAGUCHI prefecture, SW HONSHU, W JAPAN, 22 mi/35 km N of YAMAGUCHI; 34°22′N 131°11′E. Processed foods (known for Chinese-style dumplings); chicken (broilers). Includes OUMI islet, in nearby Kita-Nagato Seacoast quasi-national park. Hot springs in the vicinity.

Nagato (nah-GAH-to), town, Chisagata county, NAGANO prefecture, central HONSHU, central JAPAN, 28 mi/45 km S of NAGANO; 36°15′N 138°16′E. Enoki mushrooms, medicinal plants. Traditional papermaking.

Nagatoro (nah-GAH-to-ro), town, Chichibu county, SAITAMA prefecture, E central HONSHU, E central JAPAN, 34 mi/55 km N of URAWA; 36°06′N 139°06′E. In scenic ARAKAWA area.

Nagaur (nuh-GOUR), district (□ 6,841 sq mi/17,786.6 sq km), RAJASTHAN state, NW INDIA; ⊙ NAGAUR.

Nagaur (nuh-GOUR), town, ⊙ NAGAUR district, central RAJASTHAN state, NW INDIA, 75 mi/121 km NNE of JODHPUR. Trade center for bullocks, wool, hides, cotton; hand-loom weaving, manufacturing of camel fittings, metal utensils, ivory goods. Livestock raising nearby.

Nagavali River (NAH-gah-vah-lee), c. 150 mi/241 km long, in S ORISSA and NE ANDHRA PRADESH states, E central INDIA; rises in EASTERN GHATS, S of BHAWANIPATNA (Orissa state); flows SSE, past PALKONDA (Andhra Pradesh state), to BAY OF BENGAL 5 mi/8 km SSE of SRIKAKULAM.

Nagawa (NAH-gah-wah), town, Sannohe county, Aomori prefecture, N HONSHU, N JAPAN, 40 mi/65 km S of AOMORI; 40°25′N 141°20′E. Apples; dried chrysanthemums.

Nagawa (NAH-gah-wah), village, S Azumi county, NAGANO prefecture, central HONSHU, central JAPAN, 50 mi/80 km S of NAGANO; 36°05′N 137°41′E. Vegetable and buckwheat processing, maple and white-birch cultivation. Hot springs; skiing area. Nomugi mountain pass nearby.

Nagayeva, RUSSIA: see MAGADAN.

Nagayo (nah-GAH-yo), town, West Sonogi county, NAGASAKI prefecture, NW KYUSHU, SW JAPAN, 6 mi/10 km N of NAGASAKI; 32°49′N 129°52′E. Mandarin oranges.

Nagbhir, INDIA: see BRAMHAPURI.

Nagcarlan (nahg-kahr-LAHN), town, LAGUNA province, S LUZON, PHILIPPINES, 8 mi/12.9 km NE of SAN PABLO; 14°08′N 121°25′E. Agricultural center (rice, coconuts, sugarcane).

Nagchhu, CHINA: see NAGQU.

Nagchu, CHINA: see NAGQU.

Nagele (NAH-khuh-luh), village, S NORTH-EAST POLDER, FLEVOLAND province, central NETHERLANDS, 14 mi/23 km NE of LELYSTAD.

Nagem (NAH-kuhm), hamlet, Redange commune, W LUXEMBOURG, 2 mi/3.2 km NW of the village of REDANGE; 49°47′N 05°52′E.

Nagercoil (NAH-guhr-koil), city (2001 population 208,179), ⊙ KANNIYAKUMARI district, TAMIL NADU state, southernmost city in INDIA. Industries in automotive repair, rubber goods, and rice- and cotton-milling. Ilmenite and monazite mined from sand. An important Christian center in a predominantly Hindu area.

Nagerkurnool, INDIA: see NAGARKARNUL.

Nagfa, Eritrea: see NAKFA.

Naggar, INDIA: see NAGAR.

Nag Hammadi (NAG hah-MAD-ee), town, QENA province, Upper EGYPT, on the W bank of the NILE River, 30 mi/48 km WSW of QENA; 26°03′N 32°15′E. Center of processing agricultural products. Near the ancient town of Chenoboskion, where, in 1945, a large cache of gnostic texts in the Coptic language was discovered. The Nag Hammadi manuscripts, dating from the fourth century C.E., include twelve codices of tractates, one loose tractate, and a copy of Plato's *Republic*—making fifty-three works in all. Originally composed in Greek, they were translated (second–third century C.E.) into Coptic. Most of the texts have a strong Christian element. The presence of non-Christian elements, however, gave rise to the speculation that gnosticism, which taught salvation by knowledge, was not originally a Christian movement. Until the texts' discovery, knowledge of Christian gnosticism was confined to reports and quotations of their orthodox opponents, such as Irenaeus and Tertullian. Among the codices are apocalypses, gospels, a collection of sayings of the resurrected Jesus to his disciples, homilies, prayers, and theological treatises. Also spelled Naj' Hamadi.

Nagi (NAH-gee), town, Katsuta county, OKAYAMA prefecture, SW HONSHU, W JAPAN, 37 mi/60 km N of OKAYAMA; 35°07′N 134°10′E. Beef cattle.

Nagina (nah-GEE-nah), town, BIJNOR district, N UTTAR PRADESH state, N central INDIA, 13 mi/21 km SSE of NAJIBABAD. Trade center (rice, wheat, gram, barley, sugarcane, oilseeds); noted carved-ebony work; glass manufacturing. Sacked 1805 by Pindari dacoits.

Naginimara, INDIA: see NAZIRA.

Nagir, PAKISTAN: see NAGAR.

Nagishot (NAH-gee-shot), village, E EQUATORIA state, S SUDAN, in hilly region, near UGANDA border, 145 mi/233 km ESE of JUBA; 04°07′N 33°04′E. Also spelled Nagichot.

Nagiso (nah-gee-SO), town, Kiso county, NAGANO prefecture, central HONSHU, central JAPAN, 81 mi/130 km S of NAGANO; 35°36′N 137°36′E. *Rokuro*-style pottery; traditional hats.

Nagles Mountains (NAI-guhlz), Gaelic *Sliabh an Nóglaigh*, range extending to 10 mi/16 km E-W in NE CORK county, SW IRELAND; 52°07′N 08°28′W. Rises to 1,406 ft/429 m, 7 mi/11.3 km WSW of FERMOY.

Nago (NAH-go), city on W coast of OKINAWA island, in the RYUKYUS, OKINAWA prefecture, SW JAPAN, 31 mi/50 km N of NAHA, on inlet of EAST CHINA SEA 26°35′N 127°58′E. Pineapples. Locally called Nagu.

Nagod (NAH-god), town (2001 population 19,474), SATNA district, NE MADHYA PRADESH state, central INDIA, 16 mi/26 km W of SATNA; 24°34′N 80°34′E. Market center for millet, gram, wheat, building stone. Was capital of former princely state of Nagod of CENTRAL INDIA agency; in 1948, merged with VINDHYA PRADESH, later with Madhya Pradesh state.

Nagold (NAH-golt), town, N BLACK FOREST, BADEN-WÜRTTEMBERG, GERMANY, on the NAGOLD RIVER, and 15 mi/24 km WNW of TÜBINGEN; 48°32′N 08°39′E. Railroad junction; manufacturing of textiles, clothing, machinery; woodworking. Summer resort. Has ruins of 12th-century castle.

Nagold River (NAH-golt), 57 mi/92 km long, S GERMANY; rises in BLACK FOREST, 7 mi/11.3 km SW of BERNECK; flows E to NAGOLD, then N to the ENZ RIVER at PFORZHEIM.

Nagol'no-Tarasovka, UKRAINE: see NAHOL'NO-TARASIVKA.

Nagongera (nah-GON-ge-ruh), town, EASTERN region, SE UGANDA, on railroad, and 12 mi/19 km WNW of TORORO; 00°45′N 34°01′E. Cotton, corn, millet, sweet potatoes, bananas. Was part of former EASTERN province.

Nagoorin (nah-GOO-rin), settlement, SE QUEENSLAND, NE AUSTRALIA, 47 mi/76 km S of GLADSTONE; 24°22′S 151°18′E. Oil shale. Beef cattle; grain; dairying.

Nagor, in Thai names: see NAKHON.

Nagore (nah-GOR), town, in Nagappatinam municipality, NAGAPPATINAM Quaid-E-Milleth district, TAMIL NADU state, S INDIA, port on COROMANDEL COAST of BAY OF BENGAL, 4 mi/6.4 km N of twin port of NAGAPPATINAM, at mouth of arm of KAVERI River, at S border of PONDICHERRY Union Territory. Railroad terminus. Muslim pilgrimage center; Arabic schools. Tourist center.

Nagorn, in Thai names: see NAKHON.

Nagorno-Karabakh Autonomous Region (nuh-GAWR-nuh–kuh-ruh-BAHK), autonomous region (□ 1,699 sq mi/4,417.4 sq km), SE AZERBAIJAN, between the CAUCASUS and the KARABAKH range; ⊙ XANKANDI. The region has numerous mineral springs as well as deposits of lithographic stone, marble, and limestone. Farming and grazing are important and there are various light industries. The population of the region is mainly Armenian (76%), with Azeri (23%), Russian, and Kurdish minorities. Xankandi and SHUSHA are the chief towns. A part of Caucasian ALBANIA called Artsakh, the area was taken by ARMENIA in the 1st century C.E. and by the Arabs in the 7th century. The region was renamed Karabakh in the 13th century. In the early 17th century, it passed to the Persians, who permitted local autonomy, and in the mid-18th century the Karabakh khanate was formed. Karabakh alone was ceded to RUSSIA in 1805; the khanate passed to the Russians by the Treaty of Gulistan in 1813. In 1822 the Karabakh khanate was dissolved and the area became a Russian province. The Nagorno-Karabakh (Mountain-Karabakh) autonomous region was established 1923. In the late 1980s and early 1990s, the region became a focal point in a growing war between the republics of Armenia and Azerbaijan, as Armenian nationalists demanded the inclusion of the region into Armenia. Since ceasefire, effectively part of Armenia.

Nagornoye (nah-GOR-nuh-ye), former town, now a suburb of SAINT PETERSBURG, central LENINGRAD oblast, NW European Russia, 7 mi/11 km S of the city center and just SE of KRASNOYE SELO, in a picturesque hilly location (Dudergof Heights); 59°43′N 30°06′E. Elevation 223 ft/67 m. Site of a palace (built in 1829). Until 1944, called Dudergof.

Nagornskiy (nah-GORN-skeeyee), town (2006 population 555), E central PERM oblast, E European Russia, W URAL Mountains, on road and railroad (Nagornaya station), 5 mi/8 km S, and under administrative jurisdiction, of GUBAKHA; 58°46′N 57°32′E. Elevation 1,108 ft/337 m. In Kizel bituminous-coal basin; lumbering; coal mining. Developed during World War II. Population declined sharply following the closing of many coal mines in the post-Soviet period.

Nagornyy (nah-GOR-niyee), settlement (2006 population 145), S SAKHA REPUBLIC, S RUSSIAN FAR EAST, in the STANOVOY range, on the upper Timpton River (tributary of the ALDAN RIVER), on the administrative border with AMUR oblast, on railroad and the Yakutsk-Never highway, 180 mi/290 km S of ALDAN; 55°58′N 124°57′E. Elevation 2,680 ft/816 m. Mica mining. Formerly a gold-mining town; largely abandoned after the mine became exhausted.

Nagorsk (nah-GORSK), town (2005 population 4,930), N KIROV oblast, E central European Russia, on the VYATKA River, on highway, 65 mi/105 km NE of KIROV; 59°19′N 50°48′E. Elevation 426 ft/129 m. Lumbering; timber floating; livestock veterinary station; bakery. Formerly known as Nagorskoye.

Nagorskoye, RUSSIA: see NAGORSK.

Nagor Sridhamaraj, THAILAND: see NAKHON SI THAMMARAT.

Nagor'ye (nah-GOR-ye), village, SW YAROSLAVL oblast, central European Russia, on road, 20 mi/32 km NW of PERESLAVL'-ZALESSKIY, to which it is administratively subordinate; 56°55′N 38°15′E. Elevation 495 ft/150 m. Flax.

Nagoya (NAH-go-yah), city (2005 population 2,215,062), ⊙ AICHI prefecture, S central HONSHU, central JAPAN, on ISE BAY; 35°10′N 136°54′E. A major port, transportation hub, and industrial center, it has aircraft factories and automotive works (vehicles and machine tools). Manufacturing includes typewriters, sewing machines, synthetic fibers, food; plywood, aluminum. Produces traditional items, such as crests (*mon*), Buddhist altars, and paulownia furniture; *yuzen*-style silk printing. The world's fifth-largest tonnage port. Seat of many universities, of which Nagoya University is the most famous. Atsuta Shrine (founded in the 2nd century), where the sacred imperial sword is housed; Higashi Honganji shrine (built 1692). The Tokugawa Art Museum and Higashiyama Park are two other notable attractions. A fortress town in the 16th century, Nagoya retains a castle built by Tokyugawa Ieyasu in 1612 and reconstructed in 1959.

Nagpur (NAHG-poor), district (□ 3,819 sq mi/9,929.4 sq km), MAHARASHTRA state, central INDIA, on DECCAN PLATEAU; ⊙ NAGPUR. At S base of central SATPURA Range; partly bordered NW by WARDHA, SE by WAINGANGA RIVER; drained by KANHAN and PENCH Rivers. Mainly undulating plain; major cotton-growing tract in W, wheat area in E; famous NAGPUR oranges, mango groves, betel farms, mahua, and tamarind in fertile central river valleys; millet, oilseeds (chiefly flax). Sal and satinwood in dispersed forest areas (notably around Umrer). An important source of India's manganese; extensive mining area in vicinity of RAMTEK and between KHAPA and SAONER. Sandstone and marble quarries near KAMPTEE (trade center). NAGPUR (former MADHYA PRADESH state capital) is a major communications hub and cotton-textile center. Major transportation hub, as MUMBAI (Bombay)-CALCUTTA and CHENNAI (Madras)-DELHI railroads all pass through here, as do roads and air routes. Cotton ginning and manufacturing, oilseed milling, handicraft glassmaking and silk weaving. Pop. 86% Hindu, 7% Muslim, 6% tribal (mainly Gond). Part of a Gond dynasty in early 18th century; in 1743, became part of Maratha kingdom of Nagpur. In 1817, as a result of British victory over Marathas at fortified hill of Sitabaldi (in NAGPUR city), area passed to British and, with SE portion of province, was indirectly administered by government of India until formation (1861) of Central Provinces. Seat of university, several colleges, and technical institutions.

Nagpur (NAHG-poor), city (2001 population 2,129,500), ⊙ NAGPUR district, MAHARASHTRA state, central INDIA, on the Nag River, in the "heart" of India. Formerly capital of CENTRAL PROVINCES AND BERAR and then of MADHYA PRADESH state, it is still a seat of government for one month every winter, when the legislative assembly of the Maharashtra state holds its session here with the entire council of ministers in attendance. Transportation center and a leading Indian industrial and commercial city. Railroads, air routes, and roadways crisscross the country here. Manufacturing includes textiles, transport equipment, and ferromanganese products. Industrial estates have given rise to a variety of manufacturing units. Recent planning and development by the Municipal Corporation has improved local services and the state government has paid special attention to retaining its importance. Also the "orange capital" of the country and a center of trade and distribution of a flourishing orange crop in the vicinity. Seat of Nagpur University (established 1923) and of numerous colleges and technical institutes (e.g., medical, engineering). Religious and historical site. Has a wildlife sanctuary in the vicinity. The Gond tribals also live in the district. Founded in the 18th century as capital of the Nagpur Maratha kingdom, it passed to the British in 1853. It has expanded to include the neighboring town of KAMPTEE.

Nagqu (NAHG-CHOO), town, ⊙ Naggu county, E TIBET, SW CHINA, near headstream of SALWEEN RIVER, on main LHASA-YUSHU (QINGHAI province)-XINING trade route, and 135 mi/217 km NNE of Lhasa; 31°30′N 92°00′E. Elevation 14,580 ft/4,444 m. Hot spring just S. Also appears as Nagchu, Nagchhu, or Nakchukha, or Nagu.

Nags Head (NAGZ HED), town (□ 6 sq mi/15.6 sq km; 2006 population 3,054), DARE county, E NORTH CAROLINA, 4 mi/6.4 km NNE of MANTEO, barrier beach on OUTER BANKS, sand barrier between ATLANTIC OCEAN (E) and ROANOKE SOUND (W), SE of ALBEMARLE SOUND; 35°56′N 75°37′W. Bridge to mainland is 4 mi/6.4 km to S. Swimming, game hunting; commercial fishing. Beach resort area. Retail trade; manufacturing (crab meat processing; consumer goods). Jockey's Ridge State Park to N, largest sand dunes on E coast. Nags Head Woods, a national natural landmark (□ 2 sq mi/5.2 sq km) of forest, dunes and wetlands. Established early 1830s.

Nagu, CHINA: see NAGQU.

Nagua (NAH-gwah), town (2002 population 32,035), on ESCOCESA BAY, N DOMINICAN REPUBLIC, ⊙ MARÍA TRINIDAD SÁNCHEZ province; 19°23′N 69°50′W. Fishing port and tourist center in a rice- and cocoa-growing area. Formerly called Julia Molina or Villa Julia Molina.

Naguabo (nah-GWAH-bo), town (2006 population 24,209), E PUERTO RICO, near the coast, 6 mi/9.7 km ENE of HUMACAO. Manufacturing (electronics, apparel, fabricated metal products); dairying (milk products); livestock, hogs, poultry. Fishing. Tourism at nearby beaches. Its port, Playa de Naguabo, is 2 mi/3.2 km SE.

Naguanagua (nah-gwah-NAH-gwah), town, ⊙ Naguanagua municipio, CARABOBO state, N VENEZUELA, 5 mi/8 km N of VALENCIA; 10°15′N 68°01′W. Elevation 1,919 ft/584 m. Coffee, sugarcane, cotton, corn, fruit.

Naguilian (nah-geel-YAHN), town, ISABELA province, N LUZON, PHILIPPINES, on CAGAYAN RIVER near its junction with the MAGAT RIVER, and 8 mi/12.9 km SSW of ILAGAN; 17°01′N 121°52′E. Agricultural center (rice, corn).

Naguilian, town, LA UNION province, N central LUZON, PHILIPPINES, near W coast, 8 mi/12.9 km SE of SAN FERNANDO; 16°32′N 120°25′E. Rice-growing center.

Naguri (NAH-goo-ree), village, Iruma county, SAITAMA prefecture, E central HONSHU, E central JAPAN, 25 mi/40 km W of URAWA; 35°52′N 139°11′E.

Nagyag, ROMANIA: see SĂCĂRÎMB.

Nagyatád (NAHD-yah-tahd), city, SOMOGY county, SW HUNGARY, on Rinya River, and 22 mi/35 km SW of KAPOSVÁR; 46°13′N 17°22′E. Manufacturing (food preservatives, foods, cotton textiles).

Nagybánya, ROMANIA: see BAIA MARE.

Nagybecskerek, SERBIA: see ZRENJANIN.

Nagyberezna, UKRAINE: see VELYKYY BEREZNYY.

Nagybocsko, UKRAINE: see VELYKYY BYCHKIV.

Nagyderzsida, ROMANIA: see BOBOTA.

Nagydisznód, ROMANIA: see CISNĂDIE.

Nagyenyed, ROMANIA: see AIUD.

Nagy Fatra, SLOVAKIA: see GREATER FATRA.

Nagyfödémes, SLOVAKIA: see VELKE ULANY.

Nagyhalmágy, ROMANIA: see HĂLMAGIU.

Nagyida, SLOVAKIA: see VELKA IDA.

Nagyilonda, ROMANIA: see ILEANDA.

Nagyimánd (NAHD-ye-mahnd), village, KOMÁROM county, HUNGARY, 19 mi/31 km E of Győr; 47°38′N 18°05′E. Large plant making food formulas and additives.

Nagykálló (NAHD-yuh-kahl-lo), city, SZABOLCS-SZATMÁR county, NE HUNGARY, 8 mi/13 km SE of NYIREGYHÁZA; 47°53′N 21°51′E. Wheat, tobacco, corn, apples; cattle, hogs. Manufacturing (transportation equipment, metal structures; pasta; flour milling); tobacco warehouses. Agricultural high school.

Nagykalota, ROMANIA: see CĂLATA.

Nagykanizsa (NAHD-yuh-kah-ne-zhah), city (2001 population 52,106), ZALA county, SW HUNGARY. Wheat, corn, barley, rape; cattle, hogs. Manufacturing (brick, tile and ceramics, furniture, footwear, farinaceous foods; flour milling, meatpacking). Founded c. 1300 as a fortress, Nagykanizsa was captured by the Turks in 1600. It has a museum, an 18th-century Franciscan church, and the ruins of the old fortress. Sometimes called Kanizsa.

Nagykapos, SLOVAKIA: see VELKE KAPUSANY.

Nagykaroly, ROMANIA: see CAREI.

Nagykáta (NAHD-yuh-kah-tah), city, PEST county, N central HUNGARY, 31 mi/50 km ESE of BUDAPEST; 47°25′N 19°45′E. Manufacturing (flour milling, baking; pasta). Castle.

Nagykér, SLOVAKIA: see VELKY KYR.

Nagykőrös (NAHD-yuh-kuh-rush), city (2001 population 25,543), PEST county, central HUNGARY, 9 mi/14.5 km NNE of KECSKEMÉT; 47°02′N 19°47′E. It is the center of a grain- and fruit-growing region; manufacturing (food preservatives, pasta, dairy products); has an old Reformed church, a college, and a town hall.

Nagylak, ROMANIA: see NĂDLAC.

Nagylévárd, SLOVAKIA: see VELKE LEVARE.

Nagylomnicz, SLOVAKIA: see VELKA LOMNICA.

Nagymaros (NAHD-yuh-mah-rosh), village, PEST county, N HUNGARY, on the DANUBE River, and 22 mi/35 km NNW of BUDAPEST; 47°47′N 18°58′E. Agriculture (wine, honey; poultry); river port, summer resort. Paper mills; champagne manufacturing. A dam was built to complement the larger Slovak-Hungarian project (the GABCIKOVO DAM, which has diverted the Danube further upstream) from which Hungary withdrew amidst much political controversy. The dam here was to be dismantled during the 1990s.

Nagymaros, HUNGARY: see GABCIKOVO DAM.

Nagymegyer, SLOVAKIA: see VELKY MEDER.

Nagymihály, SLOVAKIA: see MICHALOVCE.

Nagyrákóc, UKRAINE: see VELYKYY RAKOVETS′.

Nagyróna, SLOVAKIA: see VELKE ROVNE.

Nagysajó, ROMANIA: see ŞIEU.

Nagysármás, ROMANIA: see SĂRMAŞ.

Nagysáros, SLOVAKIA: see VELKY SARIS.

Nagyselyk, ROMANIA: see ŞEICA MARE.

Nagysink, ROMANIA: see CINCU.

Nagysomkút, ROMANIA: see ŞOMCUTA MARE.

Nagyszalonta, ROMANIA: see SALONTA.

Nagyszeben, ROMANIA: see SIBIU.

Nagyszöllős, UKRAINE: see VYNOHRADIV.

Nagyszombat, SLOVAKIA: see TRNAVA.

Nagyvárad, ROMANIA: see ORADEA.

Naha (NAH-hah), city (2005 population 312,393), ⊙ Okinawa prefecture, SW JAPAN, port on the SW coast of OKINAWA island, in the RYUKYU ISLANDS; 26°12′N 127°40′E. Island's chief manufacturing center. In 1853, U.S. Commodore Perry chose Naha as his first base for the penetration of Japan. Virtually destroyed during World War II. In 1945 it became the headquarters of the U.S. military governor of the Ryukyus; prefectural capital since the island was returned to Japan in 1972. Ruins of Shuri Castle. Also spelled Nafa and Nawa.

Nahalal (nah-hah-LAHL) or **Nahalol**, moshav, ISRAEL, on the NW fringes of the PLAIN OF JEZREEL, SW of NAZARETH; 32°41′N 35°11′E. Elevation 259 ft/78 m. Predominantly agricultural (livestock; orchards); refrigerated food storage. Founded 1921 as Israel's first cooperative settlement. Home of Moshe Dayan. Home and burial place of astronaut Ilan Ramon. In biblical times, part of the inheritance of the children of Zebulun was called by this name, though the exact location of the ancient site remains in dispute. Also known as Nahallah.

Nahalein, Arab village, Bethlehem district, 7.5 mi/12 km SW of BETHLEHEM, in the JUDAEAN HIGHLANDS, WEST BANK; 31°41′N 35°07′E. Cereals; vineyards. Believed related to the biblical site of Nahlat. Some archaeological remains from the medieval era were found here.

Nahallah, ISRAEL: see NAHALAL.

Nahal Shizef, stream, ISRAEL: see HATZERIM.

Nahan (NAH-hahn), town, ⊙ Sirmaur district, S HIMACHAL PRADESH state, N INDIA, in SHIWALIK RANGE, 38 mi/61 km SSE of SHIMLA. Local market center for wheat, corn, rice, potatoes, spices; timber; ironworks (sugarcane crushers), rosin factory; hand-loom weaving, wood carving. Was capital of former PUNJAB HILL state of SIRMUR. Cantonment of SHAMSHERPUR lies 1 mi/1.6 km W.

Nahanni National Park (nuh-HAN-ee) (□ 1,840 sq mi/4,784 sq km), FORT SMITH region, NORTHWEST TER-RITORIES, N Canada, W of FORT SIMPSON, just E of the YUKON border. Extends along the lower portion of the SOUTH NAHANNI RIVER. The river's spectacular course passes through 3 deep canyons and over VIRGINIA FALLS (c.300 ft/90 m high) and numerous rapids. A wilderness area, the park has hot springs and caves and a variety of plant and animal life. Established 1972.

Nahant (nuh-HAHNT), resort and residential town (2000 population 3,632), ESSEX county, E MASSACHUSETTS, on rocky peninsula jutting S of LYNN into MASSACHUSETTS BAY; 42°25′N 70°55′W. Site of Nahant Beach. East point site of former coastal artillery World War II. Present town park and research station for Northeastern University. Settled 1630, incorporated 1853.

Nahar (NAH-huhr), village, ROHTAK district, HARYANA state, N INDIA, 35 mi/56 km SSW of ROHTAK. Local trade in millet, gram, salt, cotton.

Naharayim (nuh-huh-rah-YEEM), village, NW JORDAN, on left bank of the JORDAN River (Israeli border), at mouth of the YARMUK River, 11 mi/18 km S of TIBERIAS; 32°38′N 35°35′E. Site of former Rutenberg Works, hydroelectric power station. The station and village were abandoned during the 1948 war with ISRAEL and became a Jordanian military camp.

Nahari (nah-HAH-ree), town, Aki county, KOCHI prefecture, S SHIKOKU, W JAPAN, 31 mi/50 km SE of KOCHI; 33°25′N 134°01′E. Silk.

Nahariya (nah-hah-REE-yah) or **Naharia**, city (2006 population 50,300), UPPER GALILEE, NW ISRAEL, on MEDITERRANEAN SEA, 15 mi/24 km NNE of HAIFA; 33°00′N 35°05′E. At sea level. Majority Jewish population. Seaside resort. Building materials, textiles, high-tech products; meat processing; fruit and vegetable canning, dairying. Founded 1934.

Naharlagun, town, ⊙ PAUM PARE district, ARUNACHAL PRADESH state, NE INDIA.

Nahāvand (nah-AV-ahnd), city (2006 population 73,141), Hamadān province, W IRAN, 40 mi/64 km S of HAMADAN; 34°10′N 48°22′E. Agricultural trade center. Scene of a decisive victory of the Arabs over the Persians in 641 or 642. The name also appears as Nehavand and Nihavand.

Nahef (NAH-hef), Arab township, ISRAEL, E of AKKO (Acre) in UPPER GALILEE; 32°56′N 35°19′E. Elevation 912 ft/277 m. Kilns from Byzantine times.

Nahe River (NAH-e), 60 mi/97 km long, W GERMANY; rises 5 mi/8 km N of St. WENDEL; flows generally NE, past BAD KREUZNACH, to the RHINE RIVER at BINGEN. Receives the GLAN RIVER (right).

Nahiya, EGYPT: see NAHYA.

Nahma (NAH-muh), village, DELTA county, S UPPER PENINSULA, N MICHIGAN, 20 mi/32 km ENE of ESCANABA, on BIG BAY DE NOC; 45°50′N 86°39′W. In Hiawatha National Forest. Lumber milling. Native American settlement nearby.

Nahmakanta Lake (nah-muh-KAN-tuh), PISCATAQUIS county, central MAINE, 20 mi/32 km WNW of MILLINOCKET; 4 mi/6.4 km long. In lumbering, recreational area. Joined by stream to PEMADUMCOOK LAKE.

Nahol'no-Tarasivka (nah-HOL-no–tah-RAH-seef-kah) (Russian *Nagol'no-Tarasovka*), town, S LUHANS′K oblast, UKRAINE, in the DONBAS, 8 mi/12.9 km SE of ROVEN′KY and subordinated to its city council; 48°03′N 39°28′E. Elevation 833 ft/253 m. Coal mines. Established at the beginning of the 18th century; town since 1938.

Nahouri, province (□ 1,447 sq mi/3,762.2 sq km; 2005 population 132,798), CENTRE-SUD region, S BURKINA FASO, on GHANA (S) border; ⊙ PO; 11°15′N 01°15′W. Borders SISSILI (W, formed partly by SISSILI RIVER), ZIRO (NW), ZOUNDWÉOGO (NNE, formed partly by NAZINON RIVER), and BOULGOU (E, formed by NAZINON RIVER) provinces. Drained by Sissili and Nazinon rivers. Thermal center. Main centers are Po and TIÉBÉLÉ.

Nahr Alnil, state, SUDAN: see River Nile, state.

Cross-references are shown in SMALL CAPITALS. The pronunciation guide is shown on page xix. The sources of population figures are shown on page xvii.

Nahr an Nil, state, SUDAN: see River Nile, state.

Nahr Elnil, state, SUDAN: see River Nile, state.

Nahsholim (nahkh-sho-LEEM) or **Nachsholim**, kibbutz, NW ISRAEL, on MEDITERRANEAN SEA, 15 mi/24 km SSW of HAIFA, near railroad; 32°37'N 34°55'E. At sea level. Seaside resort; manufacturing of plastic products. Mixed farming; fruit, vegetables. Established 1948 on site of former Arab village of Tantura. Modern village on site of ancient Canaanite city-state of DOR, seat of one of Solomon's "twelve officers over Israel."

Nahualá (nah-hwah-LAH), town, SOLOLÁ department, SW central GUATEMALA, 10 mi/16 km NW of SOLOLÁ, near head of NAHUALATE RIVER (here called the Nahualá); 14°51'N 91°19'W. Elevation 8,094 ft/2,467 m. Produces woolen blankets, corn-grinding stones; corn, black beans. Quiché-speaking residents.

Nahualate River (nah-hwah-LAH-te), c.100 mi/161 km long, SW GUATEMALA; rises in W highlands near NAHUALÁ; flows S past SAN MIGUEL PANÁN, through coffee and sugarcane region, to the PACIFIC OCEAN, 7 mi/11.3 km SE of Tahuesco; 14°03'N 91°32'W.

Nahuatzén (nah-hwat-SEN), town, MICHOACÁN, central MEXICO, 17 mi/27 km NNE of URUAPAN; 19°42'N 101°50'W. Sugarcane, fruit, tobacco, corn; livestock.

Nahuelbuta (nah-wel-BOO-tah), national park, ARAUCANIA region, CHILE, 25 mi/40 km SW of ANGOL. Protects large stand of Arucaria (monkey-puzzle) trees and native wild animals. Created in 1939.

Nahuelbuta, Cordillera de (nah-wel-BOO-tah, kordee-YE-rah), pre-Andean forested range in S central CHILE, extends c.90 mi/145 km N-S in ARAUCO, BÍO-BÍO, and MALLECO provinces; rises to 4,725 ft/1,440 m.

Nahuel Huapí (nah-HWAIL hwah-PEE), village, ⊙ Los Lagos department, SW NEUQUÉN province, ARGENTINA, on E end of Lake NAHUEL HUAPÍ, at outlet of LIMAY RIVER, in Nahuel Huapí national park; 41°03'S 71°09'W. Resort; livestock-raising center. Coal deposits nearby.

Nahuel Huapí (nah-HWAIL hwah-PEE), lake (□ 210 sq mi/546 sq km), in RÍO NEGRO and NEUQUÉN provinces, W central ARGENTINA; 45 mi/72 km long. Drained NE by the LIMAY RIVER. Part of Nahuel Huapí National Park (established 1934) and is a popular resort area.

Nahuizalco (nah-wee-ZAHL-ko), town and municipality, SONSONATE department, W EL SALVADOR, on Río Grande, and 4 mi/6.4 km NNW of SONSONATE; 13°47'N 89°45'W. Mat and basket weaving, pottery making; grain, coffee, manioc. Population largely Native American.

Nahulingo (nah-oo-LEEN-go), municipality and town, SONSONATE department, EL SALVADOR, SE of SONSONATE city; 13°42'N 89°43'W.

Nahunta (nah-HUHN-tuh), city (2000 population 930), ⊙ BRANTLEY county, SE GEORGIA, 22 mi/35 km E of WAYCROSS; 31°13'N 81°59'W. Light manufacturing and vegetable processing.

Nahunta, NORTH CAROLINA: see FREMONT.

Nahya (NAH-yah), village, GIZA province, N Upper EGYPT, 7 mi/11.3 km W of CAIRO; 30°03'N 31°07'E. Flax industry. Also spelled Nahiya.

Naiak (nah-YAK), town, BAMIAN province, central AFGHANISTAN, 25 mi/40 km N of PANJAO, in the HAZARAJAT; 34°44'N 66°57'E. Hq. of YAKAOLANG (Yakkaolang) district, for which the town is sometimes named. Also spelled Nayak.

Naibandan, IRAN: see NEHBANDAN.

Naibo, village, Asia: see DOBROYE.

Naibo, RUSSIA: see DOBROYE, SAKHALIN oblast.

Naibuchi, RUSSIA: see BYKOV.

Naibuchi River, RUSSIA: see NAYBA RIVER.

Naic (NAH-eek), town, CAVITE province, S LUZON, PHILIPPINES, near MANILA BAY, 24 mi/39 km SW of MANILA; 14°18'N 120°48'E. Fishing and agricultural center (rice).

Naica (nah-EE-kah), mining settlement, CHIHUAHUA, N MEXICO, in E outliers of SIERRA MADRE OCCIDENTAL, 70 mi/113 km SE of CHIHUAHUA; 27°51'N 105°30'W.

Naicam (NAI-kuhm), town (2006 population 690), central SASKATCHEWAN, W CANADA, 30 mi/48 km ENE of HUMBOLDT; 52°25'N 104°30'W. Grain elevators.

Nai Dab Wali, INDIA: see DABWALI.

Naie (NAH-ee-e), town, Sorachi district, Hokkaido prefecture, N JAPAN, 37 mi/60 km N of SAPPORO; 43°25'N 141°53'E.

Naigawan Rebai (NEI-gah-wahn RAI-bei), former petty state of CENTRAL INDIA agency. In 1948, merged with VINDHYA PRADESH, later with MADHYA PRADESH state.

Naiguatá (nah-ee-gwah-TAH), town, VARGAS state, N VENEZUELA, minor port on the CARIBBEAN SEA, at foot of coastal range, 13 mi/21 km E of LA GUAIRA; 10°38'N 66°44'W. Fishing.

Naiguatá, Pico (nei-gwah-TAH, PEE-ko), peak (9,070 ft/2,765 m) in N Caribbean coastal range of VENEZUELA, 10 mi/16 km ENE of CARACAS; 10°33'N 66°46'W. Highest elevation in Cadena del Litoral and the entire CORDILLERA DE LA COSTA.

Naihati (NEI-hah-tee), town, NORTH 24-PARGANAS district, SE WEST BENGAL state, E INDIA, on HUGLI River (bridge), 22 mi/35 km N of CALCUTTA city center. An important railroad junction. Jute, rice, oilseeds; paper milling, paint manufacturing. Old cultural center of BENGAL. GOURIPUR is N suburb; large power station.

Naihoro, RUSSIA: see GORNOZAVODSK, SAKHALIN oblast.

Naikawa, RUSSIA: see TIKHMENEVO.

Naila (NEI-lah), town, UPPER FRANCONIA, NE BAVARIA, GERMANY, in FRANCONIAN FOREST, on small Selbitz River, and 9 mi/14.5 km W of HOF; 50°20'N 11°42'E. Manufacturing (machinery, textiles). Chartered 1454. Town since 1566.

Naila, INDIA: see JANJGIR.

Naila Janjgir, INDIA: see JANJGIR.

Nailsea (NAHL-see), town (2001 population 16,546), North Somerset, SW ENGLAND, 8 mi/12.9 km WSW of BRISTOL; 51°26'N 02°45'W. Has 15th-century church. Former coal-mining site.

Nailsworth (NAILZ-wuhth), town (2001 population 6,075), central GLOUCESTERSHIRE, W ENGLAND, 12 mi/19.2 km S of GLOUCESTER; 51°41'N 02°13'W. Previously ham and bacon curing.

Naim, ancient city, ISRAEL: see NEIN.

Naiman, KYRGYZSTAN: see NAYMAN.

Naimisharanya, INDIA: see NIMKHAR.

Nain (NAIN), town (2006 population 1,034), NE NEWFOUNDLAND AND LABRADOR, E Canada, on inlet of the Atlantic Ocean, 98 mi/158 km N of HOPEDALE; 56°33'N 61°41'W. Established as Moravian mission in 1771. Fishing port.

Nain (nah-EEN), town, Esfahān province, central IRAN, near railroad, 80 mi/129 km ENE of ESFAHAN, and on road to YAZD; 32°51'N 53°05'E. Important road junction. Grain, nuts; wool and carpet weaving.

Nain, town, SAINT ELIZABETH parish, SW JAMAICA, 8 mi/13 km SW of MANDEVILLE; 17°58'N 77°36'W. Road junction town. Airfield.

Nainital (NEI-nee-thahl) [tal=lake], district (□ 1,316 sq mi/3,421.6 sq km; 2001 population 762,909), KUMAON division, S Uttarakhand state, N central INDIA; ⊙ NAINI TAL. Crossed N by central SHIWALIK RANGE; bounded E by SARDA RIVER (Nepal border). BHABAR tract, a submontane area of gravel and shingle, in S; dense sal jungle; growing of oilseeds. Agriculture (rice, wheat, gram, corn, barley, sugarcane); a major Indian sugar-processing district. Main towns are NAINITAL and HALDWANI. Popular tourist destination; mentioned often in literature. NAINI TAL, Jim Corbett National Park, Bhim Tal, Saat Tal are attractions; water sports, hiking. In 1995, UDHAM SINGH

NAGAR district carved out of S Nainital district. During 1990s, concerns among environmentalists, scientists, and conservationists began to grow due to the increasing number of tourists and their negative effect on environment; government has since tried to implement measures reducing environmental degradation to the area.

Nainital (NEI-nee TAHL), town (2001 population 38,559), ⊙ Kumaon division and NAINITAL district, S Uttarakhand state, N central INDIA, 145 mi/233 km ENE of DELHI, in outer Kumaon Himalayas. Popular hill resort (elevation 6,346 ft/1,934 m); former summer headquarters of UTTAR PRADESH government. Seat of Kumaon University and several boarding schools, including Sherwood College, St. Joseph's College, St. Mary's Convent, and All Saints' College. Derives name from picturesque NAINI Lake [tal = lake] (4,703 ft/1,433 m long, 1,518 ft/463 m wide); yacht club; sulphur springs near S end. Suffered 1880 from severe landslide. Town founded 1841. China Peak (8,565 ft/2,611 m) is 1 mi/1.6 km NW; highest point in district. At MUKTESWAR, 13 mi/21 km NE, is Imperial Veterinary Research Institute, founded 1890 at PUNE and removed 1893 to present site. Corbett National Park 43 mi/70 km to W. Also known as Naini Tal.

Naini Tal (NEI-nee-thahl) town, NAINITAL district, S Uttarakhand state, N central INDIA; elevation 6,346 ft/1,934 m; 4,703 ft/1,433 m long, 1,518 ft/463 m wide. Situated in KUMAON foothills of outer Himalayas; surrounded by mountains, including Naini Peak (also known as China Peak), from which name is derived [tal =lake]. Popular resort area.

Nainpur, INDIA: see MANDLA.

Naintré (nan-TRAI), town (□ 9 sq mi/23.4 sq km), VIENNE department, POITOU-CHARENTES region, W FRANCE, 13 mi/ 21 km NNW of POITIERS; 46°46'N 00°29'E.

Nainwa, INDIA: see NAENWA.

Naipalganj Road, INDIA: see NANPARA.

Naira, INDONESIA: see BANDANAIRA.

Nairai (NEI-rei), volcanic island (□ 9 sq mi/23.4 sq km), FIJI, SW PACIFIC OCEAN, c.50 mi/80 km E of VITI LEVU; 5 mi/8 km long; 17°49'S 179°24'E. Copra, bananas. Once known for mats, baskets.

Nairn (NERN), town (2001 population 8,418), HIGHLAND, N Scotland, at mouth of the NAIRN RIVER, on MORAY FIRTH, and 12 mi/19.2 km NE of INVERNESS; 57°35'N 03°52'W. Tourist resort and fishing harbor. Other industries include dairy and crop farming and whiskey distilling. Granite quarrying. Cawdor Castle, legendary scene of the murder of Duncan by Macbeth, is nearby.

Nairn and Hyman (NERN, HEI-muhn), township (□ 61 sq mi/158.6 sq km; 2001 population 420), S central ONTARIO, E central Canada; 46°21'N 81°38'W. Formed 1896. Also written Nairn & Hyman.

Nairne (NERN), village, SE SOUTH AUSTRALIA state, S central AUSTRALIA, 20 mi/32 km ESE of ADELAIDE; 35°02'S 138°54'E. Dairy products; livestock.

Nairn River (NERN), 38 mi/61 km long, HIGHLAND, N Scotland; rises 15 mi/24 km NE of FORT AUGUSTUS; flows NE, past Daviot, to MORAY FIRTH at NAIRN.

Nairnshire (NERN-shir), administrative district of HIGHLAND, N Scotland; 57°30'N 03°50'W. Fishing; dairy products.

Nairo, RUSSIA: see GASTELLO.

Nairobi (nah-ee-RO-bee), city (2007 population 2,940,911), ⊙ KENYA, S Kenya, in the East AFRICA highlands. Kenya's largest city and its administrative, communications, and economic center. Trade and distribution center for a productive agricultural area specializing in coffee, tea, and cattle. Has a large industrial complex that makes motor vehicles, food products, beverages, construction materials, cigarettes, chemicals, textiles, clothing, glass, and furniture. Linked by road with the rest of Kenya and by railroad with MOMBASA (on the INDIAN OCEAN coast),

W Kenya, and UGANDA. Although Nairobi is only 90 mi/145 km S of the equator, it has a moderate climate, largely because of its high elevation (c.5,500 ft/1,680 m). Many tourists are attracted to NAIROBI NATIONAL PARK, a large wildlife sanctuary on the city's outskirts, and to nearby scenic areas. Founded in 1899 on the site of a waterhole of the pastoral Maasai as a railhead camp on the Mombasa-Uganda line. In 1905 it replaced Mombasa as capital of the British East Africa Protectorate (known as Kenya Colony, 1920–1963). Nairobi became the center of the prosperous European-dominated highlands farming area. In the 1950s the Mau Mau insurgency flared among Kikuyu people near Nairobi; there were related disturbances in the city. Seat of Nairobi University, Kenyatta University, and several medical and technical schools. The National Museum of Kenya has extensive collections of Kenya's prehistory and natural history and the Sorsbie art gallery is here. Seat of the national railroad and airway corporations. Many international organizations have their African headquarters in the area, including the UN's Environmental Program (UNEP). International airport to E, regional airport to S.

Nairobi National Park (nah-ee-RO-bee) (□ 44 sq mi/114.4 sq km), KENYA, 5 mi/8 km SE of NAIROBI. Wildlife features lion, gazelle, cheetah, leopard, giraffe, and buffalo.

Nair, Piz, SWITZERLAND: see PIZ NAIR.

Naissaar (NEI-sahr), German *Nargen*, ESTONIAN island (□ 7 sq mi/18.2 sq km), in GULF OF FINLAND, 6 mi/10 km NW of TALLINN; 5 mi/8 km long, 2 mi/3.2 km wide; 59°34′N 24°30′E. Lighthouse.

Naitauba (nei-TAHM-bah), volcanic and limestone island (□ 3 sq mi/7.8 sq km), LAU group, FIJI, SW PACIFIC OCEAN; 2 mi/3.2 km long. Copra. Sometimes spelled Naitamba or Naitaumba.

Naivasha (nah-ee-VA-shah), town (1999 population 32,222), RIFT VALLEY province, W central KENYA, on NE shore of Lake NAIVASHA, on railroad, and 50 mi/80 km NNW of NAIROBI. Elevation 6,231 ft/1,899 m. Popular resort; agriculture (wheat) and livestock (cattle, horses); dairying center. HELL'S GATE NATIONAL PARK located to S. Ostriches farmed nearby.

Naivasha (nah-ee-VA-shah), lake, RIFT VALLEY province, W central KENYA, E AFRICA, in the GREAT RIFT VALLEY, 68 mi/110 km NW of NAIROBI; 12 mi/19 km long and 9 mi/14.5 km wide. Large flower farms that supply European flower markets have been developed nearby. Lake and town of NAIVASHA are a popular resort. Waterfowl, and hippopotamus abound here. Small fish population.

Najac (nah-ZHAK), agricultural commune (□ 20 sq mi/52 sq km), AVEYRON department, MIDI-PYRÉNÉES region, S FRANCE, on height above AVEYRON RIVER, in ROUERGUE district, 9 mi/14.5 km SSW of VILLEFRANCHE-DE-ROUERGUE; 44°13′N 01°59′E. Two resort villages developed nearby. Ruins of 12th–13th-century fortress and main street lined with 13th–16th-century houses.

Najaf, Arabic *An Najaf*, province, SW IRAQ, on SAUDI ARABIA (SW) border; ☉ NAJAF; 31°59′N 44°20′E. Bordered SE by MUTHANNA, W by KARBALA, and N by BABYLON provinces. SYRIAN DESERT dominates province.

Najaf (NUH-zhuhf), Arabic, *An Najaf*, city, ☉ NAJAF province, S central IRAQ, on a lake W of the EUPHRATES River; 31°33′N 45°13′E. Scene of violent clashes between followers of Shiite cleric Muqtada al-Sadr and U.S.-led Coalition troops after the Iraq war, espeically in spring and summer of 2004. Sometimes called Mashad Ali, after the tomb of Ali, son-in-law of the prophet Muhammad. The tomb is an object of pilgrimage by Shiite Muslims and a starting point for the pilgrimage to MECCA. Also spelled Nejef.

Najafabad (nah-jah-fah-BAHD), city (2006 population 208,647), Esfahān province, W central IRAN, c.20 mi/32 km W of ESFAHAN. It is the trade center for an agricultural region noted for its pomegranates.

Najafgarh (NUH-juhf-guhr) or **Najafgarli**, town, DELHI state, N INDIA, on the outskirts of the NATIONAL CAPITAL REGION.

Najasa (nah-HAH-sah), town, CAMAGÜEY province, E central CUBA, at crossroads of secondary highways, 23 mi/37 km SSE of CAMAGÜEY city, at foot of hills with same name; 21°10′N 77°45′W.

Najasa River (nah-HAH-sah), 64 mi/104 km long, CAMAGÜEY province, E CUBA; rises SE of CAMAGÜEY; flows SW and S to the CARIBBEAN SEA, 4 mi/6 km E of SANTA CRUZ DEL SUR. Sometimes called San Juan de Najasa River. Small Sierra Najasa lies along its upper (left) course. Drains 346 sq mi/895 sq km.

Naj'd, SAUDI ARABIA: see CENTRAL REGION.

Nájera (NAH-hai-rah), town, LA RIOJA province, N SPAIN, 14 mi/23 km WSW of LOGROÑO, at foot of hill crowned by ruins of ancient castle, and on banks of NAJERILLA RIVER; 42°25′N 02°44′W. Manufacturing of furniture, wine, liquor, chocolate; tanning; food processing. Cereals, grapes, livestock in area. Romanesque monastery of Santa María la Real (founded in 11th century) has church (rebuilt 15th century) with tombs of Navarrese and Castilian kings, and fine 16th-century cloisters. Was frequently residence of kings of Navarre in 11th century. Peter IV defeated (1367) Henry of Trastamara near here.

Najerilla River (nah-hai-REE-lyah), c.40 mi/64 km long, LA RIOJA province, N SPAIN; rises on outer N edge of central plateau; flows NNE, past NÁJERA, to the EBRO RIVER, 18 mi/29 km W of LOGROÑO.

Naj' Hamadi, EGYPT: see NAG HAMMADI.

Najibabad (nuh-JEE-bah-bahd), town, BIJNOR district, N UTTAR PRADESH state, N central INDIA, 60 mi/97 km NNW of MORADABAD. Trade center (rice, wheat, gram, barley, sugarcane, oilseeds, cotton, timber); sugar processing, blanket manufacturing. Has ruins of Afghan fort, built 1775 by town's founder, Najib-ud-daula, and his tomb. Glass manufacturing 14 mi/23 km W, at BALAWALI village.

Najin (NAH-jeen), city, NORTH HAMGYONG province, NORTH KOREA, 40 mi/64 km NE of CHONGJIN, on SEA OF JAPAN; 42°10′N 130°05′E. Naval base. The commercial fishing port (ice-free) is connected by 10-mi/16-km tunnel with nearby port of UNGGI. In World War II Najin was chief Japanese naval base in Korea; captured August 13, 1945, by the Russians. Connects primary railroads. One of the three port cities included in North Korea's ZONE OF FREE ECONOMY AND TRADE.

Najran (nuhzh-RAHN), town and oasis (2004 population 246,880), ASIR hinterland, ☉ Najran province, SAUDI ARABIA, near YEMEN border; 17°25′N 44°15′E. Agriculture (dates, alfalfa, wheat, millet; livestock raising). Mainly inhabited by the Yam tribes. Reached (24 B.C.E.) by a Roman expedition from EGYPT; seat of important Christian colony (500–635); visited (1869–1870) by the French orientalist Joseph Halévy. Najran and the surrounding area were annexed by Saudi Arabia in 1934 following a war with Yemen, when its tribes accepted Saudi rule. Also spelled Nejran.

Naj Tunich Caves (nah too-NEECH), 14 mi/23 km E of POPTÚN, PETÉN department, GUATEMALA, near BELIZE border. Cavern system with Maya drawings on walls and artifacts. No road access.

Naju (NAH-JOO), city (2005 population 87,212), SOUTH CHOLLA province, SOUTH KOREA, 25 mi/40 km NE of MOKPO; 35°01′N 126°43′E. Commercial center for fruit-growing area (pears, peaches).

Naka (NAH-kah), town, Taka district, HYOGO prefecture, S HONSHU, W central JAPAN, 28 mi/45 km N of KOBE; 35°02′N 134°55′E. Textile dyeing.

Naka, town, Naka county, IBARAKI prefecture, central HONSHU, E central JAPAN, 6 mi/10 km N of MITO; 36°26′N 140°29′E. Nuclear fusion laboratory.

Nakabaru (nah-KAH-bah-roo), town, Miyaki county, SAGA prefecture, N KYUSHU, SW JAPAN, 9 mi/15 km N of SAGA; 33°20′N 130°26′E.

Nakada (nah-KAH-dah), town, Tome county, MIYAGI prefecture, N HONSHU, NE JAPAN, 34 mi/55 km N of SENDAI; 38°42′N 141°14′E.

Nakada, EGYPT: see NAQADA.

Naka-dake, JAPAN: see ASO-ZAN.

Nakadori-shima (nah-kah-DO-ree–SHEE-mah), second-largest island (□ 67 sq mi/174.2 sq km) of island group GOTO-RETTO, NAGASAKI prefecture, SW JAPAN, 25 mi/40 km W of KYUSHU; 24 mi/39 km long (with long, narrow N and S peninsulas), 10 mi/16 km wide. Deeply indented coastline. Whaling, fishing. Sometimes called Nakadori. Chief town, ARIKAWA.

Nakafurano (NAH-kah-foo-RAH-no), town, Kamikawa district, Hokkaido prefecture, N JAPAN, 62 mi/100 km E of SAPPORO; 43°24′N 142°25′E. Rice; watermelons, burdock, lavender (and lavender products), tea, coffee.

Nakagawa (nah-KAH-gah-wah), town, Chikushi county, FUKUOKA prefecture, N KYUSHU, SW JAPAN, 6 mi/10 km S of FUKUOKA; 33°29′N 130°25′E. Bullet-train testing center.

Nakagawa, town, Kamikawa district, Hokkaido prefecture, N JAPAN, 127 mi/205 km N of SAPPORO; 44°48′N 142°04′E. Ammonite. Kubinaga Ryu fossils found here.

Nakagawa, town, Naka county, TOKUSHIMA prefecture, SE SHIKOKU, W JAPAN, 9 mi/15 km S of TOKUSHIMA; 33°56′N 134°39′E.

Nakagawa (nah-KAH-gah-wah), village, Kamiina county, NAGANO prefecture, central HONSHU, central JAPAN, 71 mi/115 km S of NAGANO; 35°37′N 137°56′E.

Nakago (nah-KAH-go), village, Nakakubiki county, NIIGATA prefecture, central HONSHU, N central JAPAN, 74 mi/120 km S of NIIGATA; 36°58′N 138°13′E. Walnuts, tomatoes.

Nakagusuku (NAH-kah-GOOS-koo), village, S OKINAWA island, Nakagami county, Okinawa prefecture, SW JAPAN, 9 mi/15 km N of NAHA; 26°15′N 127°47′E.

Nakagusuku Bay (nah-kah-GOOS-koo), inlet of PHILIPPINE SEA, in SE Okinawa prefecture, OKINAWA Islands, in the RYUKYUS, SW JAPAN; 12 mi/19 km long, 5 mi/8 km wide. Best harbor of island was major Japanese naval and seaplane base in World War II. Also known as Buckner Bay since American occupation, named for General Simon Bolivar Buckner. Formerly sometimes called Mathews.

Nakahechi (nah-KAH-he-chee), town, West Muro county, WAKAYAMA prefecture, S HONSHU, W central JAPAN, 37 mi/59 km S of WAKAYAMA; 33°47′N 135°31′E. Kumano Kodo here is among the nation's best-preserved traditional streetscapes. Nonaka-no-Shimizu is nearby.

Nakai (nah-KAH-ee), town, Ashigarakami county, KANAGAWA prefecture, E central HONSHU, E central JAPAN, 25 mi/40 km S of YOKOHAMA; 35°19′N 139°13′E.

Nakaizu (nah-KAH-eez), town, Tagata county, SHIZUOKA prefecture, central HONSHU, E central JAPAN, 34 mi/55 km E of SHIZUOKA; 34°57′N 139°00′E. Wasabi, hollyhock, shiitake mushrooms.

Nakajima (nah-KAH-jee-mah), town, Onsen county, EHIME prefecture, NW SHIKOKU, W JAPAN, 12 mi/20 km N of MATSUYAMA; 33°58′N 132°37′E. Mandarin oranges, lemons.

Nakajima, town, Kashima county, ISHIKAWA prefecture, central HONSHU, central JAPAN, 40 mi/65 km N of KANAZAWA; 37°06′N 136°51′E.

Nakajima (nah-kah-JEE-mah), village, W Shirakawa county, FUKUSHIMA prefecture, N central HONSHU, NE JAPAN, 40 mi/65 km S of FUKUSHIMA city; 37°08′N 140°21′E.

Nakajo (nah-KAH-jo), town, North Kanbara county, NIIGATA prefecture, central HONSHU, N central JAPAN, 22 mi/35 km N of NIIGATA; 38°03′N 139°24′E. Pigs.

Cross-references are shown in SMALL CAPITALS. The pronunciation guide is shown on page xix. The sources of population figures are shown on page xvii.

Nakajo (nah-KAH-jo), village, Kamiminochi county, NAGANO prefecture, central HONSHU, central JAPAN, 9 mi/15 km W of NAGANO; 36°36′N 138°02′E.

Nakakawane (nah-KAH-kah-WAH-ne), town, Haibara county, SHIZUOKA prefecture, central HONSHU, E central JAPAN, 19 mi/30 km N of SHIZUOKA; 35°02′N 138°05′E. Tea.

Nakama (nah-KAH-mah), city, FUKUOKA prefecture, N KYUSHU, SW JAPAN, 25 mi/40 km N of FUKUOKA; 33°48′N 130°42′E.

Nakambe River, c.550 mi/885 km long, in BURKINA FASO and GHANA, W AFRICA; rises in Burkina Faso N of OUAHIGOUYA; flows through Lake Bam, enters Ghana NW of BAWKU; flows generally S past YAPEI, joins the MOUHOUN RIVER, 38 mi/61 km NW of Yeji, to form VOLTA RIVER. Drains most of upper and N regions of Ghana. Receives NAZINON RIVER near GAMBAGA. Formerly called White Volta River.

Nakamichi (nah-kah-MEE-chee), town, East Yatsushiro county, YAMANASHI prefecture, central HONSHU, central JAPAN, 5 mi/8 km S of KOFU; 35°35′N 138°34′E.

Nakaminato (NAH-kah-mee-NAH-to), city, IBARAKI prefecture, central HONSHU, E central JAPAN, on the PACIFIC OCEAN, 9 mi/15 km E of MITO; 36°20′N 140°36′E.

Nakamti, ETHIOPIA: see NEKEMTI.

Nakamun Park, summer village (2001 population 31), central ALBERTA, W Canada, 38 mi/61 km from EDMONTON, in Lac Sainte Anne County; 53°53′N 114°12′W. Incorporated 1966.

Nakamura (nah-KAH-moo-rah), city, KOCHI prefecture, SW SHIKOKU, W JAPAN, on Shimando River, 53 mi/85 km S of KOCHI; 32°59′N 132°56′E.

Nakanai Mountains, mountain range, WEST NEW BRITAIN province, rising to 7,169 ft/2,185 m, E PAPUA NEW GUINEA; 05°30′S 151°12′E.

Nakanida (nah-kah-NEE-dah), town, Kami county, MIYAGI prefecture, N HONSHU, NE JAPAN, 22 mi/35 km N of SENDAI; 38°34′N 140°51′E.

Nakano (nah-KAH-no), ward, TOKYO city, Tokyo prefecture, E central HONSHU, E central JAPAN, W of central Tokyo. Bordered N by NERIMA ward, E by SHINJUKU and TOSHIMA wards, S by Shibuya ward, and W by SUGINAMI ward.

Nakano (nah-KAH-no), city, NAGANO prefecture, central HONSHU, central JAPAN, 12 mi/20 km N of NAGANO; 36°44′N 138°22′E. Apples, grapes, asparagus, enoki mushrooms. Woodwork. Hot springs.

Nakanojo (nah-KAH-no-jyo), town, Agatsuma county, GUMMA prefecture, central HONSHU, N central JAPAN, 22 mi/35 km N of MAEBASHI; 36°35′N 138°50′E. Konnyaku (paste made from devil's tongue), pickles. Hot springs.

Nakanokuchi (nah-KAH-NOK-chee), village, West Kanbara county, NIIGATA prefecture, central HONSHU, N central JAPAN, 12 mi/20 km S of NIIGATA; 37°42′N 138°58′E.

Nakanoshima (nah-kah-NO-shee-mah), town, South Kanbara county, NIIGATA prefecture, central HONSHU, N central JAPAN, 25 mi/40 km S of NIIGATA; 37°32′N 138°52′E. Lotus root.

Naka-no-shima (nah-KAH–no–SHEE-mah), island (□ 13 sq mi/33.8 sq km) of DOZEN group of the OKIGUNTO, SHIMANE prefecture, W JAPAN, in SEA OF JAPAN, 7 mi/11.3 km SW of DOGO, just E of NISHI-NO-SHIMA; 6 mi/9.7 km long, and mi/6.4 km wide. Mountainous; cattle raising, fishing; raw silk.

Nakano-shima, JAPAN: see TOKARA-RETTO.

Nakapanya (nah-kah-PAH-nyah), village, RUVUMA region, S TANZANIA, 35 mi/56 km ENE of Tundura, near MUHUVESI RIVER; 10°57′S 37°52′E. Timber; livestock; subsistence crops.

Nakapiripirit, administrative district (2005 population 180,400), NORTHERN region, E UGANDA, on KENYA border (to E); ⊙ Nakapiripirit; 01°55′N 34°40′E. Elevation averages 3,281 ft/1,000 m–5,249 ft/1,600 m. As of Uganda's division into eighty districts, Nakapir-

ipirit's border districts include KAPCHORWA and SIRONKO (S), KATAKWI (W), and MOROTO (N). Primary languages are Ngakarimojong and Pokot. Fertile agricultural area (especially sorghum and maize, also including bananas, beans, groundnuts, peas, simsim, sugar cane, and sunflowers). Livestock (including goats, cattle, and donkeys). Agriculture and livestock raising are the district's major economic activities. Mount KADAM (10,063 ft/3,067 m) to SE of Nakapiripirit town. Towns include Nakapiripirit and Namalu. Formed in 2001 from S portion of former MOROTO district (current Moroto district formed from N portion).

Nakasato (nah-KAH-sah-to), town, N Tsugaru county, Aomori prefecture, N HONSHU, N JAPAN, 19 mi/30 km N of AOMORI; 40°57′N 140°26′E. Lumber; tea.

Nakasato (nah-KAH-sah-to), village, Tano county, GUMMA prefecture, central HONSHU, N central JAPAN, 28 mi/45 km S of MAEBASHI; 36°05′N 138°49′E. Fossilized dinosaur prints (first such discovery in Japan).

Nakasato, village, Nakauonuma county, NIIGATA prefecture, central HONSHU, N central JAPAN, 62 mi/100 km S of NIIGATA; 37°03′N 138°42′E. Lily of the valley bulbs; rice.

Nakasatsunai (NAH-kah-SAHTS-nah-ee), village, Tokachi district, Hokkaido prefecture, N JAPAN, 96 mi/155 km E of SAPPORO; 42°41′N 143°08′E. Beans, potatoes; poultry and meat products.

Nakaseke, administrative district, CENTRAL region, central UGANDA. As of Uganda's division into eighty districts, borders NAKASONGOLA (N), LUWERO (E), WAKISO and MITYANA (S), KIBOGA (W), and MASINDI (NW) districts. Marsh area. Primarily agricultural area; livestock also raised here. Formed in 2005 from W portion of former LUWERO district created in 1997 (current Luwero district formed from E portion).

Nakasen (nah-KAH-sen), town, Senhoku county, Akita prefecture, N HONSHU, NE JAPAN, 25 mi/40 km S of AKITA city; 39°32′N 140°32′E.

Nakashibetsu (nah-KAH-shee-BETS), town, Kushiro district, Hokkaido prefecture, N JAPAN, 186 mi/300 km E of SAPPORO; 43°32′N 144°59′E. Potatoes; dairying, cheese making.

Naka-shima (nah-KAH–shee-mah), island (□ 9 sq mi/23.4 sq km), EHIME prefecture, W JAPAN, in IYO SEA, 7 mi/11.3 km off NW coast of SHIKOKU, near MATSUYAMA; 5 mi/8 km long, 2 mi/3.2 km wide. Hilly; fertile (citrus fruit, sweet potatoes, ginger). Fishing.

Naka-shiretoko-misaki, RUSSIA: see ANIVA, CAPE.

Nakasongola, administrative district (2005 population 131,400), CENTRAL region, central UGANDA; ⊙ NAKASONGOLA; 01°20′N 32°30′E. As of Uganda's division into eighty districts, borders APAC and AMOLATAR (N, formed by VICTORIA NILE RIVER and LAKE KYOGA), KAYUNGA (E), LUWERO (S), NAKASEKE (SWW), and MASINDI (NW) districts. Agriculture (cotton, maize, bananas). Marsh area W of NAKASONGOLA town. Nakasongola town on road between KAMPALA and MASINDI towns. Created in 1997 from NE portion of former LUWERO district ([now former] new LUWERO district was created from S and W portions).

Nakasongola (na-kah-sun-GO-lah), town (2002 population 6,499), ⊙ NAKASONGOLA district, CENTRAL region, central UGANDA, 60 mi/95 km N of KAMPALA; 01°19′N 32°32′E. Agricultural center (cotton, maize, bananas). Military airfield.

Nakatane (nah-KAH-tah-ne), town, in central part of TANEGA-SHIMA island, Kumage county, KAGOSHIMA prefecture, SW JAPAN, on EAST CHINA SEA, 81 mi/130 km S of KAGOSHIMA; 30°31′N 130°57′E. Agricultural center; sugarcane, tobacco, sweet potatoes, peas; flowers.

Nakatomi (nah-kah-TO-mee), town, South Koma county, YAMANASHI prefecture, central HONSHU, central JAPAN, 16 mi/25 km S of KOFU; 35°27′N 138°26′E. Pulp.

Nakatonbetsu (nah-kah-TON-bets), town, Soyashi district, Hokkaido prefecture, N JAPAN, 140 mi/225 km N of SAPPORO; 44°58′N 142°17′E.

Nakatosa (nah-KAH-to-sah), town, Takaoka county, KOCHI prefecture, S SHIKOKU, W JAPAN, 25 mi/40 km S of KOCHI; 33°19′N 133°13′E.

Nakatsu (nah-KAHTS), city, OITA prefecture, NE KYUSHU, SW JAPAN, on the Suo Sea, at the mouth of the Yamakuni River, 34 mi/55 km N of Oita; 33°35′N 131°11′E. Commercial center and port manufacturing traditional umbrellas. Yukichi Fukuzawa (1835–1901), Japan's most famous early modern thinker, born here. Nakatsu Castle here has a moat (an unusual feature in Japan).

Nakatsu (nah-KAHTS), village, Hidaka county, WAKAYAMA prefecture, S HONSHU, W central JAPAN, 19 mi/31 km S of WAKAYAMA; 33°57′N 135°17′E. Hassaku oranges, shiitake mushrooms, taro; senryo trees; guinea fowl. Vegetable processing.

Nakatsue (nah-KAHTS-e), village, Hita county, OITA prefecture, E KYUSHU, SW JAPAN, 40 mi/65 km W of OITA; 33°07′N 130°56′E. Cryptomeria forest.

Nakatsugawa (nah-KAHTS-gah-wah), city, GIFU prefecture, central HONSHU, central JAPAN, 43 mi/70 km E of GIFU; 35°29′N 137°30′E. Electronic goods manufacturing, chestnut processing. Ena Gorge nearby.

Naka-umi (nah-KAH–OO-mee), lagoon (□ 39 sq mi/101.4 sq km) in SHIMANE prefecture, SW HONSHU, W JAPAN, connected with LAKE SHINJI (W) and inlet of SEA OF JAPAN (E); 10 mi/16 km long, 7 mi/11.3 km wide. Contains several islets; largest, DAIKON-JIMA. MATSUE is on W shore, YONAGO on SE shore.

Nakawn, in Thai names: see NAKHON.

Nakayama (nah-KAH-yah-mah), town, Iyo county, EHIME prefecture, W SHIKOKU, W JAPAN, 12 mi/20 km S of MATSUYAMA; 33°38′N 132°42′E. Chestnuts.

Nakayama, town, Saihoku county, TOTTORI prefecture, S HONSHU, W JAPAN, 37 mi/60 km W of TOTTORI; 35°31′N 133°36′E. Fruit, broccoli, rice, tobacco; livestock. Dumplings.

Nakayama, town, East Murayama county, YAMAGATA prefecture, N HONSHU, NE JAPAN, 6 mi/10 km N of YAMAGATA city; 38°19′N 140°17′E. Rice, fruit.

Nakazato (nah-KAH-zah-to), village, Shimajiri county, Okinawa prefecture, SW JAPAN, 56 mi/90 km W of NAHA; 26°20′N 126°48′E.

Nakchukha, CHINA: see NAGQU.

Nakdong River, Korean Nakdong-gang, 326 mi/525 km long, largest river in SOUTH KOREA; rises in mountains just SW of SAMCHOK; flows S, then W and generally SE past Samnangjin to KOREA STRAIT near PUSAN. Navigable 214 mi/344 km by motorboat. Lower course drains large agricultural area. Was major Allied defense line in Korean War (1950).

Nakel, POLAND: see NAKLO.

Nakfa (NAHK-fah), Italian Nacfa, town, SEMENAWI KAYIH BAHRI region, N ERITREA, on road, and 62 mi/100 km N of Keren; 16°40′N 38°30′E. In cattle-raising region. Was a major base for Eritrean liberation forces and was heavily damaged during the war with ETHIOPIA. Has airfield, technical school, trade center; hot springs nearby. Also known as Nagfa.

Nakhabino (nah-HAH-bee-nuh), town (2006 population 27,915), central MOSCOW oblast, central European Russia, on crossroads and railroad, 17 mi/27 km WNW of MOSCOW, and 5 mi/8 km W of KRASNOGORSK, to which it is administratively subordinate; 55°50′N 37°10′E. Elevation 629 ft/191 m. Summer resort; furniture, toys.

Nakhichevan (nuh-khee-che-VAHN), ancient Naxuana, city, ⊙ NAKHICHEVAN AUTONOMOUS REPUBLIC, in AZERBAIJAN. Its industries include manufacturing (furniture); electrical-engineering plant; tobacco-fermentation enterprises. Ruled by Armenians, Persians, Arabs, Mongols, and Turks; became a flourishing ARMENIAN trade center in the 15th century and was

---

Area is shown by the symbol □, and capital city or county seat by ⊙.

ceded by PERSIA to RUSSIA by the Treaty of Turkmanchai (1828). In the 19th century it was an important trading post between Persia and Russia. Nakhichevan has Greek and Roman remains and two 12th century mausoleums.

Nakhichevan Autonomous Republic (nuh-khee-che-VAHN), autonomous republic (□ 2,124 sq mi/5,522.4 sq km), SE AZERBAIJAN, on IRAN and TURKEY (both S) and ARMENIA (N) borders; ⊙ NAKHICHEVAN. ORDUBAD and Dzhulfa are the other main cities. The lowlands are irrigated and produce cotton, tobacco, rice, winter wheat, and fruits. In the foothills grapes are grown for the wine industry, and silkworms are raised. There are salt, molybdenum, lead, and zinc deposits. The republic's industries include food processing, cotton cleaning, and the bottling of mineral water. The population consists mainly of Azerbaijani Turks (82%), with Russian and Armenian minorities. The republic was founded in 1924. It has a 110-member parliament.

Nakhl (NAH-kuhl), desert station, SINAI province, EGYPT, SINAI peninsula, in the center of the plateau of Et TIH, 60 mi/97 km WNW of AQABA; 29°55′N 33°45′E. Airfield.

Nakhodka (nah-KOT-kuh), city (2006 population 146,285), S MARITIME TERRITORY, SE RUSSIAN FAR EAST, on highway junction and spur railroad, 105 mi/169 km E of VLADIVOSTOK; 42°50′N 132°53′E. A port city on the Nakhodka Bay of the Sea of JAPAN, with fewer winter ice problems than Vladivostok, Nakhodka has assumed an increasingly large share of shipping from the Russian Far East. Base for a fishing fleet and commercial passenger lines; ship repair yard; tin can factory; fisheries. Made city in 1950.

Nakhon Chai Si (NUH-KAWN CHEI SEE), village and district center, NAKHON PATHOM province, S THAILAND, on THA CHIN RIVER, on railroad, and 20 mi/32 km W of BANGKOK; 13°48′N 100°11′E. Rice and sugar milling; sugarcane plantation; fruit gardens.

Nakhon Nayok (NUH-KAWN NAH-YOK), province (□ 834 sq mi/2,168.4 sq km), Central region, THAILAND; ⊙ NAKHON NAYOK; 14°00′N 101°06′E. Rice, corn, fruit; animal husbandry. Site of the elite Chulachomklao Royal Military Academy. Contains part of KHAO YAI NATIONAL PARK.

Nakhon Nayok (NUH-KAWN NAH-YOK), town, ⊙ NAKHON NAYOK province, E THAILAND, on NAKHON NAYOK RIVER, and 66 mi/106 km NE of BANGKOK; 14°12′N 101°13′E. Rice center.

Nakhon Nayok River (NUH-KAWN NAH-YOK), 40 mi/64 km long, S THAILAND; rises in SAN KAMPHAENG RANGE (at elevation c.3,000 ft/914 m); flows S, past NAKHON NAYOK, to BANG PAKONG RIVER. Linked to PA SAK RIVER canal system.

Nakhon Pathom (NUH-KAWN PUH-TOM) province, Central region, SW THAILAND; ⊙ NAKHON PATHOM; 13°59′N 100°05′E. Rice, fruit, sugarcane; animal husbandry. Some industrial production. Transportation and communication center.

Nakhon Pathom (NUH-KAWN PUH-TOM), city, ⊙ NAKHON PATHOM province, SW THAILAND, 34 mi/55 km from BANGKOK, on the MEKONG RIVER; 13°49′N 100°03′E. A transportation and commercial center on the BANGKOK-SINGAPORE railroad. Home to Thailand's most sacred Buddhist monument, a shrine 380 ft/116 m tall, which marks the spot where Buddhism was introduced to the region 2,300 years ago.

Nakhon Phanom (NUH-KAWN PUH-NOM), province (□ 3,747 sq mi/9,742.2 sq km), Northeastern region, THAILAND, on LAOS border; ⊙ NAKHON PHANOM; 17°20′N 104°25′E. Bordered by MEKONG RIVER (Thai *Maenam Khong*). Rice, tapioca; hog and poultry raising; timber trade. Silk and rice weaving in Renu Nakhon village. Wat That Phanom temple is 33 mi/53 km S of Nakhon Phanom town; built perhaps as long as 1,500 years ago, the temple is stylistically very similar to Wat That Luang in VIENTIANE (LAOS). The chedi is about 187 ft/57 m high, and the spire is decorated with 22 lb/10 kg of gold.

Nakhon Phanom (NUH-KAWN PUH-NOM), town, ⊙ NAKHON PHANOM province, E THAILAND, on right bank of MEKONG RIVER (LAOS border), opposite THA KHAEK (LAOS), 130 mi/209 km E of Udon (linked by road; projected railroad), and 459 mi/739 km NE of BANGKOK; 17°24′N 104°47′E. Rice, hog and poultry raising; timber trade. Sometimes spelled NAGOR (or Nakhon) Pnom. Religious center THAT PHANOM is S.

Nakhon Range, THAILAND: see SITHAMMARAT RANGE.

Nakhon Ratchasima (NUH-KAWN RAHD-CHUH-SEE-MAH) province, Northeastern region, THAILAND; ⊙ NAKHON RATCHASIMA; 15°00′N 102°10′E. Transportation and communications center of NE THAILAND. Rice, corn, peanuts, soybeans, cotton, tapioca, tobacco, fruits; animal husbandry; handicrafts; some industrial production. Phimai, 37 mi/60 km NE of NAKHON RATCHASIMA town, is the site of Prasat Hin Phimai, a Khmer shrine dating back to the 10th–11th century. Another Khmer shrine, Prasat Phanomwan, is located about halfway between Phimai and Nakhon Ratchasima town.

Nakhon Ratchasima (NUH-KAWN RAHD-CHUH-SEE-MAH), city, ⊙ NAKHON RATCHASIMA province, E THAILAND, on the MUN RIVER, 155 mi/249 km NE of BANGKOK; 14°58′N 102°07′E. Strategically located near the mountain pass leading from the central plain to NE THAILAND, NAKHON RATCHASIMA is the administrative, economic, and transportation center of the KORAT PLATEAU. Copper deposits are nearby. Founded in the 17th century, the city grew rapidly after the construction (1890) of the railroad from BANGKOK. The nearby Royal Thai Air Force Base was an operation center for U.S. planes during the Vietnam War.

Nakhon Sawan (NUH-KAWN SUH-WUHN), province (□ 3,778 sq mi/9,822.8 sq km), Northern region, THAILAND; ⊙ NAKHON SAWAN; 15°42′N 100°05′E. Rice, corn, cotton; hog raising; fisheries. Sawmilling, paper manufacturing; transportation center. Large Chinese community in NAKHON SAWAN town; known for its Chinese New Year celebrations.

Nakhon Sawan (NUH-KAWN SUH-WUHN), town, ⊙ NAKHON SAWAN province, central THAILAND, on right bank of CHAO PHRAYA RIVER just below confluence of PING and NAN rivers, 140 mi/225 km NNW of BANGKOK, and on BANGKOK-CHIANG MAI railroad (PAKNAMPHO station on left bank); 15°41′N 100°07′E. Important commercial center, river-navigation hub, and a leading teak-collecting center of THAILAND. Sawmilling, paper manufacturing. Rice, corn, cotton; hog raising; fisheries. Also called PAKNAMPHO or Paknampo; sometimes spelled Nagor Svarga.

Nakhon Si Thammarat (NUH-KAWN SEE-TUHM-muh-RAHD), province (□ 3,952 sq mi/10,275.2 sq km), Southern region, SE THAILAND, on E coast of MALAY PENINSULA; ⊙ NAKHON SI THAMMARAT; 08°15′N 100°00′E. Other important towns include PAK PHANANG, port on the GULF OF THAILAND. Rice, rubber, sugarcane, coconuts, fruit; handicrafts (weaving, black and gold pottery, *yan lipao* basketry, buffalo-hide shadow puppets, dance masks). Tin mining at Ronphubin and CHA MAI, iron and lead deposits (NW). Said to be the site where Thai shadow puppetry and classical dance-drama were developed. NAKHON SI THAMMARAT town predates the subjugation of the MALAY PENINSULA in the 8th century by the Srivijaya Empire.

Nakhon Si Thammarat (NUH-KAWN SEE TUHM-muh-RAHD), town, ⊙ NAKHON SI THAMMARAT province, S THAILAND, near E coast of MALAY PENINSULA, on railroad spur, and 370 mi/595 km S of BANGKOK; 08°26′N 99°58′E. Its port on GULF OF THAILAND is PAK PHANANG. Agricultural center (rice, fruit, coconuts, and rubber plantations); handicrafts (weaving) include black and gold pottery and *yan lipao* basketry. Tin mining at Ronphibun and CHA MAI; iron and lead deposits (NW). An ancient walled town, it was until 13th century capital of a state controlling middle MALAY PENINSULA. Sometimes spelled Nagor (or Nakon) Sridhamaraj (Srithamrat or Srithamarat).

Nakhrachi, RUSSIA: see KONDINSKOYE.

Nakhtarana (nuhk-tuh-RAH-nah), town, tahsil headquarters, KACHCHH district, GUJARAT state, W central INDIA, 26 mi/42 km WNW of BHUJ.

Nakijin (nah-KEE-JEEN), village, on peninsula extending W from N OKINAWA island, Kunigami county, Okinawa prefecture, SW JAPAN, 37 mi/60 km N of NAHA; 26°40′N 127°58′E.

Nakina (nuh-KEE-nuh), unincorporated town (□ 92 sq mi/239.2 sq km; 2001 population 645), N central ONTARIO, E central Canada, near Upper Twin Lake (10 mi/16 km long), 150 mi/241 km NE of THUNDER BAY, and included in Greenstone; 50°10′N 86°42′W. Elevation 1,052 ft/321 m. Gold mining.

Nakivali (nah-KEE-vah-lee), lake (□ 14 sq mi/36.4 sq km), WESTERN region, SW UGANDA, 20 mi/32 km SE of MBARARA; maximum depth 12 ft/4 m; 00°50′S 30°54′E. Elevation 4,093 ft/1,248 m. One of five interconnected lakes that constitute the Koki lakes. Produced by tectonic forces. Was in former SOUTHERN province.

Nakkila (NAHK-ki-lah), village, TURUN JA PORIN province, SW FINLAND, on the KOKEMÄENJOKI (river), and 10 mi/16 km SE of PORI.; 61°22′N 22°00′E. Elevation 83 ft/25 m.

Naklerov (NAH-kle-RZHOF), Czech *Nakléřov*, German *Nollendorf*, village, SEVEROCESKY province, NW BOHEMIA, CZECH REPUBLIC, 5 mi/8 km NNW of ÚSTÍ NAD LABEM. Prussians defeated the French here in 1813. Naklerov Pass (2,231 ft/680 m) marks NE end of the ORE MOUNTAINS.

Naklo (NAH-klo), Polish *Nakło nad Notecią*, German *Nakel* (NAH-kel), town (2002 population 19,699), Bydgoszcz province, N central POLAND, port on NOTEC River, at W end of BYDGOSZCZ CANAL, and 17 mi/27 km W of BYDGOSZCZ. Railroad junction; manufacturing of tools, fertilizers, roofing materials; brewing, flour and beet-sugar milling, sawmilling, tanning, distilling, woodworking.

Naknek, village (2000 population 678), S ALASKA, near head of ALASKA PENINSULA, on KVICHAK BAY of BRISTOL BAY at mouth of Naknek River; 58°40′N 157°01′W. Fishing and fish processing. Formerly known as Libbyville, Pawik, Suwarof, or Suworof.

Naknek Lake, S ALASKA, near base of ALASKA PENINSULA; 40 mi/64 km long, 3 mi/4.8 km–8 mi/12.9 km wide; 58°40′N 156°12′W. Game-trout fishing. E part is in KATMAI National Monument. Drains W into KVICHAK BAY of BRISTOL BAY by Naknek River (35 mi/56 km long).

Nakodar (nah-KO-duhr), town, JALANDHAR district, central PUNJAB state, N INDIA, 15 mi/24 km SSW of JALANDHAR. Railroad junction; trade center (wheat, gram, corn, cotton); handicrafts (hookah tubes, iron jars, cotton cloth).

Nakon, in Thai names: see NAKHON.

Nakonde, ZAMBIA: see TUNDUMA.

Nakskov (NAHK-skou), city (2000 population 14,708), STORSTRØM county, SE DENMARK, a seaport at the head of Nakskov Fjord (an arm of the Langelands Bælt); 54°55′N 11°08′E. Has large sugar refineries.

Nakum (nah-KOOM), archaeological site, PETÉN department, GUATEMALA, 16 mi/26 km E of TIKAL; 17°10′N 89°26′W. A Classical-era Mayan trading center.

Nakur (nah-KOOR), town, SAHARANPUR district, N UTTAR PRADESH state, N central INDIA, 15 mi/24 km WSW of SAHARANPUR. Wheat, rice, rape, mustard, gram. Jain temple.

Nakuru (nah-KOO-roo), city (2004 population 256,300), ⊙ RIFT VALLEY province, W central KENYA;

00°16′S 36°04′E. Founded in the early 20th century as a center of European settlement, Nakuru is a growing commercial and industrial city. Manufacturing (textiles, processed food, and pyrethrum extract). Nearby is Lake Nakuru (c.35 sq mi/90 sq km), a small lake; the surrounding area has been developed as a national park (LAKE NAKURU NATIONAL PARK) noted for its flamingo haunts. Transport center.

Nakusp (nuh-KUHSP), village (□ 3 sq mi/7.8 sq km; 2001 population 1,698), SE BRITISH COLUMBIA, W Canada, on Upper Arrow Lake (COLUMBIA RIVER), and 55 mi/89 km SSE of REVELSTOKE, in Central Kootenay regional district; 50°15′N 117°48′W. Elevation 1,476 ft/443 m. Fruit; mixed farming. Tourism; Nakusp Hot Springs. Formed 1892.

Nal (NUL), village, KALAT district, KALAT division, BALUCHISTAN province, SW PAKISTAN, on HINGOL (Nal) River, and 95 mi/153 km SSW of KALAT; 27°40′N 66°12′E. Excavations here have uncovered pottery of prehistoric INDUS valley civilization.

Nala, CONGO: see ISIRO.

Nalagarh (NUH-lah-guhr), town, SOLAN district, HIMACHAL PRADESH state, N INDIA, 34 mi/55 km WSW of SHIMLA. Local market for wheat, corn, barley. Was capital of former princely state of Nalagarh of PUNJAB HILL STATES; in 1948, state (also called Hindur) merged with PATIALA AND EAST PUNJAB STATES UNION, then with PUNJAB state, finally becoming part of Himachal Pradesh state.

Nalaikha, MONGOLIA: see NALAYH.

Nalanda (nah-LUHN-dah), district (□ 914 sq mi/2,376.4 sq km; 2001 population 2,368,327), S BIHAR state, E INDIA; ⊙ BIHAR SHARIF.

Nalanda (NAH-lahn-dah), village, CENTRAL PROVINCE, SRI LANKA, in MATALE VALLEY, 25 mi/40 km N of KANDY; 07°40′N 80°38′E. Rice, coconut palms, vegetables. Has Hindu temple (8th century C.E.), now relocated near Nalanda reservoir.

Nalanda, INDIA: see BARAGAON.

Nalayh (NAH-leek), town, TÖV province, central MONGOLIA, on railroad, and 20 mi/35 km SE of ULAANBAATAR; 47°46′N 107°17′E. Coal mining, mainly for Ulaanbaatar. Mining began in 1915. Production of lignite increased greatly following completion of narrow-gauge railroad (1938–1939). Also spelled Nalaikha or Nalaykha.

Nalaykha, MONGOLIA: see NALAYH.

Nalbach (NAHL-bahkh), village, SAARLAND, GERMANY, on PRIMS RIVER, and 5 mi/8 km NNE of SAARLOUIS; 49°23′N 06°47′E. Metalworking.

Nalbari (nuhl-BAH-ree), district (□ 871 sq mi/2,264.6 sq km), ASSAM state, NE INDIA; ⊙ NALBARI.

Nalbari (nuhl-BAH-ree), town, ⊙ NALBARI district, W ASSAM state, NE INDIA, on tributary of the BRAHMAPUTRA RIVER, 27 mi/43 km NW of GUWAHATI. Road center; trades in rice, mustard, jute, cotton; silk-weaving factory.

Nal'chik (NAHL-cheek), city (2005 population 285,750), ⊙ KABARDINO-BALKAR REPUBLIC, N CAUCASUS, S European Russia, on the Nal'chik River (CHEREK RIVER basin), on the N slope of the Greater CAUCASUS Mountains, on railroad, 1,160 mi/1,867 km SSE of MOSCOW; 43°30′N 43°37′E. Elevation 1,541 ft/469 m. A health and tourist resort, it is the gateway to the Mount EL'BRUS region. Electrical machinery, semiconductors, electrical apparatus, machine tools, chemicals; molybdenum-tungsten mill. Founded in 1817 as a Russian stronghold. Made city in 1921.

Nalchiti (nawl-chee-tee), town (2001 population 35,278), Bakarganj district, S EAST BENGAL, BANGLADESH, on BISHKHALI RIVER (distributary of Arial Khan River), and 8 mi/12.9 km SW of BARISAL; 22°36′N 90°16′E. Trades in rice, oilseeds, sugarcane, jute, betel nuts; rice and oilseed milling.

Nalda (NAHL-dah), town, LA RIOJA province, N SPAIN, 10 mi/16 km SSW of LOGROÑO; 42°20′N 02°29′W. Fruit, olive oil, vegetables.

Nälden (NEL-den), village, JÄMTLAND county, NW SWEDEN, on S shore of Lake Näldsjön (10 mi/16 km long, 1 mi/1.6 km–3 mi/4.8 km wide), 16 mi/26 km NW of ÖSTERSUND; 63°21′N 14°15′E.

Naldrug (NUHL-druhg), town, OSMANABAD district, MAHARASHTRA state, W central INDIA, 29 mi/47 km SE of OSMANABAD. Millet, wheat, cotton. Has 14th-century fort (remodeled in 16th century). Was district capital until 1853. Also spelled Naldurg.

Nalepkovo, SLOVAKIA: see VONDRISEL.

Nalerigu (nah-le-REE-goo), town, NORTHERN REGION, GHANA, 5 mi/8 km E of GAMBAGA, on road to Bawki; 10°32′N 00°22′W.

Naletale National Monument, MIDLANDS province, S central ZIMBABWE, near border of MATABELELAND SOUTH province, 33 mi/53 km SSW of GWERU; 19°53′S 29°31′E. Preserves Naletale Ruins, stone walls with geometric designs; hilltop site dates to 17th century.

Nalgonda (nahl-GON-dah), district (□ 5,498 sq mi/14,294.8 sq km), ANDHRA PRADESH state, S INDIA, on DECCAN PLATEAU; ⊙ NALGONDA. Bordered S by KRISHNA RIVER; mainly lowland, drained by MUSI RIVER. Nagarjuna Sagar Reservoir lies to S. Largely sandy red soil, with alluvial soil along rivers; millet, oilseeds (chiefly peanuts, castor beans), rice. Rice and oilseed milling, cotton ginning, manufacturing of brass and copper vessels. Trade centers include BHONGIR, NALGONDA, SURIAPET. Became part of HYDERABAD during state's formation in 18th century and part of Andhra Pradesh state after independence.

Nalgonda (nahl-GON-dah), town, ⊙ NALGONDA district, ANDHRA PRADESH state, S INDIA, 55 mi/89 km ESE of HYDERABAD. Road center in agricultural area; rice and oilseed milling.

Nalhati (nahl-HAH-tee), town, tahsil headquarters, BIRBHUM district, W WEST BENGAL state, E INDIA, 35 mi/56 km NNE of SIURI. Railroad junction; rice milling, cotton weaving, metalware manufacturing; rice, pulses, wheat, sugarcane. Rice milling and cotton weaving 10 mi/16 km NNE, at MURARAI. Silk growing nearby.

Naliboki (nah-li-BO-ki), village, W MINSK oblast, BELARUS, 29 mi/47 km NE of NOVOGRUDOK, in Naliboki forest (□ 160 sq mi/414 sq km). Noted for old coniferous and deciduous species. Has old wooden church.

Naliya (NUHL-yah), town, tahsil headquarters of Abdasa tahsil, KACHCHH district, GUJARAT state, W central INDIA, 55 mi/89 km W of BHUJ. Market center for wheat, barley, salt; embroidering.

Nalkheda, INDIA: see NALKHERA.

Nalkhera (nuhl-KAI-rah) or **Nalkheda**, town (2001 population 14,201), SHAJAPUR district, W central MADHYA PRADESH state, central INDIA, 27 mi/43 km N of SHAJAPUR. Cotton, millet, wheat; cotton ginning.

Nallihan, village, central TURKEY, 50 mi/80 km NE of ESKIŞEHIR; 40°12′N 31°22′E. Grain; mohair goats. Noted for its silk needlework.

Nallur, SRI LANKA: see JAFFNA.

Nalón River (nah-LON), 80 mi/129 km long, OVIEDO province, NW SPAIN; rises in CANTABRIAN MOUNTAINS, 30 mi/48 km ENE of PAJARES PASS; flows NW, across rich coal- and iron-mining region, to BAY OF BISCAY 5 mi/8 km NNE of PRAVIA.

Nal River, PAKISTAN: see HINGOL RIVER.

Naltagua (nahl-TAH-gwah), mining settlement, SANTIAGO province, METROPOLITANA DE SANTIAGO region, central CHILE, on MAIPO RIVER, and 26 mi/42 km SW of SANTIAGO. Copper-mining and smelting center.

Nalut (NAH-lut), town, TRIPOLITANIA region, NW LIBYA, c. 12 mi/19 km from Tunisian border, on W edge of JABAL NAFUSAH plateau, 100 mi/161 km SW of ZUWARAH; 31°52′N 10°58′E. Elevation c. 2,000 ft/610 m. Road junction. BERBER and Turkish forts. Troglodyte dwellings.

Nalwadi, INDIA: see SEVAGRAM.

Nalwar, INDIA: see YADGIR.

Nam [Thai=river], for Thai names beginning thus see: under following part of the name.

Nama (NAH-mah), coral island, State of CHUUK, E CAROLINE ISLANDS, Federated States of MICRONESIA, W PACIFIC, 39 mi/63 km SE of CHUUK ISLANDS; c.½ mi/⅘ km in diameter. Rises to 20 ft/6 m.

Namaacha District, MOZAMBIQUE: see MAPUTO.

Namacurra (nah-muh-KUHR-uh), village, Zambézia province, central MOZAMBIQUE, on railroad, and 25 mi/40 km NNE of QUELIMANE; 17°29′S 37°01′E.

Namacurra District, MOZAMBIQUE: see ZAMBÉZIA.

Namadgi National Park (nuh-MA-jee) (□ 363 sq mi/943.8 sq km), S 40% of AUSTRALIAN CAPITAL TERRITORY, SE AUSTRALIA, 40 mi/64 km S of CANBERRA; 40 mi/64 km long, 19 mi/31 km wide; 35°09′S 148°57′E. Administered by Australian Capital Territory Parks and Conservation Service. Includes several mountain ranges surrounding the Cotter and Gudgenby river catchments, water supply for Canberra. KOSCIUSKO NATIONAL PARK (NEW SOUTH WALES) adjoins to W. Forests of alpine ash; snow gums; sub-alpine bogs and peatlands. Aboriginal and European heritage sites throughout. Kangaroo, wallaby, koala, dingo, glider, possum; cockatoo, emu. Camping, picnicking, hiking, mountain climbing, cross-country skiing, fishing, horse riding. Established 1984 as part of National Capital Open Space System.

Namaka, hamlet, S ALBERTA, W Canada, 8 mi/13 km from STRATHMORE, in Wheatland County; 50°57′N 113°17′W.

Namakagon River, WISCONSIN: see NAMEKAGON RIVER.

Namakan Lake (NA-muh-kuhn), ST. LOUIS county, NE MINNESOTA and Rainy Lake district, W ONTARIO, in chain of lakes on border between U.S. and CANADA, 35 mi/56 km ESE of INTERNATIONAL FALLS; 12 mi/19 km long, 3 mi/4.8 km wide; 48°30′N 92°38′W. Drains NW into RAINY LAKE, through Short Kettle Falls River, Kettle Falls Dam enlarges both lakes, is continuous with KABETOGAMA Lake (reservoir), to W. Fed from E by Namakan River, which flows across part of Ontario from Lac LA CROIX (also on border); also receives Pipestone, SEINE, and Turtle rivers from Ontario. U.S. portion in VOYAGEURS NATIONAL PARK.

Namakgale, town, LIMPOPO province, SOUTH AFRICA, 15 mi/24 km W of PHALABORWA, and 55 mi/88 km E of TZANEEN, near KRUGER NATIONAL PARK; 23°57′S 31°02′E. Elevation 1295 ft/394 m. Dormitory town for mine workers from Phalaborwa, where copper, phosphates, and vermiculite are mined.

Namakia (nah-mah-KEE-ah), village, MAHAJANGA province, NW MADAGASCAR, on left arm of MAHAVAVY RIVER estuary, 5 mi/8 km NNW of MITSINJO; 15°55′S 45°50′E.

Namakkal (NAH-muhk-kuhl), town, SALEM district, TAMIL NADU state, S INDIA, 30 mi/48 km S of SALEM. Road center in cotton area. Asbestos here and corundum to SSW.

Namak Lake (ne-MAHK), Persian Daryacheh-ye-Namak, great salt lake of N central IRAN, S of TEHRAN and E of QOM; 40 mi/64 km across. The size of the lake varies considerably depending on the season and year. Salt marshes and deposits along shore. Receives the SAVEH (QARA) River in the W and SHUR, KARAJ, and JAJ rivers in the N.

Namakwaland, region, SW Africa: see NAMAQUALAND.

Namaland, SW Africa: see NAMAQUALAND.

Namaliga (nah-muh-LEE-guh), town, CENTRAL region, S UGANDA, 18 mi/29 km N of KAMPALA. Cotton, coffee, sugar; livestock. Just W is BOMBO. Was part of former Buganda province.

Namanakula, SRI LANKA: see UVA BASIN.

Namanga, town, RIFT VALLEY province, S KENYA, 165 mi/266 km S of NAIROBI on border between Kenya and TANZANIA; 02°33′S 36°48′E. Important border-crossing point. Livestock.

Namangan (NAH-mahng-gahn), wiloyat (□ 2,400 sq mi/6,240 sq km), E UZBEKISTAN; ⊙ NAMANGAN. In N

FERGANA VALLEY; drained by NARYN River (E); bounded S by the SYR DARYA River. Extensive irrigation, especially in E; cotton growing, sericulture; wheat on mountain slopes; cattle and horses; sheep breeding in non-irrigated areas. Extensive cotton ginning, some silk milling. Antimony mine at KASANSAI. KOKAND-Namangam-ANDIJAN railroad passes through S section. Population chiefly Uzbek. Formed 1941.

Namangan (NAH-mahng-gahn), city, ⊙ NAMANGAN wiloyat, UZBEKISTAN, in the N FERGANA Basin; 41°00′N 71°40′E. A center for the production of cotton and silk, it also has food-processing plants. Captured by Russian forces in 1875.

Namapa (nuh-MA-puh), village, NAMPULA province, N MOZAMBIQUE, on road, and 110 mi/177 km NNE of NAMPULA; 17°20′S 36°28′E. Cotton, sisal.

Namapa District, MOZAMBIQUE: see NAMPULA.

Namaqualand (nah-MAHK-we-land) or **Namaland**, region (□ 150,000 sq mi/390,000 sq km), NAMIBIA. Extends from WINDHOEK (N) to the NORTHERN CAPE province (SOUTH AFRICA; S) and from the NAMIB (W) to the KALAHARI (E) deserts. The ORANGE RIVER divides the region into Great Namaqualand (in Namibia) and Little Namaqualand (in South Africa). An arid region, Namaqualand is populated chiefly by the pastoral-agricultural Nama, who speak a Khoikhoi language, and scattered commercial sheep farms. Near the ATLANTIC OCEAN are extensive alluvial diamond beds; copper is mined in Little Namaqualand. Karakul pelts are a major export of the region. Also spelled Namakwaland.

Namarrói (nuh-MAHR-oi), village, Zambézia province, central MOZAMBIQUE, 130 mi/209 km N of QUELIMANE; 15°57′S 36°51′E. Manioc, corn, beans. Also spelled Nhamarrói.

Namarroi District, MOZAMBIQUE: see ZAMBÉZIA.

Namasagali (nah-muh-suh-GAH-lee), town, KAMULI district, EASTERN region, UGANDA, on the VICTORIA NILE RIVER, and 13 mi/21 km NNW of MBULAMUTI. Elevation 3,417 ft/1,042 m. Railroad terminus and head of Victoria Nile River navigation from MASINDI PORT on LAKE KYOGA. Has river-boat repair dock. Was part of former BUSOGA province.

Namasigüe (nah-mah-SEE-gwai), town, CHOLUTECA department, S central HONDURAS, on paved road, 8 mi/12.9 km SE of CHOLUTECA; 13°12′N 87°09′W. Airfield. Small farming; grain, beans; livestock.

Namasuba (nah-mah-SOO-bah), town, WAKISO district, CENTRAL region, S UGANDA, 7 mi/11 km NE of KAMPALA; 00°16′N 32°34E. Suburb of Kampala.

Namatanai (nah-muh-tuh-NEI), town, on E coast of NEW IRELAND, PAPUA NEW GUINEA; 03°40′S 152°30′E.

Nambe (NAHM-BAI), pueblo (□ 29 sq mi/75.4 sq km), SANTA FE county, N central NEW MEXICO, 15 mi/24 km N of SANTA FE, in Nambe Pueblo land grant between SANGRE DE CRISTO MOUNTAINS and the RIO GRANDE. Elevation 6,082 ft/1,854 m. Chief activity is agriculture (grain, chilies, fruit); metal alloy pottery. Inhabitants are Mexicans and Native American Pueblo; languages spoken are Spanish, English, and Tewa.

Nambling (NAM-bling), village, EAST NEW BRITAIN province, E NEW BRITAIN island, E PAPUA NEW GUINEA, on Wide Bay of SOLOMON SEA, below E end of NAKANAI MOUNTAINS, 70 mi/113 km SSW of RABAUL; 05°12′S 152°01′E. Boat access. Cocoa, coffee; tuna; timber.

Nambondo (nam-BON-do), village, LINDI region, E TANZANIA, 35 mi/56 km NW of KILWA MASOKO; 08°42′S 39°07′E. Timber; sisal, cashews, grain; sheep, goats.

Namborn (NAHM-born), village, SAARLAND, GERMANY, 6 mi/9.7 km NNW of ST. WENDEL; 49°32′N 07°08′E. Livestock; grain.

Nambour (NAM-boor), town (2001 population 12,205), SE QUEENSLAND, NE AUSTRALIA, 55 mi/89 km N of BRISBANE; 26°38′S 152°57′E. Railroad junction; sugar

mill. Agriculture center (sugarcane, pineapples, coffee, and citrus fruit); "Big Pineapple" tourist park.

Nambroca (nahm-BRO-kah), town, TOLEDO province, central SPAIN, 6 mi/9.7 km SE of TOLEDO; 39°47′N 03°56′W. Olives, cereals; olive-oil pressing, dairying; sheep.

Nambucca Heads (nam-BUHK-uh), town, NEW SOUTH WALES, SE AUSTRALIA, 318 mi/512 km NE of SYDNEY; 30°38′S 152°59′E. Tourism; holiday, retirement center; also abbatoirs; timber; beef cattle, dairying, fishing, oyster farming.

Nambung National Park (NAM-buhng) (□ 68 sq mi/176.8 sq km), W central WESTERN AUSTRALIA state, W AUSTRALIA, 180 mi/290 km NNW of PERTH, S of CERVANTES, on INDIAN OCEAN; 15 mi/24 km long, 8 mi/13 km wide. Yellow sand dunes on foreshore, white lime dunes immediately inland; limestone Swan Coastal Plain remainder. The Pinnacles Desert at center consists of hundreds of limestone pillars jutting up from surface, ranging from finger-sized to 20 ft/6 m high, formed by down-cutting of limestone bed. Acacia thickets, tuart woodlands in small valleys; casuarina and banksia dominate in heath areas. Grey kangaroo; emu, shore birds; skink, snake, sea turtle. Chain of water holes along Nambung River support wildlife communities. Limited camping, picnicking, hiking.

Nam Ca Dinh (NAHM KAH DIN), river, N central LAOS, major tributary of MEKONG River; 18°19′N 104°00′E.

Namcha Barwa, CHINA: see NAMJAGBARWA.

Namche Bazaar (nahm-CHAI buh-ZAHR), village, central NEPAL; 27°48′N 86°43′E. Elevation 11,290 ft/3,441 m. Largest Sherpa village. Tourism. Yak herding; potatoes, barley. Headquarters for SAGARMATHA NATIONAL PARK.

Namcheju (NAHM-JE-JOO), county (□ 232 sq mi/603.2 sq km), S CHEJU province, SOUTH KOREA, on S slope of HALLA Mountain, surrounded by CHINA SEA on E, W, and S. Two inhabited and nine uninhabited islands are within its jurisdiction. Narrow fields formed along the coast produce barley, sweet potatoes, rape, beans, tangerines, and pineapples. Songsanilchulbong, Sanbang Mountain, and Andok Valley are nationally renowned scenic areas.

Namchi (NAHM-chee), town, ⊙ South district, SIKKIM state, NE INDIA, 19 mi/31 km SW of GANGTOK, in SE foothills of NEPAL HIMALAYA. Corn, rice, pulses. Noted Buddhist monasteries of SANGACHELLING (or Sangnga Chöling) and PEMIONGCHI (or Pamiongchi, Pemayangtse) are 17 mi/27 km NW, that of TASHIDING (Tassiding) 4 mi/6.4 km W.

Nam Co, Chinese *Namu-zuh* (NAH-MOO-zuh), salt lake (□ 950 sq mi/2,470 sq km), central TIBET, SW CHINA. Largest lake in Tibet; elevation 15,180 ft/4,627 m. Also appears as Namu Lake or Nam Tso.

Namdalen (NAHM-dahl-uh), valley of NAMSEN River in NORD-TRØNDELAG county, central NORWAY. Lumbering; fishing (salmon), centered at NAMSOS and GRONG.

Nam Dinh (NAHM DIN), city, ⊙ NAM HA province, N VIETNAM, on canal linking the Song Dai and SONG HONG (RED RIVER), on railroad, and 50 mi/80 km SE of HANOI; 20°25′N 106°10′E. Commercial hub. Major silk and cotton center; spinning, weaving; jute mill, distillery; salt extraction. Light manufacturing; administrative complex. Traditional Vietnamese scholastic center. Booming city center with development plans on the board. Formerly Namdinh.

Nam Dinh, VIETNAM: see NAM HA.

Namegawa (nah-ME-gah-wah), town, Hiki county, SAITAMA prefecture, E central HONSHU, E central JAPAN, 22 mi/35 km N of URAWA; 36°03′N 139°21′E.

Namekagon River (NAM-e-KAH-gun), c.95 mi/153 km long, NW WISCONSIN; rises E of Namekagon Lake, in SE BAYFIELD county; flows SW, past HAYWARD, then NW, to SAINT CROIX RIVER, 45 mi/72 km S of SU-

PERIOR. Entire length of river below Namekagon Lake is S branch of SAINT CROIX NATIONAL SCENIC RIVERWAYS. Trout fishing. Formerly spelled Namakagon.

Namen, BELGIUM: see NAMUR.

Namentenga, province (□ 2,497 sq mi/6,492.2 sq km; 2005 population 306,265), CENTRE-NORD region, E central BURKINA FASO; ⊙ BOULSA; 13°15′N 00°35′W. Borders SOUM (N), SÉNO (NE), GNAGNA (E), KOURITENGA (S), GANZOURGOU (SW), and SANMATENGA (W) provinces. Agriculture (groundnuts, rice) and livestock (sheep). Gold mining.

Namerikawa (nah-me-ree-KAH-wah), city, TOYAMA prefecture, central HONSHU, central JAPAN, 9 mi/15 km N of TOYAMA, port on SE shore of TOYAMA BAY; 36°45′N 137°20′E.

Namest nad Oslavou (NAH-mnyesht-yuh NAHD O-slah-VOU), Czech *Náměšt' nad Oslavou*, German NAMIEST AN DER OSLAWA, town (1991 population 5,115), JIHOMORAVSKY province, W MORAVIA, CZECH REPUBLIC, on railroad, on Oslava River (affluent of the JIHLAVA RIVER), and 21 mi/34 km W of Brno; 49°13′N 16°09′E. Agriculture (grain); manufacturing (textile); woodworking. First book on Czech grammar printed here in 1533. Military base.

Namestovo (nah-MES-to-VO), Slovak *Námestovo*, Hungarian *Námesztó*, town, STREDOSLOVENSKY province, N SLOVAKIA, on S slope of the BESKIDS, on W bank of ORAVA DAM, and 24 mi/39 km NNE of RUŽOMBEROK; 49°24′N 19°30′E. Summer resort; manufacturing (clothing, machinery). Has 18th-century church.

Nametil (nah-me-teel), village, NAMPULA province, NE MOZAMBIQUE, 40 mi/64 km S of NAMPULA; 15°43′S 39°20′E. Cotton, peanuts.

Nämforsen (NEM-FORSH-en), waterfalls on ÅNGERMANÄLVEN RIVER, VÄSTERNORRLAND county, NE SWEDEN, 20 mi/32 km NW of SOLLEFTEÅ. Major hydroelectric station, 113-MW capacity (1947). Rock carvings dating from c.2000 B.C. nearby.

Nam Ha, province (□ 962 sq mi/2,501.2 sq km), N VIETNAM, in RED RIVER Delta; ⊙ NAM DINH; 20°25′N 106°10′E. Bordered N by HA TAY and HAI HUNG, NE by THAI BINH, SE by VINH BAC BO, SW by NINH BINH, and W by HOA BINH provinces. Partly framed by the Red River and its distributaries and interlaced by irrigation canals, Nam Ha is an area of fertile alluvial soils and productive agriculture (wet rice cultivation, livestock raising; vegetables, mulberry, cotton, jute). Fishing and aquaculture, industries (silk weaving, jute milling, sericulture; textiles, chemicals, brick and tile; distillery, salt production; food processing). Light manufacturing; shipbuilding. Predominantly Kinh population. Future plans call for the province to be divided into two: HA NAM (capital, PHU LY) and NAM DINH (capital, Nam Dinh) provinces.

Namhaé (NAHM-HAI), county, SOUTH KYONGSANG province, SOUTH KOREA, in KOREA STRAIT, 20 mi/32 km S of CHINJU. Includes Namhae Island (fourth-largest island in South Korea) and eighty uninhabited islets. Bounded N by HADONG, E by CHUNGMU, W by YOSU provinces. Connected to S coast of Korean mainland by the Great Namhaedaegyo Bridge. Nearly divided into two parts, with three wide peninsulas. Generally low and fertile. Agriculture (rice, barley, soybeans, sweet potatoes); fishing (cod, snapper); cattle raising; raw silk. Gold, silver, and copper mines.

Nam Het (NAHM HET), river, N LAOS, major tributary of SONG Ca River (Vietnam); 20°50′N 104°01′E.

Nam Hin Boun, river, KHAMMUAN province, central LAOS, enters MEKONG River at HIN BOUN. Short river with important tin mining along its banks.

Namhkam (NAHM-kahm), village and township, SHAN STATE, MYANMAR, 40 mi/64 km SE of BHAMO, on branch of BURMA ROAD, and on left bank of the SHWELI RIVER. Trade center near CHINA border. Site of Namhkam Hospital, founded in World War II by Dr. Gordon Seagrave.

Namhsan, village and township, SHAN STATE, MYAN-MAR, 35 mi/56 km W of LASHIO. Was capital of former TAWNGPENG STATE.

Namiai (nah-mee-AH-ee), village, Shimoina county, NAGANO prefecture, central HONSHU, central JAPAN, 93 mi/150 km S of NAGANO; 35°22′N 137°41′E.

Namib (nah-MEEB), desert, SW AFRICA, along the coast of NAMIBIA; c.800 mi/1,287 km long, 30 mi/48 km–100 mi/161 km wide. Occupies a rocky platform between the ATLANTIC OCEAN and the escarpment of the interior plateau. Isolated mountains rise from the desert and some of the highest sand dunes in the world (1,000 ft/305 m) cover its S portion. It receives less than 0.5 in/1.3 cm of rain annually and is barren of vegetation, though it supports a delicate ecosystem, which receives moisture from sea mist blowing in from the cold Atlantic Ocean. Tungsten, salt, and alluvial diamonds are mined. Rock and surf angling is popular along its coastal areas.

Namibe (nah-MEE-bai), province (□ 86,087 sq mi/223,826.2 sq km), SW corner of ANGOLA, on southern ATLANTIC OCEAN coast, and on NAMIBIA (S) border; ⊙ NAMIBE. Bordered N by BENGUELA, E by HUÍLA and SE by CUNENE provinces. Drained by Curoca and CUNENE rivers. Includes IONA NATIONAL PARK in S, Mt. Negros, and Namib Desert. Major railroad connects Namibe to MENONGUE in CUANDO CUBANGO. Agriculture. Minerals include granite, marble. Main centers are Namibe, TOMBUA, Iona, BIBALA, FOZ DO CUNENE.

Namibe (nah-MEE-bai), town (2004 population 132,900), ⊙ NAMIBE province, SW ANGOLA, a port on the ATLANTIC OCEAN. International airport to S. Iron ore is the leading export; sisal, cotton, tobacco, frozen meat, hides, and skins are also significant. Also known as Mossamedes.

Namibia (nah-MIB-ee-uh), formerly South West Africa, republic (□ c.318,000 sq mi/823,620 sq km; 2004 estimated population 1,954,033; 2007 estimated population 2,055,080), SW AFRICA; ⊙ and largest city WINDHOEK.

Geography
Bordered N by ANGOLA, NE by ZAMBIA, E by BOTS-WANA, and SE and S by SOUTH AFRICA, and W by the ATLANTIC OCEAN. The ORANGE RIVER forms the S border, and the KUNENE, OKAVANGO, and ZAMBEZI rivers form parts of the N and NE borders. The country includes the CAPRIVI STRIP in the NE. Major towns include Windhoek, KEETMANSHOOP, TSUMEB, LÜ-DERITZ, GOBABIS, OTJIWARONGO, OSHAKATI, and SWAKOPMUND. The enclave of WALVIS BAY in the W, previously administered by South Africa, was re-integrated into Namibia on February 28, 1994, along with the offshore islands. The country has four main geographical regions: the arid and barren NAMIB Desert, which runs along the entire Atlantic coast with widths of 50 mi/80 km–80 mi/129 km; an extensive central plateau that averages c.3,600 ft/1,100 m in elevation; the W fringes of the KALAHARI DESERT in the E; and an alluvial plain in the N that includes the ETOSHA PAN, a large salt marsh. Namibia is divided into thirteen primary administrative divisions, called regions: CA-PRIVI, ERONGO, HARDAP, KARAS, KHOMAS, KUNENE, OHANGWENA, OKAVANGO, OMAHEKE, OMUSATI, OSH-ANA, OTJIKOTO, and OTJOZONDJUPA. The highest point is BRANDBERG Mountain (8,545 ft/2,605 m), situated in the W part of the central plateau. Rainfall varies from less than 0.5 in/1.3 cm per annum (SW) to in excess of 20 in/51 cm (NE). Climate is subtropical arid.

Population
The country has an ethnically diverse population that includes the Bantu-speaking Ovambo, Kavango, and Herero; various Nama groups; the Ova-Himba; San (Bushmen); and Europeans of South African, German, and British descent. English is the official language. Most of the population is Christian, and the rest follow traditional beliefs.

Economy
Because of inadequate rainfall, crops are not widely raised and pastoralism forms the backbone of the agricultural sector. Goats and sheep are raised mainly in the S, and cattle are herded chiefly in the central plateau N. Game animals are also raised. Agricultural income is derived mainly from Karakul pelts, livestock, and dairy products. The country's few manufactures are made up mostly of processed food. There is an extensive mining industry, run principally by foreign-owned companies. Mining accounted for approximately 73% of export earnings in 1988. The chief minerals are diamonds, uranium, copper, lead, manganese ore, zinc concentrate, salt, lithium, silver, gold, amethyst, germanium, and vanadium. Fishing fleets operate in the Atlantic. Unrestricted fishing by commercial companies has severely depleted the country's supply of certain types of fish. Fish stocks are being replenished by careful control and the strict application of the 124-mi/200-km exclusive fishing zone. The world's last remaining large-scale seal hunt takes place here. Tourism is also a major source of foreign exchange. The central part of the country is served by roads and railroad lines that are linked to the South African systems.

History: to 1890
The earliest inhabitants of Namibia were San hunters and gatherers, who lived here as early as 2,000 years ago. By c.500, Nama herders had entered the region; they have left early records of their activities in the form of cave paintings. The Herero people settled in the W and N areas of Namibia around 1600. The Ovambo migrated into Namibia from the N after c.1800. Diego Cao and Bartholomew Dias, both Portuguese navigators, landed on the coast in the early 15th century. Portuguese and Dutch expeditions explored the coastal regions, and in the late 18th century Dutch and British captains laid claim to parts of the coast. These claims, however, were disallowed by their governments. In the 18th century, English missionaries arrived, and they were followed by German missionaries in the 1840s. Britain annexed Walvis Bay in 1878. The Bremen (Germany) trading firm of F.A.E. Lüderitz gained a cession of land at Angra Pequeña (now Lüderitz) in 1883, and in 1884 the German government under Otto von Bismarck proclaimed a protectorate over this area, to which the rest of South West Africa (German *Süd-West Afrika*) was soon added.

History: 1890 to 1960
Conflicts between the indigenous population and the Europeans, mainly over control of land, led to outbreaks of violence in the 1890s and especially in the 1900s. In 1903 the Nama began a revolt, joined by the Herero in 1904. The Germans pursued an uncompromising military campaign that by 1908 had resulted in the death of about 54,000 Herero (out of a total Herero population of about 70,000), many of whom were driven E into the Kalahari Desert, where they perished; 30,000 others also died in the revolt. In 1908 diamonds were discovered near Lüderitz, and a large influx of Europeans began. During World War I the country was occupied (1915) by South African forces, and after the war South Africa began (1920) to administer it as a "C"-type Mandate under the League of Nations. In 1921–1922 the Bondelzwarts, a small Nama group, revolted against South African rule, but they were crushed by South African forces employing air power. After the founding of the UN in 1945, South Africa, unlike the other League of Nations mandatories, refused to surrender its mandate and place South West Africa under the UN trusteeship system.

History: 1960 to 1973
In 1960, Ethiopia and Liberia (both of which had been members of the League) initiated proceedings in the International Court of Justice (The HAGUE) to have the mandate declared as being in force and to have South Africa charged with failing to fulfill the terms of the mandate. The court ruled in July 1966 that Ethiopia and Liberia had not established a legal right or interest entitling them to bring the case. In frustration at this decision, the South West Africa People's Organization (SWAPO), operating in exile, undertook small-scale guerrilla warfare in South West Africa. Next, the UN General Assembly in October 1966, passed a resolution terminating the mandate, and in 1968 it resolved that the country be known as Namibia. In June 1971, the International Court of Justice reaffirmed the General Assembly's 1966 resolution and ruled that South Africa should immediately withdraw its administrators. However, the South African government maintained that the UN had no authority over South West Africa, and it proceeded with plans for establishing ten African homelands (Bantustans) in the country and for tying it more closely to South Africa itself.

History: 1973 to 1988
Political repression was met with SWAPO's extensive boycott of the Bantustan elections in OVAMBOLAND in 1973. South Africa held a constitutional conference (the Turnhalle Conference) in 1975 and delayed deciding Namibia's status. Responding to threats from the world community, the government promised Namibian independence by the end of 1978. In 1977, the government adopted a new constitution that upheld apartheid policies, restricted SWAPO participation in politics and sought to allow a continuance of South African control over foreign affairs after independence. SWAPO was able to attract popular support in spite of the regime's campaign of repression. SWAPO and other opposition groups effectively waged guerrilla warfare and gained control of areas in the N. A UN resolution in 1978 called for a cease-fire and elections to be monitored by UN civilian and military personnel. South Africa balked at elections, fearing a SWAPO-led Namibian government.

History: 1988 to Present
Ten years later, under a 1988 agreement brokered by the U.S., the withdrawal of pro-SWAPO Cuban troops from Angola was linked with the implementation of the UN plan in Namibia. Elections were held in 1989 with SWAPO winning a majority of the seats and party leader Sam Nujoma was elected president. The constitution was adopted in February 1990 and Namibia became independent on March 21, 1990. Nujoma was reelected in 1994 and 1999. In 2004 Hifikepunye Pohamba, the SWAPO candidate and Nujoma's handpicked successor, was elected president. The desire of some inhabitants of the Caprivi Strip for independence has led to clashes there. Namibia has a significant AIDS problem, with more than 40% of the population infected in some areas.

Government
Namibia is governed under the constitution of 1990. The president, who is head of state, is popularly elected for a five-year term and is eligible for a second term. The government is headed by a prime minister, who is appointed by the president. There is a bicameral legislature. The National Council has 26 seats, with two members chosen from each regional council to serve six-year terms. Members of the 72-seat National Assembly are popularly elected for five-year terms. The current head of state is President Hifikepunye Pohamba (since March 2005). Prime Minister Nahas Angula has been head of government since March 2005.

Namib-Naukluft Park, reserve (over □ 20,000 sq mi/51,800 sq km), central W NAMIBIA, from Luderitz-Aus Road N to Windhoek-Swakopmund Road. World's 4th-largest. Extends from escarpment to ATLANTIC OCEAN and includes almost 30% of the NAMIB Desert. Incorporates Sossusvlei and Kuiseb Canyon. Game includes oryx, mountain zebra, and ostrich.

Namie (nah-mee-E), town, Futaba county, FUKUSHIMA prefecture, N central HONSHU, NE JAPAN, 34 mi/55 km S of FUKUSHIMA city; 37°29′N 140°59′E. Fish; pottery.

Namiest an der Oslawa, CZECH REPUBLIC: see NAMEST NAD OSLAVOU.

Namikata (nah-MEE-kah-tah), town, Ochi county, EHIME prefecture, NW SHIKOKU, W JAPAN, 19 mi/30 km N of MATSUYAMA; 34°05′N 132°57′E.

Namikawa, RUSSIA: see TROITSKOYE, SAKHALIN oblast.

Namin (ne-MAIN), town, Ardabīl province, NW IRAN, on road, and 15 mi/24 km NE of the city of ARDEBIL, near AZERBAIJAN Republic border; 38°25′N 48°28′E.

Namino (NAH-mee-no), village, Aso county, KUMAMOTO prefecture, W KYUSHU, W JAPAN, 28 mi/45 km N of KUMAMOTO; 32°55′N 131°13′E.

Namioka (nah-mee-O-kah), town, S Tsugaru county, Aomori prefecture, N HONSHU, N JAPAN, 12 mi/20 km S of AOMORI; 40°42′N 140°35′E. Apples.

Namiquipa (nah-mee-KEE-pah), town, ⊙ Namiquipa municipio, CHIHUAHUA, N MEXICO, in SIERRA MADRE OCCIDENTAL, on SANTA MARÍA RIVER, and 90 mi/145 km WNW of CHIHUAHUA; 29°15′N 107°25′W. Elevation 5,997 ft/1,828 m. Corn, beans, fruit, cattle.

Namirembe (nah-MEE-rem-bai), town, CENTRAL region, S UGANDA, just W of KAMPALA. Has Protestant cathedral. Was part of former Buganda province.

Namiriin Usa, MONGOLIA: see HAR US NUUR.

Namjagbarwa, Chinese *Nanzhabawa* (NAHN-JAH-BAH-WAH), highest peak (25,445 ft/7,756 m) in E ASSAM HIMALAYAS, in E TIBET, SW CHINA, in bend of the BRAHMAPUTRA RIVER; 29°40′N 95°10′E. Also spelled as Namcha Barwa.

Nam Khan (NAHM KAHN), river, N LAOS, major tributary of MEKONG River; 19°55′N 102°10′E.

Namlea (NAHM-le-ah), chief town of BURU island, INDONESIA, on NE coast of island, on Kajeli Bay (small inlet of BANDA SEA), 80 mi/129 km WNW of AMBON Island; 03°16′S 127°06′E. Port and trade center; ships cajuput oil, resin, skins, rattan, timber; imports rice and dried fish.

Namling, Chinese *Nanmulin* (NAHN-MOO-LIN), Tibetan *dzong*, town, ⊙ Namling county, SE TIBET, SW CHINA, on left tributary of the BRAHMAPUTRA RIVER, and 120 mi/193 km W of LHASA; 29°40′N 89°03′E. Elevation 12,220 ft/3,725 m. Oil crops; livestock.

Nam Mae [Thai=river], for Thai names beginning thus: see under following part of the name.

Nammekon (nah-MAI-kon), village, KAYAH STATE, MYANMAR, 5 mi/8 km W of LOIKAW. Former capital of Nammekon State.

Nam Ngum (NAHM NGUHM), river, N central LAOS, major tributary of MEKONG River; 18°10′N 103°06′E. Site of NAM NGUM dam, a major producer of hydroelectric power for export.

Nam Nhiep (NAHM NEE-ahp), river, N central LAOS, major tributary of MEKONG River; 18°25′N 103°36′E.

Namoi River (NA-moi), 526 mi/846 km long, N central NEW SOUTH WALES, SE AUSTRALIA; rises as Peel River in LIVERPOOL RANGE; flows generally NW, past TAMWORTH, GUNNEDAH, and NARRABRI, to DARLING RIVER at WALGETT; 30°00′S 148°07′E. Used for irrigation.

Namoluk (NAH-mo-look), atoll, state of CHUUK, E CAROLINE ISLANDS, Federated States of MICRONESIA, W PACIFIC, 140 mi/225 km SE of CHUUK; c.3 mi/4.8 km long, 2 mi/3.2 km wide. Comprises six wooded islets on triangular reef. Formerly Hashmys Island.

Namonuito (NAH-mon-wee-to), atoll, State of CHUUK, E CAROLINE ISLANDS, Federated States of MICRONESIA, W PACIFIC, 95 mi/153 km NW of CHUUK ISLANDS; 45 mi/72 km long, 24 mi/39 km wide. Ulul (3 mi/4.8 km long) is largest islet. Formerly Los Jardines.

Namorik (NAH-mo-reek), atoll (□ 1 sq mi/2.6 sq km; 1999 population 772), RALIK CHAIN, MAJURO district, MARSHALL ISLANDS, W central PACIFIC, 210 mi/338 km S of KWAJALEIN; 5 mi/8 km long; 05°36′N 168°07′E. Comprises two islets.

Nampa (NAM-puh), village (□ 1 sq mi/2.6 sq km; 2001 population 372), central ALBERTA, W Canada, 16 mi/26 km from PEACE RIVER town, and in Northern Sunrise County; 56°02′N 117°08′W. Incorporated in 1958.

Nampa, city (2000 population 51,867), CANYON county, SW IDAHO, in the fertile Treasure Valley; 43°35′N 116°34′W. Elevation 2,484 ft/757 m. Railroad junction and the commercial, processing, and shipping center for an irrigated agricultural, orchard, and dairy region. Has food-processing plants and a large sugar factory. Consumer goods, furniture, fabricated metal products, and wood products are also manufactured. Seat of Northwest Nazaréne University; Canyon County Historical Society & Museum; Deer Flat Wildlife Refuge at LAKE LOWELL, to SW. Incorporated 1890.

Nampicuan (nahm-pee-KOO-ahn), town, NUEVA ECIJA province, central LUZON, PHILIPPINES, on railroad, and 17 mi/27 km NNE of TARLAC; 15°43′N 120°40′E. Rice-growing center.

Nampinga (nam-PEEN-gah), village, LINDI region, S central TANZANIA, 170 mi/274 km WNW of LINDI, on Mbarangadu River, in SELOUS GAME RESERVE; 09°48′S 37°07′E. Livestock.

Nampo (NAHM-PO), special city, W NORTH KOREA, on KOREA BAY; 38°43′N 125°24′E. The port city for PYONGYANG; also a leading metallurgical center. Other industries include shipbuilding, glassmaking, and electrode manufacturing. Linked by express highways to Pyongyang, Nampo, KAESONG, and WONSAN.

Nampula (nahm-POO-luh), province (2004 population 3,563,224), MOZAMBIQUE, on INDIAN OCEAN (E); ⊙ NAMPULA. Bounded N and NW by CABO DELGADO and NIASSA and S and SW by Zambézia provinces. Coastal areas dotted with mangroves and swamps. Commercial agriculture (sisal, cashews, cache nuts), concentrated in the NAMPULA area. Railroad line links the coastal port town of NACALA to the interior across to both LICHINGA and Malawai in NIASSA province. Predominately inhabited by the Muslim Macua- and Swahili-speaking people. Muslim influences indicate; was the seat of sultanate of ANGOCHE. Comprises 18 districts: ANGOCHE, ERATI, LALAUA, MALEMA, MECONTA, MEMBA, MOMA, MONGICUAL, MOGOVOLAS, MONAPO, MOSSURIL, MECUBURI, MUECATI, MURRUMPULA, NOVA VELHA, NAMAPA, NAMPULA, and RIBAUE. Has 126 habitats.

Nampula (nahm-POO-luh), city (2004 population 371,800), ⊙ NAMPULA province, NE MOZAMBIQUE; 15°09′S 39°18′E. Agricultural trade center, located on the railroad connecting the seaports of LUMBO and NACALA with MALAWI and Liehinga in NIASSA province. International airport to NE. Cement is manufactured.

Nampula District, MOZAMBIQUE: see NAMPULA.

Nampungu (nam-POON-goo), village, RUVUMA region, S TANZANIA, 95 mi/153 km ESE of SONGEA; 10°57′S 37°03′E. Timber; subsistence crops; livestock.

Namsen (NAHMS-uhn), river, 120 mi/193 km long, NORD-TRØNDELAG county, central NORWAY; issues from a lake of same name 10 mi/16 km NNE of Gjersvika; flows SW and W to NAMSFJORDEN at NAMSOS. Waterfall FISKEMFOSS is near GRONG. Nordland railroad follows its valley.

Nam Seng (NAHM SENG), river, N LAOS, major tributary of MEKONG River via Nam Suong; 19°59′N 102°14′E.

Namsfjorden (NAHMS-fyawr-uhn), inlet of the NORTH SEA, in NORD-TRØNDELAG county, central NORWAY; extends inland (SE and E) 22 mi/35 km; c.2 mi/3.2 km wide. At its head (at mouth of NAMSEN River) lies NAMSOS.

Namskoye, RUSSIA: see NAMTSY.

Namslau, POLAND: see NAMYSLOW.

Namsos (NAHM-saws), town, NORD-TRØNDELAG county, W NORWAY, a port at the mouth of the NAMSEN River, on the Namsenfjord; 64°29′N 11°30′E. Lumbering; fishing; textiles; canning; copper mining. In World War II, Namsos was the scene (1940) of heavy fighting between the British and the Germans.

Namsskogan (NAHMS-kaw-gahn), village, NORD-TRØNDELAG county, central NORWAY, on NAMSEN River, on railroad, and 25 mi/40 km NE of GRONG; 64°55′N 13°10′E. Lumbering; hunting.

Nam Tha (NAHM TAH), town, ⊙ LUANG NAM THA province, N LAOS, 20°57′N 101°25′E. Market and administrative center. Shifting cultivation, opium trade, agroforestry. Airport. Meo, Lao, Mien, Thai, Lolo, and other minorities.

Nam Tha (NAHM TAH), river, N LAOS, major tributary of MEKONG River; 20°53′N 101°09′E.

Nam Theun (NAHM TOO-ahn), river, N central LAOS, major tributary of NAM CA DINH River; 18°03′N 104°58′E. Site of proposed Nam Theum 2, a dam designed to produce hydroelectric power, mainly for export to other countries, such as THAILAND and VIETNAM.

Namti (NAHM-tee), village, MOGAUNG township, KACHIN STATE, MYANMAR, on railroad and LEDO Road, 25 mi/40 km W of MYITKYINA.

Nam Tso, CHINA: see NAM CO.

Namtsy (NAHM-tsi), village (2006 population 7,895), central SAKHA REPUBLIC, central RUSSIAN FAR EAST, on the left bank of the LENA RIVER, on road junction, 45 mi/72 km N of YAKUTSK; 62°43′N 129°37′E. Elevation 242 ft/73 m. In agricultural area. Also known as Namskoye (NAHM-skuh-ye).

Namtu (NAHM-too), town and township, SHAN STATE, MYANMAR, on MYITNGE RIVER (here called Nam Tu), and 25 mi/40 km NW of LASHIO. Smelting center for BAWDWIN mines (6 mi/9.7 km W). Population largely Indian and Chinese.

Namtumbo (nam-TOOM-bo), RUVUMA region, S TANZANIA, 32 mi/51 km ENE of SONGEA; 10°33′S 36°06′E. Road junction. Timber; subsistence crops; livestock.

Namu (NAH-moo), community, W BRITISH COLUMBIA, W Canada, in Central Coast regional district; 51°52′N 127°52′W. First Nations heritage site; historic industrial site (former fish cannery).

Namu (NAH-moo), atoll (□ 2 sq mi/5.2 sq km; 1999 population 903), RALIK CHAIN, MARSHALL ISLANDS, W central PACIFIC, 40 mi/64 km SSE of KWAJALEIN; 35 mi/56 km long; 08°11′N 167°58′E. Comprises fifty-one islets.

Namu Lake, CHINA: see NAM CO.

Namuli Mountains (na-MOO-lee), group of high peaks (7,936 ft/2,419 m) in N central MOZAMBIQUE, rising above plateau, c.180 mi/290 km N of QUELIMANE.

Namuno District, MOZAMBIQUE: see CABO DELGADO.

Namunukula, SRI LANKA: see UVA BASIN.

Namur (NAH-mur), Flemish *Namen* (NAH-men), province (□ 1,415 sq mi/3,679 sq km; 2006 population 459,904), S BELGIUM, on border of FRANCE (S); ⊙ NAMUR; 50°30′N 05°25′E. The chief cities are Namur and DINANT. The province is generally hilly; drained by the MEUSE, SAMBRE, and LESSE rivers and traversed in the S by the ARDENNES. Largely agricultural. There are also extensive marble, limestone, and granite quarries; iron mines; and glass and cutlery factories. Mainly French-speaking. Includes the former province of Namur, part of the former prince-bishopric of LIÈGE, and part of HAINAUT. A resort industry is developing along the Lesse and the S stretches of the Meuse.

Namur (NAH-mur), Flemish *Namen* (NAH-men), commune (□ 450 sq mi/1,170 sq km; 2006 population 107,411), ⊙ Namur district and of NAMUR province, S central BELGIUM, at the confluence of the MEUSE and SAMBRE rivers; 50°28′N 04°52′E. Commercial and industrial center; railroad junction. Manufacturing includes machinery, leather goods, and porcelain. It is also an episcopal center and a tourist area. Was a Merovingian fortress (first mentioned in the seventh century). Changed hands repeatedly; later shared the history of the Austrian and Spanish Netherlands. Noteworthy are the Church of St. Loup (seventeenth century), St. Aubain Cathedral (eighteenth century), and the Citadel.

Namur (nah-MYUR), village (□ 22 sq mi/57.2 sq km; 2006 population 568), Outaouais region, SW QUEBEC, E Canada, 6 mi/9 km from CHÉNÉVILLE; 45°54′N 74°56′W.

Namur, MARSHALL ISLANDS: see KWAJALEIN.

Namutumba, administrative district, EASTERN region, SE UGANDA. As of Uganda's division into eighty districts, borders PALLISA (NE), BUTALEJA (E), BUGIRI (S), IGANGA (W), and KALIRO (NW) districts. Agricultural area. Railroad between KASESE town (W Uganda) and MOMBASA (SE KENYA) runs through district. Formed in 2006 from NE portion of former IGANGA district created in 2000 (current Iganga district formed from all but NE portion).

Namwala (nah-em-WAH-lah), township, SOUTHERN province, S central ZAMBIA, 125 mi/201 km WSW of LUSAKA, on KAFUE RIVER, in Kafue Flats Swamp; 15°45′S 26°27′E. Agriculture (corn, tobacco); cattle. Tourism. KAFUE NATIONAL PARK to W. Musungwa Safari Lodge is here; Ngoma Lodge to W.

Namwendwa (nahm-WEN-dwuh), town, EASTERN region, SE central UGANDA, on railroad, and 33 mi/53 km N of JINJA. Cotton, tobacco, coffee, bananas, corn. Was part of former BUSOGA province.

Namwera (NAM-we-rah), village, MANGOCHE district, Southern region, E MALAWI, on road, and 20 mi/32 km ENE of Mangoche; 14°22′S 35°30′E. In tobacco-growing area; cotton, corn, rice. Sometimes called Mangoche. Fort Mangoche (abandoned c.1930) is 6 mi/9.7 km SSW.

Namwon (NAHM-WON), city (□ 49 sq mi/127.4 sq km; 2005 population 86,052), SE NORTH CHOLLA province, SOUTH KOREA, at center of NAMWON county, W of CHIRI Mountains. Fertile field developed in basin of Yo River, a small branch of the SOMJIN River Mountains in N and W. Economic activity involves agriculture (rice), sericulture, raw silk spinning, and tobacco processing. Important crossroad connecting Honam and Yongnam. Served by Expressway 88 and Cholla railroad. CHIRI MOUNTAIN NATIONAL PARK nearby. Kwanghanru (setting of *Chunhyangjon*—most famous ancient novel of Korea) and Maninuichong (cemetery) are here.

Namyangju (NAHM-YAHNG-JOO), county (□ 162 sq mi/421.2 sq km), central KYONGGI province, SOUTH KOREA, just E of SEOUL. Surrounded by KWANGJU mountains to the N, W, and S; the E part of the county opens into a wide field used for dry-farming. Agriculture (rice, barley, beans, nuts; poultry; dairy) and sericulture are widely practiced. Paldang, Pamsom, and Chonmasan resorts, and Kwangnung attract tourists.

Namyslow (nah-MEES-lov), Polish *Namysłów*, German *Namslau*, town, OPOLE province, SW POLAND, 30 mi/48 km E of WROCŁAW (Breslau); 51°05′N 17°43′E. Railroad junction; agricultural market (grain, sugar beets, potatoes; livestock); brewing, sawmilling. Has castle built 1360 by Emperor Charles IV. Formerly in German-administered LOWER SILESIA; since 1945, part of Poland.

Nan (NAHN), province (□ 5,730 sq mi/14,898 sq km), Northern region, THAILAND, on LAOS border; ⊙ NAN; 18°50′N 100°40′E. Rice, corn, beans, fruit, cotton, tobacco, lac. Trade in forest products, hides, horns. Salt mining at BO KLUA. There are several historic temples in and around NAN town dating back as far as 1355. Annual boat races held here during the *thot kathin* festival at the end of the Buddhist Lent (Oct.–Nov.).

Nan (NAHN), town, ⊙ NAN province, N THAILAND, in Phi Pan Nam Mountains, on NAN RIVER, 115 mi/185 km E of CHIANG MAI, and 434 mi/698 km N of BANGKOK, on road from railroad station of DEN CHAI; 18°47′N 100°47′E. Rice, cotton, tobacco, lac. Trade in teak, forest products, hides, horns. Salt mining at BO KLUA (NE). Pop. is largely LAO. Famous for annual boat races and several temples.

Nan, THAILAND: see NAN RIVER.

Nanacamilpa (nah-nah-kah-MEEL-pah) or **San José Nanacamilpa** (sahn ho-ZAI nah-nah-kah-MEEL-pah), town, ⊙ Nanacamilpa de Mariano Arista municipio, TLAXCALA, central MEXICO, 22 mi/35 km NW of TLAXCALA; 19°29′N 98°33′W. Maguey, cereals; livestock.

Nanacamilpa de Mariano Arista, MEXICO: see NANACAMILPA.

Nanae (nah-NAH-e), town, Oshima county, Hokkaido prefecture, N JAPAN, 87 mi/140 km S of SAPPORO; 41°53′N 140°41′E. Apples. Near Onuma quasi-national park.

Nana-Grébizi, economic prefecture (□ 7,718 sq mi/20,066.8 sq km; 2003 population 117,816), N central CENTRAL AFRICAN REPUBLIC; ⊙ KAGA BANDORO. Bordered N by CHAD, E by BAMINGUI-BANGORAN prefecture, S by KÉMO prefecture, W by OUHAM prefecture. Drained by GRIBINGUI RIVER. Cotton; cotton ginning. Chrome reserves. Main centers are Kaga Bandoro and Bilague. Also called Gribingui.

Nanaimo (nuh-NEI-mo), regional district (□ 786 sq mi/2,043.6 sq km; 2001 population 127,016), SW BRITISH COLUMBIA, W Canada, on Vancouver Island; 49°15′N 124°20′W. Consists of 8 unincorporated electoral areas and 4 municipalities (NANAIMO, PARKSVILLE, Lantzville, and QUALICUM BEACH). Established in 1967.

Nanaimo (nuh-NEI-mo), city (□ 34 sq mi/88.4 sq km; 2001 population 73,000), SW BRITISH COLUMBIA, W Canada, on E coast of VANCOUVER ISLAND, 69 mi/111 km N of VICTORIA, in Nanaimo regional district; 49°09′N 123°55′W. A port, the base of a commercial fishing fleet, and the trade center for a farm and lumbering region. It is the site of a federal fisheries and oceanographic research station. A tourist center, Nanaimo hosts an annual Bathtub Race across the straits to VANCOUVER.

Nanakai (nah-nah-KAH-ee), village, W Ibaraki county, IBARAKI prefecture, central HONSHU, E central JAPAN, 12 mi/20 km N of MITO; 36°27′N 140°15′E.

Nana Kru (NAH-nah KROO), town, SINOE county, SE LIBERIA, minor port on ATLANTIC OCEAN, at mouth of Nana Kru River (70 mi/113 km long; rises in NIETE MOUNTAINS), and 25 mi/40 km ESE of GREENVILLE.

Nanakuli (NAH-NAH-KOO-lee), city (2000 population 10,814), OAHU, HONOLULU county, HAWAII, on W coast, 19 mi/31 km WNW of HONOLULU; 21°23′N 158°09′W. Lualualei U.S. Naval Reservation to N. WAIANAE RANGE to E; Maili Point to NW; Nanaikapono Beach County Park to N; Kalanianaoli Beach County Park is here.

Nanam (NAH-NAHM), district of CHONGJIN (special city), NORTH KOREA, 8 mi/12.9 km SW of CHONGJIN; 41°43′N 129°42′E. Fruit-growing area producing cider and sake. Site of army base established in 1915. Became provincial capital in 1920, supplanting KYONGSONG; now part of Chongjin.

Nana-Mambéré, prefecture (□ 23,604 sq mi/61,370.4 sq km; 2003 population 233,666), W CENTRAL AFRICAN REPUBLIC, on CAMEROON (W) border; ⊙ BOUAR. Bordered N and E by OUHAM-PENDÉ, SE by OMBELLA-M'POKO, and S by MAMBÉRÉ-KADÉÏ prefectures. Drained by MAMBÉRÉ and Nana rivers. Agriculture (cattle; coffee; cotton); textiles. Main centers are Bouar, BABOUA, BAORO.

Nan'an (NAHN-AHN), town, ⊙ NAN'AN county, SE FUJIAN province, SE CHINA, 20 mi/32 km NW of Jinjiang; 24°57′N 118°23′E. Rice, sugarcane, oilseeds.

Nanango (nuh-NANG-go), town, SE QUEENSLAND, NE AUSTRALIA, 85 mi/137 km NW of BRISBANE; 26°40′S 152°00′E. Agriculture center (corn, alfalfa, beans, grapes, olives; beef cattle; dairy products); timber.

Nanao (nah-NAH-o), city, ISHIKAWA prefecture, central HONSHU, central JAPAN, on E NOTO Peninsula, port on W inlet of TOYAMA BAY, opposite, NOTO-JIMA ISLAND 37 mi/60 km N of KANAZAWA; 37°02′N 136°58′E. Buddhist altars. Hot springs.

Nan'ao (NAHN-OU), island (□ 50 sq mi/130 sq km) in SOUTH CHINA SEA, off SE CHINA coast, 20 mi/32 km E of SHANTOU; 13 mi/21 km long, 1 mi/1.6 km–6 mi/9.7 km wide; 23°26′N 117°01′E. Rises to 1,900 ft/579 m. With nearby rocky islets (SE), it forms Nan'ao county (1990 population 66,257) of GUANGDONG province; 23°25′N 117°06′E. On central W lies Nan'ao, county seat and fishing port. Manufacturing includes food processing, engineering, timber processing, textile manufacturing; saltworks.

Nanatsu-jima, JAPAN: see WAJIMA.

Nanatsuka (nah-NAHTS-kah), town, Kahoku county, ISHIKAWA prefecture, central HONSHU, central JAPAN, on SEA OF JAPAN, 12 mi/20 km N of KANAZAWA; 36°43′N 136°42′E

Nanauta (nuh-NOU-tah), town, SAHARANPUR district, N UTTAR PRADESH state, N central INDIA, 19 mi/31 km SSW of SAHARANPUR. Wheat, rice, rape and mustard, gram, sugarcane, cotton.

Nanawa, town, PRESIDENTE HAYES department, central PARAGUAY, 3 mi/4.8 km NW of and across the Paraguay River from Asunción; 25°16′S 57°40′W;

Nanayama (nah-NAH-yah-mah), village, East Matsuura county, SAGA prefecture, N KYUSHU, SW JAPAN, 19 mi/30 km N of SAGA; 33°26′N 130°07′E.

Nanay River (nah-NEI), c.200 mi/322 km long, LORETO region, NE PERU, in AMAZON basin; rises near 02°45′S 75°16′W. Flows SE, E, and NE to the Amazon River at IQUITOS. Navigable for small craft. Near its mouth below Iquitos are large sawmills (mahogany, cedar).

Nanboro (nahn-BO-ro), town, Sorachi district, Hokkaido prefecture, N JAPAN, 16 mi/25 km E of SAPPORO; 43°03′N 141°39′E.

Nanbu (NAHN-BOO), town, ⊙ Nanbu county, central SICHUAN province, CHINA, 37 mi/60 km N of NANCHONG, and on right bank of JIALING River; 31°19′N 106°02′E. Rice, oilseeds, tobacco, jute, cotton. Food industry; chemicals, textiles; engineering.

Nanbu (NAHN-boo), town, Sannohe county, Aomori prefecture, N HONSHU, N JAPAN, 40 mi/65 km S of AOMORI; 40°23′N 141°16′E.

Nanbu, town, South Koma county, YAMANASHI prefecture, central HONSHU, central JAPAN, 28 mi/45 km S of KOFU; 35°16′N 138°27′E.

Nanbu Hengkuan Kunglu, TAIWAN: see Southern Cross-Island Highway.

Nancagua (nahn-KAH-gwah), town, ⊙ Nancagua comuna, COLCHAGUA province, LIBERTADOR GENERAL BERNARDO O'HIGGINS region, central CHILE, on railroad, and 13 mi/21 km SW of SAN FERNANDO; 34°40′S 71°13′W. Agricultural center (cereals, vegetables, grapes; livestock); food processing; dairying.

Nancamilpa (nahn-kah-MEEL-pah), town, ⊙ Nancamilpa municipio, TLAXCALA, MEXICO, 22 mi/35 km WNW of TLAXCALA, 5 mi/8 km off Mexico Highway 136, on paved road, near ZOQUIAPAN NATIONAL PARK; 19°29′N 98°33′W. Elevation 8,858 ft/2,700 m. Agriculture, small farming (corn, beans, agave). Also called Mariano Arista.

Nançay (nawn-SAI), commune (□ 41 sq mi/106.6 sq km), CHER department, CENTRE administrative region, N central FRANCE, 20 mi/32 km NW of BOURGES; 47°21′N 02°12′E. Located in the lake district of sologne. Has radioastronomy observatory. Artists' studios located in old stables of local castle.

Nance, county (□ 448 sq mi/1,164.8 sq km; 2006 population 3,705), E central NEBRASKA; ⊙ FULLERTON; 41°23′N 97°59′W. Agricultural region drained by LOUP and CEDAR rivers. Cattle, hogs; dairying; corn, soybeans, sorghum. Formed 1879.

Nanchang (NAHN-CHAHNG), city (□ 238 sq mi/617 sq km; 1994 estimated urban population 1,168,700; estimated total population 1,465,400), ⊙ JIANGXI province, CHINA, on the GAN RIVER, near the S end of BOYANG Lake; 28°41′N 115°53′E. A major transportation center, it has a port, and railroad links (BEIJING-JIULONG RAILROAD) to SHANGHAI, ZHEJIANG, and HUNAN, and an airport. Large commercial and industrial center with machine shops, food-processing establishments, textile and paper mills, and plants making chemicals, motor vehicles and tractors, electric and electronic equipment, cement, tires, pharmaceuticals; shipbuilding. An old walled city, Nanchang dates from the Song (or Sung) dynasty (12th century), but it received its present name in the Ming dynasty (1368–1644). Nanchang is considered the birthplace of the People's Liberation Army (PLA). Here, in August 1927, a force of 30,000 Communist troops rose against the Kuomintang government and briefly established the First Soviet republic in China. Occupied by the Japanese (1939–1945) in World War II, Nanchang was reoccupied by the Nationalists in 1945 and then came under the Communists in 1949. Seat of an agricultural institute and a medical college. The city is well known for the ancient Tengwang Pavilion. Also called Nanjing.

Nan-ching, CHINA: see NANJING.

Nanchital de Lázaro Cárdenas del Río (nahn-shee-TAHL dai LAH-zah-ro KAHR-dai-nahs del REE-o), city, SE VERACRUZ, MEXICO, 2.5 mi/4 km S of the port of COATZACOALCOS, on the banks of the COAT-ZACOALCOS River, 4.3 mi/7 km S of Mexico Highway 180, and near railroad. Hot climate. Important because of rich petroleum deposits.

Nanchong (NAHN-CHUNG), city (□ 981 sq mi/2,550.6 sq km; 2000 population 279,178), E central SICHUAN province, CHINA, 55 mi/89 km NNW of HECHUAN, on right bank of the JIALING River; 30°54′N 106°06′E. Industry and commerce are the largest sources of income for the city. Crop growing (grain); animal husbandry (hogs). Manufacturing (food processing, oil refining; textiles, machinery).

Nanchuan (NAHN-CHWAHN), town, ⊙ Nanchuan county, SE SICHUAN province, CHINA, 50 mi/80 km SE of CHONGQING; 29°07′N 107°16′E. Rice, tobacco, jute, oilseeds, medicinal herbs; coal mining. Engineering, papermaking; food and beverages, chemicals, textiles.

Nanchuang (NAHN-CHWAHNG), village, NW TAIWAN, 14 mi/23 km S of HSINCHU. Coal mining; agriculture (tea, rice, sweet potatoes, vegetables). A Buddhist retreat is nearby, at Lion Head Mountain. Sometimes spelled Nanchwang.

Nanchwang, TAIWAN: see Nanchuang.

Nancito, PANAMA: see EL NANCITO.

Ñancorainza (nyahng-ko-RIN-sah), town and canton, LUIS CALVO province, CHUQUISACA department, SE BOLIVIA, in Serranía de AGUARAGÜE, 40 mi/64 km N of VILLA MONTES, on road; 20°40′S 63°24′W. Petroleum center. Natural gas fields of Buena Vista 2 mi/3.2 km W. Formerly called Yancorainza.

Nancowry Island, one of NICOBAR ISLANDS, ANDAMAN AND NICOBAR ISLANDS Union Territory, INDIA, in BAY OF BENGAL, 50 mi/80 km NNW of GREAT NICOBAR ISLAND; 6 mi/9.7 km long N-S. Nancowry Harbour, a landlocked anchorage, lies between Nancowry Island (S) and CAMORTA and another island (N); center of interisland trade in coconuts, canoes, pottery. Japanese naval station (1942–1945) in World War II. Sometimes spelled Nankauri.

Nancuchiname National Park (nahn-koo-chee-NAH-mai), USULUTÁN department, EL SALVADOR, W of USULUTÁN city, on RÍO LEMPA.

Nancy (nawn-SEE), city (□ 5 sq mi/13 sq km); ⊙ MEURTHE-ET-MOSELLE department, NE FRANCE, on the MEURTHE RIVER (above its junction with the MOSELLE RIVER), and on the MARNE-RHINE CANAL, 175 mi/282 km E of PARIS; 48°41′N 6°11′E. The eco-

nomic, cultural, and educational center of LORRAINE and the old provincial capital. The headquarters of the newly created administative region of LORRAINE are at METZ, however. Nancy metropolitan area population is c.330,000 (1990). Situated at the edge of the Lorraine iron fields, Nancy is surrounded by industrial suburbs with steel mills, metal works, and chemical plants lining the Meurthe Valley from LUNÉVILLE to FROUARD (at junction with Moselle). The city itself specializes in apparel, leather goods, furniture, and glassware. Nancy has a noted fine arts museum, a school of mining, a center of forestry research and training, and a cultural center (for dramatic arts) in the old tobacco factory. Seat of an important university (founded 1854). Nancy grew around a fortified 11th-century castle of the dukes of Lorraine and became the duchy's capital in the 12th century. In 1477, Charles the Bold of Burgundy was defeated and killed at the gates of Nancy by Swiss troops and the forces of René II of Lorraine. The major part of the center of Nancy, a model of urban planning and a gem of 18th-century architecture, was built during the enlightened reign of Stanislas Leszczynski, duke of Lorraine (reigned 1738–1766), father-in-law of Louis XV and ex-king of Poland. Nancy reverted to the French crown in 1766. From 1870 to 1873 it was occupied by the Germans following the Franco-Prussian War, and it was partially destroyed during violent combat in World War I. Points of interest include the ornate (fine iron and gold grillwork) *Place Stanislas*, the *Place de la Carrière* (a long square leading to the ducal palace), an 18th-century cathedral, and the 13th–16th-century ducal palace fronting on the old quarter and backed by a fine garden (known as la Pépinière) with statuary by Rodin. The palace contains the noteworthy museum of the history of Lorraine. The *Place Stanislas*, *Place de la Carrière*, and *Place d'Alliance* were together designated a UNESCO World Heritage site in 1983. The church and convent of the Cordeliers (15th century) houses the magnificent tombs of the princes of Lorraine. Begining in 1890, Nancy became a center of *art nouveau* architecture and design; it still preserves many private buildings thus influenced as well as a museum of the Nancy school of decorative design. Nancy-Tomblaine international airport 3 mi/4.8 km E of city.

Nanda Devi (NUHN-dah DAI-vee), peak (25,645 ft/7,817 m), GARHWAL district, UTTAR PRADESH state, N central INDIA, in the Himalayas; 30°23′N 79°59′E. Second highest peak in India. According to Hindu belief, Goddess Nanda (wife of the god Shiva) resides here. Nanda Kot, at an elevation of 22,538 ft/6,870 m, is said to be Nanda's "couch." The peak was scaled in 1936 by an Anglo-American expedition.

Nandaime (nahn-DEI-mai), town, (2005 population 15,866), GRANADA department, SW NICARAGUA, 14 mi/23 km SSW of GRANADA, and on INTER-AMERICAN HIGHWAY. Agriculture center (coffee, sugarcane, cacao); livestock.

Nandalur (nuhn-dah-LOOR), town, CUDDAPAH district, ANDHRA PRADESH state, S INDIA, 25 mi/40 km SE of CUDDAPAH. Rice, sugarcane, turmeric.

Nandan (NAHN-DAHN), town, ⊙ Nandan county, NW GUANGXI ZHUANG autonomous region, CHINA, near GUIZHOU province border, 70 mi/113 km WNW of YISHAN, and on railroad; 24°59′N 107°32′E. Rice, wheat, beans, potatoes; tin ore mining and dressing.

Nandan (NAHN-dahn), town, Mihara district, at S tip of AWAJI-SHIMA island, HYOGO prefecture, W central JAPAN, 38 mi/62 km S of KOBE; 34°15′N 134°43′E. Onions, Chinese cabbage; rice; chrysanthemums. Noodles; tile. O-Naruto Bridge (5,344-ft/1629-m suspension bridge) to NE SHIKOKU.

Nandasmo (nahn-DAHS-mo), town, MASAYA department, SW NICARAGUA, on road, and 3 mi/4.8 km SW of MASAYA; 11°56′N 86°07′W. Rice, coffee.

Nandayure, COSTA RICA: see CARMONA.

Nanded (nahn-DED), district (□ 4,065 sq mi/10,569 sq km), MAHARASHTRA state, W central INDIA, on DECCAN PLATEAU; ⊙ NANDED. Bordered N by PENGANGA RIVER, SE by MANJRA RIVER; mainly lowland; drained by the Godavari River. Largely in black-soil area; millet, cotton, wheat, oilseeds (chiefly peanuts, flax), rice. Cotton-milling and trade center; an important Sikh pilgrimage center because of its association with Guru Nanak. Agricultural markets at UMRI and BHAINSA. Became part of HYDERABAD during state's formation in 18th century, now part of Maharashtra state. Affected by 1993 earthquake.

Nanded (nahn-DED), city (2001 population 430,733), ⊙ NANDED district, MAHARASHTRA state, W central INDIA, on the Godavari River. A commercial center; market for livestock, grain, and cotton. Known for its fine muslin. Important Sikh pilgrimage center; one of the five sacred Sikh sites. Mogul fort nearby.

Nandgaon (NAHND-goun), former princely state of CHHATTISGARH STATES, in what is now central INDIA, W central CHHATTISGARH state. Incorporated 1948 into DURG district of MADHYA PRADESH state; in 2000, territory became part of Chhattisgarh state.

Nandgaon (NAHND-goun), town, NASHIK district, MAHARASHTRA state, W central INDIA, on railroad (workshop), 60 mi/97 km ENE of NASHIK. Agricultural market (peanuts, cotton, wheat); cotton ginning, oilseed pressing, tanning.

Nandi, town, MASVINGO province, SE ZIMBABWE, 85 mi/137 km SE of MASVINGO, near Chiredzi River, on railroad; 20°58′S 31°45′E. Irrigated agricultural area (sugarcane, wheat, cotton, corn, peanuts, soybeans); cattle, sheep, goats.

Nandi, FIJI: see NADI.

Nandi, INDIA: see CHIK BALLAPUR.

Nandidrug, INDIA: see CHIK BALLAPUR KOLAR GOLD FIELDS.

Nandigama (NAHN-dee-gah-mah), town, KRISHNA district, ANDHRA PRADESH state, S INDIA, on tributary of KRISHNA River, 27 mi/43 km NW of VIJAYAWADA. Rice, peanuts, cotton, tobacco.

Nandikotkur (nuhn-dee-KOT-koor), town, KURNOOL district, ANDHRA PRADESH state, S INDIA, 14 mi/23 km E of KURNOOL. Cotton ginning; peanuts, rice, turmeric.

Nandlstadt (NAHN-duhl-shtaht), village, UPPER BAVARIA, GERMANY, 16 mi/26 km W of LANDSHUT; 48°32′N 11°47′E. Grain; livestock.

Nandod, INDIA: see RAJPIPLA.

Nandom (NAN-duhm), town (2000 population 8,060), UPPER WEST REGION, GHANA, 15 mi/24 km NE of LAWRA; 10°51′N 02°45′W. Livestock; groundnuts (peanuts), shea-nut butter.

Nandrin (nahn-DREN), commune (2006 population 5,595), Huy district, LIÈGE province, E BELGIUM, 12 mi/19 km SW of LIÈGE; 50°30′N 05°25′E.

Nandura (nuhn-DOO-rah), town, BULDANA district, MAHARASHTRA state, W central INDIA, 36 mi/58 km WNW of AKOLA. Millet, cotton, wheat grown in the region; oilseed milling.

Nandurbar (NUHN-duhr-bahr), town, DHULE district, MAHARASHTRA state, W central INDIA, at extreme N end of WESTERN GHATS, 45 mi/72 km NW of DHULE. Road and trade center (cotton, wheat, linseed, timber); cotton ginning, oilseed milling, tanning, handloom weaving, palmarosa-oil extracting; sawmills. Noted for its grapes and melons. Prosperous town in mid-17th century.

Nandu River (NAHN-DOO), c. 170 mi/274 km long, N HAINAN province, SE CHINA, rises in mountains SE of BAISHA; flows NE, past the SONGTAO reservoir, CHENGMAI and Ding'an, to HAINAN Strait.

Nandy (nawn-DEE), town (□ 3 sq mi/7.8 sq km), SEINE-ET-MARNE department, ÎLE-DE-FRANCE region, N central FRANCE; 48°35′N 02°34′E.

Nandyal (nuhnd-YAHL), city, KURNOOL district, ANDHRA PRADESH state, S INDIA, 37 mi/60 km SE of

KURNOOL. Road and agriculture trade center; cotton ginning, oilseed milling. Handmade paper at Gazulapalle village, 10 mi/16 km ESE.

Nandydroog, INDIA: see CHIK BALLAPUR KOLAR GOLD FIELDS.

Nanfeng (NAHN-FUNG), town, ⊙ Nanfeng county, E JIANGXI province, CHINA, 28 mi/45 km SSW of Nancheng, and on XU River; 27°10′N 116°24′E. Rice, oranges; timber. Food and beverages; chemicals; pharmaceuticals; engineering.

Nangade District, MOZAMBIQUE: see CABO DELGADO.

Nanga-Eboko, town (2001 population 19,200), ⊙ HAUTE-SANAGA department, Central province, CAMEROON, on SANAGA River, and 80 mi/129 km NE of Yaoundé; 04°42′N 12°24′E. Road Junction. Rice processing. Hospital.

Nangai (NAHNG-gah-ee), village, Senhoku county, Akita prefecture, N HONSHU, NE JAPAN, 22 mi/35 km S of AKITA city; 39°26′N 140°22′E.

Nangal (NAHN-gahl), town, BILASPUR district, HIMACHAL PRADESH state, N INDIA, 22 mi/35 km WNW of BILASPUR, and 4 mi/6.4 km SW of BHAKRA, in outer W HIMALAYA. Weir constructed on SUTLEJ River (just W) to supply projected irrigation canal and hydroelectric plants. Bhakra-Nangal Project (dam) completed 1968. Was part of former Bilaspur state.

Nanga Parbat (nun-GAH PUHR-bet), peak (26,660 ft/8,126 m high), in the W terminus of the HIMALAYA, DIAMIR district, SE corner of NORTHERN AREAS, extreme NE PAKISTAN; 7th highest peak in the world; 35°15′N 74°36′E. Site of several Nazi-era German expeditions in the 1930s. A German-Austrian team led by Herman Buhl finally reached the peak in 1953. Noted for its extreme geological denudation. Easily accessed from KARAKORAM HIGHWAY along INDUS River. Locally called DIAMIR.

Nanga Parbat, Mount (nun-GAH PUHR-bet) (10,780 ft/3,286 m), on border between W ALBERTA and E BRITISH COLUMBIA, W Canada, in ROCKY MOUNTAINS, at W edge of BANFF NATIONAL PARK, 2 mi/3 km SE of Mount FRESHFIELD in the Freshfield Icefield, and 65 mi/105 km NW of BANFF; 51°43′N 116°52′W.

Nangarhar (nan-gar-HAHR), province (2005 population 1,237,800), E AFGHANISTAN, on PAKISTAN (S and E) border; ⊙ JALALABAD; 34°45′N 70°50′E. Bordered N and NE by KUNAR and LAGHMAN, W by KABUL, and SW by LOGAR and PAKTIA provinces. Watered by the KABUL RIVER and its E tributaries. Population largely Pashtun of the Khugiani, Mohmand, Shinwari, and Tirahi tribes. Mild climate permits three crops a year, primarily wheat, corn, rice; olives and citrus fruit cultivation; some timber. Darunta Dam on the Kabul River provides for irrigation and hydroelectric power. Scene of considerable fighting during the 1980s, the area was ruled by a coalition of *mujahedin* groups until captured by the Taliban. Nangarhar was part of the Eastern province until it became a separate province in 1963.

Nangbeto (nahng-BAI-to), town, PLATEAUX region, S TOGO, on MONO RIVER, 35 mi/56 km ESE of ATAKPAMÉ; 07°26′N 01°25′E. Dam, hydroelectric power plant.

Nangis (nahn-ZHEE), town (□ 9 sq mi/23.4 sq km), SEINE-ET-MARNE department, ÎLE-DE-FRANCE region, N central FRANCE, 16 mi/26 km E of MELUN; 48°34′N 03°01′E. Chemical and metalworks; sugar milling. Here Napoleon defeated Austrians in 1814. Rampillon (3 mi/4.8 km ESE) has 13th-century Gothic church with fine sculpted portal.

Nangkartse, CHINA: see NAGARZE.

Nangkatse, CHINA: see NAGARZE.

Nango (NAHNG-go), town, Tooda county, MIYAGI prefecture, N HONSHU, NE JAPAN, 22 mi/35 km N of SENDAI; 38°29′N 141°08′E.

Nango, town, S Naka county, MIYAZAKI prefecture, SE KYUSHU, SW JAPAN, 34 mi/55 km N of MIYAZAKI; 31°31′N 131°22′E.

Nango (nahng-GO), village, E Usuki county, MIYAZAKI prefecture, SE KYUSHU, SW JAPAN, on PHILIPPINE SEA, 25 mi/40 km S of MIYAZAKI; 32°22′N 131°20′E. Includes offshore islet of O-shima (2 mi/3.2 km long, c.0.5 mi/0.8 km wide).

Nango (NAHNG-go), village, Sannohe county, Aomori prefecture, N HONSHU, N JAPAN, 47 mi/75 km S of AOMORI; 40°24′N 141°26′E.

Nango, village, S Aidzu county, FUKUSHIMA prefecture, N central HONSHU, NE JAPAN, 62 mi/100 km S of FUKUSHIMA city; 37°13′N 139°32′E. Tomatoes.

Nangong (NAHN-GUNG), city (□ 330 sq mi/858 sq km; 2000 population 429,351), S HEBEI province, CHINA, 50 mi/80 km W of DEZHOU; 37°22′N 115°20′E. Agriculture is an important source of income for the city, especially crop growing (grains, cotton, fruits, oil crops); also hogs, eggs. Manufacturing (textiles and pharmaceuticals).

Nangpa La (NAHNG-pah LAH), pass (18,835 ft/5,741 m), in E NEPAL HIMALAYA, on NEPAL-CHINA (TIBET) border, 34 mi/55 km S of TINGRI (Tibet) on a Nepal-Tibet trade route; 28°05′N 86°36′E.

Nangqen (NAHNG-CHUN), town, ⊙ Nangqen county, southernmost QINGHAI province, CHINA, on trade route to CHINDU; 32°15′N 96°13′E. Livestock. Machinery; electric power generation; salt mining. Also appears as Nangqian.

Nangqian, CHINA: see NANGQEN.

Nanguneri (NAHN-guh-NAI-ree), town, NELLAI KATTABOMMAN district, TAMIL NADU state, S INDIA, 17 mi/27 km S of TIRUNELVELI. In large palmyra tract in grain-growing area; jaggery, palmyra mats. Coffee grown at TIRUKURUNGUDI, 7 mi/11.3 km SW.

Nan Hai, CHINA: see SOUTH CHINA SEA.

Nanhe (NAHN-HUH), town, ⊙ Nanhe county, SW HEBEI province, CHINA, 10 mi/16 km ESE of XINGTAI, near BEIJING-WUHAN railroad; 36°58′N 114°41′E. Cotton, wheat, kaoliang, beans, oilseeds; textiles.

Nanheng Kunglu, TAIWAN: see Southern Cross-Island Highway.

Nanhua (NAHN-HWAH), town, ⊙ Nanhua county, central YUNNAN province, CHINA, on road to MYANMAR, 20 mi/32 km NW of CHUXIONG; 25°13′N 101°21′E. Elevation 6,299 ft/1,920 m. Rice, wheat, millet, beans, tobacco; tobacco industry. Also known as Chennan.

Nanhui (NAHN-HWAI), town, ⊙ Nanhui county, CHINA, 20 mi/32 km SE of SHANGHAI, near Pudong Point on EAST CHINA SEA; 31°03′N 121°45′E. An administrative unit of Shanghai municipality. Rice, wheat, beans, cotton, oilseeds. Cotton textiles, glass products, transport equipment.

Nanisivik, town, mine (□ 64 sq mi/166.4 sq km; 2001 population 77), N BAFFIN ISLAND, BAFFIN region, NUNAVUT territory, N Canada, c.17 mi/27 km W of ARCTIC BAY, c.795 mi/1,280 km NW of IQALUIT, on S shore of Strathcona Sound; 73°02′N, 84°33′W. Wildlife, vegetation scarce. First recorded exploration was by Admiral Parry in 1820. The town developed around a silver, lead, and zinc mine that opened in 1974. Historically, the area has never been inhabited by the Inuit, but today the mine employs Inuit people as well as southern workers on a rotation basis. In 1979, Nanisivik was the site of the world's northernmost marathon, now an annual event.

Nanjangud (NUHN-juhn-good), town, MYSORE district, KARNATAKA state, S INDIA, on KABBANI RIVER, 13 mi/21 km S of MYSORE city center, on railroad spur. Cotton and paper milling; mango gardens. Large temple here has annual Hindu festival in which the image of the deity is carried around in a specially decorated chariot. Nearby BADANAVAL (or Badanval, Badanwala) village is major training center for revival of regional village handicrafts; hand spinning and weaving (khaddar, wool); handmade paper; beekeeping.

Nanjiang (NAHN-JYAHNG), town, ⊙ Nanjiang county, N SICHUAN province, CHINA, on the Nan Jiang (headstream of QU River), and 35 mi/56 km N of BAZHONG, in mountain region; 32°21′N 106°50′E. Grain, oilseeds, tobacco; food and beverages, building materials. Logging; iron ore mining.

Nanjing (NAHN-JING) [Chinese=southern capital], city (□ 366 sq mi/947 sq km; 1994 estimated urban population 2,224,200; estimated total population 2,614,900), ⊙ JIANGSU province, E central CHINA, in a bend of CHANG JIANG (Yangzi River); 32°03′N 118°47′E. A major center on Chang Jiang (Yangzi River), it has served at times in the past as capital of China. The second-largest city in the region (after SHANGHAI), Nanjing is the largest river port in interior China and at the intersection of three major railroad lines. Industry, which once centered on "nankeen" cloth (unbleached cotton goods), has been vigorously developed under the Communist government. The city now has an integrated iron-steel complex, an oil refinery, food-processing establishments, and hundreds of plants making chemicals, textiles, cement, fertilizers, machinery, weapons, optical equipment, optical instruments, photographic equipment, and trucks. Long celebrated as a literary and political center. Sixteen dynasties and regions built their power base here. Capital of China from the 3rd to the 6th centuries and again from 1368 to 1421. The Treaty of Nanjing, signed in 1842 at the end of the Opium War, opened China to foreign trade. During the Taiping Rebellion insurgents held the city from 1853 to 1864. It was captured by the revolutionists in 1911, and in 1912 it became capital of China's first president, Sun Yat-sen (Sun Zhongshan). In 1928, the city became the capital of Republic of China, led by the Kuomintang under Chiang Kai-shek (Jiang Jie-shi). In 1932, when the Japanese were threatening to attack the city, the government was temporarily removed to Luoyang, and on Nov. 21, 1937, just before Nanjing fell to the Japanese, it was moved to Chongqing. The Japanese entry into the city, which was accompanied by widespread killing and brutality, became known as the "rape of Nanjing." The Japanese established here (1938) their puppet regime. Chinese forces reoccupied the city on Sept. 5, 1945, and the capitulation of the Japanese armies in China was signed here on Sept. 9. Nanjing again came under the control of the Communists in April 1949, and from 1950 until 1952, when it became the provincial capital, Nanjing was administered as part of an autonomous region. Seat of numerous institutions of higher learning, notably Nanjing University and Nanjing Institute of Technology; also, the Nanjing Military Academy. Noted for its large library, and both its astronomical observatory and its botanical gardens are among the largest in the country. The original city wall (70 ft/21 m high), most of which still stands, dates from the Ming dynasty (1368–1644), and encircles most of the modern city. The tomb of the first Ming emperor is approached by an avenue lined with colossal images of men and animals. Also of interest are the tomb of Sun Yat-sen, a memorial to China's war dead (a steel pagoda), the Taiping Rebellion museum, and the National Zijinshan Observatory. A 4-mi/6.4-km, two-level railroad and road bridge was completed across the Chang Jiang (Yangzi River) in 1968.

Nanjing (NAHN-JING), town, ⊙ Nanjing county, S FUJIAN province, SE CHINA, 45 mi/72 km W of XIAMEN, and on tributary of LONG River; 24°31′N 117°21′E. Rice, wheat, tobacco, sugarcane; food processing, furniture and chemical manufacturing; logging. Sometimes spelled Nan-ching.

Nanjing, CHINA: see NANCHANG.

Nanjo (NAHN-jo), town, Nanjo county, FUKUI prefecture, central HONSHU, W central JAPAN, 16 mi/25 km S of FUKUI; 35°49′N 136°11′E.

Nankan (NAHN-kahn), town, Tamana county, KUMAMOTO prefecture, W KYUSHU, SW JAPAN, 22 mi/35 km N of KUMAMOTO; 33°03′N 130°32′E. Railroad terminus. Bamboo shoots; noodles.

Nan-kan, TAIWAN: see Matsu.

Nankana Sahib (nun-KAH-nuh SAH-eeb), town, SHEIKHUPURA district, E central PUNJAB province, central PAKISTAN, 24 mi/39 km SW of SHEIKHUPURA; 31°27′N 73°42′E. Local trade in cotton, wheat, millet; cotton ginning. Guru Nanak, founder of Sikh religion was born here in 1469. A sacred Sikh pilgrimage center.

Nankang (NAHN-GAHNG), suburban district of TAIPEI, N TAIWAN, on the Taipei-Keelung Highway. Formerly an industrial suburb, it is best known today as the site of the Chinese Academy of Sciences, Academia Sinica.

Nankang (NAHN-KAHNG), town, ⊙ Nankang county, SW JIANGXI province, CHINA, 17 mi/27 km SW of GANZHOU; 25°40′N 114°45′E. Rice, oilseeds, sugarcane; tobacco industry.

Nankauri, INDIA: see NANCOWRY ISLAND.

Nanko (NAHN-ko), town, Sayo district, HYOGO prefecture, S HONSHU, W central JAPAN, 48 mi/77 km N of KOBE; 34°59′N 134°23′E. Near Chikusa River.

Nankoku (NAHN-ko-koo), city, KOCHI prefecture, S SHIKOKU, W JAPAN, 6 mi/10 km E of KOCHI; 33°34′N 133°38′E. Peppers; horticulture. The practice of planting rice twice a year originated here.

Nankou (NAHN-KO), town, CHINA, on railroad, 25 mi/40 km NW of BEIJING, at E end of Nankou Pass (1,900 ft/579 m) through hills and gate of Great Wall; 40°16′N 116°05′E. An administrative unit of Beijing municipality. A site of major transport-equipment production. Nearby (E) are tombs of 13 Ming emperors.

Nanle (NAHN-LUH), town, ⊙ Nanle county, N HENAN province, CHINA, near HEBEI province border, 45 mi/72 km E of ANYANG; 36°06′N 115°12′E. Grain, cotton, oilseeds; food and timber processing.

Nanling (NAHN-LING), town, ⊙ Nanling county, SE ANHUI province, CHINA, 30 mi/48 km SSW of WUHU; 30°56′N 118°20′E. Rice, wheat, oilseeds, medicinal herbs; food processing, engineering, building materials manufacturing.

Nanling (NAHN-LING), mountain range of GUANGDONG and HUNAN provinces and GUANGXI ZHUANG autonomous region, S CHINA; rises to c. 6,900 ft/2,103 m; 25°00′N 112°00′E. The Nanling forms the geographical boundary between central and S China. It separates the CHANG JIANG (Yangzi River) and Xi drainage basins; protects S China from cold N air masses, and divides the Cantonese cultural sphere and linguistic area from that of central China. The mountain range lies in parallel ridges that hinder N-S travel.

Nanma (NAHN-MAH), town, central ZHEJIANG province, CHINA, 45 mi/72 km E of JINHUA; 29°05′N 120°15′E. Rice, medicinal herbs, sugarcane.

Nanmoku (NAHN-mok), village, Kanra county, GUMMA prefecture, central HONSHU, N central JAPAN, 31 mi/50 km S of MAEBASHI; 36°09′N 138°42′E.

Nanmulin, CHINA: see NAMLING.

Nannilam (NUHN-ni-luhm), town, tahsil headquarters, Thanjavur district, TAMIL NADU state, S INDIA, in KAVERI River delta, 17 mi/27 km WNW of Nagapattinam. Rice milling.

Nannine (na-NEEN), village, W central WESTERN AUSTRALIA state, W AUSTRALIA, on GERALDTON–WILUNA railroad, 265 mi/426 km NE of Geraldton; 26°53′S 118°20′E. Gold mining.

Nanning (NAHN-NING), city (□ 708 sq mi/1,840.8 sq km; 2000 population 1,159,099), ⊙ GUANGXI ZHUANG autonomous region, CHINA, on the XI River; 22°49′N 108°19′E. In a fertile farming area. Has a medium-size integrated iron and steel complex, a sugar refinery, other food-processing plants, and factories making fertilizer, machine tools, paper, cement, and farm machinery. The city is on the Hunan-Guangxi rail-

road to N VIETNAM, and was an important base for supplies going to North Vietnam during the Vietnam War. Seat of Guangxi University, a medical college, and an agricultural institute.

Nanning-Kunming Railroad (NAHN-NING–KUN-MING), E YUNNAN and SW GUIZHOU provinces and W GUANGXI ZHUANG autonomous region, S CHINA, running 558 mi/898 km. Provides transportation to many people living in mountainous counties along the line. Construction began in 1991; in operation since late 1996.

Nanno (NAHN-NO), town, Kaizu county, GIFU prefecture, central HONSHU, central JAPAN, 12 mi/20 km S of GIFU; 35°13′N 136°36′E.

Nannup, town, WESTERN AUSTRALIA state, W AUSTRALIA, 174 mi/280 km S of PERTH; 33°57′S 115°42′E. Timber.

Nanomana, TUVALU: see NANUMANGA.

Nanomea, TUVALU: see NANUMEA.

Nanoose Bay (na-NOOS), unincorporated village, SW BRITISH COLUMBIA, W Canada, on SE VANCOUVER ISLAND, on Strait of GEORGIA, 13 mi/21 km WNW of NANAIMO, in Nanaimo regional district; 49°15′N 124°11′W. Lumber-shipping port.

Nanortalik (na-NOKH-tah-lik), town, Nanortalik commune, SW GREENLAND, between CAPE FAREWELL and JULIANEHÅB (Qaqortoq); 60°08′N 45°13′W. Fishing center.

Nanos (nah-NOS), Italian *Monte Re*, mountain, SW SLOVENIA, 12 mi/19 km SE of Ajdovščina, and 9 mi/14.5 km E of POSTOJNA; [Suhi Vrh=*dry peak*] is highest peak (□ 10 sq mi/26 sq km). Protected in regional park (□ 10 sq mi/26 sq km).

Nanouki, GILBERT ISLANDS: see ARANUKA.

Nanpara (nahn-PAH-rah), town, BAHRAICH district, N UTTAR PRADESH state, N central INDIA, 21 mi/34 km NNW of BAHRAICH. Railroad junction, with spur to NEPALGANJ (or Naipalganj) Road, 11 mi/18 km NNE, on Nepal border. Trades in rice, wheat, corn, gram, oilseeds, and in grains, hides, and ghee from Nepal.

Nanpi (NAHN-PEE), town, ⊙ Nanpi county, SE HEBEI province, CHINA, 45 mi/72 km NNE of DEZHOU, near Tianjin-Pukou railroad; 38°02′N 116°42′E. Cotton, wheat, kaoliang, oilseeds; textile manufacturing, engineerning, food processing.

Nanping (NAHN-PING), city (□ 1,024 sq mi/2,652 sq km; 1994 estimated urban population 211,700; estimated total population 469,200), N central FUJIAN province, SE CHINA, on the upper MIN River; 26°40′N 118°07′E. Light and heavy industries (textile and electrical equipment manufacturing; non-ferrous metals) form the economic base of the city. Crop growing (grain, fruit); animal husbandry (hogs), and forestry are also important. Also called Yanping.

Nan River (NAHN), 500 mi/805 km long, one of the headstreams of the CHAO PHRAYA (Me Nam, river), N THAILAND; rises on LAOS border in N LUANG PRABANG RANGE at 19°35′N 101°10′E; flows S, past NAN, UTTARADIT (head of navigation), and PHITSANULOK, forming (in lower course) a common floodplain with YOM RIVER. The combined stream joins the PING RIVER just above NAKHON SAWAN to form the CHAO PHRAYA.

Nansei (NAHN-SAI), town, Watarai county, MIE prefecture, S HONSHU, central JAPAN, on Gokasho Bay, within Ise-shima National Park, 28 mi/45 km S of TSU; 34°20′N 136°42′E. Rias coast nearby.

Nansei-shoto, JAPAN: see RYUKYU ISLANDS.

Nansemond River (NAN-suh-muhnd), c.25 mi/40 km, SE VIRGINIA; rises in W SUFFOLK city; flows N, through Suffolk to JAMES RIVER estuary, in HAMPTON ROADS harbor; lower 10 mi/16 km widens into tidal river; 36°44′N 76°35′W.

Nansen, Cape (NUHN-sen), SE point of ALEXANDRA LAND, FRANZ JOSEF LAND, ARCHANGEL oblast, extreme N European Russia, in the ARCTIC OCEAN; 80°26′N 46°10′E. Formerly called Cape Fridtjof Nansen.

Nansen Ice Sheet, an ice shelf inland of TERRA NOVA BAY in VICTORIA LAND, EAST ANTARCTICA, adjoining the N side of the DRYGALSKI ICE TONGUE, 30 mi/50 km long, 10 mi/15 km wide; 74°58′S 163°10′E.

Nansen Island (NUHN-sen), in FRANZ JOSEF LAND, ARCHANGEL oblast, extreme N European Russia, in the ARCTIC OCEAN; 12 mi/19 km long, 6 mi/9.7 km wide; 80°30′N 54°00′E. Formerly known as Fridtjof Nansen Island.

Nansen Sound or **Fridtjof Nansen Sound**, NUNAVUT territory, N Canada, arm of the ARCTIC OCEAN, between AXEL HEIBERG (W) and NW ELLESMERE (E) islands; 90 mi/145 km long, 10 mi/16 km–35 mi/56 km wide; 81°00′N 90°00′W. Connects EUREKA SOUND and GREELY FJORD (SE) with ARCTIC OCEAN (NW). In N entrance are LANDS LOKK and Fjeldhdmen islands.

Nansha Islands, CHINA: see SPRATLY ISLANDS.

Nan Shan (NAHN SHAHN) [Mandarin=south mountain], system of parallel ranges on QINGHAI-GANSU province border, CHINA, representing an E continuation of the Altun mountains and a N branch of the KUNLUN mountains, S of the SILK ROAD through the neck of Gansu province; 22°30′N 113°54′E. The system includes the DANGHE NANSHAN (Humbolt), Qinghai Nanshan, Tulai Nanshan, and DATONG mountains; rises more than 20,000 ft/6,096 m.

Nansio (nahn-SEE-o), village, MWANZA region, NW TANZANIA, 30 mi/48 km NNE of MWANZA, on S shore of UKEREWE ISLAND, Lake VICTORIA; 02°07′S 33°03′E. Lake port; landing strip. Fish; cattle, sheep, goats; cotton, grain.

Nantachie, Lake (nuh-TACH-ee), reservoir, GRANT parish, central LOUISIANA, 8 mi/13 km NW of COLFAX; c.2 mi/3 km long; 31°37′N 92°48′W. Drains S through tributary of RED RIVER.

Nantahala Mountains (nan-tuh-HAI-luh), mostly in far SW NORTH CAROLINA, extends into NE GEORGIA, transverse range of the APPALACHIANS between GREAT SMOKY MOUNTAINS (N) and the BLUE RIDGE (S); from confluence of NANTAHALA and LITTLE TENNESSEE rivers in North Carolina, extend c.50 mi/80 km S to TALLULAH FALLS (Georgia); 35°08′N 83°33′W. Highest points are Wine Spring Bald Mountain (c.5,500 ft/1,676 m), 13 mi/21 km SW of FRANKLIN (North Carolina), and Wayah Bald Mountain (5,342 ft/1,628 m), 11 mi/18 km W of Franklin. In Nantahala (North Carolina) and Chattahoochee (Georgia) national forests. Also spelled Nantehalah.

Nantahala River (nan-tuh-HAI-luh), 40 mi/64 km long, W NORTH CAROLINA; rises in NANTAHALA MOUNTAINS 65 mi/105 km SW of ASHEVILLE, in SW MACON county at GEORGIA state line; flows NNW and N, through Nantahala Lake reservoir to LITTLE TENNESSEE River in Fontana Lake reservoir where it forms S arm, 8 mi/12.9 km W of BRYSON CITY; 35°01′N 83°30′W. Entirely within Nantahala National Forest. In lower course, traverses scenic, 8-mi/12.9-km-long Nantahala Gorge, with steep sides up to 2,000 ft/610 m high. Nantahala Dam (250 ft/76 m high, 1,042 ft/318 m long; completed 1942), privately built power dam in middle course, is c.20 mi/32 km SW of Bryson City; forms Nantahala Reservoir (5 mi/8 km long, 1 mi/1.6 km wide), sometimes known as Aquone Lake.

Nantasket Beach, beach village, of HULL town, PLYMOUTH county, E MASSACHUSETTS, 3 mi/4.8 km NNE of HINGHAM, on narrow Nantasket Peninsula (5 mi/8 km long), which extends NW into MASSACHUSETTS BAY to Point Allerton, then bends W to end at Windmill Point. State public beach.

Nantehalah Mountains, NORTH CAROLINA: see NANTAHALA MOUNTAINS.

Nanterre (nawn-TER), city (□ 9 sq mi/23.4 sq km); ⊙ HAUTS-DE-SEINE department, ÎLE-DE-FRANCE region, N central FRANCE, on the left bank of the SEINE RIVER, a WNW suburb of PARIS, 7 mi/11.3 km from Notre Dame Cathedral; 48°54′N 02°12′E. Administrative and industrial center manufacturing automotive parts, electrical equipment, machine tools, tires, paper, and rolling stock. Its man-made harbor on the Seine is part of the port of Paris. The modern district of La Défense in the outskirts of Paris (on NW extension of the CHAMPS-ÉLYSÉES) lies just E. In 1968, the Nanterre branch of the University of Paris was the scene of student protests that spread to other areas and led to a national political crisis. The Basilica of Sainte-Geneviève, with a 15th-century nave, is here; also the school of dance of the Paris Opera.

Nantes (NAHNT), Breton *Naoned*, city (□ 25 sq mi/65 sq km); ⊙ LOIRE-ATLANTIQUE department, W FRANCE, seaport on the lower LOIRE RIVER above its estuary, at influx of the ERDRE (right bank) and the SÈVRE NANTAISE (left bank) rivers, and 215 mi/346 km SW of PARIS; 47°10′N 01°35′W. It is the largest urban center of W France and the administrative capital of the region known as PAYS DE LA LOIRE, which encompasses five departments in the Loire valley, the VENDÉE coastal area S of the Loire, and parts of E BRITTANY. Due to its strategic location, Nantes serves as the major commercial center for both S Brittany and the Loire valley extending upstream past ANGERS and SAUMUR. Together with its port-oriented suburbs just downstream and suburban expansion S of the Loire (numerous bridges, including a new expressway bridge in the port area W of city center), the metropolitan area has a population of nearly 500,000, and continues to be a growth center. The port of Nantes-Saint-Nazaire extends (with interruptions) along the lower 30 mi/48 km of the Loire, SAINT-NAZAIRE being the city's outport since the mid-19th century. With repeated dredging of the river, Nantes itself can be reached by oceangoing vessels with a draft of 26 ft/8 m, depending on the tide. The city's own port facilities extend along 5 mi/8 km of waterfront, including the heavy industry zone of Cheviré and the W suburbs of Basse-Indre and COUËRON. Major products moving through the port are imported hardwoods, tropical oils, fertilizer, fresh fruit and vegetables, sugar, wine of the Loire valley, petroleum products, cereals, and special steel products. Nantes has been adding electronics and aeronautics to its traditional industrial base of food processing (canned and frozen foods, biscuits, dairy products, and sugar), metalworking (agricultural equipment), shipbuilding, and container manufacturing. Along the banks of the Erdre, N of the city, a research and educational complex known as Nantes Atlantique has established itself near the university campus. The university dates from 1460. Nantes has renewed its central area since World War II when it suffered some damage from Allied bombing. The N arm of the Loire has been filled in and paved over, thus providing space for new construction: the faculties of medicine and pharmacy of the University of Nantes, new hotels, and a civic center (Centre Neptune) containing shops and convention building. Nantes is noted for its green, open spaces, including town squares, a botanical garden, greenbelts (along the former river arm), and a park surrounding the 15th-century fortress-like castle of the dukes of Burgundy (with moat) and connecting it to the Gothic Cathedral of St. Pierre and St. Paul (begun in 1434, completed in 1893), which contains the Renaissance tomb of Francis II and his consort. The city itself is rich in museums (fine arts, housed in a late 19th century edifice; natural history; and a museum dedicated to Jules Verne, a native son and prolific 19th-century author of extrordinary travel tales). The old Feydeau island, now part of the mainland, has preserved its 18th century character, with buildings formerly housing artisans of the maritime trades. Nantes was founded by the Namnetes, a Gallic tribe, before the Roman conquest of Gaul. It became a bishopric in the 4th century. Unsuccessfully besieged by the Huns, Nantes was ravaged and occupied (843–936) by Norsemen. It fell to the powerful dukes of

Brittany in the 10th century, who resided here through the 15th century; it then passed to France (1491) after the marriage of Anne of Brittany to Louis XII. The famous Edict of Nantes was issued by Henry IV in 1598, giving the Protestants freedom of worship and other privileges. Nantes reached its greatest commercial success in the 18th century with its trade between France, W Africa, and the Antilles, including the slave trade. During the French Revolution, Nantes was nearly stormed by royalist troops of the Vendée rebellion and was the scene of bloody massacres by the leaders of the Revolution in 1793. It was a center of resistance to the German occupation in World War II. Via high-speed trains (TGV) and expressway links with Paris, Brittany, and the SW, Nantes continues to prosper as the gateway to W France. Nantes-Atlantique international airport 5 mi/8 km SW of city.

Nantes (NAHNT), village (□ 46 sq mi/119.6 sq km; 2006 population 1,443), Estrie region, S QUEBEC, E Canada, 8 mi/13 km from LAC-MÉGANTIC; 45°38′N 71°02′W. Christmas tree farms. Also called Spring Hill.

Nantes à Brest, Canal de, FRANCE: see BREST-NANTES CANAL.

Nantes-Brest Canal, FRANCE: see BREST-NANTES CANAL.

Nanteuil-le-Haudouin (nahn-TOI-luh-o-DWAN), commune (□ 8 sq mi/20.8 sq km), OISE department, PICARDIE region, N FRANCE, 11 mi/18 km ESE of SENLIS; 49°08′N 02°49′E. Agricultural equipment manufacturing. It was a pivot point of Allied offensive in first battle of the Marne (1914). Its war-damaged 13th-century church has been rebuilt.

Nanteuil-lès-Meaux (nahn-toi-lai-MO), town (□ 3 sq mi/7.8 sq km), SEINE-ET-MARNE department, ÎLE-DE-FRANCE region, N central FRANCE, on MARNE RIVER, and 3 mi/4 km SSE of MEAUX; 48°56′N 02°54′E.

Nantiat (nahn-TYAH), commune (□ 9 sq mi/23.4 sq km), HAUTE-VIENNE department, LIMOUSIN region, W central FRANCE, 13 mi/21 km NNW of LIMOGES; 46°01′N 01°11′E. Metalworks.

Nanticoke (NAN-ti-kok), former city (□ 261 sq mi/678.6 sq km; 2001 population 23,588), SE ONTARIO, E central Canada, on N shore of Lake ERIE, 36 mi/58 km SW of HAMILTON, and included in HALDIMAND; 42°48′N 80°04′W. Heavy industries including Steel Co. Canada complex, oil refinery, and coal-generated electric plant. Incorporated 1974 by almagamation of 7 municipalities, including village of Nanticoke.

Nanticoke (NAN-ti-KOK), city (2006 population 10,341), LUZERNE county, E central PENNSYLVANIA, 7 mi/11.3 km WSW of WILKES-BARRE, on the SUSQUE-HANNA RIVER (bridged); 41°12′N 76°00′W. Largely residential. Manufacturing (plastic products). Formerly the heart of a major anthracite-coal-mining region, but production has declined. State correctional institution to W; part of Lackawanna State Forest to N. Founded 1793, incorporated as a city 1926.

Nanticoke (NAN-TEE-kok), fishing village, WICOMICO county, SE MARYLAND, on the EASTERN SHORE, 18 mi/29 km WSW of SALISBURY, and on an estuary of Nanticoke River Tomato cannery; ships, seafood. The name comes from a Native American tribe known as "tide-water people."

Nanticoke River, c.50 mi/80 km long, S DELAWARE and E MARYLAND; rises in several small branches in N SUSSEX county (Delaware); flows SW, past SEAFORD (Delaware; head of navigation and tides), SHARP-TOWN and VIENNA (Md.), to N end of TANGIER SOUND near NANTICOKE; marsh-bordered estuary is 3 mi/4.8 km wide at mouth. Fishing Bay Wildlife Management Area on W shore. WICOMICO RIVER shares entrance to sound, enters from ENE. Main tributary is MARSHYHOPE CREEK, formerly called NW Fork of Nanticoke River.

Nanto (NAHN-TO), town, Watarai county, MIE prefecture, S HONSHU, central JAPAN, 31 mi/50 km S of TSU; 34°16′N 136°30′E.

Nanton, town (□ 2 sq mi/5.2 sq km; 2001 population 1,841), S ALBERTA, W Canada, 50 mi/80 km SSE of CALGARY, in Willow Creek No. 26 municipal district; 50°21′N 113°46′W. Flour and cereal foods milling; dairying; ranching; wheat; livestock. Mineral springs. Large source of bottled mineral water. Established as a village in 1903; became a town in 1907.

Nantong (NAHN-TUNG), city (□ 86 sq mi/223.6 sq km), N JIANGSU province, CHINA, on CHANG JIANG (Yangzi River), 30 mi/48 km from the coast; 32°05′N 120°51′E. Nantong is an industrial center with sizable light and heavy industries. A major port on Chang Jiang (Yangzi River), the city has ocean freight connections with more than 100 countries. Commercial agriculture (grain, oil crops, cotton, vegetables, fruits); animal husbandry (hogs, poultry, beef, lamb); eggs; fishing (aquatic products). Manufacturing includes food, textiles, apparel, paper, crafts, utilities, chemicals, pharmaceuticals, synthetic fibers, plastics, machinery, transportation equipment, electrical equipment, electronics, and precision instruments; metal making. Formerly Tongzhon. Also spelled Nantung.

Nantong (NAHN-TUNG), town, ⊙ Nantong county (2000 population 1,602,029), N JIANGSU province, CHINA, 14 mi/23 km ENE of NANTONG city; 32°06′N 121°04′E. Rice, cotton, jute, medicinal herbs; textiles, electric equipment, plastics, building materials; metalworks. Also known as Jinsha.

Nantou (NAHN-TO), town, W central TAIWAN, 16 mi/26 km S of TAICHUNG; 23°55′N 120°41′E. Marketing and agricultural center (fruits, especially oranges, bananas, betel-nut); sugar milling, pineapple canning, porcelain and pottery goods manufacturing. Sometimes spelled Nantow.

Nantou (NAHN-TO), town, S GUANGDONG province, SE CHINA, on E shore of PEARL RIVER estuary, near HONG KONG border, port for Bao'an (just NE), 22 mi/35 km NW of KOWLOON; 22°35′N 113°58′E. Commercial center; textiles; fisheries.

Nantua (nahn-too-ah), commune (□ 4 sq mi/10.4 sq km), sub-prefecture of AIN department, RHÔNE-ALPES region, E FRANCE, in the JURA Mountains, 19 mi/31 km ESE of BOURG-EN-BRESSE, at SE end of Lake of NANTUA, at entrance to a forested gorge; 46°09′N 05°36′E. Some manufacturing (furniture, eyeglasses, and plastics). Church of Saint-Michel is all that is left of a 12th-century abbey destroyed in French Revolution. Has museum of the French Resistance in World War II. The main highway from Rhône-Saône Valley to GENEVA passes through a long tunnel outside Nantua.

Nantucket (nan-TUH-ket), island (□ 303 sq mi/787.8 sq km; 2006 population 10,240), NANTUCKET county, SE MASSACHUSETTS, c.25 mi/40 km S of CAPE COD (separated by NANTUCKET SOUND) and across MUSKEGET CHANNEL from MARTHA'S VINEYARD (W); c.14 mi/23 km long, 3 mi/4.8 km–6 mi/9.7 km wide; 41°15′N 70°08′W. Airport; ferry service from WOODS HOLE and HYANNIS. Exhibiting evidence of glaciation (terminal moraine, outwash plain), Nantucket has sandy beaches and low, rolling hills composed of sand and gravel. It is sparsely vegetated; wild cranberries, beach plums, heather, and wild roses predominate. Nantucket and the small adjacent islands constitute both Nantucket town (1990 population 6,012; 2000 population 9,520) and Nantucket county. Settled by Europeans in 1659, the island was part of New York (1660–1692), when it was ceded to Massachusetts. Major whaling port until the decline of the industry (c.1850); later developed into a well-known resort and artists' colony. The village of Nantucket is the trade center of the island and is known for its many old houses and fishing port. Siasconset, or 'Sconset, is a summer resort on Atlantic coast (E). The island has a whaling museum and an eighteenth-century windmill. The first U.S. lightship station (established 1856) is located near Nantucket. Great Point forms the N tip of the island (lighthouse here). Sankaty Head and lighthouse in Siasconset on entrance.

Nantucket Sound, channel of the ATLANTIC OCEAN, off SE MASSACHUSETTS, between S shore of CAPE COD (N) and NANTUCKET Island and MARTHA'S VINEYARD (S); c.30 mi/48 km long, c.25 mi/40 km wide.

Nantung, CHINA: see NANTONG.

Nantwich (NANT-wich), town (2001 population 12,515), S CHESHIRE, W ENGLAND, on WEAVER RIVER, and 4 mi/6.4 km SW of CREWE; 53°04'N 02°31'W. Leather tanning; clothing industry; cheese market. Has brine baths and saltworks and was formerly an important center of salt industry. Has grammar school (opened 1611) and church dating from 13th and 15th century.

Nanty Glo (NAN-tee GLO), borough (2000 population 3,054), CAMBRIA county, W central PENNSYLVANIA, 10 mi/16 km NNE of JOHNSTOWN, on South Branch of Blacklick Creek; 40°28'N 78°49'W. Bituminous coal. Light manufacturing. Ebensburg Airport to E. Founded 1888.

Nant-y-Moel (nant-uh–MOIL), village (2001 population 2,322), Rhonda Cyon Taff, S Wales, on OGWR RIVER, and 8 mi/12.9 km NNE of BRIDGEND; 51°37'N 03°32'W. Just S, on Ogwr River, are districts of Price Town and Ogmore Vale. Former mining communities. Formerly in MID GLAMORGAN.

Nanu (NA-nyoo), village, WESTERN province, S central NEW GUINEA island, SW PAPUA NEW GUINEA, 40 mi/64 km NW of DARU and TORRES STRAIT; 08°53'S 142°42'E. Road access. Bananas, yams, taro, sago; cattle.

Nanuet (NA-noo-et), village (□ 5 sq mi/13 sq km; 2000 population 16,707), ROCKLAND county, SE NEW YORK, 5 mi/8 km W of NYACK; 41°06'N 74°01'W. Light manuufacturing, commercial services, and extensive retail.

Nanumanga (NAH-noo-MAHNG-ah), coral island (□ 1 sq mi/2.6 sq km; 2002 population 589), TUVALU, SW PACIFIC; 06°18'S 176°20'E. Copra. Also Nanomana. Formerly called Hudson Island

Nanumea (NAH-noo-MAI-ah), atoll (□ 1 sq mi/2.6 sq km; 2002 population 664), TUVALU, SW PACIFIC; 6 mi/9.7 km long; 09°39'S 176°08'E. Included 1915 in British Gilbert and Ellice Islands Colony; included in Tuvalu in 1978. In World War II, occupied 1943 by U.S. Formerly St. Augustine Island. Also Nanomea.

Nanu Oya, town, CENTRAL PROVINCE, SRI LANKA, on HATTON PLATEAU, on small stream, and 3 mi/4.8 km SW of NUWARA ELIYA; 06°56'N 80°44'E. Elev. 5,291 ft/1,613 m. Railroad junction; tea, rubber, vegetables. Also written Nanuoya.

Nanuque (NAH-noo-kai), city (2007 population 40,261), E central MINAS GERAIS state, BRAZIL, 99 mi/159 km E of TEÓFILO OTONI, near border with SE BAHIA state, on Rio MUCURI; 17°48'S 40°25'W.

Nanvarnarluk, village, SW ALASKA, on Bethel Bay, 60°00'N 162°00'W.

Nan Wan (NAHN WAHN) [Chin.=south bay], inlet and small village at S tip of Taiwan, W of Cape Oluanpi.

Nanwei Dao, CHINA: see SPRATLY ISLAND.

Nanxi (NAHN-SHEE), town, ⊙ Nanxi county, S SICHUAN province, CHINA, 25 mi/40 km W of LUZHOU, and on left bank of CHANG JIANG (Yangzi River); 28°52'N 104°58'E. Rice, sugarcane, jute; textiles, engineering, chemicals; food industry.

Nan Xian (NAN SHYEN), town, ⊙ Nan Xian county, N HUNAN province, CHINA, on NW shore of DONGTING Lake, 45 mi/72 km ENE of CHANGDE; 29°22'N 112°25'E. Rice, cotton, sugarcane, oilseeds; fisheries.

Nanxiang (NAHN-SHYAHNG), town, CHINA, 12 mi/19 km WNW of SHANGHAI, and on Shanghai-Nanjing railroad. An administrative unit of Shanghai municipality. Commercial center; plastic products, lighting

and electrical equipment, power transmission equipment; in rice-growing area.

Nanxiong (NAHN-SHYUNG), town, ⊙ Nanxiong county, N GUANGDONG province, CHINA, at S foot of DAYU Mountains, on BEI River (North River), 55 mi/89 km NE of SHAOGUAN, near MEILING Pass; 25°07'N 114°18'E. Rice, sugarcane, oilseeds.

Nanyamba (nah-NYAM-bah), village, MTWARA region, SE TANZANIA, 33 mi/53 km SW of MTWARA, near MOZAMBIQUE border; 10°41'S 39°50'E. Cashews, copra, bananas; livestock.

Nanyang (NAHN-YAHNG), city (□ 45 sq mi/117 sq km; 2000 population 374,600), SW HENAN province, CHINA, 160 mi/257 km SW of KAIFENG, and on the BAI River; 33°06'N 112°31'E. Road hub; light industry and commerce are the largest sectors of the city's economy. Crop growing; animal husbandry. Main industries include food processing and manufacturing of beverages, tobacco, textiles, pharmaceuticals, and electrical equipment. The Nanyang agricultural area is economically associated with LAOHEKOU and FANCHENG in the HAN River valley of HUBEI province.

Nanyo (NAHN-yo), city, YAMAGATA prefecture, N HONSHU, NE JAPAN, 19 mi/30 km S of YAMAGATA city; 38°03'N 140°09'E. Grapes.

Nanyuki (nah-NYOO-kee), town (1999 population 31,577), RIFT VALLEY province, S central KENYA, at NW foot of Mount KENYA, 90 mi/145 km NNE of NAIROBI; 00°03'N 37°02'E. Elevation 9,000 ft/2,743 m. Terminus of railroad spur from Nairobi. Resort and agricultural center (coffee, sisal, wheat, corn); dairy farming. Airfield.

Nanzhao (NAHN-ZHOU), town ⊙ Nanzhao county, W HENAN province, CHINA, in FUNIU Mountains, 40 mi/64 km NNW of NANYANG; 33°30'N 112°27'E. Wheat, beans, kaoliang; engineering, crafts; food industry; iron smelting.

Nanzhou Dao (NAHN-ZHO DOU), island, SW GUANGDONG province, SE CHINA, in GUANGZHOU Bay of SOUTH CHINA SEA, SE of DONGHAI island, forms part of ZHANJIANG muncipality.

Nao, Cape (NAH-o), ALICANTE province, E SPAIN, on the MEDITERRANEAN SEA, 50 mi/80 km NE of ALICANTE; 38°43'N 00°15'E.

Naogaon (naw-gahng), town (2001 population 124,046), RAJSHAHI district, W EAST BENGAL, BANGLADESH, on JAMUNA RIVER (tributary of the ATRAI RIVER), and 36 mi/58 km NNE of RAJSHAHI; 24°51'N 88°51'E. Ganja-growing center; silk weaving factory; trades in rice, jute, oilseeds. Also spelled Naugaon.

Naoiri (nah-O-ee-ree), town, Naoiri county, OITA prefecture, E KYUSHU, SW JAPAN, 19 mi/30 km S of OITA; 33°04'N 131°23'E. Persimmons; smelts. Serikawa Dam nearby.

Naokawa (nah-o-KAH-wah), village, South Amabe county, OITA prefecture, E KYUSHU, SW JAPAN, 25 mi/40 km S of OITA; 32°53'N 131°46'E.

Naolinco, MEXICO: see NAOLINCO DE VICTORIA.

Naolinco de Victoria (mah-o-LEEN-ko de veek-TO-ree-ah), city and township, ⊙ Naolinco municipio, VERACRUZ, E MEXICO, in SIERRA MADRE ORIENTAL, 9 mi/14.5 km NNE of XALAPA ENRÍQUEZ; elevation 5,266 ft/1,605 m; 19°39'N 96°51'W. Agricultural center (coffee, corn, sugarcane, fruit).

Naol River (NOU-UH), 250 mi/402 km long, NE HEILONGJIANG province, NE CHINA; rises in hills N of MISHAN, joined by Qixing River 25 mi/40 km W of Xiaojiahe; flows NE to USSURI River near Dong'an 100 mi/161 km SSW of KHABAROVSK (Russia).

Não-Me-Toque (nou–mee–TO-kee), city, N RIO GRANDE DO SUL state, BRAZIL, 13 mi/21 km S of CARAZINHO; 23°23'S 52°49'W. Wheat, corn, potatoes, manioc; livestock.

Naomi Wilderness Area, Mount, UTAH: see RICHMOND, town.

Naoshera (nou-SHAI-rah), cantonment town, RAJAURI district, JAMMU AND KASHMIR state, N INDIA,

on right tributary of the CHENAB River, 15 mi/24 km SSW of RAJAURI. Wheat, bajra, corn, pulses. Also spelled Nowshera, Naushahra.

Naoshima (nah-O-shee-mah), town, Kagawa county, KAGAWA prefecture, NE SHIKOKU, W JAPAN, 9 mi/15 km N of TAKAMATSU; 34°27'N 133°59'E. Yellowtail, sea bream; nori. Copper, lead, tin products; chemical products.

Naos Island (NAH-os), islet in PANAMA BAY, off PANAMA, guarding PACIFIC entrance of the PANAMA CANAL, 3 mi/4.8 km S of PANAMA city; 08°55'N 79°32'W. Mole to mainland.

Naos Point, CANARY ISLANDS: see RESTINGA POINT.

Naot Mordechai (ne-OT mor-de-KHEI) or **Neot Mordechai**, kibbutz, N ISRAEL, 3.7 mi/6 km SE of KIRYAT SHMONA, and 10 mi/16 km NE of ZEFAT (Safed) in HULA VALLEY; 33°09'N 35°35'E. Elevation 239 ft/72 m. Mixed farming; shoe and sandal factory, fruit-juice plant. Nature reserves with springs and forests of Atlantic oak nearby. Founded in 1946.

Naousa (NAH-oo-sah), Macedonian *Negush*, city (2001 population 19,870), EMATHEIA prefecture, W CENTRAL MACEDONIA department, NE GREECE, near railroad station (3 mi/4.8 km E), 10 mi/16 km NW of VÉROIA, and 46 mi/74 km W of THESSALONÍKI; 40°38'N 22°04'E. Textile-milling center (silk, wool, cotton fabrics); trades in wheat and wine. Hydroelectric plant (NE). Located on site of ancient Citium, it was formerly called Niaousta or Niausta. Also spelled Naoussa.

Napa (NA-puh), county (□ 754 sq mi/1,960.4 sq km; 2006 population 133,522), W CALIFORNIA; ⊙ NAPA; 38°30'N 122°19'W. Bounded S by SAN PABLO BAY, it is a mountainous area, in COAST RANGES. Napa Valley (wine growing) extends SE from base of MOUNT SAINT HELENA to near San Pablo Bay. S county line comes within 0.5 mi/0.8 km of San Pablo Bay. Petrified redwood forest, hot springs (resorts) near CALISTOGA. Grapes, walnuts; nursery products; dairying; cattle. Wine making (since 1850s) is principal industry, together with neighboring SONOMA county (to W); also fruit processing, some manufacturing (at Napa). Mining and quarrying (mercury, pumice, sand and gravel). Formed 1850.

Napa (NA-puh), city (□ 17 sq mi/44.2 sq km; 2000 population 72,585), ⊙ NAPA county, W CALIFORNIA, 33 mi/53 km NNE of SAN FRANCISCO, on the NAPA RIVER; 38°18'N 122°18'W. Manufacturing (wineries; medical equipment, electronic equipment, beverages, computer products, apparel, plastic products, wood products; printing and publishing. Grapes and other fruits (especially citrus fruit) are grown in the adjacent Napa valley, which is famous for the wines that have been made here since the late 1870s. Numerous wineries in the region. Growing city, almost doubled in population between 1970–1990. Seat of Napa Valley College (two-year). Napa County Airport to S. Incorporated 1872.

Napa, Cerro (NAH-pah, SER-ro), Andean peak (18,880 ft/5,755 m), on CHILE-BOLIVIA border; 20°31'S 68°41'W.

Napaimiut (nuh-PAI-mee-yoot), native village, W ALASKA, on KUSKOKWIM River, and 60 mi/97 km SE of HOLY CROSS; 61°32'N 158°48'W. Airfield. Sometimes called Napai, Napamute, or Napamiute.

Napaiskak (nuh-PEI-yi-skak) or **Napakiak**, Inuit village, W ALASKA, near mouth of KUSKOKWIM River, 11 mi/18 km SW of BETHEL.

Napajedl, CZECH REPUBLIC: see NAPAJEDLA.

Napajedla (NAH-pah-YED-lah), German *Napajedl*, town, JIHOMORAVSKY province, SE MORAVIA, CZECH REPUBLIC, on MORAVA RIVER, on railroad, and 8 mi/12.9 km WSW of ZLÍN; 49°10'N 17°32'E. Manufacturing (machinery, plastics). Has a well-known stud farm for thoroughbreds. Has an 18th century castle and a baroque church.

Napakiak, Alaska: see NAPAISKAK.

Napak, Mount (nah-PAK), mountain (8,321 ft/2,536 m), NORTHERN region, NE UGANDA, N of MT. ELGON, c.25 mi NW of Nakapiripirit and 45 mi/72 km SW of MOROTO towns; 02°03′N 34°16′E.

Napalpí (nah-pahl-PEE), town, S central Chaco province, Argentina, 18 mi/29 km SE of PRESIDENCIA ROQUE SÁENZ PEÑA; 26°54′S 60°08′W. Cotton; livestock.

Napamute, Alaska: see NAPAIMIUT.

Napanee (NA-puh-nee), former town (□ 2 sq mi/5.2 sq km; 2001 population 5,351), SE ONTARIO, E central Canada, on Napanee River, near its mouth on the Bay of QUINTE, and 23 mi/37 km W of KINGSTON; included in town of Greater Napanee; 44°15′N 76°57′W. Milling, food processing, brick and tile manufacturing; dairying.

Napanoch (NA-puh-nawk), hamlet (□ 1 sq mi/2.6 sq km; 2000 population 1,168), ULSTER county, SE NEW YORK, on RONDOUT CREEK, just W of the SHAWANGUNK range, and 2 mi/3.2 km NE of ELLENVILLE; 41°45′N 74°22′W. Rondout Reservoir is 3 mi/4.8 km NW.

Napareuli (nah-pahr-ai-OO-lee), village, E GEORGIA, in KAKHETIA, 9 mi/14 km NNE of TELAVI; 42°02′N 45°29′E. Wine-making center.

Napa River, c.50 mi/80 km long, NAPA county, W CALIFORNIA; rises in NW Napa county near MOUNT ST. HELENA; flows S; through Napa Valley in the COAST RANGES, past St. Helena and Napa (head of navigation and of tidewater), to Mars Island Strait, narrow arm of San Pablo Bay. Wine-producing district. Also has dairying; cattle; walnuts, nursery stock.

Napas (nah-PAHS), village, N TOMSK oblast, W SIBERIA, RUSSIA, on the TYM RIVER, 70 mi/113 km N of NARYM; 59°53′N 82°03′E. Elevation 206 ft/62 m. Lumbering.

Napasar (nah-PAH-sahr), town, BIKANER district, N RAJASTHAN state, NW INDIA, 15 mi/24 km ESE of BIKANER. Hand-loom woolen weaving.

Napasoq (NAH-pah-shok), fishing and seal-hunting settlement, SUKKERTOPPEN (Maniitsoq) commune, SW GREENLAND, on islet in DAVIS STRAIT, 30 mi/48 km SE of SUKKERTOPPEN; 65°03′N 52°21′W.

Napata (nah-BAH-tah), ancient city of NUBIA, just below the 4th cataract of the NILE River, in what is now N SUDAN. From about the 8th century B.C.E., Napata was capital of kingdom of KUSH. Many great temples like those of THEBES were built here by Taharka (XXV dynasty). The Kushite capital later moved (c.530 B.C.E.) to MARAWI.

Napatree Point, Rhode Island: see WATCH HILL.

Napavine, town (2006 population 1,488), LEWIS county, SW WASHINGTON, 8 mi/12.9 km S of CHEHALIS; 46°35′N 122°55′W. In agricultural, dairying area; poultry. Lewis and Clark State Park to SE.

Nape (NAH-pai), village, BOLIKHAMSAI province, central Laos, in TRUONG SON RANGE (1,935 ft/590 m), near Keonua Pass, on road, and 45 mi/72 km SW of VINH, on VIETNAM border; 18°18′N 105°06′E. Shifting cultivation; cattle raising. Pagoda and monastery. Thai, H'mong, Bo, and other minorities.

Naper (NAIP-uhr), village (2006 population 94), BOYD county, N NEBRASKA, 13 mi/21 km WNW of BUTTE, between SOUTH DAKOTA state line and KEYA PAHA RIVER; 42°57′N 99°05′W.

Naperville, city (2000 population 128,358), DU PAGE and WILL counties, NE ILLINOIS, on the DU PAGE RIVER, suburb 30 mi/48 km W of downtown CHICAGO; 41°45′N 88°09′W. It has become a major office and corporate center, marked by an economy that has grown since the early 1980s. Manufacturing (fabricated metal products, electrical equipment, food products, paper products, machinery, medical supplies, plastic products; printing). Remnant agriculture (corn, oats, vegetables). The city's population has grown fourfold between 1970 and 1990. It is the fifth

fastest-growing city in Illinois. Seat of North Central College. Settled 1831–1832, incorporated as a city 1890.

Na Phao (NAH POU), town, KHAMMUAN province, central LAOS; 17°45′N 105°45′E. Also spelled Na Hao.

Napier, city (□ 41 sq mi/106.6 sq km; 2006 population 118,404), Hawke's Bay region, E NORTH ISLAND, NEW ZEALAND, harbor on HAWKE BAY, near HASTINGS (its "twin city"); 39°29′S 176°55′E. Major center for wool, as well as meat, fruit, vegetables, and dairy- and wood-producing exports. Suffered a severe earthquake in 1931, permanently changing the harbor. Seat of an Anglican cathedral. Sunny equable climate attracts vacationers and retirees. Renowned for its Art Deco period architecture.

Napier, town, WESTERN CAPE province, SOUTH AFRICA, 9 mi/14.5 km NW of BREDASDORP, at foot of Bredasdorp mountains; 34°29′S 19°52′E. Elevation 574 ft/ 175 m. Agricultural market (sheep, wool, grain). Named for British governor Sir John Napier. Established 1840.

Napier Mountains, a group of mountains in ENDERBY LAND, EAST ANTARCTICA, 40 mi/60 km S of CAPE BATTERBEE; 66°30′S 53°40′E.

Napier Peninsula (NAIP-yuhr), N NORTHERN TERRITORY, N central AUSTRALIA, point of NE ARNHEM LAND just SE of ELCHO ISLAND across Cadell Strait; 20 mi/32 km long, 6 mi/10 km wide; 12°04′S 135°43′E. Forms NW shore of BUCKINGHAM BAY.

Napierville (NAI-pyuhr-vil), former county (□ 149 sq mi/387.4 sq km), S QUEBEC, E Canada, between the SAINT LAWRENCE RIVER and the U.S. (NEW YORK) border; the county seat was NAPIERVILLE; 45°10′N 73°30′W.

Napierville (NAI-pyuhr-vil), village (□ 1 sq mi/2.6 sq km), ⊙ Les-Jardins-de-Napierville county, Montérégie region, SW QUEBEC, E Canada, on Little Montreal River, and 24 mi/39 km SSE of MONTREAL; 45°11′N 73°24′W. Dairying; vegetables. Was seat of historic NAPIERVILLE county.

Napili Bay (NAH-PEE-lee), town, MAUI island, MAUI county, HAWAII, on NW coast of West Maui Peninsula, on Napili Bay, at mouth of Honokeana Stream. Pineapples.

Napinka (nuh-PING-kuh), unincorporated village, SW MANITOBA, W central Canada, on SOURIS River, 55 mi/89 km SW of BRANDON, and in Brenda rural municipality; 49°19′N 100°51′W. Grain elevators.

Naplate (na-PLAIT), village (2000 population 523), LA SALLE county, N ILLINOIS, just W of OTTAWA, on ILLINOIS RIVER; 41°19′N 88°52′W. Buffalo Rock State Park to W.

Naples, Italian *Napoli*, city (2001 population 1,004,500), ⊙ CAMPANIA and of NAPOLI province, S central ITALY, on the BAY OF NAPLES (an arm of the TYRRHENIAN Sea); 40°50′N 14°15′E. It is a major seaport, with shipyards, and a commercial, industrial, and tourist center. Italy's third-largest city, Naples is troubled by overpopulation, high unemployment, a low per-capita income, and income inequalities. Manufacturing includes iron and steel, petroleum, textiles, food products, chemicals, electronics, porcelain ware, and machinery. An ancient Greek colony, Naples was mentioned as Parthenope, Palaepolis, and Neapolis. It was conquered (4th century B.C.E.) by the Romans, who favored it because of its Greek culture, its scenic beauty, and its baths. The Roman poet Virgil, who often stayed here, is buried nearby. In the 6th century C.E., Naples passed under Byzantine rule; in the 8th century it became an independent duchy. In 1139 the Norman Roger II added the duchy to the kingdom of Sicily. Emperor Frederick II embellished the city and founded its university (1224). The execution (1268) of a Conradin of Hohenstaufen, the last Holy Roman Emperor to claim Naples, left Charles of Anjou (Charles I) undisputed master of the kingdom; he

transferred the capital here from PALERMO. After the Sicilian Vespers insurrection (1282), Sicily proper passed to the house of Aragón, and the Italian peninsula S of the Papal States became known as the kingdom of Naples (see NAPLES, KINGDOM OF). Naples was its capital until it fell to Garibaldi and was annexed to the kingdom of SARDINIA (1860). The city suffered severe damage in World War II. Naples is beautifully situated at the base and on the slopes of the hills enclosing the Bay of Naples. The bay, dominated by Mount VESUVIUS, extends from Cape Misena (N) to the Sorrento peninsula (S) and is dotted with towns and villas. Near its entrance are the islands of CAPRI, ISCHIA, and PROCIDA. Naples is a crowded and noisy city, famous for its songs, festivals, and gaiety. Especially interesting parts of the city are the Old Spacca Quarter (the heart of Old Naples) and the seaside Santa Lucia sector. Noteworthy structures in Naples include the Castel Nuovo (1282); the Castel dell'Ovo (rebuilt by the Angevins in 1274); the Renaissance-style Palazzo Cuomo (late 15th century); the large Carthusian Monastery of Saint Martin (remodeled in the 16th and 17th centuries); the neo-Classic Villa Floridiana, which houses a museum of porcelain, china, and Neapolitan paintings; the Church of Santa Chiara (Gothic, with 18th-cent. baroque additions), which contains the tombs of Robert the Wise and other Angevin kings; the Cathedral of Saint Januarius (14th century, with numerous later additions, including a 17th-century baroque chapel); the Royal Palace (early 17th century); and the Church of Santa Maria Donna Regina. Naples has several museums, including the National Museum, which holds the Farnese collection and most of the objects excavated at nearby POMPEII and HERCULANEUM; the picture gallery, housed in Capodimonte palace; and the aquarium. As a musical center Naples reached its greatest brilliance in the 17th and 18th centuries; Alessandro and Domenico Scarlatti, Porpora, Pergolesi, Paisiello, and Cimarosa were among the representatives of the Neapolitan style in both vocal and instrumental music. The Teatro San Carlo, a famous opera house, was opened in 1737. The city has a conservatory and several art academies. Edenlandia theme park is here. Near Naples is the Camaldulian Hermitage (founded 1585), from which there is an excellent view of the bay region.

Naples (NAI-puhlz), resort city (□ 14 sq mi/36.4 sq km; 2005 population 21,709), COLLIER county, SW FLORIDA, on the GULF OF MEXICO, just W of EAST NAPLES; 26°08′N 81°47′W. Borders the BIG CYPRESS SWAMP. EVERGLADES to E. Tourism (noted beach with pristine white sand popular year-round), fishing, and shrimp fisheries are the staples of the economy. Site of the Caribbean Gardens and Collier Seminole State Park. Many championship golf courses nearby. Rapidly growing metropolitan area. Incorporated 1927.

Naples, town (2000 population 134), SCOTT county, W central ILLINOIS, on ILLINOIS RIVER, and 11 mi/18 km WNW of WINCHESTER; 39°45′N 90°36′W. In agricultural area.

Naples, resort town, CUMBERLAND county, SW MAINE, on SEBAGO and LONG lakes, 27 mi/43 km NW of PORTLAND; 43°58′N 70°36′W. Includes part of Sebago Lake State Park.

Naples (NAI-puhlz), town (2006 population 1,414), MORRIS county, NE TEXAS, 40 mi/64 km WSW of TEXARKANA; 33°12′N 94°40′W. In agricultural area (cattle; peanuts, watermelons); timber; manufacturing (building materials; peanut processing). Incorporated 1909.

Naples, town (2006 population 1,502), UINTAH county, NE UTAH, 2 mi/3.2 km SE of VERNAL, near Ashley Creek; 40°25′N 109°29′W. Residential community.

Sheep, cattle; dairying; alfalfa, barley, wheat. Oil and natural gas.

Naples, village (2006 population 1,039), ONTARIO county, W central NEW YORK, in FINGER LAKES region, near S end of CANANDAIGUA LAKE, 20 mi/32 km SSW of CANANDAIGUA; 42°37′N 77°24′W. In grape-growing Naples valley. Incorporated 1894.

Naples, village (2006 population 22), CLARK county, E central SOUTH DAKOTA, 13 mi/21 km SE of CLARK; 44°46′N 97°30′W. In farming region.

Naples, Bay of, S ITALY, a semicircular inlet of TYRRHENIAN Sea, between gulfs of GAETA (N) and SALERNO (S), between CAPE MISENO (NW) and PUNTA DELLA CAMPANELLA (SE); c.20 mi/32 km long, 10 mi/16 km wide; 40°48′N 14°16′E. On its scenic shores are NAPLES, POZZUOLI, PORTICI, TORRE DEL GRECO, TORRE ANNUNZIATA, Castellammare di Stabia, VICO EQUENSE, and SORRENTO. VESUVIUS rises in E, with ruins of POMPEII and HERCULANEUM nearby; in SE is Monte SANT'ANGELO, highest point (4,734 ft/1,443 m) near the bay. At its entrance are islands of ISCHIA and PROCIDA (NW) and CAPRI (SE). Its NW inlet is the Gulf of Pozzuoli.

Naples, Kingdom of, former state, occupying the Italian peninsula S of the former PAPAL STATES. It comprised roughly the present regions of CAMPANIA, ABRUZZI, MOLISE, BASILICATA, APULIA, and CALABRIA. NAPLES was capital. In the 11th and 12th centuries the Normans under Robert Guiscard and his successors seized S ITALY from the Byzantines. The popes, however, claimed suzerainty over S Italy and were to play an important part in the history of Naples. In 1139, Roger II, Guiscard's nephew, was invested by Innocent II with the kingdom of SICILY, including the Norman lands in S Italy. The last Norman king designated Constance, wife of Holy Roman Emperor Henry VI, as his heir, and the kingdom passed successively to Frederick II, Conrad IV, Manfred, and Conradin of Hohenstaufen. Under them S Italy flowered, but in 1266 Charles I (Charles of Anjou), founder of the Angevin dynasty, was invested with the crown by Pope Clement IV, who wished to drive the Hohenstaufen family from Italy. Charles lost Sicily in 1282 but retained his territories on the mainland, which came to be known as the kingdom of Naples. Refusing to give up his claim to Sicily, Charles and his successors warred with the house of Aragón, which held the island, until 1373, when Queen Joanna I of Naples formally renounced her claim. During her reign began the struggle for succession between Charles of Durazzo (later Charles III of Naples) and Louis of Anjou (Louis I of Naples). The struggle was continued by their heirs. Charles' descendants, Lancelot and Joanna II, successfully defended their thrones despite papal support of their French rivals, but Joanna successively adopted as her heir Alfonso V of Aragón and Louis III and René of Anjou, and the dynastic struggle was prolonged. Alfonso defeated René and in 1442 was invested with Naples by the pope. His successor in Naples, Ferdinand I (Ferrante), suppressed (1485) a conspiracy of the powerful feudal lords. Meanwhile the Angevin claim to Naples had passed to the French crown with the death (1486) of René's nephew, Charles of Maine. Charles VIII of FRANCE pressed the claim and in 1495 briefly seized Naples, thus starting the Italian Wars between France and SPAIN. Louis XII, Charles's successor, temporarily joined forces with Spain and dethroned Frederick (1501), the last Aragonese king of Naples, but fell out with his allies, who defeated him. The Treaties of Blois (1504–1505) gave Naples and Sicily to Spain, which for two centuries ruled the two kingdoms through viceroys—one at Palermo, one at Naples. Gonzalo Fernández de Córdoba was the first viceroy of Naples. Under Spain, S Italy became one of the most backward and exploited areas in Europe. Heavy taxation (from

which the nobility and clergy were exempt) filled the Spanish treasury; agriculture suffered from the accumulation of huge estates by quarreling Italian and Spanish nobles and the church; famines were almost chronic; disease, superstition, and ignorance flourished. A popular revolt against these conditions, led by Masaniello, was crushed in 1648. In the War of the Spanish Succession the kingdom was occupied (1707) by Austria, which kept it by the terms of the Peace of Utrecht (1713). During the War of the Polish Succession, however, Don Carlos of Bourbon (later Charles III of Spain) reconquered Naples and Sicily. The Treaty of Vienna (1738) confirmed the conquest, and the two kingdoms became subsidiary to the Spanish crown, ruled in personal union by a cadet branch of the Spanish line of Bourbon. Naples then had its own dynasty, but conditions improved little. In 1798, Ferdinand IV and his queen, Marie Caroline, fled from the French Revolutionary army. The PARTHENOPEAN REPUBLIC was set up (1799), but the Bourbons returned the same year with the help of the English under Lord Nelson. Reprisals were severe; Sir John Acton, the queen's favorite, once more was supreme. In 1806 the French again drove out the royal couple, who fled to Sicily. Joseph Bonaparte, made king of Naples by Napoleon I, was replaced in 1808 by Joachim Murat. Murat's beneficent reforms were revoked after his fall and execution (1815) by Ferdinand, who was restored to the throne (Marie Caroline had died in 1814). In 1816, Ferdinand merged Sicily and Naples and styled himself Ferdinand I, king of the Two Sicilies. For the remaining history of Naples, annexed to Sardinia in 1860, see TWO SICILIES, KINGDOM OF THE.

Naples Manor (NAI-puhlz), unincorporated town (2000 population 5,186), COLLIER county, SW FLORIDA, 5 mi/8 km SE of NAPLES; 26°05′N 81°43′W.

Naples Park (NAI-puhlz), unincorporated town (□ 1 sq mi/2.6 sq km; 2000 population 6,741), COLLIER county, SW FLORIDA, 8 mi/12.9 km N of NAPLES; 26°15′N 81°49′W.

Napo (NAH-po), province, (□ 13,100 sq mi/33,929 sq km; 2001 population 79,139), ⊙ TENA, in the ORIENTE, ECUADOR. Bordered N by new SUCUMBÍOS and W by PICHINCHA, COTOPAXI, and TUNGURAHUA provinces. Relief varies from the E slopes of the ANDES Mountains (Cordillera ORIENTAL; highest point in province is SUMACO volcano; 12,550 ft/3,825 m) to the AMAZONian plain. Climate varies accordingly, from temperate to humid equatorial. The main resource is petroleum, mainly in area of COCA. Also gold panning in many rivers (COCA, TIPUTINI, Curacay). Agricultural products (bananas, cacao, coffee, some corn) and forestry products. E selva area is home to several indigenous groups (Yumbos, Cofanes, Tetetes); some tourism.

Napoklu (nah-POK-loo), village, KODAGU district, KARNATAKA state, S INDIA, on KAVERI RIVER, 8 mi/12.9 km SSW of MADIKERI. Rice (terrace farming), cardamom, oranges. Evergreen forest (W). Was in former S COORG state.

Napoleon (nuh-PO-lee-uhn), city (□ 6 sq mi/15.6 sq km; 2006 population 9,119), ⊙ HENRY county, NW OHIO, on MAUMEE RIVER, 14 mi/23 km ENE of DEFIANCE; 41°23′N 84°07′W. Market center for agricultural area; machinery, preserved fruits and vegetables, transportation equipment, consumer goods, metal products, building materials.

Napoleon, town (2000 population 238), RIPLEY county, SE INDIANA, 8 mi/12.9 km SW of BATESVILLE; 39°13′N 85°20′W. In agricultural area. Manufacturing (wood products). Laid out 1820.

Napoleon, town (2000 population 208), LAFAYETTE county, W central MISSOURI, on MISSOURI River, and 9 mi/14.5 km WSW of LEXINGTON; 39°07′N 94°05′W. Soybeans, corn, wheat; cattle.

Napoleon, town (2006 population 728), ⊙ LOGAN county, S NORTH DAKOTA, 54 mi/87 km ESE of BISMARCK; 46°30′N 99°46′W. Livestock; grain; dairy products. Beaver State Park to SE. Founded in 1886 and named for Napoleon Goodsill (1841–1887), a local realtor.

Napoléon-Vendée, FRANCE: see ROCHE-SUR-YON, LA.

Napoleonville (nuh-PO-lee-uhn-vil), village (2000 population 686), ⊙ ASSUMPTION parish, SE LOUISIANA, 30 mi/48 km SSE of BATON ROUGE, on BAYOU LAFOURCHE; 29°57′N 91°02′W. In agricultural area (sugarcane; cattle); crawfish, crabs. Food processing, manufacturing (raw sugar). Oil and gas field nearby. Has fine old church, antebellum houses. Founded c.1818, incorporated 1878.

Napoléonville, FRANCE: see PONTIVY.

Napoli (NAH-po-lee), province (□ 452 sq mi/1,175.2 sq km), CAMPANIA, S ITALY; ⊙ NAPLES; 40°53′N 14°25′E. Extends around BAY OF NAPLES; includes islands of ISCHIA, CAPRI, PROCIDA, and VENTOTENE. Smallest, yet most densely populated province of Italy. Extensive agriculture (grapes, olives, fruit, vegetables, hemp, tobacco); fishing. Pozzuolana extracting (POZZUOLI, BACOLI) and lava quarrying (TORRE DEL GRECO). Has many ports that carry on extensive trade. Chief industries include pasta manufacturing (Naples, TORRE ANNUNZIATA, Castellammare di Stabia, GRAGNANO), tomato canning, cotton milling. Its resorts (Capri, Ischia, SORRENTO), historic ruins (POMPEII, HERCULANEUM, CUMAE), and famous sights (VESUVIUS, PHLEGRAEAN FIELDS) attract many tourists annually. Included CASERTA province, 1927–1945; prior to reconstitution of Caserta province, it had an area of 1,206 sq mi/3,124 sq km.

Napoli, ITALY: see NAPLES.

Naponee (nuh-PAWN-ee), village (2006 population 123), FRANKLIN county, S NEBRASKA, 10 mi/16 km W of FRANKLIN, and on REPUBLICAN RIVER; 40°04′N 99°08′W. Harlan County Dam and Reservoir and State Recreation Area to W (upstream).

Napoopoo (NAH-PO-o-PO-o), village, HAWAII island, HAWAII county, HAWAII, 55 mi/89 km WSW of HILO, 2 mi/3.2 km SW of Captain Cook, on KEALAKEKUA BAY, KONA (W) Coast. Coffee, fruit; fish. Napoopoo Beach Park and Mokuohai Battleground to SW; Kealakekua Bay State Underwater Park to N.

Napo River (NAH-po), over 550 mi/885 km long, in NE ECUADOR and NE PERU; rises in the ANDES Mountains near the COTOPAXI volcano; flows SE, past NAPO, COCA, and NUEVO ROCAFUERTE, through tropical forests to the upper AMAZON River, 40 mi/64 km NE of IQUITOS; 03°25′S 72°43′W. Main affluents are COCA, AGUARICO, and CURARAY rivers. Has good navigability. Some cattle raising along its shores. The adjoining forests yield rubber, balata, chicle, fine timber. Site of jungle lodges and river eco-tours. First explored (1540) by Orellana and later (1638) by Teixeira.

Napostá Grande, Arroyo (nah-pos-TAH GRAHN-dai, ahr-RO-yo), river, c.50 mi/80 km long, in SW BUENOS AIRES province, ARGENTINA; rises in Sierra de la VENTANA SE of TORNQUIST (1,600 ft/488 m); flows S to BAHÍA BLANCA (bay) below BAHÍA BLANCA city.

Napoule, Golfe de la (nah-POOL, GOLF duh lah), bay of the MEDITERRANEAN SEA, off ALPES-MARITIMES department, PROVENCE-ALPES-CÔTE D'AZUR region, SE FRANCE, along French RIVIERA; 6 mi/9.7 km wide, 2 mi/3.2 km long; 43°32′N 06°59′E. CANNES is on its NE shore. Sheltered on mainland coast by barren ESTÉREL range (W), from open sea by Îles de LÉRINS. The swimming resort of MANDELIEU-LA-NAPOULE (3 beaches; marinas) lies along this coast.

Napoule, La, FRANCE: see MANDELIEU-LA-NAPOULE.

Nappanee, city (2000 population 6,710), ELKHART county, N INDIANA, 14 mi/23 km SW of GOSHEN; 41°27′N 85°59′W. In agricultural area (mint, onions,

---

Cross-references are shown in SMALL CAPITALS. The pronunciation guide is shown on page xix. The sources of population figures are shown on page xvii.

grain; livestock); manufacturing (furniture, fabricated metal products; electrical goods, transportation equipment; wood products, motor vehicles and motor homes, manufactured housing). Plotted 1874, incorporated 1926.

Napue, FINLAND: see ISOKYRÖ.

Napuka, TUAMOTU ARCHIPELAGO: see DISAPPOINTMENT ISLANDS.

Naqada (nah-KAH-dah), town, QENA province, Upper EGYPT, across the NILE RIVER from and 3 mi/4.8 km WSW of QUS; 25°54′N 32°43′E. Silk weaving, pottery making, sugar refining; cereals, sugarcane, dates. Has Coptic and Roman Catholic churches. Also spelled Nakada.

Naqadeh (nah-kah-DAI), town (2006 population 73,438), ĀZARBĀYJĀN-e GHARBI province, SW AZERBAIJAN province, IRAN, S of Lake URMIA; 36°57′N 45°23′E.

Naqb Ashtar (NUHK-ib UHSH-tahr), village, S JORDAN, in hills, 30 mi/48 km SW of MA'AN, and on main road and railroad extension between Ma'an and AQABA. Trade center; camel, sheep, and goat raising. Important station on the road from Aqaba to AMMAN.

Naqib (nuh-KEEB), township, SE YEMEN, near former border between NORTH YEMEN and SOUTH YEMEN, 18 mi/29 km N of BEIHAN AL QASAB. Was center of former BEIHAN tribal area.

Naque (NAH-kai), town (2007 population 5,885), E central MINAS GERAIS state, BRAZIL, 25 mi/40 km NE of Ipaúinga, on RIO DOCE and VITÓRIA-Minas railroad; 19°25′S 42°28′W.

Naqura, Arab village, Nablus district, 4.4 mi/7 km NW of NABLUS, in the SAMARIAN Highlands, WEST BANK. Agriculture (olives, fruits, vegetables, wheat).

Nar, town, KHEDA district, GUJARAT state, W central INDIA, 6 mi/9.7 km W of PETLAD. Agriculture market (tobacco, millet, cotton).

Nara (NAH-rah), prefecture [Japanese *ken*] (□ 1,425 sq mi/3,691 sq km; 1990 population 1,375,478), S HONSHU, W central JAPAN, partly on KII PENINSULA; ⊙ NARA. Bordered N by KYOTO, E and SE by MIE, SW by WAKAYAMA, and NW by OSAKA prefectures. Largely mountainous, particularly in S area, which is drained by KUMANO RIVER. Its population centers in and around the religious center of Nara city. Agriculture (rice, fruit), crafts, and tourism (YOSHINO-KUMANO NATIONAL PARK) are the area's main industries. Lumber and charcoal produced in S. Home industries (raw silk, sake, medicine, textiles). Some copper mined at SHIMOICHI, mercury at Uda.

Nara (NAH-rah), city (2005 population 370,102), ⊙ NARA prefecture, S HONSHU, W central JAPAN; 34°40′N 135°48′E. Sumi ink and writing brushes made here. An ancient cultural and religious center, Nara was founded in 706 by imperial decree and was modeled after Chang-an, the capital of Tang China. Nara was (710–784) the first permanent capital of Japan. The noted temple, Todai-ji, has a 53.5-ft/16.3-m-high image of Buddha, said to be one of the largest bronze figures in the world. Nara Park, the largest (1,250 acres/506 ha) city park in Japan, includes the celebrated Imperial Museum, which houses ancient art treasures and relics. Near the city is wooded Mt. Kasuga, the traditional home of the gods; its trees are never cut. Also nearby is Horyu-ji, founded in 607, the oldest Buddhist temple in Japan, with the grave of Jimmu, the first emperor. Other attractions include the ruins of Heijo Kyo palace, many other temples (e.g., Kofuku, Saidai, Hoke, and Toshodai), and the Kasuga shrine.

Nara, town, SECOND REGION/KOULIKORO, W MALI, 177 mi/295 km NNE of BAMAKO, near N border with MAURITANIA; 15°10′N 07°17′W. Agriculture; livestock.

Naracoorte (NA-ruh-koort), town, SE SOUTH AUSTRALIA state, S central AUSTRALIA, 190 mi/306 km SSE of ADELAIDE, near VICTORIA state line; 36°58′S 140°44′E. Railroad junction; agriculture and sheep center. Limestone quarries, agricultural experiment station nearby. Fossils in the Naracoorte Caves, which illustrate the evolution of Australia's fauna, are considered among the world's ten greatest. Together with Riversleigh, declared a World Heritage Area 1994.

Narada Falls (nuh-RA-duh), waterfall (168 ft/51 m), on Paradise Creek, in S MOUNT RAINIER NATIONAL PARK, LEWIS county, W central WASHINGTON.

Naradhivas, THAILAND: see NARATHIWAT.

Naraguta (nah-rah-GOO-tah), town, Plateau state, central NIGERIA, on BAUCHI Plateau, 4 mi/6.4 km N of JOS. Tin-mining center.

Naraha (nah-RAH-hah), town, Futaba county, FUKUSHIMA prefecture, N central HONSHU, NE JAPAN, 43 mi/70 km S of FUKUSHIMA city; 37°16′N 140°59′E.

Naraina (nuh-REI-nah), town, JAIPUR district, E central RAJASTHAN state, NW INDIA, 8 mi/12.9 km S of SAMBHAR. Millet, gram. Headquarters of small Hindu sect of reformers (called Dadupanthis), founded late 16th century by Dadu.

Naraini (nuh-REI-nee), town, BANDA district, S UTTAR PRADESH state, N central INDIA, 22 mi/35 km SSE of BANDA. Gram, jowar, wheat, oilseeds. Pilgrimage site and extensive Hindu shrines 23 mi/37 km ENE, at village of CHITRAKUT.

Narakawa (nah-RAH-kah-wah), village, Kiso county, NAGANO prefecture, central HONSHU, central JAPAN, 50 mi/80 km S of NAGANO; 35°58′N 137°50′E. Lacquerware, woodwork, bamboo work.

Naral (nah-rahl), town (2001 population 37,018), JESSORE district, SW EAST BENGAL, BANGLADESH, on river arm of GANGA DELTA, and 19 mi/31 km E of JESSORE; 23°08′N 89°30′E. Trades in rice, jute, linseed, tobacco. Has college.

Naramata (na-ruh-MA-tuh), unincorporated village, S BRITISH COLUMBIA, W Canada, on OKANAGAN LAKE, 7 mi/11 km N of PENTICTON, in Okanagan-Similkameen regional district; 49°36′N 119°35′W. Fruit, vegetables.

Nara Nag, INDIA: see SRINAGAR.

Narandiba (NAH-rahn-zhee-bah), town (2007 population 3,941), W SÃO PAULO state, BRAZIL, 25 mi/40 km SW of PRESIDENTE PRUDENTE; 22°24′S 51°31′W.

Narang (NAH-rung), village, SHEIKHUPURA district, E PUNJAB province, central PAKISTAN, on railroad, and 34 mi/55 km NE of SHEIKHUPURA; 32°26′N 73°50′E. Agricultural market; rice milling.

Naranja (nuh-RAN-juh), unincorporated town (□ 1 sq mi/2.6 sq km; 2000 population 4,034), MIAMI-DADE county, SE FLORIDA, 24 mi/39 km SW of MIAMI; 25°31′N 80°25′W. Adjacent to Homestead Air Force Base. Hurricane Andrew virtually destroyed Naranja in 1992.

Naranjal (nah-RAHN-hahl), town, VERACRUZ, E MEXICO, 6 mi/9.7 km S of CÓRDOBA, on Blanco River. Elevation 2,799 ft/853 m. Coffee, fruit.

Naranjal (nah-rahn-HAHL), village (2001 population 20,789), GUAYAS province, SW ECUADOR, minor river port at foot of the ANDES Mountains, 38 mi/61 km SSE of GUAYAQUIL. In agricultural region (cacao, rice, tropical fruit).

Naranjal River (nah-rahn-HAHL), c.85 mi/137 km long, S and S central ECUADOR; rises at SW foot of Cerro AYAPUNGA; flows W to inlet of the Gulf of GUAYAQUIL near mouth of GUAYAS RIVER. Its midcourse is called Cañar.

Naranjito (nah-rahn-HEE-to), town, SANTA BÁRBARA department, W HONDURAS, near JICATUYO RIVER, 29 mi/47 km W of SANTA BÁRBARA; 14°47′N 88°41′W. Commercial center in coffee and tobacco area.

Naranjito (nah-rahn-HEE-to), town (2006 population 29,918), N central PUERTO RICO, 14 mi/23 km SW of SAN JUAN. Commercial and industrial center. Agriculture (coffee, bananas, plantains; livestock). A hydroelectric plant with artificial lake is nearby on the La Plata River.

Naranjo (nah-RAHN-ho), canton, VALLEGRANDE province, SANTA CRUZ department, E central BOLIVIA, SE of MOROMORO; 18°25′S 64°16′W. Elevation 6,660 ft/2,030 m. Agriculture (corn, potatoes, yucca, bananas, soy, rye); cattle.

Naranjo (nah-RAHN-ho), city, ⊙ Naranjo canton, ALAJUELA province, W central COSTA RICA, in CENTRAL VALLEY, on INTER-AMERICAN HIGHWAY, and 14 mi/23 km NW of ALAJUELA; 10°07′N 84°24′W. Important coffee center; tobacco, corn, beans, rice, sugarcane, plantains.

Naranjo, Bahía de (nah-RAHN-ho, bah-HEE-ah dai), small sheltered inlet, on NE coast of CUBA, in HOLGUÍN province, 10 mi/16 km E of GIBARA; c.2 mi/3 km long, 1 mi/2 km wide; 21°07′N 75°52′W.

Naranjo River (nah-RAHN-ho), c.60 mi/97 km long, SW GUATEMALA; rises just N of SAN MARCOS; flows S, W, and S, past SAN PEDRO SACATEPÉQUEZ and COATEPEQUE, to the PACIFIC OCEAN at OCÓS; 13°58′N 89°21′W–15°33′N 92°11′W.

Naranjos (nah-RAHN-hos), town, ⊙ AMATLÁN TUXPAN, VERACRUZ, E MEXICO, 33 mi/53 km NW of TÚXPAM DE RODRÍGUEZ CANO; 21°22′N 97°43′W. Corn, sugar, fruit. Petroleum wells.

Naranjos Agrios (nah-RAHN-hos AH-gree-os), village, GUANACASTE province, NW COSTA RICA, near E shore of LAKE ARENAL, 5 mi/8 km N or TILARÁN; 10°08′N 84°55′W.

Narao (nah-RAH-O), town, South Matsuura county, NAGASAKI prefecture, W KYUSHU, SW JAPAN, 47 mi/75 km W of NAGASAKI; 32°50′N 129°03′E.

Naraq (nah-RAHK), village, Markazī province, N central IRAN, 40 mi/64 km S of QOM; 34°00′N 50°50′E. Grain, fruit; rug making.

Narasannapeta (nuh-ruh-SUH-nuh-pet-uh), town, SRIKAKULAM district, ANDHRA PRADESH state, S INDIA, 45 mi/72 km NE of VIZIANAGARAM. Rice, oilseeds, sugarcane, coconuts. Sometimes spelled Narsannapet.

Narasapatnam, INDIA: see NARSIPATNAM.

Narasapur, INDIA: see NARSAPUR.

Narasaraopet (nuh-ruh-suh-ROU-pet), city, GUNTUR district, ANDHRA PRADESH state, S INDIA, 26 mi/42 km W of GUNTUR. Road and agriculture trade center; rice and oilseed milling, cotton ginning. Also spelled Narasaravupet.

Narashino (nah-RAH-shee-no), city (2005 population 158,785), CHIBA prefecture, E central HONSHU, E central JAPAN, on TOKYO BAY, 5 mi/8 km N of CHIBA; 35°40′N 140°01′E. Newly developed suburb of TOKYO; manufacturing of motors.

Narasimharajapura (nuh-ruhs-im-huh-RAH-juh-puhr-uh), town, CHIKMAGALUR district, KARNATAKA state, S INDIA, 26 mi/42 km NW of CHIKMAGALUR. Terminus of railroad spur from TARIKERE junction, 20 mi/32 km ENE; serves iron mines, and coffee, cardamom, and pepper estates in BABA BUDAN RANGE (SE).

Narastan, INDIA: see AWANTIPUR.

Narathiwat (NUH-RAH-TI-WAHD), province (□ 1,613 sq mi/4,193.8 sq km), Southern region, S THAILAND, on MALAYSIA border, and on GULF OF THAILAND; ⊙ NARATHIWAT; 06°20′N 101°45′E. There are a number of beaches in NARATHIWAT. Rice, coconuts; fishing; tourism. About 4 mi/6.4 km S of NARATHIWAT town is Thaksin Palace, where King Phumiphon and Queen Sirikit reside August–October every year. Population is largely Muslim; there are numerous mosques in province, of which the Wadin Husen Mosque (built in 1769 and located in the village of Lubosawo, c.9 mi/14.5 km) NW of NARATHIWAT town, is a noteworthy example. The Narathiwat Fair is held annually in September, and includes boat racing, a singing dove contest judged by the queen, handicrafts displays, and dance and martial arts performances.

Narathiwat (NUH-RAH-TI-WAHD), town, ⊙ NARATHIWAT province, S THAILAND, port on E coast of MALAY PENINSULA, on SOUTH CHINA SEA, 100 mi/161 km SE of SONGKHLA, near MALAYSIA border, on road from railroad station of Ra Ngae; 06°26'N 101°50'E. Large coconut plantations; rice fields. Noted for Ba Cho waterfall. Entertainment center. Locally known as BANG NARA; also called NARADHIVAS.

Naraura, INDIA: see DIBAI.

Narayanganj (nah-rah-yahn-gawnj), city (2001 population 241,393), E central BANGLADESH, at the confluence of the Lakhya and DHALESWARI rivers; 23°40'N 90°02'E. It is the river port for DHAKA and is Bangladesh's busiest trade center, especially for jute. The city is also a collection center for hides and skins and a reception point for imports from and exports to CALCUTTA (INDIA). Narayanganj and DHAKA together make up the principal industrial region of Bangladesh. There are jute presses, cotton textile mills, and ship repair facilities. Nearby is the celebrated shrine of the Muslim saint Kadam Rasul 1895.

Narayanghat, NEPAL: see BHARATPUR.

Narayani (nah-rah-YE-nee), administrative zone (2001 population 2,466,138), central NEPAL. Includes the districts of BARA, CHITWAN, MAKWANPUR, PARSA, and RAUTAHAT.

Narayani River, NEPAL and INDIA: see GANDAK RIVER.

Narayanpet (NAH-rah-yuhn-pait), town, MAHBUBNAGAR district, ANDHRA PRADESH state, S INDIA, 33 mi/53 km W of MAHBUBNAGAR. Millet, oilseeds, rice; noted hand-woven silk textiles, leather goods. Railroad station of Narayanpet Road is 10 mi/16 km SW.

Narbata, ISRAEL: see MA'ANIT.

Narberth (NAHR-buhrth), town (2001 population 2,358), Pembrokeshire, SW Wales, 9 mi/14.5 km N of TENBY; 51°48'N 04°45'W. Market. Burned by Norsemen (994). Has remains of 13th-century castle, built on site of Danish and Norman castles. Formerly in DYFED, abolished 1996.

Narberth (NAHR-buhrth), borough (2006 population 4,098), MONTGOMERY county, SE PENNSYLVANIA; residential suburb 6 mi/9.7 km WNW of downtown PHILADELPHIA; 40°00'N 75°15'W. Light manufacturing. Settled 1860, incorporated 1895.

Narbethong, town, VICTORIA, SE AUSTRALIA, 40 mi/65 km NE of MELBOURNE. Timber.

Narbonne (nahr-buhn), city (□ 66 sq mi/171.6 sq km), AUDE department, LANGUEDOC-ROUSSILLON administrative region, S FRANCE, near the MEDITERRANEAN coast, and 34 mi/55 km N of PERPIGNAN, at interchange of N-S (from RHÔNE valley to Spain) with E-W (to TOULOUSE and BORDEAUX) highways; 43°05'N 02°50'E. It is linked by canal with Port-la Nouvelle on the Mediterranean (10 mi/16 km S) and with the CANAL DU MIDI (which provides a water-level route to Toulouse on the GARONNE RIVER). Commercial center for the LANGUEDOC wine region and an industrial city producing sulfur, fertilizer, pottery, and textile products A uranium-processing plant operating since 1959 at Malvési (nearby). Narbonne was the first Roman colony established in Transalpine Gaul (118 B.C.E.); it was known as *Narbo Martius*; it later became capital of the Roman province of *Gallia Narbonensis*. Narbonne was an archiepiscopal see from the 4th century until French Revolution. The city was occupied by the Visigoths in 413 and taken by the Saracens in 719 and the Franks in 759. It later became the seat of the viscounts of Narbonne, vassals of the counts of Toulouse, and was united to the French crown in 1507. Its port, silted up in 1320, brought great wealth to the city, especially during the Middle Ages. Also an important medieval Jewish center. Their expulsion (late 13th century) and the Black Death (1310), which is said to have taken 30,000 lives, were severe blows to the city's prosperity. Narbonne contains some Roman remains. Of special note are the splendid but unfinished St. Just Cathedral (13th–14th century), and an archiepiscopal palace (12th–13th century), now the town hall (with neo-Gothic addition), and the art museum. The swimming resort of Narbonne-Plage is on the open sea, 8 mi/12.9 km E.

Narborough (NAH-buh-ruh), village (2001 population 8,402), central LEICESTERSHIRE, central ENGLAND, on SOAR RIVER, and 6 mi/9.7 km SW of LEICESTER; 52°34'N 01°12'W. Paper, concrete.

Narborough Island, ECUADOR: see FERNANDINA, ISLA.

Narcea River (nahr-THAI-ah), c.60 mi/97 km long, OVIEDO province (Asturias), N SPAIN; rises near LEÓN border; flows NNE, past CANGAS DE NARCEA, to NALÓN RIVER, just S of PRAVIA.

Narcisse (nahr-SEES), hamlet, S MANITOBA, W central Canada, in Armstrong municipality; 50°41'N 97°31'W. Snake dens just N; tourist attraction.

Narcondam Island (nahr-kuhn-DAHM), small extinct volcano in ANDAMAN SEA, ANDAMAN AND NICOBAR ISLANDS Union Territory, INDIA, 80 mi/129 km E of N ANDAMAN ISLANDS; 13°25'N 94°17'E. Rises to 2,330 ft/710 m.

Nardin (NAHR-din), village, KAY county, N OKLAHOMA, 22 mi/35 km WNW of PONCA CITY; 36°48'N 97°27'W. In agricultural area.

Nardò (nahr-DO), town, LECCE province, APULIA, S ITALY, 14 mi/23 km SW of LECCE; 40°11'N 18°02'E. Agricultural trade center (wine, olive oil, wheat, tobacco); clothing, textiles; food processing. Has 13th-century cathedral (restored 19th century).

Nare, Colombia: see PUERTO NARE.

Narechenski bani, Bulgaria: see CHEPELARE.

Narela (nah-RAI-lah), town, DELHI state, N INDIA, 8 mi/12.9 km NNE of DELHI city center. Handicraft pottery and glass bangles. Part of NATIONAL CAPITAL REGION.

Narembeen (na-ruhm-BEEN), town, WESTERN AUSTRALIA state, W AUSTRALIA, 175 mi/282 km E of PERTH; 32°04'S 118°23'E. Wheat. Formerly a railroad siding for nearby Emu Hill.

Narendranagar (nuh-REHN-drah-nuh-gur), town (2001 population 4,796), TEHRI GARHWAL district, W central Uttarakhand state, N central INDIA. Former capital of TEHRI GARHWAL district.

Narenta, Canale della, CROATIA: see NERETVA CHANNEL.

Narenta River, BOSNIA AND HERZEGOVINA: see NERETVA RIVER.

Nare River (NAH-rai), c.100 mi/161 km long, ANTIOQUIA department, N central COLOMBIA; rises in Cordillera CENTRAL S of MEDELLÍN; flows E to MAGDALENA River, 25 mi/40 km SW of PUERTO BERRÍO; 06°12'N 74°35'W.

Nares Abyssal Plain, flat area of ocean floor N of the PUERTO RICO TRENCH, 20°00'N 24°00'N, one of the deepest in the N ATLANTIC, with depths of 19,029 ft/5,800 m–19,685 ft/6,000 m.

Nares, Cape, N ELLESMERE ISLAND, BAFFIN region, NUNAVUT territory, N Canada, on the ARCTIC OCEAN; 83°06'N 71°45'W.

Nares Land, region in W part of PEARY LAND, N GREENLAND, on E coast of VICTORIA FJORD; 82°15'N 46°30'W.

Narestø (NAH-ruhs-tuh), village, AUST-AGDER county, S NORWAY, on Flostaøy, a small island (□ 3 sq mi/7.8 sq km; 5 mi/8 km long), 6 mi/9.7 km NE of ARENDAL; 58°31'N 08°56'E.

Narev (NAH-rev), river, c.275 mi/443 km long, BELARUS and POLAND; rises in the Białowieza Forest, W BELARUS, near the Polish border; flows generally NW through NE Poland past Łomża, the head of navigation, then SW to the VISTULA River near WARSAW. The WESTERN BUG and the BIEBRZA rivers are the chief tributaries. Canals connect its tributaries with the NEMAN and PRIPYAT rivers. During World Wars I and II, major battles took place along the its banks.

Nargen, ESTONIA: see NAISSAAR.

Nargund (NUHR-guhnd), town, DHARWAD district, KARNATAKA state, S INDIA, 31 mi/50 km NE of DHARWAD. Market center for cotton, wheat, millet, sugarcane; cotton ginning.

Narhin (nah-REEN), town, BAGHLAN province, NE AFGHANISTAN, 23 mi/37 km SE of BAGHLAN; 36°04'N 69°08'E. Also spelled Narin.

Naria (naw-ree-ah), village (2001 population 20,058), FARIDPUR district, S central EAST BENGAL, BANGLADESH, near the PADMA RIVER, 16 mi/26 km NE of MADARIPUR; 23°18'N 90°24'E. Road terminus; rice, jute, oilseeds, sugarcane.

Naricual (nah-ree-KWAHL), town, Simón Bolívar municipio, ANZOÁTEGUI state, NE VENEZUELA, 6 mi/10 km SE of BARCELONA (connected by railroad); 10°10'N 64°35'W. Thermal resources; carboniferous deposits.

Narimanov (nah-ree-MAH-nuhf), city (2005 population 12,150), S central ASTRAKHAN oblast, S European Russia, on the W bank of the VOLGA RIVER, on the Volgograd-Astrakhan highway, 40 mi/64 km NNW of ASTRAKHAN; 46°41'N 47°51'E. Below sea level. Manufacturing of models of offshore drilling rigs. Called Nizhnevolzhsk until 1984, when made a city and renamed.

Narimanov (nah-ree-MAHN-ahv), city, TOSHKENT wiloyat, NE UZBEKISTAN.

Nariño (nah-REEN-yo), department (□ 11,548 sq mi/30,024.8 sq km), SW COLOMBIA, on the PACIFIC OCEAN, bordering ECUADOR; ⊙ PASTO; 01°30'N 78°00'W. Its indented coastal lowland rises E to volcanic ANDEAN peaks. Near Ecuador's border, the main range of the Andes splits into the CORDILLERA OCCIDENTAL and Cordillera CENTRAL, separated by the torrential PATÍA RIVER which, in N, cuts W through Cordillera Occidental to the Pacific. Climate ranges from hot and humid in lowlands to glacial conditions on Andean peaks; rain is heavy along coast throughout the year. There are oil fields and bauxite resources. The abundant forests yield tropical products and woods. Considerable cattle grazing on plateaus and in upper Patía River valley. Agricultural crops include corn, wheat, rice, coffee, cacao, sugarcane, coconuts, bananas, cotton, cassava, henequen. Pasto and IPIALES are centers for trading, consumer goods, and food-processing; TUMACO, a Pacific port, is a major outlet and terminus of a crude oil pipeline. Consists of 62 municipios, with a mainly rural population.

Nariño (nah-REEN-yo), town, ⊙ Nariño municipio, ANTIOQUIA department, NW central COLOMBIA, in the Cordillera CENTRAL, 45 mi/72 km SE of MEDELLÍN; 05°36'N 75°10'W. Elevation 4,809 ft/1,465 m. Coffee, plantains; livestock.

Nariño (nah-REEN-yo), town, ⊙ Nariño municipio, CUNDINAMARCA department, central COLOMBIA, on the MAGDALENA River, 42 mi/68 km SW of BOGOTÁ; 04°24'N 74°50'W. Elevation 1,410 ft/429 m. Sugarcane, corn; livestock.

Nari River (nah-REE), c.320 mi/515 km long, in E BALUCHISTAN, SW PAKISTAN; rises in CENTRAL BRAHUI RANGE, 14 mi/23 km N of ziarat; flows E as LORALAI RIVER, S as ANAMBAR RIVER, WSW as BEJI RIVER, and finally S as the Nari, through center of KACHHI plain, past BHAG, to NORTH WESTERN CANAL of SUKKUR BARRAGE system, 9 mi/14.5 km W of USTA MUHAMMAD. Seasonal, but its flood waters irrigate small fields (S) of wheat, millet, and cotton.

Narita (NAH-ree-tah), city, CHIBA prefecture, E central HONSHU, E central JAPAN, 9 mi/15 km N of CHIBA; 35°46'N 140°19'E. Location of NARITA AIRPORT (New Tokyo International Airport). Mount Narita Shino (Buddhist) Temple dates to 18th century, Narita New Town.

Narita Airport (NAH-ree-tah), officially known as New Tokyo International Airport, NARITA, CHIBA prefecture, E JAPAN, 37 mi/60 km E of TOKYO (road, rail links), 4 mi/7 km SE of Narita; elevation 134 ft/41 m; 35°46'N 140°23'E. Single runway; two passenger terminals; cargo facilities (2.2 million tons/2 million tonnes annually); opened in 1978; handles mainly international air traffic. Tokyo's other airport, HANEDA AIRPORT, handles mainly domestic air traffic. Airport code NRT.

Nariva (nuh-REE-vuh), county (□ 206 sq mi/535.6 sq km), E TRINIDAD, TRINIDAD AND TOBAGO, on the ATLANTIC OCEAN; 10°22′N 61°10′W. Forms, together with SAINT DAVID, SAINT ANDREW, and MAYARO, the administrative district of the Eastern Cos.

Nariva Swamp (nuh-REE-vuh), E TRINIDAD, TRINIDAD AND TOBAGO, c.35 mi/56 km SE of PORT OF SPAIN, a low area stretching from palm-lined COCOS BAY on E coast c.5 mi/8 km inland. Wetlands haven for migratory birds.

Nariwa (nah-REE-wah), town, Kawakami county, OKAYAMA prefecture, SW HONSHU, W JAPAN, 25 mi/40 km N of OKAYAMA; 34°46′N 133°32′E. Home of sacred music and dance known as "Bicchu Kagura." Historic streetscape at Fukiya.

Narka, village (2000 population 93), REPUBLIC county, N central KANSAS, 14 mi/23 km NE of BELLEVILLE, near NEBRASKA state line; 39°57′N 97°25′W. In grain region.

Narkanda (NAHR-kuhn-dah), town, SHIMLA district, central HIMACHAL PRADESH state, N INDIA, 21 mi/34 km NE of SHIMLA, in LESSER HIMALAYA, on National Highway 22. Scenic hill resort (elevation c. 9,460 ft/2,883 m).

Narkatiaganj (nuhr-kuh-TYAH-guhnj), town (2001 population 40,830), WEST CHAMPARAN district, NW BIHAR state, E INDIA, 21 mi/34 km N of BETTIAH. Railroad junction, with spur to Bhikhna Thori (NEPAL) station (just across border).

Närke (NER-ke), historic province (□ 1,712 sq mi/4,451.2 sq km), S central SWEDEN. Included in S part of ÖREBRÖ county.

Narkher (nuhr-KER), town, NAGPUR district, N MAHARASHTRA state, central INDIA, 40 mi/64 km NW of NAGPUR. Cotton ginning; millet, wheat, oilseeds, oranges; livestock market. Sometimes spelled Narkhed.

Narmada River (nahr-MAH-dah), c. 775 mi/1,247 km long, INDIA; rises in MADHYA PRADESH state, central India; flows W between the SATPURA and VINDHYA Ranges through GUJARAT state to the GULF OF KHAMBAT. Because the river is turbulent and confined between steep banks, it is unsuitable for navigation or irrigation. The NARMADA, second most sacred river (GANGA is first) to Hindus, is said to have sprung from the body of the god Shiva; a round-trip pilgrimage on foot along its entire length is highly esteemed. Many holy baths and sites line its banks; near MARBLE ROCKS GORGE, whose 100-ft/30-m-high walls bear inscriptions and sculptures, is a 12th-century temple. During 1990s was a site of controversy due to World Bank and Indian government plans to dam it (which would have displaced hundreds of thousands of local population); while progress was briefly stopped due to negative world-wide attention villagers' protests drew to the project, dam-building proceeded in 2000s.

Narmashir, IRAN: see RIGAN.

Narnaul (NAHR-noul), district (□ 650 sq mi/1,690 sq km), HARYANA state, N INDIA; ⊙ NARNAUL. Former district in PATIALA AND EAST PUNJAB STATES UNION; formerly Mahendragarh district.

Narnaul (NAHR-noul), town, ⊙ NARNAUL district, HARYANA state, N INDIA, 62 mi/100 km SW of DELHI. Trade center for cotton, grain, ghee, salt, oilseeds; hand-loom weaving, cart manufacturing.

Narni (NAHR-nee), ancient *Nequinum* and *Narnia*, town, TERNI province, UMBRIA, central ITALY, on hill above NERA River, 7 mi/11 km WSW of TERNI; 42°31′N 12°31′E. Manufacturing electric furnaces, fabricated metals, machinery, chemicals, synthetic fibers. Bishopric. Has Romanesque cathedral (consecrated 1145), old castle (1360–1370; now a prison), and 13th-century palace with picture gallery. Roman emperor Nerva was born here.

Naroch (NAH-roch), urban settlement, MINSK oblast, BELARUS, on Lake NAROCH. Fish-processing plant; tourist facilities, sanatoriums.

Naroch, Lake (NAH-roch), Polish *Narocz* (□ 32 sq mi/83 sq km), N MINSK oblast, BELARUS, 18 mi/29 km S of POSTAVY, source of Naroch' River. Noted summer resort; fisheries.

Naroda (nah-RO-dah) [Russian=people's mountain], highest peak (approximately 6,180 ft/1,884 m) of the URAL Mountains, on the border of TYUMEN oblast and KOMI REPUBLIC, NE European Russia, in the N Ural Mountains; 65°04′N 60°09′E. Also called Narodnaya Gora.

Narodichi, UKRAINE: see NARODYCHI.

Narodnaya Gora, RUSSIA: see NARODA, mountain.

Narodnoye (nah-ROD-nuh-ye) village, E VORONEZH oblast, S central European Russia, on road and railroad, 18 mi/29 km NNW of BORISOGLEBSK; 51°34′N 41°47′E. Elevation 403 ft/122 m. In agricultural area; flour mill.

Narodychi (nah-RO-di-chee) (Russian *Narodichi*), town (2004 population 2,000), NE ZHYTOMYR oblast, UKRAINE, on Uzh River, and 25 mi/40 km NE of KOROSTEN'; 51°12′N 29°05′E. Elevation 465 ft/141 m. Ribbon factory, feed and flour mills, sawmill. Polluted with radioactive materials from the 1986 explosion at CHORNOBYL'; many residents have moved out of the contamination zone, particularly families with children. Known since 1545; passed to Poland (1569), to Russia (1793); town since 1958. Jewish community since 1683, numbering 3,270 in 1925; eliminated in 1941—fewer than 100 Jews remaining in 2005.

Naro-Fominsk (NAH-ruh–fuh-MEENSK), city (2006 population 73,415), SW MOSCOW oblast, central European Russia, on the Nara River (OKA River basin), on highway junction and railroad, 43 mi/69 km SW of MOSCOW; 55°23′N 36°44′E. Elevation 518 ft/157 m. Silk combine, heat and electrical insulators, plastics, arts and educational supplies, food processing (dairy, bakery, meat packing). Nearby is a former country seat of Prince Shcherbatov, with English gardens. Arose in 1840 as a spinning factory. Became city in 1926.

Narok (NAH-rok), town (1999 population 24,091), administrative center, Narok district, RIFT VALLEY province, S KENYA, on road from KIJABE, and 65 mi/105 km NW of MAGADI; 01°06′S 35°53′E. Livestock; tea, coffee, flax, corn, wheat. Airfield.

Naro Moro (nah-ro MO-ro), town, CENTRAL province, now administrative part of Nyeri district, S central KENYA, on W slopes of Mount KENYA, on railroad, and 20 mi/32 km NNE of NYERI. Elevation 6,631 ft/2,021 m. Coffee, sisal, wheat, corn.

Naron, RUSSIA: see ALI-YURT.

Narooma (na-ROO-muh), town, SE NEW SOUTH WALES, SE AUSTRALIA, 24 mi/39 km S of BATEMAN'S BAY, on PACIFIC OCEAN and Princes Highway; 36°13′S 150°03′E. Dairying; timber; fish, oysters; tourism (fishing resort). Lighthouse on MONTAGUE ISLAND 8 mi/13 km offshore. Formerly spelled Noorooma.

Narora, INDIA: see DIBAI.

Narova River, ESTONIA and RUSSIA: see NARVA River.

Narovchat (nah-ruhf-CHAHT), village, N central PENZA oblast, E European Russia, in the MOKSHA River valley, 92 mi/148 km NW of PENZA, and 25 mi/40 km N of NIZHNIY LOMOV; 53°53′N 43°41′E. Elevation 439 ft/133 m. Highway junction. In grain and hemp area; bakery. In the 14th century, a city of the Mongol-Tatar Golden Horde stood on the site of the present settlement. Fortress constructed in 1520. Made city in 1780; reduced to status of a village in 1926.

Narovlya (nah-ROV-lyah), town, GOMEL oblast, BELARUS, ⊙ NAROVLYA region, on PRIPET RIVER, and 21 mi/34 km SSE of MOZYR. Manufacturing (construction components); logging. Also called NAROVL.

Narowal (NAH-ro-wahl), town, SIALKOT district, E PUNJAB province, central PAKISTAN, 33 mi/53 km SE of SIALKOT; 32°06′N 74°53′E.

Narpai, village, NAWOIY wiloyat, central UZBEKISTAN; 40°06′N 65°26′E. Also Narpay.

Narrabeen (NA-ruh-BEEN), town, E NEW SOUTH WALES, SE AUSTRALIA, suburb of SYDNEY, on coast; 33°39′S 151°17′E. In metropolitan area; summer resort.

Narrabri (NA-ruh-brei), municipality, N NEW SOUTH WALES, SE AUSTRALIA, on NAMOI RIVER, and 215 mi/346 km NW of NEWCASTLE; 30°19′S 149°47′E. Railroad junction. Sheep; cotton.

Narracan, village, S VICTORIA, SE AUSTRALIA, 80 mi/129 km ESE of MELBOURNE, S of MOE; 38°15′S 146°14′E.

Narragansett (na-ruh-GAN-set), town (2000 population 16,361), WASHINGTON county, S RHODE ISLAND, along W shore entrance of NARRAGANSETT BAY, N of POINT JUDITH, 27 mi/43 km S of PROVIDENCE; 41°23′N 71°29′W. Suburban settlements, resorts; fishing; agriculture. Includes resort village Narragansett Pier (1990 population 3,721; 2000 population 3,671), fishing village of Galilee, and a state reservation with Scarborough Beach. Point Judith lighthouse dates from 1816. Settled in mid-17th century, set off from South Kingstown 1888. A fire in 1900 and the 1938 hurricane did great damage. The town takes its name from a Native American tribe that once lived in the territory. In 1675, colonists engaged the Narragansetts and prevailed. Incorporated 1901.

Narragansett Bay, arm of the Atlantic Ocean, deeply indenting the state of RHODE ISLAND; 30 mi/48 km long, 3 mi/4.8 km–12 mi/19 km wide. Its many inlets provided harbors that were advantageous to Colonial trade and later to resort development. At the head of the bay is PROVIDENCE; at the SE corner of the N bay portion is NEWPORT. Conanicut and Prudence islands are also in the bay, which is spanned by Newport Bridge (built 1969; 1,600 ft/488 m), which links Newport with JAMESTOWN.

Narragansett Indian Reservation, Native American land, in CHARLESTOWN, WASHINGTON county, S RHODE ISLAND; 41°25′N 71°40′W. The Narragansetts, an Algonquian tribe of Native Americans, sold their reservation lands c.1880 and only regained them over a century later. The tribe claimed an estimated 2,100 members in the 1990s. The reservation preserves ruins illustrative of Narragansett life, as well as small burial grounds and a Native church (1859) of hewn granite.

Narragansett Pier, Rhode Island: see NARRAGANSETT.

Narraguagus River (nar-uh-GWAI-guhs), c.30 mi/48 km long, SE MAINE; rises in HANCOCK and WASHINGTON counties; flows SE to Narraguagus Bay.

Narrandera (nuh-RAN-druh), municipality, S NEW SOUTH WALES, SE AUSTRALIA, on MURRUMBIDGEE RIVER, and 150 mi/241 km WNW of CANBERRA, in RIVERINA region; 34°45′S 146°33′E. Railroad junction. Sheep; agriculture. Airport. Heritage walks; koala colony in nature reserve.

Narran Lake, AUSTRALIA: see TEREWAH, LAKE.

Narrogin (NA-ruh-jin), municipality, SW WESTERN AUSTRALIA state, W AUSTRALIA, 105 mi/169 km SE of PERTH; 32°56′S 117°10′E. Railroad junction; wheat center; flour mill. Sheep, hogs. Narrogin Farm School (1914) nearby.

Narrogin Inn, WESTERN AUSTRALIA, AUSTRALIA: see ARMADALE.

Narromine (NA-ruh-MEIN), municipality, central NEW SOUTH WALES, SE AUSTRALIA, on MACQUARIE RIVER, and 205 mi/330 km NW of SYDNEY, and 24 mi/39 km W of DUBBO; 32°09′S 148°15′E. Railroad junction; agriculture center (fruit, vegetables, wheat, cotton; lambs, wool).

Narrows (NA-roz), town (2006 population 2,180), GILES county, SW VIRGINIA, in ALLEGHENY MOUNTAINS, near WEST VIRGINIA state line, 22 mi/35 km WNW of BLACKSBURG, on NEW RIVER, at mouth of Wolf Creek; 37°19′N 80°48′W. In agricultural area (alfalfa, apples; cattle, poultry); manufacturing (cellulose products, machinery, textiles); timber; coal mining. Civil War Confederate garrison. Incorporated 1904.

Narrows Arm (NA-roz) or **Narrows Inlet**, NE arm of SEECHELT Inlet, SW BRITISH COLUMBIA, W Canada, 45 mi/72 km NW of VANCOUVER; 10 mi/16 km long, 1 mi/2 km wide; 49°41′N 123°49′W. In lumbering area.

Narrowsburg, hamlet, SULLIVAN county, SE NEW YORK, on DELAWARE RIVER (here forming PENNSYLVANIA state line), and 20 mi/32 km WSW of MONTICELLO; 41°36′N 75°03′W.

Narrows Dam, ARKANSAS: see LITTLE MISSOURI RIVER.

Narrows, The, Caribbean strait, LEEWARD ISLANDS, between St. Kitts (NW) and NEVIS (SE); c.2 mi/3.2 km wide; at c.17°13′N 62°37′W.

Narrows, The, narrow passage in the U.S. VIRGIN ISLANDS, between ST. JOHN (S) and Great Thatch and TORTOLA (N) islands (BRITISH VIRGIN ISLANDS), 12 mi/19 km ENE of CHARLOTTE AMALIE; c.1 mi/1.6 km wide; 18°22′N 64°45′W.

Narrung, village, SE SOUTH AUSTRALIA state, S central AUSTRALIA, on S shore of Lake ALEXANDRINA, and 55 mi/89 km SE of ADELAIDE; 35°31′S 139°10′E. Citrus and dried fruits.

Narsalik (NAHKH-shah-lik), abandoned fishing settlement, Paamiut (Frederikshåb) commune, SW GREENLAND, on islet in the ATLANTIC OCEAN, 25 mi/40 km SSE of FREDERIKSHAAB (Paamiut); 61°39′N 49°22′W.

Narsampet (NUHR-suhm-pait), town, tahsil headquarters, WARANGAL district, ANDHRA PRADESH state, S INDIA, 22 mi/35 km ESE of WARANGAL. Rice, oilseeds.

Narsannapet, INDIA: see NARASANNAPETA.

Narsapatnam, INDIA: see NARSIPATNAM.

Narsapur (nuhr-sah-poor), town, WEST GODAVARI district, ANDHRA PRADESH state, S INDIA, near mouth of Vasishta Godavari River, in GODAVARI delta, 45 mi/72 km ESE of ELURU. Railroad spur terminus; rice milling; tobacco, sugarcane, coconuts. Former N suburb of MADAPOLLAM (now destroyed by river erosion) was site of early 17th-century Dutch trading station; occupied 1677 by English, 1756 by French; finally ceded to English in 1759. Coconut plantations nearby. Saltworks 6 mi/9.7 km WSW, at village of MOGALTURRU. Formerly spelled Narasapur.

Narsaq (NAHKH-snahk), town, Narsaq commune, SW GREENLAND, on Tunugdliarfik Fjord, 14 mi/23 km N of JULIANEHÅB (Qaqortoq); 60°54′N 46°00′W. Fishing port.

Narsaq (NAHKH-shahk), abandoned settlement, Nuuk (Godthåb) commune, SW GREENLAND, on DAVIS STRAIT, 14 mi/23 km SSE of NUUK (Godthåb); 63°59′N 51°36′W.

Narsarsuaq (NAHKH-shahk-shwahk), locality, Narsaq commune, SW GREENLAND, at head of Tunugdliarfik Fjord, 40 mi/64 km NE of JULIANEHÅB (Qaqortoq); 61°09′N 45°24′W. Site of Bluie West 1 air base, established by U.S. in World War II. Now civil airport for South Greenland.

Narsimhapur, INDIA: see NARSINGHPUR.

Narsingdi, BANGLADESH: see NARSINGHDI.

Narsinghdi, town (2001 population 124,204), DHAKA district, E central EAST BENGAL, BANGLADESH, near MEGHNA RIVER, 25 mi/40 km NE of DHAKA; 23°55′N 90°44′E. Jute-collecting and -trading center. College. Also spelled Narsingdi.

Narsinghgarh (NAHR-sing-guhr) or **Narsingarh**, town (2001 population 27,657), RAJGARH district, N central MADHYA PRADESH state, central INDIA, 37 mi/60 km NW of BHOPAL. Market center for millet, cotton, wheat; handicraft cloth weaving. Was capital of former princely state of Narsinghgarh of Central India agency; in 1948, merged with MADHYA BHARAT and later with Madhya Pradesh state.

Narsinghpur (nuhr-SING-poor) district (□ 1,982 sq mi/5,153.2 sq km; 2001 population 957,399), S central MADHYA PRADESH state, central INDIA; ⊙ NARSINGHPUR. Soapstone, dolomite, fireclay, limestone; copper and zinc utensil manufacture in Chichali. Mostly agricultural production: oil, daal mills; manufacturing: beedi, agricultural machinery, sugar. Major cities: NARSINGHPUR, GOTEGAON, Kareli, GADARWARA, Tendukheda.

Narsinghpur (nuhr-SING-poor) or **Narsimhapur**, town (2001 population 46,120), ⊙ NARSINGHPUR district, S central MADHYA PRADESH state, central INDIA, in fertile NARMADA River valley, 95 mi/153 km E of HOSHANGABAD. Trades in wheat, millet, cotton, oilseeds; sawmilling. Was capital of former Narsinghpur district until merged in early 1930s with HOSHANGABAD district; became own district again in 1956 after Reorganization of States. Marble quarries nearby. Formerly spelled Nursingpur.

Narsinghpur, town, tahsil headquarters, CUTTACK district, E ORISSA state, E central INDIA, 50 mi/80 km W of CUTTACK. Trades in rice, oilseeds, bamboo. Was capital of former princely state of Narsinghpur in ORISSA STATES, along left bank of MAHANADI River; incorporated 1949 into Cuttack district.

Narsipatnam (NUHR-suh-puht-nuhm), town, VISHAKHAPATNAM district, ANDHRA PRADESH state, S INDIA, 45 mi/72 km W of VISHAKHAPATNAM. Road and trade center in agricultural area (oilseeds, sugarcane, rice); shipping center for timber (sal, teak), myrobalan, lac, and coffee from EASTERN GHATS (W). Graphite deposits nearby. Silk-growing farm 22 mi/35 km NW, at village of Chintapalle. Narsipatnam (also spelled Narasapatnam) Road, 17 mi/27 km SSE, is railroad station. Also spelled Narasapatnam, Narasapatam, or Narsapatnam.

Narsoba Vadi (nuhr-SO-bah VAH-dee), town, KOLHAPUR district, MAHARASHTRA state, W central INDIA, 25 mi/40 km E of KOLHAPUR. Sugarcane, millet. Large annual festival fair. Also spelled Narsobavadi, Narsubachi Vadi, or Narsobas Wadi.

Nartan (nahr-TAHN), village (2005 population 12,830), central KABARDINO-BALKAR REPUBLIC, N CAUCASUS, S European Russia, on road and near railroad, 3 mi/5 km E of NAL'CHIK, to which it is administratively subordinate; 43°30′N 43°42′E. Elevation 1,243 ft/378 m. Agriculture (horse breeding). Has a sanatorium.

Narthakion (nahr-THAH-kee-on), hill range on border of THESSALY and CENTRAL GREECE departments, just NW of the OTHRYS; rises to 3,317 ft/1,011 m in the Kassidiares (Kassidiaris), 5 mi/8 km S of PHARSALA; 39°15′N 22°29′E.

Nartkala (nahrt-kah-LAH), city (2005 population 35,460), E central KABARDINO-BALKAR REPUBLIC, N CAUCASUS, S European Russia, on road and railroad, 15 mi/24 km NE of NAL'CHIK; 43°33′N 43°51′E. Elevation 994 ft/302 m. Chemical industry. Made city in 1955. Called Dokshukino until 1967.

Nartovo, RUSSIA: see KANTYSHEVO.

Naru (NAH-roo), town, on NARU-SHIMA island in GOTO-RETTO island group, South Matsuura county, NAGASAKI prefecture, SW JAPAN, 56 mi/90 km W of NAGASAKI; 32°49′N 128°56′E.

Naru, JAPAN: see NARU-SHIMA.

Năruja (nuh-ROO-zhah), village, VRANCEA county, E ROMANIA, 22 mi/35 km NW of FOCȘANI; 45°50′N 26°47′E.

Naruko (nah-roo-KO), town, Tamatsukuri county, MIYAGI prefecture, N HONSHU, NE JAPAN, 34 mi/55 km N of SENDAI; 38°44′N 140°43′E. Traditonal dolls, lacquerware; hot-springs resort. Naruko Gorge is nearby.

Naruksovo (nah-ROOK-suh-vuh), village, S NIZHEGOROD oblast, central European Russia, on local road, 28 mi/45 km S of LUKOYANOV, and 17 mi/27 km WSW of POCHINKI, to which it is administratively subordinate; 54°37′N 44°33′E. Elevation 524 ft/159 m. Agricultural products.

Narungombe (nah-roon-GOM-be), village, LINDI region, SE Tanania, 90 mi/145 km NW of LINDI, on Narungombe River; 09°19′S 38°32′E. Grain; livestock.

Narusawa (nah-roo-SAH-wah), village, South Tsuru county, YAMANASHI prefecture, central HONSHU, central JAPAN, near MOUNT FUJI, 16 mi/25 km S of KOFU; 35°28′N 138°42′E.

Naruse (NAH-roo-se), town, Mono county, MIYAGI prefecture, N HONSHU, NE JAPAN, 19 mi/30 km N of SENDAI; 38°23′N 141°10′E. Oysters; nori.

Naru-shima (nah-ROO-shee-mah), island (□ 10 sq mi/26 sq km), of GOTO-RETTO island group, NAGASAKI prefecture, SW JAPAN, in EAST CHINA SEA, 40 mi/64 km W of KYUSHU; 5.5 mi/8.9 km long, 5 mi/8 km wide. Fishing. Sometimes called Naru.

Naruto (nah-roo-TO), city, TOKUSHIMA prefecture, SE SHIKOKU, W JAPAN, across NARUTO STRAIT from AWAJI-SHIMA island (linked by O-Naruto Bridge), 6 mi/10 km N of TOKUSHIMA; 34°10′N 134°36′E. Medicine; seaweed (*wakame*), lotus root, sweet potatoes.

Naruto (nah-roo-TO), town, Sanbu county, CHIBA prefecture, E central HONSHU, E central JAPAN, on N central BOSO, 12 mi/20 km E of CHIBA; 35°35′N 140°24′E. Onions, strawberries.

Naruto Strait (nah-roo-TO), Japanese *Naruto-kaikyo* (NAH-roo-to-KAH-ee-kyo), channel connecting HARIMA SEA (E section of INLAND SEA) with KII CHANNEL, between SHIKOKU (W) and AWAJI-SHIMA (E); c. 10 mi/16 km long, c. 1 mi/1.6 km–5 mi/8 km wide. Contains OGE-SHIMA and SHIMADA-SHIMA islands; Muya is on W shore. Known for whirlpools. Called *awa-no-naruto* in classical literature.

Narva (NAHR-vah), city, NE ESTONIA, on the left bank of the NARVA River; 59°22′N 28°11′E. A leading textile center, it also has machinery plants, sawmills, flax and jute factories, and food-processing industries. The city is in an important electric-power-production area, much of which is exported. Founded by the Danes (1223), Narva passed to the Livonian Knights in 1346 and was a member of the HANSEATIC LEAGUE. In 1492, Ivan III of Russia built the fortress IVANGOROD on the right bank of the Narva River. After the dissolution (1561) of the Livonian Order, the city was 1st seized by the Russians, then the Swedes (1581); it continued to be contested by the two nations. In 1704 the city fell to Russia, and it remained so until 1919, when it was incorporated into newly independent Estonia. German forces occupied the city in World War II. In 1945 all Estonian territory E of the Narva River, including Ivangorod fortress, was ceded to the USSR. The city is dominated by two old fortresses, and it has retained a 14th century Eastern Orthodox cathedral (originally Roman Catholic), and a 17th-century town hall and exchange buildings. Today it is inhabited mainly by ethnic Russians (90% of population).

Narva (NAHR-vah), river, c.50 mi/80 km long, ESTONIA; rises in LAKE PEIPSI (Russian, *Chudskoye*), E Estonia; flows NE past the city of NARVA into the GULF OF FINLAND. Forms the border between Estonia and RUSSIA. The falls of the river have historically supplied power to the fibers industry of Narva. Also NAROVA River.

Narvacan (nahr-VAH-kahn), town, ILOCOS SUR province, N LUZON, PHILIPPINES, 13 mi/21 km SE of VIGAN, near W coast; 17°26′N 120°29′E. Rice-growing center.

Narváez (nar-VAH-ez), canton, BURNET O'CONNOR province, TARIJA department, S central BOLIVIA, 51 mi/81 km S of TARIJA, SSW of HUAYCO, on the Tarija–VILLA MONTES road; 21°24′S 64°18′W. Elevation 4,035 ft/1,230 m. Clay, limestone, and gypsum deposits. Agriculture (potatoes, yucca, bananas, corn, barley, rye, peanuts); cattle. Also known as Paccha.

Narva-Jōesuu (NAHR-vah-YUH-e-soo) or **Narva-Yyyesu**, German *Hungerburg*, town, NE ESTONIA, outer port of NARVA, on Narva Bay (inlet of GULF OF FINLAND), a port at mouth of NARVA River (Russian border), 8 mi/13 km NW of Narva; 59°27'N 28°02'E. Summer and health resort; sawmill.

Narva-Yyyesu, ESTONIA: see NARVA-JÕESUU.

Narvik (NAHR-veek), city, NORDLAND county, N NORWAY, an ice-free port on the OFOTFJORD opposite the LOFOTEN Islands; 68°26'N 17°26'E. It was founded (1887) as the ATLANTIC port for the KIRUNA and GÄLLIVARE iron mines in SWEDEN and was known as Victoriahavn until 1898. The city is now a tourist center. In World War II, Narvik fell to the Germans when they invaded Norway on April 9, 1940. To prevent the Germans from using Narvik as a shipping base for Swedish iron ore, a British expeditionary force briefly occupied (May 28–June 9, 1940) the port.

Narwana (nuhr-WAH-nah), town, JIND district, HARYANA state, N INDIA, 50 mi/80 km SW of KURUKSHETRA. Railroad junction; agriculture market (cotton, millet, gram, wheat); cotton ginning, handloom weaving. Was in former S central PATIALA AND EAST PUNJAB STATES UNION before joining Haryana state.

Narwar (NUHR-wahr), town (2001 population 15,748), SHIVPURI district, N central MADHYA PRADESH state, central INDIA, 22 mi/35 km NE of SHIVPURI, E of SIND RIVER. Millet, gram, wheat. Said to have been capital of Raja Nala in Hindu epic *Mahabharata*; has many ruins; Narwar Fort is significant in medieval Indian history. From 12th–16th centurives, hill fort (W) held by different Rajputs; captured by Moguls in 16th century; fell to Maratha chief Sindhia in early 19th century; outside walled town are Tomar chiefs' memorial pillars. Was in N central section of former MADHYA BHARAT state prior to joining Madhya Pradesh.

Nar'yan-Mar (nah-reeyahn–MAHR), city (2006 population 17,855), ⊙ NENETS AUTONOMOUS OKRUG, administratively subordinate to ARCHANGEL oblast, NE European Russia, river and maritime port and airport on the E arm of the PECHORA River delta mouth, 40 mi/64 km inland from the PECHORA BAY on the BARENTS SEA, 1,385 mi/2,229 km NNE of MOSCOW, and 680 mi/1097 km NE of ARCHANGEL; 67°39'N 53°03'E. Sawmilling (lumber export), fish combine. Until 1935 (when it became a city), called Dzerzhinsky.

Narylkov, RUSSIA: see BELYY YAR, KHAKASS Republic.

Narym (nah-RIM), village, central TOMSK oblast, W SIBERIA, RUSSIA, on the OB' RIVER, near the mouth of the KET' River, terminus of local highway branch, 210 mi/338 km NNW of TOMSK; 58°55'N 81°35'E. Elevation 147 ft/44 m. River port; fisheries, sawmilling. Founded in 1595 on the early Siberian colonization route; important 19th-century trading town; declined following construction of the TRANS-SIBERIAN RAILROAD. Stalin (1912), Kuybyshev, and Sverdlov were exiled here.

Narym Range (nah-RIM), branch of SW ALTAI MOUNTAINS, EAST KAZAKHSTAN region, KAZAKHSTAN; extends 90 mi/145 km along NARYM RIVER W to IRTYSH RIVER; rises to 11,270 ft/3,435 m (E); forms divide between Buktarma and KURCHUM rivers.

Narym River (nah-RIM), c.60 mi/97 km long, EAST KAZAKHSTAN region, KAZAKHSTAN; rises in NARYM RANGE; flows W, past ULKENNARYN (Bolshoye Narymskoye), to upper IRTYSH RIVER.

Naryn (nah-RUN), region (□ 46,707 sq mi/121,438.2 sq km; 1999 population 249,115), central KYRGYZSTAN; ⊙ NARYN. In TIANSHAN mountain area; elevation mostly over 4,920 ft/1,500 m; bounded N by TESKEY ALA-TOO (TERSKEI ALATAU) and Kyrgyz ranges, E by FERGANA RANGE, S by Kaksha Ala-Too (KOKSHAAL-TAU; CHINA border). Drained by NARYN RIVER (hydroelectric power stations). Agriculture (wheat) in mountain valley; livestock raising (cattle, sheep, horses). Tungsten, gold mining. Crossed by BISHKEK-Naryn-Osh highway. Population consists mainly of Kyrgyz. Formed 1939 as TYAN-SHAN; renamed in 1990.

Naryn (nah-RUN), city (1999 population 40,050), ⊙ NARYN region, KYRGYZSTAN, on NARYN RIVER, on

highway, and 120 mi/193 km SE of BISHKEK; 41°27'N 76°00'E. Elevation 6,610 ft/2,015 m. Center of highmountain valley agriculture (wheat) and livestock (sheep) grazing; tanning industry, sawmilling, brickworking. Established 1927.

Naryn (nah-RIN), village, SE TUVA REPUBLIC, S SIBERIA, RUSSIA, on local highway, 118 mi/190 km SE of KYZYL, and 32 mi/51 km ESE of ERZIN; 50°13'N 95°36'E. Elevation 4,435 ft/1,351 m. In agricultural area (hardy grain; livestock).

Naryn (nah-RUN), river, c.450 mi/720 km long, dropping 9,843 ft/3000 m along its course, SW KYRGYZSTAN and SE UZBEKISTAN; rises in several branches in the TIANSHAN mountain system; flows generally W through the FERGANA VALLEY where it joins with the KARA DARYA (river) to form the SYR DARYA (river). Longest river in Kyrgyzstan. Upper course contains eight hydroelectric power plants; basis for country's export of hydropower. Lower course is used for irrigating a cotton-growing area.

Naryn, UZBEKISTAN: see KHAKKULABAD.

Narynköl (nah-rin-KOOL), village, SE ALMATY region, KAZAKHSTAN, in N TIANSHAN foothills, on CHINA border, 165 mi/266 km E of ALMATY (linked by road); 42°50'N 80°00'E. Tertiary-level administrative center. Irrigated agriculture (wheat).

Naryn-Too (nah-run–TO), Kazak *Naryn-Tau*, mountain range in the TIANSHAN mountain system, KYRGYZSTAN; extends c.80 mi/129 km E-W, S of NARYN RIVER, and E of NARYN city; rises to 13,000 ft/3,962 m.

Naryshevo, RUSSIA: see OKTYABR'SKIY, BASHKORTOSTAN Republic.

Naryshkino (nah-RISH-kee-nuh), town (2006 population 9,560), W ORËL oblast, SW European Russia, on road junction and railroad, 19 mi/30 km W of ORËL; 52°58'N 35°43'E. Elevation 708 ft/215 m. Canning.

Na Sam, VIETNAM: see VAN LANG.

Nasarawa (nah-SAH-RAH-wah), state (□ 10,469 sq mi/27,219.4 sq km; 2006 population 1,863,275), central NIGERIA; ⊙ LAFIA. Bordered by KADUNA (N), PLATEAU (NEE), TARABA (SE), BENUE (S), and KOGI (SWW) states and ABUJA FEDERAL CAPITAL TERRITORY (NW). Railroad between PORT HARCOURT and KANO runs N-S through Lafia. Carved out of S Plateau state in 1996.

Nasarawa (nah-SAH-RAH-wah), town, headquarters of Nasarawa Local Government Area, Plateau state, central NIGERIA, 80 mi/129 km NW of MAKURDI. Major tin-mining center; shea-nut processing. An important native commercial center until end of 19th century. Sometimes spelled Nassarawa.

Nasarpur (NUH-sahr-puhr), town, HYDERABAD district, S central SIND province, SE PAKISTAN, 17 mi/27 km NE of HYDERABAD; 25°31'N 68°37'E. Also spelled NASIRPUR.

Nasarpur (NUH-sahr-puhr), village, PESHAWAR district, NORTH-WEST FRONTIER PROVINCE, N PAKISTAN, 5 mi/8 km E of PESHAWAR; 34°01'N 71°41'E. Railroad station. Fruit canning.

Năsăud (nuh-suh-WOOD), Hungarian *Naszód*, town, BISTRIȚA-NĂSĂUD county, N central ROMANIA, in TRANSYLVANIA, in S foothills of the Rodna Mountains, on Somesul Mare River, on railroad, and 12 mi/19 km NNW of BISTRIȚA; 47°17'N 24°24'E. Agriculture and trading center (lumber; livestock), with brewing industry; manufacturing (furniture, textiles, and chemicals). Dating from 13th century, it was given as fief to Janos Hunyadi by Ladislaus V (15th century). Under Hungarian administration, 1940–1945.

Nasavrky (NAH-sah-VUHR-ki), German *Nassaberg*, village, VYCHODOCESKY province, E BOHEMIA, CZECH REPUBLIC, on CHRUDIMKA RIVER, 8 mi/12.9 km S of CHRUDIM. In a cotton-spinning district; agriculture (sugar beets, potatoes). Has a Gothic 14th century church and a Renaissance 16th century castle.

Näsby (NES-BEE), suburb of STOCKHOLM, STOCKHOLM county, E SWEDEN, near inlet of BALTIC SEA, 7 mi/11.3

km N of Stockholm city center. Has seventeenth-century castle.

Naschel (nahs-CHAIL), town, NE SAN LUIS province, ARGENTINA, on railroad, and 60 mi/97 km ENE of SAN LUIS; 52°55'S 65°23'W. Corn, alfalfa, wheat, flax; livestock; granite and marble quarrying. Dam nearby.

Naschitti (nah-SHI-tee), settlement, SAN JUAN county, NEW MEXICO, 39 mi/63 km N of GALLUP, on the Navajo Reservation. The site of one of the first trading posts on the E side of the CHUSKA MOUNTAINS.

Nasea, FIJI: see LABASA.

Naseby, township, Central OTAGO district, E central SOUTH ISLAND, NEW ZEALAND, 60 mi/97 km NNW of DUNEDIN; 45°02'S 170°09'E. Gold-mining nostalgia; some forest, sheep farms, orchards.

Naseby (NAIZ-bee), village (2001 population 525), NORTHAMPTONSHIRE, central ENGLAND, 6 mi/9.7 km NW of NORTHAMPTON; 52°23'N 00°59'W. Nearby, in 1645, the Parliamentarians defeated the Royalists in a decisive battle of the English Civil War.

Naselle (nai-SEL), unincorporated town, PACIFIC county, SW WASHINGTON, 12 mi/19 km N of ASTORIA (OREGON), on Naselle River, at mouth of South Naselle River. Timber; dairying. Stone quarrying. Willapa National Wildlife Refuge and Willapa Bay to NW; COLUMBIA River estuary 5 mi/8 km S.

Nash (NASH), county (□ 542 sq mi/1,409.2 sq km; 2006 population 92,312), NE central NORTH CAROLINA; ⊙ NASHVILLE; 35°58'N 77°59'W. Elevation 180 ft/54.9 m. In coastal plain area; bounded NE by FISHING CREEK; drained by TAR RIVER. Manufacturing at ROCKY MOUNT, SPRING HOPE, and Nashville; service industries; agriculture (wheat, oats, soybeans, hay, sweet potatoes, peanuts, tobacco, cotton, corn; chickens, cattle, hogs); pine timber. Formed 1777 from Edgecombe County. Named for General Francis Nash (1742–1777) who was mortally wounded fighting under General Washington at the Battle of Germantown.

Nash (NASH) town (2006 population 2,394), Bowie county, NE TEXAS, suburb 7 mi/11.3 km WNW of downtown TEXARKANA; 33°26'N 94°07'W. Agricultural area (dairying; cattle, wheat; soybeans, rice). Oil and natural gas. Manufacturing (wood products, plastic products, transportation equipment, fabricated metal products).

Nash or **Nashville**, village (2006 population 205), GRANT county, N OKLAHOMA, 21 mi/34 km NNW of ENID, near SALT FORK OF ARKANSAS RIVER; 36°40'N 98°02'W. In grain-growing and dairying area. Great Salt Plains State Park (National Wildlife Refuge) and Lake to NW.

Nash Harbor, village, W ALASKA, on N shore of NUNIVAK Island; 60°15'N 166°44'W.

Nashik, district (□ 5,996 sq mi/15,589.6 sq km), MAHARASHTRA state, central INDIA; ⊙ NASHIK. W section crossed N-S by WESTERN GHATS, with spurs extending E; drained by GODAVARI and GIRNA Rivers. Agriculture (millet, wheat, oilseeds, rice, cotton); fruit, vegetables grown in S; teak and blackwood in W forests. Handicraft cloth weaving, oilseed pressing. In late 18th century, under Maratha control (several dispersed hill forts). Original district enlarged by incorporated 1949 of former Gujarat state of Surgana.

Nashik, town (2001 population 1,152,326), ⊙ NASHIK district, MAHARASHTRA state, W central INDIA. It is a center of textile manufacturing and sugar and oil processing; military aircraft assembly. The ancient hermitage of Panchavati, is holy as the site of the exile of the epic god Rama and his wife Sita described in the *Ramayana* epic. Thousands of pilgrims visit the town annually. The Indian Security Press (printing currency notes) and Military and Police Training College are here. Nearby are ancient Jain and Buddhist caves.

Nash Island, WASHINGTON county, E MAINE, small lighthouse island just E of entrance to PLEASANT BAY.

Nashoba (na-SHO-buh), former community, SHELBY county, SW TENNESSEE, on WOLF RIVER just E of MEMPHIS. Founded 1827 by Frances Wright as a colony in which slaves were to be educated for freedom, Nashoba failed as a social experiment and by 1830 had been dissolved.

Nashotah (nuh-SHO-tuh), village (2006 population 1,370), WAUKESHA county, SE WISCONSIN, 24 mi/39 km W of MILWAUKEE, in farm and lake resort region; 43°05′N 88°24′W. Manufacturing (plastic products, cable assembly, wood products). Seat of Nashotah House, an Episcopal seminary.

Nashua (NA-shoo-wuh), city (2006 population 87,157), ⊙ HILLSBOROUGH county, S NEW HAMPSHIRE, 15 mi/24 km S of MANCHESTER, just N of MASSACHUSETTS state line, and 35 mi/56 km NW of BOSTON; 42°45′N 71°29′W. Bounded E by MERRIMACK RIVER; drained by NASHUA RIVER and Salmon Brook. Railroad junction. Nashua developed (early 19th century) as a textile mill town due to the availability of water power; the closing of these mills after World War II, however, prompted the development of diverse manufacturing. Manufacturing (machinery, rubber products, wood products, building materials, computers and computer equipment, fabricated metal products, transportation equipment, electrical and electronic goods, furniture, chemicals, shoes, paper products, airshafts, plastic products, defense systems; machining). The city has also grown as a satellite community of the Boston urbanized area and is popular for its tax-free shopping. Seat of Rivier College, New Hampshire Community Technical College, Daniel Webster College. The Federal Aviation Agency has an Air Traffic Control Center here. Federal Fish Hatchery. Municipal Airport to W. Silver Lake State Park to W, in Hollis. Settled c.1655, incorporated as a city 1853.

Nashua, town (2000 population 1,618), CHICKASAW county, NE IOWA, on CEDAR RIVER at mouth of LITTLE CEDAR RIVER, 15 mi/24 km WSW of NEW HAMPTON; 42°57′N 92°32′W. Manufacturing (machinery, metal and wood products).

Nashua (NA-shoo-wuh), village (2000 population 69), WILKIN county, W MINNESOTA, 21 mi/34 km SE of BRECKENRIDGE; 46°02′N 96°17′W. Grain, sunflowers; dairying.

Nashua (NA-shwuh), village (2000 population 325), VALLEY county, NE MONTANA, 13 mi/21 km ESE of GLASGOW, at the confluence of Porcupine Creek and MILK RIVER; 48°08′N 106°22′W. FORT PECK LAKE and Dam to S; large Fort Peck Indian Reservation to E. Sheep, cattle; irrigated agriculture. Growth stimulated by construction of dam.

Nashua River (NA-shoo-uh), c.30 mi/48 km long, MASSACHUSETTS and NEW HAMPSHIRE; formed in E WORCESTER county (Massachusetts) by junction of its N and S branches near LANCASTER; flows NNE to the MERRIMACK RIVER at NASHUA (NEW HAMPSHIRE). Water power. N branch rises W of FITCHBURG (Massachusetts), flows c.30 mi/48 km generally SE, past Fitchburg (water power), joining S branch c.5 mi/8 km below its issuance from WACHUSETT RESERVOIR.

Nashville, city (2000 population 3,147), ⊙ WASHINGTON county, SW ILLINOIS, 18 mi/29 km SW of CENTRALIA; 38°21′N 89°22′W. Manufacturing (machinery); agriculture (corn, wheat, fruit, seed; livestock, poultry). Incorporated 1853.

Nashville, city, ⊙ TENNESSEE, coextensive with DAVIDSON county, central Tennessee, on the CUMBERLAND RIVER, 195 mi/314 km NE of MEMPHIS; 36°10′N 86°46′W. In a fertile farm area, Nashville is a port of entry and an important commercial and industrial center. The city has railroad shops and diverse manufacturing. Noted for its music business; it is a major recording center, especially for country music. It also has many publishing houses producing religious materials, school annuals, magazines, and telephone directories. Several large insurance and finance companies have their headquarters here, and the country's largest healthcare conglomerate, Hospital Corporation of America, is based here. Founded (1779) by a group of pioneers under James Robertson. Fort Nashborough was built on the banks of the river, and the next year sixty families arrived to settle the area. As the N terminus of the NATCHEZ TRACE, the settlement developed early as a cotton center and river port and later as a railroad hub. It became the permanent state capital in 1843. After the fall of FORT DONELSON in February 1862, Nashville was abandoned to Union troops under D. C. Buell and became an important Union base for the remainder of the Civil War. Sometimes called the "Athens of the South," Nashville has many buildings of Classical design, including a replica of the PARTHENON, built in 1897. Among its many institutions of higher education are Vanderbilt University, Fisk University, Tennessee State University, the University of Tennessee at Nashville, Meharry Medical College, American Baptist College, Lipscomb University, Belmont University, Free Will Baptist Bible College, Aquinas College, and a state school for the blind. Nashville has many cultural amenities, including a symphony orchestra, ballet and opera companies, and several theater troupes, art galleries, and museums. Points of interest include the capitol (completed 1855) with the tomb of James K. Polk; the war memorial building; the Country Music Hall of Fame and museum; Opryland; Ryman Auditorium (original home of the Grand Ole Opry); a replica of Fort Nashborough; and several old churches and antebellum homes, including Belle Meade mansion. Nearby is The HERMITAGE, home of Andrew Jackson. Incorporated as a city 1806, merged with Davidson county 1963.

Nashville, town (2000 population 4,878), ⊙ HOWARD county, SW ARKANSAS, 38 mi/61 km NNE of TEXARKANA; 33°56′N 93°50′W. Railroad junction. Shipping point for agricultural area (poultry; grain); manufacturing (consumer goods, building materials, lumber, plastic products, machinery, poultry products, beverages).

Nashville, town (2000 population 4,697), ⊙ BERRIEN county, S GEORGIA, 26 mi/42 km N of VALDOSTA; 31°12′N 83°15′W. Tobacco market. Manufacturing includes fiberglass products, transportation equipment, clothing, building materials; food canning; lumber. Incorporated 1892.

Nashville, town (2000 population 825), ⊙ BROWN county, S central INDIANA, on SALT CREEK, and 40 mi/64 km S of INDIANAPOLIS; 39°12′N 86°14′W. Densely forested; hilly terrain. Timber. Beautiful scenery here attracts tourists especially in autumn; town has an art gallery and several resident painters. BROWN COUNTY STATE PARK to S; Yellowood State Forest to W. Laid out 1836.

Nashville (NASH-vil), town (□ 1 sq mi/7.8 sq km; 2006 population 4,501), ⊙ NASH county, NE central NORTH CAROLINA, 10 mi/16 km W of ROCKY MOUNT; 35°58′N 77°57′W. Service industries; manufacturing (feeds, mobile homes, fabricated metal products, apparel, lumber; printing and publishing); in agricultural area (cotton, tobacco, peanuts, grain; poultry, livestock; timber. Settled 1780, chartered 1815.

Nashville, village (2000 population 111), KINGMAN county, S KANSAS, 21 mi/34 km SW of KINGMAN; 37°26′N 98°25′W. In wheat area.

Nashville, village (2000 population 1,684), BARRY county, SW MICHIGAN, 10 mi/16 km ESE of HASTINGS, and on THORNAPPLE river; 42°36′N 85°05′W. In agricultural area; manufacturing tool and die. Plotted 1865, incorporated 1869.

Nashville, village (2006 population 184), HOLMES county, central OHIO, 11 mi/18 km WNW of MILLERSBURG; 40°36′N 82°07′W.

Nashwaak River (NASH-wok), 70 mi/113 km long, central New Brunswick, E Canada; issues from the small Nashwaak Lake, 30 mi/48 km NE of Woodstock; flows SE and S, past MARYSVILLE, to St. John River opposite FREDERICTON.

Nashwaaksis (NASH-wok-sis), former village, S central New Brunswick, E Canada, on St. John River, at mouth of Nashwaak River, just NW of FREDERICTON; 45°58′N 66°40′W. Amalgamated with Fredericton in 1973.

Nashwauk (NASH-wawk), town (2000 population 935), ITASCA county, NE MINNESOTA, 13 mi/21 km WSW of HIBBING, in MESABI IRON RANGE; 47°22′N 93°10′W. Iron mining; timber; manufacturing (pulpwood). Growth followed discovery of iron nearby in early 1900s. George Washington State Forest to N; Hill Annex Mine State Park to W. SWAN LAKE to S.

Nasia (nah-SEE-uh), town, NORTHERN REGION, GHANA, on main road from TAMALE to BOLGATANGA, 20 mi/32 km S of Walewale, on Nafia River; 10°09′N 00°48′W. Livestock; groundnuts, shea-nut butter.

Nasia River (nah-SEE-uh), 70 mi/113 km long, NORTHERN REGION, GHANA; rises S of Gambaga Scape, 20 mi/32 km N of Gushiago; flows W into Nakanbe River; 09°57′N 00°58′W.

Našice, town, E CROATIA, 24 mi/39 km N of SLAVONSKI BROD, at N foot of KRNDIJA mountain, in SLAVONIA; 45°30′N 18°06′E. Railroad junction; woodworking, tannin manufacturing; fishing. Castle.

Nasielsk (NAH-shelsk), Russian *Nasyelsk* or *Nasyel'sk*, town, CIECHANÓW province, E central POLAND, 26 mi/42 km NNW of WARSAW; 52°35′N 20°48′E. Railroad junction; manufacturing (buttons, cement, flour). Before World War II, population 75% Jewish.

Nasigatoka (nah-SING-ah-TO-kah), town, SW VITI LEVU, FIJI, SW PACIFIC OCEAN, 60 mi/97 km W of SUVA, near SIGATOKA mouth. Sugarcane, copra. Access to "Coral Coast" tourism. Sometimes spelled Nasingatoka.

Näsijärvi (NA-see-YAR-vee), lake, SW FINLAND, extends N from TAMPERE; 20 mi/32 km long, 2 mi/3.2 km–8 mi/12.9 km wide. Drained S into Lake PYHÄJÄRVI by Tammerkoski rapids (in Tampere).

Nasingatoka, FIJI: see NASIGATOKA.

Nasipit (nah-SEE-peet), town, AGUSAN DEL NORTE province, N MINDANAO, PHILIPPINES, on Butuan Bay of MINDANAO SEA, 12 mi/19 km W of Butuan; 08°57′N 125°19′E. Port, large timber-processing complex (plywood, wallboard).

Nasirabad (nuh-SEE-rah-bahd), town, JALGAON district, MAHARASHTRA state, W central INDIA, 5 mi/8 km E of JALGAON. Trades in cotton, millet, wheat; cotton ginning, glass-bangle manufacturing.

Nasirabad, town, AJMER district, RAJASTHAN state, NW INDIA, 11 mi/18 km SSE of AJMER. Market center for millet, corn, wheat; manufacturing (condiments, handicraft cloth); exports mica. Was in former AJMER state.

Nahr 'Umar (NA-huhr oo-MUHR), town, BASRA province, SE IRAQ, on the W bank of the SHATT AL ARAB, and 20 mi/32 km NNW of BASRA; 30°45′N 47°40′E. Oil well begun here in 1947 and began producing in 1949; it is currently not in operation. Badly damaged in the IRAN-Iraq War.

Nasiriyah, An, IRAQ: see AN NASIRIYAH.

Nasirpur, PAKISTAN: see NASARPUR.

Nasirya, Al (NAH-ser-EE-ye, el), village, TRIPOLITANIA region, NW LIBYA, on road, 25 mi/40 km SW of TRIPOLI, in JIFARAH plain. Agriculture settlement (grain, vegetables, olives, nuts, fruit). Founded 1938–1939 by Italians, who left after World War II, replaced by Libyan population. Formerly Giordani.

Naskaupi River (nas-KO-pee), 180 mi/290 km long, NEWFOUNDLAND AND LABRADOR, E Canada; issues from small lake 30 mi/48 km E of MICHIKAMAU LAKE at 54°08′N 63°05′W; flows E and SE, through GRAND LAKE (30 mi/48 km long, 2 mi/3 km wide), to W end of Lake Melville, 20 mi/32 km NE of Goose Bay air base. On its upper course are waterfalls.

Naskeag Point (NAS-keg), HANCOCK county, S MAINE, 3-mi/4.8-km peninsula in BLUE HILL BAY, S of BROOKLIN. British raid repulsed here, 1778. Naskeag village is in Brooklin town.

Nason, city (2000 population 234), JEFFERSON county, S ILLINOIS, 10 mi/16 km SSW of MOUNT VERNON; 38°10′N 88°58′W. In agricultural area (wheat, sorghum; cattle). REND LAKE nearby.

Nasonville, village, in BURRILLVILLE, PROVIDENCE county, N central RHODE ISLAND, 0.25 mi/0.40 km W of Slatersville Reservoir. Little Flower Home here.

Nasosny, AZERBAIJAN: see GADZI ZEINALABDUN.

Nasretdin-Bek, UZBEKISTAN: see BOZ.

Nasriganj (NAHS-ree-guhnj), town, ROHTAS district, SW BIHAR state, E INDIA, between SON RIVER and ARA CANAL (branch of SON CANALS system), 40 mi/64 km SSW of ARA. Rice, gram, barley; oilseed, flour, and sugar milling.

Nasrullaganj (nuhs-RUHL-lah-guhnj), town (2001 population 17,240), SEHORE district, MADHYA PRADESH state, central INDIA, 40 mi/64 km SSW of BHOPAL. Wheat, cotton, oilseeds. Was in S part of former BHOPAL state.

Nassaberg, CZECH REPUBLIC: see NASAVRKY.

Nassan (nah-SAHN), village, BAJAH province, N TUNISIA, near MILYANA RIVER, on railroad, 7 mi/11.3 km S of TUNIS. Vineyards; agriculture.

Nassarawa, NIGERIA: see NASARAWA.

Nassau (NUHS-sou), former duchy, W central GERMANY, situated N and E of the MAIN and RHINE rivers. Most of its area is now included in the state of HESSE, with a section within the state of RHINELAND-PALATINATE.

Nassau (NA-saw), county (□ 725 sq mi/1,885 sq km; 2006 population 66,707), extreme NE FLORIDA, on ATLANTIC OCEAN (E) and GEORGIA (N) state line (formed by ST. MARYS RIVER); ⊙ FERNANDINA BEACH; 30°36′N 81°46′W. Lowland area, with AMELIA ISLAND (barrier) to E. Agriculture (poultry; dairy products; corn), forestry (naval stores, lumber; paper), and fishing. County is part of the greater JACKSONVILLE metropolitan area. Formed 1824.

Nassau, county (□ 285 sq mi/741 sq km; 2006 population 1,325,662), SE NEW YORK, on W LONG ISLAND, between LONG ISLAND SOUND (N) and the Atlantic Ocean (S) and immediately E of NEW YORK CITY (QUEENS borough); ⊙ MINEOLA; 40°43′N 73°35′W. Bordered E by SUFFOLK county. Part of New York city metropolitan area; chiefly residential, with many suburban retail centers including ROOSEVELT FIELD shopping complex NW of LEVITTOWN. Deeply indented N shore has country-estate communities; yachting, fishing. The S shore has resorts; its bays, with many marshy islands, are sheltered from the Atlantic by fine barrier beaches (swimming, surf fishing) which are linked to Long Island by causeways; JONES BEACH State Park is here. County is traversed by several major highways and is served by several lines of Long Island railroad. Includes several state parks (notably Bethpage, Jones Beach), several racetracks (notably Belmont Park, in W). Pollution, dredging, and overfishing have combined to severely limit the once-major shellfish industry. Urbanization and residential tracts and industry have likewise excluded commercial agriculture, with the exception of small operations. Diversified manufacturing. Named for William of Nassau, Prince of Orange. First county to adopt charter form of government. Originally the E part of QUEENS county; created January 1, 1899, after the W part of Queens became part of New York city. Nassau Community College, rated one of the best two-year colleges in the country (and part of state university system), is in GARDEN CITY. Other educational institutions include Hofstra and Adelphi universities, SUNY campuses at OLD WESTBURY and FARMINGDALE, C.W. Post campus of Long Island University, and the U.S. Merchant Marine Academy in KINGS POINT.

Nassau, city (2000 population 210,832), ⊙ the BAHAMAS, port on NEW PROVIDENCE island; 25°05′N 77°21′W. Has a large and beautiful harbor; the commercial and social center of the country. Its warm, healthful climate and colorful atmosphere have made it a popular resort. Casino gambling at the two main resort areas in nearby PARADISE ISLAND and Cable Beach. Nassau International Airport is located here. Formerly called Charles Towne, the island was renamed Nassau in 1695. In the 18th century it was a rendezvous for pirates, among them Blackbeard. Three forts, Nassau (1697), Charlotte (1787–1794), and Fincastle (1793), were built to ward off the numerous Spanish invasions. American revolutionists in 1776 captured and held it a short time.

Nassau (NAHS-sou), town, RHINELAND-PALATINATE, W GERMANY, on the LAHN RIVER, 9 mi/14.5 km ESE of KOBLENZ; 50°19′N 07°49′E. Health resort. Has ruined ancestral castle of Nassau family. Town first mentioned 915; chartered 1348. Baron Stein was born here.

Nassau (NA-saw), village (2000 population 83), LAC QUI PARLE county, SW MINNESOTA, on SOUTH DAKOTA state line, 13 mi/21 km WNW of MADISON, near Yellow Bank River; 45°04′N 96°26′W. Grain; manufacturing (feeds, fertilizers).

Nassau, village (2006 population 1,120), RENSSELAER county, E NEW YORK, 15 mi/24 km SSE of TROY; 42°30′N 73°36′W. Light manufacturing. County fairgrounds are here.

Nassau (NA-saw), coral island, N Cook group, S PACIFIC, c.675 mi/1,085 km NW of RAROTONGA. Copra; coconut groves. Placed within NEW ZEALAND administration, 1901–1965; thereafter part of COOK ISLANDS.

Nassau Bay (NA-saw), town (2006 population 4,055), HARRIS county, SE TEXAS, residential suburb 22 mi/35 km SE of downtown HOUSTON, in area referred to as CLEAR LAKE CITY, on Clear Creek (S) and Clear Lake (E); 29°33′N 95°05′W. NASA Johnson Space Center immediately N.

Nassau Bay (NA-saw), TIERRA DEL FUEGO, extreme S CHILE, at the southernmost tip of SOUTH AMERICA, between NAVARINO (N) and WOLLASTON (S) islands; c.60 mi/97 km long, 15 mi/24 km wide.

Nassau Bay, E NEW GUINEA, MOROBE province, E central PAPUA NEW GUINEA, opens on HUON GULF, SOLOMON SEA, 10 mi/16 km S of SALAMAUA.

Nassau, Fort, NEW YORK and NEW JERSEY: see FORT NASSAU.

Nassau River (NA-saw), c.40 mi/64 km long, extreme NE FLORIDA; rises in central NASSAU county; meanders E to Nassau Sound, a small inlet of the ATLANTIC OCEAN between AMELIA (N) and TALBOT (S) islands; lower course an estuary.

Nassawadox (na-suh-WA-duks), town, NORTHAMPTON county, E VIRGINIA, 17 mi/27 km NNE of CAPE CHARLES town, in EASTERN SHORE area, between ATLANTIC OCEAN (E) and CHESAPEAKE BAY (W); 37°28′N 75°51′W. Manufacturing (seafood processing); agriculture (vegetables, grain; poultry, livestock); fish, oysters, crabs.

Nassereith (NAHS-se-reit), village, TYROL, W AUSTRIA, 27 mi/43 km NW of INNSBRUCK at the foot of FERN PASS (N); 47°19′N 10°50′E. Important road junction; seasonal tourism; castles nearby.

Nasser, Lake (NAH-suhr), (□ 1,550 sq mi/4,030 sq km), on the NILE River, SE EGYPT and N SUDAN. It is a vast reservoir for holding Nile waters created by the ASWAN HIGH DAM (built in the 1960s). Extending c.300 mi/475 km behind the dam, it is one of the largest artificial lakes in the world, submerging the second and third cataracts of the Nile. Lake Nasser averages c.6 mi/10 km in width and is 600 ft/183 m deep in places. The lake's rising waters forced over 80,000 Nubian people to relocate and submerged many historic sites; the temple of ABU-SIMBEL, built originally on land that was flooded by the lake, was

dismantled and reconstructed on higher ground. The portion of the lake that extends into Sudan is known as Lake Nubia.

Nassian (NAH-see-an), town, Zanzan region, E CÔTE D'IVOIRE, 51 mi/82 km NW of BONDOUKOU; 08°27′N 03°29′W. Agriculture (yams, bananas, maize, peanuts, coffee).

Nassīr (zul-TAN), rich oil field, CYRENAICA region, N central LIBYA, in SURT BASIN, 185 mi/298 km S of BANGHAZI, on W side of JABAL ZALTAN. Pipeline to Marsa Burayqah terminal.

Nassīr, LIBYA: see JABAL ZALTAN.

Nassiriya, IRAQ: see AN NASIRIYAH.

Nässjö (NES-SHUH), town, JÖNKÖPING county, S SWEDEN, 20 mi/32 km ESE of JÖNKÖPING; 57°39′N 14°42′E. Railroad center on STOCKHOLM-MALMÖ and GÖTEBORG-KALMAR main lines. Manufacturing (metalworking; doors, cables; printing plants). Chartered 1914.

Nassogne (nah-SON-yuh), commune (2006 population 5,008), Marche-en-Famenne district, LUXEMBOURG province, SE central BELGIUM, 7 mi/11.3 km S of MARCHE-EN-FAMENNE, on NW slope of the ARDENNES; 50°08′N 05°21′E. Lumbering.

Nass River (NAS), 236 mi/380 km long, W Canada; rises in the COAST MOUNTAINS, W BRITISH COLUMBIA; flows SW to PORTLAND INLET of the PACIFIC OCEAN; 54°59′N 129°52′W. Navigable for 25 mi/40 km. Valuable salmon fisheries.

Nastätten (NAH-shtet-tuhn), town, RHINELAND-PALATINATE, W GERMANY, 15 mi/24 km SE of KOBLENZ; 50°12′N 07°52′E.

Nasu (NAHS), town, Nasu county, TOCHIGI prefecture, central HONSHU, N central JAPAN, 34 mi/55 km N of UTSUNOMIYA; 37°00′N 140°07′E. Dairying.

Nasugbu (nah-soog-BOO), town, BATANGAS province, S LUZON, PHILIPPINES, on SOUTH CHINA SEA, 45 mi/72 km SW of MANILA; 14°06′N 120°41′E. Agricultural center (rice, sugarcane, corn, coconuts); sugar milling. Beach resorts.

Näsviken (NES-VEEK-en), village, GÄVLEBORG county, E SWEDEN, on Lake Södra Dellen (9 mi/14.5 km long, 1 mi/1.6 km–4 mi/6.4 km wide), 7 mi/11.3 km WNW of HUDIKSVAL; 61°46′N 16°53′E.

Nasworthy, Lake, Texas: see CONCHO RIVER.

Nasyelsk, POLAND: see NASIELSK.

Naszód, ROMANIA: see NĂSĂUD.

Nata, town (2001 population 4,150), CENTRAL DISTRICT, NE BOTSWANA, 110 mi/177 km from FRANCISTOWN; 20°15′S 26°12′E. At junction of roads from KASANE (N) and MAUN (E). Tourist and supply center.

Natá, town, and township, ⊙ Natá district, COCLÉ province, central PANAMA, in PACIFIC lowland, 16 mi/26 km SW of PENONOMÉ, on INTER-AMERICAN HIGHWAY. Corn, rice, beans, sugarcane; livestock raising; condensed milk factory. Founded 1520, it is one of oldest towns in Panama. Former capital of Coclé province.

Nata (NAH-tah), village, MARA region, N TANZANIA, 55 mi/89 km SE of MUSOMA, near Ruwana River, and N of SERENGETI NATIONAL PARK; 02°01′S 34°28′E. Cattle, sheep, goats; cotton, corn, wheat.

Nátaga (NAH-tah-gah), town, ⊙ Natagá municipio, HUILA department, S central COLOMBIA, 45 mi/72 km SW of NEIVA, in the Cordillera CENTRAL; 02°32′N 75°49′W. Elevation 3,809 ft/1,160 m. Coffee, plantains; livestock. Also spelled Nátaja.

Natagaima (nah-tah-GEI-mah), town, ⊙ Natagaima municipio, TOLIMA department, W central COLOMBIA, landing on MAGDALENA River, on railroad, and 50 mi/80 km SE of IBAGUÉ; 03°37′N 75°06′W. Copper-mining and agricultural center (bananas, plantains, coffee, sugarcane, sorghum, corn, rice). Gold mines nearby.

Nátaja, Colombia: see NÁTAGA.

Natal (NAH-tahl) [Portuguese=nativity], city (2007 population 774,205), ⊙ RIO GRANDE DO NORTE state,

NE BRAZIL, just above the mouth of the POTENGI RIVER; 05°47′S 35°13′W. Its port is important in the handling of coastal shipping and in the export of the state's tungsten; also some light industry in the city. Beaches; modern city that has retained its colonial flavor. Occupied by the Dutch 1633–1654 and again in 1817. Briefly the seat of a republican government until it was suppressed by imperial authorities. It grew rapidly during World War II, when an airport was built. Natal has several institutions of higher learning. Founded on Christmas Day, 1599. Formerly called Parnamirim.

Natal (nuh-TAL), unincorporated village, SE BRITISH COLUMBIA, W Canada, near ALBERTA border, in ROCKY MOUNTAINS, on Elk River, 19 mi/31 km NE of FERNIE, in East Kootenay regional district; 49°44′N 114°50′W. Elevation 3,782 ft/1,153 m. Open-pit coal mining; dairying; cattle; timber.

Natalbany (nuh-TAHL-buh-nee), unincorporated town (2000 population 1,739), TANGIPAHOA parish, SE LOUISIANA, 4 mi/6.4 km NW of HAMMOND; 30°32′N 90°29′W. In agricultural (cattle) and timber area. On Danville's map of 1732.

Natalbany River (nuh-TAHL-buh-nee), c.45 mi/72 km long, SE LOUISIANA; rises in E SAINT HELENA parish near MISSISSIPPI state line; flows S to LAKE MAUREPAS, receives TICKFAW river 2 mi/3 km above its mouth; 30°21′N, 90°29′W. Catfish, crawfish.

Natal Bay, KWAZULU-NATAL province, SOUTH AFRICA, inlet of INDIAN OCEAN; 5 mi/8 km long, 2 mi/3.2 km wide; 29°52′S 31°02′E. Durban city is on it. Narrow entrance protected by breakwaters and a bluff on the S side of the entrance to the bay. Largest natural harbor in the republic with 7.5 mi/12 km of quayside handles highest tonnage.

Natal Drakensberg Park or **Natal National Park** (□ 771.8 sq mi/1,999 sq km), W KWAZULU-NATAL province, SOUTH AFRICA, on LESOTHO and FREE STATE province borders, in DRAKENSBERG range. Comprises 10 formerly separate contiguous parks: Bushman's Nek, Cathedral Peak, Cobham, Garden Castle, Giant's Castle, Highmoor, Kamberg, Loteni, Monk's Cowl, and Vergelegen; this combined park includes some of the most spectacular mountain scenery in Africa. Also included are 2 other parks adjoining each other to the N, 55 mi/89 km WSW of LADYSMITH: Rugged Glen and Royal Natal national parks, crossed by TUGELA RIVER, which here drops 5,143 ft/1,568 m in a series of high falls, the highest is 2,010 ft/613 m, and flows through deep gorge. On W border of the Royal Natal National Park is MONT-AUX-SOURCES (10,822 ft/3,299 m), which is accessed by a 14.5-mi/23-km hiking path.

Natales, CHILE: see PUERTO NATALES.

Natalia (NUH-tahl-yuh), town (2006 population 1,847), Medina county, SW TEXAS, 25 mi/40 km SW of SAN ANTONIO; 29°11′N 98°51′W. In irrigated farm area (vegetables, peanuts; grain; cotton); manufacturing.

Natalicio Talavera (nah-tah-LEE-see-o tah-lah-VAI-rah), town, Guairá department, S PARAGUAY, 9 mi/14 km NNE of VILLARRICA; 25°37′S 56°18′W. Lumbering; agricultural (maté, sugarcane, fruit; livestock). Founded 1918.

Natal'insk (nah-TAHL-yeensk), town (2006 population 2,050), SW SVERDLOVSK oblast, W central URALS, extreme E European Russia, near the UFA RIVER, on road, 11 mi/18 km SSE of (and administratively subordinate to) KRASNOUFIMSK; 56°28′N 57°52′E. Elevation 652 ft/198 m. Glassworking. Formerly known as Natal'inskiy Zavod.

Natalio (nah-TAH-lee-o), town, ITAPUA department, SE PARAGUAY, 60 mi/97 km NE of Encarnación; 26°44′S 54°56′W. Agricultural center (soybeans, wheat).

Natal National Park, SOUTH AFRICA: see NATAL DRAKENSBERG PARK.

Natanz (nah-TAHNZ), town, Esfahān province, N central IRAN, on road, and 45 mi/72 km SE of KASHAN; 33°31′N 51°54′E. Grain, tobacco, fruit (celebrated for its pears). Woodworking, pottery manufacturing; sericulture. Coal and copper deposits nearby.

Nata River, intermittent stream, c. 200 mi/322 km long, ZIMBABWE and BOTSWANA; rises c. 30 mi/48 km SW of BULAWAYO, SW Zimbabwe; flows WNW through MATABELELAND NORTH province, entering Botswana; passes at the S tip of HWANGE NATIONAL PARK, in Makgadikgadi Pans, where it sinks into the desert landscape; no outlet. Also called Amanzamnyama River.

Natashkwan (nat-ash-KWAHN), river, 241 mi/388 km long, E Canada; rises in NEWFOUNDLAND AND LABRADOR; flows S across E QUEBEC to the Gulf of St. Lawrence; 50°06′N 61°49′W. Noted for trout and salmon fishing. Iron-bearing sands found along its banks are mined. Also spelled Natashquan.

Natasho (nah-TAH-sho), village, Onyu county, FUKUI prefecture, central HONSHU, W central JAPAN, 56 mi/90 km S of FUKUI; 35°23′N 135°41′E. Pickles. Tomb of early emperor Tshuchimikado (creator of traditional calendar) and his family.

Natashquan (nuh-TASH-kwuhn), canton (□ 75 sq mi/195 sq km; 2006 population 380), Côte-Nord region, E QUEBEC, E Canada; 50°10′N 61°45′W. Port facility.

Natazhat, Mount (13,435 ft/4,095 m), S ALASKA, in SAINT ELIAS MOUNTAINS, 150 mi/241 km NNW of YAKUTAT; 61°31′N 141°09′W.

Natchaug River (NAT-chahg), c.25 mi/40 km long, NE CONNECTICUT; rises near MASSACHUSETTS state line; flows SW, joining WILLIMANTIC RIVER to form SHETUCKET RIVER at WILLIMANTIC. Mansfield Hollow Dam (for flood control) is just above Willimantic.

Natchez (NA-chez), city (2000 population 18,464), ⊙ ADAMS county, SW MISSISSIPPI, 85 mi/137 km SW of JACKSON, on the MISSISSIPPI RIVER (bridged), opposite VIDALIA (LOUISIANA); 31°33′N 91°23′W. Railroad terminus. It is the trade, shipping, and processing center for a cotton, livestock, and timber area. Agriculture (soybeans, corn, cotton); timber; manufacturing (printing and publishing, steel fabrication, pulpwood and lumber processing; transportation equipment, machinery, building materials). One of the oldest towns on the Mississippi River, Natchez was founded in 1716 when Fort Rosalie was established; in 1729 the Native American Natchez attacked and killed the garrison members. The area passed to ENGLAND (1763), to SPAIN (1779), and to the U.S. (1798). Capital of the Mississippi Territory, 1798–1802. The city became a great river port and the cultural center of the planter aristocracy before the Civil War. Served as state capital 1817–1821. In the Civil War it was taken by Federal forces in 1863. The city has preserved its antebellum charm, and of over 500 historic homes, about twenty are available for touring and are visited during the annual festival period in March and April. NATCHEZ NATIONAL HISTORICAL PARK; St. Catherine Creek National Wildlife Refuge to SW; Homochitto National Forest to SE; SW terminus of NATCHEZ TRACE PARKWAY; Natchez National Cemetery to N; Natchez Museum of African-American History and Culture; Mostly African Market, with arts and crafts; Grand Village of the Natchez (1682–1729); Emerald Mound (c.1300 A.D.) and Natchez State Park to NE; Old South Winery. Settled 1716, incorporated 1803.

Natchez (NA-chez), village (2000 population 583), NATCHITOCHES parish, NW central LOUISIANA, 8 mi/13 km SSE of NATCHITOCHES; 31°40′N 93°03′W. Kisatchie National Forest and Red Dirt National Wildlife Refuge to S. Timber. In agricultural area. Catfish, crawfish.

Natchez National Historical Park, ADAMS county, SW MISSISSIPPI. One of the best preserved concentrations of antebellum buildings in the U.S. Interprets history of NATCHEZ area. Includes Melrose Mansion and William Johnson House. Covers 80 acres/32 ha. Authorized 1988.

Natchez Trace, historic road, in TENNESSEE, ALABAMA, MISSISSIPPI, from NASHVILLE (Tennessee) to NATCHEZ (Mississippi), of great commercial and military importance from the 1780s to the 1830s. Grew from a series of Native American trails (following trails originally traced by migrating buffalos up to 8,000 years ago) later used in the 18th century by French, English, and Spanish traders, trappers, and missionaries. Traveled originally mainly N from Natchez to Nashville; goods were later floated S to NEW ORLEANS by flatboat on CUMBERLAND, OHIO, and MISSISSIPPI rivers, and boatmen then used the Trace to trek back N. Its general use was replaced with development of steamboats, which could ply the rivers in both directions. Its heyday was in the late 1700s and early 1800s. It was made a post road in 1800 and was improved by the army. Andrew Jackson marched over the Trace to New Orleans in the War of 1812. With the coming of steamboat transportation in the 1830s, however, it passed into decline. The NATCHEZ TRACE PARKWAY and the NATCHEZ TRACE NATIONAL SCENIC TRAIL memorialize and generally follow the old Natchez Trace. Meriwether Lewis and ACKIA BATTLEGROUND national monuments were disestablished and incorporated 1961 into Natchez Trace Parkway.

Natchez Trace National Scenic Trail (NACH-ez), 445 mi/716 km long, ALABAMA and TENNESSEE. Trail extends from NASHVILLE (Tennessee), via TUPELO and JACKSON, to NATCHEZ (all in Mississippi). Parallels NATCHEZ TRACE PARKWAY, unit (□ 17 sq mi/44 sq km) of National Park System. Trail is undeveloped. Authorized 1983.

Natchez Trace Parkway, scenic 2-lane drive in central TENNESSEE, follows the route of the old NATCHEZ TRACE trade route between NASHVILLE (Tennessee) and NATCHEZ (MISSISSIPPI). Administered by the National Park Service, the route passes numerous Native American mounds, historical sites, state parks, and campgrounds. From Nashville it goes SSW, passes W of COLUMBIA (Tennessee), and FLORENCE (ALABAMA), and into Mississippi, past TUPELO, KOSCIUSKO, and JACKSON, to Natchez on Mississippi River. Total length is 445 mi/716 km, of which 400 mi/644 km are completed. Construction began in 1937 as a Depression-era relief project, yet incomplete segments still remain near Natchez and Jackson. The parkway has been designed for leisurely recreational driving rather than high speeds.

Natchitoches (NA-ki-tish), parish (□ 1,297 sq mi/3,372.2 sq km; 2006 population 38,719), NW central LOUISIANA; ⊙ NATCHITOCHES; 31°46′N, 93°06′W. Bounded on E by SALINE and RED rivers, on N in part by BLACK LAKE BAYOU and BAYOU PIERRE. Intersected by Red River and Cane River Lake; includes SALINE and BLACK lakes (N), both part of NW Louisiana State Wildlife Area. Agriculture (cotton, corn, sorghum, hay, rice, soybeans; cattle, poultry, catfish, crawfish, alligators; timber. Some manufacturing, including processing of timber, paper products, and farm products. Oil and natural gas deposits. Formed 1807. The parish courthouse has been restored. Red Dirt National Wildlife Refuge in far S, part of Peason Ridge State Wildlife Area in far S, Rebel and Los Adaes State Commemorative Areas in W, Fort Saint Jean Baptiste State Commemorative Area at center, part of Kisatchie National Forest in far NE and S, including Kisatchie Hills Wilderness in S.

Natchitoches (NA-ki-tish), city (2000 population 17,865), ⊙ NATCHITOCHES parish, NW LOUISIANA, 42 mi/68 km NW of ALEXANDRIA; 31°46′N 93°06′W. Industry centered on the production, processing, and shipping of farm products, including cotton, lumber, poultry, and cattle; also catfish, crawfish, alligators. Manufacturing (concrete products, poultry products, lumber, cottonseed oil), printing and publishing. The first permanent settlement in the Louisiana Purchase Territory, Natchitoches was founded c.1714 as a

French military and trading post. It served as the dividing line between French and Spanish territory. The city was an important port on the RED RIVER until the river changed its course in the early 1800s, leaving the 33-mi/53-km meandering riverbed known as Cane River Lake. Occupied by the Union army during the Civil War. The lasting old homes and plantations attract tourists. Northwestern State University of Louisiana is here. SIBLEY LAKE reservoir to W, National Fish Hatchery to S, Fort Saint Jean Baptiste State Commemorative Area to S (reproduction of 1730s French fort). Incorporated 1819.

Naters, commune, VALAIS canton, S SWITZERLAND, 21 mi/34 km ENE of SIERRE; 46°20′N 07°59′E.

Naters, SWITZERLAND: see BRIG.

Natewa Bay (nah-TAI-wah), E VANUA LEVU, FIJI; 35 mi/56 km long, 10 mi/16 km wide. Separates Natewa peninsula from main part of island

Nathalia (nuh-THAIL-yuh), town, N central VICTORIA, SE AUSTRALIA, 26 mi/42 km ENE of ECHUCA, on Murray Valley Highway; 36°43′S 145°12′E. In heart of MURRAY Irrigation Scheme. Serious problems have occurred resulting from salting of topsoil from subsurface. Fruit growing; dairying.

Nathdwara (naht-WAH-ruh), town, UDAIPUR district, S central RAJASTHAN state, NW INDIA, 25 mi/40 km NNE of UDAIPUR, near BANAS RIVER. Markets corn, millet, barley; handicraft jewelry. A walled town; place of Hindu pilgrimage; has noted 17th-century Vishnuite shrine. Mavli, railroad junction, is 14 mi/23 km SE.

Nathiagali (nuht-yah-GAH-lee), town, ABBOTTABAD district, NE NORTH-WEST FRONTIER PROVINCE, N PAKISTAN, 12 mi/19 km ESE of ABBOTTABAD, in extreme W PUNJAB HIMALAYA; 34°04′N 73°24′E. Popular hill resort.

Nathorst, Cape, S extremity of ELLEF RINGNES ISLAND, NUNAVUT territory, N Canada; 77°44′N 100°05′W.

Natick (NAI-dik), town, MIDDLESEX county, E MASSACHUSETTS, residential and industrial suburb of BOSTON, on Lake COCHITUATE; 42°17′N 71°21′W. Founded as a village by John Eliot in 1651, incorporated 1781. Major retailing center. Manufacturing includes electronic components, computer software, medical equipment, clocks, machine parts; food processing. Site of U.S. army research facility. Includes village of South Natick (includes historic district and park on CHARLES RIVER). Site of Cochituate State Park and Broadmoor Wildlife Sanctuary.

Natimuk (nat-i-MUHK), village, W central VICTORIA, SE AUSTRALIA, 175 mi/282 km WNW of MELBOURNE, 16 mi/25 km W of HORSHAM; 36°45′S 141°57′E. Railroad junction; wheat; sheep. Rock climbers visit the sandstone Mount Arapiles (a-ra-PI-leez), 1,168 ft/356 m high. Fishing in Lake Natimuk just N of town.

National Capital District, administrative division (2000 population 254,158), S central PAPUA NEW GUINEA, SE NEW GUINEA island, on GULF OF PAPUA (SW); ⊙ PORT MORESBY. Papua New Guinea's largest city. Bounded by CENTRAL province. Includes Port Moresby's port and suburbs. Main industrial, commercial, and service center for country; rubber plantations in area.

National Capital Region (NCR) (□ 246 sq mi/639.6 sq km), W central LUZON, PHILIPPINES, consisting of MANILA, the adjoining cities of CALOOCAN, PASAY, PASIG, and QUEZON CITY, and twelve towns. Created in 1978 to integrate the country's most populous area and deal with regional problems such as flooding, waste disposal, pollution, and traffic. Governed by a council known as the Metropolitan Manila Authority, the region is referred to as MetroManila (which is also a colloquial term for the entire Manila megalopolis, stretching from BULACAN province in the N to CAVITE province in the S and into RIZAL province in the W).

National City, city (□ 9 sq mi/23.4 sq km; 2000 population 54,260), SAN DIEGO county, S CALIFORNIA, suburb 6 mi/9.7 km SE of downtown SAN DIEGO, on E side of SAN DIEGO BAY; 32°40′N 117°06′W. Manufacturing (fabrication of structural metal, printing and publishing, shipbuilding, recycling machinery; plastics products, transportation equipment, electrical and electronic products, signs, furniture, clothing). Primarily residential, marked by a steady growth in population between 1970 and 1990. Serves as the headquarters of the Pacific Reserve Fleet. SWEETWATER RIVER to S. Incorporated 1887.

National City or **National Stockyards**, village, SAINT CLAIR county, SW ILLINOIS, adjoining EAST SAINT LOUIS (N), industrial suburb of SAINT LOUIS, near MISSISSIPPI RIVER; 38°38′N 90°09′W. Once large stockyards; periodic fires have destroyed most of facilities. Manufacturing (scrap metal, meat waste products). Gateway International Raceway nearby.

National City, village, IOSCO county, NE MICHIGAN, 11 mi/18 km WSW of TAWAS CITY; 44°14′N 83°43′W. Gypsum quarrying and processing.

National Forest System, federally owned reserves (□ 343,000 sq mi/891,800 sq km), administered by the Forest Service of the U.S. Dept. of Agriculture. The system is made up of 154 national forests and 19 national grasslands in 41 states and PUERTO RICO. The majority of reserves are found in the Western states, with ALASKA, IDAHO, and CALIFORNIA having the most extensive holdings. In the East, large national forests are in the Green, WHITE, ALLEGHENY, and BLUE RIDGE mountains. The national grasslands are found on the GREAT PLAINS. By law the reserves must be used for timber production, watershed land, wildlife preservation, livestock grazing, mining, and recreation. In 1891, Congress authorized the president to set aside forest reserves; Yellowstone Park Timber Reserve (now Shoshone National Forest) in WYOMING was the first (1891) to be established. The forest reserves were administered by the General Land Office of the Department of the Interior until 1905, when they were transferred to the Forest Service by President Teddy Roosevelt. They were designated national forests in 1907.

National Mall, landscaped park, WASHINGTON, D.C., part of the city's L'Enfant Plan. Extends from the U.S. Capitol (E) to the WASHINGTON MONUMENT (W), between Constitution and Independence avenues. The mall is the location of most Smithsonian Institute buildings Authorized 1933.

National Park (NA-shuhn-uhl PAHRK), town, TASMANIA, SE AUSTRALIA, 48 mi/77 km NW of HOBART; 42°41′S 146°44′E. Access to Mount Field National Park.

National Park, residential borough (2006 population 3,215), GLOUCESTER county, SW NEW JERSEY, on DELAWARE RIVER, and 7 mi/11.3 km SW of CAMDEN; 39°52′N 75°11′W. Monument here commemorates Revolutionary battle of Red Bank (1777), named for former locality here. Incorporated 1902.

National Park Lake Neusiedl-Seewinkel, Central EUROPE: see LAKE NEUSIEDL.

National Park of North and East Greenland, The (□ 375,289 sq mi/972,000 sq km), NE GREENLAND, from the interior parts of SCORESBYSUND in the SE, to HALL LAND in the NW. One of the world's largest national parks. Founded by decree in 1987 to conserve the wilderness of the region and simultaneously allow research and public access.

National Road, U.S. highway, built in the early 19th century. At the time of its construction, the National Road was the most ambitious road-building project ever undertaken in the U.S. It finally extended from CUMBERLAND (MARYLAND) to SAINT LOUIS and was the great highway of Western migration. Agitation for a road to the West began c.1800. Congress approved the route and appointed a committee to plan details in 1806. Contracts were given in 1811, but the War of 1812 intervened, and construction did not begin until 1815. The first section (called the Cumberland Road) was built of crushed stone. Opened in 1818, it ran from Cumberland to WHEELING (WEST VIRGINIA), following in part the Native American trail known as Nemacolin's Path. Largely through the efforts of Henry Clay it was continued (1825–1833) W through OHIO, using part of the road built by Ebenezer Zane. By this time the older part of the road was badly in need of repair. Control of the road was therefore turned over to the states through which it passed, where tolls for maintenance were collected. It was carried on to VANDALIA (ILLINOIS), and finally to Saint Louis. The old route became part of U.S. Highway 40. At points on the road copies of a statue called the *Madonna of the Trail* have been erected to honor the pioneer women who went West over the National Road.

National Stockyards, ILLINOIS: see NATIONAL CITY.

Natitingou (nah-tee-TEEN-goo), town, ⊙ ATAKORA department, NW BENIN, on road, 110 mi/177 km NW of PARAKOU; 10°19′N 01°22′E. An important link to BURKINA FASO. Shea nuts, peanuts, kapok, millet, corn.

Native States, INDIA: see BRITISH INDIA.

Natividad (nah-tee-vee-DAHD), town, in W central OAXACA, MEXICO, 31 mi/50 km NE of OAXACA DE JUÁREZ. Elevation 6,234 ft/1,900 m.

Natividade (NAH-chee-vee-dah-zhe), city (2007 population 14,925), N RIO DE JANEIRO state, BRAZIL, 11 mi/18 km NNW of ITAPERUNA; 41°03′S 41°59′W. Also called Natividade de Carangola.

Natividade (NAH-chee-vee-dah-zhee), town (2007 population 9,090), S TOCANTINS state, central BRAZIL, on a tributary of the TOCANTINS River, and 80 mi/129 km SE of PÔRTO NACIONAL; 11°43′S 47°47′W. Cattle raising; dairying; sugar, corn. Nitrate deposits.

Natividade, Brazil: see NATIVIDADE DA SERRA.

Natividade da Serra (NAH-chee-vee-dah-zhee dah SE-rah), town (2007 population 7,359), SE SÃO PAULO state, BRAZIL, in the Serra do MAR, 25 mi/40 km S of TAUBATÉ; 23°24′S 45°26′W. Brandy distilling; agriculture (sugarcane, grain, tobacco). Until 1944, Natividade.

Natividade do Carangola (nah-chee-vee-dah-zhee do kah-rahn-go-lah), city, NE MINAS GERAIS state, BRAZIL, on railroad, and 12 mi/19 km NNW of ITAPERUNA; 20°40′S 42°03′W. Coffee-growing center. Until 1944, called Catangda.

Natividade do Carangola, Brazil: see NATIVIDADE.

Natividad Island (nah-tee-vee-DAHD), islet (□ 3 sq mi/7.8 sq km) off PACIFIC OCEAN coast of BAJA CALIFORNIA SUR, NW MEXICO, at SW edge of SEBASTIÁN VIZCAÍNO BAY; 3.75 mi/6.03 km long, 0.5 mi/0.8 km– 1.5 mi/2.4 km wide; 27°50′N 115°11′W. Elevation 492 ft/ 150 m. Barren; rocky.

Natívitas, MEXICO: see SANTA MARÍA NATÍVITAS.

Natkrstac, BOSNIA AND HERZEGOVINA: see VRANICA MOUNTAINS.

Natmauk (NAHT-mouk), village and township, MAGWE division, MYANMAR, on Pyinmana-Kyaukpadaung railroad, and 34 mi/55 km ENE of MAGWE. Cotton, sesame, millet, some rice; salt extraction.

Natoena Islands, INDONESIA: see NATUNA ISLANDS.

Natogyi (nah-to-JEE), village and township, MANDALAY division, MYANMAR, on Myingyan-Mandalay railroad, and 17 mi/27 km E of MYINGYAN. Trade center in cotton-growing area.

Natoma (nuh-TO-muh), village (2000 population 367), OSBORNE county, N central KANSAS, on small affluent of SALINE RIVER, and 23 mi/37 km NNW of RUSSELL; 39°11′N 99°01′W. Grain; livestock.

Nator (nah-tor), town (2001 population 70,835), RAJSHAHI district, W EAST BENGAL, BANGLADESH, on distributary of the ATRAI RIVER, and 24 mi/39 km E of RAJSHAHI; 24°35′N 88°59′E. Silk weaving factory; rice, jute, oilseeds, wheat. Large palace of Nator rajas. Former district capital. Large sugar-processing factory 14 mi/23 km S, at Gopalpur.

Area is shown by the symbol □, and capital city or county seat by ⊙.

Natori (NAH-to-ree), city, MIYAGI prefecture, N HONSHU, NE JAPAN, 6 mi/10 km S of SENDAI; 38°10′N 140°53′E.

Natrona (nuh-TRO-nuh), county (□ 5,375 sq mi/13,975 sq km; 2006 population 70,401), central WYOMING; ⊙ CASPER; 42°58′N 106°47′W. Livestock and mining region; watered by North Platte and Sweetwater rivers, and headstreams of POWDER RIVER, ALCOVA RESERVOIR and part of PATHFINDER RESERVOIR in S. Agriculture (hay, alfalfa; cattle, sheep); oil, coal. N end of LARAMIE MOUNTAINS in SE corner; Hogaden Ski Area in SE. S end of BIGHORN MOUNTAINS in NW corner. Edness Kimball Wilkins State Park in E. Part of Pathfinder National Wildlife Refuge in S. Formed 1888.

Natrona (nah-TRO-nah), village, Harrison township, ALLEGHENY county, W PENNSYLVANIA; suburb 19 mi/31 km NE of downtown PITTSBURGH, on ALLEGHENY RIVER; 40°36′N 79°43′W. Manufacturing (fabricated metal products, chemicals). Saltworks founded here 1853.

Natrona Heights (nah-TRO-nah), unincorporated city, ALLEGHENY county, W PENNSYLVANIA; residential suburb 19 mi/31 km NE of PITTSBURGH; 40°37′N 79°43′W. Manufacturing includes commercial printing; tool and die, plastic products, building materials. Dairying; livestock, poultry; grain to N.

Natron, Lake (NAH-tron), in the GREAT RIFT VALLEY, E AFRICA, on the KENYA-TANZANIA border; in N Tanzania (ARUSHA region) and Kenya (RIFT VALLEY province); c.35 mi/56 km long and 15 mi/24 km wide. It has soda, salt, and magnesite deposits. Nkuruman Escarpment to NW; elevation 5,282 ft/1,610 m. Used by flamingoes for nesting.

Natron Lakes, EGYPT: see NATRUN, WADI EN.

Natrun, Wadi en (naht-ROON, WAH-dee ahn), valley, WESTERN DESERT province, N EGYPT, c.60 mi/97 km SSE of ALEXANDRIA, and 65 mi/105 km WNW of CAIRO; c.30 mi/48 km long, c.4 mi/6.4 km wide. Has several shallow, swampy lakes, sometimes called the Natron Lakes, as much as 25 ft/8 m below sea level, which are rich in soda deposits exploited since ancient times. A center of Coptic Christianity; several monasteries.

Nattalin (nah-tah-LIN), town and township, BAGO division, MYANMAR, 35 mi/56 km SSE of PROME, on railroad to YANGON.

Nattam (NUHT-tuhm), town, MADURAI district, TAMIL NADU state, S INDIA, 22 mi/35 km NNE of MADURAI. Road center in millet- and sesame-growing area.

Nattam Hills, INDIA: see SIRUMALAI HILLS.

Nattandiya (NAHTH-thuhn-di-yuh), town, NORTH WESTERN PROVINCE, SRI LANKA, 12 mi/19 km SSE of CHILAW; 07°25′N 79°52′E. Glass manufacturing; coconuts, rice.

Nattarasankottai (NAH-tuh-ruh-suhn-KO-tei), town, RAMANATHAPURAM district, TAMIL NADU state, S INDIA, 30 mi/48 km E of MADURAI. In cotton area; peanut-oil extraction. Also spelled Nattarasankotai.

Nattheim (NAHT-heim), village, E WÜRTTEMBERG, BADEN-WÜRTTEMBERG, SE GERMANY, 12 mi/19 km SSE of AALEN, in SWABIAN JURA; 48°42′N 10°14′E. Forestry; manufacturing (textiles).

Natuba (NAH-too-bah), city (2007 population 10,040), SE PARAÍBA state, BRAZIL, on PERNAMBUCO state border, 17 mi/27 km NE of UMBUZEIRO; 07°38′S 35°35′W.

Natukhayevskaya (nah-too-HAH-eefs-kah-yah), village (2005 population 6,460), SW KRASNODAR TERRITORY, S European Russia, on highway junction, 14 mi/23 km ESE of ANAPA, and 13 mi/21 km WNW of NOVOROSSIYSK; 44°54′N 37°34′E. Elevation 150 ft/45 m. In agricultural area (grain, fruits, grapes; livestock).

Natuna Islands (NAH-too-nah) or **Natuna Besar** (nah-TOO-nah buh-SAHR), island group, RIAU province, INDONESIA, in SOUTH CHINA SEA, between West and East MALAYSIA, and 185 mi/298 km E of ANAMBAS ISLANDS; 02°56′N–04°42′S and 102°15′E–108°00′E. Comprised of North and SOUTH NATUNA Islands. North Natuna Islands includes GREAT NATUNA (or Laut) island (40 mi/64 km long, 31 mi/50 km wide; largest in group). South Natuna Islands includes Subi, Panjang, and SERASAN islands. Generally low and wooded. Chief products are timber and coconuts; peanuts and green peas are also grown. Exploration for natural gas is going on in the waters of the adjacent South China Sea. Some basket-weave cloth is made. Also spelled Natoena Islands

Natupe, TUAMOTU ARCHIPELAGO: see REAO.

Natural Bridge, village, JEFFERSON county, N NEW YORK, 8 mi/12.9 km NE of CARTHAGE; 44°04′N 75°30′W. INDIAN RIVER has cut through limestone here to form a bridge and caverns that are a popular tourist attraction.

Natural Bridge (NA-truhl BRIJ), unincorporated village, ROCKBRIDGE county, NW VIRGINIA, in SHENANDOAH VALLEY, 12 mi/19 km SSW of LEXINGTON; 37°37′N 79°32′W. Manufacturing (industrial prototypes); agriculture (dairying; livestock; grain, apples). Nearby, over gorge of CEDAR CREEK, is famous Natural Bridge, a limestone arch 215 ft/66 m high with a span of 90 ft/27 m, once owned by Thomas Jefferson, who built a visitors' cabin and kept a guest book. A public highway now crosses the bridge. Founded 1774.

Natural Bridges National Monument (□ 12 sq mi/31 sq km), central SAN JUAN county, SE UTAH. Located in an area of colored cliffs and box canyons, the monument contains three huge natural sandstone bridges: Owachomo (also called Rock Mound), 106 ft/32 m high with a span of 180 ft/55 m; Kachina, 210 ft/64 m high with a span of 206 ft/63 m; and Sipapu, 220 ft/67 m high with a span of 268 ft/82 m. Native American Anasazi ruins under one bridge. Drained by White Canyon. Established 1908.

Natural Bridge State Resort Park (□ 3 sq mi/7.8 sq km), POWELL and WOLFE counties, E central KENTUCKY, 10 mi/16 km ESE of STANTON, just SE of SLADE, surrounded by Daniel Boone National Forest. Wooded, hilly area, with lodge and recreational facilities, nature preserve; chief attraction is Kentucky Natural Bridge, a span of Paleozoic limestone; largest natural bridge in Kentucky; arch has c.90-ft/27-m clearance, is c.75 ft/23 m wide. The cliffs and overhangs are home to VIRGINIA big-eared Bat.

Naturaliste, Cape (NA-chuhr-uh-list), SW WESTERN AUSTRALIA state, W AUSTRALIA, in INDIAN OCEAN; W headland of GEOGRAPHE BAY; 33°32′S 115°01′E. Lighthouse.

Naturaliste Channel (NA-chuhr-uh-list), small strait of INDIAN OCEAN, W WESTERN AUSTRALIA state, W AUSTRALIA, between DORRE and DIRK HARTOG islands; forms W entrance to SHARK BAY; 15 mi/24 km wide; 25°23′S 113°02′E.

Natural State, The: see ARKANSAS.

Natural Tunnel (NA-chruhl), SCOTT county, SW VIRGINIA, natural passageway on SE side of POWELL MOUNTAIN, 10 mi/16 km WNW of GATE CITY, in Natural Tunnel State Park (603 acres/244 ha); 75 ft/23 m–100 ft/30 m high, c.900 ft/274 m long; 36°42′N 82°44′W. Small stream and railroad pass through it.

Naturita, village (2000 population 635), MONTROSE county, SW COLORADO, at mouth of Naturita Creek, on SAN MIGUEL RIVER, and 40 mi/64 km SW of MONTROSE; 38°13′N 108°34′W. Elevation 5,431 ft/1,655 m. Uranium and vanadium mined here; also coal. Cattle, sheep; wheat, beans, corn. Uncompahgre National Forest to NE.

Naturno (nah-TOOR-no), German *Naturns*, village, BOLZANO province, TRENTINO-ALTO ADIGE, N ITALY, on ADIGE River, and 7 mi/11 km W of MERANO; 46°39′N 11°00′E. Agr. (fruit); organ manufacturing. Summer camping resort.

Natyrbovo (nah-TIR-buh-vuh), village, N central ADYGEY REPUBLIC, NW CAUCASUS, S European Russia, at the N foot of the Greater Caucasus Mountains, on road and near railroad, 44 mi/71 km SW of STAVROPOL; 44°44′N 40°37′E. Elevation 734 ft/223 m. Livestock; food processing.

Nau (NAW), town, central LENINOBOD region, TAJIKISTAN, on railroad (Pridonovo station), and 16 mi/26 km SW of KHUDJAND; 40°09′N 69°22′E. Cotton-growing and ginning; sericulture. Fell to Russians in 1866. Tertiary level administrative center.

Naubinway (NAW-bin-wai), village, MACKINAC county, SE UPPER PENINSULA, N MICHIGAN, 39 mi/63 km NW of SAINT IGNACE, on LAKE MICHIGAN; 46°05′N 85°26′W. Resort.

Naubise (naw-BEE-sai), village, central NEPAL, at the junction of the TRIBHUVAN RAJPATH (S to INDIA) and the MAHENDRA RAJMARG (W to POKHARA) roads; 27°44′N 85°07′E. Elevation 3,100 ft/945 m.

Naucalpan de Juárez (nah-oo-KAHL-pahn dai HWAH-res), city (2005 population 792,226) and township, ⊙ Naucalpan municipio, MEXICO state, S central MEXICO; 19°29′N 99°14′W. It is an industrial extension of MEXICO CITY.

Naucelle (no-SEL), commune (□ 8 sq mi/20.8 sq km), AVEYRON department, MIDI-PYRÉNÉES region, S FRANCE, on SÉGALA PLATEAU, 15 mi/24 km SW of RODEZ, in ROUERGUE district; 44°12′N 02°20′E. Livestock (especially sheep); cheese.

Nauchnyy (nah-OOCH-niee) [Russian=scientific], town (2004 population 5,400), S Republic of CRIMEA, UKRAINE, on ridge of the CRIMEAN MOUNTAINS, 8 mi/13 km ESE of BAKHCHYSARAY and 18 mi/29 km S of SIMFEROPOL'; 44°44′N 34°01′E. Elevation 1,610 ft/490 m. Site of the Crimean astrophysical observatory. Established in 1946; present name and town since 1957.

Naucratis (nok-RAH-tis), ancient city of EGYPT, on the former Canopic branch of the NILE River, 13 mi/21 km SE of DAMANHUR. Was probably given (seventh century B.C.E.) by Psamtik to Greek colonists from MILETUS and was the first Greek settlement in Egypt. The rise of ALEXANDRIA and the shifting of the Nile caused its decline. The site has been excavated, revealing Greek-style pottery and ruins of Greek temples. Also spelled Naukratis.

Nauders (NOU-ders), village, TYROL, W AUSTRIA, between RESCHEN and Finstermünz passes, 20 mi/32 km S of LANDECK, near Swiss and Italian borders; 46°54′N 10°30′E. Road junction, border station, resort (elevation 4,249 ft/1,295 m). Old customs point with castles (Naudersberg, 14th century; Hochfinstermünz, 13th century).

Nauen (NOU-en), town, BRANDENBURG, E GERMANY, 22 mi/35 km WNW of BERLIN; 52°37′N 12°53′E. Manufacturing of machinery, dairy and food products; sugar refining. Formerly site of major radio station, dismantled 1945–1947. Has Gothic church. Was first mentioned 981; chartered 1305.

Naugaon, BANGLADESH: see NAOGAON.

Naugaon Sadat (NOU-goun sah-DUHT), town, MORADABAD district, N central UTTAR PRADESH state, N central INDIA, 8 mi/12.9 km NNW of AMROHA. Wheat, rice, pearl millet, mustard, sugarcane. Also spelled Naugawan Sadat, Naugaon Saadat.

Naugard, POLAND: see NOWOGARD.

Naugarzan (nah-oo-gahr-ZAN), town, LENINOBOD viloyat, TAJIKISTAN, on KURAMA RANGE, on border with UZBEKISTAN, 65 mi/105 km NE of KHUDJAND; 40°56′N 70°33′E.

Naugatuck, industrial borough (2000 population 30,989), NEW HAVEN county, SW CONNECTICUT, on both sides of the NAUGATUCK River; 41°29′N 73°02′W. In 1843, Charles Goodyear established a rubber plant here, which became famous for its Goodyear tires. Other manufacturing includes machinery, chemicals, and electrical and metal (especially brass, and copper) products, candy, and medical supplies. Settled 1704, incorporated 1844.

Cross-references are shown in SMALL CAPITALS. The pronunciation guide is shown on page xix. The sources of population figures are shown on page xvii.

Naugatuck (NAW-guh-tuhk), unincorporated village, MINGO county, SW WEST VIRGINIA, 10 mi/16 km NNW of WILLIAMSON, on TUG FORK RIVER. Manufacturing (coal processing). Bituminous-coal mining. Laurel Creek Wildlife Management Area.

Naugatuck, river, 65 mi/105 km long, Connecticut; rises in NW CONNECTICUT; flows S past WATERBURY to the HOUSATONIC River at DERBY. It furnishes water power for the remaining industrial plants along its shores. Thomaston Dam (completed 1960), built for flood control, forms one reservoir on the river. Has a number of dams.

Naugawan Sadat, INDIA: see NAUGAON SADAT.

Nauhcampatépetl, MEXICO: see COFRE DE REROTE NATIONAL PARK.

Nauheim (NOUL-heim), town, S HESSE, central GERMANY, 3 mi/4.8 km NW of GROSS-GERAU; 49°57′N 08°27′E. Agriculture (fruit); manufacturing (motor vehicles; machinery).

Nauheim, Bad, GERMANY: see BAD NAUHEIM.

Naujan (NOU-hahn), town, MINDORO ORIENTAL province, near NE coast of MINDORO island, PHILIPPINES, 10 mi/16 km SE of CALAPAN; 13°15′N 121°12′E. Agricultural center (copra, abaca, rice).

Naujan, Lake (NOU-hahn) (□ 30 sq mi/78 sq km), MINDORO ORIENTAL province, NE MINDORO island, PHILIPPINES, 6 mi/9.7 km SSE of NAUJAN; 8.5 mi/13.7 km long, 4 mi/6.4 km wide; 13°10′N 121°21′E.

Naujoji Vilnia (nou-YO-yi VEEL-nyah), Polish *Nowa Wilejka*, E industrial suburb of VILNIUS, SE LITHUANIA, on right bank of the Vilnia River (small left tributary of the VILIYA RIVER), and 6 mi/10 km E of Vilnius city center; 54°42′N 25°25′E. Railroad junction (repair shops); manufacturing (machine tools, agricultural implements, cement, liquors, paper, prefabricated houses), meatpacking. In Russian Vilnius government until it passed (1921) to POLAND, later (1939) to Lithuania. Inc. 1947 into Vilnius city. Sometimes called Novo-Vilnya.

Naukluft Mountains (7,074 ft/1,965 m), S central NAMIBIA overlooking NAMIB Desert; 24°00′S 16°00′E. Rugged edge of escarpment largely comprising mountain conservation area.

Naukratis, EGYPT: see NAUCRATIS.

Naumburg (NOUM-boorg) or **Naumberg an der Saale** (NOUM-boorg ahn der SAH-le), city, SAXONY-ANHALT, E central GERMANY, on the SAALE RIVER; 51°09′N 11°48′E. Largely industrial; manufacturing includes machine tools, processed food, and textiles. Founded in the 11th century, Naumburg developed as a trade center and joined the HANSEATIC LEAGUE. It passed to SAXONY in 1564 and to PRUSSIA in 1815. The city has retained parts of its medieval walls, a 16th-century city hall; and a 12th–14th-century cathedral with some of the finest sculptures of the German Gothic period. Chartered 1028. Abbreviated as Naumburg/Saale.

Naumburg (NOUM-boorg), town, HESSE, W GERMANY, 14 mi/23 km WSW of KASSEL; 51°15′N 09°07′E. Grain.

Naumburg am Queis, POLAND: see NOWOGRODZIEC.

Naumiestis (NOU-myes-tees), Polish *Władysławów*, Russian *Vladislavov*, city, SW LITHUANIA, on the SHESHUPE, opposite KUTUZOVO (KALININGRAD oblast, RUSSIA), and 43 mi/69 km WSW of KAUNAS. Oilseed pressing, flour milling; goose raising. Passed 1795 to PRUSSIA, 1815 to Russian POLAND; in Suvalki government until 1920. Also spelled Naumiyestis.

Naungpale (NOUNG-pah-LAI), village, Demoso township, KAYAH STATE, MYANMAR, on Toungoo-Loikaw road, on and 10 mi/16 km SSW of LOIKAW. Former capital of Naungpale State.

Naunhof (NOUN-hof), town, SAXONY, E central GERMANY, on Parthe River, 10 mi/16 km ESE of LEIPZIG; 51°17′N 12°35′E. Lignite-mining region; summer resort.

Naupactus, Greece: see NAVPAKTOS.

Naupada, INDIA: see TEKKALI.

Naupaktos, Greece: see NAVPAKTOS.

Naupan (nah-OO-pahn), town, PUEBLA, central MEXICO, in SIERRA MADRE ORIENTAL, 5 mi/8 km NW of HUAUCHINANGO; 20°14′N 98°07′W. No all-weather roads. Corn, coffee, sugar.

Nauplia, Greece: see NÁVPLION.

Nauplia, Gulf of, Greece: see ARGOLIS, GULF OF.

Nauplion, Greece: see NÁVPLION.

Nauplion, Gulf of, Greece: see ARGOLIS, GULF OF.

Nauportus, SLOVENIA: see VRHNIKA.

Na'ur (NUH-OOR), township (2004 population 15,439), NW JORDAN, 8 mi/12.9 km SW of AMMAN; 31°53′N 35°50′E. Barley, wheat. Road junction.

Na'ura (NAH-uh-rah), Arab village, ISRAEL, 6 mi/10 km E of AFULA in LOWER GALILEE; 32°36′N 35°23′E. Elevation 213 ft/64 m. Founded in the 15th century, stone tools and other relics have been found from the Iron Age, the Babylonian period (6th century B.C.E.), Roman and Byzantine times.

Naurouse, Col de, FRANCE: see NAUROUZE, COL DE.

Naurouze, Col de (no-ROOZ, KOL duh), divide (634 ft/193 m), on border between AUDE and HAUTE-GARONNE departments, LANGUEDOC-ROUSSILLON region, S FRANCE, separating ATLANTIC from MEDITERRANEAN drainage, between MASSIF CENTRAL (N) and outliers of the PYRENEES MOUNTAINS (S); 43°21′N 01°51′E. It is on BORDEAUX-MARSEILLE road and railroad, and on CANAL DU MIDI (built 1660 with locks; still in use). An invasion route since Roman times. Also known as Seuil (French=threshold) de Naurouze; also spelled Naurouse.

Naurskaya (nah-OOR-skah-yah), village (2005 population 8,560), NW CHECHEN REPUBLIC, S European Russia, on railroad, on the left bank of the TEREK RIVER, opposite NADTERECHNOYE (2 mi/3.2 km to the S), near highway and railroad, 30 mi/48 km NW of GROZNYY; 43°39′N 45°19′E. Elevation 259 ft/78 m. Has a school, hospital, mosque, sports stadium. Russian military base and operational headquarters.

Naurskiy, RUSSIA: see ALPATOVO.

Nauru, republic, on a raised coral island with a shallow enclosed lagoon (c.8 sq mi/20 sq km; 2004 estimated population 12,809; 2007 estimated population 13,528), central PACIFIC, just S of the equator and W of KIRIBATI; ⊙ Yaren; 00°33′S 166°55′E. Formerly called Pleasant Island, it is one of the world's smallest independent states and is a special member of the Commonwealth.

Population

Nauruans are predominantly Micronesian with a very distinct language; there are also laborers from China, India, the Philippines, and other Pacific islands.

Economy

It has been important for its high-grade phosphate deposits, but these reserves have been depleted. Almost 2,000,000 tons/2,204,000 metric tons of phosphates were produced annually and had provided the island with a high per capita income (c.$10,000). After independence, Nauru took control of the deposits from the British Phosphate Commission. Nauru has few resources and must import virtually all goods, including food and water. Nauru has invested much of its phosphate revenue in trust funds and international real estate in order to ease the transition from phosphate mining, which has destroyed nearly 80% of the island. In an attempt to generate income, Nauru became an unregulated offshore banking center, gaining notoriety for money laundering, but left offshore banking in 2003 after threatened with crippling U.S. sanctions. By mid-2004 Nauru faced bankruptcy, and the remaining trust fund assets, mostly Australian property, were seized to pay debts.

History

It was discovered in 1798 by a British captain and annexed in 1888 by Germany. Strip mining was begun by the Germans in 1907. Occupied in 1914 by Australian forces, it was placed (1919) as a League of Nations mandate under Great Britain, Australia, and New Zealand. From August 1942 to 1944 it was occupied by the Japanese. Nauru was UN Trust Territory from 1947 to 1968, when it became independent. Australia, which controlled Nauru from 1947 to 1968, agreed to pay for the rehabilitation of the one-third of the island that was mined during that period. Nauru also has received Australia aid in exchange for its acceptance (2001–2006, 2007) of Afghan, Iraqi, and other Asian refugees that Australia refused to admit. When poor investments led to Nauru's virtual bankruptcy in mid-2004, Australian officials took charge of the country's finances. Ludwig Scotty was elected president in 2004 and reelected in 2007.

Government

Nauru is governed under the constitution of 1968. The president, who is both head of state and head of government, is elected by the unicameral Parliament for a three-year term. The eighteen members of Parliament are popularly elected, also for three-year terms. Administratively the country is divided into fourteen districts. The current head of state is Ludwig Scotty (since June 2004).

Nauset Harbor (NAW-set), SE MASSACHUSETTS, sheltered inlet of the ATLANTIC OCEAN, on E coast of CAPE COD, between EASTHAM and ORLEANS. Lighthouse and coast guard station. Part of CAPE COD NATIONAL SEASHORE.

Naushahr, IRAN: see NOW SHAHR.

Naushahr, PAKISTAN: see NOWSHERA.

Naushahra, INDIA: see NAOSHERA.

Naushahro (nuh-SHAH-ro), town, NAWABSHAH district, central SIND province, SE PAKISTAN, 44 mi/71 km NNW of NAWABSHAH; 25°50′N 68°07′E. Also called Naushahro Firoz or Naushaharo Feroze.

Naushki (NAH-oosh-kee), town (2005 population 3,490), S BURYAT REPUBLIC, S Siberia, RUSSIA, less than 4 mi/6.4 km N of the Mongolian border, on the SELENGA River, on highway and the ULAN-UDE–ULAANBAATAR railroad, 15 mi/24 km W of KYAKHTA; 50°23′N 106°06′E. Elevation 2,027 ft/617 m. Meat combine.

Naushon Island, Massachusetts: see ELIZABETH ISLANDS.

Nausori (nou-SO-ree), town, SE VITI, FIJI, SW PACIFIC OCEAN, on REWA RIVER delta. Agricultural center. Becoming industrialized as industry reaches out from high-priced Suva land. Site of Fiji's first sugar mill, 1880–1959. Fiji's main local airport.

Naus, Ras (NOUS, RAHS), cape on SW OMAN coast, at SW side of KURIA MURIA BAY of ARABIAN SEA; 17°15′N 55°18′E. Marks E limit of DHOFAR region. Also spelled Ras Naws.

Nauste (NOU-stuh), village, MØRE OG ROMSDAL county, W NORWAY, at head of ERESFJORDEN (5-mi/8-km-long SW arm of Langfjorden), and at mouth of AURA River, 30 mi/48 km E of MOLDE; 62°40′N 08°06′E. Cattle raising; river fishing. Tourist center.

Nauta (NOU-tah), town, ⊙ LORETO province, LORETO region, NE PERU, landing on MARAÑÓN RIVER near its confluence with UCAYALI River, and 65 mi/105 km SW of IQUITOS; 04°37′S 73°34′W. Sugarcane, fruit. Near RESERVA NACIONAL PACAYA-SAMIRIA.

Nautanwa (NOU-tahn-wah), town, GORAKHPUR district, E UTTAR PRADESH state, N central INDIA, 45 mi/72 km N of GORAKHPUR, on Nepal border. Railroad spur terminus; rice, wheat, barley, oilseeds. Sometimes called Nautanwan.

Nautla (nah-OO-tlah), town, VERACRUZ, E MEXICO, at mouth of small Nautla River, in GULF lowland, 40 mi/64 km SE of PAPANTLA DE OLARTE, on Mexico Highway 180; 20°15′N 96°45′W. Minor port; tobacco growing.

Nauvoo (nah-VOO), city (2000 population 1,063), HANCOCK county, W ILLINOIS, on heights overlooking the MISSISSIPPI RIVER, 9 mi/14.5 km N of KEOKUK (IOWA); 40°32′N 91°22′W. In agricultural area (fruit, corn, soybeans); manufacturing (wine, cheese). Tourism is major industry. Settled as Commerce shortly after 1830; occupied and renamed by Mormons under Joseph Smith in 1839. Population reached c.20,000 under the Mormons and was briefly the largest Illinois city; after Smith and his brother were killed (1844) by a mob in nearby CARTHAGE, the group left Illinois for UTAH (1846). The Icarians, a colony of French communists under Étienne Cabet, occupied the city, 1849–1856. Smith's house, part of an old hotel, and other old buildings are still standing. Church interests protect historic town site on flats below bluff and its square street pattern. Town damaged in 1993 floods. Nauvoo State Park nearby. Incorporated 1841.

Nauvoo (NAW-voo), town (2000 population 284), Walker co., NW Alabama, 16 mi/26 km NW of Jasper. Lumber. Originally called 'Ingle Mills,' the name was changed for a town in IL founded by Mormons in 1840. The word is derived from the Hebrew adjective for 'pleasant.' Inc. in 1906.

Nauzad (naw-ZAHD), town, HELMAND province, S central AFGHANISTAN, 40 mi/64 km N of GIRISHK, in outliers of the HINDU KUSH mountains; 32°24′N 64°28′E. Also spelled Nawzad.

Nauzontla (nah-oo-SON-tlah), town, PUEBLA, central MEXICO, 18 mi/29 km NW of TEZIUTLÁN; 19°58′N 97°35′W. Corn, fruit, vegetables, coffee.

Nava (nah-vah), town, COAHUILA, N MEXICO, on railroad, and 25 mi/40 km SW of PIEDRAS NEGRAS, on Mexico Highway 57; 28°28′N 100°45′E. Wheat, corn, cattle, istle fibers, candelilla wax.

Navabad (NAV-ah-bahd), town, TAJIKISTAN, under republic supervision, near confluence of SURKHAB and VAKHSH rivers, 84 mi/135 km NE of DUSHANBE, 12 mi/20 km W of GARM; 39°01′N 70°09′E. Area under no direct viloyat administrative division; under direct republic supervision. Called Shulmak until 1950, when it replaced Garm as oblast capital.

Navabadskii, town, W TAJIKISTAN, c.3 mi/4.8 km W of DUSHANBE. In cotton and orchard area. Formerly called Novabad. Founded in 1942.

Navacelles, Cirque de (nah-vah-SEL, SEERK duh), canyon of Vis River, in HÉRAULT department, LANGUEDOC-ROUSSILLON administrative region, S FRANCE, in the CAUSSES region of limestone tablelands, 23 mi/37 km NW of MONTPELLIER, and in the S reaches of the MASSIF CENTRAL; 44°10′N 04°15′E. The site is a spectacular steep-sided limestone gorge formed here by a cut-off meander of the Vis River. Figs are grown here.

Navacerrada (NAH-vah-the-RAH-dah), town, MADRID province, central SPAIN, on E slopes of the SIERRA DE GUADARRAMA, 28 mi/45 km NW of MADRID. Mountain resort; livestock raising, potato growing, lumbering. The NAVACERRADA PASS is 4 mi/6.4 km N.

Navacerrada Pass (c.5,905 ft/1,800 m), in the SIERRA DE GUADARRAMA, central SPAIN, on MADRID-SEGOVIA province border, at S foot of the PEÑALARA, on old Madrid-Segovia road. Royal Guadarrama hospital in vicinity. Ski resort.

Navaconcejo (NAH-vah-kon-THAI-ho), town, CÁCERES province, W SPAIN, 18 mi/29 km NE of PLASENCIA; 40°11′N 05°49′W. Olive oil, wine, fruit, pepper.

Nava de la Asunción (NAH-vah dai lah ah-soon-SYON), town, SEGOVIA province, central SPAIN, in Old Castile, on Medina del Campo–Madrid Railroad, and 24 mi/39 km NW of SEGOVIA; 41°09′N 04°29′W. Lumbering and agricultural center (cereals, carobs, chickpeas, beans, fruit, chicory, sugar beets, grapes; livestock).

Nava del Rey (NAH-vah del RAI), town, VALLADOLID province, N central SPAIN, 30 mi/48 km SW of VALLADOLID; 41°20′N 05°05′W. Wine production, cheese processing, alcohol distilling, flour milling. Agri-

cultural trade (cereals, chickpeas; livestock). Aluminum-silicate mines nearby. Parish church has fine 17th-century sculptures.

Nava de Recomalillo, La (NAH-vah dai REE-ko-mah-LEE-lyo, lah), town, TOLEDO province, central SPAIN, 23 mi/37 km SSW of TALAVERA DE LA REINA. Olive-oil pressing; grapes; livestock.

Navahermosa (NAH-vah-er-MO-sah), town, TOLEDO province, central SPAIN, on slopes of MONTES DE TOLEDO (at foot of Navahermosa Pass), 28 mi/45 km SW of TOLEDO. Agricultural center (olives, cereals, grapes; goats). Olive-oil extracting.

Navajo, county (□ 9,960 sq mi/25,896 sq km; 2006 population 111,399), E ARIZONA, on UTAH (N) state line; ⊙ HOLBROOK; 35°23′N 110°19′W. Nation's eleventh-largest county. Agriculture (sheep, cattle, hogs; alfalfa, hay, corn); Native American handicraft; tourist trade. Mountain area crossed in S by LITTLE COLORADO RIVER. BLACK MESA in far N, parts of PAINTED DESERT in N center; part of PETRIFIED FOREST NATIONAL PARK on E border, and parts of Navajo and Hopi Indian reservations cover N half of county; parts of MOGOLLON Rim escarpment crosses county in S, forms border of Apache-Sitgreaves National Forest (N) and Fort Apache Indian Reservation (S); bounded by BLACK RIVER in extreme SE. NAVAJO NATIONAL MONUMENT is in N, near Utah state line. Homolovi Ruins State Park in W center; Monument Valley, with dramatic buttes, scene in many movie westerns, in NE. Formed 1895.

Navajo (NAH-vuh-ho), unincorporated town (2000 population 2,097), MCKINLEY county, NW NEW MEXICO, 26 mi/42 km NNW of GALLUP, at ARIZONA state line, in Navajo Indian Reservation; 35°53′N 109°01′W. Cattle, sheep. Timber. Manufacturing (lumber). Reservation has four coal mines and two electric power plants. In the late 1970s to the mid-1980s some coal was converted to gas, but process then discontinued. Red and Asaayi lakes to NE. CANYON DE CHELLY NATIONAL MONUMENT (Arizona) to NW. Chuska Peak (8,795 ft/2,681 m), in CHUSKA MOUNTAINS, to E.

Navajo Lake (NAH-vuh-ho), NW KANE county, SW UTAH, in MARKAGUNT PLATEAU and Dixie National Forest, 20 mi/32 km SE of CEDAR CITY; 3 mi/4.8 km long, 0.5 mi/0.8 km wide. Elevation 9,035 ft/2,754 m. Fishing, camping. Sometimes called Duck Lake. CEDAR BREAKS National Monument 5 mi/8 km NW.

Navajo Lake (NAH-vuh-ho), reservoir, NW NEW MEXICO and SW COLORADO, on SAN JUAN RIVER, 22 mi/35 km ENE of BLOOMFIELD (New Mexico); c.25 mi/40 km long; 36°48′N 107°36′W. Maximum capacity 1,986,000 acre-ft. LOS PINOS RIVER forms 8-mi/12.9-km N arm at dam. Receives Piedra River in Colorado. Formed by Navajo Dam (402 ft/123 m high, 3,648 ft/1,112 m long), a major unit of the COLORADO RIVER storage project built (1958–1963) by the U.S. Bureau of Reclamation for irrigation and flood control. Navajo Lake State Park (New Mexico) and Navajo Reservoir State Recreation Area (Colorado) are here.

Navajo Mountain (NAH-vuh-ho) (10,388 ft/3,166 m), Laccolith formation, S UTAH, near ARIZONA state line 11 mi/18 km S of junction of San Juan and COLORADO rivers, 5 mi/8 km SE of RAINBOW BRIDGE NATIONAL MONUMENT, 38 mi/61 km ENE of PAGE (Arizona). On Navajo Indian Reservation.

Navajo National Monument, NAVAJO and COCONINO counties, NE ARIZONA. Ruins of large elaborate cliff dwellings. Betatakin Ruin Unit and headquarters 16 mi/26 km WSW of KAYENTA; Keet Seel Ruin Unit 14 mi/23 km WNW of Kayenta, both units in Navajo county. Additional acreage in Coconino county. Covers 360 acres/146 ha. Proclaimed 1909.

Naval, town, BILIRAN province, W coast of BILIRAN ISLAND, PHILIPPINES; 40 mi/64 km NNE of ORMOC; 11°35′N 124°27′E. Transportation center for island.

Naval (nah-VAHL), village, HUESCA province, NE SPAIN, 28 mi/45 km E of HUESCA; 42°12′N 00°09′E. Saltworks; lumbering; cereals, olives; sheep, goats.

Nava, La (NAH-vah, lah) or **La Nava de Santiago** (lah NAH-vah dai sahn-TYAH-go), town, BADAJOZ province, W SPAIN, 13 mi/21 km NW of MÉRIDA. Cereals, olives, vegetables; livestock.

Nava, La (NAH-vah, lah), village, HUELVA province, SW SPAIN, in the SIERRA MORENA, 11 mi/18 km WNW of ARACENA. Olives, cork, acorns, walnuts; figs, pears, apricots.

Navalacruz (nah-vah-lah-KROOTH), village, ÁVILA province, central SPAIN, 18 mi/29 km SW of ÁVILA; 40°26′N 04°56′W. Rye, vegetables, nuts.

Navalcán (nah-vahl-KAHN), town, TOLEDO province, central SPAIN, in S spur of the SIERRA DE GREDOS, 15 mi/24 km WNW of TALAVERA DE LA REINA. Grain; livestock-raising center; apiculture; dairying.

Navalcarnero (nah-vahl-kahr-NAI-ro), town, MADRID province, central SPAIN, in CASTILE-LA MANCHA, 20 mi/32 km SW of MADRID; 40°18′N 04°00′W. In fertile area (grapes, cereals, olives, carobs, chickpeas, hemp; livestock). Wine production.

Navalgund (NUHV-uhl-guhnd), town, DHARWAD district, KARNATAKA state, S INDIA, 24 mi/39 km ENE of DHARWAD. Cotton, wheat, millet; carpet manufacturing. Livestock raising nearby. Also spelled Nawalgund.

Naval Hill, locality, W FREE STATE province, SOUTH AFRICA, 2 mi/3.2 km N of MANGAUNG (Bloemfontein) city center; 29°05′E 45°00′S. Original site of Lamont-Hussey Observatory of University of Michigan. Elevation 4,888 ft/1,490 m. Begun 1927, completed 1928, closed down in 1972; the building is now a cultural center. Fossils of prehistoric giant reptiles found here 1934. Whole area incorporated into Franklin Game Reserve (established 1928).

Navalmanzano (nah-vahl-mahn-THAH-no), town, SEGOVIA province, central SPAIN, 20 mi/32 km NNW of SEGOVIA; 41°13′N 04°15′W. Cereals, carobs, chickpeas, vegetables, sugar beets, grapes; livestock.

Navalmoral (nah-vahl-mo-RAHL), town, ÁVILA province, central SPAIN, 13 mi/21 km S of ÁVILA; 40°28′N 04°46′W. Cereals, nuts, vegetables, fruit, grapes; livestock.

Navalmoral de la Mata (nah-vahl-mo-RAHL dai lah MAH-tah), city, CÁCERES province, W SPAIN, 31 mi/50 km ESE of PLASENCIA. Agricultural trade center (asparagus, cereals, grapes, olives, fruit, tobacco; livestock); wine production, oil milling, soap manufacturing, tanning.

Navalmorales, Los (nah-vahl-mo-RAH-les, los), town, TOLEDO province, central SPAIN, on N slopes of the MONTES DE TOLEDO, 20 mi/32 km SE of TALAVERA DE LA REINA. Lumbering; livestock raising (sheep, hogs, goats); agriculture (olives, cereals, grapes). Olive-oil extracting. Mineral springs.

Navalonguilla (nah-vah-long-GEE-lyah), town, ÁVILA province, central SPAIN, in W SIERRA DE GREDOS, 50 mi/80 km SSE of SALAMANCA; 40°17′N 05°30′W. Vegetables; livestock. Fishing on affluent of TORMES RIVER.

Navalosa (nah-vah-LO-sah), town, ÁVILA province, central SPAIN, 20 mi/32 km SW of ÁVILA; 40°24′N 04°55′W. Forage, cereals.

Navalperal de Pinares (nah-vahl-pai-RAHL dai pee-NAH-res), town, ÁVILA province, central SPAIN, on railroad to MADRID, and 15 mi/24 km ESE of ÁVILA; 40°35′N 04°24′W. Resort with medicinal springs. Also produces cereals, grapes, vegetables; livestock.

Navalpino (nah-vahl-PEE-no), village, CIUDAD REAL province, S central SPAIN, 40 mi/64 km WNW of CIUDAD REAL; 39°13′N 04°35′W. Cereals, olives; livestock.

Navalucillos, Los (nah-vah-loo-THEE-lyos, los), town, TOLEDO province, central SPAIN, in N MONTES DE TOLEDO, 23 mi/37 km SSE of TALAVERA DE LA REINA;

39°40′N 04°38′W. Agricultural center (olives, wheat, grapes, cork; livestock). Lumbering; olive-oil extracting, cheese processing, tanning. Lead mines.

Navaluenga (nah-vah-LWENG-gah), town, ÁVILA province, central SPAIN, on the ALBERCHE RIVER, at S foot of SIERRA DE GREDOS (crossed by Navaluenga Pass, 5,144 ft/1,568 m), and 17 mi/27 km S of ÁVILA; 40°25′N 04°42′W. Cereals, legumes, fruit, grapes, carobs, hemp; livestock.

Navalvillar de Pela (nah-vahl-vee-LYAHR dai PAI-lah), town, BADAJOZ province, W SPAIN, 19 mi/31 km ENE of VILLANUEVA DE LA SERENA; 39°06′N 05°28′W. Olive-oil pressing, tile manufacturing. Livestock raising.

Navamorcuende (NAH-vah-mor-KWEN-dai), town, TOLEDO province, central SPAIN, 12 mi/19 km N of TALAVERA DE LA REINA; 40°09′N 04°47′W. Olives, grapes, timber, wool; dairying.

Navan (NA-vuhn), Gaelic *An Úaimh*, town (2006 population 3,710), ⊙ MEATH county, NE IRELAND, at the confluence of the BOYNE and BLACKWATER rivers; 53°39′N 06°40′W. It produces woolens and has sawmills. Clothing, furniture, and carpets are also made. There are remains of the old town walls.

Navanagar (nuh-VAH-nuh-guhr), former princely state of WESTERN INDIA STATES agency, on KATHIAWAR PENINSULA. Ruled by Rajputs of Jadeja clan. Merged 1948 with SAURASHTRA; now in JAMNAGAR district, GUJARAT state. The Jam Sahib of Navanagar was a celebrated personality in English cricket. Sometimes spelled Nawanagar.

Navan Fort (NA-vuhn), Gaelic *An Eamhain*, ancient earthwork in NW ARMAGH, S Northern Ireland, just W of ARMAGH; 54°20′N 06°41′W. The large elliptical mound was the seat of kings of Ulster for several centuries. Feasting place of the heroes of the Red Branch. Place of the Isamnium of Ptolemy's *Geography*.

Navapur (nuh-VAH-poor), town, DHULE district, MAHARASHTRA state, W central INDIA, 65 mi/105 km WNW of DHULE. Local market for wheat, rice, millet, timber; sawmills; cotton ginning. Sometimes spelled Nawapur.

Navarcles (nah-VAHR-kles), town, BARCELONA province, NE SPAIN, 9 mi/14.5 km NE of MANRESA; 41°45′N 01°54′E. Sawmilling. Wine, wheat, vegetables in area.

Navarin, Cape (nah-VAH-reen), CHUKCHI AUTONOMOUS OKRUG, NE RUSSIAN FAR EAST, on the BERING SEA; 62°15′N 179°08′E. Government arctic station.

Navarino, Greece: see PYLOS.

Navarino Island (nah-vah-REE-no) (☐ 955 sq mi/2,483 sq km), TIERRA DEL FUEGO, extreme S CHILE, just S of main island of Tierra del Fuego across BEAGLE CHANNEL, SE of USHUAIA (ARGENTINA); 50 mi/80 km long, 25 mi/40 km wide; 55°05′S 67°40′W. Mountains in center rise to 3,905 ft/1,190 m. Sheep raising; fish canning; meat processing; hydroelectric plant. Chilean naval base. Airport.

Navarre (nuh-VAHR), Spanish *Navarra* (nah-VAH-rah), region, autonomous community, and former kingdom of N SPAIN and SW FRANCE, on both slopes of the W PYRENEES MOUNTAINS. The N half is dominated by the Pyrenees. Founded in 9th century; ruled from 13th century by several French dynasties until Ferdinand the Catholic seized (1512) Spanish Navarre (the larger section in S; now the Spanish province) and united it (1515) with Spain. Lower Navarre, or French Navarre (in PYRÉNÉES-ATLANTIQUES department since 1790), passed (1589) to French crown when Henry, king of Navarre, became Henry IV of France.

Navarre (nuh-VAHR), Spanish *Navarra* (nah-VAH-rah), province (2001 population 555,829), N SPAIN, bordering on FRANCE, between the W PYRENEES and the EBRO RIVER; ⊙ PAMPLONA. Forms the autonomous region of NAVARRE. The beautiful mountain slopes have extensive cattle pastures and vast forests that yield hardwoods, which are economically im-

portant. The fertile valleys produce sugar beets, cereals, and vegetables; vineyards are important in the Ebro valley. Hydroelectric energy and entrepreneurship have resulted in considerable industrialization since the 1950s. Manufacturing includes processed foods and metal parts. The population of N and W Navarre is largely of Basque stock, and the early history of the region is that of the Basques.The pass of RONCESVALLES, which leads from France to Navarre, made the region strategically important early in its history. The Basques defended themselves successfully against the Moorish invaders as well as against the Franks; the domination of Charlemagne, who conquered Navarre in 778, was short-lived. In 824 the Basque chieftain Iñigo Aritza was chosen king of Pamplona, which was expanded under his successors and became known as the kingdom of Navarre. It reached its zenith under Sancho III (reigned 1000–1035), who married the heiress of Castile and ruled over nearly all of Christian Spain. On his death the Spanish kingdoms were again divided (into Navarre, ARAGÓN, and CASTILE). The kingdom of Navarre then comprised the present province of Navarre, the BASQUE provinces (which were later lost to Castile), and, N of the Pyrenees, the district called LOWER NAVARRE, now a part of France.

Navarre passed to King Philip IV of France in 1305. Navarre stayed with the French crown until the death (1328) of Charles IV, when it passed to Charles' niece, whose son, Charles II (Charles the Bad), played an important part in the Hundred Years War and in the French civil unrest of the time. Navarre passed, through marriage, to the counts of Foix and then to the house of Albret in 1479. Ferdinand V (Ferdinand the Catholic), after defeating Jean d'Albret, annexed most of Navarre in 1515. The area N of the Pyrenees (Lower Navarre) remained an independent kingdom until it was incorporated 1589 into the French crown when Henry III of Navarre became King Henry IV of France. It was united with BÉARN into a French province. The kings of France carried the additional title king of Navarre until the French Revolution. Since the rest of Navarre was in Spanish hands, the kings of Spain also carried (until 1833) the title king of Navarre. During that period Navarre enjoyed a special status within the Spanish monarchy; it had its own cortes, taxation system, and separate customs laws. Navarre became the chief stronghold of the Carlists in 1833 but recognized Isabella II as queen in 1839. As a reward for their loyalty in the Spanish Civil War, Franco allowed the Navarrese to maintain their ancient *fueros*, charters handed down by the crown outlining a system of self-government.

Navarre (nuh-VAHR), village (☐ 2 sq mi/5.2 sq km; 2006 population 1,426), STARK county, E central OHIO, 9 mi/14 km SW of CANTON, and on TUSCARAWAS RIVER; 40°43′N 81°31′W. Agricultural trade center; makes foods and fabricated metal products. Fort Laurens State Park is nearby.

Navarre (nuh-VAHR), locality, SANTA ROSA county, NW FLORIDA, 23 mi/37 km E of PENSACOLA; 30°24′N 86°51′W. Manufacturing includes consumer goods, neon signs, and building materials.

Navarredonda de la Sierra (nah-vah-rai-DON-dah dai lah SYAI-rah), town, ÁVILA province, central SPAIN, near source of TORMES RIVER, on N slopes of the SIERRA DE GREDOS, and 31 mi/50 km SW of ÁVILA; 40°22′N 05°08′W. Resort. Point of ascension for the mountains.

Navarredondilla (nah-vah-rai-don-DEE-lyah), village, ÁVILA province, central SPAIN, 14 mi/23 km SSW of ÁVILA; 40°27′N 04°49′W. Carobs, rye, nuts, potatoes; livestock.

Navarre Française (nah-VAHR frawn-SEZ) or **Basse-Navarre** (BAHS–nah-VAHR) [French=lower Navarre], old region and kingdom of SW FRANCE, N of the W PYRENEES Mountains, included in what is now PYRÉNÉES-ATLANTIQUES department, AQUITAINE ad-

ministrative region; 42°50′N 01°40′E. Region came under French royal rule under Henry IV. Upper Navarre, S of the Pyrenees, is now a Spanish autonomous community within PAMPLONA province.

Navarrés (nah-vah-RAIS), town, VALENCIA province, E SPAIN, 13 mi/21 km NW of JÁTIVA; 39°06′N 00°41′W. Olive-oil processing, manufacturing of toys and soap; wine, vegetables.

Navarrete (nah-vah-RAI-tai), town, LA RIOJA province, N SPAIN, near ALBERCHE RIVER. Produces wine, olive oil, cereals. Dam, reservoir nearby.

Navarrevisca (nah-vah-rai-VEES-kah), village, ÁVILA province, central SPAIN, near ALBERCHE RIVER (trout fishing), 21 mi/34 km SW of ÁVILA; 40°22′N 04°53′W. Cereals, vegetables, fruit; livestock. Lumbering.

Navarro (NAH-vahr-o), county (☐ 1,086 sq mi/2,823.6 sq km; 2006 population 49,440), E central TEXAS; ⊙ CORSICANA; 32°02′N 96°28′W. Mainly rich blackland prairies; bounded NE by TRINITY RIVER; drained by Richland, Chambers and Waxahachie creeks. Diversified agriculture (cotton, corn, grains, vegetables; sorghum, wheat; herbs); livestock (beef and dairy cattle, hogs, horses, emus). Oil, natural gas wells. Manufacturing, processing of farm products and petroleum at Corsicana. NAVARRO MILLS LAKE reservoir in W (Richland Creek); large part of RICHLAND CHAMBERS Reservoir in S. Formed 1846.

Navarro (nah-VAHR-ro), town (1991 population 8,791), ⊙ Navarro district (☐ 624 sq mi/1,622.4 sq km), NE BUENOS AIRES province, ARGENTINA, 55 mi/89 km SW of BUENOS AIRES; 35°00′S 59°30′W. Railroad junction; grain; cattle; dairying.

Navarro Mills Lake (NAH-vahr-o), reservoir, NAVARRO and HILL counties, E central TEXAS, on Richland Creek, 18 mi/29 km SW of CORSICANA; c.12 mi/19 km long; 31°57′N 96°42′W. Maximum capacity 335,800 acre-ft. Formed by Navarro Mills Dam (77 ft/23 m high), built (1963) by the Army Corps of Engineers for water supply and flood control.

Navarro River, c.40 mi/64 km long, NW CALIFORNIA; rises in SE MENDOCINO county; flows NW, through Hendy Woods State Park, to the PACIFIC OCEAN 18 mi/29 km S of FORT BRAGG.

Navás (nah-VAHS), town, BARCELONA province, NE SPAIN, on the LLOBREGAT RIVER, and 13 mi/21 km NNE of MANRESA. Cotton spinning and weaving; lumbering. Wine, cereals in area.

Navas de Estena (NAH-vahs dai es-TAI-nah), village, CIUDAD REAL province, S central SPAIN, on S slopes of the MONTES DE TOLEDO, 45 mi/72 km NW of CIUDAD REAL; 40°46′N 04°20′W. Lumber, cork; sheep.

Navas de la Concepción, Las (NAH-vahs dai lah kon-thep-THYON, lahs), town, SEVILLE province, SW SPAIN, in the SIERRA MORENA, near CÓRDOBA province border, 16 mi/26 km E of CAZALLA DE LA SIERRA. Cereals, flour, olives, olive oil; livestock; timber. Manufacturing of soap and meat products.

Navas del Madroño (NAH-vahs del mah-DRO-nyo), town, CÁCERES province, W SPAIN, 18 mi/29 km NW of CÁCERES; 39°37′N 06°39′W. Olive-oil processing; wheat, potatoes, wine; livestock.

Navas del Marqués, Las (NAH-vahs del mahr-KAIS, lahs), town, ÁVILA province, central SPAIN, summer resort in SW SIERRA DE GUADARRAMA, 19 mi/31 km E of ÁVILA. Surrounded by pine forests yielding timber and naval stores. Also produces beans, cereals; livestock. Mineral springs.

Navas del Rey (NAH-vahs del RAI), town, MADRID province, central SPAIN, 30 mi/48 km W of MADRID. Grapes, cereals, olives; livestock; wine making; tiles, pottery.

Navas de Oro (NAH-vahs dai O-ro), town, SEGOVIA province, central SPAIN, 25 mi/40 km NW of SEGOVIA; 41°12′N 04°26′W. Grains, grapes; livestock. Forest industry (lumber, resins).

Navas de San Antonio (NAH-vahs dai SAHN ahn-TO-nyo), village, SEGOVIA province, central SPAIN, 17 mi/27 km SW of SEGOVIA. Grain growing; livestock raising.

Navas de San Juan, Las (NAH-vahs dai SAHN HWAHN, lahs), town, JAÉN province, S SPAIN, 12 mi/19 km NNE of ÚBEDA; 38°11′N 02°19′W. Olive-oil processing, soap manufacturing; ships olives. Cereals, beans; livestock in area. Granite quarries, lead and copper mines nearby.

Navas de Tolosa (NAH-vahs dai to-LO-sah), town, JAÉN province, S SPAIN, 2 mi/3.2 km ENE of LA CAROLINA; 38°17′N 03°35′W. Scene of decisive victory (1212) of the Christian armies under Alfonso VIII over the Moors.

Navasfrías (nah-vahs-FREE-ahs), town, SALAMANCA province, W SPAIN, near Portuguese border, 25 mi/40 km SW of CIUDAD RODRIGO; 40°18′N 06°49′W. Livestock; potatoes, rye.

Navashino (nah-VAH-shi-nuh), city (2006 population 17,470), SW NIZHEGOROD oblast, central European Russia, on the right bank of the OKA River, on highway and railroad junction, 98 mi/158 km SW of NIZHNIY NOVGOROD, and 5 mi/8 km E of MUROM; 55°31′N 42°12′E. Elevation 278 ft/84 m. Shipbuilding (since 1907); construction materials, logging and lumbering; bakery. Made city in 1957, uniting settlements of Lipnya and Morodovshchikovo.

Navasota (NAH-vah-so-tuh), town (2006 population 7,378), GRIMES county, E central TEXAS, on NAVASOTA RIVER (near its confluence with the BRAZOS RIVER), and c.60 mi/97 km NW of HOUSTON; 30°23′N 96°05′W. Cattle; timber; agribusiness (fruit); dairying; manufacturing (machinery, fabricated metal products, honey). Statue of La Salle commemorates claims that he was killed nearby. Active in Texas Revolution; laid out 1858.

Navasota River (NAH-vah-so-tuh), c.130 mi/209 km long, E central TEXAS; rises in LIMESTONE county; flows SE, through Lake Mexia and LAKE LIMESTONE reservoirs then S to the BRAZOS RIVER 5 mi/8 km SW of NAVASOTA.

Navassa (nah-VAH-suh), village (□ 2 sq mi/5.2 sq km; 2006 population 1,683), BRUNSWICK county, SE NORTH CAROLINA, 5 mi/8 km NW of WILMINGTON, on CAPE FEAR RIVER (RR bridge); 34°15′N 78°00′W. Railroad junction. Service industries; manufacturing (wood products).

Navassa Island (nuh-VA-suh), French *La Navase*, U.S.-owned CARIBBEAN islet (□ 1 sq mi/2.6 sq km), between HAITI and JAMAICA, 35 mi/56 km W of CAPE IROIS (HAITI); 18°25′N 75°02′W. Formerly yielded guano. Lighthouse.

Navatalgordo (nah-vah-TAHL-GOR-do), town, ÁVILA province, central SPAIN, near ALBERCHE RIVER, 18 mi/29 km SW of ÁVILA; 40°25′N 04°52′W. Cereals, fruit; livestock.

Navbakhor (NAHV-bah-hor), town, NAMANGAN wiloyat, NE UZBEKISTAN, in central FERGANA VALLEY, on NAMANGAN-KOKAND railroad. Also Novbakhor.

Nave (NAH-ve), town, BRESCIA province, LOMBARDY, N ITALY, 4 mi/6 km NNE of BRESCIA; 45°35′N 10°17′E. Paper-milling center; iron works. Has 16th-century churches, ancient ruins nearby.

Navegantes (NAH-ve-gahn-ches), city (2007 population 52,739), E SANTA CATARINA state, BRAZIL, 6 mi/9.7 km N of ITAJAÍ, on Atlantic Coast; 26°54′S 48°39′W.

Nävekvarn (NEV-e-KVAHRN), village, SÖDERMANLAND county, E SWEDEN, on N shore of Bråviken, 30-mi/48-km-long inlet of BALTIC SEA, 12 mi/19 km SW of NYKÖPING; 58°38′N 16°48′E.

Naver, Loch (NAI-vuhr), lake (6 mi/9.7 km long; 108 ft/33 m deep), HIGHLAND, N Scotland, 13 mi/21 km S of TONGUE; 58°17′N 04°23′W. Drained by Naver River, which issues from E end of loch and flows 19 mi/31 km N to the Atlantic Ocean 7 mi/11.3 km ENE of Tongue.

Naves (NAHV), commune (□ 13 sq mi/33.8 sq km), CORRÈZE department, SW central FRANCE, 4 mi/6.4 km N of TULLE, in LIMOUSIN region, on SW slopes of MONEDIÈRES MOUNTAINS; 45°19′N 01°46′E. Livestock raising. Woodworking. The entrenched valley of CORRÈZE RIVER is nearby.

Navesink (NA-vuh-sink), village, MONMOUTH county, E NEW JERSEY, on NAVESINK River, and 16 mi/26 km NE of FREEHOLD; 40°23′N 74°02′W. Largely residential.

Navesink Highlands (NA-vuh-sink), MONMOUTH county, E NEW JERSEY, coastal ridge (c.276 ft/84 m) between SANDY HOOK BAY (N) and NAVESINK River estuary (S). One of highest points on U.S. ATLANTIC coast; site of the Twin Towers, one of most powerful U.S. lighthouses (40°24′N 73°59′W). Often called ATLANTIC HIGHLANDS, HIGHLANDS OF NAVESINK.

Navesink River (NA-vuh-sink), estuary, c.8 mi/12.9 km long, MONMOUTH county, E NEW JERSEY; extending ENE from RED BANK (head of navigation) to junction with SHREWSBURY RIVER estuary at entrance to passage to SANDY HOOK BAY near HIGHLANDS. NAVESINK HIGHLANDS are N, on mainland.

Navgarh (nuhv-guhr), town, ☉ Sidhartnagar district, UTTAR PRADESH state, N central INDIA.

Navia (NAH-vyah), town, OVIEDO province, NW SPAIN, on NAVIA River near its mouth on BAY OF BISCAY, and 16 mi/26 km E of RIBADEO. Fishing and fish processing; cereals, potatoes; lumber; cattle.

Navia River (NAH-vyah), 65 mi/105 km long, in OVIEDO and LUGO provinces, NW SPAIN; rises in CANTABRIAN MOUNTAINS 20 mi/32 km SE of LUGO; flows NNE to BAY OF BISCAY at NAVIA. Salmon fisheries.

Navibandar (nuh-vee-BUHN-duhr), village, JUNAGADH district, GUJARAT state, W central INDIA, on ARABIAN SEA, at mouth of BHADAR RIVER, 18 mi/29 km SE of PORBANDAR. Small coastal trade (ghee, cotton, salt); fishing. Lighthouse. Was in W section of former SAURASHTRA state.

Navidad (nah-vee-DAHD), town, ☉ Navidad comuna, CARDENAL CARO province, LIBERTADOR GENERAL BERNARDO O'HIGGINS region, central CHILE, on the coast, near mouth of RAPEL RIVER, 65 mi/105 km NW of SAN FERNANDO; 33°57′S 71°50′W. Cereals; livestock.

Navidad River (NAH-vi-dahd), c.100 mi/161 km long, S TEXAS; rises in headstreams in FAYETTE county; flows generally SSE and parallel to LAVACA river, joining it 11 mi/18 km above its mouth of Lavaca Bay. Forms LAKE TEXANA 13 mi/21 km above mouth.

Navina (na-VEI-nuh), village, LOGAN county, central OKLAHOMA, 9 mi/14.5 km WSW of GUTHRIE. In agricultural area.

Naviraí (NAH-vee-rah-EE), city (2007 population 43,404), SE Mato Grosso do Sul state, BRAZIL, 31 mi/50 km SE of JUTI; 23°10′S 54°25′W. Agricultural; airstrip.

Naviti, FIJI: see YASAWA GROUP.

Naviti Levu, FIJI: see VITI LEVU.

Navlakhi (NOU-luh-kee), seaport, KACHCHH district, GUJARAT state, W central INDIA, at head of GULF OF KACHCHH, 25 mi/40 km WNW of MORBI. Railroad terminus; exports cotton, vegetable oil, wool, grain. Was in N section of former SAURASHTRA state.

Navlya (NAH-vlyah), town (2005 population 13,650), E BRYANSK oblast, central European Russia, on road and railroad, 30 mi/48 km S of BRYANSK; 52°49′N 34°30′E. Elevation 606 ft/184 m. Railroad junction. Sawmilling, woodworking; railroad ties, automotive parts and repair, dairy products, dried vegetables. Sometimes spelled—and pronounced—Novlya.

Năvodari (nuh-vo-DAHR), town, CONSTANŢA county, SE ROMANIA, on Lake Tasaul, and 10 mi/16 km N of CONSTANŢA. Stone quarrying. Petrochemical plant. Known as Caracoium under Ottomans.

Navoi, UZBEKISTAN: see NAWOIY.

Navojoa (nah-VO-ho-ah), city and township, SONORA, NW MEXICO, on lower MAYO RIVER, in coastal plain, on Gulf of CALIFORNIA, and 100 mi/161 km SE of GUAYMAS; 27°04′N 109°28′E. Railroad junction; agri-cultural center (wheat, corn, chickpeas, fruit; cattle); flour milling; native handicrafts, textiles.

Navolato (nah-vo-LAH-to), city and township, SINALOA, NW MEXICO, on CULIACÁN RIVER, and 20 mi/32 km WSW of CULIACÁN ROSALES, on railroad; 24°45′N 107°41′W. Agricultural center (sugarcane, chickpeas, corn, vegetables, fruit); lumbering.

Navoloki (NAH-vuh-luh-kee), city (2005 population 10,855), N IVANOVO oblast, central European Russia, on the VOLGA RIVER, on road and railroad spur, 65 mi/105 km NE of IVANOVO, and 7 mi/11 km WNW of KINESHMA; 57°28′N 41°57′E. Elevation 449 ft/136 m. Cotton textiles. Has a museum. Established in the 1880s; city since 1938.

Navotas (nah-BO-tahs), town (2000 population 230,403), NATIONAL CAPITAL REGION, S LUZON, PHILIPPINES, just NW of MANILA; 14°40′N 120°57′E. Largest fishing port in the Philippines. Also spelled Nabotas.

Navpaktos (NAHF-pahk-tos), Latin *Naupactus*, Italian *Lepanto*, town, AKARNANIA prefecture, WESTERN GREECE department, port on N shore of Gulf of CORINTH, near strait leading to Gulf of PATRAS, 22 mi/35 km E of MESOLONGI; 38°24′N 21°50′E. Fisheries; livestock raising; olive oil, grapes. Ancient Naupactus was the leading port of W (Ozolian) LOCRIS; identified with the legendary crossing of the Dorians to the PELOPONNESUS. Captured by ATHENS in 456 B.C.E. and assigned to the exiled Messenians following the Third Messenian War, it became the chief Athenian naval base on Gulf of Corinth. After the Peloponnesian War, it was returned (404 B.C.E.) to Locris, and passed (399) to Achaea (ACHAIA) and (338) to AETOLIA. Under Byzantine rule it was known as Epaktos or Epakhtos. Still a strategic station in the Middle Ages, it was held (1407–1499) by VENICE before passing to the Turks. Venice recaptured it briefly, 1687–1699. In independent GREECE since 1832. The naval battle of Lepanto (1571), in which the fleet of a Holy League commanded by Don John of Austria defeated the Turks, was fought at mouth of Gulf of Patras. Also Naupaktos.

Návplion (NAHF-plee-on), town, ☉ ARGOLIS prefecture, PELOPONNESE department, S mainland GREECE, fortified port on the Gulf of ARGOLIS, S of CORINTH; 37°34′N 22°48′E. Commercial center that ships tobacco, cotton, and fruits. Tourism (beaches). According to tradition, founded by Nauplius, who was the father of Palarnedes. Important during Middle Ages. Venetian fortress of Palamidi crowns city, reached by climbing 857 steps. Captured (1715) by the Ottoman Turks from Venice. Taken in 1822 by Greek insurgents and was (1830–1834) the first capital of independent Greece. The revolt (1862) against King Otto I began here. Used by British as evacuation pont during World War II. Also Nauplion, Nauplia.

Navrongo (nahr-RON-go), town, local council headquarters, UPPER EAST REGION, GHANA, near Paga on BURKINA FASO border, 105 mi/169 km NNW of TAMALE; 10°54′N 01°06′W. Road junction; shea nuts, millet, durra, yams; cattle, skins. Airfield.

Navsar, TURKEY: see SEMDINLI.

Navsari (NUHV-sah-ree), city (2001 population 232,411), VALSAD district, GUJARAT state, W central INDIA, on PURNA RIVER, 17 mi/27 km SSE of SURAT. Trades in cotton, millet, timber; cotton and silk milling, hand-loom weaving, cottonseed milling, wood carving; manufacturing of copper and brass products, glass, leather goods, oil engines, textile bobbins, soap, perfume; metalworks. Has college Headquarters of Parsi community in India. Formerly in Navsari division of BARODA state; division incorporated 1949 into SURAT district and later into Valsad district.

Navua (nah-VOO-ah), town, S VITI LEVU, FIJI, SW PACIFIC OCEAN, 18 mi/29 km WSW of SUVA; 18°13′S 178°10′E. Copra; dairying.

Cross-references are shown in SMALL CAPITALS. The pronunciation guide is shown on page xix. The sources of population figures are shown on page xvii.

Navua River (nah-VOO-ah), 40 mi/64 km long, S VITI LEVU, FIJI, SW PACIFIC OCEAN; rises in S mountain range; flows SE to Rovodrau Bay. Its delta area produces dairy foods.

Navy Board Inlet, N BAFFIN ISLAND, BAFFIN region, NUNAVUT territory, N Canada, arm of LANCASTER SOUND, between BORDEN PENINSULA of Baffin (W) and BYLOT (E) islands; 70 mi/113 km long, 6 mi/10 km–18 mi/29 km wide; 73°00′N 80°50′W.

Navyi Afon (NA-vyee–AH-fon), town, NW ABKHAZ Autonomous Republic, GEORGIA, port on BLACK SEA, 11 mi/18 km NW of SUKHUMI. Health resort; citrus fruit, tobacco. Formerly known as Psirtskha and Akhali-Afoni.

Navy Island (NAI-vee), in the NIAGARA River, just above NIAGARA FALLS, S ONTARIO, E central Canada; 43°03′N 79°00′W. Famous as the scene of the last stand made by William Lyon Mackenzie and some of his fellow rebels in the Upper Canada Rebellion of 1837.

Navy Yard City, unincorporated town (2000 population 2,638), KITSAP county, W WASHINGTON, 3 mi/4.8 km SW of BREMERTON, on Sinclair Inlet arm of PUGET SOUND; 47°33′N 122°40′W.

Nawa (NAH-wah), town, Saihaku county, TOTTORI prefecture, S HONSHU, W JAPAN, 42 mi/68 km W of TOTTORI; 35°30′N 133°30′E.

Nawa (NAH-wah), town, Der'a district, SW SYRIA, 19 mi/31 km N of Der'a; 32°53′N 36°02′E. Cereals, sesame, fodder.

Nawabashah, PAKISTAN: see NAWABSHAH.

Nawabganj (naw-uhb-gawnj), town (2001 population 152,223), RAJSHAHI district, W EAST BENGAL, BANGLADESH, on the MAHANANDA River, and 25 mi/40 km NW of RAJSHAHI; 24°37′N 88°13′E. Railroad terminus. Trades in rice, wheat, oilseed, jute, silk. Also known as Chapai Nawabganj; formerly Baragharia Nawabganj. Also spelled Nowabganj.

Nawabganj (nuh-WAHB-guhnj), town, BARA BANKI district, central UTTAR PRADESH state, N central INDIA, 18 mi/29 km ENE of LUCKNOW. Hand-loom cotton-weaving center; sugar processing; trades in rice, wheat, gram, oilseeds, barley. Founded by Shuja-ud-daula, a nawab of Oudh, in late 18th century. BARA BANKI town is just SW.

Nawabganj, town, BAREILLY district, N central UTTAR PRADESH state, N central INDIA, 18 mi/29 km NE of BAREILLY. Wheat, rice, gram, sugarcane, oilseeds.

Nawabganj, town, GONDA district, NE UTTAR PRADESH state, N central INDIA, near the GHAGHARA River, 5 mi/8 km N of FAIZABAD. Sugar processing; trades in rice, wheat, corn, gram, oilseeds. Founded 18th century by a nawab of Oudh.

Nawabshah (NWAHB-shah), district (□ 3,908 sq mi/10,160.8 sq km), central SIND province, SE PAKISTAN; ⊙ NAWABSHAH. Bounded E by THAR DESERT, W by INDUS River; irrigated by ROHRI CANAL and its branches. Flat alluvial plain. Sometimes written Nawab Shah.

Nawabshah (NWAHB-shah), town, ⊙ NAWABSHAH district, central SIND province, SE PAKISTAN, 130 mi/209 km NE of KARACHI; 26°15′N 68°25′E. Sometimes spelled NAWABASHAH.

Nawada (nuh-WAH-dah), district (□ 1,040 sq mi/2,704 sq km; 2001 population 1,809,425), S central BIHAR state, E INDIA; ⊙ NAWADA.

Nawada (nuh-WAH-dah), town (2001 population 82,291), ⊙ NAWADA district, S central BIHAR state, E INDIA, on tributary of the GANGA River, 35 mi/56 km ENE of GAYA; 24°54′N 85°33′E. Road junction; trade center (rice, gram, wheat, barley, oilseeds, corn). Mica mining nearby. Vishnuite sculpture and Hindu ruins 15 mi/24 km NNE, at AFSAR.

Nawadih, INDIA: see JAMUI.

Nawai (nuh-WEI), town, tahsil headquarters, TONK district, E RAJASTHAN state, NW INDIA, 39 mi/63 km S of JAIPUR; 26°21′N 75°55′E. Agricultural market. Sometimes spelled Niwai.

Nawakot (NAW-wuh-kot), village, N central NEPAL, near TRISULI RIVER, 17 mi/27 km NW of KATHMANDU; 27°55′N 85°10′E. Fruit growing (mango and orange orchards); rice, vegetables. Home of Thakur dynasty, ruling Kathmandu late 11th–early 14th century. Captured 1765 by Gurkha leader Prithvi Narayan Shah; treaty signed here (1792) between Chinese and Gurkhas ended CHINA-Nepal War. Winter residence of Gurkha kings until 1813.

Nawala, town, WESTERN PROVINCE, SRI LANKA; 06°53′N 79°53′E. Residential suburb of COLOMBO. Headquarters of Open University of Sri Lanka.

Nawalapitiya, town, CENTRAL PROVINCE, SRI LANKA, in SRI LANKA HILL COUNTRY, on the MAHAWELI GANGA River, on railroad, and 18 mi/29 km SSW of KANDY; 07°03′N 81°32′E. Trades in tea, rubber, rice, vegetables.

Nawalgarh (nuh-WUHL-guhr), town, JHUNJHUNUN district, NE RAJASTHAN state, NW INDIA, 18 mi/29 km NNE of SIKAR; 27°51′N 75°16′E. Trades in grain, wool, livestock, cotton; hand-loom weaving; enamel work. Commercial college.

Nawalgund, INDIA: see NAVALGUND.

Nawalparasi (NUH-wuhl-puh-rah-see), district, S central NEPAL, in LUMBINI zone; ⊙ PARASI.

Nawan (nuh-wahn), town, NAGAUR district, E central RAJASTHAN state, NW INDIA, 130 mi/209 km ENE of JODHPUR, near SAMBHAR LAKE. Salt manufacturing; hand-loom woolen weaving.

Nawanagar, INDIA: see NAVANAGAR.

Nawanshahr (nuh-WAHN-shah-huhr), town, JAMMU district, JAMMU AND KASHMIR state, N INDIA, 12 mi/19 km SW of JAMMU; 32°46′N 74°32′E. Sugar mill, alcohol distillery; wheat, rice, bajra, corn. Also called Sri Ranbirsinghpura; also written Ranbir Singhpura.

Nawanshahr, INDIA: see NAWASHAHR.

Nawapara (nuh-WAH-pah-rah), district (□ 1,316 sq mi/3,421.6 sq km), ORISSA state, E central INDIA; ⊙ NAWAPARA.

Nawapara (nuh-WAH-pah-rah), town, tahsil headquarters, ⊙ NAWAPARA district, W ORISSA state, E central INDIA, 75 mi/121 km NW of BHAWANIPATNA. Local market for rice, oilseeds, timber. Also known as Nuaparha, Nuaparha Tanwat.

Nawapur, INDIA: see NAVAPUR.

Nawarangpur (nuh-wuh-ruhng-POOR), district (□ 1,983 sq mi/5,155.8 sq km), ORISSA state, E central INDIA; ⊙ NAWARANGPUR. Formerly spelled Nowrangpur.

Nawarangpur (nuh-buh-ruhn-gah-poor), town, ⊙ NAWARANGPUR district, SW ORISSA state, E central INDIA, near INDRAVATI River, 25 mi/40 km N of JAYPUR; 19°14′N 82°33′E. Rice milling. Sal forests nearby. Carved out of KORAPUT district in 1992. Formerly spelled Nowrangpur.

Nawargaon (nuh-WUHR-goun), town, CHANDRAPUR district, MAHARASHTRA state, central INDIA, 35 mi/56 km NNE of CHANDRAPUR. Rice, flax, millet. Hematite deposits nearby. Sawmilling (teak) in forest (E). Sometimes spelled Nawergaon; sometimes called Nawargaon Buzurg.

Nawashahr (nuh-WAH-shuh-huhr), town, JALANDHAR district, central PUNJAB state, N INDIA, 35 mi/56 km SE of JALANDHAR; 31°07′N 76°08′E. Railroad junction; agricultural market (wheat, corn, cotton); hand-loom weaving. Sometimes spelled Nawanshahr.

Nawiliwili Harbor (NAH-WEE-lee-WEE-lee), SE KAUAI, KAUAI county, HAWAII, principal deep water port of island, 1 mi/1.6 km SE of LIHUE; 0.5 mi/0.8 km wide. Inner extension of Nawiliwili Bay, separated by jetty; receives Huleia Stream from NW. Port of Nawiliwili, on N side, harbors cruise ships, cargo ships.

Nawoiy (nah-vah-EE), wiloyat, central UZBEKISTAN; ⊙ NAWOIY. Includes much of E KYZYL KUM desert. Drained by the ZERAVSHAN, which runs E-W across the extreme S. Most of the population is in the S part, around Nawoiy. World's 4th-largest gold mine in N.

Nawoiy (nah-vah-EE), city, ⊙ NAWOIY wiloyat, central UZBEKISTAN; 40°09′N 65°22′E. Also NAVOI, Navoy.

Nawrangpur (nuh-ruhng-POOR), town, ⊙ NAWARANGPUR district, ORISSA state, E central INDIA.

Nawshakh, mountain (24,557 ft/7,485 m), AFGHANISTAN, SE of ISHKASHIM at entrance to Wakhan Corridor, and near PAKISTAN border (CHITRAL district); 36°25′N 71°50′E. Highest mountain in Afghanistan. Sometimes spelled Noshaq.

Nawzad, AFGHANISTAN: see NAUZAD.

Naxi (NAH-SHEE), town, ⊙ Naxi county, S SICHUAN province, CHINA, 7 mi/11.3 km SW of LUZHOU, across CHANG JIANG (Yangzi River); 28°52′N 104°58′E. Rice, wheat, sugarcane, tobacco, jute; papermaking; chemicals, food and beverages.

Naxos, earliest Greek colony of SICILY, ITALY, on Cape SCHISÒ, S of TAORMINA. Founded 735 B.C.E. from Chalcis. Its inhabitants founded CATANIA and LENTINI a few years later. Destroyed 403 B.C.E. by Dionysius. Site now occupied by lemon plantation.

Naxos (NAHK-sos), largest island (□ 169 sq mi/439.4 sq km) of the CYCLADES group, CYCLADES prefecture, SOUTH AEGEAN department, GREECE, in AEGEAN SEA, E of PÁROS; 20 mi/32 km long, 12 mi/19 km wide; 37°00′N 25°35′E. Rises to 3,284 ft/1,001 m in Mount Dryos or Drios. Produces almonds, citrus fruit, olive oil, figs, wheat, barley, and a well-known white wine; sheep, goats; cheese. Emery is a major export item; marble and granite quarries. Popular tourist area (beaches). Airport. The chief town, Naxos, is on NW shore. In Greek mythology Theseus abandoned Ariadne here. Known in antiquity for its wine and the worship of Dionysus, it was colonized by the Ionians and flourished in late 6th century B.C.E. as mistress of the Cyclades. Sacked 490 B.C.E. by the Persians, joined the Delian League, and was subjected to ATHENS after a revolt (c.450 B.C.E.). In Middle Ages, it was seat (1207–1566) of Venetian Aegean duchy and passed 1579 to Turks and 1832 to Greece. Formerly also Naxia.

Naxxar (nah-SHAR), village (2005 population 11,978), central MALTA, 4 mi/6.4 km W of VALLETTA; 35°54′N 14°26′E. Site of Malta's International Trade Fair Grounds. Just N are strange prehistoric ruts worn into the stony ground.

Naya, INDIA: see MURSAN.

Naya Dumka, INDIA: see DUMKA.

Nayaganj Hathras, INDIA: see MURSAN.

Nayagarh (nuh-YAH-guhr), district (□ 1,527 sq mi/3,970.2 sq km), ORISSA state, E central INDIA; ⊙ NAYAGARH.

Nayagarh (nuh-YAH-guhr), town, tahsil headquarters, ⊙ NAYAGARH district, E ORISSA state, E central INDIA, 55 mi/89 km WNW of PURI; 20°34′N 85°08′E. Rice, sugarcane, oilseeds. Was capital of former princely state of Nayagarh in ORISSA STATES; incorporated 1949 into PURI district.

Nayakanhatti (nuh-YAH-kahn-huh-tee), town, CHITRADURGA district, KARNATAKA state, S INDIA, 19 mi/31 km NNE of CHITRADURGA; 14°28′N 76°33′E. Rice, chilies, millet. Annual temple-festival fair. Also spelled Nayakanahatti.

Nayakhan or **Nayakan** (nah-yah-KAN), settlement, MAGADAN oblast, RUSSIAN FAR EAST, 8 mi/13 km W of EVENSK; 61°55′N 158°58′E. Elevation 104 ft/31 m. Weather station.

Nayala, province (□ 1,515 sq mi/3,939 sq km; 2005 population 154,683), BOUCLE DU MOUHOUN region, NW central BURKINA FASO; ⊙ Toma; 12°40′N 03°00′W. Bordered on N by SOUROU province, NE tip by ZONDOMA province, ENE by PASSORÉ province, ESE by SANGUIÉ province, S and W by MOUHOUN province, and NW tip by KOSSI province. Established in 1997 with fourteen other new provinces.

Nayanagar, INDIA: see BEAWAR.

Nayarit, state (□ 10,664 sq mi/27,726.4 sq km), W MEXICO, on the PACIFIC OCEAN; 22°00′N 105°00′W;

⊙ TEPIC 20°37′N 23°00′N. Mostly wild and rugged, Nayarit is broken by W spurs of the SIERRA MADRE OCCIDENTAL. In the NE are broad, tropical plains watered by the SANTIAGO RIVER, a continuation of the LERMA River. Nayarit has two large volcanoes, CEBORUCO (7,480 ft/2,280 m) and Sangangüey (7,546 ft/2,300 m). The volcanic soil, heavy rains, and elevation variations permit cultivation of a variety of products of tropical and temperate agriculture—grain, sugarcane, cotton, coffee, and tobacco. Cattle raising is also important. Forest wealth, little exploited in the past, is rapidly being developed. Mining is a significant part of the state's economy; gold has been mined in AMATLÁN DE CAÑAS municipio and in SANTA MARIA DEL ORO muncipio. Silver has been extracted in COMPOSTELA and LA YESCA (moderate); large deposits of lead, copper, silver, and gold in the state. The coastal swamps are noted bird refuges. The Nayarit region was known to the Spanish early in the sixteenth century, and one of its towns, Compostela (near Tepic), was the first capital of NUEVA GALICIA. Spain did not finally conquer the area until the early seventeenth century. Shortly afterward, Nayarit became a dependency of GUADALAJARA and, upon Mexican independence, part of JALISCO. Continued turbulence led to Nayarit's separation as a territory in 1884; it became a state in 1917. The name Nayarit is given to pre-Columbian clay figurines that are found in the vicinity.

Nayar River (nah-YAHR), c. 55 mi/89 km long (including E headstream), Garhwal district, N UTTAR PRADESH, N central INDIA; formed by junction of two headstreams in Kumaon Himalaya, 22 mi/35 km ENE of LANSDOWNE; flows SW and WNW to Ganga River 6 mi/9.7 km S of Devaprayag. Nayar Dam (650 ft/198 m high), with hydroelectric plant, is 10 mi/16 km NNW of Lansdowne, with second hydroelectric plant at river mouth; has greatly enhanced agricultural area (food grain, sugarcane, cotton).

Nayba River (NEI-bah), approximately 75 mi/121 km long, S SAKHALIN Island, SAKHALIN oblast, RUSSIAN FAR EAST; rises in the W range at 47°35′N 142°30′E; flows SE, past BYKOV, and N to the Sea of OKHOTSK near Starodubskoye. Was called Naibuchi while the area was under Japanese rule (1905–1945).

Nayba River, RUSSIA: see NAIBA RIVER.

Nay-Bourdettes (NAI–boor-DET), commune, PYRÉNÉES-ATLANTIQUES department, AQUITAINE region, SW FRANCE, on the Gave de pau, and 10 mi/16 km. SE of PAU. Produces berets. Has 15th-century church.

Naye, Rochers de, SWITZERLAND: see ROCHERS DE NAYE.

Nayland (NAI-land), town (2001 population 1,864), S SUFFOLK, E ENGLAND, on STOUR RIVER, and 6 mi/9.7 km N of COLCHESTER; 51°59′N 00°53′E. Has 15th-century church. Just W is Wissington, with sugar beet refinery; Norman church.

Naylor, city (2000 population 610), RIPLEY county, S MISSOURI, 13 mi/21 km E of DONIPHAN; 36°34′N 90°36′W. Fruit; timber; cattle; manufacturing (apparel).

Naylor, village, LOWNDES county, S GEORGIA, 13 mi/21 km ENE of VALDOSTA, near ALAPAHA RIVER; 30°55′N 83°05′W. Manufacturing of furniture.

Nayman (nei-MAHN), village, OSH region, KYRGYZSTAN, 30 mi/48 km SW of OSH, near Nayman reservoir; 40°10′N 72°23′E. Cotton. Also spelled NAIMAN.

Nayong (NAH-YUNG), town, ⊙ Nayong county, W GUIZHOU province, CHINA, 30 mi/48 km S of BIJIE; 26°50′N 105°17′E. Grain, tobacco, oilseeds; tobacco processing, food industry, papermaking. Coal mining.

Nayoro (nah-YO-ro), city, N Hokkaido prefecture, N JAPAN, on TESHIO river, and 102 mi/165 km N of SAPPORO; 44°21′N 142°28′E. Railroad junction. Asparagus, pumpkins, rice; trout. Processed fish, bath products.

Nayoshi, RUSSIA: see LESOGORSK.

Nayramadlin Orgil, MONGOLIA: see TAVAN BOGD UUL.

Nayudupeta (NUH-yoo-doo-pe-tah), town, NELLORE district, ANDHRA PRADESH state, S INDIA, 36 mi/58 km S of NELLORE; 13°54′N 79°54′E. Road center; oilseed milling; rice, millet, cashews. Also spelled Nayudupet.

Nazacara (nah-zah-KAH-rah), canton, PACAJES province, LA PAZ department, W BOLIVIA, 13 mi/20 km E of San Andres on the DESAGUADERO River, on the LA PAZ–BERENGUELA road; 17°47′S 68°58′W. Elevation 12,989 ft/3,959 m. Unmined copper deposits. Gas wells in area. Clay, gypsum, and limestone deposits. Agriculture (potatoes, yucca, bananas, rye); cattle.

Nazacara (nah-zah-KAH-rah), town and canton, INGAVI province, LA PAZ department, W BOLIVIA, in the ALTIPLANO, on DESAGUADERO River, and 25 mi/40 km NW of COROCORO; 17°47′S 68°58′W. Alpaca and sheep raising; barley, potatoes.

Naz'a, En (nahz-AH, en), town, LATTAKIA district, W SYRIA, 28 mi/45 km SSE of LATTAKIA; 35°11′N 36°04′E. Sericulture; cotton, tobacco, cereals.

Nazal Esh-Sharqiya, Arab village, Tulkarm district, 7.6 mi/12 km NW of TULKARM, in the SAMARIAN Highlands, WEST BANK. Agriculture (cereals, vegetables, olives).

Nazalt Issa, Arab village, Tulkarm district, 7 mi/1 km N of TULKARM, in the SAMARIAN Highlands, WEST BANK; 32°24′N 35°03′E. Agriculture (citrus, olives, cereals); vineyards.

Nazan, Alaska: see ATKA.

Nazaré (nah-sah-RE), city (2007 population 26,514), E BAHIA state, BRAZIL, on railroad, and 32 mi/51 km W of SALVADOR; 13°03′S 39°01′W. Ships manioc flour, coffee, sugar, rapeseed, hides; manufacturing of tobacco products, distilling, leatherworking, vegetable-oil processing. Manganese mine at Onha (5 mi/8 km W). Formerly spelled Nazareth.

Nazaré (NAH-sah-rai), town (2007 population 4,528), N TOCANTINS state, BRAZIL, 73 mi/117 km NE of PALMAS; 06°20′S 47°45′W.

Nazaré (naz-ah-RAI) or **Praia de Nazaré**, town, LEIRIA district, W central PORTUGAL, on the ATLANTIC OCEAN, 17 mi/27 km SW of LEIRIA; 39°36′N 09°04′W. Popular seaside resort and fishing port. On a height (elevation 360 ft/110 m; seasonal funicular) overlooking town is a noted pilgrimage chapel.

Nazaré, Brazil: see NAZARÉ DA MATA.

Nazaré, Brazil: see NAZARÉ PAULISTA.

Nazaré da Mata (NAH-sah-rai dah MAH-tah), city (2007 population 29,256), E PERNAMBUCO state, NE BRAZIL, on railroad, and 35 mi/56 km NW of RECIFE; 07°44′S 35°14′W; Market center; agriculture (sugar, corn, fruit, manioc). Until 1944, called Nazaré. Formerly spelled Nazareth.

Nazaré do Piauí (NAH-sah-rai do PEE-ah-oo-ee), town (2007 population 6,862), S PIAUÍ state, 28 mi/45 km SE of Floriano; 07°00′S 42°50′W.

Nazareno (NAH-sah-re-no), town (2007 population 7,716), S central MINAS GERAIS state, BRAZIL, 36 mi/58 km SW of SÃO JOÃO DEL REI; 21°20′S 44°35′W.

Nazareno Etla (nah-zah-RAI-no ET-lah), town, in central OAXACA, MEXICO, 7 mi/12 km NW of OAXACA DE JUÁREZ, off Mexico Highway 190; 17°10′N 96°49′W. Elevation 5,194 ft/1,583 m. Agriculture; irrigated by the Atoyac River.

Nazaré Paulista (NAH-sah-rai POU-lee-stah), city (2007 population 14,613), E SÃO PAULO state, BRAZIL, 30 mi/48 km NNE of SÃO PAULO; 23°11′S 46°24′W. Distilling; grain, coffee; poultry. Formerly Nazaré (old spelling, Nazareth).

Nazareth (NAH-zah-ret), agricultural commune (2006 population 10,921), Ghent district, EAST FLANDERS province, NW BELGIUM, 8 mi/12.9 km SW of GHENT; 50°58′N 03°36′E.

Nazareth, city (2006 population 64,600), N ISRAEL, in LOWER GALILEE; 32°42′N 35°17′E. Elevation 987 ft/300 m. Main Arab center of Galilee; most of Nazareth population is Christian Arab. As the home of Jesus Christ, it is a great pilgrimage and tourist center. Also the trade center for an agricultural region. Manufacturing includes processed food, cigarettes, pottery, and building materials. Mineral water is bottled here, and stone is quarried nearby. First mentioned in the New Testament, although its settlement antedates historic times. It was captured by Crusaders in 1099, taken by Saladin in 1187, and retaken by Frederick II in 1229. Muslims conquered Nazareth and massacred its Christian population in 1263. The Ottoman Empire annexed Nazareth in 1517. Adjacent to it, NAZARETH ILIT was established as a Jewish residential town. The Basilica of the Annunciation and the Mosque of Peace are here.

Nazareth (NAH-zuh-rith), town, NELLAI KATTABOMMAN district, TAMIL NADU state, S INDIA, 22 mi/35 km SE of TIRUNELVELI; 08°34′N 77°59′E. Rice, palmyra. Seat of a Christian mission.

Nazareth (NAZ-uh-reth), unincorporated village, NELSON county, KENTUCKY, 3 mi/4.8 km N of BARDSTOWN. Tobacco, grain; livestock; dairying.

Nazareth (NA-zuh-reth), village (2006 population 329), CASTRO county, NW TEXAS, 50 mi/80 km SSW of AMARILLO; 34°32′N 102°05′W. Agricultural area (cattle, sheep; wheat, cotton).

Nazareth (NA-zah-reth), borough (2006 population 6,055), NORTHAMPTON county, E PENNSYLVANIA, 6 mi/9.7 km NW of EASTON, on Shoeneck Creek; 40°44′N 75°18′W. Manufacturing (beef processing, limestone processing; building materials, textiles, consumer goods, paper products, plastic products). Agriculture (apples, soybeans, grain; livestock; dairying). Settled c.1740, incorporated 1863.

Nazareth, Brazil: see NAZARÉ.

Nazareth Ilit, city (2006 population 43,600), N ISRAEL, in LOWER GALILEE; 32°42′N 35°19′E. Elevation 1,646 ft/501 m. Founded in 1957 as a development town; became a city in 1974. Serves as administrative center in the N region and has large industrial zone, including major glass factory. The town was part of British Mandatory Palestine (1918–1948) and was captured by Israeli forces in the 1948 War of Independence. Population 90% Jewish, 10% Arabic.

Nazarezinho (NAH-sah-re-SEEN-yo), town (2007 population 6,933), W central PARAÍBA state, BRAZIL, 14 mi/23 km W of SOUSA; 06°55′S 38°20′W.

Nazarovka (nah-ZAH-ruhf-kah), village, E central RYAZAN oblast, central European Russia, on railroad, 24 mi/39 km W of SASOVO; 54°18′N 41°17′E. Elevation 544 ft/165 m. Agricultural implements.

Nazarovo (nah-ZAH-ruh-vuh), city (2005 population 54,820), SW KRASNOYARSK TERRITORY, SE SIBERIA, RUSSIA, on the CHULYM RIVER, on the Achinsk-Abakan railroad (Adadym station), 148 mi/238 km W of KRASNOYARSK, and 15 mi/24 km S of ACHINSK; 56°04′N 90°23′E. Elevation 915 ft/278 m. In lignite-mining area. Agricultural machinery; food products (dairy). Power station. Founded in 1700. Developed in the mid-1940s. Made city in 1961.

Nazas (NAH-sahs), city and township, DURANGO, N MEXICO, on NAZAS River, and 55 mi/89 km SW of TORREÓN; 25°15′N 104°06′W. Agricultural center (corn, wheat, cotton, sugar, vegetables, fruit).

Nazas (NAH-sahs), river, c.180 mi/290 km long, N MEXICO; rises in the SIERRA MADRE OCCIDENTAL, DURANGO state; flows generally E to disappear through evaporation near TORREÓN. During the wet season it usually inundates a vast desert basin and sometimes reaches Laguna de Mayran. With its control dams, it provides water for irrigating the Laguna district. Also called Sextín River. Dammed for irrigation at LAZARO CARDENAS Dam and at FRANCISCO ZARCO.

Nazca (NAHS-kah), province, extreme S ICA region, SW PERU; ⊙ NAZCA; 15°00′S 75°05′W. Site of famous Nazca lines, geometric designs on the desert floor.

Cross-references are shown in SMALL CAPITALS. The pronunciation guide is shown on page xix. The sources of population figures are shown on page xvii.

Nazca (NAHS-kah), city (2005 population 22,132), ⊙ Nazca province, ICA region, SW PERU, on Nazca River (affluent of Grande River), on PAN-AMERICAN HIGHWAY, and 84 mi/135 km ESE of ICA; 14°50'S 74°57'W. Cotton, alfalfa; livestock. Ruins of Nazca culture, known for its ceramics and for Nazca lines, geometric designs on the desert floor; tourist center. Airport.

Naze (NAH-ze), principal city and port, on N coast of AMAMI-O-SHIMA of island group AMAMI-GUNTO, in RYUKYU ISLANDS, KAGOSHIMA prefecture, SW JAPAN, 229 mi/370 km S of KAGOSHIMA; 28°22'N 129°29'E. Pongee.

Nazeing (NAIZ-ing), village (2001 population 4,675), W ESSEX, E ENGLAND, 4 mi/6.4 km NW of EPPING; 52°23'N 00°59'W. Golf course.

Naze, The (NAIZ), promontory on NORTH SEA, E extremity of ESSEX, E ENGLAND, 5 mi/8 km S of HARWICH; 51°52'N 01°17'E.

Nazik, Lake (□ 12 sq mi/31 sq km), E TURKEY, 40 mi/64 km E of MUS; 6 mi/9.7 km long, 3 mi/4.8 km wide. Elevation 6,190 ft/1,887 m. Sometimes called Nazik.

Nazilli, town (2000 population 105,665), W TURKEY, on IZMIR-DENIZLI railroad, near BÜYÜK MENDERES river, and 28 mi/45 km E of AYDIN; 37°55'N 28°20'E. Manufacturing includes cotton goods from local crop; olives, valonia, barley; antimony, emery, lignite deposits nearby.

Nazimiye, village, E central TURKEY, 40 mi/64 km SSE of ERZINCAN; 39°12'N 39°51'E. Wheat.

Nazimovo, RUSSIA: see PUTYATIN.

Nazinon River, c.200 mi/322 km long, in BURKINA FASO and GHANA, W AFRICA; rises in BURKINA FASO NW of OUAGADOUGOU; enters Ghana 28 mi/45 km WSW of BAWKU; flows SE and SSE to the NAKAMBE RIVER, 5 mi/8 km NW of GAMBAGA. Formerly called Red Volta River.

Nazira (NAH-zi-ruh), ancient *Gargaon* or *Garhgaon*, town, SIBSAGAR district, E central ASSAM state, NE INDIA, in Brahmaputra valley, on tributary of the BRAHMAPUTRA RIVER, 34 mi/55 km ENE of JORHAT; 26°55'N 94°44'E. Tea-trade center; rice, rape and mustard, sugarcane, jute; weaving, printing, and dyeing factory. Coal mining nearby. Railroad junction 2 mi/3.2 km E, at SIMALUGURI, with S spur to NAGINIMARA in Naga Hills district and N spur to MORANHAT, serving tea-garden area. As Gargaon, was capital of Ahom (Shan) kingdom from mid-16th to late 17th centuries.

Nazirhat, BANGLADESH: see SHOLASHAHAR.

Naziya (NAH-zee-yah), town (2005 population 5,720), N central LENINGRAD oblast, NW European Russia, near S shore of Lake LADOGA, on railroad and highway branch, 26 mi/42 km W of VOLKHOV, and 13 mi/21 km NE of MGA; 59°50'N 31°35'E. Elevation 144 ft/43 m. Peatworks; construction supplies.

Nazlet el Qadi (NAHZ-let el KAH-dee), village, SOHAG province, central Upper EGYPT, 3 mi/4.8 km S of TAHTA; 26°45'N 31°24'E. Cotton, cereals, dates, sugarcane. Has Coptic monastery.

Nazran' (nah-ZRAHN), city (2005 population 179,960), ⊙ INGUSH REPUBLIC, N CAUCASUS, SE European Russia, 1,190 mi/1,915 km SSE of MOSCOW, on road and railroad (Rostov-Baku line), 15 mi/24 km NNE of VLADIKAVKAZ; 43°13'N 44°47'E. Elevation 1,768 ft/538 m. Knitwear, electrical instruments, power tools; metalworks; flour milling. A Russian fortress in Caucasian wars (18th–19th centuries). Until 1944 (in Chechen-Ingush ASSR), called Nazran', then after deportations, became Kosta-Khetagurovo village within North Osset ASSR. Renamed city of Nazran' after 1967 in Chechen-Ingush ASSR, until separation of Chechen and Ingush republics in 1992. Also formerly Georgiye-Osetinskoye. Experienced significant population fluctuations because of proximity to armed hostilities in CHECHNYA.

Nazrēt or **Adama**, city (2007 population 239,525), OROMIYA state, central ETHIOPIA, on railroad, and 50 mi/80 km SE of ADDIS ABABA; 08°33'N 39°16'E. Trade center (coffee, hides, cattle, beeswax) and busy market town; citrus and fruit orchards nearby. Formerly Hadama.

Nazwa, OMAN: see NIZWA.

Nazyvayevsk (nah-zi-VYAH-eefsk), city (2006 population 12,590), W OMSK oblast, SW SIBERIA, RUSSIA, in the Ishim lowland, on road junction and the TRANS-SIBERIAN RAILROAD, 90 mi/149km NW of OMSK; 55°34'N 71°21'E. Elevation 423 ft/128 m. Railroad depots, auto shops; tannery, knitting mill. Center of agricultural area; meat packing, dairy, poultry processing. Arose as a station settlement called Sibirskiy Posad in 1911, with the construction of the Omsk-Tyumen railroad; renamed Nazyvayevskaya when a railroad station was completed in 1913. Called Novo-Nazyvayevka from 1935 to 1956, when granted city status and current name.

Nchanga, township, COPPERBELT province, N central ZAMBIA, near CONGO border and KAFUE RIVER, on railroad, 30 mi/48 km NW of KITWE; 12°31'S 27°52'E. In Copperbelt district. Open-pit mining. Produces 30% of Zambia's copper, expected to be depleted by year 2000. Agriculture (peanuts, soybeans, corn, vegetables).

Nchelenge (en-chai-LAI-eng-gai), township, LUAPULA province, N ZAMBIA, 135 mi/217 km N of MANSA, on CONGO border, and on SE shore of LAKE MWERU; 09°21'S 28°44'E. Lusenga Plain National Park to E. Agriculture (vegetables, fruit, rice); fish.

NCR, PHILIPPINES: see NATIONAL CAPITAL REGION.

Ndagaa (uhn-dah-GAH-ah), town, ZANZIBAR NORTH region, N central ZANZIBAR island, E TANZANIA, 11 mi/18 km NE of ZANZIBAR city; 06°03'S 39°16'E. Cloves, copra.

Ndala (uhn-DAH-lah), village, TABORA region, NW central TANZANIA, 35 mi/56 km NE of TABORA, near source of NJOMBE RIVER; 04°46'S 33°14'E. Timber; livestock; subsistence crops.

N'dalatando (uhn-dah-lah-TAHN-do), town, ⊙ CUANZA NORTE province, NW ANGOLA, on Luanda-Malange railroad, and 120 mi/193 km ESE of LUANDA. Elev. 2,460 ft/750 m. Agricultural center (coffee, tobacco, palms, rice, cotton). Formerly called Vila Salazar.

Ndali, town, BORGOU department, N central BENIN, 35 mi/56 km NNE of PARAKOU, on main N road to KANDI; 09°51'N 02°43'E. Cotton; livestock; shea-nut butter.

Ndanda (uhn-DAHN-dah), village, LINDI region, SE TANZANIA, 60 mi/97 km WSW of LINDI; 10°07'S 38°47'E. Cashews, peanuts, grain, bananas; sheep, goats.

Ndande (uhn-DAHN-de), village, LOUGA administrative region, W SENEGAL, on Dakar–Saint-Louis railroad, and 75 mi/121 km NE of DAKAR; 15°16'N 16°30'W. Peanut growing. Sometimes spelled N'Dande.

Ndanga, town, MASVINGO province, SE ZIMBABWE, 33 mi/53 km ESE of MASVINGO; 20°11'S 31°19'E. Lake Mtirikwe reservoir, in KYLE RECREATIONAL PARK, to W. Cattle, sheep, goats; corn. Formerly Glenlivet.

Ndareda, TANZANIA: see DAREDA.

Ndasegera (uhn-dah-sai-GAI-rah), mountain (8,291 ft/2,527 m), MARA region, N TANZANIA, 140 mi/225 km ESE of MUSOMA, near KENYA border, in NE corner of SERENGETI NATIONAL PARK; 01°58'S 35°40'E. Tourist lodge.

Ndé (uhn-DAI), department (2001 population 123,661), West province, CAMEROON; ⊙ Bangangté.

N'Délé (uhn-dai-LAI), town, ⊙ BAMINGUI-BANGORAM prefecture, N CENTRAL AFRICAN REPUBLIC, 260 mi/418 km NE of Bassangoa; 08°25'N 20°38'E. Center of local trade, caravan terminus. Cotton. Repeatedly devastated by Senussite raids in 19th and early 20th centuries. Airport.

N'dele, CONGO: see MAKABA.

Ndélélé (uhn-DAI-lai-lai), town, East province, CAMEROON, near Central Afr. Republic border; 04°03'N 14°57'E. Coffee, cocoa region.

Ndendé (uhn-DEN-dai), city, NGOUNIÉ province, S GABON, 40 mi/64 km SE of MOUILA; 02°21'S 11°21'E. Cattle breeding, poultry raising. Palm oil, food processing.

Ndhlozane River (und-HLOOZ-ahne), c.30 mi/48 km long, SOUTH AFRICA; rises at border between SWAZILAND and South Africa c.40 mi/64 km SSW of MBABANE (Swaziland); flows SE, passes briefly through South Africa; reenters SW Swaziland, joins Assegaai (Mkondo) River 50 mi/80 km S of Mbabane.

Ndian, department (2001 population 129,659), South-West province, CAMEROON; ⊙ MUNDEMBA.

Ndibi Beach (uhn-DEE-bee), town, ABIA state, SE NIGERIA, port on CROSS RIVER, 5 mi/8 km S of AFIKPO; 05°50'N 07°57'E. Palm oil and kernels, cacao, kola nuts, rice.

Ndihimbi, TANZANIA: see DIHIMBA.

N'Dikiniméki (uh-dik-IN-i-mai-kee), village, Central province, central CAMEROON, 30 mi/48 km W of BAFIA; 04°47'N 10°49'E. Also spelled Ndikiniméki.

Ndindi (uhn-DEEN-dee), town, NYANGA province, S GABON, near CONGO border, 60 mi/97 km SSE of TCHIBANGA; 03°48'S 11°12'E. Sawmill and lumber industry.

Ndiza (uhn-DEE-zah), region, RWANDA, 20 mi/32 km ENE of KIGALI. Minor administrative center.

N'djamena (uhn-jah-me-NAH), city and administrative region (2003 population 609,600), ⊙ CHAD, W Chad, on border between CHARI-BAGUIRMI and HADJER-LAMIS administrative regions, near CAMEROON border (to W), and on CHARI River; 12°06'N 15°02'E. Primarily an administrative center. Major regional market for livestock, salt, dates, and grains; meat processing is the chief industry. River port and transportation hub that lies on roads leading to NIGERIA, SUDAN, and the CENTRAL AFRICAN REPUBLIC. Has schools of administration and veterinary medicine and an international airport (N). Founded as Fort-Lamy by the French in 1900. During a period of civil war in the 1980s, Libyan forces intervened here to back Goukouni against Hissein Habre. Renamed N'djamena in 1973. Was capital of former CHARI-BAGUIRMI prefecture. When Chad's administrative divisions were reorganized from fourteen prefectures to twenty-eight departments, N'djamena became a city with department status. Became an administrative region following a decree in October 2002 that reorganized Chad's administrative divisions into eighteen regions. The region is sometimes referred to as Ville de N'djamena.

Ndjolé (uhn-JO-lai), town, MOYEN OGOOUÉ province, W central GABON, on OGOOUÉ RIVER, and 140 mi/225 km ENE of PORT-GENTIL; 00°11'S 10°44'E. Veneer production. Gold-mining and gold-trading center, terminus of navigation on the Ogooué. Sometimes spelled Njolé.

Ndogo, GABON: see SETTÉ-CAMA.

Ndola (en-DO-lah), city (2000 population 374,757), ⊙ COPPERBELT province, N central ZAMBIA, near CONGO border; 12°58'S 28°38'E. It is the main commercial and industrial center of the COPPERBELT REGION. Mining (copper, talc, limestone, emeralds); agriculture (cotton, tobacco, peanuts, grain, soybeans, flowers, coffee). Manufacturing (petroleum refining, cobalt, and copper refineries; steelworks; aluminum and bronze ingots, drilling tools, industrial and agr. chemicals, cement, processed foods, animal feeds, coffee, tea, soap, textiles, medicines, wood products). Seat of Northern Technical College (1964) and a campus of the University of Zambia (1978); Copperbelt Museum Copper mining in Ndola long predates the coming of the Europeans (c.1900).

Ndom (uhn-DAWM), town, LITTORAL province, Cameroon, 60 mi/97 km NW of Yaoundé; 04°22′N 10°51′E.

Ndop (uhn-DAWP), town (2001 population 21,200), North-West province, CAMEROON, 20 mi/32 km E of BAMENDA; 05°59′N 10°27′E. At NW corner of BAMENDJING Reservoir.

Ndoru (uhn-DO-roo), town, Rivers state, S NIGERIA, on NIGER River, 70 mi/113 km NW of PORT HARCOURT. Market center. Fish; rice, yams.

Ndoto, Mount (uhn-DO-to) (8,313 ft/2,534 m), RIFT VALLEY province, N central KENYA, N of MATTHEWS Range. Livestock.

Ndouci (uhn-DOO-see), town, Lagunes region, S CÔTE D'IVOIRE, 6 mi/10 km SE of TIASSALÉ; 05°52′N 04°46′W. Agriculture (manioc, taro, bananas, vegetables, palm oil, coffee, cacao, kola nuts).

Ndougou (uhn-DOO-goo), town, OGOOUÉ-MARITIME province, W GABON, 90 mi/145 km SE of PORT-GENTIL; 01°37′S 09°36′E.

Ndoulo (uhn-DOO-lo), village (2004 population 11,004), DIOURBEL administrative region, W SENEGAL, on Diourbel-Touba branch of Dakar-Niger railroad, 87 mi/140 km E of DAKAR; 14°44′N 16°07′W. Trading post in peanut-growing region. Also spelled N'Doulo.

N'Douo River, REPUBLIC OF THE CONGO: see KOUILOU RIVER.

Ndreketi River, FIJI: see DREKETI RIVER.

Ndu (NDOO), village, Équateur province, NW CONGO, on left bank of Mbomou River; 04°41′S 22°49′E. Elev. 1,505 ft/458 m.

Nduke, SOLOMON ISLANDS: see KOLOMBANGARA.

Ndumbwe (uhn-DOOM-bwai), village, MTWARA region, SE TANZANIA, 18 mi/29 km W of MTWARA, on Mambi River near its entrance to INDIAN OCEAN (to NE); 10°12′S 39°54′E. Cashews, peanuts, copra, bananas, sweet potatoes; sheep, goats.

Ndundu (uhn-DOON-doo), village, PWANI region, E TANZANIA, 80 mi/129 km S of DAR ES SALAAM, on RUFIJI RIVER (ferry); 08°00′S 38°57′E. Cashews, bananas, copra; livestock.

Nea Ankhialos (NAI-ah ahn-KHEE-ah-los), town, MAGNESIA prefecture, SE THESSALY department, N GREECE, 9 mi/14.5 km SW of VÓLOS, on Gulf of VÓLOS; 39°17′N 22°49′E. Tobacco, wheat, olives. Formerly Nea Anchialos.

Nea Artaki (NAI-ah ahr-TAH-kee), town, W ÉVVIA island ÉVVIA prefecture, CENTRAL GREECE department, GREECE, port on N Gulf of Évvia, 4 mi/6.4 km NNE of KHALKÍS; 38°31′N 23°38′E. Wheat, wine; fisheries. Tourism, beaches. Formerly called Vatonta or Vatonda. Formerly spelled Nea Artake.

Nea Filippias, Greece: see FILIPPIAS.

Neagari (ne-AH-gah-ree), town, Nomi county, ISHIKAWA prefecture, central HONSHU, central JAPAN, on SEA OF JAPAN, 12 mi/20 km S of KANAZAWA; 36°26′N 136°27′E.

Neagh, Lough (NAI, LAHK), Gaelic *Loch nEachach*, lake (☐ 153 sq mi/397.8 sq km), central Northern Ireland; 18 mi/29 km long, 11 mi/18 km wide; 54°37′N 06°25′W. It is the largest freshwater body in Ireland or Britain, but is not a deep lake. Fed by the UPPER BANN, BLACKWATER, and other streams and drained to the N by the Lower Bann, it is noted for pollan, trout, and eel fisheries. Mesolithic man is believed to have first appeared in Ireland (c.6000 B.C.E.) near here. Named for the horse-god Eochu, Lord of the Otherworld. According to a legend, quoted by Giraldus Cambrensis, the Norman-Welsh historian, and cited in Thomas Moore's "Let Erin Remember" (*Irish Melodies*), the lake occupies the site of a town which was flooded; buildings may sometimes be seen through the water.

Neah Bay (NEE-uh), unincorporated town (2000 population 794), CLALLAM county, NW WASHINGTON, on NEAH BAY of STRAIT OF JUAN DE FUCA, and 60 mi/97 km NW of PORT ANGELES, in (and headquarters of) Makah Indian Reservation (1990 population 1,214); 48°22′N 124°37′W. Fishing. Makah Cultural Center; Cape Flattery 4 mi/6.4 km W. N end of OLYMPIC NATIONAL PARK; coastal section to S. State's first European settlement made here by Spaniards in 1791.

Nea Ionia (NAI-ah ee-o-NEE-ah), suburb of ATHENS, ATTICA prefecture, ATTICA department, E central GREECE, 4 mi/6.4 km from city center; 38°02′N 23°45′E. Cotton textile manufacturing. Formerly called Podarades.

Nea Kokkinia, Greece: see NIKAIA.

Neales River (NEELZ), 200 mi/322 km long, N central SOUTH AUSTRALIA state, S central AUSTRALIA; rises in hills N of STUARTS RANGE; flows SE, past OODNADATTA, to Lake EYRE; 28°08′S 136°47′E. Usually dry. Sometimes spelled Neale's River.

Neale's River, AUSTRALIA: see NEALES RIVER.

Nealtican, MEXICO: see SAN BUENAVENTURA NEALTICAN.

Nea Michaniona (NAI-ah mee-kha-NYO-nah), village, THESSALONÍKI prefecture, CENTRAL MACEDONIA department, NE GREECE, 12 mi/19 km SSW of THESSALONÍKI, on Gulf of Thessaloníki; 40°28′N 22°52′E. Also spelled Nea Michaniona and Nea Mikhaniona. Formerly spelled Nea Mechaniona.

Neamţ (NAMTS), county, NE ROMANIA, in Moldavia; ⊙ PIATRA NEAMŢ; 47°00′N 26°10′E. In CARPATHIAN Mountains.

Neamţ, ROMANIA: see TÎRGU NEAMŢ.

Neanderthal (ne-AHN-der-tahl), small valley, W GERMANY, E of DÜSSELDORF; 51°14′N 06°52′E. In 1856 the remains of "Neanderthal Man" (*Homo sapiens neandertalensis*) were discovered here.

Nea Orestias, Greece: see ORESTIAS.

Nea Philippias, Greece: see FILIPPIAS.

Neapolis (nai-AH-po-lees), city, LAKONIA prefecture, PELOPONNESE department, extreme SE mainland GREECE, in S part of the easternmost peninsula, on Laconic Gulf.

Neapolis (nai-AH-po-lees), town, LASITHI prefecture, E CRETE department, 9 mi/14.5 km NW of AYIOS NIKOLAOS. Carob, raisins, olives; olive oil.

Neapolis (nee-AP-uh-lis) [Greek=new city], name of many cities in ancient Greek and Roman times, the most important being modern NAPLES, ITALY.

Neapolis, Greece: see KAVÁLLA.

Neapolis, TUNISIA: see NABUL.

Neapolis, UKRAINE: see SIMFEROPOL'.

Neapolis, WEST BANK: see NABLUS.

Nea Psara, Greece: see PSARA.

Near East: see MIDDLE EAST.

Near Islands, Alaska: see ALEUTIAN ISLANDS.

Nea Sfayia (NAI-ah sfah-YAH), SW suburb of ATHENS, ATTICA prefecture, ATTICA department, E central GREECE, 2 mi/3.2 km from city center. Meatpacking industry. Also called Tauron or Tavron, or Nea Sphageia.

Nea Smirni (NAI-ah SMIR-nee), S suburb of ATHENS, ATTICA prefecture, ATTICA department, E central GREECE, 3 mi/4.8 km from city center; 37°57′N 23°43′E. Developed following settlement (1922) of refugees from Smyrna (now IZMIR, Turkey). Formerly spelled Nea Smyrne, Nea Smyrni.

Nea Sphageia, Greece: see NEA SFAYIA.

Neath (NEETH), Welsh *Castell-nedd*, town (2001 population 18,604), Neath Port Talbot, S Wales, on the NEATH RIVER; 51°40′N 03°48′W. Metallurgical and petrochemical industries. Ruins of Neath Abbey. Formerly in WEST GLAMORGAN, abolished 1996.

Neath River (NEETH), Welsh *Afon Nedd*, 25 mi/40 km long, POWYS, S Wales; rises on BLACK MOUNTAINS 9 mi/14.5 km SW of BRECON; flows SW, past RESOLVEN and NEATH, to SWANSEA BAY at BRITON FERRY.

Néau, BELGIUM: see EUPEN.

Neauphle-le-Château (NO-fluh–luh–shah-TO), commune, YVELINES department, ÎLE-DE-FRANCE region, N central FRANCE, 11 mi/18 km W of VERSAILLES; 48°49′N 01°54′E. Liqueur distilling. Just S is 17th-century castle of JOUARS-PONTCHARTRAIN, built by Mansart.

Nea Vyssi (NAI-ah VEE-see), town, ÉVROS prefecture, EAST MACEDONIA AND THRACE department, extreme NE GREECE, on railroad, and near MARITSA (Évros) River and Turkish border, 5 mi/8 km S of EDIRNE (Turkey); 41°35′N 26°32′E. Also written Nea Vysse or Nea Vissi; formerly called Achyrochorion, Akhirokhorion, or Akhyrokhorion.

Nea Zikhni (NAI-ah ZEE-khnee), town, SÉRRAI prefecture, CENTRAL MACEDONIA department, NE GREECE, on road, and 15 mi/24 km ESE of SÉRRAI; 41°02′N 23°50′E. Cotton, tobacco, beans, potatoes. Also spelled Nea Zichni or Nea Zichna. Formerly Zeliachova or Ziliakhova.

Neba (NE-bah), village, Shimoina county, NAGANO prefecture, central HONSHU, central JAPAN, 102 mi/165 km S of NAGANO; 35°14′N 137°35′E.

Nebaj (ne-BAH), town (2002 population 14,500), QUICHÉ department, W central GUATEMALA, at E end of CUCHUMATANES MOUNTAINS, 8 mi/12.9 km NW of SACAPULAS; 15°24′N 91°08′W. Elevation 6,257 ft/1,907 m. Corn, beans; important weaving center (cotton cloth). Part of the "Ixil Triangle," an area that suffered greatly during the civil war of the 1980s. Ixil-speaking inhabitants.

Neba, Jebel, JORDAN: see NEBO, MOUNT.

Nebbi (ne-BEE), administrative district (☐ 1,270 sq mi/3,302 sq km; 2005 population 464,900), NORTHERN region, NW UGANDA, along DEMOCRATIC REPUBLIC OF THE CONGO border (to W), N OF LAKE ALBERT, on ALBERT NILE RIVER; ⊙ NEBBI; 02°25′N 31°10′E. As of Uganda's division into eighty districts, borders ARUA (N), AMURU (E, formed by Albert Nile River), and BULIISA (SE tip, on opposite shore of Lake Albert) districts. Primary inhabitants are the Alur people (part of the broader Nilotic ethnic group). Agricultural area (beans, cassava, maize, millet, simsim, sorghum, and sweet poatoes, cash crops include cotton and coffee; cattle, goats, pigs, and sheep). Fishing in Lake Albert and Albert Nile River also important. Main roads connect Nebbi town to ARUA and GULU towns, as well as Democratic Republic of the Congo.

Nebbi (ne-BEE), town (2002 population 22,741), ⊙ NEBBI district, NORTHERN region, NW UGANDA, 40 mi/64 km SSE of ARUA, W of ALBERT NILE and S of ORA rivers; 02°30′N 31°05′E. Road junction. Tobacco center. Was part of former NILE province.

Nebbio, Le (neb-YO, luh), small region of N CORSICA, FRANCE, extending inland from Saint-Florent on the gulf of same name, generally SW of BASTIA. Vineyards, olive groves, orchards, and many small villages face N toward the MEDITERRANEAN SEA.

Nebel (NE-bel), village, in SCHLESWIG-HOLSTEIN, NW GERMANY, on AMRUM ISLAND; 54°39′N 08°21′E. NORTH SEA resort.

Nebitdag, city, central BALKAN weloyat, W TURKMENISTAN, at the S foot of the Greater BALKAN range, on the TRANS-CASPIAN RAILROAD; 39°30′N 54°22′E. Industrial center of a region yielding oil and natural gas. Founded in 1933 and called Nefte-Dag until the late 1930s. Formerly spelled Nebit-dag.

Nebk, En (NE-BUHK, en), town, DAMASCUS district, SW SYRIA, on the Damascus-Homs highway, and 22 mi/35 km NE of Damascus; 34°01′N 36°42′E. Elevation 4,690 ft/1,430 m. Summer resort. Potatoes, fruits. Also spelled En NABK.

Nebo (NEE-bo), town (2000 population 220), HOPKINS county, W KENTUCKY, 9 mi/14.5 km WNW of MADISONVILLE; 37°22′N 87°38′W. In coal-mining and agricultural (tobacco, grain; livestock) area.

Nebo (NEE-bo), village (2000 population 408), PIKE county, W ILLINOIS, 12 mi/19 km S of PITTSFIELD; 39°26′N 90°47′W. Grain, fruit; livestock.

Nebo, unincorporated village, MCDOWELL county, W central NORTH CAROLINA, 5 mi/8 km ENE of MARION,

Lake JAMES reservoir (CATAWBA RIVER) to N; 35°42′N 81°55′W. Manufacturing (wood products). Lake James State Park to NE.

Nebolchi (NYE-buhl-chee), town (2006 population 2,180), N NOVGOROD oblast, NW European Russia, on the Mda River (tributary of the MSTA River), on road, 21 mi/34 km N of (and administratively subordinate to) LYUBYTINO; 59°07′N 33°21′E. Elevation 357 ft/108 m. Railroad junction; lumbering; mining of quartz sands.

Nebo, Mount (NE-bo), peak (2,644 ft/806 m), N central JORDAN, E of N end of DEAD SEA, 5 mi/8 km NW of MADABA. In the Hebrew Bible, it is the mount from where Moses viewed the Promised Land before his death. Many have believed (since early medieval times) that the site is the modern Jebel NEBA. The peaks, including the summit of Mount Pisgah (or the PISGAH), are also sometimes identified with Ras SIYAGHA a few miles W.

Nebo, Mount (NEE-bo), central UTAH, on border between UTAH and JUAB counties, 27 mi/43 km S of PROVO, in Mount Nebo Wilderness Area of Uinta National Forest. Elevation 11,877 ft/3,620 m. Mount Nebo is the southernmost extent of the middle ROCKY MOUNTAINS in the WASATCH RANGE.

Nebra (NE-brah), town, SAXONY-ANHALT, central GERMANY, on the UNSTRUT RIVER, and 14 mi/23 km NW of NAUMBURG; 51°17′N 11°35′E. Sandstone quarrying; winery. Has remains of ancient castle.

Nebraska, state (□ 77,358 sq mi/201,130.8 sq km; 2006 population 1,768,331), central UNITED STATES, in the GREAT PLAINS region, admitted as the 37th state of the union in 1867; ⊙ LINCOLN; 41°23′N 99°43′W. Nebraska is known as the "Cornhusker State" in reference to the University of Nebraska athletic teams.

Geography
The state is roughly rectangular, except in the NE and the E where the border is formed by the irregular course of the MISSOURI RIVER and in the SW where the state of COLORADO cuts out a rectangular corner. Elsewhere Nebraska is bounded W by WYOMING, N by SOUTH DAKOTA, E by IOWA and MISSOURI, and S by KANSAS. The land rises more or less gradually from 840 ft/256 m in the E to 5,426 ft/1,654 m in the W. The PLATTE RIVER, too shallow and braided for navigation, is formed in W Nebraska by the junction of the NORTH PLATTE and the SOUTH PLATTE rivers, and flows across the state from W to E to join the Missouri S of OMAHA. The river valleys have long provided routes westward, and today several of the transcontinental railroads (pioneered by the Union Pacific in the mid-1860s) and highways follow the valleys. In the far W, sandstone bedrock foundations have led to such spectacular formations as CHIMNEY ROCK and Scotts Bluff. The climate of Nebraska is continental; a low of −40°F/−40°C in the winter is not unusual, and during the short intense summers temperatures may easily reach 110°F/43°C. Rainfall is almost twice as heavy in the E as in the W.

Economy
Undulating farm lands stretch over half the state from the Missouri River westward, where the fertile silt is underlaid by deep loess soil both E and S of the Sand Hills, which occupy much of the NW interior. Nebraska's population is concentrated in the E half of the state, many being farmers who produce grains for the market or for feeding hogs and dairy cattle. Both along the Platte and along the margins of the Sand Hills, irrigation (mainly from wells) permits intensive farming, commonly with ranching and/or wheat farming, which are dependent on rainfall. Beef, both from farming and ranching, is the chief agricultural product of the state. The greatest underground water resources are found in the Great Plains. In this region are Nebraska's two major cities—Omaha, the largest and an important meat- and grain-distribution center, and Lincoln, seat of state government and an

important insurance center—as well as many of the state's larger towns. The Sand Hills of Nebraska dominate the landscape to the W and NW. Their wind-eroded contours now more or less stabilize grass coverage suitable only for cattle grazing; the low sand hills provide protection from the severe prairie winters. Mineral deposits of oil (discovered in CHEYENNE county 1949–1950), sand and gravel, and stone contribute to the state's economy, but agriculture remains the dominant occupational pursuit. To promote agriculture the University of Nebraska maintains experimental agricultural stations throughout the state. A program of soil conservation has been developed to avert a repeat of the 1930s Dust Bowl effect of overgrazing and overplowing of the fragile prairie soils. Forest plantations have been established in parts of the Sand Hills, including Nebraska and Samuel River McKelvie national forests. Nebraska's chief agricultural products are cattle, corn, hogs, wheat, and soybeans. Nebraska ranked second among the states in cattle production in 1990. Nebraska's largest industry is food processing, which derives much of its raw materials from the state's farm products. The state has diversified its industries since World War II; the manufacture of electrical machinery, primary metals, and transportation equipment is also important.

History: to 1854
The Native Americans of the plains—notably the Pawnee—were devoted to hunting the bison, which roamed in great number across the prairie, as well as the pronghorn and lesser animals. The Spanish explorer Francisco Vásquez de Coronado and his men were the first Europeans to visit the region. They probably came through Nebraska in 1541. The French also came and engaged in fur trading in the 18th century, but development began only after the area passed from France to the U.S. in the Louisiana Purchase of 1803. The Lewis and Clark expedition (1804) and the explorations of Zebulon M. Pike (1806) increased knowledge of the country, but the activities of the fur traders were more immediately valuable in terms of settlement. Manuel Lisa, a fur trader, probably established the first trading post in the Nebraska area in 1813. BELLEVUE, the first permanent settlement in Nebraska, developed originally as a trading post. Steamboating on the Missouri River, initiated in 1819, brought much business to the river ports of Omaha and BROWNVILLE. The natural highway formed by the Platte valley was used extensively by pioneers going W over the OREGON TRAIL and also the California and Mormon trails. Nebraska settlers made money supplying the wagon trains with fresh mounts and pack animals, as well as food.

History: 1854 to 1880
Nebraska became a territory after passage of the Kansas-Nebraska Act in 1854. The territory, which initially extended from latitude 40°00′N to the Canadian border, was firmly Northern and Republican in sympathy during the Civil War. The territory was reduced to its present-day size in 1863 by the creation of the territories of Dakota and Colorado. Congress passed an enabling act for statehood in 1864, but the original provision in the state constitution limiting the franchise to whites delayed statehood until 1867. In that year the Union Pacific Railroad was built across the state, and the land boom, already vigorous, became a rush. Farmers settled on free land obtained under the Homestead Act of 1862, and E Nebraska took on a settled look. The population rose from 28,841 in 1860 to 122,993 in 1870. The Pawnee were defeated in 1859, and by 1880 war with the Sioux and other Native American resistance was over. With the coming of the railroads, cow towns, such as OGALLALA and SCHUYLER, were built up as shipping points on overland cattle trails. Buffalo Bill's Wild West Shows were opened in Nebraska in 1882. Farmers had long

since been staking out homestead claims across the Sand Hills to the high plains, but ranches also prospered in the state. The ranchers, trying to preserve the open range, ruthlessly opposed the encroachment of the farmers, but the persistent farmers won. Many conservationists believe that much land was plowed under that should have been left with grass cover to prevent erosion in later dust storms. Nature was seldom kind to the people of Nebraska.

History: 1880 to World War II
The ruinous cold of the winter of 1880–1881 created problems for ranching, and farmers were plagued by insect hordes from 1856 to 1875, by prairie fires, and by the recurrent droughts of the 1890s. Many farmers joined the Granger movement in the lean 1870s and the Farmers' Alliances of the 1880s. In the 1890s many beleaguered farmers, faced with ruin and angry at the monopolistic practices of the railroads and the financiers, formed marketing and stock cooperatives and voiced their discontent by joining the Populist party. The first national convention of the Populist party was held at Omaha in 1892, and Nebraska's most famous son, William Jennings Bryan, headed the Populist and Democratic tickets in the presidential election of 1896. Populists held the governorship of the state from 1895 to 1901. Improved conditions in the early 1900s caused Populism to decline in the state, and the return of prosperous days was marked by progressive legislation, the building of highways, and conservation measures. The flush of prosperity, largely caused by the demand for foodstuffs during World War I, was almost feverish. Overexpansion of credits and overconfidence made the depression of the 1920s and 1930s all the more disastrous. Many farmers were left destitute, and many were able to survive only because of the moratorium on farm debts in 1932. They received Federal aid in the desperate years of drought in the 1930s.

History: World War II to Present
Better weather and the huge food demands of World War II renewed prosperity for Nebraskans. Since the war, efforts have continued to make the best use of the water supply, notably in such Federal plans as the Missouri River basin project (a vast dam and water-diversion scheme), but the big increase in irrigation from wells has been far more important for Nebraska. The state has attempted to diversify its economic base in order to reduce its dependence on meat processing and agriculture, industries which have destabilized the state economy for decades. These efforts have been generally successful in existing cities.

Cultural and Historic Points of Interest
Among Nebraska's noted citizens have been the pioneer and historian Julius Sterling Morton, who originated Arbor Day for tree planting, and author Willa Cather, who depicted pioneer Nebraska in her novels *My Ántonia* and *O Pioneers!* Points of interest include Father Flanagan's BOYS TOWN, near Omaha; the Fort Niobrara National Wildlife Refuge, near VALENTINE; the Oglala National Grassland in NW corner; and the HOMESTEAD MONUMENT, near BEATRICE. The pioneers' migration W over the Oregon Trail is commemorated by the Scotts Bluff National Monument and the Chimney Rock National Historic Site. AGATE FOSSIL BEDS NATIONAL MONUMENT in far W protects major Miocene mammal fossil site. The prairie, especially Sand Hills, abounds in beaver and native and migratory bird species. Hundreds of fresh and alkali lakes in the state attract sportsmen and campers. The state's leading institution of higher education is the University of Nebraska, mainly at Lincoln. The Strategic Air Command (SAC) at OFFUTT AIR FORCE BASE is S of Omaha.

Government
Nebraska's present constitution was adopted in 1875. It was amended in 1982 to ensure that rangeland and farmland could only be sold to individuals and

Area is shown by the symbol □, and capital city or county seat by ⊙.

family-farm corporations. A governor elected for a four-year term heads the executive branch. Dave Heineman is the current governor. The legislature was made unicameral by constitutional amendment in 1934, with forty-nine members elected on a nonpartisan basis for terms of four years. The state elects three representatives and two senators to the U.S. Congress; the state has five electoral votes in presidential elections.

Nebraska has ninety-three counties: ADAMS, ANTELOPE, ARTHUR, BANNER, BLAINE, BOONE, BOX BUTTE, BOYD, BROWN, BUFFALO, BURT, BUTLER, CASS, CEDAR, CHASE, CHERRY, CHEYENNE, CLAY, COLFAX, CUMING, CUSTER, DAKOTA, DAWES, DAWSON, DEUEL, DIXON, DODGE, DOUGLAS, DUNDY, FILLMORE, FRANKLIN, FRONTIER, FURNAS, GAGE, GARDEN, GARFIELD, GOSPER, GRANT, GREELEY, HALL, HAMILTON, HARLAN, HAYES, HITCHCOCK, HOLT, HOOKER, HOWARD, JEFFERSON, JOHNSON, KEARNEY, KEITH, KEYA PAHA, KIMBALL, KNOX, LANCASTER, LINCOLN, LOGAN, LOUP, MADISON, MCPHERSON, MERRICK, MORRILL, NANCE, NEMAHA, NUCKOLLS, OTOE, PAWNEE, PERKINS, PHELPS, PIERCE, PLATTE, POLK, RED WILLOW, RICHARDSON, ROCK, SALINE, SARPY, SAUNDERS, SCOTTS BLUFF, SEWARD, SHERIDAN, SHERMAN, SIOUX, STANTON, THAYER, THOMAS, THURSTON, VALLEY, WASHINGTON, WAYNE, WEBSTER, WHEELER, and YORK.

Nebraska City, city (2006 population 7,137), ⊙ OTOE county, SE NEBRASKA, 40 mi/64 km S of OMAHA, and on MISSOURI RIVER at IOWA state line; 40°40′N 95°51′W. Elevation 1,029 ft/314 m. Railroad junction. Grew as river port; has barge dock. Commercial and processing center for cattle-raising and agricultural region; dairy products; livestock; grain, apples. Manufacturing (die casting, beef processing; machinery, apparel). State school for blind. Arbor Lodge State Historical Park and Arboretum (has over 100 species of trees honoring J. Sterling Morton, founder of Arbor Day, on W edge of town). Arbor Day Lied Lodge & Conference Center designed to serve conservation groups. City is referred to as Apple Capital (by Nebraskans). Riverview Marina State Recreation Area to N. Incorporated 1855.

Nebrodi Mountains (NE-bro-dee), N SICILY, ITALY; extending 45 mi/72 km NE from MADONIE to PELORITANI mountains; 37°54′N 14°35′E. Rise to 6,060 ft/1,847 m in Mount SORI. Chestnut, oak, beech forests. Source of SIMETO and ALCANTARA rivers. Also called Monti Caronie.

Nebug (NYE-boog), settlement (2005 population 3,715), SE KRASNODAR TERRITORY, S European Russia, on the BLACK SEA coast, on road, 52 mi/84 km SSE of KRASNODAR, and 4 mi/6 km N of TUAPSE, to which it is administratively subordinate; 44°10′N 39°02′E. Elevation 692 ft/210 m. Seaside health resort; sanatoria, hotels.

Nebula (ne-BOO-lah), village, ORIENTALE province, N CONGO, 95 mi/153 km ESE of BUTA; 01°47′N 26°31′E. Gold mining.

Nebyloye (nee-bi-LO-ye), village, N VLADIMIR oblast, central European Russia, on road, 23 mi/37 km NW of VLADIMIR; 56°22′N 40°00′E. Elevation 547 ft/166 m. Agricultural products.

Necaxa, MEXICO: see NUEVO NECAXA.

Necaxa River (ne-KA-shah), c.125 mi/201 km long, in PUEBLA and VERACRUZ, MEXICO; rises NW of ZACATLÁN; flows NE, past HUAUCHINANGO, to the Gulf of MEXICO at TECOLUTLA; 20°16′N 97°27′W. Used extensively for irrigation and hydroelectric power. Its lower course in Veracruz is called Tecolutla. The large Necaxa Falls are NW of Huauchinango, near Veracruz state border.

Necedah (ne-SEE-duh), village (2006 population 857), JUNEAU county, central WISCONSIN, on YELLOW RIVER, and 28 mi/45 km SSW of WISCONSIN RAPIDS; 44°01′N 90°04′W. Railroad junction. Wood products, dairy products; manufacturing (machinery, wood products, fabricated metal products). Necedah National Wildlife Refuge to NW; Central Wisconsin Conservation Area to W; Petenwell Dam and Lake to NE; Buckhorn State Park to SE, on CASTLE ROCK LAKE RESERVOIR.

Necessity, Fort, Pennsylvania: see FORT NECESSITY.

Nechako (ni-CHAK-o), river, 287 mi/462 km long, BRITISH COLUMBIA, W Canada; rises in TETACHUCK and OOTSA lakes, W central British Columbia; flows NE, then E to the FRASER River at PRINCE GEORGE; 53°55′N 122°42′W. Kenney Dam (325 ft/99 m high; completed 1952) and Kemano Dam (320 ft/98 m high; completed 1954) are among the highest dams in Canada.

Nechanice (NE-khah-NYI-tse), German *Nechanitz*, town, VYCHODOCESKY province, NE BOHEMIA, CZECH REPUBLIC, 8 mi/12.9 km WNW of HRADEC KRÁLOVÉ; 50°14′N 15°38′E. In a sugar-beet district. Manufacturing (machinery). Has a cemetery that dates back to 1866, which holds casualties of battle of Sadová (4 mi/7 km NE).

Nechanitz, CZECH REPUBLIC: see NECHANICE.

Nechayevka (nye-CHAH-eef-kah), village, central PENZA oblast, E European Russia, on road and railroad (Simanshchina station), 22 mi/35 km WNW of PENZA; 53°16′N 44°28′E. Elevation 921 ft/280 m. In grain area; flour milling.

Nechayevka (nee-CHAH-eef-kah), village (2005 population 7,010), NW DAGESTAN REPUBLIC, SE European Russia, on road, 51 mi/82 km NW of MAKHACHKALA, and 15 mi/24 km N of KHASAVYURT; 43°16′N 46°55′E. In oil- and gas-producing region.

Neche (NECH-ee), village (2006 population 416), PEMBINA county, NE NORTH DAKOTA, near Canadian (MANITOBA) border, port of entry 14 mi/23 km NNE of CAVALIER, and on PEMBINA RIVER; 48°58′N 97°32′W. Founded in 1875 and incorporated in 1883.

Neches River (NAI-chiz), 416 mi/669 km long, E TEXAS; rises in VAN ZANDT county; flows generally SE, past BEAUMONT, to head of Sabine Lake. Has deepwater channel (an arm of SABINE-NECHES WATERWAY) from Beaumont (port) to Sabine Lake, NE of PORT ARTHUR. Plans were developed in 1947 for Rockland Dam and reservoir (for flood control, power, navigation) in N TYLER county. LAKE PALESTINE reservoir 15 mi/24 km SW of Tyler; B.P. STEINHAGEN LAKE reservoir 50 mi/80 km N of Beaumont (flows through BIG THICKET PRESERVE between lakes and Beaumont).

Nechí River (nai-CHEE), c.150 mi/241 km long, ANTIOQUIA department, N central COLOMBIA; rises in Cordillera CENTRAL SW of YARUMAL; flows NE and N, past ZARAGOZA, to CAUCA River near BOLÍVAR department border; 08°03′N 74°47′W. Main affluent, PORCE River. Navigable for small craft during rainy season. Along its course are rich gold placer mines.

Nechisar National Park (NEK-i-sar), SOUTHERN NATIONS state, S ETHIOPIA, 20 mi/32 km E of ARBA MINCH; 06°00′N 37°55′E. Includes parts of E shores of lakes CHAMO and ABAYA; game reserve with plains and forests.

Neckar (NEK-kahr), river, 231 mi/371 km long, GERMANY; rises in the BLACK FOREST, SW GERMANY; flows generally N past TÜBINGEN, STUTTGART, and HEILBRONN, then W past HEIDELBERG before joining the RHINE RIVER at MANNHEIM. The Neckar is celebrated for its scenic charm; its hilly banks are covered with fine vineyards, orchards, and woods; there are several castles and ruins along its course. Has over twenty hydroelectric power plants.

Neckar-Alb, GERMANY: see NECKAR-JURA.

Neckarau (NEK-kahr-ou), S suburb of MANNHEIM, BADEN-WÜRTTEMBERG, SW GERMANY; 49°29′N 08°28′E.

Neckarbischofsheim (NEK-kahr-BI-shofs-heim), town, LOWER NECKAR, BADEN-WÜRTTEMBERG, GERMANY, 12 mi/19 km E of WIESLOCH; 49°18′N 08°58′E. Chicory, sugar beets, fruit.

Neckargemünd (nek-kahr-ge-MYOOND), town, LOWER NECKAR, BADEN-WÜRTTEMBERG, GERMANY, on NECKAR RIVER, at mouth of the ELSENZ RIVER, and 5 mi/8 km E of HEIDELBERG; 49°23′N 08°48′E. Leatherworking; red-sandstone quarrying. Has ruined castle.

Neckar-Jura (NEK-kahr–YOO-rah), German *Neckar-Alb* (NEK-kahr–AHLB), region (□ 977 sq mi/2,540.2 sq km), BADEN-WÜRTTEMBERG, SW GERMANY. Chief town is TÜBINGEN. Bounded by region of STUTTGART (N), Danube-Iller (E), LAKE CONSTANCE–UPPER SWABIA (SE), BLACK FOREST–BAAR-HEUBERG (SW), and NORTHERN BLACK FOREST (NW).

Neckarsteinach (nek-kahr-SHTEIN-ahkh), town, S HESSE, W GERMANY, at SW foot of the ODENWALD, on the NECKAR RIVER, and 6 mi/9.7 km E of HEIDELBERG; 49°25′N 08°50′E. Woodworking. Has four castles.

Neckarsulm (nek-kahrs-ULM), town, BADEN-WÜRTTEMBERG, GERMANY, on the NECKAR RIVER, and 3 mi/4.8 km N of HEILBRONN; 49°11′N 09°14′E. Manufacturing (bicycles, motorcycles, pistons; aluminum smelting, flax and hemp weaving). Castle (built 1484–1806).

Neckartenzlingen (nek-kahr-TENTS-ling-uhn), village, STUTTGART district, BADEN-WÜRTTEMBERG, SW GERMANY, on the NECKAR RIVER, and on the E slope of Schönbuch Mountains, 13 mi/21 km S of STUTTGART; 48°35′N 09°14′E. Agriculture (vegetables); manufacturing (paper, machinery).

Neck City, city (2000 population 119), JASPER county, SW MISSOURI, on SPRING RIVER, and 12 mi/19 km N of JOPLIN; 37°15′N 94°26′W.

Necker Island, in N PACIFIC, HONOLULU county, HAWAII, c.430 mi/692 km NW of HONOLULU, 8 mi/12.9 km N of TROPIC OF CANCER; 23°34′N 164°42′W. Under jurisdiction of city and county of Honolulu; had small Hawaiian settlement. Highest elevation 277 ft/84 m. Part of Hawaiian Islands National Wildlife Refuge. From here, Necker Ocean Ridge branches to SW from the Hawaiian Ridge.

Necochea (ne-ko-CHAI-ah), city (□ 2,612 sq mi/6,791.2 sq km), ⊙ Nechochea district (□ 2,612 sq mi/6,765 sq km; 1991 pop. 84,684), S BUENOS AIRES province, ARGENTINA, port on the ATLANTIC OCEAN at mouth of QUEQUÉN GRANDE RIVER (opposite Quequén), 75 mi/121 km SW of MAR DEL PLATA; 38°15′S 59°15′W. Seaside resort and casino; cattle raising, dairying, flour milling; trout fishing. Has national college.

Neda (NAI-dah), town, LA CORUÑA province, NW SPAIN, at head of Ferrol Bay, 4 mi/6.4 km ENE of EL FERROL; 43°20′N 08°09′W. Lumber; livestock; cereals in area.

Nedan (ne-DAHN), village, LOVECH oblast, PAVLIKENI obshtina, N BULGARIA, 21 mi/34 km NNE of SEVLIEVO; 43°18′N 25°16′E. Horticulture; grain; livestock.

Neddick, Cape (NED-ik), promontory, SW MAINE, with offshore lighthouse, 3 mi/4.8 km NNE of YORK HARBOR.

Neded (nye-DYET), Hungarian *Negyed*, village, ZAPADOSLOVENSKY province, S SLOVAKIA, on VÁH RIVER, and 29 mi/47 km SSW of NITRA; 48°01′N 17°59′E. Has railroad terminus; agricultural center (sugar beets, barley, wheat). Has food processing; manufacturing of building materials. Under Hungarian rule between 1938–1945.

Nédélec (nai-dai-LEK), canton (□ 143 sq mi/371.8 sq km; 2006 population 409), Abitibi-Témiscamingue region, SW QUEBEC, E Canada, 6 mi/10 km from Notre-Dame-du-Nord; 47°40′N 79°27′W.

Nédéley (ne-dai-LAI), village, southernmost BORKOU-ENNEDI-TIBESTI administrative region, N CHAD, near borders of BATHA and KANEM administrative regions, 200 mi/322 km NE of MOUSSORO; 15°34′N 18°10′E.

Nedelino (ne-DE-lee-no), city, PLOVDIV oblast, Nedelino obshtina (1993 population 9,224), S BULGARIA, in the SE RODOPI Mountains, 6 mi/10 km N of ZLATOGRAD; 41°27′N 25°25′E. Tobacco; livestock; supplies the mining region with produce. Formerly known as Uzundere.

Nedenes (NAI-duh-nais), coastal lowland in AUST-AGDER county, S NORWAY, between the SKAGERRAK strait and the interior highlands. Nedenes was the name of Aust-Agder county until 1919.

Nederland (NE-duhr-land), city (2006 population 16,454), JEFFERSON county, SE TEXAS, suburb 6 mi/9.7 km NW of downtown PORT ARTHUR, near NECHES RIVER; 29°58′N 94°00′W. Primarily a residential suburb between BEAUMONT and PORT ARTHUR, it has two oil companies and a chemical industry. Rice is a major cash crop; soybeans; cattle. Manufacturing (electric products, barges, chemicals and chemical processing). Founded by Dutch settlers as a rice-farming community in 1897. Jefferson County Airport to SW. Incorporated 1940.

Nederland, town (2000 population 1,394), BOULDER county, N central COLORADO, on headstream of BOULDER CREEK, on W end of Barler Reservoir, in FRONT RANGE, and 30 mi/48 km WNW of DENVER; 39°57′N 105°30′W. Elevation 8,236 ft/2,510 m. Resort; tungsten mines. Lake Eldora dam serves hydroelectric plant. University of Colorado Observatory. Area surrounded by Roosevelt National Forest.

Neder Rijn, NETHERLANDS: see LOWER RHINE RIVER.

Nederweert (NAI-duhr-vert), town, LIMBURG province, SE NETHERLANDS, 3 mi/4.8 km NE of WEERT, at junction of WESSEM-NEDERWEERT (SE), ZUID-WILLEMSVAART (N, SW), and NOORDERVAART (NE) canals; 51°17′N 05°45′E. Grote Peel National Park to NW (park visitors center in town). Dairying; cattle, hogs; agriculture (grain, vegetables).

Nedon River (NE-[th]on), 14 mi/23 km long, in MESSINIA prefecture, SW PELOPONNESE department, extreme SW GREECE; rises in TAIYETOS Mountains; flows SW to Gulf of MESSENIA at KALAMATA. Unnavigable. Also Nedhon River.

Nédouncadu, INDIA: see NEDUNGADU.

Nedre Fryken, SWEDEN: see FRYKEN.

Nedrigaylov, UKRAINE: see NEDRYHAYLIV.

Nedroma (ne-dro-MAH), town, TLEMCEN wilaya, W ALGERIA, in coastal range (Trara) of the TELL ATLAS Mountains, 26 mi/42 km NW of TLEMCEN; 35°00′N 01°44′W. Trades in olives and cereals; to N are rich farming plains growing olives, cereals, fruit, and vegetables. Also, palm fiber processing and handicrafts, such as pottery, wood carving, and embroidery. Small zinc mines and marble quarries to S. Some of the finest woolen bedcovers made in weavers' workshops here.

Nedrow, hamlet, ONONDAGA county, central NEW YORK, just S of SYRACUSE; 42°59′N 76°08′W. Quarrying. Onondaga Indian Reservation is nearby.

Nedryhayliv (ne-dri-HEI-leef) (Russian *Nedrigaylov*), town, S central SUMY oblast, UKRAINE, on the SULA RIVER, and 18 mi/29 km ENE of ROMNY; 50°50′N 33°53′E. Elevation 433 ft/131 m. Raion center. Fruit cannery, dairy. Dates from the 17th century, town since 1958.

Nedungadu (NAI-duhn-guh-doo), French *Nédouncadu*, town, KARAIKAL district, PONDICHERRY Union Territory, S INDIA, 5 mi/8 km NW of KARAIKAL; 10°58′N 79°46′E.

Neduntivu Island, SRI LANKA: see DELFT.

Nedvedice (NED-vye-DYI-tse), Czech *Nedvědice*, German *Nedwieditz*, village, JIHOMORAVSKY province, N MORAVIA, CZECH REPUBLIC, on SVRATKA RIVER, on railroad, and 22 mi/35 km NW of BRNO; 49°26′N 16°20′E. Summer resort.

Nedwieditz, CZECH REPUBLIC: see NEDVEDICE.

Neebing (NEE-beeng), town, NW ONTARIO, E central Canada, 18 mi/28 km from THUNDER BAY; 48°11′N 89°28′W.

Neebing (NEE-beeng), former township (□ 337 sq mi/876.2 sq km; 2001 population 2,049), W central ONTARIO, E central Canada; 48°21′N 89°19′W. Amalgamated into city of THUNDER BAY in 1970.

Neebish Island (NEE-bish), E UPPER PENINSULA, N MICHIGAN, in SAINT MARYS RIVER, (U.S.-Can. border), just W of SAINT JOSEPH ISLAND (ONTARIO), and 15 mi/24 km SSE of SAULT SAINTE MARIE; c.8 mi/12.9 km long, 4 mi/6.4 km wide; 46°17′N 84°09′W. Resort; some agriculture. Car ferry from Barbeau, on mainland.

Needham (NEE-duhm), town (2000 population 28,911), NORFOLK county, E MASSACHUSETTS, suburb of BOSTON; 42°17′N 71°14′W. Although largely residential, paper products, electronic equipment and software, and other items are manufactured here. Founded 1680, set off from DEDHAM and incorporated 1711.

Needham Market (NEED-uhm MAH-kit), town (2001 population 4,574), central SUFFOLK, E ENGLAND, on GIPPING RIVER, and 9 mi/14.5 km NW of IPSWICH; 52°09′N 01°03′E. Bacon and ham curing. Has 15th-century church and old grammar school.

Needham's Point, headland, protecting S end of CARLISLE BAY, roadstead off BRIDGETOWN, S coast of BARBADOS. Now site of large hotel. Has fort and lighthouse.

Needle Mountain, WYOMING: see ABSAROKA RANGE.

Needles, city (□ 30 sq mi/78 sq km; 2000 population 4,830), SAN BERNARDINO county, SE CALIFORNIA, 150 mi/241 km ENE of SAN BERNARDINO, and 38 mi/61 km SW of KINGMAN (ARIZONA), on COLORADO RIVER, on upper reach of LAKE HAVASU RESERVOIR (Arizona state line); 34°49′N 114°37′W. Manufacturing (building materials). Trade center for mines, irrigated farms, and Fort Mojave Indian Reservation (Arizona, NEVADA, California) to N. Fishing; waterfowl hunting in river marshes. Temperatures here are extremely high (often over 100°F/38°C), with wide daily range and low humidity. Davis Dam (Lake Mead National Recreation Area) 22 mi/35 km to N; Lake Havasu National Wildlife Refuge (Arizona) to SE; Chemehuevi Indian Reservation to SE; SACRAMENTO MOUNTAINS to SW. Founded 1883, incorporated 1913.

Needles, The, ENGLAND: see WIGHT, ISLE OF.

Needville (NEED-vil), town (2006 population 3,448), FORT BEND county, SE TEXAS, 11 mi/18 km S of ROSENBERG; 29°23′N 95°50′W. Railroad junction; agricultural area (cattle; rice; cotton; soybeans, vegetables; nurseries); oil and gas; manufacturing (meat processing). Brazos Bend State Park to E.

Needwood Forest (NEED-wud FAH-rist), former royal forest (□ 16 sq mi/41.6 sq km), E STAFFORDSHIRE, W ENGLAND, extending along the TRENT RIVER and DERBYSHIRE border, near BURTON UPON TRENT.

Neeltje Jans (NAIL-chuh YAHNS), artificial island, SW NETHERLANDS, part of EASTERN SCHELDT storm surge barrier. Site of Delta Exposition.

Neely Henry Lake, Alabama, see: H. NEELY HENRY LAKE.

Neelyville, town (2000 population 487), BUTLER county, SE MISSOURI, near BLACK RIVER, 15 mi/24 km SSW of POPLAR BLUFF; 36°33′N 90°30′W. Cotton, rice.

Ñeembucú, department (□ 4,690 sq mi/12,194 sq km; 2002 population 76,348), S PARAGUAY; ⊙ PILAR; 27°00′S 58°00′W. A triangle formed by PARAGUAY and PARANÁ rivers, bordered W and S by ARGENTINA, NE by LAKE YPOÁ; marshy lowlands of department are intersected by TEBICUARY RIVER. Has subtropical, humid climate. Lumbering and agricultural area (oranges, sugarcane, cotton, corn, and especially cattle). Processing of agricultural products concentrated at Pilar.

Neemuch, INDIA: see NIMACH.

Neenah [Native Amer.=water], city (2006 population 24,831), WINNEBAGO county, E WISCONSIN, a suburb 7 mi/11.3 km SSW of APPLETON, on LAKE WINNEBAGO at the mouth of the FOX RIVER; 44°10′N 88°28′W. Railroad junction. Located in a dairy-farming region, Neenah is known, with its twin city MENASHA, as a center for the manufacturing of paper and paper

products. Manufacturing (rubber products, dairy products, chemicals, clothing, wood products, paper products; printing and publishing; foundries). Neenah's industrial development began c.1850 when nearby flour mills were opened. In 1865 its paper industry was established. Of interest is a replica of the home of James Duane Doty, who was the second governor of Wisconsin Territory. Hydroelectric power is generated for Neenah and Menasha by falls of the Fox River. Bergstrom Art Centre. Settled c.1835 on the site of a Winnebago village, incorporated as a city 1873.

Neepawa (NEE-puh-wah), town (2001 population 3,325), SW MANITOBA, W central Canada, on WHITEMUD RIVER, 35 mi/56 km NE of BRANDON, partly surrounded by Langford rural municipality, and partly in Rosedale rural municipality; 50°14′N 99°27′W. Woodworking, marble processing; grain; livestock. Town name is Cree for "plenty."

Neer (NER), village, LIMBURG province, SE NETHERLANDS, 5 mi/8 km N of ROERMOND; 51°15′N 05°59′E. MEUSE RIVER passes to E. Dairying; livestock; grain, vegetables.

Neerlandia, hamlet, central ALBERTA, W Canada, 14 mi/22 km W of BARRHEAD, in Barrhead County No. 11; 54°20′N 114°22′W.

Neeroeteren (nai-ROO-tuh-ruhn), agricultural village in commune of MAASEIK, Maaseik district, LIMBURG province, NE BELGIUM, 4 mi/6.4 km W of Maaseik; 51°05′N 05°42′E. Has fifteenth-century Gothic church.

Neerpelt (NER-pelt), commune (2006 population 16,140), Maaseik district, LIMBURG province, BELGIUM, 10 mi/16 km E of LOMMEL, near NETHERLANDS border; 51°13′N 05°25′E.

Neerwinden (NER-win-duhn), village in commune of LANDEN, Leuven district, BRABANT province, E BELGIUM, 5 mi/8 km SSE of TIENEN; 50°46′N 05°03′E. In the War of the Grand Alliance the FRENCH under Marshal Luxembourg defeated (1693) William III of ENGLAND here. In the French Revolutionary Wars, a French defeat (1793) here resulted in the defection of General Dumouriez to the AUSTRIANS.

Neeses (NEES-is), village (2006 population 399), ORANGEBURG county, W central SOUTH CAROLINA, 15 mi/24 km W of ORANGEBURG; 33°32′N 81°07′W. Agriculture includes livestock; grain, cotton, tobacco, pecans, peaches.

Nee Soon, town, N central Singapore island, SINGAPORE, suburb 8 mi/12.9 km N of downtown SINGAPORE; 01°24′N 103°49′E. Population is mainly Chinese. Catchment Area Nature Reserve to SW. Selatar (W) and Sungeir Selatar (E) reservoirs nearby.

Nefasit (nuh-FAH-seet), town, SEMENAWI KAYIH BAHRI region, central ERITREA, on railroad, and 10 mi/16 km E of ASMARA; 15°19′N 39°05′E. Elevation c.5,400 ft/1,646 m. Road junction; coffee, bananas. The Debre Bizen monastery (founded 1361 by Abune Filipos) is on a hill overlooking the town.

Nefas Mewcha (NAI-fahs MOO cha), town (2007 population 19,577), AMHARA state, N central ETHIOPIA, 25 mi/40 km ESE of DEBRE TABOR; 11°45′N 38°23′E. Market town in highland region.

Neffs (NEFS), unincorporated village (□ 4 sq mi/10.4 sq km; 2000 population 1,138), BELMONT county, E OHIO, 6 mi/10 km ESE of ST. CLAIRSVILLE; 40°01′N 80°49′W. In coal-mining area.

Neffsville (NEFS-vil), unincorporated town, LANCASTER county, SE PENNSYLVANIA; residential suburb 4 mi/6.4 km N of LANCASTER; 40°06′N 76°18′W. Lancaster Airport to N; Pennsylvania Farm Museum to E.

Nefteabad, town, NE LENINOBOD viloyat, TAJIKISTAN, on railroad, and 10 mi/16 km SE of KANIBADAM; 40°12′N 70°34′E. Oil field (producing since 1935).

Neftechala (nyef-te-chah-LAH), town, SE AZERBAIJAN, near KURA RIVER mouth, 80 mi/129 km SSW of BAKY; 39°36′N 49°18′E. Railroad terminus; oil fields; chemical plant (iodine production), gas refinery, petroleum stabilization plant. Developed 1931.

Neftegorsk (nyef-tye-GORSK), city (2006 population 19,595), SE SAMARA oblast, E European Russia, on the Svezhava River (tributary of the SAMARA River), on highway, 65 mi/105 km SE of SAMARA; 52°49′N 51°12′E. Elevation 160 ft/48 m. Processing of natural gas and petroleum; food industries (bakery, dairy). Founded in 1960 to exploit an oil field. Made city in 1980.

Neftegorsk (nyef-tye-GORSK), town (2005 population 4,810), SE KRASNODAR TERRITORY, S European Russia, in the NW foothills of the Greater CAUCASUS Mountains, on road and railroad spur, 25 mi/40 km SW of MAYKOP; 44°22′N 39°42′E. Elevation 1,207 ft/367 m. Petroleum refining. Developed in the 1930s.

Neftekamsk (nyef-tee-KAHMSK), city (2005 population 128,350), NW BASHKORTOSTAN Republic, E European Russia, on the E bank of the lower KAMA River before it expands into the Nizhnekamsk Reservoir, on crossroads and railroad spur, 150 mi/241 km NW of UFA; 56°05′N 54°16′E. Elevation 314 ft/95 m. Petroleum production; automobile plant; concrete; agricultural machinery; timber; leather goods. Arose in 1957 as a workers settlement of Kaseyevo, or Kasevo, with the development of the Arlansk oil field. Made city and renamed in 1963.

Neftekumsk (nyef-tye-KOOMSK), city (2006 population 28,845), E STAVROPOL TERRITORY, N CAUCASUS, SE European Russia, on the KUMA River, on road, 190 mi/306 km E of STAVROPOL, and 50 mi/80 km E of railroad (Budennovsk station); 44°40′N 44°55′E. Elevation 170 ft/51 m. Petroleum and natural gas processing. Made city in 1968.

Nefteyugansk (nyef-tee-yoo-GAHNSK), city (2005 population 114,265), central KHANTY-MANSI AUTONOMOUS OKRUG, central SIBERIA, RUSSIA, on the Yugansk channel of the OB′ RIVER, 430 mi/692 km NE of TYUMEN, 26 mi/42 km from railroad (Ostrovnaya station), and 30 mi/48 km SW of SURGUT; 61°05′N 72°42′E. Elevation 111 ft/33 m. In oil- and gas-producing area; landing on the river. Manufacture and repair of machinery and equipment for gas and oil industries; radio and television repair plant. Regional institute of oil and gas industries. Arose with the development of the Ust′-Balyk and other oil deposits; beginning of oil pipelines to Omsk and Tobol′sk. Until 1967, a rural village of Ust′-Balyk, when city status was granted.

Nefud, SAUDI ARABIA: see NAFUD.

Nefyn (NE-vin), resort (2001 population 2,819), on LLEYN PENINSULA, GWYNEDD, NW Wales, on CAERNARVON BAY, 5 mi/8 km NW of PWLLHELI; 52°56′N 04°30′W. NE, 5 mi/8 km, is peak of Eifl 1,849 ft/564 m highest point of Yr Eifl range and of LLEYN PENINSULA.

Negage (ne-GAH-gai), town, UÍGE province, ANGOLA, on major road junction, and 25 mi/40 km ESE of Uige; 07°47′S 15°27′E. Market center.

Négala (NAI-gah-lah), town, SECOND REGION/KOULIKORO, MALI, on railroad, 37 mi/60 km W of BAMAKO; 12°52′N 08°27′W.

Nega-Nega (NAI-gah–NAI-gah), township, SOUTHERN province, S central ZAMBIA, 20 mi/32 km E of MAZABUKA; 15°50′S 28°02′E. On railroad. Agriculture (tobacco, corn, cotton, sugarcane, soybeans); cattle.

Négansi, town, N BENIN, 75 mi/121 km SSW of KANDI; 10°36′N 03°48′E. Cotton; livestock; shea-nut butter.

Negaunee (nuh-GAW-nee), town (2000 population 4,576), MARQUETTE county, NW UPPER PENINSULA, N MICHIGAN, 10 mi/16 km SW of MARQUETTE, in MARQUETTE IRON RANGE; 46°30′N 87°35′W. Railroad junction for mining spurs and railroad spur to MARQUETTE. Iron mining; wood products; cattle farming. Resort. Iron was discovered here in 1844. Michigan Iron Industry Museum. Settled 1846; incorporated as village 1862, as city 1873.

Negba (NEG-bah), kibbutz, ISRAEL, 6 mi/10 km E of ASHKELON, on coastal plain; 31°40′N 34°41′E. Elevation 203 ft/61 m. Mixed farming; light industry; guest house. When founded in 1939, Negba was the southernmost Jewish settlement in Israel. Suffered from repeated Egyptian attacks in 1948 war.

Negeb, ISRAEL: see NEGEV.

Negele (ne-GAI-lai), town (2007 population 45,007), OROMIYA state, S ETHIOPIA, on plateau between DAWA and GENALE rivers, 125 mi/201 km NE of MEGA; 05°19′N 39°35′E. Road junction.

Negele (ne-GE-lai), town (2007 population 44,059), OROMIYA state, central ETHIOPIA; 07°22′N 38°41′E. Busy market town 10 mi/16 km NE of SHASHEMENE; has visitor center for ABIATA-SHALA NATIONAL PARK. Sometimes called Arsi Negele.

Negeri Sembilan (NEH-gree SEM-bee-lahn) [=nine states], state (□ 2,564 sq mi/6,666.4 sq km; 2000 population 859,924), MALAYSIA, in S MALAY PENINSULA, on the Strait of MALACCA; ⊙ SEREMBAN. Its principal rivers are the LINGGI on the W and the MUAR on the E. Rubber, pineapples, other tropical fruits, and rice are grown and exported; tin is mined and also exported. Over half the inhabitants are non-Malays (Chinese and Indians). The separate political existence of Negeri Sembilan began in the 18th century. After a considerable immigration of Minangkabaus from SUMATRA, the nine states in this area broke away (1777) from the sultanate of Riau and JOHOR to form a loose confederation. Each state was then practically independent. The British established their influence by making treaties with the separate states (1874–1889) and by reforming them into a closer federation (1895). Negeri became one of the FEDERATED MALAY STATES (1896) and in 1948 became part of the FEDERATION OF MALAYA. See MALAYSIA, FEDERATION OF.

Negev (NE-gev) [Hebrew=dry], hilly semi-desert and desert region (□ c.5,140 sq mi/13,310 sq km) of S ISRAEL, surrounded by the JUDEAN HIGHLANDS (N), the ARABAH VALLEY (E), the SINAI peninsula (W and S), and the MEDITERRANEAN coastal plain (NW); 30°30′N 34°55′E. Comprises over half of Israel's land area. The Negev receives 2 in/50 mm–8 in/200 mm of rain annually. In the Beersheba basin, NW Negev, there are fertile loess deposits, but the region's aridity prevented cultivation until irrigation was provided by the National Water Carrier Project, which taps the SEA OF GALILEE. The Negev region also has a good mineral potential; phosphates and natural gas are already commercially extracted. In ancient times there were several prosperous cities along the principal regional routes which crossed the area. In modern times the Negev was the scene of much fighting between Egyptian and Israeli forces after Israel declared independence in 1948. Many kibbutzim (collective farms) and moshavim (cooperative villages) are located here; dry farming has been attempted in some areas. Negev is home to three enormous, craterlike erosion cirques or *machteshim*, which are unique to the region: HAMACHTESH HAGADOL, HAMACHTESH HAKATAN, and MACHTESH RAMON. The major cities in the region include BEERSHEBA, DIMONA, ARAD, and EILAT. Sometimes spelled Negeb.

Negoiu (ne-GOI), peak (8,317 ft/2,535 m), central ROMANIA, NW of CÎMPULUNG; 2nd-highest peak of the TRANSYLVANIAN ALPS.

Negombo (NI-GOM-boo), town (2001 population 121,701) W SRI LANKA, at the mouth of Negombo Lagoon; 07°12′N 79°50′E. Fishing center. Trades in coconuts and rice. A busy port in medieval times. The 17th-century Portuguese fort was captured by the Dutch in 1640. Scene of Sri Lanka's only Passion play hosted by Roman Catholic fishermen on nearby island of Duwa, across the lagoon (connected by causeway).

Negoreloye (ne-go-RE-lo-ye), urban settlement SW MINSK oblast, BELARUS, 28 mi/45 km SW of MINSK.

Railroad station. Enterprises for maintenance of railroad equipment. Former Soviet border station (1921–1939) on main MOSCOW-WARSAW (POLAND) railroad.

Negotin (NEG-o-teen), town (2002 population 43,418), BOR district, E SERBIA, on railroad, and 110 mi/177 km ESE of BELGRADE, near the DANUBE River, where Serbia-Montenegro, ROMANIA, and BULGARIA meet; 44°13′N 22°32′E. Gardening; wine growing; some industry (food processing; textiles). Also called Negotin Krajinski or Negotin Krayinski.

Negotin (ne-go-teen), village, MACEDONIA, on VARDAR River, on railroad, and 50 mi/80 km SSE of SKOPJE. Trade center for wine-growing region. Also called Negotino.

Negovanoutsi (ne-go-vahn-OO-tsee), village, MONTANA oblast, NOVO SELO obshtina, BULGARIA.

Negra, Cordillera (NAI-grah, kor-dee-YAI-rah), W section of CORDILLERA OCCIDENTAL of the ANDES Mountains, ANCASH region, W central PERU, between coastal plain and Callejón de HUAYLAS; extends 110 mi/177 km SSE from SANTA RIVER to area of CHIQUIÁN; 09°25′S 77°40′W. Rises to 14,764 ft/4,500 m. Contains lead-, copper-, and silver-mining region of TICAPAMPA, and coal deposits at AIJA.

Negra, Cuchilla, Portuguese *Coxilha Negra*, hill range along BRAZIL-URUGUAY border, extending c.30 mi/48 km SW from LIVRAMENTO (Brazil) and RIVERA (Uruguay); rises to 1,000 ft/305 m.

Negrais, Cape (ne-GRAI-is), Burmese *Hainggyi*, in mouth of BASSEIN RIVER, MYANMAR, 65 mi/105 km SSW of PATHEIN; 10 mi/16 km long. Was site (1753–1759) of early British settlement in BURMA.

Negra, Laguna or **Laguna de los Difuntos**, freshwater lagoon (□ 70 sq mi/182 sq km), ROCHA department, SE URUGUAY, near the ATLANTIC ocean, 14 mi/23 km NE of Castillos; 34°03′S 53°40′W. Linked by short river with Lake MIRIM (N). Surrounded by marshes. Part of Parque Nacional Santa Teresa.

Negra, Laguna (NE-grah, lah-GOO-nah), Andean lake (□ 7 sq mi/18.2 sq km), SANTIAGO province, METROPOLITANA DE SANTIAGO region, central CHILE, near ARGENTINA border, 33 mi/53 km SE of SANTIAGO; c.3 mi/5 km long.

Négre, Cap (NAI-gruh, kahp), headland on the MEDITERRANEAN SEA, in VAR department, PROVENCE-ALPES-CÔTE D'AZUR region, SE FRANCE, at foot of the Monts des Maures massif, 16 mi/26 km E of HYÈRES; 43°08′N 06°26′E. Here the first French units landed in Allied invasion of S France (August 1944) during World War II. Other Allied forces landed on same day along coast between Hyères and SAINT-RAPHAËL.

Negreiros (ne-GRAI-ros), village, TARAPACÁ region, N CHILE, on railroad, and 30 mi/48 km SE of PISAGUA; 19°51′S 69°52′W. Former nitrate-mining center; flourished c.1900.

Négrepelisse (NAI-gruh-puh-LEES), commune (□ 19 sq mi/49.4 sq km), TARN-ET-GARONNE department, MIDI-PYRÉNÉES region, SW FRANCE, on the AVEYRON RIVER, and 9 mi/14.5 km ENE of MONTAUBAN; 44°04′N 01°31′E. Truffles and poultry shipping.

Negreşti (ne-GRESHT), town, VASLUI county, E ROMANIA, on BÎRLAD River, and 19 mi/31 km NW of VASLUI. Agriculture center; manufacturing (building materials; textiles; food processing).

Negreşti-Oaş (ne-GRESHT-o-WAHSH), Hungarian *Avasfelsőfalu*, town, SATU MARE county, NW ROMANIA, on railroad, and 26 mi/42 km NE of SATU MARE; 47°52′N 23°26′E. District is noted for colorful original folklore. Lumbering; lignite and andesite-trachyte mining; manufacturing (furniture, linen, and foodstuffs). Health resort of Băile Bixad, Hungarian *Bikszádfürdö*, with saline springs, is 5 mi/8 km NNW. Both under Hungarian rule, 1940–1945.

Negrete (nai-GRAI-tai), town, ⊙ Negrete comuna, BÍO-BÍO province, BÍO-BÍO region, S central CHILE, on railroad, and on Malchen River, 13 mi/21 km SW of

LOS ÁNGELES; 37°35′S 72°31′W. Agricultural center (grapes, cereals, vegetables; livestock); lumbering.

Negril, township and resort destination, WESTMORELAND parish, W JAMAICA, just N of SOUTH NEGRIL POINT; 18°19′N 78°20′W. Extends from Half Moon Bay in HANOVER through Long Bay S to Negril Lighthouse (18°15′N 78°20′W). The area encloses 7 mi/ 11.3 km of white sand beach protected by coral reefs. Once a quiet fishing village, Negril is now a regional tourist mecca due to its laid-back atmosphere. Prime attractions include water sports and snorkeling. Minor port ships logwood. Airport.

Negrillos (ne-GREE-yos), canton, ATAHUALLPA province, ORURO department, W central BOLIVIA, 25 mi/40 km W of HUACHACALLA, at fork on the SABAYA-ORURO road; 18°47′S 68°41′W. Elevation 12,113 ft/3,692 m. During the Colonial epoch, the Spaniards exploited the lead in the area for the silver content. Lead-bearing lode and gypsum deposits. Agriculture (potatoes, yucca, bananas); cattle.

Négrine (nai-GREEN), village and small oasis, TEBESSA wilaya, NE ALGERIA, on S slope of the Saharan ATLAS MOUNTAINS, 70 mi/113 km SSW of TEBESSA; 34°27′N 07°32′E. Date palms. Ruins of Roman military camp.

Negrito, El, HONDURAS: see EL NEGRITO.

Negritos (NAI-gree-tos), town, TALARA province, PIURA region, NW PERU; minor port on the PACIFIC OCEAN, 5 mi/8 km SSW of TALARA; 04°38′S 81°19′W. Major petroleum-production center. Petroleum is shipped via pipeline through Talara. Airport.

Negro Bay (NE-gro), inlet of INDIAN Ocean, NE SOMALIA, at mouth of NOGAL valley; 30 mi/48 km long, 5 mi/8 km wide; 08°N 49°52′E. Chief township, EIL.

Négro, Cape (nai-GRO), headland on the MEDITERRANEAN SEA, BAJAH province, N TUNISIA, 16 mi/26 km NE of TABARQAH; 37°07′N 08°59′E.

Negro, Cerro, mountain peak, ARGENTINA: see PABELLÓN, CERRO.

Negro, Cerro (NAI-gro, SER-RO), active volcano and mountain peak (3,240 ft/625 m) in CORDILLERA DE LOS MARIBIOS, W NICARAGUA, 13 mi/21 km ENE of LEÓN. Erupted last in 1992, causing damage to León city.

Negro Mountain, NW MARYLAND and SW PENNSYLVANIA, a ridge of the ALLEGHENIES, rising to 3,213 ft/ 979 m in Mount Davis, highest point on the NATIONAL ROAD; extends c.35 mi/56 km NNE from DEEP CREEK LAKE just SE of MCHENRY (Md.), to a point just S of SOMERSET (Pa.). Section of Savage River State Forest is on the slopes in Maryland. During the French and Indian War, an African-American known as Nemisis, under the command of Colonel Thomas Cresap died in a skirmish with Native Americans here; hence the name.

Negro Muerto, Cerro (NAI-gro MWER-to, SER-ro), Andean volcano (19,190 ft/5,849 m) in N CATAMARCA province, ARGENTINA, at S end of Sierra de CALALASTE, 60 mi/97 km SW of ANTOFAGASTA.

Negro Pabellón (NAI-gro pah-bai-YON), town and canton, PANTALEÓN DALENCE province, ORURO department, W BOLIVIA, in Cordillera de AZANAQUES, on road, and 18 mi/29 km SE of ORURO; 18°00′S 66°58′W. Tin mines.

Negroponte, Greece: see ÉVVIA, KHALKÍS.

Negro, Río (NAI-gro, REE-o), river, c.400 mi/644 km long, ARGENTINA; formed in central Argentina by the confluence of the NEUQUÉN and LIMAY rivers; flows E across RÍO NEGRO province (N PATAGONIA) to the ATLANTIC OCEAN. The river is used for irrigation. Site of several huge dams, and future site of several more.

Negro, Río (NAI-gro, REE-o), river, c.120 mi/193 km long, E Chaco province, ARGENTINA; rises in swamps 40 mi/64 km WNW of El Zapallar; flows SE, past PUERTO TIROL, to PARANA River at CORRIENTES.

Negro, Río, river, c.220 mi/354 km long, in SANTA CRUZ department, NE BOLIVIA; rises near Concepción;

flows NNW to Río BLANCO NNW of BAURES (BENI department); 18°02′S 64°06′W.

Negro, Rio (NEG-ro, REE-o), river, c.200 mi/322 km in SE BRAZIL; rises in the Serra do MAR on PARANÁ–SANTA CATARINA state border SE of CURITIBA; flows W along the border, past RIO NEGRO and MAFRA, to IGUASSÚ River S of SÃO MATEUS DO SUL. Navigable in small sections.

Negro, Río (NAI-gro, REE-o), river, c.1,400 mi/2,253 km long, COLOMBIA and BRAZIL; rises as the Río Guainía in E COLOMBIA; flows NE before turning S to form part of the Colombia-VENEZUELA border, then flows SE through AMAZONAS state, BRAZIL, to the AMAZON River near MANAUS; 02°03′N 67°07′W. The river is filled with islands and has many secondary channels. Its main tributary is the Río BRANCO. The Río Negro is connected with the ORINOCO basin by the CASIQUIARE, a natural canal. An important commercial channel, the Río Negro was discovered (1638) by Pedro Teixeira, a Portuguese explorer. The river was named for its black color, which results from vegetal debris, not sediment.

Negro, Río (NAI-gro, REE-o), river, c.70 mi/113 km long, in NICARAGUA and HONDURAS; rises 5 mi/8 km S of SAN MARCOS DE COLÓN (Honduras); flows SW in an arc, through CHINANDEGA department of Nicaragua, and, reentering Honduras, W to GULF OF FONSECA 18 mi/29 km SSW of CHOLUTECA.

Negro, Río, principal river of URUGUAY, c.500 mi/800 km long; rises in S BRAZIL; flows SW across central Uruguay to the URUGUAY RIVER at 33°24′S 58°22′W. It traverses a sheep-raising region; agriculture along lower course. On the river is EMBALSE DEL RÍO NEGRO (□ c.4,000 sq mi/10,360 sq km), the largest artificial lake in S America; extends 87 mi/140 km upstream from RINCÓN DEL BONETE, site of the Dr. Gabriel Terra hydroelectric dam (completed 1949) with a 128,000-kw capacity. Downstream from Bonete is RINCÓN DE BAYGORRIA (1960), a hydroelectric station with a 108,000-kw capacity. Also CONSTITUCIÓN Dam is at lower outlet of lake.

Negro, Río, GUATEMALA: see CHIXOY RIVER.

Negro, Río, HONDURAS: see SICO RIVER.

Negros (NEG-ros), island (□ 4,905 sq mi/12,753 sq km), one of the VISAYAN ISLANDS, fourth-largest of the PHILIPPINES, between PANAY and Cebu; 10°00′N 123°00′E. Area includes ninety-seven islands (twenty of which are named). Although mountainous (MOUNT CANLAON, a volcano, rises to c.8,088 ft/2,465 m), Negros has extensive arable lowlands; they are intensively cultivated and densely populated. Negros is the sugar center of the Philippines; 2⁄3 of the nation's sugarcane is grown here, and sugar processing is a major industry. In addition, paper products are made from sugarcane residue. Rice, coconuts, bananas, and corn are also grown. The island has a lumber industry and copper and coal deposits.

Negros Occidental (NEG-ros ok-see-den-TAHL), province (□ 306 sq mi/795.6 sq km), in WESTERN VISAYAS region, N and W NEGROS island, PHILIPPINES; ⊙ BACOLOD; 10°25′N 123°00′E. Population 45.7% urban, 54.3% rural. Mountainous terrain, drained by many small streams; highest peak is MOUNT CANLAON (8,088 ft/2,465 m). Major agricultural area, producing sugarcane and rice. Chief centers are cities of Bacolod, BAGO, CADIZ, LA CARLOTA, SAN CARLOS, and SILAY.

Negros Oriental (NEG-ros or-yen-TAHL), province (□ 2,086 sq mi/5,423.6 sq km), E NEGROS island, PHILIPPINES, on TAÑON STRAIT (E), and SULU SEA (S and SW); ⊙ DUMAGUETE; 09°45′N 123°00′E. Population 23.6% urban, 76.4% rural. Hills, plateaus, and mountainous terrain in interior, drained by numerous small streams. Small, fertile valleys. Much of SW interior is unexplored territory. Principal agricultural products include corn, coconuts, sugarcane, tobacco. Chief centers are cities of Dumaguete, CANLAON, BAIS.

Negru Vodă (NE-groo VO-dah), town, CONSTANȚA county, SE ROMANIA, on railroad, near border with BULGARIA, and 32 mi/51 km SW of CONSTANȚA; 43°49′N 28°12′E. Agriculture center; manufacturing of bricks and tiles. Sometimes called Cara-Omer.

Negyed, SLOVAKIA: see NEDED.

Nehalem (nuh-HAI-luhm), village (2006 population 208), TILLAMOOK county, NW OREGON, 38 mi/61 km S of ASTORIA, on NEHALEM RIVER, near mouth of North Fork; 45°43′N 123°53′W. Dairy products; cattle. Wineries. Fish hatcheries to E and NE. Oswald West State Park to NW.

Nehalem River, NW OREGON; rises in NW WASHINGTON county; flows NE past VERNONIA, curves W and SW, then flows through Clatsop and Tillamook State Forest. Receives North Fork before entering Nehalem Bay, PACIFIC OCEAN, near NEHALEM. North Fork rises in S central CLATSOP county, flows S and SW c.25 mi/40 km.

Nehalim (ne-HAH-lim), moshav, central ISRAEL, 1.9 mi/3 km S of PETAH TIKVA, near Ben-Gurion Airport; 32°03′N 34°54′E. Elevation 147 ft/44 m. Founded by religious Jews in 1948.

Nehawka, village (2006 population 227), CASS county, SE NEBRASKA, 11 mi/18 km NW of NEBRASKA CITY, and on WEEPING WATER CREEK, near MISSOURI RIVER; 40°49′N 95°59′W. Traces of prehistoric man discovered near here.

Nehbandan (nai-bahn-DAHN), town, Khorāsān province, E IRAN, 100 mi/161 km WSW of BIRJAND, between the deserts DASHT-e-Kavir (NW) and DASHT-e-Lut (SE); 31°33′N 60°04′E. Also spelled NAIBANDAN.

Nehe (NAI-HUH), town, ⊙ Nehe county, W HEILONGJIANG province, NE CHINA, on railroad, 85 mi/ 137 km NNE of QIQIHAR; 48°29′N 124°50′E. Grain, oilseeds, sugar beets; sugar refining.

Neheim-Hüsten (NE-heim–HYOOS-tuhn), district of ARNSBERG, HESSE, central GERMANY, on confluence of MÖHNE and RUHR rivers; 51°28′N 08°02′E. Electrical equipment. Established 1975.

Nehoiașu (NE-ho-YAH-shoo), village, BUZĂU county, SE central ROMANIA, on BUZĂU RIVER, and 33 mi/53 km NW of BUZĂU; 45°26′N 26°17′E. Railroad terminus, lumbering center.

Nehoiu (ne-HO-yoo), town, BUZĂU county, SE central ROMANIA, on BUZĂU RIVER, on railroad, and 32 mi/51 km NW of BUZĂU; 45°25′N 26°18′E. Summer resort and lumbering center; foundries.

Nehora (ne-HO-rah), rural service village, ISRAEL, SE of ASHKELON; 31°37′N 34°42′E. Elevation 282 ft/85 m. Founded in 1955, it is the service and educational center for surrounding settlements of the LACHISH region.

Nehuentué (nai-wen-TWAI), village, CAUTÍN province, ARAUCANIA region, S central CHILE, on IMPERIAL RIVER near its mouth on the PACIFIC OCEAN, opposite PUERTO SAAVEDRA, and 45 mi/72 km W of TEMUCO. In agricultural area (cereals, vegetables; livestock). Tourist site. Boating on river.

Neiba (ne-EE-bah), city (2002 population 18,305), ⊙ BAHORUCO province, SW DOMINICAN REPUBLIC, near E shore of Lake ENRIQUILLO, 100 mi/161 km W of SANTO DOMINGO; 18°28′N 71°23′W. Produces sugarcane and fine construction wood. Provincial capital since 1943. Rock-salt and gypsum deposits nearby. Also spelled Neyba.

Neiba Bay (ne-EE-bah), small inlet of the CARIBBEAN SEA, SW DOMINICAN REPUBLIC; c. 6 mi/9.7 km long, 6 mi/9.7 km wide; 18°15′N 71°02′W. Into it falls the YAQUE DEL SUR. BARAHONA city on SW coast.

Neiba, Sierra de (ne-EE-bah, see-ER-rah dai), range in W DOMINICAN REPUBLIC, S of the Cordillera CENTRAL, between SAN JUAN VALLEY (N) and Lake ENRIQUILLO (S), extending c. 60 mi/97 km E from HAITI border to the YAQUE DEL SUR; rises to 5,545 ft/1,690 m; 18°40′N 71°30′W.

Nei-chiang, CHINA: see NEIJIANG.

Neiden (NAI-duhn), village, FINNMARK county, NE NORWAY, on arm of VARANGERFJORDEN, 15 mi/24 km W of KIRKENES; 69°42′N 29°24′E. Nearby is Sami settlement with the only Greek Orthodox church in Norway.

Neidenburg, POLAND: see NIDZICA.

Neide River, POLAND: see WKRA RIVER.

Neige, Crêt de la (NEZH, KRAI–duh–lah), highest summit (5,660 ft/1,725 m) of the JURA MOUNTAINS, in AIN department, NE FRANCE, 11 mi/18 km WNW of GENEVA; 46°16′N 05°56′E. Panoramic views of the Swiss (E) and French (SE) Alps.

Neiges, Piton des, Réunion: see PITON DES NEIGES.

Neihart, village (2000 population 91), CASCADE county, central MONTANA, 51 mi/82 km SE of GREAT FALLS, on Belt Creek, in LITTLE BELT MOUNTAINS, in Lewis and Clark National Forest; 46°56′N 110°45′W. Elevation c.5,700 ft/1,737 m. Trading point. Silver, lead, sapphire mines. Showdown Ski Area to S.

Neihu (NAI-HOO), suburban district of TAIPEI, N TAIWAN, 8 mi/12.9 km ENE of city center. Coal mining; manufacturing (bricks and tiles, earthenware, drainage pipes, bamboo products). Rapidly urbanized area within greater Taipei.

Neihuang (NAI-HWAHNG), town, ☉ Neihuang county, N HENAN province, CHINA, on HEBEI province border, 32 mi/51 km ESE of ANYANG; 35°54′N 114°53′E. Grain, cotton; textiles, food and beverages.

Neijiang (NAI-JYAHNG), city (□ 605 sq mi/1,567 sq km; 1994 estimated urban population 295,700; estimated total population 1,333,900), central SICHUAN province, CHINA, a port city on the TUO River; 29°32′N 105°03′E. Railroad center. Agriculture and light industry (food processing; textiles, machinery; utilities) are the largest sectors of the city's economy; heavy industry and commerce are also important. Crop growing (grain, oil crops, cotton, vegetables, fruits); animal husbandry (hogs, poultry); eggs. Sometimes spelled Nei-chiang.

Neilgherry, INDIA: see NILGIRI.

Neilston (NEEL-stuhn), town (2001 population 5,168), East Renfrewshire, central Scotland, 5 mi/8 km S of PAISLEY; 55°47′N 04°25′W. Previously textile industry. Formerly in Strathclyde, abolished 1996.

Neilsville, town, ☉ CLARK county, central WISCONSIN, on BLACK RIVER, and 47 mi/76 km ESE of EAU CLAIRE, in hilly region; 44°33′N 90°35′W. Commercial center for dairying, livestock-raising, and farming area; cheese, butter; manufacturing (fabricated metal products, machinery, beverages, canned vegetables). Has hydroelectric plant. St. Mary's Mission School is here. Bruce Mound Ski Area to SW. Settled c.1844, incorporated 1882.

Nei Menggu, CHINA: see INNER MONGOLIA AUTONOMOUS REGION.

Nein (NEEN), Arab village, N ISRAEL, 3.1 mi/5 km NE of AFULA, on the N slope of Givaat Hamore, in JEZREEL VALLEY; 32°38′N 35°21′E. Elevation 685 ft/208 m. Built on the ruins of an ancient town, possibly Naim of Second Temple times (6th century B.C.E.), mentioned in the Talmud and the New Testament. Christian tradition places Jesus' revival of the dead here. A Franciscan church stands on foundations of a medieval church. Roman ruins.

Neiqiu (NAI-CHYO), town, ☉ Neiqiu county, SW HEBEI province, CHINA, 50 mi/80 km S of SHIJIAZHUANG, and on Beijing-Wuhan railroad; 37°17′N 114°31′E. Grain, cotton, oilseeds; textiles, building materials. Coal mining.

Neira (NAI-rah), town, ☉ Neira municipio, CALDAS department, W central COLOMBIA, in W Cordillera CENTRAL, and 7 mi/11 km N of MANIZALES; 05°09′N 75°32′W. Elevation 6,644 ft/2,025 m. Coffee-growing center; plantains.

Neira, INDONESIA: see BANDANAIRA.

Neiriz, IRAN: see NEYRIZ.

Neisse, Polish *Nysa*, town (2002 population 48,145), OPOLE province, SW POLAND, near CZECH border, on the GLATZER NEISSE River, and 45 mi/72 km SSE of WROCŁAW (Breslau). Railroad junction; manufacturing of chemicals, machinery; cotton and linen milling, metalworking; power station. Was (1198–1810) capital of principality held by prince-bishops of Breslau. Originally in UPPER SILESIA; captured 1741 by Frederick the Great; subsequently fortified; withstood (1758) Austrian siege. Occupied 1807–1808 by French; fortifications razed 1862. Heavily damaged in World War II. Under German rule until 1945, when it passed to Poland and was briefly spelled Nisa.

Neisse (NEIS-se), river, E GERMANY, 159 mi/256 km long; source in CZECH REPUBLIC; runs along the border between GERMANY and POLAND to ODER RIVER. Also called LAUSITZER NEISSE or GÖRLITZER NEISSE.

Neisse, two rivers of SW POLAND. The GLATZER NEISSE, Polish *Nysa Kłodzka*, c.120 mi/193 km long; rises in the SUDETES, SW POLAND, and winds generally NE past Kłodzko to the ODER River near BRZEG. A large dam at OTMUCHOW serves hydroelectric and irrigation projects. The Lausitzer Neisse or Lusatian Neisse, Czech *Lužická Nisa*, Polish *Nysa Łużycka*, c.140 mi/225 km long; rises in the Sudetes Mountains, NW CZECH REPUBLIC; flows generally N to the Oder River near GUBEN (GERMANY and Poland). Since 1945 it has formed part of the border between Germany and Poland. GÖRLITZ (Germany) is the chief city on the river. It is also known as Görlitzer Neisse.

Neisu (NAI-soo), village, ORIENTALE province, NE CONGO, on railroad, and 165 mi/266 km E of BUTA; 02°43′N 27°25′E. Elev. 2,244 ft/683 m. Palm-oil milling, soap manufacturing.

Neiva (NAI-vah), city, ☉ HUILA department, S central COLOMBIA, landing on right bank of upper MAGDALENA River, and 150 mi/241 km SW of BOGOTÁ, on railroad and highway; 02°56′N 75°18′W. Trading and processing center in agricultural region (rice, corn, cacao, coffee, plantains, rice; livestock); manufacturing of consumer goods, and food products. Due to its low elevation and proximity to the EQUATOR, Neiva has a warm climate all year. Old colonial city with park and government and national palaces. Founded 1539, rebuilt 1612 after having been destroyed by Native Americans. Phosphate rock mines nearby. Special prison for drug traffickers. Airport.

Nei-Valter, RUSSIA: see SVERDLOVO.

Neixiang (NAI-SHYAHNG), town, ☉ Neixiang county, SW HENAN province, CHINA, 40 mi/64 km W of NANYANG; 33°03′N 111°53′E. Grain, oilseeds, tobacco. Beverages, chemicals; tobacco processing, papermaking.

Nejapa (nai-HAH-pah), town and municipality, SAN SALVADOR department, W central EL SALVADOR, 7 mi/11.3 km NNW of SAN SALVADOR. Grain. Originally located at NW foot of volcano SAN SALVADOR, 4 mi/6.4 km SW of QUEZALTEPEQUE. Following its destruction in 1659 eruption, which created the large, picturesque lava field of EL PLAYÓN, Nejapa was rebuilt on present site.

Nejapa de Madero (nai-HAH-pah dai mah-DAI-ro), town, in SE OAXACA, MEXICO, 81 mi/130 km SE of OAXACA DE JUÁREZ; 16°37′N 95°59′W. Elevation 3,281 ft/1,000 m. A mountainous region with flat areas irrigated by the Grande River. Hot climate. Mostly Zapotec population. Agriculture (cereals and fruits), fine woods, lumber, mezcal; cattle and poultry raising.

Nejapa, Lake (nai-HAH-pah), small crater lake, SW NICARAGUA, 3 mi/4.8 km SW of MANAGUA. Waters are said to have medicinal value.

Nejdek (NAI-dek), German *Neudek*, town, ZAPADOCESKY province, W BOHEMIA, CZECH REPUBLIC, in ORE MOUNTAINS, on railroad, and 10 mi/16 km NW of KARLOVY VARY; 50°17′N 12°57′E. Manufacturing (machinery, spinning); woodworking. Mining settlement in 15th century.

Nejef, IRAQ: see AN NAJAF.

Nejime (NE-jee-me), town, Kimotsuki county, KAGOSHIMA prefecture, SW JAPAN, on S OSUMI PENINSULA, on SE shore of KAGOSHIMA BAY, 28 mi/45 km S of KAGOSHIMA; 31°21′N 130°46′E. Site of an old feudal castle.

Nejo (NAI-jo), township (2007 population 20,836), OROMIYA state, W central ETHIOPIA, 34 mi/55 km NW of GIMBI; 09°29′N 35°28′E. Trade center (coffee, beeswax, hides, gold).

Nejran, SAUDI ARABIA: see NAJRAN.

Neka (nai-KAH), town, Māzandarān province, N IRAN, 15 mi/25 km NE of SARI, on road and railroad to BEHSHAHR; 36°39′N 53°19′E.

Nekemte, town (2007 population 88,536), OROMIYA state, W central ETHIOPIA, on road from ADDIS ABABA to SUDAN, and 145 mi/233 km W of ADDIS ABABA; 09°05′N 36°33′E. Was capital of former WELEGA province. Elevation is c.6,880 ft/2,097 m. Trade center (coffee, hides, beeswax, honey, cereals, gold). Museum. Sometimes spelled Nekemti, Nakamti, Lakamti, or Lekemti.

Nekemti, ETHIOPIA: see NEKEMTE.

Nekhayevskaya (nee-HAH-eef-skah-yah), village (2006 population 4,720), NW VOLGOGRAD oblast, SE European Russia, on the Tishanka River near its confluence with the KHOPER River, on road junction, 27 mi/43 km SSW of URYUPINSK; 50°25′N 41°46′E. Elevation 377 ft/114 m. Margarine factory. Town status and named Nekhayevskiy since 1956; reduced to rural status and current name in 1992.

Nekheb, EGYPT: see KAB, EL.

Nekhela, EGYPT: see NIKHEILA.

Nekhvoroshcha (nek-vo-RO-shchah), village, SE POLTAVA oblast, UKRAINE, on ORIL' RIVER, and 30 mi/48 km SSE of POLTAVA; 49°09′N 34°44′E. Elevation 252 ft/76 m. Clothing manufacturing.

Neklyudovo (nee-KLYOO-duh-vuh), town (2006 population 9,000), W central NIZHEGOROD oblast, central European Russia, on railroad junction (Tolokontsevo station), 5 mi/8 km N of NIZHNIY NOVGOROD, and 4 mi/6 km NW of BOR, to which it is administratively subordinate; 56°24′N 43°59′E. Elevation 285 ft/86 m. Wool processing; felt products.

Nekmard, BANGLADESH: see THAKURGAON.

Nekoma (nuh-KO-muh), village (2006 population 44), CAVALIER county, NE NORTH DAKOTA, 13 mi/21 km S of LANGDON; 48°34′N 98°22′W. Founed in 1905 and incorporated in 1906. Name is a Chippewa Indian word for 'I promise to do something.'

Nekoosa, town (2006 population 2,521), WOOD county, central WISCONSIN, on WISCONSIN RIVER, and suburb 7 mi/11.3 km SSW of WISCONSIN RAPIDS; 44°18′N 89°54′W. In dairy area; paper products. State nursery across river to NE; large PETENWELL RESERVOIR to S on Wisconsin River. Settled 1892, incorporated 1926.

Nekrasovka (nee-KRAH-suhf-kah), village (2005 population 9,070), SW KHABAROVSK TERRITORY, RUSSIAN FAR EAST, 17 mi/27 km SE of KHABAROVSK, to which it is administratively subordinate; 48°21′N 135°14′E. Elevation 259 ft/78 m. In agricultural area. Russian air force base in the vicinity.

Nekrasovka (nee-KRAH-suhf-kah), village (2005 population 8,065), central MOSCOW oblast, W central European Russia, 13 mi/21 km SE of MOSCOW, of which it is a suburb; 55°41′N 37°55′E. Elevation 439 ft/133 m. Livestock raising.

Nekrasovskaya (nee-KRAH-suhf-skah-yah), village (2005 population 4,835), E KRASNODAR TERRITORY, S European Russia, near the confluence of the LABA and KUBAN' rivers, on road, 25 mi/40 km ENE of KRASNODAR, and 4 mi/6 km SE of UST'-LABINSK, to which it is administratively subordinate; 45°08′N 39°45′E. Elevation 200 ft/60 m. In agricultural area (grain, sunflowers, sugarbeets).

Nekrasovskiy (nee-KRAH-suhf-skeeyee), town (2006 population 9,590), central MOSCOW oblast, central

European Russia, near the MOSCOW CANAL, on road and railroad, 16 mi/26 km S, and under administrative jurisdiction, of DMITROV; 56°04′N 37°30′E. Elevation 711 ft/216 m. Manufacturing (ceramic tile, packing products).

Nekrasovskoye (nee-KRAH-suhf-skuh-ye), town (2006 population 6,310), E YAROSLAVL oblast, central European Russia, on the VOLGA RIVER, 19 mi/31 km E of YAROSLAVL; 57°40′N 40°22′E. Elevation 360 ft/109 m. Machine building, parts for agricultural machinery; milk factory. Until the 1930s, called Bol'shiye Soli.

Nekselø, DENMARK: see SEJERØ BUGT.

Neksø (NEK-soo), city and port, BORNHOLM county, DENMARK, on SE BORNHOLM island; 55°04′N 15°08′E. Fisheries; fish, fish products; standstone quarry; egg-packing plant; shipbuilding. Dueodde Lighthouse. Sometimes spelled Nexö.

Nelagoney (nel-uh-GO-nee), village, OSAGE county, N OKLAHOMA, 6 mi/9.7 km ESE of PAWHUSKA.

Nelahozeves (NE-lah-HO-ze-VES), German *Mühlhausen*, village, STREDOCESKY province, N central BOHEMIA, CZECH REPUBLIC, on VLTAVA RIVER, on railroad, and 14 mi/23 km NNW of PRAGUE; 50°16′N 14°18′E. Food processing (oil, margarine). Has a 16th century Renaissance castle with gallery, a 14th century church. Composer Antonin Dvořak was born here in 1841.

Nelamangala (nai-luh-MUHN-gah-lah), town, BANGALORE district, KARNATAKA state, S INDIA, 15 mi/24 km NW of BANGALORE; 13°06′N 77°24′E. Trades in millet, rice, tobacco.

Nelas (NAI-lahsh), town, VISEU district, N central PORTUGAL, on railroad, and 9 mi/14.5 km SSE of VISEU; 40°32′N 07°51′W. Sawmilling and woodworking; forest products. Uranium deposits nearby.

Nelaug (NAI-loug), village, AUST-AGDER county, S NORWAY, on NELAUG Lake, and on NIDELVA River, and 15 mi/24 km N of ARENDAL; 58°41′N 08°36′E. Railroad junction.

Nelchina Glacier, 22 mi/35 km long, S ALASKA; rises in CHUGACH MOUNTAINS near 61°30′N 146°50′W; flows NW, drains into TAZLINA LAKE.

Nelidovo (nee-LEE-duh-vuh), city (2006 population 25,085), SW Tver oblast, W European Russia, on the Mezha River (tributary of the WESTERN DVINA RIVER), on highway and railroad junctions, 170 mi/274 km SW of TVER, and 55 mi/89 km W of RZHEV; 56°13′N 32°46′E. Elevation 643 ft/195 m. Lignite mining; sawmilling, veneering; plastics, peat-digging machinery; hardware, hydraulic presses. Mining developed after World War II to fill Leningrad (now SAINT PETERSBURG) coal needs. Became city in 1949.

Neligh (NEE-lei), city (2006 population 1,521), ☉ ANTELOPE county, NE central NEBRASKA, 30 mi/48 km WNW of NORFOLK, and on ELKHORN RIVER; 42°07′N 98°01′W. Grain; manufacturing (flags, wood products). Neligh Mills, restored water-powered mill. Incorporated 1873.

Nel'kan (nyel-KAHN), village, central KHABAROVSK TERRITORY, E RUSSIAN FAR EAST, on the MAYA RIVER (head of navigation), on highway, 292 mi/470 km SW of OKHOTSK; 57°40′N 136°13′E. Elevation 1,089 ft/331 m. Gold and platinum mining in the vicinity.

Nellai Kattabomman, district (□ 2,618 sq mi/6,806.8 sq km), TAMIL NADU state, S INDIA; ☉ TIRUNELVELI. Also Tirunelveli Kattabomman.

Nellie (NEL-ee), village (2006 population 135), COSHOCTON county, central OHIO, 12 mi/19 km WNW of COSHOCTON, and on WALHONDING RIVER; 40°20′N 82°04′W. Mohawk Dam is nearby.

Nellieburg, unincorporated town (2000 population 1,354), LAUDERDALE county, E MISSISSIPPI, residential suburb 3 mi/4.8 km WNW of downtown MERIDIAN; 32°23′N 88°46′W.

Nellie Juan, Port, bay, S ALASKA, on E side of KENAI PENINSULA 50 mi/80 km NE of SEWARD; 30 mi/48 km long, 3 mi/4.8 km wide; opens into PRINCE WILLIAM SOUND at 60°32′N 148°17′W. Fishing, fish processing.

Nelligen, town, NEW SOUTH WALES, SE AUSTRALIA, 177 mi/285 km S of SYDNEY, 9 km from BATEMAN'S BAY, and on Clyde River; 35°39′S 150°08′E.

Nellikuppam (nail-li-kuhp-puhm), town, SOUTH ARCOT VALLALUR district, TAMIL NADU state, S INDIA, between PONNAIYAR and GADILAM Rivers, 7 mi/11.3 km WNW of CUDDALORE; 11°46′N 79°41′E. Sugar-processing center; distillery; betel farms. Sometimes spelled Nellikkuppam.

Nellis Air Force Base, NEVADA: see LAS VEGAS.

Nelliston, village (□ 1 sq mi/2.6 sq km; 2006 population 608), MONTGOMERY county, E central NEW YORK, on MOHAWK RIVER (bridged), opposite FORT PLAIN, and 22 mi/35 km W of AMSTERDAM; 42°55′N 74°36′W.

Nellore (ne-LOR), district (□ 5,049 sq mi/13,127.4 sq km), ANDHRA PRADESH state, S INDIA, on COROMANDEL COAST of BAY OF BENGAL; ☉ NELLORE. Bordered W by EASTERN GHATS; PULICAT LAKE (salt pans) on S coast; drained by Penneru River. Agriculture (millet, rice, oilseeds); extensive cashew and casuarina growing (mainly on coast). An important mica-producing area; quarries (notably near GUDUR and ATMAKUR) also yield ceramic clays and feldspar. Main towns are NELLORE, VENKATAGIRI, GUDUR.

Nellore (ne-LOR), city (2001 population 494,775), ☉ NELLORE district, ANDHRA PRADESH state, S INDIA, on the Penneru River; 14°26′N 79°58′E. Market for cotton and oilseeds. Milling and processing industries.

Nel'ma (NYEL-mah), settlement, SE KHABAROVSK TERRITORY, SE RUSSIAN FAR EAST, on the Sea of JAPAN, 100 mi/161 km S of SOVETSKAYA GAVAN'; 47°39′N 139°10′E. Fish canning.

Nelson, county (□ 424 sq mi/1,102.4 sq km; 2006 population 42,102), central KENTUCKY; ☉ BARDSTOWN; 37°47′N 85°28′W. Bounded SW by ROLLING FORK RIVER, E by BEECH FORK. Rolling agricultural area, in BLUEGRASS REGION (burley tobacco, soybeans, wheat, corn, hay, alfalfa; hogs, cattle, poultry; dairying; hardwood timber). Manufacturing at Bardstown. County is known for its whiskey distillery at Bardstown and Borton. Part of TAYLORSVILLE LAKE reservoir in NE corner; My Old Kentucky Home State Park in center; part of Bernheim Forest preserve in NW. Formed 1784.

Nelson, county (□ 995 sq mi/2,587 sq km; 2006 population 3,289), E central NORTH DAKOTA, in Pothole region; ☉ LAKOTA; 47°55′N 98°11′W. Agricultural area watered by SHEYENNE and GOOSE rivers; STUMP LAKE in W. Dairy products; cattle, poultry; wheat, oats, sunflowers. Formed 1883. Small part of DEVILS LAKE Sioux Indian Reservation in W; Johnson Lake National Wildlife Refuge on SW corner. Formed in 1883 and government formed the same year. Named for Nelson E. Nelson (1830–1913), a territorial legislator at the time.

Nelson (NEL-suhn), county (□ 474 sq mi/1,232.4 sq km; 2006 population 15,161), central VIRGINIA; ☉ LOVINGSTON; 37°47′N 78°52′W. In rolling PIEDMONT region, with BLUE RIDGE in W and NW; bounded SE by JAMES RIVER; drained by ROCKFISH RIVER. Agriculture (apples, hay, alfalfa, corn; cattle); mining (titanium ores, apatite), stone quarrying. BLUE RIDGE PARKWAY follows NW border, APPALACHIAN TRAIL crosses co. in NW, part of George Washington National Forest in NW. Formed 1807.

Nelson (NEL-suhn), city (□ 3 sq mi/7.8 sq km; 2001 population 9,298), SE BRITISH COLUMBIA, W Canada, on the Kootenay River, in Central Kootenay regional district; 49°29′N 117°17′W. Transportation and administrative center for a lumbering and farming region. Airport. Founded 1866.

Nelson, city (2000 population 212), SALINE county, central MISSOURI, on BLACKWATER RIVER, near MISSOURI River, and 12 mi/19 km SE of MARSHALL; 38°59′N 93°01′W. Corn, wheat; cattle.

Nelson, city (2006 population 529), ☉ NUCKOLLS county, S NEBRASKA, 30 mi/48 km SE of HASTINGS, and on branch of LITTLE BLUE RIVER; 40°12′N 98°04′W. Grain.

Nelson (NEL-suhn), town, SW VICTORIA, SE AUSTRALIA, 269 mi/433 km W of MELBOURNE, at mouth of GLENELG RIVER, and near SOUTH AUSTRALIA border; 38°03′S 141°01′E. Holiday destination; fishing.

Nelson (NEL-suhn), town (2001 population 28,998), LANCASHIRE, N ENGLAND, 2 mi/3.2 km SW of COLNE; 53°51′N 02°13′W. Furniture, machinery; food processing, printing. Previously textiles (cotton weaving).

Nelson, town, CHESHIRE county, SW NEW HAMPSHIRE, 8 mi/12.9 km NE of KEENE; 43°00′N 72°07′W. Agriculture (cattle, poultry; vegetables; dairying; nursery crops). Spoonwood Pond and part of Nubanusit Lake in E.

Nelson, village (2000 population 626), PICKENS and CHEROKEE counties, N GEORGIA, 32 mi/51 km W of GAINESVILLE; 34°23′N 84°22′W. Marble fabrication.

Nelson, village (2000 population 163), LEE county, N ILLINOIS, on ROCK RIVER, and 6 mi/9.7 km WSW of DIXON; 41°47′N 89°35′W. In rich agricultural area.

Nelson, village, MUHLENBERG county, W KENTUCKY, 3 mi/4.8 km N of CENTRAL CITY. In bituminous-coal mining and agricultural area.

Nelson, village (2000 population 172), DOUGLAS county, W MINNESOTA, 5 mi/8 km E of ALEXANDRIA, in lake region; 45°53′N 95°16′W. Poultry, livestock; grain; dairying; manufacturing (wood products, dairy products). Smith (to S) and Victoria (to W) lakes nearby.

Nelson, village (2006 population 373), BUFFALO county, W WISCONSIN, near the MISSISSIPPI RIVER (bridged nearby), at mouth of CHIPPEWA RIVER, 30 mi/48 km NNW of WINONA (MINNESOTA), opposite WABASHA (Minnesota); 44°25′N 92°00′W. Coulee region. Manufacturing (railroad ties, dairy products). Upper Mississippi River Wildlife and Fish Refuge on river.

Nelson (NEL-suhn), village (2001 population 4,577), CAERPHILLY, SE Wales, 2 mi/3.2 km W of YSTRAD MYNACH. Formerly in MID GLAMORGAN, abolished 1996.

Nelson (NEL-suhn), river, c.400 mi/640 km long, MANITOBA, W central Canada; issues from the NE end of Lake WINNIPEG, central Manitoba; flows NE to HUDSON BAY at PORT NELSON; 57°04′N 92°30′W. With the BOW–South Saskatchewan–SASKATCHEWAN river system, which enters NW Lake Winnipeg, the Nelson is part of a 1,600-mi/2,575-km continuous stream from W ALBERTA to HUDSON BAY. There are hydroelectric plants at Kettle Rapids, Long Spruce, and Kelsey. Nickel-mining and -refining operations at THOMPSON use electricity generated by the river. The Nelson's mouth was explored (1612) by Sir Thomas Button. The river was long followed by fur traders; from 1682 to 1957 the Hudson's Bay Company maintained a trading post at YORK FACTORY on Hudson Bay.

Nelson Bay (NEL-suhn BAI), town, NEW SOUTH WALES, SE AUSTRALIA, at mouth of Port STEPHENS, 139 mi/223 km N of SYDNEY; 32°43′S 152°09′E. Resort; many dolphins in area. Commercial fishing.

Nelson, Cape (NEL-suhn), SW VICTORIA, SE AUSTRALIA, in INDIAN OCEAN, at S tip of peninsula forming W shore of PORTLAND BAY; 38°26′S 141°33′E. Lighthouse. Cape Nelson State Park.

Nelson Forks (NEL-suhn FORKS), Hudson's Bay Company trading post, NE BRITISH COLUMBIA, W Canada, near YUKON border, on LIARD River, at mouth of FORT NELSON RIVER, and 70 mi/113 km NW of FORT NELSON; 59°30′N 124°00′W.

Nelson Island, SOUTH SHETLAND ISLANDS, off GRAHAM LAND, ANTARCTICA; 12 mi/19 km long, 7 mi/11.3 km wide; 62°18′S 59°03′W. Also known as Leipzig, O'Cain's, or Strachans Island.

Nelson Island, W ALASKA, separated from mainland (E) by narrow channel, from NUNIVAK Island (SW) by ETOLIN STRAIT; 42 mi/68 km long, 20 mi/32 km–35 mi/56 km wide; 60°37′N 164°22′W. TANUNAK village in W.

Nelson Lakes, N central SOUTH ISLAND, NEW ZEALAND, c.43 mi/70 km SW of NELSON City. Occupying the headwaters of the BULLER RIVER, including lakes ROTORUA and ROTOITI, in glacially shaped valleys flanking the Travers Range. Beech forests; a skiing area; hiking, fishing, hunting. Extends S to LEWIS PASS National Reserve.

Nelson Mandela Metropole, municipality (2001 population 1,005,776), EASTERN CAPE province, SE SOUTH AFRICA. Was created in 2001 by joining PORT ELIZABETH with UITENHAGE and DESPATCH.

Nelson Peak (NEL-suhn) (10,772 ft/3,283 m), SE BRITISH COLUMBIA, W Canada, in SELKIRK MOUNTAINS, 60 mi/97 km SW of BANFF (Alberta); 50°28′N 116°21′W.

Nelson Reservoir, NE MONTANA, in PHILLIPS county, 15 mi/24 km NE of MALTA; 10 mi/16 km long, 2 mi/3.2 km maximum width. Formed by dam on small branch of MILK River.

Nelson Strait (NEL-suhn), inlet of the PACIFIC on coast of S CHILE, N of entrance to Strait of MAGELLAN, between CAMBRIDGE Island and the ADELAIDE Islands.

Nelsonville (NEL-suhn-vil), city (2006 population 5,423), ATHENS county, SE OHIO, 12 mi/19 km NW of ATHENS, on HOCKING RIVER; 39°27′N 82°13′W. In coal- and clay-producing area; dairying, farming. Previously footwear, bricks, tile, machinery.

Nelsonville, village (☐ 1 sq mi/2.6 sq km; 2006 population 577), PUTNAM county, SE NEW YORK, 9 mi/14.5 km N of PEEKSKILL; 41°25′N 73°57′W. In dairying area.

Nelsonville, village (2006 population 178), PORTAGE county, central WISCONSIN, 12 mi/19 km E of STEVENS POINT; 44°29′N 89°18′W. Dairying area.

Nelspruit, town, ☉ MPUMALANGA province, SOUTH AFRICA, on CROCODILE RIVER, and 110 mi/177 km WNW of MAPUTO (MOZAMBIQUE); 25°29′S 30°59′E. Elevation 2,985 ft/910 m. Railroad junction; also N4 highway to KRUGER NATIONAL PARK. Agricultural center (subtropical fruit, tobacco, vegetables); tobacco processing, sawmilling. Site of government agricultural research station and center of research in citrus and tropical fruits; hydroelectric power station. Established 1892 when E railroad reached the area. Airfield.

Neluwa, village, UVA PROVINCE, SRI LANKA; 06°54′N 81°00′E. Trades in rubber and cloves.

Néma (NAI-mah), town (2000 population 13,759), ☉ Hodh Ech Chargui administrative region, SE MAURITANIA, in the SAHARA DESERT, 290 mi/467 km N of BAMAKO (MALI). Dates; livestock raising (camels, cattle, sheep, goats). Airport.

Nema (NYE-mah), village (2005 population 3,760), SE KIROV oblast, E central European Russia, on road, 20 mi/32 km E of NOLINSK; 57°30′N 50°30′E. Elevation 462 ft/140 m. Concrete products, gas pipeline service station; sawmilling, timbering, logging and lumbering; flax processing, seed inspection, bakery.

Nemacolin (NEE-ma-KO-lin), unincorporated town (2000 population 1,034), GREENE county, SW PENNSYLVANIA, 14 mi/23 km E of WAYNESBURG, on MONONGAHELA RIVER; 39°52′N 79°55′W. Hay; livestock; dairying.

Nemacolin's Path, Native American trail between the POTOMAC and the MONONGAHELA rivers, going from the site of CUMBERLAND (MARYLAND), to the mouth of Redstone Creek, where BROWNSVILLE (PENNSYLVANIA) is situated. It was blazed and cleared in 1749 or 1750 by Nemacolin, a DELAWARE chief, and Thomas Cresap, a Maryland frontiersman. The path was of military importance as the route of George Washington's first Western expedition and of General Edward Braddock's expedition in the last of the French and Indian Wars. It was known as Braddock's

Road until the Cumberland Road or NATIONAL ROAD was built on the same route.

Nemadji River (ne-MA-jee), c.20 mi/32 km long, DOUGLAS county, NW WISCONSIN; formed by joining of North Fork Nemadji and South Fork Nemadji rivers, 15 mi/24 km SSW of SUPERIOR (Wisconsin), at MINNESOTA state line; flows NE to LAKE SUPERIOR, on ALLOUEZ BAY, 4 mi/6.4 km SE of Superior at ALLOUEZ. North Fork Nemadji River, Minnesota and Wisconsin, rises in S CARLTON county, NE Minnesota, c.15 mi/24 km SSW of CARLTON; 46°24′N 92°30′W; flows generally NE c.25 mi/40 km to join South Fork Nemadji River in Douglas county, Wisconsin, at Minnesota state line. South Fork Nemadji River Minnesota and Wisconsin, formed by joining of several small creeks in S CARLTON county, NE Minnesota c.13 mi/21 km S of Carlton; flows c.15 mi/24 km to North Fork to form Nemadji River.

Nemaha (NEE-muh-hah), county (☐ 719 sq mi/1,869.4 sq km; 2006 population 10,374), NE KANSAS; ☉ SENECA; 39°46′N 96°01′W. Gently sloping terrain, bordered N by NEBRASKA; drained by South Fork Nemaha River. Cattle, hogs, poultry; corn, oats, soybeans, wheat. Food products, machinery, rubber products. Nemaha State Fishing Lake in center. Formed 1855.

Nemaha (NEE-muh-huh), county (☐ 411 sq mi/1,068.6 sq km; 2006 population 7,247), SE NEBRASKA; ☉ AUBURN; 40°23′N 95°50′W. Farming area bounded E by MISSOURI RIVER (Missouri state line); drained by LITTLE NEMAHA RIVER. Soybeans, wheat, corn, sorghum; cattle, hogs; dairy and poultry products. Cooper Station nuclear power plant (initial criticality February 21, 1974; maximum dependable capacity 778 MWe) is 23 mi/37 km S of NEBRASKA CITY; uses cooling water from Missouri River. Brownville State Recreation Area in E. Formed 1854.

Nemaha, town (2000 population 102), SAC county, W IOWA, near RACCOON RIVER, 8 mi/12.9 km NNW of SAC CITY; 42°31′N 95°05′W.

Nemaha (NEE-muh-huh), village (2006 population 183), NEMAHA county, SE NEBRASKA, 10 mi/16 km ESE of AUBURN, and on MISSOURI RIVER; 40°20′N 95°40′W. Missouri River flooded the village in 1993. Brownville State Recreation Area to N.

Neman (NYE-mahn), city (2005 population 12,440), N KALININGRAD oblast, W European Russia, on the NEMAN RIVER (Lithuanian border), on road and railroad, 80 mi/129 km NE of KALININGRAD, and 6 mi/10 km SE of SOVETSK; 55°02′N 22°01′E. Elevation 137 ft/41 m. Wood pulp and paper. Has a 15th century castle (once Teutonic Knights stronghold). Chartered in 1722. Until 1945, in German-administered EAST PRUSSIA and called Ragnit.

Neman (NE-muhn), German *Memel*, Lithuanian *Nemanus* (NE-moo-nuhs), Polish *Niemen*, river, 605 mi/974 km long, Eastern EUROPE; rises in central BELARUS, SW of MINSK; flows generally W to GRODNO, then N into LITHUANIA to KAUNAS, where it turns W and forms part of the Lithuania-RUSSIA (KALININGRAD oblast) border before entering the KURSKY ZALIV of the BALTIC SEA through a small delta W of SOVETSK (RUSSIA). Navigable c.60 mi/97 km above GRODNO. The meeting of Napoleon I and Czar Alexander I, which resulted in the Treaty of Tilsit (1807), took place on a raft in the middle of the river. Dammed at KAUNAS, forming KAUNO MARIOS reservoir (25 sq mi/65 sq km; largest in Lithuania). Drains basin of 37,915 mi/98,200 sq km (half in Lithuania) and accounts for 72 percent of Lithuania's run-off.

Nemanus River, Eastern Europe: see NEMAN.

Nemaska, Cree village (☐ 21 sq mi/54.6 sq km), Nord-du-Québec region, QUEBEC, E Canada, within Baie-James municipality. Founded 1978.

Nemausus, FRANCE: see NÎMES.

Nemawar (nai-MAH-wahr), village, DEWAS district, MADHYA PRADESH state, central INDIA, on NARMADA

RIVER, 19 mi/31 km SE of KANNOD. Annual festival fair; Siddhesvara temple.

Nembro (NEM-bro), town, BERGAMO province, LOMBARDY, N ITALY, on SERIO River, and 5 mi/8 km NE of BERGAMO; 45°45′N 09°45′E. Highly diversified secondary industrial center; foundry, paper mill; textiles.

Nemby (NEM-bee) or **San Lorenzo de la Frontera**, town (2002 population 71,909), CENTRAL department, S PARAGUAY, 11 mi/18 km SE of ASUNCIÓN; 25°22′S 57°36′W. Processing and agrilcultural center (bananas, oranges, cotton, sugarcane); sugar refining, liquor distilling. The town is part of the Asunción metropolitan area.

Nemchinovka (nyem-CHEE-nuhf-kah), settlement (2006 population 4,925), central MOSCOW oblast, central European Russia, on railroad and near highway, 5 mi/8 km NE of (and administratively subordinate to) ODINTSOVO; 55°43′N 37°21′E. Elevation 577 ft/175 m. Aluminum works.

Nemcice nad Hanou (NYEM-chi-TSE nahd HAH-nou), Czech *Nìmèice nad Hanou*, German *Niemtschitz an der Hanna*, town, JIHOMORAVSKY province, central MORAVIA, CZECH REPUBLIC, on railroad, and 9 mi/14.5 km SSE of PROSTĔJOV; 49°21′N 17°12′E. Agriculture (sugar beets); manufacturing (machinery, textiles); woodworks; sugar refinery. Has a baroque town hall and church.

Nemea (ne-MAI-ah), city of ancient GREECE, in N ARGOLIS, now in KORINTHIA prefecture, PELOPONNESE department. Nemean games, one of the four Panhellenic festivals, held at the Doric temple of Zeus here in foothills of Arkadian Mountains from 573 B.C.E.; held in the second and fourth years of each Olympiad; 37°49′N 22°40′E. Games revived in 1996. Of Pindar's odes, eleven celebrate Nemean victories. The temple and palaestra have been excavated and the latter has been restored for the games. During Corinthian War, Spartans defeated coalition forces in 394 B.C.E. here. Also the name of a river that formed the boundary of Corinth and Sicyon (near modern SIKYON).

Nemecka (nye-METS-kah), Slovak *Nemecká*, Hungarian *Németfalva*, village, STREDOSLOVENSKY province, central Slovakia, on HRON RIVER, on railroad, and 13 mi/21 km NE of BANSKÁ BYSTRICA; 48°35′N 19°27′E. Oil refinery. Baroque church.

Nemecke Jablonne, CZECH REPUBLIC: see JABLONNE V PODJESTEDI.

Nemecke Pravno, SLOVAKIA: see NITRIANSKE PRAVNO.

Nemecky Brod, CZECH REPUBLIC: see HAVLÍČKŮV BROD.

Nemencha Mountains (ne-men-SHAH), semiarid range of the Saharan ATLAS MOUNTAINS, TEBESSA wilaya, NE ALGERIA, a SE extension of the AURÈS massif, reaching out to the Tunisian border NW of GAFSA (TUNISIA). Rises to 4,800 ft/1,463 m. Important phosphate deposits, especially at Djebel ONK.

Nemenčinė (ne-MEN-shin-e), city, 15 mi/24 km NE of Vilnius, on main road to Ignalina; 54°51′N 25°29′E. Light manufacturing, food processing.

Nemёrçke (ne-MER-ke), mountain range, S ALBANIA (Epirus region), SW of VIJOSË RIVER, near Greek border; rises to 7,247 ft/2,209 m. Also spelled Nemёrçkä.

Nemeshayevo, UKRAINE: see NEMISHAYEVE.

Nemetacum, FRANCE: see ARRAS.

Németfalva, SLOVAKIA: see NEMECKA.

Nemi, Lake (NAI-mee), Latin *Nemorensis lacus*, small, picturesque crater lake, in the ALBAN HILLS, central ITALY, SE of ROME; c.1 mi/1.6 km long; 41°43′N 12°42′E. The sacred wood and the ruins of the celebrated temple of Diana are here. Two pleasure ships of the Roman emperor Caligula that were lying at the bottom of Lake Nemi for almost 2,000 years were raised (1930–1931) from an lake after its level had been lowered c.70 ft/21 m. No valuables were found, but objects interesting from an artistic and technical point of view were recovered. During World War II the ships were destroyed (1944) by the retreating German forces.

Nemirov, UKRAINE: see NEMYRIV, Vinnytsya oblast; or NEMYRIV, L'viv oblast.

Nemiscau, village, W central Quebec, E Canada, 85 mi/137 km ESE of WASKAGANISH (RUPERT HOUSE) and JAMES BAY, on NW side of Lake Nemiscau; 51°19′N 76°54′W. Located on SW margin of JAMES BAY Hydro Project. Timber.

Nemishayeve (ne-mee-SHAH-ye-ve) (Russian *Nemeshayevo*), town (2004 population 13,500), W central KIEV oblast, UKRAINE, near highway and on railroad, 20 mi/32 km NW of KIEV city center; 50°34′N 30°05′E. Elevation 465 ft/141 m. Biochemical manufacturing, Ukrainian potato research institute, exprimental farm. Established in 1900; town since 1950.

Nemmara (NE-mah-rah), town, PALGHAT district, KERALA state, S INDIA, 26 mi/42 km E of Trichur; 10°35′N 76°36′E. Trades in timber from ANAIMALAI HILLS (S); manufacturing, match manufacturing.

Nemocón (nai-mo-KON), town, ⊙ Nemocón municipio, CUNDINAMARCA department, central COLOMBIA, on railroad, and 36 mi/58 km NNE of BOGOTÁ; 05°03′N 73°53′W. Elevation 8,543 ft/2,604 m. Coffee, corn, potatoes. Salt mining. Deposits have been used since pre-Spanish days.

Nemours (nuh-MOOR), town (□ 4 sq mi/10.4 sq km), SEINE-ET-MARNE department, ÎLE-DE-FRANCE region, N central FRANCE, on the LOING RIVER (canalized), and 10 mi/16 km S of FONTAINEBLEAU; 48°16′N 02°42′E. Popular weekend resort of Parisians at S edge of Forest of Fontainebleau. Glass manufacturing (from silica quarried nearby). Has 12th-century castle (restored 15th–17th-century) containing 15th–16th-century tapestries. Was capital of medieval duchy, title to which later passed to a branch of house of Savoy and to several members of the Bourbon-ORLÉANS line. Here, in 1585, Henry III revoked concessions made to Protestants. Dupont family of Delaware originated here. There is a museum of the ancient history of Île-de-France region.

Nemrut Dag (Turkish=*Nemrut Dağı*), mountain, E TURKEY, 20 mi/32 km NNE of BITLIS, 10 mi/16 km W of LAKE VAN, in the ANTI-TAURUS MOUNTAINS. Rises to 10,010 ft/3,051 m. Site of colossal Hellenistic ruins for which Nemrut Dag was designated a UNESCO World Heritage site in 1987. Also called Mount Nemrut.

Nemsó, SLOVAKIA: see NEMSOVA.

Nemsova (nyem-SHO-vah), Slovak *Nemšová*, Hungarian *Nemsó*, town, ZAPADOSLOVENSKY province, W SLOVAKIA, on VÁH RIVER, and 5 mi/8 km NE of TRENČÍN; 48°58′N 18°07′E. Contains railroad junction; glassworks (cutting glass, bulbs); military base.

Nemunélis River, LITHUANIA: see MEMELE RIVER.

Nemuro (NE-moo-ro), city and PACIFIC OCEAN port, at easternmost point of Hokkaido prefecture, N JAPAN, on Cape NOSAPPU peninsula, 211 mi/340 km E of SAPPORO; 43°19′N 145°35′E. Marine-product processing; dairying. Lake Furen is nearby.

Nemuro Strait (NE-moo-ro), Japanese *Nemuro-kaikyo* (NE-moo-ro-KAH-ee-kyo), channel between easternmost point of HOKKAIDO, N JAPAN, and KUNASHIR ISLAND of KURIL ISLANDS, extreme E Siberian RUSSIA; connects SEA OF OKHOTSK (N) with the PACIFIC OCEAN (S); 10 mi/16 km–30 mi/48 km wide, 60 mi/97 km long. CAPE SHIRETOKO is at NW, CAPE NOSAPPU at SW side of entrance. Frozen December–March.

Nemyriv (ne-MI-reef) (Russian *Nemirov*) (Polish *Niemirów*), city, central VINNYTSYA oblast, UKRAINE, on railroad, 25 mi/40 km SE of VINNYTSYA; 48°58′N 28°51′E. Elevation 836 ft/254 m. Road junction; raion center. Fruit canning, distilling, sugar refining, flour milling, brickmaking, construction material manufacturing, sewing. Construction technical school; museum. Has a 19th-century palace, two churches

(1801, 1881). Known since 1506; passed to Poland (1569); under Turkish rule (1672–1699); to Russia (1793); city since 1985. Jewish community since 1603, developed into one of the centers of Hasidic learning in the region; reduced during the 1919 pogroms, but still numbering almost 4,200 in 1939; destroyed by the Nazis between 1941 and 1944—fewer than 100 Jews remaining in 2005.

Nemyriv (ne-MI-reef) (Russian *Nemirov*) (Polish *Niemirów*), town (2004 population 7,900), W L'VIV oblast, UKRAINE, 11 mi/18 km. SW of RAVA-RUS'KA, near the Polish border; 50°07′N 23°27′E. Elevation 905 ft/275 m. Lumbering, flour milling. Mineral water-based health resort nearby. Known since 1580, town since 1939. Jewish community since the 17th century, numbering 1,600 in 1939; wiped out during World War II.

Nenagh (NEE-nuh), Gaelic *An tAonach*, town (2006 population 7,415), ⊙ TIPPERARY county, S central IRELAND; 52°52′N 08°12′W. Agricultural market with varied manufacturing. Nenagh Castle (c.1200), a Butler stronghold, has a circular keep. Ruins of a Franciscan friary, founded in the 13th century and destroyed by Oliver Cromwell's forces in 1650, are nearby.

Nenana, village (2000 population 402), central ALASKA, on TANANA River (700-ft/213-m railroad bridge) at mouth of NENANA RIVER, and 45 mi/72 km WSW of FAIRBANKS; 64°32′N 149°05′W. Distributing and transshipment center from railroad to YUKON River tug boats and barges. Railroad museum. Scene of annual Nenana ice sweepstakes. Established 1916 as base for construction of Alaska railroad; at S end of bridge President Warren Harding opened railroad, July 1923.

Nenana River, 150 mi/241 km long, central ALASKA, S tributary of TANANA River; rises in central ALASKA RANGE, near 63°19′N 147°46′W; flows NNW to Tanana River at NENANA. River rafting. Paralleled by PARKS HIGHWAY and Alaska railroad for c.100 mi/161 km.

Nendaz (nen-DAHZ), dispersed commune, VALAIS canton, SW SWITZERLAND, S of the RHÔNE RIVER, in lower part of Val de Nendaz (Printse River), 4 mi/6.4 km SW of SION. Farming. Skiing. Consists of Basse-Nendaz (elevation 3,329 ft/1,015 m), Haute-Nendaz (elevation 4,488 ft/1,368 m), and Nendaz-Station at foot of chairlift to Pic de Nendaz (8,081 ft/2,463 m).

Nendeln, village, LIECHTENSTEIN, on railroad, and 4 mi/6.4 km NNE of VADUZ; 47°12′N 09°33′E. Manufacturing of dentures.

Nendö (NEN-duh) or **Santa Cruz**, volcanic island, SANTA CRUZ ISLANDS, SOLOMON ISLANDS, SW Pacific, 348 mi/560 km E of GUADALCANAL; 16 mi/25 km long, 11 mi/17 km wide; 10°45′S 165°55′E. Kauri pine. Graciosa Bay in NW. Formerly Ndeni.

Nene River (NEN), c.90 mi/145 km long, central ENGLAND; rises in the Northampton Uplands; flows NE past NORTHAMPTON, OUNDLE, PETERBOROUGH, and WISBECH to The WASH. It is navigable to Peterborough and drains part of the FENS. Also spelled Nen.

Nenets Autonomous Okrug (nye-NYETS), administrative division (□ 68,224 sq mi/177,382.4 sq km; 2006 population 38,110), N of ARCHANGEL oblast proper but subordinate to it, extreme NE European Russia; extending along the tundra coast of the BARENTS, WHITE, and KARA seas; ⊙ NAR'YAN-MAR, a lumber port on the PECHORA River; 67°30′N 54°00′E. The area includes the N section of the PECHORA COAL BASIN, with mines at KHALMER-YU (KOMI REPUBLIC) and along the Silova River. Until the discovery of oil and gas fields, reindeer raising, fishing, fur trapping, and seal hunting were the chief occupations. Fish canning, sawmilling, and hide processing are also important. Many of the formerly nomadic Nenets (now 12% of the population) live in agricultural settlements. Russians make up a majority (66%) of the

population. The Nenets, previously known as Samoyedes, speak a Finno-Ugric language and are either Orthodox Christians or animists. They were first mentioned in the 11th century, and became tributaries of the grand duchy of Moscow at the end of the 15th century. Nenets National Area was formed in 1929. In 1977, it changed its status from a national area to an autonomous okrug.

Nenjiang (NEN-JYAHNG), former province (□ 26,000 sq mi/67,600 sq km), NE CHINA. The capital was Qigihar. It was one of nine provinces established in 1945 by the Nationalist government in MANCHURIA. It was bordered on the S by the SONGHUA River and crossed by the NENJIANG; the soil in the valleys of these rivers is some of the most fertile in Manchuria. In 1950, Nenjiang was absorbed by HEILONGJIANG province. The name sometimes appears as Nenchiang.

Nenjiang (NEN-JYAHNG), town, ⊙ Nenjiang county, NW HEILONGJIANG province, NE CHINA, on INNER MONGOLIA border, on Nen River, on railroad, 140 mi/225 km NNE of QIQIHAR; 49°11′N 125°13′E. Grain, soybeans, oilseeds, sugar beets; sugar refining.

Nenjiang (NEN-JYAHNG), river, 740 mi/1,191 km long, NE CHINA; rises in the YILEHULI Shan (mountains), N HEILONGJIANG province, NE CHINA; flows S along the E side of the GREATER HINGGAN range to the SONGHUA River; forms part of the HEILONGJIANG-JILIN province border. Though frozen November–April, the river and its valley form an important trade artery. It is navigable for shallow-draft vessels along most of its length; a railroad follows its valley. Also known as Nonni.

Nenndorf, Bad, GERMANY: see BAD NENNDORF.

Neno (ne-no), village, MWANZA district, Southern region, S MALAWI, 38 mi/61 km NW of BLANTYRE, near MOZAMBIQUE border; 15°24′S 34°39′E. Tobacco, cotton, wheat, rice.

Nen River, ENGLAND: see NENE RIVER.

Nentón (nen-TON), town, HUEHUETENANGO department, W GUATEMALA, on W slopes of CUCHUMATANES MOUNTAINS, 38 mi/61 km NNW of HUEHUETENANGO; 15°48′N 91°45′W. Elevation 2,671 ft/814 m. Connected by all-weather road (built 1984) to INTER-AMERICAN HIGHWAY. Sugarcane, bananas, tropical fruit, grain; livestock. In Jacalteca-speaking area.

Nenzel (NEN-zil), village (2006 population 12), CHERRY county, N NEBRASKA, 28 mi/45 km W of VALENTINE, near SOUTH DAKOTA state line, and NIOBRARA RIVER; 42°55′N 101°05′W. Samuel B. McKelvie National Forest to S.

Nenzing (NEN-tsing), township, VORARLBERG, W AUSTRIA, in the WALGAU, near ILL RIVER, 6 mi/9.7 km SE of FELDKIRCH; 47°11′N 09°42′E. Manufacturing of metals (aluminum products); hydropower station in the WALGAU valley. Gothic and Roman ruins, baroque church with Carolingian foundations.

Neo (NE-o), village, Motosu county, GIFU prefecture, central HONSHU, central JAPAN, 19 mi/30 km N of GIFU; 35°37′N 136°37′E. Lumber. Neodani Fault nearby.

Neocaesarea, TURKEY: see NIKSAR.

Neoch (NECH), atoll, State of CHUUK, E CAROLINE ISLANDS, Federated States of MICRONESIA, W PACIFIC, 2 mi/3.2 km S of CHUUK ISLANDS; 11 mi/18 km long, 4 mi/6.4 km wide. Comprises four coral islets. Generally classed as one of Chuuk islands. Formerly Kuop or Royalist Island.

Neochorion, Greece: see NEOKHORIO.

Neodesha (nee-O-duh-shai), town (2000 population 2,848), WILSON county, SE KANSAS, on VERDIGRIS RIVER, near mouth of FALL RIVER, and 13 mi/21 km N of INDEPENDENCE; 37°25′N 95°40′W. Refining center in agriculture (grain; livestock) and oil area. Manufacturing (chemicals, boats, transportation equipment, hand tools, machinery, plastics products, wood products). Norman Oil Well to NE, replica of oil well W of Mississippi Museum Site of Neodesha Reservoir (for flood control on the Verdigris) is near. Incorporated 1871.

Neoga (nee-O-gah), city (2000 population 1,854), CUMBERLAND county, SE central ILLINOIS, 12 mi/19 km WNW of TOLEDO; 39°19′N 88°27′W. Corn, soybeans. Incorporated 1869.

Neokhorio (ne-o-KHO-ree-o), village, AKARNANIA prefecture, WESTERN GREECE department, GREECE, on lower AKHELÓOS River, and 9 mi/14.5 km WNW of MESOLONGI; 38°21′N 20°37′E. Railroad terminus; wheat, oats, wine. Formerly Neochorion.

Neola, town (2000 population 845), POTTAWATTAMIE county, SW IOWA, on MOSQUITO CREEK, and 18 mi/29 km NE of COUNCIL BLUFFS; 41°27′N 95°37′W. In corn, wheat, livestock area.

Neon, KENTUCKY: see FLEMING-NEON.

Neon Faliron (NE-on FAH-lee-ron) [Greek=new Phaleron], S suburb of ATHENS, ATTICA prefecture, ATTICA department, E central GREECE, in metropolitan area, just NE of PIRAEUS, on PHALERON BAY; 37°57′N 23°40′E. Silk milling; popular Athenian evening resort. Also Neon Phaleron.

Neon Karlovassion (NE-on kahr-lo-VAH-see-on), town, on N shore of SÁMOS island, SÁMOS prefecture, NORTH AEGEAN department, GREECE, 15 mi/24 km WNW of SÁMOS town; 37°47′N 26°42′E. Tanning industry; olive oil, wine, tobacco, carob; fisheries.

Neon Petrisi (NE-on pe-TREE-see), town, SÉRRAI prefecture, CENTRAL MACEDONIA department, NE GREECE, on STRUMA (Strymon) River, and 18 mi/29 km NW of SÉRRAI, on railroad, near Bulgarian border; 41°17′N 23°18′E. Barley, beans, tobacco. Also called Neon Petritsi and Neon Petritsion; formerly Vetrina.

Neon Phaleron, Greece: see NEON FALIRON.

Neopit (NEE-o-pit), village (2000 population 839), MENOMINEE county, NE WISCONSIN, 18 mi/29 km NNW of SHAWANO, on mill pond, in Menominee Indian Reservation; 44°58′N 88°49′W. Sawmilling on mill pond. Lumber.

Neópolis (NAI-o-po-lees), city (2007 population 18,829), NE SERGIPE state, NE BRAZIL, on right bank of lower SÃO FRANCISCO River (navigable here), diagonally opposite PENEDO (ALAGOAS state), and 20 mi/32 km SE of PROPRIÁ; 10°18′S 36°35′W. Textile mill; agriculture (rice, manioc). Formerly called Vila Nova.

Neosho (nee-O-sho), county (□ 578 sq mi/1,502.8 sq km; 2006 population 16,298), SE KANSAS; ⊙ ERIE; 37°33′N 95°17′W. Sloping to gently rolling area, drained by NEOSHO River. Sorghum, wheat, soybeans, hay; hogs, cattle, poultry; dairying. Textiles, plastic products, transportation equipment. Oil, gas, coal. Neosho Waterfowl Refuge and Neosho State Fishing Lake in SE. Formed 1864.

Neosho (nee-O-sho), city (2000 population 10,505), ⊙ NEWTON county, SW MISSOURI, in the OZARK MOUNTAINS, 16 mi/26 km SSE of JOPLIN; 36°50′N 94°22′W. Railroad junction. Apples, berries, vegetables; dairy products; cattle, poultry. Manufacturing (foods, furniture, apparel; poultry processing, jet engine overhauling). Tourism. Crowder College to S. Pro-Confederate Missouri state convention passed an ineffective ordinance of secession here, 1862. Thomas Hart Benton (painter and muralist) born here, 1889. National fish hatchery. Incorporated 1855.

Neosho (nee-O-sho), village (2006 population 582), DODGE county, S central WISCONSIN, on small Rubicon River, 19 mi/31 km SE of BEAVER DAM; 43°18′N 88°31′W. Dairying region.

Neosho (nee-O-sho), river, c.460 mi/740 km long, KANSAS and OKLAHOMA; rises in MORRIS county, E central Kansas; flows SE through Council Grove Lake Reservoir, past COUNCIL GROVE and EMPORIA, through John Redmond Reservoir, then past IOLA, CHANUTE, and INDEPENDENCE (all in Kansas), and MIAMI (Oklahoma) into NE Oklahoma (where it is

generally known as the GRAND RIVER) then S to join the ARKANSAS RIVER near MUSKOGEE. PENSACOLA DAM (which impounds the huge LAKE OF THE CHEROKEES) and FORT GIBSON dam and reservoir are in NE Oklahoma; there are several flood control units on the river in Kansas.

Neosho Falls (nee-O-sho), village (2000 population 179), WOODSON county, SE KANSAS, on NEOSHO River, and 12 mi/19 km NE of YATES CENTER; 38°00′N 95°33′W. Livestock; grain. Small oil fields nearby.

Neosho Rapids (nee-O-sho), village (2000 population 274), LYON county, E central KANSAS, on NEOSHO River, and 10 mi/16 km ESE of EMPORIA; 38°22′N 95°59′W. Cattle; grain.

Neos Marmaras (NE-os mahr-mah-RAHS), town, on SW coast of SITHONIA, the middle prong of the KHALKIDHIKÍ Peninsula, KHALKIDHIKÍ prefecture, CENTRAL MACEDONIA department, NE GREECE, and N of PORTO CARRAS resort; 40°06′N 23°47′E.

Neos Skopos (NE-os sko-POS), town, SÉRRAI prefecture, CENTRAL MACEDONIA department, NE GREECE, on railroad, and 5 mi/8 km SE of SÉRRAI, near STRUMA (Strymon) River; 41°01′N 23°37′E. Formerly called Kisiklik and Toumbista.

Néouvielle Massif (nai-oo-VYEL mah-SEEF), rugged granitic upland of the high PYRENEES mountains between the ADOUR and GARONNE rivers, in HAUTES-PYRÉNÉES department, MIDI-PYRÉNÉES region, SW FRANCE, culminating in the Pic de Néouvielle (10,191 ft/3,091 m); 42°51′N 00°07′E. Many features of glacial topography including tarn lakes, ice-smoothed cliffs, hanging valleys, and erratic boulders. Partly within the PYRENEES NATIONAL PARK, which extends S to the Spanish border. Accessible by road from SAINT-LARY-SOULAN on the trans-Pyrenean road, N of the BIELSA TUNNEL.

Nepal (ne-PAHL), independent kingdom (□ c.54,000 sq mi/139,860 sq km; 2004 estimated population 27,070,666; 2007 estimated population 28,901,790), central ASIA; ⊙ KATHMANDU; 28°00′N 84°00′E.

Geography

Landlocked and isolated by the HIMALAYA mountains, bordered W, S, and E by INDIA and N by the TIBET region of CHINA. Geographically, Nepal comprises three major areas: the TERAI in the S, a comparatively low region of cultivable land, swamps, and forests, valuable timber; the main section of the Himalayas, including Mount Everest (29,028 ft/8,848 m), the world's highest peak, in the N, where Nepal's major rivers, rising in Tibet, rush through deep Himalayan gorges; and Central Nepal, an area of moderately high mountains and the KATHMANDU VALLEY, the country's most densely populated region and its administrative, economic, and cultural center. Kathmandu is served by several highways linking it to India (Tribhuwan Rajpath), China (Arniko Rajmarg), POKHARA (Prithivi Rajmarg), and to E and W parts of the country (East-West Highway, Mahendra Rajmarg). Nepal has only one reliable outlet to the sea, an overland route to the Indian port of CALCUTTA, over 715 mi/1,151 km away.

Population

The population of Nepal is the result of a long intermingling of Tibeto-Burmans, who migrated from the N (especially Tibet), and peoples who came from the GANGA plain in the S. The chief ethnic group, the Newars, were probably the original inhabitants of the Kathmandu Valley. Several ethnic groups are classified together as Bhotias; among them are the Sherpas, famous for guiding mountain-climbing expeditions, and the Gurkhas, a term sometimes loosely applied to the fighting castes, who achieved fame in the British Indian army and continue to serve as mercenaries in India's army and in the British overseas forces. Nepali, the country's official language, is an Indo-European language and has similarities to Hindi. Tibeto-Burman and Munda languages and various Indo-Aryan

dialects are also spoken. Hinduism and Buddhism, particularly its Tibetan form, coexist in Nepal, where tribal and caste distinctions are still important and Brahmans (the Hindu priestly class) retain great political influence; the royal family is Hindu. Siddhartha Gautama, the Lord Buddha, was born in LUMBINI, S Nepal, c.563 B.C., living here for thirty years before forsaking his royal environs for a life of asceticism.

Economy

The overwhelming majority of Nepal's people engage in agriculture, which contributes about two-thirds of the national income. In the Terai, the main agricultural region, rice is the chief crop; other food crops include pulses, wheat, barley, and oilseeds. Jute, tobacco, cotton, and indigo are also grown in the Terai, whose forests provide sal wood and commercially valuable bamboo and rattan. In the lower mountain valleys, rice is produced during the summer, and wheat, barley, oilseeds, potatoes, and vegetables are grown in the winter. Corn, wheat, and potatoes are raised at higher altitudes, and terraced hillsides are also used for agriculture. Large quantities of medicinal herbs, grown on the Himalayan slopes, are sold worldwide. Livestock raising is second to farming in Nepal's economy; oxen predominate in the lower valleys, yaks in the higher, and sheep, goats, and poultry are plentiful everywhere. Transportation and communication difficulties have hindered the growth of industry and trade. Wood and metal handicrafts are important. BIRATNAGAR and BIRGANJ, in the Terai, are the main manufacturing towns; their products include cotton cloth, textiles, cigarettes, matches, furniture, shoes, stainless steel, sugar, processed rice, flour, oilseeds, and jute. Kathmandu is a major industrial center, with a wide variety of consumer-goods industries. Tourism is the chief source of foreign income, along with foreign aid and military pensions, carpets and garments. Nepal's trade is overwhelmingly with India, and China is also an important trading partner.

History: to 1846

By the 4th century, the Newars of the central Kathmandu Valley had apparently developed a flourishing Hindu-Buddhist culture. From the 8th to 11th century many Buddhists fled here from India, and a group of Hindu Rajput warriors set up the principality of GORKHA just W of the Kathmandu Valley. Although a Newar dynasty, the Mallas, ruled the valley 14th–18th centuries, there were internecine quarrels among local rulers. These were exploited by the Gorkha king Prithvi Narayan Shah, who conquered the Kathmandu Valley in 1768. Gorkha armies seized territories far beyond the present-day Nepal; but their invasion of Tibet, over which China claimed sovereignty, was defeated in 1792 by Chinese forces. An ensuing peace treaty forced Nepal to pay China an annual tribute, which continued until 1910. Also in 1792, Nepal first entered into treaty relations with Great Britain. Gorkha expansion into N India, however, led to a border war (1814–1816) and to British victory over the Gorkhas, who were forced by treaty to retreat into roughly the present borders of Nepal and to receive a British envoy at Kathmandu.

History: 1846 to 1959

The struggle for power among the Nepalese nobility culminated in 1846 with the rise to political dominance of the Rana family. Jung Bahadur Rana established a line of hereditary prime ministers, who controlled the government until 1950, and the Shah dynasty kings were mere figureheads. In 1854, Nepal again invaded Tibet, which was forced to pay tribute from then until 1953. Under the Ranas, Nepal was deliberately isolated from foreign influences; this policy helped to maintain independence during the colonial period but prevented economic and social modernization. Relations with Britain were cordial, however, and in 1923 a British-Nepalese treaty expressly affirmed Nepal's full sovereignty. Nepal supplied many troops for the British army in both World Wars. The successful Indian movement for independence (1947) stimulated democratic sentiment in Nepal. The newly formed Congress Party of Nepal precipitated a revolt in 1950 that forced the autocratic Ranas to share power in a new cabinet. King Tribhuvan, who sympathized with the democratic movement, took temporary refuge in India and returned as a constitutional monarch.

History: 1959 to 1961

In 1959 a democratic constitution was promulgated, and parliamentary elections gave the Congress Party a clear majority; the following year, however, King Mahendra (reigned 1952–1972) cited alleged inefficiency and corruption in government as evidence that Nepal was not ready for Western-style democracy. He dissolved the parliament, detained many political leaders, and in 1962 inaugurated a system of "basic democracy," based on the elected village council (*panchayat*) and working up to district and zonal panchayats and an indirectly elected national panchayat. Political parties were banned, and the king was advised by a council of appointed ministers. King Mahendra carried out land reform that distributed large holdings to landless families, and he instituted a law removing the legal sanctions for caste discrimination. Crown Prince Birenda succeeded to the throne (1972) upon his father's death; like previous Nepalese monarchs, he married a member of the Rana family to ensure political peace. Prior to 1989, Nepal maintained a position of nonalignment in foreign affairs, carefully balancing relationships with China, the former-USSR, the U.S., and India. The former USSR and the U.S. were major aid donors. A 1956 treaty with China recognized Chinese sovereignty over Tibet and officially terminated the centuries-old Tibetan tribute to Nepal; all Nepalese troops left Tibet in 1957.

History: 1961 to Present

The Sino-Nepalese border treaty of 1961 defined Nepal's Himalayan frontier. India's geographical proximity, cultural affinity, and substantial economic aid render it the most influential foreign power in Nepal, but India's political and economic dominance has been a constant source of worry for the government. Weeks of street protests and general strikes forced King Birenda to proclaim (November 1990) a new constitution that legalized political parties, asserted human rights, and vastly reduced the king's powers. A democratically elected government has controlled Nepal since 1992. Significant deforestation and a growing population have also greatly affected the country. In the mid-1990s, a Maoist insurgency has destabilized the country economically and politically with repeated attacks. In 2001, Crown Prince Dipendra murdered the royal family, including King Biendra and Queen Aishwaryab before attempting suicide; hospitalized and in a coma, Dipendra was declared king only to die two days later. Prince Gyanendra, brother to King Biendra and uncle to King Dipendra, assumed the throne. In 2004 Sher Bahadur Deuba became prime minister, however on January 31, 2005, King Gyanendra dismissed Nepal's government and declared a state of emergency, taking control of the kingdom for the second time in three years. After a cease-fire agreement broke down, the Maoist rebels increase attacks against government forces throughout 2004 and 2005. In January 2006, Nepal was further destabilized when political parties arranged mass protests against the king's autocratic rule and government troops opened fire on the protesters. The king agreed to reinstate parliament in April 2006, and Girija Prasad Koirala became prime minister. A cease-fire was declared and lengthy negotiations began with the rebels; the government also stripped the king of his powers and privileges. In November

2006 a peace accord was signed with the rebels, who agreed to join an interim government, which was established in April 2007. The rebels left the government in September 2007, demanding the monarchy be abolished before the planned 2007 constitutional assembly elections, which were subsequently postponed.

Government

An interim constitution adopted in January 2007 transferred the executive power of the Nepalese monarch to the prime minister, who became both head of state and head of government. It also established a 330-seat Interim Parliament, which is slated to be replaced by an elected constituent assembly in late 2007. The current head of state and head of government is Prime Minister Girija Prasad Koirala (since April 2006). Administratively, Nepal is divided into fourteen administrative zones: BAGMATI, BHERI, DHAULAGIRI, GANDAK, JANAKPUR, KARNALI, KOSI, LUMBINI, MAHAKALI, MECHI, NARAYANI, RAPTI, SAGARMATHA, and SETI. The administrative zones are divided into a total of seventy-five districts.

Nepalganj Road, INDIA: see NANPARA.

Nepalgunj (NE-pahl-gunj), city (2001 population 57,535), ⊙ BANKE district, SW Nepal, in the TERAI, near INDIA border, and 15 mi/24 km NNE of NANPARA (UTTAR PRADESH state, India); 28°03′N 81°37′E. Elevation 541 ft/165 m. Trade center (rice, corn, millet, ghee, oilseeds; hides). Railroad terminus is 3 mi/4.8 km SSW, at Nepalgunj Road, just inside India. Airport.

Nepal Himalaya, central subdivision of the HIMALAYA range, S central ASIA, roughly coextensive with Nepal, N of GANGA PLAIN; extending from KALI (SARDA) River, the border of W Nepal and N central INDIA (UTTAR PRADESH state), to MECHI River, E Nepal's border with NE India (SIKKIM state); bordered S by the SHIWALIK RANGE; 28°00′N 84°00′E. The central area is occupied by the MAHABHARAT LEKH running W-E. SINGALILA RANGE (on E Nepal-India border; E) is principal N-S spur. Toward the N is the GREAT HIMALAYA range, which contains some of highest peaks of the world, including MOUNT EVEREST (29,028 ft/8,848 m; highest), KANCHENJUNGA (28,170 ft/8,586 m; third-highest), LHOTSE (27,890 ft/8,501 m), MAKALU (27,766 ft/8,463 m), CHO OYU (26,906 ft/8,201 m), DHAULAGIRI (26,795 ft/8,172 m), MANASLU (26,781 ft/8,163 m), and ANNAPURNA I (26,545 ft/8,091 m). Divided (W-E) by KARNALI (GHAGHARA in India), NARAYANI (GANDAK in India), and SAPTA (KOSI in India) river systems into three distinct sects. Main town is KATHMANDU.

Nepaug Reservoir, CONNECTICUT: see FARMINGTON RIVER.

Nepean (ni-PEE-uhn), former city (☐ 84 sq mi/218.4 sq km; 2001 population 124,878), SE ONTARIO, E central Canada, 6 mi/10 km SSW of downtown OTTAWA, of which it is a part; formerly part of dissolved Ottawa-Carleton Regional Municipality (1991 population 678,147); 45°20′N 75°52′E. Borders OTTAWA River on NW, Rideau River on SE. Seat of Algonquin College. Rural area (mixed farming; dairying).

Nepean Bay (ni-PEE-uhn), inlet of INVESTIGATOR STRAIT of INDIAN OCEAN, on NE KANGAROO ISLAND, SOUTH AUSTRALIA state, S central AUSTRALIA; 10 mi/16 km long, 7 mi/11 km wide; 35°42′S 137°45′E. KINGSCOTE on NW shore.

Nepeña River, PERU: see SAN JACINTO.

Nephi (NEE-fei), city (2006 population 5,207), ⊙ JUAB county, central UTAH, 38 mi/61 km SSW of PROVO, in mountain region; 39°42′N 111°49′W. Elevation 5,133 ft/1,565 m. Cattle- and grain-shipping (wheat, barley, alfalfa) center; manufacturing (rubber products). Parts of Uinta National Forest to NE and SE, including Mount Nebo Wilderness Area to NE. Fountain Green Fish Hatchery to E. Settled 1851 by Mormons.

Nephin Beg (NAI-fin BEG), Gaelic *Néifinn Bheag*, mountain range (20 mi/32 km long), W MAYO county, NW IRELAND, 11 mi/18 km NNW of CASTLEBAR; 54°00′N 09°40′W. Highest peak is Nephin (2,646 ft/807 m).

Nepi (NAI-pee), ancient *Nepete* and *Nepet*, town, VITERBO province, LATIUM, central ITALY, 18 mi/29 km SE of VITERBO; 42°14′N 12°21′E. In agricultural region (cereals, grapes, olives). Bishopric. Has ancient walls (Etruscan, Roman), cathedral, town hall designed by Vignola, and ruins of 15th-century castle. Mineral baths nearby; mineral waters bottled.

Nepolokivtsi (ne-po-LO-kif-tsee) (Russian *Nepolokovtsy*), town, N central CHERNIVTSI oblast, UKRAINE, on left bank of the PRUT RIVER, on road, and on railroad spur, 15 mi/24 km WNW of CHERNIVTSI city center, and 7 mi/11 km SW of KITSMAN'; 48°23′N 25°38′E. Elevation 620 ft/188 m. Woodworking, grain milling and baking, food-flavoring manufacturing, gravel and sand extracting. Known since 1425, town since 1968.

Nepolokovtsy, UKRAINE: see NEPOLOKIVTSI.

Nepomuceno (ne-po-moo-se-no), city (2007 population 24,430), S MINAS GERAIS state, BRAZIL, near the RIO GRANDE, 25 mi/40 km S of CAMPO BELO; 21°28′S 45°15′W. Coffee, sugar, cereals. Asbestos deposits.

Nepomuk (NE-po-MUK), town, ZAPADOCESKY province, SW BOHEMIA, CZECH REPUBLIC, on USLAVA RIVER, and 14 mi/23 km NE of KLATOVY; 49°29′N 13°36′E. Railroad junction. Manufacturing (textiles, machinery); food processing. St. John of Nepomuk, Bohemia's patron saint, was born here.

Neponset (ni-PON-set), village, BUREAU county, N ILLINOIS, 18 mi/29 km WSW of PRINCETON; 41°17′N 89°47′W. In agricultural area.

Neponset River (ne-PAWN-set), c.25 mi/40 km long, E MASSACHUSETTS; rises in SW NORFOLK county; flows NE, past MILTON (head of navigation), to SW side of Boston Harbor, between BOSTON and QUINCY.

Neponsit, section of S QUEENS borough of NEW YORK city, SE NEW YORK, on ROCKAWAY PENINSULA; 40°34′N 73°52′W. Residential; beachfront community.

Neptune, township, MONMOUTH county, E NEW JERSEY, on the ATLANTIC coast, and W and S of ASBURY PARK; 40°12′N 74°02′W. Manufacturing; primarily residential. Includes OCEAN GROVE (resort). Incorporated 1879.

Neptune Beach (NEP-toon), resort city (☐ 7 sq mi/18.2 sq km; 2005 population 7,018), DUVAL county, extreme NE FLORIDA, 15 mi/24 km E of JACKSONVILLE, on the ATLANTIC OCEAN; 30°18′N 81°23′W. Incorporated 1931 (when it seceded from JACKSONVILLE BEACH).

Neptune City, resort borough (2006 population 5,150), MONMOUTH county, E NEW JERSEY, near the ATLANTIC coast, 2 mi/3.2 km SW of ASBURY PARK; 40°12′N 74°01′W. Incorporated 1881.

Neptune Islands (NEP-toon), in INDIAN OCEAN, near entrance to SPENCER GULF, 15 mi/24 km SSE of EYRE PENINSULA, SOUTH AUSTRALIA state, S central AUSTRALIA; 35°17′S 136°06′E. Comprise two small groups, North and South Neptunes. Largest island 3 mi/5 km in circumference, 1.5 mi/2.4 km long. Granite cliffs.

Neptune Range, mountain range in the PENSACOLA MOUNTAINS, EAST ANTARCTICA, located WSW of the FORRESTAL RANGE; 70 mi/113 km long; 83°30′S 56°00′W.

Nequasset, MAINE: see WOOLWICH.

Nérac (nai-RAHK), town (☐ 24 sq mi/62.4 sq km), LOT-ET-GARONNE department, AQUITAINE region, SW FRANCE, on the BAÏSE RIVER, and 15 mi/24 km WSW of AGEN; 44°10′N 00°10′E. Armagnac brandy trade. Tanneries. Has ruins of 16th-century castle of Henry IV. Petit-Nérac, the old town on right bank of Baïse River, preserves its 15th-century appearance. It was the stronghold of princes of BÉARN, who gave refuge to Calvin during Wars of Religion. Still considered the chief town of the ALBRET region, between the GARONNE RIVER and the Landes of Gascony (LANDES DE GASCOGNE NATURAL REGIONAL PARK).

Nerang (nuh-RANG), town, SE QUEENSLAND, NE AUSTRALIA, 26 mi/42 km S of BRISBANE, 9 mi/14 km W of SOUTHPORT, in GOLD COAST hinterland, on Pacific Highway, and in MCPHERSON RANGE foothills; 27°59'S 153°20'E. Hinze Dam and Advancetown lake 6 mi/10 km SW. Racetrack, landing strip 3 mi/5 km SE.

Nera River (NAI-rah), ancient *Nar*, 80 mi/129 km long, central ITALY; rises in Monti SIBILLINI 4 mi/6 km NE of NORCIA; flows SSW, past TERNI, to TIBER River 8 mi/13 km SW of NARNI. Used for hydroelectric power (Montoro). Receives VELINO River (left).

Neratovice (NE-rah-TO-vi-TSE), German *Neratowitz*, town, STREDOCESKY province, N central BOHEMIA, CZECH REPUBLIC, on ELBE RIVER, and 7 mi/11.3 km SSE of MELNIK; 50°16'N 14°31'E. Important railroad junction. Agriculture (vegetables); manufacturing (chemicals, machinery, building materials); food processing. Has a 14th century Gothic church with paintings. Paleolithic-era archaeological site.

Neratowitz, CZECH REPUBLIC: see NERATOVICE.

Néravy, INDIA: see NIRAVI.

Nerchinsk (NYER-cheensk), city (2005 population 15,240), E central CHITA oblast, S SIBERIA, RUSSIA, on the Nercha River, 4 mi/6.4 km from its confluence with the SHILKA River (AMUR River basin), on crossroads and spur of the TRANS-SIBERIAN RAILROAD, 190 mi/306 km E of CHITA; 51°59'N 116°35'E. Elevation 1,587 ft/483 m. Electromechanical works; meatpacking; forestry. Founded in 1654, the city was a Russian outpost in the area from the 17th to the 19th centuries. A Russo-Chinese border treaty signed here in 1689 granted the Transbaikalia area to Russia and left the Amur valley to China. Nerchinsk became an important customs and trade center on the caravan route from China and Mongolia to European and Far East Russia.

Nerchinskiy Zavod (NYER-cheen-skeeyee zah-VOT), village (2005 population 2,970), SE CHITA oblast, S SIBERIA, RUSSIA, near the ARGUN River, approximately 12 mi/19 km W of the Russia-China border, on highway, 270 mi/435 km E of CHITA; 51°19'N 119°36'E. Elevation 2,047 ft/623 m. Lead and zinc mining; forestry services. Founded 1704 as a silver-mining center.

Nerchinsk Range (NYER-cheensk), SE CHITA oblast, S SIBERIA, RUSSIAN FAR EAST, extends from the Mongolian border 125 mi/201 km NE to the ARGUN RIVER; rises to 4,300 ft/1,311 m; 50°50'N 117°50'E. Silver, lead, zinc. Mining centers include NERCHINSKIY ZAVOD, GAZIMURSKIY ZAVOD, ALEKSANDROVSKIY ZAVOD.

Nerdva (NYER-dvah), village, W central PERM oblast, W URAL Mountains, E European Russia, on road, 28 mi/45 km N of (and administratively subordinate to) KARAGAY; 58°41'N 55°02'E. Elevation 515 ft/156 m. In agricultural area (wheat, sunflowers; livestock).

Nerekhta (NYE-reekh-tah), city (2005 population 25,140), SW KOSTROMA oblast, central European Russia, on the Nerekhta River (VOLGA River basin), 29 mi/47 km SSW of KOSTROMA; 57°27'N 40°34'E. Elevation 344 ft/104 m. Railroad and highway junction. Main industry, formerly military spare parts, now consumer products (linen, apparel); plastics; dairy. Known from 1214. Made city in 1778.

Neresheim (NE-res-heim), town, E WÜRTTEMBERG, BADEN-WÜRTTEMBERG, GERMANY, 11 mi/18 km SE of AALEN; 48°46'N 10°20'E. Grain. Has Benedictine abbey (founded in 11th century) with splendid baroque church.

Neresnica (ner-ES-nee-tsah), village, E SERBIA, 28 mi/45 km SE of POZAREVAC; 44°27'N 21°44'E. Goldplacer mining. Also spelled Neresnitsa.

Neretva Channel (NE-ret-vah), Croatian *Neretvanski Kanal*, Ital. *Canale della Narenta*, inlet of ADRIATIC SEA, S CROATIA, bet. PELJESAC peninsula (SW) and

mainland (NE). Provides BOSNIA AND HERZEGOVINA with an outlet to the ADRIATIC via NERETVA R.

Neretva Delta Special Reserve (□ 1 sq mi/2.6 sq km), protected wetlands along lower course of NERETVA R., Dalmatia, S Croatia. Only a small portion of what was once largest wetlands in Croatia, most of which has been drained for agr. and villages over the last several decades. Important sanctuary for migratory birds en route bet. Europe and Afr.

Neretva River (ner-RET-vah), ancient *Naro*, Italian *Narenta*, 35 mi/217 km long, in DINARIC ALPS, BOSNIA AND HERZEGOVINA and DALMATIA region of CROATIA; rises 7 mi/11.3 km N of Gackol; flows NNW past KONJIC, and SSW past JABLANICA, MOSTAR, and Metković (Croatia), to ADRIATIC SEA at Ploče. Navigable for small crafts for 65 mi/105 km. Receives RAMA and BREGAVA Rivers. Followed between KONJIC and GABELA by SARAJEVO-DUBROVNIK (Croatia) railroad. Cotton grown on plain along lower Neretva River. A hydroelectric plant is on the RAMA RIVER Delta (Croatia), part of special reserve.

Nerima (ne-REE-mah), ward, in NW corner of TOKYO city, Tokyo prefecture, E central HONSHU, E central JAPAN. Bordered N by SAITAMA prefecture, NE by ITABASHI ward, E by TOSHIMA ward, S by NAKANO and SUGINAMI wards, and Mushashino city, and W by HOYA city. Has fourth-largest population in the city.

Neringa (NE-rin-guh), city, W LITHUANIA, on the COURLAND SPIT immediately S of Klaipeda; 55°22'N 21°04'E. Numerous resorts, consisting of beaches and dunes; administrative center at NIDA; also includes Joudkrante, Pervalka, and PREILA resorts and fishing communities. Museum Thomas Mann house.

Néris-les-Bains (nai-REE–lai–BAN), commune (□ 12 sq mi/31.2 sq km), spa in ALLIER department, AUVERGNE region, central FRANCE, 5 mi/8 km SE of MONTLUÇON; 46°17'N 02°40'E. Health resort with hot springs (125°F/52°C). Vestiges of Roman thermal baths (built under Augustus) have been excavated.

Neris River, Eastern EUROPE: see VILIYA RIVER.

Nerja (NER-jah), city, MÁLAGA province, S SPAIN; minor port on the MEDITERRANEAN SEA, 30 mi/48 km E of MÁLAGA; 36°44'N 03°52'W. Also fishing; livestock raising; fruit growing.

Nerl' (NYERL), village, SE TVER oblast, W central European Russia, on the NERL' RIVER, on road, 13 mi/21 km SSE of KALYAZIN; 57°03'N 37°59'E. Elevation 406 ft/123 m. Shoe manufacturing.

Nerl' (NYERL), village (2005 population 2,265), SW IVANOVO oblast, central European Russia, on road and railroad, 12 mi/19 km SSW of TEYKOVO; 56°39'N 40°23'E. Elevation 400 ft/121 m. In agricultural area (oats, rye, flax, hemp, potatoes, sunflowers).

Nerl' River (NYERL), 50 mi/80 km long, in W central European Russia; rises in PLESHCHEYEVO LAKE, S YAROSLAVL oblast; flows NW, past NERL' (TVER oblast), to the RYBINSK RESERVOIR on the VOLGA RIVER, W of RYBINSK.

Nerl' River (NYERL), approximately 120 mi/193 km long, in central European Russia; rises in the Uglich Upland in S YAROSLAVL oblast, N of PERESLAVL'-ZALESSKIY; flows E and SE, past PETROVSKIY (IVANOVO oblast), to the KLYAZ'MA RIVER at ORGTRUD (VLADIMIR oblast).

Nero, Lake (NYE-ruh), glacial lake (□ 15 sq mi/39 sq km), in SE YAROSLAVL oblast, W central European Russia; 8 mi/13 km long, 4 mi/6 km wide; 57°09'N 39°27'E. ROSTOV lies on its NW shore. Vegetable farms and orchards nearby.

Nérondes (nai-ROND), agricultural village (□ 13 sq mi/33.8 sq km), CHER department, CENTRE administrative region, central FRANCE, 21 mi/34 km ESE of BOURGES; 47°00'N 02°50'E. Poultry shipping.

Nerone, Monte (ne-RO-ne, MON-te), peak (5,006 ft/1,526 m) in Umbrian APENNINES, N central ITALY, 13 mi/21 km SW of URBINO; 43°33'N 12°31'E.

Nerópolis (ne-RO-po-lees), city (2007 population 19,374), S GOIÁS state, central BRAZIL, 40 mi/64 km NNE of GOIÂNIA; 16°18'S 49°16'W. Extension of railroad from ANÁPOLIS under construction here.

Ner River, c.70 mi/113 km long, central POLAND; rises 6 mi/9.7 km ESE of ŁÓDŹ; flows generally NW, past S suburbs of ŁÓDŹ, KONSTANTYNÓW ŁÓDZKI, and DABIE, to WARTA River 5 mi/8 km S of KOŁO.

Nersac (ner-SAHK), commune (□ 3 sq mi/7.8 sq km), CHARENTE department, POITOU-CHARENTES region; W FRANCE, on CHARENTE RIVER, and 5 mi/8 km WSW of ANGOULÊME; 45°38'N 00°03'E. Felt mill. Chocolate candy factory nearby.

Nersingen (NER-sing-uhn), village, SWABIA, BAVARIA, S GERMANY, on the Roth River near its mouth to the DANUBE RIVER, 7 mi/11.3 km ENE of ULM; 48°26'N 10°09'E. Agriculture (grain); metalworking.

Nerstrand (NUHR-strand), village (2000 population 233), RICE county, SE MINNESOTA, 11 mi/18 km ENE of FARIBAULT; 44°20'N 93°03'W. Dairying; poultry; grain, soybeans; light manufacturing. Nerstrand Woods State Park to W. Holds annual "Ring Bologna Days."

Nerva (NER-vah), town, HUELVA province, SW SPAIN, in ANDALUSIA, 35 mi/56 km NW of SEVILLE; 37°42'N 06°32'W. Copper- and pyrite-mining center; also manganese deposits. Nerva is the old Ríotinto, a name now given to the mining town 3 mi/4.8 km WNW.

Nervesa della Battaglia (ner-VAI-zah del-lah baht-TAH-lyah), village, TREVISO province, VENETO, N ITALY, on PIAVE River, and 11 mi/18 km N of TREVISO; 45°49'N 12°12'E. Fabricated metals, wood products, textiles, clothing.

Nervi (NER-vee), town, GENOVA province, LIGURIA, N ITALY, port on Riviera di Levante, 6 mi/10 km E of GENOA; 44°23'N 09°02'E. Forms E border of Greater Genoa. Cotton mills, pasta factories, distilleries (olive oil, perfume); fisheries; active trade in citrus fruit. Much-frequented winter resort.

Nerviano (ner-VYAH-no), town, MILANO province, LOMBARDY, N ITALY, on OLONA River, and 12 mi/19 km NW of MILAN; 45°33'N 08°58'E. Textile-manufacturing center (clothing); machinery, furniture, non-metallics, fabricated metals, chemicals.

Nervión River (ner-VYON), 45 mi/72 km long, VIZCAYA province, N SPAIN; rises in E spurs of the CANTABRIAN MOUNTAINS; flows N, past BILBAO (8 mi/12.9 km upstream), to BAY OF BISCAY between PORTUGALETE and LAS ARENAS; forms wide estuary (Bilbao Bay). From Bilbao to its mouth, its banks are lined with industrial towns (BARACALDO, SESTAO, ALZAGA) where Spanish metallurgical industries are concentrated. Navigable to Bilbao (inner harbor) for freighters up to 4,000 tons/4,408 metric tons; outer harbor at head of estuary.

Neryungri (nye-RYOON-gree), city (2006 population 66,300), S SAKHA REPUBLIC, S RUSSIAN FAR EAST, in the STANOVOY RANGE, in the Timpton River (tributary of the ALDAN RIVER) basin, on highway branch and terminus of N-bound spur of the BAYKAL-AMUR MAINLINE, 422 mi/679 km SSW of YAKUTSK; 56°40'N 124°43'E. Elevation 2,539 ft/773 m. Center of the South Yakut Territorial Production Complex and the Neryungri coal basin; mining and processing of bituminous coal. Regional electric-generation station; machinery repair shop; housing construction combine. Made city in 1975.

Nes, village, FRIESLAND province, N NETHERLANDS, on S side of AMELAND island, and 13 mi/21 km NW of DOKKUM, on WADDENZEE; 53°26'N 05°47'E. Sand dunes to N; car ferry to HOLWERD SE on mainland. Seaside resort; dairying; livestock; vegetables. Also known as Nes op Ameland.

Nesbit (NEZ-bit), unincorporated village, DE SOTO county, NW MISSISSIPPI, 18 mi/29 km S of MEMPHIS (TENNESSEE), and 4 mi/6.4 km N of HERNANDO. Agricultural area (cotton, corn; cattle). Manufacturing (fabricated metal products, wire racks, wood products). Site of Jerry Lee Lewis Ranch, home of the rock-'n'-roll star; has piano-shaped swimming pool.

Nesbit (NES-bit), village, HARRISON county, NE TEXAS, residential suburb 5 mi/8 km NW of downtown MARSHALL; 32°35′N 94°27′W. Located in cattle and timber area.

Nesbitt (NEZ-bit), community, SW MANITOBA, W central Canada, 8 mi/13 km W of WAWANESA, in Oakland rural municipality; 49°36′N 99°51′W.

Nesbyen (NAIS-buh-uhn), village, BUSKERUD county, S NORWAY, in the HALLINGDAL, on railroad, and 50 mi/80 km NW of HØNEFOSS; 60°34′N 09°06′E. Communications and cultural center of the Hallingdal; tourism. Has folk museum and old church. One of the warmest places in Norway during the summer season.

Nesconset, unincorporated village (□ 3 sq mi/7.8 sq km; 2000 population 11,992), SUFFOLK county, SE NEW YORK, 3.5 mi/5.6 km SE of SMITHTOWN; 40°51′N 73°09′W. Light manufacturing.

Nescopeck (NES-ko-pek), borough (2006 population 1,447), LUZERNE county, E central PENNSYLVANIA, 23 mi/37 km SW of WILKES-BARRE, on SUSQUEHANNA RIVER (bridged; opposite BERWICK to NW), and at mouth of NESCOPECK CREEK; 41°02′N 76°12′W. Manufacturing (transportation equipment, ordnance, burial vaults). Agricultural area (apples, potatoes, corn; livestock; dairying). Settled 1786, incorporated 1896.

Nescopeck Creek (NES-ko-pek), c.45 mi/72 km long, E central PENNSYLVANIA; rises in E central LUZERNE county, c.7 mi/11.3 km S of WILKES-BARRE; flows WSW, passes N of FREELAND, turns NW to SUSQUE-HANNA RIVER at NESCOPECK, opposite BERWICK; 41°05′N 75°50′W.

Nesebur (ne-SAI-buhr), ancient *Mesembria*, city, BURGAS oblast, Nesebur obshtina (1993 pop. 18,233), E BULGARIA, port on a rocky peninsula in the BLACK SEA, 17 mi/27 km NE of BURGAS; 42°39′N 27°42′E. Fisheries; vineyards. Connected with a seaside resort on mainland by isthmus 50 ft/15 m–75 ft/23 m wide. Busy summer tourism; near Slunchev Bryag resort. Has ruins of 6th–9th-century episcopal basilicas and numerous medieval churches with interesting frescoes and incrustations. Founded over 2,000 years ago; rich commercial town under Byzantine rule. Called Misivri or Missivri under Turkish rule (15th–19th century). Until 1934, known as Mesemvriya or Mesembria.

Nesflaten (NAIS-flah-tuhn), village, ROGALAND county, SW NORWAY, at NE end of SULDALSVATN Lake, and at S end of the BRATTLANDSDAL, 55 mi/89 km ENE of HAUGESUND; 59°38′N 06°48′E. Tourist center.

Neshaminy Creek (nuh-SHA-mi-nee), c.50 mi/80 km long, SE PENNSYLVANIA; rises in SW BUCKS county at joining of West and North branches; flows generally SE past CHALFANT and PARKLAND to DELAWARE RIVER at CROYDON; 40°16′N 75°12′W. West Branch (c.10 mi/16 km long) rises in NE MONTGOMERY county, flows E; North Branch (c.15 mi/24 km long) rises in N Bucks county, flows SW through Lake Galean reservoir.

Neshava, POLAND: see NIESZAWA.

Nesher (NE-sher), township, NW ISRAEL, in KISHON RIVER valley, at E foot of MOUNT CARMEL, 5 mi/8 km SE of HAIFA; 32°45′N 35°03′E. Elevation 118 ft/35 m. Cement-manufacturing center. Founded 1926.

Neshin, UKRAINE: see NIZHYN.

Neshkoro (nesh-KAW-ruh), village (2006 population 449), MARQUETTE county, central WISCONSIN, on small WHITE RIVER, and mill pond, 33 mi/53 km W of OSHKOSH; 43°58′N 89°12′W. In livestock and dairy area. Hydroelectric plant.

Neshoba (nuh-SHO-buh), county (□ 571 sq mi/1,484.6 sq km; 2006 population 30,125), E central MISSISSIPPI; ⊙ PHILADELPHIA; 32°45′N 89°07′W. Drained by

PEARL RIVER. Includes Choctaw Indian Reservation in W; also small reservation SE of Philadelphia. Agriculture (cotton, corn; poultry, cattle; dairying); timber. Manufacturing at Philadelphia. Part of Nanih Waiya Memorial State Historical Site in NE; Neshoba County Legion Lake (state lake) in SE; source of CHUNKY CREEK in E. Formed 1833.

Neskaupstaur (NES-kuhp-STAH-[th]uhr), town, extreme E ICELAND, on the Mjóifjörður, an arm of the NORWEGIAN SEA; 65°09′N 13°42′W. It is the chief town of E Iceland and a fishing port with freezing plants and fish meal factories. It was chartered in 1929.

Nesle (NEL), commune (□ 3 sq mi/7.8 sq km), SOMME department, PICARDIE region, N FRANCE, 12 mi/19 km S of PÉRONNE; 49°45′N 02°55′E. Agriculture market; manufacturing of vegetable oil. Located near main PARIS-LILLE railroad line and highway.

Neslusa (nye-SLOO-shah), Slovak *Neslušsa*, Hungarian *Neszlény*, village, STREDOSLOVENSKY province, NW Slovakia, in JAVORNIKS MOUNTAINS, and 7 mi/11.3 km N of Žilina; 49°19′N 18°45′E. Lumbering; potato growing; sheep breeding.

Nesna (NAIS-nah), village, NORDLAND county, N central NORWAY, on N shore of RANA, 30 mi/48 km WSW of Mo; 66°12′N 13°02′E. Agiculture; cattle raising; lumbering; mechanical industry.

Nesoddtangen (NAI-sawd-tahn-guh), village, AKER-SHUS county, SE NORWAY, on N tip of Nesodden peninsula at the head of OSLOFJORDEN, 3.7 mi/6 km S of OSLO. Residential area with over half of the workforce commuting to Oslo. Ferry connections (20 min.) across Oslofjorden to center of Oslo. Nearby is twelfth-century church.

Nes op Ameland, NETHERLANDS: see NES.

Nesowadnehunk Lake (ne-SOU-uh-duh-huhnk), PISCATAQUIS county, central MAINE, 32 mi/51 km NW of MILLINOCKET, near W border of Baxter State Park; 3 mi/4.8 km long, 1 mi/1.6 km wide. Source of Neso-wadnehunk Stream, which flows S to West Branch of PENOBSCOT RIVER. Formerly known as Sourdnahunk Lake.

Nespelem (nes-PEE-luhm), village (2006 population 206), OKANOGAN county, N WASHINGTON, in S end of Colville Indian Reservation, 30 mi/48 km SE of OKA-NOGAN, on Nespelem River. Indian Agency headquarters for Colville and Spokane reservations; 48°10′N 118°58′W. COLUMBIA River 4 mi/6.4 km SE. Chief Joseph Memorial. Owhi Lake reservoir to NE.

Nesquehoning (NUHS-kwa-HO-ning), borough (□ 22 sq mi/57.2 sq km; 2006 population 3,338), CARBON county, E PENNSYLVANIA, 4 mi/6.4 km W of JIM THORPE, on Nesquehoning Creek (extends from LE-HIGH RIVER to SCHUYLKILL county border); 40°51′N 75°49′W. Former anthracite-coal region. Manufacturing (chemicals, vinyl products, displays, apparel, transportation equipment, fabricated metal products). Agricultural area (potatoes, corn, hay; livestock; dairying). Mauch Chunk Lake reservoir to S.

Ness (NES), county (□ 1,075 sq mi/2,795 sq km; 2006 population 2,946), W central KANSAS, ⊙ NESS CITY; 38°28′N 99°55′W. Smoky Hills region to N, drained by WALNUT CREEK and its North and South forks, which converge at center of county, and by PAWNEE RIVER (S). Wheat, sorghum; cattle. Oil and gas extraction. Formed 1880.

Ness City (NES), town (2000 population 1,534), ⊙ NESS county, W central KANSAS, on N Fork of WALNUT CREEK, and c.60 mi/97 km W of GREAT BEND; 38°27′N 99°54′W. Elevation 2,220 ft/677 m. In agricultural area (wheat; cattle). Founded 1878, incorporated 1886.

Nesselrode, Mount (8,105 ft/2,470 m), in COAST RANGE, on border between U.S. (SE ALASKA) and CANADA (BRITISH COLUMBIA), 50 mi/80 km N of JU-NEAU (Alaska); 58°58′N 134°19′W.

Nesselsdorf, CZECH REPUBLIC: see KOPŘIVNICE.

Nesselwang (NES-sel-vahng), village, SWABIA, SW BAVARIA, GERMANY, at E foot of ALLGÄU ALPS, 5 mi/8

km NW of FÜSSEN; skiing center (2,953 ft/900 m); 47°37′N 10°30′E. Manufacturing (mathematical instruments; metalworking; brewing; dairying). Health resort, winter-sports resort.

Ness, Loch (NES), lake (22 mi/35 km long), HIGHLAND, N central Scotland, in the GREAT GLEN; 57°16′N 04°30′W. Over 700 ft/213 m deep and ice-free, it is fed by the Oich and other streams and drained by the Ness to the MORAY FIRTH. It forms part of the CALE-DONIAN CANAL. By volume, Loch Ness is the largest freshwater lake in Great Britain. As a result of the publicity surrounding a supposed "monster," Loch Ness has become a major tourist attraction.

Ness Point, ENGLAND: see ROBIN HOOD'S BAY.

Ness Ziona or **Nes Ziona** (NES tsee-YO-nah), city (2006 population 30,500), W central plain, 10 mi/16 km SSE of TEL AVIV; 31°55′N 34°48′E. Elevation 114 ft/34 m. Food, electronics, and high-tech industries; textile milling; citriculture. Founded 1883.

Ness Ziona, ISRAEL: see NES TSIYONA.

Nestani (ne-STAH-nee), town, ARKADIA prefecture, central PELOPONNESE department, S mainland GREECE, 9 mi/14.5 km NNE of TRÍPOLIS; 37°37′N 22°28′E. Livestock (goats, sheep); tobacco, wheat. Formerly called Tsipiana; formerly spelled Nestane.

Neste d'Aure, FRANCE: see NESTE RIVER.

Nestemice (NESH-tye-MI-tse), Czech *Neštěmice*, German *Nestomitz*, E district of ÚSTÍ NAD LABEM, SE-VEROCESKY province, N BOHEMIA, CZECH REPUBLIC, on left bank of ELBE RIVER; 50°40′N 14°06′E. Manufacturing (soda, chemicals, photographic equipment).

Neste River (NEST), 40 mi/64 km long, HAUTES-PYR-ÉNÉES department, MIDI-PYRÉNÉES region, SW FRANCE; formed by several headstreams descending form the high PYRENEES Mountains near Spanish border; flows through the AURE Valley, past ARREAU and Saint-Laurent-de-Neste, to the GARONNE RIVER above MONTRÉJEAU; 43°00′N 00°23′E. A canal (17 mi/27 km long) connects it with headwaters of BAÏSE, GERS, and GIMONE rivers, irrigating the LANNEMEZAN PLATEAU. Hydroelectric power stations operate along its upper valley. Also known as Neste d'Aure or Grande-Neste.

Nesterov (NYE-stee-ruhf), city (2005 population 5,080), E KALININGRAD oblast, W European Russia, on railroad, 87 mi/140 km E of KALININGRAD, and 15 mi/24 km E of GUSEV; 54°38′N 22°34′E. Elevation 288 ft/87 m. Dairying. City since 1722. Until 1945, in German-ruled EAST PRUSSIA, where it was called Stallupönen and, later (1938–1945), Ebenrode.

Nesterov, UKRAINE: see ZHOVKVA.

Nesterovskaya (NYES-tee-ruhf-skah-yah), village (2005 population 19,875), central INGUSH REPUBLIC, SE European Russia, in the foothills of the N CAU-CASUS Mountains, on road and railroad, 20 mi/32 km E of BESLAN (NORTH OSSETIAN REPUBLIC), and 5 mi/8 km S of ORDZHONIKIDZEVSKAYA; 43°14′N 45°03′E. Elevation 1,305 ft/397 m. Site of sporadic terrorist activity since the beginning of the second Russian-Chechen conflict. A school in the village was designated a backup target by the terrorists in September 2004, if the attack on a secondary school in Beslan were to fail.

Nestervar, UKRAINE: see TUL'CHYN.

Nesterville, Ontario: see NESTORVILLE.

Nestomitz, CZECH REPUBLIC: see NESTEMICE.

Neston, ENGLAND: see ELLESMERE PORT.

Nestorion (nes-TO-ree-on), town, KASTORÍA prefecture, WEST MACEDONIA department, N central GREECE, on upper ALIÁKMON River, and 13 mi/21 km SW of KASTORÍA; 40°25′N 21°04′E. Charcoal burning; wheat, beans, skins. Formerly called Nestramion.

Nestorville (NES-tuhr-vil), unincorporated town, central ONTARIO, E central Canada, on Lake HURON, 3 mi/5 km NW of THESSALON, and included in township of Huron Shores; 46°17′N 83°36′W. Dairying; mixed farming; lumbering. Also spelled Nesterville.

Area is shown by the symbol □, and capital city or county seat by ⊙.

Nestos River, SE EUROPE: see MESTA RIVER.

Nestow (NES-to), hamlet, central ALBERTA, W Canada, 12 mi/20 km from WESTLOCK, in Westlock County; 54°14′N 113°36′W.

Nestramion, Greece: see NESTORION.

Nestus River, SE EUROPE: see MESTA RIVER.

Nesvady (nes-VAH-di), village, Hungarian *Naszvad*, ZAPADOSLOVENSKY province, S SLOVAKIA, on NITRA RIVER, and 11 mi/18 km N of KOMÁRNO; 48°01′N 17°59′E. Has wheat, corn, sugar beets, fruit, vineyards. Folk art. Under Hungarian rule from 1938–1945.

Nesvetevich, UKRAINE: see LYSYCHANS′K.

Nesvitovych, UKRAINE: see LYSYCHANS′K.

Nesvizh (NES-vizh), Polish *Nieświez*, city, W MINSK oblast, BELARUS, ⊙ Nesvizh region, at the sources of the USHA RIVER (Neman Basin), and 28 mi/45 km ENE of BARANOVICHI, 53°16′N 26°40′E. Peat-processing plant. Has sixteenth-century castle and several old churches. Developed as capital of independent Polish duchy (13th–14th centuries). Chartered 1586; suffered from Swedish assaults (1654, 1706).

Nes Ziona, ISRAEL: see NES TSIYONA.

Neszlény, SLOVAKIA: see NESLUSA.

Netanya (ni-TAHN-yah), city (2006 population 173,300), W central ISRAEL, on the MEDITERRANEAN SEA; 32°20′N 34°51′E. Elevation 118 ft/35 m. It is a growing urban center within TEL AVIV orbit, and a trade and industrial center for agricultural settlements in the region. Diamond cutting and polishing (formerly Israel's major center; now shifted to Tel Aviv) and citrus packing are the chief industries. Manufacturing of food, clothing, metal, and electronics. Beach resort. Netanya, founded in 1929, was named for the U.S. philanthropist Nathan Strauss. Wingate Institute for Physical Education is on its outskirts.

Netanyeje Kamni (ne-TAHN-ye-je KUM-nee) [=oil rocks], oil town, AZERBAIJAN, built on stilts in CASPIAN SEA, 20 mi/32 km offshore. Home to 2,500 workers. Natural-gas plant.

Netawaka (net-uh-WAH-kuh), village (2000 population 170), JACKSON county, NE KANSAS, 9 mi/14.5 km N of HOLTON; 39°36′N 95°43′W. Cattle; grain.

Netcong (NET-kuhng), borough (2006 population 3,292), MORRIS county, N NEW JERSEY, near Lake MUSCONETCONG, and 13 mi/21 km NW of MORRISTOWN; 40°53′N 74°42′W. In resort area. Manufacturing and agriculture. State park nearby. Incorporated 1894; grew as residence for iron miners and iron-workers.

Nete River (NAI-tuh), French *Nèthe* (NET), 11 mi/18 km long, NE BELGIUM; formed at LIER by confluence of GROTE NETE and KLEINE NETE rivers; flows SW and W, past DUFFEL, joining DIJLE RIVER at Rumpst to form RUPEL RIVER which flows into SCHELDT RIVER.

Neteshin, UKRAINE: see NETISHYN.

Netheravon (NE-[th]uh-AI-vuhn), village (1991 population 1,217; 2001 population 1,771), WILTSHIRE, S ENGLAND, on AVON River, 4 mi/6.4 km NE of AMESBURY; 51°14′N 01°45′W. Has 18th-century dovecote. SALISBURY PLAIN to W.

Netherbury (NE-[th]uh-buh-ree), village (1991 population 1,063; 2001 population 1,890), W DORSET, SW ENGLAND, on BRIT River, 4 mi/6.4 km N of BRIDPORT, and just SSW of BEAMINSTER; 50°47′N 02°45′W. Parnham House (Tudor) adjoins to N; has 14th- and 15th-century churches. Furniture; previously wood craftsmanship.

Netherfield, ENGLAND: see CARLTON.

Nèthe River, BELGIUM: see NETE RIVER.

Netherland Island, TUVALU: see NUI.

Netherlands Antilles, federation (☐ 308 sq mi/800.8 sq km), two groups of islands, c.500 mi/805 km apart, in the WEST INDIES; ⊙ WILLEMSTAD, CURAÇAO. Autonomous part of NETHERLANDS since 1954; comprised of BONAIRE and CURAÇAO, 60 mi/97 km off the coast of VENEZUELA. In the Leewards SABA, SINT EUSTATIUS (Statia), and the S part of SAINT MARTIN (Sint Maarten). Discovered in 1499 by a Spaniard, Alonso de Ojeda, and controlled by SPAIN throughout the sixteenth century; taken by the Dutch in seventeenth century for trading post and as base for excursions against the Spanish. Economy in twentieth century prospered from the establishment of refineries on Curaçao and ARUBA after discovery of oil in Venezuela. Granted full autonomy in domestic affairs in 1954 and became a unified dependency of the Kingdom of the Netherlands. In 1986 ARUBA, W of Bonaire and Curaçao, broke away from the Antillian Federation to become a separate Netherlands dependency; plans for the other five islands to become separate dependencies by 1996 are being modified.

Netherlands, Austrian and Spanish, that part of the LOW COUNTRIES that, from 1482 until 1794, remained under the control of the imperial house of HAPSBURG. The area corresponds roughly to modern BELGIUM and LUXEMBOURG. The Low Countries passed from the house of Burgundy to that of Hapsburg through the marriage (1477) of Mary of Burgundy to Archduke Maximilian (later Emperor Maximilian I); their son Philip (later Philip I of Castile) inherited FLANDERS, BRABANT, ARTOIS, HAINAUT, the DUCHY OF LUXEMBOURG, LIMBURG, HOLLAND, and ZEELAND. His son, Emperor Charles V, added UTRECHT, GELDERLAND, OVERIJSSEL, FRIESLAND, and DRENTHE and in 1547 declared the entire Netherlands hereditary Hapsburg possessions. In 1555 he abdicated the Netherlands in favor of his son, Philip II of Spain. The provinces of the Netherlands retained their individual institutions and provincial estates, thereby limiting the powers of the Spanish governors at BRUSSELS. The harsh regime of the duke of Alba, who replaced (1567) Margaret of Parma as governor and suspended constitutional procedure, provoked the opposition of the Dutch and Flemish, led by William the Silent of Orange; Lamoral, count of Egmont; Hendrik, lord of Brederode; Marnix; and others. In 1576 the opposition united in the Pacification of GHENT. Despite the ruthless campaigns of Alba and his successors—Requesens, John of AUSTRIA, and the more diplomatic Alessandro Farnese—SPAIN recovered only the S provinces while the seven United Provinces of the Netherlands gained independence. The bloody struggle ruined the prosperous Flemish cities, particularly ANTWERP. Protestants were forced to flee the Spanish Netherlands or convert to Catholicism, which became the dominant religion of Belgium and Luxembourg. The provinces were a battleground in every major European war from the seventeenth century to World War II, but after each war their industry and commercial enterprise enabled a quick recovery. Spain lost North Brabant and part of Limburg to the United Provinces at the Peace of Westphalia (1648); Artois and parts of Hainaut and Luxembourg provinces to FRANCE at the Peace of the Pyrenees (1659); and parts of FLANDERS (including DUNKIRK and LILLE) to France in the treaties of Aix-la-Chapelle (1668) and Nijmegen (1678–1679). The remaining Spanish possessions in the Low Countries were transferred (1714) to the Austrian branch of the Hapsburgs by the Peace of UTRECHT. The bishopric of Liège, an ecclesiastic principality, was not part of the Hapsburg possessions; it fell under Spanish and (after 1714) Austrian influence. The Austrian rule of Belgium was broken by the French Revolutionary Wars. For the history of the area after its incorporation (1815) into the kingdom of the Netherlands, see BELGIUM, LUXEMBOURG (Grand Duchy).

Netherlands East Indies: see INDONESIA.

Netherlands New Guinea, INDONESIA: see IRIAN JAYA.

Netherlands, The (NE-thuhr-landz), Dutch *Nederland* or *Koninkrijk der Nederlanden*, kingdom (☐ 15,963 sq mi/41,344 sq km; 2004 estimated population 16,318,199; 2007 estimated population 16,570,613), NW EUROPE, on the NORTH SEA (N and W), and on Belgian (S) and German (E) borders. It is popularly known as HOLLAND, but strictly speaking this name only applies to two W provinces of NORTH HOLLAND and SOUTH HOLLAND. AMSTERDAM is the constitutional capital; THE HAGUE is the administrative and government capital.

Geography

The country is mostly low-lying coastal plain, much of the W part (c.27% of area) being land reclaimed from the sea since the 13th century through the construction of dikes and barrier dams and the extraction of water from shallow pockets of sea, forming new land areas (which the Dutch call polders). The Netherlands has twelve provinces: DRENTHE, FLEVOLAND, FRIESLAND, GELDERLAND, GRONINGEN, LIMBURG, NORTH BRABANT, OVERIJSSEL, UTRECHT, ZEELAND, and NORTH and SOUTH HOLLAND. Flevoland, the newest province, was created in 1986 from three existing polders in the E part of the IJSSELMEER (formerly Zuider Zee): EASTERN and SOUTHERN FLEVOLAND and the NORTH-EAST POLDER, as well as a future fourth polder called the Markerwaard (now the MARKERMEER). Below the provincial level, there are over 600 municipalities in the Netherlands. The kingdom also formerly included two overseas territories, SURINAME (now independent), in NE SOUTH AMERICA, and West New Guinea (now IRIAN JAYA, a province of INDONESIA). The NETHERLANDS ANTILLES, made up of several Caribbean islands including CURAÇAO, remain an autonomous part of the kingdom. The WADDEN (West Frisian) Islands, a line of sand barrier islands, lie off the North Sea coast of the Netherlands to N and NW, forming the WADDENZEE. Active sand dunes line the outer fringes of the islands and the fringe of the mainland coast; other areas of the country have coastal plains interrupted by areas of heath and woodland. N Brabant (the KEMPENLAND) has infertile river-deposits: sands and gravels. LIMBURG province has low, rolling hills. The highest point in the Netherlands is the VAALSEBERG (also called the Drielandenpunt), at 1,053 ft/321 m, on the SE border. The lowest point is near HOLLANDSE IJSSEL RIVER in South Holland (at 22.11 ft/6.74 m below sea level). The country is drained by the SCHELDE, MAAS, and RHINE rivers, which converge in the central and SW Netherlands. The Rhine and Maas form several channels, which enter a number of North Sea estuaries: the HARINGVLIET, GREVELINGENMEER, and the EASTERN and WESTERN SCHELDT. A network of canals connects the rivers, channels, and estuaries and extends into BELGIUM and GERMANY as well. Most of the country's estuaries have been enclosed by several dams (built since the 1960s during the Delta Project) to prevent flooding and generate electricity. Many of the delta islands are now connected to the mainland by bridges. In addition to Amsterdam and The Hague (which together coalesce in the country's urban agglomeration that forms a metropolitan center called the RANDSTAD region), important cities include ARNHEM, BREDA, DELFT, DORDRECHT, HAARLEM, LEIDEN, MAASTRICHT, ROTTERDAM, 'S-HERTOGENBOSCH, and UTRECHT.

Population

After the Netherlands obtained independence in the late 16th century, it became largely Protestant, but now there are almost as many Roman Catholics as Protestants, and freedom of worship is encouraged by the government. Linguistic conformity to Dutch, the official language, is complete except in Friesland, where Frisian is spoken in places.

Economy

The population of the Netherlands is extremely dense and highly concentrated in the cities, allowing most of the land area to be used for agriculture (mainly dairying; cattle, sheep, hogs, poultry; eggs; peas, beans, onions, potatoes, fruit, corn, wheat, oats, rye, and sugar beets). The country is known for its cheese industry; the cities of GOUDA and EDAM have given their names to varieties of cheese. The coastal margins of the two Holland provinces are known for flower raising; the tulip is a national cultural symbol. Fishing is also important, especially flounder, salmon, herring, oysters, mussels, and eels. There are large natural-gas deposits in Groningen and Friesland provinces; some petroleum is extracted in the NE. Manufacture includes textiles, machinery, electronics, transportation equipment, iron and steel, chemicals; petroleum refining, shipbuilding, and food processing. Tourism is important as well. An efficient transportation system (roads, railroads, and ferries) links the country with the rest of Europe. The Netherlands has a large foreign trade; the main exports are machinery, textiles, petroleum products, fruits and vegetables, tobacco products, and meat. The Netherlands belongs to the EU and numerous regional and global economic organizations.

History: to 1579
The Netherlands, one of the LOW COUNTRIES, did not have a unified history until the late 15th century. The region W of the Rhine formed part of the Roman province of Lower GERMANY and was inhabited by the Batavi; the area E of the Rhine was inhabited by the Frisians. Nearly the entire area was taken (4th–8th centuries) by the Franks, and in the 9th century most of it became part of the Holy Roman Empire. In the 14th and 15th centuries, Flanders, Holland, Zeeland, Gelderland, and Brabant passed to the powerful dukes of Burgundy, who controlled virtually all of the Low Countries. Most of the Dutch port towns belonged to the HANSEATIC LEAGUE in the 15th century, enjoying vast autonomous privileges. In 1555, Emperor Charles V gave the Low Countries to his son Philip II of Spain. The N provinces (i.e., the present Netherlands) began to rebel against Spanish control, in particular resisting the introduction of the Spanish Inquisition and reduction of the Low Countries to a Spanish province. The struggle for independence began (1562) in Flanders and Brabant. The N provinces succeeded in expelling the Spanish garrisons (1572–1574), but the S provinces (the Austrian Netherlands) remained in Spanish possession.

History: 1579 to 1794
The seven N provinces—Holland, Zeeland, Utrecht, Gelderland, Overijssel, Friesland, and Groningen—formed (1579) the Union of Utrecht and declared (1581) their independence. Fighting with Spain began again during the Thirty Years War (1618–1648), after which the independence of the United Provinces—as the Netherlands was then called—was recognized. Spain also ceded North Brabant, with BREDA, and part of LIMBURG, with MAASTRICHT. During the following century, the United Provinces continued its growth, particularly as a flourishing commercial and cultural center. In 1672, France invaded the country; in response, the Dutch opened their dikes and flooded the land, creating a watery barrier that was virtually impenetrable. The war, which lasted six years, devastated the provinces, but in the Treaty of NIJMEGEN (1678–1679) the Dutch won important concessions from France. In the 18th century the commercial, military, and cultural position of the United Provinces in Europe declined. The Netherlands sided against England in the American Revolution and as a result lost several colonies at the Treaty of PARIS (1783).

History: 1794 to 1945
France overran the United Provinces again (1794–1795), and in 1795 the Batavian Republic was set up. In 1806, Napoleon I established the Kingdom of Holland and made his brother Louis Bonaparte its first king. Bonaparte was deposed in 1810, and the kingdom was annexed by France. At the Congress of VIENNA (1814–1815) the United Provinces and the former Austrian Netherlands (now under Belgian control) were united. In 1830, however, the former Austrian provinces

rebelled against Dutch rule and declared independence; an agreement dividing the Dutch and Belgian provinces was reached at the LONDON Conference of 1839. During the latter half of the 19th century, the Netherlands enjoyed another period of commercial expansion, and internal development, assisted by the Industrial Revolution; the country's cultural reputation grew as well. The Netherlands was neutral in World War I. In World War II, Germany invaded (May 1940) the Netherlands, crushed Dutch resistance, and destroyed Rotterdam. Of the approximately 112,000 Jews in the Netherlands during this period, about 104,000 were deported to Poland and killed. Allied airborne landings (1944) at ARNHEM, Nijmegen, and EINDHOVEN liberated Zeeland, North Brabant, and Limburg provinces.

History: 1945 to Present
The German collapse in May 1945, was followed by the immediate restoration of the Dutch government. The Netherlands became a charter member of the UN in 1945. It also participated actively in the EC, including the Common Market, and in 1949 joined NATO. The Netherlands gave Indonesia independence in 1949, and in 1962 relinquished WEST NEW GUINEA (now part of Indonesia). The Dutch economy expanded in the 1950s and 1960s, with significant growth in industry. After a series of devastating floods in 1953, the Delta Project was begun; one of the results was that several of the islands and peninsulas were joined to the mainland and to each other and ceased to be islands. In the early 1970s the Netherlands enjoyed material prosperity and considerable influence in European affairs. Suriname was granted independence in 1975. Since World War II most Dutch governments have been coalitions led by the Christian Democrats (or its predecessors) or the Labor Party; sometimes both parties have been in the governing coalition.

Government
The Netherlands is a constitutional monarchy governed under the constitution of 1815 as amended. The hereditary monarch is the head of state; the prime minister is the head of government. There is a bicameral legislature, the States General. Members of the deliberative upper house, the seventy-five-seat First Chamber, are elected by the twelve provincial councils. Members of the more powerful lower house, the 150-seat Second Chamber, are popularly elected. All legislators serve four-year terms. The royal succession is settled on the house of Orange, which adheres to the Dutch Reformed Church. The current head of state is Queen Beatrix (since April 1980). Prime Minister Jan Peter Balkenende, a Christian Democrat, has been head of government since July 2002.

Nether Providence (NE-thur), township, DELAWARE county, SE PENNSYLVANIA; residential suburb 12 mi/19 km SW of PHILADELPHIA; 39°53′N 75°22′W.

Nether Stowey (NE-[th]uh STO-wee), village (1991 population 1,330; 2001 population 1,313), N SOMERSET, SW ENGLAND, 7 mi/11.3 km WNW of BRIDGWATER, at foot of QUANTOCK HILLS; 51°09′N 03°10′W. Has 15th-century manor house. Coleridge lived here for some time and wrote *The Rime of the Ancient Mariner*.

Netherthong, ENGLAND: see HOLMFIRTH.

Néthou, Pic de, SPAIN: see ANETO, PICO DE.

Netishyn (ne-TEE-shin) (Russian *Neteshin*), city, N KHMEL'NYTS'KYY oblast, UKRAINE, 22 mi/35 km WNW of SHEPETIVKA, on the right bank of the HORYN' River; 50°21′N 26°38′E. Elevation 685 ft/208 m. Site of the Khmel'nyts'kyy Nuclear Power Station; grain mill; professional-technical school. Mentioned in 1648, city since 1984.

Netivot (ni-tee-VOT), town (2006 population 24,700), S ISRAEL, 19 mi/30 km S of ASHKELON, in NW NEGEV; 31°25′N 34°34′E. Elevation 492 ft/149 m. Best known as the center for a religious group of Moroccan origin

that follows the Baba Sali (Rabbi Yisrael Abu-Hatzeira), who is buried here. Founded 1956.

Netley (NET-lee), village (2001 population 5,128), S HAMPSHIRE, S ENGLAND, on SOUTHAMPTON WATER, 3 mi/4.8 km SE of SOUTHAMPTON; 52°20′N 01°20′W. Has ruins of Cistercian abbey. Royal Victoria Hospital, one of the chief military hospitals in England, was built here after Crimean War, but is not in use today.

Netley Marsh (NET-lee MAHSH), village (2001 population 2,012), S HAMPSHIRE, S ENGLAND, 6 mi/9.7 km W of SOUTHAMPTON; 50°54′N 01°30′W. Has 13th-century manor house.

Netolice (NE-to-LI-tse), German *Netolitz*, town, JIHOCESKY province, S BOHEMIA, CZECH REPUBLIC, 14 mi/23 km WNW of ČESKÉ BUDEJOVICE, in moor and lake region. Railroad terminus. Manufacturing (machinery); food processing; dray-horse breeding. Has a museum.

Netolitz, CZECH REPUBLIC: see NETOLICE.

Neto River (NAI-to), 52 mi/84 km long, in CALABRIA, S ITALY; rises in LA SILA mountains on BOTTE DONATO; flows E, SE, and E to GULF OF TARANTO 9 mi/14 km N of CROTONE. The Neto and its tributaries, the ARVO and AMPOLLINO, are major hydroelectric power sources (plants near SAN GIOVANNI IN FIORE and COTRONEI).

Netphen (NET-fen), town, WESTPHALIA-LIPPE, North Rhine-Westphalia, GERMANY, on S slope of the Rothaar Mountains, near the Obernau dam, and in the source area of SIEG, LAHN, and EDER rivers, 5 mi/8 km NE of SIEGEN; 50°55′N 08°06′E. Manufacturing (machinery); forestry; tourism. Formed 1969 by union of twenty-four villages.

Netrakona (net-traw-ko-nah), town (2001 population 56,786), MYMENSINGH district, E central EAST BENGAL, BANGLADESH, 21 mi/34 km ENE of MYMENSINGH; 24°56′N 90°39′E. Rice, jute, oilseeds.

Netravati River (nai-TRAH-vuh-tee), 60 mi/97 km long, DAKSHINA KANNADA district, KARNATAKA state, S INDIA; rises in WESTERN GHATS; flows S and W past BANTVAL to MALABAR COAST of ARABIAN SEA at MANGALORE. Navigable for c. 40 mi/64 km above mouth.

Netstal (NETS-tahl), commune, GLARUS canton, E central SWITZERLAND, at junction of the Löntschbach and the LINTH RIVER, 2 mi/3.2 km N of GLARUS; 47°03′N 09°03′E. Manufacturing (textiles, paper, cement; metalworking). Löntsch hydroelectric plant is here.

Nettapakkam (NE-tuh-puh-kuhm), French *Nettapacom*, town, Pondicherry district, Pondicherry Union Territory, S INDIA, 13 mi/21 km SW of Pondicherry. Rice, plantains, sugarcane.

Nettersheim (NET-ters-heim), village, RHINELAND, North Rhine-Westphalia, W GERMANY, on N slope of the EIFEL, and on the URFT RIVER, 15 mi/24 km SSW of EUSKIRCHEN; 50°29′N 06°38′E. Forestry; grain; lime quarries.

Nettetal (NET-te-tahl), town, RHINELAND, North Rhine-Westphalia, W GERMANY, on Dutch border, 7 mi/11.3 km NW of VIERSEN; 51°19′N 06°17′E. Manufacturing (textiles, metal, cigars); tourism. Has 17th-century watermill, moated Ingenhoven castle (13th–16th centuries), moated Krickenbeck castle. Town first mentioned 1205; chartered c. 1628; incorporated Kaldenkirchen (first mentioned 1205, chartered 1268), Loberich, and three smaller villages in 1970.

Nettilling Lake (NE-chi-ling), freshwater lake (□ 1,956 sq mi/5,085.6 sq km), S BAFFIN ISLAND, BAFFIN region, NUNAVUT territory, N Canada; 66°28′N 70°19′W. One of the largest lakes entirely within Canada. Located in an arctic lowland region and fed by AMADJUAK LAKE and numerous streams that drain the tundra. It empties through the Koukdjuak River W into FOXE BASIN. Frozen most of the time.

Nett Lake (□ 11 sq mi/28 sq km), in KOOCHICHING and ST. LOUIS counties, N MINNESOTA, in Nett Lake Indian

Reservation, 35 mi/56 km SSE of INTERNATIONAL FALLS; 6 mi/9.7 km long, 4 mi/6.4 km wide; 48°07′N 93°07′W. Drains WNW through Nett Lake River to LITTLE FORK RIVER. Nett Lake village is small Native American community, on E shore of lake, at Big Point. Big Island in S.

Nettleton, town, CRAIGHEAD county, NE ARKANSAS, 2 mi/3.2 km ESE of JONESBORO; part of city of Jonesboro. Ships farm produce. Founded 1881.

Nettleton, town (2000 population 1,932), LEE and MONROE counties, E MISSISSIPPI, 13 mi/21 km SSE of TUPELO, near Chipawa River; 34°05′N 88°37′W. Agriculture (corn, cotton, wheat, soybeans; cattle; dairying), Manufacturing (furniture, apparel, valves).

Nettuno (nait-TOO-no), town, in LATIUM, central ITALY, on the TYRRHENIAN Sea; 41°27′N 12°39′E. It is an agricultural center and a seaside resort. With nearby ANZIO, it was the site of an Allied landing (Jan. 22, 1944) in World War II.

Netzahualcóyotl, MEXICO: see CIUDAD NETZAHUALCÓYOTL.

Netzarim, Jewish settlement, 2.5 mi/4 km SW of GAZA. The settlement's name is influenced by the Arab town of Nuseirat nearby.

Netze River, POLAND: see NOTEĆ River.

Netzschkau (NETSH-kou), town, SAXONY, E central GERMANY, 10 mi/16 km NE of PLAUEN; 50°37′N 12°18′E. Manufacturing (textiles, machinery); food processing. Has 15th-century castle. Chartered 1496.

Neu-Anspach (NOI–AHNS-pahkh), town, HESSE, central GERMANY, in the TAUNUS, 6 mi/9.7 km NW of BAD HOMBURG; 50°18′N 08°31′E. Manufacturing (furniture); forestry.

Neubabelsberg, GERMANY: see BABELSBERG.

Neu Bentschen, POLAND: see ZBASZYNEK.

Neuberg (NOI-berg), village, HESSE, central GERMANY, 14 mi/23 km NE of FRANKFURT; 50°12′N 08°58′E. Agriculture (fruit, grain).

Neuberg an der Mürz (NOI-berg ahn der MOOIRTS), village, NE STYRIA, E central AUSTRIA, at S foot of the SCHNEEALPE massif, on Mürz River, and 6 mi/9.7 km NNW of Mürzzuschlag; 46°40′N 15°35′E. Railroad terminus; summer resort. Abandoned Cistercian abbey, founded 1327. Medieval character still evident. Famous Neuberger Madonna here.

Neubeuern (noi-BOI-ern), village, UPPER BAVARIA, BAVARIA, SE GERMANY, on the INN RIVER; 47°46′N 12°07′E. Agriculture (fruit). Metalworking.

Neubiberg (noi-BEE-berg), suburb of MUNICH, UPPER BAVARIA, BAVARIA, SE GERMANY, 4 mi/6.4 km SE of city center; 48°06′N 11°39′E. Village first mentioned 1017.

Neu Bidschow, CZECH REPUBLIC: see NOVY BYDZOV.

Neubistritz, CZECH REPUBLIC: see NOVA BYSTRICE.

Neubourg, Le (NU-boor, luh), commune (□ 3 sq mi/7.8 sq km), EURE department, HAUTE-NORMANDIE region, NW FRANCE, 14 mi/23 km NW of ÉVREUX; 49°09′N 00°55′E. Market; manufacturing (packing materials). Has 16th-century church.

Neubrandenburg (noi-BRAHN-duhn-burg), city (2006 population 67,517), MECKLENBURG–WESTERN POMERANIA, NE GERMANY, on the TOLLENSE LAKE, 36 mi/58 km S of GREIFSWALD; 53°33′N 13°15′E. Manufacturing (paper, machinery, food products). Founded in 1248 by the margraves of BRANDENBURG, Neubrandenburg passed to Mecklenburg in 1292. Most of its medieval buildings were destroyed during World War II, and the city has generally been rebuilt along modern lines. Wall and four town gates are UNESCO monument.

Neubreisach, FRANCE: see NEUF-BRISACH.

Neubukow (noi-BU-kou), town, Mecklenburg-Western Pomerania, N GERMANY, near MECKLENBURG BAY of the BALTIC, 13 mi/21 km NE of WISMAR; 54°01′N 11°40′E. Agricultural market (grain, sugar beets, potatoes; livestock). Archaeologist Heinrich

Schliemann was born here in 1822. Town chartered c.1250.

Neubulach (noi-BOO-lahkh), town, N BLACK FOREST, BADEN-WÜRTTEMBERG, GERMANY, in BLACK FOREST, 4 mi/6.4 km SSW of CALW; 48°40′N 08°42′E. Summer resort (elevation 1,916 ft/584 m).

Neuburg am Inn (NOI-burg ahm IN), village, LOWER BAVARIA, BAVARIA, SE GERMANY, on the INN RIVER, near Austrian border, 4 mi/6.4 km S of PASSAU; 48°31′N 13°26′E. Forestry. Manufacturing (machinery, electronic equipment).

Neuburg an der Donau (NOI-burg ahn der DO-nou), city, SWABIA, central BAVARIA, GERMANY, on the DANUBE, and 11 mi/18 km W of INGOLSTADT; 48°44′N 11°10′E. Manufacturing (textiles, metalworking, printing); silica quarrying. Has castle and baroque city hall. Capital of former principality of Neuburg (1507–1685).

Neuchâtel (nuh-SHAH-tel), German *Neuenburg*, canton (□ 308 sq mi/800.8 sq km), NW SWITZERLAND, in the JURA mountains; ⊙ NEUCHÂTEL. Fields and pastures in the lowlands, forests in the uplands. Cattle are raised, and cheese is produced. Vineyards cover much of the slopes rising from LAKE NEUCHÂTEL. Watches, mainly manufactured at LE LOCLE and LA CHAUX-DE-FONDS, have been an important industrial product since the 18th century. There are rich asphalt deposits at Val de TRAVERS and a small oil refinery at Cressier. The population is mainly French-speaking and Protestant. A part of BURGUNDY by the 10th century, Neuchâtel was later governed by counts under the Holy Roman Empire. The county passed (1504) to the French house of Orléans-Longueville and in 1648 became independent. In 1707 it chose Frederick I of Prussia as its prince. In 1806 Napoleon I placed it under the rule of his chief of staff, Marshal Berthier. In 1815 the principality joined the Swiss Confederation and its government was bound simultaneously to Berthier, the King of Prussia, and the Federal Diet at BERN. Attempts to create a republic failed in 1831, but succeeded in 1848. In 1857, the king of PRUSSIA renounced his claim to the canton, but kept a courtesy title for two more decades.

Neuchâtel (nuh-SHAH-tel), district, W NEUCHÂTEL canton, SWITZERLAND, on LAKE NEUCHÂTEL; 47°00′N 06°58′E. Main town is NEUCHÂTEL; population is French-speaking and Protestant.

Neuchâtel (nuh-SHAH-tel), German *Neuenburg*, town (2000 population 32,914), ⊙ NEUCHÂTEL, NW SWITZERLAND, on N shore of LAKE NEUCHÂTEL; 47°00′N 06°58′E. Elevation 1,427 ft/435 m. Located between Lake Neuchâtel and the first ridge of the JURA, the town has industries that produce watches, tobacco, paper, and chocolate; it is home to a significant wine market. The town still retains a medieval aspect, with its numerous statues, fountains, and old structures. It has an old church (12th–13th century), a castle (12th–17th century), and a noted university (founded 1838)— Louis Agassiz, the noted naturalist and paleontologist, was its first rector.

Neuchâtel, Lake (nuh-SHAH-tel), German *Neuenburgersee* or *Langensee*, ancient *Lacus Eburodunensis*, W SWITZERLAND, bordering on cantons of NEUCHÂTEL, BERN, FRIBOURG, and VAUD; 24 mi/39 km long, about 5 mi/8 km wide (□ 83 sq mi/215 sq km); extends SW-NE; 46°54′N 06°53′E. Elevation 1,407 ft/429 m; maximum depth 1,220 ft/372 m. The JURA rise on NW shores; fine views of the ALPS to the SE. THIELLE RIVER (French *La Thielle*; German *Zihl River*) traverses lake from SW to NE and flows, as Thielle or Zihl Canal, to LAKE BIEL. LAKE MORAT (German *Murtensee*) is connected to Lake Neuchâtel by the BROYE RIVER. Several ferry lines cross the lake. Remnants of neolithic lake dwellings (see LA TÈNE). Vineyards border the lake on the NW. Main towns on lake, NEUCHÂTEL and YVERDON.

Neudamm, POLAND: see DEBNO.

Neudeck (NOO-deck), Polish *Podzamek* (pod-zah-mek), estate in EAST PRUSSIA, near Gdańsk province, N POLAND, 19 mi/31 km ESE of KWIDZYŃ. Former country estate of the Hindenburg family; Paul von Hindenburg died here. Until 1919, it was in WEST PRUSSIA.

Neudek, CZECH REPUBLIC: see NEJDEK.

Neudenau (NOI-den-ou), town, FRANCONIA, BADEN-WÜRTTEMBERG, GERMANY, on the JAGST RIVER, and 7 mi/11.3 km SE of MOSBACH; 49°18′N 09°16′E. Cucumbers, sugar beets. Chartered 1236.

Neudorf (NOO-dorf), village (population 373), SE SASKATCHEWAN, CANADA, 17 mi/27 km SW of MELVILLE; 50°43′N 103°00′W. Mixed farming, livestock.

Neudorf, FRANCE: see VILLAGE-NEUF.

Neudorf, SLOVAKIA: see SPIŠSKÁ NOVÁ VES.

Neudörfl (NOI-derfl), township, N BURGENLAND, E AUSTRIA, on the LOWER AUSTRIAN border, 2 mi/3.2 km E of WIENER NEUSTADT; 47°48′N 16°17′E. Manufacturing (furniture, printing).

Neudorf Ostra, CZECH REPUBLIC: see OSTROZSKA NOVA VES.

Neue Elde River, GERMANY: see ELDE RIVER.

Neueibau, GERMANY: see EIBAU.

Neuenahr-Ahrweiter, Bad, GERMANY: see BAD NEUE-NAHR-AHRWEITER.

Neuenbürg (NOI-en-booirg), town, N BLACK FOREST, BADEN-WÜRTTEMBERG, GERMANY, on N slope of BLACK FOREST, on the ENZ RIVER, and 6 mi/9.7 km SW of PFORZHEIM; 48°51′N 08°35′E. Summer resort. Has 16th-century castle; ruins of 12th–13th-century castle; former iron mine that can be visited.

Neuenburg, POLAND: see NOWE.

Neuenburg, SWITZERLAND: see NEUCHÂTEL.

Neuenburg am Rhein (NOI-en-burg ahm REIN), town, S UPPER RHINE, BADEN-WÜRTTEMBERG, GERMANY, on the RHINE (French border; railroad and road bridge), and 18 mi/29 km SW of FREIBURG; 47°49′N 07°33′E. Customs station.

Neuenburgersee, SWITZERLAND: see NEUCHÂTEL, LAKE.

Neuendettelsau (noi-en-DET-tels-ou), village, MIDDLE FRANCONIA, BAVARIA, S GERMANY, 9 mi/14.5 km ESE of ANSBACH; 49°17′N 10°46′E. Village first mentioned 1141.

Neuenegg (NOY-uhn-egg), commune, BERN canton, W SWITZERLAND, on SENSE RIVER, on FRIBOURG border, and 8 mi/12.9 km WSW of Bern.

Neuengamme, GERMANY: see VIERLANDE.

Neuenhagen bei Berlin (noi-en-HAH-gen bei ber-LEEN), town, BRANDENBURG, E GERMANY, 12 mi/19 km E of BERLIN; 52°32′N 13°41′E. Livestock.

Neuenhaus (NOI-en-hous), town, LOWER SAXONY, NW GERMANY, near the VECHTE, 6 mi/9.7 km NW of NORDHORN, near Dutch border; 52°30′N 06°58′E. Manufacturing (textiles).

Neuenhof, commune, AARGAU canton, N SWITZERLAND, on LIMMAT RIVER, and 2 mi/3 km SE of BADEN; 47°27′N 08°19′E.

Neuenkirch (NOY-uhn-keyrsh), commune, LUCERNE canton, central SWITZERLAND, 6 mi/9.7 km NW of LUCERNE; 47°06′N 08°12′E. Elevation 1,811 ft/552 m.

Neuenkirchen (noi-en-KIR-khun), town, WESTPHALIA-LIPPE, North Rhine-Westphalia, NW GERMANY, 4 mi/6.4 km SW of RHEINE; 52°14′N 07°22′E. Dairying.

Neuenkirchen (noi-en-KIR-khun), village, LOWER SAXONY, N GERMANY, on N foot of Teutoburger Forest, 16 mi/26 km SE of OSNABRÜCK; 52°09′N 08°24′E. Tourism. Manufacturing (textiles). Agriculture (grain, vegetables).

Neuenkirchen, village, LOWER SAXONY, N GERMANY, 7 mi/11.3 km NW of SOLTAU. Agricultural area.

Neuenkirchen-Vörden (noi-en-KIR-khun–VOER-den), village, LOWER SAXONY, N GERMANY, near small HASE RIVER, and on W foot of Dammer Mountains, 16 mi/26 km N of OSNABRÜCK; 52°31′N 08°04′E. Forestry.

Neuenrade (noi-en-RAH-de), town, WESTPHALIA-LIPPE, North Rhine-Westphalia, W GERMANY, 8 mi/12.9 km ENE of LÜDENSCHEID; 51°17′N 07°47′E. Metalworking.

Neuenstadt, SWITZERLAND: see NEUVEVILLE, LA, town.

Neuenstadt am Kocher (NOI-en-shtaht ahm KUH-kher), town, FRANCONIA, BADEN-WÜRTTEMBERG, GERMANY, on the KOCHER RIVER, and 8 mi/12.9 km NE of HEILBRONN; 49°14′N 09°20′E. Food processing. Has 16th-century castle.

Neuenstein (NOI-en-shtein), town, FRANCONIA, BADEN-WÜRTTEMBERG, GERMANY, 16 mi/26 km ENE of HEILBRONN; 49°12′N 09°35′E. Wine. Renaissance castle.

Neuerburg (NOI-er-burg), town, RHINELAND-PALATINATE, W GERMANY, in the EIFEL, 24 mi/39 km NW of TRIER, 6 mi/9.7 km E of LUXEMBOURG border; 50°01′N 06°19′E. Has ruined 13th–14th-century castle.

Neuern, CZECH REPUBLIC: see NYRSKO.

Neufahrn bei Freising (NOI-fahrn bei FREI-sing), village, UPPER BAVARIA, BAVARIA, S GERMANY, near the ISAR, 13 mi/21 km NNE of MUNICH; 48°18′N 11°39′E. Former pilgrimage location for Catholics. Manufacturing (metalworking, consumer goods). Agriculture (vegetables).

Neufahrn in Niederbayern (NOI-fahrn in NEE-der-bei-ern), town, LOWER BAVARIA, BAVARIA, SE GERMANY, on the Small Laaber, 14 mi/23 km N of LANDSHUT; 48°44′N 12°11′E. Grain, sugar beets. Castle.

Neufahrwasser, POLAND: see GDAŃSK.

Neuf-Brisach (nuf-bree-ZAHK), German, *Neubreisach* (noi-BREI-zahk), commune (2004 population 2,237), HAUT-RHIN department, ALSACE region, E FRANCE, on RHÔNE-RHINE CANAL, near the RHINE (opposite Germany's *Breisach am Rhein*), and 9 mi/14.5 km SE of COLMAR; 48°01′N 07°33′E. Aluminum industry. Built (17th century) on a grid pattern by Vauban as a frontier fortress; the remnants of a part of the MAGINOT LINE (French fortification of World War II) are just outside the 17th century moat. Here last German resistance W of the Rhine collapsed in February 1945, during World War II. Hydroelectric plant on Grand Canal of ALSACE paralleling the Rhine. The FESSENHEIM nuclear power plant is 9 mi/14.5 km S (on same canal).

Neufchâteau (NUHF-shah-TO), commune (□ 523 sq mi/1,359.8 sq km; 2006 population 6,571), ⊙ Neufchâteau district, LUXEMBOURG province, SE BELGIUM, in the ARDENNES, 20 mi/32 km NW of ARLON; 49°50′N 05°26′E. Agriculture and livestock market center.

Neufchâteau (NUF-shah-to), town (□ 9 sq mi/23.4 sq km), VOSGES department, LORRAINE region, E FRANCE, on upper MEUSE RIVER, and 32 mi/51 km SW of NANCY; 48°15′N 05°50′E. Road center. Agriculture market. Manufacturing (furniture, toys, and food products). Has two medieval churches and a 16th-century town hall with ornate stairway. Just above the town the Meuse reappears on the surface after a 2.5 mi/4 km underground flow.

Neufchâtel-en-Bray (nuf-shah-tel-ahn-brai), town (□ 25 sq mi/65 sq km), SEINE-MARITIME department, HAUTE-NORMANDIE region, N FRANCE, on the Béthune River, and 25 mi/40 km NE of ROUEN; 49°44′N 01°27′E. Market center for Bray agricultural district, known for its Neufchâtel cheese and other milk products. Has 12th–16th-century church and 16th-century wooden houses. A museum displays old cider-making equipment. The Forest of Eawy (□ 16,308 acres/6,600 ha), SW of town, is noted for its extensive old beech stands.

Neufchâtel-Hardelot, FRANCE: see HARDELOT-PLAGE.

Neufeld an der Leitha (NOI-feld ahn der LEI-tah), township, N BURGENLAND, E AUSTRIA, on the Lower Austrian border, 7 mi/11.3 km W of EISENSTADT; 47°52′N 16°23′E. Nearby lake (Neufelder See), which originates from former open-cast lignite mining.

Neufelden (noi-FEL-den), village, N UPPER AUSTRIA, AUSTRIA, in the Mühlviertel, on Grosse Mühl River

(here dammed), 18 mi/29 km NW of LINZ; 48°29′N 14°00′E. Former center of linen weaving. Gothic, renaissance, and baroque burger houses; castle Pürnstein nearby.

Neuffen (NOIF-fen), town, STUTTGART district, BADEN-WÜRTTEMBERG, GERMANY, 5 mi/8 km SSE of NÜRTINGEN; 48°33′N 09°23′E. Cattle. Has 14th-century church; also ruined castle.

Neugedein, CZECH REPUBLIC: see KDYNE.

Neugersdorf (NOI-gers-dorf), town, SAXONY, E central GERMANY, in UPPER LUSATIA, in LUSATIAN MOUNTAINS, 11 mi/18 km NW of ZITTAU; frontier station on Czech border, opposite JIRIKOV; 50°59′N 14°36′E. Linen- and cotton-milling center. SPREE RIVER rises just N.

Neuhaldensleben, GERMANY: see HALDENSLEBEN.

Neu Hannover, BISMARCK ARCHIPELAGO: see NEW HANOVER.

Neuhaus, CZECH REPUBLIC: see JINDŘICHŮV HRADEC.

Neuhaus am Rennweg (NOI-hous ahm REN-veg), town, THURINGIA, central GERMANY, in THURINGIAN FOREST, 10 mi/16 km NNW of SONNEBERG; 50°31′N 11°08′E. Manufacturing (industrial products, laboratory equipment). Climatic health and winter sports resort. Physicist Geissler was born here. Founded in 1673, chartered 1933.

Neuhäusel, SLOVAKIA: see NOVÉ ZÁMKY.

Neuhausen (noi-HOU-suhn), village, BADEN-WÜRTTEMBERG, SW GERMANY, in the BLACK FOREST, 8 mi/12.9 km SE of PFORZHEIM; 48°47′N 08°47′E.

Neuhausen, RUSSIA: see GURYEVSK, KALININGRAD oblast.

Neuhausen am Rheinfall (NOI-hah-ou-zuhn am REIN-fahl), town (2000 population 9,959), SCHAFFHAUSEN canton, N SWITZERLAND, on the N bank of the RHINE RIVER. Manufacturing center just SW of SCHAFFHAUSEN.

Neuhausen auf den Fildern (noi-HOU-suhn ouf den Fil-dern), town, STUTTGART district, BADEN-WÜRTTEMBERG, GERMANY, 4 mi/6.4 km SSW of ESSLINGEN (linked by tramway); 48°41′N 09°17′E. Manufacturing (textiles, machinery). Agriculture (grain). Stuttgart airport nearby. Two 16th-century castles here. Village first mentioned in 1153.

Neuhof (NOI-hof), village, HESSE, central GERMANY, on the Fliede River, and on W foot of the HOHE RHÖN, 8 mi/12.9 km SW of FULDA; 50°28′N 09°37′E. Dairy; industrial salt mining. First mentioned 1165; moated castle (16th–18th centuries).

Neuhofen (NOI-ho-fuhn), village, RHINELAND-PALATINATE, W GERMANY, 4 mi/6.4 km S of LUDWIGSHAFEN; 49°26′N 08°26′E. Grain, sugar beets.

Neuhofen an der Krems (noi-HO-fen ahn der KREMS), township, E central UPPER AUSTRIA, AUSTRIA, 9 mi/14.5 km E of WELS; 48°08′N 14°14′E. Diversified small industry. Agriculture (wheat; pigs). Moated castle.

Neuhrosenkau, CZECH REPUBLIC: see NOVY HROZENKOV.

Neuilly-en-Thelle (noi-YEE-ahn-tel), commune (□ 6 sq mi/15.6 sq km), OISE department, PICARDIE region, N FRANCE, 13 mi/21 km W of SENLIS; 49°14′N 02°17′E. Manufacturing (textiles).

Neuilly-Plaisance (noi-YEE-plai-ZAHNS), residential town (□ 1 sq mi/2.6 sq km), SEINE-SAINT-DENIS department, ÎLE-DE-FRANCE region, N central FRANCE, an outer E suburb of PARIS, 7 mi/11.3 km from Notre Dame Cathedral; 48°52′N 02°31′E. Diverse manufacturing.

Neuilly-Saint-Front (noi-YEE-san-fron), commune (□ 7 sq mi/18.2 sq km), AISNE department, PICARDIE region, N FRANCE, near the OURCQ, 11 mi/18 km NW of CHÂTEAU-THIERRY; 49°10′N 03°16′E.

Neuilly-sur-Marne (noi-YEE-syur-mahrn), town (□ 2 sq mi/5.2 sq km), SEINE-SAINT-DENIS department, ÎLE-DE-FRANCE region, N central FRANCE, an outer E suburb of PARIS, 8 mi/12.9 km from Notre Dame

Cathedral, on right bank of the MARNE (river-front sports); 48°51′N 02°32′E. Manufacturing (rubber, electric wire, and furniture). Two psychiatric hospitals. Has Gothic church dating from 1200.

Neuilly-sur-Seine (noi-YEE-syur-sen), innermost NW suburb (□ 1 sq mi/2.6 sq km) of PARIS, HAUTS-DE-SEINE department, ÎLE-DE-FRANCE region, N central FRANCE, just N of the BOIS DE BOULOGNE; 48°53′N 02°16′E. One of the wealthiest suburbs of Paris, Neuilly-sur-Seine has grown rapidly with high-rise residential development and office buildings. The American Hospital of Paris is here. Noted for its automotive industry. Also spelled Neuilly sur Seine.

Neuilly sur Seine, FRANCE: see NEUILLY-SUR-SEINE.

Neu-Isenburg (NOI-EE-sen-burg), town, S HESSE, GERMANY, 4 mi/6.4 km S of FRANKFURT; 50°03′N 08°32′E. Manufacturing (food products, photochemistry). Livestock. Founded 1699.

Neukieritzsch (noi-KEE-ritsh), village, SAXONY, E central GERMANY, near the PLEISSE, 13 mi/21 km S of LEIPZIG; 51°09′N 12°25′E. Grain.

Neukirchen (noi-KIR-khuhn), town, HESSE, W GERMANY, 15 mi/24 km W of BAD HERSFELD; 50°52′N 09°21′E. Textiles; lumber milling; health resort. Was first mentioned 1142; chartered 1351. Has Gothic church.

Neukirchen am Grossvenediger (noi-KIR-khuhn ahm GROS-ve-nai-di-guhr), township, SW SALZBURG, W AUSTRIA, in the PINZGAU region near the SALZACH RIVER, 25 mi/40 km WSW of ZELL AM SEE; 47°15′N, 12°17′E. Elevation 2,606 ft/794 m. Summer tourism; winter sports at Wildkogel; National Park Hohe Tauern nearby.

Neukirchen bei Heiligen Blut (noi-KIR-khuhn bei HEI-li-gen BLOOT), village, LOWER BAVARIA, GERMANY, in BOHEMIAN, at NE foot of the HOHER BOGEN, 32 mi/51 km NE of STRAUBING; 49°16′N 12°58′E. Rye, potatoes; cattle. Has Franciscan monastery (founded 1659), and early 18th-century pilgrimage church.

Neukirchen-Erzgebirge (noi-KIR-khuhn-ERTS-ge-beer-ge), village, SAXONY, E central GERMANY, 5 mi/8 km SW of CHEMNITZ; 50°47′N 12°52′E.

Neukirchen-Vluyn (noi-KIR-khuhn-FLOOIN), town, W GERMANY, North Rhine-Westphalia, in the RUHR industrial district, 2 mi/3.2 km W of MOERS, in coal-mining region; 51°27′N 06°33′E. Formed 1928 through incorporated of three neighboring villages. Church and castle from 15th century. Neukirchen first mentioned in 13th century; Vluyn first mentioned in 812.

Neukirch-Lausitz (noi-KIRKH-LOU-sits), village, SAXONY, E central GERMANY, in UPPER LUSATIA, 11 mi/18 km SW of BAUTZEN, near Czech border; 51°06′N 14°18′E. Manufacturing (leather products, ceramics, food); cotton milling, woodworking.

Neukloster (NOI-klaw-ster), town, Mecklenburg–Western Pomerania, N GERMANY, 9 mi/14.5 km ESE of WISMAR; 53°52′N 11°41′E. Agricultural market (grain, sugar beets, potatoes; livestock); tourist resort. Has 13th-century church of former Cistercian convent (founded nearby; moved here in 1219; secularized in 16th century). Has been a town since 1938.

Neukölln (noi-KOELN), workers' residential district of BERLIN, NE GERMANY, on TELTOW CANAL, and 5 mi/8 km SSE of city center; 52°29′N 13°37′E. Manufacturing (machinery, chemicals, apparel). Formerly called RIXDORF.

Neukuhren, RUSSIA: see PIONERSKIY.

Neu-Langenburg, TANZANIA: see TUKUYU.

Neu Lauenberg: see DUKE OF YORK ISLANDS.

Neulengbach (noi-LENG-bahkh), township, central LOWER AUSTRIA, AUSTRIA, at W foot of WIENERWALD (forest), 12 mi/19 km E of Sankt Pölten; 48°12′N 15°55′E. Fruit cultivation; market center; Renaissance castle in dominant position.

Neulingen (NOI-ling-uhn), village, BADEN-WÜRTTEMBERG, SW GERMANY, on N slope of BLACK FOREST, 8 mi/12.9 km N of PFORZHEIM; 48°59′N 08°42′E. Manufacturing (metalworking, electronic equipment). Agriculture (grain, fruit).

Neulussheim (noi-LUS-heim), village, BADEN-WÜRT-TEMBERG, GERMANY, near the RHINE, 4 mi/6.4 km ESE of SPEYER; 49°18′N 08°31′E. Sugar beets, strawberries.

Neum (NAI-um), village, W HERZEGOVINA, BOSNIA AND HERZEGOVINA, on Neretva Channel of ADRIATIC SEA, 9 mi/14.5 km S of Metković (CROATIA), in narrow corridor providing Herzegovina with outlet to the sea; 42°55′N 17°37′E.

Neumark, POLAND: see NOWE MIASTO LUBAWSKIE.

Neumarkt, POLAND: see NOWY TARG.

Neumarkt, POLAND: see ŚRODA SLĄSKA.

Neumarkt, ROMANIA: see TÎRGU MUREŞ.

Neumarkt am Wallersee (NOI-mahrkt ahm VAHL-ler-sai), township, N SALZBURG, W central AUSTRIA, in the FLACHGAU region, 13 mi/21 km NE of SALZ-BURG; 47°57′N 13°14′E. Manufacturing (electric equipment, furniture); summer tourism; castles.

Neumarkt im Mühlkreis (NOI-mahrkt im MOOIL-kreis), township, NE UPPER AUSTRIA, AUSTRIA, in the Mühlviertel, 6 mi/9.7 km S of FREISTADT; 48°26′N 14°29′E. Dairy farming.

Neumarkt in der Oberpfalz (NOI-mahrkt in der O-buhr-pfalts), town, UPPER PALATINATE, central BA-VARIA, GERMANY, on LUDWIG CANAL and 22 mi/35 km SW of AMBERG; 49°16′N 11°27′E. Manufacturing (synthetic materials, apparel, machinery, toys). Has late-Gothic church, 16th-century castle. Mineral springs nearby. Was first mentioned 1160; chartered 1235.

Neumarkt in Steirmark (NOI-mahrkt in SHTEI-uhr-mahrk), township, STYRIA, S central AUSTRIA, at S end of Neumarkt Pass, leading into CARINTHIA, 12 mi/19 km SW of JUDENBURG; 47°04′N 14°26′E. Resort (elevation 2,548 ft/777 m). Fortified settlement since 1224, large parts of the fortifications still standing; the castle serves as conference center. Winter sports facilities at Grebenzen.

Neumarkt-Sankt Veit (NOI-mahrkt–sahnkt FEIT), town, UPPER BAVARIA, BAVARIA, SE GERMANY, on the Roth River, 21 mi/34 km SE of LANDSHUT; 48°21′N 12°30′E. Furniture making; textile industry. Town first mentioned 1269; late-Gothic church (15th century).

Neumeyer Station (NOO-mah-yuh), ANTARCTICA, German station on EKSTRÖM ICE SHELF; 70°39′S 08°15′W.

Neumittelwalde, POLAND: see MIEDZYBORZ.

Neumühlen-Dietrichsdorf (NOI-mooi-len–DEET-rikhs-dorf), NE industrial district of KIEL, NW GER-MANY, on E bank of KIEL FIRTH; 54°20′N 10°07′E. Manufacturing (ship machinery, boats).

Neumünster (noi-MOOIN-ster), city, SCHLESWIG-HOLSTEIN, N central GERMANY, 35 mi/56 km N of HAMBURG; 54°04′N 09°59′E. It is a transportation and industrial center; manufacturing (machinery, textiles, leather, paper). Known in the 12th century, Neu-münster was chartered in 1870.

Neunburg vorm Wald (NOIN-burg form VAHLT), town, UPPER PALATINATE, E BAVARIA, GERMANY, in BOHEMIAN FOREST, 12 mi/19 km E of SCHWANDORF; 49°21′N 12°22′E. Food processing, metalworking. Has 15th-century church, 15th–16th-century castle. Town first mentioned 1017, chartered c.1300.

Neunkirchen (noin-KIR-khuhn), city, SAARLAND, SW GERMANY, 12 mi/19 km NE of SAARBRÜCKEN; 49°21′N 07°10′E. Metalworking; coal-mining area until 1968. City first mentioned 1287.

Neunkirchen (noin-KIR-khuhn), town, SE LOWER AUSTRIA, E AUSTRIA, in the STEINFELD, on the SCHWARZA RIVER; 47°44′N 16°05′E. Market center; manufacturing (concrete, construction materials, timber products, machines). Monastery, renaissance and baroque burger houses.

Neunkirchen (noin-KIR-khuhn), town, WESTPHALIA-LIPPE region, North Rhine-Westphalia, W GERMANY, 6 mi/9.7 km S of SIEGEN; 50°48′N 08°00′E. Manu-facturing (machinery, synthetic fiber, metal working, electronic equipment). Formed in 1969 by unification of six smaller villages.

Neunkirchen am Brand (noin-KIR-khuhn ahm BRAHND), village, UPPER FRANCONIA, N central BAVARIA, GERMANY, 5 mi/8 km ENE of ERLANGEN; 49°37′N 11°06′E. Brewing. Hops. Has 16th-century church.

Neunkirchen am Sand (noin-KIR-khuhn ahm SAHND), village, MIDDLE FRANCONIA, BAVARIA, central GERMANY, on the PEGNITZ, and on W slope of the FRANCONIAN JURA, 12 mi/19 km NE of NUREM-BERG; 49°31′N 11°19′E. Manufacturing (paper, electronic equipment). Agriculture (fruit).

Neu Oettingen, CZECH REPUBLIC: see NOVA VCELNICE.

Neuötting (noi-OET-ting), town, UPPER BAVARIA, GERMANY, on the INN RIVER, just NE of ALTÖTTING; 48°15′N 12°41′E. Manufacturing (chemicals). Has late-Gothic church. Was first mentioned 1231, chartered 1321.

Neupaka, CZECH REPUBLIC: see NOVA PAKA.

Neupokoyev, Cape (nye-oo-puh-KO-yeef), southernmost point of BOL'SHEVIK ISLAND, SEVERNAYA ZEM-LYA archipelago, KRASNOYARSK TERRITORY, N RUSSIA; 77°50′N 99°30′E. Government observation post.

Neu Pommern, BISMARCK ARCHIPELAGO: see NEW BRITAIN.

Neupré (nuh-PRAI), commune (2006 population 9,751), Liège district, LIÈGE province, E BELGIUM, 9 mi/14 km SW of LIÈGE.

Neuquén (ne-oo-KAIN), province (□ 36,429 sq mi/94,715.4 sq km; 2001 population 474,155), W central ARGENTINA; ⊙ Neuquén. Andean mountainous area bordering CHILE, with LIMAY RIVER on S and the Rio Colorado and BARRANCAS River on N; intersected by NEUQUÉN RIVER. In W are numerous volcanoes (e.g., Copahue, LANÍN), in SW is the Argentine lake district resort area (including lakes NAHUEL HUAPÍ and TRA-FUL), much of it in a national park. Climate, depending on elevation, is temperate and dry in populated areas. Mineral resources include gold (Cordillera del VIENTO, ANDACOLLO, CHOS MALAL), coal (Sierra de HUANTRAICÓ, Nahuel Huapí area), asphalt (Chos Malal), petroleum (Plaza Huin-cul, CHALLACO); natural gas (near Neuquén city and CUTRAL-CÓ); also lead, copper, and sulphur. Extensive pine forests in Andean foothills; also oak, cypress. Agriculture restricted to sub-Andean valleys and semiarid E plains; alfalfa, wheat, corn; in irrigated valleys near Neuquén are fruit growing and viticulture. Livestock raising (sheep, goats, cattle) in valleys and foothills. Mining, sawmilling, flour milling, wine making; dairying; paper mills (ALUMINÉ), petroleum refineries (Plaza Huincul, Challaco). Several large hydroelectric dams, also used for irrigation. Established 1884 as national territory.

Neuquén (ne-oo-KAIN), town, ⊙ NEUQUÉN province and Confluencia department, S central ARGENTINA, at confluence of NEUQUÉN and LIMAY rivers (forming the Río Negro), on railroad, and 600 mi/966 km SW of BUENOS AIRES; 38°57′S 68°05′W. Inland river port and agricultural center for irrigated area (two large dams nearby). Produces fruit, wine, and alfalfa. Manufacturing (farm equipment, construction materials); food processing. Has administrative buildings, theater. Bridge across Neuquén River.

Neuquén River (ne-oo-KAIN), c.320 mi/515 km long, NEUQUÉN province, ARGENTINA; rises in the ANDES at CHILE border near 36°30′S; flows S and SE, past AN-DACOLLO, CHOS MALAL, and AÑELO, to NEUQUÉN, where it joins the LIMAY RIVER to form the Río Negro. Lower course is dammed and used for irrigation (fruit growing). Receives the Río AGRIO.

Neureichenau (noi-REI-khen-ou), village, LOWER BA-VARIA, BAVARIA, SE GERMANY, in BAVARIAN FOREST, and near Czech and Austrian borders, 18 mi/29 km NE of PASSAU; 48°44′N 13°44′E. Forestry. Manufacturing (textiles, machinery).

Neuried (noi-REED), commune, S UPPER RHINE, BADEN-WÜRTTEMBERG, SW GERMANY, near French border, 7 mi/11.3 km W of OFFENBURG; 48°28′N 07°48′E. Agriculture (vegetables, tobacco). Metalworking. Formed by unification of several villages.

Neuried (noi-REED), residential village, UPPER BA-VARIA, BAVARIA, S GERMANY, 6 mi/9.7 km SW of MUNICH; 48°05′N 11°27′E.

Neurode, POLAND: see NOWA RUDA.

Neurohlau, CZECH REPUBLIC: see NOVA ROLE.

Neuruppin (noi-RUP-pin), town, BRANDENBURG, E GERMANY, on W shore of RUPPIN LAKE, 40 mi/64 km NW of BERLIN; 52°56′N 12°49′E. Manufacturing (foods, furniture, lithography, printing). Climatic health resort. Old town walls and 13th-century monastery church. Was capital of Ruppin co. until 1524, then passed to BRANDENBURG. Frederick the Great, as crown prince, lived here (1732–1736). Largely destroyed by fire in 1787. Dramatist Fontane and architect Schinkel was born here. Sometimes spelled Neu Ruppin. Chartered 1256.

Neusalz, POLAND: see NOWA SÓL.

Neu Sandec, POLAND: see NOWY SĄCZ.

Neusäss (noi-SEHS), town, SWABIA, BAVARIA, S GER-MANY, near the Schmutter, just E of AUGSBURG; 48°24′N 10°48′E. Manufacturing (transportation equipment, printing, paint, machinery). Was first mentioned 1178.

Neusattl, CZECH REPUBLIC: see NOVE SEDLO.

Neuschwanstein, GERMANY: see SCHWANGAU.

Neuse Forest (NOOS), unincorporated town (□ 3 sq mi/7.8 sq km; 2000 population 1,426), CRAVEN county, E NORTH CAROLINA, 10 mi/16 km SSE of NEW BERN, on NEUSE RIVER, at N edge of Croatan National Forest; 34°57′N 76°57′W. Cherry Point Marine Corps Air Station to SE. Service industries; retail trade.

Neuse River (NOOS), c.275 mi/443 km long, E NORTH CAROLINA; formed 8 mi/12.9 km NE of DURHAM by junction of small Flat (c.25 mi/40 km long), Eno, and Little rivers; flows generally SE through FALLS LAKE RESERVOIR, passes to E of RALEIGH, flows past SMITHFIELD, GOLDSBORO, KINSTON, and NEW BERN where it widens into an estuary (c.5 mi/8 km wide) extending c.40 mi/64 km SE and NE to PAMLICO SOUND; 36°05′N 78°48′W. Estimated to be two million years old; archaeological evidence indicates that the River's banks were settled c.14,000 years ago. Early Native American settlers included the Tuscarora, Coree, Neusiok, and Secotan tribes. Today the River serves as a watershed for more than 1.5 million people, and as a source of recreation and commerce.

Neuses (NOI-ses) or **Neuses bei Coburg**, NW suburb of COBURG, UPPER FRANCONIA, NE BAVARIA, GER-MANY; 50°17′N 10°56′E. Home of poet Friedrich Rückert (1820–1826).

Neusiedl am See (NOI-seedl ahm SAI), town, N BURGENLAND province, E AUSTRIA, on N shore of LAKE NEUSIEDL, 22 mi/35 km NE of EISENSTADT; 47°57′N 16°51′E. Railroad junction; market center; canning of agricultural products; vineyards; summer tourism.

Neusiedl an der Zaya (NOI-seedl ahn der TS-yah), township, NE LOWER AUSTRIA, AUSTRIA, 10 mi/16 km ENE of MISTELBACH; 48°36′N 16°47′E. Manufacturing (ski bindings); vineyards; oil production.

Neuss (NOIS), city, RHINELAND, North Rhine-West-phalia, W GERMANY, 6 mi/9.7 km WSW of DÜSSEL-DORF; 51°12′N 06°42′E. Railroad junction and canal port, near the left bank of the RHINE opposite DÜS-SELDORF. Manufacturing (machinery, paper, apparel, food products); important flower market. Horse-racing track here. Built on the site of a Roman camp called Novaesium, Neuss was chartered in the 12th century. It belonged to the archbishopric of COLOGNE until the French Revolutionary Wars. Unsuccessfully besieged (1474–1475) by Charles the Bold. Burned 1586. Passed to PRUSSIA in 1815. Noteworthy structures include the Romanesque Church of St. Quirinus (13th century), a city gate (13th century).

Neustadt (NOI-shtaht), town, HESSE, W GERMANY, 16 mi/26 km ENE of MARBURG; 50°52′N 09°07′E. Grain. Also called Neustadt-Hessen.

Neustadt (NU-stat), former village (□ 1 sq mi/2.6 sq km; 2001 population 562), S ONTARIO, E central Canada, on South Saugeen River, and 35 mi/56 km S of OWEN SOUND; 44°04′N 81°00′W. Dairying, mixed farming. Amalgamated into West Grey township.

Neustadt (NOI-shtaht), village, RHINELAND-PALATINATE, W GERMANY, in the WESTERWALD, and on the WIED, 14 mi/23 km N of NEUWIED; 50°38′N 07°26′E. Winery. Also called Neustadt-Wied.

Neustadt, CZECH REPUBLIC: see NEW TOWN.

Neustadt, POLAND: see PRUDNIK.

Neustadt, POLAND: see WEJHEROWO.

Neustadt am Rübenberge (NOI-shtaht ahm ROOI-ben-ber-ge), town, LOWER SAXONY, W GERMANY, on the LEINE, and 14 mi/23 km NW of HANOVER; 52°31′N 09°27′E. Manufacturing (machinery). Has 13th-century church.

Neustadt an der Aisch (NOI-shtaht ahn der EISH), town, MIDDLE FRANCONIA, W BAVARIA, GERMANY, on the AISCH, 18 mi/29 km W of ERLANGEN; 49°35′N 10°37′E. Hops market. Manufacturing (artists' supplies, soap, wool, toys); woodworking, printing, brewing, aluminum works. Has 14th-century church. Chartered in early 14th century.

Neustadt an der Donau (NOI-shtaht ahn der DO-nou), town, LOWER BAVARIA, GERMANY, near the DANUBE, 9 mi/14.5 km SW of KELHEIM; 48°48′N 11°44′E. Manufacturing (synthetic material); oil refining; testing course for motor vehicles. Chartered 1270.

Neustadt an der Mettau, CZECH REPUBLIC: see NOVE MESTO NAD METUJI.

Neustadt an der Saale, Bad, Germany: see BAD NEUSTADT AN DER SAALE.

Neustadt an der Tafelfichte, CZECH REPUBLIC: see NOVE MESTO POD SMRKEM.

Neustadt an der Waldnaab (NOI-shtaht ahn der VAHLD-nahb), town, UPPER PALATINATE, NE BAVARIA, GERMANY, on the Waldnaab River, and 3.5 mi/5.6 km N of WEIDEN; 49°44′N 12°09′E. Railroad junction. Manufacturing (cut-glass). Agriculture (grain; livestock). Has 18th-century church. Chartered in early 14th century.

Neustadt an der Weinstrasse (NOI-shtaht ahn der VEIN-shtrah-se), city, RHINELAND-PALATINATE, SW GERMANY; 49°21′N 07°38′E. Center of the Rhineland-Palatinate wine trade. Manufacturing (metal products, paper, wine). The city is also a tourist center. Gothic basilica 1368–1489; ruined castle from the 13th–14th centuries. City first mentioned 1235, chartered 1275.

Neustadt, Bad, Germany: see BAD NEUSTADT AN DER SAALE.

Neustadt bei Coburg (NOI-shtaht bei KAW-burg), city, UPPER FRANCONIA, N BAVARIA, GERMANY, 8 mi/12.9 km NE of COBURG, 3 mi/4.8 km SW of SONNEBERG; 50°20′N 11°07′E. Manufacturing (toys, machinery); lumber mills. Was first mentioned 1248; chartered in 1300s.

Neustadt-Dosse (NOI-shtaht–DUHS-se), town, BRANDENBURG, E GERMANY, on DOSSE, 17 mi/27 km WSW of NEURUPPIN; 52°51′N 12°27′E. Site of stud farm, established 1789. Dairying. Chartered 1664.

Neustädtel, POLAND: see NOWE MIASTECZKO.

Neustadt-Glewe (NOI-shtaht–GLEH-ve), town, MECKLENBURG–WESTERN POMERANIA, N GERMANY, on regulated ELDE RIVER, 18 mi/29 km SSE of SCHWERIN; 53°23′N 11°35′E. Manufacturing (cement, leather products). Hydroelectric power station. Has

17th-century palace; remains of old fort, town walls, and gates. Founded in 13th century.

Neustadt im Schwarzwald, Germany: see TITISEE-NEUSTADT.

Neustadt in Holstein (NOI-shtaht in HOL-shtein), town (2006 population 16,479), in SCHLESWIG-HOLSTEIN, NW GERMANY, port on LÜBECK BAY, 16 mi/26 km NNE of LÜBECK; 54°06′N 10°48′E. Manufacturing (metal products, concrete, leather products), food processing. Trade (grain, wood). BALTIC seaside resort. Has 13th-century church. Chartered 1244.

Neustadt in Sachsen (NOI-shtaht in SAHK-sen), town, SAXONY, E central GERMANY, in SAXONIAN SWITZERLAND, 22 mi/35 km E of DRESDEN, near CZECH border; 51°02′N 14°12′E. Manufacturing (agricultural machinery, artificial flowers); woodworking. Formed after 1300.

Neustadtl, CZECH REPUBLIC: see NOVE MESTO NA MORAVE.

Neustadt-Orla (NOI-shtaht–OR-lah), town, THURINGIA, central GERMANY, on ORLA RIVER, 15 mi/24 km SE of JENA; 50°44′N 11°45′E. Manufacturing (machine tools); woolen milling, metal- and woodworking. Has 16th-century church and 15th-cent. town hall. Was first mentioned 1287.

Neustettin, POLAND: see SZCZECINEK.

Neustraschnitz, CZECH REPUBLIC: see NOVE STRASECI.

Neustrelitz (noi-SHTREH-lits), city, MECKLENBURG–WESTERN POMERANIA, NE GERMANY, 16 mi/26 km SSW of NEUBRANDENBURG; 53°22′N 13°04′E. Transportation center. Metalworks, food processing, and wood mills. Founded (1733) as capital of Mecklenburg-Strelitz after the earlier ducal residence at nearby Strelitz had burned down (1712).

Neustria (NUS-tree-ah), W portion of the kingdom of the Franks in the 6th, 7th, and 8th centuries, during the rule of the Merovingians. It comprised the SEINE and LOIRE country and the region to the N; its principal towns were SOISSONS and PARIS. The realm originated with the partition of the lands of Clovis I (died 511) during 6th century. Long dynastic rivalry persisted between Neustria and E portion of Frankish kingdom, known later as Austrasia. Neustria and Austrasia were reunited briefly in 7th century. But in 687, Austrasia's Pepin of Herstal defeated his Neustrian rival, and united Austrasia and Neustria for good. His descendants, the Carolingians, continued to rule the two realms, first as mayors and after 751 as kings.

Neuteich, POLAND: see NOWY STAW.

Neutitschein, CZECH REPUBLIC: see NOVÝ JIČÍN.

Neutoggenburg (NOI-tog-uhn-boorg), district, W central ST. GALLEN canton, SWITZERLAND. Main town is WATTWIL; population is German-speaking and Roman Catholic.

Neutomischel, POLAND: see NOWY TOMYSL.

Neutra, SLOVAKIA: see NITRA.

Neutra River, SLOVAKIA: see NITRA RIVER.

Neutraubling (noi-TRAUB-ling), town, UPPER PALATINATE, BAVARIA, SE GERMANY, 5 mi/8 km SE of REGENSBURG; 48°59′N 12°10′E. Grain, sugar beets.

Neu-Ulm (NOI–ULM), city, SWABIA, W BAVARIA, GERMANY, on right bank of the DANUBE (2 bridges), at mouth of the ILLER, opposite ULM; 48°24′N 10°01′E. Railroad junction. Manufacturing (textiles); leather-, metal-, and woodworking; brewing. Founded 1810 on territory taken from ULM.

Neuves-Maisons (nuv–mai-ZON), town (□ 1 sq mi/2.6 sq km), MEURTHE-ET-MOSELLE department, LORRAINE region, NE FRANCE, on the MOSELLE and Canal de l'EST, and 7 mi/11.3 km SSW of NANCY; 48°37′N 06°06′E. Metalworking shops in old iron-mining area.

Neuveville, La (NUHVE-vil, lah), district, W BERN canton, SWITZERLAND. Main town is LA NEUVEVILLE; population is French-speaking and Protestant.

Neuveville, La (NUHVE-vil, lah), German *Neuenstadt*, town, BERN canton, W SWITZERLAND, at SW end of

LAKE BIEL, and 9 mi/14.5 km SW of BIEL, near NEUCHÂTEL border. Elevation 1,430 ft/436 m. Watches, metal products.

Neuvic (nu-VEEK), commune (□ 28 sq mi/72.8 sq km), resort village, CORRÈZE department, LIMOUSIN region, S central FRANCE, 11 mi/18 km S of USSEL on a man-made lake (hydroelectric facility); 45°23′N 02°16′E. Cattle market. Center of underground resistance against GERMANY in 1944. Has 12th–15th-century church. Also called Neuvic-d'Ussel.

Neuvic (nu-veek), commune (□ 10 sq mi/26 sq km), DORDOGNE department, AQUITAINE region, SW FRANCE, on the ISLE RIVER, and 14 mi/23 km SW of PÉRIGUEUX; 45°06′N 00°28′E. Footwear manufacturing. Has 16th-century Renaissance château. Also known as Neuvic-sur-l'Isle.

Neuvic-d'Ussel, FRANCE: see NEUVIC.

Neuvic-sur-l'Isle, FRANCE: see NEUVIC.

Neuville (nu-VEEL), village (□ 28 sq mi/72.8 sq km), Capitale-Nationale region, S QUEBEC, E Canada, on the SAINT LAWRENCE RIVER, and 20 mi/32 km WSW of QUEBEC city; 46°42′N 71°35′W. Dairying; fruit, vegetables (corn).

Neuville-aux-Bois (nu-VEEL-o-bwah), commune (□ 12 sq mi/31.2 sq km), LOIRET department, CENTRE administrative region, N central FRANCE, at N edge of Forest of ORLÉANS, 13 mi/21 km NE of ORLÉANS; 48°04′N 02°02′E. Market for cereals, cattle.

Neuville-de-Poitou (nu-VEEL-duh–PWAH-too), commune (□ 6 sq mi/15.6 sq km), VIENNE department, POITOU-CHARENTES region, W central FRANCE, 8 mi/12.9 km NW of POITIERS; 46°41′N 00°15′E. Agriculture market for wine, poultry, potatoes, and vegetables.

Neuville-en-Condroz (nuh-VEEL-AW–kaw-DROZ), village in commune of NEUPRÉ, Liège district, LIÈGE province, E BELGIUM, 9 mi/14.5 km SW of LIÈGE; 50°32′N 05°27′E. Site of U.S. World War II military cemetery.

Neuville-en-Ferrain (nu-VEEL–ahn-fuh-ran), town (□ 2 sq mi/5.2 sq km; 2004 population 9,433), NORD department, NORD-PAS-DE-CALAIS region, N FRANCE, on Belgian border, 4 mi/6.4 km N of TOURCOING; 50°45′N 03°09′E. Computer and information science center within LILLE metropolitan area.

Neuville-Saint-Rémy (nu-VEEL-san–RAI-mee), commune, NORD department, NORD-PAS-DE-CALAIS region, N FRANCE; 50°11′N 03°14′E.

Neuville-Saint-Vaast (nu-VEEL-san–vahst), village (□ 4 sq mi/10.4 sq km), PAS-DE-CALAIS department, NORD-PAS-DE-CALAIS region, N FRANCE, 5 mi/8 km N of ARRAS, just S of VIMY Ridge; 50°21′N 02°46′E. Nearby are large British, French, and German military cemeteries, and allied memorials of battle of Arras (1917) in World War I.

Neuville-sous-Montreuil, FRANCE: see MONTREUIL.

Neuville-sur-l'Escaut (nu-VEEL-syoor-les-KO), commune (□ 3 sq mi/7.8 sq km), NORD department, NORD-PAS-DE-CALAIS region, N FRANCE, on Escaut (SCHELDT) River (canalized), and 3 mi/4.8 km SW of DENAIN. Cement plant.

Neuville-sur-Saône (nu-VEEL-syoor-son), town (□ 2 sq mi/5.2 sq km), RHÔNE department, RHÔNE-ALPES region, E central FRANCE, on left bank of SAÔNE RIVER, and 14 mi/14.5 km N of LYON; 45°52′N 04°51′E. Textile milling and dyeing; chemical industry.

Neuvy-Saint-Sépulchre (nu-vee-san–sai-pul-kruh), commune (□ 13 sq mi/33.8 sq km), INDRE department, CENTRE administrative region, central FRANCE, 9 mi/14.5 km W of La CHÂTRE; 46°36′N 01°49′E. Pump manufacturing. Cattle raising. Has 11th–12th-century circular church patterned on Holy Sepulcher in Jerusalem; it is a pilgrimage center (one of the pilgrimage sites en route to the Cathedral of Santiago de Compostela, home of the shrine dedicated to St. James, the patron saint of Spain).

Neu-Walter, RUSSIA: see SVERDLOVO.

Area is shown by the symbol □, and capital city or county seat by ⊙.

Neuwarp, POLAND: see NOWE WARPNO.

Neuwedell, POLAND: see DRAWNO.

Neuwerk (NOI-verk), island of EAST FRISIAN group, NORTH SEA, belonging to HAMBURG, NW GERMANY, 8 mi/12.9 km NW of CUXHAVEN; □ 1.3 sq mi/3.4 sq km; 53°56'N 08°30'E. Ship-salvage station; some agriculture; tourism; 14th-century light tower.

Neuwied (noi-VEED), city, RHINELAND-PALATINATE, W GERMANY, a port at the confluence of the RHINE and WIED rivers, 10 mi NW of KOBLENZ; 50°26'N 07°29'E. Largely industrial; manufacturing (building materials, machinery, electronic equipment, chemicals). Neuwied developed around a palace begun by Count Frederick III of Wied in 1648. Castle from 18th-century Roman ruins nearby. Chartered 1662.

Neu Wulmstorf (NOI VULMS-dorf), village, LOWER SAXONY, NW GERMANY, in the Altes Land, 10 mi/16 km WSW of HAMBURG; 53°28'N 09°48'E. Residential area; formed after 1945 as home for 4,000 refugees. Agriculture (fruit).

Nevada, state (□ 110,566 sq mi/286,366 sq km; 2000 population 1,998,257; 1995 estimated population 1,530,108), W UNITED STATES, admitted as the 36th state of the Union in 1864; ⊙ CARSON CITY; 39°59'N 117°01'W. Nevada is called the "Silver State" because of its large silver mine industries, and the "Sage State" or "Sagebrush State" for the wild sage that grows there.

Geography

Nevada's elevation is 5,500 ft/1,676 m. LAS VEGAS is the largest city, and RENO is the second largest. Nevada is bounded on the W and SW by CALIFORNIA, on the N by OREGON and IDAHO, on the E by UTAH, and on the SE by ARIZONA (with the COLORADO RIVER marking most of the border). Most of the state lies within the Basin and Range region, or "GREAT BASIN" or "Intermountain West." The rivers in the SE belong to the Colorado River system, while those of the extreme N drain into the COLUMBIA system (SNAKE). Like the HUMBOLDT, TRUCKEE, and CARSON rivers, most Nevada rivers flow into brackish and intermittent lakes or dry sinks—which have no outlet to the oceans, where water loss due to evaporation, absorption, and human use is greater than inflow and precipitation—except where they have been diverted for irrigation and reclamation. About 500,000 acres/202,350 ha of land are being reclaimed by the HUMBOLDT project, the NEWLANDS project, and the Truckee River storage project. The alkali sinks and great arid stretches clothed with sagebrush, tule (bulrushes), and creosote bush typify Nevada's valley landscapes. There are more than 200 ranges in Nevada. Its mountain chains generally run N and S, further segmenting the state. In its angled corner on its W boundary with California, are part of the SIERRA NEVADA range, practically the only exception to the state's dry basin and range landscape. The driest state in the nation, the days and nights are generally clear, and the temperature varies with the season as well as the elevation. In the N and W the winters are extremely cold, while in parts of the S the summers approach ovenlike heat.

Economy

Many of the high plateau areas are utilized for grazing; cattle and sheep raising are important industries in the state. Because of the prevailing dryness and the steep slopes, agriculture is not highly developed, but is devoted mainly to growing forage crops, such as alfalfa and hay; however, potatoes, vegetables, cantaloupe, and barley are grown in irrigated valleys in the W and far S; dairying and poultry are also important. Much of the state's foodstuffs are imported. The population has been sparse since the Paiute and other Native American tribes eked out a living from the land and the animals. The fortune of Nevada has been not in its land but in the almost incredible wealth below the surface of the land. Mining drew people to Nevada, swelling some mining districts to 20,000 and more. Nevada is the leading producer of gold, silver, diatomite, and mercury in the U.S. Copper mining, once a major industry, is now virtually nonexistent. Sand and gravel are also mined. Petroleum was discovered in 1954, and commercial exploitation began in the 1970s. There is also some manufacturing (gaming machines and products, aerospace equipment, military supplies, lawn and garden irrigation devices, and seismic monitoring equipment).

Economy: Tourism and Gambling

Nevada's economy is now overwhelmingly based on tourism, especially the gambling (legalized in 1931), resort entertainment, and convention industries centered in Las Vegas and, to a lesser extent, RENO and Lake TAHOE. Gambling taxes are a primary source of state revenue; gambling and associated services account for half the state's employment. In addition to major gambling destinations like Las Vegas and Reno, several border towns have developed during 1970s and 1980s, some from nothing, to provide services to travelers entering and leaving the state: LAUGHLIN, MESQUITE, WEST WENDOVER, MCDERMITT, JACKPOT, and STATELINE. Liberal divorce laws made Reno "the divorce capital of the world" for many years, until other states liberalized their laws. The state has become a distribution center for the W U.S., including California. Besides Reno and Las Vegas, there are many points of interest. Hoover Dam impounds Lake MEAD, one of the largest artificial lakes in the world. LAKE MEAD NATIONAL RECREATION AREA has facilities for fishing, swimming, and boating. Other attractions include Lake TAHOE, on the Nevada-California state line, part of DEATH VALLEY NATIONAL MONUMENT (mostly in California), GREAT BASIN NATIONAL PARK (includes the former LEHMAN CAVES NATIONAL MONUMENT), and restored mining ghost towns like VIRGINIA CITY. The state has widely scattered units of two national forests; Toiyabe, in the center, W and S, and Humboldt in the N and E; part of Inyo National Forest is on the SW boundary. Indian reservations: Pyramid Lake and Walker River in the W, Summit Lake in the NW, part of Duck Valley on the N boundary (W Idaho), South Fork in the NE, part of Goshute on the E boundary (Utah), Moap River Indian Reservation in the SE.

History: to 1850

In the 1770s several Spanish explorers came near the area of present-day Nevada but it wasn't until half a century later that fur traders venturing beyond the ROCKY MOUNTAINS publicized the region. Jedediah S. Smith came across S Nevada on his way to California in 1827. The following year Peter Skene Ogden, a Hudson's Bay Company man trading out of the Oregon country, entered NE Nevada. Joseph Walker in 1833–1834 went along the Humboldt and crossed the Sierra Nevada to California. With Kit Carson, John C. Frémont had explored much of the state between 1843 and 1845, and his reports gave the Federal government its first comprehensive information on the area, which the U.S. acquired from MEXICO in the Mexican-American War. Later many wagon trains crossed Nevada on the way to California, especially during and after the gold rush of 1849. Travelers going to California over the Old Spanish Trail also crossed S Nevada, and Las Vegas became a station on the route. Reports from these sources possibly aided Brigham Young when he was shepherding the Mormons W to build a new home in Utah.

History: 1850 to 1869

When in 1850 the Federal government set up the Utah Territory, almost all of Nevada was included except the S tip, which was then part of NEW MEXICO. Non-Mormons had been averse to settling in Mormon-dominated territory, but after gold was found in 1859 non-Mormons did come into the area. A rush from California began and multiplied manyfold as news of the COMSTOCK LODE silver strike spread. Most of the newcomers preferred to consider themselves as still being within California, and a political question was added to the general upheaval. Meanwhile, miners came helter-skelter, raising camps that grew overnight into such booming and raucous places as VIRGINIA CITY. Partly to impose order on the lawless, wide-open mining towns, Congress made Nevada into a territory in 1861 as migrant prospectors and settlers poured in. The territory was then enlarged by increasing its E boundary by one degree of longitude in 1862. It was rushed into statehood in 1864, with Carson City as its capital. President Lincoln (in order to get more votes to pass the 13th Amendment) had signed the proclamation even though the territory did not actually meet the population requirement for statehood. In 1866, Nevada acquired its present-day boundaries when the S tip was added and more E land was gained from Utah.

History: 1869 to Present

Communications with the East, which had been briefly maintained by the pony express, were firmly established by the completion of the transcontinental railroad in 1869. The state continued to be dependent on its precious ores, and its fate was affected by new strikes such as the "big bonanza" (1873), which enriched the silver kings, J. W. Mackay and J. G. Fair, and the discovery (1900) of silver deposits at TONOPAH, of copper at ELY, and of gold at GOLDFIELD (1902). In the 20th century the Federal government has played an active role in Nevada, and in 1990 owned over 85% of the state's land. The Newlands Irrigation Project (1907) was the nation's first irrigation project built by the Federal government. The HOOVER DAM was completed in 1936. The U.S. Atomic Energy Commission (now the Nuclear Regulatory Commission) began nuclear tests in Nevada at FRENCHMAN FLAT and YUCCA Flat in the 1950s. In 1987, the U.S. Department of Energy named Yucca Mountain as a prospective site for the storage of high-level nuclear waste. The state bitterly opposed the decision and has continued to fight it. Nevada's population was the fastest growing in the nation during the 1980s and increased 650% from 1950 to 1990. A large influx of retired citizens has swelled populations of existing cities and has led to the creation of new communities, such as PAHRUMP and GARDNERVILLE RANCHOS, and increased demand for water to the desert environment. The state's leading institution of higher education is the University of Nevada, at Reno and at Las Vegas.

Government

Nevada's constitution was adopted in 1864. The legislature is composed of twenty senators elected for four-year terms and forty-two assemblymen elected for two-year terms. The governor is elected for a four-year term. The present governor is Kenny Gunn. The state elects two U.S. senators and two representatives; it has four electoral votes.

Nevada has sixteen counties: CHURCHILL, CLARK, DOUGLAS, ELKO, ESMERALDA, EUREKA, HUMBOLDT, LANDER, LINCOLN, LYON, MINERAL, NYE, PERSHING, STOREY, WASHOE, and WHITE PINE; and one independent city, CARSON CITY, (ORMSBY county until 1969).

Nevada (nuh-VAI-duh), county (2006 population 9,471), SW ARKANSAS; ⊙ PRESCOTT; 33°40'N 93°17'W. Bounded N by LITTLE MISSOURI RIVER and drained by Bayou DORCHEAT and Terre Rouge and Caney creeks; White Oak Lake State Park and Poison Springs Wildlife Management Area on E boundary. Agriculture (chicken); lumber milling, oil and gas. Formed 1871.

Nevada, county (□ 958 sq mi/2,490.8 sq km; 2006 population 98,764), E CALIFORNIA; ⊙ NEVADA CITY; 39°18'N 120°46'W. Narrow strip extending E across foothills to crest of SIERRA NEVADA, here crossed by DONNER PASS, bounded by NEVADA state on E. Highest point is LOLA MOUNT (9,160 ft/2,792 m). Rugged, wooded mountain country, with many beautiful lakes (e.g., DONNER LAKE). Popular recreational region (winter sports in Donner Pass area;

camping, hiking, hunting, fishing). Most of E portion in Tahoe National Forest. BEAR RIVER forms part of S boundary in SW; Middle Yuba and YUBA rivers form NW boundary. At N end of MOTHER LODE, it has been a leading California county in gold production (peak years were 1850s–1950s; also mining of silver, quarrying of sand and gravel). Agriculture (apples, nursery stock; cattle; timber). Drained by South Yuba River. Middle and South forks converge at ENGLEBRIGHT RESERVOIR in NW corner to form Yuba River. Part of Beale Air Force Base in SW; Malakoff Diggins State Historic Park in NE; Empire Mine State Historic Park in W center. GRASS VALLEY is largest city. Formed 1851.

Nevada (nuh-VAI-duh), city (2000 population 6,658), ⊙ STORY county, central IOWA, 30 mi/48 km NNE of DES MOINES; 42°01′N 93°27′W. Railroad junction; livestock shipping. Manufacturing (food processing, agricultural equipment, printing, consumer goods). Airfield here. Founded c.1853, incorporated 1869.

Nevada, city (2000 population 8,607), ⊙ VERNON county, W MISSOURI, near MARMATON RIVER, 52 mi/84 km N of JOPLIN; 37°50′N 94°20′W. Railroad junction; ships grain, livestock. Manufacturing (furniture, china plumbing fixtures, apparel, leather products); asphalt and coal mines, oil wells. Cottey College. State Southwest Missouri Mental Health Center. Evacuated and burned by Union forces in 1863, during Civil War. Osage (Indian) Village State Historic Site and Camp Clark nearby. Founded 1855.

Nevada (nuh-VA-duh), village (□ 1 sq mi/2.6 sq km; 2006 population 775), WYANDOT county, N central OHIO, 8 mi/13 km E of UPPER SANDUSKY; 40°49′N 83°08′W. Manufacturing (food products, building materials).

Nevada (NE-vah-duh), village (2006 population 680), COLLIN county, N TEXAS, 30 mi/48 km ENE of DALLAS; 33°02′N 96°22′W. Agricultural area just beyond urban fringe of Dallas–FORT WORTH area (cotton, sorghum; cattle, horses).

Nevada City, city (□ 2 sq mi/5.2 sq km; 2000 population 3,001), ⊙ NEVADA county, E central CALIFORNIA, 55 mi/89 km NE of SACRAMENTO, 5 mi/8 km NE of GRASS VALLEY, on Deer Creek, and on W slope of the SIERRA NEVADA; 39°16′N 121°01′W. Elevation c.2,500 ft/762 m. Manufacturing (broadcasting equipment, winery). Agriculture (apples, grapes, nursery stock; cattle; timber). Center of rich placer-hydraulic-mining area after 1850; lode gold is still produced. Headquarters of Tahoe National Forest (actual forest to E). Recreational region. Empire Mine State Historic Park to SW; Malakoff Diggins State Historic Park to NE. Laid out 1849, incorporated 1851.

Nevada Falls, California: see YOSEMITE NATIONAL PARK.

Nevada, Sierra: see SIERRA NEVADA.

Nevadaville, village, GILPIN county, N central COLORADO, in FRONT RANGE, 35 mi/56 km W of DENVER. Elevation c.9,150 ft/2,789 m. Former gold-mining camp, now ghost town. Nearby is Glory Hole, huge mining pit still worked for ore.

Nevado, Cerro (ne-VAH-do, SER-ro), pre-Andean peak (12,500 ft/3,810 m), S central MENDOZA province, ARGENTINA, E of Lake LLANCANELO, 70 mi/113 km S of SAN RAFAEL; 35°35′S 68°30′W.

Nevado, Cordón (ne-VAH-do kor-DON), Andean mountain range on ARGENTINA-CHILE border, N of 42°S; extends c.35 mi/56 km along the border; rises to c.7,500 ft/2,286 m.

Nevado de Toluca National Park (ne-VAH-do dai to-LOO-kah) (□ 199 sq mi/517.4 sq km), in SW MEXICO state, MEXICO, 14 mi/22 km SW of the city of TOLUCA DE LERDO. This park is named for the volcano Nevado de Toluca (or ZINANTÉCATL; 14,954 ft/4,558 m). This volcano forms part of the TRANSVERSE VOLCANIC AXIS SW of the city of Toluca. The park is served by a paved road branching S from Mexico Highway 130.

Neva River (nee-VAH), 46 mi/74 km long, LENINGRAD oblast, NW European Russia; issues from Lake LADOGA at PETROKREPOST'; flows W to the Gulf of FINLAND, forming a delta mouth at SAINT PETERSBURG. Width varies between 1,600 ft/488 m and 4,000 ft/1,219 m; canalized channel is navigable April through November. Receives Mga, TOSNA, and IZHORA rivers (left). Connected with the VOLGA RIVER by means of Mariinsk, Tikhvin, and Vyshnevolotsk canal systems, and with the WHITE SEA by means of WHITE SEA-BALTIC CANAL.

Neva River, ITALY: see ARROSCIA RIVER.

Nevasa (NAI-vah-sah), town, tahsil headquarters, AHMADNAGAR district, MAHARASHTRA state, W central INDIA, on PRAVARA RIVER, 33 mi/53 km NNE of AHMADNAGAR; 19°32′N 74°56′E. Sugarcane, millet; manufacturing of gur (jaggery). Sometimes spelled Newasa.

Nevatim (ni-vah-TEEM), cooperative settlement, S ISRAEL, 5 mi/8 km ESE of BEERSHEBA in N NEGEV; 31°13′N 34°54′E. Elevation 1,131 ft/344 m. Agriculture (flowers, fruit; poultry). Population mainly comprised of Jews from COCHIN, India. Parts of the synagogue, dismantled in Cochin and reassembled in Nevatim, date back to the 16th century. Air base. Founded 1949.

Nevdi, RUSSIA: see NEVON.

Nevdubstroy, RUSSIA: see KIROVSK, LENINGRAD oblast.

Neve Dekalim or **Neve Deqalim**, Jewish settlement, S of GAZA, in the GAZA STRIP (protected by Israeli army). Founded in 1983.

Neve Efraim (ni-VE ef-REI-yeem), **Neve Monosson** or **Neve Efraim Monosson**, town, ISRAEL, 5.5 mi/8.9 km E of TEL AVIV, and 2.5 mi/4 km NW of Ben-Gurion Airport; 32°01′N 34°52′E. Elevation 104 ft/31 m. Founded in 1953 for a group of El Al (Israel's national airline) employees and other airport workers by American philanthropist Efraim (Fred) Monosson, and grew since into a larger community, most of whose residents commute to Tel Aviv. Merged with YEHUD in 2003 to create municipality.

Neve Eitan (ni-VE e-TAHN) or **Neve Etan**, kibbutz, NE ISRAEL, in Jordan valley, 2 mi/3.2 km ESE of BEIT SHEAN; 32°29′N 35°32′E. Below sea level 728 ft/221 m. Manufacturing (construction equipment). Agriculture (mixed farming), fish breeding. Founded 1938.

Neve Ilan (ni-VE ee-LAHN), town, JUDEAN HIGHLANDS, ISRAEL, 9 mi/14.5 km W of JERUSALEM; 31°48′N 35°05′E. Elevation 1,679 ft/511 m. Founded 1946 as a kibbutz, but disbanded in 1954 to be replaced by an educational institution. Reestablished in 1971 as an industrial cooperative settlement; electronics, and some agriculture, studio and film-production center.

Neveklov (NE-vek-LOF), German *Neweklau*, village, STREDOCESKY province, S central BOHEMIA, CZECH REPUBLIC, 7 mi/11.3 km SW of BENESOV; 49°45′N 14°32′E. Agriculture (rye, potatoes). Has a Gothic church, and a 17th century synagogue.

Nevel' (NYE-veel), city (2006 population 17,600), S PSKOV oblast, W European Russia, 150 mi/241 km SE of PSKOV, and 32 mi/51 km SW of VELIKIYE LUKI; 56°02′N 29°55′E. Elevation 515 ft/156 m. Railroad and road junction. Manufacturing (communications equipment, metal works, footwear). Known since the 16th century. Passed in 1772 to Russia; chartered in 1777.

Nevele (NAI-vuh-luh), agricultural commune (2006 population 11,258), Ghent district, EAST FLANDERS province, NW BELGIUM, 8 mi/12.9 km W of GHENT; 51°02′N 03°33′E.

Nevel'sk (NYE-veelsk), city (2006 population 17,315), S SAKHALIN oblast, RUSSIAN FAR EAST, port on the TATAR STRAIT, Sea of JAPAN, on W coast railroad and highway, 37 mi/60 km WSW of YUZHNO-SAKHALINSK; 46°40′N 141°52′E. Elevation 347 ft/105 m. Fish canneries, cold-storage plant, shipyards. Base for a trawler fleet. Under Japanese rule (1905–1945), called Honto.

Nevel'skoy, Mount (nee-vyel-SKO-yee), highest point (6,604 ft/2,013 m) of SAKHALIN Island, SAKHALIN oblast, RUSSIAN FAR EAST, in the E range, 55 mi/89 km SE of ALEKSANDROVSK-SAKHALINSKIY. Also Mount Nevel'skoy.

Never (nye-VER), village, NW AMUR oblast, RUSSIAN FAR EAST, 10 mi/16 km E of SKOVORODINO on the TRANS-SIBERIAN RAILROAD (Bol'shoy Never station); 53°59′N 124°10′E. Elevation 1,568 ft/477 m. Terminus of a highway N to ALDAN gold fields and YAKUTSK; road-railroad transfer point. Truck repair and servicing. Formerly known as Larinskiy.

Neverí River (nai-vai-REE), c.55 mi/89 km long, NE VENEZUELA; rises in coastal range of SUCRE state SW of CUMANACOA; flows W, past BARCELONA (ANZOÁTEGUI state), to the CARIBBEAN 3 mi/5 km NW of Barcelona, up to which it is navigable for small craft; 10°10′N 64°42′W.

Neverkino (nee-VER-kee-nuh), village (2005 population 5,100), SE PENZA oblast, E European Russia, near highway, 25 mi/40 km SSE of KUZNETSK; 52°47′N 46°44′E. Elevation 820 ft/249 m. In agricultural area; food processing (dairy, bakery); fertilizer; agricultural machinery.

Nevers (ne-ve), city (□ 7 sq mi/18.2 sq km); ⊙ NIÈVRE department, in BURGUNDY, central FRANCE, on right bank of the LOIRE at junction with the NIÈVRE, 36 mi/58 km ESE of BOURGES, and 85 mi/137 km SE of ORLÉANS; 46°55′N 03°20′E. Commercial center of the historic NIVERNAIS region, it is noted for its pottery and chinaware indusrty. Other manufacturing (metal products, mechanical and electrical equipment, chemicals, textiles, printing). Nevers became the seat of a bishopric in 6th century and was long the capital of duchy and province of Nivernais. Of cultural interest are the ducal palace (15th–16th century), now a courthouse; the Church of Saint-Étienne (11th century), a gem of Romanesque architecture; the cathedral (13th–16th century), a vast basilica mixing architectural styles; and the Church of St. Bernadette-du-Banlay (1966), of current architectural interest. In the Convent of St. Gildard are the remains of St. Bernadette, who lived here from 1866 to 1879. The municipal museum contains a fine collection of faïence chinaware, some dating from early 17th century.

Neversink River, c.65 mi/105 km long, SE NEW YORK; rises in the CATSKILL MOUNTAINS W of ASHOKAN RESERVOIR; flows SW and S, paralleling W base of the SHAWANGUNK range in lower course, to the DELAWARE at PORT JERVIS. In upper course (called Neversink Creek) is earth-fill Neversink Dam (2,800 ft/853 m long, 200 ft/61 m high; begun 1941), impounding Neversink Reservoir. A 5-mi/8-km water tunnel extends to Rondout Reservoir, which connects with DELAWARE AQUEDUCT.

Never Summer Mountains, range of ROCKY MOUNTAINS in N COLORADO; extend c.20 mi/32 km N-S along CONTINENTAL DIVIDE; forms part of W boundary of ROCKY MOUNTAIN NATIONAL PARK, part of JACKSON county line, and Continental Divide. Prominent peaks in park and range include MOUNT CUMULUS (12,724 ft/3,878 m), MOUNT NIMBUS (12,730 ft/3,880 m), MT. CIRRUS (12,797 ft/3,901 m), Howard Mt. (12,810 ft/3,904 m), and MOUNT RICHTHOFEN (12,940 ft/3,944 m). Range sometimes included within MEDICINE BOW MOUNTAINS of WYOMING and Colorado. Includes the dramatic rock points of Nokhu Crags in N (S part of Colorado State Forest).

Nevertire (NE-vuhr-teir), township, NEW SOUTH WALES, SE AUSTRALIA, 327 mi/526 km NW of SYDNEY, 12 mi/19 km SW of WARREN; 31°52′S 147°47′E. Railway junction. Grain storage facilities.

Neves (NAIV-esh), town (2001 population 6,635), NW coast of SÃO TOMÉ island, SÃO TOMÉ AND PRÍNCIPE; 00°21′N 06°33′E. Industrial fishing; cold-storage and petroleum-storage port facilities.

Area is shown by the symbol □, and capital city or county seat by ⊙.

Neves (NE-ves), N suburb of NITERÓI, RIO DE JANEIRO state, BRAZIL, on GUANABARA BAY, opposite RIO DE JANEIRO; 22°51′S 43°06′W. Industrial center (metalworks, shipbuilding yards).

Neves, Brazil: see NEVES PAULISTA.

Neve Shaanan (ni-VE shah-ah-NAHN), SE residential suburb of HAIFA, NW ISRAEL, on NE slope of MOUNT CARMEL; 32°47′N 35°00′E. Elevation 521 ft/158 m. Israel's Technical University nearby.

Nevesinje (ne-ve-SEEN-yah), town, central HERZEGOVINA, BOSNIA AND HERZEGOVINA, 16 mi/26 km ESE of MOSTAR, on crossroad; 43°15′N 18°06′E. Local trade center. Numerous mineral springs nearby. First mentioned in 12th century. Site of anti-Turkish uprising in 1875. Also spelled Nevesinye.

Nevesinjsko Polje (ne-ve-SEENS-ko- POL-ye) or **Nevesinje Plain**, plain and historical region, upper HERZEGOVINA, BOSNIA AND HERZEGOVINA, in karst. Principal town is NEVESINJE.

Neves Paulista (NE-ves POU-lee-stah), town (2007 population 8,843), NW SÃO PAULO state, BRAZIL, 15 mi/24 km W of SÃO JOSÉ DO RIO PRÊTO; 20°49′S 49°38′W. Coffee, beans, corn, rice; cotton. Until 1944, Neves, and, 1944–1948, Iboti.

Nevestino (ne-VES-tee-no), village, SOFIA oblast, NEVESTINO obshtina (1993 population 5,599), BULGARIA; 42°15′N 22°51′E.

Nevėžis River (ne-VAI-zeehs), 132 mi/212 km long, in central LITHUANIA; rises W of ANYKSCIAI; flows NW, past PANEVEZYS, and S, past KEDAINIAI, to Nemunas River just W of KAUNAS. Navigable for 18 mi/29 km above mouth.

Neville, township, ALLEGHENY county, W PENNSYLVANIA; suburb 7 mi/11.3 km NW of PITTSBURGH; 40°30′N 80°08′W. Township is constituted entirely of Neville Island in OHIO RIVER. Manufacturing (fabricated metal products, polyester resins, chemicals, portland cement). Emsworth Lock and Dam here.

Neville (NE-vil), village (2006 population 138), CLERMONT county, SW OHIO, 18 mi/29 km S of BATAVIA, and on OHIO RIVER (here forming KENTUCKY state line); 38°48′N 84°13′W.

Nevinnomyssk (nye-veen-nuh-MISK), city (2006 population 135,125), W STAVROPOL TERRITORY, N CAUCASUS, S European Russia, on the KUBAN River, at the mouth of the Great ZELENCHUK River, at the head of the NEVINNOMYSSK Canal, 38 mi/61 km S of STAVROPOL; 44°37′N 41°56′E. Elevation 1,108 ft/337 m. Railroad center (branch line S to CHERKESSK) on the Rostov-Baku railroad; freight yards, workshops; fuel depot. Manufacturing (chemicals, electrical products); wool processing; flour mill. Made city in 1939.

Nevinnomyssk Canal (nye-veen-nuh-MISK), irrigation waterway, 32 mi/51 km long, in W STAVROPOL TERRITORY, S European Russia; extends from the KUBAN' River, at NEVINNOMYSSK, NW through Nedremannaya Mountain (3-mi/5-km tunnel), to the YEGORLYK RIVER, 15 mi/24 km WSW of STAVROPOL; 44°53′N 41°42′E. Canal supplies spring and summer floodwater from the Kuban' River to parched Yegorlyk agricultural basin. Svistukha hydroelectric station is on the canal, SW of Stavropol. Construction (begun 1935) was interrupted by World War II and completed 1948.

Nevis (NEE-vis), village (2000 population 364), HUBBARD county, central MINNESOTA, 11 mi/18 km ENE of PARK RAPIDS, at E end of ELBOW LAKE; 46°57′N 94°50′W. Agriculture (oats, barley, rye, beans, alfalfa; sheep). Manufacturing (wood products). Eighth Crow Wing Lake to E.

Nevis (NEE-vis), island (□ 36 sq mi/93.6 sq km), Federation of ST. KITTS AND NEVIS, WEST INDIES; 17°09′N 62°35′W. The smaller of two islands making up the federation. Located 2 mi/3.2 km SE of St. Kitts; linked by ferry boat service at CHARLESTOWN, the island's commercial center. Dominated by volcanic, cone-shaped Nevis Peak (3,232 ft/985 m), surrounded by

rainforests and woodlands. Columbus used the Spanish word for snow, "nieves," to name this island with cloud-shrouded peak. Alexander Hamilton b. here. Extensive damage in 1989 Hurricane Hugo. Former sugarcane estates have been converted to hotels; Canadian-based Four Seasons complex (opened in 1991) is first full-service resort. Sea island cotton is chief crop; coconuts, green lemons, and spices also grown.

Nevis, Loch (NE-vis), sea inlet, HIGHLAND, N Scotland, extending 15 mi/24 km E from the SOUND OF SLEAT at MALLAIG; 57°01′N 05°43′W.

Nevon (NYE-vuhn), settlement, N central IRKUTSK oblast, E central SIBERIA, RUSSIA, on the W bank of the ANGARA RIVER, on road, 5 mi/8 km NE of UST'-ILIMSK, to which it is administratively subordinate; 58°03′N 102°43′E. Elevation 679 ft/206 m. Logging, lumbering. Formerly known as Nevdi (NYEF-dee).

Nevrokop, Bulgaria: see GOTSE DELCHEV.

Nevsehir, town (2000 population 67,864), ⊙ Nevsehir province, central TURKEY, 45 mi/72 km N of NIGDE; 38°38′N 34°42′E. Gateway to CAPPADOCIA. Agriculture (rye, vetch, legumes, potatoes, onions). Excavated ruins nearby. Among remaining historical buildings are the Seljuk-period castle, on the city's highest point, and Kursunlu Mosque, built for the Grand Vizier Damat Ibrahim Pasha. Nevsehir Museum displays local artifacts. Town name means "new city." Formerly sometimes Nev-Shehr.

Nev-Shehr, TURKEY: see NEVSEHIR.

Nevskaya Dubrovka, RUSSIA: see DUBROVKA, LENINGRAD oblast.

Nev'yansk (neev-YAHNSK), city (2006 population 25,560), W SVERDLOVSK oblast, extreme W Siberian Russia, in the E foothills of the central URALS, on the NEYVA RIVER, on road junction and railroad, 61 mi/98 km NNW of YEKATERINBURG; 57°29′N 60°12′E. Elevation 787 ft/239 m. Manufacturing (machinery, transportation equipment, cement, construction materials); metallurgy. Mining of gold and kaolin. Founded in 1689 with the construction of a gun-manufacturing plant; developed as a metallurgical center; became city in 1917.

New Aberdeen, (A-bur-deen) coal-mining suburb of GLACE BAY, NE NOVA SCOTIA, CANADA, on CAPE BRETON ISLAND.

New Addington (NYOO AD-ing-tuhn), district (2001 population 21,527), GREATER LONDON, SE ENGLAND, 12 mi/19 km SSE of LONDON, 3 mi/4.8 km NNW of BIGGIN HILL; 51°21′N 00°01′W. Commuter belt.

Newagen, Cape (noo-WAI-guhn), LINCOLN county, SW MAINE. Has summer colony at S tip of SOUTHPORT town.

Newala (nai-WAH-lah), village, MTWARA region, SE TANZANIA, 70 mi/113 km SW of MTWARA, near MOZAMBIQUE border, on MAKONDE PLATEAU; 10°56′S 39°17′E. Airstrip. Cashews, corn, sweet potatoes; manioc; sheep, goats.

New Albany, city (2000 population 37,603), ⊙ FLOYD county, S INDIANA, near the falls of the OHIO River opposite LOUISVILLE, KENTUCKY, and at the foot of the Knobstone Escarpment, which lies to the W; 38°18′N 85°50′W. The city was a shipbuilding center in the 19th century, and the riverboats *Robert E. Lee* and *Eclipse* were built there. Manufacturing (wood products, apparel, electronic equipment, chemicals, food products, paper products, transportation equipment, sheet metal, furniture). Bridges link New Albany with Louisville. William Vaughn Moody lived in the city. Seat of Indiana University Southeast. Laid out 1813, incorporated 1819.

New Albany (AWL-buh-nee), town (2000 population 7,607), ⊙ UNION county, N MISSISSIPPI, 23 mi/37 km NW of TUPELO, on TALLAHATCHIE RIVER; 34°29′N 89°01′W. Railroad junction. Trade center for agricultural area (cotton; corn, soybeans; cattle; dairying) and manufacturing (machinery, marble tops, asphalt

and foam products, apparel, printing and publishing, furniture, concrete). Holly Springs National Forest to W. Settled c.1840.

New Albany, village (2000 population 73), WILSON county, SE KANSAS, on FALL RIVER, and 7 mi/11.3 km WNW of FREDONIA; 37°34′N 95°55′W. Livestock raising, farming.

New Albany (AWL-buh-nee), village (□ 1 sq mi/2.6 sq km; 2006 population 6,345), FRANKLIN county, central OHIO, 13 mi/21 km NE of COLUMBUS; 39°13′N 82°12′W. In agricultural area.

New Albany (NOO AHL-buh-nee), borough (2006 population 293), BRADFORD county, NE PENNSYLVANIA, 12 mi/19 km S of TOWANDA, on South Branch of Towanda Creek; 41°36′N 76°26′W. Manufacturing (feeds and supplements, machining). Agriculture (apples, corn, hay; dairying). Hatch Mountain to N.

New Albin, town (2000 population 527), ALLAMAKEE county, extreme NE IOWA, on MINNESOTA state line, 18 mi/29 km NNE of WAUKON, near MISSISSIPPI and UPPER IOWA rivers; 43°30′N 91°17′W. Lumber. Limestone quarries and Indian Fish Farm mounds to S.

New Ålesund, NORWAY: see NY-ÅLESUND.

New Alexandria (noo a-leg-ZAN-dree-uh), unincorporated town, FAIRFAX county, NE VIRGINIA, residential suburb, 2 mi/3 km S of ALEXANDRIA, and 9 mi/15 km S of WASHINGTON, D.C., on POTOMAC RIVER; 38°46′N 77°03′W. Woodrow Wilson Memorial Bridge to NE.

New Alexandria (A-leks-AN-dree-ah), village (2006 population 216), JEFFERSON county, E OHIO, 6 mi/10 km SW of STEUBENVILLE; 40°17′N 80°40′W. In coal-mining area.

New Alexandria (NOO A-leks-AN-dree-ah), borough (2006 population 581), WESTMORELAND county, SW PENNSYLVANIA, 9 mi/14.5 km NE of GREENSBURG, on LOYALHANNA CREEK; 40°23′N 79°25′W. Manufacturing (concrete, pool tables); coal, timber. Agriculture (corn, hay; livestock, dairying). Keystone State Park to SE.

Newalla (noo-AHL-uh), village, OKLAHOMA county, central OKLAHOMA, suburb 21 mi/34 km ESE of OKLAHOMA CITY. In oil-producing area; manufacturing (plumbing equipment). Part of Oklahoma City.

New Almaden, village, SANTA CLARA county, W CALIFORNIA, 9 mi/14.5 km SSE of downtown SAN JOSE, on Guadalupe Creek. Quicksilver mines (since 1845).

New Alresford (NYOO AWL-riz-fuhd) or **Alresford,** town (2001 population 5,102), central HAMPSHIRE, S ENGLAND, on ITCHEN RIVER, and 7 mi/11.3 km ENE of WINCHESTER; 51°05′N 01°10′W. Mary Mitford born here. Just N is Old Alresford. Evidence of Neolithic, Bronze and Iron Age occupation here.

New Amsterdam, town (2002 population 15,997), ⊙ EAST BERBICE–CORENTYNE district, NE GUYANA, port at mouth of BERBICE RIVER (here joined by CANJE RIVER), on the ATLANTIC OCEAN, 55 mi/89 km SE of GEORGETOWN and across from ROSIGNOL; 06°16′N 57°32′W. Trading center for fertile coastal region (sugar, rice; cattle). Rosignol, accessible by ferry, has terminus of railroad from Georgetown. Town of Dutch character, intersected by canals. Built 1740 by the Dutch as Fort Saint Andries. In 1790, seat of government was moved here from Fort Nassau (55 mi/89 km SSW on the Berbice) in 1803. Became British in 1803. Has Anglican church.

New Amsterdam, town (2000 population 1), HARRISON county, S INDIANA, on OHIO River, near mouth of INDIAN CREEK, and 11 mi/18 km SW of CORYDON; 38°06′N 86°17′W. In agricultural area. Laid out 1815.

New Amsterdam, Dutch settlement at the mouth of the HUDSON RIVER and on the S end of MANHATTAN island; est. 1624. It was capital of the colony of NEW NETHERLAND from 1626 to 1664, when it was captured by the British and renamed NEW YORK city.

New Amsterdam, FRANCE: see AMSTERDAM ISLAND.

Newark (NOO-wuhrk), city (2000 population 42,471), ALAMEDA county, W CALIFORNIA, suburb 23 mi/37 km

SSE of OAKLAND, 15 mi/24 km NNW of SAN JOSE; 37°31'N 122°02'W. Manufacturing (plastics products, furniture, feeds, semiconductors, chemicals, machine parts, paper products, food processing, gypsum products, computers). Dumbarton Bridge crosses SAN FRANCISCO BAY to PALO ALTO (W); HETCH HETCHY AQUEDUCT passes through city in S and crosses San Francisco Bay S of Dunbarton Bridge; Sky Sailing Airport to SE; Coyote Hills Regional Park to NW; San Francisco Bay National Wildlife Refuge visitors center to W.

Newark (NOO-work), city (2000 population 28,547), NEW CASTLE county, NW DELAWARE, 11 mi/18 km WSW of WILMINGTON, on White Clay Creek; 39°41'N 75°45'W. Elevation 98 ft/29 m. Near MARYLAND state line; PENNSYLVANIA state line to N. Railroad junction. The third-largest city in the state, it is the seat of the University of Delaware. Manufacturing (metal products, printing and publishing, electrical equipment, machinery, consumer goods, food products, transportation equipment, plastics products, construction materials, textiles). The only Revolutionary battle on Delaware soil was fought (September 1777) at nearby COOCH'S BRIDGE. White Clay Creek State Park to N; Lums Pond State Park to S; FAIR HILL Natural Resource Management Area and Covered Bridge to NW, in Maryland. Settled before 1700, incorporated 1852.

Newark, city (2006 population 281,402), ⊙ ESSEX county, NE NEW JERSEY, on the PASSAIC River and NEWARK BAY, 8 mi/13 km W of NEW YORK city; 40°44'N 74°10'W. It is a port of entry and the largest city in the state. Newark is a transportation, industrial, commercial, and manufacturing center. A major center for transshipment of railroad, truck, and water freight; deepwater terminal (Port Newark), developed in World War I, and NEWARK LIBERTY INTERNATIONAL AIRPORT (opened 1928) are units of Port of New York and Jersey City Authority. Its leather industry dated from the 17th century, and it was once the jewelry manufacturing capital of the UNITED STATES. Insurance businesses began in the early 19th century and remain significant, as well as banking. Light manufacturing. Newark Liberty International Airport is one of the nation's busiest, and the important seaport is operated by the Port Authority of New York and New Jersey. The city was settled (1666) by Puritans from CONNECTICUT under the leadership of Robert Treat. It was the scene of Revolutionary skirmishes. Industrial growth began after the American Revolution, aided by the development of transportation facilities. The MORRIS CANAL was opened in 1832, and the railroads arrived in 1834 and 1835. A flourishing shipping business resulted, and Newark became the industrial center of the area. In the late 19th century its industry was further developed, especially through the efforts of such men as Seth Boyden and J. W. Hyatt. Newark Port opened in 1915, and the city's shipbuilding played an important role in World War I. During the latter half of the 20th century, Newark's economy and living standards had greatly declined. Many central city residents fled to the outlying suburbs, which had been marked by a boom in corporate development, shopping center growth, and housing construction. Poverty and unemployment plagued Newark. In July 1967, the city was the scene of a major race riot, resulting in 26 deaths and more than 1,300 injuries before the National Guard restored order. By the 1980s nearly 50% of the city's residents were black. Now, as an Urban Enterprise Zone, Newark is being revitalized and there is a tremendous growth in private housing. Landmarks include Trinity Cathedral (1810, with the spire of a church built in 1743); the Sacred Heart Cathedral (begun 1898, completed 1953); the First Presbyterian Church (1791); the Newark Public Library (founded 1888); the Newark Museum (1909); and the county courthouse (1906), with Gutzon Borglum's statue of

Lincoln in front. Other points of interest include Borglum's large group *Wars of America* (1926) in Military Park (a Revolutionary War drilling ground and a Civil War tenting area) and many historic homes. Celluloid (1872) and photographic film (1887) first made here. Newark's educational institutions include Rutgers University in Newark, the New Jersey Institute of Technology, Essex County College, and a preparatory academy founded in 1774. Aaron Burr and Stephen Crane born here. Settled 1666, incorporated as a city 1836, adopted commission government 1917.

Newark (NOO-wuhrk), city (□ 20 sq mi/52 sq km; 2006 population 47,242), ⊙ LICKING county, central OHIO, on the LICKING RIVER, 33 mi/53 km ENE of COLUMBUS; 40°04'N 82°25'W. In a livestock area. Trade and processing center; transportation hub. Manufacturing includes glass, aluminum products, automobile parts, and plastics. The city's Native American mounds attract many visitors. The Newark Earthworks State Memorial includes three locations within the city's limits: the Great Circle, the Octagon Mound (with smaller mounds inside the octagon), and the Wright Earthworks. Museum of Native American Art. Newark campus of the Ohio State University. Incorporated 1826.

Newark (noo-AHRK), town (2000 population 1,219), INDEPENDENCE county, NE central ARKANSAS, 12 mi/19 km ESE of BATESVILLE; 35°42'N 91°26'W. In agricultural and timber area. Large electricity generator nearby.

Newark (NOO-ahrk), town (2000 population 100), KNOX county, NE MISSOURI, on SOUTH FABIUS RIVER, and 16 mi/26 km SE of EDINA; 39°59'N 91°58'W. Limestone quarry.

Newark, town, CALEDONIA CO., NE VERMONT, 20 mi/32 km N of St. Johnsbury; 44°42'N 71°55'W. Census of 1880 recorded the population high of 679 residents.

Newark, village (2000 population 887), KENDALL county, NE ILLINOIS, 10 mi/16 km SW of YORKVILLE; 41°32'N 88°34'W. In rich agricultural area.

Newark, village, WORCESTER county, SE MARYLAND 19 mi/31 km ESE of SALISBURY. In farming and timber area; canneries. The railroad station here is given the name Queponco to avoid confusion with other Newarks.

Newark (NOO-wuhrk), village (□ 5 sq mi/13 sq km; 2006 population 9,284), WAYNE county, W NEW YORK, on the BARGE CANAL, 29 mi/47 km ESE of ROCHESTER; 43°02'N 77°05'W. In agricultural area. Light manufacturing. Incorporated 1839.

Newark, village, MARSHALL county, NE SOUTH DAKOTA, 10 mi/16 km N of BRITTON, on NORTH DAKOTA state line; 45°55'N 97°47'W. In agricultural area.

Newark, Canada: see NIAGARA-ON-THE-LAKE.

Newark, ENGLAND: see NEWARK-ON-TRENT.

Newark Bay (NOO-uhrk), NE NEW JERSEY, estuary at confluence of PASSAIC and Hackensack rivers, between shores of NEWARK and ELIZABETH (W), JERSEY CITY and BAYONNE (E), and STATEN ISLAND (S); linked to Upper New York Bay by KILL VAN KULL, to Lower New York Bay by ARTHUR KILL; 6 mi/9.7 km long, 1 mi/1.6 km wide; 40°39'N 74°08'W. Port Newark is deepwater terminal on W shore, connected with dredged channel in bay.

Newark Liberty International Airport, NEWARK, New Jersey, serving the NEW YORK city metropolitan region (16 mi/25 km W of downtown MANHATTAN); elevation 18.3 ft/5.6 m; 40°42'N 74°10'W. Covers 2,027 acres/820 ha. Operated by the Port Authority of New York and New Jersey (see also LAGUARDIA AIRPORT and JFK (JOHN F. KENNEDY) INTERNATIONAL AIRPORT). Opened October 1928, it has five runways, three passenger terminals (over 30 million passengers annually) and five cargo centers (including centers dedicated to United Parcel Service, FedEx, and Continental and United Airlines). After the September 11, 2001 terrorist

attacks in New York and WASHINGTON, D.C., Newark International Airport added "Liberty" to the name. Airport Code EWR.

Newark-on-Trent (NYOO-wuhk–on–TRENT), town (2001 population 25,376), NOTTINGHAMSHIRE, central ENGLAND, on a branch of the TRENT RIVER, 16 mi/26 km NE of NOTTINGHAM; 53°04'N 00°49'W. Commuter town of Nottingham. Previously manufacturing (machinery and equipment). Remains of 12th-century castle, 13th-century church; grammar school founded 1529.

Newark Valley, village (2006 population 1,032), TIOGA county, S NEW YORK, on OWEGO CREEK, 17 mi/27 km NW of BINGHAMTON; 42°13'N 76°11'W. Wood products.

Newarthill (NYOO-ahr-thuhl), town (2001 population 6,849), North Lanarkshire, central Scotland, 3 mi/4.8 km NE of MOTHERWELL; 55°48'N 03°56'W. Light industry. Formerly in Strathclyde, abolished 1996.

Newasa, INDIA: see NEVASA.

New Ashford, agricultural town, BERKSHIRE county, NW MASSACHUSETTS, 11 mi/18 km N of PITTSFIELD; 42°37'N 73°14'W. Recreation; ski resort.

Newata (nai-WAH-tah), village, MTWARA region, SE TANZANIA, 25 mi/40 km SSW of MTWARA; 10°36'S 39°59'E. Cashews, peanuts, bananas, grain; livestock.

New Athens (AI-thens), village (2000 population 1,981), SAINT CLAIR county, SW ILLINOIS, on KASKASKIA RIVER, and 17 mi/27 km SSE of BELLEVILLE; 38°19'N 89°52'W. Manufacturing (stoves, enamelware); bituminous coal mines. Agriculture (corn, wheat, fruit; dairy products, livestock). Incorporated 1869.

New Athens (A-thinz), village (2006 population 349), HARRISON county, E OHIO, 6 mi/10 km S of CADIZ; 40°11'N 80°59'W. In coal-mining area.

New Auburn, village (2000 population 488), SIBLEY county, S MINNESOTA, 8 mi/12.9 km N of GAYLORD, on W shore of High Island Lake; 44°40'N 94°13'W. Livestock; grain, soybeans; dairying.

New Auburn, village (2006 population 552), CHIPPEWA county, W central WISCONSIN, 26 mi/42 km N of EAU CLAIRE; 45°12'N 91°34'W. Dairying; livestock raising. State fishery.

New Augusta (aw-GUHS-tuh), village, ⊙ PERRY county, SE MISSISSIPPI, 17 mi/27 km ESE of HATTIESBURG, on LEAF RIVER; 31°11'N 89°01'W. In N edge of De Soto National Forest. Agriculture (cotton, corn; poultry); timber. Manufacturing (telephone poles, pulpwood processing). Camp Shelby National Guard base, to W.

Newaygo (ne-WAI-go), county (□ 861 sq mi/2,238.6 sq km; 2006 population 49,840), W MICHIGAN; ⊙ WHITE CLOUD; 43°32'N 85°47'W. Drained by MUSKEGON, PERE MARQUETTE, and WHITE rivers and short Tamarack and ROGUE rivers. Agriculture (cattle, hogs, poultry; dairying; onions, carrots, asparagus, apples, cherries, peaches, grains). Some manufacturing at NEWAYGO. Resorts. Manistee National Forest, Newaygo State Park in E, White Cloud State Park at center; Manistee National Forest covers NE ⅔ of county except extreme NE corner, HARDY DAM is here. Organized 1851.

Newaygo (ne-WAI-go), town, NEWAYGO county, W central MICHIGAN, 8 mi/12.9 km S of WHITE CLOUD, and on MUSKEGON RIVER; 43°25'N 85°47'W. In farm area (fruit, vegetables); dairy products. Manufacturing (machinery). Airport. Manistee National Forest to NE. Settled 1836, incorporated as village 1867.

New Baden (BAY-den), village (2000 population 3,001), in CLINTON and SAINT CLAIR counties, SW ILLINOIS, 15 mi/24 km E of BELLEVILLE; 38°32'N 89°42'W. In agricultural (corn, wheat, poultry; dairy products) and bituminous-coal-mining area. Incorporated 1867.

New Baltimore, city (2000 population 7,405), MACOMB and SAINT CLAIR counties, E MICHIGAN, 10 mi/16 km

NE of MOUNT CLEMENS, on ANCHOR BAY of LAKE SAINT CLAIR; 42°40′N 82°44′W. Satellite community of DETROIT. Resort. Agriculture (apples; dairy, poultry). Manufacturing (thermoplastic molding, fabricated metal products). Incorporated as village 1867, as city 1931.

New Baltimore, hamlet, GREENE county, SE NEW YORK, on W bank of the HUDSON, and 15 mi/24 km. S of ALBANY; 42°25′N 73°51′W.

New Baltimore, borough (2006 population 157), SOMERSET county, SW PENNSYLVANIA, 16 mi/26 km E of SOMERSET, on Raystown Branch of JUNIATA RIVER; 39°59′N 78°46′W. Corn, hay, oats; dairying.

Newbattle (NYOO-bah-tuhl), town, MIDLOTHIAN, E Scotland, just S of DALKEITH; 55°52′N 03°04′W. Nearby is Newbattle Abbey; seat of marquis of Lothian is now a college for adult education; includes remains of Cistercian abbey founded 1140 by David I and burned down 1544 by earl of Hertford.

New Bavaria (buh-VE-ree-uh), village (2006 population 78), HENRY county, NW OHIO, 13 mi/21 km S of NAPOLEON; 41°12′N 84°10′W.

New Beaver (NOO BEE-vuhr), borough (2006 population 1,613), LAWRENCE county, W PENNSYLVANIA, 8 mi/12.9 km SSW of NEW CASTLE; 40°52′N 80°21′W. Agriculture (dairying; corn, hay).

New Bedford, city (2000 population 93,768), ⊙ BRISTOL county, SE MASSACHUSETTS, at the mouth of the ACUSHNET RIVER, on BUZZARDS BAY; 41°40′N 70°57′W. Formerly one of the world's greatest whaling ports, it has become a leading port for the fishing and scalloping industries. During the Revolution the harbor was a haven for American privateers, prompting the British to invade and burn the town in 1778. The whaling industry boomed after the Revolution, reaching a peak in the 1850s. The first cotton-textile mill in the city dates from 1846, but the textile industry declined in the 1920s. Manufacturing (apparel, textiles, electrical and electronic equipment, rubber products, medical supplies, prepared food, metal products). The Seamen's Bethel, described by Herman Melville in *Moby Dick*; the Bourne Whaling Museum; the Old Dartmouth Historical Society; Friends' Academy (1810); and the Swain School of Design are in New Bedford. The Free Public Library holds a large collection of material on whaling. A sizable Portuguese-speaking population is in the city. Settled 1640, set off from DARTMOUTH 1787, incorporated as a city 1847.

Newberg, city (2006 population 21,576), YAMHILL county, NW OREGON, 20 mi/32 km SW of PORTLAND, on WILLAMETTE RIVER; 45°18′N 122°57′W. Agriculture (wheat, barley, oats, fruit, nuts; poultry, cattle); dairy products; wineries nearby. Manufacturing (chemicals, furniture, pulp, plastic products, machinery, electronics assembly). George Fox College Bald Peak State Park to N; Champoag State Park to SE. Founded by Quakers; named, 1869, incorporated 1893.

New Berlin (BUHR-lin), city (2006 population 39,234), WAUKESHA county, SE WISCONSIN, a suburb 10 mi/16 km WSW of downtown MILWAUKEE; 42°58′N 88°07′W. It is largely residential. Manufacturing (rubber products, computer equipment, printing, wire forms, medical equipment, electrical and electronic equipment, plastic molding, transportation equipment). Founded 1840, incorporated 1959.

New Berlin (BER-lin), village (2000 population 1,030), SANGAMON county, central ILLINOIS, 15 mi/24 km WSW of SPRINGFIELD; 39°43′N 89°54′W. Bituminous-coal mining. Agriculture (corn, wheat, soybeans).

New Berlin (BUHR-lin), village (□ 1 sq mi/2.6 sq km; 2006 population 1,116), CHENANGO county, central NEW YORK, on UNADILLA RIVER, and 36 mi/58 km S of UTICA; 42°37′N 75°20′W. In dairying and farming area. Incorporated 1819.

New Berlin (BUHR-lin), borough (2006 population 824), UNION county, central PENNSYLVANIA, 11 mi/18 km SW of LEWISBURG, on PENNS CREEK; 40°52′N 76°59′W. Manufacturing (machinery, apparel). Agriculture (grain, soybeans; poultry, livestock, dairying). Shamokin Mountain ridge to N; Bald Eagle State Forest to W.

New Bern (NOO BUHRN), city (□ 27 sq mi/70.2 sq km; 2006 population 27,650), ⊙ CRAVEN county, E NORTH CAROLINA, 40 mi/64 km SE of GREENVILLE on NEUSE RIVER, at mouth of Treat River; 35°07′N 77°04′W. Railroad junction. Service industries; manufacturing (lumber, textiles, pharmaceuticals, asphalt, fabricated metal and plastic products, food processing, transportation equipment, apparel). Settled in 1710 by Swiss and German colonists under Baron Christopher de Graffenried and John Lawson, New Bern was the second town in North Carolina and an early colonial capital; in 1774 it was the seat of the first provincial convention. Notable among the old buildings is the beautiful Tryon Palace (1767–1770), which was the colonial capitol and governor's mansion; it was badly burned in 1798 and was not reconstructed until the 1950s. In the Civil War the city was captured (March 1862) by Union forces under General S. E. Burnside. Hoffman Forest to SW, Croatan National Forest to S, Cherry Point Marine Corps Air Station to SE. Tryon Palace Historical Site and Garden. Christ Episcopal Church (1752). Civil War Museum; New Bern Academy (opened 1764). First Presbyterian Church (opened 1822). Founded 1710, incorporated 1723.

Newbern, town (2000 population 231), Hale co., W central Alabama, 10 mi/16 km SE of Greenboro; 32°35′N 87°32′W. Named after Newberne, NC. Inc. in 1859.

Newbern, town (2006 population 3,117), DYER county, NW TENNESSEE, 9 mi/14 km NE of DYERSBURG; 36°07′N 89°16′W. In agricultural and light manufacturing area.

Newbern (NOO-buhrn), unincorporated village, PULASKI county, SW VIRGINIA, 5 mi/8 km ENE of PULASKI, near NEW RIVER, CLAYTOR LAKE reservoir; 37°04′N 80°41′W. Manufacturing (handmade carousel horses). Agriculture (corn; cattle, sheep; dairying). Claytor Lake State Park to E.

Newberry, county (□ 647 sq mi/1,682.2 sq km; 2006 population 37,762), NW central SOUTH CAROLINA, ⊙ NEWBERRY; 34°17′N 81°35′W. Bounded E by BROAD RIVER, S by SALUDA RIVER, N by ENOREE RIVER; part of Lake MURRAY in SE. Includes part of Sumter National Forest. Mainly agriculture (chickens, turkeys, hogs, cattle; eggs; corn, oats, soybeans, sorghum, hay); timber, granite. Manufacturing at Newberry. Formed 1785.

Newberry (NOO-ber-ee), city (□ 2 sq mi/5.2 sq km; 2005 population 3,804), ALACHUA county, N central FLORIDA, 17 mi/27 km W of GAINESVILLE; 29°38′N 82°36′W. Small-scale farming; phosphate quarries nearby.

Newberry, town (2000 population 168), GREENE county, SW INDIANA, on West Fork of WHITE RIVER, and 9 mi/14.5 km SW of BLOOMFIELD; 38°55′N 87°01′W. In agricultural area (corn, wheat; turkeys). Manufacturing (poultry products). Laid out 1822.

Newberry, town (2000 population 2,686), ⊙ LUCE county, NE UPPER PENINSULA, MICHIGAN, c.55 mi/89 km WSW of SAULT SAINTE MARIE, near TAHQUAMENON RIVER; 46°21′N 85°30′W. Logging; trade point for farm area (cattle; forage crops, oats); manufacturing (strand board). Starting point for fishermen, scenic route to Tahquamenon Falls State Park, to NE. Big Valley Ski Area to S. Settled 1882, incorporated 1886.

Newberry, industrial town (2006 population 10,874), ⊙ NEWBERRY county, NW central SOUTH CAROLINA, 45 mi/72 km WNW of COLUMBIA; 34°16′N 81°36′W. Manufacturing (textiles, wood products, printing and publishing, paper products, communication equipment); lumber, food processing; granite quarries. A fish hatchery is here. Seat of Newberry College.

New Bethlehem (BETH-luh-hem), borough (2006 population 1,005), CLARION county, W central PENNSYLVANIA, 16 mi/26 km NE of KITTANNING, on Redbank Creek; 41°00′N 79°19′W. Manufacturing (wooden products, feeds, refractories); bituminous coal, former oil and gas center. Agricultural area (grain, soybeans, potatoes; livestock; dairying). Settled 1785, laid out 1840, incorporated 1853.

Newbiggin-by-the-Sea (NYOO-big-in), town (2001 population 5,957), E NORTHUMBERLAND, NE ENGLAND, on NORTH SEA, 15 mi/24 km N of NEWCASTLE UPON TYNE; 55°10′N 01°31′W. Fishing port, seaside resort. Former coal-mining site. Cable to ARENDAL, NORWAY. To W is Woodhorn Demesne.

New Bloomfield, PENNSYLVANIA: see BLOOMFIELD.

New Bloomington (BLOOM-ing-tuhn), village (2006 population 546), MARION county, central OHIO, 10 mi/16 km W of MARION; 40°35′N 83°19′W. In agricultural area.

New Boatlift, Canal du Centre, Strépy-Thieu, HAINAUT province, SW BELGIUM, on CANAL DU CENTRE just W of LA LOUVIÈRE. Gigantic boatlift of 240 ft/73 m on Canal du Centre.

New Bohemia (noo bo-HEEM-yuh), unincorporated town, PRINCE GEORGE county, E VIRGINIA; 37°11′N 77°19′W.

Newborn, village (2000 population 520), NEWTON county, N central GEORGIA, 11 mi/18 km ESE of COVINGTON; 33°31′N 83°42′W.

Newboro (NOO-buh-ruh), former village (□ 1 sq mi/2.6 sq km; 2001 population 309), SE ONTARIO, E central Canada, on Newboro Lake (8 mi/13 km long, 3 mi/5 km wide), 30 mi/48 km NNE of KINGSTON; 44°39′N 76°19′W. Dairying, mixed farming. Amalgamated into Rideau Lakes township in 1998.

New Boston, city (2000 population 632), MERCER county, NW ILLINOIS, on the MISSISSIPPI (ferry here), and 14 mi/23 km W of ALEDO; 41°10′N 91°00′W. In agricultural area.

New Boston, town, HILLSBOROUGH county, S NEW HAMPSHIRE, 11 mi/18 km W of MANCHESTER; 42°58′N 71°41′W. Drained by South Branch of PISCATAQUOG RIVER. Joe English Hill (1,240 ft/378 m) in S. Agriculture (livestock, poultry; vegetables; dairying; timber). Manufacturing (prefabricated buildings, robotics lighting systems, casting molds).

New Boston (BAHS-tuhn), town (2006 population 4,642), BOWIE county, NE TEXAS, 22 mi/35 km W of TEXARKANA; 33°27′N 94°25′W. Trade center in timber; agricultural area (cattle; dairying; wheat, rice, vegetables). Oil and gas. Manufacturing (lumber, printing, concrete, fabricated metal products). State prison. In area settled in 1820s; town incorporated 1910.

New Boston, village, WAYNE county, SE MICHIGAN, unincorporated, suburb 23 mi/37 km SW of downtown DETROIT, and immediately S of ROMULUS, on HURON RIVER; 42°09′N 83°24′W. Manufacturing (concrete, steel, plastic products). Detroit Metropolitan Wayne County Airport to NE. Lower Huron Metropark along river.

New Boston (BAW-stuhn), village (□ 1 sq mi/2.6 sq km; 2006 population 2,182), SCIOTO county, S OHIO, on OHIO RIVER, just NE of PORTSMOUTH; 38°45′N 82°56′W. Manufacturing (steel, metal products). Founded 1891.

New Boston (BAWS-tuhn), unincorporated village, Mahanoy township, SCHUYLKILL county, E central PENNSYLVANIA, 1 mi/1.6 km SSW of MAHANOY CITY; 40°47′N 76°09′W. Anthracite-coal region.

New Boston, MASSACHUSETTS: see SANDISFIELD.

New Braintree (BRAIN-tree), town, WORCESTER county, central MASSACHUSETTS, 16 mi/26 km WNW of WORCESTER; 42°19′N 72°08′W. In agricultural and dairying area. State police academy.

New Brancepeth, ENGLAND: see BRANDON.

New Braunfels (BROUN-felz), city (2006 population 49,969), ⊙ COMAL county, S central TEXAS, 30 mi/48 km NE of SAN ANTONIO, on the GUADALUPE RIVER; 29°41′N 98°07′W. Elevation 620 ft/189 m. The economy is largely based on diversified manufacturing (portland cement, consumer goods, crushed limestone, furniture, leather products) and food processing, and agriculture (cattle, sheep, goats, hogs, exotic animals; corn, wheat, sorghum); winery; stone, lime, sand and gravel. New Braunfels was founded (1845) by Prince Carl von Solms-Braunfels and settled by thousands of German immigrants. It still retains many Germanic features. Local attractions include a historical museum; Landa Park, which contains COMAL springs, river, and lake; Natural Bridge Caverns to W; Lake McQueeney to SE. Incorporated 1847.

New Bremen (BREE-muhn), village (□ 2 sq mi/5.2 sq km; 2006 population 2,991), AUGLAIZE county, W OHIO, 13 mi/21 km SW of WAPAKONETA; 40°26′N 84°23′W. Agriculture (dairy products). Manufacturing (rubber goods, machinery). Bicycle Museum of America here. Incorporated 1833.

Newbridge, Gaelic *Droichead Nua*, town (2006 population 17,042), central KILDARE county, E central IRELAND, on the LIFFEY RIVER, and 5 mi/8 km ENE of KILDARE; 53°11′N 06°48′W. Agricultural market; rope manufacturing. Town grew around barracks built here 1816. Houses headquarters of Bórd na Móna, agency responsible for exploiting peat resources.

New Brighton, city (2000 population 22,206), RAMSEY county, SE MINNESOTA, a suburb 6 mi/9.7 km NE of MINNEAPOLIS, and 9 mi/14.5 km NW of ST. PAUL; 45°04′N 93°12′W. Drained by Rice Creek. Railroad junction. Manufacturing (metal products, machinery, leather, primary metals, printing and publishing, machining). A theological seminary is in New Brighton. Long Lake in N. Incorporated 1891.

New Brighton (BREI-tuhn), borough (2006 population 6,231), BEAVER county, W PENNSYLVANIA, 2 mi/3.2 km SSE of BEAVER FALLS, on BEAVER RIVER (bridged); 40°44′N 80°18′W. Manufacturing (chemicals, primary metals, fabricated metal products, machinery, plastic products). Agriculture (corn, hay; livestock; dairying). Merrick Art Gallery. Lapic Winery to E. Blockhouse erected here 1789. Settled c.1801, incorporated 1838.

New Brighton, section of STATEN ISLAND borough of NEW YORK city, SE NEW YORK, on N Staten Island, at junction of KILL VAN KULL and Upper NEW YORK BAY; 40°36′N 74°06′W. Residential, industrial. Sailors' Snug Harbor, a home for retired seamen, is nearby.

New Brighton, suburb, EASTERN CAPE province, SOUTH AFRICA, on ALGOA BAY of the INDIAN OCEAN; N suburb of NELSON MANDELA METROPOLE (Port Elizabeth); 33°54′S 25°36′E. Largely black residential area during apartheid rule. Many residents work in industrial area of Struandale (W) and Deal Pasty (E), either side of the suburb.

New Britain, industrial city (2000 population 71,538), HARTFORD county, central CONNECTICUT; 41°40′N 72°47′W. The tin shops and brassworks in the city were established in the 18th century. New Britain became famous as the "Hardware City" because of its tool and household hardware industry, which remains economically important. Central Connecticut State University is here. Of interest are the city hall (1884), a park designed by Frederick Law Olmsted in the center of the city, and a museum of American Art. Elihu Burritt born here. Settled c.1686, incorporated 1871.

New Britain (BRI-tuhn), borough (2006 population 2,300), BUCKS county, SE PENNSYLVANIA; residential suburb 3 mi/4.8 km WSW of DOYLESTOWN, and 23 mi/37 km N of downtown PHILADELPHIA; 40°17′N 75°10′W. Manufacturing (concrete products, rubber products). Agriculture (grain, soybeans, apples; livestock, poultry, dairying). Lake Galena reservoir to N; Peace Valley Winery to N.

New Britain, volcanic island (□ 14,600 sq mi/37,960 sq km), SW PACIFIC OCEAN, largest island of the BISMARCK ARCHIPELAGO, and part of PAPUA NEW GUINEA; 05°40′S 151°00′E. RABAUL is the chief town and port. The island is mountainous, with active volcanoes, hot springs, and peaks over 7,000 ft/2,130 m high. The major export is copra, and some copper, gold, iron, and coal are mined. Visited and named by the English explorer William Dampier in 1700, New Britain became part of German New Guinea in 1884. GERMANY called it NEU POMMERN (New Pomerania). In 1920 it was mandated to AUSTRALIA by the League of Nations and in 1947 was made a UN trust territory under Australian control.

New Brockton, town (2000 population 1,250), Coffee co., SE Alabama, 9 mi/14.5 km E of Elba; 31°22′N 85°55′W. Woodworking. Originally called 'Brock,' then 'Brockton' for Huey E. Brock, who built the first residence on the town site. 'New' was added to the name in 1907. Inc. in 1902.

Newbrook (NOO-bruk), hamlet, central ALBERTA, W Canada, 12 mi/20 km from THORHILD, in Thorhild County No. 7; 54°19′N 112°57′W.

New Brookland, SOUTH CAROLINA: see WEST COLUMBIA.

New Brunswick (NYOO BRUHNZ-wik), province (□ 28,345 sq mi/73,697 sq km; 2006 population 729,997), E CANADA; ⊙ FREDERICTON (third-largest city); 46°30′N 66°45′W. The largest city is ST. JOHN and the second-largest is MONCTON. One of the Maritime Provinces, New Brunswick is bounded on the N by CHALEUR BAY and QUEBEC province; on the E by the GULF OF SAINT LAWRENCE, NORTHUMBERLAND STRAIT, and NOVA SCOTIA; on the S by the BAY OF FUNDY; and on the W by MAINE. It is connected to the province of PRINCE EDWARD ISLAND by a bridge (completed 1997) across Northumberland Strait at Cape Tormentine. Its irregular coastline provides excellent facilities for fishing and shipping enterprises. Rivers cross the rolling countryside; they were the first means of transportation and are still important arteries of travel and commerce. The largest river, the St. John, crosses the province from NW to SE, and the Miramichi River flows NE and drains the central lowlands. Most of the roads follow the river banks.

Dairying thrives on fine pasturage, and the major crops are potatoes, hay, clover, oats, berries, and fruit. A careful conservation program maintains a supply of second-growth hardwoods and softwoods; forests cover about 90% of the total area, and lumbering is New Brunswick's most important industry. Great quantities of pulpwood and paper are produced. Manufacturing has greatly expanded since World War II; in addition to wood items and pulp and paper, products include food and beverages, ships, chemicals, refined oil, and shoes. Industry is generally run by hydroelectric power, and fuel resources include coal and much untapped water power, which is being developed. There is a nuclear reactor at Point Lepreau. Mining is an important industry, with zinc, silver, and lead the most important minerals. Other minerals include copper, bismuth, cadmium, gold, antimony, potash, oil, and natural gas. New Brunswick's fisheries, even with periodic slumps in industry, are among the most valuable in Canada, with a variety of freshwater and saltwater fish (cod, salmon, herring, and sardines) as well as shellfish (lobsters, oysters, and clams). Trade flows in and out of the ports of St. John and Moncton, facilitated by railroad connections throughout the province, E to Nova Scotia and W to Quebec. Tourism is one of New Brunswick's most important industries. Its forests are still filled with bear, deer, and moose, and the rivers abound in trout and salmon, although overfishing and acid wastes from paper mills have reduced the salmon population. Easy accessibility from the U.S. via I-95 has made WOODSTOCK the gateway to the

province. Summer residences are concentrated around PASSAMAQUODDY BAY. Natural attractions include the Grand Falls on the upper reaches of the St. John as well as the spectacular Fundy tides—the highest in the world, sometimes surging to over 50 ft/15 m. The tides in turn cause the Reversing Falls at St. John and the "Bore," a twice-daily tidal wave coming up the PETITCODIAC RIVER. They have also sculpted the famous Hopewell Rocks, another tourist attraction.

New Brunswick's first inhabitants were the Micmacs, a Native American people whose settlements stretched along the coast from Nova Scotia and Prince Edward Island to the S GASPÉ PENINSULA. The first European said to have sailed along the New Brunswick coast was a Portuguese navigator, Estevão Gomes (1525), although there is evidence of Basque fishermen at an earlier date. Jacques Cartier landed at Point Escuminac in 1534 and skirted the shores of Miramichi Bay. The first European settlement was made in 1604 at the mouth of the St. Croix River by Champlain and the sieur de Monts. During this period, while France and England made conflicting territorial claims, the present province of Nova Scotia and the coast of New Brunswick were considered one region, called ACADIA by the French and Nova Scotia by the British. British control of this region was confirmed by the Peace of Utrecht (1713–1714). Doubting the loyalty of the Acadians, the British expelled them in 1755, although many fled into the interior, which was still effectively controlled by the French. Others sought refuge in the thirteen American colonies or returned to France. (Today about 35% of the people of New Brunswick are Acadians.) Great Britain gained possession of the rest of New Brunswick when it gained all of Canada after the French and Indian Wars. When the population of New Brunswick was increased by many thousands of Loyalists who fled New England after the American Revolution, that area was organized (1784) into a separate colony. As trees were cut down for shipbuilding, the land was cleared for farming. By the middle of the 19th century lumbering and farming were extending into the interior, and St. John was a busy port and shipbuilding town. Dissatisfaction with the arbitrary rule of the provincial government resulted in the achievement of responsible (or cabinet) government in 1849. In 1867, under the British North American Act, federation with the other provinces into the dominion of Canada was somewhat reluctantly accepted. New Brunswick sends ten senators (appointed) and ten representatives (elected) to the national parliament. The province's four universities are the University of New Brunswick, Université de Moncton, Mount Allison University, and St. Thomas University.

New Brunswick, city (2006 population 50,172), ⊙ MIDDLESEX county, central NEW JERSEY, on the RARITAN River; 40°29′N 74°27′W. Originally developed as a commercial center (especially for collecting and shipping grain), New Brunswick manufactures pharmaceuticals, electrical equipment, transportation equipment, and medical and surgical supplies. Headquarters for large health-care products provider. The city is the seat of Rutgers University and New Brunswick Theological Seminary. Washington, retreating from NEW YORK, stayed a week in New Brunswick in 1776. Joyce Kilmer born here; his birthplace is an American Legion post. Adjoining New Brunswick was Camp Kilmer, a major army base during World War II and the Korean War, and now the site for part of Rutgers University campus. Settled 1681, incorporated as a city 1784.

New Buffalo, town (2000 population 2,200), BERRIEN county, extreme SW MICHIGAN, 8 mi/12.9 km NE of MICHIGAN CITY, INDIANA, on LAKE MICHIGAN; 41°47′N 86°45′W. In orchard and farm area (fruits, vegetables). Manufacturing (steel castings, plastic products, air compressors); resorts. Railroad junction to NE. Settled 1835, incorporated 1936.

New Buffalo (BUH-fah-lo), borough (2006 population 121), PERRY county, S central PENNSYLVANIA, 14 mi/23 km NNW of HARRISBURG, on SUSQUEHANNA RIVER; 40°27′N 76°58′W. Livestock, dairying.

New Bullards Bar Dam, California: see YUBA RIVER.

Newburg, city (2000 population 484), PHELPS county, central MISSOURI, in the OZARKS, on Little Piney River, 8 mi/12.9 km W of ROLLA; 37°55′N 91°54′W. Artist colony. Surrounded by Mark Twain National Forest. Settled 1823.

Newburg, village (2006 population 83), BOTTINEAU county, N NORTH DAKOTA 23 mi/37 km WSW of BOTTINEAU, near SOURIS RIVER (Mouse River); 48°42′N 100°54′W. J. Clark Salyer National Wildlife Refuge to E. Founded in 1905 and incorporated in 1906. Named for early settler, Andrew H. Newberg.

Newburg, village (2006 population 362), PRESTON county, N WEST VIRGINIA, 10 mi/16 km ENE of GRAFTON; 39°23′N 79°50′W. Agricultural and coal-mining area.

Newburg, village (2006 population 1,227), WASHINGTON county, E WISCONSIN, 7 mi/11.3 km E of WEST BEND, on branch of MILWAUKEE RIVER; 43°25′N 88°02′W. Agriculture (poultry; general farming). Manufacturing (transportation equipment, machinery).

Newburg (NOO-buhrg), borough (2000 population 81), CLEARFIELD county, central PENNSYLVANIA, 22 mi/35 km SW of CLEARFIELD, on Chest Creek; 40°50′N 41°78′W. Agriculture (corn, hay; livestock; dairying). Post office name is La Jose.

Newburg, borough (2000 population 372), CUMBERLAND county, S PENNSYLVANIA, 6 mi/9.7 km NNW of SHIPPENSBURG, near CONODOGUINET CREEK; 40°08′N 77°32′W. Manufacturing (fruit juice, potpourri). Agriculture (apples, grain; livestock; dairying).

Newburgh, city (□ 4 sq mi/10.4 sq km; 2006 population 28,345), ORANGE county, SE NEW YORK, on the W bank of the HUDSON RIVER, opposite BEACON; 41°30′N 74°01′W. The city has become an area wholesale and trucking center. Light manufacturing and commercial services. Growing Hispanic population. Newburgh has many old houses, and the streets run sharply to the river. At Hasbrouck House (1750; now a museum), George Washington made his headquarters from April 1782 to August 1783. It was in Newburgh that the Continental Army was disbanded. The first Edison power plant was located here (1915), and Newburgh was the first American city to receive electric street lighting. During the 1960s and 1970s it faced economic decline, but in recent decades it has had success in rejuvinating older areas. Mount St. Mary College is in the city. WEST POINT is located a few miles to the S. Settled 1709 by Palatines; incorporated 1800. Artist Ellsworth Kelly and politician Geraldine Ferraro hail from Newburgh.

Newburgh, town (2000 population 3,088), WARRICK county, SW INDIANA, on OHIO River, and 10 mi/16 km SW of BOONVILLE; 37°57′N 87°24′W. Suburb of EVANSVILLE. Manufacturing (aluminum ingots). Settled 1803, laid out 1818.

Newburgh, town, PENOBSCOT county, S MAINE, 13 mi/21 km WSW of BANGOR; 44°42′N 69°01′W. Includes Newburgh Center and Newburgh Village.

Newburgh (NOO-buhrg), former village (□ 6 sq mi/15.6 sq km; 2001 population 732), SE ONTARIO, E central Canada, on Napanee River, and 8 mi/12 km NE of Greater Napanee; 44°20′N 76°52′W. Paper milling. Amalgamated into Stone Mills township in 1998.

Newburgh (NYOO-buhrg), fishing village (2001 population 1,392), Aberdeenshire, NE Scotland, on YTHAN RIVER, near its mouth on the NORTH SEA, and 13 mi/21 km NNE of ABERDEEN; 57°16′N 04°30′W. Just SW is agricultural village of Foveran. Formerly in Grampian, abolished 1996.

Newburgh (NYOO-buhrg), village (2001 population 1,954), FIFE, E Scotland, on Firth of Tay, and 8 mi/12.9 km ESE of PERTH; 56°20′N 03°15′W. Fishing port. Previously linoleum manufacturing and malting. Just E are remains of Lindores Abbey, founded 1178 by David, earl of Huntingdon; David, duke of Rothesay, buried here 1402.

Newburgh Heights (NOO-buhrg, NYOO-), village (2006 population 2,197), CUYAHOGA county, N OHIO; suburb just S of CLEVELAND; 41°27′N 81°40′W. Ceramic plants.

Newburn (NYOO-buhn), town (2001 population 41,294), TYNE AND WEAR, N England, on the TYNE RIVER; 54°58′N 01°44′W. Former coal-mining and industrial center. Brewing. Has 12th-century church. Site of battle between English and Scottish armies in 1640. Includes the districts of Denton and Throckley.

New Burnside, village (2000 population 242), JOHNSON county, S ILLINOIS, 16 mi/26 km NE of VIENNA; 37°34′N 88°46′W. Surrounded by Shawnee National Forest.

Newbury (NYOO-buh-ree), town (2001 population 28,339), ⊙ West Berkshire, S central ENGLAND, 3 mi/4.8 km W of THATCHAM; 51°24′N 01°19′W. Newbury previously traded in wool, malt, and farm products. Manufacturing (paper, furniture, metal products) and high-technology industries. The town was an important textile-manufacturing center in the Middle Ages. The 16th-century cloth hall contains a museum. Civil war battles were fought here in 1643 and 1644. Racecourse to SE.

Newbury, residential town, ESSEX county, NE MASSACHUSETTS, 15 mi/24 km ENE of LAWRENCE, just S of NEWBURYPORT; 42°46′N 70°53′W. Includes sections of Byfield, South Byfield, and PLUM ISLAND. Plum Island Airport. Site of Parker River National Wildlife Refuge and Northeast Wildlife Management Area. Governor Dummer Academy preparatory school in South Byfield. Plum Island State Park nearby.

Newbury, town, MERRIMACK county, S central NEW HAMPSHIRE, 27 mi/43 km WNW of CONCORD; 43°18′N 72°01′W. Tourism; poultry; dairying. SUNAPEE LAKE on NW boundary with Mount Sunapee State Park Beach at S end of lake. Mount Sunapee Ski Area, in Mount Sunapee State Park, in W.

Newbury, town (2006 population 436), including Newbury village, ORANGE CO., E VERMONT, on the CONNECTICUT River, and 24 mi/39 km SE of BARRE; 44°06′N 72°07′W. Site of American Revolutionary War fort and E terminus of military road built 1776. Includes industrial village of Wells River. Settled 1762 on site of Indian village.

Newbury (NOO-buh-ree), village (□ 1 sq mi/2.6 sq km; 2001 population 422), S ONTARIO, E central Canada, 28 mi/45 km NE of CHATHAM; 42°41′N 81°48′W. Fruit growing. Incorporated 1873.

Newbury National Volcanic Monument, DESCHUTES county, 10 mi/16 km S of BEND, central OREGON, surrounded by Deschutes National Forest. Includes Newbury Crater, 22 mi/35 km S of Bend, with E and W lakes in its caldera, also Lava Cast Forest, Lava River Cave, and Lava Butte Geological Area.

Newbury Park, unincorporated town, VENTURA county, S CALIFORNIA, independent suburb 4 mi/6.4 km W of THOUSAND OAKS, near Arroyo Conejo. Manufacturing (metal products, consumer goods, electrical and computer equipment, communications equipment, transportation equipment, printing and publishing, machinery). Ranco Conejo Airport to N. SANTA MONICA MOUNTAINS NATIONAL RECREATION AREA to S.

Newburyport, city (2000 population 17,189), ⊙ ESSEX county, NE MASSACHUSETTS, at the mouth of the MERRIMACK RIVER. Its silverware industry dates from colonial times; textiles, scientific instruments, and electronic equipment are also made. Summer resort; antiques; fishing; whale watching. An early ship-building, whaling, and shipping center, it declined after Jefferson's embargo of 1808 and the War of 1812. Birthplace of U.S. Coast Guard and PLUM ISLAND Coast Guard Station. There are several historic and maritime museums here. Its notable old houses include the Coffin House (c.1651), the Swett-Isley House (c.1671), the Short House (c.1732), and Custom House Maritime Museum. William Lloyd Garrison and Francis Cabot Lowell were born here. National Wildlife Refuge nearby and state park. Settled 1635, set off from NEWBURY and incorporated 1764.

New Bussa (boo-SAH), French *Boussa*, town, NIGER state, W NIGERIA, on NIGER RIVER, and 50 mi/80 km NE of KAIAMA, at S end of KAINJI LAKE; 10°14′N 04°34′E. Gold-mining center; shea-nut processing, cotton weaving. Agriculture: cassava, millet, rice, and durra. Hydroelectric power station. Scottish explorer Mungo Park died here, 1806. Formerly an important town in native state of Borgu.

New Butler, WISCONSIN: see BUTLER.

New Calabar River (KAH-luh-bahr), arm of NIGER RIVER delta, Rivers state, S NIGERIA, entering GULF OF GUINEA W of BONNY.

New Caledonia (KA-luh-DON-yah), French *Nouvelle Calédonie* (noo-vel kah-lai-do-nee), overseas territory of FRANCE (□ 7,082 sq mi/18,342 sq km; 1991 population 173,300), South PACIFIC, c.700 mi/1,100 km NE of AUSTRALIA; ⊙ NOUMÉA, on New Caledonia Island; 21°30′S 165°30′E. Composed of the island of New Caledonia, the Isle Of PINES, the LOYALTY ISLANDS, WALPOLE ISLAND, and the HUON, CHESTERFIELD, and BELEP groups. New Caledonia island, the largest island of the territory (□ 6,467 sq mi/16,750 sq km), is continental in geological structure, mountainous, and subtropical in climate. It is rich in mineral resources, especially in its extensive areas of ultrabasic serpentine rocks, and yields nickel (about 25% of the world's nickel resources), iron, manganese, cobalt, gold, and silver. The island is densely forested in some places, but almost all the kauri pine that was once an important export has been cut down. The principal industries have been the mining and refining of nickel, chrome, and iron ore, and the production of coffee and copra. Cattle and poultry are raised, but many foodstuffs must still be imported from Australia. Tourism is important and the territory is largely dependent on France for economic aid. The government consists of a governor appointed by France, an elected territorial assembly, and a council. L'Union Calédonienne, a largely indigenous party, controls the assembly. The population is more than 33% Melanesian (Kanak) and 45% European (mostly French) with Polynesians in the outlying islands. Captain James Cook sighted and named the main island in 1774; the French annexed it in 1853. In the 1980s, the Kanaks pushed for independence, but a 1984 referendum rejected a proposal by French president François Mitterand. Civil strife ensued as many Kanaks revolted. A 1988 agreement between French and Melanesian delegations divided the territory into three autonomous provinces (Loyalty Islands, Nord, and Sud). A referendum for independence was held in 1998 but did not pass; a new referendum is scheduled for 2014.

New Cambria (KAIM-bree-yuh), town (2000 population 222), MACON county, N central MISSOURI, near CHARITON RIVER, 15 mi/24 km W of MACON; 39°46′N 92°45′W. Agiculture.

New Cambria (KAIM-bree-uh), village (2000 population 150), SALINE county, central KANSAS, on SALINE RIVER near its mouth on SMOKY HILL RIVER, and 6 mi/9.7 km ENE of SALINA; 38°52′N 97°30′W. Wheat; livestock.

New Canaan, town, FAIRFIELD county, SW CONNECTICUT; 41°09′N 73°30′W. It is mainly a residential town, with specialty shops and small office buildings Silvermine College of Art is located in New Canaan. Settled c.1700, incorporated 1801.

New Canada, plantation, AROOSTOOK county, NE MAINE, on FISH RIVER, and 42 mi/68 km NW of PRESQUE ISLE; 47°07′N 68°30′W. Hunting, fishing.

New Caney (CAI-nee), unincorporated town, MONTGOMERY county, SE TEXAS, 30 mi/48 km NNE of HOUSTON. Agricultural area on outer fringe of Houston urbanized area. Manufacturing (machinery); 30°09′N 95°12′W. LAKE HOUSTON State Park to SE.

New Canton, town (2000 population 417), PIKE county, W ILLINOIS, near the MISSISSIPPI, 16 mi/26 km W of PITTSFIELD; 39°38′N 91°05′W. In agricultural area.

New Carlisle (KAHR-leil), city (□ 2 sq mi/5.2 sq km; 2006 population 5,616), CLARK county, W central OHIO, 12 mi/19 km W of SPRINGFIELD, on small Honey Creek; 39°57′N 84°01′W. In agricultural area; food products. Founded 1810.

New Carlisle, town (2000 population 1,505), SAINT JOSEPH county, N INDIANA, 14 mi/23 km W of SOUTH BEND; 41°42′N 86°31′W. In agricultural area. Manufacturing (animal by-products, construction materials, snow melting systems, steel products, animal feed). Laid out 1835.

New Carlisle (noo KAHR-leil), village (□ 25 sq mi/65 sq km), ⊙ BONAVENTURE county, Gaspésie—Iles-de-la-Madeleine region, E QUEBEC, E Canada, on SE GASPÉ PENINSULA, on N side of entrance of CHALEUR BAY, 50 mi/80 km E of DALHOUSIE; 48°00′N 65°20′W. Market center in lumbering, dairying region; resort.

New Carrollton, city (2000 population 12,589), PRINCE GEORGES county, central MARYLAND, 9 mi/14.5 km NE of WASHINGTON, D.C.; 38°58′N 76°53′W. A rapidly growing suburb of Washington, it was named Carrollton by the developer, Albert Turner after Charles Carroll of Carrollton. When a post office was opened in 1965, mail was misdirected to Carrollton, a small rural village in Carroll county, which had first claim on the name. The growing suburb became New Carrollton. A major light railroad terminal for the D.C. subway system is located here. Incorporated in 1953.

Newcastel Harbour, AUSTRALIA: see HUNTER, PORT.

New Castile, SPAIN: see CASTILE.

New Castle, county (□ 493 sq mi/1,281.8 sq km; 2006 population 525,587), N DELAWARE; ⊙ WILMINGTON, state's largest city and main commercial and industrial center; 39°34′N 75°38′W. Bounded N by PENNSYLVANIA state line, W by MARYLAND state line, S in part by SMYRNA RIVER, E by DELAWARE RIVER (NEW JERSEY state line). Highest point in Delaware (442 ft/135 m; unnamed) on Pennsylvania state line 6 mi/9.7 km N of Wilmington. Agriculture (dairy products, poultry, livestock; corn, wheat); shipping at Wilmington. Crossed E-W at center by Chesapeake and Delaware Canal, part of INTRACOASTAL WATERWAY; drained by CHRISTINA RIVER, and BRANDYWINE, Red Clay, and White Clay creeks. Highly urbanized in N, especially NE around Wilmington; continuation of PHILADELPHIA metropolitan area, to NE. Silver Run and Augustine Wildlife Area in E; Canal National Wildlife Refuge follows Chesapeake and Delaware Canal, at center; BRANDYWINE CREEK, Bellevue, and Fox Point state parks in NE; White Clay State Park in NW; Lums Pond State Park in W center, N of Canal; Fort Delaware and Fort Dupont state parks in E; two small sections of Delaware, including Finn's Point, are on the New Jersey side of DELAWARE RIVER. University of Delaware is at Newark. Formed 1672.

Newcastle (NOO-ka-suhl), city (2001 population 279,975), NEW SOUTH WALES, SE AUSTRALIA, on the PACIFIC OCEAN; 32°55′S 151°45′E. It is the center of one of the largest coal-mining areas in the country and is a large port. Coal, wool, iron, steel, and wheat are exported. The city has steel mills and shipyards; a giant steelworks along the banks of HUNTER RIVER (opened in 1915) closed in 2000. The city is a commercial and transportation complex; chemicals, glass, fertilizer,

and textiles are also produced. The first permanent settlement on the site was made in 1804. The University of Newcastle is in the city.

New Castle, city (2000 population 17,780), ⊙ HENRY county, E INDIANA, 18 mi/29 km S of MUNCIE; 39°55′N 85°21′W. It is the trade center of an agricultural and farm region. Manufacturing (transportation equipment, feed, steel products, rubber products, food products, construction materials, machinery, pharmaceuticals). The city has a number of prehistoric Native American mounds. Wilbur Wright's birthplace is to the E. Laid out 1823, incorporated 1839.

New Castle (NOO KA-suhl), city (2006 population 24,732), ⊙ LAWRENCE county, W PENNSYLVANIA, 40 mi/64 km NNW of PITTSBURGH, at junction of the SHENANGO RIVER and Neshannock Creek; 41°00′N 80°20′W. Shenango River joins MAHONING RIVER 3 mi/4.8 km SW of city to form BEAVER RIVER. Fertile farm area (apples, soybeans, grain; livestock, dairying). Coal, limestone, and clay deposits found in the region contribute to the city's economy. Manufacturing (fabricated metal products, plastic products, food products, machinery, transportation equipment, printing and publishing, consumer goods). The Hoyt Institute of Fine Arts; County Historical Society; New Castle Municipal Airport to NW; McConnells Mill State Park to SE, including Hells Hollow Falls and McConnells Mill. Incorporated 1825.

Newcastle, former town, E central New Brunswick, Canada, on the Miramichi River; 47°00′N 65°34′W. In lumbering region. Sawmills and a large pulp mill. Newcastle was the birthplace of the Canadian leader Peter Mitchell and the boyhood home of Lord Beaverbrook. Amalgamated with Miramichi in 1995. Now called Miramichi West.

Newcastle (NOO-ka-suhl), unincorporated town, S ONTARIO, E central Canada, on Lake ONTARIO, 15 mi/24 km E of OSHAWA, and included in town of Clarington; 43°55′N 78°35′W. Includes large rural area. Dairying; mixed farming; apples.

Newcastle (NOO-ka-suhl), Gaelic An Caisleán Nua, town (2001 population 7,431), SE DOWN, SE Northern Ireland, on DUNDRUM BAY of the IRISH SEA, 28 mi/45 km S of BELFAST, at foot of the MOURNE MOUNTAINS; 54°12′N 05°53′W. Fishing port and seaside and golfing resort. Castle, of which there are some remains, was built 1588 on site of earlier structure.

Newcastle, town, SAINT ANDREW parish, E JAMAICA, in BLUE MOUNTAINS, 9 mi/14.5 km NE of KINGSTON; 18°04′N 76°43′W. Elevation c.3,700 ft/1,128 m. Mountain resort; military barracks.

Newcastle, town, NW KWAZULU-NATAL province, SOUTH AFRICA, on Ncandu River, a tributary of the BUFFALO and TUGELA rivers, and 160 mi/257 km NNW of DURBAN, 160 mi/257 km SE of JOHANNESBURG, at foot of DRAKENSBERG range; 27°45′S 29°58′E. Elevation 4,428 ft/1,350 m. Named for British secretary for the colonies, Duke of Newcastle. Steel center; coal mining; manufacturing (stoves, construction materials, dairy products); wool and grain market. Largest thermal power station, the Ingagane. Base of British military operations in South African War (1899–1902). Established in 1864.

New Castle, town (2000 population 4,862), NEW CASTLE county, N DELAWARE, 5 mi/8 km S of WILMINGTON, on the DELAWARE RIVER, opposite PENNSVILLE, NEW JERSEY; 39°40′N 75°34′W. Elevation 0 ft/0 m. Railroad junction. Manufacturing; sawmill. Peter Stuyvesant built a Dutch fort on the site, and the settlement was called Niew Amstel until it was renamed in 1664. The state of Delaware was formed at a convention in New Castle on September 21, 1776, and for a year the city served as state capital. The Immanuel Church (1710) is a historic landmark. Other colonial buildings are the Old Dutch House (dating from the late 1600s), the New Castle Court House

(1732), and Amstel House Museum Wilmington College (4-year) is here. DELAWARE MEMORIAL BRIDGE (Delaware River) 3 mi/4.8 km to NE. New Castle County Airport to NW. Fort Delaware State Park, on PEA PATCH ISLAND, 5 mi/8 km to S. Colonial architecture and cobblestone streets make the town an important tourist destination.

New Castle, town (2000 population 919), ⊙ HENRY county, N KENTUCKY, 23 mi/37 km NW of FRANKFORT; 38°25′N 85°10′W. In BLUEGRASS agricultural area. Drennon Springs, on Drennon Creek to E, site of annual storytelling festival (October). Established 1817.

Newcastle, town, LINCOLN county, S MAINE, on DAMARISCOTTA RIVER inlet (bridged to DAMARISCOTTA), and 6 mi/9.7 km NE of WISCASSET; 44°02′N 69°34′W.

New Castle, town, ROCKINGHAM county, SE NEW HAMPSHIRE, on small island (□ 0.8 sq mi/2 sq km) in PORTSMOUTH harbor; bridged to mainland, 12 mi/19 km SE of DOVER; 43°03′N 70°43′W. Former summer resort; now residential suburb of Portsmouth. Settlement here originally called Great Island; was pre-Revolutionary governor's seat. Ruins of Fort Constitution (then Castle William and Mary), seized by colonists in 1774. Chartered 1693.

New Castle, town (2006 population 117), MCCLAIN county, central OKLAHOMA, suburb 18 mi/29 km SSW of downtown OKLAHOMA CITY, on CANADIAN River. Agricultural area; manufacturing (consumer goods, computer equipment).

New Castle (NOO KA-suhl), town (2006 population 174), ⊙ CRAIG county, SW VIRGINIA, in ALLEGHENY MOUNTAINS, 17 mi/27 km NNW of ROANOKE, on CRAIG CREEK, in Jefferson National Forest; 37°30′N 80°06′W. Manufacturing (sand processing, apparel). Agriculture (grain; cattle, sheep).

Newcastle, town (2006 population 3,272), ⊙ WESTON county, NE WYOMING, just SW of BLACK HILLS, near SOUTH DAKOTA state line, 50 mi/80 km WSW of RAPID CITY, South Dakota; 43°51′N 104°12′W. Elevation 4,334 ft/1,321 m. Shipping point for cattle, lumber, oil products, and bentonite. Manufacturing (lumber, petroleum products). Pioneer relics at Anna Miller Museum. Scenic region of caves, canyons, lakes, and streams in nearby Black Hills. Thunder Basin National Grassland to W. Founded 1889.

Newcastle, unincorporated village, PLACER county, central CALIFORNIA, in SACRAMENTO VALLEY, 3 mi/4.8 km SW of AUBURN. Ships fruit, nuts; cattle, sheep. Manufacturing (boatbuilding). FOLSOM LAKE State Recreation Area to SE.

New Castle, village (2000 population 1,984), GARFIELD county, W COLORADO, on COLORADO RIVER, and 11 mi/18 km W of GLENWOOD SPRINGS; 39°35′N 107°31′W. Elevation c.5,552 ft/1,692 m. In irrigated agricultural region (cattle; oats, hay). Manufacturing (wood roof trusses). Coal deposits. Part of White River National Forest to N and S. Harvey Gap, Rifle Gap, and Rifle Falls state parks to NW.

Newcastle, village (2006 population 285), DIXON county, NE NEBRASKA, 10 mi/16 km NW of PONCA, near MISSOURI RIVER; 42°38′N 96°52′W. In fertile agricultural region (livestock, poultry products; grain).

Newcastle (NOO-kas-suhl), village (2006 population 573), YOUNG county, N TEXAS, near BRAZOS RIVER, c.50 mi/80 km SSW of WICHITA FALLS; 33°11′N 98°44′W. Elevation 1,126 ft/343 m. Oil fields nearby. Agriculture (livestock; cotton; wheat). Restored Fort Belknap (established 1851) is just S. Lake Graham reservoir to E. Former coal-mining center.

Newcastle, Wales: see BRIDGEND.

Newcastle Bay (NOO-ka-suhl), inlet of TORRES STRAIT, N CAPE YORK PENINSULA, N QUEENSLAND, NE AUSTRALIA, just S of Cape YORK; 11 mi/18 km long, 10 mi/16 km wide; 10°50′S 142°37′E. Shallow.

Newcastle Emlyn (NOO-kah-suhl EM-lin), town (2001 population 973), Ceredigion, W Wales, on TEIFI

RIVER, and 9 mi/14.5 km ESE of CARDIGAN; 52°02′N 04°28′W. Agriculture and processed dairy products. Formerly in DYFED, abolished 1996.

Newcastle Harbour, AUSTRALIA: see HUNTER, PORT.

New Castle Northwest (NOO KA-suhl NORTH-west), unincorporated town (2000 population 1,535), LAWRENCE county, W Pennsylvania; residential suburb 2 mi/3.2 km NW of NEW CASTLE; 41°01′N 80°21′W.

Newcastleton (NYOO-kah-suhl-tuhn), town (2001 population 772), S Scottish Borders, S Scotland, on LIDDEL WATER, and 8 mi/12.9 km ENE of LANGHOLM; 55°10′N 02°49′W. Hermitage Castle (built c.1224, restored in 19th century) is 5 mi/8 km N. Manufacturing (apparel). Newcastleton Forest (to E) is part of Border Forest Park.

Newcastle-under-Lyme (NYOO-kah-suhl–uhn-duh–LEIM), town (2001 population 74,427) and district, STAFFORDSHIRE, W central ENGLAND, on the Lyme River, 2 m/3.2 km NW of STOKE-ON-TRENT; 53°00′N 02°14′W. Manufacturing (construction materials, apparel, computers, electric motors, machinery). There are ruins of a castle built in the 12th century. Parish church dates from 13th century. Within the district, Chesterton has extensive Roman remains; Wolstanton is an industrial suburb.

Newcastle upon Tyne (NYOO-kah-suhl UHP-ahn TEIN), city (2001 population 189,863) and district, ⊙ TYNE AND WEAR, NE ENGLAND, on the TYNE RIVER; 54°59′N 01°36′W. Important shipping and trade center. Airport. The famous coal-shipping industry began in the 13th century; coal, however, was exceeded by wool exports until the 16th century. A number of heavy industries are also found, such as shipbuilding, marine machinery and equipment, defense equipment, chemicals, and pharmaceuticals. A recent industrial revival has been spearheaded by manufacturing of computer components, motor vehicles, and household appliances. The city rests on the site of the Roman military station Pons Aelii, at HADRIAN'S WALL. Later the site was occupied by the Angles until the Norman conquest. In 1080, Robert III, Duke of Normandy and eldest son of William the Conqueror, had a fortified castle built (from which Newcastle takes its name). The Cathedral of St. Nicholas dates partly from the 14th century. Several notable old buildings include Trinity Almshouse (1492) and the Royal Grammar School, founded in the 16th century. Among the many educational institutions are the University of Newcastle upon Tyne and Northumbria University. Districts of the city include Benwell, Elswick, Fenham, Heaton, Jesmond, and Walker.

Newcastle Waters (NOO-ka-suhl), settlement, N central NORTHERN TERRITORY, N central AUSTRALIA, 375 mi/603 km SSE of DARWIN; 17°20′S 133°21′E. Airport. Sheep.

Newcastle West, Gaelic *An Caisleán Nua*, town (2006 population 5,098), W LIMERICK county, SW IRELAND, near DEEL RIVER, 23 mi/37 km SW of LIMERICK; 52°27′N 09°03′W. Agricultural market. Has 12th-century castle of the Knights Templars.

New Centerville (NOO SEN-tuhr-vil), borough (2006 population 182), SOMERSET county, SW PENNSYLVANIA, 8 mi/12.9 km SW of SOMERSET; 39°56′N 79°11′W. Agriculture (corn, oats; livestock, dairying). Laurel Hill State Park to NW.

New Chicago, town (2000 population 2,063), Lake county, extreme NW INDIANA, SE of GARY, and N of HOBART; 41°34′N 87°16′W. In industrial area.

New City, suburb (□ 16 sq mi/41.6 sq km; 2000 population 34,038) of NEW YORK city, ⊙ ROCKLAND county, SE NEW YORK, 26 mi/42 km N of midtown MANHATTAN; 41°08′N 73°58′W. It is primarily residential. Movie mogul Adolph Zukor donated land for Zukor Park here.

New Cold Harbor, Virginia: see COLD HARBOR.

New Columbia (NOO kuh-LUHM-bee-ah), unincorporated village, UNION county, E central PENNSYLVANIA, 2 mi/3.2 km NNW of MILTON, on West Branch of SUSQUEHANNA RIVER (bridged to N); 41°02′N 76°52′W. Manufacturing (construction materials). Agriculture (dairying). Parts of Bald Eagle and Tiadaghton state forests to NW; Allenwood Federal Prison Camp to NW.

New Columbus (ko-LUHM-buhs), unincorporated village, NESQUEHONING borough, CARBON county, E PENNSYLVANIA, 1 mi/1.6 km NW of Nesquehoning town center; 40°52′N 75°50′W.

New Columbus (ko-LUHM-buhs), borough (2006 population 206), LUZERNE county, NE central PENNSYLVANIA, 22 mi/35 km WSW of WILKES-BARRE; 41°10′N 76°17′W. Drained by PINE CREEK. Agriculture (potatoes, corn, hay; livestock; dairying).

Newcomb, mining town, ESSEX county, NE NEW YORK, in the ADIRONDACKS, 38 mi/61 km WNW of TICONDEROGA; 44°01′N 74°08′W.

Newcomb (NOO-kuhm), suburb 2 mi/3 km SE of GEELONG, VICTORIA, SE AUSTRALIA. Formerly known as West Maloop.

Newcomerstown (NOO-kuh-muhrz-toun), village (2000 population 4,008), TUSCARAWAS county, E OHIO, 17 mi/27 km SSW of NEW PHILADELPHIA, and on TUSCARAWAS RIVER; 40°16′N 81°36′W. Manufacturing (tools, hardware, clay products). Settled 1815 on site of Delaware Indians' capital; incorporated 1838.

New Concord (KAHNG-kuhrd), village (□ 2 sq mi/5.2 sq km; 2006 population 2,669), MUSKINGUM county, central OHIO, 15 mi/24 km ENE of ZANESVILLE; 39°59′N 81°44′W. In agricultural area. Seat of Muskingum College. William Rainey Harper, first president of University of Chicago, born here. Birthplace of John Glenn, first American to orbit earth (1962). Founded c.1827.

New Corinth, Greece: see CORINTH.

New Cross, LONDON: see LEWISHAM.

New Croton Dam and Aqueduct, SE NEW YORK, water supply system for NEW YORK city and WESTCHESTER and PUTNAM counties; 41°14′N 73°52′W. A complex, integrated system of twelve dams and reservoirs, watersheds, and above- and underground tunnels, it was constructed to supplement and eventually replace the OLD CROTON DAM AND AQUEDUCT. The project began in 1885, and by 1890 additional water was being successfully delivered to the CENTRAL PARK Reservoir. New Croton Dam, begun in 1892 at a site 4 mi/6.4 km downstream from Old Croton Dam, was finished in 1907. When the new system began operations, city consumption was able to rise from 102 million gal/386.1 million liters per day to 170 million gal/643.5 million liters. Today it supplies New York city (esp. Manhattan's W Side and parts of the BRONX and Westchester county) with 300 million gal/1,135,590,000 liters of water per day, although stresses on the system's watersheds have resulted from increasing urbanization. The system still does not adequately meet New York city's water demands; the city's water supply is augmented with water from the CATSKILLS and the DELAWARE RIVER, and the city continues to look for supplementary water sources. There are trails and a park in the dam area.

New Cumberland, town (2006 population 1,030), ⊙ HANCOCK county, N WEST VIRGINIA, in NORTHERN PANHANDLE, 7 mi/11.3 km N of WEIRTON, on the OHIO RIVER (OHIO state line); 40°30′N 80°36′W. Light manufacturing. Agriculture (grain, nursery crops; fruit farms in region. TOMLINSON RUN STATE PARK to N; NEW CUMBERLAND Lock and Dam (Ohio River) are here. Plotted 1839.

New Cumberland (NOO KUHM-buhr-luhnd), borough (2006 population 7,115), CUMBERLAND county, S central PENNSYLVANIA, residential suburb 2 mi/3.2 km S of downtown HARRISBURG, on SUSQUEHANNA

RIVER (bridged to S end of Harrisburg), at mouth of Yellow Breeches Creek; 40°13′N 76°52′W. Manufacturing (machinery, electronics assembly, diverse light manufacturing). Defense Distribution Center (depot) and Central City Airport to SE. CAMP HILL State Correctional Institution to SW. Laid out c.1810, incorporated 1831.

New Cumnock (KUHM-nuhk), town (2001 population 3,165), East Ayrshire, S Scotland, on NITH RIVER, at the mouth of AFTON WATER, and 5 mi/8 km SE of CUMNOCK; 55°24′N 04°11′W. Light manufacturing. Formerly in Strathclyde, abolished 1996.

Newdale (NOO-dail), community, SW MANITOBA, W central Canada, 37 mi/59 km N of BRANDON, in Harrison rural municipality; 50°21′N 100°12′W. Original town site is just S; shifted in 1885 to be nearer rail line.

Newdale, village (2000 population 358), FREMONT county, E IDAHO, 8 mi/12.9 km SE of ST. ANTHONY; 43°54′N 111°36′W. Elevation 5,068 ft/1,545 m. Agriculture (cattle, sheep; potatoes). Teton Dam Site 5 mi/8 km ENE; on June 5, 1976, the dam collapsed, sending billions of gallons of water into the TETON RIVER valley just W of Newdale, killing 11 people.

New Dayton (NOO DAI-tuhn), hamlet, S ALBERTA, W Canada, 23 mi/37 km from MILK RIVER town, in Warner County No. 5; 49°25′N 112°23′W.

New Deal (DEEL), village (2006 population 724), LUBBOCK county, NW TEXAS, 10 mi/16 km N of LUBBOCK, on Blackwater Draw creek; 33°43′N 101°50′W. Agricultural area (cotton, wheat, sunflowers; cattle, sheep). Oil and natural gas.

New Delhi (DEL-ee), city (2001 population 294,783), ⊙ DELHI state and INDIA, N India, on the right bank of the YAMUNA RIVER; 28°37′N 77°13′E. Predominantly an administrative center, it was built (1912–1929) adjacent to OLD DELHI to replace CALCUTTA as capital of BRITISH INDIA; New Delhi was officially inaugurated in 1931. Transportation hub and trade center manufacturing electronics, motor vehicle parts, machinery, and electrical products. Designed by British architects Sir Edwin Lutyens and Herbert Baker, New Delhi has broad, symmetrically aligned streets that provide vistas of historic monuments. Between the main government buildings a broad boulevard, the Raj Path, leads E to W from India Gate, a massive war memorial arch (built 1921), through a great court to the resplendent sandstone and marble edifice Rashtrapati Bhavan (formerly the viceroy's palace; now the official residence of India's president). In the S section of the city is the Birla Bhavan, where Mahatma Gandhi was assassinated (1948). In the W are Balmiki and Lakshminarayan temples, which Gandhi frequented. Many sports stadiums were constructed on the eve of the ninth Asian Games held in 1983. Numerous hospitals, medical institutions, and scientific laboratories are here. Seat of three universities, including Jamia Milia, the Urdu University. Has an international airport.

New Delta, EGYPT: see TOSHKA CANAL.

New Denver (NOO DEN-vuhr), village (2001 population 538), SE BRITISH COLUMBIA, W Canada, on E side of SLOCAN LAKE, 35 mi/56 km N of NELSON, in Central Kootenay regional district; 49°59′N 117°22′W. Highway junction. Tourism, ski resort.

New Dongola, SUDAN: see DONGOLA.

New Dorp, section of STATEN ISLAND borough of NEW YORK city, SE NEW YORK, on E central Staten Island; 40°34′N 74°06′W. The name derives from the early Dutch *Niuwe Dorp*, "new town." Today the population is predominantly Italian-American along with Polish and Irish concentrations. Residential; some manufacturing.

New Dorp Beach, NEW YORK: see SOUTH BEACH.

New Douglas, village (2006 population 19,791), MADISON county, SW ILLINOIS, 20 mi/32 km NE of EDWARDSVILLE; 38°58′N 89°40′W. In agricultural area (corn, wheat; dairy products; poultry, livestock).

New Dundee (duhn-DEE), unincorporated village, S ONTARIO, E central Canada, 8 mi/13 km SSW of KITCHENER, and included in Wilmot; 43°21′N 80°32′W. Dairying, mixed farming.

New Durham, town, STRAFFORD county, E central NEW HAMPSHIRE, 13 mi/21 km NW of ROCHESTER; 43°28′N 71°08′W. Drained by Merrymeeting and Ela rivers. Agriculture (cattle; dairying; nursery crops; timber). Manufacturing (wire brushes). MERRYMEETING LAKE and Powder Mill Fish Hatchery in NW.

New Eagle (EE-guhl), borough (2006 population 2,246), WASHINGTON county, SW PENNSYLVANIA, 15 mi/24 km S of PITTSBURGH, and 1 mi/1.6 km W of MONONGAHELA, on MONONGAHELA RIVER; 40°12′N 79°57′W. Manufacturing (vermiculite and perlite, electronic products). Agriculture (corn, hay; livestock; dairying). Incorporated 1912.

New Echota Marker National Memorial (□ 0.92 acres/ 0.37 ha), NW GEORGIA, near CALHOUN; site of the last capital of the Cherokee tribe in Georgia. Native American newspaper was printed here in syllabary devised by Sequoyah. Central site of Cherokee removal connected with the Trail of Tears, the march that led to the Cherokee's resettlement in Oklahoma. Established 1930.

New Effington, village (2006 population 226), ROBERTS county, NE SOUTH DAKOTA, 25 mi/40 km NNE of SISSETON, in Lake Traverse (Sisseton Wahpeton) Indian Reservation; 45°51′N 96°55′W. Livestock; corn, wheat.

New Egypt, village (2000 population 2,519), OCEAN county, central NEW JERSEY, on CROSSWICKS CREEK, and 17 mi/27 km SE of TRENTON; 40°03′N 74°31′W. Agriculture and manufacturing.

Neweklau, CZECH REPUBLIC: see NEVEKLOV.

Newell (NOO-uhl) town (2000 population 887), BUENA VISTA county, NW IOWA, 10 mi/16 km ESE of STORM LAKE; 42°36′N 95°00′W. Concrete products, feed.

Newell, unincorporated town, MECKLENBURG county, S NORTH CAROLINA, residential suburb 6 mi/9.7 km NE of downtown CHARLOTTE; 35°16′N 80°44′W. University of North Carolina, Charlotte, to N; University Research Park to NW. Agriculture to E (grain; livestock).

Newell, unincorporated town (2000 population 1,602), HANCOCK county, N WEST VIRGINIA, in NORTHERN PANHANDLE, 2 mi/3.2 km SW of EAST LIVERPOOL, OHIO, across OHIO RIVER (bridged); 40°37′N 80°35′W. Manufacturing (consumer goods, electrical equipment, graphite, chemicals, bauxite), oil refining. Agriculture (grain, nursery crops).

Newell, village (2006 population 631), BUTTE county, W SOUTH DAKOTA, 22 mi/35 km E of BELLE FOURCHE; 44°43′N 103°25′W. In heart of irrigated area; diversified farming (sugar beets, grain, dairy products, honey). Nearby is a substation of a state experimental farm. BELLE FOURCHE RESERVOIR (Orman Dam) to W.

Newell (NOO-uhl), borough (2006 population 525), FAYETTE county, SW PENNSYLVANIA, 25 mi/40 km S of PITTSBURGH, in sharp bend of MONONGAHELA RIVER; 40°04′N 79°53′W. Manufacturing (chemicals). Agriculture (corn, hay; dairying).

Newell County No. 4 (NOO-uhl), municipality (□ 2,279 sq mi/5,925.4 sq km; 2001 population 7,137), S ALBERTA, W Canada; 50°36′N 111°56′W. Agriculture; oil and gas. Includes BASSANO, TILLEY, DUCHESS, Rosemary, and the hamlets of Gem, Rolling Hills, Patricia, Scandia, and Rainier. Home of Dinosaur Provincial Park. Formed as a municipal district in 1948.

New Ellenton, town (2006 population 2,252), AIKEN county, W SOUTH CAROLINA, 10 mi/16 km S of AIKEN; 33°25′N 81°41′W. Grew with the establishment of the Department of Energy's SAVANNAH RIVER Nuclear Site (S). Agriculture (livestock; poultry; grain, soybeans, corn, peanuts, peaches).

Newellton (NOO-uhl-tuhn), town, TENSAS parish, NE LOUISIANA, 29 mi/47 km SW of VICKSBURG, MISSISSIPPI, on NW end of LAKE SAINT JOSEPH; 32°04′N 91°14′W. In agricultural area (cotton, rice, soybeans, peanuts, wheat; cattle). Tensas River National Wildlife Refuge to NW.

New England, name applied to the region comprising six states of the NE U.S.—MAINE, NEW HAMPSHIRE, VERMONT, MASSACHUSETTS, RHODE ISLAND, and CONNECTICUT. The region is thought to have been so named by Captain John Smith because of its resemblance to the English coast (another source has it that Prince Charles, afterward Charles I, inserted the name on Smith's map of the country). Topographically it is partly delineated from the rest of the nation by the APPALACHIAN MOUNTAINS on the W. From the GREEN MOUNTAINS, the WHITE MOUNTAINS, and the BERKSHIRE HILLS, the land slopes gradually toward the ATLANTIC OCEAN. Many short, swift rivers furnish water power. The CONNECTICUT River is the region's longest river. Because of the generally poor soil, agriculture was never a major part of the region's economy. However, excellent harbors and nearby shallow banks teeming with fish made New England a fishing and commercial center. Shipbuilding was important until the end (mid-1800s) of the era of wooden ships. During the colonial period the region carried on a more extensive foreign commerce than the other British colonies and was therefore more affected by the passage of the British Navigation Acts. New England was the major center of the events leading up to the American Revolution, particularly after 1765, and was the scene of the opening Revolutionary engagements. The return of peace necessitated a reorganization of commerce, with the result that connections were made with the American NW and CHINA. The War of 1812 had an adverse effect on the region's trade, and opposition to the war was so great that New England threatened secession. After the war the growth of manufacturing (especially of cotton textiles) was rapid, and the region became highly industrialized. A large part of the great migration to the Old Northwest Territory originated here. Agriculture dwindled with the growth of the West. Prior to the Civil War, this region furnished many social and humanitarian leaders and movements. New England has also long been a leading literary and educational center of the country. After World War II the character of New England industry changed. Traditional industries (e.g., shoe and textile) have been superseded by more modern industries such as electronics. Tourism, long a source of income for the region, remains important, especially the ski industry in winter. There is also stone quarrying, dairying, and potato farming. BOSTON has long been the chief urban center of New England; corporate activity, however, has sprung up in many of the smaller cities and suburbs. The area is abundant in educational institutions, having some of the foremost universities in the U.S.

New England, town (2006 population 609), HETTINGER county, SW NORTH DAKOTA, 24 mi/39 km S of DICKINSON, and on CANNONBALL RIVER; 46°32′N 102°52′W. Founded in 1887 as Mayflower and changed to New England City later the same year. The name was shortened in 1894.

New England Plateau, AUSTRALIA: see NEW ENGLAND RANGE.

New England Range (NEW EEN-gluhnd), NE NEW SOUTH WALES, SE AUSTRALIA, part of GREAT DIVIDING RANGE; extends c.120 mi/193 km S from TENTERFIELD to URALLA; 30°00′S 151°50′E. Rises to 5,000 ft/1,524 m at Mount Bajimba, SSE of Tenterfield. Sapphire mines. Sometimes called New England Plateau.

New England Seamount Chain, chain of submarine volcanic mountains (6,562 ft/2,000 m–9,843 ft/3,000 m), SE of GEORGES BANK, between 36°N and 40°N, along the NE edge of the BERMUDA RISE plateau in the NW ATLANTIC Basin, and extending to a distance of c.600 mi/966 km.

Newenham, Cape, SW ALASKA, on BERING SEA, between KUSKOKWIM BAY (N) and BRISTOL BAY (S), 130 mi/209 km W of DILLINGHAM; 58°39′N 162°02′W. Extremity of small peninsula formed by rugged mountains.

Newent (NYOO-wuhnt), town (2001 population 5,073), NW GLOUCESTERSHIRE, W ENGLAND, 9 mi/14.5 km NW of GLOUCESTER; 51°56′N 02°24′W. Located near Forest of DEAN. Has 13th-century church and market house.

New Era, village (2000 population 461), OCEANA county, W MICHIGAN, 9 mi/14.5 km S of HART; 43°33′N 86°20′W. Fruit and vegetable canning.

New Fairfield, town, FAIRFIELD county, SW CONNECTICUT, between NEW YORK state line and Lake CANDLEWOOD, 6 mi/9.7 km N of DANBURY; 41°28′N 73°29′W. State park, state forest here; resorts. Manufacturing (machine tools, electrical parts). Incorporated 1740.

Newfane, town (2006 population 113), including Newfane village, ⊙ WINDHAM CO., SE VERMONT, on WEST RIVER, and 10 mi/16 km NW of BRATTLEBORO; 42°58′N 72°42′W. Resort, winter and summer recreation.

Newfane, village (□ 4 sq mi/10.4 sq km; 2000 population 3,129), NIAGARA county, W NEW YORK, near Lake ONTARIO, 25 mi/40 km NNE of BUFFALO; 43°17′N 78°42′W. In fruit-growing area; nurseries.

New Featherstone, village, MIDLANDS province, ZIMBABWE, 60 mi/97 km SSW of HARARE, and 4 mi/6.4 km NE of FEATHERSTONE; 18°42′S 30°53′E. Livestock; grain, cotton, tobacco.

New Ferry (NYOO FER-ee), village, on WIRRAL peninsula, MERSEYSIDE, NW ENGLAND, on MERSEY RIVER, and 2 mi/3.2 km SE of BIRKENHEAD; 53°21′N 03°00′W. Manufacturing (chemicals).

Newfield, town, YORK county, SW MAINE, 14 mi/23 km NW of ALFRED; 43°38′N 54°70′W. Severely damaged (1947) by forest fire.

Newfield, borough (2006 population 1,664), GLOUCESTER county, S NEW JERSEY, 10 mi/16 km N of MILLVILLE; 39°32′N 75°01′W. Manufacturing (powdered metals, glass products).

Newfields, town, ROCKINGHAM county, SE NEW HAMPSHIRE, 10 mi/16 km WSW of PORTSMOUTH; 43°02′N 70°58′W. Bounded E by Squamscott River; drained by Piscassic River. Agriculture (cattle, poultry; vegetables; dairying; nursery crops). Manufacturing (transportation equipment, screen printing).

New Florence, town (2000 population 764), MONTGOMERY county, E central MISSOURI, 5 mi/8 km SSE of MONTGOMERY CITY; 38°54′N 91°27′W. Agriculture (corn, wheat, soybeans; hogs). Manufacturing (barrel staves, wine, champagne); limestone quarries.

New Florence, borough (2006 population 741), WESTMORELAND county, SW central PENNSYLVANIA, 10 mi/16 km WNW of JOHNSTOWN, on CONEMAUGH RIVER (bridged); 40°22′N 79°04′W. Agriculture (soybeans, grain; livestock, dairying). Laurel Bridge State Park to SE.

Newfolden (noo-FOL-duhn), village (2000 population 362), MARSHALL county, NW MINNESOTA, on MIDDLE RIVER, and 18 mi/29 km NNW of THIEF RIVER FALLS; 48°21′N 96°19′W. Agriculture (potatoes, sugar beets, beans, wheat; cattle, sheep, poultry). Light manufacturing.

New Forest (NYOO FAH-rist), district (□ 145 sq mi/377 sq km; 2001 population 169,331), HAMPSHIRE, S ENGLAND. Bounded by the AVON River, The SOLENT, and SOUTHAMPTON WATER. William I organized the area in 1079 as a royal forest to provide revenue and timber. The Court of Verderers has administered the area as a public park since 1877. It consists of c.145 sq mi/ 376 sq km, of which 102 sq mi/264 sq km belong to the crown and 43 sq mi/112 sq km are wooded. The woods contain mostly oak and beech trees as well as extensive tracts of bog and heath. About 25 percent of the district's land is cultivated. Pigs, cattle, and ponies are raised, and tourism is important. An oil terminal has been constructed at FAWLEY. Other towns in the New Forest district include LYNDHURST, RINGWOOD, and NEW MILTON. Acquired national park status from the government in 1992.

New Fort Hamilton, village, W ALASKA, on arm of YUKON RIVER delta, 80 mi/129 km SW of SAINT MICHAEL. Sometimes called New Hamilton. HAMILTON village 10 mi/16 km N.

Newfound Gap, pass (5,048 ft/1,539 m), on TENNESSEE–NORTH CAROLINA state line, 8 mi/12.9 km SSE of GATLINBURG, Tennessee. Crossed by highway between Gatlinburg and CHEROKEE, North Carolina, through Great Smoky Mountains National Park.

Newfound Lake (NOO-found), resort lake in hilly region, GRAFTON county, central NEW HAMPSHIRE, 12 mi/19 km NNW of FRANKLIN; 6 mi/9.7 km long, 2.5 mi/4 km wide. Drains S through Newfound River (2.5 mi/4 km long), past Bristol (water power), to the PEMIGEWASSET RIVER.

Newfoundland, Canada: see NEWFOUNDLAND AND LABRADOR.

Newfoundland and Labrador (NYOO-fuhn-luhnd, LA-bruh-DOR), province (□ 143,048 sq mi/371,924.8 sq km), E Canada, consists of the island of Newfoundland and adjacent islands and the mainland area of Labrador and adjacent islands, ⊙ ST. JOHN'S; 52°00′N 56°00′W.

Newfoundland island lies at the mouth of the Gulf of SAINT LAWRENCE and is bounded on the N, E, and S by the Atlantic Ocean and separated on the NW from Labrador by the Strait of Belle Isle. Labrador, part of the LABRADOR-UNGAVA peninsula, forms the NE tip of the Canadian mainland. It is bounded on the E by the Atlantic Ocean down to the Strait of Belle Isle and on the S and W by Quebec. Cape Chidley, Labrador's northernmost point, is on the Hudson Strait. Newfoundland has a rocky, irregular coast, indented with numerous inlets. The major portion of the island is a plateau, with many lakes and marshes, and with forests covering less than half the area. Throughout the province the inland wilderness has an abundance of waterfowl, fish, and fur-bearing animals, while caribou graze on the tundra of the N. The cod-fishing area of the Grand Banks is probably the best in the world. Lobster, flounder, redfish, herring, and salmon are caught throughout the coastal waters; cod fishing collapsed in mid-1990s due to overfishing and failure to rebound. The province has a generally cool and moist climate. In Labrador, the cold Labrador current, bringing temperatures below freezing eight months of the year, and the lack of transportation facilities have combined to retard economic development. However, Labrador is rich in mineral resources (iron, zinc, copper, asbestos, gold, oil, natural gas), timber, and water power. Exploitation of the tremendous iron reserves in the SW lake district, begun in the 1950s, and the growth of the logging industry have brought new towns and roads. Also, what is believed to be the world's largest deposit of nickel was discovered in an 800-sq-mi/2,072-sq-km area surrounding Voisey's Bay. The area also has deposits of cobalt and copper.

There is a giant hydroelectric project at Churchill Falls. A sixty-five-year contract with Quebec (1976) gives all but a fraction of revenue from project to Hydro Quebec. This development and the collapse of the cod industry in the mid-1990s has increased pressure on province to develop offshore oil and gas fields, of which there have been significant discoveries in recent years. These discoveries are expected to have a favorable economic impact on the province's economy, although the negative impact upon coastal fishing is argued by environmentalists. Biggest concern is potential for spills and damage to oil rigs by icebergs, rough seas, and seismic activity. Mining is the main industry, and Newfoundland provides about half of Canada's iron ore. The processing of fish and the manufacturing of wood products are also important. There are large pulp and paper mills at GRAND FALLS and CORNER BROOK, both on Newfoundland.

Agriculture in the province is limited by the unfavorable soil and climate, and much of the food supply must be imported. The population is centered on the AVALON PENINSULA, which is the province's most important commercial and administrative region. Corner Brook is the second-largest city. Most of the inhabitants are of English or Irish descent, but in Labrador there are small numbers of Inuit and Montagnais-Naskapi. The Beothuk, a Native American tribe on the island of Newfoundland, died out in the 19th century, presumably of European diseases. Vikings visited the area c.1000 and briefly established a settlement on Newfoundland at L'ANSE AUX MEADOWS, the sole confirmed Viking site in North America.

After the two voyages of John Cabot to the area at the end of the 15th century, fishermen and explorers from several European countries followed. In 1535–1536, Jacques Cartier sailed through the Cabot Strait and the Strait of Belle Isle. Sir Humphrey Gilbert claimed Newfoundland for England in 1583. The first settlers arrived in 1610. After it changed hands several times, the Treaty of Paris of 1763 definitively awarded Newfoundland and Labrador (where the French had established trading posts) to Great Britain. The Peace of Utrecht in 1713 granted France the fishing rights on the NW coast of Newfoundland and awarded St. Pierre and MIQUELON to it. In 1783 the "French Shore" was redefined to include the entire W coast. In the early 19th century the Hudson's Bay Company developed the fur trade; this development and the expanding fishing industry led to increased immigration from Europe, particularly from Ireland.

Representative government was introduced in 1832 and parliamentary government in 1855. The port of HEART'S CONTENT became the W terminal of the transatlantic cable in 1866. In 1869 the voters of Newfoundland rejected union with Canada; in 1895, after a disastrous fire in St. John's and the failure of local banks, negotiations to join Canada resumed but were unsuccessful. In 1895 iron ore was discovered in the Grand Falls (now Churchill Falls) region of Labrador. As part of the Anglo-French Entente Cordiale of 1904, France abandoned the French Shore. In 1927, the British Privy Council demarcated the W boundary, enlarged the Labrador land area, and confirmed Newfoundland's title to it, but Quebec continues to dispute the border.

During the economic depression of the 1930s, Britain suspended Newfoundland's self-government and assumed administrative and financial control. Actual authority was exercised by a joint commission of Newfoundlanders and British. During World War II, U.S. and Canadian military bases were established in Labrador and on Newfoundland. After the war, Newfoundland voted to join Canada, and in 1949 it became Canada's 10th province. Joseph Smallwood, a Liberal who led the drive to join Canada, became premier and held office until 1972, when the Conservatives gained a majority under Frank Moores. Power has since moved back and forth. Newfoundland sends six senators and seven representatives to the national parliament. Memorial University of Newfoundland is in St. John's. Name changed from Newfoundland to Newfoundland and Labrador in 2001.

Newfoundland Canal, SW NEWFOUNDLAND AND LABRADOR, Canada, extends 10 mi/16 km WSW-ESE between NE end of DEER LAKE and NE end of GRAND LAKE.

Newfoundland Mountains, BOX ELDER county, NW Utah, in GREAT SALT LAKE DESERT. DESERT PEAK (6,984 ft/2,129 m) is highest point. Small range c.15 mi/24 km N-S, Newfoundland Evaporation Basin to W; Hill Air Force Firing Range and GREAT SALT LAKE to E.

New France: see CANADA.

New Franklin, city (2000 population 1,145), HOWARD county, central MISSOURI, just N of BOONVILLE across the Missouri; 39°01′N 92°44′W. Agriculture (corn, apples, wheat, soybeans). Manufacturing (concrete); lumber; limestone quarries. Laid out 1828 after the town of Franklin (established 1816) was washed away by the MISSOURI River. The SANTA FE TRAIL began at Franklin–New Franklin in the 1820s. Boonslick State Park to NW.

New Freedom (FREE-duhm), borough (2006 population 4,032), YORK county, S PENNSYLVANIA, 16 mi/26 km S of YORK on MARYLAND state line; 39°44′N 76°42′W. Railroad junction. Manufacturing (lumber, machinery, plastic molds). Agriculture (apples, soybeans, grain; poultry, livestock, dairying); timber. Incorporated 1879.

New Galilee, borough (2006 population 396), BEAVER county, W PENNSYLVANIA, 6 mi/9.7 km WSW of ELLWOOD CITY, on North Fork of Little Beaver Creek; 40°49′N 80°24′W. Manufacturing (cement products, candy); agriculture (corn, hay, alfalfa; livestock; dairying).

New Galloway (GA-lo-wai), town, DUMFRIES AND GALLOWAY, S Scotland, on Ken Water, 17 mi/27 km N of KIRKCUDBRIGHT; 55°04′N 04°09′W. Just S, on the Ken, is Loch Ken (5 mi/8 km long) with ruins of 15th–17th-century Kenmure Castle. Near S end of lake is hamlet of Little Duchrae.

Newgate, ENGLAND: see LONDON.

New Georgia, andesitic volcanic island group (□ 2,000 sq mi/5,200 sq km), SOLOMON ISLANDS, SW PACIFIC, 104 mi/167 km NW of GUADALCANAL; 08°15′S 157°30′E. Largest island, also called New Georgia, is 50 mi/80 km long, c.20 mi/32 km wide. Munda and Noro are chief centers. Produces copra, timber. In World War II Munda in SW became site of Japanese air base, now the second-busiest airfield in Solomons. Noro in W has developed with a Japanese fish cannery. Roviana Lagoon in S, Marovo Lagoon in N. Other islands of group are KOLOMBANGARA, VANGUNU, RENDOVA, GHIZO, TETEPARE, Nggatokae, and several smaller ones.

New Georgia Sound, SW PACIFIC, in SOLOMON ISLANDS, bounded N by CHOISEUL and SANTA ISABEL, S by the NEW GEORGIA island group; extends c.310 mi/499 km SE from SHORTLAND ISLANDS (off BOUGAINVILLE) to Florida and SAVO (both near GUADALCANAL). During World War II, the Japanese used (1942) the passage as supply route for reinforcement of Guadalcanal. Nicknamed "The Slot" by Americans.

New Germany, village, W NOVA SCOTIA, CANADA, on LAHAVE RIVER, 15 mi/24 km N of BRIDGEWATER; 44°32′N 64°43′W. Elevation 298 ft/90 m. Number of early settlers were German and most likely named the village for their homeland. Former and merged community names: Chesleys Corner.

New Germany, village, GARRETT county, W MARYLAND, in the ALLEGHENIES 18 mi/29 km W of CUMBERLAND. Headquarters for surrounding Savage River State Forest.

New Germany, village (2000 population 346), CARVER county, S central MINNESOTA, 36 mi/58 km W of MINNEAPOLIS; 44°52′N 93°58′W. Grain, soybeans, alfalfa; livestock; dairying; light manufacturing.

New Germany State Park, within Savage River State Forest, GARRETT county, W MARYLAND, in the ALLEGHENIES 18 mi/29 km W of CUMBERLAND. One of two state parks in the forest (the other being Big Run). The stately virgin pines here, which the NATIONAL ROAD ran through, reminded some of cathedrals. Most pioneers, fearful of Indian attack, called them "Shades of Death."

New Glarus (GLAIR-us), town (2006 population 2,070), GREEN county, S WISCONSIN, on branch of Sugar River, 23 mi/37 km SW of MADISON; 42°48′N 89°37′W. In dairying region. Manufacturing (meat products); cheese center; makes Swiss embroidery. Annual Swiss festival held here; state trail. Swiss historical village. N terminus of Sugar River State Trail. Founded by Swiss in 1845, incorporated 1901.

New Glasgow, (GLAZ-go) town (2001 population 9,432), N NOVA SCOTIA, CANADA, on East River; 45°35′N 62°38′W. Industrial town in a coal region. Manufacturing includes steel products and machinery; large pulp mill nearby. Named afer Glasgow, Scotland.

New Gloucester (GLAHS-tuhr), rural town, CUMBERLAND county, SW MAINE, 20 mi/32 km N of PORTLAND; 43°57′N 70°17′W. In it is Sabbathday Lake village, Shaker community settled 1793 near SABBATHDAY LAKE, the last living Shaker community in the U.S. (museum, visitor center). Includes village of Upper Gloucester. Settled c.1743, incorporated 1794.

New Goa, INDIA: see PANAJI.

New Goshen, village, VIGO county, W INDIANA, 9 mi/14.5 km NNW of TERRE HAUTE, between WABASH RIVER and ILLINOIS boundary. Corn, soybeans; cattle. Bituminous coal area, surface mines. Laid out 1853.

New Granada, former SPANISH colony, N SOUTH AMERICA. It included at its greatest extent present COLOMBIA, ECUADOR, PANAMA, and VENEZUELA. Between 1499 and 1510 a host of conquerors explored the CARIBBEAN coast of Panama and South America. After 1514, Pedro Arias de Ávila was successful in assuring permanent colonization of the Isthmus of PANAMA. At SANTA MARTA (1525) and CARTAGENA (1533), Spanish control of the Colombian coast was firmly established, and in the next few years the N hinterland was explored. German adventurers, notably Nikolaus Federmann, penetrated the Venezuelan and Colombian LLANOS between 1530 and 1546. By far the greatest of the conquerors was Gonzalo Jiménez de Quesada, who in 1536 ascended the MAGDALENA River, climbed the mighty Andean cordillera, where he subdued the powerful Chibcha (an advanced Indian civilization), and by 1538 had founded Santa Fé de Bogotá, later known simply as BOGOTÁ. He named the region El Nuevo Reino de Granada [=the new kingdom of Granada]. During the next 10 years the conquest was virtually completed. No civil government was established in New Granada until 1549, when an audiencia court, a body with both executive and judicial authority, was set up in Bogotá. To further stabilize colonial government, New Granada was made a presidency (an administrative and political division headed by a governor) in 1564, and the audiencia was relegated to its proper judicial functions. Loosely attached to the viceroyalty of PERU, the presidency came to include Panama, Venezuela, and most of Colombia. Disputes with—and the great distance from—LIMA led to the creation (1717) of the viceroyalty of New Granada, comprising Colombia, Ecuador, Panama, and Venezuela. Later the captaincy general of Venezuela and the presidency of QUITO were detached, creating a political division that was to survive the revolution against Spain and the efforts of Simón Bolívar to establish a republic of Greater Colombia. The struggle for independence began in 1810, and by 1830 Venezuela and Ecuador had seceded, and the remnant (Colombia and Panama) was renamed the Republic of New Granada. This became the United States of Colombia in 1863 and the Republic of Colombia in 1886, from which the present Panama seceded in 1903. This was a colony in South America, not a province of Spain.

New Grand Chain, village (2000 population 233), PULASKI county, S ILLINOIS, 17 mi/27 km NW of METROPOLIS, near OHIO River; 37°15′N 89°01′W. Grain, sorghum. Lock and Dam Number 53 to S. Shawnee Community College to NW.

Newgrange (NYOO-grainj), Gaelic *Sí An Bhrú*, national monument, MEATH county, NE IRELAND, 7 mi/11.3 km SW of DROGHEDA. Passage grave. Cairn 150 ft/46 m in diameter and 45 ft/14 m high. Surrounded by twelve pillar stones (2800 B.C.–2400 B.C.E.)

New Grove, village (2000 population 9,184), GRAND PORT district, SE MAURITIUS, 6 mi/9.7 km WSW of MAHÉBOURG. Sugarcane.

New Guinea (GI-nee), island (☐ 342,000 sq mi/889,200 sq km), SW PACIFIC OCEAN, N of AUSTRALIA; 05°00′S 140°00′E. The world's 2nd-largest island (after GREENLAND). Politically it is divided into 2 sects: the Indonesian province of IRIAN JAYA (Irian Barat or WEST NEW GUINEA; formerly Netherlands New Guinea) in W; and the independent country of PAPUA NEW GUINEA in E. The island is c.1,500 mi/2,410 km long and c.400 mi/640 km wide at the center. Largely tropical, it has vast mountain ranges such as the OWEN STANLEY and the BISMARCK mountains; JAYA PEAK (16,503 ft/5,030 m) in Irian Jaya is the highest point. The lower courses of the large rivers (the FLY, SEPIK, MAMBERAMO, and PURARI) are generally swampy, with a few grassy plains. The indigenous inhabitants of New Guinea are Melanesians, Negritos, and Papuans. The fauna consists largely of marsupials and monotremes, with venomous snakes among the reptiles. The island is known for its many unique species of butterflies and birds of paradise. There are mangrove and sandalwood forests, and the region exports copra, copal, maté, nutmeg resin, and wood. Near Tembagapura are the world's largest copper deposits. In addition to copper, gold, silver, and manganese are mined and oil is extracted. In the early 1980s, efforts began to further explore Irian Jaya's natural resources. Since that time, the area has been scoured in the hopes of discovering additional nickel, petroleum, and copper deposits. New Guinea was sighted by the Portuguese explorer Antonio d'Abreu in 1511 and was named for its resemblance to the Guinea coast of west AFRICA. During the next 2 centuries, the island was visited by Europeans from many nations. In 1828 the Dutch formally annexed the W half of the island, and in 1885 the British proclaimed a protectorate over the SE coast and the adjacent islands under the name of British New Guinea; in the same year, the Germans took possession of the NE. AUSTRALIA obtained control of British New Guinea in 1905 and renamed it the Territory of PAPUA. During World War I, Australian forces occupied the German-controlled region in the NE, which was mandated to Australia by the League of Nations in 1920. Renamed the Territory of New Guinea, this area became a UN trust territory under Australian control after World War II. The island was the scene of bitter fighting between Japanese and Allied forces. In 1949 the territories of Papua and New Guinea were merged administratively, and in 1973 they were united into a self-governing country. Full independence was gained in 1975. Netherlands New Guinea was transferred to Indonesian administration in 1963 and became the province of Irian Jaya in 1969. Since 1972, the economy of Irian Jaya has been transformed by the growth of forestry, the massive Freeport copper mine, by the "transmigration" of Javanese settlers, and by the immigration of Bugis. Christian and Muslim missionary activity has been encouraged by the government to promote "modern" lifestyles over "traditional" ones. Resentment over government cultural and economic policies and over the domination of admin. posts by non-Papuans led to the 1965 founding of the OPM (Free Papua Movement), which continues to conduct a sporadic guerrilla war against Jakarta authorities. Despite a 1986 treaty between Indonesia and Papua New Guinea, sporadic border skirmishes continue between the 2 countries, resulting from Indonesian military operations and the crossing of Papuan refugees into Papua New Guinea.

New Guinea, Territory of: see PAPUA NEW GUINEA.

Newgulf (NOO-guhlf), unincorporated village, WHARTON county, S TEXAS, c.45 mi/72 km SW of HOUSTON; 29°15′N 95°53′W. In sulphur-mining area; oil wells. Agriculture (rice; cotton).

Newhalen, village, ALASKA, 225 mi/362 km SW of ANCHORAGE, on N shore of ILIAMNA Lake, largest lake in Alaska (75 mi/121 km long, E-W; 20 mi/32 km wide); 59°43′N 154°52′W. Road connection to village of ILIAMNA and Lake CLARK, in LAKE CLARK National Park and Preserve. Fishing, hunting.

Newhall, town (2000 population 886), BENTON county, E central IOWA, 15 mi/24 km W of CEDAR RAPIDS; 41°59′N 91°58′W. Manufacturing (feed, wood and metal products).

Newham (NYOO-uhm), inner borough (☐ 14 sq mi/36.4 sq km; 2001 population 243,891) of GREATER LONDON, SE ENGLAND, on the THAMES RIVER; 51°30′N 00°05′E. Newham is residential in the NE. The Royal Docks and associated industries are in the S; chemical factories and railroad yards predominate in the NW. Few buildings in Newham are more than a century old; the area's growth stemmed largely from London's 19th-century industrial expansion. The SW especially suffered from slum conditions. Much of the borough was destroyed by bombs during World War II, but was rebuilt in the 1960s. The docks are now largely redundant. The borough includes Manor Park, East Ham, and West Ham.

New Hamburg, unincorporated community, SCOTT county, SE MISSOURI, 3 mi/4.8 km NW of BENTON. Agriculture.

New Hamburg (HAM-buhrg), unincorporated village, S ONTARIO, E central Canada, on NITH RIVER, 12 mi/19 km WSW of KITCHENER, and included in Wilmot; 43°23′N 80°42′W. Suburb of Kitchener. Within Regional Municipality of WATERLOO (formerly Waterloo county). Manufacturing (furniture); mixed farming; dairying; tobacco, peanuts.

New Hamilton, Alaska: see NEW FORT HAMILTON.

New Hampshire, state (☐ 9,350 sq mi/24,310 sq km; 2006 population 1,314,895), NE UNITED STATES, in NEW ENGLAND; 43°31′N 71°25′W; ⊙ CONCORD. One of the thirteen original U.S. states and former English colony. Known as "The Granite State."

Geography

The state is bounded on the N by CANADA (QUEBEC province), on the E by MAINE and the ATLANTIC OCEAN, on the S by MASSACHUSETTS, and on the W by the CONNECTICUT RIVER (VERMONT state line). The largest city is MANCHESTER, followed by NASHUA; Concord is third. Other important cities are ROCHESTER, DOVER, KEENE, and PORTSMOUTH. The continental ice sheet once covered the entire state and, in receding, scraped the mountains, eroded the intervening upland areas, and rerouted the water courses into precipitous streams and beautiful lakes. Across the N central part of the state the residual WHITE MOUNTAINS of the APPALACHIAN chain form ranges abruptly broken by passes (locally referred to as notches) through their rocky walls. Between the CARTER-MORIAH RANGE and the PRESIDENTIAL RANGE in the E, the ELLIS RIVER falls 80 ft/24 m on the side of PINKHAM NOTCH. W of the Presidential Range (which includes MOUNT WASHINGTON, the highest peak in NE U.S. at 6,288 ft/1,917 m), the cascading courses of the AMMONOOSUC and SACO rivers divide it from the FRANCONIA MOUNTAINS at CRAWFORD NOTCH. To the SW FRANCONIA NOTCH contained the famous Old Man of the Mountain (rock profile resembling face; collapsed May 2003), beneath which the PEMIGEWASSET RIVER tumbles on its way to join the MERRIMACK RIVER. The northernmost gap, DIXVILLE NOTCH, is surrounded by rocky pinnacles that look down upon a wild, fir-covered country abounding in lakes and streams. S of the mountains the lake and upland area is frequently interrupted by isolated peaks called "monadnocks," including MOUNT MONADNOCK in the SW.

Climate
The weather changes rapidly, and occasional high winds and violent storms roar through the narrow valleys and sweep over the rocky summits, especially at Mount Washington, which has clocked winds of 231 mi/372 km per hour. Annual precipitation ranges from 30 in/76 cm to 70 in/180 cm, with snowfall mounting to 8 ft/2.4 m in the mountain regions. The S and coastal areas, moderated by the ocean, experience less snowfall than other parts.

Economy
Agriculture is hampered by the mountainous topography and by extensive areas of unfertile and stony soil, but there is good farmland along the seacoast and along the Merrimack and Connecticut river valleys. The upper Connecticut valley (known as "Coos country") is pleasantly pastoral. Agricultural commodities include dairy products, apples, eggs; greenhouse products; cattle, poultry. Agriculture reached its peak in the 1880s and vast areas of farmland were abandoned in the 20th century. Today less than 5% of New Hampshire's land is in farms and some of the best land is being taken for residential and industrial development. Manufacturing has been predominant since the late 1800s. Based upon the percentage of population employed by industry, New Hampshire is one of the most industrialized states in the union. The textile mills and factories producing leather goods (such as shoes and boots), which once lined the state's fast-moving rivers, have moved to SE states and given way to high-technology firms, many of them migrating from the BOSTON area and its high tax rates. Moreover, many of the residents of S New Hampshire commute to the high-tech firms located in the Boston metropolitan area. Electrical and other machinery, as well as primary metals, fabricated metals, and plastics are also manufactured. Lumbering has been important since the first sawmill was built on the SALMON FALLS RIVER in 1631. Most of the timber cut is used in paper production, which is still important. The state's only port, Portsmouth, is situated on the estuary of the PISCATAQUA RIVER and serves as a commercial center in the state. Although New Hampshire has long been known as the Granite State, its large deposits of granite—used for building as early as 1623—are no longer extensively quarried, though some quarrying is still done around Concord. The use of steel and concrete in modern construction has greatly decreased the granite market. Today sand and gravel, stone, and clays are the state's leading minerals. Nevertheless, mineral production remains a minor factor in the New Hampshire economy.

Tourism
The year-round tourist trade is second only to industry in economic importance. Many visitors come to New Hampshire each year to enjoy the state's beaches, mountains, and lakes. The largest lake, beautiful LAKE WINNIPESAUKEE in the center of the state, is dotted with some 274 islands, while along the Atlantic shore 18 mi/29 km of curving beaches (many state-owned) attract vacationers. Of the rugged ISLES OF SHOALS off the coast (partly in Maine), three belong to New Hampshire. The landscape is dotted with covered bridges, especially in the N and W. Skiers flock N in the winter, and the state has responded to the increasing popularity of winter sports by greatly expanding its facilities. Ski resorts are distributed throughout the state (except in the SE); hiking and backpacking are important during other seasons. Some of New Hampshire's industries include the manufacturing of recreational equipment and hiking boots. The Appalachian Trail (APPALACHIAN NATIONAL SCENIC TRAIL) crosses state from HANOVER on

the Vermont state boundary, continuing NE through the White Mountains into Maine. Native crafts such as wood carving, weaving, and pottery making have been revived to attract the tourist market. New Hampshire has numerous state parks and forests, and the White Mountain National Forest in the N, which extends into Maine, has c.1,131 sq mi/2,930 sq km in New Hampshire.

Artistic New Hampshire
The state's scenic beauty and serenity have long inspired writers and artists. Hawthorne, Whittier, and Longfellow summered in New Hampshire. Augustus Saint-Gaudens sculpted many of his finest works at the artists' colony at CORNISH (Saint-Gaudens National Historic Site on the Connecticut River). The MacDowell Colony at PETERBOROUGH is a summer haven for musicians, artists, and writers; their ranks have included E. A. Robinson and Thornton Wilder. The state is most intimately connected with the works of Robert Frost; Frost himself once said that there was not one of his poems "but has something in it of New Hampshire."

History: to 1741
Martin Pring (1603) and Samuel de Champlain (1605) first explored the region. The Council for New England, formerly the Plymouth Company, received in 1620 a royal grant of land between latitude 40°N and 48°N. One of the council's leaders, Sir Ferdinando Gorges, formed a partnership with Captain John Mason and in 1622 obtained rights between the Merrimack and KENNEBEC rivers, then called the province of Maine. By a division Mason took (1629) the area between the Piscataqua and the Merrimack, naming it New Hampshire. Farmers and fishermen founded Portsmouth in 1630. Through claims based on a misinterpretation of its charter, Massachusetts annexed S New Hampshire between 1641 and 1643. Although New Hampshire was proclaimed a royal colony in 1679, Massachusetts continued to press land claims until the King-in-Council decided the E and S boundaries in 1741. Although they were technically independent of each other, the crown habitually appointed a single man to govern both colonies until 1741, when Benning Wentworth was made the first governor of New Hamphire alone.

History: 1741 to 1788
Wentworth was an expansionist and granted lands beyond the Connecticut River, close to the HUDSON, thereby provoking a protracted controversy with NEW YORK. Although a royal order in 1764 established the Connecticut River as the W boundary of New Hampshire, the dispute flared up again in the American Revolution and was only settled when Vermont became a state. The French and Indian Wars had prevented colonization of the inland areas, but a land rush began after the wars. Lumber camps were set up and sawmills were built along the streams. The Scots-Irish settlers had already initiated the textile industry by growing flax and weaving linen. By the time of the Revolution many of the inhabitants had tired of British rule and were eager for independence. In December 1774, a band of patriots overpowered Fort William and Mary (later Fort Constitution) and secured the arms and ammunition for their cause. New Hampshire was the first colony to declare its independence from Great Britain and to establish its own government (January 1776). New Hampshire became the ninth, and last, necessary state to ratify the new Constitution of the U.S. in 1788.

History: 1788 to 1930
New Hampshire's N boundary was fixed in 1842 when the Webster-Ashburton Treaty set the international line between Canada and the U.S. The Democrats remained in political control until their inability to take a united antislavery stand brought about their decline. When Franklin Pierce, the only U.S. president from New Hampshire (1853–1857), tried to smooth

over the slavery quarrel and unite his party, antislavery sentiment was strong enough to alienate many of his supporters. During the Civil War, New Hampshire was a strong supporter of the Northern cause and contributed many troops to the Union forces. After the war its economy began to emerge as primarily industrial, and population growth was steady although never spectacular. The production of woolen and cotton goods and the manufacture of shoes led all other enterprises. The forests were rapidly and ruthlessly exploited, but a bill was passed in 1911 to protect big rivers by creating forest reserves at their headwaters, including the White Mountain National Forest. Since that time numerous conservation measures have been enacted and large tracts of woodland have been placed under state and national ownership.

History: 1930 to Present
The Great Depression of the 1930s dislocated the state's economy severely, especially in the one-industry towns. The effort made then to broaden economic activities has been continually intensified. The recent establishment of important new industries such as electronics has successfully counterbalanced the departure to other states of older industries such as textiles. However, when Massachusetts fell into deep recession in the late 1980s and early 1990s, New Hampshire was similarly affected. Among the state's more prominent institutions of higher learning are Dartmouth College, at Hanover; the University of New Hampshire, at DURHAM; Keene State College, at KEENE; and Franklin Pierce University, at RINDGE.

Government and Politics
New Hampshire's present constitution was adopted in 1784; it is the second-oldest in the country. New Hampshire is the only state in which amendments to the constitution must be proposed by convention; once every seven years a popular vote determines the necessity for constitutional revision. The state's executive branch is headed by a governor and five powerful elected officers called councillors. The governor is elected for a two-year term and is traditionally limited to two successive terms. John Lynch, the current governor, has held office since January 2005. Perhaps the most unusual feature of New Hampshire politics is the size of its bicameral legislature (General Court); it is one of the largest representative bodies in the Western world, with twenty-four senators and from 375 to 400 representatives, all elected for two years. The state elects two senators and two representatives to the U.S. Congress and has four electoral votes. The New Hampshire presidential primary is among the first to be held in election years and has often forecast national trends or influenced important political decisions. Republicans have dominated New Hampshire politics since the Civil War.

New Hampshire has ten counties: BELKNAP, CARROLL, CHESHIRE, COOS, GRAFTON, HILLSBOROUGH, MERRIMACK, ROCKINGHAM, STRAFFORD, and SULLIVAN. As in other New England states, New Hampshire counties are subdivided into towns (townships) that are more important than the counties as the secondary level of government. Villages within towns have no government of their own. In Coos county, in far N, some mountain areas have land grants and land purchases (instead of towns) that have little or no population.

New Hampton, city (2000 population 3,692), ⊙ CHICKASAW county, NE IOWA, 18 mi/29 km E of CHARLES CITY; 43°03′N 92°18′W. Manufacturing (dairy products, egg processing, food processing, feed, beverages, building materials, motor vehicle parts). Incorporated 1873.

New Hampton, city (2000 population 349), HARRISON county, NW MISSOURI, 8 mi/12.9 km W of BETHANY; 40°16′N 94°12′W. Corn, wheat, soybeans; cattle.

New Hampton, town, BELKNAP county, central NEW HAMPSHIRE, 11 mi/18 km NW of LACONIA; 43°37′N 71°37′W. Bounded W and SW by PEMIGEWASSET

RIVER; town center near river. Manufacturing (machinery, commercial printing). Agriculture (nursery crops; cattle; poultry; dairying). Seat of New Hampton School. New Hampton Fish Hatchery, state's oldest fish hatchery (1920). Winter sports. Pemigewasset Lake in NE center. Winona Lake and Lake Waukewan in NE. Hersey Mountain (2,005 ft/611 m) on SE boundary.

New Hanover (NOO HAN-o-vuhr), county (□ 327 sq mi/850.2 sq km; 2006 population 182,591), SE NORTH CAROLINA, ⊙ WILMINGTON; 34°10′N 77°51′W. Bounded E by ATLANTIC OCEAN (Onslow Bay), W by CAPE FEAR RIVER, and NW by NORTHEAST CAPE FEAR RIVER. Forested tidewater area. Very little agriculture. Service industries; manufacturing and shipping at Wilmington; beach resorts along coast. Stone quarrying. INTRACOASTAL WATERWAY channel parallels coast S to Carolina Beach Inlet, continues S in Cape Fear River estuary. WRIGHTSVILLE BEACH resort area in E. PLEASURE ISLAND in S, including KURE BEACH and CAROLINA BEACH resort areas, also Carolina Beach State Park, FORT FISHER State Recreation Area and Historical Site. North Carolina State Aquarium. Formed 1729 from Craven County. Named for House of Hanover, royal family of England.

New Hanover, volcanic island (□ 460 sq mi/1,196 sq km), in the BISMARCK ARCHIPELAGO, NEW IRELAND province, NE PAPUA NEW GUINEA, 20 mi/32 km W of NEW IRELAND island; 02°30′S 150°15′E. Highest point (3,150 ft/960 m) in S center. New Hanover is mountainous and densely forested. Village of Taskul at E end, village of Umbukal at W end; in PACIFIC OCEAN; YSABEL CHANNEL to N, BISMARCK SEA to S. Coconuts; fish; crabs, lobsters; timber. GERMANY, which held the island from 1884 until World War I, called it NEU HANNOVER (New Hanover). This island region is known for long canoes, capable of holding 30 passengers. In New Ireland/New Hanover, the men have mastered the art of "shark calling," when sharks are actually called up from the depths, then speared or gaffed.

New Harmony, town (2000 population 916), POSEY county, SW INDIANA, on the WABASH RIVER, 14 mi/23 km N of MOUNT VERNON; 38°08′N 87°56′W. Agricultural area (grain, melons, fruit; livestock). Founded 1814 by the Harmony Society under George Rapp. In 1825 the Harmonists sold their holdings to Robert Owen and moved to ECONOMY, PENNSYLVANIA, where their sect survived for another seventy-eight years. Owen established a communistic colony in New Harmony that gained prominence as a cultural and scientific center and attracted many noted scientists, educators, and writers. Dissension arose, and in 1828 the community ceased to exist as a distinct enterprise, although the town remained an intellectual center. The nation's first kindergarten, first free public school, first free library, and first school with equal education for boys and girls were all established there. In New Harmony twenty-five old Rappite buildings remain. Harmony State Park nearby to SSW.

New Harmony, village (2006 population 193), WASHINGTON county, SW UTAH, 18 mi/29 km SW of CEDAR CITY, on Ash Creek; 37°28′N 113°18′W. Elevation 5,306 ft/1,617 m. Dixie National Forest to W; KOLOB CANYON section of ZION NATIONAL PARK to E. Ash Creek Reservoir to SE. Settled 1852.

New Hartford, resort town, LITCHFIELD county, NW CONNECTICUT, on FARMINGTON RIVER, just SE of WINSTED; 41°50′N 73°00′W. Agriculture; manufacturing of plumbing supplies, aircraft parts, electronics, blenders, springs, guitars, business forms, and plastics; skiing. Includes Pine Meadow village; state forest; part of Nepaug Reservoir. Flooded (1936) when nearby Greenwood Dam broke, and again in 1955. Sorner Dam and Lake McDonough just E. Summer tubing down Farmington River at nearby state park. Settled 1733, incorporated 1738.

New Hartford, town (2000 population 659), BUTLER county, N central IOWA, 15 mi/24 km WNW of WATERLOO; 42°34′N 92°37′W. Feeds.

New Hartford, village (2006 population 1,835), ONEIDA county, central NEW YORK, just SW of UTICA; 43°04′N 75°17′W. Light manufacturing. Settled c.1787, incorporated 1870.

New Haven, county (□ 862 sq mi/2,241.2 sq km; 2006 population 845,244), S CONNECTICUT, on LONG ISLAND SOUND; ⊙ NEW HAVEN and WATERBURY; 41°21′N 72°54′W. Manufacturing, agriculture, resort region, with industrial Waterbury, WALLINGFORD, NAUGATUCK, New Haven, ANSONIA, DERBY, and MERIDEN, producing wide variety of goods, especially metal products, hardware, firearms, rubber goods, silverware, electrical equipment and appliances, bricks, textiles; oil refining; agriculture (fruit, vegetables, dairy products; poultry); seed growing; fisheries; shore resorts. Drained by HOUSATONIC (W boundary), NAUGATUCK, QUINNIPIAC, and Hammonasset (E boundary) rivers. Constituted 1666.

New Haven, city (2000 population 123,626), NEW HAVEN county, S CONNECTICUT, port of entry where the QUINNIPIAC and other small rivers enter LONG ISLAND SOUND; 41°18′N 72°55′W. Firearms and ammunition, clocks and watches, tools, rubber and paper products, and textiles are among the manufacturing. The city is an educational center, being the seat of Yale University and its allied institutions and of Albertus Magnus College and Southern Connecticut State University. New Haven was founded in 1637–1638 by Puritans led by Theophilus Eaton and John Davenport. The city hosted the 1995 World Special Olympics, which attracted many international heads of state. It was one of the first planned communities in America and was the chief town of a colony that later included MILFORD, GUILFORD, STAMFORD, BRANFORD, and SOUTHOLD (on LONG ISLAND). Its government was theocratic; religion was a test for citizenship, and life was regulated by strict rules. In 1665 the colony was reluctantly united with Connecticut; it was joint capital with HARTFORD from 1701 to 1875. In the late 18th and early 19th centuries, New Haven was a thriving port. Manufacturing grew, and New Haven firearms, hardware, coaches, and carriages became famous products. New Haven was raided by a British and Tory force in the American Revolution, and the port was blockaded during the War of 1812. The world's first commercial telephone exchange was established there in 1879. The city centers upon a large public green, dating from 1680, on which stand three churches built between 1812 and 1816—Center and United Churches (both Congregational) and Trinity Church (Episcopal). Many old buildings have been preserved, and there is a historic district. Landmarks in the city are two traprock cliffs—West Rock, with the Judges' Cave, and East Rock. Since the 1950s, New Haven has received national attention for its pioneering urban renewal projects. The nation's first antipoverty program began there in 1962. Despite these improvements, the city suffered a serious racial riot in 1967. New Haven's manufacturing-based economy has since declined, and by 1990 manufacturing employed less than 20% of city's workforce. Noah Webster and Eli Whitney lived and are buried in the city. Incorporated 1784.

New Haven, city (2000 population 12,406), ALLEN county, NE INDIANA, on MAUMEE RIVER, suburb 6 mi/9.7 km E of FORT WAYNE; 41°04′N 85°02′W. Residential community. Manufacturing (machinery, livestock feed, steel fabricating, consumer goods). Laid out 1839.

New Haven, city (2000 population 1,867), FRANKLIN county, E central MISSOURI, on MISSOURI River, 12 mi/19 km W of WASHINGTON; 38°36′N 91°13′W. Corn, soybeans; dairying; cattle; manufacturing (grain products, apparel, fabricated metal, sporting equipment). Laid out 1856. Settled by German immigrants, 1840s.

Newhaven (NOO-ha-vuhn), township, VICTORIA, SE AUSTRALIA, 77 mi/124 km S of MELBOURNE, on PHILLIP ISLAND. Settled by Europeans 1842. Formerly known as Woody Point.

Newhaven (NYOO-hai-vuhn), town (2001 population 11,171), East SUSSEX, SE ENGLAND, on the CHANNEL at mouth of OUSE RIVER, 7 mi/11.3 km S of LEWES; 50°47′N 00°02′E. Seaside resort and seaport; terminal of container ships and cross-Channel ferries to DIEPPE, FRANCE. Church dates from 12th century. Just to ENE is residential district of Denton.

New Haven, town (2000 population 849), NELSON county, central KENTUCKY, on ROLLING FORK, 15 mi/24 km E of ELIZABETHTOWN; 37°39′N 85°35′W. Agriculture (tobacco, grain; livestock; dairying); manufacturing (furniture). Gethsemani Farms Trappist Monastery to E. Site of Knob Creek Farm, which was the Lincoln family home (1811–1816). Kentucky Railway Museum.

New Haven, town (2000 population 3,071), MACOMB county, SE MICHIGAN, 10 mi/16 km NNE of MOUNT CLEMENS; 42°43′N 82°47′W. Satellite community of DETROIT. In farm area; manufacturing (motor vehicle parts, iron foundry). Incorporated 1869.

New Haven, town, ADDISON CO., W VERMONT, on NEW HAVEN RIVER, 9 mi/14.5 km N of MIDDLEBURY. Named for NEW HAVEN, CONNECTICUT.

New Haven, town (2006 population 1,527), MASON county, W WEST VIRGINIA, near the OHIO RIVER, 14 mi/23 km NE of POINT PLEASANT; 38°59′N 81°58′W. Agriculture (grain, tobacco); livestock; dairying. Coal-mining area. Manufacturing (metal alloys, bricks, food). Railroad junction to S. Racine Lock and Dam (Ohio River) to SE.

New Haven, village (2000 population 477), GALLATIN county, SE ILLINOIS, on LITTLE WABASH RIVER and 14 mi/23 km NNE of SHAWNEETOWN; 37°53′N 88°07′W. In agricultural area.

Newhaven, Scotland: see EDINBURGH.

New Haven River, c.25 mi/40 km long, W VERMONT; rises in GREEN MOUNTAINS E of MIDDLEBURY; flows NW and SW to OTTER CREEK at WEYBRIDGE.

New Hazelton (noo HAI-zuhl-tuhn), district municipality (□ 10 sq mi/26 sq km; 2001 population 750), central BRITISH COLUMBIA, W Canada; 55°15′N 127°35′W. Forestry, tourism. Incorporated 1980.

New Hebrides: see VANUATU.

New Hebron (noo HEE-bruhn), village, LAWRENCE county, S central MISSISSIPPI, 40 mi/64 km SSE of JACKSON. Agriculture (cotton, corn; livestock); manufacturing (food, apparel).

New Hempstead, village (□ 2 sq mi/5.2 sq km; 2006 population 4,828), ROCKLAND county, SE NEW YORK, 8 mi/12.9 km NW of NYACK; 41°08′N 74°02′W. Sizable Orthodox Jewish population, along with other ethnicities.

New Hogan Reservoir, c.8 mi/12.9 km long, CALAVERAS county, central CALIFORNIA, impounded in CALAVERAS RIVER, 9 mi/14.5 km W of SAN ANDREAS. North Fork of Calaveras River enters upstream and from NE.

New Holland, village, HALL county, NE GEORGIA, 3 mi/4.8 km NE of GAINESVILLE; 34°18′N 83°48′W. Textile manufacturing.

New Holland, village (2000 population 318), LOGAN county, central ILLINOIS, 13 mi/21 km WNW of LINCOLN; 40°10′N 89°34′W. In agricultural and bituminous coal area.

New Holland (HAH-luhnd), village (□ 2 sq mi/5.2 sq km; 2006 population 787), PICKAWAY county, S central OHIO, 16 mi/26 km WSW of CIRCLEVILLE; 39°33′N 83°15′W. In livestock and general agricultural area.

New Holland, borough (2006 population 5,146), LANCASTER county, SE PENNSYLVANIA, 12 mi/19 km ENE of LANCASTER; 40°06′N 76°05′W. In Pennsylvania Dutch region. Manufacturing (feeds, wire and cable, farm equipment, furniture, clothing, wood products, concrete, lumber, printing and publishing, food products); agriculture (grain, soybeans, apples; livestock; dairying). Settled 1728, incorporated 1895.

New Holstein (HOL-steen), town (2006 population 3,181), CALUMET county, E WISCONSIN, 23 mi/37 km NW of SHEBOYGAN; 43°57′N 88°05′W. In dairying and grain-growing area. Manufacturing (construction equipment, machinery, animal feed, engines). Settled 1849, incorporated 1926.

New Hope, city (2000 population 20,873), HENNEPIN county, E MINNESOTA, suburb 7 mi/11.3 km NW of MINNEAPOLIS; 45°02′N 93°23′W. Manufacturing (lithography, electronic equipment, printing). Several small lakes in area; Medicine Lake to SW.

New Hope, unincorporated town (2000 population 1,964), LOWNDES county, E MISSISSIPPI, 6 mi/9.7 km ESE of COLUMBUS, near ALABAMA state line; 33°27′N 88°19′W. Agriculture (cotton, corn, soybeans; cattle). Lake Lowndes State Park to S.

New Hope (NOO HOP), unincorporated town, WAKE county, central NORTH CAROLINA, residential suburb 5 mi/8 km NE of RALEIGH, near NEUSE RIVER; 35°48′N 78°33′W.

New Hope, unincorporated town, WAYNE county, E central NORTH CAROLINA, residential suburb 8 mi/12.9 km ENE of GOLDSBORO; 35°22′N 77°53′W. In agricultural area. Seymour Johnson Air Force Base to SW.

New Hope, village (2000 population 2,539), Madison co., N ALABAMA, near Paint Rock River, 17 mi/27 km SE of Huntsville. Originally called 'Cloud's Town' for William Cloud, the first settler. Later known a 'Vienna,' it was changed to New Hope in 1834. Inc. in 1956.

New Hope (HOP), village (2006 population 706), COLLIN county, N TEXAS, residential suburb 32 mi/51 km NNE of DALLAS; 4 mi/6.4 km ENE of MCKINNEY, near East Fork of TRINITY RIVER; 33°12′N 96°33′W. Agricultural area in urban fringe.

New Hope, borough (2006 population 2,291), BUCKS county, SE PENNSYLVANIA, 14 mi/23 km NW of TRENTON (NEW JERSEY), on DELAWARE RIVER (bridged), opposite LAMBERTVILLE (New Jersey); 40°21′N 57°74′W. Agricultural area (grain, soybeans, apples, grapes; livestock; dairying); manufacturing (concrete, furniture, optical instruments). Artist, literary, and theatrical colony. Tourist center, shops. WASHINGTON CROSSING State Park to S, one of two places where George Washington crossed Delaware River in 1776; Bucks County Winery to W; Buckingham Valley Winery to SW. Settled c.1712, incorporated 1837.

New Houlka (HUHLK-uh), village (2000 population 710), CHICKASAW county, NE central MISSISSIPPI, 24 mi/39 km SW of TUPELO, near source of YALOBUSHA RIVER. Agriculture (cotton, corn, soybeans; dairying; timber); manufacturing (furniture). Part of Tombigbee National Forest to E. Old Houlka 1 mi/1.6 km to E. Also called Houlka.

New Hudson, village, OAKLAND county, SE MICHIGAN, unincorporated suburb 18 mi/29 km SW of PONTIAC; 42°30′N 83°36′W. In farm area; manufacturing (glass products, motor vehicle parts, machinery). Oakland Southwest Airport is here. Island Lake State Recreation Area to W.

New Hunstanton, ENGLAND: see HUNSTANTON.

New Hyde Park, village (2006 population 9,393), NASSAU county, SE NEW YORK, on LONG ISLAND; 40°43′N 73°41′W. It is a residential community with some manufacturing and a few small farms along with commercial services. To the N is North New Hyde Park. Incorporated 1927.

Newi, NIGERIA: see NNEWI.

New Iberia (ei-BIR-ee-uh), city (2000 population 32,623), ⊙ IBERIA parish, S LOUISIANA, 16 mi/26 km SSE of LAFAYETTE, on BAYOU TECHE, which is connected to the INTRACOASTAL WATERWAY by a canal; 30°01′N 91°49′W. Railroad junction. Printing and

publishing, manufacturing (oil and gas drilling equipment, fabricated steel, food products, hunting equipment, ceramics, lumber, animal feeds); especially known for pepper sauces. Acadian refugees from NOVA SCOTIA, CANADA, settled here beginning c.1765, and French is still spoken by many of the inhabitants. Base for the pirate Jean Lafitte's slave trading. Numerous antebellum houses are in the area; among them are Justine (1822) and Shadows on the Teche (1834), a classic example of Greek Revival architecture. A sugarcane festival is held in New Iberia every September. Has a gambling and red-light district. Port of New Iberia, 5 mi/8 km SSW, shallow-draft facilities. Nearby are many wildlife refuges, sheltering a multitude of migratory birds. Longfellow Evangeline State Commemorative Area to N. Incorporated 1836.

Newington (NOO-weeng-tuhn), unincorporated city, FAIRFAX county, NE VIRGINIA suburb 9 mi/15 km SW of ALEXANDRIA, 15 mi/24 km SW of WASHINGTON, D.C., on Accotink Creek; 38°44′N 77°12′W. Manufacturing (concrete, fabricated metal products, stains, commercial printing, satellites). FORT BELVOIR Military Reservation to SE, Fort Belvoir Proving Ground to NW.

Newington, town (2000 population 29,306), HARTFORD county, central CONNECTICUT, suburb of HARTFORD; 41°41′N 72°43′W. Chiefly residential. Industry includes milk processing and the manufacturing of airplane parts, ball bearings, metal fabrication, tools, and plumbing supplies. Site of children's hospital. Settled 1670, incorporated 1871.

Newington, town (2000 population 322), SCREVEN county, E GEORGIA, 16 mi/26 km SE of SYLVANIA, near SOUTH CAROLINA state line; 32°35′N 81°31′W.

Newington, town, ROCKINGHAM county, SE NEW HAMPSHIRE, 5 mi/8 km NW of PORTSMOUTH; 43°06′N 70°50′W. Bounded W by GREAT BAY, NW by Little Bay, NE by PISCATAQUA RIVER (Maine state line). Manufacturing (lab equipment, food processing, communications equipment, swimming pools); agriculture (nursery crops; cattle, poultry, seafood; dairying). Pease International Tradeport (including airport, formerly Pease Air Force Base) in E; shipping; oil and gas terminal.

New Inlet (NOO IN-luht), channel through the OUTER BANKS, DARE county, E NORTH CAROLINA, connecting PAMLICO SOUND with the ATLANTIC c.30 mi/48 km N of Cape HATTERAS. Channel closed by sand movements, linking PEA ISLAND (N) with HATTERAS ISLAND (S); 36°12′N 75°45′W. Also formerly called Trinite Harbor.

New Ipswich, town, HILLSBOROUGH county, S NEW HAMPSHIRE, 20 mi/32 km W of NASHUA; 42°44′N 71°52′W. Bounded S by MASSACHUSETTS state line; drained by SOUHEGAN RIVER. Manufacturing (concrete products, metal products, textiles); agriculture (fruit, vegetables, corn, nursery crops; poultry, cattle, hogs; dairying). State's first cotton mill built here (1803), first woolen mill (1801). Windblown Ski Touring Center in W; Annett State Forest to W in RINDGE. Incorporated 1762.

New Ireland, province (2000 population 118,350), E PAPUA NEW GUINEA, includes NEW IRELAND and NEW HANOVER islands and LIHIR, TABAR, St. Matthias, TANGA and FENI island groups; ⊙ KAVIENG, at NW end of New Ireland. Bounded on S by SOLOMON SEA, SW by SAINT GEORGE CHANNEL and BISMARCK SEA, NE by PACIFIC OCEAN. Copra, coconuts, palm oil; fish.

New Ireland, volcanic island (□ 3,340 sq mi/8,684 sq km), SW PACIFIC OCEAN, in the BISMARCK ARCHIPELAGO, part of PAPUA NEW GUINEA; 03°20′S 152°00′E. New Ireland is largely mountainous, rising to circa 4,000 ft/1,220 m. Much of the island is under cultivation, especially the E coast. KAVIENG is the chief town and port. The island was first sighted in 1616, but until 1797 it was thought to be part of NEW BRITAIN,

from which it is separated by a 20-mi/32-km channel. The island was a German protectorate from 1884 to 1914 and was called Neu Mecklenburg (New Mecklenburg) by the Germans.

New Jersey, state (□ 8,722 sq mi/22,677.2 sq km; 2006 population 8,724,560), E UNITED STATES, one of the Mid-Atlantic states and one of the original thirteen colonies; ⊙ TRENTON; 40°16′N 74°42′W. NEWARK is the largest city. New Jersey is known as the "Garden State."

Geography

Surrounded by water except along 50 mi/80 km of the N border with NEW YORK state, New Jersey is bounded on the E by the HUDSON RIVER, NEW YORK BAY, and the ATLANTIC OCEAN and on the S and W by DELAWARE BAY and the DELAWARE RIVER (which separate it from DELAWARE and PENNSYLVANIA). The N third of New Jersey lies within the APPALACHIAN Highland region, where ridges running NE and SW shelter valleys containing pleasant streams and glacial lakes. Beyond the crest of wooded slopes are long-established farms given over to dairying and field crops. KITTATINNY MOUNTAIN, with the state's highest elevations (up to 1,803 ft/550 m), stretches across the NW corner of New Jersey from the New York state line to the DELAWARE WATER GAP. SE of the Highlands lie the Triassic lowlands or PIEDMONT plains, extending from the NE border to Trenton and encompassing every major city of the state except CAMDEN and ATLANTIC CITY. The monotony of the lowlands is broken by ancient trap-rock ridges that extend to the PALISADES of the Hudson, and many commuter towns are located along the wooded slopes. E of Newark and HACKENSACK acres of tidal marshes have been converted to industrial, office, and commercial use. This area, called the Meadowlands, also contains a huge sports and entertainment complex. Drainage is provided by the state's major rivers, the PASSAIC, the RARITAN, and the HACKENSACK. The busy lowlands give way in the SE to the coastal plains, which cover more than 50% of the state. The coast itself is highly developed as a resort area. Offshore barrier islands make large harbors impractical but provide 115 mi/185 km of sheltered waterways that have made possible a superior combination of bay and ocean facilities.

Economy

Only four states are smaller in size than New Jersey, yet New Jersey ranks ninth in the nation in population, a fact indicative of its economic importance. It is a major industrial center, an important transportation terminus, and a long-established playground for summer vacationers. The state is noted for its output of chemicals and pharmaceuticals, machinery, and a host of other products, including telecommunications research and development and electronic equipment. CAMDEN is an important port. Stone, zinc, and sand and gravel are the state's only native mineral resources of consequence. There are oil refineries at LINDEN and CARTERET. The area near PRINCETON has developed into a center for high-technology industry, and New Jersey has been a leader in the research industry since Thomas Edison established a research facility at MENLO PARK in 1876. The finance industry has also become important to the state's economy, attracting many corporations from NEW YORK city. These developments have to a large extent reversed New Jersey's role as a suburban commuter area for New York city and PHILADELPHIA.

Transportation

A tremendous transportation system, concentrated in the industrial lowlands, moves products and a huge volume of interstate traffic through the state. Busy highways like the Garden State Parkway and the NEW JERSEY TURNPIKE are part of a network of toll roads and freeways. New Jersey is linked to Delaware and Pennsylvania by many bridges across the Delaware River. Traffic to and from New York is served by

---

Cross-references are shown in SMALL CAPITALS. The pronunciation guide is shown on page xix. The sources of population figures are shown on page xvii.

railroad and subway tunnels and by the facilities of the PORT AUTHORITY OF NEW YORK AND NEW JERSEY—the double-decked GEORGE WASHINGTON BRIDGE, the LINCOLN and HOLLAND vehicular tunnels, and three bridges to STATEN ISLAND. Airports are operated by many cities, and NEWARK LIBERTY INTERNATIONAL AIRPORT (controlled by the Port Authority) ranks among the nation's busiest. Shipping in New Jersey centers on the ports of the NEWARK BAY and New York Bay areas, with relatively minor seagoing traffic on the Delaware as far N as Trenton. Because of this extensive transportation network, New Jersey's ocean beaches, inland lakes, forests, and low mountain areas are the basis for the state's traditional vacation industry. Tourism is New Jersey's second-largest industry, due largely to the emergence of Atlantic City as a major gambling center.

Agriculture

In addition to being a center of industry, transportation, and tourism, New Jersey is a leading state in agricultural income per acre. The scrub pine area of the S inland region is used for cranberry and blueberry growing. N of the pine belt the soil is extremely fertile and supports a variety of crops, most notably potatoes, corn, hay, peaches, and vegetables (especially tomatoes and asparagus). Dairy products, eggs, and poultry are also important. Huge commercial and residential expansion, however, has taken over much of the state's farmland.

History: to 1681

The history of New Jersey goes back to Dutch and Swedish communities established prior to settlement by the English. Dutch claims to the Hudson and Delaware valleys were based on the voyages of Henry Hudson, who sailed into Newark Bay in 1609. Under the auspices of the Dutch West India Company patroonships were offered for settlement, and small colonies were located on the present sites of HOBOKEN, JERSEY CITY, and GLOUCESTER CITY. Swedes and Finns of NEW SWEDEN, who predominated in the Delaware Valley after 1638, were annexed by the NEW NETHERLAND colony in 1655. In 1664, New Netherland was seized by the English, but the Dutch disputed this claim. Proprietorship of lands between the Hudson (at latitude 41°N) and the northernmost point of the Delaware was granted to Lord John Berkeley and Sir George Carteret. The original grants to Berkeley and Carteret divided the region in two sections. The split was further defined in the Quintipartite Deed of 1676, which divided the province into East and West Jersey, East Jersey being held by Carteret.

History: 1681 to 1776

In 1681, William Penn and eleven other Quakers purchased East Jersey from Carteret's widow. In both Jerseys, confusion resulting from the unwieldy number of proprietors together with widespread resentment against authority caused the proprietors to surrender voluntarily their governmental powers to the crown in 1702, although they retained their land rights. New Jersey's independence from New York was recognized, but authority was vested in the governor of New York until 1738, when Lewis Morris was appointed governor of New Jersey alone. Under the royal governors the same problems persisted—land titles were in dispute and opposition to the proprietors culminated in riots in the 1740s. E Jersey was dominated by Calvinism, implanted by Scottish and New England settlers, while in W Jersey the Quakers soon developed a landed aristocracy with strong political and economic influence. Anti-British sentiment gradually spread from its stronghold in E Jersey throughout the colony and took shape in Committees of Correspondence. Although the Tory party was to prove strong enough to raise six Loyalist battalions, the patriot cause was generally accepted, and in June 1776, the provincial congress adopted a constitution and declared New Jersey a state.

History: 1776 to 1790

Because of its strategic position, New Jersey was of major concern in the American Revolution. George Washington's memorable Christmas attack on the Hessians at Trenton in 1776, followed by his victory at PRINCETON, restored the confidence of the patriots. In June 1778, Washington fought another important battle in New Jersey, at Monmouth. Altogether, about ninety engagements were fought in the state, and Washington moved his army across it four times, wintering twice at MORRISTOWN. At the Federal Constitutional Convention in 1787, the delegates from New Jersey sponsored the cause of the smaller states and carried the plan for equal representation in the Senate. New Jersey was the third state to ratify (December 1787) the U.S. Constitution. By this time New Jersey's population had grown from an estimated 15,000 in 1700 to approximately 184,000. Trenton became the state's capital in 1790.

History: 1790 to 1865

Agriculture had been supplemented by considerable mining and processing of iron and copper and by the production of lumber, leather, and glass. During the next fifty years, a period of enormous economic expansion, the dominance of the landed aristocracy gave way to industrial growth and to a more democratic state government. The important textile industry, powered by the falls of the Passaic, was initiated at PATERSON. Potteries, shoe factories, and brickworks were built. Roads were improved, the MORRIS, DELAWARE AND RARITAN canals were chartered, and the Camden and Amboy Railroad completed a line from New York city to Philadelphia with monopoly privileges. Prior to the Civil War an era of reform resulted in the framing of a new state constitution (1844) in which property qualifications for suffrage were abolished, provisions were made for the popular election of the governor and the assemblymen, and a balance of power and responsibility was established among the executive, legislative, and judicial departments. In spite of some pro-Southern sentiment, New Jersey recruited its quota of regiments in the Civil War and gave valuable financial aid to the Union. The war demands proved lucrative for commerce and industry, and the expanding labor market attracted large numbers of European immigrants.

History: 1865 to 1966

By 1865 the pattern of the state's development was molded. Population and industry showed rapid and steady growth. Large economic interests grasped control of political power, giving rise to sporadic but unsustained popular movements for reform. The Camden and Amboy Railroad was transferred by lease to the Pennsylvania Railroad in 1871, and its monopolistic power was lessened by legislation opening the state to all railroad lines and by the assessment and taxation of railroad properties. After the 1870s easy incorporation laws and low corporation tax rates attracted new trusts to incorporate through "dummy" offices in the state. There was much liberal sentiment against the power of "big business." A general reform movement sponsored by Woodrow Wilson when he was governor (1910–1912) resulted in such legislation as the direct primary, a corrupt practices act, and the "Seven Sisters" acts for the regulation of trusts (later repealed). The state voted predominantly Democratic from the Civil War until 1896. Since that time it has frequently voted Republican in national elections, and in state politics it has often divided power between Democratic governors and Republican legislatures. The powerful political machine of Frank Hague, centered in Jersey City, wielded great influence in the Democratic Party from 1913 to 1949, when it was defeated by insurgents within its ranks. In 1947 a new constitution was framed and accepted to replace the antiquated constitution of 1844. The liberal Bill of Rights was preserved and extended, governmental

departments were streamlined, the cumbersome court system was simplified, the executive power was strengthened, and labor's right to organize and bargain collectively was recognized. In 1966 another convention was called to rewrite those portions of the 1947 constitution invalidated by application of the U.S. Supreme Court's "one man, one vote" rule to state legislatures. The convention drafted sweeping revisions, which were approved by the electorate in November 1966.

History: 1967 to Present

A six-day race riot in Newark in July 1967 drew attention to the urgent need for social and political reform in many of the state's urban centers. Kenneth A. Gibson was elected as Newark's first African-American mayor in 1970. During the early 1970s the state government proposed plans for massive urban renewal and economic development projects, but the trend of movement away from central cities like Newark increased throughout the 1970s and 1980s and into the 1990s. During this period, New Jersey lost thousands of manufacturing jobs but replaced them through the dramatic development of the service and trade sectors of the economy. In 1976 the state legalized casino gambling and in 1978 the first casino opened in Atlantic City. The Meadowlands Sports Complex opened in 1976 and grew to include Giants Stadium (1977), home of the NFL's New York Giants and New York Jets football teams, and Continental Airlines Arena (1981), home of the NBA's New Jersey Nets basketball team and the NHL's New Jersey Devils hockey team, and special events. New Jersey was hard hit by recession in the early 1990s as economic growth slowed and the state suffered the effects of overdevelopment. Political scandal in 2004 prompted the resignation of Governor James McGreevey.

Government

The New Jersey legislature consists of a senate of forty members, elected to serve four-year terms, and an assembly of eighty members, elected for two-year terms. The governor serves a four-year term and may be reelected once. The current governor is Jon S. Corzine, formerly a U.S. Senator for the state, who was sworn in January 2006. New Jersey sends thirteen representatives and two senators to the U.S. Congress and has fifteen electoral votes.

New Jersey has twenty-one counties: ATLANTIC, BERGEN, BURLINGTON, CAMDEN, CAPE MAY, CUMBERLAND, ESSEX, GLOUCESTER, HUDSON, HUNTERDON, MERCER, MIDDLESEX, MONMOUTH, MORRIS, OCEAN, PASSAIC, SALEM, SOMERSET, SUSSEX, UNION, and WARREN.

New Jersey Turnpike, part of NEW JERSEY highway system extending NE from DEEPWATER on DELAWARE RIVER at DELAWARE MEMORIAL BRIDGE to RIDGEFIELD PARK, with an extension to FORT LEE and the GEORGE WASHINGTON BRIDGE. Toll road.

New Juaben, GHANA: see KOFORIDUA.

New Kensington, city (2006 population 13,935), WESTMORELAND county, SW PENNSYLVANIA, suburb 15 mi/24 km NE of PITTSBURGH, on the ALLEGHENY RIVER (bridged); 40°34′N 79°45′W. Coal-mining area. Manufacturing (ceramics, machinery, concrete, plastic products, crushed stone, chemical blending, rubber products). Pennsylvania State University (New Kensington campus). Laid out 1891 on the site of Fort Crawford (1778), incorporated as a city 1933.

New Kent (NOO KENT), county (□ 223 sq mi/579.8 sq km; 2006 population 16,852), E VIRGINIA; ⊙ NEW KENT; 37°30′N 77°00′W. Agriculture (hay, barley, wheat, corn, soybeans, sweet and white potatoes; poultry); timber. In Tidewater region, bounded S by CHICKAHOMINY RIVER, N by PAMUNKEY RIVER, and NE by YORK RIVER estuary. Formed 1654.

New Kent (NOO KENT), unincorporated village, ⊙ NEW KENT county, E VIRGINIA, 25 mi/40 km E of RICHMOND, near PAMUNKEY RIVER; 37°31′N 76°58′W.

Agriculture (potatoes, grain, soybeans; poultry). County seat since 1961.

New Kingston, resort village, DELAWARE county, S NEW YORK, in the Catskills, c.40 mi/64 km NW of KINGSTON; 42°13′N 74°41′W.

Newkirk, town (2006 population 2,141), ⊙ KAY county, N OKLAHOMA, 13 mi/21 km N of PONCA CITY; 36°52′N 97°03′W. Elevation 1,154 ft/352 m. Grain elevators; oil and gas wells; light manufacturing KAW LAKE is E. Settled c.1893.

New Knoxville (NAHKS-vil), village (2006 population 907), AUGLAIZE county, W OHIO, 8 mi/13 km SW of WAPAKONETA; 40°29′N 84°19′W. Lumber, clay products.

New Lambton (NOO LAM-tuhn), W residential suburb of NEWCASTLE, NEW SOUTH WALES, SE AUSTRALIA; 32°56′S 151°43′E.

New Lanark, Scotland: see LANARK.

Newland, village (2006 population 667), ⊙ AVERY county, NW NORTH CAROLINA, 27 mi/43 km NNW of MORGANTON, in the BLUE RIDGE MOUNTAINS, at edge of Pisgah National Forest; 36°05′N 81°55′W. North Toe River, near its source. Retail trade; agriculture (cattle; tobacco). GRANDFATHER MOUNTAIN (5,964 ft/1,818 m) to E.

Newland, ENGLAND: see REDBROOK.

Newlands (NOO-luhndz), cattle station, E central QUEENSLAND, NE AUSTRALIA, 70 mi/113 km N of MORANBAH; 33°40′S 115°52′E. Coal mine.

Newlands, suburb of CAPE TOWN, WESTERN CAPE province, SOUTH AFRICA, SSE of city center; 33°58′S 18°58′E. Site of KIRSTENBOSCH NATIONAL BOTANICAL GARDENS (□ 400 acres/162 ha), established 1913. Nearby is Groote Schuur, residence of the state president, used while Parliament in session. Oldest cricket and rugby grounds located here, known simply as "Newlands."

Newlands project, on the CARSON and TRUCKEE rivers, W NEVADA. Was the 1st irrigation project built by the U.S. Bureau of Reclamation (1903–1908), begun with Derby Dam. Nearby LAHONTAN DAM, completed in 1915, was built to produce electricity for the project.

New Lebanon (LE-buh-nuhn), village (□ 2 sq mi/5.2 sq km; 2006 population 4,160), MONTGOMERY county, W OHIO, 10 mi/16 km W of DAYTON; 39°45′N 84°24′W. In agricultural area.

New Lebanon (LE-bah-nahn), borough (2006 population 196), MERCER county, NW PENNSYLVANIA, 16 mi/26 km SSE of MEADVILLE; 41°25′N 80°04′W. Corn, hay; livestock; dairying. WILHELM LAKE reservoir, in M. K. Goddard State Park, to SW.

New Lebanon, hamlet, COLUMBIA county, SE NEW YORK, near MASSACHUSETTS state line, 8 mi/12.9 km W of PITTSFIELD (Mass.); 42°28′N 73°26′W. Site of an early Shaker community. Politician Samuel Tilden was born here.

New Leipzig, village (2006 population 240), GRANT county, S NORTH DAKOTA, 18 mi/29 km E of MOTT, near CANNONBALL RIVER; 46°22′N 101°57′W. Founded in 1910.

New Lenox (LEN-iks), village (2000 population 17,771), WILL county, NE ILLINOIS, 5 mi/8 km E of JOLIET, residential satellite community of CHICAGO; 41°31′N 87°58′W. Agriculture (corn, soybeans; dairying); manufacturing (machinery, furniture). Incorporated 1946.

New Lenox, MASSACHUSETTS: see LENOX.

New Lexington (LEK-sing-tuhn), city (□ 2 sq mi/5.2 sq km; 2006 population 4,617), ⊙ PERRY county, central OHIO, 19 mi/31 km SSW of ZANESVILLE; 39°43′N 82°12′W. Trade center and distribution point for coal-, sand-, and oil-producing area. Previously tiles, machine tools. Laid out 1817.

New Liberty, town (2000 population 121), SCOTT county, E IOWA, 20 mi/32 km NW of DAVENPORT; 41°43′N 90°52′W. In agricultural area; manufacturing.

New Limerick (LIM-uhr-ik), town, AROOSTOOK county, E MAINE, just W of HOULTON; 46°07′N 67°58′W. Agriculture.

Newlin (NOO-lin), unincorporated village, HALL county, NW TEXAS, 17 mi/27 km NW of CHILDRESS and on PRAIRIE DOG TOWN FORK of RED RIVER; 34°35′N 100°26′W. Cotton, peanuts; cattle.

New Lisbon, town (2006 population 2,361), JUNEAU county, central WISCONSIN, on LEMONWEIR RIVER, c.55 mi/89 km E of LA CROSSE; 43°52′N 90°09′W. Railroad junction. In dairying and farming region. Manufacturing (furniture, machinery). Necedah National Wildlife Refuge and Central Wisconsin Conservation Area to N. Incorporated 1889.

New Liskeard (LI-skuhrd), town (□ 3 sq mi/7.8 sq km; 2001 population 4,906), E ONTARIO, E central Canada, at N end of Lake TIMISKAMING, 5 mi/8 km NNW of HAILEYBURY; 47°30′N 79°40′W. Manufacturing (paper products), printing, pulp and lumber milling, food processing, dairying; resort. Market center for surrounding mining region. Agricultural research station (1922). Forms part of city of Temiskaming Shores.

New Llano (LAH-no), town (2000 population 2,415), VERNON parish, W LOUISIANA, 2 mi/3 km S of LEESVILLE; 31°06′N 93°16′W. In agricultural area (poultry, cattle; sweet potatoes, watermelons; dairying); timber. Llano Cooperative Colony, founded by a socialist utopian group, was established here in 1917. FORT POLK MILITARY RESERVATION to E, ANACOCO Prairie State Game and Fish Preserve to W, Kisatchie National Forest to SE. Also spelled Newllano.

New London, county (□ 771 sq mi/2,004.6 sq km; 2006 population 263,293), SE CONNECTICUT, on LONG ISLAND SOUND and RHODE ISLAND state line, bounded W by the CONNECTICUT; ⊙ NEW LONDON and NORWICH; 41°28′N 72°07′W. New London is a Coast Guard and submarine center; county has diversified manufacturing (textiles, metal products, medical supplies, machinery, printing presses, paper products, chemicals, clothing, consumer goods, silverware, boats, leather products); agriculture (dairy products; poultry; fruit, vegetables). Resorts on coast. Drained by YANTIC, SHETUCKET, THAMES, MYSTIC, QUINEBAUG, PAWCATUCK (E boundary), and Niantic rivers. Located 3.2 mi/5.1 km ENE of New London are three nuclear power plants: Millstone 1 (initial criticality October 26, 1970; maximum dependable capacity of 650 MWe), Millstone 2 (initial criticality October 17, 1975; maximum dependable capacity of 860 MWe), and Millstone 3 (initial criticality January 23, 1986; maximum dependable capacity of 1150 MWe); all use cooling water from Long Island Sound. Constituted 1666.

New London, city, NEW LONDON county, SE CONNECTICUT, on the THAMES River near its mouth on LONG ISLAND SOUND; 41°19′N 72°05′W. It is a deepwater port of entry, with manufacturing (shipbuilding; textiles; high-technology research and engineering; building materials; fishing, tourism; furniture, and paper products). New London survived a partial burning by the British under Benedict Arnold in 1781 and a British blockade during the War of 1812. The city reached the peak of its maritime prosperity in the 19th century, when it flourished as a shipping, shipbuilding, and whaling port. The excellent harbor is used by the U.S. Navy as a principal submarine base, by yachters and students of the U.S. Coast Guard Academy (located in the city), and by the Coast Guard Officers Training Command. Annual Yale-Harvard boat races are held on the Thames River. Connecticut College and Mitchell College are here. The city has a whaling museum, an art museum, and many old buildings, including the Hempsted House (1678) the old town mill (1650) and Old Fort Trumbull (1849). Laid out 1646 by John Winthrop, incorporated 1784.

New London, city (2000 population 1,001), ⊙ RALLS county, NE MISSOURI, near SALT RIVER, 8 mi/12.9 km S of HANNIBAL; 39°34′N 91°24′W. Soybeans, corn; hogs; manufacturing (consumer goods); agricultural trade center. Historic courthouse (1857–1858).

New London, town (2000 population 1,937), HENRY county, SE IOWA, 8 mi/12.9 km ESE of MOUNT PLEASANT. In livestock and grain area. Founded as Dover; incorporated 1860.

New London, town (2000 population 1,066), KANDIYOHI county, S central MINNESOTA, 14 mi/23 km NNE of WILLMAR, on CROW RIVER; 45°17′N 94°57′W. Manufacturing (machinery, concrete). Fish hatchery here. Sibley State Park, on Lake Andrew to W; Mount Tom (1,375 ft/419 m) to W; GREEN LAKE to SE, Mud Lake to N.

New London, town, MERRIMACK county, S central NEW HAMPSHIRE, 27 mi/43 km NW of CONCORD, in hilly country; 43°25′N 71°59′W. Manufacturing (printing and publishing); timber; agriculture (nursery crops, vegetables; livestock; dairying). Seat of Colby-Sawyer College. Resort and retirement area. Sunapee Lake on SW boundary. Little Squam Lake in W; Pleasant Lake in E; Norsk Ski Touring Area in SE.

New London (LUHN-duhn), town (2006 population 1,005), RUSK county, E TEXAS, 21 mi/34 km ESE of TYLER; 32°16′N 94°55′W. Oil town in E Texas field; oil and gas; cattle, horses; dairying; vegetables, watermelons; timber. A school explosion here (March 18, 1937) took the lives of hundreds of pupils and teachers.

New London, town (2006 population 7,019), on the border between WAUPACA and OUTAGAMIE counties, E central WISCONSIN, at confluence of WOLF and EMBARRASS rivers, 27 mi/43 km NNW of OSHKOSH; 44°23′N 88°44′W. Railroad junction. In agricultural area (cabbage; poultry); manufacturing (food, machinery, furniture, dairy products, wood products, brick); lumber milling. Has Carr Museum, with natural history and historical exhibits. Founded c.1853, incorporated 1877.

New London (NOO LUHN-duhn), village (2006 population 320), STANLY county, S central NORTH CAROLINA, 8 mi/12.9 km NW of ALBEMARLE; 35°26′N 80°13′W. Railroad junction to S. Manufacturing (lumber, furniture); service industries; agriculture (cotton, grain, soybeans; dairying; livestock; timber. BADIN to E, TUCKERTOWN LAKE reservoir to NE, both on Yadkin (PEE DEE) River.

New London (LUHN-duhn), village (□ 3 sq km; 2006 population 2,597), HURON county, N OHIO, 15 mi/24 km SE of NORWALK; 41°04′N 82°24′W. Trade and shipping point; manufacturing of clay and cement products. Settled 1816.

New Lothrop (LAH-thrup), village (2000 population 603), SHIAWASSEE county, S central MICHIGAN, 16 mi/26 km W of FLINT; 43°07′N 83°58′W.

Newlyn, ENGLAND: see PENZANCE.

New Lynn, S suburb within WAITAKERE CITY, a NW extension of AUCKLAND urban area, N NORTH ISLAND, NEW ZEALAND. Manufacturing of brick, tile, pottery. Incorporated into Waitakere City in 1989.

New Maas River, Dutch *Nieuwe Maas*, 13 mi/21 km long, SW NETHERLANDS; formed by junction of LEK and NOORD rivers (channels of the MEUSE RIVER), just W of KRIMPEN AAN DEN LEK; flows W, past ROTTERDAM, PERNIS, and VLAARDINGEN. Joins OLD MAAS RIVER 7 mi/11.3 km WSW of Rotterdam to form the NEW WATERWAY, main shipping channel for Rotterdam. New Maas serves as Rotterdam's harbor. Joined by HOLLANDSE IJSSEL RIVER 2 mi/3.2 km E of Rotterdam; IJSSELMONDE island on S.

Newmachar (NYOO-mah-kuhr), town (2001 population 2,318), Aberdeenshire, NE Scotland, 9 mi/14.5 km NNW of ABERDEEN; 57°16′N 02°11′W. Formerly in Grampian, abolished 1996.

Cross-references are shown in SMALL CAPITALS. The pronunciation guide is shown on page xix. The sources of population figures are shown on page xvii.

New Madison (MAD-i-suhn), village (2006 population 772), DARKE county, W OHIO, 11 mi/18 km SSW of GREENVILLE; 39°58′N 84°42′W. Canned foods; grain.

New Madrid, county (□ 679 sq mi/1,765.4 sq km; 2006 population 18,314), extreme SE MISSOURI, the MISSISSIPPI RIVER on E; ⊙ NEW MADRID; 36°35′N 89°39′W. Crossed by LITTLE RIVER and drainage channels. Land surface and drainage affected by the New Madrid earthquakes of 1811–1812. Cotton region; rice, soybeans, wheat, melons; manufacturing (aluminum processing; telecommunications shelters). Settled c.1788 by French Creoles and Americans. One of Missouri's original five counties. Formed 1812.

New Madrid, city (2000 population 3,334), ⊙ NEW MADRID county, extreme SE MISSOURI, on MISSISSIPPI RIVER and protected by high levees, and 35 mi/56 km SW of CAIRO (ILLINOIS); 36°35′N 89°32′W. Cotton, wood products; manufacturing (aluminum processing; telecommunications shelters); river port. Laid out 1789 when under Spanish rule. In Civil War, Federal troops captured city before taking (1862) nearby Island Number 10 in the Mississippi River; island, then TENNESSEE territory, has since vanished. The name has been given to the fault zone that runs SW to NE through several states. It and its branch faults have been the origin of numerous earthquakes, most notably the extremely high-energy quakes of 1811 and 1812. They reversed flow of Mississippi River, created Reelfort Lake in Tennessee, rang church bells in BOSTON, and were felt in CANADA and CHARLESTON (SOUTH CAROLINA). Original townsite is in the Mississippi River channel.

Newmains (NYOO-mainz), town (2001 population 5,329), North Lanarkshire, S SCOTLAND, 1 mi/1.6 km NE of WISHAW; 55°46′N 03°53′W. Manufacturing and residential. Formerly in Strathclyde, abolished 1996.

Newman, city (□ 1 sq mi/2.6 sq km; 2000 population 7,093), STANISLAUS county, central CALIFORNIA, in San Joaquin Valley, 22 mi/35 km S of MODESTO; 37°19′N 121°01′W. Dairying, farming (poultry; fruit, vegetables, grain, including rice); manufacturing (cheese, fabricated metal products, fruit). George J. Hatfield State Park to SE; DIABLO RANGE to SW; SAN JOAQUIN RIVER to E (mouth of MERCED RIVER). Laid out 1887, incorporated 1908.

Newman, city (2000 population 956), Douglas county, E ILLINOIS, 17 mi/27 km E of TUSCOLA; 39°47′N 87°59′W. Ships grain. Founded 1857, incorporated 1872.

Newman (NOO-muhn), town, NW WESTERN AUSTRALIA state, W AUSTRALIA, in HAMERSLEY RANGE, E PILBARA region, 240 mi/386 km SSE of PORT HEDLAND; 23°22′S 119°44′E. Planned community. Iron ore discovered S of town in 1957, starting iron-mining boom of region. Large open-cut iron mine. Terminus of railroad from iron-transfer facilities at Port Hedland.

New Manchester, unincorporated village, HANCOCK county, N WEST VIRGINIA, 9 mi N of WEIRTON, near OHIO RIVER. Agriculture (grain, nursery crops). TOMLINSON RUN STATE PARK to N; Hillcrest Wildlife Management Area.

Newman Grove, village (2006 population 765), MADISON county, NE central NEBRASKA, 18 mi/29 km WSW of MADISON, on SHELL CREEK of PLATTE RIVER, in grassy Loess Hills; 41°45′N 97°46′W. Dairy products; grain.

Newmanstown, unincorporated town (2000 population 1,536), LEBANON county, SE central PENNSYLVANIA, 12 mi/19 km E of LEBANON, near Mill Creek; 40°21′N 76°12′W. Manufacturing (furniture, apparel, steel fabrication); agriculture (grain, soybeans; livestock; dairying). Middle Creek Waterfowl Management Lake reservoir to S.

Newmarket (NOO-mahr-ket), town (□ 15 sq mi/39 sq km), York region, S ONTARIO, E central Canada, on Holland River, 27 mi/43 km N of TORONTO; 44°03′N 79°28′W. Commuter suburb of Toronto. Manufacturing (motor vehicle parts, hardware, furniture, rubber and plastic products, electrical equipment; food processing.

Newmarket (NYOO-MAH-kit), town (2001 population 14,995), SUFFOLK, E ENGLAND, 13 mi/21 km ENE of CAMBRIDGE; 52°15′N 00°25′E. Horse-racing center since early 17th century. Home of the Jockey Club. Wicken Fen, a nearby nature reserve, is one of the only remaining parts of The FENS.

Newmarket, Gaelic *Áth Trasna*, town (2006 population 949), NW CORK county, SW IRELAND, 4 mi/6.4 km NW of KANTURK; 52°13′N 09°00′W. Agricultural market.

New Market, town (2000 population 659), MONTGOMERY county, W central INDIANA, 6 mi/9.7 km S of CRAWFORDSVILLE; 39°57′N 86°55′W. In agricultural area; manufacturing (grain mixing).

New Market, town (2000 population 456), TAYLOR county, SW IOWA, 10 mi/16 km WNW of BEDFORD; 40°43′N 94°54′W. In agricultural area.

New Market, town (2000 population 427), FREDERICK county, MARYLAND, 8 mi/12.9 km ESE of FREDERICK; 39°23′N 77°16′W. Its many antiques shops are now popular with collectors. Laid out by Nicholas Hall in 1793.

Newmarket, town, ROCKINGHAM county, SE NEW HAMPSHIRE, 9 mi/14.5 km W of PORTSMOUTH; 43°04′N 70°57′W. Bounded E by GREAT BAY, drained by LAMPREY RIVER. Manufacturing (electrical products, sheet metal fabrication); agriculture (nursery crops; poultry; cattle; dairying). Set off from EXETER in 1727.

New Market, town (2006 population 1,322), JEFFERSON county, E TENNESSEE, 23 mi/37 km ENE of KNOXVILLE; 36°06′N 83°33′W. In agricultural area; recreation.

New Market (NOO MAHR-ket), town (2006 population 1,847), SHENANDOAH county, NW VIRGINIA, in SHENANDOAH VALLEY, 16 mi/26 km NE of HARRISONBURG; 38°38′N 78°40′W. Manufacturing (fabricated copper, apparel; bookbinding); agricultural area (apples, peaches, grain, soybeans; livestock; dairying). Site of Confederate Civil War victory (May 15, 1864). New Market Battlefield Park. Endless Caverns (S), Shenandoah Caverns (N).

New Market, village (2000 population 332), SCOTT county, S MINNESOTA, 28 mi/45 km S of MINNEAPOLIS; 44°34′N 93°20′W. Grain; livestock; dairying. Small lakes in area.

New Market, village, MIDDLESEX county, NE NEW JERSEY, 4 mi/6.4 km N of NEW BRUNSWICK. Structural steel plant. Settled early in 18th century; site of a Revolutionary camp. Has house built (1814) by Duncan Phyfe. In suburban area.

Newmarket, residential and industrial suburb of AUCKLAND, NORTH ISLAND, NEW ZEALAND. Incorporated into Auckland in 1989.

New Market, unincorporated rural community, PLATTE county, W MISSOURI, 10 mi/16 km N of PLATTE CITY. Settled 1830.

Newmarket-on-Fergus (NYOO-mahr-kuht–ahn–FUHR-guhs), Gaelic *Cora Chaitlín*, town (2006 population 1,542), S CLARE county, W IRELAND, 13 mi/21 km WNW of LIMERICK; 52°45′N 08°53′W. Agricultural market. Nearby is Dromond Castle, now a hotel, the birthplace of William Smith O'Brien, leader of the Young Irelanders. Shannon Airport is nearby.

New Marlboro or **New Marlborough**, town, BERKSHIRE county, SW MASSACHUSETTS, in the BERKSHIRES, 23 mi/37 km S of PITTSFIELD; 42°06′N 73°14′W. Dairy products. Mill River village is seat of town; also includes Hartsville and Southfield villages.

New Martinsville, town (2006 population 5,649), ⊙ WETZEL county, NW WEST VIRGINIA, on the OHIO RIVER (bridged), 30 mi/48 km SSW of WHEELING; 39°39′N 80°51′W. Agriculture (corn, potatoes); cattle; poultry. Manufacturing (glassware, chemicals, lumber, textiles). Oil and natural-gas wells; sandpits. HANNIBAL LOCK AND DAM to N, Lock and Dam No. 15 to S, both on Ohio River. Lewis Wetzel Wildlife Management Area to SE. Platted 1838.

New Meadows, village (2000 population 533), ADAMS county, W IDAHO, 18 mi/29 km NNE of COUNCIL, on LITTLE SALMON RIVER; 44°58′N 116°17′W. Unincorporated village of Meadows 2 mi/3.2 km E. Railroad terminus in cattle-grazing area; manufacturing (lumber); ski resort. Lost Valley Reservoir to W; Brundage Mountain Ski Area to NE, Payette Lakes Ski Area to E; parts of Payette National Forest to W and E.

New Meadows River, SW MAINE, inlet of CASCO BAY extending from S tip of SEBASCODEGAN Island. c.12 mi/19 km inland, between BRUNSWICK and BATH.

New Melones Lake, reservoir (□ 21 sq mi/54.6 sq km), on CALAVERAS-TUOLUMNE county border, E central CALIFORNIA, on STANISLAUS RIVER, on W edge of Stanislaus National Forest in MOTHER LODE region, and 35 mi/56 km NE of MODESTO; 37°57′N 120°31′W. Maximum capacity 2,870,000 acre ft. Formed by New Melones Dam (625 ft/191 m high), built (1979) by the Bureau of Reclamation for irrigation, flood control, recreation, and as a fish and wildlife pond.

New Merwede River (MER-vai-duh), Dutch *Nieuwe Merwede*, channel, 12 mi/19 km long, SW NETHERLANDS, outlet of the WAAL, Maas (MEUSE), and RHINE rivers; formed by dividing of UPPER MERWEDE RIVER into New Merwede River (SW) and LOWER MERWEDE RIVER (W) 9 mi/14.5 km W of DORDRECHT; flows SW past HARDINXVELD-GIESSENDAM and SLIEDRECHT, joins AMER RIVER (extension of BERGSE MAAS) 8 mi/12.9 km SSE of Dordrecht to form HOLLANDS DIEP channel. Biesbosch National Park on both sides of channel.

New Mexico, state (□ 121,598 sq mi/316,154.8 sq km; 2006 population 1,954,599), SW UNITED STATES ⊙ SANTA FE; 34°38′N 105°50′W. The largest city is ALBUQUERQUE. New Mexico was admitted to the Union in 1912 as the forty-seventh state. New Mexico is often referred to as the "Land of Enchantment."

Geography

The state is bounded on the N by COLORADO, on the E by OKLAHOMA and TEXAS, on the S by Texas and MEXICO (SONORA and CHIHUAHUA states), and on the W by ARIZONA. At its NW corner, Arizona, UTAH, Colorado, and New Mexico meet at right angles—the only point in the U.S. common to four states (Four Corners). New Mexico is roughly bisected by the RIO GRANDE and has an approximate mean elevation of 5,700 ft/1,737 m. The topography of the state is marked by broken mesas, wide deserts, heavily forested mountain wildernesses, and high, bare peaks. The mountain ranges, part of the ROCKY MOUNTAINS, rising to their greatest elevation (more than 13,000 ft/3,962 m) in the SANGRE DE CRISTO MOUNTAINS, are in broken groups, running N to S through central New Mexico and flanking the Rio Grande. The tumbled Gila Wilderness is in the SW. Broad, semiarid plains, particularly prominent in S New Mexico, are covered with cactus, yucca, creosote bush, sagebrush, and desert grasses. Water is rare in these near-arid regions, where the scanty rainfall is subject to rapid evaporation.

Economy

Because irrigation opportunities are few, most of the farmland is given over to grazing. There are many large ranches, and cattle and sheep graze year-round on the open range. The two notable rivers besides the Rio Grande—the PECOS and the SAN JUAN—are used for some irrigation; the Carlsbad and Fort Sumner reclamation projects are on the Pecos, and the Tucumcari project is nearby. Other projects utilize the COLORADO RIVER basin; however, the Rio Grande, harnessed by the ELEPHANT BUTTE DAM, remains the major irrigation source for the area of most extensive farming (DOÑA ANA county). In the regions that support dry farming, the major crops are hay and sorghum grains. Important are crops relating to the popular Mexican food industry, especially chiles, jalapeños, and blue corn. Onions, potatoes, and dairy products are also very important, and several crops, such as piñon nuts, pinto beans, and chilies, are especially characteristic of New Mexico. Much of the state's income is derived from its considerable mineral wealth. New Mexico is a leading producer of uranium ore, manganese ore, potash, salt, perlite, copper ore, natural gas, beryllium, and tin concentrates. Petroleum and coal are also found in smaller quantities. Silver and turquoise have been used in making Indian jewelry since long before European exploration. Navajo pottery is made in the NW and "Pueblo" pottery in the N central. The Federal government is the largest employer in the state, accounting for over one-quarter of New Mexico's jobs. A large percentage of government jobs in the state are related to the military. The climate of the state and the increasing population have aided New Mexico's effort to attract new industries; manufacturing, centered especially around Albuquerque, includes food and mineral processing and the production of chemicals, electrical equipment, and ordnance. High-technology manufacturing has also become increasingly important, much of it in the defense industry. Pinewood is the chief commercial wood.

Tourism
Millions of acres of the state are under Federal control as national forests and monuments, and, together with the attractive climate and scenery, make tourism a chief source of income. Best known of the state's attractions are the CARLSBAD CAVERNS NATIONAL PARK in the SE and the AZTEC RUINS NATIONAL MONUMENT in the NW. Thousands of tourists annually visit the WHITE SANDS, BANDELIER, CAPULIN VOLCANO, EL MORRO, EL MALPAIS, FORT UNION, GILA CLIFF DWELLINGS, and Salina Pueblo Missions (or SALINAS) national monuments and the CHACO CULTURE NATIONAL HISTORICAL PARK. National forests include Carson and Santa Fe in the N; Cibola, Gila, and part of Apache-Sitgreaves in the W; Lincoln in the S. Kiowa National Grasslands in the NE. Part of the large FORT BLISS MILITARY RESERVATION (extends N from Texas) and Holoman Air Force Base in the S. The state is a popular place for winter or year-round residence, particularly for retirees. Many writers and artists have made their homes in communities such as TAOS and Santa Fe.

Native American and Hispanic Population
The Apache, Navajo, and Ute live on Federal reservations within the state—the Navajo reservation, with over 23,166 sq mi/60,000 sq km, is the largest in the country—and the Pueblo people live in pueblos scattered throughout the N part of the state. Indian reservations in the state include Mescalero Apache in the S; Zuñi, Acoma, Laguna, Alamo Band–Navajo in the W; parts of Navajo and Ute Mountain in NW; Jicarilla Apache in the N; Isleta in the center, and several small Puelo Indian reservations and land grants in the N. Over one-third of the population today is of Hispanic origin (some are recent immigrants from Mexico) and roughly the same percentage speak Spanish fluently. The state has made a great effort to re-establish its Mexican roots.

History: to 1610
Use of the land and minerals goes back to the prehistoric time of the early Indian cultures in the Southwest that long preceded the flourishing sedentary civilization of the Pueblos that the Spanish found along the Rio Grande and its tributaries. Word of the pueblos reached the Spanish through Cabeza de Vaca, who may have wandered across S New Mexico between 1528 and 1536; Fray Marcos de Niza identified these pueblos as the fabulously rich Seven Cities of CIBOLA. New Spain dispatched a full-scale expedition

(1540–1542), under the leadership of Francisco Vásquez de Coronado, to find the cities. The treatment of Native Americans by Coronado and his men led to the long-standing hostility between the Native Americans and the Spanish, slowing Spanish conquest. Juan de Oñate founded the first regular colony at SAN JUAN in 1598. The Native Americans of ACOMA revolted against the Spanish encroachment, and were severely suppressed. Pedro de Peralta was made governor of the "Kingdom and Provinces of New Mexico" in 1609. He founded his capital at Santa Fe a year later.

History: 1610 to 1846
The little colony did not prosper greatly, although some of the missions flourished and haciendas were founded. The subjection of Native Americans to forced labor and attempts by missionaries to convert them resulted in violent revolt by the Apache in 1676 and the Pueblo in 1680. These revolts drove the Spanish entirely out of New Mexico. The Spanish did not return until the campaign of Diego de Vargas Zapata re-established their control in 1692. In the 18th century the development of ranching and of some farming and mining was more thorough, laying the foundations for the Spanish culture in New Mexico to sites near present-day EL PASO, Texas and JUAREZ, Mexico, that still persists. When Mexico achieved its independence from Spain in 1821, New Mexico became a province of Mexico, and trade was opened with the U.S. By the following year the SANTA FE TRAIL was being traveled by the wagon trains of American traders. A group of Texans embarked on an expedition in 1841 to assert Texan claims to part of New Mexico; they were captured.

History: 1846 to 1879
The Mexican-American War marked the coming of the Anglo-American culture to New Mexico. Stephen W. Kearny entered (1846) Santa Fe without opposition, and two years later the Treaty of Guadalupe Hidalgo ceded New Mexico to the U.S. The territory, which included Arizona and other territories, was enlarged by the Gadsden Purchase (1853). The Compromise of 1850, which settled the Texas boundary question in New Mexico's favor and organized New Mexico as a territory without restriction on slavery, halted a bid for statehood with an antislavery constitution. In the Civil War, Confederate troops from Texas initially occupied New Mexico, but Union forces took it over in early 1862. After the war and the withdrawal of the troops, the territory was plagued by conflict with the Apache and Navajo. The surrender of Apache chief Geronimo in 1886 ended conflict in New Mexico and Arizona (which was made a separate territory in 1863). However, there were local troubles even after that time. Already the ranchers had taken over much of the grasslands.

History: 1879 to Present
The coming of the SANTA FE RAILROAD in 1879 encouraged the great cattle boom of the 1880s. There were typical cow towns, feuds among cattlemen as well as between cattlemen and the authorities (notably the Lincoln County War), and colorful characters such as Sheriff Pat Garrett and the outlaw Billy the Kid. The cattlemen were unable to keep out the sheepherders and were overwhelmed by the homesteaders and squatters, who fenced in and plowed under the "sea of grass." Land claims gave rise to bitter quarrels among the homesteaders, the ranchers, and the old Spanish families, who made claims under the original grants. Despite overgrazing and reduction of lands, ranching survived and continues to be important together with the limited, but scientifically controlled irrigated and dry-land farming. Statehood was granted in 1912. Pancho Villa raided COLUMBUS, New Mexico, in March 1916. The U.S. government built LOS ALAMOS as a center for atomic research in 1943. The first atomic bomb exploded at the White Sands Missile Range in July 1945. The growth of

military establishments and advanced research facilities, including Sandia National Laboratory (opened 1956), has greatly contributed to the economic advance of New Mexico in recent years. Since the 1970s, high-technology industries have become prominent in the state economy. The scarcity of water, however, could slow New Mexico's impressive recent growth. The University of New Mexico, at Albuquerque, is the most prominent educational institution in the state.

Government
The legislature has a senate of forty-two members elected for four-year terms and a house of representatives with seventy members elected for two-year terms. The governor is elected for four years, and may be reelected. Bill Richardson is the current governor. The state elects two U.S. senators and three representatives and has five electoral votes. New Mexico has been generally Democratic in politics, although it joined the national trend toward conservatism in the 1980s.

The state has thirty-three counties: BERNALILLO, CATRON, CHAVES, CIBOLA, COLFAX, CURRY, DE BACA, DOÑA ANA, EDDY, GRANT, GUADALUPE, HARDING, HIDALGO, LEA, LINCOLN, LOS ALAMOS, LUNA, MCKINLEY, MORA, OTERO, QUAY, RIO ARRIBA, ROOSEVELT, SANDOVAL, SAN JUAN, SAN MIGUEL, SANTA FE, SIERRA, SOCORRO, TAOS, TORRANCE, UNION, and VALENCIA.

New Miami (mei-AM-ee), village (2006 population 2,524), BUTLER county, extreme SW OHIO, 4 mi/6 km NNE of HAMILTON, and on GREAT MIAMI RIVER; 39°26′N 84°32′W.

New Middletown, town (2000 population 77), HARRISON county, S INDIANA, 6 mi/9.7 km SE of CORYDON; 38°10′N 86°03′W. In agricultural area.

New Middletown (MID-uhl-toun), village (2000 population 1,682), MAHONING county, E OHIO, 10 mi/16 km SSE of YOUNGSTOWN, near PENNSYLVANIA state line; 40°58′N 80°33′W.

New Milford, town, LITCHFIELD county, W CONNECTICUT, on the HOUSATONIC River; 41°36′N 73°25′W. Situated in a dairy region; manufacturing includes paper products and electronic equipment. The town hall is on the homesite of Roger Sherman, a drafter and signer of the Declaration of Independence. The Canterbury School is in New Milford. Incorporated 1712.

New Milford, borough (2006 population 16,243), BERGEN county, NE NEW JERSEY, on the HACKENSACK River, 2 mi/3.2 km NNE of HACKENSACK; 40°55′N 74°01′W. A suburb of NEW YORK city, it is primarily residential. NEW MILFORD was settled in 1695 by French Huguenots. One of the original homes still stands, and there is a Huguenot cemetery in the city. In 1776, George Washington's forces crossed the Hackensack River here during their retreat from FORT LEE to TRENTON. Washington used the New Bridge Inn (still standing). Incorporated 1922.

New Milford, borough (2006 population 835), SUSQUEHANNA county, NE PENNSYLVANIA, 9 mi/14.5 km ENE of MONTROSE, on Salt Lick Creek; 41°52′N 75°43′W. Manufacturing (plastic products). Agriculture (corn, hay; dairying). Numerous small lakes in area; Salt Springs State Park to NW.

New Mill (NYOO MIL), village (2001 population 1,259), WEST YORKSHIRE, N ENGLAND, 5 mi/8 km S of HUDDERSFIELD; 53°34′N 01°47′W. Towns of Hepworth (S), Scholes (SSW), and Fulstone (NE) are nearby.

Newmill, Scotland: see KEITH.

New Millford, village (2000 population 541), WINNEBAGO county, N central ILLINOIS, residential suburb 7 mi/11.3 km S of ROCKFORD, on KISHWAUKEE RIVER; 42°10′N 89°04′W.

New Mills (NYOO MILZ), town (2001 population 9,625), NW DERBYSHIRE, central ENGLAND, on Goyt River, 8 mi/12.9 km NNW of BUXTON; 53°22′N 02°00′W. Light manufacturing. Commuter town of MANCHESTER.

Newmilns (NYOO-milnz) or **Newmilns with Green-holm**, town (2001 population 3,057), East Ayrshire, S Scotland, on the IRVINE RIVER, and 7 mi/11.3 km E of KILMARNOCK; 55°36′N 04°19′W. Previously muslin- and lace-manufacturing center. Formerly in Strathclyde, abolished 1996.

New Milton and Barton-on-Sea (NYOO MIL-tuhn and BAHR-tuhn–on–SEE), urban locality (2001 population 23,753), HAMPSHIRE, S ENGLAND, on Christchurch Bay, 4 mi/6.4 km NE of CHRISTCHURCH; 50°46′N 01°40′W.

New Minden (MEN-din), village (2000 population 204), WASHINGTON county, SW ILLINOIS, 7 mi/11.3 km N of NASHVILLE; 38°26′N 89°22′W. In agricultural area (wheat, corn, soybeans; hogs, poultry); coal; oil.

New Munich (MYOO-nik), village (2000 population 352), STEARNS county, central MINNESOTA, on SAUK RIVER, 28 mi/45 km W of ST. CLOUD; 45°37′N 94°45′W. Grain; poultry; dairying; manufacturing (concrete).

Newnan (NOO-nuhn), city (2000 population 16,242), ⊙ COWETA county, W GEORGIA, 35 mi/56 km SW of ATLANTA; 33°23′N 84°47′W. Emerging exurb of metropolitan Atlanta. Manufacturing includes clothing, plastics, paper products, consumer goods, textiles, crushed stone, electrical equipment. Many old houses and gardens. Nearby planned community of Shenandoa established in the 1970s. Several important cultural institutions including the Male Academy Museum with Civil War artifacts and the Manget Bannon Alliance for the Arts gallery and theater. Incorporated 1828.

Newnans Lake (NOO-nuhnz), ALACHUA county, N central FLORIDA, 5 mi/8 km E of GAINESVILLE; 4 mi/6.4 km long, 2 mi/3.2 km wide; 29°39′N 82°20′W.

Newnes (NYOONZ), village, E NEW SOUTH WALES, SE AUSTRALIA, 75 mi/121 km NW of SYDNEY; 33°11′S 150°15′E.

New Netherland, territory included in a commercial grant by the government of HOLLAND to the chartered Dutch West India Company in 1621. Originally extended from the ATLANTIC OCEAN for an indefinite distance W of the DELAWARE RIVER, and from Nova Francia (CANADA) on the N to VIRGINIA on the S. Colonists were settled along the HUDSON RIVER region; in 1624 the 1st permanent settlement was established at Fort Orange (now ALBANY, NEW YORK). The principal settlement in the tract after 1625 was NEW AMSTERDAM (later NEW YORK CITY) at the S end of MANHATTAN island, which was purchased from the Indians in 1626. Colonization proceeded slowly, hampered by trouble with the Indians, poor administration, and rivalry with NEW ENGLAND settlers. In 1664 the territory was taken by the English, who divided it into the 2 colonies of New York and NEW JERSEY.

Newnham (NYOO-nuhm), village (2001 population 1,285), W GLOUCESTERSHIRE, W ENGLAND, on W bank of SEVERN RIVER, 10 mi/16 km WSW of GLOUCESTER; 51°49′N 02°27′W.

New Norcia (NOO NOR-shuh), village, W WESTERN AUSTRALIA state, W AUSTRALIA, 70 mi/113 km NNE of PERTH; 30°58′S 116°13′E. Wool, wheat, wine. Benedictine monastery.

New Norfolk (NOO NOR-fuhk), town, SE TASMANIA, SE AUSTRALIA, 14 mi/23 km NW of HOBART and on DERWENT RIVER; 42°47′S 147°04′E. Agriculture center; hops, fruit, oats; sheep. Settled 1808 by *Bounty* mutineers from NORFOLK ISLAND; tourism.

New Norway (NOO NOR-wai), village (2001 population 292), central ALBERTA, W Canada, 12 mi/18 km S of CAMROSE, in Camrose County; 52°52′N 112°57′W. Established 1910.

New Orleans (OR-lee-uhnz), city (2000 population 484,674), coextensive with ORLEANS parish, SE Louisiana, between the MISSISSIPPI RIVER (SW), Lake PONTCHARTRAIN (N), and Lake BORGNE (SE), 107 mi/172 km by water from the river mouth; 30°04′N 89°56′W. Founded 1718 by the sieur de Bienville; incorporated 1805. Also known as the Crescent City and the Big Easy.

Geography

The city is c.32 mi/51 km long, reaching NE to the RIGOLETS Channel, only 4 mi/6 km W of Mississippi state line. CHEF MENTEUR PASS bisects city in E. Parts of the city/parish are undeveloped. It was built within a great bend of the Mississippi (and is therefore called the Crescent City) on subtropical lowlands, now protected from flooding by levees and by spillways located upstream, notably the MORGANZA and BONNET CARRE spillways. Level of Mississippi River is higher than the city, but levees have been built for protection. The river is crossed here by the ALGIERS Bridge (completed 1991), the Huey P. Long Bridge (completed 1935), and the Greater New Orleans Bridge (completed 1958), which is one of the largest cantilever bridges in the country. There are also 3 toll ferries operating within the city. Lake Pontchartrain is spanned by a 24-mi/39-km double causeway (opened 1957; longest bridge in the U.S.; severely damaged in the August 2005 hurricane).

Economy

The largest city in Louisiana and one of the largest in the South, New Orleans is a major U.S. port of entry. It has long been one of the busiest and most efficient international ports in the country, leading the nation in tonnage of goods conveyed. Coffee, sugar, and bananas are among its imports (the coffee and banana wharves are tourist attractions); exports include oil, petrochemicals, rice, cotton, and corn. Manufacturing (wood products, paper products, consumer goods, fabricated metal products, food and beverages, marble, granite, slate, medical equipment, boats, concrete, communication systems, building equipment, apparel, aircraft parts, printing and publishing). Coastwise traffic is heavy (the city is at the junction of the Intracoastal Waterway and the Mississippi River), and New Orleans is a major hub for railroad, highway, air, river, and ocean transportation. Louis Armstrong International Airport is 10 mi/16 km W, at Kenner. Its fine port accommodates ship and barge traffic, helping to make the New Orleans area one of the leading industrial transportation centers in the South. The region has extensive shipbuilding and repair yards as well as plants manufacturing a wide variety of products. A long corridor of oil and chemical plants lines the Mississippi River between New Orleans and BATON ROUGE.

Culture

The unusual life and history of the city have produced a literature, including the works of George W. Cable, Lafcadio Hearn, Grace Elizabeth King, Charles Gayarré, and Alcée Fortier. The picturesque French Quarter (Vieux Carré) of the old city, N of broad Canal St., is a major tourist attraction. In the heart of the quarter is Jackson Square (the former Place d'Armes); fronting on the square are the Cabildo (1795; formerly the government building, it now houses part of the Louisiana state museum); St. Louis Cathedral (1794); and other 18th- and 19th-cent. structures. Known for its music and world-famous restaurants, specializing in seafood and spicy Cajun and Creole cooking, which uphold the New Orleans tradition of good living. The annual Mardi Gras on Shrove Tuesday is perhaps the best-known festival in the U.S. Also adding to the color of the city are the many parks, museums (including a voodoo museum and the New Orleans Museum of Art), and gardens. The metropolitan area has 2 racetracks. Jean Lafitte National Historic Park includes a 33-block section of New Orleans' French Quarter, Chalmette Battlefield and Cemetery to E, and a large natural area in the swamps of Jefferson parish to S. The Superdome, home of the National Football League's New Orleans Saints, is also the site of the annual Sugar Bowl football game. Other points of interest include Louisiana Nature and Science Center, an art museum, the Aquarium of the Americas with an IMAX theater, and the fairgrounds. Algiers U.S. Naval Station and U.S. Quarantine Station are here. Lake St. Catherine is in NE, Bayou Savage State Wildlife Area is near center, Fort Pike State Commemorative Area is in NE, Fort Macomb State Commemorative Area is in SE, and Bayou Segnette State Park is to SW (Jefferson parish).

Education

New Orleans is also an educational center, the seat of Dillard University, Loyola University, Tulane University, University of New Orleans, Southern University in New Orleans, Louisiana State University Medical Center, Delgado Community College, Our Lady of Holy Cross College, New Orleans Baptist Theological Seminary and several other theological seminaries. The 1st attempts to integrate New Orleans public schools aroused a great deal of controversy in 1960. Since then, blacks have come to comprise the large majority of students and teachers in the school system, as many whites have moved from the city to the suburbs.

History: Founding to Civil War

Although most of the larger industries have been developed recently, it was soon after the sieur de Bienville had the city plotted in 1718 it took prominence as a port, and in 1722 it became the capital of the French colony. In the late 1700s thousands of French settlers arrived from Acadia, Nova Scotia, after being expelled by British. The transfer of Louisiana to Spain by the secret Treaty of Fontainebleau (1762) was confirmed by the Treaty of Paris (1763). New Orleans—deeply involved in the struggle for control of the Mississippi—was returned to French hands only briefly before passing to the U.S. with the Louisiana Purchase (1803). Nevertheless, the tone of the city's life was dominated by Creole culture until late in the 19th century, and the French influence is still seen today. After Andrew Jackson's victory over the British at New Orleans (Jan. 8, 1815) had written a postscript to the War of 1812, the W movement in the U.S. carried the "Queen City of the Mississippi" to almost fabulous heights as a port and market for cotton and slaves. New Orleans then was stamped with its lasting reputation for glamour, extravagant living, elegance, and wickedness. Then, as now, African-Americans were a large element in the population, and they contributed to the exotic flavor of the city. Jazz had its origin in the late 19th century among the black musicians of New Orleans. The quadroon balls—sumptuous affairs attended by rich whites and their quadroon mistresses—disappeared with the Civil War, but African folkways and stories of voodoo magic persist into the 20th century. The golden era ended when in the Civil War the city fell (1862) to Admiral David G. Farragut and suffered under the occupation of Union troops led by Benjamin F. Butler. New Orleans recovered from Reconstruction and passed through the end of the river-steamboat era to emerge as a modern city.

Recent History

Since the 1960s, the population of the metropolitan area has risen at a rate slightly higher than that at which the population of the city has declined, reflecting the trend toward suburbanization that has left the inner city increasingly troubled by poverty and crime. Attempts have been made at urban revitalization; in the 1970s many new buildings were erected as the city benefited from high oil prices. In the 1980s, however, the economy suffered as oil prices fell and the state's energy industry floundered. In 1983 New Orleans hosted a world's fair, but the attention it attracted and its contribution to the local economy fell far below expectations. In response to rising crime rates, the police department has introduced many innovations and encouraged the formation of citizen patrols. Gambling was legalized in 1992 and there is a casino at the end of Canal Street by the French Quarter.

Natural Disasters

Many deaths and much property damage resulted from Hurricane Camille, which swept through the region in 1969. However, the city suffered catastrophically in terms of propery damage and human suffering when several levees on Lake Pontchartrain and related canals failed in the aftermath of Hurricane Katrina in August 2005. Nearly eighty percent of the city flooded, in some places up to rooftops, and tens of thousands of people were stranded without electricity, supplies, or support for days after the storm. Recovery from Katrina has been slow and estimates at the beginning of 2006 showed that the city's population had dropped nearly 60 percent.

New Oxford, borough (2006 population 1,783), ADAMS county, S PENNSYLVANIA, 10 mi/16 km ENE of GETTYSBURG, on South Branch of Conewago Creek; 39°51′N 77°03′W. Manufacturing (feeds, fabricated metal products, paper products, iron castings, food products). Agricultural area (grain, soybeans, potatoes, apples; livestock; dairying); timber. Laid out 1792, incorporated 1874.

New Palestine, town (2000 population 1,264), HANCOCK county, central INDIANA, on SUGAR CREEK, 8 mi/12.9 km SW of GREENFIELD; 39°43′N 85°53′W. Primarily a residential area of INDIANAPOLIS. In agricultural area (grain). Laid out 1838.

New Paltz, village (2006 population 6,714), ULSTER county, SE NEW YORK, on small WALLKILL RIVER, 13 mi/21 km SSW of KINGSTON; 41°45′N 74°05′W. In agricultural area. State University of New York at New Paltz is here. Summer recreational area, known for rock climbing. Settled by Huguenots in 1677; incorporated 1887. Abolitionish Sojourner Truth lived in slavery here before escaping to freedom.

New Paris, village (2000 population 1,006), ELKHART county, N INDIANA, 5 mi/8 km S of GOSHEN, near ELKHART RIVER; 41°30′N 85°50′W. Railroad junction. Manufacturing (motor vehicles, dairy products, furniture, boats, wire products, feed). Cattle, poultry; dairying. Laid out 1838.

New Paris (PER-is), village (2006 population 1,535), PREBLE county, W OHIO, on East Fork of WHITEWATER RIVER, at Indiana state line, 12 mi/19 km NW of EATON; 39°51′N 84°48′W. Lumber. Incorporated 1832.

New Paris, borough (2006 population 204), BEDFORD county, S PENNSYLVANIA, 10 mi/16 km NW of BEDFORD, on Dunning Creek; 40°06′N 78°38′W. Agriculture (grain; livestock; dairying). Several covered bridges to NE; Shawnee State Park to S.

New Pekin, town (2000 population 1,334), WASHINGTON county, S INDIANA, 8 mi/12.9 km SSE of SALEM; 38°30′N 86°01′W. In agricultural area.

New Perlican (PUHR-li-kuhn), town (2001 population 223), SE NEWFOUNDLAND AND LABRADOR, Canada, on E side of Trinity Bay, 13 mi/21 km NNW of CARBONEAR; 47°55′N 53°22′W. Historic fishing port; lumbering. Town settled in 17th century.

New Philadelphia (FIL-uh-del-fee-ah), city (□ 8 sq mi/20.8 sq km; 2006 population 17,433), ⊙ TUSCARAWAS county, E OHIO, on TUSCARAWAS RIVER, c.40 mi/64 km S of AKRON; 40°29′N 81°26′W. In a coal and clay area. Previously foundry products, machinery, and pottery. The Tuscarawas Regional Campus of Kent State University is here. Nearby is the SCHOENBRUNN VILLAGE STATE MEMORIAL, a reconstruction of the first settlement in Ohio. Outdoor epic theater of Moravian settlement and massacre of Christian Indians in 1782. Founded 1804, incorporated 1833.

New Philadelphia, borough (2006 population 1,103), SCHUYLKILL county, E central PENNSYLVANIA, 5 mi/8 km NE of POTTSVILLE, on SCHUYLKILL RIVER, at

mouth of Silver Creek; 40°43′N 76°07′W. Manufacturing (textiles). Anthracite coal. Laid out c.1828, incorporated 1868.

New Pitsligo (PITS-lee-go), village (2001 population 927), Aberdeenshire, NE Scotland, 10 mi/16 km SW of FRASERBURGH; 57°35′N 02°11′W. It was once center of illicit whiskey distilling. Formerly in Grampian, abolished 1996.

New Plymouth, major town (2006 population 49,281), New Plymouth district (□ 853 sq mi/2,217.8 sq km; 2006 population 49,281), TARANAKI region (□ 4,880 sq mi/12,640 sq km), NORTH ISLAND, NEW ZEALAND, on the TASMAN SEA; 39°04′S 174°05′E. It is a port and servicing center for intensive dairying, including New Zealand's main cheese-producing area. Other industries include natural gas and a little oil processing and metal working. Offshore volcanic SUGARLOAF ISLANDS form a marine park.

New Plymouth, town (2000 population 1,400), PAYETTE county, SW IDAHO, 10 mi/16 km SE of PAYETTE and on PAYETTE RIVER, 5 mi/8 km upstream (SE) from its mouth (SNAKE RIVER); 43°58′N 116°49′W. Apples, peaches, plums, vegetables, sugar beets, grains; cattle, sheep; dairying.

New Plymouth, BAHAMAS: see GREEN TURTLE CAY.

New Point, town, DECATUR county, SE central INDIANA, 9 mi/14.5 km E of GREENSBURG. In agricultural area.

New Point Comfort (NOO POINT KUHM-fuhrt), low promontory, MATHEWS county, E VIRGINIA, between CHESAPEAKE BAY (E) and MOBJACK BAY (W), 8 mi/13 km SSE of MATHEWS; 37°18′N 76°16′W. Lighthouse. Villages of Bavon, New Point are nearby.

New Pomerania, PAPUA NEW GUINEA: see NEW BRITAIN.

Newport, county (□ 313 sq mi/813.8 sq km; 2006 population 82,144), SE RHODE ISLAND, on MASSACHUSETTS state line and the Atlantic, and including CONANICUT, PRUDENCE, and RHODE islands (Aquidneck) in NARRAGANSETT BAY; ⊙ NEWPORT; 41°31′N 71°16′W. Resort, agricultural area, with manufacturing. Incorporated 1703.

Newport, city (2000 population 17,048), ⊙ CAMPBELL county, N KENTUCKY, suburb 1 mi/1.6 km SE of CINCINNATI (OHIO), on the OHIO RIVER opposite Cincinnati at mouth of LICKING RIVER; 39°05′N 84°29′W. Manufacturing (wood products, food products, paper products, building equipment, oil and gas, fabricated steel products, printing, dairy products). Newport was a station on the Underground railroad, and Kentucky's only antislavery newspaper was edited here in the 1850s. Riverboat Row Riverwalk. Laid out 1791, incorporated as a city 1835.

Newport, city (2006 population 24,409), ⊙ NEWPORT county, SE RHODE ISLAND, on Aquidneck Island (also called RHODE ISLAND); 41°29′N 71°19′W. A port of entry, the city's economy revolves chiefly around tourism, educational institutions, and fishing. The manufacturing of electrical equipment is also important. Founded in 1639, Newport was united (1640) with PORTSMOUTH and then entered (1654) in a permanent federation with PROVIDENCE and WARWICK. Shipbuilding, dating from 1646, and foreign commerce brought pre-Revolutionary prosperity to Newport. In the American Revolution the British occupied the town (1776–1779); many buildings were destroyed, most of the citizens moved away, and Newport never regained its former economic prestige. It was replaced in importance by Providence, with which it was joint state capital until 1900.In the 19th century, Newport developed as a fashionable resort of the wealthy, and many palatial mansions were built. Outstanding tourist attractions are The Breakers, the former summer house of Cornelius Vanderbilt; Belcourt Castle; The Elms; Marble House; and Châteausur-Mer. Cliff Walk and Ocean Drive are known for their spectacular views of the ocean and the coastline.

Of historic interest are the Wanton-Lyman-Hazard House (c.1675; scene of a Stamp Act riot in 1765); Trinity Church (1726); Touro Synagogue (1763), oldest in the country and since 1946 a national historic site; the Redwood Library and Athenaeum (1747); and the brick market house or city hall (1762), also a national historic site. Fort Adams dates from 1776 and remained a coastal defense command post through World War II. Newport hosts yacht races, and it was the site of the America's Cup race until the early 1980s. Tennis was popularized here; the National Tennis Hall of Fame is in the Newport casino. The Newport Jazz Festival was held until 1971, but other music and dance fests continue. The city is the seat of Salve Regina College, the U.S. Naval War College (on a small island connected to the main island by a causeway), and other naval training schools. U.S. navy facilities that closed in the 1970s created significant unemployment. Pell Bridge (1969; commonly known as Newport Bridge) spans the E passage of Narragansett Bay, linking the city with Jamestown. Matthew Perry was born in Newport. Settled 1639, incorporated 1784.

Newport, city (2006 population 7,391), ⊙ COCKE county, E TENNESSEE, on PIGEON RIVER, 40 mi/64 km E of KNOXVILLE, at W foot of the GREAT SMOKY MOUNTAINS; 35°58′N 83°11′W. Light manufacturing. John Sevier Fish and Game Preserve (□ 125,000 acres/50,588 ha), and Great Smoky Mountains National Park are nearby.

Newport, city (2006 population 5,289), ⊙ ORLEANS CO., N VERMONT, on Lake MEMPHREMAGOG, 50 mi/80 km NNE of MONTPELIER, near QUEBEC (CANADA) border; 44°56′N 72°12′W. Resort, trade center, port of entry; wood products, clothing; lumber, dairy products. Just W is Newport town, including Newport Center village (lumber). Settled 1793, chartered 1803, incorporated 1917.

Newport (NOO-port), Welsh *Casnewydd*, city and county (2001 population 137,011), SE Wales, on the USK RIVER; 51°35′N 02°59′W. Previously steel works and manufacturing. Electronics, semiconductors, and chemicals. Center for civil service and government offices. Cathedral. Just to the S is Saint Woollos. Formerly in GWENT, abolished 1996.

Newport (NYOO-pawt), town (2001 population 10,850), NE Telford and Wrekin, W ENGLAND, 12 mi/19 km WSW of STAFFORD; 52°47′N 02°22′W. Former agricultural market. Has 14th-century church.

Newport (NYOO-pawt), town (2001 population 22,957), ⊙ Isle of WIGHT, S ENGLAND, 4 mi/6.4 km S of COWES; 50°42′N 01°18′W. It is also a port and the commercial center of the island, with light industries. In the 17th century, King Charles I was imprisoned in nearby Carisbrooke Castle. Parkhurst is located here. Grammar school dates from early 17th century. Remains of Roman villa.

New Port, town, S CURAÇAO, NETHERLANDS ANTILLES, minor port 7 mi/11.3 km SE of WILLEMSTAD; 12°03′N 68°49′W. Ships phosphate, which is carried by cableway from SANTA BARBARA, 2.5 mi/4 km N. Sometimes spelled Newport.

Newport, Gaelic *Baile Uí bhFiacháin*, town (2006 population 590), W MAYO county, NW IRELAND, at head of CLEW BAY, 6 mi/9.7 km NNW of WESTPORT; 53°53′N 09°33′W. Angling and deep-sea fishing resort. Nearby Rockfleet Castle was the stronghold of Gráinne Ni Mháille.

Newport, Gaelic *Tulach Sheasta*, town (2006 population 1,286), W TIPPERARY county, S central IRELAND, 10 mi/16 km ENE of LIMERICK; 52°43′N 08°24′W. Agricultural market.

Newport, town (2000 population 7,811), ⊙ JACKSON county, NE ARKANSAS, 37 mi/60 km WSW of JONESBORO and on WHITE RIVER; 35°37′N 91°14′W. Railroad and commercial center for farm area (wheat; hogs; rice, pecans). Manufacturing (machinery, lumber, railroad equipment, aluminum, motor vehicles, food). Jacksonport State Park to NW. Settled c.1873.

Newport, town (2000 population 1,122), NEW CASTLE county, N DELAWARE, and suburb 4 mi/6.4 km SW of WILMINGTON, on CHRISTINA RIVER; 39°42′N 75°36′W. Elevation 65 ft/19 m. Manufacturing (plastics products). New Castle County Airport to S; Delaware Park Horse Race Track to W. Founded in 1735. Birthplace of Oliver Evans, inventor of automatic flour-milling machinery (1785), which revolutionized grain processing.

Newport, town (2000 population 578), ⊙ VERMILLION county, W INDIANA, 28 mi/45 km N of TERRE HAUTE, on LITTLE VERMILION RIVER near its mouth on WABASH RIVER; 39°53′N 87°25′W. Manufacturing (ordnance, chemicals). Corn, wheat; cattle.

Newport, town, including Newport village, PENOBSCOT county, S central MAINE, 25 mi/40 km W of BANGOR and on SEBASTICOOK LAKE; 44°51′N 69°13′W. Manufacturing (textiles, wood products). Settled 1808, incorporated 1814.

Newport, town (2000 population 3,715), WASHINGTON county, SE MINNESOTA, on MISSISSIPPI RIVER (bridged), 7 mi/11.3 km SSE of ST. PAUL; 44°52′N 93°00′W. Manufacturing (consumer goods, wood products, steel).

Newport, town (2000 population 6,269), ⊙ SULLIVAN county, SW NEW HAMPSHIRE, 8 mi/12.9 km E of CLAREMONT; 43°22′N 72°12′W. Drained by SUGAR RIVER and its South Branch, with town center at confluence. Trade center, manufacturing (printing and publishing, concrete products; sawmilling, machining; wood products, ordnance, metal fabrication); timber; agriculture (apples, nursery crops, hay, corn; poultry, cattle). Covered bridge in NW. Includes Guild, mill village. SUNAPEE LAKE to E. Incorporated 1761.

Newport, town (□ 7 sq mi/18.2 sq km; 2006 population 3,976), CARTERET county, E NORTH CAROLINA, 10 mi/16 km WNW of MOREHEAD CITY, 7 mi/11.3 km SSE of HAVELOCK, in Croatan National Forest; 34°47′N 76°51′W. Service industries; manufacturing (apparel).

Newport, town (2006 population 9,896), LINCOLN county, W OREGON, at entrance of Yaquina Bay of the Pacific, estuary of YAQUINA River, 40 mi/64 km W of CORVALLIS; 44°37′N 124°02′W. Manufacturing (seafood processing, printing, publishing, boatbuilding); timber-shipping point; resort. Site of the Log Cabin Museum, Coast Aquarium. Part of Siuslaw National Forest to SE; numerous state parks along coast to N and S; Yaquina Bay State Park and Lighthouse in town; South Beach State Park to S. Yaquina Head Lighthouse 3 mi/4.8 km to NW. Municipal airport to S. Settled c. 1855, incorporated 1891.

Newport, town (2000 population 1,921), ⊙ PEND OREILLE county, NE WASHINGTON, near IDAHO state line opposite Old Town (Idaho), 40 mi/64 km NE of SPOKANE, on PEND OREILLE RIVER; 48°11′N 117°03′W. Railroad junction. Potatoes, hay, alfalfa; dairying; manufacturing (printing and publishing); timber. Kaniksu National Forest to N; Pend Oreille State Park to SW. County Historical Society Museum. Albeni Falls Dam 2 mi/3.2 km E, in Idaho. Settled c.1885, incorporated 1903.

Newport (NOO-port), Welsh *Trefdraeth*, town (2001 population 1,122), PEMBROKESHIRE, SW Wales, on NEWPORT BAY, 7 mi/11.3 km ENE of FISHGUARD; 52°01′N 04°50′W. Fishing port. Formerly in DYFED, abolished 1996.

Newport (NOO-port), village (□ 104 sq mi/270.4 sq km; 2006 population 739), Estrie region, S QUEBEC, E Canada; 45°23′N 71°27′W. An independent municipality, it forms part of the Cookshire-Eaton agglomeration.

Newport (NOO-port), former village, E QUEBEC, E Canada; 48°11′N 64°45′W. Amalgamated into CHANDLER in 2001.

Newport, village, DE SOTO county, NW MISSISSIPPI, 17 mi/27 km SW of MEMPHIS (TENNESSEE), near MISSISSIPPI RIVER; 34°53′N 90°13′W. Agricultural area.

Newport, village (2006 population 87), ROCK county, N NEBRASKA, 10 mi/16 km E of BASSETT; 42°36′N 99°19′W. Hay-shipping point.

Newport, village, CUMBERLAND county, SW NEW JERSEY, on short Nantuxent Creek, 9 mi/14.5 km SSE of BRIDGETON; 39°17′N 75°10′W. Sand pits.

Newport, village (2006 population 612), HERKIMER county, central NEW YORK, 13 mi/21 km NE of UTICA; 43°11′N 75°00′W.

Newport, borough (2006 population 1,470), PERRY county, central PENNSYLVANIA, 20 mi/32 km NW of HARRISBURG, on JUNIATA RIVER at mouth of Little Buffalo Creek; 40°28′N 77°07′W. Manufacturing (machinery, crushed stone, apparel, printing and publishing, flour); agriculture (corn, hay; livestock; dairying); limestone. Little Buffalo State Park to SW. Settled 1789, laid out 1814, incorporated 1840.

Newport Bay, at Newport Beach, ORANGE county, S CALIFORNIA, dredged harbor (yachting, sport and commercial fishing) 10 mi/16 km S of SANTA ANA. BALBOA (resorts) on Balboa Island and Pennisula (part of Newport Beach City); street bridge to Balboa Island, ferry to Balboa Pennisula; CORONA DEL MAR and Corona del Mar State Beach are on SE shore; entrance to Pacific on E end of bay; Balboa Island in E; Lido Island in W.

Newport Bay (NOO-port), Newport, SE Wales, N coast of BRISTOL CHANNEL; 52°02′N 04°52′W. Extends from Oxwich Point W to Pwlldu Head. Formerly in GWENT, abolished 1996.

Newport Beach, city (2000 population 70,032), ORANGE county, S CALIFORNIA, suburb 33 mi/53 km SE of LOS ANGELES, 17 mi/27 km SE of LONG BEACH, on NEWPORT BAY and the PACIFIC OCEAN; 33°37′N 117°55′W. Upper Newport Bay divides city; Peters Canyon Wash enters from N. Manufacturing (electrical equipment, computers, medical equipment, shipbuilding, adhesives, printing and publishing). It is a popular seaside resort and yachting center. John Wayne (Orange county) Airport to N. Includes Balboa Island in bay and Peninsula between bay and ocean, residential and recreational section; LIDO ISLAND in W part of bay; CORONA DEL MAR State Beach in S. Incorporated 1906.

Newport Center, VERMONT: see NEWPORT.

Newport Hills, unincorporated town, KING county, W WASHINGTON, residential suburb 7 mi/11.3 km SE of SEATTLE, on SE shore of LAKE WASHINGTON; 47°33′N 122°10′W. MERCER ISLAND is to W.

New Portland, town, SOMERSET county, W central MAINE, on the CARRABASSETT, 20 mi/32 km WNW of SKOWHEGAN; 44°53′N 70°04′W. Wood products.

Newport News (NOO-port NOOZ), independent city (□ 69 sq mi/179.4 sq km; 2006 population 178,281), SE VIRGINIA, at mouth of JAMES River, off HAMPTON ROADS, 8 mi/13 km NW of NORFOLK; 37°04′N 76°30′W. Bounded on E by independent city of HAMPTON. Port for transatlantic and intracoastal shipping; commodities handled include coal, oil, tobacco, grain, ores. One of the nation's major shipbuilding and repair centers, its shipbuilding industry began in 1886. The U.S.S. *Enterprise II*, the first nuclear-powered aircraft carrier, was constructed here. Manufacturing (office equipment, seafood processing, printing and publishing, plastic products, apparel, dairy products, shipbuilding, paper products, building materials, electronic equipment); oil refineries. Settled by Irish colonists c.1620 but did not grow appreciably until 1880, when it became E terminus of Chesapeake and Ohio railroad. In 1862 battle between ironclad ships *Monitor* and *Merrimac* took place off Newport News. Mariners Museum, War Memorial Museum of Virginia, Virginia Living Museum and Planetarium, and Victory Arch (1919, rebuilt 1962).

FORT EUSTIS Military Reservation, with Matthew Jones House (1660) on the fort's grounds in NW. Christopher Newport University and Apprentice School of Newport News. Williamsburg–Newport News International Airport in E. Monitor-Merrimac Memorial Bridge connects S to SUFFOLK, JAMES RIVER Bridge connects SW to ISLE OF WIGHT county, both bridges span James River. U.S. Naval Weapons Station to N. Incorporated 1896.

Newport-on-Tay (NYOO-port–on–TAI), town (2001 population 4,214), FIFE, E Scotland, on the Firth of Tay, opposite and 2 mi/3.2 km SE of DUNDEE; 56°26′N 02°55′W. Seaside resort and port, with ferry to Dundee. Village of Wormit is 2 mi/3.2 km SW, at S end of the Tay Bridge.

Newport Pagnell (NYOO-pawt PAG-nuhl), town (2001 population 14,739), Milton Keynes, central ENGLAND, on OUSE RIVER at mouth of Ouzel River, 12 mi/19 km N of LEIGHTON BUZZARD; 52°05′N 00°43′W. Former agricultural market. Has 14th-century church and 17th-century hospital. Town formerly noted for its lace.

New Port Richey, city (□ 5 sq mi/13 sq km; 2005 population 16,928), PASCO county, W central FLORIDA, 25 mi/40 km NW of TAMPA, near the GULF OF MEXICO; 28°15′N 82°43′W. Citrus fruit; concrete products. Major resort and retirement area; recreational fishing.

New Prague (PRAIG), town (2000 population 4,559), LE SUEUR and SCOTT counties, S MINNESOTA, 33 mi/53 km SSW of MINNEAPOLIS; 44°33′N 93°34′W. Grain, soybeans; livestock; dairying; manufacturing (coffee, flours, food products, fabricated steel, building supplies, printing and publishing). Incorporated as village 1877, as city 1891.

New Preston, CONNECTICUT: see WASHINGTON.

New Providence, town, CLARK county, SE INDIANA, 17 mi/27 km NW of JEFFERSONVILLE; 38°28′N 85°55′W. In agricultural area. Clark State Forest nearby to ENE.

New Providence, town (2000 population 227), HARDIN county, central IOWA, 7 mi/11.3 km SW of ELDORA; 42°16′N 93°10′W. In agricultural area.

New Providence, village, MONTGOMERY county, central TENNESSEE, on the CUMBERLAND River just NW of CLARKSVILLE; 36°33′N 87°23′W. Old stone blockhouse (1788) here.

New Providence, borough (2006 population 11,915), UNION county, NE NEW JERSEY; 40°42′N 74°24′W. It is largely residential but has some light industry. Originally called Turkey, its name was changed to New Providence in 1778. Settled c.1720, set off and incorporated 1899.

New Providence, island (2000 population 210,832), central BAHAMAS, 170 mi/275 km ESE of MIAMI, FLORIDA; 25°02′N 77°25′W. The most populous island in the Bahamas, with two-thirds of the nation's residents, it is the site of the capital, NASSAU. Mostly flat, with low ridges and several shallow lakes, notably Lake Killarney and Lake Cunningham. The urban center of Nassau is on the NE coast, with suburban areas to the SE, S, and W. Nassau International Airport is near the W end of the island. The island has become a major vacation resort, and the pine woodlands and swamps are giving way to golf courses, beach resorts, and residential developments. Some vegetables and tropical fruits are grown, and the rum industry is important. Settled in the late 17th century, when the British built several forts. Fort Nassau was completed in 1697, and the city laid out in 1729. During the 18th century, the Spanish and French made incursions into the area, although the island was eventually ceded by treaty to Great Britain. In the 1780s American loyalists took refuge here. The British maintained an air base on the island during WWII.

Newquay (NYOO-kee), town (2001 population 19,423), CORNWALL, SW ENGLAND, on the ATLANTIC,

11 mi/18 km N of TRURO; 50°25′N 05°05′W. Port and tourist resort; fishing. Exports kaolin and building stone.

New Quay (KEE), Welsh *Ceinewydd*, resort (2001 population 1,115), CEREDIGION, W WALES, on CARDIGAN BAY of IRISH SEA, 16 mi/26 km NE of Cardigan; 52°13′N 04°22′W. Formerly in DYFED, abolished 1996.

New Quebec, Canada: see UNGAVA.

New Radnor (RAD-nuh), village (2001 population 410), POWYS, E WALES, 10 mi/16 km E of LLANDRINDOD WELLS, at edge of RADNOR FOREST; 52°15′N 03°08′W. Remains of Norman castle. Located 3 mi/4.8 km ESE is village of Old Radnor.

New Raymer, Colorado: see RAYMER.

New Richland, town (2000 population 1,197), WASECA county, S MINNESOTA, 12 mi/19 km S of WASECA; 43°53′N 93°29′W. Grain, soybeans; livestock; dairying; manufacturing (feeds; grain processing).

New Richmond (noo RICH-muhnd), town (□ 65 sq mi/169 sq km), Gaspésie—Îles-de-la-Madeleine region, E QUEBEC, E Canada, on S GASPÉ PENINSULA, on CHALEUR BAY, at mouth of Little Cascapedia River, 26 mi/42 km ENE of DALHOUSIE; 48°10′N 65°50′W. Fishing port, in mining region.

New Richmond, town (2000 population 349), MONTGOMERY county, W central INDIANA, 11 mi/18 km NNW of CRAWFORDSVILLE; 40°11′N 86°59′W. In agricultural area; manufacturing (alfalfa).

New Richmond, town (2006 population 7,963), ST. CROIX county, W WISCONSIN, on small WILLOW RIVER, 28 mi/45 km ENE of SAINT PAUL (MINNESOTA); 45°07′N 92°32′W. Manufacturing (flour milling, food processing, machinery, wire products, printing and publishing).

New Richmond (RICH-muhnd), village (□ 4 sq mi/10.4 sq km; 2006 population 2,483), CLERMONT county, SW OHIO, 19 mi/31 km SSW of BATAVIA, and on OHIO RIVER; 38°58′N 84°16′W. In agricultural area; metal products. Birthplace of Ulysses S. Grant is nearby. Founded 1816.

New Riegel (REE-guhl), village (2006 population 217), SENECA county, N OHIO, 8 mi/13 km WSW of TIFFIN; 41°03′N 83°19′W.

New Ringgold (RING-old), borough (2006 population 281), SCHUYLKILL county, E central PENNSYLVANIA, 8 mi/12.9 km S of TAMAQUA, on Little Schuylkill River; 40°41′N 75°59′W. Manufacturing; agriculture (apples, grain; livestock; dairying). APPALACHIAN TRAIL passes to SE, on BLUE MOUNTAIN ridge.

New River (NOO RI-vuhr), unincorporated village, PULASKI county, SW VIRGINIA, 2 mi/3 km NW of RADFORD, on NEW RIVER; 37°08′N 80°35′W. Coal mining nearby.

New River, outlet of New River Lagoon, 60 mi/97 km long, N BELIZE; flows NNE, past ORANGE WALK, to CHETUMAL BAY of CARIBBEAN SEA, 3 mi/4.8 km S of COROZAL. Navigable for small craft. Sugarcane, corn; lumbering.

New River, c.200 mi/322 km long, S GUYANA; rises in the SERRA ACARAÍ near BRAZIL border; flows N, through dense forests, to the CORENTYNE RIVER at SURINAME border; 03°25′N 57°35′W.

New River (NOO RI-vuhr), c.320 mi/515 km long, NORTH CAROLINA, VIRGINIA, and WEST VIRGINIA; rising in the BLUE RIDGE MOUNTAINS, NW North Carolina, formed by joining of North and South forks on the line between ASHE and ALLEGHANY counties; flows NE into SW Virginia, through Clayton Lake reservoir past PULASKI and RADFORD. Continues NW past PEARISBURG, Virginia, and into West Virginia through BLUESTONE LAKE reservoir, past HINTON, and through NEW RIVER GORGE NATIONAL PARK (national park unit), joins GAULEY RIVER c.25 mi/40 km SE of CHARLESTON to form KANAWHA RIVER, which flows NW to OHIO RIVER, then joins with the Gauley River to form the Kanawha River; 34°42′N 77°26′W. It is used extensively to generate electricity. Bluestone

Dam (completed 1952), near Hinton, West Virginia, provides flood control and power, and its reservoir extends 36 mi/58 km upstream. The New River Gorge Bridge, built (1977) across the New River Gorge, is the largest steel-arch bridge in the U.S., with a 1,700-ft/518-m span on U.S. Highway 19 at FAYETTEVILLE. South Fork of New River rises in S WATAUGA county, NW North Carolina, near BOONE; flows NNE c.75 mi/121 km. North Fork of New River rises on Watauga-Ashe county line, NW North Carolina, near TENNESSEE state line; flows NE c.40 mi/64 km.

New River, (NOO RI-vuhr) c.50 mi/80 km long, in SE NORTH CAROLINA; rises in NW ONSLOW county; flows SSE past JACKSONVILLE where it widens into estuary c.2 mi/3.2 km wide and meanders through CAMP LEJEUNE MARINE BASE, to ONSLOW BAY, enters ATLANTIC OCEAN via New River inlet; 34°45′N 77°24′W. INTRACOASTAL WATERWAY crosses entrance behind coastal sand barrier.

New River, c.40 mi/64 km long, in S SOUTH CAROLINA; rises in N JASPER county, near RIDGELAND; flows S to the ATLANTIC OCEAN S of DAUFUSKIE ISLAND, N of mouth of SAVANNAH River; partly navigable.

New River Gorge National River (□ 97 sq mi/251 sq km), S central WEST VIRGINIA, in FAYETTE, RALEIGH, and SUMMERS counties, c.40 mi/64 km SE of CHARLESTON, on NEW RIVER, forms BLUESTONE LAKE between HINTON and VIRGINIA state line. The national river is a 52-mi/84-km-long section of one of oldest rivers in NORTH AMERICA. White-water stream flows through deep canyons; rafting. White-water rafting. Authorized 1978.

New Roads, city (2000 population 4,966), ⊙ POINTE COUPEE parish, SE central LOUISIANA, 24 mi/39 km NW of BATON ROUGE, and on False River (c.13-mi/21-km-long oxbow lake of the MISSISSIPPI RIVER); 30°42′N 91°27′W. In agricultural area (pecans, corn, sugarcane, soybeans, rice); sawmilling; fishing (crawfish), light manufacturing. Oil and natural-gas fields. Toll ferry across Mississippi River to SAINT FRANCISVILLE, 5 mi/8 km NE. Primary recreational area for people from Baton Rouge. Annual Mardi Gras parade.

New Rochelle, city (□ 13 sq mi/33.8 sq km; 2006 population 73,446), WESTCHESTER county, SE NEW YORK, on LONG ISLAND SOUND; 40°55′N 73°46′W. Although mainly a residential suburb of NEW YORK city, it has some light industry. The house where Thomas Paine lived has been preserved. The city has also been the home of several artists, authors, and other celebrities, including Norman Rockwell, Lou Gehrig, and Elia Kazan. Iona College and the College of New Rochelle are in the city. David's Island, site of former Fort Slocom, is here. Settled by Huguenots 1688, incorporated as a village 1858, incorporated as a city 1899.

New Rockford, town (2006 population 1,309), ⊙ EDDY county, E central NORTH DAKOTA, 33 mi/53 km SW of DEVILS LAKE and on JAMES RIVER; 47°40′N 99°08′W. Railroad junction; dairy products, wheat; livestock. Devils Lake Sioux Indian Reservation to NE. Founded in 1883 and incorporated 1912. Named for ROCKFORD, ILLINOIS.

New Rome (ROM), village (2000 population 60), FRANKLIN county, central OHIO, 7 mi/11 km W of COLUMBUS; 39°57′N 83°08′W.

New Romney (NYOO ROM-nee), town (2001 population 6,953), KENT, SE ENGLAND, in ROMNEY MARSH, and 12 mi/19.2 km SSE of ASHFORD; 50°59′N 00°56′E. Until the sea receded, New Romney lay on the coast and was one of the CINQUE PORTS. Cinque Ports documents are kept in town guildhall. Famous sheep fair annually in August. Partly Norman church. Old Romney is a village farther inland.

New Ross, Gaelic *Ros Mhic Thriúin*, town (2006 population 4,677), W WEXFORD county, SE IRELAND, on BARROW RIVER 2 mi/3.2 km below mouth of the NORE RIVER, 12 mi/19 km NE of WATERFORD; 52°23′N

06°56′W. Tanning, woolen milling, iron founding, brewing. There are remains of ancient fortification and of 13th-century Franciscan friary. The Earl of Pembroke founded town in 12th century, according to tradition. The town surrendered to Cromwell in 1649. United Irishmen's attack on town was repulsed in 1798. On W shore of Barrow River, in KILKENNY county, is suburb of Rosbercon, once site of important abbey.

New Ross, town (2000 population 334), MONTGOMERY county, W central INDIANA, near RACCOON CREEK, 12 mi/19 km SE of CRAWFORDSVILLE; 39°58′N 86°43′W. In agricultural area (corn, soybeans; hogs).

New Rossington (NYOO ROS-sing-tuhn), suburb (2001 population 13,255), SOUTH YORKSHIRE, N ENGLAND, 4 mi/6.4 km SE of DONCASTER; 53°29′N 01°04′W. Previously coal mining. Suburb also contains Old Rossington, on other side of East Coast Main Line (railroad).

Newry (NOO-ree), Gaelic *An Iubhar*, town (2001 population 27,300), DOWN, SE Northern Ireland, on the Clanrye River and the Newry Canal; 54°11′N 06°20′W. It has canal connections with CARLINGFORD LOUGH, the BANN RIVER, and LOUGH NEAGH. Newry is a seaport, formerly with linen mills and tobacco production. Food processing and varied manufacturing. Retail shopping. In the 12th century, Muirchertach MacLochlainn, king of Ireland, founded an abbey on the site, around which the town grew. The abbey became a collegiate church of secular priests in 1543 but was later dissolved. The town's castle was taken by Edward Bruce in 1315; the duke of Berwick burned part of Newry in his retreat before the forces of the duke of Schomberg in 1689. Newry is the seat of the Roman Catholic bishop of Dromore and contains St. Patrick's parish church (1578), the first Protestant church built in Ireland. John Mitchel, the 19th-century Republican, is buried in the Unitarian churchyard.

Newry, town, OXFORD county, W MAINE, on BEAR RIVER, 23 mi/37 km NNW of SOUTH PARIS; 44°31′N 70°49′W. In recreational area.

Newry (NOO-ree), village, OCONEE county, NW SOUTH CAROLINA, 5 mi/8 km NW of CLEMSON, on Lake KEOWEE Reservoir at LITTLE RIVER DAM, on KEOWEE RIVER.

Newry, borough (2006 population 236), BLAIR county, S central PENNSYLVANIA, 8 mi/12.9 km S of ALTOONA, near Frankstown Branch of JUNIATA RIVER; 40°23′N 26°78′W. Manufacturing (meat processing). Agriculture (grain; livestock; dairying).

New Salem, town, FRANKLIN county, central MASSACHUSETTS, 14 mi/23 km ESE of GREENFIELD, near QUABBIN RESERVOIR; 42°27′N 72°19′W. Fruit.

New Salem, town (2006 population 884), MORTON county, central NORTH DAKOTA, 25 mi/40 km W of MANDAN; 46°50′N 101°25′W. Dairy; produce; livestock; printing, concrete. Sweet Briar Dam and Reservoir to NE. To highlight the community's dairy industry, the world's largest holstein cow was built of fiberglass in 1974 and 38 feet high and 50 feet long.

New Salem, village (2000 population 136), PIKE county, W ILLINOIS, 7 mi/11.3 km NNW of PITTSFIELD; 39°42′N 90°50′W. In agricultural area (corn, wheat, soybeans; cattle, hogs).

New Salem, unincorporated village, Menallen township, FAYETTE county, SW PENNSYLVANIA, 6 mi/9.7 km WNW of UNIONTOWN, on Dunlap Creek; 39°53′N 47°76′W. Manufacturing (textiles); agriculture (corn, hay; dairying).

New Salem (NOO SAI-luhm), borough, YORK county, S PENNSYLVANIA, 5 mi/8 km SW of YORK, on Codorus Creek; 39°54′N 76°47′W. Agriculture (apples, grain; livestock; dairying).

New Salisbury, village, HARRISON county, S INDIANA, 6 mi/9.7 km N of CORYDON. Railroad junction. Manufacturing (furniture, lumber). Laid out 1830.

Cross-references are shown in SMALL CAPITALS. The pronunciation guide is shown on page xix. The sources of population figures are shown on page xvii.

New Sarepta (suh-REP-tuh), village (□ 1 sq mi/2.6 sq km; 2001 population 382), central ALBERTA, W Canada, 17 mi/28 km E of LEDUC, in Leduc County; 53°16′N 113°08′W. Settled 1904; incorporated 1960.

New Sarpy (SAHR-pee), unincorporated town, SAINT CHARLES parish, 2 mi/3.2 km E of HAHNVILLE; 29°58′N 90°23′W. On E bank (levee) of MISSISSIPPI RIVER. Name given to distinguish it from Sarpy (now called NORCO).

New Sarum, ENGLAND: see SALISBURY.

New Schwabenland (SHWAI-buhn-land), ANTARCTICA, in QUEEN MAUD LAND, large mountainous upland back of PRINCESS ASTRID COAST and PRINCESS MARTHA COAST; centers on 72°30′S 01°00′E. Was first surveyed 1939 from air by German expedition.

New Scone (SKOON), village (2001 population 4,430), PERTH AND KINROSS, E Scotland, 2 mi/3.2 km NE of PERTH; 56°25′N 03°25′W. Scone Palace (1803–1808) 2 mi/3.2 km to W. Pole Hill (945 ft/288 m) 3 mi/4.8 km to E. Formerly in TAYSIDE, abolished 1996.

New Seabury, resort village, MASHPEE, SE MASSACHUSETTS, BARNSTABLE county, W CAPE COD, on Popponesset Bay and Popponesset Beach. South Cape Beach State Park nearby.

New Sharlston (SHAHL-stuhn), village, WestYorkshire, 3 mi/4.8 km E of Wakefield.

New Sharon, town (2000 population 1,301), MAHASKA county, S central IOWA, near SKUNK RIVER, 12 mi/19 km N of OSKALOOSA; 41°28′N 92°39′W. Settled by Quakers.

New Sharon, town, FRANKLIN county, W central MAINE, on SANDY RIVER and 7 mi/11.3 km ESE of FARMINGTON; 44°38′N 70°00′W.

New Sheffield, AUSTRALIA: see MITTAGONG.

New Shildon, ENGLAND: see SHILDON.

New Shoreham, Rhode Island: see BLOCK ISLAND.

New Siberian Islands, Russian *Novosibirskiye Ostrova*, archipelago (approximately 10,900 sq mi/28,231 sq km), in the SAKHA REPUBLIC, RUSSIAN FAR EAST, in the ARCTIC OCEAN between the LAPTEV and EAST SIBERIAN seas; 75°00′N 142°00′E. The archipelago is separated into 2 groups by the SANNIKOV STRAIT. The N group, the New Siberian or Anjou islands (approximately 8,200 sq mi/21,200 sq km) includes the KOTELNY, FADDEI, NOVAYA SIBIR, BELKOVSKI, FIGURIN, and other smaller islands; the S group consists of the LYAKHOV ISLANDS. The DE LONG ISLANDS, NE of NOVAYA SIBIR, are also part of the archipelago. The islands are almost always covered by snow and ice and have a very scant tundra; ice dating from the Pleistocene Ice Age and intermingled with sediment is found there. The sparsely settled islands were sighted (1773) by Ivan Lyakhov, a Russian merchant. Mammoth fossils have been found (1870s) here by the Swedish explorer Nils A. E. Nordenskjöld, as well as by Siberian fur and ivory hunters. The islands were neglected until 1927, when meteorological stations were set up here.

New Smyrna Beach (SMUHR-nuh), city (□ 31 sq mi/80.6 sq km; 2005 population 22,356), VOLUSIA county, E central FLORIDA, 15 mi/24 km SSE of DAYTONA BEACH, on INDIAN RIVER and on PONCE DE LEON INLET of ATLANTIC OCEAN; 29°01′N 80°55′W. Resort and tourist area. Citrus-fruit packing; commercial fishing; seafood processing. Light manufacturing. Incorporated 1903.

Newsoms, town (2006 population 284), SOUTHAMPTON county, SE VIRGINIA, 12 mi/19 km WSW of FRANKLIN; 36°37′N 77°07′W. Manufacturing (lumber; meat processing); agriculture (peanuts, cotton, grain, melons; livestock); timber.

New South Wales (NOO SOUTH WAILZ), state (□ 309,443 sq mi/804,551.8 sq km), SE AUSTRALIA; ⊙ SYDNEY; 33°00′S 146°00′E. It is bounded on the E by the PACIFIC OCEAN. The other principal urban centers are NEWCASTLE, WAGGA WAGGA, LISMORE, WOLLONGONG, and BROKEN HILL. More than half the population lives in the Sydney metropolitan area. Located in the temperate zone, the state has a generally favorable climate. There are four main geographic regions: the coastal lowlands; the E highlands, culminating in Mount KOSCIUSKO, the highest peak of the AUSTRALIAN ALPS and of Australia; the W slopes; and the W plains, which cover about two-thirds of the state. The MURRAY RIVER, which forms the greater part of the S border, and its principal tributaries are important for the state's extensive irrigation systems. New South Wales is economically the most important state in Australia. The Sydney-Newcastle-Wollongong area is the greatest industrial region in the commonwealth, with steel the principal product; also financial services, tourism. Agriculture is also important: wheat, wool, and meat are produced, and there is considerable dairy farming. Tropical fruits and sugarcane are grown in the NE. The state's rich mineral resources include coal, gold, iron, copper, silver, lead, and zinc. More than 50% of the Australian Aboriginal people live in New South Wales and QUEENSLAND. The coast of Queensland was explored in 1770 by Captain James Cook, who proclaimed British sovereignty over the E coast of Australia. Sydney, the first Australian settlement, was founded in 1788 as a prison farm. During the 1820s and 1830s, the character of New South Wales changed as the wool industry grew and the importation of convicts ceased. In the early 19th century the colony included TASMANIA, SOUTH AUSTRALIA, VICTORIA, Queensland, the NORTHERN TERRITORY in Australia, and NEW ZEALAND. These territories were separated and made colonies in their own right between 1825 and 1863. In 1901, New South Wales was federated as a state of the Commonwealth of Australia. The AUSTRALIAN CAPITAL TERRITORY (site of CANBERRA, the federal capital), an enclave in New South Wales, was ceded to the commonwealth in 1911. JERVIS BAY, S of Sydney, became commonwealth territory in 1915. The nominal head of the state government is the governor; however, actual executive functions are exercised by the premier and cabinet, who are responsible to a bicameral state parliament.

New Spain: see MEXICO, republic.

New Springville, section of STATEN ISLAND borough of NEW YORK city, SE NEW YORK, on central Staten Island, 6 mi/9.7 km SW of SAINT GEORGE; 40°35′N 74°09′W.

New Square, village (2006 population 6,715), ROCKLAND county, SE NEW YORK, 2 mi/3.2 km N of SPRING VALLEY, 7 mi/11.3 km NW of NYACK; 41°08′N 74°01′W. A community of Orthodox Hasidic Jews of the Skvirer sect lives here.

New Stanton, borough (2006 population 2,086), WESTMORELAND county, SW PENNSYLVANIA, 7 mi/11.3 km SSW of Greersburg on Sewickley Creek; 40°13′N 79°36′W. Manufacturing includes plastic products, nuclear components, temperature controls, machine tools. Agriculture includes dairying; livestock; corn. Natural gas in area.

Newstead Abbey (NYOO-sted AH-bee), location, NOTTINGHAMSHIRE, central ENGLAND, on the border of SHERWOOD FOREST, between NOTTINGHAM and MANSFIELD; 53°04′N 01°10′W. It was founded (originally as an Augustinian priory) c.1170 by Henry II in atonement for the murder of Thomas à Becket. It was secularized and granted to John Byron by Henry VIII. George Byron lived at the abbey intermittently from 1806 to 1816. The Abbey was given to the Nottingham Corporation in 1931.

New Straitsville (STRAITZ-vil), village (□ 1 sq mi/2.6 sq km; 2006 population 804), PERRY county, central OHIO, 8 mi/13 km SSW of NEW LEXINGTON; 39°34′N 82°14′W. In coal-mining area. Founded 1870.

New Stuyahok (STOO-yah-hok), village (2000 population 471), SW ALASKA, c.50 mi/80 km NE of DILLINGHAM; 59°28′N 157°15′W.

New Suffolk, resort village, SUFFOLK county, SE NEW YORK, on N peninsula of E LONG ISLAND, on GREAT PECONIC BAY, 11 mi/18 km ENE of RIVERHEAD; 41°00′N 72°29′W.

New Summerfield (SUH-muhr-reeld), village (2006 population 1,038), CHEROKEE county, E TEXAS, 29 mi/47 km SSE of TYLER; 31°59′N 95°05′W. Agriculture and timber area. Manufacturing (rubber products). Lake Striker reservoir to E.

New Sweden, town, AROOSTOOK county, NE MAINE, 20 mi/32 km NNW of PRESQUE ISLE; 46°58′N 68°07′W. Settled 1870 by Swedish immigrants.

New Sweden, former Swedish colony (1638–1655), on the DELAWARE RIVER; included parts of what are now PENNSYLVANIA, NEW JERSEY, and DELAWARE. With the support of statesman Axel Oxenstierna (a Swede), Admiral Klas Fleming (a Finn), and Peter Minuit (a Dutchman), the New Sweden Company was organized in Sweden in 1633. The *Kalmar Nyckel* and the *Fogel Grip*, 2 ships commanded by Minuit, reached the Delaware River in March 1638. Minuit immediately bought land from the Native Americans and founded Fort Christina, where WILMINGTON, Delaware, stands. In 1643, TINICUM ISL. (at Philadelphia) became the colony's capital. About ½ of the colonists were Finns. Peter Stuyvesant, with a Dutch force larger than the population of New Sweden, took the little colony in 1655.

New Tecumseth (NOO te-KUHM-suhth), town (□ 106 sq mi/275.6 sq km; 2001 population 26,141), SE ONTARIO, E central Canada, 20 mi/33 km from BARRIE, with NOTTAWASAGA RIVER nearby; 44°05′N 79°45′W. Formed in 1991 from the municipalities of ALLISTON, BEETON, TOTTENHAM, and Tecumseth Township.

New Territories: see HONG KONG.

Newton, county (□ 823 sq mi/2,139.8 sq km; 2006 population 8,411), NW ARKANSAS; ⊙ JASPER; 35°55′N 93°13′W. Drained by BUFFALO RIVER and its tributaries; situated in OZARK region. Agriculture (cattle, hogs). Lead and zinc mining; lumber milling. Part of Ozark National Forest is in S, small unit to N; W part of BUFFALO NATIONAL RIVER crosses N part of county; part of Gene Bush–Buffalo River Wildlife Management Area in E. Formed 1842.

Newton, county (⊙ 279 sq mi/723 sq km; 1990 population 41,808), N central GEORGIA; ⊙ COVINGTON; 33°33′N 83°50′W. Drained by ALCOVY and YELLOW rivers; includes part of LLOYD SHOALS RESERVOIR. Textile manufacturing Piedmont agricultural area producing fruit; cattle and poultry. Formed 1821.

Newton, county (□ 403 sq mi/1,047.8 sq km; 2006 population 14,293), NW INDIANA; ⊙ KENTLAND; 40°57′N 87°24′W. Bounded W by ILLINOIS state line, N by KANKAKEE RIVER; also drained by IROQUOIS RIVER. Native prairie grasses. Agricultural area; corn, rye, oats, soybeans; cattle, hogs. Some manufacturing at Kentland. Ships seeds, grain. Limestone quarrying. Part of LaSalle State Fish and Wildlife Area in NW corner; Willow Slough State Fish and Wildlife Area in W. Formed 1859 (the last county organized in Indiana).

Newton, county (□ 579 sq mi/1,505.4 sq km; 2006 population 22,413), E central MISSISSIPPI; ⊙ DECATUR; 32°24′N 89°07′W. Drained by CHUNKY and Potterchitto creeks. Agriculture (cotton, corn; poultry, cattle; dairying); timber. Part of Bienville National Forest in SW corner. Formed 1836.

Newton, county (□ 629 sq mi/1,635.4 sq km; 2006 population 56,047), SW MISSOURI; ⊙ NEOSHO; 36°54′N 94°19′W. In the OZARKS; includes part of city of JOPLIN in NW. Vegetables, berries, hay, apples, corn, wheat; dairying; cattle, horses, poultry; oak timber. Manufacturing at Neosho, Joplin, GRANBY and SENECA. GEORGE WASHINGTON CARVER National Monument in N. Crowder College at former Camp Crowder. Formed 1854.

Newton (NOO-tuhn), county (□ 939 sq mi/2,441.4 sq km; 2006 population 14,090), E TEXAS; ⊙ NEWTON;

Area is shown by the symbol □, and capital city or county seat by ⊙.

30°50′N 93°45′W. Bounded E by SABINE RIVER (LOUISIANA state line); drained by its tributaries. In pine-forest belt; lumbering is chief industry; peaches, vegetables; oil and natural gas. TOLEDO BEND RESERVOIR in NE corner. Formed 1846.

Newton, city (2000 population 3,069), ⊙ JASPER county, SE central ILLINOIS, on EMBARRAS RIVER, 22 mi/35 km SE of EFFINGHAM; 38°59′N 88°09′W. In agricultural area; livestock; corn, wheat; dairy products. Sam Parr State Park nearby. Settled 1828, incorporated 1831.

Newton, city (2000 population 15,579), ⊙ JASPER county, central IOWA, 30 mi/48 km ENE of DES MOINES; 41°42′N 93°02′W. Manufacturing (appliances, paper products, steel fabrication, machinery, foundry products). Riverview Relenst Center (correctional facility) to S. Settled 1846, incorporated 1857.

Newton, city (2000 population 17,190), ⊙ HARVEY county, S central KANSAS, 27 mi/43 km N of WICHITA; 38°02′N 97°20′W. Elevation 1,420 ft/433 m. In an agricultural area. It is a railroad division point with railroad shops and has a large mobile home industry in addition to oil wells. Manufacturing (machinery, flour milling, motor vehicle parts, plastic products, glass, furniture, lumber, concrete). The CHISHOLM TRAIL passed through the site. In the early 1870s, German Mennonites from RUSSIA brought seed for what became the first hard winter wheat in Kansas. The city still has a large Mennonite population, and a monument to their ancestors is here. Museum here. Bethel College is in NORTH NEWTON. Incorporated 1872.

Newton, city (2000 population 83,829), MIDDLESEX county, E MASSACHUSETTS, suburb of BOSTON on the CHARLES RIVER; 42°20′N 71°13′W. Industries include publishing, chemicals, precision instruments, and computers. Newton is known as a regional education center. The city is the seat of Newton College, Mount Alvernia College, Andover Newton Theological School, Boston College, Mount Ida College, Lovell College, and Pine Manor College. Horace Mann, Nathaniel Hawthorne, Mary Baker Eddy, and Samuel Francis Smith lived in Newton. It comprises fourteen individual residential villages; including Auburndale, Eliot, Newton Centre, Newton Highlands, Newton Lower Falls, Newton Upper Falls, Newtonville, Nonantum (or Silver Lake), Riverside, Waban, West Newton, Chestnut Hill, Oak Hill, Newton Corner. Settled before 1640, incorporated as a city 1873.

Newton (NOO-tuhn), city (□ 13 sq mi/33.8 sq km; 2006 population 13,160), ⊙ CATAWBA county, W central NORTH CAROLINA, 8 mi/12.9 km SE of HICKORY 40 mi/64 km NNW of CHARLOTTE; 35°40′N 81°13′W. Railroad junction. Manufacturing (furniture, motor vehicle parts, textiles, apparel, printing, wood, paper, and metal products, flour, medical products). County Historical Museum. Hickory Motor Speedway to NW. Catawba Valley Community College to NW. Settled mid-18th century.

Newton, town (2000 population 1,708), Dale co., SE Alabama, on the Choctawhatchee, and 10 mi/16 km S of Ozark. In vegetable and pecan area. Inc. in 1887.

Newton, town (2000 population 3,699), NEWTON county, E central MISSISSIPPI, 27 mi/43 km W of MERIDIAN; 32°19′N 89°09′W. Railroad junction. Agriculture (cotton, corn; poultry, cattle; dairying); hardwood timber; manufacturing (fabricated metal products, furniture, cheese, apparel, feeds). Bienville National Forest to SW.

Newton, town, ROCKINGHAM county, SE NEW HAMPSHIRE, 9 mi/14.5 km SW of EXETER; 42°52′N 71°02′W. Bounded SE by MASSACHUSETTS state line. Manufacturing (consumer goods). Includes village of Newton Junction in W.

Newton, town (2006 population 8,337), ⊙ SUSSEX county, NW NEW JERSEY, 21 mi/34 km NW of MOR-

RISTOWN; 41°02′N 74°45′W. Revitalized residential area. Former Don Bosco College, now Sussex County Community College here. Little Flower Monastery (Benedictine) nearby. Settled c.1760, incorporated 1864.

Newton (NOO-tuhn), town (2006 population 2,321), ⊙ NEWTON county, E TEXAS, near SABINE River, c.55 mi/89 km NNE of BEAUMONT; 30°51′N 93°45′W. In agricultural, timber region; lumber milling; manufacturing (pulpwood, lumber, wood products).

Newton (NYOO-tuhn), village (2001 population 403), WARWICKSHIRE, central ENGLAND, 3 mi/4.8 km NW of RUGBY; 52°25′N 01°17′W. Gravel extraction.

Newton (NYOO-tuhn), village, South Lanarkshire, S Scotland, 6 mi/9.7 km SE of GLASGOW; 55°47′N 04°11′W. Former steel-milling center. Formerly in Strathclyde, abolished 1996.

Newton, village, SIERRA LEONE, on SIERRA LEONE PENINSULA, 3 mi/4.8 km E of WATERLOO; 08°20′N 13°00′W. Hog and poultry raising; vegetable gardens.

Newton, village (2000 population 851), ⊙ BAKER county, SW GEORGIA, 22 mi/35 km SSW of ALBANY, on FLINT RIVER; 31°19′N 84°20′W. Peanut processing.

Newton, village (2006 population 644), CACHE county, N UTAH, on CLARKSTON CREEK near its confluence with BEAR River, and 12 mi/19 km NW of LOGAN; 41°51′N 111°59′W. Irrigated agricultural area; fruit, vegetables; dairying. Elevation 4,525 ft/1,379 m. NEWTON DAM is 3 mi/4.8 km N.

Newton Abbot (NYOO-tuhn AB-uht), town (2001 population 23,580), S DEVON, SW ENGLAND, on TEIGN RIVER estuary, at mouth of Lemon River, 14 mi/23 km SSW of EXETER; 50°32′N 03°36′W. Railroad junction. Previously light manufacturing (leather goods, textiles) and quarrying (for clay pipes). At Forde House William III was proclaimed king (1688). Church dates from 14th century.

Newton Aycliffe (NYOO-tuhn AI-klif), town (2001 population 25,655), DURHAM, NE ENGLAND, 6 mi/9.7 km N of DARLINGTON; 54°37′N 01°34′W. Light industry. Designated one of the New Towns (1947).

Newton Centre, MASSACHUSETTS: see NEWTON.

Newton Corner, MASSACHUSETTS: see NEWTON.

Newton Dam and Reservoir, UTAH: see CLARKSTON CREEK.

Newton Falls (NOO-tuhn FALZ), city (□ 2 sq mi/5.2 sq km; 2006 population 4,782), TRUMBULL county, NE OHIO, 8 mi/13 km SW of WARREN, and on MAHONING RIVER; 41°11′N 80°58′W. Manufacturing of structural steel and tubing; also motor vehicles, machinery.

Newton Falls, village, ST. LAWRENCE county, N NEW YORK, near CRANBERRY LAKE, on OSWEGATCHIE RIVER (dam nearby) and 32 mi/51 km S of POTSDAM; 44°13′N 75°00′W.

Newtongrange (NYOO-tuhn-grainj), village, MIDLOTHIAN, E Scotland, on South Esk River, and 2 mi/3.2 km S of DALKEITH; 55°50′N 03°04′W. Former coal-mining community. Formerly in LOTHIAN, abolished 1996.

Newton Grove (NOO-tuhn GROV), village (□ 3 sq mi/7.8 sq km; 2006 population 627), SAMPSON county, S central NORTH CAROLINA, 22 mi/35 km SW of GOLDSBORO near Great Coharrie Creek; 35°15′N 78°21′W. Service industries; manufacturing (hog feeders; corn milling). Agricultural area (grain, cotton, peanuts, tobacco, sweet potatoes; livestock). Bentonville Battleground State Historical Site to N.

Newton Hamilton, borough (2006 population 259), MIFFLIN county, central PENNSYLVANIA, 3 mi/4.8 km E of MOUNT UNION, on JUNIATA RIVER; 40°23′N 77°50′W. Agriculture (corn, hay; livestock; dairying). Tuscarora State Forest to E; Rothrock State Forest to NW.

Newton Heath, ENGLAND: see MANCHESTER.

Newton Highlands, MASSACHUSETTS: see NEWTON.

Newtonhill (NYOO-tuhn-hil), village (2001 population 2,940), Aberdeenshire, NE Scotland, on NORTH SEA, and 3 mi/4.8 km SSW of PORTLETHEN; 57°02′N

02°09′W. Fishing. Has increasingly become "dormitory" for ABERDEEN (8 mi/12.9 km NNE). Formerly in Grampian, abolished 1996.

Newtonia, city (2000 population 231), NEWTON county, SW MISSOURI, in the OZARKS, 10 mi/16 km E of NEOSHO; 36°52′N 94°10′W. Civil War battle, September 30, 1862.

Newton-in-Makerfield, ENGLAND: see NEWTON-LE-WILLOWS.

Newton Lake, reservoir, JASPER county, SE ILLINOIS, on Weather Creek (a tributary of the BIG MUDDY RIVER), 13 mi/21 km SW of NEWTON; 38°52′N 88°17′W. Sandy Creek (N) arm is 10 mi/16 km long, Laws Creek (NNE) arm 15 mi/24 km long. Maximum capacity of 44,000 acre-ft. Formed by Newton Dam (52 ft/16 m high), built in 1974.

Newton-le-Willows (NYOO-tuhn–luh–WIL-oz), town (2001 population 21,307), MERSEYSIDE, NW ENGLAND, 5 mi/8 km E of ST. HELENS; 53°27′N 02°36′W. Previously locomotive works. Light manufacturing. Just W is town of Earlestown. Formerly called Newton-in-Makerfield.

Newton Lower Falls, MASSACHUSETTS: see NEWTON.

Newton Mearns, Scotland: see GLASGOW.

Newtonmore (NYOO-tuhn-mor), village (2001 population 982), HIGHLAND, N Scotland, on the SPEY RIVER, 3 mi/4.8 km WSW of KINGUSSIE; 57°04′N 04°08′W. Tourist resort and skiing center. Clan Macpherson Museum.

Newton, Mount, Norwegian *Newtontoppen,* highest peak (5,633 ft/1,717 m) of SPITSBERGEN, NORWAY, in NE part of island, near head of WIJDE FJORD, near 79°00′N 16°30′E.

Newton Stewart (NYOO-tuhn STOO-uhrt), town (2001 population 3,573), DUMFRIES AND GALLOWAY, S Scotland, on CREE RIVER, and 7 mi/11.3 km NNW of WIGTOWN; 54°58′N 04°29′W. Cattle market, previously with agricultural-implement works.

Newton St. Faith, ENGLAND: see HORSHAM ST. FAITH.

Newtonsville (NOO-tuhnz-vil), village (2006 population 531), CLERMONT county, SW OHIO, 10 mi/16 km NNE of BATAVIA; 39°11′N 84°05′W.

Newtontoppen, NORWAY: see SVALBARD.

Newton Upper Falls, MASSACHUSETTS: see NEWTON.

Newtonville, MASSACHUSETTS: see NEWTON.

New Town, Czech *Nové Město,* German *Neustadt,* S central district of PRAGUE, on right bank of VLTAVA RIVER, PRAGUE province, central BOHEMIA, CZECH REPUBLIC. City shopping center founded in 1348 by Charles IV in order to relieve the older part of the town of the noise from workshops and commercial activity. Located here is the noted National Theatre, National Museum, and St. Wenceslas Square (Czech *Václavské něsti*).

Newtown, town, WESTERN REGION, GHANA, on Gulf coast of Ghana and Côte d'Ivoire border, 13 mi/21 km W of Half-Assini, between Tano Lagoon and Gulf of GUINEA; 05°07′N 03°06′W. Coastal road terminus. Fishing; copra. Also called Nyekyima.

Newtown, town, FAIRFIELD county, SW CONNECTICUT, on the HOUSATONIC, 8 mi/12.9 km ENE of DANBURY; 41°23′N 73°17′W. Dairy and fruit farms are in the area. Industry includes the manufacturing of pressure gauges, plastics, paper products, and fabricated metal products. Maximum security state prison. Incorporated 1711.

Newtown, town (2000 population 162), FOUNTAIN county, W INDIANA, 7 mi/11.3 km SE of ATTICA. Agricultural area.

Newtown, town (2000 population 209), SULLIVAN county, N MISSOURI, 16 mi/26 km NW of MILAN; 40°22′N 93°19′W.

New Town, town (2006 population 1,690), MOUNTRAIL county, NW NORTH DAKOTA, 25 mi/40 km SSW of STANLEY, in Fort Berthold Indian Reservation; 47°58′N 102°29′W. Located near N end of Van Hook Arm, large bay of LAKE SAKAKAWEA. Tribal community college, tribal casino. Four Bears State Recreation Area to W. Manufacturing (printing, electronics). Founded around 1950.

Newtown, township, DELAWARE county, SE PENNSYLVANIA, residential suburb 11 mi/18 km W of PHILADELPHIA; 39°59′N 75°24′W. Includes the communities of Florida Park, NEWTOWN, and NEWTOWN SQUARE. State park to SW.

Newtown (NOO-tuhn), Welsh *Y Drenewydd*, town (2001 population 10,783), POWYS, E Wales, on SEVERN RIVER, 8 mi/12.9 km SW of MONTGOMERY; 52°30′N 03°18′W. Former woolen (flannel) milling and leather tanning. Manufacturing of electrical appliances and tools. New Town designated in 1967 extends to W. Just N is town of Llanllwchaiarn.

Newtown (NOO-toun), village (□ 2 sq mi/5.2 sq km; 2000 population 2,420), HAMILTON county, extreme SW OHIO; suburb 9 mi/14.4 km E of CINCINNATI, across Little Miami River; 39°07′N 84°21′W. Concrete products, plastics, machinery. Laid out 1801.

Newtown, borough (2006 population 2,255), BUCKS county, SE PENNSYLVANIA, suburb 22 mi/35 km NNE of PHILADELPHIA; 40°13′N 74°55′W. Diverse manufacturing; agriculture (grain, apples, soybeans; livestock; dairying). Tyler State Park to W; Bucks County Community College to W. Settled 1684, laid out 1733, incorporated 1838.

New Town (NOO TOUN), N residential suburb of HOBART, TASMANIA, SE AUSTRALIA, within city limits; 42°51′S 147°18′E.

Newtown (NOO-toun), residential suburb, E NEW SOUTH WALES, SE AUSTRALIA, 4 mi/6 km SW of SYDNEY; 37°42′S 143°39′E. In metropolitan area; light industry.

Newtownabbey (NOO-toun-A-bee), town (2001 population 62,022), DOWN, SE Northern Ireland, 2 mi/3.2 km NE of BELFAST, on BELFAST LOUGH; 54°40′N 05°52′W. Recently created to accommodate Belfast "overspill." Mainly residential.

Newtownards (NOO-toun-ahrdz), Gaelic *Baile Nua na hArda*, town (2001 population 27,795), DOWN, E Northern Ireland, near the head of STRANGFORD LOUGH; 54°35′N 05°40′W. Commuters to BELFAST. Laid out by the Montgomerys and Hamiltons on former O'Neill-owned land in 1616. Lord Castlereagh born here.

Newtownbarry (nyoo-TOUN-BA-ree), Gaelic *Bun Clóidighe*, town (2006 population 1,544), NW WEXFORD county, SE IRELAND, on SLANEY RIVER, 11 mi/18 km NNW of ENNISCORTHY, at foot of MOUNT LEINSTER. Agricultural market.

Newtown Battlefield State Park, NEW YORK: see ELMIRA.

Newtown Creek, tidal arm of EAST RIVER, within NEW YORK city, SE NEW YORK; c.4 mi/6.4 km long, partly separates BROOKLYN and QUEENS boroughs. Carries shipping for industrial NW Queens.

Newtown Grant, unincorporated town (2000 population 3,887), BUCKS county, SE PENNSYLVANIA, residential suburb 22 mi/35 km NNE of PHILADELPHIA and 1 mi/1.6 km NW of NEWTOWN; 40°15′N 74°57′W.

Newtown Hamilton (NOO-toun HAM-uhl-tuhn), town (2001 population 645), W ARMAGH, S Northern Ireland, 12 mi/19 km SSE of ARMAGH; 54°11′N 06°35′W. Market. In the heart of the Fews, former woodlands destroyed for lead smelting in the 18th century. Established 1770.

Newtown Square, unincorporated village, DELAWARE county, SE PENNSYLVANIA, suburb 11 mi/18 km W of PHILADELPHIA; 39°59′N 75°24′W. Manufacturing includes chemicals and plastics, pharmaceuticals, printing and publishing.

Newtown Stewart (NOO-toun STOO-wahrt), town, NW TYRONE, Northern Ireland, on Mourne River, 9 mi/14.5 km NNW of OMAGH; 54°43′N 07°24′W. Former plantation and linen-making. Market.

New Tredegar (tre-DEE-guhr), locality (2001 population 4,945), Caerphilly, SE Wales, on RHYMNEY RIVER, 2 mi/3.2 km N of BARGOED; 51°43′N 03°14′W. Former coal and iron producer. Formerly in MID GLAMORGAN, abolished 1996.

New Trier (TREE-yuhr), village (2000 population 116), DAKOTA county, SE MINNESOTA, 10 mi/16 km S of HASTINGS; 44°36′N 92°55′W. Agricultural area (grain; livestock; dairying). Richard J. Dorer Memorial Hardwood State Forest to E.

New Tulsa, village, WAGONER county, NE OKLAHOMA, residential suburb 16 mi/26 km ESE of TULSA; 36°06′N 95°44′W.

New Ulm (UHLM), city (2000 population 13,594), ⊙ BROWN county, S MINNESOTA, 25 mi/40 km WNW of MANKATO, on the MINNESOTA RIVER at the mouth of COTTONWOOD RIVER; 44°18′N 94°27′W. Elevation 818 ft/249 m. New Ulm is a processing and trade center for agricultural area (grain, soybeans, peas, sugar beets; livestock; dairying); manufacturing (dairy products, beer, electronics, metal fabrication, printing and publishing). It was settled in 1854 by Germans, who named it after ULM, GERMANY. In 1862, C. E. Flandrau, then a justice of the Minnesota supreme court, led the defense of the city during a Sioux uprising. Municipal Airport to W. County Fairgrounds here. Seat of Dr. Martin Luther King College. Flandrau State Park at SW edge of city, on Cottonwood River. Incorporated as a city 1876.

New Ulm (UHLM), unincorporated village, AUSTIN county, S TEXAS, 67 mi/108 km W of HOUSTON, on SAN BERNARD RIVER; 29°53′N 96°29′W. Oil and natural gas. Agricultural area.

New Underwood, village (2006 population 661), PENNINGTON county, SW central SOUTH DAKOTA, 20 mi/32 km E of RAPID CITY, and on Box Elder Creek; 44°05′N 102°49′W. In farm region (grain; livestock).

New Valley, province (2004 population 166,211), SW EGYPT, in LIBYAN DESERT; ⊙ KHARGA, in Kharga oasis. Bordered E by the NILE valley, N by MATRUH, GIZA, and MINYA provinces, W by LIBYA, S by the SUDAN. In it are the oases of Kharga and DAKHLA, which, irrigated by springs, grow palm trees, fruit, rice, cereals. Kharga is reached by railroad from NAG HAMMADI. Both oases have important archaeological remains, especially temples of the Roman period. Formerly called Southern Desert province [Arabic *Al–Sahra' al-Janubiyah*], it was the location of the failed "New Valley" development project in the 1960s, where artesian wells were to be tapped and used to irrigate desert lands. Another scheme, the "New Delta," is currently underway that involves piping water from Lake NASSER through the proposed TOSHKA CANAL to connect the region's oases.

New Vienna, town (2000 population 400), DUBUQUE county, E IOWA, 22 mi/35 km WNW of DUBUQUE; 42°32′N 91°06′W. In livestock area.

New Vienna (vee-EN-ah), village (2006 population 1,388), CLINTON county, SW OHIO, 14 mi/23 km ESE of WILMINGTON, and on East Fork of Little Miami River; 39°19′N 83°42′W.

Newville, town (2000 population 553), Henry co., SE Alabama, 12 mi/19 km SSW of Abbeville. First called 'Wells Station' for James Madison Wells, an early property owner, it was renamed in 1894. Inc. in 1903.

Newville, borough (2006 population 1,319), CUMBERLAND county, S PENNSYLVANIA, 11 mi/18 km WSW of CARLISLE, on Big Spring Creek; 40°10′N 77°24′W. Manufacturing (apparel, farm equipment). Agriculture (grain, soybeans; livestock; dairying). Big Spring to S; Colonel Denning State Park to N. Laid out 1794, incorporated 1817.

New Vineyard, town, FRANKLIN county, W central MAINE, 9 mi/14.5 km NNE of FARMINGTON; 44°48′N 70°07′W. Wood products.

New Virginia, town (2000 population 469), WARREN county, S central IOWA, 15 mi/24 km SSW of INDIANOLA and 29 mi/47 km S of DES MOINES; 41°10′N 93°43′W. In agricultural area.

New Wadi Halfa, Arabic *Shashm Al-Girba*, town, Northern state, N SUDAN, SE bank of Lake NASSER, on railroad and near Egyptian border. Main urban center within KHASHM AL-GIRBA agricultural scheme, in which the people of (Old) WADI HALFA have been resettled after their land was covered with Lake NASSER (Lake NUBIA) water. The construction of Khashm Al-Girbah Dam on ATBARA RIVER to irrigate the scheme was completed in 1966.

New Washington, town, AKLAN province, N PANAY island, PHILIPPINES, on SIBUYAN SEA, 23 mi/37 km WNW of ROXAS; 11°34′N 122°29′E. Fishing town and agricultural center (tobacco, rice). Fishing school.

New Washington, village, CLARK county, S INDIANA, 10 mi/16 km NE of CHARLESTOWN. Cattle, poultry.

New Washington (WAHSH-ing-tuhn), village (□ 1 sq mi/2.6 sq km; 2006 population 941), CRAWFORD county, N central OHIO, 12 mi/19 km NE of BUCYRUS; 40°58′N 82°51′W. Fabricated metal products, food products.

New Washington, borough (2006 population 85), CLEARFIELD county, central PENNSYLVANIA, 11 mi/18 km NNE of BARNESBORO, on Chest Creek; 40°49′N 78°42′W. Corn; livestock; dairying.

New Washoe City (wuh-SHO), unincorporated town, WASHOE county, W NEVADA, residential community 8 mi/12.9 km N of CARSON CITY, near E shore of WASHOE LAKE; 39°17′N 119°46′W. Washoe Lake State Park to S; Mount Davidson to E.

New Waterford, (WA-tur-furd) town, on NE CAPE BRETON ISLAND, NOVA SCOTIA, NE of SYDNEY; 46°15′N 60°05′W. Mine closures in the 1960s and 1970s caused steady outmigration from this traditional coal-mining center. Likely named after the Irish seaport, Waterford, by immigrants. Formerly known as Barrachois, meaning lagoon or pond.

New Waterford (WAH-tuhr-fuhrd), village (2006 population 1,382), COLUMBIANA county, E OHIO, 9 mi/14 km ENE of LISBON; 40°51′N 80°37′W. In agricultural and coal area; furniture, pottery.

New Waterway, Dutch *Nieuwe Waterweg*, canal, SOUTH HOLLAND province, SW NETHERLANDS; main shipping channel for ROTTERDAM harbor; extends 11 mi/18 km WNW from NEW MAAS RIVER, which serves as Rotterdam's inland harbor, at joining of OLD MAAS RIVER 7 mi/11.3 km WSW of Rotterdam, to NORTH SEA 2 mi/3.2 km W of HOEK VAN HOLLAND, through the HET SCHEUR channel. EUROPOORT, large port facility with oil refineries and storage facilities, on S side of canal. Caland Canal, a secondary shipping channel, parallels canal on S.

New Waverly (WAI-vuhr-lee), town (2006 population 913), SAN JACINTO county, E TEXAS, 14 mi/23 km S of HUNTSVILLE; 30°32′N 95°28′W. In timber and oil area; manufacturing (lumber, oil field equipment, tools). Settled in 1830s as plantation center. LAKE CONROE reservoir to SW; surrounded W, N, and E by Sam Houston National Forest; Huntsville State Park to N.

New Westminster (NOO WEST-min-stuhr), city (□ 6 sq mi/15.6 sq km; 2001 population 54,656), SW BRITISH COLUMBIA, W Canada, on the FRASER River, part of metropolitan VANCOUVER; 49°12′N 122°55′W. Founded in 1859 as Queensborough, it was the capital of British Columbia until VICTORIA was made the capital after the union of British Columbia and VANCOUVER ISLAND in 1866. New Westminster is a year-round port, with an excellent harbor that is the base of the Fraser River fishing fleet and a shipping point for grain, lumber, minerals, and canned goods. Among the city's industries are salmon, fruit, and vegetable canneries; distilleries and breweries; oil refineries; paper, lumber, and flour mills; and shipbuilding plants. Columbia and Saint Louis colleges are in the city, as are Anglican and Roman Catholic cathedrals.

New Weston (WES-tuhn), village (2006 population 134), DARKE county, W OHIO, 16 mi/26 km N of GREENVILLE; 40°21′N 84°39′W. In agricultural area; ceramics.

New Whiteland, town (2000 population 4,579), JOHNSON county, central INDIANA, satellite community of INDIANAPOLIS, 15 mi/24 km S of downtown, and 4 mi/6.4 km NNW of FRANKLIN; 39°34′N 86°06′W.

New Willard (WIL-uhrd), unincorporated village, POLK county, E TEXAS, 38 mi/61 km SSW of LUFKIN; 30°47′N 94°53′W. Lumbering.

New Wilmington (WIL-meeng-tuhn), borough (2006 population 2,402), LAWRENCE county, W PENNSYLVANIA, 8 mi/12.9 km N of NEW CASTLE, near Little Neshannock Creek; 41°07′N 80°19′W. Manufacturing (tools, cheese, printing and publishing). Agriculture (apples, corn, hay; dairying). Westminister College. Covered bridge to SE; Amish Town Farm to N. Incorporated 1863.

New Windsor, town (2000 population 1,303), CARROLL county, N MARYLAND, 19 mi/31 km ENE of FREDERICK, near Little Pipe River; 39°32′N 77°06′W. Called Sulphur Springs when first settled in the early 19th century, it was renamed in 1844 for WINDSOR in ENGLAND. Site of the headquarters of the Church World Service of the National Council of the Church of Christ in the U.S.

New Windsor, village, MERCER county, NW ILLINOIS, 17 mi/27 km NNW of GALESBURG; 41°12′N 90°26′W. Dairy products, corn, soybeans. Corporate name is the Village of Windsor.

New Windsor, suburban village (2000 population 9,077), ORANGE county, SE NEW YORK, on W bank of the HUDSON, just S of NEWBURGH; 41°28′N 74°07′W. Light manufacturing. Site of Stewart International Airport, an air force base opened to commercial aircraft in 1990 and now a regional airport. Industrial park located at the airport. Was home of George Clinton. De Witt Clinton was born here.

New Windsor, ENGLAND: see WINDSOR.

New Woodville, village, MARSHALL county, S OKLAHOMA, near Lake Texoma, 2 mi/3.2 km N of WOODVILLE and 11 mi/18 km, SSE of MADILL.

New World Island (□ 62 sq mi/161 sq km; 2001 population 2,402), in NOTRE DAME BAY, SE NEWFOUNDLAND AND LABRADOR, Canada; 20 mi/32 km long, 8 mi/13 km wide; 49°35′N 54°40′W. Fishing.

New Year's Islands, Spanish *Islas Año Nuevo*, small archipelago (□ 2.2 sq mi/5.7 sq km) in the South ATLANTIC OCEAN off the Argentine TIERRA DEL FUEGO, just N of ISLA DE LOS ESTADOS; three small islands, on largest of which (54°39′S 64°08′W) are lighthouse and observatory.

New Year's Point, California: see AÑO NUEVO POINT.

New York, state (□ 54,475 sq mi/141,635 sq km; 2006 population 19,306,183), NE UNITED STATES, one of the Mid-Atlantic states and one of the original thirteen colonies; ⊙ ALBANY; 42°54′N 75°40′W. New York is known as the Empire State. NEW YORK is the largest city.

Geography

The state is bounded on the N by the Canadian provinces of QUEBEC and ONTARIO, with the SAINT LAWRENCE River and Lake ONTARIO marking the Ontario border. In the NW the NIAGARA River, with scenic NIAGARA FALLS, forms the border with Ontario between Lake Ontario and Lake ERIE. Lake Erie itself and a minute portion of Pennsylvania constitute the rest of the W border. PENNSYLVANIA and NEW JERSEY are to the S, except where the state extends into the ATLANTIC OCEAN at New York city and LONG ISLAND. To the E, New York borders on CONNECTICUT, MAS-

SACHUSETTS, and VERMONT. Lake Champlain's outlet, the RICHELIEU River, stretching past the Canadian border, is part of the state line with Vermont; it is also the chief N feature of the Great Valley (including the HUDSON RIVER) that dominates all of E New York. The Hudson is noted for its beauty, as are Lake CHAMPLAIN and neighboring Lake GEORGE, which have many resorts. W of the lakes are the ADIRONDACK MOUNTAINS, including the rugged High Peaks section of S CLINTON, E ESSEX and N WARREN counties, another major vacationland, with woods in the N and sports centers like LAKE PLACID and SARANAC LAKE. Mount MARCY (5,344 ft/1,629 m), the highest point in the state, is in the Adirondacks near Lake Placid. The rest of NE New York is hilly, sloping gradually to the valleys of the Saint Lawrence and Lake Ontario. The MOHAWK RIVER, which flows from ROME to the Hudson N of ALBANY, and ONEIDA LAKE are part of the NEW YORK STATE BARGE CANAL, a major route to the GREAT LAKES and the midwestern U.S. as well as the only complete natural route across the APPALACHIAN MOUNTAINS. Most of the S part of the state is on the Appalachian plateau, which rises in the SE just W of the Hudson Valley as the CATSKILL MOUNTAINS, an area that attracts many vacationers from New York city and its environs. New York city, in turn, attracts multitudes of tourists from all over the world. The W extension of the state to Lakes Ontario and Erie contains many bodies of water, notably ONEIDA LAKE and the celebrated FINGER LAKES. The W region in rural areas has an abundance of farms (agriculture is the state's single largest industry) as well as large, traditionally industrial cities such as BUFFALO on Lake Erie, ROCHESTER on Lake Ontario, SYRACUSE, and UTICA. The W section is drained SW by the ALLEGHENY RIVER, W by CATTARAUGUS CREEK, N by the GENESEE, and S by the SUSQUEHANNA and DELAWARE RIVER systems.

Economy

The Delaware River Basin Compact, signed in 1961 by New York, New Jersey, Pennsylvania, Delaware, and the Federal government, regulates the utilization of water of the Delaware system. SCHENECTADY, Albany, and New York city, once the major industrial cities of the lower Mohawk and the Hudson, continue their longtime manufacturing decline, as have many cities in the NE U.S. Except in the mountain regions, the areas between cities are rich agriculturally. The Ontario Lake plain, from the Niagara Frontier to near OSWEGO, is a major fruit- and vegetable-producing region. New York ranks second only to Washington state in the production of apples. The Finger Lakes region, one of New York's leading regions, has extensive vineyards and is famous for its wines. Other areas of the state produce diverse crops, especially grains, vegetables, hay, and potatoes (grown in great quantity on E Long Island). New York's mineral resources include crushed stone, cement, salt, and zinc. The state has a complex system of railroads, air routes, and modern highways, notably the NEW YORK STATE THRUWAY. The rivers and the NEW YORK STATE BARGE CANAL, an improvement of the old ERIE CANAL, still carries limited freight between Albany and Syracuse, and a comprehensive plan has been instituted for the historical and recreational development of the canal. Ocean shipping is handled by the port of New York city and to a much lesser extent, by the upstate ports of Buffalo, Albany, Oswego, and OGDENSBURG. Hydroelectricity for N New York is produced by the Saint Lawrence power project and by the Niagara power project, which began producing in 1961. In spite of significant decline, New York has retained some important manufacturing industries, and, by virtue of New York city, it has strengthened its position as a commercial and financial leader. Although the largest percentage of the state's jobs lie in the

service sector, its manufacture is extremely diverse and includes apparel, food products, machinery, chemicals, paper, electrical equipment (notably at Schenectady), computer equipment, optical instruments and cameras (at Rochester), and transportation equipment. Printing and publishing, mass communications, advertising, and entertainment are among New York city's notable industries. BROOKHAVEN NATIONAL LABORATORY, an atomic energy testing and research center, is in central Long Island. Many corporate headquarters and research facilities have relocated to WESTCHESTER county, N of New York city. Fishing in Lakes Erie and Ontario is confined to sports fishing, while commercial fishing takes place in the waters around Long Island. The state has c.29,336 sq mi/75,980 sq km of forest, and forestry is a major industry. It includes the harvesting and exporting of hardwoods, harvesting pulp for timber, and use for the manufacture of a variety of hardwood and forest products, such as maple syrup, poles, pallets, and baseball and softball bats.

Culture, Learning, Places of Interest

Early in its history New York state emerged as one of the cultural leaders of the nation. In the early 19th century local authors Washington Irving, William Cullen Bryant, and James Fenimore Cooper were among the country's foremost literary figures. The natural beauty of New York inspired the noted Hudson River School of American landscape painters. With New England's decline as a literary hub, many writers came to New York city from other parts of the nation, helping to make it a literary and publishing center and the cultural heart of the country. Outside of New York city, the institutions of higher education in the state include Alfred University, Bard College, Colgate University, Cornell University, Hamilton College, Hobart College, Long Island University, Rensselaer Polytechnic Institute, Sarah Lawrence College, Skidmore College, Syracuse University, U.S. Military Academy, University of Rochester, Vassar College, and Wells College. The State University of New York has major campuses at Stony Brook, Albany, Binghamton, and Buffalo.In addition to the great forest preserves of the Adirondack and Catskill mountains, New York has many state parks, including JONES BEACH, and BUTTERMILK FALLS STATE PARK, and CHENANGO VALLEY STATE PARK. Part of FIRE ISLAND is a national seashore. The racetrack of SARATOGA SPRINGS and Saratoga Performing Arts Center makes it both a pleasure and health resort, and the THOUSAND ISLANDS are popular with summer vacationers. Among the several places of historic interest in the state under Federal administration are those at HYDE PARK, with the burial place of Eleanor and Franklin D. Roosevelt, and the VANDERBILT MANSION.

History to 1629

Before Europeans began to arrive in the 16th century, New York was inhabited mainly by Algonquian- and Iroquoian-speaking Native Americans. The Algonquians, including the Mahican, Lenni Lenape, and Wappinger, lived chiefly in the Hudson valley and on Long Island. The Iroquois, living in the central and W parts of the state, included the Cayuga, Mohawk, Oneida, Onondaga, and Seneca, who joined c.1570 to form the Iroquois Confederacy. Europeans first approached New York from both the sea and from Canada. Giovanni da Verrazano, a Florentine in the service of France, visited (1524) the excellent harbor of NEW YORK BAY but did little exploring. In 1609, French explorer Samuel de Champlain traveled S on Lake Champlain from Canada, and Henry Hudson, an Englishman in the service of the Dutch, sailed the Hudson nearly to Albany. The French, who had allied themselves with the Hurons of Ontario, continued to push into N and W New York from Canada, but met with resistance from the Iroquois Confederacy, which dominated W New York. The Dutch claimed the Hudson region early on, and the Dutch West India Company (chartered in 1621, organized in 1623) planted (1624) their colony of NEW NETHERLAND, with its chief settlements at NEW AMSTERDAM on the lower tip of present-day MANHATTAN island (purchased in 1626 from the Canarsie for goods worth about $24) and at Fort Nassau, later called Fort Orange (present-day Albany).

History: 1629 to 1765

To speed up the slow pace of colonization the Dutch set up the patroon system in 1629, thus establishing the landholding aristocracy that became the hallmark of colonial New York. The last and most able of the Dutch administrators was Peter Stuyvesant (in office 1647–1664). The English claimed the whole region on the basis of the explorations of John Cabot, and in 1664 an English fleet sailed into the harbor of New Amsterdam. Stuyvesant surrendered without a struggle, and New Netherland then became the colonies of New York and New Jersey. The popular governor Thomas Dongan (in office 1683–1688) put New York on a firm basis and began to establish the alliance of the English with the Iroquois, which later played an important part in New York history. The French threat was continuous, and New York was involved in a number of the French and Indian Wars (1689–1763). Unrest from Native Americans caused much of W New York to remain unsettled by colonists throughout the 18th century. Slowly, however, the colony, with its busy shipping and fishing fleets, and its expanding farms, was beginning to establish its own separate identity from that of England. Colonial self-assertiveness grew after the warfare with the French ended; there was considerable objection to the restrictive commercial laws, and the Navigation Acts were flouted by smugglers. When the Stamp Act was passed, New York was a leader of the opposition, and the Stamp Act Congress met (1765) in New York city.

History : 1765 to 1791

The Stamp Act occasioned considerable complaint, and unrest grew. When trouble flared into the American Revolution, New Yorkers were divided. About one-third of all the military engagements of the American Revolution took place in New York state. The first major military action in the state was the capture (May 1775) of TICONDEROGA by Ethan Allen and his Green Mountain Boys and Benedict Arnold. In August 1776, however, George Washington was unable to hold lower New York against the British and lost battles at HARLEM Heights and WHITE PLAINS. The British successfully invaded New York city; however, the state had declared independence, with KINGSTON as its capital. New York was in 1777 the key to the overall British campaign plan, which was directed toward taking the entire state and thus separating NEW ENGLAND from the S. This failed finally (October 1777) in the battles near the present-day resort of Saratoga Springs. For the rest of the war there was more or less a stalemate in New York, with the British occupying New York city and the patriots holding most of the rest of the state except for Westchester county.

History: 1791 to 1827

The E boundary of New York was established after much conflict and violence when Vermont was admitted as a state in 1791. New York city was capital of the new nation briefly and was also the state capital until 1797, when Albany succeeded it. From the 1780s increased commerce (somewhat slowed by the Embargo Act of 1807) and industry, especially textile milling, marked the turn away from the old, primarily agricultural, order. In the War of 1812, New York saw action in 1813–1814, with the British capture of FORT NIAGARA and particularly with the brilliant naval victory over the British on Lake Champlain at PLATTSBURGH. The state continued its development,

which was quickened and broadened by the building of the Erie Canal. The canal, completed in 1825, and railroad lines constructed (from 1831) parallel to it made New York the major E-W commercial route in the 19th century and helped to account for the growth and prosperity of the port of New York. Cities along the canal (Buffalo, Syracuse, Rome, Utica, and Schenectady) prospered, Albany grew, and New York city became the financial capital of the nation. New constitutions broadened the suffrage in 1821 and again in 1846; slavery was abolished in 1827.

History: 1827 to 1867

Politics was largely controlled from the 1820s to the 1840s by the Albany Regency, which favored farmers, artisans, and small businessmen. New York was a leader in numerous 19th-century reform groups, including antislavery and women's rights groups. Migrants from New England had been settling on the W frontier, and in the 1840s famine and revolution in Europe resulted in a great wave of Irish and German immigrants, whose first stop in America was New York city. In 1850, Millard Fillmore became the second New Yorker to be president of the U.S.; the first was Martin Van Buren (1837–1841). The split of the Democrats over the slavery issue helped pave the way for New York's swing to the Republicans and Abraham Lincoln in the fateful election of 1860. During the Civil War, New York state strongly favored the Union and contributed much to its cause. Industrial development was stimulated by the needs of the military, and railroads increased their capacity. New York city's newspapers had considerable national influence, and after the war the publication of periodicals and books centered more and more in the city, whose libraries expanded.

History: 1867 to 1932

From 1867 to 1869, Cornelius Vanderbilt consolidated the New York Central Railroad system. As economic growth accelerated, political corruption became rampant, particularly the Tweed Ring of New York city, headed by William Magear "Boss" Tweed. Chester A. Arthur (1881–1885) and Grover Cleveland (1885–1889, 1893–1897) were New Yorkers who served as presidents of the U.S. in the late 19th century. The inpouring after 1880 of immigrants from Ireland, Italy, and E Europe brought workers for the old industries, which were expanding, and for the new ones, including the electrical and chemical industries, which were being established. Working conditions worsened but were challenged by the growing labor movement in the state. Service as New York city's police commissioner and then as reforming governor of New York helped Theodore Roosevelt establish the national reputation that sent him to the vice presidency and then to the WHITE HOUSE (1901–1909). In the early 20th century, New York politics seesawed from the Republicans to the Democrats and back again. Reform programs, emphasizing public works, conservation, reorganization of state finances, social welfare, and extensive labor laws, continued to gain ground, however.

History: 1932 to Present

Franklin D. Roosevelt, former governor of New York, went to the White House in 1932, and then-Governor Herbert H. Lehman (1932–1942) followed the president's national New Deal program by instituting the Little New Deal in New York state. At the same time Fiorello LaGuardia, Republican mayor of New York city (1934–1945), enthusiastically supported Roosevelt's social and economic reforms. Thomas E. Dewey, elected governor in 1942, 1946, and 1950, had the immense task of coordinating state activities with national efforts in World War II, which strained New York's resources to the utmost. He also built upon the reforms of his predecessors. Nelson Rockefeller, elected governor in 1958, 1962, 1966, and 1970, increased the state's social welfare programs, greatly

expanded the State University (established 1948), and was largely responsible for the construction of a large state-office and cultural complex in Albany. New York's growth slowed considerably during the 1970s and 1980s as it lost its dominant position in manufacturing. However, New York city remains the nation's most populous city (with twice the number of inhabitants as Los Angeles, which is second). On September 11, 2001, New York city was the site (along with the PENTAGON in VIRGINIA) of a terrorist attack that destroyed the twin towers of the WORLD TRADE CENTER in lower Manhattan, killing nearly 3,000 people. Reconstruction of the site began in 2006.

New York has sixty-two counties: ALBANY, ALLEGANY, Bronx (coextensive with the BRONX borough of New York city), BROOME, CATTARAUGUS, CAYUGA, CHAUTAUQUA, CHEMUNG, CHENANGO, CLINTON, COLUMBIA, CORTLAND, DELAWARE, DUTCHESS, ERIE, ESSEX, FRANKLIN, FULTON, GENESEE, GREENE, HAMILTON, HERKIMER, JEFFERSON, Kings (coextensive with the BROOKLYN borough of New York city), LEWIS, LIVINGSTON, MADISON, MONROE, MONTGOMERY, NASSAU, NEW YORK, NIAGARA, ONEIDA, ONONDAGA, ONTARIO, ORANGE, ORLEANS, OSWEGO, OTSEGO, PUTNAM, QUEENS, RENSSELAER, RICHMOND, ROCKLAND, SAINT LAWRENCE, SARATOGA, SCHENECTADY, SCHOHARIE, SCHUYLE, SENECA, STEUBEN, SUFFOLK, SULLIVAN, TIOGA, TOMPKINS, ULSTER, WARREN, WASHINGTON, WAYNE, WESTCHESTER, WYOMING, and YATES.Eliot Spitzer is the current governor.

New York, county (□ 33 sq mi/85.8 sq km; 2006 population 1,611,581), SE NEW YORK, coextensive with MANHATTAN borough of NEW YORK city; 40°46′N 73°58′W.

New York, city (□ 309 sq mi/803.4 sq km; 2006 population 8,214,426), SE NEW YORK, often referred to as the Big Apple, Empire City or Gotham, largest city in the U.S. and one of the largest in the world, on NEW YORK BAY at the mouth of the HUDSON RIVER; 40°40′N 73°56′W.

Geography

It comprises five boroughs, each coextensive with a county: MANHATTAN (New York county), the heart of the city, an island; the BRONX (Bronx county), on the mainland, NE of Manhattan and separated from it by the HARLEM RIVER; QUEENS (Queens county), on LONG ISLAND, E of Manhattan across the EAST RIVER. BROOKLYN (Kings county), also on Long Island, on the East River adjoining Queens and on New York Bay; and STATEN ISLAND (Richmond county), on Staten Island, SW of Manhattan and separated from it by the Upper Bay. The metropolitan area (1990 estimated population 18,087,251) encompasses parts of SE New York state, NE NEW JERSEY, and SW CONNECTICUT.

Economy

The port of New York (which is now centered on the New Jersey side of the Hudson River), remains one of the leading ports in the world. The city is a vibrant center for commerce and business and one of the top three global cities (along with LONDON and TOKYO) that heavily influence world finance. Manufacturing—primarily of small but highly diverse types—accounts for a large but declining amount of employment. Clothing and other apparel, such as furs; chemicals; metal products; and processed foods are some of the principal manufactures. New York is also a major center of TV broadcasting, book publishing, advertising, and other facets of mass communications. The most celebrated newspapers are the *New York Times* and the *Wall Street Journal*. New York attracts many business and professional conventions. It was the site of two World's Fairs (1939–1940; 1964–1965); the city has recently hosted conventions by the country's two main political parties: the Democratic Party's in 1992 and the Republican Party's in 2004. With its vast cultural and educational resources, famous shops and restaurants, places of entertainment

---

Area is shown by the symbol □, and capital city or county seat by ⊙.

(including the theater district and many off-Broadway theaters), striking and diversified architecture (including the EMPIRE STATE BUILDING and, before 2001, the soaring towers of the WORLD TRADE CENTER), and parks and botanical gardens, New York draws millions of tourists every year.

Transportation

The city is served by three major airports: JFK (JOHN F. KENNEDY) INTERNATIONAL AIRPORT, and LAGUARDIA AIRPORT, both in Queens, and NEWARK LIBERTY INTERNATIONAL AIRPORT, in New Jersey. Railroads (national and commuter) and interstate highways converge upon New York from all points, giving access to outlying areas. The city's many bridges include the GEORGE WASHINGTON BRIDGE, BROOKLYN BRIDGE, HENRY HUDSON BRIDGE, TRIBOROUGH BRIDGE, BRONX-WHITESTONE BRIDGE, THROGS NECK BRIDGE, and the VERRAZANO-NARROWS BRIDGE. The HOLLAND TUNNEL (the first vehicular tunnel under the Hudson) and the LINCOLN TUNNEL link Manhattan with New Jersey. The QUEENS-MIDTOWN TUNNEL and the BROOKLYN-BATTERY TUNNEL, both under the East River, connect Manhattan with W Long Island. Islands in the East River include ROOSEVELT ISLAND (formerly Welfare Island), RIKERS ISLAND (site of a city penitentiary), and RANDALLS ISLAND (with Downing Stadium). In New York Bay are Liberty Island (featuring the STATUE OF LIBERTY NATIONAL MONUMENT); GOVERNORS ISLAND; and ELLIS ISLAND.

New York Neighborhoods and Demographics

Some of its streets and neighborhoods have become accepted symbols throughout the nation. WALL STREET means finance; BROADWAY, the theater; FIFTH AVENUE, fine shopping; MADISON AVENUE, advertising; and SOHO, art. New York city is also famous for its ethnic diversity, manifesting itself in scores of communities representing virtually every nation and people on earth, each preserving its cultural identity. LITTLE ITALY, CHINATOWN, and the LOWER EAST SIDE (once a predominantly Jewish enclave, now a melange primarily consisting of E and S Asians, Dominicans, Central Americans, and Latin Americans) date back to the mid-19th century. African-Americans from the South began to migrate to HARLEM after 1910, and in the 1940s large numbers of Puerto Ricans and other Hispanic-Americans began to settle in what is now known as Spanish Harlem. The city's population size has remained static, however, for over fifty years. In the 1980s, New York city experienced a new wave of immigration largely from Latin America (especially the Dominican Republic), Israel, E and S Asia, Jamaica, Haiti, Africa, and Russia. This new infusion, however, did little more than replace the second- and third-generation immigrants who were moving out of the city. The 2000 population of 8,008,278 is close to what it was in 1950. Among the many other places of interest are ROCKEFELLER CENTER; GREENWICH VILLAGE, with its cafés and restaurants; and TIMES SQUARE, with its lights and theaters. Of historic interest are Fraunces Tavern (built 1719), where George Washington said farewell to his officers after the American Revolution; Gracie Mansion (built late 18th century), now the official residence of the mayor; the Edgar Allen Poe Cottage; Grant's Tomb; and

Culture, Education, Hospitals, and Houses of Worship

New York city is the seat of the UNITED NATIONS HEADQUARTERS. LINCOLN CENTER FOR THE PERFORMING ARTS is a complex of buildings housing the Metropolitan Opera Company, the New York Philharmonic, the New York City Ballet, the New York City Opera, and the Juilliard School. Also in the city are Carnegie Hall and New York City Center, featuring performances by musical and theatrical companies. Among the best known of the city's many museums and scientific collections are the Metropolitan Museum of Art, the Museum of Modern Art, the Solomon R. Guggenheim Museum (designed by Frank Lloyd Wright), the Frick Collection (housed in the Frick mansion), the Whitney Museum of American Art, the Museum of the City of New York, the American Museum of Natural History (with the Rose Center for Earth and Space, featuring the Hayden Planetarium), the Pierpont Morgan Library, the Museum of Jewish Heritage (with a Holocaust memorial), the American Folk Art Museum, the Cooper-Hewitt Museum of Design, the museum and library of the New York Historical Society, and the Brooklyn Museum. The New York Public Library is one of the largest in the U.S. Major educational institutions include the City University of New York, Columbia University, Cooper Union, Fordham University, General Theological Seminary, Jewish Theological Seminary, New School University, New York University, Pace University, Yeshiva University, Pratt Institute, and Union Theological Seminary. A center for medical treatment and research, New York has more than 130 hospitals and several medical schools. Noted hospitals include Bellevue Hospital, Mount Sinai Hospital, Columbia-Presbyterian Medical Center, and New York Hospital. Among New York's noted houses of worship are Trinity Church, St. Paul's Chapel (dedicated 1776), St. Patrick's Cathedral, the Cathedral of St. John the Divine, Riverside Church, Temple Emanu-El, and New York Mosque.

Parks and Recreation

New York's parks and recreation centers include parts of GATEWAY NATIONAL RECREATION AREA, CENTRAL PARK, The BATTERY, Washington Square Park, Riverside Park, and Fort Tryon Park (with the Cloisters) in Manhattan; the New York Zoological Park (with the Bronx Zoo), the New York Botanical Garden, and Van Cortlandt Park in the Bronx; and CONEY ISLAND (with boardwalks, beaches, and an aquarium) and PROSPECT PARK in Brooklyn. Sporting events are held at MADISON SQUARE GARDEN in Manhattan, YANKEE STADIUM (American Leauge New York Yankees) in the Bronx, and SHEA STADIUM (National League New York Mets) in Queens.

History: First Settlement to New Amsterdam

Although Giovanni da Verrazano was probably the first European to explore the region and Henry Hudson certainly visited the area, it was with Dutch settlements on Manhattan and Long Island that the city truly began to emerge. In 1624 the colony of NEW NETHERLAND was established, with the town of NEW AMSTERDAM on the lower tip of Manhattan as its capital. Peter Minuit of the chartered Dutch West India Company supposedly bought the island from its Native American inhabitants for about $24 worth of merchandise (the sale was completed in 1626). Under the Dutch, schools were opened and the Dutch Reformed Church was established. The Native American population was forced out of the area of European settlement in a series of bloody battles. Charles II gave New Netherland to his brother James, Duke of York and Albany. In 1664 the duke sent a squadron of frigates to seize New Amsterdam, the seat of the colony's government, headed by Director-General Peter Stuyvesant. New Amsterdam was renamed New York city, and it became capital of the new British province of New York. The Dutch returned to power briefly (1673–1674) before the reestablishment of English rule.

History: Colonial Era and Revolution

The autocratic rule of British governors was one of the causes of an insurrection that broke out in 1689. In 1741, an alleged plot by African-American slaves to burn New York was ruthlessly suppressed. Throughout the 18th century New York was an expanding commercial and cultural center. The city's first newspaper, the New York *Gazette*, appeared in 1725, and its first institution of higher learning, Kings College (now Columbia University), was founded in 1754. New York was active in the colonial opposition to British measures after trouble in 1765 over the Stamp Act. As revolutionary sentiments increased, the New York Sons of Liberty forced (1775) Governor William Tryon and the British colonial government from the city. Although many New Yorkers were Loyalists, Continental forces commanded by George Washington tried to defend the city. After the patriot defeat in the Battle of Long Island and the succeeding actions at Harlem Heights and WHITE PLAINS, Washington gave up New York, and the British occupied the city until the end of the war for independence. Under the British occupation two mysterious fires (1776 and 1778) destroyed a large part of the city.

History: Post-Revolution to Civil War

After the Revolution, New York was briefly the first capital of the U.S. and was the state capital until 1797. New development was marked by such events as the founding (1784) of the Bank of New York and the beginning of the stock exchange (c.1790). By 1790, New York was the largest city in the U.S., with over 33,000 inhabitants; by 1800 the number had risen to 60,515. In 1811 plans were adopted for the laying out of most of Manhattan on a grid pattern. The opening of the ERIE CANAL (1825) made New York city the seaboard gateway for the Great Lakes region, ushering in another era of commercial expansion. The New York and Harlem railroad was built in 1832. In 1835 a massive fire destroyed much of Lower Manhattan, but it brought about new building laws and the construction of the OLD CROTON DAM AND AQUEDUCT water system. By 1840, New York had become the leading port of the nation. A substantial wave of Irish and German immigration after 1840 dramatically changed the character of urban life and politics in the city. The coming of the Civil War found New Yorkers unusually divided; many shared Mayor Fernando Wood's Southern sympathies, but under the leadership of Governor Horatio Seymour most supported the Union. However, in the Draft Riots (1863), which broke out in protest against the Federal Conscription Act, the rioters—many of whom were Irish and other recent immigrants—directed most of their anger against African-Americans.

History: Immigration and Industry to World War II

Extensive immigration had begun before the Civil War, and after 1865, with the acceleration of industrial development, another wave of immigration began, reaching its height in the late 19th and early 20th centuries. As a result of this immigration, which was predominantly from E and S Europe, the city's population reached 3.4 million by 1900 and 7 million by 1930. New York's many distinct neighborhoods, divided along ethnic and class lines, included such notorious slums as Five Points, CLINTON, also known as Hell's Kitchen, and the Lower East Side. They were often side-by-side with such exclusive neighborhoods as GRAMERCY PARK and BROOKLYN HEIGHTS. Until 1874, when portions of WESTCHESTER were annexed, the city's boundaries were those of present-day Manhattan. With the adoption of a new charter in 1898, New York became a city of five boroughs—New York city was split into the present Manhattan and Bronx boroughs, and the independent city of Brooklyn was annexed, as was the W portions of Queens county and Staten Island. The opening of the first subway line (1903) and other means of mass transportation spurred the growth of the outer boroughs, and this trend has continued into the 1990s. The Flatiron Building (1902) foreshadowed the skyscrapers that today give Manhattan its famed skyline. The need for regional planning resulted in the nation's first zoning legislation of 1916 and the formation of such bodies as the Port of New York Authority (1921; now the PORT AUTHORITY OF NEW YORK AND NEW JERSEY), the Regional Plan Association

---

Cross-references are shown in SMALL CAPITALS. The pronunciation guide is shown on page xix. The sources of population figures are shown on page xvii.

(1929), the Municipal Housing Authority (1934), and the City Planning Commission (1938).

History: After World War II

In the period following World War II, New York began to experience the urban problems that are common to most large U.S. cities. Urban renewal plans of the 1960s did little to alleviate the growing poverty of large sections of the city. These problems were highlighted in the near-bankruptcy of New York city in 1975. In the 1980s drug-related crime became a major concern, even while many sections of the city experienced marked economic growth. A drop in the crime rate and a stabilization of the economy marked the 1990s. On September 11, 2001, the city was the target, along with the PENTAGON, of terrorist attacks that resulted in the destruction of the famed WORLD TRADE CENTER in lower Manhattan, with the loss of nearly 3,000 lives. Clean up took place over the next few years, and construction of a new complex, including a memorial, began in 2006.

Government

The city's chief executive is a mayor and local government is the City Council, with fifty-one members led by the council Speaker. In the 20th century, New York city was served by such mayors as Seth Low, William J. Gaynor, James J. Walker (whose resignation was brought about by the Seabury investigation), Fiorello H. LaGuardia, Robert F. Wagner, Jr., John V. Lindsay, Edward I. Koch, David Dinkins (the city's first African-American mayor), and Rudolph Giuliani. The first of the city's mayors in the 21st century was Michael R. Bloomberg.

New York (YORK), unincorporated village (2006 population 2,227), HENDERSON county, E TEXAS, 12 mi/19 km E of ATHENS; 32°10′N 95°40′W. Noted for its cheesecakes, distributed throughout U.S. Lake Athens to W; LAKE PALESTINE to E.

New York Bay, arm of the ATLANTIC OCEAN at the mouth of the HUDSON RIVER, SE NEW YORK and NE NEW JERSEY, enclosed by the shores of NEW JERSEY, E STATEN ISLAND, S MANHATTAN, and W LONG ISLAND (BROOKLYN) and opening on the SE to the Atlantic Ocean between SANDY HOOK, New Jersey, and Rockaway Point, New York; 40°40′N 74°02′W. It is a sheltered deep harbor able to accommodate the largest ships. The tidal range of the bay is very small and it is ice-free. New York Bay is divided into Upper and Lower bays, which are connected by The Narrows, a strait (c.3 mi/4.8 km long; 1 mi/1.6 km wide) separating Staten Island from Brooklyn. The VERRAZANO-NARROWS BRIDGE spans the strait between Fort Wadsworth and FORT HAMILTON. Upper Bay, c.5.5 mi/8.9 km in diameter, is joined to NEWARK BAY (to the W) by KILL VAN KULL and to LONG ISLAND SOUND by the EAST RIVER. Historically the bay has been one of the world's busiest harbors with port facilities on all shores, but now the New Jersey waterfront is the most active. ELLIS and Liberty islands (both part of STATUE OF LIBERTY NATIONAL MONUMENT) and GOVERNORS ISLAND (site of Fort Jay) are in Upper Bay. Ferries cross the bay from Manhattan to Staten Island and New Jersey. The larger Lower Bay, which includes RARITAN BAY on the W and Gravesend Bay on the NE is joined to Newark Bay by ARTHUR KILL. JAMAICA BAY is an E extension of Lower Bay. Sections of Lower Bay's shoreline are part of GATEWAY NATIONAL RECREATION AREA. Ambrose Channel, federally maintained, crosses Sandy Hook bar at the bay's entrance and extends N to the piers of Upper Bay, where it is 2,000 ft/610 m wide.

New York Harbor, port of entry for NEW YORK city and the Greater New York metropolitan area, SE NEW YORK; at mouth of HUDSON RIVER on NEW YORK BAY, connects to LONG ISLAND SOUND by the EAST RIVER between the borough of MANHATTAN on the W and QUEENS and BROOKLYN on the E. From the earliest colonial days, when fur trading put NEW AMSTERDAM

on the map, to today's modern, massive container operations, the harbor has been vital to the regional and national economy. Simultaneous emergence of the jet age and containerized shipping of cargo, as well as changing trade patterns since the 1950s, have affected the harbor's maritime passenger and cargo movements; however, the modern port is still the third-largest container port in North America, as well as one of the top fifteen in the world.Harbor facilities and operations occur on both the New York and New Jersey sides. More than 400,000 motor vehicles now pass through the port each year, making it the leading U.S. port for auto importing and exporting. It is also a center for bulk and breakbulk cargoes. Planned, governed, and operated by the joint PORT AUTHORITY OF NEW YORK AND NEW JERSEY, the port serves not only the seventeen million consumers within a defined 25-mi/40-km radius of the STATUE OF LIBERTY, but also North America's largest marketplace of eighty million people, reaching into the MIDWEST, NEW ENGLAND, and E CANADA. The port district includes Upper and Lower New York bays; Raritan, Gravesend, Flushing, Newark, and Jamaica bays; Hudson (North), East, Hackensack, Passaic, Raritan, and Harlem rivers, Newtown Creek, Arthur Kill, and Kill Van Kull. Cargo facilities within the New York state confines of the harbor are the Green St. Lumber Exchange facilities (on the East River in Queens); Pier 40 (on the Hudson River side of Manhattan); the forty-acre Red Hook Terminal (along the 40-ft/13-m deep Buttermilk Channel at the S end of the East River in Brooklyn); the 2-mi/3.2-km S Brooklyn Marine Terminal; and the Howland Hook Marine Terminal (the W side of STATEN ISLAND, bordering Arthur Kill).On the New Jersey side of the harbor are two major port cargo facilities: the Auto Marine Terminal (on Upper New York Bay) and the huge (□ 3 sq mi/7.8 sq km) Port Newark–Elizabeth Marine Terminal (on Newark Bay), the largest and most versatile terminal in the harbor. The harbor's leading general cargo exports include waste paper, plastic, motor vehicles and parts, lumber, and paper and paperboard. Major bulk exports are iron and steel, fuel oils, and corn. Leading imports include alcoholic beverages, organic chemicals, clothing, machinery, footwear, and motor vehicles and parts. Leading export markets for harbor goods are Europe and SE Asia. The New York City Passenger Ship Terminal's three finger piers on Manhattan's W Side (between West 48th and West 52nd streets) are the only vestiges left of the scores of piers which once lined the Hudson to service the mainly trans-Atlantic passenger steamships. Today it primarily serves the seasonal cruise ship industry.From New York city's earliest days, the history of the port has gone hand in hand with the history of the city. Its location directly on the ATLANTIC OCEAN gave the city an early advantage over PHILADELPHIA and BALTIMORE, and the 1825 opening of the ERIE CANAL made it the seaboard gateway for the vast Midwest. Fulton's steamship *Clemont* made its first voyage (1807) on the Hudson, and in 1838 the first vessels to cross the Atlantic entirely under steam power docked in the harbor. Was one of the nation's most vital supply ports and ports of embarkation in both World Wars. The Port Authority of New York and New Jersey agreed in 1996 to undertake a large-scale dredging operation to deepen the harbor channels so larger vessels could be accommodated.

New York Mills, town (2000 population 1,158), OTTER TAIL county, W MINNESOTA, 14 mi/23 km WNW of WADENA; 46°31′N 95°22′W. Grain, sugar beets, sunflowers; poultry, livestock; dairying; manufacturing (fabricated metal products, furniture, printing and publishing). RUSH LAKE to W.

New York Mills, village (□ 1 sq mi/2.6 sq km; 2006 population 3,136), ONEIDA county, central NEW YORK, 3 mi/4.8 km W of UTICA; 43°06′N 75°17′W. Light manufacturing. Incorporated 1922.

New York State Barge Canal, waterway system, 525 mi/845 km long, traversing NEW YORK state and connecting the GREAT LAKES with the HUDSON RIVER and Lake CHAMPLAIN. The canal, a modification and improvement of the old ERIE CANAL, was authorized (1903) by public vote, begun in 1905, and completed in 1918. Its main sections are the Erie Canal, extending from TROY to TONAWANDA; the Champlain Canal, joining the Erie Canal at WATERFORD and extending N (via the Hudson as far as Fort Edward) to WHITEHALL on Lake Champlain; the Oswego Canal, connecting the Erie Canal with OSWEGO on Lake ONTARIO at Three Rivers, N of SYRACUSE; and the Cayuga and Seneca Canal, joining the Erie Canal with CAYUGA and SENECA lakes at Montezuma. It is operated by the New York State Thruway Authority's Canal Recreationway Commission, created by the state legislature in 1992. The Barge Canal (12 ft/4 m deep), with fifty-seven electrically operated locks, can accommodate 2,000-ton/1,814–metric ton vessels. With the significant decline in the commercial shipping of commodities that occurred in the 1970s, New York state has essentially abandoned the idea of modernizing the system and integrating it into the nation's extensive and commercially viable network of waterways. The state now envisions preserving the canal's history and environment, emphasizing leisuretime activities on and along the canal, and re-energizing adjacent communities through cooperative public-private projects under the canal corporation, a wholly owned subsidiary of the Canal Recreationway Commission. Numerous leisure craft use the canal. Additional uses provided by the canal system are as a supply of fresh water and a method of flood control and generating hydroelectric power.

New York State Thruway, toll expressway between NEW YORK CITY and RIPLEY in W NEW YORK (on the Pennsylvania state line). Completed in 1960, in 1964 it was officially named the Governor Thomas E. Dewey Thruway, after the state governor who sponsored legislation in 1950 for its construction. With a total length of 641 mi/1,032 km, it is the longest state expressway in the U.S. The main line originally stretched 426 mi/686 km between New York city and BUFFALO, the state's two largest cities. The main line and connections totalled 559 mi/900 km; the expressway grew another 11 mi/18 km in April 1991 with the state's acquisition of the Cross Westchester Expressway (I-287), and another 71 mi/114 km in October 1991 with the acquisition of I-84 from PORT JERVIS to the Connecticut state line. Various sects. connect to New England states, Pennsylvania, New Jersey, the Midwest, and Canada. In April 1987, a 540-ft/165-m span over SCHOHARIE CREEK collapsed, and six people were killed.

New Zealand, country (□ 104,454 sq mi/270,536 sq km; 2004 estimated population 3,993,817; 2007 estimated population 4,115,771), in the S PACIFIC OCEAN, c.1,000 mi/1,609 km SE of AUSTRALIA; WELLINGTON. New Zealand is sometimes called AOTEAROA by Maoris.

Geography

New Zealand is comprised of the NORTH ISLAND and the SOUTH ISLAND (the two principal islands), STEWART ISLAND, and the CHATHAM ISLANDS. Small outlying islands that belonging to New Zealand include the AUCKLAND ISLANDS, KERMADEC ISLANDS, CAMPBELL ISLAND, ANTIPODES ISLANDS, THREE KINGS ISLANDS, BOUNTY ISLANDS, SNARES ISLANDS, and SOLANDER ISLANDS. TOKELAU, though developing self-government, is formally a New Zealand territory, and New Zealand also has jurisdiction over the ROSS DEPENDENCY in ANTARCTICA. The COOK ISLANDS and NIUE are "free association states" with New Zealand. The North Island is known for its active volcanic mountains and its hot springs. The country's longest river, the WAIKATO, and TAUPO, the most extensive lake, are both on the North Island. On South Island,

Area is shown by the symbol □, and capital city or county seat by ⊙.

the massive SOUTHERN ALPS extend almost the length of the island, and in the SW are majestic fiords. The largest areas of virgin forest are in the W South Island and the axial (or main) ranges of the North Island. Among the unusual species native to New Zealand are the kiwi, certain species of parrot, the tuatara (the only survivor of a prehistoric order of reptiles, Rhynchocephalia), and various frogs and fish. New Zealand has no native land mammals other than bats. Large oyster beds are found in FOVEAUX STRAIT and elsewhere. Extensive areas have been set aside as national parks, including the FIORDLAND, Mount COOK, and TONGARIRO volcano. In addition to Wellington, the principal cities are AUCKLAND (the leading urban area and port), CHRISTCHURCH, DUNEDIN, HAMILTON, PALMERSTON NORTH, and INVERCARGILL.

Population

More than 85% of the population live in urban areas. New Zealand has no established official religion; the three largest denominations are Anglican, Presbyterian, and Roman Catholic. Dutch navigator Abel Tasman was the first recorded European to discover New Zealand, inhabited since at least 1000 by Polynesian Maoris. Of the present Maori population of about 435,000, most live on the North Island. There are seven universities at Auckland, Hamilton, Palmerston North, Wellington, Christchurch, Lincoln, and Dunedin.

Economy

Agriculture is the mainstay of a trading economy, although manufacturing and tertiary activities employ a larger number of people. The principal exports are wool, meat, dairy produce, fish, fruit, and timber products. Small amounts of coal, gold, iron, lignite, and natural gas are also tapped, and aluminum smelted from Australian bauxite. Food processing is the largest branch of industry; and there are a variety of small light-manufacturing industries.

History: to 1907

New Zealand was visited and charted by Captain James Cook during his three voyages (1769–1778). Between 1792 and 1840, sealing, whaling, and timber and flax trading led to European settlement. In a series of intertribal wars between 1815 and 1840, some tens of thousands of Maoris died. In 1840 the first planned settlement was made at Wellington by a group sent by the New Zealand Company, founded by Edward Gibbon Wakefield. In that year the Treaty of WAIT-ANGI secured to the Maoris the full possession of their land with future sales to be under official government (rather than private) control, for protection, in exchange for their recognition of British sovereignty. But as European settlement increased, Maori opposition to land settlement resulted in continuing conflict from 1860 to 1872. Originally part of NEW SOUTH WALES (AUSTRALIA), New Zealand became a separate colony in 1840 and received a large measure of self-government after 1852.

History: 1907 to Present

In 1907 it assumed complete self-government as the Dominion of New Zealand, but, preferring that GREAT BRITAIN handle most of its foreign affairs, did not confirm the enabling Statute of WESTMINSTER (1931) until 1947. New Zealand was a leader in progressive social legislation. It was the first country (1893) to grant women over twenty-one the right to vote, and a comprehensive social security system was begun in 1898 with the enactment of an old age pension law. During World War I and World War II, New Zealand fought on the side of the Allies, and joined the UN forces in the Korean War. New Zealand also sent troops to aid the U.S. war effort in SOUTH VIETNAM in the 1960s. In 1951, New Zealand joined in a mutual defense treaty with the U.S. and Australia (ANZAC). This pact collapsed in 1986 after New Zealand refused to let any ships with potential nuclear arms or power enter its ports. New Zealand signed the

Closer Economic Relations (CER) agreement with Australia, promoting free trade between the two countries for goods comprised of at least 50% local content. This has had a negative impact on the New Zealand auto industry because of restricted access to the larger Australian market, as well as a reduction in tariffs that permits greater imports of vehicles into New Zealand. Only four auto assembly plants remained in operation in New Zealand in 1997, compared to the seventeen that were open in 1965. In recent years, the government has sought to come to terms with longstanding Maori grievances. In 1997, Jenny Shipley of the National Party, which had been in power since 1990, became New Zealand's first woman prime minister. Helen Clark and a Labour-led coalition defeated the National Party in 1999, in 2002, and, more narrowly, in 2005.

Government

New Zealand is governed under the Constitution Act of 1986, adopted in 1987, as well as other legal documents. The monarch of Great Britain and NORTHERN IRELAND, represented by the governor general, is the head of state. The government is headed by the prime minister, who is appointed by the governor general following legislative elections. Members of the 120-seat unicameral parliament (the House of Representatives) are elected by popular vote for three-year terms using a system of mixed constituency and proportional representation. The chief political parties are the Labour Party and the National Party. The current governor general is Anand Satyanand (since 2006). The current head of government is Prime Minister Helen Clark (since 1999). Administratively, the country is divided into sixteen regions and one territory (the Chatham Islands). New Zealand is a member of the Commonwealth of Nations and the UN.

Nexö, DENMARK: see NEKSØ.

Nexon (neg-zon), commune (□ 15 sq mi/39 sq km; 2004 population 2,385), HAUTE-VIENNE department, LIMOUSIN region, W central FRANCE, 11 mi/18 km SSW of LIMOGES; 45°40′N 01°12′E. Arabian horse breeding. Has a summer training program for circus acrobats.

Nextlalpan de Felipe Sánchez Solís, town, ⊙ Nextlalpan municipio, MEXICO state, central MEXICO, 22 mi/35 km N of MEXICO CITY. Cereals; livestock.

Ney (NAI), village (2006 population 341), DEFIANCE county, NW OHIO, 10 mi/16 km NW of DEFIANCE; 41°23′N 84°31′W. Grain and lumber mills.

Neya (NYE-yah), city (2005 population 11,025), central KOSTROMA oblast, central European Russia, on the Neya River (right affluent of the UNZHA RIVER, VOLGA RIVER basin), on road and the TRANS-SIBERIAN RAILROAD, 150 mi/241 km NE of KOSTROMA, and 55 mi/89 km E of GALICH; 58°17′N 43°52′E. Elevation 380 ft/115 m. Sawmilling, woodworking; automotive repair. Agriculture (rye, oats, wheat, potatoes; livestock). Has a heritage museum. Established as a settlement around a train station in 1906; made city in 1958.

Neyagawa (ne-YAH-gah-wah), city (2005 population 241,816), OSAKA prefecture, S HONSHU, W central JAPAN, 9 mi/15 km N of OSAKA, on the Shinyodo River; 34°45′N 135°37′E. Suburb of Osaka.

Neye, GERMANY: see WIPPERFÜRTH.

Neyland (NAI-luhnd), town (2001 population 3,276), Pembrokeshire, SW WALES, on MILFORD HAVEN, 3 mi/4.8 km NW of PEMBROKE; 51°43′N 04°57′W. Formerly in DYFED, abolished 1996.

Neyriz (NAI-reez), town, Fārs province, S IRAN, 110 mi/177 km ESE of SHIRAZ, and on road to KERMAN; 29°11′N 54°19′E. Trade center; fruit, nuts, gums. Lead and iron deposits. Seat of Babist religious section Also spelled NIRIZ or NEIRIZ.

Neyriz, Lake, IRAN: see BAKHTEGAN.

Neyshabur (nai-SHAH-boor), city (2006 population 208,860), Khorāsān province, NE IRAN, 45 mi/72 km

W of MASHHAD, on TEHRAN-MASHHAD road; 36°12′N 58°47′E. It is the trade center for a farm region where cotton, fruit, and grain are grown. Manufacturing of the city include food products and leather goods; turquoise is mined nearby. Neyshabur was founded by the Sassanid ruler Shapur I in the 3rd century and was rebuilt (4th century) by Shapur II. The city became one of the foremost centers of PERSIA. Under the Seljuk Turks (11th–12th century) it was made into an important cultural center; several colleges were founded there by Nizam al-Mulk. Al-Ghazali, the noted philosopher of the 11th–12th century, studied in Neyshabur, and his famous contemporary Omar Khayyam, the poet and mathematician, was born in the city and is buried there. The tomb of Omar was rebuilt in 1934. Near Neyshabur archaeologists have made important finds of glazed pottery and stucco work from the 9th and 10th centuries. The city is also known as NISHAPUR.

Neyva River (nai-VAH), SVERDLOVSK oblast, W Siberian Russia; rises in the central URAL Mountains, 10 mi/16 km NNE of VERKH-NEYVINSKIY; flows N, past NEV'YANSK, E, past PETROKAMENSKOYE and NEYVO-SHAYTANSKIY, and generally NE, past ZYRYANOVSKIY and ALAPAYEVSK, joining the REZH RIVER 20 mi/32 km ENE of Alapayevsk to form the NITSA RIVER; 202 mi/325 km long. Has several dams holding industrial water supply.

Neyva River, RUSSIA: see NEIVA RIVER.

Neyveli, city, SOUTH ARCOT VALLALUR district, TAMIL NADU state, S INDIA; 11°32′N 79°29′E.

Neyvo-Rudyanka (NYAI-vuh–roo-DYAHN-kah), town (2006 population 2,880), W SVERDLOVSK oblast, E URAL Mountains, extreme W Siberian Russia, 5 mi/8 km SE (and under administrative jurisdiction) of KIROVGRAD; 57°20′N 60°08′E. Elevation 846 ft/257 m. Railroad junction; wood chemicals. Until 1928, called Neyvo-Rudyanskiy Zavod.

Neyvo-Rudyanka, RUSSIA: see NEIVO-RUDYANKA.

Neyvo-Shaytanskiy (NYAI-vuh–shei-TAHN-skeeye), town (2006 population 2,590), S SVERDLOVSK oblast, E URALS, W Siberian Russia, on the NEYVA RIVER (OB′ RIVER basin), on road junction and railroad spur, 18 mi/29 km SW of ALAPAYEVSK, to which it is administratively subordinate; 57°43′N 61°15′E. Elevation 465 ft/141 m. Ferrous metallurgy.

Neyvo-Shaytanskiy, RUSSIA: see NEIVO-SHAITANSKI.

Neyyattinkara (nai-YAH-tin-kah-rah), city, THIRUVANANTHAPURAM district, KERALA state, S INDIA, 10 mi/16 km SE of THIRUVANANTHAPURAM; 08°24′N 77°05′E. Manufacturing of rope and mats of coconut and palmyra fiber, copra, jaggery; cassava and cashewnut processing, hand-loom weaving.

Neyyur, INDIA: see IRANIEL.

Nezahualcóyotl, MEXICO: see CIUDAD NEZAHUALCÓYOTL.

Nezamayevskaya (nye-zah-MAH-eefs-kah-yah), village (2005 population 3,135), N central KRASNODAR TERRITORY, S European Russia, on the YEYA RIVER, on road, 22 mi/35 km NW, and under administrative jurisdiction, of NOVOPOKROVSKAYA; 46°10′N 40°16′E. Elevation 193 ft/58 m. In oil-producing region.

Nezametnyy, RUSSIA: see ALDAN.

Nezeros, Greece: see KALLIPEFKI.

Nezeros, Lake, Greece: see XINIAS, LAKE.

Nezer Sereni (NE-zer se-RE-nee) or **Netzer Sereni**, kibbutz, ISRAEL, 2 mi/3.2 km NE of REHOVOT, on coastal plain; 31°55′N 34°49′E. Elevation 183 ft/55 m. Mixed farming; manufacturing (metals and furniture). Founded in 1948 by Holocaust survivors and named for Enzo Sereni, who parachuted into Europe in World War II and was captured and executed by the Nazis. Nearby are the remains of a 19th century German Templar farm, which General Allenby made his headquarters during World War I.

Nezhinka (NYE-zhin-kah), village (2006 population 5,090), central ORENBURG oblast, SE European Russia, on the URAL River, on highway, 9 mi/14 km E of (and administratively subordinate to) ORENBURG; 51°46'N 55°22'E. Elevation 275 ft/83 m. Holiday getaway destination for residents of neighboring cities and towns; summer homes, sanatoria. Administrative offices and service units of regional geological surveys. The oblast's central capital is 3 mi/5 km to the E. Sometimes spelled Nezhenka (same pronounciation).

Nezib, ancient city in the Hebrew Bible, in what is now the SW WEST BANK, circa 6 mi/9.7 km SE of Beth Guvrin, believed to be the present site of Khirbet Beith Nasib, NW of HEBRON.

Nezib, TURKEY: see NIZIP.

Nezinscot River (ne-ZIN-skaht), c.11 mi/18 km long, W MAINE; rises in 12-mi/19-km E and W branches in OXFORD county, joining at BUCKFIELD; flows E to the ANDROSCOGGIN above AUBURN.

Nez Perce, county (□ 856 sq mi/2,225.6 sq km; 2006 population 38,324), ⊙ LEWISTON; 46°20'N 116°45'W. Livestock-raising and agricultural area (cattle; potatoes, barley, wheat, alfalfa, vegetables) bounded W by SNAKE RIVER and WASHINGTON, far S by SALMON RIVER; drained by CLEARWATER RIVER. Lumber milling at Lewiston. Lowest point in Idaho (710 ft/216 m) in NW, at point where Snake River flows into Washington state. Mountains are in S, part of Nez Perce Indian Reservation in E, covers about ½ CO. NEZ PERCE NATIONAL HISTORICAL PARK (Spalding Area Unit) in N; Soldier Meadow Reservoir in SE; Winchester Lake State Park on SE boundary; Hells Gate in W, S of Lewiston. Formed 1861.

Nezperce (NEZ-purse), village (2000 population 523), ⊙ LEWIS county, W IDAHO, 40 mi/64 km ESE of LEWISTON, in Nez Perce Indian Reservation; 46°14'N 116°14'W. Wheat, alfalfa; cattle.

Nez Perce National Historical Park (□ 3 sq mi/7.8 sq km), NEZ PERCE, IDAHO and CLEARWATER counties, NW IDAHO. Comprises 24 sites that preserve and commemorate the history and culture of the Nez Perce. Includes Spalding Area Site, Nez Perce county, which has a museum and buildings from a Presbyterian Indian Agency (1861) and church (c.1885); also East Kamiah Site, E of KAMIAH, and site S of GRANGEVILLE, both Idaho county, also some acreage in Clearwater county. Authorized 1965.

Nezpique, Bayou (nehz-PEEK, BEI-yoo), c.70 mi/113 km long, S LOUISIANA; rises in EVANGELINE parish; flows S, joining BAYOU DES CANNES near MERMENTAU to form MERMENTAU RIVER. Navigable by small shallow-draft boats for 23 mi/37 km of lower course.

N'Fiss, Oued (uhn-FEES, wahd), 70 mi/113 km long, stream of SW MOROCCO, a left tributary of the Oued TENSIFT river; rises at TIZI N'TEST pass in the High ATLAS mountains. Dam (170 ft/52 m high; completed 1935), 25 mi/40 km SSW of MARRAKECH, irrigates HAOUZ plain. Lalla Takerkoust hydroelectric plant built here 1938. Also Nfis or Nfiss.

Ngabe (uhng-gah-BAI), town, POOL region, SE Congo Republic, on CONGO River, and 100 mi/161 km NNE of BRAZZAVILLE.

Ngadi Chuli (NUH-dee CHOO-lee) or **Peak 29** (25,820 ft/7,871 m), central Nepal; 28°30'N 84°34'E.

Ngaga (NGAH-gah), village, MTWARA region, SE TANZANIA, 10 mi/16 km SW of Masai, on MBANGALA River; 10°53'S 38°40'E. Cashews, peanuts, grain; livestock; timber.

Ngahere (NAH-HER-ee), township, Grey district, WEST COAST region, SOUTH ISLAND, NEW ZEALAND, 13 mi/21 km NE of GREYMOUTH, on GREY river; 42°24'S 171°27'E. Sawmills, coal mines.

Ngajira (nga-JEE-rah), village, IRINGA region, central TANZANIA, 50 mi/80 km W of IRINGA, near GREAT RUAHA RIVER, in SE part of RUAHA NATIONAL PARK; 07°42'S 36°53'E. Airstrip; tourist lodge and facilities.

Ngala (uhng-GAH-lah), town, BORNO state, extreme NE NIGERIA, near Lake CHAD and CAMEROON border, 23 mi/37 km NE of DIKWA, on road to N'djamena

(CHAD); 12°20'N 14°11'E. Peanuts, cotton, millet; cattle.

Ngale (NGAH-lai), village, Équateur province, NW CONGO, N of LISALA; 02°27'N 21°29'E. Elev. 1,348 ft/410 m. Subsistence crops.

Ngamaba, town, POOL region, S Congo Republic, 10 mi/16 km W of BRAZZAVILLE.

Ngambé (n-GAM-bai), town, LITTORAL province, CAMEROON, 67 mi/108 km NW of Yaoundé; 04°13'N 10°39'E. Cocoa and palm oil are produced.

Ngamda (EN-DAH), town, E TIBET, SW CHINA, 25 mi/40 km WSW of QAMDO, and on road to LHASA; 31°10'N 96°44'E. Livestock.

Ngami, Lake (NGAH-mee), reedy marsh, NORTH-WEST DISTRICT, NW BOTSWANA, c.40 mi/64 km long and 4 mi/6.4 km–8 mi/12.9 km wide; 20°30'S 22°45'E. During the Pleistocene epoch, the lake covered an extensive area. Since the late 1880s, when papyrus growth blocked the mouth of its main tributary, the lake has greatly shrunk in size; it now intermittently receives water from the CUBANGO River. Nearest town is SEHITHWA, at NW end of lake.

Ngamiland District (NGAH-mee-lahnd), former administrative division (2001 population 75,070), NW BOTSWANA; the capital was MAUN. Bounded by NAMIBIA W and N (CAPRIVI STRIP), former CHOBE DISTRICT NE, CENTRAL DISTRICT E, and GHANZI DISTRICT S. Included OKAVANGO DELTA drainage basin in central part of district, with MOREMI GAME RESERVE; NXAI PAN NATIONAL PARK in E; lake NGAMI in S; Tsodilo Hills in NW and AHA HILLS in NW. Merged with Chobe District circa 2001 to form NORTH-WEST DISTRICT.

Ngamo, town, MATABELELAND NORTH province, W central ZIMBABWE, 100 mi/161 km NW of BULAWAYO, on railroad, near E boundary of HWANGE NATIONAL PARK; 19°07'S 27°29'E. Livestock.

Ngandjoek, INDONESIA: see NGANJUK.

Ngandong, INDONESIA: see TRINIL.

Ngangla Ringco (AHNG-LAH RING-ZO), salt lake, W central TIBET (CHINA), 150 mi/241 km E of GARYARSA, between ALING KANGRI and KAILAS ranges; 30 mi/48 km long; 31°40'N 83°00'E.

Nganglong Kangri, CHINA: see ALING KANGRI.

Ngangora (ngan-GO-rah), village, LINDI region, SE central TANZANIA, 115 mi/185 km W of LINDI, 5 mi/8 km S of LIWALE; 09°53'S 37°56'E. Timber; cashews, subsistence crops; livestock.

Ngangze Co (AHNG-GAHNG-ZUH ZO), lake, central TIBET (CHINA), 250 mi/402 km WNW of LHASA; 31°01'N 86°56'E. Qumingxung is on W shore.

Nganjuk (NGAHN-jok), town, NGanjuk district, Java Timur province, INDONESIA, 60 mi/97 km WSW of SURABAYA; 07°36'S 111°55'E. Trade center for agricultural area (rice, cassava, corn, peanuts). Also spelled Ngandjoek or Ngandjuk.

Ngan Son (NGAHN), village, CAO BANG province, N VIETNAM, at S end of Nganson Range, 25 mi/40 km SW of CAO BANG; 22°26'N 105°59'E. Market center. Mining of silver-lead-zinc ores. Thai, Dao, and other minority peoples. Formerly spelled Nganson.

Ngan Son Range (NGAHN) or **piaouac range**, N VIETNAM, 25 mi/40 km W of CAO BANG; 22°37'N 105°52'E. Range extends c. 40 mi/64 km N-S, rising to 6,335 ft/1,931 m. Has Vietnam's chief source of tin and tungsten; principal mines at TINH TUC. Lead, zinc, and silver mined at NGAN SON.

Ngaoui (uhn-GAH-wee), mountain (4,659 ft/1,420 m), OUHAM-PENDÉ prefecture, NW CENTRAL AFRICAN REPUBLIC, on CAMEROON border. Highest peak in the country; composed mainly of granite.

Ngaoundal (n-GAWN-dahl), town (2001 population 16,500), ADAMAOUA province, CAMEROON, 63 mi/101 km SSW of Ngaoundéré; 06°28'N 13°19'E. Some bauxite deposits are nearby. Railroad communication point.

N'Gaoundéré ([n]-GOUN-dai-rai), city (1998 estimated population 156,804; 2001 estimated population

189,800), ⊙ VINA department and ADAMAOUA province, N central CAMEROON, 275 mi/443 km NE of Yaoundé; 07°23'N 13°33'E. Elevation 3,670 ft/1,119 m. Railroad and road junction. Large local market and agr. center; cattle, horses, sheep, and goats; coffee plantations. Hospital, airport, meteorological station, experimental farms, and hydroelectric power plant. Seat of University of Ngaoundéré. Site of the Palace of the Lamido, the traditional Fulani chief. Also known as N'Gaundere; also spelled Ngaoundéré.

Ngape (uhng-gah-PAI), village, MYANMAR, at foot of the ARAKAN YOMA, 27 mi/43 km WSW of MINBU, on AN PASS road.

Ngaputaw (uhng-gah-poo-TAW), village, MYANMAR, on BASSEIN RIVER, 16 mi/26 km S of PATHEIN.

Ngara (NGAH-rah), village, KAGERA region, NW TANZANIA, 110 mi/177 km SW of BUKOBA, near upper KAGERA RIVER (Rusumo Falls to NE); 02°27'S 30°39'E. BURUNDI border to N; BURIGI Game Reserve to E. Tobacco, corn, wheat; sheep, goats. Tin deposits.

Ngarama (ngah-rah-MAH), village, NE RWANDA, 28 mi/45 km ENE of KIGALI; 01°32'S 30°14'E.

Ngare Nanyuki, TANZANIA: see ENGARE NANYUKI.

Ngari (ung-AH-ree), Chinese=A-li (AH-LEE), westernmost historical province of TIBET, SW CHINA; 32°28'N 79°48'E. Existed c. 1635–1713. Main town, GARYARSA.

Ngaruawahia (nah-ROO-ah-WAH-ee-ah), town, WAIKATO district, N NORTH ISLAND, NEW ZEALAND, 12 mi/20 km NW of HAMILTON, at junction of WAIPA RIVER with WAIKATO River, near Taupiri Gorge; 37°40'S 175°09'E. Agricultural center. Coal mine at nearby Glen Massey. Associated with Maori King movement (1858) and community activities (canoe racing).

N'Gaski (uhn-GAHS-kee), town, KEBBI state, W NIGERIA, on road, 30 mi/48 km WSW of KONTAGORA, on E shore of KAINJI reservoir; 10°25'N 04°43'E. Fishing; millet, maize.

Ngatea (NAH-tee-ah), township, in drained flats of PIAKO RIVER in HAURAKI PLAINS, Hauraki district, WAIKATO region, N NORTH ISLAND, NEW ZEALAND. S of FIRTH OF THAMES, 50 mi/80 km SE of AUCKLAND; 37°17'S 175°30'E. Dairy products and lamb.

Ngathainggyaung (uhn-gah-THEIN-jaung), town, MYANMAR, on BASSEIN RIVER (ferry), and 50 mi/80 km NNE of PATHEIN. Also spelled Ngathaingyaung.

Ngau, FIJI: see GAU.

N'gauma District, MOZAMBIQUE: see NIASSA.

N'Gaundere, CAMEROON: see N'GAOUNDÉRÉ.

Ngauruhoe, active conical volcanic peak (7,513 ft/2,290 m) in TONGARIRO NATIONAL PARK, central NORTH ISLAND, NEW ZEALAND; 39°09'S 175°38'E. Emits steam and gas regularly, ash occasionally.

Ngawi (NGAH-wee), town, ⊙ Ngawi district, Java Timur province, INDONESIA, on SOLO RIVER, 40 mi/64 km ENE of SURAKARTA; 07°24'S 111°26'E. Trade center for agricultural area (rice, corn, coffee, kapok, cassava).

Ngawun River, MYANMAR: see BASSEIN RIVER.

Ngazun (uhng-gah-ZOON), village, MYANMAR, landing on left bank of AYEYARWADY RIVER, and 27 mi/43 km WSW of MANDALAY.

Ngcheangel, atoll, Republic of Palau (Belau), W CAROLINE ISLANDS, W PACIFIC, 15 mi/24 km N of Babeldaob. Comprises four wooded islets. Formerly called Kayangel.

Ngeaur, raised coral island (□ 3 sq mi/7.8 sq km), Republic of PALAU (Belau), W CAROLINE ISLANDS, W PACIFIC; 06°53'N 134°08'E. Chalk cliffs; known for phosphate deposits; c.2.5 mi/4 km long.

Ngemelachel, volcanic island, Republic of PALAU (Belau), W CAROLINE ISLANDS, W Pacific; 07°20'N 134°27'E. Rises to 393 ft/120 m; c.0.75 mi/1.2 km in diameter. Ngemelachel Harbor is chief port of Palau (Belau) and is essentially the industrial and maritime extension of OREOR for overseas and inter-island shipping. Formerly called Malakal.

Ngerekebesang, wooded volcanic island, Republic of Palau (Belau), W CAROLINE ISLANDS, W PACIFIC, just NW of OREOR; 07°21′N 134°27′E. Island is 2 mi/3.2 km long, 0.75 mi/1.2 km wide. In World War II, site of Japanese army base. Formerly called Arakabesan.

Ngerengere (ngai-rain-GAI-raih), town, MOROGORO region, E TANZANIA, 35 mi/56 km E of MOROGORO, on railroad; 06°48′S 38°06′E. Sisal, corn, wheat, manioc; livestock.

Ngeruktabel, raised coral island, Republic of PALAU (Belau), W CAROLINE ISLANDS, W PACIFIC, 6 mi/9.7 km SW of BABELDAOB; c.10 mi/16 km long; rises to 587 ft/179 m. Formerly called Urukthapel.

Ngetik (NGE-tik), atoll, State of POHNPEI, E CAROLINE ISLANDS, Federated States of MICRONESIA, W PACIFIC, 75 mi/121 km SW of POHNPEI Island; 11 mi/18 km long, 5 mi/8 km wide; eight low islets on triangular reef. Also spelled Ngatik.

Ngezi, town, MATABELELAND SOUTH province, S central ZIMBABWE, 25 mi/40 km SE of ZVISHAVANE, on NGEZI RIVER; 20°35′S 30°24′E. Cattle, sheep, goats; corn, wheat, soybeans, cotton, tobacco.

Ngezi Recreational Park, MASHONALAND WEST province, central ZIMBABWE, bounded S by MIDLANDS province, 70 mi/113 km SSW of HARARE; 18°45′S 30°25′E. Surrounds Ngezi Reservoir, formed on Ngezi River Wildlife; recreation. Formerly called Ngezi National Park.

Ngezi River, c.70 mi/113 km long, central ZIMBABWE; rises c. 50 mi/80 km S of HARARE; flows W, through Ngezi Reservoir; joins MUNYATI RIVER 28 mi/45 km NE of KWEKWE. One of two rivers so named in Zimbabwe.

Ngezi River, c. 95 mi/153 km long, S central ZIMBABWE; rises c. 65 mi/105 km ENE of BULAWAYO, W of SOMABHULA; flows SE past NGEZI town; and joins RUNDE RIVER 30 mi/48 km SE of ZVISHAVANE. One of two rivers so named in Zimbabwe.

Nggamea, FIJI: see QUAMEA.

Ngga Pilimsit, mountain peak, INDONESIA: see PILIMSIT, MOUNT.

Nghe An (NGAI AHN), province (□ 6,323 sq mi/16,439.8 sq km), N central VIETNAM, N border with LAOS and THANH HOA province, E border with VINH BAC BO, S border with HA TINH province, W border with LAOS; ⊙ Vinh; 19°00′N 105°20′E. Drained by the Song Ca, province exhibits considerable range of environmental niches, ethnic communities, and localized economies. Once densely forested, its mountains and hills have been heavily exploited for wood products in recent decades. Diverse soils and mineral resources (limestone, tin, gold, manganese, marble, mineral water), wet rice farming, shifting cultivation, opium growing, livestock raising, forest products (gums, resins, medicinals, wood chips, paper stock, honey, beeswax, foods), agro-forestry, fishing and aquaculture, and commercial agriculture (rubber, coffee, tea, peanuts, vegetables and fruits, cinnamon). Lumbering, handicrafts, various industries (paper, cement, sericulture, vegetable oil, brick and tile, leather goods, beer and soft drinks; meat, fish, and food processing). Light manufacturing, shipbuilding. Kinh population with Thai, Kho Mu, Muong, and other minorities.

Ngidinga (ngee-DEEN-guh), village, BAS-CONGO province, W CONGO, on road, 95 mi/153 km S of KINSHASA; 05°37′S 15°17′E. Elev. 2,089 ft/636 m.

N'giri River, CONGO: see GIRI RIVER.

N'Giva, ANGOLA: see ONDJIVA.

Ngoenoet, INDONESIA: see NGUNUT.

Ngohi (uhng-GO-hee), town, BORNO state, NE NIGERIA, on road, 40 mi/64 km ENE of BIU; 10°40′N 12°42′E. Market center. Groundnuts, maize, cotton; livestock.

Ngo-Ketunjia, department (2001 population 174,173), NORTH-WEST province, CAMEROON.

N'Goko River (N GO-ko), in SE CAMEROON, along the Cameroon-Congo border; rises SE of Abong-M'Bang, formed by the confluence of the DJA and Boumba

rivers near Moloundou; flows into the SANGHA RIVER at OUESSO. Navigable year-round for small steamboats downstream (for c.80 mi/129 km) from MOLOUNDOU. Also spelled Ngoko.

Ngom, Equatorial Guinea: see MBINI RIVER.

Ngomba (NGOM-bah), village, MBEYA region, SW TANZANIA, 55 mi/89 km NW of MBEYA, on Lake RUKWA; 08°52′S 32°52′E. Road terminus. Cattle, sheep, goats; corn, wheat. Phosphate mining.

Ngombezi (ngom-BAI-zee), village, TANGA region, NE TANZANIA, 5 mi/8 km W of KOROGWE, on PANGANI RIVER; 04°11′S 38°22′E. Sisal, rice, grain, bananas; goats, sheep; timber.

Ngomedzap (NGO-med-zap), town, Central province, CAMEROON, 48 mi/77 km SSW of Yaoundé; 03°16′N 11°15′E.

Ngomeni (ngo-MAI-nee), village, TANGA region, NE TANZANIA, 15 mi/24 km WSW of TANGA, on railroad; 04°10′S 38°51′E. Road junction. USAMBARA Mountains to NW. Seat of Agricultural Research Institute (Mlingano); sisal research. Sisal, tea, corn; livestock.

Ngong (NGO-ng), town, Kajiado district, RIFT VALLEY province, S KENYA, on road, and 13 mi/21 km SW of NAIROBI; 01°23′S 36°40′E. Corn; livestock raising; slaughterhouse; coffee, wheat, corn. Ngong Hills form E escarpment of GREAT RIFT VALLEY.

Ngongo (NGON-go), village, MBEYA region, W TANZANIA, 68 mi/109 km W of MBEYA; 08°48′S 32°25′E. Cattle, sheep, goats; corn, wheat. One of two villages so named in Tanzania.

Ngongo, village, RUKWA region, W TANZANIA, 15 mi/24 km ENE of SUMBAWANGA, near Lake RUKWA; MBIZI Mountains to SW; 08°48′S 32°25′E. Sheep, goats; corn, wheat. One of two villages so named in Tanzania.

Ngong Shuen Chau, HONG KONG: see STONECUTTERS ISLAND.

Ngor (n-GOR), fishing village and beach resort (2004 population 10,810), DAKAR administrative region, W SENEGAL, on N CAPE VERDE peninsula, 6 mi/9.7 km NW of DAKAR. Also spelled N'Gor.

Ngora (n-GOR-uh), town, EASTERN region, UGANDA, 21 mi/34 km SSE of SOROTI. Cotton, peanuts, sesame; livestock. Was part of former EASTERN province.

Ngornu (uhng-GAWR-noo), town, BORNO state, extreme NE NIGERIA, 70 mi/113 km NE of MAIDUGURI, 10 mi/16 km from Lake CHAD, near CAMEROON border. Cassava, millet, gum arabic; salt; cattle, skins. An important center in Bornu kingdom until end of 19th century. Sometimes spelled N'Gornu.

Ngorongoro Crater (ngo-rong-GO-ro), caldera in Great Rift Valley, ARUSHA region, N TANZANIA, 80 mi/129 km W of ARUSHA, at center of Ngorongoro Conservation Area; 03°15′S 35°30′E. Rim is 7,900 ft/2,408 m high; floor (□ 126 sq mi/326 sq km) is 2,000 ft/610 m below rim. Inhabited by wildlife. MAGADI Lake in W center. Caldera created by collapse of large volcano. Major tourist attraction.

Ngoulemakong (n-GOO-lai-mai-kong), town, South province, CAMEROON, 55 mi/89 km SSW of Yaoundé; 3°06′N 11°26′E. Cocoa-growing region.

Ngouma (n-GOO-ma), village, FIFTH REGION/MOPTI, MALI, 56 mi/90 km NW of DOUENTZA; 15°38′N 03°22′W.

Ngoumba-Ngéoul, SENEGAL: see GUÉOUL.

Ngoumou (n-GOO-moo), town, Central province, CAMEROON, 25 mi/40 km SW of Yaoundé; 03°35′N 11°19′E.

Ngouna, VANUATU: see NGUNA.

Ngounié (n-GOON-ee-ai), province (□ 14,575 sq mi/37,895 sq km; 2002 population 100,300), S central GABON; ⊙ MOUILA (also largest city); 01°30′S 11°15′E. Bounded on S by CONGO, E by OGOOUÉ-LOLO province, N by MOYEN OGOOUÉ province, W by NYANGA and OGOOUÉ-MARITIME provinces. Coffee and palm oil are produced in this forested province.

Ngounié River (n-GOON-ee-ai), c.275 mi/443 km long, central GABON; rises 25 mi/40 km SE of MBIGOU; flows

SW and NW, past MOUILA and SINDARA, to OGOOUÉ RIVER 12 mi/19 km above LAMBARÉNÉ. Navigable most of the year for 50 mi/80 km below Sindara.

Ngouoni (n-GOO-o-nee), town, HAUT-OGOOUÉ province, SE GABON, 15 mi/24 km NE of FRANCEVILLE; 01°30′S 13°48′E.

Ngoura (uhn-goo-RAH), village, HADJER-LAMIS administrative region, W CHAD, 100 mi/161 km NE of N'DJAMENA, on road; 12°52′N 16°27′E.

N'Gouri (uhn-goo-REE), town, on border between LAC and KANEM administrative regions, W CHAD, near E shore of Lake CHAD, 40 mi/64 km N of MAO; 13°38′N 15°22′E. Former French military outpost. Was part of former LAC prefecture.

Ngourti (n-GOOR-tee), town, DIFFA province, NIGER, 80 mi/129 km N of N'GUIGMI; 15°19′N 13°12′E.

N'Goussa (nuh-goo-SAH), village and oasis, OUARGLA wilaya, S ALGERIA, in SAHARA Desert, 12 mi/19 km N of OUARGLA. Date palms.

Ngoywa (NGO-hwah), village, TABORA region, W central TANZANIA, 55 mi/89 km S of TABORA, near Wulua River; 05°55′S 32°45′E. Timber; livestock; rice.

Ngozi, province (□ 569 sq mi/1,479.4 sq km; 1999 population 601,382), N BURUNDI, on RWANDA border (to N, formed partly by Kanyaru River); ⊙ NGOZI; 02°52′S 29°55′E. Borders KIRUNDO (NE), MUYINGA (E), KARUZI (SES, formed partly by Murarangaro River), GITEGA (S), and KAYANZA (W) provinces.

Ngozi (ngo-ZHEE), village (2004 population 40,200), ⊙ NGOZI province, N BURUNDI, 36 mi/58 km NNW of GITEGA; 02°54′S 29°50′E. Market for food staples; cattle raising, manufacturing of bricks, tiles. Established 1921.

Ngudu (NGOO-doo), village, MWANZA region, NW TANZANIA, 40 mi/64 km SE of MWANZA; 02°59′S 33°19′E. Road junction. Cotton, corn, wheat, millet; goats, sheep, cattle.

Ngudwini, river, SOUTH AFRICA: see DONNYBROOKE.

N'Guigmi (n-GEE-mee), town, DIFFA province, SE NIGER, on NW shore of LAKE CHAD, on desert road, and 280 mi/451 km E of ZINDER; 14°14′N 13°08′E. Region produces millet, beans, peanuts, manioc, indigo, cotton, wheat, and henna; livestock. Meteorological station. Also spelled Nguimi.

Nguiu (wee-OO, nyoo-YOO), township, SE BATHURST ISLAND, TIWI ISLANDS, NORTHERN TERRITORY, N central AUSTRALIA, between ARAFURA and TIMOR seas; 11°45′S 130°41′E. Principal township of Bathurst Island, an Aboriginal reservation owned by the Tiwi people.

Ngulakula (ngoo-la-KOO-lah), village, PWANI region, E TANZANIA, 70 mi/113 km SSW of DAR ES SALAAM; 07°47′S 39°47′E. Rice, cashews, bananas, sweet potatoes; sheep, goats.

Ngulu (NGOO-loo), atoll, State of YAP, W CAROLINE ISLANDS, Federated States of MICRONESIA, W PACIFIC, 59 mi/95 km SSW of Yap; 19 mi/31 km long, 12 mi/19 km wide. Coconut palms.

Nguna (uhng-OO-nah), French *Ngouna*, volcanic island, VANUATU, SW PACIFIC OCEAN, 4 mi/6.4 km N of EFATE; 17°26′S 168°21′E. Island is 5 mi/8 km long, 2 mi/3.2 km wide. Formerly called Montague Island.

Ngundu, town, MASVINGO province, SE central ZIMBABWE, 50 mi/80 km S of MASVINGO, near RUNDE RIVER; 20°48′S 30°49′E. Road junction. Cattle, sheep, goats; corn, wheat, soybeans.

Ngunut (NGOO-noot), town, Java Timur province, INDONESIA, on BRANTAS RIVER, 10 mi/16 km W of BLITAR; 08°06′S 112°01′E. Trade center for agricultural area (coffee, corn, tea; cinchona bark, rubber). Also spelled Ngoenoet.

Nguru (uhng-GOO-roo), town, YOBE state, NE NIGERIA, 140 mi/225 km ENE of KANO. Railroad terminus; cotton, peanuts, millet, durra; cattle raising; saltworks.

Nguruka (ngoo-ROO-kah), village, KIGOMA region, W TANZANIA, 90 mi/145 km ESE of KIGOMA, N of Lake SAGARA; 05°07′S 31°01′E. Rice; livestock; timber.

Nguti (n-GOO-tee), town, SOUTH-WEST province, CAMEROON, 90 mi/145 km NW of DOUALA; 05°19′N 09°26′E.

Nguyen Binh (NGOO-yen BIN), town, CAO BANG province, N VIETNAM, 20 mi/32 km W of CAO BANG, at foot of NGAN SON RANGE; 22°39′N 105°56′E. Market and transportation center. Tin, tungsten mining at TINH TUC, 5 mi/8 km W. Tai, Dao, and other minorities. Formerly spelled Nguyenbinh.

Ngwale (NGWAH-laih), village, LINDI region, SE TANZANIA, 20 mi/32 km S of LIWALE; 10°05′S 37°56′E. Subsistence crops; livestock.

Ngwane Park (ung-WAHN-e), town, MANZINI district, central SWAZILAND, 20 mi/32 km SE of MBABANE, 2 mi/3.2 km S of MANZINI; 26°32′S 31°22′E. Cattle, goats, sheep, hogs; corn, vegetables, citrus.

Ngwasi (NGWAH-see), village, LINDI region, SE central TANZANIA, 110 mi/177 km of MOROGORO, on RUFIJI RIVER, below (N of) confluence of Luwego and Luhombero rivers, in SELOUS GAME RESERVE; 08°30′S 37°28′E. SIGURI FALLS to S. Livestock.

Ngwempisi River, c.75 mi/121 km long, SOUTH AFRICA and SWAZILAND; rises c.5 mi/8 km N of Amsterdam, MPUMALANGA province, South Africa; flows SE, enters Swaziland, and continues generally E; passes S of MANKAYANE, joins GREAT USUTU RIVER 12 mi/19 km S of MANZINI.

Ngwena, town, MASHONALAND WEST province, NW ZIMBABWE, 26 mi/42 km N of KWEKWE; 18°33′I 29°50′E. Livestock; tobacco, cotton, corn, soybeans.

Ngwenya (ung-WAI-nee-ya), town, HHOHHO district, NW SWAZILAND, 10 mi/16 km NW of MBABANE, at S end of Ngwenya Hills; 26°13′S 31°01′E. SOUTH AFRICA border 1 mi/1.6 km to W. Iron-ore mine closed in 1979; former terminus of railroad line from central Swaziland (track removed 1995). Airstrip. Glass manufacturing. Mining began here in 26,000 B.C.E. during the Middle Stone Age. Nearby summit of Ngwenya hills (5,998 ft/1,828 m) is second highest peak in Swaziland.

Ngwerere (eng-wai-LAI-lai), township, CENTRAL province, S central ZAMBIA, 10 mi/16 km N of LUSAKA; 15°18′S 28°19′E. On railroad. Agriculture (cotton, tobacco, peanuts, soybeans, vegetables); cattle.

Nha Be (NAH BAI), town, HO CHI MINH CITY urban region, S VIETNAM, on DONG NAI River below confluence of SAIGON RIVER, 8 mi/12.9 km SSE of Ho Chi Minh City; 10°42′N 106°44′E. Nha Be is closely linked with city center economy, yet with its own industrial zone. Presently the scene of rapid urbanization. Formerly spelled Nhabe.

Nhamatanda District, MOZAMBIQUE: see SOFALA.

Nhamundá (nee-yah-moon-DAH), city (2007 population 17,553), E AMAZONAS, BRAZIL, 28 mi/45 km N of PARINTINS, on Rio Amazonas; 02°16′S 56°46′W. Area in litigation with neighboring PARÁ.

Nha Nam (NAH NAHM), town, HA BAC province, N VIETNAM, 35 mi/56 km NNE of HANOI; 21°27′N 106°06′E. Road center. Formerly spelled Nhanam.

Nhandeara (NAHN-dai-AH-rah), town (2007 population 10,323), NW são PAULO, BRAZIL, 40 mi/64 km W of SÃO JOSÉ DO RIO PRÊTO; 20°40′S 50°02′W. Grain, cotton, coffee, fruit; cattle.

Nha Trang (NAH TRAHNG), city, ⊙ KHANH HOA province, E central S VIETNAM; 12°15′N 109°11′E. The French developed Nha Trang as a beach resort and administrative center. Luxurious villas of Emperor Bao Dai located here. It served as a major U.S. base during the Vietnam War. More recently the city has undergone considerable development, especially in mariculture, trade, and light industrial sectors. Nha Trang is a commercial center and major hub of railroad and road transport. In addition, it has two commercial ports on SOUTH CHINA SEA (BA NGOI and CAU DA ports) also neighborhood district of Chut (summer harbor). Tourism industry is a major source of revenue, with plans for extensive resort develop-

ment. The area beaches and nearby CAM RANH BAY, are major attractions. The city is also one of Vietnam's main cultural, religious, and educational centers. Key local and nearby tourist sites include Po Nagar Cham towers, Long Son pagoda, and Nha Trang cathedral. Pasteur and Oceanographic Institutes.

Nhill (NIL), town, W VICTORIA, SE AUSTRALIA, 210 mi/338 km WNW of MELBOURNE, halfway between Melbourne and ADELAIDE on Western Highway, c.50 mi/80 km E of SOUTH AUSTRALIA border; 36°20′S 141°40′E. In agricultural area (wheat, oats); flour mill.

Nhlangano (unh-lahn-GAH-no), town, SHISELWENI district, SW SWAZILAND, 45 mi/72 km S of MBABANE; 27°07′S 31°12′E. On main E-W road at junction of N road to MANZINI. Border with SOUTH AFRICA 6 mi/9.7 km to SW; airstrip to E. Cattle, goats, sheep, hogs; corn, cotton, tobacco, vegetables. Casino.

Nhommalat (no-mah-LAHT), town, KHAMMUAN province, central LAOS, NE of Thaket town, at crossroads; 17°36′N 105°12′E. Dist. capital.

Nho Quang, VIETNAM: see HOANG LONG.

Niafunké or **Niafounké** (both: NEE-a-foon-kai), town, SIXTH REGION/TIMBUKTU, S central MALI, in mid-NIGER RIVER depression, 85 mi/137 km SW of TIMBUKTU; 15°56′N 04°00′W. Market for livestock, wool, hides. Region also grows millet and rice.

Niagara, county (☐ 1,139 sq mi/2,961.4 sq km; 2006 population 216,130), W NEW YORK; ⊙ LOCKPORT; 43°19′N 78°47′W. Bounded W by NIAGARA River and Lake ERIE, N by Lake ONTARIO; crossed by the BARGE CANAL; drained by TONAWANDA CREEK. Includes NIAGARA FALLS resort area. Agricultural and extensive manufacturing area; commercial services. Some mining and oil refining. Contains Tuscarora and part of Tonawanda Indian reservations. Formed 1808.

Niagara, town (2006 population 1,780), MARINETTE county, NE WISCONSIN, at falls of MENOMINEE RIVER, 5 mi/8 km SSE of IRON MOUNTAIN (MICHIGAN); 45°46′N 88°00′W. In potato-growing and dairy area. Manufacturing (paper milling, iron and steel products). Incorporated 1914.

Niagara, village (2006 population 53), GRAND FORKS county, E NORTH DAKOTA, 40 mi/64 km WNW of GRAND FORKS; 48°00′N 97°52′W. Founded in 1883 and incorporated in 1907. Named by early settlers who came from NIAGARA County, NEW YORK.

Niagara (nei-A-gruh), region (2001 population 410,574), S ONTARIO, E central Canada, between lakes ERIE and ONTARIO; 43°02′N 79°18′W. One of 6 regional governments of Ontario; successor government to the counties of LINCOLN and Welland. Agricultural area; wine production; mining (shale, gypsum); peat and petroleum; tourism (NIAGARA FALLS). Composed of the cities of NIAGARA FALLS, PORT COLBORNE, SAINT CATHARINES, THOROLD, and WELLAND; and the towns of FORT ERIE, GRIMSBY, LINCOLN, NIAGARA-ON-THE-LAKE, PELHAM, Wainfleet, and West Lincoln.

Niagara, river, 34 mi/55 km long, NEW YORK and ONTARIO; issuing from Lake ERIE between BUFFALO, New York, and FORT ERIE, Ontario (Canada); flows N around GRAND ISLAND and over NIAGARA FALLS to Lake ONTARIO; the river forms part of the U.S.-Canada border. The upper section of the river is navigable for c.20 mi/32 km to a series of rapids above the falls; in its last 7 mi/11 km it is again navigable, from LEWISTON, New York, to Lake Ontario. The NEW YORK STATE BARGE CANAL enters the river at TONAWANDA, New York; the WELLAND SHIP CANAL, several miles W on the Ontario side, is a lake-freighter route around the falls. Hydroelectric power is generated by diverting water from the river above Niagara Falls to generating plants. Many bridges cross the Niagara River, notably Peace Bridge (1927); bridges linking Grand Island with both shores (1935); Rainbow Bridge (1941) below the falls; and American Rapids Bridge (1960), linking GOAT ISLAND with the mainland.

Niagara, Canada: see NIAGARA-ON-THE-LAKE.

Niagara Escarpment (nei-A-gruh), geological formation, S ONTARIO, E central Canada; 43°32′N 79°57′W. Ridge extending 451 mi/725 km from Queenstown, near NIAGARA FALLS, to the tip of the Bruce Peninsula at Tobermory, through both cities and farmland. Forested, with mountains, streams, valleys, and waterfalls; home to many species of birds, mammals, and fish, as well as wild orchids and the white cedar tree. Recreation opportunities, including the Bruce Trail, Canada's oldest footpath. Recognized in 1990 by UNESCO as a World Biosphere Reserve, an internationally significant ecosystem.

Niagara Falls (nei-A-gruh FAHLZ), city (☐ 81 sq mi/210.6 sq km; 2001 population 78,815), Niagara region, S ONTARIO, E central Canada, on the NIAGARA River opposite NIAGARA FALLS (NEW YORK); 43°06′N 79°03′W. Formerly called Clifton, it is a port of entry, an important industrial city, and the home of Canadian factories for many well-known U.S. firms. Electric power supplied by the falls supports industries that make chemicals, abrasives, silverware, machinery, sporting equipment, and paper products. The falls are also an international tourist attraction. Between the city and the falls and along the gorge below the falls is Queen Victoria Park. Casino Niagara, a very large gambling casino operated by the provincial government of Ontario, is 100 yd/91 m from the U.S. border. In 1963 the adjacent city of Stamford merged with Niagara Falls, increasing its population; restructured again in 1970 when CHIPPAWA, Willoughby township, and part of Crowland joined the city.

Niagara Falls (nei-A-gruh), city (☐ 16 sq mi/41.6 sq km; 2006 population 52,326), NIAGARA county, W NEW YORK, at the great falls of the NIAGARA River; 43°05′N 79°01′W. Tourism is one of its oldest industries, and many state parks are in the area, including New York State Niagara Reservation. The city is also a port of entry; its manufacturing includes abrasives, mechanical and electrochemical products, and paper and aluminum goods. This was the site of one of the world's first hydroelectric power plants; it was replaced between 1963 and 1965 by a plant capable of producing 2,100 MW. Settled by Native Americans, the site was occupied by the French in the 1680s, captured by the British in 1759, and settled by Americans in 1805. Lost to the British during the War of 1812, it was regained by the U.S. after the Treaty of Ghent in December 1814. Several bridges span the river to Canada. Niagara University and a community college are here, as well as a large casino opened by the provincial government of Ontario (Canada), 100 yd/91 m from the U.S. border. Incorporated 1892.

Niagara Falls (nei-A-gruh), in the NIAGARA River, W NEW YORK and S ONTARIO (Canada); 43°05′N 29°04′W. One of the most famous spectacles in N America, the falls are on the international border between the cities of NIAGARA FALLS, New York, and NIAGARA FALLS, Ontario. GOAT ISLAND splits the cataract into the American Falls (167 ft/51 m high and 1,060 ft/323 m wide) and the Horseshoe, or Canadian, Falls (158 ft/48 m high and 2,600 ft/792 m wide). The falls were formed c.10,000 years ago as the retreating glaciers exposed the Niagara escarpment, thus permitting the waters of Lake ERIE to flow N to Lake ONTARIO. The escarpment has been gradually eroded back toward Lake Erie, a process that has formed the Niagara Gorge (c.7 mi/11.3 km long); the Whirlpool Rapids and the Whirlpool are here. Horseshoe Falls is eroding upstream at a faster rate than the American Falls due to the greater volume of water passing over it. A great rock slide occurred (1954) at the American Falls and formed a huge talus slope at its base. Water was diverted from the American Falls for several months in 1969 by the U.S. Corps of Engineers to study the bedrock and to remove some of the talus. International agreements control the diversion of

Area is shown by the symbol ☐, and capital city or county seat by ⊙.

water for hydroelectric power; weirs divert part of the flow above the Horseshoe Falls to supplement the flow in the shallower American Falls. Hydroelectric-power developments were authorized under the Niagara Diversion Treaty (1950), which stipulated a small flow to be reserved for the falls and the equal division of the remaining flow between the U.S. and Canada. In the U.S. the project was undertaken by the Power Authority of the State of New York. Water is diverted from the river above the upper rapids into underground conduits (46 ft/14 m wide and 66 ft/20 m high). It is then conveyed overland, dropping 314 ft/96 m to a point below the lower rapids where, as it returns to the river, the water passes through turbines that power thirteen generators of the Robert Moses Niagara Power Plant (1,950-MW capacity; opened 1961). Associated with the New York hydroelectric-power project is the construction in the area of new roads, bridges, and parks. In Canada it was undertaken by the Hydro-Electric Power Commission of Ontario. Water is diverted from the river above the falls and is fed into the Sir Adam Beck Generating Stations (1,775 MW; opened 1954) by way of a series of tunnels and canals. The governments of the U.S. and Canada also control the appearance of the surrounding area, much of which has been included in parks since 1885. That the falls represented a major physical obstacle to the Great Lakes' full access to regional and ocean commerce via the SAINT LAWRENCE RIVER was recognized as early as 1829, when a canal-river bypass was constructed between Lakes Erie and Ontario. In 1839 the Canadian government assumed ownership of the canal, enlarging it in 1845 and again in 1887. In 1932, the present WELLAND SHIP CANAL, part of the SAINT LAWRENCE SEAWAY system, was opened. In 1973 $110 million was allocated to straighten and enlarge 8 mi/12.9 km of this 27-mi/43-km-long canal, located 14 mi/23 km W of Niagara Falls. The earliest written description of the falls is that of Louis Hennepin (in *Nouvelle Découverte*, 1697), who was with the expedition of French explorer Robert Cavalier, sieur de La Salle, in 1678. In the 19th century, daredevils attempted to brave the falls in barrels, boats, and rubber balls. The great Blondin performed (1859) on a tightrope over the falls, which continue to be a major center of international tourism. Historical and natural history material relating to the region is in the Niagara Falls Museum in the city of NIAGARA FALLS, New York.

Niagara, Fort, NEW YORK: see FORT NIAGARA.

Niagara-on-the-Lake or **Niagara** (nei-A-gruh), city (☐ 52 sq mi/135.2 sq km; 2001 population 13,839), NE Niagara region, S ONTARIO, E central Canada, on Lake ONTARIO at the mouth of the NIAGARA River; 43°15'N 79°04'W. It was settled (1784) by American Loyalists and in 1792 Lieutenant Governor Simcoe made the town the capital of Upper Canada, renaming it Newark. The legislature met here until 1796. Fort George, built (1796–1799) to defend the settlement, was taken in 1813 by the U.S. but Canada took it back the same year. The town, officially called Niagara-on-the-Lake to distinguish it from the Canadian and U.S. cities of Niagara Falls, is an architectural and historical treasure, with many well-preserved 19th-century buildings. It is the site of the Shaw Festival, an annual festival of plays.

Niagassola (NIR-gas-so-lah), town, Siguiri prefecture, Kankan administrative region, NE GUINEA, in Haute-Guinée geographic region, near MALI border, 70 mi/113 km N of SIGUIRI.

Niah National Park (NEE-ah) (☐ 12 sq mi/31.2 sq km), NW central SARAWAK, E MALAYSIA, 30 mi/48 km SW of MIRI; 03°58'N 113°41'E. Tropical rain forest. Has Niah Cave, one of largest in world; prehistoric paintings in Painted Cave; prehistoric skeletons discovered in Great Cave. Bats (twelve species), snakes, scorpions in cave; macaques, squirrels, flying lizards, Rajah Brooke butterflies. Accessible by road from Miri and BINTULU.

Niakaramandougou (nee-ah-kah-rah-man-DOO-goo), town, Vallée du Bandama region, central CÔTE D'IVOIRE, 39 mi/62 km NNW of Katiola; 08°40'N 05°17'W. Agriculture (corn, beans, tobacco, peanuts, cotton).

Niamati, INDIA: see NYAMATI.

Niamey (NEE-ah-mai), city (2005 population 808,346), SW NIGER, port on the NIGER RIVER; ☉ Niger and TILLABÉRY province; 13°31'N 02°07'E. Niamey is Niger's largest city and its administrative and economic center. Much of its importance stems from its location on the Niger River at the crossroads of the country's two main highways. The city is the trade center for an agricultural region that specializes in growing groundnuts. Manufacturing includes bricks, food products, beverages, ceramic goods, cement, and shoes. Niamey was a small town when the French colonized the area in the late 19th century, but it grew after it became capital of Niger in 1926. It is the site of the National School of Administration (1963), the National Museum, which has ethnological and zoological collections, and a university.

Niamtougou, town, ☉ DOUFELGOU prefecture, KARA region, N TOGO, 65 mi/105 km N of SOKODÉ; 09°48'N 01°07'E.

Niangara (nyahng-GAH-rah), town, ORIENTALE province, NE CONGO, on UELE region, on CONGO-NILE highway, 220 mi/354 km ENE of BUTA; 03°42'N 27°52'E. Elev. 2,116 ft/644 m. Cotton ginning; cotton, coffee, sesame growing. Seat of vicar apostolic. Has Roman Catholic and Protestant missions and schools, Roman Catholic seminary for priests, hospital.

Niangbo, CÔTE D'IVOIRE: see NYANGBO.

Niangua (nei-ANG-gwuh), city (2000 population 445), WEBSTER county, S central MISSOURI, in the OZARKS, at headwaters of NIANGUA RIVER, 29 mi/47 km ENE of SPRINGFIELD; 37°23'N 92°49'W. Fruit; dairying; cattle.

Niangua River (nei-ANG-gwuh), c.90 mi/145 km long, central MISSOURI; rises in the OZARKS near MARSHFIELD; flows N into the Niangua arm of LAKE OF THE OZARKS in CAMDEN county. Bennett Spring State Park along the river in DALLAS county. Canoeing, fishing.

Nia Nia (NYAH NYAH), village, ORIENTALE province, NE CONGO, near ITURI province, 125 mi/201 km W of IRUMU; 01°24'N 27°36'E. Elev. 1,853 ft/564 m. Communications point with airfield at junction of roads to SUDAN and UGANDA; repairing of motor vehicles.

Niantic (ni-ANT-ick), village (2000 population 738), MACON county, central ILLINOIS, 10 mi/16 km W of DECATUR; 39°51'N 89°10'W. Grain; livestock; dairy products.

Niantic, CONNECTICUT: see EAST LYME.

Niaouli (NYOU-lee), village, S BENIN, 35 mi/56 km NW of COTONOU; 06°44'N 02°08'E. Experiment station for coffee cultivation.

Niaousta, Greece: see NAOUSA.

Niapu (NYAH-poo), village, ORIENTALE province, N CONGO, near source of RUBI RIVER, 110 mi/177 km ESE of BUTA; 02°25'N 26°58'E. Elev. 2,034 ft/619 m. Trading post in cotton area.

Niaqornaarsuk (nyah-kor-NAHR-sook), fishing settlement, Kangaatsiaq commune, W GREENLAND, on ARFERSIORFIK fjord, near its mouth on DAVIS STRAIT, 30 mi/48 km S of EGEDESMINDE (Aasiaat); 68°14'N 52°50'W.

Niaqornat (nyah-KOR-naht), fishing and hunting settlement, Uummannaq commune, W GREENLAND, on N shore of NUUSSUAQ peninsula, on UUMMANNAQ FJORD, 35 mi/56 km WNW of UUMMANNAQ; 70°47'N 53°39'W.

Niari, region (2007 population 234,014), SW Congo Republic, bordered by GABON to the N, KOUILOU region and the Angolan enclave of CABINDA to the S, and LÉKOUMOU and BOUENZA regions to the E; ☉ LOUBOMO; 03°00'S 12°30'E. The NIARI RIVER flows

NE-SW through the center of the region. Mining (gold, lead) and agriculture (rice); also wood products, tanning, fiber processing, livestock raising, sisal plantations.

Niari River, REPUBLIC OF THE CONGO: see KOUILOU RIVER.

Nias (NEE-yahs), volcanic island (☐ 1,842 sq mi/4,771 sq km; 1990 population 588,543), INDONESIA, in the INDIAN OCEAN, 78 mi/125 km W of SUMATRA, part of Sumatra Utara province; 01°05'N 97°35'E. Highest elevation 2,907 ft/886 m. Most of the population are descended from the Niah people; their economy is largely agricultural. The chief town is GUNUNGSITOLI, capital of Nias district; airport. Native handicrafts are highly developed; megalithic shrines dot the land. The Dutch began trading here in 1669. The island is subject to severe earthquakes.

Niassa (NEE-ah-suh), province (☐ 107,328 sq mi/279,052.8 sq km; 2004 population 966,579), MOZAMBIQUE, bounded N by ROVUMA River and TANZANIA, W by MALAWI (lakes NIASSA, CHIUTA, and CHILWA form part of the border), E by CABO DELGADO province, S by NAMPULA and ZAMBEZIA provinces; ☉ LICHINGA. Rises in the W to savanna hill country dotted with isolated higher peaks. Railroad line links Lichinga to coastal port town of NACALA in NAMPULA province and to Malawi to W. Agriculture (tobacco, oilseeds, cotton, sisal). Inhabited by the Yao ethnic group and the Chihanja speaking spread over 79 demographic conurbations. Divided in 15 districts: CUAMBA, LAGO, LICHINGA, MAJUNE, MANDIMBA, MARRUPA, MAUA, MAVAGO, MECANHELAS, MECULA, METARICA, MUEMBE, N'guama, NIPEPE, and SANGA.

Niau (nee-AH-oo), atoll, N TUAMOTU ARCHIPELAGO, FRENCH POLYNESIA, S PACIFIC; 16°10'S 146°21'W. Circular, 4 mi/6.4 km in diameter. Formerly called Greig Island.

Niaux (nee-O), commune (☐ 1 sq mi/2.6 sq km), ARIÈGE department, MIDI-PYRÉNÉES region, SW FRANCE; 42°49'N 01°35'E. Cave paintings of Magdalenian age (c.12,000 B.C.E.) in an inhabited site.

Niazbatyr (nee-AHZ-bah-tir), town, ANDIJAN wiloyat, UZBEKISTAN.

Nibe (NEE-buh), city, NORDJYLLAND county, N JUTLAND, DENMARK, on LIMFJORD, 11 mi/18 km WSW of ÅLBORG; 56°58'N 09°32'E. Fisheries; cattle.

Nibley, town (2006 population 3,062), CACHE county, N UTAH, 4 mi/6.4 km S of LOGAN, on Blacksmith Creek; 41°40'N 111°50'W. Barley, wheat, vegetables; dairying; cattle. Part of Wasatch National Forest to E.

Nibo (NEE-bo), village, central SADO island, Sado county, NIIGATA prefecture, N central JAPAN, 34 mi/55 km N of NIIGATA; 38°00'N 138°25'E.

Nicaea, ASIA MINOR: see IZNIK.

Nicaea, FRANCE: see NICE.

Nicaragua (nee-kah-RAH-gwah), republic (☐ 49,579 sq mi/128,410 sq km; 2004 estimated population 5,359,759; 2007 estimated population 5,675,356), CENTRAL AMERICA; ☉ MANAGUA.

Geography

Nicaragua is bordered on the N and NW by HONDURAS, on the E by the CARIBBEAN SEA, on the S by COSTA RICA, and on the SW by the PACIFIC OCEAN. There are four main geographic areas. The NW highlands have peaks as high as 8,000 ft/2,438 m. On the Caribbean is the tropical MOSQUITO COAST, with the historic port of BLUEFIELDS. Home to Miskitos and descendants of black slaves, the inhabitants of this region have continually sought greater autonomy. A lowland belt running NW to SE contains lakes MANAGUA and NICARAGUA. The fourth region is a narrow volcanic belt squeezed between the lakes and the Pacific; in this region the productive wealth and the population (almost entirely of Spanish and Native American descent) are concentrated. CORINTO, on the Pacific, is the chief port.

Population

The population is overwhelmingly Roman Catholic. There are universities at LEÓN and Managua.

Economy

Agriculture employs about 30% of the workforce and accounts for about 30% of the GDP. The chief commercial crops are coffee and sugarcane; these, together with meat, are the largest exports. Timber and gold are also exported. The principal manufactured goods are chemicals, textiles, and processed foods.

History: to 1850

The country probably takes its name from Nicarao, the leader of an indigenous community inhabiting the shores of Lake Nicaragua which was defeated in 1522 by the Spanish under Gil González de Ávila. Under Spanish rule Nicaragua was part of the captaincy general of GUATEMALA. After declaring independence from Spain (1821), Nicaragua was briefly part of the Mexican Empire of Agustín de Iturbide and then (1825–1838) a member of the Central American Federation. Nicaraguan politics were wracked by conflict between Liberals and Conservatives, centered respectively in León and GRANADA; Managua was founded as the capital in 1855 as a compromise. British influence had been established along the E coast in the 17th century, and in 1848 the British seizure of SAN JUAN DEL NORTE opened a period of conflict over control of the Mosquito Coast. The U.S. was interested in a transisthmian canal, and its interest was heightened by the discovery of gold in California (1849).

History: 1850 to 1925

The Clayton-Bulwer Treaty (1850) settled some of the issues between Great Britain and the U.S. concerning the proposed canal, but Nicaragua remained in a state of disorder that culminated in the temporary triumph (1855–1857) of the filibuster William Walker. After Walker's defeat there was a long period of quiet under Conservative control until the Liberal leader, José Santos Zelaya, who became president in 1894. His financial dealings with Britain aroused the apprehension of the U.S. and helped bring about his downfall (1909). In 1912, U.S. marines were landed to support the provisional president, Adolfo Díaz, in a civil war. The Bryan-Chamorro Treaty, giving the U.S. exclusive rights for a Nicaraguan canal and other privileges, was ratified in 1916 (it was terminated in 1970). The Liberals opposed the U.S. intervention, and there was guerrilla warfare against the U.S.-supported regime for years.

History: 1925 to 1972

American occupation ended in 1925 but resumed the next year, when Emiliano Chamorro attempted to seize power. Augusto César Sandino was a leader of the anti-occupation forces. U.S. diplomat Henry Lake Stimson succeeded in getting most factions to agree (1927) to binding elections, although Sandino continued to fight. The marines were withdrawn in 1933; in 1936 Anastasio Somoza emerged as the strong man in Nicaragua. Nicaragua virtually became Somoza's private estate; the regime aroused much criticism among liberal groups in Latin America. Under Somoza relations with other Central American republics were poor. Somoza was assassinated in 1956, and his son Luis Somoza Debayle became president. René Schick Gutiérrez was chosen by the Somoza family to be elected president in 1963. After his death in 1966, Lorenzo Guerrero, the vice president, succeeded. Anastasio Somoza Debayle was elected president in 1967.

History: 1972 to 1990

Although Somoza resigned from office in May 1972, he retained effective control of the country as head of the armed forces, and, after the earthquake (December 1972) that devastated Managua, as director of the emergency relief operations. From this position Somoza diverted international aid to himself and his

associates, an abuse which solidified opposition to the Somoza regime among all classes. The opposition was grouped under two large factions, the Sandinista National Liberation Front (FSLN) and the Democratic Liberation Union (UDEL). In 1979, the FSLN and UDEL joined in a popular revolt which toppled the Somoza government. The more radical, left-wing FSLN took control of the government, instituting widespread social, political, and economic changes. Many economic institutions and resources were nationalized, land was redistributed, and social services such as health care and education were improved. In 1981 the U.S., politically unsupportive of the Sandinista government and suspicious of the relations it was cultivating with the Soviet Union and Cuba, began to take serious steps to weaken the government. In 1984 the U.S. illegally mined Nicaragua's principal export harbors, in 1985 it instituted a trade embargo, and throughout the decade it continued to support the anti-government "Contra" rebels. At the same time the government, although it received substantial aid from the Soviet Union, became increasingly unable to maintain the national economy. The government also curtailed civil liberties to silence popular dissatisfaction.

History: 1990 to Present

In February 1990, the FSLN was defeated by an opposition coalition in the general elections, and Violeta Barros de Chamorro became president. The U.S. lifted its trade embargo in March of that year, and in April the Contras ceased their campaign. The Chamorro administration worked toward reviving the Nicaraguan economy as well as generating a conciliatory political environment. Arnoldo Alemán was elected president in 1996. Hurricane Mitch devastated the country in 1998, damaging economic growth. In the 2001 presidential election Enrique Bolaños, of the Liberal party, won. In late 2004 Bolaños's anticorruption campaign, which had resulted in Alemán's conviction, led Alemán's and Ortega's supporters to ally against the president to reduce his powers, sparking a yearlong constitutional crisis; an agreement in October 2005 delayed the changes until after Bolaños's term. The 2006 presidential election returned Ortega to power after sixteen years.

Government

Nicaragua is governed under the constitution of 1987 as amended. Executive power is held by the president, who is both head of state and head of government. The president is popularly elected for five years and may not serve consecutive terms. Members of the unicameral ninety-two-seat National Assembly are also elected for five years. The current head of state is President Daniel Ortega Saavedra (since January 2007). Administratively, the country is divided into fifteen departments and two autonomous regions.

Nicaragua (nee-kah-RAH-gwah), sugar-mill village, HOLGUÍN province, E CUBA, on BANES BAY, 4 mi/6 km S of BANES; 20°55′N 75°58′W.

Nicaragua Canal (nee-kah-RAH-gwah), proposed waterway between the ATLANTIC and the PACIFIC oceans. Planned to be 172.8 mi/278.1 km long, it would generally follow the SAN JUAN RIVER, then go through LAKE NICARAGUA near the S shore and across the narrow isthmus of RIVAS to the Pacific Ocean. Proposed by Secretary of State Henry Clay in 1826, the route was an important factor in negotiation of the Clayton-Bulwer Treaty (1850). In later times the route has been considered as an adjunct to the PANAMA CANAL; it would shorten the water distance between NEW YORK city and SAN FRANCISCO by nearly 500 mi/805 km. Under the Bryan-Chamorro Treaty (1916), the U.S. paid NICARAGUA $3 million for an option in perpetuity and free of taxation, including ninety-nine year leases to the CORN ISLANDS and a site for a naval base on the GULF OF FONSECA. COSTA RICA protested that its rights to the San Juan River had been

infringed, and EL SALVADOR maintained that the proposed naval base affected both it and Honduras. Both protests were upheld by the Central American Court of Justice, but the court rulings were ignored by Nicaragua and the U.S. The action was bitterly criticized by Latin Americans and others as an example of U.S. imperialism.

Nicaragua, Lake (nee-kah-RAH-gwah) (□ 3,089 sq mi/ 8,031.4 sq km), SW NICARAGUA; 11°30′N 85°30′W. The largest lake of CENTRAL AMERICA (c.100 mi/161 km long and up to 45 mi/72 km wide), it is drained into the CARIBBEAN SEA by the SAN JUAN RIVER. LAKE NICARAGUA, along with LAKE MANAGUA (which drains into it from the NW), occupies part of the Nicaragua Depression, an extensive lowland region stretching across the isthmus. Once part of the sea, the lake was formed when the land rose. There are several islands in the lake (the largest is ISLA DE OMETEPE); and small volcanoes rise above its surface. The fresh water of Lake Nicaragua contains fish usually associated with salt water, including tuna and sharks, which have adapted to the environmental change. The lake is a transportation route; GRANADA is its chief port. Located only 110 ft/34 m above sea level, the lake reaches a depth of 84 ft/26 m. It was to be an important link in the proposed NICARAGUA CANAL.

Nicaro (nee-KAHR-o), town, HOLGUÍN province, E CUBA, on LEVISA BAY (ATLANTIC OCEAN), at base of small Lengua de Pájaro peninsula, 6 mi/10 km E of MAYARÍ; 20°43′N 75°32′W. Refining of nickel oxide, which is mined nearby (S) at foot of SIERRA DEL CRISTAL. Refining operations were started in 1943.

Nicastro (nee-KAHS-tro), town (2001 population 70,501), CATANZARO province, CALABRIA, S ITALY, 16 mi/26 km WNW of CATANZARO; 38°59′N 16°19′E. Commercial center (wine, olive oil, wheat, fruit). Bishopric. Has ruins of ancient castle. Frequently damaged by earthquakes.

Nicatous Lake (nik-uh-TOU-uhs), HANCOCK county, E central MAINE, 36 mi/58 km NE of BANGOR; 8.5 mi/13.7 km long. In hunting, fishing area.

Nice (NEES), Italian *Nizza*, city (□ 27 sq mi/70.2 sq km); ⊙ ALPES-MARITIMES department, PROVENCE-ALPES-CÔTE D'AZUR region, SE FRANCE, on the MEDITERRANEAN SEA, backed by the coastal ranges of the MARITIME ALPS, 420 mi/676 km SE of PARIS, 15 mi/24 km SW of the Italian border, 8 mi/12.9 km NE of MONACO; 44°00′N 07°15′E. Nice is the largest and one of the best-known resorts on the French RIVIERA (*Côte d'Azur*). Although its economy depends mainly on the tourist trade—thanks to the mild Mediterranean climate, its physical setting, and the many attractions of the Riviera—Nice is also a seaport. The old port (E of PAILLON RIVER, which bisects the city) handles commercial fishing and passenger service to CORSICA. The international airport (Nice–Côte d'Azur) is 3 mi/4.8 km SW at mouth of VAR RIVER on the Mediterranean. The city's industrial development is relatively recent, emphasizing electronics and research facilities. Perfumes, soap, furniture, musical instruments, and cotton goods are local products. Service industries are the main source of employment. Nice is known for its local food specialties, including *Salade Niçoise* (sah-lahd nee-swahz), a tomato salad with eggs, tuna or anchovies, and olives. The Carnival of Nice marks the height of the city's festival season, but there are innumerable cultural events that take place in the theater, exposition hall, and museums. Probably a Greek colony (*Nikaio*, or *Nicaea* in Latin) established 5th century B.C.E. on a hill just W of the old port, Nice became an episcopal see in the 4th century C.E. It was pillaged and burned by the Saracens in 859 and 880. In the 13th and 14th centuries the city belonged to the counts of PROVENCE and SAVOY. In 1543 the united forces of Francis I and the Turks' Barbarossa attacked and burned Nice. It was annexed to France in 1793, restored to SARDINIA in

1814, and again ceded to France in 1860 after a plebiscite. At the beginning of the French Revolution the city was a haven for Royalist émigrés. Together with its hinterland extending to the Italian border, Nice was claimed and temporarily occupied by Mussolini during World War II. The city's most prominent urban feature is the famed *Promenade des Anglais*, a wide avenue that lines 3 mi/4.8 km of waterfront on the azure Bay of the Angels. Place Masséna is the city's focal point. In the N part of Nice, in the district of *Cimiez* is the site of a Gallo-Roman community once known as *Cemenelum*. Adjacent to the excavations are the Matisse museum (tracing his development as an Impressionist painter); the archaeological museum; and an old Franciscan monastery rebuilt in 16th century. Noteworthy as well are museums of fine arts, contemporary art, and primitive art. The huge botanical garden (opened 1991) organizes its floral exhibits along climatic zones and includes an indoor exhibit of tropical plants. The Chagall museum (built 1972), near midtown, contains the largest collection of the artist's works found in one place, including his "biblical message" expressed on 18 canvas panels and some of his sculptures and mosaic tiles. Nice also has a university (with a Chagall mosaic in the law school) that is a center of Mediterranean higher education, a national school of the decorative arts, and a conservatory of music.

Nice, unincorporated town (2000 population 2,509), LAKE county, NW CALIFORNIA, 6 mi/9.7 km NE of LAKEPORT, at NW end of CLEAR LAKE (Redman Slough enters lake to W); 39°08′N 122°51′W. Cattle; grain; timber. Mendocino National Forest to N and NE. Bloody Island Massacre Historic Marker to NW.

Nice, TURKEY: see IZNIK.

Nicephorium, SYRIA: see RAQQA.

Niceville (NEIS-vil), city (□ 11 sq mi/28.6 sq km; 2005 population 12,582), OKALOOSA county, NW FLORIDA, c.45 mi/72 km ENE of PENSACOLA, on CHOCTAWHATCHEE BAY; 30°31′N 86°28′W.

Nichelino (nee-ke-LEE-no), town, TORINO province, PIEDMONT, NW ITALY, 6 mi/10 km S of TURIN; 44°59′N 07°38′E. Machinery, fabricated metals, chemicals.

Nichicun Lake or **Nichikun Lake** (both: NI-chi-kuhn) (□ 150 sq mi/390 sq km), central QUEBEC, E Canada, on SAINT LAWRENCE–HUDSON BAY watershed, at foot of the OTISH MOUNTAINS; 20 mi/32 km long, 12 mi/19 km wide; 53°05′N 71°00′W. Elevation 1,737 ft/529 m. Drained N by FORT GEORGE River. Just SE is Naokokan Lake (18 mi/29 km long, 12 mi/19 km wide).

Nichihara (nee-CHEE-hah-rah), town, Kanoashi county, SHIMANE prefecture, SW HONSHU, W JAPAN, 93 mi/150 km S of MATSUE; 34°32′N 131°50′E. Ayu. Gourd processing.

Nichinan (nee-CHEE-NAHN), city, MIYAZAKI prefecture, SE KYUSHU, SW JAPAN, on the PACIFIC OCEAN, 22 mi/35 km S of MIYAZAKI; 31°35′N 131°22′E. Agricultural center (pigs) and manufacturing center for paper pulp and paper products.

Nichinan (nee-CHEE-nahn), town, Hino county, TOTTORI prefecture, S HONSHU, W JAPAN, 57 mi/92 km S of TOTTORI; 35°09′N 133°19′E. Chrome ore. Sekka Gorge is nearby.

Nichol (NI-kuhl), former township (□ 42 sq mi/109.2 sq km; 2001 population 4,372), S ONTARIO, E central Canada, 19 mi/31 km from KITCHENER; 43°41′N 80°23′W. Amalgamated into Guelph/Eramosa township in 1999.

Nicholas, county (□ 196 sq mi/509.6 sq km; 2006 population 6,958), N KENTUCKY; ⊙ CARLISLE; 38°20′N 84°00′W. Bounded NE by LICKING RIVER. Gently rolling upland agricultural area (burley tobacco, hay, alfalfa, corn; cattle, poultry; dairying), in BLUEGRASS REGION. Includes BLUE LICKS Battlefield State Park in N; Clay Wildlife Management Area in E. Formed 1799.

Nicholas, county (□ 649 sq mi/1,687.4 sq km; 2006 population 26,446), central WEST VIRGINIA; ⊙ SUMMERSVILLE; 38°17′N 80°47′W. On ALLEGHENY PLATEAU; bounded SW by GAULEY and MEADOW rivers; drained by Gauley, Cranberry, Cherry, and Birch rivers. Agriculture (corn, potatoes, alfalfa, hay, apples); cattle, poultry. Bituminous-coal mining. Timber. CARNIFEX FERRY BATTLEFIELD STATE PARK in center; part of Monongahela National Forest in E; part of Gauley River National Recreation Area in SW; SUMMERSVILLE LAKE reservoir (Gauley River) and Wildlife Management Area in center. Formed 1818.

Nicholas Channel or **Saint Nicholas Channel,** strait off NW coast of CUBA, 90 mi/145 km E of HAVANA, 80 mi/129 km SE of KEY WEST (FLORIDA); 23°15′N 79°45′W. Extends c.100 mi/161 km E, bounded N by CAY SAL BANK (20 mi/32 km off Cuba). Continued E by OLD BAHAMA CHANNEL. Sometimes spelled Nicolas or Canal de San Nicolas.

Nicholas II Land, RUSSIA: see SEVERNAYA ZEMLYA.

Nicholasville, city (2000 population 19,680), ⊙ JESSAMINE county, central KENTUCKY, 13 mi/21 km SSW of LEXINGTON; 37°52′N 84°34′W. In BLUEGRASS region. Agricultural area (dairying; cattle, horses, hogs, poultry; burley tobacco, corn, wheat, soybeans); manufacturing (shoes). Early gristmill (1782) nearby. Jim Beam Nature Preserve and Camp Nelson National Cemetery to S. Settled 1798.

Nicholls, town (2000 population 1,008), COFFEE county, S central GEORGIA, 13 mi/21 km E of DOUGLAS; 31°31′N 82°38′W. Manufacturing (clothing and textiles).

Nicholls' Town, town, W BAHAMAS, on NE shore of ANDROS Island (40 mi/64 km W of NASSAU; 25°08′N 77°59′W; 2000 population 3,444). Fishing. Submarine caverns nearby.

Nichols, town, MUSCATINE county, SE IOWA, 13 mi/21 km WNW of MUSCATINE; 41°28′N 91°18′W. In agricultural area.

Nichols, village, MUSCATINE county, SE IOWA, 13 mi/21 km WSW of MUSCATINE. Corn, soybeans; cattle, hogs.

Nichols, village (2006 population 550), TIOGA county, S NEW YORK, near the SUSQUEHANNA, 25 mi/40 km WSW of BINGHAMTON; 42°01′N 76°22′W.

Nichols, village (2006 population 404), MARION county, E SOUTH CAROLINA, 35 mi/56 km E of FLORENCE, on LUMBER RIVER at its mouth on LITTLE PEE DEE RIVER; 34°13′N 79°09′W. Manufacturing includes furniture, fish processing, aquaculture; agricultural products include sorghum, corn, cotton, tobacco; livestock; catfishing.

Nichols, village (2006 population 278), OUTAGAMIE county, E WISCONSIN, 20 mi/32 km N of APPLETON; 44°34′N 88°28′W. Dairying; poultry; grain. Manufacturing (printing).

Nichols Hills or **Nichols Hill,** town (2006 population 3,990), OKLAHOMA county, central OKLAHOMA, residential suburb 5 mi/8 km N of OKLAHOMA CITY; 35°32′N 97°32′W. Oklahoma Museum of Art here. Lake Hefner reservoir to NW. Incorporated 1929.

Nicholson, town (2000 population 1,247), JACKSON county, NE central GEORGIA, 10 mi/16 km N of ATHENS; 34°07′N 83°26′W.

Nicholson, unincorporated village, PEARL RIVER county, SE MISSISSIPPI, 3 mi/4.8 km SSW of PICAYUNE, on PEARL RIVER. Manufacturing (fireworks, concrete products, chemicals). Stennis Space Center (NASA) to SE. Bogue Chitto National Wildlife Refuge (Louisiana) to W.

Nicholson (NI-kuhl-suhn), borough (2006 population 675), WYOMING county, NE PENNSYLVANIA, 16 mi/26 km NNW of SCRANTON, on Tunkhannock Creek; 41°37′N 75°47′W. Manufacturing (feeds); flagstone quarrying; agriculture (corn, hay; dairying). Lackawanna State Park to SE; several small lakes to SE. Incorporated 1875.

Nicholson, former trading post, NORTHWEST TERRITORIES, Canada, at head of Liverpool Bay; on inlet (30 mi/48 km long, 5 mi/8 km–10 mi/16 km wide) of the BEAUFORT SEA, at mouth of ANDERSON RIVER; 69°45′N 128°52′W. Meteorological station. Formerly called Stanton.

Nicholson River (NI-kuhl-suhn), 130 mi/209 km long, N AUSTRALIA; rises in BARKLY TABLELAND in NORTHERN TERRITORY; flows generally E to Gulf of CARPENTARIA on NW coast of QUEENSLAND, near mouth of ALBERT RIVER; 17°32′S 139°36′E. In generally arid area. Gregory River, main tributary.

Nickajack Lake, reservoir, MARION and HAMILTON counties, SE TENNESSEE, on the TENNESSEE RIVER, 18 mi/29 km WSW of CHATTANOOGA, 2 mi/3.2 km N of ALABAMA state line; 35°00′N 85°37′W. Maximum capacity 252,400 acre-ft; c.50 mi/80 km long. Formed by Nickajack Dam (67 ft/20 m high), built (1967) by the Tennessee Valley Authority for navigation, flood control, and power generation. Prentice Cooper State Forest on N shore. Former Hales Bar Dam site at Haletown (Guild), 6 mi/9.7 km upstream.

Nickelsdorf (NIK-kels-dorf), village, N BURGENLAND, E AUSTRIA, on the Heidebaden, near LEITHA RIVER and Hungarian border, 14 mi/23 km S of BRATISLAVA (SLOVAKIA); 47°56′N 17°04′E. Main border station opposite HUNGARY (on motorway and railroad between VIENNA and BUDAPEST); border commerce; vineyards.

Nickelsville (NI-kuhlz-vil), town, SCOTT county, SW VIRGINIA, 16 mi/26 km NW of BRISTOL, near Copper Creek; 36°45′N 82°25′W. Agriculture (dairying; livestock; tobacco, corn).

Nickerie (ni-KIR-ri), district (□ 2,067 sq mi/5,353 sq km; 2004 population 36,639), NW SURINAME, ⊙ NIEUW NICKERIE. Populated coastal area; agriculture (rice, bananas). Population mainly Hindustani and Javanese.

Nickerie Point (ni-KIR-ri), on ATLANTIC OCEAN coast of NW SURINAME, at mouth of NICKERIE RIVER, 3 mi/4.8 km NW of NIEUW NICKERIE; 05°58′N 57°03′W. Fishing; shrimp.

Nickerie River (ni-KIR-ri), c.200 mi/322 km long, W SURINAME; rises in outliers of the GUIANA HIGHLANDS at 04°13′N 56°54′W; flows N and WNW to the ATLANTIC OCEAN just below NIEUW NICKERIE, at mouth of CORANTIJN RIVER. Tropical forests rich in hardwood and gums along upper course; fertile alluvial land (rice grown on large scale) along lower course. Navigable for c.60 mi/97 km upstream. Linked by natural waterway with COPPENAME RIVER, wide estuary.

Nickerson, town (2000 population 1,194), RENO county, S central KANSAS, on ARKANSAS RIVER, 10 mi/16 km NW of HUTCHINSON; 38°08′N 98°05′W. In grain and livestock area. Incorporated 1879.

Nickerson, village (2006 population 433), DODGE county, E NEBRASKA, 6 mi/9.7 km N of FREMONT, near ELKHORN RIVER; 41°32′N 96°28′W. Productive terrace land.

Nickerson Ice Shelf, off the W end of the RUPPERT COAST of MARIE BYRD LAND, WEST ANTARCTICA, 35 mi/60 km wide; 75°44′S 145°00′W.

Nicoadala District, MOZAMBIQUE: see ZAMBÉZIA.

Nicobar Islands (ni-ko-BAHR), district (□ 754 sq mi/1,960.4 sq km; 2001 population 42,026), the S section of ANDAMAN AND NICOBAR ISLANDS Union Territory, INDIA; ⊙ Car Nicobar. Consists of nineteen islands, twelve of which are inhabited. The main islands are CAR NICOBAR, GREAT NICOBAR, CAMORTA, and NANCOWRY ISLANDS. Part of a great submarine mountain range extending from ARAKAN YOMA (MYANMAR) to the mountains of SUMATRA (INDONESIA). Mild, warm equatorial climate. Fish and forest products with export potential. Major ports at Car Nicobar (N) and Camorta (S).

Nicola River (NI-ko-luh), 100 mi/161 km long, S BRITISH COLUMBIA, W Canada; rises c.40 mi/64 km SSE of KAMLOOPS; flows generally W, through Nicola

Lake (14 mi/23 km long), to MERRITT, then NW to THOMPSON River just ENE of SPENCES BRIDGE; 50°26′N 121°19′W.

Nicolás Bravo (nee-ko-LAHS-BRAH-vo), town, PUEBLA, central MEXICO, 10 mi/16 km N of TEHUACÁN. Corn, sugar, fruit; livestock. Also called San Felipe Maderas.

Nicolas Channel, CUBA: see NICHOLAS CHANNEL.

Nicolás Flores (nee-ko-LAHS FLO-res), town, N central HIDALGO, MEXICO, 8 mi/83 km N of PACHUCA DE SOTO; 20°46′N 99°07′W. Agricultural area produces wheat and corn. No all-weather roads.

Nicolás Romero, MEXICO: see VILLA NICOLÁS ROMERO.

Nicolás Ruiz (nee-ko-LAHS ROO-ees), town, CHIAPAS, S MEXICO, at SW foot of Sierra de HUEYTEPEC, 23 mi/37 km SSE of SAN CRISTÓBAL DE LAS CASAS; 16°24′N 92°32′W. Isolated Tzotzil Maya community. Fruit; livestock.

Nicolás Suárez (nee-ko-LAHS SWAH-rez), province (□ 3,791 sq mi/9,856.6 sq km), PANDO department, NW BOLIVIA; ⊙ PORVENIR; 11°14′S 68°42′W. Elevation 715 ft/218 m. Agriculture (rubber, cacao, coca, yucca, bananas, coffee, tobacco, peanuts); cattle and horse raising.

Nicolet (nee-ko-LAI), former county (□ 626 sq mi/1,627.6 sq km), S QUEBEC, E Canada, on the SAINT LAWRENCE RIVER; county seat was BÉCANCOUR; 46°15′N 72°15′W.

Nicolet (nee-ko-LAI), town (□ 37 sq mi/96.2 sq km), ⊙ Nicolet-Yamaska county, Centre-du-Québec region, S QUEBEC, E Canada, on NICOLET RIVER, near its mouth on the SAINT LAWRENCE RIVER, 9 mi/14 km SSW of TROIS-RIVIÈRES; 46°13′N 72°36′W. Textile knitting, manufacturing of optical equipment, hosiery, furniture; in dairying region. Seat of Roman Catholic bishop; site of cathedral and seminary. Restructured in 2000.

Nicolet, Lake, E UPPER PENINSULA, MICHIGAN, an expansion of SAINT MARYS RIVER, 4 mi/6.4 km SE of SAULT SAINTE MARIE and W of SUGAR ISLAND; 46°25′N 84°14′W. Lake is c.13 mi/21 km long, 2 mi/3.2 km wide.

Nicolet River (nee-ko-LAI), 100 mi/161 km long, S QUEBEC, E Canada; flows in a winding course generally NW, past NICOLET, to Lake SAINT PETER; 46°15′N 72°39′W.

Nicolet-Sud (nee-ko-LAI–SOOD), former village, S QUEBEC, E Canada; 46°13′N 72°35′W. Amalgamated into NICOLET in 2000.

Nicolet-Yamaska (nee-ko-LAI–yuh-MA-skuh), county (□ 387 sq mi/1,006.2 sq km; 2006 population 22,915), Centre-du-Québec region, S QUEBEC, E Canada; ⊙ NICOLET; 46°04′N 72°50′W. Agriculture (corn, alfalfa, soy, barley, oats). Composed of 16 municipalities. Formed in 1982.

Nicollet (NI-kuh-let), county (□ 466 sq mi/1,211.6 sq km; 2006 population 31,313), S MINNESOTA; ⊙ ST. PETER; 44°20′N 94°15′W. Triangular-shaped county, bounded SW and E by MINNESOTA RIVER. Agricultural area (corn, oats, alfalfa, hay, soybeans, peas; hogs, sheep, cattle, poultry; dairying). SWAN and Middle lakes near center of county. Part of Fort Ridgely State Park in far NW corner. Formed 1853.

Nicollet (NI-kuh-let), town (2000 population 889), NICOLLET county, S MINNESOTA, 11 mi/18 km NW of MANKATO, on small stream which flows S out of Middle Lake to Nicollet Creek; 44°16′N 94°11′W. Dairying; livestock; grain, soybeans, beans; manufacturing (boating equipment, wood products). SWAN LAKE to NW, MINNESOTA RIVER to S.

Nicolosi (nee-ko-LO-zee), village, CATANIA province, E SICILY, ITALY, on S slope of Mount ETNA, 8 mi/13 km NNW of CATANIA; 37°37′N 15°01′E. Point of ascent for Mount Etna.

Nicoma Park (nuh-KO-muh), town (2006 population 2,377), OKLAHOMA county, central OKLAHOMA, residential suburb 10 mi/16 km E of OKLAHOMA CITY; 35°29′N 97°19′W. In oil-producing area.

Nicomedia, ancient city, NW ASIA MINOR, near the BOSPORUS, in present-day TURKEY; 40°47′N 29°55′E. Refounded (264 B.C.E.) by Nicomedes I of BITHYNIA to replace Astacus as his capital, it flourished for centuries. The Goths sacked the city in C.E. 258. Diocletian chose it for the E imperial capital, but it was soon superseded by BYZANTIUM (CONSTANTINOPLE). The modern city on its site is IZMIT.

Nico Pérez, URUGUAY: see JOSÉ BATLLE Y ORDÓÑEZ.

Nicopolis, ancient city in PONTUS, ASIA MINOR, on Lycus (modern KELKIT) River, 75 mi/121 km NE of modern SIVAS, TURKEY. Here Pompey defeated Mithridates, king of Pontus, in 66 B.C.E. Later it became Roman capital of Lesser Armenia.

Nicopolis, Bulgaria: see NIKOPOL.

Nicopolis, ISRAEL: see EMMAUS.

Nicopolis ad Istrum, Bulgaria: see NIKYUP.

Nicopolis ad Nestos, Bulgaria: see GOTSE DELCHEV.

Nicoreşti (nee-ko-RESHT), village, GALAŢI county, E ROMANIA, 5 mi/8 km NW of TECUCI. Wine growing.

Nicosia (nik-uh-SEE-yah), Greek *Lefkosia* or *Leukosia* (both: lef-kuh-SEE-yah), city (2001 population 191,549), ⊙ CYPRUS and LEFKOSIA district; 35°10′N 33°22′E. Located on N central CYPRUS, on PEDHIEOS RIVER, in Mesaoria lowland. Since Turkish occupation of N CYPRUS in 1974, Attila Line divides the city, the N sector (Lefkoşa) being Turkish, the remainder (NICOSIA) being Greek. Lefkoşa is capital (but as yet unrecognized) Turkish Republic of North Cyprus. In addition to being national capital, the two halves of the city also serve as the capital of the Greek LEFKOSIA and Turkish Lefkoşa districts. Both parts of the city are manufacturing centers (tobacco products, machinery, textiles, crafts, plastics, construction materials, chemicals, and food processing). Known as Ledra or LEDRAE in antiquity, the city was the residence of the Lusignan kings of CYPRUS from 1192, became a Venetian possession in 1489, and fell to the Turks in 1571. The walled city in the center, built by the Venetians, is divided in ½ E-W by the demarcation line. Notable buildings in the Greek sector include the Archbishop's Palace, churches, and several notable museums and art galleries. Places of interest in the Turkish sector include an art museum (formerly a monastery), a mosque, and a library. NICOSIA is the seat of the University of Cyprus, Cyprus College, and several other schools. Athalassa National Forest Park is 4 mi/6.4 km to SSE; Nicosia International Airport is 6 mi/9.7 km to W; Ercan International Airport (serving the Turkish sector only) is 8 mi/12.9 km to E.

Nicosia (nee-ko-ZEE-ah), town, ENNA province, N central SICILY, ITALY, near SALSO River, 14 mi/23 km NNE of ENNA; 37°45′N 14°24′E. In livestock-raising and cereal-growing region. Elevation 2,340 ft/713 m. Rock salt and sulphur mines nearby. Bishopric. Has 14th-century cathedral, ruins of Norman castle. Scene of heavy fighting (1943) in World War II.

Nicotera (nee-KO-te-rah), town, CATANZARO province, CALABRIA, S ITALY, near GULF OF GIOIA, 13 mi/21 km SW of VIBO VALENTIA; 38°33′N 15°56′E. Fishing center; wine, olive oil. Bishopric since 8th century. Has baroque cathedral and 11th-century castle (rebuilt 18th century). Just S is bathing resort of Nicotera Marina.

Nicoya (nee-KO-yah), city (2000 population 13,334), ⊙ Nicoya canton (2000 population 13,334), GUANACASTE province, NW COSTA RICA, 33 mi/53 km S of LIBERIA; 10°05′N 85°30′W. Agriculture (coffee, plantains) and livestock center. Connected by paved road with Liberia. Has nineteenth century colonial-style church. One of oldest cities of Costa Rica.

Nicoya, Gulf of (nee-KO-yah), inlet of the PACIFIC OCEAN, CENTRAL AMERICA, between the NICOYA PENINSULA and the NW mainland of COSTA RICA. The catch from the fine fishing in the gulf is canned at PUNTARENAS. The village of NICOYA on the peninsula

was probably the first Spanish settlement (c.1530) in Costa Rica.

Nicoya Peninsula (nee-KO-yah), GUANACASTE and PUNTARENAS provinces, W COSTA RICA, separating the PACIFIC OCEAN from the GULF OF NICOYA. About 75 mi/121 km long and 19 mi/31 km–37 mi/60 km wide, it is Costa Rica's largest peninsula. The Pacific Coast possesses many fine beaches and is a popular resort area, annexed voluntarily from NICARAGUA in 1824. The town of NICOYA, probably the first Spanish settlement (c.1530) in Costa Rica, is a transportation and commercial center for the peninsula. The peninsula is linked to the mainland by highway and by ferry.

Nictheroy, Brazil: see NITERÓI.

Nida (NEE-duh), village, SOHAG province, central Upper EGYPT, on E bank of the NILE River, 4 mi/6.4 km NNW of AKHMIM. Cotton, cereals, dates, sugarcane.

Nida (NEE-dah), German *Nidden*, seaside resort, W LITHUANIA, on COURLAND SPIT, 28 mi/45 km SSW of Klaipėda; 55°18′N 21°01′E. In Kursiu Nerijos National Park, in MEMEL TERRITORY, 1920–1939. Old Nida was first inhabited in the late 14th century but was covered by sand dunes and moved N to current location. Once attracted the Russian elite, now a summer vacation spot for wealthy Lithuanians and German tourists.

Nidadavole (ni-DUH-duh-vol), town, WEST GODAVARI district, ANDHRA PRADESH state, SE INDIA, in GODAVARI RIVER delta, 9 mi/14.5 km SW of RAJAHMUNDRY; 16°55′N 81°40′E. Railroad junction; rice milling; oilseeds, tobacco, sugarcane. Also spelled Nidadavolu, Nidadavol.

Nidamangalam, INDIA: see MANNARGUDI.

Nida River (NEE-dah), c.90 mi/145 km long, SE POLAND; rises 6 mi/9.7 km NE of SZCZĘKOCINY; flows E and SSE, past PIŃCZÓW, to VISTULA River 13 mi/21 km SSE of BUSKO Zdrój.

Nida River, POLAND: see WKRA RIVER.

Nidaros, NORWAY: see TRONDHEIM.

Nidau (NEE-dah-ou), commune, BERN canton, W SWITZERLAND, on Aare Canal, on NE shore of LAKE BIEL, adjoining BIEL; 47°07′N 07°14′E. Watches, metal products. Has 14th-century church, old castle.

Nidau (NEE-dah-ou), district, NW BERN canton, SWITZERLAND; 47°07′N 07°14′E. Main town is NIDAU; population is German-speaking and Protestant.

Nidda (NID-dah), town, central HESSE, GERMANY, on the NIDDA, 12 mi/19 km ENE of FRIEDBERG; 50°25′N 09°01′E. Lumber milling. Has 17th-century church and castle.

Nidda River (NID-dah), 61 mi/98 km long, HESSE, W GERMANY; rises 3 mi/4.8 km NNE of SCHOTTEN; flows generally SW to the MAIN, just W of HÖCHST.

Niddatal (NID-dah-tahl), town, HESSE, central GERMANY, on the NIDDA, 5 mi/8 km SE of FRIEDBERG; 50°18′N 08°48′E. Agriculture (grain, fruit, vegetables). Romanesque basilica (1159), monastery (18th-century), castle. Formed 1970 by unification of Assenheim, Bönstadt, Ilbenstadt, and Kaichen.

Nidden, LITHUANIA: see NIDA.

Nidderau (NID-der-ou), residential town, HESSE, central GERMANY, near the NIDDA, 11 mi/18 km ENE of BAD HOMBURG; 50°16′N 08°51′E. Machinery. Formed in 1970 by unification of Windecken and Heldenbergen; later incorporated OSTHEIM, Erbstadt, and Eichen.

Nidd River (NID), 50 mi/80 km long, NORTH YORKSHIRE, N ENGLAND; rises on GREAT WHERNSIDE; flows SE past KNARESBOROUGH to OUSE RIVER 7 mi/11.3 km NW of YORK. Valley of upper course is the Nidderdale.

Nideggen (NEE-deg-gen), town, RHINELAND, North Rhine-Westphalia, W GERMANY, near the ROER, 7 mi/11.3 km S of DÜREN; 50°41′N 06°30′E. Summer resort. Has ruined 12th-century castle. Chartered 1313.

Nidelva (NEED-el-vah), river, 100 mi/161 km long, largely in central NORWAY; rises in JÄMTLAND county

(SWEDEN), 38 mi/61 km NE of Røros, NORWAY; flows WNW through the TYDAL to SELBU Lake, then, as the NIDELVA, flows to TRONDHEIMSFJORDEN at TRONDHEIM. Forms several falls, notably HYTTEFOSSEN and LEIRFOSSEN.

Nidelva, river, 60 mi/97 km long, in S NORWAY; rises in NISSER Lake in TELEMARK county; flows S into AUSTAGDER county to the SKAGERRAK at ARENDAL. Several falls furnish hydroelectric power. Railroad follows the river.

Nidha Plain (NEE-[th]ah), plateau (□ 1 sq mi/2.6 sq km), IRÁKLION prefecture, central CRETE department, GREECE, on E slopes of Mount Psiloriti, and below IDHEAN Cave. Elev. 4,560 ft/1,390 m. Mainly pasture for sheep.

Nidhauli (ni-DOU-lee), town, ETAH district, W UTTAR PRADESH state, N central INDIA, on UPPER GANGA CANAL, 11 mi/18 km W of ETAH. Wheat, pearl millet, barley, corn, oilseeds.

Nidwalden, SWITZERLAND: see UNTERWALDEN.

Nidze (NEED-zhe), Serbo-Croatian *Nidže*, Greek *Voras*, mountain massif on Macedonian-Greek border. The name of its highest peak, the Kajmakcalan (8,280 ft/2,524 m), 15 mi/24 km NW of EDHESSA, Greece, is sometimes applied to the entire massif. Also spelled Nidzhe.

Nidzica (nee-DEE-tsah), German *Neidenburg*, town, OLSZTYN province, NE POLAND, on WKRA RIVER, and 30 mi/48 km S of OLSZTYN (Allenstein); 53°22′N 20°26′E. Grain and cattle market; sawmilling. Teutonic Knights established castle, 1382. Historian Gregorovius born here. In 1914, town heavily damaged in battle with Russians. In World War II, c.90% destroyed.

Niebla (NYAI-vlah), town, HUELVA province, SW SPAIN, on the RÍO TINTO, on railroad, and 17 mi/27 km ENE of HUELVA; 37°22′N 06°41′W. Cereals, grapes, vegetables, olives, timber; goats, sheep. Has castle and medieval walls.

Niebüll (NEE-buel), village, SCHLESWIG-HOLSTEIN, NW GERMANY, 23 mi/37 km W of FLENSBURG, in NORTH FRIESLAND; 54°47′N 08°50′E. Railroad junction. Market center for cattle region. Town since 1960.

Niedenstein (NEE-den-shtein), town, HESSE, W GERMANY, 9 mi/14.5 km SW of KASSEL; 51°14′N 09°18′E. Lumber.

Niederaltaich (nee-der-AHL-teikh), village, LOWER BAVARIA, GERMANY, on the DANUBE, and 6 mi/9.7 km SSE of DEGGENDORF; 48°46′N 13°01′E. Has Benedictine monastery founded c.741.

Niederanven (NEED-uhr-ahn-ven), commune (2001 population 5,430), central LUXEMBOURG, 5 mi/8 km NE of LUXEMBOURG city; 49°39′N 06°16′E.

Niederaula (nee-der-OU-lah), village, HESSE, central GERMANY, on the FULDA, 6 mi/9.7 km SW of BAD HERSFELD; 50°48′N 09°36′E. Forestry; manufacturing (clothes).

Niederbayern, GERMANY: see LOWER BAVARIA.

Niederbipp (NEE-duhr-beep), residential commune, BERN canton, NW SWITZERLAND, 8 mi/12.9 km NE of SOLOTHURN; 47°16′N 07°41′E. Commune of Oberbipp is nearby.

Niederbronn-les-Bains (NEE-der-bruhn-lai-ban), spa (□ 12 sq mi/31.2 sq km), BAS-RHIN department, ALSACE region, E FRANCE, in the N VOSGES MOUNTAINS, 12 mi/19 km NW of HAGUENAU; 48°58′N 07°39′E. Thermal establishment with mineral springs. Its metallurgical industry dates from 1860. Founded by Romans (48 B.C.E.), it was destroyed in 5th century, rebuilt in 16th century, and flourished in 1860s under Napoleon III. Niederbronn lies within the regional park of the N Vosges (VOSGES DU NORD NATURAL REGIONAL PARK) established 1976. The area is densely forested.

Niedercorn (NEE-duhr-KORN), German *Niederkorn*, town, DIFFERDANGE commune, SW LUXEMBOURG, 5 mi/8 km WNW of ESCH-SUR-ALZETTE; 49°32′N

05°54′E. Iron-mining center. Sometimes spelled Niederkorn.

Niedereinsiedel, CZECH REPUBLIC: see DOLNI POUSTEVNA.

Niedereschach (nee-der-ESH-ahkh), village, BLACK FOREST/BAAR-HEUBERG, BADEN-WÜRTTEMBERG, S GERMANY, in BLACK FOREST, 5 mi/8 km SW of ROTTWEIL; 48°08′N 08°32′E. Forestry; manufacturing (electronic equipment).

Niedere Tauern (NEE-de-re TOU-ern), range of Eastern Alps, central AUSTRIA, between Murtörl, SE SALZBURG, and Schaberpass in STYRIA, extending 65 mi/105 km ENE from headwaters of ENNS and MUR rivers, between their valleys; 47°03′N 13°06′E. Rises to 8,726 ft/2,660 m in the HOCHGOLLING mountain. Forested, with many small lakes and high peaks. Pastures on S slopes. Subranges are (W-E): the Radstadt Tauern, Schladming Tauern, Wölz Tauern, Rottenmann Tauern, Trieben Tauern, and Seckau Alps. The Niedere Tauern are crossed by the Tauern motorway (SALZBURG-VILLACH) and three passes: the Radstadt Tauern Pass (5,300 ft/1,615 m), the Sölk Tauern Pass (5,456 ft/1,663 m), and the Trieben Taurern Pass (3,883 ft/1,184 m). There is some mining at Hohentauern (magnesite) and TRIEBEN (graphite) and hydropower stations at Sölk and Zederhaus. The W part or the range has some prominent winter sports sites. Popular for hiking in the summer.

Niederfinow (nee-der-FIN-ou), village, BRANDENBURG, E GERMANY, on FINOW CANAL, 4 mi/6.4 km E of EBERSWALDE; 52°50′N 13°56′E. Ship elevator at nearby ODER-HAVEL CANAL (118 ft/36 m level difference), completed 1934.

Niederfischbach (nee-der-FISH-bahkh), village, RHINELAND-PALATINATE, W GERMANY, 7 mi/11.3 km W of SIEGEN; 50°51′N 07°53′E. Hilly region. Agriculture.

Niedergösgen (NEE-duhr-GES-guhn), commune, SOLOTHURN canton, N SWITZERLAND, on AARE RIVER, and 5 mi/8 km ENE of OLTEN; 47°22′N 07°58′E. Hydroelectric plant; shoes, rubber goods. With Obergösgen (W), it is known as Gösgen.

Niederhasli (NEE-dur-hahs-lee), commune, ZÜRICH canton, N SWITZERLAND, 7 mi/11.3 km NNW of ZÜRICH; 47°29′N 08°29′E. Racetrack.

Nieder-Ingelheim, GERMANY: see INGELHEIM AM RHEIN.

Niederkassel (NEE-der-kahs-sel), town, RHINELAND, North Rhine-Westphalia, W GERMANY, on right bank of the RHINE, 9 mi/14.5 km SSE of COLOGNE; 50°49′N 07°02′E. Machinery; dockyard, building industry; vegetables. International airport 4 mi/6.4 km NNE.

Niederkerschen, LUXEMBOURG: see BASCHARAGE.

Niederkorn, LUXEMBOURG: see NIEDERCORN.

Niederkrüchten (nee-der-KRUKH-ten), town, RHINELAND, North Rhine-Westphalia, W GERMANY, on the Schwalm, and near Dutch border, 10 mi/16 km W of MÖNCHENGLADBACH; 51°12′N 06°13′E. Textile industry; tourism. Late-Gothic church (15th century), military airport. Formed 1972 by unification of Elmpt (first mentioned 1203) and NIEDERKRÜCHTEN.

Niederlausitz, GERMANY and POLAND: see LUSATIA.

Nieder Lischna, CZECH REPUBLIC: see TŘINEC.

Niederlössnitz, GERMANY: see RADEBEUL.

Niedermarsberg, GERMANY: see MARSBERG.

Niedernberg (NEE-dern-berg), village, LOWER FRANCONIA, BAVARIA, S GERMANY, on left bank of the MAIN, 5 mi/8 km S of ASCHAFFENBURG; 49°56′N 09°05′E. Fruit; chemical industry.

Niedernhall (nee-dern-HAHL), town, FRANCONIA, BADEN-WÜRTTEMBERG, GERMANY, on the KOCHER, 15 mi/24 km SW of MERGENTHEIM; 48°18′N 09°37′E. Wine.

Niedernhausen (nee-dern-HOU-suhn), town, HESSE, central GERMANY, in the TAUNUS, 7 mi/9.7 km NNE of WIESBADEN; 50°10′N 08°20′E. Fruit; furniture; tourism. Was first mentioned 1233.

Nieder-Olm (NEE-der-OLM), village, RHINELAND-PALATINATE, W GERMANY, on the Selz, 7 mi/11.3 km SSW of MAINZ; 49°55′N 08°13′E. Wine, fruit; chemical industry.

Niederösterreich, AUSTRIA: see LOWER AUSTRIA.

Niederpfalz, GERMANY: see PALATINATE.

Niederrad (NEE-der-raht), SW district of FRANKFURT, HESSE, W GERMANY, on left bank of the MAIN. Racetrack.

Niedersachsen, GERMANY: see LOWER SAXONY.

Niedersachswerfen (NEE-der-sahks-VER-fuhn), village, THURINGIA, central GERMANY, at S foot of the lower HARZ, 4 mi/6.4 km NNW of NORDHAUSEN; 51°33′N 10°46′E. Gypsum quarrying.

Niederschlesien, GERMANY: see SILESIA.

Niedersimmental (NEE-duhr-ZEEM-uhn-tahl), district, S BERN canton, SWITZERLAND, in the lower SIMME valley. Main town is SPIEZ. Population is German-speaking and Protestant.

Niederstetten (NEE-der-shtet-tuhn), town, FRANCONIA, BADEN-WÜRTTEMBERG, GERMANY, 9 mi/14.5 km SE of MERGENTHEIM; 49°26′N 09°55′E. Spelt, oats. Has 16th-century castle.

Niederstotzingen (nee-der-SHTAHT-tsing-uhn), town, E WÜRTTEMBERG, BADEN-WÜRTTEMBERG, GERMANY, 10 mi/16 km SSE of HEIDENHEIM; 48°32′N 10°14′E. Grain.

Niederurnen (NEE-duhr-OOR-nuhn), commune, GLARUS canton, E central SWITZERLAND, 6 mi/9.7 km N of GLARUS, near WALENSEE and LINTH CANAL; 47°03′N 09°03′E. Textiles, cement. With commune of Oberurnen, just S, it is known as URNEN.

Niederwald (NEE-der-vahlt), mountain (1,086 ft/331 m), SW tip of the TAUNUS, W GERMANY, on a bend of the RHINE, just W of RÜDESHEIM. Vine-covered slopes. On summit, overlooking the RHINE, is the National Monument, with principal relief symbolizing the "Watch on the Rhine." Viticulture.

Niederwampach (NEE-duhr-VAHM-pahk), hamlet, WINCRANGE commune, NW LUXEMBOURG, in the ARDENNES, 5 mi/8 km NW of WILTZ, near WILTZ RIVER; 50°00′N 05°51′E. Former lead-mining center.

Niederwerrn (NEE-der-vern), village, LOWER FRANCONIA, BAVARIA, central GERMANY, 3 mi/4.8 km W of SCHWEINFURT; 50°04′N 10°11′E. Agricultural products; machinery.

Niederwiltz, LUXEMBOURG: see WILTZ.

Niederzier (NEE-der-tsir), town, RHINELAND, North Rhine-Westphalia, W GERMANY, on small Elle-Bach, 18 mi/29 km ENE of AACHEN; 50°53′N 06°28′E. Grain, sugar beets; forestry; lignite mining. Castle (13th century). Village first mentioned 922.

Niefang (NEE-fahng) or **Sevilla de Niefang**, town, CENTRO SUR province, continental EQUATORIAL GUINEA, on MBINI River, and 35 mi/56 km E of BATA; 01°53′N 10°16′E. Coffee, cacao.

Niefern-Öschelbronn (NEE-fern-OE-shel-bruhn), town, N BLACK FOREST, BADEN-WÜRTTEMBERG, GERMANY, on N slope of BLACK FOREST, on the ENZ, 4 mi/6.4 km NE of PFORZHEIM; 48°55′N 08°48′E. Metalworking, paper milling. Village first mentioned 1082. Formed 1971 by the union of Niefern and Öschelbronn.

Niegocin, Lake (nee-GO-tseen), German *Löwentin*, one of Masurian Lakes (□ 10 sq mi/26 sq km), EAST PRUSSIA, after 1945 in NE POLAND, S of GIZYCKO; 6 mi/9.7 km long, 1 mi/1.6 km–4 mi/6.4 km wide; 54°00′N 21°47′E. Drains N into Lake MAMRY.

Nieheim (NEE-heim), town, WESTPHALIA-LIPPE, North Rhine-Westphalia, NW GERMANY, 16 mi/26 km ENE of PADERBORN; 51°49′N 09°07′E. Grain.

Niel (NEEL), commune (2006 population 8,833), Antwerp district, ANTWERPEN province, N BELGIUM, near junction of RUPEL RIVER and SCHELDT RIVER, 9 mi/14.5 km SSW of ANTWERP; 51°07′N 04°20′E. Manufacturing.

Cross-references are shown in SMALL CAPITALS. The pronunciation guide is shown on page xix. The sources of population figures are shown on page xvii.

Niélé (nee-ai-LAI), town, SAVANES region, N CÔTE D'IVOIRE, 36 mi/58 km NW of OUANGOLODOUGOU; 10°12′N 05°38′W. Agriculture (sorghum, rice, millet, maize, beans, peanuts, tobacco, cotton). Also spelled Niéllé.

Nielsville (NEELZ-vil), village (2000 population 91), POLK county, NW MINNESOTA, 19 mi/31 km SW of CROOKSTON, near RED RIVER; 47°31′N 96°49′W. Grain.

Niemba (NYEM-bah), village, KATANGA province, E CONGO, on railroad, and 50 mi/80 km W of KALEMIE; 05°57′S 28°26′E. Elev. 2,165 ft/659 m. Trading post; cotton.

Niemcza (NYEM-chah), German *Nimptsch*, town in LOWER SILESIA, Wałbrzych province, SW POLAND, 8 mi/12.9 km E of DZIERŻONIÓW (Reichenbach); 50°43′N 16°51′E. Metalworking, malt processing. Has 16th century castle. First mentioned 11th century.

Niemen, BELARUS: see NEMAN.

Niemes, CZECH REPUBLIC: see MIMON.

Niemirów, UKRAINE: see NEMYRIV, Vinnytsya oblast; or NEMYRIV, L'viv oblast.

Niemodlin (nee-MOD-leen), German *Falkenberg*, town, OPOLE province, SW POLAND, 14 mi/23 km W of OPOLE (Oppeln); 50°39′N 17°37′E. Agricultural market (grain, sugar beets, potatoes; livestock). Has 16th century palace.

Niemtschitz an der Hanna, CZECH REPUBLIC: see NEMCICE NAD HANOU.

Niéna (NEE-ai-na), town, THIRD REGION/SIKASSO, MALI, 47 mi/78 km W of SIKASSO; 11°26′N 06°21′W. Agriculture and livestock.

Nienburg/Saale (NEEN-burg/SAH-le), town, SAXONY-ANHALT, central GERMANY, on the SAALE, at mouth of the BODE, and 4 mi/6.4 km NNE of BERNBURG; 51°50′N 11°45′E. In potash-mining region; manufacturing (paper bags, cement). Some agriculture. Has early-Gothic church; remains of former Benedictine monastery. Town first mentioned 961; chartered 1233.

Nienburg (Weser) (NEEN-burg, VEH-ser), city, LOWER SAXONY, W GERMANY, on right bank of the WESER, 28 mi/45 km NW of HANOVER; 52°39′N 09°13′E. Manufacturing of chemicals (phosphates, vulcanizing agents), metalworking. Glassworks. Has Gothic church (1491). Chartered 1025.

Nienhagen (NEEN-hah-gen), village, LOWER SAXONY, N GERMANY, 5 mi/8 km S of CELLE; 52°34′N 10°06′E. Oil wells.

Nienstadt (NEEN-shtaht), village, LOWER SAXONY, GERMANY, on N slope of the Weserbergland, 3 mi/4.8 km SSW of STADTHAGEN; 52°17′N 09°11′E. Agricultural products. Sometimes called Nienstädt bei Stadthagen.

Niepolomice (nye-po-lo-MEE-tse), Polish *Niepołomice*, town, Kraków province, S POLAND, near the VISTULA, 12 mi/19 km E of CRACOW; 50°02′N 20°14′E. Railroad branch terminus. Steam-power mill, tile kiln. Niepołomice Forest, Polish *Niepołomska Puszcza* or *Niepołomicka Puszcza*, lies between NIEPOŁOMICE and RABA RIVER (E).

Niepos (nee-AI-pos), town, CAJAMARCA region, NW PERU, on W slopes of CORDILLERA OCCIDENTAL, 100 mi/161 km WNW of SAN MIGUEL DE PALLAQUES; 06°56′S 79°08′W. Elevation 7,985 ft/2,434 m. Sugarcane, coffee, tobacco, potatoes, cereals.

Nieppe (NYEP), town (□ 6 sq mi/15.6 sq km), NORD department, NORD PAS-DE-CALAIS region, N FRANCE, near Belgian border, 2 mi/3.2 km NW of ARMENTIÈRES; 50°42′N 02°50′E. Textile weaving and bleaching.

Nieppe Forest, FRANCE: see MERVILLE.

Nieriko, SENEGAL: see OUASSADOU.

Niers River (NIRS), c.75 mi/121 km long, W GERMANY; rises 3 mi/4.8 km S of Wickrath; flows generally NNE, past MÖNCHENGLADBACH, crosses Dutch border 5 mi/8 km NW of GOCH, continuing to the MAAS 1 mi/1.6 km NW of GENNEP.

Nierstein (NIR-shtein), village, RHINELAND-PALATINATE, W GERMANY, on left bank of the RHINE, 9 mi/14.5 km SSE of MAINZ; 49°23′N 08°20′E. Noted for wine.

Niesky (NEES-kee), town, UPPER LUSATIA, SAXONY, E central GERMANY, 12 mi/19 km NW of GÖRLITZ; 51°18′N 14°49′E. Manufacturing railroad cars, machinery, clothing). Quitzdorf dam nearby (727 million cu ft/20.6 million cu m). Founded 1742 by Moravian Brothers. Incorporated several neighboring villages in 1929; town since 1935.

Niesky, U.S. Virgin Islands: see NISKY.

Niestetal (NEES-te-tahl), town, HESSE, central GERMANY, near the FULDA, 4 mi/6.4 km E of KASSEL; 51°19′N 09°35′E.

Nieszawa (nye-SHAH-vah), Russian *Neshava*, town, Włocławek province, central POLAND, port on the VISTULA, 14 mi/23 km NNW of WŁOCŁAWEK. Flour milling, brewing. Monastery.

Niet Ban Tinh Xa (NYET BAHN TIN SAH), temple, BA RIA-VUNG TAU province, S VIETNAM, N of VUNG TAU on the W side of small mountain. One of the country's biggest Buddhist temples. Established 1971.

Niete Mountains (nei-E-tai), SE LIBERIA, 85 mi/137 km NNW of HARPER; 05°35′N 08°10′W. Rise to 2,996 ft/913 m. Form divide between CAVALLY RIVER (N) and coastal streams (Grand Cess, Nana Kru, and SINOE rivers).

Nieul-sur-Mer (nyul–syoor–mer), town (□ 4 sq mi/10.4 sq km), CHARENTE-MARITIME department, POITOU-CHARENTES region, W FRANCE, N suburb of La ROCHELLE; 46°12′N 01°10′W.

Nieuw Amsterdam, town (pop. 4,246), ⊙ COMMEWIJNE district, N SURINAME, at confluence of COMMEWIJNE and SURINAME rivers, near the ATLANTIC OCEAN, 6 mi/9.7 km NE of PARAMARIBO; 05°52′N 55°07′W. Fertile agricultural region (sugarcane, coffee, rice). Formerly a Dutch fort.

Nieuw-Amsterdam (NEE-oo–AHM-stuhr-dahm), village, DRENTHE province, NE NETHERLANDS, 6 mi/9.7 km SSW of EMMEN; 52°43′N 06°52′E. Verlengde-Hoogeveense Canal passes to N. Railroad junction. Dairying; livestock; grain, vegetables; packing machinery. Name is Dutch for "New Amsterdam."

Nieuw-Antwerpen, CONGO: see MAKANZA.

Nieuw-Beijerland (NEE-oo–BEI-uhr-lahnt), village, SOUTH HOLLAND province, SW NETHERLANDS, part of Hoekse Waard district, on the Spui (car ferry), and 10 mi/16 km SW of ROTTERDAM; 51°49′N 04°21′E. Dairying; livestock; vegetables, sugar beets. Formerly spelled Nieuw-Beyerland.

Nieuwdonk Recreation Park, near Overmere, EAST FLANDERS province, NW BELGIUM, 9 mi/14.5 km E of GHENT.

Nieuwegein (NEE-vuh-khein), suburb of UTRECHT, UTRECHT province, W central NETHERLANDS, 5 mi/8 km SSW of city center, on Merewede Canal; 52°02′N 05°05′E. LEK RIVER passes to S; AMSTERDAM-RIJN Canal passes to NE. Dairying; cattle, poultry; vegetables, nursery stock, sugar beets.

Nieuwe Maas, NETHERLANDS: see NEW MAAS RIVER.

Nieuwe Merwede, NETHERLANDS: see NEW MERWEDE RIVER.

Nieuwenhagen, village, LIMBURG province, SE NETHERLANDS, 3.5 mi/5.6 km NE of HEERLEN, and 2 mi/3.2 km W of German border.

Nieuwenhoorn (NEE-vuhn-hawrn), village, SOUTH HOLLAND province, SW NETHERLANDS, in VOORNE region, and 16 mi/26 km WSW of ROTTERDAM; 51°51′N 04°09′E. Voorns Canal passes to SE. Dairying; cattle; grain, vegetables, sugar beets.

Nieuwe-Pekela (NEE-vuh–PAI-kuh-lah), town, GRONINGEN province, NE NETHERLANDS, 5 mi/8 km SW of WINSCHOTEN, on Pekel Aa canal; 53°05′N 06°58′E. Dairying; cattle, sheep; grain, vegetables; light manufacturing.

Nieuwerkerk (NEE-vuhr-KERK), village, ZEELAND province, SW NETHERLANDS, 4 mi/6.4 km E of ZIER-IKZEE, in DUIVELAND region; 51°39′N 04°00′E. Dairying; cattle, hogs; grain, vegetables, sugar beets.

Nieuwerkerk aan den Ijssel (NEE-vuhr-KERK ahn duhn EI-suhl), village, SOUTH HOLLAND province, W NETHERLANDS, on HOLLANDSE IJSSEL RIVER, 5 mi/8 km WSW of GOUDA, and 7 mi/11.3 km ENE of ROTTERDAM; 51°58′N 04°36′E. Lowest point in Netherlands (22.11 ft/6.74 m below sea level) is near here. Dairying; cattle, poultry; vegetables, flowers, sugar beets, fruit; food processing, manufacturing of shoes. Recreational center to SE.

Nieuwerkerken (NEE-wuhr-ker-kuhn), agricultural commune (2006 population 6,663), Hasselt district, LIMBURG province, NE BELGIUM, 5 mi/8 km N of SINT-TRUIDEN; 50°55′N 04°00′E.

Nieuweschans (NEE-vuh-SKHAHNS), village, GRONINGEN province, NE NETHERLANDS, 7 mi/11.3 km ENE of WINSCHOTEN, on Westerwoldse Aa canal; 53°11′N 07°13′E. Easternmost community in Netherlands. German border 0.5 mi/0.8 km to E; Dollard bay of Eems River estuary 4 mi/6.4 km to N. Dairying; livestock; grain, vegetables.

Nieuwe Waterweg, NETHERLANDS: see NEW WATERWAY.

Nieuwkoop (NEE-oo-kawp), village, SOUTH HOLLAND province, W NETHERLANDS, on NW side of NIEUWKOOPSE PLASSEN lakes, and 17 mi/27 km SW of AMSTERDAM; 52°09′N 04°47′E. Dairying; cattle, poultry; vegetables, fruit, flowers; food processing, manufacturing of hinges.

Nieuwkoopse Plassen (NEE-oo-KAWP-suh PLAH-suhn), lakes, SOUTH HOLLAND province, W central NETHERLANDS, c.10 mi/16 km NNE of GOUDA; 52°09′N 04°46′E. Small lakes formed from peat diggings. Town of NIEUWKOOP on NW; Kromme Mijdricht Canal passes to E.

Nieuwkuijk (NEE-oo-koik), village, NORTH BRABANT province, S NETHERLANDS, 5 mi/8 km W of 's-HERTOGENBOSCH; 51°41′N 05°11′E. Recreational center to W. Dairying; cattle, hogs; grain, vegetables. Also spelled Nieuwkuyk.

Nieuwkuyk, NETHERLANDS: see NIEUWKUIJK.

Nieuw Nickerie (new ni-KIR-ri), town, ⊙ NICKERIE district, NW SURINAME, on E bank of ⊙ CORANTIJN RIVER where it enters the ATLANTIC OCEAN; 05°55′N 57°03′W. On border with GUYANA. The second-largest town in the country; population is largely Hindustani and Javanese. Bananas, rice, shrimp. Ferry connects town to SPRINGLANDS (Guyana) across the CORANTIJN RIVER Airport.

Nieuwpoort (NEE-we-port), commune (2006 population 10,848), Veurne district, WEST FLANDERS province, W BELGIUM, on the NORTH SEA at the mouth of the YSER RIVER; 51°08′N 02°45′E. A fishing port, an industrial center, and a beach resort. Nearby is resort of Nieuwpoort-Bad.

Nieuwpoort (NEE-oo-pawrt), village, SOUTH HOLLAND province, W central NETHERLANDS, 9 mi/14.5 km NNW of GORINCHEM; 51°56′N 04°52′E. LEK RIVER (car ferry) 1 mi/1.6 km to N. Natural-gas field. Dairying; cattle; grain, vegetables, fruit.

Nieuwpoort-Bad, BELGIUM: see NIEUWPOORT.

Nieuw-Vennep (NEE-oo–VE-nuhp), town, SOUTH HOLLAND province, W NETHERLANDS, 8 mi/12.9 km S of HAARLEM, in center of HAARLEMMERMEER polder, surrounded by RINGVAART Canal; 52°16′N 04°38′E. Dairying; cattle; vegetables, flowers, nursery stock; food processing.

Nieva (NYAI-vah), village, SEGOVIA province, central SPAIN, 19 mi/31 km WNW of SEGOVIA. Cereals, grapes. Lumbering.

Nieves (nee-EH-ves), town, ⊙ Francisco River Murguía municipio, ZACATECAS, N central MEXICO, on interior plateau, 14 mi/23 km NE of RÍO GRANDE, 90 mi/145 km NNW of ZACATECAS; 23°59′N 103°01′W. Elevation 6,617 ft/2,017 m. Silver mining still moderately active. Also known as Francisco Murguía.

Nieves Ixpantepec, MEXICO: see IXPANTEPEC NIEVES.

Nieves Negras Pass (nee-AI-ves NAI-grahs) (12,500 ft/ 3,810 m–13,000 ft/3,962 m), in the ANDES, on AR-GENTINA-CHILE border, at S foot of SAN JOSÉ VOL-CANO, on road between SAN CARLOS (MENDOZA province, Argentina) and EL VOLCAN (Chile); 33°51'S 69°54'W.

Nieves, Puerto de las (ni-E-ves, PWER-to dhai lahs), small bay and landing, NW Grand Canary, CANARY ISLANDS, SPAIN, 17 mi/27 km W of LAS PALMAS.

Nièvre (nye-vruh), department (□ 2,632 sq mi/6,843.2 sq km), in old NIVERNAIS province, central FRANCE; ⊙ NEVERS; 47°05'N 03°35'E. Bounded by the MORVAN upland (E), by ALLIER and LOIRE rivers (W); occupied by Nivernais Hills (center). Drained by the Loire and its tributaries (Allier, NIÈVRE), the YONNE, and the CURE. Crossed S-N by NIVERNAIS CANAL (from the Loire to the Yonne). Department is primarily agricultural, with important forest stands and extensive livestock industry. Chief crops are green vegetables and potatoes; good cereal yields. Vineyards on Loire River slopes, especially at POUILLY-SUR-LOIRE. Iron and coal mines, once very active, are no longer worked. Kaolin quarries at Fleury-sur-Loire and SAINT-PIERRE-LE-MOÛTIER. Some metalworks in operation in Nevers area and along the Loire. Ceramics (Nevers, Saint-Amand-en-Puisaye) and woodworking are still practiced, though mostly as handicraft and specialty products. Spas at POUGUES-LES-EAUX and SAINT-HONORÉ-LES-BAINS have thermal springs. The department forms part of the Bourgogne (BUR-GUNDY) administrative region.

Nièvre River (nye-vruh), c.30 mi/48 km long, NIÈVRE department, BURGUNDY region, central FRANCE; rises SW of Varzy; flows S, past GUÉRIGNY, to the LOIRE at NEVERS; 47°07'N 03°11'E.

Nif, TURKEY: see KEMALPASA.

Nifisha (ne-FEE-shah), township, SHARQIYA province, Lower EGYPT, on ISMAILIA CANAL, on CAIRO-ISMAILIA railroad, and 2 mi/3.2 km SW of Ismailia. Here the Ismailia Canal branches off (S) at the SUEZ Freshwater Canal, and the Cairo-Ismailia railroad has a branch to SUEZ.

Nigde (nee-DAI), Turkish *Niğde*, town (2000 population 78,088), S central TURKEY, on ADANA-KAYSERI railroad, 75 mi/121 km NNW of Adana; 37°58'N 34°42'E. RISES to 4,100 ft/1,250 m. TILE MAKING; RYE, WHEAT, BARLEY, LEGUMES. SOME OF ITS BUILDINGS, SUCH AS THE OCTAGONAL HUDAVEND HATUN MAU-SOLEUM (1312), DATE FROM MIDDLE AGES. NIGDE'S ARCHAEOLOGICAL MUSEUM, AK MADRESE, IS HOUSED IN A FORMER 15TH-CENTURY RELIGIOUS SCHOOL. WAS AN IMPORTANT CITY OF THE SELJUK TURKS; FELL TO OTTOMAN TURKS IN 15TH CENTURY.

Nigel, town, GAUTENG province, SOUTH AFRICA, 30 mi/ 48 km ESE of JOHANNESBURG; 26°27'S 28°29'E. Elevation 5,478 ft/1,670 m. Gold-mining center. Increased gold mining in region has resulted in rapid growth of town. Airfield. Founded in 1909.

Niger (NEE-zher), republic (□ 489,189 sq mi/1,267,000 sq km; 2004 estimated population 11,360,538; 2007 estimated population 12,894,865), W AFRICA; ⊙ NIA-MEY; 16°00'N 10°00'E.

Geography

Borders on BURKINA FASO and MALI in the W, on ALGERIA and LIBYA in the N, on CHAD in the E, and on NIGERIA and BENIN in the S. Major cities include MARADI, TAHOUA, and ZINDER. Administratively, the country is divided into the capital city and eight regions: AGADEZ, DIFFA, DOSSO, MARADI, TAHOUA, TILLABÉRY, and ZINDER. Niger is extremely arid except along the NIGER RIVER in the SW and near the border with Nigeria in the S, where there are strips of savanna. Most of the rest of the country is either semidesert or part of the SAHARA. Located in N central Niger is the AÏR MASSIF (average elevation 3,000 ft/914 m; maximum elevation 6,627 ft/2,020 m), which

receive slightly more rainfall than the surrounding desert.

Population

The main ethnic groups are the Hausa, the Kanuri, the Songhai and Djerma (Zarma), the Fulani, and the Tuareg. The great majority of the population (80%) is rural and lives in the S. There is a significant migration of seasonal labor to Ghana, Nigeria, and Chad. The majority of the population is Muslim; less than 1% practice Christianity. The country's official language is French, and several indigenous languages as well as Arabic are also spoken.

Economy

The economy of Niger is overwhelmingly agricultural, with about 90% of the workforce engaged in farming (largely of a subsistence type). The Hausa, Kanuri, and Songhai are mainly sedentary farmers, and the Fulani and Tuareg are principally nomadic and seminomadic pastoralists. The leading crops are millet, sorghum, cassava, cowpeas, groundnuts, rice, cotton, sugarcane, and dates (grown in oases in the desert). Large numbers of poultry, goats, cattle, sheep, and camels are raised, but drought conditions have reduced the production of livestock. Most of the country's limited manufacture is confined to consumer goods such as processed food, beverages, footwear, and radios. In addition, groundnut oil, ginned cotton, and construction materials (mainly brick and cement) are produced. In the early 1970s large high-grade uranium ore deposits were discovered at ARLIT in the Aïr Massif and a uranium-ore concentrating plant was opened. Small quantities of cassiterite (tin ore), low-grade iron ore, gypsum, phosphates, coal, natron, and salt also are extracted in the country. There is a small but growing fishing industry, operating mainly in the Niger River and in LAKE CHAD (in the SE). Niger has a very limited transportation network; there is no railroad, and most of the country's all-weather roads are confined to the S and SW. A major road also runs N from Zinder, through AGADEZ, and on into ALGERIA. Niger is landlocked and has only poor access to the sea. The annual cost of Niger's imports (2002 estimate $400 million) usually is considerably higher than the value of its exports (2002 estimate $280 million). The leading imports are textiles and clothing, machinery, foodstuffs, motor vehicles, and petroleum products; the chief exports are groundnuts and groundnut products, live cattle, cotton, and uranium. The principal trade partners are European nations (especially FRANCE), Nigeria, and JAPAN. Niger is an ACP (African, Caribbean, and Pacific) member of the EU.

History: to 1926

Numerous Neolithic remains of early pastoralism have been found in the desert areas of Niger. Ptolemy wrote of Roman expeditions to the Aïr mountains (massif). In the 11th century, Tuareg migrated from the desert to the Aïr region, where they later established a state centered at Agadez, which was situated on a major trans-Saharan caravan route that connected N Africa with present-day N Nigeria. In the 14th century, the Hausa founded several city-states in S Niger. For most of the 16th century much of W and central Niger was part of the Songhai empire (centered at GAO on the Niger River in present-day Mali). In the early 19th century, the Fulani gained control of S Niger as a result of a holy war waged against the Hausa. The Conference of Berlin (1884–1885) placed the territory of Niger within the French sphere of influence, but colonization was delayed by Tuareg resistance. In 1900, Niger was made a military territory within UPPER SENEGAL–NIGER, and in 1922 it was constituted a separate colony within FRENCH WEST AFRICA.

History: 1926 to 1980

Niamey became the capital in 1926, replacing Zinder. The French generally governed through existing political structures and did not alter substantially the

institutions of the country; they undertook little economic development and provided few new educational opportunities. National political activity began in 1946, when Niger received its own assembly under the constitution establishing the FRENCH UNION. The first important political organization was the Niger Progressive Party (PPN), and Hamani Diori, leader of the PPN, became the country's first president after Niger gained full independence in August 1960; he was reelected president in 1965 and 1970. Despite its weak economy, occasional ethnic conflicts (especially between the Tuareg and the government), and sporadic campaigns of warfare waged by outlawed leftist rebels (who were later granted amnesty by Diori), Niger enjoyed political stability in the early years of independence. Close ties were retained with France, which gave Niger considerable aid. The country was severely affected by the Sahelian drought (1968–1974); much of its livestock died and crop production fell drastically. In 1974, Diori was supplanted by Lt. Colonel Seyni Kountché in a military coup.

History: 1980 to Present

The uranium boom of the early 1980s produced disparities of wealth that led to civil unrest. Kountché died in 1987 and was succeeded by General Ali Saïbou as head of state. In 1991, a 1,204-member national conference suspended the constitution and dissolved the government. A transitional government ruled until Mahamane Ousmane was elected the new civilian president of Niger in March 1993. Niger's severe economic condition and political rivalries led to another military coup in January 1996. A National Salvation Council was established with the stated aim of returning Niger to democratic civilian rule. Conflict between the government and the Tuareg subsided with the signing of a peace accord in April 1995. Some Tuareg attacks, however, continued into the 21st cent. A coup in 1999 was followed by the creation of a National Reconciliation Council that effected a transition to civilian rule. In November 1999 a retired colonel, Mamadou Tandja, was elected president of Niger; he was reelected in December 2004.

Government

Niger is governed under the constitution of 1999. The president, who is the head of state, is popularly elected for a five-year term and is eligible for a second term. The prime minister is the head of government and is appointed by the president. The unicameral National Assembly has 113 members who are popularly elected for five-year terms. Since December 1999 President Mamadou Tandja has been head of state. Prime Minister Seyni Oumarou has been head of government since June 2007.

Niger (NEI-juhr), former province, one of NORTHERN PROVINCES, W central NIGERIA; capital was Minna. Now Niger state.

Niger (NEI-juhr), state (□ 29,483 sq mi/76,655.8 sq km; 2006 population 3,950,249), W central NIGERIA; ⊙ Minna; 10°00'N 06°00'E. Bordered N by KEBBI and ZAMFARA states, E by KADUNA state and ABUJA FEDERAL CAPITAL TERRITORY, S by KWARA and KOGI states (NIGER River, border), and W by BENIN. In middle belt zone; drained by KADUNA and Niger rivers. Agriculture includes cotton, kenaf, sorghum, millet, maize, rice, yams, cassava, potatoes, sugarcane, fruits; livestock; fishing. Deposits of gold, marble, kaolin, glass sands, chalk, limestone, and copper. Main centers are Minna, Bono, Bida, Kontagora, Suleja, and Wawa. Gurara and Falls, Zuma rocks, and Shiroro dam are here. Nupe people live mainly in Niger state. They reached their political zenith between the 16th and 18th centuries.

Niger, great river of W AFRICA, c.2,600 mi/4,184 km long; rises on the FOUTA DJALLON plateau, SW Republic of GUINEA; flows NE through Guinea and into the MALI Republic. In central Mali the Niger forms its vast inland delta (□ c.30,000 sq mi/77,700 sq km), a maze of channels and shallow lakes. An irrigation project in the delta, begun by the French in the 1930s and including a large dam at SANSANDING (1941), has opened more than 100,000 acres/40,470 ha to farming, especially rice cultivation. Downstream from TIMBUKTU, Mali, the Niger begins a great bend, flowing first E and then SE out of Mali, through the Republic of NIGER (where it forms part of the border with BENIN), and into NIGERIA. At LOKOJA, central Nigeria, the BENUE RIVER, its chief tributary, joins the Niger, which then flows S, emptying through a great delta into the Gulf of GUINEA. The delta (□ c.14,000 sq mi/36,260 sq km)—the largest in Africa—is characterized by swamps, lagoons, and navigable channels. The region is a major source of palm oil and petroleum. Major towns in the delta are PORT HARCOURT and BONNY. Much of the Niger is seasonally navigable, and below Lokoja it is open to ships virtually all year. The Niger is a major source of fish, especially perch and tiger fish. A hydroelectric and irrigation project, centered around the Kainji dam (1968), is located on the Niger near JEBBA, Nigeria. The upper Niger region was an important part of the former empires of MALI and Monghai. The course of the Niger long puzzled European geographers; only from 1795 to 1797 did Mungo Park, the Scottish explorer, correctly establish the E flow of the upper Niger, and it was not until 1830 that Richard and John Lander, English explorers, found that the river emptied into the Gulf of Guinea. The water level of the Niger has been substantially lowered as a result of the long-term W African drought in the late 1960s, 1970s, and 1980s.

Niger Coast, region, NIGERIA; coastal area extending from Lagos Colony to E limits of the Niger Delta, over which GREAT BRITAIN declared a protectorate in 1885. Territory extended upstream to LOKOJA on the NIGER River and then E up the BENUE River to IBI.

Nigeria (nei-JIR-ree-uh), republic (□ 356,667 sq mi/923,768 sq km; 2006 population 140,003,542; 2007 estimated population 135,031,164), W AFRICA, bordering on the GULF OF GUINEA (an arm of the ATLANTIC OCEAN) in the S, BENIN in the W, NIGER in the NW and N, CHAD in the extreme NE, and CAMEROON in the E; ⊙ ABUJA.

Geography
LAGOS is the largest city; other major cities include ABA, ABEOKUTA, ADO, BENIN, EDE, ENUGU, IBADAN, IFE, ILESHA, ILORIN, IWO, KADUNA, KANO, MAIDUGURI, MUSHIN, OGBOMOSHO, ONITSHA, OSHOGBO, PORT HARCOURT, and ZARIA. Nigeria is divided into ABUJA FEDERAL CAPITAL TERRITORY and thirty-six states: ABIA, ADAMAWA, AKWA IBOM, ANAMBRA, BAUCHI, BENUE, BORNO, CROSS RIVER, DELTA, EDO, ENUGU, IMO, JIGAWA, KADUNA, KANO, KATSINA, KEBBI, KOGI, KWARA, LAGOS, NIGER, OGUN, ONDO, OSUN, OYO, PLATEAU, RIVERS, SOKOTO, TARABA, YOBE, and the following six, created in 1996 to make the total equal the number of years since the gaining of independence: BAYELSA, EBONYI, EKITI, GOMBE, NASARAWA, and ZAMFARA. The NIGER River and its tributaries (including the BENUE, KADUNA, and KEBBI rivers) drain most of the country. Nigeria has a 500-m/800-km coastline, for the most part made up of sandy beaches, behind which lies a belt of mangrove swamps and lagoons that averages 10 mi/16 km in width but increases to around 60 mi/100 km wide in the great Niger delta in the E. N of the coastal lowlands is a broad hilly region, with rain forest in the S and savanna in the N. Behind the hills is the great plateau of Nigeria (average elevation 2,000 ft/610 m), a region of plains covered largely with savanna but merging into scrubland in the N. Greater altitudes are attained on the BAUCHI and JOS plateaus in the center and in the Adamawa Massif (which continues into Cameroon) in the E, where Nigeria's highest point (around 6,700 ft/2,040 m) is located.

Population
Nigeria is easily the most populous nation in Africa and one of the fastest-growing on earth. The inhabitants of Nigeria are divided into around 250 ethnic groups, the largest of which are the Hausa and Fulani in the N, the Yoruba in the SW, and the Igbo in the SE. Other peoples include the Kanuri, Nupe, and Tiv of the N, the Edo of the S, and the Ibibio-Efik and Ijaw of the SE. English is the official language and each ethnic group speaks its own language. About half of the population, living mostly in the N, are Muslim; about one-third, living almost exclusively in the S, are Christian; and the rest follow traditional beliefs.

Economy
The economy of Nigeria is mainly agricultural, with more than half of the workforce engaged in farming (much of which is of a subsistence type). The chief crops are sorghum, millet, soybeans, groundnuts, cotton, maize, yams, rice, palm products, cacao, and rubber. In addition, large numbers of poultry, goats, sheep, and cattle are raised in the country. In times of drought, food production can fall below adequate levels and must be imported. Petroleum is the leading mineral produced in Nigeria; it is found in the Niger delta and offshore in the bights of BENIN and BIAFRA. Petroleum production on an appreciable scale began in the late 1950s, and by the early 1970s it was by far the leading earner of foreign exchange. The growing oil industry attracted many to urban centers, to the detriment of the agricultural sector. In the 1980s, revenues from petroleum exports fell as a result of a decline in world oil prices and motivated the government to bolster the agricultural sector. The low sulphur content of much of Nigeria's petroleum makes it especially desirable in a pollution-conscious world. Other minerals extracted include tin, limestone, columbite, tantalite, low-grade iron ore, and gold. Industry in Nigeria is largely confined to the processing of agricultural goods and to the manufacture of basic consumer goods, such as clothing, soap, tobacco products, and furniture. Other manufacturing includes refined petroleum, cement, and metal goods. In addition, numerous traditional craftsmen produce woven goods, pottery, metal objects, and carved wood and ivory. Nigeria's road and railroad systems are constructed basically along N-S lines; the country's chief seaports are Lagos, WARRI, Port Harcourt, and CALABAR. Nigeria has a developing fishing industry. Except when oil prices are low, Nigeria generally earns more from exports than it spends on imports. Other important exports include cacao, rubber, and palm products. The main imports are machinery, chemicals, textiles, motor vehicles, and manufactured consumer goods. The leading trade partners are the UNITED STATES, GREAT BRITAIN, GERMANY, CHINA, and ITALY.

History: to 16th Century
Little is known of the earliest history of Nigeria. By c. 2000 B.C.E. most of the country was sparsely inhabited by people who had a rudimentary knowledge of raising domesticated food plants and of herding animals. From c. 800 B.C.E to C.E. c. 200 the neolithic Nok culture flourished on the JOS Plateau; the people made fine terra-cotta sculptures and probably knew how to work tin and iron. The first important centralized state to influence Nigeria was Kanem-Bornu, which probably was founded in the 8th century, to the N of Lake CHAD (outside modern Nigeria). In the 11th century, by which time its rulers had been converted to Islam, Kanem-Bornu expanded S of Lake Chad into present-day Nigeria, and in the late 15th century its capital was moved there. Beginning in the 11th century seven independent Hausa city-states were founded in N Nigeria—Biram, Daura, Gobir, Kano, Katsina, Rano, and Zaria. Kano and Katsina competed for the lucrative trans-Saharan trade with Kanem-Bornu, and for a time had to pay tribute to it.

History: 16th Century to 1804
In the early 16th century all of Hausaland was briefly held by the Songhai Empire. However, in the late 16th century, Kanem-Bornu replaced Songhai as the leading power in N Nigeria, and the Hausa states regained their autonomy. In SW Nigeria two states—OYO and Benin—had developed by the 14th century; the rulers of both states traced their origins to IFE, renowned for its naturalistic terra-cotta and brass sculpture. BENIN was the leading state in the 15th century but began to decline in the 17th century. By the 18th century, OYO controlled Yorubaland and also Dahomey (now Benin). The Igbo people in the SE lived in small village communities. In the late 15th century, Portuguese navigators became the first Europeans to visit Nigeria. They soon began to purchase slaves and agricultural produce from coastal middlemen. The Portuguese were followed by British, French and Dutch traders. Among the Igbo and Ibibio, a number of city-states were established by people who had become wealthy by engaging in the slave trade; these included BONNY, Owome, and Okrika. There were major internal changes in Nigeria in the 19th century.

History: 1804 to 1886
In 1804, Usuman dan Fodio (1754–1817), a Fulani and a pious Muslim, began a jihad [Arabic=holy war] to reform the practice of Islam in the N. He soon conquered the Hausa city-states, but BORNU maintained its independence. In 1817, Usuman dan Fodio's son, Muhammad Bello (d.1837), established a state centered at SOKOTO, which controlled most of N Nigeria until the coming of the British (1900–1906). Muslim culture, and also trade, flourished in the Fulani empire. The Bornu Empire disintegrated during the reign of Umar (1835–1880). In the 19th century, many African middlemen turned from selling slaves to selling palm products, which soon became Nigeria's chief export. In 1817 a long series of civil wars began in the Oyo Empire; they lasted until 1893, by which time the empire had disintegrated completely. To stop the slave trade there, Britain annexed Lagos in 1861. Britain claimed S Nigeria at the Conference of Berlin (1884–1885), and in the following years, gained control over SW Nigeria as well, partly by signing treaties (as in the Lagos hinterland) and partly by using force (as at Benin in 1897).

History: 1886 to 1922
The Royal Niger Company, given (1886) a British royal charter to administer the Niger River and N Nigeria, antagonized Europeans and Africans alike by its monopoly of trade on the Niger; in addition, it was not sufficiently powerful to gain effective control over N Nigeria, which was also sought by the French. Therefore, in 1900 its charter was revoked and British forces under Frederick Lugard began to conquer the N, taking Sokoto in 1903. By 1906 Britain controlled Nigeria, which was divided into the Colony (i.e., Lagos) and Protectorate of Southern Nigeria and the Protectorate of Northern Nigeria. In 1914 the two regions were amalgamated and the Colony and Protectorate of Nigeria was established. The administration of Nigeria was based on a system called "indirect rule," where Britain ruled through existing political institutions headed by Africans rather than establishing a wholly new administrative network. All important decisions were made by the British governor, however, and the African rulers soon lost most of their traditional authority over their subjects. Under the British, railroads and roads were built and the production of cash crops, such as palm nuts and kernels, cacao, cotton, and groundnuts, was encouraged. The country became more urbanized as LAGOS, IBADAN, KANO, ONITSHA, and other cities grew in size and importance.

History: 1922 to 1960
From 1922, African representatives from Lagos and CALABAR were elected to the legislative council of

Southern Nigeria, but they were only a small, powerless minority. A small Western-educated elite developed in Lagos and a few other S cities. In 1947, Great Britain promulgated a constitution that gave the traditional authorities a greater voice in national affairs. The Western-educated elite was excluded, and, led by Herbert Macaulay and Nnamdi Azikiwe, its members vigorously denounced the constitution. As a result, a new constitution, providing for elected representation on a regional basis, was instituted in 1951. As a result, three major political parties emerged— the National Council of Nigeria and the Cameroons (NCNC; from 1960 known as the National Convention of Nigerian Citizens), led by Azikiwe and largely based among the Igbo; the Action Group, led by Obafemi Awolowo and with a mostly Yoruba membership; and the Northern People's Congress (NPC), led by Ahmadu Bello and based in the N. The constitution proved unworkable by 1952, and a new one, solidifying the division of Nigeria into three regions (Eastern, Western, and Northern) plus the Federal Territory of Lagos, came into force in 1954. In 1956 the Eastern and Western regions became internally self-governing, and the Northern region achieved this status in 1959. With Nigerian independence scheduled for 1960, elections were held in 1959; no party won a majority, and the NPC combined with the NCNC to form a government.

History: 1960 to 1967

Nigeria attained independence on Oct. 1, 1960, with Abubakar Tafawa Balewa of the NPC as prime minister and Azikiwe of the NCNC as governor-general; when Nigeria became a republic in 1963, Azikiwe was made president. The first years of independence were characterized by severe conflicts within and between regions. In the Western region, a bloc of the Action Group split off (1962) under S. I. Akintola to form the Nigerian National Democratic party (NNDP); in 1963 the Mid-Western region (whose population was mostly Edo) was formed from a part of the Western region. National elections late in 1964 were hotly contested, with an NPC-NNDP coalition (called the National Alliance) emerging victorious. In October 1965, elections in the Western region were marred by widespread irregularities. In January 1966, Igbo army officers staged a successful coup d'etat, which resulted in the deaths of Federal Prime Minister Balewa, Western Prime Minister Akintola, and Northern Prime Minister Ahmadu Bello. Major-General Johnson T. U. Aguiyi-Ironsi, an Igbo, became head of a military government and suspended the national and regional constitutions. His attempts to abolish the regions and establish a centralized government brought a violent reaction in the N. In July 1966, a coup led by Hausa army officers ousted Ironsi (who was killed) and placed Lt. Col. Yakubu Gowon at the head of a new military regime. Gowon attempted to start Nigeria along the road to civilian government but met determined resistance from the Igbo.

History: 1967 to 1974

In May 1967, the Eastern parliament gave Lt. Col. Chukwuemeka O. Ojukwu, the region's leader, authority to declare the region an independent republic. Gowon proclaimed a state of emergency, and, as a gesture to the Igbos, redivided Nigeria into twelve states (including one, the East-Central state, that comprised most of the Igbo people). However, on May 30, Ojukwu proclaimed the independent Republic of BIAFRA, and in July fighting broke out between Biafra and Nigeria. Biafra made some advances early in the war, but soon federal forces gained the initiative. After much suffering, Biafra capitulated on Jan. 15, 1970, and the secession ended. The early 1970s were marked by reconstruction in areas that were formerly part of Biafra, by the gradual reintegration of the Igbo into national life, and by a slow return to civilian rule. Spurred by the booming petroleum in-

dustry, the Nigerian economy quickly recovered from the effects of the civil war and made impressive advances, although inflation, high unemployment, a decline in the world price of groundnuts and cocoa, and drought caused problems. Nigeria joined OPEC in 1971. The prolonged drought that desiccated the SAHEL region of Africa began to have a profound effect on semiarid N Nigeria in late 1973; thousands of head of livestock died, farming suffered, and the fishing industry on Lake Chad collapsed as the lake shrank in size. As a result, there was a migration of peoples into the less arid areas and into the cities of the S. The oil boom of the early 1970s brought a great deal of revenue to Nigeria but also led to government corruption and uneven distribution of wealth, which continues to this day.

History: 1974 to 1990

In 1974 Gowon broke his promise to introduce civilian rule. His regime was overthrown in 1975 by General Murtala Muhammad and a group of officers who formed the Supreme Military Council and pledged a return to civilian rule. In the mid-1970s the capital was moved from Lagos to Abuja, a move that drained the national economy. A major dams project was executed in the 1970s, producing Kaangimi, Zaria, Kainji and Jebba dams along the Niger River, producing hydroelectricity, agricultural lands (irrigation, floodplains), and regional stability (against flooding and droughts). Muhammad was assassinated in an attempted coup one year after taking office and was succeeded by General Obasanjo. The government banned all student organizations, restricted public opposition to the regime, controlled union activity, nationalized land, and increased oil industry regulation. Nigeria sought Western support under Obasanjo while supporting African nationalist movements. In 1979, elections were held under the new constitution (1978), bringing Alhayi Shehu Shagari to the presidency. In 1983, he was reelected but subsequently overthrown after only a few months in office. In 1985 there was still another coup, this one led by Army Chief of Staff and Supreme Military Council (SMC) member Major-General Ibrahim Babangida, who became head of state and commander in chief of the Armed Forces and promoted himself to the rank of general.

History: 1990 to Present

General Babangida promised to return the country to civilian rule in 1990, later shifting the date to 1992 and again to Aug. 23, 1993. In June 1993, a presidential election was held, but the military annulled the results and instituted an interim national government, led by General Sani Abacha. Abacha proclaimed himself head of state and commander in chief of the Armed Forces. In 1998, he handed over power to an elected president, Olusegun Obasanjo, a former military officer who had ruled Nigeria from 1976 to 1979. President Obasanjo reelected in 2003, but the election was marred by vote rigging. Ethnic and economic conflict in the oil-rich Niger delta became a significant and ongoing problem in Obasanjo's second term. The dispute with Cameroon over the location of the border in the BAKASSI Peninsula was decided in Cameroon's favor by the International Court of Justice in 2002, but Nigeria did not begin turning the disputed region over until 2006. In 2005–2006 Obasanjo unsuccessfully maneuvered to secure the chance to run for a third term. The presidential and state elections of April 2007 were marked by blatant fraud and intimidation; Umaru Yar'Adua, Obasanjo's hand-picked candidate, was declared with winner with 70% of the vote. Nigeria increasingly exercises its influence as a regional power, taking the lead in sending multinational forces to stabilize situations in SIERRA LEONE and LIBERIA.

Government

Nigeria is governed under the constitution of 1999. The president, who is both head of state and head of

government, is popularly elected for a four-year term and is eligible for a second term. The bicameral legislature, the National Assembly, consists of the 109-seat Senate and a 360-seat House of Representatives; all legislators are elected by popular vote for four-year terms. The current head of state is President Umaru Musa Yar'Adua (since May 2007).

Nigg (NIG), village, EAST BERBICE–CORENTYNE district, NE GUYANA; 06°15′N 57°23′W. Located in ATLANTIC OCEAN coastland, 10 mi/16 km E of NEW AMSTERDAM. Rice-growing area.

Nigg (NIG), village, HIGHLAND, N Scotland, on CROMARTY FIRTH, 3 mi/4.8 km NNE of CROMARTY; 57°43′N 04°00′W. Oil refinery.

Nighasan (nig-HAH-suhn), town, KHERI district, N UTTAR PRADESH state, N central INDIA, 20 mi/32 km NNE of LAKHIMPUR; 28°14′N 80°52′E. Rice, wheat, gram, sugarcane.

Nighrita, Greece: see NIGRITA.

Nighthawk, village, OKANOGAN county, N WASHINGTON, port of entry at BRITISH COLUMBIA border, on OKANOGAN RIVER, and 38 mi/61 km N of OKANOGAN.

Night Hawk Lake (NEIT HAWK), E ONTARIO, E central Canada, 15 mi/24 km E of TIMMINS; 13 mi/21 km long, 6 mi/10 km wide; 48°28′N 80°58′W. Drains N into ABITIBI RIVER.

Nightingale (NEI-tuhn-gail), hamlet, S ALBERTA, W Canada, 8 mi/13 km N of STRATHMORE, in Wheatland County; 51°10′N 113°20′W.

Nightingale Islands, group of 3 uninhabited islets in S ATLANTIC, 13 mi/21 km SSW of TRISTAN DA CUNHA, with which it is a dependency of SAINT HELENA; largest island is 1 mi/1.6 km long, 0.75 mi/1.2 km wide; 37°25′S 12°30′W. It was discovered in 1760 by a British naval officer. Tiny Stoltenhoff and Middle islands are just NNW of Nightingale Island. Dependency of St. Helena since 1938.

Nightmute, village (2000 population 208), SW ALASKA, on NELSON ISLAND, 70 mi/113 km W of BETHEL; 60°27′N 164°48′W.

Nigrita (nee-YREE-tah), town, SÉRRAI prefecture, CENTRAL MACEDONIA department, NE GREECE, on road, and 13 mi/21 km S of SÉRRAI; 40°54′N 23°30′E. Trading center for grain, cotton, sesame, tobacco, corn, barley. Noted sulphur springs nearby. Also spelled Nighrita.

Nigüelas (nee-GWAI-lahs), town, GRANADA province, S SPAIN, 14 mi/23 km SSE of GRANADA; 36°59′N 03°32′W. Olive-oil processing; wine, cereals, potatoes.

Nihama (NEE-hah-mah), city, EHIME prefecture, N SHIKOKU, W JAPAN, 31 mi/50 km E of MATSUYAMA; 33°57′N 133°17′E. Commercial port and manufacturing and mining center, producing copper and chemical products. World leader in supply of resin for semiconductor manufacturing. Ruins of old Besshi copper mine.

Nihari (NEE-hah-ree), village, Nihari county, IBARAKI prefecture, central HONSHU, E central JAPAN, 25 mi/40 km S of MITO; 36°07′N 140°09′E.

Niharu (nee-HAH-roo), village, Tone county, GUMMA prefecture, central HONSHU, N central JAPAN, 25 mi/40 km N of MAEBASHI; 36°40′N 138°55′E. Carp.

Nihing River, PAKISTAN: see DASHT River.

Nihoa (nee-HO-ah), uninhabited island, HONOLULU county, HAWAII, 130 mi/209 km NW of NIIHAU, 270 mi/435 km NW of HONOLULU, 20 mi/32 km S of TROPIC OF CANCER. Highest point 910 ft/277 m; covers 155 acres/63 ha. Under jurisdiction of city and county of Honolulu. Once had small Hawaiian settlement. Most SE point of NW Hawaiian Islands. Part of Hawaiian Islands National Wildlife Refuge. Formerly Bird Island.

Nihonmatsu (nee-HON-mahts), city, FUKUSHIMA prefecture, N central HONSHU, NE JAPAN, on ABUKUMA RIVER, 12 mi/20 km S of FUKUSHIMA city; 37°35′N 140°25′E. Cucumbers. Known for Lantern and

---

Cross-references are shown in SMALL CAPITALS. The pronunciation guide is shown on page xix. The sources of population figures are shown on page xvii.

Chrysanthemum-doll festivals. Hot springs nearby. Sometimes spelled Nihonmatu.

Nihtaur (NEE-tour), town, BIJNOR district, N UTTAR PRADESH state, N central INDIA, 16 mi/26 km ESE of BIJNOR; 29°20′N 78°23′E. Road junction. Sugar refining; rice, wheat, gram, sugarcane.

Nihuil (nee-hoo-EEL), village, central MENDOZA province, ARGENTINA, on ATUEL RIVER, and 40 mi/64 km SW of SAN RAFAEL; 35°03′S 68°36′W. Irrigation dam nearby.

Niigata (NEE-gah-tah), prefecture [Japanese *ken*] (□ 4,856 sq mi/12,577 sq km; 1990 population 2,474,602), central and N HONSHU, N central JAPAN, on Sea of JAPAN (W); ⊙ NIIGATA, its port. Bordered N by YAMAGATA prefecture, NE by FUKUSHIMA prefecture, SE by TOCHIGI and GUMMA prefectures, SW by NAGANO prefecture, and W by TOYAMA prefecture; includes SADO Island. Partly mountainous terrain; SHINANO RIVER drains large fertile plain producing rice, tea, and other farm products. Oil field in W coast area. Fishing, gold and silver mining (on Sado Island), raw-silk culture, manufacturing of machinery, textiles, and lacquerware. Principal centers include Niigata, KASHIWAZAKI (oil refining), NAGAOKA (heavy industry), Takada, and SANJO.

Niigata (NEE-gah-tah), city (2005 population 813,847), ⊙ NIIGATA prefecture, N HONSHU, N central JAPAN, on the SEA OF JAPAN at the mouth of the SHINANO RIVER; 37°54′N 139°02′E. Main port (at Niigata Higashi) for the W coast of Honshu; important point for oil storage and the importing of liquid natural gas. Watermelon; rice. Manufacturing includes iron molds; also, Buddhist altars, lacquerware, and sake. Traversed by many canals. Site of one of the largest flower farms in E ASIA. Opened to foreign trade in 1869.

Niihau (NEE-ee-HOU), island (□ 70 sq mi/182 sq km), KAUAI county, HAWAII, N end 13 mi/21 km WSW of Mana Point; 18 mi/29 km long and 6 mi/9.7 km wide. It is mostly semiarid lowland, rising to 1,281 ft/390 m at Paniau Mountain on NE coast. Administered as a district of Kauai county. Villages of Puuwai and Nonopapa on W coast; Kii Landing, at NE tip, main transfer point to adjacent islands. Main industries are cattle raising, manufacturing of shell leis. The island is inhabited mainly by pure-blooded Hawaiians who maintain their cultural heritage. Halulu and Halalii, intermittent natural lakes, in S center. Small LEHUA Island off N end.

Nii-jima (NEE-jee-mah), volcanic island (□ 11 sq mi/ 28.6 sq km) of island group Izushichito, Tokyo prefecture, SE JAPAN, in PHILIPPINE SEA, 19 mi/31 km SSW of O-SHIMA; 7 mi/11.3 km long, 2 mi/3.2 km wide. Hilly; rises to 1,404 ft/428 m. Produces camellia oil, charcoal, grain; livestock raising, dairying. Honmura, on W coast, is chief settlement. Nearby Shikinejima (2 mi/3.2 km long, 1 mi/1.6 km wide) has hot springs.

Niirala (NEE-rah-lah), village, POHJOIS-KARJALAN province, SE FINLAND, 40 mi/64 km SE of JOENSUU; 62°10′N 30°37′E. Elevation 248 ft/75 m. Frontier station on RUSSIAN border, opposite VYARTSILYA.

Níjar (NEE-hahr), city, ALMERÍA province, S SPAIN, 17 mi/27 km NE of ALMERÍA; 36°58′N 02°12′W. Olive-oil processing. Esparto, cereals, almonds; lumber. Silver-bearing lead, gold, garnet mines, and clay quarries in vicinity.

Nijima (NEE-jee-mah), village, on NII-JIMA island, Oshima district, Tokyo prefecture, SE JAPAN, 16 mi/25 km S of TOKYO; 34°22′N 139°15′E.

Nijkerk (NEI-kerk), town, GELDERLAND province, central NETHERLANDS, 6 mi/9.7 km NE of AMERSFOORT; 52°13′N 05°30′E. Nijkerksluis (locks) 3 mi/4.8 km to N, connecting Nijkerkernauw and NULDERNAUW channels, which separate FLEVOLAND polder from mainland. Dairying; cattle, hogs, poultry, eggs; grain, vegetables, sugar beets, fruit; manufacturing (food processing, heat exchangers). Town since 1413. Castle to N. Formerly spelled Nykerk.

Nijlen (NEI-luhn), commune (2006 population 20,910), Mechelen district, ANTWERPEN province, N BELGIUM, 13 mi/21 km ESE of ANTWERP; 51°10′N 04°39′E. Agriculture market.

Nijmegen (NEI-mai-khuhn), city, GELDERLAND province, E NETHERLANDS, on the WAAL RIVER (bridge), 37 mi/60 km ESE of AMSTERDAM; 51°50′N 05°51′E. N entrance to MAAS-WAAL CANAL, on Waal River, 3 mi/ 4.8 km to NW; German border 3 mi/4.8 km to E. Railroad junction. Dairying; cattle, hogs, poultry; vegetables, grain; manufacturing (construction materials, machinery, fabricated metal products, clothing, dairy products; food processing, printing and publishing). Among the oldest cities in the Netherlands, Nijmegen was founded in Roman times and flourished under Charlemagne (eighth–early ninth century). It was chartered in 1184, became a free imperial city, and later joined the Hanseatic League. It subscribed (1579) to the Union of Utrecht, formed as a defensive measure against Philip II of Spain. The treaties of Nijmegen (1678–1679), which ended the Dutch War (1672–1678) of Louis XIV of France, were signed here. In World War II, Allied airborne troops wrested (Sept. 1944) Nijmegen from the Germans. In February 1944 the entire center of the city was destroyed when the American Air Force bombed Nijmegen, mistaking the city for the German city of KLEVE. Nijmegen has a thirteenth-century church (the St. Stevens Kerk), a sixteenth-century town hall, a seventeenth-century weighhouse, and the remains of a palace built (c.777) by Charlemagne and rebuilt by Frederick Barbarossa in 1165. It is the seat of the Catholic University of Nijmegen (founded 1923), and has several museums. The city is known in French as Nimègue and in German as Nimwegen. Sometimes spelled Nymegen.

Nijo (NEE-jo), town, Itoshima county, FUKUOKA prefecture, N KYUSHU, SW JAPAN, 12 mi/25 km W of FUKUOKA; 33°30′N 130°08′E.

Nijvel, BELGIUM: see NIVELLES.

Nijverdal (NEI-vuhr-dahl), town, OVERIJSSEL province, E NETHERLANDS, 8 mi/12.9 km W of ALMELO; 52°22′N 06°28′E. Regge River passes to E. Dairying; cattle, hogs, poultry; grain, vegetables; manufacturing (fabrics, piping). Formerly spelled Nyverdal.

Nikaho (nee-KAH-ho), town, Yuri county, Akita prefecture, N HONSHU, NE JAPAN, 31 mi/50 km S of AKITA city; 39°17′N 139°57′E. Condensers.

Nikaia (NEE-kai-ah), SW suburb of ATHENS, ATTICA prefecture, ATTICA department, E central GREECE, 5 mi/8 km from city center; 39°34′N 22°28′E. Also Nea Kokkinia.

Nikaria, Greece: see IKARÍA.

Nikatsubo (NEE-kahts-boo), town, Hidaka district, Hokkaido prefecture, N JAPAN, 68 mi/110 km S of SAPPORO; 42°21′N 142°18′E. Lumber; horses.

Nikaweratiya, village, NORTH WESTERN PROVINCE, SRI LANKA; 07°45′N 80°07′E. Timber production. Trades in rice and coconuts.

Nike (NEE-kai), town, ENUGU state, S central NIGERIA, on road, 10 mi/16 km N of ENUGU; 08°43′N 07°54′E. Market town. Cassava, rice, fruits, vegetables.

Nikel' (NEE-kyel), town (2006 population 15,295), NW MURMANSK oblast, NW European RUSSIA, on road and railroad near, on Lake Kuets-Yarvi, 5 mi/8 km SE of SALMIYARVI; 69°24′N 30°13′E. Elevation 459 ft/139 m. Nickel mining; developed after 1935. Called Kolosjoki until ceded (1944) by FINLAND to USSR.

Nikeltau (nee-kel-TOU), village, N AKTÖBE region, KAZAKHSTAN, on railroad, 45 mi/72 km ENE of AKTÖBE (Aktyubinsk); 50°24′N 58°09′E. In nickel-mining area. Junction of railroad spur to DONSKOYE chromium mine.

Nikesiane, Greece: see NIKISIANI.

Nikhab, EGYPT: see KAB, EL.

Nikheila (ne-KAI-luh), village, ASYUT province, central Upper EGYPT, on W bank of the NILE, on railroad, 2 mi/3.2 km S of ABU TIG. Pottery making, wood carving; cereals, dates, sugarcane. Also spelled Nekhela.

Niki (NEE-kee), town, Shiribeshi district, Hokkaido prefecture, N JAPAN, 31 mi/50 km W of SAPPORO; 43°09′N 140°46′E. Cherries.

Nikiforovka (nee-KEE-fuh-ruhf-kah), village, W TAMBOV oblast, S central European RUSSIA, on the Lesnoy Voronezh River (right headstream of the VORONEZH RIVER), on road and railroad, 10 mi/16 km E of MICHURINSK; 52°53′N 40°46′E. Elevation 429 ft/130 m. Paper products.

Nikiforovo, RUSSIA: see BOLOGOYE.

Nikisiani (nee-KEE-syah-nee), town, KAVÁLLA prefecture, EAST MACEDONIA AND THRACE department, NE GREECE, at E foot of Mount PANGAION, 14 mi/23 km W of KAVÁLLA; 40°57′N 24°09′E. Tobacco, corn. Also Nikesiane.

Nikitinka (ni-ki-TEEN-kuh), village, W EAST KAZAKHSTAN region, KAZAKHSTAN, in KALBA RANGE, 35 mi/56 km S of UST-KAMENOGORSK; 49°34′N 82°27′E. Tertiary-level (raion) administrative center. Gold mining.

Nikitinskiy (nee-KEE-teen-skeeyee), town (2005 population 2,400), W central KEMEROVO oblast, S central SIBERIA, RUSSIA, on highway and near railroad, 5 mi/8 km SW of LENINSK-KUZNETSKIY; 54°35′N 86°06′E. Elevation 757 ft/230 m. In a coal-mining and -processing region.

Nikitovka (nee-KEE-tuhf-kah), village (2004 population 2,920), central BELGOROD oblast, SW European RUSSIA; on the OSKOL RIVER; on road, 6 mi/10 km N of ROSSOSH; 50°22′N 38°24′E. Elevation 511 ft/155 m. Fisheries.

Nikitovka, UKRAINE: see MYKYTIVKA.

Nikitsch (NI-kich), Croatian *Filež* (FEE-lezh), village, central BURGENLAND, E AUSTRIA, 11 mi/18 km SSE of SOPRON, HUNGARY; 47°32′N 16°40′E. Sugar beets; poultry. Croatian community.

Nikki (NEE-kee), town, BORGOU department, E BENIN, on road, 55 mi/89 km NE of PARAKOU, and 15 mi/24 km W of NIGERIA border; 09°56′N 03°12′E. Ginning of kapok and cotton. Historic capital of the Baatonu people.

Nikko (NEEK-ko), city, TOCHIGI prefecture, central HONSHU, N central JAPAN, in NIKKO NATIONAL PARK, near Mount Nantai and Lake Chuzenji, 22 mi/35 km N of UTSUNOMIYA; 36°44′N 139°36′E. Copper products. Tourist resort and religious center, famous for its ornate temples and shrines, dating from the Edo period (1600–1868), and notable for rich coloring. Within the shrine of Tokugawa Ieyasu is the Yomeimon (Gate of Sunlight), perhaps the most beautiful gate in Japan. The park is noted for its mountain scenery, waterfalls (Kegon Falls), and cryptomeria forests. Has Buddhist as well as Shinto shrines. Tokugawa Ieyasu is buried in Nikko Toshogo Shrine.

Nikko National Park (NEEK-ko) (□ 220 sq mi/572 sq km), in TOCHIGI and GUMMA prefectures, central HONSHU, N central JAPAN. In mountain area with cryptomeria forests, scenic waterfalls, hot springs. Lake Chuzenji is a well-known attraction. Includes the historic city of NIKKO.

Nikla el 'Inab (nik-LAL i-NAB), village, BEHEIRA province, Lower EGYPT, on RASHID branch of the NILE River, and 6 mi/9.7 km ENE of ITYAI EL BARUD; 30°55′N 30°46′E. Cotton, rice, cereals.

Niklasdorf (NIK-lahs-dorf), village, central STYRIA, SE central AUSTRIA, on MUR RIVER, and 3 mi/4.8 km E of LEOBEN; 47°24′N 15°10′W. Paper products; technology park.

Nikolaevo (nee-ko-LAI-vo), city, HASKOVO oblast, SVILENGRAD obshtina, BULGARIA; 42°38′N 25°48′E. Fishing area; coal mines; manufacturing (porcelain).

Nikolaevo (nee-ko-LAI-vo), village, LOVECH oblast, PLEVEN obshtina, BULGARIA; 43°15′N 24°37′E.

Nikolai, village (2000 population 100), S central ALASKA, in upper reaches of KUSKOKWIM River, 35 mi/56 km E of MCGRATH; 63°00′N 154°23′W.

Nikolai, POLAND: see MIKOLOW.

Nikolaiken, POLAND: see MIKOLAJKI.

Nikolainkaupunki, FINLAND: see VAASA.

Nikolaistad, FINLAND: see VAASA.

Nikolaital, SWITZERLAND: see VISPA RIVER.

Nikola-Kozlevo (nee-KO-lah–koz-LE-vo), village, VARNA oblast, Nikola-Kozlevo obshtina (1993 population 8,294), BULGARIA; 43°35′N 27°14′E.

Nikolayev, UKRAINE: see MYKOLAYIV, Mykolayiv oblast; MYKOLAYIV, L'viv oblast; or MYKOLAYIV, Khmel'nyts'kyy oblast.

Nikolayevka (nee-kuh-LAH-eef-kah), town (2005 population 7,335), NE JEWISH AUTONOMOUS OBLAST, S RUSSIAN FAR EAST, on the TUNGUSKA River (tributary of the AMUR River), on highway and the TRANS-SIBERIAN railroad, 13 mi/21 km W of KHABAROVSK; 48°34′N 134°46′E. Elevation 121 ft/36 m. Woodworking combine. Also known as Nikolayevsk.

Nikolayevka (nee-kuh-LAH-eef-kah), town (2006 population 6,400), SW ULYANOVSK oblast, E central European Russia, on the Kanadey River (VOLGA RIVER basin), on road junction and railroad, 19 mi/31 km W of NOVOSPASSKOYE; 53°07′N 47°12′E. Elevation 570 ft/173 m. In agricultural area; creamery.

Nikolayevka (nee-kuh-LAH-eef-kah), village (2006 population 5,045), central MORDVA REPUBLIC, central European Russia, on the INSAR River, on road and railroad, 3 mi/5 km S of (and administratively subordinate to) SARANSK; 54°09′N 45°08′E. Elevation 495 ft/150 m. In agricultural area; produce processing, dairying.

Nikolayevka (nee-kuh-LAH-eef-kah), settlement (2006 population 4,020), S MARITIME TERRITORY, SE RUSSIAN FAR EAST, on the Suchan River (which feeds into the Sea of JAPAN), on road and railroad spur, 3 mi/5 km NE of PARTIZANSK (across the river), to which it is administratively subordinate; 43°05′N 133°12′E. Elevation 505 ft/153 m. Russian military airfield in the vicinity.

Nikolayevka, UKRAINE: see MYKOLAYIVKA, Odessa Oblast; MYKOLAYIVKA, Dnipropetrovs'k oblast; MYKOLAYIVKA, Luhans'k oblast; or MYKOLAYIVKA, Donets'k oblast.

Nikolayevka, UKRAINE: see NOVOVORONTSOVKA.

Nikolayevka Vtoraya, UKRAINE: see MYKOLAYIVKA, Odessa oblast.

Nikolayevka-Vyrevskaya, UKRAINE: see ZHOVTNEVE, Sumy oblast.

Nikolayev oblast, UKRAINE: see MYKOLAYIV oblast.

Nikolayevsk (nee-kuh-LAH-eefsk), city (2006 population 16,020), NE VOLGOGRAD oblast, SE European Russia, on the E shore of the Volgograd Reservoir of the VOLGA RIVER, on road, 120 mi/190 km NE of VOLGOGRAD, and 5 mi/8 km SSE of (across the river from) KAMYSHIN; 50°01′N 45°26′E. Elevation 150 ft/45 m. Dairying, sunflower oil extraction, canning. Formerly called Nikolayevskaya and Nikolayevskiy. Made city in 1967.

Nikolayevsk, RUSSIA: see NIKOLAYEVKA, JEWISH AUTONOMOUS Oblast.

Nikolayevsk, RUSSIA: see PUGACHÉV.

Nikolayevskaya (nee-kuh-LAH-eef-skah-yah), village (2006 population 3,615), central ROSTOV oblast, S European Russia, on the right bank of the DON River, terminus of local highway branch, 60 mi/97 km E of SHAKHTY; 47°37′N 41°30′E. In agricultural area; produce processing.

Nikolayevskaya, RUSSIA: see NIKOLAYEVSK.

Nikolayevskaya oblast, UKRAINE: see MYKOLAYIV oblast.

Nikolayevskiy (nee-kuh-LAH-eef-skeeyee), village (2004 population 950), central BURYAT REPUBLIC, S Siberian Russia, 13 mi/21 km SSE of ULAN-UDE; 51°39′N 107°48′E. Elevation 2,286 ft/696 m. Pork processing.

Nikolayevsk-na-Amure (nee-kuh-LAH-eefsk–nah-ah-MOO-rye), city (2005 population 26,715), E KHABAROVSK TERRITORY, RUSSIAN FAR EAST, on local highway, 600 mi/966 km NE of KHABAROVSK, port on the AMUR River, 25 mi/40 km above its mouth; 53°08′N 140°44′E. Fishing, shipbuilding and repair, fish canning. Important port of Russian Far East, although frozen 6.5 months annually. Founded in 1850; leading Russian city in Russian Far East until the rise of Khabarovsk after 1880; nearly destroyed (1920) during the Russian civil war. Naval base was transferred in the 1920s to SOVETSKAYA GAVAN'. Made city in 1856.

Nikolo-Arkhangel'skoye, RUSSIA: see NIKOL'SKO-ARKHANGEL'SKIY.

Nikolo-Berëzovka (nee-KO-luh–bee-RYO-zuhf-kah), town (2005 population 6,000), NW BASHKORTOSTAN Republic, E European Russia, on the W bank of the KAMA River (landing), on highway branch, 70 mi/113 km NW of BIRSK; 56°08′N 54°09′E. Elevation 219 ft/66 m. Flour milling, sawmilling. Formerly Nikolo-Berezovskoye.

Nikolo-Berezovskoye, RUSSIA: see NIKOLO-BERËZOVKA.

Nikologory (nee-kuh-luh-GO-ri), town (2006 population 6,195), E VLADIMIR oblast, central European Russia, on road junction, 10 mi/16 km SW of VYAZNIKI; 56°08′N 41°59′E. Elevation 436 ft/132 m. Linen textiles and clothing.

Nikolo-Moshenskoye, RUSSIA: see MOSHENSKOYE.

Nikolo-Pavdinskiy Zavod, RUSSIA: see PAVDA.

Nikolo-Pavlovskoye (nee-KO-luh–PAHV-luhf-skuh-ye), village (2006 population 4,695), W SVERDLOVSK oblast, E URALS, extreme W Siberian Russia, on the TAGIL RIVER, on road junction and railroad, 10 mi/16 km S of NIZHNIY TAGIL, and 7 mi/11 km E of CHER-NOISTOCHINSK; 57°47′N 60°03′E. Elevation 708 ft/215 m. Founded as an ironworks settlement.

Nikolovo (nee-KO-lo-vo), village, RUSE oblast, RUSE obshtina, BULGARIA; 43°52′N 26°06′E. Park and recreation.

Nikolo-Vsekhsvyatskoye, RUSSIA: see NIKOL'SKIY.

Nikolsburg, CZECH REPUBLIC: see MIKULOV.

Nikol'sk (nee-KOLSK), city (2006 population 8,660), SE VOLOGDA oblast, N central European Russia, on the YUG RIVER (head of navigation), on the Sharya-Kotlas road, 270 mi/435 km E of VOLOGDA, and 90 mi/145 km SSW of VELIKIY USTYUG; 59°30′N 45°27′E. Elevation 488 ft/148 m. In agricultural area; meat packing, flax processing, lumbering, dairying. Chartered in 1780.

Nikol'sk (NEE-kuhlsk), city (2005 population 23,400), NE PENZA oblast, S central European Russia, on railroad, 75 mi/120 km NE of PENZA; 53°42′N 46°05′E. Elevation 875 ft/266 m. Highway junction; local transshipment point. In forested grain and hemp area; forestry; glassworking; food processing (bakery, creamery); lighting fixtures; printing house. Founded in the 17th century. Until 1954, called Nikol'skaya Pestravka.

Nikol'sk, RUSSIA: see USSURIYSK.

Nikol'skaya Pestravka, RUSSIA: see NIKOL'SK, PENZA oblast.

Nikol's'ke, UKRAINE: see VOLODARS'KE.

Nikolski, village (2000 population 39), SW UMNAK Island, ALEUTIAN ISLANDS, SW ALASKA, 80 mi/129 km SW of KASHEGA; 52°55′N 168°47′W. Has church and school.

Nikol'skiy (NEE-kuhl-skeeyee), town (2005 population 2,910), NE LENINGRAD oblast, NW European Russia, on the SVIR' River, on railroad, 22 mi/35 km NE of LODEYNOYE POLE, and 3 mi/5 km W of PODPOROZHYE; 60°55′N 34°04′E. Sawmilling, woodworking.

Nikol'skiy (nee-KOL-skeeyee), former town, central MOSCOW oblast, RUSSIA, now a suburb of MOSCOW, near the MOSCOW CANAL, 7 mi/11 km NW of the city center; 55°50′N 37°29′E. Elevation 544 ft/165 m. Until 1938, called Nikolo-Vsekhsvyatskoye.

Nikol'skiy Khutor, RUSSIA: see SURSK.

Nikol'sko-Arkhangelskiy (nee-KOL-skuh–ahr-HAN-gyel-skeeyee), town (2006 population 18,440), central MOSCOW oblast, central European Russia, 12 mi/19 km E of MOSCOW, and 4 mi/6 km SSW of BALASHIKHA, to which it is administratively subordinate; 55°45′N 37°56′E. Elevation 436 ft/132 m. Formerly a village of Nikolo-Arkhangel'skoye.

Nikol'skoye (NEE-kuhl-skuh-ye), city (2005 population 17,435), W central LENINGRAD oblast, NW European Russia, on the TOSNA RIVER (tributary of the NEVA RIVER), on railroad, 30 mi/48 km SE of SAINT PETERSBURG; 59°30′N 34°58′E. Elevation 498 ft/151 m. Manufacturing (ceramic products). Made city in 1990.

Nikol'skoye (nee-KOL-skuh-ye), town, N ASTRAKHAN oblast, S European Russia, on the E bank of the VOLGA River (landing), near highway, 120 mi/193 km NNW of ASTRAKHAN; 47°46′N 46°24′E. Below sea level. In agricultural area producing fruit, wheat; cattle, and sheep.

Nikol'skoye (nee-KOL-skuh-ye), village (2005 population 750) and administrative center of KOMANDORSKIY ISLANDS, extreme E KAMCHATKA oblast, RUSSIAN FAR EAST, on NW BERING ISLAND; 55°11′N 165°59′E. Seismological station. Has an Aleut cultural museum.

Nikol'skoye (NEE-kuhl-skuh-ye), village (2005 population 4,850), SW KOSTROMA oblast, central European Russia, near the confluence of the KOSTROMA and VOLGA rivers, near highway, 6 mi/10 km WNW, and under administrative jurisdiction, of KOSTROMA; 57°50′N 40°45′E. Elevation 259 ft/78 m. Popular holiday getaway spot; summer homes.

Nikol'skoye (nee-KOL-skuh-ye), village, N central ORENBURG oblast, in the S URALS, SE European Russia, on the SAKMARA RIVER (tributary of the URAL River), on road, 30 mi/48 km NE of ORENBURG; 52°02′N 55°43′E. Elevation 505 ft/153 m. In agricultural area producing wheat and livestock.

Nikol'skoye (nee-KOL-skuh-ye), village (2006 population 5,285), NW VORONEZH oblast, S central European Russia, on the VORONEZH RIVER near its confluence with the DON River, near highway and railroad, 8 mi/13 km SSE of VORONEZH; 51°35′N 39°18′E. Elevation 360 ft/109 m. In agricultural area; food processing.

Nikol'skoye, RUSSIA: see KALININO, AMUR oblast.

Nikol'skoye, RUSSIA: see SHEKSNA, town.

Nikol'skoye, RUSSIA: see USSURIYSK.

Nikol'skoye, RUSSIA: see VOLODARS'KE.

Nikol'skoye-na-Cheremshane (nee-KOL-skuh-ye–nah-chyee-reem-SHAH-nye), town (2006 population 1,960), NE ULYANOVSK oblast, E central European Russia, on the GREATER CHEREMSHAN RIVER, 38 mi/61 km SE of ULYANOVSK, and 13 mi/21 km SW of DIMITROVGRAD, to which it is administratively subordinate; 54°03′N 49°14′E. Elevation 190 ft/57 m. Knitwear. Fisheries.

Nikol'sk-Ussuriyskiy, RUSSIA: see USSURIYSK.

Nikonga River (nee-KON-gah), NW TANZANIA; rises c. 60 mi/97 km NW of MWANZA, near boundary of MWANZA and SHINYANGA regions; flows W, roughly following regional boundary, then S; joins Moyowosi River in marshy area 105 mi/169 km WNW of TABORA.

Nikopol (nee-KO-pol), city, LOVECH oblast, NIKOPOL obshtina (1993 population 15,497), N BULGARIA, a port on the DANUBE River bordering ROMANIA; 43°43′N 24°54′E. Farming, viticulture, and fishing are the chief occupations. Electric equipment; wine making. Paper products nearby in Cherkovitsa village. Founded in 629 by Byzantine emperor Heraclius, Nikopol (then Nicopolis) became a flourishing trade and cultural center of the second Bulgarian kingdom. In 1396 at Nikopol, the Ottoman Turks under Bayazid I defeated an army of Crusaders led by King Sigismund of Hungary (later Emperor Sigismund). The Turkish victory removed the last serious obstacle to a Turkish advance on Christian Europe. However, when Timur defeated Bayazid (1402), Europe gained a respite. The Turks strongly fortified Nikopol, which was strategically important during the Russo-Turkish wars (18th–19th century), but the city later declined.

Nikopol' (NEE-ko-pol) (in Ukrainian also **Nykopil'**), city (2001 population 136,280), S DNIPROPETROVS'K oblast, UKRAINE, on the N shore of the KAKHOVKA RESERVOIR; 47°34′N 34°24′E. Raion center, railroad terminus, river park, and the industrial center of one of the world's richest manganese-mining areas. The city has metallurgical plants, pipe making and machine tool factories, and food-processing and brewing industries. Nikopol' stands on the site of one of the earliest trade routes and strategic crossing points (Mykytyn Rih) over the DNIEPER (Ukrainian *Dnipro*) River. In the early 17th century, a portage settlement, Mykytyne, was established here; site of Zaporozhian (Mykytyn) Sich (1636–1652); after destruction of Zaporozhian Sich by the Russian Army (1775), fort Slavyansk (Ukrainian *Slov'yans'k*) was built nearby; it was renamed Nikopol' (1782). Manganese production at Nikopol' became important in the 19th century. Jewish community since 1782, numbering 2,700 in 1939, mainly craftsmen and professionals; largely evacuated E with the industrial complex during World War II, the rest destroyed by the Nazis. The community reestablished after the war, with close to 10,000 Jews in Nikopol' in 2005.

Nikopolis (nee-KO-po-lees) [Greek=city of victory], ancient city, in what is now EPIRUS department, NW GREECE, on peninsula between IONIAN SEA and AMBRACIAN GULF, near PREVEZA; 40°53′N 23°11′E. Founded by Octavian (later Augustus) to celebrate the victory (31 B.C.E.) at nearby ACTIUM. Largely eclipsed Ambracia. Mentioned in the New Testament by St. Paul (Titus 3:12). Formerly spelled Nicopolis. Also Nikopol.

Nikopolis ad Nestum (nee-KO-po-lees ahd NE-stum), Roman ruins near GOTSE DELCHEV, SW BULGARIA.

Niksar, ancient *Neocaesarea*, town (2000 population 44,808), N central TURKEY, on KELKIT RIVER, and 28 mi/45 km NE of TOKAT; 40°35′N 36°59′E. Grain. Ancient Neocaesarea was important town of PONTUS. There are Roman and Byzantine remains here.

Nikshahr (neek-SHAHR), town, Sīstān va Balūchestān province, SE IRAN, 35 mi/56 km W of QASR-e-Qand; 26°13′N 60°12′E.

Nikšić (NEEK-sheech), city (2003 population 58,212), S central MONTENEGRO; 42°46′N 18°56′E. The commercial center of an agricultural region, it is also an important industrial city with a steel mill, sawmill, ironworks, and a hydroelectric plant. Bauxite deposits nearby. Founded in the early Middle Ages, it was under Turkish rule until 1878, when it passed to Montenegro. The town has a Serbian Orthodox cathedral (the gift of Nicholas II of RUSSIA) and a Roman bridge.

Nikunau (nee-koo-NOU), coral island (□ 7 sq mi/18.2 sq km; 2005 population 1,912), N GILBERT ISLANDS, KIRIBATI, W central PACIFIC OCEAN; 01°23′S 176°26′E. Discovered 1765 by Byron. Also called Nukunau; formerly Byron Island.

Nikyup (nee-KYUP), village, LOVECH oblast, VELIKO TURNOVO obshtina, N BULGARIA, 9 mi/15 km NNW of GORNA ORYAHOVITSA; 43°14′N 25°35′E. Flour milling; livestock raising. Site of ruins of an ancient Roman town Nicopolis ad Istrum, which existed up to 600 A.D.

Nilakkottai (ni-LUHK-ko-tei), town, Nagapattinam Quaid-E-Milleth district, TAMIL NADU state, S INDIA, 25 mi/40 km NW of MADURAI; 10°10′N 77°52′E. Grain, sesame.

Nilambur, INDIA: see SHORANUR.

Niland, unincorporated town (2000 population 1,143), IMPERIAL county, S CALIFORNIA, 18 mi/29 km N of BRAWLEY, at N end of IMPERIAL VALLEY, 5 mi/8 km E of SALTON SEA; 33°14′N 115°31′W. Citrus fruit, vegetables, tomatoes, dates, melons, sugar beets; cotton; cattle, sheep. Coachella Canal and CHOCOLATE MOUNTAINS to NE.

Nilandhe Atoll, central group of MALDIVES, in INDIAN OCEAN, between 02°40′N 72°48′E and 03°21′N 73°05′E. Consists of two neighboring physical atolls (North and South Nilandhe). Coconuts.

Nilanga (nee-LUHN-gah), town, LATUR district, MAHARASHTRA state, W central INDIA, 50 mi/80 km WNW of BIDAR; 18°06′N 76°46′E. Millet, cotton, rice, sugarcane.

Nilaveli, village, EASTERN PROVINCE, SRI LANKA, on NE coast, 8 mi/12.9 km N of TRINCOMALEE; 08°41′N 81°12′E. Coconut-palm plantations, vegetable gardens. Tourism.

Nil Desperandum Mine, ZIMBABWE: see ZVISHAVANE.

Nile, state, SUDAN: see RIVER NILE, state.

Nile, former province (□ 6,071 sq mi/15,784.6 sq km), NW UGANDA; capital was ARUA. Was bordered N by SUDAN, W and S by Zaire (now DEMOCRATIC REPUBLIC OF THE CONGO), E by NORTHERN province. Drained by ALBERT NILE, ORA, and Ola rivers. Consisted of ZOKA forest. Tobacco, groundnuts, cassava; livestock-raising. Main centers were Arua, MOYO, NEBBI, PAKWACH.

Nile, longest river in the world, c. 4,160 mi/6,695 km long from its remotest headstream, the Luvironza River in BURUNDI, central AFRICA, to its delta on the MEDITERRANEAN SEA, NE EGYPT. The Nile flows N and drains c. 1,100,000 sq mi/2,850,000 sq km, about 10% of Africa, including parts of Egypt, SUDAN, ETHIOPIA, KENYA, UGANDA, RWANDA, Burundi, and CONGO. Its waters support 98% of all agriculture in Egypt, furnish water for more than 20% of Sudan's total crop area, and are widely used throughout the basin for navigation and hydroelectric power. The trunk stream of the Nile is formed at KHARTOUM, Sudan, 1,857 mi/2,988 km from the sea, by the junction of the BLUE NILE (c. 1,000 mi/1,610 km long) and the WHITE NILE (c. 2,300 mi/3,700 km long). The Blue Nile rises in the headwaters of Lake Tana, NW Ethiopia, a region of heavy summer rains, and is the source of floodwaters that reach Egypt in July and peak in September; the Blue Nile contributes almost 60% of all Nile waters throughout the year. During floodtime it also carries great quantities of silt from the highlands of Ethiopia; these now collect in LAKE NASSER behind the ASWAN HIGH DAM, but for centuries they were left on the floodplain after the floods and helped replenish the fertility of Egypt's soil. The White Nile (known in various sections as the Bahr Al-Abyad, Bahr Al-Gabal, ALBERT NILE, and VICTORIA NILE) rises in the headwaters of Lake Victoria in a region of heavy, year-round rainfall; unlike the Blue Nile, it has a constant flow, owing in part to its source area and in part to the regulating effects of its passage through lakes Victoria and Albert and the SUDD swamps of Sudan. Other important tributaries of the Nile are the ATBARA and SOBAT rivers. The GEZIRA, or "island," formed between the Blue Nile and the White Nile before they come together at Khartoum is Sudan's principal agricultural area and the only large tract of land outside Egypt irrigated with Nile waters. From Khartoum to the Egyptian border at WADI HALFA and on to ASWAN in Egypt, the Nile occupies a narrow entrenched valley with little floodplain for cultivation; in this stretch it is interrupted by six cataracts (rapids). From Aswan the river flows 550 mi/885 km N to CAIRO, bordered by a floodplain that gradually widens S of Cairo to c. 12 mi/20 km; irrigated by the river, this intensively cultivated valley contrasts with the barren desert on either side. N of Cairo is the great Nile delta (c. 100 mi/160 km long and up to 115 mi/185 km wide), which contains 60% of Egypt's cultivated land and extensive areas of swamps and shallow lakes. Two distributaries, the Damietta (DUMYAT) on the E and the ROSETTA (RASHID) on the W, each c. 150 mi/240 km long, carry the river's remaining water (after irrigation) to the MEDITERRANEAN SEA. Regular steamship service is maintained on the Nile between ALEXANDRIA (reached by canal) and Aswan; the Blue Nile is navigable June–December from SUKI (above Sennar Dam) to Al-Roseires Dam; the White Nile is navigable all year between KHARTOUM and JUBA in Sudan, and between NIMULE and MURCHISON FALLS on the Victoria Nile. The use of the Nile for irrigation, now regulated by the Nile Waters Treaty of 1959, dates back to at least 4000 B.C.E. in Egypt. The traditional system of basin irrigation—in which Nile floods were trapped in shallow basins and a cool-season crop of wheat or barley was grown in soaked and silt-replenished soil—has been replaced in the last 150 years by a system of perennial irrigation and the production of two or three crops a year, including cotton, sugarcane, and peanuts. The delta barrages, just below CAIRO, channel water into a system of feeder canals for the delta, and other barrages at ISNA, ASYUT, and NAG HAMMADI keep the level of the Nile high enough all year for perennial irrigation in the valley of Upper Egypt; the Idfina Barrage on the Rosetta prevents infiltration by the sea into the river. Nile water is also used for irrigation in the Fayyum Basin. The Aswan Dam (completed 1902 and raised twice since then) was the first dam built on the Nile to store part of the autumn flood for later use; it has a storage capacity of 167 billion cu ft/5 billion cu m and is now supplemented by the Aswan High Dam (completed 1971), 5 mi/8 km upstream, with a storage capacity of 5,724 billion cu ft/162 billion cu m, sufficient (with existing dams) to hold back the entire flood for later use. Construction of the Aswan High Dam has added c.1,800,000 acres/728,500 ha of irrigated land to Egypt's cultivable area and converted c. 730,000 acres/295,400 ha from basin to perennial irrigation. Lake Nasser, created by the Aswan High Dam, has experienced problems with silting. There has been a reduction of soil replenishment downstream and a reduction of nutrients that once fed the E Mediterranean Sea. In addition, major delta distributaries are threatened by reduced water flow due to Egypt's extensive irrigation schemes. Other important storage dams, all outside Egypt, but built with Egypt's help or cooperation, are the OWEN FALLS DAM (1954) in Uganda and Jebel Aulia Dam (1937) in Sudan on the White Nile; the Sennar (1927) and Al-Roseires (1966) on the Blue Nile (Sudan); and the Kashm-el-Girba Dam (1964) on the ATBARA RIVER (Sudan). The source of the Nile and its life-giving floods was a mystery for centuries. Ptolemy held that the source was the "Mountains of the Moon," and the search for these and for the origin of the Nile attracted much attention in the 18th and 19th centuries. James Bruce, the Scottish explorer, identified (1770) Lake Tana as the source of the Blue Nile, and John Speke, the British explorer, is credited with the identification (1861–1862) of Lake Victoria and RIPON FALLS as the source of the White Nile.

Niles, city, BERRIEN county, SW MICHIGAN, 10 mi/16 km N of SOUTH BEND, Indiana; 41°49′N 86°15′W. Part of urbanized area, on the ST. JOSEPH RIVER. In a farm and fruit area. Manufacturing includes paper products, transportation equipment, fabricated metal products, and machinery. It was the site of a Jesuit Mission (1690) and of Fort Saint Joseph, built by the French (1697). The fort fell to the British (1761), to the Native Americans (Pontiac's Rebellion, 1763), and to the Spanish and Native Americans (1780, 1781). Permanent settlement began in 1827, and as a station on the stagecoach route between DETROIT and CHICAGO, Niles grew as a commercial and industrial center. Airport to E. Ring Lardner was born here. Over 1,000-acres/405-ha Botanic Garden here. Incorporated 1829.

Niles (NEI-uhlz), city (□ 9 sq mi/23.4 sq km; 2006 population 19,824), TRUMBULL county, NE OHIO, on the MAHONING RIVER, 4 mi/6.4 km SE of WARREN; 41°11′N 80°45′W. Produces steel, building materials, and lathes. There is a memorial to President William McKinley, who was born here. Settled 1806, incorporated as a city 1895.

Niles, unincorporated town, ALAMEDA county, W CALIFORNIA, suburb 23 mi/37 km SE of OAKLAND and 2 mi/3.2 km E of FREMONT, on Alameda Creek. Railroad junction. Niles Canyon (E) is recreational area.

Niles, village (2000 population 30,068), COOK county, NE ILLINOIS, a residential suburb just NW of CHICAGO, on the N branch of CHICAGO RIVER; 42°01′N 87°48′W. Settled 1832, incorporated 1899. The village has a replica (half size) of the leaning tower of Pisa. Niles College of Loyola University is here.

Nileswaram, INDIA: see KASARAGOD.

Nilgiri (NEEL-gi-ree), former princely state in ORISSA STATES, INDIA. Incorporated 1949 into Balasore district.

Nilgiri (NEEL-gi-ree), Sanskrit and Tamil *Nilagiri* [=blue hills], district (□ 984 sq mi/2,558.4 sq km), TAMIL NADU state, S INDIA; ⊙ UDAGAMANDALAM (or Ooty). Consists mainly of NILGIRI HILLS (E), with SE portion of the Wynaad in NW. Numerous mountain streams (trout, mahseer, carp) drain into MOYAR (N) and BHAVANI (on SE border) Rivers. Temperate-zone flower and vegetable gardens and experimental farms; millet, wheat, barley, pyrethrum. Important cinchona plantations (main quinine factory at NADUVATTAM); numerous eucalyptus, tea, and coffee estates. Headworks of hydroelectric system at PYKARA. Main towns are Udagamandalam (or Ooty), COONOOR, KOTAGIRI. Formerly spelled Neilgherry.

Nilgiri (NEEL-gee-ree), mountain peak (23,170 ft/7,061 m), central NEPAL; 28°41′N 83°44′E. Dominates KALI GANDAKI RIVER valley. Climbed by Dutch team in 1962.

Nilgiri Hills, INDIA: see GHATS DODABETTA.

Nilikluguk, village, SW ALASKA, on Bethel Bay, 80 mi/129 km WSW of BETHEL.

Ni'lin, Arab village, Ramallah district, WEST BANK, 15 mi/24 km W of RAMALLAH, on the W slopes of the SAMARIAN Highlands, at the boundary of the Judean Wilderness; 31°57′N 35°01′E. Agriculture (fruit, olives, cereal).

Nilo (NEE-lo), town, ⊙ Nilo municipio, CUNDINAMARCA department, central COLOMBIA, 30 mi/48 km SW of BOGOTÁ; 04°19′N 74°38′W. Elevation 1,456 ft/443 m. Coffee, sugarcane; livestock.

Nilo Peçanha (NEE-lo pe-SAHN-yah), city (2007 population 12,961), BAHIA state, central Atlantic coast of BRAZIL, 17 mi/27 km S of VALENÇA; 13°37′S 39°07′W.

Nilópolis (NEE-lo-po-lees), suburb (2007 population 152,223) of RIO DE JANEIRO, RIO DE JANEIRO state, BRAZIL, 15 mi/24 km from city center; 22°49′S 43°25′W. Metallurgical plants. Orange groves.

Nilphamari (neel-fah-mah-ree), town (2001 population 40,084), RANGPUR district, N EAST BENGAL, BANGLADESH, 29 mi/47 km WNW of RANGPUR; 25°57′N 88°52′E. Trades in rice, jute, tobacco, oilseeds; hosiery manufacturing.

Niltepec, MEXICO: see SANTIAGO NILTEPEC.

Nilufer River (Turkish=*Nilüfer*), 60 mi/97 km long, NW TURKEY; rises on ULU DAG 11 mi/18 km SE of BURSA; flows N and W, passing near Bursa to SIMAV RIVER 7 mi/11.3 km NE of KARACABEY.

Nilus, AFRICA: see NILE, river.

Nilvange (neel-VAHNJ), town (□ 1 sq mi/2.6 sq km), MOSELLE department, LORRAINE region, NE FRANCE, 5 mi/8 km WSW of THIONVILLE; 49°21′N 06°03′E. Metallurgical works.

Nilwood, town (2000 population 284), MACOUPIN county, SW central ILLINOIS, 9 mi/14.5 km NNE of CARLINVILLE; 39°23′N 89°48′W. In agricultural and bituminous-coal area.

Nima (NEE-mah), town, Nima county, SHIMANE prefecture, SW HONSHU, W JAPAN, on SEA OF JAPAN, 43 mi/69 km S of MATSUE; 35°08′N 132°24′E. Tile manufacturing; fish processing.

Nimach, district, NW INDIA (□ 1,490 sq mi/3,874 sq km; 2001 population 725,457). District formed out of N MANDSAUR district in 1998; 290 mi/462 km NW of BHOPAL. Mostly agriculture: wheat, maize, peanuts, gram, garlic, soyabeans, mustard. Famous for opium production: has government opium and alkaloid factory under central government control. Regional folklore states that the district's Bhadwa Matha Temple's water can cure paralysis. The Gomabai Nethralaya and Research Center, one of the leading ophthalmic centers in Asia, is located here; Nimach is nicknamed the "eye-donation capital of India" due to its large percentage of locals who donate their eyes.

Nimach (NEE-much) or **Neemuch**, town (2001 population 107,496), NIMACH district, SW MADHYA PRADESH state, central INDIA, 30 mi/48 km NNW of MANDSAUR; 24°44′N 72°26′E. Trades in wheat, cotton, opium, building stone; hand-loom weaving. Airport. In 1857, sepoys stationed here rebelled. Formerly in Gwalior state, later in MADHYA BHARAT.

Nimaima (nee-MEI-mah), town, ⊙ Nimaima municipio, CUNDINAMARCA department, central COLOMBIA, 30 mi/48 km NW of BOGOTÁ; 05°06′N 74°24′W. Elevation 3,904 ft/1,189 m. Coffee, sugarcane; livestock.

Nimaj (NEE-mahj), town, PALI district, central RAJASTHAN state, NW INDIA, 60 mi/97 km ESE of JODHPUR; 26°10′N 74°00′E. Millet, wheat, gram, oilseeds.

Nimar, INDIA: see EAST NIMAR WEST NIMAR.

Nimba (NIM-bah), county (□ 4,650 sq mi/12,090 sq km; 1999 population 338,887), LIBERIA; ⊙ SANOQUELLI; 06°45′N 08°45′W. Borders GUINEA (N), CÔTE D'IVOIRE (E), GRAND GEDEH county (SE), SINOE county (S), and RIVERCESS county (W). A highland area, including the NIMBA MOUNTAINS in the N. Much of the land is forested. Iron and gem deposits exist in the N. Main centers include YEKEPA and GANTA. Nimba is divided into six districts: Sanniquelleh-Mahn (N), Gbehlageh (NE), Zoegeh (E), Tappita (SES), Yarwein-Mehnsohnneh (SW), and Saclepea (W).

Nimbahera (NEEM-buh-HAI-ruh), town, CHITTAURGARH district, S RAJASTHAN state, NW INDIA, 25 mi/40 km E of CHITTAURGARH; 24°37′N 74°41′E. Agricultural market (millet, corn, wheat, oilseeds); cotton ginning; handicraft metal utensils. Sandstone quarries SW.

Nimba Mountains (NIM-buh), in W AFRICA bulge along LIBERIA–CÔTE D'IVOIRE–GUINEA border. Highest elevation 6,000 ft/1,830 m. Iron ore mining at nearby SANOQUELLI.

Nimbi (NEM-be), town, Rivers state, S NIGERIA, in NIGER River delta region, 20 mi/32 km NE of BRASS; 04°32′N 06°24′E. Market center. Fish; cassava, rice, yams. Also Nembe.

Nimbin (NIM-bin), town, NEW SOUTH WALES, SE AUSTRALIA, 488 mi/785 km N of SYDNEY, 16 mi/25 km N of LISMORE. Rainforest at nearby Nightcap National Park. Fruit growing; permaculture agricultural practices.

Nimbschen, GERMANY: see GRIMMA.

Nimburg, CZECH REPUBLIC: see NYMBURK.

Nimbus, Mount, Colorado: see NEVER SUMMER MOUNTAINS.

Nimègue, NETHERLANDS: see NIJMEGEN.

Nîmes (NEEM), ancient *Nemausus*, city; ⊙ GARD department, LANGUEDOC-ROUSSILLON region, S FRANCE, 60 mi/97 km NW of MARSEILLE, in the MEDITERRANEAN coastal plain at the edge of limestone hills, and just NW of the RHÔNE river delta; 43°55′N 04°23′E. On the expressway junction and railroad line running SW from the Rhône valley and from Marseille to the Spanish border. An important marketplace, Nîmes is

noted for its remarkable collection of splendidly preserved Roman buildings and as France's chief bullfighting and bull-running center. It has a long manufacturing tradition (clothing, footwear, fruit canning in syrup) and conducts an active trade in wine of the Languedoc region. It has several gastronomic specialties (almond biscuits, candied olives, and crisp bakery products). The leading points of interest are the perfectly preserved Roman arena (1st century A.D.), still in use and able to accomodate 24,000 spectators; and the Maison Carrée [French=square house], a remarkable Corinthian temple from the 1st or 2nd century C.E., which now houses a museum of antiquities. The Tour Magne, a watchtower built c.50 B.C.E. overlooks Nîmes from atop a hill. At its foot is an 18th-century park (Jardin de la Fontaine) which contains a temple of Diana (2nd century C.E.) and the spring (named Nemausus by the Romans) to which the city owes its name. There are several museums, including the recently modernized fine art museum. The famous Pont du Gard Roman aqueduct crossing the GARD RIVER, a major tourist attraction, is 12 mi/19 km NE. Founded by Augustus, Nîmes soon became one of the most important urban centers of the Roman Empire as capital of Narbonensis province. Attached to the possessions of the counts of Toulouse in 1185, it was united with the French crown in 1229. It later became a stronghold of the Huguenots and, as a consequence, suffered severely from the revocation (1685) of the Edict of Nantes, when many skilled and affluent Protestants left France. It is now a bishopric. Alphonse Daudet, the popular French author of the 19th century, was born here. Nîmes-Garons International Airport 6 mi/9 km SSE of the city.

Nimfaion, Cape, Greece: see AKRATHOS.

Nimi (NEE-MEE), city, OKAYAMA prefecture, SW HONSHU, W JAPAN, 34 mi/55 km NW of OKAYAMA; 34°58′N 133°28′E. Railroad junction. Tobacco. Cool geyser nearby.

Nimity Bell, AUSTRALIA: see NIMMITABEL.

Nimitz Glacier, in the central ELLSWORTH MOUNTAINS, WEST ANTARCTICA, flowing between the SENTINEL and Bastien Ranges into MINNESOTA GLACIER; 40 mi/65 km long, 5 mi/8 km wide; 78°55′S 85°10′W.

Nim-ka-Thana (NEEM–kah–TAH-nah), town, SIKAR district, E RAJASTHAN state, NW INDIA, 55 mi/89 km N of JAIPUR; 27°44′N 75°48′E. Millet, barley.

Nimkhar (NEEM-kahr), town, SITAPUR district, central UTTAR PRADESH state, N central INDIA, on the GOMATI RIVER, 19 mi/31 km SW of SITAPUR. Wheat, rice, gram, barley. Pilgrimage site. Since 1948 officially Naimisharanya.

Nimla (NEEM-lah), village, NANGARHAR province, E AFGHANISTAN, 26 mi/42 km WSW of JALALABAD, on highway to KABUL; 34°18′N 70°06′E. Royal cypress garden.

Nimmitabel (NI-mi-tuh-bel), village, NEW SOUTH WALES, SE AUSTRALIA, 278 mi/447 km SW of SYDNEY, on GREAT DIVIDING RANGE; 36°31′S 149°16′E. Formerly known as Nimity Bell.

Nimmons (NIM-uhnz), village (2000 population 100), CLAY county, extreme NE ARKANSAS, 28 mi/45 km NE of PARAGOULD, near SAINT FRANCIS River; 36°18′N 90°05′W.

Nimpkish River (NIMP-kish), 80 mi/129 km long, SW BRITISH COLUMBIA, W Canada, on VANCOUVER ISLAND; rises near VICTORIA PEAK; flows NW, through Nimpkish Lake (□ 12 sq mi/31 sq km), to JOHNSTONE STRAIT opposite ALERT BAY; 50°34′N 126°59′W.

Nimptsch, POLAND: see NIEMCZA.

Nimrod, village (2000 population 75), WADENA county, central MINNESOTA, on CROW WING RIVER, 19 mi/31 km NE of WADENA; 46°38′N 94°52′E. Dairying. Lyons State Forest to S; Huntersville State Forest to N.

Nimrod Glacier, a major outlet glacier in EAST ANTARCTICA, flowing from the polar plateau through the

TRANSANTARCTIC MOUNTAINS into the ROSS ICE SHELF at the SHACKLETON COAST; 85 mi/140 km long; 82°22′S 163°00′E.

Nimrod Lake, reservoir (□ 6 sq mi/15.6 sq km), YELL county, W central ARKANSAS, on FOURCHE LA FAVE RIVER, in Ouachita National Forest, 35 mi/56 km NNE of HOT SPRINGS; 34°57′N 93°00′W. Extends E-W. Maximum capacity 336,000 acre-ft. Formed by Nimrod Dam (100 ft/30 m high), built (1942) by Army Corps of Engineers for power generation, recreation, and as a fish and wildlife pond.

Nimrud, IRAQ: see CALAH.

Nimruz (NEEM-rooz), province (2005 population 135,900), SE AFGHANISTAN, ⊙ Zaranj; 30°30′N 62°00′E. Borders on IRAN (W); BALUCHISTAN province, PAKISTAN (S); HELMAND province (E); GHOR province (NE); and HERAT province (N). Drained by KHASH RUD River; HELMAND River crosses E-W in center. Highway from Zaranj N to Herat. Agriculture in Helmand River valley; animal husbandry. Baluch nomads seasonally graze their herds in the deserts. Population primarily Baluch and some Tajiks. Strong winds prevail for c.120 days of the year; countless windmills transform these winds into power for grain milling and water pumping. Ruins testify to a once flourishing civilization that was destroyed during the Mongol invasions.

Nimtita, INDIA: see DHULIAN.

Nimule (NEE-moo-lai), town, central EQUATORIA state, southernmost SUDAN, on UGANDA border, on right bank of the BAHR AL-GABAL (here also called ALBERT NILE), the upper course of the WHITE NILE, and 95 mi/153 km SSE of JUBA (linked by road); 03°36′N 32°03′E. N limit of navigation on ALBERT NILE (from Lake ALBERT); steamer-road transfer point for bypass to JUBA. Customs station. A dam here is planned for long-range storage of Lake ALBERT waters.

Nimwegen, NETHERLANDS: see NIJMEGEN.

Nimy (nee-MEE), village in commune of MONS, Mons district, HAINAUT province, SW BELGIUM, 2 mi/3.2 km N of Mons, near CANAL DU CENTRE; 50°28′N 03°57′E.

Nin (NEEN), Italian *Nona*, ancient *Aenona*, village, S CROATIA, on ADRIATIC SEA, 9 mi/14.5 km N of ZADAR, in DALMATIA. Ruins of Roman town and 8th–9th century Croatian churches. Medieval bishopric. Center of RAVNI KOTARI, the most important region of the medieval Croatian state prior to the Turkish invasions.

Ninacaca (nee-nah-KAH-kah), town, PASCO province, PASCO region, central PERU, in Cordillera CENTRAL of the ANDES, on railroad, and 15 mi/24 km SE of CERRO DE PASCO; 10°51′S 76°07′W. Elevation 13,930 ft/4,246 m. Cereals, potatoes; livestock.

Nina Rodrigues (NEE-nah ROD-ree-ges), town (2007 population 10,006), NE MARANHÃO state, BRAZIL, 6 mi/10 km N of VARGEM GRANDE; 03°28′S 43°52′W.

Nindirí (neen-dee-REE), town (2005 population 17,313), MASAYA department, SW NICARAGUA, 3 mi/4.8 km NW of MASAYA; 12°00′N 86°07′W. Agriculture (tobacco, coffee, rice, sugarcane). Has archaeological museum. An important Indian center in pre-colonial times.

Nine (NEE-nai), town, BRAGA district, N PORTUGAL, 8 mi/12.9 km SW of BRAGA; 41°25′N 08°28′W. Railroad junction.

Nine Cantons, former name of original component units of W Aden Br. Protectorate. The Nine Cantons included tribal dists. of Abdali, Amiri, Haushabi, Fadhli, Yafa, Subeihi, Aqrabi, Aulaqi, and Alawi. This followed an agreement bet. Britain and Turkey on Yemen's boundaries. Under treaties bet. Britain (Aden Colony) and local rulers, concluded bet. 1839 and 1904, 9 local tribal entities accepted Br. suzerainty.

Nine Degree Channel, channel of the Lakshadweep (Laccadive), LAKSHADWEEP Union Territory, INDIA, in ARABIAN SEA, at 09°00′N, between MINICOY ISLAND (S) and the Lakshadweep proper (N).

Nine Mile Falls, unincorporated village, SPOKANE county, E WASHINGTON, suburb 9 mi/14.5 km NNW of downtown SPOKANE, on SPOKANE RIVER, at Nine Mile Dam. Manufacturing (furniture, wood products). Parts of Riverside State Park to SW and N.

Ninemile Peak, NEVADA: see ANTELOPE RANGE.

Nine Mile Point Nuclear Station, two nuclear power–generating stations, OSWEGO county, central NEW YORK, on S shore of Lake ONTARIO, 7 mi/11 km NE of OSWEGO; 43°31′N 76°25′E. This is one of three nuclear power sites in New York state. Nine Mile Point Unit 1 opened in 1969 and has a 610-MW capacity; Unit 2 opened in 1988 and has a 1,100-MW capacity. Both are run by Constellation Energy Group, which owns Unit 1 in its entirety and holds an 82% share in Unit 2. In 2006 the federal Nuclear Regulatory Commission granted twenty-year license extensions for both units.

Ninette (ni-NET), unincorporated village, SW MANITOBA, W central Canada, 15 mi/24 km N of KILLARNEY, at NW end of Pelican Lake, and in Riverside rural municipality; 49°24′N 99°38′W. Grain; livestock.

Ninety Mile Beach Marine National Park (NEIN-tee MEIL BEECH) (□ 11 sq mi/28.6 sq km), 3-mi/5-km-long coast and marine sanctuary, 19 mi/30 km S of SALE, VICTORIA, SE AUSTRALIA, adjacent Gippsland Lakes Coastal Park; 38°13′S 147°23′E. Situated on a long, narrow sand dune, the beach area has many species, especially small sand-dwelling creatures; sponges, soft coral, and other invertebrates inhabit the calcarenite reefs covered in red seaweeds. Also written Ninety-Mile Beach.

Ninety Six, town (2006 population 1,922), GREENWOOD county, W SOUTH CAROLINA, 7 mi/11.3 km E of GREENWOOD. Livestock, poultry; dairying; grain; knitted cloth, cotton fabrics, bricks. SALUDA RIVER (E) forms like GREENWOOD reservoir (hydroelectric power); Lake Greenwood state park here. Town settled around nearby trading post built c.1730; moved to present site with railroad's arrival 1855.

Ninety Six Historic Site (□ 1 sq mi/2.6 sq km), NW SOUTH CAROLINA, located 2 mi/3.2 km S of NINETY SIX town and 7 mi/11.3 km ESE of GREENWOOD; 34°10′N 82°01′W. A frontier trading post after 1769 and Revolutionary War stronghold. Authorized 1976.

Nineveh, Arabic *Ninawa*, province (□ 14,555 sq mi/37,698 sq km; 2007 estimated population 2,700,000; 1987 population 1,507,000), N IRAQ; ⊙ MOSUL; 36°10′N 42°35′E. SYRIA to W, and traversed by TIGRIS RIVER. Ancient city of NINEVEH, capital of Assyrian empire, near E bank of the Tigris, is N of Mosul. Chiefly Sunni Arab population. Formerly called Mosul.

Nineveh (nin-NIV-eh), ancient city, capital of the Assyrian Empire, near E bank of the TIGRIS RIVER, N of the site of modern MOSUL, IRAQ. A shaft dug at Nineveh has yielded a pottery sequence that can be equated with the earliest cultural development in N MESOPOTAMIA. The old capital, ASHUR, was replaced by CALAH, which seems to have been replaced by Nineveh. Nineveh was thereafter generally the capital, although Sargon built Dur Sharrukin (KHORSABAD) as his capital. Nineveh reached its full glory under Sennacherib and Assurbanipal. It continued to be the leader of the ancient world until it fell to a coalition of Babylonians, Medes, and Scythians in 612 B.C.E. and the Assyrian Empire came to an end. Excavations, begun in the middle of the 19th century, have revealed an Assyrian city wall with a perimeter of c.7.5 mi/12.1 km. The palaces of Sennacherib and Assurbanipal, containing magnificent sculptures, have been discovered, as well as Assurbanipal's library, including over 20,000 cuneiform tablets. The city is mentioned often in the Bible.

Nineveh, unincorporated hamlet, BROOME county, S NEW YORK, 16 mi/26 km NE of BINGHAMTON; 42°12′N 75°37′W. Home of cartoonist Johnny Hart (*B.C.* strip).

Ninga (NIN-guh), unincorporated village, SW MANITOBA, W central Canada, 11 mi/18 km WNW of KILLARNEY, and in Turtle Mountain rural municipality; 49°13′N 99°53′W. Grain; livestock.

Ningaloo Marine Park (NIN-jah-loo) (□ 1,660 sq mi/4,316 sq km), NW WESTERN AUSTRALIA state, W AUSTRALIA, 745 mi/1,199 km N of PERTH, in INDIAN OCEAN, adjacent to CAPE RANGE NATIONAL PARK. Includes waters around North West Cape of Cape Range Peninsula and extends S 160 mi/257 km to Amherst Point. Averages 12 mi/19 km wide. Only coral reef on W side of a continent. Declared a UNESCO World Heritage Area 1994. Milyerang Visitors' Centre serves Ningaloo and Cape Range parks. Whales, dugongs, dolphins; sea turtles; sea birds; coral-eating butterfly fish. Shipwrecks at Point Cloates and Point Murat. Snorkeling, scuba diving, swimming, fishing, whale watching. Vlamingh Head Lighthouse W of North West Cape. Established 1987.

Ning'an (NING-AN), town, ⊙ Ning'an county, SE HEILONGJIANG province, NE CHINA, on railroad, on left bank of MUDAN River, and 17 mi/27 km S of MUDANJIANG; 44°23′N 129°26′E. Grain, tobacco, sugar beets; sugar refining.

Ningbo (NING-BO), city (□ 399 sq mi/1,037.4 sq km; 2000 population 3,350,851), NE ZHEJIANG province, SE CHINA, at the confluence of the YONG (or Ningbo) and Yao rivers; 29°53′N 121°33′E. Situated at the terminus of the E Zhejiang railroad, it is an industrial center and one of China's leading seaports. It was designated an "open" city in 1984 to stimulate foreign trade and investment. Its ports and economic and technological development zone are at Beilun. Ningbo has a variety of heavy and light industries, including ship building, food processing, textile mills, and the manufacturing of machinery. It is a transportation center with canal, road, and railroad links, and steamer services to places such as SHANGHAI. Long a center of culture, religion, and foreign exchange, Ningbo has many temples, Buddhist monasteries, and ancient buildings. Most notable are the Tianyi Pavilion, Tianting Temple, and Ayiwang Temple. The present site of Ningbo has been occupied since the 8th century C.E., and during the Ming dynasty (1368–1644), it was known as Qingyuan. The Tashan Dam, built in 833, is still in use for irrigation and flood control. From 1433 to 1549 it served as the port of entry for Japanese missions to the Chinese court. The Portuguese, who had established a trading settlement here in the 16th century, called the city Liampo. In the Opium War (1841), British forces occupied the city. The Treaty of Nanjing (1842) made Ningbo a treaty port. The city was known as Ningxian from 1911 to 1949. Also spelled Ningpo.

Ningcheng (NING-CHENG), town, ⊙ Ningcheng county, S central INNER MONGOLIA AUTONOMOUS REGION, N CHINA, 36 mi/58 km S of CHIFENG; 41°34′N 119°20′E. Grain, oilseeds, sugar beets; sugar refining.

Ningde (NING-DUH), city (□ 576 sq mi/1,497.6 sq km; 2000 population 361,223), NE FUJIAN province, CHINA, 40 mi/64 km NNE of FUZHOU, on EAST CHINA SEA; 26°40′N 119°30′E. Agriculture constitutes the largest sector of the city's economy. Grain; hogs; fishing; food and beverages.

Ningdu (NING-DU), town, ⊙ Ningdu county, S JIANGXI province, CHINA, 65 mi/105 km NE of GANZHOU; 26°22′N 115°48′E. Rice, oilseeds, tobacco, jute; logging; timber processing; furniture.

Ninggang (NING-GANG), town, ⊙ Ninggang county, W JIANGXI province, CHINA, 65 mi/105 km WSW of Ji'an, near HUNAN border; 26°45′N 113°58′E. Logging; furniture, building materials.

Ningguo (NING-GU-uh), town, ⊙ Ningguo county, SE ANHUI province, CHINA, near ZHEJIANG border, on railroad, 65 mi/105 km SE of WUHU; 30°38′N 118°58′E. Rice, wheat; cement.

Ninghai (NING-HEI), town, ⊙ Ninghai county, E ZHEJIANG province, CHINA, 40 mi/64 km S of NINGBO,

Area is shown by the symbol □, and capital city or county seat by ⊙.

near SANMEN BAY of EAST CHINA SEA; 29°18′N 121°25′E. Rice, wheat, oilseeds, cotton; fisheries; food and beverages, textiles, rubber and plastics; machinery and equipment manufacturing.

Ninghe (NING-HUH), town, ⊙ Ninghe county, CHINA, 40 mi/64 km NE of TIANJIN; 39°20′N 117°48′E. An administrative unit of TIANJIN municipality. Grain, cotton; food products, textiles, building materials; machinery and equipment manufacturing, papermaking.

Ninghsia Hui Autonomous Region, CHINA: see NING-XIA HUI AUTONOMOUS REGION.

Ninghua (NING-HUAH), town, ⊙ Ninghua county, W FUJIAN province, CHINA, on JIANGXI border, 35 mi/56 km NNE of CHANGTING, near source of SHA or Ninghua River (SW tributary of MIN River); 26°14′N 116°36′E. Rice, tobacco, jute; logging; building materials, machinery, chemicals.

Ningjin (NING-JIN), town, ⊙ Ningjin county, SW HEBEI province, CHINA, 35 mi/56 km SE of SHIJIAZHUANG; 37°37′N 114°55′E. Cotton, wheat, kaoliang, beans; textiles, chemicals; food processing, machinery and equipment manufacturing.

Ninglang (NING-LANG), town, ⊙ Ninglang county, NW YUNNAN province, CHINA, 40 mi/64 km NE of LIJIANG; 27°19′N 100°53′E. Timber, rice, millet; food processing.

Ningling (NING-LING), town, ⊙ Ningling county, NE HENAN province, CHINA, 65 mi/105 km SE of KAIFENG; 34°27′N 115°24′E. Grain, cotton, oilseeds; beverage manufacturing.

Ningming (NING-MING), town, ⊙ Ningming county, SW GUANGXI ZHUANG AUTONOMOUS REGION, CHINA, on railroad, 75 mi/121 km SW of NANNING; 22°12′N 107°05′E. Rice, wheat, oilseeds; sugarcane.

Ningnan (NING-NAN), town, ⊙ Ningnan county, SW SICHUAN province, CHINA, 55 mi/89 km SSE of XICHANG; 27°03′N 102°46′E. Grain, sugarcane, tobacco; food processing.

Ningpo, CHINA: see NINGBO.

Ningqiang (NING-CHIANG), town, ⊙ Ningqiang county, SW SHAANXI province, CHINA, near SICHUAN border, 45 mi/72 km WSW of Nanzheng; 32°50′N 106°15′E. Medicinal herbs, potatoes, rice, wheat, kaoliang, oilseeds; building materials, chemicals; food processing, iron ore mining, non-ferrous ore mining.

Ningshan (NING-SHAN), town, ⊙ Ningshan county, S SHAANXI province, CHINA, 70 mi/113 km SSW of Xi'an; 33°19′N 108°19′E. In mountain region. Grain; logging, timber processing.

Ningwu (NING-WU), town, ⊙ Ningwu county, N SHANXI province, CHINA, near S section of GREAT WALL, on railroad, 80 mi/129 km N of TAIYUAN; 39°01′N 112°18′E. Grain, oilseeds; coal mining.

Ningxia Hui Autonomous Region (NING-SIAH HUAI), region (□ 25,600 sq mi/66,560 sq km; 2000 population 5,486,393), N CHINA; ⊙ YINCHUAN; 37°00′N 106°00′E. Ningxia is part of the Inner Mongolian plateau, and desert and grazing land make up most of the area. But extensive land-reclamation and irrigation projects have increased cultivation, pushing the nomadic herdsmen N or forcing them to change their lifestyles. The N section, through which the HUANG HE (Yellow River) flows, is the best agricultural land. Irrigation canals and aqueducts have been built since the Qin (221 B.C.–206 B.C.E.), Han (C.E. 206–220), and Tang (618–907) dynasties. The Communist government has built dams, notably a major project at QINGTONGXIA, to regulate irrigation flow and to generate hydroelectric power. Now there are c.1,260 sq mi/3,266 sq km of drained land and c.77 sq mi/200 sq km of irrigated land. Wheat, kaoliang, rice, beans, fruit, and vegetables are grown. Wools, furs, and hides are exported. Ningxia has rich coal reserves, chiefly mined in the HELAN mountains. Desert lakes yield salt and soda. Major industries include coal mining, machinery, metals, power generation, che-

micals, and light industries. Major export products are coal, machines, rugs, knitting, flannel, and porcelain. The chief cities—YINCHUAN, WUZHONG, QINGTONGXIA, and SHIZUISHAN—are all on the Huang He. The main railroads travel N-S, with W spurs. The Batou-Lanzhou railroad, completed in 1958, forms a part of railroad trunk in N China. The Baoji-Zhongwei railroad, completed in 1994, connects Ningxia with GANSU and SHAANXI provinces. A highway system centered on Yinchuan connects to various settlements and mines. The Ningxia part of the Hunag He is navigable in most sections. The Chinese population is by far the largest; other ethnic groups include the Hui (Muslim), Mongols, Tibetans, and Manchus. Formerly a province, Ningxia was incorporated into Gansu in 1954 but was detached and reconstituted as an autonomous region for the Hui people in 1958. Ningxia received a part of the INNER MONGOLIA AUTONOMOUS REGION in 1969, but this area was returned in 1979. Ningxia University is in Yinchuan. The name sometimes appears as Ninghsia Hui; also known as Ning.

Ning Xian (NING SIAN), town, ⊙ Ning Xian county, SE GANSU province, CHINA, on Malian River, 34 mi/55 km NE of JINGCHUAN; 35°27′N 107°50′E. Grain, oilseeds, tobacco; textiles; food processing.

Ningxiang (NING-SIANG), town, ⊙ Ningxiang county, NE HUNAN province, CHINA, 28 mi/45 km WNW of CHANGSHA; 28°15′N 112°33′E. Rice, tea, oilseeds, jute. Hometown of Liu Shaoqi, former chairman of People's Republic of China. Liu's memorial hall was built here in 1985.

Ningyang (NING-YANG), town, ⊙ Ningyang county, W SHANDONG province, CHINA, 15 mi/24 km N of Yanzhou; 35°46′N 116°47′E. Grain, oilseeds, cotton, jute; food processing, machinery and equipment manufacturing, coal mining; chemicals, textiles, clothing.

Ningyuan, town, ⊙ Ningyuan county, S HUNAN province, CHINA, 50 mi/80 km SSE of Lingling (YONGZHOU); 25°36′N 111°54′E. Rice, wheat, corn, tobacco; logging; papermaking; building materials, food and beverages.

Ninh Binh (NIN BIN), province (□ 536 sq mi/1,393.6 sq km), N VIETNAM, in RED RIVER Delta; ⊙ NINH BINH; 20°15′N 106°05′E. NW border with HOA BINH province, E border with NAM HA province, S border on VINH BAC BO, SW border with THANH HOA province. Partly framed by distributaries of the Red River and interlaced with irrigation canals, province is an area of fertile alluvial soils and productive agriculture (livestock raising; wet rice cultivation; vegetables, cotton, coffee, mulberry). Fishing and aquaculture, industries (sericulture; silk weaving, food processing); light manufacturing. Predominantly Kinh population.

Ninh Binh (NIN BIN), city, ⊙ NINH BINH province, N VIETNAM, on the Song Dai, and 60 mi/97 km S of HANOI; 20°15′N 109°59′E. Market and administrative center. Transportation hub (railroad, road, and water) in cotton- and coffee-growing region; light manufacturing; cattle-trading center; sericulture. Formerly Ninhbinh.

Ninh Giang (NIN YAHNG), town, HAI HUNG province, N VIETNAM, on the SONG CAU, 45 mi/72 km ESE of HANOI; 20°44′N 106°24′E. Market hub; rice-growing and -trading center. Formerly Ninhgiang.

Ninh Hoa (NIN HWAH), town, KHANH HOA province, S central VIETNAM, transportation hub (RR and road) on railroad, and 18 mi/29 km N of NHA TRANG; 12°29′N 109°08′E. Trading center and outlet for BUON MA THUOT (linked by highway); served by the port of BA NGOI on CAM RANH BAY, 8 mi/12.9 km NE. Formerly Ninhhoa.

Ninh Thuan (NIN TOO-uhn), province (□ 1,324 sq mi/3,442.4 sq km), S central VIETNAM; ⊙ PHAN RANG; 11°40′N 108°50′E. N border with KHANH HOA province, E border on the SOUTH CHINA SEA, S border with

BINH THUAN province, W border with LAM DONG province. Including narrow coastal lowlands, hilly midlands, and a mountainous interior, province has a range of habitats and localized economies. Its once heavily forested mountains and hills have suffered widespread deforestation in recent decades. Parts of province have suffered much defoliation and devastating bombing during the Vietnam War and still bear scars from the conflict. Diverse soils and mineral resources (granite, kaolin, construction sands), wet rice farming, shifting cultivation, forest products (resins, medicinals, bamboo, foraged foods), fishing and aquaculture, agro-forestry, livestock raising, and commercial agriculture (coconut, cassava, tobacco, sugarcane, castor beans, tea, vegetables, fruits). Sawmilling, handicrafts, various industries (sugar refining, brick and tile, salt extraction, construction materials, machinery, fish curing, food processing); light manufacturing; boat building. International and domestic tourism focused on ancient Cham ruins. Predominantly Kinh population with Raglai, Churu, Cham, Chinese, and other minorities.

Ninhue (NEEN-wai), town, ⊙ Ninhue comuna, ÑUBLE province, BÍO-BÍO region, S central CHILE, 22 mi/35 km NW of CHILLÁN; 36°24′S 72°24′W. In agricultural area (grapes, cereals, vegetables, fruit; livestock); lumbering. Artisan center.

Ninigo Group, coral islands, MANUS province, NW PAPUA NEW GUINEA, BISMARCK ARCHIPELAGO, SW PACIFIC OCEAN, 150 mi/241 km W of ADMIRALTY ISLANDS; 01°16′S 144°19′E. Manu Island to W, HERMIT ISLANDS to E, BISMARCK SEA to S, Pacific Ocean to N. Comprise numerous islets and atolls, including NINIGO Island, Heina Island, and Awin island. Copra; trepang.

Ninigret Pond, Rhode Island: see CHARLESTOWN.

Ninilchik, village on W coast of KENAI PENINSULA, S ALASKA, on COOK INLET 75 mi/121 km W of SEWARD, on Kenai-Homer Highway; 60°15′N 152°25′W. Fishing, fish processing. Orthodox church. Ninilchik Recreation Area nearby. Founded c.1830 by employees of Russian American Company.

Ninnescah River (nin-ES-kuh), 49 mi/79 km long, S KANSAS; formed by confluence of its North Fork (87 mi/140 km long; forms Cherry Reservoir N of Cherry) and South Fork (92 mi/148 km long) in SEDGWICK county 6 mi/9.7 km SE of CHENEY; flows SE to ARKANSAS RIVER 19 mi/31 km NNW of ARKANSAS CITY.

Ninnis Glacier, a large, steep, heavily crevassed glacier on the GEORGE V COAST of WILKES LAND, EAST ANTARCTICA; 68°22′S 147°00′E. Flows into NINNIS GLACIER TONGUE, its floating extension.

Ninnis Glacier Tongue, seaward extension of NINNIS GLACIER, ANTARCTICA, off GEORGE V COAST; 68°07′S 147°51′E. Dimensions vary with episodic breakoffs of icebergs. Discovered 1913 by Sir Douglas Mawson.

Ninohe (nee-NO-he), city, IWATE prefecture, N HONSHU, NE JAPAN, 40 mi/65 km N of MORIOKA; 40°16′N 141°18′E. Tobacco, soybeans.

Ninomiya (nee-NO-mee-yah), town, Naka county, KANAGAWA prefecture, E central HONSHU, E central JAPAN, on N shore of SAGAMI Bay, 25 mi/40 km S of YOKOHAMA; 35°17′N 139°15′E. Peanuts.

Ninomiya, town, Haga county, TOCHIGI prefecture, central HONSHU, N central JAPAN, 12 mi/20 km S of UTSUNOMIYA; 36°22′N 139°58′E. Strawberries.

Niño Perdido (NEE-nyo per-DHEE-dho), suburb of VILLARREAL, Castellón de la Plana province, E SPAIN, 7 mi/11.3 km SW of CASTELLÓN DE LA PLANA; 39°53′N 00°06′W. In orange-growing district; ships fruit, vegetables.

Ninotsminda (NEE-nots-MIN-dah), urban settlement and administrative center of Bogdanovka region, S GEORGIA, on Akhalkalaki-Leninkan highway, and 70 mi/113 km SW of TBILISI; 41°16′N 43°35′E. Butter and cheese processing; garment manufacturing; machine shop. Formerly Bogdanovka.

Ninove (NEE-no-vuh), commune (2006 population 35,738), AALST district, EAST FLANDERS province, W central BELGIUM, on DENDER RIVER, and 15 mi/24 km W of BRUSSELS; 50°50′N 04°01′E. Textile center (cotton, artificial silk).

Ninovka (NEE-nuhf-kah), village (2004 population 2,900), central BELGOROD oblast, SW European Russia on the OSKOL RIVER; on road, 14 mi/23 km S of CHERNYANKA; 50°44′N 37°49′E. Elevation 383 ft/116 m. Furniture manufacturing.

Nio (NEE-o), town, Mitoyo county, KAGAWA prefecture, N SHIKOKU, W JAPAN, on HIUCHI SEA, 25 mi/40 km W of TAKAMATSU; 34°11′N 133°38′E.

Nioaque (NEE-o-ah-kai), city, S Mato Grosso do Sol, BRAZIL, on upper RIO MIRANDA, 90 mi/145 km SW of CAMPO GRANDE; 21°12′S 55°47′W. Cattle.

Niobrara (nei-uh-BRAR-uh), county (□ 2,627 sq mi/6,830.2 sq km; 2006 population 2,253), E WYOMING; ⊙ LUSK; 43°02′N 104°28′W. Grain, livestock region, bordering on SOUTH DAKOTA and NEBRASKA; watered by South Fork of Cheyenne River and its branches. Agriculture (wheat, oats; sheep); petroleum, natural gas. Small part of Thunder Basin National Grassland near N boundary. Formed 1911.

Niobrara (nee-o-BRUH-ruh), village (2006 population 352), KNOX county, NE NEBRASKA, on MISSOURI RIVER (LEWIS AND CLARK LAKE reservoir) at mouth of NIOBRARA RIVER, at SOUTH DAKOTA state line, and 20 mi/32 km NNW of CREIGHTON; 42°45′N 98°01′W. Dairying; livestock; grain. Resort. Santee Indian and Ponca Indian agencies here. Ponca tribal powwow in mid-August. Santee Indian Reservation to E. Niobrara State Park to W.

Niobrara (nei-uh-BRAR-uh), river, c. 430 mi/692 km long; rises in the High Plains in S NIOBRARA county, E WYOMING, flows E as an intermittent stream past LUSK, Wyoming, into NW NEBRASKA, through AGATE FOSSIL BEDS NATIONAL MONUMENT, through Box Butte Reservoir (completed 1946), past VALENTINE and BUTTE, entering Missouri River in KNOX county, NE Nebraska, 4 mi/6.4 km NNW of Niobrara.

Nioka (NYO-kah), village, ORIENTALE province, NE CONGO, 75 mi/121 km NE of IRUMU; 02°10′N 30°39′E. Elev. 5,885 ft/1,793 m. Center of agricultural research, with model livestock-breeding farm, plantations of coffee, tea, pyrethrum, cinchona, tobacco, aleurites, essential-oil plants, African staples; palm-oil milling. KWANDRUMA, 7 mi/11.3 km E, also known as RHETI or RETHY, has Protestant mission, schools. Nioka vicinity is noted as an area of European agricultural settlement, supplying KISANGANI with vegetables.

Nioki (NYO-kee), village, BANDUNDU province, SW CONGO, on right bank of FIMI RIVER, and 70 mi/113 km SW of INONGO; 02°43′S 17°41′E. Elev. 1,082 ft/329 m. Lumbering center; sawmilling, woodworking; rubber plantations. Airport.

Niokolo Koba National Park (NEE-o-ko-lo KO-bah), French *Parc National du Niokolo Koba*, game reserve (□ 3,525 sq mi/9,165 sq km), TAMBACOUNDA administrative region, SE SENEGAL, 62 mi/100 km SE of TAMBACOUNDA; 13°00′N 13°00′W. Senegal's largest national park. Hippopotamuses, lions, and elephants live here in ever-dwindling numbers. Tourist attraction.

Niono (NEE-o-no), township, FOURTH REGION/SÉGOU, S MALI, in mid-NIGER RIVER depression (irrigation), 57 mi/95 km NNE of SÉGOU; 14°15′N 06°00′W. Cotton growing. Manufacturing of vegetable oil, soap; sugar factory in nearby Siribala.

Nioro (NEE-or-o), town, FIRST REGION/KAYES, W MALI, 130 mi/209 km NE of KAYES; 15°15′N 09°35′W. Gum, peanuts, kapok; also millet, corn, manioc, sweet potatoes, melons. Animal husbandry. Rural Development Operation (ODIK). High school. Also known as Nioro du Sahel.

Nioro-du-Rip (NEE-or-o–doo–rip), town (2007 population 19,721), KAOLACK administrative region, W

SENEGAL, near GAMBIA border, 34 mi/55 km SE of KAOLACK; 13°45′N 15°48′W. Cotton, millet, peanuts. Senegalese Agricultural Research Institute station nearby.

Niort (nee-OR), city (□ 26 sq mi/67.6 sq km); ⊙ DEUX-SÈVRES department, POITOU-CHARENTES region, W FRANCE, port on SÈVRE NIORTAISE RIVER, and 34 mi/55 km ENE of LA ROCHELLE; 46°20′N 00°20′W. Commercial and transportation center with a long history of producing leather goods (especially gloves); manufacturing of wood products, fabricated metal products, electronic equipment, chemicals, and clothing. Niort has also become an important insurance center. Niort preserves two towers of a 12th–13th-century fortress, several Renaissance buildings, old houses, and museums of fine arts and natural history. It was a Huguenot stronghold in the 16th–17th century. Madame de Maintenon (the secret spouse of Louis XIV) born in a local jail. Niort holds an annual fair exhibiting its products. It is an entry point to the regional park of the nearby Poitevin Marshes (MARAIS POITEVIN, VAL DE SÈVRE ET DE LA VENDÉE NATURAL REGIONAL PARK).

Nios, Greece: see IOS.

Niota (nei-O-tuh), city (2006 population 802), MCMINN county, SE TENNESSEE, 45 mi/72 km SW of KNOXVILLE; 35°30′N 84°32′W.

Niotaze (NEE-o-taiz), village (2000 population 122), CHAUTAUQUA county, SE KANSAS, 5 mi/8 km NW of CANEY, near OKLAHOMA state line; 37°04′N 96°00′W. In livestock and grain area. Oil fields nearby.

Nioumachoua (nee-oo-MAH-shwah), town, Mwali island and district, SW Comoros Republic, 6 mi/9.7 km S of Fomboni, on S coast of island; 12°21′S 43°44′E. Fish; livestock; ylang-ylang, vanilla, cloves, bananas, coconuts. The Marine Reserve, off coast to S, has several islets including Chissioua Ouenefou island, former leper colony.

Nipani (nee-PAH-nee), town, BELGAUM district, KARNATAKA state, S INDIA, 37 mi/60 km NNW of BELGAUM; 16°24′N 74°23′E. Tobacco market; trades in chili, jaggery, corn; biri manufacturing. Seat of a college.

Nipas (NEE-pahs), town, ⊙ Ranquil comuna, ÑUBLE province, BÍO-BÍO region, central CHILE, 22 mi/35 km W of CHILLÁN; 36°36′S 72°32′W. On railroad. Fruit, vegetables.

Nipawin (NI-puh-win), town (2006 population 4,061), E central SASKATCHEWAN, CANADA, on SASKATCHEWAN RIVER, and 75 mi/121 km E of PRINCE ALBERT; 53°22′N 104°01′W. Flour milling, dairying.

Nipe Bay (NEE-pai), sheltered ATLANTIC inlet, HOLGUÍN province, NE CUBA, 50 mi/80 km N of SANTIAGO DE CUBA, and linked with sea by narrows; 14 mi/23 km long, 8 mi/13 km wide; 20°45′N 75°45′W. Receives MAYARÍ RIVER and small Nipe River. PRESTON is on E, ANTILLA on N shore.

Nipepe District, MOZAMBIQUE: see NIASSA.

Nipe, Sierra de (NEE-pai, see-ER-ah dai), range, HOLGUÍN and SANTIAGO DE CUBA provinces, E CUBA, extending c.25 mi/40 km S from NIPE BAY towards SANTIAGO, forming divide between Nipe (W) and MAYARÍ (E) rivers. Rises to 3,264 ft/995 m at Loma de Mensura. Iron mined in N. Also called Los Pinales.

Niphad (nee-PAHD), town, tahsil headquarters, NASHIK district, MAHARASHTRA state, W central INDIA, 22 mi/35 km ENE of NASHIK; 20°05′N 74°07′E. Market center for millet, wheat, gur, grapes; oilseed (peanuts) pressing.

Nipigon (NI-pi-gahn), town (□ 42 sq mi/109.2 sq km; 2001 population 1,964), NW central ONTARIO, E central Canada, on Nipigon Bay of Lake SUPERIOR, at mouth of Nipigon River, 60 mi/97 km NE of THUNDER BAY; 49°01′N 88°15′W. Lumbering. Gemstones (amethyst) found in area. Orient Bay Rock ice-climbing area nearby.

Nipigon, Lake (NI-pi-gahn) (□ 1,870 sq mi/4,862 sq km), central ONTARIO, E central Canada; 48°59′N 88°20′W. Has many islands. Its outlet, the Nipigon River (40 mi/64 km long) flows S, past the logging town of NIPIGON, into Lake SUPERIOR.

Nipisiguit River (ni-PI-zi-gwit), c.100 mi/161 km long, N New Brunswick, Canada; rises at foot of Mt. Carleton; flows E, turning NNE to Nipisiguit Bay, inlet of Chaleur Bay, at BATHURST. Grand Falls, 20 mi/32 km SSW of Bathurst, are four falls with total height of 140 ft/43 m.

Nipissing (NI-pi-sing), district (□ 6,589 sq mi/17,131.4 sq km; 2001 population 82,910), NE ONTARIO, E central Canada, on Lake NIPISSING; ⊙ NORTH BAY; 46°15′N 79°00′W.

Nipissing (NI-pi-seeng), township (□ 149 sq mi/387.4 sq km; 2001 population 1,553), S ONTARIO, E central Canada, 16 mi/26 km from NORTH BAY; 46°05′N 79°31′W.

Nipissing, Lake (NIP-i-sing) (□ 350 sq mi/910 sq km), S ONTARIO, E central Canada, between the OTTAWA River and GEORGIAN BAY. It extends W from the city of NORTH BAY and is drained SW by the FRENCH RIVER c.50 mi/80 km to Georgian Bay; 46°16′N 79°47′W.

Nipomo, unincorporated town (2000 population 12,626), SAN LUIS OBISPO county, SW CALIFORNIA 5 mi/8 km NNW of SANTA MARIA, 10 mi/16 km E of PACIFIC OCEAN; 35°02′N 120°29′W. Apples, strawberries, avocados, vegetables, grain, nursery products; cattle. SANTA MARIA RIVER to S; TWITCHELL RESERVOIR to E; Los Padre National Forest, in SIERRA MADRE to E.

Nippersink Lake, ILLINOIS: see CHAIN O' Lakes.

Nipple Mountain (12,199 ft/3,718 m), in SANGRE DE CRISTO MOUNTAINS, S central COLORADO. Boundaries of FREMONT, CUSTER, and SAGUACHE counties converge at summit. San Isabel National Forest to NE; Rio Grande National Forest to SW.

Nipple Top, peak (4,620 ft/1,408 m) of the High Peaks section of the ADIRONDACKS, in ESSEX county, NE NEW YORK, c.6 mi/9.7 km ESE of MOUNT MARCY, and 16 mi/26 km SE of LAKE PLACID village; 44°05′N 73°50′W.

Nippon: see JAPAN.

Nippur (nip-POOR), ancient city of BABYLONIA, a N Sumerian settlement in the QADISSIYA province of modern IRAQ, NE of DIWANIYA. It was the seat of the important cult of the god Enlil, or Bel. Excavations at Nippur have yielded the remains of several temples that date from the middle of the 3rd millennium B.C.E. and were later rebuilt and restored many times. Over 40,000 clay tablets found there serve as a primary source of information on Sumerian civilization. Assurbanipal erected a ziggurat in Nippur. Relics of the Persian and Parthian periods have also been unearthed at the site.

Nique, Alturas de, DARIÉN province, PANAMA, highest point (5,144 ft/1,568 m) on Panama-Colombia frontier, in DARIÉN NATIONAL PARK; 07°40′N 77°45′W.

Niquelândia (nee-kai-LAHN-zhee-ah), city (2007 population 38,517), central GOIÁS, central BRAZIL, 125 mi/201 km N of ANÁPOLIS; 14°27′S 48°28′W. Nickel mining; electric smelter under construction. Extensive nickel, cobalt, and copper deposits in region. Until 1944, called São José do Tocantins.

Niquero (nee-KER-o), town (2002 population 21,636), HOLGUÍN province, E CUBA, port on SE Guacanayabo Bay, 35 mi/56 km SW of MANZANILLO; 20°30′N 75°40′W. In fertile region (sugarcane, fruit; livestock; timber). Airfield.

Niquinohomo (nee-kee-no-HO-mo) or **La Victoria**, town, MASAYA department, SW NICARAGUA, 5 mi/8 km S of MASAYA; 11°54′N 86°06′W. Health resort; coffee, rice, corn. Scene of government victory (1860) over rebel forces.

Niquitao (nee-kee-TOU), town, TRUJILLO state, W VENEZUELA, at E foot of Teta de Niquitao (13,143 ft/

4,006 m), 19 mi/31 km S of TRUJILLO. Elevation 6,355 ft/1,937 m. Wheat, corn, potatoes.

Nir Am (NIR AHM), kibbutz, SW ISRAEL, on coastal plain, at NW edge of the NEGEV, 6 mi/9.7 km E of GAZA; 31°31′N 34°34′E. Elevation 331 ft/100 m. Manufacturing of cutlery; food processing. Agriculture (dairying; vegetables, fruit). Founded 1943.

Nira River (NEE-rah), 115 mi/185 km long, MAHARASHTRA state, W central INDIA; rises in WESTERN GHATS S of PUNE; flows ESE to BHIMA RIVER 12 mi/19 km SE of INDAPUR. LLOYD DAM (1.5 mi/2.4 km N of BHOR) supplies canal irrigation system, with headworks 16 mi/26 km E. Right (S) canal extends 106 mi/171 km ESE, past PHALTAN, to point 5 mi/8 km SE of PANDHARPUR; left (N) canal extends 100 mi/161 km E, past BARAMATI to point 4 mi/6.4 km W of Indapur. System irrigates large sugarcane area.

Nirasaki (nee-RAH-sah-kee), city, YAMANASHI prefecture, central HONSHU, central JAPAN, 9 mi/15 km N of KOFU; 35°42′N 138°26′E. Silk cocoons; peaches, grapes.

Niravi (NEE-rah-vee), French *Néravy*, town, KARAIKAL district, in former FRENCH INDIA, now in Pondicherry Union Territory, S INDIA, 2 mi/3.2 km W of KARAIKAL.

Nirayama (nee-RAH-yah-mah), town, Tagata county, SHIZUOKA prefecture, central HONSHU, E central JAPAN, 34 mi/55 km E of SHIZUOKA; 35°03′N 138°57′E. Strawberries. Hojo Masako, political figure of the early 13th century, born here.

Nir David (NIR dah-VEED) or **Tel Amal**, kibbutz, ISRAEL, at NE foot of MOUNT GILBOA, 3 mi/4.8 km W of BEIT SHEAN; 32°30′N 35°27′E. Below sea level 364 ft/110 m. Textile and metal industries; mixed farming, banana growing; fowl raising, fish breeding. There are remains of Iron Age buildings, including a factory for weaving and dying, on a *tel* on the grounds. The ancient settlement was destroyed in the 10th century B.C.E., possibly by Shishak. Founded 1936.

Ñirehuau (nyee-RE-wou), village, AISÉN province, AÍSEN DEL GENERAL CARLOS IBAÑEZ DEL CAMPO region, S CHILE, in the ANDES, 40 mi/64 km NE of PUERTO AISÉN; sheep-raising center. Also spelled Ñirihuau, Nirehuao, or Ñirihuao.

Nir Ezion (NIR e-tsee-YON), cooperative, ISRAEL, 8 mi/12.9 km S of HAIFA, on SW slopes of CARMEL range; 32°41′N 34°59′E. Elevation 853 ft/259 m. Mixed farming; guest house. Founded by religious Jews in 1950.

Nirgua (NEER-gwah), town, ⊙ Nirgua municipio, YARACUY state, N VENEZUELA, in coastal range, 39 mi/63 km W of VALENCIA (CARABOBO state); 10°09′N 68°34′W. Elevation 2,867 ft/873 m. Agricultural center (coffee, tobacco, cacao, sugarcane; cotton; cattle); liquor distilling, sugarmilling.

Ñirihuau, CHILE: see ÑIREHUAU.

Nirim (nee-REEM), kibbutz, SW ISRAEL, in the NW NEGEV, 12 mi/19 km SE of GAZA; 31°20′N 34°24′E. Elevation 318 ft/96 m. Mixed farming. Produces electronic-irrigation management equipment. Founded 1946.

Nirit (nee-REET), kibbutz, ISRAEL, 8 mi/12.9 km NE of PETAH TIKVA, in E fringes of the SHARON plain. Founded 1982.

Nirivilo (nee-ree-VEE-lo), village, MAULE region, S central CHILE, 32 mi/51 km NE of CAUQUENES; 35°33′S 72°05′W. Grain, potatoes, lentils, sheep; lumbering.

Niriz, IRAN: see NEYRIZ.

Nirmal (NIR-muhl), town, ADILABAD district, ANDHRA PRADESH state, central INDIA, 40 mi/64 km SSW of ADILABAD; 19°06′N 78°21′E. Road center in agricultural area; cotton ginning, rice milling; noted toy handicrafts.

Niroemoar, INDONESIA: see KUMAMBA ISLANDS.

Nirsa (NIR-sah), village, BARDDHAMAN district, WEST BENGAL state, E INDIA, in W RANIGANJ coalfield, 39 mi/63 km NE of Puruliya. Coal mining.

Nirumoar, INDONESIA: see KUMAMBA ISLANDS.

Nir Yizhak (NIR yeets-KHAHK) or **Yitzhaq**, kibbutz, S ISRAEL, 25 mi/40 km W of BEERSHEBA, in NW NEGEV; 31°14′N 34°22′E. Elevation 324 ft/98 m. Mixed farming; light industry. Founded 1949.

Niš (NEESH), city, in Serbia, on the Nišava River; 43°19′N 21°54′E. An important railroad and industrial center; manufacturing (textiles, electronics, spirits, and locomotives). Airport. The Roman *Naissus*, it was the site of a victory (269) of Claudius II over the Ostrogoths and was the birthplace of Constantine I (Constantine the Great). In 441 it was destroyed by the Huns but was rebuilt (6th century) by Emperor Justinian I. In the Middle Ages the city passed back and forth between the Bulgarian and Serb empires. The Turks captured it c.1386, were defeated here in 1443 by John Hunyadi, and recaptured it again in 1456. It became (until 1878) their most important military stronghold in the Balkans. It passed to Serbia in 1878. The city retains a medieval fortress that dominates the S Morava valley. The Tower of Skulls (Serbian *Cele Kula*) was built to commemorate the Serbs massacred by the Turks in the uprising of 1809. Also spelled Nish.

Nisa (NEE-zah) or **Niza**, town, PORTALEGRE district, central PORTUGAL, c.20 mi/32 km NW of PORTALEGRE; 39°31′N 07°39′W. Set in olive-growing area; town has market; some 18th-century buildings. Thermal waters nearby.

Nisa, POLAND: see NEISSE, town.

Nisa, POLAND: see NEISSE, RIVER.

Nisab (nee-SAHB), township, S YEMEN, 70 mi/113 km NNW of AHWAR; 14°31′N 46°30′E. Center of cotton-growing area; cotton weaving (shawls, turbans). Was main center of the former Upper AULAQI sultanate.

Nisa River, POLAND and CZECH REPUBLIC: see NEISSE.

Nisato, town, Shimohei county, IWATE prefecture, N HONSHU, NE JAPAN, 34 mi/55 km E of MORIOKA; 39°37′N 141°47′E.

Nisato (NEE-sah-to), village, Seta county, GUMMA prefecture, central HONSHU, N central JAPAN, 12 mi/20 km E of MAEBASHI; 36°24′N 139°14′E. Cucumbers.

Nišava, Serbian *Nišavski Okrug*, district (□ 1,054 sq mi/2,740.4 sq km; 2002 population 381,757), ⊙ NIŠ, SE central SERBIA; 43°16′N 21°58′E. Includes municipalities (*opštinas*) of ALEKSINAC, Doljevac, GADZIN HAN, Merosina, Nis, Razanj, and SVRLJIG. Electronics, textiles; tobacco.

Nisava River, BULGARIA and SERBIA: see NISHAVA RIVER.

Niscemi (nee-SHAI-mee), town, CALTANISSETTA province, S SICILY, 9 mi/14 km NE of GELA; 37°09′N 14°23′E. In cork-growing region; wine; food processing. Panoramic coastal view.

Niseko (nee-SE-ko), town, Shiribeshi district, Hokkaido prefecture, N JAPAN, 37 mi/60 km W of SAPPORO; 42°48′N 140°41′E. Potatoes. Skiing area.

Nish, SERBIA: see NIŠ.

Nishapur, IRAN: see NEYSHABUR.

Nishava River (nee-SAH-vah), Serbo-Croatian *Nišava*, c.100 mi/161 km long, NW BULGARIA and SE SERBIA; rises in the BERKOVITSA Mountains at the S foot of KOM peak; flows WNW, into SERBIA, past DIMITROVGRAD, PIROT, BELA PALANKA, Nishka Banja, and NIS, to SOUTHERN MORAVA River 7 mi/11 km WNW of Nis (43°00′N 23°56′E). Sofia-Nis railroad follows its course. Sometimes spelled Nisava River.

Nishiaidzu (nee-shee-AH-eedz), town, Yama county, FUKUSHIMA prefecture, N central HONSHU, NE JAPAN, 47 mi/75 km W of FUKUSHIMA city; 37°35′N 139°39′E.

Nishiarie (nee-SHEE-AH-ree-eh), town, South Takaki county, NAGASAKI prefecture, W KYUSHU, SW JAPAN, on SE SHIMABARA PENINSULA, 40 mi/64 km E of NAGASAKI, near ARIE; 32°39′N 130°18′E.

Nishiarita (nee-SHEE-AH-ree-tah), town, West Matsuura county, SAGA prefecture, N KYUSHU, SW JAPAN, 28 mi/45 km W of SAGA; 33°12′N 129°51′E.

Nishiawakura (nee-SHEE-ah-WAHK-rah), village, Aida county, OKAYAMA prefecture, SW HONSHU, W JAPAN, 43 mi/70 km N of OKAYAMA; 35°10′N 134°20′E. Cryptomeria forest.

Nishiazai (nee-shee-ah-ZAH-ee), town, Ika county, SHIGA prefecture, S HONSHU, central JAPAN, 37 mi/60 km N of OTSU; 35°29′N 136°07′E.

Nishibiwajima (nee-SHEE-bee-wah-JEE-mah), town, West Kasugai county, AICHI prefecture, S central HONSHU, central JAPAN, 3.1 mi/5 km N of NAGOYA; 35°12′N 136°52′E.

Nishigoshi (nee-shee-GO-shee), town, Kikuchi county, KUMAMOTO prefecture, W KYUSHU, SW JAPAN, 6 mi/10 km N of KUMAMOTO; 32°54′N 130°44′E. Watermelon.

Nishihara (nee-SHEE-hah-rah), town, on S OKINAWA island, Nakagami county, Okinawa prefecture, SW JAPAN, 5.6 mi/9 km N of NAHA; 26°13′N 127°46′E.

Nishihara (nee-SHEE-hah-rah), village, Aso county, KUMAMOTO prefecture, W KYUSHU, SW JAPAN, near Mount Tawara, 9 mi/15 km E of KUMAMOTO; 32°49′N 130°54′E. Sweet potatoes.

Nishiharu (nee-SHEE-HAH-roo), town, West Kasugai county, AICHI prefecture, S central HONSHU, central JAPAN, 5 mi/8 km N of NAGOYA; 35°14′N 136°52′E. Traditional dolls. Enzymatic industries.

Nishiiyayama (nee-SHEE-ee-yah-YAH-mah), village, Miyoshi county, TOKUSHIMA prefecture, SE SHIKOKU, W JAPAN, 43 mi/70 km W of TOKUSHIMA; 33°53′N 133°49′E. Nearby attractions include Iya (spanned by Iya-no-kazura Bridge) and Oboke gorges.

Nishiizu (nee-shee-EEZ), town, Kamo county, SHIZUOKA prefecture, central HONSHU, E central JAPAN, 25 mi/40 km S of SHIZUOKA; 34°46′N 138°46′E. Marguerites. Dogashima Cave is nearby.

Nishikata (nee-shee-KAH-tah), village, Kamikatsuga county, TOCHIGI prefecture, central HONSHU, N central JAPAN, 9 mi/15 km S of UTSUNOMIYA; 36°27′N 139°44′E.

Nishikatsura (nee-shee-KAHTS-rah), town, South Tsuru county, YAMANASHI prefecture, central HONSHU, central JAPAN, 19 mi/30 km S of KOFU; 35°31′N 138°51′E.

Nishikawa (nee-shee-KAH-wah), town, West Kanbara county, NIIGATA prefecture, central HONSHU, N central JAPAN, 9 mi/15 km S of NIIGATA; 37°47′N 138°55′E.

Nishikawa, town, West Murayama county, YAMAGATA prefecture, N HONSHU, NE JAPAN, 19 mi/30 km N of YAMAGATA city; 38°25′N 140°09′E.

Nishiki (NEE-shee-kee), town, Taki district, HYOGO prefecture, W central JAPAN, 27 mi/44 km N of KOBE; 35°05′N 135°10′E.

Nishiki, town, Kuma county, KUMAMOTO prefecture, W KYUSHU, SW JAPAN, 40 mi/65 km S of KUMAMOTO; 32°11′N 130°50′E.

Nishiki, town, Kuga county, YAMAGUCHI prefecture, SW HONSHU, W JAPAN, near Mount Rakan, 31 mi/50 km E of YAMAGUCHI; 34°15′N 131°57′E. Nishiki (carp) and Jakuchi rivers nearby.

Nishiki (nee-shee-KEE), village, Senhoku county, Akita prefecture, N HONSHU, NE JAPAN, near LAKE TAZAWA, 25 mi/40 km E of AKITA city; 39°39′N 140°34′E. Chestnuts.

Nishiko (nee-shee-KO), village, W Shirakawa county, FUKUSHIMA prefecture, N central HONSHU, NE JAPAN, 47 mi/75 km S of FUKUSHIMA city; 37°08′N 140°09′E. Fish; shiitake mushrooms, potatoes; cork leather. Hot springs nearby.

Nishime (nee-shee-ME), town, Yuri county, Akita prefecture, N HONSHU, NE JAPAN, 25 mi/40 km S of AKITA city; 39°20′N 140°01′E.

Nishimera (nee-SHEE-me-rah), village, Koyu county, MIYAZAKI prefecture, SE KYUSHU, SW JAPAN, 28 mi/45 km N of MIYAZAKI; 32°13′N 131°09′E. Citrons and citron processing.

Nishimeya (nee-shee-ME-yah), village, Nakatsugaru county, Aomori prefecture, N HONSHU, N JAPAN, 28 mi/45 km S of AOMORI; 40°34′N 140°17′E. Lumber.

Nishinasuno (nee-shee-NAHS-no), town, Nasu county, TOCHIGI prefecture, central HONSHU, N central JAPAN, 22 mi/35 km N of UTSUNOMIYA; 36°53′N 139°58′E.

Nishine (nee-shee-NE), town, Iwate county, IWATE prefecture, N HONSHU, NE JAPAN, near Mount Iwate, 16 mi/25 km N of MORIOKA; 39°54′N 141°06′E. Nearby Yakibashiri has protected remains of lava flow.

Nishinomiya (nee-shee-NO-mee-yah), city (2005 population 465,337), HYOGO prefecture, S HONSHU, W central JAPAN, third-largest city on OSAKA BAY, 11 mi/17 km E of KOBE; 34°44′N 135°20′E. Famous sake produced here. Resort and site of several temples that were founded in the 7th and 8th century. Seat of Kobe Women's College. Koshien Baseball Stadium is loated here.

Nishinoomote (nee-SHEE-no-O-mo-te), chief city and port on W coast of TANEGA-SHIMA island, KAGOSHIMA prefecture, SW JAPAN, on OSUMI STRAIT, 65 mi/105 km S of KAGOSHIMA; 30°43′N 130°59′E. Scissors.

Nishinoshima (nee-shee-NO-shee-mah), town, NISHI-NO-SHIMA island, DOZEN island group, Oki county, SHIMANE prefecture, W JAPAN, near Mount Takuhi, 44 mi/71 km N of MATSUE; 36°05′N 132°59′E.

Nishi-no-shima (nee-SHEE-no-SHEE-mah), largest island (□ 33 sq mi/85.8 sq km) and town of DOZEN group of the OKI-GUNTO, SHIMANE prefecture, W JAPAN, 8 mi/12.9 km SW of DOGO; 12 mi/19 km long, c.0.25 mi/0.4 km–4 mi/6.4 km wide. Mountainous, forested; lumber, charcoal, rice. Sometimes called Nishi Shima.

Nishi-notoro-saki, RUSSIA: see CRILLON, CAPE.

Nishinoura, Cape (nee-SHEE-no-OO-rah), Japanese *Nishiura-misaki* (nee-SHEE-no-OO-rah-mee-SAH-kee), SAGA prefecture, N KYUSHU, SW JAPAN, forms W side of entrance to HAKATA BAY (inlet of GENKAI SEA); 33°40′N 130°13′E.

Nishio (nee-SHEE-o), city, AICHI prefecture, S central HONSHU, central JAPAN, 22 mi/35 km S of NAGOYA; 34°51′N 137°03′E. Automotive parts; tea. Tenteko Festival held here.

Nishiokoppe (NEE-shee-o-KOP-pe), village, Abashiri district, Hokkaido prefecture, N JAPAN, 118 mi/190 km N of SAPPORO; 44°19′N 142°56′E.

Nishi-sakutan, RUSSIA: see BOSHNYAKOVO.

Nishisenhoku (nee-shee-SEN-ho-koo), town, Senhoku county, Akita prefecture, N HONSHU, NE JAPAN, 19 mi/30 km S of AKITA city; 39°32′N 140°22′E.

Nishitosa (nee-SHEE-to-sah), village, Hata county, KOCHI prefecture, S SHIKOKU, W JAPAN, 50 mi/80 km S of KOCHI; 33°10′N 132°47′E. Lumber.

Nishiumi (nee-SHEE-oo-mee), town, S Uwa county, EHIME prefecture, NW SHIKOKU, W JAPAN, 65 mi/105 km S of MATSUYAMA; 32°56′N 132°30′E. Fish and fish processing.

Nishiwaki (nee-SHEE-wah-kee), city, HYOGO prefecture, S HONSHU, W central JAPAN, 24 mi/38 km N of KOBE; 34°59′N 134°58′E. Textiles.

Nishiyama (nee-shee-YAH-mah), town, Kariwa county, NIIGATA prefecture, central HONSHU, N central JAPAN, 37 mi/60 km S of NIIGATA; 37°27′N 138°40′E.

Nishiyoshino (NEE-shee-YO-shee-no), village, Yoshino district, NARA prefecture, S HONSHU, W central JAPAN, 29 mi/46 km S of NARA; 34°16′N 135°45′E. Fruits; persimmon vinegar. Hot springs nearby.

Nishnabotna River, c.100 mi/161 km long, in SW IOWA and NW MISSOURI; formed by EAST and WEST NISHNABOTNA rivers; rises in CARROLL county, IOWA; both flow S and W, then join near HAMBURG to form Nishnabotna River; flows c.12 mi/19 km S to MISSOURI RIVER 2 mi/3.2 km W of WATSON in extreme NW Missouri. Used for hydroelectric power. Tributaries and main stream are canalized for flood control. Because of record rains, serious flooding occurred along the river in 1993.

Nísia Floresta (NEE-see-ah FLO-re-stah), city (2007 population 22,919), E RIO GRANDE DO NORTE state, NE BRAZIL, near the Atlantic Ocean, 19 mi/31 km S of NATAL; 06°04′S 35°08′W. Cotton, corn, manioc; cattle. Until 1948, called Papari.

Nisib, TURKEY: see NIZIP.

Nisibis, TURKEY: see NUSAYBIN.

Nisida (NEE-zee-dah), ancient *Nesis*, rocky islet, S ITALY, in BAY OF NAPLES, 3 mi/5 km SE of POZZUOLI; 40°47′N 14°11′E. Site of penal institution. Brutus had a villa here.

Nisiro, Greece: see NISYROS.

Niška Banja (NEESH-kah BAHN-yah), village (2002 population 15,359), E SERBIA, on railroad, and 6 mi/9.7 km ESE of NIS, near NISAVA RIVER; 43°18′N 22°01′E. Health resort. Also spelled Nishka Banja.

Niskayuna, hamlet (□ 1 sq mi/2.6 sq km; 2000 population 4,892), SCHENECTADY county, E NEW YORK, on MOHAWK RIVER, and 5 mi/8 km E of SCHENECTADY; 42°47′N 73°51′W.

Nisko (NEE-sko), town, Tarnobrzeg province, SE POLAND, on SAN RIVER, on railroad, and 33 mi/53 km N of RZESZOW; 50°32′N 22°09′E. Brewing, lumber milling; brickworks.

Nisku, hamlet, central ALBERTA, W Canada, 5 mi/7 km N of LEDUC, in Leduc County; 53°20′N 113°32′W.

Nisky (NES-kee), village, S ST. THOMAS Island, U.S. VIRGIN ISLANDS, 1.5 mi/2.4 km W of CHARLOTTE AMALIE. Site of a Moravian mission (founded 1755), in picturesque setting. Sometimes spelled Niesky.

Nisland, village (2006 population 210), BUTTE county, W SOUTH DAKOTA, 15 mi/24 km E of BELLE FOURCHE, and on BELLE FOURCHE RIVER; 44°40′N 103°32′W. County fair takes place here.

Níspero, El, HONDURAS: see EL NÍSPERO.

Nisporeni (nis-po-REN), village (2004 population 12,105), W MOLDOVA, 30 mi/48 km WNW of CHISINAU (Kishinev); 47°04′N 28°10′E. Corn, wheat, fruit, wine. Formerly spelled Nisporeny.

Nisqually River (nuh-SKWAH-lee), 81 mi/130 km long, W central WASHINGTON; fed by Nisqually Glacier on S slopes of MOUNT RAINIER, MOUNT RAINIER NATIONAL PARK, SE PIERCE county; flows SW, then W through Alder Lake reservoir, then NW past YELM through part of Fort Lewis Military Reservation, and past Nisqually Indian Reservation to PUGET SOUND 10 mi/16 km ENE of OLYMPIA. ALDER DAM (completed 1944; 330 ft/101m high, 1,600 ft/488m long) and LA GRANDE DAM (completed 1945; 215 ft/66m high, 710 ft/216m long) furnish power. Nisqually National Wildlife Refuge on W side of mouth.

Nisramont Lake and Dam (nee-zrah-MAW), nature reserve near NISRAMONT, LUXEMBOURG Province, SE BELGIUM, 10 mi/16 km NNW of BASTOGNE.

Nissan, atoll, N SOLOMON ISLANDS, PAPUA NEW GUINEA, SW PACIFIC OCEAN, 100 mi/161 km E of NEW IRELAND; 04°31′S 154°11′E. Comprises 4 islands (NISSAN, Barahun, Sirot, Han) on reef 10 mi/16 km long, 5 mi/8 km wide. Coconuts. In World War II, Allied capture (1944) of island completed Solomons campaign. Sometimes called GREEN ISLAND.

Nissan (NIS-sahn), river, 100 mi/161 km long, SW SWEDEN; rises SW of JÖNKÖPING; flows SW, past GISLAVED and OSKARSTRÖM, to KATTEGATT at HALMSTAD.

Nissan-lez-Anserune, FRANCE: see NISSAN-LEZ-ENSÉRUNE.

Nissan-lez-Ensérune (nee-sahn–lai–zahn-sai-ROON), commune (□ 12 sq mi/31.2 sq km), HÉRAULT department, LANGUEDOC-ROUSSILLON region, S FRANCE, near the CANAL DU MIDI, 6 mi/9.7 km SW of BÉZIERS; 43°17′N 03°08′E. Vineyards. Nearby are the remains (and museum) of ancient fort of Ensérune built in 6th century B.C.E., later occupied by Greek mariners, then by Romans who founded their colony at nearby NARBONNE in 118 B.C.E. The fort was probably abandoned in 1st century C.E. when peace prevailed in the Roman Empire. Also spelled Nissan-lez-Anserune.

Nissedal (NIS-suh-dahl), village, TELEMARK county, S NORWAY, on NISSER Lake, 40 mi/64 km W of SKIEN. Lumbering, fishing. From near Treungen (administration center), 11 mi/18 km S of the village, is a railroad line to GRIMSTAD and ARENDAL.

Nissequogue (NIS-uh-kwahg), residential village (□ 5 sq mi/13 sq km; 2006 population 1,457), SUFFOLK county, SE NEW YORK, on N shore of LONG ISLAND, near mouth of NISSEQUOGUE RIVER, 12 mi/19 km E of HUNTINGTON; 40°54′N 73°11′W. Residential and recreational area.

Nissequogue River (NIS-uh-kwahg), c.8 mi/12.9 km long, SE NEW YORK; rises in small lake just S of SMITHTOWN BRANCH on LONG ISLAND; flows SW and NE, past SMITHTOWN, to SMITHTOWN BAY 4 mi/6.4 km W of STONY BROOK.

Nisser (NIS-suhr), lake, TELEMARK county, S NORWAY, 49 mi/79 km W of SKIEN; 22 mi/35 km long, 2 mi/3.2 km wide. Elev. 794 ft/242 m. Outlet: NIDELVA River.

Nissin (NEE-seen), town, Aichi county, AICHIprefecture, S central HONSHU, central JAPAN, 9 mi/15 km E of NAGOYA; 35°07′N 137°02′E.

Nissum Fjord (NI-soom), inlet, c.10 mi/16 km long, of NORTH SEA, W JUTLAND, DENMARK, S of Lim Fjord. STORA River flows into it.

Nisswa (NIS-wah), town (2000 population 1,953), CROW WING county, central MINNESOTA, 13 mi/21 km NNW of BRAINERD; 46°30′N 94°17′W. Resort area. Dairying; poultry; oats, alfalfa; manufacturing (boat lifts and docks, labels). Cullen Lake to N, GULL LAKE RESERVOIR to SW, PELICAN LAKE to NE; Pillsbury State Forest to SW.

Nistelrode (NIS-tuhl-raw-duh), village, NORTH BRABANT province, E central NETHERLANDS, 11 mi/18 km E of 's-HERTOGENBOSCH; 51°42′N 05°34′E. Dairying; livestock; vegetables, fruit, grain; manufacturing (toys).

Nisutlin River S, 150 mi/241 km long, central YUKON, Canada, headstream of Yukon River; rises in SE part of PELLY MOUNTAINS; flows in wide arc W and S to TESLIN LAKE.

Nisyros (NEE-see-ros), Italian *Nisiro*, volcanic island (□ 18 sq mi/46.8 sq km), DODECANESEprefecture, SOUTH AEGEAN department, GREECE, off RESADIYE Peninsula (SW Turkey), and S of KÓS; 4 mi/6.4 km long, 4 mi/6.4 km wide; 36°35′N 27°10′E. Rises to 2,270 ft/692 m. Has dormant volcano (sulphur deposits; hot springs). Produces barley, olive oil, figs, almonds, wheat. Mandraki (main town) is on NW shore. Also spelled Nisiros.

Nita (NEE-TAH), town, Nita county, SHIMANE prefecture, SW HONSHU, W JAPAN, 20 mi/33 km S of MATSUE; 35°11′N 133°00′E. Beef cattle; rice. Traditional iron manufacturing; abacus making.

Nitchequon, unincorporated village, E central QUEBEC, E Canada, on NE shore of NICHICUN LAKE, where La Grande River exits and flows N from lake, and included in Baie-James; included in Baie-James municipality; 53°12′N 70°55′W. Within JAMES BAY Hydro Project. Populated by Cree Indians. Iron ore and timber reserves in area. Fishing, hunting.

Niterói (NEE-te-roi), city (2007 population 474,002), SE BRAZIL, on GUANABARA BAY opposite the city of RIO DE JANEIRO; 22°53′S 43°07′W. Residential suburb of Rio, and many of its citizens commute to work across the bay. A highway bridge (opened in 1974) connects the two cities across Guanabara Bay. Shipbuilding center; foodstuffs, transportation equipment, textiles, pharmaceuticals, and metals. The area was settled by native Americans in 1573 on land granted by the king of Portugal. The city's name derives from the Native American word *Nyteroi* meaning "hidden water." By 1819 the indigenous community was extinct, and in 1835 Niterói became the provincial capital until 1974 when Guanabara was absorbed into Rio de Janeiro state, and the capital was moved to Rio de Janeiro city. Also spelled Nichtheroy.

Nith River (NITH), c.70 mi/113 km, SW ONTARIO, E central Canada; rises SE of LISTOWEL; flows SE, past NEW HAMBURG, to THAMES River 9 mi/14 km SW of Galt; 43°11′N 80°22′W. Receives Millbank River.

Nith River (NITH), 80 mi/129 km long, East Ayrshire and DUMFRIES and GALLOWAY, S Scotland; rises 4 mi/6.4 km E of DALMELLINGTON; flows NE to New Cumnock, E to Sanquhar, and SE past Thornhill and Dumfries to SOLWAY FIRTH 11 mi/18 km S of Dumfries. Its valley is called Nithsdale. Receives AFTON WATER at New Cumnock.

Nitinat Lake (NI-tuh-nat), (□ 10 sq mi/26 sq km), SW BRITISH COLUMBIA, W Canada, on SW VANCOUVER ISLAND, 70 mi/113 km WNW of VICTORIA, in lumbering area; 13 mi/21 km long, 1 mi/2 km wide; 48°49′N 124°40′W.

Niti Pass (NEE-TEE), pass (16,627 ft/5,068 m), in SE ZASKAR RANGE of Kumaun Himalayas, SW TIBET, SW CHINA, near INDIA (border undefined), 60 mi/97 km SSW of GARYARSA, at foot of Peak KAMET (25,447 ft/7,756 m); 30°57′N 79°53′E. Also called KIUNGLANG.

Nitmiluk National Park or **Katherine Gorge National Park** (KATH-rin GORJ) (□ 696 sq mi/1,809.6 sq km), N central NORTHERN TERRITORY, N central AUSTRALIA, 210 mi/338 km SE of DARWIN, 20 mi/32 km NE of KATHERINE; 14°15′S 132°30′E. Length is 45 mi/72 km, width 18 mi/29 km. Administered by Northern Territory Conservation Commission; owned by Aboriginal people. Deep sandstone gorge of Katherine River; waterfalls. Kangaroos, euros, echidnas, dingoes; freshwater crocodiles, goannas; emus, brolga cranes, hooded parrots, great bowerbird. Pandanus, freshwater mangrove, sand palms. Visitor center, kiosk. Camping, picnicking, hiking, boating, swimming. Established 1980.

Niton Junction, hamlet, central ALBERTA, W Canada, 27 mi/44 km from EDSON, in Yellowhead County; 53°37′N 115°46′W.

Nitra (nyit-RAH) German *Neutra*, Hungarian *Nyitra*, city (2000 population 87,285), ZAPADOSLOVENSKY province, SLOVAKIA, on the NITRA RIVER, a tributary of the DANUBE RIVER; 48°19′N 18°05′E. Agricultural market center; extensive food-processing industries; manufacturing (machinery, chemicals, textiles). Dating from Roman times, Nitra was important from the 9th century onward as a religious center and fortress. It became a free city by royal decree in 1248, and was made a Roman Catholic bishopric in 1288. The bishopric church and a castle (founded c.830) are the oldest structures in Slovakia. Seat of an agrigultural college; has an agricultural museum. Also contains a military base.

Nitra River (nyit-RAH), GERMAN *Neutra*, HUNGARIAN *Nyitra*, c.122 mi/196 km long, SW SLOVAKIA; rises on SE slope of the LESSER FATRA, 13 mi/21 km N of PRIEVIDZA; flows generally S, past NITRA and NOVÉ ZÁMKY, to VÁH River just N of KOMÁRNO.

Nitrianske Pravno (nyit-RYAHN-ske prahf-NO), Hungarian *Nyitrapróna*, village, STREDOSLOVENSKY province, W central SLOVAKIA, 6 mi/9.7 km N of PRIEVIDZA; 48°52′N 18°38′E. Contains railroad terminus, food processing, and clothing manufacturing. Neolithic-era archaeological site. Until 1946, was called NEMECKE PRAVNO (nye-METS-ke prah-VE-nets), Slovak *Nemecké Pravno*, German *Deutschproben*. Large furniture factory just S, in PRAVENEC (prah-VE-nets).

Nitrianske Rudno (nyit-RYAHN-ske rud-NO), Hungarian *Nyitrarudnó*, village, STREDOSLOVENSKY province, W central SLOVAKIA, 6 mi/9.7 km WNW of PRIEVIDZA; 48°48′N 18°29′E. Clothing manufacturing; noted summer resort (river, dam). Has 17th-century UHROVSKY castle (u-HROU-skee), Slovak *Uhrovský*, 5 mi/8 km SW.

Nitriansky Hradok, SLOVAKIA: see BANOV.

Nitro (NEI-tro), town (2006 population 6,739), PUTNAM and KANAWHA counties, W central WEST VIRGINIA, on KANAWHA RIVER (bridged), suburb 11 mi/18 km WNW of CHARLESTON; 38°25′N 81°49′W. Agriculture (corn, tobacco); cattle; poultry. Coal-producing area. Manufacturing (chemicals, glass, screens, protective castings; printing and publishing). A boomtown in World War I, government explosives plant here was later abandoned. Incorporated 1932.

Nitsanim, ISRAEL: see NITZANIM.

Nitsa River (NEE-tsah), 100 mi/161 km long, SVERDLOVSK oblast, W Siberian Russia; formed 20 mi/32 km ENE of ALAPAYEVSK by the confluence of the Nieva and REZH rivers; flows generally ESE, past IRBIT and YELAN′, to the TURA RIVER 10 mi/16 km SSE of TURINSKAYA SLOBODA. Navigable below Irbit; timber floating. Receives the Irbit River.

Nitsu (NEETS), city, NIIGATA prefecture, central HONSHU, N central JAPAN, 9 mi/15 km S of NIIGATA; 37°47′N 139°07′E. Important transportation hub and commercial center.

Nitsuru (NEETS-roo), village, Onuma county, FUKUSHIMA prefecture, N central HONSHU, NE JAPAN, 37 mi/60 km S of FUKUSHIMA city; 37°30′N 139°50′E. Medicinal plants.

Nitta (NEET-TAH), town, Nitta county, GUMMA prefecture, central HONSHU, N central JAPAN, 16 mi/25 km S of MAEBASHI; 36°17′N 139°18′E. Beef cattle.

Nittenau (NIT-ten-ou), town, UPPER PALATINATE, E BAVARIA, GERMANY, in BOHEMIAN FOREST, on the REGEN, 15 mi/24 km NE of REGENSBURG; 49°12′N 12°16′E. Manufacturing includes machinery, clothing, rubber products. Town since 1953.

Nittendorf (NIT-ten-dorf), village, UPPER PALATINATE, BAVARIA, GERMANY, on the NAAB near its mouth to the DANUBE, 5 mi/8 km WNW of REGENSBURG; 49°02′N 11°59′E. Dairy products, machinery.

Nitzana (nee-tzah-NAH), crossing point between ISRAEL and EGYPT in NE SINAI on the Beersheba-Ismailiya road; 30°53′N 34°25′E. Elevation 744 ft/226 m.

Nitzanim (nee-tzah-NEEM) or **Nitsanim**, kibbutz, W ISRAEL, in coastal plain, near MEDITERRANEAN, 16 mi/26 km SW of REHOVOT; 31°43′N 34°37′E. Elevation 75 ft/22 m. Produces chairs; mixed farming. Adjoining it is a youth village and agriculture school and training center; beach. Founded 1943; captured and destroyed (1948) by Egyptians; shortly afterwards retaken by Israeli forces and rebuilt.

Niuafo′ou (nee-OO-ah-FAW-o), island, northernmost island of TONGA, S PACIFIC OCEAN; 3.5 mi/5.6 km long, 3 mi/4.8 km wide; 15°45′S 175°45′W. Volcanic; large crater lake, hot springs. Island was evacuated in 1946 because of periodic eruptions, and later reoccupied. Also spelled Niuafoo; popularly called TIN CAN ISLAND. Formerly called Proby Island.

Niuatoputapu (NYOO-ah-TAW-poo-TAH-poo), volcanic island, extinct, N TONGA, S PACIFIC OCEAN, 150 mi/241 km N of VAVA′U; c.3 mi/4.8 km long, 1 mi/1.6 km wide; 16°00′S 173°48′W. Rises to 350 ft/107 m. Coral airstrip. Also spelled Niuatobutabu; also called Keppel Island.

Niue (nee-OO-ee), coral island (□ 100 sq mi/260 sq km), S PACIFIC OCEAN, 300 mi/483 km E of TONGA ⊙ ′ALOFI; 19°02′S 169°55′W. Exports coconut milk, honey, fruit juices, and reconstituted milk. Since 1974, Niue is a self-governing state in free association with NEW ZEALAND, with New Zealand citizenship. In 1991, Niue's population was outnumbered more than fivefold by the ethnic Niueans living in NEW ZEALAND. Formerly called Savage Island.

Niulakita (NYOO-lah-KEE-tah), atoll (2002 population 35), TUVALU, SW PACIFIC; 10°45′S 179°30′E; 1 mi/1.6 km long, 0.5 mi/0.8 km wide. Covers 104 acres/42 ha. Also called Nurakita; formerly called Sophia Island and Rocky.

Niutao (NEE-oo-TAH-o), coral island (2002 population 663), TUVALU, SW PACIFIC; 06°06′S 177°16′E.

Covers 625 acres/253 ha. Copra. Formerly called Speiden Island.

Niuxiatai (NOO-SI-AH-TEI), town, central LIAONING province, NE CHINA, on railroad spur, 8 mi/12.9 km NE of BENXI. Grain, oilseeds; coal and alunite mining center.

Niuzhuang (NIU-JUANG), town, SW LIAONING province, NE CHINA, 25 mi/40 km NE of YINGKOU, and on arm of lower LIAO River; 40°56′N 122°32′E. The oldest port for the Liao valley, the town flourished throughout 18th century. Changes in the lower Liao course and subsequent silting caused its gradual decline (19th century).

Nivala (NI-vah-lah), village, OULUN province, W FINLAND, near the KALAJOKI (river), and 55 mi/89 km E of KOKKOLA; 63°55′N 24°58′E. Elevation 297 ft/90 m.

Niva River (nee-VAH), 22 mi/35 km long, on SW KOLA PENINSULA, MURMANSK oblast, extreme NW European Russia; leaves Lake IMANDRA (elevation 420 ft/128 m) at ZASHEYEK; flows S, past NIVSKIY (site of the Niva hydroelectric plant), to KANDALAKSHA BAY at KANDALAKSHA. Its strong current keeps it from freezing during the winter.

Nivelle, Mount (NI-vuhl) (10,620 ft/3,237 m), SE BRITISH COLUMBIA, W Canada, near ALBERTA border, in ROCKY MOUNTAINS, 50 mi/80 km SSE of BANFF (Alberta); 50°31′N 115°11′W.

Nivelles (nee-VEL), Flemish *Nijvel*, commune (□ 421 sq mi/1,094.6 sq km; 2006 population 24,378), ⊙ Nivelles district, BRABANT province, central BELGIUM; 50°36′N 04°20′E. Industrial center and a railroad junction. Manufacturing includes machinery, linen, cotton goods, and lace. Of note are a seventh-century convent and a Romanesque church (eleventh century; rebuilt in the eighteenth century). The Nivelles district is predominantly French-speaking.

Nivenskoye (NEE-vyen-skuh-ye), settlement (2005 population 7,900), W KALININGRAD oblast, W European Russia, on road and railroad, 9 mi/14 km SSE of KALININGRAD; 54°35′N 20°34′E. Elevation 108 ft/32 m. Russian military air base in the vicinity. Until 1945, in EAST PRUSSIA and called Wittenberg.

Nive River (NEEV), 45 mi/72 km long, PYRÉNÉES-ATLANTIQUES department, AQUITAINE region, SW FRANCE; formed by confluence of several headstreams near SAINT-JEAN-PIED-DE-PORT; flows NNW, past CAMBO-LES-BAINS and USTARITZ (head of shallow navigation), to the ADOUR at BAYONNE; 43°31′N 01°30′W. Its narrow valley, in BASQUE country, is followed by a railroad.

Nivernais (nee-ver-NAI), region and former province, central FRANCE; historic capital was NEVERS. It roughly coincides with NIÈVRE department, in BURGUNDY. Drained by the LOIRE and the YONNE rivers, it is chiefly a series of hills, rising to the MORVAN massif (outlier of the MASSIF CENTRAL) in the E and bounded by the Loire in the W. A county after the 10th century, it passed (1384) through inheritance to Philip the Bold of Burgundy, and later, as a duchy, it passed through a complicated succession to the house of Gonzaga in 1601. Cardinal Mazarin bought (1659) the title, which remained with his family even after Louis XIV incorporated (1669) Nivernais into the royal domain.

Nivernais Canal (nee-ver-NAI), French *Canal du Nivernais* (kah-nahl dyoo nee-ver-NAI), 108 mi/174 km long, YONNE and NIÈVRE departments, BURGUNDY region, central FRANCE, connects DECIZE (on the LOIRE) with AUXERRE (on the YONNE), paralleling the small ARON to its source; 47°40′N 03°40′E. It crosses the W slopes of the MORVAN near Châtillon-en-Bazois, and enters Yonne River valley S of CORBIGNY. It then runs alongside Yonne River past CLAMECY, merging with it above Auxerre. Used for lumber and wood product shipments.

Niverville, town (□ 3 sq mi/7.8 sq km; 2001 population 1,921), SE MANITOBA, W central Canada, 20 mi/32 km

---

Cross-references are shown in SMALL CAPITALS. The pronunciation guide is shown on page xix. The sources of population figures are shown on page xvii.

S of WINNIPEG; 49°36′N 97°02′W. Grain elevator. Founded as a village in 1877; became a town in 1993.

Nivnice (NYIV-nyi-TSE), German *niwnitz*, village, JI-HOMORAVSKY province, SE MORAVIA, CZECH REPUBLIC, 10 mi/16 km SE of UHERSKÉ HRADIŠTĚ; 48°59′N 17°38′E. Food processing and distilling. Has folk architecture, an 18th-century baroque church. It is the most probable birthplace of Czech scholar and founder of modern pedagogy Jan Amos Komenský-Comenius (1592–1670).

Nivskiy (NEEF-skeeye), town, SW MURMANSK oblast, NW European Russia, on the NIVA RIVER, on highway and the Murmansk railroad, 9 mi/14 km N, and under administrative jurisdiction, of KANDALAKSHA; 67°18′N 32°29′E. Elevation 324 ft/98 m. Hydroelectric station.

Niwai, INDIA: see NAWAI.

Niwas (nee-WUHS), town, MANDLA district, SE MADHYA PRADESH state, central INDIA, 31 mi/50 km N of MANDLA; 23°03′N 80°26′E. Rice, wheat, oilseeds. Lac cultivation in nearby dense sal forests.

Niwnitz, CZECH REPUBLIC: see NIVNICE.

Niwot, unincorporated town (2000 population 4,160), BOULDER county, N central COLORADO, suburb 8 mi/12.9 km NE of BOULDER, and 5 mi/8 km SW of LONGMONT; 40°06′N 105°09′W. Elevation 5,090 ft/1,551 m. Manufacturing of concrete, plastic parts; publishing; IBM plant.

Nixa (NIK-suh), town (2000 population 12,124), CHRISTIAN county, SW MISSOURI, in the OZARKS, near JAMES RIVER, 12 mi/19 km S of downtown SPRINGFIELD; 37°02′N 93°17′W. Fruit; dairying; manufacturing (plastic molding, deodorants, lamps, fixtures, aluminum and sand castings, furniture, pottery).

Nixdorf, CZECH REPUBLIC: see MIKULASOVICE.

Nixon, unincorporated town (2000 population 1,404), BUTLER county, W PENNSYLVANIA, residential community 6 mi/9.7 km SSW of BUTLER; 40°46′N 79°55′W. Butler county airport to W.

Nixon (NIK-suhn), town (2006 population 2,241), GONZALES county, S central TEXAS, c.45 mi/72 km ESE of SAN ANTONIO; 29°16′N 97°45′W. Poultry-packing and -shipping center. Agriculture (peanuts; corn, wheat; cattle, poultry).

Nixon, unincorporated village, WASHOE county, W NEVADA, 33 mi/53 km NE of RENO, on TRUCKEE River, 5 mi/8 km SE of its entrance to PYRAMID LAKE, in SE part of Pyramid Lake Indian Reservation. Cattle, sheep. Pyramid Lake State Park to NW; Anaho Island National Wildlife Reserve to N.

Niya, CHINA: see MINGFENG.

Niyodo (nee-YO-DO), village, Takaoka county, KOCHI prefecture, S SHIKOKU, W JAPAN, 22 mi/35 km W of KOCHI; 33°31′N 133°07′E. Tea.

Niyodo River (nee-YO-do), Japanese *Niyodo-gawa* (nee-YO-do-GAH-wah), 82 mi/132 km long, KOCHI prefecture, SW SHIKOKU, W JAPAN; rises in mountains near KUMA; flows generally SE, through forested area, past OCHI, INO, and Takaoka, to TOSA BAY 7 mi/11.3 km SW of KOCHI.

Niza (NEE-ZAH), city (2005 population 153,305), SAITAMA prefecture, E central HONSHU, E central JAPAN, on the Yonase River, 6 mi/10 km S of URAWA; 35°47′N 139°34′E. Suburb of TOKYO. Printing.

Niza, PORTUGAL: see NISA.

Nizamabad (nee-ZAH-mah-bahd), district (□ 3,072 sq mi/7,987.2 sq km), ANDHRA PRADESH state, SE INDIA, on DECCAN PLATEAU; ⊙ NIZAMABAD. Bordered N by GODAVARI RIVER; NIZAM SAGAR (reservoir) in SW. Mainly lowland (largely sandy red soil), drained by tributaries of the Godavari River; rice, millet, oilseeds (especially peanuts, sesame), cotton, tobacco; sugarcane grown along Nizam Sagar Canal. Rice and oilseed milling, cotton ginning, biri manufacturing, silk weaving. Main towns: Nizamabad (trade center), BODHAN. Part of HYDERABAD from beginning (early

18th century) of state's formation; part of Andhra Pradesh state since 1948.

Nizamabad (nee-ZAH-mah-bahd), city (2001 population 288,722), ⊙ NIZAMABAD district, ANDHRA PRADESH state, SE INDIA, 90 mi/145 km NNW of HYDERABAD; 18°40′N 78°07′E. A district administrative center. A road and agricultural trade center in sugarcane area; market for grain, sugar, and vegetable oil. Rice milling, cotton ginning. Industrial school. The district is irrigated by the NIZAM SAGAR hydroelectric project.

Nizamabad (ni-ZAHM-ah-bahd), village, GUJRANWALA district, E PUNJAB province, central PAKISTAN, 1 mi/1.6 km S of WAZIRABAD; 32°26′N 74°07′E.

Nizampatam (nee-ZAHM-puh-tuhm), village, GUNTUR district, ANDHRA PRADESH state, SE INDIA, 35 mi/56 km SW of MACHILIPATNAM. Was English trading station, founded 1611 and called Pedapalle; soon outgrown by MACHILIPATNAM.

Nizam Sagar (nee-zahm sah-guhr), reservoir in MEDAK district, ANDHRA PRADESH state, SE INDIA, NW of HYDERABAD; 15 mi/24 km long, 10 mi/16 km wide; 18°10′N 77°55′E. Impounded by 115-ft/35-m-high dam across MANJRA RIVER, 32 mi/51 km S of BODHAN; constructed in 1930s. Feeds Nizam Sagar Canal, which extends 70 mi/113 km N and NE, past RUDRUR (sugarcane experimental farm) and NIZAMABAD, to just S of ARMUR.

Nizankowice, UKRAINE: see NYZHANKOVYCHI.

Nizao (nee-ZOU), town, PERAVIA province, S DOMINICAN REPUBLIC, on the coast, 26 mi/42 km SW of SANTO DOMINGO; 18°15′N 70°13′W. Agricultural (coffee, rice, bananas.)

Nizhankovichi, UKRAINE: see NYZHANKOVYCHI.

Nizhegorod (NEE-zhi-guh-ruht), Russian *Nizhegorodskaya*, oblast (□ 29,626 sq mi/77,027.6 sq km; 2006 population 3,445,670), central European Russia; ⊙ NIZHNIY NOVGOROD. Drained by the VOLGA and lower Oka rivers. Nizhniy Novgorod and satellite cities, as well as industrialized lower OKA River valley (SW), have large urban centers; metal handicraft industries (PAVLOVO), and metallurgical works (VYKSA, TASHINO, KULEBAKI). Manufacturing (transportation equipment, motor vehicles, paper-making machinery, tools). Lumbering, with wood cracking (VAKHTAN) and paper milling (PRAVDINSK). Chemical industry (DZERZHINSK) uses phosphorites, peat, limestone, and gypsum. Also petroleum refining at KSTOVO, shoe and leather industry at BOGORODSK, hemp milling (ropes, fishnets) at RESHETIKHA and GORBATOV. Volga and Oka rivers carry half of all freight, with good railroad network (S); wooded steppe (SE) has grain agriculture; extensive forested region (N) produces flax and potatoes. Formed in 1929 as a territory (Russian *kray*) that included Chuvash Autonomous SSR, Mari and Udmurt autonomous oblasts; became an oblast in 1936. Called Nizhegorod (a contracted form of Nizhniy Novgorod) until 1932. Renamed Gor'kiy after writer Maxim Gor'kiy. Changed back to Nizhegorod in 1992.

Nizhne– [Russian combining form=lower], in Russian names: see also NIZHNEYE, NIZHNI, NIZHNIYE, or NIZHNYAYA.

Nizhneangarsk (neezh-nye-ahn-GAHRSK), town, NW BURYAT REPUBLIC, S central SIBERIA, RUSSIA, on the N end of Lake BAYKAL, at the mouth of the UPPER ANGARA RIVER, on highway branch and railroad, 285 mi/459 km NNE of ULAN-UDE; 55°47′N 109°33′E. Elevation 1,948 ft/593 m. Fish canneries, lumbering, gold mining. Until 1938, called Kozlovo.

Nizhnebakanskiy (neezh-nye-bah-KAHN-skeeye), settlement (2005 population 8,690), SW KRASNODAR TERRITORY, S European Russia, in the extreme NW outliers of the Greater CAUCASUS Mountains, on road and railroad, 26 mi/42 km E of ANAPA, and 9 mi/14 km N of NOVOROSSIYSK; 44°51′N 37°52′E. Elevation 396 ft/120 m. Railroad depots. Population largely Meskhetian Turks.

Nizhne-Chirskaya, RUSSIA: see NIZHNIY CHIR.

Nizhnedevitsk (neezh-nye-dee-VEETSK), village (2006 population 6,295), W VORONEZH oblast, S central European Russia, on road junction, 53 mi/85 km WSW of VORONEZH; 51°32′N 38°22′E. Elevation 547 ft/166 m. In agricultural area (livestock). Made city in 1779; reduced to status of a village in 1928.

Nizhnegniloye, RUSSIA: see SOSNOVKA, KURSK oblast.

Nizhnegorskiy, UKRAINE: see NYZHN'OHIRS'KYY.

Nizhneilimsk (neezh-nye-ee-LEEMSK), village, W central IRKUTSK oblast, E central SIBERIA, RUSSIA, on the ILIM RIVER, 210 mi/338 km NNE of TULUN; 57°11′N 103°16′E. Elevation 826 ft/251 m. Iron-ore deposits nearby. Also known as Nizhne-Ilimskoye.

Nizhne-Ilimskoye, RUSSIA: see NIZHNEILIMSK.

Nizhneisetskiy (neezh-nye-ee-SYETS-keeye), town, central SVERDLOVSK oblast, W Siberian Russia, in residential suburbs of YEKATERINBURG (less than 9 mi/14 km SE of city center), connected by highway branch and railroad spur; 56°45′N 60°41′E. Elevation 774 ft/235 m.

Nizhneivkino (neezh-nye-EEF-kee-nuh), resort town (2005 population 2,210), central KIROV oblast, E central European Russia, on the Ivkika River (left tributary of the BYSTRITSA River, VYATKA River basin), on highway, 37 mi/60 km S of KIROV; 58°11′N 49°31′E. Elevation 482 ft/146 m. Local tourism and vacation spot (mineral springs).

Nizhnekamchatsk (neezh-nye-kahm-CHAHTSK), settlement, NE KAMCHATKA oblast, RUSSIAN FAR EAST, on E KAMCHATKA PENINSULA, on the lower KAMCHATKA RIVER, 255 mi/410 km N of PETROPAVLOVSK-KAMCHATSKIY; 56°21′N 162°12′E. Elevation 187 ft/56 m. Founded in 1700. One of the ports of call for Capt. Vitus Bering during his explorative eastward expeditions between 1727 and 1743.

Nizhnekamsk (neezh-nye-KAHMSK), city (2006 population 237,390), NE TATARSTAN Republic, E European Russia, on the KAMA River, terminus of railroad spur, 150 mi/241 km E of KAZAN′, and 20 mi/32 km of SW of NABEREZHNYYE CHELNY; 55°38′N 51°49′E. Elevation 396 ft/120 m. Oil refining, manufacturing (petroleum chemicals, concrete chemicals, auto tires, low-voltage cables, bricks, plastics, furniture, textiles, garments), food industries (macaroni plant, bakery). Has institutes of business management, economics, and chemical technologies; drama theater. Offices of Tatarstan Ministry of Ecology are located here. Regional airport is 10 mi/16 km to the ESE, at Begishevo. Made city in 1966.

Nizhnekamsk Reservoir (neezh-nye-KAHMSK), Russian *Nizhnekamskoye Vodokhranilishche* (□ 1,042 sq mi/2,709.2 sq km), in NE TATARSTAN, NW BASHKORTOSTAN, and SE UDMURT REPUBLIC, E European Russia; 55°53′N 52°48′E. Formed by the confluence of the KAMA, BYSTRYY TANYP, BELAYA, and a number of smaller rivers. Hydroelectric power plant in the vicinity of NIZHNEKAMSK. Gravel extracted from its bottom is used for production of building materials.

Nizhnekartli Plain (NIZH-nai-KART-lee), between the spurs of Trialeti and Somkheti ranges, mostly extending along the Khrami and Algeti rivers; 40 mi/65 km long, 22 mi/35 km wide. Elevation 919 ft/280 m–1,312 ft/400 m long.

Nizhne-Kolosovskoye, RUSSIA: see KOLOSOVKA.

Nizhnekolymsk (neezh-nye-kuh-LIMSK), settlement, NE SAKHA REPUBLIC, N RUSSIAN FAR EAST, on the KOLYMA River at the mouth of the ANYUY River, 200 mi/322 km NE of SREDNEKOLYMSK; 68°32′N 160°56′E. River port; reindeer raising. Founded in 1644; early center of Siberian exploration.

Nizhnepavlovka (neezh-nye-PAHV-luhf-kah), village (2006 population 3,685), central ORENBURG oblast, SE European Russia, on the URAL River, on highway, 11 mi/18 km ESE of ORENBURG; 51°43′N 54°48′E. Elevation 314 ft/95 m. Agricultural products. Also called Nizhnyaya Pavlovka.

Nizhne-Saraninskiy, RUSSIA: see SARANA.

Nizhneshilovka (neezh-nye-SHI-luhf-kah), village (2005 population 5,115), SE KRASNODAR TERRITORY, S European Russia, on the S slopes of the NW Greater CAUCASUS Mountains, on the RUSSIA-GEORGIA border, 7 mi/11 km NNE of (and administratively subordinate to) SOCHI; 43°28′N 40°03′E. Elevation 1,204 ft/366 m. Livestock raising (sheep, pigs). Skiing resort in the vicinity. Also known as Nizhneshilovskoye.

Nizhneshilovskoye, Russia: see NIZHNESHILOVKA.

Nizhne-Teploye, UKRAINE: see TEPLE.

Nizhnetroitskiy (neezh-nye-TRO-eets-keeyee), town (2005 population 3,790), W BASHKORTOSTAN Republic, E European Russia, 95 mi/153 km WSW of UFA, and 21 mi/33 km S of TUYMAZY, to which it is administratively subordinate; 54°20′N 53°41′E. Elevation 702 ft/213 m. Woolen milling. Until 1928, called Nizhnetroitskiy Zavod.

Nizhnetroitskiy Zavod, RUSSIA: see NIZHNETROITSKIY.

Nizhneudinsk (neezh-nye-OO-deensk), city (2005 population 38,430), SW IRKUTSK oblast, E central SIBERIA, RUSSIA, on the Trans-Siberian Railroad, on the UDA RIVER (landing), 310 mi/506 km NW of IRKUTSK; 54°54′N 99°02′E. Elevation 1,368 ft/416 m. Highway and railroad junction; locomotive depots. Aircraft manufacturing, mica factory, lumbering, food processing (meat packing, bakery, confectionery, brewery). Founded in 1648, made city in 1783.

Nizhne-Ufaleyskiy Zavod, RUSSIA: see NIZHNIY UFALEY. **Nizhne-Uvel′skoye**, RUSSIA: see YUZHNOURAL′SK.

Nizhnev, UKRAINE: see NYZHNIV.

Nizhnevartovsk (neezh-nye-VAHR-tuhfsk), city (2005 population 246,765), W central KHANTY-MANSI AUTONOMOUS OKRUG, TYUMEN oblast, central SIBERIA, RUSSIA, on the OB′ RIVER, on railroad (long line from Tyumen), 583 mi/938 km NE of Tyumen; 60°56′N 76°34′E. Elevation 147 ft/44 m. The discovery of a huge oilfield at nearby Lake Samotlor in 1965 quickly transformed the small village of Nizhnevartovskaya into a bustling center of an oil and gas field. Hundreds of thousands of oil workers poured into the region to drill and maintain oil wells, process oil and gas, and construct and maintain oil pipelines and pumping stations. Other industries include an automaking plant, brewery. Has a pedagogical institute. Made city in 1972.

Nizhnevolzhsk, RUSSIA: see NARIMANOV.

Nizhneye, UKRAINE: see NYZHNYE.

Nizhni [Russian=lower], in Russian names: see also NIZHNE- [Russian combining form], NIZHNIYE, NIZHNI or NIZHNYAYA.

Nizhniy Baskunchak (NEEZH-neeyee–bahs-koon-CHAHK), town (2005 population 3,150), NE ASTRAKHAN oblast, SE European Russia, on Lake BASKUNCHAK, less than 12 mi/19 km S of the RUSSIA-KAZAKHSTAN border, on road and railroad, 140 mi/225 km NNW of ASTRAKHAN; 48°13′N 46°50′E. Below sea level. Major salt-extracting center; gypsum works.

Nizhniy Cherek (NEEZH-neeyee CHYE-reek) [Russian=lower Cherek], village (2005 population 3,130), E central KABARDINO-BALKAR REPUBLIC, N CAUCASUS, S European Russia, on the CHEREK RIVER, on highway, 13 mi/21 km E of NAL′CHIK; 43°31′N 43°55′E. Elevation 980 ft/298 m. Logging, lumbering. Hydroelectric power station.

Nizhniy Chir (NEEZH-neeyee CHEER), town (2006 population 4,770), SW VOLGOGRAD oblast, SE European Russia, on the Tsimlyansk Reservoir of the DON River, just below the mouth of the CHIR River, on highway, 65 mi/105 km WSW of VOLGOGRAD, and 18 mi/29 km SE of SUROVIKINO, to which it is administratively subordinate; 48°22′N 43°05′E. Food and fish combines; woodworking. Population largely Cossack. Formerly called Nizhne-Chirskaya.

Nizhniy Dzhengutay (NEEZH-neeyee jen-goo-TEI) [Russian=lower Dzhengutay], village (2005 population 7,630), central DAGESTAN REPUBLIC, NE CAUCASUS, SE European Russia, on road, 19 mi/31 km SW of MAKHACHKALA; 42°42′N 47°14′E. Elevation 2,171 ft/661 m. Highway junction. Agriculture (livestock); dairy processing.

Nizhniy Dzherakh, RUSSIA: see KAMBILEYEVSKOYE.

Nizhniye [Russian=lower], in Russian names: see also NIZHNE- [Russian combining form], NIZHNEYE, NIZHNI, or NIZHNYAYA.

Nizhniye Achaluki (NEEZH-nee-ye ah-chah-LOO-kee), village (2005 population 5,435), W INGUSH REPUBLIC, extreme N CAUCASUS, S European Russia, on road, 13 mi/21 km N of NAZRAN; 43°24′N 44°45′E. Elevation 1,564 ft/476 m. Has a hospital, mosque.

Nizhniye Kresty, RUSSIA: see CHERSKIY.

Nizhniye Kropachi, RUSSIA: see STULOVO.

Nizhniye Sergi (NEEZH-nee-ye syer-GEE), city (2006 population 11,920), SW SVERDLOVSK oblast, W central URALS, extreme E European Russia, on small lake, on road junction and railroad, 75 mi/120 km WSW of YEKATERINBURG; 56°39′N 59°18′E. Elevation 1,249 ft/380 m. Metallurgical works (pig iron, steel), lumbering; bakery. Health resort (baths). Arose in 1743 as a village of Nizhnesergiyevskaya, later Nizhneserginskiy Zavod (until 1928). Made city in 1943.

Nizhniye Serogozy, UKRAINE: see NYZHNI SIROHOZY.

Nizhniye Ustriki, POLAND: see USTRZYKI DOLNE.

Nizhniye Vorota, UKRAINE: see NYZHNI VOROTA.

Nizhniye Vyazovyye (NEEZH-nee-ye VYAH-zuh-vi-ye), village (2006 population 8,300), W TATARSTAN Republic, E European Russia, on the right bank of the VOLGA RIVER, on railroad spur, across the river from ZELENODOL′SK (3 mi/5 km to the N), to which it is administratively subordinate; 55°48′N 48°31′E. Elevation 193 ft/58 m. Woodworking. Site of a federal penal colony.

Nizhniy Ingash (NEEZH-neeyee een-GAHSH) [Russian=lower Ingash], town (2005 population 8,415), SE KRASNOYARSK TERRITORY, SE SIBERIA, RUSSIA, on highway junction and the TRANS-SIBERIAN RAILROAD (Ingashskaya station), 40 mi/64 km E of KANSK; 56°12′N 96°32′E. Elevation 813 ft/247 m. Lumber milling; creamery. Found in 1899 with the construction of the railroad; town since 1961.

Nizhniy Kislyay (NEEZH-neeyee kees-LYEI), village (2006 population 4,375), central VORONEZH oblast, S central European Russia, on the BITYUG RIVER, on highway junction and railroad spur, 19 mi/31 km WSW of BUTURLINOVKA; 50°51′N 40°10′E. Elevation 370 ft/112 m. In agricultural and woodworking area. Popular holiday getaway spot for residents of neighboring cities.

Nizhniy Kundysh, RUSSIA: see LESSER KUNDYSH RIVER.

Nizhniy Kurkuzhin (NEEZH-neeyee koor-KOO-zhin) [Russian=lower Kurkuzhin], village (2005 population 3,715), N central KABARDINO-BALKAR REPUBLIC, N CAUCASUS, S European Russia, on highway, 18 mi/29 km NW of NAL′CHIK; 43°45′N 43°21′E. Elevation 2,125 ft/647 m. Agricultural products.

Nizhniy Lomov (NEEZH-neeyee–LO-muhf), city (2005 population 23,700), N PENZA oblast, E European Russia, 75 mi/120 km WNW of PENZA; on the Lomovka River (OKA River basin); 53°31′N 43°40′E. Elevation 501 ft/152 m. Railroad spur terminus; road hub; plywood factory; distillery; bakery; electromechanical plant. Founded in 1636; made city in 1780.

Nizhniy Mamon (NEEZH-neeyee MAH-muhn), village (2006 population 3,895), S VORONEZH oblast, S central European Russia, on the DON River, on road junction, 4 mi/6 km E of VERKHNIY MAMON; 50°11′N 40°30′E. Elevation 275 ft/83 m. In agricultural area; produce processing, fishing.

Nizhniy Nagol′chik, UKRAINE: see NYZHNIY NAHOL′CHYK.

Nizhniy Naur, RUSSIA: see NADTERECHNOYE.

Nizhniy Novgorod (NEEZH-neeyee NOV-guh-ruht), city (2006 population 1,275,320), ⊙ NIZHEGOROD oblast, central European Russia, at the confluence of the VOLGA and OKA rivers, 270 mi/435 km E of MOSCOW; 56°20′N 44°00′E. Elevation 442 ft/134 m. A major river port and a railroad and air center, it is one of the chief industrial cities of the Russian Federation. Heavy machinery, motor vehicles, transportation equipment, machine tools, communication equipment, textiles, ferrous and nonferrous metallurgy and petroleum chemicals are produced. The city is the site of one of the largest automobile factories in Russia. Nizhniy Novgorod stretches along the Volga and Oka rivers and is surrounded by large satellite towns such as BALAKHNA, BOR, PRAVDINSK, and KSTOVO. In 1221, a prince of Vladimir founded the city as a frontier post against the Volga Bulgars and Mordvinians. It became a major trading point for Russia and the East. In 1350, it became the capital of the Suzdal′-Nizhniy Novgorod principality and was annexed in 1392 by Moscow. Nizhniy Novgorod was famous for its annual trade fairs, held from 1817 to 1930. Many architectural monuments. Nizhniy Novgorod was named Gor′kiy from 1932 to 1991 for Maxim Gor′kiy, a popular Soviet-era writer, who was born here.

Nizhniy Odes (NEEZH-neeyee O-dees), town (2005 population 10,980), central KOMI REPUBLIC, NE European Russia, on road, 34 mi/55 km E of SOSNOGORSK; 63°38′N 54°52′E. Elevation 590 ft/179 m. Oil and gas production; sawmilling, timbering.

Nizhniy Pyandzh (NEEZH-neeyee pee-YAHNJ) [Russian=lower Panj], town, SW KHATLON viloyat, TAJIKISTAN, on PANJ RIVER (AFGHANISTAN border), and 45 mi/72 km S of QURGHONTEPPA (connected by narrow-gauge railroad); 37°12′N 68°35′E. Sheep, goats. Head of navigation on the AMU DARYA River system.

Nizhniy Tagil (NEEZH-neeyee tah-GEEL), city (2006 population 378,030), W SVERDLOVSK oblast, E URALS, extreme W Siberian Russia, on the TAGIL RIVER (OB′ RIVER basin), on highway and railroad junctions, 92 mi/148 km N of YEKATERINBURG; 57°55′N 60°00′E. Elevation 823 ft/250 m. A leading center of ferrous metallurgy and heavy industry, it uses the iron ore from deposits at KACHKANAR and on Mount VYSOKAYA. Manufacturing (RR cars, machinery, boilers and radiators, thermal insulation, medical instruments, ferro-concrete, asbestos and cement, construction ceramics, chemicals, and plastics); food processing (flour mill, meat-packing plant, confectionery, brewery, distillery). Founded in 1725; made city in 1917.

Nizhniy Tsasuchey (NEEZH-neeyee tsah-soo-CHAI), village (2005 population 3,230), S CHITA oblast, S SIBERIA, Russia, on the ONON RIVER, on road, 125 mi/201 km SE of CHITA; 50°31′N 115°08′E. Elevation 2,076 ft/632 m. In agricultural area (wheat; livestock); forestry.

Nizhniy Ufaley (NEE-zhneeyee oo-fah-LYAI), town, NW CHELYABINSK oblast, S URALS, RUSSIA, on the Ufaley River, 2 mi/3.2 km above its confluence with the upper UFA RIVER, on road and railroad spur, 12 mi/19 km SW (under jurisdiction) of VERKHNIY UFALEY; 55°55′N 59°58′E. Elevation 1,194 ft/363 m. Repair of metallurgical equipment. Until 1928, called Nizhne-Ufaleyskiy Zavod.

Nizhniy Uslon (NEEZH-neeyee oo-SLON), village, NW TATARSTAN Republic, E European Russia, on the right bank of the VOLGA RIVER, near highway branch, 5 mi/8 km SW of KAZAN′; 55°42′N 48°58′E. Elevation 177 ft/53 m. In agricultural area; cannery.

Nizhnyaya [Russian=lower], in Russian names: see also NIZHNE- [Russian combining form], NIZHNEYE, NIZHNI, or NIZHNIYE.

Nizhnyaya Arpa, RUSSIA: see URAY.

Nizhnyaya Berëzovka (NEEZH-nyah-yah bee-RYO-zuhf-kah), W suburb of ULAN-UDE, W central BURYAT REPUBLIC, S SIBERIA, RUSSIA; 58°43′N 47°18′E.

Nizhnyaya Dobrinka, RUSSIA: see DOBRINKA, VOLGO-GRAD oblast.

Nizhnyaya Duvanka, UKRAINE: see NYZHNYA DU-VANKA.

Nizhnyaya Krynka, UKRAINE: see NYZHNYA KRYNKA.

Nizhnyaya Maktama (NEEZH-nee-yah mahk-tah-MAH), village (2006 population 10,610), SE TATAR-STAN Republic, E European Russia, on the ZAY RIVER (left tributary of the KAMA RIVER), on railroad spur and near highway, less than 3 mi/5 km SSE of (and administratively subordinate to) AL'MET'YEVSK; 54°51′N 52°22′E. Elevation 465 ft/141 m. Agricultural products.

Nizhnyaya Omka (NEEZH-nyah-yah OM-kah), village (2006 population 5,225), E OMSK oblast, SW SIBERIA, RUSSIA, on highway, 30 mi/48 km NE of KALACHINSK; 55°26′N 74°55′E. Elevation 364 ft/110 m. In agricultural area; butter factory.

Nizhnyaya Pavlovka, RUSSIA: see NIZHNEPAVLOVKA.

Nizhnyaya Pesha (NEEZH-nye-yah PYE-shah), village, W NENETS AUTONOMOUS OKRUG, ARCHANGEL oblast, NE European Russia, on the small Pesha River, near the CHESHA BAY of the BARENTS SEA, 115 mi/185 km NE of MEZEN; 66°46′N 47°47′E. Trading post.

Nizhnyaya Poyma (NEEZH-nyah-yah PO-yee-mah), town (2005 population 9,510), SE KRASNOYARSK TERRITORY, SE Siberian Russia, on highway and railroad junction (Reshoty station on the TRANS-SI-BERIAN RAILROAD), 21 mi/34 km W of NIZHNIY IN-GASH, to which it is administratively subordinate; 56°09′N 97°11′E. Elevation 882 ft/268 m. Building materials; food processing. Founded in 1899, in conjunction with the railroad's construction; town since 1961.

Nizhnyaya Salda (NEEZH-nyah-yah sahl-DAH), city (2006 population 17,280), W central SVERDLOVSK oblast, extreme W Siberian Russia, in the E foothills of the central URALS, on the Salda River (right tributary of the TURA RIVER), on highway junction and on railroad, 125 mi/205 km N of YEKATERINBURG, and 30 mi/48 km ENE of NIZHNIY TAGIL; 58°04′N 60°43′E. Elevation 613 ft/186 m. Old metallurgical center (pig iron, steel); metalworking, machine building. Founded in 1760 with the construction of metallurgical works. Became city in 1938. Formerly called Nizhne-Saldinskiy Zavod.

Nizhnyaya Sarana, RUSSIA: see SARANA.

Nizhnyaya Shakhtama (NEEZH-nyah-yah shahkh-tah-MAH), village (2004 population 525), SE CHITA oblast, S SIBERIA, RUSSIA, near highway, 60 mi/97 km S of SRETENSK, and 21 mi/34 km SW of GAZIMURSKIY ZAVOD; 51°24′N 117°40′E. Elevation 2,601 ft/792 m. Gold mines; graphite deposits.

Nizhnyaya Suyetka (NEEZH-nye-yah soo-YET-kah), village, NW ALTAI TERRITORY, S SIBERIA, RUSSIA, on the KULUNDA STEPPE, on local road junction, 50 mi/80 km ENE of SLAVGOROD; 53°14′N 79°54′E. Elevation 364 ft/110 m. In dairy-farming area.

Nizhnyaya Tavda (NEEZH-nye-yah tahv-DAH), settlement (2006 population 6,700), SW TYUMEN oblast, W SIBERIA, RUSSIA, on the TAVDA RIVER, on road junction, 40 mi/64 km NE of TYUMEN; 57°40′N 66°10′E. Elevation 203 ft/61 m. In agricultural area (grain; livestock); dairy plant. Town status granted in 1965, revoked in 1992.

Nizhnyaya Tunguska River, RUSSIA: see LOWER TUN-GUSKA RIVER.

Nizhnyaya Tura (NEEZH-nyah-yah too-RAH), city (2006 population 23,700), W SVERDLOVSK oblast, extreme W Siberian Russia, in the E foothills of the central URALS, on small lake formed by the TURA RIVER, on highway and railroad spur, 150 mi/243 km N of YEKATERINBURG; 58°37′N 59°51′E. Elevation 590 ft/179 m. Old metallurgical center (pig iron, steel); manufacturing (machinery, electric hardware); food industries (sausage factory, fish farm). Power station.

Founded in 1754. Made city in 1949. Until 1929, called Nizhne-Turinskiy Zavod.

Nizhnyaya Veduga (NEEZH-nye-yah VYE-doo-gah), village, W VORONEZH oblast, S central European Russia, on road junction and near railroad, 23 mi/37 km WNW of VORONEZH; 51°44′N 38°39′E. Elevation 633 ft/192 m. In wheat-growing area.

Nizhyn (NEE-zhin) (Russian *Neshin*), city (2001 population 76,625), W CHERNIHIV oblast, UKRAINE, on the OSTER RIVER, 56 mi/90 km N of KIEV; 51°03′N 31°53′E. Elevation 380 ft/115 m. It is a railroad junction on the main Moscow-Kiev line, a raion center, and an agricultural (notably cucumbers) trade center. Industries include engineering, food processing, and the manufacture of farm machinery and railroad cars. Known since 1147 as Unenezh; renamed Nizhen or Nizhyn in 1514 while under Lithuanian rule; passed to Poland in 1618; granted the rights of Magdeburg law in 1625; center of the Nizhyn Ukrainian Cossack regiment from 1648 to 1782. It became an important trading center in the 17th and 18th centuries after Greek merchants received permission from Hetman Khmel'nyts'kyy to settle there in 1657; in 1820, Nizhyn Lyceum was established (now a pedagogical institute); presently it also has a medical school, an agricultural mechanization technical school, vocational technical schools, drama theater and several museums. Architectural monuments include three cathedrals (1668, 1702, 1778), several 18th-century churches, and the lyceum building (1807–1820). Jewish community since 1792, had grown into one of the centers for Talmudic learning by the end of the 19th century; greatly reduced by the pogroms of 1881 (and subsequent emigration), 1905, and 1919; numbering 6,130 in 1926; destroyed by the Nazis in 1941—fewer than 1,000 Jews remaining in 2005.

Nizi (NEE-zee), village, ORIENTALE province, NE CONGO, 4 mi/6.4 km SE of KILO-MINES; 01°45′N 30°18′E. Elev. 4,940 ft/1505 m. Gold-mining and trading center; gold processing. Machine shops.

Nizip, town (2000 population 71,629), S TURKEY, 23 mi/37 km ESE of GAZIANTEP; 37°02′N 37°47′E. Olives, pistachios, wheat. Here in 1839 Turks were defeated by Egyptians. On the edge of Nizip are the ruins of the ancient city Belkis (Zeugnia or Zeugma), on the EU-PHRATES, where excavations were completed in 2000, in anticipation of sections of the region being covered by the waters of the BIRECIK Dam. Formerly spelled Nezib, Nisib.

Nízke Tatry, SLOVAKIA: see LOW TATRAS.

Nizna (nyizh-NAH), Slovak *Nižná*, Hungarian *Nizsna*, village, STREDOSLOVENSKÝ province, N SLOVAKIA, on railroad, on ORAVA RIVER, and 9.94 mi/16 km NNW of LIPTOVSKÝ MIKULAS; 49°19′N 19°31′E. MANU-FACTURING (television sets).

Nizne Ruzbachy, SLOVAKIA: see VYSNE RUZBACHY.

Nizniów, UKRAINE: see NYZHNIV.

Nizny Svidnik, SLOVAKIA: see SVIDNIK.

Nizwa (NIZ-wuh), town (2003 population 68,785), in highland OMAN, 20 mi/32 km W of IZKI, at S foot of the JABAL AKHDAR; 22°56′N 57°32′E. Strategic and commercial center; fortress. Copper and brass working; leather, pottery, and weaving handicrafts. Noted *halwa* production. Was long the capital of Oman; also the center of armed opposition to the sultan of Oman. The seat of the religious leader (Mufti) who headed the rebellion. Center of the Ibadite (Muslim) sect, which resisted the rule of the sultan and actually ruled much of the interior of Oman. Sometimes spelled Nazwa.

Nizza, FRANCE: see NICE.

Nizza Monferrato, town, ASTI province, PIEDMONT, NW ITALY, on BELBO River, and 12 mi/19 km SE of ASTI; 44°46′N 08°21′E. In grape- and vegetable-growing region; railroad junction; sulphur refining; artisan industries (wine making, wood- and metalworking).

Njakwa (njah-kuah), village, Northern region, N MA-LAWI, S of NYIKA PLATEAU, 35 mi/56 km SSW of LI-

VINGSTONIA; 11°03′S 33°53′E. Road junction. Cassava, corn.

Njala (uhn-JAH-lah), village, SOUTHERN province, S central SIERRA LEONE, on JONG RIVER, and 4 mi/6.4 km N of MANO; 08°07′N 12°05′W. University of Sierra Leone, experimental farm, headquarters government (Deptartment of Agriculture).

Njarasa, Lake, TANZANIA: see EYASI, LAKE.

Njazidja (nuh-jah-ZEE-jah), island and administrative district (□ 443 sq mi/1,151.8 sq km; 2003 population 296,177), NW Comoros Republic, one of three main islands comprising the republic, most NW of four main islands of Comoros islands, in Mozambique Channel, Indian Ocean, 360 mi/579 km WNW of Madagascar; ⊙ Moroni; 11°40′S 43°20′E. Other towns include Mitsamiouli, Dembeni, and Foumbouni. Mwali island to SE. The island is volcanic; rises to 7,746 ft/2,361 m at Mt. Karthala in SE. Porous rock accounts for the lack of rivers; most of its residents live on relatively level W side. Timber; vanilla, ylang-ylang, rice, cloves, cassava, sweet potatoes, coconuts, bananas; livestock; fish. International airport N of Moroni, on W coast. Formerly composed of seven independent sultanates; placed under French protectorate 1886; became part of independent Comoros in 1974. Called Grande Comore under French rule. Sometimes spelled Ngazidja.

Njegos (NYE-gos), Serbian *Njegoš* (NYE-gosh), mountain of DINARIC ALPS, W MONTENEGRO (elevation 5,658 ft/1,725 m), 14 mi/23 km NW of NIKSIC; 42°53′N 18°45′E. Also spelled Nyegosh.

Njegusi (NYE-goo-see), Serbian *Njeguši* (NYE-goo-shee), village, S MONTENEGRO, at NE foot of LOVCEN mountain, 5 mi/8 km NW of CETINJE. Road junction. Also spelled Nyegushi.

Njeru (NJAI-roo), town (2002 population 51,236), MU-KONO district, CENTRAL region, S UGANDA, 5 mi/8 kms SW of Jinja; 0°27′N 33°11′E. Dairy center across the VICTORIA NILE RIVER and OWEN FALLS DAM from Jinja. Dam power plants are here.

Njila (NJEE-lah), village, MBEYA region, SW TANZANIA, 60 mi/97 km NW of MBEYA, and 5 mi/8 km NW of NGOMBA, on Lake RUKWA; 08°18′S 32°46′E. Cattle, sheep, goats; corn, wheat.

Njinjo (NJEEN-jo), village, LINDI region, SE TANZANIA, 45 mi/72 km WNW of KILWA MASOKO, on MATANDU RIVER; 08°52′S 38°50′E. Sheep, goats; corn, wheat; timber. SELOUS GAME RESERVE to W.

Njolé, GABON: see NDJOLE.

Njombe (NJOM-bai), town, IRINGA region, S TANZA-NIA, 90 mi/145 km ESE of MBEYA; LIVINGSTONE MOUNTAINS to SW; 09°20′S 34°46′E. Wheat, corn, pyrethrum; livestock.

Njombe River (NJOM-bai), c. 135 mi/217 km long, central TANZANIA; rises in N MBEYA region c. 65 mi/105 km NNE of MBEYA; flows NE, forms boundary between RUAHA NATIONAL PARK (SE) and RUNGWA GAME RESERVE (NW); enters KISIGO RIVER 80 mi/129 km NW of IRINGA.

Njoro (NJO-ro), town, RIFT VALLEY province, W KENYA, on railroad, and 10 mi/16 km WSW of NAKURU; 00°20′S 35°57′E. Elevation 7,113 ft/2,168 m. Corn, wheat, flax; dairy farming. Agricultural school.

Njoro (NJO-ro), village, MANYARA region, N central TANZANIA, 83 mi/134 km NE of DODOMA, in S part of MASAI STEPPE; 05°14′S 36°30′E. Cattle, sheep, goats; wheat, corn.

Njoro Neganga (NJO-ro nai-GAN-gah), village, MAN-YARA region, N central TANZANIA, 80 mi/129 km NE of DODOMA, on Kitwei Plain; 05°30′S 36°42′E. Cattle, sheep, goats; wheat, corn. Lol Lanok Well to S.

Njurundabommen (NYOO-RUND-ah-BOM-men), village, VÄSTERNORRLAND county, NE SWEDEN, on LJUNGAN RIVER, 8 mi/12.9 km SSE of SUNDSVALL; 62°16′N 17°23′E. Has old church.

Nkalago (uhn-KAH-lah-GO), town, ENUGU state, SE NIGERIA, 20 mi/32 km E of ENUGU. Silver and lead-zinc mining; limestone, salt deposits. Cement manufacturing.

Nkam, department (2001 population 66,979), LITTORAL province, CAMEROON, ⊙ YABASSI.

Nkambé (NKAM-bai), town (2001 population 32,900),⊙ Donga-Mantung department, North-West province, CAMEROON, 55 mi/89 km NE of BAMENDA; 06°35'N 10°42'E. Livestock raising. Hospital.

Nkana, township, COPPERBELT province, N central ZAMBIA, just W of KITWE, 33 mi/53 km WNW of NDOLA; 12°50'S 28°12'E. Copper-mining center; smelter, electrolytic refinery. Also produces cobalt and gold, with silver and selenium as by-products; emeralds. Also called Nkana Mine.

Nkawkaw (NKOR-kor), town, EASTERN REGION, GHANA, on railroad, 45 mi/72 km NW of KOFORIDUA, and just S of MPRAESO; 06°33'N 00°46'W. Cacao market.

Nkayi, town (2007 population 59,169), BOUENZA region, S Congo Republic, 35 mi/56 km E of LOUBOMO. Processing of agricultural products (sugarcane). Airfield.

Nkayi, town, MATABELELAND NORTH province, W central ZIMBABWE, 60 mi/97 km W of KWEKWE, on SHANGANI RIVER; 19°00'S 28°54'E. Cattle, sheep, goats; grain, cotton. MBASE PAN SANCTUARY to N.

Nkenkaasu, GHANA: see NKENKASO.

Nkenkaso (nken-KAH-soo), town, ASHANTI region, GHANA, 40 mi/64 km SE of Techiman; 07°18'N 01°53'W. Road junction. Cocoa, coffee, timber; tourism. Formerly spelled NKENKAASU or NKINKASO.

Nkhata Bay (NKAH-tah), district and administrative center (2007 population 199,083), Northern region, MALAWI, port on W shore of Lake Malawi (LAKE NYASA), 22 mi/35 km NNE of CHINTHECHE. Fishing; cassava, corn. Tourist resort.

Nkhoma (NKO-mah), village, central MALAWI, 25 mi/40 km E of LILONGWE; 12°43'S 33°24'E. Tobacco.

Nkhotakota, district (2007 population 301,604), Central region, central MALAWI; ⊙ NKHOTAKOTA. Bordered by SALIMA (S), Ntchisi (SW), KASUNGU (W), MZIMBA (NW), and NKHATA BAY (N) districts and LAKE MALAWI (E).

Nkhotakota (nko-TAH-ko-tah), town, ⊙ Nkhotakota district, Central region, central MALAWI, port on Lake Malawi (LAKE NYASA), 80 mi/129 km NNE of LILONGWE; 12°55'S 34°18'E. Rice-growing center; also tobacco, cotton, corn; fishing. Political center (seat of representative of Sultan of ZANZIBAR) in Arab slave-trade days (1880s, 1890s).

Nkhota-kota, MALAWI: see NKHOTAKOTA.

Nkinkaso, GHANA: see NKENKASO.

Nkolabona (nko-lo-BO-nuh), town, WOLEU-NTEM province, N GABON, 50 mi/80 km N of MITZIC; 01°12'N 11°43'E. Cocoa-producing region.

Nkomi, GABON: see OMBOUÉ.

Nkondjock (NKAWNG-jok), town, LITTORAL province, CAMEROON, 66 mi/106 km NE of DOUALA; 04°49'N 10°14'E. Agr. (coffee, cocoa) nearby; bauxite deposits. Also spelled Nkongjok.

N'Kongsamba (uhn-kawng-SAHM-buh), city (1998 estimated population 104,908; 2001 estimated population 110,600), ⊙ MOUNGO department, LITTORAL province, CAMEROON, 140 mi/225 km NW of Yaoundé; 04°58'N 09°58'E. Elevation 2,900 ft/884 m. Commercial and tourist center; terminus of railroad from DOUALA. Sawmilling; palm-oil and coffee plantations, food processing industry, and repair shops. Bauxite deposits nearby. Has meterological station, junior college Airfield. Hospital. Also spelled Nkongsamba.

Nkoteng (uhn-KO-taing), town (2001 population 17,900), Central province, CAMEROON, 55 mi/89 km NE of Yaoundé; 04°28'N 12°05'E.

Nkubu (NKOO-boo), town, Meru district, EASTERN province, KENYA, on road, and 6 mi/10 km S of MERU; 00°03'S 37°44'E. Market and trading center.

Nkululu River (nkoo-LOO-loo), c. 50 mi/80 km long, central TANZANIA; rises in W SINGIDA region c. 115 mi/185 km SW of SINGIDA; flows SW joins Mungu River 150 mi/241 km SW of Singida to form SHAMA RIVER.

Nkwanta, GHANA: see DUAYAW NKWANTA.

Nmai River (uhn-MEI), Burmese *Nmai Kka*, Chinese *En-mei-k'ai*, left (main) headstream of AYEYARWADY RIVER, c. 300 mi/480 km long, N MYANMAR; rises in high mountains on CHINA (YUNNAN) border, N of PUTAO; flows S, joining the MALI RIVER 25 mi/40 km N of MYITKYINA to form AYEYARWADY RIVER.

Nnewi (uhn-NOO-wee), town, ANAMBRA state, S central NIGERIA, 12 mi/19 km SSE of ONITSHA; 06°01'N 06°55'E. Large market rivals that of Onitsha. Manufacturing (transportation equipment, tools, and textiles); palm oil and kernels, kola nuts. Sometimes spelled Newi.

No (NO), town, West Kubiki county, NIIGATA prefecture, central HONSHU, N central JAPAN, 81 mi/130 km S of NIIGATA; 37°05'N 137°59'E. Crabs.

No, lake (max. □ c.40 sq mi/100 sq km), S central SUDAN, in the swampy SUDD region; 09°30'N 30°28'E. It is formed by the floodwaters of the WHITE NILE River, near its confluence with BAHR AL-GHAZAL, and varies in size seasonally. Papyrus grows in the lake.

No. 1, UKRAINE: see ORDZHONIKIDZE, Dnipropetrovs'k oblast.

Noailles (no-EI), commune (□ 3 sq mi/7.8 sq km), OISE department, PICARDIE region, N FRANCE, 9 mi/14.5 km SE of BEAUVAIS; 49°20'N 02°12'E. Manufacturing of brushes, wood products.

Noakhali (no-ah-kah-lee), district (□ 1,658 sq mi/4,310.8 sq km), EAST BENGAL, BANGLADESH; ⊙ NOAKHALI. Bounded NE by TRIPURA state (INDIA), W by MEGHNA RIVER; GANGA DELTA in S part of district; includes SANDWIP (SE) and HATIA (SW) islands, separating MEGHNA RIVER mouths from BAY OF BENGAL. Mainly alluvial soil; agriculture (rice, jute, pulse, chili, oilseeds, sugarcane); extensive coconut and betal palm groves; hill tract NE (sal trees). Major cotton-weaving center at Chaumuhani; brassware manufacturing at NOAKHALI, perfume manufacturing at FENNY; coconut-oil milling. District is subject to severe cyclones (over half of HATIA ISLAND population was destroyed in 1876). Became part of Bengal under Afghan governors in 1353; figured in 17th-century struggle between Moguls and Portuguese; was passed in 1765 to the English. SANDWIP ISLAND was a 17th-century Portuguese and Arakanese pirate stronghold. Present district formed in 1822. Part of former British Bengal province in INDIA until 1947, when it was incorporated into the new Pakistani province of East Pakistan. Formerly called Bhulua.

Noakhali (no-ah-kah-lee), town (2001 population 75,956), ⊙ NOAKHALI district, SE EAST BENGAL, BANGLADESH, on MEGHNA RIVER, and 75 mi/121 km SE of DHAKA; 22°51'N 90°48'E. Trades in rice, jute, oilseeds; brassware manufacturing. Formerly called Sudharam.

Noale (no-AH-le), town, VENEZIA province, VENETO, N ITALY, 15 mi/24 km NW of VENICE; 45°33'N 12°04'E. Diversified secondary industrial center; manufacturing (agricultural machinery, stoves, beads).

Noalejo (no-ah-LAI-ho), town, JAÉN province, S SPAIN, 18 mi/29 km SSE of JAÉN; 37°32'N 03°39'W. Olive oil, cereals, vegetables; livestock.

Noamundi (NO-ah-muhn-dee), town (2001 population 16,228), WEST SINGHBHUM district, S JHARKHAND state, E INDIA, on railroad, 65 mi/105 km SW of JAMSHEDPUR; 22°09'N 85°32'E. Major hematite-mining center for Jamshedpur iron- and steelworks. Hematite mining near GUA (10 mi/16 km WNW) and BARABIL (9 mi/14.5 km WSW), both railroad spur termini.

Noarlunga (nor-LUHN-guh), village, SE SOUTH AUSTRALIA state, S central AUSTRALIA, 19 mi/31 km SSW of ADELAIDE, of which it is a suburb; 35°11'S 138°30'E.

Fruit-growing and dairying center. Port Noarlunga is nearby on Gulf SAINT VINCENT.

Noatak, Inuit village (2000 population 428), NW ALASKA, on NOATAK RIVER, and 50 mi/80 km NNW of KOTZEBUE; 67°34'N 163°00'W. Just W is CAPE KRUSENSTERN National Monument.

Noatak Preserve (□ 10,272 sq mi/26,604 sq km), ALASKA; authorized 1978. Mountain-ringed river basin.

Noatak River, c.400 mi/644 km long, NW ALASKA; rises in BROOKS RANGE, near 67°28'N 154°50'W; flows W to KOTZEBUE SOUND opposite KOTZEBUE.

Nobber, Gaelic *An Obair*, town (2006 population 233), N MEATH county, NE IRELAND, on DEE RIVER, and 13 mi/21 km NNW of NAVAN; 53°49'N 06°56'W. Agricultural market. Turlough O'Carolan, the harper, born here.

Nobel' (NO-bel), village (2004 population 2,400), NW RIVNE oblast, UKRAINE, on Lake Nobel' and Pripet (Ukrainian *Prypyat'*), 15 mi/24 km WNW of ZARICHNE; 51°52'N 25°46'E. Elevation 459 ft/139 m. Site of a settlement dating back to Mesolithic/early Neolithic, thus is the oldest known settlement in the PRIPYAT MARSHES.

Nobeoka (no-BE-o-kah), city, MIYAZAKI prefecture, SE KYUSHU, SW JAPAN, on the mouth of the GOKASE RIVER, 47 mi/75 km N of MIYAZAKI; 32°34'N 131°40'E. Commercial and fishing port and a production center for foodstuffs (kombu rolls, dried fish, pickles) and textiles (crepe). Also synthetic fibers, vinyl chloride bulbs, pipes, inkstones. Shiitake mushrooms. Traditional masks, sword parts, and folk art. Nippo coast is nearby.

Noble, county (□ 417 sq mi/1,084.2 sq km; 2006 population 47,918), NE INDIANA; ⊙ ALBION; 41°24'N 85°25'W. Agricultural area (hogs, cattle, poultry; fruit, corn, soybeans, produce; dairy). Manufacturing (dairy- and farm-product processing) at KENDALLVILLE, Albion, LIGONIER. Gravel pits. Drained by ELKHART RIVER. Chain O' Lakes State Park in S central; part of Tri-County Fish and Wildlife Area in SW corner. Gene Stratton Porter State Memorial in N. About twenty small lakes, glacial in origin, scattered throughout county; largest is Sylvan Lake near ROME CITY in N. Formed 1836.

Noble (NO-buhl), county (□ 404 sq mi/1,050.4 sq km; 2006 population 14,165), E OHIO; ⊙ CALDWELL; 39°50'N 81°28'W. Drained by WILLS CREEK and small Duck and Seneca creeks. Includes SENECAVILLE LAKE reservoir. In the Unglaciated Plain physiographic region. Agriculture (livestock); grain, tobacco; dairy products); manufacturing (lumber and wood products, motor vehicle parts and accessories, electrometallurgical products); coal mines, clay pits, limestone quarries. Formed 1851.

Noble, county (□ 742 sq mi/1,929.2 sq km; 2006 population 11,152), N OKLAHOMA; ⊙ PERRY; 36°23'N 97°14'W. Bounded NE by ARKANSAS RIVER in extreme NE. Drained by small BLACK BEAR and RED ROCK creeks. Agricultural area (wheat, oats, barley; cattle). Manufacturing (metal products, construction machinery) at Perry. Oil and natural gas wells. Includes Tonkawa and Ponca Otoe Indian national headquarters. and tribal lands. Lake McMurray is S; Sooner Lake reservoir on E boundary. Formed 1893.

Noble, town (2006 population 5,591), CLEVELAND county, central OKLAHOMA, suburb 6 mi/9.7 km SSE of NORMAN, and 22 mi/35 km SSE of downtown OKLAHOMA CITY, on CANADIAN River; 35°08'N 97°22'W. In agricultural area; manufacturing (belt buckles, figurines).

Noble, village (2000 population 746), RICHLAND county, SE ILLINOIS, 7 mi/11.3 km WSW of OLNEY; 38°42'N 88°13'W. In agricultural area (corn, wheat, apples; livestock).

Noble, village (2000 population 259), SABINE parish, W LOUISIANA, 57 mi/92 km S of SHREVEPORT; 31°42'N

93°41′W. Agriculture; timber; recreation. Oil wells nearby. Large TOLEDO BEND RESERVOIR (SABINE RIVER; TEXAS boundary) to W.

Noble, DOMINICAN REPUBLIC: see VICENTE NOBLE.

Nobleboro, town, LINCOLN county, S MAINE, on DAMARISCOTTA LAKE, and 11 mi/18 km NE of WISCASSET; 44°06′N 69°28′W. Sometimes called Nobleborough.

Nobleford (NO-buhl-fuhrd), village (□ 1 sq mi/2.6 sq km; 2001 population 615), S ALBERTA, W Canada, 17 mi/27 km NW of LETHBRIDGE, in Lethridge County; 49°53′N 113°03′W. Wheat; dairying. Incorporated 1918.

Noblejas (no-VLAI-hahs), town, TOLEDO province, central SPAIN, on railroad, and 33 mi/53 km SSE of MADRID; 39°59′N 03°26′W. In agricultural region (cereals, olives, grapes, esparto). Olive-oil pressing, plaster manufacturing. Mineral springs.

Noble Park (NO-buhl PAHRK), suburb 10 mi/16 km SE of MELBOURNE, VICTORIA, SE AUSTRALIA; 37°58′S 145°10′E. Ethnically diverse area.

Nobles, county (□ 722 sq mi/1,877.2 sq km; 2006 population 20,445), SW MINNESOTA; ⊙ WORTHINGTON; 43°40′N 95°45′W. Bordered by IOWA on S and West Branch of LITTLE ROCK RIVER. Agricultural area (corn, oats, soybeans, alfalfa; hogs, sheep, cattle, poultry; dairying). Includes part of COTEAU DES PRAIRIES; several small natural lakes of glacial origin in NE and SE, includes OCHEDA and OKABENA lakes (SE), West and East Graham lakes (NE). Formed 1857. Ocheda Lake is in SE.

Noblestown, unincorporated village, North Fayette township, ALLEGHENY county, W PENNSYLVANIA, suburb 11 mi/18 km WSW of downtown PITTSBURGH, on Robinson Run; 40°23′N 80°11′W. Gas and oil field. Pennsylvania Motor Speedway is here.

Noblesville, city (2000 population 28,590), ⊙ HAMILTON county, central INDIANA, on WHITE RIVER, and 20 mi/32 km NNE of INDIANAPOLIS; 40°04′N 86°02′W. Suburb of Indianapolis. Railroad junction. In livestock and grain area; manufacturing (concrete and molded products, ductile iron castings, plastics, medical products, air springs, food products, agricultural products, consumer goods; printing and publishing); horse breeding. MORSE RESERVOIR nearby to the NW. Laid out 1823; incorporated as town in 1851, as city in 1887.

Nobleton (NO-buhl-tuhn), unincorporated village, York region, S ONTARIO, E central Canada, 18 mi/29 km from TORONTO, and included in the township of KING; 43°54′N 79°39′W.

Noboribetsu (no-BO-ree-bets), city, Hokkaido prefecture, N JAPAN, 47 mi/75 km S of SAPPORO; 42°24′N 141°06′E.

Nobres, city (2007 population 14,825), S central MATO GROSSO, BRAZIL, 93 mi/150 km N of CUIABÁ; 14°42′S 56°28′W.

Nobsa (NOB-sah), town, ⊙ Nobsa municipio, BOYACÁ department, central COLOMBIA, 30 mi/48 km NE of TUNJA; 05°46′N 72°57′W. Elevation 8,500 ft/2,590 m. Coffee, corn, sugarcane; livestock.

Nobska Point (NAWB-skuh), SW CAPE COD, FALMOUTH, MASSACHUSETTS, low promotory just SE of WOODS HOLE; extends into VINEYARD SOUND. Lighthouse (41°31′N 70°39′W).

Nocaima (no-KEI-mah), town, ⊙ Nocaima municipio, CUNDINAMARCA department, central COLOMBIA, 26 mi/42 km NW of BOGOTÁ; 05°04′N 74°23′W. Elevation 3,589 ft/1,093. Coffee, sugarcane; livestock.

Nocatee (NAHK-uh-tee), village, DESOTO county, central FLORIDA, near PEACE RIVER, 4 mi/6.4 km S of ARCADIA; 27°10′N 81°52′W. Large crate factory.

Noccalula Falls (nah-kuh-LOO-luh), in small Black Creek, just NW of GADSDEN, ALABAMA, drop of c.100 ft/30 m over limestone ridge of LOOKOUT MOUNTAIN; c.15 mi/24 km long.

Nocera Inferiore (no-CHAI-rah een-fe-ree-O-re), ancient *Nuceria Alfaterna*, town, Salerno province, CAMPANIA, S ITALY, 8 mi/13 km NW of SALERNO;

40°44′N 14°38′E. Railroad junction; lumber mills; fabricated metals, machinery, textiles, apparel; food processing. Bishopric. Has ruins of Angevin castle. Destroyed by Hannibal (216 B.C.E.); rebuilt by Augustus.

Nocera Umbra (no-CHAI-rah OOM-brah), village, PERUGIA province, UMBRIA, central ITALY, 9 mi/14 km ENE of ASSISI; 43°05′N 12°47′E. Bishopric. Mineral baths nearby.

Noce River (NO-che), TRENTINO-ALTO ADIGE, N ITALY; rises in glaciers on Cima dei TRE SIGNORI in ORTLES group; flows ENE, through VAL DI SOLE, and S to ADIGE RIVER 6 mi/10 km N of TRENT; 50 mi/80 km long. Used for hydroelectric power (MEZZOCORONA).

Noceto (no-CHAI-to), town, PARMA province, EMILIA-ROMAGNA, N central ITALY, 7 mi/11 km W of PARMA; 44°48′N 10°11′E. Fabricated metals; food processing.

Nochistlán, MEXICO: see ASUNCIÓN NOCHIXTLÁN, OAXACA.

Nochistlán, MEXICO: see NOCHISTLÁN DE MEJÍA.

Nochistlán de Mejía (no-chees-TLAHN dai me-HEE-ah), city and township, ZACATECAS, N central MEXICO, on interior plateau, 50 mi/80 km SW of AGUASCALIENTES; 21°21′N 102°50′W. Elevation 6,332 ft/1,930 m. Agrcultural center (grain, sugarcane, beans, fruit; livestock).

Nochrich (NO-kreek), German *Leschkirch*, Hungarian *Újegyház*, village, SIBIU county, central ROMANIA, on railroad, and 15 mi/24 km NE of SIBIU; 45°54′N 24°27′E. Agriculture center.

Noci (NO-chee), town, BARI province, APULIA, S ITALY, 14 mi/23 km SSW of MONOPOLI; 40°48′N 17°07′E. Diversified small industrial center; olive oil, pasta.

Nockamixon, Lake, reservoir (□ 2 sq mi/5.2 sq km), BUCKS county, E central PENNSYLVANIA, on Tohickon Creek, in Nockamixon State Park, 16 mi/25 km SE of BETHLEHEM; 40°28′N 75°11′W. Maximum capacity 71,000 acre-ft. Formed by Nockamixon Dam (102 ft/31 m high) built (1973) for recreation.

Nocona (NAH-ko-nah), town (2006 population 3,279), MONTAGUE county, N TEXAS, c.45 mi/72 km E of WICHITA FALLS; 33°46′N 97°43′W. Elevation 988 ft/301 m. In agricultural, ranching, oil-producing area; a leather goods manufacturing center (belts, accessories, tack). Lake Nocona to NE.

Nocupétaro de Morelos (no-koo-PE-tah ro dai moh-RE-los), town, ⊙ Nocupétaro municipio, MICHOACÁN, central MEXICO, 45 mi/73 km S of MORELIA; 18°50′N 101°02′W. Rice, sugarcane, fruit.

Noda (no-DAH), city (2005 population 151,240), CHIBA prefecture, E central HONSHU, E central JAPAN, on the Edo River, 19 mi/30 km N of CHIBA; 35°56′N 19°51′E. Commercial and industrial center known for its soy sauce factories; also rice crackers.

Noda (no-DAH), town, Izumi county, KAGOSHIMA prefecture, SW KYUSHU, SW JAPAN, 37 mi/60 km N of KAGOSHIMA; 32°03′N 130°16′E.

Noda (no-DAH), village, Kunohe county, IWATE prefecture, N HONSHU, NE JAPAN, 47 mi/75 km N of MORIOKA; 40°06′N 141°49′E. Manganese. Tofugaura Beach is nearby.

Noda, RUSSIA: see CHEKHOV, SAKHALIN oblast.

Nodagawa (no-DAH-gah-wah), town, Yosa county, KYOTO prefecture, S HONSHU, W central JAPAN, 50 mi/80 km N of KYOTO; 35°31′N 135°06′E. *Tango* crepe.

Nodaway (NAH-duh-wai), county (□ 877 sq mi/2,280.2 sq km; 2006 population 21,660), NW MISSOURI; ⊙ MARYVILLE; 40°21′N 94°52′W. Drained by NODAWAY RIVER, LITTLE PLATTE RIVER, and ONE HUNDRED AND TWO RIVER. Agricultural region (corn, wheat, oats, soybeans; cattle, hogs); manufacturing at Maryville. NW Missouri State University at Maryville. Formed 1845.

Nodaway, town, ADAMS county, SW IOWA, on East Nodaway River, and 9 mi/14.5 km WSW of CORNING.

Nodaway River, 188 mi/303 km long (including MIDDLE NODAWAY RIVER), in SW IOWA and NW MISSOURI;

formed near VILLISCA in MONTGOMERY county (Iowa) by junction of MIDDLE and WEST NODAWAY rivers; flows to MISSOURI RIVER above ST. JOSEPH; 40°56′N 94°54′W. Receives East Nodaway River (c.60 mi/97 km long) near SHAMBAUGH.

Nodushan (nuh-DOO-shahn), village, YAZD province, central IRAN, 50 mi/80 km WNW of YAZD, in sheep-grazing area. Grain, dairy products (especially cheese). Also spelled Nadushon or NUDUSHAN.

Noel (NO-wuhl), town (2000 population 1,480), MCDONALD county, extreme SW MISSOURI, in the OZARKS, on ELK RIVER, and 9 mi/14.5 km SSW of ANDERSON; 36°32′N 94°29′W. Sport fishing; berries, chickens, dairying; manufacturing (poultry processing, wood products). Bluff Dwellers Cave to S. Plotted 1891.

Noëlville (no-EL-vil), settlement, SE central ONTARIO, E central Canada; included in town of French River; 46°08′N 80°25′W. Formed c.1900. Agriculture, forestry, tourism. Formerly called Cosby.

Noemfoor, INDONESIA: see NUMFOOR.

Noépé (no-AI-pai), village, MARITIME region, S TOGO, 15 mi/24 km NW of LOMÉ, on railroad and near GHANA border; 06°16′N 01°02′E. Palm oil and kernels, copra. Customhouse.

Noesa, in Indonesian names beginning thus: see under NUSA.

Noetinger (no-e-TEEN-ger), town, E CÓRDOBA province, ARGENTINA, 55 mi/89 km E of VILLA MARÍA; 32°22′S 62°19′W. Wheat, flax, soybeans, corn, alfalfa; livestock.

Noeux-les-Mines (no-u–lai–meen), town (□ 3 sq mi/7.8 sq km), PAS-DE-CALAIS department, NORD-PAS-DE-CALAIS region, N FRANCE, 4 mi/6.4 km SSE of BÉTHUNE; 50°29′N 02°40′E. Textile industry, metalworking. Has museum of mining. Coal mines were closed down by 1990.

Noevtsi (NOIV-tsee), village, SOFIA oblast, BREZNIK obshtina, BULGARIA; 42°42′N 22°53′E.

Noez (no-ETH), town, TOLEDO province, central SPAIN, 12 mi/19 km SW of TOLEDO; 39°45′N 04°11′W. Wheat, barley, carobs, chick peas, olives, grapes; livestock.

Nofit (no-FEET), suburb, ISRAEL, 9 mi/14.5 km SE of SW foothills of LOWER GALILEE. Founded 1987. Also known as Pi Ner.

Nogaisk, UKRAINE: see PRYMORS′K.

Nogales (no-GAH-les), city and township, VERACRUZ, E MEXICO, in valley of SIERRA MADRE ORIENTAL, at S foot of PICO DE ORIZABA, on railroad, and 4 mi/6.4 km WSW of ORIZABA; 18°50′N 97°12′W. Textile-milling and agricultural center (coffee, tobacco, sugarcane, fruit).

Nogales (NO-gah-les), city (2000 population 20,878), SANTA CRUZ county, S ARIZONA, 60 mi/97 km S of TUCSON, on the Mexican (SONORA) border opposite NOGALES, MEXICO; 31°21′N 110°55′W. Elevation 3,865 ft/1,178 m. Railroad terminus (transfer point to Mexican railroad system). There are copper, silver, and lead mines. Skirmishes occurred in Nogales against Pancho Villa in 1916. Industrial development, primarily in the 1980s, in the Mexican city of Nogales has resulted in a growth of maquiladoras (manufacturing plants operated by low-cost labor for the production of U.S. trading goods). Parts of Coronado National Forest to E and W; Patagonia Lake State Park to NE; Tubac Presidio State Historic Park and TUMACACORI National Monument to N.

Nogales (no-GAH-les), town, ⊙ Nogales comuna, QUILLOTA province, VALPARAISO region, central CHILE, on railroad, and PAN-AMERICAN HIGHWAY, 33 mi/53 km NE of VALPARAISO. Agricultural center (cereals, grapes, fruit; livestock).

Nogales (no-GAH-les), town, BADAJOZ province, W SPAIN, 23 mi/37 km SE of BADAJOZ. Cereals, vegetables, olives; livestock.

Nogales, MEXICO: see HEROICA NOGALES.

Nogal Peak, NEW MEXICO: see SIERRA BLANCA.

Nogal Valley (noo-GAHL), NE central SOMALIA, extends 150 mi/241 km ESE to NEGRO Bay of INDIAN Ocean; intermittent stream in lower section below KALIS.

Nogamut (NO-guh-moot), Indian village, W ALASKA, on Holitna River, and 75 mi/121 km SE of ANIAK. Sometimes spelled Nugammute and Nogamiut.

Nogara (no-GAH-rah), village, VERONA province, VENETO, N ITALY, 19 mi/31 km S of VERONA; 45°11'N 11°04'E. Railroad junction; diversified secondary industrial center.

Nogaro (no-gah-ro), commune (□ 4 sq mi/10.4 sq km), GERS department, MIDI-PYRÉNÉES region, SW FRANCE, 25 mi/40 km SW of CONDOM; 43°46'N 00°02'W. Vineyards; ARMAGNAC brandy distilling. Has 11th-century church.

Nogata (NO-gah-tah), city, FUKUOKA prefecture, N KYUSHU, SW JAPAN, 22 mi/35 km N of FUKUOKA; 33°44'N 130°43'E. Railroad junction. Dumplings.

Nogatino (nah-GAH-tyi-nuh), former town, central MOSCOW oblast, RUSSIA, on the S bank of the S port of MOSCOW (SE). Now incorporated into Moscow city.

Nogat River (NO-gaht), 35 mi/56 km long, N POLAND, E arm of VISTULA River delta estuary; leaves the Vistula 11 mi/18 km SW of MALBORK (Marienburg); flows NE, past Malbork, to VISTULA LAGOON 9 mi/14.5 km NW of Elbląg. From 1919 to 1939, formed border between territory of Free City of GDAŃSK and EAST PRUSSIA.

Nogaysk, UKRAINE: see PRYMORS'K.

Nogent (no-ZHAHN), town (□ 21 sq mi/54.6 sq km), HAUTE-MARNE department, CHAMPAGNE-ARDENNE region, NE FRANCE, 11 mi/18 km SE of CHAUMONT; 46°46'N 03°33'E. Noted cutlery-manufacturing center since 18th century. Once known as Nogent-en-Bassigny.

Nogent-en-Bassigny, FRANCE: see NOGENT.

Nogent-l'Abbesse (no-ZHAHN–lah-BES), village (□ 3 sq mi/7.8 sq km), MARNE department, CHAMPAGNE-ARDENNE region, N FRANCE, 6 mi/9.7 km E of REIMS; 49°15'N 04°09'E. Vineyards (champagne). Its fort and Fort de la Pompelle (3 mi/4.8 km SSW) were key defense points against German attacks during the battle of Champagne in World War I.

Nogent-le-Roi (no-zhahn–luh–rwah), commune (□ 5 sq mi/13 sq km; 2004 population 4,067), EURE-ET-LOIR department, CENTRE administrative region, NW central FRANCE, on the EURE, and 10 mi/16 km SE of DREUX; 48°39'N 01°32'E. In rich agricultural area; flour milling, flask manufacturing. Has 15th–17th-century church of Saint-Sulpice.

Nogent-le-Rotrou (no-zhahn–luh–ro-troo), town (□ 9 sq mi/23.4 sq km), EURE-ET-LOIR department, CENTRE administrative region, NW central FRANCE, on the HUISNE, and 32 mi/51 km WSW of CHARTRES; 48°20'N 01°00'E. Horse-breeding center (Percheron horses), and agriculture market for cider and poultry. Industries include meat-slaughtering for PARIS market, electronic equipment, pharmaceuticals, and plastics. Has ruins of 11th–16th-century castle of the counts of Perche and the tomb of Sully, treasurer of Henry IV. The town was the capital of historic PERCHE district.

Nogent-sur-Marne (no-zhahn–syoor–mahrn), town (□ 1 sq mi/2.6 sq km), VAL-DE-MARNE department, ÎLE-DE-FRANCE region, N central FRANCE, E residential suburb of PARIS, 6 mi/9.7 km from Notre Dame Cathedral, on hill above right bank of the MARNE, and at E edge of BOIS DE VINCENNES; 48°50'N 02°29'E. Diverse manufacturing. Has 13th-century church and a curving railroad viaduct, 2,500 ft/762 m long. Pleasant waterfront development.

Nogent-sur-Oise (no-zhahn–syoor–wahz), town (□ 2 sq mi/5.2 sq km), OISE department, PICARDIE region, N FRANCE, near OISE River, 7 mi/11.3 km NW of SENLIS; 49°16'N 02°28'E. Smelters (copper, brass), metalworks in industrial district of CREIL.

Nogent-sur-Seine (no-zhahn–syoor–sen), town (□ 7 sq mi/18.2 sq km), AUBE department, NE central FRANCE, in CHAMPAGNE-ARDENNE region, 11 mi/18 km ESE of PROVINS, and 28 mi/45 km NW of TROYES; 48°25'N 03°45'E. Flour mill; agricultural silos; dairy products. On farm (4 mi/6.4 km ESE) once stood abbey of the Paraclete (French Le Paraclet), founded 1123 by Abélard for Héloïse, its first abbess. The Nogent nuclear power plant (capacity 2600 MW) is 2 mi/3.2 km NNE on right bank of SEINE RIVER; it has two cooling towers, to recirculate condenser cooling water taken from the river.

Nogent-sur-Vernisson (no-zhahn–syoor–ver-nee-son), commune (□ 12 sq mi/31.2 sq km; 2004 population 2,538), LOIRET department, CENTRE administrative region, N central FRANCE, in the GÂTINAIS district, 11 mi/18 km S of MONTARGIS; 47°51'N 02°45'E. Manufacturing of automobile accessories. Has school of forestry and water management. Arboretum.

Nogi (no-GEE), town, Shimotsuga county, TOCHIGI prefecture, central HONSHU, N central JAPAN, 22 mi/35 km S of UTSUNOMIYA; 36°14'N 139°43'E.

Noginsk (nuh-GEENSK), city (2006 population 115,480), E MOSCOW oblast, central European Russia, on the KLYAZ'MA RIVER, on spur railroad and highway junction, 42 mi/68 km E of MOSCOW; 55°51'N 38°26'E. Elevation 524 ft/159 m. It is a major textile center, processing cotton, silk, and wool; fuel equipment, roofing and water proofing materials, cutting tools, plastics, rubber goods, garments, food processing (dairy, bakery, meat packing). Founded in the 16th century as Rogozhi, the city was later called Bogorodsk until 1930. Made city in 1781.

Noginsk (nuh-GEENSK), village, W EVENKI AUTONOMOUS OKRUG, KRASNOYARSK TERRITORY, N SIBERIA, RUSSIA, on the LOWER TUNGUSKA RIVER, on highway, 260 mi/418 km W of TURA; 64°29'N 91°14'E. Elevation 826 ft/251 m. Graphite mines nearby.

Nogir (nuh-GEER), village (2006 population 11,543), E NORTH OSSETIAN REPUBLIC, RUSSIA, on the slopes of the N central CAUCASUS Mountains, on the TEREK RIVER, near railroad, 3 mi/5 km N of VLADIKAVKAZ; 43°05'N 44°38'E. Elevation 2,057 ft/626 m. Agricultural market; light manufacturing.

Nogliki (NO-glee-kee), town (2006 population 9,895), N SAKHALIN oblast, RUSSIAN FAR EAST, on the Sea of OKHOTSK (E coast of SAKHALIN Island), at the mouth of the TYM' RIVER, on railroad and road, 75 mi/121 km NE of ALEKSANDROVSK-SAKHALINSKIY; 51°50'N 143°10'E. Lumbering, fishing.

Nogoa River (nuh-GO-uh), 180 mi/290 km long, E central QUEENSLAND, NE AUSTRALIA; rises in GREAT DIVIDING RANGE SW of SPRINGSURE; flows NE, joining COMET RIVER near COMET to form MACKENZIE RIVER; 23°33'S 148°32'E. Used for irrigation.

Nogoyá (no-go-YAH), town (1991 population 18,968), ⊙ Nogoyá department (□ 1,670 sq mi/4,342 sq km), S central ENTRE RÍOS province, ARGENTINA, near Nogoyá River, 60 mi/97 km SE of PARANÁ; 32°23'S 59°47'W. Railroad junction. Agricultural area (flax, wheat, corn, grapes; livestock); flour milling. Airport.

Nógrád (NOG-rahd), county (1,090 sq mi/2,823 sq km), N HUNGARY; ⊙ SALGÓTARJÁN; 48°00'N 19°35'E. Heavily forested Börzsöny Mountains in W, CSERHÁT MOUNTAINS in S; drained by ZAGYVA RIVER; bounded W and N by the IPOLY RIVER. Grain, potatoes; hogs, sheep. Coal and lignite mines; manufacturing (chemicals, plastics, farinaceous foods; baking, meat-packing). Cities include SALGÓTARJÁN and BALASSAGYARMAT. In the 1990s, Nógrád county was among the worst affected by economic restructuring. Its highly uneconomical lignite mines are mostly shut down; other industries contracted sharply. Unattractive to foreign investment; and unemployment is among top five Hungarian counties. Has many ruins of castles and fortresses, some built in the first three centuries of the Hungarian Kingdom. It was a frontier region for the Ottoman Empire with frequent warfare. During the adjoining region of SLOVAKIA is the home of the Palóc people, a Hungarian subgroup that has preserved to this day many of its folk customs. The village of Hollókö is designated a national treasure and is on UNESCO's World Heritage List. The famous Hungarian novelist Kálmán Mikszáth was born in the region of BALASSAGYARMAT; the Palóc and the Slovak peoples often figure significantly in his writings.

Nógrád (NOG-rahd), village, NÓGRÁD county, N HUNGARY, 25 mi/40 km N of BUDAPEST; 47°55'N 19°03'E. Fortress here, now in ruins, dates back to Árpád rulers.

Nógrádveröce (NOG-rahd-ver-uh-tse), village, NÓGRÁD county, N HUNGARY, on the DANUBE River, and 5 mi/8 km NW of VÁC. Summer resort; brickworks.

Nogueira, Serra de (no-GAI-rah, SER-rah dai), range in BRAGANÇA district, N PORTUGAL, SW of BRAGANÇA; 41°42'N 06°52'W. Rises to 4,324 ft/1,318 m.

Noguera Pallaresa River (no-GAI-rah pah-lyah-RAI-sah), c.100 mi/161 km long, LÉRIDA province, NE SPAIN; rises N of SORT, near French border in the central PYRENEES; flows S to the SEGRE RIVER above CAMARASA. Forms the Tremp reservoir (with 300-ft/91-m dam), feeding two hydroelectric plants. The FLAMISELL RIVER joins it at POBLA DE SEGUR.

Noguera Ribagorzana River (no-GAI-rah ree-vah-gor-THAH-nah), 80 mi/129 km long, NE SPAIN; rises in the mountains of VALLE DE ARÁN in the central PYRENEES; flows S, forming CATALONIA-ARAGÓN border as far as ALFARRÁS, to the SEGRE River 6 mi/9.7 km above LÉRIDA.

Noguères (no-GER), commune, PYRÉNÉES-ATLANTIQUES department, AQUITAINE region, SW FRANCE, on GARONNE River 12 mi/19 km, NW of PAU, and 1.5 mi/2.4 km E of MOURENX; 43°22'N 00°36'W. Large aluminum reduction plant, part of the LACQ natural gas field's industrial complex.

Nohant-Vic (no-ahn–veek), rural district (□ 8 sq mi/20.8 sq km), INDRE department, CENTRE administrative region, central FRANCE, near INDRE RIVER, 3 mi/4.8 km NNW of La CHÂTRE; 46°38'N 01°58'E. The 18th-century Château of Nohant, once the home of George Sand, is now a museum. At Vic there is a Romanesque church with a group of wall paintings. Also spelled Nohant-Vicq.

Nohant-Vicq, FRANCE: see NOHANT-VIC.

Nohar (NO-huhr), town, GANGANAGAR district, N RAJASTHAN state, NW INDIA, 74 mi/120 km SW of GANGANAGAR; 29°11'N 74°46'E. Millet, gram; livestock raising; hand-loom weaving.

Nohays'ke, UKRAINE: see PRYMORS'K.

Noheji (NO-he-jee), town, Kamikita county, Aomori prefecture, N HONSHU, N JAPAN, on S MUTSU bay, 22 mi/35 km E of AOMORI; 40°51'N 141°07'E. Scallops.

Nohfelden (NO-fel-den), town, SAARLAND, W GERMANY, 9 mi/14.5 km N of ST. WENDEL; 49°35'N 07°09'E. Furniture; metalworking; tourism. Formed 1973 by unification of thirteen villages.

Nohmul, archaeological site, ORANGE WALK district, BELIZE, midway between ORANGE WALK and COROZAL W of Northern Highway; 18°14'N 88°34'W. A large and important classical-era Mayan site. On private property.

NoHo, section of Lower MANHATTAN, NEW YORK city, SE NEW YORK, bounded on S by East Houston Street, on N by East 8th Street, on E by the Bowery, and on W by Mercer where it abuts GREENWICH VILLAGE; 40°43'N 73°59'W. Name is an emblematic acronym referring to the gentrification process that has occurred in the area "north of Houston Street" (similar to the way the name SoHo stands for "south of Houston Street"). Like SOHO, its more glamorous counterpart, NoHo is still zoned for light manufacturing, with joint living and working quarters for artists. In 1976, artists' request that the city re-zone the district in the same manner as SoHo—that is, as an historical, industrial manufacturing area—was granted. Many of the former warehouses have been officially certified by New York City Dept. of Cultural Affairs as residences of artists. For young, less affluent artists, NoHo has developed as a cheaper alternative to SoHo, which is now characterized by skyrocketing rents and lack of space.

Noichi (NO-ee-chee), town, Kami county, KOCHI prefecture, S SHIKOKU, W JAPAN, 9 mi/15 km E of KOCHI; 33°33′N 133°42′E.

Noida, city (2001 population 305,058), GHAZIABAD district, UTTAR PRADESH state, N central INDIA. Residential and industrial suburb of DELHI.

Noire, Montagne, FRANCE: see MONTAGNE NOIRE.

Noire, Rivière, VIETNAM: see BLACK RIVER.

Noires, Montagnes, FRANCE: see MONTAGNES NOIRES.

Noirétable (nwahr-ai-tah-bluh), commune (□ 15 sq mi/39 sq km), resort in LOIRE department, RHÔNE-ALPES region, SE central FRANCE, in Monts du Forez of the MASSIF CENTRAL, 11 mi/18 km ESE of THIERS; 45°49′N 03°45′E. Cutlery manufacturing; plastics, molds; woodworking, embroidering of military uniforms. Has 15th-century Flamboyant Gothic church. The regional park of LIVRADOIS-FOREZ lies SW.

Noir Island, in SW TIERRA DEL FUEGO, CHILE, 30 mi/48 km S of SANTA INÉS ISLAND; 7 mi/11 km long. Elevation 600 ft/183 m. Terminates SW in rocky Cape Noir, 54°30′S 73°06′W.

Noir, Lac, FRANCE: see ORBEY.

Noirmoutier-en-l'Île (nwahr-moo-tyai–ahn–leel), chief town (□ 7 sq mi/18.2 sq km), Île de NOIRMOUTIER, VENDÉE department, PAYS DE LA LOIRE region, W FRANCE, 19 mi/31 km S of SAINT-NAZAIRE; 47°00′N 02°15′W. A bathing resort and small fishing port on NE shore. Saltworks. Church of abbey (founded c.680) contains empty crypt of Saint-Philibert (8th century). Here, in 1794, the leader of the Vendée rebellion was captured and shot. The 15th-century chateau, with older towers, is now a museum.

Noirmoutier, Île de (nwahr-moo-tyai, eel duh), island (□ 22 sq mi/57.2 sq km), in BAY OF BISCAY, off VENDÉE department, PAYS DE LA LOIRE region, W France, c.18 mi/29 km S of SAINT-NAZAIRE; 12 mi/19 km long, 1 mi/1.6 km to 4 mi/6.4 km wide, and separated from mainland by Fosse de Fromentine, a shallow strait which drains at low tide and is crossed by bridge since 1971; 47°00′N 02°15′W. Island consists of low, barren tracts, sand dunes, and salt marshes; it produces salt, fruits, potatoes, and flowers. Oyster beds. Port of NOIRMOUTIER-EN-L'ÎLE on NE shore, l'Épine has a beach on W shore. There are many windmills on island. Fromentine is nearest village on mainland.

Noisiel (nwah-zyel), town (□ 1 sq mi/2.6 sq km), SEINE-ET-MARNE department, ÎLE-DE-FRANCE region, N central FRANCE, on left bank of the MARNE, and 13 mi/21 km E of PARIS; 48°51′N 02°38′E. Chocolate factory. It forms part of the "new town" of MARNE-LA-VALLÉE.

Noissy-le-Roi (nwah-see–luh–rwah), town, YVELINES department, ÎLE-DE-FRANCE region, N central FRANCE; 48°51′N 02°04′E. Just S of the Forêt de Marly, and 4 mi/6 km S of SAINT-GERMAIN-EN-LAYE.

Noisy-le-Grand (nwah-zee–luh–grahn), city (□ 5 mi/13 sq km), SEINE-SAINT-DENIS department, ÎLE-DE-FRANCE region, N central FRANCE, an outer E suburb of PARIS, 9 mi/14.5 km from Notre Dame Cathedral, on left bank of the MARNE, opposite NEUILLY-SUR-MARNE; 48°51′N 02°34′E. Manufacturing (rubber, paints, and surgical instruments).

Noisy-le-Sec (nwah-zee–luh–sek), industrial town (□ 1 sq mi/2.6 sq km), SEINE-SAINT-DENIS department, ÎLE-DE-FRANCE region, N central FRANCE, a NE suburb of PARIS, 6 mi/9.7 km from Notre Dame Cathedral, on OURCQ CANAL; 48°53′N 02°28′E. Railroad yards; metalworks, cement products.

Nojima, Cape (no-JEE-MAH), Japanese *Nojima-saki* (no-JEE-MAH-SAH-kee), CHIBA prefecture, E central

HONSHU, E central JAPAN, on S BOSO PENINSULA, in the PACIFIC OCEAN, near entrance to TOKYO BAY; 34°54′N 139°53′E. Lighthouse.

Nojiri (NO-jee-ree), town, W Morokata county, MIYAZAKI prefecture, SE KYUSHU, SW JAPAN, 19 mi/30 km W of MIYAZAKI; 31°57′N 131°06′E.

Nokami (no-KAH-mee), town, Kaiso county, WAKAYAMA prefecture, S HONSHU, W central JAPAN, on W KII PENINSULA, 9 mi/15 km S of WAKAYAMA; 34°09′N 135°18′E. Traditional housewares.

Nokia (NO-kee-ah), town, HÄMEEN province, SW FINLAND; 61°28′N 23°30′E. Elevation 281 ft/85 m. On NÄSIJÄRVI (lake). Industrial center for wood and rubber products.

Nokilalaki, Mount (no-KEE-lah-lah-kee) mountain (10,863 ft/3,311 m), Sulawesi Tengah province, INDONESIA, 45 mi/72 km NNW of POSO; 01°13′S 128°08′E.

Nok Kundi (NOK KUHN-dee), town, CHAGAI district, NW BALUCHISTAN province, SW PAKISTAN, on railroad, and 205 mi/330 km WSW of NUSHKI; 28°48′N 62°46′E. Sulphur-ore, gypsum, limestone deposits nearby; potential iron ore reserves, mini-steel plant scheduled to be built with Chinese assistence. Also called KUNDI, Nokkundi, KONDI, Nok Kundi.

Nokomis, city (2000 population 2,389), MONTGOMERY county, S central ILLINOIS, 16 mi/26 km NE of HILLSBORO; 39°17′N 89°17′W. In agricultural area; dairy products. Incorporated 1867.

Nokomis (no-KO-mis), town (2006 population 404), S central SASKATCHEWAN, CANADA, 22 mi/35 km SE of WATROUS; 51°31′N 105°01′W. Railroad junction, grain elevators, lumbering.

Nokomis (nuh-KO-mis), unincorporated town (□ 2 sq mi/5.2 sq km; 2000 population 3,334), SARASOTA county, W central FLORIDA, 17 mi/27 km SE of SARASOTA; 27°07′N 82°26′W. Manufacturing includes motor vehicles, printing and publishing, machining, and aluminum fabricating.

Nokou (no-KOO), town, KANEM administrative region, W CHAD, 200 mi/322 km N of N'DJAMENA; 14°35′N 14°47′E.

Nokoué, Lake (no-KWAI) (□ 30 sq mi/78 sq km), coastal lagoon, S BENIN; 12 mi/19 km long, 5 mi/8 km wide; 06°26′N 02°27′E. Located N of COTONOU; linked to adjoining PORTO-NOVO lagoon. Main arm of OUÉMÉ RIVER empties into it. Tourism, fishing, recreation.

Nokrek Peak (NOK-raik), highest point (c. 4,630 ft/ 1,411 m) in GARO HILLS, WEST GARO HILLS district, MEGHALAYA state, NE INDIA, 7 mi/11.3 km SE of TURA.

Nol (NOOL), town, ÄLVSBORG county, SW SWEDEN, on GÖTA ÄLV RIVER, 6 mi/9.7 km NE of KUNGÄLV; 57°55′N 12°05′E.

Nola (NO-lah), town, in CAMPANIA, S ITALY. Agricultural center and highly diversified secondary industrial center, including food-processing industries. An Etruscan stronghold as early as 500 B.C.E., Nola flourished after passing (c.316 B.C.E.) to Rome and was an important center of early Christianity. Nearby are Roman ruins (an amphitheater and tombs) and an old cemetery where Christian martyrs are buried. Augustus died here in C.E. 14.

Nola (no-LAH), village, ⊙ SANGHA-MBAÉRÉ economic prefecture, SW CENTRAL AFRICAN REPUBLIC, at confluence of headstreams of SANGHA RIVER, 55 mi/89 km SSE of BERBÉRATI; 03°28′N 16°08′E. Coffee plantations, diamond mining.

Nolan (NO-luhn), county (□ 914 sq mi/2,376.4 sq km; 2006 population 14,812), W TEXAS; ⊙ SWEETWATER. Drained by Sweetwater Creek and other tributaries of the COLORADO and the BRAZOS rivers; 32°18′N 100°24′W. Elevation 2,000 ft/610 m–2,700 ft/823 m. Ranching, agriculture; cattle, sheep, goats, hogs; cotton, sorghum, wheat, hay; oil and natural gas production; gypsum quarries, stone, sand and gravel, clay. Recreation areas at SWEETWATER and Oak Creek lakes. Formed 1876.

Nolanville (NO-luhn-vil), town (2006 population 2,361), BELL county, central TEXAS, residential suburb 7 mi/11.3 km E of KILLEEN, and 14 mi/23 km WSW of TEMPLE, on Nolan Creek; 31°04′N 97°36′W. Agricultural area. Limestone. Large FORT HOOD Military Reservation to N, STILLHOUSE HOLLOW LAKE reservoir to S.

Nolay (no-lai), commune (□ 5 sq mi/13 sq km; 2004 population 1,490), CÔTE-D'OR department, in BURGUNDY, E central FRANCE, on S slope of the CÔTE D'OR hills, 11 mi/18 km SW of BEAUNE; 46°57′N 04°38′E. Burgundy wines. Meat preserving. Lazare Carnot born here.

Noli (NO-lee), village, SAVONA province, LIGURIA, NW ITALY, port on GULF OF GENOA, and 8 mi/13 km SSW of SAVONA; 44°12′N 08°26′E. Fishing center; canneries. Bishopric. Has 12th-century church. Just SE is Cape Noli; quartz mines. Tourist, bathing center.

Noli (NO-lee), village, LINDI region, SE TANZANIA, 93 mi/150 km WSW of LINDI, and 25 mi/40 km NW of NACHINGWEA; 10°25′S 38°17′E. One of the centers of failed groundnut-growing scheme of late 1940s and early 1950s. Subsistence crops; goats, sheep.

Nolichucky (NAHL-uh-chuck-ee), river, c.150 mi/240 km long, W NORTH CAROLINA; rises in the BLUE RIDGE; flows NW and W to the FRENCH BROAD RIVER W of GREENEVILLE, TENNESSEE A power dam impounds Davy Crockett Lake SW of Greeneville. The first settlement on the river was made in 1772.

Nolin Lake (NO-lin), reservoir (□ 9 sq mi/23.4 sq km), EDMONSON county, central KENTUCKY, on NOLIN RIVER, 24 mi/39 km NNE of BOWLING GREEN; 37°17′N 86°15′W. Maximum capacity 609,400 acre-ft. Formed by Nolin Lake Dam (146 ft/45 m high), built (1963) by Army Corps of Engineers for flood control; also used for recreation. MAMMOTH CAVE NATIONAL PARK to S. Also known as Nolin River Lake.

Nolin River (NO-lin), 105 mi/169 km long, central KENTUCKY; rises in E LARUE county, flows W past HODGENVILLE, and SSW through Nolin River Lake reservoir and W end of MAMMOTH CAVE NATIONAL PARK, to GREEN RIVER 3 mi/4.8 km NE of BROWNSVILLE.

Nolinsk (nuh-LEENSK), city (2005 population 10,225), S KIROV oblast, E central European Russia, on the VOYA River (tributary of the VYATKA River), on crossroads, 90 mi/145 km S of KIROV; 57°33′N 49°56′E. Metalworking; sawmilling, timbering, woodworking; agriculture (seed nursery, bakery, poultry factory, fur farm, veterinary station and laboratory). Founded in 1668; chartered in 1780. Called Molotovsk, 1940–1957.

Nollendorf, CZECH REPUBLIC: see NAKLEROV.

Nólsoy, Danish *Nolsø*, island (□ 4 sq mi/10.4 sq km) of the E FAEROE ISLANDS, separated from Stremoy by Nólsoyafjørdður, 3 mi/4.8 km E of TÓRSHAVN; 6 mi/9.7 km long, 1.5 mi/2.4 km wide. Accessible by mail boat. Has one village near N end of island, by the same name. N part is flat; highest point 1,217 ft/371 m. Fishing; sheep. Puffin colonies on island's coastal cliffs. Remains of medieval settlement S of Nólsoy isthmus. Lighthouse at S end.

Noma, Cape (NO-mah), Japanese *Noma-misaki* (no-MAH-mee-SAH-kee), KAGOSHIMA prefecture, SW KYUSHU, SW JAPAN, on SW SATSUMA Peninsula, on EAST CHINA SEA; 31°25′N 130°07′E.

No Mans Land, SE MASSACHUSETTS, island in the AT-LANTIC OCEAN, 6 mi/9.7 km S of GAY HEAD; c.2 mi/3.2 km long, 0.5 mi/0.8 km–1 mi/1.6 km wide. The island is a National Wildlife Refuge. Formerly U.S. Navy bombing range.

Nombela (nom-BAI-lah), town, TOLEDO province, central SPAIN, 21 mi/34 km NE of TALAVERA DE LA REINA; 40°09′N 04°30′W. Grapes, cereals; livestock. Lumbering. Mineral springs.

Nombre de Dios (NOM-brai dai DEE-os), city and township, DURANGO, N MEXICO, on interior plateau, 30 mi/48 km ESE of VICTORIA DE DURANGO; 23°50′N

104°15'W. Grain, cotton, vegetables, livestock. Silver mines nearby. First Spanish settlement in Durango. Also called LA VILLITA.

Nombre de Dios, village and minor civil division of Santa Isabel district, COLÓN province, central PANAMA, minor port on CARIBBEAN SEA, and 20 mi/32 km NE of COLÓN. In banana area; also abacá, cacao, coconuts; livestock. Manganese deposits. Important port in sixteenth century for shipment of Spanish colonial riches.

Nombre de Dios, Cordillera de (NOM-brai de dee-OS kor-dee-YE-rah dai), mountain range of N HONDURAS, on ATLÁNTIDA-YORO department border, S of TELA; 15°40'N 86°40'W. Rises to over 7,700 ft/2,347 m.

Nombre de Jesus (NOM-brai dai hai-SOOS), municipality and town, CHALATENANGO department, EL SALVADOR, E of CHALATENANGO city in far E of department on RIVER LEMPA; 14°00'N 88°44'W.

Nome, city (2000 population 3,505), W ALASKA, on the S side of SEWARD PENINSULA, on NORTON SOUND; 64°30'N 165°24'W. Founded c.1898, when gold was discovered on the beach here. It is the commercial, government, and supply center for NW Alaska, with an airport. Major economic mainstays are mining, tourism fishing, and government The city is also a center of Eskimo handicrafts. Nome was a gold rush town from 1899 to 1903; its population swelled to 30,000, but many died or left because of the hardships. Dredging, which replaced older methods of mining, ceased in 1962, renewed in 1980s. The city is the scene of an annual Midnight Sun Festival (late June), Iditarod Dog Sled Race (late February, from ANCHORAGE) and the All-Alaska Championship Dog Race. Cape NOME lies to SE. Road system connects with TAYLOR (N), Keller (NW), and OPHIR (NE).

Nome (NOM), village (2006 population 65), BARNES county, SE NORTH DAKOTA, 20 mi/32 km SE of VALLEY CITY; 46°40'N 97°49'W. Railroad junction to NE at Lucca. Founded in 1901 and incorporated in 1907. Named for NOME, ALASKA.

Nome (NOM), village (2006 population 495), JEFFERSON county, SE TEXAS, 18 mi/29 km W of BEAUMONT; 30°01'N 94°25'W. Oil and natural gas. Cattle; rice, soybeans.

Nome, Cape, W ALASKA, S SEWARD PENINSULA, on N shore of NORTON SOUND, 12 mi/19 km ESE of NOME; 64°27'N 165°00'W.

Nomeland (NAW-muh-lahn), village, AUST-AGDER county, S NORWAY, on OTRA RIVER, and 70 mi/113 km NNW of KRISTIANSAND; 58°21'N 07°54'E.

Nomelandsfoss (NAW-muh-lahns-FAWS), waterfall (66 ft/20 m) on OTRA RIVER, VEST-AGDER county, S NORWAY, 15 mi/24 km N of KRISTIANSAND. Hydroelectric plant.

No-men-han, MONGOLIA: see NOMONHAAN.

Nomexy (nom-SEE), commune (□ 3 sq mi/7.8 sq km), VOSGES department, LORRAINE region, NE FRANCE, port on MOSELLE RIVER, and Canal de l'Est, 10 mi/16 km NNW of ÉPINAL; 48°19'N 06°22'E. Cotton spinning.

Nomi (NO-mee), town, central NOMI-SHIMA island, Saeki county, HIROSHIMA prefecture, W JAPAN, 12 mi/20 km S of HIROSHIMA; 34°13'N 132°26'E. Oysters; fish processing. Flowers.

Nominingue (no-meen-ANG), village (□ 119 sq mi/309.4 sq km; 2006 population 2,227), Laurentides region, SW QUEBEC, E Canada, on Lake Nominingue (6 mi/10 km long, 3 mi/5 km wide), in the LAURENTIANS, 45 mi/72 km NW of SAINTE-AGATHE-DES-MONTS; 46°23'N 75°01'W. Dairying.

Nomi-shima (no-MEE-shee-MAH), island (□ 39 sq mi/101.4 sq km), HIROSHIMA prefecture, W JAPAN, in HIROSHIMA BAY, just S of HIROSHIMA and W of KURE; 10 mi/16 km long, 3 mi/4.8 km wide; has peninsula 8 mi/12.9 km long, 0.5 mi/0.8 km–4 mi/6.4 km wide. S part is called Higashi-nomi-shima, NW part, Nishi-nomi-shima, and NE part, Etajima. Mountainous; fertile. Rice, tea, citrus fruit; livestock. Fishing.

Nõmme (NUH-mai), S residential district of single-family homes in TALLINN, ESTONIA, within city limits.

Nomo, Cape (NO-mo), Japanese *Nomo-misaki* (no-MO–mee-SAH-kee), S extremity of NOMO PENINSULA, NAGASAKI prefecture, NW KYUSHU, SW JAPAN; forms NW side of entrance to AMAKUSA SEA; 32°34'N 129°44'E. Kaba-shima, islet, is nearby.

Nomoi Islands or MORTLOCK ISLANDS, State of CHUUK, E CAROLINE ISLANDS, W Pacific, 150 mi/241 km SE of Chuuk. Include ETAL, LUKUNOR, and SATAWAN atolls.

Nomonhaan (no-mon-HAHN), Chinese *No-men-han*), hill on NW MANCHURIA (CHINA)-MONGOLIA border, 120 mi/190 km SSW of HAILAR, at HALHIN GOL (Khalka River); 47°46'N 118°45'E. Scene of Soviet-Japanese border incident (May-September 1939).

Nomo Peninsula (NO-mo), Japanese *Nomo-hanto* (no-MO–HAHN-to), NAGASAKI prefecture, NW KYUSHU, SW JAPAN, S projection of HIZEN PENINSULA, between EAST CHINA (W) and AMAKUSA (E) seas; 17 mi/27 km long, 5 mi/8 km wide; terminates (S) at CAPE NOMO. NAGASAKI is at NW corner of base, where peninsula joins SONOGI.

Nomozaki (no-MO-zah-kee), town, West Sonogi county, NAGASAKI prefecture, NW KYUSHU, SW JAPAN, 12 mi/20 km S of NAGASAKI; 32°34'N 129°45'E. Fish and fish processing; loquats, pineapples. Big eel habitat.

Nomura (no-MOO-rah), town, E Uwa county, EHIME prefecture, NW SHIKOKU, W JAPAN, 34 mi/55 km S of MATSUYAMA; 33°22'N 132°38'E.

Nomwin, atoll, HALL ISLANDS, State of CHUUK, E CAROLINE ISLANDS, Federated States of MICRONESIA, W PACIFIC, 45 mi/72 km N of Chuuk; 15 mi/24 km long, 9 mi/14.5 km wide; nine wooded islets.

Nona, CROATIA: see NIN.

Nonacho Lake (□ 305 sq mi/790 sq km), NORTHWEST TERRITORIES, Canada, E of GREAT SLAVE LAKE; 61°42'N 109°40'W; 30 mi/48 km long, 1 mi/2 km–20 mi/32 km wide. Drained W into Great Slave Lake by TALTSON RIVER.

No-Name City, GERMANY: see POING.

Nonancourt (no-nahn-KOOR), agricultural commune (□ 2 sq mi/5.2 sq km), EURE department, HAUTE-NORMANDIE region, N central FRANCE, on the AVRE, and 8 mi/12.9 km WNW of DREUX; 48°46'N 01°13'E. Cattle and horse raising, sawmilling. Has 16th-century Gothic church.

Nonantola (no-NAHN-to-lah), town, MODENA province, EMILIA-ROMAGNA, N central ITALY, near PANARO River, 6 mi/10 km ENE of MODENA; 44°41'N 11°02'E. Small and medium firms provide economic resources of the region. Tourism. Has Romanesque abbey of San Silvestro (founded c.753).

Nonantum, MASSACHUSETTS: see NEWTON.

Nondaburi, THAILAND: see NONTHABURI.

Nondalton, village (2000 population 221), S ALASKA, at SW end of Lake CLARK, 190 mi/306 km WSW of ANCHORAGE; 59°59'N 154°50'W. Fishing resort.

None (NO-ne), village, TORINO province, PIEDMONT, NW ITALY, 12 mi/19 km SSW of TURIN; 44°56'N 07°32'E. Fabricated metals, machinery; food processing.

Nonesuch River, SW MAINE, tidal stream; follows 12-mi/19-km semi-circular course, NE above SACO, then SE and S to the ATLANTIC at SCARBOROUGH.

Nong'an (NUNG-AN), town, ⊙ Nong'an county, NW JILIN province, NORTHEAST province, CHINA, 38 mi/61 km N of CHANGCHUN, on railroad; 44°26'N 125°11'E. Agriculture (livestock; soybeans, sugar beets, oilseeds, tobacco, grain); manufacturing (sugar refining, food and beverages, forage processing, engineering, chemicals, pharmaceuticals).

Nong Bua Lamphu (NAWNG bu-ah LUHM-POO), province, NE THAILAND; ⊙ NONG BUA LAMPHU; 17°10'N 102°15'E. Formed from the W part of CHANGWAT UDON THANI in 1993. Rice, corn, tapioca; animal husbandry.

Nong Bua Lamphu (NAWNG BU-ah LUHM-POO), town, NE THAILAND, 377 mi/607 km NNE of BANGKOK; ⊙ NONG BUA LAMPHU province; 17°11'N 102°25'E.

Nong Et (NAWNG ET), town, Xieng Khwang province, central LAOS, 115 mi/185 km ESE of Luang Prabang, on highway to VINH; 19°29'N 103°59'E. Market center; shifting cultivation, agroforestry. Thai, H'mong, and other minorities. Also called Nong Het.

Nong Het, LAOS: see NONG ET.

Nong Khai (NAWNG KEI), province (□ 2,907 sq mi/7,558.2 sq km), NE THAILAND, along MAENAM KHONG [Thai=MEKONG RIVER] and the border with LAOS; ⊙ NONG KHAI; 17°55'N 103°05'E. Rice, beans, tobacco, tapioca.

Nong Khai (NAWNG KEI), town, ⊙ NONG KHAI province, NE THAILAND, on right bank of MEKONG RIVER (LAOS line), near VIENTIANE, 30 mi/48 km N of Udon (linked by road and projected railroad), and 382 mi/615 km NNE of BANGKOK; rice, beans, tobacco. Sometimes spelled Nongkay or Nongkhay.

Nongkhlaw (NAWNG-klou), village, East Hills district, MEGHALAYA state, NE INDIA, on SHILLONG PLATEAU, 17 mi/27 km WNW of SHILLONG. Rice, sesame, cotton.

Nongson, VIETNAM: see QUE TRUNG.

Nongstoin (NAWNG-stoin), town, ⊙ WEST KHASI HILLS district, MEGHALAYA state, NE INDIA, on SHILLONG PLATEAU, 39 mi/63 km W of SHILLONG; 25°31'N 91°16'E. Rice, sesame, cotton.

Nonni, CHINA: see NENJIANG, river.

Nonnweiler (nuhn-VEI-luhr), village, N SAARLAND, W GERMANY, on PRIMS RIVER, 6 mi/9.7 km NE of WADERN, opposite HERMESKEIL; 49°38'N 06°58'E. Cattle; grain. Health and winter sports resorts. Just N is Prinstal Dam, largest reservoir of SAARLAND (37/100 sq mi/96/259 sq km; 706 million cu ft/ 20 million cu m).

Nono (NO-no), village, W CÓRDOBA province, ARGENTINA, at foot of Sierra GRANDE, 55 mi/89 km SW of CÓRDOBA; 31°46'S 65°00'W. Resort in livestock and fruit area.

Nonoai (NO-no-ah-ee), city (2007 population 12,327), N RIO GRANDE DO SUL state, BRAZIL, 48 mi/77 km NE of PALMEIRA DAS MISSÕES; 27°21'S 52°47'W. Wheat, corn, soybeans, manioc; livestock.

Nonoava (no-no-AH-vah), town, CHIHUAHUA, N MEXICO, on tributary of CONCHOS RIVER, and 90 mi/145 km SW of CHIHUAHUA; 27°30'N 106°41'W. Corn, cotton, beans, livestock, lumbering.

Nonoc Island (NO-nok) (□ 19 sq mi/49.4 sq km), SURIGAO DEL NORTE province, PHILIPPINES, just off NE tip of MINDANAO, at S end of DINAGAT ISLAND; 09°51'N 125°37'E. Major nickel mine and processing facility (world's largest nickel deposits here). Coconuts; fishing.

Nonogasta (no-no-GAHS-tah), town, central LA RIOJA province, ARGENTINA, in FAMATINA valley, on railroad, and 40 mi/64 km W of LA RIOJA; 29°18'S 67°30'W. Fruit-growing and wine-producing center.

Nonoichi (no-NO-ee-chee), town, Ishikawa county, ISHIKAWA prefecture, central HONSHU, central JAPAN, 3.1 mi/5 km S of KANAZAWA; 36°31'N 136°37'E.

Nonouti (no-no-OO-tee), atoll (□ 10 sq mi/26 sq km; 2005 population 3,179), S GILBERT ISLANDS, KIRIBATI, W central PACIFIC OCEAN; 00°41'S 174°23'E. Reef is 24 mi/39 km long. Produces copra. Formerly Sydenham Island.

Nonquitt, MASSACHUSETTS: see DARTMOUTH.

Nonsan (NON-SAHN), town (□ 21 sq mi/54.6 sq km; 2005 population 125,332), SOUTH CHUNGCHONG province, SOUTH KOREA, 21 sq mi/54 sq km SW of TAEJON. Rice-growing center. Has 11th century Buddhist temple with 86 ft/26 m tall granite Buddha.

Nonthaburi (NON-TUH-BU-REE), province (□ 250 sq mi/650 sq km), central THAILAND; ⊙ NONTHABURI. Rice, fruit, vegetables; 13°45'N 100°30'E. Famous for its durian.

Cross-references are shown in SMALL CAPITALS. The pronunciation guide is shown on page xix. The sources of population figures are shown on page xvii.

Nonthaburi (NON-TUH-BU-REE), town, ⊙ NON-THABURI province, S THAILAND, on CHAO PHRAYA RIVER, and 12 mi/19 km N of BANGKOK; 13°50′N 100°29′E. Rice- and fruit-growing area. Sometimes spelled NONDABURI.

Nontron (non-tron), commune (□ 9 sq mi/23.4 sq km), DORDOGNE department, AQUITAINE region, SW FRANCE, on height overlooking BANDIAT RIVER, and 24 mi/39 km N of PÉRIGUEUX; 45°32′N 00°40′E. Manufacturing of footwear and pocket-knives; food processing (goose-liver paté and truffles); lumber trade.

Noodle (NOO-duhl), unincorporated village, JONES county, W central TEXAS, 20 mi/32 km WNW of ABILENE; 32°36′N 100°03′W. Agricultural area (cotton, wheat; cattle).

Nooitgedacht, locality, GAUTENG province, SOUTH AFRICA, at foot of MAGALIESBERG MOUNTAIN RANGE in Magaliesberg Nature Area, 20 mi/32 km NW of KRUGERSDORP; 25°51′S 27°33′E. Elevation 5,412 ft/1,650 m. Scene of Boer War battle (1900) in which Boers under Delarey defeated British force.

Noojee, town, VICTORIA, SE AUSTRALIA, 66 mi/107 km E of MELBOURNE; 37°55′S 146°00′E.

Nooksack (NOOK-sak), village (2006 population 909), WHATCOM county, NW WASHINGTON, 15 mi/24 km NE of BELLINGHAM, and on Sumas River (NOOKSACK RIVER is 2 mi/3.2 km W at EVERSON); 48°56′N 122°19′W. In agricultural area; grain, vegetables, berries; dairying; cattle, poultry.

Nooksack River, c.25 mi/40 km long, NW WASHINGTON; rises NE of Mount Baker as NORTH FORK (c.50 mi/80 km long); flows W and SW through Nooksack Falls (power plant), first joining MIDDLE FORK, then SOUTH FORK; main stream flows NW, and SW again to Bellingham Bay at Lummi Indian Reservation, E WHATCOM county; receives South Fork (c.40 mi/64 km long) and Middle Fork (c.20 mi/32 km long). Has power plant at falls near source.

Noonan, village (2006 population 146), DIVIDE county, NW NORTH DAKOTA, port of entry 7 mi/11.3 km N of village of CROSBY near CANADIAN border; 48°53′N 103°00′W. Lignite mining; dairy products, wheat, corn. Founded in 1907 and incorporated in 1929. Named for the local Noonan family.

Noonen Reservoir, ARAPAHOE county, W COLORADO, on Deer Trail Creek, 53 mi/85 km E of DENVER; 1 mi/1.6 km long; 39°43′N 103°58′W. Formed by Noonen Dam (42 ft/13 m high), built (1970) by private interests for flood control.

Noonkanbah, aboriginal reserve, NE WESTERN AUSTRALIA state, W AUSTRALIA, 40 mi/64 km W of FITZROY CROSSING, 90 mi/145 km long, average 30 mi/48 km wide; 18°33′S 124°51′E. Bisected by FITZROY RIVER. Beef cattle.

Noonmark, peak (elevation 3,552 ft/1,083 m) of the High Peaks section of the ADIRONDACKS, in ESSEX county, NE NEW YORK, c.8 mi/12.9 km E of Mount Marcy, and 16 mi/26 km SE of LAKE PLACID village; 44°08′N 73°47′W.

Noord-Beveland, NETHERLANDS: see NORTH BEVELAND.

Noord-Brabant, NETHERLANDS: see NORTH BRABANT.

Noordervaart, canal, LIMBURG province, SE NETHERLANDS; extends 10 mi/16 km NE from convergence of WESSEM-NEDERWEERT CANAL and ZUID-WILLEMSVAART canal 3 mi/4.8 km NE of WEERT.

Noord-Holland, NETHERLANDS: see NORTH HOLLAND.

Noordholland Kanaal, NETHERLANDS: see NORTH HOLLAND CANAL.

Noordoewer (nort-o-ver), village, S NAMIBIA, on bridge over ORANGE RIVER, 150 mi/241 km S of KEETMANSHOOP; 28°45′S 17°37′E. Customs post into SOUTH AFRICA on route to CAPE TOWN.

Noord-Oost Polder, NETHERLANDS: see NORTH-EAST POLDER.

Noord River (NAWRT), 6 mi/9.7 km long, SW NETHERLANDS; formed by dividing of LOWER MERWEDE RIVER into Noord River, and OLD MAAS RIVER 1 mi/1.6 km N of DORDRECHT; flows NNW past ALBLASSERDAM and KINDERDIJK, joining LEK RIVER 0.5 mi/0.8 km W of KRIMPEN AAN DEN LEK to form NEW MAAS RIVER. Entire length navigable.

Noordwijk (NAWRT-veik), town, SOUTH HOLLAND province, W NETHERLANDS, on NORTH SEA, and 7 mi/11.3 km NNW of LEIDEN; 52°15′N 04°26′E. Seaside resort; cattle; flowers, nursery stock, vegetables, fruit; fishing. Site of European Space and Technology Centre. Lighthouse. Formerly spelled Noordwyk aan Zee.

Noordwijk-Binnen (NAWRT-veik—BI-nuhn), town, SOUTH HOLLAND province, W NETHERLANDS, 6 mi/9.7 km NNW of LEIDEN; 52°14′N 04°26′E. NORTH SEA 1 mi/1.6 km to W. Resort area; agriculture (dairying; cattle, poultry; flowers, vegetables, fruit).

Noordwijkerhout (NAWRT-veik-uhr-hout), town, SOUTH HOLLAND province, W NETHERLANDS, 8 mi/12.9 km N of LEIDEN; 52°16′N 04°30′E. NORTH SEA 2 mi/3.2 km to W. Agriculture (dairying; cattle; flowers, nursery stock, vegetables); manufacturing (fences, PVC fittings, airlocks).

Noord-Willems Kanaal, NETHERLANDS: see NORTH WILLEMS CANAL.

Noordwolde (nawrt-VAWL-duh), village, FRIESLAND province, N NETHERLANDS, 10 mi/16 km ESE of HEERENVEEN; 52°54′N 06°09′E. National Park Het Drents-Friese Woud to E; recreational center to S. Dairying; cattle; vegetables, grain.

Noordzee Kanaal, NETHERLANDS: see NORTH SEA CANAL.

Noormarkku (NAWR-MAHR-koo), Swedish *Noormark*, village, TURUN JA PORIN province, SW FINLAND, 8 mi/12.9 km NNE of PORI; 61°35′N 21°52′E. Elevation 99 ft/30 m. Lumbering; machine shops.

Noorooma, AUSTRALIA; see NAROOMA.

Noorvik, village (2000 population 634), NW ALASKA, on KOBUK RIVER, and 45 mi/72 km E of KOTZEBUE; 66°49′N 161°02′W. Within Noorvik Indian Reservation.

Noosa Heads (NOO-suh HEDZ) or **Noosa**, resort village, QUEENSLAND, NE AUSTRALIA, 111 mi/178 km N of BRISBANE, and at N end of Sunshine Coast; 26°23′S 153°09′E. Noosa National Park (□ 9 sq mi/23 sq km).

Nooseneck, Rhode Island: see WEST GREENWICH.

Nootka Island (NOOT-kuh) (□ 206 sq mi/535.6 sq km), SW BRITISH COLUMBIA, W Canada, in the PACIFIC, off W central coast of VANCOUVER ISLAND, 45 mi/72 km NW of TOFINO; 21 mi/34 km long, 16 mi/26 km wide; center near 49°45′N 126°45′W. Fishing, lumbering. Nootka village on SE coast.

Nootka Sound (NOOT-kuh), inlet of the PACIFIC OCEAN and natural harbor on the W coast of VANCOUVER ISLAND, SW BRITISH COLUMBIA, W Canada, lying between the mainland and NOOTKA ISLAND (206 sq mi/534 sq km); 49°36′N 126°34′W. The mouth of the sound was sighted (1774) by Juan Pérez, the Spanish explorer. The sound itself was visited by Captain James Cook (1778), who was the first European to land in that region. John Meares, the British explorer, established a trading post on Nootka Sound in 1788. Its seizure by Spaniards in 1789 became the subject of a controversy between Spain and England over claims in the region. The Nootka Convention (1790) resolved the dispute and opened the N Pacific coast to British settlement.

Nopala (no-PAH-lah), town, ⊙ Nopala de Villagran municipio, HIDALGO, central MEXICO, near railroad, and 25 mi/40 km SE of SAN JUAN DEL RÍO. Corn, beans, maguey, livestock.

Nopaltepec (no-PAHL-te-pek), town, MEXICO state, central MEXICO, on railroad, and 35 mi/56 km NE of MEXICO CITY, in the ZONA METROPOLITANA DE LA CIUDAD DE MÉXICO; 19°41′N 98°25′W. Elevation 8,202 ft/2,500 m. Cereals, maguey.

Nopalucan de la Granja, town, ⊙ Nopalucan municipio, PUEBLA, central MEXICO, 28 mi/45 km NE of PUEBLA; 19°13′N 97°49′W. Cereals, beans, maguey.

Ñoquera (nyo-KE-rah), canton, AVILÉS province, TARIJA department, S central BOLIVIA, 6 mi/10 km NE of YUNCHARA; 21°40′S 65°06′W. Elevation 5,600 ft/1,707 m. Lead, zinc, silver, and copper deposits in area (N); clay, limestone deposits. Agriculture (potatoes, yucca, bananas, corn, sweet potatoes); cattle.

Noqui (NO-kee), town, ZAIRE province, NW ANGOLA, river port on the Congo (ZAÏRE border), 4 mi/6.4 km S of MATADI.

Nora (NOO-rah), town, ÖREBRO county, S central SWEDEN, on Lake Norasjön (3 mi/4.8 km long), 18 mi/29 km NNW of ÖREBRO; 59°31′N 15°01′E. Manufacturing (metalworking; explosives and drills); trade center for BERGSLAGEN mining region (N) until eighteenth century. Chartered 1643.

Nora, village (2000 population 118), JO DAVIESS county, NW ILLINOIS, 20 mi/32 km NW of FREEPORT; 42°27′N 89°56′W. In agricultural area. Apple River Canyon State Park is nearby.

Nora, village (2006 population 19), NUCKOLLS county, S NEBRASKA, 10 mi/16 km NNE of SUPERIOR; 40°09′N 97°58′W.

Nora (NOR-uh), unincorporated village, DICKENSON county, SW VIRGINIA, in CUMBERLAND Mountains, 18 mi/29 km NE of NORTON; 37°04′N 82°20′W. Agriculture (cattle; tobacco); timber.

Nor Achin, urban settlement, ARMENIA, 40°18′N 44°35′E. Stone production for construction and industry. Arzni Hydroelectric Plant.

Norak (noor-YEK), city (2000 population 19,000), W TAJIKISTAN, on VAKHSH RIVER, at bottom of reservoir with same name, and 32 mi/51 km ESE of DUSHANBE; 38°23′N 69°21′W. Area is under no direct viloyat administrative division; under direct republic supervision. Wheat; salt deposits, gold placers. Called Norak until 1937, then Nurek, and now Norak again.

Noranda, Canada: see ROUYN-NORANDA.

Norangdal (NAW-rahng-dahl), valley in MØRE OG ROMSDAL county, W NORWAY; extends from Øye on Hjørundfjorden, SE of Ålesund, c.10 mi/16 km SE to Hellesylt. Popular scenic region. A landslide in 1908 dammed the valley, forming a lake which submerged the surrounding farms.

Norashen, AZERBAIJAN: see IL'ICHEYSK.

Nora Springs, town (2000 population 1,532), FLOYD county, N IOWA, on SHELL ROCK RIVER, and 17 mi/27 km WNW of CHARLES CITY; 43°08′N 93°00′W. Concrete blocks. Limestone quarry, sand and gravel pits nearby. Incorporated 1875.

Norberg (NOOR-BER-ye), town, VÄSTMANLAND county, central SWEDEN, in BERGSLAGEN region, 6 mi/9.7 km NE of FAGERSTA; 60°05′N 15°56′E. Among Sweden's oldest iron- and lead-mining centers; mine reconstructed in Nya Lapphyttan. Has thirteenth-century church; mining museum.

Norberto de la Riestra (nor-BER-to dai lah ree-AIS-trah), town, N central BUENOS AIRES province, ARGENTINA, 25 mi/40 km ENE of VEINTICINCO DE MAYO; 35°15′S 59°46′W. Cattle- and hog-raising center; seed cultivating; tanning.

Norbertville (nor-ber-VEEL), village (□ 1 sq mi/2.6 sq km), Centre-du-Québec region, S QUEBEC, E Canada, 25 mi/40 km W of THETFORD MINES; 46°06′N 71°49′W. Dairying; cattle; wheat, potatoes.

Norbestos (nor-BE-stuhs), unincorporated asbestos mining village, S QUEBEC, E Canada, 5 mi/8 km NE of ASBESTOS, and included in Saint-Rémi-de-Tingwick; 45°49′N 71°50′W.

Norborne (NAHR-buhrn), city (2000 population 805), CARROLL county, NW central MISSOURI, near MISSOURI River, 10 mi/16 km WSW of CARROLLTON; 39°17′N 93°40′W. Wheat, corn, soybeans; hogs, cattle. Laid out 1868.

Area is shown by the symbol □, and capital city or county seat by ⊙.

Norcatur (nor-KAI-tuhr), village (2000 population 169), DECATUR county, NW KANSAS, 18 mi/29 km E of OBERLIN; 39°49′N 100°11′W. Grain; cattle.

Nor Chichas (nor CHICH-chahs), province (□ 3,467 sq mi/9,014.2 sq km), POTOSÍ department, W central BOLIVIA; ⊙ Santiago de COTAGAITA; 20°50′S 65°37′W. Elevation 8,596 ft/2,620 m. Terrain is primarily mountainous. Antimony mining at Mina Churquini. Agriculture (potatoes, cassava, bananas); cattle. Created in August 1893 by Governor José María de Achá.

Norcia (NOR-chah), ancient *Nursia*, village, PERUGIA province, UMBRIA, central ITALY, near Monti SIBILLINI, 19 mi/31 km ENE of SPOLETO; 42°48′N 13°05′E. Railroad terminus; manufacturing (woolen textiles, metal furniture). Bishopric. Has 14th-century walls and a castle (1554–1563). St. Benedict was born here.

Nor Cinti, province, CHUQUISACA department, SE BOLIVIA; ⊙ CAMARGO; 20°20′S 65°00′W.

Norco, city (2000 population 24,157), RIVERSIDE county, SE CALIFORNIA, suburb 35 mi/56 km ESE of downtown LOS ANGELES, 10 mi/16 km W of RIVERSIDE, on SANTA ANA RIVER; 33°56′N 117°33′W. Diversified agriculture being displaced by rapid urban development from Los Angeles. Manufacturing (lubricating oils). California Institute for Women (prison) to W, California Institute for Men (prison) and Chino Aiport to NW, Corona Municipal Airport to SW. CHINO HILLS State Park to W; Prado Reservoir to SW. Incorporated 1964.

Norco (NOOR-ko), unincorporated town (2000 population 3,579), SAINT CHARLES parish, SE LOUISIANA, near E bank (levee) of the MISSISSIPPI RIVER, c.20 mi/32 km W of NEW ORLEANS; 30°00′N 90°25′W. Railroad junction; oil refining; manufacturing (gas and liquid oxygen, gasoline and other fuels, fabricated metals). BONNET CARRE SPILLWAY to NW. LAKE PONTCHARTRAIN shore 4 mi/6 km NNE.

Norcross, town (2000 population 8,410), GWINNETT county, N central GEORGIA, 15 mi/24 km NE of ATLANTA; 33°56′N 84°13′W. Emerging center of high-tech innovation and production in this N Atlanta suburb. Many high-tech business and industrial parks house domestic and multinational. firms. Manufacturing (chemicals, computer hardware, concrete blocks, ventilation equipment, particleboard, consumer goods, telecommunications equipment, apparel; sheet metal fabrication, printing and publishing).

Norcross, village (2000 population 59), GRANT county, W MINNESOTA, on MUSTINKA RIVER, and 14 mi/23 km SW of ELBOW LAKE town; 45°52′N 96°12′W. Dairying; grain.

Nord, region (□ 6,255 sq mi/16,263 sq km; 2005 population 1,152,402), N central BURKINA FASO; ⊙ OUAHIGOUYA. Borders SAHEL (NE), CENTRE-NORD (E), PLATEAU CENTRAL and CENTRE-OUEST (S), and BOUCLE DU MOUHOUN (W) regions and MALI (N). Composed of LOROUM, PASSORÉ, YATENGA, and ZONDOMA provinces.

Nord (NOR), department (□ 2,217 sq mi/5,764.2 sq km), in French FLANDERS and HAINAUT (Belgium), historic provinces, N FRANCE; ⊙ LILLE; 50°30′N 03°30′E. The northernmost French department, it is bounded by Belgium (N and E), has narrow frontage on NORTH SEA, and abuts the ARDENNES foothills (SE). Generally level (low and swampy near coast) it is drained by SAMBRE, Escaut (SCHELDT), SCARPE, LYS, DEÛLE, and YSER rivers, interconnected by a dense network of canals (collectively known as Flanders canals). A leading agricultural department, with soils skillfully irrigated and cultivated, its chief crops are wheat, oats, sugar beets, flax, hops, chicory, tobacco, potatoes, vegetables. Cattle pastures. S central part of department is crossed by old-established Franco-Belgian coal basin, whose mines are virtually closed down due to high cost of extraction, depleted seams, dangerous

conditions for underground miners, and foreign competition. Nord is one of France's chief urban-industrial regions, with textile and metallurgical plants, and a good start towards a diversified modern industrial base. Leading textile centers are Lille (fabrics), ROUBAIX, TOURCOING, and FOURMIES (all wool-manufacturing centers), ARMENTIÈRES (linen), CAMBRAI (fine textiles). Metalworks are heavily concentrated in Lille metropolitan area (locomotive works), DENAIN (steel), and DOUAI. Other products of the region are glass, paper, chemicals, and processed foods (sugar, flour products, chocolates). The principal port is DUNKERQUE, completely rebuilt and expanded since its virtual destruction in World War II. Department is first in population among the 96 departments that constitute metropolitan France. It has attracted many foreigners (Poles, Belgians), especially in early 20th century, when coal mines were booming. Long a European battleground, Nord was ravaged in World War I. In 1940, more than 300,000 Allied (mostly British) troops, cut off by German drive to the ENGLISH CHANNEL, were evacuated to ENGLAND from Dunkerque. Nord, and the Lille urban complex in particular, are well served by a dense railnet connecting all the countries of the EU, now that the UNITED KINGDOM is also linked via the CHANNEL TUNNEL with the continent's road and railroad net. Lille has become the hub of the TGV (high-speed train) network between PARIS, LONDON, BRUSSELS, and points E and S. Nord department forms part (together with PAS-DE-CALAIS department) of the administrative region known as NORD-PAS-DE-CALAIS, with its administrative headquarters at Lille.

Nord (NOR), department (2003 population 773,546), N HAITI, bounded on the W by NORD-OUEST and ARTIBONITE departments, S by ARTIBONITE and CENTRE departments, E by NORD-EST department; ⊙ CAP-HAÏTIEN (Haiti's second-largest city); 19°36′N 72°18′W. Agriculture (sugarcane, coffee, citrus fruits); copper deposits.

Nord, CAMEROON: see NORTH.

Nordaustlandet, SPITSBERGEN: see NORTHEAST LAND.

Nordborg (NOR-bor), city (2000 population 7,652), SØNDERJYLLAND county, DENMARK, on N ALS island, 10 mi/16 km N of SØNDERBORG; 55°02′N 09°50′E. Founded around twelfth-century castle.

Norddeich (NORD-deikh), seaside resort, LOWER SAXONY, in EAST FRIESLAND, NW GERMANY, on NORTH SEA, 3 mi/4.8 km NW of NORDEN; 53°35′N 07°10′E. Railroad terminus; ferry connection to islands of JUIST, and NORDERNEY. Commune is called DITHMARSCHEN.

Nord-du-Québec (NOR-dyoo-kai-BEK), region (□ 277,309 sq mi/721,003.4 sq km; 2005 population 40,246), N QUEBEC, E Canada; 56°10′N 74°25′W. Mining, forestry. Composed of 51 municipalities, centered on CHIBOUGAMAU. Cree, Inuit, and James Bay Nation cultural communities. Sparsely populated.

Nordegg (NUHR-deg) or **Brazeau** (BRA-zo), unincorporated town, SW ALBERTA, W Canada, in ROCKY MOUNTAINS, near North SASKATCHEWAN River, 130 mi/209 km SW of EDMONTON, in Clearwater County; 52°28′N 116°05′W. Elevation 4,471 ft/1,363 m. Hydroelectricity; cattle; timber.

Norden (NOR-den), town, LOWER SAXONY, NW GERMANY, in EAST FRIESLAND, near NORTH SEA, 15 mi/24 km N of EMDEN; 53°35′N 07°13′E. In suburb of NORDDEICH is a port with ferry connection to JUIST, Norderny, and BALTRUM islands. Linked by canal with EMS RIVER estuary; railroad junction; metalworking, distilling (brandy); tea. Tourism. Has 15th-century church, 16th-century town hall. Oldest town of EAST FRIESLAND (first mentioned 1255).

Norden, unincorporated village, NEVADA county, E central CALIFORNIA, in the SIERRA NEVADA, 35 mi/56 km SW of RENO (Nevada), on SOUTH YUBA RIVER, at Lake Van Norden. Elevation 6,880 ft/2,097 m. DON-

NER PASS to E (Interstate Highway 80) and Norden Tunnel (railroad; c.10,000 ft/3,048 m long).

Nordenham (NOR-den-hahm), town, OLDENBURG, LOWER SAXONY, NW GERMANY, port on left bank of the WESER (just above its estuary) and 30 mi/48 km below BREMEN, nearly opposite BREMERHAVEN; 53°29′N 08°29′E. Shipbuilding; lead and zinc refining; manufacturing of transoceanic cables, phosphates. Dispatches deep-sea fishing fleet. Chartered 1908.

Nordenskjöld Archipelago (nor-deens-kee-YOLD), Russian *Arkhipelag Nordenshel'da*, in KARA SEA of ARCTIC OCEAN, 10 mi/16 km–15 mi/24 km off N TAYMYR PENINSULA, in KRASNOYARSK TERRITORY, N RUSSIA; includes Vilkitskiy, Tsivolka, PAKHTUSOV, and Litke island groups; 76°45′N 96°00′E. Northernmost island, RUSSKIY, has a government observation post. Named for the 19th century Swedish explorer Nils Adolf Nordenskjöld.

Nordenskjöld Basin (NOR-duhns-kyeld), ANTARCTICA, submarine basin in westernmost ROSS SEA, along N SCOTT COAST, S of DRYGALSKI BASIN; 76°S 164°30′E.

Nordenskjöld Bay (nor-deens-kee-YOLD), on SW coast of ALEXANDRA LAND, FRANZ JOSEF LAND, ARCHANGEL oblast, extreme N European Russia, in the ARCTIC OCEAN.

Nordenskjöld Coast (NOR-duhns-kyeld), ANTARCTICA, NE coast of ANTARCTIC PENINSULA between CAPE LONGING (64°30′S) and Cape Fairweather (65°S).

Nordenskjöld Glacier, GREENLAND: see ARFERSIORFIK.

Nordenskjöld Sea, RUSSIA: see LAPTEV SEA.

Norden Tunnel, California: see NORDEN.

Norderney (NOR-der-nai), town, LOWER SAXONY, NW GERMANY, on EAST FRISIAN island NORDERNEY, 23 mi/37 km N of EMDEN; 53°42′N 07°09′E. Ferry connection with NORDDEICH. Casino; tourism; fishing museum. Town first mentioned 1398, chartered 1948; oldest German NORTH SEA health resort (since 1797). Small airport, windmill (1862), lighthouse (1873).

Norderney (NOR-der-nai), NORTH SEA island (□ 9 sq mi/23.4 sq km) of EAST FRISIAN group, GERMANY, 5 mi/8 km N of NORDDEICH (ferry connection); 8 mi/12.9 km long (E-W), c.1 mi/1.6 km wide (N-S); 53°42′N 07°15′E. The town of NORDERNEY (S tip) is Germany's oldest and most popular NORTH SEA resort (1797).

Nord-Est (NOR-EST), department (2003 population 300,493), NE HAITI. Bounded on N by ATLANTIC OCEAN, E by DOMINICAN REPUBLIC, S by CENTRE department, W by NORD department; ⊙ FORT LIBERTÉ; 19°32′N 71°42′W. Sugarcane, fruit, sisal growing; gold deposits.

Nordeste (NORD-esht), town, Ponta Delgada district, E Azores, on NE shore of São Miguel Island, 29 mi/47 km ENE of Ponta Delgada; 37°50′N 25°09′W. Tea growing; beekeeping; woodworking.

Nordestína (NOR-des-CHEE-nah), town (2007 population 12,171), NE BAHIA, BRAZIL, 16 mi/25 km NE of QUEIMADAS; 10°49′S 39°26′W.

Nordfjord (NAWR-fyawr), inlet, c.50 mi/80 km long, SOGN OG FJORDANE county, SW NORWAY. To the S, between Nordfjord and SOGNEFJORDEN, is JOSTEDALSBREEN glacier. The Nordfjord's several branches, cutting deeply into the mountains and celebrated for their scenery, are favored tourist spots.

Nordfjordeid (NAWR-fyawr-AID), village, SOGN OG FJORDANE county, W NORWAY, at head of a N branch of NORDFJORD, 38 mi/61 km NE of FLORØ; 61°54′N 06°00′E. Fishing; lumbering; cattle raising; foodstuff industries. County hospital here. Regional center for Nordfjord area.

Nordfriesland, GERMANY: see NORTH FRIESLAND.

Nordhausen (nord-HOU-suhn), city, THURINGIA, central GERMANY, at the S foot of the HARZ MOUNTAINS; 51°30′N 10°48′E. Industrial center and railroad junction. Manufacturing (clothing, beer and liquor, chewing tobacco, wood products, heavy machinery); potash mines nearby. Known since 927. Nordhausen was chartered in the 13th century and was a free imperial city from 1220 to 1803. In 1815 it passed to PRUSSIA. Severely damaged in World War II (seventy-five percent). Noteworthy buildings include the cathedral (14th century) and the late-Gothic city hall from 1360.

Nordheim (NORD-heim), village, FRANCONIA, BADEN-WÜRTTEMBERG, SW GERMANY, 5 mi/8 km SW of HEILBRONN, on left bank of the NECKAR; 49°06′N 09°08′E. Grain, vegetables; food processing.

Nordheim (NORD-heim), village (2006 population 327), DE WITT county, S TEXAS, 38 mi/61 km WNW of VICTORIA; 28°55′N 97°36′W. Cotton, corn; cattle, poultry, dairying; oil and gas.

Nordholz (NORD-holts), village, LOWER SAXONY, NW GERMANY, 17 mi/27 km N of BREMERHAVEN, near NORTH SEA; 53°46′N 08°37′E. Military airport nearby.

Nordhorn, town, LOWER SAXONY, NW GERMANY, on the Vechte, at junction of ALMELO-NORDHORN, EMS-VECHTE, and SÜD-NORD canals, 11 mi/18 km WSW of LINGEN (EMS), near Dutch border; 52°26′N 07°04′E. Textile center (cotton spinning and weaving, manufacturing of artificial fiber, bleaching, dyeing). Has 15th-century church. Town first mentioned c.900. Chartered 1379.

Nordiya (nor-dee-YAH), moshav, ISRAEL, 1 mi/1.6 km SE of NETANYA on SHARON plain; 32°19′N 34°54′E. Elevation 144 ft/43 m. Mixed farming. Light industry. Founded in 1948 by members of the Zionist group Betar.

Nordjylland (NOR-YUH-lan), county (□ 2,386 sq mi/6,203.6 sq km), N JYLLAND, N DENMARK, includes island of LAESO; ⊙ ALBORG; 88 mi/142 km long, 57 mi/92 km wide; 57°00′N 09°50′E. Bounded on NW by SKAGERRAK, E by KATTEGAT strait, S by ARHUS county, SW by VIBORG county. Bisected E-W by LIMFJORDEN and Langerak waterways. Sand dunes along Skagerrak coast, ending at Skagen Odde (The Skaw), northernmost tip of Denmark proper. Rolling farmland with remnant forest; heath covered hills in S. Ferry crossings to NORWAY and SWEDEN. Dairying, meatpacking; cattle, hogs; rye; fishing. Manufacturing, shipbuilding, port facilities at Alborg and FREDERIKSHAVN. Tourism.

Nordkapp, NORWAY: see NORTH CAPE.

Nordkette (NORD-ket-te), range of TYROLEAN-BAVARIAN LIMESTONE and the KARWENDEL mountain range in TYROL, W AUSTRIA, forming N wall (c. 20 mi/32 km long) of Inn River Valley at INNSBRUCK. Highest peak, Solstein (8,038 ft/2,450 m). Attractive for mountaineering; views on Innsbruck and the Central alps. The MARTINSWAND, a very steep rock between Innsbruck and ZIRL, divides the Upper from the Lower Inn Valley.

Nordkirchen (nord-KIR-khuhn), village, WESTPHALIA-LIPPE, North Rhine-Westphalia, W GERMANY, 15 mi/24 km SSW of MÜNSTER; 51°44′N 07°31′E. Agricultural products. Baroque castle (built 1703–1734 by Plettenberg dynasty).

Nord-Kivu, province (2004 population 4,626,000), NE DEMOCRATIC REPUBLIC OF THE CONGO; ⊙ GOMA. Bordered NWN by ORIENTALE province (part of W portion of border formed by LINDI RIVER and small part of central portion formed by Lenda River), E by UGANDA, SE by RWANDA, S by SUD-KIVU province, and W by MANIEMA province. Lenda, Ibina, and SEMLIKI rivers in N; LINDI and Oso rivers in central part of province; LOWA and Luhoho rivers in S. E portion of province falls in the GREAT RIFT VALLEY. Part of the Mitumba Mountains in E of province. MOUNT MARGHERITA and part of RUWENZORI mountain range in NE. Part of LAKE EDWARD in E (on border of UGANDA). VIRUNGA NATIONA PARK. NORD-KIVU, SUD-KIVU, and MANIEMA form the region of KIVU.

Nord-Kivu, CONGO: see KIVU.

Nordkyn, Cape (NAWR-chun) or **Kinnarodden**, northernmost point of the European mainland, FINNMARK county N NORWAY, E of NORTH CAPE, at latitutde 71°08′N.

Nordland (NAWR-lahn), county (□ 14,798 sq mi/38,327 sq km), 2007 estimated population 235,502), N central NORWAY, bordering on the NORWEGIAN SEA in the W and on SWEDEN in the E; ⊙ BODØ. Chief towns are Mo and NARVIK. The county has many fjords and approximately 600 islands, including the LOFOTEN AND VESTERÅLEN groups. Fishing, mining (iron, lead, and zinc), livestock raising, and farming are the main occupations. In addition, copper is mined and processed at SULITJELMA.

Nördlingen (NYOORD-ling-uhn), town, SWABIA, BAVARIA, S central GERMANY; 48°51′N 10°30′E. Manufacturing center and a railroad junction, with industries in metalworking, synthetics, and precision instruments. Home to an annual horse race, the oldest in Germany. Historically a Swabian town, Nördlingen was founded in the 9th century and was a free imperial city c.1217–1803. In the Thirty Years War an imperial army under Gallas defeated (1634) troops here led by Duke Bernhard of Saxe-Weimar; the victory by the imperial side was a major reason for France's entry into the war in 1635. In 1645 the town was the scene of a German defeat at the hands of French troops under Condé. Passed to BAVARIA in 1803. The picturesque town retains its walls (14th–17th century), a town hall (14th century), the late-Gothic Church of St. George (1427–1505), and numerous 16th- and 17th-century houses.

Nordmaling (NORD-MAHL-eeng), town, VÄSTER-BOTTEN county, N SWEDEN, on small inlet of GULF OF BOTHNIA, 30 mi/48 km SW of UMEÅ; 63°34′N 19°31′E. Medieval church.

Nordman, unincorporated village, BONNER county, N IDAHO, 30 mi/48 km NW of SANDPOINT, near WASHINGTON state line, in Kaniksu National Forest; 48°38′N 116°56′W. Timber; tourism. N terminus of State Highway 57. PRIEST LAKE to E; Upper Priest Lake Scenic Area to N.

Nordmark (NORD-mahrk), village, VÄRMLAND county, W SWEDEN, in BERGSLAGEN region, 8 mi/12.9 km N of FILIPSTAD; 59°50′N 14°07′E.

Nordmarsch-Langeness (NORD-mahrsh-LAHNG-ge-nes) or **Langeness**, NORTH SEA island (□ 5 sq mi/13 sq km) of NORTH FRISIAN group, NW GERMANY, in HALLIG ISLANDS, 6 mi/9.7 km off SCHLESWIG-HOL-STEIN coast (connected by dike); 5.5 mi/8.9 km long (E-W), 1 mi/1.6 km wide (N-S); 54°39′N 08°36′E. Grazing.

Nord, Massif du (NOR, mah-SEEF dyoo), range in N HAITI, between the PLAINE-DU-NORD (N) and Central Plain (S), continuing the CORDILLERA CENTRAL of the DOMINICAN REPUBLIC; extends c.80 mi/129 km SE from PORT-DE-PAIX; 19°72′15′W. Rises to 2,525 ft/770 m. The PLAINE-DU-NORD continues the CIBAO region of the DOMINICAN REPUBLIC; agriculture (sugarcane, coffee, cacao, tobacco, bananas, citrus fruit, pineapples, sisal, fine wood); copper deposits.

Nordostlandet, NORWAY: see NORTHEAST LAND.

Nordost Rundingen, GREENLAND: see NORTHEAST FORELAND.

Nord-Ostsee-Kanal, GERMANY: see KIEL CANAL.

Nord-Ouest (NOR-WEST), department (□ 1,060 sq mi/2,756 sq km; 2003 population 445,080), NW HAITI; 19°45′N 73°05′W. To the N is the ATLANTIC OCEAN, to the S the GULF OF GONAÏVES and ARTIBONITE department, to the E NORD department; ⊙ PORT-DE-PAIX. Coffee, rice, sisal, cocoa growing; beekeeping; has copper deposits.

Nord-Ouest, Cameroon: see NORTH-WEST.

Nordøyane (NAWR-uh-ah-nuh) [Norwegian=north islands], island group in NORTH SEA, MØRE OG ROMSDAL county, W NORWAY, paralleling the coast,

and extending from mouth of MOLDEFJORD 30 mi/48 km SW to ÅLESUND. Include islands of HARØYA, and HARAMSØYA.

Nord-Pas-de-Calais (nor-pah–duh–kah-lai), administrative region (□ 4,793 sq mi/12,461.8 sq km) of N FRANCE that includes the departments of NORD and PAS-DE-CALAIS; 50°30′N 02°30′E. Within this region lies the urban complex of LILLE, ROUBAIX, and TOURCOING, which is benefiting from the extensive road net leading N, E, and S from the French terminus of the CHANNEL TUNNEL.

Nord-Pas-de-Calais-Audomarois Natural Regional Park (nor-pah–duh–kah-lai-o-do-mah-rwah), PAS-DE-CALAIS department, NORD-PAS-DE-CALAIS region, N FRANCE. It includes many municipal forests in the Saint-Omer region.

Nord-Pas-de-Calais Plaine de la Scarpe et de l'Escaut Natural Regional Park, FRANCE: see VALENCIENNES.

Nordre Isortoq (NOR-druh i-SOKH-tok), Greek *Sismiut Isortuat Fjord*, 85 mi/137 km long, 1 mi/1.6 km–3 mi/4.8 km wide, of DAVIS STRAIT, SW GREENLAND, 17 mi/27 km N of HOLSTEINSBORG (Sisimiut); 67°15′N 52°30′W. Receives several glacier streams from inland ice.

Nordre Strom Fjord (NOR-druh STRUM FYOR), Danish *Nordre Strømfjord*, Greenlandic *Nassuttooq*, inlet of DAVIS STRAIT, W GREENLAND, 60 mi/97 km S of EGEDESMINDE (Aasiaat); 110 mi/177 km long, 1 mi/1.6 km–16 mi/26 km wide; 67°40′N 52°W. Extends inland to edge of the inland ice.

Nordrhein-Westfalen, GERMANY: see NORTH RHINE-WESTPHALIA.

Nordseebad Sankt Peter, GERMANY: see SANKT PETER-ORDING.

Nordstemmen (NORD-shtem-men), town, LOWER SAXONY, NW GERMANY, 7 mi/11.3 km W of HILDE-SHEIM; 52°10′N 09°47′E. Metal- and woodworking, sugar refining.

Nordstrand (NAWR-strahn), residential suburb of OSLO, SE NORWAY, on Bonnefjorden (arm of OSLOF-JORDEN), 4 mi/6.4 km SSE of city center. Until 1948, in AKERSHUS county.

Nordstrand (NORD-shtrahnd), NORTH SEA island (□ 19 sq mi/49.4 sq km) of NORTH FRISIAN group, NW GERMANY, in HALLIG ISLANDS, 2 mi/3.2 km off SCHLESWIG-HOLSTEIN coast (connected by dike); 5.7 mi/9.2 km wide (E-W), 4.5 mi/7.2 km long (N-S); 54°30′N 08°53′E. Grazing. Nordstrand (1994 population 2,400) is the main village. It is remainder of larger island, which was partly flooded in 1634.

Nord-Trøndelag (NAWR–TRUHN-nu-lahg), county (□ 8,647 sq mi/22,396 sq km; 2007 estimated population 128,928), central NORWAY, N of TROND-HEIMSFJORDEN and bordering on the ATLANTIC OCEAN in the W; ⊙ STEINKJER. The chief towns are NAMSOS and LEVANGER. The economy is based on farming, fishing, mining, and the manufacturing of forest products

Nordur-Isafjardarsýsla (NAW-[th]uhr–EE-sahf-yahr-[th]ahrs-EES-lah), county (1993 population 118), Westfjords region, NW ICELAND⊙ and chief town, ISAFJÖRÐUR. Fishing villages at BOLUNGARVIK and Hnifsdajur. DENMARK STRAIT coastline is dominated by the large fjord, ÍSAFJARÐARDJÚP, and its several branch fjords, the Jokulfirdir being the largest. DRANGAJOKULL is on NE boundary. Mountainous and rugged coastline. N part is sparsely inhabited; Hornstrandi Landscape Reserve protects large peninsula 7 mi/11.3 km from ARCTIC CIRCLE. Fishing (cod, haddock), shrimp, lobster.

Nordur-Múlasýsla (NAW-[th]ur–MOO-lahs-EES-lah), county (1993 population 905), NE ICELAND ⊙ Seydisfjörður, a city in, but independent of, the county. Extends along NE coast of Iceland between LANGANES peninsula (N) and Seydisfjörður (S). Drained by JO-KULSAÁ Brú and LAGARFLJOT rivers. Fishing.

Nordur-Thingeyjarsýsla (NAW-[th]ur–THEENG-gai-yahrs-EES-lah), county (1993 population 506), NE ICELAND ⊙ Kopaskar. Other villages include Raufarhofn and THORSHOFN. Its GREENLAND SEA coastline extends from the W shore of Axarjörður E to the NW shore of Bakkafloi Bay; includes THISTILFJÖRÐUR, the large MELRAKKASLETTA Peninsula (lighthouses at Raudinupur and RIFSTANGI), which comes within 1 mi/1.6 km of the ARCTIC, mainland Iceland's northernmost point, and the smaller LANGANES Peninsula (lighthouse at Fontur). Includes large inland area with active volcanic activity. JOKULSA A Fjollum River in W has DETTIFOSS waterfall and JOKULSARGLJUFUR NATIONAL PARK. Fish; sheep.

Norduz River, TURKEY: see BOTAN RIVER.

Nordvik-Ugol'naya (NOR-tveek OO-guhl-nah-yah), locality, E TAYMYR (DOLGAN-NENETS) AUTONOMOUS OKRUG, in KRASNOYARSK TERRITORY, N SIBERIA, RUSSIA, N of the ARCTIC CIRCLE, on the W shore of Nordvik Bay of the KHATANGA GULF, part of the LAPTEV SEA, 235 mi/378 km NE of KHATANGA; 74°02'N 111°32'E. Elevation 150 ft/45 m. Lignite deposits.

Nordwalde (nord-VAHL-de), village, WESTPHALIA-LIPPE, NORTH RHINE-WESTPHALIA, NW GERMANY, 7 mi/11.3 km SE of STEINFURT; 52°05'N 07°29'E. Dairying.

Noreña (no-RAI-nyah), town, OVIEDO province (ASTURIAS), NW SPAIN, 7 mi/11.3 km ENE of OVIEDO; 43°24'N 05°42'W. Meat-processing center; agriculture (apples, corn; cattle, hogs).

Nörenberg, POLAND: see INSKO.

Nore River, Gaelic *An Fheoir,* c.70 mi/113 km long; rises in NE TIPPERARY county, S central IRELAND; flows NE, then SE through a rich agricultural region to the BARROW RIVER near NEW ROSS. The Nore valley is noted for its scenic beauty.

Noresund (NAW-ruh-soon), village, BUSKERUD county, SE NORWAY, on the KRØDEREN Lake, 20 mi/32 km W of HØNEFOSS; 60°10'N 09°37'E. Tourist center noted for its scenery. Agriculture, lumbering. Norefjell Mountains rise to 4,810 ft/1,466 m 10 mi/16 km NW.

Nore, The (NAW), sandbank in the THAMES estuary, SE ENGLAND, 4.8 km E of SHEERNESS, KENT; 51°26'N 00°46'W. At E end is Nore lightship, the first of its kind in the world (1732). The name is also applied to part of the Thames estuary (famous anchorage). A mutiny in the British fleet here, shortly after the SPITHEAD mutiny in 1797, was suppressed.

Norfolk (NAW-fuhk), county (□ 2,054 sq mi/5,340.4 sq km; 2001 population 796,728), E ENGLAND; ⊙ NORWICH; 52°40'N 01°00'E. The region is one of flat, fertile farmlands, with a long, low coast bordering on the NORTH SEA and The WASH. The principal rivers are the OUSE, the BURE, the YARE and its tributary the WENSUM, and the WAVENEY. A series of connected shallow lakes, known as The BROADS, occupies the NE portion of the county and are popular for boating. Norfolk produces cereal and root crops and supports extensive breeding of cattle and poultry. Fishing; light industries also important. Numerous vestiges of habitation dating from prehistoric times remain. After the Anglo-Saxon invasion of England, Norfolk became a part of the kingdom of East ANGLIA, the home of the "north folk" of that region (thus its name). Towns of significance include GREAT YARMOUTH, KING'S LYNN, CROMER, and THETFORD.

Norfolk, county (□ 443 sq mi/1,151.8 sq km; 2006 population 654,753), E MASSACHUSETTS; ⊙ DEDHAM, S of BOSTON, with QUINCY and HINGHAM bays on BOSTON BAY; bounded SW by RHODE ISLAND; 42°10'N 71°11'W. Drained by CHARLES and NEPONSET rivers. NE area is thickly populated, with residential suburbs of Boston (e.g., BROOKLINE, MILTON). Manufacturing (textiles, metal and wood products, building materials, machinery, paper products), printing and publishing;

former granite quarrying in its industrial towns, notably QUINCY. Agiculture in SW. Formed 1793. State Prison.

Norfolk (NOR-fuhk), former county, SE VIRGINIA, is now the independent cities of PORTSMOUTH and CHESAPEAKE.

Norfolk, city (2006 population 23,896), MADISON county, NE NEBRASKA, on ELKHORN RIVER, c.95 mi/152 km NW of OMAHA; 42°01'N 97°25'W. Elevation 1,527 ft/465 m. Railroad terminus. Trade and railroad center in a fertile farming region; livestock market. Manufacturing (animal feeds, food and beverages, electronics; printing and publishing, construction materials). Northeast Community College; Veterans' Home. Karl Stefan Memorial Airport. Incorporated 1881.

Norfolk (NOR-fuhk), independent city (□ 66 sq mi/171.6 sq km; 2006 population 229,112) SE VIRGINIA, bounded N by CHESAPEAKE BAY, NW by HAMPTON ROADS harbor, W by ELIZABETH RIVER (PORTSMOUTH opposite), S by CHESAPEAKE city, E by VIRGINIA BEACH city; 36°55'N 76°14'W. Lafayette River estuary indents city from Elizabeth River in center, South Branch Elizabeth River in S. Railroad junction and terminus; port of entry; major commercial, industrial, shipping, and distribution center. Manufacturing (lumber, steel and sheet metal fabrication, leather products, agricultural equipment, motor vehicles, metal doors, textiles, shipbuilding, transportation equipment, furniture; machining, meat processing, printing and publishing). With Portsmouth and NEWPORT NEWS, it forms the Port of Hampton Roads, one of world's best natural harbors. Waterfront (50 mi/80 km), extensive maritime trade; quantities of coal, grain, tobacco, seafood, farm products exported. Major military center; with Portsmouth the city forms an extensive naval complex. A rallying point for Tory forces at the start of the American Revolution, Norfolk was attacked (1776) by Americans and in the ensuing battle caught fire and was nearly destroyed. Civil War Confederate naval base until the Union takeover (May 1862); battle between the *Monitor* and the *Merrimack* fought in Hampton Roads. St. Paul's Church (1738; only building to survive the burning of 1776); Fort Norfolk (1794); General Douglas MacArthur Memorial, where the general is buried; Virginia Zoological Park; Norfolk Botanical Gardens; Hermitage Foundation Museum; many historic homes. Old Dominion University, Norfolk State University, Eastern Virginia Medical School; Virginia Wesleyan College on boundary with Virginia Beach. CHESAPEAKE BAY BRIDGE-TUNNEL and railroad barge link Norfolk from Virginia Beach with Delmarva Peninsula (CAPE CHARLES) and Hampton Roads Bridge-Tunnel with HAMPTON, Virginia. Norfolk International Terminal (shipping port) on Elizabeth River; Norfolk International Airport in E, on Virginia Beach boundary. Norfolk Naval Base in NW. INTRACOASTAL WATERWAY passes through Elizabeth River. Founded 1682, incorporated as a city 1845.

Norfolk, resort town, LITCHFIELD county, NW CONNECTICUT, in LITCHFIELD HILLS, near MASSACHUSETTS line, 8 mi/12.9 km NW of WINSTED; 41°58'N 73°12'W. Agriculture; manufacturing wooden and metal products. Recreational area, with state parks; Annual Litchfield county choral concerts began here, 1899. William Henry Welch born here. Settled 1755, incorporated 1758.

Norfolk, agricultural town, NORFOLK county, E MASSACHUSETTS, 22 mi/35 km SW of BOSTON; 42°07'N 71°20'W. State correctional institute here. Settled 1795, incorporated 1870.

Norfolk, village (□ 1 sq mi/2.6 sq km; 2000 population 1,334), ST. LAWRENCE county, N NEW YORK, on RAQUETTE RIVER, and 10 mi/16 km SSW of MASSENA; 44°47'N 74°59'W. Settled in 1809. Amish people moved into the area in the mid-1970s and opened a cheese mill, cheese factories, sawmills, and gristmills;

they are all gone now, though. William P. Rogers (U.S. Attorney General and Secretary of State), grew up here.

Norfolk Bay, AUSTRALIA: see FREDERICK HENRY BAY.

Norfolk County (NOR-fuhk), city (□ 634 sq mi/1,648.4 sq km; 2001 population 60,847), S ONTARIO, E central Canada, on Lake ERIE; 42°50'N 80°02'W. Retains pre-1973 borders. Since 2001, forms W half of former Haldimand-Norfolk regional municipality.

Norfolk Island (NOR-fuhk), island (□ 13 sq mi/33.8 sq km), South PACIFIC, a territory of AUSTRALIA, c.1,035 mi/1,670 km NE of SYDNEY; 29°02'S 167°57'E. A resort, Norfolk has luxuriant vegetation and is known for its pine trees. Explored in 1774 by Captain James Cook, the island was claimed by GREAT BRITAIN in the hope that the pines would provide masts for the navy. When the wood proved unsatisfactory, Norfolk was made into a prison island (1788–1855). In 1856 the prisoners were removed and some of the descendants of the *Bounty* mutineers were moved to Norfolk from PITCAIRN ISLAND. Norfolk Island was annexed to TASMANIA in 1844, became a dependency of NEW SOUTH WALES. in 1896, and was transferred to the Commonwealth of Australia in 1913. Many of the old prison colony buildings have been restored and contribute to the main industry: tourism. The main languages are English and Tahitian.

Norfolk Island National Park (NOR-fuhk) (□ 3 sq mi/7.8 sq km), N end of NORFOLK ISLAND, AUSTRALIA, PACIFIC OCEAN, 4 mi/6 km NNW of KINGSTON; 2 mi/3 km long, 1.5 mi/2.4 km wide, 1.5 mi/2.4 km shoreline; 29°01'S 167°57'E. Highest peaks include Mount Bates (1,043 ft/318 m) and Mount Pitt (1,037 ft/316 m). Captain Cook Monument overlooks Duncombe Bay on N. Palm forests, stands of Norfolk pine. White terns, red-tailed tropicbirds, grey fantails. Rare green parrot, Boobook owl. Walking tracks. Established 1986.

Norfork, village (2000 population 484), BAXTER county, N ARKANSAS, c.45 mi/72 km E of HARRISON, at junction of WHITE RIVER and the North Fork; 36°12'N 92°16'W. Norfork Dam is to NE on North Fork (White) River backs up the North Fork River in NORFORK LAKE.

Norfork Lake, reservoir (□ 34 sq mi/88.4 sq km), mainly in extreme N central ARKANSAS (BAXTER county) and MISSOURI, on North Fork of WHITE RIVER, 4 mi/6.4 km W of MOUNTAIN HOME; 36°15'N 92°14'W. Maximum capacity 1,983,000 acre-ft. Extends N-S. Formed by Norfork Dam (216 ft/66 m high), built (1944) by Army Corps of Engineers for power generation, recreation, water supply and as a fish and wildlife pond.

Norg (NAWRKH), town, DRENTHE province, N NETHERLANDS, 7 mi/11.3 km NW of ASSEN; 53°04'N 06°28'E. Dairying; livestock; grain, fruit, vegetables. Ancient tombs to N and E.

Norge, unincorporated village, JAMES CITY county, SE VIRGINIA, 7 mi/11 km NNW of WILLIAMSBURG; 37°22'N 76°46'W. Agriculture (tobacco, grain, soybeans; cattle; dairying).

Norge: see NORWAY.

Norglenwold, summer village (2001 population 267), S central ALBERTA, W Canada, 14 mi/22 km from RED DEER, in Red Deer County; 52°19'N 114°07'W.

Norheimsund (NAWR-haim-SOON), village, HORDALAND county, SW NORWAY, on HARDANGERFJORDEN, 10 mi/16 km WSW of ÅVIK; 60°22'N 06°08'E. Tourism.

Noria de Ángeles (NO-ree-ah dai AHN-he-les), town, ZACATECAS, N central MEXICO, 50 mi/80 km SE of ZACATECAS; 22°26'N 101°54'W. Maguey, fruit; livestock; silver and copper deposits.

Noria de San Pantaleón (NO-ree-ah dai sahn pahn-tah-le-ON), mining settlement, Sombrerete municipio, ZACATECAS, N central MEXICO, 10 mi/16 km ENE of SOMBRERETE; 23°40'N 103°46'W. Silver, gold, lead, copper deposits.

Noria, La, CHILE: see LA NORIA.

Noric Alps (NO-rik), German *Norische Alpen*, central Alps in S AUSTRIA, mainly along STYRIA-CARINTHIA line; between KATSCHBERG PASS in the W and Bacher, SLOVENIA in the SE; bounded N by upper MUR RIVER, S by DRAU RIVER valley. Consists of the GURKTAL ALPS (W) rising to 7,440 ft/2,268 m in the EISENHUT mountains, the Sectal Alps, Saualpe, Koralpe, Possruck, and Bacher (SE in Slovenia). Sometimes called Styrian-Carinthian Alps.

Noricum (NO-ri-kum), province of the Roman Empire. It corresponded roughly to modern AUSTRIA S of the DANUBE, W of VIENNA and E of TYROL. It was bordered on the W by RHAETIA and on the E by PANNONIA. Noricum was incorporated into the Roman Empire in 16 B.C.E. It prospered for centsuries, then declined and was overrun by Germanic tribes in the 5th century.

Norikura, Mount (no-REE-koo-rah), Japanese *Norikura-dake* (no-ree-KOO-rah–DAH-ke), peak (9,925 ft/3,025 m), central HONSHU, central JAPAN, on GIFU-NAGANO prefecture border, in Chubu-Sangaku National Park, 17 mi/27 km E of TAKAYAMA.

Noril'sk (nuh-REELSK), city (2006 population 125,130), W central TAYMYR (DOLGAN-NENETS) AUTONOMOUS OKRUG, administratively in KRASNOYARSK TERRITORY, NE Siberian Russia, approximately 930 mi/1,497 km N of KRASNOYARSK; 69°20′N 88°13′E. Elevation 374 ft/113 m. The northernmost major city of Russia and the world's second-largest city (after MURMANSK) N of the ARCTIC CIRCLE, Noril'sk is the center of a mining region (nickel, copper, cobalt, precious metals, platinum, gold, silver, palladium, iridium, and coal). A railroad links Noril'sk with the YENISEY port of DUDINKA, from where ores are shipped via the NORTHERN SEA ROUTE to European Russia. Hydroelectric plants are nearby. Natural-gas deposits are in the region. Has a drama theater. Founded in 1935, Noril'sk was the site of forced labor camps during the Stalin era.

Norische Alpen, AUSTRIA: see NORIC ALPS.

Nor Kharberd, urban settlement, ARMENIA; 40°05′N 44°28′E. Combine producing articles from semiprecious stones. SE of YEREVAN.

Norlane (NOR-lain), suburb just N of GEELONG, VICTORIA, SE AUSTRALIA; 38°06′S 144°21′E.

Norlina (nor-LEIN-uh), town (□ 1 sq mi/2.6 sq km; 2006 population 1,034), WARREN county, N NORTH CAROLINA, 14 mi/23 km NE of HENDERSON, 8 mi/12.9 km S of JOHN H. KERR and GASTON lakes reservoirs; 36°27′N 78°12′W. Manufacturing (textiles); agriculture (grain, soybeans, tobacco; poultry, livestock; dairying). Founded 1900. Formerly called Ridgeway Junction, Woodyard.

Nor Lípez, province, POTOSÍ department, W central BOLIVIA, ⊙ Colchaca Villa Martín; 20°30′S 67°50′W.

Normal, town, Madison co., N ALABAMA, 3 mi/4.8 km N of Huntsville. Named for the state Normal and Industrial School, which is now the Alabama Agr. and Mechanical University.

Normal, town (2000 population 45,386), MCLEAN county, central ILLINOIS; 40°31′N 89°00′W. It is the center of a productive farming area. Manufacturing (motor vehicles). The town originally grew around Illinois State University (1857; formerly called Illinois State Normal University, the first public University in Illinois), which remains a major contributor to its economy. Incorporated 1867.

Norman, county (□ 877 sq mi/2,280.2 sq km; 2006 population 6,850), NW MINNESOTA; ⊙ ADA. Bounded W by RED RIVER of the North (NORTH DAKOTA state line) and drained by WILD RICE RIVER; 47°20′N 96°28′W. Agricultural area (wheat, oats, barley, sugar beets, beans, sunflowers, soybeans, alfalfa, hay). Formed 1881.

Norman, city (2006 population 102,827), ⊙ CLEVELAND county, central OKLAHOMA, suburb 18 mi/29 km S of OKLAHOMA CITY, on CANADIAN River; 35°13′N 97°20′W. Elevation 1,170 ft/357 m. It is the center of a livestock region. Oil wells; manufacturing (machinery, communication equipment, bakery products, nutritional products, consumer goods; food processing, publishing and printing). Norman is the seat of the University of Oklahoma. Max Westheimer Field airport here. Lake Thunderbird reservoir and Little River State Park in E part of city. The city has been marked by a population increase of more than 55% between 1970 and 1990. Incorporated 1891.

Norman, village (2000 population 423), MONTGOMERY county, W ARKANSAS, 35 mi/56 km W of HOT SPRINGS, in Ouachita National Forest; 34°27′N 93°40′W. Lumbering.

Norman, village (2006 population 49), KEARNEY county, S NEBRASKA, 7 mi/11.3 km E of MINDEN; 40°28′N 98°47′W.

Norman (NOR-muhn), village (2006 population 73), RICHMOND county, S NORTH CAROLINA, 16 mi/26 km NNE of ROCKINGHAM; 35°10′N 79°43′W. Retail trade; agriculture (tobacco, grain, cotton; livestock). Railroad terminus. JOHN H. KERR RESERVOIR to NW, Lake GASTON reservoir to NE.

Normanby (NOR-muhn-bee), former township (□ 109 sq mi/283.4 sq km; 2001 population 2,679), S ONTARIO, E central Canada, 34 mi/55 km from OWEN SOUND; 44°04′N 80°53′W. Amalgamated into West Grey township.

Normanby Island, volcanic island (□ 400 sq mi/1,040 sq km), D'ENTRECASTEAUX ISLANDS, MILNE BAY province, SE PAPUA NEW GUINEA, SW PACIFIC OCEAN, 10 mi/16 km NE of SE tip of NEW GUINEA, across Goschen Stait; 45 mi/72 km long, 15 mi/24 km wide; 10°00′S 151°00′E. Separated on NW from FERGUSSON ISLAND by Dobu Passage; village of Esa'ala at NW end; Bwaslalal at S end. Gold.

Norman, Cape, promontory at N extremity of Newfoundland, NEWFOUNDLAND AND LABRADOR, Canada, on Strait of Belle Isle; 51°38′N 55°54′W. Lighthouse.

Normandia (NOR-mahn-zhee-ah), city (2007 population 7,782), NE RORAIMA state, BRAZIL, 123 mi/198 km NE of BOA VISTA; 04°08′S 59°51′W. Named after Frenchman who escaped from Devil's Island. Located near GUYANA border in proposed Macuxi Indian reservation.

Normandie, FRANCE: see NORMANDY.

Normandie, Basse, FRANCE: see BASSE-NORMANDIE.

Normandie, Haute, FRANCE: see HAUTE-NORMANDIE.

Normandie-Maine Natural Regional Park (normahn-dee–men), a regional park of NW FRANCE, within the historic provinces of NORMANDY and MAINE. Covers 578,206 acres/234,000 ha; occupies parts of the departments of MANCHE, MAYENNE, ORNE, and SARTHE, in BASSE-NORMANDIE region; 48°30′N 00°10′W. Established 1975, the park serves both recreational and redevelopment objectives, especially among the smaller communities experiencing out-migration to the cities. It aims to encourage the revival of handicraft arts, sport activities, folklore festivals and exhibits, native costumes, and the local production of cider. The largest nearby city is ALENçON; park headquarters is at Carrouges (12 mi/19 km NW of Alençon).

Normandie-Maine Natural Regional Park (NOR-mahn-dee–ME-nuh), on the borders of MANCHE, MAYENNE, SARTHE, and ORNE departments (□ 888 sq mi/2,300 sq km), W FRANCE.

Normandin (nor-mahn-DAN), town (□ 82 sq mi/213.2 sq km), Saguenay—Lac-Saint-Jean region, S central QUEBEC, E Canada, 26 mi/42 km NW of ROBERVAL; 48°50′N 72°31′W. Lumbering; food processing.

Normandy (NOR-muhn-dee), French *Normandie* (nor-mahn-dee), region and former province, NW FRANCE, bordering on the ENGLISH CHANNEL; 49°00′N 00°00′W. In 1790 it was divided into five departments: CALVADOS, EURE, MANCHE, ORNE, and SEINE-MARITIME. Normandy is a region of lowland, gentle hills, and farmland which is especially known for the hedgerows that partition farmers' fields and orchards. Its climate is humid, suitable for livestock raising (and supplying milk and cheese to the markets of the PARIS BASIN). Normandy is also known for its diverse districts or "pays" (French=distinct country areas), such as the COTENTIN (the large peninsula extending into the Channel), the BOCAGE NORMAN (fields and hedgerows), the Bessin, the "pays" of CAUX (underlain by massive limestone and bounded by white cliffs along the coast), the Norman VEXIN N of the SEINE, and others. The lower Seine River, which meanders in its course to an estuary on the Channel, roughly divides Upper (N) from Lower (SW) Normandy, though the department of Eure (administratively in Upper Normandy) actually straddles the Seine valley. Most of the regions' industry is found along the lower course of the Seine, with oil refineries, metal and chemical plants and textile mills concentrated in the urban ribbon between ROUEN, the historic capital of Normandy, and Le Havre (see HAVRE, LE), one of France's leading seaports. CHERBOURG, at the tip of the Cotentin Peninsula, is also an important port for passenger and freight traffic and a shipbuilding center, while CAEN still makes steel and alloys. There are many beach resorts along Normandy's Channel coast, ranging from the luxurious and animated (DEAUVILLE, TROUVILLE, ÉTRETAT, FÉCAMP) to the smaller family resorts. Along the Normandy coast three nuclear power plants have been established and a nuclear reprocessing center formerly operated at La HAGUE near Cherbourg. Tourism has become an important industry, with the Normandy landing sites of the Allied Forces in June 1944 (World War II) of BAYEUX and the famed MONT-SAINT-MICHEL island (where BRITTANY and Normandy meet) among the principal attractions. Part of ancient GAUL, the region was conquered by Julius Caesar and became part of the province of Lugdunensis. It was Christianized in the 3rd century and conquered by the Franks in the 5th century. Repeatedly devastated (9th century) by the Norsemen, it finally was ceded (911) to their chief, Rollo, first duke of Normandy, by Charles III (Charles the Simple) of France. The Norsemen (or Normans), for whom the region was named, soon accepted Christianity. Rollo's successors acquired neighboring territories in a series of wars. In 1066, Duke William (William the Conqueror), son of Robert I, invaded England, where he became king as William I. The succession disputed among William's sons (Robert II of Normandy, William II, and Henry I of England), passed to England after the battle of Tinchebrai (1106), in which Henry defeated Robert. In 1144, Geoffrey IV of Anjou conquered Normandy; his son, Henry Plantagenet (later Henry II of England), was invested (1151) with the duchy by King Stephen of England. It was by this series of events that branches of the Angevin dynasty came to rule England, as well as vast territories in France, Sicily, and S Italy, where the Normans had begun to establish colonies in the 11th century Normandy was joined to France in 1204 after the invasion and conquest by Philip II. It was again devastated during the Hundred Years War (1337–1453). The Treaty of Brétigny (1360) confirmed Normandy as a French possession, but Henry V of England invaded the region and conquered it once more. With the exception of the larger Channel Islands, Normandy was permanently restored to France in 1450, and in 1499, Louis XII established a provincial *parlement* for Normandy at Rouen. The Protestants made great headway in Normandy in the 16th century, and there were bitter battles between Catholics and Huguenots. Louis XIV sought to complete the assimilation of Normandy into France, and in 1654 the provincial estates were suppressed. The revocation of the Edict of NANTES (1685) led to a mass emigration of

Huguenots from Normandy and a grave economic setback for the region. In the 18th century, however, prosperity returned. In 1790 the province, with others in France, was abolished and replaced by the present-day departments. In the 1870s a large reverse migration of Protestants (and others) took place from German-occupied ALSACE. The region was the scene of the Allied invasion (1944) of continental Europe in World War II.

Normandy, town (2000 population 5,153), SAINT LOUIS county, E MISSOURI, residential suburb 8 mi/12.9 km NW of downtown ST. LOUIS; 38°42′N 90°17′W. Served by Metrolink light railroad; University of Missouri campus in adjacent BELLERIVE. Incorporated since 1940.

Normandy, town (2006 population 151), BEDFORD county, central TENNESSEE, on DUCK RIVER, and 13 mi/21 km SE of SHELBYVILLE; 35°27′N 86°16′W. Normandy Dam and Fish Hatchery nearby.

Normandy Beaches (NOR-muhn-dee), CALVADOS department, BASSE-NORMANDIE region, NW FRANCE, on ENGLISH CHANNEL, N and NW of city of CAEN. Here Allied forces established a beachhead on the continent on June 6, 1944, the first step in the reconquest of W Europe from Hitler's Germany. The American landings took place on OMAHA BEACH (NW of BAYEUX) and UTAH BEACH (near SAINTE-MÈRE-ÉGLISE, at SE base of COTENTIN PENINSULA). British and Canadian forces landed at Gold, Juno, and Sword beaches, between the artificial harbor established at ARROMANCHES-LES-BAINS and the mouth of the ORNE RIVER N of Caen. There are several museums commemorating the invasion, and eleven military cemeteries, testimonials to the bloody fighting. At first visited mainly by British, Canadian, and American tourists, they have become a focus for German visitors as well, and increasingly for many French groups (especially schoolchildren and their teachers).

Normandy Hills (NOR-muhn-dee), in NORMANDY, NW FRANCE, extend E-W in several irregular chains from FALAISE area (CALVADOS department) to vicinity of COUTANCES (at base of COTENTIN PENINSULA, MANCHE department). Average elevation 600 ft/183 m. Paralleled S by outlying hill-range extending from ALENÇON area (E) to the ENGLISH CHANNEL near AVRANCHES (W). They are covered by wooded tracts and hedgerows, and include the small regions of the SUISSE NORMANDE and the BOCAGE. Scene of fighting during Normandy campaign of World War II.

Normandy Park, town (2006 population 6,228), KING county, W WASHINGTON, residential suburb 10 mi/16 km S of downtown SEATTLE, on PUGET SOUND; 47°26′N 122°22′W. Three Tree Point to W. Seattle-Tacoma (SEATAC) International Airport to E.

Normangee (NOR-man-jee), village (2006 population 772), LEON and MADISON counties, E central TEXAS, 29 mi/47 km NE of BRYAN; 31°01′N 96°07′W. Trade point in farm area (vegetables, watermelons; cattle, hogs). State park nearby.

Norman G. Wilder Wildlife Area, DELAWARE: see CAMDEN.

Norman Island, islet, BRITISH VIRGIN ISLANDS, 3 mi/4.8 km E of SAINT JOHN; 18°20′N 64°37′W. Has old pirates' caves.

Norman Isles, United Kingdom: see CHANNEL ISLANDS.

Norman, Lake (NOR-muhn), reservoir, on LINCOLN-MECKLENBURG county border and in IREDELL and CATAWBA counties, SW NORTH CAROLINA on CATAWBA RIVER, 18 mi/29 km NNW of CHARLOTTE; c.25 mi/40 km long; 35°28′N 80°58′W. Maximum capacity 1,092,429 acre-ft. Formed by COWANS FORD DAM (115 ft/35 m), built (1963) by the Duke Power Company for power generation. Duke Power State Park is on the NE shore. Site of McGuire One and Two nuclear power plants.

Norman Park, city (2000 population 849), COLQUITT county, S GEORGIA, 9 mi/14.5 km NE of MOULTRIE; 31°16′N 83°41′W.

Norman River (NOR-muhn), 260 mi/418 km long, N QUEENSLAND, NE AUSTRALIA; rises in SE GREGORY RANGE; flows WNW and NNW, past NORMANTON, to Gulf of CARPENTARIA; 17°28′S 140°49′E. Navigable 30 mi/48 km below Normanton by small craft. Yappar and Clara rivers, main tributaries.

Normans, MARYLAND: see KENT ISLAND.

Norman's Woe, NE MASSACHUSETTS, rocky headland on W side of entrance to Gloucester Harbor. Just offshore, to NE, is **Norman's Woe Rock**, islet with surrounding reefs, which figures in Longfellow's poem "The Wreck of the Hesperus."

Normanton (NAW-muhn-tuhn), town (2001 population 14,958), WEST YORKSHIRE, N ENGLAND, 4 mi/6.4 km ENE of WAKEFIELD; 53°41′N 01°25′W. Former railroad junction. Previously coal mining. Has 16th-century grammar school. Commuter town of LEEDS.

Normanton (NOR-muhn-tuhn), village and river port, N QUEENSLAND, NE AUSTRALIA, on NORMAN RIVER, 30 mi/48 km from its mouth on Gulf of CARPENTARIA, and 390 mi/628 km NNW of TOWNSVILLE; 17°40′S 141°05′E. Terminus of railroad from CROYDON; exports cattle; tourism.

Normantown, town, TOOMBS county, E central GEORGIA, 7 mi/11.3 km N of VIDALIA; 32°18′N 82°22′W.

Norman Wells, village, INUVIK region, NORTHWEST TERRITORIES, Canada, on the MACKENZIE RIVER, W of GREAT BEAR LAKE; 65°17′N 126°51′W. River port and center of an oil region. In 1985 a pipeline to the Zama, Alberta oil field (S) was completed, continues on to EDMONTON refineries. Radio, TV, scheduled air service. Originated as oil-producing town. Once the terminus of Canol Road (1942–1958) from Yukon (now Canol Heritage Trail); proposed extension of MACKENZIE HIGHWAY (now a winter road).

Normétal (nor-mai-TAHL), village (□ 22 sq mi/57.2 sq km), Abitibi-Témiscamingue region, W QUEBEC, E Canada, 55 mi/89 km NNW of ROUYN-NORANDA, near ONTARIO border; 49°00′N 79°22′W. Gold, copper, zinc mining.

Nor Oriental del Marañón (NOR o-ree-en-TAHL del mah-rahn-YON), former planning region (□ 22,202 sq mi/57,725.2 sq km), NW PERU. Contained AMAZONAS, CAJAMARCA, and LAMBAYEQUE departments (today regions) and their twenty-three provinces and 237 districts. Agriculture (rice, corn, wheat, beans, potatoes, coffee, cotton, soy, sugarcane, barley, garlic, yucca, lemons, plantains); meat (lamb, fowl, pork, beef) and dairying. Mineral resources (silver, lead, copper, gold, carbon, salt, zinc, limestone). Established in 1988 as part of Peru's 1988 regionalization program. These regions never quite caught on and were were abandoned.

Nor-Oriental del Marañón, PERU: see MARAÑÓN.

Norphlet (NOR-flit), village (2000 population 822), UNION county, S ARKANSAS, 7 mi/11.3 km N of EL DORADO; 33°19′N 92°39′W.

Norquay (NOR-kwai), town (2006 population 412), E SASKATCHEWAN, CANADA, near MANITOBA border, 23 mi/37 km NNW of KAMSACK; 51°52′N 101°59′W. Mixed farming.

Ñorquin, ARGENTINA: see EL HUECÚ.

Ñorquincó (nyor-keen-KO), village, ⊙ Ñorquincó department, SW río negro province, ARGENTINA, on railroad, and 55 mi/89 km SSE of SAN CARLOS DE BARILOCHE; 41°51′S 70°54′W. Livestock raising (sheep, cattle, goats). Coal deposits nearby.

Norrahammar (NOR-rah-HAHM-mahr), town, JÖNKÖPING county, S SWEDEN, incorporated in JÖNKÖPING; 57°42′N 14°07′E.

Norrbotten (NOR-BOT-ten), northernmost and largest county (□ 40,750 sq mi/105,950 sq km) of SWEDEN; ⊙ LULEÅ; 67°00′N 20°00′E. In Lappland, bounded by NORWAY (W and NW), FINLAND (NE and E), and GULF OF BOTHNIA (SE); comprises Norrbotten province (□ 10,703 sq mi/27,721 sq km; 1995 population 200,920) and N part of Lappland province. Over half of county is N of ARCTIC CIRCLE. Population includes c.4,400 Lapps and c.40,000 Finns. Fertile lowland (E part), rising toward W, becoming mountainous on Norwegian border. Near border is KEBNEKAISE (6,965 ft/2,123 m), Sweden's highest peak. Drained by PITE ÄLVEN, LULEÄLVEN, TORNEÄLVEN, and KALIX ÄLVEN rivers; several large lakes, including TORNETRÄSK, STORA LULEVATTEN, HORNAVAN, and UDDJAURE lakes. Manufacturing (lumbering, sawmilling); research industries; agriculture (sheep, cattle); vast iron deposits at KIRUNA, GÄLLIVARE, and MALMBERGET, ore shipped from LULEÅ and NARVIK, Norway. Hydroelectric power at PORJUS and HARSPRÅNGET. Important towns include Luleå, BODEN, Kiruna, HAPARANDA, and PITEÅ.

Norrbyn (NOR-BEEN), village, VÄSTERBOTTEN county, N SWEDEN, on GULF OF BOTHNIA, 20 mi/32 km SW of UMEÅ; 63°34′N 19°50′E.

Nørre Åby (NU-ruh O-buh), town, FYN county, DENMARK, on NW FYN island, and 21 mi/34 km WNW of ODENSE; 55°28′N 9°52′E. Manufacturing (marzipan).

Nørre Alslev (NOO-ruh AHL-slev), town, STORSTRØM county, DENMARK, on N FALSTER Island, and 17 mi/27 km NE of MARIBO; 54°53′N 11°53′E. Agriculture (sugar beets, barley); manufacturing (candles).

Nørre Sundby (NUH-ruh SOON-buh), city and port, NORDJYLLAND county, N JUTLAND, DENMARK, on LIMFJORD (here bridged), opposite ÅLBORG, 27 mi/43 km S of HJØRRING; 57°04′N 09°55′E. Manufacturing (cement, chemicals, textiles, plastic products); fish, seafood.

Norrhult-Klavreström (NOR-HULT–KLAHV-re-STRUHM), village, KRONOBERG county, S SWEDEN, 20 mi/32 km NE of VÄXJÖ; 57°08′N 15°10′E.

Norridge, village (2000 population 14,582), COOK county, NE ILLINOIS, near W edge of CHICAGO; 41°58′N 87°49′W. Incorporated since 1948.

Norridgewock (NOR-ij-wahk), town, SOMERSET county, W central MAINE, on the KENNEBEC, and 5 mi/8 km SW of SKOWHEGAN; 44°43′N 69°48′W. Light manufacturing (shoes). In seventeenth century Norridgewock Indians here were visited by Jesuit missionaries; monument commemorates chapel of Sébastien Rasles who, with the Indians, was killed by English in 1724. Incorporated 1788.

Norris, city (2006 population 1,465), ANDERSON county, E TENNESSEE, on CLINCH River, and 16 mi/26 km NW of KNOXVILLE; 36°12′N 84°04′W. Built originally as residential community for construction workers on nearby Norris Dam. Nearby is Norris Park (resort) on NORRIS Reservoir, and the Museum of Appalachia, a living history museum. Incorporated 1949.

Norris, town (2006 population 844), PICKENS county, NW SOUTH CAROLINA, c. 9 mi/14.5 km SSW of PICKENS; 34°45′N 82°45′W. Manufacturing of clothing; agriculture includes dairying, poultry, hogs; corn.

Norris, village (2000 population 194), FULTON county, W central ILLINOIS, 18 mi/29 km NNE of Lewiston; 40°37′N 90°01′W. In agricultural and bituminous-coal area.

Norris, village, MADISON county, SW MONTANA, at the confluence of Burnt Creek and Hot Springs Creek, a branch of MADISON River, and 32 mi/51 km WSW of BOZEMAN. Corundum of gem quality mined here, coal outcroppings; cattle. Norris Hot Springs. TOBACCO ROOT MOUNTAINS to W; unit of Lee Metcalf Wilderness Area to SE. Units of Deerlodge and Beaverhead national forests to W, Gallatin National Forest to SE.

Norris, reservoir (□ 23 sq mi/59.8 sq km), E TENNESSEE, on CLINCH River, 25 mi/40 km N of KNOXVILLE; 36°13′N 84°05′W. Maximum capacity 255,200 acre-ft. Has 2 main branches; W branch, on Yellow and POWELL rivers extends NE from dam through CAMPBELL, UNION and CLAIBORNE counties; E branch on CLINCH River extends E between Union and Claiborne

(N) and GRAINGER (S) counties. Formed by Norris Dam (265 ft/81 m high), built (1936) by Tennessee Valley Authority for flood control; also used for power generation, navigation, and recreation. Chuck Swan State Forest between W and E branches; Big Ridge Rustic Park situated directly S of reservoir.

Norris Beach (NAH-ris BEECH), summer village (2001 population 29), S central ALBERTA, W Canada, 25 mi/39 km from MILLET, in Wetaskiwin County No. 10; 52°58′N 114°01′W. Established 1988.

Norris City, village (2000 population 1,057), WHITE county, SE ILLINOIS, 13 mi/21 km SW of CARMI; 37°58′N 88°19′W. Trade center in agricultural and bituminous-coal, oil field area; wheat, corn, soybeans; livestock. Incorporated 1901.

Norristown, borough (2006 population 30,337), ⊙ MONTGOMERY county, SE PENNSYLVANIA, suburb 15 mi/24 km NW of downtown PHILADELPHIA, on the SCHUYLKILL RIVER; 40°07′N 75°20′W. Manufacturing (textiles, medical equipment, machinery, fabricated metal products, petroleum products, explosive devices, furniture, food products, consumer goods, windows and doors; printing and publishing). The borough is named for Isaac Norris (1671–1735), a Quaker merchant and a mayor of Philadelphia, who in 1704 bought a large tract of land here from his friend William Penn. General Winfield Scott Hancock, a commander during the Civil War and Democratic candidate for president in 1880, was born in Norristown and is buried here. Norristown State Hospital is here; Wings Field Airport to E; Elmwood Park Zoo; Fort Washington State Park to E; EVANSBURG STATE PARK to N; VALLEY FORGE NATIONAL HISTORICAL PARK to W.

Norrköping (NOR-SHUHP-eeng), town, ÖSTERGÖT-LAND county, SE SWEDEN, seaport at head of Bråvi-ken, a narrow inlet of BALTIC SEA; 58°35′N 16°11′E. Commercial, industrial, and transportation center; manufacturing (pulp and paper, telecommunication equipment, electrical goods). Former major engineering and textile center. International airport to NE. Burned by Russians in Northern War (1719). Many historic buildings, including Hedvig's Church (seventeenth century). Founded fourteenth century.

Norrland: see SWEDEN.

Norsälven (NORSH-ELV-en), river, central SWEDEN; rises in NORWAY; flows S in FRYKSDALEN valley to LAKE VÄNERN W of KARLSTAD.

Norseman (NORS-muhn), town, WESTERN AUSTRALIA state, W AUSTRALIA, 451 mi/726 km E of PERTH; 32°15′S 121°47′E. Mining (gold, quartz); tourism. Last town before road travel E across NULLARBOR PLAIN.

Norsjø (NAWR-shuh), lake, (□ 23 sq mi/59.8 sq km), TELEMARK county, S NORWAY, just W and NW of SKIEN; 18 mi/29 km long, 1 mi/1.6 km–3 mi/4.8 km wide, 577 ft/176 m deep. Elev. 49 ft/15 m. Boats from Skien (via SKIEN River) traverse the lake to ULEFOSS, the E terminus of the BANDAK-NORSJØ CANAL.

Norte de Santander (NOR-tai dai sahn-tahn-DER), department (□ 8,297 sq mi/21,572.2 sq km), N CO-LOMBIA; ⊙ CÚCUTA; 08°00′N 73°00′W. Bounded E by VENEZUELA (TÁCHIRA RIVER), it is located in deeply dissected Cordillera ORIENTAL and includes a narrow strip of MARACAIBO lowlands. Watered by ZULIA RIVER, affluent of the CATATUMBO. Tropical to temperate climate, depending on elevation. Mineral resources include coal, kaolin, sulphur, gypsum, iron, talc, copper, tin, lead; rich Petrólea oil fields of the Barco concession are linked by pipeline with COVEÑAS (SUCRE department) on CARIBBEAN coast. Forests yield fine lumber, balsam, resins, tagua nuts. Around Cúcuta is a rich coffee growing region; other products are cacao, sugarcane, tobacco, cotton, corn, wheat, fique (sisal) fibers; livestock. Cúcuta, PAMPLONA, and OCAÑA are its trading and processing centers. Contains 39 municipios (2004); most people live in municipio centers and other urban places such as PAMPLONA, OCAÑA, and VILLA DEL ROSARIO.

Nörten-Hardenberg (NYOOR-ten–HAHR-ten-berg), village, LOWER SAXONY, W GERMANY, on the LEINE RIVER, 7 mi/11.3 km N of GÖTTINGEN; 51°38′N 09°56′E. Sugar refining, distilling, woodworking.

Norte, Sierra del (NOR-te, see-YER-rah del), pampean mountain range of Sierra de CÓRDOBA system, NW CÓRDOBA province, ARGENTINA, extends c.35 mi/56 km N from DEÁN FUNES; rises to c.3,000 ft/914 m.

North, province, N RWANDA. Bordered by EAST (E), KIGALI (SES), SOUTH (SSW), and West (W) provinces, as well as DEMOCRATIC REPUBLIC OF THE CONGO (NW) and UGANDA (N). Created in 2006 following a reorganization that replaced Rwanda's twelve provinces with five provinces.

North, district (□ 1,632 sq mi/4,243.2 sq km), SIKKIM state, NE INDIA; ⊙ MANGAN.

North, town (2006 population 782), ORANGEBURG county, W central SOUTH CAROLINA, 16 mi/26 km NW of ORANGEBURG; 33°37′N 81°05′W. Manufacturing includes lumber, knife cases, screen doors, apparel. Agriculture (timber; poultry, livestock; grain, cotton, peanuts, tobacco, pecans, peaches).

North, for names beginning thus and not found here: see under Northern. In Russian names: see also Severnaya, Severniye, Severnoye, Severny, or Severo-.

North 24-Parganas, district, WEST BENGAL state, E INDIA, in GANGA DELTA; ⊙ Alipur. Bounded W by HUGLI River, and Calcutta Municipal Corporation, E by BANGLADESH; drained by river arms of the Ganga Delta. Alluvial soil (rice, jute, pulses, potatoes, betel leaves, mustard, sugarcane, beets); extensive swamps. Heavily industrialized area extends along the Hugli River the length of the W edge of district. Major Indian jute-milling and chemical-manufacturing center; cotton and rice milling; manufacturing of matches, glass, soap, cement, silk cloth. Main centers include TITA-GARH and BARAKPUR, which figured in Sepoy rebellions of 1824 and 1857. Ruins of 18th-century fort at GARULIA.

North Abington, MASSACHUSETTS: see ABINGTON.

North Adams, city (2000 population 14,681), BERK-SHIRE county, NW MASSACHUSETTS, in the BERK-SHIRE HILLS, on the HOOSIC RIVER near the VERMONT border; 42°42′N 73°07′W. The city is located in a summer resort and winter ski area. Manufacturing includes electronic and electrical components, paper products, textiles, and machinery. Airport. North Adams State College was a normal school in 1894. New art museum in city. Mount Greylock State Reservation. Settled c.1737, set off from ADAMS and incorporated 1878.

North Adams, village (2000 population 514), HILLSDALE county, S MICHIGAN, 6 mi/9.7 km NE of HILLSDALE; 41°58′N 84°31′W. In agricultural area; manufacturing of screw machine products, metal stampings.

North Aegean (e-JEE-ahn), department (□ 1,481 sq mi/3,850.6 sq km), comprising the N part of the CYCLADES Islands in the AEGEAN SEA, Greece; ⊙ MITILÍNI (LES-BOS island). Includes the islands of LÍMNOS, Lesbos, SÁMOS, KHÍOS, and IKARÍA. Divided into Khíos, Les-bos, and Sámos prefectures. Chief towns include Mitilíni, Khíos (Khíos island), and Sámos (Sámos island).

North Albanian Alps, Serbo-Croatian *Prokletije*, mountain group at S end of DINARIC ALPS, on borders of ALBANIA and SERBIA-MONTENEGRO in Serbia's KOSOVO province, extends c.60 mi/97 km in an alpine dome between Lake SHKODËR and WHITE DRIN river. Rises to 8,839 ft/2,694 m in the JEZERCE. A wild, precipitous highland region, more easily accessible from N (Montenegro and Serbia) slopes.

North Algona-Wilberforce (NORTH al-GO-nuh-WIL-buhr-fors), township (□ 146 sq mi/379.6 sq km; 2001 population 2,729), SE ONTARIO, E central Canada, 15 mi/24 km from PEMBROKE; 45°37′N 77°13′W.

Northallerton (nawth-AL-uh-tuhn), town (2001 population 15,517), NORTH YORKSHIRE, N ENGLAND, 15 mi/24 km N of RIPON, in valley of the Wiske; 54°20′N 01°26′W. Administrative and trade center, with dairying and leather-tanning industries. Has 12th-13th-century church, remains of Roman station and of 12th-century palace. Standard Hill, 2 mi/3.2 km N, was site of defeat (1138) of Scots by English in "Battle of the Standard."

Northam (NOR-thuhm), municipality, SW WESTERN AUSTRALIA state, W AUSTRALIA, 50 mi/80 km ENE of PERTH, and on Avon River; 31°40′S 116°40′E. Railroad junction; wheat center; flour mill. Muresk agricultural college nearby.

Northam (NAWTH-uhm), town (2001 population 11,604), N DEVON, SW ENGLAND, on TORRIDGE RIVER, near its mouth on BARNSTAPLE BAY of the BRISTOL CHANNEL, and 2 mi/3.2 km N of BIDEFORD; 51°02′N 04°13′W. Scene of 9th-century battle between Saxons and Danes. Has 14th-century church. Urban district includes APPLEDORE and WESTWARD HO!

North America, third largest continent (□ c. 9,400,000 sq mi/24,346,000 sq km; 1990 estimated pop. 365,000,000; 2004 estimated pop. 440,600,000), the N of the two continents of the Western Hemisphere.

Boundaries

North America includes all of the mainland and related offshore islands lying N of the ISTHMUS OF PA-NAMA (which connects it with SOUTH AMERICA). The term "Anglo-America" is frequently used in reference to CANADA and the UNITED STATES combined, while the term "Middle America" is used to describe the region including MEXICO, the republics of CEN-TRAL AMERICA, and the CARIBBEAN. The continent is bounded on the N by the ARCTIC OCEAN, on the W by the PACIFIC OCEAN and the BERING SEA, and on the E by the ATLANTIC OCEAN and the GULF OF MEXICO. Its coastline is long and irregular. With the exception of the Gulf of Mexico, HUDSON BAY is by far the largest body of water indenting the continent; others include the GULF OF SAINT LAWRENCE and the GULF OF CA-LIFORNIA (SEA OF CORTÈS). There are numerous islands off the continent's coasts; the ARCTIC ARCHIPELAGO, the GREATER and LESSER ANTILLES, the ALEXANDER ARCHIPELAGO, and the ALEUTIAN IS-LANDS are the principal groups.

Mountains - Rivers - Lakes

MOUNT MCKINLEY (also known as Denali; 20,320 ft/6,194 m), ALASKA, is the highest point on the continent; the lowest point (282 ft/86 m below sea level) is in DEATH VALLEY, CALIFORNIA. The MISSOURI-MIS-SISSIPPI river system (c. 3,740 mi/6,020 km long) is the longest of North America. Together with the OHIO RIVER and numerous other tributaries, it drains most of S central North America and forms the world's greatest inland waterway system. Other major rivers include the COLORADO, COLUMBIA, DELAWARE, MACKENZIE, NELSON, RIO GRANDE, SAINT LAWRENCE, SUSQUEHANNA, and YUKON. LAKE SUPERIOR (□ 31,820 sq mi/82,414 sq km), the westernmost of the GREAT LAKES, is the continent's largest lake. The SAINT LAWRENCE SEAWAY, which utilizes the Saint Lawrence River and the Great Lakes, enables ocean-going vessels to penetrate into the heart of North America.

Regions

Physiographically, the Anglo-American section of the continent may be divided into five major regions: the CANADIAN SHIELD, a geologically stable area of ancient rock that occupies most of the NE quadrant, including GREENLAND; the APPALACHIAN MOUN-TAINS, a geologically old and eroded system that extends from NEWFOUNDLAND to ALABAMA; the Atlantic-Gulf Coastal Plain, a belt of lowlands widening to the S that extends from S NEW ENGLAND to Mexico; the Interior Lowlands, which extend down the middle of the continent from the Mackenzie valley to the Gulf Coastal Plain and includes the GREAT

PLAINS on the W and the agriculturally productive Interior Plains on the E; and the North American Cordillera, a complex belt of geologically young mountains and associated plateaus and basins, which extend from Alaska into Mexico and include two orogenic belts—the PACIFIC MARGIN on the W and the ROCKY MOUNTAINS on the E—separated by a system of intermontane plateaus and basins. The Coastal Plain and the main belts of the North American Cordillera continue S into Mexico (where the Mexican Plateau, bordered by the SIERRA MADRE ORIENTAL and the SIERRA MADRE OCCIDENTAL, is considered a continuation of the intermontane system) to join the Transverse Volcanic Range, a zone of high and active volcanic peaks S of MEXICO CITY.

Climate

North America, extending to within 10° of latitude of both the Equator and the NORTH POLE, embraces every climatic zone, from tropical rain forest and savanna on the lowlands of Central America to areas of permanent ice cap in central Greenland. Subarctic and tundra climates prevail in N Canada and N Alaska, and desert and semiarid conditions are found in interior regions cut off by high mountains from rain-bearing W winds. However, a high proportion of the continent has temperate climates very favorable to settlement and agriculture. During the Ice Age of the late Cenozoic era, a continental ice sheet, centered W of Hudson Bay (the floor of which is slowly rebounding after being depressed by the great weight of the ice), covered most of N North America; glaciers descended the slopes of the Rocky Mountains and those of the Pacific Margin. Extensive glacial lakes, such as BONNEVILLE, LAHONTAN, AGASSIZ, and ALGONQUIN, were formed by glacial meltwater; their remnants are still visible in the GREAT BASIN and along the edge of the CANADIAN SHIELD in the form of the GREAT SALT LAKE, the Great Lakes, and the large lakes of W central Canada.

Population

The first human inhabitants of North America are believed to be of Asian origin; they crossed over to Alaska from NE ASIA roughly 20,000 years ago, and then moved S through the Mackenzie River valley. European discovery and settlement of North America dates from the 10th century, when Norsemen settled (986) in Greenland. Although evidence is fragmentary, they probably reached E Canada c. 1000 at the latest. Of greater impact on the subsequent history of the continent were Christopher Columbus's exploration of the BAHAMAS in 1492 and later landings in the WEST INDIES and Central America, and John Cabot's explorations of E Canada (1497), which established English claims to the continent. Spanish and French expeditions also explored much of North America.Although the population of Canada and the U.S. is still largely of European origin, it is growing increasingly diverse with substantial immigration from Asia, LATIN AMERICA, and AFRICA; it is also highly urbanized (about 74% live in urban areas); much of the population is centered in large conurbations and coalescing urban belts along the S margin of Canada and in the NE midwestern quadrant of the U.S. around the Great Lakes and along the Atlantic coast, and in California. Mexico's population, about 60% mestizo (of European and Native American descent), is increasingly urbanized (about 72%). People of European descent are a minority in most Central American and Caribbean countries, and the population outside the major cities is largely rural. The largest urban agglomerations on the continent are Mexico City, NEW YORK CITY, LOS ANGELES, and CHICAGO. For more detailed information, please go to other entries in the Gazetteer.

Economy

The extensive agricultural lands (especially in Canada and the U.S.) are a result of the interrelationship of favorable climatic conditions, fertile soils, and technology. Irrigation has turned certain arid and semi-arid regions into productive oases. North America produces most of the world's corn, meat, cotton, soybeans, tobacco, and wheat, along with a variety of other food and industrial raw material crops. Mineral resources are also abundant; the large variety includes coal, iron ore, bauxite, copper, natural gas, petroleum, mercury, nickel, potash, and silver. The manufacture that provided a high standard of living for the people of Canada and the U.S. has significantly declined, and formerly abundant factory jobs are increasingly replaced by those in the service sector. Some of this manufacture has moved to Mexico (especially in the border zone adjoining the U.S.), which offers a large and inexpensive labor force. In 1993 and 1994 the U.S., Canada, and Mexico signed and ratified NAFTA, which sharply reduced customs and other trade barriers among the three countries. For more detailed information, please go to other entries in the Gazetteer.

North Amherst, MASSACHUSETTS: see AMHERST.

North Amity (AM-uh-tee), town, AROOSTOOK county, E MAINE, 15 mi/24 km S of HOULTON, near NEW BRUNSWICK border (CANADA). Agriculture; lumbering.

Northampton (north-HAMP-tuhn), county (□ 550 sq mi/1,430 sq km; 2006 population 21,247), NE NORTH CAROLINA; ⊙ Jackson; 36°25′N 77°24′W. In PIEDMONT region, bounded N by VIRGINIA state line, SW by ROANOKE RIVER (forms ROANOKE RAPIDS and GASTON reservoirs in extreme W); drained by MEHERRIN RIVER, forms part of NE boundary. Manufacturing (textiles, apparel, lumber, chemicals, boats, sheet metal, paper products); service industries; agriculture (peanuts, cotton, corn, tobacco, wheat, soybeans, hay; hogs, cattle, poultry); timber. Formed 1741 from Bertie County. Named for James Compton, Earl of Northampton (1687–1754).

Northampton, county (□ 377 sq mi/980.2 sq km; 2006 population 291,306), E PENNSYLVANIA; ⊙ EASTON; 40°45′N 75°18′W. Industrial region; bounded E by DELAWARE RIVER (forms NEW JERSEY state boundary), NW by LEHIGH RIVER (drained in S by Lehigh River); BLUE MOUNTAIN and KITTATINNY MOUNTAIN (both ridges) along N boundary; APPALACHIAN TRAIL follows N boundary. Manufacturing at BETHLEHEM, NORTHAMPTON, and NAZARETH; slate quarrying. Agriculture (corn, wheat, oats, barley, hay, alfalfa, soybeans, potatoes, apples; sheep, hogs, cattle, poultry; dairying). Small part of DELAWARE WATER GAP NATIONAL RECREATION AREA in NE corner; Jacobsburg State Park in center. Settled first by English and Scotch-Irish, later by Germans. Formed 1752.

Northampton (north-HAMP-tuhn), county (□ 795 sq mi/2,067 sq km; 2006 population 13,609), E VIRGINIA; ⊙ EASTVILLE; 37°17′N 75°55′W. At S end of EASTERN SHORE (DELMARVA) peninsula; CAPE CHARLES at S tip, on N side of entrance to CHESAPEAKE BAY; CHESAPEAKE BAY BRIDGE-TUNNEL (18 mi/29 km long) crosses mouth of bay. Many barrier and bay islands lie off ATLANTIC coast. Fertile coastal plain. Manufacturing (several canneries process seafood, farm produce); agriculture (known for white and sweet potatoes; also strawberries, corn, barley, wheat, soybeans; poultry, hogs); pine timber; limestone; crabs, oysters. Cape Charles, resort town, is a port and a railroad and ferry terminus. Kiptopeke State Park in S, WRECK ISLAND State Natural Area in E, FISHERMANS ISLAND National Wildlife Refuge in S. INTRACOASTAL WATERWAY passes through bay to W. Formed 1663.

Northampton (nawth-AMP-tuhn), city (2001 population 194,458) and district, ⊙ NORTHAMPTONSHIRE, central ENGLAND, on the NENE River; 52°15′N 00°53′W. Shoemaking was the chief industry, but ceased in the 20th century. Some engineering and manufacturing. The city was an important settlement of the Angles and of the Danes, and its Norman castle was the scene of sieges as well as parliaments from the 12th to the 14th centuries. Roman and ancient British relics are in the vicinity. The Church of St. Giles has a Norman doorway; All Saints' has a 14th-century tower; St. Peter's (12th century) has a Norman interior; and there is a Roman Catholic cathedral designed by A. W. Pugin. The 12th-century St. Sepulchre's is one of the four round churches in England. St. John's Hospital was founded in 1138. One of the few remaining Eleanor Crosses is near Northampton, at Hardingstone.

Northampton (nor-THAM-tuhn), city (2000 population 28,978), ⊙ HAMPSHIRE county, W MASSACHUSETTS, on the CONNECTICUT River; 42°20′N 72°41′W. Includes village of Leeds. Brushes, wire, optical devices, and plastic products are made in Northampton. It is the seat of Smith College and has Clarke School for the Deaf. President Calvin Coolidge was a former mayor of Northampton; his papers and mementos are preserved in the Forbes Library. Jonathan Edwards was pastor here, and Sylvester Graham lived and is buried in the city. Historic Deerfield is nearby. Dinosaur Footprint Park nearby. Incorporated as a town 1656; as a city 1883.

Northampton (NOR-THAMP-tuhn), town, W WESTERN AUSTRALIA state, W AUSTRALIA, 30 mi/48 km N of GERALDTON; 28°21′S 114°37′E. Lead-mining center of state; citrus fruits; flower cultivation; fishing.

Northampton (north-HAMP-tuhn), borough (2006 population 9,765), NORTHAMPTON county, E PENNSYLVANIA, suburb 5 mi/8 km N of ALLENTOWN, on LEHIGH RIVER; 40°41′N 75°29′W. Railroad junction. Manufacturing (machinery, metal castings, men's suits, flow meters, apparel), agriculture (grain, apples, soybeans, potatoes; livestock; dairying). Allentown State Farm to E; Mary Immaculate Missionary College to N; Lehigh Valley International Airport to SE. Settled c.1763; incorporated 1901.

Northamptonshire (nawth-AMP-tuhn-shir) or **Northants,** county (□ 914 sq mi/2,376.4 sq km; 2001 population 629,676), central ENGLAND; ⊙ NORTHAMPTON; 52°20′N 00°55′W. The terrain is undulating, and is comprised mainly of pasture and forests. The principal river is the NENE. The iron and steel industry, which flourished on the basis of local ore from the start of the 20th century, has dwindled in significance. The county's specialization in boot and shoe production has also withered in the face of competition. The Roman roads ERMINE and WATLING streets crossed the county. In Anglo-Saxon times the area was part of the kingdom of MERCIA and was probably organized as a shire in Danish times. Other significant towns include CORBY, KETTERING, and WELLINGBOROUGH.

North Andover (nor-THAN-do-vuhr), town, ESSEX county, NE MASSACHUSETTS, on the MERRIMACK RIVER; 42°40′N 71°05′W. In a dairy and farm area. A former textile town, its manufacturing includes telephone equipment, chemicals, paper products, valves, prepared foods, and plastics. It is the seat of Merrimack College, Brooks Preparatory School, and a Boston University theology center; textile museum. The scenic spring-fed Lake Cochichewick is nearby. Boston Hill ski resort; Harold Parker State Forest. Settled c.1644, set off from ANDOVER, and incorporated 1855.

North Anna 1 and 2 Nuclear Power Plants, Virginia: see LOUISA, county.

North Anna Dam, Virginia: see ANNA, LAKE.

North Anna River (NORTH A-nuh), c.70 mi/113 km long, central VIRGINIA; rises in PIEDMONT region in W ORANGE county (38°09′N 78°05′W); flows SE, through LAKE ANNA reservoir; joins SOUTH ANNA RIVER c.20 mi/32 km N of RICHMOND to form PAMUNKEY RIVER.

Northants, ENGLAND: see NORTHAMPTONSHIRE.

North Apollo, borough (2006 population 1,343), ARMSTRONG county, W central PENNSYLVANIA, 1 mi/1.6 km N of APOLLO, and 15 mi/24 km S of KITTANNING, on KISKIMINETAS RIVER opposite VANDERGRIFT; 40°35'N 79°33'W. Manufacturing (fabricated metal products); agriculture (corn, hay; dairying). Incorporated 1930.

North Arcot Ambedkar (AHR-kot AHM-bed-kahr), Tamil *Vada Arkadu*, district (□ 4,736 sq mi/12,313.6 sq km), TAMIL NADU state, S INDIA; ⊙ VELLORE. Bordered NW by S EASTERN GHATS. Mainly lowland, except for JAVADI HILLS (W; sandalwood, hemp) and numerous isolated granite outcrops (quarries); drained by PALAR RIVER and its tributary, the CHEYYAR. Largely red ferruginous soil; alluvial soil along rivers. Mainly agricultural, including rice, peanuts (intensive cultivation), sesame, sugarcane, cotton, tobacco; mango and orange groves. Rice and sugar milling, peanut- and sesame-oil extraction. Cotton-milling center at GUDIYATTAM; tanneries at AMBUR and VANIYAMBADI; railroad workshops at ARAKKONAM. Medical (hospital) and psychiatric research center at VELLORE. Under Chola kingdom and successive Indian powers, until it became main arena of 18th-century struggle among French, English, Mysore sultans, and nawabs of Arcot in larger Anglo-French contest for supremacy in India (decisive British victory, 1760, at Wandiwash). Ceded to English 1801 by nawab of Arcot.

North Arlington, borough (2006 population 15,077), BERGEN county, NE NEW JERSEY, 2 mi/3.2 km NNE of NEWARK; 40°47'N 74°07'W. A residential and industrial suburb of Newark, on the PASSAIC River. Settled 1700s, incorporated 1896.

North Atlanta, town, DEKALB county, N central GEORGIA, suburb of ATLANTA; 33°51'N 84°20'W.

North Atlantic Coast Autonomous Region, ZELAYA department, NICARAGUA, occupies the N half of the department, created by the constitution of 1987 to provide for autonomous rule of Miskito, Rama, and Sumo Indian communities. Miskito, Spanish, and English are all widely spoken in this region.

North Atlantic Current, warm ocean current in the N part of the ATLANTIC OCEAN. It is a continuation of the GULF STREAM, the merging point being at 40°00'N 60°00'W. Off the BRITISH ISLES it splits off a branch going S (the CANARY CURRENT), while it continues N along the coast of W and N EUROPE, where it exerts considerable influence upon the climate as far as NW Europe.

North Atlantic Gyre, circuit of mid-latitude currents, centered at c. 30°N–40°N, 45°W, around periphery of North Atlantic Ocean. Consists of GULF STREAM, NORTH ATLANTIC CURRENT, CANARY CURRENT, NORTH EQUATORIAL CURRENT, and ANTILLES CURRENT.

North Attleboro, industrial town, BRISTOL county, SE MASSACHUSETTS, near the RHODE ISLAND line, 20 mi/32 km NNW of FALL RIVER; 41°58'N 71°20'W. Jewelry (since 1807); other manufacturing includes boxes, metal products, eyewear; fish hatchery nearby. The Woodcock tavern dates from 1670. Settled 1669, set off from ATTLEBORO and incorporated 1887.

North Auburn, unincorporated city (2000 population 11,847), PLACER county, central CALIFORNIA, residential suburb 2 mi/3.2 km N of AUBURN, on North Fork AMERICAN RIVER, opposite mouth of RUBICON RIVER; 38°56'N 121°05'W. Cattle, sheep; apiary products; fruit, nuts.

North Auckland, NEW ZEALAND: see NORTHLAND.

North Augusta, city (2006 population 19,926), AIKEN county, SW SOUTH CAROLINA, on the SAVANNAH River. Suburb of AUGUSTA, GEORGIA; 33°31'N 81°57'W. Located in an agricultural region, it is mostly residential. Manufacturing includes reconditioned railroad equipment, building materials, machinery, chemicals, furniture; printing and publishing. Agriculture (cotton, grain, peanuts, peaches, soybeans; livestock, poultry). Many local residents are employed at the Atomic Energy Commission's nearby SA-VANNAH RIVER Nuclear site and in Augusta, Georgia Settled c.1860, incorporated 1906.

North Augusta (NORTH uh-GUHS-tuh), unincorporated village, SE ONTARIO, E central Canada, 12 mi/19 km W of BROCKVILLE, and included in Augusta township; 44°45'N 75°44'W. Dairying, mixed farming.

North Aurora (aw-ROR-ah), village (2000 population 10,585), KANE county, NE ILLINOIS, on FOX RIVER (bridged here), just N of AURORA; 41°47'N 88°19'W.

North Australia, AUSTRALIA: see NORTHERN TERRITORY.

North Ayrshire (ER-shire), county (□ 339 sq mi/881.4 sq km; 2003 population 135,820), SW Scotland, on the Firth of Clyde, includes GREAT CUMBRAE and LITTLE CUMBRAE islands and ARRAN, ⊙ IRVINE; 55°50'N 04°50'W. Includes SKELMORLIE, LARGS, FAIRLIE, Kilbernie, Beith, Dalry, Kilwinning, Steventson, Saltcoats, Ardrossan, and WEST KILBRIDE. Deepwater port and nuclear power station at Hunterston. Tourism, livestock raising, oil refining, and chemicals; 6 mi from Glasgow Prestwick International Airport. Formerly in STRATHCLYDE, abolished in 1996.

North Bačka, Serbian *Severno-Bački Okrug*, district (□ 689 sq mi/1,791.4 sq km; 2002 population 200,140), ⊙ SUBOTICA, VOJVODINA province, N SERBIA; 46°02'N 19°39'E. Includes the municipalities (*opštinas*) of BAČKA TOPOLA, MALI IDJOŠ, and Subotica. Industry (food processing and food products, meatpacking) and agriculture (corn, wheat, sunflowers). Population is ethnically and religiously mixed, including Orthodox and Roman Catholic and other small religious communities. Sometimes called Bačka North.

North Baddesley (BADZ-lee), village (2001 population 12,878), S HAMPSHIRE, S ENGLAND, 3 mi/4.8 km E of ROMSEY; 50°58'N 01°26'W. Residential.

North Bahr Al-Ghazal, Arabic *Shamal Bahr al Ghazal*, state, BAHR AL-GHAZAL region, SW SUDAN; ⊙ AWEIL. Bordered by S DARFUR (N), S KURDOFAN (NE), Warrap (E), and WEST BAHR AL-GHAZAL (S and W) states. Created in 1996 following a reorganization of administrative divisions. Railroad between WAU (W BAHR AL-GHAZAL state) and W KURDOFAN state travels N-S through AWEIL. The area making up N Bahr Al-Ghazal state was part of former BAHR AL-GHAZAL province. Also called Northern Bahr al Ghazal.

North Baldwin, hamlet of Baldwin township, CUMBERLAND county, SW MAINE, c.28 mi/45 km WNW of PORTLAND; 43°50'N 70°41'W. Saddleback Hills nearby.

North Baldy, NEW MEXICO: see MAGDALENA MOUNTAINS.

North Baltimore (BAHL-tuh-mor), village (□ 2 sq mi/5.2 sq km; 2006 population 3,326), Wood county, NW OHIO, 13 mi/21 km S of BOWLING GREEN; 41°11'N 83°40'W. In diversified farming area (corn, wheat). Previously leather goods, machine-shop products, rubber products. Stone quarries nearby. Settled 1834.

North Banat, Serbian *Severno-Banatski Okrug*, district (□ 1,257 sq mi/3,268.2 sq km; 2002 population 165,881), ⊙ KIKINDA, VOJVODINA province, N SERBIA; 45°50'N 20°27'E. Includes the municipalities (*opštinas*) of ADA, Coka, KANJIŽA, KIKINDA, NOVI KNEŽEVAC, and SENTA. Industry (construction material (clay), metal fabrication, agricultural equipment and machinery). Noted for production of clay.

North Bank, administrative division coextensive with KEREWAN local government area (□ 871 sq mi/2,264.6 sq km; 2003 population 172,806), NW THE GAMBIA, along N bank of the GAMBIA RIVER; ⊙ KEREWAN; 13°30'N 16°00'W. Bordered N by SENEGAL, S by WESTERN and LOWER RIVER divisions (the Gambia River, border), E by CENTRAL RIVER division, and W by the ATLANTIC OCEAN. Extends over 80 mi/129 km up the Gambia River from the ATLANTIC. Peanuts and rice are grown and fishing is important. FARAFENNI is the largest town. The W part is linked by ferry to BANJUL; the Trans-Gambian Highway runs N-S through here. Comprises six smaller administrative divisions called districts: Central Baddibu, Jokadu, Lower Baddibu, Lower Niumi, Upper Baddibu, and Upper Niumi.

North Barakpur (bah-ruhk-poor), city, NORTH 24-PARGANAS district, WEST BENGAL state, E INDIA.

North Barrington, village (2000 population 2,918), Lake county, NE ILLINOIS, residential suburb 27 mi/43 km NW of downtown CHICAGO, 2 mi/3.2 km NW of LAKE ZURICH; 42°12'N 88°07'W. Manufacturing greeting cards.

North Bass Island, Ohio; see BASS ISLANDS.

North Battleford (BAT-uhl-fuhrd), city (2006 population 13,190), W SASKATCHEWAN, CANADA, at the confluence of the NORTH SASKATCHEWAN and Battle rivers, opposite BATTLEFORD; 52°46'N 108°17'W. It is the service and distributing center for NW Saskatchewan, which has rich farming, lumbering, and fishing.

North Bay (NORTH BAI), city (□ 122 sq mi/317.2 sq km; 2001 population 52,771), NE ONTARIO, E central Canada, on Lake NIPISSING; 46°18'N 79°27'W. It is the transportation and commercial center of lumbering and mining districts and a popular summer resort. Mining equipment is manufacturing here. It is the site of a large air base.

North Bay, village (2006 population 252), RACINE county, SE WISCONSIN, residential suburb 3 mi/4.8 km NNE of RACINE, on LAKE MICHIGAN; 42°45'N 87°46'W. Urban growth area.

North Beach, summer resort town (2000 population 1,880), CALVERT county, S MARYLAND, on CHESAPEAKE BAY, 19 mi/31 km S of ANNAPOLIS; 38°43'N 76°32'W. Declined in popularity since opening of Bay Bridge (Atlantic beaches). Fires have destroyed a boardwalk, two piers and hotels; hurricanes have eroded the beaches.

North Belle Vernon, borough (2006 population 2,001), WESTMORELAND county, SW PENNSYLVANIA, 22 mi/35 km S of PITTSBURGH, and 2 mi/3.2 km SE of MONESSEN, near MONONGAHELA RIVER; 40°07'N 79°52'W. Agriculture (corn, hay; dairying). Incorporated 1876.

North Bellmore, unincorporated town (2000 population 20,079), NASSAU county, SE NEW YORK, on LONG ISLAND; 40°41'N 73°32'W. Chiefly residential.

North Bellport, village (2000 population 9,007), SUFFOLK county, SE NEW YORK, just E of PATCHOGUE; 40°45'N 72°58'W. Pop. figure includes Hagerman.

North Belmont or **Belmont Junction** (BEL-mahnt), village, GASTON county, S NORTH CAROLINA, a suburb of CHARLOTTE, 2 mi/3.2 km N of BELMONT; 35°16'N 81°02'W.

North Bend, town (2006 population 1,220), DODGE county, E NEBRASKA, 15 mi/24 km W of FREMONT, and on PLATTE RIVER; 41°28'N 96°46'W. Livestock, poultry; grain; manufacturing (fertilizer, pallets).

North Bend, town (2006 population 9,846), COOS county, SW OREGON, on inside of large curve of COOS BAY estuary, c.70 mi/113 km SW of EUGENE, adjoins city of COOS BAY on S side of bay; 43°24'N 124°14'W. Wood preserving; paper mills. Fisheries. Simpson Wayside State Park in NE part of town. North Bend Municipal Airport in NW part of town. Oregon Dunes National Recreation Area to N. Settled 1853, incorporated 1903.

North Bend, unincorporated town, Chapman township, CLINTON county, N central PENNSYLVANIA, 3 mi/4.8 km ENE of RENOVO, on West Branch of SUSQUEHANNA RIVER; 41°21'N 77°42'W. Area surrounded by Sproul State Forest. Hyner Run and Hyner View state parks to E.

North Bend, town (2006 population 4,621), KING county, W central WASHINGTON, suburb 25 mi/40 km ESE of SEATTLE, and on South Fork Snoqualmie River, 2 mi/3.2 km S of junction of South, Middle and North forks to form main stream; 47°30'N 121°48'W.

Farm and dairy products; lumber. Chester Morse Lake, formed by Masonry Dam, to S; Mount Si (4,167 ft/1,270 m) to E. Olallie State Park to SE; Mount Baker–Snoqualmie National Forest to S and E; SNO-QUALMIE PASS (3,022 ft/921 m) and Ski Area to E.

North Bend, village (2006 population 604), HAMILTON county, extreme SW OHIO, on the OHIO RIVER, and 11 mi/18 km W of CINCINNATI; 39°08'N 84°45'W. Benjamin Harrison was born here; William Henry Harrison Memorial State Park nearby. Founded 1789.

North Bennington, village (2006 population 1,356), of BENNINGTON town, BENNINGTON CO., SW VERMONT, 4 mi/6.4 km N of Benington; 42°55'N 73°14'W. Wood products, paper.

North Bergen, residential suburban township, HUDSON county, NE NEW JERSEY, NE of JERSEY CITY; 40°47'N 74°01'W. Light manufacturing. Incorporated 1861.

North Berwick, town, including North Berwick village, YORK county, SW MAINE, on GREAT WORKS RIVER, and 12 mi/19 km S of ALFRED; 43°20'N 70°46'W. Manufacturing (wood and metal products, textiles). Settled c.1630, set off from BERWICK 1831.

North Bessemer, unincorporated village, ALLEGHENY county, W PENNSYLVANIA, 12 mi/19 km ENE of downtown PITTSBURGH, now part of PENN HILLS borough; 40°29'N 79°47'W.

North Beveland (BAI-vuh-land), Dutch *Noord-Beve-land*, region, ZEELAND province, SW NETHERLANDS, bounded N and E by EASTERN SCHELDT estuary, S and W by VEERSE MEER Lake. Connected to SOUTH BEVELAND (S) by ZANDKREEK DAM, to WALCHEREN region (W) by VEERSEGAT DAM, to DUIVELAND region (NE) by Zeelandbrug (bridge), to SCHOUWEN region (NW) by OOSTERSCHELDE DAM. Wissenkerke (near N shore), whose name derives from a beautiful seventeenth-century church, is the chief town. Cattle; sugar beets, vegetables, potatoes, grain.

North Bimini, BAHAMAS : see BIMINI ISLANDS.

North Bonneville, village (2006 population 750), SKA-MANIA county, SW WASHINGTON, 30 mi/48 km E of VANCOUVER, Washington and on COLUMBIA RIVER, opposite BONNEVILLE, OREGON, no bridge connection; 45°38'N 121°58'W. Downstream (1 mi/1.6 km W) from BONNEVILLE DAM. Beacon Rock State Park to W; Gifford Pinchot National Forest to N.

North Borden Island, Canada: see BORDEN ISLANDS.

North Borneo, Malaysia: see SABAH.

Northboro, town (2000 population 60), PAGE county, SW IOWA, near MISSOURI line, 16 mi/26 km SW of CLARINDA; 40°36'N 95°17'W. In agricultural region.

Northborough, town, including Northborough village, WORCESTER county, E central MASSACHUSETTS, 10 mi/16 km ENE of WORCESTER; 42°19'N 71°39'W. Manufacturing (fiberoptics, electronics equipment, rubber products). Settled c.1672; incorporated 1766.

North Bosque River, Texas: see BOSQUE RIVER.

Northbourne (NAWTH-buhn), village (2001 population 793), E KENT, SE ENGLAND, 3 mi/4.8 km W of DEAL; 51°13'N 01°21'E. Former coal-mining area. Has 12th-century church and 14th-century Northbourne Court.

North Brabant (brah-BAHNT), Du *Noord-Brabant*, province (c.1,920 sq mi/4,970 sq km; 2007 estimated population 2,419,042), S NETHERLANDS, bordering on BELGIUM in the S and on LIMBURG province, and GERMANY in the E; ⊙ 's-HERTOGENBOSCH (Den Bosch). Other cities include TILBURG, EINDHOVEN, BERGEN OP ZOOM, and BREDA. Bounded N and NE by MEUSE RIVER, NW by LOWER MERWEDE, NEW MER-WEDE, HOLLANDS DIEP, VOLKERAK, and KRAMMER river channels; bounded W by SCHELDE-RIJN CANAL and E end of EASTERN SCHELDT estuary. German border 3 mi/4.8 km to E; province surrounds small Belgian enclave of BAARLE-HERTOG 13 mi/21 km S of Tilburg. The province has fertile soil in river margins in N and W, but elsewhere it has mainly infertile sandy soils. Dairying; cattle, hogs, poultry; wheat,

barley, oats, beans, peas, onions, beets, soybeans, potatoes, and sugar beets are grown. The history of the province was that of Brabant until the late sixteenth century, when the Dutch revolted against the harsh Spanish rule. As a result of the Spanish reconquest of the larger part of the duchy, Brabant was divided by the Peace of Westphalia (1648) between the Spanish (later Austrian) Netherlands and the United Provsinces of the Netherlands. North Brabant, the smaller part occupied by the United Provinces, remained Catholic. It was administered by the United Provinces as a territory and was not granted a seat in the States-General. In 1795, North Brabant became a province of the Netherlands During the nineteenth century most heathlands were reclaimed and settled; after 1900, settlement subsided. Nearly 50% of the population has since become urban.

North Braddock (BRA-duhk), borough (2006 population 5,915), ALLEGHENY county, W PENNSYLVANIA, residential suburb 8 mi/12.9 km ESE of downtown PITTSBURGH, near the MONONGAHELA RIVER; 40°23'N 79°50'W. Andrew Carnegie's first steel plant was built here in 1875. The borough was the site of General Edward Braddock's defeat in the last conflict of the French and Indian wars and of a mass meeting of farmers instituting the Whiskey Rebellion. Incorporated 1897.

North Branch, town (2000 population 8,023), CHISAGO county, E MINNESOTA, 39 mi/63 km N of ST. PAUL, on North Branch Sunrise River, surrounded by municipality of BRANCH; 45°30'N 92°58'W. Grain; cattle, poultry; dairying; manufacturing (metal stampings, feed ingredients, concrete, mops, tool and die, physical training devices; grain mill).

North Branch, village (2000 population 1,027), LAPEER county, E MICHIGAN, 14 mi/23 km NNE of LAPEER; 43°13'N 83°11'W. In area producing livestock; grain, apples, soybeans; dairying; manufacturing (screw machine products, rolled thread and nuts).

North Branford, town, NEW HAVEN county, S CON-NECTICUT, on the Branford River; 41°21'N 72°46'W. A large traprock quarry is here, as is some light industry. Settled c.1680, incorporated 1831.

North Brentwood, town (2000 population 469), PRINCE GEORGES county, central MARYLAND, just NE of WASHINGTON, D.C.; 38°57'N 76°57'W. Originally called Highlands, the name was changed after the Civil War. Incorporated in 1912, it is adjacent to BRENTWOOD. Incorporated in 1924.

Northbridge, town, WORCESTER county, S MASSACHU-SETTS, on the BLACKSTONE RIVER; 42°08'N 71°39'W. It includes the villages of Whitinsville (1990 population 5,639) and Linwood. Manufacturing (wood furniture, paper products, stereo components); former textile-machinery manufacturing. Settled 1704; set off from UXBRIDGE and incorporated 1772.

North Bridgton, MAINE: see BRIDGTON.

North Brisbane, suburb of BRISBANE, SE QUEENSLAND, NE AUSTRALIA.

Northbrook, village (2000 population 33,435), COOK county, NE ILLINOIS, a suburb NW of CHICAGO; 42°07'N 87°49'W. It was incorporated as Shermerville in 1901 and was reincorporated as Northbrook in 1923. Largely residential, it has some industry and research laboratories and is an insurance center. Once a farming community, Northbrook developed industry after the coming of a railroad in 1871. Botanical gardens and a forest preserve are just E of the village. Settled 1836.

North Brookfield, town, including North Brookfield village, WORCESTER county, central MASSACHUSETTS, 14 mi/23 km W of WORCESTER; 42°16'N 72°05'W. Rubber goods, beverages, poultry feed; dairying; poultry; fruit. Settled 1664, incorporated 1812.

Northbrook Island, in SW FRANZ JOSEF LAND, ARCH-ANGEL oblast, extreme N European Russia, in the ARCTIC OCEAN; 20 mi/32 km long, 12 mi/19 km wide.

80°00'N 51°00'E. Terminates W in Cape FLORA. Discovered in 1880 by the British explorer Leigh Smith.

North Brother Island, SE NEW YORK, in EAST RIVER, off S shore of BRONX borough of NEW YORK city; 40°48'N 74°00'W. Covers 13 acres/5 ha.

North Brother Mountain (4,143 ft/1,263 m), PISCATA-QUIS county, N central MAINE, 25 mi/40 km NW of MILLINOCKET, in Katahdin State Game Preserve.

North Broughton Island, Canada: see BROUGHTON ISLAND.

North Brunswick, township, MIDDLESEX county, N NEW JERSEY, 3 mi/4.8 km S of NEW BRUNSWICK; 40°27'N 74°28'W. Incorporated 1798.

North Bruny, AUSTRALIA: see BRUNY.

North Buena Vista, town (2000 population 124), CLAYTON county, NE IOWA, on the MISSISSIPPI, and 26 mi/42 km ESE of ELKADER; 42°40'N 90°57'W. In agricultural and dairying region.

North Buganda (boo-GAHN-duh), former province (□ 13,028 sq mi/33,872.8 sq km), central UGANDA; capital was BOMBO. Was bordered N by LAKE KWANIA and LAKE KYOGA, W by WESTERN province, S by CENTRAL province and LAKE VICTORIA, E by BUSOGA province. Drained by LUGOGO, MAYANJA, Nabakazi rivers. Consisted of Lake Wamala, Mt. Kakade. Agriculture included cotton; cattle; groundnuts, cassava, sugarcane, banana. Tungsten, tantalite deposits. Main centers were Bombo, MUBENDE, LUGAZI, Mukoro.

North Cachar Hill (KAH-chahr), district (□ 1,887 sq mi/4,906.2 sq km), S central ASSAM state, NE INDIA; ⊙ HAFLONG. Largely coextensive with BARAIL RANGE; bordered E by NAGALAND and MANIPUR states, W by MEGHALAYA state, S by CACHAR district, from which it was separated in 1950.

North Caicos (□ 54 sq mi/140.4 sq km), TURKS AND CAICOS ISLANDS, crown colony of GREAT BRIT-AIN, WEST INDIES, NW of MIDDLE CAICOS island; c.12 mi/19 km long; 21°50'N 72°0'W. Most populous of the group, with KEW and Bottle Creek villages. Tourism, swimming, scuba diving, sport fishing. Noted for its flamingos. Formerly site of many plantations, now in ruins.

North Caldwell, residential township (2006 population 7,207), ESSEX county, NE NEW JERSEY, 6 mi/9.7 km SW of PATERSON; 40°51'N 74°15'W. Incorporated 1898.

North Canaan, town, LITCHFIELD county, NW CON-NECTICUT, in TACONIC MOUNTAINS, on MASSACHU-SETTS state line, and 17 mi/27 km NW of TORRINGTON; 42°01'N 73°17'W. Magnesium mining and reduction; agriculture; limestone quarrying. Includes villages of CANAAN (S) and East Canaan.

North Canadian River, 760 mi/1,223 km long, NEW MEXICO, TEXAS, OKLAHOMA; rises in NW central UNION county (extreme NE New Mexico), E of Des Moines; flows E near N boundaries of Kiowa National Grassland (New Mexico) and Rita Blanca National Grassland (Oklahoma), through far W part of Oklahoma Panhandle. Drops S briefly in passing through extreme N Texas, continuing E in Oklahoma, past Guymon and Beaver, leaving Panhandle and turning SE past Woodward (roughly paralleling CANADIAN RIVER for remainder of its course). River continues through CANTON LAKE reservoir, past Watonga, through El Reno and Oklahoma City, past Shawnee and passing S of Henryetta, joining Canadian River in EUFAULA LAKE reservoir, forming large N arm of lake. Eufaula Lake also receives Deep Fork Canadian River from NW, which forms its own arm. Federal dams and reservoirs on the river and its tributary, WOLF CREEK, are part of the ARKANSAS RIVER basin project for flood control and other purposes.

North Canara, INDIA: see UTTAR KANNAD.

North Canton (KAN-tuhn), city (□ 6 sq mi/15.6 sq km; 2006 population 16,755), STARK county, NE OHIO; suburb 3 mi/4.8 km NNW of CANTON; 40°52'N

81°23'W. Diversified light manufacturing. Walsh University is here. Settled c.1815, incorporated as a city 1961.

North Canton (KAN-tuhn), village, CHEROKEE county, NW GEORGIA, near CANTON and ALLATOONA Reservoir; 34°14'N 84°29'W.

North Cape, Far North district, NORTH ISLAND, northernmost point of NEW ZEALAND; 34°25'S 173°05'E.

North, Cape, NE extremity of CAPE BRETON ISLAND, NE NOVA SCOTIA, CANADA, on Cabot Strait, 60 mi/97 km NNW of SYDNEY; 47°2'N 60°25'W. Elevation 984 ft/ 300 m. Communications linkpoint, first undersea telegraph cable from NEWFOUNDLAND (Canada) 1856.

North Cape, promontory, rising steeply c.1,000 ft/305 m from the ARCTIC OCEAN, near but not at the N end of Magerøya island, FINNMARK co., N Norway. Although Magerøya is separated by a narrow channel from the mainland, North Cape, at lat. 71° 10'N, is considered to be the northernmost important point of the European continent. The northernmost point actually situated on the mainland is Cape Nordkyn. The North Cape is a traditional stop for tourist steamers. Also known as Nordkapp.

North Cape, RUSSIA: see SHMIDT, CAPE.

North Caribou Lake (KA-ri-boo), NW ONTARIO, E central Canada, in NW PATRICIA district; 22 mi/35 km long, 14 mi/23 km wide; 52°49'N 90°46'W. Elevation 1,060 ft/323 m. Drains N into Severn River.

North Carolina (NORTH ka-ro-LEI-nuh), state (□ 53,821 sq mi/139,397 sq km; 1995 estimated population 7,195,138; 2000 population 8,049,313), SE UNITED STATES, one of the original 13 U.S. states; ⊙ RALEIGH; the largest city is CHARLOTTE; other major cities include GREENSBORO, WINSTON-SALEM, DURHAM, RALEIGH, and ASHEVILLE; 35°28'N 79°10'W. North Carolina is known as the "Old North State" and as the "Tar Heel State."

Geography

The state is bounded on the N by VIRGINIA, on the E by the ATLANTIC OCEAN, on the S by SOUTH CAROLINA and GEORGIA, and on the W by TENNESSEE. Serving as a buffer against the Atlantic is a long chain of sand barrier islands and peninsulas referred to as the OUTER BANKS, with constantly shifting sand dunes, which open and close passages, or inlets, through the banks. Prominent capes (three)—HATTERAS, LOOKOUT, and FEAR—point SE from the island chain and lie prone to the rigors of Atlantic storms. Between the Banks and the mainland are large lagoons—ALBEMARLE SOUND and PAMLICO SOUND being the largest—that receive the CHOWAN, ROANOKE, TAR, and NEUSE rivers, each forming estuaries which protrude inland; the CAPE FEAR RIVER in the SE and its estuary open directly into the ocean. WILMINGTON, North Carolina's chief port, lies at the head of Cape Fear River estuary. Most of North Carolina's cities lie at least 100 mi/161 km inland from the coast. The INTRACOASTAL WATERWAY follows the Atlantic coastline through a channel less than 1 mi/1.6 km inland in the S, and through canals in the N linking the sounds and estuaries behind the Outer Banks.

Geography: The Tidewater

The mainland bordering the sounds is low, flat tidewater country, often swampy, including the (Great) DISMAL SWAMP in the NE (Virginia-North Carolina), East Dismal Swamp in the E, and Green Swamp in the SE. In the upper coastal plain the land rises gradually from the tidewater level, reaching 500 ft/152 m at the fall line.

Geography: The Piedmont

There begins the PIEDMONT region, a rolling hill country with many swift streams such as the BROAD, CATAWBA, and PEE DEE or Great Pee Dee rivers. Power from the Piedmont fall line contributed greatly to the development of the textile mill industry in both Carolinas and the relocation of the industry from

New England by the 1960s. The hydroelectric power these rivers generate made this an important manufacturing area. The Piedmont supports most of the state's population and has its largest cities. At the W edge of the Piedmont, the land rises abruptly in the BLUE RIDGE, and in the S, dips down to several basins, and rises again in the GREAT SMOKY MOUNTAINS. Asheville is the leading urban center of this mountain region, with Mount MITCHELL (6,684 ft/2,037 m) the highest peak E of the MISSISSIPPI RIVER. The FRENCH BROAD RIVER, the WATAUGA, and other rivers rising W of the Blue Ridge flow into the Mississippi system, almost all via the TENNESSEE RIVER. Several reservoirs, including B. EVERETT JORDAN LAKE (Jordan Lake) (center; on HAW and New Hope rivers), FALLS LAKE RESERVOIR (center; Neuse River), Lake NORMAN (W center; Catawba River), part of John H. Keen Reservoir (N state line with Virginia; on the ROANOKE RIVER). North Carolina, in the warm temperate zone, has a mild, generally uniform climate, and the rainfall is abundant and well distributed, although summer dry spells are common.

Economy

The state leads the nation in the production of tobacco and is a major producer of textiles and furniture. While it grows 40% of all U.S. tobacco, it is also the country's second-largest producer of poultry and hogs. The continuing trend toward diversification has moved poultry and hogs ahead of tobacco in total value. Cattle; dairying; corn, soybeans, cotton, peanuts, and eggs are also important. Plentiful forests supply the thriving paper products and lumber industries. North Carolina has long been a major textile manufacturer, producing cotton, knit, and synthetic goods. Other leading manufacturing are electrical machinery, computers, and chemicals. The state also has mineral resources: it leads the nation in the production of feldspar, mica (primary source during World War II), and lithium materials and produces substantial quantities of olivine, crushed granite, talc, clays, and phosphate rock. There are valuable coastal fisheries, with shrimp, menhaden, and crabs the principal catches.

Tourism

North Carolina's congenial climate, its many miles of beaches, and its beautiful mountains attract large numbers of visitors and vacationers each year. Chief among the tourist attractions are the Cape HATTERAS and Cape LOOKOUT national seashores in the E, part of the BLUE RIDGE PARKWAY in the NW, and part of the GREAT SMOKY MOUNTAINS National Park in the W. Wildlife abounds in the national forests, Nantahala and Pisgah (W), Uwharrie (center), Croatan (E); also Blade Lakes State Forest (SE). The Appalachian Trail (or APPALACHIAN NATIONAL SCENIC TRAIL) closely follows most of Tennessee-North Carolina state boundary then crosses the SW end of the state to Georgia. The main range of the Appalachians on the W boundary remains a formidable barrier to the W states. Places of historic interest include FORT RALEIGH NATIONAL HISTORIC SITE, on ROANOKE ISLAND and the WRIGHT BROTHERS NATIONAL MEMORIAL, at KITTY HAWK, both in the E; the CARL SANDBURG HOME National Historic Site, at FLAT ROCK, in the SW; and GUILFORD COURTHOUSE (N center) and MOORES CREEK (SE) national battlefields. Among the largest military reservations in the nation are FORT BRAGG, near FAYETTEVILLE in the SE center, and CAMP LEJEUNE Marine base, near JACKSONVILLE, in the SE, at the mouth of the NEW RIVER. North Carolina has one Indian reservation, the Eastern Cherokee (also called Qualla Boundary) Indian Reservation in the W, bordering on the Great Smoky Mountains National Park.

History to 1711

The treacherous coast was explored by Verrazano in 1524, and possibly by some Spanish navigators. In the 1580s, Sir Walter Raleigh attempted unsuccessfully to

establish a colony on one of the islands (Roanoke Island). The first permanent settlements were made (c.1653) around Albemarle Sound by colonials from Virginia. Meanwhile, Charles I of England had granted (1629) the territory S of Virginia between the 36th and 31st parallels (named Carolina in the king's honor) to Sir Robert Heath. Heath did not exploit his grant, and it was declared void in 1663. Charles II reassigned the territory to eight court favorites, who became the "true and absolute Lords Proprietors" of Carolina. In 1664, Sir William Berkeley, governor of Virginia and one of the proprietors, appointed a governor for the province, which after 1691 was known as North Carolina. By 1700 there were only some 4,000 people, predominantly of English stock, along Albemarle Sound. There, with the help of indentured servants and African and Native American slaves, they raised tobacco, corn, and livestock, mostly on small farms. The people were semi-isolated; only vessels of light draft could negotiate the narrow and shallow passages through the island barriers, and communication by land was almost impossible, except with Virginia, and even then swamps and forests made it difficult.

History: 1711 to 1771

There was some trade (primarily with Virginia, NEW ENGLAND, and BERMUDA), and in 1711, North Carolina was made a separate colony. The destructive war with Native Americans of the Tuscarora tribe broke out that year. The Tuscarora were defeated, and in 1714 the remnants of the tribe moved N to join the Iroquois Confederacy. A long, bitter boundary dispute with Virginia was partially settled in 1728 when a joint commission ran the state line 240 mi-386 km inland. The British government made North Carolina a royal colony in 1729. Thereafter the region developed more rapidly. The Native Americans were gradually pushed back over the Appalachians as the Piedmont was increasingly occupied. Germans and Scotch-Irish followed the valleys down from PENNSYLVANIA, and Highland Scots settled along the Cape Fear River. These varied racial and national elements, in addition to smaller groups of Swiss, French, and Welsh who had migrated to the region earlier in the century, gradually amalgamated. In 1768, the back-country farmers, justifiably enraged by the excessive taxes imposed by a legislature dominated by the E aristocracy, organized the Regulator Movement in an attempt to effect reforms.

History: 1771 to 1784

The insurgents were suppressed at ALAMANCE in 1771 by the provincial militia led by Governor William Tryon, who executed seven of the Regulators. After the outbreak of the American Revolution, royal authority collapsed. A provisional government was set up, the disputed Mecklenburg Declaration of Independence was allegedly promulgated (May 1775), and the provincial congress instructed (April 12, 1776) the colony's delegates to the Continental Congress to support complete independence from Britain. Most Loyalists, including Highland Scots, fled North Carolina after their defeat (February 27, 1776) at the battle of Moores Creek Bridge near WILMINGTON. The British, however, did not give up hope of Tory assistance in the state until their failure in the Carolina Campaign (1780–1781). One explanation for the designation of North Carolinians as "Tar Heels" is said to have originated during that campaign when patriotic citizens poured tar into a stream across which Cornwallis's men retreated, and the British emerged with the substance sticking to their heels. Settlements had been established beyond the mountains before the Revolution and were increased after the war.

History: 1784 to 1839

In 1784, North Carolina ceded its W lands to the U.S., spurring the transmontane people to organize a new, short-lived government (see FRANKLIN, a former

Area is shown by the symbol □, and capital city or county seat by ⊙.

state). Within the year North Carolina repealed the act ceding the land; however, the cession was re-enacted in 1789, and that territory became (1796) the state of Tennessee. North Carolina opposed a strong central government and did not ratify the Constitution until November 1789, months after the new U.S. government had begun to function. Little social and economic progress was made under the state's undemocratic constitution (framed in 1776), which largely served the interests of the politically dominant, tidewater planter aristocracy, and North Carolina appeared to be on the verge of revolution. In 1835, however, the W part of the state, now its most populous section, finally succeeded in enacting a constitution that abolished the property and religious qualifications for voting and holding office (except for Jews) and provided for the popular election of governors. In the same year began the final forced removal of most of the Cherokee; but to check the steady, voluntary migration of whites, internal improvements, especially the building of railroads and plank roads, were effected.

History: 1839 to 1865

The Public School Law (1839) inaugurated free education, and other important reforms were instituted. The period of progress continued until the Civil War. Few North Carolinians held slaves, and considerable antislavery sentiment existed until the 1830s, when the organized agitation of Northern abolitionists began, provoking a defensive reaction that North Carolinians shared with most Southerners. Yet it was a native of the state, Hinton Rowan Helper, who made the most notable Southern contribution to antislavery literature. Not until President Lincoln's call for troops after the firing on FORT SUMTER did the state secede and join (May 1861) the Confederacy. The coast was ideal for blockade-running, and the last important Confederate port to fall (January 1865) was Wilmington (FORT FISHER). Governor Zebulon B. Vance zealously defended the state's rights against what he considered encroachments by the Confederate government. Although many small engagements were fought on North Carolina soil, the state was not seriously invaded until almost the end of the war when General William Sherman and his huge army moved N from Georgia.

History: 1865 to 1898

After engagements at Averasboro and BENTONVILLE in March, 1865, Confederate General J. E. Johnston surrendered (April 26, 1865) to Sherman near Durham; next to Lee's capitulation at APPOMATTOX, it was the largest (and almost the last) surrender of the war. In May, 1865, President Johnson applied his plan of Reconstruction to the state. The radical Republicans in Congress, however, adopted their own scheme in 1867, and the Carolinas, organized as the second military district, were again occupied by Federal troops. The Reconstruction constitution of 1868 abolished slavery, removed all religious tests for holding office, and provided for the popular election of all state and county officials. In 1871 the legislature, with conservatives again in control, impeached and convicted Governor William H. Holden. The often maligned period of Reconstruction actually saw the beginning of the modern state, with a tremendous rise in industry in the Piedmont. Increased use of tobacco in the Civil War stimulated the growth of tobacco manufacturing, first centered at Durham, and the introduction of the cigarette-making machine in the early 1880s was an immense boon to the industry, creating tobacco barons such as James B. Duke and River J. Reynolds. Agriculture, however, was in a critical condition. The old plantation system was replaced by farm tenancy, which long remained the dominant system of holding land. Much farm property was destroyed, credit was lacking, and transportation broke down. The nation-wide agrarian revolt

reached North Carolina in the Granger Movement (1875), the Farmers' Alliance (1887), and the Populist Party, which united with the Republicans to carry the state elections in 1894 and 1896.

History: 1898 to 1954

However, the Fusionists were blamed for the rise of black control in many tidewater towns and counties, and in the election of 1898, when the Red Shirts, like the Ku Klux Klan of Reconstruction days, were active, the Democrats regained control. The turn of the century marked the beginning of a new progressive era, typified by the successful airplane experiments of the Wright Brothers near Kitty Hawk. The crusade for public education for both races led by Governor Charles B. Aycock, elected in 1900, achieved wide results, and new interest was created in developing the state's agricultural and industrial resources. But one old pattern was strengthened when a suffrage amendment, the "grandfather clause" assuring white supremacy, was added (1900) to the state constitution. Since World War I the state government has increasingly followed a policy of consolidation and centralization, taking over the public school system and the supervision of county finance and roads. A huge highway development program, begun by the counties in 1921, was assumed by the state (1931) when the counties could no longer meet the costs. Expenditures for higher education were greatly increased, and the three major state educational institutions were merged into a greater University of North Carolina.

History: 1954 to Present

North Carolina, more than many other Southern states, was able to make a peaceful adjustment to integration in the public schools following the Supreme Court's desegregation ruling in 1954. Industrialization burgeoned after World War II, and in the 1950s the value of manufactured goods surpassed that of agriculture for the first time. In that period, and especially in the administration of Governor Luther H. Hodges, over $1 billion in new industry was established, making North Carolina the leading industrial state of the South. This industrialization continued during the 1960s and early 1970s, increasing at a rate unmatched by any other Southern state. The Charlotte-Douglas and Raleigh-Durham airports were both transformed into major air traffic hubs during the 1980s, reflecting the tremendous growth (most of it suburban) in those metropolitan areas, and an Interstate Highway system criss-crosses the state. In addition, North Carolina has been able recently to shift from its traditional, low-skill industries to high-technology industries. With the cooperation of Duke University, the University of North Carolina, and North Carolina State University, Governor Hodges was instrumental in establishing the RESEARCH TRIANGLE PARK, between Raleigh and Durham in the late 1950s, which helped boost North Carolina's standing in technological research and development. North Carolina experienced substantial population growth in the 1980s, becoming the tenth-largest state in the nation after the 1990 census. The state's other notable institutions of higher learning include East Carolina University, at GREENVILLE; Appalachian State University, at BOONE; and Wake Forest University, at Winston-Salem.

Government

North Carolina's first constitution was adopted in 1776. Its present constitution dates from 1868 but was thoroughly revised in 1875–1876 as a result of Reconstruction experiences; it has been amended many times since. The state's executive branch is headed by a governor elected for a four-year term. The current governor is Mike Easley. North Carolina's bicameral general assembly has a senate with fifty members and a house with 120 members, all elected for two-year terms. The state elects two Senators and twelve Re-

presentatives to the U.S. Congress and has fourteen electoral votes.

North Carolina has 100 counties: ALAMANCE, ALEXANDER, ALLEGHANY, ANSON, ASHE, AVERY, BEAUFORT, BERTIE, BLADEN, BRUNSWICK, BUNCOMBE, BURKE, CABARRUS, CALDWELL, CAMDEN, CARTERET, CASWELL, CATAWBA, CHATHAM, CHEROKEE, CHOWAN, CLAY, CLEVELAND, COLUMBUS, CRAVEN, CUMBERLAND, CURRITUCK, DARE, DAVIDSON, DAVIE, DUPLIN, DURHAM, EDGECOMBE, FORSYTH, FRANKLIN, GASTON, GATES, GRAHAM, GRANVILLE, GREENE, GUILFORD, HALIFAX, HARNETT, HAYWOOD, HENDERSON, HERTFORD, HOKE, HYDE, IREDELL, JACKSON, JOHNSTON, JONES, LEE, LENOIR, LINCOLN, MCDOWELL, MACON, MADISON, MARTIN, MECKLENBURG, MITCHELL, MONTGOMERY, MOORE, NASH, NEW HANOVER, NORTHAMPTON, ONSLOW, ORANGE, PAMLICO, PASQUOTANK, PENDER, PERQUIMANS, PERSON, PITT, POLK, RANDOLPH, RICHMOND, ROBESON, ROCKINGHAM, ROWAN, RUTHERFORD, SAMPSON, SCOTLAND, STANLY, STOKES, SURRY, SWAIN, TRANSYLVANIA, TYRRELL, UNION, VANCE, WAKE, WARREN, WASHINGTON, WATAUGA, WAYNE, WILKES, WILSON, YADKIN, and YANCEY.

North Carrollton, village (2000 population 499), CARROLL county, central MISSISSIPPI, 15 mi/24 km E of GREENWOOD, and 1 mi/1.6 km N of CARROLLTON; 33°31′N 89°55′W. Agriculture (cotton, corn; cattle); manufacturing (furniture).

North Carver, MASSACHUSETTS: see CARVER.

North Cascades National Park, SKAGIT and CHELAN counties, N WASHINGTON. Bounded by Mount Baker–Snoqualmie National Forest, E, W, and SW; by Wenatchee National Forest, S; by CHELAN NATIONAL RECREATION AREA in SE; N unit bounded by CANADA (BRITISH COLUMBIA) to N. Covers 504,781 acres/204,285 ha. Area of great alpine scenery in the CASCADE RANGE; North and South units separated by ROSS LAKE NATIONAL RECREATION AREA. Authorized 1968.

North Caspian Basin, KAZAKHSTAN: see EMBA BASIN.

North Castle, town, WESTCHESTER county, SE NEW YORK, abutting NW half of New York state border with SW CONNECTICUT, c.35 mi/56 km of NEW YORK city; 41°10′N 73°43′W. Exclusive, suburban residential area (including the hamlets of ARMONK, Banksville, East Patent, Middle Patent, North White Plains, Payne's Corners, and West Patent); site of IBM's international headquarters.

North Catasauqua (KA-tah-SAW-kwah), borough (2006 population 2,855), NORTHAMPTON and LEHIGH counties, E PENNSYLVANIA, residential suburb 4 mi/6.4 km N of downtown ALLENTOWN, on LEHIGH RIVER; 40°39′N 75°28′W. Lehigh Valley International Airport to E. Incorporated 1908.

North Caucasia, Russian *Kavkaz,* region and mountain system, SE European Russia, in KRASNODAR (including the ADYGEY REPUBLIC) and STAVROPOL territories, ROSTOV oblast, NORTH OSSETIA, CHECHEN REPUBLIC, INGUSH REPUBLIC, the Republic of KALMYKIA-KHALMG-TANGEH, and the KARACHEVO-CHERKESS, KABARDINO-BALKAR, and DAGESTAN republics. Composed mainly of plain (steppe) areas, begins at the Manych Depression and rises to the S, where it runs into the main mountain range, the Caucasus Mountains. The mountain system in this region is a series of chains running NW-SE, including Mount Elbrus (18,481 ft/5,633 m), the Dykh-Tau (17,050 ft/5,197 m), the Koshtan-Tau (16,850 ft/5,134 m), and Mount Kazbek (16,541 ft/5,042 m). The Caucasus are crossed by several passes, notably the Mamison and the Daryal, among others. Home to over 100 nationalities, the largest being Russians, Chechens, Ukrainians, Ossetians, Karbards, Circassians, Dagestani; also Armenians, Ingush, Avars, Darghins, Kumyks, and Lezghinians. Oil has been the major product of the region, with fields at GROZNYY (Chechnya) and MAYKOP (Adygey Republic) and pipelines from Groznyy to the port of MAKHACHKALA (Dagestan Republic) and to ROSTOV-NA-DONU (Rostov oblast). On the mountain slopes, which are covered by pine and deciduous trees, livestock raising. In the valleys, grain (corn, wheat), sugar beets, sunflowers, rice, hemp, vegetables, fruits, and livestock are raised. Along the BLACK SEA coast, in Krasnodar Territory between ANAPA and SOCHI, there are many resorts and summer homes. PYATIGORSK (Stavropol Territory), MINERAL'NYYE VODY, and KISLOVODSK are among the best known health and mineral resorts. Major cities are Groznyy, VLADIKAVKAZ (formerly known as Ordzhonikidze, North Ossetia), Krasnodar, Stavropol, Sochi, NAL'CHIK, Makhachkala, and NOVOROSSIYSK. The Adyg (their own term) is a general name for the once-numerous Circassians and includes present-day Adygeians and Kabards. Seminomadic Sarmatians inhabited the area in the 1st century B.C.E., when Alani, or Alans, whose language was related to Iranian, broke away and gradually adopted an agricultural, livestock-raising way of life. Their feudal state arose after the collapse of the Jewish Khazar kingdom (7th–10th century). Defeated by the Mongols in the 13th century, they migrated to the central Caucasus hills and mixed with the local population, to become the ancestors of the modern Ossetians.

North Caucasus Economic Region (□ 137,104 sq mi/ 355,100 sq km; 1995 population 15,000,000), SE European Russia, in KRASNODAR (including the ADYGEY REPUBLIC) and STAVROPOL territories, ROSTOV oblast, NORTH OSSETIA, CHECHNYA, INGUSHETIA, the Republic of KALMYKIA-KHALMG-TANGEH, and the KARACHEVO-CHERKESS, KABARDINO-BALKAR, and DAGESTAN republics. Divided into 3 areas: the plains (steppes), foothills (home to most of population and industry), and mountains. Elevation ranges from 92 ft/28 m below sea level along the CASPIAN SEA coastline up to 18,510 ft/5,642 m on Mount Elbrus. Foreign trade via BLACK SEA ports of NOVOROSSIYSK and TUAPSE (Krasnodar Territory). Extensive railroad network (2 trunk lines intersect at TIKHORETSK, Krasnodar Territory); ferry across the Kerch' Strait links Caucasus railroad with CRIMEA (Ukraine). Volga-Don Ship Canal for water transport. Extensive reserves of natural gas and oil (GROZNYY, Chechnya; NEVINNOMYSSK, Stavropol Territory; MAYKOP, Krasnodar Territory; and VOLGODONSK and KAMENSK-SHAKHTINSKIY, Rostov oblast; refined locally and transported via vast pipeline system) and coal (Donets Coal Basin); also, lead and zinc ores (SADON, Chechnya and Vladikavkaz), tungsten and molybdenum (TYRNYAUZ, Kabardino-Balkar Republic), rock salt (Shedok, Krasnodar Territory), calcareous raw materials for the chemical and cement industries (near Novorossiysk) and building materials. Rostov oblast (N), including East Donbass, most developed industrially. Food processing (esp. canning, oil, wine, groats), machine building (esp. for agr.), ferrous and nonferrous metallurgy; also, wool and leather products; nitrogen fertilizers, chemicals, synthetic fats. Black Sea coast was an important recreational and resort area under the soviet rule. A major food-producing area, growing grains, oil plants, fruits, vegetables, tea, and tobacco; also, meat and wool. Energy resources include natural gas, coal, petroleum, and water power. Growing population due to both a high natural increase and a steady influx of people from other regions. The 6 major cities are TAGANROG and ROSTOV-NA-DONU (Rostov oblast), KRASNODAR and SOCHI (Krasnodar Territory), VLADIKAVKAZ (North Ossetia), and GROZNYY (Chechnya).

North Central Province, administrative division (□ 3,768 sq mi/9,796.8 sq km), N SRI LANKA; ⊙ ANURADHAPURA; 08°20′N 80°30′E. Undulating, jungle-covered plain, with a few isolated peaks and short hill ranges extending NE-SW. Served by extensive irriga-

tion tanks near POLONNARUWA, MINNERIYA, KALAWEWA, and Anuradhapura; agriculture (rice, vegetables). Archaeological landmarks at Anuradhapura, Polonnaruwa, MIHINTALE. Created 1886.

North Channel (NOHRTH CHAN-uhl), strait, c.75 mi/ 121 km long, between Northern Ireland and Scotland, connecting the IRISH SEA with the ATLANTIC OCEAN. It is 13 mi/21 km across at its narrowest point.

North Channel (NORTH CHA-nuhl), central ONTARIO, E central Canada, N arm (120 mi/193 km long, 1 mi/2 km–20 mi/32 km wide) of Lake HURON, between N shore of lake and MANITOULIN ISLANDS; connects SAINT MARYS RIVER (W) and GEORGIAN BAY (E). Crossed by road and railroad bridge to Manitoulin Island at LITTLE CURRENT.

North Charleroi (SHAHR-luh-roi), borough (2006 population 1,337), WASHINGTON county, SW PENNSYLVANIA, suburb 1 mi/1.6 km NW of CHARLEROI, on MONONGAHELA RIVER (bridged; Lock No. 4 to SE); 40°08′N 79°54′W. Agriculture (grain; livestock; dairying). Incorporated 1894.

North Charleston, city (2006 population 87,482), CHARLESTON county, SE SOUTH CAROLINA, suburb 5 mi/8 km NNW of downtown CHARLESTON, between COOPER and ASHLEY rivers; 32°54′N 80°02′W. Railroad junction. Manufacturing includes military avionics, construction materials, consumer goods, food and beverage processing, paper bags, ship conversion and repair, textiles, fabricated metal products, machinery, steel drums, plastic products, ferrochrome alloys, therapy units, electronics. Charleston Naval Base and U.S. Army Depot to SE, Charleston Airforce Base to NW; Charleston International Airport to W.

North Chatham, MASSACHUSETTS: see CHATHAM.

North Cheek, ENGLAND: see ROBIN HOOD'S BAY.

North Chelmsford, MASSACHUSETTS: see CHELMSFORD.

North Chicago, industrial city (2000 population 35,918), Lake county, NE ILLINOIS; 42°19′N 87°51′W. Its economy is closely intertwined with the neighboring city of WAUKEGAN, which has a harbor on Lake MICHIGAN. Pharmaceuticals, medical diagnostic testing equipment, chemicals, steel, and automobile parts are among the many manufactures. A sit-down strike at a steel plant here in 1937 led to a U.S. Supreme Court decision (1939) ruling sit-down strikes illegal. Adjacent to the city is the GREAT LAKES NAVAL Training Center. Incorporate 1895.

North Chili (CHEI-LEI), hamlet, MONROE county, W NEW YORK, 10 mi/16 km WSW of ROCHESTER; 43°07′N 77°48′W. Seat of Roberts Wesleyan College.

North China Plain, Mandarin *Hua Bei Ping Yuan* (HAH-BAI-PING-YUAN), N CHINA, a combination of alluvial plains (□ 127,000 sq mi/330,200 sq km) of several major rivers in N China, including the Hai River (to the N), HUANG HE (Yellow River; in the center), and the HUAI River (to the S). The plain is bounded by the TAIHANG Mountains and FUNIU Mountains (W), BOHAI and the YELLOW Sea (E), the GREAT WALL (N), and the QINLING Mountains, and HUAI River (S). Sometimes the North China Plain refers specifically to the Hai River and Huang He alluvial plains. The area that includes the Hai River, Huang He (Yellow River), and Hai River is sometimes called the Huang-Huai-Hai Plain. The plain is of low elevation (generally lower than 165 ft/50 m). The massive lower reaches of the Huang He alluvial plain comprise the middle section of the Huang-Huai-Hai plain. The river carries a massive silt load from flowing through the easily eroded LOESS Plateau. The silting has caused numerous shifts of the Huang He channel throughout history and significantly affected the Huai and Hai systems as well. In HENAN province, the Huang He riverbed is several ft above the surrounding ground surface level. To the N of the Huang He plain, the Hai river plain inclines from the Taihang Mountains to Bohai Bay, leaving three distinct subsections: the piedmont, alluvial plain, and the coastal

plains and river delta. To the S of the Huang He plain, the Huai river system is characterized by low land and dense river and lake systems. The Huang-Huai-Hai region has a dry, windy spring and a warm, rainy summer. Monsoon season comes in late July and early August, bringing most of the year's precipitation. The seasonally fluctuating rainfall, silting, and flatness contribute to frequent drought and flooding. The region has been cultivated for 7,000–8,000 years. The major crops include wheat, corn, kaoliang, millet, beans, cotton, and peanuts. In areas with irrigation, rice is also grown. Major agriculture problems in the region are spring droughts, autumn flooding, and salinization. The ancient Dongyi ethnic group lived in the lower reaches of Huang He and along the N China coast, and eventually blended with other ethnic groups to form the Han Chinese. ZHENGZHOU, KAIFENG, ANYANG, and BEIJING have functioned as political centers for various imperial dynasties.

North Cholla, Korean, *Cholla-pukdo*, province (□ 3,104 sq mi/8,070.4 sq km), SOUTH KOREA, bounded W by YELLOW SEA, E by SOUTH KYONGSANG province, N by CHUNGCHONG province, S by SOUTH CHOLLA province; ⊙ CHONJU; 35°45′N 127°15′E. Chief port is KUNSAN. Partly mountainous, with fertile valleys drained by KUM RIVER and several small rivers. The Noryong Mountains runs centrally through the province; NE part becomes CHINAN plateau, the largest plateau in South Korea. Kum and SOMJIN rivers originate in the E; Honam plain located in W. Extensive coastline (277 mi/446 km), including more than seventy islands, provides yellow fish and shrimp. The Cholla railroad links SEOUL-TAEJON-CHONJU-YOSU and the Honam expressway connects Seoul-Chonju. Manufacturing areas include: Chonju (textiles, paper, food processing); IKSAN (jewerly, electronics); KUNSAN (sake brewing, plywoods). Primarily agriculture, the province produces rice, barley, soy beans, cotton, ramie, tobacco, ginseng, paper mulberry, and persimmons. Silk-cocoon production and lumbering are widespread. Gold, silver, molybdenum, and iron mines. Handicraft is important.

North Chungchong (CHOONG-CHUHNG), Korean *Chung-chong-pukdo*, inland province (□ 2,871 sq mi/ 7,464.6 sq km), bounded by the provinces of KYONGGI and KANGWON (N), KYONGSANG (E), NORTH CHOLLA (S), and SOUTH CHUNGCHONG (W), SOUTH KOREA; ⊙ CHONGJU; 36°45′N 128°00′E. Largely mountainous terrain, drained by KUM and HAN rivers. Agricultural products include rice, barley, hemp, tobacco; gold, tungsten, iron, coal, and graphite mined. Other products include silk cocoons, honey, textiles. Large cement factories in CHECHON and TANYANG. Mount Songui in the Sobaek Mountains designated a national park.

North City, village (2000 population 630), FRANKLIN county, S ILLINOIS, 8 mi/12.9 km W of BENTON; 37°59′N 89°03′W. In bituminous-coal and agricultural area.

Northcliffe (NORTH-klif), town, WESTERN AUSTRALIA state, W AUSTRALIA, 225 mi/362 km S of PERTH; 34°36′S 116°04′E. Timber (karri). Northcliffe Forest Park.

North Cohasset, MASSACHUSETTS: see COHASSET.

North College Hill, city (□ 2 sq mi/5.2 sq km; 2006 population 9,157), HAMILTON county, SW OHIO; suburb 10 mi/16 km N of CINCINNATI; 39°13′N 84°33′W. Mostly residential. Clovernook Home for the Blind has a braille printing shop. Revolutionary War cemetery is in the city. Incorporated as a city 1941.

North Collins, village (2006 population 1,019), ERIE county, W NEW YORK, 22 mi/35 km S of BUFFALO; 42°35′N 78°56′W. In fruit-growing region; light manufacturing; natural-gas wells. Settled c.1810, incorporated 1911.

Area is shown by the symbol □, and capital city or county seat by ⊙.

North Conway, village (2000 population 2,069), CAR-ROLL county, E NEW HAMPSHIRE, 28 mi/45 km S of BERLIN, and 5 mi/8 km N of Conway village, in town of CONWAY, on SACO RIVER; 44°02′N 71°07′W. Year-round resort area; agriculture (livestock; dairying; timber); manufacturing (furniture). White Mountain National Forest to W and N; Mount Cranmore Ski Area to NE; Echo Lake State Park to W.

North Cooking Lake (NORTH KU-keeng), hamlet (2006 population 50), central ALBERTA, W Canada, 15 mi/24 km SE of Sherwood Park, near shores of Cooking Lake, and included in municipality of Strathcona County. Residential.

North Corbin, unincorporated town (2000 population 1,662), LAUREL county, SE KENTUCKY, in Cumberland foothills on LAUREL RIVER, 1 mi/1.6 km N of CORBIN; 36°58′N 84°05′W. LAUREL RIVER LAKE reservoir to W.

North Cotabato (ko-tah-BAH-to), province (□ 2,535 sq mi/6,591 sq km), in CENTRAL MINDANAO region, central MINDANAO, PHILIPPINES; ⊙ KIDAPAWAN; 07°01′N 125°05′E. Population 18.1% urban, 81.9% rural; in 1991, 100% of urban and 50% of rural settlements had electricity. Central plain ringed by mountains to NW and E. Volcanic soil. Agriculture (rice, corn, fruit as food crops; rubber, coconuts, coffee as cash crops). Also known as Cotabato province.

Northcote (NORTH-kot), residential suburb, S VICTORIA, SE AUSTRALIA, 4 mi/6 km NNE of MEL-BOURNE; 37°46′S 145°00′E. Retail; manufacturing (apparel), printing, paper. Has merged with PRESTON to become city of Darebin.

Northcote (NORTH-kot), residential suburb within NORTH SHORE CITY, NORTH ISLAND, NEW ZEALAND, at N end of AUCKLAND Harbour bridge, WAITEMATA HARBOUR. Incorporated 1989 into North Shore City.

North Country, region of NEW YORK state that includes the ADIRONDACKS, Lake CHAMPLAIN lowlands, and the SAINT LAWRENCE River. Key summer and winter tourist area; some farming in lowlands; also lumbering. Hydroelectric production. Counties include CLINTON, ESSEX, FRANKLIN, JEFFERSON, LEWIS, and ST. LAWRENCE. PLATTSBURGH, MASSENA, OGDENSBURG, LAKE PLACID, and WATERTOWN are main centers.

North Cousin Island, one of the SEYCHELLES, in INDIAN OCEAN, off SW coast of PRASLIN ISLAND; 0.33 mi/0.53 km long, 0.33 mi/0.53 km wide; 04°20′S 55°43′E. Granite formation. Since 1973 a private bird sanctuary.

North Cowichan (KOU-i-chuhn), district municipality (□ 75 sq mi/195 sq km; 2001 population 26,148), SW BRITISH COLUMBIA, W Canada, on VANCOUVER ISLAND, in Cowichan Valley regional district; 48°50′N 123°41′W. Forestry.

North Creek, hamlet, WARREN county, E NEW YORK, in the High Peaks section of the ADIRONDACK MOUN-TAINS, on the HUDSON RIVER, and 32 mi/51 km NNW of GLENS FALLS; 43°42′N 74°00′W. Some manufacturing and gem mining. Just SW is Gore Mountain (elevation 3,595 ft/1,096 m), skiing center.

Northcrest (NORTH-krest), town, MCLENNAN county, E central TEXAS, residential suburb 7 mi/11.3 km NE of downtown WACO; 31°38′N 97°05′W. Agricultural area (cattle; dairying; cotton). Seat of Texas State Tech. College; airport.

North Crimean Canal (Ukrainian *Pivnichno-Kryms'-kyy Kanal*) (Russian *Severo-Krymskiy Kanal*), in S KHERSON oblast and NE CRIMEA, UKRAINE, extending from TAVRIYS'K, above the dam forming the KA-KHOVKA Reservoir of the DNIEPER R., generally S and then ESE, past KALANCHAK, over the Isthmus of PEREKOP past ARM'YANS'K and KRASNOPEREKOPS'K, across the North Crimean Lowland past DZHANKOY, AZOVS'KE, NYZHN'OHIRS'KYY, SOVYETS'KYY, and then E, along KERCH PENINSULA, towards KERCH, to a water reservoir at Zelenyy Yar, 23 mi/37 km W of Kerch. The canal provides Dnieper water for irrigation in the dry steppes of S Ukraine and for municipal and industrial needs of SIMFEROPOL', SEVASTOPOL',

and Kerch. Following a false start in 1951, construction began in 1957. Current length of this mainline canal is 249 mi/401 km. Pumping stations (four) are used to send water this distance. Area irrigated by water supplied via the North Crimean Canal in Crimea has increased from 464,000 acres/188,000 ha (1976) to 66,0001 acres/268,000 ha (1986) and 882,000 acres/357,000 ha (1994). Crops grown under irrigation in the North Crimean Plain include wheat, rice, vegetables; orchards and vineyards.

North Crimean Plain, UKRAINE: see CRIMEAN LOW-LAND.

North Crosby (KRAHZ-bee), former township (□ 74 sq mi/192.4 sq km; 2001 population 1,024), SE ONTARIO, E central Canada, 28 mi/44 km from KINGSTON; 44°42′N 76°23′W. Amalgamated into Rideau Lakes township in 1998.

North Crows Nest, town (2000 population 42), MAR-ION county, central INDIANA, on the WHITE RIVER, and 7 mi/11.3 km N of downtown INDIANAPOLIS. A suburb of and part of the city of Indianapolis.

North Curry (CUHR-ee), village (2001 population 1,740), central SOMERSET, SW ENGLAND, 6 mi/9.7 km E of TAUNTON; 51°01′N 02°58′W. Has 13th–15th-century church.

North Cypress (NORTH SEI-pres), rural municipality (□ 465 sq mi/1,209 sq km; 2001 population 1,853), S MANITOBA, W central Canada, 27 mi/44 km from BRANDON; 49°55′N 99°21′W. Agriculture (potatoes, canola, flax). Includes CARBERRY, Brookdale, WELL-WOOD, and Edrans.

North Dakota, state (□ 70,665 sq mi/183,022 sq km; 1995 estimated population 641,367; 2000 population 642,200) N central UNITED STATES, admitted to the Union in 1889 simultaneously with SOUTH DAKOTA (they are the 39th and 40th states). ⊙ BISMARCK, on the E bank of the MISSOURI River; FARGO is the largest city. North Dakota is nicknamed the "Peace Garden State" and the "Flickertail State," which refers to the Richardson ground squirrels which are abundant.

Geography
North Dakota is bounded on the N by the Canadian provinces of SASKATCHEWAN and MANITOBA, on the E by the RED RIVER of the N and BOIS DE SIOUX RIVER which separates it from MINNESOTA, on the S by South Dakota, and on the W by MONTANA. North Dakota is drained by the Missouri, JAMES, SHEYENNE and SOURIS (also called Mouse) rivers. Situated in the geographical center of NORTH AMERICA, North Dakota is subject to the extremes of a continental climate. Semiarid conditions prevail in the W half of the state, but in the E an average annual rainfall of 22 in/56 cm, much of it falling in the crop-growing months of spring and summer, enables the rich soil to yield abundantly.

Economy
North Dakota is one of the most rural states in the nation; the cities and towns supply the needs of neighboring farms, and industry is largely devoted to the processing of agricultural products and agricultural machinery. The E half of the state is in the central lowlands; agriculture there includes flax, grain, and sunflowers. E of the Manitoba Escarpment there is a wedge of land, c.40 mi/64 km wide at the Canadian border and tapering to 10 mi/16 km in the S, that is the floor of the former glacial Lake AGASSIZ. Treeless, except along the riversides, and relatively void of rocks, this flat land was transformed into the bonanza wheat fields of the 1870s and 1880s, with farms ranging in size from 3,000 acres/1,214 ha to 65,000 acres/26,306 ha. Today the average farm in the Red River valley is about 1,009 acres/408 ha; (the state average is about 1,267 acres/513 ha), and its major crop, wheat, is varied with such crops as sugar beets, soybeans, pinto beans, canola, flax and seed potatoes. To the W of the valley a series of escarpments rises some 300 ft/91 m to meet the drift prairies, where rolling hills, scattered lakes,

and occasional moraines form a pleasant and fertile countryside. The productivity of the soil makes North Dakota a leader in spring wheat and durum wheat (ranking second in the nation), hay, barley, sugarbeets, oats, soybeans, and sunflowers. However, cattle and cattle products exceed all the crops except wheat in income earned. In the W part of the state a combination of unfavorable topography and scant rainfall precludes intensive cultivation except in the river valleys. In the NW area of the state oil was discovered in 1951, and petroleum is now the state's leading mineral product, ahead of sand and gravel, lime, and salt. There are also natural gas fields. Underlying the W counties are lignite coal reserves estimated at 350 billion tons/386 billion metric tons, which are mixed in several areas. In close proximity to the lignite beds are fine deposits of clay of such varied types that they serve as both construction and pottery materials. Despite mineral production and some manufacturing, however, agriculture continues to be the state's principal pursuit, and the processing of grain, meat, and dairy products is vital to such cities as Fargo, GRAND FORKS, MINOT, and Bismarck. The Missouri and the Red River, once the major transportation routes, are more important now for their irrigation potential. Several dams have been built, notably GARRISON DAM, and a number of Federal reclamation projects have been completed as part of the Missouri River basin project. There has also been reforestation. With such attractions as the Badlands, the INTERNATIONAL PEACE GARDEN on the Canadian border, and recreational facilities provided by reservoirs (resulting from dam building in the 1950s), tourism has become North Dakota's 5th-ranking source of income, behind agriculture, Federal activities (2nd), mineral production (3rd), manufacturing (4th).

The Badlands
An area some 50 mi/80 km E of the Missouri River is a farm and grazing belt, divided from the drift prairies by the Missouri escarpment. W from the Missouri rolls an irregular plateau, covered with short prairie grasses and cut by deep coulees. Where wind and rain have eroded the hillsides there are unusual formations of sand and clay, glowing in yellows, reds, browns, and grays. Along the Little Missouri this section is called the BADLANDS, so named because the region (once described as "hell with the fires out") was difficult to traverse in early days. Here are the three units of the THEODORE ROOSEVELT NATIONAL PARK where from 1883 to 1886 the young Theodore Roosevelt spent part of each year ranching. Also in this area is the Little Missouri National Grassland. On the plateau cattle graze, finding shelter in the many ravines, and large ranges are an economic necessity.

History to 1832
The first farmers in the region of whom there is definite knowledge were Native Americans of the Mandan tribe. Other agricultural tribes were the Arikara and the Hidatsa. Seminomadic and nomadic tribes were the Cheyenne, Cree, Sioux, Assiniboine, Crow, and Ojibwa (Chippewa). With the Louisiana Purchase of 1803 the two-thirds of North Dakota became part of the U.S. The E one-third was acquired from Great Britain in 1818 when the international line with CA-NADA was fixed at the 49th parallel. Earlier the Lewis and Clark expedition had wintered (1804–1805) with the Mandan along the Missouri River and the North West Company and the Hudson's Bay Company had established trading posts in the Red River valley. These ventures introduced an industry that dominated the region for more than half a century. Within that era the buffalo vanished from the plains and the beaver from the rivers. From its post at FORT UNION (established 1828), John Jacob Astor's American Fur Company gradually gained monopolistic control for a time over the trade of the region. Supply and transport were greatly facilitated when a paddlewheel

steamer, the *Yellowstone*, inaugurated steamboat travel on the turbulent upper Missouri in 1832.

History: 1832 to 1876
Additional transportation was provided by the supply caravans of Red River carts, which went W across the Minnesota prairies and returned to the MISSISSIPPI loaded with valuable pelts. In 1837, the introduction of smallpox by settlers decimated the Mandan and Hidatsa tribes. The first attempt at agricultural colonization was made at PEMBINA in 1812, but the first permanent farming community was not established until 1851 when another group settled at Pembina. This was still the only farm settlement in the future state when Dakota Territory was organized in 1861 to include what eventually became present-day North Dakota, South Dakota, Montana, and WYOMING. Several military posts had been established starting in 1857 to protect travelers and railroad workers. Even when free land was opened in 1863 and the Northern Pacific Railroad was chartered in 1864 a preoccupation with the Civil War and the eruption of open warfare with the Native Americans prevented any appreciable settlement. General Alfred H. Sully joined General Henry H. Sibley of Minnesota in campaigns against the Sioux in 1863–1866. A treaty was signed in 1868.

History: 1876 to 1892
In 1876, after gold was discovered on Native American land in the BLACK HILLS, the unwillingness of the whites to respect treaty agreements led to further war, and the force of George A. Custer was annihilated at the battle of the LITTLE BIGHORN in present-day Montana. Ultimately, however, the Sioux under Chief Sitting Bull fled to Canada, where they surrendered voluntarily; they were returned to reservations in the U.S. The first cattle ranch in North Dakota was established 1878. With the construction of railroads in the 1870s and 1880s, thousands of European immigrants, principally Scandinavians and Germans (notably Germans from Russia) arrived. They worked the land on their own homesteads or on the large Eastern-financed bonanza wheat fields of the low central prairies. Borrowing the idea from Europe, they founded agricultural cooperatives. Local politics was rapidly reduced to a struggle between the agrarian groups and the corporate interests. Alexander McKenzie of the Northern Pacific was for many years the most important figure in the state. Republicans held the elective offices.

History: 1892 to 1921
Agrarian groups formed the Farmers' Alliance and in 1892, three years after North Dakota had achieved statehood, the Farmers' Alliance combined with the Democrats and Populists to elect Eli Shortridge, a Populist, as governor. Later, when the success of the La Follette Progressives in WISCONSIN encouraged the growth of the Republican Progressive movement in North Dakota, a fusion with the Democrats elected "Honest John" Burke as governor for three terms (1906–1912). Much of the agrarian discontent was focused on marketing practices of the large grain interests. Although many small cooperative grain elevators were established, they did not prove effective, and the farmers pressed for state-owned grain elevators. When this movement failed in the legislature of 1915, the Nonpartisan League, directed in North Dakota by Arthur C. Townley, was organized on a platform that included state ownership of terminal elevators and flour mills, state inspection of grain and grain dockage, relief of farm improvements from taxation, and rural credit banks operated at cost. Working primarily with the Republican Party because it was the majority party in North Dakota, the league captured the state legislature in 1919 and proceeded to enact virtually its entire platform. This included the establishment of an industrial commission to manage state-owned enterprises and the creation of the Bank of North Dakota to handle public funds and provide

low-cost rural credit. The right of recall was also enacted, by which voters could remove an elected official. The reforms were disappointing in operation. Dissension arose within the league, and the Independent Voters Association was organized to represent the conservative Republican position.

History: 1921 to Present
The industrial commission was accused of maladministration, and the provision of recall was exercised three times, the first against Governor Lake J. Frazier in 1921. William Langer, who had been active with both the Nonpartisan League and the Independent Voters Association, was elected governor in 1932 running as a Nonpartisan. Langer was convicted on a Federal charge of misconduct in office in 1934, although the conviction was later reversed. Langer again became governor in 1936, running as an individual candidate and not on the ticket of either party; subsequently he was elected to the U.S. Senate four times. In the 1950s and 1960s, a large number of military installations were built in North Dakota. The state's heavy dependence on wheat and petroleum has made it extremely vulnerable to the fluctuations of those markets. In 1991, North Dakota sought alternative sources of revenue through the legalization of some Indian tribal casino gambling. In recent years North Dakota has become more urbanized, but its overall population declined 2.1% in the 1980s. The state's institutions of higher education include the University of North Dakota, at Grand Forks; North Dakota State University, at Fargo; Jamestown College, at JAMESTOWN; and several other colleges.

Government
The state is governed under the 1889 constitution. The legislature consists of fifty-three senators elected to four-year terms and 106 representatives elected to two-year terms. The governor is elected for a four-year term. The current governor is John Hoeven. North Dakota elects two U.S. senators and one representative; it has three electoral votes.

North Dakota has fifty-three counties: ADAMS, BARNES, BENSON, BILLINGS, BOTTINEAU, BOWMAN, BURKE, BURLEIGH, CASS, CAVALIER, DICKEY, DIVIDE, DUNN, EDDY, EMMONS, FOSTER, GOLDEN VALLEY, GRAND FORKS, GRANT, GRIGGS, HETTINGER, KIDDER, LA MOURE, LOGAN, MCHENRY, MCINTOSH, MCKENZIE, MCLEAN, MERCER, MORTON, MOUNTRAIL, NELSON, OLIVER, PEMBINA, PIERCE, RAMSEY, RANSOM, RENVILLE, RICHLAND, ROLETTE, SARGENT, SHERIDAN, SIOUX, SLOPE, STARK, STEELE, STUTSMAN, TOWNER, TRAILL, WALSH, WARD, WELLS, and WILLIAMS.

North Darfur, Arabic *Shamal Darfur*, state (2005 population 1,709,000), DARFUR region, NW SUDAN; ⊙ AL-FASHER; 16°00′N 26°00′E. Bordered by Northern (NNE), N KURDOFAN (E), S DARFUR (S), and W DARFUR (SW) states, as well as CHAD (W) and LIBYA (NW). Created in 1996 following a reorganization of administrative divisions. Roads located mainly in S, with AL-FASHER as hub; main road runs through AL-FASHER W through W DARFUR state into CHAD (including ABÉCHÉ) and E to AL-NAHUD (in N KURDOFAN state). Regional airport at AL-FASHER. The area making up N Darfur state was part of former DARFUR province. Also called Northern Darfur.

North Darley, ENGLAND: see DARLEY BRIDGE.
North Dartmouth, MASSACHUSETTS: see DARTMOUTH.
North Dighton, MASSACHUSETTS: see DIGHTON.
North District, AUSTRALIA: see COSSACK.
North Dome, peak (elevation 3,593 ft/1,095 m) in GREENE county, SE NEW YORK, in the CATSKILL MOUNTAINS, 20 mi/32 km WNW of SAUGERTIES; 42°10′N 74°21′W.
North Downs (DOUNZ) and **South Downs**, ranges of chalk hills, SE ENGLAND; 51°15′N 00°05′E. Rise to 965 ft/294 m at LEITH HILL. The North Downs range, extending c.100 mi/161 km through SURREY and KENT, is

cut by the WEY, MOLE, DARENT, MEDWAY, and GREAT STOUR rivers. It is separated from The WEALD from the South Downs (c.65 mi/105 km) in West SUSSEX and HAMPSHIRE, which are cut by the ARUN, ADUR, and OUSE rivers. The Downs provide excellent pasturage for sheep; Southdown sheep are well known.
North Duffield (DUHF-feeld), village (2001 population 1,771), NORTH YORKSHIRE, N ENGLAND, 4 mi/6.4 km NE of SELBY; 53°49′N 00°57′W. DERWENT RIVER 1 mi/1.6 km to E.
North Dumdum, city, within Dumdum, suburban Calcutta, NORTH 24-PARGANAS district, WEST BENGAL state, E INDIA. Contains several large rural enclaves. Also spelled North Dum Dum.
North Dumfries (NORTH DUHM-freez), township (□ 72 sq mi/187.2 sq km; 2001 population 8,769), WATERLOO region, SW ONTARIO, E central Canada, 8 mi/12 km from KITCHENER; 43°19′N 80°23′W. Local airport.
North Dundas (NORTH DUHN-duhs), township (□ 194 sq mi/504.4 sq km; 2001 population 11,014), SE ONTARIO, E central Canada, 33 mi/52 km from CORNWALL, and c.19 mi/30 km S of OTTAWA; 45°05′N 75°23′W.
North East, town (2000 population 2,733), CECIL county, NE MARYLAND, at head of navigation on Northeast River (c.6 mi/9.7 km long), and 24 mi/39 km WSW of WILMINGTON, DELAWARE; 39°36′N 75°56′W. Trade center for hay and grain area; sand, gravel pits; manufacturing (fireworks, firebrick). Once nicknamed Herrington because of the importance of commercial fishing. The Georgian bell tower of St. Anne's Protestant Episcopal Church (c.1742) was added in 1904 by Robert Brookings, founder of the Brookings Institute in memory of his parents. The churchyard is older than the church and contains several Indian graves as well as those of colonists dating from the 1600s. Sandy Cove (summer resort) and ELK NECK (site of state forest, state park) are nearby.
North East, borough (2006 population 4,302), ERIE county, NW PENNSYLVANIA, 13 mi/21 km NE of ERIE, on Sixteenmile Creek; LAKE ERIE 1 mi/1.6 km to NW; 42°12′N 79°49′W. NEW YORK state line 3 mi/4.8 km to E. Manufacturing (frozen foods, powder coating, metal castings, electrical parts, wineries, juices and jams); agriculture (cherries, tomatoes, apples, grapes, grain; livestock; dairying). Lake Shore Railroad Museum here. Settled c.1800, incorporated 1834.
Northeast, Mandarin *Dongbei*, region (□ 308,880 sq mi/803,088 sq km), NE CHINA. It is separated from RUSSIA largely by the AMUR, ERGUN, and USSURI rivers, from NORTH KOREA by the YALU and Tumen rivers, and from MONGOLIA by the GREATER HINGGAN mountains. It included territory now in RUSSIAN FAR EAST until 1860. Provincial divisions have changed frequently, but Northeast has comprised JILIN, HEILONGJIANG, and LIAONING provinces since 1956. Much of the region's borderland is hilly to mountainous. The Great Hinggan mountains in the W, Lesser Hinggan mountains in the N, and the CHANGBAI mountains in the E are the greatest ranges. Northeast's vast forestry is the largest timber reserve in China. The region is the most important heavy-industry base in China. Mineral resources, chiefly coal and iron, are concentrated in the SW; there is a large colliery at FUSHUN and a large steel mill at ANSHAN. Magnesite, copper, lead, and zinc are also important, and there is a large oilfield at DAQING, NW of HARBIN. Uranium deposits have also been found. The great Northeast Plain (average elevation c.1,000 ft/300 m), crossed by the LIAO and SONGHUA rivers, is among the largest plains in China. Fertile and densely populated, it is a major manufacturing and agricultural center of China. One of the few areas in the country suitable for large-scale mechanized agriculture, it has numerous collective farms. Long, severe winters limit harvests to

one a year, but considerable quantities of soybeans are produced. Sweet potatoes, beans, and cereals (including rice, wheat, millet, and kaoling) are also grown, and cotton, flax, and sugar beets are raised as industrial crops. The processing of soybeans into oil, animal feed, and fertilizer is centered in cities in or near the plain, notably CHANGCHUN, HARBIN, and SHENYANG. Livestock are raised in the N and the W, and fishing is important off the YELLOW SEA coast. The chief commercial port is DALIAN, and LUDA is a leading naval base. All rivers are navigable but only the Songhua is significant for heavy traffic. When the rivers freeze, they are used as roads. An extensive railroad system connects the hinterland with the coastal ports; major lines are the South Liaoning railroad and the Northeast railroad. The construction of the railroad (after 1896) spurred industrial development. Today Northeast is a great industrial hub, with huge coal mines, iron- and steel-works, aluminum-reduction plants, paper mills, and factories making heavy machinery, tractors, locomotives, aircraft, and chemicals.

Northeast is traditionally the homeland of peoples that have invaded and sometimes ruled N China. Among the most important of these tribes were the Tungus, Eastern Turks, Khitan, and Jurchen. The region was formerly called MANCHURIA because it was the home of the Manchu nomads. The Manchus conquered China in the mid-1600s and established the Qing dynasty (1644–1911). The Manchu rulers tried to keep Northeast an imperial preserve by limiting Chinese immigration. Emigration to Northeast from the adjacent provinces has been heavy in the last c.100 years, and the population is now predominantly Chinese. JAPAN and Russia long struggled for control of this rich, strategically-important region. Japan tried to seize the Liaodong (Liao-tung) peninsula in 1895, but was forestalled by the Triple Intervention. Russia was dominant from 1898 to 1904. As a result of a Russo-Chinese alliance against Japan, the Russians built Harbin, the naval base at PORT ARTHUR, and the Chinese Eastern railroad (currently called Northeast railroad). Japan, after victory in the Russo-Japanese War (1904–1905), took control of Port Arthur and the S half of Northeast (see LIAONING), limiting Russian influence to the N.

Chiefly through the South Liaoning railroad, Japan developed the region's modern economy base. The warlords Zhang Zuo-lin and Zhang Xue-liang controlled Chinese military power in Northeast from 1918 to 1931. Japan occupied Northeast in 1931–1932, when Chinese military resistance, sapped by civil war, was weak. The seizure of Northeast was, in effect, an unofficial declaration of war on China. Northeast was a base for Japanese aggression in N China and a buffer region for Japanese-controlled Korea. In 1932, under the aegis of Japan, Northeast with REHE province was constituted Manzhoukuo (Japanese *Manchukuo*), a nominally independent state with capital in present-day Changchun (Jilin province). During World War II, the Japanese developed the Dalian, Anshan, Fushun, Shenyang, and Harbin areas into a huge industrial complex of metallurgical, coal, petroleum, and chemical industries. Soviet forces, which occupied Northeast from July 1945 to May 1946, dismantled and removed more than half of the Northeastern industrial plant. At the end of the war the Chinese Communists were strongly established in Northeast and had captured the major cities by 1948. During the 1950s, with the help of Soviet technicians, the Communists rapidly restored Northeast's large industrial capacity. After the Sino-Soviet rift in the 1960s there was a massive Soviet military buildup along the border and several border incidents occurred. With the breakup of the Soviet Union, these incidents have subsided. China's changing economic policies have led to renewed investment in the region since 1978.

Northeast Cape Fear River (NORTH-eest KAIP FEER RI-vuhr), SE NORTH CAROLINA; rises SE WAYNE county at DUPLIN county line; flows 130 mi/209 km S to CAPE FEAR RIVER at WILMINGTON; lower c.50 mi/80 km tidal; 35°10′N 78°02′W.

North East Carry, resort village, PISCATAQUIS county, W central MAINE, on MOOSEHEAD LAKE, and 28 mi/45 km N of GREENVILLE. Hunting, fishing.

North-East District, administrative division (2001 population 49,399), NE BOTSWANA; ⊙ FRANCISTOWN. Bordered by ZAMBIA N and E; CENTRAL DISTRICT S and W. SHASHE River serves as S boundary. Crossed by road and railroad. Headquarters of Bakalanga tribe. Rich farming area.

Northeastern Manitoulin and The Islands (ma-ni-TOO-lin), town (□ 192 sq mi/499.2 sq km; 2001 population 2,531), ONTARIO, E central Canada, 18 mi/29 km from GORE BAY; 45°52′N 82°06′W. Includes the port of LITTLE CURRENT, the rural Howland area (including Sheguiandah, Honora hamlets), McGregor Bay, and Bay of Islands.

North Eastern Province, province (□ 48,720 sq mi/126,185 sq km; 2003 estimated population 1,187,767; 2007 estimated population 1,368,785), N KENYA; ⊙ GARISSA. Bordered by ETHIOPIA (N), EASTERN province (W), and SOMALIA (E). Comparatively waterless, it consists of a vast undulating plain sloping upwards from the INDIAN OCEAN; in NW is section of GREAT RIFT VALLEY (here occupied by Lake TURKANA) flanked by escarpments. Most rivers that drain this semiarid region are not perennial. Has mainly scrub grasslands. A nomadic livestock region (cattle, sheep and goats); some agriculture (peanuts, sesame, corn, wild coffee). Main centers are Garissa, wajir, and Mado Gashi.

Northeastern Railroad, Mandarin *Dongbei Teilu* (DUNG-BAI-TAI-LU), NORTHEAST province, CHINA, 8,700 mi/14,000 km. Originally known by various names such as Manchuria railroad, Chinese Eastern railroad, or Chinese Changchun railroad. The earliest sections, called N Manchuria railroad, were built by the Russians to provide a short route to VLADIVOSTOK and an exit to an ice-free port (DALIAN). Northern Manchuria railroad extended from Manzhouli to Suifenhe (built 1879–1903) via HARBIN, from which city a branch (built 1903–1905) extended S past CHANGCHUN to Dalian. After the Russo-Japanese war (1904–1905), the Dalian-Changchun section passed to Japanese control and became the Southern Manchuria railroad. Manchukuo bought (1935) the rest of Northern Manchuria railroad (or the Chinese Eastern railroad) from the USSR. After World War II, the entire Manchuria railroad system (renamed Chinese Changchun railroad) passed to joint Soviet-Chinese control. Called Northeastern railroad since 1949, the system has expanded and consists of one-third of China's total railroad system. The two backbone tracks are the E-W Manzhouli-Harbin-Suifenhe section system and the Trans-Siberian railroad in Russia. In addition, the Shenyang-Shanhaiguan-Bejing railroad connects N China, and the Shenyang-Dandong-Pyongyang railroad links with NORTH KOREA. There are also many branch lines and forest railroads in the system.

Northeast Foreland, Danish *Nordostrundingen* (NOR-OST-ROO-ning-uhn), cape, NE extremity of GREENLAND, on GREENLAND SEA, at E tip of CROWN PRINCE CHRISTIAN LAND peninsula; 81°21′N 11°40′W.

North-East Frontier Agency, INDIA: see ASSAM.

North Eastham, MASSACHUSETTS: see EASTHAM.

Northeast Harbor, MAINE: see MOUNT DESERT.

North Easthope (NORTH EEST-hop), former township (□ 68 sq mi/176.8 sq km; 2001 population 2,188), SW ONTARIO, E central Canada; 43°36′N 80°52′W. Amalgamated into Perth East in 1998.

Northeast Land [Norwegian *Nordaustlandet*], formerly *Nordaustlandet*, island (□ 5,710 sq mi/14,846 sq km) of SVALBARD group, in BARENTS SEA of ARCTIC OCEAN, NE of SPITSBERGEN, from which it is separated by HINLOPEN STRAIT; 79°10′–80°31′N 17°35′–27°10′E. Island is 110 mi/177 km long (E-W), 40 mi/64 km–90 mi/145 km wide. Almost wholly glaciated, its inland ice rises to over 2,000 ft/610 m. Coastline is irregular and deeply indented, especially on N and W coasts. Discovered in early 17th century; first charted (1707) by Van Keulen. Island explored (1873) by Nordenskjöld; completely charted (1935–1936) by British expedition. Off N coast are the SJUOYANE, [Norwegian *Sjuøyane*], a group of five small mountainous islands.

North East Lincolnshire (LING-kuhn-shir), county (□ 74 sq mi/192.4 sq km; 2001 population 157), E ENGLAND, on the HUMBER RIVER estuary of the North Sea; ⊙ GRIMSBY; 53°35′N 00°05′W. Created in 1996 when Humberside was split into North Lincolnshire, North East Lincolnshire, Kingston upon Hull, and East Riding of Yorkshire. Port at Immingham Dock; fishing, food processing, and chemical plants.

North Easton, MASSACHUSETTS: see EASTON.

Northeast Passage, water route along the N coast of Europe and Asia, between the ATLANTIC and PACIFIC oceans. Beginning in the 15th century, efforts were made to find a new all-water route from Europe to India and China. Most of these attempts were directed at seeking a NORTHWEST PASSAGE. However, English, Dutch, and Russian navigators did try to seek a NE route by sailing along the N coast of Russia and far into the Arctic seas. In the 1550s, English ships made the first attempt to find the passage. Willem Barentz, the Dutch navigator, made several futile voyages in the 1590s, as did Henry Hudson in the early 17th century. The decline of Dutch shipping in the 1700s left the exploration mainly to the Russians; among the men sent out was Vitus Bering, who explored the E part of the passage. The Russian Great Northern Expedition (1733–1743) explored most of the coast of N Siberia. The Northeast Passage was not, however, traversed by anyone until Nils A. E. Nordenskjöld of Sweden accomplished the feat in 1878–1879. In the early 1900s, icebreakers sailed through the passage, and in the 1930s the Northern Sea Route, a shipping lane, was established by the USSR. Since World War II, the Soviet Union—and now Russia—has maintained a regular highway for shipping along this passage through the development of new ports and the exploitation of resources in the interior. A fleet of Russian icebreakers, aided by aerial reconnaissance and by radio weather stations, keeps the route navigable from June to October (all year between MURMANSK and the mouth of the YENISEY RIVER). The Northern Sea Route halves the distance between Russian Atlantic and Pacific ports.

North East Point, SAINT THOMAS parish, cape at NE extremity of JAMAICA, on JAMAICA CHANNEL, 33 mi/53 km ENE of KINGSTON; 18°09′N 76°20′W.

North-East Polder, Dutch *Noord-Oost Polder*, region (□ 185 sq mi/481 sq km), FLEVOLAND province, N central NETHERLANDS. Formerly part of OVERIJSSEL province until the Flevoland province was established 1986 (17 mi/27 km long, 16 mi/26 km wide). Bounded W by IJSSELMEER, S by KETELMEER channel, SE by ZWARTEMEER channel; joined to mainland on E and N; the Linde River flows from NE into polder's canal system. EMMELOORD is in center, Urkaine in SW; entire polder is in Emmeloord municipality. Cattle; vegetables, grain, fruit, rapeseed.

Northeast Providence Channel, Atlantic strait, BAHAMAS, NE of the GREAT BAHAMA BANK, bounded by NEW PROVIDENCE Island and ELEUTHERA Island (S) and GREAT ABACO ISLAND (N); c.110 mi/177 km long SWNNE. Leads through NW PROVIDENCE CHANNEL to Straits of FLORIDA.

North Edwards, unincorporated town (2000 population 1,227), KERN county, S central CALIFORNIA, 44 mi/

71 km W of BARSTOW, in MOJAVE DESERT; 35°01′N 117°50′W. EDWARDS AIR FORCE BASE and Rogue Lake (dry) to S. Service center for air base. Cattle; grain.

North Egremont, MASSACHUSETTS: see EGREMONT.

Northeim (NORT-heim), city, LOWER SAXONY, W GERMANY, on the LEINE, 12 mi/19 km N of GÖTTINGEN; 51°42′N 10°00′E. Food processing (flour products, canned goods, beer, spirits); machinery. Has late-Gothic church; remains of old fortifications. Was member of HANSEATIC LEAGUE.

North Elba, town (☐ 156 sq mi/405.6 sq km), ESSEX county, NE NEW YORK, in the ADIRONDACKS, just SE of LAKE PLACID village; 44°15′N 74°00′W. Here are the farm, home (now a museum), and grave of abolitionist John Brown.

North Elmham, village (2001 population 1,428), central NORFOLK, ENGLAND, 5 mi/8 km N of East Dereham. Site of naval training school. Has church dating from 13th–15th century. It was seat of a Saxon bishopric and traces of its defenses remain. Also called Elmham.

North El Monte, unincorporated town, LOS ANGELES county, S CALIFORNIA, residential suburb 13 mi/21 km ENE of downtown LOS ANGELES, and 2 mi/3.2 km NE of EL MONTE; drained by SAN GABRIEL RIVER; 34°06′N 118°01′W.

North Elmsall (ELMZ-uhl), town (2001 population 4,093), WEST YORKSHIRE, N ENGLAND, 6 mi/9.7 km SSE of PONTEFRACT; 53°37′N 01°16′W.

North English, town (2000 population 991), IOWA county, E central IOWA, 19 mi/31 km S of MARENGO; 41°31′N 92°04′W.

North Enid (EEN-id), town (2006 population 837), GARFIELD county, N OKLAHOMA, residential suburb 2 mi/3.2 km NNE of ENID; 36°27′N 97°51′W.

North Equatorial Current, shallow (660 ft/201 m deep) but broad, W-flowing ocean currents, generally between 15°N and 20°N, driven by NE trade winds blowing over the tropical oceans of the Northern Hemisphere. In the ATLANTIC OCEAN it is known as the ATLANTIC NORTH EQUATORIAL.

Northern or **oro**, province (2000 population 133,065), SE PAPUA NEW GUINEA, E NEW GUINEA island; ⊙ POPONDETTA. Bounded on N by MOROBE province, SW by CENTRAL province, SE by MILNE BAY province, NE by SOLOMON SEA. Includes Mambare, Kumuis, and Musa rivers and NE slopes of OWEN STANLEY RANGE. Mount LAMINGTON (5,200 ft/1,585 m), volcano, erupted in 1951 killing 3,000 at HIGATURU. Palm oil, coconuts; prawns, tuna. Noted for tapa cloth made from paper mulberry tree. Province is home of Alexandra birdwing, largest butterfly in world.

Northern, province (☐ 13,875 sq mi/36,075 sq km; 2004 population 1,745,553), N SIERRA LEONE; ⊙ MAKENI; 09°15′N 11°45′W. The province covers most of the N half of the country. Bordered N and E by GUINEA, SE by Eastern province, S by SOUTHERN province, SWW by WESTERN AREA, and W by ATLANTIC OCEAN. International airport at LUNGI (N of FREETOWN). Northern province is further divided into five secondary administrative divisions, called districts: BOMBALI (see Makeni town), Kambia (see KAMBIA town), Koinadugu (see KABALA town), Port Loko (see PORT LOKO town), and Tonkolili (see MAGBURAKA town).

Northern, Arabic *Ash Shamaliyah*, state (2005 population 634,000), Northern region, N SUDAN; ⊙ DONGOLA; 19°00′N 30°00′E. Bordered by River Nile (E), KHARTOUM (SE tip), North Kurdofan (S), and North Darfur (W) states, as well as LIBYA (NW) and EGYPT (N). Created in 1996 following a reorganization of administrative divisions. S portion of Lake NASSER reservoir extends from EGYPT into NE Northern state (where it is called Lake NUBIA). NILE River flows SW out of River Nile state, turns N near AD-DEBBA, and flows into EGYPT at Lake NASSER. SAHARA Desert extends through Northern state. Road runs parallel to NILE RIVER; a main road also runs from S shore of Lake NASSER through River Nile state to KHARTOUM. Re-

gional airports at AD-DEBBA, DONGOLA, and Merowe. The area making up Northern state was part of former Northern province.

Northern, former province (☐ 16,058 sq mi/41,750.8 sq km), N central UGANDA; capital was LIRA. Was bordered N by SUDAN, W by NILE province, S by VICTORIA NILE RIVER, E by KARAMOJA and EASTERN provinces. Drained by ACHWA, PAGER, Aguga rivers. Consisted of MT. ROM, Mt. Ladwong. Agriculture included cotton, cassava, millet; cattle. Gold and mica deposits. Main centers were KITGUM, Lira, GULU, Atiak.

Northern, province (☐ 57,076 sq mi/148,397.6 sq km; 2000 population 1,258,696), NE ZAMBIA; ⊙ KASAMA; 11°00′S 31°00′E. Bounded on N by CONGO and TANZANIA, on E by MALAWI, on E and SE by EASTERN province, on SW by CENTRAL province, on W by LUAPULA province. Drained by CHAMBESHI RIVER; LUANGWA RIVER forms SE border. LAKE BANGWEULU and Bangweulu Swamps on W border; LAKE TANGANYIKA on N border at corner of Tanzania and Congo, port of MPULUNGU (Zambia's only port) at S end of lake; LAKE MWERU. Wantipa in NW. Part of MUCHINGA MOUNTAINS extend from SW border to center; part of NYIKA PLATEAU in NE, includes Zambia's highest point (7,120 ft/2,170 m), on Malawi border. NSUMBU and part of MWERU WANTIPA national parks in NW, NORTH LUANGWA and part of SOUTH LUANGWA national parks in SE, Isangano National Park in S center, Lavushi Manda National Park in S. Tazara railroad, oil pipeline from NDOLA, and Great North Road (or "Hell Road") cross province from SW to Tanzania in NE. Agriculture in N (coffee, tobacco, fruits); cattle. Fish from lakes TANGANYIKA and Bangweulu. Tourism at national parks.

Northern, geographic region (☐ 32,970 sq mi/85,722 sq km; 2005 population 6,091,600), N UGANDA, borders DEMOCRATIC REPUBLIC OF THE CONGO (W), SUDAN (N), and KENYA (E). Includes LAKE KWANIA in S. VICTORIA NILE RIVER forms part of SSW border, also includes ALBERT NILE, ACHWA, OKERE, OKOK, and Ora rivers. MT. MORUNGOLE (9,020 ft/2,749 m) in NE. Principle town is GULU. Composed of ABIM, ADJUMANI, AMOLATAR, AMURU, APAC, ARUA, DOKOLO, GULU, KAABONG, KITGUM, KOBOKO, KOTIDO, LIRA, MARACHA-TEREGO, MOROTO, MOYO, NAKAPIRIPIRIT, NEBBI, OYAM, PADER, and YUMBE districts.

Northern, region (2005 population 1,624,000), N SUDAN. Bordered by Eastern (E), KHARTOUM (SE), KURDOFAN (S), and DARFUR (W) regions, as well as LIBYA (NW) and EGYPT (N). Current Northern region covers approximately the same area as former Northern province. Following the reorganization of administrative divisions in 1996, Northern region is composed of River Nile and Northern states.

Northern, region (2007 population 1,550,966), N MALAWI; ⊙ MZUZU. Bordered N by TANZANIA, E by LAKE NYASA (with Tanzania and MOZAMBIQUE on the opposite shore), S by Central region (portion of border formed by Rupashe River), and W by ZAMBIA. South Rukuru River in W; LIKOMA ISLAND off E coast in Lake Nyasa; Vipliya Mountains in center of region; NYIKA PLATEAU (8,291 ft/2,527 m), located in NYIKA NATIONAL PARK, in N central; Vwaza Game Reserve in W central. Towns include CHILUMBA, CHINTHECHE, EKWENDENI, Karonga, LIVINGSTONIA, Mzuzu, and RUARWE. Roads connect the region to the rest of Malawi, as well as Zambia and Tanzania. Airports at Karonga and Mzuzu. Includes the districts of CHITIPA (N), KARONGA (NE), Likoma (E, in Lake Nyasa), RUMPHI (N central), NKHATA BAY (SE), and MZIMBA (SW).

Northern, for names beginning thus and not found here: see under NORTH. In Russian names: see also SEVERNAYA, SEVERNIYE, SEVERNOYE, SEVERNY, or SEVERO-.

Northern Anyuy Range, RUSSIA: see ANYUY.

Northern Areas, federally administered territory, comprising GILGIT and BALTISTAN dists., extreme NE

Pakistan, bordering far W China (Xinjiang) and far N India (Jammu and Kashmir states). Areas once subject to nominal Br. control. Gilgit agency was under Br. suzerainty with nominal representation of Hindu maharajah of state of Jammu and Kashmir at Partition in 1947. Baltistan, like Ladakh in present-day India, once came under direct control of Jammu. Northern Areas have no direct political representation in federal govt. but have federal administration officials in Gilgit and Sakardu. Both Gilgit and Baltistan once contained several petty princely states now all incorporated into civil administration. Major military presence with military airfields or landing strips at GILGIT, Passu, and Sakardu. With one exception, Shimshal, all villages connected to vehicle tracks and metalled roads. Electric power available almost everywhere. Completion of military-political Karakoram Highway in 1979 consolidated federal power throughout Northern Areas.

Northern Bahr al Ghazal, state, SUDAN: see North BAHR AL-GHAZAL.

Northern Black Forest, German *Nord-Schwarzwald* (NORT–SCHWAHRTS-vahlt), region (☐ 903 sq mi/2,347.8 sq km), BADEN-WÜRTTEMBERG, SW GERMANY. Chief town is PFORZHEIM. Bounded by MIDDLE UPPER RHINE (NW), FRANCONIA (N), Region of STUTTGART (E), NECKAR-JURA (SE), BLACK FOREST/BAAR-HEUBERG (S), and SOUTHERN UPPER RHINE (SW).

Northern Bruce Peninsula (NOR-thuhrn BROOS), township (☐ 302 sq mi/785.2 sq km; 2001 population 3,599), SW ONTARIO, E central Canada; 45°05′N 81°24′W. Formed in 1999 by the merger of Saint Edmunds, Lindsay, Eastnor, and LION'S HEAD; also includes Tobermory.

Northern Cape, province (☐ 139,691 sq mi/361,800 sq km; 2004 estimated population 1,078,634; 2007 estimated population 1,102,200), SOUTH AFRICA, NW areas from Sout River N to ORANGE RIVER (NAMIBIA border) up to NOSSOB RIVER, then SE around BOTSWANA border to Phephane River, then SE to FREE STATE province border near SCHWEIZER-RENEKE, S around E Free State province border to Norvals Pont, and W along watershed of S escarpment to Sout River; ⊙ KIMBERLEY; 29°30′S 22°00′E. Includes NAMAQUALAND, LITTLE NAMAQUALAND, Bushmanland, upper Karoo, Gordonia, half of former BOPHUTHATSWANA and central Cape. Mostly arid and semiarid KAROO scrubland.

Northern Circars, INDIA: see CIRCARS, NORTHERN.

Northern Cross-Island Highway or **Peipu Hengkuan Kunglu** (BAI-BOO HENG-GUAN GUNG-LOO), highway, N TAIWAN; connects taoyuan on NW coast with ILAN on NE coast, over mountains. Highway branch at Chilan connects with CENTRAL CROSS-ISLAND HIGHWAY at LISHAN.

Northern Darfur, state, SUDAN: see North DARFUR.

Northern District, BELIZE: see COROZAL.

Northern Donets River, UKRAINE: see DONETS.

Northern Dvina, RUSSIA: see DVINA, river.

Northern Dvina Canal (dvee-NAH), Russian *Severo-Dvinskiy Kanal*, system of lakes, canals, and canalized rivers 35 mi/56 km long, in W central VOLOGDA oblast, European Russia; 59°45′N 38°22′E. Links SHEKSNA River (W) with KUBENO LAKE (E); E section formed by canalized POROZOVITSA River. Joins Maryinsk canal system with basin of the NORTHERN DVINA River via the SUKHONA River. Formerly called Württemberg Canal.

Northern Green Belt, CHINA: see NORTHERN PROTECTION BELT.

Northern Ireland (EI-uhr-luhnd), division (☐ 5,462 sq mi/14,201.2 sq km; 2001 population 1,685,267) of the UNITED KINGDOM OF GREAT BRITAIN AND NORTHERN IRELAND, NE IRELAND; ⊙ BELFAST; 54°40′N 06°45′W. It is located in ULSTER province and comprises six counties—ANTRIM, ARMAGH, DOWN, FERMANAGH, LONDONDERRY, and TYRONE—divided into twenty-six districts. There are currently plans to abolish the twenty-six districts in 2009. The land is mountainous and has few natural resources. (For a detailed description of Northern Ireland's physical geography, see Ireland.) Farming (livestock; dairy products; cereals). Northern Ireland was once famous for its fine linens. Heavy industry is concentrated in and around Belfast, one of the chief ports of the BRITISH ISLES. Shipbuilding has diminished in significance, as has machinery and equipment manufacturing, food processing, and the manufacturing of textiles. Previously papermaking, furniture manufacturing.

Northern Ireland has seventeen representatives in the British Parliament and three representatives to the EU. English is the official language. Nearly 43% of the population is Catholic; the majority is Protestant. Besides Belfast, the chief towns are BANGOR, LISBURN, NEWTOWNABBEY, PORTADOWN, ARMAGH, BALLYMENA, COLERAINE, DERRY, and OMAGH.

Northern Ireland's relatively distinct history began in the early 17th century, when, after the suppression of an Irish rebellion, much land was confiscated by the British crown and "planted" with Scottish and English settlers. Ulster took on a Protestant character as compared with the rest of Ireland; but there was no question of political separation until the late 19th century, when British prime minister William Gladstone presented (1886) his first proposal for home rule for Ireland. The largely Protestant population of the N feared domination under home rule by the Catholic majority in the S. In addition, industrial Ulster was bound economically more to England than to the rest of Ireland. Successive schemes for home rule widened the rift, so that by the outbreak of World War I civil war in Ireland was an immediate danger. The Government of Ireland Act of 1920 attempted to solve the problem by enacting home rule separately for the two parts of Ireland, thus creating the province of Northern Ireland.

However, the Irish Free State (established 1922), which would become the Republic of Ireland, refused to recognize the finality of the partition; and violence erupted frequently on both sides of the border. The late 1960s marked a new stage in the region's troubled history. The Catholic minority, which suffered economic and political discrimination, had grown steadily. In 1968 civil rights protests by Catholics led to widespread violence. Prime Minister Terence O'Neill had sought to end anti-Catholic bias as part of his policy of fostering closer ties between Ulster and the Irish Republic, but opponents within his ruling Unionist Party forced his resignation in April 1969. His successor, James Chichester-Clark, was unable to restrain the growing unrest and in August called in British troops to help restore order.

At the end of 1969 a split occurred in the Irish Republican Army (IRA), which is the illegal military arm of the Sinn Fein Party; the new "provisional" wing of the IRA was made up of radical nationalists from Northern Ireland. Brian Faulkner became leader of the Unionist Party and prime minister of Northern Ireland in March 1971, and began a policy of arresting and imprisoning Catholic suspects; however, this policy did not deter terrorist groups such as the IRA. On March 30, 1972, British Prime Minister Edward Heath suspended the government. An assembly was formed in June 1972, with the Unionist Party, a moderate pro-British group, in the majority. In November the Unionist Party formed a coalition with the Social Democratic Labour Party (SDLP), the major Catholic group, and the nonsectarian Alliance Party. An eleven-person Northern Ireland Executive was formed to exercise day-to-day administration, while LONDON retained responsibility for security, foreign relations, justice, and some financial matters. WESTMINSTER's direct rule over the province was renewed in March 1973.

In late 1973, the British prime minister, the head of the Executive, and the Irish Republic's prime minister agreed to form a Council of Ireland to promote closer cooperation between Ulster and the Republic. However, sentiments in Ulster were extremely factional; many Protestant activists, finding power-sharing with the Catholics untenable, threatened to destroy the Executive and the Council. The IRA denounced the SDLP for collaboration with the Protestants and also vowed to destroy the Executive. In 1974, hard-line Ulster Protestants won eleven of the province's twelve seats in the British House of Commons, becoming the fourth-largest group in that branch of Parliament; they pledged to renegotiate Ulster's constitution to terminate the Protestant-Catholic coalition and halt progress toward a Council of Ireland.

The most dramatic attempt to upset Ulster's new governmental framework came in May 1974, when militant Protestants sponsored a general strike in the province. Under pressure from the fourteen-day strike, the Northern Ireland Executive collapsed on May 28. The British government then took direct control of the province with the passage of the Northern Ireland Act of 1974. Meanwhile, bombings and other terrorist activities had spread to DUBLIN and London. The IRA assassinated Lord Mountbatten in 1979, and protests broke out in Belfast in 1981 over the death by hunger strike of Bobby Sands, an IRA member of Parliament. In 1985, the Anglo-Irish accord sought to lay the groundwork for talks between Northern Ireland and the Republic of Ireland. Dublin agreed not to contest Northern Ireland's allegiance to Great Britain in exchange for British acknowledgment of the Republic's interest in how Northern Ireland is run. An assembly formed in 1982 to propose plans for strengthening legislative and executive autonomy in Northern Ireland was dissolved in 1986 due to its lack of progress.

By early 1990, more than 2,700 people had perished in Northern Ireland's sectarian strife, and British troop strength in Ulster remained strong, sometimes reaching 20,000. In the early 1990s, terrorist violence by the IRA and Protestant groups remained a problem to varying degrees, as talks continued between Britain, Ireland, and the political parties of Northern Ireland. The IRA broke the uneasy cease-fire in 1996, and the Orange Order marches of that summer provoked continuing unrest. The Northern Ireland Act 1998 repealed the Government of Ireland Act 1920, and the Belfast Agreement, or "Good Friday Agreement," of 1998 marked a major step in the peace process. The subsequent decade was marked by fitful progress, with home rule suspended at times as the disarmament process proceeded despite setbacks and historical distrust. In May 2007, however, following an election earlier in the year, Northern Ireland's major Catholic and Protestant parties formed a power-sharing government with Democratic Unionist Party leader Ian Paisley as the Northern Ireland assembly's first minister and Martin McGuinness of Sein Fein as deputy first minister.

Northern Karoo, NORTHERN CAPE, NORTH-WEST and SW FREE STATE provinces, SOUTH AFRICA. With HIGH VELD, forms innermost and highest of the republic's plateau regions, extends N from GREAT KAROO, bounded by NAMAQUALAND (W), KOMSBERG and ROGGEVELD escarpments (SW), and merges with the High Veld of Free State and Transvaal provinces Up to 4,000 ft/1,220 m high in Northern Cape province, it rises to c.6,000 ft/1,830 m in Transvaal High Veld. Forms the lower W half of the central escarpment.

Northern Kordofan, state, SUDAN: see North KURDOFAN.

Northern Land, RUSSIA: see SEVERNAYA ZEMLYA.

Northern Lights No. 22 (NOR-thurn LEITS), municipal district (□ 8,009 sq mi/20,823.4 sq km; 2001 population 4,217), NW ALBERTA, W Canada; 57°28′N 117°59′W. Includes MANNING and the hamlets of

Deadwood, Notikewin, Dixonville, and North Star. Formed as an improvement district in 1913; incorporated as a municipal district in 1995.

Northern Mariana Islands, commonwealth (□ 185 sq mi/481 sq km) associated with the U.S., comprising sixteen islands (six regularly inhabited) of the Marianas (all except GUAM), in the W PACIFIC OCEAN; ⊙ SAIPAN; formerly part of the U.S. TRUST TERRITORY OF THE PACIFIC ISLANDS. They lie E of the PHILIPPINES and S of JAPAN and extend 350 mi/563 km N-S. The most important islands are Saipan, ROTA, and TINIAN. The chain forms the crest of a submerged ridge of andesitic continental-type volcanoes, with the surface of the N islands composed of volcanic rock, the S islands of madrepore limestone covering a volcanic base. All the islands are mountainous, with the highest peak (3,166 ft/965 m) on AGRIHAN. Sugarcane, coffee, and coconuts are the chief products. There are deposits of phosphate, sulfur, and manganese ore. The Northern Marianas receive substantial financial assistance from the U.S. Tourism, especially from Japan, is a major industry, employing roughly 10% of the workforce. Most of the inhabitants are Chamorros (mixed Spanish, Filipino, and Micronesian descent), but there are some Carolinians and other Micronesians. The islands were explored in 1521 by Ferdinand Magellan, who named them the Ladrones Islands (Thieves Islands). They were renamed the Marianas by Spanish Jesuits who arrived in 1668. Nominally a possession of SPAIN until 1898, the islands were sold to GERMANY in 1899, except for Guam, which was ceded to the U.S. The islands belonging to Germany were seized by Japan in 1914 and were mandated to Japan by the League of Nations in 1920. U.S. forces occupied the Marianas (1944) during World War II, and in 1947 the group (exclusive of Guam) was included in the U.S. Trust Territory of the Pacific Islands. Voters approved separate status for the islands as a U.S. commonwealth in 1975; these became internally self-governing under U.S. military protection in 1978. The islands are governed by an elected governor and lieutenant governor, both serving four-year terms, and a bicameral legislature consisting of a nine-member Senate and a fifteen-member House of Representatives. Residents are U.S. citizens but do not vote in U.S. Presidential elections.

Northern Mindanao (MEEN-dah-nou), region (□ 5,418 sq mi/14,086.8 sq km), N central MINDANAO, PHILIPPINES, consisting of CAMIGUIN, BUKIDNON, MISAMIS OCCIDENTAL, and Misamis Oriental provinces. Population 44.9% urban and 55.1% rural. Mixed agriculture (corn, coconut, pineapple); mining (chromite, maganese, limestone); fishing; and some manufacturing (wood products; food processing, smelting). AGUSAN DEL NORTE, AGUSAN DEL SUR, and SURIGAO DEL NORTE provinces, originally part of region, were transferred to CARAGA region c.1996. Regional center, CAGAYAN DE ORO. Also known as Region 10. Created 1975.

Northern Nafud, SAUDI ARABIA: see NAFUD.

Northern Neck (NOR-thurn NEK), E VIRGINIA, Tidewater region peninsula between POTOMAC RIVER (N) and RAPPAHANNOCK RIVER (S) estuaries, bounded E by CHESAPEAKE BAY; extends c.65 mi/105 km SE from PORT ROYAL and KING GEORGE. Includes Westmoreland, RICHMOND, NORTHUMBERLAND, and LANCASTER counties, and part of KING GEORGE county. NE shore (except for inlets) forms MARYLAND-Virginia state line. Ferries from E end to SMITH and TANGIER islands.

Northern Ossetian Preserve (uh-SYE-tee-ahn), wildlife refuge (□ 100 sq mi/260 sq km), NORTH OSSETIAN REPUBLIC, SE European Russia. Established in 1967. Has ecotourism and alpine sports camps.

Northern Panhandle, NW WEST VIRGINIA (□ 584 sq mi/1,513 sq km), narrow arm of state, extending N between PENNSYLVANIA (E) and OHIO (W and N); OHIO RIVER forms N and W boundary; includes

MARSHALL, OHIO, BROOKE, and HANCOCK counties. Largely industrial, due to proximity to PITTSBURGH steel district (to E) and to region's abundant natural resources (especially coal, clay, natural gas, glass sand, salt). WEIRTON (steel) and WHEELING (diversified industries) are principal manufacturing centers.

Northern Protection Belt or **Northern Green Belt**, CHINA, a major green-belt project designed to improve the ecological environment in semi-arid N areas of the country. Begun in 1978, the project is expected to be completed in 2050. The chief goal is to reforest a geographic belt that extends from BIN XIAN in HEILONGJIANG province in the E to the Uzibeli Pass of Xinjiang in the W. In the belt are 551 counties in BEIJING, TIANJIN, and 11 other provinces and autonomous regions in NE, N, and NW China. Major regions involved are the W Northeast Plain, the NORTH CHINA PLAIN, the HETAO area of the HUANG HE (Yellow River), the HEXI Corridor, and Xinjiang. The total projected forestation area is 137,394 sq mi/355,850 sq km, and it is expected to increase forest cover from 5 percent in 1977 to 15 percent in 2050. Also called Sanbei Protection Belt or Three Northern Regions Protection Belt because it traverses three N regions of China (NW, N, and NE China).

Northern Province, administrative division (□ 2,736 sq mi/7,113.6 sq km) of N SRI LANKA; ⊙ JAFFNA; 09°00′N 80°30′E. Comprises two natural division, JAFFNA Peninsula (N), with group of islands lying off W coast, and mainland (dense jungle area in interior known as the WANNI); lagoons along E coast. Agriculture (rice, tobacco, chilis, yams, onions, millet); extensive coconut- and palmyra-palm plantations. Fishing (Jaffna Lagoon, MANNAR Island); pearl banks in Gulf of Mannar. Main towns include Jaffna, CHAVAKACHCHERI, Mannar, MULLAITIVU, and VAVUNIYA. Large salterns at Jaffna and ELEPHANT PASS. Most of the province has been under the control of the Liberation Tigers, a Tamil militant group, since c.1986. A Sri Lankan government offensive recovered parts of the Jaffna Peninsula in 1996. Created 1833.

Northern Province, former administrative division and province (□ 236,200 sq mi/614,120 sq km), N SUDAN; ⊙ AD-DAMER. Bordered by EGYPT (N) and LIBYA (NW), it was in S part of LIBYAN DESERT and W part of NUBIAN DESERT, which are separated by the NILE. A strip 0.5 mi/0.8 km–3 mi/4.8 km wide on either side of river is cultivable. The N part of the Nile Valley within region was flooded by Lake NASSER (Lake NUBIA). Agriculture includes cotton, wheat, barley, corn, and fruits; nomadic grazing (camels, goats, and sheep). The LIBYAN DESERT (W of the NILE) is uninhabited and waterless with the exception of a few oases. The NUBIAN DESERT is crossed here by railroad from ABU HAMED to WADI HALFA, cutting off great bend of the NILE. There are numerous ancient ruins (MARAWI, NAPATA). Chief towns are ATBARAH, BERBER, SHENDI, WADI HALFA, and AD-DAMER. Province was formed in 1930s by amalgamation of Halfa, DONGOLA, and Berber provinces. Since 1996, Northern province was divided into the River Nile, the Red Sea, and N states and part of North Darfur state.

Northern Province, SOUTH AFRICA: see LIMPOPO PROVINCE.

Northern Provinces, former major administrative division of N NIGERIA; (⊙ Kaduna until 1967). Included what are now the states of ADAMAWA, BAUCHI, BENUE, BORNO, KWARA, KABBA, KANO, KATSINA, NIGER, Plateau, SOKOTO, and KADUNA. Territories came (c. 1885) under British protection and were first administered by Royal Niger Co. Passed to the crown (1900); known as protectorate of N Nigeria until amalgamated (1914) with colony and protectorate of S Nigeria. Most of the names of present-day states in N Nigeria are derived from the old Northern province names.

Northern Region (□ 27,175 sq mi/70,384 sq km; 1990 estimated population 1,472,900; 2000 population 1,820,806), N central GHANA; ⊙ TAMALE. Borders Côte d'IVOIRE to W, BRONG-AHAFO REGION to S, VOLTA region to SE, UPPER WEST REGION to NW, UPPER EAST REGION to NE, and TOGO to E. The largest of Ghana's ten regions, covering ⅓ of the country's total area, it consists of mixed savannah woodland. Main towns are Tamale, YENDI, Bembilla, and NALERIGU. Agricultural (maize, rice, sorghum, yams, tomatoes, shea nuts, and kapole); livestock. Cotton and groundnuts are commercial crops.

Northern Region, SAUDI ARABIA; geographical region that includes Northern Frontier, HAIL, and Tabouk provinces. One of five such regions of Saudi Arabia.

Northern Rhodesia: see ZAMBIA.

Northern Rockies (NOR-thurn RAH-keez), regional district (□ 32,877 sq mi/85,480.2 sq km; 2001 population 5,720), NE BRITISH COLUMBIA, W Canada. Includes the town of FORT NELSON. Forestry; agriculture; oil and gas; tourism. In the area are Muncho Lake, Stone Mountain, and Liard River (springs) provincial parks.

Northern Samar (SAH-mahr), province (□ 1,351 sq mi/3,512.6 sq km), in EASTERN VISAYAS region, N SAMAR, PHILIPPINES; ⊙ CATARMAN; 12°20′N 124°40′E. Population 35.4% urban, 64.6% rural; in 1991, 58% of urban and 23% of rural settlements had electricity. W is chiefly wooded mountains with narrow coastal plain; lowland river valleys more extensive in center and E. In typhoon belt, with year-round rain; heaviest November–February. Agriculture (rice and corn as food crops; abaca and coconuts as cash crops); province imports much of its food. Fishing; mining (coal, silicates, white clay). Airport at Catarman. Fishing and trade port at San Jose; ferry at ALLEN.

Northern Sea Route, RUSSIA: see NORTHEAST PASSAGE.

Northern Shan State, MYANMAR: see SHAN STATE.

Northern Sosva River, RUSSIA: see SOSVA RIVER.

Northern Sporades, Greece: see SPORADES.

Northern Sunrise County (NOR-thurn SUHN-reiz), municipality, N central ALBERTA, W Canada; 56°38′N 116°04′W. Oil and gas; forestry; agriculture; tourism. Includes Nampa and the hamlets of Cadotte Lake, Reno, Little Buffalo, Saint Isidore, and Marie Reine. Formed as an improvement district in 1913; incorporated as a municipal district in 1994.

Northern Territories, former British protectorate (□ 30,600 sq mi/79,560 sq km), N GHANA; capital was TAMALE; between 08°00′N and 11°00′N. Bounded W by CÔTE D'IVOIRE, N by Upper Volta (BURKINA FASO), E by TOGOLAND. Included Black Volta (MOUHOUN) (W and S borders) and White Volta (NAKAMBE) rivers, which join to form VOLTA RIVER. Hills (N and W) slope gently toward Volta basin. Was divided into tribal districts (Dagomba, Gonja, Mamprusi, Krachi, Wa) named for principal tribes. N section of British Togoland was administered as part of Northern Territories. Region came under British influence in 1897; made a protectorate in 1901. No longer recognized as a separate region; comprises sections of UPPER WEST, UPPER EAST, and NORTHERN regions.

Northern Territory (NOR-thurn), territory (□ 520,280 sq mi/1,347,525 sq km; 2006 pop. 192,898), N central AUSTRALIA; ⊙ DARWIN; 15°00′S 133°00′E. It is bounded on the N by the TIMOR SEA, the ARAFURA SEA, and the Gulf of CARPENTARIA. In the N are lowlands, in the SE are low plains sloping toward the Lake Eyre depression, and in the SW are the MACDONNELL RANGES. The main rivers are the VICTORIA, DALY, ADELAIDE, and ROPER, all of which drain into the N seas. The climate in the N is tropical, with a monsoon season; the S becomes cooler and drier as the elevation rises. Most of the population lives in the Darwin and ALICE SPRINGS metropolitan areas. More than a quarter of the Northern Territory's people identify themselves as Aboriginal or TORRES STRAIT Islander. Indigenous Australians own the land of fifteen reservations with a total area of 94,000 sq mi/243,460 sq km; the ARNHEM LAND preserve is the largest. Much of this land is extremely important to the uranium-mining and tourist industries. The territory's economic development has been accelerating in recent years. Gold is worked to a very small extent; uranium, bauxite, manganese, iron, lead, and zinc deposits are increasingly exploited. Livestock breeding, encouraged by government-development projects, is the major rural activity. There is very little farming in the territory. Peanuts, pearl shell, and trepang are the principal exports. The Northern Territory's first settlement was established at Port Essington in 1824 in an attempt to forestall French colonization. The settlement failed, and permanent settlement did not resume until 1869. The Northern Territory was part of NEW SOUTH WALES from 1825 to 1863 and of SOUTH AUSTRALIA from 1863 to 1911. Transferred to direct rule by the commonwealth in 1911, it was divided into two territories in 1926 but was reunited in 1931. In 1974, the commonwealth-appointed Legislative Council was replaced by the fully elected Legislative Assembly. In 1978, full self-government was granted to the territory. The Northern Territory elects a member with full voting rights to the commonwealth House of Representatives.

Northern Uvals (oo-VAHLS), Russian *Severnyye Uvaly*, one of the moraine uplands [Russian *uvaly*], in VOLOGDA, KOSTROMA, KIROV, and ARCHANGEL oblasts and KOMI REPUBLIC, N European Russia; form watershed between NORTHERN DVINA and Volga-Kama river basins; extend from area of SOLIGALICH (Kostroma oblast) approximately 450 mi/724 km E to the upper VYCHEGDA River; rise to 695 ft/212 m.

Northern Vosges, FRANCE: see VOSGES, NORTHERN.

North Esk River (NORTH ESK), 45 mi/72 km long, TASMANIA, SE AUSTRALIA; rises in mountains 15 mi/24 km S of SCOTTSDALE; flows S, W, and NW to LAUNCESTON, here joining SOUTH ESK RIVER to form TAMAR RIVER; 41°26′S 147°08′E.

North Fabius River, MISSOURI and IOWA: see FABIUS RIVER.

North Fairfield (FER-feeld), village (2006 population 574), HURON county, N OHIO, 10 mi/16 km S of NORWALK, and on East Branch of HURON RIVER; 41°06′N 82°37′W.

North Fair Oaks, unincorporated city (2000 population 15,440), SAN MATEO county, W CALIFORNIA, residential suburb 24 mi/39 km SSE of downtown SAN FRANCISCO, and 3 mi/4.8 km SW of SAN FRANCISCO BAY; 37°28′N 122°12′W. HETCH HETCHY AQUEDUCT passes to NW.

North Falmouth, MASSACHUSETTS: see FALMOUTH.

North Ferriby (FE-rib-ee), village (2001 population 3,819), East Riding of Yorkshire, E ENGLAND, on HUMBER RIVER, and 7 mi/11.3 km WSW of HULL; 53°43′N 00°30′W. Previously cement works.

Northfield, city (2000 population 17,147), RICE county, SE MINNESOTA, 35 mi/56 km S of MINNEAPOLIS, on the CANNON RIVER; 44°27′N 93°10′W. Railroad junction and trade center for farming region (corn, oats, soybeans; livestock, poultry; dairying); manufacturing (printed circuit boards, toys, feeds and seeds, cereals). On September 7, 1876, Jesse and Frank James and their bandit gang attempted a bank robbery here, which failed and resulted in the deaths of two Northfield citizens. Each September, Northfield holds a festival that reenacts the robbery attempt. Carleton College, St. Olaf College, and the Laura Baker School for mentally challenged children are in the city. Historical Society Museum. Nestrand Woods State Park to SE. Incorporated 1875.

Northfield, city (2006 population 8,003), ATLANTIC county, SE NEW JERSEY, 6 mi/9.7 km W of ATLANTIC CITY; 39°22′N 74°32′W. Incorporated 1905.

Northfield, town (2000 population 970), JEFFERSON county, N KENTUCKY; residential suburb 6 mi/9.7 km ENE of downtown LOUISVILLE, near OHIO RIVER;

38°17′N 85°38′W. Zachary Taylor National Cemetery to SW.

Northfield, town, WASHINGTON county, E MAINE, 9 mi/14.5 km NW of MACHIAS; 44°49′N 67°36′W.

Northfield, town, FRANKLIN county, N MASSACHUSETTS, on CONNECTICUT River, and 10 mi/16 km NE of GREENFIELD, near NEW HAMPSHIRE state line; 42°41′N 72°27′W. A beautiful NEW ENGLAND town. Dwight Lake Moody born here, founded Northfield Seminary (now Northfield School) for girls and Mount Hermon School for boys. Headquarters of American Youth Hostels; first hostel opened here 1934. Town first settled 1672; permanently settled 1714; incorporated 1723. Includes villages of East Northfield, South Vernon, and Mount Hermon.

Northfield, town, MERRIMACK county, S central NEW HAMPSHIRE, 4 mi/6.4 km S of FRANKLIN; 43°24′N 71°34′W. Bounded W by MERRIMACK RIVER, N by Winnipesaukee River. Agriculture (livestock, poultry; vegetables, apples; dairying; nursery crops; timber); manufacturing (boxes, batteries). Covered bridges in N.

Northfield, town (2006 population 3,144), including Northfield village, WASHINGTON co., central VERMONT, 10 mi/16 km SW of MONTPELIER; 44°08′N 72°41′W. Winter sports. Home to Norwich University since 1866. Chartered 1781; settled 1785; organized 1794. Probably named for NORTHFIELD, MASSACHUSETTS.

Northfield (NORTH-feeld), former village, SW QUEBEC, E Canada; 46°04′N 75°57′W. Amalgamated into GRACEFIELD in 2002.

Northfield, village (2000 population 5,389), COOK county, NE ILLINOIS, N suburb of CHICAGO, just W of WINNETKA; 42°06′N 87°46′W. Food processing.

Northfield, village (2006 population 3,715), SUMMIT county, NE OHIO, 15 mi/24 km N of downtown AKRON, near CUYAHOGA RIVER; 41°20′N 81°31′W.

North Field, natural gas reserve (□ 6,000 sq mi/15,600 sq km), in PERSIAN GULF, NE of QATAR peninsula. Discovered in 1971, Qatar's NORTH FIELD (about half the area of the country as a whole) is the world's largest gas field. The total reserve is c.500 trillion sq ft/46 trillion sq m, making Qatar one of the top five gas-rich nations in the world. The field is serviced by Ras Lafan port. Marketing plans have begun, and joint ventures with JAPAN, SOUTH KOREA, TAIWAN, TURKEY, and INDIA are underway to expand production. It has been estimated that Qatar's natural gas supply from the North Field will last for over 200 years.

Northfleet (NAWTH-fleet), town (2001 population 27,384), NW KENT, SE ENGLAND, 3 mi/4.8 km W of GRAVESEND; 51°26′N 00°20′E. Shipbuilding and the production of cables and paper are the main industries. In the center of town is a Roman Catholic church designed by George Gilbert Scott. Huggens College is located here.

North Fond du Lac (fawn doo LAK), town (2006 population 4,908), FOND DU LAC county, E WISCONSIN, on LAKE WINNEBAGO, suburb 2 mi/3.2 km NW of FOND DU LAC; 43°48′N 88°28′W. Railroad junction. In farm area. Light manufacturing; railroad shops. Incorporated 1903.

North Foreland (NAWTH FAW-luhnd), promontory at NE tip of KENT, SE ENGLAND, on Isle of THANET, on S shore of THAMES estuary, just N of BROADSTAIRS at N end of The DOWNS roadstead. Site of lighthouse; 51°22′N 01°27′E.

Northfork, village (2006 population 446), MCDOWELL county, S WEST VIRGINIA, on TUG FORK RIVER, 8 mi/12.9 km E of WELCH; 37°25′N 81°25′W. Trade center for semibituminous-coal-mining region.

North Fork or **North Fork Malheur River**, 55 mi/89 km long, E OREGON; rises in SW corner of BAKER county; flows S through Beulah Reservoir to MALHEUR RIVER at JUNTURA. Agency Valley Dam (55 mi/89 km high, 1,850 ft/564 m long; completed 1935), 12 mi/19 km

NNW of Juntura, forms Beulah Reservoir (2.5 mi/4 km long, 1.5 mi/2.4 km wide; capacity 59,900 acre-ft). Unit in VALE Irrigation Project.

North Fork Blackfoot River, c.40 mi/64 km long, W MONTANA; rises in NW LEWIS AND CLARK county at CONTINENTAL DIVIDE; flows SSW to join BLACKFOOT RIVER 2 mi/3.2 km S of OVANDO.

North Fork Musselshell River (MUH-suhl-SHEL), c.35 mi/56 km long, central MONTANA; rises in LITTLE BELT MOUNTAINS, NE MEAGHER county; flows S through Bair Reservoir, then SE, joins main MUSSELSHELL RIVER just W of WHEATLAND county line.

North Fork of Red River (RED), c.195 mi/314 km long, N TEXAS and SW OKLAHOMA; rises in the LLANO ESTACADO W of LEFORS, GRAY county, N Texas; flows generally E to SAYRE, Oklahoma, then SSE through ALTUS LAKE reservoir and S to RED RIVER 10 mi/16 km WSW of FREDERICK, Oklahoma. Elm Fork (c.90 mi/145 km long) rises in SW WHEELER county, Texas as an intermittent stream; flows ESE; enters from W, c.10 mi/16 km E of MANGUM, Oklahoma.

North Fork of Wichita River, Texas: see WICHITA RIVER.

North Fork Republican River, c.30 mi/48 km long, in COLORADO and NEBRASKA; rises W of WRAY, in YUMA county, NE Colorado; flows E, past Wray, Colorado, into DUNDY county, SW Nebraska, receiving ARIKAREE RIVER from SW at HAIGLER, then joining SOUTH FORK REPUBLICAN RIVER at BENKELMAN, forming REPUBLICAN RIVER.

North Fork River, c.100 mi/161 km long; rises in DOUGLAS county in the OZARKS, S MISSOURI; flows S, into N ARKANSAS, to the WHITE RIVER. Near its mouth is Norfork Dam (completed 1944), which impounds NORFORK LAKE and has a power plant. Popular canoeing, floating, and fishing stream above lake. Many springs feed it. Commonly called Norfork River North Fork of the White River.

North Fork Virgin River, 40 mi/64 km long, SW UTAH; rises in SE IRON county; flows through Zion Canyon, the main focal point of ZION NATIONAL PARK, joining EAST FORK VIRGIN RIVER at SPRINGDALE, just S of park, to form VIRGIN RIVER.

North Fort Myers (MEI-uhrz), city (□ 23 sq mi/59.8 sq km; 2000 population 40,214), LEE county, SW FLORIDA, 3 mi/4.8 km NNW of FORT MYERS; 26°42′N 81°53′W. Manufacturing includes signs, concrete products, shell products, and coil winding.

North Fox Island, MICHIGAN: see FOX ISLANDS.

North Franklin Mountain, Texas: see FRANKLIN MOUNTAINS.

North Freedom, village (2006 population 598), SAUK county, S central WISCONSIN, on BARABOO RIVER, and 6 mi/9.7 km W of BARABOO; 43°27′N 89°51′W. In dairy and livestock area. Railroad museum. Natural Bridge State Park to SW.

North Friesland (FREES-lahnd), German *Nordfriesland*, region (□ 658 sq mi/1,710.8 sq km), SCHLESWIG-HOLSTEIN, NW GERMANY, on NORTH SEA, between Danish border and EIDERSTEDT peninsula; includes most of NORTH FRISIAN islands. Low, level region, with reclaimed land along shore; noted cattle raising. Main town: HUSUM. Region came (15th century) to duchy of SCHLESWIG.

North Frisian Islands, DENMARK and GERMANY: see FRISIAN ISLANDS.

North Frontenac (FRAHN-tuh-nak), township (□ 439 sq mi/1,141.4 sq km; 2001 population 1,801), SE ONTARIO, E central Canada; 45°00′N 76°54′W.

North Garden (NORTH GAHRD-uhn), unincorporated village, ALBEMARLE county, N central VIRGINIA, 10 mi/16 km SW of CHARLOTTESVILLE; 37°56′N 78°38′W. Manufacturing (lumber, concrete, apple brandy); agriculture (dairying, livestock; grain, apples).

Northgate (NORTH-gait), village, SE SASKATCHEWAN, CANADA, frontier station on NORTH DAKOTA

border, 35 mi/56 km ESE of ESTEVAN; 49°00′N 102°16′W.

Northgate, village, BURKE county, NW NORTH DAKOTA, port of entry at CANADIAN border, 14 mi/23 km N of BOWBELLS; 48°59′N 102°15′W. Das Lacs National Wildlife Refuge to E. Founded in 1914. Post office closed in 1985.

Northgate, unincorporated village (2000 population 8,016), HAMILTON county, extreme SW OHIO, c.6 mi/10 km NW of CINCINNATI; 39°15′N 84°35′W.

North Geomagnetic Pole, the point on the North Hemisphere surface toward which the Earth's magnetic field lines in space converge; it is the effective North magnetic pole at large distances from the Earth. It lies near ETAH on the NW coast of GREENLAND.

North Girard, Pennsylvania: see LAKE CITY.

North Glengarry (NORTH glen-GA-ree), township (□ 248 sq mi/644.8 sq km; 2001 population 10,589), SE ONTARIO, E central Canada, 24 mi/39 km from CORNWALL; 45°22′N 74°40′W. Has Scottish heritage. Includes the communities of ALEXANDRIA, Apple Hill, Dalkeith, Dominionville, Dunvegan, Fassifern, Glen Robertson, Greenfield, Laggan, Lochiel, MAXVILLE, and Saint Elmo.

Northglenn, city (2000 population 31,575), ADAMS county, N central COLORADO, suburb 10 mi/16 km N of downtown DENVER, near SOUTH PLATTE RIVER; 39°54′N 104°58′W. Elevation 5,460 ft/1,664 m. Manufacturing of commercial greenhouses.

North Goa, district (□ 670 sq mi/1,742 sq km), GOA state, SW INDIA; ⊙ PANAJI.

North Gower (GOR), unincorporated village, SE ONTARIO, E central Canada, 18 mi/29 km S of OTTAWA, in which it is officially included; 45°07′N 75°42′W. Dairying; mixed farming.

North Grafton, MASSACHUSETTS: see GRAFTON.

North Grenville (NORTH GREN-vil), township (□ 135 sq mi/351 sq km; 2001 population 13,581), SE ONTARIO, E central Canada; 44°58′N 75°39′W.

North Grosvenor Dale, CONNECTICUT: see THOMPSON.

North Grove, village, MIAMI county, N central INDIANA, 12 mi/19 km SSE of PERU. In agricultural area.

North Guilford, town, NEW HAVEN county, S CONNECTICUT, 12 mi/19 km NE of NEW HAVEN on Coginchaug River.

North Gulfport, unincorporated town, HARRISON county, SE MISSISSIPPI, residential suburb 4 mi/6.4 km N of downtown GULFPORT, on Bernard Bayou; 30°24′N 89°05′W.

North Haledon (HAIL-duhn), borough (2006 population 9,039), PASSAIC county, NE NEW JERSEY, 3 mi/4.8 km N of PATERSON; 40°57′N 74°11′W. Incorporated 1901. Largely residential.

North Hamgyong (HAHM-GYOUNG), Korean, *Hamgyong-pukdo*, northernmost province, NORTH KOREA; ⊙ and major port is CHONGJIN. Bounded E by SEA OF JAPAN, W by TUMAN RIVER (on Manchurian and Russian borders). Extremely mountainous interior rises to 8,337 ft/2,541 m (Mount Kwanmo). There are several hot-springs resorts, the best-known being Chuuronjang. Rivers (other than the Tuman) are small; they drain extensive lumbering and coal-mining area. Iron, graphite, and gold are mined in scattered areas. Fishing is an important industry. There is some livestock raising; agriculture is generally limited to the cultivation of soybeans, millet, and potatoes. Industrial centers are Chongjin and Songjin; NAJIN is a naval base. The Kimchaek Mining Complex is one of the leading producers of high-grade electrolytic zinc and lead nationally.

North Hampton, town, ROCKINGHAM county, SE NEW HAMPSHIRE, 7 mi/11.3 km SSW of PORTSMOUTH, and 2 mi/3.2 km N of HAMPTON; 42°58′N 70°49′W. Bounded E by ATLANTIC OCEAN. Agriculture (poultry; dairying; nursery crops); manufacturing (freight elevators, software). Fuller Gardens and Hampton Beach Casino in E. Set off from Hampton 1742.

North Hampton (HAMP-tuhn), village (2006 population 360), CLARK county, W central OHIO, 8 mi/13 km NW of SPRINGFIELD; 39°59′N 83°57′W.

North Hampton (NORTH HAMP-tuhn), unincorporated suburb of HAMPTON, SE VIRGINIA.

North Hangay, MONGOLIA: see ARHANGAY.

North Hanover, township, BURLINGTON county, S NEW JERSEY, 8 mi/12.9 km N of FORT DIX; 40°04′N 74°35′W. Manufacturing. Incorporated 1905.

North Hartsville, unincorporated town (2000 population 3,136), DARLINGTON county, NE SOUTH CAROLINA, 1 mi/1.6 km N of HARTSVILLE; 34°23′N 80°04′W.

North Harwich, MASSACHUSETTS: see HARWICH.

North Hatley (north HAT-lee), village (□ 1 sq mi/2.6 sq km), Estrie region, S QUEBEC, E Canada, on Lake MASSAWIPPI, 9 mi/14 km SSW of SHERBROOKE; 45°25′N 71°54′W. Dairying; mixed farming. Resort.

North Haven, town (2006 population 23,035), NEW HAVEN county, S CONNECTICUT, on the QUINNIPIAC RIVER; settled c.1650, set off from NEW HAVEN in 1786; 41°22′N 72°51′W. Chiefly residential, it has some manufactures, such as aircraft parts, tools, chemicals, and machinery. Several 18th-century houses still stand in the town.

North Haven, resort town, KNOX county, S MAINE, on NORTH HAVEN Island (8 mi/12.9 km long, 0.5 mi/0.8 km–2 mi/3.2 km wide), in PENOBSCOT BAY, and 12 mi/19 km ENE of ROCKLAND; 44°09′N 68°53′W. Includes Pulpit Harbor village.

North Haven, resort village (□ 2 sq mi/5.2 sq km; 2006 population 868), SUFFOLK county, SE NEW YORK, on E LONG ISLAND, just N of SAG HARBOR, on peninsula just E of NOYACK BAY; 41°01′N 72°19′W.

North Henderson (NORTH HEN-duhr-suhn), unincorporated village, VANCE county, N NORTH CAROLINA, a suburb 1 mi/1.6 km N of HENDERSON; 36°20′N 78°23′W.

North Hero, island and town, ⊙ GRAND ISLE CO., NW VERMONT, in Lake CHAMPLAIN, 10 mi/16 km W of St. Albans. Island (c.12 mi/19 km long) is bridged to GRAND ISLE (S) and ALBURG (N); 44°50′N 73°16′W. North Hero and Knight Point state parks are here.

North Hickory (NORTH HI-kuh-ree), unincorporated town, CATAWBA county, W central NORTH CAROLINA, residential suburb 3 mi/4.8 km NNE of Hickory Lake; 35°46′N 81°19′W. Hickory reservoir (CATAWBA RIVER) to N and NW.

North Highlands, unincorporated city (2000 population 44,187), SACRAMENTO county, N central CALIFORNIA, a residential suburb 9 mi/14.5 km NE of downtown SACRAMENTO, in the SACRAMENTO VALLEY; 38°40′N 121°23′W. Grew dramatically 1970s–1990s. Agriculture (citrus, tomatoes, grain; nursery products; dairying; poultry); manufacturing (draperies, concrete products, lumber; printing and publishing). MCCLELLAN AIR FORCE BASE to SW.

North High Shoals, town (2000 population 439), OCONEE county, N central GEORGIA; 33°50′N 83°30′W.

North Hills, town (2006 population 863), WOOD county, NW WEST VIRGINIA, residential suburb 4 mi/6.4 km NE of PARKERSBURG; 39°18′N 81°30′W.

North Hills, residential village (□ 2 sq mi/5.2 sq km; 2006 population 4,406), NASSAU county, SE NEW YORK, on W LONG ISLAND, c.2 mi/3.2 km N of MINEOLA; 40°46′N 40°73′W.

North Himsworth, Canada: see CALLANDER.

North Holland, Dutch Noord-Holland, province (□ c.1,080 sq mi/2,800 sq km; 2007 estimated population 2,613,070), W and NW NETHERLANDS, mostly a peninsula bounded W by NORTH SEA, N by WADDENZEE, NE by IJSSELMEER, and E by MARKERMEER; includes TEXEL island to N, most SW of Wadden (East FRISIAN) Islands; ⊙ is HAARLEM. Other cities include AMSTERDAM (royal capital of Netherlands and its largest city), HILVERSUM, IJMUIDEN, DEN HELDER, and ZAANSTAD. AFSLUITDIJK barrier dam (19 mi/31 km long) extends SE from E center of peninsula to form Markermeer. North Holland largely comprises low-lying fenland, most of it reclaimed from the North Sea, with large areas below sea level. Dunes line the W coast and include the National Park De Kennemerduinen and Noordholland Dune Reserve, both in W center. It is crossed by the NOORDZEE, NORTH HOLLAND, and AMSTERDAM-RHINE canals, as well as many smaller ones. Fishing; dairying; sheep, cattle; vegetables, fruit, nursery stock. Most of the Netherlands's flower industry is centered along the province's coast, and manufacturing is concentrated at Amsterdam and in ZAANSTREEK industrial area, to NW. There are many picturesque drawbridges, windmills, and tulip fields. For the history of the province, see HOLLAND.

North Holland Canal, Dutch Noordholland Kanaal, NORTH HOLLAND province, NW NETHERLANDS, extends 50 mi/80 km N-S, from E entrance of MARSDIEP strait on WADDENZEE, arm of NORTH SEA, at DEN HELDER S past ALKMAAR, turns E, skirts to N of ALKMAARDERMEER lake, turns S at PURMEREND, and joins HET IJ channel, AMSTERDAM's inland port. Built 1819–1824 to give Amsterdam sea access (eventually proved inadequate and was superseded by the NORTH SEA CANAL).

North Hollywood, suburban section of LOS ANGELES, LOS ANGELES county, S CALIFORNIA, in SAN FERNANDO VALLEY, 12 mi/19 km NW of downtown Los Angeles, W of BURBANK. Manufacturing (wood containers, stationery, electronic capacitors, baked goods, musical instruments, metal products). Hollywood Burbank Airport to N; Universal Studios to S. Formerly called Lankershim.

North Hornell, village (2006 population 834), STEUBEN county, S NEW YORK, on CANISTEO RIVER, opposite HORNELL; 42°21′N 77°39′W.

North Horr, village, EASTERN province, N KENYA, on road, and 210 mi/338 km NNW of ISIOLO, E of Lake TURKANA; 03°19′N 37°04′E. Livestock raising; peanuts, sesame, corn. Airfield.

North Hsenwi State (shen-WEE), former state (□ 6,422 sq mi/16,697.2 sq km), MYANMAR; the capital was Hsenwi. Bounded E and N by China (YUNNAN prov.) and NW by SHWELI R., it is astride THANLWIN R. Rice in valleys, cotton, sesame, tea, licorice root, corn, opium in hills. Served by Mandalay-Lashio RR and Burma Road. Pop. consists of Kachins and Chinese (N), Shans and Palaungs (S).

North Hudson, town (2006 population 3,762), SAINT CROIX county, W WISCONSIN, on SAINT CROIX RIVER (LOWER SAINT CROIX NATIONAL SCENIC RIVERWAY), 1 mi/1.6 km N of HUDSON; 45°00′N 92°45′W. Dairying. Willow River State Park to NE.

North Huntingdon, township, WESTMORELAND county, SW PENNSYLVANIA, suburb 14 mi/23 km SE of PITTSBURGH; 40°23′N 79°44′W. Manufacturing (machinery, tools, temperature controls); agriculture (dairying; livestock; corn); however, agriculture is losing ground to urban growth. Population is centered in NW in Circleville, Sunset Valley; and NE in WESTMORELAND CITY. Shuster Cellars Winery to S.

North Huron (NORTH HYU-ruhn), township (□ 69 sq mi/179.4 sq km; 2001 population 4,984), S ONTARIO, E central Canada, 16 mi/25 km from GODERICH; 43°49′N 81°24′W. Agriculture, manufacturing, tourism. Formed in 2001 from BLYTH, East Wawanosh, and WINGHAM.

North Hwanghae (HWAHNG-HAE), province, NORTH KOREA, bounded by PYONGYANG city on S, KANGWON province on E, and SOUTH HWANGHAE on W and S; ⊙ SARIWON. Popular areas are SARIWON city, SONGNIM city, and HWANGJU co. Cement and textile industries.

North Hyde Park, VERMONT: see HYDE PARK.

North Hykeham (HEI-kuhm), town (2001 population 11,538), LINCOLNSHIRE, E ENGLAND, 3 mi/4.8 km SSW of LINCOLN; 53°11′N 00°35′W. WITHAM RIVER to E.

Northill (nawth-HIL), agricultural village (2001 population 2,288), central BEDFORDSHIRE, central ENGLAND, 3 mi/4.8 km WNW of BIGGLESWADE; 52°06′N 00°19′W. Has 14th-century church.

North Irwin (UHR-win), borough (2006 population 855), WESTMORELAND county, SW PENNSYLVANIA, residential suburb 16 mi/26 km SE of PITTSBURGH, and 1 mi/1.6 km NW of IRWIN, on Brush Creek; 40°20′N 79°42′W. Incorporated 1894.

North Island, part of MAHÉ GROUP in the SEYCHELLES, in INDIAN OCEAN, 20 mi/32 km NNW of VICTORIA; 04°23′S 55°15′E. Granite formation; covers 525 acres/212 ha.

North Island, SOUTH CAROLINA: see WINYAH BAY.

North Island Naval Air Station, California: see SAN DIEGO.

North Island, The (□ 44,702 sq mi/115,778 sq km; 2006 population 3,059,418), NEW ZEALAND; WELLINGTON. Other principal cities include AUCKLAND. Separated from the SOUTH ISLAND by COOK STRAIT, the North Island is irregularly shaped due to the conjunction of two basic geological structures. The long NW peninsula follows a pattern extending SE from North Caledonia. The hard-rock Main Ranges extending SW from EAST CAPE continue the alignment of the South Island ranges. The two structures meet in the central volcanic plateau. The highest volcanoes are MOUNT RUAPEHU (9,175 ft/2,797 m) and MOUNT EGMONT (TARANAKI) (8,260 ft/2,518 m) offset to the W. TAUPO is New Zealand's largest lake (hot springs resorts here), drained by the longest river, the WAIKATO, important for hydroelectricity. Although it is the smaller of the two principal islands, it has almost three-fourths of New Zealand's population. The island produces most of New Zealand's dairy goods, lamb, warm climate fruits and wines. Natural gas, coal, and iron facilitate its consumer-based manufacturing.

North Jay, MAINE: see JAY.

North Johns or **Johns**, town (2000 population 142), Jefferson co., N central Alabama, 20 mi/32 km SW of Birmingham.

North Judson, town (2000 population 1,675), STARKE county, NW INDIANA, 10 mi/16 km SW of KNOX; 41°13′N 86°47′W. Agriculture (grain; livestock); manufacturing (dried foliage, meatpacking, furnaces and air conditioners). Laid out 1861.

North Kanara, INDIA: see UTTAR KANNAD.

North Kansas City, city (2000 population 4,714), CLAY county, W MISSOURI, on N bank of MISSOURI River, industrial suburb 3 mi/4.8 km N of downtown KANSAS CITY; 39°08′N 94°33′W. Manufacturing (foods, chemicals, metal products, food processing, metal containers, fabricated structured steel, machinery, lighting fixtures, auto batteries, lamp bases, gypsum board; printing). Kansas City Downtown Airport on W side. Founded 1912.

North Kawartha (NORTH kuh-WOR-thuh), township (□ 295 sq mi/767 sq km; 2001 population 2,144), S ONTARIO, E central Canada, 33 mi/53 km from PETERBOROUGH; 44°45′N 78°08′W.

North Kazakhstan (kah-zahk-STAHN), Kaz. Soltustyk Kazakhstan, Russian Severo-Kazakhstan, region (□ 17,100 sq mi/44,460 sq km), N Kazakhstan; ⊙ PETROPAVLOVSK. On WEST SIBERIAN PLAIN; drained by ISHIM RIVER. Sharply continental climate. Largely agricultural (wheat, millet, oats); cattle and sheep raising. Manufacturing (centered at Petropavlovsk) includes consumer goods, vehicle engines and parts; food processing. TRANS-SIBERIAN RAILROAD crosses E-W, Trans-Kazakhstan railroad N-S. Pop. consists of Russians, Ukrainians, Kazaks. Formed in 1936 as North Kazakhstan oblast of Kazakh SSR.

North Khangai, MONGOLIA: see ARHANGAY.

North Kingstown, town (2000 population 26,326), WASHINGTON county, S central RHODE ISLAND, on NARRAGANSETT BAY; 41°34'N 71°27'W. Includes QUONSET POINT (former naval air station, now a state airport and industrial park) and the villages of Allentown, Davisville, Hamilton, Lafayette, Saunderstown, and Wickford. Manufacturing (machine tools, primary metals, printing, chemicals, plastics, and textiles).The site of North Kingstown was settled in 1641 by Roger Williams, founder of Providence, Rhode Island. It is a regional trade center and minor fishing port and attracts many tourists. Of interest are Smith's Castle (1678); Casey House (1725), which retains bullet holes made during skirmishes in the Revolutionary War; the birthplace (now a museum) of Gilbert Stuart (1755–1828), the portrait painter. Narragansett Bay is used for recreational boating and fishing. Incorporated 1674 as Kings Towne, divided into North Kingstown and South Kingstown 1723.

North Kingsville, village (2006 population 2,623), ASHTABULA county, extreme NE OHIO, on LAKE ERIE, 6 mi/10 km ENE of ASHTABULA; 41°55'N 80°40'W. Wood products.

North Kitui National Reserve, (□ 288 sq mi/748.8 sq km), EASTERN province, S central KENYA; 00°17'N 38°29'E. Designated a national reserve in 1979.

North Koel River (KO-ail), 140 mi/225 km long, W Jharkhand state, E INDIA; rises on CHOTA NAGPUR PLATEAU c. 25 mi/40 km NE of JASHPURNAGAR; flows NNW past DALTONGANJ to Son River 32 mi/51 km SSW of SASARAM.

North Kordofan, state, SUDAN: see N KURDOFAN.

North Korea or **Democratic People's Republic of Korea,** republic (□ 46,540 sq mi/120,538 sq km; 2004 estimated population 22,697,553; 2007 estimated population 23,301,725), E ASIA, N half of Korean peninsula, separating SEA OF JAPAN from YELLOW SEA; ⊙ PYONGYANG. Founded on May 1, 1948; Pyongyang is the largest city. For information on history prior to 1948 and topography of the Korean peninsula, see KOREA.

Population
A serious population loss, resulting from the exodus of several million people to the S, was somewhat offset by Chinese colonists and Koreans from MANCHURIA and JAPAN. Educational facilities have expanded enormously, emphasizing specialized and technical education. There are many technical colleges, and the major university, Kim Il Sung University, is on the outskirts of Pyongyang. The North Korean government has actively suppressed religion as contrary to Marxist belief.

Economy
After the Korean War, the Communist government of North Korea, under the leadership of Kim Il Sung, used the region's rich mineral and power resources as the basis for an ambitious program of industrialization and rehabilitation. With Chinese and Soviet aid, railroads, industrial plants, and power facilities were rebuilt. Farms were collectivized, and industries were nationalized. In a series of three-year, five-year, six-year, and seven-year economic development plans, the coal, iron, and steel industries were greatly expanded, new industries were introduced, and the mechanization of agriculture was pushed.

Economy - Lumber
Excessive cutting during the Japanese occupation (1910–1945) weakened North Korea's timber resources, once quite large. Predominant trees are larch, oak, alder, pine, spruce, and fir. Intensive government conservation and reforestation programs have increased the supply, and timber occasionally appears on North Korean export lists.

Economy - Mining
Modern mining methods have been instituted, and minerals and metals account for a significant portion of the country's export revenue. North Korea is es-pecially rich in iron and coal and has some two hundred different kinds of minerals of economic value. Some of the other more important minerals produced include copper, lead, zinc, uranium, manganese, gold, silver, and tungsten.

Economy - Agriculture
The S was the peninsula's agricultural center prior to the Korean War, and as a result agricultural self-sufficiency has become a major goal of the North Korean government. The establishment of highly mechanized state farms has been a step in that direction. The government has recently expanded irrigation facilities; numerous dams are being constructed, and land reclamation projects are in progress. Both North and South Korea must still import grain, however, in order to meet their food demands. Livestock plays only a minor role in North Korean agriculture; the steep and often barren hills are unsuitable for large-scale grazing. Fish remains the chief source of protein in the Korean diet. North Korea experienced food shortages leading to famine since the mid-1990s.

Economy - Industry
The North Korean economy was shattered by the war of 1950–1953. Postwar reconstruction was abetted by enormous amounts of foreign aid (from communist countries) and intensive government economic development programs. Economic development throughout the peninsula has been uneven, however, with the S showing significantly greater gains. By the early 1990s the GNP of South Korea was more than five times that of North Korea (per capita GNP in 1994 was $8,196). North Korea has changed from a predominantly agricultural society (in 1946) to an industrial one; 75% of its national product is now derived from manufacturing and mining. Other major products include iron, steel, and other metals, machinery, military products, textiles (synthetics, wool, cotton, and silk), and chemicals. Industrialization has been accompanied by improved transportation. During the Korean War the railroad and highway systems had been heavily damaged. The railroads have been extensively rebuilt. There is domestic air service, and an international airport located at Pyongyang.

History
North Korea maintained close relations with the Soviet Union and communist China (military aid treaties were signed with both countries in 1961), but did seek to retain a degree of independence; the Sino-Soviet rift facilitated this. With the dissolution of the Soviet Union, North Korea now looks to China as its most important ally. Relations with the U.S. have remained tense because of the U.S. military presence and economic assistance to South Korea. In 1968, North Korea seized the U.S. intelligence ship Pueblo and imprisoned its crew for eleven months, and in 1969 it shot down an American plane. Since then, the U.S. imposed sanctions (1988) on North Korea for its alleged terrorist activity and expressed concern over reports that North Korea was building a nuclear weapons plant. New tensions mounted on the peninsula in 1994 after confirming that the country had developed a nuclear program. After months of crisis, North and South Korea agreed to have a historic summit meeting, the first ever between the leaders of the two countries, to secure peace on the peninsula, a peace free of nuclear weapons. The two leaders failed to meet due to the death of Kim Il Sung in July 1994, who had been premier since the country's inception in 1948. Kim's son, Kim Jong Il, had assumed the day-to-day management of the government before his father's death, and succeeded him as North Korea's premier. After direct talks with the U.S., North Korea agreed to freeze its nuclear program in exchange for aid to its energy problem and improved diplomatic and economic relations. It opened its nuclear facilities near Yongbyon to international delegates inspecting for weapons-grade materials. To assist North Korea's energy problem the Korean Peninsula Energy Development Organization (KEDO)—a multinational consortium including the U.S., South Korea, China, Japan, and others—was formed in 1995. KEDO provided technology at an estimated project cost of US$5.5 billion (heavily contributed by North Korea) for building two light-water nuclear reactors in Shinpo area, and supply 500,000 tons/453,721 metric tons of heavy oil a year until the completion of reactors in 2003. To counter the Olympics held in Seoul in 1988, North Korea sponsored the 13th World Youth and Student Festival, which hosted over 25,000 students and was the largest international gathering ever held in North Korea. The leaders of North and South Korea met in a historic summit in 2000, but relations between the two countries improved only gradually. In 2002 oil shipments under the 1994 agreement were halted after revelations that the North had a nuclear weapons program. North Korea ended UN supervision of its nuclear facilities and made other moves toward the development of nuclear weapons, leading to a rise in tensions in the region. In 2005 the North publicly claimed to have nuclear weapons. Talks on various nuclear issues with the North have produced no concrete results. Also in 2005 the U.S. imposed sanctions on a Macao bank accused of laundering counterfeit U.S. currency produced by the North; the move, which led other banks to limit transactions with North Korea, appeared to have been undertaken in part to force North Korea to make nuclear concessions. North Korean missile tests in July 2006, despite a 1999 agreement to suspended such tests, provoked limited UN sanctions. The North then conducted a nuclear-test explosion in October 2006, which brought wide condemnation and stronger, though largely military, UN sanctions. Both China and South Korea, however, continued most trade with the North, concerned about the possibility of instability in, and confrontation with, the country. Talks in 2007 resulted (February) in an agreement that called for the North to shut down its main reactor in exchange for aid; the UN confirmed the reactor shutdown in July. A second North-South summit meeting in October 2007 ended with a joint call for talks on replacing the 1953 Korean War armistice with a permanent peace treaty.

Government
North Korea is governed under the constitution of 1972. The president is head of state, but the title of president was reserved for Kim Il Sung after his death. The chairman of the National Defense Commission is now regarded as the nation's highest administrative office. The premier, who is the head of government, is elected, unopposed, by the Supreme People's Assembly. The unicameral legislature consists of the 687-seat Supreme People's Assembly, whose members are popularly elected to five-year terms. Although nominally a republic governed by the Supreme People's Assembly, North Korea is actually ruled by the Communist Party (known in Korea as the Korea Workers' Party). The ruling party approves a list of candidates who are generally elected without opposition. The current head of state and government is Premier Kim Jong Il (since July 1994). Kim Yong Nam, the president of the Supreme People's Assembly Presidium, has the responsibility of representing the state and receiving diplomatic credentials.

In 1988 the North Korean government implemented an administrative system of nine provinces (Korean, -do) and five municipalities (Korean, -jikhalshi) having a status equivalent to that of a province. Each province is further divided into counties (Korean,-gun). The municipalities are PYONGYANG, KAESONG, NAMPO, and Najin Sonbong. The provinces are NORTH PYONGAN, SOUTH PYONGAN, CHAGANG, YANGGANG, NORTH HAMGYONG, SOUTH HAMGYONG, NORTH HWANGHAE, SOUTH HWANGHAE, and KANGWON.

North Kurdofan, Arabic *Shamal Kurdufan*, state (2005 population 1,602,000), KURDOFAN region, central SUDAN; ⊙ AL UBAYYID; 15°00'N 30°00'E. Bordered by Northern (N), KHARTOUM (NE), White Nile (E), S KURDOFAN (S), S DARFUR (SW tip), and N DARFUR (W) states. Created in 1996 following a reorganization of administrative divisions. Railroad travels S from AL UBAYYID to join railroad running W into S KURDOFAN state (with branches then extending W to NYALA in S DARFUR state and S to WAU in W BAHR AL-GHAZAL state) and E through White Nile into Sinnar states. Roads run primarily through central and S portions of state (especially around AL UBAYYID and AL-NAHUD) connecting it to surrounding SUDAN. Airport at AL UBAYYID. Absorbed N half of former W KORDOFAN state c.2007. The area making up N Kurdofan state was part of former KURDOFAN province. Also spelled North Kordofan and Northern Kordofan.

North Kyongsang (KEE-ONG-SAHNG), Korean, *Kyongsang-pukdo*, province [Korean *do*] (□ 7,344 sq mi/19,094.4 sq km), SOUTH KOREA, bounded by SEA OF JAPAN (E), and the provinces of NORTH CHUNGCHONG and KANGWON (N), SOUTH KYONGSANG (S), NORTH CHOLLA and SOUTH CHUNGCHONG (W); ⊙ TAEGU; 36°20'N 128°45'E. Largest province in South Korea. Partly mountainous and forested, with fertile lowlands drained by NAKDONG RIVER. The province is a major producer of raw silk. Also important are livestock raising and agriculture (barley, cotton, tobacco, pepper, ginseng, silk cocoon, fruit). Taegu's locally grown apples are of high quality and known throughout South Korea. There are many fishing ports and scattered salt fields. Gold, silver, coal, graphite, molybdenum, and tungsten are mined. Heavy industry complex, including iron and steel mills, was built in POHANG; electronics and textiles industries in KUMI. Historically, province was center of the Silla kingdom; relics in KYONGJU area, the kingdom's capital.

North La Junta, unincorporated village, OTERO county, SE COLORADO, suburb just N of LA JUNTA, between city and railroad yards, on ARKANSAS RIVER. Home of the Koshare Indian Dancers.

Northlake, city (2000 population 11,878), COOK county, NE ILLINOIS, a suburb of CHICAGO; 41°54'N 87°54'W. It has various manufactures. Saint John Vianney Roman Catholic Church, which is shaped like a fish, has the largest mosaic-tile mural in the Western Hemisphere. Incorporated 1949.

Northlake, unincorporated town (2000 population 3,659), ANDERSON county, NW SOUTH CAROLINA, residential suburb 4 mi/6.4 km NNW of ANDERSON, on NE arm of HARTWELL LAKE reservoir; 34°34'N 82°40'W.

North Lake, town, WAUKESHA county, SE WISCONSIN, on small North Lake, 12 mi/19 km NW of MILWAUKEE. In farm and lake-resort region. Manufacturing (lubricants).

North Lake, village, MARQUETTE county, NW UPPER PENINSULA, MICHIGAN, 2 mi/3.2 km NW of ISHPEMING; 46°29'N 87°43'W.

North Lake, village, GREENE county, SE NEW YORK, in the CATSKILLS, 2 mi/3.2 km E of HAINES FALLS; c.0.5 mi/0.8 km long; 42°12'N 74°02'W. State campsite.

Northlakes, unincorporated town (□ 1 sq mi/2.6 sq km; 2000 population 1,390), CALDWELL county, W central NORTH CAROLINA, residential suburb 5 mi/8 km NW of HICKORY, near CATAWBA RIVER, forms Hickory Lake reservoir to E and S; Lake Rhodhiss reservoir to W; 35°46'N 81°22'W. Recreation area.

North Lakhimpur (luhk-HEEM-poor), town, ⊙ LAKHIMPUR district, E central ASSAM state, NE INDIA, in Brahmaputra valley, 50 mi/80 km WSW of DIBRUGARH; 27°14'N 94°07'E. Road center; tea, rice, jute, sugarcane, rape, and mustard. Tea processing; extensive tea gardens nearby.

North Lancaster (NORTH LANG-ki-stuhr), unincorporated village, SE ONTARIO, E central Canada, 19 mi/30 km from CORNWALL, and included in South Glengarry township; 45°14'N 74°30'W.

North Lancing, ENGLAND: see SOUTH LANCING.

Northland, region (□ 23,247 sq mi/60,442.2 sq km), NEW ZEALAND; since 1989, administratively defined as the N two-thirds of the Northland or NORTH AUCKLAND peninsula, inclusive of the districts of Kaipara, WHANGAREI, and the Far North. This region is climatically relatively uniform, with warm temperate or subtropical conditions, but landscapes are variegated. The W coast is straightened by lines of sand dunes, and its inlets are bar bound. The E coast is open to the sea with intricate embayments, islands, and alluvial flats. Soils vary from rich basaltic loams to podzolized gumlands. Vegetation includes remnant kauri and shoreline mangroves, with much scrubland. Agriculture (warm-climate citrus, kiwifruit, and vines); dairying; beef cattle and sheep. An early "Cradle of New Zealand" with the sites of WAITANGI and mission stations, it preserves elements of older Maori lifestyles. Its pastures have been upgraded by modern techniques, and Whangarei has emerged as a regional center, with fertilizer, cement, glass, and especially petroleum industries. The attraction of AUCKLAND as a source of employment and vacationers is strong toward the S.

North Land, RUSSIA: see SEVERNAYA ZEMLYA.

North Laramie River (LAR-uh-mee), 69 mi/111 km, SE WYOMING; rises in LARAMIE MOUNTAINS in N ALBANY county; flows first S then mainly E to LARAMIE RIVER 5 mi/8 km N of WHEATLAND.

North Las Vegas, city (□ 60 sq mi/156 sq km; 2006 population 197,567), CLARK county, SE NEVADA, a suburb 3 mi/4.8 km N of LAS VEGAS; 36°16'N 115°08'W. Tourism and military expediture are the economic mainstays of this growing suburb. Manufacturing (furniture, food products, chemicals, sanitary food containers, septic tanks, fertilizers, gaming equipment). Community College of Southern Nevada. The Garden of Cities there features trees and plants from cities throughout the U.S. Nellis Air Force Base to E and NE. Sunrise Mountain Natural Area to SE. Las Vegas Dunes Recreation Area to NE. Desert National Wildlife Range to N. Part of Toiyabe National Forest, including Mount Charleston, to W. Incorporated 1946.

North Lauderdale (LAW-duhr-DAIL), city (□ 4 sq mi/10.4 sq km; 2005 population 42,262), BROWARD county, SE FLORIDA, 6 mi/9.7 km W of POMPANO BEACH; 26°12'N 80°13'W. Manufacturing includes cabinets and commercial printing.

Northleach (NAWTH-leech), village (2001 population 1,861), E GLOUCESTERSHIRE, W central ENGLAND, 11 mi/18 km ESE of CHELTENHAM; 51°51'N 01°50'W. Has 15th-century church. Near FOSSE WAY.

North Leigh (LEE), village (2001 population 1,919), W OXFORDSHIRE, S central ENGLAND, 3 mi/4.8 km NE of WITNEY; 51°48'N 01°26'W. Previously limestone quarrying. Church has Saxon tower.

North Leverton (LEV-uh-tuhn), village (2001 population 926), NOTTINGHAMSHIRE, central ENGLAND, 5 mi/8 km E of RETFORD; 53°20'N 00°49'W. Has working windmill (1813). Also called North Leverton with Habblesthorpe.

North Lewisburg (LOO-uhs-buhrg), village (2006 population 1,594), CHAMPAIGN county, W central OHIO, 13 mi/21 km NE of URBANA, near DARBY CREEK; 40°13'N 83°33'W. In agricultural area.

North Liberty, town (2000 population 1,402), SAINT JOSEPH county, N INDIANA, 13 mi/21 km SW of SOUTH BEND; 41°32'N 86°26'W. Dairy products, grain. Manufacturing (aluminum extrusions). Potato Creek State Park is nearby to the E. Laid out 1836.

North Liberty, town (2000 population 5,367), JOHNSON county, E IOWA, 7 mi/11.3 km NNW of IOWA CITY; 41°44'N 91°36'W. State park nearby.

North Lincolnshire (LING-kuhn-shir), county (□ 328 sq mi/852.8 sq km; 2001 population 152,849), E ENGLAND, on the HUMBER estuary, ⊙ SCUNTHORPE; 53°35'N 00°40'W. Created in 1996 when Humberside was divided into North Lincolnshire, North East Lincolnshire, Kingston Upon Hull, and East Riding of Yorkshire. Agriculture includes cerals, sugar beets, and vegetables; oil refineries, food processing, and a steel mill at Scunthorpe.

North Little Rock, city (2000 population 60,433), PULASKI county, central ARKANSAS, on the ARKANSAS RIVER (MURRAY LOCK AND DAM to W), opposite LITTLE ROCK; 34°47'N 92°15'W. Railroad center. North Little Rock lies in a cotton, rice, soybean, dairy-cattle, and produce area. Manufacturing (consumer goods, food and food products, fiberglass products, printing, building materials, hospital garments, bakery products, feed, furniture, fertilizers, electronic products, chemicals). In the early 19th century the discovery of a small silver vein drew settlers to the area, which was then called Silver City. Most of the area later became part of Little Rock, but in 1903 local citizens pushed a bill through the Arkansas legislature permitting a part of Little Rock to secede and join the small village of North Little Rock. Nearby is Camp Joseph T. Robinson National Guard Training Area, to N of city. Settled c.1856, incorporated as a city 1903.

North Logan, town (2006 population 7,558), CACHE county, N UTAH, suburb 2 mi/3.2 km N of LOGAN, W of Bear River Range; 41°46'N 111°48'W. Elevation 4,700 ft/1,433 m. Mount NAOMI WILDERNESS AREA of Wasatch National Forest to E. Settled 1878.

North Long Lake, CROW WING county, central MINNESOTA, (□ 10 sq mi/26 sq km), 4 mi/6.4 km N of BRAINERD; 6.5 mi/10.5 km long, 2 mi/3.2 km wide in middle. Drains NE to Edward Lake Resorts. Distinct bays at E and W ends, connected to lake by narrow channels. GULL LAKE to W, Round Lake to NW, MISSISSIPPI RIVER to SE.

North Loup (LOOP), village (2006 population 313), VALLEY county, central NEBRASKA, 10 mi/16 km SE of Ord, and on NORTH LOUP RIVER; 41°29'N 98°46'W. Dairying, grain; manufacturing (popcorn processing). Chalkmine State Wayside Area to SE (Greeley county).

North Loup River, Nebraska: see LOUP RIVER.

North Luangwa National Park (□ 1,788 sq mi/4,648.8 sq km), NORTHERN province, E central ZAMBIA, 90 mi/145 km N of CHIPATA. Bounded on E by LUANGWA RIVER, on W by Muchinga Escarpment. More remote than SOUTH LUANGWA NATIONAL PARK; mainly woodland, drained by MWALESHI RIVER, which descends from escarpment in series of waterfalls. Wildlife (buffalo, elephants, wildebeest, impalas, zebras, baboons, hippos, crocodiles, antelope, lions, hyenas).

North Lumberton (NORTH LUHM-buhr-tuhn), unincorporated village, ROBESON county, S NORTH CAROLINA, a suburb 1 mi/1.6 km N of LUMBERTON.

North Madison, village, JEFFERSON county, SE INDIANA, just NW of MADISON. In agricultural area.

North Magnetic Pole, the N point on the Earth's surface where the Earth's magnetic field points directly downward. Location moves over time; it currently is drifting across the Canadian Arctic toward SIBERIA. North magnetic pole was discovered in 1831 by the English explorer, Sir James C. Ross.

North Manchester, town (2000 population 6,260), WABASH county, NE central INDIANA, on EEL RIVER, and 14 mi/23 km N of WABASH; 41°00'N 85°46'W. In agricultural area (livestock; grain); manufacturing (transportation equipment, animal feed, mobile home axles, battery cables, agricultural equipment, appliances, iron castings). Seat of Manchester College. Laid out 1837.

North Manistique Lake, MICHIGAN: see MANISTIQUE LAKE.

North Manitou Island, MICHIGAN: see MANITOU ISLANDS.

North Mankato (man-KAI-do), city (2000 population 11,798), NICOLLET county, S MINNESOTA, suburb 1 mi/1.6 km NW of MANKATO, on N side of MINNESOTA RIVER; 44°10′N 94°01′W. Agriculture (grain, soybeans, peas; livestock, poultry; dairying); manufacturing (awards, printers, electronics, trusses, food and beverage processing, lumber brake systems). North Mankato Vocational Technical School. Incorporated 1922.

North Maroon Peak (14,014 ft/4,271 m), in ELK MOUNTAINS, PITKIN county, near GUNNISON county boundary, W central COLORADO, 13 mi/21 km SW of ASPEN, and 1 mi/1.6 km N of MAROON PEAK.

North Melbourne (NORTH MEL-buhrn), suburb just NW of MELBOURNE, VICTORIA, SE AUSTRALIA; 37°48′S 144°58′E.

North Merrick, unincorporated village (2000 population 11,844), NASSAU county, SE NEW YORK, on LONG ISLAND; 40°42′N 73°34′W. It is chiefly residential.

North Miami (mei-AM-ee), city (□ 10 sq mi/26 sq km; 2005 population 57,654), MIAMI-DADE county, SE FLORIDA; suburb 6 mi/9.7 km N of MIAMI, on BISCAYNE BAY; 25°53′N 80°01′W. Mainly residential. Manufacturing includes boats, wooden furniture, and aluminum products; substantial retail development since the 1980s. Film, video, and recording industry. Large Haitian-American population. Incorporated 1926.

North Miami, village (2006 population 433), OTTAWA county, extreme NE OKLAHOMA, suburb 3 mi/4.8 km N of MIAMI; 36°55′N 94°52′W. Manufacturing (precast concrete structures).

North Miami Beach (mei-AM-ee), city (□ 5 sq mi/13 sq km; 2005 population 39,442), MIAMI DADE county, SE FLORIDA, 10 mi/16 km N of MIAMI near the ATLANTIC coast; 25°55′N 80°10′W. Major office and retail area. Incorporated 1931.

North Middlesex (NORTH MI-duhl-seks), township (□ 231 sq mi/600.6 sq km; 2001 population 6,901), SW ONTARIO, E central Canada, 25 mi/40 km from LONDON; 43°10′N 81°37′W. Established 2001.

North Middletown, village (2000 population 562), BOURBON county, N central KENTUCKY, 9 mi/14.5 km ESE of PARIS; 38°08′N 84°06′W. In BLUEGRASS region. Agriculture (horses, cattle; burley tobacco, grain; dairying).

North Mountain (NORTH) (elevation c.2,000 ft/610 m–3,300 ft/1,006 m high), in VIRGINIA and WEST VIRGINIA, ridge of ALLEGHENY MOUNTAINS, extends c.50 mi/80 km SW-NE, partly along border between HARDY county, West Virginia, and SHENANDOAH county, Virginia; in George Washington National Forest (both states). SHENANDOAH VALLEY to E. Also called Great North Mountain.

North Mullins, unincorporated village, MARION county, E SOUTH CAROLINA, 28 mi/45 km E of FLORENCE.

North Muskegon (mus-KEE-guhn), town (2000 population 4,031), MUSKEGON county, SW MICHIGAN, on Muskegon Lake, suburb 1 mi/1.6 km N of MUSKEGON; 43°15′N 86°16′W. Railroad junction. Manufacturing (agricultural chemicals); fruit and vegetable farming. Muskegon State Park to W of LAKE MICHIGAN; Manistee National Forest to NE; Duck Lake State Park to NW on Lake Michigan. Incorporated as village 1881, as city 1891.

North Myrtle Beach, town (2006 population 14,972), HORRY county, E SOUTH CAROLINA, 15 mi/24 km NE of MYRTLE BEACH, on ATLANTIC OCEAN, in Grand Strand Beach resort area; 33°49′N 78°40′W. Manufacturing includes wire harnesses, food processing. INTRACOASTAL WATERWAY to NW, bridged to N.

North Naples (NAI-puhlz), town (□ 7 sq mi/18.2 sq km), COLLIER county, SW FLORIDA, 4 mi/6.4 km N of NAPLES; 26°11′N 81°47′W.

North Negril Point, HANOVER parish, cape at W end of JAMAICA, 29 mi/47 km WSW of MONTEGO BAY; 18°21′N 78°22′W. SOUTH NEGRIL POINT, westernmost cape of the island, is 6 mi/9.7 km S.

North Newton, town (2000 population 1,522), HARVEY county, S central KANSAS, suburb 2 mi/3.2 km N of NEWTON; 38°04′N 97°20′W. Seat of Bethel College.

North Norfolk (NORTH NOR-fuhk), rural municipality (□ 448 sq mi/1,164.8 sq km; 2001 population 2,941), S MANITOBA, W central Canada, 18 mi/30 km E of Sprucewoods Provincial Park; 49°54′N 98°50′W. Agriculture; manufacturing (candles, ventilation systems). MACGREGOR is the main center; other communities include Austin, Sydney, Bagot, and Rossendale. Incorporated 1882.

North Oaks, unincorporated town, LOS ANGELES county, S CALIFORNIA, residential suburb 30 mi/48 km NNW of downtown LOS ANGELES, on SANTA CLARA RIVER. SAN GABRIEL MOUNTAINS to S; Bouquet Canyon to N. Dairying; cattle, poultry; bedding plants, peaches. LOS ANGELES AQUEDUCT passes N-S through area.

North Oaks, town (2000 population 3,883), RAMSEY county, E MINNESOTA, residential suburb 9 mi/14.5 km N of downtown ST. PAUL, and 12 mi/19 km NE of downtown MINNEAPOLIS; 45°06′N 93°05′W. Numerous small lakes in area; Pleasant Lake in center.

North Ogden, city (2006 population 16,798), WEBER county, N UTAH; residential suburb 5 mi/8 km N of OGDEN; 41°18′N 111°57′W. Elevation 4,275 ft/1,303 m. Vegetables, cherries, apples, wheat, barley; dairying; cattle. WASATCH RANGE and Wasatch National Forest to E. Settled 1851; incorporated 1934.

North Okanagan (o-kuh-NAH-guhn), regional district (□ 2,901 sq mi/7,542.6 sq km; 2001 population 73,227), S BRITISH COLUMBIA, W Canada; 50°25′N 118°45′W. Consists of 5 electoral areas and 6 member municipalities (ENDERBY, ARMSTRONG, Spallumcheen, VERNON, COLDSTREAM, and LUMBY).

North Okkalapa (OK-kah-lah-PAH), suburb of YANGON, MYANMAR. E of Yangon Airport.

North Olmsted (AHM-sted), city (□ 11 sq mi/28.6 sq km; 2006 population 32,126), CUYAHOGA county, NE OHIO; suburb 13 mi/21 km SW of CLEVELAND; 41°25′N 81°55′W. Mainly residential. Printed materials. First U.S. municipal bus line began operations in North Olmsted in 1931. Incorporated as a city 1951.

Northolt, ENGLAND: see EALING.

Northome (NORTH-hom), village (2000 population 230), KOOCHICHING county, N MINNESOTA, c.40 mi/64 km NE of BEMIDJI; 47°52′N 94°16′W. Livestock; alfalfa; timber; manufacturing (meat processing, printing); logging. ISLAND LAKE and Chippewa National Forest to SE; Pine Island State Forest to N and E.

Northop (NOR-thop), village (2001 population 2,983), Flintshire, NE Wales, 3 mi/4.8 km S of FLINT; 53°12′N 03°08′W. Market. Formerly CLWYD, abolished 1996.

North Ossetian Republic (o-SE-shyun) or **North Ossetia** (o-SE-shyuh), Russian *Respublika Severnaya Ossetia*, constituent republic (□ 3,071 sq mi/7,984.6 sq km; 2006 population 727,540), SE European Russia, on the N slopes of the Greater CAUCASUS Mountains and on the adjoining plains; ⊙ VLADIKAVKAZ (formerly Ordzhonikidze). Forests cover almost one-quarter of the land. Glavnyy (Vodorazdel'nyi) and BOKOVOY ranges in the S; highest peaks include Mounts DZHIMARA (15,682 ft/4,780 m), Uilpata (15,253 ft/4,649 m), and Tepli (14,537 ft/4,431 m). The Zakinsk, Zaramag, and Verkhniaia Digora basins divide the ranges. Drained mostly by the TEREK RIVER and its tributaries, the Urukh, Ardon, Fiagdon, and Gizel'don rivers (all flowing N); also, SUNZHA (rises in extreme W), Ursdon, and Kambileevka rivers. Glaciers cover 66 sq mi/172 sq km; Karaiyom and Tseyskiy are the

largest. The sloping Ossetian Plain (forest steppe converted to farmland) is in the center; the low SUNZHA and TEREK ranges to the N, with the Mozdok Plain (former grass steppe, now agricultural fields) beyond. Fauna include wild goat, chamois, wildcat, boar, deer, and bear. Crossed by the Northern Caucasus railroad (ROSTOV-NA-DONU, Russia-BAKY, Azerbaijan). Georgian and Ossetian Military Roads (S to Transcaucasia) begin here. Fruit, wine, grain (especially wheat and corn), and cotton produced in valleys; lumbering and livestock (dairy and beef cattle, hog, sheep) and poultry raising are important in the mountains. Moderately continental climate, with mild temperatures; dry winds and arid climate on the Mozdok Plain. Plains are frost-free for over 6 months. Mineral resources include lead, silver, zinc, and boron deposits, dolomites, and mineral water springs. Mining, nonferrous metallurgical, oil-extracting, machine-building, electrical engineering, and food-processing (canning) industries; also manufacturing (woodworking, apparel, footwear, textiles, and glassware). Population mainly Ossetian, or Ossets (62.7%), with Russian (23.2%), Ingush (3%), Armenian (2.4%), Kumyk (1.8%), Georgian (1.5%), Ukrainian (0.7%), and Chechen (0.5%) minorities, as well as a number of poorly represented nationalities. Population is mainly urban, with 65.5% living in cities and urban-type settlements. Ossetians, an Iranian-speaking people, are mainly Sunni Muslims here. In 1924, North Ossetia became an autonomous region in RSFSR and then an autonomous republic in 1936. North Ossetia was a signatory to the March 31, 1992, treaty that created the RUSSIAN FEDERATION.

North Oxford, MASSACHUSETTS: see OXFORD.

North Pacific Current, warm ocean current in North PACIFIC OCEAN, formed at c. 38°N by convergence of warm JAPAN and cooler OKHOTSK CURRENTS; flows E to c. 150°W before turning around toward S in the general clockwise circulation of the North Pacific. Also known as North Pacific Drift.

North Pagi, INDONESIA: see PAGAI ISLANDS.

North Palisade (north pa-luh-SAID), mountain (14,242 ft/4,328 m), in FRESNO and INYO counties, E central CALIFORNIA; 37°06′N 118°31′W. Lies in both KING'S CANYON NATIONAL PARK and Inyo National Forest.

North Palm Beach, village (□ 6 sq mi/15.6 sq km; 2005 population 12,633), PALM BEACH county, SE FLORIDA, 10 mi/16 km N of WEST PALM BEACH; 26°49′N 80°03′W. Residence of golfer Jack Nicklaus.

North Pass, La: see PLAQUEMINES.

North Patchogue, unincorporated town, SUFFOLK county, SE NEW YORK, 20 mi/32 km SW of RIVERHEAD; 40°47′N 72°56′W. Residential area. Pop. figure includes Patchogue Highlands.

North Peak (north peek), peak (19,470 ft/5,934 m), in the ALASKA RANGE, S central ALASKA; one of the two main summits of Mount MCKINLEY; 63°06′N 151°00′W. The other summit, SOUTH PEAK is synonymous with McKinley.

North Peak, NEVADA: see BATTLE MOUNTAIN.

North Pease River, Texas: see PEASE RIVER.

North Pekin (PEE-kin), village (2000 population 1,574), TAZEWELL county, central ILLINOIS, suburb of PEORIA, near ILLINOIS RIVER; 40°36′N 89°37′W. In agricultural area; manufacturing (grinding wheels, machinery). Next to Pekin Lake Conservation Area. Incorporated 1948.

North Pelham (NORTH PEL-uhm), unincorporated village, Niagara region, S ONTARIO, E central Canada, and included in the town of PELHAM; 43°03′N 79°21′W.

North Perry (PER-ee), village (□ 4 sq mi/10.4 sq km; 2006 population 949), Lake county, NE OHIO, on LAKE ERIE, 8 mi/13 km NE of PAINESVILLE; 41°48′N 81°07′W.

North Perth (NORTH PUHRTH), town (□ 190 sq mi/ 494 sq km; 2001 population 12,055), SW ONTARIO, E central Canada; 43°43′N 80°58′W. Formed in 1998 from Wallace, Elma, and LISTOWEL.

North Petherton (PE[TH]-uh-tuhn), village (1991 population 4,736; 2001 population 5,065), central SOMERSET, SW ENGLAND, 3 mi/4.8 km SSW of BRIDGWATER; 51°06′N 03°02′W. Located near QUANTOCK HILLS in dairying region. Has 14th–15th-century church.

North Plainfield, residential borough (2006 population 21,738), SOMERSET county, NE NEW JERSEY. Settled 1736, incorporated 1885. A Revolutionary War cemetery is here.

North Plains, town (2006 population 1,806), WASHINGTON county, NW OREGON, 15 mi/24 km WNW of downtown PORTLAND, on MCKAY CREEK; 45°36′N 123°00′W. Dairying; poultry; berries, fruit, vegetables.

North Platte (PLAT), city (2006 population 24,386), ⊙ LINCOLN county, W central NEBRASKA, at the confluence of the NORTH PLATTE and SOUTH PLATTE rivers, on point below the two rivers, and 95 mi/152 km WNW of KEARNEY; 41°07′N 100°46′W. It is a processing and shipping point for grain and livestock. It has meatpacking plants, a fish hatchery, and a large railroad-repair shop and clarification yard. Manufacturing (concrete blocks, sheet-metal fabrication, wooden doors, horse meat, zoo districts, lawn ornaments, soft drinks, aprons; printing). Scout's Rest Ranch (formerly a home of "Buffalo" Bill Cody, who lived in North Platte for thirty years) and Fort McPherson National Cemetery are nearby. A nightly rodeo is held in the city during the summer at the Rodeo Arena at Scout's Rest Ranch. Mid-Plains Community College. Maloney Reservoir (on Tri-County Supply Canal) and Lake Maloney State Recreation Area to S. Old OREGON TRAIL, once guarded by Fort McPherson, follows S side Platte–South Platte rivers. U.S. Senator Chuck Hagel born here. Incorporated 1873.

North Platte, river, c.680 mi/1,094 km long; rises in the PARK RANGE, SW JACKSON county, N COLORADO, at joining of Grizzly and Little Grizzly creeks, near CONTINENTAL DIVIDE, 23 mi/37 km SSW of WALDEN, Colorado; flows N into WYOMING, past SARATOGA, through Seminos, PATHFINDER, and ALCOVA reservoirs, NE to CASPER, then E, past N end of LARAMIE MOUNTAINS, SE past DOUGLAS and TORRINGTON, Wyoming, SCOTTSBLUFF, BRIDGEPORT, and OSHKOSH, Nebraska, through Lake C.W. McConaughy Reservoir (N of OGALLALA), then joining SOUTH PLATTE at city of North Platte to form Platte River, lower 80 mi/129 km nearly parallels South Platte River. The North Platte project and the Kendrick project utilize the North Platte's water for power and irrigation. KINGSLEY DAM (170 ft/52 m high and 3.4 mi/5.5 km long; completed 1942) near Ogallala, Nebraska, is one of many dams on the river, and there are also large reservoirs. The valley of the North Platte is a chief route used by W-bound pioneers. FORT LARAMIE NATIONAL HISTORIC SITE is on the river near the mouth of the LARAMIE RIVER.

North Platte Project (PLAT), unit of the U.S. Bureau of Reclamation, in the NORTH PLATTE RIVER valley, W NEBRASKA and E WYOMING. It supplies hydroelectric power to many towns and industries and provides irrigation for land extending along the valley from GUERNSEY, Wyoming, to below BRIDGEPORT, Nebraska. Among the project's many dams and reservoirs are GUERNSEY RESERVOIR, formed by Guernsey Dam (completed 1927), and PATHFINDER RESERVOIR, created by Pathfinder Dam (completed 1909). There are also several large dams on the branches of the North Platte. The power system of the project has been integrated with the MISSOURI RIVER basin project.

North Pleasureville, village, HENRY county, N KENTUCKY, 18 mi/29 km NW of FRANKFORT, in BLUEGRASS region.

North Plymouth, MASSACHUSETTS: see PLYMOUTH.

North Point, suburban village, BALTIMORE county, central MARYLAND, on North Point, extending into CHESAPEAKE BAY at N side of PATAPSCO River mouth (lighthouse), 12 mi/19 km SE of downtown BALTIMORE. Agriculture; steel plant. Fort Howard Veterans Hospital is here. Site of minor victory over British (Sept. 12, 1814).

North Point, NW extremity of Prince Edward Island, Canada, on the Gulf of St. Lawrence, 7 mi/11 km NNE of TIGNISH; 47°04′N 63°59′W. Lighthouse.

North Polar Sea: see ARCTIC OCEAN.

North Pole, N end of the earth's axis; lat. 90°N. It is distinguished from the N magnetic pole. U.S. explorer Robert E. Peary is traditionally acknowledged as having been the first to reach (1909) the North Pole. See also ARCTIC REGIONS.

Northport, city (2000 population 19,435), Tuscaloosa co., W central Alabama, on Black Warrior River opposite Tuscaloosa. Lumber; manufacturing (apparel, fixtures). Originally called 'Kentuck,' it was also known as 'North Tuscaloosa.' Renamed to Northport in 1832, it was incorporated in 1852.

North Port, city (□ 75 sq mi/195 sq km; 2000 population 22,797), SARASOTA county, W central FLORIDA, 12 mi/19 km E of VENICE; 27°03′N 82°11′W. Contains Little Salt Spring archeological site. Rapidly-growing population.

Northport, resort town, WALDO county, S MAINE, on PENOBSCOT BAY, and just S of BELFAST; 44°21′N 68°59′W. Temple Heights Spiritualist camp of the National Spiritual Association of Churches, founded 1882.

Northport, village (2000 population 648), LEELANAU county, NW MICHIGAN, 25 mi/40 km N of TRAVERSE CITY, on W shore of GRAND TRAVERSE BAY; 45°07′N 85°37′W. Railroad terminus. In cherry and apple producing area. Light manufacturing; fisheries. Lighthouse; Leelanau State Park to NE.

Northport, residential village (□ 2 sq mi/5.2 sq km; 2006 population 7,494), SUFFOLK county, SE NEW YORK, on N shore of W LONG ISLAND, on Northport Bay, 5 mi/8 km ENE of HUNTINGTON; 40°53′N 73°20′W. Tourism (park, waterfront retail); quarries. Settled c.1683, incorporated 1894.

Northport, village (2006 population 339), STEVENS county, NE WASHINGTON, 27 mi/43 km NNE of COLVILLE, and 5 mi/8 km S of Canadian border (BRITISH COLUMBIA) on COLUMBIA River; 48°55′N 117°47′W. Ports of entry to NW and NE. Manufacturing (concrete). Parts of Colville National Forest to E and W. Upper reach of FRANKLIN D. ROOSEVELT LAKE, formed by GRAND COULEE DAM, 150 mi/241 km downstream.

North Portal (POR-tuhl), village (2006 population 123), SE SASKATCHEWAN, CANADA, frontier station on NORTH DAKOTA border, 23 mi/37 km ESE of ESTEVAN; 49°00′N 102°33′W. Mixed farming.

Northport Bay, 3 mi/4.8 km long NNS, 2.5 mi/4 km wide, an arm of LONG ISLAND SOUND indenting N shore of W LONG ISLAND, SE NEW YORK, opens on E side of HUNTINGTON BAY, 2 mi/3.2 km NE of HUNTINGTON; 40°55′N 73°24′W. Northport is on its E shore.

North Powder, village (2006 population 481), UNION county, NE OREGON, 22 mi/35 km SSE of LA GRANDE, on POWDER RIVER, W of the confluence with Powder River; 45°01′N 117°55′W. Elevation 3,256 ft/992 m. Agriculture (wheat, potatoes, cherries, apples; sheep, cattle); timber. Parts of Wallowa-Whitman National Forest to W and E.

North Prairie, town (2006 population 1,997), WAUKESHA county, SE WISCONSIN, 26 mi/42 km WSW of MILWAUKEE; 42°56′N 88°24′W. In dairying region. Dairy products; manufacturing (wood pallets). Kettle Moraine State Forest (S unit) to W.

North Princeton, village, MERCER county, W NEW JERSEY, near PRINCETON; 40°21′N 74°39′W.

North Providence, town (2000 population 32,411), PROVIDENCE county, NE RHODE ISLAND, Set off from PROVIDENCE; 41°52′N 71°28′W. Once a large textile town, it is now mainly a residential suburb. A major portion of Rhode Island College is within the town's limits. Incorporated 1765.

North Puyallup (pyoo-A-luhp), unincorporated town, PIERCE county, W WASHINGTON, residential suburb 8 mi/12.9 km SE of downtown TACOMA, and 1 mi/1.6 km N of PUYALLUP, on PUYALLUP RIVER; 47°13′N 122°19′W.

North Pyongan (PYOUNG-AHN), Korean *Pyonhanpukdo*, 2nd-largest province, NORTH KOREA, ⊙ SINUIJU. Includes SINMI ISLAND and several islets. Bounded N by YALU RIVER (forming Manchurian border), S by CHONGCHON RIVER, W by KOREA BAY (arm of YELLOW). Largely mountainous, the province is sparsely settled except for NW corner. Extensive lumbering in the interior; gold and some coal are mined in W area. Agriculture is limited to growing of potatoes, soybeans, and millet.

North Randall (RAN-duhl), village (2006 population 850), CUYAHOGA county, NE OHIO; suburb 8 mi/13 km SE of CLEVELAND; 41°26′N 81°32′W. Site of Randall Park Mall, one of the largest shopping centers in the U.S. Extensive retailing; thoroughbred racing at Randall Park.

North Reading (RE-deeng), residential town, MIDDLESEX county, NE MASSACHUSETTS, on the IPSWICH RIVER, 14 mi/23 km N of BOSTON; 42°35′N 71°05′W. Manufacturing (athletic footwear). Settled 1651, set off from READING and incorporated 1853.

North Redwood, village, REDWOOD county, SW MINNESOTA, 2 mi/3.2 km NE of REDWOOD FALLS, on REDWOOD RIVER at its mouth on MINNESOTA RIVER; 44°33′N 95°05′W. Corn, oats, soybeans; dairying; livestock.

North Rhine–Westphalia (REIN–west-FAI-lee-uh), German *Nordrhein-Westfalen*, state (□ 13,155 sq mi/ 34,203 sq km; 2005 population 18,058,105), W central GERMANY; ⊙ DÜSSELDORF. Bounded by BELGIUM and the NETHERLANDS in the W; LOWER SAXONY in the N and E; HESSE in the SE and RHINELAND-PALATINATE in the S. Situated in the lower RHINE plain, North Rhine-Westphalia includes the TEUTOBURG FOREST and the ROTHAARGEBIRGE. Drained by the Rhine, RUHR, WUPPER, LIPPE, and EMS rivers. It contains the largest industrial concentration in EUROPE (see RUHR district), with one of the largest mining and energy-producing regions on the continent. It has excellent transportation facilities, including superhighways, electrified railroad, river transport, and two large airports. Manufacturing includes chemicals, machines, processed foods, textiles, clothing, and iron and steel. More than half of the state's total land is occupied with commercial farming as well as gardens and orchards, although these enterprises amount to only a small portion of the area's GNP. North Rhine-Westphalia is also the most populous state in Germany and has numerous large cities, including AACHEN, COLOGNE, Düsseldorf, DUISBURG, ESSEN, DORTMUND, REMSCHEID, OBERHAUSEN, and WUPPERTAL. There are universities at BIELEFELD, BOCHUM, BONN, Dortmund, Düsseldorf, Cologne, and MÜNSTER. The region, which had no historic unity, was constituted as a new state (1946) in British occupation zone from union of former Prussian province of WESTPHALIA, N part of former Prussian RHINE PROVINCE, and former state of LIPPE. Significant cultural differences exist among the various peoples in the state; this diversity has been enlarged by substantial immigration from other European countries to cities throughout the region.

North Richland Hills (RICH-land), city (2006 population 8,076), TARRANT county, N TEXAS, residential suburb 9 mi/14.5 km NE of downtown FORT WORTH; 32°51′N 97°13′W. Elevation 650 ft/198 m. Drained by

Big Fossil (S) and Little Bear creeks (N). Manufacturing (corrugated boxes, food products, textiles). Its population more than doubled between 1970 and 1990 as a result of the economic development of the N Texas area. City is now surrounded by neighboring municipalities. Tarrant County Junior College NE campus 1 mi/1.6 km E in HURST. Incorporated 1953.

Northridge, unincorporated village (2000 population 6,853), CLARK county, W central OHIO, 3 mi/5 km NE of SPRINGFIELD; 39°59′N 83°46′W.

Northridge, unincorporated village (□ 2 sq mi/5.2 sq km; 2000 population 8,487), MONTGOMERY county, SW OHIO, 3 mi/5 km N of DAYTON; 39°48′N 84°11′W. Former village absorbed into suburban development.

Northridge, suburban section of LOS ANGELES, LOS ANGELES county, S CALIFORNIA, in SAN FERNANDO VALLEY, 20 mi/32 km NW of downtown Los Angeles. Manufacturing (computer equipment, electronic equipment, pharmaceuticals, machinery). Seat of California State University, Northridge. Northridge Fashion Center, one of the largest shopping centers in the UNITED STATES, is here. Van Nuys Airport to SE. SANTA SUSANA MOUNTAINS to N. Center of 1994 Northridge Earthquake.

North Ridgeville (RIJ-vil), city (□ 23 sq mi/59.8 sq km; 2006 population 27,197), LORAIN county, N OHIO, 18 mi/29 km WSW of downtown CLEVELAND; 41°23′N 82°01′W.

North River (NORTH), 75 mi/121 km long, SW QUEBEC, E Canada; flows S and SW, past SAINTE-AGATHE-DES-MONTS, SAINT-JÉRÔME, and LACHUTE, to W end of Lake of the TWO MOUNTAINS.

North River, 200 mi/322 km long, GUANGDONG province, CHINA; formed in NANLING Mountains at SHAOGUAN by union of CHENG (left) and WU (right) rivers; flows S, past YINGDE, QINGYUAN, and SANSHUI, to PEARL RIVER DELTA. At Sanshui it joins the XI River (West River) to form PEARL RIVER. Navigable in entire course. Receives Linzhou River and an arm of SUI River (right). Formerly, North River was an important water route between S and N China. The name North River is sometimes also applied to its two headstreams. Sometimes spelled Pei Kiang or Pei Jiang. Also appears as Bei Jiang or Bei River.

North River, 55 mi/89 km long, in W ALABAMA; formed by confluence of two headstreams E of FAYETTE; flows S through LAKE TUSCALOOSA, to BLACK WARRIOR RIVER NE of TUSCALOOSA.

North River, c.70 mi/113 km long, in S central IOWA; rises near MENLO in GUTHRIE county; flows E to DES MOINES RIVER 10 mi/16 km SE of DES MOINES.

North River, c.25 mi/40 km long, in E MASSACHUSETTS; rises in N PLYMOUTH county; flows NE, past HANOVER, to MASSACHUSETTS BAY c.2 mi/3.2 km S of SCITUATE. Wildlife sanctuary.

North River, c.70 mi/113 km, NE MISSOURI; rises in KNOX county; flows SE and E to the MISSISSIPPI below QUINCY, ILLINOIS.

North River (NORTH), c.25 mi/40 km long, NW VIRGINIA; rises in BLUE RIDGE in NW AUGUSTA county; flows generally SE to SW of HARRISONBURG, joins SOUTH River near PORT REPUBLIC to form South Fork of SHENANDOAH RIVER; 38°27′N 79°15′W. Receives MIDDLE RIVER 2 mi/3 km W of its mouth.

North River, Virginia: see MAURY RIVER.

North Riverside, village (2000 population 6,688), COOK county, NE ILLINOIS, residential suburb 11 mi/18 km W of downtown CHICAGO; 41°51′N 87°49′W. Incorporated 1923. Manufacturing trigger pumps. On Des Plaines River.

North Robinson (RAHB-uhn-suhn), village (2006 population 198), CRAWFORD county, N central OHIO, 6 mi/10 km E of BUCYRUS; 40°48′N 82°51′W.

North Rohachyk Irrigation System (ro-HAH-chik) (Ukrainian *Pivnichno-Rohachyts'ka zroshuval'na systema*), an irrigation system in W ZAPORIZHZHYA ob-

last, UKRAINE, on the NE part of the BLACK SEA LOWLAND. Extends S from the E end of the KAKHOVKA RESERVOIR for 26 mi/42 km, and W from the town of MYKHAYLIVKA 12 mi/19 km S of VASYLIVKA for 32 mi/51 km, some 8 mi/13 km short of the town of VERKHNIY ROHACHYK (in NE KHERSON oblast). Water is pumped from the Kakhovka Reservoir at a point 6 mi/10 km W of DNIPRORUDNE and conveyed by a mainline canal 19 mi/31 km SE to near Mykhaylivka, and then 20 mi/32 km WSW, and distributed by branch canals W, E, and S. Started in 1968, by 1990 the system provided irrigation on 263,000 acres/106,300 ha and drained 80,000 acres/32,300 ha, to grow winter wheat, corn for grain, vegetables and feed crops, like alfalfa.

Northrop, village (2000 population 262), MARTIN county, S MINNESOTA, 6 mi/9.7 km N of FAIRMONT; 43°44′N 94°26′W. Grain, soybeans; livestock. MIDDLE CHAIN OF LAKES to W, includes Martin, High, and Charlotte lakes.

North Royalton (ROI-uhl-tuhn), city (□ 21 sq mi/54.6 sq km; 2006 population 29,465), CUYAHOGA county, NE OHIO; growing suburb c.10 mi/16 km SSW of CLEVELAND; 41°19′N 81°45′W. Dairy-processing and sawmilling center in the 19th century, North Royalton has since developed a variety of light industries. Settled 1811, incorporated as a city 1961.

North Saanich (SA-nich), district municipality (□ 14 sq mi/36.4 sq km; 2001 population 10,436), SW BRITISH COLUMBIA, W Canada, 16 mi/26 km N of VICTORIA, on S VANCOUVER ISLAND, Saanich Peninsula, in Capital Regional District; 48°40′N 123°25′W. Agricultural area. Victoria International Airport and Swartz Bay Ferry Terminal located here. Established 1965.

North Sacramento, city, SACRAMENTO county, central CALIFORNIA; suburb of SACRAMENTO, across AMERICAN RIVER. A community college is here. Packed meat, brick, machinery. Incorporated 1924.

North Saint Paul, city (2006 population 11,293), RAMSEY county, SE MINNESOTA, a suburb 6 mi/9.7 km NE of downtown ST. PAUL; 45°00′N 93°00′W. Manufacturing (electronic equipment, concrete products, furniture, roofing materials, arrowheads [for archery], vending machines; millwork, printing and publishing). Silver Lake in NE corner; WHITE BEAR LAKE to N. Incorporated 1888.

North Salem, town (2000 population 591), HENDRICKS county, central INDIANA, 8 mi/12.9 km NW of DANVILLE. In agricultural area (corn, soybeans; hogs). Laid out 1835.

North Salem, town, NE WESTCHESTER county, SE NEW YORK, 50 mi/80 km N of MANHATTAN; 41°20′N 73°35′W. Acknowledged as one of the most beautiful and underdeveloped areas in New York city's metropolitan area. Open fields and rugged hills surrounding Titicus Reservoir; horseback riding is popular, as is the raising of horses for hunting and jumping. DeLancey Town Hall, a restored 18th-century Georgian manor, is on the National Register of Historic Places; museum, Japanese Stroll Garden also here. Along Titicus Road, there is a 60-ton/54–metric ton granite glacial erratic that has inspired media attention and scientific speculation. The residence of a number of notable New Yorkers.

North Salem, village, ROCKINGHAM county, SE NEW HAMPSHIRE; suburb in town of SALEM, 3 mi/4.8 km N of Salem center, at N end of Arlington Mill Reservoir. Manufacturing (radar, outboard motor accessories). America's Stonehenge, Mystery Hill, prehistoric stone structures here; Robert Frost Farm to NE.

North Salt Lake, town (2006 population 11,598), DAVIS county, N central UTAH, suburb 4 mi/6.4 km N of downtown SALT LAKE CITY, at mouth of JORDAN RIVER and GREAT SALT LAKE; 40°50′N 111°55′W. Elevation 4,300 ft/1,311 m. Gravel; manufacturing (containers and luggage, medical equipment, lighting

fixtures; petroleum refining, meatpacking). WASATCH RANGE and National Forest to E.

North Saskatchewan, Canada: see SASKATCHEWAN, river.

North Scituate, MASSACHUSETTS: see SCITUATE.

North Sea, arm of the ATLANTIC OCEAN (□ 222,000 sq mi/577,200 sq km), located between 51°00′N and 62°00′N; c. 600 mi/966 km long and c. 400 mi/644 km wide, NW of central EUROPE. It washes the shores of GREAT BRITAIN, NORWAY, DENMARK, GERMANY, the NETHERLANDS, BELGIUM, and the N tip of FRANCE. In the S, the STRAIT OF DOVER connects it with the ENGLISH CHANNEL. The North Sea is deepest (c. 2,165 ft/660 m) along the coast of Norway and contains several shallows, the largest of which is the DOGGER BANK, midway between ENGLAND and Denmark. The cod and herring fisheries of the North Sea are economically important. In 1970 significant deposits of oil and natural gas were discovered under the sea floor. They are exploited by Great Britain, Norway, WEST GERMANY, Denmark, and the Netherlands.

North Sea–Baltic Canal, Germany: see KIEL CANAL.

North Sea Canal, Dutch *Noordzee Kanaal*, North Holland province, W NETHERLANDS, extends 18 mi/29 km E-W between the HET IJ (Amsterdam's inland harbor and channel channel of MARKERMEER lake) and North Sea at IJMUIDEN. Entrance jetties extend 2 mi/3.2 km into North Sea. Canal gives Amsterdam direct ship access to sea.

North Sewickley (SUH-wik-lee), unincorporated village, Franklin township, BEAVER county, W PENNSYLVANIA, residential suburb 2 mi/3.2 km S of ELLWOOD CITY, on CONNOQUENESSING CREEK; 40°50′N 80°16′W.

North Shields, ENGLAND: see TYNEMOUTH.

Norths Highland, a broad ice-covered peninsula upland on the BANZARE COAST, WILKES LAND, ANTARCTICA, on INDIAN OCEAN; 66°40′S 126°00′E. Discovered 1840 by Charles Wilkes, U.S. explorer. Formerly known as Norths Coast.

North Shore (NORTH SHOR), residential and industrial suburb of GEELONG, VICTORIA, SE AUSTRALIA, overlooking CORIO BAY; 38°06′S 144°22′E. Railway station. Motor vehicle factory, oil refinery.

North Shore City, former NE suburbs of AUCKLAND urban area (□ 50 sq mi/130 sq km; 2001 population 219,936), including BIRKENHEAD, DEVONPORT, TAKAPUNA, NORTHCOTE, AND OTHERS. INCORPORATED IN 1989 AS A SEPARATE CITY.

North Sinai, province (□ 10,646 sq mi/27,679.6 sq km; 2004 population 302,077), NE EGYPT, ☉ Al 'ARISH; 31°30′N 34°00′E. Borders the MEDITERRANEAN SEA (N), ISRAEL (E), SOUTH SINAI province (S), and the SUEZ, ISMAILIA, and PORT SAID governates (W). Occupying the N half of the SINAI Peninsula, it is largely a plateau which slopes upwards as it extends inland from the Mediterranean coast, where some vegetables and grains are grown. Al 'Arish and Rummānah are the largest towns.

North Sioux City, city (2006 population 2,511), UNION county, extreme SE SOUTH DAKOTA; suburb 6 mi/9.7 km WNW of downtown SIOUX CITY, IOWA, on BIG SIOUX RIVER; 42°32′N 96°30′W. Agricultural area (corn; cattle, hogs, poultry). Manufacturing of computers, pet food, bakery goods, and hydraulic pumps.

North Skunk River, IOWA: see SKUNK RIVER.

North Slope, borough (□ 87,861 sq mi/228,438.6 sq km; 2006 population 6,608), N ALASKA. Bounded on N by ARCTIC OCEAN; E by YUKON TERRITORY (CANADA); includes most of the N slope of the BROOKS RANGE in S. Main town is BARROW. Part of the GATES OF THE ARCTIC National Park and Preserve and the NOATAK National Preserve to S; part of the Arctic National Wildlife Reservation to the East Alaska Pipeline crosses the borough N-S in center, terminating at PRUDHOE BAY in NE. Fishing. Caribou hunting. Oil and natural gas extraction.

North Slope, Alaska: see ALASKA NORTH SLOPE.

North Smithfield, industrial town (2000 population 10,618), PROVIDENCE county, N RHODE ISLAND, on MASSACHUSETTS line, on BRANCH RIVER, and 13 mi/21 km NW of PROVIDENCE; 41°59′N 71°33′W. Formerly an important textile center. Has several old inns. Includes villages of Forestdale and Slatersville (population center). Incorporated 1871.

North Solomons, province (□ 450,000 sq mi/1,170,000 sq km; 2000 population 175,160), E PAPUA NEW GUINEA; ⊙ KIETA, at SE end of BOUGAINVILLE. Once the district capital, Kieta was re-established as capital after SOHANO, on small Sohano Island, off N end of Bougainville, after political unrest, late 1980s. Includes Bougainville Island, also BUKA island and Outer Islands, all part of SOLOMON ISLANDS group. Kieta and ARAWA, on Bougainville, and Hutjena, on Buka, and SOHANO, on Sohano Island, are main towns. Bounded on SW by SOLOMON SEA, on NE by Pacific Ocean. Country of Solomon Islands, independent of BRITAIN since 1978, situated to immediate SE. Hutjena is main town. Independence sentiments led to violence and closing of Panguna copper mine in 1989. Bananas, coconuts, rice, yams; fish. Bougainville has 19 separate languages.

North Somerset (SUHM-uh-set), county (□ 145 sq mi/377 sq km; 2006 population 193,000), SW ENGLAND, borders Bristol, Bath and North East Somerset, and Somerset counties; ⊙ WESTON-SUPER-MARE; 51°23′N 02°48′W. The County of AVON was abolished in April 1996 and split into Bath and North East Somerset, Bristol, North Somerset, and South Gloucestershire. The county borders the BRISTOL CHANNEL with 27 mi of coastline. Most of the population resides in Weston-super-Mare, PORTISHEAD, CLEVEDON, and NAILSEA. Has Bristol International Airport.

North Springfield, unincorporated town, LANE county, W OREGON, residential suburb 3 mi/4.8 km ENE of downtown SPRINGFIELD, on MCKENZIE RIVER; 44°04′N 123°00′W.

North Springfield, unincorporated town, FAIRFAX county, residential suburb 8 mi/12.9 km W of ALEXANDRIA, 11 mi/18 km SW of WASHINGTON, D.C.; 38°47′N 77°12′W. Lake ACCOTINK Park to SW.

North Stack, Wales: see HOLYHEAD.

North Star (NORTH STAHR), hamlet, NW ALBERTA, W Canada, 5 mi/7 km S of MANNING, in Northern Lights No. 22 municipal district; 56°51′N 117°38′W.

North Star Lake, ITASCA county, N central MINNESOTA, 22 mi/35 km E of GRAND RAPIDS, in Chippewa National Forest; 3 mi/4.8 km long, 1 mi/1.6 km wide; 47°33′N 93°39′W. Drained from N end by Potato Creek. Resort area. Sometimes called Potato Lake.

North Stifford (STIF-uhd), village (2001 population 5,618), THURROCK, SE ENGLAND, 2 mi/3.2 km N of GRAYS THURROCK; 51°30′N 00°18′E.

North Stonington, town, NEW LONDON county, SE CONNECTICUT, on RHODE ISLAND state line, 12 mi/19 km SE of NORWICH; 41°28′N 71°52′W. Rural agricultural town. State forest here. Incorporated 1807.

North Stormont (NORTH STOR-muhnt), township (□ 199 sq mi/517.4 sq km; 2001 population 6,855), SE ONTARIO, E central Canada, 19 mi/31 km from CORNWALL; 45°13′N 75°01′W. Composed of the communities of Avonmore, Berwick, CRYSLER, FINCH, Monkland, and MOOSE CREEK. Formed 1998.

North Stradbroke Island (NORTH STRAD-brok) (□ 123 sq mi/319.8 sq km), in PACIFIC OCEAN, just off SE coast of QUEENSLAND, NE AUSTRALIA, S of MORETON ISLAND; 27°35′S 153°28′E. Forms E shore of MORETON BAY; 24.5 mi/39.4 km long, 7 mi/11 km wide; rises to 700 ft/213 m. Sandy. Hardwood timber; tourism. Sometimes called Stradbroke Island.

North Stratford, NEW HAMPSHIRE: see STRATFORD.

North Swansea, MASSACHUSETTS: see SWANSEA.

North Sydney (SID-nee) town, NE CAPE BRETON ISLAND, NOVA SCOTIA, on SYDNEY HARBOUR; 46°12′N 60°15′W. Elevation 0 ft/0 m. It is the coal-shipping port for the nearby SYDNEY MINES and a winter base for the Cape Breton fisheries. Named after the Hon. Thomas Townshend, the first Viscount Sydney, and its location on N side of Sydney Harbour.

North Sydney (NORTH SYD-nee), suburb, E NEW SOUTH WALES, SE AUSTRALIA, 0.5 mi/0.8 km N of SYDNEY, on N shore of PORT JACKSON; 33°51′S 151°13′E. Linked to Sydney central business district by Sydney Harbor Bridge and Tunnel. "Second downtown" to Sydney; large commercial and high-rise office center. Manufacturing (insulation, plasterboard, safety equipment); information technology, media industries; shipyards.

North Syracuse, suburb (□ 1 sq mi/2.6 sq km; 2006 population 6,694) of SYRACUSE, ONONDAGA county, central NEW YORK, 6 mi/9.7 km N of city center. Hancock International Airport; 43°07′N 76°07′W. Inc. 1925.

North Taranaki Bight (ta-ruh-NAK-ee), embayment of TASMAN SEA, TARANAKI region, NORTH ISLAND, NEW ZEALAND, separated from SOUTH TARANAKI BIGHT by hard-lava promontory of CAPE EGMONT. Receives MOKAU and Waitara River. Near S extremity is NEW PLYMOUTH.

North Tarryall Peak, Colorado: see TARRYALL MOUNTAINS.

North Tawton (TAW-tuhn), village (2001 population 1,570), central DEVON, SW ENGLAND, on TAW RIVER, and 7 mi/11.3 km NE of OKEHAMPTON; 50°48′N 03°53′W. Light industry. Previously scene of annual horse and cattle fair. Has 13th-century church and many old houses.

North Tazewell (TAZ-wel), former town, SW VIRGINIA, part of TAZEWELL county, with separate post office, in ALLEGHENY MOUNTAINS, 1 mi/2 km N of downtown TAZEWELL; 37°07′N 81°31′W. In agricultural area; (hay, corn; cattle, sheep; dairying) manufacturing (lumber, commercial printing); timber; coal-mining.

North Tidworth (TID-wuhth), town (2001 population 5,991), E WILTSHIRE, S ENGLAND, 13 mi/21 km NNE of SALISBURY; 51°14′N 01°40′W. Important military center (usually called Tidworth). Has 15th-century church.

North Tisbury, MASSACHUSETTS: see WEST TISBURY.

North Tiverton, neighborhood, in TIVERTON town, on MOUNT HOPE BAY, NEWPORT county, SE RHODE ISLAND, 17 mi/27 km SE of PROVIDENCE.

North Tonawanda, industrial and commercial city (□ 10 sq mi/26 sq km; 2006 population 31,770), NIAGARA county, W NEW YORK, on the NIAGARA River at the terminus of the BARGE CANAL; 43°02′N 78°52′W. It is a port of entry and has a variety of manufacturing, including furniture, chemicals, plastics, and castings. Settled c.1802, incorporated as a city 1897.

North Tons River, c. 150 mi/241 km long, N (left) tributary of the Ganga River, in UTTAR PRADESH state, N central INDIA; formed on GANGA PLAIN by confluence of two headstreams 5 mi/8 km WNW of AKBARPUR; flows SE, past Akbarpur, JALALPUR, AZAMGARH, MUHAMMADABAD, and MAU, to the GANGA just W of BALLIA.

North Tripura, district (□ 1,089 sq mi/2,831.4 sq km), TRIPURA state, NE INDIA; ⊙ KAILASHAHAR. Bordered N by BANGLADESH, E by ASSAM state.

North Troy, VERMONT: see TROY.

North Truchas Peak, New Mexican: see TRUCHAS.

North Truro, MASSACHUSETTS: see TRURO.

North Tunica (TOO-nik-uh), unincorporated town (2000 population 1,450), TUNICA county, NW MISSISSIPPI, residential suburb 2 mi/3.2 km N of TUNICA; 34°42′N 90°22′W. Cotton, corn, rice, soybeans, sorghum; cattle; catfish.

North Turtle Lake, OTTER TAIL county, W MINNESOTA, 10 mi/16 km E of FERGUS FALLS; 5 mi/8 km long, 1 mi/1.6 km wide at E end; 46°18′N 95°48′W. Village of UNDERWOOD is near W end of lake. Smaller South Turtle Lake (2 mi/3.2 km long, 0.5 mi/0.8 km wide) 1 mi/1.6 km to S.

North Twin Lake (□ 16 sq mi/41 sq km), S NEWFOUNDLAND AND LABRADOR, 25 mi/40 km WNW of BOTWOOD; 8 mi/13 km long, 4 mi/6 km wide. Drains into EXPLOITS RIVER.

North Twin Lake, VILAS county, N WISCONSIN, 7 mi/11.3 km NE of EAGLE RIVER city; c.5 mi/8 km long, 1.5 mi/2.4 km wide. PHELPS is at E end. Connected to the smaller South Twin Lake (2 mi/3.2 km wide, 2 mi/3.2 km long) at W end by narrow channel.

North Twin Lakes Reservoir (□ 28 sq mi/73 sq km), PENOBSCOT county, N MAINE, on West Branch PENOBSCOT RIVER, 16 mi/96 km N of BANGOR; 45°38′N 68°47′W. Max. capacity 392,680 acre-ft. Formed by North Twin Dam (35 ft/11 m high), built (1934) for power generation; also used for flood control and recreation.

North Twin Mountain, NEW HAMPSHIRE: see FRANCONIA MOUNTAINS.

North Tyrol Limestone Alps, AUSTRIA: see TYROLEAN-BAVARIAN LIMESTONE ALPS.

North Ubian Island (OOB-yahn) (□ 6 sq mi/15.6 sq km), SULU province, PHILIPPINES, in PANGUTARAN GROUP of SULU archipelago, 30 mi/48 km WNW of JOLO Island; 06°09′N 120°26′E.

Northumberland (nor-THUHM-buhr-luhnd), county (□ 4,671 sq mi/12,144.6 sq km; 2001 population 50,817), N central New Brunswick, Canada, on the Gulf of St. Lawrence; ⊙ MIRAMICHI.

Northumberland (nor-THUHM-buhr-luhnd), county (□ 735 sq mi/1,911 sq km; 2001 population 77,497), SE ONTARIO, E central Canada, on Lake ONTARIO and on RICE LAKE; ⊙ COBOURG; 44°07′N 78°00′W. Includes the municipalities of BRIGHTON, Cobourg, PORT HOPE, Trent Hills, Alnwick/Haldimand, Cramahe, and HAMILTON.

Northumberland (nawth-UHM-buh-luhnd), county (□ 1,943 sq mi/5,051.8 sq km; 2001 population 307,190), NE ENGLAND; ⊙ MORPETH; 55°15′N 02°00′W. Northernmost of the English counties, it is separated from SCOTLAND by the CHEVIOT HILLS and the TWEED RIVER, and borders on the NORTH SEA. The terrain is level along the rugged coastline and hilly in the interior, where high moorlands alternate with fertile valleys. Other rivers are the TYNE, the DERWENT, the Wansbeck, the TILL, and the COQUET. The SE of the county is heavily populated and industrialized. In the past the economy was dominated by coal mining, shipping, shipbuilding and repairing, and the production of heavy electrical machinery. These industries suffered a heavy decline, however. Sheep and cattle are raised. HADRIAN'S WALL was built in Roman times. In the 6th century the Angles established themselves in the region, which later became the kingdom of NORTHUMBRIA. The Viking influence in history is still strong in speech and culture. The area suffered severely during the border wars between England and Scotland. The chief towns of the county (in addition to Morpeth) are ALNWICK, ASHINGTON, BERWICK-UPON-TWEED, BLYTH, and HEXHAM. Has its own national park.

Northumberland (north-UHM-buhr-luhnd), county (□ 477 sq mi/1,240.2 sq km; 2006 population 91,654), E central PENNSYLVANIA; ⊙ SUNBURY; 40°51′N 42°76′W. Anthracite-coal-mining region; drained by SUSQUEHANNA RIVER and its West Branch, joining at NORTHUMBERLAND; mountainous in S. Anthracite coal; manufacturing at Sunbury, Northumberland, and MILTON; limestone. Agriculture (corn, wheat, oats, barley, alfalfa, vegetables, soybeans, potatoes, apples; poultry, sheep, hogs, cattle; dairying). Milton State Park in NW; part of Shikellamy State Park in NW. Formed 1772.

Northumberland (north-UHM-buhr-luhnd), county (□ 285 sq mi/741 sq km; 2006 population 12,820), E

VIRGINIA; ⊙ HEATHSVILLE; 37°51'N 76°22'W. On NE part of NORTHERN NECK peninsula; bounded by POTOMAC RIVER estuary (NE) and CHESAPEAKE BAY (E); coast has many bays and inlets; indented from E by Great WICOMICO RIVER estuary. Reedville (fish-processing center) is a port. Mainly agriculture (hay, barley, wheat, corn, soybeans, tomatoes; cattle, poultry); oysters, crabs, herring, fish. Shore resorts (fishing, bathing). Formed 1648. Originally called Chickacoan.

Northumberland, town, COOS county, NW NEW HAMPSHIRE, 5 mi/8 km N of LANCASTER; 44°34'N 71°31'W. Bounded W by CONNECTICUT RIVER (Vermont state line); drained by UPPER AMMONOOSUC RIVER. Includes village of Groveton (1990 population 1,255) in NE (covered bridge). Agriculture (cattle, poultry; vegetables; dairying; nursery crops; timber); manufacturing (pulp and paper). Cape Horn State Forest at center. Settled 1767, incorporated 1779.

Northumberland (north-UHM-buhr-luhnd), borough (2006 population 3,541), NORTHUMBERLAND county, E central PENNSYLVANIA, 2 mi/3.2 km N of SUNBURY, on SUSQUEHANNA RIVER, at mouth of West Branch Susquehanna River (both rivers bridged); 40°53'N 76°47'W. Manufacturing (transportation equipment, food, wood products). Dr. Joseph Priestley lived here 1794–1804. Shikellamy State Park to SW and SE, partly on Packers Island; Sunbury Airport to E on Packers Island. Laid out c.1775, incorporated 1828.

Northumberland, Cape, SE ALASKA, S extremity of DUKE ISLAND, 35 mi/56 km SSE of KETCHIKAN; 54°52'N 131°21'W.

Northumberland, Cape (NOR-thuhm-buhr-luhnd), extreme SE SOUTH AUSTRALIA state, S central AUSTRALIA, in INDIAN OCEAN; forms W end of DISCOVERY BAY (site of PORT MACDONNELL); 38°04'S 140°40'E.

Northumberland Islands (NOR-thuhm-buhr-luhnd), coral group in CORAL SEA, between GREAT BARRIER REEF and SHOALWATER BAY, off E coast of QUEENSLAND, NE AUSTRALIA; 21°40'S 150°00'E. Comprise 80-mi/129-km chain of 40 rocky islands and scattered islets; largest, Prudhoe Island (2.3 mi/3.7 km long, 1 mi/1.6 km wide; rising to 1,074 ft/327 m); generally wooded. Tourist resorts.

Northumberland Strait, c.200 mi/320 km long and from 9 mi/14 km to 30 mi/48 km wide, arm of the Gulf of St. Lawrence, separating Prince Edward Island from NEW BRUNSWICK and NOVA SCOTIA, Canada. A bridge across the strait (completed 1997) connects Cape Tormentine, New Brunswick, to Prince Edward Island

Northumbria, Kingdom of (nawth-UHM-bree-uh), one of the Anglo-Saxon kingdoms in ENGLAND. It was originally composed of two independent kingdoms divided by the TEES RIVER—BERNICIA (including modern E SCOTLAND, the Borders region, E NORTHUMBERLAND, and DURHAM) and Deira (NORTH YORKSHIRE). Both kingdoms were settled by invading Angles c.500. Sparse records tell of a King Ida of Bernicia and a King Ælli (or Ælle) of Deira in the middle of the 6th century. Æthelfrith of Bernicia (593–616) united the kingdoms to form Northumbria and added Scottish and Welsh territory. He was defeated by Edwin of Deira (616–632), who accepted (627) Roman Christianity and established Northumbrian supremacy in England. Oswald of Bernicia (633–641) brought in St. Aidan to introduce Celtic Christianity. Under Oswiu (641–670) and Ecgfrith (670–685), Northumbria's power declined gradually as that of MERCIA increased. Oswiu, however, established the Roman Church over the Celtic Church at the Synod of WHITBY (664). The Danes occupied S Northumbria, and the Angles were able to keep only a small kingdom stretching from the Tees N to the Firth of FORTH. The conquering Canute (1015) and his successors installed Danish earls, of whom Siward (d. 1055) was the last and most powerful. The North-

umbrians expelled his successor, Tostig, in 1065. Tostig was replaced by Morcar, the brother of Edwin, Earl of Mercia. Tostig returned with Harold Hardrada of NORWAY the next year and defeated Morcar and Edwin at Fulford. Harold II of England, however, came N to defeat the Danes at the battle of STAMFORD BRIDGE (1066), an event that, in effect, ended the Viking era in England.

North Utica, village (2000 population 977), LA SALLE county, N central ILLINOIS, 4 mi/6.4 km E of LA SALLE, near ILLINOIS RIVER. On Illinois and Michigan State Trail; 41°20'N 89°00'W. STARVED ROCK and Mathiessen state parks to S.

Northvale, borough (2006 population 4,562), BERGEN county, extreme NE NEW JERSEY, 13 mi/21 km NE of PATERSON, at NEW YORK line; 41°00'N 73°57'W. Largely residential. Incorporated 1916.

North Valley Stream, unincorporated town (□ 1 sq mi/2.6 sq km; 2000 population 15,789), NASSAU county, SE NEW YORK, on LONG ISLAND; 40°40'N 73°42'W. It is chiefly residential.

North Vancouver (van-KOO-vuhr), district municipality (□ 62 sq mi/161.2 sq km; 2001 population 82,310), SW BRITISH COLUMBIA, W Canada, across BURRARD INLET from VANCOUVER, and at base of COAST MOUNTAINS; 49°22'N 123°04'W. Parks, resorts; Capilano Suspension Bridge (built 1889) extends 450 ft/137 m across and 230 ft/70 m above CAPILANO RIVER.

North Vancouver (NORTH van-KOO-vuhr), city (□ 5 sq mi/13 sq km; 2001 population 44,303), SW BRITISH COLUMBIA, on BURRARD INLET of the Strait of GEORGIA, opposite VANCOUVER, of which it is a suburb; 49°19'N 123°04'W. Shipbuilding, woodworking, and the shipping of grain, lumber, and ore are the chief industries.

North Vandergrift, unincorporated town, ARMSTRONG county, W PENNSYLVANIA, 25 mi/40 km NE of PITTSBURGH, on KISKIMINETAS RIVER, opposite (1 mi/1.6 km NE of) EAST VANDERGRIFT; 40°36'N 79°33'W. Agriculture (grain; livestock; dairying). Listed as North Vandergrift–Pleasant View (1990 population 1,431) by Census Bureau; Pleasant View 1 mi/1.6 km to NW.

North Vernon, city (2000 population 6,515), JENNINGS county, SE INDIANA, 2 mi/3.2 km NW of VERNON; 39°01'N 85°38'W. Plastic injection moldings, wire harnesses, processed eggs, concrete, steel forgings, motor-vehicle fuel tubes, fixtures. Timber; limestone quarries. Selmier State Forest is nearby to the NE. Plotted 1854.

North Versilles (VER-sils), township, ALLEGHENY county, W PENNSYLVANIA, residential suburb 12 mi/19 km ESE of PITTSBURGH; 40°22'N 79°48'W. Manufacturing of nylon and polyester bags, fabric printing, plastics molding.

North Vietnam: see VIETNAM.

Northville, town, WAYNE and OAKLAND counties, SE MICHIGAN, suburb 23 mi/37 km NW of downtown DETROIT, on upper reach of Middle RIVER ROUGE; 42°26'N 83°29'W. Remnant farming; manufacturing (glass products, chemicals, computer equipment, transportation equipment, furniture). Mayberry State Park to W. Incorporated 1867.

Northville, village (□ 1 sq mi/2.6 sq km; 2000 population 1,139), FULTON county, E central NEW YORK, on NW arm of SACANDAGA Reservoir, 15 mi/24 km NE of GLOVERSVILLE; 43°13'N 74°10'W. S terminus of 133-mi/214-km Northville–Lake Placid Trail. Incorporated 1873.

Northville, village, SUFFOLK county, SE NEW YORK, on NE LONG ISLAND, 4 mi/6.4 km NE of RIVERHEAD; 40°58'N 72°37'W. In summer-recreation area.

Northville, village (2006 population 113), SPINK county, NE central SOUTH DAKOTA, 20 mi/32 km N of REDFIELD, on Snake Creek; 45°08'N 98°34'W. Trading point for agricultural area.

North Vineland, village, CUMBERLAND county, S NEW JERSEY, 3 mi/4.8 km N of VINELAND.

North Vizagapatam, INDIA: see VISHAKHAPATNAM or VIZIANAGARAM.

North Wales, borough (2006 population 3,260), MONTGOMERY county, SE PENNSYLVANIA, suburb 19 mi/31 km NNW of downtown PHILADELPHIA; 40°12'N 75°16'W. Manufacturing (paperboard boxes, power supplies, corrugated boxes, food-service equipment, pumps and valves, commercial printing, precision metal tubing, cutting blades, metal finishing, textile finishing); some agriculture. Settled 1850, laid out 1867, incorporated 1869.

North Walpole, NEW HAMPSHIRE: see WALPOLE.

North Walsham (WAWL-shuhm), town (2001 population 11,998), NE NORFOLK, E ENGLAND, 14 mi/23 km NNE of NORWICH; 52°49'N 01°23'E. Has 14th-century church, remains of Saxon church tower, and 16th-century grammar school attended by Lord Nelson.

North Warren, unincorporated town, WARREN county, NW PENNSYLVANIA, residential suburb 2 mi/3.2 km N of WARREN, on CONEWANGO CREEK; 41°52'N 79°09'W. Manufacturing (wood products). Agricultural area (corn, hay; dairying). Large Warren State Hospital complex to N.

North Washington, town (2000 population 118), CHICKASAW county, NE IOWA, 6 mi/9.7 km NW of NEW HAMPTON; 43°07'N 92°24'W.

Northway, village, E ALASKA, near YUKON border, on NABESNA RIVER, and 110 mi/177 km SW of DAWSON; 62°57'N 141°57'W. Just off ALASKA HIGHWAY.

North Waziristan, PAKISTAN: see WAZIRISTAN.

North Weald Bassett (WEELD BAH-sit), village (2001 population 6,039), W ESSEX, SE ENGLAND, 3 mi/4.8 km N of EPPING; 51°43'N 00°10'W. Nearby is airfield that was important base during battle of Britain.

North Webster, town (2000 population 1,067), KOSCIUSKO county, N INDIANA, 10 mi/16 km NE of WARSAW; 41°20'N 85°42'W. In agricultural area; manufacturing (heaters). Recreation area. Tri-County State Fish and Wildlife Area is nearby to the NE.

North Weissport, unincorporated town, CARBON county, E PENNSYLVANIA, residential suburb 1 mi/1.6 km N of WEISSPORT, and 1 mi/1.6 km E of LEHIGHTON, on LEHIGH RIVER; 40°50'N 75°41'W.

North-West, province (□ 6,876 sq mi/17,877.6 sq km; 2004 population 1,989,000), NW CAMEROON; ⊙ BAMENDA (also largest city); 06°00'N 10°30'E. Bounded on S by West province, E by ADAMAOUA province, N by Nigeria, W by South-West province. Highland area where millet, manioc, and coffee are grown; livestock. Includes the departments of Boyo, BUI, DONGAMANTUNG, MENCHUM, MEZAM, MOMO, and Ngo-Ketunjia.

Northwest Angle, MINNESOTA: see LAKE OF THE WOODS.

Northwest Arctic, borough (□ 35,863 sq mi/93,243.8 sq km; 2006 population 7,511), NW ALASKA, bounded on W by CHUKCHI SEA, indented by KOTZEBUE SOUND and HOTHAM INLET, drained by KOBUK RIVER. Main town is KOTZEBUE. Part of BERING LAND BRIDGE National Preserve to SW; part of NOATAK National Preserve to N; part of GATES OF THE ARCTIC National Park and Preserve to E; CAPE KRUSENSTERN National Monument to NW; KOBUK VALLEY NATIONAL PARK in center; Selawik National Wildlife Reserve to S. ARCTIC CIRCLE crosses borough at its middle. Fishing. Caribou and seal hunting.

Northwest Cape (NORTH-WEST), NW WESTERN AUSTRALIA state, W AUSTRALIA, in INDIAN OCEAN, on peninsula forming W shore of EXMOUTH GULF; 21°47'S 114°10'E. U.S. defense communications base.

North Westchester, village, NEW LONDON county, E central CONNECTICUT, 16 mi/26 km WNW of NORWICH, at confluence of SALMON and Pine rivers.

North-West District, administrative division (2001 population 142,970), N BOTSWANA. Bounded by

NAMIBIA W and N (CAPRIVI STRIP and CHOBE RIVER), NE by small section of ZAMBIA (ZAMBEZI RIVER) and ZIMBABWE, ESE by CENTRAL DISTRICT, S by GHANZI DISTRICT. Formed circa 2001 by combining NGAMILAND (W portion of NORTH-WEST DISTRICT) and CHOBE districts (NE portion of North-West District). Includes the towns of KASANE, MAUN, SEHITHWA, and SHAKAWE and the village of TOTENG. OKAVANGO DELTA drainage basin is in central part of district, which includes MOREMI GAME RESERVE; CHOBE NATIONAL PARK in NE; NXAI PAN NATIONAL PARK in SE; LAKE NGAMI in S; AHA HILLS in W; and Tsodilo Hills in NW. Airport at Maun.

North-Western, province (□ 48,582 sq mi/126, 313.2 sq km), NW ZAMBIA; ⊙ SOLWEZI; 13°00′S 25°00′E. Bounded on N by CONGO REPUBLIC, on W by ANGOLA, on S by WESTERN and CENTRAL provinces, on E by Central and COPPERBELT provinces. Drained by ZAMBEZI, Kabompo, and Lunga rivers; source of Zambezi River in far NW. Part of KAFUE NATIONAL PARK in S, W. Lunga National Park in center. Agriculture (honey, beeswax, pineapples, sorghum, corn, peanuts; cattle); timber. Created from Copperbelt province. Formerly called Kaonde-Lunda when Zambia was Northern Rhodesia.

North Western Canal, irrigation canal of SUKKUR BARRAGE system, N SIND province, SE PAKISTAN; from right bank of INDUS River at SUKKUR runs c.120 mi/193 km W, through SUKKUR and SW SIBI (BALUCHISTAN province) districts. Waters crops of wheat, millet.

North Western Province, administrative division (□ 2,898 sq mi/7,534.8 sq km), W SRI LANKA; ⊙ KURUNEGALA; 07°45′N 80°10′E. Bounded W by INDIAN OCEAN and Gulf of MANNAR; mainly lowland, with W foothills of SRI LANKA HILL COUNTRY in SE; drained by the DEDURU OYA River. Largely agricultural (coconut palms, rice, vegetables, cacao, rubber, fruit, tobacco). Important government salterns (PUTTALAM, PALAVI); graphite mines (DODANGASLANDA, RAGEDARA). Main towns include Kurunegala, Puttalam, CHILAW. Archaeological remains at YAPAHUWA and DAMBADENIYA. Sri Lankan Christian pilgrimage center at St. Anna's Church. Created 1844.

North-Western Provinces, name given to former British presidency of AGRA, now part of UTTAR PRADESH state, N central INDIA. Created 1833 partly out of Mogul Empire's Agra province. Named North-Western provinces in 1835.

Northwest Fork of Nanticoke River, DELAWARE and MARYLAND: see MARSHYHOPE CREEK.

North-West Frontier Province (□ 41,000 sq mi/106,600 sq km), NW PAKISTAN; ⊙ PESHAWAR. Bounded N and W by AFGHANISTAN. An area of high, barren mountains dissected by fertile valleys. Several notable peaks including TIRICH MIR (25,263 ft/7,700 m), SIKARAM (15,620 ft/4,761 m), and TAKHT-I-SULAIMAN (11,290 ft/3,441 m). On W borders are noted KHYBER and GUMAL passes. Predominantly agricultural, wheat is the chief crop; barley, sugarcane, tobacco, cotton, and fruit trees are also cultivated, and livestock is raised. Irrigation works, notably the Warsak project on the KABUL RIVER, and the SWAT RIVER canal, supplement the scanty rainfall. Mineral resources include marble, rock salt, gypsum, and limestone. Extensive forests in the N. Food processing, cotton and wool milling, papermaking, and the production of cigarettes, textiles, chemicals, and fertilizer are the main industries; handicrafts also flourish. Some Pashtun tribes inhabit the province in the plains and lower hills. Kohestani ethnic groups dominate the N. Main centers include PESHAWAR, NOWSHERA, DERA ISMAIL KHAN, KOHAT, BANNU, MARDAN, and ABBOTTABAD. The region has been historically and strategically important due to secondary passes leading into INDIA, through which came invaders from the KABUL area, and central ASIA. Alexander the Great travelled

through here c.326 B.C.E., but his garrisons were unable to hold the region. In the early century, Kanishka and his Kushan dynasty ruled the area. The Pashtuns arrived from the QUETTA region in the 7th century, and by the 10th century conquerors from AFGHANISTAN had made Islam the dominant religion in the S, but even today 3,000 people of pagan Kalasha remain in S CHITRAL district. Under local Pashtun rule from the late 12th century until Babur annexed it to his Mogul empire, the region paid nominal allegiance to the Moguls in the 16th and 17th centuries. After Nadir Shah's invasion (1738), it became a feudatory of the Kabul Durrani kingdom. The Sikhs later conquered the area, which passed to Great Britain in 1849. The British maintained large military forces and entered into treaties and paid heavy subsidies to control the Pashtun tribesmen against the intrusion of Russian influence in the INDIAN SUBCONTINENT. After 1880 until 1896 the British demarcated the border between BRITISH INDIA and the new state of AFGHANISTAN, known as the Durand Line. This border remains in effect even though it divides several Pashtun tribes. Britain separated the region from the PUNJAB of INDIA in 1901 and constituted the North-West Frontier province, whose people voted to join newly independent PAKISTAN in 1947. From 1955 to 1970 the North-West Frontier province was a section of the consolidated province of West PAKISTAN. In 1970, the region was once again granted provincial status. The Afghan civil war, starting in 1973, caused over three million refugees to flee here. PESHAWAR became the military and political center of the feuding Afghan factions and warlords. All Afghan tented refugee camps were closed by 1995. Rapid urban growth has been stimulated by unsettled conditions in AFGHANISTAN.

Northwest Gander River, S NEWFOUNDLAND AND LABRADOR, Canada, headstream of GANDER RIVER.

Northwest Harborcreek, unincorporated town (2000 population 8,658), ERIE county, NW PENNSYLVANIA, residential suburb 6 mi/9.7 km ENE of downtown ERIE, on LAKE ERIE at mouths of Sixmile and Sevenmile Creeks; 42°08′N 79°59′W.

North Westminster, VERMONT: see WESTMINSTER.

Northwest Passage, water routes through the ARCTIC ARCHIPELAGO, N CANADA, and along the N coast of ALASKA between the PACIFIC and ATLANTIC oceans. Even though the explorers of the 16th century demonstrated that the American continents were a true barrier to a short route to East Asia, there still remained hope that a natural passage would be found leading directly through the barrier. During the same period, the idea of reaching CHINA and INDIA by sailing over the NORTH POLE or by sailing through a passage N of EUROPE and ASIA—the NORTHEAST PASSAGE—also became popular. However, the Northwest Passage remained the most important goal, and the search for the passage continued even though at that time such a route had no commercial value. Sir Martin Frobisher, the English explorer, was the first European to explore (1576–1578) the eastern approaches of the passage. John Davis also explored (1585–1587) this area, and in 1610 Henry Hudson sailed north and visited HUDSON BAY while seeking a short route to the Orient. Soon afterward, William Baffin, an English explorer, visited (1616) BAFFIN BAY, through which the passage was finally found. English statesmen and merchants, anxious to have the passage found, encouraged exploration. Luke Fox and Thomas James made (1631–1632) voyages into Hudson Bay. Although one of the avowed goals of Hudson's Bay Company was to find the Northwest Passage, little was accomplished until a century after its charter, when Samuel Hearne, a British explorer with the company, went overland as far west as the Coppermine River (1771–1772) and demonstrated that there was no short passage to the sea.

The British government offered prizes for achievements in northern exploration, and Captain James Cook was inspired to make the first attempt at navigating the passage from the west. He died before he could accomplish anything. However, the British, Spanish, and Americans pushed explorations on the Pacific coast; and the explorations of the Russians about KAMCHATKA and ALASKA, together with the voyages of Alexander Mackenzie, the Canadian explorer, and the expedition of the Americans Lewis and Clark, had now shown the contours of the continental barrier. Wars between Britain and France interrupted the search for the Northwest Passage, and when resumed after the wars the explorations were made in the interests of science, not commerce. The desire to extend human knowledge was the chief motive in arctic exploration after the expeditions of British explorers John Ross and David Buchan were sent out in 1818. Ross's later voyages, and those of Sir William Edward Parry, F. W. Beechey, Sir George Back, Thomas Simpson, and Sir John Franklin pushed forward the knowledge of arctic regions and of the Northwest Passage. The last tragic expedition of Franklin indirectly had more effect than any other voyage, because of the many expeditions sent out to discover his fate. In his expedition (1850–1854), Robert J. Le M. McClure penetrated the passage from the W along the N coast of the continent and by a land expedition reached Viscount Melville Sound, which had been reached (1819–1820) by Parry from the E.

The actual existence of the Northwest Passage had been proved, and the long search was over. It was many years, however, before a transit of the passage was made. This feat, which had been attempted by so many men, was first accomplished (1903–1906) by the Norwegian explorer Roald Amundsen. Interest in the Northwest Passage slackened until the 1960s, when oil was discovered in N Alaska and there was a desire for a short water route to transport oil to the E coast of the U.S. In 1969, the SS *Manhattan*, an ice-breaking tanker, became the first commercial ship to transit the Northwest Passage. In 1988 the U.S. and Canada addressed the question of arctic sovereignty, agreeing that U.S. icebreakers could cross arctic waters, but only after approval on a case-by-case basis. In 2007, extreme loss to the arctic ice pack effectively made the passage fully navigable, and three ships successfully completed the passage that year.

North West Point, cape in HANOVER parish, NW JAMAICA, 19 mi/31 km W of MONTEGO BAY; 18°28′N 78°14′W. Also known as Pedro Point.

North Westport, MASSACHUSETTS: see WESTPORT.

Northwest Providence Channel, c.130 mi/209 km long, WNW-NESE, in BAHAMAS, just E of WEST PALM BEACH, FLORIDA, between LITTLE BAHAMA BANK (N) and GREAT BAHAMA BANK (S), linking Straits of FLORIDA (W) with NORTHEAST PROVIDENCE CHANNEL (E).

North-West Province (□ 44,861 sq mi/116,190 sq km; 2004 estimated population 3,313,959; 2007 estimated population 3,394,200), N SOUTH AFRICA, W area of former TRANSVAAL province, bordered to SW by NORTHERN CAPE province, SE by FREE STATE province, E by GAUTENG province, NE by LIMPOPO province, and N by BOTSWANA; ⊙ MAFIKENG; 26°30′S 26°00′E. Largely semiarid to temperate grassland located on highveld. Extremely rich in variety of mineral deposits. Includes much of former BOPHUTHATSWANA. Its area is 9.5% of the republic's total land area. Capital was MMABATHO before it was switched to Mafikeng.

North West River, town (□ 1 sq mi/2.6 sq km; 2006 population 492), E NEWFOUNDLAND AND LABRADOR, Canada, at W end of Lake Melville, at mouth of NASKAUPI RIVER, 10 mi/16 km NE of GOOSE BAY; 53°32′N 60°09′W. Established in 1743 by fur trader Louis Fornel. Lumbering.

Area is shown by the symbol □, and capital city or county seat by ⊙.

Northwest Territories, territory (□ 440,584 sq mi/ 1,145,518.4 sq km; 2006 population 41,464), NW CANADA; ⊙ YELLOWKNIFE; 65°00′N 118°00′W. The Northwest Territories lie W of NUNAVUT territory, N of latitude 60°N, and E of the YUKON territory. Until 1999, when the Northwest Territories were divided and the E portion became Nunavut, the region occupied more than 33% of Canada's area. Yellowknife has been the territorial capital since 1967; before 1967 the government was conducted from OTTAWA. Nunavut's establishment split the Northwest Territories along a zigzag line running from the SASKATCHEWAN-MANITOBA border through the ARCTIC ARCHIPELAGO to the NORTH POLE. Geographically, the region is largely S of the tree line, which runs roughly NW to SE, from the MACKENZIE RIVER delta in the ARCTIC OCEAN to the SE corner of the territory. Tundra characterizes the land N of the tree line; there the Inuit and Native Americans derive their small incomes from hunting, trapping fur, and making arts and crafts, and obtain many necessities from fish, seals, reindeer, and caribou. Most of the development of the Northwest Territories has taken place in the FORT SMITH and INUVIK regions, areas well covered with softwoods and rich in minerals. In these regions are two of the largest lakes in the world, GREAT SLAVE and GREAT BEAR. The geology of the region E of these lakes is that of the CANADIAN SHIELD. Great Slave and Great Bear lakes are linked to the ARCTIC OCEAN by one of the world's longest rivers, the Mackenzie, which runs 2,635 mi/4,241 km from its source in BRITISH COLUMBIA.

Agriculture in the Northwest Territories is virtually impossible except for limited cultivation S of the Mackenzie River region. Government agricultural experimental stations, however, are at Fort Simpson and Yellowknife. Trapping, the region's oldest industry, ranks second after mining. A thriving commercial fishing industry, based on whitefish and lake trout, is centered on the town of HAY RIVER, on Great Slave Lake. Oil is pumped and refined at TULITA and NORMAN WELLS on the Mackenzie River; gold is being produced in increasing quantities at Yellowknife and on the Snare River; copper is extracted on the COPPERMINE RIVER. The region also has tungsten, silver, cadmium, and nickel deposits. Important hydroelectric developments are on the Talston and Snare rivers.

Transportation and communication in the Northwest Territories are difficult. Long winters close the rivers to navigation for all but two months of the year, but allow the opening of winter roads for transport of goods. During World War II, an oil pipeline from Norman Wells to Zama, Alberta, was built. Despite the Great Slave railroad and the MACKENZIE HIGHWAY system, which links ALBERTA to the Great Slave area, most commerce, supply, and travel continue to be airborne. An extensive N roads program, first announced in 1966, was expected to help open up the area when completed, and there are now extensive telecommunications services. The Liard Highway, opened in 1984, ties FORT SIMPSON to the ALASKA HIGHWAY.

Original Native American inhabitants were the hunting and fishing Inuit and Dene (de-NAI). Vikings from GREENLAND may have been the first Europeans to venture into the E portion of the Northwest Territories, now Nunavut. Sir Martin Frobisher was the first in a long line of explorers to seek a NORTHWEST PASSAGE, but it was Henry Hudson who discovered the gateway to the NW (Hudson Bay) in 1610. For several decades the Hudson's Bay Company sent out trader-explorers through the N sea lanes and coast, and in 1771 Samuel Hearne descended the Coppermine River. In 1789, Alexander Mackenzie, exploring for the North West Company, journeyed to the mouth of the Mackenzie River. Sir John Franklin made scientific expeditions to the Arctic NW in the early 19th century, obtaining valuable geographical data. The area that is now the

Northwest Territories and Nunavut was part of the vast lands sold by the Hudson's Bay Company to the new Canadian confederation in 1870. Some of those lands were added to the provinces of QUEBEC and ONTARIO. The province of MANITOBA was carved from them in 1870, and Alberta and SASKATCHEWAN in 1905.

The present boundaries of the Northwest Territories were set in 1912 and remained fixed until the creation of Nunavut in 1999. Since the patriation of the Canadian constitution, several land claims by Native peoples have been making their way through the courts and the federal government. In 1982 the Northwest Territories voted to split the territory. The new territory of Nunavut, which is controlled by the Inuit, was established April 1, 1999. This action split the Northwest Territories along a zigzag line running from the Saskatchewan-Manitoba border to the North Pole. Other Native groups making claims are the Dene, the Métis, and the Inuvialuit. Today the territory is administered by a commissioner and a nineteen-member legislative assembly in Yellowknife, backed up by the Royal Canadian Mounted Police. The government operates a school system and, with the aid of scattered missions throughout the vast region, provides extensive health and welfare services. The Northwest Territories sends one senator and one representative to the national parliament. With its valuable minerals and their strategic importance, enhanced by transpolar navigation, this Canadian territory continues to present major challenges in every field of human endeavor. The Northwest Territories are the site of the NAHANNI and Baffin Islands national parks (both established 1972).

Northwest Territory, first possession of the U.S., comprising the region known as the Old Northwest, S and W of the GREAT LAKES, NW of the OHIO RIVER, and E of the MISSISSIPPI RIVER, including the present states of OHIO, INDIANA, ILLINOIS, MICHIGAN, WISCONSIN, and part of MINNESOTA. Men from New France began to penetrate this rich fur country in the 17th century; in 1634, the French explorer Jean Nicolet became the first to enter the region. He was followed by explorers and traders: Radisson and Groseilliers, Duluth, La Salle, Jolliet, Perrot, and Cadillac, as well as by missionaries such as Jogues, Dablon, and Marquette. The Great Lakes region was controlled by a few widely scattered French posts, such as KASKASKIA, VINCENNES, PRAIRIE DU CHIEN, and GREEN BAY; links were established between the Northwest settlements and those in French Louisiana (St. Louis, New Orleans). The two chief posts of the Old Northwest were DETROIT and MACKINAC (Michilimackinac), but French influence spread among the Native American groups E to the Iroquois country.

In the 18th century the Northwest was coveted not only by the British colonists in CANADA, but also by those in the American seaboard colonies, who organized the Ohio Company in 1747 for the purpose of extending the Virginia settlements westward. At the same time, the French sought to strengthen their hold on the Northwest by building forts. The clash of British and French interests culminated in the expedition led by George Washington that resulted in the loss of FORT NECESSITY and the outbreak of the last of the French and Indian wars. The wars ended in 1763 with the Treaty of Paris, by which GREAT BRITAIN obtained Canada and the Old Northwest. Almost immediately after the British took over, Pontiac, an Ottawa chief, led an uprising against them. The Ottawa were somewhat appeased by the British Proclamation of 1763 that closed the region W of the ALLEGHENY MOUNTAINS to white settlement in an attempt to protect the Native American fur trade and lands; yet this action caused resentment among the American frontiersmen and contributed to the American Revolution.

The mysterious machinations of Robert Rogers, an American frontiersman, further endangered the British

hold on the Old Northwest. During the Revolutionary War, an expedition led by the American general George Rogers Clark penetrated deep into the region in 1778–1779, in one of the most daring and valuable exploits of the war. The Old Northwest, which became U.S. territory in 1783 by the Treaty of Paris ending the Revolution, soon became one of the most pressing problems before the U.S. Congress. The four so-called landed states—VIRGINIA, MASSACHUSETTS, NEW YORK, and CONNECTICUT—claimed portions of the Old Northwest, while states with no W land claims, especially MARYLAND, argued that if the claims of the landed states were recognized, the wealth and population of the other states would be attracted to the W lands. The final solution was the cession of all the lands to the U.S. government, which was thus greatly strengthened; New York made its cession in 1780, Virginia in 1784, Massachusetts in 1785, and Connecticut in 1786. Two reserves were kept, the Virginia Military District and the Connecticut Western Reserve in Ohio.

The Ordinance of 1785 established the Township System for surveying, which used a rectangular grid system in order to divide the land. The Ordinance of 1787 set up the machinery for the organization of territories and the admission of states. Its terms prohibited slavery in the Northwest Territory, encouraged free public education, and guaranteed religious freedom and trial by jury. The Ohio Company of Associates, the most active force in early colonization, was followed by later companies that brought settlers into the territory. British traders, however, opposed American expansion, and the Native Americans were also hostile to their encroachment. A series of campaigns against the Native Americans culminated in 1794, when General Anthony Wayne won an American victory at Fallen Timbers; his victory was solidified by the Greenville Treaty of 1795. Meanwhile, Jay's Treaty and subsequent negotiations smoothed out some of the British-American difficulties. The Northwest posts were transferred to the Americans in 1796, although British influence remained strong among the Native Americans.

Settlers poured into the S part of the territory, and a legislature was organized in 1799. The W part was split off as Indiana Territory in 1800, and by 1802, the E portion was populated enough to seek admission as a state; it was admitted as Ohio in 1803. Other territories were then formed—Michigan in 1805, Illinois in 1809, and Wisconsin in 1836. The British traders, however, wanted the Northwest set aside as Native American land. Unrest led Tecumseh and Shawnee Prophet to seek a permanent foothold for the Native Americans. Some W Americans, meanwhile, sought to extend the Northwest to Canada. The quarrel over the Northwest was a major cause of the War of 1812. The Treaty of Ghent, which ended the war, solved the problem of the Northwest. Despite opposition from British merchants in the region, Great Britain irrevocably gave the Northwest to the U.S.

North Weymouth, MASSACHUSETTS: see WEYMOUTH.
Northwich (NAWTH-wich), town (2001 population 39,568), CHESHIRE, W central ENGLAND, at the confluence of the WEAVER and DANE rivers, 20 mi/32 km S of MANCHESTER; 53°16′N 02°31′W. Northwich was once the center of England's salt production; however, the manufacturing of chemicals has become its leading occupation. It was the site of a Roman station, Condate. The rock-salt fields were used even before the Christian era. Noteworthy is the Northwich salt museum. The town includes Winnington.
North Wilbraham, MASSACHUSETTS: see WILBRAHAM.
North Wildwood, resort city (2006 population 4,803), CAPE MAY county, S NEW JERSEY, on barrier island (bridged to mainland) off CAPE MAY peninsula, and 32 mi/51 km SW of ATLANTIC CITY; 39°00′N 74°47′W. Incorporated 1917.

North Wilkesboro, town (□ 5 sq mi/13 sq km; 2006 population 4,184), WILKES county, NW NORTH CAROLINA, 50 mi/80 km W of WINSTON-SALEM, 1 mi/1.6 km N of WILKESBORO, across Yadkin (PEE DEE) River; 36°10′N 81°08′W. Manufacturing (glass mirrors, apparel, furniture, machinery, lumber, crushed stone, consumer goods; printing and publishing); retail trade. Rendezvous Mountain Educational State Forest to NW; West Scott Kerr (Wilksboro) reservoir to W; Stone Mountain State Park to N. North Wilkesboro Speedway. Founded 1890; incorporated 1891.

North Willems Canal, Dutch *Noord-Willems Kanaal*, DRENTHE and GRONINGEN provinces, N NETHERLANDS; links S with DRENTSE HOOFDVAART canal at ASSEN; connects Assen (S) with GRONINGEN (N).

North Windham, village, WINDHAM county, NE CONNECTICUT, 4 mi/6.4 km NE of WILLIMANTIC.

North Wingfield (WING-feeld), village (2001 population 6,318), NE DERBYSHIRE, central ENGLAND, 4 mi/6.4 km SE of CHESTERFIELD; 53°11′N 01°22′W. Has 14th–15th-century church.

Northwood, town (2000 population 2,050), N IOWA, near MINNESOTA state line, on SHELL ROCK RIVER; and 20 mi/32 km N of MASON CITY; ☉ WORTH county; 43°26′N 93°13′W. Manufacturing (dairy products, food, furniture, chemicals). Settled 1853, incorporated 1875.

Northwood, town, ROCKINGHAM county, SE NEW HAMPSHIRE, 11 mi/18 km SW of ROCHESTER; 43°12′N 71°12′W. Agriculture (cattle, poultry; vegetables; dairying; nursery crops; timber); manufacturing (lumber, fabricated metal products). Source of LAMPREY RIVER. Northwood Lake in W; Jenness Pond in NW.

Northwood, town (2006 population 864), GRAND FORKS county, E NORTH DAKOTA, 28 mi/45 km SW of GRAND FORKS; 47°44′N 97°34′W. Printing; dairy products; wheat, flax, potatoes. Settled 1879, incorporated 1892. Named for NORTHWOOD, IOWA, the hometown of most of the first settlers.

Northwood (NAWTH-wud), village and parish (2001 population 3,679), on Isle of WIGHT, S ENGLAND, 2 mi/3.2 km N of COWES; 50°44′N 01°19′W. Church dates from 13th century.

Northwood, ENGLAND: see HILLINGDON.

Northwoods, town (2000 population 4,643), SAINT LOUIS county, E MISSOURI, residential suburb 7 mi/11.3 km NW of downtown ST. LOUIS; 38°42′N 90°16′W. Incorporated 1939.

North Woodstock, NEW HAMPSHIRE: see WOODSTOCK.

North Yarmouth (YAHR-muhth), town, CUMBERLAND county, SW MAINE, on ROYAL RIVER, and 15 mi/24 km NNE of PORTLAND; 43°51′N 70°14′W. Incorporated 1680, Yarmouth set off 1849.

North Yemen, former republic, officially the Yemen Arab Republic (□ 97,300 sq mi/252,980 sq km). Its early history is that of YEMEN. In 1967, after years of civil war bet. the ruling imam and republican forces, the area of Yemen was split into 2 sects.: the Yemen Arab Republic and the People's Republic of Yemen (South Yemen); the strait Bab el Mandeb was the dividing line. The 2 Yemens immediately began having violent conflicts over border disputes, which lasted throughout the 1970s and eventually led to all-out war. A peace agreement unifying the 2 sides was devised in 1979, and North and South Yemen were officially reunited in May 1990. Conflicts bet. the 2 sides have continued to the present day.

North York (NORTH YORK), former city and borough (□ 68 sq mi/176.8 sq km; 2001 population 608,288) of Metropolitan Toronto, S central ONTARIO, E central Canada, 7 mi/11 km N of downtown TORONTO; 43°45′N 79°24′W. Humber River on W boundary; E and W branches of Don River in E part. MacDonald-Cartier Freeway (Highway 401) runs E–W. Major commercial and office complexes along Freeway and Yonge Street. Ontario Science Centre, Black Creek Pioneer Village,

Civic Gardens, North York Performing Arts Centre; Downsview Canadian Air Force Base. York University campus here. Incorporated 1979. Amalgamated into TORONTO in 1998.

North York, borough (2006 population 1,657), YORK county, S PENNSYLVANIA, residential suburb 1 mi/1.6 km N of YORK; 39°58′N 76°43′W.

North Yorkshire (NAWTH YAWK-shir), county (□ 3,104 sq mi/8,070.4 sq km; 2001 population 569,660), N ENGLAND; ☉ NORTHALLERTON; 54°13′N 01°25′W. The county, the largest in terms of square miles in England, comprises the districts of Craven, Hambleton, Harrogate, Richmondshire, Ryedale, Scarborough, and Selby. The district of York was previously part of the county, but it became an independent unitary authority in April 1996. North Yorkshire consists of two upland areas; the PENNINES and deep valleys dominate the W regions, while in the E are the limestone and sandstone CLEVELAND HILLS. The uplands are separated by the Vale of York, a broad lowland consisting of clay soil. The economy is agricultural, but there is also some light industry, such as food processing and light manufacturing. Around SELBY new deep pits have been sunk since the 1970s, expanding the industrial population. The area was occupied by the Roman military until the 7th century. York flourished under the Anglians in the 8th century until invasions led to occupation by Scandinavians. William the Conqueror destroyed many settlements here. During the Middle Ages the county was governed by wealthy landowners. It was later ravaged during the 15th century and in the mid-17th century English Civil War. Many castle ruins remain. The Yorkshire Dales and the North York Moors national parks located within the county attract a growing number of tourists. The principal towns are HARROGATE, Northallerton, SCARBOROUGH, Selby, and SKIPTON.

North Yukon National Park (□ 3,926 sq mi/10,168 sq km), NW YUKON TERRITORY, Canada. Bounded on W by Alaska, on NE by BEAUFORT SEA, on SE by Babbage River, S boundary follows divide to Alaska boundary. Excludes Herschel Island. Traversed by Malcolm, Firth, and Trail rivers. British Mountains in S, Arctic Coastal Plain in N. Tundra; forested pockets in valleys. Snow geese, eagles, Arctic and red fox, Dall sheep, grizzly bears; caribou migration route and calving grounds. Established 1984. Ranger station; no facilities. Backpacking, rafting, primitive camping.

North Zanesville (ZAINZ-vil), unincorporated village (□ 4 sq mi/10.4 sq km; 2000 population 3,013), MUSKINGUM county, central OHIO, just N of ZANESVILLE; 39°59′N 82°00′W.

North Zulch (ZUHLCH), unincorporated village, MADISON county, E central TEXAS, 22 mi/35 km NE of BRYAN, near NAVASOTA RIVER; 30°55′N 96°06′W. In agricultural area, mainly cattle, horses, hogs; oil and gas.

Norton, county (□ 881 sq mi/2,290.6 sq km; 2006 population 5,584), NW KANSAS; ☉ NORTON; 39°46′N 99°54′W. Rolling plain region, bordering N on NEBRASKA; watered by BEAVER, PRAIRIE DOG creeks and North Fork SOLOMON RIVER. Wheat, barley, oats, corn, sorghum; cattle, hogs; furniture. Prairie Dog State Park and Keith Sebelius Lake Reservoir in W center. Formed 1872.

Norton, city (2000 population 3,012), NW KANSAS, on PRAIRIE DOG CREEK, and c.70 mi/113 km NNW of HAYS; ☉ NORTON county; 39°50′N 99°53′W. Elevation 2,260 ft/689 m. Processing center for agricultural region; dairying. Manufacturing (fabricated metal products). Prairie Dog State Park and Keith Sebelius Lake reservoir to SW. Incorporated 1885.

Norton (NORT-uhn), independent city (□ 7 sq mi/18.2 sq km; 2006 population 3,643), SW VIRGINIA, separate from surrounding WISE county, in CUMBERLAND Mountains, 33 mi/53 km NW of BRISTOL, on Guest River; 36°55′N 82°37′W. Manufacturing (apparel, ex-

plosives, printing and publishing, crushed stone, beverages); trade, processing and shipping center in bituminous-coal and agricultural area; bituminous-coal mining. Settled 1787; incorporated 1894.

Norton (NAW-tuhn), town (2001 population 6,943), NORTH YORKSHIRE, N ENGLAND, on DERWENT RIVER just SE of MALTON; 54°09′N 00°47′W. Previously steel milling. Racehorse training center.

Norton, town, BRISTOL county, SE MASSACHUSETTS, 11 mi/18 km SW of BROCKTON; 41°58′N 71°11′W. Light manufacturing. Wheaton College, House in the Pines School here. Settled 1669, set off from TAUNTON 1711.

Norton, town, ESSEX CO., NE VERMONT, on QUEBEC (CANADA) border, port of entry; 44°58′N 71°48′W. Norton Pond (c.3 mi/4.8 km long; fishing) is S. Part of Averill village is in Norton town.

Norton, town, MASHONALAND WEST province, central ZIMBABWE, 25 mi/40 km NE of HARARE; 17°53′S 30°42′E. Elevation 4,499 ft/1,371 m. On railroad. Lake CHIVERO reservoir (recreation park) to E, Lake MANYAME reservoir (recreation park) to N. Agriculture (tobacco, cotton, peanuts, citrus fruit, coffee, tea, vegetables); cattle, sheep, goats, hogs; dairying. Chibero College is here.

Norton, village, CENTRAL PROVINCE, SRI LANKA, on HATTON PLATEAU, on right headstream of the KELANI GANGA River, and 5 mi/8 km WNW of HATTON; 06°54′N 80°31′E. Hydroelectric project.

Norton Air Force Base, California: see SAN BERNARDINO, city.

Norton Bay, W ALASKA, NE arm of NORTON SOUND, on S side of base of SEWARD PENINSULA; 30 mi/48 km long, 20 mi/32 km wide; 64°40′N 161°27′W. Receives KOYUK RIVER.

Norton Canes (NAW-tuhn CAINZ), village (2001 population 6,394), STAFFORDSHIRE, W ENGLAND, 1 mi/1.6 km SE of CANNOCK; 52°41′N 01°58′W. Farming.

Norton-Radstock (NAW-tuhn–RAD-stahk), district (2001 population 21,325), Bath and North East Somerset, SW ENGLAND; 51°17′N 02°26′W. Includes town of Radstock, 8 mi/12.9 km SW of BATH. To W is former coal-mining town of Midsomer Norton, with paper and shoe industries.

Norton Reservoir (□ 10 sq mi/26 sq km), NORTON county, NW KANSAS, on PRAIRIE DOG CREEK, 70 mi/113 km NNW of HAYS; 39°49′N 99°56′W. Maximum capacity 193,023 acre-ft. Formed by Norton Dam (131 ft/40 m high), built (1964) by the Bureau of Reclamation for irrigation; also used for flood control and recreation. Also known as Keith Sebelius Reservoir.

Norton Shores, city (2000 population 22,527), MUSKEGON county, W MICHIGAN, on LAKE MICHIGAN, residential suburb of MUSKEGON. 43°09′N 86°15′W. Airport is here. P. F. Hoffmaster State Park in SW part of city, on Lake Michigan.

Norton Sound (NOR-tuhn), inlet of the BERING SEA, W ALASKA, S of the SEWARD PENINSULA; c.150 mi/240 km long and 125 mi/200 km across at its widest point; 63°50′N 164°16′W. NORTON BAY is its NE arm. NOME is on the N shore and the YUKON River flows into the sound from the SE. It is navigable from May to October.

Nortonville, town (2000 population 1,264), HOPKINS county, W KENTUCKY, 10 mi/16 km SSE of MADISONVILLE; 37°11′N 87°27′W. Railroad junction. Coal mining; timber; agriculture (burley tobacco, grain; hogs, cattle).

Nortonville, village (2000 population 620), JEFFERSON county, NE KANSAS, 15 mi/24 km SW of ATCHISON; 39°25′N 95°19′W. Shipping point in grain, dairy, and poultry area.

Nortorf (NOR-torf), town, SCHLESWIG-HOLSTEIN, NW GERMANY, 14 mi/23 km SW of KIEL; 54°10′N 09°51′E. Manufacturing (apparel, leather, food).

Nort-sur-Erdre (nor–syur–er-druh), town (□ 25 sq mi/65 sq km), LOIRE-ATLANTIQUE department, PAYS DE

LA LOIRE region, W FRANCE, on ERDRE RIVER, 16 mi/
26 km N of NANTES, and near NANTES-BREST CANAL;
47°26′N 01°30′W. Tanning, shoe manufacturing, and
flour milling.

Norusovo, RUSSIA: see KALININO, CHUVASH Republic.

Norvegia, Cape, located at the end of the peninsula that
separates the RIISER-LARSEN and QUAR ice shelves on
the PRINCESS MARTHA COAST of QUEEN MAUD LAND,
EAST ANTARCTICA; 71°25′S 12°18′W.

Norvelt (NOR-velt), unincorporated town, MOUNT
PLEASANT township, WESTMORELAND county, SW
PENNSYLVANIA, 7 mi/11.3 km SSE of GREENSBURG,
near Sewickley Creek; 40°12′N 79°29′W. Manufacturing (ceramics, apparel); agriculture (corn, hay,
dairying).

Nörvenich (NYOOR-ve-nikh), town, RHINELAND,
NORTH RHINE-WESTPHALIA, W GERMANY, on small
Neffel-Bach River, 6 mi/9.7 km E of MERZENICH;
50°49′N 06°39′E. Agricultural products; food processing. Has ruined castle.

Norville, La (nor-VEEL, lah), commune (□ 1 sq mi/2.6
sq km), ESSONNE department, ÎLE-DE-FRANCE region,
N central FRANCE.

Norwalk, city (2000 population 103,298), LOS ANGELES
county, S CALIFORNIA, suburb 12 mi/19 km SE of
downtown LOS ANGELES and 10 mi/16 km NE of LONG
BEACH; 33°55′N 118°05′W. Bounded by SAN GABRIEL
RIVER in W. Manufacturing (fabricated metal products, plastic products, computer equipment, furniture). With the arrival (1875) of the Southern Pacific
railroad, it became a center for the dairy and logging
industries. Norwalk's main growth occurred after
World War II, with rapid industrialization. Seat of
Cerritos College (two year). The city also holds an
annual Space, Science, and Technology Show. Settled
in the 1850s, incorporated 1957.

Norwalk, city (2000 population 82,951), FAIRFIELD
county, SW CONNECTICUT, at the mouth of the
NORWALK RIVER, on LONG ISLAND SOUND; 41°05′N
73°25′W. An oyster center; manufacturing of apparel,
electronic and electrical equipment, machinery, chemicals; aircraft research. Norwalk was burned by the
British in the American Revolution. It has a community college, Norwalk Community Technical College, and an amateur symphony orchestra. The city
includes numerous small islands in the harbor and the
village of Silvermine, an artists' colony. The Maritime
Center at Norwalk, an aquarium and marine-life
education center, is here. Settled 1640, incorporated
1913.

Norwalk (NOR-wahk), city (□ 9 sq mi/23.4 sq km; 2006
population 16,576), ⊙ HURON county, N OHIO, c.50
mi/80 km WSW of CLEVELAND; 41°14′N 82°37′W.
Trade and processing center for a farm area, with
factories that make furniture, rubber products, fabricated metal products, and machinery. City was settled (c.1817) by "Fire Sufferers" from NORWALK,
Connecticut, whose homes had been burned by the
British in the American Revolution. Incorporated
1881.

Norwalk, town (2000 population 6,884), WARREN
county, S central IOWA, near NORTH RIVER, 9 mi/14.5
km S of DES MOINES, satellite community of Des
Moines; 41°30′N 93°40′W. Agricultural and bituminous-coal area; manufacturing of food, concrete.

Norwalk, village (2006 population 617), MONROE
county, W central WISCONSIN, on a branch of KICK-
APOO RIVER, and 30 mi/48 km E of LA CROSSE;
43°49′N 90°37′W. In farm and dairy area. On La
Crosse State Trail.

Norwalk River, c.30 mi/48 km long, SW CONNECTICUT;
rises just E of RIDGEFIELD; flows N, then S, past
GEORGETOWN and WILTON, to LONG ISLAND SOUND at
NORWALK.

Norway, Norwegian *Norge* or *Noreg*, kingdom
(□ 240,451 sq mi/386,958 sq km; 2004 estimated population 4,574,560; 2007 estimated population

4,627,926), N EUROPE, occupying the W part of the
Scandinavian peninsula; ⊙ OSLO.

Geography
Extending from the SKAGERRAK c.1,100 mi/1,770 km
NE to NORTH CAPE on the ARCTIC OCEAN, the country
forms a narrow mountainous strip along the NORTH
SEA and the ATLANTIC OCEAN, which in Norway is also
called the NORWEGIAN SEA. It has a long land frontier
with SWEDEN and in the north borders on FINLAND
and RUSSIA. The coastline, c.1,700 mi/2,740 km long,
is fringed with islands (notably the LOFOTEN AND
VESTERÅLEN islands) and is deeply indented by numerous fjords. SOGNEFJORDEN, HARDANGERFJORDEN,
NORDFJORD, and OSLOFJORDEN are among the largest
and best known. From the coast the land rises sharply
to high plateaus such as DOVREFJELL and the HARD-
ANGERVIDDA. GALDHØPIGGEN, in the JOTUNHEIMEN
range, is the high point (8,100 ft/2,469 m); west of
it lies JOSTEDALSBREEN, the largest glacier field in
Europe. The mountains and plateaus are intersected
by fertile valleys, such as GUDBRANDSDALEN, and
by rapid rivers, which furnish hydroelectric power
and are used for logging. The GLOMMA, in the south,
is the longest (372 mi/598 km) and most important river. Most of the population is concentrated
along the southern coast and valleys, where the chief
cities—Oslo, BERGEN, STAVANGER, KRISTIANSAND,
and DRAMMEN—are located. Farther north along
the coast are TRONDHEIM, and in the extreme north are
NARVIK, TROMSØ, and HAMMERFEST. The beautiful
Norwegian fjords and the midnight sun of the far
north attract many tourists. Because of the North
Atlantic Drift, Norway has a mild and humid climate
for a northern country. The outlying possessions of
Norway are SVALBARD and JAN MAYEN in the Arctic
Ocean and BOUVET and PETER I island in the south
Atlantic; Norway has claims in Antarctica. Russians
have the right to exploit resources in SVALBARD
equally with Norway, and in late 1940s failed to gain
Norwegian agreement to place a military garrison
there.

Population
The majority of Norwegians are of Scandinavian
stock, but in the northern counties of FINNMARK,
Sami and Finns form a significant minority. The literary language of Norway for many years was Danish,
from which Bokmål, an official idiom of Norway, is
derived. Landsmål (officially Nynorsk), the other official idiom, is similar. Frequent spelling reforms account for the variation in Norwegian place names.
The Lutheran Church is the state church, but all other
religions enjoy freedom of worship. The king nomi-
nates the nine bishops and other clergy of the Lutheran Church. The educational level in Norway is
very high; the leading universities are in Oslo (founded 1811) and Bergen (founded 1946).

Economy
Almost three-quarters of the land is unproductive; less
than 4% is under cultivation. The vast mountain
pastures are used for the grazing of cattle and sheep,
and, in the north, for reindeer raising. The chief industries are petroleum and natural-gas production,
which are exploited from its offshore fields in the
North Sea, and shipping, and trading. The great
Norwegian merchant fleet carries a large part of the
world's trade; in 1992, it was the sixth largest, and if
measured by capita, it was number one. Fishing (notably of cod, herring, and mackerel) is important, and
fresh, canned, and salted fish from Norway are exported to the entire world. The food manufacturing,
pulp and paper, electrochemical, electrometallurgical, and printing and publishing industries are important to the economy. Mineral resources include
pyrites, copper, titanium, and iron ore, which are
heavily mined, and some coal, zinc, and lead. Nickel,
aluminum, ferroalloys, and semifinished steel are
produced. Formerly an exporter of heavy water for

nuclear reactors, Norway ceased production in the
late 1980s.

History: to 1035
The history of Norway before the age of the Vikings
is indistinct from that of the rest of SCANDINAVIA. In
the 9th century the country was still divided among
the numerous petty kings of the *fylker*. Harald I, of the
Yngling or Scilfing dynasty (which claimed descent
from one of the old Norse gods), defeated the petty
kings (c.900) and conquered the SHETLAND and the
ORKNEY islands, but failed to establish permanent
unity. Harald's campaigns drove many nobles and
their followers to settle in ICELAND and FRANCE. In the
next two centuries Norsemen raided widely in W
Europe and established the Norse duchy of NOR-
MANDY. Harald himself concentrated on developing a
dynasty; before he died (c.935) the country was divided among his sons, but one of them, Håkon I,
defeated (c.935) his brothers and temporarily reunited
the kingdom. Christianity, brought by English missionaries, gained a foothold under Olav I and was
established by Olav II (reigned 1015–1028). Olav II was
driven out of Norway by King Knut of England and
Denmark, in league with discontented Norwegian
nobles; however, his son, Magnus I, was restored
(1035) to the Norwegian throne.

History: 1035 to 1535
Both Magnus and his successor, Harald III, played a
vital part in the complex events then taking place in
ENGLAND and DENMARK. After Harald died while invading England (1066), Norway entered a period of
decline and civil war, precipitated by conflicting
claims to the throne. Among the major events of
12th-century Norwegian history were the mission of
Nicholas Breakspear (later Pope Adrian IV), who
organized the Norwegian hierarchy, and the rule of
Sverre, who created a new nobility grounded in
commerce and, with the help of the popular party, the
Birkebeiners, consolidated the royal power. His
grandson, Haakon IV, was put on the throne by the
Birkebeiners in 1217; under him and under Magnus VI
(reigned 1263–1280) medieval Norway reached its
greatest flowering and enjoyed peace and prosperity.
During this time Iceland and GREENLAND recognized
Norwegian rule. The separate development of Norway
was halted by the accession (1319) of Magnus VII,
who was also king of Sweden. He was unpopular in
Norway, which he was compelled to surrender (1343)
to his son, Haakon VI, husband of Margaret of
Denmark. Margaret subsequently united the rule of
Norway, Sweden, and Denmark in her person and had
the Kalmar Union drawn up in 1397. Although the
union was strictly a personal one, Norway virtually
ceased to exist as a separate kingdom and was ruled by
Danish governors for the following four centuries. Its
power had greatly declined even before Margaret's
accession, however, and its trade had been taken over
by the HANSEATIC LEAGUE, which maintained its chief
northern office at Bergen.

History: 1535 to 1884
Norway's political history became essentially that of
DENMARK. Christian III of Denmark (1535–1559) introduced Lutheranism as the state religion. Under
Danish rule Norway lost territory to Sweden but developed economically. The fishing industry flourished
(late 17th century), lumbering became an important
industry (18th century), the merchant class grew, and
Norway became a naval power. During the Napoleonic Wars, Norway was blockaded by the British. In
1814, Denmark, which had sided with France, was
obliged to consent to the Treaty of KIEL, by which it
ceded Norway to the Swedish crown in exchange for
W POMERANIA. The Norwegians, however, attempted
to set up a separate kingdom, with a liberal constitution and a parliament, under Prince Christian (later
King Christian VIII of Denmark). A Swedish army
obliged Norway to accept Karl XIII of Sweden, but the

---

act of union of 1814 recognized Norway as an independent kingdom, in personal union with Sweden, with its own constitution and parliament. Despite some Swedish concessions to growing Norwegian nationalism, Swedish-Norwegian relations were strained throughout the 19th century.

History: 1884 to 1930

Johan Sverdrup, the Liberal leader, succeeded in making the ministry responsible to parliament despite royal opposition (1884), but other problems remained. The Norwegian interest in obtaining greater participation in foreign policy came to a crisis in the late 19th century over the issue of a separate Norwegian consular service, justified by the spectacular growth of Norwegian shipping and commercial interests. Finally, in 1905, the Storting declared the dissolution of the union and the deposition of Oscar II. Sweden acquiesced after a plebiscite showed Norwegians nearly unanimously in favor of separation; in a second vote Norway chose to become a monarchy, and parliament elected the second son of Frederick VIII of Denmark king of Norway as Håkon VII. Two important features in Norwegian history of the late 19th and early 20th centuries were the large-scale emigration to the U.S. and the great arctic and antarctic explorations by such notable men as Fridtjof Nansen and Roald Amundsen. Three outstanding cultural figures of the period were Edvard Grieg, Henrik Ibsen, and Edvard Munch. In World War I, Norway remained neutral. The industrial development of Norway, spurred by the harnessing of water power, contributed to the rise of the Labor (socialist) Party, which has predominated in Norwegian politics since 1927.

History: 1930 to 1971

In the 1930s much social-welfare legislation was passed, including public health and housing measures, old-age pensions, aid to the disabled, and unemployment insurance. Norway attempted to remain neutral in World War II, but in April 1940, German troops invaded, and in a short time nearly the whole country was in German hands. King Haakon and his cabinet set up a government in exile in London, and the Norwegian merchant fleet was of vital assistance to the Allies throughout the war. Despite the attempts of Vidkun Quisling to promote collaboration with the Germans, the people of Norway defied the occupation forces. German troops remained in Norway until the war ended in May 1945. Although half of the Norwegian fleet was sunk during the war, Norway quickly recovered its commercial position. Postwar economic policy included a degree of socialization and measures such as permanent price, interest, and dividend controls. Norway was one of the original members of the UN (the Norwegian Trygve Lie was the first UN Secretary-General), and it became a member of NATO in 1949. King Olav V succeeded to the throne in 1957 as the second king of independent Norway. Norway joined the EFTA in 1959. Norwegian voters rejected membership in the EEC in 1972, but trade agreements with the market were made the next year. Between 1965 and 1971 the Labor Party was out of power for the first time since 1936. Since the discovery of petroleum in the EKOFISK field in 1969, the petroleum and natural-gas industries have become increasingly important for Norway's economy, bringing increased employment, but also increased inflation and a vulnerability to fluctuations in the world petroleum market.

History: 1971 to Present

The Labor Party returned to power in 1971 under the leadership of Trygve Bratteli, whose government resigned but was restored to power in the 1973 elections. Bratteli was succeeded as prime minister by Odvar Nordli in 1976, who was quickly succeeded (1977) by Gro Harlem Brundtland, Norway's first woman prime minister. Brundtland was defeated by Conservative Kåre Willoch in the 1981 election, but she returned to

the prime ministership in 1986. She was again defeated in 1989, by Conservative Jan P. Syse, but again returned as prime minister from 1990 to 1996, when she was replaced by Thorbjørn Jagland. On January 21, 1991, Harald V succeeded his father Olav V as king of Norway. In 1992 Norway refused to conform to the International Whaling Treaty. Once again, in 1994, Norwegian voters, in a referendum, rejected membership in the EU. Following elections in 1997, Jagland resigned and Christian Democrat Kjell Magne Bondevik became prime minister. Bondevik resigned in 2000 after losing a key parliamentary vote; Labor party leader Jens Stoltenberg formed a new government. Stoltenberg lost power after the 2001 elections, and Bondevik was again prime minister until Stoltenberg led Labor's coalition to a majority in 2005. Norway is a member of NATO, the UN, the EFTA, and the European Council.

Government

Norway is a constitutional monarchy governed under the constitution of 1814 as amended. The hereditary monarch is the head of state. The prime minister, who is the head of government, is appointed by the monarch with the approval of Parliament, as is the cabinet. Members of the 169-seat unicameral Parliament (*Storting*) are popularly elected to four-year terms. Administratively, the country is divided into nineteen counties (Norwegian *fylker*). The current head of state is King Harald V (since January 1991). The current head of government is Prime Minister Jens Stoltenberg (since October 2005).

Norway, town (2000 population 601), BENTON county, E central IOWA, 14 mi/23 km WSW of CEDAR RAPIDS; 41°53′N 91°55′W. In agricultural area; light manufacturing.

Norway, town, including Norway village, OXFORD county, W MAINE, 17 mi/27 km NW of AUBURN, and on LAKE PENNESSEEWASSEE; 44°13′N 70°36′W. Manufacturing (wood products), resort area. Settled 1786, incorporated 1797.

Norway, town (2000 population 2,959), DICKINSON county, SW UPPER PENINSULA, MICHIGAN, 8 mi/12.9 km ESE of IRON MOUNTAIN city; 45°47′N 87°54′W. In lumbering and farming region (potatoes; cattle). Manufacturing (printing, transportation equipment). Norway Spring, artesian spring created in 1903 by 1,904 ft/580 m drill-hole made in search of iron deposits. Iron Mountain Iron Mine is here (tourist attraction). Vulcan U.S.A. Ski Area to E. Trout hatchery nearby. Settled c.1879, incorporated 1891.

Norway, village (2006 population 365), ORANGEBURG county, W central SOUTH CAROLINA, 15 mi/24 km WSW of ORANGEBURG; 33°27′N 81°07′W. Agriculture includes poultry, livestock; grain, tobacco, peanuts, cotton, watermelons, peaches.

Norway Islands, VIETNAM: see LONG CHAU ISLANDS.

Norwegian Bay, NUNAVUT territory, Canada, arm (c.100 mi/161 km long, 90 mi/145 km wide) of the ARCTIC OCEAN, off SW ELLESMERE ISLAND; 77°00′N 90°00′W.

Norwegian Current, terminal branch of NORTH ATLANTIC CURRENT, in NORWEGIAN SEA, flowing N along coast of NORWAY before branching into BARENTS SEA and past SPITSBERGEN. Has great moderating effect on climate of N EUROPE.

Norwegian Sea, part of the ATLANTIC OCEAN, NW of NORWAY, extending from 62°00′N to c.70°00′N, between the GREENLAND SEA and the NORTH SEA. It is separated from the Atlantic by a submarine ridge linking Iceland and the FAEROE ISLANDS (to the E of ICELAND), by the Greenland-Iceland Rise (to the W of Iceland), and from the ARCTIC OCEAN by the Jan Mayen Ridge. The warm Norwegian Current gives the sea generally ice-free conditions.

Norwell (NOR-wel), town, PLYMOUTH county, E MASSACHUSETTS, on NORTH RIVER, near coast, and 20 mi/32 km SE of BOSTON; 42°10′N 70°50′W. Light

manufacturing In suburbanizing area. Settled 1634, set off from SCITUATE 1888.

Norwich (NAH-rich), city (2001 population 121,550) and district, E ENGLAND, on the WENSUM RIVER, just above its confluence with the YARE; ⊙ NORFOLK county; 52°38′N 01°18′E. Industries include food processing and insurance. It is also a center for shopping and entertainment, as well as administration. Airport. Known to the Romans as *Venta Icenorum*, Norwich has been a leading provincial city since the 11th century. It was sacked by the Danes in the 11th century and scourged by the Black Death in 1349. Norwich was the scene of events in Wat Tyler's rebellion of 1381 and in the uprising under Robert Kett in 1549. There are many medieval churches as well as a cathedral founded in 1096 by the first bishop of Norwich. Norwich Castle, part of which dates from Norman times, was made (1894) into a museum for collections of natural history and local antiquities. It also houses paintings of the 18th- and 19th-century Norwich school of artists. Other old buildings include St. Giles's Hospital (13th century), Suckling House (14th century), Strangers Hall (15th century; now a museum), the guildhall (15th century), and St. Andrew's Hall (15th century; formerly a Dominican church). The Maddermarket Theatre, a reconstruction of a Shakespearean theater, has a permanent amateur company. The Norwich grammar school dates from the 13th century. The city is also the cultural center of the county; triennial music festivals have been held here since 1824. It is the seat of the University of East Anglia (1963). The writer Harriet Martineau born here. The city includes the districts of Catton and Lakenham.

Norwich, industrial city (2000 population 36,117), SE CONNECTICUT, on hilly ground, where the YANTIC and SHETUCKET rivers form the THAMES; ⊙ NEW LONDON county; 41°32′N 72°05′W. Chemicals, plastics, paper products. The last great battle between the Mohegans and Narragansetts took place on this site in 1643, and the tribal chiefs are buried here. Norwich was a leading colonial industrial city; Thomas Danforth began making pewterware here in 1733. The many historic structures include the Leffingwell Inn (1675); the birthplace and home of Benedict Arnold; and the home of Samuel Huntington. Location of Three Rivers Community College and a state hospital Settled 1659, incorporated 1784, town and city consolidated 1952.

Norwich, city (□ 2 sq mi/5.2 sq km; 2006 population 7,203), central NEW YORK, on CHENANGO RIVER, and 35 mi/56 km NE of BINGHAMTON; ⊙ CHENANGO county; 42°31′N 75°31′W. Manufacturing (pharmaceuticals, aerospace electrical systems, wood and metal products, apparel), in dairying and farming area. Settled 1788. Incorporated 1915.

Norwich (NOR-wich, NAH-rich), township (□ 166 sq mi/431.6 sq km; 2001 population 10,478), SW ONTARIO, E central Canada, 12 mi/19 km SE of WOODSTOCK; 42°59′N 80°36′W. Agricultural area. Lumbering, food processing.

Norwich, town, WINDSOR co., E Vermont, on the CONNECTICUT River, opposite Hanover, NEW HAMPSHIRE, and 12 mi/19 km NE of Woodstock; 43°44′N 72°19′W. Includes Pompanoosuc village. Settled before 1775. Named for NORWICH, CONNECTICUT.

Norwich, village (2000 population 551), KINGMAN county, S KANSAS, 20 mi/32 km SE of KINGMAN; 37°27′N 97°50′W. Railroad junction. Wheat-shipping point.

Norwich (NOR-hwich), village (2006 population 118), MUSKINGUM county, central OHIO, 12 mi/19 km ENE of ZANESVILLE; 39°59′N 81°48′W. In agricultural area.

Norwich Park (NO-rich), railroad siding, open-cut coal mine, E central QUEENSLAND, NE AUSTRALIA, 65 mi/105 km NNW of BLACKWATER, 15 mi/24 km S of DYSART; 22°41′S 148°25′E.

Norwood (NOR-wuhd), city (□ 3 sq mi/7.8 sq km; 2006 population 19,532), HAMILTON county, extreme SW OHIO; suburb 3 mi/4.8 km E of CINCINNATI; 39°09′N 84°27′W. Manufacturing includes machinery, printing and publishing. Settled early 1800s, incorporated 1888.

Norwood (NOR-wud), town (2000 population 299), WARREN county, E GEORGIA, 12 mi/19 km W of THOMSON; 33°28′N 82°43′W.

Norwood, town (2000 population 28,587), NORFOLK county, E MASSACHUSETTS; 42°11′N 71°12′W. Chiefly residential; local industries include printing and publishing, plastics, apparel, computer software, electronic equipment. Settled 1678, set off from DED-HAM and WALPOLE and incorporated 1872.

Norwood, town, CARVER county, S MINNESOTA, 34 mi/55 km WSW of MINNEAPOLIS; 44°46′N 93°55′W. Railroad junction. Agricultural area (grain, soybeans, alfalfa; livestock, poultry; dairying); light manufacturing.

Norwood, town, WRIGHT county, S central MISSOURI, in the OZARKS, 9 mi/14.5 km W of MOUNTAIN GROVE; 37°06′N 92°25′W. Dairy; cattle, poultry.

Norwood (NOR-wud), town (□ 2 sq mi/5.2 sq km; 2006 population 2,159), STANLY county, S central NORTH CAROLINA, 10 mi/16 km SSE of Albemarle Lake; 35°13′N 80°07′W. TILLERY reservoir to E on Yadkin (PEE DEE) River. Railroad junction. Manufacturing (textiles, apparel, office and store fixtures, transportation equipment); agricultural area (cotton, grain, soybeans; poultry, livestock; dairying). Yadkin and ROCKY rivers join 5 mi/8 km to SE to form Pee Dee (or Great Pee Dee) River. Settled c.1800; incorporated 1882.

Norwood (NOR-wud), unincorporated village (□ 1 sq mi/2.6 sq km; 2001 population 1,346), SW ONTARIO, E central Canada, on Ouse River, 22 mi/35 km E of PETERBOROUGH, and included in Asphodel-Norwood township; 44°23′N 77°59′W. Dairying, feed milling, woodworking.

Norwood, village (2000 population 438), SAN MIGUEL county, SW COLORADO, near SAN MIGUEL river, 30 mi/48 km NW of TELLURIDE; 38°07′N 108°17′W. Elevation 7,014 ft/2,138 m. In diversified farming area; wheat, corn; cattle, sheep. SAN MIGUEL MOUNTAINS just S. Parts of Uncompahgre National Forest to S and NE; Miramoric State Wildlife Area to S.

Norwood, village (2000 population 473), PEORIA county, central ILLINOIS, residential suburb 5 mi/8 km W of downtown PEORIA; 40°42′N 89°42′W. Greater Peoria Airport to S. Wildlife Prairie Park to W.

Norwood, village, EAST FELICIANA parish, SE central LOUISIANA, near MISSISSIPPI state line to N, 35 mi/56 km N of BATON ROUGE; 30°58′N 91°07′W. Light manufacturing.

Norwood, village (□ 2 sq mi/5.2 sq km; 2006 population 1,620), ST. LAWRENCE county, N NEW YORK, on RA-QUETTE RIVER, and 13 mi/21 km SSW of MASSENA; 44°45′N 75°00′W. Originally an Adirondack–St. Lawrence Valley railroad center and industrial village, but now it is purely a residential community. Incorporated 1871.

Norwood, borough (2006 population 6,267), BERGEN county, extreme NE NEW JERSEY, 12 mi/19 km NE of PATERSON, near NEW YORK state line; 40°59′N 73°57′W. Largely residential. Incorporated 1905.

Norwood, borough (2006 population 5,834), DELAWARE county, SE PENNSYLVANIA, residential suburb 9 mi/14.5 km SW of downtown PHILADELPHIA, on Darby Creek; 39°53′N 75°17′W. Incorporated 1893.

Norwood, AUSTRALIA: see KENSINGTON AND NORWOOD.

Norwood, ENGLAND: see CROYDON.

Norwood Court, town (2000 population 1,061), SAINT LOUIS county, E MISSOURI, residential suburb 8 mi/12.9 km NW of downtown ST. LOUIS, adjacent to NORMANDY; 38°43′N 90°17′W.

Norwoodville, unincorporated village, POLK county, central IOWA, suburb 5 mi/8 km NW of downtown DES MOINES, in residential area.

Norwottock, Mount, MASSACHUSETTS: see HOLYOKE RANGE.

Nor Yungas, province, LA PAZ department, W BOLIVIA, ⊙ COROICO; 16°00′S 67°30′W.

Norzagaray (nor-SAH-gah-rei), town, BULACAN province, S central LUZON, PHILIPPINES, 21 mi/34 km NNE of MANILA; 14°55′N 121°11′E. Trade center for rice-growing and gold-mining area.

Nos (NOS), village, SANTIAGO province, ME-TROPOLITANA DE SANTIAGO region, central CHILE, on railroad, and 13 mi/21 km S of SANTIAGO; 33°39′S 70°43′W. Black-powder plant just W.

Nosaka (NO-sah-kah), town, Sosa county, CHIBA prefecture, E central HONSHU, E central JAPAN, 19 mi/30 km E of CHIBA; 35°39′N 140°34′E.

Nosappu, Cape (no-SAHP), easternmost point of HOKKAIDO island, N JAPAN, on narrow Nemuro Peninsula, in GOYOMAI STRAIT; 43°23′N 145°49′W. Lighthouse.

Nosara (no-SAH-rah), resort, NICOYA PENINSULA, COSTA RICA, W of CARMONA, and near OSTIONAL WILDLIFE REFUGE; 09°59′N 85°39′W. A favored retirement area for foreign residents of Costa Rica. Beach and hotel facilities.

Nose (NO-se), town, Toyono county, OSAKA prefecture, S HONSHU, W central JAPAN, 22 mi/35 km N of OSAKA; 34°58′N 135°25′E.

Nosegawa (no-SE-gah-wah), village, Yoshino district, NARA prefecture, S HONSHU, W central JAPAN, 38 mi/61 km S of NARA; 34°09′N 135°38′E.

Noshappu, Cape (no-SHAHP), one of two capes at the N tip of HOKKAIDO island, extreme N JAPAN, just W of Cape SOYA (Japan's northernmost point), in LA PÉROUSE STRAIT; 45°27′N 141°39′E. Lighthouse.

Noshaq, AFGHANISTAN: see NAWSHAKH.

Noshiro (no-SHEE-ro), city, Akita prefecture, N HON-SHU, NE JAPAN, port on SEA OF JAPAN, 34 mi/55 km N of AKITA city; 40°12′N 140°01′E. Vegetation has been planted along a 9-mi/14-km stretch of beach to prevent sand erosion. Sometimes spelled Nosiro.

Nosi Bewar, INDONESIA: see PEUNASOE.

Nosivka (NO-seef-kah), (Russian Nosovka), city, SW CHERNIHIV oblast, UKRAINE, 15 mi/24 km. SW of NIZHYN; 50°56′N 31°35′E. Elevation 390 ft/118 m. Raion center; manufacturing of food products, furniture. Agricultural research station; historical museum. Known since 1147; city since 1960.

Nosovka, UKRAINE: see NOSIVKA.

Noss (NAWS), uninhabited island (2 mi/3.2 km long, 1 mi/1.6 km wide; rises to 592 ft/180 m), of the SHET-LAND ISLANDS, extreme N Scotland, just E of BRESSAY island, and 5 mi/8 km E of LERWICK. Ponies are bred here and island is home of vast numbers of sea birds during breeding season.

Nossa Senhora da Glória (NO-sah SEN-yo-rah dah GLO-ree-ah), city (2007 population 29,545), W SER-GIPE, NE BRAZIL, 60 mi/97 km NW of ARACAJU; 10°14′S 37°25′W. Livestock; corn, manioc.

Nossa Senhora das Dores (NO-sah SEN-yo-rah dahs DO-res), city (2007 population 23,815), central SER-GIPE, NE BRAZIL, 32 mi/51 km NNW of ARACAJU; 10°29′S 37°13′W. Agricultural trade center (corn, manioc, fruit, cattle).

Nossa Senhora das Graças (NO-sah SEN-yo-rah dahs GRAH-sahs), town (2007 population 3,904), NW PARANÁ state, BRAZIL, 34 mi/55 km NNE of MARINGÁ; 22°51′S 51°46′W. Coffee, corn, rice, cotton; livestock.

Nossa Senhora de Lourdes (NO-sah SEN-yo-rah zhe LOOR-zhes), city (2007 population 6,296), N SERGIPE state, BRAZIL, 59 mi/95 km N of ARACAJU; 10°05′S 37°03′W. Manioc; livestock.

Nossa Senhora do Amparo (NO-sah SEN-yo-rah do AHM-pah-ro), city, W RIO DE JANEIRO state, BRAZIL, 16 mi/26 km N of BARRA MANSA; 22°22′S 44°05′W.

Nossa Senhora do Livramento (no-sah sen-yo-rah do lee-vrah-men-to), city (2007 population 12,302), central MATO GROSSO, BRAZIL, 23 mi/37 km SW of CUIABÁ; 15°37′S 56°30′W. Sugar, medicinal plants; livestock. Until 1943, called Livramento; 1944–1948, called São José dos Cocais.

Nossa Senhora do Ó (NO-sah SEN-yo-rah do O), city, extreme E PERNAMBUCO state, BRAZIL, 25 mi/40 km S of RECIFE, on Atlantic coast; 08°26′S 35°01′W.

Nossa Senhora dos Remedios (NO-sah SEN-yo-rah dos RE-me-zhee-os), town (2007 population 8,338), NW PIAUÍ state, BRAZIL, 78 mi/126 km N of Teresina; 04°00′S 42°48′W.

Nossebro (NOS-se-BROO), village, SKARABORG county, SW SWEDEN, 17 mi/27 km ESE of TROLL-HÄTTAN; 58°11′N 12°43′E.

Nossen (NAHS-sen), town, SAXONY, E central GER-MANY, on the FREIBERGER MULDE, 11 mi/18 km SW of MEISSEN; 51°03′N 13°18′E. Metalworking, paper mill-ing; manufacturing of machinery, electronic equipment, leather products. Towered over by large castle (16th–17th century). Nearby are remains of 12th-century Cistercian monastery of Altzella.

Nossi-Bé, MADAGASCAR: see NOSY-BÉ.

Nossob River, c.500 mi/805 km long, in NORTHERN CAPE province, SOUTH AFRICA; rises in central Na-mibia NNE of WINDHOEK near Steinhausen; flows SW, forms border between BOTSWANA and NORTHERN CAPE province (South Africa), through Kalahari Gemsbok Park, to MOLOPO RIVER at 26°55′S 20°41′E.

Nossob River (No-SOB), c.500 mi/805 km long, SW AFRICA, in SE NAMIBIA, SW BOTSWANA; rises 25 mi/40 km S of GOBABIS; flows SE across KALAHARI, crossing into BOTSWANA and forming border between SOUTH AFRICA and Botswana; receives (right) AUOB RIVER; joins MOLOPO RIVER at 26°55′S 20°41′E. Feeds artesian basin in Kalahari; flows into GEMSBOK NATIONAL PARK (Botswana). Called Nosop River in Botswana.

Nossombougou, village, SECOND REGION/KOULIKORO, MALI, 33 mi/55 km S of KOLOKANI; 13°06′N 07°56′W.

Nossop River, BOTSWANA: see NOSSOB RIVER.

Nosy Bé (NOOS-ee BAI), island (□ 115 sq mi/299 sq km), in MOZAMBIQUE CHANNEL just off coast of ANTSIRANANA province, NW MADAGASCAR; 13°20′S 48°15′E. Produces sugar, vanilla, ylang-ylang, pepper, fruit. Shrimp exports. Sugar and rum manufacturing. Major tourist center with beach hotels (especially in AMBATOLOAKA and AMBILOBE), nature reserve at Lo-kobe. Commercial center is Hellville. International airport. Formerly Nossi-Bé.

Nosy Boraha (NOOS-ee boo-RAH) (□ 77 sq mi/200.2 sq km), in INDIAN OCEAN just off TOAMASINA prov-ince, E coast of MADAGASCAR; 35 mi/56 km long, 5 mi/8 km wide; 16°55′S 49°52′E. Cloves, vanilla. Tourism (beach hotels). Airport. Main center is AMBODIFO-TATRA, on W shore. The island, long a pirate haven, was claimed by FRANCE in 18th century. Attached to Madagascar, 1896. Formerly Ste-Marie Island.

Nosy Iranja (NOOS-ee ee-RAHNJ), island, Antser-anana province, N MADAGASCAR, 3.7 mi/6 km off the coast of Saikanosin Ampasindava peninsula. Actually two islands connected by a sandbar.

Nosy Komba (NOOS-ee KUM-bah), island, ANTSIR-ANANA province, MADAGASCAR, between NOSY BÉ and mainland. Volcanic origin. Famous black lemurs; tourism. Also known as Nosy Ambarivorato.

Nosy Mitsio (NOOS-ee mit-SEE-o), archipelago, AN-TSIRANANA province, 34 mi/55 km NE of NOSY BÉ off NW side of MADAGASCAR. A dozen islands of which Mitsio is largest. Fishing, diving; beaches. Privately owned.

Nosy Varika (NOOS-ee VAHR-eek), town, FIANAR-ANTSOA province, E MADAGASCAR, on INDIAN OCEAN coast and Canal des PANGALANES, 115 mi/185 km NE of FIANARANTSOA; 20°35′S 48°32′E. Coffee center.

Nosy Ve (NOOS-ee VAI), island, Toliara province, SW MADAGASCAR, 19 mi/30 km from Toliara near mouth of ONILAHY RIVER. Haunt of pirates. Reefs make it a diving site.

Notasulga (nah-tuh-SUL-ga), town (2000 population 916), Lee and Macon counties, E Alabama, 10 mi/16 km N of Tuskegee. Lumber. First known as 'Moore's Crossroads,' the present name is a combination of Creek Indian words for plants. Inc. in 1893.

Noteborg, RUSSIA: see SHLISSELBURG.

Noteć (NO-tech), river, c.270 mi/435 km long, NW POLAND; rises S of INOWROCŁAW; flows generally W into the WARTA River near Gorzów WIELKOPOLSKI. The Noteć is connected by the BYDGOSZCZ CANAL with the VISTULA River and is navigable for almost its entire length.

Notikewin, hamlet, NW ALBERTA, W Canada, 5 mi/7 km from MANNING, in Northern Lights No. 22 municipal district; 56°59′N 117°38′W.

Notium, ancient town on AEGEAN coast of ASIA MINOR, just S of COLOPHON, for which it was the port, NE of SÁMOS island. Here in the Peloponnesian War, the Athenian fleet suffered a disastrous defeat (407 or 406 B.C.E.) by the Spartan Lysander.

Noto (NO-to), town, SIRACUSA province, SE SICILY, ITALY, 17 mi/27 km SW of SYRACUSE; 36°53′N 15°04′E. Citrus fruits, almonds, wine, olive oil, hemp; cattle. Limestone and lignite mined nearby. Bishopric. Has churches, palaces, museum. Founded 1703, replacing ancient Netum, or Neetum (5 mi/8 km NW), largely destroyed by earthquake of 1693. Bathing at Noto Marina nearby.

Noto (NO-to), town, Fugeshi county, ISHIKAWA prefecture, central HONSHU, central JAPAN, on the SE coast of NOTO Peninsula, 59 mi/95 km N of KANAZAWA; 37°18′N 137°09′E. Prehistoric Jomon-era ruins.

Noto (NO-to), Japanese *Noto-hanto* (no-TO-HAHN-to), peninsula, c.45 mi/72 km long and 6 mi/9.7 km–17 mi/27 km wide, ISHIKAWA prefecture, W central HONSHU, central JAPAN, between the SEA OF JAPAN and TOYAMA BAY. The rugged peninsula has a deeply indented E coast. Farming, lumbering, and fishing are major economic activities.

Notodden (NAWT-AWD-duhn), town, TELEMARK county, SE NORWAY; 59°34′N 09°17′E. The world's first nitrate factory was built here in 1905. Today, the town also has paper mills, iron foundries, and several large hydroelectric stations. Nearby at Heddal is the largest Norwegian stave church (c.1250).

Notogawa (no-TO-gah-wah), town, Kanzaki county, SHIGA prefecture, S HONSHU, central JAPAN, on E inlet of LAKE BIWA, 22 mi/35 km N of OTSU; 35°10′N 136°09′E. Watermelon.

Notojima (no-TO-jee-mah), town, on NOTO-JIMA island, Kashima county, ISHIKAWA prefecture, off central HONSHU, central JAPAN, 43 mi/70 km N of KANAZAWA; 37°07′N 136°59′E. Pickles; *matsutake* mushrooms. Light manufacturing.

Noto-jima (no-TO-jee-mah), island (□ 19 sq mi/49.4 sq km), ISHIKAWA prefecture, off central HONSHU, central JAPAN, in W inlet of TOYAMA BAY, N of NANAO; 7 mi/11.3 km long, 0.5 mi/0.8 km–3 mi/4.8 km wide. Flat, fertile. Rice growing; fishing.

Notranjska (no-TRAHN-ye-skah), historical region, S SLOVENIA, part of medieval Hapsburg province of KRANJSKA (German *Krain*, ancient *Carniola*); in Austria-Hungary until 1918. Encompasses approximate watershed of LJUBLJANICA and PIVKA rivers, S of Ljubljana valley, surrounded by NANOS, Hrušica, Javorniki and Snežnik mountains Karst topography (high plateaus, dry valleys, karst caverns); large forests. Principal towns are LOGATEC, POSTOJNA. Major tourist attractions are Postojna jame (POSTOJNA caverns) and Škocjanske jame (Škocjan caverns).

Notre-Dame (no-truh–dahm), village (2001 population 597), E New Brunswick, Canada, on Cocagne River, and 16 mi/26 km N of MONCTON; 46°18′N 64°43′W. Mixed farming.

Notre Dame, village (2000 population 3,615), PAMPLEMOUSSES district, MAURITIUS, 8 mi/12.8 km NE of PORT LOUIS. Sugarcane, fruits, and vegetables.

Notre Dame, INDIANA: see SOUTH BEND.

Notre-Dame-Auxiliatrice-de-Buckland (no-truh–DAHM–og-zee-lyah-TREES–duh–BUHK-luhnd), parish (□ 37 sq mi/96.2 sq km; 2006 population 806), Chaudière-Appalaches region, S QUEBEC, E Canada, 9 mi/15 km from ARMAGH; 46°37′N 70°33′W.

Notre Dame Bay, arm of the Atlantic Ocean, c.40 mi/60 km long and 50 mi/80 km wide, SE NEWFOUNDLAND AND LABRADOR, Canada. The EXPLOITS RIVER empties into it. The bay has an irregular shoreline and contains many islands; FOGO ISLAND is E of the bay. BOTWOOD is the chief town and port.

Notre-Dame-de-Bellecombe (no-truh–dahm–duh-bel-KONB), wintersport resort (□ 8 sq mi/20.8 sq km), SAVOIE department, RHÔNE-ALPES region, E FRANCE, in the Savoy Alps (ALPES FRANÇAISES), 10 mi/16 km NNE of ALBERTVILLE, and 6 mi/9.7 km SW of MEGÈVE; 45°48′N 06°31′E. It lies on a high ledge above the ARLY RIVER valley, with ski terrain at elevation of 3,700 ft/1,128 m–6,700 ft/2,042 m, with distant view of MONT BLANC.

Notre-Dame-de-Bondeville (no-truh–dahm–duh-bond-VEEL), town (□ 2 sq mi/5.2 sq km), SEINE-MARITIME department, HAUTE-NORMANDIE region, N FRANCE, 4 mi/6.4 km NW of ROUEN; 49°29′N 01°03′E. Light manufacturing.

Notre-Dame-de-Bonsecours (no-truh–DAHM–duh-bon-suh-KOOR), municipality (□ 103 sq mi/267.8 sq km; 2006 population 295), Outaouais region, SW QUEBEC, E Canada, 5 mi/7 km from PAPINEAUVILLE; 45°39′N 74°56′W.

Notre-Dame-de-Gravenchon (no-truh–dahm–duh-grah-vahn-SHON), town (□ 7 sq mi/18.2 sq km), SEINE-MARITIME department, HAUTE-NORMANDIE region, N central FRANCE, high above the lower SEINE RIVER, 19 mi/31 km E of Le HAVRE; 49°29′N 00°35′E. Has interesting modern church. Petroleum refinery on the Seine (W).

Notre-Dame-de-Ham (no-truh–DAHM–duh–AHM), village (□ 12 sq mi/31.2 sq km; 2006 population 427), Centre-du-Québec region, S QUEBEC, E Canada, 26 mi/41 km from RICHMOND; 45°54′N 71°44′W.

Notre-Dame-de-la-Merci (no-truh–DAHM–duh–lah-mer-SEE), village (□ 97 sq mi/252.2 sq km; 2006 population 836), Lanaudière region, S QUEBEC, E Canada, 10 mi/16 km from Saint-Donat; 46°14′N 74°03′W.

Notre-Dame-de-la-Paix (no-truh–DAHM–duh-lah-PAI), village (□ 41 sq mi/106.6 sq km; 2006 population 708), Outaouais region, SW QUEBEC, E Canada, 6 mi/10 km from CHÉNÉVILLE; 45°49′N 74°58′W.

Notre-Dame-de-la-Salette (no-truh–DAHM–duh-lah-sah-LET), village (□ 46 sq mi/119.6 sq km; 2006 population 719), Outaouais region, SW QUEBEC, E Canada, 11 mi/18 km from Bowman; 45°46′N 75°35′W.

Notre-Dame-de-la-Salette, FRANCE: see CORPS.

Notre Dame de la Trappe, FRANCE: see SOLIGNY-LA-TRAPPE.

Notre-Dame-de-l'Île-Perrot (NO-truh–DAHM–duh-LEEL–pe-RO), village (□ 11 sq mi/28.6 sq km; 2006 population 9,570), Montérégie region, S QUEBEC, E Canada, 17 mi/28 km from MONTREAL; 45°33′N 73°55′W. Part of the Metropolitan Community of Montreal (*Communauté Metropolitaine de Montréal*).

Notre-Dame-de-Lorette (no-truh–DAHM–duh-lo-RET), village (□ 87 sq mi/226.2 sq km; 2006 population 210), Saguenay—Lac-Saint-Jean region, S central QUEBEC, E Canada, 11 mi/17 km from Girardville; 49°05′N 72°21′W.

Notre-Dame-de-Lorette (no-truh–dahm–duh-lo-RET), hill (541 ft/165 m), PAS-DE-CALAIS department, NORD-PAS-DE-CALAIS region, N FRANCE, 8 mi/12.9 km NNW of ARRAS, just NW of VIMY RIDGE. Captured at great cost by French in May 1915. Site of large French military cemetery of World War I dead.

Notre Dame de Lourdes (no-truh dahm duh LOORD), village (□ 1 sq mi/2.6 sq km; 2001 population 619), S MANITOBA, W central Canada, 20 mi/33 km N of MANITOU; 49°31′N 98°33′W. Agriculture (wheat, oats, barley, oilseed; livestock; dairying). Incorporated 1963.

Notre-Dame-de-Lourdes (no-truh–DAHM–duh-LOORD), village (□ 14 sq mi/36.4 sq km; 2006 population 2,300), Lanaudière region, S QUEBEC, E Canada, 6 mi/9 km from JOLIETTE; 46°06′N 73°26′W.

Notre-Dame-de-Lourdes (no-truh–DAHM–duh-LOORD), parish (□ 32 sq mi/83.2 sq km; 2006 population 731), Centre-du-Québec region, S QUEBEC, E Canada, 22 mi/35 km from LOTBINIÈRE; 46°19′N 71°49′W.

Notre-Dame-de-Montauban (no-truh–DAHM–duh-mon-to-BAHN), village (□ 63 sq mi/163.8 sq km; 2006 population 776), Mauricie region, S QUEBEC, E Canada; 46°52′N 72°20′W. Formed in 1976.

Notre-Dame-de-Monts (no-truh–dahm–duh–mon), village (□ 8 sq mi/20.8 sq km), VENDÉE department, PAYS DE LA LOIRE region, W FRANCE, near BAY OF BISCAY, 29 mi/47 km NNW of Les SABLES-D'OLONNE, and just S of Île de NOIRMOUTIER; 46°50′N 02°09′W. Has bathing beach and pine forests.

Notre-Dame-de-Pierreville (nuh-truh–DAHM–duh-pyer-VEEL), former village, S QUEBEC, E Canada; 46°06′N 72°52′W. Amalgamated into PIERREVILLE in 2001.

Notre-Dame-de-Pontmain (no-truh–DAHM–duh-pon-MAN), village (□ 104 sq mi/270.4 sq km; 2006 population 669), Laurentides region, S QUEBEC, E Canada, 7 mi/10 km from Lac-du-Cerf; 46°17′N 75°38′W.

Notre Dame de Portneuf, Canada: see PORTNEUF.

Notre-Dame-de-Saint-Hyacinthe (NUH-truh–DAHM–duh-sant-yuh-SANT), former village, S QUEBEC, E Canada; 45°35′N 72°58′W. Amalgamated into SAINT-HYACINTHE in 2001.

Notre-Dame-de-Salette (no-truh–DAHM–duh-sah-LET), village (□ 46 sq mi/119.6 sq km; 2006 population 719), Outaouais region, SW QUEBEC, E Canada; 45°46′N 75°35′W.

Notre-Dame-des-Anges (no-truh–DAHM–daiz-AHNZH), parish (2006 population 456), Capitale-Nationale region, S QUEBEC, E Canada, just N of QUEBEC city; 46°49′N 71°14′W.

Notre-Dame-des-Bois (no-truh–DAHM–dai-BWAH), village (□ 74 sq mi/192.4 sq km; 2006 population 816), Estrie region, S QUEBEC, E Canada, 16 mi/25 km from LAC-MÉGANTIC; 45°24′N 71°04′W.

Notre-Dame-des-Monts (no-truh–DAHM–dai-MON), village (□ 22 sq mi/57.2 sq km; 2006 population 846), Capitale-Nationale region, S QUEBEC, E Canada, 7 mi/11 km from CLERMONT; 47°40′N 70°23′W.

Notre-Dame-des-Neiges (no-truh–DAHM–dai-NEZH), village (□ 36 sq mi/93.6 sq km; 2006 population 1,223), Bas-Saint-Laurent region, S QUEBEC, E Canada; 48°07′N 69°10′W.

Notre-Dame-des-Pins (no-truh–DAHM–dai–PAN), parish (□ 10 sq mi/26 sq km; 2006 population 1,071), Chaudière-Appalaches region, S QUEBEC, E Canada, 3 mi/5 km from BEAUCEVILLE; 46°11′N 70°43′W.

Notre-Dame-des-Prairies (no-truh–DAHM–dai-pre-REE), village (□ 7 sq mi/18.2 sq km; 2006 population 7,860), Lanaudière region, S QUEBEC, E Canada, 2 mi/4 km from JOLIETTE; 46°03′N 73°26′W.

Notre-Dame-des-Sept-Douleurs (no-truh–DAHM–dai-SET–doo-LUHR), parish (□ 4 sq mi/10.4 sq km; 2006 population 44), Bas-Saint-Laurent region, S QUEBEC, E Canada, 12 mi/20 km from RIVIÈRE-DU-LOUP; 48°00′N 69°27′W.

Notre-Dame-de-Stanbridge (no-truh–DAHM–duh-STAN-brij), parish (□ 17 sq mi/44.2 sq km; 2006 population 755), Montérégie region, S QUEBEC, E Canada, 4 mi/7 km from BEDFORD; 45°10′N 73°02′W.

Area is shown by the symbol □, and capital city or county seat by ⊙.

Notre-Dame-du-Bon-Conseil (no-truh–DAHM–dyoo–bon–kon-SE-yuh), village (□ 2 sq mi/5.2 sq km; 2006 population 1,455), Centre-du-Québec region, S QUEBEC, E Canada, 10 mi/17 km from DRUMMOND-VILLE; 46°00′N 72°21′W.

Notre-Dame-du-Bon-Conseil (no-truh–DAHM–dyoo–BON–kon-SE-yuh), parish (□ 33 sq mi/85.8 sq km; 2006 population 985), Centre-du-Québec region, S QUEBEC, E Canada, 10 mi/17 km from DRUMMOND-VILLE; 46°00′N 72°21′W.

Notre-Dame-du-Lac (no-truh–DAHM–dyoo–LAHK), town (□ 41 sq mi/106.6 sq km), ⊙ TÉMISCOUATA county, Bas-Saint-Laurent region, SE QUEBEC, E Canada, on Lake TEMISCOUATA, 40 mi/64 km ESE of RIVIÈRE-DU-LOUP; 47°36′N 68°48′W. Dairying, lumbering.

Notre-Dame-du-Laus, village (□ 334 sq mi/868.4 sq km; 2006 population 1,448), Laurentides region, S QUEBEC, E Canada, 12 mi/19 km from Bowman; 46°05′N 75°37′W.

Notre-Dame-du-Laus, FRANCE: see GAP.

Notre-Dame-du-Mont-Carmel (no-truh–DAHM–dyoo–mon–kahr-MEL), former village, S QUEBEC, E Canada; 46°29′N 72°39′W. Amalgamated into LACOLLE in 2001.

Notre-Dame-du-Mont-Carmel (no-truh–DAHM–dyoo–mon–kahr-MEL), parish (□ 49 sq mi/127.4 sq km; 2006 population 5,300), Mauricie region, S QUEBEC, E Canada, 6 mi/9 km from Saint-Maurice; 46°29′N 72°39′W.

Notre-Dame-du-Nord (no-truh–DAHM–dyoo–NOR), village (□ 40 sq mi/104 sq km; 2006 population 1,107), Abitibi-Témiscamingue region, SW QUEBEC, E Canada, near ONTARIO border, and 6 mi/10 km from Nédelec; 47°36′N 79°29′W. Trucking, trailer building.

Notre-Dame-du-Portage (no-truh–DAHM–dyoo–por-TAZH), village (□ 15 sq mi/39 sq km; 2006 population 1,224), Bas-Saint-Laurent region, S QUEBEC, E Canada, 6 mi/10 km from RIVIÈRE-DU-LOUP; 47°46′N 69°37′W. Established 1856.

Notre-Dame-du-Rosaire (no-truh–DAHM–dyoo–ro-ZER), village (□ 61 sq mi/158.6 sq km; 2006 population 399), Chaudière-Appalaches region, S QUEBEC, E Canada, 13 mi/20 km from MONTMAGNY; 46°50′N 70°24′W.

Notre-Dame-du-Sacré-Coeur-d'Issoudun (no-truh–DAHM–dyoo–sah-krai–KUHR–dee-soo-DUHN), parish (□ 24 sq mi/62.4 sq km; 2006 population 786), Chaudière-Appalaches region, S QUEBEC, E Canada, 15 mi/24 km from LOTBINIÈRE; 46°35′N 71°37′W.

Notre Dame Mountains (NO-tri DAIM), section of the Appalachian system, extending c.500 mi/805 km from the GREEN MOUNTAINS of VERMONT into the GASPÉ PENINSULA, CANADA. Worn low by erosion, the ancient mountains have an average elevation of c.2,000 ft/610 m. Named by the French explorer, Samuel de Champlain.

Notrees (NO-treez), unincorporated village, ECTOR county, W TEXAS, 23 mi/37 km WNW of ODESSA; 31°55′N 102°45′W. Agricultural area (cattle, horses; pecans). Oil and natural gas.

Notsé, village, ⊙ HAHO prefecture, PLATEAUX region, S TOGO, 55 mi/89 km NNE of LOMÉ; 06°55′N 01°10′E. On railroad. Cacao, palm oil and kernels; cotton gin. Also spelled Nouatia. Formerly spelled Nuatja.

Notsuharu (no-TSOO-hah-roo), town, Oita county, OITA prefecture, E KYUSHU, SW JAPAN, 6 mi/10 km S of OITA; 33°09′N 131°31′E. Shiitake mushrooms; miso.

Nottawa (NAH-tuh-wuh), unincorporated village, S ONTARIO, E central Canada, 3 mi/5 km S of COLLINGWOOD, and included in Clearview township; 44°27′N 80°12′W. Dairying, mixed farming.

Nottawasaga (nah-tuh-wuh-SAH-guh), former township, S ONTARIO, E central Canada, 24 mi/39 km from BARRIE; 44°23′N 80°10′W. Amalgamated into Clearview township in 1994.

Nottawasaga River (nah-tuh-wuh-SAH-guh), 60 mi/97 km long, S ONTARIO, Canada; rises NNE of TORONTO, flows N of Nottawasaga Bay, S arm of GEORGIAN BAY, 11 mi/18 km ENE of COLLINGWOOD; 44°32′N 80°00′W.

Nottaway (NAHT-uh-WAI), river, c.140 mi/230 km long; issuing from MATTAGAMI LAKE, W QUEBEC, E Canada; flows NW into S JAMES BAY; 51°22′N 78°55′W. The Waswanipi River (c.195 mi/310 km long) is its chief headstream.

Nottely River (NAHT-lee), c.40 mi/64 km long, in N GEORGIA and W NORTH CAROLINA; rises in BLUE RIDGE MOUNTAINS 17 mi/27 km NE of DAHLONEGA; flows NW into CHEROKEE county, North Carolina, then NE to HIWASSEE Reservoir (in HIWASSEE RIVER) near MURPHY; 34°45′N 83°50′W. NOTTELY Dam, in Georgia, 9 mi/14.5 km NW of BLAIRSVILLE, is major Tennessee Valley Authority dam (184 ft/56 m high, 2,300 ft/701 m long) completed 1942; used for flood control. Forms Nottely Reservoir (6.7 sq mi/17.4 sq km; 20 mi/32 km long, 1 mi/1.6 km–3 mi/4.8 km wide; capacity 184,400 acre-ft.) in UNION county, Georgia.

Nøtterøy (NUHT-tuhr-uh-oo), island (□ 17 sq mi/44.2 sq km), SE NORWAY, in OSLOFJORDEN just S of TØNSBERG (bridge); 7 mi/11.3 km long, 4 mi/6.4 km wide. Shipyards; fishing. Small section at N tip is part of Tønsberg city, to which the rest of the island is a suburb.

Nottingham (NAHT-ing-uhm), city and county (□ 29 sq mi/75.4 sq km; 2001 population 266,988), central ENGLAND, on the TRENT RIVER; ⊙ Nottingham; 52°58′N 01°10′W. A center of railroad and road transportation. Previous industries included the manufacturing of lace, hosiery, and silk. Science sector. In the 9th century, it was one of the Danish FIVE BOROUGHS. A fire destroyed much of the city in the 12th century. In 1642, Nottingham was the scene of Charles I unfurling his banner, marking the beginning of the English Civil War. Early in the 19th century, Luddites were active in the city. The 17th-century castle overlooking the Trent River was burned in 1831 during Reform Bill riots. It was restored in 1878 and now houses an art museum. The earlier Norman castle on the same site was once the prison of David II of Scotland and the headquarters of Richard III before the battle of BOSWORTH FIELD. Other features of interest are the council house in the market place, a Roman Catholic cathedral (designed by A. W. Pugin), the 16th-century grammar school (now a high school), the University of Nottingham (1948), Trent University (1992), and St. Peter's Church, part of which dates from the 12th century. According to tradition, Robin Hood born here. William Booth, founder of the Salvation Army, born here in 1829. Districts of the city include Daybrook, Mapperley, Basford, Bulwell, and Radford. Formerly part of Nottinghamshire.

Nottingham (NAW-deeng-ham), town, ROCKINGHAM county, SE NEW HAMPSHIRE, 14 mi/23 km SW of DOVER; 43°07′N 71°07′W. Drained by Pawtuckaway River (flows out of Pawtuckaway Lake in S), North River, and Black Creek. Agriculture (cattle, poultry; vegetables; dairying; timber); manufacturing (plastic products, lumber, fabricated metal products). Part of Pawtuckaway State Park in SW.

Nottingham (NAH-ting-ham), unincorporated village, CHESTER county, SE PENNSYLVANIA, 5 mi/8 km SW of OXFORD near MARYLAND state line; 39°45′N 76°00′W. Manufacturing of food products; agriculture includes dairying, livestock, poultry; grain, mushrooms. Nottingham Park to S.

Nottingham Island (□ 441 sq mi/1,142 sq km), NUNAVUT territory, Canada, in HUDSON STRAIT, between Ungava Peninsula (S) and SOUTHAMPTON ISLAND (WNW); 38 mi/61 km long, 11 mi/18 km–22 mi/35 km wide; 63°20′N 78°00′W.

Nottinghamshire (NAHT-ing-uhm-shir) or **Notts**, county (□ 834 sq mi/2,168.4 sq km; 2001 population 1,015,498), central ENGLAND; ⊙ West BRIDGFORD (previously NOTTINGHAM); 53°03′N 01°00′W. The land, partially reclaimed fenland, is low-lying and fertile. A S area of moors devoted to pasturage is known as the WOLDS. The principal river is the TRENT. The FOSSE WAY passes through the county. SHERWOOD FOREST, with its legends of Robin Hood, includes the Dukeries, a district noted for its opulent estates. Cereal crops and sugar beets are grown. Dairying is extensive, and sheep are also raised. The remaining Nottinghamshire coal fields extend along the W border; Nottingham, MANSFIELD, and WORKSOP are the chief mining centers. There are small oil fields at Egmanton and Bothamsall. The mineral wealth also includes limestone, sandstone, and gravel. Previously manufacturing of products such as hosiery, bicycles, and lace. The county was a part of the Anglo-Saxon kingdom of MERCIA. SCROOBY, the home of William Brewster, was the cradle of the Pilgrims.

Notting Hill (NAH-teeng HIL), suburb 11 mi/18 km SE of MELBOURNE, VICTORIA, SE AUSTRALIA; 37°54′S 145°07′E.

Nottoway (NAWT-uh-wai), county (□ 316 sq mi/821.6 sq km; 2006 population 15,572), central VIRGINIA; ⊙ NOTTOWAY COURT HOUSE; 37°08′N 78°02′W. Bounded S by NOTTOWAY RIVER; source of Stony Creek in SE. Manufacturing at BLACKSTONE; agriculture (mainly tobacco; also fruit, hay, barley, wheat, corn, soybeans; cattle, sheep, poultry; dairying); timber; granite quarrying. Part of Fort Pickett Military Reservation in SE. Formed 1788.

Nottoway Court House (NAWT-uh-wai), unincorporated village, central VIRGINIA, 37 mi/60 km WSW of PETERSBURG, Lake Nottoway to NE; ⊙ NOTTOWAY county. Agriculture (tobacco, grain; livestock; dairying). Also called Nottoway.

Nottoway River (NAWT-uh-wai), c.170 mi/274 km long, rises in S PRINCE EDWARD county, c.7 mi/11 km ENE of KEYSVILLE flows generally SE past COURTLAND; joins BLACKWATER RIVER 9 mi/15 km S of FRANKLIN, at NORTH CAROLINA state line to form CHOWAN RIVER; 37°06′N 78°19′W.

Notts, ENGLAND: see NOTTINGHAMSHIRE.

Nottuln (NAHT-tuln), town, WESTPHALIA-LIPPE, NORTH RHINE-WESTPHALIA, NW GERMANY, 12 mi/19 km WSW of MÜNSTER; 51°56′N 07°21′E. Manufacturing of carpets, building materials; dairying. Has late-Gothic church from 15th century.

Notus (NAH-tus), village (2000 population 458), CANYON county, SW IDAHO, 8 mi/12.9 km NW of CALDWELL, and on BOISE RIVER; 43°44′N 116°48′W. In dairying and agricultural area served by BOISE IRRIGATION PROJECT.

Nouadhibou (nwah-dee-BOO), town (2000 population 72,337), ⊙ Dakhlet Nouâdhibou administrative region, W MAURITANIA, minor ATLANTIC OCEAN port on narrow peninsula flanking DAKHLET NOUÂDHIBOU inlet (E), 4 mi/6.4 km N of Agüera (WESTERN SAHARA), 330 mi/531 km N of SAINT-LOUIS (SENEGAL); 20°55′N 17°03′W. Important industrial and economic center. RAS NOUADHIBU, bisected by Mauritania–Western Sahara border, is just S. Base for important coastal fishing grounds. Has fish salting and canning, iron, and steel industries. International airport to N. Roadstead can be entered by ships drawing up to 22 ft/7 m. Formerly named Port Étienne.

Nouakchott (nwahk-SHAWT), city and capital district (2000 population 558,195), W MAURITANIA, on the ATLANTIC OCEAN; ⊙ Islamic Republic of Mauritania and its Nouakchott district; 18°06′N 15°57′W. Nouakchott was a small village until 1957, when it was chosen as the capital of Mauritania. A large-scale construction program began in 1958. Today Nouakchott is Mauritania's largest city and its administrative center. Its ocean port, which is c.4 mi/6.4 km from the city proper, has modern storage facilities,

---

Cross-references are shown in SMALL CAPITALS. The pronunciation guide is shown on page xix. The sources of population figures are shown on page xvii.

especially for petroleum. Handicrafts are made, and light industry is carried on in the city. Power plant (oil, 32.9 MW). Gypsum and salt deposits to the N. Nouakchott is located on a major highway (NOUAKCHOTT-NÉMA HIGHWAY) and has an international airport to the E. Some historians believe that nearby stood the ribat (monastery) from which the Muslim Almoravids set out on their conquests of AFRICA and SPAIN in the 11th century.

Nouakchott-Nema Highway, S MAURITANIA, extending from NOUAKCHOTT on the ATLANTIC OCEAN coast to NÉMA in the E. Built in the 1970s to carry the agricultural products of the E to urban market centers in the W and SW. Called *Tariq Al-Amal* (Arabic=Road of Hope).

Nouamghar (nwahm-GWAHR), hamlet (2000 population 4,151), Dakhlet Nouâdhibou administrative region, W MAURITANIA, at RÂS TIMIRIST (cape), S of NOUADHIBOU; 19°22′N 16°28′W. Fishing.

Noumbiel, province (□ 1,056 sq mi/2,745.6 sq km; 2005 population 57,974), SUD-OUEST region, extreme S BURKINA FASO; ⊙ BATIÉ; 09°50′N 03°00′W. Bordered on N by PONI province, E by GHANA, and S and W by CÔTE D'IVOIRE. Established in 1997 with fourteen other new provinces.

Nouméa (noo-mai-ah), town, chief port and ⊙ of the French overseas territory NEW CALEDONIA, on New Caledonia island, S PACIFIC OCEAN; 22°16′S 166°26′E. It was the site of a U.S. airfield in World War II. Nouméa is the seat of the Secretariat of the Pacific Community (SPC; formerly the South Pacific Commission), an international body formed in 1947 to promote the economic and social welfare of Pacific island people. The town was a French penal colony from 1864 to 1897. Though focused largely on the nearby DONIAMBO nickel smelter, its industries include cementworks, foods and beverages, agriculture processing, and apparel. Road extension has improved island contacts, while international connections have recently been enhanced through major improvements at Tontouta airport and harbor installations, including a causeway joining former Nou island to the city. Ducos is the main industrial suburb. Educational and technological growth are joined with cosmopolitan characteristics. Energy is supplied by YATÉ dam and Doniambo nickel works, and watered by a dam near DUMBÉA. Tjibaou Cultural Center, designed by Renzo Piano, is here.

Noun (NOON), department (2001 population 434,542), West province, CAMEROON; ⊙ FOUMBAN.

Nouna (NOO-nah), village (2005 population 23,166), ⊙ KOSSI province, BOUCLE DU MOUHOUN region, NW BURKINA FASO, 60 mi/97 km WSW of TOUGAN on Yondo Plain, 30 mi/48 km W of MALI border; 12°44′N 03°54′W. Peanuts, cotton, sesame; cattle; cheese making. Hospital.

Noupoort, town, NORTHERN CAPE province, SOUTH AFRICA, 65 mi/105 km SE of DE AAR; 31°11′S 24°58′E. Elevation 5,445 ft/1,660 m. Major railroad junction with EASTERN CAPE and FREE STATE provinces on N9 highway. Sheep; wool, wheat, feed crops, fruit. Region was scene (1899–1900) of cavalry operations in South African War. Formerly spelled Naauwpoort. Large recent residential development called Kwazamuxolo NW of the town.

Nouveau Québec, Canada: see UNGAVA.

Nouvelle (noo-VEL), village (□ 89 sq mi/231.4 sq km; 2006 population 1,948), Gaspésie–Îles-de-la-Madeleine region, SE QUEBEC, E Canada, 8 mi/13 km from Escuminac; 48°08′N 66°19′W.

Nouvelle-Amsterdam, FRANCE: see AMSTERDAM ISLAND.

Nouvelle-Anvers, CONGO: see MAKANZA.

Nouvelle-Beauce (noo-VEL–BOS), county (□ 347 sq mi/902.2 sq km; 2006 population 31,860), Chaudière-Appalaches region, S QUEBEC, E Canada; ⊙ SAINTE-MARIE; 46°27′N 71°01′W. Composed of 11 municipalities. Formed in 1982.

Nouvelle Calédonie, FRANCE: see NEW CALEDONIA.

Nouvelle France, village (2000 population 6,691), GRAND PORT district, MAURITIUS, 19 mi/30.4 km SSE of PORT LOUIS, and 3.1 mi/5 km NW of ROSE BELLE. Sugarcane, tea.

Nouvion-en-Thiérache, Le (noo-vyon–ahn–tee-ai-rahsh, luh), commune (□ 18 sq mi/46.8 sq km), AISNE department, PICARDIE region, N FRANCE, on the SAMBRE, and 25 mi/40 km NE of SAINT-QUENTIN; 50°01′N 03°47′E. Glassworks; dairying for PARIS region's market; specialty cheese products.

Nouvion-sur-Meuse (noo-vyon–syur–muz), commune (□ 3 sq mi/7.8 sq km; 2004 population 2,235), ARDENNES department, CHAMPAGNE-ARDENNE region, N FRANCE, on right bank of the MEUSE (canalized), and 5 mi/8 km SE of CHARLEVILLE-MÉZIÈRES; 49°42′N 04°48′E. Metalworks.

Nouzonville (noo-zon-VEEL), town (□ 4 sq mi/10.4 sq km), ARDENNES department, CHAMPAGNE-ARDENNE region, N FRANCE, in entrenched MEUSE RIVER valley, 4 mi/6.4 km NNE of CHARLEVILLE-MÉZIÈRES; 49°49′N 04°45′E. Metallurgical center (stamping and sheet-metal works). Has long tradition of manufacturing nails.

Nouzov, CZECH REPUBLIC: see UNHOST.

Nova Aliança (NO-vah AH-lee-AHN-sah), town (2007 population 4,905), NW SÃO PAULO, BRAZIL, 17 mi/27 km SSW of SÃO JOSÉ DO RIO PRÊTO; 21°02′S 49°30′W. Coffee, rice, corn, cotton, beans; cattle.

Nova Almeida (no-vah ahl-MAI-dah), town, central ESPÍRITO SANTO, BRAZIL, on the Atlantic, 22 mi/35 km NNE of VITÓRIA; 20°05′S 40°15′W. Founded by Jesuits in 16th century.

Nova Alvorada do Sul (NO-vah AHL-vo-rah-dah do SOOL), town (2007 population 3,058), central Mato Grosso do Sul, BRAZIL, 30 mi/48 km NNE of Rio Brilhante; 21°29′S 54°24′W.

Nova América da Colina (NO-vah AH-me-ree-kah dah KO-lee-nah), town (2007 population 3,298), N PARANÁ state, BRAZIL, 27 mi/43 km E of LONDRINA; 23°20′S 50°42′W. Coffee, cotton, corn, rice; livestock.

Nova Andradina (NO-vah AHN-drah-zhee-nah), city (2007 population 43,508), SE Mato Grosso do Sul, BRAZIL, 58 mi/93 km E of DEODÁPOLIS; 22°15′S 53°21′W. Airport.

Nova Araçá (NO-vah AH-rah-su), town (2007 population 3,775), E RIO GRANDE DO SUL state, BRAZIL, 48 mi/77 km SE of PASSO FUNDO; 28°40′S 41°45′W. Grapes, wheat, corn, potatoes; livestock.

Nova Astrakhan' (NO-vah ahs-trah-KHAHN) (Russian *Novaya Astrakhan'*), village (2004 population 3,660), central LUHANS'K oblast, UKRAINE, 16 mi/26 km SW of STAROBIL's'K; 49°07′N 38°36′E. Elevation 246 ft/74 m. Wheat, sunflowers.

Nova Aurora (NO-vah OU-ro-rah), town (2007 population 11,734), SE GOIÁS, BRAZIL, 25 mi/40 km E of CORUMBAÍBA; 18°07′S 48°19′W.

Nova Bana (no-VAH bah-NYAH), Slovak *Nová Baňa*, HUNGARIAN *Ujbánya*, town, STREDOSLOVENSKY province, S central Slovakia, on HRON RIVER, on railroad, and 32 mi/51 km SW of BANSKÁ BYSTRICA; 48°26′N 18°39′E. Woodworking, textile manufacturing, basalt processing. Former mining town. Has 14th-century church; museum.

Nova Basan' (no-VAH bah-SAHN) (Russian *Novaya Basan'*), village, SW CHERNIHIV oblast, UKRAINE, 37 mi/60 km SSW of NIZHYN, and 13 mi/21 km SSE of BOBROVYTSYA; 50°34′N 31°31′E. Elevation 380 ft/115 m. Sugar beets.

Nova Bassano (NO-vah BAH-sah-no), town (2007 population 8,683), E RIO GRANDE DO SUL state, BRAZIL, 55 mi/89 km SE of PASSO FUNDO; 28°44′S 51°42′W. Grapes, wheat, corn, potatoes; livestock.

Nova Borova (no-VAH bo-ro-VAH) (Russian *Novaya Borovaya*), town, central ZHYTOMYR oblast, UKRAINE, on the Irsha River (tributary of the TETERIV RIVER),

and on highway and railroad, 30 mi/48 km N of ZHYTOMYR; 50°42′N 28°38′E. Elevation 590 ft/179 m. Fruit-drying plant; reinforced concrete products. Established in the late 17th century; town since 1957.

Nova Brasilândia (NO-vah BRAH-see-LAHN-zhee-ah), city (2007 population 4,877), S central MATO GROSSO, BRAZIL, NE of CUIABÁ; 14°55′S 55°02′W.

Nova Brasília (NO-vah BRAH-see-lee-ah), city, E SANTA CATARINA state, BRAZIL, 25 km/16 mi NE of TUBARÃO; 28°15′S 47°48′W.

Nova Brécia (NO-vah BRE-see-ah), town, E RIO GRANDE DO SUL state, BRAZIL, 39 km/24 mi NW of MONTENEGRO; 29°12′S 52°02′W. Grapes, corn, wheat, manioc; livestock.

Nova Bystrice (NO-vah BIS-trzhi-TSE), Czech *Nová Bystřice*, German *neubistritz*, town, JIHOCESKY province, SE BOHEMIA, CZECH REPUBLIC, near Austrian border, 10 mi/16 km SSE of JINDŘICHŮV HRADEC; 49°01′N 15°06′E. Railroad terminus. Bathing resort; manufacturing (cotton goods). Has a Gothic castle. Fish ponds in vicinity.

Nova Canaã (NO-vah KAH-nah-AH), city (2007 population 18,973), SE BAHIA, BRAZIL, in Serra Da Ouricana, 78 mi/125 km W of ITABUNA; 14°47′S 40°08′W.

Nova Canaã do Norte (NO-vah kah-nah-AH do NOR-chee), city(1996 population 10,752), N central Mato Grosso do Sul state, 124 mi/200 km NNW of Sinop; 11°45′S55°45′W.

Nova Chaves, ANGOLA: see MULONDA.

Novachene (no-VAHCH-en-e), village, LOVECH oblast, NIKOPOL obshtina, N BULGARIA, on the OSUM RIVER, 9 mi/15 km S of NIKOPOL; 43°33′N 24°58′E. Grain; livestock; vegetables.

Nova Cherna (NO-vah CHER-nah), village, RUSE oblast, TUTRAKAN obshtina, BULGARIA; 44°00′N 26°28′E.

Novaci (NO-vahch), town, GORJ county, SW ROMANIA, 20 mi/32 km NE of TÎRGU JIU. Summer resort (elev. 2,231 ft/680 m) in S foothills of the TRANSYLVANIAN ALPS. Asbestos production, cheese making; manufacturing (wood products). Includes *Novacii-Romăni* and *Novacii-Strini*. Polovragi summer resort (elev. 2,161 ft/659 m), with 17th-century monastery, is 5 mi/8 km E.

Nova Crixás (NO-vah kree-SHAHS), city (2007 population 12,639), NW GOIÁS, BRAZIL, 62 mi/100 km E of PEIXE; 14°09′S 50°23′W.

Nova Cruz (No-vah KROOS), city, extreme E PERNAMBUCO state, BRAZIL, 19 mi/31 km N of RECIFE, on ITAMARACÁ Island; 07°51′S 35°51′W.

Nova Cruz (NO-vah kroos), city (2007 population 35,241), E RIO GRANDE DO NORTE, NE BRAZIL, on PARAÍBA border, on NATAL–JOÃO PESSOA railroad, and 50 mi/80 km SSW of Natal; 06°28′S 35°26′W. Cotton, corn, manioc; hides.

Nova Dantzig, Brazil: see CAMBÉ.

Nova Dubnica (no-VAH dub-NYI-tsah), Slovak *Nová Dubnica*,Hungarian *Újtölgyes*, town, STREDOSLOVENSKY province, W SLOVAKIA, 4 mi/6.4 km NE of TRENČÍN; 48°57′N 18°09′E. Contains railroad junction; textile manufacturing. Town founded in 1953 as a settlement for industrial workers of nearby DUBNICA NAD VAHOM.

Nova Era (no-vah E-rah), city (2007 population 17,932), E central MINAS GERAIS, BRAZIL, on RIO PIRACICABA, and 60 mi/97 km ENE of BELO HORIZONTE; 19°39′S 43°03′W. Railroad spur from ITABIRA iron mine (15 mi/24 km NW) joins main Belo Horizonte-VITÓRIA line. Monlevade steel mill is 9 mi/14.5 km SW. Until 1930s, called São José da Lagoa; and, until 1944, Presidente Vargas.

Nova Erechim, Brazil: see NOVA EREXIM.

Nova Erexim (NO-vah E-re-cheen), town, W SANTA CATARINA state, BRAZIL, 16 mi/26 km NW of CHAPECÓ; 26°56′S 52°58′W. Wheat; livestock. Also spelled Nova Erechim.

Nova Esperança (NO-vah E-spe-rahn-sah), city (2007 population 25,719), NW PARANÁ state, BRAZIL, 23 mi/37 km NW of MARINGÁ; 25°57′S 53°48′W. Coffee, rice, corn; livestock.

Nova Europa (NO-vah YOO-ro-pah), town (2007 population 9,047), central SÃO PAULO state, BRAZIL, 25 mi/40 km W of ARARAQUARA; 21°46′S 48°33′W.

Nova Fatima (NO-vah FAH-chee-mah), city (2007 population 8,049), BAHIA, BRAZIL.

Nova Fátima (NO-vah FAH-chee-mah), town, N PARANÁ state, BRAZIL, 31 mi/50 km SE of LONDRINA; 23°29′S 50°33′W. Coffee, cotton, rice; livestock.

Nova Floresta (NO-vah FLO-re-stah), town (2007 population 10,033), N central PARAÍBA, Brazil, near border with RIO GRANDE DO NORTE, 4.3 mi/6.9 km NW of CUITÉ; 06°27′S 36°13′W.

Nova Friburgo (NO-vah FREE-boor-go), city (2007 population 177,376), RIO DE JANEIRO state, BRAZIL, in the Serra do MAR, on railroad, and 60 mi/97 km NE of RIO DE JANEIRO; 22°16′S 42°32′W. Elevation 2,779 ft/847 m. Popular mountain resort built in Alpine style by Swiss who first settled here in 1818. Textile milling, stone cutting, soap manufacturing. In fertile agricultural district growing potatoes, wine, fruit, and flowers. Locally called Friburgo.

Nova Goa, INDIA: see PANAJI.

Nova Gorica (NO-vah go-REE-tsah), town (2002 population 13,203), W SLOVENIA, adjoining (E of) GORIZIA (ITALY); 45°56′N 13°38′E. Furniture, textiles, and plastics manufacturing Developed after Italo-Yugoslav border settlement in 1947, which ceded land to the former Yugoslavia and divided the city of Gorizia between Nova Gorica and the Italian city of Gorizia.

Nova Gradiška (NO-vah GRAH-deesh-kah), Hungarian *Ujgradiska*, town, E CROATIA, on railroad, and 32 mi/51 km W of SLAVONSKI BROD, at S foot of PSUNJ Mountain, in SLAVONIA; 45°16′N 17°24′E. Trade center for plum-growing region; brewing, furniture manufacturing. Petroleum and lignite deposits in vicinity.

Nova Granada (NO-vah GRAH-nah-dah), city (2007 population 17,739), NW SÃO PAULO, BRAZIL, 20 mi/32 km N of SÃO JOSÉ DO RIO PRÊTO; 20°29′S 49°19′W. Pottery manufacturing, cotton ginning, coffee and rice processing.

Nova Guataporanga (NO-vah GWAH-tah-po-rahn-gah), town (2007 population 2,101), W SÃO PAULO state, BRAZIL, 16 mi/26 km NW of DRACENA; 21°20′S 51°38′W. Coffee growing.

Nova Haleshchyna (no-VAH hah-LE-shchi-nah) (Russian *Novaya Galeshchina*), town (2004 population 3,440), S POLTAVA oblast, UKRAINE, on railroad and near highway, 16 mi/26 km ENE of KREMENCHUK and 5 mi/8 km SW of KOZEL′SHCHYNA; 49°09′N 33°45′E. Elevation 242 ft/73 m. Machinery manufacturing; biological factory; grain receiving depot. Established in 1958, town since 1968.

Nova Ibiá (NO-vah EE-bee-AH), city (2007 population 7,068), BAHIA, BRAZIL.

Nova Iguaçu (NO-vah EE-gwah-soo), industrial suburb (2007 population 830,672) of RIO DE JANEIRO, BRAZIL, in RIO DE JANEIRO state, 18 mi/29 km NW of city center; 22°45′S 43°27′W. Fruit-preserving (marmalades) and vegetable-canning center. Has metalworks, paper mills, rubber factory. Also manufacturing of chemicals, pharmaceuticals, soft drinks; coffee drying, cutting of semiprecious stones.

Nova Independência (NO-vah EEN-de-pen-den-see-ah), town (2007 population 2,480), NW SÃO PAULO state, BRAZIL, 14 mi/23 km SW of ANDRADINA; 21°06′S 51°29′W. Coffee growing.

Nova Iorque (no-vah ee-or-kai), town (2007 population 4,882), E MARANHÃO, BRAZIL, on left bank of RIO PARNAÍBA (PIAUÍ border), and 140 mi/225 km SW of TERESINA; 06°35′S 44°06′W. Ships cacao, medicinal plants, babassu nuts, lumber. Airfield. Formerly spelled Nova York.

Nova Itália, Brazil: see SEVERIANO DE ALMEIDA.

Nova Itapirema (NO-vah EE-tah-pee-re-mah), city, N SÃO PAULO state, BRAZIL, 17 mi/27 km SW of SÃO JOSÉ DO RIO PRETO; 21°06′S 49°32′W. Coffee growing.

Nova Itarana (NO-vah EE-tah-RAH-nah), town (2007 population 7,511), E central BAHIA, BRAZIL, 16 mi/25 km SW of MILAGRES; 13°01′S 40°04′W.

Nova Ivanivka (no-VAH ee-VAH-nif-kah) (Russian *Novaya Ivanovka*) (Romanian *Ivanestii-Noui*), village (2004 population 4,370), SW ODESSA oblast, UKRAINE, 40 mi/64 km NNE of IZMAYIL; 45°55′N 29°05′E. Elevation 265 ft/80 m. Agricultural center.

Nova Kakhovka (no-VAH ka-KOV-kah) (Russian *Novaya Kakhovka*), city (2001 population 52,137), central KHERSON oblast, UKRAINE, 37 mi/60 km ENE of KHERSON, on the left bank of the DNIEPER (Ukrainian *Dnipro*) River, at the site of the Kakhovka hydroelectric station; 46°45′N 33°23′E. Manufacturing of machinery, construction materials, furniture; dairying, canneries, wine making. Technical schools in water melioration, agricultural mechanization, machine construction, and three other professional-technical schools; heritage museum.

Nova Kanjiza, SERBIA: see NOVI KNEZEVAC.

Nova Khrestivka, UKRAINE: see KIROVS′KE, Donets′k oblast.

Novakovo (no-va-KO-vo), village, PLOVDIV oblast, ASENOVGRAD obshtina, BULGARIA; 41°53′N 24°54′E.

Nova Lamego, GUINEA-BISSAU: see GABÚ.

Nova Laranjeiras (NO-vah LAH-rahn-zhai-rahs), city (2007 population 11,561), W PARANÁ state, BRAZIL, 62 mi/100 km SE of CASCAVEL; 25°18′S 52°32′W. Timber.

Nova Lima (no-vah lee-mah), city (2007 population 72,207), S central MINAS GERAIS, BRAZIL, near the RIO DAS VELHAS, on SE slope of the Serra do CURRAL, 7 mi/11.3 km SE of BELO HORIZONTE; 20°00′S 43°49′W. Elevation 2,760 ft/841 m. Brazil's leading gold-mining center (producing 90% of total output), with one of world's deepest mines (8,000 ft/2,438 m), worked since 1834 and called Morro Velho [Portuguese=old hill]. Arsenic and silver are by-products. Blast furnaces and ironworks at RIO ACIMA, 9 mi/14.5 km SSE. Bauxite exploited at Motuca mine nearby. Tile-roofed colonial houses; city is surrounded by eucalyptus groves.

Nova Lisboa, ANGOLA: see HUAMBO.

Novallas (no-VAH-lyahs), village, ZARAGOZA province, NE SPAIN, 4 mi/6.4 km NNE of TARAZONA; 41°57′N 01°42′W. Olive-oil processing; wine, sugar beets, cherries, apricots.

Nova Londrina (NO-vah LON-dree-nah), city (2007 population 12,626), far NW PARANÁ state, BRAZIL; 22°25′S 53°00′W. Coffee, cotton, corn, rice; livestock.

Nova Lusitânia, MOZAMBIQUE: see BÚZI.

Nova Mahala (NO-vah ma-ha-LAH), village, PLOVDIV oblast, BATAK obshtina, BULGARIA; 41°56′N 24°16′E.

Nova Mayachka (no-VAH mah-YAHCH-kah) (Russian *Novaya Mayachka*), town (2004 population 6,320), SW central KHERSON oblast, UKRAINE, 28 mi/45 km E of KHERSON; 46°36′N 33°14′E. Sanatorium for children. Established in the early 19th century, town since 1957.

Nova Mutum (NO-vah kah-nah-AH do NOR-chee), town (2007 population 24,368), MATO GROSSO, BRAZIL.

Nova Nadezhda (NO-vo na-DEZH-da), village, HASKOVO oblast, HASKOVO obshtina, BULGARIA; 42°01′N 25°43′E.

Nova Odesa (no-VAH o-DE-sah) (Russian *Novaya Odessa*), city, central MYKOLAYIV oblast, UKRAINE, on the Southern BUH, and 25 mi/40 km NNW of MYKOLAYIV on highway; 47°19′N 31°47′E. Elevation 180 ft/54 m. Raion center; food processing; manufacturing of furniture, construction materials. Vocational technical school; heritage museum. Established in 1776 as the military settlement of Fedorivka (Russian *Fedorovka*); renamed Nova Odesa in 1832; city since 1976.

Nova Olinda (NO-vah O-leen-dah), city (2007 population 12,974), S central CEARÁ, Brazil, in Serra dos Haveres, 24 mi/39 km NW of CRATO; 07°10′S 39°40′W.

Nova Olinda (NO-vah O-leen-dah), town (2007 population 6,293), SW PARAÍBA, BRAZIL, 6 mi/9.7 km SE of PEDRA BRANCA; 07°28′S 38°03′W.

Nova Olinda, town (2007 population 10,518), N TOCANTINS state, BRAZIL, 31 mi/50 km S of PALMAS on BR 153 Highway; 07°50′S 48°30′W.

Nova Olinda do Maranhão (NO-vah O-leen-dah do MAH-rahn-youn), city (2007 population 17,419), NW MARANHÃO state, BRAZIL, on BR 316 Highway; 02°43′S 45°45′W.

Nova Olinda do Norte (NO-vah O-leen-dah do NOR-te), city (2007 population 29,184), E central AMAZONAS, BRAZIL, on Rio MADEIRA, 40 mi/65 km from confluence with Rio Amazonas; 03°02′S 59°04′W.

Nova Paka (No-vah PAH-kah), Czech *Nová Paka*, German *neupaka*, town, VYCHODOCESKY province, NE BOHEMIA, CZECH REPUBLIC, on railroad, and 23 mi/37 km NNW of HRADEC KRÁLOVÉ; 50°29′N 15°31′E. Manufacturing (textiles, consumer goods); brewery (established 1872). Museum of precious stones. Village of STARA PAKA, Czech *Stará Paka* (STAH-rah PAH-kah) (railroad junction), is 2 mi/3.2 km NNW; 13th century castle of PECKA is 4 mi/6.4 km SE.

Nova Palma (NO-vah PAHL-mah), town (2007 population 6,444), central RIO GRANDE DO SUL state, BRAZIL, 20 mi/32 km NE of SANTA MARIA; 29°28′S 53°28′W. Livestock.

Nova Palmeira (NO-vah PAHL-mai-rah), town (2007 population 3,934), N central PARAÍBA, BRAZIL, 7 mi/11.3 km NE of PEDRA LAVRADA; 06°41′S 36°25′W.

Nova Ponte (NO-vah PON-chee), town (2007 population 11,609), W central MINAS GERAIS, BRAZIL, 51 mi/82 km SE of UBERLÂNDIA, in TRIÂNGULO MINEIRO, on Rio Araquari; 19°11′S 47°44′W.

Nova Praha (no-VAH PRAH-hah) (Russian *Novaya Praga*), town, E KIROVOHRAD oblast, UKRAINE, on road, and 29 mi/47 km E of KIROVOHRAD; 48°34′N 32°54′E. Elevation 433 ft/131 m. Cannery, cheese factory, sewing factory, asphalt plant. Established in 1730 by a Zaporozhian Cossack P. Petryk and called Petrykivka; renamed Nova Praha in 1821; town since 1957.

Nova Prata (NO-vah PRAH-tah), city (2007 population 22,257), NE RIO GRANDE DO SUL, BRAZIL, in the Serra GERAL, 38 mi/61 km NW of CAXIAS DO SUL; 28°47′S 51°36′W. Wine growing; wheat, corn. Until 1944, Prata.

Novar (NO-vahr), unincorporated village, S ONTARIO, E central Canada, 40 mi/64 km E of PARRY SOUND, and included in town of HUNTSVILLE; 45°27′N 79°15′W. Diatomite mining.

Novara (no-VAH-rah), province (□ 1,393 sq mi/3,621.8 sq km), PIEDMONT, N ITALY; ⊙ NOVARA; 45°58′N 08°24′E. Bordered N by SWITZERLAND; drained by TOCE, AGOGNA, and ANZA rivers. Contains LAKE of ORTA and part of Lago MAGGIORE. Mountainous terrain in N, including LEPONTINE ALPS, which rise to over 9,500 ft/2,896 m. Highly irrigated plains in S, crossed by CAVOUR Canal. A major rice region of Piedmont, producing also fodder crops and grapes; cattle raising. Quarries (granite, marble, limestone); several gold mines in MONTE ROSA region (Pestarena). Industry at Novara, DOMODOSSOLA, INTRA, PALLANZA, and OMEGNA. Important rice market at Novara. Area reduced in 1927 to form VERCELLI province.

Novara (no-VAH-rah), city (2001 population 100,910), ⊙ NOVARA province, PIEDMONT, N ITALY; 45°58′N 08°24′E. It is an agricultural and industrial center and a railroad junction. Manufacturing (textiles, clothing, chemicals, machinery, metals, processed food, paper, and printed materials). It is a major market for rice. Several battles were fought (1500, 1513) near Novara during the Italian Wars. At Novara, in March 1849, the Austrians under Radetzky defeated the Piedmontese under Charles Albert. The Church of San Gaudenzio (16th–17th century) has an impressive campanile (19th century).

Nova Redenção (NO-vah RE-den-SOUN), city (2007 population 8,947), BAHIA, BRAZIL.

Nova Resende (NO-vah RE-sen-dai), city (2007 population 14,156), SW MINAS GERAIS, BRAZIL, 38 mi/61 km NE of GUAXUPÉ; 21°15′S 46°28′W.

Nova Role (NO-vah RO-le), Czech *Nová Role*, German *neurohlau*, town, ZAPADOCESKY province, W BOHEMIA, CZECH REPUBLIC, 4 mi/6.4 km NW of KARLOVY VARY; 50°16′N 12°48′E. Railroad junction. Manufacturing (porcelain); kaolin mining in vicinity. Has a 13th century church.

Nova Russas (no-vah ROO-sahs), city (2007 population 30,632), W CEARÁ, BRAZIL, on CAMOCIM-CRATEÚS railroad, and 32 mi/51 km N of Crateús, near PIAUÍ border; 04°40′S 40°32′W. Livestock raising. Graphite deposits.

Nova Scotia (NO-vuh SKO-shuh) [Latin=New Scotland], province (□ 21,425 sq mi/55,705 sq km; 2006 population 913,462), E CANADA; ⊙ HALIFAX; 45°00′N 63°00′W. One of the MARITIME PROVINCES, it is comprised of a mainland peninsula and the adjacent CAPE BRETON ISLAND. In addition to Halifax, important cities are DARTMOUTH, SYDNEY, GLACE BAY, TRURO, and NEW GLASGOW. Nova Scotia is bounded on the N by the Gulf of ST. LAWRENCE, on the E and S by the ATLANTIC OCEAN, and on the W by NEW BRUNSWICK, from which it is largely separated by the Bay of FUNDY. The climate is moderate and the rainfall abundant. The E coast is rocky, with numerous bays and coves, and is dotted with many charming fishing villages. Off the beautiful S shore is SABLE ISLAND, called the graveyard of the Atlantic; on the W coast huge Fundy tides wash the shores.There is still some mining activity in Nova Scotia. Coal is mined principally in the SYDNEY–GLACE BAY area of CAPE BRETON ISLAND. Gypsum, barite, and salt are also mined. Fishing is important in Nova Scotia with its 200 miles of coastline. Fleets operate on the continental shelf edging the coast and also move out to the GRAND BANKS. Certain types of fishing, including cod, have been curtailed in the 1990s and 2000s, but lobster, scallops, and haddock continue to be important. The moratoria, however, have resulted in a smaller fishing fleet and a loss of fish processing jobs. Inland, the forests yield spruce lumber, and the province's industries produce much pulp and paper. In the NW there is dairying, the most important sector of Nova Scotia's agricultural economy, and the region of ANNAPOLIS and Cornwallis supports valuable apple orchards. There are also important hay, grain, fruit, and vegetable crops. The bay lowlands, reclaimed by dikes in the 17th century, are very productive. Manufacturing is one of the largest production sectors of Nova Scotia's economy. In addition to the iron and steel produced at Sydney, the province's manufacturing includes processed food, motor vehicles, tires, sugar, and construction materials. In addition to its all-year port facilities, Halifax is a railroad terminus. The rivers of Nova Scotia have a number of small hydroelectric stations that help support the economy. Tourism is another major contributor, employing some 30,000.The charms of the rural and coastal countryside and an abundance of historical sites attract over one million tourists per year. Frequently visited historical spots include the Alexander Graham Bell Museum at BADDECK, the Shrine of Evangeline at GRAND PRÉ, and the town of ANNAPOLIS ROYAL, site of the first permanent Canadian settlement (1610). Cape Breton Island (established 1936) and KEJIMKUJIK (established 1968) national parks are in Nova Scotia. Sportsmen are attracted by abundant game and all types of fishing, and some of the best sailing on the continent. Two Algonquian tribes, the

Abnaki and the Micmac, inhabited the area before Europeans arrived. John Cabot may have landed (1497) on the tip of Cape Breton Island; European fishermen were already making regular stops during their yearly expeditions. An unsuccessful French settlement was made in 1605 at Port Royal (now Annapolis Royal). In 1610 the French succeeded at the same site. For the next 150 years FRANCE and ENGLAND contested bitterly for colonial rights to ACADIA, which included present-day Nova Scotia, NEW BRUNSWICK, and PRINCE EDWARD ISLAND. In 1621 Sir William Alexander obtained a patent from James I for the colonization of Acadia. Control alternated between France and England through several wars and treaties. Under the Peace of Utrecht (1713–1714), the Nova Scotia peninsula was restored to England, although Cape Breton Island was retained by the French. Hostilities were renewed in 1744.During the French and Indian War (1755–1763), the French Acadians were expelled—an event which loosely inspired Henry Wadsworth Longfellow's *Evangeline*. The Treaty of Paris (1763) gave all of French North America to England. Prince Edward Island, joined to Nova Scotia in 1763, became separate in 1769. With the influx (c.1784) of United Empire Loyalists, additional settlement occurred. In 1784 New Brunswick and Cape Breton also became separate colonies; Cape Breton rejoined Nova Scotia in 1820. During the early 19th century, thousands of Scots and Irish emigrated to Nova Scotia. Under the leadership of Joseph Howe, Nova Scotia became the first colony to achieve (1848) responsible (or cabinet) government. It acceded to the Canadian confederation as one of the four original members in 1867 after considerable difficulty over economic arrangements.In recent years Nova Scotia has struggled to stabilize its economy. Federal government programs to develop secondary industries or to discover offshore oil or natural-gas deposits have been largely unsuccessful. The province sends ten senators and eleven representatives to the national parliament. Nova Scotia has pioneered in Canadian history with the first newspaper (*Halifax Gazette*, 1752), the first printing press (1751), and the first university (King's College, Windsor, 1788–1789). Additional educational institutions include Dalhousie University, St. Francis Xavier University, Saint Mary's, Mount Saint Vincent, University College of Cape Breton, Sainte-Anne University, and the Technical University of Nova Scotia.

Nova Sedlica, SLOVAKIA: see STAKCIN.

Nova Serrana (NO-vah SER-rah-nah), city (2007 population 36,596), central MINAS GERAIS, BRAZIL, 31 mi/50 km W of PARÁ DE MINAS; 19°40′S 45°10′W.

Nova Sintra, CAPE VERDE ISLANDS: see VILA NOVA SINTRA.

Nova Sofala (NO-ve so-FAH-luh), city, SE MOZAMBIQUE, on MOZAMBIQUE CHANNEL. An early Arab trading post, it was settled by the Portuguese in 1505, when a fort was built. Nova Sofala was the starting point for expeditions into the mineral-rich hinterland. Formerly Nova Sofala.

Nova Sofala, MOZAMBIQUE: see SOFALA.

Nova Soure (NO-vah SO-rai), city (2007 population 25,412), NE BAHIA, BRAZIL, near border with SERGIPE, 6 mi/10 km S of CIPÓ; 11°15′S 38°30′W.

Novate Milanese (no-VAH-te mee-lah-NAI-ze), town, MILANO province, LOMBARDY, N ITALY, 5 mi/8 km NNW of MILAN; 45°32′N 09°08′E. In gardening region; fabricated metals, machinery, textiles, paper.

Nova Timboteua (no-vah cheem-bo-tai-oo-ah), city (2007 population 12,042), E PARÁ, BRAZIL, near BELÉM-BRAGANÇA railroad, 80 mi/129 km ENE of Belém; 01°12′S 47°21′W. Rubber, cereals.

Novato, city (2000 population 47,630), MARIN county, W CALIFORNIA, suburb 24 mi/39 km NNW of downtown SAN FRANCISCO, on Novata Creek, 4 mi/6.4 km W of SAN PABLO BAY; 38°05′N 122°34′W.

Dairying; poultry, lambs; fruit, nuts. Manufacturing (cosmetics, fabricated metal products, telephone apparatus, lumber, wiring devices). Its population has increased along with the economic development of N California. HAMILTON AIR FORCE BASE to SE, a major West Coast installation, is situated in Novato, and the county airport is just N of the city. City is surrounded by dairy farming. Annual celebration called Western Weekend is here. Petaluma Adobe State Historical Park to NE; POINT REYES NATIONAL SEASHORE, on PACIFIC OCEAN, to SW. Incorporated 1960.

Novator (nuh-VAH-tuhr), town, NE VOLOGDA oblast, central European Russia, on the NORTHERN DVINA River, on road and railroad, 7 mi/11 km SW of VELIKIY USTYUG; 60°44′N 46°12′E. Elevation 311 ft/94 m. Plywood factory. Formerly known as Galuzino.

Nova Trento (NO-vah TREN-to), town (2007 population 11,325), E SANTA CATARINA, BRAZIL, 30 mi/48 km NW of FLORIANÓPOLIS; 27°17′S 48°55′W. Textile mill; dairying, wine growing. Settled after 1876 by Italians.

Nova Ushytsya (oo-SHI-tsya) (Russian *Novaya Ushitsa*), town (2004 population 5,250), SE KHMEL'-NYTS'KYY oblast, UKRAINE, on the KAL'MIUS RIVER, 33 mi/53 km ENE of KAMYANETS'-PODIL'S'KYY; 48°50′N 27°17′E. Elevation 836 ft/254 m. Raion center; fruit canning, flour and feed milling, asphalt and brick making. Agricultural mechanization technical school. Known since 1939 as Litvintsi, when it passed to Poland; passed to Russia in 1793; promoted to status of a company center and renamed Nova Ushytsya in 1829; town since 1924. Site of battles between the Ukrainian National Republic Army and Red Army (1918, 1920).

Nova Varoš (NO-vah var-osh), village (2002 population 19,982), W SERBIA, 26 mi/42 km S of UZICE, in the Sanjak region, at N end of Lake Zlatarsko; 43°28′N 19°49′E. Also spelled Nova Varosh.

Nova Vcelnice (NO-vah FCHEL-nyi-TSE), Czech *Nová Včelnice*, German *neu oettingen*, village, JIHOCESKY province, SE BOHEMIA, CZECH REPUBLIC, on railroad, and 7 mi/11.3 km NNE of JINDŘICHŮV HRADEC; 49°14′N 15°04′E. Manufacturing (textiles [woolens]); wine making; agriculture (wheat, barley); cattle. Has an 18th century baroque castle.

Nova Velha District, MOZAMBIQUE: see NAMPULA.

Nova Venécia (no-vah ve-NE-see-ah), town (2007 population 43,926), N ESPÍRITO SANTO, BRAZIL, 35 mi/56 km NW of SÃO MATEUS (linked by railroad); 18°36′S 40°28′W. Coffee, oranges, bananas.

Nova Veneza (NO-vah VE-ne-sah), town (2007 population 12,657), SE SANTA CATARINA state, BRAZIL, 11 mi/18 km W of CRICIÚMA; 29°39′S 49°30′W.

Nova Vicenza, Brazil: see FARROUPILHA.

Nova Vicosa (NO-vah VEE-so-sah), city (2007 population 34,792), BAHIA state, SE BRAZIL, on ATLANTIC coast, 16 mi/25 km S of CARAVELAS; 17°55′S 39°23′W.

Nova Vida (NO-vah VEE-dah), city, N RONDÔNIA state, BRAZIL, 123 mi/198 km SE of PÔRTO VELHO; 10°11′S 62°42′W.

Nova Vodolaha (no-VAH vo-do-LAH-hah) (Russian *Novaya Vodolaga*), town, W KHARKIV oblast, UKRAINE, on railroad, and 23 mi/37 km SW of KHARKIV; 49°43′N 35°52′E. Elevation 449 ft/136 m. Raion center; food processing (flour mill, dairy, fruit and vegetable cannery), construction materials and asphalt making. Established in 1675 as a fortified settlement on the Murava Road; developed as a silk trading center in the 18th century, when danger of Tatar attacks declined; town since 1938.

Nova Xavantina (NO-vah SHAH-vahn-chee-nah), city (2007 population 18,657), SE MATO GROSSO, BRAZIL, on RIO DAS MORTES; 14°45′S 52°13′W.

Novaya [Russian=new], in Russian names: see also NOVO- [Russian combining form], NOVOYE, NOVY, or NOVYE.

Novaya Aleksandriya, POLAND: see PULAWY.

Novaya Astrakhan', UKRAINE: see NOVA ASTRAKHAN'.

Novaya Basan', UKRAINE: see NOVA BASAN'.

Novaya Borovaya, UKRAINE: see NOVA BOROVA.

Novaya Bryan' (NO-vah-yah BRYAHN), village (2005 population 4,830), central BURYAT REPUBLIC, S Siberian Russia, on highway branch and railroad spur, 31 mi/50 km ESE of ULAN-UDE, and 7 mi/11 km S of ZAIGRAYEVO; 51°43′N 108°17′E. Elevation 1,965 ft/598 m. Has an Old Believers Orthodox church.

Novaya Bukhtarma, KAZAKHSTAN: see ZHANA BUKTYRMA.

Novaya Chara (NO-vah-yah CHAH-rah), village, N CHITA oblast, S Siberian Russia, on the Chara River, on road and the BAYKAL-AMUR MAINLINE, 6 mi/10 km S of CHARA; 56°49′N 118°18′E. Elevation 2,411 ft/734 m. Marble mines in the vicinity. Heritage museum. Formerly part of the Gulag prison camp system.

Novaya Chigla (NO-vah-yah chee-GLAH), village (2006 population 2,965), central VORONEZH oblast, S central European Russia, on local road junction, 21 mi/34 km ENE of BOBROV; 51°13′N 40°28′E. Elevation 314 ft/95 m. In agricultural area (wheat, oats, potatoes, vegetables).

Novaya Derevnya, RUSSIA: see NOVOSIBIRSK.

Novaya Eushta, RUSSIA: see TIMIRYAZEVSKIY.

Novaya Galeshchina, UKRAINE: see NOVA HALESHCHYNA.

Novaya Igirma (NO-vah-yah ee-geer-MAH), town (2005 population 10,590), central IRKUTSK oblast, E central Siberian Russia, on the ILIM RIVER, on road and UST'-ILIMSK-bound branch of the BAYKAL-AMUR MAINLINE, 37 mi/60 km N of ZHELEZNOGORSK-ILIMSKIY; 57°08′N 103°54′E. Elevation 1,040 ft/316 m. Sawmilling, woodworking.

Novaya Ivanovka, UKRAINE: see NOVA IVANIVKA.

Novaya Kakhovka, UKRAINE: see NOVA KAKHOVKA.

Novaya Kalitva (NO-vah-yah kah-leet-VAH), village, S VORONEZH oblast, S central European Russia, on the DON River, near local road junction, 22 mi/35 km ESE, and under administrative jurisdiction, of ROSSOSH; 50°05′N 39°59′E. Elevation 360 ft/109 m. In agricultural area; produce processing.

Novaya Kazanka (NO-vuh-yuh kuh-ZAHN-kuh), village, W WEST KAZAKHSTAN region, KAZAKHSTAN, in Caspian Lowland, 170 mi/274 km SW of ORAL (Uralsk); 49°00′N 49°36′E. Millet, wheat; camel and cattle breeding. Natural-gas wells nearby. Greater and Lesser Uzun rivers disappear into salt lakes (E).

Novaya Krestovka, UKRAINE: see KIROVS'KE, Donets'k oblast.

Novaya Ladoga (NO-vah-yah LAH-duh-gah), city (2005 population 9,600), N LENINGRAD oblast, NW European Russia, port on the S shore of Lake LADOGA, at the mouth of the VOLKHOV RIVER, near highway, 85 mi/141 km ENE of SAINT PETERSBURG; 60°06′N 32°18′E. Ship repair yard; fish canning; textiles; leather goods and haberdashery. Chartered in 1704.

Novaya Lyalya (NO-vah-yah LYAH-lyah), city (2006 population 14,140), W central SVERDLOVSK oblast, W SIBERIA, RUSSIA, on the LYALYA RIVER, on road junction and railroad, 150 mi/243 km N of YEKATERINBURG, and 80 mi/129 km NNE of NIZHNIY TAGIL; 59°03′N 60°36′E. Elevation 298 ft/90 m. Wood pulp and paper; sawmilling. Until 1928, called Novo-Lyalinskiy Zavod. Made city in 1938.

Novaya Maka (NO-vah-yah mah-KAH) [Russian=new Maka], village (2005 population 4,655), SE DAGESTAN REPUBLIC, in pre-Caspian valley just E of the foothills of the E Greater CAUCASUS Mountains, approximately 7 mi/11 km NW of the Russia-AZERBAIJAN border, on road, 80 mi/129 km SSE of MAKHACHKALA; 41°46′N 48°22′E. Elevation 495 ft/150 m. Agriculture (grain, fruits, vegetables). Formerly called Yanankala.

Novaya Malykla (NO-vah-yah mah-lik-LAH), village (2006 population 3,285), NE ULYANOVSK oblast, E central European Russia, on highway and railroad, 12 mi/19 km E of DIMITROVGRAD; 54°12′N 49°57′E. Elevation 282 ft/85 m. In agricultural area (wheat, sun-

flowers; livestock); creamery. Sometimes called Novomalykla.

Novaya Mayachka, UKRAINE: see NOVA MAYACHKA.

Novaya Mayna (NO-vah-yah MEI-nah), urban settlement (2006 population 6,430), E ULYANOVSK oblast, E central European Russia, in the GREATER CHEREMSHAN RIVER basin, on highway, 7 mi/11 km SE of (and administratively subordinate to) DIMITROVGRAD; 54°09′N 49°45′E. Elevation 265 ft/80 m. Lumbering; fisheries.

Novaya Mel'nitsa (NO-vah-yah MYEL-nee-tsah) [Russian=new mill], settlement, W central NOVGOROD oblast, NW European Russia, in the VOLKHOV RIVER basin, 3 mi/5 km W of NOVGOROD, to which it is administratively subordinate; 58°32′N 31°12′E. Sawmilling, lumbering.

Novaya Mysh (NO-vah-yah MYSH), Polish *Nowa Mysz*, village, S BREST oblast, BELARUS, 5 mi/8 km W of BARANOVICHI. Manufacturing (flour milling, pitch processing; bricks).

Novaya Odessa, UKRAINE: see NOVA ODESA.

Novaya Pis'myanka, RUSSIA: see LENINOGORSK.

Novaya Pokrovka, RUSSIA: see LISKI.

Novaya Praga, UKRAINE: see NOVA PRAHA.

Novaya Sibir' Island (NO-vah-yah see-BEER) [Russian=new Siberia], easternmost of the Anjou group of NEW SIBERIAN ISLANDS, off the coast of SAKHA REPUBLIC, RUSSIAN FAR EAST; 75 mi/121 km long, 35 mi/56 km wide; 75°00′N 149°00′E. Rises to 330 ft/101 m. Discovered in 1806.

Novaya Slobodka (NO-vah-yah sluh-BOT-kah), former town, NE IVANOVO oblast, central European Russia, on the VOLGA RIVER, at the mouth of the UNZHA RIVER, opposite YURYEVETS. Sawmilling center for timber floated down Unzha River. Emtpied of residents and flooded during the construction of the Gorkiy reservoir.

Novaya Tavolzhanka (NO-vah-yah tah-vuhl-ZHAHN-kah), village (2004 population 2,700), S BELGOROD oblast, SW European Russia; in the SEVERSKIY DONETS river basin, approximately 3 mi/5 km N of the Ukrainian border; near railroad; 19 mi/30 km S of BELGOROD and 5 mi/8 km SW of SHEBEKINO; 50°21′N 36°50′E. Elevation 416 ft/126 m. Sugar refinery. Also known as Novotavolzhanka.

Novaya Ushitsa, UKRAINE: see NOVA USHYTSYA.

Novaya Usman' (NO-vah-yah OOS-mahn), village (2006 population 22,070), W central VORONEZH oblast, S central European Russia, on highway junction, 8 mi/13 km ESE of VORONEZH; 51°38′N 39°25′E. Elevation 328 ft/99 m. Asphalt plant. Has a 19th century Trinity church. Formerly known as Sobakino.

Novaya Vodolaga, UKRAINE: see NOVA VODOLAHA.

Novaya Zaimka (NO-vah-yah zah-EEM-kah), settlement, SW TYUMEN oblast, W SIBERIA, RUSSIA, on road and the TRANS-SIBERIAN RAILROAD, 25 mi/40 km SE of YALUTOROVSK; 56°29′N 66°55′E. Elevation 344 ft/104 m. Flour milling, dairying; fodder plant.

Novaya Zemlya, archipelago (□ c.35,000 sq mi/90,650 sq km), in the ARCTIC OCEAN between the BARENTS and KARA seas, NW RUSSIA; 74°00′N 57°00′E. It consists of two main islands (separated by MATOCHKIN Strait) and many smaller ones. The mountains of Novaya Zemlya, rising to c.3,500 ft/1,067 m, are a continuation of the URALS. In the N the archipelago is glaciated and is covered with arctic desert; the S part is tundra. Copper, lead, zinc, and asphaltite are found here. Fishing, sealing, and trapping are the chief occupations of the region's small population, which lives mainly along the W coast. The islands have been used by the Russians for thermonuclear testing. Explored by Novgorodians in the 11th or 12th centuries, the islands were sighted by explorers searching for the Northeast Passage in the 1500s. Since the mid-1800s Russians have built settlements and scientific stations.

Novaya Zhizn', RUSSIA: see GELDAGANA.

Novaya Zhizn', RUSSIA: see KAZINKA.

Nova York, Brazil: see NOVA IORQUE.

Nova Zagora (NO-vah zah-GO-rah), city, BURGAS oblast, NOVA ZAGORA obshtina (1993 population 48,941), E central BULGARIA, near a branch of the SAZLIIKA RIVER, 19 mi/31 km ENE of Stara Zagora; 42°30′N 26°00′E. Railroad junction; agricultural center (grain, cotton); livestock; produces textiles, tobacco; agricultural machinery building and repair, canning. Health resort BANYA is 9 mi/15 km to the N, near the TUNDZHA RIVER.

Novbaran, IRAN: see NOWBARAN.

Nove (no-VE) [Ukrainian = new] (Russian *Novoye*), town, central KIROVOHRAD oblast, UKRAINE, near highway and railroad, 6 mi/10 km WNW of the city center and subordinated to KIROVOHRAD; 48°33′N 32°10′E. Elevation 531 ft/161 m. Cast-iron foundry, asphalt and cement plant, pre-fabricated apartment building fabrication, farm-building construction. Established in 1977.

Nove (NO-ve), village, VICENZA province, VENETO, N ITALY, near BRENTA River, 14 mi/23 km NE of VICENZA; 45°43′N 11°40′E. Hydroelectric station. Noted for its ceramic industry.

Nove Benatky, CZECH REPUBLIC: see BENATKY NAD JIZEROU.

Novegradi, CROATIA: see NOVIGRAD.

Nove Hrady (NO-ve HRAH-di), Czech *Nové Hrady*, German *gratzen*, town, JIHOCESKY province, S BOHEMIA, CZECH REPUBLIC, on railroad, and 21 mi/34 km E of CESKY KRUMLOV, near Austrian border; 48°47′N 14°47′E. Peat extraction from nearby marshes. Founded in the 13th century (Gothic church and castle, museum).

Novelda (no-VEL-dah), city, ALICANTE province, E SPAIN, in VALENCIA, 16 mi/26 km WNW of ALICANTE; 38°23′N 00°46′W. Saffron, almonds, grapes, cereals, olives; wine production, oil milling, vegetable canning; stones for construction. Sandstone and marble quarries.

Novellara (no-vel-LAH-rah), town, REGGIO NELL'EMILIA province, EMILIA-ROMAGNA, N central ITALY, 11 mi/18 km NNE of REGGIO NELL'EMILIA; 44°51′N 10°44′E. Motors, generators, fabricated metals, textiles. Has a Gonzaga castle.

Novelty, town (2000 population 119), KNOX county, NE MISSOURI, on NORTH river, and 11 mi/18 km S of EDINA; 40°00′N 92°12′W. Lumber.

Nové Město, CZECH REPUBLIC: see NEW TOWN.

Nove Mesto nad Metuji (NO-ve MNYES-to NAHD ME-tu-YEE), Czech *Nové Město nad Metují*, German *neustadt an der mettau*, town, VYCHODOCESKY province, NE BOHEMIA, CZECH REPUBLIC, in foothills of the Eagle Mountains, on railroad, and 17 mi/27 km NE of HRADEC KRÁLOVÉ; 50°20′N 16°10′E. Summer resort; manufacturing (textiles, watches); distilling; food processing. Has a picturesque castle, a 16th century church, and a museum of ANTARCTICA.

Nové Mesto nad Váhom (no-VAI mes-TO NAHD vah-HOM), German *Neustadtl an der Waag*, Hungarian *Vágújhely*, city (2000 population 21,327), ZAPADOSLOVENSKY province, W SLOVAKIA, on VÁH RIVER; 48°45′N 17°50′E. Contains railroad junction; manufacturing (machinery, textiles; food); brewing, distilling. Has 15th-century Gothic church. Paleolithic-era archaeological site. Military base.

Nove Mesto na Morave (NO-ve MNYES-to NAH MO-rah-VYE), Czech *Nové Město na Moravě*, German *neustadtl*, town, JIHOMORAVSKY province, W MORAVIA, CZECH REPUBLIC, in BOHEMIAN-MORAVIAN HEIGHTS, on railroad, and 24 mi/39 km NE of JIHLAVA; 49°34′N 16°05′E. Excursion center; manufacturing (skis, medical instruments, machinery). Has a 16th century castle, and a museum.

Nove Mesto pod Smrkem (NO-ve MNYES-to POT smuhr-KEM), Czech *Nové Město pod Smrkem*, German *neustadt an der tafelfichte*, town, SEVEROCESKY province, N BOHEMIA, CZECH REPUBLIC, at NW foot of SMRK peak, on railroad, and 13 mi/21 km NNE of LIBEREC; 50°56′N 15°15′E. Agriculture (oats, potatoes); manufacturing (textiles, metal products). Has a Gothic 14th century church.

Nove Misto (no-VE MEES-to) [Ukrainian=new city] (Russian *Novyy Gorod* or *Novoye Misto*), town (2004 population 7,000), W L'VIV oblast, UKRAINE, on Vyrva River (tributary of Vyar River) and San (Ukrainian *Syan*) River, on road and railroad, 14 mi/23 km NW of STARYY SAMBIR, and 4 mi/7 km NE of DOBROMYL'; 49°37′N 22°52′E. Elevation 938 ft/285 m. Has old churches (1512, 1529); site of old settlement, with Roman and Hungarian coins among archeological finds; site of battles (1919) between Ukrainian and Polish armies.

Nove Myasto, POLAND: see NOWE MIASTO.

Novena, Paso de, SWITZERLAND: see NÜFENEN PASS.

Noventa Vicentina (vee-chen-TEE-nah), village, VICENZA province, VENETO, N ITALY, 18 mi/29 km S of VICENZA; 45°17′N 11°32′E. In cereal-growing region; liquor, agricultural machinery, clothing. Main plaza, porticoed cathedral, villas.

Noves (NOV), town (□ 10 sq mi/26 sq km), BOUCHES-DU-RHÔNE department, PROVENCE-ALPES-CÔTE D'AZUR region, SE FRANCE, on left bank of the DURANCE, and 7 mi/11.3 km SE of AVIGNON; 43°52′N 04°54′E. Olive- and grape-growing, soap manufacturing. Petrarch's Laura born here. An old place with narrow streets and a 12th-century church.

Novés (no-VAIS), town, TOLEDO province, central SPAIN, 19 mi/31 km NW of TOLEDO; 42°36′N 00°37′W. Cereals, potatoes; sheep, hogs; sawmilling.

Nove Sady (no-VAI sah-DI), Slovak *Nové Sady*, village, ZAPADOSLOVENSKY province, SW SLOVAKIA, on railroad, and 8 mi/12.9 km NW of NITRA; 48°29′N 17°59′E. Manufacturing of ceramics (electric insulators). Has 17th-century Renaissance castle. Until 1948 was called ASAKERT (ah-SHAH-kert-yuh), Slovak *Ašakert'*, HUNGARIAN *Alsókürth*.

Nove Sedlo (NO-ve SED-lo), Czech *Nové Sedlo*, German *neusattl*, town, ZAPADOCESKY province, W BOHEMIA, CZECH REPUBLIC, 5 mi/8 km WSW of KARLOVY VARY; 50°13′N 12°44′E. Railroad junction. Glassworks and manufacture of porcelain; in lignite-mining area.

Nove Selo (no-VE se-LO) (Russian *Novoye Selo*), (Polish *Nowe Siolo*), village (2004 population 5,750), E TERNOPIL' oblast, UKRAINE, 13 mi/21 km E of ZBARAZH; 49°39′N 26°04′E. Elevation 1000 ft/304 m. Wheat, barley; beekeeping. Village has two churches.

Nove Straseci (NO-ve STRAH-she-TSEE), Czech *Nové Strašecí*, German *neustraschnitz*, town, STREDOCESKY province, W central BOHEMIA, CZECH REPUBLIC, on railroad, and 23 mi/37 km WNW of PRAGUE; 50°09′N 13°55′E. Agriculture (wheat, sugar-beets); manufacturing (electronics); woodworking. Has a museum. Coal mining nearby. Brewery (established 1581) in KRUSOVICE, Czech *Krušovice* (KRU-sho-VI-tse), 5 mi/8 km WNW.

Nové Zámky (no-VAI zahm-KI), Hungarian *Érsekújvár*, German *Neuhäusel*, city (2000 population 42,262), ZAPADOSLOVENSKY province, S SLOVAKIA, on NITRA RIVER; 47°59′N 18°10′E. Contains railroad junction; manufacturing (consumer goods, food). Intensive tobacco growing in vicinity. Former Hungarian fortress, founded in 1561; played important part in TURKISH wars. Heavily damaged by BRITISH and AMERICAN air raid in March of 1945. Held by Hungary between 1938–1945. Majority of population is Hungarian. Has gallery, museum, and Neolithic-era archaeological site.

Novgorod (NOV-guh-ruht), oblast (□ 20,809 sq mi/54,103.4 sq km; 2006 population 673,500) in NW European Russia; ⊙ NOVGOROD. Includes N VALDAY HILLS (E) and Lake IL'MEN depression (W). Drained by MSTA, LOVAT', and VOLKHOV rivers; extensive forests (half of total area) and marshes, clays and

sandy soils. Mineral resources around BOROVICHI (lignite at KOMAROVO and ZARUBINO, refractory clays, limestone). Chief crop is flax; wheat and potatoes in same areas. Dairy cattle raised extensively. Lumber industry is important: sawmilling, paper milling (PARAKHINO-PODDUB'YE), match manufacturing (CHUDOVO, GRUZINO), veneering (STARAYA RUSSA, PARFINO). Glassworking (MALAYA VISHERA, BOL'SHAYA VISHERA, PROLETARIY) and porcelain manufacturing (KRASNOFARFORNYY). Main industrial centers include Borovichi, Novgorod, Chudovo, Staraya Russa. More than 70% of the population live in cities and other urban settlements. Formed in 1944 out of Leningrad oblast.

Novgorod (NOV-guh-ruht) or **Velikiy Novgorod** (vee-LEE-keeyee NOV-guh-ruht), city (2006 population 214,460), ⊙ NOVGOROD oblast, NW European Russia, on the VOLKHOV RIVER near the point where it leaves Lake IL'MEN, on railroad, 375 mi/603 km NW of MOSCOW; 58°31′N 31°17′E. Novgorod's industries produce machinery, electrical goods, radios, chemicals, fertilizer, and glass, wood, and food products. It has a major tourism industry. One of the oldest Russian cities, it was a major commercial and cultural center of medieval Europe. Rurik, who is said to have founded the Russian state in 862, was invited by the inhabitants of Novgorod to rule them. The capital was transferred to Kiev by Oleg in 886, but Novgorod remained the chief center of foreign trade, obtaining self-government status in 997 and achieving independence from Kiev in 1136, when it became the capital of an independent republic—Sovereign Great Novgorod, an area that embraced the whole of N European Russia to the Urals. Novgorod was governed by a popular assembly, or *veche*. The strength of the republic rested on its economic prosperity. Situated on the great trade route to the Volga valley, it became, with London, Bruges, and Bergen, one of the four chief trade centers of the Hanseatic League. Furs, hides, wax, honey, flax, and tar were the chief exports. Cloth and metals were imported from Europe and wheat from central Russia. The citizens of Novgorod repulsed the attacks of the Teutonic Knights and Livonian Knights and of the Swedes from the W and escaped the Mongol invasion from the E. At its height, in the 14th century, its population rose to nearly 400,000. Its splendor during that period, its hundreds of churches, its great shops and arsenals, its huge fairs, have all furnished rich themes for later Russian art and folklore. However, the 14th century also witnessed the start of Novgorod's long decline with the rise of Moscow. Novgorod retained its commercial position until SAINT PETERSBURG was built in 1703. The magnificent architectural monuments of Novgorod earned it the official designation of "museum city." In World War II, it suffered great damage.

Novgorodka, UKRAINE: see NOVHORODKA.

Novgorod-Severskiy, UKRAINE: see NOVHOROD-SIVERS'KYY.

Novgorodsk, RUSSIA: see POSYET.

Novgorodskoye, UKRAINE: see NOVHORODS'KE.

Novgradets, Bulgaria: see SUVOROVO.

Novhorodka (nov-ho-ROD-kah) (Russian *Novgorodka*), town, central KIROVOHRAD oblast, UKRAINE, on highway 20 mi/32 km SE of KIROVOHRAD; 48°22′N 32°39′E. Elevation 469 ft/142 m. Raion center; mineral-water-bottling plants, food processing (including dairy), feed mill, granite quarry. Vocational technical school; regional museum. Known as Kutsivka (Russian *Kutsovka*) since 1770 on Zaporozhian Sich territory; renamed Novhorodka in 1822.

Novhorod-Sivers'ke Polissya, UKRAINE: see POLISSYA, UKRAINIAN.

Novhorod-Sivers'kyy (NOV-ho-rod–SEE-ver-skiee) (Russian *Novgorod-Severskiy*), city, NE CHERNIHIV oblast, UKRAINE, 89 mi/143 km NE of CHERNIHIV, on the right bank of the DESNA River; 52°00′N 33°16′E.

Elevation 547 ft/166 m. Raion center; asphalt, construction materials, cheese making and other food processing, flax processing, cotton-textile factory; medical school, heritage museum. Established around 980, mentioned as city in 1044. Historic sites include the Spaso-Preobrazhens'kyy (=Transfiguration of our Savior) monastery (15th–19th century), the Uspens'kyy (Assumption) cathedral (1671–1715), Triumphal arch (1786–1787), market (18th–19th centuries).

Novhorods'ke (nov-ho-ROD-ske) (Russian *Novgorodskoye*), town, central DONETS'K oblast, UKRAINE, in the DONBAS, on Kryvyy Torets' River, on railroad, and 4 mi/6 km S of DZERZHYNS'K (and under its jurisdiction); 48°20′N 37°50′E. Elevation 367 ft/111 m. Phenol manufacturing; machine construction; brickworks; food industries. Established at the end of the 18th century and was called N'yu-York (New York) until 1951; located on the left bank of the river, it absorbed the adjoining town of Zalizne (Russian *Zheleznoye*) on the right bank after 1954.

Novi (NO-vee), city (2000 population 47,386), OAKLAND county, SE MICHIGAN, 24 mi/39 km WNW of downtown DETROIT and 13 mi/21 km SW of PONTIAC; 42°28′N 83°29′W. Drained in SE by Ingersoll Creek. Heavy manufacturing. WALLED LAKE to N. Maybury State Park to SW (WAYNE county). Oakland Southwest Airport to NW. Borders Wayne county on S.

Novi Bilokorovychi (no-VEE bee-lo-ko-RO-wi-chee) (Russian *Novyye Belokorovichi*), town, NW ZHYTOMYR oblast, UKRAINE, in the POLISSYA, on road and on railroad junction, 29 mi/46 km WNW of KOROSTEN'; 51°07′N 28°03′E. Elevation 547 ft/166 m. Rock-crushing plant, lumber mill, drainage administration. Established in 1901, town since 1961.

Novi Bilyary (no-VEE bee-lya-RI) (Russian *Novyye Belyary*), town (2004 population 4,400), E ODESSA oblast, UKRAINE, on W bank of Adzhalyk Liman, on road and on railroad spur, 16 mi/26 km NE of ODESSA city center and 13 mi/21 km SSE of KOMINTERNIVS'KE; 46°38′N 31°00′E. Manufacturing (construction materials). Established in the beginning of the 20th century as Anantal'; re-named in 1914; town since 1974.

Novice (NO-vis), village (2006 population 137), Coleman county, central TEXAS, 15 mi/24 km NW of COLEMAN; 31°59′N 99°37′W. Agricultural area. Lake Coleman reservoir to NE.

Novichikha (nuh-vee-CHEE-hah), village (2006 population 4,450), SW ALTAI TERRITORY, S SIBERIA, RUSSIA, on the E shore of shallow Gor'koye Lake, on highway, 50 mi/80 km N of RUBTSOVSK; 52°13′N 81°24′E. Elevation 826 ft/251 m. In agricultural area; dairy plant, woodworking.

Novi di Modena (NO-vee dee MO-de-nah), town, MODENA province, EMILIA-ROMAGNA, N central ITALY, 8 mi/13 km W of MIRANDOLA; 44°54′N 10°54′E. Wine making, food processing.

Novi Futog, SERBIA: see FUTOG.

Novigrad, medieval *Neapoli*, town, W CROATIA, on W coast of ISTRIA, 9 mi/15 km N of POREC. Sea resort and marina. St. Anastazija church (11th century); St. Pelagija church has wall paintings and Roman sarcophagi. Well-preserved Gothic palaces and town wall.

Novigrad (NO-vee-grahd), Italian *Novegradi*, village, S CROATIA, on NOVIGRAD SEA (inlet of ADRIATIC SEA), 15 mi/24 km ENE of ZADAR, in N DALMATIA. Bauxite deposits; tourism.

Novigrad Sea, Croatian *Novigradsko more*, bay in the ADRIATIC SEA, S CROATIA, in DALMATIA, NE of ZADAR. Connected by narrow canal with Adriatic Sea in near VELEBIT Mountain. Represents the lower course of the ZRMANJA RIVER that was flooded during the last Ice Age, when the Adriatic level rose by 328 ft/100 m.

Novi Han (NO-vee HAHN), village, SOFIA oblast, ELIN PELIN obshtina, BULGARIA; 42°36′N 23°36′E.

Novi Iskur (NO-vee EES-kuhr), city, GRAD SOFIA oblast, Novi Iskur obshtina, BULGARIA, 12 mi/19 km N of

SOFIA, on the ISKUR River; 42°51′N 23°23′E. Scrap nonferrous metallurgy, power generation; textiles, synthetic rubber. Comprises former villages of Kurilo, Alexander Voikov, and Gnilayne.

Novi Kneževac (NO-vee KNEZH-e-vahts), Hungarian *Törökkanizsa*, village (2002 population 12,975), VOJVODINA, N SERBIA, on TISZA RIVER, on railroad, and 22 mi/35 km E of SUBOTICA, in the BANAT region; 46°03′N 20°06′E. Raw silk manufacturing. Also spelled Novi Knezhevats. Formerly called Nova Kanjiza.

Novikovo (NO-vee-kuh-vuh), settlement (2006 population 785), S SAKHALIN oblast, RUSSIAN FAR EAST, in the SE part of SAKHALIN Island, on road junction, 41 mi/66 km SE of YUZHNO-SAKHALINSK; 46°22′N 143°21′E. Fisheries. Base camp for ecotourists. Under Japanese rule (1905–1945), called Shiretoko.

Novikovo (NO-vee-kuh-vuh), settlement, SW RYAZAN oblast, central European Russia, on highway and near railroad, 2 mi/3.2 km W, and under administrative jurisdiction, of SKOPIN; 53°49′N 39°31′E. Elevation 419 ft/127 m. Transportation and shipping enterprises.

Novi Krichim, Bulgaria: see STAMBOLIISKI, city.

Novi Ligure (NO-vee LEE-goo-re), town, ALESSANDRIA province, PIEDMONT, N ITALY, 14 mi/23 km SE of ALESSANDRIA; 44°46′N 08°47′E. Railroad junction; important industrial center with diverse consumer industries; steel mill, textile mills (silk, cotton). Austro-Russian army under Suvarov defeated French here, 1799.

Novi Marof (NO-vee MAH-rof), village, central CROATIA, 10 mi/16 km S of Varaždin, in HRVATSKO ZAGORJE; 46°10′N 16°20′E. In lignite area. Local trade center; lumbering. Castle, castle ruins; old mines.

Novinger (NAH-vin-juhr), city (2000 population 534), ADAIR county, N MISSOURI, on CHARITON RIVER, and 6 mi/9.7 km W of KIRKSVILLE; 40°13′N 92°42′W. Corn, soybeans; cattle, hogs, poultry. Settled 1830s; platted 1888. Former coal-mining town; population peaked c.1900 at c.5,000 inhabitants.

Noviomagus, GERMANY: see SPEYER.

Novi Pazar (NO-vee pah-ZAHR) [Bulgarian=new market], city, VARNA oblast, Novi Pazar obshtina (1993 population 20,978), E BULGARIA, on a headstream of the PROVADIISKA RIVER, 15 mi/24 km ENE of SHUMEN; 43°20′N 27°12′E. Agricultural center (grain; vineyards), wine making; glassworks, manufacturing (porcelain, construction ceramics, scales, parts for ships). Kaolin quarried nearby. Called Yeni Pazar under Turkish rule.

Novi Pazar (NO-vee PAH-zhah), city (2002 population 85,996), SW SERBIA, on the RASKA RIVER, 50 mi/80 km NW of PRISTINA; 43°08′N 20°31′E. It is an agricultural trading center with a well-developed textile industry. Known as Raška or Rashka in the 9th century, it was the capital of Serbia from the 12th century to the 14th century. It was captured by the Turks in 1456 and became an important trade center and the seat of the Turkish *sanjak* [=district] of Novibazar (an older spelling). The *sanjak* of Novibazar remained under Turkish civil administration until 1913, when it passed to Serbia. It became part of the former Yugoslavia after World War I. The city retains much of its Turkish architecture.

Novi Sad (NO-vee SAHD), German *Neusatz*, Hungarian *Újvidék*, city (2002 population 299,294), N SERBIA, on the DANUBE River, 70 mi/113 km NW of BELGRADE; ⊙ VOJVODINA province; 45°15′N 19°49′E. An industrial center and port, its industries produce processed foods, textiles, electrical equipment, and munitions. It is the site of a major oil refinery. Known in the 16th century, it rapidly developed as a commercial center, became an Orthodox episcopal see, and was made (1748) a royal free city of Austria-Hungary. In the 18th and early 19th centuries Novi Sad was the center of the Serbian literary revival. It was incorporated into the former Yugoslavia in 1918. The city has Serbian Orthodox churches, a university, and

numerous cultural facilities. Nearby, across the DANUBE River, is PETROVARADIN (German *Peterwardein*), the largest fortress of Austria-Hungary.

Novi Sad-Mali Stapar Canal (NO-vee SAHD–MAHlee STAH-pahr), 43 mi/69 km long, VOJVODINA province, N SERBIA; runs between the DANUBE River at NOVI SAD and DANUBE-TISZA CANAL at MALI STAPAR, in the BACKA region. Until 1945, called Franz Joseph Canal in Hungary, King Alexander Canal in the former Yugoslavia.

Novi Sanzhary (no-VEE sahn-ZHAH-ri) (Russian *Novyye Sanzhary*), town, SE POLTAVA oblast, UKRAINE, on the VORSKLA RIVER, near railroad, and 20 mi/32 km SSW of POLTAVA; 49°20′N 34°20′E. Elevation 252 ft/76 m. Raion center; deep-drilling laboratory, leather tanning, food processing (flour mill, dairy), clothing manufacturing, woodworking, furniture making; sanatorium. Established in the early 17th century; town since 1925.

Novi Strilyshcha (no-VEE STRI-lish-chah) (Russian *Novyye Strelishcha*), (Polish *Strzeliska Nowe*), town (2004 population 7,700), L′VIV oblast, UKRAINE, on road, and 27 mi/43 km SE of L′VIV; 49°31′N 24°24′E. Elevation 977 ft/297 m. Heritage museum. Known since 1513; town since 1940. Jewish community since the 17th century, numbering 1,150 in 1931; destroyed during World War II.

Nóvita (NO-vee-tah), village, ⊙ Nóvita municipio, CHOCÓ department, W COLOMBIA, 50 mi/80 km S of QUIBDÓ; 04°57′N 76°34′W. Plantains, sugarcane, corn. Gold- and platinum-placer mines.

Novi Vinodolski (NO-vee VEE-no-dol-skee), town, W CROATIA, small port on ADRIATIC SEA, 23 mi/37 km SE of RIJEKA, opposite KRK Island Seaside resort; trade center for fruit-growing region. Founded in 1288.

Novka, town, N central VITEBSK oblast, BELARUS, 20 mi/32 km NNE of VITEBSK, 55°26′N 30°24′E. Glassworks.

Novlenskoye (NOV-leen-skuh-ye), town, central VOLOGDA oblast, central European Russia, on the W shore of KUBENO LAKE, on road, 31 mi/50 km NW of VOLOGDA; 59°37′N 39°20′E. Elevation 419 ft/127 m. Linens.

Novo- [Russian combining form=new], in Russian names: see also NOVAOYA, NOVOYE, NOVY, and NOVYE.

Novo Acordo (NO-vo AH-kor-do), town (2007 population 4,063), S TOCANTINS state, BRAZIL, 121 mi/195 km SE of PALMAS; 10°00′S 47°50′W.

Novoagansk (no-vuh-ah-GAHNSK), town (2005 population 9,975), E central KHANTY-MANSI AUTONOMOUS OKRUG, TYUMEN oblast, central SIBERIA, RUSSIA, on the AGAN River (OB′ RIVER basin), 78 mi/125 km N of NIZHNEVARTOVSK, to which it is administratively subordinate; 61°58′N 76°38′E. Elevation 226 ft/68 m. In oil- and gas-producing area; river port facilities. Made town in 1978.

Novo-Aidar, UKRAINE: see NOVOAYDAR.

Novo Airão (NO-vo ei-ROUN), city (2007 population 14,630), N central AMAZONAS, BRAZIL, 120 mi/200 km NW of MANAUS, on Rio Negro; 01°57′S 61°23′W.

Novo-Aleksandrovka, UKRAINE: see MELITOPOL′.

Novoaleksandrovka, UKRAINE: see NOVOOLEKSANDRIVKA.

Novoaleksandrovsk (nuh-vuh-ah-leek-SAHNdruhfsk), city (2006 population 27,715), NW STAVROPOL TERRITORY, N CAUCASUS, S European Russia, on the Rasshevatka River (W MANYCH River basin), on highway junction and railroad, 70 mi/113 km NW of STAVROPOL; 45°29′N 41°13′E. Elevation 423 ft/128 m. Agricultural center in wheat and livestock area; meat packing, flour milling, brewing. Founded in 1804. Until 1971, a village called Novo-Aleksandrovskaya.

Novoaleksandrovsk (no-vuh-ah-leek-SAHNdruhfsk), town, S SAKHALIN oblast, RUSSIAN FAR EAST, on the E coastal highway and railroad, 6 mi/10 km N of YUZHNO-SAKHALINSK; 47°03′N 142°44′E. Elevation 177 ft/53 m. In agricultural area; railroad

junction (branch to Sinegorsk coal mines). Town status since 1989. Under Japanese rule (1905–1945), called Konuma.

Novoaleksandrovsk, LITHUANIA: see ZARASAI.

Novoalekseyevka (NO-vuh-ah-lek-se-YEF-kuh), village, NW AKTÖBE region, KAZAKHSTAN, 70 mi/113 km W of AKTÖBE (Aktyubinsk); 50°09′N 55°39′E. Tertiary-level (raion) administrative center. Millet, wheat.

Novoalekseyevka, UKRAINE: see NOVOOLEKSIYIVKA.

Novo-Alekseyevka, UKRAINE: see NOVOOLEKSIYIVKA.

Novoalekseyevskaya (no-vuh-ah-leek-SYE-eefs-kahyah), village (2005 population 3,595), E KRASNODAR TERRITORY, S European Russia, on the slopes of the NW CAUCASUS Mountains, in the Sinyukha River (tributary of the LABA RIVER) basin, on highway, 17 mi/27 km SW of ARMAVIR; 44°49′N 40°56′E. Elevation 830 ft/252 m. In oil- and gas-producing and -processing region.

Novoaltaysk (no-vuh-ahl-TEISK), city (2005 population 61,000), NE ALTAI TERRITORY, S central SIBERIA, RUSSIA, 5 mi/8 km E of BARNAUL; 53°24′N 83°57′E. Elevation 482 ft/146 m. Joined by a bridge across the OB′ RIVER, at a junction (Altayskaya station) of the TURK-SIB railroad and branch line to BIYSK. Center of agricultural area; paper and cardboard mill; metal goods; agricultural processing; flour milling. City since 1942. Developed during World War II. Until 1962, called Chesnokovka.

Novoamvrosiyevskoye, UKRAINE: see NOVOAMVROSIYIVS′KE.

Novoamvrosiyivs′ke (no-vo-ahm-VRO-see-yeef-ske) (Russian *Novoamvrosiyevskoye*), town (2004 population 4,100), SE DONETS′K oblast, UKRAINE, in the DONBAS, on the right bank of the Krynka River (right tributary of the MIUS RIVER), 4 mi/6 km NNE of AMVROSIYIVKA; 47°50′N 38°30′E. Elevation 482 ft/146 m. Town has two cement plants. Established in the late 19th century; town since 1956.

Novoannenskaya, RUSSIA: see NOVOANNINSKIY.

Novo-Annenskiy, RUSSIA: see NOVOANNINSKIY.

Novoanninskiy (nuh-vuh-AH-neen-skeeyee), city (2006 population 19,155), NW VOLGOGRAD oblast, SE European Russia, on railroad (Filonovo station), on the BUZULUK RIVER (tributary of the KHOPER River, DON River basin), 157 mi/254 km NW of VOLGOGRAD; 50°32′N 42°41′E. Elevation 357 ft/108 m. Major flour-milling center; electric medical equipment; foundry. Also livestock breeding and meat packing. Formerly called Novoannenskaya, or Novo-Annenskiy. Made city in 1956.

Novoardonskoye, RUSSIA: see EKAZHEVO.

Novo Aripuanã (NO-vo AH-ree-poo-an-AHN), city (2007 population 18,067), E central AMAZONAS, BRAZIL, 78 mi/125 km downstream from MANICORÉ, on Rio MADEIRA; 05°08′S 60°22′W.

Novoarkhangel′sk, UKRAINE: see NOVOARKHANHEL′S′K.

Novo-Arkhangel′sk, UKRAINE: see NOVOARKHANHEL′S′K.

Novoarkhanhel′s′k (no-vo-ahr-KHAHN-helsk) (Russian *Novoarkhangel′sk*), town, W KIROVOHRAD oblast, UKRAINE, on the SYNYUKHA RIVER and on road, and 33 mi/53 km E of UMAN′; 48°39′N 30°49′E. Elevation 396 ft/120 m. Raion center; food processing (feed and flour mills, cheese factory), asphalt and brickworks; small hydroelectric power plant. Established in 1742 as Arkhanhel-Horodok; renamed in 1764; town since 1957. Formerly spelled, in Russian, *Novo-Arkhangel′sk*.

Novoasbest (nuh-vuh-ahz-BYEST), town (2006 population 2,380), W central SVERDLOVSK oblast, extreme W Siberian Russia, in the E outliers of the central URALS, on road, 18 mi/29 km SE of NIZHNIY TAGIL; 57°44′N 60°17′E. Elevation 813 ft/247 m. Asbestos mining and processing. Developed in the 1930s. Called Krasnouralskiy Rudnik until 1933.

Novoaydar (no-vo-ei-DAHR), town, central LUHANS'K oblast, UKRAINE, on Aydar River, on railroad and road, and 22 mi/35 km S of STAROBIL's'K; 48°57′N 39°01′E. Elevation 262 ft/79 m. Raion center; food processing (elevator, feed mill, broiler factories, food-flavoring factory). Vocational technical school. Established in the 17th century; town since 1957. Formerly spelled Novo-Aydar, also Novo-Aidar.

Novo-Aydar, UKRAINE: see NOVOAYDAR.

Novoazovs'k (no-vo-ah-ZOVSK) (Russian *Novoazovsk*), city, S DONETS'K oblast, UKRAINE, on Sea of AZOV, 26 mi/42 km E of MARIUPOL'; 47°07′N 38°05′E. Raion center; grain mill, feed mill, asphalt-making, brick-making, poultry-processing, and food-flavoring plants. Established in 1849 as Novomykolayivs'ka Stanytsya (Russian *Novo-Nikolayeveskaya Stanitsa*), then simply Novomykolayivka (Russian *Novo-Nikolayevka*), until 1923 when it was renamed Bud'onnivka (Russian *Budënnovka*); in 1959 it was renamed Novoazovs'ke (Russian *Novoazovskoye*); promoted to city status in 1966.

Novoazovskoye, UKRAINE: see NOVOAZOVS'K.

Novobelokatay (no-vuh-bye-luh-kah-TEI), village (2005 population 5,930), NE BASHKORTOSTAN Republic, in the NW foothills of the S URALS, E European Russia, on road, 45 mi/72 km NNW of ZLATOUST; 55°42′N 58°57′E. Elevation 836 ft/254 m. Grain; livestock; lumbering.

Novobeysugskaya (no-vuh-byai-SOOGS-kah-yah), village (2005 population 3,620), E central KRASNODAR TERRITORY, S European Russia, on road, 39 mi/63 km NE of KRASNODAR; 45°28′N 39°53′E. Elevation 196 ft/59 m. In agricultural area (wheat, flax, sunflowers, castor beans).

Novobirilyussy (no-vuh-bee-ree-LYOO-si), settlement (2005 population 4,255), SW KRASNOYARSK TERRITORY, SE SIBERIA, RUSSIA, on road, 5 mi/8 km S of NAZAROVO; 56°57′N 90°40′E. Elevation 564 ft/171 m. Sawmilling, lumbering. Formerly known as Chepysheva.

Novobiryusinskiy (no-vuh-bee-ryoo-SEEN-skeeyee), industrial settlement (2005 population 4,815), W IRKUTSK oblast, E central SIBERIA, RUSSIA, on crossroads and spur of the BAYKAL-AMUR MAINLINE, 51 mi/82 km NNW of SHITKINO; 56°27′N 97°43′E. Elevation 705 ft/214 m. Logging, lumbering, sawmilling.

Novobogatinskoye (NO-vuh-bo-guh-TEEN-sko-ye), town, N ATYRAU region, KAZAKHSTAN, on arm of URAL RIVER delta mouth, 35 mi/56 km NW of Atyrau; 47°14′N 51°00′E. In petroleum area. Tertiary-level (raion) administrative center.

Novo Brasil (NO-vo brah-SEEL), town (2007 population 3,455), W central GOIÁS, BRAZIL, 75 mi/121 km SW of Goiás Velho; 16°03′S 50°39′W.

Novobrattsevskiy (no-vuh-BRAHT-tsif-skeeyee), rural settlement, central MOSCOW oblast, central European Russia, on the Skhodna River (MOSKVA River basin), on highway, 12 mi/20 km NW of MOSCOW, and 3 mi/5 km NE of KRASNOGORSK; 55°51′N 37°23′E. Elevation 462 ft/140 m. Textile production; poultry plant. Town status since 1928, revoked in 1990.

Novo Brdo (NO-vo BUHRT-do), mine, KOSOVO province, S SERBIA, 14 mi/23 km ESE of PRISTINA. Produces lead and zinc ores with some silver and gold. A noted silver-mining center in 12th–15th century Serbia.

Novobureyskiy (no-vuh-boo-RYAI-skeeyee), town, S AMUR oblast, SE SIBERIA, RUSSIAN FAR EAST, on the BUREYA RIVER, on road and the TRANS-SIBERIAN RAILROAD, 26 mi/42 km NW of ARKHARA; 49°47′N 129°53′E. Elevation 328 ft/99 m. In a protected old-growth forest area. Cargo crane manufacturing.

Novoburino (no-vuh-BOO-ree-nuh), village (2004 population 1,240), NE CHELYABINSK oblast, SW SIBERIA, RUSSIA, on road, 32 mi/51 km S of KAMENSK-URALSKIY; 55°59′N 61°38′E. Elevation 521 ft/158 m. Near railroad; grain; livestock. Peat digging nearby (S). Formerly called Burino.

Novocheboksarsk (nuh-vuh-chi-buhk-SAHRSK), city (2005 population 129,370), N CHUVASH REPUBLIC, central European Russia, 3 mi/5 km E of CHEBOKSARY, on S shore of the Cheboksary Reservoir on the VOLGA RIVER; 56°07′N 47°30′E. Elevation 236 ft/71 m. Port; manufacturing (building materials, pipes; cotton spinning); food processing (dairy, bakery, brewery). Cheboksary hydroelectric power station nearby. Arose in 1960 as a satellite of Cheboksary for the construction of the power station. Made city in 1965.

Novocheremshansk (nuh-vuh-chye-reem-SHAHNSK), town (2006 population 3,285), NE ULYANOVSK oblast, E central European Russia, on the GREATER CHEREMSHAN RIVER, on railroad spur, 25 mi/40 km ENE of DIMITROVGRAD, and 14 mi/23 km NE of NOVAYA MALYKLA, to which it is administratively subordinate; 54°21′N 50°10′E. Elevation 278 ft/84 m. Tanning extract, cardboard; lumbering. Formerly called Staryy Salavan.

Novocherkassk (nuh-vuh-chir-KAHSK), city (2006 population 165,725), W ROSTOV oblast, S European Russia, on the AKSAY RIVER (right tributary of the DON River), on highway junction and railroad, 32 mi/51 km NE of ROSTOV-NA-DONU; 47°25′N 40°05′E. Elevation 298 ft/90 m. Electric locomotives, machine tools, petroleum machinery, chemicals, synthetic products. Novocherkassk regional electric power station nearby. Founded in 1805; administrative center of the Don Cossacks until 1920; site of the hetman's palace. Has a Don Cossack historical museum. Cossack Cadet Academy, established in 1883, closed by the Soviets, reopened here in 1991 as part of Cossack revival.

Novochernorechenskiy (no-vuh-chyer-nuh-RYE-chinskeeyee), town (2005 population 3,705), SW KRASNOYARSK TERRITORY, SE SIBERIA, Russia, on highway and the TRANS-SIBERIAN RAILROAD (Chernorechenskaya station), 89 mi/144 km W of KRASNOYARSK, and 9 mi/14 km WNW of KOZUL'KA, to which it is administratively subordinate; 56°16′N 91°05′E. Elevation 912 ft/277 m. Railroad shops; lumbering. Railroad station and railroad workers' settlement built in 1898; town status granted in 1949.

Novo Cruzeiro (NO-voh KROO-sai-ro), city (2007 population 30,199), NE MINAS GERAIS, BRAZIL, 45 mi/72 km WNW of TEÓFILO OTONI; 17°32′S 41°50′W.

Novo-Dar'yevka, UKRAINE: see NOVODARYIVKA.

Novodar'yevka, UKRAINE: see NOVODARYIVKA.

Novodaryivka (no-vo-DAHR-yeef-kah) (Russian *Novodar'yevka*), town (2004 population 7,650), S LUHANS'K oblast, UKRAINE, in the DONBAS, 8 mi/13 km S, and under jurisdiction, of ROVEN'KY; 47°58′N 39°23′E. Elevation 429 ft/130 m. Coal mining; gas compressor station; asphalt making, and stone crushing. Established in the 1870s, town since 1938. Formerly spelled, in Russian, *Novo-Dar'yevka*.

Novoderevyankovskaya (no-vuh-dye-ree-VYAHN-kuhf-skah-yah), village (2005 population 6,745), NW KRASNODAR TERRITORY, S European Russia, on the Albasi River, on highway junction, 83 mi/134 km N of KRASNODAR, and 9 mi/14 km W of NOVOMINSKAYA; 46°19′N 38°45′E. In agricultural region (wheat, corn, sunflowers, vegetables, fruits); produce processing.

Novodevichye (nuh-vuh-DYE-veech-ye), village, W SAMARA oblast, E European Russia, grain port on the right bank of the VOLGA RIVER, on local highway branch, 35 mi/56 km NNE of SYZRAN'; 53°36′N 48°50′E. Elevation 347 ft/105 m. In agricultural area (grain, sunflowers). Chalk and cement-rock quarrying nearby.

Novodmitriyevskaya (no-vuh-DMEE-tree-eef-skah-yah), village (2005 population 5,255), S central KRASNODAR TERRITORY, S European Russia, in the KUBAN' River basin, on highway junction, 17 mi/27 km SSW of KRASNODAR; 44°50′N 38°52′E. Elevation 170 ft/51 m. In oil- and gas-refining region.

Novodnestrovsk, UKRAINE: see NOVODNISTROVS'K.

Novodnistrovs'k (no-vo-dnees-TROVSK) (Russian *Novodnestrovsk*), city, E CHERNIVTSI oblast, UKRAINE, on the high right bank of the DNIESTER (Ukrainian *Dnister*) River, 75 mi/121 km ENE of CHERNIVTSI and 9 mi/15 km N of SOKYRYANY; 48°36′N 27°25′E. Elevation 764 ft/232 m. Site of the Dniester hydroelectric station, its building and administration; grain-milling; technical school. Established in 1975; city since 1993.

Novodonets'ke (no-vo-do-NETS-ke) (Russian *Novodonetskoye*), town (2004 population 8,400), NW DONETS'K oblast, UKRAINE, on road and railroad, 12 mi/19 km NNW of, and subordinated to, DOBROPILLYA; 48°38′N 37°00′E. Elevation 508 ft/154 m. Coal mining; furniture making. Established in 1956, town since 1960.

Novodonetskoye, UKRAINE: see NOVODONETS'KE.

Novodruzhes'k (no-vo-DROO-zhesk), city, W LUHANS'K oblast, UKRAINE, in the DONBAS, 6 mi/10 km N, and under jurisdiction, of LYSYCHANS'K; 48°58′N 38°21′E. Elevation 600 ft/182 m. Coal mines; brewery; cement factory. Established in 1935; city since 1963, when its name was altered from Novo-Druzhes'ke (Russian *Novo-Druzheskoye*).

Novo-Druzhes'ke, UKRAINE: see NOVODRUZHES'K.

Novo-Druzheskoye, UKRAINE: see NOVODRUZHES'K.

Novodugino (nuh-vuh-DOO-gee-nuh), town (2006 population 3,860), NE SMOLENSK oblast, W European Russia, on road and railroad, 13 mi/21 km S of SYCHËVKA; 55°37′N 34°17′E. Elevation 754 ft/229 m. Agricultural products, lumbering, agricultural construction.

Novodvinsk (nuh-vuh-DVEENSK), city (2006 population 41,545), N ARCHANGEL oblast, N European Russia, on the Northern DVINA River, on highway and railroad, 15 mi/24 km SE of ARCHANGEL, and 6 mi/10 km E of ISAKOGORKA; 64°26′N 40°50′E. Building materials, plywood, logging and lumbering, film, furs; geological survey enterprise. Formerly known, at different times, as Mechkastroy, Pervomayskiy, and Voroshilovskiy.

Novodvinsk (nuh-vuh-DVEENSK), city (2005 population 43,100), N ARCHANGEL oblast, N European Russia, on the N DVINA River, on road and railroad spur (Isakogorka station), 6 mi/10 km S of ARCHANGEL; 64°26′N 40°50′E. Wood-pulp combine and paper mill; plywood factory. Established as a settlement of Pervomayskiy; city status and renamed Voroshilovskiy in 1977; current name since 1989.

Novodzhereliyevskaya (no-vuh-je-rye-LEE-eefs-kah-yah), village (2005 population 5,390), central KRASNODAR TERRITORY, S European Russia, on road and near railroad, 48 mi/77 km NNW of KRASNODAR, and 13 mi/21 km NW of TIMASHEVSK; 45°46′N 38°40′E. In natural gas-producing area.

Novo-Ekonomicheskiy Rudnik, UKRAINE: see NOVOEKONOMICHNE.

Novoekonomicheskoye, UKRAINE: see NOVOEKONOMICHNE.

Novo-Ekonomicheskoye, UKRAINE: see NOVOEKONOMICHNE.

Novoekonomichne (no-vo-e-ko-no-MEECH-ne) (Russian *Novoekonomicheskoye*), town, W DONETS'K oblast, UKRAINE, in the Donbas, on the Kazennyy Torets' River, 8 mi/12 km ENE of KRASNOARMIYS'K and 4 mi/6 km ENE of DYMYTROV; 48°18′N 37°15′E. Elevation 646 ft/196 m. Abandoned coal mines. Residents work in surrounding coal mines. Formerly spelled, in Russian, *Novo-Ekonomicheskoye*; formerly called Novoekonomichnyy Rudnyk (Russian *Novo-Ekonomicheskiy Rudnik*), and before that, Karakove (Russian *Karakovo*). Established at the end of the 18th century; town since 1956.

Novoekonomichne, UKRAINE: see DYMYTROV.

Novoekonomichnyy Rudnyk, UKRAINE: see NOVOEKONOMICHNE.

Novogagatli (no-vuh-gah-gah-TLEE), village (2005 population 4,730), NW DAGESTAN REPUBLIC, SE European Russia, on highway, 14 mi/23 km NNW of KHASAVYURT; 43°27′N 46°29′E. In oil-producing area.

Novogeorgievsk, POLAND: see MODLIN.

Novo-Georgiyevsk, UKRAINE: see SVITLOVODS′K.

Novo-Glukhov, UKRAINE: see KREMINNA, Luhans′k oblast.

Novogordino, RUSSIA: see KRASNOYE EKHO.

Novograd-Volynskiy, UKRAINE: see NOVOHRAD-VO-LYNS′KYY.

Novogremyachenskoye (nuh-vuh-gree-MYAH-chin-skuh-ye), village, NW VORONEZH oblast, S central European Russia, at a confluence of the DON and VORONEZH rivers, on highway, 11 mi/18 km E of (and administratively subordinate to) KHOKHOL′SKIY, and 8 mi/13 km SSW of VORONEZH; 51°31′N 39°03′E. Elevation 278 ft/84 m. Winery.

Novogrod, POLAND: see NOWOGROD.

Novogrodovka, UKRAINE: see NOVOHRODIVKA.

Novogroznenskiy, RUSSIA: see OYSKHARA.

Novogrudok (no-vo-GROO-dok), Polish *Nowogródek*, town, GRODNO oblast, BELARUS; ⊙ NOVOGRUDOK. Has ruins of thirteenth-century castle, several old churches, and house museum of Polish poet Adam Mickiewicz (b. here 1798). Founded 1116; belonged to LITHUANIA in mid-thirteenth century. In c.1795 into RUSSIA; in 1920 became part of POLAND; 1939 reunited with Belarus. Industry plants (gas apparatus, metal articles; butter, dried vegetables; bread-baking combine, clothing factory).

Novogrudok Upland (no-vo-GROO-dok), in GRODNO oblast, BELARUS, between upper NEMAN River and its tributaries, Shshara and Servech rivers. Elev. 1,060 ft/ 323 m. Has been extensively plowed; some forests have been preserved.

Novogurovskiy (nuh-vuh-GOO-ruhf-skeeye), settlement (2006 population 3,665), N TULA oblast, W central European Russia, on local railroad spur, 11 mi/ 18 km E of ALEKSIN, to which it is administratively subordinate; 54°28′N 37°20′E. Elevation 679 ft/206 m. Concrete manufacturing.

Novoguyvinskoye, UKRAINE: see NOVOHUYVYNS′KE.

Novo Hamburgo (NO-vah AHM-boor-go), city (2007 population 253,067), E RIO GRANDE DO SUL, BRAZIL, on railroad, and 25 mi/40 km N of PÔRTO ALEGRE; 29°41′S 51°08′W. Livestock slaughtering; tanning; manufacturing (musical instruments); agricultural trade. Settled by Germans in 19th century.

Novoheorhiyivs′ke, UKRAINE: see SVITLOVODS′K.

Novohlukhiv, UKRAINE: see KREMINNA, Luhans′k oblast.

Novo Horizonte (NO-vo O-ree-son-zhee), city (2007 population 10,535), BAHIA, BRAZIL.

Novo Horizonte (NO-vah O-ree-son-che), city (2007 population 34,264), N central SÃO PAULO, BRAZIL, near TIETÊ River and Promissão Reservoir, 37 mi/60 km ENE of LINS; 21°28′S 49°13′W. Coffee, rice, and cotton processing; pottery manufacturing; cattle raising.

Novo Horizonte do Norte (NO-vo O-ree-son-chee do NOR-chee), town (2007 population 3,815), W central MATO GROSSO, BRAZIL, 12 mi/19 km NE of PORTO DOS GAUCHOS; 11°28′S 57°20′W.

Novohrad-Volyns′kyy (no-vo-HRAHD-vo-LIN-skee) (Russian *Novograd-Volynskiy*), city (2001 population 56,259), W ZHYTOMYR oblast, UKRAINE, on SLUCH RIVER, and 50 mi/80 km WNW of ZHYTOMYR; 50°36′N 27°37′E. Elevation 652 ft/198 m. Railroad and highway junction; raion center; building and machine repair shops; stone crushing; building materials; asphalt making; food processing (flour, cheese, meat); non-alcoholic beverages; brewery; sewing; furniture manufacturing. Machine-construction technical school, vocational-technical school. Known since 1257 as Vozvyahel′ in the Kiev principality; passed to Lithuania in the mid-14th century; then renamed Zv′yahel′ in the 15th century; passed to Poland in 1569;

to Russia in 1793; renamed Novohrad-Volyns′kyy (Russian *Novograd-Volynsk*) in 1795; Russian name modified to conform more closely to Ukrainian usage in 1928. Native town of the Ukrainian poetess, Lesya Ukrayinka, in whose honor was established a memorial museum. Remains of K. Ostroz′kyy′s castle. Jewish community since 1488, population 40% Jewish until World War II (6,500 in 1939); eliminated by the Nazis between 1941 and 1943—fewer than 1,000 Jews remaining in 2005.

Novohrodivka (no-vo-HRO-dif-kah) (Russian *Novogrodovka*), city (2004 population 20,800), W central DONETS′K oblast, UKRAINE, in the DONBAS, on railroad, 9 mi/14 km SE of KRASNOARMIYS′K, and 4 mi/6 km NNE of SELYDOVE; 48°12′N 37°21′E. Elevation 711 ft/216 m. Subordinated to Selydove city council. Four bituminous-coal mines, enrichment plant, two professional-technical schools. Established in 1939, city since 1958.

Novohryhorivka, UKRAINE: see VERKHN′ODNIPROVS′K.

Novohuyvyns′ke (no-vo-HOOY-vin-ske) (Russian *Novoguyvinskoye)*, town, S ZHYTOMYR oblast, UKRAINE, on right bank of the TETERIV RIVER at the mouth of the Huyva River, and near highway, 4 mi/6 km S of ZHYTOMYR city center; 50°11′N 28°40′E. Elevation 679 ft/206 m. Repair depot. Established in 1932, town since 1973.

Novoil′insk (no-vuh-eel-YEENSK), village, E central BURYAT REPUBLIC, S SIBERIA, RUSSIA, on the Khudan River (left tributary of the UDA RIVER), on highway and spur of the TRANS-SIBERIAN RAILROAD, 47 mi/76 km E of ULAN-UDE; 51°41′N 108°40′E. Elevation 2,211 ft/673 m. Logging, lumbering, timbering.

Novoil′inskiy (nuh-vuh-eel-YEEN-skeeye), town (2006 population 3,630), S central PERM oblast, W URALS, E European Russia, on the right bank of the KAMA River, near railroad, 5 mi/8 km SE of NYTVA; 57°54′N 55°28′E. Elevation 321 ft/97 m. Petroleum deposits; pre-fabricated houses.

Novoizborsk (no-vuh-ess-BORSK), Estonian *Uus-Ir-boska*, town, W PSKOV oblast, W European Russia, on road, 20 mi/32 km WSW of PSKOV; 57°46′N 27°58′E. Elevation 187 ft/56 m. Gypsum-quarrying center. Castle ruins. On main highway between SAINT PETERSBURG and Rīga. Gypsum works are 6 mi/10 km NE of the town. In Russian Pskov gubernia until 1920; in Estonia until 1945.

Novokaolinovyy (no-vuh-kah-uh-LEE-nuh-viyee), settlement (2004 population 900), S central CHELYABINSK oblast, SW Siberian Russia, 51 mi/82 km SSW of CHESMA; 53°05′N 59°58′E. Elevation 1,427 ft/434 m. Calcite deposites in the vicinity.

Novokashirsk (no-vuh-kah-SHIRSK), town, S MOSCOW oblast, central European Russia, on the Oka River, on road and railroad spur, 35 mi/56 km SSE of MOSCOW, and 2 mi/3.2 km E of KASHIRA, to which it is administratively subordinate; 54°51′N 38°15′E. Elevation 570 ft/173 m. Site of the Kashira power plant. Formerly known as Ternovsk, then, from the mid-1930s to 1958, Kaganovich.

Novokastornoye (no-vuh-kahs-TOR-nuh-ye), village (2005 population 2,435), NE KURSK oblast, SW European Russia, on the Olym River (tributary of the SOSNA RIVER), on road and railroad, 4 mi/6 km S, and under administrative jurisdiction, of KASTORNOYE; 51°47′N 38°08′E. In agricultural area; produce processing.

Novokayakent (no-vuh-kah-yah-KYENT), village (2005 population 6,135), E DAGESTAN REPUBLIC, on the W coast of the CASPIAN SEA, on coastal railroad and highway, 38 mi/61 km SSE of MAKHACHKALA; 42°23′N 47°59′E. Below sea level. In oil- and gas-producing region; port facilities; fisheries.

Novokharitonovo (no-vuh-hah-ree-TO-nuh-vuh), town, E central MOSCOW oblast, central European Russia, on road and railroad, 10 mi/16 km ENE of (and administratively subordinate to) RAMENSKOYE;

55°35′N 38°31′E. Elevation 482 ft/146 m. Industrial ceramic products.

Novokhopërsk (nuh-vuh-huh-PYORSK), city (2006 population 7,475), E VORONEZH oblast, S central European Russia, on the KHOPER River, on highway junction and near railroad, 167 mi/270 km SE of VORONEZH, and 25 mi/40 km SW of BORISOGLEBSK; 51°06′N 41°37′E. Elevation 393 ft/119 m. Mechanical-repair shops; stone quarry; food and wood industries. Beaver reserve nearby. Novokhopërsk railroad station, with adjoining Novokhopërskiy town, lies 3 mi/5 km SW. City dates from 1716; chartered in 1762; site of shipyards under Catherine the Great.

Novokhopërskiy (nuh-vuh-huh-PYOR-skeeye), settlement (2006 population 7,635), E VORONEZH oblast, S central European Russia, on the KHOPER River, on highway and railroad, 3 mi/5 km W of NOVOKHOPÈRSK; 51°04′N 41°34′E. Elevation 380 ft/115 m. In agricultural area; sunflowerseed oil pressing.

Novokiyevsk, RUSSIA: see KRASKINO, MARITIME Territory.

Novokiyevskiy Uval (no-vuh-KEE-yeef-skeeye-oo-VAHL), village (2005 population 4,500), S AMUR oblast, SE SIBERIA, RUSSIAN FAR EAST, near the confluence of the ZEYA and SELEMDZHA rivers, on road, 35 mi/56 km NE of SVOBODNYY; 51°41′N 128°56′E. Elevation 551 ft/167 m. In agricultural area (wheat; livestock).

Novokizhinginsk (no-vuh-kee-ZHIN-geensk), village (2005 population 2,200), E BURYAT REPUBLIC, S SIBERIA, RUSSIA, on highway, 33 mi/53 km km NE of TARBAGATAY; 51°33′N 109°37′E. Elevation 2,939 ft/895 m. Has a church, Buddhist temple.

Novokorsunskaya (no-vuh-KOR-soons-kah-yah), village (2005 population 5,155), central KRASNODAR TERRITORY, S European Russia, on highway, 36 mi/58 km N of KRASNODAR; 45°38′N 39°09′E. In agricultural area; produce processing.

Novokruchininskiy (no-vuh-kroo-CHEE-neen-skeeye), town (2005 population 9,250), central CHITA oblast, S SIBERIA, RUSSIA, on the INGODA RIVER (AMUR River basin) near the influx of the Kruchina River, on road and the TRANS-SIBERIAN RAILROAD (Novaya station), 16 mi/25 km SE of CHITA; 51°47′N 113°46′E. Elevation 2,119 ft/645 m. Logging; structural components.

Novokubansk (no-vuh-koo-BAHNSK), city (2005 population 36,760), E KRASNODAR TERRITORY, S European Russia, on the KUBAN′ River, on highway branch and railroad, 116 mi/187 km E of KRASNODAR, and 7 mi/11 km NW of ARMAVIR; 45°06′N 41°03′E. Elevation 498 ft/151 m. Center of agricultural area; sugar factory; distillery; construction materials; food products. Made city in 1966, when absorbed a nearby town of Khutorok. Formerly called Novo-Kubanskaya and Novokubanskiy.

Novo-Kubanskaya, RUSSIA: see NOVOKUBANSK.

Novokubanskiy, RUSSIA: see NOVOKUBANSK.

Novokurovka (nuh-vuh-KOO-ruhf-kah), village, S KHABAROVSK TERRITORY, SE RUSSIAN FAR EAST, on the KUR RIVER, 40 mi/64 km NW of KHABAROVSK; 48°51′N 134°20′E. Elevation 209 ft/63 m. In lumbering area.

Novokuybyshevsk (nuh-vuh-KOO-yee-bi-shifsk), city (2006 population 112,260), central SAMARA oblast, E European Russia, in the VOLGA RIVER valley, on railroad, 12 mi/19 km SW of SAMARA; 53°06′N 49°55′E. Elevation 305 ft/92 m. Petroleum refining, petrochemicals, concrete structures, metal industries; garment factory, knitted goods factory, food processing. Made city in 1952.

Novokuznetsk (no-vuh-kooz-NYETSK), city (2005 population 536,255), W KEMEROVO oblast, S central SIBERIA, RUSSIA, on the TOM′ River (head of navigation), 190 mi/306 km S of KEMEROVO; 53°45′N 87°06′E. Elevation 820 ft/249 m. Highway and railroad junction. Industrial center of the KUZNETSK BASIN,

---

Cross-references are shown in SMALL CAPITALS. The pronunciation guide is shown on page xix. The sources of population figures are shown on page xvii.

the most important area for heavy industry in Siberia. Steel, mining equipment, agricultural machinery, chemicals, and aluminum. Coal mining and processing. Founded as the town of Kuznetsk by the Cossacks in 1617 and was a trading center until the 20th century. Developed from 1929 as iron and steel center. Called Kuznetsk-Sibirskiy until 1931; Stalinsk, 1932–1961. Made city in 1931. Experienced considerable economic turmoil in the post-Soviet switch to free market, especially within the mining industry, resulting in closing of a number of enterprises and subsequent emigration of workers to other locales.

Novolabinskaya (no-vuh-LAH-been-skah-yah), village (2005 population 3,750), E KRASNODAR TERRITORY, S European Russia, on the LABA RIVER where it serves as an administrative border with ADYGEY REPUBLIC, on road, 31 mi/50 km ENE of Krasnodar; 45°06′N 39°54′E. Elevation 314 ft/95 m. In agricultural area (wheat, flax, fruits, castor beans; livestock). Has a large Balkar minority population.

Novolakskoye (no-vuh-LAHK-skuh-ye), village (2005 population 4,570), W central DAGESTAN REPUBLIC, in the NE foothills of the Greater CAUCASUS Mountains, SE European Russia, just E of the administrative border with CHECHEN REPUBLIC, on road, 8 mi/13 km SSW of KHASAVYURT; 43°06′N 46°28′E. Agriculture (grain, livestock). Until 1944 (in Chechen-Ingush ASSR), called Kishen'-Aukh. Site of intense battles between Chechen fighters and Russian army units at the onset of the second Russian-Chechen conflict in August–September 1999.

Novolazarevskaya Station (no-vo-LAH-zuh-REV-skah-yuh), ANTARCTICA, Russian scientific station on PRINCESS ASTRID COAST; 70°46′S 11°50′E. Opened February 18, 1961.

Novoleushkovskaya (no-vuh-lye-oosh-KOFS-kah-yah), village (2005 population 6,885), N central KRASNODAR TERRITORY, S European Russia, on road and near railroad, 68 mi/109 km NNE of KRASNODAR, and 12 mi/19 km NNW of TIKHORETSK; 46°00′N 40°02′E. Elevation 219 ft/66 m. In agricultural area (wheat, rye, corn, castor bean; livestock); produce processing.

Novoli (NO-vo-lee), town, LECCE province, APULIA, S ITALY, 7 mi/11 km WNW of LECCE; 40°23′N 18°03′E. Wine-making center. 15th-century castle.

Novo Lino (NO-vo LEEN-o), city (2007 population 11,903), NE ALAGOAS state, BRAZIL, near PERNAMBUCO border, on highway to RECIFE; 08°55′S 35°40′W.

Novo-Lisino, RUSSIA: see FORNOSOVO.

Novolugovoye (no-vuh-loo-guh-VO-ye), village (2006 population 3,295), E NOVOSIBIRSK oblast, SW SIBERIA, RUSSIA, in the OB' RIVER basin, on road and railroad, 7 mi/11 km SE of (and administratively subordinate to) NOVOSIBIRSK; 54°58′N 83°07′E. Elevation 383 ft/116 m. Agricultural products.

Novolukoml (no-vo-loo-KOML), town, VITEBSK oblast, BELARUS. Regional power plant; bread baking; milk; pre-fabricated-housing plants.

Novol'vovsk (nuh-vuh-LVOFSK), settlement (2006 population 1,955), E TULA oblast, W central European Russia, just W of the administrative border with RYAZAN oblast, on local highway branch and railroad, 10 mi/16 km SE of KIMOVSK; 53°55′N 38°47′E. Elevation 662 ft/201 m. In agricultural and lumbering area.

Novo-Lyubino, RUSSIA: see LYUBINSK.

Novomaklakovo, RUSSIA: see LESOSIBIRSK.

Novomalorossiyskaya (no-vuh-mah-luh-ruh-SEEYEE-skah-yah), village (2005 population 5,895), E KRASNODAR TERRITORY, S European Russia, on road and 5 mi/8 km S of railroad (Beysug station), 47 mi/76 km NE of KRASNODAR; 45°38′N 39°54′E. Elevation 193 ft/58 m. Agricultural products.

Novomalykla, RUSSIA: see NOVAYA MALYKLA.

Novo-Mariinsk, RUSSIA: see ANADYR'.

Novo Mesto (NO-vo MES-to), town (2002 population 22,368), SE SLOVENIA, on KRKA RIVER, on railroad, and 36 mi/58 km SE of LJUBLJANA; 45°47′N 15°09′E.

Terminus of railroad spur to DOLENJSKE; motor vehicles, pharmaceuticals, textiles, and electronics manufacturing. Hydroelectric plant. Founded 1365 by Duke Rudolf of Austria.

Novomichurinsk (no-vuh-mee-CHOO-reensk), city (2006 population 20,860), W central RYAZAN oblast, central European Russia, on the Pronya River (OKA River basin), on railroad and near highway, 50 mi/80 km S of RYAZAN; 54°02′N 39°47′E. Elevation 374 ft/113 m. Food industries (fish processing, butter making, bakery). Regional electric power station. Made city in 1981.

Novomikhaylovskiy (no-vuh-mee-HEI-luhf-skeeyee), town (2005 population 10,420), SE KRASNODAR TERRITORY, S European Russia, on the BLACK SEA coast, in the W foothills of the NW Greater CAUCASUS Mountains, on the Tuapse-Novorossiysk highway, 48 mi/77 km S of KRASNODAR, and 25 mi/41 km NW of TUAPSE; 44°15′N 38°51′E. Elevation 383 ft/116 m. Health resort.

Novomikhaylovskiy (no-vuh-mee-HEI-luhf-skeeyee), settlement (2005 population 930), W BASHKORTOSTAN Republic, E European Russia, on road, 14 mi/23 km S of BELEBEY; 53°54′N 54°08′E. Elevation 1,085 ft/330 m. In oil-producing area.

Novo Minsk, POLAND: see MINSK MAZOWIECKI.

Novominskaya (no-vuh-MEENS-kah-yah), village (2005 population 12,105), NW KRASNODAR TERRITORY, S European Russia, on the Albasi River, on road and railroad, 40 mi/64 km SE of YEYSK; 46°19′N 38°57′E. In agricultural area (wheat, sunflowers); flour mill.

Novomirgorod, UKRAINE: see NOVOMYRHOROD.

Novo-Mirgorod, UKRAINE: see NOVOMYRHOROD.

Novomoskovsk (nuh-vuh-muhs-KOFSK), city (2006 population 129,990), E TULA oblast, W central European Russia, at the source of the DON River and SHAT' River, on highway junction and railroad, 40 mi/64 km ESE of TULA; 54°05′N 38°13′E. Elevation 685 ft/208 m. An industrial center in the Moscow lignite basin; railroad depots, gypsum quarrying, manufacturing (chemicals, concrete products, fireproof materials, ceramics, electrical instruments, machinery), agricultural machinery repair, food processing (flour mill, bakery); power station. Coal mining, once the chief industry, has lost much of its importance in the post-Soviet market economy. Founded in 1929 as Bobriki with the construction of a chemical plant; renamed Stalinogorsk in 1934; current name since 1961.

Novomoskovs'k (no-vo-mos-KOVSK) (Russian *Novomoskovsk*), city (2001 population 72,439), central DNIPROPETROVS'K oblast, UKRAINE, 12 mi/19 km NE of DNIPROPETROVS'K, on the SAMARA RIVER; 48°37′N 35°12′E. Elevation 170 ft/51 m. Raion center; steel-pipe manufacturing, sleeper-tie (beam) engineering, food processing, sewing, furniture making. Trade, veterinary schools. Established in the 18th century as Novoselytsya sloboda on the site of a Cossack settlement, Samarchyk; city since 1784, when it was renamed Novomoskovs'k, after the Moscow regiment stationed here. *Troyitskyy sobor* (Trinity cathedral) is an outstanding example of Ukrainian wooden architecture, built 1773–1778 by Yakov Pohrebnyak.

Novomoskovsk, UKRAINE: see NOVOMOSKOVS'K.

Novomoskovsk, UKRAINE: see YELANETS'.

Novomoskovs'k, UKRAINE: see YELANETS'.

Novomykolayivka (no-vo-mi-ko-LAH-yif-kah) (Russian *Novonikolayevka*), town, N ZAPORIZHZHYA oblast, UKRAINE, on the Verkhnya Tersa River, 35 mi/56 km ENE of ZAPORIZHZHYA; 47°58′N 35°54′E. Elevation 351 ft/106 m. Raion center; food processing (including food mill, dairy, broiler factory). Heritage museum. Established in the late 18th century as Kocherezhky until 1813; town since 1957. Also spelled, in Russian, *Novo-Nikolayevka*.

Novomykolayivka (no-wo-mi-ko-LAH-yif-kah), (Russian *Novonikolayevka*), town (2004 population 6,700), central DNIPROPETROVS'K oblast, UKRAINE, on DNIEPER UPLAND near DNIPRODZERZHYNS'K RESERVOIR, on highway and railroad, 7 mi/11 km SSE of VERKHN'ODNIPROVS'K; 48°34′N 34°22′E. Elevation 567 ft/172 m. Poultry-breeding state farm. Established in 1917, town since 1966.

Novo-Mykolayivka, UKRAINE: see CHKALOVE.

Novomykolayivka, UKRAINE: see NOVOAZOVS'K.

Novomyrhorod (no-vo-MIR-ho-rod) (Russian *Novomirgorod*), city, NW KIROVOHRAD oblast, UKRAINE, on Velyka Vys' River, 34 mi/55 km NW of KIROVOHRAD; 48°47′N 31°39′E. Elevation 469 ft/142 m. Raion center; lignite mine, construction-materials manufacturing, leather curing, feed milling, preserve canning, furniture making. Veterinary school; historical museum. Established in 1740 by the Zaporozhian Cossacks of the Myrohorod regiment; center of New Serbia (1752–1764); city since 1960. Previously spelled, in Russian, *Novo-Mirgorod*.

Novomyshastovskaya (no-vuh-mi-SHAHS-tuhf-skah-yah), village (2005 population 9,965), S central KRASNODAR TERRITORY, S European Russia, in the KUBAN' River basin, on road, 26 mi/42 km WNW of KRASNODAR; 45°11′N 38°53′E. In agricultural area (wheat, corn, sunflowers, fruit orchards; livestock).

Novo-Nazyvayevka, RUSSIA: see NAZYVAYEVSK.

Novo-Nikolayevka, UKRAINE: see CHKALOVE.

Novo-Nikolayevka, UKRAINE: see NOVOAZOVS'K.

Novonikolayevka, UKRAINE: see NOVOAZOVS'K.

Novo-Nikolayevka, UKRAINE: see NOVOMYKOLAYIVKA, Zaporizhzhya oblast.

Novonikolayevsk, RUSSIA: see NOVOSIBIRSK, city.

Novonikolayevskaya (no-vuh-nee-kuh-LAH-eefs-kah-yah), village (2005 population 3,770), central KRASNODAR TERRITORY, S European Russia, on the Protoka River (N KUBAN' River delta), on road, 47 mi/76 km NW of KRASNODAR; 45°35′N 38°22′E. In agricultural area (wheat, corn, castor beans, fruits, grapes; livestock).

Novonikolayevskiy (no-vuh-nee-kah-LAH-eef-skeeyee), town (2006 population 10,260), NW VOLGOGRAD oblast, SE European Russia, on highway junction, 30 mi/48 km SSE of BORISOGLEBSK; 50°58′N 42°22′E. Elevation 485 ft/147 m. Railroad junction (Aleksikovo station; branch SW to URYUPINSK); auto repair; dairy products; poultry packing, flour milling.

Novonikolayevskiy, RUSSIA: see NOVOSIBIRSK.

Novonikol'sk (no-vuh-nee-KOLSK), settlement (2006 population 4,035), SW MARITIME TERRITORY, SE RUSSIAN FAR EAST, on the Suifen River (tributary of the USSURI River), on road, 9 mi/14 km NW of USSURIYSK; 43°51′N 131°52′E. Elevation 101 ft/30 m. In agricultural area (rice, soybeans, vegetables; livestock).

Novonukutskiy (no-vuh-noo-KOOT-skeeyee), industrial settlement (2005 population 3,115), NW UST-ORDYN-BURYAT AUTONOMOUS OKRUG, in IRKUTSK oblast, E central Siberian Russia, in the ANGARA river basin, on short spur of the TRANS-SIBERIAN RAILROAD, 13 mi/21 km NE of ZALARI; 53°42′N 102°42′E. Elevation 1,384 ft/421 m. Coal and anhydrite mining; sawmilling and woodworking. Has a riverside sanatorium. Formerly known as Takhtal-Ongoy (takh-TAHL–uhn-GO-yee).

Novooleksandrivka (no-vo-o-lek-SAHN-dreef-kah), (Russian *Novoaleksandrovka*), town (2004 population 2,700), SE LUHANS'K oblast, UKRAINE, in the DONBAS, on the Velyka Kam'yanka River, 2 mi/3 km W of KRASNODON; 49°22′N 37°55′E. Elevation 557 ft/169 m. Former coal mines. Established in the 17th century; town since 1938. Formerly spelled, in Russian, *Novo-Aleksandrovka*.

Novooleksandrivka, UKRAINE: see MELITOPOL'.

Novooleksiyivka (no-vo-o-lek-SEE-yif-kah) (Russian *Novoalekseyevka*), town, SE KHERSON oblast, UKRAINE,

55 mi/89 km SW of MELITOPOL', and 10 mi/16 km WNW of HENICHES'K; 46°13′N 34°38′E. Railroad junction (spur to Heniches'k) and crossroads. Fruit and vegetable canning; bakery; grape processing. Corn-research station. Established in the 1890s during railroad construction; town since 1938. Formerly (Russian *Novo-Alekseyevka*).

Novoomskiy (nuh-vuh-OM-skeeye), settlement (2006 population 4,450), S OMSK oblast, SW SIBERIA, RUSSIA, on the IRTYSH River, on highway and local railroad branch, 10 mi/16 km SSW of OMSK; 54°50′N 73°18′E. Elevation 282 ft/85 m. In agricultural area; food processing.

Novo Oriente (NO-voh O-ree-en-chee), city (2007 population 27,497), E central CEARÁ, BRAZIL, in Serra GRANDE near border with PIAUÍ, 27 mi/43 km SW of CRATÉUS; 05°30′S 40°50′W.

Novoorlovsk (no-vuh-uhr-LOFSK), town (2005 population 2,800), central AGIN-BURYAT AUTONOMOUS OKRUG, S central CHITA oblast, approximately 22 mi/35 km SE of AGINSKOYE; 51°05′N 114°40′E. Wolfram ore processing and enrichment plant.

Novoorsk (nuh-vuh-ORSK), town (2006 population 11,195), E ORENBURG oblast, in the S URALS, extreme SE European Russia, on the Kumak River (left tributary of the URAL River), on crossroads and railroad, 25 mi/40 km NE of ORSK; 51°22′N 58°59′E. Elevation 672 ft/204 m. Non-ferrous-ore combine.

Novoorzhitskoye, UKRAINE: see NOVOORZHYTS'KE.

Novoorzhyts'ke (no-vo-OR-zhits-ke) (Russian *Novoorzhitskoye*), town (2004 population 3,400), NW POLTAVA oblast, UKRAINE, 47 mi/76 km W of CHERKASY and 15 mi/24 km WNW of LUBNY; 50°05′N 32°39′E. Elevation 321 ft/97 m. Sugar refinery. Established in 1979 on the site of the village Lazirky (Russian *Lazorki*).

Novoozerne (no-vo-o-ZER-ne) (Russian *Novoozernoye*), town, W Republic of CRIMEA, UKRAINE, on the E shore of the DONUZLAV LAGOON, 17 mi/28 km NW of, and subordinated to, YEVPATORIYA. Youth camp nearby. Established in 1969; town since 1977.

Novoozernoye, UKRAINE: see NOVOOZERNE.

Novoozeryanka (no-vo-o-zer-YAHN-kah), town, NW ZHYTOMYR oblast, UKRAINE, in Polissya, on road 15 mi/24 km E of OLEVS'K, and 6 mi/10 km N of NOVI BILOKOROVYCHI. Reinforced-concrete-culvert fabrication. Established in 1963, town since 1972.

Novopashiyskiy, RUSSIA: see GORNOZAVODSK, PERM oblast.

Novopavlivka, UKRAINE: see NOVYY BUH.

Novopavlovka (no-vuh-PAHV-luhf-kah), town (2005 population 4,040), SW CHITA oblast, S SIBERIA, RUSSIA, on the KHILOK River, on road and the TRANS-SIBERIAN RAILROAD (near Tolbaga station), 20 mi/32 km E of PETROVSK-ZABAYKALSKIY, and 6 mi/10 km E of TARBAGATAY; 51°13′N 109°14′E. Elevation 2,391 ft/728 m. Coal mining, lumbering; furniture.

Novopavlovka, UKRAINE: see NOVYY BUH.

Novopavlovsk (no-vuh-PAHV-luhfsk), city (2006 population 24,820), S STAVROPOL TERRITORY, N CAUCASUS, S European Russia, on the Kura River, on railroad (Apollonskaya station), 165 mi/268 km SE of STAVROPOL, and 15 mi/24 km SE of GEORGIYEVSK; 43°57′N 43°38′E. Elevation 1,082 ft/329 m. Center of agricultural area; food industries. A village of Novopavlovskaya until 1981, when granted city status.

Novopetrovskoye (no-vuh-pye-TROF-skuh-ye), village (2006 population 5,170), W central MOSCOW oblast, central European Russia, on road and railroad (Ustinovka station), 15 mi/24 km WNW of ISTRA; 55°59′N 36°28′E. Elevation 875 ft/266 m. Garments; poultry.

Novopistsovo (no-vuh-pees-TSO-vuh), town (2005 population 3,030), NE IVANOVO oblast, central European Russia, near highway, 9 mi/14 km NNW of VICHUGA; 57°19′N 41°50′E. Elevation 387 ft/117 m. Flour milling.

Novoplatnirovskaya (no-vuh-PLAHT-nee-ruhfs-kah-yah), village (2005 population 4,005), N central KRASNODAR TERRITORY, S European Russia, on the Chelbas River, on road, 64 mi/103 km NNE of KRASNODAR, and 5 mi/8 km E of KRYLOVSKAYA; 46°06′N 39°24′E. Elevation 108 ft/32 m. In agricultural area (wheat, corn, flax, sunflowers, vegetables). Founded by the Kuban' Cossacks.

Novopodrezkovo (no-vuh-puhd-RYES-kuh-vuh), town (2006 population 10,245), central MOSCOW oblast, central European Russia, on road and railroad, 5 mi/8 km NW, and under administrative jurisdiction, of KHIMKI; 55°56′N 37°20′E. Elevation 626 ft/190 m. Automotive repair.

Novopokrovka (no-vo-pok-ROV-kah) (Russian *Novo-Pokrovka*), town (2004 population 2,700), central DNIPROPETROVS'K oblast, UKRAINE, on Komyshuvata Sura River, and 33 mi/53 km SSW of DNIPROPETROVS'K; 48°03′N 34°37′E. Elevation 275 ft/83 m. Vegetables.

Novopokrovka (no-vo-po-KROV-kah) (Russian *Novo-Pokrovka*), town (2004 population 4,500), N central KHARKIV oblast, UKRAINE, on railroad, and 6 mi/10 km W of CHUHUYIV, in KHARKIV metropolitan area; 49°50′N 36°33′E. Elevation 334 ft/101 m. Flour milling, parts manufacturing, equipment repair; poultry breeding. Established in 1880; town since 1938.

Novopokrovka (NO-vuh-po-KROF-kah), village, N SEMEY region, KAZAKHSTAN, on TURK-SIB RAILROAD, 15 mi/24 km NE of SEMEY (Semipalatinsk); 50°40′N 80°29′E. Livestock breeding.

Novopokrovka (no-vuh-puh-KROF-kah), village (2006 population 3,915), W MARITIME TERRITORY, SE RUSSIAN FAR EAST, on the IMAN' RIVER, on road junction, 36 mi/58 km E of DAL'NERECHENSK; 45°51′N 134°30′E. Elevation 278 ft/84 m. In agricultural area (grain, soybeans).

Novopokrovka (nuh-vuh-puh-KROF-kah), village, E central ORENBURG oblast, S URALS, SE European Russia, on road, 37 mi/60 km NW of ORSK; 51°31′N 57°52′E. Elevation 1,473 ft/448 m. In Orsk-Khalilovo industrial district; flour milling. Phosphorite deposits (S).

Novopokrovka (no-vuh-puh-KROF-kah), village, N SARATOV oblast, SE European Russia, on the Tereshka River (right tributary of the VOLGA RIVER), on railroad, 29 mi/47 km SW of KHVALYNSK; 52°24′N 47°25′E. Elevation 269 ft/81 m. In agricultural area; produce processing.

Novopokrovka, RUSSIA: see LISKI.

Novo-Pokrovka, UKRAINE: see NOVOPOKROVKA, Dnipropetrovs'k oblast; or NOVOPOKROVKA, Kharkiv oblast.

Novopokrovskaya (no-vuh-puh-KROF-skah-yah), village (2005 population 19,435), E KRASNODAR TERRITORY, S European Russia, on the YEYA RIVER, on road and railroad, 27 mi/43 km ENE of TIKHORETSK; 45°57′N 40°42′E. Elevation 246 ft/74 m. In wheat- and sunflower-growing area.

Novopolotsk (no-vo-PO-lotsk), city, VITEBSK oblast, BELARUS, on the left bank of the Zapadnaia Dvina River; 55°32′N 28°39′E. Chemical combine, oil refinery, reinforced-concrete-components plant, motor vehicle, railroad, and pipeline enterprises. Heat and electric power plant.

Novopskov (no-vo-PSKOV), town, N LUHANS'K oblast, UKRAINE, on the Aydar River and highway, and 10 mi/16 km NNE of STAROBIL's'K; 49°32′N 39°07′E. Elevation 275 ft/83 m. Raion center; gas-transmission administrative office; food processing (including feed mills, butter creamery, flour mill, food-flavoring factory). Sanatorium; heritage museum. Established in the mid-17th century; town since 1957. In Russian, also hyphenated, *Novo-Pskov*.

Novo-Pskov, UKRAINE: see NOVOPSKOV.

Novoradomsk, POLAND: see RADOMSKO.

Novo Redondo, ANGOLA: see SUMBE.

Novorepnoye (nuh-vuh-RYEP-nuh-ye), village, E SARATOV oblast, SE European Russia, on the Bol'shoy Uzen' River, on road junction, 20 mi/32 km SSE of YERSHOV; 51°04′N 48°25′E. Elevation 131 ft/39 m. In wheat and cattle area.

Novorossiskoye (NO-vuh-ruhs-SEES-ko-ye), village, N AKTÖBE region, KAZAKHSTAN, 35 mi/56 km E of AKTÖBE; 50°20′N 58°30′E. Millet, wheat.

Novorossiysk (no-vuh-ruh-SEEYEESK), city (2005 population 245,340), S KRASNODAR TERRITORY, S European Russia, on the ice-free Novorororossiysk Bay on the NE coast of the BLACK SEA, on road and railroad, 84 mi/135 km SW of KRASNODAR; 44°43′N 37°46′E. The major Russian port on the Black Sea and a naval base, it has shipyards, and is a major center of the Russian cement industry. It also has machinery and metalworking, woodworking (furniture), and fish-processing industries. Has an oil terminal for pipelines from central Asia. The city stands on the site of a Genoese colony (13th–14th century) and of a Turkish fortress, captured by the Russians in 1808. The present city was founded in 1838, and the first cement factory was opened in 1882. Before 1914, it was one of the important grain-exporting cities of Russia.

Novorozhdestvenskaya (no-vuh-ruhzh-DYES-tveenskah-yah), village (2005 population 6,810), central KRASNODAR TERRITORY, S European Russia, on the Chelbas River, on road and near railroad, 59 mi/95 km NE of KRASNODAR, and 7 mi/11 km W of TIKHORETSK; 45°51′N 39°57′E. Elevation 154 ft/46 m. In agricultural area (wheat, flax, sunflowers, castor beans; livestock); produce processing.

Novorudnyy (nuh-vuh-ROOD-niyee), settlement (2006 population 1,925), E ORENBURG oblast, extreme SE European Russia, in the S URALS, on road and terminus of local railroad spur, 21 mi/34 km NW of ORSK; 51°30′N 58°11′E. Elevation 1,302 ft/396 m. Mining of non-ferrous ores.

Novoryazhsk (nuh-vuh-RYASHSK), former town, S central RYAZAN oblast, central European Russia, now a suburb of RYAZHSK, 2 mi/3.2 km E of the city center; 53°42′N 40°06′E. Elevation 462 ft/140 m. Railroad junction.

Novorzhev (nuh-vuh-RZHEF), city (2006 population 3,945), S PSKOV oblast, W European Russia, on road junction, 90 mi/144 km SE of PSKOV, and 45 mi/72 km SE of OSTROV; 57°02′N 29°20′E. Elevation 200 ft/60 m. Flax retting; dairy products. Dolomite quarries nearby. Chartered in 1777.

Novosadovoye, RUSSIA: see AVTURY.

Novo São Joaquim (NO-vo SOUN ZHWAH-keen), town (2007 population 6,880), MATO GROSSO, BRAZIL.

Novoselenginsk (no-vuh-see-lyen-GEENSK), village, SE BURYAT REPUBLIC, S SIBERIA, RUSSIA, on the SELENGA River (landing), just N of the CHIKOY RIVER mouth, 65 mi/105 km SW of ULAN-UDE; 51°06′N 106°37′E. Elevation 1,797 ft/547 m. Highway junction. Old Russian colonization center of Selenginsk founded nearby in 1666.

Novoseleznëvo (no-vuh-sye-leez-NYO-vuh), village (2006 population 3,370), SE TYUMEN oblast, SW Siberian Russia, on the ISHIM River, on road, 38 mi/61 km SSW of ISHIM; 55°41′N 69°12′E. Elevation 206 ft/62 m. In agricultural area (grain; livestock).

Novoselitsa, UKRAINE: see NOVOSELYTSYA.

Novoselitskoye (nuh-vuh-SYE-leets-kuh-ye), village (2006 population 8,780), central STAVROPOL TERRITORY, N CAUCASUS, S European Russia, on road, 25 mi/40 km S of BLAGODARNYY; 44°45′N 43°26′E. Elevation 597 ft/181 m. In agricultural area (wheat, sunflowers); flour mill.

Novoselitsy (nuh-vuh-SYE-lee-tsi), settlement, W central NOVGOROD oblast, NW European Russia, on the MSTA River, near highway, 14 mi/23 km E, and under administrative jurisdiction, of NOVGOROD; 58°31′N 31°42′E. Bread-baking plant.

Novoselivka (no-vo-SE-leef-kah) (Russian *Novoselovka*), town (2004 population 2,500), N DONETS'K oblast, UKRAINE, 8 mi/13 km NW of KRASNYY LYMAN; 49°04'N 37°41'E. Elevation 242 ft/73 m. Established in the 17th century as Khutir Zabolochans'kyy; renamed Novoselivka in 1808; town since 1938.

Novoselivka, UKRAINE: see FRUNZE.

Novoselivs'ke (no-vo-SE-leevs-ke) (Russian *Novoselovskoye*), town, W central Republic of CRIMEA, UKRAINE, 19 mi/31 km NE of YEVPATORIYA; 45°26'N 33°36'E. Elevation 331 ft/100 m. Winery; irrigation administration. Vocational school. A Jewish settlement from about 1925 to World War II. Until 1944, called Fraidorf. Town since 1965.

Novo Selo (NO-vo SE-lo), village, MONTANA oblast, Novo Selo obshtina (1993 population 5,050), NW BULGARIA, on the DANUBE, 11 mi/18 km NNW of VIDIN; 44°09'N 22°48'E. Grain; livestock; produce; vineyards; fisheries; wine making for export.

Novo Selo (NO-vo SE-lo), Hungarian *Révaújfalu*, village, VOJVODINA, NE SERBIA, 10 mi/16 km NE of PANCEVO, in the BANAT region. Also called Banatsko Novo Selo.

Novo-selo, Bulgaria: see DULGOPOL.

Novoselovka, UKRAINE: see FRUNZE.

Novoselovka, UKRAINE: see NOVOSELIVKA.

Novosëlovo (no-vuh-SYO-luh-vuh), village (2005 population 6,220), SW KRASNOYARSK TERRITORY, SE SIBERIA, RUSSIA, on the YENISEY RIVER, on highway, 90 mi/145 km SW of KRASNOYARSK; 55°00'N 90°57'E. Elevation 984 ft/299 m. In dairy-farming area. Town status granted in 1962, revoked in 1991.

Novoselovskoye, UKRAINE: see NOVOSELIVS'KE.

Novosel'skiy (no-vuh-SYEL-skeeyee), settlement, W central NOVGOROD oblast, NW European Russia, on the Porus'ya River (tributary of the LOVAT' RIVER), on highway, 9 mi/14 km S of (and administratively subordinate to) STARAYA RUSSA; 57°51'N 31°22'E. Automobile repair plant.

Novoselskoye, RUSSIA: see ACHKHOY-MARTAN.

Novosel'tsevo (nuh-vuh-SYEL-tsi-vuh), town, central MOSCOW oblast, central European Russia, on the MOSCOW CANAL, on local road, 8 mi/13 km NW of (and administratively subordinate to) MYTISHCHI; 56°00'N 37°35'E. Elevation 570 ft/173 m. Electric fixtures manufacturing.

Novoseltsi, Bulgaria: see ELIN PELIN.

Novosel'ye (nuh-vuh-SYEL-ye), settlement, N central PSKOV oblast, W European Russia, on road junction and railroad, 28 mi/45 km NE of PSKOV; 58°06'N 28°53'E. Elevation 485 ft/147 m. Flax processing.

Novoselytsya (no-vo-SE-li-tsyah) (Russian *Novoselitsa*) (Romanian *Nová Suliţa* or *Suliţa*), city (2004 population 10,800), S central CHERNIVTSI oblast, UKRAINE, in North Bukovina, 19 mi/31 km SE of CHERNIVTSI, on the left bank of the PRUT RIVER; 48°13'N 26°17'E. Elevation 469 ft/142 m. Raion center; food processing (cheese, cannery, distillery and liquer, meat, grain, food flavors), reinforced-concrete fabrication, furniture making; medical school. Known since 1456 as Shyshkivtsi; in 1617 it was renamed Novoselytsya, in Moldavia; divided border town between Austria (W) and Turkey (E) in 1774; in 1812, the Turkish part (E) of town was ceded, along with the Khotyn Company, to the Russian Empire; between world wars (1918–1940), became part of Romania; in 1940, became part of Ukrainian SSR; since 1991, in independent Ukraine.

Novosemeykino (nuh-vuh-see-MYAI-kee-nuh), urban settlement (2006 population 9,625), central SAMARA oblast, E European Russia, near railroad, 18 mi/29 km NNE of SAMARA, and 7 mi/11 km S of KRASNYY YAR, to which it is administratively subordinate; 53°23'N 50°22'E. Elevation 662 ft/201 m. Manufacturing (chemical catalysts); mayonnaise plant.

Novosergiyevka (nuh-vuh-SYER-gee-yef-kah), town (2006 population 13,170), W central ORENBURG oblast, SE European Russia, on the SAMARA RIVER, on crossroads and railroad, 30 mi/48 km SE of SOROCHINSK; 52°06'N 53°39'E. Elevation 406 ft/123 m. Mechanical shops; food industries.

Novoshakhtinsk (nuh-vuh-SHAHKH-teensk), city (2006 population 98,950), W ROSTOV oblast, S European Russia, near the border with UKRAINE, on highway, 65 mi/105 km NNE of ROSTOV-NA-DONU, and 13 mi/21 km W of SHAKHTY (linked by railroad spur); 47°45'N 39°56'E. Elevation 419 ft/127 m. Major anthracite-mining center in the E DONETS BASIN. Founded in 1863. As a mining settlement, called Imeni III Internatsionala (Russian=named after the Third International) in the 1920s and later (1929–1939), Komintern (short for *Kommunisticheskiy Internatsional*, Russian=Communist International), until made a city and renamed.

Novoshakhtinskiy (no-vuh-SHAHKH-teen-skeeyee), town (2006 population 8,025), SW MARITIME TERRITORY, SE RUSSIAN FAR EAST, on road and local spur of the TRANS-SIBERIAN RAILROAD, 9 mi/14 km NNE of (and administratively subordinate to) USSURIYSK; 43°56'N 132°00'E. Elevation 121 ft/36 m. Coal mining. Military airbase in the vicinity.

Novoshcherbinovskaya (no-vuh-shchyer-BEE-nuhf-skah-yah), village (2005 population 6,445), NW KRASNODAR TERRITORY, S European Russia, on the YASENI River, on crossroads, 17 mi/27 km SE of YEYSK, and 11 mi/18 km S of STAROSHCHERBINOVSKAYA; 46°29'N 38°38'E. In agricultural area (fruits, sunflowers, grapes; livestock); produce processing.

Novo-Shepelichi, UKRAINE: see NOVOSHEPELYCHI.

Novoshepelichi, UKRAINE: see NOVOSHEPELYCHI.

Novoshepelychi (no-vo-she-pe-LI-chee) (Russian *Novoshepelichi*), former village, N KIEV oblast, UKRAINE, on PRIPET (Ukrainian *Prypyat'*) River, 70 mi/113 km NNW of KIEV and 1 mi/2 km NW of PRYP'YAT'; 51°25'N 30°02'E. Elevation 367 ft/111 m. Abandoned after the accident at the Chornobyl' Nuclear Power Station in Prypyat'; no static population, but approximately 2,100 workers and scientists occupy the village on a daily basis. In Russian, also hyphenated, *Novo-Shepelichi*.

Novosheshminsk (nuh-vuh-shish-MEENSK), village (2006 population 4,590), central TATARSTAN Republic, E European Russia, on the SHESHMA RIVER, on road junction, 32 mi/51 km SE of CHISTOPOL'; 55°04'N 51°14'E. Elevation 318 ft/96 m. In wheat and livestock area; dry and skim milk plant.

Novosibirsk (nuh-vuh-see-BEERSK), oblast (☐ 68,988 sq mi/179,368.8 sq km; 2006 population 2,646,960) in S SIBERIA, in the geographic center of RUSSIA; ⊙ NOVOSIBIRSK. Drained by middle Ob' (E), Om' and Tara rivers. Lake CHANY is in W. Ob' Sea reservoir spans the region. Includes low, marshy BARABA STEPPE (W) and hilly forest steppe (E of the OB' RIVER), with continental climate. Novosibirsk is a key crosspoint of the main transportation routes and the TRANS-SIBERIAN RAILROAD, which serves the largest towns (Novosibirsk, BARABINSK, TATARSK), traversing oblast E to W. Except for highly industrialized Novosibirsk and its suburbs, economy is chiefly agricultural (half of land), with emphasis on dairy products, wheat, and livestock breeding. The region still, however, boasts the highest concentration of industry between the URAL Mountains and the Pacific Ocean; 80% concentrated around Novosibirsk. Metallurgy, tin smelting, food, and Siberian timber processing, and the manufacturing of electrodes, plastics and paint, chemicals, machines, construction materials, pharmaceuticals, and apparel. Population consists chiefly of Russians (94.4%), with some Germans (1.8%), Ukrainians (1.3%), Tatars (1.1%) and other smaller nationalities. Created in 1937 out of West Siberian Territory; until 1943–1944, it included areas now in TOMSK and KEMEROVO oblasts. SE of Novosibirsk is Akademgorodok (Academic City), a secluded settlement of scientific institutes. The oblast is subdivided into 30 districts and contains 14 cities and 19 towns.

Novosibirsk (nuh-vuh-see-BEERSK), city (2006 population 1,417,105), ⊙ NOVOSIBIRSK oblast, SW SIBERIA, RUSSIA, on the OB' RIVER and on the TRANS-SIBERIAN RAILROAD, 1,978 mi/3,183 km E of MOSCOW; 55°02'N 82°56'E. Elevation 475 ft/144 m. It is a large river, railroad, and air transportation hub (international airport is 10 mi/16 km W of city center) and is the leading scientific, industrial, and cultural center of Siberia. Novosibirsk has machinery (heavy, agricultural, electrical), machine tools, instruments, food, textile, chemical, and metallurgical industries. The Siberian branch of the Russian Academy of Sciences is in AKADEMGORODOK, S of Novosibirsk. There is a hydroelectric power station on the Ob' above the city. Founded in 1893, during the construction of the Trans-Siberian Railroad, at the bridge over the Ob' River, and called Novaya Derevnya (Russian=new village), it grew as a trade center. It was renamed, in quick succession, Aleksandrovskiy (1894–1895) and Novoaleksandrovskiy (1895–1903). The coal mining and industrial KUZNETSK BASIN, to the E, has been an important factor in its growth. Made a city in 1903 and renamed Novonikolayevsk; current name since 1925.

Novosibirskiye Ostrova, RUSSIA: see NEW SIBERIAN ISLANDS.

Novosibirsk Reservoir (nuh-vuh-see-BEERSK), Russian *Novosibirskoye Vodokhranilishche*, artificial lake (surface ☐ 421 sq mi/1,090 sq km) created in 1956 by expanding 103 mi/166 km of the natural course of the OB' RIVER; this was achieved by damming the river near NOVOSIBIRSK (site of a hydroelectric power station). Less than one-quarter of it (the S part) lies in the ALTAI TERRITORY, and the rest stretches in the NE direction across SE and E NOVOSIBIRSK oblast, SW SIBERIA, Russia. Mean depth is 30 ft/9 m (deepest point is 82 ft/25 m); maximum width is 12 mi/19 km. Ice-free May through October. Informally called the Ob' Sea. Its shores are popular summer vacation destination for local residents (beaches, sailing, non-commercial fishing).

Novosil' (nuh-vuh-SEEL), city (2006 population 3,980), N central ORËL oblast, SW European Russia, on road junction, 50 mi/80 km E of ORËL; 52°58'N 37°02'E. Elevation 757 ft/230 m. Agricultural products. Known since 1155. Made city in 1777.

Novosilikatnyy (no-vuh-see-lee-KAHT-niyee), town (2005 population 15,240), N central ALTAI TERRITORY, S central SIBERIA, RUSSIA, 11 mi/18 km W of BARNAUL, to which it is administratively subordinate; 53°19'N 83°38'E. Elevation 633 ft/192 m. Manufacturing (construction materials, ceramic concrete, large panels, porous concrete).

Novosineglazovskiy (no-vuh-see-nee-GLAH-zuhf-skeeyee), town (2005 population 14,590), E central CHELYABINSK oblast, SW SIBERIA, RUSSIA, on railroad (Sineglazovo station), 9 mi/15 km S of CHELYABINSK, of which it is a suburb and to which it is administratively subordinate; 55°02'N 61°22'E. Elevation 784 ft/238 m. Manufacturing (construction materials, silicate bricks).

Novosmolinskiy (no-vuh-SMO-leen-skeeyee), town (2006 population 5,100), W NIZHEGOROD oblast, central European Russia, near highway, 31 mi/50 km W of NIZHNIY NOVGOROD, and 4 mi/6 km NW of VOLODARSK, to which it is administratively subordinate; 56°16'N 43°06'E. Elevation 285 ft/86 m. Peat works. Formerly known as Smolino Vtoroye (Russian=Smolino the Second).

Novosokol'niki (nuh-vuh-suh-KOL-nee-kee), city (2006 population 9,495), S PSKOV oblast, W European Russia, on railroad junction, 178 mi/287 km SE of PSKOV, and 12 mi/19 km W of VELIKIYE LUKI; 56°20'N 30°09'E. Elevation 423 ft/128 m. Railroad enterprises; produce processing. City since 1925.

Novospasskoye (nuh-vuh-SPAHS-kuh-ye), town (2006 population 10,910), S ULYANOVSK oblast, E central European Russia, on the SYZRAN' RIVER, on highway junction and railroad, 28 mi/45 km W of SYZRAN'; 53°08'N 47°45'E. Elevation 344 ft/104 m. Logging, woodworking; petroleum products.

Novostroyevo (no-vuh-STRO-ee-vuh), settlement (2005 population 1,800), S KALININGRAD oblast, W European Russia, less than 8 mi/13 km N of the border with POLAND, on road, 7 mi/11 km WNW of OZËRSK; 54°26'N 21°50'E. Elevation 403 ft/122 m. Until 1945, in EAST PRUSSIA and called Trempen.

Novosuvorovskaya, RUSSIA: see VYSELKI.

Novo-Sventsyany, LITHUANIA: see ŠVENČIONĖLIAI.

Novosvetlovka, UKRAINE: see NOVOSVITLIVKA.

Novo-Svetlovka, UKRAINE: see NOVOSVITLIVKA.

Novosvitlivka (no-vo-SVEET-leef-kah) (Russian *Novosvetlovka*), town, SE LUHANS'K oblast, UKRAINE, in the DONBAS, on Luhanchyk River, and on highway 10 mi/16 km SE of LUHANS'K; 48°30'N 39°30'E. Elevation 367 ft/111 m. Asphalt and cement building-materials manufacturing. Established in the 1860s; town since 1961. In Russian, also hyphenated, *Novo-Svetlovka*.

Novosysoyevka (no-vuh-si-SO-eef-kah), village (2006 population 4,865), S central MARITIME TERRITORY, SE RUSSIAN FAR EAST, in the USSURI River basin, on road and railroad, 6 mi/10 km NNE of ARSEN'YEV; 44°14'N 133°23'E. Elevation 528 ft/160 m. In agricultural area (grains, soybeans, potatoes, vegetables). Formerly the site of a nuclear ballistic missile base, shut down in the 1990s.

Novotalitsy (no-vuh-TAH-lee-tsi), town (2005 population 8,445), central IVANOVO oblast, central European Russia, on highway and near railroad, less than 5 mi/8 km W, and under administrative jurisdiction, of IVANOVO; 57°01'N 40°52'E. Elevation 400 ft/121 m. Sand and gravel quarrying.

Novotavolzhanka, RUSSIA: see NOVAYA TAVOLZHANKA.

Novotitarovskaya (no-vuh-tee-TAH-ruhf-skah-yah), village (2005 population 22,740), central KRASNODAR TERRITORY, S European Russia, on road and railroad, 15 mi/24 km N of KRASNODAR; 45°14'N 38°58'E. Elevation 134 ft/40 m. Flour mill; dairying; wheat, sunflowers. Has an airfield.

Novo Triunfo (NO-vo TREE-yoon-fo), city (2007 population 14,821), BAHIA, BRAZIL.

Novotroitsk (no-vuh-TRO-eetsk), city (2006 population 106,165), SE ORENBURG oblast, extreme SE European Russia, in the foothills of the S URALS, on the right bank of the URAL River, 4 mi/6 km N of the border with KAZAKHSTAN, on highway and railroad spur, 195 mi/314 km SE of ORENBURG, and 10 mi/16 km W of ORSK; 51°12'N 58°19'E. Elevation 682 ft/207 m. In Orsk-Khalilovo industrial district. Metallurgical center (high-quality steels, chromium, and other ferroalloys); cement works. Limonite shipped from Khalilovo (15 mi/24 km NW), limestone from Izvestnyaki (just WSW; on railroad spur). Mining of chromium, nickel, and cobalt ores. Construction of the metallurgical plant began in 1942. City status granted in 1945.

Novo-Troitskoy, RUSSIA: see BALEY.

Novotroitskoye (no-vuh-TRO-eets-kuh-ye), village, W KIROV oblast, E central European Russia, on road, 45 mi/72 km WNW of KOTEL'NICH; 58°27'N 47°05'E. Elevation 495 ft/150 m. Gas pipeline service station. Flax processing; logging and lumbering.

Novotroitskoye (nuh-vuh-TRO-eets-kuh-ye), village, S central TATARSTAN oblast, E European Russia, on road, 31 mi/50 km WNW of (and administratively subordinate to) AL'MET'YEVSK, and 12 mi/19 km ESE of NOVOSHESHMINSK; 55°01'N 51°52'E. Elevation 223 ft/67 m. In agricultural area (grain); distillery.

Novotroitskoye (nuh-vuh-TRO-eets-kuh-ye), settlement (2006 population 7,960), W STAVROPOL TERRITORY, S European Russia, on the YEGORLYK RIVER (dam), on highway, 33 mi/53 km NW of STAVROPOL; 45°19'N 41°31'E. Elevation 442 ft/134 m. In grain area; flour mill. Dam (0.75 mi/1.2 km long) backs up main reservoir (□ 12 sq mi/31 sq km) of the Kuban-Yegorlyk irrigation system; feeds right and left Yegorlyk lateral canals. Also called Novotroitskaya.

Novo-Troitskoye, UKRAINE: see NOVOTROYITS'KE, Kherson oblast; or NOVOTROYITS'KE, Donets'k oblast.

Novotroitskoye, UKRAINE: see NOVOTROYITS'KE, Kherson oblast; or NOVOTROYITS'KE, Donets'k oblast.

Novotroyits'ka, UKRAINE: see TROYITS'KE, town.

Novotroyits'ke (no-vo-TRO-yeets-ke) (Russian *Novotroitskoye*), town, SE KHERSON oblast, UKRAINE, on road and irrigation canal, and 25 mi/40 km NW of HENICHES'K; 46°21'N 34°20'E. Raion center; feed milling, broiler factory, dairy, meat and bone processing, food-flavoring factory. Established in the 1860s, town since 1958. In Russian, also hyphenated, *Novo-Troitskoye*.

Novotroyits'ke (no-vo-TRO-yee-tske) (Russian *Novotroitskoye*), town, central DONETS'K oblast, UKRAINE, in the DONBAS on Sukha Volnovakha River, 21 mi/34 km SW of DONETS'K; 47°43'N 37°35'E. Elevation 597 ft/181 m. Limestone and dolomite quarries, reinforced-concrete construction materials manufacturing. Established in 1773, town since 1938. In Russian, also hyphenated, *Novo-Troitskoye*.

Novotul'skiy (nuh-vuh-TOOL-skeeyee), former town, central TULA oblast, W central European Russia, on the UPA RIVER, 3 mi/5 km SE of TULA, of which it is administratively a suburb; 54°10'N 37°40'E. Elevation 629 ft/191 m. Metallurgical center; iron- and steel-works.

Novoturukhansk, RUSSIA: see TURUKHANSK.

Novougol'nyy (nuh-vuh-OO-guhl-niyee), town (2006 population 7,005), E TULA oblast, W central European Russia, on local highway, 6 mi/10 km SE of NOVOMOSKOVSK, and 3 mi/5 km N of DONSKOY, to which it is administratively subordinate; 54°01'N 38°17'E. Elevation 767 ft/233 m. Lignite mining; furniture.

Novoukrainka, UKRAINE: see NOVOUKRAYINKA.

Novo-Ukraina, UKRAINE: see NOVOUKRAYINKA.

Novoukrainskiy (no-vuh-oo-krah-EEN-skeeyee), settlement (2005 population 3,590), S KRASNODAR TERRITORY, S European Russia, on road and railroad, 30 mi/48 km ENE of ANAPA, 5 mi/8 km W of ABINSK, and 4 mi/6 km E of KRYMSK, to which it is administratively subordinate; 44°54'N 38°03'E. Elevation 104 ft/31 m. Agricultural products.

Novoukrainskoye (no-vuh-oo-krah-EEN-skuh-ye), village (2005 population 5,955), E KRASNODAR TERRITORY, S European Russia, on the KUBAN' River, on highway, 4 mi/6 km S of KROPOTKIN; 45°22'N 40°32'E. Elevation 269 ft/81 m. In agricultural area (wheat, sunflowers, castor beans). An oil pipeline runs in the vicinity.

Novoukrayinka (no-vo-ook-rah-YEEN-kah) (Russian *Novoukrainka* or *Novo-Ukraina*), city, SW KIROVOHRAD oblast, UKRAINE, on Chornyy Tashlyk River and railroad, 36 mi/58 km WSW of KIROVOHRAD; 48°19'N 31°31'E. Elevation 433 ft/131 m. Raion center; flour-milling center; sugar refining; dairying; stone crushing, reinforced-concrete-materials manufacturing, brickworks; furniture factory. Vocational school; regional museum. Established in 1743 as Fort Pavlivs'k (Russian *Pavlovskaya krepost'*) until 1764; Novopavlivs'k (Russian *Novo-Pavlovsk*) until 1773; Posad Pavlivs'kyy (Russian *Pavlovskiy Posad*) until 1830; city since 1938. Jewish community since the 19th century, numbering 2,800 in 1939; eliminated by the Nazis in 1941–42.

Novoul'yanovsk (nuh-vuh-ool-YAH-nuhfsk), city (2006 population 17,770), E central ULYANOVSK oblast, E central European Russia, on the Kuybyshev Reservoir of the VOLGA RIVER, on road and near railroad (Belyy Klyuch station), 12 mi/20 km S, and under administrative jurisdiction, of ULYANOVSK;

54°10'N 48°23'E. Elevation 173 ft/52 m. Cement, chrome, roof tiles; bakery. Made city in 1967.

Novoural'sk (nuh-vuh-oo-RAHLSK), city (2006 population 93,355), W SVERDLOVSK oblast, E central URALS, extreme W Siberian Russia, on highway junction and railroad, 11 mi/18 km S of KIROVGRAD; 57°15'N 60°05'E. Elevation 1,085 ft/330 m. Automobile engines, construction materials, boilers; research and design enterprises in electronics and chemical industries; uranium enrichment for nuclear reactors. A closed city; until 1991, part of the Soviet Union's nuclear military complex and known under its secret designation of Sverdlovsk-44.

Novoutkinsk (nuh-vuh-OOT-keensk), town (2006 population 5,480), SW SVERDLOVSK oblast, E central URALS, extreme W Siberian Russia, near the CHUSOVAYA River, on railroad (Kourovka station), 9 mi/14 km WNW of BILIMBAY; 56°59'N 59°33'E. Elevation 964 ft/293 m. Electric welding machines and apparatus, sawmilling, school furniture.

Novouzensk (nuh-vuh-oo-ZYENSK), city (2006 population 16,955), SE SARATOV oblast, SE European Russia, on the Bol'shoy Uzen' River, on road junction and railroad, 125 mi/201 km SE of SARATOV; 50°27'N 48°08'E. Center of agricultural area; food industries. Chartered in 1835.

Novovarshavka (nuh-vuh-vahr-SHAF-kah), village (2006 population 6,465), SE OMSK oblast, SW SIBERIA, RUSSIA, on road, 5 mi/8 km W of CHERLAK (across the IRTYSH River); 54°11'N 74°42'E. Elevation 259 ft/78 m. Regional branches of state agricultural and pedagogical universities.

Novovasilyevka, UKRAINE: see NOVOVASYLIVKA.

Novo-Vasilyevka, UKRAINE: see NOVOVASYLIVKA.

Novovasylivka (no-vo-vah-SI-leef-kah) (Russian *Novovasilevka*), town, S ZAPORIZHZHYA oblast, UKRAINE, on AZOV LOWLAND, on road intersection 18 mi/29 km E of MELITOPOL' and 9 mi/14 km NE of PRYAZOVS'KE; 46°50'N 35°44'E. Elevation 190 ft/57 m. Food processing, stone crushing, brick making, granite quarry; war museum. Established in 1823 on the site of the Nogay settlement Apanly; town since 1957.

Novovasylivka (no-vo-va-SI-leef-kah) (Russian *Novovasilyevka*), village, S ZAPORIZHZHYA oblast, UKRAINE, 19 mi/31 km E of MELITOPOL'; 46°51'N 36°47'E. Food processing, construction materials (granite quarry; stone crushing, cement and asphalt making). Heritage museum. Established in 1823 on the site of the Nogay settlement Apanly; town since 1957. Also, in Russian, hyphenated, *Novo-Vasilyevka*.

Novovelichkovskaya (no-vuh-vye-LEECH-kuhf-skah-yah), village (2005 population 8,815), central KRASNODAR TERRITORY, S European Russia, on crossroads, 19 mi/31 km N of KRASNODAR; 45°16'N 38°50'E. In agricultural area; produce processing. Archaeological works in the vicinity has unearthed artifacts from the 2nd millenium B.C.E.

Novo-Vilnya, LITHUANIA: see NAUJOJI VILNIA.

Novovolyns'k (no-vo-vo-LINSK) (Polish *Nowowolynsk*) (Russian *Novovolynsk*), city (2001 population 53,838), SW VOLYN' oblast, UKRAINE, 12 mi/19 km SW of VOLODYMYR-VOLYNS'KYY; 50°44'N 24°10'E. Elevation 761 ft/231 m. Coal-mining center of the L'VIV-VOLHYNIA COAL BASIN, including six coal mines, machine construction, woodworking, fabrication of reinforced-concrete construction materials, cotton-fiber spinning, shoe-making, and food processing; electromechanical technical school, three professional-technical schools. Established in 1950, city since 1957.

Novovolynsk, UKRAINE: see NOVOVOLYNS'K.

Novovoronezh (nuh-vuh-vuh-RO-neesh), city (2006 population 37,540), central VORONEZH oblast, S central European Russia, on the DON River, on railroad spur and near highway, 25 mi/40 km S of VORONEZH; 51°19'N 39°13'E. Elevation 442 ft/134 m. Developed with construction of the Novovoronezhskiy nuclear power plant in 1958. Made city in 1987.

Novovorontsovka (no-vo-vo-ron-TSOV-kah) (Russian *Novo-Vorontsovka*), town, N KHERSON oblast, UKRAINE, on the W bank of the KAKHOVKA RESERVOIR, 20 mi/32 km WSW of NIKOPOL'; 47°30′N 33°56′E. Raion center; food processing (flour mill, dairy, food-flavor factory). Heritage museum. Established around 1795 by the Zaporozhzhian Cossacks and named Mykolayivka (Russian *Nikolayevka*); acquired in 1821 by Prince Mikhail Vorontsov and renamed Vorontsovka; town since 1956.

Novo-Vorontsovka, UKRAINE: see NOVOVORONTSOVKA.

Novovoznesenovka (no-vo-vahz-ni-SE-nuhv-kuh), village, NE ISSYK-KOL region, KYRGYZSTAN, 20 mi/32 km ENE of KARAKOL; 42°38′N 78°45′E. Wheat.

Novovyatsk (no-vuh-VYAHTSK), former city, central KIROV oblast, E central European Russia, on the VYATKA River and on railroad, now incorporated into KIROV, 5 mi/8 km SSE of city center; 58°30′N 49°43′E. Elevation 456 ft/138 m. Sawmilling; building materials, skis, mechanical shops. Formed in 1955 from the union of Lesozavodskiy and Vyatskiy. Until 1939, called Grukhi.

Novovyazniki (nuh-vuh-VYAHZ-nee-kee), town (2006 population 5,570), NE VLADIMIR oblast, central European Russia, on railroad, 3 mi/5 km S of VYAZNIKI; 56°12′N 42°10′E. Elevation 436 ft/132 m. Railroad depots; linen textiles.

Novoyavorivs'ke (no-vo-YAH-vo-rif-ske) (Russian *Novoyavorovskoye*), city, W L'VIV oblast, UKRAINE, 21 mi/34 km WNW of L'VIV; 49°55′N 23°44′E. Elevation 869 ft/264 m. City is E of the Yavoriv sulfur mine; fabrication of reinforced-concrete building materials, professional-technical school. Established in 1965, city since 1986.

Novoyavorovskoye, UKRAINE: see NOVOYAVORIVS'KE.

Novoye [Russian=new], in Russian names: see also NOVAYA, NOVO- [Russian combining form], NOVY, and NOVYE.

Novoye, UKRAINE: see NOVE.

Novoye Atlashevo (NO-vuh-ye AHT-lah-shi-vuh), settlement (2005 population 3,505), N CHUVASH REPUBLIC, central European Russia, on the W bank of the TSIVIL' River, 8 mi/13 km ESE of CHEBOKSARY; 56°01′N 47°33′E. Elevation 203 ft/61 m. In agricultural area; fruit and vegetable canning.

Novoye Churilino (NO-vuh-ye choo-REE-lee-nuh), village, NW TATARSTAN Republic, E European Russia, on railroad and near highway, 15 mi/24 km ENE of ARSK; 56°08′N 50°14′E. Elevation 606 ft/184 m. Holiday getaway spot; summer homes.

Novoye Devyatkino (NO-vuh-ye dye-VYAHT-kee-nuh), town (2005 population 8,790), W central LENINGRAD oblast, NW European Russia, on the S KARELIAN ISTHMUS, on railroad, 19 mi/31 km NNE of SAINT PETERSBURG; 60°03′N 30°28′E.

Novoyegor'yevskoye (nuh-vuh-yee-GOR-yeef-skuh-ye), village, SW ALTAI TERRITORY, S SIBERIA, RUSSIA, on small Lake Peresheyechnoye, on road junction, 20 mi/32 km NW of RUBTSOVSK; 51°46′N 80°53′E. Elevation 764 ft/232 m. In logging and woodworking area.

Novoye Leushino (NO-vuh-ye lee-OO-shi-nuh), town (2005 population 1,545), central IVANOVO oblast, central European Russia, near railroad (Sakhtysh station), 4 mi/6 km S of TEYKOVO, to which it is administratively subordinate; 56°48′N 40°32′E. Elevation 459 ft/139 m. Peat works.

Novoyelnya (no-vo-YEL-nyah), Polish *Nowojelnia*, urban settlement, S GRODNO oblast, BELARUS, 15 mi/24 km SW of NOVOGRUDOK. Railroad junction; flour milling.

Novoye Misto, UKRAINE: see NOVE MISTO.

Novoye Selo, UKRAINE: see NOVE SELO.

Novozavidovskiy (nuh-vuh-zah-VEE-duhf-skeeyee), town (2006 population 7,560), S TVER oblast, W European Russia, on the Ivankov Reservoir of the VOLGA RIVER, on highway and railroad, 27 mi/43 km SE of

TVER; 56°33′N 36°26′E. Elevation 505 ft/153 m. Sanitation and hygienic equipment; felt products; winery.

Novozybkov (no-vuh-ZIP-kuhf), city (2005 population 42,420), SW BRYANSK oblast, central European Russia, on road and railroad, 128 mi/207 km SW of BRYANSK; 52°32′N 31°56′E. Elevation 544 ft/165 m. Railroad and highway junction; local transshipment center. Machine tools, clothing, knit goods, electrothermal equipment; meat and poultry packing. Known since the 17th century. Chartered in 1809. Since 1963, the official seat of the Archbishop of the Russian Old Orthodox Church, the Old Believers branch that split from the mainstream Eastern Orthodox Christianity in the 17th century.

Novshahr, IRAN: see NOW SHAHR.

Novska (NOV-skah), town, E CROATIA, 20 mi/32 km SSW of DARUVAR, in SLAVONIA; 45°20′N 17°00′E. Railroad junction; local trade center; metal and lumber industries. Heavily damaged in 1991 conflict.

Novy [Russian=new], in Russian names: see also NOVAYA, NOVO- [Russian combining form], NOVOYE, NOVYE.

Novy Bohumin, CZECH REPUBLIC: see BOHUMÍN.

Novy Bor (NO-vee BOR), Czech *Nový Bor*, German *haida*, town, SEVEROCESKY province, N BOHEMIA, CZECH REPUBLIC, on railroad, and 24 mi/39 km ENE of ÚSTÍ NAD LABEM; 50°46′N 14°35′E. Noted glass industry (mainly artistic glassware); manufacturing (machinery); woodworking. Has a glass museum. The town today also includes commune of ARNULTOVICE, (German *arnsdorf*). Until 1946, the town was called BOR U CESKE LIPY.

Novy Bydzov (NO-vee BID-zhof), Czech *Nový Bydžov*, German *neu bidschow*, town, VYCHODOCESKY province, NE BOHEMIA, CZECH REPUBLIC, on railroad, on Cidlina River, 15 mi/24 km W of HRADEC KRÁLOVÉ; 50°15′N 15°30′E. Agriculture (sugar beets, potatoes); manufacturing (glass, costume jewelry, textiles); food processing; sugar refining.

Novy Dvor, POLAND: see NOWY DWOR.

Novye [Russian=new], in Russian names: see also NOVAYA, NOVO- [Russian combining form], NOVOYE, NOVY.

Novy Hrozenkov (NO-vee HRO-zen-KOF), Czech *Nový Hrozenkov*, German NEUHROSENKAU, village, SEVEROMORAVSKY province, E MORAVIA, CZECH REPUBLIC, on the HORNI BECVA (headstream of Becva River), on railroad, and 25 mi/40 km ENE of ZLÍN, in the BESKIDS; 49°21′N 18°13′E. Woodworking. Summer and winter resort, noted for skiing facilities, picturesque wooden cottages, and regional costumes.

Nový Jičín (NO-vee YI-cheen), German *Neutitschein*, city (2001 population 26,970), SEVEROMORAVSKY province, NE MORAVIA, CZECH REPUBLIC, on railroad, and 21 mi/34 km SW of OSTRAVA; 49°36′N 18°01′E. Manufacturing (known for men's hats [museum]; machinery; woolen textiles; wood products; food; tobacco. HODSLAVICE, birthplace of F. Palacky, is 4 mi/6.4 km S.

Novy Margelan, UZBEKISTAN: see FERGANA.

Novy Smokovec, SLOVAKIA: see STARY SMOKOVEC.

Novy Svet, CZECH REPUBLIC: see HARRACHOV.

Novyy Ardon, RUSSIA: see EKAZHEVO.

Novyy Bug, UKRAINE: see NOVYY BUH.

Novyy Buh (no-VEE BOO) (Russian *Novyy Bug*), city, N MYKOLAYIV oblast, UKRAINE, 56 mi/90 km NNE of MYKOLAYIV; 47°41′N 32°30′E. Elevation 282 ft/85 m. On highway intersection; raion center; food processing (flour and products, cheese, canned goods), sewing, footwear, furniture manufacturing, brickworks. Agricultural technical school, teachers' college; heritage museum. Established in the late 18th century as Kutsa Balka after the Cossak Yakov Kutsyy; merged with other settlements in the early 19th century to form a town called Semenivka (Russian *Semenovka*), then Novopavlivka (Russian *Novopavlovka*) and, in 1832, Novyy Buh; city since 1961.

Novyy Buyan (NO-viyee boo-YAHN), village (2006 population 3,500), W SAMARA oblast, E European Russia, on road, 35 mi/56 km N of SAMARA; 53°42′N 50°02′E. Elevation 446 ft/135 m. Agricultural products.

Novyy Bykiv (no-VIEE BI-keef) (Russian *Novyy Bykov*), town (2004 population 3,430), S CHERNIHIV oblast, UKRAINE, on the Supiy River, on highway and railroad spur, 31 mi/50 km W of PRYLUKY and 15 mi/24 km SE of BOBROVYTSYA; 50°36′N 31°40′E. Elevation 400 ft/121 m. Sugar refinery, non-alcoholic beverage plant, cattle feedlot. Known since the 16th century; town since 1964.

Novyy Bykov, UKRAINE: see NOVYY BYKIV.

Novyy Byt (NO-viyee BIT), village, S MOSCOW oblast, central European Russia, near road, 13 mi/21 km NE of SERPUKHOV, and 11 mi/18 km SE of CHEKHOV, to which it is administratively subordinate; 55°03′N 37°36′E. Elevation 613 ft/186 m. Testing and certification center for agricultural and cargo machinery.

Novyy Chirkey (NO-viyee cheer-KYAI), village (2005 population 7,735), central DAGESTAN REPUBLIC, SE European Russia, on road and railroad, 33 mi/53 km NW of MAKHACHKALA; 43°09′N 47°03′E. In oil- and gas-producing region.

Novyy Donbas, UKRAINE: see DYMYTROV.

Novyye Aldy, RUSSIA: see CHERNORECHYE.

Novyye Atagi (NO-vi-ye ah-TAH-gee) [Russian=new Atagi], village (2005 population 8,770), central CHECHEN REPUBLIC, S European Russia, on road, 11 mi/18 km SSE of GROZNYY; 43°08′N 45°46′E. Elevation 882 ft/268 m. Agriculture (vegetables, meat and dairy livestock). Has a hospital. Since 1994, besieged on numerous occasions by Russian troops fighting Chechen separatists. Formerly known as Mayskoye (1944–1959).

Novyye Belokorovichi, UKRAINE: see NOVI BILOKOROVYCHI.

Novyye Belyary, UKRAINE: see NOVI BILYARY.

Novyye Burasy (NO-vi-ye boo-RAH-si), town (2006 population 6,455), N SARATOV oblast, SE European Russia, on road junction and 5 mi/8 km from railroad (Burasy station), 40 mi/64 km N of SARATOV; 52°08′N 46°05′E. Elevation 839 ft/255 m. Dairy products.

Novyye Gorki (NO-vi-ye GOR-kee), village (2005 population 2,845), S IVANOVO oblast, central European Russia, on road and railroad, 13 mi/21 km SW of SHUYA; 56°43′N 41°03′E. Elevation 357 ft/108 m. In agricultural area (grain, flax, hemp, sunflowers, sugarbeets, soybeans, potatoes; livestock).

Novyye Lapsary (NO-vi-ye lahp-SAH-ri), town (2005 population 7,560), N CHUVASH REPUBLIC, central European Russia, near highway, 4 mi/6 km S, and under administrative jurisdiction, of CHEBOKSARY (connected by railroad); 56°04′N 47°12′E. Elevation 698 ft/212 m. Power plant equipment supplies, maintenance, and repair.

Novyy El'ton, RUSSIA: see EL'TON.

Novyye Petushki, RUSSIA: see PETUSHKI.

Novyye Sanzhary, UKRAINE: see NOVI SANZHARY.

Novyye Strelishcha, UKRAINE: see NOVI STRILYSHCHA.

Novyye Vysli (NO-vi-ye VIS-lee), village (2005 population 3,885), E CHUVASH REPUBLIC, central European Russia, near highway, 56 mi/90 km SSE of CHEBOKSARY; 55°09′N 47°27′E. Elevation 751 ft/228 m. In agricultural area (wheat, vegetables, hemp); molasses factory.

Novyy Gorod, UKRAINE: see NOVE MISTO.

Novyy Gorodok (NO-viyee guh-ruh-DOK), town (2005 population 16,215), W central KEMEROVO oblast, S central SIBERIA, RUSSIA. Town formed in 1988 in the vicinity of, and administratively subordinate to, BELOVO city; in the KUZNETSK coal basin.

Novyy Karachay (NO-viyee kah-rah-CHEI), village (2005 population 2,175), central KARACHEVO-CHERKESS REPUBLIC, S European Russia, in the NW CAUCASUS Mountains, on the right bank of the KUBAN' River, on road, 3 mi/5 km N of KARACHAYEVSK, to which it is administratively subordinate; 43°49′N 41°54′E. Elevation 2,893 ft/881 m. In agricultural area. Formerly known as Pravoberezhnoye (prah-vuh-bee-RYEZH-nuh-ye).

Novyy Khushet (NO-viyee hoo-SHET), village (2005 population 8,100), E central DAGESTAN REPUBLIC, in the Caspian lowlands, on branch of coastal railroad, 15 mi/24 km S of MAKHACHKALA, and 4 mi/6 km WNW of KASPIYSK; 42°54′N 47°33′E. Below sea level. In oil- and gas-producing region.

Novyy Kiner (NO-viyee kee-NYER), village, NW TATARSTAN Republic, E European Russia, 22 mi/35 km NNW of (and administratively subordinate to) ARSK; 56°24′N 49°44′E. Elevation 564 ft/171 m. In agricultural area (grain; livestock); woodworking, vegetable drying.

Novyy Kocherdyk, RUSSIA: see TSELINNOYE, KURGAN oblast.

Novyy Kostek (NO-viyee kuhs-TYEK) [Russian=new Kostek], village (2005 population 4,095), W central DAGESTAN REPUBLIC, SE European Russia, in fertile pre-Caspian valley, on highway, 47 mi/76 km NW of MAKHACHKALA, and 13 mi/21 km NE of KHASAVYURT; 43°20′N 46°49′E. Agriculture (grain, fruits, grapes).

Novyy Milyatin, UKRAINE: see NOVYY MYLYATYN.

Novyy Mylyatyn (no-VEE mil-YAH-tin) (Russian *Novyy Milyatin*) (Polish *Milatyn Nowy*), village (2004 population 6,000), central L′VIV oblast, UKRAINE, 6 mi/10 km W of BUS′K; 49°58′N 24°30′E. Elevation 725 ft/220 m. Grain, potatoes, flax.

Novyy Nekouz (NO-viyee nee-kuh-OOS), village (2006 population 3,605), W YAROSLAVL oblast, central European Russia, on highway and railroad, 14 mi/23 km WSW of VOLGA; 57°54′N 38°04′E. Elevation 439 ft/133 m. Agricultural machinery and equipment servicing, woodworking, cheese making.

Novyy Oskol (NO-viyee uh-SKOL) [Russian=new Oskol], city (2005 population 21,100), central BELGOROD oblast, SW European Russia, on the OSKOL RIVER, on road and railroad, 95 mi/153 km SW of VORONEZH, and 57 mi/92 km E of BELGOROD; 50°45′N 37°52′E. Elevation 442 ft/134 m. Food industries (sunflower-oil extraction, flour milling, canning; meat packing); electronic equipment; concrete structural elements. Founded in 1637. Made city in 1799.

Novyy Port (NO-viyee PORT), settlement, N YAMALO-NENETS AUTONOMOUS OKRUG, TYUMEN oblast, NW SIBERIA, RUSSIA, port on the OB′ BAY, 190 mi/306 km ENE of SALEKHARD; 67°41′N 72°56′E. Fish cannery; airport; supply port on the Arctic sea route.

Novyy Put′, RUSSIA: see SPASSKAYA GUBA.

Novyy Redant (NO-viyee ree-DAHNT), village (2005 population 5,170), NW INGUSH REPUBLIC, extreme N CAUCASUS, S European Russia, on highway, 21 mi/34 km NNW of NAZRAN′; 43°28′N 44°48′E. Elevation 1,095 ft/333 m. Agriculture (grain, livestock).

Novyy Rogachik (NO-viyee ruh-GAH-cheek), village (2006 population 7,585), S VOLGOGRAD oblast, SE European Russia, on the VOLGA-DON CANAL, on highway and railroad, 23 mi/37 km SW, and under administrative jurisdiction, of GORODISHCHE; 48°40′N 44°03′E. Elevation 173 ft/52 m. Horse breeding, poultry farm, grain processing.

Novyy Rozdil (no-VEE roz-DEEL) (Russian *Novyy Rozdol*), city, SE L′VIV oblast, UKRAINE, 25 mi/40 km SSE of L′VIV; 49°28′N 24°08′E. Elevation 875 ft/266 m. Sulfur mining, fabrication of building materials, machine building; technical and professional-technical schools. Established in 1953, city since 1965.

Novyy Rozdol, UKRAINE: see NOVYY ROZDIL.

Novyy Sulak (NO-viyee soo-LAHK) [Russian=new Sulak], village (2005 population 5,585), central DAGESTAN REPUBLIC, SE European Russia, on road and railroad, 43 mi/69 km WNW of MAKHACHKALA, and 11 mi/18 km E of KHASAVYURT; 43°12′N 46°49′E. Elevation 173 ft/52 m. In oil- and gas-producing region.

Novyy Svet, RUSSIA: see KRASNYY LUCH.

Novyy Svet, UKRAINE: see NOVYY SVIT.

Novyy Svit (no-VIEE SVEET) (Russian *Novyy Svet*) [Ukrainian=new world; Russian=new light], town, central SE DONETS′K oblast, UKRAINE, 19 mi/30 km SE of DONETS′K city center on the KAL′MIUS RIVER at the Starobesheve Reservoir; 47°48′N 38°01′E. Elevation 469 ft/142 m. Site of the Starobesheve Regional Thermal Electric Power Station; construction materials manufacturing, food processing. Established in 1954, town since 1956.

Novyy Toryal (NO-viyee tuhr-YAHL), town (2006 population 7,150), NE MARI EL REPUBLIC, E central European Russia, on the Nemda River (tributary of the VYATKA River), on road, 40 mi/64 km NE of YOSHKAR-OLA; 57°02′N 48°44′E. Elevation 400 ft/121 m. Flax processing.

Novyy Uoyan (NO-viyee oo-uh-YAHN), village (2005 population 4,840), NW BURYAT REPUBLIC, S central SIBERIA, RUSSIA, in the UPPER ANGARA RIVER basin, on the BAYKAL-AMUR MAINLINE, 281 mi/452 km NNE of KHORINSK; 56°09′N 111°42′E. Elevation 1,683 ft/512 m. Logging and lumbering.

Novyy Urengoy (NO-viyee oo-reen-GO-yee), city (2006 population 94,155), central YAMALO-NENETS AUTONOMOUS OKRUG, TYUMEN oblast, NW SIBERIA, RUSSIA, on the Yelovakha River (tributary of the Pur River, which flows into TAZ BAY), 280 mi/451 km E of SALEKHARD; 66°05′N 76°38′E. Elevation 144 ft/43 m. Terminus of a long railroad line extending 430 mi/692 km NE from SURGUT; airport; in major natural-gas region; long natural-gas pipelines extend W, SW, and S to central European Russia (the Northern Lights), the Urals, and W Siberia. Made a city in 1980. Arose with development of the Urengoy natural-gas field.

Novyy Urgal (NO-viyee oor-GAHL), settlement (2005 population 6,990), W KHABAROVSK TERRITORY, RUSSIAN FAR EAST, on the URGAL River, on road and branch connecting the BAYKAL-AMUR MAINLINE at Chegdomyn with the TRANS-SIBERIAN RAILROAD at BIRAKAN (JEWISH AUTONOMOUS OBLAST), 149 mi/240 km NW of KHABAROVSK; 51°04′N 132°34′E. Elevation 954 ft/290 m. Railroad depots.

Novyy Vasyugan (NO-viyee vah-syoo-GAHN), village, W TOMSK oblast, central SIBERIA, RUSSIA, on the W bank of the VASYUGAN RIVER, on highway, 135 mi/217 km SSW of NIZHNEVARTOVSK; 58°34′N 76°29′E. Elevation 252 ft/76 m. In oil- and natural gas-producing region.

Novyy Yarychev, UKRAINE: see NOVYY YARYCHIV.

Novyy Yarychiv (no-VIEE YAH-ri-chif) (Russian *Novyy Yarychev*), town, central L′VIV oblast, UKRAINE, 13 mi/21 km ENE of L′VIV; 49°55′N 24°18′E. Elevation 715 ft/217 m. Weaving factory. Known since 1340; part of Poland (1349–1772, 1919–1939), when it was called, until 1940, *Jaryczow Nowy*. Town since 1940.

Novyy Zay, RUSSIA: see ZAINSK.

Nowabganj, BANGLADESH: see NAWABGANJ.

Nowa Huta, POLAND: see CRACOW.

Nowa Ruda (NO-vah ROO-dah), German *Neurode*, town (2002 population 25,071), in LOWER SILESIA, Wałbrzych province, SW POLAND, near CZECH border, at S foot of the EULENGEBIRGE, 13 mi/21 km NNW of KŁODZKO (Glatz). Coal and clay mining, cotton milling, slate quarrying. Town has 13th century monastery.

Nowa Sol (NO-vah sol), Polish *Nowa Sól*, German *Neusalz*, town (2002 population 41,176) in LOWER SILESIA, after 1945 in Zielona Góra province, W POLAND, port on the ODER, and 13 mi/21 km SE of ZIELONA GÓRA (Grünberg); 51°48′N 15°43′E. Textile mills, foundries; oil refining, manufacturing of chemicals, glue, paper-milling machinery.

Nowata (no-WAH-tuh), county (□ 580 sq mi/1,508 sq km; 2006 population 10,785), NE OKLAHOMA; ⊙ NOWATA; 36°47′N 95°37′W. Bounded N by KANSAS state line; drained by VERDIGRIS RIVER Cattle; agri-

culture (corn, wheat, oats, sorghum). Oil and natural-gas fields, limestone quarries; refineries. Manufacturing at Nowata. Timber. Part of Oologah Lake in SE. Formed 1907.

Nowata (no-WAH-tuh), town (2006 population 4,005), ⊙ NOWATA county, NE OKLAHOMA, 20 mi/32 km E of BARTLESVILLE; 36°42′N 95°38′W. Elevation 708 ft/216 m. In agricultural and oil- and natural gas–producing area; manufacturing (metal industries, lithographic printing; crushed limestone); timber. N end of Oologah Lake reservoir (VERDIGRIS RIVER) to SE. Has county historical museum. Settled 1888, incorporated 1895.

Nowawes, GERMANY: see BABELSBERG.

Nowa Wilejka, LITHUANIA: see NAUJOJI VILNIA.

Nowbaran (nah-ou-bah-RAHN), village, Markazī province, W central IRAN, 35 mi/56 km WNW of SAVEH and on the SAVEH RIVER; 35°07′N 49°42′E. Cotton, wheat, fruit.

Nowe (NO-ve), German *Neuenburg*, town, Bydgoszcz province, N POLAND, port on the VISTULA, and 11 mi/18 km N of GRUDZIADZ. Railroad terminus; manufacturing of agricultural machinery, cement; sawmilling; distilling; flour milling.

Nowe Miasteczko (NO-ve mee-STECH-ko), German *Neustädtel*, town, Zielona Góra province, W POLAND, 8 mi/12.9 km S of NOWA SÓL; 51°41′N 15°44′E. Agricultural market (vegetables, grain, potatoes; livestock). City has 14th century church, 17th century city hall.

Nowe Miasto Lubawskie (NO-ve mee-AH-sto), German *Neumark*, town, Toruń province, N POLAND, on DRWĘCA RIVER, and 12 mi/19 km S of IŁAWA; 53°25′N 19°36′E. Railroad junction; manufacturing of bricks, furniture; flour milling; sawmilling. City has 14th century walls; museum.

Nowe Miasto nad Pilicą (NO-ve mee-AH-sto nahd pee-LEE-tso), Russian *Nove Myasto*, town, Radom province, central POLAND, on PILICA RIVER, opposite Drzewiczka River mouth, and 25 mi/40 km E of TOMASZOW MAZOWIECKI; 51°38′N 20°35′E. Railroad spur terminus; summer resort.

Nowe Siolo, UKRAINE: see NOVE SELO.

Nowe Warpno (NO-ve VAHRP-no), German *Neuwarp*, town, Szczecin province, NW POLAND, on small inlet of S STETTIN LAGOON, 25 mi/40 km NW of SZCZECIN, on German border; 53°44′N 14°18′E. Fishing port; fish smoking; yacht port.

Nowgaon, INDIA: see NOWGONG.

Nowgong (NOU-gou) or **Nowgaon**, town (2001 population 33,024), CHHATARPUR district, N central MADHYA PRADESH state, central INDIA; 26°21′N 92°40′E. Market center for grain, cloth, fabrics, sugar, oilseeds; chemical and pharmaceutical works; distillery. Formerly a large British cantonment. Sepoys rebelled here in Indian Mutiny (1857). Headquarters of former Bundelkhand Agency.

Nowgong, INDIA: see NAGAON.

Nowimburk, POLAND: see NOWOGRODZIEC.

Nowitna River, 250 mi/402 km long, SW ALASKA; rises N of upper KUSKOKWIM River, near 63°26′N 155°02′W; flows in a winding course generally N to YUKON River at 64°56′N 154°20′W.

Nowogard (no-VO-gahrd), German *Naugard*, town, Szczecin province, NW POLAND, 30 mi/48 km NE of STETTIN; 53°39′N 15°07′E. Flour milling, distilling, starch manufacturing. In World War II, partially destroyed. Has 14th century church; ruins of 14th century city walls.

Nowogrod (no-VO-grood), Polish *Nowogród*, Russian *Novogrod*, town, Łomża province, NE POLAND, on the Narew, and 9 mi/14.5 km WNW of ŁOMŻA. Folk museum.

Nowogrodek (no-vo-GRO-dek), Polish *Nowogródek*, former province (□ 8,945 sq mi/23,257 sq km) of E POLAND; ⊙ NOVOGRUDOK. Formed 1921 out of Russian governments; occupied 1939 by USSR and became part of BARANOVICHI and VILEIKA (after 1944, MO-LODECHNO) oblasts of the then Belorussian SSR.

Nowogrodziec (no-vo-GRO-jets), German *Naumburg am Queis*, town, in LOWER SILESIA, Jelenia Góra province, SW POLAND, on KWISA (Queis) River, and 18 mi/29 km E of GÖRLITZ; 51°12′N 15°24′E. Ceramic manufacturing. Chartered 1233. Heavily damaged in World War II. After 1945, briefly called Nowimburk.

Noworadomsk, POLAND: see RADOMSKO.

Nowo-Swięciany, LITHUANIA: see SVENCIONELIAI.

Nowowolynsk, UKRAINE: see NOVOVOLYNS'K.

Nowra (NOU-ruh), municipality, E NEW SOUTH WALES, SE AUSTRALIA, on SHOALHAVEN RIVER, and 75 mi/121 km SSW of SYDNEY; 34°53′S 150°36′E. Railroad terminus; tourist center; dairy products; hogs.

Now Shahr (nah-OU SHAHR), town, Māzanderān province, N IRAN, port on CASPIAN SEA, 37 mi/60 km ESE of Tanekabon; 36°37′N 51°30′E. Developed after 1935; held by USSR during World War II. Also Nowshahr or NAUSHAHR.

Nowshera (NOU-shai-rah), town, ABBOTTABAD district, NE NORTH-WEST FRONTIER PROVINCE, N PAKISTAN, 2 mi/3.2 km ENE of ABBOTTABAD; 34°10′N 73°15′E. Agricultural market center (corn, wheat). Nearby munitions factory.

Nowshera, town, PESHAWAR district, central NORTH-WEST FRONTIER PROVINCE, N PAKISTAN, on KABUL RIVER, and 23 mi/37 km E of PESHAWAR; 34°01′N 71°59′E. Also spelled NAUSHAHR.

Nowshera, INDIA: see NAOSHERA.

Nowy Bytom (NO-vee BEE-tom), German *Morgenroth*, town, Katowice province, S POLAND, 4 mi/6.4 km E of ZABRZE (Hindenburg). Railroad junction; coal mines; zinc and chemical works. Also called Ruda Śląska.

Nowy Dwor (NO-vee dvoor), Polish *Nowy Dwór Gdański*, German *Tiegenhof*, town, Gdańsk province, N POLAND, in VISTULA River delta, 22 mi/35 km ESE of GDAŃSK. Railroad junction.

Nowy Dwor Mazowiecki (NO-vee dvoor mahz-zo-VEETS-kee), Russian *Novy Dvor* or *Novyy Dvor*, town (2002 population 27,361), Warszawa province, E central POLAND, port on VISTULA River, at Narew River mouth, and 18 mi/29 km NW of WARSAW; 52°26′N 20°43′E. Naval base; brewing, sawmilling, flour milling; manufacturing of earthenware, soap, syrup. Before World War II, population 60% Jewish.

Nowy Port, POLAND: see GDAŃSK.

Nowy Sącz (NO-vee sonch), German *Neu Sandec*, city (2002 population 84,477), SE POLAND, on the DUNAJEC; 49°38′N 20°43′E. It is a railroad junction and an administrative and economic center. There are deposits of lignite and petroleum in the vicinity. Chartered in 1298, it passed to AUSTRIA in 1772 and was included in Poland in 1919. The city has several old churches; its 14th century palace was destroyed in World War II.

Nowy Staw (NO-vee stahf), German *Neuteich*, town, Elbląg province, N POLAND, in VISTULA River delta, 21 mi/34 km SE of GDAŃSK. Railroad junction. After 1945, briefly called Nytych.

Nowy Targ (NO-vee tahrg), town (2002 population 33,034), Nowy Sącz province, S POLAND, on DUNAJEC RIVER, and 40 mi/64 km S of CRACOW, in the POD-HALE. Railroad junction; airport; lumbering, tanning, flour milling; stone quarries; salmon hatchery. Hydroelectric plant. Founded 13th century. In Austrian Poland (until World War I), called Neumarkt.

Nowy Tomysl (NO-vee TO-mee-seel), Polish *Nowy Tomyśl*, German *Neutomischel*, town, Poznań province, W POLAND, 34 mi/55 km WSW of POZNAŃ; 52°19′N 16°09′E. Railroad junction; brewing, flour milling, sawmilling; trades in hops.

Noxapater (nahks-uh-PAI-tuhr), village (2000 population 419), WINSTON county, E central MISSISSIPPI, 9 mi/14.5 km S of LOUISVILLE; 32°59′N 89°03′W. Agriculture (cotton, corn; poultry, cattle; dairying);

manufacturing (gloves, furniture, cabinets). Nanih Waiya Historical Site, Choctaw burial mounds, to SE.

Noxen, unincorporated village, NOXEN township, WYOMING county, NE PENNSYLVANIA, 21 mi/34 km W of SCRANTON, on Bowman Creek; 41°25′N 76°03′W.

Noxon (NAHKS-uhn), village, SANDERS county, NW MONTANA, 37 mi/60 km NW of THOMPSON FALLS, on the CLARK FORK River, at upstream (SE) end of CABINET GORGE RESERVOIR, near IDAHO line. Railroad. Trapping; mining; logging; berries; hay; livestock. Noxon Rapids Dam (NOXON RESERVOIR), 3 mi/4.8 km SE on Clark Fork. Area surrounded by Kaniksu National Forest; CABINET MOUNTAINS Wilderness Area to E.

Noxon Reservoir (NAHKS-uhn), SANDERS county, NW MONTANA, on CLARK FORK river, 30 mi/48 km SSW of LIBBY, near IDAHO state line; c.30 mi/48 km long; 47°57′N 115°57′W. Maximum capacity 800,000 acre-ft. Formed by Noxon Rapids Dam (gravity and earth; 176 ft/54 m high), built (1960) by the Washington Power Company for power generation, flood control, and irrigation.

Noxubee (NAHKS-uh-bee), county (□ 700 sq mi/1,820 sq km; 2006 population 12,051), E MISSISSIPPI, bordering E on ALABAMA; ⊙ MACON; 33°06′N 88°34′W. Drained by NOXUBEE RIVER. Agriculture (cotton, corn, wheat, soybeans; cattle; dairying); timber. Part of Aliceville Lake reservoir on NE, on W boundary; part of Noxubee National Wildlife Refuge in NW corner. Formed 1833.

Noxubee River (NAHKS-uh-bee), c.140 mi/225 km long, in E MISSISSIPPI and W ALABAMA; rises in SE CHOCTAW county, E central Mississippi; flows first E through Tombigbee National Forest (Choctaw Unit) and Noxubee National Wildlife Refuge, then SE, past MACON, into SUMTER county, Alabama; enters TOM-BIGBEE River near GAINESVILLE.

Noya (NOI-ah), city, LA CORUÑA province, NW SPAIN, in GALICIA, fishing port on Noya Bay (inlet of the Atlantic), 18 mi/29 km WSW of SANTIAGO DE COM-POSTELA; 42°47′N 08°53′W. Shellfish and sardine fishing; boatbuilding, sawmilling. Has Gothic church. Tin and wolfram deposits nearby.

Noyabr'sk (nuh-YAHBRSK), city (2006 population 98,940), S YAMALO-NENETS AUTONOMOUS OKRUG, TYUMEN oblast, central SIBERIA, RUSSIA, on the Tyumen-Novyy Urengoy railroad line, 660 mi/1,062 km NE of TYUMEN, and 150 mi/241 km N of SURGUT; 63°10′N 75°37′E. Elevation 423 ft/128 m. Center for development of gas products and petroleum. Dairy plant. Regional airport. Made city in 1982.

Noyack Bay, inlet indenting N shore of S peninsula of LONG ISLAND, SE NEW YORK, c.1 mi/1.6 km W of SAG HARBOR; c.2 mi/3.2 km in diameter; 41°01′N 72°20′W. In summer-resort area; yachting; fishing.

Noyal-sur-Vilaine (nwah-yahl–syoor–vee-len), town (□ 11 sq mi/28.6 sq km; 2004 population 4,794), ILLE-ET-VILAINE department, BRITTANY, NW FRANCE, 6 mi/10 km E of RENNES; 48°07′N 01°31′W.

Noyan (nwah-YON), village (□ 17 sq mi/44.2 sq km; 2006 population 1,159), Montérégie region, S QUEBEC, E Canada, 3 mi/6 km from LACOLLE; 45°04′N 73°18′W. Agriculture (corn); dairying. Established 1855.

Noyan, MONGOLIA: see HÖVÜÜN.

Noyelles-Godault (nwah-yel-go-do), town (□ 2 sq mi/5.2 sq km), PAS-DE-CALAIS department, NORD PAS-DE-CALAIS region, N FRANCE, 5 mi/8 km NW of DOUAI; 50°25′N 02°59′E. Zinc and lead smelters.

Noyelles-sous-Lens (nwah-YEL–soo–LAHNS), E residential suburb (□ 1 sq mi/2.6 sq km) of LENS, PAS-DE-CALAIS department, NORD-PAS-DE-CALAIS region, N FRANCE; 50°26′N 02°52′E. In abandoned coal-mining district.

Noyemberyan (nuh-yem-buhr-YAHN), village, NE ARMENIA, 19 mi/31 km ENE of ALAVERDI, near AZERBAIJAN border; 41°10′N 44°59′E. Wheat. Also called Noyember.

Noyen, MONGOLIA: see HÖVÜÜN.

Noyen-sur-Sarthe (nwah-yahn–syur–sahrt), commune (□ 16 sq mi/41.6 sq km), SARTHE department, PAYS DE LA LOIRE region, W FRANCE, on the SARTHE, and 16 mi/26 km SW of Le MANS; 47°52′N 00°06′W. Agriculture market (cereals, cabbage, and potatoes).

Noyes (NOIZ), village, KITTSON county, extreme NW MINNESOTA, port of entry on CANADA (MANITOBA) border, 20 mi/32 km NW of HALLOCK, 2 mi/3.2 km NNE of ST. VINCENT, and 1 mi/1.6 km E of RED RIVER (NORTH DAKOTA state line); 48°59′N 97°12′W.

Noyil River (NOI-uhl), 100 mi/161 km long, KERALA state, S INDIA; rises on S spur of NILGIRI HILLS; flows E, past COIMBATORE (TAMIL NADU state), SINGA-NALLUR, and TIRUPPUR, to KAVERI RIVER, 24 mi/39 km SSE of ERODE. Supplies many irrigation channels in the W part of COIMBATORE district.

Noyo, unincorporated village, MENDOCINO county, NW CALIFORNIA, on the PACIFIC OCEAN, 35 mi/56 km NW of UKIAH, 2 mi/3.2 km S of FORT BRAGG and mouth of NOYO RIVER. Fruit; nursery stock; timber; fish, urchins.

Noyon (nwah-YON), town (□ 6 sq mi/15.6 sq km), OISE department, PICARDIE region, N FRANCE, near OISE RIVER and its lateral canal, 14 mi/23 km NE of COM-PIÈGNE; 49°35′N 03°00′E. Agricultural trade (cereals, cherries, black currants, artichokes, and sugar beets) and diversified manufacturing center (plumbing fixtures, metal furniture; woodworking, printing). Cookies and biscuits are a local specialty. There are grain silos and tree nurseries. In 768 Charlemagne was crowned king of the Franks here, and in 987 Hugh Capet was chosen king of France. Treaty creating alliance between France and Spain was signed here in 1516. Occupied by Germans 1914–1917, and devastated in both World Wars. The 12th–13th-century early Gothic cathedral was damaged but survived. The 16th-century town hall and birthplace of Calvin (now housing museum) restored after World War I. The partly ruined abbey of Ourscamps, dating from 12th century, is 4 mi/6.4 km S.

Noyo River, c.30 mi/48 km long, NW CALIFORNIA; rises in central MENDOCINO county, 18 mi/29 km N of Ukiah, in COAST RANGES; flows W to the PACIFIC OCEAN at Fort Bragg.

Nozawaonsen (no-ZAH-wah-ON-sen), village, Shimotakai county, NAGANO prefecture, central HONSHU, central JAPAN, 25 mi/40 km N of NAGANO; 36°55′N 138°26′E.

Nozay (no-zai), commune (□ 22 sq mi/57.2 sq km; 2004 population 4,732), LOIRE-ATLANTIQUE department, PAYS DE LA LOIRE region, W FRANCE, 14 mi/23 km SW of CHÂTEAUBRIANT; 47°34′N 01°37′W. Cider making. Iron once mined nearby.

Nozhay-Yurt (nuh-ZHEI–YOORT), village, E CHE-CHEN REPUBLIC, NE CAUCASUS, S European Russia, near highway, 69 mi/111 km ESE of GROZNYY, and 15 mi/24 km SW of KHASAVYURT; 43°05′N 46°22′E. Elevation 1,719 ft/523 m. In wheat and livestock area, near the border of DAGESTAN REPUBLIC. Successively in Chechen-Ingush Republic, Dagestan, and Chechen Republic. Called Andalaly in the 1940s and 1950s, when in Dagestan, following deportation of the Chechens for collaboration with the German army during World War II.

Nozu (NOTS), town, Ono county, OITA prefecture, E KYUSHU, SW JAPAN, 16 mi/25 km S of OITA; 33°02′N 131°41′E. Peppers. Furen limestone cave nearby.

Npologu (uhn-PO-lo-GOO), town, ENUGU state, S central NIGERIA, on road, 30 mi/48 km NW of ENUGU. Market town. Yams, cassava, cashews.

Nsanakang, town, SW CAMEROON, on CROSS River, and 25 mi/40 km WNW of MAMFE. Cacao, bananas, palm oil, and kernels; hardwood, rubber; salt deposits.

Nsanje (n-SAN-je), administrative center and district (2007 population 239,972), Southern region, MALAWI, on SHIRE RIVER, on railroad, and 105 mi/169 km S of

---

Area is shown by the symbol □, and capital city or county seat by ⊙.

ZOMBA; 16°55′S 35°16′E. Customs station on MO-ZAMBIQUE border; cotton-growing center; cotton ginning; tobacco, rice, corn. Airfield. Coal deposits (W). To immediate S and straddling the border is the Ndindi marsh, formed by the Shire River.

Nsawam (n-SAH-wahm), town, local council headquarters, EASTERN REGION, GHANA, 20 mi/32 km NNW of ACCRA; 05°18′N 00°44′W. On railroad and road junction; major cacao center.

Nsawkaw (n-SOR-kor), town, BRONG-AHAFO REGION, GHANA, 25 mi/40 km NW of Techiman near WENCHI; 07°52′N 02°19′W.

Nsiamfumu (nsee-ahm-FOO-moo), town, BAS-CONGO province, W CONGO, 15 mi/24 km N of MUANDA; 05°52′S 12°17′E. Formerly known as VISTA.

Nsimsim, GHANA: see NSINIM.

Nsinim (n-sin-eem), town, WESTERN region, GHANA; 06°29′N 02°53′W. On BIA RIVER, 40 mi/64 km W of BIBIANI. Timber, bauxite, cocoa, kola nuts. Formerly called Nsimsim.

Nsiza, town, MATABELELAND SOUTH province, SW central ZIMBABWE, 45 mi/72 km NE of BULAWAYO, at source of Nsiza River; 19°47′S 29°11′E. Elevation 4,640 ft/1,414 m. DHLODHLO NATIONAL MONUMENT (ruins) to SE. Copper mining; gold, asbestos mined in region. Cattle, sheep, goats; tobacco, corn, cotton, soybeans. Also spelled Insiza.

Nsoc, Equatorial Guinea: see NSOK.

Nsok (N-sok), town, WELE-NZAS province, continental Equatorial Guinea, near GABON border, 120 mi/193 km SE of BATA; 01°08′N 11°17′E. Coffee, cacao, hardwoods. Also spelled Nsoc.

Nsoko (un-soo-koo), town, LUBOMBO district, SE SWAZILAND, 68 mi/109 km SE of MBABANE, on railroad; 27°02′S 31°56′E. Road junction. Airstrip, clinic, and postal center. MOZAMBIQUE border 3 mi/4.8 km to E. Cattle, goats, sheep, hogs; sugarcane, pineapples, citrus, corn.

Nsontin (nson-TEEN), village, BANDUNDU province, W CONGO, 15 mi/24 km E of BANDUNDU; 03°09′S 18°00′E. Elev. 1,184 ft/360 m. Also known as SONTIN.

Nsukka (uhn-SOO-kah), town, ENUGU state, S NIGERIA, 30 mi/48 km N of ENUGU; 06°52′N 07°23′E. Agricultural trade center; palm oil and kernels, yams, cassava, corn. University town. Coal deposits at OBOLO (E).

Nsumbu National Park (□ 779 sq mi/2,025.4 sq km), NORTHERN province, N ZAMBIA, 55 mi/89 km W of MBALA (Abercorn), on shore of LAKE TANGANYIKA. Near CONGO border; hilly area with 50 mi/80 km of lake shoreline, drained by Lufubu River. Nkamba Lodge is here. Rift Valley Escarpment has granite balanced-rock formations. Abundant wildlife (elephants, bushbucks, warthogs, antelope); marine reserve has freshwater tropical fish, including goliath tiger fish.

Nsuta (n-SOO-tuh), town, WESTERN REGION, GHANA, 4 mi/6.4 km S of TARKWA; 05°16′N 01°59′W. On railroad. One of the world's leading manganese-mining centers; ore is exported via TAKORADI (130 mi/209 km SSE). Sometimes spelled INSUTA.

Ntabazinduna, historic site, MATABELELAND NORTH province, SW ZIMBABWE, 16 mi/ 26 km NW of BULAWAYO; 20°03′S 28°43′E. Site where King Mzilikazi of the Matabele tribe ordered the execution of his chiefs for hastily crowning his son when he was presumed dead during the settlement of Southern Zimbabwe in the 1830s by the Matabeles.

Ntcheu (n-che-uh), administrative center and district (2007 population 493,207), Central region, MALAWI, near MOZAMBIQUE border, 60 mi/97 km NE of ZOMBA; 14°49′S 34°38′E. Tobacco processing; wheat, corn, peanuts, potatoes.

Ntchisi, district (2007 population 231,165), Central region, central MALAWI; ⊙ Ntchisi. Bordered by Nkhotakota (NEE), SALIMA (SE tip), DOWA (S and W), and KASUNGU (NWN) districts.

Ntem, department, South province, CAMEROON: see MVILA and VALLEE DU NTEM departments.

Ntem River (n-TEM), c.250 mi/402 km long, tributary of the GULF OF GUINEA in N GABON and along CAMEROON-Equatorial GUINEA frontier; rises 40 mi/64 km SE of MINVOUL; flows NW and W to the ATLANTIC at Campo. Forms several large islands in its middle and upper course. Formerly Campo River.

Ntonso (n-TON-so), town, ASHANTI region, GHANA, 15 mi/24 km NE of KUMASI, near the edge of the Kwahu Plateau scarp; 06°50′N 01°31′W. Cacao, coffee, timber.

Ntoum (n-TOOM), town, ESTUAIRE province, NW GABON, 22 mi/35 km E of LIBREVILLE; 00°23′N 09°46′E.

N'tsaoueni (uhn-tsah-WE-nee), village, Njazidja island and district, NW Comoros Republic, 15 mi/24 km N of Moroni, on W coast of island; 11°30′S 43°16′E. Fish; livestock; ylang-ylang. Hahaya Intl. Airport to S.

Ntui (n-TOO-ee), town, Central province, CAMEROON, 38 mi/61 km NNE of Yaoundé; 04°26′N 11°39′E.

Ntungamo (n-TOON-gah-mo), administrative district (2005 population 407,400), WESTERN region, SW UGANDA; ⊙ NTUNGAMO; 00°55′S 30°15′E. Elevation ranges from 1,969 ft/600 m–6,562 ft/2,000 m. As of Uganda's division into eighty districts, borders KABALE (SW), RUKUNGIRI (NW), BUSHENYI (N), MBARARA (NE), and ISINGIRO (E) districts, and RWANDA (S). Vegetation primarily divided between grassland savannah and swampland; contains volcanic lakes. Agriculture (including beans, bananas, and matooke, coffee is main cash crop); cattle raising is important (also goats, rabbits, and sheep); tin deposits. Roads from Ntungamo lead to DEMOCRATIC REPUBLIC OF THE CONGO, Rwanda, and TANZANIA, as well as surrounding Uganda. Created in 1993 from parts of Bushenyi and Mbarara districts.

Ntungamo (n-TOON-gah-mo), town (2002 population 13,320), ⊙ NTUNGAMO district, WESTERN region, SW UGANDA, 30 mi/50 km SW of MBARARA; 00°53′S 30°12′E. Road junction.

Ntwetwe Pan, NE BOTSWANA, E of MAKGADIKGADI PANS NATIONAL PARK, SW of NATA. Vast non-perennial pan once fed by the OKAVANGO. Now dry and dusty except in rainy season (December–April).

Nuages, Col des, VIETNAM: see CLOUDS, PASS OF THE.

Nuanetsi, ZIMBABWE, MWENEZI.

Nuanetsi River, ZIMBABWE and MOZAMBIQUE: see MWENEZI RIVER.

Nuangola (NOO-an-GO-lah), borough (2006 population 658), LUZERNE county, NE central PENNSYLVANIA, 7 mi/11.3 km SW of WILKES-BARRE, on Nuangola Lake reservoir; 41°09′N 75°58′W.

Nu'ariya (noo-AH-ree-yeh), settlement (2004 population 20,964), AL AHSA region, SAUDI ARABIA, 130 mi/209 km NW of DHAHRAN, on the highway from the PERSIAN GULF to JORDAN and SYRIA; 27°30′N 28°20′E. Oil-pumping station on pipeline from ABQAIQ to SAIDA, LEBANON.

Nuba, Arab village, Hebron district, 7 mi/11 km N of HEBRON, WEST BANK; 31°36′N 35°02′E. Cereal; vineyards.

Nuba Mountains (NOO-bah), group of isolated hills in central SUDAN, S of AL UBAYYID, rising to 4,780 ft/1,457 m; 12°00′N 30°45′E. Remnants of a geologically old and much-eroded mountain system. Main centers are DALANG, RASHAD, TALODI, and KADUGLI.

Nubeena (nyoo-BEE-nuh), village, TASMANIA, SE AUSTRALIA, 70 mi/113 km SE of HOBART, on TASMAN PENINSULA; 43°06′S 147°45′E. Holiday destination; fishing. Former convict community.

Nubia (NOO-bee-ah), ancient state of NE AFRICA. At the height of its political power, Nubia extended N from the First Cataract of the NILE (near ASWAN, EGYPT) S to KHARTOUM, in the SUDAN. It came under the influence of the pharaohs early, and in the 20th century B.C.E. Seti I completed the occupation of the area. Many centuries later Egypt itself was ruled (8th–

7th century B.C.E.) by conquering Nubians of the Kush kingdom. Later, after the Assyrians expelled (c.667 B.C.E.) Tirhakah from Egypt, the Kushite capital was moved (c.530) from NAPATA to MARAWI. Marawi fell (c.350) to the Ethiopians and was abandoned. The region then came under the sway of the Nobatae, an ethnic group that mixed with the indigenous stock and formed a powerful kingdom with its capital at DONGOLA. The kingdom was converted to Christianity in the 6th century C.E. Joined with the Christian kingdom of Ethiopia, it long resisted Muslim encroachment, but in the 14th century it finally collapsed. NUBIA was then broken up into many petty states. Muhammad Ali of Egypt conquered (1820–1822) Nubia, and in the late 19th century much of the area was held by supporters of the Mahdi. The present area is commonly known as Nubia, and Nubians still inhabit the land near and around LAKE NASSER.

Nubia, Lake, SUDAN: see Lake NASSER.

Nubian Desert (NYOO-bee-uhn), E extension (□ 157,000 sq mi/408,200 sq km) of the SAHARA Desert, NE SUDAN, NE Afr., between the NILE River and the RED SEA. The arid region, largely a sandstone plateau, has numerous wadis flowing toward (but never reaching) the NILE, whose great bends are entrenched in the W margin of the region. Is crossed in a N-S direction by a railroad and road connecting EGYPT with SUDAN.

Ñuble (NYOO-blai), northernmost province (□ 5,487 sq mi/14,266.2 sq km) of BÍO-BÍO region, S central CHILE, between the ANDES and the PACIFIC OCEAN; ⊙ CHILLÁN; 36°35′S 72°00′W. Watered by ITATA and Ñuble rivers, it occupies part of the fertile central valley, has temperate climate. Mainly agricultural (grapes, cereals, vegetables; livestock). Lumbering, food processing.

Ñuble River (NYOO-blai), c.100 mi/160 km long, Ñuble province, BÍO-BÍO region, S central CHILE; rises at foot of NEVADOS DE CHILLÁN at ARGENTINA border; flows W, through irrigated valley, to ITATA RIVER 20 mi/32 km W of CHILLÁN.

Nubra River (noob-rah), 50 mi/80 km long, in JAMMU AND KASHMIR state, extreme N INDIA and NE PAKISTAN; flows from S end of SIACHEN Glacier in NORTHERN AREAS (Pakistan) S to SHYOK River just S of Liakzun in LEH district (India).

Nucet (noo-CHET), town, BIHOR county, ROMANIA, 52 mi/84 km SE of ORADEA; 46°28′N 22°35′E. Extraction of marble; manufacturing (textiles, foodstuffs).

Nuchek, Indian fishing village, S ALASKA, on W coast of HINCHINBROOK ISLAND, 35 mi/56 km WSW of CORDOVA. A nineteenth-century Russian and American trading post.

Nucía, La (lah-noo-THEE-ah), town, ALICANTE province, E SPAIN, 25 mi/40 km ESE of ALCOY. Olive pressing; wine, almonds, citrus, and other fruit.

Nuckolls, county (□ 576 sq mi/1,497.6 sq km; 2006 population 4,650), S NEBRASKA; ⊙ NELSON; 40°10′N 98°02′W. Agricultural region bounded S by KANSAS; drained by REPUBLICAN and LITTLE BLUE rivers. Cattle; dairying; hogs; corn, wheat, sorghum. Historic OREGON TRAIL in NE. Formed 1871.

Nucla, village (2000 population 734), MONTROSE county, SW COLORADO, near SAN MIGUEL RIVER, 39 mi/63 km WSW of MONTROSE; 38°16′N 108°32′W. Elevation 5,862 ft/1,787 m. In irrigated cattle, sheep, fruit, and vegetable area. Nearby deposits of carnotite yield vanadium and uranium. Coal. Uncompahgre National Forest to NE.

Nüdlingen (NYOOD-ling-uhn), village, LOWER FRANCONIA, BAVARIA, central GERMANY, 3 mi/4.8 km NE of BAD KISSINGEN, on E slope of the RHÖN; 50°13′N 10°08′E. Grain, vegetables. Has ruined castle.

Nudushan, IRAN: see NODUSHAN.

Nueces (NOO-ai-sis), county (□ 1,166 sq mi/3,031.6 sq km; 2006 population 321,457), S TEXAS; ⊙ CORPUS CHRISTI; 27°44′N 97°31′W. Deepwater port, industrial, commercial center. On Gulf plains; bounded E by GULF OF MEXICO, N by NUECES RIVER and Nueces, and E by CORPUS CHRISTI BAY; LAGUNA MADRE separates MUSTANG and PADRE Islands from mainland. A leading Texas oil and natural gas-producing county; also lime, sand, and gravel; diversified irrigated agriculture (cotton, grain sorghum, corn, wheat; some citrus); cattle. Resort beaches; fishing. Oil refining, diversified manufacturing (especially chemicals) centered in city of Corpus Christi, also processing of farm products, seafood. Mustang Island State Park on Gulf Coast in E; part of Laguna Largo, inland lake, in SE; S fringe of KING RANCH. Formed 1846.

Nueces Bay, Texas: see CORPUS CHRISTI BAY.

Nueces River (NOO-ai-sis), 315 mi/507 km long, TEXAS; rises on EDWARDS PLATEAU in REAL county; flows generally S past CRYSTAL CITY then SE past COTULLA, NE to THREE RIVERS, then again SE to Nueces Bay, a NW arm of CORPUS CHRISTI BAY. Receives FRIO RIVER at Three Rivers (ATASCOSA RIVER enters Frio 5 mi/8 km N). Dam 32 mi/51 km WNW of Corpus Christi impounds Lake Corpus Christi. West Nueces River rises in N EDWARDS county, flows S and SE c.85 mi/137 km to Nueces River, 10 mi/16 km NW of UVALDE.

Nueltin Lake (noo-EHL-tin) (□ 336 sq mi/873.6 sq km), NORTHWEST TERRITORIES, and NW MANITOBA, Canada, W of HUDSON BAY; 85 mi/137 km long, 3 mi/5 km–19 mi/31 km wide; 60°N 100°W. Drained NE by Thlewiaza River into Hudson Bay.

Nuenen (NUH-nuhn), town, NORTH BRABANT province, SE NETHERLANDS, 4 mi/6.4 km NE of EINDHOVEN; 51°28′N 05°33′E. Dairying; cattle, hogs; grain, vegetables, potatoes; manufacturing (food processing, leather products).

Nuervo, unincorporated town, RIVERSIDE county, S CALIFORNIA, 18 mi/29 km SE of RIVERSIDE, S of COLORADO RIVER AQUEDUCT, on San Jacinto River.

Nuestra Senora de la Soledad, Mission, California: see SOLEDAD.

Nueva Arcadia (NWAI-vah ar-KAH-dee-ah), town, COPÁN department, NW HONDURAS, on highway, 27 mi/43 km N of SANTA ROSA DE COPÁN; 15°02′N 88°45′W. Elevation 5,502 ft/1,677 m. Small farming; grain, tobacco.

Nueva Arica (NWAI-vah ah-REE-kah), village, CHICLAYO province, LAMBAYEQUE region, NW PERU, in W foothills of CORDILLERA OCCIDENTAL, on SAÑA RIVER, and 35 mi/56 km E of CHICLAYO; 06°53′S 79°21′W. Rice, corn, alfalfa, cotton.

Nueva Armenia (NWAI-vah ahr-ME-nee-ah), town, FRANCISCO MORAZÁN department, S central HONDURAS, 26 mi/42 km S of TEGUCIGALPA, on unpaved road; 13°44′N 87°10′W. Grain; livestock.

Nueva Asunción, department, former department along PARAGUAY's border with BOLIVIA; 21°00′S 61°00′W. This large, sparsely populated department was merged with Boquerón department in 1992.

Nueva Australia, PARAGUAY: see COLONIA NUEVA AUSTRALIA.

Nueva Bilbao, CHILE: see CONSTITUCÍON.

Nueva Caceres, PHILIPPINES: see NAGA.

Nueva Carteya (NWAI-vah kahr-TAI-ah), town, CÓRDOBA province, S SPAIN, 8 mi/12.9 km W of BAENA; 37°35′N 04°28′W. Agricultural trade center (wheat, grapes, vegetables); olive-oil processing, soap manufacturing.

Nueva Chicago (NWAI-vah chee-KAH-go), W industrial section of BUENOS AIRES, ARGENTINA.

Nueva Ciudad Guerrero, city and township, ⊙ GUERRERO municipio, TAMAULIPAS, N MEXICO, near Falcón Dam on RIO GRANDE (TEXAS border), and 50 mi/80 km SSE of NUEVO LAREDO, on Mexico Highway 30; 26°49′N 99°20′W. Agricultural center (cotton, sugarcane, corn; cattle).

Nueva Colombia (NWAI-vah ko-LOM-bee-ah), town, La Cordillera department, S central PARAGUAY, 20 mi/

32 km NE of Asunción; 25°11′S 57°21′W. In agricultural area (fruit, tobacco; livestock). Sometimes Colonia Nueva Colombia.

Nueva Concepción (NWAI-vah kon-sep-see-ON), city and municipality, CHALATENANGO department, W central EL SALVADOR, 18 mi/29 km E of SANTA ANA. Agriculture; livestock raising.

Nueva Concepcion (NWAI-vah kon-sep-see-ON), town (2002 population 10,500), ESCUINTLA department, GUATEMALA, 34 mi/55 km WSW of ESCUINTLA on the coastal plain; 14°12′N 91°18′W. Elevation 164 ft/50 m. Formed from a large agricultural reform colony created during the 1960s; agriculture (maize, sorghum, bananas, sugarcane, lemongrass); cattle.

Nueva Ecija (NWAI-vah ai-SEE-hah), province (□ 2,040 sq mi/5,304 sq km), in CENTRAL LUZON region, central LUZON Island, PHILIPPINES, drained by PAMPANGA RIVER; ⊙ CABANATUAN; 15°35′N 121°00′E. Mountainous, with fertile valleys. Principal crop is rice; corn, tobacco, abaca, and sugarcane are also grown. After World War II, the province was chief center of disturbances by rebellious Hukbalahaps. Unrest continued into the 1960s, but the situation is calmer at present.

Nueva España (NWAI-vah es-PAHN-yah), town, S central TUCUMÁN province, ARGENTINA, on railroad, 32 mi/51 km S of TUCUMÁN. Agricultural center (tobacco, sugarcane; livestock); 27°16′S 65°12′W.

Nueva Esparta (NWAI-vah es-PAHR-tah), state (□ 440 sq mi/1,144 sq km; 2001 population 373,851), N VENEZUELA; ⊙ LA ASUNCIÓN; 11°00′N 64°00′W. Consists primarily of one large and two small islands: MARGARITA, COCHE, and CUBAGUA. Little agriculture; some grazing of livestock. Mineral resources include asbestos, salt, gypsum. Main towns include PORLAMAR, SANTA ANA, PAMPATAR, JUAN GRIEGO, Boca de Rio, PARTA DE PIEDRA. Popular tourist site, especially for Venezuelans. Site of Parques Nacionales LAGUNA DE LA RESTINGA and Cerro El Copay, as well as Monumentos Naturales Laguna de Las Marites, Las Tetas de María Guevara, and Cerro Matasiete and Guayamurí.

Nueva Esparta (NWAI-vah es-PAHR-tah), municipality and town, LA UNIÓN department, EL SALVADOR, N of SANTA ROSA DE LIMA; 13°47′N 87°50′W.

Nueva Esperanza (NWAI-vah es-pe-RAHN-zah), town (1991 population 2,689), ⊙ Pellegrini department (□ 2,500 sq mi/6,500 sq km), NW SANTIAGO DEL ESTERO province, ARGENTINA, 75 mi/121 km NE of TUCUMÁN; 26°12′S 64°16′W. Agricultural center (corn, alfalfa; livestock); lumbering.

Nueva Esperanza de Machacamarca (NWAI-vah es-pe-RAHN-zah dai mah-chah-kah-MAHR-kah), canton, AROMA province, LA PAZ department, W BOLIVIA, 16°56′S 68°16′W. Elevation 12,851 ft/3,917 m. Gas wells in area. Salt extraction. Copper, gypsum, limestone, clay deposits. Agriculture (potatoes, yucca, bananas, rye); cattle.

Nueva Galicia, Spanish colonial administrative region, W MEXICO, comprising roughly the present states of JALISCO and NAYARIT with S SINALOA. Conquered (1529–1531) by Nuño de Guzmán and later governed by Francisco Vásquez de Coronado, the territory was the scene of the Mixtón War in 1541. In 1548 it was given its own audiencia at GUADALAJARA. Nominally subject to the viceroy of NEW SPAIN, it was essentially a separate administration controlled from Spain, and it came to be known after the creation (1563) of a presidential office of its own as the presidency of Nueva Galicia. Its independent character, however, declined as colonial-era authority was more and more centralized in MEXICO CITY.

Nueva Germania (NWAI-vah jer-MAH-nee-ah), town, San Pedro department, central PARAGUAY, 32 mi/51 km NE of San Pedro; 23°53′S 56°34′W. Maté-growing center.

Nueva Gerona (NWAI-vah hai-RON-ah), principal town of ISLA DE LA JUVENTUD (2002 population

46,923), SW CUBA, on small Las Casas River, and 90 mi/145 km SSW of HAVANA; 21°53′N 82°49′W. Resort, trading, and agricultural center (citrus fruit, tobacco, potatoes, winter vegetables). Served by airline and linked through nearby landing (N) with Surgidero de BATABANÓ on main island. A clean town of modern character with a customhouse and industrial art center. Fine beaches (fishing, bathing) and caves in picturesque surrounding region. Presidio Modelo penitentiary is 2 mi/3 km E. Marble, copper, iron, and gold deposits in vicinity.

Nueva Granada (NWAI-vah grah-NAH-dah), municipality and town, USULUTÁN department, EL SALVADOR, N of INTER-AMERICAN HIGHWAY, between MERCEDES UMAÑA and TRIUNFO; 13°36′N 88°27′W.

Nueva Guadalupe (NWAI-vah gwah-dah-LOO-pai), municipality and town, SAN MIGUEL department, EL SALVADOR, S of INTER-AMERICAN HIGHWAY in W part of department.

Nueva Guinea (NWAI-vah gee-NE-ah), town (2005 population 25,585), SOUTH ATLANTIC COAST AUTONOMOUS REGION, ZELAYA department, NICARAGUA, 38 mi/60 km SW of RAMA (connected to Rama highway by unpaved road), on E slope of Cordillera Chontalena; 11°41′N 84°27′W. English is widely spoken.

Nueva Helvecia (NWAI-vah el-VAI-syah), city (2004 population 10,002), COLONIA department, SW URUGUAY, on railroad, and 37 mi/60 km ENE of Colonia; 34°18′S 57°14′W. In agricultural region (grain, fruit; livestock). Serves as urban nucleus of COLONIA SUIZA agricultural settlement.

Nueva Imperial (NWAI-vah eem-pai-RYAHL), city, ⊙ Nueva Imperial comuna, CAUTÍN province, ARAUCANIA region, S central CHILE, on IMPERIAL RIVER, on railroad, and 20 mi/32 km W of TEMUCO; 38°44′S 72°57′W. Inland river port, trading and agricultural center (cereals; livestock); food processing; lumbering.

Nueva Island (NWAI-vah) (□ 45 sq mi/117 sq km), in TIERRA DEL FUEGO, at mouth of BEAGLE CHANNEL on the ATLANTIC, 8 mi/13 km S of main island of the archipelago; 9 mi/15 km long; 55°15′S 66°35′W. One of Beagle Channel Islands awarded to CHILE by Vatican intervention in 1977 and again in 1984. Natural resources (petroleum, mineral deposits, and krill) and strategic position continues to be disputed by Chile and Argentina.

Nueva Italia de Ruíz (NWAI-vah ee-TAH-lee-ah dai roo-EEZ), town, ⊙ Múgica municipio, MICHOACÁN, W MEXICO, 35 mi/56 km SSW of URUAPAN on Mexico Highway 37; 19°01′N 102°06′W. Cereals, rice, sugar, fruit.

Nueva Loja (NWAI-vah LO-hah), town, ⊙ SUCUMBÍOS province, NE ECUADOR near AGUARICO RIVER; 00°06′N 76°52′W. Rapidly growing town on the main petroleum-producing area. Airport. Formerly Lago Agrio.

Nueva Lubecka (NWAI-vah loo-BAI-kah), village, W CHUBUT province, ARGENTINA, on GENOA RIVER, 100 mi/161 km NW of SARMIENTO; 44°32′S 70°24′W. Alfalfa; sheep, horses, goats.

Nueva Ocotepeque (NWAI-vah o-ko-tai-PAI-kai), city, ⊙ OCOTEPEQUE department, W HONDURAS, in Senseti Valley, on Lempa River, and 33 mi/53 km SW of SANTA ROSA DE COPÁN, near EL SALVADOR border; 14°26′N 89°11′W. Elevation 8,957 ft/2,730 m. Commercial center in agricultural area; coffee, sugarcane, wheat, tobacco; livestock. Linked by paved road with SAN PEDRO SULA, CHIQUIMULA (GUATEMALA), and SAN SALVADOR (El Salvador). Founded just NE of site of Ocotepeque, destroyed 1935 by flood of Marchala River (small left affluent of the Lempa).

Nueva Palmira (NWAI-vah pahl-MEE-rah), city, COLONIA department, SW URUGUAY, port on URUGUAY RIVER (ARGENTINA border) above its mouth on the Rio de la PLATA, 50 mi/80 km NW of Colonia and 50 mi/80 km N of BUENOS AIRES; 33°53′S 58°28′W. In rich agricultural region. Ships grain, livestock; flour milling; steel mill.

Nueva Paz (NWAI-vah PAHZ), town, LA HABANA province, W CUBA, 45 mi/72 km SE of HAVANA, on major highway; 22°46′N 81°45′W. Rice, sugarcane, vegetables, cattle. Rice mills. Manuel Isla sugar mill 4 mi/6 km NNE.

Nueva Pompeya (NWAI-vah pom-PAI-yah), village, N Chaco province, ARGENTINA, on old bed of BERMEJO RIVER, and 24 mi/39 km SW of EL PINTADO; 24°51′S 61°30′W. Desert outpost with Franciscan mission.

Nueva Rosita (NWAI-vah ro-SEE-tah), mining city and township, ⊙ San Juan de Sabinas municipio, COAHUILA, N MEXICO, in semiarid country, in SABINAS coal district, 170 mi/274 km NW of Monterreyon; Mexico Highway 53; 27°55′N 101°17′W. Elevation 1,411 ft/430 m. Connected by railroad with SALTILLO, MONTERREY, PIEDRAS NEGRAS. Developed as prominent industrial city of N Mexico in 1930s. Coal mining; zinc smelter; manufacturing (zinc sulfate and sulfuric acid). Sometimes called simply Rosita.

Nueva San Salvador (NWAI-vah SAHN sahl-vah-DOOR), city and municipality, ⊙ LA LIBERTAD department, central EL SALVADOR; 13°41′N 89°17′W. It was founded in 1854 after the capital, SAN SALVADOR, was destroyed in an earthquake. San Salvador, 9 mi/14.5 km away, was rebuilt, and Nueva San Salvador became a wealthy suburb, now part of the San Salvador metropolitan area. It is situated among coffee farms. The city is also called SANTA TECLA.

Nueva Santa Rosa (NWAI-vah SAHN-tah RO-sah), town, SANTA ROSA department, S GUATEMALA, in Pacific piedmont, 1 mi/1.6 km ESE of SANTA ROSA DE LIMA, 8 mi/12.9 km N of CUILAPA; 14°23′N 90°17′W. Elevation 3,284 ft/1,001 m. Coffee, corn, beans.

Nueva Segovia (NWAI-vah se-GO-vee-ah), department (□ 1,595 sq mi/4,147 sq km), NW NICARAGUA, on HONDURAS border; ⊙ OCOTAL. Separated from Honduras by CORDILLERA DE DIPILTO and JALAPA; bounded SE by COCO RIVER. Gold and silver mining at EL JÍCARO. Tobacco, sugarcane, coffee; livestock (N); cacao, corn, vegetables (S). Main centers: OCOTAL, El Jícaro.

Nuevas Grandes (NWAI-vahs GRAHN-dais), narrow inlet off N coast of CUBA, on LAS TUNAS-CAMAGÜEY province border, 20 mi/32 km ESE of NUEVITAS; 8 mi/13 km long; 21°23′N 77°00′W. Receives small Cabreras River.

Nueva Tolten (NWAI-vah tol-TAIN), town, ⊙ Tolten comuna, CAUTÍN province, ARAUCANIA region, S CHILE, 40 mi/64 km SW of TEMUCO; 39°13′S 73°14′W. Grapes, vegetables, fruit; livestock.

Nueva Trinidad (NWAI-vah tree-nee-DAHD), municipality and town, CHALATENANGO department, EL SALVADOR, ENE of CHALATENANGO city; 14°04′N 88°48′W.

Nueva Villa de Padilla (NWAI-vah VEE-yah dai pah-DEE-yah), town, ⊙ Padilla municipio, TAMAULIPAS, NE MEXICO, 32 mi/51 km NE of CIUDAD VICTORIA, on Mexico Highway 101; 24°00′N 98°47′W. Cereals, sugarcane, fruit; livestock.

Nueva Vizcaya (NWAI-vah vees-KAH-yah), province (□ 1,507 sq mi/3,918.2 sq km), in CAGAYAN VALLEY region, central LUZON, PHILIPPINES; ⊙ BAYOMBONG; 16°20′N 121°20′E. Population 27.3% urban, 72.7% rural. Drained by the CAGAYAN and its tributary MAGAT RIVER. Mountainous, with fertile valleys. Agriculture (rice, corn, tobacco), livestock raising.

Nueva Vizcaya, MEXICO: see DURANGO.

Nueve de Abril (NWAI-vai dai ah-BREEL), canton, CERCADO province, ORURO department, W central BOLIVIA. Elevation 12,372 ft/3,771 m. Antimony-bearing lode; copper and clay deposits. Agriculture (potatoes, yucca, bananas, barley, oats); cattle.

Nueve de Julio or **9 de Julio** (NWAI-vai dai HOO-lee-o), town (□ 1,650 sq mi/4,290 sq km), ⊙ Nueve de Julio district (□ 1,650 sq mi/4,274 sq km; 1991 population 44,015), N central BUENOS AIRES province, ARGENTINA, 60 mi/97 km SW of CHIVILCOY. Commercial center in livestock-raising area; meatpacking, dairying.

Nueve de Julio, town (1991 population 2,417), ⊙ Nueve de Julio department, S SAN JUAN province, ARGENTINA, in SAN JUAN RIVER valley (irrigation area), 12 mi/19 km SSE of SAN JUAN. In wine grape– and fruit-growing area.

Nueve de Julio, ARGENTINA: see SIERRA COLORADA.

Nueve de Julio, Argentina: see TOSTADO.

Nuevitas (nwai-VEE-tahs), city (2002 population 38,995), CAMAGÜEY province, E CUBA, on the Guincho peninsula on the N coast; 21°33′N 75°17′W. Once the largest world port for sugar shipping, its salience declined in 1970s due to growth elsewhere. Nuevitas is sheltered by a huge harbor with an entrance through a twisted, rocky channel, has two auxiliary ports, and remains a major shipping point for Cuban sugar as well as other products from the surrounding agricultural region. Ports handle about 6% of total sugar exports, serving five sugar mills from Camagüey and CIEGO DE ÁVILA provinces, and have cargo facilities for bulk grain handling. Thermoelectric plant, 10 de Octubre, has 442 MW capacity. Fertilizer and cement plants nearby. It is connected to CAMAGÜEY city by railroad and a two-lane highway. The large bay was sighted by Columbus in 1492. Founded in 1775, the city was moved to its present site in 1828.

Nuevo Casas Grandes (NWAI-vo KAH-sahs GRAHN-des), city and township, CHIHUAHUA, N MEXICO, on arid plateau, on CASAS GRANDES RIVER (irrigation), and 130 mi/209 km SW of CIUDAD JUÁREZ; 30°24′N 107°55′W. Elevation 4,833 ft/1,473 m. Railroad junction; agricultural center (cotton, corn, beans, tobacco, cattle); tanning; flour milling; lumbering.

Nuevo Celilac (NWAI-vo se-LEE-lak), town, SANTA BÁRBARA department, NW HONDURAS, 6 mi/10 km NW of SANTA BÁRBARA; 14°56′N 88°19′W. No road. Small farming; grain; livestock.

Nuevo Chagres (NWAI-vo CHAH-grais), town, ⊙ Chagres district, COLÓN province, central PANAMA, minor port on CARIBBEAN SEA, and 14 mi/23 km SW of COLÓN; 09°14′N 80°05′W. Corn, beans, rice, coconuts; livestock. At mouth of CHAGRES RIVER 8 mi/12.9 km NE, in CANAL ZONE, are remains of old town of Chagres, which flourished until rise of Colón.

Nuevo Coahuayana, MEXICO: see COAHUAYANA DE HIDALGO.

Nuevo Colón (NWAI-vo ko-LON), town, ⊙ Nuevo Colón municipio, BOYACÁ department, central COLOMBIA, 15 mi/24 km SSW of TUNJA; 05°21′N 73°28′W. Elevation 7,808 ft/2,379 m. Coffee, sugarcane, corn; livestock.

Nuevo Cristóbal (NWAI-vo krees-TO-bahl), elite residential section of COLÓN, COLÓN province, central PANAMA; faces MANZANILLO BAY.

Nuevo Cuscatlán (NWAI-vo koos-kaht-LAHN), municipality and town, LA LIBERTAD department, EL SALVADOR, E of NUEVA SAN SALVADOR; 13°39′N 89°16′W.

Nuevo Eden de San Juan (NWAI-vo AI-den dai sahn hwahn), municipality and town, SAN MIGUEL department, EL SALVADOR, far NW of department, near the HONDURAS border; 13°49′N 88°29′W.

Nuevo Emperador (NWAI-vo em-per-ah-DOR), village and minor civil division of Arraiján district, PANAMA province, central PANAMA, 13 mi/21 km NW of PANAMA city; 09°00′N 79°44′W. Bananas, oranges; livestock. Formerly Paja.

Nuevo, Golfo (NWAI-vo, GOL-fo), inlet of S ATLANTIC OCEAN in E CHUBUT province, ARGENTINA, S of VALDÉS PENINSULA; 40 mi/64 km W-E, 30 mi/48 km N-S. Port of PUERTO MADRYN on W shore, PUERTO PIRÁMIDES (on VALDÉS PENINSULA) on NE.

Nuevo Ideal (NWAI-vo ee-dai-AHL), city, in central DURANGO, MEXICO, 37 mi/60 km N of VICTORIA DE DURANGO; 24°53′N 105°02′W.

Nuevo Laredo (NWAI-vo lah-RAI-do), city (2005 population 348,387) and township, Nuevo Laredo municipio, TAMAULIPAS state, NE MEXICO, across the RIO GRANDE from LAREDO, TEXAS; 27°30′N 99°31′W. Linked with the U.S. by road and railroad bridges, Nuevo Laredo is the N terminus of the national railroad and the INTER-AMERICAN HIGHWAY, as well as an important point of entry for U.S. tourists driving to Mexico. It is also a center of international trade and the distribution point for an agriculture (mainly cotton) and livestock-raising area; commerce; tourism industry. Nuevo Laredo has been one of the many Mexican cities affected by an influx of foreign capital, primarily due to the establishment of foreign-owned industrial plants, known as maquiladoras. Has developed into a transportation–trans-shipment center since NAFTA (1993). Founded in 1755, the city was part of Laredo until the end of the Mexican-American War in 1848. Nuevo Laredo played a role in the Mexican Revolution of 1910 and was burned extensively in 1914.

Nuevo León (NWAI-vo lai-ON), state (□ 25,136 sq mi/65,353.6 sq km), N MEXICO; ⊙ MONTERREY; 21°10′N 98°27′W. The S and W parts of the state are traversed by the SIERRA MADRE ORIENTAL, but some of the extreme W portions lie within the vast, semiarid basin lands of N Mexico, which are cultivable under irrigation. Much of the N is arid, but to the E, where the plains sweep down toward the lowlands of TAMAULIPAS and are crossed by several large rivers, the land is suitable for rain-fed agriculture. Grains and citrus fruits are grown. Nuevo León has an extremely diversified industrial structure which includes oil refining and extensive heavy and light manufacturing. The growth of *maquiladoras*, foreign-owned industrial plants that produce goods for export to the UNITED STATES, has become important. The area is also a leading national producer of iron, steel, and chemicals. Road and railroad connections within the state are excellent, and Nuevo León enjoys one of the highest living standards in Mexico. The area was explored and settled by the Spanish in the late sixteenth century. Nuevo León became a state in 1824.

Nuevo Manoa (NWAI-vo mah-NO-ah), port, GENERAL FEDERICO ROMÁN province, PANDO department, NW BOLIVIA, 6 mi/10 km N of FORTALEZA; 09°42′S 65°42′W. Elevation 367 ft/112 m. Agriculture (yucca, bananas, cacao, coffee, tobacco, coca, rubber, peanuts); cattle and horse raising.

Nuevo Morelos (NWAI-vo mo-RAI-los), town, TAMAULIPAS, NE MEXICO, in E foothills of SIERRA MADRE ORIENTAL, 90 mi/145 km WNW of TAMPICO on Mexico Highway 80; 22°34′N 99°17′W. Agave, fruit; livestock.

Nuevo Necaxa (NWAI-vo nai-KA-shah), town, ⊙ Juan Galindo municipio, PUEBLA, central MEXICO, in SIERRA MADRE ORIENTAL, on NECAXA RIVER, and 4 mi/6.4 km NE of HUAUCHINANGO. Sugarcane, coffee, fruit. Hydroelectric plant, on Necaxa Falls (c.540 ft/165 m high) nearby, supplies MEXICO CITY. Sometimes called Necaxa.

Nuevo Parangaricutiro, MEXICO: see NUEVO SAN JUAN PARANGARICUTIRO.

Nuevo Progreso (NWAI-vo pro-GRE-so), town, SAN MARCOS department, SW GUATEMALA, in Pacific piedmont, 5 mi/8 km NNW of COATEPEQUE; 14°48′N 91°55′W. Elevation 2,165 ft/660 m. Coffee, sugarcane, grain. In Mam-speaking area.

Nuevo, Río (NWAI-vo, REE-o) or **González River**, c.75 mi/121 km long, TABASCO, SE MEXICO, slough draining a portion of lower GRIJALVA, river marshes; formed 7 mi/11.3 km ENE of CÁRDENAS; flows E, N, and NW, to Gulf of CAMPECHE 6 mi/9.7 km ENE of PARAÍSO.

Nuevo Rocafuerte (NWAI-vo ro-kah-FWER-tai), village, NAPO province, E ECUADOR, landing on NAPO RIVER at mouth of the YASUNÍ, on PERU border, and 20 mi/32 km W of PANTOJA (Peru), 230 mi/370 km ESE of QUITO; 00°56′S 75°24′W. Cattle raising; collecting of rubber, balata, chicle, furs. Founded after 1942 boundary settlement, replacing ROCA-FUERTE. Nearby on Napo River are Puerto Miranda (military base; 10 mi/16 km NW) and Tiputini airfield (at mouth of TIPUTINI RIVER; 20 mi/32 km NW).

Nuevo Rocafuerte (NWAI-vo ro-kah-FWER-tai), NAPO province, former Ecuadorian military post, in AMAZON basin, on NAPO RIVER at mouth of the AGUARICO RIVER, opposite PANTOJA (PERU); 00°58′S 75°10′W. Following 1942 border settlement, ROCA-FUERTE was replaced by Nuevo Rocafuerte, 20 mi/32 km.

Nuevo San Carlos (NWAI-vo SAHN KAHR-los), town, RETALHULEU department, SW GUATEMALA, in Pacific piedmont, on branch of Coyote River, and 4 mi/6.4 km N of RETALHULEU; 14°36′N 91°42′W. Sugarcane, corn, rice. In Quiché-speaking area.

Nuevo San Juan Parangaricutiro (NWAI-vo sahn hwahn pah-rahn-gah-ree-koo-TEE-ro), town, ☉ Nuevo Parangaricutiro municipio, MICHOACÁN, central MEXICO, in Sierra de los TARASCOS, 20 mi/32 km WNW of URUAPAN; 19°24′N 102°08′W. Elevation 7,415 ft/2,260 m. In agricultural area. After eruption (1943) of PARÍCUTIN volcano, 5 mi/8 km SSW, original town was largely engulfed by lava flow, residents moved to new site.

Nuevo Santo Tomás de los Plátanos (NWAI-vo SAHN-to to-MAHS dai los PLAH-tah-nos), town, MEXICO state, central MEXICO, on MICHOACÁN border, 40 mi/64 km WSW of TOLUCA DE LERDO; 19°09′N 100°12′W. Sugarcane, fruit, corn.

Nuevo Soyaltepec (NWAI-vo so-YAHL-tai-pek), town, N OAXACA, MEXICO, 23 mi/37 km NW of TUX-TEPEC, near PRESIDENTE MIGUEL ALÉMAN Dam and within TEMASCAL LAKE NATURAL PARK. Elevation 1,640 ft/500 m. Hot climate. Agriculture (corn, beans, coffee, sugarcane, tropical fruits); wood. Temascal hydroelectric plant (generating capacity 154,080 kw) at Temascal Dam. Former town name is Temascal. A Mazatec-speaking community.

Nuevo Urecho (noo-E-vo oo-RE-cho), town, MI-CHOACÁN, central MEXICO, 22 mi/35 km SSE of UR-UAPAN. Fruit, cereals. Near railroad, poor roads.

Nuevo Zoquiapam (NWAI-vo zo-kee-AH-pahm), town, central OAXACA, MEXICO, 16 mi/26 km NE of OAXACA DE JUÁREZ; 17°18′N 96°36′W. Mountainous terrain in the sierra of IXTLÁN on a tributary of the RÍO GRANDE. Temperate climate although there is a slight variation from the E to the W. Farming (corn, beans; livestock), Zapotec Indian community. Also SAN-TIAGO ZOQUIÁPAM.

Nüfenen Pass (NYOO-fuh-nen), Italian *Passo della Novena* (8,130 ft/2,478 m), S central SWITZERLAND, between upper RHÔNE valley and Val Ticino, on border of VALAIS and TICINO cantons.

Nuffield Radio Astronomy Observatory, ENGLAND: see JODRELL BANK.

Ñuflo de Chávez, province, SANTA CRUZ department, E central BOLIVIA, ☉ CONCEPCIÓN; 15°55′S 62°30′W.

Nufringen (NUF-ring-uhn), village, region of STUTT-GART, BADEN-WÜRTTEMBERG, S GERMANY, 17 mi/27 km SW of STUTTGART, on W slope of Schönbuch; 48°37′N 08°53′E. Vegetables.

Nufud, ARABIA: see NAFUD.

Nugaal, region, NE SOMALIA; ☉ GAROE. Desert coast-line along INDIAN Ocean to the E. Some grazing and cultivation (sorghum, maize, cassava, millet).

Nugammute, Alaska: see NOGAMUT.

Nugegoda (NU-GAI-go-duh), SE suburb, WESTERN PROVINCE, SRI LANKA, 5 mi/8 km SE of COLOMBO city center; 06°52′N 79°52′E. Residential area. Handicrafts (lace, pottery); trades in vegetables, rice, coconuts, rubber. Part of KOTTE urban council.

Nugent (NYOO-juhnt), village, SE TASMANIA, SE AUSTRALIA, 25 mi/40 km NE of HOBART; 42°42′S 147°45′E.

Nuguria Islands (noo-goo-REE-ah), coral group (□ 2 sq mi/5.2 sq km) of Malum Islands, NEW IRELAND province, BISMARCK ARCHIPELAGO, PAPUA NEW GUI-NEA, SW PACIFIC OCEAN, circa 135 mi/217 km E of NEW IRELAND; 03°20′S 154°45′E. Consists of 2 atolls of circa 50 islets. Privately owned; coconut plantations; fishing.

Nuh (NOO), town, tahsil headquarters, GURGAON district, HARYANA state, N INDIA, 24 mi/39 km S of GURGAON; 28°07′N 77°01′E. Millet, gram, wheat; handicraft cloth weaving. Stone (used for road metal) quarried in hills (SW).

Nuhu Chut, INDONESIA: see NUHU CUT.

Nuhu Cut (NOO-hoo), largest island (□ 241 sq mi/626.6 sq km) of KAI ISLANDS, S Maluku province, IN-DONESIA, in BANDA SEA, 180 mi/290 km SE of SERAM; 65 mi/105 km long, 8 mi/12.9 km wide; 05°37′S 133°02′E. Wooded and mountainous, rising to 2,500 ft/762 m. Fishing; copra. Also called Noehoe Tjoet, Nuhu Tjut, or Great Kai; formerly spelled Nuhu Chut.

Nuhu Tjut, INDONESIA: see NUHU CUT.

Nui (NOO-ee), crescent-shaped atoll (2002 population 548), TUVALU, SW PACIFIC; 07°16′S 177°10′E. Consists of 500 acres/202 ha. Copra. Formerly Netherland Island.

Nuits-Saint-Georges (nwee–san–zhorzh), town (□ 8 sq mi/20.8 sq km; 2004 population 5,335), CÔTE-D'OR department, BURGUNDY region, E central FRANCE, on E slope of the Côte d'Or hill range (here called CÔTE de Nuits), 13 mi/21 km SSW of DIJON; 47°08′N 04°57′E. Renowned Burgundy vineyards and wine-distribution center, whose fame dates form the time of Louis XIV. The celebrated abbey of CÎTEAUX (founded 1098), mother house of the Cistercian monastic order, is 7 mi/11.3 km E.

Nu Jiang, CHINA: see SALWEEN.

Nuka Bay, S ALASKA, on SE coast of KENAI PENINSULA, opens into Gulf of ALASKA; 20 mi/32 km long, 1 mi/1.6 km–7 mi/11.3 km wide; 59°23′N 150°31′W.

Nuka Island, S ALASKA, in Gulf of ALASKA, off SE KENAI PENINSULA, 65 mi/105 km SW of SEWARD; 8 mi/12.9 km long, 3 mi/4.8 km wide; 59°22′N 150°41′W. Entire island used as fox farm.

Nukata (noo-KAH-tah), town, Nukata county, AICHI prefecture, S central HONSHU, central JAPAN, 28 mi/45 km S of NAGOYA; 34°54′N 137°17′E.

Nukatl' Range (noo-KAH-tuhl), N spur of the E Greater CAUCASUS Mountains, in S central DAGESTAN REPUBLIC, SE European Russia; extends in arc 30 mi/48 km N, forming right watershed of the AVAR KOISU River; 42°15′N 46°38′E. Rises to 12,860 ft/3,920 m.

Nu Kiang, CHINA: see SALWEEN.

Nuku'alofa (NOO-koo-ah-LO-fah), town (2000 population 21,538) ☉ chief port of TONGA, on the N coast of TONGATAPU island. The town has a reasonable, reef-protected harbor exporting copra, bananas, vanilla, and handicrafts. The royal palace and government buildings are here.

Nukufetau (NOO-koo-fai-TAH-oo), atoll (□ 1 sq mi/2.6 sq km; 2002 population 586), central TUVALU, SW PACIFIC; 08°00′S 178°28′E. Many islets on reef (circumference 24 mi/39 km); airfield. Formerly De Peyster Group or De Peyster's Island.

Nuku Hiva (NOO-koo HEE-vah), volcanic island (□ 127 sq mi/330.2 sq km), South PACIFIC, largest of the MARQUESAS ISLANDS, FRENCH POLYNESIA; 08°54′S 140°06′W. The island is fertile but rugged, with well-watered valleys; its highest point is c.3,999 ft/1,219 m. There are eight harbors, including Taiohae Bay on the S coast. Copra is the chief export. The village of TAIOHAE is the administrative center of the Marquesas Islands.

Nukulaelae (NOO-koo-LEI-lei), atoll (2002 population 393), TUVALU, SW PACIFIC; 09°23′S 179°51′E. Copra. Covers an area of 449 acres/182 ha. Also spelled Nukulailai. Formerly Mitchell Island.

Nukumanu (NOO-koo-MAH-noo), atoll, N SOLOMON ISLANDS, PAPUA NEW GUINEA, SW PACIFIC OCEAN, circa 250 mi/402 km ENE of BOUGAINVILLE; 11 mi/18 km long, 7 mi/11.3 km wide; 04°32′S 159°26′E. Consists of circa 40 small islands on reef. Largest island is NUKUMANU (□ 1 sq mi/2.6 sq km). Sometimes called TASMAN GROUP.

Nukunau, KIRIBATI: see NIKUNAU.

Nukunonu (NOO-koo-NAH-noo), central atoll (□ 2 sq mi/5.2 sq km), TOKELAU, NEW ZEALAND, S PACIFIC, 35 mi/56 km WNW of FAKAOFO; twenty-four islets. Village on SW island; produces copra. Once termed DUKE OF CLARENCE ISLAND.

Nukuoro (NOO-kuhr-o), atoll, state of POHNPEI, E CAROLINE ISLANDS, MICRONESIA, PACIFIC, 115 mi/185 km SE of SATAWAN; c.4 mi/6.4 km in diameter; 03°51′N 154°58′E. Rises to 12 ft/4 m; comprises forty-eight low islets. Polynesian inhabitants. Formerly Nuguor.

Nukus, city, ☉ KARAKALPAK REPUBLIC, W UZBEKISTAN, in the KHORAZM region and on the AMU DARYA River; 42°50′N 59°29′E. It has alfalfa, food-processing, and various light industries.

Nukuty (noo-KOO-ti), village, NW UST-ORDYN-BUR-YAT AUTONOMOUS OKRUG, in Irkutsk oblast, E central SIBERIA, RUSSIA, 40 mi/64 km NNW of CHER-EMKHOVO; 53°43′N 102°48′E. Elevation 1,315 ft/400 m. In agricultural area.

Nulato (noo-LAH-do), native village, on W bank of YUKON River, W ALASKA, 200 mi/322 km E of NOME; 64°43′N 158°07′W. Airfield. Russian blockhouse built here 1838.

Nuldernauw (NUHL-duhr-nou), inlet, central NETH-ERLANDS, between S. FLEVOLAND polder and GEL-DERLAND province coast, SE IJSSELMEER.

Nules (NOO-les), town, Castellón de la Plana province, E SPAIN, on irrigated plain near the MEDITERRANEAN SEA, 12 mi/19 km NW of CASTELLÓN DE LA PLANA; 39°51′N 00°09′W. Agricultural trade center (olive oil, almonds, rice, oranges, wine); sawmills. Hot mineral springs nearby. Severely damaged during Spanish civil war (1936–1939); now rebuilt.

Nulhegan River (NUHL-i-gin), c.15 mi/24 km long, NE VERMONT; rises in Averil; flows E to the CONNECTICUT River at BLOOMFIELD.

Nullagine, village, N WESTERN AUSTRALIA state, W AUSTRALIA, 140 mi/225 km SE of PORT HEDLAND; 21°53′S 120°06′E.

Nullarbor National Park (NUH-luhr-bor) (□ 2,290 sq mi/5,954 sq km), SW SOUTH AUSTRALIA state, S central AUSTRALIA, 400 mi/644 km W of PORT AU-GUSTA, 60 mi/97 km W of WESTERN AUSTRALIA border; 31°30′S 130°30′E. Name derived from Aboriginal term meaning "no tree." Occupies 100 mi/161 km of 1,000-ft/305-m coastal cliffs of INDIAN (Southern) Ocean, at S edge of NULLARBOR PLAIN. Park area extends an average of 20 mi/32 km N of coast. Large subsurface cavern systems make this one of the largest karst areas in world. High winds, crumbling rock make cliffs dangerous. Kangaroos, wombats, dingoes; emus. Right whales off coast. Eyre Highway follows coast length of park. No facilities. Established 1979.

Nullarbor Plain (NUH-luhr-BOR), vast limestone plateau of S AUSTRALIA, extending c.300 mi/483 km W from OOLDEA, South Australia, into WESTERN AUS-TRALIA, between VICTORIA DESERT and GREAT AUS-TRALIAN BIGHT; rises to 1,000 ft/305 m; 30°45′S 129°00′E. Traversed by Trans-Australian railroad. Sand dunes; sparse vegetation; few sheep; caves, caverns; tourism. Name simply means "no trees."

Nulvan (nool-VAHN), village, W BADAKHSHAN AU-TONOMOUS VILOYAT, S central TAJIKISTAN, on PANJ RIVER (AFGHANISTAN border), and 55 mi/89 km S of GARM; 38°16′N 70°33′E. Gold placers; goats.

Ñum, town, Guairá department, S PARAGUAY, 16 mi/26 km SE of VILLARRICA, on road and railroad; 25°56′S 56°19′W. Sugarcane, fruit; livestock.

Numa, town (2000 population 109), APPANOOSE county, S IOWA, 6 mi/9.7 km WSW of CENTERVILLE; 40°41′N 92°58′W.

Numai (NOO-mei), town, KOGI state, central NIGERIA, at confluence of NIGER and BENUE rivers, 5 mi/8 km NE of LOKOJA; 07°49′N 06°46′E. Market and fishing town.

Numakuma (noo-MAH-koo-mah), town, Numakuma county, HIROSHIMA prefecture, SW HONSHU, W JAPAN, 50 mi/80 km E of HIROSHIMA; 34°23′N 133°19′E. Shipbuilding.

Numan (noo-MAWNG), town, ADAMAWA state, E NI-GERIA, on BENUE River, at mouth of the GONGOLA River, 40 mi/64 km NW of YOLA; 09°28′N 12°02′E. Agricultural trade center; peanuts, millet, durra, cassava, yams; cattle raising; saltworks.

Numana (noo-MAH-nah), village, ANCONA province, The MARCHES, central ITALY, port on the ADRIATIC, 9 mi/14 km SE of ANCONA; 43°31′N 13°37′E. Bishopric.

Numantia (noo-MAN-shuh), ancient settlement, SPAIN, near the DUERO RIVER and N of modern SORIA. Numantia played a central role in the Celt-Iberian resistance to Roman conquest. Its inhabitants withstood repeated Roman attacks from the time of Cato the Elder's campaign (195 B.C.E.) until Scipio Aemilianus finally took the city in 133 B.C.E., after an eight-month blockade, thus completing the conquest of Spain. Archaeologists have uncovered the remains of Roman camps and evidence of settlement dating back to the Bronze Age.

Numarán (noo-mah-RAHN), town, MICHOACÁN, central MEXICO, on LERMA River, and 7 mi/11.3 km SE of LA PIEDAD DE CABADAS; 20°15′N 101°56′W. Cereals, fruit; livestock.

Numas (NUHM-es), large village (2004 population 23,780), N ASIR, MAKKA province, SAUDI ARABIA, in highlands, 80 mi/129 km ESE of QUNFIDHA. Grain (sorghum), vegetables, fruit. Formerly in Asir. Also An-Nimas.

Numata (noo-MAH-tah), city, GUMMA prefecture, central HONSHU, N central JAPAN, 16 mi/25 km N of MAEBASHI; 36°38′N 139°02′E. *Konnyaku* (paste made from devil's tongue). Fuel.

Numata (noo-MAH-tah), town, Sorachi district, Hokkaido prefecture, N JAPAN, 59 mi/95 km N of SAPPORO; 43°48′N 141°56′E.

Numazu (NOO-mahz), city (2005 population 208,005), SHIZUOKA prefecture, central HONSHU, central JAPAN, port on NE shore of SURUGA BAY, 28 mi/45 km N of SHIZUOKA; 35°05′N 138°51′E. Manufacturing includes transformers and computers; fish processing. Mandarin oranges; tea.

Nümbrecht (NYOOM-brekht), town, RHINELAND, NORTH RHINE-WESTPHALIA, W GERMANY, 9 mi/14.5 km S of GUMMERSBACH; 50°55′N 07°33′E. Forestry.

Numbur (NOOM-boor), mountain peak (22,830 ft/6,957 m), central Nepal, in SOLU region S of KHUMBU; 27°45′N 86°34′E. Region's major peak. Climbed by Japanese team in 1963.

Numedal (NOO-muh-dahl), valley of LÅGEN River, SE NORWAY, extends from E slope of the HARDANGERVIDDA, c.60 mi/97 km SSE to KONGSBERG. Agriculture; lumbering; fishing. Folk customs and dialect resemble those of Hallingdalen. Several hydroelectric power plants on the Lågen River.

Numfoor (num-FOOR) or **Noemfoor**, island, MISORE ISLANDS, IRIAN JAYA province, INDONESIA, off NW coast of NEW GUINEA island at entrance to CENDERAWASIH BAY, 50 mi/80 km E of MANOKWARI; 14 mi/23 km long, 12 mi/19 km wide; 01°03′S 134°54′E. Fishing. Airport.

Numidia (noo-mee-DYAH), ancient country of NW AFRICA, very roughly the area occupied by modern ALGERIA. It was part of the Carthaginian empire until Masinissa, ruler of E Numidia, allied himself (c.206 B.C.E.) with Rome in the Punic Wars. After the Roman victory over Carthage led to peace in 201 B.C.E., Masinissa was awarded rule of all Numidia. This began a flourishing period, both culturally and politically. Numidia's encroachments on reviving Carthage furnished Rome with a pretext for the Third Punic War (149 B.C.–146 B.C.E.). Masinissa's successor was Micipsa (148 B.C.–118 B.C.E.), one of whose heirs, Jugurtha, brought on a fatal war with Rome. Later, in the Roman civil war, King Juba I sided with Pompey, and Numidia lost (46 B.C.E.) all independence with Julius Caesar's victory. Juba II was favored by the Romans as a subject prince, and the region subsequently flourished for several centuries. Numidia was invaded by the Vandals in the 5th century and by the Arabs in the 8th century. The main urban centers of ancient Numidia were Cirta (now CONSTANTINE) and Hippo Regius (now ANNABA).

Num, Mios, INDONESIA: see YAPEN ISLANDS.

Nu Mountains (NU), Chinese *Nu Shan* (NU-SHAN), outlier of Tibetan highlands, NW YUNNAN province, CHINA, extends c. 250 mi/402 km N-S between MEKONG (E) and SALWEEN (W) rivers, an extension of the Taniantaweng mountains and a part of the Hengduan mountains; 27°00′N 99°00′E. Rise to 15,800 ft/4,816 m.

Numurkah (nuh-MUHR-kuh), town, N VICTORIA, SE AUSTRALIA, 120 mi/193 km N of MELBOURNE, in GOULBURN RIVER valley; 36°06′S 145°26′E. Railroad junction in agricultural area (tobacco, wheat, and oats); dairying; fruit trees.

Nunachuak (noo-NAH-choo-wak), village, SW ALASKA, on NUSHAGAK RIVER, and 60 mi/97 km NE of DILLINGHAM.

Nunapitchuk, village (2000 population 466), SW ALASKA, 30 mi/48 km SW of BETHEL, near KUSKOKWIM River; 60°30′N 162°25′W.

Nunarssuit (noo-NAHKH-shwit), island, just off SW GREENLAND, near S side of mouth of KOBBERMINE BAY, 55 mi/89 km W of JULIANEHÅB (Qaqortoq); 20 mi/32 km long, 3 mi/4.8 km–9 mi/14.5 km wide; 60°47′N 48°W. Rises to 2,610 ft/795 m. On W coast is Cape Desolation (60°44′N 48°11′W), landmark for early explorers. Formerly called Desolation Island.

Nunavik or **Nunavik-Quebec** (NOO-nuh-vik), region (□ 196 sq mi/509.6 sq km; 2006 population 10,000), N QUEBEC, E Canada, also includes N tip of LABRADOR N of Ramah, and NUNAVUT territory islands of KILLINIQ and AKPATOK (UNGAVA BAY), SALISBURY and NOTTINGHAM (HUDSON STRAIT), MANSEL, OTTAWA, Smith, and Gilmour (HUDSON BAY), North TWIN and South TWIN (JAMES BAY). The Quebec portion, also called "Arctic Quebec," includes land N of 56°30′N, including land around Ungava Bay and all of Ungava Peninsula; also Hudson Bay and James Bay shore lands S to 53°N. Terrain of tundra, mountains, taiga forest, and lakes and rivers. Includes 14 coastal Inuit villages; also Naskapi and Cree inhabitants. Inuit village of CHISASIBI, at mouth of La Grande River. Name means "great land" in Inuit (Inuktitut). Subsistence by fishing, hunting, and other traditional activities. Not a political unit, the affairs and services of region are operated by an Inuit corporation which draws much of its income from agreements over the use of lands to SE of region for the JAMES BAY PROJECT (hydroelectric scheme). The corporation operates Air Inuit, which serves region and adjacent areas.

Nunavut (noo-NAH-vut), territory (□ 772,260 sq mi/2,007,876 sq km; 2006 population 29,474), NE CANADA, ⊙ IQALUIT. Established April 1, 1999. In a May 4, 1992 plebiscite, the electorate of the NORTHWEST TERRITORIES approved the boundary that divides the territory along a zigzag line running from the SASKATCHEWAN-MANITOBA border through the ARCTIC ARCHIPELAGO at 60°N, almost to the NORTH POLE. Constitutes the E portion of the current Northwest Territories: the former BAFFIN, Keewatin (now KIVALLIQ), and KITIKMEOT regions. Its W boundary runs N to the THELON RIVER, W to just above the GREAT BEAR LAKE, and then N again to bisect VICTORIA and MELVILLE islands. Spans 3 time zones, and includes one-fifth of Canada's total land area. The establishment of Nunavut created a Canadian "Four Corners" where SW Nunavut, SW Northwest Territories, NE Saskatchewan, and NW Manitoba meet. The region includes the islands of ELLESMERE, BAFFIN, DEVON, PRINCE OF WALES, SOUTHAMPTON, and COATS, among others. Rises to 8,583 ft/2,616 km at Barbeau Peak on Ellesmere Island. The territory is largely on the CANADIAN SHIELD, and almost entirely N of the tree line; the landscape is dominated by tundra, rock, snow, and ice. The territory's capital and largest town is IQALUIT (1999 population c.6,000), on Baffin Island at FROBISHER BAY. Mount THOR, rising to 4,920 ft/1,500 m in Nunavut, is the tallest uninterrupted cliff face on Earth. The territory, Canada's third after the YUKON and the remaining Northwest Territories, is effectively controlled by the Inuit, who, with 17,500 people, comprise the majority of the area's population. Nunavut's population density is just 1 person per 27 sq mi/70 sq km; its physical isolation accounts for its having Canada's highest cost of living. Most of the richest and most well-developed parts of the current Northwest Territories, which lie along the MACKENZIE RIVER, are not included in Nunavut. To sustain itself economically, the new territory relies on the development of its mineral resources, in addition to hunting, fishing, fur trapping, sealing, and the production of arts and crafts. As a result of the Nunavut Land Claims Agreement with the federal government, the largest native land-claim settlement in Canadian history, the Inuit hold outright title to about 20% of Nunavut, including 13,896 sq mi/36,000 sq km of subsurface mineral rights. Nunavut has three official languages: Inuktitut, English, and French. In Inuktitut, the Inuit language, Nunavut means "our land."

Nunawading (nuhn-uh-WAH-deeng), suburb 11 mi/18 km E of MELBOURNE, VICTORIA, SE AUSTRALIA; 37°49′S 145°10′E.

Nun, Cap (NOON), headland on ATLANTIC OCEAN, in what was once IFNI territory, SW MOROCCO, NW AFRICA, 10 mi/16 km SW of SIDI IFNI; 29°15′N 10°18′W. Formerly confused with Cape DRA, 60 mi/97 km SW.

Nunchía (noon-CHEE-ah), town, ⊙ Nunchía municipio, CASANARE department, S COLOMBIA, 25 mi/40 km NE of YOPAL, in the Oriente; 05°38′N 72°12′W. Elevation 1,807 ft/550 m. Agriculture includes sugarcane, corn, plantains; livestock.

Nünchritz (NYUNKH-rits), village, SAXONY, E GERMANY, 22 mi/35 km NNW of DRESDEN, on right bank of the ELBE RIVER; 51°18′N 13°24′E. Grain, sugar beets.

Nunda (NUN-day), village (2006 population 1,270), LIVINGSTON county, W central NEW YORK, 23 mi/37 km NW of HORNELL; 42°34′N 77°56′W. Light manufacturing and agriculture. Letchworth State Park along GENESEE River is NW. Incorporated 1839.

Nunda, village (2006 population 46), Lake county, E SOUTH DAKOTA, 12 mi/19 km NNE of MADISON; 44°09′N 97°01′W.

Nundle (NUN-duhl), village, E NEW SOUTH WALES, SE AUSTRALIA, 110 mi/177 km NNW of NEWCASTLE, adjacent Peel (NAMOI) River, in GREAT DIVIDING RANGE foothills; 31°28′S 151°08′E. Sheep, cattle; wheat. Former gold-mining center; visitors still fossick for gold and such gemstones as quartz crystals, sapphires, zircon, and green jasper.

Nuneaton and Bedworth (nuhn-EET-uhn and BED-wuhth), district (2001 population 119,132), WARWICKSHIRE, central ENGLAND; 52°31′N 01°28′W. Primarily comprises the city of Nuneaton and the town of Bedworth. The district's growth was based on coal mining. Electronics, distribution services, and manufacturing of textiles. Nuneaton is located at a railroad junction, which adds to the growth of industry. There are remains of the 12th-century nunnery that gave Nuneaton its name. Writer George Eliot born within the district at Arbury. Other settlements include Chilvers Coton, Hartshill, and Stockingford.

Nunez (noo-NEZ), town (2000 population 131), EMANUEL county, E central GEORGIA, 7 mi/11.3 km S of SWAINSBORO; 32°29′N 82°21′W.

Nuñez (noon-YAIZ), NE residential section of BUENOS AIRES, ARGENTINA, bordering on the Rio de la Plata.

Nunez, Rio, GUINEA: see RIO NUNEZ.

Nungarin (nuhn-GA-ruhn), town, WESTERN AUSTRALIA state, W AUSTRALIA, 173 mi/278 km NE of PERTH; 31°11′S 118°06′E. Wheat; sheep, pigs, cattle; gypsum mining.

Nungwa (NOON-gwuh), town (2000 population 62,902), GREATER ACCRA region, GHANA, on Gulf of GUINEA, 11 mi/18 km ENE of ACCRA; 05°36′N 00°04′W. Fishing; coconuts, cassava, corn.

Nungwe (NOON-gwai), town, MWANZA region, NW TANZANIA, 60 mi/97 km WSW of MWANZA, at S end of NUNGWE BAY, Lake VICTORIA; 02°46′S 32°00′E. Road terminus, lake port. Fish; cattle, sheep, goats; cotton, corn, wheat, millet.

Nungwe Bay (NOON-gwai), inlet, arm of EMIN PASHA GULF, Lake VICTORIA, NW TANZANIA; 13 mi/21 km long, 3 mi/4.8 km wide. Town of NUNGWE at S end.

Nunivak, island (□ 1,700 sq mi/4,420 sq km), off W ALASKA, in the BERING SEA. It is the second-largest island in the Bering Sea. Fogbound most of the year, Nunivak is covered with low vegetation and has a small Inuit population engaged in hunting and fishing. Reindeer and musk oxen have been introduced as part of a national wildlife refuge here.

Nunkun (noon-koon), peak (23,410 ft/7,135 m) in GREATER HIMALAYAS, JAMMU AND KASHMIR state, extreme N INDIA, 40 mi/64 km E of Pahlgam, at 34°N 76°E. Also written Nun Kun.

Nunn, village (2000 population 471), WELD county, N COLORADO, 20 mi/32 km N of GREELEY; near Lone Tree Creek; 40°42′N 104°46′W. Elevation 5,186 ft/1,581 m. Farm center in cattle-grazing area, also wheat, corn, sugar beets, sunflowers. Pawnee National Grassland to E.

Nuñoa (noon-YO-ah), town, MELGAR province, PUNO region, SE PERU, on the ALTIPLANO, 28 mi/45 km N of AYAVIRI; 14°28′S 70°38′W. Elevation 13,222 ft/4,030 m. Cereals; livestock.

Nun River (NOON), a main outlet of NIGER RIVER, S NIGERIA, in middle section of delta; enters GULF OF GUINEA at AKASSA.

Nun's Island (NUHNZ), S QUEBEC, E Canada, in the SAINT LAWRENCE RIVER, opposite MONTREAL, near the LACHINE RAPIDS; 2 mi/3 km long, 1 mi/2 km wide; 45°28′N 73°32′W.

Nunspeet (NUN-spait), town, GELDERLAND province, N central NETHERLANDS, 13 mi/21 km NW of APELDOORN; 52°23′N 05°48′E. VELUWEMEER channel 2 mi/3.2 km to NW; VELUWE nature area to S. Dairying; cattle, hogs, poultry; fruit, vegetables, grain; manufacturing (ladders; food processing, construction equipment).

Nunthorpe (NUHN-thawp), village (2001 population 4,705), Middlesbrough, NE ENGLAND, 4 mi/6.4 km S of MIDDLESBROUGH; 54°31′N 01°10′W. CLEVELAND HILLS to E and S.

Nuoro (NWO-ro), province (□ 2,808 sq mi/7,300.8 sq km), central SARDINIA, ITALY; ⊙ NUORO; 40°10′N 09°20′E. Least populous, most mountainous region of Sardinia, with Monti del GENNARGENTU and Catena del MARGHINE; drained by TIRSO and FLUMENDOSA rivers. Deposits of chalcopyrite, anthracite, talc and steatite, granite, barite, lignite. Fisheries. Forestry (cork plantations in Tirso valley, wood carving in Gennargentu villages). Livestock raising in E and NW. Corn, barley, potatoes, vineyards in E and NW; olives and fruit in NW; industry at MACOMER and BOSA. Served by main N-S railroad, forking at Macomer to Bosa and Nuoro. Nuraghi in NW. Formed 1927 from SASSARI and CAGLIARI provinces.

Nuoro (NWO-ro), town, ⊙ NUORO province, E central SARDINIA, ITALY, 50 mi/80 km SE of SASSARI; 40°19′N 09°20′E. Railroad terminus; bishopric. Prehistoric rock tombs (E).

Nuporanga (NOO-po-rahn-gah), town (2007 population 6,629), NE SÃO PAULO, BRAZIL, 30 mi/48 km N of RIBEIRÃO PRÊTO; 20°44′S 47°45′W. Coffee, rice, cotton, grain.

Nuptse (NUHP-tse), mountain peak (25,770 ft/7,855 m), E Nepal, SW of MOUNT EVEREST and WESTERN CWM; 27°58′N 86°53′E. First ascent by British team in 1961.

Nuqui (NOO-kee), town, ⊙ Nuqui municipio, CHOCÓ department, W COLOMBIA, minor port on the PACIFIC, 40 mi/64 km W of QUIBDÓ; 05°43′N 77°16′W. Rice, corn, coconuts, plantains, fruit; fishing.

Nur (NOOR), town, YAZD province, IRAN, 45 mi/72 km S of YAZD; 31°25′N 54°20′E. Also spelled Nir.

Nur Abad (NOOR AH-bahd), town (2006 population 52,597), FĀRS province, SW IRAN, NW of SHIRAZ; 30°48′N 51°27′E. Wheat cultivation.

Nūrābād (NOOR-ah-BAHD), town (2006 population 56,530), Lorestān province, W IRAN, 60 mi/95 km NW of KHORRAMABAD; 34°12′N 47°48′E.

Nurabad (NOOR-ah-baht), town, TOSHKENT wiloyat, NE UZBEKISTAN.

Nuradilovo (noo-rah-DEE-luh-vuh), village (2005 population 3,450), W central DAGESTAN REPUBLIC, SE European Russia, near the administrative border with CHECHEN REPUBLIC, on road and railroad, 5 mi/8 km W, and under administrative jurisdiction, of KHASAVYURT; 43°17′N 46°27′E. Elevation 321 ft/97 m. Agriculture (fruits; livestock). Until 1944, in Chechen-Ingush ASSR and called Daud-Otar (dah-OOD uh-TAHR). Site of battles between encroaching Chechen units and Russian troops in August–September 1999, at the beginning of the second Russian-Chechen conflict.

Nurakita, TUVALU: see NIULAKITA.

Nura River (noo-RAH), 445 mi/716 km long, in N KARAGANDA and S AKMOLA regions, KAZAKHSTAN; rises in KAZAK HILLS SE of KARAGANDA; flows NW, past TEMIRTAU (site of reservoir and power plant) and Kurgaldzhinoskoye, through Lake Kurgaldzhin, to TENGIZ (lake); intermittent connection with ISHIM RIVER SW of Akmola (Akmolinsk).

Nurata, city, NAWOIY viloyat, UZBEKISTAN, in W foothills of the AK-TAU range, 55 mi/89 km NNW of KATTAKURGAN; 40°33′N 65°41′E. Wheat; sheep.

Nuratau (NOOR-ah-taw), range in N SAMARKAND wiloyat, UZBEKISTAN; extends 100 mi/161 km E-W. Elevation 6,300 ft/1,920 m. Also Nura-Tau.

Nürburg (NYOOR-boorg), village, RHINELAND-PALATINATE, W GERMANY, in the EIFEL, 12 mi/19 km W of MAYEN; 50°23′N 06°57′E. Scene of auto races (track opened 1927).

Nuremberg (NYUHR-em-berg), German *Nürnberg*, city, MIDDLE FRANCONIA, BAVARIA, S GERMANY, 41 mi/66 km SSW of BAYREUTH, on the PEGNITZ RIVER; 49°27′N 11°04′E. Important commercial, industrial, and transportation center. Manufacturing includes electrical equipment, mechanical and optical products, motor vehicles, chemicals, textiles, fancy metal-

work, and printed materials. Homemade toys and fine gingerbread (Ger. *Lebkuchen*) are traditional export items. It was first mentioned in 1050. Nuremberg received a charter in 1219, and was made a free imperial city by the end of the 13th century. Nuremberg soon became, with AUGSBURG, one of the two great trade centers on the route from ITALY to N EUROPE. Center of the German Renaissance (15th–16th century). Dürer, Kraft, Stoss, Vischer, and Wolgemut adorned the city with their works; Hans Sachs and other Meistersingers competed in contests. The city was also an early center of humanism, science, printing, and mechanical invention. The scholars Pirkheimer and Celtes lectured in the city, Koberger set up a printing press and Regiomontanus an observatory, and the first pocket watches, known as Nuremberg eggs, were made here c.1510 by Peter Henlein. Accepted the Reformation 1534. The city declined after the Thirty Years War and recovered its importance only in the late 19th century, when it grew as an industrial center. In 1806, Nuremberg passed to BAVARIA. The first German railroad, from Nuremberg to nearby FÜRTH, was opened in 1835. National Socialists (Nazis) made the city a national shrine and held their annual party congresses nearby (1933–1938); at 1935 congress the violently anti-Semitic Nuremberg Laws were promulgated. Until 1945, Nuremberg was the site of roughly half the total German production of airplane, submarine, and tank engines; as a consequence, the city was heavily bombed by the Allies during World War II and was largely destroyed. After the war, Nuremberg was the seat of the international tribunal for war crimes (the Nuremberg Trials). Since 1945 much of the city's architectural beauty has been restored. Among the historic buildings are the churches of St. Sebald (1225–1273), St. Lorenz (13th–14th century), and Our Lady (c.1350); the Hohenzollern castle (11th–16th century); the old city hall (1616–1622); and the house (now a museum) where Albrecht Dürer lived from 1509 to 1528. A large portion of the city walls (14th–17th century) still stands. Nuremberg is the site of the German National Museum (founded 1852), a part of the University of Erlangen-Nuremberg, and a museum of transportation.

Nure River (NOO-re), 40 mi/64 km long, N central ITALY; rises in LIGURIAN APENNINES on Monte MAGGIORASCA; flows NNE, past Ferriere, BETTOLA, and PONTE DELL'OLIO, to PO RIVER 5 mi/8 km E of PIACENZA.

Nuri (NOO-ree), town, Northern state, SUDAN, on the NILE River, just above KAREIMA, near ancient NAPATA; 18°33′N 31°54′E. Site of pyramid field with tombs of Ethiopian kings (after 7th century B.C.E.).

Nuriootpa (noo-ree-OOT-puh), town, SE SOUTH AUSTRALIA, 40 mi/64 km NNE of ADELAIDE; 34°29′S 139°00′E. Fruit; livestock; wheat; dairying; manufacturing (alcohol, electrical components, ceramic tiles, and steel fabrication); vineyards, Barossa Valley wine center. Marble quarry nearby.

Nuristan (noo-REE-stahn) (Persian, *land of light or the enlightened*), region on the S slopes of the HINDU KUSH, NANGARHAR province, NE AFGHANISTAN, bordered on the E by PAKISTAN; 35°30′N 70°45′E. Formerly called Kafiristan (Persian, *land of the infidels*). A remote highland NE of KABUL—wooded, mountainous, and remote—consisting of a series of deep, narrow valleys, separated by rugged outliers of the Hindukush; drained by the left-bank tributaries of KABUL RIVER. The region's ASMAR FOREST furnishes 80% of Afghanistan's timber production. It is inhabited by an ethnically distinctive people (numbering about 100,000), who practiced animism until their forcible conversion to Islam in 1895–1896. Agriculture (wheat, barley, millet, peas, wine grapes, fruit); livestock (chiefly goats). A special artisan caste specializes in woodcarving, pottery making, weaving, and metalwork. The Nuristanis, divided into several tribes, speak Dardic dialects (often mutually unintelligible) belonging to a distinct branch of the Indo-European language family. Was the scene of some of the heaviest guerrilla fighting during the 1979–1989 invasion and occupation of Afghanistan by Soviet forces.

Nuristan, province (2005 population 123,300), E AFGHANISTAN; ⊙ Nuristan. Borders BADAKHSHAN province (N), PAKISTAN (E), KUNAR and LAGHMAN provinces (S), and PANJSHIR province (W).

Nurlat (noor-LAHT), city (2006 population 35,135), S TATARSTAN Republic, E European Russia, on the KONDURCHA RIVER (VOLGA RIVER basin), on highway junction and railroad, 165 mi/268 km SE of KAZAN', and 65 mi/105 km S of CHISTOPOL'; 54°26′N 50°46′E. Elevation 442 ft/134 m. Oil and gas processing, asphalt factory; food processing (sugar, meat, bread, dairy products, alcoholic beverages). Made city in 1961.

Nurlaty (noor-LAH-ti), village, NW TATARSTAN Republic, E European Russia, near the SVIYAGA RIVER, near highway and railroad, 63 mi/101 km SW of KAZAN'; 55°02′N 48°10′E. Elevation 462 ft/140 m. In grain- and livestock-producing area.

Nurma (noor-MAH), settlement (2005 population 3,250), S central LENINGRAD oblast, NW European Russia, on road branch and railroad spur, 32 mi/51 km SE of SAINT PETERSBURG, and 4 mi/6 km E of TOSNO, to which it is administratively subordinate; 59°33′N 31°01′E. Agricultural products. Peat works.

Nurmahal (NOOR-mah-uhl), town, JALANDHAR district, central PUNJAB state, N INDIA, 16 mi/26 km S of JALANDHAR; 31°06′N 75°36′E. Wheat, cotton, grain; handicrafts (cotton cloth, shell buttons).

Nurmes (NOOR-mes), town, POHJOIS-KARJALAN province, E FINLAND, at NW end of LAKE PIELINEN, 60 mi/97 km NE of KUOPIO; 63°33′N 29°07′E. Elevation 297 ft/90 m. Lumber milling, woodworking. Area is known for its birch trees; old section of town, called Puu-Nurmes, is known for its wooden houses.

Nur Mountains, TURKEY: see AMANOS MOUNTAINS.

Nürnberg, GERMANY: see NUREMBERG.

Nurpur (NOOR-poor), town, KANGRA district, HIMACHAL PRADESH state, N INDIA, 26 mi/42 km WNW of DHARMSHALA. Market for grain, fiber; hand-woven woolen shawls.

Nurpur (NUHR-puhr), town, SARGODHA district, W central PUNJAB province, central PAKISTAN, in THAL region, 50 mi/80 km WSW of SARGODHA; 31°53′N 71°54′E.

Nurpur (NUHR-puhr), village, JHELUM district, N PUNJAB province, central PAKISTAN, in SALT RANGE, 65 mi/105 km WSW of JHELUM; 32°40′N 72°45′E. Several rock-salt mines are just S.

Nürschan, CZECH REPUBLIC: see NYRANY.

Nursingpur, INDIA: see NARSINGHPUR, MADHYA PRADESH state.

Nürtingen (NYOOR-ting-uhn), town, STUTTGART district, BADEN-WÜRTTEMBERG, GERMANY, on the NECKAR RIVER, and 8 mi/12.9 km S of ESSLINGEN; 48°37′N 09°20′E. Manufacturing (machinery, machine tools, textiles, cement). Has late-Gothic church. Chartered in 13th century. Was first mentioned 1046.

Nuruhak Dag (Turkish=*Nuruhak Dağ*) peak (9,850 ft/3,002 m), S central TURKEY, in MALATYA MOUNTAINS, 50 mi/80 km WSW of MALATYA.

Nurzec River (NOO-zets), Russian *Nurzhets*, Bialystok province, E POLAND; rises SW of KLESZCZELE, near RUSSIA border; flows WNW past BRANSK, and SSW past CIECHANOWIEC to BUG RIVER 7 mi/11.3 km SSW of Ciechanowiec; 64 mi/103 km long.

Nurzhets River, POLAND: see NURZEC RIVER.

Nusa Barung (NOO-sah bah-ROONG), uninhabited island of INDONESIA, in the INDIAN OCEAN, off SE coast of JAVA, 55 mi/89 km ESE of MALANG; 10 mi/16 km long, 4 mi/6.4 km wide; 08°28′S 113°21′E. Rocky, hilly, with steep coastal cliffs. Edible birds' nests are collected here. Population was removed in 1776 to

Java, and the island remains uninhabited. Also spelled Noesa Baroeng.

Nusa Besi, INDONESIA: see JACO.

Nusa Dua (NOO-sah DOO-wah), resort area, S BALI, INDONESIA, 6 mi/10 km SE of KUTA; 08°48′S 115°14′E. Luxury tourist resort on E coast of BUKIT PENINSULA.

Nusagandi (□ 159 sq mi/412 sq km), forest reserve and park, PANAMA, reserve established and operated by Kuna Indians within the San Blas Comarca. Accessible by motor vehicle via the El Lano–Soledad highway; lodging, trails, and guides are available. The park is noted for rain-forest vegetation and wildlife.

Nusa Laut, INDONESIA: see LAUT, NUSA.

Nusantara: see INDONESIA.

Nusa Penida (NOO-sah PUH-nee-dah), island, Bali province, INDONESIA, in LOMBOK STRAIT, between BALI (W, across Badnuna Strait) and LOMBOK (E, across Lombok Strait); 12 mi/19 km long, 8 mi/12.9 km wide; 08°44′S 115°32′E. Rice, vegetables; fishing. Main town is Sampalan on NE coast. Mondi Mountain (1,736 ft/529 m) is highest peak on island. Also called Penide.

Nusa Tenggara, INDONESIA: see SUNDA ISLANDS.

Nusatenggara, INDONESIA: see SUNDA ISLANDS.

Nusaybin, town (2000 population 74,110), SE TURKEY, near the Syrian border; 37°05′N 41°11′E. It is a commercial and transportation center. It has ruins of the ancient Nisibis, the residence of early (2nd century B.C.E.–1st century C.E.) Armenian kings. In early Christian times it was a center of Nestorianism, which held that Jesus was two distinct persons, closely and inseparably united. In 1839, the Egyptians defeated the Turks at Nusaybin. A substantial number of the population are members of the Syrian Orthodox Church.

Nuseirat, El, town and Palestinian refugee camp, 6.2 mi/10 km SW of GAZA, in the GAZA STRIP. Grows mainly citrus and vegetables.

Nushagak, village, SW ALASKA, on NUSHAGAK BAY, inlet of BRISTOL BAY, 8 mi/12.9 km S of DILLINGHAM. Fishing; cannery.

Nushagak Bay, SW ALASKA, on N shore of BRISTOL BAY; 50 mi/80 km long, 4 mi/6.4 km–20 mi/32 km wide; 58°37′N 158°31′E. Receives NUSHAGAK RIVER. Fishing and canning region. DILLINGHAM village, N.

Nushagak River, 280 mi/451 km long, SW ALASKA; rises in ALASKA RANGE near 60°50′N 154°W; flows SW, past NUNACHUAK and EKWOK, to NUSHAGAK BAY, inlet of BRISTOL BAY, just E of DILLINGHAM. Salmon stream. Upper course called MULCHATNA RIVER.

Nu Shan, CHINA: see NU MOUNTAINS.

Nushki (NOOSH-kee), village, ⊙ CHAGAI district, N BALUCHISTAN province, SW PAKISTAN, 75 mi/121 km SW of QUETTA; 29°33′N 66°01′E. Road center.

Nusle (NUS-le), SSE district of PRAGUE, PRAGUE-CITY province, central BOHEMIA, CZECH REPUBLIC. Food processing. Housing estate.

Nussdorf am Attersee (NUS-dorf ahm AHT-ter-sai), village, S UPPER AUSTRIA, on W shore of the ATTERSEE lake, and 12 mi/19 km W of GMUNDEN; 47°53′N 13°31′E. Elevation 1,630 ft/497 m. Summer resort.

Nussloch (NUS-lokh), town, LOWER NECKAR, BADEN-WÜRTTEMBERG, GERMANY, 2 mi/3.2 km N of WIESLOCH; 49°19′N 08°42′E. Limestone quarrying.

Nutak, abandoned locality, N NEWFOUNDLAND AND LABRADOR, NE Canada, at SE end of Okak Island, off Labrador Sea coast. It and village of Okak, at NW end of island, were sites of Moravian missions, established in the 1830s. December 1918, supply ship *Harmony* brought influenza to area, nearly wiping out Inuit population.

Nutarawit Lake (noo-tuh-RO-wit) (□ 350 sq mi/910 sq km), NUNAVUT territory, Canada, just NW of YATHKYED LAKE; 62°55′N 98°50′W. Drains SE into KAZAN RIVER through Yathkyed Lake.

Nutfield (NUHT-feeld), village (2001 population 2,728), E central SURREY, SE ENGLAND, 3 mi/4.8 km E

of REIGATE; 51°13′N 00°07′W. Quarries for fuller's earth. Has 15th-century church.

Nuth (NUHT), town, LIMBURG province, SE NETHERLANDS, 10 mi/16 km NE of MAASTRICHT, near Geleen River; 50°55′N 05°52′E. Dairying; cattle, hogs; vegetables, grain, fruit; manufacturing (radiators, pulleys, rubber products; food processing.

Nutley, town (2000 population 27,362), ESSEX county, NE NEW JERSEY, a residential suburb of NEWARK, 5 mi/8 km N of Newark, on the PASSAIC River; 40°49′N 74°09′W. Settled 1680, incorporated 1902. Light manufacturing. After the Civil War the town was a center for writers and artists. Annie Oakley lived here.

Nutrias, VENEZUELA: see CIUDAD DE NUTRIAS.

Nutter Fort, town (2006 population 1,645), HARRISON county, N WEST VIRGINIA, 2 mi/3.2 km SE of CLARKSBURG; 39°15′N 80°19′W. Light manufacturing. Agriculture (corn); cattle; poultry. Settled 1770; incorporated 1924.

Nutzotin Mountains (nood-ZO-tin), E ALASKA and SW YUKON, NW extension of SAINT ELIAS MOUNTAINS, between WRANGELL MOUNTAINS (SW) and upper TANANA River (NE); extend 75 mi/121 km between upper NABESNA River (NW) and upper WHITE River, YUKON (SE); 62°15′N 142°08′W. Rise to c.9,000 ft/2,743 m; continued NW by MENTASTA MOUNTAINS.

Nuuanu Pali, sheer cliff and mountain pass (1,200 ft/366 m), KOOLAU RANGE, SE OAHU island, HONOLULU county, HAWAII, 5 mi/8 km NE of HONOLULU. One of three highway routes between Honolulu and E Oahu. State highway 61 Freeway runs through pass and tunnel.

Nuugaatsiaq, fishing and hunting settlement, Uummannaq commune, W GREENLAND, in S Qeqertarsuaq Island (16 mi/26 km long, 3 mi/4.8 km–10 mi/16 km wide), on Karrats Fjord, 65 mi/105 km NNW of Umanak; 71°33′N 53°12′W.

Nuuk [Danish=*Godthåb*], town Nuuk (Godthåb) commune, ⊙ GREENLAND, on the GODTHÅBSFJORD. The largest town of Greenland, it is the seat of the national council and of the supreme court, and it has foreign consulates. Nuuk also has radio stations and newspapers, is a fishing center, and has oil and liquid-gas storage facilities. An airfield was built here in the late 1970s. The town was founded in 1728 by Hans Egede, a Norwegian missionary. The Godthåbsfjord region has fine pastures and supports reindeer herds. At the head of the fjord are the remains of VESTERBYGD, a tenth- to fourteenth-century Norse settlement.

Nu'umamiya, town, WASIT province, E central IRAQ, NW of KUT, on W bank of the TIGRIS. Center of fertile agricultural region.

Nuussuaq, peninsula, W GREENLAND, extends from inland ice cap NW into DAVIS STRAIT, between UUMMANNAQ FJORD (N) and the VAIGAT and head of DISKO BAY; 110 mi/177 km long, 18 mi/29 km–30 mi/48 km wide; 70°20′N 52°30′W. Mountainous and partly glaciated, it rises to 7,340 ft/2,237 m, 12 mi/19 km SW of Umanak. In center, between high mountains, is TASERSUAQ LAKE (26 mi/42 km long, 1 mi/1.6 km–2 mi/3.2 km wide; elevation c.885 ft/270 m). Peninsula has lignite deposits and petrified flora. Sometimes spelled Nugsuak or Nûgssuak.

Nuussuaq, peninsula, NW GREENLAND, on BAFFIN BAY; 30 mi/48 km long, 1 mi/1.6 km–4 mi/6.4 km wide; 74°12′N 56°35′W. At base is Peary Lodge (74°19′N 56°13′W), N base of University of Michigan expedition, 1932–1933. KRAULSHAVN, Greenlandic *Nutaarmiut* (1995 population 63), fishing outpost, on S coast.

Nuutajärvi (NOO-tah-YAR-vee), village, HÄMEEN province, SW FINLAND, in lake region, 30 mi/48 km SSW of TAMPERE; 61°03′N 23°26′E. Elevation 413 ft/125 m. Glassworks (Finland's oldest).

Nuwakot (noo-WAH-kot), town, central Nepal, in BAGMATI zone; ⊙ BIDUR.

Nuwara Eliya, district (□ 659 sq mi/1,713.4 sq km; 2001 population 703,610), CENTRAL PROVINCE, SRI LANKA; ⊙ NUWARA ELIYA; 07°00′N 80°45′E.

Nuwara Eliya (NOO-wuh-rah E-LI-yuh), town (2001 population 25,388), ⊙ NUWARA ELIYA district, CENTRAL PROVINCE, S SRI LANKA; 06°58′N 80°46′E. Hill resort (elev. 6,194 ft/1,889 m) and health center. Tea-growing area, which also produces potatoes, carrots, cabbages, and strawberries. First settled in 1827 by the British. Today it is a world-class resort known as "Little England." According to legend, Sita, the wife of Rama, hero of the Sanskrit epic *Ramayana*, was imprisoned nearby by Ravana, a native king.

Nuweiba, EGYPT: see SINAI.

Nuweveld Range, on boundary between WESTERN CAPE and NORTHERN CAPE provinces, SOUTH AFRICA; at NW edge of the GREAT KAROO and on S side of the NORTHERN KAROO. Extends 120 mi/193 km W from BEAUFORT WEST; rises to 6,450 ft/1,966 m on Klaverfontein mountain, 20 mi/32 km N of Beaufort West. Continued W by KOMSBERG ESCARPMENT (5,645 ft/1,721 m). KAROO NATIONAL PARK is nearby.

Nuyts Archipelago (NOOTS), in GREAT AUSTRALIAN BIGHT, 5 mi/8 km off S coast of SOUTH AUSTRALIA; 32°35′S 133°17′E. Shelters DENIAL and SMOKY bays. It is a 40-mi-/64-km chain comprising SAINT PETER ISLAND (largest), FRANKLIN ISLANDS (easternmost), Isles of SAINT FRANCIS, LACY ISLES, PURDIE ISLANDS (westernmost), and scattered islets. Sandy, hilly. Petrels, geese.

Nuzi (NOO-zee), site SW of KIRKUK, N IRAQ. Thousands of clay tablets unearthed here bear inscriptions said to have been made by the Horims (or Horites) of the Bible. The tablets, which are in Akkadian, reveal much about ancient laws and customs.

Nuzvid (NOOZ-veed), town, KRISHNA district, ANDHRA PRADESH state, SE INDIA, 24 mi/39 km NE of VIJAYAWADA; 16°47′N 80°51′E. Rice milling; hand-loom woolen blankets; coconut and mango groves. Nuzvid railroad station 12 mi/19 km SE.

Nwa (NWAH), town, NORTH-WEST PROVINCE, CAMEROON, near NIGERIA border; 06°27′N 11°15′E.

Nxai Pan National Park, park (□ 999 sq mi/2,597.4 sq km), SE NORTH-WEST DISTRICT, N central BOTSWANA. Includes NXAI PANS and famous Baines Baobab trees; lies N of main NATA-MAUN road, on border with CENTRAL DISTRICT. Small game animals.

Nyabarongo River (nyah-BAH-ron-go), c.250 mi/402 km long, RWANDA; rises E of LAKE KIVU; flows generally E, joining the RUVUBU at TANZANIA border to form KAGERA RIVER. Formerly considered the remotest headstream of the NILE, this distinction is now given to the Ruvyironza, the longest branch of the Ruvubu.

Nyac (NEI-yak), village, W ALASKA, 70 mi/113 km ENE of BETHEL.

Nyack (NEI-yak), residential village (□ 1 sq mi/2.6 sq km; 2006 population 6,706), ROCKLAND county, SE NEW YORK, on W bank of TAPPAN ZEE (widening of the Hudson River; bridge), opposite TARRYTOWN; 41°05′N 73°55′W. Manufacturing of clothing, leather goods, optical goods. It was a 19th-century health resort, port, and boat-building center. Birthplace of artists Edward Hopper and Joseph Cornell; actor Helen Hayes is buried here. Hook Mountain Park (□ c.650 acres/263 ha; a section of Palisades Interstate Park) is just N. Settled 1684, incorporated 1782.

Nyadire River, c. 105 mi/169 km long, NE ZIMBABWE; rises c. 45 mi/72 km E of HARARE; flows NE; joins MAZOWE RIVER 120 mi/193 km NE of Harare.

Nyadiri, town, MASHONALAND EAST province, NE central ZIMBABWE, 20 mi/32 km NW of HARARE; 17°40′S 30°48′E. Lake Mayame reservoir (recreational park) to S. Gold mining to N. Dairying; cattle, sheep, goats; tobacco, wheat, corn, citrus fruit, soybeans.

Nyadri, administrative district, UGANDA: see MARACHA-TEREGO.

Nyagan' (NYAH-gahn), city, NW KHANTY-MANSI AUTONOMOUS OKRUG, TYUMEN oblast, W central SIBERIA, on railroad, 143 mi/230 km N of KHANTY-MANSIYSK, and 29 mi/47 km S of Priob'ye; 62°08′N 65°23′E. Elevation 134 ft/40 m. Center of oil production in the Krasnoleninsk oil field. Regional airport and highway junction. Food processing (bakery). Originally a settlement of Nyakh; made city and renamed in 1985.

Nyah (NEI-uh), village, NW VICTORIA, SE AUSTRALIA, on MURRAY RIVER, and 200 mi/322 km NNW of MELBOURNE, on NEW SOUTH WALES border; 35°10′S 145°23′E. Irrigation farming.

Nyahanga (nyah-HAHN-gah), village, MWANZA region, NW TANZANIA, 45 mi/72 km ENE of MWANZA, near SPEKE GULF, Lake VICTORIA; 02°25′S 33°33′E. Highway junction. Fish; cattle, sheep, goats; cotton, sugarcane, corn, wheat.

Nyahokwe, ZIMBABWE: see DZIVA NATIONAL MONUMENT.

Nyahua (nyah-HOO-ah), village, TABORA region, NW central TANZANIA, 42 mi/68 km SE of TABORA, near Nyahua River, on railroad; 05°25′S 33°58′E. Timber; livestock; subsistence crops.

Nyahururu (nyah-hoo-ROO-roo), town (1999 population 24,751), RIFT VALLEY province, W central KENYA, on Laikipia Plateau E of GREAT RIFT VALLEY, 30 mi/48 km NNE of NAKURU; 00°01′N 36°23′E. Elevation 7,680 ft/2,341 m. Railroad terminus and resort, in farming district (coffee, wheat, tea, corn). Falls are 243 ft/74 m high. Also known as Thompson's Falls or Thomson's Falls.

Nyaingentanglha (NEN-CHIN-TANG-GOO-LAH), mountain range of the SE TRANS-HIMALAYAS, in SE TIBET, SW CHINA, N of the BRAHMAPUTRA River; extends c. 550 mi/885 km E-W; 30°10′N 90°00′E. Highest point, Nyaingentanglha peak (23,255 ft/7,088 m), is 60 mi/97 km NW of LHASA. S central area drained by the LHASA River. Also spelled as Nyenchen Tanglha.

Nyakabindi (nyah-ka-BEEN-dee), village, SHINYANGA region, NW TANZANIA, 75 mi/121 km E of MWANZA. Maswa Game Reserve and SERENGETI NATIONAL PARK to E; 02°36′S 33°59′E. Cattle, sheep, goats; corn, wheat, cotton.

Nyakahura (nyah-kah-HOO-rah), village, KAGERA region, NW TANZANIA, 20 mi/32 km SW of BIHARAMULO, on Moyowosi River, S of BURIGI Game Reserve; 02°49′S 30°59′E. Livestock; subsistence crops.

Nyaka Kangaga (nyah-kah kan-GAH-gah), village, KIGOMA region, NW TANZANIA, 75 mi/121 km NE of KIGOMA, near MALAGARASI RIVER; 04°10′S 30°30′E. Timber; livestock; grain, tobacco.

Nyakanyasi (nyah-kah-NYAH-see), village, KAGERA region, NW TANZANIA, 43 mi/69 km WNW of BUKOBA, on KAGERA RIVER; 01°12′S 31°12′E. UGANDA border 15 mi/24 km to N. Livestock; subsistence crops.

Nyakh, RUSSIA: see NYAGAN'.

Nyakrom (NEE-ah-krahm), town, CENTRAL REGION, GHANA, located 20 mi/32 km NNW of WINNEBA; 05°37′N 00°47′W. Cacao, palm oil and kernels, cassava. Sometimes spelled Nyakrum or Nyaakrom.

Nyaksimvol' (nyek-SEEM-vuhl), village, W KHANTY-MANSI AUTONOMOUS OKRUG, TYUMEN oblast, W SIBERIA, RUSSIA, on the NORTHERN SOS'VA RIVER (head of navigation) near its confluence with the Nyays (left tributary) and Lep'lya (right tributary) rivers, on local unpaved highway, 280 mi/451 km NW of KHANTY-MANSIYSK; 62°25′N 60°52′E. Elevation 170 ft/51 m. Logging, lumbering; fishing.

Nyala (NYAH-lah), town, ⊙ S DARFUR state, W SUDAN, 115 mi/185 km SSW of AL-FASHER; 12°03′N 24°53′E. Terminal of railroad from KOSTI; road junction and trade center (gum arabic). Regional airport.

Nyalam (NAH-LAH-MOO), town, S TIBET, SW CHINA, in N NEPAL Himalayas, on KATMANDU-LHASA trade route, and 50 mi/80 km SW of TINGRI; 28°09′N 85°58′E. Livestock.

Ny-Ålesund (NUH–AW-luh-soon) Norwegian *New Ålesund*, town, on KONGSFJORDEN, NW SPITSBERGEN island, SVALBARD. It is a coal-mining settlement. Ny-Ålesund was (1926) the base of polar flights by Richard Byrd, in an airplane, and by Roald Amundsen, who, with Lincoln Ellsworth and Umberto Nobile, flew from here to ALASKA in the dirigible *Norge*. Polar-research station since 1968.

Nyalikungu (nyah-lee-KOON-goo), village, SHINYANGA region, N TANZANIA, 40 mi/64 km NE of SHINYANGA; 03°12′S 33°47′E. Road junction. Cattle, sheep, goats; cotton, wheat, corn.

Nyamandhlovu (en-YAH-mah-en-DLO-voo), town, MATABELELAND NORTH province, SW ZIMBABWE, 30 mi/48 km NW of BULAWAYO, near KHAMI RIVER, on railroad; 19°52′S 28°16′E. Elevation 3,942 ft/1,202 m. Cattle, sheep, goats; tobacco, peanuts, corn, soybeans, cottton.

Nyamapanda (en-YAH-mah-PAH-en-dah), village, MASHONALAND EAST province, NE ZIMBABWE, 130 mi/209 km NE of HARARE, near MUDZI RIVER, at MOZAMBIQUE border (crossing); 16°52′S 32°51′E. Airstrip to W. Livestock; subsistence crops.

Nyamaropa (en-YAH-mah-RO-pah), village, MANICALAND province, E ZIMBABWE, 70 mi/113 km NNE of MUTARE, on GAIREZI RIVER (MOZAMBIQUE border); 17°58′S 32°57′E. NYANGA MOUNTAINS to W. Coffee, tea, fruit, macadamia nuts; cattle, sheep, goats.

Nyamati (NYAH-muh-tee), town, SHIMOGA district, KARNATAKA state, SW INDIA, 15 mi/24 km N of SHIMOGA; 14°09′N 75°35′E. Local trade in grain, betel nuts, jaggery. Also spelled Nyamti; formerly Niamati.

Nyambeni (nyahm-BAI-nee), mountain range (8,244 ft/2,513 m), EASTERN province, central KENYA, E of ISIOLO. Market area; livestock. Also spelled Nyambene.

Nyambiti (nyahm-BEE-tee), village, MWANZA region, NW TANZANIA, 42 mi/68 km SE of MWANZA, on railroad; 02°52′S 33°26′E. Cattle, sheep, goats; corn, wheat, millet.

Nyamina (NEE-a-mee-na), township, SECOND REGION/KOULIKORO, S MALI, on the NIGER RIVER, and 81 mi/130 km NE of BAMAKO; 13°19′N 06°59′W. Peanuts, shea nuts; livestock. Market. Also spelled Niamina.

Nyamira, town, ⊙ Nyamira district, NYANZA province, SW KENYA; 00°34′N 34°57′E. Tea, bananas; food processing. Market and trade.

Nyamirembe (nyah-mee-RAIM-bai), village, KAGERA region, NW TANZANIA, 80 mi/129 km S of BUKOBA, on EMIN PASHA GULF, Lake VICTORIA, BIHARAMULO GAME RESERVE to W, RUBONDO (Island) National Park to NE; 02°35′S 31°41′E. Fish; livestock; grain.

Nyamlagira (nyahm-lah-GEE-rah), active volcano (c.10,000 ft/3,048 m) of the VIRUNGA range, E CONGO, N of LAKE KIVU, and 15 mi/24 km NE of SAKE, in S part of VIRUNGA NATIONAL PARK, and 30 mi/48 km N of GOMA; 01°25′S 29°12′E. Lava from Nyamlagira covers 579 sq mi/1,500 sq km of East African Rift. Since 1882, it has erupted 30 times. Before 1938, main point of volcanic activity was the central crater, a large lake of incandescent lava. In 1938, further eruption opened new outlet on SW slopes, resulting in lava lake drying out. Streams of lava reached LAKE KIVU and closed Sake Bay to NW of the lake. Quiet since the 1980s, it erupted in 1996, setting fire to forests lining its crater.

Nyamtukusa (nyahm-too-KOO-sah), village, MWANZA region, NW TANZANIA, 35 mi/56 km SSW of MWANZA, near head of Mwanza Gulf (inlet), Lake VICTORIA; 03°01′S 32°44′E. Fish; cattle; sugarcane, cotton, corn, wheat.

Nyandoma (NYAHN-duh-mah), city (2005 population 22,000), SW ARCHANGEL oblast, N European Russia, on road and railroad, 210 mi/338 km S of ARCHANGEL; 61°40′N 40°12′E. Elevation 721 ft/219 m. Railroad junction; railroad establishments; wood and food industries. Arose in 1896 with the construction of a railroad. City since 1939.

Nyanga (NEE-ahn-gah), province (□ 8,218 sq mi/ 21,366.8 sq km; 2002 population 50,800), SW GABON; ⊙ TCHIBANGA; 03°00′S 11°00′E. Bounded on E and S by CONGO, N by NGOUNIÉ and OGOOUÉ-MARITIME, W by GULF OF GUINEA. Rice is a major crop grown in this coastal lowland region. Some iron deposits are near the coast.

Nyanga, town, MANICALAND province, E ZIMBABWE, 45 mi/72 km NE of RUSAPE, in NYANGA MOUNTAINS; 18°14′S 32°44′E. Elevation 5,514 ft/1,681 m. Nyanga National Park and Nyangombe Falls to S. Fruit-growing center (pears, apples, citrus); cattle, sheep, goats; dairying. Rhodes Inyanga Estate, tourist resort, is 6 mi/9.7 km to S. Dziva Ruins to NW. Formerly called Inyanga.

Nyangadzi River, c. 75 mi/121 km long, NE ZIMBABWE; rises c. 30 mi/48 km E of MARONDERA, near UMFESERI; flows NE to join MWARAZI RIVER 30 mi/48 km ESE of MTOKO to form RUENYA (Luenha) River.

Nyanga Mountains, MANICALAND province, E ZIMBABWE, N of MUTARE, extending c.50 mi/80 km N–S to W MOZAMBIQUE border; rise to 8,508 ft/2,593 m at Mount Nyangani (18°18′S 32°50′E), highest point of Zimbabwe. S part of range in Nyanga (formerly RHODES-INYANGA) National Park. Formerly Inyanga Mountains.

Nyangani, Mount, ZIMBABWE: see NYANGA MOUNTAINS.

Nyanga River (NEE-ahn-guh), 240 mi/386 km long, SW GABON; rises 50 mi/80 km E of M'Bigou; flows SW, NW, and WSW, past TCHIBANGA, to the ATLANTIC 40 mi/64 km NW of MAYUMBA. Many rapids.

Nyangbo (nee-AHNG-bo), town, N CÔTE D'IVOIRE, 14 mi/23 km NE of NIAKARAMANDOUGOU, on railroad; 08°48′N 05°10′W. Agriculture (manioc, rice, corn, beans, peanuts, tobacco, cotton). Also spelled Niangbo.

Nyang Chu, CHINA: see NYANG QU.

Nyange (NYAHN-gai), village, KIGOMA region, NW TANZANIA, 68 mi/109 km NE of KIGOMA, near BURUNDI border; 04°07′S 30°20′E. Tobacco, subsistence crops; livestock; timber.

Nyang Qu (NANG CHU), river, SE TIBET, SW CHINA; right tributary of upper BRAHMAPUTRA RIVER; rises in small lakes Gala and Bam (elevation c. 14,800 ft/4,511 m) in NW ASSAM HIMALAYAS, 21 mi/34 km SW of KANGMAR; flows c. 110 mi/177 km N and NW, along main INDIA-Tibet trade route, past GYANGZE, Bainang, and XIGAZE, to the Brahmaputra 2 mi/3.2 km NE of Xigaze. Sometimes spelled Nyang Chu.

Nyanguge (nyah-goo-GAI), village, MWANZA region, NW TANZANIA, 18 mi/29 km E of MWANZA, near Lake VICTORIA; 02°34′S 33°10′E. Cattle, sheep, goats; sugarcane, cotton, grain.

Nyangwe (NYAHNG-gwai), village, MANIEMA province, E CONGO, on right bank of the LUALABA, and 30 mi/48 km NW of KASONGO; 04°13′S 26°11′E. Elev. 1,666 ft/507 m. Trading center, in cotton-growing area. Former center of Arab slave-traders in center of Africa, established 1863. It was reached by David Livingstone and V. Lake Cameron in 1870. Stanley began the first upstream voyage on CONGO RIVER here in 1874. Nyangwe fell to the Belgians in 1893. Also known as PENE BWEBWE.

Nyantakara (nyah-NTAH-kah-rah), village, KAGERA region, NW TANZANIA, 35 mi/56 km SSE of BIHARAMULO; 02°49′S 31°23′E. Livestock; subsistence crops.

Nyanza (NYAHN-zah), province (□ 6,239 sq mi/16,159 sq km, including 1,457 mi/2,345 km of lake area; 2003 estimated population 4,804,078; 2007 estimated population 5,039,776), W KENYA; ⊙ KISUMU. Bordered W by Lake VICTORIA, N by WESTERN province, and S by TANZANIA. Consists of savanna plateau sloping toward the lake. Agriculture (cotton, peanuts, sesame,

sugarcane, corn); livestock raising. Sorghum and millet are chief food crops. Dairy and poultry farming; lake fisheries. Gold mining in KAKAMEGA area. Main centers are Kisumu and KISII.

Nyanza (NYAN-zha), town (pop. 13,500), RWANDA, 35 mi/56 km SW of KIGALI. Local trade center and residence of *mwami* (sultan) of Rwanda; cinchona plantations; dairying. Roman Catholic mission.

Nyanza [Bantu=lake], generic name sometimes applied to lakes VICTORIA, ALBERT, and EDWARD, in E central AFRICA.

Nyanza Lac (nyahn-ZHAH lahk), village, MAKAMBA province, S BURUNDI, on NE shore of Lake TANGANYIKA, near TANZANIA border, 65 mi/105 km SSW of Kitega; 03°00′S 30°23′E. Small port exporting palm kernels and food staples. Sometimes called Nyanza-Lac.

Nyárádszereda, ROMANIA: see MIERCUREA-NIRAJ.

Nyasa, Lake (NYAH-sah) or **Lake Malawi** (□ c.11,600 sq mi/30,044 sq km), MALAWI, MOZAMBIQUE, and TANZANIA, in GREAT RIFT VALLEY, E AFRICA; c.360 mi/ 579 km long and from 15 mi/24 km to 50 mi/80 km wide. The third-largest lake in Africa, is bounded by steep mountains, except in S. Its main tributary is the RUHUHU RIVER, which enters from Tanzania in NE; the SHIRÉ, a tributary of the ZAMBEZI, exits S end of lake. Shipping lines ply the length of the lake. Malawi-Mozambique border runs through center of S part of lake, except LIKOMA and CHIZUMULU islands, which are part of Malawi; Malawi-Tanzania border follows NE (Tanzania) shore. Lake Nyasa [*nyasa*=lake] was named by David Livingstone in 1859. Ports include CHIPOKA and NKHATA BAY (Malawi), Cobue (Mozambique), and ITUNGI (Tanzania). Called Lake Niassa in Mozambique, according to Portuguese spelling, and Lake Malawi in Malawi.

Nyasaland, former state, central AFRICA: see MALAWI.

Nyasoso, village, SW CAMEROON, 23 mi/37 km NE of KUMBA. Cacao, bananas, palm oil, and kernels.

Nyaunglebin (nyoung-LAI-bin), town and township, MYANMAR, on Yangon-Mandalay railroad, near SITTANG RIVER, and 90 mi/145 km NNE of YANGON. In rice-growing area.

Nyaungu (NYOUNG-OO), town, MYANMAR, on left bank of AYEYARWADY RIVER, and 15 mi/24 km SW of PAKOKKU, immediately upstream of PAGAN. Lacquer works. Also spelled Nyaung-u.

Nyazepetrovsk (nyah-zye-pee-TROFSK), city (2005 population 12,500), NW CHELYABINSK oblast, SW Siberian Russia, in the central URALS, on the Nyazya River (tributary of the UFA RIVER), on road and railroad, 210 mi/338 km NW of CHELYABINSK, and 65 mi/105 km SW of SVERDLOVSK; 56°03′N 59°36′E. Elevation 1,161 ft/353 m. Railroad repair shops; crushing machinery, construction cranes; woodworking. Arose in 1747 as a workers settlement of Knyazepetrovskiy for the nearby ironworks. Became city in 1944.

Nyazura, town, MANICALAND province, E ZIMBABWE, 12 mi/19 km SSE of RUSAPE, on Nyazura River, on railroad; 18°43′S 32°05′E. Elevation 3,994 ft/1,217 m. Cattle, sheep, goats; dairying; tobacco, corn, coffee, tea.

Nyazwidzi River, c.100 mi/161 km long, E central ZIMBABWE; rises c. 5 mi/8 km SE of Chivu, in E MIDLANDS province; flows SE through Ruti Reservoir; joins DEVURE RIVER 25 mi/40 km WNW of BIRCHENOUGH BRIDGE.

Nyborg (NUH-bor), city (2000 population 15,792), FYN county, central DENMARK, a seaport at the head of Nybord Fjord (an arm of the STORE BÆLT); 55°20′N 10°45′E. It is an industrial center, with shipyards and plants manufacturing textiles and tobacco products. In Nyborg castle, built to control the Store Bælt (c.1170; now in ruins) Eric V granted (1282) the first Danish constitution. It was a royal capital (1200–1430).

Nybro (NEE-BROO), town, KALMAR county, SE SWEDEN, 19 mi/31 km WNW of KALMAR; 56°45′N 15°54′E. Railroad junction; manufacturing (glass, crystal; paper milling, woodworking). Incorporated 1932.

Nyby bruk (NEE-BEE BROOK), village, SÖDERMANLAND county, E SWEDEN, 4 mi/6.4 km NNW of ESKILSTUNA.

Nyda (NI-dah), village, central YAMALO-NENETS AUTONOMOUS OKRUG, TYUMEN oblast, NW SIBERIA, RUSSIA, just N of the ARCTIC CIRCLE, on the S shore of OB' BAY, at the mouth of the short Nyda River, 170 mi/ 274 km E of SALEKHARD; 66°37′N 72°55′E. Fish cannery; reindeer raising.

Nydalen (NUH-dah-luhn), suburb of OSLO, SE NORWAY, 4 mi/6.4 km ENE of city center. Metalworking. Until 1948, in AKERSHUS county.

Nye, county (□ 18,159 sq mi/47,213.4 sq km; 2006 population 42,693), S and central NEVADA; ⊙ TONOPAH; 38°02′N 116°27′W. Mountain region bordering on CALIFORNIA on SW: 4 large sections of Toiyabe National Forest are in N, in SHOSHONE, TOIYABE, TOQUIMA and MONITOR ranges; small part of Toiyabe National Forest in Spring Mountains in far E. PAHUTE MESA, AMARGOSA DESERT, and parts of DEATH VALLEY NATIONAL MONUMENT (main part in California) are in S. Mining (silver, gold, clay, magnesium, sand and gravel); cattle, sheep; ranching. Formed 1863. Berlin-Ichthyosaur State Park in NW. Nevada Test Site, and part of large Nellis Air Force Bombing and Gunnery Range in S center. Ash Meadows Wildlife Management Area in S. National Wildhorse Management Area in SE center. Nevada's only winery at PAHRUMP, a growing retirement community in extreme SE. Two parts of Humboldt National Forest in NE, in WHITE PINE and GRANT ranges. Railroad Valley and Wayne E. Kirch Wildlife Management Areas in NE. Duckwater Indian Reservation in NE. Lunar Crater in NE center. Largest county (land area) in Nevada, 3rd-largest co. in U.S. Drained by AMARGOSA RIVER (S), REESE RIVER (NW), White River (NE).

Nyegosh, MONTENEGRO: see NJEGOS.

Nyegushi, MONTENEGRO: see NJEGUSI.

Nye Mountains, a group of mountains in ENDERBY LAND, EAST ANTARCTICA, located E of RAYNER GLACIER; 30 mi/50 km long, 10 mi/16 km–15 mi/24 km wide; 68°10′S 49°00′E.

Nyenchen Tanglha, CHINA: see NYAINGENTANGLHA.

Nyergesújfalu (NYAR-gesh-uy-fah-loo), Hungarian *Nyergesújfalu*, city, KOMAROM county, N HUNGARY, on the DANUBE River, and 9 mi/14 km W of ESZTERGOM; 47°46′N 18°33′E. Silk manufacturing, cement- and brickworks.

Nyeri (NYE-ree), town (1999 population 46,969), ⊙ CENTRAL province, S central KENYA, between ABERDARE RANGE and Mount KENYA, on road, and 60 mi/97 km N of NAIROBI; 00°25′S 36°57′E. Elevation 5,900 ft/1,798 m. Tourist resort and agricultural trade center; coffee, sisal, wheat, corn; dairying. Airfield. Roman Catholic mission. Nyeri railroad station is 5 mi/8 km ENE.

Nyhammar (NEE-hahm-mahr), village, KOPPARBERG county, central SWEDEN, in BERGSLAGEN region, 11 mi/ 18 km NW of LUDVIKA; 60°17′N 14°58′E.

Nyika (NYI-kah), arid steppe in E KENYA, between tropical coastal lowland and central highlands. Elevation 500 ft/152 m–3,000 ft/914 m. Watered by TANA River.

Nyika National Park (□ 31 sq mi/80.6 sq km), EASTERN province, NE ZAMBIA, 160 mi/257 km ESE of KASAMA, on NYIKA PLATEAU. On MALAWI border, adjoins Malawi's larger Nyika National Park to E. Elevation generally exceeds 6,500 ft/1,981 m; Zambia's highest point, 7,120 ft/2,170 m, is N of park, on Malawi border. Variety of alpine flora and fauna, including orchids.

Nyika National Park, MALAWI, encloses most of the NYIKA PLATEAU in Northern region, 84 mi/135 km

NE of MZUZU. The country's first national park. A remote park with large herds of zebra, eland, roan antelope, reedbuck, bushbuck, and warthog.

Nyika Plateau (NYI-kah), grassy highland, Northern region, N MALAWI, W of LIVINGSTONIA; rises to 7,000 ft/2,134 m–8,000 ft/2,438 m. Wild game. Chelinda visitor center here.

Nyimba, village, EASTERN province, SE ZAMBIA, on Great East Road, 180 mi/290 km ENE of LUSAKA, near MOZAMBIQUE. Agriculture (tobacco, corn).

Nyinahin, town, ☉ ASHANTI region, GHANA, 40 mi/64 km W of KUMASI; 06°37′N 02°07′W. Timber; cocoa, kola nuts, coffee, rice. Bauxite mining.

Nyiragongo (nyee-rah-GAWNG-go), active volcano (c.11,400 ft/3,475 m) of the VIRUNGA range in E CONGO, near RWANDA border, NE of LAKE KIVU and 12 mi/19 km N of GOMA, in S part of VIRUNGA NATIONAL PARK; 01°31′S 29°15′E. Central crater is c.1.3 mi/2.1 km in diameter, 800 ft/244 m deep. There are several old craters, easily accessible and noted for their picturesque scenery and flora. In 1948, some cones opened at WSW base and lava flows issued, one reaching LAKE KIVU and obstructing SAKE-GOMA road. Erupted in 1977, covering hundreds of villages and parts of GOMA. During the last major eruption on January 17, 2002, lava flows reached the city of GOMA, forcing the displacement of c.500,000 people. Native legends associate the volcano with the place of expiation of guilty souls. Also known as TSHANINAGONGO.

Nyirbátor (NYER-bah-tor), Hungarian *Nyírbátor*, city, SZABOLCS-SZATMÁR county, NE HUNGARY, 21 mi/34 km ESE of NYIREGYHÁZA; 47°50′N 22°08′E. Railroad junction; grain, apples, tobacco, alfalfa, rye; cattle; manufacturing (vegetable oil, soap, footwear, paint, furniture, apparel, flour, pasta); tobacco warehouses.

Nyirbéltek (NYER-bel-tek), Hungarian *Nyírbéltek*, village, SZABOLCS-SZATMÁR county, NE HUNGARY, 25 mi/40 km ENE of DEBRECEN; 47°42′N 22°08′E. Rye, wheat, cattle; manufacturing (tiles); apple orchards.

Nyirbogdány (NYER-bog-dah-nyuh), Hungarian *Nyírbogdány*, village SZABOLCS-SZATMÁR county, NE HUNGARY, 10 mi/16 km NE of NYIREGYHÁZA; 48°03′N 21°53′E. Small oil refinery.

Nyíregyháza (NYE-redy-hah-zah), city (2001 population 118,795), NE HUNGARY. It is a county administrative center, a road and railroad junction, and the market for an extensive agricultural region. Industries include textiles, synthetic fibers, resins, apparel, paper and wood products, footwear, rubber tires, and natural-gas processing. It has secondary and higher education institutes (pedogical and technical college). It has specialized health-care and hospital facilities. Nyíregyháza is the seat of the Greek Catholic church in Hungary. Known in the 13th century, the city was destroyed during the Turkish occupation (16th century) of Hungary but was rebuilt in the 18th century. Its museum contains gold relics dating from Avar times.

Nyírség (NYER-shaig), Hungarian *Nyírség*, section of the Alföld, NE HUNGARY. Main town is NYIREGYHÁZA. The Nyirség occupies the W ⅔ of SZABOLCS-SZATMÁR county. With its sandy, acidic soil, it also receives much less sunshine than its sandy counterpart between the DANUBE and TISZA rivers. Apples are grown extensively, but other fruits are rare. Some wheat, corn, sugar beets, and tomatoes are grown, but potatoes, tobacco, and rye suit the environment more. Historically a poor region, it always had a labor surplus. Economic development during 1970–1990, including the large apple plantations, depended overwhelmingly on exports to the USSR. With the disintegration of the Soviet market and price liberalization, economic opportunities collapsed. Unemployment in the mid-1990s rose to between 70% and 75% above the country's average and almost four times higher than in the Budapest area.

Nyitra, SLOVAKIA: see NITRA.

Nyitrapróna, SLOVAKIA: see NITRIANSKE PRAVNO.

Nyitrarudnó, SLOVAKIA: see NITRIANSKE RUDNO.

Nykarleby (NUH-KAHRL-uh-BUH), Finnish *Uusikaarlepyy*, city, VAASAN province, W FINLAND, 40 mi/64 km NE of VAASA; 63°32′N 22°32′E. Elevation 66 ft/20 m. Located near mouth of Lapuanjoki (100 mi/161 km long), on GULF OF BOTHNIA. Trade center in lumbering, woodworking region. Population is largely Swedish-speaking. Swedish teachers' seminary; Saint Bridget church (1708). Poet Zacharias Topelius born here.

Nykerk, NETHERLANDS: see NIJKERK.

Nÿkerkernauw (NEI-kuhr-kuhr-nou), inlet, central NETHERLANDS between S FLEVOLAND polder and UTRECHT province coast, S IJSSELMEER.

Nykirke, NORWAY: see BORRE.

Nykøbing (NUH-KOH-bing), city, ☉ STORSTRØM county, SE DENMARK, on FALSTER ISLAND and on the GULDBORG SUND, connected by bridge with LOLLAND; 54°46′N 11°53′E. It is a seaport and has sugar refineries, breweries, shipyards, and textile mills. Of note are a Gothic church (until 1532 a Franciscan monastery) and the ruins of a twelfth-century castle. Jesperhus Blomster Park, largest gardens in Scandinavia, are here.

Nykopil', UKRAINE: see NIKOPOL'.

Nyköping (NEE-SHUHP-eeng), town, ☉ SÖDERMANLAND county, SE SWEDEN, port on BALTIC SEA; 58°45′N 17°01′E. Commercial and industrial center; manufacturing (motor-vehicle parts, electrical equipment, plastic and metal goods). Airport. Atomic research center at Studsvik. Founded thirteenth century on site of former trading town. Destroyed by fire (1665), rebuilt; sacked by Russians (1719). St. Nicholas Church (thirteenth–eighteenth centuries); City Hall (seventeenth century); Nyköpingshus castle (thirteenth century) ruins. Annual Renaissance fair.

Nykroppa (NEE-krop-pah), village, VÄRMLAND county, W SWEDEN, in BERGSLAGEN region, on Lake Östersjön (5 mi/8 km long), 7 mi/11.3 km SE of FILIPSTAD; 59°37′N 14°18′E. Railroad junction.

Nykvarn (NEE-KVAHRN), town, STOCKHOLM county, E SWEDEN, 6 mi/9.7 km W of SÖDERTÄLJE; 59°11′N 17°26′E.

Nyland (NEE-lahnd), village, VÄSTERNORRLAND county, NE SWEDEN, on ÅNGERMANÄLVEN RIVER estuary, 25 mi/40 km N of HÄRNÖSAND; 63°01′N 17°46′E.

Nyland, FINLAND: see UUDENMAAN.

Nylga (nil-GAH), village, S central UDMURT REPUBLIC, E European Russia, on crossroads, 30 mi/48 km W of IZHEVSK, and 15 mi/24 km SSE of UVA, to which it is administratively subordinate; 56°45′N 52°22′E. Elevation 374 ft/113 m. Flax processing.

Nylstroom, SOUTH AFRICA: see MODIMOLLE.

Nymagee, village, central NEW SOUTH WALES, SE AUSTRALIA, 285 mi/459 km E of BROKEN HILL; 32°04′S 146°20′E.

Nymburk (NIM-buhrk), German NIMBURG, town, STREDOCESKY province, E central BOHEMIA, CZECH REPUBLIC, on ELBE RIVER, and 19 mi/14.5 km NNW of KUTNÁ HORA; 50°11′N 15°03′E. Railroad junction (large railroad workshops); river port. Manufacturing (machinery; food processing). Founded in Middle Ages, the town was destroyed during Hussite wars. Has a 13th-century church, and remains of fortifications. Brewery was established in 1865.

Nymegen, NETHERLANDS: see NIJMEGEN.

Nymphenburg (NIMPF-en-burg), group of châteaus and a large park, near MUNICH, BAVARIA, S GERMANY. The main building is the Nymphenburg château (built 1663–1728), which belonged to the dukes (later kings) of BAVARIA; 48°08′N 11°30′E. Also noteworthy is Amalienburg (1734–1739), a small baroque hunting château designed by François de Cuvilliés. A famous porcelain factory was founded here in 1747. By the Treaty of Nymphenburg (1741) SPAIN promised Charles Albert of BAVARIA its support in his attempt to secure the imperial election.

Nymphio, TURKEY: see KEMALPASA.

Nynäshamn (NEE-NES-HAHMN), town, STOCKHOLM county, E SWEDEN, on BALTIC SEA, 25 mi/40 km S of STOCKHOLM; 58°54′N 17°56′E. Seaport, terminus of VISBY steamers; manufacturing (petrochemicals, telecommunication equipment). Railroad construction (1901) stimulated development. Has fifteenth-century manor house nearby.

Nyngan (NING-guhn), municipality, central NEW SOUTH WALES, on BOGAN RIVER, and 290 mi/467 km NW of SYDNEY; 31°34′S 147°11′E. Wool; wheat; cattle. Former mining center (copper, gold). Copper mining in area. Bird sanctuary at nearby Macquarie Marshes. Extensively damaged in 1990 by floods; flood museum at old railway station.

Nyombé (nee-YOM-bai), city, LITTORALprovince, W CAMEROON, 36 mi/58 km NNW of DOUALA; 04°35′N 09°40′E. Location of experimental farms producing tropical fruits. Railroad station and Roman Catholic mission.

Nyombi, village, MASVINGO province, SE central ZIMBABWE, 57 mi/92 km E of GWERU, near SHASHE RIVER; 19°27′S 30°42′E. Livestock; grain.

Nyon (nee-YON), district, SW VAUD canton, SWITZERLAND, on LAKE GENEVA; 46°23′N 06°15′E. Main town is NYON; population is French-speaking and Protestant, but many Roman Catholics live here.

Nyon (nee-YON), ancient *Noviodunum* or *Civitas Julia Equestris*, town (2000 population 16,182), VAUD canton, SW SWITZERLAND, on LAKE GENEVA, and 13 mi/21 km NNE of GENEVA; 46°23′N 06°15′E. Elevation 1,345 ft/410 m. Resort town; chemicals, leather products; metalworking. Has 16th-century castle with museum.

Nyong-et-Kelle (nee-YAWNG–ai–khel-ai), department (2001 population 145,181), Central province, CAMEROON; ☉ Eséka.

Nyong-et-Mfoumou (nee-YAWNG–ai–MFOO-moo), department (2001 population 130,321), Central province, CAMEROON; ☉ AKONOLINGA.

Nyong-et-Soo (nee-YAWNG–ai–so), department (2001 population 142,907), Central province, CAMEROON; ☉ Mbalmayo. Also spelled Nyong-et-So.

Nyong River (nee-yawng), c.400 mi/644 km long, central and W CAMEROON; rises c.25 mi/40 km E of Abong-M'Bang; flows generally W, past Abong-M'Bang, AYOS, Akonolinga, M'Balmayo, and Eséka, to the ATLANTIC 40 mi/64 km SSW of Edéa. Its middle course (Abong-M'Bang to M'Balmayo) is navigable for small powerboats April–Nov.

Nyons (nee-YON), town (□ 9 sq mi/23.4 sq km), DRÔME department, RHÔNE-ALPES region, SE FRANCE, on the Eygues River, and 33 mi/53 km NNE of AVIGNON; 44°20′N 05°10′E. Agricultural trade center (wines, truffles, lavender essence, black olives, almonds, and honey); fruit and olive processing and shipping. Colorful weekly market in PROVENCE setting attracts tourists. Has 14th-century arched bridge and a picturesque old quarter.

Nyord (NUH-or), island (□ 2 sq mi/5.2 sq km), DENMARK, between Møn and SJÆLLAND islands, between STEGE BUGT (SE) and FAKSE BUGT (NW); 55°03′N 12°13′E.

Nyoriya Husenpur, town, PILIBHIT district, N UTTAR PRADESH state, N central INDIA, 9 mi/14.5 km NE of PILIBHIT. Trades in rice, wheat, gram, sugarcane. Also called Neoria, Neoriya, Neoria Husainpur.

Nyos, Lake (nee-OS), lake in NORTH-WEST PROVINCE, CAMEROON, 15 mi/24 km ENE of WUM; 06°25′N 10°19′E. Site of August 21, 1986, carbon-dioxide-gas eruption which killed 1,700 people in nearby villages.

Nyrany (NEE-rzhah-NI), Czech *Nýřany*, German *nürschan*, town, ZAPADOCESKY province, W BOHEMIA, CZECH REPUBLIC, 8 mi/12.9 km WSW of PLZEŇ; 49°43′N 13°12′E. Railroad junction. Manufacturing (electronics, machinery, construction materials); food processing.

Nyrob (NI-ruhp), town (2006 population 7,255), N PERM oblast, W URALS, E European Russia, near the KOLVA RIVER, on road, 25 mi/40 km NNE of CHERDYN'; 60°43′N 56°42′E. Elevation 469 ft/142 m. Lumbering.

Nyrsko (NIR-sko), Czech *Nýrsko*, German *neuern*, town, ZAPADOCESKY province, SW BOHEMIA, CZECH REPUBLIC, in foothills of BOHEMIAN FOREST, on railroad, on UHLAVA RIVER, and 10 mi/16 km SW of KLATOVY, near German border; 49°18′N 13°09′E. Manufacturing (goggles and glasses, textiles); wood and food processing.

Nysa, POLAND: see NEISSE.

Nysa Klodzka River, POLAND: see NEISSE.

Nysa Luzycka River, POLAND: see NEISSE.

Nysa River, POLAND: see NEISSE.

Nyslott, FINLAND: see SAVONLINNA.

Nyssa (NIS-uh), town (2006 population 3,053), MALHEUR county, E OREGON, on Snake River, 5 mi/8 km NNE of mouth of Owyhee River, 8 mi/12.9 km S of ONTARIO; 43°52′N 117°00′W. Elevation 2,178 ft/664 m. Railroad junction. Sugar-beet processing. The town is a market for irrigation projects on Owyhee and BOISE rivers and for the Vale Irrigation Project. Site of an agricultural museum. Owyhee Dam 20 mi/32 km to SW. Lake Owyhee State Park to SW. Incorporated 1903.

Nyssa, name of several ancient cities devoted to the worship of Dionysus. The best known of them is a town of CAPPADOCIA, ASIA MINOR, near the Halys (now the KIZIL IRMAK) river. It was the residence of St. Gregory of Nyssa.

Nystad, FINLAND: see UUSIKAUPUNKI.

Nysted (NUH-stedh), city and port, STORSTRØM county, DENMARK, on LOLLAND island, and 12 mi/19 km SE of MARIBO, on BALTIC SEA; 54°42′N 11°40′E. Agriculture (barley, sugar beets); shipbuilding; fishing (flounder, herring).

Nytva (NIT-vah), city (2006 population 20,480), W central PERM oblast, W URALS, E European Russia, on the Nytva River near its confluence with the KAMA River (city protected by embankment on the river), on highway branch and railroad spur, 52 mi/84 km WSW of PERM; 57°56′N 55°20′E. Elevation 321 ft/97 m. Ferrous metallurgy and metal processing (quality steels); furniture; creamery. Arose in 1756 at the site of ironworks. Became city in 1942. Called Nytvinskiy prior to World War I. Landing at Ust'-Nytva, 5 mi/8 km S, on the left bank of the Nytva River.

Nytych, POLAND: see NOWY STAW.

Nyudo, Cape (NYOO-do), Japanese *Nyudo-misaki* (nyoo-DO—mee-SAH-kee), Akita prefecture, N HONSHU, NE JAPAN, at NW tip of OGA PENINSULA, in SEA OF JAPAN; 40°00′N 139°42′E.

Nyukawa (NYOO-kah-wah), village, Ono county, GIFU prefecture, central HONSHU, central JAPAN, 62 mi/100 km N of GIFU; 36°10′N 137°18′E.

Nyuksenitsa (nyoo-ksee-NEE-tsah), village (2006 population 4,295), NE VOLOGDA oblast, N central European Russia, on the SUKHONA River, on road, 75 mi/121 km WSW of VELIKIY USTYUG; 60°25′N 44°14′E. Elevation 288 ft/87 m. Flax and dairy processing; logging and lumbering.

Nyulikunga (nyoo-lee-KOON-gah), village, MWANZA region, NW TANZANIA, 35 mi/56 km E of MWANZA, on SIMIYU RIVER, near its entrance to Lake VICTORIA; 02°37′S 33°25′E. Cattle, sheep, goats; cotton, sugarcane, corn, wheat.

Nyundo, RWANDA: see GISENYI.

Nyungwe National Forest (nyoon-GWAI), RWANDA, located W of GIKONGORO and E of CYANGUGU. Protected equatorial montane forest.

Nyurba (NYOOR-bah), town (2006 population 9,865), W SAKHA REPUBLIC, W RUSSIAN FAR EAST, on VILYUY River, on highway, 100 mi/161 km WSW of VILYUYSK; 63°17′N 118°20′E. Elevation 393 ft/119 m. Diamond mining. Has a drama theater.

Nyustya, SLOVAKIA: see HNUSTA.

Nyuvchim (NYOOV-chim), town, SW KOMI REPUBLIC, NE European Russia, on the SYSOLA RIVER (tributary of the VYCHEGDA River), 31 mi/50 km S of SYKTYVKAR; 61°24′N 50°44′E. Elevation 364 ft/110 m. Iron mines; iron foundry dating from 1756 was reconstructed in the 1930s. Lumbering.

N'yu-York, UKRAINE: see NOVHORODS'KE.

Nyuzen (NYOO-ZEN), town, Nakanikawa county, TOYAMA prefecture, central HONSHU, central JAPAN, on SEA OF JAPAN, 22 mi/35 km N of TOYAMA; 36°55′N 137°30′E. Watermelons; tulips.

Nyverdal, NETHERLANDS: see NIJVERDAL.

Nyzhankovychi (ni-zhahn-KO-vee-chee) (Russian *Nizhankovichi*) (Polish *Nizankowice*, town), L'VIV oblast, UKRAINE, on Polish border, 7 mi/11 km N of DOBROMYL'; 49°41′N 22°49′E. Sawmilling, furniture making. Known since the 13th century as Zahumenka; renamed Nyzhankovychi in the 15th century; town since 1940.

Nyzhni Sirohozy (NIZH-nee see-ro-HO-zi) (Russian *Nizhniye Serogozy*), town, E KHERSON oblast, UKRAINE, 46 mi/74 km W of MELITOPOL'; 46°50′N 34°23′E. Elevation 131 ft/39 m. Raion center; flour mill, food-flavoring plant. Regional museum.

Nyzhniv (NIZH-neef) (Polish *Niznió́w*) (Russian *Nizhnev*), village (2004 population 7,900), E IVANO-FRANKIVS'K oblast, UKRAINE, on the DNIESTER River, and 17 mi/27 km E of IVANO-FRANKIVS'K; 48°57′N 25°05′E. Elevation 918 ft/279 m. Lime processing, basketmaking, flour milling.

Nyzhni Vorota (NIZH-nee vo-ro-TAH) [Ukrainian= lower gates] (Russian *Nizhniye Vorota*), town (2004 population 5,000), N TRANSCARPATHIAN oblast, UKRAINE, in the EASTERN BESKYDS, on the upper reaches of LATORYTSYA RIVER, on highway near the VERETS'KYY PASS, 16 mi/26 km NNE of SVALYAVA, and 6 mi/10 km NW of VOLOVETS'; 48°46′N 23°06′E. Elevation 1,909 ft/581 m. Woodworking factory, agricultural research station. Known since the 12th century; town since 1971.

Nyzhniy Nahol'chyk (NIZH-nee nah-HOL-chik) (Russian *Nizhniy Nagol'chik*), town, S LUHANS'K oblast, UKRAINE, in the DONBAS, 6 mi/10 km S of ANTRATSYT; 48°01′N 39°04′E. Elevation 396 ft/120 m. Greenhouse vegetable production; former silver, lead, and coal mines. Established in 1801; town since 1938.

Nyzhn'odnistrovs'ka zroshuval'na systema, UKRAINE: see LOWER DNIESTER IRRIGATION SYSTEM.

Nyzhn'ohirs'kyy (nizh-nyo-HEER-skie) (Russian *Nizhnegorskiy*), town, NE Republic of CRIMEA, UKRAINE, on railroad, on the SALHYR River, and 23 mi/37 km SE of DZHANKOY; 45°27′N 34°44′E. Flour mill. Until 1944, known as Seitler.

Nyzhn'oteple, UKRAINE: see TEPLE.

Nyzhnya Duvanka (NIZH-nyah doo-VAHN-kah) (Russian *Nizhnyaya Duvanka*), town (2004 population 2,750), NW LUHANS'K oblast, UKRAINE, 40 mi/64 km NW of STAROBIL'S'K; 49°35′N 38°09′E. Elevation 295 ft/89 m. Dairy. Established in 1732; town since 1960.

Nyzhnya Krynka (NIZH-nyah KRIN-kah) (Russian *Nizhnyaya Krynka*), town, E DONETS'K oblast, UKRAINE, in the Donbas, on the Krynka River (tributary of the MIUS RIVER), 6 mi/10 km N of KHARTSYZ'K, and 11 mi/18 km ENE of (and subordinated to) MAKIYIVKA; 48°07′N 38°10′E. Elevation 396 ft/120 m. Coal mine. Established in 1788; town since 1938.

Nyzhnye (NIZH-nye) (Russian *Nizhneye*), town, W LUHANS'K oblast, UKRAINE, in the DONBAS, on the right bank of the DONETS River, 13 mi/21 km SE of LYSYCHANS'K, and 12 mi/19 km NE of (and subordinated to) PERVOMAYS'K; 48°46′N 38°37′E. Elevation 318 ft/96 m. Coal mines. Established in 1754; town since 1983.

Nzebela (ZAI-bai-lah), town, Macenta prefecture, N'Zérékoré administrative region, SE GUINEA, in

Guinée-Forestière geographic region, along the LOFA RIVER, 40 mi/64 km SE of MACENTA; 08°05′N 09°06′W. Forested region.

Nzega (NZAI-gai), town, TABORA region, NW central TANZANIA, 60 mi/97 km NNE of TABORA; 04°14′S 33°10′E. Important highway junction. Agricultural trade center (grain; livestock).

N'Zérékoré, administrative region, SE Guinea; ⊙ N'ZÉRÉKORÉ. Bordered NWN by Faranah administrative region, N by Kankan administrative region (central portion of border formed by Dion River), ESE by CÔTE D'IVOIRE (part of central portion of border formed by Férédougouba River), SSW by LIBERIA, and W by SIERRA LEONE. Dion River in NNE, Kouraï River in NE, Férédougouba and Gouan rivers in E, Oulé River in S, SAINT PAUL RIVER rises in S on Liberian border, Lawa (later called LOFA) and Via rivers in W central, and MOA RIVER in NW. Part of Fon Going ridge in N. Pic de Tibé mountain (elevation 4,934 ft/1,504 m) in N central, Kourandu mountain (elevation 4,055 ft/1,236 m) in NE, Mount Tétini (elevation 4,124 ft/1,257 m) in E, and Ziama mountain (elevation 4,551 ft/1,387 m) in W central. Towns include BEYLA, GUÉKÉDOU, LOLA, MACENTA, and Yomou. Several main roads branch out of N'Zérékoré town: one running N through Gouéké, BOOLA, and Beyla towns before entering Kankan administrative region; one running E through Lola and Nzoo towns before entering Côte d'Ivoire; one running S through Diéké before entering Liberia; and one running NW through NZEBELA, Yirié, SÉRÉDOU, Macenta, and Guéckédou towns; another main road also runs N out of Guéckédou town into Faranah administrative region. Smaller roads also run through the region. Airport at N'Zérékoré. Includes the prefectures of Beyla in NE, Lola in SE, N'Zérékoré in S central, and Yomou in SW, Macenta in W central, and Guéckédou in extreme NW. All of N'Zérékoré administrative region is in Guinée-Forestière geographic region.

N'Zérékoré, prefecture, N'Zérékoré administrative region, SE GUINEA, in Guinée-Forestière geographic region; ⊙ N'ZÉRÉKORÉ. Bordered N by Beyla prefecture, E by Lola prefecture, S by LIBERIA, SWW by Yomou prefecture, and WNW by Macenta prefecture. Oulé River runs through center of prefecture. Towns include Gouéké, Koulé, and N'Zérékoré. Roads branch out of N'Zérékoré town to all parts of the prefecture and administrative region, as well as NE Guinea, CÔTE D'IVOIRE, and Liberia. Airport at N'Zérékoré town.

N'zérékoré (uhn-ze-re-KO-rai), town (1999 population 116,300), ⊙ N'Zérékoré prefecture and administrative region and Guinée-Forestière geographic region, SE Guinea, near Liberia border, 165 mi/266 km NE of Monrovia (road); 07°45′N 08°49′W. Agr. center (rice, palm kernels, coffee, tobacco, pepper, kola nuts, pimento; sheep, goats); timber industry. Manufacturing of cigars. Gold and iron deposits in vicinity. Churches.

N'zeto (n-ZAI-to), town, ZAIRE province, NW ANGOLA, small port on the ATLANTIC OCEAN, on road to UÍGE, 110 mi/177 km N of LUANDA; 07°15′S 12°55′E. Ships cotton, oilseeds; bananas, corn; fisheries. Airfield. Formerly Ambrizete (ahm-bree-ZE-tuh).

N'zi-Comoé, region (□ 7,550 sq mi/19,630 sq km; 2002 population 909,800), E central CÔTE D'IVOIRE; ⊙ DIMBOKRO; 07°15′N 04°10′W. Bordered NW by Vallée du Bandama region, E by KOMOÉ RIVER (NE by by Zanzan region, E by Moyen-Comoé region), SES by Agnéby region, SW by Lagunes region, and W by Lacs region. NZI RIVER in W. Towns include ARRAH, BOCANDA, BONGOUANO, DAOUKRO, DIMBOKRO, GROUMANIA, MBAHIAKRO, and PRIKRO. Railroad runs N-S through Dimbokro and W part of the region, from Lagunes and Agnéby into Lacs.

Nzinge (NZEEN-gai), village, DODOMA region, central TANZANIA, 10 mi/16 km W of DODOMA; 06°07′S 35°35′E. Cattle, sheep, goats; peanuts, corn, wheat.

Cross-references are shown in SMALL CAPITALS. The pronunciation guide is shown on page xix. The sources of population figures are shown on page xvii.

Nzi River (uhn-ZEE), c.280 mi/451 km long, CÔTE D'IVOIRE; rises SE of FERKÉSSÉDOUGOU; flows in meandering course generally S to BANDAMA RIVER near TIASSALÉ at 05°57′N 04°50′W.

Nzoia River (NZO-ee-ah), c.150 mi/241 km long, SW KENYA; rises in CHERANGANY HILLS; flows SW, past MUMIAS, to Lake VICTORIA just S of PORT VICTORIA. Its right affluents (from Mount ELGON) drain the Trans-Nzoia region (chief town, KITALE).

Nzo River (uhn-ZO), W CÔTE D'IVOIRE; rises in NIMBA MOUNTAINS on border of GUINEA and LIBERIA; flows SSE to join SASSANDRA RIVER at BUYO.

Nzoro, CONGO: see WATSA.

Nzubuka (nzoo-BOO-kah), village, TABORA region, NW central, 17 mi/27 km N of TABORA, on railroad; 04°44′S 32°49′E. Timber; sheep, goats; subsistence crops.

Nzwani (nuh-ZWAH-nee) or **Johanna Island**, island and administrative district (□ 164 sq mi/426.4 sq km; 2003 population 243,732), one of three main islands of Comoros Republic, one of four main islands of Comoros Islands, in Mozambique Channel, Indian Ocean c.300 mi/483 km W of N tip of Madagascar; ⊙ MUTSAMUDU; 12°10′S 44°20′E. Mayotte island (French) to E; Njazidja (Grand Comore) island to NW; Mwazi (Moheli) island to W. Major towns include Domoni, Sima, and Moya. Timber; vanilla, ylang-ylang, rice, cassava, sweet potatoes, bananas, coconuts, sugarcane, cloves; livestock; fish. Nzwani is 20 mi/32 km long, 18 mi/29 km wide. The island is of volcanic origin; it rises to 5,233 ft/1,595 m at Mt. Ntingui (Tingue) near center. Formerly the economic center of the Comoros Islands, Nzwani played a major role in the slave market and was subject to strong Arab influence from the 15th century. Became part of French protectorate in 1885; broke away from French to become part of Comoros Republic in 1974. Called Anjouan under French rule. Economically depressed conditions on the island (unemployment hovers around 90%) led to bitter resentment against the republic, and rebels took control of the island in 1997, declaring its secession from the republic and asserting their desire to become a French colony once again; they repelled a government invasion seeking to put down the rebellion.

Oacoma (O-uh-KO-muh), village (2006 population 418), LYMAN county, S central SOUTH DAKOTA, 4 mi/6.4 km W of CHAMBERLAIN, across MISSOURI RIVER; 43°47′N 99°22′W. Manganese deposits nearby. Site of Fort Kiowa (1822) to N.

Oadby (OD-bee), town (2001 population 22,679), central LEICESTERSHIRE, central ENGLAND, 3 mi/4.8 km SE of LEICESTER; 52°36′N 01°05′W. Hosiery, knitwear, shoes. Has 13th-century church. Site of racecourse.

Oadweina, SOMALIA: see ODUEINE.

Oahe, Lake (O-AW-hee), reservoir (c.250 mi/402 km long), SOUTH DAKOTA and NORTH DAKOTA, on MISSOURI RIVER, 6 mi/9.7 km N of PIERRE; 44°28′N 100°24′W. Has three arms (all in South Dakota), formed by CHEYENNE, MOREAU, and GRAND rivers. Formed by Oahe Dam (242 ft/74 m high), built (1948–1963) by the Army Corps of Engineers for power generation, flood control, irrigation, and recreation. Cheyenne River (South Dakota) and Standing Rock (North Dakota) Indian reservations on W shore, as is Fort Rice Historic Site (North Dakota); West Whitlock and Swan Creek state recreation areas (South Dakota) on E shore.

Oahu (o-AH-hoo), island (□ 593 sq mi/1,541.8 sq km), third-largest and chief island of HAWAII, comprising (with offshore islands) HONOLULU county, separated from KAUAI island (NW) by Kauai Pass, from MOLOKAI island (E) by KAIWI CHANNEL. Oahu is composed of 2 parallel mountain ranges, WAIANAE (W) and KOOLAU (E), separated by a broad, rolling plain dissected by deep gorges. Mount KAALA (4,046 ft/1,233 m) is the island's highest peak. Oahu has no active volcanoes, but there are many extinct craters, among them DIAMOND HEAD, KOKO HEAD, and PUNCHBOWL. PEARL HARBOR indents the island's S coast. HONOLULU, the state capital and the economic center of Hawaii, and capital of Honolulu county, is on the highly urbanized S coast of Oahu. Honolulu is the site of the University of Hawaii (at Manoa), Chaminade University, Hawaii Pacific University, and Honolulu and Kapiolani community colleges. The island is an important defense area that includes the headquarters of the U.S. Marine Corps Pacific Command and the Pearl Harbor naval base and Hickam Air Force Base. Other military installations include Kaneohe Marine Corps Base (on Mokapu Peninsula), Barbers Point Naval Air Station (SW), Lualualei Naval Reservation (W), Schofield Barracks Military Reservation and Wheeler Army Air Field (center), Kaena Military Reservation (NW). There are many swimming beaches (including WAIKIKI at Honolulu). Large pineapple plantations persist in central rural areas of the island, but sugarcane is no longer grown. Dairy farming, fruit, vegetables, and nursery crops and fishing are important activities, but services, tourism and military operations are the principal economic mainstay of Oahu. Honolulu is main manufacturing center of Oahu and state; new industrial centers include AIEA, WAIPAHU, and KAPOLEI. Urban development has spread W, N and NE from Honolulu; 3 interstate highways extend in these directions. Numerous forest reserves protect mountain areas, including Waianae Kai Forest Reserve (W) and KAHUKU, EWA, Waiahole and Honolulu Watershed (E). Kahana Valley and Sacred Falls state parks and Malaekahana State Recreational Area in NE; Kuaiwa Heiau State Recreational Area in S; Kaena Point State Park in NW; Honolulu Stadium State Park and Diamond Head State Monument at Honolulu.

Oak, village (2006 population 56), NUCKOLLS county, S NEBRASKA, 17 mi/27 km NNE of SUPERIOR, and on LITTLE BLUE RIVER; 40°14′N 97°54′W.

Oakbank (OK-bank), community, SE MANITOBA, W central Canada, 4 mi/6 km from DUGALD, 14 mi/23 km from WINNIPEG, and in Springfield rural municipality; 49°56′N 96°50′W.

Oak Bay (OK BAI), district municipality, SE suburb (□ 4 sq mi/10.4 sq km; 2001 population 17,798) of VICTORIA, SW BRITISH COLUMBIA, W Canada, at SE extremity of VANCOUVER ISLAND, on JUAN DE FUCA Strait, in Capital regional district; 48°27′N 123°18′W. Established 1906.

Oak Bluff (OK BLUHF), community, S MANITOBA, W central Canada, 11 mi/18 km from WINNIPEG, and in Macdonald rural municipality; 49°46′N 97°19′W.

Oak Bluffs, town, DUKES county, SE MASSACHUSETTS, on NE MARTHA'S VINEYARD, 23 mi/37 km SE of NEW BEDFORD; 41°26′N 70°35′W. Resort. Passenger ferry connections with WOODS HOLE. Location of famous Methodist campground dating from 1870, Victorian Gingerbread cottages. Oldest operating carousel in U.S. Lighthouse at entrance. Oak Bluffs State Beach Park. Settled 1642, set off from EDGARTOWN as Cottage City in 1880, renamed 1907.

Oakboro, village (□ 2 sq mi/5.2 sq km; 2006 population 1,180), STANLY county, S central NORTH CAROLINA, 11 mi/18 km SW of ALBEMARLE, near ROCKY RIVER; 35°13′N 80°19′W. Manufacturing (textiles, apparel, roller tubing, lumber, machinery); service industries; agriculture (cotton, grain, soybeans; poultry, livestock; dairying). Incorporated 1915.

Oak Brae (OK BRAI), community, S MANITOBA, W central Canada, 14 mi/23 km from WINNIPEGOSIS, in Mossey River rural municipality; 51°26′N 99°52′W.

Oak Brook, village (2000 population 8,702), DU PAGE county, NE ILLINOIS, suburb 16 mi/26 km W of CHICAGO; 41°50′N 87°57′W. High-rise office complexes; shopping district. Manufacturing (consumer goods, transportation equipment, coated paper and film, detergents, machinery). Site of Oakbrook Shopping Center, one of the largest malls in the U.S.

Oakbrook Terrace, city (2000 population 2,300), DU PAGE county, NE ILLINOIS, residential suburb 17 mi/27 km W of downtown CHICAGO, and NW of OAK BROOK; 41°51′N 87°58′W.

Oakburn (OK-buhrn), community, SW MANITOBA, W central Canada, 9 mi/14 km N of town of SHOAL LAKE, in Shoal Lake rural municipality; 50°33′N 100°34′W.

Oak City (OK SIT-ee), village (2006 population 352), MARTIN county, E NORTH CAROLINA, 28 mi/45 km E of ROCKY MOUNT; 35°57′N 77°17′W. Manufacturing (apparel); agriculture (tobacco, cotton, peanuts, grain; poultry, hogs, cattle). Established in the early 1880s as a small, rural trading center. Originally called Goose Neck.

Oak City, village (2006 population 624), MILLARD county, W central UTAH, 12 mi/19 km E of DELTA; 39°22′N 112°20′W. Elevation 5,105 ft/1,556 m. Alfalfa; dairying; cattle. Fishlake National Forest to S and E; Fools Creek Reservoir to N. Settled 1860, originally named Oak Creek.

Oak Creek, city (2006 population 32,341), MILWAUKEE county, SE WISCONSIN, a suburb 8 mi/12.9 km S of downtown MILWAUKEE, on LAKE MICHIGAN; 42°52′N 87°54′W. Manufacturing (machinery, electronic products, plastic products, computers, paper products, chemicals, transportation equipment, fabricated metal products, concrete products; metal fabricating). Small farms dot the city's surrounding region. Incorporated 1955.

Oak Creek, village (2000 population 849), ROUTT county, NW COLORADO, Oak county, in SW foothills of PARK RANGE, and 15 mi/24 km SSW of STEAMBOAT SPRINGS; 40°16′N 106°57′W. Elevation 7,414 ft/2,260 m. Oil and gas. Cattle, sheep; wheat, hay, barley. Coal mines in vicinity. Parts of Routt National Forest to E

and SW; Stagecoach State Park to E. Incorporated 1907.

Oakdale, city (2000 population 15,503), STANISLAUS county, central CALIFORNIA, in San Joaquin Valley, 25 mi/40 km ESE of STOCKTON, and on STANISLAUS RIVER; 37°46′N 120°51′W. Dairying; irrigated farming (fruit, almonds, vegetables, grain, melons), horticulture; poultry. Manufacturing (wood products, machinery, chocolate; printing and publishing). Annual rodeo. HETCH HETCHY AQUEDUCT to SE, delivers water from SIERRA NEVADA range to SAN FRANCISCO. Woodward and Farmington reservoirs to N. Incorporated 1906.

Oakdale, city, ALLEN parish, SW central LOUISIANA, 52 mi/84 km NE of LAKE CHARLES city, on CALCASIEU RIVER; 30°49′N 92°40′W. In lumber and agricultural area (soybeans, vegetables, peaches; cattle); manufacturing (chemicals, plywood and wood products, apparel). Laid out in 1886. West Bay State Wildlife Area to W.

Oakdale, city, WASHINGTON county, E MINNESOTA, residential suburb 7 mi/11.3 km ENE of downtown ST. PAUL, in area of small natural lakes; 44°59′N 92°58′W. Manufacturing (machinery, pressure vessels, plastic products, polyethylene tubing).

Oakdale, unincorporated town, MCCRACKEN county, W KENTUCKY; residential suburb 3 mi/4.8 km SE of downtown PADUCAH, on OHIO RIVER, at mouths of TENNESSEE and CLARKS rivers. Includes villages of Woodlawn and Tyler, both names duplicated in other parts of Kentucky.

Oakdale, town (2006 population 237), MORGAN county, NE central TENNESSEE, 7 mi/11 km SE of WARTBURG, in the CUMBERLAND MOUNTAINS; 35°59′N 84°33′W.

Oakdale, village, within MONTVILLE, NEW LONDON county, SE CONNECTICUT.

Oakdale, village (2006 population 316), ANTELOPE county, NE central NEBRASKA, 5 mi/8 km SE of NELIGH, and on ELKHORN RIVER; 42°04′N 97°58′W. Livestock; dairy products; grain.

Oakdale, unincorporated village (□ 3 sq mi/7.8 sq km; 2000 population 8,075), SUFFOLK county, SE NEW YORK, on S shore of LONG ISLAND, 2 mi/3.2 km NW of SAYVILLE; 40°44′N 73°07′W. Seat of Dowling College. HECKSCHER STATE PARK is nearby.

Oakdale, borough (2006 population 1,454), ALLEGHENY county, SW PENNSYLVANIA, suburb 11 mi/18 km WSW of downtown PITTSBURGH, on Robinson Run; 40°23′N 80°11′W. Diversified manufacturing; oil. Settlers Cabin Park to NE. Incorporated 1872.

Oakdale, MASSACHUSETTS: see WEST BOYLSTON.

Oakengates (O-kuhn-GAITZ), town (2001 population 8,507), Telford and Wrekin, W ENGLAND, 4 mi/6.4 km E of WELLINGTON; 52°42′N 02°27′W. Adjacent to TELFORD, a New Town.

Oakes, town (2006 population 1,819), DICKEY county, SE NORTH DAKOTA, 27 mi/43 km ENE of ELLENDALE, near JAMES RIVER; 46°08′N 98°05′W. Railroad junction, agricultural shipping center. Manufacturing (machinery); dairy products; grain. Founded in 1886, incorporated in 1888. Named for Thomas Fletcher Oakes, president of the Northern Pacific Railroad.

Oakesdale, village (2006 population 384), WHITMAN county, SE WASHINGTON, 18 mi/29 km NNE of COLFAX, near Pine Creek, in the Palouse Hills; 47°08′N 117°15′W. Wheat, oats, peas, barley; manufacturing (flour milling). Steptoe Butte State Park to S. Historic flour mill; historical museum.

Oakes Field, airport, just E of NASSAU, NEW PROVIDENCE ISLAND, N central BAHAMAS.

Oakey (O-kee), town, SE QUEENSLAND, NE AUSTRALIA, 80 mi/129 km W of BRISBANE, 18 mi/29 km NW of TOOWOOMBA; 27°26′S 151°43′E. Agriculture center (wheat, corn); dairy, beef cattle.

Oakey Creek (O-kee KREEK), railroad siding, coal mine, E central QUEENSLAND, NE AUSTRALIA, 45 mi/72 km NW of BLACKWATER; 31°37′S 149°42′E.

Oakfield, town, WORTH COUNTY, S central GEORGIA, 20 mi/32 km NE of ALBANY, near FLINT RIVER; 31°46′N 83°58′W.

Oakfield, agricultural town, AROOSTOOK county, E MAINE, 15 mi/24 km W of HOULTON; 46°04′N 68°04′W. Incorporated 1897.

Oakfield, town (2006 population 1,009), FOND DU LAC county, E central WISCONSIN, 8 mi/12.9 km SSW of FOND DU LAC; 43°40′N 88°32′W. Railroad spur terminus. In dairy and farm region. Manufacturing (limestone products). Horicon National Wildlife Refuge to SW.

Oakfield, village (□ 17 sq mi/44.2 sq km; 2006 population 1,703), GENESEE county, W NEW YORK, 7 mi/11.3 km NW of BATAVIA; 43°03′N 78°16′W. In wheat area. Food processing; quarrying. Iroquois National Wildlife Refuge (□ 17 sq mi/44 sq km), featuring migratory waterfowl of E N. America, occupies much of Orchard Swamp, 5 mi/8 km NW of village center. Settled 1850, incorporated 1858.

Oakford, village (2000 population 309), MENARD county, central ILLINOIS, near SANGAMON RIVER, 10 mi/16 km NW of PETERSBURG; 40°06′N 89°58′W. In agricultural area.

Oak Forest, city (2000 population 28,051), COOK county, NE ILLINOIS, a residential suburb 20 mi/32 km SSW of downtown CHICAGO; 41°36′N 87°45′W. Residential development interspersed with remnant agriculture (grain; livestock); manufacturing (chemicals; commercial printing). Incorporate 1947.

Oak Grove, city (2000 population 5,535), JACKSON county, W MISSOURI, 25 mi/40 km ESE of KANSAS CITY; 39°00′N 94°07′W. Corn, wheat, sorghum; cattle. Satellite community of Kansas City.

Oak Grove, unincorporated city (2000 population 12,808), CLACKAMAS county, NW OREGON, residential suburb 7 mi/11.3 km S of PORTLAND, on WILLAMETTE RIVER; 45°24′N 122°38′W. Manufacturing (machinery).

Oak Grove, town (2000 population 7,064), CHRISTIAN county, SW KENTUCKY, 15 mi/24 km SSE of HOPKINSVILLE, at TENNESSEE state line; 36°40′N 87°25′W. Agricultural area (dark and burley tobacco, soybeans, grain; livestock; dairying). Limestone. Manufacturing (concrete). Fort Campbell Military Reservation to W.

Oak Grove, town (2000 population 2,174), ⊙ WEST CARROLL parish, NE LOUISIANA, near ARKANSAS state line, 50 mi/80 km NE of MONROE; 32°52′N 91°23′W. In agricultural area (cotton, corn, vegetables; cattle); manufacturing (canned vegetables, building components). Incorporated as village in 1908, as town in 1928. BAYOU MACON State Wildlife Area to E.

Oak Grove, unincorporated town, LEXINGTON county, central SOUTH CAROLINA, residential suburb 6 mi/9.7 km SW of downtown COLUMBIA. Columbia Municipal Airport to SE; 33°58′N 81°08′W.

Oak Grove, town (2000 population 4,072), WASHINGTON county, NE TENNESSEE, 8 mi/13 km NNE of JOHNSON CITY; 36°24′N 82°25′W.

Oak Grove, village (2000 population 1,318), ROCK ISLAND county, NW ILLINOIS, 5 mi/8 km S of ROCK ISLAND; 41°24′N 90°34′W. Residential suburb in agriculture area (corn, soybeans; livestock).

Oakham (O-kuhm), town (2001 population 9,975), ⊙RUTLAND, central ENGLAND, 17 mi/27 km E of LEICESTER; 52°40′N 00°44′W. Rutland Water, one of the largest man-made lakes in EUROPE, to E. Manufacturing of hosiery, plastics, electrical equipment, and machinery. Has public school (founded 1584), 12th-century castle, and 14th-century church.

Oakham (O-kuhm), town, WORCESTER county, central MASSACHUSETTS, 13 mi/21 km WNW of WORCESTER; 42°21′N 72°03′W. Agriculture.

Oak Harbor, city (2006 population 22,713), Island county, NW WASHINGTON, on WHIDBEY ISLAND and 30 mi/48 km NW of EVERETT. Dairying; wheat; poultry; manufacturing (concrete products, lab instruments). WHIDBEY ISLAND Naval Air Station to N on Saratoga Passage, arm of PUGET SOUND. Joseph Whidbey (W) and DECEPTION PASS (W) state parks nearby.

Oak Harbor, village (□ 2 sq mi/5.2 sq km; 2006 population 2,819), OTTAWA county, N OHIO, 11 mi/18 km W of PORT CLINTON, on PORTAGE RIVER; 41°31′N 83°08′W. Baskets, food products, building materials, barrels; ships fruit. Sometimes spelled Oakharbor.

Oakhaven, village (2000 population 54), HEMPSTEAD county, SW ARKANSAS, 4 mi/6.4 km N of HOPE; 33°43′N 93°37′W. Hope Wildlife Management Area to N.

Oak Hill, city (2006 population 4,825), DAVIDSON county, central TENNESSEE, residential suburb 6 mi/10 km S of NASHVILLE; 36°04′N 86°47′W. Radnor Lake State Natural Area is nearby. Incorporated 1952.

Oak Hill, town (2000 population 37), Wilcox co., SW central Alabama, 14 mi/23 km SE of Camden.

Oak Hill, town (2006 population 7,272), FAYETTE county, S central WEST VIRGINIA, 6 mi/9.7 km S of FAYETTEVILLE; 37°58′N 81°09′W. Railroad junction. Bituminous-coal mines. Manufacturing (building materials, machinery). Agriculture (grain, alfalfa); cattle. Timber. NEW RIVER GORGE NATIONAL RIVER to E; Plum Orchard Wildlife Management Area to SW. Settled 1820.

Oak Hill, village (2000 population 35), CLAY county, N central KANSAS, 14 mi/23 km SW of CLAY CENTER; 39°15′N 97°20′W. In grain and livestock area. Also spelled Oakhill.

Oak Hill, village (2006 population 1,647), JACKSON county, S OHIO, 12 mi/19 km SSE of JACKSON; 33°54′N 82°34′W. Forest area. Welsh-American Heritage Museum.

Oak Hill, hamlet, GREENE county, SE NEW YORK, in CATSKILL MOUNTAINS, on CATSKILL CREEK, and 20 mi/32 km NW of CATSKILL; 42°25′N 74°09′W.

Oak Hill, MASSACHUSETTS: see NEWTON.

Oak Hills, unincorporated town (2000 population 2,335), BUTLER county, W PENNSYLVANIA, residential suburb 3 mi/4.8 km SSW of BUTLER; 40°49′N 79°54′W.

Oakhurst, unincorporated town (2000 population 2,868), MADERA county, central CALIFORNIA, 38 mi/61 km NNE of FRESNO, on FRESNO RIVER, in W foothills of SIERRA NEVADA; 37°20′N 119°39′W. Cattle, poultry; dairying; nuts, fruit, grain. Manufacturing (power valves; printing and publishing). Yosemite National Forest to N and E.

Oakhurst, village (2000 population 4,152), MONMOUTH county, E NEW JERSEY, near the ATLANTIC coast, 3 mi/4.8 km N of ASBURY PARK; 40°15′N 74°01′W. Residential.

Oakhurst (OK-huhrst), village (2000 population 230), SAN JACINTO county, E TEXAS, 14 mi/23 km E of HUNTSVILLE, in N edge of Sam Houston National Forest; 30°44′N 95°18′W. In timber, agricultural area. Lake Livingston reservoir to E.

Oak Island, islet in MAHONE BAY, S NOVA SCOTIA, E CANADA, 4 mi/6 km SW of CHESTER; 44°31′N 64°18′W. The treasure of Captain Kidd is reputedly hidden here; many unsuccessful attempts have been made to recover it. Named for the oak trees on island.

Oak Island, one of the THOUSAND ISLANDS, ST. LAWRENCE county, N NEW YORK, in the SAINT LAWRENCE River, near Canadian (ONTARIO) border, just SW of CHIPPEWA BAY; c.2 mi/3.2 km long, max. width c.1 mi/1.6 km; 44°26′N 75°48′W.

Oak Lake (OK), town (□ 1 sq mi/2.6 sq km; 2001 population 359), SW MANITOBA, W central Canada, near ASSINIBOINE RIVER, 30 mi/48 km W of BRANDON, and surrounded by Sifton rural municipality; 49°46′N 100°38′W. Mixed farming; livestock; muskrat farms. Incorporated 1907.

Oakland (OK-luhnd), rural municipality (□ 222 sq mi/577.2 sq km; 2001 population 1,111), SW MANITOBA, W central Canada, traversed in its SW area by SOURIS River; 49°37′N 99°50′W. Agriculture (canola, wheat, barley) and related businesses. Nesbitt is the main center; other communities include Rounthwaite, Carroll, Banting, and WAWANESA. Incorporated 1883.

Oakland, county (□ 908 sq mi/2,360.8 sq km; 2006 population 1,214,255), SE MICHIGAN; 42°39′N 83°22′W; ⊙ PONTIAC. Drained by SHIAWASSEE, HURON, and CLINTON rivers, and by the RIVER ROUGE. Part of DETROIT metropolitan area. Fruit growing (apples); agriculture in N (vegetables, wheat, oats, corn; cattle, hogs, sheep, poultry; dairy products; nurseries). Manufacturing at Pontiac, FERNDALE, and ROYAL OAK. Highly urbanized in S ½ from neighboring city of Detroit. Many small lakes (resorts, especially in W). County has ten state parks and state recreational areas and 4 ski areas. Organized 1820.

Oakland (OK-luhnd), city (2000 population 399,484), ⊙ ALAMEDA county, W CALIFORNIA, 6 mi/9.7 km E of SAN FRANCISCO, on the E side of SAN FRANCISCO BAY; 37°46′N 122°13′W. Together with San Francisco and SAN JOSE, the city comprises the fourth-largest metropolitan area in the U.S. A containerized shipping port and a major railroad terminus, Oakland has shipyards, chemical plants, glassworks, food-processing establishments, and an iron foundry. Manufacturing includes foods, cleaners, electronic goods, canvas products, and fabricated metal products. The SAN FRANCISCO–OAKLAND BAY BRIDGE was opened in 1936 and connects Oakland with other nearby cities. Oakland is the headquarters and hub of the Bay Area Rapid Transit (BART), a three-county rapid transit system connected to San Francisco that began operation in 1972. Oakland International Airport, at Bay Farm Island (peninsula) in S.

An extremely severe earthquake on October 17, 1989, which struck during a World Series baseball game in the San Francisco Bay area, resulted in great damage to Oakland as well as to the San Francisco–Oakland Bay Bridge. About 1 mi/1.6 km of Interstate 880 highway, in Oakland, collapsed. The earthquake's toll took sixty-two lives and injured thousands; major repair and reconstruction efforts immediately ensued. Other principal redevelopment since the 1970s has focused on Oakland's waterfront area.

Of interest are the Oakland Museum, Chabot Observatory, the Morcom Rose Garden, and Jack London Square. The city has a symphony orchestra, notable parks, a state arboretum, a children's amusement park, and a zoo. Seat of Mills College, Holy Names University, California College of the Arts, Laney College, and Merritt College; University of California at Berkeley is to N. The large U.S. Naval Supply Depot and Oakland Army Base are in W part of city, on bay; Alameda Naval Air Station to SW, in ALAMEDA; Oak Knoll Naval Medical Center in SE. McAfee Coliseum is the home of the Oakland Athletics professional baseball team and the Oakland Raiders professional football team. Anthony Chabot and Redwood regional parks to E; Upper San Leandro Reservoir to E, LAKE CHABOT reservoir to SE, Lake Merritt reservoir E of downtown; Knowland State Park and Arboretum in SE. Incorporated 1852.

Oakland, city (2000 population 996), COLES county, E ILLINOIS, near EMBARRAS RIVER, 14 mi/23 km NNE of CHARLESTON; 39°39′N 88°01′W. In rich agricultural area (corn, wheat, soybeans; livestock). Incorporated 1855.

Oakland, town (2000 population 1,487), POTTAWATTAMIE county, SW IOWA, on WEST NISHNABOTNA RIVER, and 25 mi/40 km E of COUNCIL BLUFFS; 41°18′N 95°24′W. In livestock, grain, poultry area. Fabricated metal products; rendering works. Incorporated 1882.

Oakland, town, including Oakland village, KENNEBEC county, S MAINE, 16 mi/26 km N of AUGUSTA, and on MESSALONSKEE LAKE; 44°33′N 43°69′W. Wood products. Set off from WATERVILLE 1873.

Oakland, town, ⊙ GARRETT county, extreme W MARYLAND, in the ALLEGHENIES near WEST VIRGINIA state line, c.40 mi/64 km WSW of CUMBERLAND; 39°25′N 79°54′W. Trade center in resort and agricultural area (dairy products; vegetables, grain); wood products, maple sugar. Swallow Falls State Forest is just NW. William Armstrong, the first permanent settler, built a cabin in 1806 here at a ford where a packhorse trail crossed the YOUGHIOGHENY River. Resort hotels were built here in "Switzerland of America" after the BALTIMORE AND OHIO RAILROAD came through in 1849. The railroad station (c.1851) is a National Historic Landmark. Buffalo Bill Cody came to the funeral in Crook's Crest of General George Crook, who fought the Sioux and Apache after the Civil War, and retired here in 1890. Joseph E. Harwood, the former owner of Proudfoot's Drugstore, wrote a classic botany book, *Wildflowers of the Appalachians*. Deer Park Spring water, popular among late-19th-century presidents, is still available, although the famed Deer Park Hotel was razed in 1944.

Oakland, town (2000 population 1,540), SAINT LOUIS county, E MISSOURI, residential suburb 10 mi/16 km WSW of downtown ST. LOUIS; 38°34′N 90°22′W.

Oakland, town (2006 population 1,281), BURT county, E NEBRASKA, 13 mi/21 km W of TEKAMAH, and on LOGAN CREEK, near MISSOURI RIVER; 41°50′N 96°28′W. Trade and shipping point for rich agricultural area; livestock; grain. Printing. Settled 1863, incorporated 1881.

Oakland, town (2006 population 975), DOUGLAS county, SW OREGON, 15 mi/24 km N of ROSEBURG, on Calapooya Creek; 43°25′N 123°17′W. Wood products; dairy products; poultry, sheep, cattle.

Oakland, unincorporated town (2000 population 1,516), Union township, LAWRENCE county, W PENNSYLVANIA, residential suburb 1 mi/1.6 km SW of NEW CASTLE; 40°59′N 80°22′W.

Oakland, unincorporated town (2000 population 1,272), SUMTER county, central SOUTH CAROLINA, residential suburb 10 mi/16 km WNW of downtown SUMTER, at NW edge of SHAW AIR FORCE BASE; 33°59′N 80°30′W.

Oakland, town (2006 population 3,306), FAYETTE county, SW TENNESSEE, 9 mi/14 km W of SOMERVILLE; 35°14′N 89°31′W.

Oakland, village (2000 population 260), WARREN county, S KENTUCKY, 12 mi/19 km ENE of BOWLING GREEN; 37°02′N 86°15′W. In agricultural area (tobacco, grain; livestock; dairying); manufacturing (glass products).

Oakland, village (2000 population 586), YALOBUSHA county, N central MISSISSIPPI, 20 mi/32 km NNW of GRENADA; 34°02′N 89°54′W. Agriculture (cotton, corn; cattle, poultry); timber; manufacturing (lumber). Section of Holly Springs National Forest to SE; ENID LAKE reservoir and George Payne Cossar State Park to NE.

Oakland, village (2006 population 750), MARSHALL county, S OKLAHOMA, 2 mi/3.2 km W of MADILL, and 20 mi/32 km ESE of ARDMORE; 34°06′N 96°47′W. In farm area.

Oakland, village, in BURRILLVILLE town, PROVIDENCE county, NW RHODE ISLAND, 10 mi/16 km WSW of WOONSOCKET; 41°57′N 71°38′W.

Oakland, residential borough (2006 population 13,558), BERGEN county, NE NEW JERSEY, near Ramapo River, 8 mi/12.9 km NW of PATERSON; 41°01′N 74°14′W. Ramapo State Forest to W.

Oakland, borough (2006 population 593), SUSQUEHANNA county, NE PENNSYLVANIA, 37 mi/60 km N of SCRANTON, on SUSQUEHANNA RIVER, opposite (N of) SUSQUEHANNA DEPOT; 41°57′N 75°36′W. Agriculture (grain; livestock; dairying).

Oakland Acres, village (2000 population 166), JASPER county, central IOWA, 5 mi/8 km WSW of GRINNELL; 41°43′N 92°49′W. Rock Creek State Park to N. Corn; cattle, hogs, sheep.

Oakland Beach, neighborhood, WARWICK, KENT county, central RHODE ISLAND.

Oakland Beach, NEW YORK: see SOUTH BEACH.

Oakland City, city (2000 population 2,588), GIBSON county, SW INDIANA, 11 mi/18 km E of PRINCETON; 38°20′N 87°21′W. Grain farms. Bituminous coal, surface mines. Seat of Oakland City College. Laid out 1856.

Oakland Park (OK-luhnd), city (□ 7 sq mi/18.2 sq km; 2005 population 31,713), BROWARD county, SE FLORIDA, on ATLANTIC coast, 3 mi/4.8 km N of FORT LAUDERDALE; 26°10′N 80°09′W.

Oak Lawn, village (2000 population 55,245), COOK county, NE ILLINOIS, a suburb SE of CHICAGO; 41°43′N 87°45′W. It is chiefly residential with some light manufacturing industries. Products include metalwork, wood products, and school supplies. Incorporated 1909.

Oakleigh (OK-lee), residential and light industrial suburb, S VICTORIA, SE AUSTRALIA, 9 mi/14 km SE of MELBOURNE; 37°54′S 145°06′E. In metropolitan area. Monash University nearby.

Oakley, unincorporated city (2000 population 25,619), CONTRA COSTA county, W CALIFORNIA, suburb 34 mi/55 km NE of downtown OAKLAND, and 6 mi/9.7 km E of ANTIOCH, near SAN JOAQUIN RIVER; 38°00′N 121°43′W. Fruit, vegetables (especially asparagus); grain, nuts.

Oakley (OK-lee), town (2001 population 4,123), FIFE, E Scotland, 4 mi/6.4 km W of DUNFERMLINE; 56°04′N 03°33′W.

Oakley, town (2000 population 2,173), ⊙ LOGAN county, NW KANSAS, 21 mi/34 km SSE of COLBY; 39°07′N 100°50′W. Railroad junction. Trading point in grain, livestock region; dairying. Located in extreme NE corner of county, unusual for a county seat. Fick Fossil Museum. Incorporated 1887.

Oakley (OK-lee), village (2001 population 5,322), HAMPSHIRE, S ENGLAND, 5 mi/8 km W of BASINGSTOKE; 51°15′N 01°11′W. Farming. Also known as Church Oakley.

Oakley, village (2000 population 668), CASSIA county, S IDAHO, on GOOSE CREEK, and 20 mi/32 km SSW of BURLEY; 42°14′N 113°53′W. Elevation 4,191 ft/1,277 m. Railroad terminus, in agricultural area (potatoes, sugar beets, wheat; cattle, sheep); flour milling. Oakley Dam (145 ft/44 m high, 1,025 ft/312 m long) is on Goose Creek 4 mi/6.4 km SSW; forms Lower Goose Creek Reservoir (6 mi/9.7 km long, 0.5 mi/0.8 km wide), used for irrigation. Parts of Sawtooth National Forest to E and W. CITY OF ROCKS NATIONAL HISTORIC LANDMARK to SE; granite columns (c.600 ft/183 m high) were important landmark for pioneers.

Oakley, village (2000 population 339), SAGINAW county, E central MICHIGAN, 22 mi/35 km SW of SAGINAW, near SHIAWASSEE RIVER; 43°08′N 84°10′W. Lumber.

Oakley (OK-lee), unincorporated village, PITT county, E NORTH CAROLINA, 11 mi/18 km NNE of GREENVILLE; 35°45′N 77°17′W.

Oakley, village (2006 population 1,299), SUMMIT county, N UTAH, 15 mi/24 km SSE of COALVILLE, and on WEBER RIVER; 40°43′N 111°17′W. Elevation 6,517 ft/1,986 m. Rockport Reservoir and State Park to NW. UINTA MOUNTAINS, in part of Wasatch National Forest, to E. Originally named Oak Creek.

Oaklyn, residential borough (2006 population 4,080), CAMDEN county, SW NEW JERSEY, 4 mi/6.4 km SE of CAMDEN; 39°53′N 75°04′W. Settled 1682 by Friends, laid out c.1890, incorporated 1905.

Oakman, town (2000 population 944), Walker co., NW central ALABAMA, 10 mi/16 km SW of Jasper. Coal mining. Originally known a 'Day's Gap' for W. B. Day, the first settler, the name was changed to 'York' before being changed to its current name, in honor of W. G. Oakman, director of the Sloss-Sheffield Iron and Steel Co.

Oakman, town, GORDON county, NW GEORGIA, 15 mi/24 km ENE of CALHOUN; 34°33′N 84°42′W.

Oakmont, borough (2006 population 6,505), ALLEGHENY county, SW PENNSYLVANIA, suburb 10 mi/16 km NE of downtown PITTSBURGH, on ALLEGHENY RIVER (bridged); 40°31′N 79°50′W. Manufacturing (wire, seals, plastic products, fabricated metal products, furniture, machinery, building materials, electronic goods). Incorporated 1889.

Oakmont, Pennsylvania: see HAVERFORD.

Oakmulgee Creek (ok-MUL-gee), c.40 mi/64 km long, central Alabama; rises in SE Bibb co.; flows S to Cahaba River 9 mi/14.5 km NW of Selma. Its tributary Little Oakmulgee Creek rises in Chilton County and flows between Perry County and Dallas County into the larger stream. The name is derived from Hitchiti, and means 'bubbling water.'

Oakner (OK-nuhr), community, SW MANITOBA, W central Canada, 33 mi/53 km from BRANDON, in Hamiota rural municipality; 50°05′N 100°35′W.

Oak Orchard Creek, c.60 mi/97 km long, W NEW YORK; rises in N central GENESEE county; flows N and W, then N and NE, crossing the BARGE CANAL at MEDINA, to Lake ONTARIO 9 mi/14.5 km N of ALBION.

Oak Park, city (2000 population 52,524), COOK county, NE ILLINOIS, a residential suburb adjacent to CHICAGO; 41°53′N 87°47′W. Some 25 houses in the village were designed by Frank Lloyd Wright, who lived here. Ernest Hemingway was born here. Emmaus Bible School is here. Settled 1833, incorporated 1901.

Oak Park, city (2000 population 29,793), OAKLAND county, SE MICHIGAN, a suburb 10 mi/16 km NW of downtown DETROIT, and just N of Detroit city; 42°27′N 83°10′W. It is chiefly residential, but there is some industry. Manufacturing (laser cutting, sheet metal forming; fabricated metal products, foods, textiles). Marian Sandweiss born here. Detroit Zoological Park to NE. Incorporated 1927.

Oak Park, unincorporated town (2000 population 2,320), VENTURA county, SW CALIFORNIA, residential suburb 6 mi/9.7 km ESE of THOUSAND OAKS, on Media Creek; 34°10′N 118°46′W. Simi Peak to N.

Oak Park, town (2000 population 366), EMANUEL county, E central GEORGIA, 16 mi/26 km S of SWAINSBORO; 32°22′N 82°19′W.

Oak Park (OK PAHRK), suburb 7 mi/11 km N of MELBOURNE, VICTORIA, SE AUSTRALIA; 37°43′S 144°55′E. Railway station.

Oak Park Heights, town, WASHINGTON county, E MINNESOTA, suburb 16 mi/26 km ENE of downtown ST. PAUL, and 2 mi/3.2 km S of STILLWATER, on Lake St. Croix, natural lake on ST. CROIX River (WISCONSIN state line); 45°01′N 92°48′W. Light manufacturing; agriculture (cattle, sheep; corn, oats, alfalfa, soybeans). Minnesota Correctional facilities are here and in adjacent BAYPORT (to S). LOWER ST. CROIX NATIONAL SCENIC RIVERWAY on St. Croix River.

Oak Point (OK), unincorporated village, S MANITOBA, W central Canada, on Lake MANITOBA, 55 mi/89 km NW of WINNIPEG, in St. Laurent rural municipality; 50°30′N 98°02′W. Fishing.

Oak Point, hamlet, ST. LAWRENCE county, N NEW YORK, on the SAINT LAWRENCE River, and 18 mi/29 km SW of OGDENSBURG; 44°31′N 75°45′W.

Oak Ridge, city (2006 population 27,638), ANDERSON and ROANE counties, E TENNESSEE, on Black Oak Ridge and the CLINCH River, 17 mi/27 km WNW of KNOXVILLE; 36°02′N 84°12′W. Many activities in the fields of atomic energy and nuclear physics are pursued; manufacturing includes complex nuclear instruments, electronic products, radioactive pharmaceuticals, and nuclear fuel. More recently, the mission has broadened to include research into variety of energy technologies and strategies. The site was chosen (1942) for what was then called the Clinton Engineer Works, and the city was built by the Federal government to house the workers who developed the uranium-235 and plutonium-239 for the atomic bomb. The existence and purpose of the community were kept secret from most of the country until the summer of 1945. The project was under the control of the Atomic Energy Commission, but the city has since (1955–1959) been turned over to its residents. The former Clinton National Laboratory for nuclear research became (1948) the Oak Ridge National Laboratory. The Oak Ridge Institute of Nuclear Studies (1948), composed of many sponsoring educational institutions, and the University of Tennessee Biomedical Science graduate school are also here. Tourist attractions include the American Museum of Atomic Energy; a nearby nuclear graphite reactor; the K-25 overlook, from which can be seen the Oak Ridge gaseous diffusion plant; and an arboretum. Founded by the U.S. government 1942, incorporated as an independent city 1959.

Oak Ridge, town, ST. CLAIR county, central ALABAMA, 3 mi/4.8 km N of PELL CITY.

Oak Ridge, town (2000 population 202), CAPE GIRARDEAU county, SE MISSOURI, between MISSISSIPPI and WHITEWATER rivers, 9 mi/14.5 km NNW of JACKSON; 37°30'N 89°43'W.

Oak Ridge (OK RIJ), unincorporated town (□ 14 sq mi/36.4 sq km; 2006 population 4,295), GUILFORD county, N central NORTH CAROLINA, 12 mi/19 km NW of downtown GREENSBORO, near HAW RIVER; 36°10'N 79°59'W. Manufacturing (wooden roof and floor trusses); agricultural area (tobacco, grain; poultry, livestock; dairying).

Oakridge, town (2006 population 3,132), LANE county, W OREGON, 38 mi/61 km SE of EUGENE, on Middle Fork of WILLAMETTE RIVER; 43°45'N 122°28'W. Concrete, traffic counters. Timber. Fish hatchery to E. WILLAMETTE PASS to SE. Surrounded by WILLAMETTE National Forest. Umpqua National Forest to SW; Waldo Lake Wilderness Area to E; HILLS CREEK Reservoir to SE. Incorporated 1934.

Oak Ridge, village (2000 population 142), MOREHOUSE parish, NE LOUISIANA, 14 mi/23 km SSE of BASTROP; 32°37'N 91°46'W. An agricultural area (cotton, rice, soybeans, vegetables; cattle). Coulee State Game Refuge to W.

Oak Ridge, HONDURAS: see JOSÉ SANTOS GUARDIOLA.

Oak Ridge North (OK RIJ) town (2006 population 3,381), MONTGOMERY county, SE TEXAS, suburb 30 mi/48 km N of downtown HOUSTON metropolitan area; 30°09'N 95°26'W. Agriculture (cattle, ratites [ostriches, emus]; nursery and greenhouse crops).

Oak Ridge Reservoir (1 sq mi/2.6 sq km), MORRIS county, N NEW JERSEY, on PEQUANNOCK River, 14 mi/22 km NW of PARSIPPANY; 41°02'N 74°31'W. Maximum capacity 15,000 acre-ft. Formed by Oak Ridge Reservoir Dam (60 ft/18 m high), built (1892) for water supply.

Oak River (OK), unincorporated village, SW MANITOBA, W central Canada, on Oak River, 30 mi/48 km NW of BRANDON, and in Blanshard rural municipality; 50°08'N 100°25'W. Grain; livestock.

Oaks, unincorporated town, MONTGOMERY county, SE PENNSYLVANIA, 20 mi/32 km WNW of PHILADELPHIA, on SCHUYLKILL RIVER; 40°07'N 75°27'W. Diverse manufacturing. Agriculture includes dairying; grain, apples.

Oaks, village, DELAWARE county, NE OKLAHOMA, 18 mi/29 km NNE of TAHLEQUAH; 36°10'N 94°50'W. Agricultural area. Rocky Ford State Park to S.

Oaks Goldfield, The: see KIDSTON.

Oakton (OK-tuhn), unincorporated city, FAIRFAX county, N VIRGINIA, 13 mi/21 km W of WASHINGTON, D.C., 2 mi/3 km N of FAIRFAX; 38°53'N 77°17'W. DULLES INTERNATIONAL AIRPORT to W.

Oaktown, town (2006 population 607), KNOX county, SW INDIANA, 14 mi/23 km NNE of VINCENNES; 38°52'N 87°26'W. Fruit (especially watermelons), wheat, corn. Oil wells. Laid out 1867.

Oak Vale, village, LAWRENCE county, S central MISSISSIPPI, 15 mi/24 km NNW of COLUMBIA, near PEARL RIVER. Also spelled Oakvale.

Oakvale, village (2006 population 138), MERCER county, S WEST VIRGINIA, 8 mi/12.9 km ESE of PRINCETON, near VIRGINIA state line; 37°19'N 80°58'W. Coal-mining area. State fish hatchery.

Oak View, unincorporated town (2000 population 4,199), VENTURA county, S CALIFORNIA, 8 mi/12.9 km N of VENTURA, on Ventura River; 34°24'N 119°18'W. Avocados, citrus, vegetables; flowers, nursery products. Lake Casitas reservoir to W; Los Padres National Forest to N; Emma Wood State Beach to S; SAN BUENAVENTURA MISSION to N.

Oakville (OK-vil), town (□ 54 sq mi/140.4 sq km), HALTON region, S ONTARIO, E central Canada, on Lake ONTARIO, between TORONTO and HAMILTON; 43°26'N 79°40'W. A major component of the local economy is the Ford Motor Company plant, one of the largest motor vehicle plants in Canada. Originally inhabited by the Mississauga, Oakville became a shipbuilding center in the 19th century due to its good harbor. Founded 1827.

Oakville, town (2000 population 439), LOUISA county, SE IOWA, on IOWA RIVER, and 10 mi/16 km SE of WAPELLO; 41°06'N 91°02'W.

Oakville (OK-vil), unincorporated village, S MANITOBA, W central Canada, 14 mi/23 km ESE of city of PORTAGE LA PRAIRIE, and in Portage la Prairie rural municipality; 49°55'N 98°00'W. Grain; livestock; dairying.

Oakville, unincorporated village, NAPA county, W CALIFORNIA, in NAPA RIVER valley, 15 mi/24 km E of SANTA ROSA. Winery; grapes, walnuts; dairying; cattle; nursery products. U.S. Department of Agriculture experimental vineyard here.

Oakville, village, section of LITCHFIELD county, W CONNECTICUT.

Oakville, village (2006 population 723), GRAYS HARBOR county, W WASHINGTON, 20 mi/32 km SW of OLYMPIA, and on CHEHALIS RIVER Manufacturing (wood products, lumber). Chehalis Indian Reservation to SE; Black Hills to N.

Oakwood (OK-wuhd), city (□ 3 sq mi/7.8 sq km; 2006 population 8,611), MONTGOMERY county, SW OHIO, 3 mi/4.8 km S of DAYTON; 39°43'N 84°10'W. Incorporated 1908.

Oakwood, town (2000 population 2,689), HALL county, NE GEORGIA, 5 mi/8 km SW of GAINESVILLE; 34°14'N 83°53'W. Manufacturing of animal feeds, paints, plastics, transportation equipment; commercial printing.

Oakwood, unincorporated town (2000 population 2,249), Union township, LAWRENCE county, W PENNSYLVANIA, residential suburb 1 mi/1.6 km W of NEW CASTLE, on SHENANGO RIVER; 41°00'N 80°22'W. New Castle Municipal Airport to NW.

Oakwood, unincorporated village, Madison co., Alabama.

Oakwood, village (2000 population 1,502), VERMILION county, E ILLINOIS, 7 mi/11.3 km W of DANVILLE; 40°06'N 87°46'W. In agricultural and bituminous-coal area. Kickapoo State Park is nearby.

Oakwood (OK-wuhd), village (2006 population 566), PAULDING county, NW OHIO, on AUGLAIZE RIVER, 11 mi/18 km ESE of PAULDING; 41°05'N 84°22'W. In agricultural area.

Oakwood, village (2006 population 68), DEWEY county, W OKLAHOMA, 17 mi/27 km WNW of WATONGA; 35°55'N 98°42'W. In agricultural area (grain; cattle).

Oakwood (OK-wood), village (2006 population 502), LEON county, E central TEXAS, 17 mi/27 km SW of PALESTINE, near TRINITY RIVER; 31°34'N 95°50'W. Lumbering center and agricultural area.

Oakwood (OK-wud), unincorporated village, BUCHANAN county, SW VIRGINIA, 40 mi/64 km ENE of NORTON; 37°12'N 82°00'W. Manufacturing (concrete products, machinery); agriculture (tobacco, potatoes; cattle); bituminous coal.

Oakwood Hills, village (2000 population 2,194), MCHENRY county, NE ILLINOIS, residential suburb 4 mi/6.4 km ENE of CRYSTAL LAKE; 40°06'N 87°46'W.

Oamaru (om-ah-ROO), town (2006 population 12,681), WAITAKI district (□ 2,785 sq mi/7,241 sq km; 2006 population 12,681), OTAGO region, SOUTH ISLAND, NEW ZEALAND, 71 mi/115 km NE of DUNEDIN; 45°05'N 170°59'E. Service center for N Otago area, producing grain, wool, lambs, and Oamaru limestone. Center also for WAITAKI RIVER dams for hydroelectricity and some irrigation. Has small harbor with concrete breakwater. Oamaru is within a coastal section of Waitaki district; the interior with dams is associated with the CANTERBURY region.

Oamishirasato (O-ah-mee-shee-rah-SAH-to), town, Sanbu county, CHIBA prefecture, E central HONSHU, E central JAPAN, 9 mi/15 km E of CHIBA; 35°31'N 140°19'E.

Oarai (O-ah-rah-ee), town, E Ibaraki county, IBARAKI prefecture, central HONSHU, E central JAPAN, 6 mi/10 km S of MITO; 36°18'N 140°34'E.

Oas (O-wahs), town, ALBAY province, SE LUZON, PHILIPPINES, on railroad, and 8 mi/12.9 km WNW of LEGASPI; 13°09'N 123°21'E. Abaca-growing center.

Oasa (O-ah-sah), town, Yamagata county, HIROSHIMA prefecture, SW HONSHU, W JAPAN, 25 mi/40 km N of HIROSHIMA; 34°46'N 132°27'E.

Oates Bank, ANTARCTICA, bank off PENNELL COAST; 70°15'S 165°00'E.

Oates Coast, between PENNELL and GEORGE V coasts, PACIFIC coast of N VICTORIA LAND, ANTARCTICA; 154°00'E–164°00'E. Discovered by Pennell on February 22, 1911, and named for Laurence Oates. Sometimes called Oates Land.

Oatlands (OT-luhndz), town, E central TASMANIA, SE AUSTRALIA, 40 mi/64 km N of HOBART; 42°18'S 147°22'E. Merino wool; cattle; turnips, oats, wheat, forage crops; sport fishing. Lake Dulverton wildlife sanctuary nearby.

Oatman, unincorporated village, MOHAVE county, W ARIZONA, 20 mi/32 km SW of KINGMAN. Gold-mining town in W foothills of BLACK MOUNTAINS. Mount Nutt (NE) and Warm Springs (SE) wilderness areas nearby. Fort Mojave Indian Reservation to W.

Oaussou, town, GUINEA: see also WASSOU.

Oaxaca (wah-HAH-kah), state (□ 36,375 sq mi/94,575 sq km), S MEXICO, on the PACIFIC OCEAN and its arm, the Gulf of TEHUANTEPEC; ⊙ OAXACA DE JUÁREZ 16°45'N 94°10'W. The N part of the state is dominated by the Sierra Madre de OAXACA; there are deep, narrow valleys in the S and broad, open, semiarid valleys and plateaus in the N. Except on the W and the N the periphery of the state is tropical. The climate of the interior is generally temperate. Fertile valleys are farmed; agriculture is the principal economic activity. Sugarcane, coffee (of which Oaxaca is a leading national producer), tobacco, cereals, and tropical and semitropical fruits are grown; livestock is raised. Oaxaca's reportedly extensive mineral deposits remain largely unexploited. The state's limited industrial activity centers around oil refining, beverage and paper manufacturing, and sugar and flour milling. Oaxaca is also known for its handicrafts, especially handwoven textiles, pottery, and leather goods. Despite the existence of several highways, inadequate communications remain the chief barrier to the state's development. There are famous archaeological sites at MITLA and MONTE ALBÁN. Native Americans predominate here, as in few other states (notably YUCATÁN and CHIAPAS), with Mixtecs dominating in the W highlands and Zapotecs elsewhere. Beach resorts are under development at HUATULCO BAYS and elsewhere on the S coast, which should augment the already important contribution of tourism to the state's economy. Porfirio Díaz and Benito Juárez were born here.

Oaxaca de Juárez (wah-HAH-kah dai HWAH-res), city (2005 population 258,008) and township,

⊙ OAXACA state, S MEXICO; 17°03′N 96°43′W. Commercial and tourist center situated in the Valley of OAXACA. The church and monastery of Santo Domingo is a national monument and UN World Heritage Site. Noted for hand-wrought gold and silver filigree, pottery, and woven goods that rank among the finest in MEXICO. The chief city of S MEXICO, Oaxaca is linked with the federal capital by railroad and the Inter-American Highway (190). Subject to severe earthquakes. According to Aztec tradition, Oaxaca was founded as Huasyacac in 1486, during the brief ascendancy of the Aztecs over the Mixtecs and Zapotecs. Prominent in the Mexican Revolution against SPAIN, the city also joined in the War of the Reform and in resistance to the French intervention. Known as Antequera in colonial period.

Oaxaca, Sierra Madre de (wah-HAH-kah, see-ER-rah MAH-drai dai), mountain range, OAXACA, PUEBLA, and VERACRUZ states, central MEXICO; extending roughly 164 mi/200 km NW-SE and paralleling the Gulf of MEXICO coast from CÓRDOBA to the Isthmus of TEHUANTEPEC; 18°30′N 97°00′W. A part of the SIERRA MADRE DEL SUR physiographic province. Divided into several poorly demarcated ranges, including Sierra ZONGOLICA (Puebla, Veracruz), Sierra MAZATECA (Puebla, Oaxaca), and Sierra de JUÁREZ (Oaxaca).

Oaxaca, Valley of (wah-HAH-kah), region, OAXACA state, S MEXICO, largest valley of SIERRA MADRE DEL SUR, surrounding the city of OAXACA DE JUÁREZ. Broken into three arms: ETLA, which extends N of Oaxaca de Juárez; TLACOLULA, the E extension; and the Valle Grande, the S branch of the valley. Site of extensive pre-Columbian settlement at MONTE ALBÁN, MITLA, and other sites; still, the heart of one of the country's most Indian regions, where those of non-European descent predominate.

Ob' (OP), city (2006 population 24,165), E NOVOSIBIRSK oblast, SW SIBERIA, RUSSIA, on road and the TRANS-SIBERIAN RAILROAD, 10 mi/16 km W of NOVOSIBIRSK, to which it is administratively subordinate; 54°59′N 82°42′E. Elevation 305 ft/92 m. Manufacturing of sanitary equipment; headquarters of regional airline. Tolmachevo international airport is 3 mi/5 km to the WNW. Railroad station since 1896; town since 1947; made city in 1969.

Oba (AW-bah), town, ONDO state, SW NIGERIA, on road, 50 mi/80 km ENE of AKURE. Market town. Kola nuts, millet, maize.

Oba (O-buh), unincorporated village, central ONTARIO, E central Canada, on Oba Lake (12 mi/19 km long, 2 mi/3 km wide), 100 mi/161 km SW of KAPUSKASING, and included in the unorganized N part of Algoma; 49°04′N 84°06′W. Gold mining.

Obabika Lake (o-buh-BEE-kuh), SE central ONTARIO, E central Canada, 50 mi/80 km NE of SUDBURY; 15 mi/24 km long, 2 mi/3 km wide; 47°02′N 80°15′W. In gold-mining region. Drains E into Lake TEMAGAMI.

Obagan River (o-bah-GAHN), c.200 mi/322 km long, E KOSTANAI region, KAZAKHSTAN; rises in W KAZAK HILLS; flows N, through Lake KUSHMURUN (formerly Lake Obagan), to TOBOL RIVER near ZVERINOGOLO-VSKOYE (Russia). Also spelled Ubagan River.

Obah, AFGHANISTAN: see OBEH.

Obal, YEMEN: see UBAL.

Obala (O-bah-lah), town (2001 population 19,300), Central province, CAMEROON, 20 mi/32 km N of Yaoundé; 04°09′N 11°33′E.

Obalski Lake (o-BAHL-skee), W QUEBEC, E Canada, on HARRICANAW RIVER, 11 mi/18 km NE of AMOS; 7 mi/11 km long, 2 mi/3 km wide; 48°44′N 77°57′W. On shore are gold and copper deposits; airfield.

Obama (o-BAH-mah), city, FUKUI prefecture, central HONSHU, W central JAPAN, port on WAKASA BAY, 47 mi/75 km S of FUKUI; 35°29′N 135°44′E. Flatfish. Lacquerware manufacturing, fish pickling, traditional papermaking; agate handicrafts. Sotomo Cave nearby.

Obama (o-BAH-mah), town, South Takaki county, NAGASAKI prefecture, NW KYUSHU, SW JAPAN, on W coast of SHIMABARA PENINSULA, 19 mi/30 km E of NAGASAKI across TACHIBANA BAY; 32°43′N 130°12′E. Potatoes; sardines and sardine processing; rice crackers. UNZEN-AMAKUSA NATIONAL PARK (Unzen Fugendake mountain) is nearby. Hot springs.

Oban (O-ban), township and harbor, NE STEWART ISLAND, NEW ZEALAND, 22 mi/35 km SSW of BLUFF, SOUTH ISLAND, across FOVEAUX STRAIT, N of PATERSON INLET (8 mi/12.9 km E-W, 5 mi/8 km N-S). Now usually called HALF MOON BAY.

Oban (AW-bahn), town, CROSS RIVER state, extreme SE NIGERIA, near CAMEROON border, 30 mi/48 km NE of CALABAR; 05°19′N 08°34′E. Agricultural trade center; hardwood, rubber, palm oil and kernels, cacao, kola nuts. OBAN Hills are N.

Oban (O-buhn), town (2001 population 8,120), Argyll and Bute, W Scotland, on the Firth of Lorn, and 26 mi/41.6 km N of LOCHGILPHEAD; 56°25′N 05°28′W. A port and seaside resort, its circular bay makes a fine yacht basin. Ferry services to the HEBRIDES. Annual Highland Gathering Games are held here each August. Distillery, glass and tweed factories. Near the ruins of Dunollie Castle, a 12th-century fortress, is the Dog Stone, to which Fingal chained his dog Bran. Formerly in Strathclyde, abolished 1996.

Obanazawa (o-BAH-nah-zah-wah), city, YAMAGATA prefecture, N HONSHU, NE JAPAN, 25 mi/40 km N of YAMAGATA city; 38°35′N 140°24′E. Hops; watermelon. Traditional dolls; origin of traditional *hanagasa* dance. Hot springs nearby.

Obando (o-BAHN-do), town, ⊙ Obando municipio, VALLE DEL CAUCA department, W COLOMBIA, on railroad, and near the CAUCA River, 170 mi/274 km NNE of CALI; 04°34′N 75°58′W. Corn, sugarcane, sorghum, soybeans; livestock.

Obando (o-BAHN-do), town (2000 population 52,906), BULACAN province, S central LUZON, PHILIPPINES, 7 mi/11.3 km NNW of MANILA, near MANILA BAY; 14°43′N 120°56′E. Rice growing.

Oban Hills (AW-bahn), in SE NIGERIA, N of OBAN; 30 mi/48 km long (N-S); rise to around 3,500 ft/1,067 m. CALABAR River rises on W slopes.

Obara (o-BAH-rah), town, KANAGAWA prefecture, E central HONSHU, E central JAPAN, 5 mi/8 km WSW of HACHIOJI. Mineral springs.

Obara (O-bah-rah), village, West Kamo county, AICHI prefecture, S central HONSHU, central JAPAN, 25 mi/40 km S of NAGOYA; 35°13′N 137°17′E. Paper.

Obata (O-bah-tah), town, Watarai county, MIE prefecture, S HONSHU, central JAPAN, 19 mi/30 km S of TSU; 34°30′N 136°40′E. Also Kobata.

Obatake (o-BAH-tah-ke), town, Kuga county, SE YAMAGUCHI prefecture, SW HONSHU, W JAPAN, on IYO SEA across from YASHIRO JIMA island (linked by Oshima bridge), and 43 mi/70 km E of YAMAGUCHI; 33°57′N 132°10′E.

Ob' Bay (OB, pronounced with a soft 'B'), Russian *Obskaya Guba*, shallow estuary of the OB' RIVER, TYUMEN oblast, NW Siberian Russia; forms an inlet of the KARA SEA, between Yamal and Gydan peninsulas; 500 mi/805 km long, 35 mi/56 km–50 mi/80 km wide; 69°00′N 73°00′E. Fresh water, low shores. W bend called the Nadym Bay. Receives Ob', Nadym, and Nyda rivers (S). Has an E arm called the Taz Bay. Abounds in fish. Chief port, NOVY PORT. Also called the Ob' Gulf.

Obbe, Scotland: see LEVERBURGH.

Obbia (oo-BI-yah), township, E central SOMALIA, port on INDIAN OCEAN, 360 mi/579 km NE of MOGADISHO; 05°21′N 48°32′E. Road junction; fishing, weaving (mats, baskets). Has old native fort. Also spelled A'bbia, HOBIO.

Obbola (OB-boo-lah), town (☐ 8 sq mi/20.8 sq km), VÄSTERBOTTEN county, N SWEDEN, on Obbola island (☐ 8.5 sq mi/22 sq km), in GULF OF BOTHNIA, at

mouth of UMEÄLVEN RIVER, 9 mi/14.5 km SSE of UMEÅ; 63°42′N 20°19′E.

Obbrovazzo, CROATIA: see OBROVAC.

Obdach (OB-dahkh), township, STYRIA, central AUSTRIA, at N foot of Obdacher Sattel (Obdach Pass; 2,908 ft/886 m) and near Carinthian border; 7 mi/11.3 km S of JUDENBURG; 47°04′N 14°42′E. Elevation 2,652 ft/808 m. Manufacturing of machines, clothes. Ensemble of fine burgher houses.

Obdorsk, RUSSIA: see SALEKHARD.

Obedinenie (o-be-dee-NEN-ee), agricultural village, LOVECH oblast, POLSKI TRUMBESH obshtina, N BULGARIA, 20 mi/32 km NNW of VELIKO TURNOVO; 43°20′N 25°30′E. Formerly called Knyaz Simeonovo (1941–1978).

Obedjiwan, unorganized territory, S QUEBEC, E Canada; 48°40′N 74°56′W. Attikameks First Nations community reservation. Sawmill. Forms part of La TUQUE agglomeration.

Obedovo, RUSSIA: see OKTYABR′SKIY, IVANOVO oblast.

Obed Wild and Scenic River and Riverway (☐ 7 sq mi/18 sq km), E central TENNESSEE, on the CUMBERLAND PLATEAU. Includes portions of Daddy's and Clear creeks and Emory and Obed rivers. Rugged scenery. Authorized 1976.

Obeh (O-buh), town, HERAT province, NW AFGHANISTAN, 55 mi/89 km E of HERAT, on the HARI RUD, at E end of its irrigated valley; 34°22′N 63°10′E. Livestock-raising center; manufacturing (felt and coarse woolen goods). Old walled town. Also spelled Obah.

Obeid, Tell el (o-BAID, TEL ahl), excavation site, 8 mi/13 km from W bank of the lower EUPHRATES River, in THI-GAR province, SE IRAQ, c.15 mi/24 km WSW of NASIRIYA, c.4 mi/6 km WNW of site of ancient UR. Finds of Sumerian kings have been made.

Obejo (o-VAI-ho), town, CÓRDOBA province, S SPAIN, 17 mi/27 km N of CÓRDOBA; 38°08′N 04°48′W. Olive-oil processing; cereals, peas; livestock.

Obelya (o-BEL-yah), neighborhood, SOFIA city, SOFIA oblast, W BULGARIA; 43°45′N 23°18′E. Former village; now residential quarter.

Oberá (o-be-RAH), town, S MISIONES province, ARGENTINA, 50 mi/80 km E of POSADAS; 27°29′S 55°08′W. Lumbering and farming center (maté, tobacco, hardwood); corn and rice milling, maté processing; sawmills. Formerly Yerbal Viejo.

Oberaar Glacier, SWITZERLAND: see OBERAARHORN.

Oberaarhorn (o-buhr-AHR-horn), peak (11,936 ft/3,638 m) in BERNESE ALPS, BERN canton, S central SWITZERLAND, 9 mi/14.5 km WSW of GRIMSEL PASS, headwall of the OBERAAR GLACIER (which feeds the AARE RIVER).

Oberägeri, SWITZERLAND: see ÄGERISEE.

Oberalp (O-buhr-AHLP), Alpine pass (6,706 ft/2,044 m), between the Vorderhein in GRISONS canton and the REUSS valley in Uri canton, S central SWITZERLAND. Watershed between Reuss River and UPPER RHINE drainage. The Oberalpsee (10,919 ft/3,328 m high) is 8 mi/12.9 km NE of the pass, on border between GLARUS and Uri cantons.

Oberalpstock, SWITZERLAND: see OBERALP.

Oberalsbach (o-buhr-AHLS-bahkh), village, MIDDLE FRANCONIA, BAVARIA, GERMANY, 6 mi/9.7 km WSW of NUREMBERG; 49°25′N 10°52′E. Metalworking.

Oberammergau (o-buhr-AHM-mer-gou), village, UPPER BAVARIA, S GERMANY, in the Bavarian Alps; 48°35′N 11°03′E. It has been a noted center of wood-carving since the 12th century. Famous for the Passion Play performed here every ten years (last in 1990), originally (1634) in fulfillment of a vow made during a plague in 1633. Tourism is the town's major source of income. Numerous fresco-decorated houses and a rococo church are here.

Oberaudorf (o-buhr-OU-dorf), village, UPPER BAVARIA, BAVARIA, GERMANY, on E slope of the Bavarian Alps, near the INN RIVER, 5 mi/8 km N of KUFSTEIN, near Austrian border; 47°39′N 12°08′E. Textiles; summer and winter resort (elevation 1,585 ft/483 m).

Oberaula (o-buhr-OU-lah), village, HESSE, central GERMANY, on the Aula River, and in the Knüll, 10 mi/16 km W of BAD HERSFELD; 50°52′N 09°28′E. Grain.

Oberaurach (o-buhr-OU-rahkh), village, LOWER FRANCONIA, BAVARIA, S GERMANY, on the Aurach, 10 mi/16 km WNW of BAMBERG; 49°55′N 10°40′E. Viticulture, food processing; grain; fruit.

Oberbayern, GERMANY: see UPPER BAVARIA.

Oberbipp, SWITZERLAND: see NIEDERBIPP.

Oberboihingen (o-buhr-BOI-hing-uhn), village, STUTTGART region, BADEN-WÜRTTEMBERG, S GERMANY, 12 mi/19 km SE of STUTTGART, on the NECKAR RIVER; 48°38′N 09°21′E. Viticulture; grain; synthetics.

Ober-Briz, CZECH REPUBLIC: see HORNI BRIZA.

Oberburg (O-buhr-boorg), commune, BERN canton, NW central SWITZERLAND, 1 mi/1.6 km S of BURGDORF; 47°02′N 07°37′E. Linen textiles, tiles; metal- and woodworking.

Obercorn (O-buhr-KORN), German *Oberkorn*, town, DIFFERDANGE commune, SW LUXEMBOURG, 4 mi/6.4 km W of ESCH-SUR-ALZETTE; 49°31′N 05°54′E. Iron-mining center.

Oberdan (O-ber-DAHN), village, CYRENAICA region, NE LIBYA, 17 mi/27 km NE of Al MARJ, on plateau. Agricultural settlement (grain, olives, fruit). Founded 1938–1939 by Italians, who left after World War II, replaced by Libyan population.

Oberderdingen (o-buhr-DER-ding-uhn), village, MIDDLE UPPER RHINE, BADEN-WÜRTTEMBERG, S GERMANY, 19 mi/30 km ENE of KARLSRUHE; 49°04′N 08°49′E. Fruit.

Oberding (O-buhr-ding), village, UPPER BAVARIA, BAVARIA, S GERMANY, 18 mi/29 km NE of MUNICH, near Middle Isar Canal; 48°19′N 11°50′E.

Oberdrauburg (o-buhr-DROU-boorg), township, CARINTHIA, S AUSTRIA, on DRAU RIVER, and near TYROL border, 11 mi/18 km SE of LIENZ; 46°45′N, 12°58′E. Elevation 1,920 ft/585 m. Road junction; summer tourism. Ruins of a castle, fine burgher houses.

Oberegg (O-buhr-eg), commune, APPENZELL Inner Rhoden half-canton, NE SWITZERLAND, in exclave between Ausser Rhoden and ST. GALLEN; 47°25′N 09°33′E. Embroideries, textiles.

Oberengstringen (o-buhr-EENG-string-gen), commune, ZÜRICH canton, N SWITZERLAND, 5 mi/8 km NW of ZÜRICH; 47°24′N 08°27′E.

Oberentfelden (O-buhr-uhnt-fel-den), commune, AARGAU canton, N SWITZERLAND, on SUHRE RIVER, and 3 mi/4.8 km S of AARAU; 47°21′N 08°02′E. Shoes, rubber products; woodworking. Unterentfelden is just N.

Oberer Hauenstein Pass, SWITZERLAND: see HAUENSTEIN.

Oberfeulen, LUXEMBOURG: see FEULEN.

Oberfranken, GERMANY: see UPPER FRANCONIA.

Obergabelhorn (O-buhr-gah-buhl-horn), peak (13,330 ft/4,063 m) in PENNINE ALPS, VALAIS canton, S SWITZERLAND, 4 mi/6.4 km WNW of ZERMATT, NW of the MATTERHORN.

Oberglogau, POLAND: see GLOGOWEK.

Obergösgen, SWITZERLAND: see NIEDERGÖSGEN.

Ober-Grafendorf (O-buhr-GRAH-fen-dorf), township, central LOWER AUSTRIA, on Rielach River, 5 mi/8 km SW of Sankt Pölten; 48°09′N 15°33′E. Railroad junction; metal manufacturing.

Obergünzburg (o-buhr-GYOONTS-boorg), village, SWABIA, SW BAVARIA, GERMANY, 10 mi/16 km NE of KEMPTEN; 47°51′N 10°25′E. Textile manufacturing; dairying. Summer resort (elevation 2,418 ft/737 m).

Obergurgl (o-buhr-GUHR-guhl), village, TYROL, W AUSTRIA, in Ötztal Alps, near the Italian border, on the Ötztaler Ache, and 31 mi/50 km SW of INNSBRUCK; 46°52′N 11°02′E. Highest village (6,320 ft/1,926 m) in Austria; winter-sports center. Cable cars to Festkogel at an elevation of 9,260 ft/2,822 m. N end of road

across Timmelsjoch at an elevation of 7,541 ft/2,298 m. to MERANO (South Tyrol).

Oberhaching (o-buhr-HUH-khing), suburb of MUNICH, UPPER BAVARIA, BAVARIA, S GERMANY, 8 mi/12.9 km S of city center; 48°01′N 11°34′E.

Oberhaid (O-buhr-heit), village, UPPER FRANCONIA, NE BAVARIA, GERMANY, 5 mi/8 km NW of BAMBERG, near the MAIN RIVER; 49°56′N 10°46′E. Fruit.

Oberhaid, CZECH REPUBLIC: see HORNI DVORISTE.

Oberhalbstein (O-buhr-HAHLB-shtein), Romansh *Sursès*, valley of the Julia River, GRISONS canton, E SWITZERLAND. Road leads from ALBULA RIVER in N to Julier Pass.

Oberhasli (o-buhr-HAHS-lee), district, S BERN canton, SWITZERLAND, in the AARE valley. Main town is MEIRINGEN; population is German-speaking and Protestant.

Oberhausen (o-buhr-HOU-suhn), city, NORTH RHINE-WESTPHALIA, W GERMANY, 6 mi/9.7 km W of ESSEN, on the Emscher River; 51°29′N 06°52′E. An industrial center of the RUHR district, port on the RHINE-HERNE CANAL, and a railroad junction. Has railroad workshops and a large thermoelectric plant. Manufacturing includes iron and steel, coal, refined zinc, dyes, steam boilers, wire rope, glass, machinery, and chemicals. Chartered 1874; OSTERFELD and STERKRADE were incorporated in 1929.

Oberhausen-Rheinhausen (o-buhr-HOU-suhn–rein-HOU-suhn), village, MIDDLE UPPER RHINE, BADEN-WÜRTTEMBERG, GERMANY, on an arm of the RHINE RIVER, and 4.5 mi/7.2 km SE of SPEYER; 49°16′N 08°30′E. Asparagus. Sugar refinery. Has 15th century church.

Oberhof (o-buhr-hof), village, THURINGIA, central GERMANY, in THURINGIAN FOREST, 17 mi/27 km S of GOTHA; 50°42′N 10°43′E. Popular health and winter-sports resort.

Oberhofen am Thunersee (o-buhr-HOF-en um TOO-nuhr-zai), commune, BERN canton, central SWITZERLAND, on LAKE THUN, and 2 mi/3.2 km SSE of THUN. Elevation 1,847 ft/563 m. Resort. Old castle.

Oberhomburg, FRANCE: see HOMBOURG-HAUT.

Ober-Ingelheim, GERMANY: see INGELHEIM AM RHEIN.

Oberkassel (O-buhr-kah-sel), suburb of DÜSSELDORF, NORTH RHINE-WESTPHALIA, W GERMANY.

Oberkirch (O-buhr-kirkh), town, S UPPER RHINE, BADEN-WÜRTTEMBERG, GERMANY, at W foot of BLACK FOREST, 7 mi/11.3 km NE of OFFENBURG; 48°32′N 08°05′E. Noted for its wine. Fruit trade; largest strawberry market in GERMANY. Summer resort.

Oberklettgau (o-buhr-KLAT-gou), district, in SW SCHAFFHAUSEN canton, SWITZERLAND. Population is German-speaking and Protestant.

Oberkochen (o-buhr-KUH-khuhn), town, E WÜRTTEMBERG, BADEN-WÜRTTEMBERG, S GERMANY, 4 mi/6.4 km S of AALEN, in the SWABIAN JURA; 48°47′N 10°06′E. Manufacturing (machinery, tools, paper, optics); optic museum. Strawberry market. First mentioned in Middle Ages; chartered 1968.

Oberkorn, LUXEMBOURG: see OBERCORN.

Oberkotzau (o-buhr-KOT-sou) or **Oberkotzau an der Saale**, village, UPPER FRANCONIA, NE BAVARIA, GERMANY, on the SAALE RIVER, at mouth of small Schwesnitz River, 4 mi/6.4 km S of HOF. Textile manufacturing, glass- and metal-working, brewing.

Oberlaa (O-buhr-lah), part of VIENNA, AUSTRIA, 5 mi/8 km SE of city center. Spa with sulphurous thermal water.

Oberland, Bernese, SWITZERLAND: see BERNESE ALPS.

Oberländischer Kanal, POLAND: see WARMIA CANAL.

Oberlandquart (O-buhr-lahnd-kwahrt), district, in N GRISONS canton, SWITZERLAND. Main town is DAVOS; population is German-speaking and Protestant, though many Roman Catholics live in Davos.

Oberleutensdorf, CZECH REPUBLIC: see LITVÍNOV.

Oberlin (O-buhr-lin), city (□ 4 sq mi/10.4 sq km; 2006 population 8,239), LORAIN county, N OHIO, 7 mi/11 km

SW of ELYRIA; 41°17′N 82°13′W. In vegetable, dairy, and poultry area. Seat of Oberlin College.

Oberlin (O-buhr-lin), town (2000 population 1,994), ⊙ DECATUR county, NW KANSAS, on SAPPA CREEK, and 32 mi/51 km W of NORTON; 39°49′N 100°31′W. Railroad terminus. Trading point in corn, barley, wheat, rye, cattle area; dairying. Manufacturing (wood products, foods). Site of last Native American raid on Kansas soil 1878. Laid out 1878, incorporated 1885.

Oberlin, town (2000 population 1,853), ⊙ ALLEN parish, SW central LOUISIANA, 38 mi/61 km NE of LAKE CHARLES city, near CALCASIEU RIVER; 30°37′N 92°46′W. In agricultural area (rice, soybeans, peaches; cattle; dairying); timber. West Bay State Wildlife Area to N.

Oberlössnitz, GERMANY: see RADEBEUL.

Oberlungwitz (o-buhr-LUNG-vits), town, SAXONY, E central GERMANY, 9 mi/14.5 km WSW of CHEMNITZ; 50°47′N 12°44′E. Textile milling, hosiery knitting, metalworking. Founded 1273; town since 1936.

Obermarsberg, GERMANY: see MARSBERG.

Obermarschendorf, CZECH REPUBLIC: see HORNI MARSOV.

Ober-Mörlen (O-buhr–MUHR-len), village, HESSE, central GERMANY, 17 mi/27 km N of FRANKFURT AM MAIN, on E foot of the TAUNUS; 50°22′N 08°40′E. Grain, vegetables.

Obernai (o-ber-NAI), town (□ 10 sq mi/26 sq km), BAS-RHIN department, ALSACE region, E FRANCE, at E foot of the VOSGES MOUNTAINS, 15 mi/24 km SW of STRASBOURG; 48°28′N 07°29′E. Wine-producing and marketing center of Alsatian wines. Also has a brewery and hosiery manufacturing plants. A tourist center noted for its picturesque market square with 14th–17th-century town hall, historical museum, and many old houses and commercial buildings. The ramparts of Obernai are well-preserved. Sainte Odile, patron saint of Alsace, born here in 7th century. Seat of early dukes of Alsace. Mont SAINTE-ODILE, in the Vosges nearby, rises to 2,500 ft/762 m above the Alsatian plain; spectacular panorama enjoyed by visitors who also make the pilgrimage to the restored convent in a dense fir forest nearby.

Obernberg am Inn (O-buhrn-berg ahm IN), township, W UPPER AUSTRIA, W central AUSTRIA, in the INNVIERTEL region, on Inn River, and on German border, 10 mi/16 km NW of RIED IM INNKREIS; 48°19′N, 13°20′E. Elevation 1,052 ft/321 m. Border station; hydropower station. Bavarian-type market square, castle.

Obernburg am Main (O-buhrn-boorg ahm MEIN), town, LOWER FRANCONIA, NW BAVARIA, GERMANY, on the MAIN RIVER at the mouth of MÜMLING RIVER, 9 mi/14.5 km S of ASCHAFFENBURG; 49°51′N 09°08′E. Manufacturing of synthetic fiber, clothing. Built on site of Roman camp; chartered 1313. Portions of medieval wall still surround town; former Roman fort.

Oberndorf am Neckar (o-buhrn-dorf ahm NEK-kahr), town, BLACK FOREST, Barr-Henberg, BADEN-WÜRTTEMBERG, SW GERMANY, at E foot of BLACK FOREST, on the NECKAR RIVER, 9 mi/14.5 km NNW of ROTTWEIL; 48°18′N 08°35′E. Manufacturing (office equipment, machinery). Site (until 1945) of well-known arms factory. Chartered in mid-13th century.

Oberndorf bei Salzburg (O-buhrn-dorf bei SAHLTS-boorg), township, SALZBURG, W AUSTRIA, on Salzburg River in the FLACHGAU region, 11 mi/18 km NNW of SALZBURG; 47°56′N 12°56′E. Former suburb of LAUFEN (Germany). In 1818, the first performance of the Christmas song *Stille Nacht* (*Silent Night, Holy Night*) took place in the former parish church here.

Oberneukirchen (O-buhr-NOI-kir-khuhn), township, N UPPER AUSTRIA, 11 mi/18 km NNW of LINZ; 48°28′N 14°14′E. Dairy farming.

Obernick, POLAND: see OBORNIKI.

Area is shown by the symbol □, and capital city or county seat by ⊙.

Obernigk, POLAND: see OBORNIKI SLASKIE.

Obernkirchen (o-buhrn-KIR-khuhn), town, LOWER SAXONY, W GERMANY, 9 mi/14.5 km E of MINDEN; 52°16′N 09°08′E. Health resort in suburb of STADTHAGEN. Glass working; machinery; limestone quarries. Has a 12th-century church. Was first mentioned 1167; chartered 1181; town since 1615.

Obernzell (o-buhrn-TSEL), village, LOWER BAVARIA, GERMANY, on the DANUBE RIVER, 8 mi/12.9 km E of PASSAU; 48°33′N 13°38′E. Manufacturing of melting pots, tanning, woodworking. Graphite mining in area. Chartered c.1300. Has been metalworking center since 16th century. Formerly called HAFNERZELL.

Ober-Olm (O-buhr–OLM), village, RHINELAND-PALATINATE, W GERMANY, 7 mi/11.3 km SE of MAINZ; 49°58′N 08°14′E. Viticulture; winery, distillery.

Oberon (O-buh-rahn), village, E central NEW SOUTH WALES, SE AUSTRALIA, 80 mi/129 km WNW of SYDNEY; 33°43′S 149°52′E. Agriculture; beef cattle; nuts; timber; granite quarries. Railroad terminus. Jenolan (juh-NO-luhn) limestone caves 12 mi/19 km SE in BLUE MOUNTAINS; tourism.

Oberon (O-buhr-ahn), village (2006 population 82), BENSON county, N central NORTH DAKOTA, 21 mi/34 km SW of DEVILS LAKE; 47°55′N 99°12′W. Junction of railroad spur to MINNEWAUKAN. Devils Lake Sioux Indian Reservation just E; Fort Totten Historic Site to NE. Founded in 1886.

Oberösterreich, AUSTRIA: see UPPER AUSTRIA.

Oberpahlen, ESTONIA: see PÕLTSAMAA.

Oberpfalz, GERMANY: see UPPER PALATINATE.

Oberplan, CZECH REPUBLIC: see HORNI PLANA.

Oberpullendorf (o-buhr-PUL-len-dorf), Hungarian *Felöpulya*, town, central BURGENLAND, E AUSTRIA, 14 mi/23 km SSW of SOPRON (HUNGARY); 47°30′N 16°30′E. Railroad terminus; market center; electro-technical industry; sugar beets. Large Hungarian population in the town and the general vicinity.

Ober-Ramstadt (O-buhr–RAHM-shtaht), town, HESSE, GERMANY, 5 mi/8 km SE of DARMSTADT; 49°50′N 08°45′E. Paints; foundry; flour milling, printing and publishing.

Oberrheintal (O-buhr-REIN-tahl), district in NE ST. GALLEN canton, SWITZERLAND. Main town is ALTSTÄTTEN; population is German-speaking and Protestant.

Oberriet (O-buhr-reet), commune, ST. GALLEN canton, NE SWITZERLAND, 7 mi/11.3 km E of APPENZELL, near the RHINE RIVER, and the Austrian border; 47°19′N 09°34′E. Elevation 1,381 ft/421 m. Embroidery; metalworking.

Oberschaeffolsheim (o-ber-SHE-folz-HEIM), commune (□ 3 sq mi/7.8 sq km; 2004 population 2,052), BAS-RHIN department, ALSACE region, NE FRANCE, 3 mi/ 5 km WNW of STRASBOURG; 48°35′N 07°38′E. Agricultural village.

Oberschleissheim (o-buhr-SHLEIS-heim), suburb of MUNICH, UPPER BAVARIA, BAVARIA, S GERMANY, 8 mi/12.9 km N of city center; 48°15′N 11°33′E. Neues Schloss [Ger.=New Castle] to SE.

Obersee, SWITZERLAND: see LAKE CONSTANCE.

Obersee, SWITZERLAND: see ZÜRICH, LAKE.

Obershausen (o-buhrs-HOU-suhn), town, HESSE, central GERMANY, 4 mi/6.4 km ESE of OFFENBACH, on the Rodau River; 50°05′N 08°52′E. Machinery; metalworking, leatherware. Formed 1977 by unification of Hausen and Obertshausen (both first mentioned 865); chartered 1979.

Obersiggenthal (o-buhr-ZEE-guhn-tahl), commune, AARGAU canton, N SWITZERLAND, near LIMMAT RIVER, 2 mi/3.2 km NW of BADEN; 47°30′N 08°19′E. Resort. Metal products. UNTERSIGGENTHAL is NW.

Obersimmental (o-buhr-ZEE-muhn-tahl), district, SW BERN canton, SWITZERLAND. Population is German-speaking and Protestant.

Obersontheim (o-buhr-SONT-heim), village, FRANCONIA, BADEN-WÜRTTEMBERG, S GERMANY, 8 mi/ 12.9 km ESE of SCHWÄBISCH HALL; 49°03′N 09°53′E. Forestry.

Oberstaufen (o-buhr-SHTOU-fuhn), village, SWABIA, SW BAVARIA, S GERMANY, 17 mi/27 km SW of KEMPTEN, on the WEISSACH, and in the Bavarian Alps; 47°33′N 10°01′E. Tourism; health resort. Chartered 1243; passed 1805 to PRUSSIA. Cable railroad onto Hochgrat (elevation 6,014 ft/1,833 m); chapel (c.1440).

Oberstdorf (O-buhrst-dorf), town, SWABIA, SW BAVARIA, GERMANY, in ALLGÄU ALPS, 7 mi/11.3 km S of SONTHOFEN; 47°25′N 10°17′E. Railroad terminus. Well-known climatic health and winter sports resort (elevation 2,731 ft/832 m). Chartered 1495.

Oberstein, GERMANY: see IDAR-OBERSTEIN.

Oberstenfeld (O-buhr-sten-feld), village, STUTTGART region, BADEN-WÜRTTEMBERG, S GERMANY, 10 mi/16 km SSE of HEILBRONN, on W slope of the Swabian-Franconian Forest; 49°01′N 09°20′E. Viticulture; manufacturing (metalworking; machinery).

Ober Suchau, CZECH REPUBLIC: see HORNI SUCHA.

Obersulm (O-buhrs-ulm), town, FRANCONIA, BADEN-WÜRTTEMBERG, GERMANY, 7 mi/11.3 km E of HEILBRONN, on W slope of Swabian-Franconian Forest; 49°08′N 09°23′E. Railroad junction. Viticulture; grain, fruit; steelwork.

Obert, village (2006 population 45), CEDAR county, NE NEBRASKA, 12 mi/19 km ENE of HARTINGTON, near MISSOURI RIVER; 42°41′N 97°01′W.

Obertauern (O-buhr-tou-ern), village, SE SALZBURG, W central AUSTRIA, on Radstädter Tauern Pass, 17 mi/27 km ESE of SANKT JOHANN IM PONGAU; 47°15′N 13°32′E. Elevation 5,072 ft/1,546 m. Winter-sports center; cable cars to Zehnerkar (elevation 7,257 ft/2,212 m).

Oberteuringen (o-buhr-TOI-ring-uhn), village, LAKE CONSTANCE–UPPER SWABIA, BADEN-WÜRTTEMBERG, S GERMANY, 6 mi/9.7 km N of FRIEDRICHSHAFEN; 47°43′N 09°29′E. In fruit-growing region.

Oberthal (O-buhr-tahl), village, SAARLAND, W GERMANY, 5 mi/8 km NW of ST. WENDEL, on the BLIES RIVER; 49°31′N 07°06′E. Steelworking.

Oberthulba (O-buhr-tul-bah), village, LOWER FRANCONIA, NW BAVARIA, GERMANY, on small Thulba River, and 4 mi/6.4 km W of BAD KISSINGEN; 50°12′N 09°58′E. Rye, peas, lentils; wine growing.

Obertin, UKRAINE: see OBERTYN.

Obertoggenburg (o-buhr-TO-guhn-boorg), district, S central ST. GALLEN canton, SWITZERLAND; 47°10′N 09°15′E. Main town is EBNAT-KAPPEL; population is German-speaking and mixed Roman Catholic and Protestant.

Obertraubling (o-buhr-TROUB-ling), village, UPPER PALATINATE, BAVARIA, GERMANY, 4 mi/6.4 km SE of REGENSBURG; 48°58′N 12°08′E. Agricultural products; machinery.

Obertraun (o-buhr-troun), village, S UPPER AUSTRIA, in the SALZKAMMERGUT, near E shore of LAKE OF HALLSTATT, 11 mi/18 km SSE of BAD ISCHL; 47°33′N 13°41′E. Tourist center (elevation 1,579 ft/481 m). Cable railroad to the DACHSTEIN massif (Krippenstein, elevation 6,435 ft/1,961 m); caves.

Obertrum am See (O-buhr-trum ahm SAI), village, N SALZBURG, W central AUSTRIA, in the FLACHGAU, and on S tip of Obertrumer See (Obertrum Lake, 1.9 mi/3.1 km long, 1.29 km wide, maximum depth 107 ft/ 33 m), 9 mi/14.5 km N of SALZBURG; 47°56′N 13°05′E. Elevation 1,548 ft/472 m. Manufacturing of building materials; brewery. Summer tourism.

Obertyn (o-ber-TIN), (Russian *Obertin*), town (2004 population 8,700), E IVANO-FRANKIVS'K oblast, UKRAINE, on the left tributary of the PRUT RIVER, and on crossroads, 13 mi/21 km NNE of KOLOMYYA; 48°42′N 25°10′E. Elevation 1,036 ft/315 m. Flour milling; two brickworks, can making (for a nearby cannery). Site of the defeat (1531) of the Walachians by Polish hetman Jan Tarnowski. Known since 1416; town since 1940. Jewish community since the 18th century; received full citizenship rights under the Austro-Hungarian rule in

1867; numbered 1,130 in 1939; destroyed during World War II.

Oberursel (o-buhr-UHR-sel), town, HESSE, W GERMANY, on S slope of the TAUNUS, 7 mi/11.3 km NNW of FRANKFURT (tramway connection); 50°10′N 08°35′E. Manufacturing of jewelry, machinery. Has 14th–15th-century Gothic church. First mentioned 791; town hall dates to 1659; passed to PRUSSIA 1866. Belonged to electors of MAINZ (1535–1581 and 1632–1635).

Oberuzwil (O-buhr-OOTS-veel), commune, ST. GALLEN canton, NE SWITZERLAND, 12 mi/19 km W of ST. GALLEN; 47°26′N 09°07′E. Textiles, leather, shoes.

Obervellach (O-buhr-vel-lahkh), township, CARINTHIA, S AUSTRIA, on Möll River, and 16 mi/26 km NW of SPITTAL AN DER DRAU; 46°56′N 13°12′E. Hydroelectric station; summer resort (elevation 2,250 ft/ 686 m). Castles in the area.

Oberviechtach (O-buhr-VEEKH-tahkh), town, UPPER PALATINATE, E BAVARIA, GERMANY, in BOHEMIAN FOREST, 17 mi/27 km NE of SCHWANDORF; 49°28′N 12°24′E. Machinery; woodworking. Health resort; tourism. Has rococo church (1771–1776); ruined castle. First mentioned 1130. Passed to Bavaria 1272. Chartered 1337; town since 1952.

Oberwampach (O-buhr-VAHM-pahkh), village, WINCRANGE commune, NW LUXEMBOURG, in the ARDENNES Mountains, 5 mi/8 km NW of WILTZ; 50°01′N 05°52′E. Former lead-mining village.

Oberwart (O-buhr-vahrt), town, S BURGENLAND, E AUSTRIA, 20 mi/32 km WNW of SZOMBATHELY (HUNGARY); 47°17′N 16°13′E. Railroad terminus. Educational center for linguistic minorities. Manufacturing of textiles and steel equipment. Cattle auctions; fairs.

Oberwesel (O-buhr-VE-sel), town, RHINELAND-PALATINATE, W GERMANY, on left bank of the RHINE RIVER (landing), and 10 mi/16 km SE of BOPPARD; 50°07′N 07°43′E. Wine; tourism. Has 14th-century walls and towers, several Gothic churches. Modern castle Schönburg and ruins of 12th-century castle Schönburg tower above town. Chartered 1216.

Oberweser (O-buhr-VE-ser), village, HESSE, central GERMANY, 19 mi/30 km N of KASSEL, on left bank of the WESER RIVER, in the Weserbergland; 51°36′N 09°38′E. Forestry.

Oberwiesenthal (o-buhr-VEE-sen-tahl) or **kurort**, town, SAXONY, E central GERMANY, in the Erzgebirge, 11 mi/18 km S of ANNABERG, frontier point on CZECH border, 5 mi/8 km NE of JACHYMOV (Czech Republic); 50°26′N 12°58′E. Ribbon and glove manufacturing; wood carving. Winter-sports center at foot of the FICHTELBERG (funicular railroad). Tourism. Chartered 1530; passed to Saxony 1559; combined with UNTERWIESENTHAL in 1921.

Oberwil (O-buhr-veel), commune (2000 population 9,363), Basel-Land half-canton, N SWITZERLAND, on BIRS RIVER, and 4 mi/6.4 km SSW of BASEL. Tiles; woodworking.

Oberwölz Stadt (O-buhr-wuhlts SHTAHT), town, STYRIA, S central AUSTRIA, on S foot of Wölzer Tauern, 19 mi/31 km W of JUDENBURG; 47°12′N 14°17′E. Elevation 2,545 ft/776 m. Town before 1305; many medieval fortifications and buildings preserved; late-Romanesque parish church; impressive castle Rothenfels.

Oberzeiring (o-buhr-TSEI-ring), township, STYRIA, S central AUSTRIA, at E foot of Wölzer Tauern, 9 mi/14.5 km NW of JUDENBUR; 47°15′N 14°29′E. Elevation 2,835 ft/864 m. Formerly an important silver mine; mining museum; asthma spa. Old burgher houses.

Obetz (O-bets), village (□ 4 sq mi/10.4 sq km; 2006 population 4,064), FRANKLIN county, central OHIO; suburb 5 mi/8 km SSE of COLUMBUS; 39°52′N 82°57′W.

Obey River (O-bee), 58 mi/93 km long, N TENNESSEE; rises in E PUTNAM county; flows NW and W to CUMBERLAND RIVER at CELINA; Dale Hollow Dam, 7 mi/11 km above mouth, impounds 51-mi/82-km-long DALE HOLLOW Reservoir; 36°33′N 85°30′W.

Óbidos (O-bee-dos), city, W PARÁ, BRAZIL, on heights above left bank of the lower AMAZON (which here narrows down to 1.25 mi/2.01 km), and 65 mi/105 km NW of SANTARÉM; 01°45′S 55°30′W. Port of call (river steamers, hydroplanes). Exports (tobacco, cacao, coffee, sugar, lumber, oil seeds, cattle); manufacturing (chocolate); fruit preserving, jute processing. Establishede 1697, and fortified in the colonial period, the city occupies a strategic site on the Amazon just below influx of TROMBETAS River. Formerly also spelled Obydos.

Óbidos (O-bee-doosh), town, LEIRIA district, W central PORTUGAL, on railroad, and 3 mi/4.8 km SSW of CALDAS DA RAINHA. Manufacturing (chemical fertilizer, wax candles), resin processing. Apple orchards. Surrounded by walls of well-preserved 14th-century castle. Recaptured from Moors 1148.

Obi-Dzhuk, TAJIKISTAN: see VARZOB.

Obigarm (o-bee-GAHRM), town, W TAJIKISTAN, on VAKHSH RIVER, and 40 mi/64 km SW of GARM; 38°43′N 69°42′W. Under direct republic supervision (i.e., not part of any viloyat). Wheat; cattle.

Obihiro (o-bee-HEE-ro), city (2005 population 170,580), S central Hokkaido prefecture, N JAPAN, 93 mi/150 km E of SAPPORO, on TOKACHI RIVER; 42°55′N 143°12′E. Railroad junction. Commercial center for agriculture (beef cattle; dairy products; beans, sugar beets, potatoes, sweet corn, cinnamon vine; wheat).

Obi Islands (O-bee) or **Ombi Islands**, island group, N Maluku province, INDONESIA, in SERAM SEA, between HALMAHERA (N) and SERAM (S); 01°20′S 127°38′E. Comprise Obir, or Obira (52 mi/84 km long, 28 mi/45 km wide), and several small offshore islands. Obir is largely mountainous, rising to 4,321 ft/1,317 m, with narrow coastal plain; chief village is Laiwui, on N coast. Other islands of group are Bisa (N; 15 mi/24 km long), Obi Latu (W; 8 mi/12.9 km long), Tobalai (E; 7 mi/11.3 km long). Principal products of group are resin and sago. Obi Islands became a Dutch possession in 1682.

Obi-Khingou River (o-bee–KEEN-goo), c.120 mi/193 km long, central TAJIKISTAN; rises in several branches in DARVAZA RANGE; flows W, past SANGVOR and Tavil-Dara, joining SURKHAB RIVER near KOMSOMOLABAD to form the VAKHSH RIVER.

Obikiik (o-bee-kee-EEK), village, KHATLON viloyat, TAJIKISTAN, in Vakhsh valley, 12 mi/20 km NNW of QURGHONTEPPA; 38°08′N 68°40′E. Tertiary level administrative center.

Obilić (O-bee-leech), village, KOSOVO province, SW SERBIA, on railroad, and 5 mi/8 km WNW of PRISTINA. Manufacturing of explosives. Also spelled Obilich.

Obing (O-bing), village, UPPER BAVARIA, BAVARIA, SE GERMANY, 16 mi/26 km NE of ROSENHEIM, on Obinger Lake; 48°00′N 12°23′E.

Obion (o-BEI-uhn), county (□ 550 sq mi/1,430 sq km; 2006 population 32,184), NW TENNESSEE, on KENTUCKY (N) state line; ⊙ UNION CITY; 36°22′N 89°09′W. Bounded NW by REELFOOT LAKE; drained by OBION RIVER and its tributaries. Fertile farm region; diverse manufacturing. Formed 1823. Includes Gooch Wildlife Management Area and part of Reelfoot National Wildlife Refuge.

Obion, town (2006 population 1,104), OBION county, NW TENNESSEE, on OBION RIVER, and 13 mi/21 km 13 mi/21 km SSW of UNION CITY; 36°16′N 89°12′W. In fertile farm area. Future site of ethanol distribution center for NW Tennessee.

Obion Creek, c.50 mi/80 km long, SW KENTUCKY; rises in S GRAVES county, c.15 mi/24 km S of MAYFIELD; flows NW through Obion Creek Wildlife Management Area, then SW to MISSISSIPPI RIVER 2 mi/3.2 km NE of HICKMAN.

Obion River, c.50 mi/80 km long, W TENNESSEE; formed in OBION county by confluence of canalized North (c.45 mi/72 km long), South (c.55 mi/89 km long), and Rutherford (c.50 mi/80 km long) forks; flows SW past OBION, to MISSISSIPPI RIVER 13 mi/21 km NW of RIPLEY; 35°54′N 89°38′W. Receives FORKED DEER RIVER.

Obir, AUSTRIA: see HOCHOBIR.

Obir or **Obira**, INDONESIA: see OBI ISLANDS.

Obira (O-bee-rah), town, Rumoi district, Hokkaido prefecture, N JAPAN, 68 mi/110 km N of SAPPORO; 44°00′N 141°39′E.

Obiralovka, RUSSIA: see ZHELEZNODOROZHNYY, MOSCOW oblast.

Öbisfelde, GERMANY: see OEBISFELDE.

Obispos (o-BEES-pos), town, ⊙ Obispos municipio, BARINAS state, W VENEZUELA, in LLANOS, 7.0 mi/11.3 km E of BARINAS; 08°36′N 70°06′W. Livestock.

Obispo Santiesteban, province, SANTA CRUZ department, E central BOLIVIA, ⊙ MONTERO; 16°30′S 63°30′W.

Obispo Trejo (o-BEES-po TRAI-ho), village, N CÓRDOBA province, ARGENTINA, on railroad, 65 mi/105 km NE of CÓRDOBA; 30°46′S 63°25′W. Flax; livestock; lumber.

Obitel (o-BEE-tel), village, RUSE oblast, OMURTAG obshtina, BULGARIA; 43°04′N 26°34′E.

Obitochnaya kosa, UKRAINE: see OBYTICHNA SPIT.

Obitochnyy zaliv, UKRAINE: see OBYTICHNA BAY.

Objat (ob-ZHAH), commune (□ 3 sq mi/7.8 sq km), CORRÈZE department, LIMOUSIN region, S central FRANCE, 9 mi/14.5 km NW of BRIVE-LA-GAILLARDE; 45°16′N 01°25′E. Agricultural trade (fruit, vegetables, truffles, and chestnuts; hogs). Manufacturing of rattan furniture, food processing.

Öblarn (OB-lahrn), township, NW STYRIA, central AUSTRIA, on ENNS RIVER, 13 mi/21 km SW of LIEZEN; 47°28′N 14°00′E. Elevation 2,073 ft/632 m. Old silver and copper mine. The poet Paula Grogger lived here.

Oblatos, Barranca de, MEXICO: see BARRANCA DE OBLATOS.

Obligado (O-ble-GAH-do), town, SE PARAGUAY, Itapúa department, 25 mi/40 km NE of Encarnación; 27°02′S 55°38′W. In an agricultural area (soybeans, wheat, fruits, yerba maté).

Oblivskaya (uh-BLEEF-skah-yah), village (2006 population 9,525), E ROSTOV oblast, S European Russia, on the CHIR' RIVER, on road junction and railroad, 35 mi/56 km ENE of MOROZOVSK; 48°32′N 42°29′E. Elevation 160 ft/48 m. In agricultural area (wheat, sunflowers; livestock); flour mill, dairy; logging and lumbering.

Oblong (OB-long), village (2000 population 1,580), CRAWFORD county, SE ILLINOIS, 9 mi/14.5 km W of ROBINSON; 39°00′N 87°54′W. Oil, natural-gas wells. Agriculture (livestock, poultry; corn, wheat, soybeans). Incorporated 1883.

Obluchye (uh-BLOO-cheye), city (2005 population 10,685), NW JEWISH AUTONOMOUS OBLAST, S RUSSIAN FAR EAST, on the Khingan River (left tributary of the AMUR River), on highway and the TRANS-SIBERIAN RAILROAD, 100 mi/159 km W of BIROBIDZHAN; 49°00′N 131°00′E. Elevation 1,079 ft/328 m. Railroad establishments; food industries (dairy, bakery). Founded in 1909–1911 as a railroad station; made city in 1938.

Obninsk (OB-neensk), city (2005 population 107,965), NW KALUGA oblast, central European Russia, on road and railroad, on the PROTVA RIVER (OKA River basin), 66 mi/106 km SW of MOSCOW, and 46 mi/74 km NW of KALUGA; 55°06′N 36°36′E. Elevation 570 ft/173 m. Scientific center with research institutes of meteorology, geology, medical radiology, nuclear energy, and energy physics. First nuclear power station in the former Soviet Union (began operation in 1954); research reactor; enriched uranium storage site. Production of scientific and medical instruments; prosthetic-orthopedic equipment, food-processing machinery, plumbing supplies. Food processing industries (dairy, bakery, sausage making, meat packing). Made city in 1956.

Obnova (ob-NO-vah), village, LOVECH oblast, LEVSKI obshtina, N BULGARIA, 9 mi/15 km ENE of Levski; 43°25′N 25°00′E. Flour milling; livestock, poultry; vegetables.

Obo (O-BO), town, ⊙ HAUTE-M′BOMOU prefecture, SE CENTRAL AFRICAN REPUBLIC, near SUDAN and CONGO borders, 260 mi/418 km ENE of BANGASSOU; 05°18′N 26°28′E. Trade center.

Obock, district (2006 population 19,460), NE DJIBOUTI; ⊙ OBOCK. Bordered NWN by ERITREA, NE by the STRAITS OF BAB EL MANDEB, E and S by the GULF OF ADEN, and SWW by TADJOURAH district. Towns include Khozor Anghar, Moulhoulè, and OBOCK. Obock district is divided into two secondary administrative divisions: Alaili Dadda in the NNW and Obock in the SSE.

Obock (o-BOK), town, ⊙ OBOCK district, DJIBOUTI, port on N shore of GULF OF TADJOURA, 26 mi/42 km NNE of DJIBOUTI; 11°58′N 43°17′E. Fisheries; hot springs. Airport. Ceded 1862 to FRANCE; was capital of FRENCH SOMALILAND between 1884–1892; then transferred to Djibouti. Also spelled Obok.

Obodivka (o-bo-DEEF-kah), (Russian *Obodovka*), village (2004 population 6,640), SE VINNYTSYA oblast, UKRAINE, at road junction, 25 mi/40 km SE of TUL′-CHYN; 48°24′N 29°15′E. Elevation 741 ft/225 m. Sugar mill. Lignite deposits. Jewish community since the 16th century, numbering close to 400 at the beginning of the 20th century; wiped out almost completely during the 1919 civil war pogroms; site of a ghetto for the Jews from Savran' and other towns in Bukovyna and Bessarabia from 1941 until 1944, when the ghetto was liquidated by the retreating Germans.

Obodovka, UKRAINE: see OBODIVKA.

Obogu (o-BO-goo), town, ASHANTI region, GHANA, 5 mi/8 km S of JUASO; 06°31′N 01°07′W. Cacao, coffee; timber.

Obok, DJIBOUTI: see OBOCK.

Obokum (o-BAW-koom), town, CROSS RIVER state, extreme SE NIGERIA, on CROSS RIVER (CAMEROON border), 12 mi/19 km ESE of IKOM; 05°54′N 08°54′E. Hardwood, rubber; palm oil and kernels, cacao.

Obol' (o-BOL), urban settlement, VITEBSK oblast, BELARUS, on Obol' River railroad station; brick and pressed-peat insulation plants.

Oboldino (uh-buhl-dee-NO), urban settlement, central MOSCOW oblast, central European Russia, near railroad, 4 mi/6 km WSW of (and administratively subordinate to) SHCHËLKOVO; 55°53′N 37°56′E. Elevation 505 ft/153 m. Paints, packing materials.

Obolo (o-BO-lo), town, ENUGU state, S central NIGERIA, on road, 40 mi/64 km NNE of ENUGU, just E of NSUKKA. Market town. Cassava, rice, cashews; coal deposits.

Obolon' (o-bo-LON), village (2004 population 3,000), W central POLTAVA oblast, UKRAINE, 28 mi/45 km S of LUBNY; 49°36′N 32°53′E. Elevation 272 ft/82 m. Wheat, flax, mint.

Obol' River (o-BOL), 92 mi/148 km long, in VITEBSK oblast, NE BELARUS; issuing from Lake EZERISHCHE in extreme NE corner of BELARUS, on Russian border; flows generally SW to ULLA RIVER just NW of ULLA.

Obón (o-VON), village, TERUEL province, E SPAIN, 25 mi/51 km WSW of ALCAÑIZ; 40°54′N 00°43′W. Cereals, olive oil, saffron; sheep.

Obora, town (2007 population 7,142), OROMIYA state, ETHIOPIA, near SHAMBU.

Oborishte (o-BOR-eesh-te), village, PLOVDIV oblast, PANAGYURISHTE obshtina, BULGARIA; 42°32′N 24°04′E.

Oborniki (o-bor-NEE-kee), German *Obernick*, town, Poznań province, W POLAND, on WARTA River, at WEŁNA River mouth, and 17 mi/27 km N of POZNAŃ; 52°39′N 16°49′E. Railroad junction; manufacturing of wooden shoes; flour milling, sawmilling; cementworks. Tuberculosis sanatorium.

Oborniki Slaskie (o-bor-NEE-kee SLON-skye), Polish *Oborniki Śląskie*, German *Obernigk*, town, Wrocław

---

Area is shown by the symbol □, and capital city or county seat by ⊙.

province, SW POLAND, 14 mi/23 km NNW of WROCŁAW (Breslau); 51°18′N 16°55′E. Agricultural market (grain, sugar beets, potatoes; livestock). In German-ruled LOWER SILESIA prior to 1945, when it passed to Poland. Heavily damaged in World War II.

Oborona, RUSSIA: see MORDOVO.

Oboyan′ (uh-buh-YAHN′), city (2005 population 14,375), central KURSK oblast, SW European Russia, on the PSEL RIVER, 35 mi/56 km S of KURSK; 51°12′N 36°17′E. Elevation 679 ft/206 m. Railroad terminus; highway hub. Automotive repair shop; wallboard; food industry (meatpacking, flour milling, fruit and vegetable canning). Founded in 1639, made city in 1779.

Obozërskiy (uh-buh-ZYOR-skeeyee), town (2006 population 3,535), W ARCHANGEL oblast, N European Russia, on upper reaches of the Baymuga River (Northern DVINA River basin), on road, 75 mi/121 km S of ARCHANGEL; 63°26′N 40°19′E. Elevation 344 ft/104 m. Near the junction of railroad line to ONEGA and BELOMORSK. Limeworks.

Obra River (OB-rah), c.150 mi/241 km long, W POLAND; rises 3 mi/4.8 km N of Koźmin; flows generally WNW past KOŚCIAN, W and N past ZBĄSZYN and MIĘDZYRZECZ to WARTA River at Skierzyna. Just below KOŚCIAN, its W course divides into 3 parallel canals; the longest (N) canal (Obra Canal, Polish *Kanal Obry*), extends E to Warta River near MOSINA; canals drain Obra Marshes, Polish *Legi Oberskie*.

Obraztsovo-Travino (uh-brahs-TSO-vuh–TRAH-veenuh), village, E ASTRAKHAN oblast, SE European Russia, in the VOLGA RIVER delta, on local highway branch, 25 mi/40 km S of ASTRAKHAN; 45°58′N 48°00′E. Below sea level. Fisheries; fruit. Formerly called Obraztsovo.

Obregón, Cañadas de, MEXICO: see CAÑADAS DE OBREGÓN.

Obregón, Ciudad, MEXICO: see CIUDAD OBREGÓN.

Obrenovac (o-BRE-no-vahts), city (2002 population 70,975), N central SERBIA, on KOLUBARA RIVER, and 17 mi/27 km SW of BELGRADE, near the SAVA River; 44°39′N 20°12′E. Railroad junction; manufacturing (food processing). Also spelled Obrenovats.

O′Brien, county (□ 575 sq mi/1,495 sq km; 2000 population 15,102), NW IOWA; ⊙ PRIMGHAR; 43°04′N 95°37′W. Prairie agricultural area (hogs, cattle, sheep, poultry; corn, oats, barley) drained by Little Sioux and FLOYD rivers. Mill Creek State Park in S. Field and stream flooding occurred in 1993. Formed 1851.

O′Brien, ARGENTINA: see GENERAL O′BRIEN.

O′Brien Island (o-BREI-uhn), TIERRA DEL FUEGO, CHILE, between the main island and LONDONDERRY ISLAND; 54°50′S 70°40′W.

Obrigheim (O-brig-heim), village, Lower NECKAR, BADEN-Württemberg, S GERMANY, 20 mi/32 km ESE of HEIDELBERG, on left bank of the NECKAR River, in the ODENWALD; 49°21′N 09°05′E. Agriculture products; tourism. Nuclear power station.

Ob′ River (OP), approximately 2,300 mi/3,701 km long, W Siberian Russia. With the IRTYSH River, its chief tributary, it is approximately 3,460 mi/5,568 km long and the world's fourth-longest river. Formed by the junction of the BIYA and KATUN rivers (both of which rise in the Altai Mountains) SW of BIYSK (Altai Territory), the upper Ob′ flows NW, then NE past BARNAUL and NOVOSIBIRSK (Novosibirsk oblast) through the W Siberian lowlands to be joined by the Tom′ River in Tomsk oblast. The middle Ob′ flows NW through the swampy forests in the TOMSK and NARYM regions and then is joined by the CHULYM, KET, and Irtysh rivers. The lower Ob′ consists of the Great Ob′ and the Small Ob′ and flows N, then E into OB′ BAY, an estuary and shallow arm (approximately 500 mi/ 800 km long, 35 mi/56 km–50 mi/80 km wide) of the ARCTIC OCEAN between the YAMAL and GYDAN peninsulas (Tyumen oblast). The width of the Ob′ increases downstream to about 25 mi/40 km near its mouth. The valley of the middle Ob′ is subject to

flooding each spring as the thaw occurs in the upper Ob′ basin before the ice in the lower course of the Ob′ has melted. Although frozen 5–6 months of the year, the Ob′ is an important trade and transport route; Novosibirsk, Barnaul, KAMEN-NA-OBI (Altai Territory), and MOGOCHIN (Tomsk oblast) are the chief ports. There is a large hydroelectric power station at Novosibirsk. The largest oil and gas deposits in Russia are found in the basin of the middle and lower Ob′. Severe pollution in the lower Ob′ has damaged the river's formerly famous fisheries.

Obrochishte (o-BRO-cheesh-te), village, VARNA oblast, BALCHIK obshtina, on the BLACK SEA; 42°31′N 24°05′E.

Obrovac (OB-ro-vahts), Italian *Obbrovazzo*, village, S CROATIA, on ZRMANJA RIVER, and 22 mi/35 km ENE of ZADAR, in N DALMATIA. Former bauxite-manufacturing area; aluminum plant closed due to insufficient bauxite supply. Part of Srpska (Serb) Krajina 1991–1995.

Obruchishte (ob-ROO-chee-shte), village, HASKOVO oblast, GULUBOVO obshtina, BULGARIA; 42°09′N 25°56′E.

Obry, Kanal, POLAND: see OBRA RIVER.

Ob′ Sea, RUSSIA: see NOVOSIBIRSK RESERVOIR.

Observatory Inlet (uhb-ZUHR-vuh-tor-ee), bay, W BRITISH COLUMBIA, W Canada, long narrow arm of PORTLAND INLET (arm of DIXON ENTRANCE), extending inland near and parallel to the S tip of the ALASKA panhandle; 45 mi/72 km long, 1 mi/2 km–4 mi/6 km wide; 55°15′N 129°49′W. Alice Arm extends E to ALICE ARM village. On W side of Observatory Inlet is copper mining center of ANYOX.

Obsharovka (uhb-SHAH-ruhf-kah), village (2006 population 5,655), SW SAMARA oblast, E European Russia, in the VOLGA RIVER basin, on railroad and near highway, 7 mi/11 km SE of OKTYABR′SK; 53°07′N 48°51′E. Elevation 170 ft/51 m. Manufacturing (abrasive instruments). In agricultural area; grain elevator.

Obshchiy Syrt (OP-shcheeyee SIRT), SW foothills of the S URAL Mountains, in ORENBURG, SAMARA, and SARATOV oblasts, SE European Russia, N of the URAL River; rise to approximately 1,400 ft/427 m.

Obstruction Mountain (uhb-STRUHKT-shuhn), (10,394 ft/3,168 m), SW ALBERTA, W Canada, in ROCKY MOUNTAINS, near SE edge of JASPER NATIONAL PARK, 60 mi/97 km SE of JASPER; 52°23′N 116°53′W.

Obu (O-boo), city, AICHI prefecture, S central HONSHU, central JAPAN, 12 mi/20 km S of NAGOYA; 35°00′N 136°57′E.

Obuasi (o-BWAH-see), town (2002 population 122,600), ASHANTI, GHANA, on road and railroad link; 06°12′N 01°41′W. Highly concentrated gold ore is mined, and there are gold-extraction plants. Gold was mined in Obuasi by indigenous peoples as early as the 17th century. From the late 1890s it was developed by Europeans into a modern mining town. Operating since 1965, it remains Ghana's only privately owned mine. In rice-growing area.

Obubra (aw-boo-BRAH), town, CROSS RIVER state, SE NIGERIA, port on CROSS RIVER, 25 mi/40 km SE of ABAKALIKI; 06°05′N 08°19′E. Palm oil and kernels, cacao, kola nuts.

Óbuda, HUNGARY: see BUDAPEST.

Obudu (aw-boo-DOO), town, CROSS RIVER state, SE NIGERIA, 29 mi/47 km E of OGOJA; 06°40′N 09°10′E. Shea nuts, sesame.

Obukhiv (o-BOO-kheef), (Russian *Obukhov*), city, central KIEV oblast, UKRAINE, on highway 23 mi/37 km S of KIEV; 50°06′N 30°38′E. Elevation 593 ft/180 m. Raion center. Carton-paper mill, dairy, biochemical plant; building materials (silicate walls, bricks). Literary museum in honor of the poet, A. Malyshko. Known since the 14th century as Lukavytsya; renamed Obukhiv at the end of the 16th century; company center of Kiev regiment in the 17th–18th centuries.

Obukhivka, UKRAINE: see KIROVS′KE, Dnipropetrovs′k oblast.

Obukhov, UKRAINE: see OBUKHIV.

Obukhovka, UKRAINE: see KIROVS′KE, Dnipropetrovs′k oblast.

Obukhovo (O-boo-huh-vuh), town, E LENINGRAD oblast, NW European Russia, on the VOLKHOV RIVER, 4 mi/6 km N of VOLKHOV; 59°58′E 32°20′E. Manufacturing (construction materials and structures; electronic and telecommunication equipment).

Obukhovo (O-boo-huh-vuh), town (2006 population 10,630), E central MOSCOW oblast, central European Russia, on the KLYAZ′MA RIVER (OKA River basin), on highway junction, 7 mi/11 km WSW of NOGINSK, to which it is administratively subordinate; 55°50′N 38°16′E. Elevation 433 ft/131 m. Carpets and rugs; plastic polymers.

Obuse (o-BOO-se), town, Kamitakai county, NAGANO prefecture, central HONSHU, central JAPAN, 9 mi/15 km N of NAGANO; 36°41′N 138°18′E.

Obwalden, SWITZERLAND: see UNTERWALDEN.

Obyachevo (uhb-YAH-chee-vuh), village (2005 population 5,455), SW KOMI REPUBLIC, NE European Russia, on the LUZA RIVER, on road and local railroad spur, 120 mi/193 km N of KIROV (KIROV oblast); 60°20′N 49°37′E. Elevation 439 ft/133 m. In flax and potato area.

Obydos, Brazil: see ÓBIDOS.

Ob′-Yenisey Canal System (OP–yee-nee-SYAI), in TOMSK oblast and KRASNOYARSK TERRITORY, W Siberian Russia, 70-mi/113-km waterway joining Ob′ and YENISEY rivers; 59°10′N 88°30′E. Kas River (tributary of the Yenisey) and Ket River (tributary of the Ob′) are joined by means of their dredged branches (with locks) and a 5-mi/8-km canal across the watershed. Built in 1882–1891 for small craft; fell into disuse.

Obytichna Bay (o-bi-TEECH-nah), (Ukrainian *Obytichna zatoka*), (Russian *Obitochnyy zaliv*), shallow N embayment of the Sea of AZOV, in S ZAPORIZHZHYA oblast, UKRAINE. Partly separated form the Sea of Azov (SE) by the OBYTICHNA SPIT (Ukrainian *Obytichna kosa*); 19 mi/31 km long, maximum depth 26 ft/ 8 m; receives Obytichna (NE) and Lozuvatka (NW) rivers. Fishing villages.

Obytichna kosa, UKRAINE: see OBYTICHNA SPIT.

Obytichna Kosa Landshaftnyy zakazynk, UKRAINE: see OBYTICHNA SPIT.

Obytichna Spit (o-bi-TEECH-nah), (Ukrainian *Obytichna Kosa*), (Russian *Obitochnaya Kosa*), a spit on the N coast of the Sea of AZOV, extending from 5 mi/8 km to 24 mi/38 km SW of PRYMORS′K, S ZAPORIZHZHYA oblast, UKRAINE. It partially separates OBYTICHNA BAY from the Sea of Azov (S) and BERDYANS′K BAY (E); has an indented NW coast, smooth SE coast, rises up to 6 ft/2 m above sea level, consists of sand and shells; covered by halophytic steppe vegetation, aquatic emergent plants; site of nesting and migration resting point of waterfowl, spawning of marine fishes. Site of the Obytichna Spit Landscape Reserve (Ukrainian *Obytichna Kosa Landshaftnyy zakaznyk*) (□ 34 sq mi/88 sq km).

Obytichna Spit Landscape Reserve, UKRAINE: see OBYTICHNA SPIT.

Obytichna zatoka, UKRAINE: see OBYTICHNA BAY.

Obzor (ob-ZOR), city, VARNA oblast, NESEBUR obshtina, BULGARIA, on the BLACK SEA; 42°49′N 27°53′E. Resort area; vineyards. Formerly known as Gyozeken.

Obzor, Bulgaria: see BYALA, VARNA oblast.

Ocala (o-KA-lah), city (□ 39 sq mi/101.4 sq km; 2005 population 49,745), ⊙ MARION county, N central FLORIDA; 29°11′N 82°07′W. Trade and processing center for citrus fruit, vegetables, and other agricultural goods. The surrounding region is known for its thoroughbred horses, cattle, lumber, and phosphates. The cattle-raising industry here has grown significantly since 1970. Tourism is also important to the city; fish and game abound in the many nearby lakes and streams and in Ocala National Forest. Incorporated 1868.

Ocallí (o-kah-YEE), town, LUYA province, AMAZONAS region, N PERU, on E ANDEAN slopes, 32 mi/51 km W of CHACHAPOYAS; 06°09'S 78°10'W. Cacao, coca, coffee, cereals, potatoes.

Ocamonte (o-kah-MON-tai), town ⊙ Ocamonte municipio, SANTANDER department, N central COLOMBIA, 46 mi/74 km S of BUCARAMANGA; 06°20'N 73°07'W. Coffee, corn; livestock.

Oca, Montes de (O-kuh, MON-tais dai), ANDEAN range on COLOMBIA-VENEZUELA border, an offshoot of Cordillera ORIENTAL, extending c.30 mi/48 km NE from Serranía de VALLEDUPAR; rises to c.3,500 ft/1,067 m; 11°07'N 72°25'W.

Oca, Montes de, low range, BURGOS province, N SPAIN, 10 mi/16 km E of BURGOS; extends c.30 mi/48 km NW-SE; rises to c.3,250 ft/991 m.

Ocampo (o-KAHM-po), city and township, GUANAJUATO, central MEXICO, in SIERRA MADRE OCCIDENTAL, 52 mi/84 km NW of DOLORES HIDALGO, on Mexico Highway 51; 21°38'N 101°24'W. Beans, wheat, corn, mescal, livestock.

Ocampo, city and township, TAMAULIPAS, NE MEXICO, in E foothills of SIERRA MADRE ORIENTAL, 33 mi/54 km WNW of CIUDAD MANTE, on Mexico Highway 70; 22°50'N 99°20'W. Cereals, agave; livestock.

Ocampo, town, COAHUILA, N MEXICO, on plateau E of SIERRA MADRE ORIENTAL, 66 mi/107 km WNW of MONCLOVA; 27°20'N 102°24'W. Elevation 3,773 ft/1,150 m. Silver and lead mining. Also known as VILLA OCAMPO.

Ocampo, town, MICHOACÁN, central MEXICO, on railroad, and 10 mi/16.5 km N of ZITÁCUARO. Corn; livestock.

Ocampo (o-KAHM-po), mining settlement, CHIHUAHUA, N MEXICO, in SIERRA MADRE OCCIDENTAL, 92 mi/148 km WSW of Cuantemoc on unpaved road; 28°12'N 108°24'W. Elevation 5,689 ft/1,734 m. Some silver, gold, lead, copper mining. Airfield.

Ocampo, ARGENTINA: see VILLA OCAMPO.

Ocampo, MEXICO: see VILLA OCAMPO.

Ocampo, San Pedro, MEXICO: see MELCHOR OCAMPO.

Ocampo, Villa, MEXICO: see VILLA OCAMPO.

Ocaña (o-KAHN-yah), city, ⊙ Ocaña municipio, NORTE DE SANTANDER department, N COLOMBIA, in W valley of Cordillera ORIENTAL, 65 mi/105 km WNW of CÚCUTA; 08°14'N 73°21'W. Elevation 3,940 ft/1,201 m. Trading and agricultural center (coffee, sugarcane, plantains, rice; livestock); consumer goods. Old colonial town, founded 1573, formerly a cinchona-growing center. A triumphal arch commemorates an independence convention held here 1828.

Ocaña, TOLEDO province, central SPAIN, in CASTILE-LA MANCHA, on railroad, and 8 mi/12.9 km SE of ARANJUEZ; 39°56'N 03°31'W. Cereals, grapes, olives, anise; livestock. Olive-oil pressing, alcohol and liquor distilling, wine making, tanning; manufacturing of ceramics. Site of French victory (November 1809) in Peninsular War.

Ocara (O-kah-RAH), city (2007 population 23,340), E central CEARÁ state, BRAZIL, 24 mi/39 km SSW of PACAJUS; 04°30'S 38°40'W.

Ocauçu (O-kah-oo-koo), town (2007 population 4,180), central SÃO PAULO state, BRAZIL, 16 km/10 mi S of MARÍLIA; 22°26'S 49°56'W.

Occidental, unincorporated town (2000 population 1,272), SONOMA county, W CALIFORNIA, 6 mi/9.7 km W of SEBASTOPOL; 38°24'N 122°57'W. Apples, grapes, nursery products; dairying; poultry, cattle.

Occidental, Pico, VENEZUELA: see LA SILLA DE CARACAS.

Occoquan (UH-kuh-kwahn), town (2006 population 809), PRINCE WILLIAM county, NE VIRGINIA, 20 mi/32 km SW of WASHINGTON D.C., on Occoquan Creek, 3 mi/5 km NW of its mouth, on POTOMAC RIVER; 38°40'N 77°15'W. Occoquan Regional Park (FAIRFAX county) to NW.

Occoquan River (UH-kuh-kwahn), c.20 mi/32 km long, NE VIRGINIA; formed 4 mi/6 km S of MANASSAS by joining of Broad Run and Cedar Run creeks; flows ENE through LAKE JACKSON (Lake Jackson Dam) and Occoquan reservoirs, receives Bull Run creek from NW, turns SE, flows past OCCOQUAN to Occoquan Bay, arm of POTOMAC RIVER, at WOODBRIDGE; 38°41'N 77°29'W. Occoquan Dam 5 mi/8 km above mouth, forming Occoquan Reservoirs.

Océan, department (2001 population 133,062), South province, CAMEROON; ⊙ KRIBI.

Ocean, county (□ 915 sq mi/2,379 sq km; 2006 population 562,335), E NEW JERSEY, on BARNEGAT BAY (E); ⊙ TOMS RIVER; 39°52'N 74°15'W. LONG BEACH ISLAND and ISLAND BEACH peninsula, between Barnegat Bay and the ATLANTIC OCEAN, have many popular summer resorts and fisheries. Inland agricultural area; varied manufacturing. Part of county is in PINE BARRENS region (timber; cranberries, blueberries; here including Lebanon State Forest. Drained by TOMS and METEDECONK rivers and Cedar Creek. Formed 1850. OYSTER CREEK NUCLEAR POWER PLANT, began operation on May 3, 1969 (nation's third-oldest nuclear plant), in FORKED RIVER, 9 mi/14.5 km S of Toms River; uses cooling water from Barnegat Bay, and has a maximum dependable capacity of 605 MWe.

Ocean, township, MONMOUTH county, E central NE NEW JERSEY, 1 mi/1.6 km NW ASBURY PARK; 40°15'N 74°02'W. Incorporated 1849.

Oceana (o-shee-AN-ah), county (□ 1,306 sq mi/3,395.6 sq km; 2006 population 28,639), W MICHIGAN, on LAKE MICHIGAN (W); ⊙ HART; 43°39'N 86°31'W. Drained by WHITE RIVER and short Pentwater River. Fruit (apples, cherries, peaches) and vegetables; cattle, hogs, poultry; dairying. Some manufacturing at Hart, SHELBY, and PENTWATER. Fisheries. Resort. Includes part of Manistee National Forest in E and far N; several lakes, mainly in E and NW; Charles Mears (NW) and Silver Lake (W) state parks, both on Lake Michigan. Organized 1855.

Oceana (o-shee-AN-uh), town (2006 population 1,457), WYOMING county, S WEST VIRGINIA, on Clear Fork, 9 mi/14.5 km NW of PINEVILLE; 37°41'N 81°37'W. Agriculture (corn, potatoes); cattle. River D. Bailey Lake reservoir and Wildlife Management Area to SW.

Ocean Beach, suburban section of SAN DIEGO city, SAN DIEGO county, S CALIFORNIA, 6 mi/9.7 km WNW of downtown San Diego, on PACIFIC OCEAN. MISSION BAY and mouth of SAN DIEGO RIVER to N. POINT LOMA Naval Reservation and CABRILLO National Monument to S. Point Loma College and National University are in area. San Diego International Airport to E. Residential and beach resort area.

Ocean Beach, NEW YORK: see FIRE ISLAND.

Ocean Bluff, MASSACHUSETTS: see MARSHFIELD.

Ocean Cape, SE ALASKA, on Gulf of ALASKA, on E shore of entrance to YAKUTAT BAY; 59°32'N 139°52'W. Nearby are a Native fishing area and an air strip built in World War II.

Ocean City, city (2006 population 15,124), CAPE MAY county, SE NEW JERSEY, resort on the ATLANTIC coast, on an 8-mi/13-km-long island between the Atlantic Ocean and Great Egg Harbor Bay; linked to the mainland by a 2-mi/3.2-km causeway; 39°16'N 74°35'W. Its boardwalk, amusement rides, and proximity to other New Jersey beaches make it a popular summer vacation spot. Incorporated 1897.

Ocean City (o-SHUHN), unincorporated town (□ 2 sq mi/5.2 sq km; 2000 population 5,594), OKALOOSA county, NW FLORIDA, directly N of FORT WALTON BEACH across village of CINCO BAYOU; 30°26'N 86°36'W.

Ocean City, town (2000 population 7,173), WORCESTER county, SE MARYLAND, 28 mi/45 km E of SALISBURY, and extends 10 mi/16 km along a barrier beach; 38°23'N 75°02'W. Largest ocean resort in the state. Originally patented in the 1870s as "The Ladies Resort to the Sea." Tourism is its economic mainstay and the population greatly increases during the summer. Storm tides here, particularly those of March 6–8, 1962, were the most devastating on the East Coast. A major center for deep-sea fishing. The Coast Guard Station here, established on August 4, 1790, is one of the oldest in the country. The First Maryland-DELAWARE Boundary Marker, at the foot of the Fenwick Island Lighthouse (no longer in operation), was erected in 1751. Incorporated 1880.

Ocean Drive Beach, unincorporated village, HORRY county, E SOUTH CAROLINA, 20 mi/32 km NE of MYRTLE BEACH, on ATLANTIC OCEAN, in Grand Strand Beach resort area.

Ocean Falls (O-shuhn FAHLZ), community, unincorporated town, W BRITISH COLUMBIA, W Canada, on Cousins Inlet of the PACIFIC OCEAN, 300 mi/483 km NW of VANCOUVER, in Central Coast regional district; 52°22'N 127°41'W. Former paper-milling and -shipping center; hydroelectric power. Heavy annual rainfall.

Ocean Falls, Canada: see CENTRAL COAST regional district.

Ocean Gate, borough (2006 population 2,130), OCEAN county, E NEW JERSEY, on TOMS RIVER, and 4 mi/6.4 km SE of TOMS RIVER; 39°55'N 74°08'W. In fishing and resort area.

Ocean Grove, town and resort (2000 population 4,256), part of NEPTUNE township (1990 population 4,818; 2000 population 4,256), MONMOUTH county, E NEW JERSEY, on the ATLANTIC coast, just S of ASBURY PARK. Founded 1869 by Methodist camp meeting association as a tent city for summer camp meetings, with an auditorium (seating 7,000) designed according to biblical rules. Noted for its Victorian architecture. Blue laws kept cars off the roads, and clothes off the lines on Sundays. Roads in and out of here were roped off all day to keep the traffic out. The New Jersey Supreme Court struck down the laws in 1979, thereby opening the streets to Sunday traffic.

Ocean Grove, MASSACHUSETTS: see SWANSEA.

Oceania or **Oceanica**, collective name for the seas and islands of the PACIFIC OCEAN, usually excluding such nontropical areas as the RYUKYU and ALEUTIAN islands and JAPAN, as well as TAIWAN, INDONESIA, and the PHILIPPINES, whose populations are more closely related to those of mainland ASIA. Generally considered synonymous with either the South Sea Islands, divided ethnologically into MELANESIA, MICRONESIA, and POLYNESIA; or with AUSTRALIA, NEW ZEALAND, and the PACIFIC ISLANDS.

Ocean Island, HAWAII: see KURE ATOLL.

Ocean Isle Beach (O-shuhn EIL BEECH), village (□ 4 sq mi/10.4 sq km; 2006 population 522), BRUNSWICK county, SE NORTH CAROLINA, 40 mi/64 km SW of WILMINGTON, on ATLANTIC OCEAN; 33°53'N 78°25'W. Tubbs Inlet to W; INTRACOASTAL WATERWAY canal passes to N. Beach resort area. Incorporated 1959.

Oceanlake, unincorporated town, LINCOLN county, W OREGON, on PACIFIC OCEAN, 2 mi/3.2 km N of LINCOLN CITY. Resort area. Roads End Beach State Wayside to N. Devils Lake to SE.

Oceano, unincorporated town (2000 population 7,260), SAN LUIS OBISPO county, SW CALIFORNIA, on PACIFIC OCEAN, 12 mi/19 km S of SAN LUIS OBISPO; 35°06'N 120°37'W. Flowers, nursery stock; vegetables, apples, strawberries, avocados, grain; cattle. Beach resort; tourism. PISMO STATE BEACH is here.

Ocean Park, unincorporated town (2000 population 1,459), PACIFIC county, SW WASHINGTON, 20 mi/32 km SW of RAYMOND, on North Beach Peninsula, on PACIFIC OCEAN; 46°30'N 124°02'W. Beach resort; tourism. Manufacturing (processed seafood). Willapa Bay 1 mi/1.6 km E. Leadbetter State Park, near entrance to Willapa Bay, to N. Parts of Willapa National Wildlife Refuge to N and SE.

Ocean Pond, lake, E NEWFOUNDLAND AND LABRADOR, E Canada, 30 mi/48 km SW of BONAVISTA; 4 mi/6 km long, 2 mi/3 km wide; 48°19′N 53°39′W. Drains into TRINITY BAY.

Oceanport, borough (2006 population 5,751), MONMOUTH county, E NEW JERSEY, on SHREWSBURY RIVER estuary (head of navigation), and 3 mi/4.8 km NW of LONG BRANCH; 40°19′N 74°01′W. Largely residential. Computer manufacturing. Fort MONMOUTH and Monmouth Park Racetrack nearby. Incorporated 1920.

Ocean Ridge, town (□ 1 sq mi/2.6 sq km; 2005 population 1,699), PALM BEACH county, SE FLORIDA, 1 mi/1.6 km E of BOYNTON BEACH, on the ATLANTIC OCEAN; 26°31′N 80°02′W.

Ocean Shores, town (2006 population 4,658), GRAYS HARBOR county, W WASHINGTON, 14 mi/23 km W of ABERDEEN, on peninsula between GRAYS HARBOR and PACIFIC OCEAN; 46°58′N 124°09′W. Entrance to harbor and Point Brown to S. Ferry to Westport at S side of harbor. Ocean City State Park to N. Beach resort.

Oceanside, city (2000 population 161,029), SAN DIEGO county, S CALIFORNIA, suburb 36 mi/58 km NNW of downtown SAN DIEGO, on the GULF OF SANTA CATALINA; 33°14′N 117°19′W. Railroad junction. Commercial and trading center for an inland farm area and for Camp Pendleton Marine Corps Base to N. Mainly residential, the city produces rubber goods, electronic components, hardware, motors, and clothing; also has a large citrus, flower, and bulb industry. Deep-sea fishing and tourism, owing to its seaside location, are also important. Oceanside is one of the fastest-growing U.S. cities, marked by a population increase of over 67% between 1980 and 1990. Seat of Mira Costa College (two year). Nearby is SAN LUIS REY Mission (founded 1798). CARLSBAD (S) and San Onofre (NW) state beaches nearby. Incorporated 1888.

Oceanside, unincorporated town (□ 5 sq mi/13 sq km; 2000 population 32,733), NASSAU county, SE NEW YORK, on the S shore of LONG ISLAND; 40°37′N 73°38′W.

Ocean Springs, city (2000 population 17,225), JACKSON county, extreme SE MISSISSIPPI, 3 mi/4.8 km ENE of BILOXI, at entrance to BILOXI BAY (bridged), on Gulf of MEXICO; 30°24′N 88°47′W. Manufacturing (electronics products, plastics, optical goods, pottery, apparel, consumer goods, boats, wood products; printing and publishing, shrimp processing); fish, shrimp. Seat of marine research laboratory (1948). Walter Anderson Museum of Art; The Doll House, doll museum; Fort Maurepas (replica 1 mi/1.6 km from original 1699 site); Mississippi Sandhill Crane National Wildlife Refuge to NE; GULF ISLANDS National Seashore Visitors Center, on mainland, to E; De Soto National Forest to N; Gulf Marine State Park is here; Deer Island to S. Town established on site of Old Biloxi, founded in 1699 by Iberville as first European settlement in lower Mississippi valley.

Ocean State: see RHODE ISLAND.

Ocean View, village (2000 population 1,006), SUSSEX county, SE DELAWARE, 16 mi/26 km SSE of LEWES, on Assawoman Canal; 38°32′N 75°05′W. Elevation 13 ft/3 m. In poultry-raising region. Atlantic Coast 2 mi/3.2 km to E; INDIAN RIVER BAY to N; Assawoman Wildlife Area to S. Gateway to BETHANY BEACH resort area.

Oc Eo (OK AI-O), historic site, KIEN GIANG province, S VIETNAM, just outside what is now RACH GIA; 10°00′N 105°06′E. Ancient trade city of FUNAN empire (1st–6th century C.E.).

O. C. Fisher Lake (FISH-uhr), reservoir, TOM GREEN county, W central TEXAS, on North CONCHO RIVER, 3 mi/4.8 km W of downtown SAN ANGELO; 5 mi/8 km long; 31°30′N 100°28′W. Maximum capacity 696,300 acre-ft. Formed by San Angelo Dam (122 ft/37 m

high), built (1952) by the Army Corps of Engineers for water supply and flood control. City parks on S and SW shores.

Ochagavía (o-chah-gah-VEE-ah), town, NAVARRE province, N SPAIN, 30 mi/48 km ENE of PAMPLONA; 42°55′N 01°05′W. Sawmilling; wheat, potatoes; cattle, sheep.

Ochakiv (o-CHAH-kif) (Russian *Ochakov*), city, S MYKOLAYIV oblast, UKRAINE, on the DNIEPER (Ukrainian *Dnipro*) River estuary opening to the BLACK SEA, 30 mi/48 km S of MYKOLAYIV; 46°37′N 13°33′E. Raion center; center of an agricultural district (flour milling, juice making; wineries), with sewing factories; seaport with fishing industries. In the 7th and 6th centuries B.C.E., there were several Greek colonies in the area, and Ochakiv is on the site of the ancient Greek city of Alektor. In the first half of the 14th century, a Lithuanian grand duke, Vytautas, built a fortress here called Dashiv. Passed to the Crimean Tatar control in 1492, and the Crimean khan built a fortress called Kara-Kermen here. When the Turks took control of it, they renamed it Achi-Kale (origin of the modern name). In the 16th and 17th centuries, the Ukrainian Cossacks attacked the Turks here. The city fell to the Russians (1788) during the Russo-Turkish War of 1787–1792. In the Crimean War, it was occupied (1855) by the allies. Near the site of the ancient Greek colony of Olbia.

Ochakov, UKRAINE: see OCHAKIV.

Ochandiano (o-chahn-DYAH-no), village, VIZCAYA province, N SPAIN, 20 mi/32 km SE of BILBAO; 42°55′N 01°05′W. Destroyed (April 1937) Spanish civil war fighting.

Oche, Greece: see OKHI.

Ocheda Lake (o-CHEE-duh), NOBLES county, SW MINNESOTA, 3 mi/4.8 km S of WORTHINGTON; 7 mi/11.3 km long, 5 mi/8 km wide; 43°32′N 95°38′W. Fed from N by short stream from OKABENA LAKE. Drains S into OCHEYEDAN RIVER. Lake is divided into three distinct sections, connected by narrow channels.

Ochelata (och-uh-LAIT-uh), village (2006 population 502), WASHINGTON county, NE OKLAHOMA, 9 mi/14.5 km S of BARTLESVILLE, near CANEY RIVER; 36°36′N 95°58′W. In farm and ranch area.

Ochemchiri (o-chem-CHEE-ree) or **Ochamchira**, city, S ABKHAZ Autonomous Republic, GEORGIA, port on BLACK SEA, 30 mi/48 km SE of SUKHUMI. Administrative center. Railroad center and station. Oil-extraction plant, cannery, tea factories, industrial combine; poultry hatchery. Citrus fruit. Port for TKVARCHELI coalfields (linked by railroad spur, 78 mi/126 km long). Formerly spelled Ochemchiry.

Ocher (O-chyer), city (2006 population 15,340), W PERM oblast, W URAL Mountains, E European Russia, on the Ocher River (short right tributary of the KAMA River), on highway junction and railroad, 78 mi/126 km WSW of PERM; 57°53′N 54°42′E. Elevation 390 ft/118 m. Agricultural machinery manufacturing, machinery and metal industries (pipe-laying machines); food processing. Until 1929, called Ocherskiy Zavod.

Ocheretino, UKRAINE: see OCHERETYNE.

Ocheretyne (o-che-RE-ti-ne) (Russian *Ocheretino*), town (2004 population 3,600), central DONETS′K oblast, UKRAINE, in the DONBAS, on road and railroad, 19 mi/30 km NNW of DONETS′K city center, and 9 mi/15 km NW of AVDIYIVKA; 48°14′N 37°37′E. Elevation 784 ft/238 m. Building materials manufacturing, especially for the S leg of the DNIEPER-DONBAS CANAL; grain-collection point. Established in the 1880s, in conjunction with railroad construction; town since 1957.

Ocherskaya, RUSSIA: see VERESHCHAGINO.

Ocheyedan (o-CHEE-duhn), town (2000 population 536), OSCEOLA county, NW IOWA, near OCHEYEDAN RIVER, 11 mi/18 km E of SIBLEY; 43°25′N 95°32′W. Makes popcorn. Nearby is Ocheyedan Mound, highest point (1,675 ft/511 m) in Iowa.

Ocheyedan River, 58 mi/93 km long, MINNESOTA and IOWA; rises in NOBLES county, SW Minnesota, in OCHEDA LAKE, 3 mi/4.8 km S of WORTHINGTON; flows S through Lake Bella reservoir then SE into NW Iowa to Little Sioux River at SPENCER.

Ochi (O-chee), town, Okawa county, KAGAWA prefecture, NE SHIKOKU, W JAPAN, 19 mi/30 km S of TAKAMATSU; 34°14′N 134°20′E.

Ochi, town, Takaoka county, KOCHI prefecture, S SHIKOKU, W JAPAN, on NIYODO RIVER, and 16 mi/25 km W of KOCHI; 33°31′N 133°15′E.

Ochi, town, East Matsuura county, SAGA prefecture, NW KYUSHU, SW JAPAN, on N HIZEN PENINSULA, 21 mi/34 km NE of SASEBO and 19 mi/30 km N of SAGA; 33°20′N 130°00′E.

Ochi, town, Ochi county, SHIMANE prefecture, SW HONSHU, W JAPAN, 38 mi/62 km S of MATSUE; 35°04′N 132°35′E.

Ochiai (O-chee-ah-ee), town, Maniwa county, OKAYAMA prefecture, SW HONSHU, W JAPAN, 28 mi/45 km N of OKAYAMA; 35°00′N 133°45′E. Sometimes spelled Otiai.

Ochiai, RUSSIA: see DOLINSK.

Ochil Hills (OK-hil), range of hills of volcanic origin in PERTH AND KINROSS and Clackmannanshire, central Scotland, extends 25 mi/40 km ENE-WSW between STIRLING and the Firth of Tay; 56°14′N 03°40′W. Mineral deposits; silver formerly mined here. Highest points are Ben Cleuch (2,363 ft/720 m), 3 mi/4.8 km N of TILLICOULTRY, and King's Seat (2,111 ft/643 m), 2 mi/3.2 km NNE of Tillicoultry. DEVON RIVER rises here. Formerly in Tayside and Central, abolished 1996.

Ochiltree (OK-uhl-tree), county (□ 918 sq mi/2,386.8 sq km; 2006 population 9,550), extreme N TEXAS, on OKLAHOMA (N) state line; ⊙ PERRYTON; 36°16′N 100°48′W. On high plains of the PANHANDLE; elevation c.2,600 ft/792 m–3,100 ft/945 m. Drained by WOLF and Kiowa creeks, tributaries of the North CANADIAN RIVER. A leading wheat-producing county of U.S.; also cattle ranching; sheep, hogs, horses; cotton, wheat, sorghum; oil and gas, gypsum, sand and gravel, clay. Formed 1876.

Ochiltree (OK-hil-tree), agricultural village (2001 population 693), East Ayrshire, S Scotland, on LUGAR WATER, and 4 mi/6.4 km W of CUMNOCK; 55°27′N 04°24′W. Formerly in Strathclyde, abolished 1996.

Ochkhamuri (och-hah-MOO-ree), urban settlement, KOBULETI region, GEORGIA; 41°51′N 41°50′E. Railroad station. Tung oil manufacturing; tea factory.

Ochlocknee (ok-LAHK-nee), town (2000 population 605), THOMAS county, S GEORGIA, 10 mi/16 km NNW of THOMASVILLE; 30°59′N 84°03′W. Also spelled Ochlockonee.

Ochlockonee River (ok-LAHK-uh-nee), c.150 mi/241 km long, in GEORGIA and FLORIDA; rises SW of SYLVESTER in SW GEORGIA; flows generally S, across NW Florida, into W end of APALACHEE BAY of the Gulf of Mexico, 14 mi/23 km SSW of CRAWFORDVILLE; 31°27′N 83°55′W. W of TALLAHASSEE, a dam forms Lake TALQUIN (c.14 mi/23 km long, 1 mi/1.6 km–4 mi/6.4 km wide).

Ochoco Creek (O-chuh-ko), c.40 mi/64 km long, OREGON; rises in mt. region of N CROOK county, central Oregon; flows W to CROOKED RIVER at PRINEVILLE. Ochoco Reservoir (4 mi/6.4 km long), 8 mi/12.9 km E of Prineville, is formed by small dam in upper course. Used for irrigation.

Ochomogo Pass (o-cho-MO-go) (5,138 ft/1,566 m), on CONTINENTAL DIVIDE, in central COSTA RICA, connecting SAN JOSÉ and CARTAGO (GUARCO VALLEY) sections of central plateau. Site of the deciding battle of the revolution of 1948 (which established Costa Rica's reform government). Used by railroad between San José and LIMÓN and by INTER-AMERICAN HIGHWAY.

Ocho Rios, town, SAINT ANN parish, NE JAMAICA, on the CARIBBEAN SEA; 18°25'N 77°07'W. Major tourist center noted for its rivers and waterfalls, as well as a commercial port that exports mainly bauxite, logwood, and sugar.

Ochozská Cave, CZECH REPUBLIC: see MOKRA-HORAKOV.

Ochre River (O-kuhr), unincorporated village, W MANITOBA, W central Canada, on Ochre River, near DAUPHIN LAKE, 14 mi/23 km ESE of DAUPHIN, and in Ochre River rural municipality; 51°03'N 99°47'W. Grain; mixed farming.

Ochre River (O-kuhr), rural municipality (□ 207 sq mi/538.2 sq km; 2001 population 952), S MANITOBA, W central Canada, 20 mi/31 km from DAUPHIN, and S of Dauphin Lake; 51°00'N 99°40'W. Agriculture (wheat, barley, canola; livestock). Includes the communities of OCHRE RIVER village, Makinak.

Ochrid, MACEDONIA: see OHRID, LAKE.

Ochrida, MACEDONIA: see OHRID, town, and OHRID, LAKE.

Ochsenfurt (OK-sen-foort), town, Lower FRANCONIA, W BAVARIA, GERMANY, on the Main River, and 10 mi/16 km SE of Würzburg; 49°40'N 10°04'E. Manufacturing (textiles, chemicals, food, wood products, metal products), brewing, sugar refining; dairy. Partly surrounded by 14th-century walls. Has 14th-century church; late-Gothic chapel; town hall (1497–1513). Former monastery (founded c.1138).

Ochsenhausen (OK-sen-HOU-suhn), town, DANUBE-Iller, SE BADEN-Württemberg, S GERMANY, 24 mi/38 km S of ULM, on the Rottum River; 48°04'N 09°58'E. Manufacturing (synthetic fiber, furniture). Grew around a Benedictine monastery; first mentioned 1137; secularized 1803; passed 1825 to Württemberg; chartered 1950. Has 14th-century Gothic church (rebuilt in baroque style 1725–1732), monastery buildings (1583–1791).

Ochsenkopf (OK-sen-kopf), peak (3,356 ft/1,023 m) of the FICHTELGEBIRGE, BAVARIA, GERMANY, 8 mi/12.9 km W of WUNSIEDEL; 50°02'N 11°55'E. Separated from neighboring (NE) SCHNEEBERG by a moor.

Ochtendung (OKHT-en-dung), village, RHINELAND-PALATINATE, W GERMANY, 9 mi/14.5 km W of KOBLENZ, near the Nette-Bach; 50°21'N 07°24'E. In wine- and fruit-growing region.

Ochtrup (OKH-trup), town, WESTPHALIA-LIPPE, North Rhine–WESTPHALIA, NW GERMANY, 7.5 mi/12.1 km NW of STEINFURT; 52°12'N 07°12'E. Cattle. Moated castle from 14th century. Chartered 1593.

Ocilla (o-SIL-uh), town (2000 population 3,270; ⊙ IRWIN county, S central GEORGIA, 8 mi/12.9 km S of FITZGERALD; 31°36'N 83°15'W. Agricultural trade center; manufacturing includes chemicals, roasted nuts, clothing. Incorporated 1897.

Ockbrook (AHK-bruk), village (2001 population 7,331), SE DERBYSHIRE, central ENGLAND, 4 mi/6.4 km E of DERBY; 52°55'N 01°22'W. Site of 18th-century Moravian settlement and school. Has church with 12th-century tower. Evidence of Neolithic and Iron Age activity. Part of the kingdom of MERCIA.

Ockelbo (OK-el-BOO), town, GÄVLEBORG county, E SWEDEN, 20 mi/32 km NW of GÄVLE; 60°54'N 16°43'E. Railroad junction.

Öckerö (UHK-er-UH), fishing town, GÖTEBORG OCH BOHUS county, SW SWEDEN, on islet of same name in SKAGERRAK, 12 mi/19 km W of GÖTEBORG; 57°43'N 11°39'E. Has eighteenth-century church.

Ocker River, GERMANY: see OKER RIVER.

Ocland (OK-lahnd), Hungarian *Oklánd*, village, HARGHITA county, central ROMANIA, 11 mi/18 km SE of ODORHEIU-SECUIESC; 46°10'N 25°25'E. Manufacturing of bricks and tiles. Under Hungarian administration, 1940–1945.

Ocmulgee (ok-MUHL-gee), river, c.255 mi/410 km long, central GEORGIA; formed SE of ATLANTA, NW Georgia, by the confluence of the YELLOW, SOUTH,

and ALCOVY rivers; 33°19'N 83°50'W; flows SE past MACON to join the OCONEE River and form the ALTAMAHA RIVER near LUMBER CITY. Major river of middle Georgia. The river passes the remains of prehistoric Native American villages, preserved in OCMULGEE NATIONAL MONUMENT in Macon near Fort Hawkins. Macon's cotton market flourished due to its location at the head of the river's navigation.

Ocmulgee National Monument (ok-MUHL-gee), (□ 1 sq mi/1.6 sq km), central GEORGIA, on OCMULGEE River, just E of MACON; 32°48'N 83°35'W. Mounds, pyramids, artifacts, and village remains of early Mississippian civilization; reconstructed earth lodge on one of eight grass-covered mounds. Authorized 1934; established 1936.

Ocna, UKRAINE: see VIKNO.

Ocna-Dejului, ROMANIA: see DEJ.

Ocna Mureş (OK-nah MOO-resh), Hungarian *Marosújvár*, town, ALBA county, central ROMANIA, in TRANSYLVANIA, on MUREŞ RIVER, and 24 mi/39 km NNE of ALBA IULIA; 46°23'N 23°51'E. Salt-production center and health resort with saline springs and baths; extensive vineyards and tobacco growing in vicinity.

Ocna Sibiului (OK-nah see-BEE-woo-loo-i), German *Salzburg*, Hungarian *Vízakna*, town, SIBIU county, central ROMANIA, on railroad, and 7 mi/11.3 km NW of SIBIU; 45°53'N 24°03'E. Health resort with saline lakes; saltworking, manufacturing of cutlery and steel products. Has two old churches. Known as spa in Roman times. Also spelled Ocna-Sibiu.

Ocna Şugatag (OK-nah soo-gah-TAHG), Hungarian *Aknasugatag*, village, MARAMUREŞ county, NW ROMANIA, on W slopes of the CARPATHIANS, 20 mi/32 km NE of BAIA MARE; 47°47'N 23°56'E. Salt-mining center; also a summer resort (elev. 1,608 ft/490 m) with alkaline springs. Under Hungarian administration, 1940–1945.

Ocnele Mari (ok-NE-le MAHR), town, VÎLCEA county, S central ROMANIA, in WALACHIA, on railroad, and 3 mi/4.8 km SW of RÎMNICU VILCEA; 45°05'N 24°19'E. Major salt-production center and health resort on salt lake. Has 16th-century church. Tîrgul de la Ocne, Ocna cea Mare, Ocna, Ocnele de la Rîmnic bathing resorts just W.

Ocniţa (OK-nee-tsah), town (2004 population 9,300), N MOLDOVA, 16 mi/26 km WSW of Mogilev-Podolski (Ukraine), near Ukrainian border; 48°22'N 27°25'E. Flour and oilseed milling. Phosphorite deposits nearby. OCNITA railroad junction is 2 mi/3.2 km NE. Formerly spelled Oknitsa.

Ocoa Bay (o-KO-ah), inlet of the CARIBBEAN SEA, S DOMINICAN REPUBLIC, 45 mi/72 km W of SANTO DOMINGO; 18°22'N 70°39'W.

Ocobamba (o-ko-BAHM-bah), town, LA CONVENCIÓN province, CUSCO region, S central PERU, on affluent of URUBAMBA RIVER, and 35 mi/56 km NNW of URUBAMBA; 13°30'S 73°34'W. Potatoes, cereals.

Ocobaya (o-ko-BAH-yah), canton, SUD YUNGAS province, LA PAZ department, W BOLIVIA, 32 mi/52 km W of LA PAZ, and 19 mi/30 km SW of CHULUMANI; 16°25'S 67°31'W. Elevation 5,686 ft/1,733 m. Tin-bearing lode; tungsten mining at Mina Reconquistada, Mina CHOJLLA, Mina Bolsa Negra; clay and gypsum deposits. Agriculture (potatoes, yucca, bananas, rye; important coffee center); cattle.

Ococingo, MEXICO: see OCOSINGO.

Ocoee (o-KO-ee), city (□ 14 sq mi/36.4 sq km; 2005 population 29,849), ORANGE county, central FLORIDA, 10 mi/16 km W of ORLANDO, near LAKE APOPKA; 28°34'N 81°31'W. Ships citrus fruit and general produce.

Ocoee River (o-KO-ee), in N GEORGIA and SE TENNESSEE, formed in BLUE RIDGE MOUNTAINS 9 mi/14 km SE of BLUE RIDGE city, Georgia, by confluence of two headstreams; flows c.80 mi/129 km NW through FANNIN county (here called TOCCOA RIVER), into Tennessee, past Copper Hill, through POLK county,

and N to HIWASSEE RIVER 2 mi/3 km N of BENTON; 90 mi/145 km long; 35°12'N 84°30'W. Drains forest area in both states. Drainage area is 639 sq mi/1,655 sq km. Dammed in upper course by BLUE RIDGE DAM, forming Lake TOCCOA. Three other dams are in the Tennessee Ocoee. Among the three dams, Ocoee # 1 (135 ft/41 m high, 840 ft/256 m long; completed 1912) is in lower course, 6 mi/10 km S of Benton; forms Lake Ocoee (7 mi/11 km long, 0.5 mi/0.8 km wide; also known as PARKSVILLE RESERVOIR). Ocoee # 2 (30 ft/9 m high, 450 ft/137 m long; completed 1913) is further upstream, 6 mi/10 km NW of COPPERHILL. Ocoee # 3 (110 ft/34 m high, 612 ft/187 m long; completed 1943) is at Copperhill, just N of Georgia line. Major whitewater recreation area and scheduled site of the 1996 Summer Olympics kayaking events.

Ocoña (o-KON-yah), town, CAMANÁ province, AREQUIPA region, S PERU; minor PACIFIC port, at mouth of OCOÑA RIVER, on PAN-AMERICAN HIGHWAY, and 100 mi/161 km W of AREQUIPA; 16°26'S 73°07'W. Cotton, sugarcane, olives, grapes, cereals.

Ocoña River (o-KON-yah), c.110 mi/177 km long, AREQUIPA region, S PERU; rises as COTAHUASI RIVER at S foot of Cordillera de Huanzo on APURÍMAC region border; flows S to the PACIFIC at OCOÑA (16°34'S 73°07'W). Used for irrigation.

Oconee (o-KO-nee), county (□ 186 sq mi/483.6 sq km; 2006 population 30,858), NE central GEORGIA; ⊙ WATKINSVILLE; 33°50'N 83°26'W. Piedmont area drained by APALACHEE and OCONEE rivers. Manufacturing includes furniture, fixtures, wholesale goods; agriculture includes cotton, soybeans, fruit; cattle, poultry, hogs. Formed 1875.

Oconee, county (□ 673 sq mi/1,749.8 sq km; 2006 population 70,567), extreme NW SOUTH CAROLINA; ⊙ WALHALLA; 34°45'N 83°04'W. Bounded NW by CHATTOOGA RIVER, SW by TUGALOO RIVER, E by KEOWEE and SENECA rivers. Includes much of South Carolina part of the BLUE RIDGE MOUNTAINS; summer-resort area, with part of SUMTER National Forest. Manufacturing includes granite and sand. Agricultural products include chickens, eggs, hogs, cattle, dairying, corn, wheat, soybeans, sorghum, hay, apples. Formed 1868. Nuclear power plants OCONEE 1 (initial criticality April 19, 1973), Oconee 2 (initial criticality November 11, 1973), and Oconee 3 (initial criticality September 5, 1974) are 30 mi/48 km W of GREENVILLE; they use cooling water from Lake KEOWEE, and each has a max. dependable capacity of 846 MWe.

Oconee, village (2000 population 202), SHELBY county, central ILLINOIS, 19 mi/31 km SW of SHELBYVILLE; 39°17'N 89°06'W. In agricultural area.

Oconee (o-KO-nee), river, 282 mi/454 km long; rising in the APPALACHIAN MOUNTAINS, N GEORGIA; flowing SE to the OCMULGEE River to form the ALTAMAHA RIVER; 34°23'N 83°39'W. SINCLAIR Dam (completed 1953) and Furman Shoals Dam (completed 1953) are on the Oconee.

Oconee 1, 2, and 3 Nuclear Power Plants, SOUTH CAROLINA: see OCONEE county.

Oconee, Lake (o-KO-nee), reservoir (□ 30 sq mi/78 sq km), PUTNAM, MORGAN, and GREENE counties, N central GEORGIA, on OCONEE River, 18 mi/29 km NNE of MILLEDGEVILLE; 33°21'N 83°09'W. Maximum capacity 370,000 acre-ft. Formed by WALLACE DAM (117 ft/36 m high) built (1980) for power generation. Oconee National Forest at N end.

Oconee State Park, SOUTH CAROLINA: see WALHALLA.

Ocongate (o-kon-GAH-tai), town, S QUISPICANCHI province, CUSCO region, central PERU, in Cordillera de VILCANOTA, 40 mi/64 km E of CUSCO, on road from URCOS to PUERTO MALDONADO; 13°37'S 71°23'W. Elevation 12,349 ft/3,763 m. Potatoes, cereals, alfalfa.

Oconi (o-KO-nee), canton, MANUEL MARÍA CABALLERO province, SANTA CRUZ department, E central BOLIVIA, 12 mi/20 km SE of COMARAPA, on the COCHABAMBA

Area is shown by the symbol □, and capital city or county seat by ⊙.

border; 17°56′S 64°41′W. Elevation 5,955 ft/1,815 m. Undrilled gas deposits in area. Clay and limestone deposits. Agriculture (potatoes, yucca, bananas, corn, wheat, sweet potatoes, peanuts, sugarcane); cattle.

O'Connor (o-KAH-nuhr), township (□ 42 sq mi/109.2 sq km; 2001 population 724), NW ONTARIO, E central Canada, 20 mi/32 km from city of THUNDER BAY; 48°21′N 89°42′W. Organized 1907.

Oconomowoc (o-KAW-nuh-mo-wawk), city (2006 population 14,167), WAUKESHA county, SE WISCONSIN, on Oconomowoc River (tributary of ROCK RIVER), between small Fowler and LA BELLE lakes, and 13 mi/ 21 km ESE of WATERTOWN; 43°06′N 88°30′W. Manufacturing (dairying and baking products, wheelchairs, labels, electronics, fabricated metal products; food processing); resort with mineral springs. Annual winter-sports carnival. Incorporated 1875.

Oconto (o-KAWN-to), county (□ 1,149 sq mi/2,987.4 sq km; 2006 population 37,958), NE WISCONSIN; ☉ OCONTO; 44°59′N 88°13′W. Primarily a dairying and lumbering area; agriculture (barley, oats, wheat, brans, hay; cattle, hogs, poultry). Bounded E by GREEN BAY; drained by OCONTO RIVER. Menominee Indian Reservation W of county; large section of Nicolet National Forest in NW; Copper Culture Mound State Park in E. Formed 1851.

Oconto (o-KAWN-to), town (2006 population 4,715), ☉ OCONTO county, NE WISCONSIN, at mouth of OCONTO RIVER, on W shore of GREEN BAY, 28 mi/45 km NNE of GREEN BAY city; 44°53′N 87°52′W. Railroad junction. Commercial center for lumbering and dairying area. Manufacturing (wood products, textiles, beer, food processing, pleasure boats, machinery); fisheries. Father Allouez founded a mission here in 1669; the first Christian Science church was built here in 1886. Was important lumbering center. Copper Culture Mound State Park to SW. Incorporated 1869.

Oconto, village (2006 population 136), CUSTER county, central NEBRASKA, 20 mi/32 km SSW of BROKEN BOW, and on WOOD RIVER; 41°08′N 99°45′W.

Oconto Falls (o-KAWN-to), town (2006 population 2,855), OCONTO county, NE WISCONSIN, on OCONTO RIVER, and 26 mi/42 km NNW of GREEN BAY city; 44°52′N 88°09′W. Railroad terminus. Manufacturing (paper milling, feeds and fertilizer, magnetic printing cylinders, paper products). Incorporated 1919.

Oconto River (o-KAWN-to), 87 mi/140 km long, NE WISCONSIN; rises in several small lakes in SE FOREST county; flows S and E, past OCONTO FALLS, to GREEN BAY at OCONTO city.

Ocós (o-KOS), town, SAN MARCOS department, SW GUATEMALA, minor port on PACIFIC OCEAN coast, 25 mi/40 km SW of COATEPEQUE, at mouth of NARANJO RIVER; 14°30′N 92°11′W. Railroad terminus. Saltworks nearby. Flourished (19th century) in indigo and cochineal trade; declined after rise of PUERTO BARRIOS. Has given name to pre-Olmec Mesoamerican Ocós culture, which occupied the area c.1,500 B.C.E.

Ocosingo (o-ko-SEEN-go), city and township, CHIAPAS, S MEXICO, on N plateau of SIERRA MADRE, 36 mi/ 58 km NE of SAN CRISTÓBAL DE LAS CASAS; 16°54′N 92°05′W. Corn, rice, mangoes, oranges. Ancient Maya ruins of great interest and artistic merit nearby. Sometimes spelled OCOCINGO. Population is largely Tzectal-speaking Maya Indian.

Ocosito River (o-ko-SEE-to), QUEZALTENANGO and RETALHULEU departments, GUATEMALA; 14°30′N 92°11′W. Drains SIETE OREJAS volcano; flows across coastal plain, forms part of boundary between Quezaltenango and Retalhuleu departments on coastal plain. Sometimes called Tilapa River in its lower course.

Ocotal (o-ko-TAHL), city and township (2005 population 34,190), ☉ NUEVA SEGOVIA department, NW NICARAGUA, on branch of INTER-AMERICAN HIGHWAY, and 105 mi/169 km NNW of MANAGUA, near

COCO RIVER; 13°38′N 86°29′W. Commercial center; manufacturing (shoes, furniture, beverages), agriculture processing (coffee, sugarcane, tobacco).

Ocotepec (o-KO-te-pek), town, CHIAPAS, S MEXICO, in N spur of SIERRA MADRE, 30 mi/48 km N of TUXTLA GUTIÉRREZ; 17°13′N 93°09′W. Elevation 4,629 ft/1,411 m. Cereals, sugarcane, fruit. No paved roads. A Zoque Indian town.

Ocotepec, town, PUEBLA, central MEXICO, 7 mi/11 km ENE of LIBRES 1 mi/2 km E of Mexico Highway 129. Corn, maguey.

Ocotepeque (o-ko-te-PE-kai), department (□ 859 sq mi/2,233.4 sq km; 2001 population 108,029), W HONDURAS, on EL SALVADOR and GUATEMALA borders; ☉ NUEVA OCOTEPEQUE; 14°30′N 89°00′W. Largely mountainous (CORDILLERA DEL MERENDÓN); drained by LEMPA RIVER (W) and upper JICATUYO RIVER (E). Agriculture (wheat, corn, coffee, tobacco, rice, fruit); cattle, hogs. Local industries (flour milling, mat weaving, cigar manufacturing). Main centers: Nueva Ocotepeque and SAN MARCOS. Formed 1906 from COPÁN department.

Ocotepeque, HONDURAS: see NUEVA OCOTEPEQUE.

Ocotillo (o-ko-TEE-yo), unincorporated town, MARICOPA county, central ARIZONA, residential suburb 16 mi/26 km SE of downtown PHOENIX. Gila River Indian Reservation to W and S. CHANDLER Municipal Airport to NE.

Ocotlán (o-ko-TLAHN), city and township, JALISCO, central MEXICO, on NE shore of Lake CHAPALA, at outlet of Santiago (LERMA) River, and 45 mi/72 km SE of GUADALAJARA; 20°21′N 102°42′W. Railroad junction; processing and agricultural center (grain, vegetables, fruit; livestock); milk canneries, rayon-yarn plants. Point of departure for Lake Chapala resort district.

Ocotlán de Morelos (o-ko-TLAHN dai mo-RE-los), city and township, OAXACA, S MEXICO, in SIERRA MADRE DEL SUR, on railroad, and 19 mi/31 km S of OAXACA DE JUÁREZ on Mexico Highway 175; 16°49′N 96°40′W. Agricultural center (cereals, sugarcane, coffee, fruit; livestock; timber); silver and gold deposits.

Ocoyoacac (o-ko-yo-ah-KAHK), town, MEXICO state, central MEXICO, 12 mi/19 km E of TOLUCA DE LERDO, off Mexico Highway 15; 19°16′N 99°26′W. Agricultural center (cereals; livestock); dairying.

Ocoyucan, MEXICO: see SANTA CLARA OCOYUCAN.

Ocozocoautla de Espinosa (o-ko-zo-ko-ah-OO-tlah), city and township, CHIAPAS, S MEXICO, in SIERRA MADRE, 18 mi/29 km W of TUXTLA; 16°46′N 93°22′W. Agricultural center (corn, beans, sugarcane, coffee, tobacco, fruit). A Zoque Indian town.

Ocracoke (O-kruh-kok), unincorporated village (□ 9 sq mi/23.4 sq km; 2000 population 769), HYDE county, E NORTH CAROLINA, 27 mi/43 km WSW of Cape HATTERAS, on PAMLICO SOUND, near SW end of Ocracoke Island (c.12 mi/19 km long, 1 mi/1.6 km wide), in OUTER BANKS, sand barrier between the ATLANTIC OCEAN and Pamlico Sound in CAPE HATTERAS NATIONAL SEASHORE; 35°06′N 75°58′W. Service industries; retail trade. An important port before Civil War. British Cemetery, graves of four Royal Navy crewmen who died here (1942). OCRACOKE INLET 3 mi/4.8 km to W. Ocracoke Lighthouse (1823), oldest operating lighthouse in North Carolina, at center of village. Toll ferries to SWAN QUARTER, on mainland (NW) and CEDAR ISLAND (SW, carries State Highway 12 traffic).

Ocracoke Inlet (O-kruh-kok IN-let), E NORTH CAROLINA, passage connecting PAMLICO SOUND (NW) with the ATLANTIC OCEAN (SE) between OCRACOKE Island (CAPE HATTERAS NATIONAL SEASHORE) to NE and PORTSMOUTH Island (CAPE LOOKOUT NATIONAL SEASHORE) to SW. HYDE-CARTERET county line passes through.

Ocros (O-kros), province, ANCASH region, W central PERU; ☉ OCROS.

Ocros (O-kros), town, ☉ Ocros province, ANCASH region, W central PERU, in CORDILLERA OCCIDENTAL, 60 mi/97 km S of HUARÁZ; 10°24′S 77°24′W. Cereals; livestock.

Ócsa (O-chah), village, PEST county, N central HUNGARY, 15 mi/24 km SSE of BUDAPEST. Has Romanesque church.

Ocsad, SLOVAKIA: see OSCADNICA.

Octa (AHK-tuh), village (2006 population 81), FAYETTE county, S central OHIO, 10 mi/16 km WNW of WASHINGTON COURT HOUSE; 39°37′N 83°36′W.

Octavia (awk-TAI-vee-uh), village (2006 population 140), BUTLER county, E NEBRASKA, 7 mi/11.3 km NNE of DAVID CITY, near PLATTE RIVER; 41°21′N 97°03′W.

Octay, CHILE: see PUERTO OCTAY.

Octeville (OK-tuh-veel), SW suburb (□ 2 sq mi/5.2 sq km) of CHERBOURG, MANCHE department, BASSE-NORMANDIE region, NW FRANCE; 49°37′N 01°39′W.

October Revolution Island, RUSSIA: see OKTYABR'S-KAYA REVOLYUTSIYA ISLAND.

Octoraro Creek (AHK-to-RER-o), SE PENNSYLVANIA and NE MARYLAND; formed by joining of West and East branches in Octoraro Lake reservoir, 4 mi/6.4 km WNW of OXFORD, Pennsylvania; flows c.20 mi/32 km SE in Maryland to SUSQUEHANNA RIVER at 8 mi/12.9 km NNW of HAVRE DE GRACE; 39°48′N 76°02′W. East Branch rises in NE LANCASTER county, near CHESTER county line, flows c.20 mi/32 km. South West Branch rises in E Lancaster county, flows c.20 mi/32 km SSW and SE.

Ocú (o-KOO), town, ☉ Ocú district, HERRERA province, S central PANAMA, 18 mi/29 km WSW of CHITRÉ. Corn, rice, beans; livestock.

Ocucaje (o-koo-KAH-hai), village, ICA region, ICA province, SW PERU, on ICA RIVER, and 20 mi/32 km S of ICA; 14°21′S 75°41′W. Cotton; vineyards.

Ocuilan de Arteaga (o-KWEE-lahn dai ahr-te-AH-gah), town, ☉ Ocuilan municipio, MEXICO state, central MEXICO, 12 mi/20 km WSW of CUERNAVACA; 18°58′N 99°25′W. Cereals, sugarcane, vegetables; livestock.

Ocuituco (o-kwee-TOO-ko), town, MORELOS, central MEXICO, 12 mi/19 km NE of CUAUTLA; 18°52′N 98°46′W. Grain, sugar, fruit; livestock.

Ocumare de la Costa (o-koo-MAH-rai dai lah KOS-tah), town, ARAGUA state, N VENEZUELA, at N foot of coastal range, in narrow lowland, on small river, and 16 mi/26 km E of PUERTO CABELLO (CARABOBO state); 10°28′N 67°45′W. Cacao and fruit. Its port is 3 mi/5 km N on the CARIBBEAN.

Ocumare del Tuy (o-koo-MAH-rai del TOO-ee) or **Ocumare**, town, ☉ Lander municipio, MIRANDA state, N VENEZUELA, in valley of coastal range, on TUY RIVER, and 28 mi/45 km SSE of CARACAS (linked by railroad); 10°08′N 66°46′W. Agricultural center (coffee, cacao, sugarcane, cereals).

Ocuri (o-koo-REE), canton, NOR CINTI province, CHUQUISACA department, SE BOLIVIA, 12 mi/20 km E of Padcoyo; 20°18′S 65°05′W. Elevation 7,894 ft/2,406 m. Limestone deposits. Agriculture (potatoes, yucca, bananas, corn, wheat, rye, peanuts); cattle and hog raising.

Ocuri (o-KOO-ree), town and canton, CHAYANTA province, POTOSÍ department, W central BOLIVIA, on road, and 16 mi/26 km SE of COLQUECHACA; 18°50′S 65°50′W. Corn, barley, potatoes; alpaca, sheep.

Ocussi, INDONESIA: see AMBENO.

Oda (O-dah), city, SHIMANE prefecture, SW HONSHU, W JAPAN, 37 mi/59 km S of MATSUE; 35°11′N 132°30′E. Livestock. Ruins of Iwami silver mine nearby.

Oda (O-dah), town, local council headquarters, EASTERN REGION, GHANA, on railroad, and 45 mi/72 km NNW of WINNEBA; 05°55′N 00°59′W. Diamond-mining center.

Oda (o-DAH), town, Kamiukena county, EHIME prefecture, NW SHIKOKU, W JAPAN, 19 mi/30 km S of MATSUYAMA; 33°33′N 132°48′E. Leaf tobacco.

Cross-references are shown in SMALL CAPITALS. The pronunciation guide is shown on page xix. The sources of population figures are shown on page xvii.

Oda (O-dah), town, Nyu county, FUKUI prefecture, central HONSHU, W central JAPAN, 12 mi/20 km S of FUKUI; 35°57′N 136°03′E.

Odae Mountain National Park, Korean, *Odae-san Kungnip Kongwon*, (□ 115 sq mi/299 sq km), E KANGWON province, SOUTH KOREA, 15 mi/24 km W of KANGNUNG. Established 1975. Many lofty peaks including Piro-bong (5,128 ft/1,563 m), Hwangbyong-san (4,616 ft/1,407 m), and Noin-bong (4,390 ft/1,338 m). The stretch between two Buddhist temples (Woljong-sa and Sangwon-sa) embraces many valleys, waterfalls, streams, and thick fir tree forests and shrubs. Sokumgang Valley in Chonghak-dong comprises Kumgang-sa temple, Changgundae peak, many unusual rock formations (Seshimdae, Chongshimdae and Manmulsang), Kuryong and Kunja waterfalls, and Shipcha-so pool. Pang a Dari mineral spring is known to contain some thirty different minerals (including carbonic acid and iron). Lodges, campgrounds, hiking trails.

Odai (O-dah-ee), town, Taki county, MIE prefecture, S HONSHU, central JAPAN, 22 mi/35 km S of TSU; 34°23′N 136°24′E. Tea. Trade in wooden food containers.

Odaid, Khor al (O-deed, KOR el), inlet of PERSIAN GULF, at SE base of QATAR peninsula, separating Qatar from ABU DHABI emirate (UAE); 24°35′N 51°25′E. Also spelled Khawr al-Odayd.

Odaka (O-dah-kah), town, Soma county, FUKUSHIMA prefecture, N central HONSHU, NE JAPAN, on the PACIFIC OCEAN, 31 mi/50 km S of FUKUSHIMA city; 37°33′N 140°59′E. Also spelled Otaka.

Ödåkra (UHD-OK-rah), town, SKÅNE county, SW SWEDEN, 5 mi/8 km NE of HELSINGBORG; 56°6′N 12°45′E. Distilling.

Odanah (o-DAN-uh), rural municipality (□ 147 sq mi/382.2 sq km; 2001 population 505), S MANITOBA, W central Canada, between BRANDON and RIDING MOUNTAIN NATIONAL PARK (N); 50°08′N 99°45′W. Agriculture and related businesses. Includes communities of Cordova, Moore Park.

Odanah (o-DAN-uh), village (2000 population 254), ASHLAND county, extreme N WISCONSIN, 9 mi/14.5 km E of ASHLAND, on BAD RIVER, in Bad River Reservation (for Chippewa Indians), in barren, marshy area; 46°36′N 90°40′W. Wild rice. Near LAKE SUPERIOR.

Oda River (O-dah), tributary of OFIN RIVER, GHANA; rises NE of KUMASI in Moinsi Hills. Unnavigable.

Odate (O-dah-te), city, Akita prefecture, N HONSHU, NE JAPAN, on RIVER, 41 km E of NOSHIRO, and 43 mi/70 km S of AKITA; 40°16′N 140°34′E. Cryptomeria [Japanese=*akitasugi*].

Odawara (o-DAH-wah-rah), city (2005 population 198,741), KANAGAWA prefecture, E central HONSHU, E central JAPAN, on NW shore of SAGAMI Bay, 34 mi/55 km S of YOKOHAMA; 35°15′N 139°09′E. In agricultural area (mandarin oranges); tourist resort (hot springs). Produces computers; also, paper lanterns, wooden-inlay work; fish and dried-apricot processing). Has large Zen Buddhist temple founded in the 15th century. Site of 16th century castle of Hojo clan.

Odawara Bay, JAPAN: see SAGAMI SEA.

Odayd, Khawr al-, Qatar: see ODAID, KHOR AL.

Odda, NORWAY: see HARDANGERFJORDEN.

Oddar Mean Chey, province, CAMBODIA: see OTDÂR MÉAN CHEAY.

Odde, DENMARK: see FANØ.

Odder (O-duhr), city (2000 population 9,955), ÅRHUS county, E JUTLAND, DENMARK, 8 mi/13 km S of ÅRHUS; 55°55′N 10°10′E. Agriculture (barley, fruit and fruit juice; dairying; hogs); barley; manufacturing (coffee, furniture).

Odderøya (AWD-duhr-uh-yah), island (□ c.1 sq mi/2.6 sq km) in an inlet of the SKAGERRAK strait, VEST-AGDER county, S NORWAY, part of the city of KRISTIANSAND. Has fortifications for Kristiansand.

Oddur, SOMALIA: see HUDDUR.

Ode (o-DAI), town, OGUN state, SW NIGERIA, 20 mi/32 km WNW of IJEBU-ODE; 06°58′N 03°41′E. Road center; cacao industry; cotton weaving, indigo dyeing; palm oil and kernels, rice. Also Ode Remo.

Odebolt (O-duh-bolt), town (2000 population 1,153), SAC county, W IOWA, 15 mi/24 km SW of SAC CITY; 42°18′N 95°15′W. Popcorn. Incorporated 1877.

Odell (o-DEL), town, COOS county, N NEW HAMPSHIRE, 20 mi/32 km NW of BERLIN. Wilderness area in Nash Stream State Forest. Drained by Nash Stream. Timber.

Odell (o-DEL), village (2000 population 1,014), LIVINGSTON county, NE central ILLINOIS, 10 mi/16 km NNE of PONTIAC; 41°00′N 88°31′W. In agricultural area; manufacturing of clothing.

Odell, village (2006 population 336), GAGE county, SE NEBRASKA, 15 mi/24 km S of BEATRICE, and on branch of BIG BLUE RIVER, near KANSAS state line; 40°02′N 96°47′W. Grain; livestock; dairy and poultry products; animal crates and cribs.

Odell (o-DEL), village, WILBARGER county, N TEXAS, 15 mi/24 km NNW of VERNON, near RED RIVER (OKLAHOMA state line); 34°20′N 99°25′W. In agriculture area.

Odelzhausen (o-delts-HOU-suhn), village, UPPER BAVARIA, BAVARIA, S GERMANY, 14 mi/23 km ESE of AUGSBURG, on the Glonn River; 48°18′N 11°10′E.

Odem (o-DEM), town (2006 population 2,547), SAN PATRICIO county, S TEXAS, 15 mi/24 km NW of CORPUS CHRISTI, near NUECES RIVER; 27°56′N 97°35′W. Railroad junction in farm area (cotton, sorghum; cattle). Oil and natural gas. W end of Nueces Bay to SE. Incorporated as city 1929.

Odemira (o-dai-MEE-rah), town, BEJA district, S PORTUGAL, river port on the MIRA RIVER, 12 mi/19 km from its mouth, on railroad, and 52 mi/84 km SW of BEJA. Cork-producing center; grain milling.

Odemis, Turkish *Ödemiş*, town, W TURKEY, railroad terminus from IZMIR, 45 mi/72 km ESE of Izmir; 38°11′N 27°58′E. Emery, mercury, antimony, arsenic, iron. Olives, figs, valonia, tobacco, barley, potatoes. Odemis Archaeological Museum displaying regional artifacts is c. 37 mi/60 km E of Izmir.

Oden (O-duhn), village (2000 population 220), MONTGOMERY county, W ARKANSAS, c.40 mi/64 km WNW of HOT SPRINGS, on OUACHITA RIVER, in Ouachita National Forest; 34°37′N 93°47′W.

Oden, village, EMMET county, NW MICHIGAN, on CROOKED LAKE, and 6 mi/9.7 km NE of PETOSKEY; 45°25′N 84°49′W. Has large state fish hatchery. Resort area.

Ödenburg, HUNGARY: see SOPRON.

Odendaalsrus, town, W FREE STATE province, SOUTH AFRICA, 35 mi/56 km WSW of KROONSTAD, 11 mi/18 km NNE of WELKOM, and 90 mi/145 km NNE of MANGAUNG (BLOEMFONTEIN); 27°52′S 26°40′E. Elevation 4,420 ft/1,347 m. Established 1899 as a church center. Center of goldfield discovered 1946. Maize and cattle raising. Airfield.

Odenkirchen (o-den-KIR-khuhn), S suburb of Mönchengladbach, North RHINE—WESTPHALIA, W GERMANY, on the NIERS; 51°06′N 06°27′E.

Odenpäh, ESTONIA: see OTEPÄÄ.

Odense (O-duhn-suh), city (2000 population 156,923), ⊙ FYN county, S central DENMARK, a seaport linked by canal with the ODENSE FJORD (an arm of the KATTE-GAT strait); 55°23′N 10°23′E. Denmark's third-largest city, it is an important commercial, industrial, and cultural center and a railroad junction. There are large shipyards and plants manufacturing metal goods, motor vehicles, machinery, textiles, and processed food. Founded in the tenth century, Odense is one of the oldest cities in N EUROPE. It has been an episcopal see since 1020. Of note in the city are a twelfth-century church and the thirteenth-century Cathedral of Saint Knud, one of the finest examples of Danish Gothic architecture. Odense has several colleges and a university (1964). The house of the writer Hans Christian Andersen, who was b. here in 1805, is now a museum.

Odense Å (O-duhn-suh o), largest river on FYN island, c.40 mi/64 km long, DENMARK; rises in Arreskov Lake; flows N, past ODENSE, to ODENSE FJORD.

Odense Fjord, inlet of the KATTEGAT strait, N FYN island, DENMARK; connects with ODENSE through ODENSE Å river and Odense Canal; c.10 mi/16 km long.

Odenthal (O-den-tahl), town, RHINELAND, North Rhine—WESTPHALIA, W GERMANY, 3 mi/4.8 km NNW of BERGISCH-GLADBACH; 51°02′N 07°08′E.

Odenton (O-duhn-tuhn), village (2000 population 20,534), ANNE ARUNDEL county, central MARYLAND, 14 mi/23 km NW of ANNAPOLIS; 39°04′N 76°42′W. In vegetable-farming area; plastics, furniture and corrugated box plant. Named for Oden Bowie, who became governor of Maryland in 1869. The town caters to Fort George G. Meade, just NW, and has fast-food restaurants lining the road.

Odenville, town (2000 population 1,131), St. Clair co., NE central Alabama, c.9 mi/14.5 km NW of Pell City.

Odenwald (O-den-vahlt), hilly, forested region, S central GERMANY, bordering on the NECKAR and MAIN rivers and the Rhine plain. Its highest point (2,055 ft/626 m) is the KATZENBUCKEL. Fruit and grapes are grown in the W and S, and there are porphyry quarries. The region is a popular tourist area, its landscape dotted with castles and medieval ruins.

Oder, Czech and Polish *Odra*, river, 562 mi/904 km long; the second-longest river of POLAND; rises in the E SUDETES Mountains, NE CZECH REPUBLIC; flows generally NW through SW Poland, then N along the border between Poland and GERMANY to the BALTIC SEA N of SZCZECIN (Poland). The WARTA and the LAUSITZER NEISSE rivers are its chief tributaries. There are power dams on the Oder's headwaters in Czech Republic. Navigable from RACIBÓRZ (Poland), the Oder is an important waterway of central and E EUROPE, connecting the industrial region of SILESIA with the sea. Barges on the river carry iron, coal, and coke. The Oder is linked by canals with the SPREE and ELBE rivers; the Warta connects it with the VISTULA River, WROCŁAW (Poland), FRANKFURT AN DER ODER (Germany), and Szczecin are the chief cities on the Oder.

Öderan, GERMANY: see OEDERAN.

Oderberg, CZECH REPUBLIC: see BOHUMÍN.

Oderbruch, GERMANY: see ODER MARSHES.

Oder-Danube Canal, POLAND and CZECH REPUBLIC: see DANUBE-ODER CANAL.

Ode Remo, NIGERIA: see ODE.

Odergebirge, CZECH REPUBLIC: see ODER MOUNTAINS.

Oder-Havel Canal (O-der–HAHV-el), GERMANY, waterway linking BERLIN with the ODER RIVER; starts at the HAVEL River, N of ORANIENBURG (52°45′N 13°15′E) and continues for 52 mi/83 km to the Oder River at the Polish border, 45 mi/72 km NE of Berlin (52°50′N 14°06′E). The FINOW, a minor, 25-mi/40-km-long canal, runs parallel to it. Formerly known as the Hohenzollernkanal, it used to be part of the Berlin-STETTIN (now Szczecin) Canal, a connection between Berlin and SZCZECIN which today is of little economic importance.

Oderin Island (o-DIR-in), SE NEWFOUNDLAND AND LABRADOR, E Canada, in Placentia Bay, 40 mi/64 km W of Argentia; 2 mi/3 km long; 47°18′N 54°49′W. Residents of fishing town of Oderin resettled in 1966.

Oder Marshes, German *Oderbruch*, Polish *Bagna Odry*, in E GERMANY, extend c.30 mi/48 km N-S along W bank of the lower Oder River, between Küstrin (KOSTRZYN, POLAND) and ODERBERG (Germany). The old ODER (German *Alte Oder*, Polish *Odrzyca*) arm of the river runs along W edge to HOHENZOLLERN CANAL. The marshes were drained by Frederick the Great. Partly owing to its role as a very quiet frontier between the former East Germany and Poland, the region became ecologicallly valuable and is now prized and partially protected.

Oder Mountains, Czech *Oderské vrchy* (O-der-SKE VUHR-khi), German *odergebirge*, easternmost range of the SUDETES, SEVEROMORAVSKY province, N central MORAVIA, CZECH REPUBLIC; extend c.30 mi/48 km W-E, between OLOMOUC and ODRY. Highest point (2,231 ft/680 m is 10 mi/16 km E of Olomouc. ODRA RIVER rises in S.

Oder-Neisse line (O-der–NEIS-se), border established 1945 between E GERMANY and W POLAND; it follows the ODER and W NEISSE rivers from the BALTIC SEA to the CZECH border. Sanctioned informally at Yalta (Feb. 1945); recognized, pending a peace treaty with Germany, at POTSDAM (Aug. 1945). Despite the absence of such a treaty, the German Democratic Republic (former East Germany) and Poland acknowledged the line as the inviolable Polish-German border in 1950. Recognized by West German government in 1970; by reunified government in 1990.

Oderské vrchy, CZECH REPUBLIC: see ODER MOUNTAINS.

Oder-Spree Canal (O-der–SHPRAI), E GERMANY, extends 52 mi/84 km E-W between the Oder River at Fürstenberg and SEDDIN LAKE, where it is connected, via DAHME RIVER, with the SPREE River at Köpenick. Has seven locks; navigable for ships up to 1,000 tons/907 metric tons. Utilizes 13-mi/21-km-long section of Spree River near Fürstenwalde. Opened 1891. Sometimes called Spree-Oder Canal.

Oderzo (o-DER-tso), ancient *Opitergium*, town, TREVISO province, VENETO, N ITALY, 15 mi/24 km ENE of TREVISO, between PIAVE and LIVENZA rivers; 45°46′N 12°29′E. In livestock-raising and silk- and tobacco-growing region. Wine center; alcohol distillery, bicycle factory, lime- and cementworks. Has church dating partly from 10th century.

Odesa, UKRAINE: see ODESSA.

Ödeshög (UHD-es-HUHG), village, ÖSTERGÖTLAND county, S SWEDEN, near W shore of LAKE VÄTTERN, 18 mi/29 km WSW of MJÖLBY; 58°14′N 14°39′E.

Odes'ka oblast, UKRAINE: see ODESSA oblast.

Odessa (o-DES-suh), (Ukrainian *Odes'ka*), (Russian *Odesskaya*), oblast (□ 12,861 sq mi/33,438.6 sq km; 2001 population 2,469,057), S UKRAINE; ⊙ ODESSA. In BLACK SEA LOWLAND, rising (N) into PODOLIAN UPLAND; bounded SE by BLACK SEA, SW by DANUBE River and its Kiliya Estuary, W by Moldova. Agricultural steppe region with sugar beets (extreme N), wheat, corn, barley, and sunflowers (center and S), rice (S), vegetables (near Odessa); lesser crops include grapes and fruit (W), castor beans. Chief industrial center and port is Odessa, associated outports at ILLICHIVS'K and YUZHNE; manufacturing also at IZMAYIL, BALTA, KOTOVS'K, ANAN'YIV, BILYAYIVKA, BILHOROD-DNISTROVS'KYY, KILIYA, ARTSYZ, VYLKOVE, TATARBUNARY. Sugar refining, flour milling, dairying, wine making. Formed in 1932. In 1954, was enlarged to SW by incorporation of Izmayil oblast, which was formed in 1940 out of Romanian departments of Izmail and Cetatea-Alba of S BESSARABIA. Has (1993) seventeen cities, thirty three towns, twenty six rural raions.

Odessa (o-DE-sah), (Ukrainian *Odesa*), city (2001 population 1,029,049), ⊙ ODESSA oblast, in UKRAINE, a port on Odessa Bay of the BLACK SEA; 46°28′N 30°44′E. Elevation 134 ft/40 m. The third-largest Ukrainian city after KIEV and KHARKIV, Odessa is an important railroad junction and highway hub and is a major industrial, cultural, scientific, and resort center. Grain, sugar, machinery, coal, petroleum products, cement, metals, jute, and timber are the chief items of trade at the port, which was the leading Soviet Black Sea port and remains such for independent Ukraine. Also a naval base and home port of fishing and antarctic whaling fleets. Shipbuilding, oil refining, machine building, metalworking, food processing, and the manufacturing of chemicals, machine tools, clothing, and products made of wood, jute, and silk. Large and numerous health resorts located both within and outside city limits. The most posh resort

area, Arcadia, is dubbed "the French Riviera of Eastern Europe." Ukrainians, Russians, Jews, Bulgarians and Greeks predominate in Odessa's cosmopolitan population, with large Tatar and Gypsy minorities. Named after an ancient Miletian Greek colony (Odessos, Ordyssos, or Ordas) that was thought to have occupied the site and then disappeared between the 3rd and 4th centuries C.E. In the 14th century, the site was occupied by a settlement, Kotsyubiyiv, which was fortified by the Lithuanian grand duke Vytautas in the 15th century. In 1480, the fort was captured by the Crimean Tatars and renamed Khadzhy-Bey (Ukrainian *Kachybey*). In 1764, it passed to the Turks, who renamed it Gadzhy-Bey and built a fortress (Yenu-Duniya) to protect the harbor. Captured by the Russians in 1789. By the Treaty of Jassy in 1792, Turkey ceded the region between the DNIESTER and BUH rivers (including Odessa) to Russia, which rebuilt Odessa as a fort, commercial port, and naval base. The city that developed around the fort grew rapidly as the chief grain-exporting center of Ukraine; its importance was further enhanced with the coming of the railroad in the second half of the 19th century. It was a free port from 1819 to 1849, and in 1866 it was linked by railroad with Kiev, Kharkiv, and JASSY (Romania). Industrialization began in the latter part of the 19th century. Odessa was a center of emigré Greek and Bulgarian patriots, of the Ukrainian cultural and national movement, of Jewish, Zionist, and literary culture, and of the labor movement and social democracy. The city's first workers' organization was founded in 1875. Scene in 1905 of a workers' uprising led by sailors from the battleship *Potemkin*. From the 1880s to the 1920s, Odessa's Jewish community was second largest in all of Russian Empire (next to Warsaw). Over one-third of the population was Jewish in the 1930s, including retailers, craftsmen, and professionals. When Turkey closed the Dardanelles to the Allies in World War I, the port of Odessa was also closed and was later bombarded by the Turkish fleet. Following the 1917 Bolshevik Revolution, the city was successively occupied by the Central Powers, the French, the Reds, and the Whites until the Red Army decisively took it from General Denikin in 1920 and united it with the Ukrainian SSR. Odessa suffered greatly in the famine of 1921–1922 after the Russian civil war. Despite a heroic defense during World War II, the city fell to German and Romanian forces in October 1941. It was under Romanian administration as the capital of Transnistria until its liberation (April 1944) by the Soviet Army. Many buildings were ruined, and approximately 280,000 civilians (mostly Jews) were massacred or deported during the Axis occupation. Seat of a university (established 1865), opera and ballet theater (1809), historical museum (1825), municipal library (1830), astronomical observatory (1871), opera house (1883–1887), and art gallery (1898).

Odessa, city (2000 population 4,818), LAFAYETTE county, W central MISSOURI, 33 mi/53 km E of KANSAS CITY; 39°00′N 93°57′W. Processes and ships farm products. Corn, wheat, sorghum; cattle, hogs; manufacturing (consumer goods). Plotted 1878.

Odessa (o-DE-suh), city (2006 population 95,163), ⊙ ECTOR county, extends E into MIDLAND county, W TEXAS, 18 mi/29 km SW of MIDLAND, and 123 mi/198 km SSW of LUBBOCK; 31°52′N 102°20′W. Elevation 2,891 ft/881 m. Great oil deposits just to the S changed Odessa from a small ranch town into a large and growing oil center with refineries and plants that produce fuels, carbon black, chemicals, plastics, synthetic rubber, industrial gas, and machinery. The region has potash, salt, and limestone deposits. Schlemeyer Field airport to N, Midland Airport (to E) handles many commuters to and from DALLAS and HOUSTON. Coliseum. Seat of Odessa College (two-year) and University of Texas of the Permian Basin. Founded 1881, incorporated 1927.

Odessa (o-DES-suh), town (2006 population 934), LINCOLN county, E WASHINGTON, 32 mi/51 km SW of DAVENPORT, on Crab Creek, at mouth of Duck Creek; 47°20′N 118°42′W. In COLUMBIA basin agricultural region; ships barley, oats, rye, potatoes, wheat, cattle. Coffeepot Lake and Twin Lakes to NE; Pacific Lake to N. Town settled by Germans and Russians.

Odessa (o-DE-suh), unincorporated village, SE ONTARIO, E central Canada, 12 mi/19 km WNW of KINGSTON, and included in Loyalist township; 44°17′N 76°43′W. Cheese making.

Odessa (O-de-sah), village (2006 population 201), S SASKATCHEWAN, Canada, 40 mi/64 km ESE of REGINA. Wheat; dairying.

Odessa (o-DES-suh), village (2000 population 286), NEW CASTLE county, N central DELAWARE, 20 mi/32 km S of WILMINGTON; 39°27′N 75°39′W. Elevation 42 ft/12 m. In agricultural region. Has 18th century homes; Friends meeting house dates from 1783. Winterthur Museum. Silver Run Wildlife Area to NE. Named after ODESSA, UKRAINE (then RUSSIA). Formerly called Cantwell's Bridge. A busy port for grain shipments until 1855, when the new railroad bypassed the town, leaving it to serve as a small local trade center.

Odessa, village (2000 population 113), BIG STONE county, SW MINNESOTA, 7 mi/11.3 km ESE of ORTONVILLE, on MINNESOTA RIVER, near SOUTH DAKOTA state line; 45°15′N 96°19′W. Grain. Lac qui Parle Wildlife Area to SE.

Odessa, village (□ 7 sq mi/18.2 sq km; 2006 population 607), SCHUYLER county, W central NEW YORK, in FINGER LAKES region, 15 mi/24 km WSW of ITHACA; 42°18′N 76°49′W.

Odessadale (o-DES-uh-dail), town, MERIWETHER county, W GEORGIA, 12 mi/19 km ESE of LA GRANGE; 33°00′N 84°48′W.

Odesskaya oblast, UKRAINE: see ODESSA oblast.

Odesskoye (uh-DYES-kuh-ye), village (2006 population 6,000), S OMSK oblast, SW SIBERIA, RUSSIA, approximately 6 mi/10 km N of the KAZAKHSTAN border, on road junction, 55 mi/89 km S of OMSK; 54°13′N 72°58′E. Elevation 393 ft/119 m. In agricultural area.

Odet River (o-DAI), 35 mi/56 km long, in FINISTÈRE department, BRITTANY, W FRANCE; rises in MONTAGNES NOIRES (a Breton hill range), 5 mi/8 km SE of CHÂTEAUNEUF-DU-FAOU; flows SW and S, past QUIMPER (head of navigation), through a drowned valley estuary to the ATLANTIC OCEAN 5 mi/8 km E of PONT-L'ABBÉ; 47°53′N 04°06′W. The lower course is scenic with meanders enclosed by wooded escarpments. The small resort of BÉNODET is at the estuary's mouth.

Odeypore, INDIA: see UDAIPUR.

Odia Island, MARSHALL ISLANDS: see AILINGLAPALAP.

Odiel River (o-DYEL), c.90 mi/145 km long, HUELVA province, SW SPAIN; rises in the SIERRA MORENA 4 mi/6.4 km ESE of ARACENA; flows SW and S in wide estuary to the ATLANTIC OCEAN. Joined by the RÍO TINTO 3 mi/4.8 km S of HUELVA. Only the estuary is navigable.

Odienné (o-dyen-NAI), town, ⊙ Denguélé region, NW CÔTE D'IVOIRE, road junction and market 130 mi/209 km ESE of KANKAN communicating with Guinea and Burkina Faso; 09°30′N 07°34′W. Mining (copper, molybdenum) nearby. Climatological and medical stations. Airport.

Odiham (OD-i-uhm), village (2001 population 4,959), NE HAMPSHIRE, S ENGLAND, 7 mi/11.3 km E of BASINGSTOKE; 51°15′N 00°57′W. Has 13th-century church. Nearby is Odiham Castle, with 12th-century keep.

Odin, village (2000 population 1,122), MARION county, S central ILLINOIS, 5 mi/8 km W of SALEM; 38°37′N 89°02′W. In agricultural and oil-producing area. Incorporated 1865.

Odin, village (2000 population 125), WATONWAN county, S MINNESOTA, 10 mi/16 km SW of ST. JAMES,

---

on South Fork of WATONWAN RIVER; 43°52′N 94°44′W. Grain, soybeans; livestock; manufacturing (agricultural equipment parts, feeds). Small lakes in area.

Odin, Mount (9,751 ft/2,972 m), SE BRITISH COLUMBIA, W Canada, in MONASHEE MOUNTAINS, 30 mi/48 km S of REVELSTOKE; 50°33′N 118°08′W.

Odintsovo (uh-deen-TSO-vuh), city (2006 population 137,810), central MOSCOW oblast, central European Russia, on road and railroad, 14 mi/23 km WSW of MOSCOW; 55°40′N 37°16′E. Elevation 603 ft/183 m. Steel works, machinery, furniture, chemicals, paint and varnish, plastic products; refractories, brick works. Made city in 1957.

Odioñgan (o-JONG-ahn), town, ROMBLON province, PHILIPPINES, on W TABLAS ISLAND, on TABLAS STRAIT; 12°24′N 122°02′E. Agricultural center (rice, coconuts).

Odiorne's Point (O-dee-yuhr), small peninsula, SE NEW HAMPSHIRE, near mouth of the PISCATAQUA RIVER, between PORTSMOUTH and RYE. Odiorne State Park. Site of state's first European settlement (1623).

Odivelas (o-dee-VAI-lahs), suburb of LISBON, LISBOA district, W PORTUGAL, 6 mi/9.7 km NNW of city center; 38°47′N 09°11′W.

Odoben (o-DO-ben), town, EASTERN REGION, GHANA, 5 mi/8 km W of NYAKROM; 05°55′N 00°59′W. Cacao, palm oil and kernels, cassava.

Odobeşti (o-do-BESHT), town, VRANCEA county, E ROMANIA, on railroad, and 7 mi/11.3 km NW of FOCŞANI. Wine center, with school of viticulture. Has 18th-century church.

Odolanow (o-do-LAH-noov), Polish *Odolanów*, German *Adelnau*, town, Kalisz province, W POLAND, on BARYCZ RIVER, and 22 mi/35 km SW of KALISZ; 51°34′N 17°42′E. Flour milling, sawmilling.

Odolena Voda (O-do-LE-nah VO-dah), German *wodolka*, village, STREDOCESKY province, N central BO-HEMIA, CZECH REPUBLIC, 10 mi/16 km N of PRAGUE; 50°14′N 14°25′E. Manufacturing (machinery; food processing); agriculture (sugar beets, vegetables). Has an 18th century baroque church. Important manufacturing of aircraft just SSW, in VODOCHODY.

Odon, town (2000 population 1,376), DAVIESS county, SW INDIANA, 15 mi/24 km NE of WASHINGTON; 38°50′N 86°59′W. In agricultural area (corn, soybeans; dairying); manufacturing (agricultural and industrial supplies; concrete burial vaults, lumber). Laid out 1846.

Odonkawkrom (o-duhn-KOR-kruhm), town, district headquarters, EASTERN REGION, GHANA, in Agram Plains, 10 mi/16 km W of VOLTA Lake; 07°03′N 00°06′W. Tobacco; fish.

O'Donnell (O DAHN-uhl) town, LYNN and DAWSON counties, NW TEXAS, on the LLANO ESTACADO 45 mi/72 km S of LUBBOCK; 32°58′N 101°49′W. Elevation 3,110 ft/948 m. In irrigated agricultural area (wheat, cotton; livestock); manufacturing (farm tools). Settled 1908, incorporated 1923.

Odon River (o-DON), 25 mi/40 km long, CALVADOS department, BASSE-NORMANDIE region, NW FRANCE; rises in NORMANDY HILLS above AUNAY-SUR-ODON; flows NE to the ORNE RIVER at CAEN. Heavy fighting along its banks in World War II (Normandy campaign).

Odorheiu-Secuiesc (o-dor-HAH-yoo–se-koo-YESK), Hungarian *Székelyudvarhely*, city, HARGHITA county, central ROMANIA, in TRANSYLVANIA, on railroad, and 39 mi/63 km SE of TÎRGU MUREŞ; 46°18′N 25°18′E. Distilling, brewing, tanning, manufacturing of bricks and tiles; trade in lumber and livestock. Has ruins of Roman and medieval citadels, 17th-century Franciscan monastery, and Roman Catholic church. Mineral springs nearby (SW). Over 90% population are Magyars (Szeklers). Part of HUNGARY, 1940–1945.

Odoyev (uh-DO-eef), town (2006 population 6,325), W TULA oblast, W central European Russia, on the UPA RIVER (tributary of the OKA River), on road junction,

40 mi/64 km SW of TULA; 53°56′N 36°41′E. Elevation 692 ft/210 m. Bakery, butter factory. Known since the second half of the 14th century. Became city in 1777; reduced to status of a village in 1926; made town in 1959.

Odra, E EUROPE: see ODER.

Odra Port (OD-rah), Polish *Odra-Warszów*, German *Ostswine*, village, Szczecin province, NW POLAND, at W end of WOLIN island, on Swine mouth of the ODER River, opposite SWINEMÜNDE (GERMANY), and 35 mi/56 km NNW of SZCZECIN. Fishing port. Terminus since 1948 of railroad ferry across BALTIC SEA to TRELLEBORG (SWEDEN). Formerly in German-administered POMERANIA; passed to Poland in 1945.

Odra River, c.25 mi/40 km long, central CROATIA; formed by several headstreams joining 13 mi/21 km SE of ZAGREB; flows SE to KUPA RIVER just NW of SISAK.

Odrau, CZECH REPUBLIC: see ODRY.

Odra-Warszów, POLAND: see ODRA PORT.

Odry (OD-ri), German *odrau*, town, SEVEROMORAVSKY province, S central SILESIA, CZECH REPUBLIC, on railroad, on ODRA RIVER, and 26 mi/42 km ENE of OLOMOUC; 49°40′N 17°50′E. Agriculture (oats); manufacturing (rubber, balls for sports); food and wood processing.

Odrzyca River, POLAND: see ODER MARSHES.

Odueine (ood-WAIN-ye), village, NW SOMALIA, in OGO highland, on road, and 35 mi/56 km WSW of BURAO; 09°24′N 45°04′E. Camels, sheep, goats. Formerly spelled OADWEINA, Odweina.

Odum (O-duhm), village (2000 population 414), WAYNE county, SE GEORGIA, 10 mi/16 km NW of JESUP, on LITTLE SATILLA Creek; 31°40′N 82°02′W. Manufacturing includes textile printing and meat processing.

Odurne (O-duhr-ne), village, LOVECH oblast, PORDIM obshtina, BULGARIA; 43°20′N 24°56′E.

Odžaci (OD-jah-tsee), Hungarian *Hódság*, village (2002 population 35,582), VOJVODINA, NW SERBIA, 31 mi/50 km NW of NOVI SAD, in the BACKA region; 45°30′N 19°17′E. Railroad junction; hemp processing. Also spelled Odzhatsi.

Odžak (O-zhahk), town, N BOSNIA, BOSNIA AND HER-ZEGOVINA, on BOSNA RIVER, and 8 mi/12.9 km WSW of BOSANSKI Šamac. Also spelled Odzhak.

Odzhatsi, SERBIA: see ODZACI.

Odzi, town, MANICALAND province, E ZIMBABWE, 20 mi/32 km W of MUTARE, on ODZI RIVER, on railroad; 18°58′S 32°23′E. Coffee, tea; citrus fruit, macadamia nuts, corn, wheat; dairying; cattle, sheep, goats, hogs, poultry.

Odzi River, c.120 mi/193 km long, MANICALAND province, E ZIMBABWE; rises c. 40 mi/64 km N of MUTARE, in NYANGA MOUNTAINS; flows SSW, forming Lake OSBORNE (Osborne Dam), past ODZI and HOT SPRINGS towns; joins SAVE RIVER 55 mi/89 km SSW of MUTARE.

Oe, town, Kasa county, KYOTO prefecture, S HONSHU, W central JAPAN, near Mount Oe, 43 mi/70 km N of KYOTO; 35°23′N 135°08′E.

Oe, town, West Murayama county, YAMAGATA prefecture, N HONSHU, NE JAPAN, 12 mi/20 km N of YA-MAGATA city; 38°22′N 140°12′E. Tea; trays. Bentonite.

Oea, LIBYA: see TRIPOLI.

Oeanthea, Greece: see GALAXIDHI.

Oebisfelde (e-bis-FEL-de), town, SAXONY-ANHALT, central GERMANY, near Spetze River, on the ALLER RIVER, 20 mi/32 km NW of HALDENSLEBEN, 8 mi/12.9 km E of WOLFSBURG; 52°26′N 10°59′E. Agriculture market (grain, sugar beets, vegetables, livestock); food canning. Was first mentioned 1226; passed 1680 to BRANDENBURG. Has late-Gothic parish church; ruined castle (13th–14th century); half-timbered houses (15th–17th century). Sometimes spelled Öbisfelde.

Oebroeg, INDONESIA: see UBRUG.

Oe-Cusse, INDONESIA: see AMBENO.

Oecussi, EAST TIMOR: see AMBENO.

Oedelem (OO-duh-lem), agricultural village, in commune of BEERNEM, Bruges district, WEST FLANDERS

province, NW BELGIUM, 6 mi/9.7 km ESE of BRUGES; 51°10′N 03°20′E. Brick manufacturing.

Oederan (UH-de-rahn), town, SAXONY, E central GERMANY, in the Erzgebirge, 11 mi/18 km E of CHEMNITZ; 50°52′N 13°07′E. Produces textiles (hosiery, cotton, wool, carpets), furniture; metalworking. Sometimes spelled Öderan.

Oedheim (UHD-heim), village, FRANCONIA, BADEN-Württemberg, S GERMANY, 8 mi/12.9 km N of HEIL-BRONN, on the KOCHER RIVER; 49°14′N 09°16′E. Fruit, grain.

Oedjoengbatoe, INDONESIA: see BANYAK ISLANDS.

Oegstgeest (OOKHST-khaist), town, SOUTH HOLLAND province, W NETHERLANDS, 2 mi/3.2 km NW of LEI-DEN; 52°11′N 04°28′E. NORTH SEA 3 mi/4.8 km to NW. Cattle; flowers, nursery stock, vegetables, fruit.

Oehringen, GERMANY: see ÖHRINGEN.

Oeiras (o-AI-rahs), city (2007 population 35,322), central PIAUÍ state, BRAZIL, near CANINDÉ River, 140 mi/225 km SSE of TERESINA; 07°01′S 42°08′W. Road junction and cattle-raising center. Ships carnauba wax, rice. Airfield. Was capital of Piauí province until 1852.

Oeiras (o-ER-ahsh), town, LISBOA district, W central PORTUGAL, on N shore of TAGUS RIVER estuary at its influx into the ATLANTIC OCEAN, 9 mi/14.5 km W of LISBON; 38°41′N 09°19′W. Fishing port and swimming resort; manufacturing (paper, biscuits). Important military base nearby.

Oeiras do Para (O-ai-rahs do pah-RAH), city (2007 population 25,336), N central PARÁ state, BRAZIL, on RIO PARÁ, across river from CAMETÁ, 100 mi/161 km WSW of BELÉM; 01°58′S 49°52′W. Airport.

Oel (O-ail), town, KHERI district, N UTTAR PRADESH state, N central INDIA, 8 mi/12.9 km SSW of LA-KHIMPUR. Sugar milling; rice, wheat, gram, corn, oilseeds.

Oelde (UHL-de), town, WESTPHALIA-LIPPE, NORTH RHINE–Westphalia, NW GERMANY, 11 mi/18 km ENE of AHLEN; 51°50′N 08°09′E. Dairying, hog raising; metalworking; pumpernickel, machinery, furniture.

Oeleëlheuë, INDONESIA: see ULEELHUE.

Oeliaser Islands, INDONESIA: see ULIASERS.

Oella (O-el-lah), suburban village, BALTIMORE county, central MARYLAND, on PATAPSCO River, and 10 mi/16 km W of downtown BALTIMORE. The large cotton mill established here in 1845 is said to have been named for the first woman in America to spin cotton. Once exclusively a workers' town, it is becoming gentrified. Benjamin Banneker, "the free man of color" who helped lay out WASHINGTON, D.C. for Thomas Jefferson, was associated with the Mount Gillboa Church. Patapsco Valley State Park is nearby and BALTIMORE AND OHIO RAILROAD Museum. The stone mill is now a community center, gallery, and crafts shop.

Oelrichs, village (2006 population 143), FALL RIVER county, SW SOUTH DAKOTA, 20 mi/32 km SE of HOT SPRINGS, and on Horsehead Creek; 43°10′N 103°13′W. Buffalo Gap National Grassland to E, S, and W.

Oels, CZECH REPUBLIC: see OLESNICE.

Oels, POLAND: see OLESNICA.

Oelsnitz (UHLS-nits), town, SAXONY, E central GER-MANY, at N foot of the Erzgebirge, 9 mi/14.5 km E of ZWICKAU; coal-mining center; 50°43′N 12°43′E. Power station.

Oelsnitz, town, SAXONY, E central GERMANY, at N foot of the Erzgebirge, on the White ELSTER River, and 6 mi/9.7 km SSE of PLAUEN, in the VOGTLAND; 50°25′N 12°11′E. Textile milling (cotton, wool, carpets); manufacturing of machinery; construction. Former coal-mining center (1943–1970). Once divided into three villages, united 1839. Baroque church (1724), Gothic church from 16th century; castle Voigtsberg from 1270–1405. Town since 1923. Sometimes spelled Ölsnitz.

Oelwein, city (2000 population 6,692), FAYETTE county, NE IOWA, 20 mi/32 km S of WEST UNION; 42°40′N 91°54′W. Railroad center; poultry-packing plant; feed, beverages. Incorporated 1888.

Area is shown by the symbol □, and capital city or county seat by ⊙.

Oembilin, INDONESIA: see UMBILIN.

Oenaoena, INDONESIA: see TOGIAN ISLANDS.

Oene (OO-nuh), town, GELDERLAND province, E central NETHERLANDS, 10 mi/16 km NNE of APELDOORN; 52°21′N 06°03′E. Grote Wetering River to E, Apeldoorns Canal to W. Dairying; livestock; grain, vegetables.

Oengaran, INDONESIA: see UNGARAN.

Oeno Island, South Pacific: see PITCAIRN ISLAND.

Oensingen (UHN-sing-uhn), commune, SOLOTHURN canton, N SWITZERLAND, 10 mi/16 km ENE of SOLOTHURN; 47°17′N 07°44′E.

Oenussae Islands, Greece: see OINOUSSAI ISLANDS.

Oer-Erkenschwick (UHR–ER-ken-shvik), town, WESTPHALIA-LIPPE, NORTH RHINE–Westphalia, W GERMANY, in the RUHR district, 3 mi/4.8 km NE of RECKLINGHAUSEN; 51°39′N 07°15′E. Coal mining; nitrogen production.

Oerle, BELGIUM: see OREYE.

Oerlenbach (UHR-len-bahkh), village, Lower FRANCONIA, BAVARIA, central GERMANY, 9 mi/14.5 km NNW of SCHWEINFURT, on E slope of the Rhön; 50°09′N 10°08′E. Grain; machinery.

Oerlikon (UHR-lee-kon), N suburb of ZÜRICH, N SWITZERLAND. Noted for its production of firearms and antiaircraft guns.

Oerlinghausen (uhr-ling-HOU-suhn), town, WESTPHALIA-LIPPE, NORTH RHINE–Westphalia, NW GERMANY, in TEUTOBURG FORREST, 9 mi/14.5 km WNW of DETMOLD; 51°58′N 08°40′E. Woodworking. Sometimes spelled Örlinghausen. Chartered 1926.

Oescus, Bulgaria: see BORIL, GIGEN.

Oesterdam, NETHERLANDS: see EASTERN SCHELDT.

Oestrich-Winkel (UHS-strikh–VING-kel), town, HESSE, central GERMANY, 11 mi/18 km WSW of WIESBADEN, on right bank of the RHINE River; 50°01′N 08°03′E. Tourism; viticulture; fine fabrics industry. Formed 1972 by unification of OESTRICH, WINKEL, and Mittelheim. Late-Gothic parish church (1508), town hall (1684), remains of former moated castle (14th century).

Oeta, Mount, Greece: see OITI.

Oetling (o-AIT-leeng), town, S Chaco province, ARGENTINA, 50 mi/80 km SSE of PRESIDENCIA ROQUE; 27°24′S 60°01′W. Railroad terminus; agricultural center (corn, cotton; livestock).

Oetrange (uh-TRAWNZH), village, CONTERN commune, SE LUXEMBOURG, on SYRE RIVER, and 6 mi/9.7 km E of LUXEMBOURG city; 49°36′N 06°16′E. Fruit (plums, cherries).

Oettingen, GERMANY: see OTTINGEN IM BAYERN.

Oetz or **Ötz** (both: OO-ETS), village, TYROL, W AUSTRIA, on Ötztaler Ache River, and 23 mi/37 km WSW of INNSBRUCK; 47°12′N 10°54′E. Summer and winter resort. Waterfalls (Stribenfall) nearby.

Oever, Den, NETHERLANDS: see DEN OEVER.

Oeynhausen, Bad, GERMANY: see BAD OEYNHAUSEN.

Of, village, NE TURKEY, on BLACK SEA, 28 mi/45 km E of TRABZON; 40°57′N 40°17′E. Corn, beans, vetch.

Ofakim (o-fah-KEEM) or **Ofaqim**, town (2006 population 24,400), S ISRAEL, 15 mi/24 km NW of BEERSHEBA, in NW NEGEV; 31°17′N 34°37′E. Elevation 577 ft/175 m. Textiles. Service center for surrounding rural settlements. Nahal Ofakim (Ofakim stream) nature reserve, which includes some archaeological ruins, is 1.2 mi/2 km SE. Founded 1955.

O'Fallon, city (2000 population 21,910), SAINT CLAIR county, SW ILLINOIS, suburb of SAINT LOUIS, 12 mi/19 km E of EAST SAINT LOUIS; 38°35′N 89°54′W. Bituminous coal in area; agricultural area (corn, soybeans, wheat; hogs); flour mill, dairying. SCOTT AIR FORCE BASE is SE. City derives population and revenue from base personnel. Incorporated 1865.

O'Fallon, city (2000 population 46,169), SAINT CHARLES county, E MISSOURI, residential and commercial suburb 25 mi/40 km WNW of ST. LOUIS; 38°46′N 90°42′W. Roman Catholic convent. Agriculture re-

mains mainly to N, along MISSISSIPPI RIVER (soybeans, corn, vegetables). Manufacturing (paper products, fabricated metal products, transportation equipment, consumer goods). Rapid urban growth since 1970s.

O'Fallon Creek (O-FAH-luhn), c.90 mi/145 km long, SE MONTANA; rises in NW CARTER county, flows N, past ISMAY, to YELLOWSTONE RIVER 7 mi/11.3 km ENE of TERRY.

Ofanto River (o-FAHN-to), ancient *Aufidus*, 83 mi/134 km long, S ITALY; rises in the APENNINES near Torella de' Lombardi; flows E, N, and NE to ADRIATIC Sea 4 mi/6 km NW of BARLETTA. Southernmost river on E coast; below it are only a few small streams. Forms border between CAMPANIA, BASILICATA, and APULIA regions in upper course, between BARI and FOGGIA provinces in lower course.

Ofen (OF-en), Romansch *Pass dal Fuorn*, Italian *Passo del Fuarno*, alpine pass (7,051 ft/2,149 m), GRISONS canton, E SWITZERLAND. The Ofen Pass Road links the ENGADINE VALLEY to the Italian TYROL.

Ofen, HUNGARY: see BUDAPEST.

Offa (aw-FAH), town, KWARA state, SW NIGERIA, on railroad, 25 mi/40 km SSE of ILORIN. Agricultural trading center; shea-nut processing, cotton weaving; cassava, yams, corn; cattle.

Offaly (AH-fuh-lee), Gaelic *Uíbh Fhailí*, county (□ 768 sq mi/1,996.8 sq km; 2006 population 70,868), central IRELAND; ⊙ TULLAMORE. Borders ROSCOMMON, WESTMEATH, and MEATH counties to N, KILDARE county to E, LAOIGHIS and TIPPERARY counties to S, and GALWAY county to W. A part of the central plain of Ireland, the county is mostly flat with some sections covered by the BOG OF ALLEN. The SLIEVE BLOOM mountains are on the SE border. The SHANNON, the chief river, forms much of the W border. The BROSNA RIVER flows through the NW part of the county. Agriculture is the chief occupation; cattle, pigs, and poultry are bred in considerable quantity. Grains and potatoes are grown. Distilling. With adjacent areas, the region formed the kingdom of Offaly in ancient Ireland. It was known as King's County until the establishment of the Irish Free State. At CLONMACNOISE are the ruins of one of the principal religious centers of early Ireland. Other towns include BANAGHER, BIRR, and EDENDERRY.

Offa's Dyke (AW-fahz DEIK), ancient entrenchment of W ENGLAND and E WALES, from the DEE estuary to near the WYE estuary. It was built in the 8th century by Offa, king of MERCIA, as a barrier against the Welsh, specifically those residing in the kingdom of POWYS, and lies mainly along the border of England and Wales. Watt's Dyke, a similar work, roughly parallels a section of Offa's at a distance of c.2 mi/3.2 km. Parts of the dikes are well preserved.

Offemont (of-uh-MON), town (□ 2 sq mi/5.2 sq km), Territoire de BELFORT department, FRANCHE-COMTÉ region, NE FRANCE; 47°40′N 06°53′E. NE suburb of BELFORT.

Offenbach am Main (OF-fuhn-bahkh ahm MEIN), city, HESSE, S central GERMANY, 4 mi/6.4 km E of FRANKFURT, port on the Main River; 50°06′N 08°46′E. Industrial center long famous for the manufacturing of leather goods; chemicals, electrical products, textiles, and machinery. First mentioned in 977; it passed to the counts of Isenburg in the 15th century, and was annexed by Hesse-Darmstadt in 1815. Renaissance-style palace (1564–1578); museum of leathercraft, typography, and graphics.

Offenbach an der Quelch (OF-fuhn-bahkh ahn der KVELKH), village, RHINELAND-PALATINATE, W GERMANY, 4 mi/6.4 km E of Landau in der Pfalz, on the Quelch River; 49°12′N 08°12′E. Viticulture.

Offenburg (OF-fuhn-boorg), town, S UPPER RHINE, BADEN-Württemberg, GERMANY, at W foot of BLACK, on the KINZIG RIVER, 24 mi/39 km SSW of BADEN-BADEN; 48°29′N 07°57′E. Railroad junction. Market

center for farming (fruit, vegetables) and winegrowing region. Manufacturing of machinery, metal and woodworking. Has remnants of old fortifications; 18th-century town hall. Town first mentioned in early 12th century. Created free imperial city in 1235. Destroyed by French in 1689. Passed to Baden 1803.

Offerle (AWF-uhr-lee), village (2000 population 220), EDWARDS county, S central KANSAS, 8 mi/12.9 km WSW of KINSLEY; 37°53′N 99°33′W. In wheat region.

Officer River (AH-fi-suhr), 100 mi/161 km long, N SOUTH AUSTRALIA, S central AUSTRALIA; rises in SE MUSGRAVE RANGES; flows S and SE. Disappearing in arid central area; usually dry.

Offida (of-FEE-dah), town, ASCOLI PICENO province, The MARCHES, central ITALY, 8 mi/13 km NE of ASCOLI PICENO; 42°56′N 13°41′E. Wine-making center; woolen mill. Known for lace-making. Has 14th-century church, picturesque piazza.

Offingen (OF-fing-uhn), village, SWABIA, BAVARIA, S GERMANY, 4 mi/6.4 km ENE of Günzburg, near the DANUBE River; 48°29′N 10°23′E.

Offranville (o-frahn-VEEL), commune (□ 6 sq mi/15.6 sq km), SEINE-MARITIME department, HAUTE-NORMANDIE region, N FRANCE, 4 mi/6.4 km SSW of DIEPPE; 49°52′N 01°03′E. Manufacturing (plastics, cables).

Offutt Air Force Base, U.S. military installation (□ 3 sq mi/7.8 sq km), SARPY county, E NEBRASKA, c.11 mi/18 km SSE of OMAHA; 41°06′N 95°55′W. Established 1896 as Fort Crook, an army base. Converted to an air base in the early 1900s and renamed in 1924, it is now the headquarters of the U.S. Strategic Command. The Strategic Air Command (SAC) Museum previously on the base (moveed to ASHLAND 1998). President George W. Bush conducted one of the first major strategy sessions for the U.S. response to the September 11, 2001, terrorist attacks from this base on the day of the attack.

Ofidousa (o-fee-[TH]OO-sah), Italian *Ofidusa*, westernmost island of DODECANESE prefecture, SOUTH AEGEAN department, GREECE; 36°33′N 26°10′E. Also Ophidousa.

Ofin River (O-feen), 150 mi/241 km long, GHANA; rises N of KUMASI in ASHANTI uplands 20 mi/32 km NW of MAMPONG; flows SW, S, and E, past DUNKWA, to PRA RIVER, 20 mi/32 km ESE of Dunkwa. Its drainage basin is a timber and mining area; lumber mill at Dunkwa.

Ofotfjorden (AW-fawt-fyawr-uhn), inlet of VESTFJORDEN, NORDLAND county, N NORWAY, opposite LOFOTEN AND VESTERÅLEN islands; 45 mi/72 km long, 3 mi/4.8 km–6 mi/9.7 km wide. NARVIK is near its head. In World War II, the fjord and its E arm, Rombakkenfjorden, were scene (April–June 1940) of repeated naval action for control of NARVIK.

Ofra, Jewish settlement, 4 mi/7 km NE of RAMALLAH, on the border between Judaea and Samaria, WEST BANK. Orchards, tea factory, light industry. Center for Cave Research, run by the Society for the Protection of Nature in ISRAEL, is located here. Founded in 1975, it was the first of the settlements established by the right-wing Gush Emunim group of Jewish settlers on the ideologically extreme fringe.

Ofterdingen (OF-ter-ding-uhn), village, NECKAR-JURA, BADEN-Württemberg, S GERMANY, 8 mi/12.9 km S of Tübingen, on W slope of the SWABIAN JURA; 48°25′N 09°02′E. Fruit. Minstrel Heinrich von Ofterdingen lived here.

Oftersheim (OF-ters-heim), town, Lower NECKAR, BADEN-Württemberg, SW GERMANY, 1.5 mi/2.4 km SSE of SCHWETZINGEN; 49°22′N 08°36′E. Asparagus.

Oftringen (OFT-ring-uhn), commune (2000 population 10,305), AARGAU canton, N SWITZERLAND, 2 mi/3.2 km SSE of OLTEN; 47°19′N 07°55′E. Elevation 1,316 ft/401 m. Paper, watches.

Ofu (O-foo), volcanic island and county, Manua district, AMERICAN SAMOA, S PACIFIC OCEAN, c.60 mi/90 km from TUTUILA. Rises to c.1,585 ft/483 m. Thick vegetation.

Ofun (o-FAHNG), town, CROSS RIVER state, SE NIGERIA, on CROSS RIVER, 85 mi/137 km N of CALABAR. Market and fishing town.

Ofunato (OF-NAH-to), city, IWATE prefecture, N HONSHU, NE JAPAN, on the PACIFIC OCEAN, 53 mi/85 km S of MORIOKA; 39°04′N 141°42′E. Fishing (salmon); oysters, seaweed (*wakame*). Cement industry.

Ófutak, SERBIA: see FUTOG.

Oga (O-gah), city, Akita prefecture, N HONSHU, NE JAPAN, on OGA PENINSULA, 19 mi/30 km S of AKITA city; 39°53′N 139°51′E. Traditional woodcarving (masks). *Hatahata* sushi. Setting of Namahage Festival. Oga quasi-national park and Kojaku-no-Iwaya Cavern nearby.

Ogachi (O-gah-chee), town, Ogachi county, Akita prefecture, N HONSHU, NE JAPAN, 50 mi/80 km S of AKITA city; 39°03′N 140°27′E. Paulownia; strawberries; silver. Kurikoma quasi-national park and hot springs nearby. Poet Onono Komachi born here, c. C.E. 833.

Ogaden (O-gah-den), geographical region, SOMALI state, SE ETHIOPIA; 08°00′N 44°00′E. This is an arid plateau (1,500 ft/457 m–3,000 ft/914 m) intermittently watered by the FAFEN and JERER rivers. Borders on SOMALIA, and is inhabited mainly by Somali pastoral nomads. The region was conquered by Menelik II of Ethiopia in 1891. A clash (December 5, 1934) between Italian and Ethiopian troops at the watering hole of WELWEL was used as a pretext by ITALY to begin a war (1935–1936) against ETHIOPIA. Since 1960, Somali nationalists have demanded its union with SOMALIA, and there have been violent clashes over the region's precise borders. In the late 1970s, the SOVIET UNION, CUBA, SOUTH YEMEN, and LIBYA backed Ethiopian interests in the region, and SOMALIA withdrew its troops. However, fighting continued intermittently until 1988, when SOMALIA and ETHIOPIA signed a non-aggression pact. The war and devastating drought conditions have resulted in millions of refugees and acute resettlement problems.

Ogaki (O-gah-kee), city (2005 population 162,070), GIFU prefecture, central HONSHU, central JAPAN, on IBI RIVER, and 6 mi/10 km W of GIFU; 35°21′N 136°36′E. Cotton thread. Lime, marble. Has feudal castle. Bombed in World War II (1945).

Ogaki (O-gah-kee), town, on S coast of NOMI-SHIMA, HIROSHIMA prefecture, W JAPAN, 16 mi/25 km S of HIROSHIMA; 34°10′N 132°27′E. Oysters. Carnations; mandarin oranges.

Ogallala (o-guh-LUH-luh), city (2006 population 4,649), ⊙ KEITH county, SW central NEBRASKA, 50 mi/80 km W of NORTH PLATTE city, and on SOUTH PLATTE RIVER; 41°07′N 101°43′W. Manufacturing (machinery, electronic products, building materials, textiles, foods, machine products; printing). Livestock (cattle); wheat, corn, beans; dairy and poultry products. Kingsley Dam and LAKE C. W. MCCONAUGHY (Kingsley Reservoir) near by, on NORTH PLATTE RIVER. Old OREGON TRAIL follows S side of river. Lake McConaughy State Recreation Area to N and NW (seven units scattered along S sides of lake); Lake Ogallala State Recreation Area to N, S of Kingsley Dam. Front Street offers a reconstruction and reenactment of Old West town life. City laid out 1875, incorporated 1930.

Ogallala Aquifer, the main source of irrigation water in NEBRASKA and the remainder of the LLANO ESTACADO, of which it is the chief component. The Ogallala, of Tertiary age, contains thick permeable materials, consisting largely of sand and gravel, and underlies the Nebraska Sand Hills, extending beyond, especially at the S and E, where it merges with Quaternary sand and gravel. In places it is overlain by recent alluvium, as along the PLATTE RIVER.

Ogano (o-GAH-no), town, Chichibu county, SAITAMA prefecture, E central HONSHU, E central JAPAN, 37 mi/60 km W of URAWA; 36°00′N 139°00′E. Cucumbers.

Oga Peninsula (O-gah), Japanese *Oga-hanto* (O-gah–HAHN-to), AKITA prefecture, N HONSHU, NE JAPAN, in Sea of JAPAN; 13 mi/21 km long, 8 mi/12.9 km wide. Mountainous, with rocky shores. At its base is HACHIRO-GATA lagoon; Funakawaminato is on S shore.

Ogarëvka (uh-gah-RYOF-kah), town (2006 population 3,780), central TULA oblast, W central European Russia, on railroad spur and near highway, 21 mi/34 km S of TULA; 53°55′N 37°32′E. Elevation 679 ft/206 m. Iron and lignite mining.

Ogasa (o-GAH-sah), town, Ogasa county, SHIZUOKA prefecture, central HONSHU, E central JAPAN, 25 mi/40 km S of SHIZUOKA; 34°41′N 138°06′E.

Ogasawara (o-GAH-sah-WAH-rah), village, on CHICHI-JIMA island, BONIN Islands, Ogasawara district, Tokyo prefecture, SE JAPAN, 25 mi/40 km S of SHINJUKU; 27°05′N 142°11′E.

Ogasawara-gunto, JAPAN: see BONIN ISLANDS.

Ogata (O-gah-tah), town, Hata county, KOCHI prefecture, SW SHIKOKU, W JAPAN, 50 mi/80 km S of KOCHI; 33°01′N 133°00′E. Flowers. Formed 1943 by combining former villages of Nanasato, Irino, and Tanokuchi.

Ogata, town, Nakakubiki county, NIIGATA prefecture, central HONSHU, N central JAPAN, 59 mi/95 km S of NIIGATA; 37°13′N 138°19′E.

Ogata, town, Ono county, OITA prefecture, E KYUSHU, SW JAPAN, 19 mi/30 km S of OITA; 32°58′N 131°28′E.

Ogata (O-gah-tah), village, South Akita county, Akita prefecture, N HONSHU, NE JAPAN, 22 mi/35 km N of AKITA city; 40°00′N 139°57′E. Reclaimed land at HACHIROGATA, an agricultural development project.

Ogatsu (O-gahts), town, Mono county, MIYAGI prefecture, N HONSHU, NE JAPAN, 37 mi/60 km N of SENDAI; 38°31′N 141°28′E. Fishery.

Ogawa (o-GAH-wah), town, E Ibaraki county, IBARAKI prefecture, central HONSHU, E central JAPAN, 16 mi/25 km S of MITO; 36°10′N 140°22′E. Tobacco; chicken.

Ogawa, town, Shimomashiki county, KUMAMOTO prefecture, W KYUSHU, SW JAPAN, 12 mi/20 km S of KUMAMOTO; 32°35′N 130°42′E.

Ogawa, town, Hiki county, SAITAMA prefecture, E central HONSHU, E central JAPAN, 25 mi/40 km N of URAWA; 36°03′N 139°15′E.

Ogawa, town, Nasu county, TOCHIGI prefecture, central HONSHU, N central JAPAN, 19 mi/30 km N of UTSUNOMIYA; 36°45′N 140°08′E.

Ogawa (o-GAH-wah), village, Naka county, IBARAKI prefecture, central HONSHU, E central JAPAN, 19 mi/30 km N of MITO; 36°36′N 140°19′E.

Ogawa, village, Kamiminochi county, NAGANO prefecture, central HONSHU, central JAPAN, 12 mi/20 km W of NAGANO; 36°36′N 137°58′E.

Ogawara (O-gah-wah-rah), town, Shibata county, MIYAGI prefecture, N HONSHU, N JAPAN, 16 mi/25 km SW of SENDAI; 38°02′N 140°44′E.

Ogawara-numa (o-gah-WAH-rah–NOO-mah), lagoon (□ 25 sq mi/65 sq km), Aomori prefecture, N HONSHU, N JAPAN, 16 mi/26 km NNW of HACHINOHE, near E coast; 9 mi/14.5 km long, 3 mi/4.8 km wide; stream connects with the PACIFIC OCEAN.

Ogbomosho (og-BO-mo-sho), city, headquarters of Local Government Area, OYO state, SW NIGERIA; 08°08′N 04°16′E. It is the trade center for a farming region, producing cotton (for weaving cloth), yams, cassava, maize, and tobacco. Founded in the 17th century. It resisted Fulani invasions in the early 19th century and grew by absorbing refugees from towns destroyed by the Fulani. Seat of a teachers college.

Ogdem, TURKEY: see YUSUFELI.

Ogden, city (2006 population 78,086), ⊙ WEBER county, N UTAH, at the confluence of the OGDEN and WEBER rivers; 41°13′N 111°58′W. Aerospace industries; HILL AIR FORCE BASE (Utah's largest employer, with 17,000 employees) and the Ogden Defense Depot to W are the major employers. The site of a trading post in the 1820s, the area was settled by Mormons in 1847.

Seat of Weber State University and a state industrial school; has a Mormon temple and tabernacle. Surrounded by mountains, and major ski resorts in the area include Snow Basin to E (elevation 4,300 ft/1,311 m), Nordic Valley, and Powder Mountain. One of the sites of the annual Sundance Film Festival. Important railroad center. Manufacturing (paper mill; weapons design; transportation equipment, machinery, fabricated metal products, foods, building materials, carbon fiber products, chemicals, computer products, electrical and electronic goods, consumer goods, apparel, motor vehicles, paper goods; printing and publishing, sugar beet refining). WASATCH RANGE and National Forest to E. Mount OGDEN 5 mi/8 km to SE. Fort Buenaventura State Historical Park to W. Ogden Bay Waterfowl Management Area to W. Incorporated 1851.

Ogden, town (2000 population 2,023), BOONE county, central IOWA, 7 mi/11.3 km W of BOONE; 42°02′N 94°01′W. In agricultural area. Incorporated 1878.

Ogden, town (2006 population 1,762), RILEY county, NE central KANSAS, on KANSAS River, and 9 mi/14.5 km SW of MANHATTAN; 39°06′N 96°42′W. Grazing; agriculture. At entrance to Fort Riley Military Reserve.

Ogden (AHG-duhn), unincorporated town (□ 4 sq mi/10.4 sq km; 2000 population 5,481), NEW HANOVER county, SE NORTH CAROLINA, residential suburb 7 mi/11.3 km ENE of WILMINGTON, near Middle Sound (INTRACOASTAL WATERWAY); 34°16′N 77°47′W. ATLANTIC OCEAN 4 mi/6.4 km to SE. Service industries; retail trade.

Ogden (AHG-den), village (□ 29 sq mi/75.4 sq km; 2006 population 793), Estrie region, S QUEBEC, E Canada, 3 mi/5 km from STANSTEAD; 45°03′N 72°08′W.

Ogden, village (2000 population 214), LITTLE RIVER county, extreme SW ARKANSAS, 11 mi/18 km N of TEXARKANA, and near RED RIVER; 33°34′N 94°02′W.

Ogden, village (2000 population 743), CHAMPAIGN county, E ILLINOIS, 14 mi/23 km E of URBANA; 40°07′N 87°57′W. In agricultural area.

Ogden Dunes, town (2000 population 1,313), PORTER county, NW INDIANA, on Lake MICHIGAN, 7 mi/11.3 km E of downtown GARY; 41°37′N 87°11′W. Residential suburb. INDIANA DUNES State Park nearby to E.

Ogden Island, ST. LAWRENCE county, N NEW YORK, in the SAINT LAWRENCE River (rapids here), at Canadian (ONTARIO) border, just N of WADDINGTON; c.2.75 mi/4.43 km long, 0.5 mi/0.8 km–1.5 mi/2.4 km wide; 44°52′N 75°13′W. Connected to New York shore by bridge.

Ogden, Mount, on border between U.S. (ALASKA) and CANADA (BRITISH COLUMBIA), in COAST RANGE, 40 mi/64 km ENE of JUNEAU (Alaska); 58°26′N 133°23′W.

Ogden, Mount, (9,572 ft/2,918 m), N central UTAH, 5 mi/8 km SE of OGDEN, in WASATCH Mountains, WEBER county, in Wasatch National Forest. Snow Basin Ski Resort to NE.

Ogden River, 35 mi/56 km long, UTAH; formed in the WASATCH RANGE, WEBER county, N Utah, at the convergence of its North, Middle, and South forks at Pineview Reservoir near HUNTSVILLE; flows SW through Wasatch National Forest and city of OGDEN to join the WEBER RIVER just W of Ogden. The river has been used for irrigation for nearly a century. The Ogden-Brigham Canal (c.20 mi/30 km long) carries water N to BRIGHAM CITY. Pineview Dam, completed by the Bureau of Reclamation in 1937, is at the head of Ogden Canyon. The headwater region of the Ogden is a winter sports area. South Fork formed at Causey Reservoir in E Weber county by three forks, flows W c.20 mi/32 km; Middle Fork rises in NE Weber county flows SW c.15 mi/24 km; North Fork rises in N Weber county, flows SSE c.15 mi/24 km.

Ogdensburg (AHG-duhnz-buhrg), city (□ 8 sq mi/20.8 sq km; 2006 population 11,346), ST. LAWRENCE county, N NEW YORK, on the SAINT LAWRENCE RIVER at the mouth of the OSWEGATCHIE RIVER, in a resort area, opposite PRESCOTT (ONTARIO); connected by Ogdensburg-Prescott International Bridge); 44°42'N 75°28'W. Seaport. A variety of light industrial products are made here, including shade rollers and blinds. Settled by French missionaries and trappers 1749; was strategically important in the War of 1812. Seat of Mater Dei, and Wadhams Hall Seminary and College and a museum with works of Frederic Remington, who lived here. Rhoda Fox Graves was the first woman to serve the state's Assembly and Senate. Incorporated as a city 1868.

Ogdensburg, village (2006 population 211), WAUPACA county, central WISCONSIN, 7 mi/11.3 km N of WAUPACA; 44°27'N 89°01'W. In farm area. Manufacturing (fiberglass products).

Ogdensburg (AHG-duhnz-buhrg), borough (2006 population 2,623), SUSSEX county, NW NEW JERSEY, on WALLKILL RIVER, and 8 mi/12.9 km ENE of NEWTON; 41°04'N 74°35'W. Incorporated 1914.

Ogeechee River (o-GEE-chee), c.250 mi/402 km long, E GEORGIA; rises E of GREENSBORO; flows SE, past LOUISVILLE and MILLEN, to the ATLANTIC OCEAN through OSSABAW Sound, 15 mi/24 km S of SAVANNAH; 33°31'N 82°54'W. Rice plantations formerly located along lower reaches.

Ogema (O-ge-muh), village (2000 population 143), BECKER county, W MINNESOTA, 20 mi/32 km N of DETROIT LAKES, in SW part of White Earth Indian Reservation, near WILD RICE RIVER; 47°06'N 95°55'W. Dairying; grain, wild rice. Numerous small lakes to E, including WHITE EARTH LAKE.

Ogema, village, PRICE county, N WISCONSIN, 28 mi/45 km W of TOMAHAWK. In forested area. Manufacturing (wood products, machinery). Timms Hill (1951.5 ft/594.8 m) highest point in Wisconsin, 5 mi/8 km to E.

Ogemaw (AH-ge-maw), county (□ 574 sq mi/1,492.4 sq km; 2006 population 21,665), NE central MICHIGAN; ⊙ WEST BRANCH; 44°19'N 84°07'W. Drained by AU GRES and RIFLE rivers. Agriculture (cattle, hogs; corn, wheat, oats, barley; dairy products). Part of Huron. National Forest in NE corner; state forest, game refuge, and many small lakes, especially in E (resorts), are in county. Rifle River State Park in NE. Organized 1875.

Oger, LATVIA: see OGRE.

Oge-shima (o-GE-shee-MAH), island (□ 4 sq mi/10.4 sq km), TOKUSHIMA prefecture, W JAPAN, just off NE coast of SHIKOKU, in NARUTO STRAIT opposite SW coast of AWAJI-SHIMA island; 3.5 mi/5.6 km long, 2.5 mi/4 km wide. Deeply indented W coast. Hilly; fertile. Produces salt, rice. Fishing.

Oggersheim (OG-gers-heim), W district of LUDWIGSHAFEN, RHINELAND-PALATINATE, W GERMANY; 49°30'N 08°23'E. Manufacturing of machinery, cotton; malting, brewing, distilling. Home of German chancellor Helmut Kohl. Has 18th-century church.

Oggevatn (AWG-guh-VAH-tuhn), lake (□ 4 sq mi/10.4 sq km), AUST-AGDER county, S NORWAY, 18 mi/29 km N of KRISTIANSAND. Skirted by railroad.

Oggiono (od-JYO-no), town, COMO province, LOMBARDY, N ITALY, 13 mi/21 km E of COMO, on small lake; 45°47'N 09°20'E. Manufacturing (silk textiles, bedspreads, machinery). Octagonal baptistery (11th–12th century).

Ogi (O-gee), town, on S SADO island, Sado county, NIIGATA prefecture, N central JAPAN, 40 mi/65 km W of NIIGATA; 37°48'N 138°16'E. Traditional boats.

Ogi, town, Naoiri county, OITA prefecture, E KYUSHU, SW JAPAN, 28 mi/45 km S of OITA; 32°55'N 131°18'E.

Ogi, town, Ogi county, SAGA prefecture, NW KYUSHU, SW JAPAN, 6 mi/10 km N of SAGA; 33°17'N 130°12'E.

Ogidi (aw-gee-DEE), town, ANAMBRA state, S NIGERIA, 5 mi/8 km E of ONITSHA. Road junction; palm oil and kernels, kola nuts. Iyi-Enu Mission Hospital.

Ogíjares (o-HEE-hah-res), S suburb of GRANADA, GRANADA province, S SPAIN; 37°07'N 03°36'W. Olive oil, wine, vegetables, hemp, tobacco.

Ogilvie (O-guhl-vee), community, S MANITOBA, W central Canada, 42 mi/67 km from PORTAGE LA PRAIRIE, and in Westbourne rural municipality; 50°17'N 99°04'W.

Ogilvie, village (2000 population 474), KANABEC county, E MINNESOTA, 7 mi/11.3 km WSW of MORA, on Groundhouse River; 45°49'N 93°25'W. Grain, potatoes; livestock, poultry; dairying; manufacturing. Fish Lake to E.

Ogilvie Range, E YUKON, NW Canada, at NW end of MACKENZIE MOUNTAINS, extending c.150 mi/241 km WNW from NORTHWEST TERRITORIES border.

Ogimi (O-gee-mee), village, Kunigami county, Okinawa prefecture, SW JAPAN, 43 mi/70 km N of NAHA; 26°41'N 128°07'E. Yanbaru water railroad located here.

Oginski Canal (o-GEEN-skee), Polish *Kanal Ogiński-kiego*, SW BELARUS, in PRIPET Marshes, between YASELDA RIVER (S) and VYGONOVO LAKE (N); 34 mi/55 km long. Built 1777. TELEKHANY on W shore. Part of DNIEPER-NEMAN waterway.

Oglanli (og-lahn-LEE), oil town, W BALKAN weloyat, W TURKMENISTAN, at N foot of GREATER BALKAN RANGE, 25 mi/40 km N of NEBITDAG; 39°52'N 54°30'E. Bentonite deposits.

Ogle (O-guhl), county (□ 763 sq mi/1,983.8 sq km; 2006 population 54,826), N ILLINOIS; ⊙ OREGON; 42°02'N 89°19'W. Agriculture (cattle, hogs; dairy products; corn, soybeans, hay). Food processing plants; manufacturing (textile finishing, die-casting; paper products). Also other manufacturing at Oregon and ROCHELLE. Drained by ROCK, LEAF, and KYTE rivers, and by Kilbuck and ELKHORN creeks. Includes White Pines Forest, Lowden, and Castle Rock state parks and Lowder-Miller State Forest. John Deere home at GRAND DETOUR (S). Nuclear power plants BYRON 1 (initial criticality February 2, 1985) and Byron 2 (initial criticality January 9, 1987) are 17 mi/27 km SW of ROCKFORD; use cooling water from the Rock River and each has a maximum dependable capacity of 1,105 MWe. Formed 1836.

Oglesby (O-guhlz-bee), city (2000 population 3,647), LA SALLE county, N ILLINOIS, on ILLINOIS RIVER, near mouth of VERMILION RIVER, 3 mi/4.8 km SE of LA SALLE; 41°17'N 89°03'W. Manufacturing (consumer goods, building materials); agriculture (dairying; corn, wheat, soybeans; cattle, hogs). Incorporated 1902. Illinois Valley Community College 1 mi/1.6 km W.

Oglesby (AHG-uhlz-bee), village (2006 population 424), CORYELL county, central TEXAS, near LEON RIVER, 23 mi/37 km WSW of WACO; 31°25'N 97°30'W. In cattle, farm area; manufacturing (machining). Beelton Lake reservoir and Mother Neff State Park to S.

Oglethorpe (O-guhl-thorp), county (□ 442 sq mi/1,149.2 sq km; 2006 population 13,997), NE GEORGIA; ⊙ LEXINGTON; 33°53'N 83°05'W. Piedmont agriculture (wheat, fruit); cattle, hogs, poultry; in timber area. Formed 1793.

Oglethorpe (O-guhl-thorp), town (2000 population 1,200), ⊙ MACON county, W central GEORGIA, c.45 mi/72 km SW of MACON, and on FLINT RIVER opposite MONTEZUMA; 32°17'N 84°04'W. Manufacturing includes animal feeds, wood and paper products. Incorporated 1849.

Oglethorpe, Mount (O-guhl-thorp) (3,290 ft/1,003 m), PICKENS county, N GEORGIA, 6 mi/9.7 km NE of JASPER, at S end of the BLUE RIDGE, at SW terminus of APPALACHIAN Trail; 34°29'N 84°19'W. A white marble shaft has been erected at its summit as a memorial to General James E. Oglethorpe, founder of the Georgia colony. Called Grassy Mountain until 1929.

Oglio River (O-lyo), ancient *Ollius*, 175 mi/282 km long, LOMBARDY, N ITALY; formed at PONTE DI LEGNO by union of several glacial streams rising in S ORTLES mountain group; flows SSW, past EDOLO, through VAL CAMONICA and LAGO D'ISEO, S, past PALAZZOLO SULL'OGLIO, and SE, across Lombard plain, to PO RIVER 10 mi/16 km SW of MANTUA. Receives MELLA and CHIESE rivers (left) and waters from ADAMELLO mountain glaciers. Navigable for 20 mi/32 km in lower course; forms border between BRESCIA and CREMONA provinces. Extensively used for hydroelectric power in Val Camonica and for irrigation in Lombard plain.

Ogmore River, Wales: see OGWR, AFON.

Ogmore Vale, Wales: see NANT-Y-MOEL.

Ognon River (on-YON), 115 mi/185 km long, E FRANCE; rises in the S VOSGES MOUNTAINS near the Balloon (summit) de SERVANCE; flows SW, past Montbozon, Marnay, and Pesmes, to the SAÔNE RIVER above Pontailler-sur-Saône. Through most of its course it forms border between HAUTE-SAÔNE and DOUBS departments. A favorite stream for fishermen and kayak enthusiasts.

Ognyanovo (og-NYAH-no-vo), village, SOFIA oblast, GURMEN obshtina, BULGARIA; 41°37'N 23°46'E. Mineral spa resort.

Ognyanovo (og-NYAH-no-vo), village, PLOVDIV oblast, PAZARDZHIK obshtina, BULGARIA; 42°09'N 24°24'E.

Ogo (O-go), town, Seta county, GUMMA prefecture, central HONSHU, N central JAPAN, 6 mi/10 km N of MAEBASHI; 36°24'N 139°09'E.

Ogo (oo-GOU), highland of N SOMALIA, rising steeply from Gulf of ADEN and merging in S with the HAUD plateau. Rises to 7,900 ft/2,408 m.

Ogoe (O-go-e), town, Tamura county, FUKUSHIMA prefecture, N central HONSHU, NE JAPAN, 28 mi/45 km S of FUKUSHIMA city; 37°22'N 140°37'E.

Ogoja (aw-GAW-jah), former province, one of the former Eastern Provinces, SE NIGERIA; capital was OGOJA. Much of the area now is in CROSS RIVER state.

Ogoja (aw-GAW-jah), town, headquarters of Local Government Area, CROSS RIVER state, SE NIGERIA, 85 mi/137 km ENE of ENUGU; 06°39'N 08°43'E. Trade center; shea nuts, palm oil and kernels, sesame, yams, cassava, corn.

Ogoki (o-GO-kee), river, c.300 mi/480 km long, ONTARIO, E central Canada; rises in lakes W of Lake NIPIGON, W central Ontario; flows NE to the ALBANY RIVER; 51°38'N 85°57'W. A dam at Waboose Rapids forms a reservoir (45 mi/72 km long), which drains to the S into Lake Nipigon.

Ogoño, Cape (o-GO-nyo), VIZCAYA province, N SPAIN, on BAY OF BISCAY, 19 mi/31 km NE of BILBAO; 43°26'N 02°38'W.

Ogooué (o-go-WAI), river, circa 560 mi/901 km long, central Africa; rises on the Batéké Plateau, Congo Republic; flows NW and W, past FRANCEVILLE and LAMBARÉNÉ and across SE and central GABON to the GULF OF GUINEA, near PORT-GENTIL (Gabon), where it forms a large delta. Navigable for most of its length, the river is the chief economic artery of Gabon. Also spelled Ogowe.

Ogooué-Ivindo (OO-goo-ai–EE-vin-do), province (□ 17,790 sq mi/46,254 sq km; 2002 population 63,000), NE GABON, on CONGO (E) border; ⊙ MAKOKOU; 00°48'N 13°00'E. Bounded S by HAUT-OGOOUÉ and OGOOUÉ-LOLO, N by WOLEU-NTEM, and W by MOYEN OGOOUÉ and NGOUNIÉ provinces. Forested, but also grows important coffee and cocoa crops. Large iron deposits are present.

Ogooué-Lolo (OO-goo-ai–LO-lo), province (□ 9,799 sq mi/25,477.4 sq km; 2002 population 56,600), S GABON, on CONGO (S) border; ⊙ KOULAMOUTOU; 00°50'S 12°37'E. Bounded E by HAUT-OGOOUÉ, N by OGOOUÉ-IVINDO, and W by NGOUNIÉ provinces. Cocoa and coffee grow in this forested region. Some gold deposits exist.

Ogooué-Maritime (OO-goo-ai–MER-i-teem), province (□ 8,838 sq mi/22,978.8 sq km; 2002 population 126,200), W GABON, on GULF OF GUINEA (W); ⊙ PORT-GENTIL; 01°21'S 09°33'E. Bounded on S by NYANGA, E

Cross-references are shown in SMALL CAPITALS. The pronunciation guide is shown on page xix. The sources of population figures are shown on page xvii.

by MOYEN OGOOUÉ and NGOUNIÉ, and N by ESTUAIRE provinces. Heavily forested coastal lowland where sawmills and the lumber industry dominate many small villages. Oil deposits exist in the area. Port-Gentil, the only large city, contains many industrial and commercial businesses.

Ogori (o-GO-ree), city, FUKUOKA prefecture, N KYUSHU, SW JAPAN, 16 mi/25 km S of FUKUOKA; 33°23′N 130°33′E.

Ogori (o-GO-ree), town, Yoshiki county, YAMAGUCHI prefecture, SW HONSHU, W JAPAN, 6 mi/10 km S of YAMAGUCHI; 34°05′N 131°23′E.

Ogose (O-go-se), town, Iruma county, SAITAMA prefecture, E central HONSHU, E central JAPAN, 22 mi/35 km W of URAWA; 35°57′N 139°17′E.

Ogosta River (o-GO-stah), 91 mi/146 km long, NW BULGARIA; rises W of BERKOVITSA in the CHIPROVSKA MOUNTAINS; flows generally NE, past MONTANA, to the DANUBE River 5 mi/8 km W of ORYAHOVO; 43°36′N 23°40′E. Receives the Botunya River (right).

Ogou, prefecture (2005 population 264,915), PLATEAUX region, S TOGO ⊙ ATAKPAMÉ; 07°30′N 01°20′E.

Ogowe, river, central AFRICA: see OGOOUÉ.

Ogradea (o-GRAH-da), village, MEHEDINŢI county, SW ROMANIA, on the DANUBE River, and 5 mi/8 km SW of ORŞOVA; 44°40′N 22°19′E. Chromium and nickel mining. Includes Ogradena–Nou (Hungarian Újasszonyrét) and Ogradena–Veche (Hungarian Óasszonyrét).

Ógradiska, CROATIA: see STARA GRADIŠKA.

Ograzden (o-grahzh-DEN), Bulgarian Ograzhden, Serbo-Croatian Ogražden, mountain, on Macedonian-Bulgarian border, between STRUMA (E) and STRUMICA (S, W) rivers; 41°33′N 22°53′E. Heavily forested. Highest peak (5,717 ft/1,743 m) is 14 mi/23 km NE of STRUMICA (MACEDONIA).

Ogre (O-grai), German Oger, city (2000 population 26,573), central LATVIA, in VIDZEME, on right bank of the DVINA (DAUGAVA) River, at mouth of the Ogre River, and 20 mi/32 km ESE of RĪGA; 56°49′N 24°36′E. Health and summer resort; sawmilling, food processing. Kegums hydroelectric plant is 6 mi/10 km SE.

Ogreyak, SE EUROPE: see KADIITSA.

Ogualik Island, Canada: see COD ISLAND.

Oguchi (O-goo-chee), town, Niwa county, AICHI prefecture, S central HONSHU, central JAPAN, 9 mi/15 km N of NAGOYA; 35°19′N 136°54′E.

Oguchi (O-goo-chee), village, Ishikawa county, ISHIKAWA prefecture, central HONSHU, central JAPAN, 22 mi/35 km S of KANAZAWA; 36°16′N 136°39′E. Hot springs; skiing areas.

Ogulin (O-guh-leen), town, W CROATIA, on DOBRA RIVER (which here disappears underground), and 20 mi/32 km SW of KARLOVAC, in the GORSKI KOTAR region; 45°16′N 15°14′E. Railroad junction; local trade center; lumber and wood processing, appliance manufacturing. Has 15th-century castle.

Ogumali, NIGERIA: see IGUMALE.

Ogun (o-GOON), state (□ 6,471 sq mi/16,824.6 sq km; 2006 population 3,728,098), SW NIGERIA, on the Bight of BENIN; ⊙ ABEOKUTA; 07°00′N 03°35′E. Bordered N by OYO and OSUN states, E by ONDO state, S by LAGOS state, and W by BENIN. In tropical rain forest zone; drained by OGUN RIVER. Agriculture includes cocao, kola nuts, rubber, oil palms; timber, livestock. Manufacturing of ceramics, clay brick, carpets, and clothing; breweries. Limestone and chalk phosphate deposits. Main centers are Abeokuta, OTTA, and SHAGAMU. Olumo Rock (a religious site in Abeokuta) and Birikisu Sungho Shrine are here.

Oguni (O-goo-nee), town, Aso county, KUMAMOTO prefecture, N central KYUSHU, SW JAPAN, 28 mi/45 km N of KUMAMOTO; 33°06′N 131°04′E. Dairying. Aso-Kuju National Park nearby.

Oguni, town, Kariwa county, NIIGATA prefecture, central HONSHU, N central JAPAN, 43 mi/70 km N of NIIGATA; 37°18′N 138°42′E.

Oguni, town, West Okitama county, YAMAGATA prefecture, N HONSHU, NE JAPAN, 37 mi/60 km S of YAMAGATA city; 38°03′N 139°44′E. Bandai-Asahi National Park nearby.

Ogunquit (o-GUHN-kwit), village, in WELLS town, YORK county, SW MAINE, on the ATLANTIC coast, 17 mi/27 km SSE of ALFRED; 43°15′N 70°36′W. Summer resort frequented especially by artists; summer playhouse.

Ogun River (o-GOON), 200 mi/322 km long, SW NIGERIA; rises in OYO state NW of Igboho; flows S into OGUN state, past OLOKEMEJI and ABEOKUTA, emptying into LAGOS Lagoon, across from LAGOS (LAGOS state).

Oguta (aw-goo-TAH), town, IMO state, S NIGERIA, on arm of NIGER River, near head of delta, 25 mi/40 km NW of OWERRI; 05°42′N 06°48′E. Market town. Cocoyams, cassava, rice, palm oil and kernels, kola nuts. Lignite deposits. Fish from area lake.

Oguz (uh-GOOZ), town, ⊙ Oguz region, N AZERBAIJAN, on S slope of Greater CAUCASUS Mountains, 15 mi/24 km SE of Nukha; 41°06′N 47°28′E. Tobacco, rice; asphalt. Formerly Vartashen.

Ogwashi Uku (og-WAH-shee OO-koo), town, DELTA state, S NIGERIA, 60 mi/97 km E of BENIN; 06°10′N 06°31′W. Palm oil and kernels, hardwood, rubber, kola nuts, yams, cassava, corn, plantains. Lignite deposits. Also Ugwashi Uku.

Ogwr, Afon (O-goor) or **Ogmore River**, 16 mi/26 km long, Rhondda Cynon Taff, S Wales; rises just N of NANT-Y-MOEL; flows S and SW, past Bridgend and Newcastle, to BRISTOL CHANNEL 3 mi/4.8 km ESE of PORTHCAWL. Receives Garw River 3 mi/4.8 km N of Bridgend. Formerly in MID GLAMORGAN, abolished 1996.

Ogwu (OG-woo), town, IMO state, S NIGERIA, on road, 20 mi/32 km NW of OWERRI; 05°26′N 07°43′E. Market center. Bananas, maize, cassava.

Ógyalla, SLOVAKIA: see HURBANOVO.

Ohaaki, site of geothermal power station, at Broadlands, 19 mi/30 km NE of TAUPO, Volcanic Plateau, NORTH ISLAND. Site of NEW ZEALAND's second geothermal power station (and world's newest, 1989) drawing steam and water from 0.6 mi/1 km down to drive turbines, and pump water and geothermal fluids back into ground.

Ohai (O-hei), town, SOUTHLAND region, SOUTH ISLAND, NEW ZEALAND, 48 mi/77 km N of INVERCARGILL; 45°55′S 167°57′E. Significant coal-mining center, at railroad terminus from Invercargill.

Ohakune (O-ha-KOO-nee), town, Ruapehu District, central NORTH ISLAND, NEW ZEALAND, near Main Trunk railroad midpoint between WELLINGTON (S) and AUCKLAND (N); 39°25′S 175°25′E. Once railroad camp, now serving area with livestock, sawmills, some market gardening, and tourism linked to adjacent volcanic peak, RUAPEHU.

Ohalatva, RUSSIA: see AGALATOVO.

Ohanes (o-AH-nes), town, ALMERÍA province, S SPAIN, 20 mi/32 km NW of ALMERÍA; 37°02′N 02°44′W. Cherries, grapes, wine.

Ohangwena Region (o-hahn-GWAI-nuh), administrative division (2001 population 228,384), N NAMIBIA, stretches in narrow strip along Angolan border from OSHAKATI to Mpungu; 17°20′S 16°40′E. It is the 2nd-smallest political region, densely populated and tribally split by border with ANGOLA.

Ohara (O-hah-rah), town, Isumi county, CHIBA prefecture, E central HONSHU, E central JAPAN, on E coast of BOSO PENINSULA, 22 mi/35 km S of CHIBA; 35°15′N 140°23′E.

Ohara, town, Aida county, OKAYAMA prefecture, SW HONSHU, W JAPAN, 40 mi/65 km N of OKAYAMA; 35°07′N 134°19′E.

O'Hara (o-HER-ah), township, ALLEGHENY county, W PENNSYLVANIA, residential suburb 6 mi/9.7 km NE of PITTSBURGH; 40°30′N 79°54′W. ALLEGHENY RIVER is township's S boundary.

O'Hare International Airport, serving CHICAGO, ILLINOIS, (20 mi/32 km northwest of the city center) has an of elevation 669 ft/204 m. The airport is served by the Chicago Transit Authority's Blue Line providing 24-hour service between the airport and downtown. The world's second-busiest airport, O'Hare has seven runways and four passenger terminals, handling over 76 million passengers annually. Its fifteen cargo terminals handle over 1.7 million tonnes of cargo annually. The airport also has a light rail intra-airport transit system with service between terminals.

Developed as Orchard Field in 1945, in 1949 the site was renamed O'Hare International Airport for the U.S. Congressional Medal of Honor winner, Lieutenant Commander Edward "Butch" O'Hare. There is a replica of the World War II F3F-4 flighter plane that he flew in terminal 2. In 1955, O'Hare's first official year of operation, it served 176,902 passengers.

Oharu (O-hah-roo), town, Ama county, AICHI prefecture, S central HONSHU, central JAPAN, 5 mi/8 km W of NAGOYA; 35°10′N 136°49′E.

Ohasama (o-HAH-sah-mah), town, Hienuki county, IWATE prefecture, N HONSHU, NE JAPAN, 11 mi/18 km SSE of MORIOKA; 39°27′N 141°17′E.

Ohata (O-hah-tah), town, Shimokita county, Aomori prefecture, N HONSHU, N JAPAN, on TSUGARU Strait, 47 mi/75 km N of AOMORI; 41°24′N 141°10′E. Hot springs nearby.

Ohatchee (o-HAT-chee), town (2000 population 1,215), Calhoun co., E Alabama, 12 mi/19 km NW of Anniston, between Ohatchee Creek and Neely Henry Lake Manufacturing (paint and children's clothes). Ohatchee is the incorporated name for Ohatchie. The Ohatchee Creek Ranch, which houses more than 50 different animal species, is located just E. Inc. in 1956.

Ohaton (o-HA-tuhn), unincorporated village, central ALBERTA, W Canada, 9 mi/14 km ESE of CAMROSE, in Camrose County No. 22; 52°58′N 112°40′W. Coal mining; oil and gas; wheat, oats, barley.

Ohau, Lake (O-ou) (□ 23 sq mi/59.8 sq km), upper WAITAKI valley, SOUTH ISLAND, NEW ZEALAND, 65 mi/105 km W of TIMARU; 44°15′S 169°51′E. Glacially shaped valley; 10 mi/16 km long, 3 mi/4.8 km wide. Outlet: Ohau River and Ohau Hydro-Electricity Canal, which flows E to LAKE BENMORE. One of three lakes which occupy glacially shaped valleys on the E side of the SOUTHERN ALPS in the upper Waitaki and MACKENZIE districts.

Oheteroa, FRENCH POLYNESIA: see RURUTU.

Ohey, commune (2006 population 4,298), Namur district, NAMUR province, S central BELGIUM, 12 mi/19 km E of NAMUR; 50°26′N 05°08′E.

O'Higgins, CHILE: see LIBERTADOR GENERAL BERNARDO O'HIGGINS.

O'Higgins Land, ANTARCTICA: see ANTARCTIC PENINSULA.

Ohinemuri (o-hi-NEE-moo-ree), stream, NEW ZEALAND. Enters WAIHOU RIVER at PAEROA from former gold-mining area.

Ohio (o-HEI-o), state (□ 44,828 mi/116,105 sq km; 1995 estimated population 11,150,506; 2000 population 11,353,140), N U.S., in the GREAT LAKES region of the MIDWEST, admitted as the seventeenth state of the Union in 1803; ⊙ COLUMBUS (the largest city). Ohio is nicknamed the "Buckeye State" because of the many Buckeye trees that once covered its hills and plains.

Geography

Other major cities are CLEVELAND, CINCINNATI, TOLEDO, and AKRON. The OHIO RIVER, from which the state takes its name, separates it in the SE from WEST VIRGINIA and in the S from Kentucky. Ohio is bounded on the W by INDIANA, on the N by MICHIGAN and LAKE ERIE, and on the E (N of the Ohio River) by PENNSYLVANIA. From the dunes on Lake Erie to the gorge-cut plateau along the Ohio River, the land is fairly flat, with some pleasant rolling country and, in the SE, rugged little hills leading to the mountains of

Area is shown by the symbol □, and capital city or county seat by ⊙.

West Virginia. Before the coming of settlers to Ohio, it was covered with many square miles of virgin forest, but today only vestiges of the trees that helped to build the many cities remain.

Economy

The state is highly industrialized. Yet it continues to draw from the earth; it leads the nation in the production of lime, is second in the production of clays, and ranks high in the production of salt. Also produced are sand and gravel, stone, and coal. The land supports rich farms, especially where the soil was improved ages ago by glacier-ground limestone. Although most of the state's income is derived from commerce and manufacturing, Ohio has extensive farms, and produces large amounts of corn, soybeans, hay, wheat, cattle, hogs, and dairy items. Railroads and highways crisscross the state, bearing raw materials and manufactured goods. The Lake Erie ports, Toledo and Cleveland, handle much iron and copper ore, coal, oil, and finished materials (including steel and automobile parts). In spite of a massive decline in manufacturing employment since the 1970s, the state is still a major manufacturing center. Heavy industry is still significant but there is more employment in polymer industries now than in the steel industry at its peak. Leading products include polymers, transportation equipment, primary and fabricated metals, and machinery.

History - 1600s to 1776

In prehistoric times Ohio was inhabited by the Mound Builders, many of whose mounds are preserved in state parks and in the HOPEWELL CULTURE NATIONAL HISTORICAL PARK (previously Mound City Group National Monument). Before the arrival of Europeans, E Ohio was the scene of warfare between the Iroquois and the Erie, resulting in the extermination of the Erie. In addition to the Iroquois, other Native American tribes soon prominent in the region were the Miami, the Shawnee, and the Ottawa. La Salle began his explorations of the Ohio valley in 1669 and claimed the entire area for France. The Ohio River became a magnet for fur traders and landseekers, and the British, moving in, hotly contested the French claims. Rivalry for control of the forks of the Ohio River (PITTSBURGH, Pennsylvania) led to the outbreak (1754) of the last of the French and Indian Wars. The defeat of the French caused the transfer of the land to the British, but British possession was disturbed by Pontiac's Rebellion. The British government issued a proclamation in 1763 forbidding settlement W of the APPALACHIAN MOUNTAINS. Then in 1774, with the Quebec Act, the British placed the region between the Ohio River and the Great Lakes within the boundaries of CANADA. The colonists' resentment over these acts contributed to the discontent that led to the American Revolution. In that war military operations were conducted in the Ohio country.

History - 1783 to 1802

Ohio was part of the vast area ceded to the U.S. by the Treaty of Paris (1783). Conflicting claims to land in that area made by Connecticut, Massachusetts, and Virginia were settled by relinquishment of almost all of the claims and the organization of the Old Northwest (formerly NORTHWEST TERRITORY) by the Ordinance of 1787. Ohio was the first region developed under the provisions of that ordinance, with the activities of the Ohio Company of Associates promoted by Rufus Putnam and Manasseh Cutler. MARIETTA, founded in 1788, was the first permanent American settlement in the Old Northwest. In the years that followed, various land companies were formed, and settlers poured in from the E, down the Ohio on flatboats and barges, or across the mountains by wagon and pack horse—their numbers varying with conditions but steadily expanding the population. The Native Americans, supported by the British, resisted American settlement. A coalition of Indian tribes

successfully opposed campaigns led by Josiah Harmar and Arthur St. Clair but were decisively defeated by Anthony Wayne in the battle of Fallen Timbers (1794). The British thereafter (1796) withdrew their outposts from the Northwest under the terms of Jay's Treaty, and the area was pacified. Ohio became a territory in 1799. General St. Clair, as the first governor, ruled in an arbitrary fashion that made Ohio natives distrustful of all government for many years afterward.

History - 1802 to 1837

A state convention drafted a constitution in 1802, and Ohio entered the Union in 1803, with CHILLICOTHE as its capital. Columbus became the permanent capital in 1816. In the War of 1812 the Americans lost many of the first battles of the war in the Old Northwest, and their military frontier was pushed back to the Ohio River. The two British attacks on Ohio soil were successfully resisted: one against FORT MEIGS at the mouth of the MAUMEE RIVER and another against Fort Stephenson on the SANDUSKY RIVER. The area was further secured with Oliver Hazard Perry's naval victory on Lake Erie near PUT-IN-BAY, and William Henry Harrison's victory in the battle of the THAMES on Canadian soil. After the war, Ohio's growth was spurred by the building of the ERIE CANAL, other canals, and toll roads. The Ohio National Road facilitated settlement and was a vital commercial artery. Settlement of the Western Reserve by New Englanders (especially those from Connecticut) gave NE Ohio a decidedly New England cultural landscape. Before the opening of the Erie Canal (Buffalo to Albany), Ohio's society of small farms exported their produce down the Ohio and MISSISSIPPI rivers to ST. LOUIS and NEW ORLEANS.

History - 1837 to 1865

In 1837, Ohio won a territorial struggle with Michigan usually called the Toledo War. The Loan Law, adopted in the Panic of 1837, encouraged railroad and industrial development. Railroads gradually succeeded canals, preparing the way for the industrial expansion that followed the Civil War. In the war most Ohio natives were sympathetic with the Union, and the state contributed many soldiers to the Union army. Native sons such as Joshua R. Giddings, Salmon P. Chase, and Edwin M. Stanton had long been prominent opponents of slavery. Nevertheless, the Peace Democrats, the Knights of the Golden Circle, and the Copperheads were very active; Clement L. Vallandigham, leader of the Copperheads, drew many votes in the gubernatorial election of 1863. Ohio was the scene of the northernmost penetration of Confederate forces in the war—the famous raid (1863) of John Hunt Morgan, which terrorized the people of the countryside until Morgan and most of his men were finally captured in the SE corner of the state.

History - 1865 to 1913

After the Civil War, industrial development increased rapidly when the shipment of ore from the upper Great Lakes region was intensified and the development of the petroleum industry in NE Ohio shifted the center of economic activity from the banks of the Ohio River to the shores of Lake Erie, particularly around Cleveland. Immigrants began to swell the population, and huge fortunes were made. Ohio became very important politically. The state contributed seven American presidents: Ulysses S. Grant, Rutherford B. Hayes, James A. Garfield, Benjamin Harrison, William McKinley, William Howard Taft, and Warren G. Harding. Big business and politics became entwined as in the relations of Marcus A. Hanna and William McKinley. City bosses such as Cincinnati's George B. Cox are consistent with this pattern. The state as a whole was for many years steadily Republican despite the rise of organized labor in the late 19th century and considerable labor strife. In the 1890s the reforming mayor of Toledo, Samuel "Golden Rule" Jones, won national fame for his espousal of the city ownership of municipal utilities.

History - 1913 to 1970

Floods in the many rivers flowing to the Ohio and in the Ohio River itself have long been a problem; a devastating flood in 1913 led to the establishment of the Miami valley conservation project. Continuing long-term state and Federal projects have improved locks and dams along the entire length of the Ohio and its tributaries, for navigation as well as flood-control purposes. The Muskingum Watershed Conservancy District maintains dams and lakes to control floodwaters on upstream tributaries, and for fishing and other recreational purposes. Both farms and industries in Ohio were hard hit by the Great Depression that began in 1929. In the 1930s the state was wracked by major strikes such as the sit-down strikes in Akron (1935–1936) and the so-called Little Steel strike (1937). World War II brought great prosperity to Ohio, but labor strife was later resumed in the steel strikes of 1949 and 1959.

History - 1970 to Present

Four Kent State University students were killed on May 4, 1970 when national guardsmen fired on a group of Vietnam War protesters. Ohio's economy went into massive decline in the 1970s and 1980s as the automobile, steel, and coal industries virtually collapsed, causing unemployment to soar. Akron, once world famous as a major rubber center, stopped producing rubber products altogether by the mid-1980s. During this period, the state's N industrial centers were especially hard hit and lost much of their population. Since then, Ohio has concentrated on diversifying its economy and transforming its manufacturing base by incorporating advanced manufacturing techniques. The state achieved this objective through the attraction of Japanese automobile manufacturers and parts suppliers, the expansion of the service sector, and the widespread expansion of polymer coatings and plastic products, including composite materials. Ohio has become an important center for the health care industry with the opening of the Cleveland Clinic. Industrial research is also important, with Nela Park near Cleveland and Battelle Memorial Institute in Columbus and many Edison Technology Centers: two for advanced manufacturing, two for materials, and two for biotechnologies being among the more notable research centers. There are still important polymer research laboratories in Akron and the Liquid Crystal Institute in KENT.

Education and Culture

Ohio's cultural development has been marked by the fame achieved by the Cleveland Orchestra and by the unusually large number of institutions of higher learning located in the state. Among them are Antioch College and Antioch University McGregor at YELLOW SPRINGS; Bowling Green State University, at BOWLING GREEN; Case Western Reserve University (formerly Western Reserve University and Case Institute of Technology), at Cleveland; Kent State University, at Kent; Kenyon College, at GAMBIER; Miami University, at OXFORD; Muskingum College, at NEW CONCORD; Denison University, at GRANVILLE; Oberlin College, at OBERLIN; the Ohio State University, at Columbus; Ohio University, at ATHENS; Ohio Wesleyan University, at DELAWARE; University of Cincinnati; University of Toledo; and a large number of other state and private institutions.

Government

Ohio's present constitution was adopted in 1851. It has been amended many times, most notably in 1912 after a constitutional convention and the adoption of thirty-three changes, including many progressive labor provisions and such measures as initiative, referendum, and the direct primary. The state's executive branch is headed by a governor elected for a four-year term, and permitted two successive terms. Ted Strickland is the current governor. Ohio's bicameral general assembly has a senate with thirty-three members, elected for four-year terms and a house with ninety-nine members elected for two-year terms. The state elects two Senators and eighteen Representatives to the U.S. Congress and has twenty electoral votes in presidential elections.

Ohio has eighty-eight counties: ADAMS, ALLEN, ASHLAND, ASHTABULA, ATHENS, AUGLAIZE, BELMONT, BROWN, BUTLER, CARROLL, CHAMPAIGN, CLARK, CLERMONT, CLINTON, COLUMBIANA, COSHOCTON, CRAWFORD, CUYAHOGA, DARKE, DEFIANCE, DELAWARE, ERIE, FAIRFIELD, FAYETTE, FRANKLIN, FULTON, GALLIA, GEAUGA, GREENE, GUERNSEY, HAMILTON, HANCOCK, HARDIN, HARRISON, HENRY, HIGHLAND, HOCKING, HOLMES, HURON, JACKSON, JEFFERSON, KNOX, LAKE, LAWRENCE, LICKING, LOGAN, LORAIN, LUCAS, MADISON, MAHONING, MARION, MEDINA, MEIGS, MERCER, MIAMI, MONROE, MONTGOMERY, MORGAN, MORROW, MUSKINGUM, NOBLE, OTTAWA, PAULDING, PERRY, PICKAWAY, PIKE, PORTAGE, PREBLE, PUTNAM, RICHLAND, ROSS, SANDUSKY, SCIOTO, SENECA, SHELBY, STARK, SUMMIT, TRUMBULL, TUSCARAWAS, UNION, VAN WERT, VINTON, WARREN, WASHINGTON, WAYNE, WILLIAMS, WOOD, and WYANDOT.

Ohio, county (□ 87 sq mi/226.2 sq km; 2006 population 5,826), SE INDIANA; ☉ RISING SUN; 38°57′N 84°58′W. Bounded E by KENTUCKY state line, here formed by OHIO River; drained by small LAUGHERY CREEK. Agricultural area (hogs, cattle, tobacco); manufacturing at Rising Sun. Smallest county in Indiana (both population and area). Formed 1844.

Ohio, county (□ 596 sq mi/1,549.6 sq km; 2006 population 23,844), W KENTUCKY; ☉ HARTFORD; 37°28′N 86°50′W. Bounded SW and SE (in part) by GREEN RIVER; drained by ROUGH RIVER and South Fork of PANTHER CREEK. Rolling agricultural area (burley and dark tobacco, hay, alfalfa, soybeans, wheat, corn; hogs, cattle); bituminous coal mines, limestone quarries; timber. Formed 1798.

Ohio, county (□ 109 sq mi/283.4 sq km; 2006 population 44,662), N WEST VIRGINIA, in NORTHERN PANHANDLE; ☉ WHEELING; 40°06′N 80°37′W. Industrial and commercial center of the Panhandle. Bounded W by OHIO RIVER (OHIO state line), E by PENNSYLVANIA; drained by Wheeling Creek. Coal mines, natural-gas wells. Iron, steel, and metal-working industries and other manufacturing at Wheeling. Agriculture (corn, wheat, oats, barley, hay); cattle; dairying; hogs. Oglebay Park in W; Bear Rocks Lakes Wildlife Management Area in E; part of Castleman Run Wildlife Management Area in NE. Formed 1776.

Ohio, village, GUNNISON county, W central COLORADO, just W of SAWATCH MOUNTAINS, 17 mi/27 km E of GUNNISON, on Quartz Creek, at mouth of Gold Creek. Elevation 8,560 ft/2,609 m. Surrounded by Gunnison National Forest.

Ohio, village (2000 population 540), BUREAU county, N ILLINOIS, 14 mi/23 km N of PRINCETON; 41°33′N 89°27′W. In agricultural area; dairy products.

Ohio, river, 981 mi/1,579 km long; formed by the confluence of the ALLEGHENY and MONONGAHELA rivers in SW PENNSYLVANIA, in downtown PITTSBURGH (40°26′N 80°00′W); first flows NW to MONACA, where it receives BEAVER RIVER from N, then generally SW, forms OHIO-WEST VIRGINIA state line and flows past STEUBENVILLE, Ohio, WHEELING and HUNTINGTON, West Virginia, continuing WNW forming Ohio-KENTUCKY state line, past ASHLAND, Kentucky, and PORTSMOUTH and CINCINNATI, Ohio, turns WSW, forms INDIANA-Kentucky and ILLINOIS-Kentucky state line, flowing past LOUISVILLE and OWENSBORO, Kentucky, EVANSVILLE, Indiana, and PADUCAH, Kentucky, entering MISSISSIPPI RIVER at CAIRO, Illinois, opposite MISSOURI. Receives KENTUCKY RIVER from S, 40 mi/64 km NE of Louisville; receives WABASH RIVER from N, 28 mi/45 km WSW of Evansville; receives CUMBERLAND RIVER and Kentucky River 12 mi/19 km

ENE and 3 mi/4.8 km ESE of Paducah, respectively. The Ohio is navigable for its entire length; a series of locks and dams improves its navigability and controls flooding. The Ohio's course follows a portion of the S edge of the region covered by continental ice during the late Cenozoic era; glacial meltwater probably cut its original channel. The river is a major tributary of the Mississippi and supplies more water to it than does the Missouri River. The Ohio River basin covers c.204,000 sq mi/528,360 sq km; the chief tributaries are the TENNESSEE, Cumberland, Wabash, and Kentucky. The Ohio is prone to spring flooding, and extensive flood-control and protection devices have been constructed along the river and its tributaries. These devices also improve the river's navigability; a 9-ft/2.7-m channel is maintained along its entire length. A system of modern locks and dams, constructed since 1955 to replace older structures, speeds the transit of barges and leisure craft. A canal (first opened in 1830) at Louisville bypasses the Falls of the Ohio, a 2.25-mi/3.62-km-long series of rapids having a 24 ft/7 m drop. Oil and steel account for most of the cargoes moved on the river. The principal river ports are Cincinnati, Louisville, and Pittsburgh. The Ohio River basin is one of the most populated and industrialized regions of the U.S. Eight states (Illinois, Indiana, Kentucky, New York, Ohio, Pennsylvania, VIRGINIA, and West Virginia) affected by the river's industrial pollution ratified (1948) the Ohio River Valley Sanitation Compact. Some results of their cleanup efforts have become discernible, and the river now supports marinas and recreational facilities. The French explorer La Salle reportedly reached the Ohio River in 1669, but there was no significant interest in the valley until the French and the British began to struggle for control of the river in the 1750s. An early settlement was established at the forks of the Ohio (modern Pittsburgh) by the Ohio Company of Virginia in 1749, but it was captured by the French in 1754, and the unfinished Fort Prince George was renamed FORT DUQUESNE; it was recaptured by the British and renamed Fort Pitt in 1758. At the end of the French and Indian Wars, BRITAIN gained control of the river by the treaty of 1763, but settlement of the area was prohibited. Britain ceded the region to the U.S. at the end of the Revolutionary War (1783), and it was opened to settlement by the Ordinance of 1787, which established the NORTHWEST TERRITORY. Until the opening of the ERIE CANAL in 1825, the Ohio River was the main route to the newly opened West and the principal means of market transportation of the region's growing farm output. Traffic declined on the river after the railroad was built in the mid-1800s, although it revived after World War II. The Ohio River remains a vital link in the river transportation system of the Midwest. Most shipping is done with barges pushed by towboats.

Ohio and Erie Canal, former waterway of OHIO, 307 mi/494 km long, between LAKE ERIE at CLEVELAND and the OHIO RIVER at PORTSMOUTH; built 1825–1832. It utilized part of the courses of the CUYAHOGA, MUSKINGUM, and SCIOTO rivers and had forty-nine locks. The canal flourished as a means of transporting freight until the advent of the railroad era in the 1850s. It was responsible for the growth of cities along its route, especially Cleveland, AKRON, and COLUMBUS. Badly damaged by 1913 flood. Parts of the canal are maintained, to this day, as a water supply for local industries.

Ohio Caverns, Ohio: see WEST LIBERTY.

Ohio City, village (2006 population 775), VAN WERT county, W OHIO, 7 mi/11 km SSW of VAN WERT; 40°46′N 84°37′W. Dairy products, canned foods, grain products.

Ohiopyle (o-HEI-yo-pai-uhl), borough (□ 29 sq mi/75.4 sq km; 2006 population 75), FAYETTE county, SW PENNSYLVANIA, 12 mi/19 km E of UNIONTOWN, on the YOUGHIOGHENY RIVER, at center of Ohiopyle State

Park (□ 29 sq mi/75 sq km); 39°52′N 79°29′W. A recreation area including part of Potomac National Scenic Trail. Tourism. Fallingwater, a home designed by Frank Lloyd Wright in 1936, is to N.

Ohio Range, NE end of the HORLICK MOUNTAINS, in the TRANSANTARCTIC MOUNTAINS of EAST ANTARCTICA, E of the WISCONSIN RANGE; 30 mi/50 km long and 10 mi/16 km wide; 84°45′S 114°00′W.

Ohioville (o-HEI-yo-vil), borough (2006 population 3,666), BEAVER county, W PENNSYLVANIA, 10 mi/16 km W of BEAVER on OHIO state line; 40°40′N 80°28′W. Agriculture includes dairying, corn, hay.

Ohiowa, village (2006 population 132), FILLMORE county, SE NEBRASKA, 10 mi/16 km SE of GENEVA; 40°24′N 97°27′W.

Ohira (O-hee-rah), town, Shimotsuga county, TOCHIGI prefecture, central HONSHU, N central JAPAN, 19 mi/30 km S of UTSUNOMIYA; 36°20′N 139°42′E. Appliance manufacturing (air conditioners, refrigerators). Grapes, strawberries.

Ohira (O-hee-rah), village, Kurokawa county, MIYAGI prefecture, N HONSHU, NE JAPAN, 12 mi/20 km N of SENDAI; 38°27′N 140°52′E.

Ohito (O-hee-to), town, Tagata county, SHIZUOKA prefecture, central HONSHU, E central JAPAN, on N central IZU PENINSULA, 31 mi/50 km E of SHIZUOKA; 35°00′N 138°56′E. Cash registers.

O. H. Ivie Lake (EI-vee), reservoir (□ 30 sq mi/78 sq km), COLEMAN, CONCHO, and RUNNELS counties, central TEXAS, on COLORADO RIVER, 40 mi/64 km WSW of BROWNWOOD; 31°30′N 99°40′W. Also fed by Concho River Formed by Simon Freese Stacy Dam (148 ft/45 m high), built (1989) for water supply; also used for recreation.

Ohiya (O-HI-yuh), town, UVA PROVINCE, S central SRI LANKA, in SRI LANKA HILL COUNTRY, near HORTON PLAINS, 34 mi/55 km SSE of KANDY; 06°49′N 80°50′E.

Ohlau, POLAND: see OLAWA.

Ohle River, POLAND: see OLAWA RIVER.

Ohligs (AW-liks), outer district of SOLINGEN, RHINELAND, North RHINE-Westphalia, W GERMANY, 3.5 mi/5.6 km W of city center; 51°10′N 07°05′E.

Ohlsdorf (AWLS-dorf), district of HAMBURG, NW GERMANY, on left bank of the ALSTER, 5 mi/8 km NNE of city center; 53°37′N 10°03′E. Site of Hamburg's main cemetery (founded 1877).

Ohm River (OM), 35 mi/56 km long, HESSE, W GERMANY; rises 6 mi/9.7 km NE of SCHOTTEN in Vogels Mountains; flows generally NW to the LAHN, 2 mi/3.2 km N of Marburg.

Öhningen (O-ning-uhn), village, BADEN-Württemberg, S GERMANY, 13 mi/21 km W of KONSTANZ, on Swiss border, and on right bank of the Rhine; 47°39′N 08°54′E.

Ohogamiut (o-HO-gah-mee-yoot), village, W ALASKA, on lower YUKON River and 55 mi/89 km N of BETHEL; 61°35′N 161°56′W. Sometimes spelled Ohogamut. Also called Iguak.

Ohoka (o-HO-kuh), township, Waimakariri district, CANTERBURY region, NEW ZEALAND, 13 mi/21 km NNW of CHRISTCHURCH; 43°22′S 172°34′E. In agricultural area.

Ohoopee (o-HOOP-ee), village, TOOMBS county, E central GEORGIA, 10 mi/16 km E of VIDALIA; 32°10′N 82°13′W.

Ohoopee River (o-HOOP-ee), c.100 mi/161 km long, E central GEORGIA; rises S of SANDERSVILLE; flows SE to ALTAMAHA RIVER 13 mi/21 km S of REIDSVILLE; 32°56′N 82°47′W. Receives Little Ohoopee River (c.45 mi/72 km long).

Ohrdruf (OR-druf), town, THURINGIA, central GERMANY, at N foot of THURINGIAN FOREST, 9 mi/14.5 km S of GOTHA; 50°50′N 10°44′E. Manufacturing of toys, porcelain, metal and woodworking. Has 18th-century church on site of first Christian chapel in THURINGIA (built 724 by St. Boniface); 16th-century Ehrenstein castle, on site of early monastery. J.S. Bach attended school here. Town was site of notorious concentration camp and of extensive subterranean German army headquarters during World War II. Chartered 1348.

Ohre River (O-re), 65 mi/105 km long, central GERMANY; rises 6 mi/9.7 km NW of GARDELEGEN; flows generally SE, past Calvörde, HALDENSLEBEN, and WOLMIRSTEDT, to the Elbe 5 mi/8 km NW of Burg.

Ohrid (O-reed), town, in MACEDONIA, on a rock above Lake OHRID, on the Albanian border; 41°07′N 20°48′E. Macedonia's chief resort, it is a tourist and commercial center, as well as a railroad terminus. Fishing and farming are the chief occupations. Ohrid stands on or near the site of the Greek colony of Lychnidos, founded in the third century B.C.E. It was captured by the Romans in C.E. 168 and became a major trade center and an early episcopal see. In the 9th century Ohrid was incorporated into the first Bulgarian empire, and in the 10th century it became the seat of the Bulgarian patriarchate and flourished as the political and cultural center of Bulgaria. Traditionally a Slavic cultural center, Ohrid served as a conduit of Christianity into other Slav-inhabited areas. After Ohrid's reconquest in 1018 by the BYZANTINE EMPIRE, the patriarchate was abolished; but the town remained a metropolitan see. Ohrid was captured by the Serbs in 1334 and fell to the Turks in 1394. It was briefly reconquered by the Albanian hero Scanderbeg in the 15th century. During World War I, Ohrid was taken by Serb troops; after the war, it was joined to the former YUGOSLAVIA. Bulgarian forces held the town during World War II, but it was then restored to Yugoslavia. Ohrid's numerous ancient churches and other historical relics include the cathedrals of St. Sophia (9th century) and St. Clement (1299), both with medieval frescoes; two 14th-century churches; and the walls and towers of the former Turkish citadel. The town is also noted for its museum, galleries, fishing institute, and other educational facilities. Airport. Also spelled Ochrida or Okhrida.

Ohrid, Lake (O-reed), Serbo-Croatian *Ohridsko Jezero*, Albanian *Liqen i Ohrit*, deepest lake (□ 134 sq mi/348.4 sq km) in the Balkans, on Macedonian-Albanian border, W of Galicica Mountain, MACEDONIA; 1,017 ft/310 m deep; 41°00′N 20°45′E. Elevation 2,280 ft/695 m. Drained by BLACK DRIN RIVER. Connected with Lake PRESPA by underground channels. Water, received from coastal and bottom springs, is unusually transparent (sometimes down to 72 ft/22 m). Fishing (carp, eel, trout) is important source of local livelihood; fish hatchery. Noted for its beauty, the lake is surrounded by beaches and historic buildings (e.g., monastery at Sveti (St.) Naum). Coastal towns (STRUGA, OHRID, POGRADEC) are connected by boat lines. Outlet is the Black Drin River. Sometimes spelled Ochrid, Ochrida, or Okhrid.

Ohrigstad, town, MPUMALANGA, SOUTH AFRICA, on OHRIGSTAD RIVER and 25 mi/40 km NNE of LYDENBURG; 24°45′S 30°33′E. Elevation 5,199 ft/1,585 m. Cotton, corn, tobacco, sunflower seeds, and deciduous fruit. Founded 1845 by Boer trekkers, who established *Volksraad* (people's council) here; abandoned 1850 because of malaria outbreak, later resettled in 1923 after mosquitoes were brought under control. Railroad connection.

Öhringen (O-ring-uhn), town, FRANCONIA, BADEN-Württemberg, SW GERMANY, 13 mi/21 km ENE of HEILBRONN; 49°12′N 09°30′E. Manufacturing of agriculture machinery, precision mechanics, tinware, furniture, shoes, apparel; leather and woodworking. Has late-Gothic church (1454–1501); Renaissance castle (1611–1616) used today as the town hall. Was Roman settlement. Town first mentioned 1073. Chartered 1253. Sometimes spelled Oehringen.

Ohura (o-HUHR-ruh), township, Ruapehu District, W NORTH ISLAND, NEW ZEALAND, 50 mi/80 km ENE of NEW PLYMOUTH; 38°50′S 174°59′E. Sawmills, some coal.

Oi (O-ee), town, Oi county, FUKUI prefecture, HONSHU, W central JAPAN, 53 mi/85 km S of FUKUI; 35°28′N 135°37′E. Nuclear power plant located here.

Oi, town, Ashigarakami county, KANAGAWA prefecture, E central HONSHU, E central JAPAN, 28 mi/45 km S of YOKOHAMA; 35°19′N 139°09′E.

Oi, town, Iruma county, SAITAMA prefecture, E central HONSHU, E central JAPAN, 6 mi/10 km S of URAWA; 35°51′N 139°30′E.

Oia (EE-ah), village, on NW tip of SANTORINI island, CYCLADES prefecture, SOUTH AEGEAN department, GREECE, 8 mi/12.9 km N of Fira and c.500 ft/152 m above the caldera; 36°28′N 25°22′E. Long the commercial center of the island, it was all but destroyed in the 1956 earthquake. Now a popular tourist center, as many of its old buildings have been restored and many homes built into the cliff side. Also Ia.

Oiapoque (o-ee-ah-po-kai), town (2007 population 19,181), extreme N AMAPÁ, N BRAZIL, on right bank of OIAPOQUE (Oyapock) River (FRENCH GUIANA border), and 85 mi/137 km SSE of CAYENNE, French Guiana; 03°48′N 51°41′W. Gold found in area.

Oiapoque River, Brazil and French Guiana: see OYAPOCK RIVER.

Oiba (O-ee-bah), town, ⊙ Oiba municipio, SANTANDER department, N central COLOMBIA, 50 mi/80 km SSW of BUCARAMANGA; 06°15′N 73°17′W. Coffee, sugarcane, corn, livestock.

Oich, Loch (OIK), lake in HIGHLAND, N Scotland, extending 4 mi/6.4 km NE-SW along the GREAT GLEN of Scotland between LOCH NESS and LOCH LOCHY, forming highest section of the CALEDONIAN CANAL; 57°04′N 04°08′W. Drained by Oich River, which flows 7 mi/11.3 km NE to Loch Ness at FORT AUGUSTUS.

Oies, Île aux (ZWAH, ee-lo) or **Island of Geese**, in the SAINT LAWRENCE RIVER, S QUEBEC, E Canada, 40 mi/64 km ENE of QUEBEC city; 6 mi/10 km long, 1 mi/2 km wide; 47°05′N 70°30′W. Just SW is Île aux GRUES.

Oigawa (O-ee-GAH-wah), town, Shida county, SHIZUOKA prefecture, SE HONSHU, E central JAPAN, 12 mi/20 km S of SHIZUOKA; 34°48′N 138°17′E. Known for *sakura* [=cherry blossom] shrimp.

Oignies (WAHN-yee), town (□ 2 sq mi/5.2 sq km), PAS-DE-CALAIS department, NORD-PAS-DE-CALAIS region, N FRANCE, 8 mi/12.9 km NW of DOUAI; 50°28′N 02°59′E. A former coal-mining center, seeking a new industrial base in coal by-products and charcoal. A center recounting the history of mining and railroads in this coal basin is located here.

Oil Center, unincorporated village, LEA county, extreme SE NEW MEXICO, 22 mi/35 km SSW of HOBBS. In oil and natural-gas producing area of LLANO ESTACADO. Cattle, sheep, cotton, grain.

Oil City, city (2006 population 10,849), VENANGO county, NW PENNSYLVANIA, 70 mi/113 km NNE of PITTSBURGH, on the ALLEGHENY RIVER, at mouth of OIL CREEK; 41°25′N 79°42′W. Manufacturing (continuous casting equipment, plastic products, printing and publishing, aluminum castings, concrete production, motor oil, machinery, antioxidants, steel tubing). Agricultural area (corn, hay; livestock, dairying). The city was founded after Edwin Lake Drake struck oil nearby in TITUSVILLE in 1859. It is a refining and shipping center for the state's oil industry and a producer of oil-field equipment. Clarion College (Venango campus) to SW; Oil Creek State Park to N. Incorporated 1871.

Oil City, town (2000 population 1,219), CADDO parish, extreme NW LOUISIANA, 20 mi/32 km NW of SHREVEPORT, near CADDO LAKE. Enter of oil field discovered in 1906; 32°45′N 93°58′W. Caddo-Pine Island Oil and Historical Society museum here. Caddo Lake reservoir to SW.

Oil City (OIL), unincorporated village, S ONTARIO, E central Canada, 5 mi/8 km SSE of PETROLIA, and included in Enniskillen; 42°48′N 82°07′W. Oil production.

Oil Creek, 7 mi/11.3 km long, TELLER county, central COLORADO; rises W of PIKES PEAK; flows SSW then NNW, joining FOURMILE CREEK at Mueller State Park.

Oil Creek, c.45 mi/72 km long, NW PENNSYLVANIA; rises in headstreams in N CRAWFORD county, in Canadohta Lake; flows SE, past TITUSVILLE, and S, through Oil Creek State Park, through oil-producing region to ALLEGHENY RIVER at OIL CITY; 41°48′N 79°50′W. On its banks near Titusville, first successful oil well in U.S. was drilled in 1859.

Oildale, unincorporated city (2000 population 27,885), KERN county, S central CALIFORNIA, a residential suburb 1 mi/1.6 km N of BAKERSFIELD, across KERN RIVER; 35°25′N 119°02′W. Railroad junction. Oil field center. Irrigated agricultural area, cotton; dairying; cattle; grain, melons, pumpkins, fruit, nuts, grapes, sugar beets, beans, vegetables. Bakerfield Airport to N. Greenhouse Mountains and Sequoia National Forest to NE.

Oil Rivers, region, NIGERIA, distributaries of the Niger Delta, and long known to slave traders. This region of various creeks possessed no organized form of government until the mid-19th century. Was part of the NIGER COAST British Protectorate.

Oil Springs (OIL SPREENGZ), village (□ 3 sq mi/7.8 sq km; 2001 population 758), SW ONTARIO, E central Canada, 7 mi/11 km SSE of PETROLIA; 42°47′N 82°06′W. Oil production. Originally called Black Creek. Noted as the site of the first commerical oil well in North America. Incorporated 1865.

Oilton (OIL-tuhn), town (2006 population 1,124), CREEK county, central OKLAHOMA, 34 mi/55 km W of TULSA, and 10 mi/16 km NE of CUSHING, on CIMARRON RIVER; 36°04′N 96°34′W. In petroleum and agricultural area (grain, corn); manufacturing (horse and stock trailers); oil and natural gas wells. Founded c.1915.

Oilton (OIL-tuhn), unincorporated village, WEBB county, S TEXAS, 31 mi/50 km E of LAREDO; 27°27′N 98°58′W. Oil and natural gas. Cattle.

Oil Trough, village (2000 population 218), INDEPENDENCE county, NE central ARKANSAS, 14 mi/23 km SE of BATESVILLE and on WHITE RIVER; 35°37′N 91°27′W. In agricultural area.

Oilville (OIL-vil), unincorporated village, GOOCHLAND county, central VIRGINIA, 22 mi/35 km WNW of RICHMOND; 37°42′N 77°47′W. Manufacturing (tobacco processing, carrying cases, machining); agriculture (grain, tobacco, soybeans; cattle).

Oinoussa (ee-NOO-sah), island (□ 6 sq mi/15.6 sq km), in AEGEAN SEA, largest of the OINOUSSAI group, in KHÍOS prefecture, NORTH AEGEAN department, GREECE, between NE KHÍOS island and KARABURUN Peninsula (Turkey); 5 mi/8 km long, 1.5 mi/2.4 km wide; 38°32′N 26°15′E. Livestock, fisheries. Also called Agnousa and Ignusi.

Oinoussai Islands (ee-NOO-se), Latin *Oenussae*, archipelago in IONIAN SEA, off Cape AKRITAS, MESSENIA prefecture, SW PELOPONNESE department, off extreme SW coast of GREECE; 36°45′N 21°46′E. Constitute shipping hazard. Include islands of SKHIZA and SAPIENDZA.

Oio (OI-o), province (2004 population 179,048), N central GUINEA-BISSAU; ⊙ FARIM; 12°15′N 15°15′W. Bounded on N by SENEGAL, E by BAFATÁ PROVINCE, S by QUINARA PROVINCE, and W by BISSAU and CACHEU provinces. Phosphate deposits; peanuts and tropical fruits.

Oiron (WAH-ron), commune (□ 14 sq mi/36.4 sq km), DEUX-SÈVRES department, POITOU-CHARENTES region, W FRANCE, 20 mi/32 km S of SAUMUR; 46°57′N 00°05′W. It has a fine Renaissance collegiate church built in mid-16th century, and a castle that exhibits two centuries (16th and 17th) of formal French architecture; interesting wall paintings.

Oirot Autonomous Region, RUSSIA: see ALTAI REPUBLIC.

Oirschot (AWR-skhawt), town, NORTH BRABANT province, S NETHERLANDS, on WILHELMINA CANAL

Cross-references are shown in SMALL CAPITALS. The pronunciation guide is shown on page xix. The sources of population figures are shown on page xvii.

and 11 mi/18 km ESE of TILBURG; 51°30′N 05°19′E. Dairying; cattle, hogs; grain, vegetables. Also spelled Oorschot.

Oisans (wah-zahn), valley of ROMANCHE and Vénéon rivers, HAUTES-ALPES and ISÈRE departments, SE FRANCE, in the DAUPHINÉ ALPS, extending from Col du LAUTARET (E) to Le BOURG-D'OISANS (W) and the ÉCRINS Massif (S). Numerous dams (including huge CHAMBON DAM) and hydroelectric plants power its electrochemical and aluminum industries. Alpinism is popular (ascent of Mont PELVOUX and other peaks). Winter sports have grown in importance with development of such ski centers as Les DEUX-ALPES, L'ALPE-D'HUEZ, La GRAVE (at base of MEIJE glacier). ÉCRINS NATIONAL PARK (French *Parc National des Écrins*) extends S of the Romanche valley. Oisans is the new name of the Alpine mountain bloc formerly described as Massif du Pelvoux.

Oise (WAHZ), department (□ 2,263 sq mi/5,883.8 sq km), N FRANCE, occupying parts of historic ÎLE-DE-FRANCE and PICARDY; ⊙ BEAUVAIS. Generally level, plateau country, with large wooded tracts (forests of COMPIÈGNE, CHANTILLY) in S; 49°06′N 02°45′E. Drained NE-SW by the OISE, which receives the AISNE, its principal tributary, near Compiègne. A leading agricultural district (wheat, sugar beets, oats, feed crops, apples, cherries, black currants, vegetables) with an important cattle-raising and dairying industry. Numerous clay and sandstone quarries. Principal industries: metallurgy (in CREIL area), woolen milling (noted tapestry production formerly at Beauvais), woodworking (in Compiègne and SENLIS area), glass manufacturing and food processing (sugar refining, cider distilling, fruit and vegetable preserving for PARIS market). Chief towns: Beauvais (large cathedral), Compiègne, Creil, Senlis, and Chantilly (horse racing). Population growth benefits from proximity of expanding Greater Paris. The department forms part of the administrative region of PICARDIE.

Oise (WOI-zuh), river, 186 mi/299 km long, S BELGIUM; rising in the ARDENNES mountains; flowing through N FRANCE generally SW past COMPIÈGNE to join the SEINE RIVER near PONTOISE. Navigable for most of its length, the Oise is an important transportation route; canals link it with the AISNE, SAMBRE, and THÉRAIN rivers.

Oise-Aisne Canal (wahz–en kah-nahl), c.40 mi/64 km long, AISNE department, PICARDIE region, N FRANCE; connects OISE RIVER (below CHAUNY) with AISNE RIVER (above VAILLY-SUR-AISNE); 49°36′N 03°11′E. Follows course of AILETTE RIVER, pierces the CHEMIN DES DAMES ridge in a tunnel. Also called Canal de l'Oise à l'Aisne.

Oise à l'Aisne, Canal de l', FRANCE: see OISE-AISNE CANAL.

Oise River (WAHZ), 186 mi/299 km long, in N FRANCE; rises in the ARDENNES region of S BELGIUM, enters France N of HIRSON; flows generally SW, past La FÈRE, COMPIÈGNE, CREIL, and PONTOISE, to the SEINE at CONFLANS-SAINTE-HONORINE; 49°00′N 02°04′E. Navigable upstream to Compiègne, then paralleled by lateral canal. Connected with SOMME and Escaut (SCHELDT) rivers by SAINT-QUENTIN CANAL, and with SAMBRE RIVER by OISE-SAMBRE CANAL, both beginning above CHAUNY. The AISNE is its chief tributary. The Oise River Valley was the scene of bitter battles (1918) in World War I and has frequently been an invasion route aimed at the PARIS BASIN.

Oise-Sambre Canal (wahz–sahn-bruh kah-nahl), 50 mi/80 km long, AISNE and NORD departments, N FRANCE; begins at CHAUNY; follows the OISE upstream, past Fargniers (where it is joined by SAINT-QUENTIN CANAL), La FÈRE, and RIBEMONT, crosses watershed W of Wassigny and follows upper SAMBRE RIVER to LANDRECIES; 49°39′N 03°20′E. In battle line during World War I. Also called Canal de la Sambre à l'Oise.

Oishida (O-ee-shee-dah), town, North Murayama county, YAMAGATA prefecture, N HONSHU, NE JAPAN, on MOGAMI RIVER and 25 mi/40 km N of YAMAGATA city; 38°35′N 140°22′E. Traditional straw rainwear.

Oisina Point (OI-see-nah), westernmost point of TIMOR, in ROTI STRAIT, opposite SE end of SEMAU ISLAND; 10°21′S 123°28′E. Sometimes spelled OL SINA.

Oismäe (OIS-ma-ai), suburban district within TALLINN, ESTONIA. A high-rise community of approximately 50,000 residents.

Oiso (O-ee-so), town, Naka county, KANAGAWA prefecture, E central HONSHU, E central JAPAN, on N shore of SAGAMI Bay, just W of HIRATSUKA and 22 mi/35 km S of YOKOHAMA; 35°18′N 139°18′E.

Ois River, AUSTRIA: see YBBS RIVER.

Oissel (wah-sel), town (□ 8 sq mi/20.8 sq km), SEINE-MARITIME department, HAUTE-NORMANDIE region, N FRANCE, small port on left bank of the SEINE and 5 mi/8 km S of ROUEN; 49°20′N 01°06′E. An industrial center with chemical and textile plants.

Oisterwijk (AWST-uhr-veik), town, NORTH BRABANT province, S NETHERLANDS, 6 mi/9.7 km ENE of TILBURG; 51°35′N 05°13′E. Dairying; cattle, hogs; grain, vegetables. Castle to NE. Sometimes spelled Oosterwijk.

Oistins (oi-STINZ), town, S BARBADOS, BRITISH WEST INDIES, 5 mi/8 km ESE of BRIDGETOWN. Fisheries, modern fish market. Sometimes Oistin's or Oistin's Town.

Oita (O-EE-tah), prefecture [Japanese *ken*] (□ 2,447 sq mi/6,338 sq km; 1990 population 1,236,924), NE KYUSHU, SW JAPAN; chief port and ⊙ OITA. Bounded N by SUO SEA (W section of INLAND SEA), NE by IYO SEA (SW section of INLAND SEA), SE by HOYO STRAIT. FUKUOKA prefecture to N, MIYAZAKI prefecture to S, and KUMAMOTO prefecture to W. Mountainous terrain; rises to 5,850 ft/1,783 m in MOUNT KUJU, highest peak of Kyushu. KUNISAKI PENINSULA is NE projection. Drained by CHIKUGO RIVER (largest river of island). Numerous hot springs at BEPPU and vicinity. Volcanic area around Mount Kuju is part of Aso National Park. Densely forested (cedar, pine, bamboo). Gold, silver, tin mined in S, with refineries at SAGANOSEKI. Rice, wheat, soybeans grown in coastal region and in valleys of interior; known for village-specific agricultural product specialization. Widespread production of raw silk, charcoal, lumber. Thriving fishing industry. Produces silk textiles, paper, sake. Manufacturing centers include Oita (E), NAKATSU (N).

Oita (O-EE-tah), city (2005 population 462,317), ⊙ OITA prefecture, NE KYUSHU, SW JAPAN, a port on BEPPU BAY; 33°14′N 131°36′E. Oil and iron manufacturing; computer components. An important castle town in the 16th century; traded with the Portuguese. Tourist center; Mount Takasaki Zoo is known for its large collection of Japanese monkeys.

Oital (oi-TAHL), town, SW ZHAMBYL region, KAZAKHSTAN, on branch of TURK-SIB RAILROAD (MERKE station), 90 mi/145 km E of ZHAMBYL; 42°53′N 73°15′E. Beet-sugar refinery. Also spelled Oytal.

Oiti (EE-tee), massif, in PHOCIS and FTHIOTIDA prefectures, CENTRAL GREECE department, 15 mi/24 km S of LAMÍA; Pírgos [Greek=tower] is highest peak (7,057 ft/2,151 m); 38°49′N 22°17′E. Forests of Greek fir, flowery meadows, tablelands, springs, and vistas. In classical times it was known as Mount Oeta, where Hercules died on a pyre after being poisoned by Nessus' robe. Also Iti.

Oiticica (O-ee-chee-see-kah), village, N central PIAUÍ, BRAZIL, near CEARÁ border, 33 mi/53 km W of CRATEÚS; 05°14′S 41°48′W. Present terminus of railroad from FORTALEZA and Crateús.

Oituz Pass (oi-TOOZ), in the Moldavian Carpathians, E central ROMANIA, 15 mi/24 km NE of TÎRGU SECUIESC. Highway corridor between Moldavia and TRANSYLVANIA. Elev. 2,837 ft/865 m.

Oiwake (OI-wah-ke), town, Iburi district, Hokkaido prefecture, N JAPAN, 28 mi/45 km S of SAPPORO; 42°52′N 141°49′E.

Oizumi (O-ee-zoo-mee), town, Oura county, GUMMA prefecture, central HONSHU, N central JAPAN, 25 mi/40 km S of MAEBASHI; 36°14′N 139°24′E. Refrigerators.

Oizumi (O-ee-ZOO-mee), village, North Koma county, YAMANASHI prefecture, central HONSHU, central JAPAN, 19 mi/30 km N of KOFU; 35°51′N 138°23′E.

Ojai, city (2000 population 7,862), VENTURA county, S CALIFORNIA, 12 mi/19 km N of VENTURA, near Ojai Creek; 34°27′N 119°15′W. In fertile Ojai Valley in COAST RANGES. Railroad terminus. Vegetables, citrus, avocados; flowers, nursery products; manufacturing (cutting tools, printing and publishing). Seat of several private schools. Year-round resort. Los Padres National Forest to N; Pine Mountain (7,510 ft/2,289 m) to N. Incorporated 1921.

Ojcowski National Park (oi-tsov-skee) (□ 10 sq mi/26 sq km), POLAND, NW of KRAKÓW. Part of Kraków-Częstochowa Upland, established 1956, the smallest national park in Poland, caves and impressive rock formations, two medieval castles, wide variety of plant life, tourist attraction is the Trail of the Eagles' Nests.

Ojén (o-HEN), town, MÁLAGA province, S SPAIN, in coastal spur of the CORDILLERA PENIBÉTICA, 27 mi/43 km WSW of MÁLAGA; 36°34′N 04°51′W. In picturesque setting known for its Ojén liqueur, containing anise. Also has iron mines and a nickel foundry. Among the agricultural produce of the region are oranges, carob beans, olives, corn; also livestock.

Oji (O-JEE), town, North Katsuragi district, NARA prefecture, S HONSHU, W central JAPAN, on NW KII PENINSULA, 9 mi/15 km S of NARA; 34°35′N 135°42′E.

Ojibwa (o-JIB-wah), village, SAWYER county, N WISCONSIN, on CHIPPEWA RIVER, and 24 mi/39 km SE of HAYWARD. Lumbering, dairying. On Tuscobia State Trail. Relics of early logging days are exhibited here. Former Ojibwa State Roadside Park nearby (now a county park).

Ojika (o-JEE-kah), town, on OJIKA-SHIMA island, North Matsuura county, NAGASAKI prefecture, SW JAPAN, 56 mi/90 km N of NAGASAKI; 33°11′N 129°03′E.

Ojika Peninsula (o-JEE-KAH), Japanese *Ojika-hanto*, MIYAGI prefecture, N HONSHU, NE JAPAN, between ISHINOMAKI BAY (W) and the PACIFIC OCEAN (E); 12 mi/19 km long, 1 mi/1.6 km–6 mi/9.7 km wide. Mountainous. Chief town, ONAGAWA. Just off SE tip of peninsula is sacred Kinkazan, a small wooded site with popular shrine and lighthouse.

Ojika-shima (o-jee-KAH–shee-mah), island (□ 10 sq mi/26 sq km) of island group GOTO-RETTO, NAGASAKI prefecture, W JAPAN, in EAST CHINA SEA, 26 mi/42 km W of KYUSHU; 3.5 mi/5.6 km long, 2.5 mi/4 km wide. Hilly. Fishing.

Ojima (o-JEE-mah), town, Nitta county, GUMMA prefecture, central HONSHU, N central JAPAN, 19 mi/30 km S of MAEBASHI; 36°15′N 139°19′E. Onions.

Ojinaga (o-hee-NAH-gah), city and township, CHIHUAHUA, N MEXICO, on the RIO GRANDE, opposite PRESIDIO (TEXAS), at mouth of CONCHOS RIVER, on railroad, terminus and Mexico Highway 16, 120 mi/193 km NE of CHIHUAHUA; 29°35′N 104°26′W. Border trade; cotton and cattle center.

Ojiya (O-jee-yah), city, NIIGATA prefecture, central HONSHU, N central JAPAN, 43 mi/70 km S of NIIGATA; 37°18′N 138°47′E. Computer components, precision machines and equipment, textiles, pongee; noodles, Buddhist altars. Carp; rice. Natural gas.

Ojocaliente (o-ho-kah-lee-EN-tai), city and township, ⊙ Ojocaliente muncipio, ZACATECAS, N central MEXICO, on central plateau, 23 mi/37 km SE of ZACATECAS; 22°34′N 102°15′W. Elevation 6,936 ft/2,114 m. Railroad terminus. Gold, silver mining; sulfur plant.

Ojo Caliente (O-ho kah-lee-EN-tai), unincorporated village, RIO ARRIBA and TAOS counties, N NEW MEXICO, on Rio Tusas, and 26 mi/42 km N of ESPAÑOLA.

Area is shown by the symbol □, and capital city or county seat by ⊙.

Elevation 6,213 ft/1,894 m. Health resort with hot mineral springs. Manufacturing (signs). Cattle, sheep, alfalfa; chilies. Pueblo ruins in area; Carson National Forest surrounds village, except in S.

Ojo Caliente, CHIHUAHUA, Mexico: see CAMARGO.

Ojo de Agua (O-ho dai AH-gwah) or **Villa Ojo de Agua**, town (□ 2,485 sq mi/6,461 sq km), ⊙ Ojo de Agua department (□ 2,485 sq mi/6,436 sq km; 1991 population 11,806), S SANTIAGO DEL ESTERO province, ARGENTINA, at NE slopes of Sierra de Córdoba, 120 mi/193 km; 28°28′S 65°27′W. SSE of SANTIAGO DEL ESTERO. Agricultural center (livestock).

Ojo de Agua, COSTA RICA: see SAN RAFAEL, village.

Ojo del Toro (O-ho del TO-ro), peak (1,749 ft/533 m), GRANMA province, E CUBA, near W end of the SIERRA MAESTRA, 22 mi/35 km E of the CAPE CRUZ; 22°38′N 77°27′W.

Ojojona (o-ho-HO-nah), town, FRANCISCO MORAZÁN department, S central HONDURAS, in SIERRA DE LEPATERIQUE, 13 mi/21 km SSW of TEGUCIGALPA; 13°56′N 87°18′W. Elevation 4,378 ft/1,334 m. Summer resort; pottery and ropemaking; grain, coffee; livestock. A gold- and silver-mining center in colonial times.

Ojos de Agua (O-hos dai AH-gwah), municipality and town, CHALATENANGO department, EL SALVADOR, NNE of CHALATENANGO city.

Ojos de Agua (O-hos dai AH-gwah), town, COMAYAGUA department, W central HONDURAS, on the HUMUYA RIVER, 22 mi/33 km N of COMAYAGUA, on unpaved road; 14°43′N 87°39′W. Small farming; grain, beans; livestock. In area flooded by El Cajón Dam.

Ojos del Guadiana (O-hos dhel gwahdh-YAH-nah), marshes and ponds, CIUDAD REAL province, S central SPAIN, 20 mi/32 km NE of CIUDAD REAL, in low basin into which flow the GIGÜELA, ZÁNCARA, and AZUER rivers, and from which emerges the GUADIANA RIVER.

Ojos del Salado (O-hos del sah-LAH-do), peak (22,638 ft/6,900 m), in the ANDES, on the border between ARGENTINA and CHILE; 27°07′S 68°32′W. It is the second highest of the Andean peaks. In 1956 a Chilean expedition reported its summit elevation to be 23,239 ft/7,083 m, thus making it greater than ACONCAGUA and therefore the tallest peak in the Western Hemisphere, but subsequent measurments (including the 2000 Shuttle Radar Topography Mission [SRTM]), though still varied, put the peak in the elevation range cited above; elevation sometimes cited betweeen 22,520 ft/6,864 m and 22,654 ft/6,905 m.

Ojos Negros (NAI-gros), village, TERUEL province, E SPAIN, 35 mi/56 km NW of TERUEL; 40°44′N 01°30′W. Cereals, grapes, saffron; livestock; wine making; salt and coal mines.

Ojstro, SLOVENIA: see HRASTNIK.

Ojuelos de Jalisco (o-HWAI-los dai hah-LEES-ko), town, Ojuelos de Jalisco muncipio, JALISCO, central MEXICO, in SIERRA MADRE OCCIDENTAL, near ZACATECAS border, 45 mi/72 km SW of SAN LUIS POTOSÍ, at junction of Mexico Highways 51, 70, and 80; 21°52′N 101°40′W. A major highway hub; elevation 7,493 ft/ 2,284 m. Corn, beans, chilies; livestock.

Ojus (O-juhs), city (□ 3 sq mi/7.8 sq km; 2000 population 16,642), MIAMI DADE county, SE FLORIDA, 12 mi/19 km NNE of MIAMI; 25°57′N 80°09′W. Dairy products; limestone quarries.

Ok (AWK), glacier, W ICELAND, 45 mi/72 km NE of REYKJAVIK; rises to 3,743 ft/1,141 m at 64°36′N 20°53′W.

Oka (aw-KAH), town, ONDO state, S NIGERIA, 25 mi/40 km NE of owo and 40 mi/64 km ENE of AKURE. Cacao, palm oil and kernels, kola nuts, cassava, maize.

Oka (O-kuh), village (□ 26 sq mi/67.6 sq km; 2006 population 4,678), Laurentides region, S QUEBEC, SE Canada, on the N shore of the Lake of the TWO MOUNTAINS (a widening of the OTTAWA RIVER) and 21 mi/34 km SW of MONTREAL; 45°29′N 74°07′W. It is

noted as the site of a Trappist monastery and farm (established 1881), where Oka cheese is made. An agricultural institute here is affiliated with the University of Montreal. About 600 Native Americans live at Oka. In 1982, Oka Crisis was a native clash resulting in death and a lengthy stand-off with the Quebec government. Village is part of the Metropolitan Community of Montreal (*Communauté Metropolitaine de Montréal*).

Oka (uh-KAH), river, approximately 925 mi/1,489 km long, rising S of ORËL (ORËL oblast), central European Russia; flows N past Orël, through W TULA oblast and into KALUGA oblast, past KALUGA, E past SERPUKHOV (S MOSCOW oblast) and KOLOMNA, forming part of the Tula oblast border with Kaluga and Moscow oblasts, passes E to RYAZAN (RYAZAN oblast), and then NE past MUROM (VLADIMIR oblast) to join the VOLGA RIVER at NIZHNIY NOVGOROD (NIZHEGOROD oblast). It is navigable by large vessels below Kolomna, approximately 550 mi/885 km upstream, and traverses densely populated agricultural and industrial areas. Among its tributaries are the Moskva, Klyazma, and Moksha rivers. Forms most of the border between Vladimir and Nizhegorod oblasts.

Oka (uh-KAH), river, approximately 600 mi/966 km long, RUSSIA; rises in the SAYAN MOUNTAINS, SW BURYAT REPUBLIC, S Siberian Russia; flows N through IRKUTSK oblast to join the ANGARA RIVER below BRATSK. The lower Oka valley is flooded by waters contained behind the Bratsk Dam.

Okaba (O-kah-bah), town, IRIAN JAYA, INDONESIA, 78 mi/125 km NE of MERAUKE; 08°06′S 139°42′E.

Okabe (O-kah-be), town, Shida county, SHIZUOKA prefecture, central HONSHU, E Central JAPAN, 9 mi/15 km S of SHIZUOKA; 34°54′N 138°16′E.

Okabena (o-kuh-BEE-nuh), village (2000 population 185), JACKSON county, SW MINNESOTA, 19 mi/31 km WNW of JACKSON; 43°44′N 95°19′W. Grain; livestock. HERON Lake to NE, South Heron Lake to E.

Okabena Lake (o-kuh-BEE-nuh), NOBLES county, SW MINNESOTA; 2 mi/3.2 km long, 1 mi/1.6 km wide. Drains into OCHEDA LAKE, 3 mi/4.8 km S through OCHEYEDAN RIVER. WORTHINGTON is at NE end and on N shore. Man-made channel links lake to Okabena Creek to N (no natural link).

Okagaki (o-KAH-gah-kee), town, Onga county, FUKUOKA prefecture, N KYUSHU, SW JAPAN, 22 mi/35 km N of FUKUOKA; 33°51′N 130°36′E.

Okahandja (o-ke-HAHN-juh), town, N central NAMIBIA, 40 mi/64 km NNW of WINDHOEK; 21°59′S 16°55′E. Elev. 4,398 ft/1,341 m. Livestock-raising and craft center; site of National Institute for Educational Development.

Okaharu (o-KAH-hah-roo), village, Kuma county, KUMAMOTO prefecture, W KYUSHU, SW JAPAN, 40 mi/ 65 km S of KUMAMOTO; 32°13′N 130°56′E.

Okahukura (O-kah-hoo-KOOR-ah), village, W NORTH ISLAND, NEW ZEALAND, c.6 mi/10 km N of Taumaranui, and c.168 mi/270 km S of AUCKLAND; 38°47′S 175°13′E. On Main Trunk railroad Auckland-WELLINGTON.

Okaihau (o-KEI-ou), township, Far North district, NEW ZEALAND, 120 mi/193 km NNW of AUCKLAND, between BAY OF ISLANDS and HOKIANGA HARBOUR, near Lake Omapere; 35°19′S 173°47′E. Northernmost railroad terminus; sawmills, limonite quarries.

Okakarara (o-ke-kuh-RAH-ruh), town, N central NAMIBIA, 60 mi/97 km ESE of OTJIWARONGO; 20°35′S 17°26′E. End of paved road. Supplies needs of local farming community. Hosp. and technical institute here.

Okak Islands (O-kak), group of two adjoining islands, NE NEWFOUNDLAND AND LABRADOR, NE Canada, at entrance of Okak Bay on the Atlantic; each island is 10 mi/16 km long, 10 mi/16 km wide. On NW island is abandoned fishing settlement of NUTAK. Adjacent are several islets.

Okaloosa (o-kuh-LOOS-suh), county (□ 1,082 sq mi/ 2,813.2 sq km; 2006 population 180,291), NW FLORIDA, bounded by ALABAMA state line (N) and GULF OF MEXICO (S); ⊙ CRESTVIEW; 30°40′N 86°35′W. Rolling agricultural area (corn, peanuts, cotton; hogs, cattle, poultry) drained by BLACKWATER, YELLOW, and SHOAL rivers. Also forestry (lumber, naval stores) and some fishing. Includes part of Choctawhatchee National Forest and CHOCTAWHATCHEE BAY. Formed 1915.

Okamanpeedan Lake (O-kuh-MAN-pee-dan), MARTIN county, S MINNESOTA, and EMMET county, NW IOWA, 10 mi/16 km SSW of FAIRMONT, Minnesota; 7 mi/11.3 km long, 1 mi/1.6 km wide; 43°30′N 94°34′E. Fed and drained by East Fork DES MOINES RIVER. Okamanpeedan State Park is on S shore of lake, in Iowa. Also called Tuttle Lake.

Okanagan Lake (o-kuh-NAH-guhn), S BRITISH COLUMBIA, W Canada; 69 mi/111 km long and from 2 mi/ 3 km to 4 mi/6 km wide; 49°45′N 119°44′W. It drains S through the Okanagan (OKANOGAN) River. The lake is in a prosperous fruit-growing region.

Okanagan Landing (o-kuh-NAH-guhn), former village, S BRITISH COLUMBIA, W Canada, on OKANAGAN LAKE; 50°14′N 119°22′W. Fruit, vegetables. Amalgamated into city of VERNON in 1993.

Okanagan River, U.S. and Canada: see OKANOGAN RIVER.

Okanagan-Similkameen (o-kuh-NAH-guhn–si-MILkuh-meen), regional district (□ 4,021 sq mi/10,454.6 sq km; 2001 population 76,635), S BRITISH COLUMBIA, W Canada; 49°25′N 120°00′W. Manning Provincial Park to W. Consists of 8 electoral areas and 6 municipalities (PENTICTON, SUMMERLAND, OLIVER, OSOYOOS, KEREMEOS). Agriculture; manufacturing; forestry; tourism, retirement center. Incorporated 1966.

Okanda National Park, game reserve, OGOOUÉ-IVINDO province, GABON, 150 mi/250 km ESE of LIBREVILLE; 00°30′S 11°40′E. Also called Lopé-Okanda National Park.

Okanogan (o-kuh-NAH-guhn), county (□ 5,281 sq mi/ 13,730.6 sq km; 2006 population 40,040), N WASHINGTON, bounded by CANADA (BRITISH COLUMBIA.) to N; ⊙ OKANOGAN; 48°33′N 119°45′W. Bounded S by COLUMBIA River, forms LAKE PATEROS (Wells Dam) SW, RUFUS WOODS LAKE (Chief Joseph Dam) S and SE; large GRAND COULEE DAM in SE corner; drained by OKANOGAN and METHOW rivers. CASCADE MOUNTAINS in W. Timber; apples, barley, oats, alfalfa, wheat; cattle, sheep; gold, silver, copper; tourism and recreation, several ski resorts. Includes parts of Chelan and Colville national forests and Colville Indian Reservation. Pearrygin Lake State Park in W; Osoyos Lake State Park on N, lake extends N into British Columbia; Conconally State Park in center; Alta Lake, Fort Okanogan, and Bridgeport state parks in S; nearly all of Okanogan National Forest is in W, including Pasayten and part of Lake Chelan-Sawtooth wilderness areas; several smaller sections of Okanogan National Forest in NE; part of large Colville Indian Reservation (roughly half) in SE. Largest county in land area in Washington (ranks fifty-fifth in U.S.). Formed 1888.

Okanogan (o-kuh-NAH-guhn), town (2006 population 2,391), ⊙ OKANOGAN county, N WASHINGTON, 73 mi/117 km NNE of WENATCHEE and on OKANOGAN RIVER, at mouth of SALMON CREEK; center of Okanogan irrigation project; 48°22′N 119°35′W. Settled 1886, c.20 mi/32 km from Fort Okanogan, which was first American settlement in Washington; incorporated 1907. Silver, copper; timber, apples (in irrigated Okanogan River valley), wheat; manufacturing (concrete). Gateway to Chelan National Forest (NW). Loup Loup Ski Area, at Loup Loup Pass, to W; Conconally State Park to NW. County Historical Museum, Okanogan National Forest to W.

Okanogan River (o-kuh-NAH-guhn), 115 mi/185 km long, N WASHINGTON and S BRITISH COLUMBIA,

---

CANADA; issues from S end of OKANAGAN LAKE at PENTICTON, British Columbia; flows S, through Skaha Lake, past OLIVER, through Osoyoos Lake, in which it crosses British Columbia-Washington border, past OROVILLE, OMAK and OKANOGAN, to COLUMBIA River 2 mi/3.2 km E of BREWSTER. Receives SIMILKAMEEN RIVER from W at Oroville.

Okara (o-KAH-rah), city, SAHIWAL district, MULTAN division, PUNJAB province, N central PAKISTAN; 30°49′N 73°27′E. Major market for food grains, oilseed, and cotton; manufacturing textiles, hosiery, metal boxes, and carpets; cotton milling and ginning. Has a national dairy farm (5 mi/8 km W) and a livestock research center. Technical school.

Okarche (o-KAHR-chee), town (2006 population 1,155), on KINGFISHER-CANADIAN county line, central OKLAHOMA, 30 mi/48 km WNW of OKLAHOMA CITY; 35°43′N 97°58′W. In grain and livestock area; manufacturing (fiber processing, concrete, commercial and industrial heating and air conditioning); oil and natural gas in region (declining).

Okarem (o-ka-REM), town, BALKAN weloyat, TURKMENISTAN, on CASPIAN SEA, 105 mi/175 km SSW of NEBITDAG, and 39 mi/63 km N of ESENGULY. Oil production nearby. Also Oqarem.

Okatibbee Creek (o-kuh-TIB-ee), E MISSISSIPPI; rises in W KEMPER county; flows through Okatibbee Lake reservoir and past MERIDIAN to W, joins CHUNKY CREEK at ENTERPRISE to form CHICKASAWHAY RIVER.

Okauchee (o-KAW-chee), village, WAUKESHA county, SE WISCONSIN, on Okauchee Lake (c.3 mi/4.8 km long), 13 mi/21 km NW of WAUKESHA. Concrete.

Okavango Delta (o-KAH-vahn-go), NORTH-WEST DISTRICT, N central BOTSWANA. Covers more than 5,000 sq mi/12,950 sq km. This is one of the largest inland drainage deltas in AFRICA. This swampy area occupies a depression that contained a large prehistoric lake. The delta is formed by the CUBANGO River that enters the country from ANGOLA, crossing the CAPRIVI STRIP, and subdivides into many distributaries that are blocked by a low ridge forming the S boundary of the depression. The N part of the swamp has papyrus growth and is wet throughout the year; the rest of the swamp fills with water as the seasonal cycle progresses. Supports a wide variety of wildlife.

Okavango Region (o-kuh-VAHN-go), administrative division (2001 population 202,694), NE NAMIBIA extending from Kavangor S to Omatakor; 18°20′S 19°30′E. Political region centering around RUNDU; forest savanna; major timber area; traditional woodcarving.

Okavango River, SW AFRICA: see CUBANGO, river.

Okawa (O-kah-wah), city, FUKUOKA prefecture, NW KYUSHU, SW JAPAN, on CHIKUGO RIVER and 28 mi/45 km S of FUKUOKA; 33°12′N 130°23′E. Rush (handmade mats). Furniture. *Etsu* fish live exclusively in Chikugo River.

Okawa (O-kah-wah), town, Okawa county, KAGAWA prefecture, NE SHIKOKU, W JAPAN, 12 mi/20 km S of TAKAMATSU; 34°15′N 134°14′E.

Okawa (O-kah-wah), village, Tosa county, KOCHI prefecture, S SHIKOKU, W JAPAN, 16 mi/25 km N of KOCHI; 33°46′N 133°28′E.

Okawachi (O-kah-WAH-chee), town, Kanzaki district, HYOGO prefecture, S HONSHU, W central JAPAN, 35 mi/56 km N of KOBE; 35°03′N 134°44′E.

Okawville, village (2000 population 1,355), WASHINGTON county, SW ILLINOIS, 12 mi/19 km WNW of NASHVILLE; 38°25′N 89°32′W. Health resort, with mineral springs.

Okay, village (2006 population 595), WAGONER county, E OKLAHOMA, suburb 8 mi/12.9 km SSE of WAGONER and 7 mi/11.3 km NNE of MUSKOGEE, on VERDIGRIS RIVER near its confluence with ARKANSAS RIVER; 35°51′N 95°18′W. In agricultural area. FORT GIBSON LAKE reservoir (NEOSHO River) to NE, including Bay State Park.

Okaya (O-kah-yah), city, NAGANO prefecture, central HONSHU, central JAPAN, on W shore of Lake SUWA, near efflux of TENRYU RIVER, 40 mi/65 km S of NAGANO; 36°03′N 138°03′E. Manufacturing includes cameras, machine tools, optical equipment, timepieces. Raw-silk center. Carnations; miso, *koya* tofu, sake; fish processing.

Okayama (o-KAH-yah-mah), prefecture [Japanese *ken*] (□ 2,721 sq mi/7,047 sq km; 1990 population 1,925,913), SW HONSHU, W JAPAN, on HIUCHI and HARIMA seas (S; central and E sections of INLAND SEA); ⊙ OKAYAMA. Bounded N by TOTTORI prefecture, E by HYOGO prefecture, and W by HIROSHIMA prefecture. Includes many offshore islands; largest, KONO-SHIMA and KITAGI-SHIMA. Mountainous terrain, drained by many small streams. Mainly agricultural, with chief products being rice, wheat, peppermint, fruit (persimmons, pears, watermelons, peaches). Major production area for tatami mats. Widespread livestock raising; saltmaking on shores of Inland Sea. Manufacturing of motor vehicles, steel, petrochemicals, textiles, pottery; oil refining. Chief centers are Okayama, KURASHIKI, TSUYAMA, TAMANO.

Okayama (o-KAH-yah-mah), city (2005 population 674,746), ⊙ OKAYAMA prefecture, SW HONSHU, W JAPAN, on an inlet of the INLAND SEA; 34°39′N 133°55′E. Railroad hub and industrial and marketing center. Manufacturing of synthetic fibers; nori. Fruits. Okayama Castle dates to the 16th century and Korakuen Garden is an 18th century park. Seat of Okayama University, which has a famous medical college Sadaiji Eyo Festival is held here.

Okazaki (o-KAH-zah-kee), city (2005 population 363,807), AICHI prefecture, S central HONSHU, central JAPAN, 22 mi/35 km S of NAGOYA; 34°57′N 137°10′E. Motor vehicles, synthetic fibers. Strawberries. Granite. Has feudal castle. Matsudaira Takechiyo, later Tokugawa Ieyasu (founder of Tokugawa shogunate) born here.

Okchon, county (□ 207 sq mi/538.2 sq km), S NORTH CHUNGCHONG province, SOUTH KOREA. Located between Sobaek and Noryong mountains. KUM RIVER flows N, creating agricultural basins (rice, barley, beans, potatoes, sweet potatoes, tobacco). Sericulture; animal husbandry. Kyongbu railroad and Expressway. Resorts.

O'Kean (o-KEEN), village, RANDOLPH county, NE ARKANSAS, 20 mi/32 km WNW of PARAGOULD; 36°10′N 90°49′W.

Okecie (o-KEN-tsee), Polish *Okęcie*, residential suburb of WARSAW, Warszawa province, E central POLAND, 4 mi/6.4 km SW of city center. Principal Warsaw airport.

Okeechobee (o-kee-CHO-bee), county (□ 891 sq mi/2,316.6 sq km; 2006 population 40,406), central FLORIDA, bounded W by KISSIMMEE RIVER and S by LAKE OKEECHOBEE; ⊙ OKEECHOBEE; 27°23′N 80°53′W. Cattle-raising area of grassy plains with many small lakes and some swamps; also poultry and vegetable farming. Formed 1917.

Okeechobee (o-kee-CHO-bee), city (□ 4 sq mi/10.4 sq km; 2005 population 5,900), ⊙ OKEECHOBEE county, central FLORIDA, 35 mi/56 km SW of FORT PIERCE, near N end of LAKE OKEECHOBEE; 27°14′N 80°49′W. Shipping center for vegetables, poultry, fish, frog legs, and palm fronds. Established 1915.

Okeechobee, Lake (o-kee-CHO-bee) (□ 700 sq mi/1,820 sq km), S central FLORIDA, N of the EVERGLADES; 26°56′N 80°48′W. Third-largest freshwater lake and fourth-largest lake wholly within the U.S. It is c.35 mi/56 km long and up to 25 mi/40 km wide, with a maximum depth of 15 ft/5 m. The KISSIMMEE RIVER is its chief source and the CALOOSAHATCHEE RIVER its main outlet. In reclaiming the Everglades and adjacent lands, many canals were built extending from the S part of the lake, itself a link in the OKEECHOBEE WATERWAY. A levee, built after the disastrous hurricane of 1926, encircles the lake's shores and protects the region from flood waters. The levees and canals have impeded the flow of water from the lake into the Everglades, which now suffers from water shortages. The drained lands bordering the lake produce vegetables and sugarcane.

Okeechobee Waterway (o-kee-CHO-bee) or **Cross-Florida Waterway**, 155 mi/249 km long, across S central FLORIDA, from STUART on the ATLANTIC OCEAN to FORT MYERS on the GULF OF MEXICO. Its main segments are the ST. LUCIE CANAL, LAKE OKEECHOBEE, Lake Hicpochee, and CALOOSAHATCHEE RIVER. The shallow (6 ft/1.8 m) waterway has five locks and is used by small commercial and leisure craft. It is also an outlet for the flood waters of Lake Okeechobee.

Okeene (o-KEEN), town (2006 population 1,183), BLAINE county, W central OKLAHOMA, 31 mi/50 km SW of ENID, near CIMARRON RIVER; 36°07′N 98°19′W. Wheat, oats; cattle; manufacturing (wheat by-prods., gypsum, flour).

Okefenokee Swamp (o-kee-fuh-NO-kee) (□ c.600 sq mi/1,554 sq km), SE GEORGIA, extending into N FLORIDA; c.40 mi/64 km long and averaging 20 mi/32 km in width; 00°40′N 82°20′W. It is a saucer-shaped depression with low ridges and small islands rising above the water and vegetation cover. Water depth averages 6 ft/2 m, but is as deep as 24 ft/8 m in places. It abounds in varied wildlife, and is drained by the SUWANEE and St. Marys rivers. In Georgia the swamp makes up most of the OKEFENOKEE National Wildlife Refuge and Wilderness Area (established 1937). Extensive timbering of cypress trees in the early 1990s. Many carnivorous plants; wading birds, fish, and other waterfowl found here.

Okegawa (o-KE-gah-wah), city, SAITAMA prefecture, E central HONSHU, E central JAPAN, 9 mi/15 km N of URAWA; 35°59′N 139°33′E.

Okehampton (OK-HAMP-tuhn) town (2001 population 5,846), W central DEVON, SW ENGLAND, on OKEMENT RIVER, at N edge of DARTMOOR, and 15 mi/24 km NNE of TAVISTOCK; 50°44′N 04°00′W. Former agricultural market; manufacturing (shoes, agricultural machinery, fertilizer); limestone and slate quarries. Has remains of Norman castle. Artillery ranges on moor to S.

Okemah (o-KEE-muh), town (2006 population 2,973), ⊙ OKFUSKEE county, central OKLAHOMA, 22 mi/35 km SW of OKMULGEE; 35°25′N 96°17′W. Elevation 913 ft/278 m. In agricultural area (grain, pecans, peanuts, corn). Manufacturing (flex circuits, printing, aerospace parts, jeans). Was home of Woody Guthrie. Settled 1902.

Okement River (OK-muhnt), 13 mi/21 km long, DEVON, SW ENGLAND; rises in two headstreams near center of DARTMOOR, 5 mi/8 km S of OKEHAMPTON; flows N to TORRIDGE RIVER 2 mi/3.2 km NNE of HATHERLEIGH.

Okene (o-KE-ne), town, headquarters of Local Government Area, KOGI state, S central NIGERIA, 40 mi/64 km WSW of LOKOJA. Agricultural trade center; sheanut processing, cotton weaving, sack making; palm oil and kernels, durra, corn, plantains.

Okere (O-ke-rai), 20 mi/32 km long, river, NORTHERN region, NE UGANDA; extends from the KOTIPE RIVER; flows to meet the OKOK RIVER in N EASTERN region.

Oker River (O-ker), 65 mi/105 km long, NW GERMANY; rises in the upper HARZ S of GOSLAR; flows N, past BRUNSWICK, to the ALLER 8 mi/12.9 km WNW of GIFHORN. Dammed in the Harz (reservoir holds 1,679 million cu ft/47.4 million cu m). Formerly also spelled Ocker.

Oketo (O-ke-to), town, Abashiri district, Hokkaido prefecture, N JAPAN, 121 mi/195 km E of SAPPORO; 43°42′N 143°35′E. Lumber.

Oketo (o-KEE-to), village (2000 population 87), MARSHALL county, NE KANSAS, on BIG BLUE RIVER, and 9 mi/14.5 km NNE of MARYSVILLE, near NEBRASKA state line; 39°57′N 96°35′W. Grain.

Area is shown by the symbol □, and capital city or county seat by ⊙.

Okfuskee (ok-FUHS-kee), county (□ 628 sq mi/1,632.8 sq km; 2006 population 11,370), central OKLAHOMA; ☉ OKEMAH; 35°28′N 96°19′W. Intersected by NORTH CANADIAN RIVER (forms county boundary in SW) and DEEP FORK of Canadian River (in N). Agriculture (melons, fruit, grain, pecans, peanuts); cattle; dairying; manufacturing (machinery, apparel). Formed 1907.

Okha (uh-HAH), city (2006 population 26,100), N SAKHALIN oblast, RUSSIAN FAR EAST, on the Sea of OKHOTSK (NE coast of SAKHALIN Island), on railroad, 660 mi/1,062 km N of YUZHNO-SAKHALINSK; 53°34′N 142°57′E. Major petroleum-producing center, linked by pipeline to KOMSOMOL'SK-NA-AMURE; sawmilling; asphalt deposits; electric power station. Became city in 1938.

Okha (uhk-HAH), port, JAMNAGAR district, GUJARAT state, W central INDIA, on W end of KATHIAWAR PENINSULA, at mouth of GULF OF KACHCHH, 17 mi/27 km NNE of DWARKA; 22°28′N 69°05′E. Railroad terminus; trade center; fishing, salt manufacturing; motor vehicle assembly plant. Lighthouse (NE). Large chemical works (manufacturing of soda ash, bleaching powder) 5 mi/8 km SW, at Mithapur.

Okhaldhunga (ahk-hul-DOONG-gah), district, E Nepal, in SAGARMATHA zone; ☉ OKHALDHUNGA.

Okhaldhunga (ahk-hul-DOONG-gah), town, ☉ OKHALDHUNGA district, E central Nepal, 80 mi/129 km ESE of KATHMANDU; 27°19′N 86°30′E. Elevation 6,066 ft/1,849 m.

Okhansk (uh-HAHNSK), city (2006 population 7,895), W central PERM oblast, W URALS, E European Russia, port on the right bank of the KAMA River, on highway, 75 mi/119 km SW of PERM; 57°43′N 55°23′E. Elevation 383 ft/116 m. Clothing factory; logging; dairying. Founded in the 17th century, city since 1781.

Okhi (O-khee), Latin *Ocha*, mountain on SE Évvia island, Évvia prefecture, CENTRAL GREECE department, extends c.10 mi/16 km from KARISTOS NE to Cape KAFIREOS; 38°04′N 24°28′E. Rises to 4,585 ft/1,398 m at the Ayios, or Hagios, Elias [=St. Elias], 4 mi/6.4 km NNE of KARISTOS. Also Oche.

Okhota River (uh-HO-tah), 220 mi/354 km long, RUSSIAN FAR EAST; rises in highland S of OYMYAKON PLATEAU, near the border of NE KHABAROVSK TERRITORY and SE SAKHA REPUBLIC; flows S across Khabarovsk Territory to the Sea of OKHOTSK near OKHOTSK.

Okhotnikovoye, UKRAINE: see ROKYTNE, Rivne oblast.

Okhotnykove, UKRAINE: see ROKYTNE, Rivne oblast.

Okhotsk (uh-HOTSK), town (2005 population 5,515), NE KHABAROVSK TERRITORY, RUSSIAN FAR EAST, maritime fishing port on the NW shore of the Sea of OKHOTSK, near the mouth of the OKHOTKA RIVER, on highway, 1040 mi/1677 km N of KHABAROVSK, and 440 mi/708 km N of NIKOLAYEVSK-NA-AMURE; 59°23′N 143°18′E. Fish combine; ship repair. One of the oldest Russian settlements in the Far East and an administrative center of the region until the early-20th century. Founded in 1649. Made city in 1783; naval port until 1850; reduced to status of a village, 1927–1949; made town in 1949.

Okhotsk Current, current, Pacific Ocean: see OYASHIO.

Okhotsk, Sea of (ah-KHOTSK), Russian *Okhotskoye More* (□ 590,000 sq mi/1,528,100 sq km), NW arm of the PACIFIC OCEAN, W of the KAMCHATKA PENINSULA and the KURIL Islands, off far E coast of SIBERIA, Russian Far East, bounded by KHABAROVSK TERRITORY (NW) and MAGADAN (N), SAKHALIN (S), and KAMCHATKA (NE) oblasts. It is connected with the Sea of JAPAN by the TATAR and LA PÉROUSE straits and with the Pacific Ocean by passages through the Kuril Islands. The sea is generally less than 5,000 ft/1,524 m deep; its deepest point, near the Kuriles, is 11,033 ft/3,363 m. The sea is icebound from November to June and has frequent heavy fogs. Very rich fishing and crabbing area. MAGADAN and KORSAKOV are the largest ports.

Okhrid, MACEDONIA: see OHRID.

Okhtyrka (o-HTIR-kah) (Russian *Akhtyrka*), city (2001 population 50,399), SE SUMY oblast, UKRAINE, near the VORSKLA RIVER, 40 mi/64 km S of SUMY; 50°18′N 34°53′E. Elevation 396 ft/120 m. Railroad terminus; raion center; food processing, agricultural and other machine building; furniture-making; gas extraction office; technical college, heritage museum. Founded by the Ukrainian Cossacks in 1641; center of the Okhtyrka regiment (1655–1765) of the Sloboda Ukraine, a S frontier of Muscovy, then Russia.

Oki (O-KEE), town, Mizuma county, FUKUOKA prefecture, NE KYUSHU, SW JAPAN, 25 mi/40 km S of FUKUOKA; 33°12′N 130°26′E. Rush products (mats).

Okiep, town, NAMAQUALAND, NORTHERN CAPE province, SOUTH AFRICA, 300 mi/483 km N of CAPE TOWN, 5 mi/8 km N of SPRINGBOK, on main route N7 to NAMIBIA; 29°35′S 17°52′E. Elevation 3,690 ft/1,725 m. Original site of copper mining, now moved to NABABEEP 7 mi/12 km E where large smelter is located.

Oki-gunto (o-kee–GUN-to), island group (□ 145 sq mi/377 sq km), SHIMANE prefecture, W JAPAN, in SEA OF JAPAN, 35 mi/56 km N of MATSUE, off SW HONSHU. Includes DOGO (largest island) and DOZEN (group of three islands). Generally mountainous and forested. Cattle raising, lumbering, fishing. Produces rice, raw silk. Emperor Gotoba exiled here in 1239. SAIGO (on Dogo), chief town and port. Anglicized as Oki Islands; sometimes called Oki-no-shima.

Okigwi (aw-KEE-gwee), town, IMO state, S NIGERIA, 35 mi/56 km NE of OWERRI. Trade center; palm-oil milling; kola nuts. Has hospital.

Okikeska Lake (o-ki-KE-skuh), W QUEBEC, E Canada, on HARRICANAW RIVER, 22 mi/35 km NW of VAL-D'OR; 7 mi/11 km long, 3 mi/5 km wide. In gold-mining region.

Okimi (O-kee-mee), town, Saeki county, HIROSHIMA prefecture, SW HONSHU, SW JAPAN, 9 mi/15 km S of HIROSHIMA; 34°12′N 132°24′E. Oysters.

Okinawa (o-kee-NAH-wah), city (2005 population 126,400), S OKINAWA island, Okinawa prefecture, SW JAPAN, 12 mi/20 km N of NAHA; 26°19′N 127°48′E. Kadena military base is here.

Okinawa (o-kee-NAH-wah), main island (□ 454 sq mi/1,180.4 sq km), of Okinawa prefecture, W PACIFIC OCEAN, SW of KYUSHU, SW JAPAN. It is the largest of the OKINAWA ISLANDS in the RYUKYU ISLANDS archipelago. NAHA is the largest city and chief port. Long, narrow, irregularly-shaped island of volcanic origin with coral formations in the S part. The N part is mountainous, rising to 1,657 ft/505 m, and has a dense vegetation cover. Most of the population is located in the S. Humid subtropical climate. Sugarcane, sweet potatoes, and rice are grown, sugar is refined, cattle are raised, and fishing is important. There is some light industry in Naha. Major domestic tourist destination. Scene of the last great U.S. amphibious campaign in World War II. U.S. army and marine forces landed here on April 1, 1945 and fought one of the bloodiest campaigns of the war, while the navy offshore suffered heavy damage in resisting attacks by kamikaze planes. The Japanese garrison, having lost 103,000 of its 120,000 men and ⅓ of its civilian population, ended organized resistance on June 21, 1945. U.S. casualties were 48,000, one-quarter listed as dead. Placed in August 1945 under a U.S. military governor and remained under U.S. control until May 1972, when it was returned to Japan. U.S. military bases (including the large Kadena Air Force Base), which occupy 20% of the island's land area, were allowed to remain despite growing opposition by native Okinawans. As of 1996, 29,000 American troops (two-thirds of those stationed in Japan) were based here.

Okinawa Islands (o-kee-NAH-wah), Japanese *Okinawa-gunto* (o-kee-NAH-wah–GUN-to), central group (□ 579 sq mi/1,505.4 sq km) of RYUKYU ISLANDS, Okinawa prefecture, SW JAPAN, between EAST CHINA (W) and PHILIPPINE(E) seas, 325 mi/523 km S of KYUSHU; 26°04′N 127°40′S–27°03′N 127°59′E. Its 70-mi/113-km chain comprises volcanic islands of OKINAWA(largest), KERAMA-RETTO, IE-JIMA, KUME-SHIMA, DAITO-JIMA, IHEYA-SHOTO, and scattered coral islets. Generally mountainous and fertile; in typhoon zone. Pine and oak forests. Agriculture (sweet potatoes, sugarcane, bananas, pineapples); livestock-breeding. Produces textiles, lacquerware, pottery. Tourism. NAHA on Okinawa is chief center.

Okino-daito-shima, JAPAN: see DAITO-JIMA.

Okinoerabu-shima (o-kee-NO-e-RAH-boo–SHEE-mah), island (□ 37 sq mi/96.2 sq km) of island group AMAMI-GUNTO, in RYUKYU ISLANDS, Oshima county, KAGOSHIMA prefecture, SW JAPAN, between EAST CHINA (W) and PHILIPPINE (E) seas, 35 mi/56 km N of OKINAWA; 12 mi/19 km long, 6 mi/9.7 km wide. Hilly; fertile (sugarcane, sweet potatoes). Also called Erabu-shima, Okierabu-shima. Chief town, WADO-MARI.

Okino-shima (o-KEE-no–SHEE-mah), island (□ 5 sq mi/13 sq km), SUKUMO city, KOCHI prefecture, W JAPAN, in HOYO STRAIT just off SW coast of SHIKOKU; 3 mi/4.8 km long, 2 mi/3.2 km wide. Hilly, fertile. Agriculture (barley, sweet potatoes); raw silk, ornamental coral. Fishing.

Oki-no-shima, JAPAN: see OKI-GUNTO.

Okitipupa (o-kee-chee-POO-pah), town, ONDO state, S NIGERIA, 40 mi/64 km S of ONDO. Agricultural trading center; cacao, palm oil and kernels, rubber, timber. Sawmilling.

Okku (OK-GOO), county (□ 114 sq mi/296.4 sq km), NW NORTH CHOLLA province, SOUTH KOREA, on KUNSAN peninsula, adjacent to IKSAN city on E, KIMJE over Mankyong on S, SOUTH CHUNGCHONG province over KUM RIVER on N, and facing YELLOW SEA on W. Alluvial valley covers most of county, low mountains in N. Agriculture (rice, fruit); fishery (croaker, shrimp, pike, skate); sericulture; animal husbandry. Known for rush mats. Estuary dike on Kum River; Sonyudo Beach.

Oklahoma, state (□ 69,902 sq mi/181,046 sq km; 1995 estimated population 3,277,687; 2000 population 3,450,654), SW UNITED STATES; ☉ OKLAHOMA CITY. Admitted as the forty-sixth state of the Union in 1907.

Geography

The state is bounded on the N by COLORADO and KANSAS and on the E by MISSOURI and ARKANSAS; the RED RIVER marks the S border with TEXAS; Texas also bounds the state on the W and on the S of the OKLAHOMA Panhandle, a 34-mi/55-km-wide strip of land that extends 166 mi/267 km W from the NW corner of the state, bordering NEW MEXICO on its W end. Oklahoma City and TULSA are important cities. Oklahoma is a land of climatic transition. The OUACHITA MOUNTAINS of the SE average more than 50 in/126 cm of precipitation per year while BLACK MESA averages less than 16 in/41 cm. Consequently, dense forests were the original cover for most of E Oklahoma while short grasslands dominated the W. The high, short-grass plains of W Oklahoma are part of the GREAT PLAINS and, like the rest of that area, are chilled by N winds in the winter and baked by intense heat in the summer. There are extensive grazing lands and wheat fields. The plains are broken here and there, notably by Black Mesa in the Panhandle and by the WICHITA MOUNTAINS in the SW, but the general slope is downward to the E, and central and E Oklahoma is mostly prairie, rising in the NE to the OZARK and BOSTON MOUNTAINS and in the SE to the Ouachita Mountains. Lesser ranges include the ARBUCKLE MOUNTAINS in S. The rivers that flow W-E across the state—the ARKANSAS, and its tributaries, the CIMARRON and the CANADIAN (with the NORTH CANADIAN) in the N; the Red River with the WASHITA and other tributaries in the S—are much more prominent in the E.

Economy

Formerly the major crop of Oklahoma was cotton, but now wheat is the leading cash crop; however, income from livestock (especially cattle) exceeds that from crops. Other important crops are peanuts and sorghums. Also sheep, poultry, exotic fowl [emu, ostrich] gained popularity, in the 1980s, being raised for their meat. Many minerals are found in the state, including coal, but the resource that has given the state its wealth is oil. After the first well was drilled in 1888, the petroleum industry grew by fits and starts to enormous proportions, and Oklahoma City and Tulsa were among the great natural gas and petroleum centers of the world. Oklahoma remains a major—but declining—oil-producing state. Many of Oklahoma's factories process raw materials found in the state and its chief industry includes non-electrical machinery and fabricated metal products.

History: to 1819

Oklahoma has a rich Native American heritage. The Native American population is the largest in the nation; the 1990 census reported 252,420 Native Americans in Oklahoma (c.8% of total state population). Several Native American cultures existed in the area before the first European visited here in 1541. Francisco Coronado almost certainly crossed Oklahoma in that year, and Hernando De Soto may have visited E Oklahoma. Later Juan de Oñate passed through W Oklahoma, and some other Spanish explorers and traders and French traders from LOUISIANA visited the region, but there was no development of the area. Native Americans dominated the landscape, tribes of the Plains cultures—Osage, Kiowa, Comanche, and Apache—in the W, and the Wichita and other relatively sedentary tribes farther E. It is asserted that the first European trading post was established at SALINA by the Chouteau family of SAINT LOUIS before the territory was transferred to the U.S. by the Louisiana Purchase in 1803, but the land remained in control of the sparse and nomadic Native American population. For the most part only traders, official explorers (notably Stephen H. Long), and scientific and curious travelers (among them Washington Irving and George Catlin) came into the present-day state.

History: 1819 to 1860

In 1819 the Adams-Onís Treaty with Spain defined Oklahoma as the SW border of the U.S. After the War of 1812 the U.S. government invited the Cherokee of GEORGIA and TENNESSEE to move into the area, and a few had come to settle before intense white pressure for their lands, with the approval of President Andrew Jackson, forced the Cherokee and the others of the Five Civilized Tribes (the Choctaw, the Chickasaw, the Creek, and the Seminole) to abandon their old homes E of the Mississippi and to take up residence in what was to become the INDIAN TERRITORY. Their tragic removal is known as the TRAIL OF TEARS. They settled on the hills and little prairies of the E section and built separate organized states and communities. The Cherokee particularly had a highly Europeanized culture, with a written language, invented by their great leader Sequoyah, and highly developed institutions. Some of the Cherokee were slaveholders and ran their agriculture in the traditional Southern plantation pattern; others were small farmers. The Five Civilized Tribes clashed briefly with the Plains Indians, particularly the Osage, but they were for a time free from white interference, and they were able to establish a civilization that strongly affected the whole history of the region.

History: 1860 to 1870

The troubles of the whites did not, however, long escape them, and the Civil War was a major disaster. Although no major battle of the war was fought in present-day Oklahoma, there were innumerable skirmishes. Most Native Americans allied themselves with the Confederacy, but Unionist disaffection was widespread, and individual violence was so prevalent that many fled, leaving their farms to desolation. As a punishment for taking the Confederate side the Five Civilized Tribes lost the W part of the Indian Territory, and the Federal government began assigning lands there to such landless Eastern tribes as the Delaware and the Shawnee, as well as to nomadic Plains tribes, who caused much trouble before they were subdued and settled on reservations. The territory was victimized by lawlessness and served as a hideout for white outlaws. After the establishment of a Federal court at FORT SMITH, Isaac Parker became famous as the hanging judge. Immediately after the Civil War the long drives of cattle from Texas to the Kansas railroad began to cross Oklahoma, traveling over the cattle trails that became part of Western folklore. The best known is the CHISHOLM TRAIL. The cattle were fattened on the virgin ranges of Oklahoma, and cattlemen began to look on the grasslands with speculative and covetous eyes.

History: 1870 to 1890

The first railroad to cross Oklahoma was built between 1870 and 1872, and thereafter it was not possible to keep white settlers out. They came despite laws and treaties with the Native Americans, and by the 1880s there was a strong admixture of whites. Ranches were developed, too, nominally owned by Native Americans, but actually controlled by white cattlemen and their cowboys; the region took on a tinge of the Old West of the cattle frontier, a tinge that it has never wholly lost. In the 1880s, land-hungry frontier farmers, the boomers, agitated to obtain the "unassigned" lands in the central section—the lands not given to any Native American tribe. The agitation succeeded, and a large strip was opened for settlement in 1889. On April 22, 1889, prospective settlers lined up on the territorial border, and at noon, at the sound of a pistol shot, were allowed to run into the state to compete for the best lands. Some settlers who illegally entered ahead of the set time were referred to as the Sooners, hence the state's nickname, the Sooner State. Later other strips of territory were opened, and settlers poured in from the Midwest and the South. The W section of what is now the state of Oklahoma became the Oklahoma Territory in 1890; it included the Panhandle, the narrow strip of territory that, taken from Texas by the Compromise of 1850, had become a no-man's-land where settlers came in undisturbed.

History: 1890 to World War II

In 1893 the Dawes Commission was appointed to implement a policy of dividing the tribal lands into individual holdings; the Native Americans resisted, but the policy was finally enforced in 1906. The wide lands of the INDIAN TERRITORY were thus made available to whites. The Civilized Tribes made the best of a poor bargain, and the Indian Territory and Oklahoma Territory were united in 1907 to form the state of Oklahoma, with a constitution that included provision for initiative and referendum. Already the oil boom had reached major proportions, and the young state was on the verge of great economic development. At the same time, cotton, wheat, and corn were major money crops, and cattle-land holdings, although shrinking, were still enormous. In World War I the great demand for farm products brought an agricultural boom to the state, but in the 1920s the state fell upon hard times. Recurrent drought burned the wheat in the fields, and overplanting, overgrazing, and unscientific cropping aided the weather in making Oklahoma part of the Dust Bowl of the 1930s. Farm tenancy increased in the 1920s, and in both the E and W the farms tended more and more to be held by large interests and to be consolidated in large blocks. A great number of tenant farmers were compelled to leave their dust-stricken farms and went W as migrant laborers; the tragic plight of these Okies, many of whom took Route 66 (the Highway of the Okies) to

CALIFORNIA, is the theme of John Steinbeck's *The Grapes of Wrath*. A larger migration, however, took place within the state as rural residents moved to the cites. With the return of rains, however, and with increasing care in selecting crops and in conserving and utilizing water and soil resources, much of the Dust Bowl was again made into productive farm land.

History: World War II to Present

The demands for food in World War II and Federal price supports for agricultural products after the war aided farm prosperity. Large state and Federal programs for conserving the water of rivers and for supplying irrigation have resulted in the construction of many large dams and reservoirs, such as the reservoir impounded by Kerr Dam on the Arkansas River, resulting in extension of barge navigation on the Arkansas River Navigation System to the Tulsa area in 1971, improved agricultural conditions and new recreation areas. (For more detailed information on irrigation projects, see separate articles on the rivers of Oklahoma.) Oklahoma experienced a boom in its economy during the late 1970s when oil prices rose dramatically. In the mid-1980s, Oklahoma's economy was hurt (as it had been in the 1930s) by dependence on a single industry as oil prices fell rapidly. Oklahoma has increased its industrial diversity and has moved, along with Texas, into the apparel industry (due to the availability of cotton and low-cost labor). Also important is the state's aircraft and rocket industries. The bombing of the Murrah Federal Building in Oklahoma City, on April 19, 1995, killed 166 people and interrupted the state's usual tranquillity. During the 1920s two governors, John C. Walton and Henry S. Johnston, were impeached. Prohibition, in effect since statehood, was repealed in 1959. The most important institutions of higher learning in the state are the University of Oklahoma, Oklahoma State University, and the University of Tulsa.

Government

The original 1907 constitution is still in effect. The Cheyenne and Arapacho tribes are currently suing the state government, claiming that their tribal lands were illegally seized in 1883 and 1948, including a 12-sq-mi/ 31-sq-km piece of land (formerly the Army base of Fort Reno) with unmarked graves of their people and ritual dance grounds, as well as significant oil and gas reserves. They are asking for the land's return; as yet, the issue has not been resolved. Oklahoma has a legislature of forty-eight senators and 101 representatives, elected for four- and two-year terms, respectively. The governor is elected for a four-year term. The current governor is Brad Henry. The state elects two U.S. senators and six representatives and has eight electoral votes.

Oklahoma has seventy-seven counties: ADAIR, ALFALFA, ATOKA, BEAVER, BECKHAM, BLAINE, BRYAN, CADDO, CANADIAN, CARTER, CHEROKEE, CHOCTAW, CIMARRON, CLEVELAND, COAL, COMANCHE, COTTON, CRAIG, CREEK, CUSTER, DELAWARE, DEWEY, ELLIS, GARFIELD, GARVIN, GRADY, GRANT, GREER, HARMON, HARPER, HASKELL, HUGHES, JACKSON, JEFFERSON, JOHNSTON, KAY, KINGFISHER, KIOWA, LATIMER, LE FLORE, LINCOLN, LOGAN, LOVE, McCLAIN, MCCURTAIN, MCINTOSH, MAJOR, MARSHALL, MAYES, MURRAY, MUSKOGEE, NOBLE, NOWATA, OKFUSKEE, OKLAHOMA, OKMULGEE, OSAGE, OTTAWA, PAWNEE, PAYNE, PITTSBURG, PONTOTOC, POTTAWATOMIE, PUSHMATAHA, ROGER MILLS, ROGERS, SEMINOLE, SEQUOYAH, STEPHENS, TEXAS, TILLMAN, TULSA, WAGONER, WASHINGTON, WASHITA, WOODS, and WOODWARD.

Oklahoma, county (□ 718 sq mi/1,866.8 sq km; 2006 population 691,266), central OKLAHOMA; ⊙ OKLAHOMA CITY, which extends into three adjacent counties; 35°32′N 97°24′W. Other important cities include EDMOND, SPENCER, and MIDWEST CITY. Intersected by NORTH CANADIAN River and the DEEP FORK of Ca-

Area is shown by the symbol □, and capital city or county seat by ⊙.

nadian River. Dairying; cattle; wheat; agriculture mostly in E and NW parts. Varied manufacturing at Oklahoma City and Edmond. Oil and natural-gas fields; sand, granite; refineries. Most urbanized county in state. Formed 1890.

Oklahoma, borough (2006 population 863), WESTMORELAND county, W central PENNSYLVANIA, 24 mi/ 39 km ENE of PITTSBURGH and 1 mi/1.6 km S of VANDERGRIFT, on KISKIMINETAS RIVER; 40°34′N 34°79′W.

Oklahoma City, city (2006 population 537,734), ⊙ state and of OKLAHOMA county, central Oklahoma, on the NORTH CANADIAN River; 35°28′N 97°30′W. The largest city in the state, it is an important livestock market, a wholesale, distributing, industrial, and financial center, and a farm trade and processing point. Oil is a major product; the city is situated in the middle of an oil field (opened 1928), with oil derricks even on the capitol grounds. Diversified light and heavy manufacturing. In SE part of city, Tinker Air Force Base, a logistics center with one of the world's largest air depots, has the largest employment at one site. Oklahoma City was quickly settled in a land rush after the area was opened to homesteaders on April 22, 1889. It became the state capital in 1910. One of the largest U.S. cities in terms of land area (604 sq mi/1,564 sq km), it extends into three counties and has many parks. Of interest are the capitol, the state historical museum, the National Cowboy Hall of Fame and Western Heritage Center, sports stadium, the state fairgrounds, the civic center buildings and monuments, a theater complex, a convention center, the state library, the Oklahoma Health Sciences Center, and a zoo. Educational institutions include Oklahoma City University, Oklahoma Christian University of Science and Arts, an Oklahoma State University branch campus, the medical school of the University of Oklahoma, Southern Nazarene University (in nearby Bethany), Oklahoma City Community College, Oklahoma State University (two-year branch), University of Central Oklahoma in adjacent Edmond (N), University of Oklahoma at NORMAN (S). The city also has a symphony orchestra. Reservoirs for recreation and water supply include Lake Stanley Druper and Shawnee Reservoir (SE), Arcadia Lake (N), Lake Hefner and Lake Overholder (W). Softball Hall of Fame; Firefighters Museum, Kirkpatrick Center Museum Complex. Will Rogers World Airport in SW part of city; Wiley Post Airport in NW part. On April 19, 1995, Oklahoma City was the victim of one of the worst terrorist attacks in U.S. history. A bomb explosion destroyed an entire side of the nine-story Murrah Federal Building; total death toll 166. Timothy McVeigh and Terry Nichols were convicted on Federal charges in association with the bombing. Incorporated 1890.

Oklánd, ROMANIA: see OCLAND.

Oklaunion (O-kluh-yoon-yuhn), unincorporated village, WILBARGER county, N TEXAS, near OKLAHOMA state line, c.40 mi/64 km WNW of WICHITA FALLS; 34°07′N 99°08′W. In agricultural area.

Oklawaha River (ahk-lah-WAH-hah), c.140 mi/225 km long, central and N central FLORIDA; rises in extensive central Florida lake system (with LAKE APOPKA at its upper end); flows N, receiving outlets of SILVER SPRINGS and ORANGE LAKE, then turns E, converging with ST. JOHNS RIVER near Welaka. The upper course, extremely tortuous, is dredged to LEESBURG.

Oklee, village (2000 population 396), RED LAKE county, NW MINNESOTA, on Lost River, and 19 mi/31 km E of RED LAKE; 47°50′N 95°50′W. Falls. Grain, sugar beets, sunflowers, potatoes; cattle; manufacturing (fertilizers, feeds).

Okmulgee (ok-MUHL-gee), county (□ 702 sq mi/ 1,825.2 sq km; 2006 population 39,670), E central OKLAHOMA; ⊙ OKMULGEE; 35°38′N 95°58′W. Intersected by the DEEP FORK; includes Okmulgee and Henryetta

lakes (recreation). Agriculture (peanuts, corn; cattle; dairy products). Manufacturing (glass containers, metal stampings, chemicals) at Okmulgee and Henryetta. Oil and natural-gas fields; timber. Coal mining. State recreation area. Lake OKMULGEE reservoir and Okmulgee State Park in W. Formed 1907.

Okmulgee (ok-MUHL-gee), city (2006 population 12,829), ⊙ OKMULGEE county, E central OKLAHOMA, 31 mi/50 km S of TULSA; 35°37′N 95°57′W. Elevation 670 ft/204 m. In an oil and farm area. An agricultural processing center, it has oil and gas wells; manufacturing (glass containers, marker boards, carbonated beverages, publishing and printing). It was founded on the site of the Creek capital (1868–1907) and boomed with the discovery of oil in 1907. An old Creek (Muscogee) Nation Council House (1878) is on the town square. Yucchi Tribal Headquarters and Oklahoma State University Technical Branch (two-year) here. Lake OKMULGEE and Okmulgee State Park to W; EUFAULA LAKE reservoir to SE. Incorporated 1900.

Okmulgee, Lake (ok-MUHL-gee), reservoir, OKMULGEE county, E central OKLAHOMA, in Okmulgee State Park, 6 mi/9.7 km WSW of OKMULGEE; c.5 mi/8 km long; 35°36′N 96°03′W. Built for Okmulgee water supply.

Okna, UKRAINE: see VIKNO.

Oknitsa, MOLDOVA: see OCNIȚA.

Okno, UKRAINE: see VIKNO.

Okny, UKRAINE: see KRASNI OKNY.

Oko, RUSSIA: see YASNOMORSKIY.

Okoboji, town (2000 population 820), DICKINSON county, NW IOWA, just S of SPIRIT LAKE, between lakes EAST OKOBOJI (c.5 mi/8 km long) and WEST OKOBOJI (c.6 mi/9.7 km long); 43°23′N 95°08′W. Situated in popular resort area containing several state parks on West Okoboji Lake, and fish hatchery on BIG SPIRIT LAKE to N.

Okodongwe (o-ko-DAWNG-gwai), village, ORIENTALE province, NE CONGO, 230 mi/370 km ENE of BUTA; 03°17′N 28°13′E. Elev. 2,857 ft/870 m. Cotton ginning.

Okok (O-kok), river, c.70 mi/113 km long, NORTHERN and EASTERN regions, NE UGANDA, extension of Dopeth River; rises near MT. MORUNGOLE; flows S to marshes between LAKE KYOGA and LAKE BISINA in Eastern region.

Okola, town, Central province, CAMEROON, 13 mi/21 km NW of Yaoundé; 04°01′N 11°24′E.

Okolona (o-kuh-LO-nuh), unincorporated city (2000 population 17,807), JEFFERSON county, N KENTUCKY; residential suburb 10 mi/16 km SSE of downtown LOUISVILLE; 38°08′N 85°41′W. Agriculture to SE; motor-vehicle plant to NW.

Okolona (o-kuh-LO-nuh), town (2000 population 3,056), ⊙ CHICKASAW county, NE central MISSISSIPPI, 18 mi/29 km S of TUPELO; 34°00′N 88°45′W. In agricultural, dairying, livestock, and timber area; manufacturing (furniture, apparel, corrugated boxes). Seat of Okolona College. Founded 1848; burned in Civil War. Has a Confederate cemetery. Nearby is a U.S. game preserve. Part of Tombigbee National Forest to W; NATCHEZ TRACE PARKWAY passes to W.

Okolona (o-kuh-LO-nuh), village (2000 population 160), CLARK county, S central ARKANSAS, 18 mi/29 km WSW of ARKADELPHIA; 34°00′N 93°20′W. In agricultural area, near ANTOINE RIVER.

Okomfukrom (o-kuhm-FOO-kruhm), town, VOLTA region, GHANA, 15 mi/24 km W of JASIKAN; 07°22′N 00°35′E. Road terminus on Lake VOLTA. Fishing; cacao.

Okondja (o-KON-juh), town, HAUT-OGOOUÉ province, E GABON, 75 mi/121 km NNE of FRANCEVILLE; 00°41′S 13°48′E. Coffee growing. Until 1946, in Middle CONGO colony.

Okonek (o-KO-nek), German *Ratzebuhr*, town in POMERANIA, Piła province, NW POLAND, 14 mi/23 km SSE of SZCZECINEK; 53°32′N 16°51′E. Woolen milling,

steel; dairying. Until 1938, in former Prussian province of Grenzmark Posen-Westpreussen.

Okoneshnikovo (uh-kuh-NYESH-nee-kuh-vuh), village (2006 population 5,480), SE OMSK oblast, SW SIBERIA, RUSSIA, on crossroads, 23 mi/37 km SE of KALACHINSK; 54°50′N 75°05′E. Elevation 360 ft/109 m. Flour milling.

Okoppe (o-KOP-pe), town, Abashiri district, Hokkaido prefecture, N JAPAN, 130 mi/210 km N of SAPPORO; 44°28′N 143°07′E. Dairying.

Okorsh (o-KORSH), village, RUSE oblast, DULOVO obshtina, BULGARIA; 43°45′N 27°01′E.

Okoruro (o-ko-ROO-ro), canton, PACAJES province, LA PAZ department, W BOLIVIA; 17°53′S 68°43′W. Elevation 12,989 ft/3,959 m. Unmined copper deposits; gas wells in area. Clay, gypsum, and limestone deposits. Agriculture (potatoes, yucca, bananas, rye); cattle.

Okotoks (O-kuh-tahks), town (□ 7 sq mi/18.2 sq km; 2001 population 11,664), SW ALBERTA, W Canada, at foot of ROCKY MOUNTAINS, satellite community 23 mi/37 km S of CALGARY, in Foothills No. 31 municipal district; 50°44′N 113°59′W. Elevation 3,448 ft/1,051 m. Oil-shipping center for TURNER VALLEY field; natural-gas and oil production, oil refining, pharmaceuticals manufacturing; wheat, flax; cattle. Dormitory town for Calgary. Established as a village in 1899; became a town in 1904.

Okpara (ok-PAH-rah), town, DELTA state, S central NIGERIA, on road, 15 mi/24 km NE of WARRI. Market town. Fish; cassava, maize, bananas.

Okpara River, c.160 mi/257 km long, E BENIN and SW NIGERIA; rises in Benin N of PARAKOU; flows S (c.75 mi/121 km) along Nigerian border to OUÉMÉ RIVER, 25 mi/40 km NNE of Zagnando (Benin).

Okrilla, GERMANY: see OTTENDORF-OKRILLA.

Oksarka, RUSSIA: see AKSARKA.

Øksendal (UHK-suhn-dahl), village, MØRE OG ROMSDAL county, W NORWAY, 5 mi/8 km from head of SUNNDALSFJORDEN, 40 mi/64 km E of MOLDE. Cattle raising in mountains nearby.

Oksfjordhamn (AWKS-fyawr-HAHM-uhn), fishing village, TROMS county, N NORWAY, on Nordreisa (small fjord ofNORWEGIAN SEA), 60 mi/97 km ENE of TROMSØ; 69°55′N 21°15′E.

Oksino (AWK-see-nuh), settlement, central NENETS AUTONOMOUS OKRUG, ARCHANGEL oblast, NE European Russia, on the W arm of the PECHORA River delta mouth, 22 mi/35 km WSW of NARYAN-MAR; 67°36′N 52°12′E. Trading post; fisheries; reindeer raising.

Oksøy (AWKS-uh-u), small island in the SKAGERRAK approach to KRISTIANSAND, VEST-AGDER county, S NORWAY. Beacon light.

Oktaha, village (2006 population 335), MUSKOGEE county, E OKLAHOMA, 13 mi/21 km SSW of MUSKOGEE; 35°34′N 95°28′W. In agricultural area.

Ok Tedi (AHK te-dee), mine complex, WESTERN province, central NEW GUINEA island, W central PAPUA NEW GUINEA, 300 mi/483 km NW of DARU, 15 mi/24 km E of Indonesian border. Located in Star Mountains. Copper, gold. Mining begun in 1984; town of Tabubil constructed 15 mi/24 km S as service center. Road built to KIUNGA. Operations helped save Papua New Guinea economy during late 1980s, but environmental damage has affected indigenous peoples. Permission granted in 1989 to discharge 150,000 tons/136,116 metric tons of waste into FLY RIVER, polluting river and TORRES STRAIT. More gold reserves at nearby Frieda River.

Oktemberyan (uhk-tem-bir-YAHN), city, W ARMENIA, in irrigated ARAS RIVER valley, on railroad and 27 mi/ 43 km W of YEREVAN; 40°08′N 44°00′E. Cotton ginning, fruit canning, metalworking, tanning; geranium-oil press. Orchards, vineyards; cotton, sugar beets, tobacco; manufacturing includes machine parts, furniture, glass products; wine combine; dairy processing; cannery. Until c.1935, Sardarabad; sometimes called Oktember. Developed in 1930s, in irrigated desert zone.

Cross-references are shown in SMALL CAPITALS. The pronunciation guide is shown on page xix. The sources of population figures are shown on page xvii.

Oktiabrsk (ok-TYAH-buhrsk), town, DASHHOWUZ weloyat, TURKMENISTAN, in Amu Darya delta, c.62 mi/100 km NW of DASHHOWUZ. Tertiary-level administrative center. Also Oktiabrsk.

Oktibbeha (ok-TIB-uh-hah), county (☐ 462 sq mi/1,201.2 sq km; 2006 population 41,633), E MISSISSIPPI; ⊙ STARKVILLE; 33°25′N 88°52′W. Drained by NOXUBEE and Oktibbeha rivers. Agriculture (cotton, corn, hay, soybeans, honey; cattle; dairying); timber. Starr Forest Wildlife Management Area and part of Noxubee National Wildlife Refuge in S; small part of Tombigbee National Forest in SW. Formed 1833.

Oktwin (OUK-TWIN), village, MYANMAR, on railroad, and 8 mi/12.9 km S of TOUNGOO.

Oktyabrsk (uhk-TYAH-buhrsk), city, N central AKTÖBE region, KAZAKHSTAN, on TRANS-CASPIAN RAILROAD, on ILEK RIVER, 100 mi/161 km S of AKTÖBE (Aktyubinsk); 49°30′N 57°22′E. Tertiary-level (raion) administrative center. In phosphorite area; junction for railroad to ATYRAU (SW) and ORSK (N). Formerly Kandagach or Kandagash.

Oktyabr'sk (uhk-TYAH-buhrsk), city (2006 population 24,530), SW SAMARA oblast, E European Russia, on the W bank of the SARATOV Reservoir on the VOLGA RIVER, on railroad junction, 95 mi/153 km W of SAMARA; 53°10′N 48°42′E. Freight transshipping; building materials, industrial materials; food industries. Made city in 1956 out of Batraki and other nearby villages.

Oktyabrsk (uhk-TYAH-buhrsk), town, KHATLON viloyat, SW TAJIKISTAN, in Vakhsh valley, 5 mi/8 km SSE of QURGHONTEPPA; long-staple cotton. Formerly called Chichka.

Oktyabr'skaya (uhk-TYAHBR-skah-yah), village (2005 population 11,665), N KRASNODAR TERRITORY, S European Russia, on road and railroad, 9 mi/14 km N of PAVLOVSKAYA; 46°16′N 39°48′E. Elevation 206 ft/62 m. In agricultural area (wheat, flax, sunflowers; livestock); food processing.

Oktyabr'skaya, UKRAINE: see POKROVS'KE, Luhans'k oblast.

Oktyabr'skaya Revolyutsiya Island (uhk-TYAH-buhrs-kah-yah rye-vuh-LYOO-tsi-yah) [Russian= October revolution], central island (☐ 5,400 sq mi/14,040 sq km) of SEVERNAYA ZEMLYA archipelago, in the ARCTIC OCEAN, off KRASNOYARSK TERRITORY, N Siberian Russia. Glaciers cover 45% of the land.

Oktyabrs'ke (uhk-TYABR-ske) (Russian Oktyabrskoye), town, central Autonomous Republic of CRIMEA, UKRAINE, on railroad (Elevatornaya station), 22 mi/35 km N of SIMFEROPOL; 45°17′N 34°08′E. Elevation 206 ft/62 m. Wine, milk, and flour factories. Population largely Tatar prior to World War II. Until 1945, called Biyuk-Onlar.

Oktyabrski (ahk-TYAH-buhr-skee), suburb of TOKMOK, CHÜY region, KYRGYZSTAN, on CHU RIVER; 42°46′N 75°16′E. Beet-sugar refining.

Oktyabrskii, town, W TAJIKISTAN, on railroad (station at Cheptura), and 25 mi/40 km W of DUSHANBE. Cotton. Also spelled Oktyabrskiy.

Oktyabrskii (ahk-TYAH-buhr-skee), urban settlement, GOMEL oblast, BELARUS; ⊙ OKTYABRSKIY region. Dried milk plant, distillery.

Oktyabr'skiy (uhk-TYAHBR-skeeyee), city (2005 population 109,800), W BASHKORTOSTAN Republic, E European Russia, on the IK RIVER (tributary of the KAMA River), on railroad, 120 mi/195 km WSW of UFA; 54°28′N 53°28′E. Elevation 574 ft/174 m. Major petroleum center in the Tuimazy oil field, with pipelines. Manufacturing of equipment for oil industry; machinery and metal working, electrical apparatus; ceramics; furniture; tanning. Arose in 1937 with the discovery of the oil field, and developed in the 1940s through urbanization of Naryshevo and nearby villages; became city in 1946.

Oktyabr'skiy (uhk-TYAHBR-skeeyee), town, NW AMUR oblast, SE RUSSIAN FAR EAST, on road junction,

110 mi/177 km N of ZEYA; 53°01′N 128°37′E. Elevation 1,115 ft/339 m. Gold mining.

Oktyabr'skiy (uhk-TYABR-skeeyee), town (2005 population 9,600), S ARCHANGEL oblast, N European Russia, on the Ust'ya River (NORTHERN DVINA River basin), on road and near railroad (Kostylëvo station on the Konosha-Kotlas line), 47 mi/75 km E of VEL'SK; 61°05′N 43°11′E. Elevation 278 ft/84 m. Timber transshipment base. Also known as Pervomayskiy.

Oktyabr'skiy (uhk-TYABR-skeeyee), former town, W CHELYABINSK oblast, SE URALS, RUSSIA, now a suburb of MIASS, less than 3 mi/5 km SE of the city center; 55°03′N 60°09′E. Elevation 1,866 ft/568 m. In gold-mining region.

Oktyabr'skiy (uhk-TYABR-skeeyee), town (2005 population 1,215), NW IVANOVO oblast, central European Russia, on railroad spur, 24 mi/39 km WNW of IVANOVO; 57°08′N 40°20′E. Elevation 492 ft/149 m. Peat works. Until 1941, called Obedovo.

Oktyabr'skiy (uhk-TYABR-skeeyee), town (2005 population 7,020), SW BELGOROD oblast, S central European Russia, on the Lopan' River (N DONETS River basin), on railroad, 14 mi/23 km SW of BELGOROD; 50°26′N 36°21′E. Elevation 623 ft/189 m. Sugar refinery; dairy products. Originally called Voskresenovka village, in 1938 made a town and called Mikoyanovka until 1960.

Oktyabr'skiy (uhk-TYAH-buhr-skeeyee), town (2005 population 2,160), SW KAMCHATKA oblast, RUSSIAN FAR EAST, on the Sea of OKHOTSK, on SW KAMCHATKA PENINSULA, near the Bystraya River, 20 mi/32 km S of UST'-BOL'SHERETSK; 52°40′N 156°14′E. Fish cannery. Formerly known as Mikoyanovskiy.

Oktyabr'skiy (uhk-TYABR-skeeyee), town (2005 population 6,270), E KHABAROVSK TERRITORY, RUSSIAN FAR EAST, on the TATAR STRAIT, on local road and cargo railroad spur, 5 mi/8 km N of SOVETSKAYA GAVAN' (across the bay); 49°03′N 140°17′E. Elevation 137 ft/41 m. Fisheries. Platinum mining in the vicinity.

Oktyabr'skiy (uhk-TYAHBR-skeeyee), town (2005 population 1,905), NW KIROV oblast, E central European Russia, on railroad, 4 mi/6 km NNW of, and under administrative jurisdiction, of MURASHI; 59°26′N 48°54′E. Elevation 774 ft/235 m. Lumber milling.

Oktyabr'skiy (uhk-TYAHBR-skeeyee), town (2005 population 615), NE KOMI REPUBLIC, NE European Russia, on the Vorkuta River (PECHORA River basin), on railroad, 5 mi/8 km NW of VORKUTA, to which it is administratively subordinate; 67°34′N 64°08′E. Elevation 643 ft/195 m. Before the fall of the USSR, a busy mining town; closing of many coal mines in the area after the switch to market economy in 1992 led to drastic population decrease.

Oktyabr'skiy (uhk-TYAHBR-skeeyee), town, central KOSTROMA oblast, central European Russia, on railroad, 70 mi/113 km E of GALICH; 58°19′N 44°19′E. Elevation 459 ft/139 m. Sawmilling. Until 1939, known as Brantovka.

Oktyabr'skiy (uhk-TYAHBR-skeeyee), town (2006 estimate population 10,025), central MOSCOW oblast, central European Russia, near the MOSKVA River, on road and railroad spur, 6 mi/10 km SE of (and administratively subordinate to) LYUBERTSY; 55°36′N 37°58′E. Elevation 393 ft/119 m. Cotton textiles.

Oktyabr'skiy (uhk-TYAHBR-skeeyee), town (2006 population 7,190), W NIZHEGOROD oblast, central European Russia, on the VOLGA RIVER, on road, 6 mi/10 km SE, and under administrative jurisdiction, of BOR; 56°17′N 44°11′E. Elevation 249 ft/75 m. Shipyards. Earlier called Imeni Molotova.

Oktyabr'skiy (uhk-TYAHBR-skeeyee), town (2006 population 9,710), SE PERM oblast, W central URALS, E European Russia, on road junction and railroad (Chad station), 20 mi/32 km WSW of KRASNOUFIMSK; 56°30′N 57°12′E. Elevation 1,092 ft/332 m. Lumbering; road building; creamery.

Oktyabr'skiy (uhk-TYAHBR-skeeyee), former town, now a suburb of SHAKHTY, SW ROSTOV oblast, S European Russia, 10 mi/16 km E of city center. Anthracite mining.

Oktyabr'skiy (uhk-TYAHBR-skeeyee), town (2006 population 4,775), SW RYAZAN oblast, central European Russia, on railroad, 4 mi/6 km SW of SKOPIN, to which it is administratively subordinate; 53°47′N 39°29′E. Elevation 567 ft/172 m. In the Moscow coal basin; lignite mining.

Oktyabr'skiy (uhk-TYAHBR-skeeyee), town (2006 population 6,495), W RYAZAN oblast, central European Russia, on the Pronya River (tributary of the OKA River), on highway branch and railroad (Tsementnaya station), 5 mi/8 km W, and under administrative jurisdiction, of MIKHAYLOV; 54°13′N 38°53′E. Elevation 567 ft/172 m. Cement-making; limestone quarry. Until 1927, called Sapronovo.

Oktyabr'skiy (uhk-TYAHBR-skeeyee), former town (2006 population 115), S SVERDLOVSK oblast, E central URALS, extreme W Siberian Russia, 11 mi/18 km NNE of YEKATERINBURG; 57°02′N 60°44′E. Elevation 885 ft/269 m. Lumbering.

Oktyabr'skiy (uhk-TYAHBR-skeeyee), town (2005 population 6,080), E KRASNOYARSK TERRITORY, SE Siberian Russia, just NW of the administrative border with IRKUTSK oblast, in the CHUNA RIVER basin, on road and railroad, 112 mi/180 km WNW of TASEYEVO, and 57 mi/92 km S of BOGUCHANY; 57°26′N 97°31′E. Elevation 551 ft/167 m. Sawmilling, lumbering; food processing. Formerly called Maleyeva.

Oktyabrskiy, town, NE KASHKADARYO wiloyat, UZBEKISTAN, near SHAKHRISABZ; 40°28′N 68°47′E. In cotton area.

Oktyabr'skiy (uhk-TYABR-skeeyee), village (2005 population 2,080), SE CHITA oblast, S SIBERIA, RUSSIA, near the ARGUN' RIVER, on road, 7 mi/11 km E of KRASNOKAMENSK; 50°05′N 118°11′E. Elevation 2,345 ft/714 m. Local airfield.

Oktyabr'skiy (uhk-TYABR-skeeyee), village (2005 population 5,665), W IRKUTSK oblast, E central SIBERIA, RUSSIA, on the CHUNA RIVER, on the administrative border with KRASNOYARSK TERRITORY, on highway and BOGUCHANY-bound spur of the BAYKALAMUR MAINLINE, 80 mi/129 km WSW of BRATSK; 56°04′N 99°25′E. Elevation 875 ft/266 m. Logging, lumbering, sawmilling, woodworking.

Oktyabr'skiy (uhk-TYABR-skeeyee), settlement, central UDMURT REPUBLIC, E European Russia, near railroad, 5 mi/8 km E of IZHEVSK, and less than 3 mi/5 km N of ZAVYALOVO, to which it is administratively subordinate; 56°50′N 53°22′E. Elevation 360 ft/109 m. In agricultural area (poultry raising). Automotive repair.

Oktyabrskiy (ahk-TYAH-buhr-skee), urban settlement, VITEBSK oblast, BELARUS.

Oktyabr'skiy, RUSSIA: see BUDËNNOVSKIY.

Oktyabr'skiy, RUSSIA: see TAKHTAMUKAY.

Oktyabr'skoe (ahk-TYAH-buhr-skaw-ye), village, SW JALAL-ABAD region, KYRGYZSTAN, near railroad (Bagysh station), 11 mi/18 km NE of JALAL-ABAD; 40°59′N 73°05′E. Walnut processing.

Oktyabr'skoye (uhk-TYABR-skuh-ye), village (2004 population 1,350), E CHELYABINSK oblast, SW SIBERIA, RUSSIA, approximately 17 mi/27 km N of the KAZAKHSTAN border, on crossroads, 50 mi/80 km NE of TROITSK; 54°24′N 62°43′E. Elevation 587 ft/178 m. In wheat and livestock area.

Oktyabr'skoye (uhk-TYABR-skuh-ye), village (2004 population 4,700), NE CHUVASH REPUBLIC, central European Russia, on road, 15 mi/24 km SSE of MARIINSKIY POSAD, and 19 mi/30 km E of CHEBOKSARY; 55°54′N 47°49′E. Elevation 393 ft/119 m. In grain-growing area. Until 1939, known as Ismeli.

Oktyabr'skoye (uhk-TYAHBR-skuh-ye), village, SE LIPETSK oblast, S central European Russia, on road and near railroad, 23 mi/37 km SSE of LIPETSK, and 18 mi/29 km N of USMAN', to which it is administratively

subordinate; 52°18′N 39°43′E. Elevation 521 ft/158 m. In agricultural area; dairying. Until 1920, known as Dryazgi; between 1938 and 1957, called Molotovo.

Oktyabr'skoye (uhk-TYAH-buhr-skuh-ye), village (2006 population 10,830), E NORTH OSSETIAN REPUBLIC, in the N CAUCASUS Mountains, RUSSIA, near highway, 3 mi/5 km E of VLADIKAVKAZ; 43°04′N 44°44′E. Elevation 2,112 ft/643 m. Fruit and vegetable processing, woodworking. Until 1944 (in Chechen-Ingush ASSR), called Sholkhi; later called Kartsa.

Oktyabr'skoye (uhk-TYAHBR-skuh-ye), village (2006 population 7,590), N ORENBURG oblast, S URALS, SE European Russia, in the SAKMARA RIVER basin, on road and railroad, 45 mi/72 km SSW of MELEUZ, and 32 mi/51 km N of ORENBURG; 52°20′N 55°30′E. Elevation 452 ft/137 m. In wheat and livestock area. Originally called Isayevo-Dedovo, and from 1923 until 1937, Kashirinskoye.

Oktyabr'skoye (uhk-TYAHBR-skuh-ye), settlement (2005 population 3,960), NW KHANTY-MANSI AUTONOMOUS OKRUG, TYUMEN oblast, W central Siberian Russia, on the OB' RIVER, 25 mi/40 km N of NYAGAN'; 62°27′N 66°02′E. Fisheries. Seat of local regional administration; has a hospital, hotel, museum of indigenous arts and crafts. Formerly known as Kondinskoye (KON-deen-skuh-ye).

Oktyabrskoye, UKRAINE: see OKTYABRS'KE.

Oktyabr'skoye, RUSSIA: see TSOTSIN-YURT.

Oktyabrskoye, UKRAINE: see ZHOVTEN'.

Oktyabrskoye, UKRAINE: see ZHOVTNEVE.

Oku (O-koo), town, Oku county, OKAYAMA prefecture, SW HONSHU, W JAPAN, 9 mi/15 km E of OKAYAMA; 34°39′N 134°05′E. Oysters; grapes; pottery.

Okučani (O-kuh-chah-nee), village, E CROATIA, on ZAGREB–SLAVONSKI BROD railroad, and 20 mi/32 km S of DARUVAR, at SW foot of the PSUNJ, in W SLAVONIA. Occupied by Serbs 1991–1995.

Okuchi (O-koo-chee), city, KAGOSHIMA prefecture, W central KYUSHU, extreme SW JAPAN, 31 mi/50 km N of KAGOSHIMA; 32°03′N 130°36′E. Pigs. Site of feudal castle.

Okulovka (uh-KOO-luhf-kah), city (2006 population 13,655), E central NOVGOROD oblast, NW European Russia, on highway, 87 mi/140 km E of NOVGOROD, and 22 mi/35 km W of BOROVICHI; 58°22′N 33°18′E. Elevation 528 ft/160 m. Railroad junction. Pulp and paper, woodworking; radios, furniture, food enterprises (bakery). Made city in 1965.

Okulovskiy, RUSSIA: see KAMENKA, ARCHANGEL oblast.

Okuma (O-koo-mah), town, Futaba county, FUKUSHIMA prefecture, N central HONSHU, NE JAPAN, 37 mi/60 km S of FUKUSHIMA city; 37°24′N 140°59′E.

Okuma Bay, a bay into the E end of the ROSS ICE SHELF front, ANTARCTICA, at its junction with EDWARD VII PENINSULA; 77°48′S 159°20′W.

Okuni (aw-KOO-nee), town, CROSS RIVER state, extreme SE NIGERIA, on main road, 70 mi/113 km NNE of CALABAR. Market center. Bananas, maize, cocoyams.

Okura (O-koo-rah), village, Mogami county, YAMAGATA prefecture, N HONSHU, NE JAPAN, 31 mi/50 km N of YAMAGATA city; 38°42′N 140°14′E.

Okushiri (o-koo-SHEE-ree), town, on OKUSHIRI-TO island, Hiyama district, SW Hokkaido prefecture, N JAPAN, 112 mi/180 km SW of SAPPORO; 42°10′N 139°31′E. Abalone. Nabetsuru Rock is nearby.

Okushiri-to (o-KOO-shee-ree-TO), island (□ 56 sq mi/145.6 sq km), in Hiyama district, Hokkaido prefecture, N JAPAN, in SEA OF JAPAN, just W of SW peninsula of HOKKAIDO island; 15 mi/24 km long, 5 mi/8 km wide; 42°09′N 139°28′E. Mountainous, with fertile coastal area; hot springs. Rice growing, lumbering, fishing. Sometimes spelled Okushiri-sima, Okujirijima. Earthquake 1994.

Okusi, INDONESIA: see AMBENO.

Okuta (aw-KOO-tah), town, KWARA state, W NIGERIA, 65 mi/105 km NW of KAIAMA. Road center; customs depot on BENIN border. Shea-nut processing, cotton weaving; cassava, yams, corn.

Okutama (o-koo-TAH-mah), town, West Tama county, Tokyo prefecture, E central HONSHU, E central JAPAN, 12 mi/20 km W of SHINJUKU; 35°48′N 139°06′E. Wasabi. Nearby attractions include Chichibu-Tama National Park, Okutama Lake, Ogochi Dam, and Nippara Cave (limestone).

Okutsu (O-koot-soo), town, Tomata county, OKAYAMA prefecture, SW HONSHU, W JAPAN, 34 mi/55 km N of OKAYAMA; 35°08′N 133°53′E. Okutsu Gorge and hot springs nearby.

Okuwa (O-koo-wah), village, Kiso county, NAGANO prefecture, central HONSHU, central JAPAN, 74 mi/120 km S of NAGANO; 35°40′N 137°40′E. Hinoki cypress trees.

Okwoga (ok-WO-gah), town, KOGI state, S central NIGERIA, on road, 90 mi/145 km SE of LOKOJA. Market town. Yams, cotton, rice.

Okwuzi (ok-WOO-zee), town, IMO state, S NIGERIA, in NIGER River delta region, 25 mi/40 km NW of OWERRI. Market center. Cashews, soybeans, bananas.

Olá (o-LAH), town, ⊙ Olá district, COCLÉ province, central PANAMA, 15 mi/24 km WSW of PENONOMÉ; 08°25′N 80°39′W. Corn, rice, beans, sugarcane; livestock.

Ola (O-lah), town (2006 population 6,185), S MAGADAN oblast, E RUSSIAN FAR EAST, at the mouth of the Ola River in the Tauysk Bay on the Sea of OKHOTSK, on highway, 22 mi/35 km E of MAGADAN; 59°35′N 151°17′E. Fisheries.

Ola (O-luh), town (2000 population 1,204), YELL county, W central ARKANSAS, 18 mi/29 km SSW of RUSSELLVILLE; 35°01′N 93°13′W. Manufacturing (lumber). Petit Jean Wildlife Management Area to N; Ouachita National Forest to W.

Olaa, HAWAII: see KEAAU.

Ólafsfjörður (AW-lahfs-FYUHR-[th]ur) or **Olafsfjordhur**, city (2000 population 1,038), in, but independent of, EYJAFJARÐARSÝSLA county, N ICELAND, on SW arm of EYJAFJÖRÐUR, 160 mi/257 km NE of REYKJAVIK; 66°04′N 18°39′W. Fishing port.

Ólafsvík (AW-lahfs-VEEK), village (2000 population 989), Snæfellsnes county, W ICELAND, NW Snæfellsnes peninsula, on BREIÐAFJÖRÐUR; 64°53′N 23°43′W. Cod and haddock fisheries; shrimp, lobster. Iceland's oldest established trading town (1687 charter).

Olaine (O-lei-nai), city (2000 population 12,952), LATVIA, 12 mi/19 km SW of Rīga; 56°48′N 23°56′E. Center of Latvia's chemical industry.

Olamon, MAINE: see GREENBUSH.

Olancha Peak (12,123 ft/3,695 m), on INYO-TULARE county line, E CALIFORNIA, in the SIERRA NEVADA, 24 mi/39 km N of LONE PINE. In Inyo National Forest. Pacific Crest Trail passes peak.

Olanchito (o-lahn-CHEE-to), city (2001 population 22,856), YORO department, N HONDURAS, in Olanchito valley, near AGUÁN RIVER, 27 mi/43 km SE of LA CEIBA; 15°30′N 86°34′W. On railroad and linked by road with ports of La Ceiba and TRUJILLO. Commercial center in banana area; dairy farming. Airfield. Founded in early 17th century.

Olancho (o-LAHN-cho), department (□ 12,986 sq mi/33,763.6 sq km; 2001 population 419,561), E central HONDURAS; ⊙ JUTICALPA; 14°50′N 86°00′W. Bounded SE by Oro River and NICARAGUA; largely mountainous (Sierra de La Esperanza, Sierra de Agalta, Montañas de Colón); drained by PATUCA RIVER and its two headstreams (Guayapa and Guayambre rivers) and by SICO RIVER. Livestock raising and dairying, mainly in CATACAMAS VALLEY; agriculture (coffee, tobacco, sugarcane, rice). Gold placers in principal rivers. Population is concentrated in W half of department; E half is covered by tropical, hardwood, and pine forests; hardwood lumbering is primary activity. Tanning, sugar milling, cigar manufacturing are lesser industries. Liquidambar oil is produced. Main centers (linked by paved road with TEGUCIGALPA): Juticalpa and CATACAMAS. Formed 1825.

Öland (UH-lahnd), province and narrow island (□ 520 sq mi/1,352 sq km), KALMAR county, SE SWEDEN, in BALTIC SEA, separated from mainland Sweden by KALMAR SOUND. Manufacturing (cement), quarrying; agriculture (sugar beets, cereals, vegetables; cattle); fishing. Many summer resorts. BORGHOLM is chief town. Numerous Stone Age monuments; first mentioned eighth century. Frequent battleground during wars among Scandinavian countries.

Olănești (o-luh-NESHT), village, SE MOLDOVA, on DNIESTER (Nistru) River, and 31 mi/50 km SE of TIGHINA (Bender); 46°29′N 29°55′E. Red wine-making center.

Olangbecho (O-lahng-BE-cho), town, KOGI state, S central NIGERIA, on road, 95 mi/153 km ESE of LOKOJA. Market town. Maize, millet.

Olanpi, Cape, TAIWAN: see OLUANPI, Cape.

Olan, Pic d' (do-lahn, peek), peak (11,739 ft/3,578 m) of the ÉCRINS Massif, DAUPHINÉ ALPS, SE FRANCE, on ISÈRE–HAUTES-ALPES department border, 10 mi/16 km NE of Saint-Firmin, overlooking VALGODEMAR valley (S). Glacier. Difficult ascent. It lies entirely within the ÉCRINS NATIONAL PARK.

Olanta (o-LANT-uh), village (2006 population 626), FLORENCE county, E central SOUTH CAROLINA, 22 mi/35 km SSW of FLORENCE; 33°56′N 79°55′W. Manufacturing of yarn for motor vehicle industry. Agriculture includes tobacco, grain, cotton; hogs, chickens.

Olar (O-luhr), village (2006 population 218), BAMBERG county, W central SOUTH CAROLINA, 29 mi/47 km SW of ORANGEBURG; 33°10′N 81°11′W. Manufacturing of hardwood veneers, fence posts.

Olary, village, E SOUTH AUSTRALIA state, S central AUSTRALIA, 140 mi/225 km ENE of PORT PIRIE; 32°17′S 140°19′E. On Port Pirie–BROKEN HILL railroad. Sheep. Uranium at nearby Radium Hill.

Olascoaga (o-lahs-ko-AH-gah), town, central BUENOS AIRES province, ARGENTINA, 10 mi/16 km SW of BRAGADO; 35°12′S 60°36′W. Railroad junction and agricultural center (corn, wheat, hogs, cattle).

Olathe (o-LAITH-uh), city (2000 population 92,962), ⊙ JOHNSON county, NE KANSAS, suburb 19 mi/31 km SW of downtown KANSAS CITY, Kansas; 38°53′N 94°48′W. In an area of livestock farms. Agriculture to S and W. Manufacturing. Olathe has been growing at a fast rate; its population more than tripled between 1970 and 1990. The city's name is derived from the Shawnee word for *beautiful*. Kansas State School for the Deaf, Mid-American Nazarene College, and Johnson Community Vocational Technical School are located here. Incorporated 1858.

Olathe, town (2000 population 1,573), MONTROSE county, W COLORADO, on Uncompahgre River and 45 mi/72 km SE of GRAND JUNCTION; 38°36′N 107°58′W. Elevation 5,346 ft/1,629 m. Shipping point in cattle, sheep region; potatoes, onions, beans. BLACK CANYON OF THE GUNNISON NATIONAL MONUMENT to E; Switzer Lake State Park to N.

Olavakkot, INDIA: see PALAKKAD.

Olavarría (o-lah-vahr-REE-ah), city (1991 population 72,821), ⊙ Olavarría district (□ 2,959 sq mi/7,693.4 sq km), S central BUENOS AIRES province, ARGENTINA, on the Arroyo TAPALQUÉ, and 175 mi/282 km SW of BUENOS; 36°45′S 60°45′W. Railroad junction and agricultural center; sheep and cattle raising; flour milling, dairying, meatpacking.

Olawa (o-LAH-vah), Polish *Oława*, German *Ohlau*, town (2002 population 31,154), Wrocław province, SW POLAND, port between ODER and OLAWA rivers (here linked by canal), 17 mi/27 km SE of WROCŁAW (Breslau); 50°56′N 17°18′E. Chemical manufacturing, metalworking, zinc and lead refining, sawmilling. Gothic church, 16th century castle (former seat of Polish Piast princes). Town suffered heavily in Thirty

---

Cross-references are shown in SMALL CAPITALS. The pronunciation guide is shown on page xix. The sources of population figures are shown on page xvii.

Years War. Has 14th century castle; 13th century church.

Olawa River (o-LAH-vah), 62 mi/100 km long, German *Ohle*, SW POLAND; rises E of ZĄBKOWICE; flows N past ZIĘBICE and STRZELIN, NNE past WIAZÓW and OLAWA (here linked by canal to the ODER), and NW to the Oder at WROCLAW (Breslau).

Olaya (o-LEI-yah), town, ⊙ Olaya municipio, ANTIOQUIA department, NW central COLOMBIA, in CAUCA valley, 25 mi/40 km NW of MEDELLÍN; 06°37′N 75°49′W. Elevation 2,440 ft/743 m. Coffee, cacao, plantains; livestock.

Olazagutía (o-lahth-ah-goo-TEE-ah), town, NAVARRE province, N SPAIN, 28 mi/45 km W of PAMPLONA; 42°53′N 02°12′W. Cement manufacturing, sawmilling; livestock raising.

Olbernhau (OL-bern-hou), town, SAXONY, E central GERMANY, in the ERZGEBIRGE, 23 mi/37 km SE of CHEMNITZ, near CZECH border; 50°40′N 13°20′E. Toy manufacturing center; woodworking; also manufacturing of electrical equipment.

Olbersdorf (OL-bers-dorf), village, SAXONY, E central GERMANY, in UPPER LUSATIA, at foot of LUSATIAN MOUNTAINS, 2 mi/3.2 km SW of ZITTAU, near CZECH border; 50°52′N 14°46′E. Lignite mining; manufacturing of steel products.

Olbersdorf, CZECH REPUBLIC: see MESTO ALBRECHTICE.

Olbia (ol-BEE-ah), town (2001 population 45,366), SASSARI province, NE SARDINIA, ITALY, port on Gulf of TERRANOVA, and 50 mi/80 km NE of SASSARI; 40°55′N 09°31′E. Sardine fishing, corkworking. Passenger port (main service to mainland at CIVITAVECCHIA). Has Pisan church of San Simplicio. Originally Greek colony of Olbia. Called Terranova Pausania until early 1940s.

Olbia, UKRAINE: see MYKOLAYIV.

Olca, Cerro (OL-kah, SER-ro), Andean peak (17,420 ft/5,310 m) on CHILE-BOLIVIA border; 20°54′S.

Olching (OL-khing), town, UPPER BAVARIA, BAVARIA, GERMANY, on the AMPER, and 12 mi/19 km NW of MUNICH; 48°13′N 11°19′E. Manufacturing of metalware. Mainly a residential village.

Olcott, hamlet (□ 5 sq mi/13 sq km; 2000 population 1,156), NIAGARA county, W NEW YORK, on Lake ONTARIO, 30 mi/48 km NNE of BUFFALO; 43°20′N 78°42′W. Manufacturing (industrial chemicals).

Old Aberdeen, Scotland: see ABERDEEN.

Old Alresford, ENGLAND: see NEW ALRESFORD.

Old Andreafsky (an-dree-YAF-skee), village, W ALASKA, on YUKON River, and 100 mi/161 km NW of BETHEL. Formerly important trading center (salmon fishing and packing) and port with dock installations now destroyed. Has Russian Orthodox church. Trading post established by Russians, c.1853.

Old Appleton, town (2000 population 82), CAPE GIRARDEAU county, SE MISSOURI, on Apple Creek, near MISSISSIPPI RIVER, 13 mi/21 km SE of PERRYVILLE; 37°35′N 89°42′W. Settled 1808. Formerly Appleton. Old Appleton since 1917.

Old Bahama Channel (old buh-HAH-muh), strait off N coast of CUBA and CAMAGÜEY ARCHIPELAGO, S of GREAT BAHAMA BANK; c.100 mi/161 km long, 15 mi/24 km wide.

Old Baldy, Arizona: see WRIGHTSON MOUNT.

Old Baldy, California: see SAN ANTONIO, MOUNT.

Old Bennington, VERMONT: see BENNINGTON.

Old Bight, town, central Bahamas, on SW Cat Island, 140 mi/225 km SE of NASSAU; 24°22′N 75°22′W. Sisal, fruit.

Old Bridge, township, MIDDLESEX county, E NEW JERSEY, on SOUTH RIVER and 7 mi/11.3 km SE of NEW BRUNSWICK; 40°23′N 74°18′W. Concrete blocks. Largely residential.

Oldbridge, agricultural village, NE MEATH county, NE IRELAND, on the BOYNE, and 3 mi/4.8 km W of DROGHEDA; 53°03′N 06°17′W. Scene (1690) of the battle of the Boyne, in which William of Orange and Schom-

berg defeated army of James II. Schomberg was killed here; James II was forced to flee to France.

Old Bridgeport, coal mining village, NE NOVA SCOTIA, CANADA, on NE coast of CAPE BRETON ISLAND, 9 mi/14 km NE of SYDNEY; 46°13′N 60°01′W. Elevation 108 ft/32 m.

Old Brookville, village (□ 3 sq mi/7.8 sq km; 2006 population 2,244), NASSAU county, SE NEW YORK, on W LONG ISLAND, 4 mi/6.4 km SE of GLEN COVE; 40°49′N 73°35′W.

Oldbury, ENGLAND: see BRIDGNORTH.

Oldbury, ENGLAND: see WARLEY.

Oldbury and Smethwick (OLD-buhr-EE and SMETH-wik), towns (2001 population 21,834) and district, WEST MIDLANDS, central ENGLAND, 5 mi/8 km W of BIRMINGHAM; 52°30′N 02°00′W. Some manufacturing.

Old Cairo, EGYPT: see FUSTAT, EL.

Old Calabar, NIGERIA: see CALABAR.

Old Castile, SPAIN: see CASTILE.

Oldcastle, Gaelic *Sean-Chaisleán an Fhásaigh*, town (2006 population 1,316), NW MEATH county, NE IRELAND, 21 mi/34 km WNW of NAVAN; 53°46′N 07°09′W. Agricultural market. Many ancient cairns and tumuli excavated nearby.

Old Chitambo, ZAMBIA: see CHITAMBO.

Old Cleeve (CLEEV), village (2001 population 2,306), W SOMERSET, SW ENGLAND, 2 mi/3.2 km WSW of WATCHET; 51°11′N 03°22′W. Has 15th-century church. Ruins of Cleeve Abbey.

Old Corinth, Greece: see CORINTH.

Old Croton Dam and Aqueduct, WESTCHESTER and PUTNAM counties, SE NEW YORK, a 50-ft/15-m-high, granite-faced, timber-rock dam, and a 33-mi/53-km-long, brick-lined aqueduct that was the main source of NEW YORK CITY's water supply 1842–1907. The first supply of public water to New York city was a 21,000-gal/79,491-liter elevated tank, but it proved less than adequate. In 1832, Governor DeWitt Clinton asked the state legislature for plans for new water sources. In 1834 the legislature approved the construction of a dam on CROTON RIVER, which would produce an estimated yield of 20 million gals/75.7 million liters per day. The dam project started in 1837 and was completed in 1842. The aqueduct from Croton crossed over HARLEM RIVER at High Bridge, then to CENTRAL PARK reservoir, then to Bryant Park reservoir, an Egyptian-style enclosure with 50-ft/15-m-high, 25-ft/7.5-m-thick walls. That reservoir was used until 1890, and in 1897 it was razed to make way for the New York Public Library at 42nd Street and 5th Avenue. Repairs to the system in 1881 strengthened it, increasing the yield to 95 million gals/359.6 million liters per day. The Old Croton System was used until 1907, when it was replaced by the NEW CROTON DAM AND AQUEDUCT. The Old Croton Aqueduct is listed on the National Register of Historic Places. Today the aqueduct route is part of the 173-acre/70-ha Old Croton Trailways State Park; it stretches 20 miles/32 km from the dam along the public right-of-way following the HUDSON RIVER to Van Cortlandt Park in the BRONX, a distance of 20 mi/32 km.

Old Crow, village (□ 5 sq mi/13 sq km; 2006 population 253), N YUKON, on Porcupine River, at mouth of Old Crow River; and 250 mi/402 km N of DAWSON; 67°35′N 139°50′W. Trading post, Royal Canadian Mounted Police station. Controversial archaeological finds of early occupation of Americas. Home of the Vuntut Gwitchin First Nation; only Yukon community north of the Arctic Circle. No road access.

Old Dailly, Scotland: see DAILLY.

Old Deer or **Deer** (DIR), agricultural village, Aberdeenshire, NE Scotland, 10 mi/16 km W of PETERHEAD; 57°31′N 02°02′W. Distilling, brewing; granite and limestone quarrying. Of monastery founded here by St. Columba in 6th century, the *Book of Deer*, now in Cambridge University Library, is sole relic. There are ruins of 13th-century Cistercian abbey. Village of New Deer is 6 mi/9.7 km W.

Old Delhi, INDIA: see MAHRAULI.

Old Dominion State; see VIRGINIA.

Old Dongola, SUDAN: see DONGOLA.

Oldeani (ol-dai-AH-nee), village, ARUSHA region, N TANZANIA, 75 mi/121 km W of ARUSHA, S of NGORONGORO CRATER and OLDUVAI GORGE; 03°22′S 35°33′E. Several tourist lodges in area. Subsistence crops; livestock.

Oldebroek (AWL-duh-brook), village, GELDERLAND province, central NETHERLANDS, 9 mi/14.5 km E of ZWOLLE; 52°28′N 05°56′E. Dairying; livestock; grain, vegetables, sugar beets. Nature area to SE.

Oldehove (AWL-duh-haw-vuh), village, GRONINGEN province, N NETHERLANDS, 9 mi/14.5 km NW of GRONINGEN; 53°18′N 06°24′E. Reitdiep canal to N. Dairying; cattle, sheep; vegetables, grain. Natural gas field. Also spelled Oldenhove.

Oldenburg (OL-den-burg), former state, NW GERMANY, now an administrative division (Ger. *Verwaltungsbezirk*) of LOWER SAXONY.

Oldenburg (OL-den-burg), city, LOWER, NW GERMANY, 23 mi/37 km W of BREMEN, on the HUNTE RIVER, the Küstenkanal (Coast Canal); 53°09′N 08°14′E. It is a railroad junction, transshipment point, agriculture market, and industrial center. Manufacturing includes ships, glass, and textiles. It was the seat of the counts of Oldenburg until 1667, when it passed, with the county, to DENMARK. From 1777 to 1918 it served as the residence of the dukes (later grand dukes) of Oldenburg. Noteworthy buildings include the former ducal palace (17th–18th century) and the Gothic Lambertikirche, a church built in the 13th century (rebuilt 18th–19th century). Oldenburg was first mentioned in 1108 and chartered in 1345. Also spelled Oedenburg.

Oldenburg (OL-den-burg) or **Oldenburg in holstein**, town, SCHLESWIG-HOLSTEIN, NW GERMANY, 15 mi/24 km NE of EUTIN; 54°17′N 10°53′E. Grain; flour milling. Has 13th-century church. Was seat (948–1160) of bishop. Chartered 1235.

Oldenburg, town (2000 population 647), FRANKLIN county, SE INDIANA, 11 mi/18 km SW of BROOKVILLE; 39°20′N 85°12′W. In agricultural area. Seat of Convent of the Immaculate Conception and Academy (Oldenburg Franciscan Community of Sisters). Laid out 1837.

Oldendorf, Hessisch, GERMANY: see HESSISCH OLDENDORF.

Oldendorf, Preussisch, GERMANY: see PREUSSISCH OLDENDORF.

Old England, town, MANCHESTER parish, S JAMAICA, 6 mi/9.7 km SSE of MANDEVILLE; 17°59′N 77°27′W. In agricultural region (tropical spices and citrus; livestock).

Oldenhorn (OL-duhn-horn), peak (10,246 ft/3,123 m) in the DIABLERETS of the BERNESE ALPS, SW SWITZERLAND, 9 mi/14.5 km NW of SION, at juncture of VAUD, FRIBOURG, and BERN cantons and just S of COL DU PILLON.

Oldenhove, NETHERLANDS: see OLDEHOVE.

Oldenzaal (AWL-duhn-zahl), city, OVERIJSSEL province, E NETHERLANDS, on branch of TWENTE CANAL, 7 mi/11.3 km ENE of HENGELO; 52°12′N 06°56′E. German border 4 mi/6.4 km to SE; arboretum to E (at De Lutte); recreational park to W. Dairying; cattle, hogs, poultry; grain, vegetables; manufacturing (hydraulic systems, elevators, clothing; food processing). Has tenth-eleventh-century church (Plechelmuskerk).

Oldesloe, Bad, GERMANY: see BAD OLDESLOE.

Oldevatnet (AWL-duh-VAHT-nuht), lake (□ 3 sq mi/7.8 sq km) in a fissure of the glacier JOSTEDALSBREEN, SOGN OG FJORDANE county, W NORWAY; extends 8 mi/12.9 km S from Olden village, 60 mi/97 km ENE of FLORØ. Tourist center.

Old Factory River (OLD FAK-tuh-ree), village, NW QUEBEC, E Canada, on Old Factory Bay, on E side of

JAMES BAY; 52°36′N 78°43′W. Hudson's Bay Company trading post.

Old Faithful, WYOMING: see YELLOWSTONE NATIONAL PARK.

Old Field, village (□ 2 sq mi/5.2 sq km; 2006 population 988), SUFFOLK county, SE NEW YORK, on N shore of LONG ISLAND, overlooking SMITHTOWN BAY, and 5 mi/8 km NW of PORT JEFFERSON; 40°57′N 73°07′W. Lighthouse at OLD FIELD POINT.

Old Field Point, promontory on N shore of W LONG ISLAND, SE NEW YORK, at W side of entrance to PORT JEFFERSON HARBOR. Lighthouse (40°59′N 73°07′W).

Old Fletton (FLE-tuhn), suburb (2006 population 9,200), Peterborough (previously part of CAMBRIDGESHIRE), E ENGLAND, just S of PETERBOROUGH; 52°33′N 00°14′W. Brick-making center. Nearby are towns of Fletton (with Saxon and Norman church), Woodston, and Stanground.

Old Forge, hamlet, HERKIMER county, N central NEW YORK, in the ADIRONDACK MOUNTAINS, on a lake of FULTON CHAIN OF LAKES, c.45 mi/72 km NNE of UTICA; 43°43′N 74°58′W. Winter sports (snowmobiling) and summer recreation (camping, canoeing, fishing). First place in U.S. to employ mail boat for postal service.

Old Forge, borough (2006 population 8,569), LACKAWANNA county, NE PENNSYLVANIA, suburb 5 mi/8 km SW of SCRANTON, on LACKAWANNA RIVER; 41°22′N 75°44′W. Anthracite coal. Manufacturing (food products, winery, machinery, printing, apparel, consumer goods). Preate Winery is here. Settled 1798, incorporated 1895.

Old Fort (OLD FORT), town (□ 1 sq mi/2.6 sq km; 2006 population 963), MCDOWELL county, W central NORTH CAROLINA, 21 mi/34 km E of ASHEVILLE on CATAWBA RIVER, near its source; 35°37′N 82°10′W. Manufacturing (motor vehicle parts, textiles, machine parts); agricultural area (corn, apples; poultry, cattle). Timber. Pisgah National Forest to N.

Old Fort Harrod State Park, in HARRODSBURG, MERCER county, central KENTUCKY. It commemorates old fort pioneer settlement W of the Alleghenies, made here in 1774 by Captain James Harrod. On site of old Fort Harrod, park includes reproduction of fort and pioneer cabins, George Rogers Clark memorial, cabin in which Nancy Hanks and Thomas Lincoln (Abraham Lincoln's parents) were married, pioneer cemetery, and a museum. Formerly called Pioneer Memorial State Park. Established 1934.

Old Fort Henry, Canada: see Fort HENRY.

Old Glory (GLOR-ee), unincorporated village, STONEWALL county, NW central TEXAS, c.50 mi/80 km NNW of ABILENE; 33°07′N 100°03′W. Cattle; wheat, peanuts, hay.

Old Goa, INDIA: see GOA.

Old Grand Port or **Vieux Grand Port**, village (2000 population 2,779), GRAND PORT district, SE MAURITIUS, on the Grand Port (inlet of INDIAN OCEAN), on road, 3 mi/4.8 km NNE of MAHÉBOURG. Fishing; sugarcane; lime industry, ceramics. Dutch landed in 1810.

Old Greenwich, CONNECTICUT: see GREENWICH.

Old Gumbiro (goom-BEE-ro), village, RUVUMA region, S TANZANIA, 45 mi/72 km N of SONGEA; 10°00′S 35°23′E. Subsistence crops; livestock.

Oldham (OLD-uhm), county (□ 196 sq mi/509.6 sq km; 2006 population 55,285), N KENTUCKY; ⊙ LA GRANGE; 38°23′N 85°26′W. In N part of BLUEGRASS region. Bounded W and NW by OHIO RIVER (INDIANA state line); drained by FLOYDS FORK river and Harrods Creek. Rolling upland agricultural area (burley tobacco, corn, wheat, hay, alfalfa, soybeans; hogs, cattle, poultry; dairying); limestone. Urbanized in SW; extends from LOUISVILLE. Formed 1823.

Oldham (OLD-uhm), county (□ 1,501 sq mi/3,902.6 sq km; 2006 population 2,133), extreme N TEXAS; ⊙ VEGA, in high plains of PANHANDLE; 35°23′N 102°35′W. Elevation 3,200 ft/975 m–4,100 ft/1,250 m. Wheat, sorghum; cattle ranching; oil and gas, sand

and gravel, stone. Drained by CANADIAN RIVER and RITA BLANCA CREEK. Formed 1876.

Oldham (OLD-uhm), town (2001 population 103,544), GREATER MANCHESTER, W ENGLAND, 7 m/11.2 km NE of MANCHESTER; 53°33′N 02°07′W. This former coal-mining town in the PENNINES previously manufactured textiles and had a dominant cotton industry. The town hall, art gallery, museum, and Alexandra Park are noteworthy. There is also a 17th-century grammar school and Oldham College. Winston Churchill was the member of Parliament for Oldham. William Walton born here.

Oldham, village (2006 population 185), KINGSBURY county, E SOUTH DAKOTA, 16 mi/26 km SE of DE SMET; 44°13′N 97°18′W. In agricultural area.

Old Hamilton, Alaska: see HAMILTON.

Old Harbor, village (2000 population 237), ALASKA, on SE KODIAK ISLAND, 60 mi/97 km SW of KODIAK; 57°15′N 153°22′W. Fishing, fish processing. Site of THREE SAINTS BAY, Russian settlement in Alaska, established 1784.

Old Harbour, town, SAINT CATHERINE parish, S JAMAICA, in coastal lowland, on railroad, 20 mi/32 km W of KINGSTON; 17°55′N 77°06′W. Sugarcane, tropical fruit, coffee. The minor port OLD HARBOUR BAY (2 mi/3.2 km S), on bay of the same name, was once noted for shipbuilding.

Old Harbour Bay, town, SAINT CATHERINE parish, S central JAMAICA, 11 mi/18 km SW of SPANISH; 17°54′N 77°06′W. Fishing center on S coast of Jamaica.

Old Harbour Bay, CARIBBEAN inlet, SAINT CATHERINE parish, S JAMAICA; bounded by CLARENDON and Saint Catherine parishes; c.20 mi/32 km SW of KINGSTON; 17°55′N 77°03′W. At its SW gate is PORTLAND RIDGE peninsula. The bay is dotted with many islets, such as GREAT GOAT Island and PIGEON Island, which, together with sections of the shoreline, were leased to the UNITED STATES for naval bases in 1940. The minor port of OLD HARBOUR BAY is at its head. Sometimes called Portland Bight.

Old Head, cape on S shore of CLEW BAY, SW MAYO county, NW IRELAND, 2 mi/3.2 km NE of LOUISBURGH; 53°47′N 09°48′W. Elevation 348 ft/106 m.

Old Head of Kinsale, Gaelic *An Seancheann*, Atlantic cape, S CORK county, SW IRELAND, 7 mi/11.3 km S of KINSALE. Lighthouse (51°36′N 08°31′W).

Old Hickory, village, DAVIDSON county, N central TENNESSEE, on OLD HICKORY Reservoir (CUMBERLAND RIVER), 10 mi/16 km NNE of NASHVILLE; 36°15′N 86°39′W. Formerly owned by E. I. du Pont de Nemours and Company, whose plant here was founded to make munitions during World War I. Light manufacturing. Until 1923, called Jacksonville.

Old Hickory Lake, reservoir, on border of DAVIDSON and SUMNER counties and in WILSON, TROUSDALE, and SMITH counties, central TENNESSEE, on CUMBERLAND RIVER, 12 mi/19 km NE of NASHVILLE; 36°17′N 86°39′W. Serpentine; c.65 mi/105 km long. Cordell Hull Dam at E tip. Formed by Old Hickory Dam, built (1950s) by Army Corps of Engineers. Bledsoe Creek State Park on N shore.

Oldhorn Mountain (OLD-horn) (10,125 ft/3,086 m), W ALBERTA, W Canada, near BRITISH COLUMBIA border, in ROCKY MOUNTAINS, in JASPER NATIONAL PARK, 11 mi/18 km W of JASPER; 52°46′N 118°13′W.

Old IJssel River (EIS-sel) or **Old Yssel River**, Dutch *Oude IJssel*, in NW GERMANY and E NETHERLANDS; rises 3 mi/4.8 km NW of BORKEN; flows 45 mi/72 km generally NE, past Bacholt, forms border between Germany and the Netherlands (for 2 mi/3.2 km), enters the Netherlands just SSE of ULFT (head of navigation), continuing, past TERBORG, DOETINCHEM, and DOESBURG, to IJSSEL RIVER just SSW of Doesburg.

Old Iliamna, Alaska: see ILIAMNA.

Old Kasaan National Monument, SE ALASKA, on E PRINCE OF WALES ISLAND, 30 mi/48 km WNW of KETCHIKAN. Area is 38 acres/15 ha. Includes historic

ruins of abandoned Haida Indian village, with grave houses and totem poles. Established 1916.

Old Kilpatrick or **West Kilpatrick** (KIL-pah-trik), town, West Dunbartonshire, W Scotland, on the CLYDE RIVER, 5 mi/8 km ESE of DUMBARTON; 55°55′N 04°27′W. Previously shipbuilding, ironworking, distilling. ANTONINE WALL (Roman) passes by the town. Old Kilpatrick is one of places claiming to be birthplace of St. Patrick. Formerly in Strathclyde, abolished 1996.

Old Leighlin (LEE-lin) or **Leighlinbridge**, Gaelic *Sean-Leithghlin*, town (2006 population 674), W CARLOW county, SE IRELAND, 8 mi/12.9 km SSW of CARLOW; 52°44′N 06°59′W. St. Laserian founded monastery here in 7th century. Site of Protestant cathedral (founded c.1185, rebuilt 1529–1549), seat of bishop of Ossory, Ferns, and Leighlin. Town is also seat of Roman Catholic bishop of Kildare and Leighlin. Sometimes called Leighlin.

Old Line State; see MARYLAND.

Old Lyme, residential and resort town, NEW LONDON county, SE CONNECTICUT, on LONG ISLAND SOUND, on the E side of the mouth of the CONNECTICUT River; 41°19′N 72°17′W. Its noteworthy old houses built by sea captains have attracted many artists. The Congregational Church (1817; burned in 1909, but carefully restored) has been portrayed by Childe Hassam. Chiefly a residential community that is home to art galleries, art museums, and the Lyme Art Academy (Fine Arts). High Hopes Therapeutic Riding stables hosted equestrian events for 1995 World Special Olympics. Location of the initial discovery of Lyme disease. Seasonal tourism centered around the town's beaches, boating activities, and scenic beauty. Includes villages of South Lyme, Laysville, and Blackhall. Settled c.1655, incorporated 1855.

Old Maas River, Dutch *Oude Maas*, SW NETHERLANDS, river channel, formed by dividing of NEW MERWEDE RIVER into NOORD RIVER and Old Maas River 1 mi/1.6 km N of DORDRECHT; flows 19 mi/31 km WNW, joining NEW MAAS RIVER 7 mi/11.3 km WSW of ROTTERDAM to form the NEW WATERWAY, shipping channel to NORTH SEA 11 mi/18 km to WNW for Rotterdam's inland harbor. Dortsche Kil River branches S from it 1 mi/1.6 km SW of Dordrecht; SPUI RIVER branches SW from it 6 mi/9.7 km SSW of Rotterdam. BEIJERLAND and PUTTEN regions on S; IJsselmonde island on N. Entire length navigable.

Old Malda, town, MALDAH district, N WEST BENGAL state, E INDIA, on the MAHANANDA RIVER, 2 mi/3.2 km N of Ingraj Bazar. Rice, wheat, oilseeds, jute. Has 16th-century mosque. Site of former Dutch and French factories. Also spelled Malda.

Old Malton, ENGLAND: see MALTON.

Oldman (OLD-muhn), river, c.250 mi/400 km long; rises in the ROCKY MOUNTAINS, SW ALBERTA, W Canada; 49°57′N 111°42′W. Flows generally E past LETHBRIDGE to join the BOW RIVER W of MEDICINE HAT and form the South SASKATCHEWAN River. The BELLY RIVER is its chief tributary. The Oldman flows through a farming region; wheat and sugar beets are the main crops.

Old Man of Coniston, ENGLAND: see CONISTON.

Old Man of Hoy, Scotland: see HOY.

Old Man of the Mountain, former cliff and landmark, Franconia Notch State Park, GRAFTON county, NEW HAMPSHIRE; 44°10′N 71°41′W. Also known as "the Profile" or "the Great Stone Face," which inspired the Nathaniel Hawthorne story of the same name. Formed by a series of cliffs on CANNON Mountain, the formation resembled the face of a man in profile and became the official state trademark. The formation collapsed in May, 2003.

Oldmans Creek, c.25 mi/40 km long, SW NEW JERSEY; rises SW of GLASSBORO; flows generally NW, forming

GLOUCESTER-SALEM county line, to DELAWARE RIVER above PENNS GROVE. Navigable for c.11 mi/18 km above mouth.

Old Man's Pond, lake (8 mi/13 km long, 1 mi/2 km wide), SW NEWFOUNDLAND AND LABRADOR, Canada, 10 mi/16 km NNE of CORNER BROOK.

Oldmeldrum (OLD-mel-druhm), town (2001 population 2,003), Aberdeenshire, NE Scotland, 15 mi/24 km NNW of ABERDEEN; 57°19′N 02°19′W. Previously cotton milling. Just S is Barra Hill (634 ft/193 m), with prehistoric (Iron Age) fort. Formerly in Grampian, abolished 1996.

Old Mill Creek, village (2000 population 251), Lake county, NE ILLINOIS, residential suburb 44 mi/71 km NNW of CHICAGO, 11 mi/18 km NW of WAUKEGAN, near WISCONSIN state line; 42°25′N 87°58′W. Drained by North Mill Creek.

Old Mission Peninsula, MICHIGAN: see GRAND TRAVERSE BAY.

Old Mobeetie, Texas: see MOBEETIE.

Old Monroe, town (2000 population 250), LINCOLN county, E MISSOURI, on CUIVRE RIVER, near the MISSOURI, and 13 mi/21 km ESE of TROY; 38°55′N 90°45′W. Lumber; fertilizers and feed.

Old Mulkey Meeting House State Park, KENTUCKY: see TOMPKINSVILLE.

Old Mystic, NEW LONDON county, SE CONNECTICUT, 8 mi/12.9 km ENE of NEW LONDON, on Whitford River. Many old houses.

Old North State; see NORTH CAROLINA.

Old Northwest: see NORTHWEST TERRITORY.

Old Ocean (O-shuhn), unincorporated town, BRAZORIA county, SE TEXAS, 24 mi/39 km WNW of FREEPORT, near Linville Bayou. Oil and natural gas region. Agricultural (rice, cotton, soybeans). Manufacturing (petro-chemicals).

Old Orchard Beach, town (2000 population 8,856), YORK county, SW MAINE, on the ATLANTIC coast; 43°31′N 70°23′W. For many years a popular summer resort, it has a beach and amusement facilities. A trading post was located nearby before 1630. Settled c.1631, incorporated 1883.

Old Oyo (aw-YAW), town, OYO state, SW NIGERIA, 38 mi/61 km WSW of JEBBA. Cotton weaving, shea-nut processing; cattle raising. Was capital of dispersal of Yoruba tribe in 17th and 18th centuries. Sometimes called Katunga.

Old Paphos, CYPRUS: see KOUKLIA.

Old Perlican (PUHR-li-kuhn), town (2001 population 714), SE NEWFOUNDLAND AND LABRADOR, Canada, on E side of Trinity Bay, 25 mi/40 km NNE of CARBONEAR; 48°05′N 53°01′W. Fishing. Just offshore is PERLICAN ISLAND.

Old Point Comfort (OLD POINT KUHM-fuhrt), resort area, HAMPTON independent city, SE VIRGINIA, on small peninsula (Old Point Comfort), on CHESAPEAKE BAY, on N side of entrance to HAMPTON ROADS harbor, 3 mi/5 km E of HAMPTON; 37°00′N 76°18′W. Bathing, fishing resort since 1830s. U.S. Fort Monroe (long known as Fortress Monroe), built 1819–1834 on site of 17th-century fortifications. Held by Union throughout Civil War; Jefferson Davis imprisoned here, 1865–1867. Long a coast artillery post, it became (1946) headquarters of U.S. Army field forces. Now FORT MONROE Military Reservation; historic Jefferson Davis Casement. Hampton Roads Bridge-Tunnel 1 mi/2 km to SW.

Old Radnor, Wales: see NEW RADNOR.

Old Rampart, Indian village, NE ALASKA, near YUKON (CANADA) border, on PORCUPINE River, 110 mi/177 km ENE of FORT YUKON; 67°10′N 141°40′W. Established 1869 as trading post of Hudson's Bay Company.

Old Rhine River, Dutch *Oude Rijn*, arm of the RHINE RIVER, 42 mi/68 km long, central and W NETHERLANDS, a continuation of CROOKED RHINE RIVER at UTRECHT. Flows W past WOERDEN, BODEGRAVEN, ALPHEN AAN DEN RIJN, and LEIDEN, to NORTH SEA at Katwijk aan Zee. Crossed by AMSTERDAM-RHINE CANAL 1 mi/1.6 km WSW of Utrecht. Waters from the main Rhine River channel in E Netherlands have been diverted into the country's canal system and to the channels in the Delta Project of SW Netherlands.

Old Ripley, village (2000 population 127), BOND county, SW central ILLINOIS, 8 mi/12.9 km W of GREENVILLE; 38°53′N 89°34′W. In agricultural area (corn, wheat; dairy products; livestock).

Old River, stream, 7 mi/11km long, S CONCORDIA parish, LOUISIANA; connects the MISSISSIPPI RIVER with the RED and ATCHAFALAYA rivers; 31°04′N 91°30′W. The OLD RIVER CONTROL STRUCTURE, which prevents the Mississippi River from being captured by the Atchafalaya River, is here.

Old River, GUATEMALA and BELIZE: see BELIZE RIVER.

Old River Control Structure, LOUISIANA: see ATCHAFALAYA RIVER.

Old River-Winfree (WIN-free), unincorporated town (2006 population 1,710), CHAMBERS county, SE TEXAS, residential suburb 30 mi/48 km ENE of HOUSTON, on Old River, near TRINITY RIVER; 29°52′N 94°49′W. Agricultural area on urban fringe.

Old Road, village, W St. Kitts, ST. KITTS AND NEVIS, WEST INDIES, 5 mi/8 km WNW of BASSETERRE. Here landed (1623) Sir Thomas Warner with British settlers. Sometimes called Old Road Town.

Olds (OLDZ), town (□ 4 sq mi/10.4 sq km; 2001 population 6,607), S ALBERTA, W Canada, 51 mi/82 km N of CALGARY, in Mountain View County. In a livestock-raising, dairy-producing, and wheat-farming region; 51°47′N 114°06′W. It has grain elevators and is the seat of a provincial agricultural school. Established as a village in 1896; became a town in 1905.

Olds, town (2000 population 249), HENRY county, SE IOWA, 12 mi/19 km N of MOUNT PLEASANT; 41°07′N 91°32′W. In livestock area.

Old San Juan (SAHN WAHN), neighborhood of SAN JUAN, NE PUERTO RICO, W portion of SAN JUAN ISLET, facing San Juan Bay and the ATLANTIC OCEAN. A walled city within the city of San Juan, Old San Juan is characterized by its extraordinary Spanish colonial architecture and its cobblestone streets. Important government offices are here, including La Fortaleza, where all governors of the island have lived since Juan Ponce de León. Administrative, commercial, banking, recreational, and tourist center of Puerto Rico.

Old Sarum (SER-uhm), site of a former city, SE WILTSHIRE, S ENGLAND, just N of SALISBURY (New Sarum); 51°06′N 01°48′W. Excavations in the old settlement's mound have revealed remains of an ancient British camp, the Roman station *Sorviodunum*, and a later Saxon (then Norman) town. The bishopric, moved to Old Sarum from SHERBORNE in 1075, was transferred to Salisbury in 1220. Old Sarum's cathedral was torn down and parts of it were used in the construction of the cathedral at Salisbury. At Old Sarum the Use of Sarum, a ritual adopted in S England, was compiled. Old Sarum was an important city until strife between the men of the castle and garrison and the men of the religious institution arose. That turmoil led to the cathedral's removal and eventually to the decay of the old city; water shortage and harsh winds may also have been causes of its decline. The "rotten borough" of Old Sarum continued to be represented in Parliament until the Reform Bill of 1832 was passed.

Old Saybrook, resort town, MIDDLESEX county, S CONNECTICUT, on LONG ISLAND SOUND, at mouth of the CONNECTICUT River (here bridged), 28 mi/45 km E of NEW HAVEN; 41°17′N 72°22′W. Agriculture; fishing; tourism; manufacturing (hardware, boats, food products, electronics, printing, metal fabrication). Includes Fenwick (resort borough), Old Saybrook village, and Saybrook Point (summer colony). Collegiate School of America here (early 18th century) was nucleus of Yale University. Settled 1635 as Saybrook colony, incorporated 1854 as Old Saybrook. A town formerly named Saybrook (now Deep River) is N.

Old Scone, Scotland: see SCONE.

Old Sennar, SUDAN: see SENNAR.

Old Shawneetown, village (2000 population 278), GALLATIN county, S ILLINOIS, on OHIO River, 23 mi/37 km E of HARRISBURG; 37°42′N 88°08′W. Highway bridge to KENTUCKY. Shawneetown State Historical Site is here. Shawnee National Forest to SW. Gallatin County Courthouse and most residents moved to (new) Shawneetown in late 1930s after disastrous Ohio River flood of 1937. One of state's oldest settlements; important area in the early nineteenth century. The first state chartered bank opened here.

Oldsmar (OLDZ-mahr), city (□ 10 sq mi/26 sq km; 2005 population 13,552), PINELLAS county, W central FLORIDA, 10 mi/16 km ENE of CLEARWATER; 28°03′N 82°40′W.

Old Sodbury (SAHD-buh-ree), agricultural village (2006 population 4,900), South Gloucestershire, SW ENGLAND, 12 mi/19 km NE of BRISTOL, in the COTSWOLD HILLS; 51°32′N 02°21′W. Has 12th-century church. Site of ancient British camp, later one of largest Roman camps in Britain. Just N is agricultural village and parish of Little Sodbury, with manor house in which William Tyndale was tutor (1521) to family of Sir John Walsh and where he translated Erasmus' *Enchiridion* and began translation of New Testament.

Old Spec Mountain (4,180 ft/1,274 m), OXFORD county, W MAINE, near NEW HAMPSHIRE state line, 20 mi/32 km N of RUMFORD.

Old Tappan, borough (2006 population 6,013), BERGEN county, extreme NE NEW JERSEY, 12 mi/19 km NE of PATERSON, near HACKENSACK River and NEW YORK state line; 41°01′N 73°58′W. Largely residential.

Old Town, Czech *Staré Město*, German ALTSTADT, central district (1991 population 13,040) of PRAGUE, PRAGUE-CITY province, central BOHEMIA, CZECH REPUBLIC, on right bank of VLTAVA RIVER. From the 11th century, trade routes intersected here (marketplace). Today the district is noted for its picturesque narrow streets, Old Town Square (Czech *Staroměstské náměstí*), and Old Town Hall with calendar-clock, which was made in 1410.

Old Town, city (2000 population 8,130), PENOBSCOT county, S central MAINE, on the PENOBSCOT, and 13 mi/21 km above BANGOR; 44°57′N 68°44′W. Manufacturing (canoes, pulp, paper; lumber mills). On small island N of Old Town village is a reservation for Penobscot Indians. Maine's first railroad (1836) connected Old Town with Bangor. Includes villages of Stillwater and Great Works. Settled 1774, incorporated 1840, city chartered 1891.

Oldtown, village (2000 population 190), BONNER county, N IDAHO, at WASHINGTON state line adjacent to NEWPORT (Washington); 48°11′N 117°02′W. Manufacturing (sawmill products, lumber).

Oldtown, unincorporated village, GREENUP county, NE KENTUCKY, 8 mi/12.9 km S of GREENUP, on LITTLE SANDY RIVER. Tobacco, alfalfa, soybeans, corn; cattle. Oldtown Covered Bridge (1880) spans river. Greenbo State Resort Park to N.

Oldtown, village, ALLEGANY county, W MARYLAND, on North Branch of the POTOMAC (bridged), 11 mi/18 km SE of CUMBERLAND. It was originally a ford called Skipton settled in 1741 by College Thomas Cresap on an old Indian trail known as "The Warriors' Path." The Six Nations of the Iroquois Confederacy of the N and the Cherokees, Catawbas, and Shawnees of the S raided each other over the path. Private toll bridge across Potomac to GREEN SPRINGS (WEST VIRGINIA); closed to motor vehicle traffic in 1995.

Old Trafford, ENGLAND: see MANCHESTER.

Old Uppsala, SWEDEN: see GAMLA UPPSALA.

Olduvai Gorge (ol-doo-VAH-ee), in GREAT RIFT VALLEY, ARUSHA region, N TANZANIA, 80 mi/129 km

Area is shown by the symbol □, and capital city or county seat by ⊙.

WNW of ARUSHA, in NGORONGORO Conservation Area; 02°57′S 35°14′E. Erosional processes have exposed geological strata in the gorge dating to the lower Pleistocene epoch, 600,000 to 1.8 million years ago. The site was made famous by the numerous hominid fossils excavated by Louis Leakey and his wife, Mary Leakey, as well as by later researchers including their son Richard Leakey. Examples of at least 3 species of hominids have been found at Olduvai, including *Australopithecus boisei*, *Homo habilis*, and *Homo erectus*. In addition, the two earliest stone tool traditions, Oldowan and Acheulian, have been found along with fossil remains. Both the fossils and the tools have been important lines of evidence in understanding human evolution, including the social and dietary adaptations of early hominids. SERENGETI PLAIN to W.

Old Westbury, residential village (□ 8 sq mi/20.8 sq km; 2006 population 5,187), NASSAU county, SE NEW YORK, on W LONG ISLAND, 3 mi/4.8 km NE of MINEOLA; 40°47′N 73°35′W. Old Westbury Gardens and manor of former Phipps estate rivals formal gardens of Europe. Seat of New York Institute of Technology's central campus; State University of New York at Old Westbury. Incorporated 1924.

Old Windsor (WIN-zuhr), town (2001 population 4,775), Windsor and Maidenhead, SE ENGLAND, on the THAMES, just SE of WINDSOR; 51°27′N 00°35′W. Has 13th-century church. Edward the Confessor had castle here.

Old Wives Lake (OLD WEIVZ) (□ 123 sq mi/319.8 sq km), S SASKATCHEWAN, Canada, 22 mi/35 km SW of MOOSE JAW; 20 mi/32 km long, 12 mi/19 km wide; 50°06′N 106°00′W.

Old Yssel River, NETHERLANDS and GERMANY: see OLD IJSSEL RIVER.

Olean (O-lee-AN), city (□ 6 sq mi/15.6 sq km; 2006 population 14,584), CATTARAUGUS county, W NEW YORK, on the ALLEGHENY RIVER near the PENNSYLVANIA state line; 42°04′N 78°25′W. An important commercial center of the region. Once an oil-based economy emanating from nearby "Pennsylvania" oil fields, manufacturing include turbines and compressors for the oil industry, electrical items, cutlery, and dairy products. St. Bonaventure University and Allegheny State Park are nearby. Major outfitting post for settlers moving W down the Allegheny and OHIO rivers in early 1800s. In 1972 a severe flood associated with Hurricane Agnes flooded large areas and damaged more than 2,900 homes. Settled 1804, incorporated 1893.

Olean (O-lee-AN), town (2000 population 157), MILLER county, central MISSOURI, 5 mi/8 km N of ELDON; 38°24′N 92°31′W.

Olean Creek, NEW YORK: see ISCHUA CREEK.

Olecko (o-LET-sko), German *Treuburg*, town in EAST PRUSSIA, Suwałki province, NE POLAND, in Masurian Lakes region, on small lake, 18 mi/29 km W of SUWAŁKI; 54°02′N 22°31′E. Railroad junction; grain and cattle market. In World War I, important flank position of Russian lines in first stage (September 1914) of battle of Masurian Lakes. In World War II, c.90% destroyed. Founded 1560. Until 1928, called Marggrabowa.

Oledo (o-LAI-doo), town, CASTELO BRANCO district, central PORTUGAL, 15 mi/24 km NE of CASTELO BRANCO; 39°58′N 07°18′W. Agriculture includes grain, corn, beans. Oak woods.

Oleggio (o-LED-jo), town, NOVARA province, PIEDMONT, N ITALY, 11 mi/18 km N of NOVARA; 45°36′N 08°38′E. Railroad junction; cotton and hosiery mills. Has Romanesque church with 10th-century frescoes.

Oleh (O-le), former tribal confederation of S Yemen, astride the Kaur al Audhilla. Although centered on MUDIA plain of Dathina dist., it also includes tribes on N slopes of the KAUR range in UPPER AULAQI country. The confederation ceased to exist after SOUTH YEMEN won its independence.

Olehleh, INDONESIA: see ULEELHUE.

Oleiros (o-LAI-roosh), town, CASTELO BRANCO district, central PORTUGAL, 23 mi/37 km WNW of CASTELO BRANCO. Lumbering, resin and pitch extracting. Has 16th–17th-century churches.

Olëkma (uh-LYOK-mah), river, approximately 820 mi/1,320 km long; rises in the YABLONOVY RANGE, CHITA oblast, SE Siberian Russia; flows N through NW AMUR oblast and the SAKHA REPUBLIC to the LENA RIVER below OLËKMINSK.

Olëkminsk (uh-LYOK-meensk), city (2006 population 9,535), SW SAKHA REPUBLIC, W RUSSIAN FAR EAST, on the LENA RIVER just W of its confluence with the Olekma River, on local highway, 400 mi/651 km SW of YAKUTSK; 60°22′N 120°25′E. Elevation 511 ft/155 m. Center of agricultural, cattle-raising area; lumbering, woodworking, food industries, gypsum. Founded in 1635 at the mouth of the Olekma River, later moved 7 mi/12 km upstream on the Lena River.

Oleksandrivka (o-lek-SAHN-dreef-kah) (Russian *Aleksandrovka*), town (2004 population 5,575), central DONETS'K oblast, UKRAINE, in the DONBAS, 15 mi/24 km SSW of DONETS'K; 49°07′N 37°35′E. Elevation 524 ft/159 m. Established in 1841 as Kreminna (Russian *Kremennaya*); renamed in 1903; town since 1938; it is now a commuter suburb of Donets'k.

Oleksandrivka (o-lek-SAHN-dreef-kah) (Russian *Aleksandrovka*), town (2004 population 12,100), NW MYKOLAYIV oblast, UKRAINE, 9 mi/14 km NNW of VOZNESENS'K and 37 mi/59.5 km E of MYKOLAYIV; 46°50′N 32°47′E. Elevation 101 ft/30 m. Prefabricated building components plant, reinforced concrete fabrication, silicate brick, granite quarry, asphalt plant, food processing.

Oleksandrivka (o-lek-SAHN-dreef-kah) (Russian *Aleksandrovka*), town, N KIROVOHRAD oblast, UKRAINE, 30 mi/48 km N of KIROVOHRAD; 48°58′N 32°14′E. Elevation 406 ft/123 m. Raion center; sugar mill, gas pipeline-manufacturing plant. Established in the early 17th century, town since 1957.

Oleksandrivka (o-lek-SAHN-dreev-kah) (Russian *Aleksandrovka*), town, NW DONETS'K oblast, UKRAINE, on the SAMARA RIVER, 15 mi/24 km SSW of BARVINKOVE; 47°48′N 38°03′E. Elevation 479 ft/145 m. Raion center; flour mill; river fishery research and demonstration center and processing plant.

Oleksandrivka, UKRAINE: see LOZNO-OLEKSANDRIVKA.

Oleksandrivka, UKRAINE: see ORDZHONIKIDZE, Dnipropetrovs'k oblast.

Oleksandrivs'k (o-lek-SAHN-dreefsk) (Russian *Aleksandrovsk*), city, S central LUHANS'K oblast, UKRAINE, on the LUHAN' RIVER, and 4 mi/6 km W of LUHANS'K and subordinated to its city council; 48°35′N 39°11′E. Elevation 167 ft/50 m. Woodworking, vegetable processing. Established in the 1770s, city since 1961, when its name was changed from Oleksandrivka (Russian *Aleksandrovka*).

Oleksandrivs'k, UKRAINE: see ZAPORIZHZHYA, city.

Oleksandriya (o-lek-sahn-DREE-yah) (Russian *Aleksandriya*), city (2001 population 93,357), KIROVOHRAD oblast, UKRAINE, on the INHULETS' River, 40 mi/64 km ENE of KIROVOHRAD; 48°40′N 33°07′E. Elevation 354 ft/107 m. Raion center; center of lignite-mining area (developed after 1945); manufacturing (mining machinery, excavators, chemicals), woodworking, clothing factories, meatpacking; flour milling. Granite quarries. Technical and professional schools. Dates from the mid-18th century; originally an agricultural trade center. In World War II, held by the Germans (1941–1943).

Oleksandriya (o-lek-SAHN-dree-yah) (Polish *Aleksandrja*) (Russian *Aleksandriya*), village, central RIVNE oblast, UKRAINE, on the HORYN' River, 8 mi/13 km NNE of RIVNE; 50°44′N 26°21′E. Elevation 538 ft/163 m. Flour milling, sawmilling, tanning, agricul-

Oleksandro-Hryhoryivka, UKRAINE: see DONETS'K.

Oleksiyeve-Leonove, UKRAINE: see TOREZ.

Oleksiyeve-Orlivka, UKRAINE: see SHAKHTARS'K.

Oleksiyivka (o-lek-SEE-yif-kah) (Russian *Alekseyevka*), village, central KHARKIV oblast, UKRAINE, 38 mi/61 km S of KHARKIV; 49°24′N 36°17′E. Elevation 442 ft/134 m. Dairying. Former Russian name was Alekseyevskoye.

Olen (O-luhn), commune (2006 population 11,382), Turnhout district, ANTWERPEN province, N BELGIUM, 3 mi/4.8 km SSE of HERENTALS; 51°09′N 04°51′E. Formerly spelled Oolen.

Ølen (UH-luhn), village, HORDALAND county, SW NORWAY, on a S inlet of HARDANGERFJORDEN, 24 mi/39 km ENE of HAUGESUND. Agriculture; canneries; forestry; mechanical works.

Olenegorsk (uh-lye-nee-GORSK), city (2006 population 23,175), central MURMANSK oblast, NW European Russia, on road and railroad (Olenya station), 70 mi/113 km S of MURMANSK, and 40 mi/64 km N of APATITY; 68°09′N 33°18′E. Elevation 495 ft/150 m. Mining and concentrating of iron ore from the Olenegorsk ore deposit; mechanical plant. Became city in 1957.

Olenëk (uh-lee-NYOK), village, NW SAKHA REPUBLIC, NW RUSSIAN FAR EAST, on the OLENËK River, on local highway, 400 mi/644 km NNW of NYURBA; 68°33′N 112°18′E. Trading post; local airport. Formerly called Olenëkskaya Kul'tbaza (Russian=Olenëk cultural base).

Olenëk (uh-lee-NYOK), river, approximately 1,350 mi/2,173 km long, E Siberian Russia; rises in the CENTRAL SIBERIAN PLATEAU, KRASNOYARSK TERRITORY; winds E then N through NW SAKHA REPUBLIC to the LAPTEV SEA. It is navigable for about 600 mi/966 km upstream and abounds in fish.

Olenino (uh-LYE-nee-nuh), town (2006 population 5,025), SW TVER oblast, W European Russia, on road and railroad junctions, 30 mi/48 km W of RZHEV; 56°12′N 33°29′E. Elevation 823 ft/250 m. Lumbering; dairy products, clothing.

Olenivka (o-LE-nif-kah) (Russian *Yelenovka*), town, E DONETS'K oblast, UKRAINE, in the DONBAS, 3 mi/5 km NE (under jurisdiction) of YENAKIYEVE; 47°50′N 37°40′E. Elevation 656 ft/199 m.

Olenivka, UKRAINE: see ZORYNS'K.

Oleniv'ski Kar'yery, UKRAINE: see DOKUCHAYEVS'K.

Olenovka (o-LEN-of-kah) (Russian *Yelenovka*), town, S central DONETS'K oblast, UKRAINE, in the DONBAS, on road and railroad, 13 mi/21 km SW of DONETS'K; 48°15′N 38°19′E. Elevation 613 ft/186 m. Railroad car depot; bakery; fireproof clays. Limestone quarries at OLENIV'SKI KAR'YERY (Russian *Yelenovskiye Karyery*), 6 mi/10 km S. Est. 1840; town since 1938.

Olentangy River (O-luhn-TAN-jee), c.75 mi/121 km long, central OHIO; rises near Galion in CRAWFORD county; flows NW, then S, past Delaware and Worthington, to SCIOTO RIVER at COLUMBUS; 39°57′N 83°01′W.

Óleo (O-lee-o), town, W central SÃO PAULO state, BRAZIL, 35 mi/56 km E of OURINHOS; 22°58′S 49°21′W. Coffee and cotton processing, sawmilling, pottery manufacturing.

Oléron, Île d' (do-lai-ron, eel), largest island (□ 68 sq mi/176.8 sq km) in Bay of BISCAY, part of CHARENTE-MARITIME department, POITOU-CHARENTES region, W FRANCE; 45°55′N 01°16′W. Extends NW-SE opposite mouth of CHARENTE RIVER; separated from mainland by the Pertuis (French=strait) de MAUMUSSON (1 mi/1.6 km wide) and from Île de Ré by the Pertuis d'ANTIOCHE. Island is 18 mi/29 km long, 4 mi/6.4 km to 6 mi/9.7 km wide; generally level and sandy with pine-covered dunes in S. Produces early vegetables, forage crops, and wine; fisheries, aquaculture, and oyster beds. The tourist trade is the island's main economic activity. Principal localities are SAINT-PIERRE-D'OLÉRON, SAINT-TROJAN-LES-BAINS; the main beach is on W shore. Lighthouse at N end of island. A bridge-viaduct links island with mainland since 1966.

The island was a stronghold of Protestantism in the 16th century.

Olesa de Montserrat (o-LAI-sah dhai mon-se-RAHT), city, BARCELONA province, NE SPAIN, in CATALONIA, on the LLOBREGAT RIVER, 17 mi/27 km NW of BARCELONA; 41°33′N 01°54′E. Manufacturing of cotton and wool textiles, felt, cement, dyes; olive-oil processing; lumbering. Livestock; wine; cereals, fruit in area. Warm sulphur springs nearby.

Oleshky, UKRAINE: see TSYURUPYNS'K.

Oleshky Sich, UKRAINE: see TSYURUPYNS'K.

Oleshya, UKRAINE: see TSYURUPYNS'K.

Oles'ko (o-LES-ko), town (2004 population 6,000), E L'VIV oblast, UKRAINE, 12 mi/19 km N of ZOLOCHIV; 49°58′N 24°53′E. Elevation 754 ft/229 m. Mineral water-bottling plant. Vocational-technical school. Has a restored 17th-century castle, which houses a public art gallery, a 15th-century church, and an 18th-century monastery. Known since 1366, after it came under Polish rule; granted Magdeburg law in 1441; since it was located on the Kuchman Trail, it was often attacked by the Tatars in the 16th century; Polish king Jan II Sobieski built the castle (in the 1680s) to revive the town's economy. Jewish community since 1500, reaching its peak in the second half of the 19th century, when the town became one of the centers of Hasidic Orthodox learning; suffered through pogroms of the beginning of the 20th century and reduced to 740 in 1935; destroyed during World War II.

Olesko, UKRAINE: see OLES'KO.

Olesnica (o-le-SNEE-tsah), Polish *Oleśnica*, German *Oels* or *Öls*, town (2002 population 37,276), Wrocław province, SW POLAND, 17 mi/27 km ENE of WROCŁAW (Breslau). Linen and paper milling, manufacturing of shoes, chemicals, furniture, food products; brewing, distilling, flour milling, sawmilling. Mentioned 1214 as trade center; chartered 1255. Was (1312–1492) capital of duchy under branch of Piast dynasty. Passed 1884 to PRUSSIA. Considerably damaged in World War II; 13th century tower, medieval gate, and remains of old town walls survived. In LOWER SILESIA until 1945.

Olesnice (O-lesh-NYI-tse), Czech *Olešnice*, German OELS, village, JIHOMORAVSKÝ province, W central MORAVIA, CZECH REPUBLIC, in E part of BOHEMIAN-MORAVIAN HEIGHTS, 25 mi/40 km NNW of BRNO. Agriculture (potatoes); cattle; meat processing; manufacturing (artificial flowers).

Olesno, German *Rosenberg*, town in UPPER SILESIA, Częstochowa province, S POLAND, 25 mi/40 km NE of OPOLE (Oppeln). Textiles; agricultural market (grain, potatoes; livestock). Old church is object of pilgrimage.

Oletta (o-le-TAH), village (□ 10 sq mi/26 sq km; 2004 population 1,245), N CORSICA, HAUTE-CORSE department, FRANCE, 7 mi/11.3 km SW of BASTIA; 42°38′N 09°21′E. Olive groves overlook resort of Saint-Florent and the MEDITERRANEAN.

Oletta, ETHIOPIA: see HOLETA GENET.

Oleum, oil-refining and -shipping point, CONTRA COSTA county, W CALIFORNIA, on SAN PABLO BAY, 9 mi/14.5 km NE of RICHMOND. Tanker docks.

Olevs'k (O-levsk) (Russian *Olevsk*) (Polish *Olwsk*), town, NW ZHYTOMYR oblast, UKRAINE, on Ubort' River (right tributary of the PRIPET RIVER), 45 mi/72 km WNW of KOROSTEN'; 51°13′N 27°39′E. Elevation 554 ft/168 m. Raion center; ceramics; food processing, flax processing, sewing, lumbering. Known since 1488 when under Lithuania; passed to Poland in 1569; granted the rights of Magdeburg Law in 1641; passed to Russia in 1793; served as USSR-Poland frontier station (1921–1939); center of Ukrainian insurgents (the Polisian Sich of Bulba-Borovets') during World War II. Also spelled Olevs'ke. Jewish community since 1721, numbering 2,900 in 1926; destroyed during World War II—fewer than 100 Jews remaining in 2005.

Olevsk, UKRAINE: see OLEVS'K.

Olevs'ke, UKRAINE: see OLEVS'K.

Oley (O-lee), unincorporated town, BERKS county, SE central PENNSYLVANIA, 7 mi/11.3 km NE of READING, on Little Manatawny Creek; 40°23′N 75°47′W. Light manufacturing; agriculture (grain; livestock; dairying).

Olfen (OL-fen), town, WESTPHALIA-LIPPE, North RHINE–Westphalia, NW GERMANY, on DORTMUND-EMS CANAL, 5 mi/8 km SW of Lüdinghausen; 51°43′N 07°20′E. Grain; cattle, hogs.

Ölfusá (UHL-vuh-sou), wide river, 18 mi/29 km long, SW ICELAND; formed 10 mi/16 km NE of EYRARBAKKI by confluence of HVITA and SOG rivers; flows SW to ATLANTIC OCEAN, 2 mi/3 km WNW of Eyrarbakki. Noted for its salmon.

Ol'ga (OL-gah), town (2006 population 4,315), SE MARITIME TERRITORY, SE RUSSIAN FAR EAST, port on the small Olga Gulf of the Sea of JAPAN, on road junction, 175 mi/282 km ENE of VLADIVOSTOK; 43°44′N 135°17′E. Elevation 101 ft/30 m. Timber shipping. Formerly known as Permskaya.

Olgiate Comasco (ol-JAH-te ko-MAH-sko), town, COMO province, LOMBARDY, N ITALY, 6 mi/10 km WSW of COMO; 45°48′N 08°58′E. Silk mills; clothing, machinery, chemicals.

Olgiate Olona (o-LO-nah), town, VARESE province, LOMBARDY, N ITALY, on OLONA River, 2 mi/3 km NE of BUSTO ARSIZIO; 45°38′N 08°53′E. Manufacturing (textile machinery, textiles, fabricated metals, alcohol, candy).

Ol'ginka, UKRAINE: see OL'HYNKA.

Olginka, UKRAINE: see OL'HYNKA.

Ol'gino (OL-gee-nuh), town, W central LENINGRAD oblast, NW European Russia, near the SE shore of the Gulf of FINLAND, on highway, 13 mi/21 km WSW of SAINT PETERSBURG, of which it is a suburb; 59°50′N 29°55′E. Construction materials; tobacco products. Also known as Olino (O-lee-nuh).

Ol'gino, RUSSIA: see MOSKALENKI.

Ol'ginskaya (OL-geen-skah-yah), village (2005 population 4,460), W KRASNODAR TERRITORY, S European Russia, on road junction and railroad, 60 mi/97 km NW of KRASNODAR, and 19 mi/31 km NW of ROGOVSKAYA; 45°56′N 38°33′E. In agricultural area (wheat, hemp, corn, sunflowers).

Ol'ginskoye (OL-geen-skuh-ye), village (2006 population 3,200), E NORTH OSSETIAN REPUBLIC, RUSSIA, in the N CAUCASUS Mountains, on road and near railroad, 8 mi/13 km N of VLADIKAVKAZ; 43°09′N 44°41′E. Elevation 1,863 ft/567 m. Agriculture (grain; livestock); produce processing.

Ölgiy (UH-luh-gai), town (2000 population 25,791), ⊙ BAYAN ÖLGIY province, W MONGOLIA, on HOVD GOL (Kobdo River), and 100 mi/160 km NW of HOVD (Kobdo); 48°56′N 89°57′E. Elevation 5,610 ft/1710 m. Also spelled Ölögey, Ulegei, or Ulegey.

Ølgod (UL-godh), town, RIBE county, SW JUTLAND, DENMARK, 24 mi/39 km N of ESBJERG; 55°45′N 08°40′E. Agriculture (cattle; potatoes); manufacturing (wood cabinets).

Ol'gopol', UKRAINE: see OL'HOPIL'.

Olhão (ol-YOU), town, FARO district, S PORTUGAL, port on the ATLANTIC OCEAN (S coast), on railroad, 5 mi/8 km E of FARO; 37°02′N 07°50′W. Major fishing and fish-canning center (sardines and tuna); saltworks; pottery manufacturing. Tourism.

Ölheim, GERMANY: see EDEMISSEN.

Olho d'Água (OL-yo DAH-gwah), town, W central PARAÍBA, BRAZIL, 14 mi/23 km E of PIANCÓ; 07°13′S 37°45′W.

Olho d' Água das Cunhãs (OL-yo DAH-gwah dahs KOON-yahs), city, N central MARANHÃO state, BRAZIL, 55 km/34 mi NW of BACABAL; 03°55′S 45°12′W.

Olho d'Agua das Flores (OL-yo dah-GWAH das FLO-res), city, W central ALAGOAS state, BRAZIL, at base of Serra São Joaquim; 09°34′S 37°17′W.

Olho d'Água do Casado (OL-yo dah-GWAH do KAH-sah-do), town, W ALAGOAS state, BRAZIL, 12 mi/19 km NW of PIRANHAS; 09°33′S 37°48′W.

Olho d'Agua do Seco (OL-yo dah-GWAH do SE-ko), village, W central BAHIA, BRAZIL, 43 mi/70 km W of PALMEIRAS; 12°34′S 42°08′W.

Olho d'Agua Grande (OL-yo dah-GWAH GRAHN-zhee), town, SE ALAGOAS state, BRAZIL, 12 mi/19 km N of PORTO REAL DO COLEGIO, on SALVADOR-RECIFE railroad; 10°05′S 36°50′W.

Ôlho Marinho (ol-oo mah-EEN-yoo), village, LEIRIA district, W central PORTUGAL, 8 mi/12.9 km SW of CALDAS DA RAINHA. Wine, beans.

Ol'hopil' (OL-ho-peel) (Russian *Ol'gopol'*), village (2004 population 4,400), SE VINNYTSYA oblast, UKRAINE, 12 mi/19 km S of BERSHAD'; 48°12′N 29°30′E. Elevation 629 ft/191 m. Formerly town with small metalworks (1950s). In the 18th century, belonged to Lubomirski family and called Rohuzka-Chechel'-nyts'ka. Passed to Russia (1793) and renamed (1795) in honor of Catherine II's granddaughter, Olga. Became company center and grew in the 19th century to 10,000, including a sizeable Jewish community (numbering 1,660 in 1939). The Jews were destroyed by the Nazis and the town largely depopulated during World War II.

Ol'hynka (OL-hin-kah) (Russian *Ol'ginka*), town (2004 population 6,600), central DONETS'K oblast, UKRAINE, in the DONBAS, 25 mi/40 km SW of DO-NETS'K; 47°42′N 37°31′E. Elevation 669 ft/203 m. Manufacturing containers for railroad cars, building materials; feed milling. Incubator research station; forestry research station. Known since 1779; town since 1938.

Oliana (o-LYAH-nah), town, LÉRIDA province, NE SPAIN, on SEGRE RIVER (dam), 20 mi/32 km SSW of SEO DE URGEL; 42°04′N 01°19′E. Livestock; agricultural products.

Oliarus, Greece: see ANTIPAROS.

Olías del Rey (o-LEE-ahs dhel RAI), town, TOLEDO province, central SPAIN, 6 mi/9.7 km NNE of TOLEDO; 39°57′N 03°58′W. Cereals, grapes, olives, olive oil, cherries.

Olib Island (O-leeb), Italian *Ulbo*, Dalmatian island in ADRIATIC SEA, S CROATIA, 29 mi/47 km NNW of ZADAR; c.5 mi/8 km long, c.2 mi/3.2 km wide. Chief village, Olib.

Olicana, ENGLAND: see ILKLEY.

Oliete (o-LYAI-tai), town, TERUEL province, E SPAIN, 28 mi/45 km W of ALCAÑIZ; 41°00′N 00°40′W. Olive-oil processing, flour milling; saffron, beans, potatoes. Irrigation reservoir nearby.

Olifants River, Portuguese *Rio dos Elefantes*, 350 mi/563 km long, SOUTH AFRICA and MOZAMBIQUE; rises in S of LIMPOPO province (South Africa) W of WITBANK; flows in a wide arc NE and E, crossing into MOZAMBIQUE at 23°39′S 31°57′E, to meet LIMPOPO R. 130 mi/209 km N of MAPUTO.

Olifants River, c.144 mi/230 km long, SOUTH AFRICA; rises in EASTERN CAPE province, formed by the confluence of Traka and Sand rivers (which flow S through the Toorwater Poort in the SWARTBERG mountain range into Western Cape province, then E between Swartberg and Kammanassie mountains S of OUDTSHOORN), to confluence with GAMKA RIVER 10 mi/16 km S of CALITZDORP in the Rooiberge range. The GOURITZ RIVER is formed, which eventually flows into the INDIAN OCEAN 19 mi/30 km SW of MOSSEL BAY.

Olifants River, 170 mi/274 km long, in WESTERN CAPE province, SOUTH AFRICA; rises NNW of WORCESTER, in basin between Skurweberg and Witsenberg ranges; flows NNW through Beaverlac Nature Reserve, past CITRUSDAL and then past CLANWILLIAM (dammed here for irrigation), to the Atlantic along winding lower course to it mouth, 30 mi/48 km NNW of LAMBERTS BAY at Papendorp (31°43′S 18°10′E). On lower course are extensive irrigation works.

Olifants River (o-LEE-fahnts), NAMIBIA; rises in AUAS MOUNTAINS, E of WINDHOEK; flows SE to join AUOB

Area is shown by the symbol □, and capital city or county seat by ⊙.

OLIVER 2793

RIVER; tributary of Auob River, also feeds irrigation areas around STAMPRIET and Gochas.

Olimarao (O-lee-mah-ROU), uninhabited atoll, State of YAP, W CAROLINE ISLANDS, Federated States of MICRONESIA, W PACIFIC, 21 mi/34 km WNW of ELATO; 2.3 mi/3.7 km long, 1.5 mi/2.4 km wide; 07°41′N 145°52′E. Comprised of two wooded islets.

Olimar River (O-leem-MAHR), c.100 mi/161 km long, TREINTA Y TRES department, E central URUGUAY; rises in the CUCHILLA GRANDE PRINCIPAL SW of Santa Clara; flows E, past TREINTA Y TRES, to the CEBOLLATÍ, 6 mi/9.7 km SW of General Enrique Martínez; 33°16′S 53°52′W.

Olimbos (O-leem-bos), village, on NW coast of KÁR-PATHOS island, DODECANESE prefecture, SOUTH AEGEAN department, GREECE, on flank of Profitis Ilias; 35°44′N 27°11′E. Reputed to be the island's oldest settlement, possibly inhabited by direct descendants of the ancient Dorian Greeks. The local dialect has some ancient elements. Local population wears traditional costumes. Also spelled Olimpos.

Ólimbos, Greece: see OLYMPOS.

Olímpia (O-leem-pee-ah), city (2007 population 48,020), N SÃO PAULO, BRAZIL, 30 mi/48 km ENE of SÃO JOSÉ DO RIO PRÊTO; 20°44′S 48°54′W. Livestock market; butter, beverages, macaroni products, coffee. Formerly Olympia.

Olimpo, PARAGUAY: see FUERTE OLIMPO.

Olimpos, Greece: see OLIMBOS.

Olin, town (2000 population 716), JONES county, E IOWA, 10 mi/16 km SE of ANAMOSA; 42°00′N 91°08′W.

Olinalá (o-LEE-nah-LAH), town, GUERRERO, SW MEXICO, in SIERRA MADRE DEL SUR, 19 mi/30 km NNW of TLAPA DE COMONFORT on unpaved roads; 17°48′N 98°44′W. Cereals, sugarcane, coffee, fruit, forest products (resin, vanilla).

Olinda (O-leen-dah), city (2007 population 391,433), PERNAMBUCO state, E BRAZIL, on the Atlantic Ocean; 08°01′S 34°51′W. Established 1537, it was captured by the Dutch in the 1630s and burned to the ground. The rebuilt city served as a provincial capital until 1827. Its reputation as a center of learning dates from 1796, when a Jesuit seminary was founded there. Teacher's college, colonial architecture, bedroom community for RECIFE. Crafts industry (small).

Olinda, unincorporated village, ORANGE CO., S CALIFORNIA, 12 mi/19 km NNE of SANTA ANA. Oil field here (since 1890s). Annexed by BREA.

Olinda, settlement, VICTORIA, SE AUSTRALIA, 27 mi/44 km E of MELBOURNE, in DANDENONG RANGES; 37°51′S 145°22′E. Arts and crafts galleries. National Rhododendron Gardens.

Olindina (O-leen-ZHEE-nah), city (2007 population 25,705), NE BAHIA, BRAZIL, 56 mi/90 km N of ALAGOINHAS; 11°21′S 38°20′W.

Olino, RUSSIA: see OL′GINO.

Olintepeque (o-leen-te-PE-kai), town, QUEZALTENANGO department, SW GUATEMALA, on headstream of SAMALÁ RIVER, 3 mi/4.8 km N of QUEZALTENANGO; 14°53′N 91°31′W. Elevation 7,680 ft/2,341 m. Cotton weaving; corn, potatoes, wheat, fodder grasses; livestock. Quiché-speaking town.

Olintla (o-LEEN-tlah), town, PUEBLA, central MEXICO, 23 mi/37 km ESE of HUAUCHINANGO; 20°06′N 97°41′W. An isolated community in the S SIERRA MADRE ORIENTAL with Totonac Indian population. Sugar, coffee, fruit. Also called San José Olintla.

Olishevka, UKRAINE: see OLYSHIVKA.

Olisipo, PORTUGAL: see LISBON.

Olita, LITHUANIA: see ALYTUS.

Olite (o-LEE-tai), town, NAVARRE province, N SPAIN, 23 mi/37 km SE of PAMPLONA; 42°29′N 01°39′W. In wine-growing area; brandy and chocolate manufacturing. Has 15th-century castle (temporary residence of kings of Navarre) and two Gothic churches (built 12th and 13th centuries).

Olitsikas, Greece: see TOMAROS.

Oliva (o-LEE-vah), Polish *Oliwa*, residential district of Gdańsk, Gdańsk province, N POLAND, near GULF OF GDAŃSK of BALTIC SEA, 5 mi/8 km NW of city center. Has former abbatial church (begun c.1175; completed 1836); now serves as cathedral of Roman Catholic bishop of GDAŃSK. At Peace of Oliva (1660), John II of Poland renounced claim of his line to Swedish throne and confirmed Swedish possession of N LIVONIA. Frederick William, elector of BRANDENBURG, was recognized in full sovereignty over EAST PRUSSIA and in turn confirmed Polish possession of WEST PRUSSIA.

Oliva (o-LEE-vah), city, VALENCIA province, E SPAIN, near the MEDITERRANEAN SEA, 5 mi/8 km SE of GANDÍA, 42 mi/68 km SSE of VALENCIA; 38°55′N 00°07′W. Agricultural trade center in rich farming area; manufacturing of colored tiles, soap; sawmilling, olive-oil processing; hog raising. Mineral springs.

Oliva (o-LEE-vah), town (1991 population 10,696), ⊙ Tercero Arriba department (☐ 1,800 sq mi/4,680 sq km), central CÓRDOBA province, ARGENTINA, 90 mi/145 km SE of CÓRDOBA; 32°03′S 63°34′W. Dairy products (casein, cheese); trading in grain. Corn, wheat, flax, soybeans, peanuts; livestock.

Oliva or **Villa Oliva** (both: O-lee-vah), town, Ñeembucú department, S PARAGUAY, on Paraguay River (Argentina border), 50 mi/80 km SSW of Asunción; 26°00′S 57°53′W. Lumbering; cattle raising.

Oliva de la Frontera (o-LEE-vah dhai lah fron-TAI-rah) or **Oliva de Jerez** (o-LEE-vah dhai he-RETH), town, BADAJOZ province, W SPAIN, in EXTREMADURA, in mountain country near Portuguese border, 8 mi/12.9 km WSW of JEREZ DE LOS CABALLEROS; 38°16′N 06°55′W. Livestock-raising and agricultural center (cereals, acorns, olives, cork). Galena deposits nearby.

Oliva de Mérida (o-LEE-vah dhai MAI-ree-dhah), town, BADAJOZ province, W SPAIN, 15 mi/24 km SE of MÉRIDA; 38°47′N 06°07′W. Olives; livestock; manufacturing of tiles, pottery.

Oliva, La (o-LEE-vah, lah), town, FUERTEVENTURA island, CANARY ISLANDS, SPAIN, 9 mi/14.5 km NNW of PUERTO DE CABRAS. Barley, chickpeas, cereals; sheep, goats, camels. Jasper quarrying; limekilns.

Olivar Alto (o-lee-VAHR AHL-to), village, ⊙ Olivar comuna, CACHAPOAL province, LIBERTADOR GENERAL BERNARDO O'HIGGINS region, central CHILE, on CACHAPOAL RIVER, 5 mi/8 km SW of RANCAGUA; 34°12′S 70°49′W. In agricultural area (cereals, vegetables, grapes; livestock.

Olivares (o-lee-VAH-res), town, SEVILLE province, SW SPAIN, on Seville-Huelva Railroad, 9 mi/14.5 km W of SEVILLE. Agricultural center (olives, grain, grapes, fruit; livestock). Its church is noted for paintings by Roelas, who died here (1625). Has palace of counts of Olivares.

Olivares, Cerro de (o-lee-VAH-res, SER-ro dai), Andean peak (20,512 ft/6,252 m) on ARGENTINA-CHILE border, 40 mi/64 km WSW of RODEO, Argentina; 30°25′S 69°48′W.

Olivares de Júcar (o-lee-VAH-res dai HOO-kahr), town, CUENCA province, E central SPAIN, 25 mi/40 km SSW of CUENCA; 39°45′N 02°21′W. Olives, saffron, grapes, cereals, potatoes, produce; sheep; apiculture; lumbering.

Olive Branch, town (2000 population 21,054), DE SOTO county, extreme NW MISSISSIPPI, near TENNESSEE state line, suburb 15 mi/24 km SE of MEMPHIS (Tennessee); 34°57′N 89°49′W. In agricultural area (cotton, grain; cattle; dairying); manufacturing (aluminum and metal products, printing and publishing; textiles, machinery, furniture).

Olive Branch, unincorporated village, ALEXANDER county, S ILLINOIS, 13 mi/21 km SE of CAPE GIRARDEAU (MISSOURI), near MISSISSIPPI RIVER. Shawnee National Forest to N; 37°10′N 89°21′W. HORSESHOE LAKE State Conservation Area to S.

Olivebridge, hamlet, ULSTER county, town of Olive, SE NEW YORK, in the CATSKILLS, near ASHOKAN RE-SERVOIR, 11 mi/18 km W of KINGSTON; 41°56′N 74°13′W. Also spelled Olive Bridge.

Olive Bridge Dam, NEW YORK: see ASHOKAN DAM.

Olivebridge Dam, NEW YORK: see ASHOKAN DAM.

Olive Cove, fishing village, SE ALASKA, on N shore of ETOLIN ISLAND, 20 mi/32 km S of Wrangell; 56°11′N 132°19′W.

Olivedos (O-lee-ve-dos), town (2007 population 3,489), central PARAÍBA, BRAZIL, 16 mi/26 km NE of SOLEDADE; 06°59′S 36°15′W.

Olive Hill, town (2000 population 1,813), CARTER county, NE KENTUCKY, 12 mi/19 km W of GRAYSON, on TYGARTS CREEK; 38°17′N 83°10′W. Agriculture, clay, sand and gravel, coal; manufacturing (limestone processing, lumber, apparel, pumps, consumer goods). Northeastern Kentucky Museum to NE, near Carter Caves State Park. Carter Caves State Resort Park to NE; Cascade Caverns State Nature Preserve to NE; Grayson Lake reservoir and State Park to SE; Tygarts State Forest to N; several caves and natural bridges in area.

Olivehurst, unincorporated town (2000 population 11,061), YUBA county, N central CALIFORNIA, 3 mi/4.8 km SSE of MARYSVILLE; 39°05′N 121°33′W. Peaches, prunes, olives, walnuts, kiwi; dairying; cattle; manufacturing concrete. Yuba County Airport is here. Beale Air Force Base to E.

Oliveira (o-lee-vai-rah), old city (2007 population 37,848), S central MINAS GERAIS, BRAZIL, on railroad, 45 mi/72 km NW of SÃO JOÃO DEL REI; 20°36′S 44°46′W. Elevation 3,150 ft/960 m. Manufacturing (textiles, dried meat, lard); coffee and rice processing; dairying.

Oliveira dos Brejinhos (O-lee-VAI-rah do BRAI-zheen-yos), city (2007 population 21,747), W central BAHIA, BRAZIL, near Rio SÃO FRANCISCO in Serra da Macaúbas; 12°21′S 42°56′W.

Oliveira Fortes (o-lee-vai-rah for-chees), town (2007 population 1,934), S MINAS GERAIS, BRAZIL, on railroad, 20 mi/32 km SE of BARBACENA; 21°30′S 43°36′W. Rutile and nickel deposits. Until 1944, called Livramento.

Olivença (O-lee-VEN-sah), city (2007 population 10,549), W central ALAGOAS state, BRAZIL, 11 mi/18 km N of BATALHA; 09°32′S 37°12′W.

Olivença (O-lee-VEN-sah), town on Atlantic coast, BAHIA, BRAZIL, 11 mi/18 km S of ILHÉUS; 14°57′S 39°02′W.

Olivenza (o-lee-VEN-thah), city, BADAJOZ province, W SPAIN, in Estremadura, near Portuguese border, 15 mi/24 km SW of BADAJOZ; 38°41′N 07°06′W. Processing and agricultural center on fertile plain (olives, grapes, cereals, acorns; livestock). Manufacturing of olive oil, flour, liquor, beverages, meat products, tiles, soap. Has notable church. Formerly fortified, it belonged to Portugal until 1801. Local language is a Portuguese dialect.

Oliver, county (☐ 720 sq mi/1,872 sq km; 2006 population 1,808), central NORTH DAKOTA, ⊙ CENTER; 47°06′N 101°20′W. Agricultural area watered by Square Butte and Otter Creek; bounded E by MISSOURI RIVER (on border between Mountain and Central time zones). Lignite mines; farming; cattle; wheat, barley, potatoes. Nelson Lake SE of center; Cross Ranch State Park in E; Fort Clark Historic Site on N boundary. Formed 1885 and government organized the same year. Named for Harry S. Oliver (1855–1909), a Republican politician. County seat was Sanger from 1885–1902.

Oliver (AH-li-vuhr), town (☐ 2 sq mi/5.2 sq km; 2001 population 4,224), S BRITISH COLUMBIA, W Canada, near U.S. (WASHINGTON) border, on OKANAGAN RIVER, 22 mi/35 km S of PENTICTON, in Okanagan-Similkameen regional district; 49°11′N 119°33′W. In irrigated farming region; fruit and vegetable packing, canning; tourism, retirement center. Municipal airport. Incorporated 1945.

Cross-references are shown in SMALL CAPITALS. The pronunciation guide is shown on page xix. The sources of population figures are shown on page xvii.

Oliver (AH-li-vuhr), unincorporated town (2000 population 2,925), North Union township, FAYETTE county, SW PENNSYLVANIA, residential suburb 2 mi/3.2 km N of UNIONTOWN; 39°55′N 79°43′W. Railroad junction.

Oliver, village, SCREVEN county, E GEORGIA, 15 mi/24 km SSE of SYLVANIA; 32°31′N 81°32′W. Manufacturing (wood processing).

Oliver, village (2006 population 423), DOUGLAS county, extreme NW WISCONSIN, 4 mi/6.4 km SW of SUPERIOR, on SAINT LOUIS RIVER (MINNESOTA state line); 46°38′N 92°11′W. Railroad terminus. Grain growing. Mont Du Lac Ski Area to SW.

Oliverea (AHL-iv-ree), hamlet, ULSTER county, town of Shandaken, SE NEW YORK, in the CATSKILL MOUNTAINS, 26 mi/42 km WNW of KINGSTON; 42°04′N 74°28′W.

Oliveria de Azeméis (o-lee-vai-REE-ah dai ah-zai-MAISH), town, AVEIRO district, N central PORTUGAL, 16 mi/26 km NNE of AVEIRO; 40°50′N 08°29′W. Museum.

Oliveria de Frades, town, VISEU district, N central PORTUGAL, on VOUGA RIVER, 14 mi/23 km WNW of VISEU; 40°44′N 08°11′W. Resin extracting.

Oliveria do Bairro (o-lee-vai-REE-ah doo BEI-roo), town, AVEIRO district, N central PORTUGAL, on railroad and 12 mi/19 km SE of AVEIRO; 40°31′N 08°30′W. Vineyards; textile milling, brick manufacturing.

Oliveria do Conde (o-lee-vai-REE-ah doo KOND), village, VISEU district, N central PORTUGAL, on railroad, and 16 mi/26 km S of VISEU; 40°26′N 07°58′W. Sawmilling.

Oliveria do Douro (o-lee-vai-REE-ah doo DO-roo), town, PÔRTO district, N PORTUGAL, SE suburb of OPORTO, on left bank of DUERO RIVER. Footwear, hosiery, paper.

Oliveria do Hospital (o-lee-vai-REE-ah), town, COIMBRA district, N central PORTUGAL, on N slope of SERRA DA ESTRÊLA, 31 mi/50 km ENE of COIMBRA; 40°21′N 07°52′W. Cheese manufacturing. Has 13th–14th-century chapel.

Oliveros (o-lee-VAI-ros), town, SE SANTA FE province, ARGENTINA, 30 mi/48 km NNW of ROSARIO; 32°34′N 60°51′W. Agricultural area (corn, wheat, flax, soybeans; livestock).

Oliver Paipoonge, township (□ 135 sq mi/351 sq km; 2001 population 5,862), NW ONTARIO, E central Canada, 12 mi/19 km from THUNDER BAY; 48°24′N 89°31′W. Agriculture (livestock); meat, cheese processing; fertilizer mixing and soybean roasting plants.

Oliver Springs, town (2006 population 3,319), MORGAN, ROANE, and ANDERSON counties, E TENNESSEE, 23 mi/37 km WNW of KNOXVILLE; 36°03′N 84°20′W. Recreation. TVA windmill farm here.

Olives, Mount of or **Olivet**, ridge, E of JERUSALEM, ISRAEL, mentioned in the Old Testament as the scene of David's flight from the city, Ezekiel's theophany, and Zechariah's prophecy, and in the New Testament as a frequent resort of Jesus and the scene of His Ascension. The principal hill of the mountain is often called "the Ascension." Bethany and Bethphage lie near its foot, and the garden of GETHSEMANE is on the W slope. Mentioned in 2 Samuel 15:30; Ezekiel 11:23; Zechariah 14:4; Matthew 21:1; and Acts of the Apostles 1:12. Most of the W and S slopes are taken up by Jewish cemeteries. Numerous churches, monasteries, and religious institutions.

Olivet (ah-li-VET), town (2000 population 1,758), EATON county, S central MICHIGAN, 16 mi/26 km NE of BATTLE CREEK; 42°27′N 84°55′W. Manufacturing (fabricated metal products). Seat of Olivet College.

Olivet, town (2006 population 65), ⊙ HUTCHINSON county, SE SOUTH DAKOTA, 30 mi/48 km NNW of YANKTON; 43°14′N 97°40′W.

Olivet (ah-li-VET), village (2000 population 64), OSAGE county, E central KANSAS, near MARAIS DES CYGNES

River, 9 mi/14.5 km S of LYNDON, on S side of Melvern Lake Reservoir; 38°28′N 95°45′W. In livestock and grain area.

Olivet (o-lee-ve), S residential suburb (□ 9 sq mi/23.4 sq km) of ORLÉANS, LOIRET department, CENTRE administrative region, N central FRANCE, 3 mi/4.8 km S of city center, on left bank of LOIRET RIVER whose source rises in a small park amidst an array of flower beds; 47°52′N 01°52′E. Tree nurseries and flower gardens.

Olivette, town (2000 population 7,438), SAINT LOUIS county, E MISSOURI, suburb 10 mi/16 km WNW of ST. LOUIS; 38°40′N 90°22′W. Manufacturing (paper products, pharmaceuticals, cosmetics, glass products printing and publishing, building materials, furniture); distillery (makes Southern Comfort). Incorporated 1930.

Olive View, suburban section of LOS ANGELES, LOS ANGELES county, S CALIFORNIA, in SAN FERNANDO VALLEY, just N of SAN FERNANDO, in foothills of SAN GABRIEL MOUNTAINS. Veterans Administration Hospital here. Pacoima Reservoir to NE; Angeles National Forest to N and E.

Olivia, village (2000 population 3,646), FLACQ district, MAURITIUS, 26 mi/41.6 km E of PORT LOUIS. Sugarcane, vegetables, and fruits.

Olivia, village (2000 population 2,570), ⊙ RENVILLE county, SW MINNESOTA, 28 mi/45 km E of GRANITE FALLS; elevation 1,086 ft/331 m; 44°46′N 95°00′W. Agricultural area (grain, soybeans, beans, sugar beets; livestock, poultry; dairying); manufacturing (agricultural products, canned food, food products). Plotted 1878, incorporated 1881.

Olivos (o-LEE-vos), city in Greater BUENOS AIRES, ARGENTINA, on the Rio de la Plata, and 10 mi/16 km NW of Buenos Aires; 34°30′S 58°28′W. Residential suburb, popular beach resort, and diverse manufacturing center. Fishing and yachting ground, with small port facilities. Theater. Country residence of Argentine president.

Oliwa, POLAND: see OLIVA.

Ol Kalou (ol kah-LAH-oo), town, ⊙ Nyandarua district, RIFT VALLEY province, W central KENYA, on railroad spur, and 20 mi/32 km E of NAKURU; 00°17′S 36°23′E. Coffee, wheat, corn.

Olkhivchyk, UKRAINE: see Shakhtars'k.

Ol'khon Island (uhl-HAWN), largest island in Lake BAYKAL, E IRKUTSK oblast, E central Siberian Russia; 46 mi/74 km long, 7 mi/11.3 km wide; 53°09′N 107°24′E. Highest point, Izhimey (4,254 ft/1,297 m). Manganese deposits.

Ol'khovatka (uhl-huh-VAHT-kah), town (2006 population 4,345), S VORONEZH oblast, S central European Russia, on road junction and terminus of railroad spur, 12 mi/19 km WNW of ROSSOSH; 50°17′N 39°16′E. Elevation 364 ft/110 m. Sugar refinery, flour mill.

Ol'khovatka (ol-ko-VAHT-kah), town, E DONETS'K oblast, UKRAINE, in the DONBAS, 9 mi/14.5 km E and under jurisdiction of YENAKIYEVE; 48°15′N 38°25′E. Elevation 807 ft/245 m. Coal mine. Established in early 18th century; town since 1938.

Ol'khovatka, UKRAINE: see VIL'KHUVATKA.

Olkhovchik, UKRAINE: see SHAKHTARS'K.

Ol'khovka (uhl-HOF-kah), village, NW KURGAN oblast, SW SIBERIA, RUSSIA, 20 mi/32 km N of SHADRINSK; 56°21′N 63°46′E. Elevation 446 ft/135 m. Flour mill, dairy plant.

Ol'khovka (uhl-HOF-kah), village (2006 population 5,320), N central VOLGOGRAD oblast, SE European Russia, on the ILOVLYA RIVER (DON River basin), on road and near railroad, 40 mi/64 km SW of KAMYSHIN; 49°52′N 44°34′E. Elevation 252 ft/76 m. In agricultural area (wheat); lumbering and woodworking, road making and repair.

Ol'khovka, UKRAINE: see USPENKA.

Ol'khovskiy, RUSSIA: see ARTËMOVSK.

Olkiluoto (OL-kee-loo-o-to), island, TURUN JA PORIN province, FINLAND, 9 mi/14.5 km N of RAUMA (Nor-

way), 22 mi/35 km S of PORI; 61°14′N 21°29′E. Located in mouth of Eurajoki. Site of nuclear power plant providing c.twenty percent of Finland's electricity; the two reactors (on-line in 1979 and 1982) have combined 1,470 MW capacity; Finland's two other reactors are in LOVISA. Suitability of Olkiluoto, Äänekoski, and Kannonkoski as storage sites for spent nuclear fuel is being studied.

Olkusz (OL-koosh), Russian *Olkush*, town (2002 population 38,173), Katowice province, S POLAND, on railroad, 22 mi/35 km NW of CRACOW, at W foot of CRACOW JURA; 50°17′N 19°34′E. Iron-ore mining; sawmilling, manufacturing of enameled products. Center of one of oldest Polish mining districts; mining of lead ore (containing some silver) flourished in 14th century; site of silver mint in 17th century. Mining declined in 18th century; redeveloped in 19th century following building of local railroad; included zinc mines. At present, lead and zinc mining concentrated near BOLESLAW, Polish *Boleslaw*, 4 mi/6.4 km W of town. Dolomite and lime deposits (used as flux and for cement manufacturing) in vicinity. Castle ruins just NE. In World War II, under German rule, called Ilkenau.

Olla (AH-luh), town (2000 population 1,417), LA SALLE parish, central LOUISIANA, c.45 mi/72 km NNE of ALEXANDRIA; 31°54′N 92°14′W. In agricultural area (cotton, soybeans; cattle, hogs); manufacturing (oil storage tanks, lumber).

Ollachea (o-yah-CHAI-ah), town, CARABAYA province, PUNO region, SE PERU, on N slopes of Cordillera ORIENTAL, on road to Lanlacuni Bajo, 17 mi/27 km NNW of MACUSANI; 13°49′S 70°29′W. Elevation 8,940 ft/2,725 m. Copper and silver mining; potatoes, corn, coca; livestock.

Ollagüe (o-yah-GWAI), village, ⊙ Ollagüe comuna, EL LOA province, ANTOFAGASTA region, N CHILE, at NW foot of Ollagüe Volcano and at the edge of the Salar de Ollagüe, in the ANDES, last Chilean stop on railroad to BOLIVIA, and 95 mi/153 km NE of CALAMA; 21°14′S 68°16′W. Elevation 12,125 ft/3,696 m. Sulphur mining and refining.

Ollagüe, Cerro (o-yah-GWAI, SER-ro), Andean peak (19,260 ft/5,870 m) on CHILE-BOLIVIA border; 21°18′S.

Ollainville (o-lan-VEEL), town (□ 4 sq mi/10.4 sq km), ESSONNE department, ÎLE-DE-FRANCE region, N central FRANCE; 48°35′N 02°13′E.

Ollantaytambo (o-yahn-tei-TAHM-bo) or **Ollantaitambo** (o-yahn-tei-TAHM-bo), town, URUBAMBA province, CUSCO region, S central PERU, in a canyon on URUBAMBA RIVER, on railroad from CUSCO to MACHUPICCHU, and 32 mi/51 km NW of CUSCO; 13°16′S 72°16′W. In agricultural region (cereals, potatoes); silver, gold, copper deposits. Has remains of ancient Inca fortress, famed for its massive structure.

Olleria (o-lyai-REE-ah), town, VALENCIA province, E SPAIN, 6 mi/9.7 km SSW of JÁTIVA; 38°55′N 00°33′W. Manufacturing of glass and crystal, baskets; meat processing; olive oil, wine, cereals, vegetables.

Ollerton (AHL-uh-tuhn), town (2001 population 6,619), central NOTTINGHAMSHIRE, central ENGLAND, on Maun River, and 8 mi/12.9 km NE of MANSFIELD; 53°13′N 01°01′W. Previously coal mines. SHERWOOD FOREST to S.

Ollie, town (2000 population 224), KEOKUK county, SE IOWA, 11 mi/18 km SE of SIGOURNEY; 41°12′N 92°05′W. Limestone quarries nearby.

Ollioules (ol-YOOL), residential town (□ 7 sq mi/18.2 sq km), VAR department, PROVENCE-ALPES-CÔTE D'AZUR region, SE FRANCE, 3 mi/4.8 km WNW of TOULON; 43°08′N 05°51′E. In flower-growing district that exports cut flowers.

Ollon (o-LON), commune, VAUD canton, SW SWITZERLAND, 2 mi/3.2 km SE of AIGLE; 46°18′N 07°00′E. Vineyards and orchards.

Ollur (UH-loor), town, THRISSUR district, KERALA state, S INDIA, suburb (3 mi/4.8 km SW) of THRISSUR

10°28'N 76°14'E. Timber-trade center; sawmilling, rice and oilseed milling, tile manufacturing.

Olmaliq, UZBEKISTAN: see ALMALYK.

Olmeda del Rey (ol-MAI-dhah dhel RAI), village, CUENCA province, E central SPAIN, 18 mi/29 km S of CUENCA; 39°48'N 02°05'W. Cereals, chickpeas, grapes; sheep, goats. Lumbering (pine).

Olmedo (ol-MAI-dho), town, VALLADOLID province, N central SPAIN, 25 mi/40 km S of VALLADOLID; 41°17'N 04°41'W. Manufacturing of consumer goods, footwear; sawmilling. Wine, cereals, fruit in area. Has ancient walls and towers and several churches and former convents. Played notable role in history of Castile and was seat of powerful noble families in Middle Ages.

Olmito (OL-mi-to), unincorporated village, CAMERON county, extreme S TEXAS, 8 mi/12.9 km N of BROWNSVILLE; 26°01'N 97°32'W. In irrigated agriculture area of lower RIO GRANDE valley; manufacturing (radio equipment, boats).

Olmitz (OL-mits), village (2000 population 138), BARTON county, central KANSAS, 14 mi/23 km NW of GREAT BEND; 38°31'N 98°56'W. In wheat region. Gas and oil fields nearby.

Olmos (OL-mos), town, LAMBAYEQUE region, NW PERU, in W foothills of CORDILLERA OCCIDENTAL, at E edge of OLMOS DESERT, on PAN-AMERICAN HIGHWAY, 50 mi/80 km NNE of LAMBAYEQUE; 05°59'S 79°46'W. Corn, cotton, sugarcane; livestock.

Olmos Dam, Texas: see OLMOS PARK.

Olmos Desert (OL-mos), LAMBAYEQUE region, NW PERU, in W foothills of CORDILLERA OCCIDENTAL, N of LA LECHE RIVER, and just E of OLMOS; 25 mi/40 km wide, 35 mi/56 km long. It is continued by SECHURA DESERT (NW) and MÓRROPE DESERT (SW). Some agriculture (corn, cotton) and cattle raising in irrigated areas.

Olmos, Lake (□ 31 sq mi/80.6 sq km), SE CÓRDOBA province, ARGENTINA, formed by the Río CUARTO, 10 mi/16 km E of LA CARLOTA; c.25 mi/40 km long. Saltwater lake.

Olmos Park (OL-mos), city (2006 population 2,303), BEXAR county, S central TEXAS, residential suburb 3 mi/4.8 km N of SAN ANTONIO, near SAN ANTONIO RIVER; 29°28'N 98°29'W. Olmos Dam (flood control) built here across small Olmos Creek (tributary of San Antonio River) after 1921 flood. Trinity University in S. Incorporated 1939.

Olmsted, county (□ 654 sq mi/1,700.4 sq km; 2006 population 137,521), SE MINNESOTA; ⊙ ROCHESTER; 44°00'N 92°24'W. Drained by ROOT RIVER and South and Middle forks of ZUMBRO RIVER. Agricultural area (corn, oats, soybeans, peas, hay, alfalfa; sheep, hogs, cattle, poultry; dairying); manufacturing at Rochester. Mayo Clinic at Rochester. Part of Whitewater Wildlife Area in NE corner; parts of Richard J. Dorer Memorial Hardwood State Forest in NW and S. Formed 1855.

Olmsted, village (2000 population 299), PULASKI county, extreme S ILLINOIS, 13 mi/21 km NNE of CAIRO; 37°10'N 89°04'W. In agricultural area. Near Lock and Dam Number 53 on OHIO River.

Olmsted Air Force Base, Pennsylvania: see MIDDLETOWN.

Olmsted Falls, city (□ 4 sq mi/10.4 sq km; 2006 population 8,333), CUYAHOGA county, N OHIO, 14 mi/23 km SW of CLEVELAND, and on ROCKY RIVER; 41°22'N 81°54'W. Manufacturing.

Olmstedville, hamlet, ESSEX county, NE NEW YORK, in the ADIRONDACK MOUNTAINS, 35 mi/56 km NNW of GLENS FALLS; 43°46'N 73°56'W.

Olmue (ol-MOO-ai), city, ⊙ Olmue comuna, QUILLOTA province, VALPARAISO region, N central CHILE, 12 mi/19 km SSE of QUILLOTA city. Fruit, grapes, vegetables. Originally a *reduccion* of Indians. Resort. Fruit; livestock.

Olmué (ol-moo-AI), town, VALPARAISO province, central CHILE, 25 mi/40 km ENE of VALPARAISO. In

agricultural area. Parque Nacional LA CAMPANA nearby.

Olmütz, CZECH REPUBLIC: see OLOMOUC.

Olna Firth, Scotland: see SWARBACKS MINN.

Olne (AWL-nuh), commune (2006 population 3,783), Verviers district, LIÈGE province, E BELGIUM, 9 mi/14.5 km ESE of LIÈGE; 50°35'N 05°45'E. Textile industry; agriculture.

Olney (AWL-nee), city (2000 population 8,631), ⊙ RICHLAND county, SE ILLINOIS, 30 mi/48 km W of VINCENNES (INDIANA); 38°43'N 88°05'W. Trade and shipping center in agriculture (corn, wheat, apples; livestock); manufacturing (dairy products, vinegar). Olney Central College near here. Famous for white squirrels. Incorporated 1841.

Olney (AWL-nee), town (2001 population 6,032), Milton Keynes, central ENGLAND, on OUSE RIVER, 17 mi/27 km N of LEIGHTON BUZZARD; 52°09'N 00°42'W. Leather tanning, flour milling. Limestone quarries nearby. Home of poet William Cowper, whose house is now museum. Has 14th-century church. Site of Roman building. Annual pancake race on Shrove Tuesday.

Olney (OL-nee), town (2000 population 3,336), YOUNG county, N TEXAS, 40 mi/64 km S of WICHITA FALLS; 33°21'N 98°45'W. Elevation 1,184 ft/361 m. Commercial center for oil and gas; livestock (cattle, sheep, goats, hogs). Agricultural area (wheat, cotton, pecans, nursery crops), oil refining; manufacturing (aircraft components, building materials, clothing, fabricated aluminum). Settled 1891, incorporated 1909.

Olney, residential village (2000 population 31,438), MONTGOMERY county, central MARYLAND, 17 mi/27 km N of WASHINGTON, D.C. 39°09'N 77°05'W. Site of county hospital. It is said to have been named after a town in ENGLAND which was the home of poet William Cowper. The summer theater here (popular with Washington residents) dates back to 1875.

Olney (AWL-nee), N residential section of PHILADELPHIA, PENNSYLVANIA; 40°02'N 75°07'W.

Olney Springs, village (2000 population 389), CROWLEY county, SE central COLORADO, near ARKANSAS RIVER, 10 mi/16 km WSW of ORDWAY; 38°10'N 103°56'W. Elevation 4,391 ft/1,338 m. In irrigated sugar-beet region.

Olocuilta (o-lo-KWEEL-tah), city and municipality, LA PAZ department, S EL SALVADOR, on road, and 11 mi/18 km SE of SAN SALVADOR, on PACIFIC slope of coastal range; 13°34'N 89°07'W. Market center; manufacturing (hats, baskets); grain. Has colonial Spanish church.

Olodio (o-LO-dee-o), town, Bas-Sassandra region, SW CÔTE D'IVOIRE, 22 mi/35 km NNW of TABOU; 04°43'N 07°28'W. Agriculture (manioc, bananas, rice, palm oil).

Olofström (OOL-of-STRUHM), town, BLEKINGE county, S SWEDEN, 20 mi/32 km NE of KRISTIANSTAD; 56°17'N 14°32'E. Manufacturing (motor vehicle bodies, hardware). Power station.

Ölögey, MONGOLIA: see ÖLGIY.

Olokemeji (aw-LO-kai-me-jee), town, OGUN state, SW NIGERIA, on railroad, on Ogun River, 20 mi/32 km NNE of ABEOKUTA. Cotton weaving, indigo dyeing; cacao, palm oil and kernels, cotton. Has agricultural experiment station.

Olombrada (o-lom-BRAH-dhah), town, SEGOVIA province, central SPAIN, 33 mi/53 km N of SEGOVIA; 41°25'N 04°09'W. Cereals, grapes, chickpeas, vetch.

Olomega, Lake (o-lo-MAI-gah) or **Lake Camalotal** (□ c.15 sq mi/39 sq km), SAN MIGUEL department, SE EL SALVADOR, 12 mi/19 km SSE of SAN MIGUEL; 5 mi/8 km long, 2 mi/3.2 km wide; 13°19'N 88°04'W.

Olomouc (O-lo-MOUTS), German *Olmütz*, city (2001 population 102,607), SEVEROMORAVSKY province, CZECH REPUBLIC, N central MORAVIA, on the MORAVA RIVER; 49°35'N 17°15'E. Center of HANA region. Industrial city; manufacturing (machinery, appliances, and food products, especially candy and chocolate). An ancient town, it was once the leading city of

Moravia and was strongly fortified. In 1242, Wenceslaus II of BOHEMIA defeated the Mongol invaders here. Here, in 1469, Matthias Corvinus, king of HUNGARY, had himself crowned king of Bohemia. The city was later held by the Swedes from 1642 to 1650. In 1758, Frederick II besieged it unsuccessfully. An agreement between AUSTRIA and PRUSSIA was signed here in 1850, which dissolved the German Union under Prussia's presidency and restored the GERMAN CONFEDERATION, headed by Austria. Prussia suffered from the "humiliation of Olmütz" until 1866, when it defeated Austria in war. Present-day landmarks include the Cathedral of St. Wenceslaus (begun in the 12th century), the city hall (rebuilt 13th century), and two Gothic churches. Also in the city are the Cyril-Methodius theological faculty (a university, which was founded in 1566), and several libraries. The Marquis de Lafayette was once imprisoned in Olomouc's fortress. Today there are lovely parks and gardens where the fortress formerly stood. Military base.

Olona River (o-LO-nah), c.65 mi/105 km long, LOMBARDY, N ITALY; rises 4 mi/6 km N of VARESE; flows SSE, past LEGNANO and MILAN, to LAMBRO River at Sant' Angelo Lodigiano. At Milan part of its waters are canalized and flow S to Pavia, with a branch past Cortedona, continuing to the PO 4 mi/6 km NE of STRADELLA.

Olonets (uh-LO-nyets), Finnish *Aunus*, city (2005 population 9,930), S Republic of KARELIA, NW European Russia, on the OLONETS ISTHMUS, on the Olonka River, near Lake LADOGA, on railroad branch, 100 mi/161 km SW of PETROZAVODSK, and 90 mi/145 km SE of SORTAVALA; 60°59'N 32°58'E. Center of agricultural area; lumbering, building materials, food industries. Known since 1137. Chartered in 1648.

Olonets Isthmus (uh-luh-NYETS), in Republic of KARELIA, NW European Russia, between lakes LADOGA (W) and ONEGA (E); 100 mi/161 km long, 80 mi/129 km–100 mi/161 km wide; bounded S by the SVIR' River. OLONETS city in the SW portion.

Olongapo (o-LON gah-po), city (□ 40 sq mi/104 sq km; 2000 population 194,260), ZAMBALES province, central LUZON, PHILIPPINES, on E shore of SUBIC BAY, at base of BATAAN Peninsula, 50 mi/80 km WNW of MANILA; 14°50'N 120°16'E. Former site of U.S. naval base, the loss of which has resulted in economic depression in the area.

Olonki (uh-LON-kee), village, W central UST-ORDYN-BURYAT AUTONOMOUS OKRUG, in IRKUTSK oblast, E central SIBERIA, RUSSIA, on the ANGARA RIVER, on highway, 45 mi/72 km SW of IRKUTSK, and 32 mi/51 km SE of CHEREMKHOVO; 52°35'N 103°42'E. Elevation 1,394 ft/424 m. Sawmilling; wood products.

Olonkomi, town (2007 population 8,924), OROMIYA state, central ETHIOPIA, 35 mi/56 km W of ADDIS ABABA; 09°00'N 38°15'E.

Olonne-sur-Mer (ol-luhn–syur–mer), town (□ 17 sq mi/44.2 sq km), VENDÉE department, PAYS DE LA LOIRE region, W FRANCE, near BAY OF BISCAY, 3 mi/4.8 km N of Les SABLES-D'OLONNE; 46°32'N 01°47'W. Saltworks, limekilns. Its port is no longer active.

Ólonos, Mount, Greece: see ERIMANTHOS.

Olopa (o-LO-pah), town, CHIQUIMULA department, GUATEMALA, 16 mi/26 km SE of CHIQUIMULA; 14°41'N 89°21'W. Elevation 4,429 ft/1,350 m. Indian population is Chortí-speaking. Subsistence farming (corn, beans; livestock).

Oloron, Gave d' (do-lo-RON, gahv), river of the W PYRENEES, 75 mi/121 km long, SW FRANCE; formed by the confluence of the Gave d'ASPE and Gave d'OSSAU at OLORON-SAINTE-MARIE; flows NW, past Navarrenx and SAUVETERRE-DE-BÉARN, to the Gave de pau above PEYREHORADE; 43°35'N 01°05'W. Receives SAISON RIVER (left). Its flow is often torrential in the upper reaches where its valley is very narrow.

Oloron-Sainte-Marie (o-lo-RON–sant–mah-ree), town (□ 26 sq mi/67.6 sq km), PYRÉNÉES-ATLANTIQUES department, AQUITAINE region, SW FRANCE, at junction of Gave d'ossau and Gave d'aspe rivers, 14 mi/23 km SW of PAU; 43°10′N 00°45′W. Road center and N terminus of trans-Pyrenean highway to Spain via SOMPORT PASS. Has modern facility supporting aircraft industry of TOULOUSE region. There is also a small textile industry (berets, blankets, and slippers) and a chocolate factory. Town has former 11th–14th-century Gothic cathedral (with fine Romanesque portal). It was the seat of a bishop, 4th–18th century.

Olosega (o-lo-SENG-ah), volcanic island and county, Manua district, AMERICAN SAMOA, S PACIFIC OCEAN, c.60 mi/90 km from TUTUILA. Rises to 2,095 ft/639 m. Mountainous, fertile. Sometimes spelled Olosenga.

Olot (o-LOT), city, GERONA province, NE SPAIN, in CATALONIA, 22 mi/35 km NW of GERONA; 42°11′N 02°29′E. Road center; wool and cotton spinning, tanning, paper- and sawmilling, meat processing, hat manufacturing; woodcarving. Trades in livestock, cereals, potatoes. In region of dormant volcanoes. Completely destroyed (1427) by earthquake. A prosperous textile center until 18th century. Has art school and museum. Carlist headquarters in last Carlist War (19th century).

Olovi (O-lo-VEE), Czech *Olový*, German *bleistadt*, village, ZAPADOCESKY province, W BOHEMIA, CZECH REPUBLIC, in ORE MOUNTAINS, on railroad, 12 mi/19 km WNW of KARLOVY VARY; 50°15′N 12°33′E. Glassworks; lignite mining in vicinity.

Olovyannaya (uh-luh-VYAHN-nah-yah) [Russian= made of tin], town (2005 population 8,130), SE CHITA oblast, S SIBERIA, RUSSIA, on the ONON RIVER (AMUR River basin), on road and branch of the TRANS-SIBERIAN RAILROAD, 120 mi/193 km SE of CHITA; 50°57′N 115°34′E. Elevation 1,929 ft/587 m. In tin-ore area. Lifting equipment, building materials; forestry.

Olpad (OL-pahd), town, SURAT district, GUJARAT state, W central INDIA, 11 mi/18 km NNW of SURAT; 21°20′N 72°45′E. Rice; fishing (pomfrets), cotton ginning.

Olpe (AWL-pe), town, WESTPHALIA-LIPPE, North RHINE-Westphalia, W GERMANY, 15 mi/24 km SE of Lüdenscheid; 51°02′N 06°51′E. Ironworks; copper refining. Lister reservoir 3 mi/4.8 km NNW.

Olpe (OL-pee), village (2000 population 504), LYON county, E central KANSAS, 10 mi/16 km S of EMPORIA; 38°15′N 96°10′W. Cattle; grain.

Olperer (AWL-pe-rer), peak (10,595 ft/3,229 m) of ZILLERTAL ALPS, in TYROL, W AUSTRIA, 8 mi/12.9 km ENE of BRENNER PASS; 47°03′N 11°40′W.

Öls, POLAND: see OLESNICA.

Olsany, CZECH REPUBLIC: see POSTRELMOV.

Olsberg, town, WESTPHALIA-LIPPE, North RHINE-Westphalia, W GERMANY, 9 mi/14.5 km E of MESCHEDE, in the SAUERLAND, and on the RUHR RIVER; 51°21′N 08°30′E. Manufacturing (metal and woodworking, ovens, plastics); health and winter sports resort. Mining museum. Formed 1969 by unification of Bigge and OLSBERG.

Olsburg, village (2000 population 192), POTTAWATOMIE county, NE KANSAS, 15 mi/24 km N of MANHATTAN, E of Tuttle Creek Reservoir; 39°25′N 96°37′W.

Olše River, CZECH REPUBLIC and POLAND: see OLZA RIVER.

Ol'shana, UKRAINE: see VIL'SHANA.

Ol'shanka, UKRAINE: see VIL'SHANKA.

Ol'shans'ke (ol-SHAHN-ske) (Russian *Ol'shanskoye*), town, S MYKOLAYIV oblast, UKRAINE, on the right bank of the Southern BUH RIVER, on road and on railroad, 19 mi/30 km NW of MYKOLAYIV city center and 6 mi/10 km S of NOVA ODESA; 47°12′N 31°47′E. Cement, reinforced concrete products manufacturing, yeast-making, railroad transport shop. Established in 1957, town since 1968.

Ol'shanskoye, UKRAINE: see OL'SHANS'KE.

Ol'shany, UKRAINE: see VIL'SHANY.

Ol Sina, INDONESIA: see OISINA POINT.

Ölsnitz, GERMANY: see OELSNITZ.

Olst, town, OVERIJSSEL province, E NETHERLANDS, on IJSSEL RIVER (car ferry), 6 mi/9.7 km NNW of DEVENTER; 52°21′N 06°07′E. Dairying; cattle; grain, vegetables; manufacturing (brick-making machines).

Olsztyn or **Olsztyn**, city, ☉ OLSZTYN province, NE POLAND, on LYNA (Alle) River, 65 mi/105 km S of KALININGRAD; 53°47′N 20°29′E. Railroad junction; commercial center; popular health resort; cattle market; sawmilling, woodworking; power station. Seat of Roman Catholic bishop of WARMIA. Has fairs; agricultural college. In World War II, c.45% of the city was destroyed. Town hall now houses part of Torun university. Founded 1348 by Teutonic Knights, who built castle here; chartered 1353. Passed 1466 to Poland, 1772 to PRUSSIA; retained by GERMANY after 1920 plebiscite. Entirely resettled by Poles after World War II. Copernicus lived here for some time. In EAST PRUSSIA until 1945.

Olsztyn (OLSH-teen), province (☐ 8,106 sq mi/21,075.6 sq km), NE POLAND; ☉ OLSZTYN (Allenstein). Borders N on RUSSIA, NW on the BALTIC SEA. Agricultural region of low rolling hills, partly forested, and dotted with numerous small lakes. In E part of extensive Masurian Lakes region. Drained by PASLEKA (Passarge), LYNA (Alle), and ANGERAPP rivers. Principal crops are rye, potatoes, oats, barley; livestock. Sawmilling and woodworking are chief industries; fishing. Principal cities are Allenstein, KETRZYN (Rastenburg), DZIALDOWO. In EAST PRUSSIA until 1945; German population was expelled and replaced by Poles. Subsequently briefly called Mazury. In 1950 province was enlarged by small sections of Bydgoszcz and Warszawa provinces.

Olsztyn (OLSH-teen), German *Allenstein,* city (2002 population 173,102), N POLAND. It is a trade, manufacturing, and railroad center. Founded (1348) by the Teutonic Knights, who built its impressive castle, it was ceded to Poland in 1466 and to PRUSSIA in 1772. The city was retained by Germany after a plebiscite in 1920. It suffered heavy damage in World War II and reverted to Poland in 1945.

Olsztynek (OL-shtee-neck), German *Hohenstein*, town, OLSZTYN province, NE POLAND, in Masurian Lakes region, 17 mi/27 km SSW of OLSZTYN (Allenstein); 53°35′N 20°18′E. Grain and cattle market. In 14th century, Teutonic Knights built castle here. In World War I, heavily damaged during battle of TANNENBERG. In EAST PRUSSIA until 1945.

Olt, county, S ROMANIA, in WALACHIA on Bulgarian border; ☉ SLATINA; 44°20′N 24°30′E. Flat terrain; DANUBE River on S border; lakes. Agriculture.

Olta, town (1991 population 2,874), ☉ General Belgrano department (☐ 888 sq mi/2,308.8 sq km), SE LA RIOJA province, ARGENTINA, at E foot of Sierra de los LLANOS, 55 mi/89 km W of SERREZUELA (CÓRDOBA province); 30°37′S 66°16′W. Agricultural center (livestock).

Olten, district, NE SOLOTHURN canton, SWITZERLAND; 47°21′N 07°55′E. Population is largely German-speaking and Roman Catholic, though many Protestants live in main town of OLTEN.

Olten, town (2000 population 16,757), SOLOTHURN canton, N SWITZERLAND, on the AARE RIVER; 47°21′N 07°55′E. Elevation 1,299 ft/396 m. It is an important railroad center at S end of Unter Hauenstein Tunnel, and has manufacturing of machinery, motor vehicles, textiles, and shoes.

Oltenia (ol-TE-nyah), the W part (☐ 9,305 sq mi/24,193 sq km) of WALACHIA, ROMANIA.

Oltenița (ol-te-NEE-tsah), town, CĂLĂRAȘI county, S ROMANIA, on DANUBE River (BULGARIA border) opposite TUTRAKAN, at mouth of ARGEȘ RIVER, 35 mi/56 km SE of BUCHAREST; 44°05′N 26°38′E. Railroad terminus and inland port, trading in fish, fowl, eggs, cheese, and wool; textiles, woodworking, flour milling, tanning, and brick making.

Olte, Sierra de (OL-te, see-YER-rah dai), subandean range in W central CHUBUT province, ARGENTINA, 30 mi/48 km WNW of Paso de los Indios; c.35 mi/56 km N-S; rises to c.3,500 ft/1,067 m.

Olton (OL-tuhn), town (2006 population 2,238), LAMB county, NW TEXAS, on the LLANO ESTACADO, 25 mi/40 km W of PLAINVIEW; 34°10′N 102°08′W. Cattle; sheep; cotton; oil and gas; manufacturing (building materials). Was the capital of Lamb county until 1946.

Oltovsk (uhl-TOFSK), city, S KRASNOYARSK TERRITORY, central SIBERIA, RUSSIA, on right bank of the YENISEY RIVER, 55 mi/89 km N of KRASNOYARSK, on a local spur of the TRANS-SIBERIAN RAILROAD. Industrial center supplying nearby Yenisey goldfields with machinery and industrial equipment. Shipped refined gold and silver. Mining college. Founded in 1929 on site of a 16th-century Russian fur-trading post destroyed by the Tatars, Oltovsk boomed in the 1930s and became a leading economic center of S Siberia. Originally called Slovetsky until 1931, then Kyz-Kurt until 1939. After World War II, the existence and activities of settlements along the Yenisey River N of Krasnoyarsk were shrouded in military secrecy.

Olt River, German *Aluta*, 348 mi/560 km long, in TRANSYLVANIA and WALACHIA, central and S ROMANIA; rises on W slopes of the Moldavian Carpathians, 7 mi/11.3 km E of GHEORGHENI; flows S past MIERCUREA-CIUC and SFÎNTU GHEORGHE, then wends its way W past FĂGĂRAS and cuts S across the TRANSYLVANIAN ALPS through TURNU ROȘU PASS, passing RÎMNICU VÎLCEA and joining the DANUBE just S of TURNU MĂGURELE, opposite NIKOPOL (BULGARIA). Used for logging in upper and middle courses; navigable for small craft in lower course, below Slatina.

Oltu, township, NE TURKEY, on OLTU RIVER, 60 mi/97 km NE of ERZURUM; 40°34′N 41°59′E. Coal; grain.

Oltu River, 80 mi/129 km long, NE TURKEY; rises in Kargapazari Mountains 8 mi/12.9 km SE of TORTUM; flows NE and WNW, past OLTU, to CORUH RIVER 12 mi/19 km SE of YUSUFELI.

Oluanpi, Cape (O-LWAHN-BEE), also called *Nan Chia* (NAN-JIAH), meaning *South Cape* in Chinese, southernmost point of TAIWAN; 21°54′N 120°51′E. Lighthouse (1882), tourist beach. KENTING NATIONAL PARK nearby. Sometimes spelled Olwanpi or Olanpi.

Ólubló, SLOVAKIA: see STARA LUBOVNA.

Olula del Río (o-LOO-lah dhel REE-o), town, ALMERÍA province, S SPAIN, on ALMANZORA RIVER, 18 mi/29 km WSW of HUÉRCAL-OVERA; 37°21′N 02°18′W. Marble cutting and shipping. Olive oil, cereals, esparto.

Olustee (o-LUHS-tee), village (2006 population 635), JACKSON county, SW OKLAHOMA, 8 mi/12.9 km SW of ALTUS, near SALT FORK OF RED RIVER; 34°32′N 99°25′W. In cotton and grain area.

Oluta (o-LOO-tah), town, VERACRUZ, SE MEXICO, on Isthmus of TEHUANTEPEC, 2 mi/3.2 km SE of ACAYUCAN; 17°55′N 94°54′W. Rice, coffee, fruit; livestock. A Popoluca Indian town.

Olutanga Island (o-loo-TAHN-gah) (☐ 78 sq mi/202.8 sq km), ZAMBOANGA DEL SUR province, PHILIPPINES, just off W coast of MINDANAO, across SIBUGUEY BAY from ZAMBOANGA Peninsula; 07°20′N 122°51′E.

Ólvega (OL-vai-gah), town, SORIA province, N central SPAIN, 25 mi/40 km E of SORIA. Livestock raising; meatpacking.

Olvera (ol-VAI-rah), town, CÁDIZ province, SW SPAIN, in spur of the CORDILLERA PENIBÉTICA, 15 mi/24 km NNW of RONDA; 36°56′N 05°16′W. Lumbering and agricultural center (cereals, vegetables, olives, acorns; livestock). Olive-oil pressing. Sulphur springs. Taken (1327) by Alfonso XI.

Olveston (AHL-vuhz-tuhn), village (2001 population 2,021), South Gloucestershire, SW ENGLAND, 9 mi/14.5 km N of BRISTOL; 51°35′N 02°34′W. Former agricultural market. Church dates from 12th century.

Ol'viopol, UKRAINE: see PERVOMAYS'K, Mykolayiv oblast.

Olwanpi, Cape, TAIWAN: see OLUANPI, Cape.

Olwsk, UKRAINE: see OLEVS'K.

Olyka (o-LI-kah) (Polish *Ołyka*), town, E VOLYN' oblast, UKRAINE, 22 mi/35 km E of LUTS'K; 50°43′N 25°49′E. Elevation 590 ft/179 m. Lumber- and grain-trading center; agricultural processing (cereals, vegetable oils, hops), sawmilling, brick manufacturing. Has ruins of a town hall, 16th-century castle. An old Ukrainian settlement; became (16th century) residence of Polish gentry; sacked (1648) by the Cossacks. Passed to Russia in 1795; reverted to Poland in 1921; annexed to Ukrainian SSR in 1939. Jewish community since the 17th century, representing the majority of the population by the beginning of the 20th century; reduced during the civil war pogroms, but still numbering 2,500 in 1939; wiped out by the Nazis in 1941.

Olymbos, Greece: see OLYMPOS.

Olympia, city (2006 population 44,645), ⊙ of WASHINGTON and of THURSTON county, W Washington, 26 mi/42 km SW of TACOMA, a S extremity of PUGET SOUND, on Budd Inlet; 47°02′N 122°54′W. Railroad junction. A port of entry, it ships lumber products and agricultural produce. Oyster fisheries are here; manufacturing (printing and publishing, explosives, consumer goods, sports equipment, plastic bottles, metal cans, plastic products, paper products, veneer, furniture, cheese, steel, aircraft engines, porcelain enamel); logging. State Capital Museum. Settled in 1846, it was made capital of the newly created Washington Territory in 1853. Of interest are the state library, the old capitol building (1893), and the newer, imposing group of white sandstone capitol buildings. Olympia Municipal Airport to S. Fort Lewis Military Reservation to E. Nisqually National Wildlife Reservation to E; Nisqually Indian Reservation to NE. Millersylvania State Park to S; Tolmie State Park to NE. A local attraction is the annual salmon run from Budd Inlet into Capitol Lake St. Martin's College (at Lacey, to E), South Puget Sound Community College and Evergreen State College are in Olympia. The OLYMPIC MOUNTAINS can be seen to the N, and MOUNT RAINIER to the NE. Incorporated 1859.

Olympia (o-LIM-pee-ah), Greek *Olimbéa*, important center of the worship of Zeus in ancient GREECE, in ancient ELIS region on N bank of the Alpheus River, in what is now ILIA prefecture, WESTERN GREECE department, NW PELOPONNESUS, S mainland Greece. Scene of the Olympic Games from 776 B.C.E.–C.E. 393, when Theodosius I banned them; in 426 Theodosius II ordered total destruction of the sanctuary's temples. Modern Olympic Games held throughout the world, revived in 1896 with ATHENS Olympics. The great temple of Zeus was especially celebrated for the ivory, gold-adorned statue of Zeus by Phidias—one of the Seven Wonders of the World. Excavation, which revealed the great temple, also uncovered the Hermes of Praxiteles, several other temples within the sacred enclosure (called the Altis), and the enormous stadium. Museum contains artifacts from Neolithic to Roman periods.

Olympia, Brazil: see OLÍMPIA.

Olympia Fields, village (2000 population 4,732), COOK county, NE ILLINOIS, S suburb of CHICAGO, just W of CHICAGO HEIGHTS; 41°31′N 87°41′W. One of few upscale suburbs S of the city.

Olympian Village, village (2000 population 669), JEFFERSON county, E MISSOURI, residential community 5 mi/8 km E of DE SOTO; 38°07′N 90°27′W.

Olympic Hot Springs, CLALLAM county, NW WASHINGTON, in OLYMPIC MOUNTAINS, OLYMPIC NATIONAL PARK, N of Mount Olympus and 13 mi/21 km SW of PORT ANGELES, 4 mi/6.4 km W of Lake Mills, on Boulder Creek, branch of ELWHA RIVER. Elevation c.2,000 ft/610 m. Former resort.

Olympic Mountains, range, highest part of the COAST RANGES, on the OLYMPIC PENINSULA, NW WASHINGTON. Mount Olympus (7,965 ft/2,427 m) is the highest point in the mountains (has permanent snowcap), which are composed mainly of sedimentary rock. The W side of the mountains is in one of the areas of greatest precipitation in the U.S., with an annual rainfall of c.130 in/330 cm; the NE side, in the rain shadow, is in one of the driest areas on the West Coast. On the upper slopes are about 60 small glaciers fed by heavy winter snows. The greater part of the Olympic Mountains is included in OLYMPIC NATIONAL PARK (□ 922,654 acres/373,398 ha); established 1938. Rugged mountains, alpine meadows, coniferous rain forests, glaciers, lakes, and streams characterize this area. The national park includes a separate 50-mi/80-km stretch of shoreline along the PACIFIC OCEAN that contains scenic seascapes and wildlife. Center of park 50 mi/80 km W of SEATTLE. Temperate rain forests in valleys extending W toward Pacific Ocean; bounded on all sides by parts of Olympic National Forest. Mount Olympus at center of park. JEFFERSON and CLALLAM counties, extends S into MASON and GRAYS HARBOR counties.

Olympic National Park (□ 1,441 sq mi/3,732 sq km), NW WASHINGTON. Rain forests and glaciers in the OLYMPIC MOUNTAINS. Authorized 1938.

Olympic Peninsula, NW WASHINGTON, bounded E by PUGET SOUND and HOOD CANAL, W by PACIFIC OCEAN, N by STRAIT OF JUAN DE FUCA. Includes OLYMPIC MOUNTAINS, OLYMPIC NATIONAL PARK, and Olympic National Forest. CLALLAM, JEFFERSON, and parts of MASON and GRAYS HARBOR counties. Kitsap Peninsula to E, across narrow Hood Canal; city of PORT ANGELES on N shore, main community on peninsula. Makah Indian Reservation at Cape Flattery, NW point; Quinault Indian Reservation in SW.

Olympos (o-LIM-pos), Greek *Ólimbos*, range, N GREECE, on the border of THESSALY and central MACEDONIA departments, near the AEGEAN coast, c.25 mi/40 km long; 40°05′N 22°21′E. Rises to c.9,570 ft/2,917 m at Mount Olympos, the highest point in Greece. Was first ascended in 1913. In Greek mythology the summit, shut from the sight of men on earth by clouds, was the home of the Olympian gods. Later the name was given to the remote heavenly palace of the gods. Also spelled Olymbos. Formerly spelled Olympus.

Olympus, Mount (uh-LIM-puhs) (10,132 ft/3,088 m), SW ALBERTA, W Canada, near BRITISH COLUMBIA border, in ROCKY MOUNTAINS, in JASPER NATIONAL PARK, 50 mi/80 km SE of JASPER; 52°28′N 117°02′W.

Olympus, Mount, peak (6,401 ft/1,951 m), W central CYPRUS, in TROODOS MOUNTAINS, on boundary of LEFKOSIA and LIMASSOL districts, 35 mi/56 km SW of NICOSIA; 34°56′N 32°52′E. Highest point in CYPRUS. Popular mountain resort area; ski resorts and spas, popular walking trails encircle mountain. Several Greek monasteries in area.

Olympus, Mount, AUSTRALIA: see SAINT CLAIR, LAKE.

Olympus, Mount, Greece: see OLYMPOS.

Olympus, Mount, TURKEY: see ULU DAG.

Olym River (uh-LIM), approximately 72 mi/116 km long, in S central European Russia. Rises 10 mi/16 km E of GORSHECHNOYE (KURSK oblast) and flows generally N, serving for the first 12 mi/19 km of its course as a natural border between Kursk and VORONEZH oblasts; continues through NE Kursk oblast, past Olymskiy, Novokastornoye, and KASTORNOYE, and into SW Lipetsk oblast, past Naberezhnoye. For approximately 13 mi/21 km past Pokrovskoye, it serves as a natural border between ORËL and Lipetsk oblasts; feeds into the SOSNA RIVER in W Lipetsk oblast, approximately 21 mi/34 km WSW of YELETS. Used mainly for irrigation and fishing.

Olymskiy (uh-LIM-skeey), settlement (2005 population 3,105), NE KURSK oblast, SW European Russia, on the Olym River (tributary of the SOSNA RIVER), on road and railroad, 19 mi/31 km ESE of KSHENSKIY; 51°46′N 38°07′E. Elevation 639 ft/194 m. In agricultural area (wheat, rye, oats, potatoes, sugar beets).

Olynthos (O-leen-thos), ancient city of GREECE, on the KHALKIDHIKÍ Peninsula, NE of POTIDAIA, in what is now KHALKIDHIKÍ prefecture, CENTRAL MACEDONIA department, NE Greece, at the head of the Gulf of ORFANI between SITHONIA and KASSANDRA peninsulas. A league of Chalcidic cities grew up in the late 5th century B.C.E., and Olynthus, as the head of this Chalcidian League, vigorously opposed the threats of ATHENS and SPARTA. Athens captured the city and held it for a brief time. In 379 B.C.E., Sparta defeated Olynthus and dissolved the league, which was, however, re-formed after the fall of Sparta. Olynthus had been allied with Philip II of Macedon against Athens, but, fearing Philip's power, sought Athenian aid. Philip attacked, and Demosthenes in his Olynthiac orations eloquently urged his fellow Athenians to help the threatened city. Although they did give aid to the city, Philip destroyed it (348 B.C.E.). Excavations at Olynthus have revealed the city's layout.

Olyphant (O-li-fant), borough (2006 population 4,910), LACKAWANNA county, NE PENNSYLVANIA, suburb 6 mi/9.7 km NE of SCRANTON, on LACKAWANNA RIVER; 41°27′N 75°34′W. Manufacturing (lumber, wood products, pharmaceuticals, consumer goods). Archbald Pothole State Park to NE; Moosic Mountains to SE; Bell Mountain to NW. Settled 1858, incorporated 1877.

Olyshivka (o-LI-sheef-kah) (Russian *Olishevka*), town (2004 population 2,830), W central CHERNIHIV oblast, UKRAINE, on the Smolyanka River, tributary of the DESNA River, 18 mi/29 km S of CHERNIHIV; 51°13′N 31°20′E. Elevation 357 ft/108 m. Flax, potatoes; forestry; brickworks. Nearby is the Sosyns'ky forest preserve. Established at the beginning of the 16th century; town since 1967.

Olytsikas, Greece: see TOMAROS.

Olyutorskiy (uh-LYOO-tuhr-skeey), village, E KORYAK AUTONOMOUS OKRUG, RUSSIAN FAR EAST, on the OLYUTORSKIY Gulf of the BERING SEA, at the mouth of the Apuka River, 350 mi/563 km NE of PALANA; 60°26′N 169°40′E. Fishing port. Fisheries; canning plant.

Olyutorskiy Gulf (uh-LYOO-tuhr-skeey), a bay of the BERING SEA, on the NE side of the KAMCHATKA PENINSULA, off the E coast of KORYAK AUTONOMOUS OKRUG. Rich in salmon, herring, and other stock fish species; fishing industry is well developed along its shores, with centers in Pakhachi, OLYUTORSKIY, and Kavacha.

Olza River, Czech *Olše*, c.40 mi/64 km long, S POLAND and SILESIA, CZECH REPUBLIC; rises in Poland in the W BESKIDS, 8 mi/12.9 km E of JABLUNKOV (Czech Republic); flows E and NNW, partly along Czech-Polish border, past CIESZYN, CESKY TESIN, and Frystat, to ODER River just NW of NOVY BOHUMIN. Supplies most of water power for metallurgical industry of OSTRAVA area.

Ol'zheras, RUSSIA: see MEZHDURECHENSK.

Oma (O-mah), town, Shimokita county, Aomori prefecture, at northernmost tip of HONSHU, N JAPAN, 50 mi/80 km N of AOMORI, on E TSUGARU Strait; 41°31′N 140°54′E.

Oma, Cape (O-mah), Japanese *Oma-zaki* (o-MAH-zah-kee), northernmost point of HONSHU, Aomori prefecture, N JAPAN, in TSUGARU Strait, opposite Hakodate peninsula of HOKKAIDO; 41°32′N 140°55′E.

Omachi (O-mah-chee), city, NAGANO prefecture, central HONSHU, central JAPAN, 19 mi/30 km S of NAGANO; 36°29′N 137°51′E. Noodles. Apples, blueberries. Wine. Nishina lakes (three), hot spring, and ski area nearby.

Omachi (O-mah-chee), town, Kishima county, SAGA prefecture, NW KYUSHU, SW JAPAN, 12 mi/20 km W of SAGA, on E HIZEN PENINSULA; 33°12′N 130°07′E.

Cross-references are shown in SMALL CAPITALS. The pronunciation guide is shown on page xix. The sources of population figures are shown on page xvii.

Omaezaki (o-MAH-e-zah-kee), town, Haibara county, SHIBARA prefecture, central HONSHU, E central JAPAN, on Cape Omaezaki, 28 mi/45 km S of SHI-ZUOKA; 34°36′N 138°12′E. Bonito and bonito processing; seaweed (*arame*). Melons; peanuts; baby's breath. Protected sea turtle habitat.

Omagari (O-mah-GAH-ree), city, Akita prefecture, N HONSHU, NE JAPAN, 28 mi/45 km S of AKITA; 39°27′N 140°28′E. Ham and sausage.

Om Ager, Eritrea: see OMHAJER.

Omagh (O-muh), Gaelic *An Ómaigh*, town (2001 population 19,836), TYRONE, W Northern Ireland, on the Strule River; 54°36′N 07°18′W. Market. Retail center. Bogs here are the site of undeveloped gold and silver veins.

Omaha, city (2006 population 419,545), ⊙ DOUGLAS county, E NEBRASKA, on the W bank of the MISSOURI RIVER, across the river from COUNCIL BLUFFS, Iowa, and 50 mi/80 km NE of LINCOLN; 41°15′N 95°56′W. The largest city in the state, it is a busy port of entry and a major transportation center. It is also one of the largest livestock markets in the world and a market for agricultural products. Much of the city's industry is devoted to food processing, although it is a diverse manufacturing center. Omaha is the home of many insurance companies, telecommunications firms, and a center for medical treatment and research.

Founded when the Nebraska Territory was opened to settlement in 1854, it grew as a supply point for westward migration and became a thriving transportation and industrial center after the arrival of the Union Pacific Railroad in 1865, and its transcontinental link-up with the Central Pacific in 1869. It was the territorial capital from 1855 to 1867. The Trans-Mississippi and International Exhibition, a world's fair, was held here in 1898. Fort Omaha (built 1868) serves as headquarters of the naval reserve and marine reserve training command.

BOYS TOWN is 9 mi/14.5 km to the W, and OFFUTT AIR FORCE BASE, headquarters of the Strategic Command, is to the S. Some river and local flash flooding occurred during floods of 1993. President Gerald Ford's birthplace, including his home and rose garden, is here. The city has noted park and school systems and is the seat of Creighton University, the University of Nebraska at Omaha, University of Nebraska Medical Center, Nebraska Christian College, and the College of St. Mary. Of interest are the Joslyn Art Museum, Durham Western Heritage Museum, an aerospace museum, a Mormon cemetery, Fontenelle Forest, Henry Doorly Zoo, and Rosenblatt Stadium (baseball facility which has hosted the College World Series since 1950). Omaha Home for Boys, Nebraska School for the Deaf. Eppley Airfield (main airport) 4 mi/6.4 km NE of downtown. Omaha Livestock Market in SE part of city (Union Stockyards). Barge dock on Missouri River. Incorporated 1857.

Omaha (O-mah-ha), town (2006 population 984), MORRIS county, NE TEXAS, c.45 mi/72 km WSW of TEXARKANA; 33°10′N 94°44′W. Watermelons; peanuts; poultry; cattle; timber.

Omaha, village (2000 population 165), BOONE county, N ARKANSAS, 15 mi/24 km NNW of HARRISON, near MISSOURI state line; 36°27′N 93°11′W. Manufacturing (charcoal). Recreation area. Arm of TABLE ROCK reservoir to W; BULL SHOALS LAKE to E.

Omaha, village, STEWART county, SW GEORGIA, 14 mi/23 km WNW of LUMPKIN, near CHATTAHOOCHEE RIVER; 32°09′N 85°01′W.

Omaha, village (2000 population 263), GALLATIN county, SE ILLINOIS, 16 mi/26 km NNW of SHAWNEETOWN; 37°53′N 88°17′W. In agricultural area.

Omaha Beach, name given to stretch of NORMANDY coast, CALVADOS department, BASSE-NORMANDIE region, NW FRANCE, between GRANDCAMP-MAISY (W) and PORT-EN-BESSIN (E), NW of BAYEUX, where units

of American First Army landed on June 6, 1944, in invasion of France during World War II. Heavy fighting especially at COLLEVILLE-SUR-MER (monument to engineer brigade) and SAINT-LAURENT-SUR-MER (U.S. military cemetery); 49°22′N.

Omaha Reservation (1990 population c.1,500), mainly in SE THURSTON county, NE NEBRASKA. A patchwork of tribal and allotted land comprises the reservation. In 1855, the Omaha ceded extensive hunting grounds in exchange for a reservation. In 1865, the N half was sold to the Winnebago tribe through the U.S. government. Following authorization by President Chester Arthur in 1882, c.118 sq mi/308 sq km were allotted and 86 sq mi/224 sq km were reserved for future generations. The remainder was given to whites. Subsequently, Omaha land has been much reduced, with little allotted land remaining (most of which is leased to non-Indians). Commonly the leases permit the Omaha family to live in a house on the land, and there is some acquired tribal land. The majority of the Omaha population live in villages, especially MACY, where the Omaha agency is located. Many more live off the reservation, mainly in cities, some distant. The Omaha hold an annual powwow in Macy in late August.

Omaheke Region (o-muh-HAI-kai), administrative division (2001 population 68,039), E central NAMIBIA, stretching to BOTSWANA border, centered around GOBABIS; 22°00′S 19°10′E. Rich cattle country and game-ranching area in NE end of the KALAHARI DESERT.

Omai (o-MEI), gold mine, CUYUNI-MAZARUNI district, N central GUYANA, on left bank of ESSEQUIBO River, 70 mi/113 km S of ROCKSTONE; 05°26′N 58°45′W. Largest open-pit mine in SOUTH AMERICA. One of Guyana's largest sources of revenue. Owned by Cambior (CANADA) and Golden Star Resources (COLORADO).

Omak, town (2006 population 4,751), OKANOGAN county, N WASHINGTON, on OKANOGAN RIVER at mouth of Omak Creek, 4 mi/6.4 km NE of OKANO-GAN; 48°25′N 119°32′W. Apples; lumber, wood products; manufacturing (millwork, printing and publishing); logging. Fish hatchery. Omak Lake (on No Name Creek) to SE; large Colville Indian Reservation to E, across river. Saint Mary's Mission to SE. Incorporated 1911.

Omaké, town, BENIN; 09°30′N 01°35′E. Cotton; livestock; shea-nut butter.

Omalur (O-muh-loor), town, SALEM district, TAMIL NADU state, S INDIA, 10 mi/16 km NW of SALEM, on railroad branch to Mettur Dam; 11°44′N 78°04′E. Cotton weaving; steatite culinary vessels. Magnesite and chromite mines in CHALK HILLS (SE); magnetite, magnesite, and chromite mines in KANJAMALAI hill (S).

Omama (O-mah-mah), town, Yamada county, GUMMA prefecture, central HONSHU, N central JAPAN, 12 mi/20 km E of MAEBASHI; 36°25′N 139°16′E.

Oman, ISRAEL: see EVEN SHMUEL.

Oman (o-MAHN), sultanate (□ c.119,498 sq mi/309,500 sq km; 2003 population 2,340,815; 2007 estimated population 3,204,897), SE Arabian peninsula, on the GULF OF OMAN and the ARABIAN SEA; ⊙ MUSCAT. It was formerly known as Muscat and Oman; in ancient times known as Muzoon or Mugan.

Geography
It is bordered on the SW by YEMEN, W and SW by SAUDI ARABIA, and on the NW by the UNITED ARAB EMIRATES, which separate the major portion of the sultanate from a small area on the Strait of HORMUZ. The largest city is Muscat. Oman fronts on the Gulf of Oman between the OMAN PROMONTORY (MUSANDAM Peninsula) and the cape RAS AL HADD and on the Arabian Sea from the cape Ras al Hadd and the YEMEN border; its shoreline is 1,056 mi/1,700 km long. Physiographically, the country is dominated by the

HAJAR hill region, which extends along the entire coast of the Gulf of Oman and rises to 9,900 ft/3,000 m in the JABAL AKHDAR. Between the Western Hajar and the Gulf of Oman lies the fertile and populous BATINA plain, where most of the towns are located; the EASTERN HAJAR hills (186 mi/300 km long; 9 mi/15 km–11 mi/18 km wide) approach closer to the coast SE of Muscat. At the N tip of the Oman Promontory is the detached region of MUSANDAM, separated from the rest of the sultanate by the United Arab Emirates. On the landward side of the Hajar lies a sandy tableland (elevation 4,000 ft/1,219 m) extending to the margins of the desert RUB' AL KHALI. Off the desolate Arabian Sea coast are MASIRA ISLAND and the KURIA MURIA ISLANDS. In the extreme S, the DHOFAR region extends 249 mi/400 km between Ras NAUS and Ras DHARBAT 'ALI, on the Yemen border.

Climate
The climate of the gulf coastal belt is relatively pleasant November–March, when the temperature falls to 60°F/15.5°C–70°F/21°C, but high humidity is coupled with high temperatures (to 110°F/43°C) during the rest of the year. On the Dhofar coast, the temperature is pleasant, and the Hajar hill country experiences a fresher, temperate climate. Between June and September is the monsoon season. Rainfall averages 5 in/12.7 cm annually, and agriculture depends on irrigation.

Population and Education
The inhabitants are mostly Arabs; there are also minorities of Pakistanis, Indians, black Africans, Baluchis, and migrant workers of varied ethnicities. The main language is Arabic-Swahili, spoken by Omanis who emigrated from ZANZIBAR; English is widely spoken by the Comazari tribe (1995 population 3000) living in a secluded area in Musandam Cape near the STRAIT OF HORMUZ. Since 1970, immigration has increased, although the number of non-Omanis (27.7% in 1993) is still less that other Gulf countries. Sultan Qabus University is in Muscat, and technical and science institutes are scattered around the sultanate.

Economy
In the extreme N, dates, limes, nuts, and vegetables are cultivated, and in the SW there is an abundance of cattle and other livestock. Fishing is an important industry. There is a small, but slowly growing, banking sector, and tourism is flourishing. The major product, however, is oil, which was discovered in A 'dakilia governate in Oman in 1964 and first exported in 1967. Natural-gas production and small copper mines developed in the early 1980s and are a part of Oman's growing industries.

History: to 1981
Much of the coast of Oman was controlled by POR-TUGAL from 1508 until their expulsion by the Ya'ariba tribe in the 17th century. In 1659 the Ottoman Empire took possession. The Ottoman Turks were driven out in 1741 by Ahmad ibn Said of Yemen, who founded the present royal line. In the late 18th century, Oman began its close ties with Great Britain, which still continue. In the early 19th century, Oman was the most powerful state in ARABIA, controlling ZANZIBAR and neighboring areas of the E African coast and much of the coast of IRAN and BALUCHISTAN. Zanzibar was lost in 1856, and the last Omani hold on the Baluchistan coast, GWADAR, was ceded to Pakistan in 1958. The sultan of Oman has had frequent clashes with the imams (leaders) of the interior ethnic groups. In 1957 the groups revolted but were suppressed with British aid. In 1965 the UN called for the elimination of British influence in Oman. In 1970, Sultan Said ibn Timur was deposed by his son, Qabus bin Said, who promised to use oil revenues for modernization. Rebel activity continued, however, particularly in Dhofar, in the S, where a Chinese-aided liberation front was strong. Oman joined the UN and the Arab League in 1971, but

it did not become part of the Organization of Petroleum Exporting Countries (OPEC).

History: 1981 to Present

In May 1981, Oman joined Arab Persian Gulf nations and Saudi Arabia in founding the Gulf Cooperation Council (GCC). Relations between Oman and the U.S. have been particularly close since the 1970s. As a result of the Iraqi invasion of KUWAIT in August 1990, the country opened its bases to international coalition forces against Iraq in the 1991 Persian Gulf War and to U.S. forces mounting strikes against Afghanistan and Osama bin Laden in 2001. In 1996 the sultan announced a new basic law that provided for a legislature with limited powers and guaranteed basic civil liberties for Omani citizens. The government has ambitious environmental programs; many wildlife reserves exist (including Jedat Al Harsees protected area, Ras Al Hadd reserve, and Al Demanat reserve); recycling has started in some towns.

Government

Oman does not have a constitution, but the Basic Law, which was promulgated by royal decree in 1996, is considered by the government to be a constitution. The monarch is both head of state and head of government. The bicameral legislature, the Majlis Oman, consists of the 58-seat Majlis al-Dawla, or upper house, whose members are appointed by the monarch, and the 84-seat Majlis al-Shura, or lower house, whose members are popularly elected to serve four-year terms. The legislature is mainly an advisory body. The head of state (since 1970) is Sultan and Prime Minister Qaboos bin Said Al Said. Administratively, the country is divided into five regions (Al BATINA, A'DAKHLIYAH, A'SHARQIYAH, Al WUSTA, and Az Zahirah) and four governates (Al Buryam, Dhofar, Musandam, and Muscat).

Oman, Gulf of (o-MAHN), NW arm of ARABIAN SEA, between OMAN section of Arabian Peninsula and MAKRAN coast of IRAN; 350 mi/563 km long, 200 mi/322 km wide between Ras al HADD and Gwatar Bay. Ports of MUSCAT and MATRAH are on S coast, JASK and CHAHBAHAR on N coast. Connects NW with PERSIAN GULF via Strait of HORMUZ.

Oman Promontory (o-MAHN), N projection of Oman region of Arabian Peninsula, between the PERSIAN GULF and Gulf of OMAN, terminating in rocky Cape MUSANDAM on Strait of HORMUZ. Except for MUSANDAM (at N point), a detached region of OMAN sultanate, Oman Promontory belongs to the emirates of the UNITED ARAB EMIRATES. SHARJA and DUBAI are on W coast.

Oman Proper (o-MAHN), former district of OMAN, on interior side of the JABAL AKHDAR. Main towns are NIZWA, BAHLA, and IZKI. Until transfer of sultan's residence to MUSCAT and coastal belt in 1741, Oman Proper was principal seat of political power and most prosperous region in Oman.

Omao (O-MAH-o), town (2000 population 1,221), KAUAI island, KAUAI county, HAWAII, 8 mi/12.9 km WSW of LIHUE, 2 mi/3.2 km inland from S coast, on Omao Stream, on Kaumualii Highway; 21°55′N 159°29′W. Sugarcane, fruit. Lihue-Koloa Forest Reserve to N.

Omar, unincorporated town, LOGAN county, SW WEST VIRGINIA, 7 mi/11.3 km S of LOGAN. In coal region.

Omaruru (o-mah-ROO-roo), town, NW NAMIBIA, on OMARURU RIVER, and 110 mi/177 km NW of WINDHOEK; 21°26′S 15°56′E. Mining region (tin, gold). During Herero campaign German forces were besieged here (1904).

Omaruru River (o-mah-ROO-roo), central NAMIBIA; rises near Omatako Mountain, 50 mi/80 km S of OTJIWARONGO; flows WSW to Henties Bay. Episodic river providing water to Omaruru and Henties Bay. Name means "sour milk" in Otjiherero.

Omasuyos, province, LA PAZ department, W BOLIVIA, ⊙ ACHACACHI; 15°55′S 68°50′W.

Omate (o-MAH-tai), town, ⊙ GENERAL SÁNCHEZ CERRO province, MOQUEGUA region, S PERU, at S foot of the Huaina Putina or Omate volcano, on affluent of TAMBO RIVER, and 30 mi/48 km NNE of MOQUEGUA, on road to AREQUIPA; 16°41′S 70°59′W. Elevation 7,086 ft/2,160 m. Cereals, fruit; livestock.

Omba, VANUATU: see AOBA, island.

Ombai, INDONESIA: see ALOR ISLANDS.

Ombai Strait (om-BEI), channel, connecting BANDA SEA (NE) and SAVU SEA (SE), between TIMOR (S) and ALOR ISLANDS (N); 20 mi/32 km–50 mi/80 km wide; 08°30′S 125°00′E. PANTE MAKASAR is on S shore. Also called Matua Strait.

Ombalantu (OM-buh-LAHN-too), NW NAMIBIA, 50 mi/80 km ENE of OSHAKATI, 10 mi/16 km from ANGOLA border; 17°30′S 15°00′E. Trading center on paved road to Rualana, junction with S road from Tsandi.

Ombella-M'Poko, prefecture (□ 12,288 sq mi/31,948.8 sq km; 2003 population 356,725), SW CENTRAL AFRICAN REPUBLIC; ⊙ BANGUI. Bordered N by OUHAM prefecture, E by KÉMO prefecture, SE by Democratic Republic of the CONGO, S by LOBAYE prefecture, and W by MAMBÉRÉ-KADÉÏ and NANA-MAMBÉRÉ prefectures. Drained by UBANGI, M'bali, and M'POKO rivers. Agriculture (coffee, cotton); manufacturing (textiles); copper and iron deposits. Main centers are Bangui, Bodanga.

Ombella River (om-be-LAH), c.100 mi/161 km long, in E OMBELLA-M'POKO prefecture, S CENTRAL AFRICAN REPUBLIC; rises 40 mi/64 km WSW of SIBUT; flows generally SE to UBANGI RIVER 20 mi/32 km WSW of POSSEL. Also called Yambéré.

Ombessa (awm-BES-uh), town, Central province, CAMEROON, 55 mi/89 km NNW of Yaoundé; 04°35′N 11°17′E.

Ombi Islands, INDONESIA: see OBI ISLANDS.

Ombilin, INDONESIA: see UMBILIN.

Ombo (AWM-baw), island (□ 22 sq mi/57.2 sq km) in BOKNAFJORDEN, ROGALAND county, SW NORWAY, at mouth of Jøsenfjorden, 25 mi/40 km ESE of HAUGESUND; 6 mi/9.7 km long (WSW-ESE), 4 mi/6.4 km wide. Fishing; fruit.

Ombos, ancient city, S EGYPT, on the E bank of the NILE River, in ASWAN province, just N of KOM OMBO. It was strategically located on top of a hill. The city attained great importance under the Ptolemies, who built there a mighty temple complex dedicated to the crocodile-headed god Suchos and the falcon-headed Haroeris.

Omboué (OM-boo-ai), town,, OGOOUÉ-MARITIME province, W GABON, on the ATLANTIC, on Fernan-Vaz lagoon (also called Nkomi lagoon), 70 mi/113 km SSE of PORT-GENTIL; 01°35′S 09°15′E. In lumbering area; noted for its mahogany and *okumé* wood; palm-oil milling. Formerly called Fernan-Vaz.

Ombrone River (om-BRO-ne), 100 mi/161 km long, TUSCANY, central ITALY; rises in S Monti CHIANTI 10 mi/16 km ENE of SIENA; flows generally SW, past BUONCONVENTO, to TYRRHENIAN Sea 10 mi/16 km SW of GROSSETO. Chief tributaries include Orcia River (left) and Arbia River (right).

Ombúes de Lavalle (om-BOO-es dai lah-VAH-yai), town, COLONIA department, SW URUGUAY, 38 mi/61 km N of Colonia; 33°55′S 57°47′W. Agricultural center (livestock; wheat, corn, tobacco, fruit). Founded 1890.

Omdurman (om-duhr-MAN), Arabic *Umm Durman*, city, central SUDAN, on the NILE River near the confluence of the WHITE NILE and BLUE NILE rivers opposite KHARTOUM; 15°39′N 32°29′E. It is the largest city and chief commercial center of SUDAN and part of a tri-city metropolitan area (with KHARTOUM and KHARTOUM NORTH) that forms the country's industrial and cultural heart. Industries include textile, clothing, leather tanning, furniture and pottery making, food and agricultural products processing. In 1884 the Mahdi made his military headquarters at the village of Omdurman. After the Mahdist forces destroyed KHARTOUM (1885), the Mahdi's successor, Khalifa Addallah, made Omdurman his capital, and the city grew rapidly as the site of the Mahdi's tomb. The battle of Karari, which took place (1898) near Omdurman, marked the defeat of the Mahdist state in the SUDAN by the Anglo-Egyptian army of Lord Kitchener. Although most of the city was destroyed after the battle, the Mahdi's tomb has been restored and embellished. The Khalifa's former residence is now a museum.

Omealca (o-me-AHL-kah), town, VERACRUZ, E MEXICO, in SIERRA MADRE ORIENTAL foothills, on Blanco River, 14 mi/23 km SE of CÓRDOBA; 18°45′N 96°46′W. Coffee, sugarcane, fruit.

Omega (o-MAI-guh), town (2000 population 1,340), TIFT county, S central GEORGIA, 9 mi/14.5 km SW of TIFTON; 31°20′N 83°36′W. In farm area.

Omegna (o-MAI-nyah), town, NOVARA province, PIEDMONT, N ITALY, port at N end of Lake of ORTA, 17 mi/27 km SSE of DOMODOSSOLA; 45°53′N 08°24′E. Metalworks (iron, steel), aluminum factory, tanneries, cotton and paper mills.

Omei, CHINA: see EMEI.

Omemee (o-MEE-mee), former village (□ 1 sq mi/2.6 sq km; 2001 population 1,319), S ONTARIO, S central Canada, near S end of PIGEON LAKE, 12 mi/19 km W of PETERBOROUGH; 44°18′N 78°33′W. Dairying; mixed farming. Amalgamated into Kawartha Lakes when that city was created in 2001.

Omemee (o-MEM-ee), village, BOTTINEAU county, N NORTH DAKOTA, 9 mi/14.5 km SSE of BOTTINEAU; 48°41′N 100°20′W. Railroad junction. Founded 1890 and incorporated in 1902. Named for OMEMEE, CANADA. Post office closed in 1967.

Omena (o-MEE-nah), village, LEELANAU county, NW MICHIGAN, 10 mi/16 km N of TRAVERSE CITY, on NW shore of GRAND TRAVERSE BAY; 45°03′N 85°35′W. Has Native American cemetery. Resorts. Winery.

Omeo (O-mee-o), village, E central VICTORIA, SE AUSTRALIA, 150 mi/241 km ENE of MELBOURNE; 37°06′S 147°36′E. Sheep, cattle; access to high plains grazing; timber. Old gold-mining center; tourism.

Omer (O-mer), township, ISRAEL, 3.7 mi/6 km NE of BEERSHEBA; 31°17′N 34°51′E. Elevation 1,112 ft/338 m. Developed from a communal settlement that was founded in 1949 into an upper-middle-class neighborhood of Beersheba. Many residents are professionals who work at the nearby city's university and hospital. High-tech industrial park.

Omer (AH-mer), village (2000 population 337), ARENAC county, E MICHIGAN, 7 mi/11.3 km NE of STANDISH, and on RIFLE RIVER near its mouth on SAGINAW BAY; 44°02′N 83°51′W. In farm area; fishing; manufacturing (aluminum doors, concrete). Settled 1873, incorporated as city 1903.

Omereque (o-me-RE-kai), canton, CAMPERO province, COCHABAMBA department, central BOLIVIA, 25 mi/40 km SE of MIZQUE, on the N bank of the MIZQUE RIVER; 18°06′S 64°53′W. Elevation 7,356 ft/2,242 m. Clay, limestone, and gypsum deposits. Agriculture (potatoes, yucca, bananas, corn, rye, peanuts); cattle for dairy products and meat.

Omerville (o-mer-VEEL), former town, S QUEBEC, E Canada; 45°17′N 72°06′W. Amalgamated into MAGOG in 2002.

Ometepec (o-ME-te-pek), city and township, ⊙ Ometepec municipio, GUERRERO, SW MEXICO, in PACIFIC lowlands, 100 mi/161 km E of ACAPULCO DE JUÁREZ; 16°41′N 98°24′W. Agricultural center (cereals, sugarcane, tobacco, cotton, fruit, livestock).

Ometepec Bay (o-ME-te-pek), small inlet of GULF OF CALIFORNIA, on NE coast of LOWER CALIFORNIA, NW MEXICO, near mouth of COLORADO RIVER, 85 mi/137 km SE of MEXICALI. Large deposits of sea salt.

Ometepe Island (o-mai-TAI-pai), SW NICARAGUA, in LAKE NICARAGUA, 5 mi/8 km offshore, 8 mi/12.9 km

ENE of RIVAS; 11°30'N 85°35'W. Consists of two islands connected by 2–mi/3.2–km isthmus. Larger (N) is 12 mi/19 km long, 10 mi/16 km wide; smaller (S) is circular (7 mi/11.3 km across). Rises to 5,066 ft/1,544 m in volcano CONCEPCIÓN (N), to 4,350 ft/1,326 m in volcano MADERA (S). Coffee, tobacco. Main centers: ALTAGRACIA, MOYOGALPA.

Omhajer (oom-HAH-juhr), Italian *Om Ager* or *Om Hager*, village, GASH-BARKA region, W ERITREA, near Ethiopian and Sudanese borders, above SETIT (Tekeze) River, and 50 mi/80 km S of Tessenei; 14°19'N 36°42'E. Elevation c.1,800 ft/549 m. Border crossing point; cotton ginning. Also spelled Om Hājer or Umm Hajar.

Om Hājer, Eritrea: see OMHAJER.

Omi, former province in S HONSHU, JAPAN; now SHIGA prefecture.

Omi (O-mee), town, Sakata county, SHIGA prefecture, S HONSHU, central JAPAN, 34 mi/55 km N of OTSU; 35°20'N 136°18'E. Floss.

Omi (O-mee), village, E Chikuma county, NAGANO prefecture, central HONSHU, central JAPAN, 16 mi/25 km S of NAGANO; 36°27'N 138°02'E. Chinese quinces.

Omīdīyeh (uhm-eed-ee-YAI), town (2006 population 58,616), Khuzestān province, IRAN 8 mi/13 km W of AGHA JARI, in oil-producing region; 30°45'N 49°44'E.

Omigawa (o-MEE-gah-wah), town, Katori county, CHIBA prefecture, E central HONSHU, E central JAPAN, on TONE RIVER, 22 mi/35 km N of CHIBA; 35°50'N 140°36'E. Sweet potatoes. Also spelled Omikawa.

Omihachiman (O-mee-hah-CHEE-mahn), city, SHIGA prefecture, S HONSHU, central JAPAN, near E shore of LAKE BIWA, 16 mi/25 km N of OTSU; 35°07'N 136°06'E. Rice; pearl farming and processing. Tile, pottery, reed products (screens). *Konnyaku* (paste made from devil's tongue). Traditional streetscape at town of OMI.

Omin, CHINA: see EMIN.

Omiš (O-meesh), Italian *Almissa*, town, S CROATIA, in DALMATIA, port on ADRIATIC SEA, at CETINA RIVER mouth, 14 mi/23 km ESE of SPLIT; 43°28'N 16°42'E. Seaside resort; electrochemical industries; manufacturing (phosphates, artificial fertilizers, flour products); wine trade. DUGI RAT, village, is 2 mi/3.2 km W; manufacturing (calcium carbide, cyanamid). Hydroelectric power plant (built 1961) at Zakučac, serveral miles upstream, on Cetina River.

Omišalj (O-mee-shahl-ye), Italian *Castelmuschio*, village, W CROATIA, on KRK Island, on ADRIATIC SEA, 13 mi/21 km N of KRK. Seaside resort. Has terminal for oil refinery in RIJEKA (Fiume). Has old church, castle ruins. Rijeka International Airport nearby.

Omishima (O-MEE-shee-mah), town, Ochi county, EHIME prefecture, W SHIKOKU, W JAPAN, 31 mi/50 km N of MATSUYAMA; 34°14'N 132°59'E. Mandarin oranges.

Omi-shima (o-MEE–shee-mah), island (□ 26 sq mi/67.6 sq km), EHIME prefecture, W JAPAN, in W HIUCHI SEA off N coast of SHIKOKU, N of Cape OSUMI, 10 mi/16 km N of IMABARI; 6 mi/9.7 km long, 3.5 mi/5.6 km wide; broad peninsula in SW. Mountainous; rises to 1,430 ft/436 m. Fishing. Site of ancient Shinto shrine.

Omi-shima, island (□ 7 sq mi/18.2 sq km), YAMAGUCHI prefecture, W JAPAN, in SEA OF JAPAN, just off SW HONSHU, N of Senzaki; 5 mi/8 km long, 2 mi/3.2 km wide; long, narrow E peninsula. Mountainous; scenic. Produces citrus fruit; fishing. Known for ancient temple. Also called Aomi-shima.

Omitlán de Juárez (o-mee-TLAHN dai HWAH-res), town, HIDALGO, central MEXICO, 6 mi/9.7 km NE of PACHUCA de SOTO on Mexican Highway 105; 29°19'N 99°37'W. Corn, maguey, beans, livestock.

Omiya (O-MEE-YAH), city, SAITAMA prefecture, E central HONSHU, E central JAPAN, 3.1 mi/5 km N of URAWA; 35°54'N 139°37'E. Medicines; cameras; bonsai plants. Known for ancient Shinto shrine dedicated to Susano-wo (brother of the sun goddess).

Omiya (O-mee-yah), town, Naka county, IBARAKI prefecture, central HONSHU, E central JAPAN, 12 mi/20 km N of MITO; 36°32'N 140°24'E.

Omiya, town, Naka county, KYOTO prefecture, S HONSHU, W central JAPAN, 53 mi/85 km N of KYOTO; 35°34'N 135°05'E. *Tango* crepe.

Omiya, town, Watarai county, MIE prefecture, S HONSHU, central JAPAN, 25 mi/40 km S of TSU; 34°21'N 136°25'E.

Omkarji, INDIA: see GODARPURA.

Omlouj, SAUDI ARABIA: see UMM LAJJ.

Ommaney, Cape, SE ALASKA, S tip of BARANOF ISLAND, on W shore of S entrance to CHRISTIAN SOUND, 7 mi/11.3 km S of PORT ALEXANDER; 56°10'N 134°40'W.

Ommen (O-muhn), town, OVERIJSSEL province, E NETHERLANDS, on VECHT RIVER, and 14 mi/23 km E of ZWOLLE; 52°31'N 06°26'E. Regge River enters 1 mi/1.6 km to N. Agriculture (dairying; cattle; grain, vegetables); manufacturing (trailers). Established 1248.

Ömnögovï [Mongolian=south Gobi], province (□ 63,700 sq mi/165,620 sq km; 2000 population 46,858), S MONGOLIA; ⊙ DALANDZADGAD; 43°00'N 104°00'E. Bounded S by CHINA's INNER MONGOLIA AUTONOMOUS REGION, it lies in the GOBI DESERT and is the mostly sparsely settled province of Mongolia. Also spelled Umuni Gobi or Umnegov.

Omo (O-mo), town, ENUGU state, S NIGERIA, on road, 30 mi/48 km WNW of ENUGU; 06°31'N 06°58'E. Market town. Cashews, cocoyams, cassava, rice.

Omø (O-muh), island (□ 2 sq mi/5.2 sq km), DENMARK, in SMÅLANDSFARVAND strait, just SW of SJÆLLAND island; 55°09'N 11°10'E.

Omoa (o-MO-ah), town, CORTÉS department, NW HONDURAS; minor port on OMOA BAY of GULF OF HONDURAS, 8 mi/12.9 km WSW of PUERTO CORTÉS; 15°46'N 88°02'W. Coastal trade; coconuts, rice, fruit; livestock. Site of San Fernando fort (completed 1795; now a state prison). Founded 1752; flourished until rise of Puerto Cortés (1870).

Omoa, Scotland: see CLELAND.

Omoa Bay (o-MO-ah), NW HONDURAS, S inlet of GULF OF HONDURAS, SW of PUERTO CORTÉS; 12 mi/19 km wide between MOTAGUA RIVER mouth (W) and OMOA (E); 6 mi/9.7 km long; 15°45'N 88°10'W.

Omoa, Sierra de (o-MO-ah, see-ER-rah dai), N section of CORDILLERA DEL MERENDÓN system, NW HONDURAS; extends c.30 mi/48 km from SANTA BÁRBARA department border NE to GULF OF HONDURAS near PUERTO CORTÉS; forms left watershed of CHAMELECÓN RIVER; 15°35'N 88°12'W. Rises to 7,310 ft/2,228 m at Cerro San Ildefonso (in CUSUCO NATIONAL PARK).

Omodeo, Lake, ITALY: see TIRSO, LAKE.

Omodhos (o-MO-[th]os), village, LIMASSOL district, SW central CYPRUS, 14 mi/23 km NW of LIMASSOL, in S foothills of TROODOS MOUNTAINS; 34°51'N 32°48'E. Wine production; fruit, nuts; livestock. Mountain resort area.

Omogo (o-MO-go), village, Kamiukena county, EHIME prefecture, NW SHIKOKU, W JAPAN, 19 mi/30 km S of MATSUYAMA; 33°41'N 133°02'E. Plums.

Omoko (o-MO-ko), town, Rivers state, S NIGERIA, in NIGER RIVER delta region, 55 mi/89 km NW of PORT HARCOURT; 05°21'N 06°39'E. Market center. Fish; oil palms, rice.

Ómoldova, ROMANIA: see MOLDOVA-VECHE.

Omolon River (uh-muh-LON), 715 mi/1,151 km long, N MAGADAN oblast, W CHUKCHI AUTONOMOUS OKRUG, NE SAKHA REPUBLIC, RUSSIAN FAR EAST; rises in the KOLYMA RANGE; flows N to the KOLYMA River, 60 mi/97 km SW of NIZHNE-KOLYMSK (Sakha Republic).

O Mon (O MON), town, CAN THO province, S VIETNAM, near Song Hau, in MEKONG delta, 12 mi/19 km NW of CAN THO; 10°07'N 105°38'E. Market center; rice-growing area. Formerly Omon.

Omo Nada, village (2007 population 3,756), OROMIYA state, SW ETHIOPIA; 07°38'N 37°15'E. Near OMO RIVER, at the foot of MOUNT MAI GUDO, 40 mi/64 km ESE of JIMMA. Also called Nada.

Omo National Park (O-mo), park, SOUTHERN NATIONS state, SW ETHIOPIA, 120 mi/193 km W of ARBA MINCH; 06°00'N 35°50'E. Known for large African game.

Omonogawa (o-MO-no-GAH-wah), town, Hiraka county, Akita prefecture, N HONSHU, NE JAPAN, 34 mi/55 km S of AKITA city; 39°17'N 140°25'E. Watermelon.

Omori (O-mo-ree), town, Hiraka county, Akita prefecture, N HONSHU, NE JAPAN, 31 mi/50 km S of AKITA city; 39°20'N 140°26'E.

Omo River, c.500 mi/805 km long, central and S ETHIOPIA; rises in highlands S of MOUNT GOROCH'AN 40 mi/64 km ENE of NEKEMTE; flows generally S to LAKE TURKANA. Receives GOJEB and GIBE rivers.

Omorphita, CYPRUS: see KAIMAKLI.

Omotego (o-mo-te-GO), village, W Shirakawa county, FUKUSHIMA prefecture, N central HONSHU, NE JAPAN, 50 mi/80 km S of FUKUSHIMA city; 37°03'N 140°18'E.

Ompompanoosuc River (AHM-pahm-pan-oo-suhk), c.25 mi/40 km long, E VERMONT; rises in E ORANGE co.; flows S to the CONNECTICUT River near NORWICH.

Om' River (OM), 475 mi/764 km long, N NOVOSIBIRSK and SE OMSK oblasts, Siberian Russia; rises in VASYUGANYE marshes, near border of Novosibirsk and TOMSK oblasts; flows W, past KUYBYSHEV, UST'-TARKA, and KALACHINSK (Omsk oblast), to the IRTYSH River at OMSK. Navigable for 200 mi/322 km above its mouth.

Omro (AHM-ro), town (2006 population 3,298), WINNEBAGO county, E central WISCONSIN, on FOX RIVER, and 10 mi/16 km W of OSHKOSH; 44°02'N 88°44'W. Manufacturing (business forms, polyethylene tubing). Settled 1845; incorporated as village in 1857, as city in 1944.

Omsk, oblast (□ 54,473 sq mi/141,629.8 sq km; 2006 population 2,046,515) in SW SIBERIAN Russia; ⊙ OMSK. Forest-tayga in the N, forest-steppe with fertile black-earth soil in the S. Drained by the middle IRTYSH River, which flows S-N through the oblast and is navigable throughout. Continental climate, with warm summers and snowy winters; average temperature −2°F/−19°C (January) and 66°F/19°C (July). Annual precipitation varies between 12–16 in and 30–41 cm. Most advanced agro-industrial sector in Siberia. Wheat on S steppe, dairy farming on central wooded steppe; flax grown around TARA; also, vegetables, sunflower seeds. The TRANS-SIBERIAN RAILROAD crosses the oblast E-W (two branches join at Omsk). Lumbering in N forests. Manufacturing (concentrated at Omsk) includes oil refining and petrochemicals, producing tires, synthetic rubber, polystyrene; also machines, metal goods. Plans developed to convert local military enterprises to civilian use. Urban centers include Omsk (largest), ISILKUL, KALACHINSK, Nazyvaevsk, Tara, Tukalinsk; 67% of the population is urban. Formed 1934 out of West Siberian Territory. Major portion of original area organized into TYUMEN oblast (1944).

Omsk, city (2006 population 1,127,865), ⊙ OMSK oblast, SW SIBERIA, RUSSIA, at the confluence of the IRTYSH and Om' rivers, on the TRANS-SIBERIAN RAILROAD, 1,580 mi/2,543 km E of MOSCOW; 55°00'N 73°24'E. Elevation 265 ft/80 m. The second-largest city in Siberia (after NOVOSIBIRSK). It is a major river port and produces agricultural machinery, television sets, tape recorders, gas equipment, motors, rocket engines, electric appliances, and railroad equipment. There are also oil refineries supplied by pipelines from the West Siberian basin, and woodworking, rubber, and plastic industries. Factories in Omsk also produce footwear, clothing, carpets. Educational and cultural center, with numerous vocational and higher-education institutions, theaters, and museums. Architectural landmarks include the Nikol'skiy Cossack cathedral (built in 1840), former palace of governor-general (1861), former Cadet pavilion (1826, renovated in 1879), and hotel "Rossiya" (1906). Founded as a fortress in 1716, Omsk became city in 1782, and a major transportation and administrative center in the 19th century and a place of detention for political exiles. Industries developed rapidly after the relocation of a large number of war plants to the city from the European part of the country in 1941, ahead of advancing German armies.

Omsukchan (uhm-sook-CHAHN), town (2006 population 4,095), central MAGADAN oblast, E RUSSIAN FAR EAST, in the Omsukchan Mountains (N extension of the KOLYMA RANGE), on the Sugoy River (tributary of the KOLYMA River), terminus of a highway connecting it to MAGADAN, 260 mi/418 km to the SW; 62°32′N 155°48′E. Elevation 1,709 ft/520 m. Mining center (gold, silver, iron ore, quartz).

Omu (O-moo), town, Abashiri district, Hokkaido prefecture, N JAPAN, 133 mi/215 km N of SAPPORO; 44°34′N 142°57′E. Dairying; salmon, crab.

Omulew River (o-MOO-lev), Russian *Omulev*, 65 mi/105 km, NE POLAND; rises in Lake OMULEW 11 mi/18 km NNE of NIDZICA; flows SE to Narew River opposite OSTROLEKA.

Omu Peak (O-moo), highest mountain (elev. 8,236 ft/2,510 m) in the BUCEGI group, S central ROMANIA, 6 mi/9.7 km NW of SINAIA.

Omura (O-moo-rah), city, NAGASAKI prefecture, W KYUSHU, SW JAPAN, port on OMURA BAY, 12 mi/20 km N of NAGASAKI; 32°53′N 129°57′E. Nagasaki Airport (world's first marine airport) located here. Aircraft plant here during World War II; bombed 1944.

Omura Bay (O-moo-rah), Japanese *Omura-wan* (o-MOO-rah-WAHN), landlocked gulf of EAST CHINA SEA, W HIZEN Peninsula, NAGASAKI prefecture, W KYUSHU, SW JAPAN; sheltered W by SONOGI Peninsula; 20 mi/32 km long, 8 mi/12.9 km wide. Connected N to SASEBO harbor by two narrow channels.

Omurtag (o-muhr-TAHG), city (1993 population 8,865), RUSE oblast, Omurtag obshtina, E central BULGARIA, on the N slope of the LISA MOUNTAINS, 28 mi/45 km WSW of SHUMEN; 43°06′N 26°25′E. Market center; tobacco; sawmilling, canning, manufacturing (parquet, metal construction parts, motor vehicle parts, textiles). Until 1934, called Osman Pazar.

Omusati Region (o-moo-SAH-tee), administrative division (2001 population 228,842), N NAMIBIA north of ETOSHA NATIONAL PARK to Angolan border W of KUNENE REGION; 18°00′S 15°05′E. Third-smallest political region with high population density. Mopane savanna grassland.

Omuta (O-moo-tah), city, FUKUOKA prefecture, W KYUSHU, SW JAPAN, a port on the AMAKUSA SEA, 40 mi/65 km S of FUKUOKA; 33°01′N 130°26′E. Chemical products (agricultural chemicals), medicine, dyestuff; coal, aluminum. Mandarin oranges. Nori; fish (*mutsugoro*). Also spelled Omuda.

Omutinskiy (uh-moo-TEEN-skeeye), town (2006 population 9,580), S TYUMEN oblast, W SIBERIA, RUSSIA, on the VAGAY River (tributary of the IRTYSH River), on road junction and the TRANS-SIBERIAN RAILROAD (Omutinskaya station), 105 mi/169 km SE of TYUMEN; 56°28′N 67°40′E. Elevation 321 ft/97 m. Oil and gas processing, woodworking, auto repair; agricultural products. Founded in 1913 in conjunction with railroad construction; town since 1959.

Omutninsk (uh-moot-NEENSK), city (2005 population 25,135), E KIROV oblast, E European Russia, on spur railroad (Stal'naya station), and Omutnaya River near its confluence with VYATKA River, 120 mi/193 km E of KIROV; 58°40′N 52°11′E. Elevation 646 ft/196 m. Chemical and metallurgical center, producing high-grade steels; woodworking, timbering; food processing (bakery, meat packing, produce preserving; livestock veterinary station). The metallurgical plant and the settlement were founded in 1773. Made city in 1921.

On, EGYPT: see HELIOPOLIS.

Oña (O-nyah), town, BURGOS province, N SPAIN, in the EBRO RIVER valley, 31 mi/50 km NNE of BURGOS; 42°44′N 03°24′W. Fruit, vegetables; livestock. Sawmilling; manufacturing of resins, turpentine. Known for Jesuit San Salvador College, founded as a monastery in 10th century by Benedictines; one of the most famed in Spain in Middle Ages.

Oña (ON-yah), village, AZUAY province, S ECUADOR, in the ANDES Mountains, on PAN-AMERICAN HIGHWAY, 10 mi/16 km NNE of SARAGURO, and 60 mi/97 km S of CUENCA; 03°27′S 79°10′W. Elev. 9,580 ft/2,920 m. Cereals, potatoes; livestock.

Onaga (o-NAIG-uh), village (2000 population 704), POTTAWATOMIE county, NE KANSAS, on Vermillion Creek, 14 mi/23 km NE of WESTMORELAND; 39°29′N 96°10′W. Trade and shipping point in livestock and grain region. Manufacturing (clothing).

Onagawa (o-nah-GAH-wah), town, Oshika county, MIYAGI prefecture, N HONSHU, NE JAPAN, at base of Ojika Peninsula, on the PACIFIC OCEAN, 34 mi/55 km N of SENDAI; 38°26′N 141°26′E. Railroad terminus. Oysters.

Onakawana (o-nah-kuh-WAH-nuh), unincorporated village, NE ONTARIO, E central Canada, on ABITIBI RIVER, and 110 mi/177 km NNW of COCHRANE; 50°35′N 81°26′W. Lignite deposits.

Onalaska (AHN-uh-LAS-kah), city (2006 population 16,186), LA CROSSE county, W WISCONSIN, on MISSISSIPPI RIVER (MINNESOTA state line), suburb 6 mi/9.7 km N of downtown LA CROSSE; 43°53′N 91°13′W. In farm and dairy area. Dairy products; manufacturing (consumer goods, printing, printed circuit boards, machinery); summer resort. Hamlin Garland born nearby. Lock and Dam No. 7 and La Crosse Airport to E. Interstate 90 bridge to Minnesota. Settled 1854, incorporated 1887.

Onalaska (O-nuh-las-kuh), village (2006 population 1,533), POLK county, E TEXAS, 45 mi/72 km SSW of LUFKIN, on NE shore of LAKE LIVINGSTON (TRINITY RIVER); 30°48′N 95°06′W. Residential and recreational community in timber- and oil-producing area.

Onamia (o-NAI-mee-yuh), village (2000 population 847), MILLE LACS county, E central MINNESOTA, 30 mi/48 km SE of BRAINERD, on RUM RIVER, at its exit from Lake Onamia, 4 mi/6.4 km S of MILLE LACS LAKE; 46°04′N 93°40′W. Agriculture (grain; livestock, poultry; dairying); manufacturing (lumber and wood chips, custom circuit hybrid manufacturing); timber. Mille Lacs Kathio State Park and part of Mille Lacs Indian Reservation to NW; part of Rum River Wildlife Area to W; Mille Lacs Wildlife Area to SE.

Onancock (o-NAN-kawk), town (2006 population 1,448), ACCOMACK county, E VIRGINIA, 4 mi/6 km W of ACCOMAC, on Onancock River estuary, arm of CHESAPEAKE BAY; 37°42′N 75°44′W. Manufacturing (seafood processing, commercial printing); trade center for farming region (vegetables, grain, fruit; poultry, livestock); timber (pine). Ferry to TANGIER ISLAND. Founded 1680.

Onangué, Lake (O-nahn-gai), lake, MOYEN OGOOUÉ province, W GABON, 20 mi/32 km SW of LAMBARÉNÉ; 00°58′S 10°05′E. About 15 mi/24 km wide, 20 mi/32 km long; drains into the OGOOUE River.

Onanole, community, SW MANITOBA, W central Canada, just S of RIDING MOUNTAIN NATIONAL PARK, and in Park rural municipality; 50°37′N 99°58′W.

Onaping Lake (O-nuh-ping), SE central ONTARIO, E central Canada, 40 mi/64 km NW of SUDBURY; 35 mi/56 km long, 3 mi/5 km wide; 47°08′N 81°31′W. Drains S into WANAPITEI LAKE.

Onaqui Mountains (O-nuh-kee), range (6,000 ft/1,829 m–9,000 ft/2,743 m) in TOOELE county, W UTAH; forms S extension of STANSBURY MOUNTAINS. The ranges (two) separated by Johnsons Pass (6,237 ft/1,901 m). RUSH VALLEY to E.

Onarga (o-NAHR-gah), village (2000 population 1,438), IROQUOIS county, E ILLINOIS, 15 mi/24 km WSW of WATSEKA; 40°42′N 88°00′W. In rich agricultural area (corn, wheat, soybeans; livestock; nursery stock); canned foods. Incorporated 1867.

Ona River, RUSSIA: see BIRYUSA RIVER.

Oñate (on-YAH-tai), city, GUIPÚZCOA province, N SPAIN, 18 mi/29 km SW of TOLOSA; 43°02′N 02°24′W. Metalworking (wire, electrical equipment); other manufacturing (dyes, furniture). Formerly seat of university. Was headquarters of Don Carlos in Carlist Wars (19th century). Aranzanzu monastery is 4 mi/6.4 km S.

Onavas (o-NAH-vahs), town, SONORA, NW MEXICO, on YAQUI RIVER, and 100 mi/161 km ESE of HERMOSILLO; 28°28′N 109°32′W. Cereals, fruit; livestock.

Onawa, city (2000 population 3,091), ☉ MONONA county, W IOWA, near the MISSISSIPPI RIVER, 37 mi/60 km SSE of SIOUX CITY; 42°01′N 96°05′W. Trade center for grain. Agriculture (livestock; dairying); manufacturing (farm equipment, leather goods). Lewis and Clark State Park is W. Incorporated 1858.

Onawa (AHN-uh-wah), village, PISCATAQUIS county, MAINE, 14 mi/23 km NNW of DOVER-FOXCROFT, on Onawa Lake.

Onaway (AHN-ah-wai), town (2000 population 993), PRESQUE ISLE county, NE MICHIGAN, 20 mi/32 km WSW of ROGERS CITY; 45°21′N 84°13′W. In lake-resort and farm area (livestock; grains, potatoes; dairy products). Fishing on nearby BLACK LAKE. State forest nearby and Onaway State Park, on E end of Black Lake, to N. Settled c.1880; incorporated as village 1899, as city 1903.

Onaway, village (2000 population 230), LATAH county, N IDAHO, 16 mi/26 km NNE of MOSCOW, borders POTLATCH, near PALOUSE RIVER; 46°56′N 116°54′W. Timber; cattle, sheep. St. Joe National Forest to N; Coeur d'Alene Indian Reservation to N.

Onbetsu (ON-bets), town, Kushiro district, Hokkaido prefecture, N JAPAN, 130 mi/210 km E of SAPPORO; 42°53′N 143°56′E.

Oncativo (on-kah-TEE-vo), town, central CÓRDOBA province, ARGENTINA, 50 mi/80 km SE of CÓRDOBA; 31°55′S 63°40′W. Manufacturing (food processing, flour milling); agriculture (grain, flax, soybeans, peanuts, potatoes; livestock).

Onda (ON-dah), city, Castellón de la Plana province, E SPAIN, near MIJARES RIVER, 13 mi/21 km WSW of CASTELLÓN DE LA PLANA (linked by branch railroad); 39°58′N 00°15′W. Important ceramics center. Rice, oranges, wine, olive oil in area. Mineral springs. Clay quarries and bituminous schists nearby. Irrigation reservoir on Mijares River.

Ondal, INDIA: see ANDAL.

Ondangwa (on-dahn-gwuh), town, traditional; ☉ OVAMBOLAND, N NAMIBIA, near ANGOLA border; 17°58′S 16°01′E. Administrative center of OSHANA REGION. Center of cattle-raising, grain-growing, trading region. Airfield and large hospital.

Ondara (on-DAH-rah), town, ALICANTE province, E SPAIN, 15 mi/24 km SE of GANDÍA; 38°50′N 00°01′E. Olive-oil processing; raisins, oranges, cereals.

Ondárroa (on-DAH-ro-ah), town, VIZCAYA province, N SPAIN; fishing port on BAY OF BISCAY, 26 mi/42 km ENE of BILBAO; 43°19′N 02°25′W. Fish processing (anchovies), manufacturing of fishing supplies, boatbuilding, cider distilling. Bathing resort.

Ondava River (on-DAH-vah), 91 mi/146 km long, VYCHODOSLOVENSKY province, E SLOVAKIA; rises on S slope of the BESKIDS, 6 mi/9.7 km NE of BARDEJOV; flows S, joins LATORICA river in TISA LOWLANDS, 11 mi/18 km SSE of TREBIŠOV, to form BODROG RIVER; receives TOPLA RIVER Domasa Dam (Slovak *Domaša*) 5 mi/8 km S of STROPKOV.

Onda Verde (ON-dah VER-zhe), town (2007 population 3,736), N SÃO PAULO state, BRAZIL, 17 mi/27 km N of são josé do rio preto; 20°37′S 49°18′W. Coffee growing.

Onderneeming (awn-duhr-NEE-meeng), village, po-meroon-supenaam district, N guyana; 07°06′N 58°29′W. Located on coast at mouth of essequibo River, 30 mi/48 km NW of georgetown, and 1 mi/1.6 km S of suddie.

Onderstepoort, N suburb of tshwane, gauteng province, south africa; 25°35′S 28°11′E. Site of Veterinary Research Institute; established 1908 on 17,290-acre/7,000-ha farm. N of Bon Accord Dam. Largest veterinary school in Africa.

Ondjiva (on-JEE-vuh), town, provincial capital of cu-nene, SSW angola, near namibia border, 210 mi/338 km SE of Lubango. Formerly called Vila Pereira de Eça (VEE-lah pei-REI-rah de ES-sah), Vila Pereira d'Eça, and Ngiva (or N'Giva).

Ondo (ON-do), one of former western provinces, S nigeria; now part of ondo state.

Ondo (ON-do), state (□ 5,639 sq mi/14,661.4 sq km; 2006 population 3,441,024), SW nigeria, on the Bight of benin; ⊙ akure; 07°10′N 05°15′E. Bordered N by ekiti state, NE by kogi state, E by edo state, and SE tip by delta state, S by Bight of Benin, W by ogun state, and NW by osun state. In rain forest zone. Idanre and Oka hills and Ebomi lake are here. Agriculture (cashews, coffee, rubber, kola nuts, maize, cocoyams, cassava, rice, and beans; livestock; manufacturing (glass, electronics, beverages, brass and bronze work; palm oil mills; cocoa processing, wood processing and carving). Main centers are akure, Ayede-Ogebese, Igbokoda, Ile-Oluji, and okitipupa. Ikogosi Warm Springs is in Ondo state. Ekiti state carved out of N section in 1996.

Ondo (ON-do), city, ondo state, SW nigeria. Market center for a cacao- and timber-producing region and has rice mills and sawmills. Ondo formerly was a capital of a Yoruba kingdom.

Ondo (ON-do), town on kurahashi-jima island, Aki county, hiroshima prefecture, W japan, opposite kure (linked by Ondo Bridge), 16 mi/25 km S of hiroshima; 34°11′N 132°32′E. Oysters; fish processing. Crepe; fishnet parts.

Ondores (on-DO-res), town, junín province, junín region, central peru, in Cordillera central of the andes, on Lake junín, and 12 mi/19 km WNW of junín; 11°04′S 76°08′W. Cereals, potatoes; livestock. Salt mining nearby. Near reserva nacional junín.

Öndörhaan (UHN-duhr-KHAHN), town (2000 population 18,003), ⊙ hentiy province, E central mongolia, on kerulen River, and 180 mi/290 km E of ulaanbaatar; 47°19′N 110°39′E. Elevation 3,370 ft/1,027 m. Highway and trading center; coal mining. Until 1931, called Tsetsen Khan.

Ondozero (uhn-DO-zee-ruh), Finnish Ontajärvi, lake (□ 65 sq mi/169 sq km) in central Republic of karelia, NW European Russia, 60 mi/97 km SW of belomorsk; 15 mi/24 km long, 5 mi/8 km wide; 63°48′N 33°20′E. Frozen November through April; outlet is the Onda River (E), an affluent of vyg river.

O'Neals, unincorporated village, madera county, central california, 26 mi/42 km NNE of fresno. Dairying; cattle, poultry; nuts, fruit, honey, grain. Manufacturing (concrete). Sierra National Forest to NE; millerton lake reservoir and State Recreation Area to S.

One and a Half Degree Channel, seaway of indian ocean, roughly along 01°30′N, between hadhdhun-mathi (N) and huvadhu (S) atolls of maldive islands; c.60 mi/97 km wide.

Oneco (O-ni-ko), village, windham county, E connecticut, 20 mi/32 km NE of norwich, near rhode island border. A postal section of sterling.

Onega (uh-NYE-gah), city (2005 population 22,500), NW archangel oblast, N European Russia, maritime lumber-shipping port on the white sea, at the mouth of the onega River, on crossroads and railroad

spur, 165 mi/266 km WSW of archangel; 63°55′N 38°04′E. Sawmilling; plywood, wood chemicals. Known since the 16th century. Chartered in 1780.

Onega (uh-NYE-gah), river, approximately 260 mi/418 km long; rises in Lake lacha, archangel oblast, NW European Russia; flows N into the onega bay of the white sea, SW of archangel. It is navigable (May–November) except for the rapids in its middle course.

Onega Bay (uh-NYE-gah), S inlet of the white sea, W of the onega peninsula, off archangel oblast and Republic of karelia, in NW European russia; 30 mi/48 km–50 mi/80 km wide, 100 mi/161 km long; receives kem, vyg, and onega rivers; 64°20′N 36°30′E. Main ports: kem, belomorsk (N end of White Sea–Baltic Canal), onega. Solovetskiye Islands are at its entrance.

Onega, Lake (uh-NYE-gah), Finnish *Aäninen*, Russian *Onezhskoye Ozero*, lake (surface □ 3,819 sq mi/9,890 sq km), NW European Russia, in Republic of karelia, between Lake ladoga and the white sea; 61°40′N 35°30′E. Elevation 114 ft/35 m. The second-largest lake in Europe, it is 150 mi/241 km long with a maximum width of 60 mi/97 km and a maximum depth of 394 ft/120 m. The lake is located at the S edge of the heavily glaciated Baltic Shield. Its shores are low and sandy in the S, rocky, indented, and covered with forest in the N. It is frozen from November to May. The lake receives the Vytegra and the Vodla rivers and drains SW through the Svir' River into Lake Ladoga. The baltic-white sea canal has its S terminus at Povenets on the lake's N shore. petrozavodsk is the chief city and port on Lake Onega. Parallel to the S shore of the lake runs the Onega Canal, 45 mi/72 km long, which joins the Svir and Vytegra rivers and forms part of the volga-baltic waterway. Fresh-water salmon population in the lake has been restored to the levels to allow licensed private fishing again beginning in 2005. More than 800 prehistoric drawings have been discovered along the lake's shores, pointing at its being an area of early human settlement.

Onega Peninsula (uh-NYE-gah), N archangel oblast, NW European Russia, on the white sea, between onega and dvina bays; 15 mi/24 km–60 mi/97 km wide, 80 mi/129 km long. Marshy lowland. Fisheries along the coast.

One Hundred and Two River, c.90 mi/145 km long, in SW iowa and NW missouri; rises in S iowa (S adams county); flows S to little platte river 6 mi/9.7 km E of saint joseph. Also called Hundred and Two.

One Hundred Mile House or **100 Mile House** (wuhn HUHN-dred MEIL HOUS), district (□ 20 sq mi/52 sq km; 2001 population 1,739), S central british columbia, W Canada, 185 mi/297 km NE of vancouver, in Cariboo Regional District; 51°39′N 121°17′W. The area's many lakes and wetlands are part of the Fraser Basin system. Named for its location 100 miles from Lillooet (mile o) of the Cariboo Wagon Road.

Onehunga (o-nee-HUNG-ah), S suburb, and minor tasman sea port of auckland City, north island, new zealand, on NE shore of shallow, bar-bound manukau harbour. Woolen mills, tanneries, sawmills; fishing.

Oneida (o-NEI-duh), county (□ 1,201 sq mi/3,122.6 sq km; 2006 population 4,176), SE idaho; ⊙ malad city; 42°13′N 112°31′W. Mountain region bordering on utah and crossed in SE by malad river and tributaries. Cattle; agriculture (wheat, alfalfa, barley). Part of Caribou National Forest in E; part of Sawtooth National Forest in NW corner; part of Curlew National Grassland in center of county. Formed 1864.

Oneida, county (□ 1,257 sq mi/3,268.2 sq km; 2006 population 233,954), central new york; ⊙ utica; 43°14′N 75°26′W. Partly bounded W by oneida lake; rises to the adirondack mountains in E and NE; drained by mohawk and black rivers, and by Oneida, oriskany, and west canada creeks. Tra-

versed by the barge canal. Extensive manufacturing (variety of industrial wire products, fishing tackle, ornamental ironwork, tools, dies and machinery) at Rome, Utica; dairying, livestock-raising, farming; limestone quarries and lime and lime products. Historically known for silverware production. Casino gambling in verona. Year-round recreation on lakes and in Adirondacks. Formed 1798.

Oneida, county (□ 1,236 sq mi/3,213.6 sq km; 2006 population 36,779), N wisconsin; ⊙ rhinelander; 45°42′N 89°31′W. Largely a resort area with numerous lakes; contains part of American Legion State Forest in N and a section of Nicolet National Forest in E corner. Lumbering; dairying and farming (mostly brans) on cutover forest land. Bearskin State Trail (N-S) in W part. Drained by wisconsin river. Formed 1885.

Oneida, city (2000 population 752), knox county, NW central illinois, 10 mi/16 km NE of galesburg; 41°04′N 90°13′W. In agricultural area.

Oneida, city (□ 22 sq mi/57.2 sq km; 2006 population 10,935), madison county, central new york; 43°04′N 75°39′W. Silverware is its best-known product; factories also make cutlery, industrial wire and cable, paper and plastic goods. Nearby was the Oneida Community, a religious society of Perfectionists that was established in 1848 by John Humphrey Noyes. Members of the sect held all property in common and practiced complex marriage and common care of the children. The community prospered by making steel traps and silverware. In 1881 it was reorganized as a joint stock company, and the social experiments were abandoned. Incorporated 1901.

Oneida (o-NEI-duh), town, Delaware co., E iowa, 6 mi/9.7 km NE of Manchester; 42°32′N 91°20′W.

Oneida, town (2006 population 3,682), scott county, N tennessee, 50 mi/80 km NW of knoxville, in the cumberland mountains; 44°30′N 88°12′W. Agriculture; natural resources; manufacturing. big south fork national river and recreation area here. Settled 1868; incorporated 1914.

Oneida (o-NEED-uh), village (2000 population 70), nemaha county, NE kansas, 7 mi/11.3 km E of seneca; 39°52′N 95°56′W. Livestock; grain.

Oneida (o-NEI-duh), village, butler county, extreme SW ohio; 39°28′N 84°23′W.

Oneida Castle, village (2006 population 611), oneida county, central new york, on Oneida Creek and just SE of oneida; 43°04′N 75°37′W. Center of oneida territory, site of chief settlement of Oneida people.

Oneida Lake (□ c.80 sq mi/207 sq km), central new york, NE of syracuse; 22 mi/35 km long, 1–5 mi/1.6–8 km wide; 43°12′N 75°55′W. The new york state barge canal links the E end of the lake with mohawk river, and also follows part of the Oneida River, which flows from the W end of the lake c.20 mi/32 km into the oswego river.

Oneida Territory, parcel of scrubby land (□ 5 sq mi/13 sq km), oneida county, central new york, adjacent to Route 46 near verona. Site of Turning Point gambling casino, this region is all that remains of millions of acres of land once owned by the Oneida, part of the six Nations of Iroquois Confederacy. In the late 18th century, the U.S. government led by President George Washington, tried to settle land disputes with the entire Iroquois Confederacy, including the Oneida. In 1785 and 1788, however, before an agreement could be reached with the Federal government, New York state signed treaties with the Oneida that resulted in the tribe surrendering all rights to most of its land. In 1978, several Oneida groups filed suit, claiming that the state did not have the authority to make those treaties, and that they were invalid since the Federal government had not ratified them. They laid claim to 9,375 sq mi/24,281 sq km of land (equal to one fifth of the state) bounded on the W by a meridonal line just E of syracuse and extending from the SAINT LAWRENCE River to the New York-Pennsylvania border and on the E by another meridonal line. They also demanded fair rental payments from the state for the use of their property for the last one hundred years. The claim affected all or part of twelve counties, the cities of BINGHAMTON, NORWICH, and WATERTOWN, numerous municipal corporations, and land owned by 60,000 individuals. New York state offered the Oneida $8 million and title to Sampson State Park in SENECA and CAYUGA counties, but they refused. In 1980 the case went to court, and a year later it was decided in favor of the state. The U.S. Court of Appeals upheld the decision in a 1988 hearing, and the Supreme Court refused to hear a further appeal. A 1998 suit involving "nonpossessory claims" was, however, allowed to proceed under a Court of Appeals ruling.

O'Neill, city (2000 population 3,733), ⊙ HOLT county, N NEBRASKA, 70 mi/113 km NW of NORFOLK, and on ELKHORN RIVER; 42°27′N 98°39′W. Elevation 2,000 ft/610 m. Railroad terminus. Shipping point for grain and livestock region. Printing; concrete. Founded 1874.

Onejime (O-NE-jee-me), town, Kimotsuki county, KAGOSHIMA prefecture, S KYUSHU, extreme SW JAPAN, on W OSUMI PENINSULA, port on E shore of KAGOSHIMA BAY, 28 mi/45 km S of KAGOSHIMA; 31°14′N 130°47′E. Produces ponka liquor (made from ponkan variety of mandarin oranges).

Onekama (AHN-ah-KAH-mah), village (2000 population 647), MANISTEE county, NW MICHIGAN, 9 mi/14.5 km NNE of MANISTEE, on PORTAGE LAKE, and 3 mi/4.8 km E of LAKE MICHIGAN; 44°22′N 86°12′W. Manufacturing (electrical wire harnesses); resort.

Onekotan Island (uh-nee-kuh-TAHN) (□ 121 sq mi/314.6 sq km), one of the N main KURIL ISLANDS group, SAKHALIN oblast, extreme E SIBERIA, RUSSIAN FAR EAST; separated from PARAMUSHIR ISLAND (N) by Fourth Kuril Strait, from KHARIMKOTAN ISLAND (S) by Sixth Kuril Strait; 27 mi/43 km long, 7 mi/11.3 km wide; 49°25′N 154°45′E. Krenitsyn Peak [Japanese *Kuroishi-yama*], secondary volcanic cone in the S part of the island, rises to 4,350 ft/1,326 m in a circular crater lake. Small population engaged in fishing. Also spelled Onnekotan Island.

Onemen Gulf, RUSSIA: see ANADYR' BAY.

Onemen River, RUSSIA: see BOL'SHAYA RIVER.

Oneonta (o-nee-AHN-tah), city (□ 4 sq mi/10.4 sq km; 2006 population 13,238), OTSEGO county, E central NEW YORK, W of the CATSKILLS, on the SUSQUEHANNA River; 42°27′N 75°04′W. Light manufacturing, commercial services, agricultural products. Oneonta grew after the coming of the railroad in 1865; however, no vestiges of importance as a railroad center exist today. Brotherhood of Railroad Brakemen founded here in 1883, which was renamed Brotherhood of Railroad Trainmen in 1889. Seat of the State University of New York at Oneonta; Hartwick College. Regional medical center. National Soccer Hall of Fame. Settled c.1780, incorporated as a city 1909.

Oneonta (o-nee-AHN-tah), town (2000 population 5,576), ⊙ Blount co., N central Alabama, 35 mi/56 km NE of Birmingham. Vegetable area; clothing; millwork, meat packaging; tire rims; motor vehicle customizing. Named by William Newbold for his former home in NY. Inc. in 1891.

Oneşti (o-NESHT), city, BACĂU county, ROMANIA, 23 mi/37 km SSW of BACĂU; 46°15′N 26°45′E. Power station; large chemical and petrochemical plant. Gymnastics training center. Formerly called Gheorghe Gheorghiu-Dej.

Onet-le-Château (o-ne-luh–shah-TO), town (□ 15 sq mi/39 sq km), AVEYRON department, MIDI-PYRÉNÉES region, S FRANCE; 44°23′N 02°32′E. NW suburb of RODEZ.

One Tree Hill, residential suburb within AUCKLAND City, NORTH ISLAND, NEW ZEALAND. Built around a reserve with conspicuous volcanic crater, early Maori earthworks, prehistoric sites, and prominent monument to Maori people consisting of an obelisk and a single tree. Parks and hospitals, including New Zealand National Women's Hospital.

Onex, town (2000 population 16,419), GENEVA canton, SW SWITZERLAND, SW suburb of GENEVA; 46°11′N 06°06′E.

Ong, village (2006 population 64), CLAY county, S NEBRASKA, 14 mi/23 km SE of CLAY CENTER; 40°23′N 97°50′W.

Onga (AHN-guh), town, HAUT-OGOOUÉ province, E GABON, 75 mi/121 km NE of FRANCEVILLE; 00°46′S 14°20′E.

Onga (ON-gah), town, Onga county, FUKUOKA prefecture, N KYUSHU, SW JAPAN, 25 mi/40 km N of FUKUOKA; 33°50′N 130°40′E.

Ongeri (on-GER-ree), town, KASAI-OCCIDENTAL province, SW CONGO, 20 mi/32 km E of LUSAMBO; 04°37′S 25°10′E. Elev. 2,001 ft/609 m.

Ongerup, town, WESTERN AUSTRALIA state, W AUSTRALIA, 230 mi/370 km SE of PERTH; 33°59′S 118°28′E. In mixed farming area (grain; sheep, cattle). Ecotourism. Museum.

Ongin Gol, MONGOLIA: see ONGIYN GOL.

Ongiyn Gol (AWNG-geen GOL), river, 270 mi/435 km long, central MONGOLIA; rises on SE slopes of HANGAYN NURUU (Khangai Mountains); flows SE to ULAAN NUUR (lake) in the GOBI DESERT at 44°33′N 103°42′E. Intermittent in lower course.

Ongjin (AWNG-JEEN), county, SOUTH HWANGHAE province, S NORTH KOREA, on small peninsula in YELLOW SEA, 90 mi/145 km WNW of SEOUL. Gold-mining center.

Ongjin (AWNG-JEEN), administrative district (□ 79 sq mi/205.4 sq km) of INCHON city, SOUTH KOREA. Consists of thirty-seven inhabited and nine uninhabited islands scattered throughout KYONGGI Bay. Originally, Ongjin was made up of Ongjin peninsula and nearby islands, most of which were allocated to NORTH KOREA at the division of Korean peninsula along the 38th Parallel in 1945, and were not recovered. Staples such as rice and barley, potatoes and sweet potatoes are main agricultural products, and croaker fishing near Yonpyong Island is on the decline. Sopori Beach is well known. Passenger ships in service from Inchon harbor.

Ongniud Qi, CHINA: see WUDAN.

Ongole (uhn-GOL), city, PRAKASAM district, ANDHRA PRADESH state, SE INDIA, 60 mi/97 km SSW of GUNTUR; 15°30′N 80°03′E. Road junction; tobacco-curing center; rice and oilseed milling, ghee processing; leather goods. Salt refining at KANUPARTI, 12 mi/19 km NE.

Onguday (uhn-goo-DEI), village, central ALTAI REPUBLIC, S central SIBERIA, RUSSIA, on the Chuya mountain highway (Russian *Chuyskiy Trakt*), 85 mi/137 km S of GORNO-ALTAYSK; 50°45′N 86°09′E. Elevation 2,687 ft/819 m. In the protected old-growth forest area. Weather station.

Ongurën (uhn-goo-RYON), village, SE IRKUTSK oblast, E central SIBERIA, RUSSIA, on the W shore of Lake BAYKAL, 70 mi/113 km ESE of KACHUG; 53°38′N 107°35′E. Elevation 1,604 ft/488 m. Fisheries.

Onhaye, commune (2006 population 3,129), Dinant district, NAMUR province, S central BELGIUM, 4 mi/6 km W of DINANT; 50°15′N 04°50′E.

Oni (O-nee), city and administrative center of Oni region, N GEORGIA, on RIONI River, on OSSETIAN MILITARY ROAD, and 40 mi/64 km NE of KUTAISI; 42°34′N 43°26′E. Creamery; winery; clothing manufacturing. Has medieval Imeretian castle.

Onida, city (2006 population 664), ⊙ SULLY county, central SOUTH DAKOTA, 27 mi/43 km NNE of PIERRE; 44°42′N 100°04′W. Ranching area.

Onikshty, LITHUANIA: see ANYKŠČIAI.

Onil (o-NEEL), town, ALICANTE province, E SPAIN, 12 mi/19 km SWS of ALCOY; 38°38′N 00°40′W. Toy and footwear manufacturing, olive-oil processing; cereals, wine, almonds. Has palace of dukes of Dos Aguas.

Onilahy River (oo-nee-LAH-hee), c.250 mi/402 km long, TOLIARY province, SW MADAGASCAR; rises c.20 mi/32 km S of BETROKA (23°26′S 46°05′E); flows generally W to ST. AUGUSTIN BAY on MOZAMBIQUE CHANNEL, just N of SOALARA. Navigable year-round for shallow-draught boats in its lower course. Used for irrigation.

Onishi (O-nee-shee), town, Ochi county, EHIME prefecture, W SHIKOKU, W JAPAN, 16 mi/25 km N of MATSUYAMA; 34°03′N 132°55′E. Shipbuilding.

Onishi, town, Tano county, GUMMA prefecture, central HONSHU, N central JAPAN, 19 mi/30 km S of MAEBASHI; 36°09′N 139°03′E. Hot-springs resort. Sometimes called Oniishi.

Onitsha (aw-nee-CHAH), one of former EASTERN PROVINCES, S NIGERIA; capital was ONITSHA. Now part of ANAMBRA state.

Onitsha, city, ANAMBRA state, SE NIGERIA, a port on the NIGER River; 06°10′N 06°47′E. Has the largest market E of the Niger. Manufacturing includes textiles, beverages, shoes, lumber, phonograph records, and printed materials; cotton mills, breweries. Fishing and canoe-building are traditional local industries. Onitsha is the N limit of year-round navigation on the Niger and is an important entrepôt linking traders from the Niger delta with the upper Niger and BENUE rivers and with a wide region of E Nigeria. A road bridge (built 1965) across the Niger here is a vital link between E and W Nigeria. Probably founded in the 16th century by immigrants from Benin. In the 17th century it became the capital of an Igbo kingdom. In 1857 a British trading station and a Christian mission were established here, and in 1884 Onitsha came under British protection.

Onival, FRANCE: see AULT.

Onjuku (ON-jyoo-koo), town, Isumi county, CHIBA prefecture, E central HONSHU, E central JAPAN, on E BOSO PENINSULA, 22 mi/35 km S of CHIBA; 35°11′N 140°21′E. Abalone, turban shells; seaweed (*wakame*). Known for its *ama* female pearl divers.

Onk, Djebel (ONGK, JE-bel), mining town, TEBESSA wilaya, E ALGERIA, 57 mi/92 km S of TEBESSA, near Tunisian border; 34°52′N 07°57′E. In a major phosphate ore-producing region; the ore is transported to ANNABA.

Onk, Djebel (ONGK, JE-bel), mountain (4,390 ft/1,338 m), NE ALGERIA, near TUNISIA border, 50 mi/80 km SSW of TEBESSA. Large phosphate deposits.

Onkhor, SOMALIA: see ANKHOR.

Onley (ON-lee), town (2006 population 491), ACCOMACK county, E VIRGINIA, 3 mi/5 km SW of ACCOMAC, on DELMARVA PENINSULA; 37°41′N 75°43′W. Manufacturing (printing and publishing, concrete); agricultural area (vegetables, grain; poultry, livestock).

Onna (ON-NAH), village, on W coast of central OKINAWA island, Kunigami county, Okinawa prefecture, SW JAPAN, 22 mi/35 km N of NAHA; 26°29′N 127°51′E. Okinawa Seacoast quasi-national park nearby.

Onnaing (o-NANG), town (□ 5 sq mi/13 sq km), NORD department, NORD-PAS-DE-CALAIS region, N FRANCE, 4 mi/6.4 km NE of VALENCIENNES, near Belgian border; 50°23′N 03°36′E. Engineering works. Breweries.

Onne (AW-ne), town, Rivers state, SE NIGERIA, in NIGER River delta region, 10 mi/16 km W of PORT HARCOURT; 04°43′N 07°09′E. Market center. Fish; oil palms, cassava, rice, yams.

Onnekotan Island, RUSSIA: see ONEKOTAN ISLAND.

Onnekotan-kaikyo, RUSSIA: see KURILE STRAIT.

Önnestad (UHN-ne-STAHD), village, SKÅNE county, S SWEDEN, 5 mi/8 km WNW of KRISTIANSTAD; 56°03′N 14°02′E.

Ono (O-no), city, FUKUI prefecture, central HONSHU, W central JAPAN, 16 mi/25 km E of FUKUI; 35°58′N 136°29′E. Traditional string handicrafts (*mizuhiki saiku*). Skiing.

Ono (o-NO), city, HYOGO prefecture, S HONSHU, W central JAPAN, 17 mi/28 km N of KOBE; 34°51′N 134°56′E. Soybeans. Abacus making, spinning industries, woodworking; manufacturing of automotive parts, carpenter's tools, and cutlery.

Ono (o-NO), town, Tamura county, FUKUSHIMA prefecture, N central HONSHU, NE JAPAN, 34 mi/55 km S of FUKUSHIMA city; 37°17′N 140°37′E. Tobacco, asparagus; stone quarrying.

Ono (O-no), town, Ibi county, GIFU prefecture, central HONSHU, central JAPAN, 6 mi/10 km N of GIFU; 35°25′N 136°42′E. Persimmons.

Ono, town, Saeki county, HIROSHIMA prefecture, SW HONSHU, W JAPAN, 12 mi/20 km S of HIROSHIMA; 34°16′N 132°16′E.

Ono, town, Oshima county, Hokkaido prefecture, N JAPAN, 90 mi/145 km S of SAPPORO; 41°52′N 140°38′E.

Ono, town, Ono county, OITA prefecture, E KYUSHU, SW JAPAN, 16 mi/25 km S of OITA; 33°02′N 131°30′E.

Ono (O-no), village, Kashima county, IBARAKI prefecture, central HONSHU, E central JAPAN, 25 mi/40 km S of MITO; 36°02′N 140°36′E.

Ono, village, Kunohe county, IWATE prefecture, N HONSHU, NE JAPAN, 47 mi/75 km N of MORIOKA; 40°16′N 141°37′E. Woodworking.

Ono (O-no), volcanic island (□ 12 sq mi/31.2 sq km), FIJI, SW PACIFIC OCEAN, c.4 mi/6.4 km NE of KADAVU; 4.5 mi/7.2 km long, 3.5 mi/5.6 km wide; 18°54′S 178°29′E. Rises to 1,160 ft/354 m. Copra.

Onoda (O-no-dah), city, YAMAGUCHI prefecture, SW HONSHU, W JAPAN, port on SUO SEA, 22 mi/35 km S of YAMAGUCHI; 34°00′N 131°11′E. Railroad terminus.

Onoda (O-no-dah), town, Kami county, MIYAGI prefecture, N HONSHU, NE JAPAN, 22 mi/35 km N of SENDAI; 38°34′N 140°47′E. Vegetables; fish. Pongee.

Onoe (o-NO-e), town, S Tsugaru county, Aomori prefecture, N HONSHU, N JAPAN, 16 mi/25 km S of AOMORI; 40°36′N 140°34′E.

Onogami (o-NO-gah-mee), village, N Gumma county, GUMMA prefecture, central HONSHU, N central JAPAN, 16 mi/25 km N of MAEBASHI; 36°32′N 138°56′E.

Onohara (O-no-HAH-rah), town, Mitoyo county, KAGAWA prefecture, NE SHIKOKU, W JAPAN, 28 mi/45 km S of TAKAMATSU; 34°04′N 133°40′E. Onions.

Onojo (o-NO-jo), city, FUKUOKA prefecture, NE KYUSHU, SW JAPAN, 6 mi/10 km S of FUKUOKA; 33°31′N 130°28′E.

Onokhino (uh-NO-hee-nuh), village (2006 population 3,005), SW TYUMEN oblast, SW Siberian Russia, on the PYSHMA RIVER, on road, 12 mi/19 km S of (and administratively subordinate to) TYUMEN; 56°56′N 65°32′E. Elevation 160 ft/48 m. In agricultural area; poultry factory.

Onokhoy (uh-nuh-HO-yee), town (2005 population 10,415), central BURYAT REPUBLIC, S SIBERIA, RUSSIA, on the UDA RIVER (tributary of the SELENGA River), on road and railroad, 25 mi/40 km E of ULAN-UDE; 51°55′N 108°02′E. Elevation 1,781 ft/542 m. Lumber shipping and processing.

Onomi (O-no-mee), village, Takaoka county, KOCHI prefecture, S SHIKOKU, W JAPAN, 28 mi/45 km S of KOCHI; 33°20′N 133°08′E.

Onomichi (o-NO-mee-chee), city (2005 population 150,225), HIROSHIMA prefecture, SW HONSHU, W JAPAN, 43 mi/70 km E of HIROSHIMA; 34°24′N 133°12′E. Rush mats for tatami. Baked sea bream is a local specialty. Site of several Buddhist temples, notably that of Senko-ji (10th century).

Onondaga, county (□ 805 sq mi/2,093 sq km; 2006 population 456,777), central NEW YORK; ⊙ SYRACUSE; 43°00′N 76°12′W. Situated in FINGER LAKES region; drained by SENECA and OSWEGO rivers and by Onondaga and small Chittenango creeks; crossed by the NEW YORK STATE BARGE CANAL. Includes Onondaga Indian Reservation. Resorts on ONEIDA and Skaneateles lakes. Dairying and farming area (poultry; corn, potatoes, hay), with extensive manufacturing. Formed 1794.

Onondaga (ahn-uhn-DO-guh), unincorporated town (□ 34 sq mi/88.4 sq km; 2001 population 1,758), SE ONTARIO, E central Canada, 8 mi/13 km E of BRANTFORD, and included in BRANT; 43°07′N 80°06′W. Mohawk Indian reserve here.

Onondaga Lake, brackish lake, central NEW YORK, NW of SYRACUSE; 5 mi/8 km long and 1 mi/1.6 km wide; 43°05′N 76°12′W. In 1654, Father LeMoyne, a missionary, was taken to salt springs along the lakeshore by the Onondagas. He showed them how to obtain salt from the water by boiling it. In 1795 the lake was purchased from the Native Americans by New York state for its salt resources. The Salt Museum on the lakeshore near LIVERPOOL contains relics of the early salt industry, which thrived in the mid-19th century. Due to industry pollution (including chemical manufacturing), all normal lake recreational activities are precluded (with the exception of boating). Ranks as one of the nation's most polluted bodies of water.

Onon Gol, MONGOLIA and RUSSIA: see ONON RIVER.

Onon River (O-nuhn), Mongolian *Onon Gol*, 640 mi/1,032 km long, NE MONGOLIA and Russian SIBERIA; rises in the HENTIYN NURUU (Kentei Mountains) in Mongolia at 48°N 109°E; flows E and NE, crossing into RUSSIA, and past OLOVYANNAYA (in CHITA oblast), joining the INGODA RIVER 15 mi/24 km NW of SHILKA to form navigable SHILKA River (51°42′N 115°51′E). Used for logging. Lower course forms S and E border of AGIN-BURYAT AUTONOMOUS OKRUG.

Onota, Lake (uh-NO-duh), reservoir, BERKSHIRE county, W MASSACHUSETTS, on branch of HOUSATONIC River, in the BERKSHIRE Mountains, 3 mi/4.8 km NW of PITTSFIELD; c.3 mi/4.8 km long; 42°28′N 73°16′W.

Onoto (o-NO-to), town, ⊙ Juan Manuel Cajigal municipio, ANZOÁTEGUI state, NE VENEZUELA, landing on UNARE RIVER (navigable in rainy season), and 50 mi/80 km SW of BARCELONA; 09°35′N 65°12′W. Cotton, sugarcane, fruit; livestock.

Onoto (o-NO-to), town, TRUJILLO state, W VENEZUELA, 13 mi/21 km NE of VALERA; 09°41′N 70°11′W. Elevation 4,783 ft/1,457 m.

Onotoa (o-no-TO-ah), atoll (□ 5 sq mi/13 sq km; 2005 population 1,644), KINGSMILL GROUP, S GILBERT ISLANDS, W central PACIFIC OCEAN; 01°50′S 175°33′E. Copra. Formerly Clerk Island.

Onoway (AH-no-wai), town (□ 1 sq mi/2.6 sq km; 2001 population 847), central ALBERTA, W Canada, 30 mi/48 km WNW of EDMONTON, in Lac Sainte Anne County; 53°42′N 114°11′W. Agriculture. Formed as a village in 1923; became a town in 2005.

Onsen (ON-SEN), town, Mikata district, HYOGO prefecture, S HONSHU, W central JAPAN, 70 mi/113 km N of KOBE; 35°33′N 134°29′E. *Tajima ushi* variety of beef cattle. Hot-springs resort; skiing area. Sometimes called Yumura.

Onset, MASSACHUSETTS: see WAREHAM.

Onslow (AHNZ-lo), county (□ 908 sq mi/2,360.8 sq km; 2006 population 150,673), E NORTH CAROLINA, on the ATLANTIC; ⊙ JACKSONVILLE; 34°42′N 77°24′W. Bounded S by Atlantic Ocean, NE by White Oak River. INTRACOASTAL WATERWAY channel parallels coast (SE). Drained by NEW RIVER; New River estuary indents county from SE. Heavily forested and partly swampy tidewater area. Service industries; retail trade; some manufacturing at Jacksonville. Fishing. Agricultural area (tobacco, corn, wheat, cotton, soybeans, hay). U.S. Marine Corps Base CAMP LEJEUNE in SE. Catherine Lake in NW; Hammocks Beach State Park on Atlantic in E; part of Hoffmann Forest in N; part of Topsail Island, beach resort area in S. Formed 1734 from New Hanover County. Named for Arthur Onslow (1691–1768), member of the British Parliament and Speaker of the House of Commons.

Onslow (AHNZ-lo), town and port, NW WESTERN AUSTRALIA, NW AUSTRALIA, 490 mi/789 km N of GERALDTON, at mouth of ASHBURTON RIVER, near NE entrance to EXMOUTH GULF; 21°39′S 115°06′E. Airport; exports gold; livestock; near oil reserves.

Onslow, town (2000 population 223), JONES county, E IOWA, 14 mi/23 km E of ANAMOSA; 42°06′N 91°00′W. In livestock and grain area; fertilizers; feeds.

Onslow Bay (AHNZ-lo BAI), SE NORTH CAROLINA, bay of the ATLANTIC OCEAN from CAPE LOOKOUT (NE) to CAPE FEAR (SW); c.100 mi/161 km long, c.25 mi/40 km wide; 34°35′N 77°13′W. Long curvature in Atlantic coastline.

Onsong (ON-SUHNG), county, NORTH HAMGYONG province, NORTH KOREA, near Manchurian border, 45 mi/72 km NNW of UNGGI. Coal-mining and livestock-raising area.

Onsøy (AWNS-uh-u), village, Østfold county, SE NORWAY, part of FREDRIKSTAD. Center of stone-quarrying region (granite, paving stone); construction industry; agriculture, strawberries. Summer resorts with tourism.

Onsted (AHN-sted), village (2000 population 813), LENAWEE county, SE MICHIGAN, 12 mi/19 km NW of ADRIAN; 42°00′N 84°11′W. In farm area. Manufacturing (hardware).

Onstwedde (AWNST-ve-duh), village, GRONINGEN province, NE NETHERLANDS, 5 mi/8 km ENE of STADSKANAAL city; 53°02′N 07°03′E. Dairying; livestock; vegetables, grain.

Ontajärvi, RUSSIA: see ONDOZERO.

On-take (ON-tah-kai), peak (10,108 ft/3,081 m), on GIFU-NAGANO prefecture border, central HONSHU, central JAPAN, 36 mi/58 km SW of MATSUMOTO. On its summit are a 14th century Shinto shrine and several small lakes.

Ontario (awn-TER-ree-o), province (□ 412,582 sq mi/1,072,713.2 sq km; 2006 population 12,160,282), E central Canada; ⊙ TORONTO, the largest metropolitan area in Canada; 50°00′N 86°00′W. Other important cities are OTTAWA, HAMILTON, KITCHENER, LONDON, WINDSOR, THUNDER BAY, and SAINT CATHARINES. Ontario, the second-largest Canadian province, is the most populous and the most important in terms of mineral, industrial, and agricultural output and in terms of financial and other services. It is bounded on the N by HUDSON BAY and JAMES BAY; on the E by QUEBEC; on the S by the SAINT LAWRENCE RIVER, lakes ONTARIO, ERIE, HURON, and SUPERIOR, and by the U.S.; and on the W by MANITOBA. The province has three main geographic regions. In the W and central portion is the CANADIAN SHIELD, a region of mineral-rich rock covered with forests and broken by a labyrinth of rivers and lakes. In the N is the Hudson Bay Lowlands bordering on Hudson and James bays, an area consisting mainly of marshes, swampland, and forest. In the S and E is the GREAT LAKES–Saint Lawrence lowlands, where 90% of the population lives and where industry and agriculture are concentrated. Climate varies among the regions. The far N has subarctic conditions, while the W has a temperate climate. Around the Great Lakes the weather is moderate and summers are longer than in other parts of the province. The Saint Lawrence River gives Ontario access to the ATLANTIC. Other important rivers are the OTTAWA (which forms part of the boundary with Quebec), the SAINT CLAIR, the DETROIT, and the SAINT MARYS RIVER. Several of the province's rivers are used to generate hydroelectric power, among them the NIAGARA, with its famous falls.

The most important economic activity in Ontario is manufacturing, and the Toronto-Hamilton region is the most highly industrialized section of the country. The area from OSHAWA around to NIAGARA FALLS is known as the "Golden Horseshoe." Major industrial products include motor vehicles and parts, iron, steel, and other metal products, foods and beverages, electrical goods, machinery, chemicals, petroleum and coal products, and paper products. Ontario has attracted many high technology industries, especially around Ottawa, and its service industries are second in importance only to manufacturing. Agriculture is also important, with cattle, dairy products, and hogs producing the most income. Other major crops are corn, wheat, potatoes, and soybeans. On the shores of the E Great Lakes are orchards and tobacco plantations. Mining is important in the Canadian Shield region, where iron ore, copper, zinc, gold, silver, and uranium are found. The area around SUDBURY is particularly rich in copper and nickel; Ontario produces half the world's nickel. Ontario is also a major producer of lumber and pulp and paper.

With steady immigration by people from ITALY, GERMANY, PORTUGAL, the WEST INDIES, INDIA, and East ASIA, Ontario's ethnic composition is rapidly diversifying. People of British ancestry make up about half the province's population, and 10% are French. More than 80% of Ontario's residents live in urban centers. Ontario has four national parks and numerous tourist attractions, three of the most notable being Niagara Falls, the annual Shakespeare Festival at Stratford, and the annual George Bernard Shaw Festival at Niagara-on-the-Lake. Among the province's institutions of higher education are the University of Toronto, the University of Ottawa, McMaster University, Queen's University, the University of Western Ontario, Brock University, Carleton University, Trent University, University of Waterloo, Wilfred Laurier University, York University, and Ryerson Polytech University.

Before the arrival of Europeans the area of Ontario was inhabited by several Algonquian (Ojibwa, Cree, and Algonquin) and Iroquoian (Iroquois, Huron, Petun, Neutral, Erie, and Susquehannock) tribes. Étienne Brulé explored S Ontario in 1610–1612. Henry Hudson sailed into Hudson Bay in 1611 and claimed the region for ENGLAND. Within a few years Samuel de Champlain reached (1615) the E shores of Lake Huron, and French explorers, missionaries, and trappers had established posts at several points. However, settlement was long hindered by the presence of the Iroquois. In the late 17th century the British established trading posts in the Hudson Bay area, and the Anglo-French struggle for control of Ontario began. The conflict was resolved by the Treaty of Paris of 1763, which gave GREAT BRITAIN all of France's mainland N American territory. In 1774 the British attached Ontario to Quebec, which had a predominantly French culture. When many pro-British Loyalists migrated to Ontario after the American Revolution, the desire for institutions and a government separate from those of Quebec grew. The Constitutional Act of 1791 split Quebec into Lower Canada (present-day Quebec) and Upper Canada (present-day Ontario), with the Ottawa River as the dividing line. During the War of 1812, Americans raided Upper Canada and burned Toronto (1813). After the war many English, Scottish, and Irish settlers came to the colony. Conflict developed between the conservative, aristocratic governing group, known as the Family Compact, and the reformers and radicals led by William Lyon Mackenzie. The radicals staged an armed uprising in 1837 but were easily suppressed. However, the rebellion occurred at the same time as a revolt in Lower Canada, and the British government sent over Lord Durham to study the situation in the N. American colonies. He recommended the reunion of the 2 colonies (to place the French of Quebec in a minority) and the granting of self-government. Accordingly Upper and Lower Canada were joined in 1841 and became known, respectively, as Canada West and Canada East. Parliamentary self-govt. was not granted until 1849. However, conflict between French and English made the united province unworkable, and in 1867 when the confederation of Canada was formed, Ontario and Quebec became separate provinces. With the construction of the transcontinental railroad in the 1880s, settlement increased in W Canada and Ontario's commerce and industry flourished. The exploitation of the minerals in the Canadian Shield region began in the early 20th century. Ontario sends twenty-four senators and ninety-nine representatives to the national parliament. As part of a national trend beginning in the 1990s, many of Ontario's municipalities have merged, resulting in a more than 40% reduction of their numbers, from 815 to 447, between 1996 and 2001.

Ontario (ahn-TE-ree-o), former county (□ 853 sq mi/ 2,217.8 sq km), S ONTARIO, E central Canada, on Lake ONTARIO and Lake SIMCOE; its ⊙ WHITBY; 44°15′N 79°05′W. Along with Durham county, it was superseded in 1973 by the regional municipality of Durham.

Ontario, county (□ 662 sq mi/1,721.2 sq km; 2006 population 104,353), W central NEW YORK; ⊙ CANANDAIGUA; 42°51′N 77°17′W. Situated in FINGER LAKES region, and partly bounded E by SENECA Lake; includes CANANDAIGUA, HONEOYE, and CANADICE lakes. Drained by small Honeoye, Mud, and Flint creeks, and Canandaigua Outlet. Diversified manufacturing, especially at GENEVA, Canandaigua, NAPLES; fruit-growing area; nurseries; agricultural products. Formed 1789.

Ontario, city (2000 population 158,007), SAN BERNARDINO county, S CALIFORNIA, suburb 35 mi/56 km E of downtown LOS ANGELES, and 20 mi/32 km W of SAN BERNARDINO; 34°03′N 117°37′W. In a region of vineyards. Also important to the economy are the manufacturing of aircraft and aircraft parts, aerospace vehicle parts, sporting goods, leather goods, electrical equipment, and plastics. The growing number of high-technological industries in the S California area have added to Ontario's development. Founded in 1882, the city is the site of Chaffey College, Ontario International Airport, and Ontario Motor Speedway in E part of city; California Institutes for Men and Women (separate prison facilities) to S. It is also one of the fastest-growing U.S. cities, marked by a population increase of 50% between 1980 and 1990. SAN GABRIEL MOUNTAINS and San Bernardino National Forest to N; Angeles National Forest to NW. Incorporated 1891.

Ontario, city (2006 population 11,093), MALHEUR county, E OREGON, on Snake River, at mouth of MALHEUR RIVER, 45 mi/72 km NW of BOISE, IDAHO; 44°01′N 116°58′W. Elevation 2,154 ft/657 m. Railroad, trade, and highway center for large irrigated area in Owyhee River project. Agriculture (sugar beets, potatoes, grain); dairy products. Manufacturing (printing, publishing; concrete). Site of Treasure Valley Community College. Ontario State Park to N. Incorporated 1914.

Ontario (awn-TER-ree-o), village (2006 population 462), VERNON county, SW WISCONSIN, on KICKAPOO RIVER, and 32 mi/51 km ESE of LA CROSSE; 43°43′N 90°35′W. In farm and dairy area. Wildcat Mountain State Park to S.

Ontario, hamlet, WAYNE county, W NEW YORK, 16 mi/26 km ENE of ROCHESTER; 43°15′N 77°19′W. In agricultural area.

Ontario, Lake (□ 7,540 sq mi/19,529 sq km), between SE ONTARIO, CANADA, and NW NEW YORK; 193 mi/311 km long and 53 mi/85 km at its greatest width; smallest and lowest of the Great Lakes, it has a surface elevation of 246 ft/75 m above sea level and a maximum depth of 778 ft/237 m. Lake Ontario is fed chiefly by the waters of Lake ERIE by way of the NIAGARA River; other tributaries are the GENESEE, OSWEGO, and BLACK rivers in New York and the TRENT RIVER in Ontario. The lake is drained to the NE by the ST. LAWRENCE River. Oceangoing vessels reach the lake through the SAINT. LAWRENCE SEAWAY and use the WELLAND SHIP CANAL to bypass NIAGARA FALLS and reach Lake Erie; smaller craft (mostly pleasure boats) can travel the RIDEAU CANAL between KINGSTON and OTTAWA, and the TRENT CANAL between the Bay of QUINTE and GEORGIAN BAY. Navigation on the lake is not usually impeded by ice in winter.

The chief Canadian lakeshore cities are ST. CATHARINES, HAMILTON, TORONTO, OSHAWA, and KINGSTON; on the S shore are ROCHESTER and OSWEGO, New York. Commercial fishing is important, but pollution has been a problem. A U.S.-Canadian pact (1972) established that water quality would be improved and further pollution ended. Recreational facilities are provided at state and provincial parks. The first European to see (1615) Lake Ontario was Étienne Brulé, the French explorer; later that year Samuel de Champlain visited it.

Onteniente (on-ten-YEN-tai), city, VALENCIA province, E SPAIN, 14 mi/23 km SSW of JÁTIVA; 38°49′N 00°37′W. Textile center; also manufacturing of barrels, chairs, paper, knit goods; olive-oil processing, wine making. Cereals, grapes, olives, carobs in area.

Ontiñena (on-tee-NYAI-nah), village, HUESCA province, NE SPAIN, 17 mi/27 km NW of FRAGA; 41°40′N 00°05′E. Wine; olive oil; figs; livestock.

Ontonagon (ahn-TAHN-ah-guhn), county (□ 3,741 sq mi/9,726.6 sq km; 2006 population 7,202), NW UPPER PENINSULA, MICHIGAN; ⊙ ONTONAGON. Bounded N by LAKE SUPERIOR; drained by ONTONAGON RIVER and by small IRON and Firesteel rivers; 46°58′N 89°16′W. Lumber and agricultural area (cattle; oats; hay; dairy products); fishing; resorts. Manufacturing in Ontonagon. Boundary with GOGEBIC county (S) and IRON county (SE) coterminous with Central and Eastern time zone boundary. Ontonagon county westernmost extension 89°50′W of Eastern time zone (extends further W in ONTARIO, CANADA to 90°W). PORCUPINE MOUNTAINS are in NW, part of Porcupine Mountains State Park in W; part of Ottawa National Forest in S ½ of county and part of Gogebic Lake on S boundary; Ontonagon Indian Reservation in NE. Organized 1848.

Ontonagon (ahn-TAHN-ah-guhn), town (2000 population 1,769), ⊙ ONTONAGON county, NW UPPER PENINSULA, MICHIGAN, 40 mi/64 km SW of HOUGHTON, on LAKE SUPERIOR at mouth of ONTONAGON RIVER; 46°52′N 89°18′W. Railroad terminus; railroad/ship transfer point. Manufacturing (paperboard); shipping center; fishing. Resort. The Ontonagon boulder, a huge copper mass, was found near the river, and was moved to the Smithsonian Institute. Ontonagon Indian Reservation to NE; Porcupine Mountains State Park and Ski Area to W; Ottawa National Forest to S. Established on site of Native American village. Incorporated 1885.

Ontonagon River (ahn-TAHN-ah-guhn), c.22 mi/35 km long, W UPPER PENINSULA, MICHIGAN; formed by several branches uniting in ONTONAGON county, c.6 mi/9.7 km SW of Massachusetts; flows NNW to LAKE SUPERIOR at ONTONAGON. East Branch rises in Kunzie Lake, SE HOUGHTON county; flows c.35 mi/56 km NNW. Middle Branch rises in E GOGEBIC county, flows generally c.40 mi/64 km N past WATERSMEET. South Branch rises at WISCONSIN state line, flows c.55 mi/89 km NNE, past EWEN; 46°41′N 89°09′W. VICTORIA DAM (power) is on the South Branch just below the influx of West Branch (c.20 mi/32 km long), which flows from N end of GOGEBIC LAKE. All generally flow through Ottawa National Forest.

Ontong Java Atoll (AWN-tawng JAH-vah), SOLOMON ISLANDS, SW PACIFIC, c.160 mi/260 km N of SANTA ISABEL; 05°25′S 159°30′E; four islands on reef 30 mi/48 km long, 20 mi/32 km wide. Coconuts. Polynesian inhabitants. Also called Luangiua after largest settlement; Lord Howe Atoll (or Island).

Ontur (on-TOOR), town, ALBACETE province, SE central SPAIN, 14 mi/23 km NE of HELLÍN; 38°38′N 01°29′W. Olive-oil processing; cereals, wine, fruit.

Onufriyevka, UKRAINE: see ONUFRIYIVKA.

Onufriyivka (o-NOO-free-yif-kah) (Russian *Onufriyevka*), town, E KIROVOHRAD oblast, UKRAINE, 12 mi/19 km S of KREMENCHUK; 48°54′N 33°27′E. Elevation 416 ft/126 m. Raion center; radio assembly; horse breeding. Museum of horse-breeding, heritage museum. Established at the beginning of the 17th century; town since 1968.

Onverdacht (AWN-vuhr-dahkht), village, Para district, N SURINAME, on SURINAME RIVER just N of PARANAM, and 15 mi/24 km SSE of PARAMARIBO; 05°36′N 55°12′W. Commonly called BILLITON (BIL-LI-tohn), for the bauxite plant (built 1942) here.

Onverwacht (AWN-vuhr-vahkht), village, ⊙ Para district, N SURINAME, 17 mi/27 km S of PARAMARIBO; 05°35′N 55°15′W, Paramaribo's international airport to immediate S at Sanderij. Coffee, rice, fruit.

Onward, town (2000 population 81), CASS county, N central INDIANA, 10 mi/16 km ESE of LOGANSPORT; 40°41′N 86°12′W. In agricultural area.

Onyang (ON-YAHNG), city, N SOUTH CHUNGCHONG province, SOUTH KOREA, in center of ASAN CO.; 35°34′N 129°07′E. Low mountains S of city, rivers in N form wide field in their basin. Famed hot springs. Historic and scenic sites (Onyang Folk Museum, Hyonchung Temple, artificial lakes). Tourism; fruit farming. CHANGHANG railroad passes to N.

Onzaga (on-SAH-gah), town, ⊙ Onzaga municipio, SANTANDER department, N central COLOMBIA, in Cordillera ORIENTAL, 51 mi/82 km SE of BUCARAMANGA; 06°21′N 72°50′W. Elevation 6,673 ft/2,034 m. Coffee, cassava, corn.

Onzain (on-zan), commune (□ 11 sq mi/28.6 sq km), LOIR-ET-CHER department, CENTRE administrative region, N central FRANCE, 10 mi/16 km SW of BLOIS; 47°30′N 01°11′E. Vineyards.

Onze Lieve Vrouw-Waver (ON-zuh LEE-vuh vrou-WAH-vuhr), French *Wavre-Notre Dame*, village in commune of SINT-KATELIJNE-WAVER, Mechelen district, ANTWERPEN province, N BELGIUM, 5 mi/8 km ENE of MECHELEN; 51°04′N 04°34′E. Agricultural market.

Oô (O), village (□ 12 sq mi/31.2 sq km), HAUTE-GARONNE department, MIDI-PYRÉNÉES region, S FRANCE, in central PYRENEES Mountains, 5 mi/8 km W of BAGNÈRES-DE-LUCHON, near Spanish border; 42°48′N 00°30′E. Nearby Lake of Oô (elevation 4,912 ft/1,497 m), fed by waterfall (850 ft/259 m high), has hydroelectric station; water level may fluctuate. Splendid mountain setting.

Oodeypore, INDIA: see UDAIPUR, city, RAJASTHAN state.

Oodla Wirra (OOD-luh WIR-uh), village, S SOUTH AUSTRALIA state, S central AUSTRALIA, 65 mi/105 km ENE of PORT PIRIE; 32°53′S 139°04′E. On Port Pirie–BROKEN HILL railroad; wool, some wheat. Sometimes spelled Oodlawirra.

Oodlawirra, AUSTRALIA: see OODLA WIRRA.

Oodnadatta (OOD-nuh-DA-tuh), village, N central SOUTH AUSTRALIA state, S central AUSTRALIA, 405 mi/652 km NNW of PORT PIRIE and on NEALES RIVER; 27°33′S 135°28′E. Formerly on Port Pirie–ALICE SPRINGS railroad line, which closed 1981; railway museum. Cattle.

Ooka (O-kah), village, Sarashina county, NAGANO prefecture, central HONSHU, central JAPAN, 16 mi/25 km S of NAGANO; 36°30′N 137°59′E. Lumber.

Ookala (o-O-KAH-lah), village, NE HAWAII island, HAWAII county, HAWAII, 23 mi/37 km NW of HILO on Hamakua Coast. Manowaialee Forest Reserve to S; Hamakua Forest Reserve to W; Hilo Forest Reserve to SE.

Ooldea (ool-DEE-uh), settlement, S SOUTH AUSTRALIA state, S central AUSTRALIA, at E edge of NULLARBOR PLAIN, 410 mi/660 km NW of PORT PIRIE; 30°30′S 131°45′E. On Trans-Australian Railway; railroad station. Near a permanent water source in arid area.

Oolen, BELGIUM: see OLEN.

Oolitic, town, LAWRENCE county, S INDIANA, on SALT CREEK, and 4 mi/6.4 km NNW of BEDFORD. Limestone quarries.

Oologah (OOL-uh-guh) or **Oolagah**, town, ROGERS county, NE OKLAHOMA, 25 mi/40 km NE of TULSA. In stock-raising and agricultural area. Will Rogers' birthplace to E. To S, on VERDIGRIS RIVER, is site of OOLOGAH LAKE Dam, for flood control, hydroelectric power.

Oologah Lake, reservoir, ROGERS and NOWATA counties, NE OKLAHOMA, on VERDIGRIS RIVER, 26 mi/42 km NE of TULSA; 36°25′N 95°41′W. Maximum capacity 1,519,000 acre-ft. Formed by dam (137 ft/42 m high), built (1963) by Army Corps of Engineers for flood control; also used for water supply and navigation. Will Rogers' Home near dam.

Ooltewah (OOL-te-wah), town (2000 population 5,681), HAMILTON county, SE TENNESSEE, 12 mi/19 km E of CHATTANOOGA, 35°04′N 85°03′W.

Ooltgensplaat (AWLT-khens-plaht), village, SOUTH HOLLAND province, SW NETHERLANDS, at E end of OVERFLAKKEE island, 11 mi/18 km NNW of ROOSENDAAL; 51°41′N 04°21′E. HARINGVLIET Bridge and VOLKERAK DAM to N; pumping station to W. Dairying; livestock; grain, vegetables, sugar beets.

Oorgaum, INDIA: see KOLAR GOLD FIELDS.

Oorschot, NETHERLANDS: see OIRSCHOT.

Ooru, INDIA: see KHARAGHODA.

Oos (AWS), NW suburb of BADEN-BADEN, BADEN-Württemberg, S GERMANY, on the OOS; 48°47′N 08°12′E. Airport.

Oostakker (OS-tah-kuhr), village, EAST FLANDERS province, NW BELGIUM, 3 mi/4.8 km NNE of GHENT, near GHENT-TERNEUZEN CANAL; 51°06′N 03°46′E.

Oostanaula River (oo-stuh-NAW-luh), c.45 mi/72 km long, NW GEORGIA; formed by confluence of CONASAUGA and COOSAWATTEE rivers 3 mi/4.8 km NE of CALHOUN; flows SSW to ROME, where it joins with ETOWAH River to form the COOSA; 34°32′N 84°51′W.

Oostburg (AWST-buhrg), town, ZEELAND province, SW NETHERLANDS, 10 mi/16 km SW of VLISSINGEN; 51°19′N 03°29′E. Belgian border 3 mi/4.8 km to S and 5 mi/8 km to W; NORTH SEA 5 mi/8 km to N. Dairying; cattle, hogs; vegetables, grain, sugar beets. Town chartered in thirteenth century; reached its zenith as trading center in Middle Ages.

Oostburg (OOST-buhrg), town (2006 population 2,806), SHEBOYGAN county, E WISCONSIN, 10 mi/16 km SSW of SHEBOYGAN; 43°37′N 87°47′W. In dairy and farm area; manufacturing (dairy products, canned vegetables, plastic products); fishing. Kohler-Andrae State Park (on LAKE MICHIGAN) to NE (formerly called Terry-Andrae).

Oostcamp, BELGIUM: see OOSTKAMP.

Oostduinkerke (ost-DOIN-ker-kuh), village in Veurne district, WEST FLANDERS province, W BELGIUM, near NORTH SEA, 3 mi/4.8 km W of NIEUWPOORT; 51°07′N 02°41′E. Just NNW, on NORTH SEA, is seaside resort of Oostduinkerke-Bad.

Oostende (os-TEN-duh), French *Ostende*, commune (□ 113 sq mi/293.8 sq km; 2006 population 68,898), ⊙ Oostende district, WEST FLANDERS province, W BELGIUM, on the NORTH SEA; 51°13′N 02°55′E. It is a major commercial and fishing port, connected by canals with BRUGES and GHENT. Oostende is also an industrial and railroad center, as well as a seaside resort. It has a ferry terminal that connects the city with ENGLAND. Manufacturing includes processed food, ships, soap, tobacco, and chemicals. A port by the time of the First Crusade (eleventh century), Oostende was fortified (1583) by William the Silent and played a leading role in the Dutch struggle for independence. The city was taken (1604) by the Spanish under Ambrogio Spinola after a three-year siege in which it was almost totally destroyed, was sacked again in 1745 by the French, and suffered heavy Allied bombardment in World War II. From the mid-nineteenth century to World War I it was one of Europe's most fashionable social centers. The city is also known as Ostend.

Oosterbeek (AWST-uhr-baik), suburb of ARNHEM, GELDERLAND province, central NETHERLANDS, on LOWER RHINE RIVER, and 3 mi/4.8 km W of city center; 51°58′N 05°51′E. Dairying; cattle, poultry; fruit, vegetables, grain. Allied airborne landing here (1944), and subsequent battle. Doorwerth Castle (with Airborne Museum) 2 mi/3.2 km to SW.

Oosterhout (AWST-uhr-hout), city, NORTH BRABANT province, SW NETHERLANDS, 6 mi/9.7 km NE of BREDA, at junction of WILHELMINA and MARK canals; 51°38′N 04°52′E. Dairying; cattle, hogs; vegetables, potted plants, grain; manufacturing (doors, piping, chemicals, furniture, food processing).

Oosterland (AWST-uhr-lahnt), village, ZEELAND province, DUIVELAND region, SW NETHERLANDS, 5 mi/8 km E of ZIERIKZEE; 51°39′N 04°02′E. Mastgaat strait (car ferry) 2.5 mi/4 km to E; GREVELINGEN Dam 2.5 mi/4 km to ENE. Dairying; livestock; vegetables, sugar beets.

Oosterschelde, NETHERLANDS: see EASTERN SCHELDT.

Oosterschelde Dam (AWST-uhr-SKHEL-duh), ZEELAND province, SW NETHERLANDS, 10 mi/16 km NNE of Mindelburg; 51°38′N 03°43′E. Serpentine barrier dam 6 mi/9.7 km long at NORTH SEA entrance to EASTERN SCHELDT estuary, built as part of the Delta Project. Connects NORTH BEVELAND island (S) with SCHOUWEN-DUIVELAND island (N). Road crosses dam; Delta Expo at N center of dam.

Oosterwijk, NETHERLANDS: see OISTERWIJK.

Oosterwolde (AWST-uhr-VAWL-duh), town, FRIESLAND province, N central NETHERLANDS, 11 mi/18 km W of ASSEN, on Tjonger Canal; 52°59′N 06°18′E. Het Drents-Friese Woud (national park) to S. Dairying; cattle; grain, vegetables, fruit; manufacturing (glass, food processing).

Oosterzele (OS-tuhr-zai-luh), agricultural commune (2006 population 13,162), Ghent district, EAST FLANDERS province, NW BELGIUM, 8 mi/12.9 km SSE of GHENT; 50°57′N 03°48′E.

Oostkamp (OST-kahmp), commune (2006 population 21,831), Bruges district, WEST FLANDERS province, NW BELGIUM, 4 mi/6.4 km S of BRUGES; 51°09′N 03°14′E. Agriculture, lumbering. Formerly spelled Oostcamp.

Oostmahorn (awst-MAH-hawrn), village, FRIESLAND province, N NETHERLANDS, on LAUWERSMEER, and 8 mi/12.9 km ENE of DOKKUM; 53°23′N 06°10′E. Barrier dam on WADDENZEE 3 mi/4.8 km to N. Dairying; cattle, sheep; grain, potatoes.

Oostrozebeke (ost-RO-zuh-bai-kuh), commune (2006 population 7,439), Tielt district, WEST FLANDERS province, W BELGIUM, 7 mi/11.3 km NNE of KORTRIJK; 50°55′N 03°20′E. Textile industry; agricultural market.

Oost-Vlaanderen, BELGIUM: see EAST FLANDERS.

Oost-Vlieland, NETHERLANDS: see VLIELAND.

Oostvoorne (AWST-fawr-nuh), town, SOUTH HOLLAND province, SW NETHERLANDS, 17 mi/27 km W of ROTTERDAM; 51°55′N 04°06′E. On BRIELSE MEER, former channel of MEUSE RIVER, in Voone region. Formerly on NORTH SEA coast, town is now 3 mi/4.8 km E of coastline; EUROPOORT, large port facility, to NW. Dairying; cattle; vegetables, sugar beets, potatoes. Lighthouse 5 mi/8 km NW at entrance to NEW WATERWAY.

Ootacamund, INDIA: see UDAGAMANDALAM.

Ootmarsum (awt-MAHR-sum), village, OVERIJSSEL province, E NETHERLANDS, 11 mi/18 km ENE of ALMELO; 52°25′N 06°54′E. ALMELO-NORDHORN canal passes to S; German border 3 mi/4.8 km to N. Resort area; dairying; livestock; grain, vegetables.

Ootsa Lake (OOT-suh) (□ 50 sq mi/130 sq km), W central BRITISH COLUMBIA, W Canada, on N boundary of TWEEDSMUIR PARK, 120 mi/193 km WSW of PRINCE GEORGE; 53°47′N 126°15′W. Drains ESE into NECHAKO River by short Ootsa River.

Ooty, INDIA: see UDAGAMANDALAM.

Opaka (o-PAH-kuh), city (1993 population 3,592), RUSE oblast, Opaka obshtina, NE BULGARIA, on the RUSENSKI LOM RIVER, 7 mi/11 km NNW of POPOVO;

43°27′N 26°11′E. Grain; poultry, livestock, oil bearing plants, vineyards; tailoring.

Opal (O-puhl), village (2006 population 97), LINCOLN county, SW WYOMING, on HAMS FORK, and 11 mi/18 km E of KEMMERER; 41°46′N 110°19′W. Elevation c. 6,668 ft/2,032 m. Shipping point for sheep and wool. Agriculture (barley, alfalfa, hay; cattle).

Opal (O-puhl), hamlet, central ALBERTA, W Canada, 13 mi/21 km from THORHILD, in Thorhild County No. 7; 53°59′N 113°13′W.

Opala (o-PAH-lah), village, ORIENTALE province, central CONGO, on left bank of LOMAMI RIVER, and 110 mi/177 km SW of KISANGANI; 00°37′S 24°21′E. Elev. 1,453 ft/442 m. Steamboat landing in rice- and fiber-growing area. Roman Catholic mission.

Opalaca, Cordillera de (o-pah-LAH-kah, kor-dee-YER-rah dai), section of Continental Divide, in IN-TIBUCÁ department, W HONDURAS; extends c.20 mi/32 km NW-SE of LA ESPERANZA; 14°00′N 87°58′W–14°30′N 88°22′W. Elevation 7,298 ft/2,225 m. Forms divide between ULÚA RIVER (N) and LEMPA RIVER basins (S).

Opal Cliffs, unincorporated town (2000 population 6,458), SANTA CRUZ county, SW CALIFORNIA, residential suburb 3 mi/4.8 km E of downtown SANTA CRUZ, on W side of Soquel Cove, MONTEREY BAY, PACIFIC OCEAN; 36°58′N 121°58′W. CAPITOLA and New Brighton state beaches to E.

Opalenica (o-pah-le-NEE-tsah), German *Opalenitza*, town, Poznań province, W POLAND, 23 mi/37 km WSW of POZNAŃ; 52°18′N 16°26′E. Railroad junction. Beet sugar and flour milling; tanning.

Opalescent River, NEW YORK: see HUDSON RIVER.

Opalikha (uh-PAH-lee-hah), village (2006 population 5,070), central MOSCOW oblast, central European Russia, on road and railroad, 3 mi/5 km W, and under administrative jurisdiction, of KRASNOGORSK; 55°49′N 37°15′E. Elevation 643 ft/195 m. Holiday getaway spot for residents of Moscow and neighboring cities; hotels, summer homes. Has an airfield.

Opa-locka (o-puh–LAHK-uh), city (□ 4 sq mi/10.4 sq km; 2005 population 15,763), MIAMI-DADE county, SE FLORIDA, 14 mi/23 km NNW of MIAMI; 25°53′N 80°15′W. There is some diverse light industry, and a large county airport. A U.S. Coast Guard station is located at the airport. Opa-locka's city hall is patterned after a Moorish castle, and other buildings are of Arabic architecture. Incorporated 1926.

Opan (o-PAHN), village (1993 population 512), HASKOVO oblast, Opan obshtina, BULGARIA; 42°13′N 25°41′E.

Opanayake (O-pah-NAH-yah-kuh), town, SABAR-AGAMUWA PROVINCE, S central SRI LANKA, 16 mi/26 km ESE of RATNAPURA. Railroad terminus; rubber, tea, vegetables, rice. Also spelled Opanayaka.

Opanets (o-pah-NETS), village, LOVECH oblast, PLEVEN obshtina, BULGARIA; 43°25′N 23°34′E.

Opari (o-PAH-ree), village, E EQUATORIA state, S SUDAN, near UGANDA border, on road, and 70 mi/113 km SSE of JUBA; 03°56′N 32°03′E. Cotton, corn, and durra; livestock.

Oparino (uh-PAH-ree-nuh), town (2005 population 4,935), NW KIROV oblast, E central European Russia, on road and railroad, 100 mi/161 km NNW of KIROV; 59°51′N 48°17′E. Highway and railroad junction; local transshipment point. Elevation 613 ft/186 m. Dairy, bakery, produce processing; seed inspection, veterinary clinic; logging, lumbering.

Oparo, FRENCH POLYNESIA: see RAPA, island.

Opasatika, township (□ 127 sq mi/330.2 sq km; 2001 population 325), central ONTARIO, E central Canada; 49°07′N 83°07′W.

Opatija (O-pah-tee-yah), Italian *Abbazia*, town, W CROATIA, on KVARNER GULF of ADRIATIC SEA, 7 mi/11.3 km W of RIJEKA (Fiume), in NE ISTRIA region, at NE foot of Učka Mountain; 45°20′N 14°16′E. Noted summer and winter resort (Optaija Riviera). Includes

suburbs of LOVRAN (S), and VOLOSKO, Italian *Volosca* (N).

Opatoro (o-pah-TO-ro), town, LA PAZ department, SW HONDURAS, 21 mi/34 km SW of LA PAZ; 14°05′N 87°54′W. Agriculture (coffee, wheat); manufacturing (palm hats, rope, pottery).

Opatow (o-PAH-toov), Polish *Opatów*, Russian *Opatov*, town, Tarnobrzeg province, E POLAND, 35 mi/56 km ESE of KIELCE. Manufacturing (chemicals, cement product brushes). Known from 1040; flourished at end of 14th century, trading with GREECE, PERSIA, ARMENIA, NUREMBERG, and HOLLAND. Has old church and town gate.

Opava (O-pah-VAH), German *Troppau*, city (2001 population 61,382), SEVEROMORAVSKY province, CZECH REPUBLIC, central SILESIA, on the OPAVA RIVER and near the Polish border; 49°57′N 17°55′E. A prosperous market center in a fertile agricultural region, it has food-processing plants and industries producing clothing and machinery. Road and railroad hub. Founded in the 12th century, it later became the capital of Austrian Silesia. In 1820 representatives of the great European powers met here at the Congress of Troppau to discuss problems arising after the settlement of the Napoleonic Wars. City landmarks include a 15th century cathedral built by the Teutonic Knights, the 15th century Church of St. George, and a 17th century Jesuit foundation.

Opava River (O-pah-VAH), German *Oppa*, Polish *opawa*, 81 mi/130 km long, SEVEROMORAVSKY province, central SILESIA, CZECH REPUBLIC; formed by three headstreams joining at VRBNO POD PRADEDEM; flows SE, then NE past Krnov and SE along the Czech/Polish border for c.12 mi/19 km, past OPAVA, to ODER RIVER, at W outskirts of OSTRAVA. Receives the MORAVICE RIVER (right).

Opawa River, CZECH REPUBLIC: see OPAVA RIVER.

Opbrakel (AHP-brah-kuhl), agricultural village in commune of BRAKEL, Oudenaarde district, EAST FLANDERS province, W central BELGIUM, 7 mi/11.3 km ESE of OUDENAARDE; 50°47′N 03°45′E.

Opechee Bay (O-puh-chee), S extension (2 mi/3.2 km long) of LAKE WINNIPESAUKEE, BELKNAP county, central NEW HAMPSHIRE; continuation of larger PAUGUS BAY to NE, in city of LACONIA, and connected by narrow channel to E side of Opechee Bay. Drains SW through Winnipesaukee River, 1 mi/1.6 km to WINNISQUAM LAKE. Sometimes called Opechee Lake.

Opechenskiy Posad (uh-PYE-chin-skeeyee puh-SAHT), village, E NOVGOROD oblast, NW European Russia, on the MSTA River, on highway, 10 mi/16 km SE of BOROVICHI; 58°16′N 34°07′E. Elevation 472 ft/143 m. In agricultural area (flax).

Opelika (ah-puh-LEE-kuh), city (2000 population 23,498), ⊙ Lee co., E Alabama, c.15 mi/24 km W of Chattahoochee River (Ga. line), and c.23 mi/37 km NW of Phenix City. Trade center, with textile, lumber, paperboard (box), tires, sporting goods, magnetic recording tape, drilling equipment, and metallurgical industries. Factory outlet mall here. Inc. 1854. Founded in 1830 as Opelikan, it was changed in 1850 to Opelika, the name of a town in Coosa County. Chosen as county seat in 1866.

Opelousas (ah-puh-LOO-suhs), city, ⊙ SAINT LANDRY parish, S central LOUISIANA, 20 mi/32 km N of LAFAYETTE; 30°31′N 92°05′W. Railroad junction. Industries based chiefly on the agricultural products and livestock of the surrounding region; printing and publishing, manufacturing (water treating units, plasma, glass, cooking oil, sausage and food production, apparel, pipe fittings). Oil and gas nearby. Opelousas still retains some of its early French and Spanish flavor, and many antebellum structures remain. It was founded c.1765 by French traders and served (1863) as state capital for a period during the Civil War. Boyhood home of James Bowie here. Antebellum homes in area. Incorporated 1821.

Open Door, town, NE BUENOS AIRES province, ARGENTINA, 5 mi/8 km NNE of LUJÁN; 34°30′S 59°05′W. Agricultural center (alfalfa, grain, flax; livestock).

Open Lake, TENNESSEE, see: RIPLEY.

Openshaw, ENGLAND: see MANCHESTER.

Opeongo Lake (o-pee-AHNG-go), S ONTARIO, E central Canada, in ALGONQUIN PROVINCIAL PARK, 60 mi/97 km W of PEMBROKE; 20 mi/32 km long, 5 mi/8 km wide; 45°42′N 78°23′W. Elevation 1,323 ft/403 m. Drained by MADAWASKA RIVER.

Opfikon, town (2000 population 12,062), ZÜRICH canton, N SWITZERLAND, NE suburb of ZÜRICH; 47°26′N 08°35′E.

Opglabbeek (AHP-glah-baik), commune (2006 population 9,630), Hasselt district, LIMBURG province, NE BELGIUM, 7 mi/11 km NNE of GENK; 51°03′N 05°35′E.

Opheim, village (2000 population 111), VALLEY county, NE MONTANA, near CANADA (SASKATCHEWAN) border, 48 mi/77 km N of GLASGOW; 48°52′N 106°25′W. Opheim port of entry 10 mi/16 km N. Agriculture (wheat, barley, oats, hay; cattle, sheep, hogs); manufacturing potato lefse (pancakes).

Ophel, hill in ancient JERUSALEM, E ISRAEL. In the Hebrew Bible, it was the site of original Jebusite settlement captured by David. Fortified by Jotham and Manasseh.

Opheusden (awp-HUHZ-duhn), village, GELDERLAND province, central NETHERLANDS, 9 mi/14.5 km ENE of TIEL; 51°56′N 05°37′E. LOWER RHINE RIVER 0.5 mi/0.8 km to N, WAAL RIVER 2.5 mi/4 km to S. Agriculture (dairying; cattle, poultry; grain, vegetables); fruit, tree nurseries.

Ophidousa, Greece: see OFIDOUSA.

Ophir (O-fuhr), village, SW central ALASKA, on INNOKO RIVER, and 30 mi/48 km WNW of MCGRATH. Trapping; placer gold mining. Airstrip.

Ophir, village (2000 population 113), SAN MIGUEL county, SW COLORADO, in SAN JUAN MOUNTAINS, 5 mi/8 km S of TELLURIDE, on Howard Fork Creek; 37°51′N 107°49′W. Elevation c.9,800 ft/2,987 m. Surrounded by Uncompahgre National Forest. In goldmining region. Ophir Loop (elevation 9,280 ft/2,829 m), 2 mi/3.2 km W, turnoff to Ophir.

Ophir, village (2006 population 27), TOOELE county, N central UTAH, 12 mi/19 km S of TOOELE, in OQUIRRH MOUNTAINS; 40°22′N 112°15′W. Elevation 6,498 ft/1,981 m. Silver, gold, lead deposits; lead and silver still mined. Old mining town founded 1874, once had population of 6,000, population in 1959 was 199; semi-ghost town with general store.

Ophir, settlement, E central NEW SOUTH WALES, SE AUSTRALIA, 18 mi/29 km NE of ORANGE; 33°10′S 149°14′E. Site of first payable gold fields in Australia (1851). Open to fossicking (gold and gem hunting); is also a nature reserve.

Ophir, Mount, peak (elev. 9,554 ft/2,912 m), N Sumatra Barat province, INDONESIA, in PADANG HIGHLANDS of BARISAN MOUNTAINS, 75 mi/121 km NNW of PADANG; 00°04′N 99°55′E.

Ophir, Mount (O-fuhr), Malay *Gunong Ledang* (GOO-nong LEH-dahng), highest peak (4,187 ft/1,276 m) in Johore, S MALAYSIA, 100 mi/161 km NW of Johore Bharu, near MALACCA line; 02°21′N 102°39′E.

Opi (o-PEE), town, ENUGU state, SE NIGERIA, on road, 30 mi/48 km N of ENUGU; 06°47′N 07°26′E. Market town. Cashews, rice.

Opichén (o-pee-CHEN), town, YUCATÁN, SE MEXICO on Mexico Highway 184, 33 mi/53 km SSW of MÉRIDA; 20°56′N 89°41′W. Henequen, sugarcane, fruit.

Opicina (o-pee-CHEE-nah), Slovenian *Opčine* (OP-chee-nai), N suburb of TRIESTE, NE ITALY, on the karst above the city, near SLOVENIAN border. Railroad junction (Poggioreale Campagna) and frontier station; fine views.

Opico, EL SALVADOR: see SAN JUAN OPICO.

Opienge (o-PYENG-gai), village, ORIENTALE province, E CONGO, 125 mi/201 km ESE of KISANGANI; 00°12′N

27°30′E. Elev. 2,677 ft/815 m. Trading post with Protestant mission; rice processing.

Opillya, W UKRAINE: see PODOLIAN UPLAND.

Opishnya (o-PEESH-nyah) (Russian *Oposhnya*), town, NE POLTAVA oblast, UKRAINE, on the VORSKLA RIVER, and 25 mi/40 km N of POLTAVA; 49°58′N 34°37′E. Elevation 482 ft/146 m. Weaving, pottery, ceramic industry; flour mill. Known since the 17th century; company town of Ukrainian Cossack (Hadyach, then Poltava, Zinkiv and again Hadyach) regiments; known for handicrafts since the 18th century; town since 1925.

Opiskoteo Lake, E central QUEBEC, E Canada, near NEWFOUNDLAND AND LABRADOR border; 22 mi/35 km long, 15 mi/24 km wide; 53°10′N 68°10′W. Elevation 2,025 ft/617 m.

Oploca (o-PLO-kah), town and canton, SUD CHICHAS province, POTOSÍ department, SW BOLIVIA, on Tupiza River, and 10 mi/16 km NNW of TUPIZA, on VILLA-ZÓN–UYUNI railroad; 21°19′S 65°46′W. In agricultural area; corn, potatoes; orchards. Oploca tin mines lie 45 mi/72 km NW, near CHOCAYA, on W slopes of Cordillera de CHICHAS.

Opobo (o-PO-bo), town, Rivers state, SE NIGERIA, in the NIGER RIVER delta; 04°31′N 07°32′E. It is a palm oil collection center and has fishing and boatbuilding industries. Founded in 1869 by a group of immigrants from nearby BONNY led by Jaja, a middleman in the palm oil trade with Europeans. Opobo prospered, but Jaja antagonized the Europeans by hampering their trade. Jaja was deported by the British in 1887, after which Opobo declined.

Opochka (uh-POCH-kah), city (2006 population 13,410), SW PSKOV oblast, W European Russia, on the VELIKAYA RIVER, 80 mi/130 km S of PSKOV, and 65 mi/105 km NW of NEVEL′; 56°42′N 28°43′E. Elevation 311 ft/94 m. Road center; woodworking, flax and food processing. Founded in 1412 as a fortress. Made city in 1777.

Opochno, POLAND: see OPOCZNO.

Opocno (O-poch-NO), Czech *Opočno*, German *opotschno*, town, VYCHODOCESKY province, E BOHEMIA, CZECH REPUBLIC, 13 mi/21 km ENE of HRADEC KRÁLOVÉ; 50°16′N 16°07′E. Railroad junction. Manufacturing (electronics); food processing. Noted Renaissance castle.

Opoco (o-PO-ko), canton, ANTONIO QUIJARRO province, POTOSÍ department, W central BOLIVIA, 9 mi/15 km SE of RÍO Mulato, on the Río Mulato–UYUNI road; 19°50′S 66°42′W. Elevation 12,024 ft/3,665 m. Antimony mining in area. Clay and limestone deposits. Potatoes, yucca, bananas; cattle.

Opoczno (o-POCH-no), Russian *Opochno*, town (2002 population 22,589), Piotrków province, central POLAND, on Drzewiczka River, on railroad, and 45 mi/72 km SE of ŁÓDŹ; 51°22′N 20°17′E. Clay pit, limekiln; ironworks; glass, porcelain; flour milling, brewing. Small electric plant.

Opodepe (o-po-DAI-pai), town, SONORA, NW MEXICO, on San Miguel de Horcasitas River, and 60 mi/97 km NNE of HERMOSILLO; 29°56′N 110°38′W. Corn, beans; cattle.

Opol (O-pol), town, Misamis Oriental province, N MINDANAO, PHILIPPINES, near MACAJALAR BAY, 6 mi/9.7 km NW of CAGAYAN DE ORO; 08°25′N 124°30′E. Agricultural center (corn, coconuts). Fishing port. Chromite deposits.

Opole (o-PO-le), province (□ 3,633 sq mi/9,445.8 sq km), S POLAND; ⊙ OPOLE (Oppeln); 51°22′N 20°17′E. Borders S on CZECH REPUBLIC. Surface rises from fertile Oder valley SW to the SUDETES range, which slopes down (SE) toward Moravian Gap. Drained by ODER and GLATZER NEISSE (Nysa Klodzka) rivers. Principal industries: leather-goods, manufacturing, tanning, textile milling, sugar refining. Heavy industry (manufacturing of machinery, electrical equipment) concentrated at RACIBORZ (Ratibor). Iron, galena, tetrahedryte are worked. Chief crops: rye, potatoes,

oats, wheat, sugar beets, flax. Principal cities: Opole, Raciborz, NEISSE (Nysa), Brieg (BRZEG), and KOZLE (Cosel). Province created 1950 when Slask (Polish *Śląsk*) province, established 1945 and briefly called SLASK DABROWSKI (Polish *Śląsk Dąbrowski*), was divided between Opole province (W) and Katowice province (E). It also received small part (W) of Wrocław province. German population expelled after 1945 and replaced by Poles.

Opole (o-PO-le), German *Oppeln*, city (2002 population 129,946), S POLAND, on the ODER River. A river port and railroad junction, it is also an important trade center, with manufactures of cement, metals, and furniture. Originally a Slavic settlement, it was the seat (1163–1532) of the dukes of Opole of the Piast dynasty. The duchy passed (1532) to the house of HAPSBURG and (1742) to PRUSSIA, and was incorporated into Poland in 1945. It was the capital (1919–1945) of the Prussian province of UPPER SILESIA. In the city are the churches of Saint Adalbert (10th century) and of the Holy Cross (14th century).

Opon, PHILIPPINES: see LAPU-LAPU.

Opoqueri (o-po-KE-ree), canton, CARANGAS province, ORURO department, W central BOLIVIA, 25 mi/40 km SE of CHOQUECOTA, and 18 mi/30 km SW of CORQUE; 18°30′S 67°58′W. Elevation 12,448 ft/3,794 m. Clay, limestone, and gypsum deposits. Agriculture (potatoes, yucca, bananas); cattle.

Oporapa (o-po-RAH-pah), town, ⊙ Oporapa municipio, HUILA department, S central COLOMBIA, 80 mi/129 km SW of NEIVA; 02°02′N 75°58′W. Elevation 3,661 ft/1,115 m. Coffee, plantains; livestock.

Opornyi (o-POR-nee), town, MANGYSTAU region, KAZAKHSTAN, on highway and railroad, near ATYRAU region border, 60 mi/100 km E of CASPIAN SEA, and 50 mi/80 km NNW of BEYNEU; 46°13′N 54°27′E. Oil and gas pipelines from MANGYSHLAK fields to S. Oil and gas industry support; some seasonal agriculture.

Oporto (o-PORT-oo), Portuguese *Pôrto*, city, ⊙ PÔRTO district and DOURO LITORAL, NW PORTUGAL, near the mouth of the DUERO RIVER. It is the second-largest city of Portugal, after LISBON, and an important ATLANTIC port. Its outer harbor is at LEIXÕES. Most famous export is port wine, to which the city gives its name. Cork, fruits, olive oil, and building materials are also exported. Manufacturing includes cotton, silk, and wool textiles, wood and leather goods; fish and other food processing. International airport to NW. The ancient settlement, probably of pre-Roman origin, was known as Cale and later as Portus Cale. Captured by the Moors in C.E. 716 and retaken by Christians in 1092. The centuries of war depopulated the town. Henry of Burgundy secured the title of duke of Portucalense in the 11th century, and Oporto thus gave its name to the state that became a kingdom. For some time became the chief city, although not the capital, of little Portugal. Wine exports increased after the Methuen Treaty (1703) with ENGLAND. The creation by the marquês de POMBAL of a wine monopoly brought on the "tipplers' revolt" (1757) in Oporto. After the French conquest of Portugal in the Peninsular War, Oporto was the first city to revolt (1808). Retaken by the French but liberated (1809) by Wellington. In 1832, in the Miguelist Wars, Dom Pedro I of BRAZIL long withstood a siege of the city by his brother, Dom Miguel. Later a center of republican thought; in 1891 an abortive republican government was set up here. The city's most conspicuous landmark is the Torre dos Clérigos, a baroque tower; also noteworthy are the Romanesque cathedral, the two-storied Dom Luis bridge (1881–1887), which spans the Douro, and the Crystal Palace (1865). Seat of a university. Several housing projects have been built since World War II, but overcrowding is still a problem, particularly in the slums of the old city.

Oposhnya, UKRAINE: see OPISHNYA.

Opotiki (o-po-TEE-kee), town, OPOTIKI district (□ 1,198 sq mi/3,114.8 sq km), NORTH ISLAND, NEW ZEALAND, on BAY OF PLENTY flats, where direct road through Waioeka River gorge from GISBORNE (37 mi/60 km to SE) meets longer EAST CAPE route. Agricultural center for corn-raising; dairy products; sheep.

Opotschno, CZECH REPUBLIC: see OPOCNO.

Opp (AHP), city (2000 population 6,607), Covington co., S Alabama, 15 mi/24 km E of Andalusia. Clothing, textiles; food processing; lumber. Douglas MacArthur State Technical College. Inc. 1902. Named for Henry Opp, a lawyer and the settlement's principal promoter.

Oppa River, CZECH REPUBLIC and POLAND: see OPAVA RIVER.

Oppau (OP-ou), N district (since 1938) of LUDWIG-SHAFEN, GERMANY; 49°33′N 08°24′E. Manufacturing of chemicals (chemicals, textiles).

Oppdal (AWP-duhl), village, SØR-TRØNDELAG county, central NORWAY, on DRIVA River, on railroad, and 40 mi/64 km NNE of DOMBÅS. Slate quarry; tourism and skiing center in the DOVREFJELL. The Dovrefjell National Park is nearby.

Oppdøl (AWP-duhl), village, MØRE OG ROMSDAL county, W NORWAY, on SUNNDALSFJORDEN, 45 mi/72 km E of MOLDE. Tourist center, starting point for trips to TROLLHEIMEN mountains.

Oppegård (AWP-puh-gawr), residential village (2007 population 23,993), AKERSHUS county, SE NORWAY, on E shore of Bundefjord (SE arm of OSLOFJORD), 9 mi/14.5 km S of OSLO; 59°46′N 10°45′E.

Oppeln, POLAND: see OPOLE.

Oppenau (OP-pen-ou), town, S UPPER RHINE, BADEN-Württemberg, SW GERMANY, in BLACK FOREST, 10 mi/16 km E of OFFENBURG; 48°29′N 08°07′E. Lumber milling, woodworking. Climatic health resort (elevation 879 ft/268 m) and tourist center.

Oppenheim (OP-pen-heim), town, RHINELAND-PALATINATE, W GERMANY, on left bank of the Rhine, and 11 mi/18 km SE of MAINZ; 49°52′N 08°21′E. Noted for its wine; textiles. Ruins of former imperial fortress Landskron (destroyed 1689 by French) tower above town. On site of Roman settlement; was free imperial city. Gothic church (12th–15th century). Was first mentioned 765, chartered 1226.

Oppenweiler (op-pen-VEI-luhr), village, Region of STUTTGART, BADEN-Württemberg, S GERMANY, 13 mi/21 km ENE of LUDWIGSBURG, on the MURR; 48°59′N 09°28′E.

Oppido Mamertina, town, REGGIO DI CALABRIA province, CALABRIA, S ITALY, on N slope of the ASPRO-MONTE, 9 mi/14 km ESE of PALMI; 38°17′N 18°59′E. In lumbering and livestock-raising region. Bishopric. Rebuilt after earthquakes of 1783 and 1908.

Oppland (AWP-lahn), county (□ 9,720 sq mi/25,174 sq km; 2007 estimated population 182,972), S central NORWAY; ⊙ LILLEHAMMER. The chief town is GJØVIK. The county is traversed from NW to SE by the GUD-BRANDSDALEN valley. Farming and lumbering are the main occupations, and tourism is an important industry. The N section is mountainous.

Opponitz (AWP-puh-nits), village, SW LOWER AUSTRIA, AUSTRIA, on YBBS RIVER, and 6 mi/9.7 km SSE of Waidhofen; 47°53′N 14°49′E. Hydroelectric plant.

Opportunity, unincorporated city (2000 population 25,065), SPOKANE county, E WASHINGTON, a residential suburb 7 mi/11.3 km E of downtown SPOKANE; 47°39′N 117°14′W. It is a growing residential town.

Opportunity, locality, DEER LODGE county, SW MONTANA, within city of ANACONDA (city and county boundaries are coterminous), 6 mi/9.7 km ESE of Anaconda proper, near Silver Bow Creek, which becomes CLARK FORK river 5 mi/8 km to NE. Former copper mining area, closed 1970. Tailings pond to N. Anaconda Stack State Park to W, 585 ft/178 m smokestack.

Opportunity No. 17 (ah-por-TOO-ni-tee), municipal district (□ 11,267 sq mi/29,294.2 sq km; 2001 population 3,436), N ALBERTA, W Canada; 56°14′N 113°47′W. Oil and gas; logging; diamond excavation fields; tourism. Includes the hamlets of Calling Lake, Wabasca-Desmarais, Red Earth Creek, and Sandy Lake; and the settlements of Chipewyan Lake, Loon Lake, Peerless Lake, and Trout Lake. First formed as an improvement district in 1913.

Opryland USA, theme park, in DAVIDSON county, central TENNESSEE, on the CUMBERLAND RIVER, 6 mi/10 km NE of NASHVILLE; 36°12′N 86°41′W. Covers 120-acre/49-ha complex including an amusement park, a hotel, and performance theaters, including the Opry House. The Grand Ole Opry, a live country music program founded in 1925, is broadcast weekly from the Opry House.

Optikogorsk, RUSSIA: see KRASNOGORSK, MOSCOW oblast.

Optima (AWP-ti-muh), village, TEXAS county, central OKLAHOMA PANHANDLE, 10 mi/16 km NE of GUYMON, near NORTH CANADIAN RIVER (Beaver River); 36°45′N 101°20′W. In wheat-growing area. Nearby is site of Optima Lake reservoir, for flood control, irrigation; includes Optima National Wildlife Refuge around lake.

Opua (o-POO-ah), township, NORTH ISLAND, Far North district, NEW ZEALAND, 110 mi/177 km N of AUCKLAND, and on S shore of BAY OF ISLANDS; 35°19′S 174°07′E. Railroad terminus; in agriculture and resort area. Deep-sea fishing.

Opunake (o-poo-NAK-ee), town, W NORTH ISLAND, in S TARANAKI district, NEW ZEALAND, 30 mi/48 km SSW of NEW PLYMOUTH; 39°27′S 173°51′E. Intensive dairying center where Taranaki Ring Plain meets sea.

Opuntian Locris, Greece: see LOCRIS.

Opuwo (o-poo-wo), town, NW NAMIBIA, 100 mi/161 km NNW of KHORIXAS; 18°04′S 13°51′E. Trading center for region. High school and clinic center of N KUNENE REGION.

Opuzen (O-puh-zen), village, S CROATIA, on NERETVA RIVER, near its mouth, on railroad, and 4 mi/6.4 km SW of METKOVIC, in DALMATIA. Former Austrian strongpoint called FORT OPUS.

Opwijk (AHP-weik), commune (2006 population 12,319), Halle-Vilvoorde district, BRABANT province, central BELGIUM, 11 mi/18 km NW of BRUSSELS; 50°58′N 04°11′E. Textile industry; agricultural market.

Opzullik, BELGIUM: see SILLY.

Oqqŭrghon, UZBEKISTAN: see AKKURGAN.

Oquaga Lake (ok-WAH-gah), resort village, Broome county, S NEW YORK, near PENNSYLVANIA state line, 25 mi/40 km ESE of BINGHAMTON, on small Oquaga Lake; 42°02′N 75°27′W. Golf course.

Oquawka (ok-WAW-kah), village (2000 population 1,539), ⊙ HENDERSON county, W ILLINOIS, on the MISSISSIPPI, and 11 mi/18 km NNE of BURLINGTON, IOWA; 40°56′N 90°57′W. Summer resort; agriculture (produce; livestock); limestone quarries.

Oquirrh Mountains (O-kuhr), NW UTAH, extend c.30 mi/48 km S from GREAT SALT LAKE, SALT LAKE, TOOELE, and UTAH counties. Rise to c.11,000 ft/3,353 m. Copper, lead, zinc, gold and silver mined near BINGHAM CANYON; Kennecott Corporation Open Cut Copper Mine, largest in world. On boundary of Tooele (W) and Salt Lake and Utah counties (E).

Oquitoa (o-kee-TO-ah), town, SONORA, NW MEXICO, on Altar River, and 60 mi/97 km SW of NOGALES; 30°46′N 111°42′W. Wheat, corn, cotton, beans; cattle.

Oquossoc, MAINE: see RANGELEY, village.

Ora (OR-uh), river, 70 mi/113 km long, NEBBI and ARUA districts, NORTHERN region, NW UGANDA; rises from the N sections of DEMOCRATIC REPUBLIC OF THE CONGO's Mt. Bleus range and flows E to the ALBERT NILE RIVER.

Ora Banda (O-ruh BAN-duh), locality, WESTERN AUSTRALIA state, W AUSTRALIA, 41 mi/66 km NE of KALGOORLIE; 30°27′S 121°04′E. Goldfields. Virtual ghost town.

Oracabessa, town, SAINT MARY parish, N JAMAICA, banana port on fine CARIBBEAN bay, 4 mi/6.4 km WNW of PORT MARIA; 18°28′N 76°56′W.

Oracle, unincorporated town (2000 population 3,563), PINAL county, S central ARIZONA, 31 mi/50 km NNE of TUCSON. Cattle, sheep; alfalfa. BLACK HILLS to N; SANTA CATALINA MOUNTAINS, in part of Coronado National Forest, to S. Site of Biosphere 2 environmental living experiment to S, at Mount LEMMON.

Oradea (o-RAH-dyah), Hungarian *Nagyvárad*, German *Grosswardein*, city, W ROMANIA, in TRANSYLVANIA, 66 mi/106 km NNE of ARAD, near the Hungarian border; ⊙ BIHOR county; 47°04′N 21°56′E. It is the marketing and shipping center for a livestock and agricultural region. Oradea is also an important industrial city with manufacturing of machinery, mining equipment, and chemicals; food processing. Airport. There are health resorts nearby. The city was made (1083) the seat of a River C. bishop by King Ladislaus I of HUNGARY. Destroyed (1241) by the Mongols, it was rebuilt in the 15th century. Oradea was held by the Turks from 1660 to 1692. From mid-19th century until World War II, 20% to 25% of the population was Jewish. It was a early center for Reform (Neolog) as well as Orthodox Judaism. Hungary ceded it (1919) to Romania after World War I, but Hungarian forces occupied the city during World War II. The Jewish community was mostly destroyed in the Holocaust; the few who survived later emigrated to Israel. About ½ the population is Hungarian. Most of the city's architecture is baroque, dating from the reign of Maria Theresa.

Oradell (or-RUH-del), residential borough (2006 population 7,957), BERGEN county, NE NEW JERSEY, on HACKENSACK River, at SW end of Oradell Reservoir, and 5 mi/8 km N of HACKENSACK; 40°57′N 74°01′W. Settled by Dutch before the Revolution, incorporated 1894.

Oradna, ROMANIA: see RODNA.

Oradour-sur-Glane (o-rah-DOOR–syor–glahn), commune (□ 14 sq mi/36.4 sq km), HAUTE-VIENNE department, LIMOUSIN region, W central FRANCE, 12 mi/19 km NW of LIMOGES; 45°56′N 01°03′E. It was burnt down and the entire population (642 people) massacred by a German S.S. detachment, June 10, 1944, in World War II. The horror of this event is compounded by the fact that the village was chosen for its innocence and thus, according to the S.S., the massacre was more apt to terrorize the rest of the country. Village was rebuilt next door to ruins, and a memorial was built to commemorate the event.

Öræfajökull (UHR-ree-vah-YUH-ku-tuhl), mountain, SE ICELAND, rising from the VATNAJÖKULL glacier. ÖRÆFAJÖKULL is an ice-covered, three-peaked volcano. The largest of its four recorded eruptions occurred in 1362; all have been destructive, owing to the devastating floods caused by melting ice. Its highest peak, Hvannadalshnukur (6,950 ft/2,118 m), is also the highest point in Iceland.

Orahovac (or-AH-ho-vahts), village, KOSOVO province, SW SERBIA, 14 mi/23 km N of PRIZREN, in the METOHIJA valley; 42°24′N 20°40′E. Also spelled ORAKHOVATS.

Orahovica (O-rah-ho-vee-tsah), village, E CROATIA, 31 mi/50 km SE of VIROVITICA, at E foot of the PAPUK, in SLAVONIA; 45°32′N 17°54′E. Railroad terminus; local trade center. Castle ruins. First mentioned in 1228.

Orai (O-rei), town, S UTTAR PRADESH state, N central INDIA, 65 mi/105 km SW of KANPUR; ⊙ JALAUN district; 25°59′N 79°28′E. Road center; trades in gram, wheat, oilseeds, jowar.

Oraibi (o-REI-bei), pueblo, NAVAJO county, N ARIZONA, on a mesa 55 mi/89 km N of WINSLOW, in Hopi Indian Reservation. It was built 1100 and was discovered in 1540 by Pedro de Tovar, a lieutenant of Coronado. The mission of San Francisco, established on the site in 1629, was destroyed in the Pueblo revolt of 1680. Oraibi was long the most important pueblo of the Hopi Indians, but because of economic disturbances and internal dissension many of the inhabitants left in 1907 to form the pueblos of HOTEVILLA and Bacavi. Also called Old Oraibi.

Oraison (o-rai-ZON), commune (□ 14 sq mi/36.4 sq km), ALPES-DE-HAUTE-PROVENCE department, PROVENCE-ALPES-CÔTE D'AZUR region, SE FRANCE, near left bank of DURANCE RIVER, 20 mi/32 km SW of DIGNE-LES-BAINS, in Provence (MARITIME) Alps; 43°55′N 05°55′E. Market for olives and fruit. Has hydroelectric plant on Durance River canal.

Orakhovats, SERBIA: see ORAHOVAC.

Or Akiva (OR ah-KEE-vah) or **Or Aqiva**, township, ISRAEL, 1.2 mi/2 km E of ancient CAESAREA and N of HADERA; 32°30′N 34°55′E. Elevation 88 ft/26 m. Industrial zone with large factories for polymers, carpets, upholstery, and synthetic furs. Founded 1951 and named for Rabbi Akiva, who was slain in the ancient Roman attack on Caesarea.

Oral (o-RAHL), city (1999 population 195,459), NW KAZAKHSTAN, on the URAL RIVER; ⊙ WEST KAZAKHSTAN region; 51°19′N 51°20′E. Manufacturing of food; grain processing; tanning. Founded (1613) as Yaitskii Gorodok by the Ural Cossacks, who fought with Stenka Razin in the uprising of 1667 and against the Bolsheviks in 1918–1919. Renamed (1775) Uralsk, it was an important trade center on the border of European Russia and Kazakhstan. Current name adopted after Kazakhstan became independent in 1991. Also known as Uralsk.

Oran (o-RAWNG), Arabic *Wahran*, wilaya or province, W ALGERIA, on the MEDITERRANEAN SEA; ⊙ ORAN. Includes the city of Oran, where ¾ of the wilaya population lives, and a few other smaller urban centers in the vicinity, such as ARZEW, MERS EL KEBIR, and ESSENIA. One of the country's major industrial poles, with hydrocarbon and chemical industries located at Arzew and BETHIOUA. Also, an oil and gas terminal at Arzew. Has some rich farmland. Under French rule, Oran department covered all of NW Algeria, but most of its former territory has been divided among other wilaya since independence.

Oran (o-RAWNG), city, NW ALGERIA, a port on the MEDITERRANEAN SEA; ⊙ ORAN wilaya; 35°45′N 00°38′W. The second-largest city and second-most important city (after ALGIERS), Oran is built mainly on a plateau rising steeply from the shore. Manufacturing includes steel, metal, and glass; also, textile manufacturing and food processing. International airport. During the medieval heyday of TLEMCEN, Oran served as its Mediterranean port. Later, almost 3 centuries of Spanish occupation cut Oran off from its hinterland. SPAIN had envisioned Oran to be the capital of a new Spanish empire in N. AFRICA. Under French rule (from 1831), the city expanded considerably and attracted a large settler community, becoming the most European of Algerian cities. The French also developed a naval base here. Oran has an old Arab-Spanish section and a growing, modern part, where the writer Albert Camus lived. Toward the end of the War of Independence (1954–1962), the city witnessed some of the most violent and bloody terrorist actions by the European settlers' secret army (Organisation Armée Secrète, or OAS), which wanted to maintain French rule. After independence, 200,000 Europeans left the city, replaced by native Muslims. Oran's population subsequently grew rapidly, leading to the development of new residential areas and modern apartment towers. Growth has caused such problems as traffic congestion and, above all, water shortages.

Oran (o-RAN), city (2000 population 1,264), SCOTT county, SE MISSOURI, in MISSISSIPPI alluvial plain, 16 mi/26 km SSW of CAPE GIRARDEAU; 37°05′N 89°39′W. Corn, soybeans; livestock; some manufacturing. Platted 1869.

Cross-references are shown in SMALL CAPITALS. The pronunciation guide is shown on page xix. The sources of population figures are shown on page xvii.

Orán (o-RAHN), town, N SALTA province, ARGENTINA, 95 mi/153 km NE of JUJUY; ⊙ Orán department (□ 10,660 sq mi/27,716 sq km); 27°10′S 65°33′W. Railroad; farming, lumbering, and mining center. Oil wells, sawmills, woodworking plants, grain mills. Citrus fruit, watermelons, papayas. Gold, silver, copper, bismuth, kaolin, and nickel deposits nearby.

Orang (UH-RAHNG), county, NORTH HAMGYONG province, NORTH KOREA, 10 mi/16 km SW of CHONGJIN and just S of NANAM. Agricultural center (raw silk, tobacco). Has agriculture school. Provincial capital until 1920, preceding NANAM. Previously called KYONGSONG.

Orange, county (□ 790 sq mi/2,054 sq km; 2006 population 3,002,048), S CALIFORNIA; ⊙ SANTA ANA; 33°40′N 117°47′W. Coastal plain and foothill region, drained by SANTA ANA RIVER; rises to SANTA ANA MOUNTAINS (over 5,000 ft/1,524 m) along E border. SAN GABRIEL RIVER forms part of W boundary. Although statistically counted separately from LOS ANGELES, Orange county is a direct continuation of the Los Angeles urbanized area. Urbanization in the county increased after World War II, starting in NW, near Los Angeles, and encroaching SE. Agriculture (beans, celery, avocados, artichokes, peppers, tomatoes, strawberries, oranges; livestock). Extensive petroleum and natural gas fields in N and offshore. Large packing, canning, and processing industries; oil refining. Coast has HUNTINGTON BEACH, NEWPORT BEACH (includes BALBOA), LAGUNA BEACH, resort areas that have become extensions of the Los Angeles urbanized area. Disneyland theme park at ANAHEIM, in NW. Along coast, Bolsa Chica, Huntington, Doheny, CORONA DEL MAR, and SAN CLEMENTE state beaches, lagoons and marshes; part of Cleveland National Forest, in SANTA ANA MOUNTAINS, in NE. County declared bankruptcy in 1995 as result of imprudent investments. Formed 1889.

Orange (OR-uhnj), county (□ 1,004 sq mi/2,610.4 sq km; 2006 population 1,043,500), central FLORIDA, bounded E by ST. JOHNS RIVER; ⊙ ORLANDO; 28°30′N 81°19′W. Heavily urbanized as recent growth of Orlando metropolitan area. Hilly lake region in W includes part of LAKE APOPKA; lowland in E. Previously a major citrus fruit-growing area; severe winters in the 1980s destroyed many orange groves. Also small-scale farming, dairying, and poultry raising. Manufacturing of high-tech electronics and food products. Formed as Mosquito county in 1824; renamed Orange county in 1845.

Orange, county (□ 408 sq mi/1,060.8 sq km; 2006 population 19,659), S INDIANA; ⊙ PAOLI; 38°32′N 86°30′W. Drained by LOST and PATOKA rivers and Lick Creek. Mainly agricultural (corn, fruit; cattle, hogs, poultry; dairy products); stone quarrying. Mineral springs (notably FRENCH LICK) are resorts. Manufacturing at Paoli. Lost River runs E to W below surface in limestone karst topography c.18 mi/29 km in N part of county. Springs Valley State Fish and Wildlife Area, Tillery Hill and Jackson State Recreation Areas and PATOKA LAKE all in SW part of county. Hoosier National Forest in S, W, and NW. S and W parts of county are hilly and heavily forested; N and E parts of county are in karst plain. Formed 1816.

Orange, county (□ 838 sq mi/2,178.8 sq km; 2006 population 376,392), SE NEW YORK; ⊙ GOSHEN; 41°23′N 74°18′W. Bounded E by the HUDSON RIVER, SW by NEW JERSEY and PENNSYLVANIA borders and DELAWARE River; includes parts of the Hudson highlands, the RAMAPOS, and the SHAWANGUNK range. Drained by small WALLKILL and Ramapo rivers and by Shawangunk Kill. Now largely suburban with several major shopping malls, was once an important dairying region; farming (fruit, hay; poultry). Large farm migratory worker population, mainly Hispanic. Many mountains and lakes. Includes WEST POINT military reservation and part of PALISADES Interstate Park.

Manufacturing at NEWBURGH, MIDDLETOWN. Formed 1683.

Orange, county (□ 401 sq mi/1,042.6 sq km; 2006 population 120,100), N central NORTH CAROLINA; ⊙ HILLSBOROUGH; 36°03′N 79°07′W. Bounded by HAW RIVER, in SW corner, drained by Eno River. PIEDMONT area; Service industries; manufacturing at Hillsborough (textiles, furniture); agriculture (tobacco, corn, wheat, barley, soybeans, hay; cattle, hogs; dairying); timber (pine, oak); stone quarrying. Eno River State Park on E boundary. University Lake reservoir in S. Formed 1753 from Johnston, Bladen, and Granville counties. Most likely named for William V of Orange whose grandfather was George II of England.

Orange (OR-uhnj), county (2006 population 84,243), SE Texas; ⊙ ORANGE; 30°07′N 93°53′W. Deepwater port, industrial center. Bounded E by SABINE RIVER (here the LOUISIANA line), W and SW by NECHES RIVER, S by Sabine Lake. Gulf coastal plains in S; wooded (chiefly pine; timber) in N. Livestock (cattle, horses, hogs); fish farming, aquaculture; honey. Hunting, fishing. Oil, natural gas wells; sand and gravel, clay. Part of Big Thicket National Preserve in W. Formed 1852.

Orange, county (□ 691 sq mi/1,796.6 sq km; 2006 population 29,440), central and E VERMONT, bounded E by the Connecticut; ⊙ Chelsea; 44°00′N 72°22′W. Dairying, lumbering; manufacturing (machinery, wood products, paper); maple sugar; winter and summer resorts. Drained by WHITE, WAITS, and OMPOMPANOOSUC rivers. Organized 1781.

Orange (AHR-uhnj), county (□ 343 sq mi/891.8 sq km; 2006 population 31,740), N central VIRGINIA; ⊙ ORANGE; 38°14′N 78°00′W. In the PIEDMONT region; bounded N by RAPIDAN RIVER, S by NORTH ANNA RIVER (head of LAKE ANNA reservoir in SE). Agriculture (barley, wheat, corn, soybeans, grapes, fruit, legumes, sweet potatoes, tobacco, hay, alfalfa; cattle, sheep, poultry; dairying); oak timber. Many historic estates, including "Montpelier," SW of Orange in W center. Lake of the Woods reservoir in NE. Formed 1734.

Orange, city (2000 population 128,821), ORANGE county, S CALIFORNIA, suburb 26 mi/42 km SE of downtown LOS ANGELES and 4 mi/6.4 km ESE of ANAHEIM; 33°49′N 117°49′W. City drained by SANTA ANA RIVER and Santiago Creek. Railroad junction. Citrus fruits and nuts are packed, processed, and shipped; rubber products, electronic components, plastics products, aircraft parts, and industrial furnaces are manufactured in Orange. The growing city has been marked by a population increase of about 35% between 1970 and 1990. Orange was founded as Richland until it was renamed in 1875. Chapman College is there. Anaheim Stadium to W. Santiago Reservoir (Irvine Lake) to E; SANTA ANA MOUNTAINS and part of Cleveland National Forest to E; large CHINO HILLS State Park to NE. Incorporated 1888.

Orange, city (2000 population 32,868), ESSEX county, NE NEW JERSEY; 40°46′N 74°13′W. Orange and the surrounding municipalities of EAST ORANGE, WEST ORANGE, SOUTH ORANGE, and MAPLEWOOD are known as "The Oranges," a single suburb of NEWARK and NEW YORK city. Although chiefly residential, Orange has some manufacturing. Settled c.1675, set off from Newark 1806, incorporated as a city 1872.

Orange (OR-uhnj), city (2006 population 17,891), SE TEXAS, 18 mi/29 km NE of PORT ARTHUR and 22 mi/35 km E of BEAUMONT; ⊙ Orange county; 30°06′N 93°45′W. A deepwater port on the SABINE RIVER (LOUISIANA boundary). The INTRACOASTAL WATERWAY follows Sabine Lake and River from Texas Coast (SW) to point 5 mi/8 km SSE of ORANGE, where it continues E across Louisiana marsh, 22 mi/35 km inland. In the wet, lush country of the Gulf Coast, it is a port of entry, with shipyards, oil and gas wells, and

major petrochemical plants. It also has facilities for processing paper, lumber, and rice. Cattle, hogs; honey; winery. Lamar University–Orange Campus (two-year). The U.S. navy has a "mothball fleet" there. Settled c.1800, incorporated 1858.

Orange (o-RAHNZH), town (□ 28 sq mi/72.8 sq km), VAUCLUSE department, PROVENCE-ALPES-CÔTE D'AZUR region, SE FRANCE, near the lower RHÔNE RIVER and 13 mi/21 km N of Avignon; 44°08′N 04°48′E. An agricultural market center for wines, early fruit, and truffles. The town produces refined sugar, pâtés, preserves, chemicals, woolens, and apparel. Orange lies next to a major highway interchange where expressway leading S from LYON splits to MARSEILLE (SE) and Spain (SW). Town is also on France's main N-S railroad line. A nuclear power plant operates at Marcoule on Rhône River (5 mi/8 km W). A military airbase is nearby. Tourism is important at Orange due to well-preserved Roman monuments dating from 35 B.C.E., when the town had become a center of Roman colonization. It was destroyed by the Visigoths in C.E. 412. Orange later became an earldom probably founded by Charlemagne. It became the capital of a principality (12th century) and was passed from family to family and through inheritance (in 1544), to William the Silent of the house of Nassau. Among William's descendants were William III of England and the ruling family of the Netherlands. Orange was conquered (1672) by Louis XIV and confirmed in French possession by the Treaty of Ryswick (1697) and the Peace of Utrecht (1713), although the title remained with the Dutch princes of Orange. The town's noted Roman ruins include a triumphal arch (1st century C.E.) and an amphitheater (c. C.E. 120) so well preserved that it is still in use. The ancient theater and its surroundings, along with the triumphal arch, were designated a UNESCO World Heritage site in 1981. Atop the isolated hill of Saint-Eutrope, just S of the amphitheater, was the castle (demolished in 17th century) of the Princes of Orange. The municipal museum contains a Roman plan of the town drawn upon marble slabs.

Orange, town (2000 population 13,233), NEW HAVEN county, SW CONNECTICUT; 41°16′N 73°01′W. A residential suburb of NEW HAVEN; set off from MILFORD 1822. It is a major retail center. Manufacturing of fabricated metal products, furniture, electronic equipment, food, transportation equipment. A major retail shopping area has developed along U.S. Route 1 N of Interstate 95. The town's first house (1720) still stands. Settled 1720, incorporated 1921.

Orange, town, including Orange village, FRANKLIN county, N MASSACHUSETTS, on MILLERS RIVER and 15 mi/24 km E of GREENFIELD; 42°37′N 72°18′W. Fabricated metal products. Settled c.1746, incorporated 1810.

Orange, town, GRAFTON county, W central NEW HAMPSHIRE, 15 mi/24 km SW of PLYMOUTH; 43°38′N 71°57′W. Manufacturing (lumber); agriculture (vegetables, apples, nursery crops; cattle, poultry; dairying); chalk, ochre deposits. Includes solitary Mount Cardigan (3,121 ft/951 m) in Cardigan State Forest to E.

Orange, town, ORANGE CO., E central VERMONT, 5 mi/8 km SE of BARRE; 44°09′N 72°23′W. Largest population recorded as just over 1000 in the 1830 census.

Orange (AHR-uhnj), town (2006 population 4,536), N central VIRGINIA, 25 mi/40 km NE of CHARLOTTESVILLE; ⊙ ORANGE county; 38°15′N 78°06′W. Railroad junction; manufacturing (fixtures, printing and publishing, apparel, wood products); trade center for agricultural area (grain, soybeans, tobacco; livestock; dairying). "Montpelier," home (built c.1760) of James Madison, is 4 mi/6 km W; Madison and his wife are buried nearby. James Madison Museum to SW. Settled c.1810, incorporated 1856.

Orange (AH-ruhnj), municipality (2001 population 31,970), E central NEW SOUTH WALES, SE AUSTRALIA,

130 mi/209 km WNW of SYDNEY; 33°17′S 149°06′E. Railroad junction. Apple orchards, grapes, olives, berries; lambs, cattle, hogs. Former gold-mining center. Agricultural research center.

Orange (OR-uhnj), village (□ 4 sq mi/10.4 sq km; 2006 population 3,319), CUYAHOGA county, NE OHIO; suburb 12 mi/19.2 km ESE of CLEVELAND; 41°26′N 81°28′W.

Orange Bay, small inlet of Nassau Gulf, in SE coast of HARDY PENINSULA (HOSTE ISLAND), TIERRA DEL FUEGO, CHILE, 45 mi/72 km NW of Cape HORN; 55°30′S 68°02′W. Excellent harbor.

Orange Beach, town (2000 population 3,784), Baldwin co., SW Alabama, 8 mi/12.9 km E of Gulf Shores, near Gulf StatePark; 30°17′N 87°35′W. Manhole cover manufacturing. On Perdido Bay, justoff Gulf of Mexico.

Orangeburg, county (□ 1,127 sq mi/2,919 sq km; 1990 population 84,803; 2000 population 91,582), S central SOUTH CAROLINA; ⊙ ORANGEBURG; 33°26′N 80°47′W. Bounded SW by South Fork of EDISTO RIVER, NE by Lake MARION reservoir; drained by North Fork of the Edisto. One of leading agricultural counties in the South. Sand, clay, limestone; apparel, textiles, wood products, asphalt, chemicals, cement, concrete, food, lumber, printing and publishing. Agriculture includes poultry products, hogs, cattle, corn, wheat, rye, tobacco, sorghum, peanuts, pecans, cotton, watermelons, peaches. Formed 1785.

Orangeburg, city (2006 population 13,563), central SOUTH CAROLINA, 36 mi/58 km SSE of COLUMBIA on the North Fork of EDISTO RIVER; ⊙ ORANGEBURG county; 33°29′N 80°52′W. It is the trade and processing center of a cotton and agricultural area with large textile and apparel industries and a variety of light industry including printing and publishing, food, asphalt, cement, and concrete products, chemicals. Orangeburg was a planned settlement established by German-Swiss immigrants who had free grants of land. It is the seat of South Carolina State University Orangeburg-Calhoun Technical College and Claflin College. Of interest are the Edisto gardens and the Donald Bruce House (c.1735). Settled 1732, incorporated as a city 1883.

Orangeburg, hamlet (□ 3 sq mi/7.8 sq km; 2000 population 3,388), ROCKLAND county, SE NEW YORK, 3 mi/4.8 km SSW of NYACK; 41°02′N 73°57′W. Some manufacturing. Site of Rockland Psychiatric Center.

Orange, Cape, headland at N tip of main island of TIERRA DEL FUEGO, CHILE, on STRAIT OF MAGELLAN 3 mi/5 km SE of POINT ANEGADA; 52°28′S 69°24′W.

Orange City (OR-uhnj), city (□ 6 sq mi/15.6 sq km; 2005 population 7,862), VOLUSIA county, E central FLORIDA, 24 mi/39 km SW of DAYTONA BEACH; 28°56′N 81°17′W. Known for its pure water, bottled here and shipped all over Florida.

Orange City, town (2000 population 5,582), NW IOWA, 37 mi/60 km NNE of SIOUX CITY; ⊙ SIOUX county; 43°00′N 96°03′W. Manufacturing (aircraft, chemicals, apparel, printing and publishing, concrete, feeds, fertilizers). Sand and gravel pits nearby. Northwestern College is here. Holds annual Tulip Day. Founded 1869 by Dutch settlers; incorporated 1884.

Orange Cove, city (2000 population 7,722), FRESNO county, central CALIFORNIA, 27 mi/43 km ESE of FRESNO, in sheltered valley of SIERRA NEVADA foothills; 36°37′N 119°19′W. Citrus fruit, avocados, olives, grapes, almonds, vegetables; dairying. Sequoia National Forest to NE, SEQUOIA NATIONAL PARK to E. Incorporated 1948.

Orange Free State, SOUTH AFRICA: see FREE STATE.

Orange Grove, unincorporated city, HARRISON county, SE MISSISSIPPI, residential suburb 4 mi/9.7 km N of GULFPORT, near Bernard Bayou; 30°26′N 89°05′W. Cotton corn, citrus, pecans. Gulfport-Biloxi Regional Airport to S.

Orange Grove (OR-uhnj), town (2006 population 1,408), JIM WELLS county, S TEXAS, 35 mi/56 km WNW of CORPUS CHRISTI, near NUECES RIVER; 27°57′N 97°56′W. In area producing oil, natural gas, vegetables; cotton; cattle; light manufacturing.

Orange Lake, village, ORANGE county, SE NEW YORK, 6.5 mi/10.5 km SW of NEWBURGH on Orange Lake; 41°27′N 74°07′W.

Orange Lake, village, ORANGE county, SE NEW YORK, 6.8 mi/11 km NW of NEWBURGH on Orange Lake; 41°32′N 74°05′W.

Orange Lake (OR-uhnj), c.16 mi/26 km long, ALACHUA and MARION counties, N central FLORIDA, 10 mi/16 km SE of GAINESVILLE; drains into OKLAWAHA RIVER (E) via Orange Creek (c.6 mi/9.7 km long); 29°27′N 82°10′W.

Orange Park (OR-uhnj), town (□ 6 sq mi/15.6 sq km; 2005 population 9,205), CLAY county, NE FLORIDA, on ST. JOHNS RIVER, and 12 mi/19 km SSW of JACKSONVILLE; 30°10′N 81°42′W.

Orange Range, Dutch *Oranje*, central NEW GUINEA, forms E section of SNOW MOUNTAINS, extends circa 200 mi/322 km E from Nassau Range to border of Australian territories. Rises to circa 15,585 ft/4,750 m in Mount WILHELMINA.

Orange River, chief river of SOUTH AFRICA, c.1,300 mi/2,090 km long; rises in MALUTI MOUNTAINS, N LESOTHO; flows SW through Lesotho, then NW and W through central South Africa forming the boundary first between FREE STATE and Lesotho, and then Free State and EASTERN CAPE provinces, as well as part (c.194 mi/310 km) of the South Africa–NAMIBIA border before entering the ATLANTIC OCEAN at ALEXANDER BAY. The VAAL RIVER is its chief tributary. The lower Orange River flows through the S part of the KALAHARI and NAMIB deserts; in very dry years it does not reach the sea. At the mouth of the river are rich alluvial diamond beds. Shoals, falls (AUGHRABIES FALLS is 400 ft/122 m high), irregular flow, and a sandbar at its mouth limit navigation even in its lower course, but the river is used extensively for irrigation. The Orange River Project, a thirty-year scheme begun in the early 1960s, was planned to bring c.750,000 acres/305,000 hectares of land under irrigation and provide hydroelectric power and municipal water supplies. The GARIEP DAM (completed 1972) is the project's principal unit and largest body of fresh water in South Africa (60 mi/97 km long). Dams and tunnels in the upper Orange basin will divert water to the FISH and SUNDAY rivers. Major towns along its course from E to W include ALIWAL NORTH, Frieska, and UPINGTON.

Orange River Colony, SOUTH AFRICA: see FREE STATE.

Orangetown, Dutch *Oranjestad*, village, NETHERLANDS ANTILLES, WEST INDIES, overlooking small bay; ⊙ SINT EUSTATIUS Island; 17°28′N 62°59′W. Snorkeling, ruins of fort.

Orangeville, unincorporated city, SACRAMENTO county, central CALIFORNIA, residential suburb 16 mi/26 km NE of downtown SACRAMENTO. Citrus, grain, nursery products, dairying, poultry. Manufacturing (machinery). Near AMERICAN RIVER, FOLSOM LAKE reservoir and State Recreation Area to NE.

Orangeville (AH-ruhnj-vil), town (□ 6 sq mi/15.6 sq km; 2001 population 25,248), S ONTARIO, E central Canada, on CREDIT RIVER and 40 mi/64 km WNW of TORONTO; ⊙ DUFFERIN county; 43°55′N 80°05′W. Milling, dairying.

Orangeville, town (2006 population 1,344), EMERY county, central UTAH, 2 mi/3.2 km NW of CASTLE DALE; 39°13′N 111°03′W. Elevation 5,790 ft/1,765 m. Alfalfa; cattle; coal mining. Manti-La Sal National Forest to W. Settled 1877 on Cottonwood Creek, forms Joes Valley Reservoir to NW.

Orangeville, village (2000 population 751), STEPHENSON county, N ILLINOIS, near WISCONSIN line, 11 mi/18 km N of FREEPORT; 42°28′N 89°38′W. In agricultural area (dairying; corn, oats; cattle; hogs).

Orangeville, village, ORANGE county, S INDIANA, near LOST RIVER and 6 mi/9.7 km WSW of ORLEANS. In agricultural area (cattle). Near rise of Lost River. Laid out 1849.

Orangeville (OR-uhnj-vil), village (2006 population 184), TRUMBULL county, NE OHIO, 17 mi/27 km NE of WARREN, and on PYMATUNING CREEK, at PENNSYLVANIA line; 41°20′N 80°31′W.

Orangeville, borough (2006 population 482), COLUMBIA county, E central PENNSYLVANIA, 6 mi/9.7 km NNE of BLOOMSBURG; 41°04′N 76°24′W. Light manufacturing; agriculture (grain; livestock; dairying). Covered bridges in area.

Orange Walk, district (□ 4,636 sq mi/12,053.6 sq km; 2005 population 44,900), BELIZE; ⊙ ORANGE WALK; 17°50′N 88°50′W. Agricultural area, producing sugarcane, citrus fruits, rum, and dairy products; cattle. Mennonite areas with important Maya sites.

Orange Walk, town (2000 population 13,483), BELIZE, on NEW RIVER and 50 mi/80 km NNW of BELIZE CITY; ⊙ ORANGE WALK district; 18°06′N 88°33′W. Citrus; sugar and rum manufacturing; commercial and farming center. Site of bloody conflicts between Maya and Creole settlers in 1860s and 1870s.

Orango, GUINEA-BISSAU: see BIJAGOS ISLANDS.

Orani (o-RAH-nee), town, Bataan province, S LUZON, PHILIPPINES, on MANILA BAY, at base of BATAAN peninsula, 33 mi/53 km WNW of MANILA; 14°48′N 120°30′E. Fishing port. Sheltered by small Tubutubu Island. On route of Bataan death march (1942).

Oranienbaum (o-RAH-nee-en-boum), town, SAXONY-ANHALT, central GERMANY, 9 mi/14.5 km ESE of DESSAU; 51°48′N 12°24′E. Distilling, glass manufacturing. Chinese teahouse (1794–1797) and 17th century palace here.

Oranienbaum, RUSSIA: see LOMONOSOV.

Oranienburg (o-RAH-nee-en-burg), town, BRANDENBURG, NE GERMANY, on the HAVEL River, and ODER-HAVEL CANAL, 15 mi/24 km N of BERLIN; 52°45′N 13°14′E. Center of a fruit-growing region. Manufacturing includes chemicals and steel products. Oranienburg was the site of one of the earliest concentration camps (1933) set up by the National Socialists. Camp moved (1936) to SACHSENHAUSEN, just N. Town first mentioned 1216. Baroque castle from 17th century.

Oranit (o-rah-NEET), urban Jewish settlement, 6 mi/10 km NE of PETAH TIKVA (and 5.6 mi/9 km SE of KFAR SABA) on the E fringe of the SHARON PLAIN, in the WEST BANK, near border with Israel; 32°07′N 34°58′E. Elevation 524 ft/159 m. Founded 1985.

Oranje Mountains (O-rahn-zhe), rise to 2,430 ft/741 m, in S SURINAME, NE spur of the GUIANA HIGHLANDS; extend c.60 mi/97 km NW from TUMUC–HUMAC Mountains (BRAZIL line).

Oranjemund (o-rahn-ye-munt), coastal town, on S border of NAMIBIA, at mouth of ORANGE RIVER; 28°33′S 16°26′E. Diamond-mining town with strict security; almost self-contained. Alluvial coastal diggings.

Oranje Range, NEW GUINEA: see ORANGE RANGE.

Oranjestad (aw-RAH-nyuh-STAHT), town, DUTCH WEST INDIES, port on ARUBÁS W coast, 80 mi/129 km WNW of WILLEMSTAD, CURAÇAO; ⊙ Aruba; 12°32′N 70°03′W. Only 20 mi/32 km N of the VENEZUELA coast, it is an important petroleum trans-shipping and refining point, at which most of the island's population is concentrated. The oil refineries are in NW outskirts.

Oranjestad (aw-RUH-nyuh-STAHT) or **Orange Town**, principal settlement of SAINT EUSTATIUS, NETHERLANDS ANTILLES, in the LEEWARD ISLANDS, 8 mi/12.9 km NW of ST. KITTS, c.530 mi/853 km NE of WILLEMSTAD, CURAÇAO; 17°29′N 62°58′W. Yams, cotton, corn, livestock raised in vicinity. Has two old forts.

Oranje Vrystaat, SOUTH AFRICA: see FREE STATE.

Oran Sebkha (o-RAHN seb-KAH), salt flat in ORAN and AÏN TÉMOUCHENT wilaya, NW ALGERIA, just S of ORAN city; 25 mi/40 km long, 5 mi/8 km wide. Filled with shallow lake after rainfall.

Cross-references are shown in SMALL CAPITALS. The pronunciation guide is shown on page xix. The sources of population figures are shown on page xvii.

Oranzherei (uh-rahn-zhi-RYE-ee), town (2005 population 4,360), SW ASTRAKHAN oblast, S European Russia, in the VOLGA RIVER delta, 31 mi/50 km SSW of ASTRAKHAN; 45°51'N 47°34'E. Below sea level. Fish processing.

Oraoumba Boka, CÔTE D'IVOIRE: see ORUMBO BOKA.

Orapa (o-RAH-pah), town (2001 population 9,151), CENTRAL DISTRICT, BOTSWANA, 138 mi/222 km W of FRANCISTOWN; 21°25'S 25°25'E. Diamond mining. Airfield. Restricted area.

Oras (O-rahs), town, EASTERN SAMAR province, E SAMAR island, PHILIPPINES, on small inlet of PHILIPPINE SEA, 45 mi/72 km NE of CATBALOGAN; 12°09'N 125°24'E. Agricultural center (rice, corn, coconuts).

Orasac (o-RAHSH-ahts), village, central SERBIA, 33 mi/53 km S of BELGRADE, in the SUMADIJA region. Lignite mine. Also spelled Orashats.

Orasje (o-RAHSH), Serbo-Croatian *Orašje*, village, NW MACEDONIA, in outlier of SAR MOUNTAINS, on left bank of VARDAR River and 13 mi/21 km WNW of SKOPJE; 42°08'N 21°08'E. Railroad terminus for RADUSA chromium mine, 2 mi/3.2 km SE across the Vardar River. Also spelled Orashye.

Orăștie (o-ruhsh-TEE-ye), German *Broos*, Hungarian *Szászváros*, town, HUNEDOARA county, W central ROMANIA, on railroad and 13 mi/21 km E of DEVA; 45°50'N 23°12'E. Manufacturing of chemicals, bricks; clay mining; woodworking; fur and medicinal herb processing. Trade in livestock, grain, charcoal, fruit, and wine. Has large cemetery of World War I. Founded by German colonists (13th century). Early printing works established here (1582) published first church books in Romanian.

Orașul-Stalin, ROMANIA: see BRAȘOV.

Orativ (o-RAH-tif) (Russian *Oratov*), town (2004 population 3,400), E VINNYTSYA oblast, UKRAINE, 45 mi/72 km E of VINNYTSYA; 49°12'N 29°32'E. Elevation 767 ft/233 m. Raion center; dairy. Known since 1545; town since 1984.

Oratorio (o-rah-TO-ree-o), town, SANTA ROSA department, S GUATEMALA, in Pacific piedmont, near INTER-AMERICAN HIGHWAY, 8 mi/12.9 km ESE of CUILAPA; 14°14'N 90°10'W. Coffee, sugarcane; livestock.

Oratorio de Concepción (o-rah-TO-ree-o dai kon-sep-see-ON), municipality and town, CUSCATLÁN department, EL SALVADOR, NW of COJUTEPEQUE in W part of department; 13°49'N 89°04'W.

Oratov, UKRAINE: see ORATIV.

Orava (o-RAH-vah), HUNGARIAN *Árva*, historical region (□ 642 sq mi/1,669.2 sq km), STREDOSLOVENSKY province, N SLOVAKIA, on POLISH border, in MAGURA (Slovak *Oravská Magura*) (o-RAHF-skah mah-GU-rah) Mountains and in valley of ORAVA RIVER (and BIELA Orava River) between BESKIDS and HIGH TATRAS, in today's border of DOLNY KUBIN district. Until 1920, Orava was also included as part of POLAND (□ 138 sq mi/357 sq km, between MOUNT BABIA GORA and JABŁONKA). It has always been an agricultural region (cattle breeding) with lumbering (fir forests). Its original center was in ORAVSKY PODZAMOK, then in VELICNA (ve-LICH-nah), Slovak *Veličná*, just WSW of DOLNY KUBIN, and finally, since 1683, in Dolny Kubin (machinery). Other important towns are by ORAVA DAM (NAMESTOVO, TRSTENA, and TVRDOSIN).

Orava Dam (o-RAH-vah), STREDOSLOVENSKY province, N SLOVAKIA, on POLISH border, 20 mi/32 km N of LIPTOVKÝ MIKULAS, on cofluence of Biela Orava and Cierna Orava rivers. The largest Slovak dam built in 1953 (□ 13.5 sq mi/35 sq km); has a maximum depth of 90 ft/27 m; power plant (total output 22 MW). Many summer resorts on S banks (recreation).

Oravais (O-rah-veis), Finnish *Oravainen*, village, VAASAN province, W FINLAND, on inlet of GULF OF BOTHNIA, 30 mi/48 km NE of VAASA; 63°18'N 22°23'E. Elevation 66 ft/20 m. Woolen mills.

Orava River (o-RAH-vah), HUNGARIAN *Árva*, POLISH *Orawa*, c.70 mi/113 km long, STREDOSLOVENSKY province, N SLOVAKIA; formed by junction of its headstreams (Biela ORAVA and Cierna Orava rivers) on S slope of the BESKIDS, 2 mi/3.2 km NW of TRSTENA; flows SW, past ORAVSKY PODZAMOK, DOLNY KUBIN, to VAH RIVER at KRALOVANY. Major reservoir (ORAVA DAM) at conflux of its two headstreams. Part of territory N and NE of Trstena. The Orava River was appropriated by Poland (1938) after the MUNICH Pact and was returned by GERMANY to Slovakia after partition (1939) of Polish state.

Oravița (o-rah-VEE-tsah), Hungarian *Oravicabánya*, German *Orawitza*, town, CARAȘ-SEVERIN county, SW ROMANIA, in BANAT, 21 mi/34 km SSW of REȘIȚA; 45°02'N 21°40'E. Railroad junction; iron mining, iron working, flour milling, manufacturing of apparel. Formerly had a large German population. Known as mining center since Roman times (notably for gold, silver, and iron). Sometimes spelled Oravitsa.

Oravska Magura Mountains, SLOVAKIA: see ORAVA.

Oravska Polhora (o-RAHF-skah pol-HO-rah), Slovak *Oravská Polhora*, Hungarian *Arvapolhora*, village, STREDOSLOVENSKY province, northernmost village of Slovakia, on SW slope of BESKIDS, by Polish border and 30 mi/48 km NNW of LIPTOVSKÝ MIKULAS; 49°31'N 19°27'E. Potato growing, sheep breeding. Traditional log cabins.

Oravsky Podzamok (o-RAHF-skee pod-ZAH-mok), Slovak *Oravský Podzámok*, Hungarian *Árvaváralja*, village, STREDOSLOVENSKY province, N SLOVAKIA, on right bank of ORAVA RIVER, on railroad, and 28 mi/45 km ENE of Žilina; 49°16'N 19°22'E. Manufacturing (wood products, fabricated metal products). Tourist center at foot of Orava Castle. The castle, built on 360-ft/110-ft-high cliff in 13th century and enlarged in 15th and 16th century, dominates the river; has extensive natural science collections, Gothic chapel. Original center of ORAVA region.

Orawa River, SLOVAKIA: see ORAVA RIVER.

Orawitza, ROMANIA: see ORAVIȚA.

Orbassano (or-bahs-SAH-no), town, TORINO province, PIEDMONT, NW ITALY, 8 mi/13 km SW of TURIN; 45°01'N 07°32'E. Fabricated metal products, machinery, plastics.

Orb, Bad, GERMANY: see BAD ORB.

Orbe (ORB), commune, VAUD canton, W SWITZERLAND, on ORBE RIVER and 15 mi/24 km NNW of LAUSANNE; 46°43'N 06°32'E. Elevation 1,572 ft/479 m. Known in Roman times (*Urba*), some ancient architecture still stands 1 mi/1.6 km N.

Orbe (ORB), district, N VAUD canton, SWITZERLAND; 46°43'N 06°32'E. Main town is ORBE; population is largely French-speaking and Protestant.

Orbec (or-BEK), commune (□ 3 sq mi/7.8 sq km), CALVADOS department, BASSE-NORMANDIE region, NW FRANCE, 12 mi/19 km SE of LISIEUX; 49°01'N 00°25'E. Cider distilling, metalworking; light manufacturing. Has 15th- and 16th-century houses, including one housing the municipal museum. Also called Orbec-en-Auge.

Orbe River (ORB), French *l'Orbe*, 35 mi/56 km long, W SWITZERLAND; rises in small lake (Lac des Rousses) in the JURA, just inside French line; flows NE into VAUD canton, entering SW end of LAC DE JOUX and leaving N end of Lac des Brenets, then to LAKE NEUCHÂTEL at YVERDON (here canalized and called THIÈLE RIVER). Hydroelectric plant, N of Lac des Brenets, is on the Orbe.

Orbetello (or-bet-TEL-lo), town (□ 10 sq mi/26 sq km), GROSSETO province, TUSCANY, central ITALY, on spit in lagoon of ORBETELLO (□ 10 sq mi/26 sq km; 5 mi/8 km long, 3 mi/5 km wide) formed by two tombolos connecting Monte ARGENTARIO with mainland; 42°27'N 11°13'E. Bathing resort; fishing center; manufacturing of food, explosives, machinery. Has school of aerial navigation. From its seaplane base Italo Balbo led mass transatlantic flights to RIO DE JANEIRO (1931) and CHICAGO (1933). Badly damaged by air bombing (1944) in World War II.

Orbey (or-BAI), commune (□ 17 sq mi/44.2 sq km), HAUT-RHIN department, ALSACE region, E FRANCE, in resort area in the VOSGES, 10 mi/16 km WNW of COLMAR; 48°08'N 07°10'E. Elevation 1,620 ft/494 m. Dairying. Scenic lakes Blanc (□ 0.8 sq mi/2.1 sq km; elevation 3,458 ft/1,054 m) and Noir (□ 0.5 sq mi/1.3 sq km; elevation 3,117 ft/950 m; hydroelectric station) are 3 mi/4.8 km W, near crest of the Vosges, followed by crest highway, the *Route des Crêtes*.

Órbigo River (OR-vee-go), 67 mi/108 km long, LEÓN and ZAMORA provinces, NW SPAIN; rises on S slopes of the CANTABRIAN MOUNTAINS, 8 mi/12.9 km SE of VILLABLINO; flows SSE, past LA BAÑEZA, to the ESLA RIVER near BENAVENTE.

Orbisonia (OR-bi-SO-nee-ah), borough (2006 population 407), HUNTINGDON county, S central PENNSYLVANIA, 10 mi/16 km S of MOUNT UNION, on Blacklog Creek, adjoining ROCKHILL FURNACE to W; 40°14'N 77°53'W. Manufacturing (printing and publishing); agriculture (corn, hay, alfalfa; livestock, dairying). BLACKLOG MOUNTAIN, and Blacklog Gap, to E.

Orbó (or-VO), mining village, PALENCIA province, N central SPAIN, on S slopes of the CANTABRIAN MOUNTAINS, 44 mi/71 km SW of SANTANDER; 42°53'N 04°15'W. Anthracite and bituminous coal mines.

Orbost (OR-bahst), town, VICTORIA, SE AUSTRALIA, 190 mi/306 km E of MELBOURNE, near SE coast and SNOWY MOUNTAINS; 37°42'S 148°27'E. Railroad terminus in mixed agricultural region. Rainforest information center.

Orb River (ORB), 71 mi/114 km long, in HÉRAULT department, LANGUEDOC-ROUSSILLON administrative region, S FRANCE; rises in the CÉVENNES Mountains near Le Caylar; flows S across LANGUEDOC region, past BÉDARIEUX, to the Gulf of LION (MEDITERRANEAN SEA) 9 mi/14.5 km below BÉZIERS; 43°15'N 03°18'E.

Örbyhus (UHR-BEE-HOOS), village, UPPSALA county, E SWEDEN, on FYRISÅN RIVER, 25 mi/40 km N of UPPSALA; 60°14'N 17°42'E. Manufacturing (wood- and metalworking). Medieval castle, rebuilt seventeenth century.

Orcadas Station (or-KAH-dahz), ANTARCTICA, Argentinian scientific station on LAURIE ISLAND, South Orkney Islands; 60°44'S 44°44'W.

Orcas Island (OR-kuhs), SAN JUAN county, NW WASHINGTON, in NE part of SAN JUAN ISLANDS, 14 mi/23 km long; 8 mi/12.9 km wide. Bounded E by ROSARIO STRAIT, N by Strait of Georgia, W by President and San Juan channels, S by Pole Pass, Harney Channel, Lopez Sound, and Obstruction Pass. Island is deeply indented from S in 3 places: East Sound splits island into 2 main lobes, with town of EASTSOUND at N end, on isthmus, West Sound is in W, and Deer Harbor is at far W end. Moran State Park is in E; eight other state parks on small offshore islands. Mount Constitution (2,408 ft/734 m) in E, highest point in San Juans. Other communities: West Sound, Deer Harbor, Olga. Fishing. Tourism.

Orce (OR-thai), town, GRANADA province, S SPAIN, 23 mi/37 km NE of BAZA; 37°43'N 02°28'W. Ships hemp, esparto. Livestock raising, lumbering; wine, cereals, sugar beets.

Orcera (or-THAI-rah), town, JAÉN province, S SPAIN, in mountainous region, near GUADALIMAR RIVER, 27 mi/43 km NE of VILLACARRILLO; 38°19'N 02°39'W. Olive-oil processing, flour milling and sawmilling; light manufacturing. Produce; lumber. Gypsum quarries.

Orchanie, Bulgaria: see BOTEVGRAD.

Orchard, town (2000 population 88), MITCHELL county, N IOWA, near CEDAR RIVER, 4 mi/6.4 km SSE of

OSAGE; 43°13′N 92°46′W. Limestone quarries, sand and gravel pits in vicinity.

Orchard, village (2006 population 354), ANTELOPE county, NE NEBRASKA, 17 mi/27 km NW of NELIGH; 42°20′N 98°14′W. Dairy center; grain, livestock; light manufacturing.

Orchard (OR-chuhrd), village (2006 population 489), FORT BEND county, SE TEXAS, 38 mi/61 km WSW of HOUSTON, near BRAZOS RIVER; 29°36′N 95°58′W. Oil and natural gas; agricultural area.

Orchard Beach, village, ANNE ARUNDEL county, central MARYLAND, 10 mi/16 km SE of downtown BALTIMORE, near CHESAPEAKE BAY. Colonel Thomas Cresap built a fortified home here before pushing on to OLDTOWN.

Orchard Beach, NEW YORK: see PELHAM BAY PARK.

Orchard City, town (2000 population 2,880), DELTA county, COLORADO, on branch of GUNNISON RIVER and 9 mi/14.5 km NE of DELTA; 38°48′N 107°57′W. Elevation c.5,800 ft/1,768 m. In dairying area. Light manufacturing. Post office name is Eckert. Grand Mesa National Forest to N.

Orchard Hill, village (2000 population 230), SPALDING county, W central GEORGIA, 6 mi/9.7 km SE of GRIFFIN; 33°11′N 84°13′W. Light manufacturing.

Orchard Lake, village, OAKLAND county, SE MICHIGAN, suburb 1 mi/1.6 km SW of PONTIAC; 42°34′N 83°22′W. In lake-resort and farm area. Seat of Saint Mary's College. Surrounds ORCHARD LAKE, Upper Straits Lake in W, borders CASS LAKE on N, other lakes in area.

Orchard Park, residential village (□ 1 sq mi/2.6 sq km; 2006 population 3,105) in Orchard Park town (1990 population 24,632), ERIE county, W NEW YORK, 10 mi/16 km SE of BUFFALO; 42°45′N 78°45′W. Some manufacturing. Seat of Erie County Community College (S campus). NFL franchise at Rich Stadium (capacity 80,290), 2.5 mi/4 km to NW. Until 1934, called E. Hamburg. Incorporated 1921.

Orchards, unincorporated town, CLARK county, SW WASHINGTON, suburb 6 mi/9.7 km NE of VANCOUVER, Washington. Dairying; poultry; fruit, nuts, vegetables. Clark County Airport to NE.

Orchha (OR-chuh), former princely state of CENTRAL INDIA AGENCY. A Rajput state, founded c. 1500. In 1948, merged with VINDHYA PRADESH; with MADHYA PRADESH state in 1956; today territory comprises TIKAMGARH district.

Orchha (OR-chuh), village, TIKAMGARH district, N central MADHYA PRADESH state, central INDIA, on BETWA RIVER, 7 mi/11.3 km SE of JHANSI. Archeological as well as popular tourist site, due to its several notable large fortresses, palaces, and temples.

Orchies (or-SHEE), town (□ 4 sq mi/10.4 sq km), NORD department, NORD-PAS-DE-CALAIS region, N FRANCE, 10 mi/16 km NE of DOUAI; 50°28′N 03°14′E. Agriculture-growing and shipping-center (especially sugar beets and potatoes).

Orchilla, VENEZUELA: see LA ORCHILA, ISLA.

Orchilla, Cape (or-CHEE-lyah), westernmost point of HIERRO island and of CANARY ISLANDS, SPAIN, 16 mi/26 km WSW of VALVERDE; 27°42′N 18°10′W. Prime meridian was once drawn through it.

Orchomenos (or-kho-me-NOS), ancient city of BOEOTIA, in what is now Boeotia prefecture, CENTRAL GREECE department, NW of Lake COPAIS; 38°30′N 22°59′E. After 1600 B.C.E. it was an important center of the Mycenaean civilization. Later eclipsed by THEBES. Nearby, Roman general Sulla won (85 B.C.E.) a significant victory over Archelaus, general of Mithradates VI. Excavations here have been extensive. There was another Orchomenus in Arcadia, NW of MANTINEA (now in Peloponnese department). Also Orkhomenos.

Orchy River (OR-kee), 16 mi/26 km long, Argyll and Bute, W Scotland; issues from Loch Tulla (2.5 mi/4 km long, 1 mi/1.6 km wide); flows SW past DALMALLY to head of LOCH AWE. Formerly in Strathclyde, abolished 1996.

Orcières (or-SYER), commune (□ 37 sq mi/96.2 sq km), resort in HAUTES-ALPES department, PROVENCE-ALPES-CÔTE D'AZUR region, SE FRANCE, 15 mi/24 km NE of GAP, in the CHAMPSAUR Valley (headwaters of DRAC RIVER) at S edge of the ÉCRINS NATIONAL PARK; 44°41′N 06°20′E. Park headquarters located here. Orcières-Merlette (1 mi/1.6 km N) is a ski resort established 1962, with cross-country and downhill trails at elevation of 5,906 ft/1,800 m to 8,530 ft/2,600 m. The station has a modernistic multi-sports palace amidst ski chalets.

Orcival (or-see-VAHL), commune (□ 10 sq mi/26 sq km), PUY-DE-DÔME department, AUVERGNE region, central FRANCE, in volcanic AUVERGNE MOUNTAINS, 13 mi/21 km WSW of CLERMONT-FERRAND; 45°41′N 02°51′E. Produces religious articles. Has beautiful 12th-century Romanesque piligrimage church, built of volcanic rock. The Château of Cordès (1.5 mi/2.4 km N) dates from 13th–15th century, restored in 17th century.

Orcoma (or-KO-mah), canton, CAPINOTA province, COCHABAMBA department, central BOLIVIA, near the ORURO-COCHABAMBA railroad; 17°44′S 66°20′W. Elevation 7,815 ft/2,382 m. Clay, limestone, and an extensive 22-mi/35-km gypsum lode. Agriculture (potatoes, yucca, bananas, corn, rye, sweet potatoes, soy, coffee); cattle; dairy products.

Orco River (OR-ko), 50 mi/80 km long, in PIEDMONT, NW ITALY; rises in glaciers of GRAN PARADISO region, 7 mi/11 km NW of CERESOLE REALE; flows SE past CUORGNÈ, to PO RIVER 1 mi/1.6 km W of CHIVASSO. Used for hydroelectric power and irrigation.

Orcotuna (or-ko-TOO-nah), town, CONCEPCIÓN province, JUNÍN region, central PERU, in MANTARO RIVER valley, on road, and 16 mi/26 km SE of JAUJA; 11°58′S 75°20′W. Cereals; livestock.

Orcutt, unincorporated village, SANTA BARBARA county, SW CALIFORNIA, 6 mi/9.7 km S of SANTA MARIA. In oil field. Cattle; grain, fruit, vegetables.

Ord, city (2006 population 2,116), ⊙ VALLEY county, central NEBRASKA, 55 mi/89 km NW of GRAND ISLAND, and on NORTH LOUP RIVER; 41°36′N 98°55′W. Railroad terminus. Livestock, dairy, poultry, grain. Manufacturing (printing and publishing). Fort Hartsuff State Historical Park to NW. Surveyed 1874.

Orda (or-DAH), village, W WEST KAZAKHSTAN region, KAZAKHSTAN, 130 mi/209 km E of VOLGOGRAD (Russia); 48°44′N 47°30′E. In livestock-breeding area (cattle, camels, horses). Was capital of former Bukei government (1917–1925). Also spelled Urda.

Orda (uhr-DAH), village (2006 population 5,225), SE PERM oblast, W central URALS, E European Russia, on road, 16 mi/26 km S of KUNGUR; 57°12′N 56°54′E. Elevation 419 ft/127 m. Agricultural products. Gypsum quarries in the vicinity. Population mostly Tatar and Bashkir.

Ordelos, Greece and Bulgaria: see ORVILOS.

Órdenes (OR-dhe-nes), town, LA CORUÑA province, NW SPAIN, 20 mi/32 km SSW of LA CORUÑA. Tanning; lumbering, stock raising; cereals, potatoes.

Orderville, village (2006 population 606), KANE county, SW UTAH, 18 mi/29 km NNW of KANAB and near E Fork of VIRGIN RIVER; 37°16′N 112°37′W. Elevation 5,280 ft/1,609 m. Apples; cattle. Dixie National Forest to N; ZION NATIONAL PARK to W. Founded 1874 as Mormon economic experiment by United Order communal society; settled 1875.

Ordesa y Monte Perdido National Park (□ 60 sq mi/155 sq km), N HUESCA province, NE SPAIN, in ARAGÓN, within MONTE PERDIDO, near French border; 42°40′N 00°10′W. Elevation 3,576 ft/1,090 m–11,001 ft/3,353 m. Rugged, mountainous area. Includes special protection zone for birds. Established 1918.

Ord, Fort, California: see MONTEREY, city.

Ordhilos, Greece and Bulgaria: see ORVILOS.

Ordinary (OR-di-ne-ree), unincorporated town, GLOUCESTER county, E VIRGINIA, 21 mi/34 km N of NEWPORT NEWS, 4 mi/6 km N of GLOUCESTER POINT; 37°18′N 76°30′W. Light manufacturing; agriculture (cattle; grain, soybeans). Coleman Swamp to SE.

Ordino, town, ANDORRA, on Valira del Nord River, 4 mi/6.4 km N of ANDORRA LA VELLA; 42°33′N 01°32′E. Iron and lead mining. Tourism. Moorish ruins.

Ord, Mount (7,155 ft/2,181 m), on MARICOPA/GILA county line, central ARIZONA, in MAZATZAL MOUNTAINS, c.50 mi/80 km NE of PHOENIX, in Tonto National Forest.

Ordos Plateau (OR-DUHS), sandy desert plateau (□ 35,000 sq mi/91,000 sq km), where the Mu Us desert and Hobg desert are located, INNER MONGOLIA Autonomous Region, N CHINA. Almost encircled by the great N bend of the HUANG HE (Yellow River) in W and N. The GREAT WALL of China forms the S boundary and separates the Ordos from the fertile loess land to the S and E. The desert receives less than 10 in/25 cm of rain annually, mainly in the form of thunderstorms. The region has many salt lakes and intermittent streams. Large soda deposits are mined. The alkaline soil supports some grasslands, and nomadic Mongol herdsmen raise sheep and goats; some oasis-farming is also practiced.

Ord River (ORD), 300 mi/483 km long, in NE WESTERN AUSTRALIA state, W AUSTRALIA; rises SE of HALLS CREEK; flows N through mountainous area, to CAMBRIDGE GULF near WYNDHAM; 15°30′S 128°21′E. Dammed for major irrigation and hydroelectric scheme (1972), still controversial. KUNUNURRA Dam, built in 1963, and Lake Argyle Dam, 25 mi/40 km upstream, built in 1972. Irrigation for fruit and sugarcane production; recreation.

Ordsall, ENGLAND: see SALFORD.

Ordu, city (2000 population 112,525), N TURKEY, a port on the BLACK SEA; 41°00′N 37°52′E. At the foot of a forested hill. Copper and iron are exported. Hazelnut production. Site of Cotyora, founded by Greek colonists c.500 B.C.E.; evidence of habitation c.3000 B.C.E.

Ordu, TURKEY: see YAYLADAGI.

Ordubad (uhr-doo-BAHD), city, SE NAKHICHEVAN Autonomous Republic, AZERBAIJAN, near ARAS River, 38 mi/61 km SE of NAKHICHEVAN; 38°54′N 46°01′E. Manufacturing (food, textiles, construction materials). Silk-cocoon collecting station. Until c.1940, spelled Ordubat.

Orduña (or-DHOO-nyah), town, VIZCAYA province, N SPAIN, in an enclave between ÁLAVA and BURGOS provinces, on NERVIÓN RIVER, and 19 mi/31 km SSW of BILBAO; 42°59′N 03°00′W. Cement manufacturing, tanning. Cereals, beans, potatoes in area. Mineral springs nearby.

Ordway, town (2000 population 1,248), SE central COLORADO, near Bob Creek, 45 mi/72 km E of PUEBLO; ⊙ CROWLEY county; 38°13′N 103°45′W. Elevation 4,312 ft/1,314 m. Melon shipping point in agricultural region; alfalfa meal, dairy products, sugar beets; turkeys. Small lake nearby. LAKE MEREDITH RESERVOIR to SE; Lake Henry Reservoir to NE. Incorporated 1900.

Ordway, village, BROWN county, N SOUTH DAKOTA, 10 mi/16 km NNE of ABERDEEN, near Elm Creek; 45°34′N 98°24′W.

Ordynskoye (uhr-DIN-skuh-ye), town (2006 population 10,270), SE NOVOSIBIRSK oblast, SW SIBERIA, RUSSIA, on the NOVOSIBIRSK Reservoir of the OB' RIVER, on highway, 55 mi/89 km SW of NOVOSIBIRSK; 54°22′N 81°54′E. Elevation 396 ft/120 m. Dairy products, fish processing, metalworks.

Ordzhonikidze (ord-zho-nee-KEED-ze), city (2004 population 43,780), S DNIPROPETROVS'K oblast, UKRAINE, on Bazavluk River, on road and 17 mi/27 km WNW of NIKOPOL'; 47°40′N 34°03′E. Elevation 193 ft/58 m. Manganese mining; construction materials manufacturing. Heritage museum. Amalgamated with Kalinin and No. 1 settlements and Oleksandrivka (Russian *Aleksandrovka*) village to form a city. Established in 1934.

Cross-references are shown in SMALL CAPITALS. The pronunciation guide is shown on page xix. The sources of population figures are shown on page xvii.

Ordzhonikidze (uhrd-zhoo-NEE-ki-deez), town, S AZERBAIJAN, on railroad (Dashburun station) and 50 mi/80 km SW of SABIRABAD. Cotton-ginning center; shipping point for Beilagansky cotton district.

Ordzhonikidze, agricultural town, N AZERBAIJAN, 22 mi/35 km NE of YEVLAKH. Livestock grazing.

Ordzhonikidze (or-jo-ni-KEED-ze), town, KOSTANAI region, KAZAKHSTAN, on railroad and highway, 90 mi/150 km SW of KOSTANAI; 52°27′N 61°39′E. Tertiary-level (raion) administrative center. Grains; meat.

Ordzhonikidze (ord-zho-nee-KEED-ze), town (2004 population 7,000), SE Republic of CRIMEA, UKRAINE, on the BLACK SEA coast, 8 mi/13 km SW and under jurisdiction of FEODOSIYA; 44°55′N 35°20′E. Trass (hydraulic cement rock) for Novorossiysk mills.

Ordzhonikidze, GEORGIA: see HARAGAULI.

Ordzhonikidze, RUSSIA: see VLADIKAVKAZ.

Ordzhonikidzeabad (or-jawn-i-KEED-je-band), city, W TAJIKISTAN, on KAFIRNIGAN RIVER, and 12 mi/19 km E of DUSHANBE (linked by railroad); 38°34′N 69°01′E. Area is under no direct viloyat administrative division; under direct republic supervision. Cotton, sericulture; flour milling, metalworking. Until 1937, called Yangi-Bazar.

Ordzhonikidzegrad, RUSSIA: see BEZHITSA.

Ordzhonikidzevskaya (uhr-juh-nee-KEE-dzeef-skah-yah), village (2005 population 74,425), E central IN-GUSH REPUBLIC, in the foothills of the N CAUCASUS Mountains, S European Russia, on road and railroad, on the left bank of the SUNZHA RIVER, 32 mi/51 km W of GROZNYY (CHECHEN REPUBLIC); 43°19′N 45°03′E. Elevation 1,040 ft/316 m. Junction of a lumber railroad to GALASHKI (S); in agricultural area. Until 1939, called Sleptsovskaya. More than half of Ordzhoni-kidzevskaya's current population are Chechen refugees, living in two impromptu settlements within the village limits.

Ordzhonikidzevskiy (uhr-juh-nee-KEE-dzyef-skeeyee), town, NW KHAKASS REPUBLIC, S central SIBERIA, RUSSIA, in the foothills of the KUZNETSK ALATAU, 110 mi/177 km NW of ABAKAN; 54°46′N 88°58′E. Elevation 2,171 ft/661 m. Highway junction. Woodworking combine; in area of gold deposits.

Ordzhonikidzevskiy (uhr-juh-nee-KEE-dzif-skeeyee), town (2006 population 3,250), central KARACHEVO-CHERKESS REPUBLIC, N CAUCASUS, RUSSIA, in the W Greater Caucasus Mountains, on the KUBAN' River, on mountain highway, 33 mi/53 km SSW of CHER-KESSK, and 4 mi/6 km N of KARACHAYEVSK, to which it is administratively subordinate; 43°50′N 41°54′E. Elevation 3,005 ft/915 m. Coal mining.

Ordzhonikidzevskiy (uhr-juh-nee-KEE-dzif-skeeyee), settlement (2005 population 3,250), central KAR-ACHEVO-CHERKESS REPUBLIC, S European Russia, in the NW CAUCASUS, on the KUBAN' River, 4 mi/6 km N of KARACHAYEVSK; 43°50′N 41°54′E. Elevation 3,005 ft/915 m. In agricultural area.

Orea (o-RAI-ah), village, GUADALAJARA province, central SPAIN, near TERUEL province border, 21 mi/34 km SE of MOLINA; 40°34′N 01°43′W. Wheat, potatoes; livestock.

Orealla (or-ree-AH-lah), village, EAST BERBICE-COR-ENTYNE district, E GUYANA; 05°19′N 57°20′W. Located on CORENTYNE RIVER (SURINAME border) and 65 mi/105 km SSE of NEW AMSTERDAM.

Oreamuno, COSTA RICA: see SAN RAFAEL, CARTAGO.

Oreamuno, San Rafael de, COSTA RICA: see SAN RA-FAEL, CARTAGO province.

Oreana, village (2000 population 892), MACON county, central ILLINOIS, suburb 7 mi/11.3 km NE of down-town DECATUR; 39°56′N 88°52′W. Agricultural area; light manufacturing.

Orebić (O-re-beech), Italian *Sabbioncello*, village, S CROATIA, on ADRIATIC SEA, on S coast of Pelješac peninsula, 3 mi/4.8 km NE of Korčula, in DALMATIA. Seaside resort; center of olive- and wine-growing re-gion. Park (cypress forest).

Örebro (UH-re-BROO), county (□ 3,559 sq mi/9,253.4 sq km), S central SWEDEN; ☉ ÖREBRO; 59°30′N 15°00′E. Includes NÄRKE, SE VÄRMLAND, and E VÄSTMANLAND provinces. Low and level with fertile soil (S), becomes undulating highland in N. Drained by ARBOGAÅN, LETÄLVEN, and Svartån rivers; dotted with numerous small lakes. Manufacturing (steel, paper, pulp, lumber miling). Fomerly important iron, zinc, and copper mines, and timber stands. Important towns are Örebro, KARLSKOGA, KUMLA, LINDESBERG, and NORA.

Örebro (UH-re-BROO), town, ☉ ÖREBRO county, S central SWEDEN, W of LAKE HJÄLMAREN; 59°17′N 15°12′E. Commercial, industrial, and transportation center; manufacturing (machinery, metal goods, paper, processed food). International airport to W. Known since eleventh century; site of fifteen national diets, especially that which brought Reformation to Sweden (1529), and the one electing Bernadotte (later Charles XIV) crown prince of Sweden and NORWAY (1810).

Ore City (OR), town (2004 population 1,161), UPSHUR county, NE TEXAS, 27 mi/43 km NW of MARSHALL; 32°47′N 94°43′W. Dairying; poultry; vegetables; tim-ber; manufacturing (food, lumber). LAKE O' THE PINES to NE (Big Cypress Creek).

Oredezh (O-ree-dyesh), settlement, SW LENINGRAD oblast, NW European Russia, on the Oredezh River (left tributary of the LUGA RIVER), on road and rail-road, 19 mi/31 km ENE of LUGA; 58°49′N 30°20′E. Elevation 190 ft/57 m. In agricultural area; produce processing, livestock.

Orefield (OR-feeld), unincorporated village, LEHIGH county, E PENNSYLVANIA, 6 mi/9.7 km WNW of AL-LENTOWN on Jordan Creek; 40°38′N 75°35′W. Man-ufacturing includes printing and publishing, food, fabricated metal products, electronic equipment. Agriculture includes dairying, livestock, poultry; grain, apples. Covered bridges in area.

Oregon, state (□ 98,386 sq mi/255,803.6 sq km; 2006 population 3,700,758); NW UNITED STATES, in the Pacific Northwest; ☉ SALEM, the third largest city; 44°07′N 120°21′W. The largest city is PORTLAND, fol-lowed by EUGENE. Oregon is nicknamed the "Beaver State." Admitted 1859 as the thirty-third state.

Geography

Oregon is bounded on the N by WASHINGTON; most of the N boundary, from the N center W to the PACIFIC OCEAN, is formed by the COLUMBIA River; on the E by IDAHO, with the SNAKE RIVER forming the boundary in the NE; on the S by NEVADA and CALIFORNIA; and on the W by the Pacific Ocean. The state's contrasting physical features are characterized by great forested mountain slopes and treeless basins, rushing rivers and barren playas, lush valleys and extensive waste-lands. The major determinants for these unusual cli-matic differences are the CASCADE RANGE, a rugged mountain chain running N to S c.100 mi/160 km in-land, and the COAST RANGE. As the moist, warm E-moving air masses that are cooled by the California Ocean Current and blocked by mountain ranges rise, cool precipitation occurs over the W one-third of Oregon, especially autumn through spring. A dry continental climate prevails over the two-thirds of the state, receiving little of the rainfall from the W. The W shoreline (c.300 mi/483 km) is bordered by narrow coastal plains of sandy beaches, luxuriant pastures, and occasional jutting promontories. About 25 mi/40 km inland, the rugged Coast Range rises to heights of 4,000 ft/1,219 m. The WILLAMETTE River, which flows N, in NW, is bounded on the W by the Coast Range and the E by the Cascade Range. Its valley is the main artery of Oregon's economy and population. Most of its agriculture and manufacturing is there. Portland, whose metropolitan area contains nearly half the state's population, straddles the Willamette near its junction with the Columbia and extends to the Co-lumbia River. Salem and Eugene lie S in the valley,

which rises in the S in the low range of the Calapooia Mountains. The snowcapped volcanic peaks of the Cascades include Mount HOOD in the N of range, rising to the state's highest elevation (11,235 ft/3,424 m). Much of Oregon's timber resources are in several national forests: Ochoco and Malheur in the E center; Umatilla and Wallow-Whitman in the NE; Siskiyou and Siuslaw in the W; Umpqua, Rogue River, Winema, Deschutes, and Fremont in the S center; Willamette and Mount Hood in the N center. Two large state forests, Clatsop and Tillamook, are in the NW.

Eastward the Cascades level out into high plateaus drained in the N by the DESCHUTES and the JOHN DAY rivers. The area just E of the Cascades is referred to as the High Desert Region. To the S a variegated pattern of marshland, intermittent lakes, and mountains merges in the E into the semiarid Harney Basin, part of the larger GREAT BASIN, a large area, covering part of six states, that has no outlet to the oceans. There, little vegetation grows, and the absence of potable water makes habitation difficult. NE of this area rise the pine-covered BLUE MOUNTAINS and Wallowa Moun-tains, which in some places extend to the Snake River to form precipitous gorges. Part of HELLS CANYON (Grand Canyon of the Snake) National Recreational Area in the NE, extends into Idaho. Other parts of the region where the Snake cuts through the plateau are more level and have been made productive through irrigation. Oregon's irrigation projects include the Deschutes, the UMATILLA, and the Vale; the KLAMATH, shared with California; and the BOISE and the OWY-HEE, shared with Idaho.

Economy

The state's major sources of farm income are cattle and sheep, with large herds grazing on the plateaus E of the Cascades, dairying, and poultry. Fur farms are in the NW. Chief crops in terms of value are hay, wheat, pears, and onions. Oregon is one of the na-tion's leading producers of snap beans, peppermint, sweet cherries (orchards are particularly numerous in the N Willamette valley), broccoli, strawberries, and grass seed. Oregon has developed an important and growing wine industry (NW) since 1980. The state's c.30,700,000 acres/12,400,000 ha of rich forestlands (almost half the area of the state) comprise the country's greatest reserves of standing timber; large areas have been set aside for conservation. Oregon has been the nation's foremost lumber state since 1950, producing much of the nation's lumber and most of its plywood. It exports lumber in increasing amounts to Japan. Wood processing is Oregon's major indus-try. Douglas fir predominates in the Cascades and Ponderosa pine in the eastern regions. Other major products are food, paper and paper items, machinery, and fabricated metal products. Printing and pub-lishing are important businesses. Oregon is also home to many high-technology computer and electronics companies. Abundant, cheap electric power is sup-plied by numerous dams, most notably those on the Columbia River—BONNEVILLE DAM, THE DALLES DAM, and MCNARY Dam. The JOHN DAY Dam is one of the largest hydroelectric generators in the world. The many dams provide locks for navigation. The Bon-neville Dam, in the steep gorge through which the Columbia River pierces the Cascades, enables large vessels to travel far inland, and although river traffic is not as vital as it once was, the Columbia River cities still serve as transport centers for a vast hinterland to the E. Oregon's river resources are one of its greatest assets. Although commercial fishing has declined in recent decades, it remains an important part of the economy of Oregon's coastal counties. Lincoln county, on the middle coast (around NEWPORT), is Oregon's largest fish producer; salmon, tuna, halibut, crabs, clams, and shrimp are important. Although mining is still relatively underdeveloped, Oregon leads the nation in the production of nickel. Sand and gravel, stone, and cement are also major sources of mineral income.

Tourism

Oregon's beautiful, long ocean beaches with dramatic rock formations on and off the coastline, and its lakes and mountains, draw thousands of visitors annually, making tourism one of the state's largest industries. Major attractions are the OREGON CAVES NATIONAL MONUMENT (SW), FORT CLATSOP HISTORIC SITE (NW), and MCLOUGHLIN HOUSE NATIONAL HISTORIC SITE; CRATER LAKE NATIONAL PARK (SW center) is a vacationer's paradise. There are thirteen national forests, one national grassland, and more than 220 state parks.

History: to 1792

Initial European interest in the region was aroused by the search for the Northwest Passage. Spanish seamen skirted the Pacific coast from the 16th to the 18th century, hoping to claim the area. The English may first have arrived in the person of Sir Francis Drake, who sailed along the coast in 1579, possibly as far as Oregon. Much later, in 1778, Captain James Cook, seeking the award of £20,000 for the discovery of the NORTHWEST PASSAGE, charted some of the coastline. By this time the Russians were pushing S from posts in ALASKA and the British fur companies were exploring the West. Oregon's furs promised to become an important factor in the rapidly expanding China trade, and the Oregon coast was soon active with the vessels of several nations engaged in fur trade with the Native Americans. British captains, among them John Meares and George Vancouver, made the coastal area known, but it was an American, Robert Gray, who first sailed up the Columbia River (1792), thus establishing U.S. claim to the areas that it drained.

History: 1792 to 1846

Canadian traders of the North West Company were approaching the Columbia River country when the overland Lewis and Clark expedition arrived in 1805. David Thompson was already making his way to the lower river when John Jacob Astor's agents (in the Pacific Fur Company) founded ASTORIA, the first permanent settlement in the Oregon country. In the War of 1812 the post was sold (1813) to the North West Company, but in 1818 a treaty provided for ten years of joint rights for the U.S. and Great Britain in Oregon (i.e., the whole Columbia River area). This agreement was later extended. The North West Company merged with the Hudson's Bay Company in 1821, and soon the region was dominated by John McLoughlin at FORT VANCOUVER. In 1842 and 1843 enormous wagon trains began the "great migration" westward over the Oregon Trail. Trouble between the settlers and the British followed. The Americans set out to form their own government, and demanded the British be removed from the whole of the Columbia River country up to latitude 54°40′N; one of the slogans of the 1844 election was "Fifty-four forty or fight." War with Britain was a threat momentarily, but diplomacy prevailed.

History: 1846 to 1880

In 1846 the boundary was set at the line of latitude 49°N. Soon afterward the Oregon Territory was created in 1848, embracing the area W of the Rockies from the 42nd to the 49th parallel. The area was reduced with the creation of the Washington Territory in 1853, and Oregon became a state in 1859 with a constitution that prohibited slaveholding, but which also forbade free blacks from entering the state. Although the California gold rush caused a temporary exodus of settlers, it also brought a new market for Oregon's goods, and the Oregon gold strike that followed attracted some permanent settlement to the E hills and valleys. Wheat farming prospered and in 1867–1868 a surplus crop was shipped to England, the beginning of Oregon's great wheat export trade. Cattle and sheep were driven up from California to graze on the tall grass of the semiarid plateaus, and

soon cattle barons, such as Henry Miller, acquired large herds. They dominated the industry until the late 19th century, when sheepmen and homesteaders succeeded in reducing the cattle range. The 1850s, 1860s, and 1870s were plagued by Native American uprisings, but by 1880 Native American troubles were over, and the next few decades brought increasing settlement and internal improvements.

History: 1880 to 1930

During the 1880s, and largely under the management of Henry Villard of the Northern Pacific Railroad, transcontinental railroad lines were completed to the coast and down the Willamette valley into California, bringing new trade and stimulating the beginnings of manufacturing. Lumbering, which had long been important, became a leading industry. Seemingly almost overnight logging camps and sawmills were built in the W foothills. The huge stands of Douglas fir and cedar brought fortunes to the lumbering kings, and the threat to natural resources led ultimately to the creation of national forests. By the time of the Lewis and Clark Centennial Exposition at Portland in 1905, less than fifty years after statehood had been gained, the frontier era had passed. Most of the feuding on the E plateaus was over.

Oregon has been a leader in social, environmental, and political reforms. It was the first state, for example, to institute initiative, referendum, and recall; and to initiate a ban against nonrecyclable containers. Several issues have divided conservative and liberal thought sharply. One has been the question of minority groups. In the 1880s the influx of Chinese threatened the labor market and brought violent anti-Chinese sentiment, and in the 20th century there was opposition to the Japanese. Feeling against minorities has never been universal, however, and large groups have vigorously opposed it.

History: 1930 to Present

In the 1930s one of the most disputed issues was the question of public or private development of power. Today, however, it has to be recognized that the Federal power and irrigation projects have had a profound effect on the economy of the entire Pacific Northwest. Many acres have been opened to irrigated farming, and the tremendous industrial expansion of World War II was to a large extent dependent on Bonneville power. Environmental issues have dominated since the 1970s. The state's numerous hydroelectric dams are killing Pacific salmon, which in 2001 led to protests over restrictions on water use in the Klamath Basin designed to protect the coho salmon. A major controversy was raised in the late 1980s over the spotted owl, which has been endangered as its forest habitat is cut down. The tension between preserving nature and maintaining the state's economic growth continues as many high-technology electronic and computer firms and individuals seeking to escape from crowding, migrate from California to Oregon. Among the state's more prominent institutions of higher learning are the University of Oregon at Eugene; Oregon State University at CORVALLIS; Reed College and Portland State University at Portland; and Willamette University at Salem.

Government

Oregon still operates under its original constitution, drawn and ratified in 1857. Its executive branch is headed by a governor elected for a four-year term. The current governor is Ted Kulongoski. Its bicameral legislature has a senate with thirty members elected for four-year terms and an assembly with sixty members elected for two years. The state elects two senators and five representatives to the U.S. Congress and has seven electoral votes.

Oregon has thirty-six counties: BAKER, BENTON, CLACKAMAS, CLATSOP, COLUMBIA, COOS, CROOK, CURRY, DESCHUTES, DOUGLAS, GILLIAM, GRANT, HARNEY, HOOD RIVER, JACKSON, JEFFERSON, JOSE-

PHINE, KLAMATH, LAKE, LANE, LINCOLN, LINN, MALHEUR, MARION, MORROW, MULTNOMAH, POLK, SHERMAN, TILLAMOOK, UMATILLA, UNION, WALLOWA, WASCO, WASHINGTON, WHEELER, and YAMHILL.

Oregon, county (□ 784 sq mi/2,038.4 sq km; 2006 population 10,407), S MISSOURI; ⊙ ALTON; 36°41′N 91°24′W. In the OZARKS, borders ARKANSAS on S, drained by ELEVEN POINT RIVER and SPRING RIVER. Agriculture, hay, notably livestock (cattle, hogs); dairying; oak, hickory, walnut timber; manufacturing at THAYER and Alton. Grand Gulf State Park in SW; Part of Mark Twain National Forest in NE. Formed 1845.

Oregon, city (2000 population 4,060), N ILLINOIS, on ROCK RIVER (bridged) and 21 mi/34 km SW of ROCKFORD; ⊙ OGLE county; 42°00′N 89°20′W. Trade and industrial center in rich agricultural area (corn, soybeans, cattle, hogs; dairying); manufacturing (machinery, glass). Nearby is Eagle's Nest Art Colony, founded 1898 by Lorado Taft and others; Taft's soldiers' monument is in the city, and his great *Black Hawk* statue is on bluff overlooking Rock River. White Pines State Forest and Lowden and Castle Rock state parks are nearby. Incorporated 1843.

Oregon, city (2000 population 935), NW MISSOURI, near MISSOURI River, 22 mi/35 km NW of SAINT JOSEPH; ⊙ HOLT county, 39°58′N 95°08′W. Corn, apples, wheat; hogs, cattle. Big Lake State Park and Squaw Creek National Wildlife Refuge to NW. Platted 1841.

Oregon (OR-uh-gahn), city (□ 28 sq mi/72.8 sq km; 2006 population 19,110), LUCAS county, NW OHIO; suburb adjacent (SE) to TOLEDO, on LAKE ERIE; 41°39′N 83°27′W. It is a port with railroad-owned and -operated docks. The city has industries producing oil, chemicals, and fabricated metal products. The majority of the city's area is open farmland, where tomatoes, soybeans, greenhouse vegetables, fruits, and grains are grown. Incorporated 1958.

Oregon, town (2006 population 8,810), DANE county, S WISCONSIN, suburb 10 mi/16 km S of MADISON; 42°55′N 89°22′W. In dairying and farming area (hogs; tobacco); creamery; light manufacturing. Settled 1842, incorporated 1883.

Oregon Caves National Monument (488 acres/197 ha), S JOSEPHINE county, SW OREGON, 32 mi/51 km SW of MEDFORD, 12 mi/19 km ESE of CAVE JUNCTION (its headquarters), in SISKIYOU MOUNTAINS, bounded on all sides by Siskiyou National Forest, 5 mi/8 km N of CALIFORNIA state boundary. Authorized by Presidential Proclamation, 1909. Intricate cave formations in marble bedrock.

Oregon City, city (2006 population 30,667), NW OREGON, suburb 11 mi/18 km S of downtown PORTLAND, at falls (c.40 ft/12 m high), bypassed by locks of WILLAMETTE RIVER, at mouth of CLACKAMAS RIVER; ⊙ CLACKAMAS county; 45°20′N 122°35′W. Railroad junction. Manufacturing (food, lumber, printing, publishing, fabricated metal products). End of the OREGON TRAIL. Territorial capital until late in 1851. Had first newspaper (*Oregon Spectator*, 1846) W of the Missouri River. Edwin Markham was born here (1852). MCLOUGHLIN HOUSE NATIONAL HISTORIC SITE (established 1941; affiliated area administered but not owned by National Park Service) preserves the home (1846–1857) of Dr. John McLoughlin, a prominent figure in the early development of the Pacific Northwest, who plotted the city and is considered to be the "Father of Oregon." Site of Clackamas Community College. Laid out in 1842, incorporated 1849.

Oregon Inlet (AH-re-guhn IN-let), channel, DARE county, E NORTH CAROLINA (c. 1 mi/1.6 km wide) through the OUTER BANKS, 12 mi/19 km SSE of MANTEO, c.40 mi/64 km N of Cape HATTERAS and BODIE ISLAND to N (lighthouse), PEA ISLAND to S, including Peas Island National Wildlife Refuge and CAPE HATTERAS NATIONAL SEASHORE; connects PAMLICO SOUND with the ATLANTIC OCEAN.

Cross-references are shown in SMALL CAPITALS. The pronunciation guide is shown on page xix. The sources of population figures are shown on page xvii.

Oregon River, U.S. and Canada: see COLUMBIA RIVER.

Oregon Trail, national historic trail, runs through MISSOURI, KANSAS, NEBRASKA, WYOMING, IDAHO, OREGON, and WASHINGTON. It is an affiliated area of the National Park Service, whose role is to implement and manage the plan to preserve this primary route of the Oregon Trail (2,000 mi/3,219 km long), used by pioneer settlers during the 1840s, and 125 other associated historic sites.

Oregon Trail, overland emigrant route in the United States from the Missouri River to the Columbia River country (all of which was then called Oregon) in the 1840s and 1850s. The pioneers by wagon train did not, however, follow any single narrow route. In open country the different trains might spread out over a large area, only to converge again for river crossings, mt. passes, and other natural constrictions. In time many cutoffs and alternate routes also developed. They originated at various places on the Missouri, although Independence and Westport (now part of Kansas City, Missouri) were favorite starting points, and St. Joseph had some popularity. Those starting from Independence followed the same route as the Santa fe Trail for some 40 mi/64 km, then turned NW to the Platte and generally followed that river to the junction of the North Platte and the South Platte. Crossing the South Platte, the main trail followed the North Platte to Fort Laramie, then to the present Casper, Wyoming, and through the mountains by the broad, level South Pass to the basin of the Colorado River. The travelers then went SW to Fort Bridger, from which the Mormon Trail continued SW to the Great Salt Lake, while the Oregon Trail went NW across a divide to Fort Hall, on the Snake River. It then went along the Snake River. The California Trail branched off to the southwest, but the Oregon Trail continued to Fort Boise. From that point the travelers had to make the hard climb over the Blue Mountains. Once the mountains were crossed, paths diverged somewhat; many went to Fort Walla Walla before proceeding down the south bank of the Columbia River, traversing the Columbia's gorge where it passes through the Cascade Mountains to the Willamette valley, where the early settlement centered. The end of the trail shifted as settlement spread. The Mountain Men were chiefly responsible for making the route known, and Thomas Fitzpatrick and James Bridger were renowned as guides. Capt. Benjamin de Bonneville 1st took wagons over South Pass in 1832. The 1st genuine emigrant train was that led by John Bidwell in 1841, half of which went to California, the rest proceeding from Fort Hall to Oregon on horses and mules. The 1st train of emigrants to reach Oregon was that led by Elijah White in 1842. In 1843 occurred the "great emigration" of more than 900 persons and more than 1,000 head of livestock. Four trains made the journey in 1844, and by 1845 the emigrants reached a total of over 3,000. Although it took the average emigrant train six months to traverse the c.2,000-mi/3,200-km route, the trail continued in use for many years. Travel upon the trail gradually declined with the coming of the railroad, and it was abandoned in the 1870s.

Öregrund (UH-re-GRUND), town, UPPSALA county, E SWEDEN, on Öregrundsgrepen, 20-mi/32-km-long strait of GULF OF BOTHNIA, opposite GRÄSÖ island, 45 mi/72 km NE of UPPSALA; 60°20′N 18°27′E. Portuguese seaside resort. Former shipping center for UPPLAND province iron mines. Chartered 1491.

Orehovitsa (o-RE-ho-veet-sah), village, LOVECH oblast, DOLNA MITROPOLIYA obshtina, N BULGARIA, on the ISKUR River, 17 mi/27 km NW of PLEVEN; 43°35′N 24°32′E. Wheat, corn; livestock. Sometimes spelled Oryehovitsa.

Orehovo, Bulgaria: see ORYAHOVO.

Orei (o-re-EE), village, NW Évvia island, Évvia prefecture, CENTRAL GREECE department, port on OREOS Channel, 45 mi/72 km NW of KHALKÍS; 38°57′N 23°06′E. Wheat, wine; fisheries. Also spelled Oreoi.

Orekhov, UKRAINE: see ORIKHIV.

Orekhovo (uh-RYE-huh-vuh), village, W KOSTROMA oblast, central European Russia, on road and near railroad, 17 mi/27 km E of BUY; 58°29′N 41°59′E. Elevation 482 ft/146 m. In flax-growing area.

Orekhovo, Bulgaria: see ORYAHOVO.

Orekhovo, UKRAINE: see TOREZ.

Orekhovo-Zuyevo (uh-RYE-huh-vuh–ZOO-ee-vuh), city (2006 population 119,285), E MOSCOW oblast, central European Russia, on the KLYAZ'MA RIVER, on road and railroad junction, 55 mi/89 km ENE of MOSCOW; 55°49′N 38°59′E. Elevation 406 ft/123 m. One of the oldest Russian centers of the cotton textile industry (dating from the 18th century); machinery, ceramics; woodworking; food processing (bakery).

Orekhovsk (o-RE-hofsk), settlement, SE VITEBSK oblast, BELARUS, 14 mi/23 km NNE of ORSHA, 54°40′N 30°30′E. Manufacturing (food). Large peat-fed power plant (*Belgres*) here. Until 1946, Orekhi-Vydritsa.

Orël (uh-RYOL), oblast (□ 9,472 sq mi/24,627.2 sq km; 2006 population 845,235), in SW European Russia; ⊙ ORËL. In CENTRAL RUSSIAN UPLAND; drained by the SOSNA and upper OKA rivers; black-earth, wooded steppe. Basic crops include hemp, potatoes, wheat, coarse grain, sugar beets; orchard products, vegetable gardens. Hogs extensively raised. Rural industries include hemp processing and milling, distilling, starch making. Metalworking, food processing (Orël, YELETS, LIVNY). Orël province was first established at the end of the reign of Peter the Great (early 18th century), and later received the status of guberniya, occupying much of the territories of current BRYANSK and LIPETSK oblasts. Following the Russian civil war of 1918–1921, the guberniya was split up and seized to exist as administrative entity by 1934, with its territory going to both old and newly formed oblasts. Reformed in September 1937 out of Western oblast and parts of Kursk oblast; within its current borders since 1954. Divided into 24 districts (Russian *rayony*) and contains seven cities and thirteen towns. Sometimes spelled Orel (pronunciation does not change).

Orël (uh-RYOL), city (population 337,000), SW European Russia, on the OKA River, 235 mi/378 km S of MOSCOW; 52°58′N 36°05′E. Elevation 646 ft/196 m. ⊙ ORËL oblast. Railroad junction; many kinds of machinery, metalworking, glassworks, ceramics, toys. Like Moscow, its main streets are rings and radii moving outward from a central core. It was founded in 1566 by Ivan IV as a fortified settlement to protect the southern border of Muscovy from Crimean Tatar attacks. In the 18th and 19th centuries, it was a large agricultural trade center.

Orël (uh-RYOL), town (2006 population 1,855), N central PERM oblast, W URALS, E European Russia, port on the E bank of the Kama Reservoir on the KAMA River, opposite YAYVA RIVER mouth, 8 mi/13 km SW of BEREZNIKI; 59°20′N 56°35′E. Elevation 341 ft/103 m. Logging, lumbering. Founded 1564 as a Russian stronghold of Kergedan.

Oreland (OR-luhnd), unincorporated town (2000 population 5,509), MONTGOMERY county, SE PENNSYLVANIA, residential suburb 12 mi/19 km N of PHILADELPHIA; 40°06′N 75°10′W. Manufacturing includes printing and publishing, fabricated metal products, industrial glass products.

Orel-Izumrud (uh-RYEL–ee-zoom-ROOT), settlement (2005 population 4,635), SE KRASNODAR TERRITORY, S European Russia, on the E shore of the BLACK SEA, near the RUSSIA-GEORGIA border, on road, 12 mi/19 km SE of SOCHI, and 2 mi/3.2 km N of ADLER; 43°27′N 39°55′E. Elevation 278 ft/84 m. Holiday and vacation getaway spot; spas, summer homes. Airport in the vicinity.

Orel'ka, UKRAINE: see ORIL'KA.

Orellana (o-rai-YAH-nah), town, LORETO region, E central PERU; landing on UCAYALI River and 29 mi/47 km NNW of CONTAMANA; 06°54′S 75°10′W. Sugarcane, fruit. Also known as Francisco de Orellana.

Orellana (o-re-YAH-nah), village, AMAZONAS region, N PERU, landing on upper MARAÑON RIVER, and 20 mi/32 km WSW of SANTA MARÍA DE NIEVA; 04°38′S 78°08′W. Bananas, plantains, yucca. Tropical woods.

Orellana la Sierra (o-re-LYAH-nah lah SYE-rah) or **Orellanita** (o-re-lyah-NEE-tah), village, BADAJOZ province, W SPAIN, 16 mi/26 km E of VILLANUEVA DE LA SIERRA; 39°02′N 05°30′W. Cereals, olives, oranges, grapes.

Orellana la Vieja (o-re-LYAH-nah lah VYAI-hah), town, BADAJOZ province, W SPAIN, 14 mi/23 km E of VILLANUEVA DE LA SERENA; 39°00′N 05°32′W. Cereals, grapes, olives, livestock; olive-oil pressing. Lead mines.

Orellanita, SPAIN: see ORELLANA LA SIERRA.

Orel' River, UKRAINE: see ORIL' RIVER.

Orem, city (2006 population 90,857), UTAH county, N central UTAH, suburb 4 mi/6.4 km N of PROVO, 36 mi/58 km SSE of SALT LAKE CITY, near UTAH LAKE. Orem is located in an irrigated vegetable and fruit-growing area; 40°17′N 111°42′W. It has a large steel mill, and a variety of light industrial products are manufactured here (chemicals, food, furniture, fabricated metal products, printing and publishing, electronic equipment, steel, computer software); motion picture production. Unta National Forest, Bridal Veil Falls, and Sundance Ski Area to NE. Utah Valley Community College. The city has been marked by growth, and its population more than doubled between 1970 and 1990. Originally named Provo Beach. Settled 1861, incorporated 1919.

Ore Mountains [German=berg mountain], Czech *Krušné hory* (KRUSH-ne HO-ri), German *erzgebirge*, mountain range, along the Czech-German border, extending c.95 mi/150 km from the FICHTELGEBIRGE in the SW to the ELBE river in the NE. It reaches its highest point (4,080 ft/1,244 m) in KLINOVEC (German *keilberg*) CZECH REPUBLIC. The OHŘE and BILINA rivers drain most of the range. Silver and iron were mined here extensively in the 14th–19th centuries, notably at JACHYMOV. In the 20th century, the mineral products included uranium, lead, and zinc. There is no ore mining at present. Coal and lignite mines are also exploited. Air pollution has resulted in serious environmental problems, including the loss of large areas of forest. Has many famous mineral springs (notably at KARLOVY VARY and TEPLICE, in the CZECH REPUBLIC) and is an important industrial area, manufacturing especially chemicals, machinery and textiles. Embroidering and toy making have long been traditional home industries. In 1938 the Czech part of the mountain was transferred to GERMANY by the Munich Pact, but it was later restored to CZECHOSLOVAKIA in 1945.

Orenburg (uh-ryen-BOORG), oblast (□ 48,008 sq mi/124,820.8 sq km; 2006 population 2,164,190) in SE European Russia; ⊙ ORENBURG; 52°00′N 56°00′E. In SW foothills of the S URAL Mountains (E) and OBSHCHIY SYRT hills (W); drained by URAL, SAKMARA, SAMARA, GREATER KINEL, and ILEK rivers. Borders KAZAKHSTAN in the S and SE, SAMARA oblast in the W, TATARSTAN Republic in the NW, BASHKORTOSTAN Republic in the N, and CHELYABINSK oblast in the NE. Humid continental (N; short summers) and steppe (S) climate. Minerals chiefly mined in Orsk-Khalilovo industrial district include limonite, hematite, nickel, cobalt, copper, chromite, lignite, jasper, and limestone; salt (SOL'-ILETSK), petroleum and asphalt (BUGURUSLAN, SARAKTASH), gold (AIDYRLINSKIY), oil shale, phosphorite, gypsum, and fireproof clay (S, SE). Large natural gas deposit near Orenburg prompted construction of a gas pipeline to countries in E Europe. Extensive agriculture (wheat, millet, sunflowers) and cattle raising; sheep and goats (S); vegetable gardening (around ORSK and Orenburg). Lightly forested (NE,

NW). Industries based on mining (metallurgy in Novo-Troitsk, nickel refining in Orsk, copper and sulphur works in MEDNOGORSK, saltworks in Sol'-Iletsk) and agriculture (flour milling, meatpacking, distilling, food processing and preserving). Oil cracking (Orsk, Buguruslan), machine manufacturing (Orsk, BUZULUK, Orenburg), light manufacturing (Orenburg). Rural industries include wool weaving, sheepskin processing. Orenburg, Orsk, Buzuluk, and Buguruslan are the main urban centers. Well-developed railroad network; occasional spring navigation on the Ural River below Orenburg. Formed in 1934 out of Middle Volga Territory, which had absorbed Orenburg guberniya in 1928. Gained industrial importance following the development of Orsk-Khalilovo district before World War II. For a while Orenburg oblast and city were named Chkalov after a Russian aviator who flew from Orenburg to Vancouver via the North Pole.

Orenburg (uh-ryen-BOORK), city (2006 population 550,550), S URALS, SE European Russia, on the URAL River, 915 mi/1,473 km SE of MOSCOW; ☉ ORENBURG oblast; 51°47′N 55°06′E. Elevation 452 ft/137 m. Major railroad junction where the TRANS-CASPIAN RAILROAD, the Central Asian Railroad, and the Samara-Orenburg line meet. Manufacturing of machinery, leather products, chemicals, apparel, food, and silk. Natural gas processing. Vocational schools and institutes of higher learning proliferate. Has a museum of Russian literary classics; theater of musical comedy. Founded in 1735 as a fortress and chartered in 1743, it became a center for Russian trade with KAZAKHSTAN and central ASIA. Called Chkalov between 1938 and 1957. Entered a period of rapid urban development and explosive population growth in the early 1940s, as many essential industries were evacuated here from the western part of the Soviet Union ahead of the German advance during World War II. Discovery of natural gas deposits in the vicinity in 1966 contributed to continuous growth.

Orense (o-RAIN-sai), province (☐ 2,691 sq mi/6,996.6 sq km; 2001 population 338,446), NW SPAIN, in GALICIA; ☉ ORENSE. Bounded S by PORTUGAL; crossed by GALICIAN MOUNTAINS, here reaching highest point (CABEZA DE MANZANEDA, 5,833 ft/1,778 m). Drained by the Miño and its tributaries (e.g., the SIL), and by LIMA and TÁMEGA rivers, flowing into Portugal. Tin and tungsten mines (richest in Spain), noted since ancient times, are now exploited somewhat. Numerous mineral springs. Principally agricultural, with extensive forests (lumbering) and pastures (cattle and hog raising); fertile valleys, favored by sufficient rainfall; vineyards, orchards, vegetable gardens; also cereals, potatoes, flax, honey. Industries limited to domestic type. Population widely scattered in small hamlets. Major towns include CELANOVA and RIBADAVIA. Also called Ourense.

Orense (o-RAIN-sai) or **Ourense**, city (2001 population 107,510), ☉ ORENSE province, NW SPAIN, in GALICIA, on the Miño River, 40 mi/64 km ESE of PONTEVEDRA; 42°20′N 07°51′W. It is the center of an agricultural region and has some light industry. A Roman settlement, it reached its greatest importance as the capital of the kings of the Suebi (5th–6th century). It has a fine 12th-century bridge and a Gothic cathedral, frequently restored. There are hot sulphur springs, known since Roman times.

Orense (o-RAIN-sai), town, BUENOS AIRES province, ARGENTINA, 36 mi/58 km SE of TRES ARROYOS; 38°40′S 59°47′W. Wheat, oats, sheep, cattle; dairying. Beach resort on ATLANTIC OCEAN coast 9 mi/14.5 km S.

Oreoi, Greece: see OREI.

Oreor, volcanic island (☐ c.3 sq mi/7.8 sq km); c.3 mi/4.8 km long; rises to 459 ft/140 m; W CAROLINE ISLANDS, W PACIFIC, 1 mi/1.6 km SW of Babeldaob; ☉ Republic of PALAU (Belau); 07°20′N 134°30′E. Formerly Koror or Corrora.

Oreos Channel (o-RE-os), arm of AEGEAN SEA, CENTRAL GREECE department, between the NW end of Évvia island, Évvia prefecture (SE) and NE Fthiotida prefecture on Greek mainland (NW); joins Trikeri Channel (NE) and Maliakos Gulf (SW); 20 mi/32 km long, 2 mi/3.2 km–4 mi/6.4 km wide. Village of OREI on S shore.

Orepuki (o-re-POOK-ee), township, SOUTHLAND district, SOUTH ISLAND, NEW ZEALAND, 30 mi/48 km W of INVERCARGILL; 46°17′S 167°44′E. Agricultural center.

Ore River, ENGLAND: see ALDE RIVER.

Oresh (o-RESH), village, LOVECH oblast, SVISHTOV obshtina, N BULGARIA, on Svishtov Lake, 5 mi/8 km WSW of Svishtov; 43°34′N 25°15′E. Railroad junction; grain, livestock.

Oreshak (o-re-SHAHK), village, LOVECH oblast, TROYAN obshtina, BULGARIA. Craft center, Troyanski Manastir nearby; 42°53′N 24°47′E. Arts and crafts center.

Oreshek, RUSSIA: see SHLISSELBURG.

Oreshki (uh-RYESH-kee), town, W MOSCOW oblast, central European Russia, near highway, 5 mi/8 km ENE of RUZA, to which it is administratively subordinate; 55°43′N 36°21′E. Elevation 708 ft/215 m. Building materials.

Oressa River (o-RE-sah) or *Rassa*, 94 mi/151 km long, in MINSK and GOMEL oblasts, BELARUS. Tributary of PTICH RIVER. Used to float timber.

Orestes (aw-RES-tuhs), town (2000 population 334), MADISON county, E central INDIANA, 2 mi/3.2 km W of ALEXANDRIA; 40°16′N 85°44′W. Agricultural area; manufacturing (food).

Orestia, TURKEY: see EDIRNE.

Orestias (o-re-stee-AHS), town (2001 population 15,246), Évros prefecture, EAST MACEDONIA AND THRACE department, extreme NE GREECE, on railroad and 12 mi/19 km S of EDIRNE (formerly Adrianople), TURKEY, near MARITSA (Évros) River (Turkish border); 41°30′N 26°31′E. Trade center for fertile agricultural lowland; silk, wheat, cotton, rice, sesame; dairy products. Named for Byzantine name of Adrianople; sometimes called Nea Orestias.

Orestias, Lake, Greece: see KASTORÍA, LAKE.

Øresund, SWEDEN and DENMARK: see ORESUND.

Öresund, SWEDEN and DENMARK: see ORESUND.

Oresund (UH-re-SUND), Swedish Öresund, Danish Øresund, strait (c.45 mi/72 km long) between Danish island of SJAELLAND and SWEDEN, connecting KATTEGATT with BALTIC SEA, to which it is deepest channel. Between HELSINGBORG and HELSINGØR, DENMARK; 2.5 mi/4 km wide. COPENHAGEN, Denmark, and MALMÖ, Sweden, are on Oresund. Strategic passage long contested between Denmark and Sweden. Also known as The Sound.

Oretana, Cordillera, SPAIN: see TOLEDO, MONTES DE.

Oreti River (o-REET-ee), estuary, 105 mi/169 km long, S SOUTH ISLAND, NEW ZEALAND; rises S of LAKE WAKATIPU; flows SE past LUMSDEN, to FOVEAUX STRAIT on S coast, 3 mi/4.8 km S of INVERCARGILL.

Orewa (o-REE-wah), township, NORTH ISLAND, NEW ZEALAND, c.31 mi/50 km N of AUCKLAND, between estuary of Orewa River and Whangaparoa Bay, at base of Whangaparoa peninsula, a dog-leg promontory extending c.10 mi/16 km eastward into HAURAKI GULF; RODNEY district (☐ 5,776 sq mi/14,960 sq km); 36°34′S 174°42′E. A retirement and vacation center.

Oreye (aw-RAI), Flemish *Oerle*, commune (2006 population 3,476), Waremme district, LIÈGE province, E BELGIUM, 12 mi/19 km NW of LIÈGE; 50°44′N 05°22′E. Beet sugar refining.

Orfani Gulf (or-fah-NEE), 14 mi/23 km wide, 17 mi/27 km long inlet of N AEGEAN SEA, off CENTRAL MACEDONIA department, on E side of KHALKIDIKÍ Peninsula, NE GREECE. Receives STRUMA (Strymon) River on N shore and outlet of Lake VOLVI (W). In a wider sense, it is applied to the entire section of the Aegean Sea between island of THÁSOS and Acti

(Athos) arm of Khalkhidikí Peninsula. In ancient times, STAGIRA and AMPHIPOLIS were on its shores. Also called Gulf of Orphane (Orfani) for small village on N shore; also Rendina and Contessa. Formerly Strymonic Gulf.

Orford (OR-fuhrd), canton (☐ 52 sq mi/135.2 sq km; 2006 population 3,206), Estrie region, S QUEBEC, E Canada, 8 mi/14 km from MAGOG; 45°23′N 72°12′W.

Orford (OR-fuhrd), township, TASMANIA, SE AUSTRALIA, 50 mi/81 km E of HOBART, and on Prosser River; 42°34′S 147°51′E. Holiday, fishing resort.

Orford (OR-fuhrd), town, GRAFTON county, W NEW HAMPSHIRE, 19 mi/31 km NNE of LEBANON; 43°53′N 72°04′W. Bounded W by CONNECTICUT RIVER, opposite FAIRLEE, Vermont; drained by Jacobs Brook. Orford Ridge, site of fine colonial and federal houses, in Orford Village. APPALACHIAN TRAIL crosses E part. SMARTS MOUNTAIN (3,240 ft/988 m) in SE corner.

Orford, ENGLAND: see WARRINGTON.

Orfordville, town (2006 population 1,380), ROCK county, S WISCONSIN, 12 mi/19 km WSW of JANESVILLE; 42°37′N 89°15′W. In tobacco, dairying, and grain area.

Organ, unincorporated village, DOÑA ANA county, S NEW MEXICO, 13 mi/21 km ENE of LAS CRUCES, W of SAN AUGUSTIN PASS (5,719 ft/1,743 m). Cattle, sheep. Light manufacturing. ORGAN MOUNTAINS to S; SAN ANDRES MOUNTAINS to N. At SW boundary of WHITE SANDS MISSILE RANGE. Aguirre Springs National Recreation Area to SE.

Organ Cave, WEST VIRGINIA: see RONCEVERTE.

Organ Mountains, S NEW MEXICO, DOÑA ANA county, just E of RIO GRANDE and LAS CRUCES. Prominent points: Organ Peak (8,872 ft/2,704 m) and Organ Needle (8,990 ft/2,740 m). Just N is SAN AUGUSTIN PASS, separating range from SAN ANDRES MOUNTAINS and village of ORGAN, W of pass. Aguirre Springs National Recreation Area in N; WHITE SANDS MISSILE RANGE to NE.

Organ Mountains, Brazil: see ORGÃOS, SERRA DOS.

Órganos, Sierra de los (OR-gah-nos, see-ER-rah de los), mountain range, PINAR DEL RÍO province, W CUBA, extending c.60 mi/97 km NE from MANTUA to the SIERRA DEL ROSARIO; rises to c.1,350 ft/411 m. Rich in copper and lumber. Sometimes considered to include the Sierra del Rosario.

Organ Pipe Cactus National Monument (☐ 516 sq mi/1,336 sq km), PIMA county, S ARIZONA. Organ pipe cactus and other unique desert growth. Bounded by MEXICO (SONORA) on S; by Ajo Range and Tohono O'odham (Papago) Indian Reservation on E; by Cabeza Prieta National Wildlife Refuge on W; border village of LUKEVILLE in S; Quitobaquito Springs in SW. Authorized 1937.

Organ Pipes National Park (OR-guhn PEIPS), VICTORIA, SE AUSTRALIA. In deep KEILOR plains gorge. Features hexagonal columns of basalt, one of the world's largest lava flows, formed c.one milion years ago, rising 65 ft/20 m above Jackson Creek; also Rosette Rock, a radial arrangement of basalt columns. Approximately 145 species of native vegetation; birdlife (waterbirds, blue wrens, cockatoos, rosellas, magpies, wedge-tailed eagles). Established 1972.

Orgãos, Serra dos (OR-gouns, SE-rah dos) [Portuguese=Organ Mountains], range in central RIO DE JANEIRO state, BRAZIL, forming part of the great coastal escarpment (Serra do MAR) overlooking GUANABARA BAY, and extending E from PETRÓPOLIS to beyond TERESÓPOLIS. Highest peaks are Pedra do Sino (7,365 ft/2,245 m) and Pedra Aqu (7,265 ft/2,214 m). One of its many serrate heights, the sheer Dedo de Deus [Portuguese=finger of God] (5,561 ft/1,695 m) is a landmark seen from RIO (30 mi/48 km SSW) as a national park (☐ 9 sq mi/23.3 sq km).

Orgaz or **Orgaz con Arísgotas** (or-GAHTH kon ah-REEZ-go-tahs), town, TOLEDO province, central

SPAIN, in CASTILE-LA MANCHA, 16 mi/26 km SSE of TOLEDO; 39°39′N 03°54′W. Agricultural center in picturesque mountain terrain; olives, grapes, cereals, goats, sheep. Wool and tanning industry; also olive-oil extracting, sawmilling. Unexploited copper mine. Agricultural station. Has noteworthy Santo Tomás church and ruins of castle.

Orgeval (or-zhuh-vahl), town (□ 5 sq mi/13 sq km), YVELINES department, ÎLE-DE-FRANCE region, N central FRANCE; near SEINE RIVER and 6 mi/10 km WNW of SAINT-GERMAIN-EN-LAYE; 48°54′N 01°59′E.

Orgeyev, MOLDOVA: see ORHEI.

Órgiva, SPAIN: see ÓRJIVA.

Orgon (or-GON), commune (□ 13 sq mi/33.8 sq km; 2004 population 2,913), BOUCHES-DU-RHÔNE department, PROVENCE-ALPES-CÔTE D'AZUR region, SE FRANCE, on left bank of the DURANCE at E foot of the ALPILLES hill range and 16 mi/26 km SE of AVIGNON; 43°47′N 05°02′E. Olive groves amidst limestone hills. Has 14th-century church.

Orgosolo (or-GO-zo-lo), village, NUORO province, E central SARDINIA, 8 mi/13 km S of NUORO; 40°12′N 09°21′E. Village of sheepherders who maintain traditional culture and values. Known for producing traditional regional women's costumes. Nuraghe nearby.

Orgtrud (ork-TROOT), town (2006 population 4,360), N central VLADIMIR oblast, central European Russia, on the KLYAZ'MA RIVER (OKA River basin), at the mouth of the NERL' RIVER, near highway, 8 mi/13 km NE of VLADIMIR; 56°11′N 40°36′E. Elevation 406 ft/123 m. Cotton milling; lumbering. Called Lemeshenskiy (1927–1940).

Orgun, Afghanistan: see URGUN.

Orhaneli, village, NW TURKEY, near Orhaneli River, 20 mi/32 km SSW of BURSA; 39°52′N 28°59′E. Cereals; rich chromium deposits. Formerly Atranos.

Orhangazi, township, NW TURKEY, near W end of Lake IZNIK, 25 mi/40 km NE of BURSA; 40°30′N 29°18′E. Olives, potatoes, vetch, cereals. Formerly Pazarkoy.

Orhei, city (2004 population 25,641), central MOLDOVA, on RUT RIVER and 25 mi/40 km N of CHISINAU (Kishinev); 47°22′N 28°49′E. In rich agricultural district (orchards, vineyards, tobacco plantations); flour milling, fruit and tobacco processing, brewing; light manufacturing. Has 17th-century church. An ancient settlement, it flourished under Moldavian and Turkish rule (15th–18th century), trading with N Moldavia and Tatar lands across the DNIESTER (Nistru) River. While in Romania (1918–1940; 1941–1944), it was capital of Orhei department. Formerly called Orgeyev.

Orhon Gol (AWR-kuhn), river, c. 700 mi/1,124 km long, MONGOLIA and RUSSIA; rises in the HANGAYN NURUU (Khangai mountains), N central Mongolia; flows E, then N, past the site of ancient KARAKORUM, and then NE to join the SELENGA River just S of the Russian border at 50°14′N 106°08′E. It is navigable for shallow-draft vessels only during July and August. The Orhon Inscriptions, discovered in 1889 by the Russian explorer N. M. Yadrinstev near the site of ancient Karakorum, date from the 8th century and are the oldest known material in a Turkic language.

Orhon River, MONGOLIA: see ORHON GOL.

Oria (O-ree-ah), ancient *Uria*, town, BRINDISI province, APULIA, S ITALY, 19 mi/31 km SW of BRINDISI. Agricultural center (olives, grapes, figs, vegetables). Bishopric. Has cathedral, castle (1227–1233; restored), several palaces.

Oria (O-ryah), town, ALMERÍA province, S SPAIN, 18 mi/29 km SW of VÉLEZ RUBIO. Lumbering, livestock raising; olive oil, cereals, esparto, fruit.

Orica (O-ree-kah), town, FRANCISCO MORAZÁN department, central HONDURAS, 45 mi/72 km NE of TEGUCIGALPA; 14°42′N 86°57′W. Corn, wheat, coffee.

Orichi (O-ree-chee), town (2005 population 7,725), central KIROV oblast, E central European Russia, on road and the TRANS-SIBERIAN RAILROAD, 25 mi/40 km

WSW of KIROV; 58°24′N 49°03′E. Elevation 505 ft/153 m. Railroad depots. Furniture; food industries (bakery, produce preserves); livestock veterinary station; seed nursery. Established in 1906 with the construction of the railroad; town since 1960.

Orient, city (2000 population 296), FRANKLIN county, S ILLINOIS, 8 mi/12.9 km SSW of BENTON; 37°55′N 88°58′W. In bituminous coal-mining and agricultural area.

Orient, town (2000 population 402), ADAIR county, SW IOWA, 11 mi/18 km N of CRESTON; 41°12′N 94°25′W. In agricultural region.

Orient, agricultural town, AROOSTOOK county, E MAINE, on GRAND LAKE and 22 mi/35 km S of HOULTON; 45°49′N 67°51′W. In recreational area.

Orient, village (2006 population 49), FAULK county, N central SOUTH DAKOTA, 10 mi/16 km S of FAULKTON; 44°53′N 99°05′W. Lake Faulkton State Lakeside Use Area to N.

Orient, hamlet (□ 1 sq mi/2.6 sq km), SUFFOLK county, SE NEW YORK, NE LONG ISLAND, overlooking Orient Harbor (inlet of GARDINERS BAY), 24 mi/39 km NE of RIVERHEAD; 41°09′N 72°15′W. To E are Orient Point resort village, with ferry connections with NEW LONDON, CONNECTICUT; Orient Point promontory at tip of N peninsula ("North Fork") of Long Island; and Orient Beach State Park (□ 342 acres/138 ha).

Oriental, town, PUEBLA, central MEXICO, on central plateau, 45 mi/72 km NE of PUEBLA; 19°24′N 97°40′W. Corn, maguey. Railroad junction.

Oriental (or-ee-EN-tuhl), town (□ 1 sq mi/2.6 sq km; 2006 population 830), PAMLICO county, E NORTH CAROLINA, on NEUSE RIVER and 20 mi/32 km SE of NEW BERN; 35°01′N 76°40′W. Service industries; manufacturing (food); agriculture inland (potatoes, grain, cotton; hogs). Settled in the 1870s, incorporated 1899.

Oriental, administrative region (2004 population 1,918,094), E MOROCCO; ⊙ OUJDA. Bordered by Meknès-Tafilalet (SW), Fes-Boulemane (W), and Taza-Al Hoceima-Taounate (NW) administrative regions, as well as MEDITERRANEAN SEA (N) and ALGERIA (E and S). Part of ATLAS MOUNTAINS here. Railroad runs W out of Oujda, through TAOURIRT, to MEKNES; railroad also runs E from Oujda into Algeria; secondary railroad runs N from BOUARFA, joining the Meknes-Oujda railroad W of Oujda. Roads runs through the region, especially in N, connecting it to rest of Morocco and Algeria. International airport at Oujda. Further divided into six secondary administrative divisions called prefectures and provinces: Berkane, Figuig, Jerada, Nador, Oujda-Angad, and Taourirt.

Oriental, Cordillera (o-ree-en-TAHL, kor-dee-YAI-rah) [Spanish=eastern cordillera], name applied to several mountain ranges in Spanish-speaking countries, notably in the ANDES Mountains; 14°00′S 71°00′W. In BOLIVIA, term refers to a branch range of the Andes, bounding the ALTIPLANO on E and following an arc from S central to W central Bolivia; its highest section is known as Cordillera Real or Cordillera de la Paz. In PERU, it is the easternmost range between Nudo de VILCANOTA and Nudo de PASCO and beyond. In ECUADOR, it forms the E of 2 ranges enclosing the central tableland. In COLOMBIA, it is again the easternmost of the 3 main cordilleras which fan out into roughly parallel chains near the Ecuador border N to the CARIBBEAN SEA, running through the center of the country.

Orientale, province (□ 204,164 sq mi/530,826.4 sq km; 2004 population 6,886,000), NE and N CONGO; ⊙ KISANGANI. Borders NE on SUDAN, E on UGANDA (along SEMLIKI RIVER and LAKE ALBERT), N on central AFRICAN REPUBLIC (along BOMU RIVER). Drained by UELE-KIBALI, ARUWIMI-ITURI, and CONGO-LUALABA rivers. Covered by equatorial rain forest in center and

S, by savannas in N. The nation's leading gold-mining region (KILO-MOTO); also produces cotton, coffee, rice, fibers, peanuts, sesame, rubber, cacao, palm kernels, and palm oil. Several hydro-electric plants provide power for the region. In NE region of high plateaus (to 8,070 ft/2,460 m), an area of agricultural settlements, vegetables, potatoes, and essential-oil plants are grown and cattle raised. Experimental plantations of cinchona, tea, and tobacco. Main navigable waterways are the CONGO-LUALABA (except in the sector of BOYOMA FALLS), LOMAMI, LAKE ALBERT, and lower ITIMBIRI. Railroads include the Ubundi-KISANGANI section and a secondary system in N. Chief towns are BUTA, IRUMU, and BUNIA. YANGAMBI and NIOKA are centers of tropical agricultural research. Region was called STANLEYVILLE, 1935–1947. Was the political stronghold of Patrice Lumumba. In 1960, Lumumba's followers, centered at KISANGANI (then STANLEYVILLE), attempted to establish a government to rival the central government at KINSHASA (then LEOPOLDVILLE). The STANLEYVILLE regime controlled most of the region until the central government re-established control in 1962. There were further rebellions throughout the 1960s. Formerly called HAUT-ZAÏRE, changed name c.1998.

Oriental, Pico, VENEZUELA: see LA SILLA DE CARACAS.

Orient Beach State Park, New York: see ORIENT.

Oriente (or-ee-AIN-tai), region and former province, E CUBA. Was the largest and most populous province of the island, occupying its easternmost section. Now subdivided into GRANMA, GUANTÁNAMO, HOLGUÍN, LAS TUNAS, and SANTIAGO DE CUBA provinces.

Oriente (or-YEN-tai), town, S BUENOS AIRES province, ARGENTINA, on QUEQUÉN SALADO RIVER (hydroelectric station) and 37 mi/60 km E of CORONEL DORREGO; 38°44′S 60°37′W. In wheat-growing and livestock district.

Oriente (O-ree-en-che), town (2007 population 6,079), W central SÃO PAULO, BRAZIL, on railroad, 10 mi/16 km WNW of MARÍLIA; 22°09′S 50°11′W. Agricultural zone (cotton, sugarcane, fruit, coffee); sugar milling.

Oriente, El (o-ree-EN-tai, el), traditional name of the area of ECUADOR E of the ANDES, characterized by forested hills and alluvial plains dissected by rivers flowing toward the AMAZON. Originally the largest part of the Ecuadorian territory, it was reduced by the Rio de Janeiro Protocol of 1942 and it represents now about 50% of the country's land. Although increasingly attracting colonization beacuse of oil exploration and exploitation, it had only about 5% of the country's population in 1990. Originally it was a single province; in 1925 it was subdivided into NAPO-PASTAZA and SANTIAGO-ZAMORA provinces. It is now divided into five provinces.

Orient Point, NEW YORK: see ORIENT.

Orihuela (o-ree-WAI-lah), city, ALICANTE province, E SPAIN, in VALENCIA, on SEGURA RIVER, and 32 mi/51 km SW of ALICANTE; 38°05′N 00°57′W. Oil milling; canning. Irrigation of surrounding garden region (hemp, pimientos, fruit, cereals, olives) dates from Moorish times. Livestock. Marble quarries. Mineral springs. Bishopric; and formerly seat (1568–1835) of university. On side of hill above city is large seminary. Has cathedral (14th–15th century; restored), several old mansions, and 16th-century churches of Santiago and Santo Domingo (with adjoining 17th-century former university building). Was held by Romans, Visigoths, and Moors; liberated (1264) by Christians. Partly destroyed (1829) by earthquake.

Orijärvi (O-ri-YAR-vee), village, TURUN JA PORIN province, SW FINLAND, 45 mi/72 km ESE of TURKU, 17 mi/27 km NNE of EKENÄS; 60°13′N 23°33′E. Elevation 165 ft/50 m. In lake region; copper, zinc, lead, and silver mines.

Orikhiv (o-REE-khif) (Russian *Orekhov*), city, N ZAPORIZHZHYA oblast, UKRAINE, on KINS'KA (Russian

Konka) River, 30 mi/48 km SE of ZAPORIZHZHYA; 47°34′N 35°47′E. Elevation 193 ft/58 m. Railroad and highway hub; raion center; machinery, construction materials; food processing (flour, vegetable canning, dairy products). Agricultural technical school, vocational school, heritage museum. Established around 1783; city since 1938.

Orikhove, UKRAINE: see TOREZ.

Orillia (o-RIL-ee-uh), city (□ 11 sq mi/28.6 sq km; 2001 population 29,121), SE ONTARIO, E central Canada, on COUCHICHING Lake; 44°36′N 79°25′W. Manufacturing includes machinery, consumer goods, rubber products. It is also a summer resort. A monument to Champlain, erected in 1925, commemorates his explorations. Canadian humorist Stephen Leacock had a summer home here.

Orillia (o-RIL-ee-uh), former township, S ONTARIO, E central Canada, 26 mi/41 km from BARRIE; 44°43′N 79°29′W. Amalgamated into Severn township in 1994.

Oril' River (o-REEL) (Russian *Orel'*), 215 mi/346 km long, in E UKRAINE; rises 37 mi/60 km S of KHARKIV in Dnieper-Donets divide, in KHARKIV oblast; then flows in S-shaped course W past PERESHCHEPYNE and NEKHVOROSHCHA, where it forms the N boundary of DNIPROPETROVS'K oblast, and then SW into the DNIPRODZERZHYNS'K RESERVOIR on the DNIEPER (Ukrainian *Dnipro*) River, N of VERKHN'ODNIPROVS'K.

Orimattila (O-ri-MAHT-ti-lah), village, UUDENMAAN province, S FINLAND, 13 mi/21 km SSE of LAHTI; 60°48′N 25°45′E. Elevation 215 ft/65 m.

Orinda, city (2000 population 17,599), CONTRA COSTA county, W CALIFORNIA, residential suburb 5 mi/8 km NE of downtown OAKLAND; 37°53′N 122°11′W. Manufacturing (computer equipment, light manufacturing). Caldecott Tunnel to Oakland (Mount Diablo Boulevard) to SW. Mokelumne Aqueduct passes to N, enters SAN PABLO Reservoir to NW, BRIONES RESERVOIR to N, Lafayette Reservoir to E; BERKELEY HILLS to SW; Briones Regional Park to NE.

Orindiúva (O-reen-dee-oo-vah), town (2007 population 4,916), N SÃO PAULO state, BRAZIL, 47 mi/76 km N of SÃO JOSÉ DO RIO PRETO; 20°12′S 49°22′W. Coffee growing.

Orinin, UKRAINE: see ORYNYN.

Orinoca (o-ree-NO-kah), canton, SUD CARANGAS province, ORURO department, W central BOLIVIA, 3 mi/5 km W of Lake POOPÓ, on the UYUNI-ORURO road; 18°56′S 67°15′W. Elevation 12,959 ft/3,950 m. Gas wells to S; clay, limestone deposits. Agriculture (potatoes, yucca, bananas, rye); cattle.

Orinoco River (o-ree-NO-ko), estimated to be from 1,500 mi/2,414 km to 1,700 mi/2,736 km long, in VENEZUELA; rising near Mount Delgado Chalbaud in the GUIANA HIGHLANDS, S Venezuela; flows in a wide arc through tropical rain forests and savannas (LLANOS), forming part of the VENEZUELA-COLOMBIA border, and enters the ATLANTIC OCEAN through a large marshy delta (□ c.7,800 sq mi/20,200 sq km) in NE Venezuela; 08°35′S 61°50′W. One of SOUTH AMERICA'S longest rivers, it and its branches drain an extensive basin; the APURE RIVER is its chief tributary. The Orinoco is joined to the AMAZON system by CASIQUIARE, a natural canal. The huge flow of the Orinoco varies markedly with the season. Divided into upper and lower courses by the ATURE and MAIPURES cataracts, the river is navigable for most of its length. Dredging permits oceangoing vessels to reach CIUDAD BOLÍVAR, c.270 mi/435 km upstream; a high suspension bridge crosses the river there. The major cities on the river are Ciudad Bolívar and CIUDAD GUAYANA, which developed in an industrial zone in the late 1960s. Christopher Columbus probably discovered the mouth of the Orinoco in 1498, and Lope de Aguirre, the Spanish adventurer, seems to have traveled most of its length in 1560. In 1799, Alexander von Humboldt, the German geographer and naturalist, explored the upper reaches, but it was not until 1944 that an aerial expedition sighted the source area in the remote highlands. Further explorations in 1951 and 1956 located 2 rivulets now considered the headwaters.

Orio (O-ryo), town, GUIPÚZCOA province, N SPAIN, 8 mi/12.9 km WSW of SAN SEBASTIÁN; 43°16′N 02°08′W. Manufacturing (ceramics, chemicals, textiles) corn, apples; cattle.

Oriomo (o-ree-O-mo), village, WESTERN province, S central NEW GUINEA island, SW PAPUA NEW GUINEA, 10 mi/16 km N of DARU; 08°52′S 143°11′E. Access by walking track. Palm oil, sago; cattle.

Orion (OR-yon), town, Bataan province, S LUZON, PHILIPPINES, on E BATAAN Peninsula, on MANILA BAY, 28 mi/45 km W of MANILA; 14°35′N 120°32′E. Agricultural center (sugarcane, rice).

Orion, town, OAKLAND county, MICHIGAN, suburb 10 mi/16 km W of PONTIAC. Manufacturing (electronic equipment, transportation equipment, machinery).

Orion, village (2000 population 1,713), HENRY county, NW ILLINOIS, 12 mi/19 km WNW of CAMBRIDGE; 41°21′N 90°22′W. In agricultural area.

Orion (o-REI-uhn), hamlet, SE ALBERTA, W Canada, 38 mi/61 km from BOW ISLAND town, in Forty Mile County No. 8; 49°27′N 110°49′W.

Oriska (or-IS-kuh), village (2006 population 116), BARNES county, E NORTH DAKOTA, 10 mi/16 km E of VALLEY CITY; 46°55′N 97°47′W. Founded in 1872 as Fourth Siding and then name was changed in 1879 to Carleton and changed again in 1881 to Oriska. Incorporated in 1912.

Oriskany (ah-RIS-kuh-nee), village (2006 population 1,428), ONEIDA county, central NEW YORK, near mouth of ORISKANY CREEK on MOHAWK River, 7 mi/11.3 km NW of UTICA; 43°09′N 75°19′W. Light manufacturing. Obelisk at Oriskany Battlefield (NW) marks site of an engagement (Aug. 6, 1777) of the Saratoga campaign, one of the bloodiest battles of the American Revolution. Incorporated 1914.

Oriskany Creek (ah-RIS-kuh-nee), c.30 mi/48 km long, central NEW YORK; rises in MADISON county S of ONEIDA; flows S and NNE to MOHAWK River 6 mi/9.7 km NW of UTICA.

Oriskany Falls (ah-RIS-kuh-nee), village (2006 population 679), ONEIDA county, central NEW YORK, on ORISKANY CREEK and 16 mi/26 km SW of UTICA; 42°56′N 75°27′W. Light manufacturing.

Orissa (o-RI-suh), state (□ 60,118 sq mi/156,306.8 sq km; 2001 population 36,804,660), E central INDIA, on the BAY OF BENGAL; ⊙ BHUBANESHWAR. The relatively unindented coastline (c. 200 mi/320 km long) lacks good ports save for the newly constructed deepwater facility at Paradwip, The narrow, level coastal strip, including the delta of the MAHANADI River, is exceedingly fertile. Rainfall is heavy and regular, and two crops of rice (by far the most important food) are grown annually. In the S are the EASTERN GHATS, which yield valuable timber. Supporting a major industrial zone, including iron and steel manufacturing, in the N are deposits of iron, manganese, coal, and mica. A canal system links the Mahanadi River with the HUGLI River in WEST BENGAL state. CUTTACK is the principal city. The temple-dotted cities of Konarka, PURI, and Bhubaneshwar support a thriving tourist industry. The dense population, concentrated on the coastal alluvial plain, is Oriya-speaking. The interior of Orissa, inhabited largely by Munda-speaking aborigines, is hilly and mountainous. In ancient times the region of Orissa was the center of the strong Kalinga kingdom, although it was temporarily conquered (c. 250 B.C.E.) by Asoka and held for almost a century by the Mauryas. With the gradual decline of Kalinga, several Hindu dynasties arose and built temples at Bhubaneshwar, Puri, and Konarka. After long resistance to the Muslims, the region was finally overcome (1568) by Afghan invaders and passed to the Mogul empire. After the fall of the Moguls, Orissa was divided between the nawabs of BENGAL and the Marathas. In 1803 it was conquered by the British. The coastal section, which was made (1912) part of Bihar and Orissa Province, became in 1936 the separate province of Orissa. In 1948 and 1949 the area of Orissa was almost doubled and the population was increased by one third with the addition of twenty-four former princely states. In 1950, Orissa became a constituent state of India. It is governed by a chief minister and cabinet responsible to an elected unicameral legislature and by a governor appointed by the president of India.

Orissa States (o-RI-suh), subordinate agency of former EASTERN STATES Agency, INDIA. Comprised princely states of Athgarh, Athmallik, Bamra, Baramba, Baudh, Bonai, Daspalla, Dhenkanal, Gangpur, Hindol, Keonjhar, Khandpara, Kharsawan, Narsinghpur, Nayagarh, Nilgiri, Pal Lahara, Rirakhol, Ranpur, Saraikela, Sonepur, Talcher, and Tigiria. Kharsawan and Saraikela were incorporated 1948 into BIHAR state, while rest of states merged with ORISSA state.

Oristano (o-rees-TAH-no), town, Oristano province, W SARDINIA, near mouth of TIRSO River, 55 mi/89 km NNW of CAGLIARI, in the CAMPIDANO; 39°54′N 08°36′E. Pottery; canneries (fish, fruit), olive oil and pasta factories, flour mills; domestic embroidery. Archbishopric. Has 18th-century cathedral (replacing one built in 13th century; two of its towers remain), medieval fortifications. Pisan church of Santa Giusta (12th century) is 1 mi/1.6 km S.

Oristano, Gulf of, W SARDINIA, inlet of MEDITERRANEAN Sea, between CAPE SAN MARCO (N) and CAPE FRASCA (S); 13 mi/21 km long, 6 mi/10 km wide. Fisheries (tunny, lobster). Receives TIRSO River.

Öriszentpéter (UH-re-sant-pai-tar), village, VAS county, HUNGARY, 19 mi/31 km W of ZALAEGERSZEG, on upper course of ZALA RIVER. Alfalfa, silage corn, clover; dairy; cattle. Manufacturing (footwear, food products).

Orito (o-REE-to), town, PUTUMAYO department, SW COLOMBIA, 140 mi/225 km E of TUMACO; 00°41′N 76°52′W. Elevation 2,240 ft/682 m. Site of oil refinery.

Orituco River (o-ree-TOO-ko), c.150 mi/241 km long, GUÁRICO state, central VENEZUELA; rises on S slopes of coastal range near MIRANDA state border; flows SSW and SW past ALTAGRACIA DE ORITUCO, to GUÁRICO RIVER, 13 mi/21 km S of CALABOZO; 08°44′N 67°24′W.

Orivesi (O-ree-VAI-see), village, HÄMEEN province, SW FINLAND, on small Lake Orivesi, 25 mi/40 km ENE of TAMPERE; 61°41′N 24°21′E. Elevation 495 ft/150 m. In agricultural region (grain, potatoes; livestock). Quarry provides special stones for Finnish saunas.

Oriximiná (o-ree-shee-mee-NAH), city (2007 population 55,034), W PARÁ, BRAZIL, on Rio TROMBETAS and 25 mi/40 km NW of ÓBIDOS, in marshy flood area; 01°35′S 55°50′W. Rubber, Brazil nuts, medicinal plants.

Orizaba (o-ree-SAH-bah), city and township, VERACRUZ state, E central MEXICO 18°51′N 97°08′W. It is the commercial center of a prosperous coffee and sugar growing region on Mexico Highway 150. The development of water power has stimulated manufacturing industries, especially cotton and wool textile factories. Site of large brewery. Orizaba is a popular vacation spot. Mineral springs are nearby, and the majestic cone of PICO DE ORIZABA (Citlaltépetl) rises in the distance. The city is a cultural center noted for its fine arts institute. The federal school in Orizaba houses murals by José Clemente Orozco. In 1862, Benito Juárez, seeking to forestall foreign intervention in Mexican affairs, called a conference at Orizaba of foreign representatives; his efforts failed. French for-

---

Cross-references are shown in SMALL CAPITALS. The pronunciation guide is shown on page xix. The sources of population figures are shown on page xvii.

ces subsequently used the city as a base for their invasion of Mexico.

Orizatlán (o-ree-sah-TLAHN), town, HIDALGO, central MEXICO, in foothills of SIERRA MADRE ORIENTAL, 13 mi/21 km WNW of HUEJUTLA DE REYES; ⊙ SAN FELIPE ORIZATLÁN municipio; 21°10′N 98°36′W. Corn, rice, tobacco, coffee, fruit, cigars.

Orizona (o-ree-so-nah), city (2007 population 14,378), S GOIÁS, central BRAZIL, 60 mi/97 km SE of ANÁPOLIS; 17°01′S 48°20′W. Lard, cereals. Rutile deposits. Until 1944, Campo Formoso.

Orjen (OR-yen), mountain (6,216 ft/1,895 m), in DINARIC ALPS, on the MONTENEGRO–BOSNIA and HERZEGOVINA border, 10 mi/16 km N of HERCEG NOVI; 42°34′N 18°32′E. Also spelled Oryen.

Órjiva or **Órgiva** (both: OR-hee-vah), town, GRANADA province, S SPAIN; chief center of ALPUJARRAS district, 22 mi/35 km SSE of GRANADA. Olive-oil and cheese processing, brandy distilling. Oranges, cereals; livestock. Silver-bearing lead mines nearby. Starting point for the ascent of the Mulhacén.

Orkanger (AWRK-ahng-uhr), village, SØR-TRØNDELAG county, central NORWAY, at head of Orkdalsfjorden (inlet of TRONDHEIMSFJORDEN), at mouth of ORKLA River, on railroad and 20 mi/32 km SW of TRONDHEIM; 63°19′N 09°52′E. Industrial center; ferrosilisium.

Örkelljunga (UHR-kel-YUNG-ah), town, SKÅNE county, SW SWEDEN, 16 mi/26 km ENE of ÄNGELHOLM; 56°17′N 13°17′E.

Orkhomenos, Greece: see ORCHOMENOS.

Orkhon Gol, MONGOLIA: see ORHON GOL.

Orkhon River, MONGOLIA: see ORHON GOL.

Orkla (AWR-klah), river, 100 mi/161 km long, central NORWAY; rises in the NE DOVREFJELL of SØR-TRØNDELAG county 40 mi/64 km NE of DOMBÅS; flows E to KVIKNE in HEDMARK county, then NW to SØr-Trøndelag county, past RENNEBU, MELDAL, Orkland, and Orkdal, to an inlet of TRONDHEIMSFJORDEN at ORKANGER.

Orkney, town, GAUTENG, SOUTH AFRICA, on Schoonspruit River near its confluence with the VAAL River border off FREE STATE, 6 mi/9.7 km S of KLERKSDORP. Elevation 4,275 ft/1,303 m. Railroad junction; gold mining at Vaal Reef mine. Major disaster in mine, May 1995, killed 106 people. Airfield. Formerly Eastleigh.

Orkney Islands (ORK-nee), county (□ 376 sq mi/977.6 sq km; 2001 population 19,245), N Scotland, consisting of an archipelago of about seventy islands (about twenty of which are inhabited) in the ATLANTIC OCEAN and NORTH SEA, N of HIGHLAND across PENTLAND FIRTH; ⊙ KIRKWALL; 59°00′N 03°00′W. MAINLAND (Pomona) has Kirkwall and STROMNESS. Other large islands are HOY, SOUTH RONALDSAY, Stronsay, SANDAY, WESTRAY, and ROUSAY. The climate is mild, windy, and wet. Beef cattle, sheep, and pigs are raised. Some fishing, mainly for lobster, is carried on in SCAPA FLOW and in the N. The discovery of North Sea oil in the early 1970s provided employment for many of Orkney's inhabitants, with an oil terminal on FLOTTA and service bases at Car Ness and Stromness, Mainland; and at Lyness, Hoy. Airport at Grimsetter, near Kirkwall. Orkney was settled by Picts. There were Viking invasions in the 8th century. From 875 to 1231 it was a Viking earldom under the Norwegian crown. In 1231, Orkney passed to the Scottish earls of Angus on the death of the last Viking earl. It became a possession of the Scottish crown in 1472 in trust for the undelivered dowry of Margaret of Norway on her marriage to James III (1469), but the long Norse occupation left marked Scandinavian traces in the people and their culture. James V visited Kirkwall in 1540 and made Orkney a county. Orkney has many prehistoric relics. Stone Age villages have been unearthed at SKARA BRAE on Mainland and a broch (prehistoric fort) at Rinyo on Rousay. Other relics are the burial chambers at Maeshowe and the standing stones at STENNESS.

Orkney Springs (ORK-nee), unincorporated village, SHENANDOAH county, NW VIRGINIA, near WEST VIRGINIA state line, 25 mi/40 km N of HARRISONBURG, in ALLEGHENY MOUNTAINS foothills, in George Washington National Forest; 38°47′N 78°48′W. Mineral springs; resort. Boyce Resort to N.

Orland, city (2000 population 6,281), GLENN county, N central CALIFORNIA, in SACRAMENTO VALLEY, 18 mi/29 km W of CHICO; 39°45′N 122°11′W. Shipping, processing center; fruit, nuts, olives, rice, wheat, corn; nursery livestock; cattle, sheep; dairying; dried fruits and nuts. Headquarters of U.S. Bureau of Reclamation irrigation project begun in 1910. BLACK BUTTE LAKE reservoir to W, Stony Gorge Reservoir to SW, both on STONY CREEK. Tehama Colusa Canal passes to E and S. Orland Buttes (1,038 ft/316 m) to W. Founded 1881, incorporated 1909.

Orland, town (2000 population 341), STEUBEN county, NE INDIANA, near MICHIGAN line, 11 mi/18 km NW of ANGOLA; 41°44′N 85°10′W. Manufacturing (plastic products, wire products, food, transportation equipment, fabricated metal products). Fawn River State Fish Hatchery nearby to NW. Laid out 1838.

Orland, town, HANCOCK county, S MAINE, near BUCKSPORT, 16 mi/26 km W of ELLSWORTH; 44°34′N 68°40′W. Settled 1764.

Orland Hills, village (2000 population 6,779), COOK county, NE ILLINOIS, residential suburb 23 mi/37 km SSW of downtown CHICAGO, 2 mi/3.2 km S of ORLAND PARK; 41°35′N 87°50′W.

Orlândia (OR-lahn-zhee-ah), city (2007 population 36,162), NE SÃO PAULO, BRAZIL, on railroad, 31 mi/50 km N of RIBEIRÃO PRÊTO; 20°43′S 47°53′W. Sugarcane, coffee, cotton, corn, rice; manufacturing of food.

Orlando (OR-lan-DO), city (□ 101 sq mi/262.6 sq km; 2005 population 213,223), ⊙ ORANGE county, central FLORIDA, in a lake region, and c.80 mi/128 km NE of TAMPA; 28°31′N 81°22′W. It is a growing center of a high-tech manufacturing region as well as one of the most visited vacation spots in the world. Orlando International Airport S of downtown area. The city's economy focuses on the aerospace and electronics industries, but tourism brings in the largest revenues. Located 15 mi/24 km SW of Orlando is the famous Walt Disney World (opened 1971), with its many theme parks, which draw millions of visitors annually to the Orlando area. Hotels, restaurants, and tourist facilities abound in and around Orlando, which is noted for its subtropical climate. The film and television industries are becoming increasingly important in Orlando, which is often called "Hollywood East." The University of Central Florida and Rollins College are in the metropolitan area. Site of Amway Arena, home of the Orlando Magic basketball team.

Orlando, residential town, GAUTENG, SOUTH AFRICA, on WITWATERSRAND, 8 mi/12.9 km WSW of JOHANNESBURG; 26°15′S 28°57′E. Elevation 5,379 ft/1,640 m. Under jurisdiction of Johannesburg city; almost entire population consists of black residents, as a result of former apartheid policy of separate development based on ethnicity. Extension of SOWETO to the E. On main railroad to SW.

Orlando, village (2006 population 215), LOGAN county, central OKLAHOMA, 19 mi/31 km N of GUTHRIE; 36°08′N 97°22′W. Trade center for farming area. Lake Carl Blackwell reservoir to E.

Orlandovtsi (or-LAHN-dov-tsee), village, SOFIA district, W BULGARIA, just N of SOFIA; now part of Sofia city; 42°35′N 23°00′E.

Orland Park, village (2000 population 51,077), COOK county, NE ILLINOIS, suburb 24 mi/39 km SW of downtown CHICAGO, 12 mi/19 km NE of JOLIET; 41°36′N 87°50′W. Despite rapid urban growth, area still has some agriculture (corn; dairying); light manufacturing.

Orla River (OR-lah), 25 mi/40 km long, central GERMANY; rises ESE of Tripti; flows W past TRIPTIS and NEUSTADT, to the SAALE at Orlamünde.

Orlau, CZECH REPUBLIC: see ORLOVÁ.

Orle (OR-le), mountain (4,966 ft/1,514 m), MACEDONIA, along left bank of the lower CRNA REKA, 16 mi/27 km E of PRILEP; 41°21′N 21°53′E.

Orleães (OR-lai-eins), city, SE SANTA CATARINA, BRAZIL, on railroad and 35 mi/56 km WNW of LAGUNA, in coal-mining area; 28°21′S 49°18′W. Livestock. Also spelled Orleans.

Orléanais (or-lai-ah-NAI), region and former province, CENTRE administrative region, N central FRANCE, on both sides of the middle LOIRE RIVER; 47°50′N 02°00′E. ORLÉANS, the historic capital, CHARTRES, MONTARGIS, and BLOIS are the chief centers. The region includes LOIRET, LOIR-ET-CHER, and a part of EURE-ET-LOIR department. The BEAUCE in the N, the GÂTINE and Vendômois in the W, and GÂTINAIS district in the E are rich agricultural districts; the large ancient forest of ORLÉANS (NE of the city) occupies the center of the region. The fertile Loire valley yields fruits, vegetables, and horticultural products. It contains many fine châteaux, notably Blois and CHAMBORD, CHEVERNY, CHAUMONT. S of the Loire bend is the lake-studded SOLOGNE district, now improved by drainage. The nucleus of the Orléanais has been part of the royal domain since the time of Hugh Capet (10th century). But during several periods it was a duchy ruled by the powerful Orléans family. It definitively became part of the French crown in 1661. There are ruins of fortresses and churches from the Carolingian period (c.7th century). Orléans is considered part of the *pays de la loire* (the Loire country).

Orléans (or-lai-AHN), community, SE ONTARIO, E central Canada, and included in city of OTTAWA; formerly part of dissolved Ottawa-Carleton Regional Municipality (1991 population 678,147), on S bank of OTTAWA River; 45°28′N 75°31′W. Residential area. Mixed farming.

Orleans, parish (□ 199 sq mi/517.4 sq km; 2006 population 223,388), SE LOUISIANA, coextensive with city of NEW ORLEANS; 30°04′N, 89°56′W. Home gardens, oysters, shrimp, crabs, finfish. Formed 1805.

Orleans, county (□ 817 sq mi/2,124.2 sq km; 2006 population 43,213), W NEW YORK; ⊙ ALBION; 43°22′N 78°13′W. Bounded N by Lake ONTARIO; crossed by the NEW YORK STATE BARGE CANAL; drained by Oak Orchard Creek. Diversified manufacturing at Albion, MEDINA, LYNDONVILLE. Fruit-growing area; also dairy products. Formed 1924.

Orleans, county (□ 720 sq mi/1,872 sq km; 2006 population 27,718), N VERMONT, on Canadian (QUEBEC) line; ⊙ Newport; 44°49′N 72°15′W. Dairying, lumbering; wood products, furniture; maple sugar. Resorts on Lake MEMPHREMAGOG, Lake WILLOUGHBY, and smaller lakes; winter sports. Drained by BARTON, MISSISQUOI, Black, and CLYDE rivers. Organized 1792.

Orléans (or-lai-AHN), city (□ 10 sq mi/26 sq km), N central FRANCE, on right bank of the LOIRE RIVER, and 70 mi/113 km SSW of PARIS; ⊙ LOIRET department and CENTRE region; 47°55′N 02°05′E. A commercial, administrative, and transportation center; manufacturing of food, machinery, electrical equipment, pharmaceuticals, and textiles. The main industrial growth zone is at CHAPELLE-SAINT-MESMIN. The railroad station at FLEURY-LES-AUBRAIS (1.5 mi/2.4 km N of city center) is a junction of lines fanning out in many directions. The old city, hugging the banks of the Loire, is surrounded by extensive suburbs where most of the growth has occurred since World War II, when Orléans was severely damaged. Orléans-la-Source (4.5 mi/7.2 km SSE of city center) is a "new town" that has absorbed much residential growth. The urban area's population exceeds 240,000. Orléans was first known as *Genabum*, a commercial center of the Carnutes, a Celtic tribe. It revolted against Julius Caesar (52 B.C.E.), was burned, and rebuilt as *Aurelianum*. Unsuccessfully attacked by Attila the Hun (451), it was taken by Clovis I (498). Under the Capetians, the first kings of France, the city became (10th century), after Paris, the principal residence of the French kings. Orléans, with the surrounding province, known as Orléanais, constituted part of the originally small nucleus of the royal domain, and it was several times given in appanage as a duchy to the eldest brother of the king of France and to his descendants, known as the Orléans family. The siege of Orléans (1428–1429) by the English threatened to bring all of France under England's rule, and its lifting by Joan of Arc turned the tide of the Hundred Years War (1337–1453). In the Wars of Religion (16th century) the city was briefly the headquarters of the Huguenots and was besieged in 1563 by Catholic forces under the Duc de Guise, who was assassinated under its walls. Orléans remained in Catholic hands until the Edict of NANTES (1598). During the 17th and 18th century the city was a prosperous center, and its university (founded 14th century) became famous throughout Europe. Despite repeated wartime damage, several fine structures remain, including the Cathedral of Sainte-Croix, rebuilt (17th–19th century) after its destruction by the Huguenots in 1568; and the Renaissance town hall (restored in 19th century), where Francis II died in 1560. The museum of fine arts has a collection of paintings and sculptures spanning 6 centuries. The feast of Joan of Arc has been celebrated in Orléans with particular splendor each May since 1435. Orléans, the "city of roses," prides itself on being the gateway to the "garden of France," the lower valley of the Loire.

Orleans, town (2000 population 2,273), ORANGE county, S INDIANA, near LOST RIVER, 7 mi/11.3 km N of PAOLI; 38°40′N 86°27′W. In agricultural area (corn, cattle); manufacturing (lumber, food, furniture). Karst topography. Settled 1815.

Orleans, resort town (2000 population 583), DICKINSON county, NW IOWA, on Spirit Lake, 2 mi/3.2 km NE of SPIRIT LAKE; 43°27′N 95°05′W. Fish hatchery and Marble Beach State Park here.

Orleans, town, BARNSTABLE county, SE MASSACHUSETTS, near elbow of CAPE COD, 17 mi/27 km ENE of BARNSTABLE; 41°46′N 69°58′W. Summer resort. Transatlantic cable to BREST, FRANCE. Includes villages of East Orleans, Rock Harbor, South Orleans. CAPE COD NATIONAL SEASHORE just E; Nickerson State Park and Point of Rocks Beach nearby. Settled 1693, set off from EASTHAM 1797.

Orleans, village (2006 population 378), HARLAN county, S NEBRASKA, 6 mi/9.7 km WNW of ALMA, and on REPUBLICAN RIVER, near KANSAS state line; 40°07′N 99°27′W. Railroad junction. Grain.

Orleans, village (2006 population 836) in Barton town, ORLEANS CO., N VERMONT, 9 mi/14.5 km S of Newport and on BARTON RIVER; 44°48′N 72°12′W. Settled c.1821.

Orléans Canal (or-lai-ahn kah-nahl), 45 mi/72 km long, in LOIRET department, CENTRE administrative region, N central FRANCE, one of the waterways connecting the LOIRE and SEINE rivers; 47°55′N 01°55′E. It begins just above ORLÉANS, traverses the Forest of ORLÉANS, and joins the BRIARE CANAL below MONTARGIS, whence both are continued by the LOING CANAL to the Seine.

Orléans, Forest of (or-lai-AHN) (□ 150 sq mi/390 sq km), one of largest forests in FRANCE, extends 40 mi/64 km E and ESE of ORLÉANS, LOIRET department, CENTRE region, N of the LOIRE RIVER; 47°55′N 02°15′E. Consists of oak, birch, and hornbeam stands.

Orléans, Île d' (dor-lai-AHN, eel) or **Orléans Island** (or-lai-AHN), 20 mi/32 km long and 5 mi/8 km wide, S QUEBEC, E Canada, in the SAINT LAWRENCE RIVER NE of QUEBEC city; 46°55′N 70°58′W. It is connected with the mainland by a highway bridge. Settled (1651) by the French, it was the site of one of Wolfe's camps in his attack on Quebec in 1759. It is a popular tourist attraction and a residential community for Quebec city. Potatoes, strawberries, cheese, and poultry are the chief products.

Orléans-la-Source, FRANCE: see ORLÉANS.

Orléans Strait, 10 mi/15 km wide and 20 mi/30 km long; separates TRINITY ISLAND from the DAVIS COAST of GRAHAM LAND, ANTARCTIC PENINSULA; 63°50′S 60°20′W.

Orléans Strait, 10 mi/15 km wide and 20 mi/30 km long; separates TRINITY ISLAND from the DAVIS COAST of GRAHAM LAND, ANTARCTIC PENINSULA; 63°50′S 60°20′W.

Orlice River (OR-li-TSE), German *adler*, 83 mi/134 km long, VYCHODOCESKY province, E BOHEMIA, CZECH REPUBLIC; formed by junction of two headstreams, DIVOCHA ORLICE RIVER (right) and TICHA ORLICE RIVER (left), 1 mi/1.6 km above TYNISTE NAD ORLICI; flows c.20 mi/32 km NW to ELBE RIVER at HRADEC KRÁLOVÉ.

Orlik (OR-leek), village (2005 population 1,995), SW BURYAT REPUBLIC, S SIBERIA, RUSSIA, in the EASTERN SAYAN MOUNTAINS, on the OKA River, on highway, 130 mi/209 km S of TULUN (IRKUTSK oblast); 52°30′N 99°49′E. Elevation 4,550 ft/1,386 m. Gold mining.

Orlik Dam (OR-leek), Czech *Orlík*, JIHOCESKY and STREDOCESKY provinces, S central BOHEMIA, CZECH REPUBLIC, on VLTAVA R. Built during 1954–1962; impounds reservoir (□ 10.6 sq mi/27.5 sq km; maximum depth 243 ft/74 m); power plant; summer resort (fishing, water sports). Nearby is a 13th century Gothic castle Orlik (Czech *Orlík*).

Orlinghausen, GERMANY: see OERLINGHAUSEN.

Orlinoye, UKRAINE: see ORLYNE.

Orljava River (OR-lyah-vah), c.50 mi/80 km long, E CROATIA, in SLAVONIA; rises on NE slope of the PSUNJ; flows SE past Požega and PLETERNICA, and SSW to Sava River 15 mi/24 km WSW of SLAVONSKI BROD. Receives Londža River.

Orlov (uhr-LOF), city (2005 population 8,125), central KIROV oblast, E European Russia, on the VYATKA River, on local road, 48 mi/77 km W of KIROV; 58°32′N 48°54′E. Elevation 341 ft/103 m. Dairy, bakery; poultry factory, veterinary station; light manufacturing. First mentioned in 1459; chartered in 1627; important trading center until the 19th century. Called Khalturin between 1923 and 1992.

Orlová (OR-lo-VAH), German *Orlau*, city (2001 population 34,856), SEVEROMORAVSKY province, NE SILESIA, CZECH REPUBLIC; 49°51′N 18°26′E. Coal-mining center of Ostrava-Karvina coal basin; manufacturing (electronics, construction materials), tanning, food processing. Founded in the 13th century; coal mining dates to 1817.

Orlovets (or-LO-vets), peak (8,753 ft/2,668 m), in the PIRIN MOUNTAINS, SW BULGARIA.

Orlovista (or-luh-VIS-tuh), unincorporated town (□ 2 sq mi/5.2 sq km; 2000 population 6,047), ORANGE county, central FLORIDA, 5 mi/8 km WSW of ORLANDO; 28°32′N 81°27′W.

Orlovka (ahr-LAWV-kah), town, CHÜY region, KYRGYZSTAN, at E edge of CHU valley, 10 mi/16 km SW of KEMIN; 42°44′N 75°36′E. Gold mining in nearby Kyrgyz Ala-Too (ALATAU) mountains.

Orlovka (uhr-LOF-kah), settlement, NE TATARSTAN Republic, E European Russia, on the KAMA River, 12 mi/19 km E of YELABUGA, and 4 mi/6 km NE of NABEREZHNYYE CHELNY, to which it is administratively subordinate; 55°44′N 52°23′E. Elevation 400 ft/121 m. Metalworking, construction materials.

Orlovo (uhr-LO-vuh), village (2006 population 4,030), N VORONEZH oblast, S central European Russia, in the VORONEZH RIVER basin, on road junction and railroad, 17 mi/27 km ENE of VORONEZH; 51°45′N 39°35′E. Elevation 341 ft/103 m. In agricultural area producing wheat, corn, sunflowers.

Orlovo (uhr-LO-vuh), settlement, SW SAKHALIN oblast, RUSSIAN FAR EAST, on the TATAR STRAIT, near coastal highway, 15 mi/24 km SSW of UGLEGORSK; 48°51′N 141°56′E. Fishing. Under Japanese rule (1905–1945), called Ushiro.

Orlovo-Rovenets'ka, UKRAINE: see ROVEN'KY.
Orlovo-Rovenetskaya, UKRAINE: see ROVEN'KY.

Orlovskiy (uhr-LOF-skeeyee), town (2005 population 2,300), central AGIN-BURYAT AUTONOMOUS OKRUG, S central CHITA oblast, 16 mi/26 km S of MOGOYTUY; 51°02′N 114°50′E. Elevation 2,680 ft/816 m. Tungsten and beryl ore mining and processing. Created as a workers settlement after the discovery of the Spokoyninskoye tungsten deposit in the area in the early 1970s, and originally called Spokoynyy (Russian= calm). Post-Soviet transition to market economy resulted in a severe economic depression, closing of some of the mines and processing plants, and steady population outflow since the mid-1990s.

Orlovskiy (uhr-LOF-skeeyee), town (2006 population 19,320), S ROSTOV oblast, S European Russia, on highway junction and railroad (Dvoynaya station), 38 mi/61 km NE of SAL'SK; 46°52′N 42°03′E. Elevation 396 ft/120 m. Construction materials, agricultural machinery; flour mill, dairy.

Orlovskiy (uhr-LOF-skeeyee), village (2005 population 2,300), E central AGIN-BURYAT AUTONOMOUS OKRUG, S central CHITA oblast, 16 mi/26 km S of MOGOYTUY; 51°02′N 114°50′E. Elevation 2,680 ft/816 m. Ore mining and processing. Established as a workers settlement for the Spokoyninskiy ore deposit, discovered in the late 1960s, and originally called Spokoynyy.

Orlu (AW-loo), town, IMO state, S NIGERIA, 20 mi/32 km N of OWERRI; 05°47′N 07°02′E. Palm oil and kernels, kola nuts. Also Awlu.

Orly (or-LEE), SSE suburb (□ 2 sq mi/5.2 sq km) of PARIS, VAL-DE-MARNE department, ÎLE-DE-FRANCE region, N central FRANCE, 9 mi/14.5 km from Notre Dame Cathedral; 48°45′N 02°24′E. It is the site of Orly airport, one of the two international and domestic commercial airports of PARIS. It has two terminals, linked to the city by limited access roads and by railroad. There is a large industrial and warehouse zone N of the runways. The other main airport, Charles de Gaulle, is at ROISSY-EN-FRANCE, 14 mi/23 km NE of central Paris.

Orlyak (orl-YAHK), village, VARNA oblast, TERVEL obshtina, BULGARIA; 43°38′N 27°22′E.

Orlyne (or-LI-ne) (Russian *Orlinoye*), village, S Republic of CRIMEA, UKRAINE, in Baydar Valley, at W end of the CRIMEAN MOUNTAINS, 15 mi/24 km SE, and under jurisdiction, of SEVASTOPOL; 44°27′N 33°46′E. Elevation 915 ft/278 m. Tourist excursion center. In a pass (2.5 mi/4 km S) is the Baidar Gate, a stone structure built 1848, offering fine view of S Crimean coast. Originally known as Baydary, renamed around 1950.

Ormara (or-MAH-rah), village, LAS BELA district, KALAT division, BALUCHISTAN province, SW PAKISTAN, on promontory in ARABIAN SEA, 150 mi/241 km WNW of KARACHI; 25°12′N 64°38′E. Fishing; some coastal trade.

Orme (orm), town (2006 population 119), MARION county, S TENNESSEE, near ALABAMA line, 29 mi/47 km W of CHATTANOOGA; 35°00′N 85°48′W.

Ormenion (or-ME-nee-on), village, Évros prefecture, EAST MACEDONIA AND THRACE department, extreme NE GREECE, on railroad and 18 mi/29 km WNW of EDIRNE (Adrianople), Turkey, on MARITSA (Évros) River (Bulgarian border); 41°43′N 26°13′E. Border post opposite SVILENGRAD (Bulgaria).

Ormesby, ENGLAND: see MIDDLESBROUGH.

Ormesson-sur-Marne (or-me-son–syor–mahrn), town (□ 1 sq mi/2.6 sq km), VAL-DE-MARNE department, ÎLE-DE-FRANCE region, N central FRANCE, on left bank of the MARNE and 10 mi/16 km SE of PARIS; 48°47′N 02°33′E. Orthopedic center; pharmaceuticals. Has large golf course. Its 17th–18th-century château with garden was laid out by Le Nôtre.

Örmezö, SLOVAKIA: see STRAZSKE.

Ormidhia (or-MEE-[th]yah), town, LARNACA district, SE CYPRUS, 10 mi/16 km ENE of LARNACA, in enclave surrounded by DHEKELIA U.K. SOVEREIGN BASE AREA; 34°59′N 33°47′E. Larnaca Bay, MEDITERRANEAN SEA 1 mi/1.6 km to S. Base service area.

Ormiston (OR-mis-tuhn), village, S SASKATCHEWAN, Canada, near small Shoe Lake, 45 mi/72 km S of MOOSE JAW; 49°44′N 105°22′W. Light manufacturing.

Ormiston (OR-mis-tuhn), village (2001 population 2,079), East Lothian, E SCOTLAND, on TYNE RIVER, and 2 mi/3.2 km S of TRANENT; 55°54′N 02°56′W. Just NW is village of Elphinstone, with remains of Elphinstone Tower, a 14th-century border fortress. Formerly in LOTHIAN, abolished 1996.

Ormoc (OR-mok), city (□ 179 sq mi/465.4 sq km; 2000 population 154,297), LEYTE province, W LEYTE, PHILIPPINES, 32 mi/51 km SW of TACLOBAN; 11°00′N 124°37′E. Chief port and commercial center for W coast of Leyte. Exports rice, sugar. Sugar milling. Tongonan geothermal power station (with 100 MW capacity) at Tongonan Hot Springs National Park, 12 mi/19 km N. During World War II, city was principal Japanese supply port on Leyte; taken December 1944, by U.S. forces after a bitter struggle. Chartered as city 1947. Severely damaged by typhoon (with more than 4,000 people dead) November 1991.

Ormoc Bay (OR-mok), inlet of CAMOTES SEA, W LEYTE, PHILIPPINES; 13 mi/21 km E-W, 16 mi/26 km N-S. ORMOC city is at head of bay.

Ormond, suburb 7 mi/12 km SE of MELBOURNE, VICTORIA, SE AUSTRALIA; 37°54′S 145°02′E.

Ormond Beach (OR-muhnd), resort and residential city (□ 29 sq mi/75.4 sq km; 2005 population 38,613), VOLUSIA county, E central FLORIDA, 5 mi/8 km NNW of DAYTONA BEACH, on HALIFAX RIVER (a lagoon) and the ATLANTIC OCEAN; 29°17′N 81°05′W. Incorporated 1880.

Ormond-by-the-Sea (OR-muhnd), unincorporated town (□ 2 sq mi/5.2 sq km; 2000 population 8,430), VOLUSIA county, E central FLORIDA, 12 mi/19 km NNW of DAYTONA BEACH; 29°20′N 81°04′W.

Or, Mont d' (dor, mon), summit (4,800 ft/1,463 m) of the E JURA Mountains, near Franco-Swiss border, 11 mi/18 km S of PONTARLIER (FRANCE), overlooking VALLORBE (Switzerland) and the Swiss lakes; 46°44′N 06°22′E. Pierced by international railroad tunnel (4 mi/6.4 km long).

Ormož (OR-mozh), village, NE SLOVENIA, on Drava River, and 23 mi/37 km SE of MARIBOR, on CROATIA border; 46°24′N 16°09′E. In lignite-mining and wine-growing region. Railroad junction; tuberculosis sanatorium.

Ormsby, former county (□ 141 sq mi/366.6 sq km), W NEVADA. In July 1969 CARSON CITY, ⊙ state and county, consolidated with the county and replaced it, becoming an independent city. Formed 1854.

Ormsby, village (2000 population 154), MARTIN and WATONWAN counties, S MINNESOTA, 18 mi/29 km NW of FAIRMONT; 43°51′N 94°42′W. Agriculture (grain, soybeans; livestock). Manufacturing (fertilizers, lumber). Small lakes in area.

Ormskirk (AWMZ-kuhk), town (2001 population 23,392), LANCASHIRE, N ENGLAND, 12 mi/19.6 km NNE of LIVERPOOL; 53°34′N 02°54′W. Electrical equipment manufacturing. The church, with an embattled tower, contains the burial chapel of the earls of Derby. Nearby are ruins of Burscough Abbey (12th century). Edge Hill University, a teacher training college, is here.

Ormstown (ORMZ-toun), village (□ 55 sq mi/143 sq km), Montérégie region, S QUEBEC, E Canada, on Châteauguay River, and 12 mi/19 km SE of SALABERRY-DE-VALLEYFIELD; 45°08′N 74°00′W. Dairying.

Ormuz, IRAN: see HORMUZ, STRAIT OF.

Ornain River (or-nan), 75 mi/121 km long, in MEUSE and MARNE departments, CHAMPAGNE-ARDENNE region, NE central FRANCE; rises above Gondrecourt-le-Château; flows NW, past LIGNY-EN-BARROIS, BAR-LE-DUC, and REVIGNY-SUR-ORNAIN, to the SAULX at PARGNY-SUR-SAULX; the Saulx River, in turn, flows W

to the MARNE at VITRY-LE-FRANÇOIS; 48°45′N 04°47′E. Paralleled along most of its course by MARNE-RHINE CANAL.

Ornans (or-NAHN), town (□ 12 sq mi/31.2 sq km), DOUBS department, FRANCHE-COMTÉ region, E FRANCE, 11 mi/18 km SE of BESANÇON, in the JURA Mountains; 47°06′N 06°09′E. Manufacturing (electric motors). Museum for painter Gustave Courbet, born here. The old bridge and a row of old houses are reflected in the calm waters of the LOUE RIVER.

Orne (ORN), department (□ 2,356 sq mi/6,125.6 sq km), in NORMANDY, NW FRANCE; ⊙ ALENÇON; 48°40′N 00°05′E. Rolling area traversed by PERCHE hills (E), with wooded heights (Forest of ÉCOUVES) and Alençon plain (center), and hedgerow country (W). Watershed between LOIRE RIVER tributaries flowing S (MAYENNE, SARTHE, HUISNE), and streams (ORNE, DIVES, TOUQUES, RISLE) draining N into ENGLISH CHANNEL. Agriculture (cereals, potatoes, apples, and pears). Butter and CAMEMBERT cheese making (especially in VIMOUTIERS area). Percheron horses reared mainly in Perche district; cattle fattened and poultry shipped to PARIS market. Limited iron mining in DOMFRONT area. Traditional small hardware manufacturing is waning; metalworks at Alençon. Textile milling and lace making are also in decline. BAGNOLES-DE-L'ORNE is a thermal resort. No major urban centers beyond Alençon. Heavy fighting occurred in department during battle of ARGENTAN-FALAISE pocket (August 1944) in World War II. The regional park of NORMANDIE-MAINE occupies S part of department, between Alençon and Domfront. Orne department is part of the administrative region of BASSE-NORMANDIE.

Ornelas, MEXICO: see SAN JOSÉ DE GRACIA, MICHOACÁN.

Orne River (ORN), 95 mi/153 km long, in ORNE and CALVADOS departments, BASSE-NORMANDIE region, NW FRANCE; rises just E of SÉES; flows NW and N in an arc, past ARGENTAN, and CAEN, to the ENGLISH CHANNEL at OUISTREHAM. Crosses NORMANDY HILLS in gorge-like valley. Receives the ODON (left). Below Caen it is paralleled by a ship canal to its mouth. Along its banks heavy fighting took place in Normandy campaign (June–July 1944) of World War II.

Ornes (AWR-nais), village, SOGN OG FJORDANE county, W NORWAY, on E shore of Lusterfjorden (an arm of SOGNEFJORDEN), 9 mi/14.5 km NE of SOGNDAL; 61°18′N 07°19′E. Norway's oldest stave church, built c.1090, is well preserved.

Orneta (or-NE-tah), German *Wormditt*, town in EAST PRUSSIA, Elbląg province, NE POLAND, near PASLEKA RIVER, 30 mi/48 km E of ELBLĄG; 54°07′N 20°08′E. Grain and cattle market. Founded by Silesian colonists; chartered 1312.

Ornovgay, RUSSIA: see ESSO.

Örnsköldsvik (UHRN-SHUHLDS-VEEK), town, VÄSTERNORRLAND county, NE SWEDEN, on GULF OF BOTHNIA, at mouth of ÅNGERMANÄLVEN RIVER, 75 mi/121 km NE of SUNDSVALL; 63°18′N 18°44′E. Includes village of Gullänget. Airport. Seaport (icebound in winter), shipping timber and wood products. Manufacturing (machinery, motor vehicles, wallboard). Has museum. Incorporated 1894.

Orø, DENMARK: see ISE FJORD.

Oro, PAPUA NEW GUINEA: see NORTHERN, province.

Oro Bay, inlet, arm of SOLOMON SEA, NORTHERN province, SE PAPUA NEW GUINEA, SE NEW GUINEA island, circa 20 mi/32 km S of BUNA. Musa River enters from S; Cape NELSON to E. Also called DYKE ACKLAND BAY.

Orobayaya (o-ro-bei-AH-yah), town and canton, ITÉNEZ province, BENI department, NE BOLIVIA, on Río BLANCO, and 27 mi/43 km NE of MAGDALENA; 13°21′S 63°45′W. Rice, fruits (mangoes, bananas). On unpaved road.

Orobó (O-ro-bo), city (2007 population 21,746), NE PERNAMBUCO state, BRAZIL, 55 mi/89 km NW of RE-

CIFE; 07°44′S 35°36′W. Manioc, aloe, corn, fruit, potatoes.

Orochen Pervyy (uh-RO-chin PYER-viyee), village, SAKHA REPUBLIC, S central RUSSIAN FAR EAST, in the Aldan Hills (N central extension of the STANOVOY RANGE), on the Yakutsk-Never highway, 10 mi/16 km S of ALDAN; 58°29′N 125°25′E. Elevation 2,775 ft/845 m. Gold mines.

Orocovis (o-ro-KO-vis), town (2006 population 24,654), central PUERTO RICO, in Cordillera Central, 23 mi/37 km SW of SAN JUAN. Elevation 1,430 ft/436 m. Industrial and commercial area; light manufacturing (electric items, apparel, furniture). Coffee-growing and processing. Artisanry.

Orocué (o-ro-KWAI), town, ⊙ Orocué municipio, CASANARE department, E COLOMBIA, landing on navigable META RIVER, and 90 mi/145 km ESE of YOPAL; 01°50′N 72°44′W. Livestock; corn, sugarcane.

Orocuina (o-ro-KWEE-nah), town, CHOLUTECA department, S HONDURAS, 12 mi/19 km NE of CHOLUTECA; 13°29′N 87°06′W. Corn, beans; livestock.

Orodaro, town (2005 population 21,798), ⊙ KÉNÉDOUGOU province, HAUTS-BASSINS region, W BURKINA FASO, on road, 50 mi/80 km WSW of BOBO-DIOULASSO; 10°56′N 04°56′W. Sorghum, vegetables; cotton.

Oro, El, ECUADOR: see EL ORO.

Oro, El, MEXICO: see EL ORO DE HIDALGO.

Orofino [Span.=pure gold], city (2000 population 3,247), ⊙ CLEARWATER county, N IDAHO, on CLEARWATER RIVER, near mouth of East Fork Clearwater River, and 40 mi/64 km E of LEWISTON; 46°29′N 116°16′W. Elevation 1,027 ft/313 m. Railroad junction. Lumber-milling point and gateway to one of country's largest stands of white pine. Agriculture (cattle; wheat, barley, alfalfa). Manufacturing (wood products, lumber). Limestone quarries nearby. In NE part of Nez Perce Indian Reservation, main center for reservation. DWORSHAK RESERVOIR, Dam, and State Park to N, Dworshak National Fish Hatchery 1 mi/1.6 km N. Founded 1898 (c.25 mi/40 km from original site, established in gold rush of 1860), incorporated 1906.

Orog Nuur (O-rokh NOOR), lake (□ 54 sq mi/140.4 sq km) in SW central MONGOLIA, in GOBI DESERT, 170 mi/275 km SSW of TSETSERLEG, at N base of IH BOGD UUL (mountain peak) in the GOBI ALTAY mountain range; 17 mi/27 km long, 5 mi/8 km wide; 45°03′N 100°42′E. Elevation 3,930 ft/1,200 m. Has maximum depth 15 ft/4.5 m. Receives the TÜYN GOL (river), which flows S from S slopes of HANGAYN NURUU (mountain range). The size of the lake is variable. In wet years the lake is fresh, in dry years saline and divided into pools; contains abundant fish and water fowl.

Oro Grande, unincorporated village, SAN BERNARDINO county, S CALIFORNIA, 34 mi/55 km N of SAN BERNARDINO, in MOJAVE DESERT, on MOJAVE RIVER. Manufacturing asphalt; limestone quarrying in area.

Orohena, Mount (o-ro-HAI-nah), peak (7,352 ft/2241 m), W TAHITI, SOCIETY ISLANDS, S PACIFIC; 17°37′S 149°28′W. Highest mountain of island group.

Oro Ingenio (O-ro een-HE-nee-o), canton, SUD CHICHAS province, POTOSÍ department, W central BOLIVIA, 18 mi/30 km NW of TUPIZA, on the UYUNI-VILLAZÓN railroad and highway; 21°17′S 65°55′W. Zinc and lead (at Mina Tatasi), antimony (at Mina Chilcobija), and iron mining; clay, limestone deposits. Agriculture (potatoes, yucca, bananas, oats); cattle.

Orolaunum, BELGIUM: see ARLON.

Oroluk (OR-ah-look), atoll, State of POHNPEI, E CAROLINE ISLANDS, Federated States of MICRONESIA, W PACIFIC, 165 mi/266 km WNW of POHNPEI; 18 mi/29 km in diameter; 07°38′N 155°10′E.

Oro-Medonte, township (□ 227 sq mi/590.2 sq km; 2001 population 18,315), S ONTARIO, E central Canada, 15 mi/24 km from BARRIE; 44°34′N 79°36′W.

Oromiya (or-o-MEE-yah), state (2007 population 27,304,000), central and S central ETHIOPIA; 08°00′N

40°00′E. Has narrow border with KENYA (S), and borders SOUTHERN NATIONS and GAMBELA states (SW), BENISHANGUL-GUMUZ state (W), AMHARA and AFAR states (N), SOMALI state (E). Includes areas largely populated by the Oromo people. Comprises the former provinces of SHEWA, WELEGA, ARSSI, and parts of the BALE, SIDAMO, HARERGE, ILUBABOR, and KEFA provinces. ADDIS ABABA is within Oromiya, but is considered a separate political entity. The W and central sects are dominated by highlands and mountain ranges, and the GREAT RIFT VALLEY bisects the state. The MENDEBO MOUNTAINS are in S central area. Major rivers include the GIBE, AWASH, WABĒ SHEBELĒ, OMO, and BLUE NILE (Abay). There is a variety of agricultural products grown, including coffee in W, sugarcane in central areas near the AWASH RIVER, and various cereal crops throughout. Livestock raising is also important. Larger cities include NEKEMTE, JIMMA, NAZRĒT, DEBREZIET, SHASHEMENE, ASELA, and GOBA.

Oromocto (o-ro-MAHK-to), town (2001 population 8,843), S central New Brunswick, Canada, on the St. John River; 45°51′N 66°28′W. The town developed because of its proximity to CAMP GAGETOWN, the largest (436 sq mi/1,129 sq km) military camp in Canada.

Oromocto Lake (o-ro-MAHK-to) (□ 16 sq mi/41.6 sq km), SW New Brunswick, Canada, 16 mi/26 km WSW of FREDERICTON; 8 mi/13 km long, 4 mi/6 km wide.

Oromocto River (o-ro-MAHK-to), 20 mi/32 km long, SW New Brunswick, Canada; rises in two branches, one issuing from Oromocoto Lake, the other from South Oromocto Lake, joining 21 mi/34 km S of FREDERICTON; flows N to St. John River at Oromocto.

Oron (o-RO), district, W central VAUD canton, SWITZERLAND. Population is largely French-speaking and Protestant.

Oron (O-rawn), town, AKWA IBOM state, extreme SE NIGERIA, port at mouth of CROSS RIVER, 12 mi/19 km SSW of CALABAR (linked by ferry); 04°50′N 08°14′E. Palm oil and kernels; hardwood, rubber. Fisheries. Also Idua Oron.

Orona (o-RAW-nah), lozenge-shaped atoll (□ 2 sq mi/5.2 sq km), centrally placed in the PHOENIX ISLANDS, S PACIFIC, 110 mi/177 km SW of KANTON Island; on reef 4.5 mi/7.2 km long, 2.5 mi/4 km wide; 04°30′S 172°11′W. Archaeological remains indicate ancient Polynesian population, and inhabitants when found by Wilkes indicate Tahitian contacts. In 1889 it was (following coconut planting) annexed by British, included in British GILBERT AND ELLICE ISLANDS Colony in 1937, and in 1979 incorporated in independent KIRIBATI. Gilbertese settlers occupied Orona 1938–1963, after which they were resettled in Waghena in the SOLOMON ISLANDS. The United States has since built a radar station there. Formerly Hull Island.

Orono (OR-uh-no), town, including Orono village, PENOBSCOT county, S MAINE, on the PENOBSCOT RIVER, and 8 mi/12.9 km above BANGOR; 44°52′N 68°42′W. Seat of main campus of University of Maine. Settled c.1775, called Stillwater until incorporated 1806, included Old Town until 1840.

Orono (OR-no), town (2000 population 7,538), HENNEPIN county, E MINNESOTA, residential suburb 15 mi/24 km W of downtown MINNEAPOLIS, on N shore of Lake MINNETONKA, includes Big Island and several landlocked bays, N extension of lake; 44°58′N 93°35′W. Long Lake in NE. Morris T. Baker Park Reserve on N boundary, in NW.

Orono (AH-ro-no), unincorporated village, S ONTARIO, E central Canada, 15 mi/24 km ENE of OSHAWA, and included in town of Clarington; 43°58′N 78°37′W. Food processing.

Oronoco (or-ruh-NO-ko), village (2000 population 883), OLMSTED county, SE MINNESOTA, 10 mi/16 km NNW of ROCHESTER; 44°09′N 92°32′W. In agricultural area (grain, soybeans; livestock, poultry; dairy-

ing). Manufacturing (signs). Parts of Richard J. Dorer Memorial Hardwood State Forest to NE and SW.

Oronogo (o-ruh-NO-go), city (2000 population 976), JASPER county, SW MISSOURI, near SPRING RIVER, suburb 8 mi/12.9 km N of JOPLIN; 37°11′N 94°28′W. Agriculture (wheat, soybeans; cattle). Former lead, zinc mines.

Oronsay (OR-uhn-sai), barely inhabited island (□ 3 sq mi/7.8 sq km), one of the INNER HEBRIDES, Argyll and Bute, W Scotland; 56°01′N 06°16′W. Separated from COLONSAY by a shallow sound uncovered at low tide. Formerly in Strathclyde, abolished 1996.

Oronsay (OR-uhn-sai), uninhabited island (2 mi/3.2 km long, 1 mi/1.6 km wide) at mouth of LOCH SUNART, SW HIGHLAND (LOCHABER district), N Scotland, 25 mi/40 km NW of OBAN; 56°39′N 05°53′W.

Oronsay (OR-uhn-sai), uninhabited island (0.5 mi/0.8 km long), OUTER HEBRIDES, Eilean Siar, NW Scotland, just E of N peninsula of BARRA; 57°02′N 07°20′W. On E side of Northbay airfield.

Orontes River (O-RUHN-tees), Arabic *Assi* or *Asi*, c.250 mi/402 km long, LEBANON, SYRIA, and TURKEY; rises in the N part of the Al Beqa'a valley, Lebanon; flows generally N through Syria, then W into S Turkey, and into the MEDITERRANEAN SEA; celebrated because of the antiquity of settlement along its banks. The river is unnavigable but is important for irrigation, especially in Syria. Marshes on its middle course have been drained and the land reclaimed for farming. On its lower course, the river has cut below the surrounding plain, and it was noted for remarkable water wheels, 20 ft/6 m–70 ft/21 m in diameter, at HOMS and Hamah; the wheels, turned by the current, used to lift water onto the plains. In addition to irrigation projects, there are several dams along the river with hydroelectric power stations. In June 2002, the Syrian dam Zeyzoun, near Hama and completed in 1996, collapsed and the ensuing flood destroyed several villages.

Oropesa (o-ro-PAI-sah), town, ANTABAMBA province, CUSCO region, S central PERU, on CUSCO-PUNO railroad, and 16 mi/26 km ESE of CUSCO; 13°35′S 71°46′W. Elevation 10,751 ft/3,276 m. Cereals, potatoes, vegetables; alfalfa. Archaeological remains nearby.

Oropesa y Corchuela (ee kor-CHWAI-lah), town, TOLEDO province, central SPAIN, on railroad and highway to CÁCERES, and 18 mi/29 km W of TALAVERA DE LA REINA. Agriculture (cereals, acorns, olives, grapes; sheep, hogs). Olive-oil extracting, sawmilling, charcoal burning, tile manufacturing; lime kilns. Town of medieval character.

Oropeza, province, CHUQUISACA department, SE BOLIVIA, ⊙ YOTALA; 18°40′S 65°01′W.

Oropolí (o-ro-po-LEE), town, EL PARAÍSO department, SW central HONDURAS, on unpaved road, 20 mi/32 km SW of DANLÍ; 13°48′N 86°48′W. Small farming, grain; livestock.

Oropus (o-ro-POS), ancient town of ATTICA, in what is now ATTICA department, ATTICA department, E central GREECE, on ASOPOS River, and 24 mi/39 km E of THEBES, near S arm of Gulf of Évvia; 38°18′N 23°45′E. Frequently mentioned in frontier wars between Athenians and Boeotians. Nearby was oracle of Amphiaraüs. On site is modern village of Oropos (lignite mining), and, just NE, on the gulf, is its harbor, Skala Oropou.

Oroquieta (o-ro-KYE-tah), city (□ 75 sq mi/195 sq km; 2000 population 59,843), ⊙ MISAMIS OCCIDENTAL province, W MINDANAO, PHILIPPINES, port on ILIGAN BAY, 50 mi/80 km NE of PAGADIAN; 08°29′N 123°48′E. Agricultural center (corn, coconuts); processes and ships copra.

Orós (o-ROS), city (2007 population 21,305), SE Ceará, Brazil, on Rio Jaguaribé, 19 mi/31 km NW of Icó; 06°20′S 38°52′W. Site of Orós dam and reservoir. Airport.

Orós (or-OS), town, SW CEARÁ, BRAZIL, 28 mi/45 km ENE of IGUATU; 06°16′S 38°50′W. Terminus of rail-

road spur from ALENCAR. Magnesite deposits. Site of dam impounding waters of RIO JAGUARIBE for irrigation and flood control.

Orosei (o-ro-ZAI), village, NUORO province, E SARDINIA, near GULF OF OROSEI, 20 mi/32 km ENE of NUORO; 40°23′N 09°42′E. Known for marble caves. Has medieval castle. Port (Marina di Orosei) is 2 mi/3.2 km E.

Orosei, Gulf of, inlet of TYRRHENIAN Sea, E SARDINIA, between Cape COMINO and Cape MONTE SANTO; 31 mi/50 km wide, 8 mi/13 km deep. Fisheries. Chief port, OROSEI.

Orosh (o-ROSH), village, Mirditë district, N ALBANIA, 32 mi/51 km ESE of SHKODËR, in the MIRDITË tribal region; 41°50′N 20°05′E. Also spelled Oroshi.

Orosháza (O-rosh-hah-zah), city (2001 population 31,764), BÉKÉS county, SE HUNGARY, 22 mi/35 km WSW of BÉKÉSCSABA; 46°34′N 20°40′E. Railroad, agricultural center (sugar beets, grain; sunflowers, tobacco, vineyards; cattle). Manufacturing (food processing, dairy products, paper, textiles); flour milling; largest glassworks in Hungary. One of Hungary's largest natural gas fields nearby. German investment in early 1990s. Resort of Gyopárosfürdő is nearby.

Orosi, unincorporated town (□ 2 sq mi/5.2 sq km; 2000 population 7,318), TULARE county, S central CALIFORNIA, in San Joaquin Valley, 15 mi/24 km N of Visalia; 36°33′N 119°17′W. Citrus, peaches, apples, plums, olives, nuts, grapes; cotton, nursery products; cattle, hogs, poultry.

Orosi (o-RO-see), village, CARTAGO province, central COSTA RICA, on REVENTAZÓN RIVER, and 3 mi/4.8 km SSE of PARAÍSO; 09°48′N 83°51′W. Agriculture (coffee, chayote squash, sugarcane; livestock). Has eighteenth century colonial church. Founded by Franciscans; an old mission town.

Orosi (o-RO-see), extinct volcano (5,056 ft/1,541 m) in NW COSTA RICA, at NW end of the CORDILLERA DE GUANACASTE, 12 mi/19 km SE of LA CRUZ. Slopes are covered with dense rain forest.

Oroska, SLOVAKIA: see POHRONSKY RUSKOV.

Oroszka, SLOVAKIA: see POHRONSKY RUSKOV.

Oroszlány (O-ros-lah-nyuh), city (2001 population 20,280), KOMÁROM county, HUNGARY, 29 mi/47 km W of BUDAPEST; 47°29′N 18°19′E. One of the socialist mining cities built in the 1950s on site of small village. Lignite; large electric power station. Manufacturing (footwear, handicrafts); employing women, built to supply Soviet market. Small plant for electric appliances. All industries contracted sharply in 1990. Unemployment well above average.

Oroszvár, SLOVAKIA: see RUSOVCE.

Orota (o-RO-tah), village, ANSEBA region, N ERITREA, 93 mi/150 km NW of ASMARA; 16°14′N 37°55′E. Supply base and operation center for Eritrean liberation forces during war with ETHIOPIA; now largely abandoned.

Orotava, La (o-ro-TAH-vah, lah), city, SANTA CRUZ DE TENERIFE province, CANARY ISLANDS, SPAIN, 18 mi/29 km WSW of SANTA CRUZ DE TENERIFE, in Orotava Valley, famed for its luxuriant vegetation; 28°23′N 16°31′W. Served by its port, PUERTO DE LA CRUZ, 2 mi/3.2 km NW. A noted health resort and among archipelago's most prosperous town. Famous garden founded in 18th century. Region produces chiefly bananas; also tomatoes, cereals, grapes, tobacco.

Orotina (o-ro-TEE-nah), city, ALAJUELA province, ⊙ Orotina canton, W central COSTA RICA, in Tárcoles valley, 22 mi/35 km ESE of ALAJUELA; 09°54′N 84°34′W. Agriculture (grain, fruit). Developed in twentieth century with construction of railroad (no longer working).

Orotukan (uh-ruh-too-KAHN), town (2006 population 2,495), central MAGADAN oblast, E RUSSIAN FAR EAST, on the Orotukan River (short right tributary of the KOLYMA River), 400 mi/644 km N of Magadan (linked by highway); 62°16′N 151°42′E. Elevation 1,525 ft/464 m. Drilling equipment; gold mines. Formerly called Urutukan.

Orovada (or-uh-VAH-duh), unincorporated village, HUMBOLDT county, N NEVADA, 40 mi/64 km N of WINNEMUCCA, 30 mi/48 km S of OREGON state boundary, E of Quinn River. Mercury, gold, silver. Cattle, sheep. SANTA ROSA MOUNTAINS and Santa Rosa Peak (9,701 ft/2,957 m), in section of Humboldt National Forest, to E. Parts of Fort McDermitt Indian Reservation to N and SW.

Oro Valley, town (2000 population 29,700), PIMA county, S ARIZONA, residential suburb 12 mi/19 km N of TUCSON, on Oro River; 32°25′N 110°57′W. Cattle. Coronado National Forest to E; Catalina State Park to NE.

Oroville, city (□ 10 sq mi/26 sq km; 2000 population 13,004); ⊙ BUTTE county, N central CALIFORNIA, 63 mi/101 km N of SACRAMENTO, on FEATHER RIVER, in Sacramento Valley, at W base of the SIERRA NEVADA; 39°30′N 121°34′W. Diverse manufacturing. Agriculture (citrus, olives, kiwi fruit, plums, peaches, walnuts, almonds, grain, nursery stock; cattle); timber. LAKE OROVILLE reservoir and State Recreation Area to E (Feather River); Feather River Canyon and FEATHER FALLS are to E; Plumas National Forest to NE. Settled 1849 as gold camp (Ophir City), incorporated 1857.

Oroville (OR-o-vil), town (2006 population 1,597), OKANOGAN county, N WASHINGTON, port of entry 4 mi/6.4 km S of CANADA (BRITISH COLUMBIA) border, and 6 mi/9.7 km S of osoyoos (British Columbia) 40 mi/64 km N of OKANOGAN, and on OKANOGAN RIVER at mouth of Similkamern River (from W) and at S end of Osoyoos Lake (extends into Canada); 48°57′N 119°26′W. Railroad terminus. Gold, silver; Agriculture (timber, apples). Manufacturing (wood products). Osoyoos Lake State Park N of town; small section of Okanogan National Forest to SE. Incorporated 1908.

Oroville East, unincorporated town (2000 population 8,860), BUTTE county, N central CALIFORNIA, residential suburb 1 mi/1.6 km E of OROVILLE, near FEATHER RIVER. LAKE OROVILLE reservoir, and State Recreation Area to NE; 39°31′N 121°29′W. Citrus, kiwi fruit, olives, almonds, grain, nursery stock; cattle.

Oroville, Lake, reservoir (c.17 mi/27 km long), BUTTE county, N central CALIFORNIA, on FEATHER RIVER (largest unit), 5 mi/8 km ENE of OROVILLE; 39°31′N 121°18′W. Elevation 899 ft/274 m. West, North (longest), Middle, and South Forks of the river form four large arms extending N and E. Formed by Oroville Dam (770 ft/235 m high, 7,600 ft/2,316 m long) built (1957–1968) to provide electric power, drinking water, and irrigation. Plumas National Forest to NE; surrounded by Lake Oroville State Recreation Area.

Oroza (o-RO-zah), canton, ANICETO ARCE province, TARIJA department, S central BOLIVIA, 3 mi/5 km E of PADCAYA; 21°52′S 64°43′W. Elevation 6,549 ft/1,996 m. Abundant gas wells to E. Limestone deposits. Agriculture (potatoes, yucca, bananas, corn); cattle.

Orphane, Gulf of, Greece: see ORFANI GULF.

Orpheus Island (OR-fee-uhs), island (□ 5 sq mi/13 sq km) and national park, in GREAT BARRIER REEF, in CORAL SEA, off central QUEENSLAND, NE AUSTRALIA, 50 mi/80 km NE of TOWNSVILLE, 12 mi/20 km E of INGHAM, and 7 mi/11 km long, 0.6 mi/1 km wide; 18°37′S 146°30′E. Volcanic in origin. Administered by the state, except for a private resort and a marine research station. Beaches, sheltered bays, protected fringing reefs, significant birdlife; walking tracks.

Orpington, ENGLAND: see BROMLEY.

Orp-Jauche (awr–ZO-shuh), commune (2006 population 7,921), Nivelles district, BRABANT province, central BELGIUM, 8 mi/12.9 km SSE of TIENEN. Cement.

Orr, village (2000 population 249), ST. LOUIS county, NE MINNESOTA, 43 mi/69 km N of HIBBING, on E end of PELICAN LAKE, Pelican River exits to S, in Kabetogama State Forest area; 48°03′N 92°49′W. Resort

area; timber. Manufacturing (wood products). Nett Lake Indian Reservation to W; Superior National Forest to E; VOYAGEURS NATIONAL PARK to N.

Orrefors (OR-re-FORSH), village, KRONOBERG county, SE SWEDEN; 56°50′N 15°45′E. Manufacturing (fine crystal and glassware). Simon Gate and Edward Hald, outstanding glassware engravers, based here; their work represented in numerous collections. Museum of glass.

Orrell (AH-ruhl), town (2001 population 11,203), GREATER MANCHESTER, W ENGLAND, 3 mi/4.8 km WSW of WIGAN; 53°32′N 02°42′W. Previously cotton milling. Includes village of Kit Green.

Orres, Les (OR, laiz), commune (□ 28 sq mi/72.8 sq km), Alpine ski resort in HAUTES-ALPES department, PROVENCE-ALPES-CÔTE D'AZUR region, SE FRANCE, 5 mi/8 km ESE of EMBRUN. Elevation 5,200 ft/1,585 m. Perched above the DURANCE RIVER valley upstream from SERRE-PONÇON Lake, it has become a major sports center of the S ALPS since its establishment in 1970.

Orrhoe, TURKEY: see EDESSA.

Orrick (O-rik), city (2000 population 889), RAY county, NW MISSOURI, near MISSOURI River, 9 mi/14.5 km SW of RICHMOND; 39°12′N 94°07′W. Agriculture (corn, wheat, soybeans; hogs). Manufacturing (plastic molding). Plotted 1873.

Orrington, town, PENOBSCOT county, S MAINE, on the PENOBSCOT RIVER, and 8 mi/12.9 km below BANGOR; 44°43′N 68°46′W. Settled 1770, incorporated 1788.

Or River, 206 mi/332 km long, in NW KAZAKHSTAN and RUSSIAN FEDERATION (SE ORENBURG oblast); rises in central MUGODZHAR HILLS; flows generally N and W to URAL RIVER at ORSK. Unnavigable.

Orroroo, village, S SOUTH AUSTRALIA state, S central AUSTRALIA, 45 mi/72 km NE of PORT PIRIE, against S FLINDERS RANGES; 32°44′S 138°37′E. Wheat, wool; dairying; wine.

Orrs Island, MAINE: see HARPSWELL.

Orrstown (ORS-toun), borough (2006 population 232), FRANKLIN county, S PENNSYLVANIA, 5 mi/8 km W of SHIPPENSBURG, on CONODOGUINET CREEK; 40°03′N 36°77′W. Agriculture (grain; livestock; dairying). Light manufacturing. Letterkenny Army Depot Directorate of Supply and Transportation to SW. BLUE MOUNTAIN ridge to NW.

Orrum (OR-uhm), village (2006 population 103), ROBESON county, SE NORTH CAROLINA, 10 mi/16 km S of LUMBERTON; 34°28′N 79°00′W. Manufacturing; agriculture (grain, tobacco, flowers; livestock). Lumber River State Park to SE.

Orrville (OR-vil), city (□ 5 sq mi/13 sq km; 2006 population 8,466), WAYNE county, N central OHIO, 20 mi/32 km W of CANTON; 40°51′N 81°46′W. In agricultural area; food and dairy products. Manufacturing (transportation equipment, machinery, chemicals, wood products, leather goods, mattresses). Settled c.1850, incorporated 1864.

Orrville, town (2000 population 230), Dallas co., S central ALABAMA, 15 mi/24 km SW of SELMA. Manufacturing (wooden church pews). Named after James F. Orr, operater of Orr's Mills and the first postmaster. Inc. in 1854.

Orsa (OOR-shah), town, KOPPARBERG county, central SWEDEN, on N shore of Lake Orsasjön (7 mi/11.3 km long, 2 mi/3.2 km–4 mi/6.4 km wide), 45 mi/72 km NW of FALUN; 61°07′N 14°37′E. Railroad junction. Manufacturing (iron links). Has fourteenth-century church.

Orsaro, Monte (or-SAH-ro, MON-te), peak (6,004 ft/1,830 m) in ETRUSCAN APENNINES, N central ITALY, on PARMA–MASSA CARRARA province border, 6 mi/10 km ENE of PONTREMOLI.

Orsay (OR-sai), town (□ 3 sq mi/7.8 sq km), ESSONNE department, ÎLE-DE-FRANCE region, N central FRANCE, on small Yvette River, and 13 mi/21 km SSW of PARIS; 48°42′N 02°11′E. Center of advanced science

education (campus of Institute of Higher Social Studies of University of Paris; technical education facilities in optics and energy technology; French Atomic Energy Commission headquarters). Psychiatric hospital. Town is reached by Parisian suburban railroad. Other major research facilities nearby.

Orsay (OR-sai), island (1 mi/1.6 km long), of the INNER HEBRIDES, Argyll and Bute, W Scotland, off SW end of ISLAY, just SW of Portnahaven; 55°40′N 06°29′W. Lighthouse. Formerly in Strathclyde, abolished 1996.

Orsett (AW-sit), village (2001 population 5,627), THURROCK, SE ENGLAND, 4 mi/6.4 km N of TILBURY; 51°31′N 00°23′E. Has church of Norman origin.

Orsha (or-SHA), city, ⊙ Orsha region, VITEBSK oblast, BELARUS; a port at the confluence of the DNIEPER and Orshitsa rivers. One of Belarus's leading railroad and water transport junctions and industrial centers. The city's industries include textiles, machinery, consumer goods, silicates, construction materials, food processing, transportation equipment. First mentioned as Rsha in 1067, the city passed to LITHUANIA in the thirteenth-century. It was an important Polish fortress and trade center from the sixteenth-century until its annexation by RUSSIA in 1772, during the first partition of POLAND.

Orsha (uhr-SHAH), village (2006 population 2,250), SE TVER oblast, W central European Russia, on local railroad spur, 11 mi/18 km ENE, and under administrative jurisdiction, of TVER; 56°55′N 36°14′E. Elevation 462 ft/140 m. In agricultural area; produce processing.

Orshanka (uhr-SHAN-kah), town (2006 population 6,725), N MARI EL REPUBLIC, E central European Russia, in the LESSER KOKSHAGA RIVER basin, on crossroads, 18 mi/29 km N of YOSHKAR-OLA; 56°55′N 47°53′E. Elevation 334 ft/101 m. Flax processing.

Orsha Upland (or-SHA), VITEBSK oblast, BELARUS. Elevation up to 860 ft/262 m. Large areas under cultivation; some forests preserved.

Orshevtsy, UKRAINE: see ORSHIVTSI.

Orshivtsi (OR-shif-tsee) (Russian *Orshevtsy*) (Romanian *Oraçseni*), town (2004 population 11,900), W CHERNIVTSI oblast, UKRAINE, in N Bukovyna, on the PRUT RIVER, and 15 mi/24 km WNW of CHERNIVTSI; 48°25′N 25°37′E. Elevation 705 ft/214 m. Railroad junction; sawmilling. Polish-Romanian frontier station (1921–1939).

Orshütz River, POLAND: see ORZYC RIVER.

Orsières (or-SEE-er), commune, VALAIS canton, SW SWITZERLAND, on DRANCE RIVER in Val d'Entremont, and 6 mi/9.7 km SSE of MARTIGNY, N of GREAT ST. BERNARD PASS; 46°02′N 07°09′E. Elevation 2,956 ft/901 m. Railroad terminus; hydroelectric plant.

Orsk (ORSK), city (2006 population 245,490), SE ORENBURG oblast, in the foothills of the S URALS, extreme SE European Russia, on the URAL River, on highway, 200 mi/322 km SE of ORENBURG, and 10 mi/16 km E of NOVOTROITSK; 51°13′N 58°34′E. Elevation 656 ft/199 m. Railroad junction. Center of the Orsk-Khalilovo industrial area, which has rich iron, copper, nickel, and coal deposits. Nonferrous metallurgical, chemical works, and metalworking plants; manufacturing of machinery, oil refinery. Founded in 1735. Made city in 1865.

Orşova (OR-sho-vah), town, MEHEDINŢI county, SW ROMANIA, in BANAT, on DANUBE River near the IRON GATE and opposite TEKIJA (SERBIA), on railroad and 15 mi/24 km NW of DROBETA-TURNU SEVERIN. Transshipment point; geographical research station; hydroelectric station; petroleum refining; mineral processing; manufacturing of woolen textiles, apparel, furniture. Built on the ruins of a Roman citadel, it played a prominent part in the Turco-Aus. wars (17th–18th centuries).

Ørstavik (UHR-stah-vik), village, MØRE OG ROMSDAL county, W NORWAY, at head of Ørstafjorden (6-mi/9.7-km-long inlet of NORTH SEA), 18 mi/29 km S of ÅESUND. Furniture manufacturing. A mountaineering center, with a regional airport at HOVDEN, 3.1 mi/5 km SW.

Ortaklar, village, SW TURKEY, 11 mi/18 km NNE of SOKE; 37°53′N 27°30′E. On the IZMIR–AYDIN railroad.

Orta-koi, Bulgaria: see IVAILOVGRAD, city.

Orta, Lake of (OR-tah) (□ 7 sq mi/18.2 sq km), NOVARA province, PIEDMONT, N ITALY, 24 mi/39 km NNW of NOVARA; 8 mi/13 km long, 1.5 mi/2.4 km wide. Elevation 951 ft/290 m, maximum depth 469 ft/143 m. Discharges into Lago MAGGIORE, 7 mi/11 km E, through outlet (N) to TOCE River. Furnishes water to industries of OMEGNA, on N shore. Contains islet of SAN GIULIO. Sometimes called Lake Cusio.

Orta Nova (OR-tah NO-vah), town, FOGGIA province, APULIA, S ITALY, 13 mi/21 km SE of FOGGIA; 41°19′N 15°42′E. Paper.

Orta Novarese (no-vah-RAI-ze), village, NOVARA province, PIEDMONT, N ITALY, port on E shore of Lake of ORTA, opposite islet of SAN GIULIO, 23 mi/37 km SSE of DOMODOSSOLA. Resort; glove manufacturing. Has palace (1582). On nearby hill is sanctuary (started 1590) comprising twenty chapels with frescoes and terra cottas of the 16th–17th centuries.

Orte (OR-te), town, VITERBO province, LATIUM, central ITALY, on the TIBER River, and 15 mi/24 km E of VITERBO; 42°27′N 12°23′E. Railroad junction. In cereal- and grape-growing region. Travertine quarries nearby.

Ortega (or-TAI-gah), town, ⊙ Ortega municipio, TOLIMA department, W central COLOMBIA, in E foothills of Cordillera CENTRAL, 36 mi/58 km S of IBAGUÉ; 03°56′N 75°13′W. Elevation 1,774 ft/540 m. Agricultural center (coffee, plantains, rice, sorghum; livestock).

Orteguaza River (or-tai-GWAH-sah), c.100 mi/161 km long, CAQUETÁ department, S COLOMBIA; rises in Cordillera ORIENTAL; flows SSE, past FLORENCIA, to CAQUETÁ River at TRES ESQUINAS; 00°43′N 75°16′W.

Ortelsburg, POLAND: see SZCZYTNO.

Ortenberg (OR-ten-berg), town, S UPPER RHINE, BADEN-Württemberg, SW GERMANY, at W foot of BLACK FOREST, near the KINZIG, and 1.5 mi/2.4 km SW of OFFENBURG; 48°27′N 07°53′E. Noted for its wine. Castle (12th–13th century), destroyed by French in 1668. Remains of town wall (13th century) here. Founded c.1266.

Ortenburg (OR-ten-burg), village, Lower BAVARIA, GERMANY, 10 mi/16 km WSW of PASSAU; 48°33′N 13°13′E. Brewing. Agriculture (wheat; cattle, horses). Has 14th century Gothic church and 16th century castle.

Orth an der Donau (ORT ahn der DO-nou), township, E LOWER AUSTRIA, E AUSTRIA, in the MARCHFELD plain on left bank of Danube, 16 mi/26 km ESE of VIENNA; 48°09′N 16°42′E. Elevation 442 ft/135 m. Manufacturing (pharmaceuticals). Moated castle, medieval core, reconstructed 1550. National Park Danube. Flood Plains nearby.

Orthez (or-TEZ), town (□ 17 sq mi/44.2 sq km), PYRÉNÉES-ATLANTIQUES department, AQUITAINE region, SW FRANCE, on the Gave de PAU, and 24 mi/39 km NW of PAU; 43°29′N 00°46′W. Market (trade in hams) and center of chemical industry. Has 13th–14th-century bridge, the tower of a medieval castle, and picturesque old houses. Was capital of BÉARN region until 1460. Here Wellington defeated the French in 1814.

Orthon River, BOLIVIA: see ORTON RIVER.

Ortiga, Cordillera de la (or-TEE-gah, kor-dee-YER-rah dai lah), Andean range in NW SAN JUAN province, ARGENTINA, near CHILE border; extends c.30 mi/48 km N-S; rises to c.18,860 ft/5,749 m at 29°13′S.

Ortigueira (or-tee-GAI-rah) or **Santa Marta de Ortigueira** (SAHN-tah MAHR-tah dhai or-tee-GAI-rah), town, LA CORUÑA province, NW SPAIN; fishing port on inlet of BAY OF BISCAY, SE of Cape Ortegal, and 24 mi/39 km NE of EL FERROL; 43°41′N 07°50′W. Fish processing (sardines), boatbuilding; lumbering. Livestock raising.

Orting, town (2006 population 5,462), PIERCE COUNTY, W central WASHINGTON, 17 mi/27 km SE of TACOMA, and on PUYALLUP RIVER at mouth of Carbon River to N; 47°05′N 122°13′W. Railroad junction. Agriculture (bulbs, timber, fruit; dairy products). Manufacturing (machinery, corrugated boxes). A fish hatchery is here. Lake Kapowsin to S. Incorporated 1889.

Ortisei (or-tee-SAI), German *Sankt Ulrich*, town, BOLZANO province, TRENTINO–ALTO ADIGE, N ITALY, in Val GARDENA, 16 mi/26 km NE of BOLZANO; 46°34′N 11°40′E. Resort (elevation 4,050 ft/1,234 m); woodcarving center (toys, religious articles).

Ortiz (or-TEES), town, ⊙ Ortiz municipio, GUÁRICO state, N central VENEZUELA, 22 mi/35 km SSE of SAN JUAN DE LOS MORROS; 09°37′N 67°16′W. Livestock; coffee, tobacco.

Ortiz, village, CONEJOS county, S COLORADO, on branch of CONEJOS River, in SE foothills of SAN JUAN MOUNTAINS, at NEW MEXICO state line, and 5 mi/8 km SSW of ANTONITO. Elevation c.8,000 ft/2,438 m. Rio Grande National Forest to W; Carson National Forest (New Mexico) to SW.

Ortiz Mountains (or-TEEZ), range, N central NEW MEXICO, SANTA FE county, E of RIO GRANDE, 25 mi/40 km SSW of SANTA FE. Highest point: Placer Peak (8,897 ft/2,712 m). Coal mines.

Ortles (ORT-lais), range of the Ötztal Alps, in TRENTINO–ALTO ADIGE, N ITALY; 46°30′N 10°40′E. It has many glaciers. Ortles peak, 12,792 ft/3,899 m high, the highest peak, was first ascended in 1804.

Ortley, village (2006 population 54), ROBERTS county, NE SOUTH DAKOTA, 24 mi/39 km SSW of SISSETON, in Lake Traverse (Sisseton Wahpeton) Indian Reservation; 45°20′N 97°12′W.

Ortoire River (AWR-twahr), 31 mi/50 km long, SE TRINIDAD, TRINIDAD AND TOBAGO; flows E to the ATLANTIC OCEAN; 10°20′N 61°00′W. Not navigable. Formerly called Guataro River.

Orton (OR-tuhn), unincorporated village, S ONTARIO, E central Canada, 17 mi/28 km from GUELPH, and included in town of Erin; 43°47′N 80°13′W.

Ortona (or-TO-nah), town, ABRUZZI, central ITALY, on the ADRIATIC Sea; 42°21′N 14°24′E. Now a small fishing port and a seaside resort. Manufacturing (fabricated metals, apparel, plastics). It was a major port from the 11th century to 1447, when its fleet and arsenal were destroyed by the Venetians. The 12th-century cathedral (now restored) and the Aragonese castle (15th century) were heavily damaged in World War II.

Orton River (OR-ton), 120 mi/193 km long, PANDO department, N BOLIVIA; formed by confluence of TAHUAMANU and MANURIPI rivers at PUERTO RICO; flows E, through tropical forest, past INGAVI and HUMAITÁ, to BENI River 10 mi/16 km N of RIBERALTA; 10°50′S 66°04′W. Navigable for entire course. Also spelled Orthon.

Ortonville, town (2000 population 2,158), ⊙ BIG STONE county, W MINNESOTA, 67 mi/108 km WNW of WILLMAR, on SOUTH DAKOTA state line, at S end of BIG STONE LAKE (at outlet of MINNESOTA RIVER), opposite BIG STONE CITY, South Dakota; 45°17′N 96°26′W. Elevation 1,021 ft/311 m. Manufacturing (fertilizer, soft drinks, printing and publishing); granite quarries nearby. Big Stone Lake State Park to NW. Settled 1872, laid out 1873.

Ortonville, village (2000 population 1,535), OAKLAND county, SE MICHIGAN, 17 mi/27 km NNW of PONTIAC; 42°51′N 83°26′W. In farm area. Diverse manufacturing. Ortonville State Recreational Area to N; Mount Hilly and Pine Knob ski areas to SW.

Orto-Tokoy (awr-to–to-KOI), town, ISSYK-KOL region, KYRGYZSTAN, adjacent to ORTO-TOKOY reservoir, 12 mi/20 km W of ISSYK-KOL Lake.

Orto-Tokoy (awr-to–to-KOI), reservoir (□ 15 sq mi/39 sq km), ISSYK-KOL region, KYRGYZSTAN, on CHU RIVER, between Kyrgyz Range and KÜNGEY ALA-TOO (KUNGEI ALATAU) mountains, 15 mi/24 km W of Balakchy; 7 mi/11.3 km long; 42°18′N 75°57′E. Irrigates thousands of acres in Kyrgyzstan and KAZAKHSTAN. Constructed in late 1940s.

Ortuella (or-TWE-lyah), town, VIZCAYA province, N SPAIN, 8 mi/12.9 km NW of BILBAO; 43°19′N 03°04′W. In iron-mining region.

Ortulu, Turkish *Örtülü*, village, NE TURKEY, 75 mi/121 km NE of ERZURUM. Grain. Known for its handicrafts.

Oru (o-ROO), town, OGUN state, SW NIGERIA, 10 mi/16 km N of IJEBU-ODE. Road center; cacao and hardwood industry; rubber, palm oil and kernels, rice.

Oruch-gaazi, Bulgaria: see YUNAK.

Orumbo Boka (o-ROOM-bo BO-kah), sacred mountain (1,729 ft/529 m), S central CÔTE D'IVOIRE, 17 mi/28 km SE of TOUMODI. Also spelled ORAOUMBA BOKA.

Oruro (o-ROO-ro), department (□ 20,690 sq mi/53,794 sq km; 2005 population 433,481), W BOLIVIA; ⊙ ORURO, in CERCADO province; 18°40′S 67°30′W. Includes Lake POOPÓ (E) and Salar de COIPASA (SW). W Cordillera of the ANDES Mountains (W) separates department from CHILE, Cordillera de AZANAQUES (part of the E Cordillera; E) from POTOSÍ department. The ALTIPLANO (elevation over 12,000 ft/3,658 m), between the two ranges, is drained by DESAGUADERO and LAUCA rivers. Severe climate and arid soil conditions permit raising only of potatoes, oca, and quinoa; alpaca, sheep, and llama breeding. In the E mountain ranges is one of Bolivia's richest mineral districts, with main tin mines at HUANUNI, MOROCOCALA, AVICAYA, and ANTEQUERA. Salt, saltpeter, and sulphur deposits near SALINAS DE GARCI MENDOZA. Also copper. Manufacturing centered at Oruro. E part of department served (N-S) by Oruro-uyuni railroad. Created September 5, 1826, by Antonio José de Sucre.

Oruro, city (2001 population 201,230), ⊙ ORURO department, W BOLIVIA; 17°59′S 67°09′W. It is Bolivia's railroad center. Oruro's economy is based on exploitation of the region's tin, tungsten, silver, and copper mining (also copper smelting); textiles. Because of its high elevation (12,146 ft/3,702 m), agriculture is almost nonexistent. Oruro was founded in 1595 to exploit the rich silver deposits nearby. When silver production declined in the nineteenth century, it became almost a ghost town. It expanded with the development of other mineral resources. Also known as "Folkloric Capital of Bolivia."

Orusco (o-ROO-sko), town, MADRID province, central SPAIN, on TAJUÑA RIVER, on railroad, and 27 mi/43 km ESE of MADRID; 40°17′N 03°13′W. Manufacturing (woolen goods).

Orust (OOR-UST), island (□ 133 sq mi/345.8 sq km), GÖTEBORG OCH BOHUS county, SW SWEDEN, in SKAGERRAK, 25 mi/40 km N of GÖTEBORG; 18 mi/29 km long, 7 mi/11.3 km–9 mi/14.5 km wide. Separated from mainland by channel 1 mi/1.6 km–3 mi/4.8 km wide. Hallevikstrand (W) chief fishing village; Henån municipality center (1990 population 1,659).

Oruzgan (uh-ROOZ-gahn), province (2005 population 291,500), central AFGHANISTAN, ⊙ TARIN KOT; 33°15′N 66°00′E. Bordered by DAYKUNDI (N), GHAZNI (NE), ZABUL (ESE), KANDAHAR (S), and HELMAND (W) provinces. Part of the HAZARAJAT region, heartland of the Hazara people, which was in 1963 divided into several provinces. Mountainous area, accessible only from the S over valleys at 3,000 ft/914 m–7,600 ft/2,316 m. Traversed by the HELMAND River and its tributaries, which provide water for irrigation. Largely agricultural (cereal grains); major handicrafts is production of woven carpets, called *gelim*. In 2004, N part of Oruzgan province was excised and formed into Daykundi province.

Orvault (or-vo), town (□ 10 sq mi/26 sq km), LOIRE-ATLANTIQUE department, PAYS DE LA LOIRE region, W FRANCE, an outer NW suburb of NANTES, 6 mi/9.7 km from city center; 47°16′N 01°37′W. Manufacturing (telephone and related communications equipment).

Cross-references are shown in SMALL CAPITALS. The pronunciation guide is shown on page xix. The sources of population figures are shown on page xvii.

Orvieto (or-VYE-to), town, in UMBRIA, central ITALY, on the Poglia River; 42°43′N 12°07′E. Situated at the top of a rocky hill, it is a tourist and pilgrimage center. A commercial and agricultural center producing wines and olive oil. The wine of the same name is particularly prized. Orvieto is probably located on the site of the Etruscan town of VOLSINII (sacked by the Romans in 280 B.C.E.), which was later rebuilt as Urbs Vetus. It became a free commune by the 12th century, but was later at the mercy of indigenous and foreign tyrants until it passed to the popes in 1448. There are notable Romanesque, Gothic, and Renaissance buildings in Orvieto, but the fame of the town is due mainly to its beautiful cathedral (begun in 1290). The cathedral's white- and black-marble facade is decorated with delicate sculptures and colorful mosaics, and the Chapel of San Brizio, inside, has frescoes by Fra Angelico and by Luca Signorelli, whose powerful scenes of the Apocalypse inspired Michelangelo. The town also has a well (200 ft/61 m deep) dug in rock (completed 1537).

Örviken (UHR-VEEK-en), village, VÄSTERBOTTEN county, N SWEDEN, on islet in GULF OF BOTHNIA, at mouth of SKELLEFTE ÄLVEN RIVER, 8 mi/12.9 km SE of SKELLEFTEÅ; 64°41′N 21°12′E.

Orville Coast, forms part of the N coast of the RONNE ICE SHELF, along southeasternmost ELLSWORTH LAND, ANTARCTICA; 75°45′S 65°30′W.

Orvilos (OR-vee-los), Bulgarian *Slavyanka*, massif, in CENTRAL MACEDONIA and EAST MACEDONIA AND THRACE departments, NE GREECE, and SW BULGARIA, at S end of PIRIN MOUNTAINS, 25 mi/40 km N of both SÉRRAI and DRÁMA (Greece) and 8 mi/12.9 km SSW of GOTSE DELCHEV (Bulgaria); 41°23′N 23°37′E. Karstlike formations; sparse vegetation. Highest peak (7,258 ft/ 2,212 m) known to Greeks as Ali Boutous, to Bulgarians as Gocev Vrah. Also Ordhilos, Ordelos. Formerly called Alibotush Mountains.

Orvin Mountains, a large group of mountains in NEW SCHWABENLAND, EAST ANTARCTICA, located between the WOHLTHAT MOUNTAINS and the MÜHLIG-HOFMANN MOUNTAINS, 65 mi/105 km long; 72°00′S 09°00′E.

Orwell, town, ADDISON CO., W VERMONT, on Lake CHAMPLAIN, 21 mi/34 km NW of RUTLAND, and 15 mi/ 24 km SSW of MIDDLEBURY; 43°48′N 73°17′W. Named for Francis Vernon (1715–1783), first Baron Orwell. Contains Mount Independence which once had a large fort built by 12,000 soldiers in 1776 to defend against British attack during the American Revolutionary War. Well preserved ruins of the fort remain.

Orwell (OR-wel), village (2006 population 1,505), ASHTABULA county, extreme NE OHIO, 25 mi/40 km S of ASHTABULA; 41°32′N 80°51′W. In dairy, grain, and poultry area.

Orwell, hamlet, OSWEGO county, N central NEW YORK, 27 mi/43 km NE of OSWEGO; 43°32′N 75°57′W. Also known as Orwell Corners.

Orwell Bay, inlet S PRINCE EDWARD ISLAND, CANADA, opening SE from HILLSBOROUGH BAY; 10 mi/16 km long, mouth 4 mi/6 km wide.

Orwell River (AW-wel), 21 mi/34 km long, SUFFOLK, E ENGLAND. Rises as Gipping River in several headstreams near STOWMARKET; flows SE, past NEEDHAM MARKET and IPSWICH (where it becomes Orwell River), to STOUR RIVER estuary opposite HARWICH.

Orwigsburg (OR-wigs-buhrg), borough (2006 population 2,985), SCHUYLKILL county, E central PENN-SYLVANIA, 5 mi/8 km ESE of POTTSVILLE; 40°38′N 76°05′W. Manufacturing (fabricated metal products, pens and markers, machinery, wood products, food products, textiles, apparel). Agriculture area (potatoes, corn, apples; livestock; dairying). Hawk Mountain Sanctuary and APPALACHIAN TRAIL to E. Settled 1747, laid out 1796, incorporated 1813.

Oryahovo (or-YA-ko-vo), city (1993 population 6,712), MONTANA oblast, Oryahovo obshtina, N BULGARIA,

port on the right bank of the DANUBE (Romanian border), opposite the mouth of the JIU RIVER, 40 mi/ 64 km NNE of VRATSA; 43°44′N 23°56′E. Railroad terminus. Agricultural center (grain and fruit exports). Manufacturing (metal products, machinery), wood processing. Has ruins of Roman castles. Scene of a defeat of Turks by Stephen the Great (1475). Formerly called Rakhovo (later spelled Rahovo), and Orekhovo or Oryekhovo until 1945.

Or Yehuda (OR yi-hoo-DAH), city (2006 population 31,100), central ISRAEL, 4.3 mi/6.9 km SE of TEL AVIV on coastal plain; 32°01′N 34°50′E. Elevation 82 ft/24 m. Site of an ancient *tel*—the Ono from the times of Ezra and Nehemiah. An important city in period of Second Temple and later, possibly until the Arab conquest of the 7th century. Founded in 1950, it was granted city status in 1988.

Oryekhovo, Bulgaria: see ORYAHOVO.

Oryen, MONTENEGRO: see ORJEN.

Orynyn (o-RI-nin) (Russian *Orinin*), village (2004 population 5,800), SW KHMEL′NYTS′KYY oblast, UKRAINE, on the Zhvanchyk, tributary of the DNIESTER, 9 mi/15 km NW of KAMYANETS′-PODIL′S′KYY; 48°46′N 26°24′E. Elevation 813 ft/247 m. Clothing, metal products.

Orzhev, UKRAINE: see ORZHIV.

Orzhitsa, UKRAINE: see ORZHYTSYA.

Orzhits River, POLAND: see ORZYC RIVER.

Orzhiv (OR-zhif) (Russian *Orzhev*), town (2004 population 6,800), W central RIVNE oblast, UKRAINE, on the left bank of the HORYN′ River, and on railroad 10 mi/16 km NW of RIVNE; 50°45′N 26°07′E. Elevation 538 ft/163 m. Woodworking mill. Established at the beginning of the 16th century; town since 1959.

Orzhytsya (OR-zhi-tsyah) (Russian *Orzhitsa*), town, W central POLTAVA oblast, UKRAINE, on the Orzhytsya River, tributary of the SULA, and 20 mi/32 km SW of LUBNY; 49°48′N 32°42′E. Elevation 262 ft/79 m. Raion center. Manufacturing (feed, flour, food products), peat working. Known since 1630; company center of Ukrainian Cossack regiments (17th–18th century); town since 1968.

Orzinuovi (or-tsee-NWO-vee), town, BRESCIA province, LOMBARDY, N ITALY, near OGLIO River, 17 mi/27 km SW of BRESCIA; 45°24′N 09°55′E. Manufacturing (foundries; metal fabricating, food processing; apparel). Built 1193 by Brescia as fortress against CREMONA.

Orzu (or-TSOO), town, KHATLON viloyat, SW TAJIKI-STAN, in Vakhsh valley, 16 mi/25 km S of KURGAN-TYUBE, and 19 mi/30 km N of AFGHANISTAN border; 37°36′N 68°49′E.

Orzyc River (O-zeets), German *Orshütz*, Russian *Orzhits*, 84 mi/135 km long, N POLAND; rises 4 mi/6.4 km ENE of MŁAWA; flows NNE and SSE, past CHORZELE and MAKÓW Mazowiecki, to Narew River 7 mi/11.3 km NE of PUŁTUSK.

Orzysz (O-zeesh), German *Arys*, town, in EAST PRUS-SIA, Suwałki province, NE POLAND, in Masurian Lakes region, 17 mi/27 km W of EŁK; 53°49′N 21°57′E. Grain, cattle; sawmilling. Tourist resort. Chartered 1726.

Os (AWS), village, HORDALAND county, SW NORWAY, on the BERGEN Peninsula, port on Bjørnafjorden, 15 mi/24 km SSE of Bergen. Furniture. This suburb of Bergen has a conference center.

Osa (uh-SAH), city (2006 population 23,240), SW PERM oblast, W URALS, E European Russia, port on the VOTKINSK Reservoir of the KAMA River, on highway, 90 mi/146 km W of PERM; 57°17′N 55°26′E. Elevation 295 ft/89 m. Manufacturing (machinery), woodworking, sawmilling; food industries. Oil fields in the vicinity. An old Khanty (Ostyak) village; became Russian town of Nikol′skaya Sloboda in 1557; fortified and renamed Osa in 1737. Became city in 1781.

Osa (O-SAH), town, Atetsu county, OKAYAMA prefecture, SW HONSHU, W JAPAN, 37 mi/60 km N of OKAYAMA; 35°04′N 133°34′E.

Osa (uh-SAH), village (2006 population 4,445), central UST-ORDYN-BURYAT AUTONOMOUS OKRUG, in IR-KUTSK oblast, E central SIBERIA, RUSSIA, 35 mi/56 km NE of CHEREMKHOVO; 53°23′N 103°52′E. Elevation 1,473 ft/448 m. Highway junction. In dairy-farming area.

Osa, canton, COSTA RICA: see CIUDAD CORTÉS.

Osa de la Vega (O-sah dhai lah VAI-gah), town, CUENCA province, E central SPAIN, in LA MANCHA, 45 mi/72 km SW of CUENCA; 39°40′N 02°45′W. Cereals, grapes, olives; sheep, goats.

Osafune (o-SAHF-ne), town, Oku county, OKAYAMA prefecture, SW HONSHU, W JAPAN, 9 mi/15 km E of OKAYAMA; 34°41′N 134°06′E. Traditional sword making.

Osage (o-SAIJ), county (□ 719 sq mi/1,869.4 sq km; 2006 population 16,958), E KANSAS; ☉ LYNDON; 38°38′N 95°43′W. Gently rolling plains area, drained by MARAIS DES CYGNES River and Dragoon Creek. Agriculture (hogs, cattle; soybeans, wheat, apples; sorghum, hay). Manufacturing (paper products, metal products). Osage State Fishing Lake in N. Pomona Lake Reservoir and Pomona State Park in E; Melvern Lake Reservoir and Melvern State Park in SW. Formed 1859.

Osage (O-saij), county (□ 601 sq mi/1,562.6 sq km; 2006 population 13,498), central MISSOURI; ☉ LINN; 38°27′N 91°51′W. In OZARK region, on MISSOURI (N) and OSAGE (W) rivers; drained by GASCONADE and Maries rivers. Agriculture (corn, wheat, vegetables, hay; cattle, poultry; dairying; wineries). Fire-clay. Known for picturesque villages. Settled largely by German immigrants beginning in the 1830s. Formed 1841.

Osage (o-SAIJ), county (□ 2,303 sq mi/5,987.8 sq km; 2006 population 45,549), contiguous with Osage Indian Reservation, N OKLAHOMA; ☉ Pawhuska, Osage agency headquarters; 36°37′N 96°24′W. Largest county (land area) in Oklahoma; bounded N by KANSAS state line, SW by ARKANSAS RIVER (forms KAW LAKE in NW, KEYSTONE LAKE in SE); drained by CANEY RIVER, Salt, Hominy, and BIRD creeks; hilly in W. County includes part of city of BARTLESVILLE in E, and part of city of TULSA in SE corner. Cattle-ranching and oil- and natural gas–producing area, with some agriculture (soybeans, sorghum, hay; cattle, horses). Manufacturing (apparel, machinery, petroleum and coal products, process control instruments). Includes Osage Hills State Park in E center; Bluestem Lake reservoir in center; SKIATOOK LAKE reservoir in SE; Birch Lake reservoir in E; Hulah Lake reservoir and Wah-Sha-She State Park in NE; Walnut Creek State Park in S. Formed 1907.

Osage, city (2000 population 3,451), ☉ MITCHELL county, N IOWA, near CEDAR RIVER, and 22 mi/35 km ENE of MASON CITY; 43°16′N 92°48′W. Manufacturing (textiles, apparel, paper products, wood products, lime, feed, food products). Limestone quarries, sand and gravel pits nearby. Settled 1853, incorporated 1871.

Osage (o-SAIJ) or **Osage City**, village (2006 population 192), OSAGE county, N OKLAHOMA, 25 mi/40 km WNW of TULSA, and on ARKANSAS RIVER (KEYSTONE LAKE); 36°17′N 96°25′W. In agricultural area. Walnut Creek State Park to SE.

Osage, village, MONONGALIA county, N WEST VIRGINIA, 3 mi/4.8 km NW of MORGANTOWN, on Scotts Run creek, near MONONGAHELA RIVER; 39°39′N 80°00′W. Livestock; grain.

Osage, village, WESTON county, NE WYOMING, 14 mi/23 km NW of NEWCASTLE. Elevation c. 4,300 ft/1,311 m. Oil refining. Thunder Basin National Grassland to N and W.

Osage Beach, town (2000 population 3,662), MILLER county, central MISSOURI, 35 mi/56 km SW of JEF-FERSON CITY, at E end of the LAKE OF THE OZARKS, 5 mi/8 km S of Bagnell Dam (OSAGE RIVER). Lake of the Ozarks State Park adjoins town on SE. Resort area; tourism. Manufacturing (printing, medical products, transportation equipment, concrete, limestone products, cut stone); sand and gravel processing.

Osage City (o-SAIJ), town (2000 population 3,034), OSAGE county, E KANSAS, 25 mi/40 km NE of EMPORIA; 38°37′N 95°49′W. Railroad junction. Trade center for livestock, grain area. Manufacturing (prefabricated wood buildings). Melvern Lake reservoir and Melvern State Park to S. Settled 1865 near Santa Fe Trail incorporated 1872.

Osage River, c.360 mi/579 km long (including lakes), MISSOURI; formed SE of RICH HILL, W Missouri, by junction of MARAIS DES CYGNES River and LITTLE OSAGE RIVER; flows SE and E, into the OZARKS, widening into Truman Lake impounded by Harry S. Truman Dam at WARSAW, then into the LAKE OF THE OZARKS impounded by Bagnell Dam (near BAGNELL), then NE to MISSOURI River E of JEFFERSON CITY. Power plants at the dams at Bagnell and Warsaw.

Osa Gulf, COSTA RICA: see DULCE, GOLFO.

Osaka (o-SAH-kah), prefecture (□ 730 sq mi/1,898 sq km), S HONSHU, W central JAPAN, on OSAKA BAY (SW; E arm of INLAND SEA); ⊙ OSAKA, its principal port. Bounded N by KYOTO prefecture, E by NARA prefecture, S by WAKAYAMA prefecture, and NW by HYOGO prefecture. Interior has rugged terrain; coastal area is generally flat and fertile, with most of urban centers on wide Osaka plain. Manufacturing (machinery and equipment, chemicals, textiles, rubber and plastic products); flour milling. Important home industries are weaving, woodworking. Agriculture (rice, wheat, fruit; poultry); lumbering, raw silk production. Chief centers include Osaka, SAKAI, FUSE, KISHIWADA, SUITA, TOYONAKA, KAIZUKA, IKEDA, TAKATSUKI.

Osaka (o-SAH-kah), city (2005 population 2,628,811), ⊙ OSAKA prefecture, S HONSHU, W central JAPAN, on OSAKA BAY, at the mouth of the YODO RIVER; 34°41′N 135°30′E. One of Japan's largest cities and principal industrial and commercial centers, Osaka is the focal point of a chain of industrial cities (within the *Hanshin* or *Kinki*) stretching to KOBE, Osaka's port. Manufacturing (steel, transportation equipment, metal products, appliances, pharmaceuticals, apparel). Also a major port, transportation hub, and financial and media center. An educational and cultural center known for its *bunraku* puppet and other theaters and for Osaka and Kansai universities. Noted parks and gardens. Landmarks include the Buddhist temple of Shitennoji, founded in 593; Temmangu, a Shinto shrine founded in 949; and ruins of Naniwakyu. As Naniwa, the city was the site of imperial palaces as early as the 4th century. Its importance as a commercial center dates from the 16th century, when it became the seat of ruler Hideyoshi Toyotomi and grew to be Japan's leading trade center. Hideyoshi's huge castle, reconstructed in 1931, still dominates the city.

Osaka (O-sah-kah), town, Mashita county, GIFU prefecture, central HONSHU, central JAPAN, near ON-TAKE mountain, 50 mi/80 km N of GIFU; 35°56′N 137°15′E. Lumber and woodworking. Trout. Hot springs.

Osaka Bay (o-SAH-kah), E arm of INLAND SEA, off OSAKA prefecture, W central JAPAN, between AWAJI-SHIMA (W) and S coast of HONSHU; merges with HARIMA SEA (W); connected with PHILIPPINE SEA (S) by KII CHANNEL; 35 mi/56 km long, 20 mi/32 km wide. OSAKA is on NE shore. Part of Kii Peninsula forms E shore. Three ports (KOBE, Osaka, Amagasaki-Nishinomiya). Within bay is Port Island, a huge landfill with port facilities and factories.

Osakarovka (o-suh-kuh-ROF-kuh), village, N KARAGANDA region, KAZAKHSTAN, on railroad, 50 mi/80 km NNW of KARAGANDA; 50°35′N 73°20′E. Tertiary-level (raion) administrative center. In cattle area.

Osakasayama (O-sah-kah-SAH-yah-mah), city, OSAKA prefecture, S HONSHU, W central JAPAN, 12 mi/20 km S of OSAKA; 34°30′N 135°33′E.

Osaki (O-sah-kee), town, Toyota county, HIROSHIMA prefecture, SW HONSHU, W JAPAN, 28 mi/45 km E of HIROSHIMA; 34°14′N 132°53′E. Mandarin oranges.

Osaki, town, Soo county, KAGOSHIMA prefecture, S KYUSHU, extreme SW JAPAN, on E OSUMI PENINSULA, 28 mi/45 km S of KAGOSHIMA; 31°25′N 131°00′E. Sweet potatoes; chicken, beef cattle.

Osaki-kami-shima (o-SAH-kee–kah-MEE–shee-mah), island (□ 16 sq mi/41.6 sq km), HIROSHIMA prefecture, W JAPAN, in HIUCHI SEA, 3 mi/4.8 km S of TAKEHARA on SW HONSHU, just W of OMI-SHIMA; 5 mi/8 km long, 4 mi/6.4 km wide. Mountainous, fertile (fruit growing). KINOE (E) is chief town.

Osakis (o-SAI-kuhs), town (2000 population 1,567), DOUGLAS and TODD counties, W MINNESOTA, at S end of Lake OSAKIS, 11 mi/18 km E of ALEXANDRIA; 45°52′N 95°09′W. Agriculture (grain; poultry; dairying). Manufacturing (construction materials, diversified light manufacturing).

Osaki-shimo-shima (o-SAH-kee–shee-MO–shee-mah), island (□ 6 sq mi/15.6 sq km), HIROSHIMA prefecture, W JAPAN, in HIUCHI SEA, just SW of OSAKI-KAMI-SHIMA, 10 mi/16 km SE of KURE; 4 mi/6.4 km long, 2 mi/3.2 km wide. Hilly, fertile (fruit growing). Fishing.

Osakis, Lake (□ 10 sq mi/26 sq km), in TODD and DOUGLAS counties, W MINNESOTA, 12 mi/19 km E of ALEXANDRIA; 8 mi/12.9 km long, 2 mi/3.2 km wide. Elevation 1,320 ft/402 m. Drains into small affluent of LONG PRAIRIE RIVER. Fishing, boating, bathing.

Osan (O-SAHN), city (2005 population 132,532), SW KYONGGI province, SOUTH KOREA. Hilly in NW, plain in the basin of the Osan River to the E. Agriculture and stockbreeding on decline. Rapid industrialization, caused by the city's proximity to large urban centers. Large-scale manufacturing (textiles, chemistry, paper, electronics). Located on Kyongbu Expressway and railroad between SEOUL and PUSAN.

Osa Península (O-sah), on the PACIFIC in S COSTA RICA; 35 mi/56 km long, 10 mi/16 km–15 mi/24 km wide. Separated from the mainland by the GOLFO DULCE (E). Site of Corcovado National Park, over 100,000 acres/40,470 ha of tropical forest and hundreds of plant and animal species. Placer gold mining is carried out legally near the park and illegally in it.

Osasco (O-sah-SKO), W suburb of SÃO PAULO city (2007 population 701,012), SE SÃO PAULO state, BRAZIL, on TIETÊ RIVER; 23°32′S 46°46′W. On railroad to SOROCABA. Manufacturing (food products, transportation equipment).

Osato (O-sah-to), town, Kurokawa county, MIYAGI prefecture, N HONSHU, NE JAPAN, 12 mi/20 km N of SENDAI; 38°25′N 141°00′E.

Osato (O-sah-to), village, Osato county, SAITAMA prefecture, E central HONSHU, E central JAPAN, 22 mi/35 km N of URAWA; 36°05′N 139°24′E.

Osavulivka, UKRAINE: see YESAULIVKA.

Osawano (O-sah-wah-no), town, Kaminikawa county, TOYAMA prefecture, central HONSHU, central JAPAN, 9 mi/15 km S of TOYAMA; 36°34′N 137°12′E.

Osawatomie (o-suh-WAHT-uh-mee), town (2000 population 4,645), MIAMI county, E KANSAS, on the MARAIS DES CYGNES River, 7 mi/11.3 km SW of PAOLA; 38°30′N 94°57′W. Railroad junction. The town, once a station on the Underground Railroad, has a memorial park that contains the cabin where John Brown lived in 1856; John Brown Museum Founded 1855 by the New England Emigrant Aid Company, incorporated 1883.

Osborn, Ohio: see FAIRBORN.

Osborne (AHZ-born), community, S MANITOBA, W central Canada, 26 mi/41 km from WINNIPEG, in Macdonald rural municipality; 49°32′N 97°22′W.

Osborne (OZ-bawn), district of COWES (2001 population 2,071), on Isle of Wight, S ENGLAND, just W of the town; 50°45′N 01°17′W. Site of Osborne House (1846), summer residence and place of death of Queen Victoria. Presented to nation by Edward VII. The House is now restored to its former splendor as the summer palace of the queen.

Osborne (AHZ-born), county (□ 894 sq mi/2,324.4 sq km; 2006 population 3,978), N central KANSAS; ⊙ OSBORNE; 39°21′N 98°46′W. Smoky Hills region, drained by North and South forks of SOLOMON RIVER. Cattle, hogs; wheat, rye, barley, sorghum. Manufacturing (farm equipment); food processing. N arm of Waconda Lake Reservoir in NE. Geodetic Center of U.S in SE (Meades Ranch). Formed 1871.

Osborne (AHZ-born), town (2000 population 1,607), ⊙ OSBORNE county, N KANSAS, on South Fork of SOLOMON RIVER, and 39 mi/63 km NNE of RUSSELL; 39°26′N 98°42′W. Elevation 1,500 ft/457 m. Railroad junction. Shipping center for livestock and grain area. Manufacturing (feeds, farm machinery, wood products). Founded 1871, incorporated 1879.

Osborne (AHS-born), borough (2006 population 530), ALLEGHENY county, SW PENNSYLVANIA, residential suburb 11 mi/18 km NW of downtown PITTSBURGH, on OHIO RIVER; 40°31′N 80°10′W.

Osborne, Lake, reservoir, MANICALAND province, E ZIMBABWE, formed on ODZI RIVER by Osborne Dam 17 mi/27 km NW of MUTARE; c.10 mi/16 km long; 18°50′S 32°28′E. Nyatanda River enters from N forming N arm. Irrigation.

Osborne Park (AHZ-buhrn PAHRK), town, SW WESTERN AUSTRALIA state, W AUSTRALIA, N residential suburb of PERTH. Manufacturing (paper products, consumer goods).

Osbornville, village, OCEAN county, E NEW JERSEY, 7 mi/11.3 km NNE of TOMS RIVER.

Osbourne Store, town, CLARENDON parish, S central JAMAICA, 5 mi/8 km W of MAY PEN; 17°58′N 77°20′W. A highway junction along main road W of May Pen.

Osburn, town (2000 population 1,545), SHOSHONE county, N IDAHO, 2 mi/3.2 km NW of WALLACE, across highway (I-90) from SILVERTON; 47°31′N 116°00′W. Manufacturing (concrete); silver mining. St. Joe National Forest to S; Coeur d'Alene National Forest to N.

Osby (OOS-BEE), town, SKÅNE county, S SWEDEN, on N side of Osbysjön Lake (3 mi/4.8 km long), 18 mi/29 km NNE of HÄSSLEHOLM; 56°23′N 14°00′E. Manufacturing (consumer goods, furniture). Tourist resort. Seat of agricultural college. Danish fortifications built 1611.

Oscadnica (osh-CHAHD-nyi-TSAH), Slovak *Oščadnica*, Hungarian *Ocsad*, village, STREDOSLOVENSKY province, NW SLOVAKIA, in BESKIDS, and 16 mi/26 km NNE of ŽILINA; 49°26′N 18°53′E. Woodworking. Tourist resort; folk architecture and costumes, museum.

Oscar Bressane (O-skahr BRE-sah-ne), town (2007 population 2,483), W SÃO PAULO state, BRAZIL, 22 mi/35 km SW of MARÍLIA; 22°18′S 50°17′W.

Oscar II Coast, on the E coast of GRAHAM LAND, ANTARCTIC PENINSULTA, between CAPE FAIRWEATHER and CAPE ALEXANDER; 65°45′S 62°30′W. Originally called KING OSCAR LAND, later redefined and renamed.

Oscarsborg (AWS-kahrs-BAWR), fortress on islet at N of DRØBAK, AKERSHUS county, SE NORWAY, 18 mi/29 km S of OSLO. During German invasion (April 1940), its guns sank cruiser *Blücher*.

Oscarville, village (2000 population 61), SW ALASKA, near BETHEL; 60°42′N 161°45′W.

Oscawana, hamlet, PUTNAM county, SE NEW YORK, on Oscawana Lake (2 mi/3.2 km long, c.1 mi/1.6 km wide), 8 mi/12.9 km NNE of PEEKSKILL; 41°13′N 73°56′W. In dairying and farming area. Also known as Oscawana Corners. Post office in Putnam Valley.

Osceola (ah-see-O-luh), county (□ 1,506 sq mi/3,915.6 sq km; 2006 population 244,045), central FLORIDA; ⊙ KISSIMMEE; 28°03′N 81°09′W. Lowland area with many lakes (notably KISSIMMEE, TOHOPEKALIGA, and East Tohopekaliga), and streams of KISSIMMEE RIVER system. Agriculture (citrus fruit, vegetables; cattle);

lumber. County is sparsely settled except for its NE corner in suburban ORLANDO. Formed 1887.

Osceola, county (□ 399 sq mi/1,037.4 sq km; 2006 population 6,629), NW IOWA, on MINNESOTA line; ⊙ SIBLEY; 43°22′N 95°37′W. Rises to highest elevation (1,670 ft/509 m) in Iowa 6 mi/9.7 km NE of Sibley. Drained by OCHEYEDAN RIVER and Otter Creek. Prairie agricultural area (hogs, cattle, poultry; corn, oats, soybeans). Iowa and Rush lakes in NE. Formed 1851.

Osceola (O-see-O-lah), county (□ 573 sq mi/1,489.8 sq km; 2006 population 23,584), central MICHIGAN; ⊙ REED CITY; 43°58′N 85°19′W. Intersected by MUSKEGON RIVER and drained by PINE RIVER. Agriculture (forage crops; cattle, hogs, poultry; wheat, oats, corn; dairy products). Some manufacturing at REED CITY. Resorts. Part of Manistee National Forest in W and immediately beyond NW and SW corners of county. Organized 1869.

Osceola (oh-see-OLE-uh), city (2000 population 4,659), ⊙ CLARKE county, S IOWA, near WHITEBREAST CREEK, c.40 mi/64 km SSW of DES MOINES; 41°01′N 93°46′W. Manufacturing (electrical equipment, metal products, transportation equipment, apparel, dairy equipment and products). Stephens State Forest to SE. Settled 1850, incorporated 1859.

Osceola (o-see-O-luh), city (2000 population 835), ⊙ SAINT CLAIR county, W MISSOURI, on OSAGE RIVER, arm of Truman Lake (Harry S. Truman Reservoir), and 85 mi/137 km SSE of KANSAS CITY; 38°02′N 93°42′W. Agriculture (cattle; wheat, sorghum). Recreational activities.

Osceola (aw-see-O-luh), city (2006 population 888), ⊙ POLK county, E central NEBRASKA, 20 mi/32 km SW of COLUMBUS, and on branch of BIG BLUE RIVER; 41°10′N 97°32′W. Livestock, poultry; grain. Founded c.1872.

Osceola (o-see-O-luh), town, shares ⊙ functions with BLYTHEVILLE, MISSISSIPPI county, NE ARKANSAS, 15 mi/24 km S of Blytheville, and on MISSISSIPPI RIVER; 35°42′N 89°59′W. In rich cotton and soybean area. Manufacturing (consumer goods, construction materials, furniture, apparel, food products, metal products). Incorporated 1838.

Osceola, town (2000 population 1,859), SAINT JOSEPH county, N INDIANA, 10 mi/16 km E of SOUTH BEND, and near SAINT JOSEPH RIVER; 41°40′N 86°04′W. Light manufacturing.

Osceola (ahs-ee-O-lah), town (2006 population 2,685), POLK county, W WISCONSIN, on SAINT CROIX RIVER (MINNESOTA state line), and 19 mi/31 km NNE of STILLWATER (Minnesota); 45°19′N 92°42′W. In dairying and stock-raising area. Manufacturing (wooden products, metal products, electrical equipment, publishing, cheese). Saint Croix Falls and Osceola Fish Hatcheries to NE.

Osceola Mills (AH-see-O-lah MILS), borough (2006 population 1,179), CLEARFIELD county, central PENNSYLVANIA, 25 mi/40 km NNE of ALTOONA, on Moshannon Creek; 40°51′N 78°16′W. Manufacturing (apparel); surface bituminous coal, clay, timber in area. Agriculture (grain; livestock, dairying). Laid out c.1857, incorporated 1864.

Osceola, Mount (aw-see-YO-luh), peak (4,326 ft/1,319 m) in WHITE MOUNTAINS, GRAFTON county, central NEW HAMPSHIRE, 6 mi/9.7 km ESE of North Woodstock, in White Mountain National Forest.

Oschatz (O-shats), town, SAXONY, E central GERMANY, 32 mi/51 km E of LEIPZIG; 51°18′N 13°07′E. Manufacturing (shoes); woolen and felt milling, sugar refining. Was first mentioned 1246. Has 15th century church, 16th century town hall.

Oschersleben-Bode (O-shers-leh-ben–BAW-de), town, SAXONY-ANHALT, central GERMANY, on the BODE, 19 mi/31 km WSW of MAGDEBURG; 52°02′N 11°14′E. Manufacturing (light metals, agriculture machinery, chemicals, cigars, food products); sugar re-

fining, malting. Obtained town status 1235; castle here from 1545.

Oscoda (ah-SKO-duh), county (□ 571 sq mi/1,484.6 sq km; 2006 population 9,140), NE central MICHIGAN; ⊙ MIO; 44°40′N 84°07′W. Intersected by AU SABLE RIVER, and drained by Upper South Branch of THUNDER BAY RIVER. Agriculture (grain; cattle, sheep; dairy products, poultry). Manufacturing (wood products, metal products). Recreational area. Includes part of Huron National Forest in S and several small lakes, especially N and S center; Mio Mountain Ski Area in center. Organized 1881.

Oscoda (ah-SKO-duh), village (2000 population 992), IOSCO county, NE MICHIGAN, c.45 mi/72 km SE of ALPENA, on LAKE HURON at mouth of AU SABLE RIVER; 44°25′N 83°19′W. Trade center for resort and farm area (potatoes; cattle); fisheries. Manufacturing (transportation equipment, vinyl floor coverings). Paul B. Wurstmith Air Force Base to NW.

Oscura Peak (os-KOO-ruh) (9,650 ft/2,941 m), highest point in SIERRA OSCURA, on LINCOLN-SOCORRO county line, S central NEW MEXICO, 29 mi/47 km W of CARRIZOZO. In Lincoln National Forest.

Oscuro, Cerro (os-KOO-ro SE-ro), peak (c.7,000 ft/ 2,134 m) in continental divide, on GUATEMALA-HONDURAS border, 10 mi/16 km E of ESQUIPULAS (Guatemala); 14°35′N 89°12′W.

Oseberg (AW-suh-BAR), village, VESTFOLD county, SE NORWAY, 3 mi/4.8 km NE of TØNSBERG; 59°19′N 10°27′E. Scene (1903) of excavation of well-preserved ninth-century Viking ship, now in OSLO museum.

Osek (O-sek), German *osseg*, town, SEVEROCESKY province, NW BOHEMIA, CZECH REPUBLIC, in NE foothills of the ORE MOUNTAINS, on railroad, and 6 mi/9.7 km WSW of TEPLICE; 50°37′N 13°42′E. Manufacturing (chemicals); coal mining in vicinity. Has famous Cistercian abbey, which was founded in 12th century.

Osek, GERMANY: see HOHER BOGEN.

Ösel, ESTONIA: see SAAREMAA.

Oseras, Altos de las (o-SAI-rahs, AHL-tos dai lahs), ANDEAN massif (12,565 ft/3,830 m), W central COLOMBIA, in Cordillera ORIENTAL, 60 mi/97 km SSW of BOGOTÁ; S point of Páramo de SUMAPAZ; 03°45′N 74°30′W.

Oseredok, Mount, UKRAINE: see POKUTIAN-BUKOVINIAN CARPATHIANS.

Osetia, GEORGIA and RUSSIA: see OSSETIA.

Oseto (o-SE-to), town, W Sonogi county, NAGASAKI prefecture, W KYUSHU, SW JAPAN, 19 mi/30 km N of NAGASAKI; 32°55′N 129°38′E. Loquats.

Osgood (town (2000 population 1,669), RIPLEY county, SE INDIANA, 5 mi/8 km N of VERSAILLES; 39°08′N 85°17′W. Farm trading center, with some manufacturing (cement products, shoes); grain milling, timber; limestone quarries. Laid out 1857.

Osgood, town (2000 population 51), SULLIVAN county, N MISSOURI, 12 mi/19 km W of MILAN; 40°12′N 93°20′W.

Osgood (AHS-guhd), village (2006 population 253), DARKE county, W OHIO, 18 mi/29 km NNE of GREENVILLE; 40°20′N 84°29′W. In agricultural area.

Osgoode (AHZ-gud), former township (□ 147 sq mi/ 382.2 sq km; 2001 population 17,607), SE ONTARIO, E central Canada, near Rideau River and RIDEAU CANAL, and included in city of OTTAWA; 45°14′N 75°30′W.

Osgoode (AHZ-gud) or **Osgoode Station**, unincorporated village, SE ONTARIO, E central Canada, near Rideau River and RIDEAU CANAL, and 20 mi/32 km S of OTTAWA, of which it is a part; 45°08′N 75°36′W. Dairying, mixed farming.

Osgood Mountains, N NEVADA, in E HUMBOLDT county Adam Peak (8,678 ft/2,645 m) 25 mi/40 km NE of WINNEMUCCA, is highest point. Gold mining, tungsten deposits.

Osh (AWSH), region (□ 17,834 sq mi/46,368.4 sq km; 1999 population 1,175,998), SW KYRGYZSTAN; ⊙ OSH.

On N slopes of ALAY and TURKESTAN ranges, except for Alay Valley, which is on S slope of Alay Range; includes S fringe of FERGANA VALLEY (N; site of cotton area). Wheat and livestock raised in mountain valleys. Extensive coal mining (SÜLÜKTÜ, KYZYL-KYYA, UZGEN); antimony and mercury (FRUNZE, KHAYDARKAN, CHAUVAY); sericulture near Osh. Population consists of Kyrgyz, Uzbeks. Formed 1939; divided between Osh and JALAL-ABAD regions in 1991.

Osh (AWSH), city (1999 population 208,520), ⊙ OSH region, S KYRGYZSTAN, in the FERGANA VALLEY; 40°25′N 72°50′E. Country's second-largest city; terminus of the Osh-KHOROG (KAZAKHSTAN) highway. City has silk, cotton, textile, and food-processing industries and is one of central ASIA's largest silk markets. Currently a center for opium trade, based on poppy growing in nearby countries; main production center in KHORUGH (TAJIKISTAN). One of the oldest settlements of Central Asia (c.3,000 years), Osh was for centuries a major silk-production center, strategically situated on a trade route to India. The old city adjoins the larger modern section. Seat of two universities (one is country's oldest); airport. The Tash-Sulayman [=Solomon's throne], an odd-shaped rock, was once a place of Muslim pilgrimage. Osh was the scene of several violent Uzbek-Kyrgyz ethnic clashes in 1990.

Oshakati (o-shuh-kah-tee), town (2001 population 28,255), N central NAMIBIA, 30 mi/48 km from Angolan border, 120 mi/193 km NW of TSUMEB. Largest town in OVAMBOLAND, located in OSHANA REGION. Trade center and popular tourist center. Largest teachers' college 6 mi/9.7 km S at Ongwediva.

Oshamanbe (o-shah-MAHN-be), town, Oshima county, Hokkaido prefecture, N JAPAN, on NW shore of UCHIURA BAY, 62 mi/100 km S of SAPPORO; 42°30′N 140°23′E. Scallops, crabs.

Oshana Region (o-SHAH-nuh), administrative division (2001 population 161,916), directly N of Etosha in the center of OVAMBOLAND, NAMIBIA; 18°10′S 15°45′E. Smallest, most densely populated area, housing main towns of OSHAKATI and ONDANGWA.

Osha River (uh-SHAH), approximately 150 mi/241 km long, central OMSK oblast, RUSSIA; rises in small Lake Saltaim, 25 mi/40 km NW of TYUKALINSK; flows ENE, past STAROSOLDATSKOYE and KOLOSOVKA, and N to the IRTYSH River below ZNAMENSKOYE.

O'Shaughnessy Dam, California: see HETCH HETCHY VALLEY.

Oshawa (AHSH-uh-wuh), city (□ 56 sq mi/145.6 sq km; 2001 population 139,051), DURHAM region, SE ONTARIO, E central Canada, on Lake ONTARIO; 43°54′N 78°52′W. The production of motor vehicles, begun in 1907, is the leading industry, since Oshawa is the home site of General Motors of Canada, although plant operations will be downscaled by 2008. Other products include metals, glass, plastics, machine parts, furniture, leather and woolen goods, and electrical products. Automobile museum. On the site of a French trading post, established 1750; incorporated as a city in 1924. Name from the Aboriginal meaning "crossing of a stream" or "carrying place."

Oshchepkovo, RUSSIA: see PYSHMA, town.

Oshika (O-shee-kah), town, Oshika county, MIYAGI prefecture, N HONSHU, NE JAPAN, near Mount Kinka, 34 mi/55 km E of SENDAI; 38°17′N 141°30′E. *Wakame* (seaweed).

Oshika (O-shee-kah), village, Shimoina county, NAGANO prefecture, central HONSHU, central JAPAN, 74 mi/120 km S of NAGANO; 35°34′N 138°02′E.

Oshikango (o-shi-kahn-go), village, NAMIBIA, N border town in OVAMBOLAND, 40 mi/64 km N of Ondangua; 17°24′S 15°53′E. At end of paved road. Border post into ANGOLA. Communal livestock-raising area.

Oshima (O-shee-mah), town, W Sonogi county, NAGASAKI prefecture, W KYUSHU, SW JAPAN, 25 mi/40 km N of NAGASAKI; 33°01′N 129°37′E. Shipbuilding, *shochu* distilling.

Oshima, town, TOKYO municipality, off SE HONSHU, SE JAPAN, 19 mi/30 km S of SHINJUKU; 34°44′N 139°21′E. Confections. Camellia flowers. Mount Mihara volcano and port of Habu nearby.

Oshima, town, Imizu county, TOYAMA prefecture, central HONSHU, central JAPAN, 9 mi/15 km N of Toyama; 36°43′N 137°03′E.

Oshima, town, on Oshima island, Oshima county, YAMAGUCHI prefecture, off SW HONSHU, W JAPAN, in IYO SEA, 47 mi/75 km E of YAMAGUCHI; 33°55′N 132°11′E. *Wakame* (seaweed), mandarin oranges.

Oshima (O-shee-mah), village, Munakata county, FUKUOKA prefecture, NE KYUSHU, SW JAPAN, 22 mi/35 km N of FUKUOKA; 33°53′N 130°26′E.

Oshima, village, N Matsuura county, NAGASAKI prefecture, W KYUSHU, SW JAPAN, 53 mi/85 km N of NAGASAKI; 33°28′N 129°33′E.

Oshima, village, E Kubiki county, NIIGATA prefecture, central HONSHU, N central JAPAN, 59 mi/95 km S of NIIGATA; 37°08′N 138°30′E. Processed edible wild plants, pickles.

O-shima (O–shee-mah), island (□ 35 sq mi/91 sq km), TOKYO prefecture, SE JAPAN, near the entrance to TOKYO BAY. The largest and most northerly of the IZUSHICHITO group, it is the site of volcanic Mount Mihara (2,477 ft/7,550 m), which last erupted in 1986. Agriculture and fishing are chief activities here. It is a recreation area for Tokyo. Visited (17th century) by Maarten Vries, the Dutch navigator.

O-shima, island (□ 17 sq mi/44.2 sq km), EHIME prefecture, W JAPAN, in HIUCHI SEA, just off N coast of SHIKOKU, NE of IMABARI; 6 mi/9.7 km long, 3 mi/4.8 km wide. Hilly, fertile (rice, wheat, fruit). Sake brewery, fishery.

O-shima, island (□ 6 sq mi/15.6 sq km), NAGASAKI prefecture, SW JAPAN, off NW KYUSHU, in EAST CHINA SEA, 22 mi/35 km NNW of SASEBO, N of HIRADO-SHIMA island; 5 mi/8 km long, 3 mi/4.8 km wide; 33°28′N 129°33′E. Fishing. Sometimes called Okoshima.

O-shima, island (□ 5 sq mi/13 sq km), NAGASAKI prefecture, SW JAPAN, in EAST CHINA SEA, just off NNW coast of SONOGI PENINSULA, NW KYUSHU, 9 mi/14.5 km SW of SASEBO; 3.5 mi/5.6 km long, 1.5 mi/2.4 km wide; 33°03′N 129°37′E. Fishing.

O-Shima (O–shee-ma), island (□ 2 sq mi/5.2 sq km), WAKAYAMA prefecture, W central JAPAN, in KUMANO SEA, just E of SHIO POINT on S KII PENINSULA in S HONSHU; shelters KUSHIMOTO harbor; 3 mi/4.8 km long, 1 mi/1.6 km wide. Tourist resort.

Oshimizu (o-SHEE-meez), town, Hakui county, ISHIKAWA prefecture, central HONSHU, central JAPAN, 19 mi/30 km N of KANAZAWA; 36°49′N 136°45′E.

Oshino (O-shee-no), village, S Tsuru county, YAMANASHI prefecture, central HONSHU, central JAPAN, 22 mi/35 km S of KOFU; 35°27′N 138°51′E. Robots.

Oshkosh (AWSH-kawsh), city (2006 population 64,084), ☉ WINNEBAGO county, E WISCONSIN, on LAKE WINNEBAGO, where the Upper FOX RIVER enters; 44°01′N 88°32′W. Manufacturing (apparel, transportation equipment, wood products, electrical equipment, machinery). Summer resort. Father Allouez visited the site in 1670; French explorers traveled there in the 18th century; and a French fur-trading post was set up in the early 19th century. Oshkosh grew as a lumber town. The downtown area was destroyed by fire in 1875. A branch of the University of Wisconsin is here. Paine Art Center and Arboretum, Winnebago Mental Health Institute to N, Experimental Aircraft Association Museum and annual Fly-In in July at Whitman Field. Incorporated 1846.

Oshkosh (AWSH-kawsh), town (2006 population 762), ☉ GARDEN county, W NEBRASKA, 37 mi/60 km NW of OGALLALA, and on NORTH PLATTE RIVER; 41°24′N 102°20′W. In irrigated sugar-beet region; livestock, grain. Manufacturing (wood furniture). OREGON TRAIL follows opposite side of river.

Oshmyany (osh-MYA-nee), Polish *Oszmiana*, city, ☉ Oshmyany region, GRODNO oblast, BELARUS, 28 mi/45 km ESE of VILNA. Creamery, flax-processing and bread-baking plants, meat combine, dairy industry. Has ruins of seventeenth-century churches. Old Russian settlement; successively captured by Lithuanians, Teutonic Knights (1384), and Poles. Passed (1795) from POLAND to RUSSIA; reverted (1921) to Poland; ceded to USSR in 1945.

Oshmyany Upland (osh-MYA-nee), between Vilija and NEMAN rivers, W BELARUS. Elevation up to 1,050 ft/320 m. Most of upland under cultivation; small tracts of forest have been preserved.

Oshnuiyeh, IRAN: see USHNUIYEH.

Oshogbo (o-SHO-bo), city, ☉ OSUN state, SW NIGERIA, on the Oshun River; 07°46′N 04°34′E. Primarily a farming and commercial city; manufacturing (cotton, steel, textiles, cigarettes, food processing). Road and railroad junction; airport. Oshogbo probably was founded in the 17th century as a town in the Yoruba kingdom of Ijesha. In 1839 it was the site of a decisive battle in which IBADAN, a Yoruba city-state, defeated ILORIN, an expansionist Fulani state, thus halting Ilorin's S advance. An influx of refugees after the battle swelled Oshogbo's population.

Oshtorinan (ush-TUHR-ee-NAHN), town, Lorestān, SW IRAN, 10 mi/16 km NW of BORUJERD; 33°58′N 48°38′E.

Oshun, state, SW NIGERIA: see OSUN.

Oshwe (OSH-wai), village, BANDUNDU province, W CONGO, on left bank of LUKENIE RIVER, and 130 mi/209 km SE of INONGO; 03°24′S 19°30′E. Elev. 1,210 ft/368 m. Steamboat landing and trading center; fibers, copal.

Osicala (o-see-KAH-lah), city and municipality, MORAZÁN department, E EL SALVADOR, 9 mi/14.5 km NNW of SAN FRANCISCO GOTERA; 13°48′N 88°09′W. Agriculture (henequen, sugarcane, coffee; livestock raising). Deptartment capital 1875–1887.

Osieczna (o-SEE-chnah), German *Storchnest*, town, Leszno province, W POLAND, 6 mi/9.7 km NE of LESZNO. Cement manufacturing, flour milling; trades in horses, pigs. Historic sites include a 17th century monastery and 16th century castle.

Osierfield (O-zhuhr-feeld), town, IRWIN county, S central GEORGIA, 8 mi/12.9 km ESE of FITZGERALD; 31°40′N 83°06′W.

Osijek, Ger. *Esseg*, Hung. *Eszék*, city, E CROATIA, on the DRAVA R.; 45°35′N 18°40′E. The chief city of the historic region of SLAVONIA, it is a river port and industrial center. Foods (chocolate), chemicals (soap and detergents), textiles, leather goods, and agr. machinery are among its industrial prods. International airport. Grew around a castle built (1091) on the site of the Roman colony and fortress of Mursa. Became an early episcopal see; under Turk. rule from 1526 to 1687. Later part of Austria-Hungary; part of the former Yugoslavia in 1918. Damaged by Serb artillery in 1991.

Osimo (O-zee-mo), ancient *Auximum*, town, ANCONA province, The MARCHES, central ITALY, 9 mi/14 km S of ANCONA; 43°29′N 13°29′E. Manufacturing (silk textiles, metal products, machinery, plastics); food processing. Bishopric. Has cathedral and remains of Roman walls.

Osinki (uh-SEEN-kee), settlement (2006 population 3,095), S central SAMARA oblast, E European Russia, on an irrigation canal from the Chagra River (tributary of the VOLGA RIVER), on highway junction, 10 mi/16 km SW of CHAPAYEVSK; 52°51′N 49°31′E. Elevation 370 ft/112 m. In agricultural and manufacturing (chemicals) region.

Osinniki (uh-SEEN-nee-kee), city (2005 population 48,220), SW KEMEROVO oblast, S central SIBERIA, RUSSIA, on branch of the TRANS-SIBERIAN RAILROAD (Kandalep station), at the confluence of the Kandalep River and Kondoma River, 215 mi/350 km S of KEMEROVO, and 14 mi/23 km SE of NOVOKUZNETSK;

53°37′N 87°21′E. Elevation 1,089 ft/331 m. Coal-mining center in the KUZNETSK BASIN. Mining equipment repair; bakery. Developed in the 1930s. Until 1938, called Osinovka. Made city in 1938. Large S Kuzbas coal-fed electric power plant is nearby.

Osinovka (uh-SEE-nuhf-kah), town, W central IRKUTSK oblast, E central SIBERIA, RUSSIA, on the E bank of the ANGARA RIVER, on road and the BAYKAL-AMUR MAINLINE (Gidrostroitel' station), at a dam on the Angara River backing up Bratsk reservoir, just SE of city of BRATSK, to which it is administratively subordinate; 56°17′N 101°53′E. Elevation 1,204 ft/366 m. Sawmilling, woodworking.

Osinovka (uh-SEE-nuhf-kah), village, N KEMEROVO oblast, S central Siberian Russia, on highway, 9 mi/14 km E of KEMEROVO; 55°23′N 86°19′E. Elevation 554 ft/168 m. Poultry processing.

Osinovka, RUSSIA: see OSINNIKI.

Osinovo (uh-SEE-nuh-vuh), rural settlement (2006 population 7,835), W TATARSTAN Republic, E European Russia, 12 mi/19 km NW of KAZAN'; 55°53′N 48°53′E. Elevation 403 ft/122 m. In agricultural area; poultry farm.

Osintorf (o-sin-TORF), urban settlement, SE VITEBSK oblast, BELARUS, 13 mi/21 km NE of ORSHA; 54°40′N 30°39′E. Peat works supply power plant at OREKHOVSK (W).

Osipenko, UKRAINE: see BERDYANS'K.

Osipovichi (o-si-PO-vi-chi), city, S MOGILEV oblast, BELARUS, ☉ Osipovichi region, near SVISLOCH river, 60 mi/97 km SE of MINSK; 53°19′N 28°36′E. Railroad junction (repair shops). Manufacturing (cardboard, construction materials, concrete products); food processing and canning, flour milling.

Osiraq, nuclear facility, IRAQ, reputed to be for weapons production; bombed and destroyed by Israeli warplanes in 1981. While fissionable material remained intact after the bombing, the Osiraq-type reactor does not seem to have been replaced.

Osisko Lake (o-SI-sko) W QUEBEC, E Canada; 2 mi/3 km long, 2 mi/3 km wide. On W shore is mining center ROUYN-NORANDA; 48°15′N 78°59′W.

Osječenica (os-YECH-neet-sah), mountain in DINARIC ALPS, W BOSNIA, BOSNIA AND HERZEGOVINA, along lower right bank of UNAC RIVER; highest point, Velika Osječenica (5,888 ft/1,795 m), is 7 mi/11.3 km WSW of BOSANSKI PETROVAC. Also spelled Osyechenitsa.

Oskaloosa, city (2000 population 10,938), ☉ MAHASKA county, SE IOWA, on the NORTH and South Skunk rivers; 41°17′N 92°38′W. It is the trade and processing center of a rich farm and livestock area. Manufacturing (machinery, feeds, egg products); steel fabrication. Coal has been mined here for over 100 years. A small fort was established here in 1835, and it became a post on a much-traveled W trail. Pioneer Farm and Craft Museum, William Penn College, and Vennard College are there. Lake Keomah State Park to E. The city was settled (1844) by Quakers; incorporated 1852.

Oskaloosa (ahs-kuh-LOO-suh), town (2000 population 1,165), ☉ JEFFERSON county, NE KANSAS, 23 mi/37 km NE of TOPEKA; 39°13′N 95°18′W. Trading point in grain, livestock, and dairy region. Perry Reservoir to W.

Oskarsborg, NORWAY: see OSCARSBORG.

Oskarshamn, town, KALMAR county, SE SWEDEN, seaport on KALMAR SOUND (arm of BALTIC SEA); 57°16′N 16°27′E. Manufacturing (transportation equipment, electrical equipment). Nuclear plant in Simpevarp. Chartered 1856.

Oskarström (OS-kahr-STRUHM), town, HALLAND county, SW SWEDEN, on NISSAN RIVER, 10 mi/16 km NNE of HALMSTAD; 56°48′N 12°58′E.

Öskemen, KAZAKHSTAN: see UST-KAMENOGORSK.

Oskol River (uhs-KOL), Ukrainian *Oskil*, 285 mi/459 km long, in SW European Russia and NE UKRAINE; rises ESE of KURSK (Kursk oblast), in CENTRAL RUSSIAN UPLAND; flows S through BELGOROD oblast, past

STARY OSKOL, NOVYY OSKOL, Valyuki, and KU-P'YANS'K (Ukraine), to the NORTHERN DONETS River E of IZYUM. High right bank; forms the CHERVO-NOOSKIL RESERVOIR below Kupyans'k to the E of Izyum. Timber floating in spring.

Osku, IRAN: see USKU.

Oslavany (OS-lah-VAH-ni), German *oslawan*, town, JIHOCESKY province, S MORAVIA, CZECH REPUBLIC, 13 mi/21 km WSW of BRNO; 49°08′N 16°20′E. Railroad terminus. Manufacturing (machinery); power plant. Coal mining in vicinity. Renaissance castle. Paleolithic-era archaeological site.

Oslawan, CZECH REPUBLIC: see OSLAVANY.

Osler (OS-luhr, OZ-), town (2006 population 926), central SASKATCHEWAN, Canada, 18 mi/29 km NNE of SASKATOON; 52°22′N 106°32′W. Mixed farming, dairying.

Oslo (AWS-law), city (2007 estimated population 548,617), ⊙ NORWAY, and of Oslo county (□ 175 sq mi/455 sq km; 2007 population 548,617), also ⊙ AKERSHUS county, which physically surrounds Oslo on three sides, SE Norway, at the head of the OSLOFJORDEN (a deep inlet of the SKAGERRAK); 59°56′N 10°45′E. Oslo is Norway's largest city, its main port, and its chief commercial, industrial, communication, and transportation center. Manufacturing (food, textiles, forest products, machines). It has a significant electrotechnical, graphics, and printing industry. Founded c.1050 by Harald III, Oslo became (1299) the national capital. In the 14th century it came under the dominance of the Hanseatic League. After a great fire (1624), the city was rebuilt by Christian IV and was renamed Christiania (or Kristiania); in 1925 the name Oslo again became official. The city's modern growth dates from the late 19th century, when it also replaced BERGEN as the main city in Norway. In World War II, Oslo fell (April 9, 1940) to the Germans, and it was occupied until the surrender (May 8, 1945) of the German forces in Norway. The neighboring industrial commune of Aker was incorporated into Oslo in 1948. Today, Oslo is a modern city in design and construction, and its government has fostered contemporary art in a number of impressive public projects. Among these are the 150 sculptural groups by Gustav Vigeland in the famous Frogner Park. Oslo's major suburbs are Bærum and Asker to the W, and Oppegård and Lørenskog to the S and SE. The city's chief public buildings include the royal palace (1848), the Storting (parliament), and the city hall (1950), which was decorated by many Norwegian artists. Surviving medieval structures include the Akers kirke (12th century) and the Akershus fortress (13th century), and there are ruins of the Cathedral of St. Hallvard, the first cathedral of Oslo. The University of Oslo (founded 1811), the national theater (1899), the national gallery, the Nobel Institute, and a college of architecture are among the city's cultural institutions. In addition, the Folk Museum has reconstructions of old Norwegian timber houses and of a 12th-century stave church, and the Kon-Tiki Museum has mementos of Thor Heyerdahl's trip (1947) across the Pacific Ocean. The forested hills surrounding Oslo are popular excursion points; the annual Holmenkollen ski meet nearby attracts an international group of skiers. The 1952 winter Olympic games were held at Oslo. GARDERMOEN international airport at GARDERMOEN, 29 mi/47 km to the N, opened in 1998 (rail link).

Oslo (AWS-lo), village (2000 population 347), MARSHALL county, NW MINNESOTA, 20 mi/32 km N of GRAND FORKS, NORTH DAKOTA, on RED RIVER (North Dakota state line); 48°11′N 97°07′W. Agriculture (grain, potatoes, beans, sunflowers, sugar beets; livestock, poultry). Manufacturing (fertilizer).

Oslob (os-LOB), town, CEBU province, S Cebu island, PHILIPPINES, on BOHOL STRAIT, 19 mi/31 km E of TANJAY; 09°31′N 123°24′E. Agricultural center (corn, coconuts).

Oslofjorden or **Olsofjord**, SE Norway, c.62 mi/100 km long, deep arm of the Skagerrak strait, with OSLO at its N end. Site of battles during the World War II invasion of Norway by German troops. Formerly known as Kristianiafjord.

Osma (O-zmah), town, SORIA province, N central SPAIN, in CASTILE-LEÓN, near the DUERO RIVER, 35 mi/56 km WSW of SORIA, 1 mi/1.6 km W of the cathedral town El Burgo de Osma. In agricultural region; sugar refining, flour milling. Was ancient Celtiberian settlement *Uxama*. Heavily disputed during Moorish wars. Adjoined by ruins and Mudejar watchtower.

Osma Gradishte, Bulgaria: see GRADISHTE.

Osma Kalugerovo, Bulgaria: see ASENOVTSI.

Osmanabad (OS-mah-nah-BAHD), district (□ 2,922 sq mi/7,597.2 sq km), MAHARASHTRA state, W central INDIA, on DECCAN PLATEAU; ⊙ OSMANABAD. Bordered N by MANJRA RIVER; highland (NE); lowland (SW). In black-soil area; millet, wheat, cotton, oilseeds (chiefly peanuts). Cotton ginning, flour and oilseed milling, road-metal quarrying. Main trade center is Osmanabad. Became part of HYDERABAD during state's formation in 18th century. Called Naldrug for its former capital until c. 1900. Sometimes spelled Usmanabad.

Osmanabad (OS-mah-nah-BAHD), town, ⊙ OSMANABAD district, MAHARASHTRA state, W central INDIA, 35 mi/56 km N of SOLAPUR; 18°10′N 76°03′E. Trade center for grain (chiefly millet, wheat, rice), cotton, oilseeds (chiefly peanuts). Ancient Jain and Vishnuite caves nearby. Sometimes spelled Usmanabad; formerly called Dharaseo.

Osmancik, township, N central TURKEY, on the W bank of the KIZIL IRMAK, and 30 mi/48 km N of ÇORUM; 40°58′N 34°50′E. Grain, cotton, mohair goats. Sometimes spelled Osmanjik.

Osmaneli, village, NW TURKEY, near W bank of the SAKARYA RIVER, on IZMIT-BILECIK railroad, and 15 mi/24 km N of Bilecik; 40°22′N 30°01′E. Cereals.

Osmaniye, town (2000 population 173,977), S TURKEY, near CEYHAN RIVER, on railroad, 50 mi/80 km E of ADANA; 37°04′N 36°15′E. Wheat, cotton. Formerly Cebelibereket (Jebel-Bereket).

Osmanjik, TURKEY: see OSMANCIK.

Osmannagar, INDIA: see SULTANABAD.

Osman Pazar, Bulgaria: see OMURTAG.

Osma River, Bulgaria: see OSUM RIVER.

Osmond (AWZ-muhnd), village (2006 population 738), PIERCE county, NE NEBRASKA, 10 mi/16 km NNW of PIERCE, and on branch of ELKHORN RIVER; 42°21′N 97°35′W. In grain and livestock area; dairy and poultry products. Manufacturing (agricultural equipment, irrigation systems).

Osmore River, PERU: see MOQUEGUA RIVER.

Osnabrock (AHZ-nuh-brahk), village (2006 population 154), CAVALIER county, NE NORTH DAKOTA, 12 mi/19 km SE of LANGDON; 48°40′N 98°09′W. Founded in 1883 and incorporated in 1903. Named for Osnabrock, CANADA.

Osnabrück (os-nah-BROOIK), city, Lower SAXONY, NW GERMANY, on the HASE RIVER, 27 mi/43 km NW of BIELEFELD, linked by canal with the MIDLAND CANAL; 52°16′N 08°03′E. Inland port, railroad junction, and industrial center, with iron and steel mills, machinery plants, and factories manufacturing textiles, paper, and machinery. Located on the site of an ancient Saxon settlement, Osnabrück was made (780) an episcopal see by Charlemagne. The city became a member of the Hanseatic League and a center of the linen trade. It accepted the Reformation in 1543; however, the cathedral remained Catholic, and under the Peace of Westphalia—one of whose treaties was signed (Aug. 1648) in the Osnabrück city hall—the see was occupied alternately by Catholic and Lutheran bishops. The bishopric of Osnabrück was secularized in 1803, and the city passed (1815) to HANOVER at the Congress of Vienna. Noteworthy buildings include

the three-towered cathedral (begun 11th century; burned down 1254; rebuilt in Romanesque style with Gothic additions); the Gothic Church of St. Mary (c.1300); the city hall (1487–1512); and a baroque castle (1667–1690). Osnabrück also contains a teachers college (housed in a 17th-century palace), university founded 1970, and a museum. Airport serves Münster and Osnabrück.

Osnaburgh, Scotland: see DAIRSIE.

Osnaburgh House (AHZ-nuh-buhrg), unincorporated village, NW ONTARIO, E central Canada, in PATRICIA district, on Lake SAINT JOSEPH, 23 mi/37 km SSW of PICKLE LAKE, 100 mi/161 km NE of SIOUX LOOKOUT, and included in unorganized part of Kenora; 51°08′N 90°16′W. Gold mining. Nearby is hydroelectric station.

Osno (OS-no), Polish *Ośno Lubuskie*, German *Drossen*, town, BRANDENBURG, Gorzów province, W POLAND, 16 mi/26 km ENE of FRANKFURT. In lignite-mining region; vegetable and flower market. Has 13th–15th century churches, remains of 15th century town walls. Founded c.1150 by bishops of Lebus.

Osny (os-NEE), residential town (□ 5 sq mi/13 sq km), VAL-D'OISE department, ÎLE-DE-FRANCE region, N central FRANCE, 2 mi/3.2 km NW of PONTOISE; 49°04′N 02°04′E.

Osoblaha (O-so-BLAH-hah), German *hotzenplotz*, village, SEVEROMORAVSKY province, N SILESIA, CZECH REPUBLIC, 13 mi/21 km N of KRNOV, near Polish border; 50°17′N 17°43′E. Railroad terminus. Produces smoked meats. Has a 17th century Jewish cemetery.

Osogov Mountains (o-so-GOV), Bulgarian and Serbo-Croatian *Osogovska Planina*, on the Macedonian-Bulgarian border; form a divide between the VARDAR (W) and STRUMA (E) rivers and between the KRIVA REKA (N) and BREGALNICA (S) rivers; 42°10′N 22°30′E. Rise to c.7,390 ft/2,252 m at Rujen peak. Crossed by the VELBUZHDKI PASS at GYUESHEVO, BULGARIA. Scattered copper, lead, zinc, silver, gold, and iron deposits (mined at KRATOVO and ZLETOVO, both in MACEDONIA). Heavily forested. Also called Osogovo Mountains.

Oso, Mount, peak (c.12,925 ft/3,940 m), in ROCKY MOUNTAINS, LA PLATA county, SW COLORADO, 25 mi/40 km NE of DURANGO.

Osoppo (o-ZOP-po), village, UDINE province, FRIULI–VENEZIA GIULIA, NE ITALY, on TAGLIAMENTO River, and 15 mi/24 km NNW of UDINE; 46°15′N 13°04′E. Manufacturing (transportation equipment, furniture, soap). Has fort (damaged in World War II), in which Friulians resisted Austrians for six months in 1848.

Osório (O-so-ree-o), city, E RIO GRANDE DO SUL, BRAZIL, near the Atlantic, amid coastal lagoons, 60 mi/97 km ENE of PÔRTO ALEGRE; 29°54′S 50°16′W. Hydroelectric plant nearby. Brandy distilling, fish drying, livestock slaughtering, wool processing. Airfield. Formerly called Conceição do Arroio.

Osorno (o-SOR-no), N province (□ 3,507 sq mi/9,118.2 sq km) of LOS LAGOS region, S central CHILE; ⊙ OSORNO; 40°40′S 73°05′W. Situated between the ANDES and the PACIFIC, it includes part of the Chilean lake district, notably lakes RUPANCO and PUYEHUE and N tip of Lake LLANQUIHUE. The volcanoes OSORNO and PUNTIAGUDO are in SE. Has a temperate, wet climate. Contains an agricultural valley (of which Osorno city is the center), engaging in wheat growing and cattle and sheep raising. Also rich in timber. Freshwater fishing in Andean lakes.

Osorno (o-SOR-no), city (2002 population 132,245), ⊙ Osorno comuna and province, LOS LAGOS, S central CHILE, in the heart of the lake district, on railroad, and PAN-AMERICAN HIGHWAY, at the confluence of Rahue and Damas rivers; 40°34′S 73°09′W. Osorno is chiefly an agricultural processing and distributing center. Founded in 1558, it was later destroyed by the Araucanian and was re-established in 1796 by order of Ambrosio O'Higgins. An influx of immigrants in the latter half of the 19th century has given Osorno the atmosphere of a German town. It has an active tourist trade, based on such nearby attractions as OSORNO VOLCANO, the lake district, and hot springs. Airport.

Osorno, town, PALENCIA province, N central SPAIN, 30 mi/48 km NNE of PALENCIA; 42°24′N 04°22′W. Cereals, wine, vegetables.

Osorno Volcano (o-SOR-no), Andean peak (8,725 ft/2,659 m), on OSORNO-Llanquihue province border, LOS LAGOS region, S central CHILE, between LAKE LLANQUIHUE (W) and LAKE TODOS LOS SANTOS (E), in Chilean lake district; 41°07′S 72°30′W. Winter sports. Thermal baths nearby. Known for its symmetrical cone.

Osov, CZECH REPUBLIC: see HOSTOMICE.

Osoyoos (O-soo-yoos), town (□ 4 sq mi/10.4 sq km; 2001 population 4,295), S BRITISH COLUMBIA, W Canada, near U.S. (WASHINGTON) border, on Osoyoos Lake (12 mi/19 km long), 30 mi/48 km S of PENTICTON, in Okanagan-Similkameen regional district; 49°02′N 119°28′W. Semi-arid climate. In irrigated farming region (fruit, vegetables); tourism. Incorporated 1946.

Ospedaletti (os-pe-dah-LET-tee), village, IMPERIA province, LIGURIA, NW ITALY, port on GULF OF GENOA, and 3 mi/5 km WSW of SAN REMO; 43°48′N 07°43′E. In flower-growing region; flower market; fisheries. Winter resort.

Ospina (os-PEE-nah), town, ⊙ Ospina municipio, NARIÑO department, SW COLOMBIA, 20 mi/32 km WSW of PASTO, in the ANDES; 01°03′N 77°34′W. Elevation 9,248 ft/2,818 m. Wheat, sugarcane, vegetables; livestock.

Ospina Pérez (os-PEE-nah PAI-res), town, ⊙ Ospina Pérez municipio, CUNDINAMARCA department, central COLOMBIA, in valley of Sumapaz River, 36 mi/58 km SSW of BOGOTÁ; 04°05′N 74°28′W. Elevation 6,190 ft/1,886 m. Sugarcane, coffee; livestock.

Ospino (os-PEE-no), town, ⊙ Ospino district, PORTUGUESA state, W VENEZUELA, in llanos, 27 mi/43 km NE of GUANARE; 09°18′N 69°27′W. Agricultural region with extensive land dedicated to cotton, sorghum, corn, rice, sugarcane; livestock.

Ospitaletto (os-pee-tah-LET-to), town, BRESCIA province, LOMBARDY, N ITALY, 7 mi/11 km WNW of BRESCIA. Manufacturing (hosiery, fabricated metals, machinery, wood products).

Osprey (AWS-pree), former township (□ 114 sq mi/296.4 sq km; 2001 population 2,466), S ONTARIO, E central Canada, 34 mi/54 km from OWEN SOUND; 44°18′N 80°20′W. Amalgamated into Grey Highlands in 2001.

Osprey (AHS-spree), unincorporated town (□ 6 sq mi/15.6 sq km; 2000 population 4,143), SARASOTA county, W central FLORIDA, 10 mi/16 km S of SARASOTA; 27°11′N 82°29′W. Plastic products, light manufacturing.

Ospringe, unincorporated village, S ONTARIO, E central Canada, 12 mi/20 km from GUELPH, and included in town of Erin; 43°41′N 80°08′W.

Osroene (ahs-ruh-WEE-nee), ancient kingdom of NW MESOPOTAMIA, in present-day SE TURKEY and NE SYRIA. EDESSA (now SANLIURFA) was capital. It broke away (2nd century B.C.E.) from the Seleucid empire and formed a separate kingdom. It came under Roman rule late in the 2nd century C.E.

Oss (AWS), city, NORTH BRABANT province, S NETHERLANDS, 12 mi/19 km ENE of ʼs-HERTOGENBOSCH; 51°45′N 05°32′E. MEUSE RIVER (car ferry, bridge) 4 mi/6.4 km to N; recreational center to W. Agriculture (dairying; cattle, hogs, poultry; vegetables, grain, fruit). Manufacturing (machinery, hardware, pharmaceuticals, carpeting); food processing. Chartered 1399.

Ossa (O-sah), peak (c.6,490 ft/1,980 m), NE THESSALY department, N GREECE, NE of LÁRISSA; 39°49′N 22°40′E. According to legend, the Aloadae piled Mount PELION on Ossa when they stormed Mount OLYMPOS.

Ossabaw Island (AH-suh-baw), one of the SEA ISLANDS (c.9 mi/14.5 km long, 7 mi/11.3 km wide), just off SE GEORGIA coast, in CHATHAM county, 15 mi/24 km S of SAVANNAH; 31°48′N 81°05′W. Marshy. OSSABAW Sound (c.5 mi/8 km long and wide), at N end of island, receives Ogeechee River. Extensive agriculture in nineteenth-century, much of which has reverted to a natural state. Used by Creek Indians as hunting ground. Now administered by the Georgia Department of Natural Resources.

Ossa de Montiel (O-sah dhai mon-TYEL), town, ALBACETE province, SE central SPAIN, 22 mi/35 km SSW of VILLARROBLEDO; 38°58′N 02°45′W. Cereals, wine, honey. Gypsum quarries. One of the Guadiana's headstreams originates from chain of lagoons SW of here.

Ossa, Serra de (OS-sah, SER-rah dai), hills in S central PORTUGAL, in NE ÉVORA district, between ÉVORA and ELVAS; 38°43′N 07°36′W. Rise to 2,150 ft/655 m. Iron deposits on N slopes.

Ossau, Gave d' (do-so, gahv), river, 40 mi/64 km long, in PYRÉNÉES-ATLANTIQUES department, AQUITAINE region, SW FRANCE; rises in the PYRENEES NATIONAL PARK near POURTALET pass (Spanish border), flows NW, through picturesque Ossau valley (resort area), past GABAS, Les EAUX-CHAUDES, LARUNS, and ARUDY, joining the Gave d'aspe at OLORON-SAINTE-MARIE to form the Gave d'OLORON; 43°12′N 00°36′W. Used for hydroelectric power.

Ossau, Pic du Midi d', FRANCE: see PIC DU MIDI D'OSSAU.

Osseg, CZECH REPUBLIC: see OSEK.

Osseo (AH-see-yo), town (2000 population 2,434), HENNEPIN county, E MINNESOTA, industrial suburb 12 mi/19 km NNW of downtown MINNEAPOLIS, near MISSISSIPPI RIVER; 45°07′N 93°24′W. Manufacturing (diversified manufacturing, printing and publishing, machining). Elm Creek Park Reserve to NW.

Osseo (AH-see-o), town (2006 population 1,641), TREMPEALEAU county, W WISCONSIN, on BUFFALO RIVER, and 21 mi/34 km SE of EAU CLAIRE; 44°34′N 91°13′W. In dairy, livestock, and poultry area; cheese. Manufacturing (paint spray bottles, furniture). On Buffalo River State Trail. Settled 1851; incorporated as village in 1893, as city in 1941.

Ossetia (o-SE-tee-yuh), region of the central Caucasus Mountains, divided between the Republic of GEORGIA and the RUSSIAN FEDERATION. On the N slope is the NORTH OSSETIAN REPUBLIC, in SE European RUSSIA. This region extends N beyond the TEREK RIVER. On the S slope is the South Ossetian Autonomous Region (□ 1,500 sq mi/3,885 sq km; 1990 est. population 100,000), in GEORGIA; ⊙ TSKHINVALI. The region extends S almost to the KURA River. Both sections of Ossetia have valleys that produce fruit, wine, grain, and cotton. Lumbering and livestock raising are important in the mountains. The Ossetians, a Farsi-speaking people, are Eastern Orthodox Christians in the S, where Georgian culture prevails. They are descendents from medieval Alan. During the 17th century the North Ossetians were subjects of Karbada princelings. From the 18th century they came under strong Russian influence, and between 1801 and 1806 all of Ossetian territory was annexed to Russia. Ossetian artwork includes wood, stone, and silver carving. In March 1918, the entire area was declared an Autonomous Soviet Republic, and in January 1920 was renamed the Mountain Autonomous Republic. For further history of North Ossetia, see NORTH OSSETIAN REPUBLIC. In 1922, South Ossetia was made part of Georgia. It lost its autonomous-region status by an act of the Georgian Supreme Soviet in 1990. Following Georgia's independence from the USSR, South Ossetian nationalists demanded either independence from Georgia or incorporation into the North Ossetian Republic. In April 1992, the South Ossetian Autonomous Region was reestablished in Georgia.

Ossetian Military Road (uh-SYE-tee-yahn), highway, approximately 170 mi/274 km long, across the CAUCASUS Mountains, S European Russia and GEORGIA, linking KUTAISI (GEORGIA) with ALAGIR (NORTH OSSETIAN REPUBLIC, RUSSIA). One of the two main routes over the Great Caucusus, it crosses the Caucasian crest through the pass at MAMISON. Construction of the road began in 1888.

Ossett (AH-sit), town (2001 population 15,788), WEST YORKSHIRE, N ENGLAND, 8 mi/13 km S of LEEDS; 53°40′N 01°34′W. Manufacturing. Former coal-mining site.

Ossiach (AWS-si-ahkh), village, CARINTHIA, S AUSTRIA on SE shore of OSSIACHER SEE (lake), 8 mi/12.9 km NE of VILLACH; 46°41′N 13°59′E. Elevation 1,548 ft/472 m. Former Benedictine abbey (founded before 1028) with baroquised church, grave of the Polish king Boleslaw II. Site of a music festival (Carinthian Summer).

Ossiacher See (AWS-si-ah-kher SAI), lake (□ 4 sq mi/10.4 sq km), in CARINTHIA, S AUSTRIA, 3 mi/4.8 km NE of VILLACH; 7 mi/11.3 km long, 1 mi/1.6 km wide, maximum depth of 158 ft/48 m; 46°40′N 13°57′E. Site of tourist resorts (Ossiach, Steindorf, Annenheim, Sattendorf, Bodensdorf). Cable cars from Annenheim to Kanzelhöhe at an elevation of 4,645 ft/1,416 m.

Ossian (OSH-en), town (2000 population 2,943), WELLS county, E INDIANA, on small Longlois Creek, and 10 mi/16 km N of BLUFFTON; 40°53′N 85°10′W. Agricultural area (soybeans, corn). Manufacturing (food products, transportation equipment); meat processing. Laid out 1850.

Ossian, town (2000 population 853), WINNESHIEK county, NE IOWA, 11 mi/18 km S of DECORAH; 43°08′N 91°45′W. In grain and dairy area. Limestone quarries nearby.

Ossineke (ah-SIN-e-kee), village (2000 population 1,059), ALPENA county, NE MICHIGAN, on Thunder Bay of LAKE HURON, 10 mi/16 km S of ALPENA; 44°54′N 83°25′W. Michigan Islands National Wildlife Refuge to ESE; Negwegon State Park to SE (ALCONA county).

Ossining, village (□ 6 sq mi/15.6 sq km; 2006 population 23,578), WESTCHESTER county, SE NEW YORK, on the HUDSON RIVER; 41°09′N 73°52′W. Mainly residential; some manufacturing (medical instruments, pharmaceuticals). Ossining is the site of Sing Sing state prison (built 1825–1828). This prison was long known for its extreme discipline, but under Thomas Mott Osborne and Lewis Edward Lawes, notable reforms were introduced. By end of the 19th century, Ossining was the second-largest industrial center in Westchester. Brickyards produced bricks for OLD CROTON Aqueduct. Aqueduct Bridge in Ossining, one of most impressive parts of the aqueduct, is now part of the Osinning Urban Cultural Park. MARYKNOLL, the headquarters of the Catholic Foreign Mission Society, is nearby. Settled c.1750, incorporated 1813 as Sing Sing, renamed 1901.

Ossipee (AW-si-pee), town (2000 population 4,211), ⊙ CARROLL county, E NEW HAMPSHIRE, 28 mi/45 km N of ROCHESTER; 43°42′N 71°08′W. Drained by Ossipee, Pine, Chocorua, and Lovell rivers. Railroad terminus at West Ossipee. Manufacturing (plastic products, printing and publishing, asphalt, lumber); timber; major source of sand and gravel for BOSTON. Agriculture (nursery crops, vegetables; livestock, poultry; dairying). Covered bridge at village of West Ossipee, in NW (summer home of poet John Greenleaf Whittier). Resort area. OSSIPEE LAKE in NE; Ossipee Lake and Heath Pond natural areas at S end of lake. Deer Cap Ski Area in E. Mount Shaw (2,975 ft/907 m) on SW boundary, in Ossipee Mountains. Village of Center Ossipee in N. Incorporated 1785.

Ossipee (AH-suh-pee), unincorporated village (□ 2 sq mi/5.2 sq km; 2006 population 514), ALAMANCE county, N central NORTH CAROLINA, near HAW RIVER, 7 mi/11.3 km NNW of BURLINGTON; 36°10′N 79°30′W.

Manufacturing; agriculture (tobacco, grain; chicken). Reported as ALTAMAHAW-OSSIPEE by Census Bureau (2000 population 996).

Ossipee Lake (AW-si-pee), 3.5 mi/5.6 km long, 2 mi/3.2 km wide, E NEW HAMPSHIRE, fed by Pine River (S), Lovell River (W), and Bearcamp River (NW); 43°07′N 71°08′W. Has 3-mi/4.8-km long E arm (Broad Bay), from which OSSIPEE RIVER drains into MAINE. Ossipee Lake and Heath Bog natural areas on S end.

Ossipee Mountain (AHS-uh-pee) (1,058 ft/322 m), YORK county, SW MAINE, 7 mi/11.3 km N of ALFRED. In WATERBORO resort area.

Ossipee River (AW-si-pee), c.20 mi/32 km long, E NEW HAMPSHIRE; rises in OSSIPEE LAKE, E central CARROLL county; flows E through Broad Bay, past village of Effingham Falls, into SW MAINE, past Kezar Falls (dam) to SACO RIVER, 15 mi/24 km WNW of PORTLAND, Maine; 43°48′N 71°09′W.

Ossora (uhs-SO-rah), town (2005 population 2,365), S KORYAK AUTONOMOUS OKRUG, RUSSIAN FAR EAST, on the neck of the E coast of KAMCHATKA PENINSULA, on the Ossora Bay of the BERING SEA, 110 mi/177 km E of PALANA; 59°14′N 163°04′E. Fish canning. Administrative offices for the okrug's department of forestry services.

Ossory (AH-suhr-ree), ancient kingdom of IRELAND, the borders of which are now largely traced by those of the Catholic episcopal see of Ossory, including KILKENNY county and parts of OFFALY and LAOIGHIS counties. An independent state on the borders of LEINSTER and MUNSTER provinces, its overlordship was long disputed. It became part of Leinster under the Normans in the 12th century, and became part of the earldom of Ormonde, held by the Butler family, by the middle of the 14th century.

Ossu (AW-soo), town, EAST TIMOR, in E TIMOR ISLAND, 55 mi/89 km ESE of DILI; 08°44′S 126°22′E. Palm oil, copra.

Ossun (o-suhn), commune (□ 10 sq mi/26 sq km), HAUTES-PYRÉNÉES department, MIDI-PYRÉNÉES region, SW FRANCE, 6 mi/9.7 km SW of TARBES; 43°11′N 00°02′W. Airport, aircraft industry. Tarbes-Ossun-Lardes international airport.

Ostaf′yevo (uhs-TAH-feeye-vuh), village, central MOSCOW oblast, central European Russia, on railroad spur, 17 mi/27 km S of MOSCOW; 55°30′N 37°31′E. Elevation 534 ft/162 m. Agriculture (grain, vegetables).

Östansjö (UHS-tahn-SHUH), village, ÖREBRO county, S central SWEDEN, 17 mi/27 km SSW of ÖREBRO; 59°03′N 14°58′E.

Ostashevo (uh-stah-SHO-vuh), village, W MOSCOW oblast, central European Russia, on the Ruza River (left tributary of the MOSKVA River), on highway, 12 mi/19 km S of VOLOKOLAMSK; 55°51′N 35°52′E. Elevation 574 ft/174 m. Hog farming. Also spelled Ostashëvo or Ostashovo (same pronounciation).

Ostashkov (uhs-TAHSH-kuhf), city (2006 population 19,205), W TVER oblast, central European Russia, in the VALDAY HILLS, on Lake SELIGER (fisheries), on highway and railroad, 118 mi/190 km WNW of TVER; 57°09′N 33°06′E. Elevation 708 ft/215 m. Leather (from 1730), apparel, food industries. Known since the 16th century; chartered in 1770.

Ostbevern (ost-BE-vern), suburb of Münster, WESTPHALIA-LIPPE, North RHINE—Westphalia, W GERMANY, 11 mi/18 km ENE of city center, in the Münsterland; 52°03′N 07°50′E.

Ostenburg, POLAND: see PULTUSK.

Ostend-Bruges Canal (OS-ten-duh—BROO-zhuh), Flemish *Groot Zwin*, 16 mi/26 km long, NW BELGIUM, runs E-W, between BRUGES and NORTH SEA at OSTENDE.

Ostende, BELGIUM: see OOSTENDE.

Österås (UHS-ter-OS), residential village, VÄSTERNORRLAND county, NE SWEDEN, on ÅNGERMANÄLV RIVER, at mouth of Faxälven River, 4 mi/6.4 km NW of SOLLEFTEÅ; 63°13′N 17°10′E.

Osterburg/Altmark (OS-ter-burg/AHLT-mahrk), town, SAXONY-ANHALT, central GERMANY, 13 mi/21 km NNW of STENDAL; 52°47′N 11°45′E. Manufacturing (lenses, eyeglass frames, bricks). Has church rebuilt in 13th century.

Osterburken (OS-ter-bur-ken), town, Lower NECKAR, BADEN-Württemberg, SW GERMANY, 16 mi/26 km WSW of Mergentheim; 49°26′N 09°26′E. Railroad junction. Agriculture (fruit, wheat). Was Roman castrum.

Österbybruk (UHS-ter-BEE-BROOK), town, UPPSALA county, E SWEDEN, 20 mi/32 km NNE of UPPSALA; 60°12′N 17°54′E. Former ironworks founded seventeenth century by Dutch industrialists who employed workers originally brought here from LOW COUNTRIES. Museum.

Ostercappeln (os-ter-KAHP-peln), village, LOWER SAXONY, NW GERMANY, 9 mi/14.5 km NE of Osnabrück, in the WIEHEN MOUNTAINS; 52°21′N 08°14′E. Tourism.

Østerdalen (UHS-tuhr-dahl-uhn), valley (100 mi/161 km long) of GLOMMA River, HEDMARK county, E NORWAY, extends from E slope of the DOVREFJELL SSE to ELVERUM. FOLLDAL is a tributary valley (NW). Lumbering, agriculture, livestock raising. Valley is traversed by secondary railroad linking OSLO and TRONDHEIM.

Osterdock, town (2000 population 50), CLAYTON county, NE IOWA, on TURKEY RIVER, and 15 mi/24 km ESE of ELKADER; 42°43′N 91°09′W. In dairying area.

Osterfeld (OS-ter-felt), industrial district of OBERHAUSEN, North RHINE—Westphalia, W GERMANY, N of RHINE-HERNE CANAL, 2 mi/3.2 km E of city center; 51°28′N 06°53′E.

Osterfjorden (AWS-tuhr-FYAWR-uhn), inlet of the NORTH SEA in HORDALAND county, SW NORWAY; c.20 mi/32 km long, 1 mi/1.6 km wide. Extends NE from coast, separating the island OSTERØY from mainland; joins with SØRFJORDEN at head, 25 mi/40 km NNE of BERGEN.

Östergötland (UHS-ter-YUHT-lahnd), county (□ 4,266 sq mi/11,091.6 sq km), SE SWEDEN; ⊙ LINKÖPING; 58°25′N 15°45′E. Comprises Östergötland province (□ 4,239 sq mi/10,979 sq km; 1995 population 412,710) and small section of SÖDERMANLAND province, between BALTIC SEA and LAKE VÄTTERN. Fertile plain dotted with lakes (ROXEN, Glan, Sommen); drained by MOTALA STRÖM, Svartån, and STÅNGÅN rivers; partly wooded. Manufacturing (aircraft, motor vehicles, electronics, pulp and paper, processed food); agriculture (grain, oil plants, fodder crops; livestock; dairying). Important towns are NORRKÖPING, LINKÖPING, MOTALA, MJÖLBY (railroad center), VADSTENA, SÖDERKÖPING, and SKÄNNINGE.

Osterhofen (OS-ter-HOH-fuhn), town, Lower BAVARIA, BAVARIA, GERMANY, near the DANUBE River, 9 mi/14.5 km SE of PLATTLING; 48°42′N 13°00′E. Manufacturing (textiles).

Osterholz-Scharmbeck (OS-ter-holts-SHAHRM-bek), town, LOWER SAXONY, GERMANY, 9 mi/14.5 km N of BREMEN; 53°13′N 08°48′E. Manufacturing (tobacco products, machinery, wood products); food processing. Romanesque 12th century church.

Oste River (OS-te), c.90 mi/145 km long, NW GERMANY; formed 7 mi/11.3 km SW of ZEVEN, flows N, past Bremervörde, to Elbe River estuary 3 mi/4.8 km N of Neuhaus. Navigable in lower course.

Ostermundigen (ahs-tur-MOON-di-gen), commune (2000 population 15,452), BERN canton, E suburb of BERN, W SWITZERLAND, 3 mi/4.8 km NE of BERN; 46°48′N 07°30′E. Some industries created from Bolligen commune in 1983.

Östermyra, FINLAND: see SEINÄJOKI.

Osterode, POLAND: see OSTRODA.

Osterode am Harz (os-tuh-RO-de ahm HAHRTS), town, LOWER SAXONY, W GERMANY, at W foot of the HARZ, 19 mi/31 km NE of Göttingen; 51°44′N 10°15′E.

Manufacturing (textiles, machinery); food processing. Gypsum quarries nearby. Tourism. Has 16th century church and town hall. Was first mentioned 1136; chartered 1293.

Osterøy (AWS-tuhr-uh-u), island (□ 127 sq mi/330.2 sq km), HORDALAND county, SW NORWAY, NE of BERGEN; 19 mi/31 km long (N-S), 12 mi/19 km wide. Separated from mainland by very narrow SØRFJORDEN on SW, S, and E, and by OSTERFJORDEN on NW.

Österreich: see AUSTRIA.

Oster River (os-TER), 120 mi/193 km long, in CHERNIHIV oblast, UKRAINE; rises 12 mi/19 km SSW of BAKHMACH; flows generally WSW, past NIZHYN and KOZELETS′, to the DESNA River at OSTER.

Osterrönfeld (os-ter-RAWN-felt), village, SCHLESWIG-HOLSTEIN, N GERMANY, just ESE of RENDSBURG, on the NORD-OSTSEE Canal; 54°17′N 09°42′E.

Östersund (UHS-ter-SUND), town, ⊙ JÄMTLAND county, central SWEDEN, on LAKE STORSJÖN; 63°11′N 14°38′E. Commercial, industrial, agricultural, tourist, and transportation center. Airport. Founded 1786.

Osterville, MASSACHUSETTS: see BARNSTABLE, town.

Osterwieck (OS-ter-veek), town, SAXONY-ANHALT, central GERMANY, at N foot of the upper HARZ, 10 mi/16 km NNW of WERNIGERODE; 51°59′N 10°40′E. Has 16th century church and town hall; many half-timbered houses.

Ostfildern (ost-FIL-dern), town, region of STUTTGART, BADEN-Württemberg, S GERMANY, 5 mi/8 km SE of STUTTGART, near the FILS; 48°43′N 09°14′E. Manufacturing (machinery, textiles, precision mechanics, electronic equipment). Church with Romanesque tower (1120); castle (1777). International airport is 2 mi/3.2 km SW. Formed 1975 by several villages; chartered 1976.

Østfold (UHST-fawl), county (□ 1,615 sq mi/4,183 sq km; 2007 estimated population 262,543), SE NORWAY, between the OSLOFJORD in the W and the SWEDEN border in the E; ⊙ SARPSBORG. Grain, vegetables; dairying. The area around the lower GLOMMA River is one of the important industrial centers in Norway, producing ships, chemicals, rubber, and processed food. Timber is floated down the Glomma from the ØSTERDALEN valley for processing. The region has been important in the cultural and military history of Norway and includes numerous prehistoric remains.

Ostfriesland, GERMANY: see EAST FRIESLAND.

Ostgeim, UKRAINE: see TEL′MANOVE.

Östhammar (UHST-hahm-mahr), town, UPPSALA county, E SWEDEN, on Granfjärden, 12-mi/19-km-long inlet of GULF OF BOTHNIA, 35 mi/56 km NE of UPPSALA; 60°15′N 18°22′E. Port; seaside resort. Has eighteenth-century town hall; museum. Chartered fourteenth century; destroyed by Russians (1719).

Ostheim, UKRAINE: see TEL′MANOVE.

Ostheim vor der Rhön (OST-heim for der ROEN), town, LOWER FRANCONIA, N BAVARIA, GERMANY, 11 mi/18 km SW of MEININGEN; 50°28′N 10°14′E. Manufacturing of organs; cherries; brewery. Was in Thuringian exclave between 1920 and 1945. Church and castle from 15th century. Was first mentioned 840; chartered 1586.

Osthofen (OST-hoh-fuhn), town, RHINELAND-PALATINATE, W GERMANY, 5 mi/8 km NNW of WORMS; 49°42′N 08°19′E. Manufacturing of malt, food processing. Vineyards. Town since 1970.

Ostia (OS-tee-uh), ancient city of ITALY, at the mouth of the TIBER. Founded (4th century B.C.E.) as a protection for Rome, then developed (from the 1st century B.C.E.) as a Roman port, rivaling PUTEOLI. Augustus, Claudius I, Trajan, and Hadrian expanded the city and harbor. From the 3rd century C.E. the city began to decline. The ruins, of great archaeological interest, rival those of POMPEII in showing the layout of an ancient Italian city. In the late 1980s it served as a refugee camp for processing Jewish émigrés from the SOVIET UNION. Lido di Ostia is a popular bathing area.

Area is shown by the symbol □, and capital city or county seat by ⊙.

Ostional, NICARAGUA: see OSTIONAL, EL.

Ostional, El (os-tee-o-NAHL), village, RIVAS department, SW NICARAGUA, 22 mi/35 km S of RIVAS, on the PACIFIC. Livestock; coffee.

Ostional Wildlife Refuge, on shore of Guanacaste Peninsula, COSTA RICA, on PACIFIC coast. Small (400 acres/162 ha) reserve protects beach (1 of 4 in the world) to which rare sea turtles return for their annual nesting.

Ostpreussen, GERMANY: see EAST PRUSSIA.

Ostrach (OST-rahkh), village, LAKE CONSTANCE-UPPER SWABIA, BADEN-Württemberg, S GERMANY, 12 mi/19 km SE of SIGMARINGEN, on the OSTRACH; 47°57′N 09°23′E.

Ostrander (AWS-tran-duhr), village (2000 population 212), FILLMORE county, SE MINNESOTA, 28 mi/45 km S of ROCHESTER, near IOWA state line; 43°36′N 92°25′W. Grain, soybeans; livestock, poultry; dairying. Lake Louise State Park to SW.

Ostrander (AHS-tran-duhr), village (2006 population 542), DELAWARE county, central OHIO, 7 mi/11 km WSW of DELAWARE; 40°16′N 83°13′W. Leather products.

Ostrau, CZECH REPUBLIC: see OSTRAVA.

Ostrava (OS-trah-VAH), German *Ostrau*, city (2001 population 316,744), SEVEROMORAVSKY province, on SILESIA-MORAVIA border, N central CZECH REPUBLIC, near the junction of the ODER and OSTRAVICE rivers; 49°50′N 18°17′E. It is the heart of the Ostrava-Karviná industrial and mining region, the most heavily industrialized area of the Czech Republic. Anthracite and bituminous coal, iron and steel, railroad cars, and chemicals are the major products. It is the Czech Republic's third-largest city, a regional administrative center, a road and railroad hub, and the site of several hydroelectric stations. Although a small town in the Middle Ages, it was well known; it later became important because of its strategic location guarding the MORAVIAN GATE, the entrance to the Moravian lowlands. The city's industrial prominence dates from the late 19th century, after the opening of its first coal mine and the coming of the railroad. German forces occupied Ostrava 1939–1945. It is a cultural and educational center, noted especially for its college of mining and metallurgy.

Ostravice River (OS-trah-VI-tse), German *Ostrawitz*, c.40 mi/64 km long, SEVEROMORAVSKY province, E SILESIA, CZECH REPUBLIC; formed in the BESKIDS, by two headstreams joining 22 mi/35 km E of VALAŠSKÉ MEZIŘÍČÍ; flows NNW, past FRYDLANT NAD OSTRAVICI, forming part of the MORAVIA-Silesia border, past FRÝDEK-MISTEK and OSTRAVA, to ODER River at N part of Ostrava. SANCE Dam (Czech ŠANCE) impounds reservoir (830 acres/336 ha) 11 mi/18 km SSE of Frýdek-Místek.

Ostredok (o-STRE-dok), highest mountain (5,223 ft/1,592 m) of the GREATER FATRA; STREDOSLOVENSKY province, W central SLOVAKIA, 16 mi/26 km SW of RUŽOMBEROK; 48°54′N 19°05′E.

Ostrhauderfehn (ost-ROU-der-fehn), village, LOWER SAXONY, NW GERMANY, 9 mi/14.5 km SSE of LEER, in RHAUDERFEHN region, and on the Leda; 53°08′N 07°37′E. Vegetables.

Ostrica (AH-strik-uh), village, PLAQUEMINES parish, extreme SE LOUISIANA, on E bank (levee) of the MISSISSIPPI RIVER, and c.50 mi/80 km SE of NEW ORLEANS, in the delta. Oyster culture; hunting, fishing. Ostrica Canal, with lock through levee here, connects the Mississippi with Breton Sound (E). Oil and natural-gas wells in vicinity.

Östrich-Winkel (OE-strich–VING-kel), town, HESSE, GERMANY, in the RHEINGAU, on right bank of the Rhine, 11 mi/18 km WSW of WIESBADEN; 50°01′N 08°02′E. Manufacturing of chemicals. Has 17th century castle; formed 1972 by the union of three villages.

Ostricourt (os-tree-KOOR), town (□ 2 sq mi/5.2 sq km), NORD department, NORD-PAS-DE-CALAIS region,

N FRANCE, 6 mi/9.7 km NNW of DOUAI; 50°27′N 03°02′E. Abandoned coal mines. Miners' villages are being redeveloped into residential subdivisions.

Ostrih, UKRAINE: see OSTROH.

Östringen (OE-string-uhn), town, MIDDLE UPPER RHINE, BADEN-Württemberg, SW GERMANY, 9 mi/14.5 km NE of BRUCHSAL; 49°13′N 08°42′E. Manufacturing of cigars and cigarettes.

Ostrivyans'ke Lake, UKRAINE: see SHATS'K LAKES.

Ostroda (os-TROO-dah), Polish *Ostróda*, German *Osterode*, town (2002 population 33,769) in EAST PRUSSIA, after 1945 in OLSZTYN province, NE POLAND, in Masurian Lakes region, on DRWECA RIVER, and 20 mi/32 km WSW of OLSZTYN (Allenstein); 53°42′N 19°59′E. Railroad junction; trade center; grain and cattle market; sawmilling. In World War I, German army headquarters here during battle of TANNENBERG. In World War II, c.50% destroyed.

Ostrog, UKRAINE: see OSTROH.

Ostrogozhsk (uh-struh-GOSHSK), city (2006 population 34,355), W VORONEZH oblast, S central European Russia, on the TIKHAYA SOSNA RIVER (DON River basin), on highway and railroad, 88 mi/142 km S of VORONEZH; 50°52′N 39°05′E. Elevation 295 ft/89 m. Fruit canning, dairying, flour milling, tanning; auto-repair shops. Founded 1652 as a fortress.

Ostroh (ost-RO) (Russian *Ostrog*) (Polish *Ostróg*), city, SE RIVNE oblast, UKRAINE, on HORYN' River and 22 mi/35 km SSE of RIVNE; 50°20′N 26°31′E. Elevation 577 ft/175 m. Raion center; agricultural and manufacturing center; food processing (sugar, cereals, meat, milk, fruit, vegetables); brewing, mineral water bottling, sewing and leather industries, sawmilling, furniture making, brick and asphalt making. Lignite deposits nearby. Has ruins of a 14th-century castle. Known since 1100; was capital of an independent principality until the 17th century. Passed (1793) from Poland to Russia; following a struggle for Ukrainian independence, it reverted (1921) to Poland; annexed by USSR in 1939 as part of Ukrainian SSR; independent Ukraine since 1991.

Ostrołęka (os-tro-WEN-kah), Russian *Ostrolenka*, town (2002 population 54,238), NE POLAND, on the Narew River. A railroad junction and a manufacturing center where pulp and paper, lumber, and foodstuffs are produced. Chartered in 1427, it passed to PRUSSIA in 1795 and to RUSSIA in 1815. It reverted to Poland in 1920. The town has several churches built in the 14th and 17th centuries.

Ostropil' (ost-ro-PEEL) (Russian *Ostropol'*), village (2004 population 5,900), E KHMEL'NYTS'KYY oblast, UKRAINE, on the SLUCH RIVER, and 15 mi/24 km ENE of STAROKOSTYANTYNIV; 49°48′N 27°34′E. Elevation 839 ft/255 m. Flour mill. Formerly a town.

Ostropol', UKRAINE: see OSTROPIL'.

Ostrorog (o-STRO-rog), Polish *Ostroróg*, German *Scharfenort*, town, Poznań province, W POLAND, on railroad, and 25 mi/40 km NW of POZNAŃ; 52°38′N 16°28′E. Flour milling; cattle trade.

Ostrov (OS-truhf), city (2006 population 24,115), SW PSKOV oblast, W European Russia, on the VELIKAYA RIVER, on railroad, 32 mi/51 km S of PSKOV; 57°20′N 28°22′E. Elevation 177 ft/53 m. Highway hub in flax-producing region; electrical equipment, flax processing, food industries. Founded in 1342 as an outpost of Pskov principality; later border fortress of MOSCOW domain. City since 1777.

Ostrov (OS-trof), German *schlackenwerth*, town, ZAPADOCESKY province, W BOHEMIA, CZECH REPUBLIC, on railroad, 6 mi/9.7 km NNE of KARLOVY VARY. Manufacturing (trolley buses, machinery, textiles); woodworking. Old mining settlement. Has a Romanesque church (13th century) and Baroque castle (17th century). Health resort (elevation 1,175 ft/358 m) of KYSELKA (mineral water) is 4 mi/6 km SE, on OHRE RIVER.

Ostrov (os-TROV), village, MONTANA oblast, ORYAHOVO obshtina, N BULGARIA, near the DANUBE, 12

mi/19 km ESE of ORYAHOVO; 43°41′N 24°10′E. Vineyards; livestock; fisheries.

Ostrov (OS-trov), village, CONSTANŢA county, SE ROMANIA, in DOBRUJA, on DANUBE River, and 7 mi/11.3 km S of CĂLĂRAŞI. Trades in grain, livestock, animal products, fish, grapes, and wine.

Ostrov, POLAND: see OSTROW, Lubin Province.

Ostrovets (o-strov-VETS), Polish *Ostrowiec*, urban settlement, N GRODNO oblast, ⊙ OSTROVETS region, 13 mi/21 km N of OSHMYANY. Lumbering, food processing.

Ostrov Genriyetta, RUSSIA: see HENRIETTA ISLAND.

Ostrovnoy (uhs-truhv-NO-yee), town (2006 population 4,655), NE MURMANSK oblast, NW European Russia, on the BARENTS SEA shore, on local road, 126 mi/203 km ESE of SEVEROMORSK; 68°03′N 39°31′E. Fleet refueling facilities; nuclear fuel storage. Naval base for nuclear submarines of the Russian Navy's Northern Fleet and known as Yokanga, or Murmansk-140, during the Soviet rule.

Ostrovnoye (uh-struhv-NO-ye), village, NW CHUKCHI AUTONOMOUS OKRUG, NE RUSSIAN FAR EAST, on the LESSER ANYUY RIVER, on local road junction, 425 mi/684 km NW of ANADYR'; 68°08′N 164°12′E. Elevation 272 ft/82 m. Reindeer farms; trading post, airfield.

Ostrovo (os-TRO-vo), village, RUSE oblast, ZAVET obshtina, NE BULGARIA, 12 mi/19 km NNE of RAZGRAD; 43°42′N 26°37′E. Wheat, rye, sunflowers. Formerly known as Golyama-ada.

Ostrovon, Greece: see ARNISSA.

Ostrovon, Lake, Greece: see VEGORITIS, LAKE.

Ostrov Russkiy, RUSSIA: see RUSSIAN ISLAND.

Ostrovskoye (uhs-TROF-skuh-ye), village (2005 population 5,095), S KOSTROMA oblast, central European Russia, on a short tributary of the VOLGA RIVER, on road and near railroad, 49 mi/79 km E of KOSTROMA; 57°48′N 42°14′E. Elevation 406 ft/123 m. Agricultural machinery. Formerly known as Semënovskoye (see MYO-nuhf-skuh-ye).

Ostrovtsy (OS-truhf-tsi), village (2006 population 3,675), central MOSCOW oblast, central European Russia, on the MOSKVA River, on highway, 7 mi/11 km W, and under administrative jurisdiction of RAMENSKOYE; 55°35′N 38°00′E. Elevation 354 ft/107 m. Agricultural products.

Ostrov u Macochy, CZECH REPUBLIC: see MORAVIAN KARST.

Ostrov Vrangelya, RUSSIA: see WRANGEL ISLAND.

Ostrov Zhannetta, RUSSIA: see JEANNETTE ISLAND.

Ostrow (O-stroov), Polish *Ostrów Wielkopolski*, German *Ostrowo*, city (2002 population 73,568), Poznań province, W central POLAND, 65 mi/105 km SSE of POZNAŃ. Railroad junction; trade center; manufacturing of railroad cars, agricultural machinery, chemicals, furniture, bricks; brewing, flour milling, sawmilling. Passed (1793, 1815) to PRUSSIA; returned to Poland in 1919.

Ostrow (O-stroov), Polish *Ostrów Lubelski*, Russian *Ostrov*, town, Lublin province, E POLAND, on TYŚMIENICA River, and 21 mi/34 km NNE of LUBLIN. Tanning, flour and groat milling; brickworks. Sometimes called Ostrów Siedlecki.

Ostrowiec (o-STRO-veets), Polish *Ostrowiec Swietokrzyski*, Russian *Ostrovets*, town (2002 population 75,639), KIELCE province, E POLAND, on KAMIENNA RIVER, on railroad, and 33 mi/53 km ENE of KIELCE. In lignite and iron-ore mining region; linked by pipeline with JASLO-KROSNO natural-gas field. Ironworks; manufacturing of munitions, firebricks, cement products; food processing, sawmilling, tanning, brewing. Development in 1920s because of existence of local iron ore. Before World War II, population was Jewish.

Ostrów Mazowiecka (O-stroov mah-zo-VEET-skah), town (2002 population 22,431), Ostrołęka province, E central POLAND, on railroad, and 55 mi/89 km NE of WARSAW. Manufacturing of bricks, cement, knit goods, hosiery; brewing, flour milling. Before World War II, population 50% Jewish. Formerly Ostrowia.

Ostrozhets (ost-ro-ZHETS) (Polish *Ostroż ec*), village (2004 population 4,700), W RIVNE oblast, UKRAINE, 11 mi/18 km SE of LUTS′K; 50°40′N 25°33′E. Elevation 669 ft/203 m. Rye, wheat.

Ostrozska Nova Ves (OS-trozh-SKAH NO-vah VES), Czech *Ostrožská Nová Ves*, German NEUDORF OSTRA, village, JIHOMORAVSKY province, CZECH REPUBLIC, on railroad, and 4 mi/6.4 km SSW of UHERSKÉ HRADIŠTĚ; 49°01′N 17°26′E. Manufacturing (building materials); agriculture (fruit, vegetables). Health resort (elevation 577 ft/176 m).

Ostryhon Corners, hamlet, Niagara region, S ONTARIO, E central Canada; included in the township of Wainfleet; 42°53′N 79°21′W.

Ostryna (o-STRYA-nah), urban settlement, N GRODNO oblast, BELARUS, 29 mi/47 km E of GRODNO. Manufacturing of tiles; food processing.

Ostrzeszow (ost-SHE-shoov), Polish *Ostrzeszów*, German *Schildberg*, town, Kalisz province, SW central POLAND, on railroad, and 25 mi/40 km SSW of KALISZ; 51°25′N 17°57′E. Manufacturing (bricks, machinery, chemicals, cement, flour).

Ostseebad Kühlungsborn, Germany: see KÜHLUNGS-BORN.

Oststenbek (ost-STEN-bek), suburb of HAMBURG, SCHLESWIG-HOLSTEIN, N GERMANY, 7 mi/11.3 km E of city center; 53°33′N 10°19′E.

Ostswine, POLAND: see ODRA PORT.

Osttirol, AUSTRIA: see TYROL.

Ostuacán (os-too-ah-KAHN), town, CHIAPAS, S MEXICO, 45 mi/72 km NW of TUXTLA GUTIÉRREZ; 17°25′N 93°18′W. Cacao, rice. Sometimes called Usumacinta or Nuevo Usumacinta.

Ostuncalco (os-toon-KAHL-ko), town (2002 population 11,500), QUEZALTENANGO department, SW GUATEMALA, 6 mi/9.7 km WNW of QUEZALTENANGO; 14°52′N 91°37′W. Elevation 8,200 ft/2,499 m. In agricultural region; market center; grain; livestock. In Mam-speaking area. Also called San Juan Ostuncalco (SAHN HWAHN os-toon-KAHL-ko).

Ostuni (os-TOO-nee), town, BRINDISI province, APULIA, S ITALY, 20 mi/32 km WNW of BRINDISI; 40°44′N 17°35′E. Agricultural center (wine, olive oil, fruit, tobacco, vegetables). Bishopric. Has cathedral.

Ostwald (UHST-vahld), town (□ 2 sq mi/5.2 sq km), outer SSW suburb of STRASBOURG, BAS-RHIN department, ALSACE region, E FRANCE, on the Illinois Road junction; 48°33′N 07°43′E.

Ost-Württemberg, Germany: see EAST WURTTEMBERG.

Ostyak-Vogul′sk, RUSSIA: see KHANTY-MANSIYSK.

Ostyak-Vogul National Okrug, RUSSIA: see KHANTY-MANSI AUTONOMOUS OKRUG.

Osuka (OS-kah), town, Ogasa county, SHIZUOKA prefecture, HONSHU, E central JAPAN, 31 mi/50 km S of SHIZUOKA; 34°40′N 137°59′E.

O'Sullivan Dam, WASHINGTON: see COLUMBIA BASIN PROJECT.

Osumacinta, town, CHIAPAS, S MEXICO, 10 mi/16 km NNE of TUXTLA; 16°54′N 93°05′W. Elevation 3,281 ft/1,000 m. Corn, fruit, livestock.

Osumi, former province in S Kyushu, JAPAN; now part of KAGOSHIMA prefecture.

Osumi (O-soo-mee), town, Soo county, KAGOSHIMA prefecture, S KYUSHU, extreme SW JAPAN, 25 mi/40 km E of KAGOSHIMA; 31°35′N 130°59′E.

Osumi, Cape (O-soo-mee), Japanese *Osumi-misaki* (o-SOO-mee-mee-SAH-kee), EHIME prefecture, N SHIKOKU, W JAPAN, in HIUCHI SEA (central section of INLAND SEA), W of O-SHIMA; 34°08′N 132°57′E. Lighthouse.

Osumi Peninsula (O-soo-mee), Japanese *Osumi-hanto* (o-SOO-mee–HAHN-to), KAGOSHIMA prefecture, S KYUSHU, SW JAPAN, between KAGOSHIMA (W) and ARIAKE (E) bays; 45 mi/72 km long (terminating S at

Cape Sata), 20 mi/32 km wide. SAKURA-JIMA (formerly an island) is NW projection.

Osumi Strait (O-soo-mee), Japanese *Osumi-kaikyo* (o-SOO-mee–KAH-ee-kyo), channel connecting EAST CHINA (W) and PHILIPPINE (E) seas, between S coast of KYUSHU and TANEGA-SHIMA island, KAGOSHIMA prefecture, SW JAPAN; 35 mi/56 km long, 20 mi/32 km wide. Cape SATA of OSUMI Peninsula forms NW side. Formerly sometimes called Van Diemen Strait.

Osum River (O-sum), c.70 mi/113 km long, S central ALBANIA; rises in the GRAMOS mountains (Greek border) near ERSEKË; flows W and NW, past BERAT, joining DEVOLL RIVER 8 mi/12.9 km NW of Berat to form Seman River. Also spelled Osumi River.

Osum River (O-sum), 207 mi/333 km long, N BULGARIA; formed N of TROYAN by a confluence of the Beli Osum and Cherni Osum rivers, which rise in the Troyan Mountains; flows N, past LOVECH, NE, and NW, past LEVSKI, to the DANUBE 3 mi/5 km W of NIKOPOL; 43°30′N 25°00′E. Sometimes called Osma River.

Osun (aw-SHOON), state (□ 3,571 sq mi/9,284.6 sq km; 2006 population 3,423,536), SW NIGERIA; ⊙ OSHOGBO. Bordered N by KWARA state, NEE by EKITI state, ESE by ONDO state, S by OGUN state, and W by OYO state. Agriculture includes cocoa, oil palms, maize, yams, millet, cassava, plantains; livestock. Forestry and fishing. Deposits of gold, clay, limestone, granite. Main centers are OSHOGBO, EDE, ILESHA, IFE, ILA, IWO, ILOBU, and IKIRUN. Carved out of Oyo state in 1991. Also spelled Oshun.

Osuna (o-SOO-nah), city, SEVILLE province, SW SPAIN, on fertile ANDALUSIAN plain, on railroad, and 50 mi/80 km E of SEVILLE; 37°14′N 05°07′W. Trading and processing center for agricultural region (olives, cereals, grapes; livestock. Vegetable-oil processing, wine making; manufacturing of esparto goods, linen and wool textiles, soap, hats. An ancient town with many historic relics. Above it is a noted 16th-century collegiate church containing a *Crucifixion* by Ribera. A restored crypt contains graves of the dukes of Osuna. The university, founded in 1549 and dissolved in early 19th century, was a noted cultural center. Osuna was an important Roman colony, alternatively called *Urso, Gemina Urbanorum,* or *Orsona.* Taken in 711 by the Moors (who spelled it Oxuna), it was recaptured in 1239.

Osvaldo Cruz (OS-vahl-do KROOS), city, N RIO GRANDE DO SUL state, BRAZIL, 34 mi/55 km N of PALMEIRA DAS MISSÕES; 27°25′S 53°25′W. Wheat, soybeans, corn, potatoes; livestock.

Osvaldo Cruz (OS-vahl-do kroos), city (2007 population 30,150), W SÃO PAULO state, BRAZIL, 60 mi/97 km WNW of MARÍLIA; 21°47′S 50°50′W. Agriculture (cotton, coffee, corn, rice, bananas, peanuts, manioc).

Osveia Lake (o-SVE-iyah) or **Osveya** (□ 20.4 sq mi/52.8 sq km), VITEBSK oblast, BELARUS; effluence flows through the Dytiarevka River into Lake LISKO.

Osveya (o-SVE-iyah), urban settlement, VITEBSK oblast, BELARUS, on Osveya Lake, 45 mi/72 km NW of POLOTSK. Manufacturing (food processing, peat briquettes, bricks).

Oswaldtwistle (AHZ-wuhl-twis-uhl), town (2001 population 12,527), E LANCASHIRE, NW ENGLAND, adjoining ACCRINGTON, 3 mi/4.8 km E of BLACKBURN; 53°44′N 02°25′W. Previously manufacturing of textiles.

Oswayo (ahs-WAI-yo), borough (2006 population 147), POTTER county, N PENNSYLVANIA, on Oswayo River, and 10 mi/16 km N of COUDERSPORT, on Oswayo Creek; 41°55′N 78°01′W. Agriculture (corn, hay; dairying); timber. State fish hatchery.

Oswegatchie River (aks-wugh-GAH-chee), c.150 mi/241 km long, N NEW YORK; rises in small lakes of the W ADIRONDACK MOUNTAINS; flows N, entering and issuing from CRANBERRY LAKE, thence generally NW,

past GOUVERNEUR, beyond which it turns sharply SW and again NE in a bend, then continues NE and N to the SAINT LAWRENCE at OGDENSBURG. West Branch (c.45 mi/72 km long) enters from S 7 mi/11.3 km SE of Gouverneur.

Oswego (ahs-WEE-go), county (□ 1,312 sq mi/3,411.2 sq km; 2006 population 123,077), N central NEW YORK; ⊙ OSWEGO; 43°28′N 76°12′W. Bounded NW by Lake ONTARIO, S by ONEIDA Lake and Oneida River; crossed by the NEW YORK STATE BARGE CANAL; drained by OSWEGO and SALMON rivers. Diversified manufacturing, especially at Oswego. Dairying area; farming (strawberries); poultry, dairy cows. Limited shipping uses Barge Canal access to Lake Ontario. Intensive recreational use of lakes and lakeshores. Formed 1816.

Oswego (ahs-WEE-go), city (□ 11 sq mi/28.6 sq km; 2006 population 17,638), ⊙ OSWEGO county, N central NEW YORK, on Lake ONTARIO and the OSWEGO River; 43°27′N 76°30′W. The largest U.S. port on Lake Ontario, it is a port of entry and a N terminus of the BARGE CANAL. The city's manufacturing includes metal processing. A trading post established here after the English founded Oswego (1722) became a vital outlet for the ALBANY fur trade. The strategic location prompted the building of Fort Oswego (1727), Fort George (1755), and Fort Ontario (1755; an active U.S. army post until 1946, and a state historic site since 1951). These fortifications were much contested in the colonial wars. The city's importance as a lake port came with the completion of the Barge Canal (1917) and the SAINT LAWRENCE SEAWAY (1959). It is the seat of State University of New York at Oswego. Located NE of Oswego are three nuclear power plants: James A. Fitzpatrick (maximum dependable capacity of 800 MW), 8 mi/12.9 km NE; and Nine Mile Point 1 (maximum dependable capacity of 610 MW), and Nine Mile Point 2 (maximum dependable capacity of 1,080 MW), both 6 mi/9.7 km NE. James Fenimore Cooper's novel *The Pathfinder* is set in the Oswego River valley. Founded 1722, incorporated as a city 1848.

Oswego (ahs-WEE-go), town (2000 population 2,046), ⊙ LABETTE county, extreme SE KANSAS, on NEOSHO River, and 14 mi/23 km SSE of PARSONS; 37°10′N 95°06′W. Railroad junction. Trading point in livestock, grain, and poultry area. Manufacturing (pottery, construction machinery, cultured marble products, truck and bus bodies). Sawmill. Founded 1865, incorporated 1870.

Oswego (os-WEE-go), village (2000 population 13,326), KENDALL county, NE ILLINOIS, on FOX RIVER (bridged here), and 4 mi/6.4 km SSW of AURORA, satellite community of CHICAGO; 41°42′N 88°19′W. Initial stages of urban growth occurring here. Dairying; manufacturing (consumer goods, water softeners, telecommunications equipment).

Oswego River (ahs-WEE-go), 23 mi/37 km long, formed by the confluence of the Oneida and the SENECA rivers (referred to as Three Rivers), central NEW YORK, NW of SYRACUSE and flowing NW to Lake ONTARIO at OSWEGO. It is a part of the state's canal-transportation system. As with ROCHESTER and BUFFALO, Oswego's GREAT LAKES port function has declined significantly with improvements in the SAINT LAWRENCE SEAWAY, development of the nation's system of superhighways and subsequent growth in trucking and, in Oswego's case, drop in demand for coal as an energy source.

Oswestry (AHZ-wuhs-tree), town (2001 population 15,613) and district, SHROPSHIRE, W central ENGLAND; 52°51′N 03°03′W. Some industries (plastics, clothing). The area is named for St. Oswald, a king of NORTHUMBRIA who was killed here in a battle (7th century) against King Penda of MERCIA. The poet Wilfred Owen was born here. OFFA'S DYKE to the W.

Oświęcim (o-see-VEEN-tseem), German *Auschwitz* (OUSH-vits), town (1992 estimated population

45,100), SE POLAND; 50°02'N 19°14'E. It is a railroad junction and industrial center producing chemicals, leather, and agricultural implements; machining, paper. There are coal deposits in the vicinity. In World War II, the Germans organized a concentration- and death-camp system here, consisting of three main camps and thirty forced-labor camps. At the BRZEZINKA (German *Birkenau*) extermination camp, as many as 4,000,000 prisoners, mostly Jews, were exterminated. Museum and memorial.

Osyechenitsa, BOSNIA AND HERZEGOVINA: see OSJEČENICA.

Osyka (o-SAHK-ee), village (2000 population 481), PIKE county, SW MISSISSIPPI, on TANGIPAHOA RIVER, and 16 mi/26 km S of MCCOMB, at LOUISIANA state line; 31°00'N 90°28'W. Agriculture (cotton, corn; cattle; dairying); manufacturing (transportation equipment).

Osynovyy Rovenok, UKRAINE: see ROVEN'KY.

Osypenko, UKRAINE: see BERDYANS'K.

Ota (O-tah), southernmost ward of TOKYO city, Tokyo prefecture, E central HONSHU, E central JAPAN, on TOKYO BAY (E). Bordered NW by SETAGAYA and MEGURO wards, NE by SHINAGAWA ward, and S by KANAGAWA prefecture. Ward has second highest population in the city. Haneda Airport is in SE corner.

Ota (O-tah), city (2005 population 213,299), GUMMA prefecture, central HONSHU, N central JAPAN, 22 mi/35 km E of MAEBASHI; 36°17'N 139°22'E. Motor vehicles. Silk cocoons.

Ota (O-tah), town, Senhoku county, Akita prefecture, N HONSHU, NE JAPAN, 31 mi/50 km S of AKITA city; 39°30'N 140°36'E.

Ota (o-TAH), village, W Kunisaki county, OITA prefecture, E KYUSHU, SW JAPAN, 19 mi/30 km N of OITA; 33°29'N 131°33'E.

Otaci (o-TAH-chi), village (2004 population 8,400), N MOLDOVA, on DNIESTER (Nistru) River (UKRAINE border), opposite Mogilev-Podolski, Ukraine; 48°26'N 27°46'E. On railroad; agricultural center; flour and oilseed milling. Until World War II, population largely Jewish. Formely known as Ataki.

Otacílio Costa (O-tah-see-lee-o KO-stah), city (2007 population 15,693), S central SANTA CATARINA state, BRAZIL, 23 mi/37 km NE of LAJES; 27°34'S 50°07'W. Corn, fruit; livestock.

Otáez (o-TAH-es) or **santa maría de otáez**, town, DURANGO, N MEXICO, in W foothills of SIERRA MADRE OCCIDENTAL, 45 mi/72 km SW of SANTIAGO PAPASQUIARO; 24°42'N 105°59'W. Corn, cotton, chickpeas.

Otago (oh-TA-go), region (□ 14,918 sq mi/38,786.8 sq km), corresponding (as defined in 1989) with the former province and land district, NEW ZEALAND, but without the Upper WAITAKI district (assigned to the CANTERBURY region), some marginal N and W areas, and particularly the SOUTHLAND region (which previously become a province 1861–1870 and therafter rejoined Otago province, though remaining a land district). The region now includes the Queenstown-Lakes district, which is oriented toward tourism and sheep ranching in the area of lakes WAKATIPU, WANAKA, and HAWEA. The Central Otago district is characterized by a dry climate, grassland vegetation, the harnessing of river sites for generation of hydroelectricity and irrigation of apricots (as well as other fruit), sheep farming, and a tendency toward continentality evinced in seasonal change. A gold-mining past and vistas of range and basin topography also create a regional mystique for this district. In the coastal Waitaki district, or N Otago, plains and downlands support a mixed crop and livestock farming system continuous with that of S Canterbury. Southward, in the CLUTHA district, a transition to somewhat wetter and cooler climate is accompanied by greener pastures and emphasis on fodder crops such as oats, grown for prime lamb, beef, and dairy cattle. Native forest sur-

vives in the CATLINS area (divided with Southland region), but intensification of agriculture appears near DUNEDIN, as in the TAIERI PLAIN. Dunedin, with associated PORT CHALMERS, remains the commercial and educational center of Otago, but the 1989 expansion to be New Zealand's largest city (in terms of sq mileage) reflects adminstrative extension to rural surroundings, not urban growth.

Otago Harbour (oh-TA-go), inlet from S PACIFIC, OTAGO region, SOUTH ISLAND, NEW ZEALAND; formed between volcanic hills; 13 mi/21 km long, 1 mi/1.6 km wide at mouth; PORT CHALMERS on W shore. Dredged channel enables ships to reach Dunedin's commercial area. Flanked to E by OTAGO PENINSULA, running 16 mi/25 km NE; DUNEDIN spreads around S end of harbor.

Otago Peninsula (o-TA-go), NEW ZEALAND, extends 16 mi/25 km to NE, to albatross rookery at Taiaroa Head, and CAPE SAUNDERS with lighthouse. Popular resort sites.

Otahuhu (o-TAH-hoo), SE suburb of AUCKLAND city, NW NORTH ISLAND, NEW ZEALAND, on narrow isthmus between TAMAKI RIVER and MANUKAU HARBOUR; 36°57'S 174°51'E. Metallurgical and chemical plants, soap and candle factories, abattoirs.

Otake (O-tah-ke), city, HIROSHIMA prefecture, SW HONSHU, W JAPAN, on HIROSHIMA BAY, 19 mi/30 km S of HIROSHIMA; 34°13'N 132°13'E. Synthetic fibers.

Otaki (O-tah-kee), town, Isumi county, CHIBA prefecture, E central HONSHU, E central JAPAN, on central BOSO Peninsula, 19 mi/30 km S of CHIBA; 35°16'N 140°14'E.

Otaki (o-TAK-ee), town (2001 population 5,643), S NORTH ISLAND, NEW ZEALAND, 40 mi/64 km NNE of WELLINGTON; 40°45'S 175°09'E. On narrow coastal plain. Dairying, vegetable, fruit center. Resort beach nearby.

Otaki (O-tah-kee), village, Iburi district, HOKKAIDO prefecture, N JAPAN, 31 mi/50 km S of SAPPORO; 42°40'N 141°04'E.

Otaki, village, Kiso county, NAGANO prefecture, central HONSHU, central JAPAN, 62 mi/110 km S of NAGANO; 35°48'N 137°33'E. Medical products. Hinoki cypress trees. Ontake mountain (skiing area) nearby.

Otaki, village, Chichibu county, SAITAMA prefecture, E central HONSHU, E central JAPAN, 40 mi/65 km W of URAWA; 35°56'N 138°56'E.

Otama (O-tah-mah), town, Adachi county, FUKUSHIMA prefecture, N central HONSHU, NE JAPAN, 16 mi/25 km S of FUKUSHIMA city; 37°31'N 140°22'E.

Otanche (o-TAHN-chai), town, ⊙ Otanche municipio, BOYACÁ department, central COLOMBIA, 60 mi/97 km W of TUNJA; 05°39'N 74°11'W. Elevation 3,766 ft/1,147 m. Coffee, corn, sugarcane; livestock.

Otani, RUSSIA: see SOKOL, SAKHALIN oblast.

Otar (o-TAHR), town, E ZHAMBYL region, KAZAKHSTAN, on TURKISH-SIBERIAN RAILROAD, 90 mi/145 km W of ALMATY; 43°30'N 75°13'E.

Otari (O-tah-ree), village, North Azumi county, NAGANO prefecture, central HONSHU, central JAPAN, 19 mi/30 km N of NAGANO; 36°46'N 137°54'E. Grapes and wine. Alpine plants. Skiing area. On Shio-no-michi, ancient salt road from Sea of Japan coast.

Ota River, JAPAN: see HIROSHIMA, city.

Otaru (O-tah-roo), city (2005 population 142,161), Hokkaido prefecture, N JAPAN, 22 mi/35 km N of SAPPORO, port on Otaru Canal and ISHIKARI BAY; 43°11'N 140°59'E. Manufacturing includes marine-product processing; crafts (glass, pottery); wine. Tourism; ancient writing preserved in the Temiya caves. Site of the Asarigawa Spa and the Otaru Aquarium.

Otatitlán (o-tah-tee-TLAHN), town, VERACRUZ, SE MEXICO, on Papaloapan River, and 9 mi/14 km NE of TUXTEPEC; 18°12'N 96°02'W. Sugarcane, coffee, fruit.

Otautau (o-TOO-too), township, SOUTHLAND district, SOUTH ISLAND, NEW ZEALAND, 27 mi/43 km NW of INVERCARGILL, where Aparima River Plain impinges against forested Longwood Range; 46°09'S 168°00'E. Sawmill, dairy plant.

Otava (O-tah-vah), village, MIKKELIN province, SE FINLAND, 7 mi/11.3 km WSW of MIKKELI; 61°39'N 27°04'E. Elevation 330 ft/100 m. In SAIMAA lake region; sawmills; agricultural school.

Otavalo (o-tah-VAH-lo), town (2001 population 30,965), IMBABURA province, N ECUADOR, on PANAMERICAN HIGHWAY, 35 mi/56 km NNE of QUITO, in magnificent setting, surrounded by Andean peaks; 00°14'N 78°16'W. Elev. 8,400 ft/2,560 m. Health resort (thermal springs), textile and agricultural center (vegetables, coffee, cotton, sugarcane, tobacco, cereals, potatoes, fruit; livestock). Manufacturing of woolen goods (ponchos, carpets), cotton textiles, leather goods. Indians predominate in the region. Internationally known for its Saturday craft market, featuring Otavalo Indian weaving.

Otava River (O-tah-VAH), German *wottawa*, c.55 mi/89 km long, ZAPADOCESKY and JIHOCESKY provinces, S BOHEMIA, CZECH REPUBLIC; formed 4 mi/6.4 km SW of KASPERSKE HORY by junction of two headstreams rising in foothills of BOHEMIAN FOREST; flows NNE, past SUSICE, E, past STRAKONICE, and N, past PÍSEK, to VLTAVA RIVER at historic ruins of ZVIKOV.

Otavi (o-TAH-vee), canton, JOSÉ MARÍA LINARES province, POTOSÍ department, W central BOLIVIA, 3 mi/5 km SE of PUNA, on the BETANZOS-VILLAZÓN road, and 22 mi/35 km S of the POTOSÍ-CAMARGO railroad.

Otavi (o-tah-vee), town, N NAMIBIA, 50 mi/80 km WSW of GROOTFONTEIN; 19°39'S 17°20'E. Elev. 4,658 ft/1,420 m. Railroad junction in cattle-raising, copper- and lead-mining region; also maize and sunflower production area.

Otawara (o-TAH-wah-rah), city, TOCHIGI prefecture, central HONSHU, N central JAPAN, 22 mi/35 km N of UTSUNOMIYA; 36°52'N 140°01'E. X-ray equipment. Soybeans. Also Otahara.

Otay, unincorporated town, SAN DIEGO county, S CALIFORNIA, residential suburb 12 mi/19 km SSE of downtown SAN DIEGO, on OTAY RIVER, S of CHULA VISTA. S end of SAN DIEGO BAY to NW; Tijuana Mountains 4 mi/6.4 km to S.

Otay Mesa, unincorporated town, SAN DIEGO county, S CALIFORNIA, 16 mi/26 km SE of downtown SAN DIEGO and 4 mi/6.4 km NE of TIJUANA, MEXICO. Brown Field Airport and U.S. Space Surveillance Station are here. Tijuana International Airport (Mexico) to S.

Otay River, intermittent stream (c.25 mi/40 km long), S CALIFORNIA; rises in SW SAN DIEGO county, 15 mi/24 km E of downtown SAN DIEGO, at San Miguel Mountain; flows first S through Upper and Lower Otay reservoirs, then W through San Ysidro section of San Diego, 3 mi/4.8 km N of Mexican border, to S end of SAN DIEGO BAY. Savage Dam (172 ft/52 m high, 750 ft/229 m long; completed 1919) impounds a water-supply reservoir for San Diego.

Otchinjau, town, CUNENE province, ANGOLA, on road, and 130 mi/209 km WNW of ONDJIVA; 16°30'S 13°57'E. Market center.

Otdâr Méan Cheay, province (□ 2,377 sq mi/6,180.2 sq km; 2007 population 102,835), NW CAMBODIA, bordering THAILAND (to N); ⊙ SAMRONG. Borders PREAH VIHEAR (E), SIEM REAP (S), and BANTEAY MEAN CHEAY (SWW) provinces. Also spelled Oddar Mean Chey.

Otdykh, RUSSIA: see ZHUKOVSKIY.

Oteapan (o-te-AH-pan), town, VERACRUZ, SE MEXICO, on Isthmus of TEHUANTEPEC, 8 mi/12.9 km W of MINATITLÁN, on railroad; 18°00'N 94°39'W. Fruit; livestock; in petroleum-producing area.

Oteen (O-teen), unincorporated town, BUNCOMBE county, W NORTH CAROLINA, residential suburb 3 mi/

4.8 km E of downtown ASHEVILLE, near Swannanoa River; 35°35′N 82°29′W. BLUE RIDGE PARKWAY passes through town. Folk Art Center.

Otego (o-TEE-go), village (□ 1 sq mi/2.6 sq km; 2006 population 999), OTSEGO county, central NEW YORK, on the SUSQUEHANNA River, and 7 mi/11.3 km SW of ONEONTA; 42°23′N 75°10′W. In dairying and grain-growing area.

Oteiza (o-TAI-thah), town, NAVARRE province, N SPAIN, 5 mi/8 km SE of ESTELLA. Grain, wine.

Otélé (o-TAI-lai), village, Central province, SW central CAMEROON, 30 mi/48 km SW of Yaoundé; 03°37′N 11°15′E. Railroad junction and market; cacao and coffee plantations in vicinity.

Oţelu Roşu (O-tse-loo RO-shoo), town, CARAŞ-SEVERIN county, ROMANIA, 10 mi/16 km NE of CARANSEBEŞ; 45°32′N 22°22′E. Metallurgy.

Otepää (O-tai-pah), German *Odenpäh*, city, SE ESTONIA, 24 mi/39 km SSW of TARTU, in Otepää hills (rising to 804 ft/245 m); 58°03′N 26°29′E. Agricultural market; oats; dairy farming; auto reconstruction. Dates from 13th century.

Otero (o-TER-o), county (□ 1,269 sq mi/3,299.4 sq km; 2006 population 19,452), SE COLORADO, ⊙ LA JUNTA; 37°53′N 103°42′W. Irrigated agricultural area, drained by PURGATOIRE, APISHAPA, and ARKANSAS (forms part of N boundary in W) rivers, also HORSE and TIMPAS creeks. Cattle, poultry; wheat, hay, beans, corn, melons, vegetables, cantaloupes. Large part of Comanche National Grassland in S; HORSE CREEK RESERVOIR on W boundary, in NE. BENT'S OLD FORT NATIONAL HISTORIC SITE in E. Formed 1889.

Otero, county (□ 6,627 sq mi/17,230.2 sq km; 2006 population 62,744), S NEW MEXICO, ⊙ ALAMOGORDO; 32°37′N 105°43′W. Livestock-grazing area bordering TEXAS on S; small part of state line in SE corner of county is Central-Mountain time zone boundary (New Mexico and HUDSPETH and EL PASO counties, Texas, are in Mountain time zone). Gold, silver; cattle, sheep; fruit, nuts, pecans, hay, alfalfa, timber, some wheat. Part of large FORT BLISS MILITARY RESERVATION in S half of county extends N from EL PASO, Texas; Mescalero Apache Indian Reservation in N; SACRAMENTO MOUNTAINS in N; part of GUADALUPE MOUNTAINS in SE; parts of Lincoln National Forest in SE and center; part of WHITE SANDS NATIONAL MONUMENT in W; Tularosa Valley, desert plain, in W, including White Sands Desert; Three Rivers National Recreation Site in N; Oliver Lee State Park in N center. Formed in 1899.

Otero de Herreros (o-TAI-ro dhai e-RAI-ros), town, SEGOVIA province, central SPAIN, 9 mi/14.5 km SW of SEGOVIA; 40°49′N 04°12′W. Wheat, rye, barley, oats; livestock.

Otford (AHT-fuhd), village (2001 population 3,258), W KENT, SE ENGLAND, on DARENT RIVER, and 3 mi/4.8 km N of SEVENOAKS; 51°19′N 00°12′E. Former agricultural market. Has 14th-century church and remains of palace of archbishops of CANTERBURY.

Othello, town (2006 population 6,293), ADAMS county, SE WASHINGTON, 65 mi/105 km ENE of YAKIMA; 46°49′N 119°10′W. In COLUMBIA basin agricultural region; wheat, barley, oats, alfalfa, fruits, vegetables, peppermint; dairying; cattle, hogs; manufacturing (fertilizers, food products, printing and publishing). Scootenay Reservoir to SE; Eagle Lakes to S; large POTHOLES RESERVOIR and O'SULLIVAN DAM to NW. Columbia National Wildlife Refuge to NW. U.S. Dept. of Energy Hanford Site to SW.

Othe, Pays d' (DOT, pai), forested upland SW of TROYES in the PARIS BASIN, CHAMPAGNE-ARDENNE region, FRANCE.

Othis (o-TEE), town (□ 5 sq mi/13 sq km), SEINE-ET-MARNE department, ÎLE-DE-FRANCE region, N central FRANCE, just N of Charles de Gaulle airport and 20 mi/32 km NE of PARIS; 49°05′N 02°41′E.

Othmaniya, SAUDI ARABIA: see ITHMANIYA.

Otho, village (2000 population 571), WEBSTER county, central IOWA, near DES MOINES RIVER, 5 mi/8 km S of FORT DODGE; 42°25′N 94°09′W.

Othoni (o-tho-NEE), island (□ 3 sq mi/7.8 sq km) in IONIAN Sea, in KÉRKIRA prefecture, IONIAN ISLANDS department, off W coast of GREECE, and 14 mi/23 km NW of KÉRKIRA (Corfu) island; 3.5 mi/5.6 km long, 1.5 mi/2.4 km wide; 39°50′N 19°26′E. Largely mountainous. Fisheries. Also called Fano, Fanos, or Phanos.

Othrys (O-threes), mountain massif in N CENTRAL GREECE department, on border of S THESSALY department, between MALIAKOS GULF (S) and Gulf of VÓLOS (NE); 39°02′N 22°37′E. Rises to 5,773 ft/1,760 m in the Gerakovouni (Yerakovouni), 17 mi/27 km ENE of LAMIÁ. Copper deposits. Also called Orthrys or Orthris; also spelled Othris.

Oti, prefecture (2005 population 138,919), SAVANES region, N TOGO ⊙ SANSANNÉ-MANGO; 10°20′N 00°35′E.

Otibanda (O-tee-BAN-dah), village, MOROBE province, E NEW GUINEA island, NE PAPUA NEW GUINEA, 10 mi/16 km W of BULOLO; 07°18′S 146°31′E. On NE flank of Bowutu Mountains. Road access. Coffee, tea; cattle; timber.

Ötigheim (OE-tikh-heim), village, MIDDLE UPPER RHINE, BADEN-Wřurttemberg, SW GERMANY, 3 mi/4.8 km NNE of RASTATT; 48°54′N 08°13′E. Corn, horse-radish.

Otira (o-TEER-ah), township, WESTLAND district, SOUTH ISLAND, NEW ZEALAND, 70 mi/113 km NW of CHRISTCHURCH, at W end of steep electrified OTIRA tunnel (5.3 mi/8.5 km long), at Arthur's Pass National Park; 42°50′S 171°33′E. Tunnel through SOUTHERN ALPS is only railroad route between E and W coasts. Winter sports.

Oti River (O-tee), 320 mi/515 km long, TOGO and GHANA, maVoltain E tributary of VOLTA RIVER; rises W of ATAKORA MOUNTAINS near TOGO–BURKINA FASO border; flows SSE in meandering course, past SANSANNÉ–MANGO, forming 80 mi/129 km of TOGO-GHANA border before entering Lake VOLTA. Canoe transport; fishing. Unnavigable outside of Lake Volta region.

Otis (O-tis), township, HANCOCK county, S MAINE, c.10 mi/16 km N of ELLSWORTH; 44°42′N 68°28′W. In recreational area.

Otis (O-dis), town, BERKSHIRE county, SW MASSACHUSETTS, 19 mi/31 km SSE of PITTSFIELD; 42°13′N 73°05′W. Agriculture; resort. Otis Ridge Ski Area nearby.

Otis (O-tis), village (2000 population 534), WASHINGTON county, NE COLORADO, 13 mi/21 km E of AKRON; 40°08′N 102°57′W. Elevation 4,335 ft/1,321 m. Cattle; wheat, sunflowers; manufacturing (construction materials).

Otis, village (2000 population 325), RUSH county, central KANSAS, 14 mi/23 km E of LA CROSSE; 38°31′N 99°02′W. In wheat area.

Otisco Lake (o-TIS-ko), easternmost of the eleven FINGER LAKES, ONONDAGA county, central NEW YORK, 14 mi/23 km SW of SYRACUSE; 6 mi/9.7 km long, 0.5 mi/0.8 km–1 mi/1.6 km wide; 42°52′N 76°17′W. Drains NE to ONONDAGA Lake via Ninemile Creek. Some minor pollution from fertilizer runoff; invasive zebra muscles. Part of lake is man-made.

Otisfield, town, CUMBERLAND county, SW MAINE, between CROOKED RIVER and THOMPSON LAKE, c. 30 mi/48 km NNW of PORTLAND; 44°05′N 70°32′W.

Ötisheim (OE-tis-heim), village, N BLACK FOREST, BADEN-Württemberg, S GERMANY, 7 mi/11.3 km NNE of PFORZHEIM; 48°58′N 08°48′E. Fruit; metal-working.

Otish Mountains, range in central QUEBEC, E Canada, on N side of the LAURENTIAN PLATEAU; c.50 mi/80 km long; 52°20′N 70°40′W. Rises to 3,700 ft/1,128 m. EASTMAIN, PERIBONCA, and OUTARDES rivers rise here. Range is part of the SAINT LAWRENCE–HUDSON BAY watershed.

Otis Orchards, unincorporated town, SPOKANE county, E WASHINGTON, 14 mi/23 km ENE of SPOKANE, and 15 mi/24 km W of COEUR D'ALENE, Idaho, 3 mi/4.8 km W of IDAHO boundry. Agriculture area; timber. Manufacturing (wood products, mining services). Newman Lake reservoir to N; LIBERTY LAKE reservoir to S.

Otisville, village (2000 population 882), GENESEE county, SE central MICHIGAN, 14 mi/23 km NE of FLINT; 43°10′N 83°31′W. In farm area. Manufacturing (fishing tackle).

Otisville, village (2006 population 1,091), ORANGE county, SE NEW YORK, 7 mi/11.3 km WNW of MIDDLETOWN, in the SHAWANGUNK range; 41°28′N 74°32′W.

Otívar (o-tee-VAHR), town, GRANADA province, S SPAIN, 10 mi/16 km WNW of MOTRIL; 36°49′N 03°41′W. Brandy distilling, olive-oil processing. Sugarcane, fruit, raisins, esparto.

Otjikoto Lake (ot-jee-ko-to), NAMIBIA, 10 mi/16 km NW of TSUMEB and 25 mi/40 km E of ETOSHA NATIONAL PARK; 19°11′S 17°33′E. Natural sinkhole in limestone topography. The circular lake (actual depth unknown) maintains a constant level all year. Tourist attraction.

Otjikoto Region (ot-JEE-ko-to), administrative division (2001 population 161,007), SE part of OVAMBOLAND, NAMIBIA, from ONDANGWA SE to TSUMEB in N. Largest region in Ovamboland extending around ETOSHA NATIONAL PARK. Woodland and mountain savanna.

Otjiwarongo (ot-jee-wah-RON-go), town (2001 population 19,614), N NAMIBIA, 150 mi/241 km NNW of WINDHOEK. Elev. 4,774 ft/1,455 m. Railroad and road junction; distributing center for N part of country, in cattle-raising region. Fluorspar deposits. Traditional headquarters of Herero 60 mi E at OKAKARARA. Education center.

Otjozondjupa Region (ot-JO-zon-JOO-puh), administrative division (2001 population 135,384), central NAMIBIA, S of OKAVANGO REGION E from OTJIWARONGO to BOTSWANA border, S to include OKAHANDJA; 20°20′S 18°00′E. Political region includes most of Hereroland E and the rich farmlands of GROOTFONTEIN/OTAVI.

Otley (AHT-lee), town (2001 population 14,124), WEST YORKSHIRE, N ENGLAND, on WHARFE RIVER, and 10 mi/16 km NW of LEEDS; 53°54′N 01°43′W. Previously woolen and paper milling. Manufacturing of leather, biscuits, and printing machinery. Has church of 15th-century origin.

Otley, village, MARION county, S central IOWA, 10 mi/16 km NNE of KNOXVILLE. In agricultural area.

Otmuchow (ot-MOO-hoov), Polish *Otmuchów*, German *Ottmachau*, town (□ 8 sq mi/20.8 sq km) in UPPER SILESIA, after 1945 in OPOLE province, S POLAND, near CZECH border, at N foot of REICHENSTEIN MOUNTAINS, on the GLATZER NEISSE, and 8 mi/12.9 km W of NEISSE (Nysa); 50°28′N 17°09′E. Textile and paper milling, manufacturing of glass, furniture. Just W, on the Glatzer Neisse, is dam and irrigation reservoir (□ 8 sq mi/20.7 sq km; 5 mi/8 km long, 3 mi/4.8 km wide), completed 1933. Hydroelectric station. Town was residence of prince-bishops of BRESLAU. Has 17th century church, remains of 15th century castle.

Oto (O-TO), town, Tagawa county, FUKUOKA prefecture, NE KYUSHU, SW JAPAN, 25 mi/40 km E of FUKUOKA; 33°36′N 130°51′E.

Oto, town (2000 population 145), WOODBURY county, W IOWA, on Little Sioux River, and 31 mi/50 km ESE of SIOUX CITY; 42°16′N 95°53′W. In livestock and grain area.

Oto (O-to), village, Yoshino district, NARA prefecture, HONSHU, W central JAPAN, 36 mi/58 km S of NARA; 34°10′N 135°45′E. Lumber.

Oto, village, W Muro county, WAKAYAMA prefecture, HONSHU, W central JAPAN, 39 mi/63 km S of WAKAYAMA; 33°43′N 135°29′E.

Area is shown by the symbol □, and capital city or county seat by ⊙.

Otobe (O-to-be), town, Hiyama district, Hokkaido prefecture, N JAPAN, 96 mi/155 km S of SAPPORO; 41°57′N 140°08′E. Lily roots; wine. Vegetable canning.

Otočac (o-TO-chahts), town, W CROATIA, on GACKA RIVER, and 19 mi/31 km ESE of SENJ, in the LIKA region. Local trade center; leather, textile, and lumber industries. First mentioned in 12th century.

Otoe (AW-toe), county (□ 619 sq mi/1,609.4 sq km; 2006 population 15,747), SE NEBRASKA; ☉ NEBRASKA CITY; 40°38′N 96°08′W. Agriculture and commercial region bounded E by MISSOURI RIVER, forming IOWA and MISSOURI boundaries; drained by LITTLE NEMAHA RIVER and its branches. Manufacturing at Nebraska City. Dairying; cattle, hogs; apples, corn, wheat, soybeans, sorghum. Arbor Lodge State Historical Park and Riverview Marina State Recreation Area at Nebraska City in E. Formed 1855.

Otoe, village (2006 population 216), OTOE county, SE NEBRASKA, 13 mi/21 km W of NEBRASKA CITY, and on branch of LITTLE NEMAHA RIVER; 40°43′N 96°07′W.

Otofuke (o-TOF-ke), town, Tokachi district, Hokkaido prefecture, N JAPAN, 93 mi/150 km E of SAPPORO; 42°59′N 143°12′E. Beans, potatoes, sugar beets, wheat; dairying; livestock raising. Hot springs. Large flower clock.

Otoineppu (o-TO-ee-NEP), village, Kamikawa district, Hokkaido prefecture, N JAPAN, 124 mi/200 km N of SAPPORO; 44°43′N 142°15′E. Woodworking.

Otola, town, COLLINES department, W central BENIN, 30 mi/48 km WNW of SAVALOU, on TOGO border; 08°09′N 01°39′E. Groundnuts, cotton.

Otomari, RUSSIA: see KORSAKOV.

Oton (O-ton), town, ILOILO province, SE PANAY island, PHILIPPINES, near PANAY GULF, 6 mi/9.7 km W of Iloilo; 10°44′N 122°28′E. Rice-growing center.

Otonabee-South Monaghan (o-TAH-nuh-bee–south MAH-nuh-huhn), township (□ 135 sq mi/351 sq km; 2001 population 6,669), S ONTARIO, E central Canada, on RICE LAKE, and 6 mi/10 km from PETERBOROUGH; 44°15′N 78°13′W.

Otone (o-TO-ne), town, N Saitama county, SAITAMA prefecture, E central HONSHU, E central JAPAN, 19 mi/30 km N of URAWA; 36°08′N 139°40′E.

Otopeni (o-to-PEN), village, ILFOV AGRICULTURAL SECTOR, S ROMANIA, 7 mi/11.3 km N of BUCHAREST; 44°33′N 26°04′E. Corn, wheat. International airport. Includes Otopeni-de-Sus and Otopeni-de-Jos. Also spelled Otopenii.

Otoque Island, in BAY OF PANAMA, arm of the PACIFIC OCEAN, PANAMÁ province, E PANAMA, 24 mi/39 km SSW of PANAMA city; 2 mi/3.2 km long, 1 mi/1.6 km wide; 08°36′N 79°36′W.

Otoque Occidente, village and minor civil division of Taboga district, PANAMÁ province, PANAMA, ONOTOQUE ISLAND in PANAMA BAY; 08°36′N 79°37′W.

Otoque Oriente, village and minor civil division of Taboga district, PANAMÁ province, PANAMA, on OTOQUE ISLAND in PANAMA BAY; 08°36′N 79°36′W.

Otorohanga (o-to-ro-HANG-ah), township (□ 797 sq mi/2,072.2 sq km), W central NORTH ISLAND, NEW ZEALAND, 37 mi/60 km S of HAMILTON; 38°11′S 175°12′E. Main trunk railroad and road route. Servicing center for dairy, prime lamb area. Nearest to WAITOMO limestone caves.

Otoro Valley (o-TO-ro), INTIBUCÁ department, SW HONDURAS, between JESÚS DE OTORO and MASAGUARA; 14°26′N 88°01′W. Drained by Rio Grande de Otoro. Livestock; tobacco, sugarcane.

Otowa (o-TO-wah), town, Hoi county, AICHI prefecture, S central HONSHU, central JAPAN, 31 mi/50 km S of NAGOYA; 34°51′N 137°18′E.

Otoyo (O-to-yo), town, Nagaoka county, KOCHI prefecture, S SHIKOKU, W JAPAN, 16 mi/25 km N of KOCHI; 33°45′N 133°40′E.

Ot, Piz, SWITZERLAND: see PIZ OT.

Otpor, RUSSIA: see ZABAYKAL'SK.

Otradinskiy (uh-TRAH-deen-skeeyee), settlement (2006 population 3,010), N central ORËL oblast, SW European Russia, near the OKA River, on highway and railroad, 25 mi/40 km NE of ORËL, and 5 mi/8 km SW of MTSENSK, to which it is administratively subordinate; 53°15′N 36°24′E. Elevation 751 ft/228 m. Natural gas processing; sugar refining.

Otradnaya (uh-TRAH-dnah-yah), village (2005 population 23,095), SE KRASNODAR TERRITORY, S European Russia, in the NW foothills of the Greater CAUCASUS Mountains, on the URUP RIVER, on crossroads, 45 mi/72 km SSE of ARMAVIR; 44°23′N 41°31′E. Elevation 1,427 ft/434 m. Flour mill, brewery; bast-fiber processing (hemp).

Otradnoye (uh-TRAHD-nuh-ye), city (2005 population 21,340), central LENINGRAD oblast, NW European Russia, on the NEVA RIVER, on highway branch and railroad (Pella station), 25 mi/40 km E of SAINT PETERSBURG; 59°47′N 30°49′E. Port facilities; ship construction and repair; plastic pipes manufacturing. Made city in 1970.

Otradnoye (uh-TRAHD-nuh-ye), village (2006 population 3,200), NW VORONEZH oblast, S central European Russia, on the VORONEZH RIVER, near highway, 6 mi/10 km E of VORONEZH; 51°40′N 39°21′E. Elevation 318 ft/96 m. In agricultural area; fishing, woodworking. Weekend getaway spot for residents of Voronezh; summer homes.

Otradnoye (uh-TRAHD-nuh-ye), settlement, W KALININGRAD oblast, W European Russia, on coastal railroad, 2 mi/3.2 km W of SVETLOGORSK, to which it is administratively subordinate; 54°56′N 20°06′E. Elevation 167 ft/50 m. Baltic seaside resort. Home of Herman Brachert, a German sculptor, whose house is now a museum.

Otradnyy (uh-TRAHD-niyee), city (2006 population 50,170), E SAMARA oblast, E European Russia, near the SAMARA River, on road and railroad (Novoootradnaya station), 47 mi/75 km E of SAMARA; 53°22′N 51°21′E. Elevation 246 ft/74 m. Developed with the exploitation of the Mukhanovsk oil fields from 1952. Oil and gas production; manufacturing (polymer building materials, petroleum refining, metalworks, electrical and mechanical workshop). Made city in 1956.

Otrado-Kubanskoye (uh-TRAH-duh-koo-BAHN-skuh-ye), settlement (2005 population 3,715), E KRASNODAR TERRITORY, S European Russia, in the KUBAN' River basin, on crossroads and railroad, 7 mi/11 km S of GUL'KEVICHI, and 6 mi/10 km SE of KAVKAZSKAYA; 45°14′N 40°50′E. Elevation 521 ft/158 m. In agricultural area (wheat, corn, hemp, sunflowers).

Otrado-Ol'ginskoye (uh-TRAH-duh–OL-geen-skuh-ye), settlement (2005 population 3,380), E KRASNODAR TERRITORY, S European Russia, on the left bank of the KUBAN' River, on highway junction, 12 mi/19 km ESE of GUL'KEVICHI; 45°18′N 40°57′E. Elevation 370 ft/112 m. In agricultural area (wheat, corn, hemp, sunflowers).

Otranto (o-TRAHN-to), town, LECCE province, APULIA, extreme S ITALY, on the STRAIT OF OTRANTO, which links the ADRIATIC and IONIAN seas; 40°09′N 18°30′E. It is a small fishing port and a seaside resort. Originally a Greek settlement, Otranto became an important port under the Romans. Later ruled by the Byzantines and the Normans, it never recovered from its devastation (1480) by the Turks. Of note are an 11th-century cathedral (restored 17th–18th centuries), with a fine mosaic floor (12th century), and the ruins of an imposing Aragonese castle (15th century) that provided the setting for Horace Walpole's Gothic novel, *The Castle of Otranto*.

Otranto, Cape, easternmost point of ITALY, in APULIA, on "heel" of Italian peninsula, at W end of STRAIT OF OTRANTO, 3 mi/5 km SSE of OTRANTO; 40°07′N 18°31′E.

Otranto, Strait of, connects ADRIATIC Sea (N) with IONIAN Sea (S); extends c.43 mi/69 km between Cape OTRANTO (S ITALY) and Cape Linguetta (ALBANIA).

Otrar, KAZAKHSTAN: see SHAULDER.

Otra River (AWT-rah), 150 mi/241 km long, AUST-AGDER county, S NORWAY; rises in lakes in the Bykle Mountains, flows S through the SETESDAL and the BYGLANDSFJORD to the SKAGERRAK at KRISTIANSAND. Below the Byglandsfjord it is sometimes called the Torridal.

Otricoli (o-TREE-ko-lee), village, TERNI province, UMBRIA, central ITALY, near the TIBER, 13 mi/21 km SW of TERNI; 42°25′N 12°29′E. Nearby are remains of ancient Ocriculum.

Otrokovice (O-tro-KO-vi-TSE), German *Otrokowitz*, city, JIHOMORAVSKY province, E MORAVIA, CZECH REPUBLIC, 6 mi/9.7 km E of Zlín; 49°12′N 17°32′E. Railroad junction; tanning center producing mainly semi-finished goods for ZLÍN workshops; manufacturing (rubber, tires, chemicals, aircrafts). Municipal airport of Zlín area here. Founded in the 12th century. Part of Zlín 1949–1954.

Otrozhka (uh-TRO-shkah), industrial suburb of VORONEZH (5 mi/8 km NE of the city center), NW VORONEZH oblast, S central European Russia, on the left bank of the VORONEZH RIVER (tributary of the DON River), on highway and near railroad (Otrozhka junction); 51°44′N 39°16′E. Elevation 321 ft/97 m. Manufacturing (locomotives, railroad cars). Incorporated into Voronezh in 1940.

Ötscher (OET-cher), isolated mountain peak (5,770 ft/1,759 m) of the Styrian–Lower Austrian Limestone Alps, in the EISENWURZEN REGION, SW LOWER AUSTRIA, central AUSTRIA, 31 mi/50 km SW of Sankt Pölten; 47°52′N 15°12′E. Many caves; gorge at S foot; ski area (Lackenhof) at NW foot.

Otsego (aht-SEE-go), county (□ 526 sq mi/1,367.6 sq km; 2006 population 24,711), N MICHIGAN; ☉ GAYLORD; 45°01′N 84°36′W. Drained by STURGEON and BLACK rivers, and by North Branch of AU SABLE RIVER. Agriculture (cattle, hogs; potatoes, wheat, oats; dairy products); manufacturing (wood products, electronic equipment). Year-round resort area. Includes many small lakes and a state forest. Otsego State Park in SW; Tyrolean Hills Ski Area in E; Sylvan and Michaway ski areas in center. Organized 1875.

Otsego, county (□ 1,013 sq mi/2,633.8 sq km; 2006 population 62,583), central NEW YORK; ☉ COOPERSTOWN; 42°37′N 75°02′W. Bounded W by UNADILLA RIVER; drained by the SUSQUEHANNA River, issuing here from OTSEGO Lake. Manufacturing at ONEONTA, UNADILLA, WORCESTER. Dairying area; also livestock. Canadarago and 14-mi/23-km-long Otsego lakes are physically the easternmost of the FINGER LAKES. Mineral springs at Richfield Springs. Attractive region with many downstate residents' second homes. Formed 1791.

Otsego (aht-SEE-go), town (2000 population 3,933), ALLEGAN county, SW MICHIGAN, 9 mi/14.5 km ESE of ALLEGAN, and on KALAMAZOO RIVER (water power); 42°27′N 85°42′W. In agricultural area; manufacturing (plastic, paper, and metal products; machinery). Settled 1832; incorporated as village 1865, as city 1918.

Otsego (aht-SEE-go), town, WRIGHT county, E MINNESOTA, 27 mi/43 km NW of MINNEAPOLIS, on MISSISSIPPI RIVER, NE of mouth of CROW RIVER; 45°17′N 93°36′W. Residential community on NW fringe of Minneapolis–ST. PAUL (Twin Cities) urban area. Agriculture includes dairying; poultry, livestock; grain, soybeans.

Otsego Lake, OTSEGO county, N MICHIGAN; 4 mi/6.4 km long; 44°57′N 84°41′W. At Send, 7 mi/11.3 km S of GAYLORD, is Otsego Lake village (1990 population c.50). Resort with fine beaches. Otsego Lake State Park is on E side.

Otsego Lake (aht-SEE-go), E central NEW YORK, SE of UTICA, in a resort region; c.9 mi/14.5 km long; 42°45′N 74°54′W. The SUSQUEHANNA River originates here at lake outlet from its S end at COOPERSTOWN. State University of New York at Oneonta's Biological Field

Station on the W shore is a major limnological research facility. The lake is the Glimmerglass of James Fenimore Cooper's tales and physically easternmost of a series of glacially derived "finger" lakes in New York state.

Otselic River (aht-SIL-ik), c.45 mi/72 km long, central NEW YORK; rises in S MADISON county; flows SW, past CINCINNATUS, to TIOUGHNIOGA River at WHITNEY POINT.

Otse Peak (oo-TSAI) (5,000 ft/1,525 m), SE BOTSWANA, between RAMOTSWA and KANYE. Highest point in country. Formed by an outrider of the Magalieberg River from SOUTH AFRICA.

Otsu (OTS), city (2005 population 323,719), ⊙ SHIGA prefecture, S HONSHU, central JAPAN, on Lake BIWA; 35°00′N 135°51′E. Tourist center and port for pleasure boats on Lake Biwa. Manufacturing includes appliances, computer components, and plastic goods. A former imperial seat (2nd and 7th centuries), Otsu is the site of Mount Hiei Enryaku and the 7th-century Buddhist Midera temples. Ruins of a Tokugawa-era toll barrier. The poet Basho is buried here.

Otsuchi (OTS-chee), town, Kamihei county, IWATE prefecture, N HONSHU, NE JAPAN, 47 mi/75 km S of MORIOKA; 39°21′N 141°54′E. Seaweed (*wakame*, tangle).

Otsuki (O-tsuh-kee), city, YAMANASHI prefecture, central HONSHU, central JAPAN, near FUJIYAMA, 19 mi/30 km S of KOFU; 35°36′N 138°56′E. Textiles. Saruhashi bridge is located here.

Otsuki (OTS-kee), town, Hata county, KOCHI prefecture, S SHIKOKU, W JAPAN, 68 mi/110 km S of KOCHI; 32°50′N 132°42′E.

Otta (AW-tah), town, OGUN state, extreme SW NIGERIA, 20 mi/32 km SE of ILARO. Road junction. Cotton weaving, indigo dyeing; cacao, palm oil and kernels. Phosphate deposits.

Otta (AWT-tah), village, OPPLAND county, S central NORWAY, in GUDBRANDSDALEN, on LÅGEN River at mouth of OTTA River, on railroad, and 60 mi/97 km NW of LILLEHAMMER; 61°46′N 09°32′E. Slate and steatite quarries; livestock raising. Tourist resort. Nearby monument celebrates defeat (1612) of Scottish mercenaries in Swedish service by Otta peasants. In World War II, scene (April, 1940) of engagement between Anglo-Norwegian and German forces.

Ottakring (AWT-tah-kring), outer W district (□ 3 sq mi/7.8 sq km) of VIENNA, AUSTRIA; 48°13′N 16°19′E. Traditional workers' district with a high share of guest workers; large brewery; Castle Wilhelminerberg.

Ottappalam (o-tuh-PAH-luhm), town, PALAKKAD district, KERALA state, S INDIA, on PONNANI river, 50 mi/80 km SE of KOZHIKODE; 10°46′N 76°23′E. Rice, cassava, pepper, ginger. Sometimes spelled Ottapalam.

Otta River, 70 mi/113 km long, OPPLAND county, S central NORWAY; rises on NW slope of JOTUNHEIMEN Mountains near head of GEIRANGERFJORDEN; flows generally ESE through expansion (25 mi/40 km long) of Vågåvatn, past VÅGÅMO, to LÅGEN River at OTTA.

Ottauquechee River (AWT-ah-KWEE-chee), c.40 mi/64 km long, E VERMONT; rises in GREEN MOUNTAINS NE of RUTLAND, flows generally E, past WOODSTOCK (dam here), to the CONNECTICUT River S of White River Junction. Formerly Quechee River.

Ottaviano (OT-tah-VYAH-no), town, NAPOLI province, CAMPANIA, S ITALY, at NE foot of VESUVIUS, 11 mi/18 km E of NAPLES; 40°51′N 14°28′E. Wine center, known for famous "Lacrima Christi." Food processing, textile manufacturing (cotton, linen). Formerly Ottaiano.

Ottawa (AH-tah-wah), county (□ 721 sq mi/1,874.6 sq km; 2006 population 6,168), N central KANSAS; ⊙ MINNEAPOLIS; 39°07′N 97°39′W. Rolling prairie region, drained by SOLOMON RIVER and Salt Creek and in SW by SALINE RIVER. Railroad junction. Wheat, sorghum, alfalfa; cattle. Diversified manufacturing. Ottawa State Fishing Lake near center of

county. Rock sandstone formations in W center. Formed 1866.

Ottawa, county (□ 1,632 sq mi/4,243.2 sq km; 2006 population 257,671), SW MICHIGAN; ⊙ GRAND HAVEN; 42°55′N 86°13′W. Bounded W by LAKE MICHIGAN; drained by GRAND and BLACK rivers. Hogs, cattle, sheep, poultry; apples, cherries, onions, asparagus, grain; dairy products; tulip growing. Manufacturing at GRAND HAVEN, HOLLAND, and ZEELAND. Fisheries. Tourism; has resorts. SPRING LAKE, large backwater lake of Lake Michigan, receives GRAND RIVER in NW corner of county. Grand Haven Ski Bowl and Grand Haven State Park in NW; Holland State Park in SW. Organized 1837.

Ottawa (AH-tah-wah), county (□ 263 sq mi/683.8 sq km; 2006 population 41,331), N OHIO; ⊙ PORT CLINTON; 41°32′N 83°06′W. Bounded NE by LAKE ERIE; drained by PORTAGE RIVER and small Toussaint and Packer creeks. Includes PERRY'S VICTORY AND INTERNATIONAL PEACE MEMORIAL National Monument. In the Lake Plains physiographic region. Agricultural area (corn, soybeans; hogs); manufacturing at GENOA, Oakharbor, PORT CLINTON (rubber and plastic products, machinery). Limestone quarries; fisheries. BASS ISLANDS are resorts. Davis-Besse Nuclear Power Plant, 21 mi/34 km ESE of TOLEDO, uses cooling water from Lake Erie, and has a maximum dependable capacity of 918 Mwe. The first electricity was generated on August 28, 1977. Plant provides power to the city of Toledo and surrounding areas. The city was formed in 1840.

Ottawa, county (□ 484 sq mi/1,258.4 sq km; 2006 population 33,026), extreme NE OKLAHOMA; ⊙ MIAMI; 36°50′N 94°48′W. Bounded N by KANSAS state line, E by MISSOURI state line; part of the OZARKS are in E. Drained by NEOSHO (headwaters of LAKE OF THE CHEROKEES) and Spring rivers; includes section of Lake of the Cherokees (recreation). Contains part of lead- and zinc-mining region (centering on Picker) extending into Kansas and Missouri. Also livestock raising, dairying, agriculture (corn, wheat, hay, sorghum, soybeans); manufacturing at Miami; timber. Spring River State Park in NE; Twin Bridges State Park near center. Formed 1907.

Ottawa (AHT-uh-wuh), city (□ 1,073 sq mi/2,789.8 sq km; 2006 population 812,129), ⊙ CANADA, SE ONTARIO, at the confluence of the OTTAWA and Rideau rivers; 45°25′N 75°42′W. HULL, QUEBEC, just across the Ottawa at the mouth of the GATINEAU River, forms part of the metropolitan area. The RIDEAU CANAL separates the city into upper and lower towns; along its banks and those of the rivers are many landscaped drives as well as much of the city's land area, which totals 1,500 acres/607 ha. Ottawa is not primarily an industrial center; however, it has industries that produce, among other goods, paper and paper products, printed materials, telecommunications equipment, and electronics. In the 1980s, Ottawa and its suburbs NEPEAN and KANATA became, with the growth of high-technology industries, Canada's "Silicon Valley." The area's industries utilize the hydroelectric power of the Ottawa (CHAUDIÈRE FALLS) and Gatineau valleys. Since 1940, the largest employer in Ottawa has been the federal government. The National Capital Commission, a developer of public works, has done much to redevelop the core of the city, removing old railroad lines and building new parks (Confederation Square) and national buildings (National Arts Center, National Defence Building, Bank of Canada Building). Development of the national capital plan began with commissioning of French architect Jacques Greber, 1937–1939. The Greber plan, which included broad boulevards, green spaces, and a regional district that took in both sides of Ottawa River and the city of Hull, Quebec, was adopted in 1950, and led to the establishment of the National Capital Region (□ 4,662 sq mi/12,075 sq km; 2,719.5 sq mi/7,043.5

sq km in Ontario, 1,942.5 sq mi/5,031.1 sq km in Quebec). Gatineau National Capital Commission Park, Quebec, the only federal park not managed by Parks Canada, was developed as part of the plan. Several federal government offices and the Canadian Museum of Civilization were relocated to Hull in 1970s and 1980s, as well as the National Gallery of Canada. In part due to these development projects, tourism has become Ottawa's second-largest industry, attracting about 4 million people annually. Ottawa proper was founded in 1827 by John By, an engineer in charge of construction of the Rideau Canal. At first called Bytown, it was named after the Ottawa, an Algonquian-speaking people, in 1854. In 1858, Ottawa was chosen by Queen Victoria to be the capital of the United Provinces of Canada, and in 1867 it became the capital of the Dominion of Canada. The government buildings, built between 1859 and 1865, were burned in 1916 but immediately rebuilt on an enlarged scale. Other notable buildings are Rideau Hall, the residence of the Governor General; the Anglican and Roman Catholic cathedrals; the Bytown Museum; the Canadian Museum of Nature; the National Gallery; the National Arts Centre; the National Aviation Museum; the National Library and Public Archives Building; the National Museum of Science and Technology; the Dominion Observatory; the Royal Mint; and the Rideau Centre complex. The University of Ottawa, St. Paul University, and Carleton University are in the city. Ottawa is largely a bilingual city because federal government employees are required to know both English and French. In 2001 the former regional municipality of Ottawa-Carleton amalgamated into the city of Ottawa, significantly increasing its population. Name from the Aboriginal meaning "to trade."

Ottawa (AHT-uh-wah), city (2000 population 18,307), ⊙ LA SALLE county, N central ILLINOIS, at the confluence of the FOX and ILLINOIS rivers; 41°21′N 88°50′W. In a fertile farm area, the city has grain-oriented agriculture and manufactures glass, tools, building materials, and automobile parts. Points of interest include the site of the first Lincoln-Douglas debate (1858) and Fort Johnson (1832). Several state parks are in the area, and scenic attractions along the rivers draw many visitors. Incorporated as a city 1853.

Ottawa, city (2000 population 11,921), ⊙ FRANKLIN county, E KANSAS, on the MARAIS DES CYGNES River; 38°36′N 95°16′W. The railroad and industrial center of a farm area, it has a variety of light industries. The city is named for the Ottawa, who moved there (1832) after ceding their OHIO lands to the U.S.; they were subsequently removed (1867) to OKLAHOMA. Ottawa University is here. Incorporated 1867.

Ottawa (AH-tah-wah), village (2006 population 4,448), ⊙ PUTNAM county, NW OHIO, 18 mi/29 km N of LIMA, and on BLANCHARD River; 41°01′N 84°02′W. Manufacturing of television equipment, food and dairy products, clay and wood products. Oil wells, stone quarries. Founded 1833. Severe flooding in August 2007 destroyed much of the village.

Ottawa (AHT-uh-wuh), river, c.700 mi/1,130 km long, largest tributary of the SAINT LAWRENCE RIVER, Canada; rises in the Laurentian Highlands, SW QUEBEC, and flows generally W through La Vérendrye Provincial Park to Lake TIMISKAMING, then SE forming part of the Quebec-ONTARIO border, past OTTAWA, and into the Saint Lawrence River near MONTREAL; 45°19′N 73°55′W. Its lower course has several expansions, known as the ALLUMETTE, CHATS, and DESCHÊNES lakes and Lake of the TWO MOUNTAINS. Among its chief tributaries are the GATINEAU, LIÈVRE, and COULONGE rivers. Hydroelectric power stations at La Cave, DES JOACHIMS, BRYSON, Chenaux, Chats, CHAUDIÈRE FALLS, and CARILLON have a combined generating capacity of about 1.5 million kw. The river is navigable for large vessels as far as Ottawa; it is connected with Lake ONTARIO by the RIDEAU CANAL

Area is shown by the symbol □, and capital city or county seat by ⊙.

system. There is some farming in the valley below PEMBROKE, but lumbering is the chief industry along the lower river. Samuel de Champlain, the French explorer, was the first European to visit (1613–1615) the valley; the river, known then as the Grand River, later became an important highway for fur traders and missionaries.

Ottawa-Carleton, Canada: see OTTAWA, CARLETON.

Ottawa Hills (AH-tah-wah), residential village (2006 population 4,620), LUCAS county, NW OHIO, just W of TOLEDO; 41°40′N 83°39′W. Settled 1916, incorporated 1924.

Ottawa International Airport (AH-tuh-wuh), French, *Aéroport international d'Ottawa* (ai-ro-POR an-ter-na-syo-NAHL DAH-tuh-wuh) or **Ottawa Macdonald-Cartier International Airport** (AH-tuh-wuh muhk-DAH-nuhld–kahr-TYAI), SE ONTARIO, E central Canada, 6 mi/10 km S of OTTAWA city center, and just E of Rideau River. Handles c.3 million passengers annually. One of several Canadian airports with U.S. border pre-clearance (*prédédouanement*). Built c.1920; formerly a Canadian Armed Forces base. Airport code: YOW.

Ottawa Islands, NUNAVUT territory, Canada, group of 24 small islands and islets in HUDSON BAY, off NW Ungava Peninsula; c.70 mi/113 km long, 50 mi/80 km wide; 60°00′N 80°00′W. Main islands are Booth, Bronson, Gilmour (rises to over 1,800 ft/549 m), Perley, J. Gordon, Paltee, and Eddy.

Ottawa River (AH-tah-wah), c.50 mi/80 km long, W OHIO; rises in HARDIN county; flows NW and W, past LIMA, then N, past ELIDA, to AUGLAIZE RIVER in PUTNAM county; 40°59′N 84°14′W.

Ottendorf-Okrilla (OT-ten-dorf–o-KRIL-lah), village, SAXONY, E central GERMANY, 12 mi/19 km NNE of DRESDEN; 51°12′N 13°49′E. Manufacturing (glass, automotive parts).

Otterbach (OT-ter-bahkh), village, RHINELAND-PALATINATE, W GERMANY, on the LAUTER, and 3 mi/4.8 km NNW of KAISERSLAUTERN; 49°29′N 07°45′E. Grain, potatoes.

Otterbein, town (2000 population 1,312), BENTON county, W INDIANA, 15 mi/24 km SE of FOWLER; 40°29′N 87°05′W. Manufacturing (building components, transportation equipment). Agricultural area (corn, soybeans; livestock). Laid out 1872.

Otterberg (OT-ter-berg), town, RHINELAND-PALATINATE, W GERMANY, on the LAUTER, 4 mi/6.4 km N of KAISERSLAUTERN; 49°31′N 07°46′E. Automotive accessories; agriculture (rye, oats, potatoes). Has 13th century church.

Otterbourne (AHT-uh-bawn), village (2001 population 1,466), HAMPSHIRE, S ENGLAND, near ITCHEN RIVER, 5 mi/8 km SW of WINCHESTER; 51°00′N 01°20′W.

Otterburn (AHT-uh-buhn), village (2001 population 550), NORTHUMBERLAND, NE ENGLAND, 16 m/25.6 km WNW of MORPETH; 55°13′N 02°10′W. It was the scene of the Battle of Chevy Chase, a victory (1388) of the Scots over the English. The engagement, in which Sir Henry Percy was taken captive, is the subject of the English ballad "Chevy Chase" and the Scots ballad "Otterburn." Military camp 2 mi/3.2 km N. Artillery ranges on moors to N.

Otterburne (AH-tuhr-buhrn), unincorporated village, SE MANITOBA, W central Canada, on Rat River, 27 mi/43 km S of WINNIPEG, and in De Salaberry rural municipality; 49°30′N 97°03′W. Grain; dairying.

Otterburn Park (AH-tuhr-buhrn PAHRK), city (□ 2 sq mi/5.2 sq km; 2006 population 8,468), Montérégie region, S QUEBEC, E Canada, 21 mi/34 km from MONTREAL; 45°31′N 73°13′W. Established 1969. Part of the Metropolitan Community of Montreal (*Communauté Metropolitaine de Montréal*).

Otter Creek, 30 mi/48 km long, S central UTAH; rises in FISH LAKE PLATEAU in central SEVIER county; flows SSW, through SEVIER PLATEAU, through Koosharem Reservoir, past KOOSHAREM to Otter Creek Reservoir

(5.5 mi/8.9 km long), formed by dam on East Fork of SEVIER RIVER. Creek is dammed near source to SEVIER RIVER 13 mi/21 km ESE of JUNCTION.

Otter Creek, c.100 mi/161 km long, W VERMONT; rises in GREEN MOUNTAINS near DORSET, flows generally NW, past RUTLAND, PROCTOR, MIDDLEBURY, and VERGENNES, to Lake CHAMPLAIN near FERRISBURG. In upper course, flows between the Taconics (W) and GREEN MOUNTAINS.

Otter Creek Recreational Area, KENTUCKY: see MEADE.

Otterfing (OT-ter-fing), village, Upper Bavaria, BAVARIA, S GERMANY, 12 mi/19 km NNE of BAD Tölz; 47°54′N 11°39′E.

Otter Lake (AH-tuhr LAIK), municipality (□ 192 sq mi/499.2 sq km; 2006 population 867), Outaouais region, SW QUEBEC, E Canada, 24 mi/38 km from Pontiac; 45°51′N 76°26′W.

Otter Lake, village (2000 population 437), LAPEER and GENESEE counties, E MICHIGAN, 18 mi/29 km NE of FLINT; 43°12′N 83°27′W. In farm and lake region. Manufacturing (transportation equipment, chemicals).

Otter Lake, lake resort hamlet, ONEIDA county, central NEW YORK, in the W ADIRONDACK MOUNTAINS, on small Otter Lake, 35 mi/56 km N of Utica; 43°36′N 75°06′W.

Otter Lake, reservoir (□ 1 sq mi/2.6 sq km), MACOUPIN county, central ILLINOIS, on West Fork of Otter Creek, 30 mi/48 km NW of SPRINGFIELD; 39°25′N 89°55′W. Maximum capacity of 24,708 acre-ft. Formed by Otter Lake Dam (71 ft/22 m high), built (1969) for water supply; also used for recreation.

Otterndorf (OT-tern-dorf), town, LOWER SAXONY, NW GERMANY, near Elbe estuary, 9 mi/14.5 km ESE of CUXHAVEN; 53°48′N 08°54′E. Agricultural market center; food processing (flour and dairy products, canned goods, spirits); feed. Was first mentioned 1261; chartered 1400.

Otterøya (AWT-tuhr-uh-yah), island (□ 29 sq mi/75.4 sq km) at mouth of MOLDEFJORDEN, MØRE OG ROMSDAL county, W NORWAY, 7 mi/11.3 km W of MOLDE; 10 mi/16 km long, 4 mi/6.4 km wide. Fisheries.

Otter, Peaks of (AH-tuhr) (Flat Top: 4,004 ft/1,220 m; Sharp Top: 3,870 ft/1,180 m), twin peaks, BEDFORD county, W central VIRGINIA, 8 mi/13 km NW of BEDFORD, in BLUE RIDGE, in Jefferson National Forest, immediately S of BLUE RIDGE NATIONAL PARKWAY; 37°26′N 79°35′W.

Otter River (AHT-uhr), 24 mi/39 km long, DEVON, SW ENGLAND; rises near SOMERSET border; flows SW, past HONITON, to the CHANNEL at BUDLEIGH SALTERTON.

Otter River, MASSACHUSETTS: see GARDNER.

Ottersberg (OT-ters-berg), town, LOWER SAXONY, NW GERMANY, 14 mi/23 km E of BREMEN; 53°07′N 10°36′E. Sawmilling.

Ottersweier (ot-ters-VEI-uhr), village, MIDDLE UPPER RHINE, W BADEN-Württemberg, SW GERMANY, 9 mi/14.5 km SW of BADEN-BADEN; 48°40′N 08°07′E. Viticulture, fruit.

Otter Tail, county (□ 2,225 sq mi/5,785 sq km; 2006 population 57,817), W MINNESOTA; ⊙ FERGUS FALLS; 46°24′N 95°42′W. Extensively watered agricultural area drained by PELICAN, POMME DE TERRE, and OTTER TAIL rivers, and numerous lakes. Alfalfa, hay, wheat, corn, oats, barley, soybeans, sugar beets, beans, sunflowers; hogs, sheep, cattle, poultry; dairying. Chief lakes are Lake LIDA in NW, and DEAD LAKE and OTTER TAIL LAKE in center. Orwell Wildlife Area in SW on Orwell Reservoir; Maplewood State Park in NW center; Inspiration Peak (1,750 ft/533 m) in State Park in S. Formed 1858.

Ottertail, village (2000 population 451), OTTER TAIL county, W MINNESOTA, just E of OTTER TAIL LAKE, 27 mi/43 km NE of FERGUS FALLS; 46°25′N 95°33′W. Dairying; poultry; grain; manufacturing (fur processing; concrete). Numerous lakes in area. Sometimes Otter Tail.

Otter Tail Lake (□ 23 sq mi/60 sq km), OTTER TAIL county, W MINNESOTA, 17 mi/27 km ENE of FERGUS FALLS; 9 mi/14.5 km long, 3 mi/4.8 km wide; 46°24′N 95°39′W. Elevation 1,320 ft/402 m. Fed from NW and drained to SW by OTTER TAIL RIVER. Fishing resorts.

Otter Tail River, 150 mi/241 km long, W central MINNESOTA; rises in S part of CLEARWATER county, c.35 mi/56 km SW of BEMIDJI; flows S, through ELBOW, Many Point, and Round lakes, through Tamarac National Wildlife Refuge, past FRAZEE, through Rose and LONG lakes, E through MUD, Little Pine, and PINE lakes, S through Pine, RUSH LAKE, and OTTER TAIL lakes, then W, past FERGUS FALLS through Orwell Reservoir, to BRECKENRIDGE (opposite WAHPETON, NORTH DAKOTA), where it joins BOIS DE SIOUX RIVER to form RED RIVER of the North; 47°01′N 95°32′W.

Otterup (O-tuh-roop), town, FYN county, DENMARK, on N FYN island, and 8 mi/12.9 km N of ODENSE; 55°30′N 10°20′E. Manufacturing (foundry; feed production).

Otterville, town (2000 population 120), JERSEY county, W ILLINOIS, 8 mi/12.9 km SSW of JERSEYVILLE; 39°02′N 90°24′W. In apple-growing area.

Otterville, town (2000 population 476), COOPER county, central MISSOURI, near LAMINE RIVER, 12 mi/19 km E of SEDALIA; 38°42′N 93°00′W. Cattle; corn, wheat; lumber.

Otterville (AH-tuhr-vil), unincorporated village, S ONTARIO, E central Canada, on Otter Creek, 16 mi/26 km SE of WOODSTOCK, and included in NORWICH township; 42°55′N 80°36′W. Dairy products; tobacco (declining), fruit (cherries, peaches, apples, strawberries), peanuts, vegetables.

Ottery St. Mary (AHT-uh-ree suhnt MER-ee), town (2001 population 8,219), E DEVON, SW ENGLAND, on OTTER RIVER, and 11 mi/18 km E of EXETER; 50°45′N 03°16′W. Has 13th-century church and Tudor mansion. Coleridge born here.

Ottignies-Louvain-la-Neuve (AH-teen-yee–loo-VAN–lah–NUHV), commune (2006 population 29,669), Nivelles district, BRABANT province, central BELGIUM, on DYLE RIVER, and 15 mi/24 km SE of BRUSSELS. Agricultural market. Bois-des-Rêves Provincial Estate nearby.

Ottilien River, PAPUA NEW GUINEA: see RAMU RIVER.

Ottine (O-tein), unincorporated village, GONZALES county, S central TEXAS, on SAN MARCOS RIVER, and 10 mi/16 km NW of Gonzales; 29°35′N 97°35′W. Health resort, with mineral springs. Site of Palmetto State Park, known for its semi-tropical plants.

Öttingen in Bayern (OT-ting-uhn in BEI-ern) or **Oettingen,** town, SWABIA, W BAVARIA, SE GERMANY, on the Wörnitz, 9 mi/14.5 km NNE of NÖRDLINGEN; 48°57′N 10°36′E. Tourism; manufacturing of organs, woodworking, brewing. Potatoes; ducks, hogs. Situated on N edge of Ries crater, which resulted from a meteor impact fifteen million years ago. Geological teaching path. Has 17th century castle and 14th century church. Chartered 12th century.

Ottmachau, POLAND: see OTMUCHOW.

Ottmarsheim (UHT-mahrz-heim), commune (□ 10 sq mi/26 sq km), HAUT-RHIN department, ALSACE region, E FRANCE; 47°47′N 07°31′E. Hydroelectric plant (built 1948–1952) on Grand Canal of ALSACE, parallel to RHINE RIVER. River port with major chemical complex. Metallurgy. Octagonal church (11th century). City of MULHOUSE is 8 mi/12.9 km W. Major bridge crosses Rhine into GERMANY.

Ottnang am Hausruck (AWT-nahng ahm HOUS-ruk), township, central UPPER AUSTRIA, in the HAUSRUCK mountains, 6 mi/9.7 km N of Vöcklabruck. Manufacturing of wood products, metals.

Otto Beit Bridge, ZIMBABWE: see CHIRUNDU.

Ottobeuren (ot-tuh-BOI-ren), village, SWABIA, SW BAVARIA, GERMANY, 6 mi/9.7 km ESE of MEMMINGEN; 47°57′N 10°18′E. Metal and woodworking, brewing,

---

Cross-references are shown in SMALL CAPITALS. The pronunciation guide is shown on page xix. The sources of population figures are shown on page xvii.

meat preserving, tanning. Church of former Benedictine abbey (1764–1802) is one of outstanding baroque buildings of GERMANY. Kneipp born here.

Ottobrunn (ot-tuh-BRUN), suburb of MUNICH, BAVARIA, S GERMANY, 6 mi/9.7 km SE of city center; 48°04′N 11°39′E. Aircraft industry.

Otto, Cerro (OT-to, SER-ro), Andean peak (4,600 ft/ 1,402 m), SW RÍO NEGRO province, ARGENTINA, in Nahuel Huapí national park, 5 mi/8 km WSW of SAN CARLOS DE BARILOCHE; 41°09′S 71°23′W. Popular skiing area.

Ottoman Empire, vast state founded in the late 13th century by Turkish tribes in ANATOLIA and ruled by the descendants of Osman I until its dissolution in 1918. Modern TURKEY formed only part of the empire, but the terms "Turkey" and "Ottoman Empire" were often used interchangeably. The Ottoman state began as one of many small Turkish states that emerged in ASIA MINOR during the breakdown of the empire of the Seljuk Turks. The Ottoman Turks began to absorb the other states, and during the reign (1451–1481) of Muhammad II they ended all other local Turkish dynasties. The early phase of Ottoman expansion took place under Osman I, Murad Orkhan I, and Beyazid I at the expense of the BYZANTINE EMPIRE, BULGARIA, and SERBIA. BURSA fell in 1326 and Adrianople (the modern EDIRNE) in 1361; each in turn became the capital of the empire. The great Ottoman victories of Kossovo (1389) and Nikopol (1396) placed large parts of the BALKAN PENINSULA under Ottoman rule and awakened Europe to the Ottoman danger. The Ottoman siege of CONSTANTINOPLE was lifted at the appearance of Tamerlane, who defeated and captured Beyazid in 1402. The empire, reunited by Muhammad I, expanded victoriously under Muhammad's successors Murad II and Muhammad II. The victory (1444) at VARNA over a crusading army led by Ladislaus III of POLAND was followed in 1453 by the capture of Constantinople. Within a century the Ottomans had changed from a nomadic horde to the heirs of the most ancient surviving empire of Europe. Turkish expansion reached its peak in the 16th century under Selim I and Sulayman I (Sulayman the Magnificent). The Hungarian defeat (1526) at Mohacs prepared the way for the capture (1541) of Buda and the absorption of the major part of HUNGARY by the Ottoman Empire; TRANSYLVANIA became a tributary principality, as did WALACHIA and Moldavia. The Asiatic borders of the empire were pushed deep into PERSIA and ARABIA. Selim I defeated the Mamluks of EGYPT and SYRIA, took CAIRO in 1517, and assumed the succession to the Caliphate. ALGIERS was taken in 1518. Most of the Venetian and other Latin possessions in Greece also fell to the sultans. During the reign of Sulayman I began (1535) the traditional friendship between FRANCE and TURKEY, directed against Hapsburg AUSTRIA and SPAIN. The first serious blow by Europe to the empire was the naval defeat of Lepanto (1571), inflicted on the fleet of Selim II by the Spanish and Venetians under John of Austria. However, Murad IV in the 17th century temporarily restored Turkish military prestige by his victory (1638) over Persia. CRETE was conquered from VENICE, and in 1683 a huge Turkish army under Grand Vizier Kara Mustafa surrounded VIENNA. The relief of Vienna by John III of Poland and subsequent campaigns cost Turkey Hungary and other territories. Economically, socially, and militarily, Turkey remained a medieval state, unaffected by the developments in the rest of Europe. Turkish domination over the northern part of Africa (except TRIPOLI and Egypt) was never well defined or effective, and the E border was inconstant, shifting according to frequent wars with Persia. A positive feature in Ottoman administration was the religious toleration generally extended to all non-Muslims. This, however, did not prevent occasional massacres

and discriminatory fiscal practices. In Constantinople the Greeks and Armenians held a privileged status and were very influential in commerce and politics. The Russo-Turkish war of 1828–1829 and the war with Muhammad Ali of Egypt resulted in the loss of Greece and Egypt, the protectorate of Russia over Moldavia and Walachia, and the semi-independence of Serbia. By the 19th century Turkey was known as the Sick Man of Europe. Through a series of treaties of capitulation from the 16th to the 18th century, the Ottoman Empire gradually lost its economic independence. As a result of the Russo-Turkish war of 1877–1878, in which Turkey was defeated despite its surprisingly vigorous stand, Romania (i.e., Walachia and Moldavia), Serbia, and Montenegro were declared fully independent, and BOSNIA AND HERZEGOVINA passed under Austrian administration. Bulgaria, made a virtually independent principality, annexed (1885) EASTERN RUMELIA with impunity. The Armenian massacres of the late 19th century turned world public opinion against Turkey. In 1908 the rebellious Young Turks forced the restoration of the constitution of 1876, and in 1909 the parliament deposed the sultan and put Muhammad V on the throne. In the two successive Balkan wars (1912–1913), Turkey lost nearly its entire territory in Europe to Bulgaria, Serbia, Greece, and newly independent ALBANIA. The outbreak of World War I found Turkey solidly lined up with the Central Powers. Although the Turkish troops were successful against the Allies in the Gallipoli (GALIBOLU) campaign (1915), Arabia rose against Turkish rule when British forces occupied (1917) BAGHDAD and JERUSALEM. In 1918, Turkish resistance collapsed both in Asia and Europe. An armistice was concluded in October, and the Ottoman Empire came to an end. The Treaty of Sèvres confirmed its dissolution. With the victory of the Turkish nationalists, who had refused to accept the peace terms and who overthrew the sultan in 1922, the history of modern Turkey began.

Ottosen, town (2000 population 61), HUMBOLDT county, N central IOWA, 15 mi/24 km NW of DAKOTA CITY; 42°53′N 94°22′W.

Ottoville (AH-to-vil), village (2006 population 862), PUTNAM county, NW OHIO, 18 mi/29 km NW of LIMA, and on AUGLAIZE RIVER; 40°56′N 84°20′W.

Ottumwa, city (2000 population 24,998), ⊙ WAPELLO county, SE IOWA, on both banks of the DES MOINES RIVER; 41°01′N 92°25′W. A commercial and industrial center in a farm and coal area, Ottumwa's economy is based on its meat-packing plant and a farm-machinery industry, although it has a diverse variety of other manufacturing. In the center of the city is Ottumwa Park, which was developed from a reclaimed river bottom. Indian Hills Community College extension is 6 mi/9.7 km N, near airport. Incorporated 1851.

Ottweiler (OT-vei-luhr), town, E SAARLAND, GERMANY, on BLIES RIVER, 15 mi/24 km NNE of Saarbrücken; 49°24′N 07°10′E. Coal mining; iron and steel industry. Renaissance and Baroque buildings.

Ottynia, UKRAINE: see OTYNIYA.

Otumba de Gómez Farías (o-TOOM-bah dai GO-mes fah-REE-ahs), city and township, ⊙ Otumba municipio, MEXICO state, central MEXICO, 30 mi/48 km NE of MEXICO CITY; 19°42′N 98°45′W. Maguey, corn; livestock. On nearby plains Cortés fought a fierce battle (1520) in which 20,000 Indians are said to have been slain.

Otuquis, Bañados del (o-TOO-kees, bah-NYAH-dos del), marshy area in SANTA CRUZ department, E BOLIVIA, in the CHACO, c.35 mi/56 km WSW of PUERTO SUÁREZ; c.45 mi/72 km long, 20 mi/32 km wide; 19°20′S 58°30′W. Formed by OTUQUIS River.

Otuquis River, c.100 mi/161 km long, SANTA CRUZ department, E BOLIVIA; formed by confluence of TUCAVACA and SAN RAFAEL rivers c.3 mi/4.8 km S of Tucavaca; flows SSE, across the NE CHACO, through

the Bañados del OTUQUIS to the PARAGUAY near BAHÍA NEGRA (PARAGUAY); 20°10′S 58°10′W.

Otura (o-TOO-rah), town, GRANADA province, S SPAIN, 6 mi/9.7 km S of GRANADA. Cereals, olive oil, wine.

Ótura, SLOVAKIA: see STARA TURA.

Oturkpo (aw-TUHRK-po), town, BENUE state, S central NIGERIA, on railroad, 45 mi/72 km SW of MAKURDI; 07°13′N 08°09′E. Shea-nut processing; sesame, cassava, durra.

Otuyo (o-TOO-yo), canton, CORNELIO SAAVEDRA province, POTOSÍ department, W central BOLIVIA, 6 mi/10 km S of MILLARES, on the POTOSÍ-SUCRE highway; 19°29′S 65°15′W. Elevation 10,886 ft/3,318 m. Limestone, phosphate, and gypsum deposits. Agriculture (potatoes, yucca, bananas, wheat, barley); cattle.

Otuzco (o-TOOS-ko), province (□ 1,343 sq mi/3,491.8 sq km), LA LIBERTAD region, NW PERU; ⊙ OTUZCO; 07°50′S 78°35′W. N central province of La Libertad region. N border on CAJAMARCA region.

Otuzco (o-TOOS-ko), city, ⊙ OTUZCO province, LA LIBERTAD region, NW PERU, in CORDILLERA OCCIDENTAL of the ANDES, 34 mi/55 km ENE of TRUJILLO; 07°54′S 78°35′W. Elevation 8,645 ft/2,635 m. Cotton, rice, cereals; livestock.

Otuzy, UKRAINE: see SHCHEBETOVKA.

Otvazhnyy, RUSSIA: see ZHIGULEVSK.

Otvotsk, POLAND: see OTWOCK.

Otway (AHT-wai), village (2006 population 84), SCIOTO county, S OHIO, 13 mi/21 km WNW of PORTSMOUTH; 38°52′N 83°11′W. Sawmills.

Otway Bay (OT-wai), inlet of the PACIFIC in TIERRA DEL FUEGO, CHILE, between SANTA INÉS Island and DESOLATION Island.

Otway, Cape (OT-wai), S VICTORIA, SE AUSTRALIA, on N side of W approach to BASS STRAIT, on GREAT OCEAN ROAD, 138 mi/222 km SW of MELBOURNE; 38°51′S 143°31′E. Lighthouse. OTWAY NATIONAL PARK, Port Campbell National Park.

Otway National Park (OT-wai) (□ 49 sq mi/127.4 sq km), S central VICTORIA, SE AUSTRALIA, 125 mi/201 km SW of MELBOURNE at Cape OTWAY; 12 mi/19 km long, 9 mi/14 km wide. BASS STRAIT to SE, INDIAN (Southern) Ocean to SW. S end of OTWAY RANGES, covered in mountain ash, myrtle beech, tree ferns. Wallabies, ringtail possums; bowerbirds, parrots, bristlebirds, goshawks. Camping, picnicking, surfing, fishing, hiking. Established 1981.

Otway Ranges (AHT-wai), mountain range, S central VICTORIA, SE AUSTRALIA, beginning 50 mi/80 km SW of MELBOURNE, extending 50 mi/80 km, from N of ANGLESEA on NE to Cape OTWAY on SW; 38°40′S 143°35′E. Parallels coast of BASS STRAIT of Southern (INDIAN) Ocean. Average 10 mi/16 km wide. Highest point Mount Sabine (1,912 ft/583 m) near center. OTWAY NATIONAL PARK at SW end of range. Timber; dairying.

Otway Sound (OT-wai), large inlet of Strait of MAGELLAN, in S MAGALLANES province, CHILE, between BRUNSWICK PENINSULA and RIESCO Island; 50 mi/80 km long, 12 mi/19 km—20 mi/32 km wide.

Otwell, village, PIKE county, SW INDIANA, 9 mi/14.5 km WNW of JASPER. Agriculture (wheat, corn; cattle, hogs). Bituminous coal, old surface mines. Pike State Forest to SW.

Otwock (OT-votsk), Russian *Otvotsk*, town (2002 population 42,621), Warszawa province, E central POLAND, near the VISTULA, on railroad, and 14 mi/23 km SE of WARSAW; 52°08′N 21°19′E. Cement manufacturing, flour milling; health resort (notably for tubercular patients). Meteorological station.

Otyniya (o-TI-nee-yah) (Russian *Otynya*) (Polish *Ottynia*), town (2004 population 10,400), central IVANO-FRANKIVS'K oblast, W UKRAINE, on the right tributary of the BYSTRYTSYA RIVER, on road and railroad, and 15 mi/24 km SSE of IVANO-FRANKIVS'K; 48°44′N 24°51′E.

Elevation 859 ft/261 m. Furniture factory, bakery; building materials, pottery, bricks; vegetable-oil extracting. Known since the 13th century; town since 1940. Jewish community since the 18th century; received full citizenship rights under Austro-Hungarian rule in 1867; numbering 1,730 in 1939; destroyed during World War II.

Otynya, UKRAINE: see OTYNIYA.

Ötz, AUSTRIA: see OETZ.

Otzberg (OTS-berg), commune, HESSE, central GERMANY, 12 mi/19 km ESE of DARMSTADT; 49°52′N 08°54′E. Sugar beets. Formed by several villages. Has ruined castle.

Otzoloapan (ot-so-LO-AH-pahn), town, in SW MEXICO state, MEXICO, 16 mi/25 km NW of TEJUPILCO DE HIDALGO; 19°14′N 99°56′W. Elevation 5,906 ft/1,800 m. Agriculture (sugarcane, peanuts), cattle raising.

Otzolotepec, MEXICO: see VILLA CUAUHTÉMOC.

Ötztal (OOETS-tahl), valley of TYROL, W AUSTRIA, extending 30 mi/48 km S from Inn River along the Ötztaler Ache, which rises in the Ötztal Alps; 47°14′N 10°51′E. Flanked by peaks and glaciers of the Ötztal Alps (SW) and STUBAI ALPS (NE). The valley is known for its scenic beauty. Important tourist region in summer and winter. The resorts of Oetz, Umgauser, Löngerfeld, Sölden, and Obergurgl are located here. Panoramic pass road across Timmelsjoch at an elevation of 7,541 ft/2,298 m to South Tyrol.

Ötztal Alps (OOETS-tahl), mountain group of the Central Alps, in the TYROL province, W AUSTRIA, S of the Inn River, and extending into N Italy; 46°45′N 10°55′E. Rises to 11,485 ft/3,501 m in the Wildspitze, which is noted for glaciers, of which Gepatschfoner is the largest. Glacier skiing takes place here in the summer. The village of Obergurgl is a skiing resort and a starting point for mt. climbing. In 1991 at the Tisenjoch, a mummified body of man from the Stone Age, dating back to 3000 B.C.E. was found here.

Ouachita (WAHSH-uh-tah), county (□ 739 sq mi/1,921.4 sq km; 2006 population 26,710), S ARKANSAS; ⊙ CAMDEN; 33°35′N 92°52′W. Drained by OUACHITA RIVER (forms S part of E boundary); LITTLE MISSOURI RIVER forms W part of N boundary, Two Bayou forms part of E boundary, Smackover Creek forms part of S boundary. Agriculture (cattle, hogs). Manufacturing at Camden. Oil and gas wells; timber; gravel, asphalt; lumbering. Poison Spring Battleground State Historical Monument in W center; part of Poison Springs Wildlife Management Area and White Oak Lake State Park in NW. Formed 1844.

Ouachita (waw-shi-TAW), parish (□ 642 sq mi/1,669.2 sq km; 2006 population 149,259), NE central LOUISIANA; ⊙ MONROE; 32°31′N 92°05′W. Bounded E by BAYOU LAFOURCHE; intersected by OUACHITA RIVER and BAYOU D'ARBONNE. Agriculture (cotton, rice, soybeans, blueberries, peaches, sorghum, corn, home gardens, nursery crops, vegetables; cattle, horses, exotic fowl, alligators; dairying). Large natural-gas fields, gas pipelines. Logging, varied manufacturing, including food products, apparel, paper products, plastic products, chemicals, electronic equipment, industrial machinery, metal products. Explored by Hernando de Soto in 1541. Formed 1807. Part of D'Arbonne National Wildlife Refuge in N, Cheniere Brake Fish Preserve and lake at center, Ouachita and part of Russell Sage State Wildlife areas in E.

Ouachita, Lake (WAHSH-uh-tah), reservoir (□ 63 sq mi/163.8 sq km), W central ARKANSAS, on OUACHITA RIVER, mostly within Ouachita National Forest, 7 mi/11.3 km NW of HOT SPRINGS; 34°34′N 93°11′W. Maximum capacity 3,760,000 acre-ft. Extends E-W into GARLAND and MONTGOMERY counties. Formed by BLAKELY MOUNTAIN DAM (240 ft/73 m high), built (1953) by Army Corps of Engineers for power generation, flood control, and recreation.

Ouachita Mountains (WAHSH-uh-tah), range of E-W ridges between the ARKANSAS and RED rivers, extending c.200 mi/320 km from central ARKANSAS into SE OKLAHOMA. MAGAZINE MOUNTAIN (c.2,800 ft/850 m high) is the tallest peak. The Ouachita Mountains are geologically considered an outlier of the APPALACHIAN MOUNTAINS. They are composed of strongly folded and faulted sedimentary rocks. A whetstone industry is near HOT SPRINGS, Arkansas. Mineral springs, lakes, and extensive wooded areas attract tourists. Several parts of the region have been set aside as public parks or forest reservations.

Ouachita Mountains (WAH-shuh-tuh), LEFLORE, MCCURTAIN, PUSHMATAHA, ATOKA, and LATIMER counties, S Oklahoma Range, formed by folding and faulting of resistant sedimentary rocks. Long, parallel ridges that run generally E-W. Ridges are as much as 1,500 ft/460 m above the adjacent valleys; greatest elevation is 2,666 ft/810 m. Farming in valleys; timber and grazing on slopes.

Ouachita River (WAHSH-uh-tah), c.600 mi/966 km long; rising in the OUACHITA MOUNTAINS, POLK county, W ARKANSAS, N of MENA; flows E, SE, and S through the Ouachita Mountains and the cotton-producing region of S Arkansas and NE LOUISIANA and into the RED RIVER system. It is joined by the TENSAS RIVER at JONESVILLE, Louisiana, to form the BLACK RIVER. HOT SPRINGS, Arkansas, and MONROE, Louisiana, are the largest cities on the river; also important are ARKADELPHIA and CAMDEN, Arkansas The river is navigable for shallow-draft vessels below Arkadelphia. Three dams in the river near Hot Springs—Remmel (completed 1925), Carpenter (1931), and Blakeley Mountain (1955)—impound respectively Lake CATHERINE, Lake HAMILTON, and Lake OUACHITA (63 sq mi/101 sq km, Arkansas's largest). Flows through Felsenthal National Wildlife Refuge in S Arkansas and Upper Ouachita National Wildlife Refuge in N Louisiana. There is a hydroelectric power plant at Blakeley Mountain Dam. Felsenthal Dam in S Arkansas impounds Lake JACK LEE. The lakes, part of a Federal flood-control project, are the center of a popular recreation area.

Ouachita River, Texas and Oklahoma: see WASHITA river.

Ouad Amlil (wahd ahm-LEEL), village, Taza province, Taza-Al Hoceima-Taounate administrative region, MOROCCO, 16 mi/25 km W of TAZA.

Ouadâne (wah-dah-NE), hamlet (2000 population 3,695), Adrar administrative region, N central MAURITANIA, close to GUELB ER RICHAT mountain, c.99 mi/159 km ENE of ATAR; 20°56′N 11°37′W.

Ouadda (wah-DAH), village, HAUTE-KOTTO prefecture, E CENTRAL AFRICAN REPUBLIC, 260 mi/418 km N of BANGASSOU; 08°09′N 22°20′E. Trading post. Diamond deposits. Airport.

Ouaddaï (wah-DEI), administrative region (2000 population 651,100), E Chad; ⊙ ABÉCHÉ; 13°00′N 21°00′E. Borders SALAMAT (S), GUERA and BATHA (W), and WADI FIRA (N) administrative regions, as well as SUDAN (E) and CENTRAL AFRICAN REPUBLIC (SE). Drained by BATHA River. Ouaddaï was a prefecture (with the same capital) prior to Chad's administrative division reorganization from fourteen prefectures to twenty-eight departments. It was recreated as a region following a decree in October 2002 that reorganized Chad's administrative divisions into eighteen regions.

Ouadi-Rimé (WAH-dee–ree-MAH), village oasis, BATHA administrative region, central CHAD, 55 mi/89 km N of ATI, on caravan road.

Ouad Laou (wahd lahw), village, Tétouan province, Tanger-Tétouan administrative region, MOROCCO, on the MEDITERRANEAN SEA at mouth of the Ouad LAOU river, 19 mi/30 km E of TETOUAN. Beach resort and fishing village.

Ouâd Naga (WAHD nah-GAH), village (2000 population 10,291), Trarza administrative region, SW MAURITANIA, 31 mi/50 km SE of NOUAKCHOTT, on NOUAKCHOTT-NÉMA HIGHWAY; 17°55′N 15°29′E. Livestock raising (cattle, goats, sheep, camels); gum arabic.

Ouad Zem (wahd ZEM), town, Khouribga province, Chaouia-Ouardigha administrative region, W central MOROCCO, 80 mi/129 km SE of CASABLANCA. Railroad terminus; agricultural trade center (grain; livestock); flour milling. Phosphates in KHOURIBGA area (W). Also spelled Oued Zem.

Ouagadougou (wah-ga-DOO-goo), city (2005 population 1,029,297), ⊙ Centre region, KADIOGO province, and BURKINA FASO; 12°20′N 01°40′W. The nation's largest city and its administrative, communications, and economic center, due in part to its central location. Also the trade and distribution center for an agricultural region whose main crop is peanuts. The city's industry is limited to textiles, handicrafts, and the processing of food and beverages. It has an international airport, railroad to ABIDJAN and CÔTE D'IVOIRE, and road links with NIAMEY, NIGER. Founded in the late 11th century as the capital of a MOSSI empire ruled by the *Moro Naba* [=ruler of the world]. It remained a center of Mossi power until 1896, when FRENCH forces captured it. Formerly spelled Wagadugu.

Ouagbo (WAHG-bo), village, S BENIN, on railroad, and 37 mi/60 km NW of PORTO-NOVO; 06°48′N 02°10′E. Palm kernels, palm oil.

Ouahigouya (wah-hee-GOO-yah), town (2005 population 62,934), ⊙ YATENGA province and NORD region, N BURKINA FASO, 100 mi/161 km NW of OUAGADOUGOU, near MALI border; 13°31′N 02°20′W. Largest town in N central Burkina Faso. Livestock-raising center. Grows chiefly shea nuts and cotton for export; also millet, corn, beans, manioc, potatoes, and sesame. Has daily market and hospital.

Ouaka (wah-KAH), prefecture (□ 19,261 sq mi/50,078.6 sq km; 2003 population 276,710), S central CENTRAL AFRICAN REPUBLIC; ⊙ BAMBARI. Bordered N by BAMINGUI-BANGORAN prefecture, NE by HAUTE-KOTTO prefecture, SE by BASSE-KOTTO prefecture, S by Democratic Republic of the CONGO, SW by KÉMO prefecture, and NW by NANA-GRÉBIZI prefecture. Drained by Kandjia and OUAKA rivers. Agriculture (cotton, coffee); manufacturing (textiles); nickel deposits. Main centers are Bambari, Battinga, DEMBA, IPPY, and KOUANGO.

Ouakam (WAH-kahm), town (2004 population 45,286), DAKAR administrative region, W SENEGAL, on CAPE VERDE peninsula, 4.5 mi/7.2 km NW of DAKAR; 14°43′N 17°29′W. Suburban residential community of Dakar. Site of military air base. Adjoined on W by Mamelles lighthouse.

Ouaka River (wah-KAH), 225 mi/362 km long, OUAKA prefecture, central and S CENTRAL AFRICAN REPUBLIC; rises 30 mi/48 km W of BRIA; flows SW and S, past BAMBARI, to UBANGI RIVER at KOUANGO. Also known as Kouango in its middle and lower courses.

Oualata (wah-LAH-tah), village (2000 population 11,779), Hodh Ech Chargui administrative region, SE MAURITANIA, in Oualata range of the SAHARA DESERT, 320 mi/515 km N of BAMAKO (MALI), linked by road; 17°18′N 07°02′W. Dates; livestock. Prehistoric remains nearby.

Oualia (wa-LEE-a), village, FIRST REGION/KAYES, MALI, 84 mi/135 km SE of KAYES. Station on the DAKAR-NIGER railroad.

Ouallam (OO-ah-luhm), town, TILLABÉRY province, SW NIGER, administrative center 59 mi/95 km N of NIAMEY; 14°19′N 02°05′E. Livestock raising.

Ouanaminthe (wah-nah-MANGT), town, NORD-EST department, NE HAITI, on MASSACRE RIVER (DOMINICAN REPUBLIC border), opposite DAJABÓN, 34 mi/55 km ESE of CAP-HAÏTIEN; 19°33′N 71°44′W. Citrus fruits, sugarcane; sugar processing. Gold deposits nearby.

Ouanary (wahn-ah-REE), town, NE FRENCH GUIANA; 04°13′N 51°40′W. Located near mouth of small Ouanary River, 70 mi/113 km SE of CAYENNE.

Ouanda Djallé (WAHN-dah jah-LAI), town, VAKAGA prefecture, N CENTRAL AFRICAN REPUBLIC, 85 mi/137 km S of BIRAO; 08°55′N 22°53′E. Market town.

Ouango (wahng-GO), village, M'BOMOU prefecture, SSE CENTRAL AFRICAN REPUBLIC, on BOMU RIVER (CONGO border), and 30 mi/48 km SSE of BANGASSOU; 04°19′N 22°30′E. Customs station, trading center; coffee plantations.

Ouango Fétini (WAN-go fai-TEE-nee), town, Zanzan region, NE CÔTE D'IVOIRE, 24 mi/38 km W of TÉHINI; 09°34′N 04°03′W. Airport.

Ouangolodougou (wahn-go-lo-DOO-goo), town, Savanes region, N CÔTE D'IVOIRE, 26 mi/42 km N of FERKÉSSÉDOUGOU; 09°58′N 05°09′W. Agriculture (corn, millet, peanuts, beans, cotton).

Ouaninou (WAHN-nee-noo), town, Bafing region, W CÔTE D'IVOIRE, 13 mi/21 km WSW of TOUBA; 08°14′N 07°52′W. Agriculture (manioc, rice, corn, beans, peanuts, tobacco, cotton). Also spelled WANINOU.

Ouaouizaght (wah-WEE-zarht), village, Azilal province, Tadla-Azilal administrative region, central MOROCCO, at SW extremity of the Middle ATLAS mountains, 30 mi/48 km E of KASBA TADLA; 31°30′N 07°24′W. Oak forests. Sometimes spelled Ouaouizarhte.

Ouara (wah-RAH), village, OUADDAÏ administrative region, E CHAD, 25 mi/40 km N of ABÉCHÉ; 14°14′N 20°40′E. Former capital of the kingdom of Ouaddaï (WADAI); burial site of its sultans. Also spelled Wara.

Ouargla (wahr-GLAH), wilaya, S ALGERIA, in the N half of the SAHARA Desert; ⊙ OUARGLA; 30°30′N 06°10′E. Stretching to the SE of the country, it includes Algeria's largest hydrocarbon reserves and production sites at HASSI MESSAOUD and El Borma. It also features the large date-production oases of Ouargla and TOUGGOURT. The region is renowned for its architectural style and such handicrafts as tapestry and leather work. There are burgeoning industries in the larger cities of Ouargla and Touggourt. One of Algeria's largest provinces until 1984, when its S ½ became ILLIZI wilaya.

Ouargla (wahr-GLAH), town and oasis, ⊙ OUARGLA wilaya, S ALGERIA, on one of the Trans-Saharan roads from N Algeria, 99 mi/160 km S of TOUGGOURT and 200 mi/322 km S of BISKRA; 32°00′N 05°16′E. The center of a vast Saharan region, its palm oasis may be the country's largest and is watered by wells from the subterranean Oued Mya. It has a morning market and trades in sheep, cereals, dates, wool, handicrafts, locally made carpets, and garments. Also, international camel trading. International airport. Has one of the earliest centers for administrative studies (1966) and an institute specializing in Saharan agriculture. Ouargla has many wide avenues and large public buildings. Settled in the 11th century by the Muslim Kharidjite sect, it remained more or less independent until occupied by the French in 1853. Because of the oil boom in the 1960s and 1970s, the town grew considerably, which caused water-supply problems. Le Musée Saharien (museum) displays crafts from all desert regions and prehistoric artifacts.

Ouarra (wah-RAH), river, 170 mi/274 km long, HAUTE-M'BOMOU prefecture, SE CENTRAL AFRICAN REPUBLIC; rises in SE ranges of the region; flows SW to join Mbomou River (CONGO border). Lower course forms border between Haute-M'Bomou and M'BOMOU prefectures.

Ouarsenis Massif (wahr-SNEES ma-SEEF), section of the TELL ATLAS, in N central ALGERIA, extending c.120 mi/193 km E-W between Ksar BOUKHARI and RELIZANE. Bounded by the CHÉLIFF valley (N and E), by

the Oued MINA (W), and by the High Plateaus (S). Rises to 6,512 ft/1,985 m in the KEF SIDI AMAR (also called Djebel OUARSENIS). Well-watered N slopes are covered with oak and pine trees. Around TÉNIET EL HAÂD is a well-known cedar forest.

Ouarzazate (WAHR-zuh-zaht), town, Ouarzazate province, Souss-Massa-Draâ administrative region, SW MOROCCO, oasis on S slope of the High ATLAS mountains, 80 mi/129 km SE of MARRAKECH (linked by road via TIZI N'TICHKA pass), at junction between main routes from DRA and DADES valleys; 30°56′N 06°55′W. Military center of region not "pacified" by French until 1930. Date palms. Noted *kasbah*. Tourism. High-grade but modest manganese deposits along Imini (c.30 mi/48 km NW) and Iriri or Irhir (20 mi/32 km W) stream beds. Important BOU AZZER cobalt mine 27 mi/43 km S. International airport to W.

Ouassadou (WAH-sah-doo), village, TAMBACOUNDA administrative region, S SENEGAL, on GAMBIA RIVER, and 35 mi/56 km SE of TAMBACOUNDA; 13°14′N 13°49′W. Sisal growing. Sometimes called Ouassadou-Nieriko. Also spelled Wassadou.

Ouatcha (OO-ah-chuh), town, ZINDER province, NIGER, 40 mi/64 km SE of ZINDER; 13°32′N 09°18′E.

Oubari, LIBYA: see AWBARI.

Oubatche (oo-BAHCH), coastal village, NEW CALEDONIA, on NE coast, 185 mi/298 km NW of NOUMÉA, adjacent to MOUNT PANIÉ; 20°26′S 164°38′E. Agricultural products.

Oubritenga, province (□ 1,073 sq mi/2,789.8 sq km; 2005 population 249,898), PLATEAU CENTRAL region, central BURKINA FASO; ⊙ Ziniaré; 12°35′N 01°25′W. Borders PASSORÉ (N), SANMATENGA (N and E), GANZOURGOU (SES), KADIOGO (S and W), and KOURWÉOGO (NW) provinces. Drained by NAKAMBE RIVER. Agriculture includes groundnuts. A portion of this province was excised in 1997 when fifteen additional provinces were formed.

Ouche (OOSH), old district of NORMANDY, NW FRANCE, along the RISLE RIVER valley, SW of ROUEN. Heavily forested.

Ouche River (OOSH), c.60 mi/97 km long, CÔTE-D'OR department, BURGUNDY region, E central FRANCE; rises W of the CÔTE D'OR hills, flows E in a great arc, past DIJON, to the SAÔNE above SAINT-JEAN-DE-LOSNE; 47°05′N 05°16′E. Above Dijon it is accompanied by a section of the BURGUNDY CANAL.

Ouchi (O-oo-chee), town, Yuri county, Akita prefecture, N HONSHU, NE JAPAN, 19 mi/30 km S of AKITA city; 39°26′N 140°05′E.

Ouchiyama (O-oo-chee-YAH-mah), village, Watarai county, MIE prefecture, S HONSHU, central JAPAN, 31 mi/50 km S of TSU; 34°16′N 136°21′E.

Ouchy (oo-SHE), port, VAUD canton, W SWITZERLAND, on LAKE GENEVA. A former fishing hamlet, it has become the port and lakefront of LAUSANNE. It is a resort and was at times the residence of Shelley and Byron, the latter of whom wrote *The Prisoner of Chillon* here.

Ouda (O-oo-dah), town, Uda district, NARA prefecture, S HONSHU, W central JAPAN, 16 mi/25 km SSE of NARA; 34°28′N 135°55′E. Arrowroot; *Kazaguruma so* flowers (N limit). Formed in early 1940s by combining former town of Matsuyama and three former villages, largest being Kambe.

Oudalan, province (□ 3,796 sq mi/9,869.6 sq km; 2005 population 169,102), SAHEL region, NE BURKINA FASO; ⊙ GOROM GOROM; 14°40′N 00°20′W. Borders SÉNO (S) and SOUM (W) provinces and MALI (N) and NIGER (E). Shea nuts; manganese, limestone. Main centers are GOROM GOROM, Tambao, and Markoye. Arid and sparsely populated province.

Oudamxay, LAOS: see UDOMSAI.

Oud-Beijerland (OUD–BEI-uhr-lahnt), town, SOUTH HOLLAND province, SW NETHERLANDS, in BEIJERLAND region, 8 mi/12.9 km SSW of ROTTERDAM;

51°49′N 04°25′E. On SPUI RIVER channel, 1 mi/1.6 km SW of its exit from the NEW MAAS RIVER. Dairying; cattle, poultry; vegetables, sugar beets; manufacturing (automobile parts, food processing). Formerly spelled Oud-Beyerland.

Oudegem (OU-duh-gem), village in Dendermonde district, EAST FLANDERS province, W central BELGIUM, 3 mi/4.8 km SW of DENDERMONDE; 51°01′N 04°04′E.

Oude IJssel, NETHERLANDS: see OLD IJSSEL RIVER.

Oude Maas, NETHERLANDS: see OLD MAAS RIVER.

Oudenaarde (OU-duh-nahr-duh), French *Audenarde,* commune (□ 162 sq mi/421.2 sq km; 2006 population 28,690), ⊙ of Oudenaarde district, EAST FLANDERS province, W BELGIUM, on the SCHELDT RIVER; 50°51′N 03°36′E. A textile center and railroad junction. At Oudenaarde, in 1708, the allies under Marlborough and Eugene of Savoy defeated the French under the dukes of BURGUNDY and of VENDÔME in the War of the Spanish Succession. Oudenaarde has a Gothic town hall (sixteenth century).

Oudenbosch (OU-duhn-baws), town, NORTH BRABANT province, SW NETHERLANDS, 5 mi/8 km NE of ROOSENDAAL, near DINTEL River; 51°35′N 04°32′E. Airport to SE. Dairying; cattle, hogs; grain, vegetables; manufacturing (cable, sweets).

Oudenburg (OU-duhn-buhkh), commune (2006 population 8,964), Ostend district, WEST FLANDERS province, NW BELGIUM, 5 mi/8 km SE of OSTENDE; 51°11′N 03°00′E. Fruit, vegetable, and flower market for Ostende. Textile center in Middle Ages.

Oud en Nieuw Gastel (OUD uhn NEE-oo KHAS-tuhl), village, NORTH BRABANT province, SW NETHERLANDS, 4 mi/6.4 km N of ROOSENDAAL; 51°35′N 04°28′E. Roosendaalse River passes to W. Dairying; cattle, hogs; grain, sugar beets, vegetables; manufacturing (dairy products).

Oude-Pekela (OU-duh–PAI-kuh-lah), town, GRONINGEN province, NE NETHERLANDS, 3 mi/4.8 km SSW of WINSCHOTEN, on Pekel Aa canal; 53°06′N 07°01′E. Dairying; cattle, sheep; grain, vegetables; manufacturing (strawboard).

Oudergem, BELGIUM: see AUDERGHEM.

Oude Rijn, NETHERLANDS: see OLD RHINE RIVER.

Ouderkerk (OU-duhr-kerk), village, SOUTH HOLLAND province, NETHERLANDS, on HOLLANDSE IJSSEL RIVER, and 7 mi/11.3 km E of ROTTERDAM; 51°56′N 04°38′E. Dairying; cattle, poultry; grain, vegetables.

Oudewater (OU-duh-vah-tuhr), town, UTRECHT province, W NETHERLANDS, on HOLLANDSE IJSSEL RIVER, and 7 mi/11.3 km E of GOUDA; 52°01′N 04°53′E. Dairying; cattle, hogs, poultry; vegetables, sugar beets, nursery stock; light manufacturing Many sixteenth- and seventeenth-century houses. Site (1575) of massacre by Spaniards.

Oudh (OUD), historic region of N central INDIA, now part of UTTAR PRADESH state. Its early history centers around the ancient kingdom of KOSALA (with capital at AYODHYA) of the legendary Rama. The region passed under Gupta rule in the 4th century C.E. and later became (11th–12th centuries) the center of the Rajput state of KANAUJ. In the 13th century it was conquered by the DELHI SULTANATE. It became (16th century) a province of the Mogul empire, and was subsequently governed by the nawabs of OUDH from their capitals of FAIZABAD (1724–1775) and LUCKNOW (1775–1856). The annexation (1856) of Oudh as a British province was a cause of the Indian Mutiny (1857–1858). In 1877, Oudh was joined with the presidency of AGRA to form the United Provinces of AGRA AND OUDH, now the Indian state of Uttar Pradesh.

Oudh, INDIA: see AYODHYA.

Oud Heverlee (OUD HAI-vuhr-lai), commune, Leuven district, BRABANT province, central BELGIUM, 4 mi/6 km S of LEUVEN; 50°50′N 04°40′E.

Oudjda, MOROCCO: see OUJDA.

Oudong (U-DONG), town, KOMPONG SPEU province, central CAMBODIA, near TONLÉ SAP (lake), 20 mi/32 km NNW of PHNOM PENH. Rice-growing area; fisheries. Tourism potential. Khmer with Cham minority. Was ancient capital of Cambodia (1618–1866), site of monasteries, pagodas, and 16th-century ruins. The ruins of Phonom Chet Ath Roeus, a vihara built in 1911 by King Sisowath, were destroyed by the Khmer Rouge along with other nearby viharas and stupas. Mass graves and memorial to victims of Khmer Rouge located nearby. Formerly spelled Udong.

Oudon River (OO-don), 40 mi/64 km long, MAYENNE and MAINE-ET-LOIRE departments, PAYS DE LA LOIRE region, W FRANCE; rises W of LAVAL; flows SE, past CRAON and SEGRÉ, to the MAYENNE 1 mi/1.6 km SE of Le LION-D'ANGERS; 47°37′N 00°42′W.

Oudtshoorn, town, WESTERN CAPE province, SOUTH AFRICA, on Grabbelaaro River, a tributary of OLIFANTS RIVER, 200 mi/322 km W of NELSON MANDELA METROPOLE (PORT ELIZABETH), 40 mi/64 km N of MOSSEL BAY, and 31 mi/50 km NW of GEORGE over the OUTENIQUA RANGE, at foot of GREAT SWARTBERG range; 33°36′S 22°12′E. Elevation 984 ft/300 m. Agriculture center (tobacco, grain, fruit, vegetables; dairying; wine and brandy making; fruit drying; ostrich and crocodile farming). Home of C. J. Langenhoven, champion of Afrikaans language. Has teachers college. Airfield. The CANGO CAVES are 15 mi/24 km N. Founded in 1847 and became a municipality in 1887; was the ostrich-feather capital in the late 1800s, with many examples of fine Victorian mansions.

Oud Turnhout (OUD TURN-hout), commune, Turnhout district, ANTWERPEN province, N BELGIUM, 1 mi/1.6 km E of TURNHOUT; 51°19′N 04°59′E.

Oued [Fr. transliteration of Arabic word for "stream"], for names in FRENCH NORTH AFRICA beginning thus and not found here, see under following part of the name, or under "Wadi."

Oued Amizour (WED ah-mee-ZOOR), village, BEJAÏA wilaya, NE ALGERIA, in Oued SOUMMAM valley, 13 mi/21 km SW of BEJAÏA. Olives, citrus fruit.

Oued Athménia (WED aht-mai-NYAH), village, MILA wilaya, NE ALGERIA, on railroad, and 20 mi/32 km SW of CONSTANTINE. In cereal-growing region.

Oued Chorfa (WED shor-FAH), village, AÏN DEFLA wilaya, N central ALGERIA, on the CHÉLIFF River, and 13 mi/21 km SW of MÉDÉA. Ghrib Dam is just S. Formerly Dollfusville.

Oued Eddahab-Lagouira, administrative region (2004 population 99,367), MOROCCO; ⊙ DAKHLA. Lies entirely in S WESTERN SAHARA, which is claimed and administered by Morocco (though the territory's status is considered unresolved by much of the world, including the U.S. and UN). Bordered N by Laayoune-Boujdour-Sakia El Hamra administrative region, E and S by MAURITANIA, and W by ATLANTIC OCEAN. Main road runs N along the coast to LAAYOUNE, with smaller roads branching off into central and E areas of the region. Airport at Dakhla. Further divided into two secondary administrative divisions called provinces: Aousserd and Oued Eddahab.

Oued, El (WED, el), wilaya, in SAHARA Desert, E central ALGERIA; ⊙ El Oued; 33°10′N 07°15′E. Bordering on TUNISIA (E). Population is concentrated in oasis towns of GUÉMAR, El M'ghaier, and Djamaa, whose economic activity focuses on date production. The centers are also marketplaces for wool, carpets, tools, cereals, and livestock. Created in 1984 from El Oued and El Maghair districts, formerly part of the large Saharan portion of BISKRA wilaya.

Oued, El (el WED) [Arab.=the river], town and oasis, ⊙ El OUED wilaya, in SAHARA Desert, S ALGERIA, near the Tunisian border, traditional capital of the SOUF region, 50 mi/80 km NE of TOUGGOURT and 137 mi/220 km E of BISKRA; 33°22′N 06°52′E. The center of a cluster of oases, it is the largest oasis of the Souf region and is surrounded by sand dunes of the Eastern ERG. Tem-

peratures can reach 131°F/55°C. Despite the name, there is no river in the area. The local economy is based on the cultivation and processing of dates; the region produces large amounts of the Deglet Nour variety. Also, tobacco cultivation, livestock rearing, and wool (much used in local crafts) production; carpet manufacturing. Airport. El Oued's building style includes domes, which are the dominant feature, even in the market. The museum displays the geology and fauna of the Souf, along with coins, costumes, and crafts. Like other oases, El Oued has expanded in recent years. There are still remnants of nomadic life in the region.

Oued El Abtal (WED el ahb-TAHL), agricultural village, MASCARA wilaya, NW ALGERIA, near the Oued MINA, 31 mi/50 km ENE of MASCARA; 35°20′N 00°42′E. railroad junction. Formerly Uzès-le-Duc.

Oued El Alleug (WED el ahl-LUHG), village, BLIDA wilaya, N central ALGERIA, in the MITIDJA plain, 6 mi/9.7 km NNW of BLIDA. Processing of essential oils. Citrus-tree nursery.

Oued Fodda (WED fo-DAH), town (□ 116 sq mi/301.6 sq km), CHLEF wilaya, E central ALGERIA. OUED FODDA Dam (292 ft/89 m high) on the CHÉLIFF River is in the vicinity. The dam, which survived the earthquakes of 1954 and 1980, has a capacity of 10,212 million cu ft/289 million cu m and can irrigate 116 sq mi/300 sq km.

Oued Imbert (WED ang-BER), village, SIDI BEL ABBÈS wilaya, NW ALGERIA, on NE slope of TESSALA MOUNTAINS, on railroad, and 14 mi/23 km NNE of SIDI BEL ABBÈS. Wheat, wine.

Oued Ksob Dam, ALGERIA: see M'SILA, WILAYA.

Oued Messelmoun (WED me-sel-MOON), locality in TIPAZA wilaya, N central ALGERIA, 6 mi/10 km W of CHERCHELL. In a nearby farm during World War II, American General Clark and representatives of the local North African resistance leaders met on October 24, 1942 and organized the Allied landing of November 12.

Oued Rhiou (WED ree-OO), village, RELIZANE wilaya, N ALGERIA, near the CHÉLIFF River, on railroad, and 25 mi/40 km NE of RELIZANE; 35°55′N 00°56′E. Trade center in irrigated valley growing cotton and citrus fruit. Grain and cattle market. Formerly Inkermann.

Oued R'Hir (WED GIR), region, El OUED wilaya, E ALGERIA, in the SAHARA Desert, extending N from TEMACINE and TOUGGOURT to S edge of the Chott MEROUANE. It is what remains of a river formed thousands of years ago. According to legend, the Oued R'Hir was once as large as the NILE or NIGER rivers. It has about 2,000,000 palm trees, of which a large proportion is devoted to the production of Deglet Nour dates. The oases are organized in self-managed cooperatives. Djamaa has Algeria's largest factory for the conditioning of dates. Also Oued Righ.

Oued Taria (WED tahr-YAH), village, MASCARA wilaya, NW ALGERIA, on railroad, and 20 mi/32 km S of MASCARA; 35°00′N 00°05′E. Cereals, wine.

Oued Tlélat (WED tlai-LAHT), village, ORAN department, NW ALGERIA, 15 mi/24 km SE of ORAN; 35°34′N 00°25′W. Railroad and road junction. Wine, olives; horse raising. Formerly Ste-Barbe-du-Tlélat.

Oued Zem, MOROCCO: see OUAD ZEM.

Oued Zenati (WED ze-nah-TEE), town, GUELMA wilaya, NE ALGERIA, on the High Plateaus, on railroad, and 31 mi/50 km E of CONSTANTINE. Agricultural market in cereal-growing region; manufacturing of flour products.

Ouégoa (wai-go-AH), village, NE NEW CALEDONIA, French Polynesia, S PACIFIC OCEAN, 185 mi/298 km NW of NOUMÉA, lying within Diahot River valley (largest river valley in New Caledonia); 20°20′S 164°21′E. Agriculture, livestock district.

Ouellette (oo-LET), settlement, SE central ONTARIO, E central Canada, 5 mi/8 km from Noëlville, and included in town of French River; 46°06′N 80°31′W.

Ouémé, department (□ 1,095 sq mi/2,847 sq km; 2002 population 730,772), SE corner of BENIN; ⊙ PORTO-

NOVO; 06°40′N 02°30′E. Bordered on N by ZOU department, E by PLATEAU department, SE by NIGERIA, S by BIGHT OF BENIN, W by ATLANTIQUE department. Includes Benin's capital city, Porto-Novo, and the town of ADJOHON. OUÉMÉ RIVER flows from N to S through W central part of the department. In 1999 the NNE portion of Ouémé was separated and established as Plateau department.

Ouémé (WAI-mai), village, S BENIN, 100 mi/161 km N of PORTO-NOVO; 08°01′N 02°23′E. On OUÉMÉ RIVER (crossed here by railroad to PARAKOU); railroad station. Cotton.

Ouémé River (WAI-mai), c.300 mi/483 km long, BENIN, W AFRICA; rises in ATAKORA MOUNTAINS at about 10°N just N of DJOUGOU; flows S, past Betero, crossed by railroad and past ADJOHON, to the Gulf of GUINEA in delta near COTONOU, emptying mostly into Lake Nokoué. Impeded by rapids, though partly navigable during rainy season. Sometimes spelled Weme. OKPARA to the E is its main affluent.

Ouénou, town, BORGOU department, N central BENIN, 20 mi/32 km N of PARAKOU; 09°47′N 02°38′E. Cotton; livestock; shea-nut butter. Formerly called Wénou.

Ouenza (wen-ZAH), town, TEBESSA wilaya, E ALGERIA, near TUNISIA border, 37 mi/60 km N of TEBESSA; 36°19′N 07°10′E. Has Algeria's largest iron mine, which produces ⅔ of the country's iron ore (which is of very high quality). Direct railroad connection to El HADJAR steel complex near ANNABA.

Ouenzerig, LIBYA: see WANZARIK.

Ouenzeriq, LIBYA: see WANZARIK.

Ouergha, Oued (wahd WER-ghuh), stream of N MOROCCO; rises in RIF mountains; flows c.120 mi/193 km WSW to the OUED SEBOU river in the GHARB lowland. Navigable in lower course. Also spelled Ouerrha.

Ouessant (we-sahn), English *Ushant*, fog-bound island (year-round population 1,080), in the ATLANTIC off W tip of BRITTANY, westernmost point of FRANCE, 26 mi/42 km WNW of BREST; 48°28′N 05°03′W. Forming part of FINISTÈRE department, 5 mi/8 km long, 2 mi/3.2 km wide, it is rockbound and barren. Sheep raising. Fishing port of Lampaul is on a SW embayment. Lighthouses mark entrance to ENGLISH CHANNEL. Near island, French and British fleets fought engagements in 1778 and 1794. Ferry service to Brest.

Ouèssè, town, COLLINES department, E central BENIN, 40 mi/64 km N of SAVÉ; 08°24′N 02°37′E. Cotton, groundnuts.

Ouesso (we-SO), town (2007 population 26,072), ⊙ SANGHA region, N Congo Republic, on SANGHA RIVER, at mouth of NGOKO RIVER, on CAMEROON border, and 400 mi/644 km NNE of BRAZZAVILLE; 01°36′S 16°03′E. Trading center, terminus of year-round stream navigation on the Sangha; palm products, hides, copal, rubber, African mahogany. Customs, meteorological stations. Founded 1891. Airfield.

Ouest (WEST), department (2003 population 3,093,699), S central HAITI, bounded on N by ARTIBONITE and CENTRE departments, E by DOMINICAN REPUBLIC, W by GULF OF GONAÏVES, S by SUD-EST department; ⊙ PORT-AU-PRINCE, both national and departmental; 18°40′N 72°20′W. Central part is the fertile CUL-DE-SAC plain, bounded in N and S by mountains. Produces sugarcane, cotton, fruits, tobacco, sisal; has bauxite and manganese deposits.

Ouest, CAMEROON: see WEST.

Ouezzane (WEZ-zan), city, Sidi Kacem province, Gharb-Chrarda-Beni Hssen administrative region, N MOROCCO, on SW slope of the RIF mountains, 60 mi/97 km NNW of FES. Olives, fruits, carpets. Sacred pilgrimage place for Muslims of Taibia brotherhood, founded here in early 18th century. Also center for Jewish pilgrims to shrine of Rabbi Ba Amrane.

Ouffet (oo-FAI), commune (2006 population 2,522), Huy district, LIÈGE province, E central BELGIUM, 11 mi/18 km ESE of HUY; 50°26′N 05°28′E. Granite quarrying.

Oughterard (oo-tuhr-RAHRD), Gaelic *Uachtarárd*, town (2006 population 1,305), W GALWAY county, W IRELAND, on W shore of LOUGH CORRIB, 16 mi/26 km NW of GALWAY; 53°25′N 09°20′W. Agricultural market; fishing resort convenient to CONNEMARA and JOYCE'S COUNTRY. Nearby, in Lough Corrib, is IN-CHAGOILL island.

Oughter, Lough (OU-tuhr, LAHK), shallow lake (7 mi/11.3 km long, 4 mi/6.4 km wide), W CAVAN county, N central IRELAND, on ERNE RIVER, NW of CAVAN; 54°00′N 07°29′W. An 11th-century round tower of Cloughoughter Castle of the O'Reillys is on one of the many islands. Lake receives ANNALEE RIVER at N end; town of KILLESHANDRA on W shore.

Oughtibridge (OO-ti-brij), village (2001 population 3,401), SOUTH YORKSHIRE, N ENGLAND, on DON RIVER, and 5 mi/8 km NW of SHEFFIELD; 53°26′N 01°33′W. Steel and paper milling.

Ougrée (oo-GRAI), town in Liège district, LIÈGE province, E BELGIUM, on the MEUSE RIVER, a suburb of LIÈGE; 50°36′N 05°32′E. It is a center of heavy industry.

Ouham (oo-AHM), prefecture (□ 19,396 sq mi/50,429.6 sq km; 2003 population 369,220), NW CENTRAL AFRICAN REPUBLIC; ☉ BOSSANGOA. Bordered N by CHAD, E by NANA-GRÉBIZI prefecture, SE by KÉMO prefecture, S by OMBELLA-M'POKO prefecture, and W by OUHAM-PENDÉ prefecture. Drained by Bakassa, Ouham, FAFA rivers. Agriculture (cotton, ground-nuts); manufacturing (textiles, cotton-ginning); copper deposits nearby. Main centers are Bossangoa, BATANGAFO, and Kouki. Includes La Nane Barya and most of Bridingui-Bamingui Animal Reserve.

Ouham-Pendé (oo-AHM–pen-DAI), prefecture (□ 12,391 sq mi/32,216.6 sq km; 2003 population 430,506), W CENTRAL AFRICAN REPUBLIC; ☉ BOZOUM. Bordered N by CHAD, E by OUHAM prefecture, SE by OMBELLA-M'POKO prefecture, S by NANA-MAMBÉRÉ prefecture, and W by CAMEROON. Drained by Ouham, PENDÉ, and Lia rivers. Agriculture (cotton, ground-nuts) and livestock (cattle); manufacturing (textiles, cotton-ginning). Main centers are Bozoum, BOCAR-ANGA, and PAOUA.

Ouham River, CENTRAL AFRICAN REPUBLIC: see BAHR SARA.

Ouiatchouanish River (wee-uhch-WAH-nish), S central QUEBEC, E Canada; outlet of COMMISSIONERS LAKE; flows N past LAC BOUCHETTE to Lake SAINT JOHN at VAL JALBERT. Just above its mouth are 236 ft/72 m falls.

Ouidah (WEE-dah), town, ☉ ATLANTIQUE department, S BENIN, 25 mi/40 km W of COTONOU; 06°22′N 02°05′E. A port on the GULF OF GUINEA, linked to Cotonou by road and railroad. Agriculture (palm oil, peanuts); fishing. Was capital of small state founded about the 16th century. From the early 17th century, Portuguese, French, and Dutch traders were inter-mittently active at Ouidah. The name was derived by Europeans from a nearby Portuguese fort called St. John of Adjuda. In the 18th and early 19th centuries, Ouidah was an important export point for slaves from the SLAVE COAST (1.5 million were exported from here). In the 1840s the French established a substantial trade with Ouidah, exchanging textiles, guns, and gunpowder for palm oil and ivory. The town was annexed by FRANCE in 1886. Center for Voudou, an-cient religion that spread to AMERICA with slavery. Also spelled Whydah or Wida.

Ouistreham (wees-TRAHM), town (□ 4 sq mi/10.4 sq km), CALVADOS department, BASSE-NORMANDIE re-gion, NW FRANCE, on ENGLISH CHANNEL, at mouth of ORNE RIVER and of Caen maritime canal, and 8 mi/12.9 km NE of CAEN; 49°17′N 00°15′E. Small fishing port and bathing resort favored by yachtsmen. Allied forces landed here in June 1944. Coastal Riva-Bella has

a fine beach. Car ferry to PORTSMOUTH, England. Coastal Riva-Bella has a fine church.

Oujda (UZH-duh), city, ☉ ORIENTAL administrative region, NE MOROCCO, near ALGERIA border; 34°41′N 01°54′W. Railroad junction, agricultural market, and commercial center. Connected by railroad to ORAN (Algeria). International airport to N. Seat of a public university. It was occupied by the French in 1844, 1859, and 1907. Was part of former Oujda province prior to reorganization of administrative divisions in 1997. Also spelled Oudjda.

Oujé-Bougoumou, village, Nord-du-Québec region, central QUEBEC, E Canada, on N shore of Lake Opé-misca, 10 mi/16 km N of Chapais, and within Baie-James municipality; 49°55′N 74°49′W. Planned Cree community established 1989 to put end to 75 years of relocations of people by mining interests. Road ter-minus. Cultural center.

Oujeft (oo-JEFT), village, Adrar administrative region, W central MAURITANIA, 36 mi/58 km S of ATAR, and c.248 mi/399 km NE NOUAKCHOTT; 20°02′N 13°03′W. Dates, subsistence crops; livestock.

Oukaimden, MOROCCO: see ASNI.

Oulad Ayad (OO-lahd ai-YAHD), city, Béni Mellal province, Tadla-Azilal administrative region, MOR-OCCO, new agricultural center in the TADLA plain, 31 mi/50 km WSW of BENI MELLAL; 32°54′N 07°22′W.

Oulad El Haj, MOROCCO: see OUTAT EL HAJ.

Oulad Naïl Mountains, ALGERIA: see OULED NAÏL MOUNTAINS.

Oulad Saïd (weld sah-EED), village, Settat province, Chaouia-Ouardigha administrative region, W MOR-OCCO, 12 mi/19 km WSW of SETTAT; 32°58′N 07°49′W. Wheat, barley.

Oulad Teima (OO-lahd TAY-muh), city, Taroudannt province, Souss-Massa-Draâ administrative region, MOROCCO, 28 mi/45 km E of AGADIR, on road between Agadir area and TAROUDANNT; 30°24′N 09°13′W. Center of greenhouse-grown vegetables, exported to EUROPE.

Oulainen (O-LEI-nuhn), Swedish *Oulais*, village, OULUN province, W FINLAND, on PYHÄJOKI (river), and 30 mi/48 km SSE of RAAHE; 64°16′N 24°48′E. Elevation 248 ft/75 m. Light manufacturing.

Oulanka National Park (O-lahng-kah) (□ 80 sq mi/208 sq km), OULUN and LAPIN provinces, NE FIN-LAND; centered at 66°22′N 29°22′E. Located on the Oulanjoki near Suomutunturi (mountain, elevation 1,338 ft/408 m). Boreal coniferous forest zone. Ou-lanka Biological Station is field station of University of Oulu. Hiking, fishing; one of many recreational sites around Ruka, the park hosts hiking and fishing.

Ould Yenje, village (2000 population 4,935), GUIDI-MAKHA administrative region, S MAURITANIA, near KARAKORO RIVER and MALI border, 78 mi/125 km NW of KAYES (Mali); 15°33′N 11°44′W. Livestock; dates, millet, sorghum.

Ouled Djellal (oo-LED jel-LAHL), town and Saharan oasis, BISKRA wilaya, N central ALGERIA, on the Oued DJEDI, 50 mi/80 km SW of BISKRA; 34°25′N 05°02′E. Dates; wool. Handicraft blankets.

Ouled Fayet (oo-LED fah-YET), village, TIPAZA wilaya, N central ALGERIA, 7 mi/11.3 km SW of ALGIERS.

Ouled Mimoun (oo-LED mee-MOON), village, TLEM-CEN wilaya, NW ALGERIA, in TLEMCEN MOUNTAINS, on railroad, and 15 mi/24 km E of TLEMCEN; 34°54′N 01°03′W. Agricultural market (olive oil, wine, vege-tables, cereals). Formerly called Lamoricière.

Ouled Naïl Mountains (oo-LED nah-EEL), range of the Saharan ATLAS in N central ALGERIA, stretching from the ATLANTIC OCEAN to the AURÈS Mountains, and separating the High Plateaus from the SAHARA Desert. Bordered N by the HODNA Massif, E by the Zibans, and SE by Djebel AMOUR, the Ouled Naïl Mountains are occupied by nomadic and seminomadic populations practicing intensive livestock rearing. Region renowned for its dances and carpets.

The GREEN BARRAGE is being planted at the foot of these mountains.

Ouled Rahmoun (oo-LED rah-MOON), village, CON-STANTINE wilaya, NE ALGERIA, 14 mi/23 km SSE of CONSTANTINE; 36°12′N 06°43′E. Railroad junction. Cereals.

Oulgaret (ool-gah-RET), town, formerly in FRENCH INDIA, now in PONDICHERRY district, Pondicherry Union Territory, S INDIA, suburb of Pondicherry, 3 mi/4.8 km W of city center. Hand-loom cotton weaving; pottery.

Oullins (oo-LAN), residential SSW suburb (□ 2 sq mi/5.2 sq km), of LYON, RHÔNE department, RHÔNE-ALPES region, E central FRANCE, near right bank of the RHÔNE; 45°43′N 04°48′E. Metalworks, textiles.

Oulmes (WOOL-mes), village, Khémisset province, Rabat-Salé-Zemmour-Zaër administrative region, W MOROCCO, on a high tableland (elevation 4,100 ft/1,250 m), 40 mi/64 km SW of MEKNES; 33°26′N 06°01′W. Tin and antimony mining. The mineral waters of nearby Oulmès-les-Thermes are bottled. Fluorite mine at El-Hammam nearby (NW).

Oulu (O-loo), Swedish *Uleåborg*, city, ☉ OULUN prov-ince, W central FINLAND; 65°01′N 25°28′E. Elevation 50 ft/15 m. Located at mouth of the OULUJOKI (river) on the GULF OF BOTHNIA. Seaport; metal shops, leather products, and wood-processing, and other manu-facturing; power station. The city grew around a castle founded in 1590, was chartered in 1610, and became an important commercial center in the nineteenth cen-tury. It was destroyed by fire in 1882 and then rebuilt. The University of Oulu was founded in 1959. It is also the site, since 1889, of the world's oldest cross-country ski race. Seat of Lutheran bishop. Airport.

Oulujärvi (O-loo-YAR-vee), Swedish *Uleträsk*, lake (□ 387 sq mi/1,006.2 sq km), OULUN province, N central FINLAND, 50 mi/80 km SE of OULU; 40 mi/64 km long (SE-NW), 2 mi/3.2 km–18 mi/29 km wide. Contains several islands. Drained (NW) by the OU-LUJOKI. KAJAANI city is on SE arm of lake. A chain of small lakes, connected with Oulujärvi, extends E to Russian border.

Oulujoki (O-loo-YO-kee), Swedish *Ule älv*, river, 60 mi/97 km long, N central FINLAND; rises from NW end of OULUJÄRVI (lake); flows NW, over Pyhäkoski falls (105 ft/32 m high; has hydroelectric plant), 20 mi/32 km SE of OULU, to GULF OF BOTHNIA at Oulu.

Oulun (O-loon), Swedish *Uleåborg*, province (□ 23,777 sq mi/61,820.2 sq km), N central FINLAND; ☉ OULU. Located between GULF OF BOTHNIA (W) and RUSSIA border (E). Low and level in coastal region, becoming hilly in E part of province; drained by IIJOKI, OULU-JOKI, KALAJOKI, PYHÄJOKI, and several smaller rivers. Of the many lakes in SE part of province, OULUJÄRVI is largest. Manufacturing (building materials; bobbin and spindle milling; fertilizers; paper); agriculture (grain; cattle); lumbering. Minerals include iron, nickel, zinc, granite, kaolin, and feldspar. Major cities are OULU, RAAHE, and KAJAANI. Province included all of Finnish LAPLAND until a separate province (LAPIN) was created in 1938.

Oum Chalouba (OOM shah-loo-BAH), town, BOR-KOU-ENNEDI-TIBESTI administrative region, E CHAD, 125 mi/201 km SW of FADA, on border with WADI FIRA administrative region; 15°48′N 20°46′E.

Oume (O-oo-me), city, Tokyo prefecture, E central HONSHU, E central JAPAN, 22 mi/35 km NW of SHINJUKU; 35°47′N 139°16′E. Computers. Chichibu-Tama National Park and Mitake Gorge nearby.

Oumé (OO-mai), village, Fromager region, S central CÔTE D'IVOIRE, 120 mi/193 km NW of ABIDJAN; 06°23′N 05°25′W. Agriculture (coffee, cacao, palm ker-nels, manioc, taro, bananas, corn, peanuts). Airport.

Oum El Adrou (OOM el ah-DROO), village, CHLEF wilaya, N central ALGERIA, on the CHÉLIFF River, on railroad, and 4 mi/6.4 km NE of CHLEF. Hydroelectric plant. Formerly Pontéba.

Area is shown by the symbol □, and capital city or county seat by ☉.

Oum El Bouaghi (OOM el bwah-GEE), wilaya, E AL-
GERIA; ⊙ OUM EL BOUAGHI; 35°50′N 07°05′E. Recent
industrial developments; cereals; sheep.

Oum El Bouaghi (OOM el bwah-GEE), village, ⊙ OUM
EL BOUAGHI wilaya, E ALGERIA, on the edge of salt
lakes and N of the AURÈS Mountains; 35°51′N 07°09′E.
The establishment of new industries and a university,
as well as its position as wilaya capital, have given the
town far more importance than it used to have, much
of it at the expense of AÏN BEÏDA, its larger neighbor.
Formerly called Canrobert.

Oum er Rbia, Oued (wahd oom er-BEE-yuh), 345 mi/
555 km long, chief river of W MOROCCO; rises in the
Middle ATLAS mountains 20 mi/32 km NE of KHE-
NIFRA; flows SW, past KASBA TADLA and DAR OULD
ZIDOUH, then NW in entrenched meanders to the
ATLANTIC OCEAN just below AZEMMOUR. Harnessed
for hydroelectric power and irrigation at Al Massira
dam (completed 1979, near Mechra Benabbou), Kasba
Zidania dam (11 mi/18 km SW of KASBA TADLA;
completed 1936), Imefout dam (360 ft/110 m high; 12
mi/19 km NW of Mechra Benabbou; completed 1944,
improved after World War II), Daourat dam (20 mi/
32 km below Imefout dam; completed 1950), and Sidi
Said Machou dam (20 mi/32 km above river mouth;
completed 1929). On the Oued EL-ABID (left tributary), Bin el Ouidane dam was completed 1953.
Sometimes spelled Oum er Rebia.

Oum-Hadjer (OOM–hah-JER), village, BATHA administrative region, central CHAD, on BATHA RIVER, and
90 mi/145 km E of ATI; 13°18′N 19°41′E. Trading post
on caravan road to DARFUR (SUDAN); vegetables.

Oumi (O-oo-mee), town, W Kubiki county, NIIGATA
prefecture, central HONSHU, N central JAPAN, 93 mi/150
km S of NIIGATA; 37°01′N 137°48′E. Carbide, chloro-
prene gum, cement-mixing materials, limestone.

Oum Toub (OOM TOOB), town, SKIKDA wilaya, E
ALGERIA, 25 mi/40 km SW of SKIKDA; 36°41′N
06°34′E. Regroupment center set up by the French
army in the 1950s; now a local administrative center.

Ou, Nam (OO, NAHM), river in N LAOS, c.280 mi/451
km long; rises on Chinese frontier N of Ou Tay; flows
S in valley (rice fields) and deep gorge to MEKONG
River 10 mi/16 km N of Luang Prabang. Navigable up
to HATSA (port of PHONGSALY).

Ounasjoki (O-nas-YO-kee), river, 210 mi/338 km long,
Lapin province, N Finland; rises on Norwegian border
SW of Lake Inari, near 68°40′N 24°00′E; flows in
winding course generally S to the Kemijoki (river) at
Rovaniemi.

Oundle (OUN-duhl), town (2001 population 5,345), NE
NORTHAMPTONSHIRE, central ENGLAND, on NENE
RIVER, and 12 mi/19 km SW of PETERBOROUGH;
52°29′N 00°28′W. Agricultural market. Has public
school founded 1556, 14th-century church, several old
inns, and 16th-century bridge.

Ou Neua, Laos: see MUONG OU NUA.

Ountivou (oon-tee-VOO), town, PLATEAUX region, SE
TOGO, 30 mi/48 km SSE of ATAKPAMÉ, on secondary
road near BENIN border; 07°21′N 01°34′E. Subsistence
farming of manioc, maize, and millet.

Oupeye (oo-PEI), commune (2006 population 23,591),
Liège district, LIÈGE province, E BELGIUM, 5 mi/8 km
N of LIÈGE, on MEUSE RIVER; 50°42′N 05°39′E.

Oupu (O-PU), town, N HEILONGJIANG province, NE
CHINA, 140 mi/225 km NNW of HEIHE, and on AMUR
River (Russian border); 52°47′N 106°01′E. Logging.

Oura (O-oo-rah), town, Oura county, GUMMA prefecture, central HONSHU, N central JAPAN, 28 mi/45 km S
of MAEBASHI; 36°15′N 139°28′E.

Oura, town, Kawanabe county, KAGOSHIMA prefecture,
SW KYUSHU, SW JAPAN, 25 mi/40 km S of KA-
GOSHIMA; 31°22′N 130°13′E.

Ourafane (OOR-uh-fahn), town, NIGER, 80 mi/129 km
NE of MARADI.

Ourâne (oo-RAH-ne), dunes, Adrar administrative
region, central MAURITANIA, range of dune formation

running NE, c.233 mi/375 km NE of ATAR; approximate coordinates: 20°50′N 07°00′W.

Ouranopolis (oo-rah-NO-po-lees), village, KHALK-
IDHIKÍ prefecture, CENTRAL MACEDONIA department,
NE GREECE, on NW coast of easternmost prong of
KHALKIDHIKÍ Peninsula; 40°20′N 23°59′E. Fishing
village established after 1922 by Greek refugees from
Turkey. Shortly thereafter, they were helped eco-
nomically by a Scottish-Australian couple, Sydney
and Joyce Loch, who began marketing the refugees'
knotted rugs with distinctive Byzantine patterns. The
Lochs lived in the nearby 12th-century Tower of
Profirion for several decades. Also called Prosfori and
Pírgos.

Ouray (YUR-ai), county (□ 542 sq mi/1,409.2 sq km;
2006 population 4,307), SW central COLORADO;
⊙ OURAY; 38°09′N 107°46′W. Sheep-grazing and
mining region, drained by UNCOMPAHGR RIVER. Gold,
silver, lead, copper. Includes parts of Uncompahgre
National Forest in NW corner and S, small part of San
Juan Forest on S boundary, and ranges of Rocky
Mountains Ridgway State Park in center. Formed 1883.

Ouray (YUR-aī), town (2000 population 813), ⊙ OURAY
county, SW central COLORADO, on UNCOMPAHGRE
RIVER, in SAN JUAN MOUNTAINS, and 85 mi/137 km SE
of GRAND JUNCTION; 38°01′N 107°40′W. Elevation
7,706 ft/2,349 m. Health resort (Ouray Hot Springs)
with mineral hot springs. Some agriculture (grain,
potatoes; livestock). Manufacturing (printing and
publishing). Gold, silver, lead, and copper mines in
vicinity. Surrounded by Uncompahgre National Forest. Wetterhorn Peak to E. Settled 1875; incorporated
1884.

Ouray Peak (13,955 ft/4,253 m), CHAFFEE and SA-
GUACHE counties, central COLORADO, in S tip of SA-
WATCH MOUNTAINS, 18 mi/29 km SW of SALIDA, E of
CONTINENTAL DIVIDE.

Ource River (OORS), 55 mi/89 km long, in CÔTE-D'OR
and AUBE departments, N central FRANCE; rises in the
Plateau of LANGRES; flows NW, past Essoyes, to the
SEINE above BAR-SUR-SEINE.

Ourcq Canal (OORK), French, *Canal de l'Ourcq* (kah-
nahl duh loork), canal, ÎLE-DE-FRANCE, N central
FRANCE; 48°51′N 02°22′E. The canal, c.68 mi/109 km
long, has many locks and runs parallel to the OURCQ
RIVER in its lower course. Populated places along the
canal include BONDY, CLAYE-SOUILLY, MEAUX, NOISY-
LE-SEC PANTIN, SEVRAN VAUJOURS, and VILLEPARISIS.
Timber trade through the canal. A slab-and-girder
bridge for high-speed train travel (*train à grande vi-
tesse*, or TGV), crosses the canal at Ocquerre.

Ourcq River (OORK), 50 mi/80 km long, in AISNE and
SEINE-ET-MARNE departments, NE central FRANCE;
rises NNW of DORMANS; flows SW in an arc across
World War I battlefield of the MARNE, past FÈRE-EN-
TARDENOIS and La FERTÉ-MILON, to the Marne just
below LIZY-SUR-OURCQ. Paralleled in lower course by
OURCQ CANAL.

Ourém (o-RAIN), city (2007 population 15,172), extreme E PARÁ state, BRAZIL, head of navigation on RIO
GUAMÁ, and 105 mi/169 km ESE of BELÉM; 01°33′S
47°08′W. Tobacco, manioc, cotton; hides.

Ouricangas (O-ree-SAHN-gahs), town (2007 population 8,055), NE BAHIA state, BRAZIL, 19 mi/31 km NW
of ALAGOINHAS; 12°01′S 38°38′W.

Ouricia, El (oo-ree-SYAH, el), village, SÉTIF wilaya, NE
ALGERIA, 7 mi/11.3 km N of SÉTIF. Wheat.

Ouricuri (O-ree-koo-ree), city (2007 population
62,367), W PERNAMBUCO state, NE BRAZIL, 110 mi/177
km NNE of JUÀZEIRO (BAHIA); 08°32′S 36°16′W.
Onions, corn; livestock. Formerly spelled Ouricury.

Ourilândia do Norte (oo-ree-LAHN-zhee-ah do NOR-
che), city (2007 population 20,417), S central PARÁ
state, BRAZIL, 50 mi/80 km ESE of Tucumá; 06°45′S
50°28′W.

Ourinhos (O-reen-yos), city (2007 population 99,008),
W central SÃO PAULO state, BRAZIL, near PARA-

NAPANEMA River (PARANÁ border), 200 mi/322 km
WNW of SÃO PAULO; 22°59′S 49°52′W. Important
railroad junction with line to agricultural district of N
Paraná. Food-processing center (processed foods),
with trade in coffee, cotton, alfalfa, fruits, livestock,
and timber.

Ourique (o-REEK), town, BEJA district, S PORTUGAL, in
BAIXO ALENTEJO; 37°39′N 08°13′W. Although tradition says Alfonso I defeated the Moors here in 1139,
the battle of Ourique was actually fought at some
undetermined place nearby.

Ourizona (O-ree-so-nah), town (2007 population
3,296), NW PARANÁ state, BRAZIL, 16 mi/26 km W of
MARINGÁ; 24°12′S 50°55′W. Coffee, cotton, rice, corn;
livestock.

Ourlal (oor-LAHL), village, BISKRA wilaya, NE AL-
GERIA, oasis in the ZIBAN region of the N SAHARA, on
railroad spur, and 18 mi/29 km SW of BISKRA; 34°38′N
05°25′E. Date palms.

Ouro (O-ro), town (2007 population 7,095), W SANTA
CATARINA state, BRAZIL, 17 mi/27 km SW of JOAÇABA;
27°22′S 51°41′W. Wheat; livestock.

Ouro Branco (O-ro BRAHN-ko), city (2007 population
33,530), S central MINAS GERAIS state, BRAZIL, 11 mi/18
km E of CONGONHAS in Iron Quadralateral (metals
mining and manufacturing area); 20°35′S 43°40′W.
Site of Brazil's third-largest steel mill, Açiminas.

Ouro Branco (O-ro BRAHN-do), town (2007 population 11,049), W ALAGOAS state, BRAZIL, on BR 423
Highway; 09°08′S 37°18′W.

Ouro Branco (O-ro BRAHN-ko), town (2007 population
4,959), S RIO GRANDE DO NORTE state, BRAZIL, 45
mi/72 km SW of CURRAIS NOVOS, on PARAÍBA border;
06°42′S 36°57′W.

Ouro Fino (o-ro fee-no), city (2007 population 31,160),
SW MINAS GERAIS state, BRAZIL, near SÃO PAULO
border, on railroad, and 35 mi/56 km SSE of POÇOS DE
CALDAS; 22°17′S 46°26′W. Agricultural trade center;
coffee and tea growing.

Ourolândia (O-ro-LAHN-zhee-ah), city (2007 population 16,301), BAHIA state, BRAZIL.

Ouro Prêto (o-ro PRE-to) [Portuguese=black gold],
city (2007 population 67,405), MINAS GERAIS state, E
BRAZIL; 20°30′S 43°35′W. A national historic site
(1933), especially the colonial mint and treasury, governor's mansion, and old churches. Most notably the
Church of São Francisco, decorated with the magnificent carvings of Aleijadinho; also the oldest theater in
South America. Founded as Vila Rica in the gold rush
near the end of the 17th century, it became a prosperous 18th-century mining town, a cultural center,
and the chief seat of the abortive move for independence led by Tiradentes. The city declined as the mines
lost importance but remained capital of Minas until
1897, when it was superseded by BELO HORIZONTE. A
mining school (established 1875) is in Ouro Prêto.

Ouro Velho (O-ro VEL-yo), town (2007 population
2,974), S central PARAÍBA state, BRAZIL, near border
with PERNAMBUCO; 07°38′S 37°10′W.

Ouro Verde, Brazil: see CANOINHAS.

Our River (OOR), 50 mi/80 km long, E BELGIUM and E
LUXEMBOURG; rises 10 mi/16 km NE of SAINT–VITH
(Belgium); flows S, forming Luxembourg-GERMANY
border for greater part of its course, past VIANDEN
(Luxembourg), to SÛRE RIVER 6 mi/9.7 km E of
DIEKIRCH (Luxembourg).

Ourscamps, Abbey, FRANCE: see NOYON.

Ourthe River (OOR-tuh), c.100 mi/160 km long, E
BELGIUM; rising in the ARDENNES mountains, near the
N tip of LUXEMBOURG; flowing first W then N to join
the MEUSE RIVER at LIÈGE. Its valley provides one of
the few passages through the Ardennes.

Ous (O-oos), settlement (2006 population 1,630), N
SVERDLOVSK oblast, W Siberian Russia, on road and
railroad junction, 40 mi/64 km ENE, and under administrative jurisdiction, of IVDEL'; 60°54′N 61°31′E.
Elevation 298 ft/90 m. Railroad depots.

Cross-references are shown in SMALL CAPITALS. The pronunciation guide is shown on page xix. The sources of population figures are shown on page xvii.

Ousa, INDIA: see AUSA.

Ouse (OOZ), town, central TASMANIA, SE AUSTRALIA, 40 mi/64 km NW of HOBART, and on OUSE RIVER; 42°28′S 146°43′E. Cattle, sheep. Has two hydroelectric power stations.

Ouse River (OOZ), 62 mi/100 km long, central TASMANIA, SE AUSTRALIA; rises in small lakes WNW of GREAT LAKE; flows SE, past WADDAMANA (hydroelectric plant) and OUSE, to DERWENT RIVER just S of OUSE.

Ouse River (OOZ), 30 mi/48 km long, in West SUSSEX, SE ENGLAND; rises 6 mi/9.7 km ESE of HORSHAM in West Sussex; flows E and S, past LEWES, to the CHANNEL at NEWHAVEN, East SUSSEX. Navigable in its lower course.

Ouse River (OOZ), c.60 mi/97 km long, NE ENGLAND; formed by the confluence of the URE and SWALE rivers near BOROUGHBRIDGE, NORTH YORKSHIRE. Flows generally SE past YORK to join with the TRENT RIVER and form the HUMBER RIVER. All of its chief tributaries (Wharfe, Nidd, Derwent, Aire, and Don) rise in the PENNINES. Navigable to York, the Ouse is an important commercial waterway.

Oussouye (OO-swee), town (2004 population 4,457), ZIGUINCHOR administrative region, SW SENEGAL, in CASAMANCE River delta, and 20 mi/32 km W of ZIGUINCHOR; 12°29′N 16°31′W. Peanuts, rice, hardwoods.

Oust River (OOST), 80 mi/129 km long, BRITTANY, W FRANCE; rises in ARMORICAN MASSIF 4 mi/6.4 km NE of Corlay (CÔTES-D'ARMOR department); flows SE into MORBIHAN department, past Rohan, JOSSELIN, and MALESTROIT, to VILAINE RIVER 1 mi/1.6 km S of REDON; 47°35′N 02°05′W. Its course through Morbihan department forms part of BREST-NANTES CANAL.

Outagamie (out-a-GAIM-ee), county (□ 644 sq mi/1,674.4 sq km; 2006 population 172,734), E WISCONSIN; ⊙ APPLETON; 44°24′N 88°27′W. Drained by WOLF, FOX, and EMBARRASS rivers. Dairying and paper milling are principal industries. Some livestock raising (cattle, hogs, sheep, poultry) and farming (barley, oats, wheat, corn, soybeans, peas, beans, alfalfa, hay). Manufacturing at Appleton, KAUKAUNA, KIMBERLY. Urbanized in S along Fox River. Part of Oneida Indian Reservation in E. Formed 1851.

Outaouais (oo-tah-WAIZ), suburban region (□ 11,778 sq mi/30,622.8 sq km; 2005 population 341,752), SW QUEBEC, E Canada, opposite OTTAWA, ONTARIO, on OTTAWA (Outaouais) River; 46°16′N 76°19′W. Includes 75 municipalities, centered on GATINEAU. GATINEAU River crosses region from N to enter OTTAWA River. Purpose of the National Capital Commission region was to coordinate government activites; maintain quality of development, including limits to construction in vicinity of Parliament buildings; and establish greenways and parks. Federal government office buildings and the Canadian Museum of Civilization were built in Hull, now part of Gatineau, in early 1980s. Inclusion of Quebec portion of plan was seen as an attempt to appease separatist sentiments in the French-speaking province.

Outardes, Rivière aux (zoo-TAHRD, reev-YER o), 300 mi/483 km central QUEBEC, E Canada, rises in OTISH MOUNTAINS, flows S, through PLETIPI LAKE, to the SAINT LAWRENCE RIVER 18 mi/29 km SW of BAIE COMEAU; 49°04′N 68°28′W. Several waterfalls.

Outat El Haj (OO-tah tel HAZH), village, Boulemane province, Fes-Boulemane administrative region, MOROCCO, agricultural and pastoral center on the Oued MOULOUYA river, 81 mi/130 km NE of MIDELT, on road; 33°21′N 03°42′W. Also called Outat Oulad el Haj.

Outeniqua Mountains, WESTERN CAPE province, SOUTH AFRICA, extend 170 mi/274 km E from GOURITZ RIVER valley and E end of LANGEBERG range S of CALITZDORP to W side of PLETTENBERG BAY. Range parallels INDIAN OCEAN coast; rises to 5,179 ft/1,579 m

on George Peak, 5 mi/8 km N of GEORGE. W part of range crossed by railroad and gravel road on MONTAGU PASS (2,348 ft/716 m), and N9 highway on Outeniqua Pass, 5 mi/8 km N of George.

Outer Banks or **the Banks** (OU-tuhr BANKS), E NORTH CAROLINA and SE VIRGINIA, chain of sand barrier islands and peninsulas paralleling the ATLANTIC coast enclosing several saltwater bodies called sounds, marked by three prominent capes, Cape HATTERAS (E), Cape LOOKOUT (SE), and Cape FEAR (S). State Highway 12 runs length of banks in N and center. Known for shifting shoals and changing water depths, hazardous to shipping; also vulnerable to hurricanes. Banks are generally 1 mi/1.6 km wide or less. Includes CAPE LOOKOUT NATIONAL SEASHORE and CAPE HATTERAS NATIONAL SEASHORE in E, Pea Island National Wildlife Refuge in NE. In S, banks are separated only by channel of INTRACOASTAL WATERWAY, part natural, part man-made; farther N, the channel widens into BOGUE and Core sounds (both 3 mi/4.8 km–5 mi/8 km wide); in center coast, channel widens further in PAMLICO SOUND (20 mi/32 km–30 mi/48 km wide), which is connected to ALBEMARLE SOUND in N by parallel CROATAN (W) and ROANOKE (E) sounds, then narrows in CURRITUCK SOUND, which extends into SE Virginia, as BACK BAY; several large estuaries extend inland, especially NEUSE RIVER, PAMLICO RIVER, and CHOWAN RIVER. N part of banks is attached to mainland at VIRGINIA BEACH, Virginia.

Outer Barrier, series of sandy barrier islands or offshore bars, extending c.75 mi/121 km along the S shore of LONG ISLAND, SE NEW YORK, from Rockaway Beach (see ROCKAWAY PENINSULA) at the W to E end of SHINNECOCK BAY, and separating a series of lagoons (GREAT SOUTH BAY, Moriches Bay, Mecox Bay, and Shinnecock Bay) from the ATLANTIC OCEAN. Inlets at EAST ROCKAWAY, JONES BEACH, FIRE ISLAND, MORICHES, Mecox Bay, and Shinnecock Bay pierce the barrier, forming the narrow, sandy islands. Includes the resort communities of ATLANTIC BEACH, LONG BEACH, and WESTHAMPTON Beach, and Fire Island National Seashore; Jones Beach State Park and other recreational areas are found here. The sparsely settled and largely undeveloped low-lying islands suffer from wave erosion. During storms they are sometimes inundated and occasionally pierced. Strong littoral drift requires constant dredging of inlets for navigation by fishing and pleasure boats. Unlike the term OUTER BANKS (North Carolina), the name *Outer Barrier* is not in wide circulation, perhaps because most residents and visitors refer to specific locations (e.g., Fire Island).

Outer Brewster Island, MASSACHUSETTS: see BREWSTER ISLANDS.

Outerbridge Crossing, NEW YORK and NEW JERSEY, highway bridge across ARTHUR KILL, between TOTTENVILLE, STATEN ISLAND, and PERTH AMBOY, New Jersey; 40°31′N 74°15′W. Completed 1928 by Port of New York Authority (see PORT AUTHORITY OF NEW YORK AND NEW JERSEY), it is a cantilever structure; total length of truss spans is 2,100 ft/640 m, and main span is 750 ft/229 m long and 142 ft/43 m above water. Although it is the outermost bridge in NEW YORK city, its name comes from Eugenius H. Outerbridge, first chairman of the Port Authority. Outerbridge Reach, a section of the channel lying 800 ft/244 m S of the bridge, figures in a novel of the same name by Robert Stone.

Outer Hebrides, Scotland: see THE HEBRIDES.

Outer Himalaya, PAKISTAN, INDIA, CHINA (TIBET), and NEPAL: see HIMALAYA.

Outer Mongolia: see MONGOLIA.

Outer Pintades, MAURITIUS: see PINTADES ISLAND.

Outjo (ot-YO), town, NW Namiba,180 mi/290 km NNW of Windhoek,40 mi/64 km NW of Otjiwarongo; 20°07′S 16°09′E. Elev. 4,139 ft/1,262 m. Railroad terminusin cattle-raising region. Airfield.

Outlook (OUT-luhk), town (2006 population 1,938), S central SASKATCHEWAN, Canada, on SOUTH SASKATCHEWAN RIVER, and 50 mi/80 km SSW of SASKATOON; 51°30′N 107°03′W. Wheat.

Outlook, village (2000 population 82), SHERIDAN county, NE MONTANA, 13 mi/21 km NW of PLENTYWOOD; CANADA (SASKATCHEWAN) border, 7 mi/11.3 km N; 48°54′N 104°47′W. Grain, hay, safflower, sunflower; sheep, hogs, cattle, exotic and game animals.

Outokumpu (O-to-KOOM-poo), village, POHJOIS-KARJALAN province, SE FINLAND, 25 mi/40 km WNW of JOENSUU; 62°44′N 29°01′E. Elevation 330 ft/100 m. In SAIMAA lake region. Former copper-mining and smelting center; mines closed in 1989.

Outreau (oo-TRO), S industrial suburb (□ 2 sq mi/5.2 km) of BOULOGNE-SUR-MER, PAS-DE-CALAIS department, NORD-PAS-DE-CALAIS region, N FRANCE; 50°42′N 01°36′E. Metalworking plants and cement works.

Outremont (oo-truh-MON), former city, borough (French *arrondissement*) of MONTREAL (□ 2 sq mi/5.2 sq km), S QUEBEC, E Canada, on central MONTREAL ISLAND; 45°31′N 73°37′W. Bounded on W by city of MONT-ROYAL. Mont-Royal Park to E; University of Montreal campus to S.

Out Skerries (OUT SKER-eez), group of small islands, easternmost of the SHETLAND ISLANDS, extreme N Scotland; 60°25′N 00°43′W. Largest island, Housay, 22 mi/35 km NE of LERWICK, is 2 mi/3.2 km long, 1 mi/1.6 km wide. Just E of Housay are islets of Bruray and Grunay or Gruna. Just E of Grunay is Bound islet, site of the Out Skerries lighthouse. Fishing is chief occupation.

Ouvéa (oo-VAI-ah), atoll, northernmost of LOYALTY ISLANDS, NEW CALEDONIA, SW PACIFIC OCEAN, 30 mi/48 km NW of LIFOU; 20°32′S 166°35′E; 23 mi/37 km long, 3.5 mi/5.6 km wide. Lagoon formed by W coast and two rows of islets. Most fertile of group. Coconuts. Produces copra. Chief town, Fayahoué. Formerly called Halgan. Also spelled Uvea (not to be confused with 'Uvea).

Ouvèze River (oo-VEZ), c.40 mi/64 km long, DRÔME and VAUCLUSE departments, SE FRANCE; rises near BUIS-LES-BARONNIES in SW outliers of the ALPS; flows generally SW, past VAISON-LA-ROMAINE and BÉDARRIDES, to the RHÔNE 5 mi/8 km above AVIGNON; 44°00′N 04°51′E. Receives the SORGUE DE VAUCLUSE (left).

Ouyen (O-yuhn), town, NW VICTORIA, SE AUSTRALIA, 240 mi/386 km NW of MELBOURNE; 35°04′S 142°19′E. Railroad junction; agricultural center (wheat, oats). In an area of extensive growth of mallee (low-growing eucalyptus).

Ouzinkie (oo-ZEEN-kee), village (2000 population 225), S ALASKA, on SPRUCE ISLAND, in Gulf of ALASKA, 9 mi/14.5 km NNW of KODIAK; 57°56′N 152°27′W. Salmon fishing and canning. Has an Orthodox church. Shipyard est. here by Russians c.1800. Also spelled Uzinki.

Ouzouer-sur-Loire (oo-ZWAI–syoor–LWAHR), commune (□ 13 sq mi/33.8 sq km), LOIRET department, CENTRE administrative region, N central FRANCE, near right bank of LOIRE RIVER, 9 mi/14.5 km NW of GIEN; 47°46′N 02°29′E. Sawmilling, lumber, trade. The nearby Forest of ORLÉANS attracts hunters.

Ovacik, village, E central TURKEY, on MUNZUR RIVER, and 29 mi/47 km SW of ERZINCAN; c.39°38′N 38°24′E. Grain. Formerly called Zerenik and Marasalcakmak.

Ovada (o-VAH-dah), town, ALESSANDRIA province, PIEDMONT, N ITALY, 19 mi/31 km S of ALESSANDRIA; 44°38′N 08°38′E. Railroad junction; wine market; diversified small industrial center including fabricated metals, machinery, textiles.

Ovadne (o-VAHD-ne), (Russian *Ovadno*), (Polish *Owadno*), village (2004 population 3,100), W VOLYN' oblast, UKRAINE, on railroad, and 5 mi/8 km NNE of VOLODYMYR-VOLYNS'KYY; 50°55′N 24°24′E. Elevation 669 ft/203 m. Wheat, barley; livestock.

Ovadno, UKRAINE: see OVADNE.

Ovalau (O-vah-LOU), volcanic island (□ 39 sq mi/101.4 sq km), FIJI, SW PACIFIC OCEAN, 10 mi/16 km E of VITI LEVU; 8 mi/12.9 km long, 5 mi/8 km wide, within Viti Levu barrier reef; rises to 2,054 ft/626 m; 17°40′S 178°48′E. Bananas, pineapples; commercial fishing and freezing. Site of Levuka, former capital of colony.

Ovalle (o-VAH-yai), city, ⊙ Ovalle comuna (2002 population 66,405) and LIMARÍ province, COQUIMBO region, N central CHILE, on LIMARÍ RIVER (irrigation), on railroad, and 50 mi/80 km S of LA SERENA; 30°35′S 71°12′W. Fruit-growing (pisco) and sheep-raising center. Copper mining; consumer goods. Airport. Near Monumento Arqueológico Valle del Encanto. Founded 1831 and named for Vice-President Ovalle.

Ovamboland (o-vam-BO-land) or **Owambo**, region (□ 20,000 sq mi/52,000 sq km), N NAMIBIA, extends W from OKAVANGO RIVER, along ANGOLA border; ⊙ ONDANGWA. Traditional home of Oshivambo-speaking people and several other tribes. Cattle- and goat-raising and grain-growing on communal lands are chief occupations. In S part of district is ETOSHA PAN, saltwater-filled depression. Fish caught in the shallow Oshanas during the rainy season (Dec.–April) to supplement local diet.

Ovan (O-vahn), town, OGOOUÉ-IVINDO province, central GABON, 47 mi/76 km SW of MAKOKOU; 00°20′N 12°12′E.

Ovando (uh-VAN-do), village, POWELL county, W MONTANA, on Warren Creek, near BLACKFOOT RIVER, 41 mi/66 km ENE of MISSOULA. Timber, hay; cattle, sheep. Salmon Lake State Park to W; Lolo National Forest to N; Helena National Forest and Lincoln State Forest to E; Lolo National Forest and Clearwater State Forest to W.

Ovar (oo-VAHR), fishing town (2001 population 33,022), AVEIRO district, N central PORTUGAL, on railroad, and 15 mi/24 km N of AVEIRO, at N end of Aveiro lagoon. Manufacturing (hardware, paper, felt, pottery); sardine processing.

Ovčar (OV-chahr), mountain (3,231 ft/985 m), in DINARIC ALPS, W SERBIA, 7 mi/11.3 km W of Cacak; 43°53′N 20°12′E. Ovcar Banja (or Ovcarska Banja) health resort is at W foot, on railroad. Also spelled Ovchar.

Ovce Polje (OV-che POL-ye), Serbo-Croatian *Ovče Polje*, wide valley, N MACEDONIA; extends c.15 mi/24 km NW from lower BREGALNICA RIVER. Sheep. Chief village, SVETI NIKOLA. Also spelled Ovche Polye.

Ovcha Mogila (OV-chah mo-GEE-lah), village, LOVECH oblast, SVISHTOV obshtina, N BULGARIA, 12 mi/19 km SSW of Svishtov; 43°26′N 25°16′E. Grain; livestock.

Ovcharets (ov-CHAH-rets), peak (9,085 ft/2,769 m), E RILA MOUNTAINS, W BULGARIA, 11 mi/18 km S of SAMOKOV; 42°20′N 24°24′E. Formerly called Yur-ushki-chal.

Ovcharitsa (ov-CHAH-reet-sah), reservoir, BURGAS oblast, SE BULGARIA, on the OVCHARITSA River; tributary of the SAZLIIKA RIVER; 42°15′N 26°10′E.

Ovcharitsa (ov-CHAH-reet-sah), river, BURGAS oblast, BULGARIA; 42°12′N 26°00′E. Tributary of the SAZLIIKA RIVER.

Ovche Polye, MACEDONIA: see OVCE POLJE.

Oveja, Cerro (o-VAI-hah, SER-ro), peak (7,237 ft/2,206 m), W CÓRDOBA province, ARGENTINA, 50 mi/80 km WSW of RÍO TERCERO, in N Sierra de COMECHINGONES, a range of the Sierra de CÓRDOBA.

Ovejas (o-VAI-hahs), town, ⊙ Ovejas municipio, SUCRE department, N central COLOMBIA, 15 mi/24 km NE of SINCELEJO; 09°31′N 75°14′W. Tobacco-growing center; sorghum; livestock.

Ovejería (o-ve-he-REE-ah), town, OSORNO province, LOS LAGOS region, S central CHILE, 7 mi/11 km SE of OSORNO.

Ovelgönne (O-vel-goen-ne), village, LOWER SAXONY, NW GERMANY, 3 mi/4.8 km WNW of BRAKE; 53°20′N 08°25′E. Wool textiles.

Ovens Peninsula (3 mi/5 km long, 1 mi/2 km wide), SW NOVA SCOTIA, CANADA, 4 mi/6 km SSE of LUNENBURG. Former gold-mining area.

Ovens River (UH-vuhnz), 110 mi/177 km long, NE VICTORIA, SE AUSTRALIA; rises in AUSTRALIAN ALPS near Mount HOTHAM; flows generally NW, past BRIGHT, MYRTLEFORD, and WANGARATTA (hydro-electric plant), to MURRAY RIVER 14 mi/23 km E of YARRAWONGA; 36°03′S 146°11′E. King River, main tributary; used for irrigation.

Overath (O-ve-raht), town, RHINELAND, NORTH RHINE-WESTPHALIA, W GERMANY, 14 mi/23 km E of COLOGNE; 50°56′N 07°17′E. railroad junction. Manufacturing of precision instruments, electrical equipment.

Overbrook, village (2000 population 947), OSAGE county, E KANSAS, 20 mi/32 km SSE of TOPEKA. Livestock; grain, feeds.

Overflakkee (AW-vuhr-flah-kai), region, SOUTH HOLLAND province, SW NETHERLANDS, former island joined by land on NW to GOEREE (also former island) to form Goeree-Overflakkee island. Bounded by the GREVELINGENMEER (SW), the KRAMMER (S), the VOLKERAK (SE, E), and the HARINGVLIET (N, NE). Main city is Middelharnis, in N. Connected to SCHOUWEN-DUIVELAND island to NW by Grevelingen Dam, to Voorne region to N by Haringvliet Dam, to mainland to E by Volkerak Dam, and to BEIJERLAND region to NE by Haringvliet Bridge, which joins Volkerak Dam in mid-channel. Dairying; cattle; vegetables, grain, sugar beets.

Overgaard, unincorporated town, NAVAJO county, E central ARIZONA, 43 mi/69 km S of WINSLOW, and 3 mi/4.8 km SE of HEBER, in Apache-Sitgreaves National Forest. Elevation c.6,000 ft/1,829 m. Timber. Fort Apache Indian Reservation and MOGOLLON Rim Escarpment to S.

Overijse (O-ve-rei-suh), commune (2006 population 24,078), Halle-Vilvoorde district, BRABANT province, central BELGIUM, 9 mi/14.5 km SE of BRUSSELS; 50°46′N 04°32′E. Market center for grape-growing area.

Overijssel (AW-vuhr-REI-suhl), province (□ c.1,500 sq mi/3,340 sq km; 2007 estimated population 1,116,374), E and E central NETHERLANDS, bounded on E by GERMANY, on NW by NORTH-EAST POLDER (formerly part of Overijssel province; became part of FLEVOLAND polder in 1986) and KETELMEER and Vossemeer channels, on N by Reest River; ⊙ ZWOLLE. Drained by VECHT and ZWARTE WATER rivers; IJSSEL River forms part of W boundary. Other cities include ALMELO, DEVENTER, ENSCHEDE, KAMPEN, and ZUTPHEN. The province is generally sandy with vegetated sand dunes but supports extensive livestock raising, agriculture (oats, wheat, rye, peas, beans, onions, potatoes, fruits), and dairying. The lordship of Overijssel belonged in the Middle Ages to the bishop of Utrecht but was sold (1527) to Emperor Charles V. It joined (1579) the Union of Utrecht and became one of the United Provinces of the Netherlands.

Overijssel Canal, Dutch *Overijssels Kanaal*, network of canals from ZWARTE WATER river, mainly in OVERIJSSEL province, NE NETHERLANDS. Connects towns of ZWOLLE, RAALTE, ALMELO, VRIEZENVEEN, and COEVORDEN. Total length, 73 mi/117 km; 25 mi/40 km branch to Vroomshoop, 18 mi/29 km branch to COEVORDEN, 8 mi/12.9 km branch to Almelo.

Overijssels Kanaal, NETHERLANDS: see OVERIJSSEL CANAL.

Överkalix (UHV-er-KAH-liks), village, NORRBOTTEN county, N SWEDEN, on KALIX ÄLVEN RIVER, 50 mi/80 km NW of HAPARANDA; 66°19′N 22°51′E. Market center in lumbering region.

Overland, residential city (2000 population 16,838), SAINT LOUIS county, E MISSOURI, residential, business, and industrial suburb 10 mi/16 km WNW of downtown ST. LOUIS; 38°42′N 90°22′W. Manufactur-ing (machinery, chemicals, printing, signs, processed foods). U.S. military records center. Incorporated 1939.

Overland Park, city (2000 population 149,080), fourth-largest city in KANSAS, JOHNSON county, NE Kansas, a residential suburb 8 mi/12.9 km SSW of downtown KANSAS CITY, Kansas; 38°54′N 94°40′W. Manufacturing (printing and publishing, apparel, aircraft parts, cement, prepared foods, salt, chemicals, marine accessories, signs). It has profited from the regional growth and development in the burgeoning Kansas City area. OVERLAND PARK has become one of the fastest-growing U.S. cities, marked by a population increase of nearly 37 percent between 1980 and 1990. Incorporated 1960.

Overland Trail, any of several trails of W migration in the U.S. The term is sometimes used to mean all the trails W from the MISSOURI to the PACIFIC and sometimes for the central trails only. Particularly, the term has been applied to a S alternate route of the OREGON TRAIL. It branched from the parent trail at the junction of the North PLATTE and South Platte rivers and followed the South Platte to present JULESBURG, COLORADO, where it left the river and W overland to the North Platte, rejoining the parent trail east of FORT LARAMIE, WYOMING. The term is also particularly applied to a route to CALIFORNIA that went W from FORT BRIDGER to the GREAT SALT LAKE (thus duplicating in part the Mormon Trail), then on to Sutter's Fort in California; it was much used by immigrants bound for California.

Overlea (OH-ver-lee), village (2000 population 12,148), BALTIMORE county, central MARYLAND, 6 mi/9.7 km NE of downtown BALTIMORE; 39°22′N 76°31′W. One of Baltimore's earliest suburbs and mostly rowhouses, it (along with nearby PARKVILLE) was once called Lavender Hill.

Overmountain Victory National Historic Trail (O-vuhr-mount-uhn VIK-tuh-ree), NORTH CAROLINA, SOUTH CAROLINA, TENNESSEE, VIRGINIA, follows the path of Revolutionary patriots. It is 272 mi/438 km long and terminates in S at KINGS MOUNTAIN NATIONAL MILITARY PARK, South Carolina. Authorized 1980.

Overo, Cerro (o-VAI-ro, SER-ro), Andean volcano (15,630 ft/4,764 m), MENDOZA province, ARGENTINA, near CHILE border, 45 mi/72 km SE of RANCAGUA (Chile); 34°35′S 70°02′W. Sulfur deposits.

Øverød, DENMARK: see HOLTE.

Overpelt (O-vuhr-pelt), commune (2006 population 13,472), Maaseik district, LIMBURG province, NE BELGIUM, 20 mi/32 km N of HASSELT; 51°13′N 05°25′E. Town of NEERPELT is just N, on SCHELDT-MEUSE JUNCTION CANAL.

Overschie (AW-vuh-SKHEE), suburb of ROTTERDAM, SOUTH HOLLAND province, W NETHERLANDS, 4 mi/6.4 km NW of city center, on Delftse Schie canal; 51°57′N 04°25′E. N of NEW MAAS RIVER (Rotterdam harbor), Rotterdam International Airport to N.

Overseal (O-vuh-seel), village (2001 population 2,049), S DERBYSHIRE, central ENGLAND, 6 mi/9.7 km SE of BURTON UPON TRENT; 52°44′N 01°34′W.

Overseas Highway (O-VUHR-seez), 100-mi/161-km segment of U.S. 1 that links the FLORIDA KEYS between KEY LARGO in the NE and KEY WEST in the SW. Follows the route of Henry Flagler's "overseas railway," which reached Key West from MIAMI in 1912. When the great hurricane of 1935 destroyed the railroad, its right-of-way and numerous bridges were rebuilt as a two-lane highway that carries U.S. 1 to Key West. It opened 1938 and was rebuilt substantially in the 1980s.

Overstrand (O-vuh-strand), village (2001 population 1,405), N NORFOLK, E ENGLAND, on NORTH SEA, 2 mi/3.2 km ESE of CROMER; 52°55′N 01°21′E. Seaside resort.

Overton, county (□ 442 sq mi/1,149.2 sq km; 2006 population 20,740), N TENNESSEE; ⊙ LIVINGSTON, 36°20′N 85°17′W. In the CUMBERLAND MOUNTAINS;

drained by affluents of OBEY and CUMBERLAND rivers. Includes part of DALE HOLLOW Reservoir. Formerly relied on lumbering and extractive industries; tourism now represents significant source of income. Standing Stone State Park here. First commercial oil well in Tennessee was sunk here in 1866. Formed 1806.

Overton (O-vuhr-tuhn), town, CLARK county, SE NEVADA, on MUDDY RIVER, near its mouth in N arm (VIRGIN RIVER) of Lake MEAD (LAKE MEAD NATIONAL RECREATION AREA), and c.45 mi/72 km NE of LAS VEGAS. 36°32′N 114°26′W. Overton Beach Marina, Lake Mead 10 mi/16 km to SE. Cattle, poultry; hay. Lost City Museum has Pueblo Indian relics. Moapa Indian Reservation to W. Valley of Fire State Park to SW.

Overton (O-vuhr-tuhn), town (2006 population 2,339), RUSK county, E TEXAS, 19 mi/31 km ESE of TYLER; an oil center in E Texas field; 32°16′N 94°58′W. Railroad junction; cattle; dairying; vegetables, watermelons; manufacturing (natural-gas processing, metal fabrication, wood products). Boomed after oil discovery (1930).

Overton (O-vuh-tuhn), village (2001 population 3,948), N HAMPSHIRE, S ENGLAND, on TEST RIVER, and 8 mi/12.9 km W of BASINGSTOKE; 51°15′N 01°15′W. Paper milling (for Bank of England notes). A sheep and lamb fair is held here. Jane Austen born at village of Steventon, 2 mi/3.2 km ESE.

Overton (O-vuh-tuhn), village (2001 population 1,015), LANCASHIRE, NW ENGLAND, on LUNE RIVER, 6 mi/9.7 km S of MORECAMBE; 54°01′N 02°52′W. IRISH SEA 2 mi/3.2 km to W. Has Norman church.

Overton, village (2006 population 663), DAWSON county, S central NEBRASKA, 15 mi/24 km ESE of LEXINGTON, near PLATTE RIVER; 40°44′N 99°32′W. Grain; livestock.

Overton (O-vuhr-tuhn), village (2001 population 1,276), WREXHAM, NE Wales, near the DEE, 6 mi/9.7 km SE of Wrexham; 52°58′N 02°56′W. Formerly in CLWYD, abolished 1996.

Övertorneå (UHV-er-toorn-e-O) or **Matarengi**, town, NORRBOTTEN county, N SWEDEN, on TORNEÄLVEN RIVER (FINNISH border), 40 mi/64 km NNW of HAPARANDA; 66°24′N 23°39′E.

Overtown (O-vuhr-tuhn), village (2001 population 2,371), NORTH LANARKSHIRE, central Scotland, 2 mi/3.2 km S of WISHAW; 55°45′N 03°56′W. Formerly in STRATHCLYDE, abolished 1996.

Överum (UHV-er-OOM), village, KALMAR county, SE SWEDEN, 20 mi/32 km NW of VÄSTERVIK; 57°59′N 16°19′E.

Overveen (AW-vuhr-vain), suburb of HAARLEM, NORTH HOLLAND province, W NETHERLANDS, 1 mi/1.6 km NW of city center; 52°24′N 04°38′E. NORTH SEA 3 mi/4.8 km to W; National Park De Kennemerduinen to W. Dairying; cattle; vegetables, flowers, fruits.

Ovett (o-VET), unincorporated village, JONES county, SE MISSISSIPPI, 16 mi/26 km SSE of LAUREL. Agriculture (cotton, corn; cattle, poultry); timber; manufacturing (pressure tanks). Oil-producing area. At W edge of De Soto National Forest.

Ovetum, SPAIN: see OVIEDO, city.

Ovezberdy Kulieva (uh-vez-ber-DEE koo-lee-YE-vah), town, AHAL weloyat, TURKMENISTAN, 6 mi/10 km NW of ASHGABAT. Also spelled Ovezberdy Kuliyeva.

Ovid, town (2006 population 1,514), CLINTON county, S central MICHIGAN, 10 mi/16 km E of SAINT JOHNS, and on MAPLE RIVER; 43°00′N 84°22′W. In agricultural area; manufacturing (metal-forming equipment, dairy products). Plotted 1857; incorporated 1869.

Ovid, village (2000 population 330), SEDGWICK county, NE COLORADO, on SOUTH PLATTE RIVER, near NEBRASKA state line, 6 mi/9.7 km WSW of JULESBURG; 40°57′N 102°23′W. Elevation 3,521 ft/1,073 m. Trade point in grain, cattle region; beet sugar. Manufacturing (prepared foods).

Ovid, unincorporated village, BEAR LAKE county, SE IDAHO, 5 mi/8 km WSW of MONTPELIER; elevation 7,424 ft/2,263 m. Timber; cattle. Lumber. Cache National Forest to W.

Ovid, village (2006 population 606), ⊙ SENECA county, W central NEW YORK, 23 mi/37 km NW of ITHACA, between SENECA and CAYUGA lakes; 42°40′N 76°49′W. In agricultural area (vineyards; fruits, grain); dairy products. Willard Correctional Facility is to W.

Ovidiopil', UKRAINE: see OVIDIOPOL'.

Ovidiopol' (o-vee-dee-O-pol), town, central ODESSA oblast, UKRAINE, on the DNIESTER LIMAN, 18 mi/29 km SW of ODESSA (linked by railroad); 46°16′N 30°26′E. Raion center; automatic-press building plant, cannery, dairy products, feed mill, flour mill, brickworks. Once identified with site of ancient Tomis, where Ovid was exiled. Established in 1793 on the site of a Turkish fortress, Khadzhy-Dere; town since 1970. Also spelled, in Ukrainian, Ovidiopil'. Sizeable Jewish community since the mid-19th century, decimated during the pogroms of 1918–1920; the majority of survivors emigrated, with only 700 Jews remaining in 1939; community completely destroyed during World War II.

Ovidiu (o-VEE-dyoo), town, CONSTANȚA county, ROMANIA, 7 mi/11.3 km NW of CONSTANȚA; 44°16′N 28°34′E.

Oviedo (o-VYAI-dho), province (□ 4,207 sq mi/ 10,938.2 sq km; 2001 population 1,062,998), NW SPAIN, on BAY OF BISCAY; ⊙ OVIEDO. Coextensive with the autonomous region of ASTURIAS.

Oviedo (o-VYAI-dho), ancient *Ovetum*, city (2001 population 201,154), ⊙ OVIEDO province and ASTURIAS autonomous region, NW SPAIN, near NALÓN RIVER (S), 230 mi/370 km NW of MADRID, and 15 mi/24 km SW of GIJÓN; 43°21′N 05°50′W. Important industrial center in mining region N of CANTABRIAN MOUNTAINS; coal and iron mining is an important industry; manufacturing includes chemicals, explosives, ordnance, firearms, textiles, food, and a variety of industrial and consumer goods. Bishopric; University of Oviedo (founded 1604). Has 14th-century Gothic cathedral (damaged in Spanish civil war) with tombs of Asturian kings and Cámara Santa, which housed the cathedral's sacred relics and treasures. Founded c.760, the city flourished (9th century) as the capital of the Asturian kings, but declined after the capital was moved to León (early 10th century). Was plundered (1809) by the French, and suffered severely during miners' revolt (1934) and Spanish civil war (1936–1939). A new quarter has been built on the SW side of the city.

Oviedo (o-VEE-do), city (□ 15 sq mi/39 sq km; 2005 population 29,848), SEMINOLE county, central FLORIDA, 14 mi/23 km NE of ORLANDO, near LAKE JESSUP; 28°39′N 81°10′W. Incorporated 1925.

Ovilla (O-vi-luh), town (2006 population 3,858), ELLIS and DALLAS counties, N TEXAS, residential suburb 17 mi/27 km SSW of downtown DALLAS; 32°32′N 96°53′W. Agricultural area on fringe of large Dallas–Fort Worth urban area (cotton, nursery crops; cattle; dairying).

Övörhangay (UH-buhr-KHAHNG-gei) [Mongolian= south Khangai], province (□ 24,300 sq mi/63,180 sq km; 2000 population 111,420), central MONGOLIA; ⊙ ARVAYHEER; 46°00′N 102°00′E. Situated on SE slopes of HANGAYN NURUU (Khangai Mountains), the province is traversed by the ONGIYN GOL (river) and extends from forested mountain slopes (N) to the GOBI DESERT (S). Livestock grazing. Also spelled Ubur Khangai, Ubur Hangay, or Uverkhangay.

Øvrebø (UHV-ruh-buh), village, VEST-AGDER county, S NORWAY, 12 mi/19 km NW of KRISTIANSAND. At Mushom, nearby, the oldest ski in SCANDINAVIA was found, said to date from 2000 B.C.E.

Øvre Fryken, SWEDEN: see FRYKEN.

Ovruch (OV-rooch), city, N ZHYTOMYR oblast, UKRAINE, on S edge of Prypyat' Marshes, and on Noryn' River, tributary of the UZH, 25 mi/40 km NNE of KOROSTEN'; 51°19′N 28°48′E. Elevation 479 ft/145 m. Raion center; railroad junction; lumber center; furniture manufacturing, food processing (dairy, cannery, bakery), flax processing, instrument manufacturing, auto and railroad servicing. Dates from the 9th century; originally called Vruchyy; passed to Lithuania (1362), Poland (1569), Russia (1793); city since 1796. Burial site of Prince Oleh (977); St. Basil's Church (1190, restored 1909).

Ovsyanka (uhf-SYAN-kah), village, central AMUR oblast, SE SIBERIA, RUSSIAN FAR EAST, on road, 14 mi/23 km SW of ZEYA; 53°35′N 126°53′E. Elevation 652 ft/198 m. In agricultural area. Has a wooden church dating back to the late 19th - early 20th century. Home and final resting place of a renowned Russian classical writer Viktor Astafyev - his house turned into a memorial museum in 2004.

Ovsyanka (uhf-SYAHN-kah), settlement (2005 population 3,705), SW KRASNOYARSK TERRITORY, SE SIBERIA, RUSSIA, on the YENISEY RIVER, 21 mi/34 km SW of KRASNOYARSK; 55°57′N 92°33′E. Elevation 856 ft/260 m. In agricultural area; grain processing.

Owadno, UKRAINE: see OVADNE.

Owando, town, ⊙ CUVETTE region, central Congo Republic, 260 mi/418 km N of BRAZZAVILLE; 00°29′S 15°54′E. Palm-oil products.

Owaneco (OH-wan-e-co), village (2000 population 256), CHRISTIAN county, central ILLINOIS, 7 mi/11.3 km SE of TAYLORVILLE; 39°28′N 89°11′W. In agricultural area. Near Taylorville Correctional Center.

Owani (O-wah-nee), town, S Tsugaru county, Aomori prefecture, N HONSHU, N JAPAN, 22 mi/35 km S of AOMORI; 40°31′N 140°34′E. Bean sprouts. Nearby is Mount AJARA (3,200 ft/975 m); ski meet held annually.

Owariasahi (o-WAH-ree-AH-sah-hee), city, Aichi prefecture, S central HONSHU, central JAPAN, 9 mi/15 km N of NAGOYA; 35°12′N 137°02′E. Computers. Traditional ceramic zodiac ornaments.

Owasa, town (2000 population 38), HARDIN county, central IOWA, 7 mi/11.3 km NW of ELDORA; 42°25′N 93°12′W. In agricultural area.

Owasco (o-WAS-ko), hamlet, CAYUGA county, W central NEW YORK, in FINGER LAKES region, between OWASCO and SKANEATELES lakes, 8 mi/12.9 km SE of AUBURN; 42°53′N 76°29′W.

Owasco Lake (o-WAS-ko), one of the FINGER LAKES, CAYUGA county, W central NEW YORK, between CAYUGA Lake (W) and SKANEATELES Lake (E); extends c.11 mi/18 km SSE from AUBURN, at its outlet; 42°50′N 76°31′W. FILLMORE GLEN STATE PARK is near S end.

Owase (O-wah-se), city, MIE prefecture, S HONSHU, central JAPAN, port on KUMANO SEA, on E KII PENINSULA, 50 mi/80 km S of TSU; 34°04′N 136°11′E. Hinoki; lumber. Marine products.

Owasso (o-WAH-so), city (2006 population 24,938), TULSA county, NE OKLAHOMA, suburb 11 mi/18 km NE of downtown TULSA, near BIRD CREEK; 36°16′N 95°50′W. Manufacturing (machining, plastics fabrication, electrical panels). Tulsa International Airport to S.

Owatonna (o-wuh-TAWN-uh), city (2000 population 22,434), ⊙ STEELE county, SE MINNESOTA, 62 mi/100 km S of MINNEAPOLIS, on STRAIGHT RIVER; 44°05′N 93°13′W. Railroad junction. Manufacturing (furniture, consumer goods, electronic equipment, apparel, machinery, printing and publishing). Agriculture (corn, oats, soybeans, peas, alfalfa; poultry, livestock; dairying). Rice Lake State Park to E.

Oweekeno or **Oweekeno Nation**, First Nations community, unincorporated village, W BRITISH COLUMBIA, W Canada, on Whannock River, in Central Coast regional district; 51°40′N 127°13′W. Commercial fishing.

Owego (o-WEE-go), village (□ 2 sq mi/5.2 sq km; 2006 population 3,778), ⊙ TIOGA county, S NEW YORK, on the SUSQUEHANNA River at mouth of OWEGO CREEK, and 18 mi/29 km W of BINGHAMTON; 42°06′N

76°15'W. Railroad junction; diverse manufacturing; in agricultural area. Politician Thomas C. Platt (leader in the creation of greater NEW YORK city) born here. Settled 1787 and incorporated 1827 on site of Native American village destroyed (1779) in Sullivan campaign.

Owego Creek, c.35 mi/56 km long, S NEW YORK; rises in S CORTLAND county; flows generally SSW to the SUSQUEHANNA River at OWEGO.

Owel, Lough (O-wuhl, LAHK) lake (4 mi/6.4 km long, 2 mi/3.2 km wide), central WESTMEATH county, central IRELAND, 2 mi/3.2 km NNW of MULLINGAR; 53°34'N 07°24'W. Reservoir for ROYAL CANAL.

Owen, county (□ 387 sq mi/1,006.2 sq km; 2006 population 22,741), SW central INDIANA; ⊙ SPENCER; 39°19'N 86°50'W. Agriculture (soybeans, corn, wheat, fruits; hogs; cattle; dairying). Manufacturing at Spencer. Lumber milling; limestone quarrying; timber. Drained by West Fork of WHITE RIVER and MILL CREEK. McCormick's Creek State Park in E; Owen-Putnam State Forest and part of Lieber State Recreation Area on Cataract Lake in N. Formed 1818.

Owen, county (□ 354 sq mi/920.4 sq km; 2006 population 11,428), N KENTUCKY; ⊙ OWENTON; 38°31'N 84°49'W. Bounded SW by KENTUCKY RIVER, NW by Eagle Creek; drained in E by Eagle Creek. Rolling upland agricultural area (burley tobacco, corn, hay, alfalfa; cattle, poultry; dairying), in BLUEGRASS region. Lead and zinc mines, limestone quarries. Twin Eagle Wildlife Management Area in NW, Kleber Wildlife Management Area in S; small Elmer Davis Lake and Elk Lake reservoirs in center of county. Formed 1819.

Owen (O-ven), town, STUTTGART district, BADEN-Württemberg, SW GERMANY, at W foot of the TECK, and on the Lauters, 4 mi/6.4 km S of KIRCHHEIM; 48°35'N 09°27'E. Grain. Has Gothic church with tombs of dukes of Teck. Chartered 1241.

Owen, town (2006 population 885), CLARK county, central WISCONSIN, on small Poplar River, and 45 mi/72 km W of WAUSAU; 44°57'N 90°34'W. Railroad junction. In dairying, livestock, and lumbering area; manufacturing (cheese, wooden products, drafting supplies). Settled c.1890; incorporated as village 1904, as city 1925.

Owen (O-wuhn), village, SE SOUTH AUSTRALIA state, S central AUSTRALIA, 45 mi/72 km N of ADELAIDE; 34°16'S 138°33'E. Agricultural center.

Owendale, village (2000 population 296), HURON county, E MICHIGAN, 14 mi/23 km WSW of BAD AXE, near SAGINAW BAY; 43°43'N 83°16'W.

Owendo (o-VEN-do), village, ESTUAIRE province, NW GABON, on N shore of GABON RIVER, part of and 5 mi/8 km SSE of LIBREVILLE; 00°20'N 09°30'E. Deepwater port facilities. Industrial plants. Ships hardwoods (notably *okume* wood).

Owen Falls, dam, located along VICTORIA NILE RIVER, SE UGANDA, 5 mi/8 km NW of Jinja; 00°27'N 33°11'E. Hydroelectric station at Jinja has installed capacity of 180 MW with plans to expand to 380 MW.

Owen, Mount, WYOMING: see GRAND TETON NATIONAL PARK.

Owensboro, city (2000 population 54,067), ⊙ DAVIESS county, W KENTUCKY, 82 mi/132 km WSW of LOUISVILLE, on the OHIO RIVER (bridged to INDIANA); 37°45'N 87°07'W. Elevation 401 ft/122 m. It is an important tobacco market and a shipping point for a farm and oil region. Manufacturing (limestone, whiskey, food processing, electrical motors, rubber gaskets, vinyl windows, coal and steel processing, office furniture, fertilizers, tobacco products, metal stamping). Confederates attacked and burned part of the city, including the courthouse, in 1864. Owensboro-Daviess County Airport in SW part of city. Owensboro is home to Kentucky Wesleyan College, Brescia University, and Owensboro Community College; the Museum of Science and Industry, Museum of Fine Arts, Owensboro Symphony, Owensboro Ice Arena. Carpenter and

Kingfisher lakes (natural) to NE; Ben Hawes State Park to W; Lock and Dam No. 45 is here, on Ohio River. Settled 1798; incorporated as city 1866.

Owens Lake, California: see OWENS RIVER.

Owen Sound (O-wen SOUND), city (□ 9 sq mi/23.4 sq km; 2001 population 21,431), SE ONTARIO, E central Canada, on Owen Sound; 44°34'N 80°56'W. Port and railroad terminal in a farming region, with large grain elevators; manufacturing (tractor trailers, glass, paper, printing); fish processing.

Owens River, c.120 mi/193 km long, E CALIFORNIA; rises in the SIERRA NEVADA, SW MONO county, c. 25 mi/40 km ESE of Yosemite Village, S of JUNE LAKE; flows SSE through Lake Crowley reservoir, past Bishop, Independence, and Lone Pine; enters Owens Lake (dry), 25 mi/40 km SE of Mount Whitney. LOS ANGELES AQUEDUCT (c.15 mi/24 km N of Independence and parallels river on W) diverts tributary streams and river's water to LOS ANGELES at a point c.45 mi/72 km NW of Owens Lake. Evaporation and ground-absorption rates in Owens Lake exceed in flow from Owens River and other streams.

Owen Stanley Range, mountain chain, circa 300 mi/480 km long, SE PAPUA NEW GUINEA, on NEW GUINEA island; 09°20'S 148°00'E. Rises to Mount VICTORIA (13,363 ft/4,073 m). The region, drained by several small rivers, is largely jungle.

Owens Valley, California: see OWENS RIVER.

Owensville, city (2000 population 2,500), GASCONADE county, E central MISSOURI, 40 mi/64 km SE of JEFFERSON CITY; 38°21'N 91°30'W. Corn; cattle, poultry; manufacturing (printing, signs, store fixtures, hats); clay. Incorporated 1900.

Owensville, town (2000 population 1,322), GIBSON county, SW INDIANA, 9 mi/14.5 km SW of PRINCETON; 38°16'N 87°41'W. Agriculture; gas and oil wells, bituminous-coal; timber; feed mill. Laid out 1817; incorporated 1881.

Owensville (o-WENZ-vil), village (2006 population 837), CLERMONT county, SW OHIO, 20 mi/32 km E of CINCINNATI; 39°07'N 84°08'W. Limestone quarry.

Owenton, city (2000 population 1,387), ⊙ OWEN county, N KENTUCKY, 24 mi/39 km N of FRANKFORT; 38°32'N 84°50'W. In BLUEGRASS agricultural area (burley tobacco, corn); manufacturing (gas regulators, printing and publishing, lumber). Fishing nearby. Elmer Davis Lake reservoir to S; Elk Lake reservoir to SE; Twin Eagle Wildlife Management Area to W; Kleber Wildlife Management Area to S.

Owerri (aw-WAI-ree), one of former EASTERN PROVINCES, S NIGERIA; capital was PORT HARCOURT. Included E part of NIGER River delta. Now within IMO state.

Owerri (aw-WAI-ree), town, ⊙ IMO state, S NIGERIA, 35 mi/56 km NW of ABA. Trade center; palm oil and kernels; kola nuts. Animal feed mill. Has a hospital, a university, and a college of education. Former capital Owerri province.

Owey (O-wee), island (□ 1 sq mi/2.6 sq km), off NW DONEGAL county, N IRELAND, 4 mi/6.4 km NE of Aran Island; 55°03'N 08°27'W.

Owingen (O-ving-uhn), village, LAKE CONSTANCE, BADEN-Württemberg, S GERMANY, 9 mi/14.5 km N of CONSTANCE, near the Überlinger Lake (a branch of LAKE CONSTANCE); 47°47'N 09°11'E. Fruits.

Owings, village, LAURENS county, NW SOUTH CAROLINA, 12 mi/19 km NW of LAURENS.

Owings Mills, village (2000 population 20,193), BALTIMORE county, N MARYLAND, 12 mi/19 km NW of downtown BALTIMORE; 39°25'N 76°47'W. Named after Samuel Owings, who established three grist mills in the mid-18th century, it is now dominated by Rosewood State Hospital. He called the mills upper, lower, and middle and used the acronym for his house, Ulm, which has been restored as a restaurant. Some of the Owings family are buried in the yard of St. Thomas Church (c.1742) in nearby GARRISON.

Owingsville, city (2000 population 1,488), ⊙ BATH county, NE KENTUCKY, 28 mi/45 km ESE of PARIS, in BLUEGRASS region; 38°08'N 83°45'W. Manufacturing (apparel, printed circuit boards, wiring harnesses, fertilizer). Ruins of Slate Creek iron furnace (built 1790) to E; to SE are Olympia Springs (used since 1791) in Daniel Boone National Forest, once-famous mineral springs and baths for which county is named. CAVE RUN LAKE reservoir to E.

Owl Creek Mountains, (highest point, 9,665 ft/2,946 m), central WYOMING, a range of the ROCKIES, between Bridger Mountains (E), ABSAROKA RANGE (NW); bounds BIGHORN BASIN (N); range forms part of boundary between HOT SPRINGS and FREMONT counties, SW of THERMOPOLIS. Entirely within Wind River Indian Reservation.

Owl Head Mountain (OUL HED), (3,425 ft/1,044 m), S QUEBEC, E Canada, on W side of Lake MEMPHRÉMAGOG, near U.S. (VERMONT) border, 30 mi/48 km SE of GRANBY.

Owls Head, resort and fishing town, KNOX county, S MAINE, just S of ROCKLAND on Owls Head peninsula; includes Ash Point village; 44°04'N 69°04'W. Lighthouse (1826) nearby.

Owo, city, ONDO state, S NIGERIA. It is primarily a farming and commercial city, in an area producing cacao, cotton, and timber. Owo was capital of the Yoruba state of the same name that was capital of the founded in the 14th century. The city came under British rule in 1893.

Owode (aw-WO-dai), town, OGUN state, SW NIGERIA, 18 mi/29 km SSE of ABEOKUTA. Cotton weaving, indigo dyeing; cacao, palm oil and kernels.

Owosso (ah-WO-so), city (2000 population 15,713), SHIAWASSEE county, S MICHIGAN, 3 mi/4.8 km WNW of CORUNNA, and on the SHIAWASSEE RIVER, 24 mi/39 km NE of LANSING, 27 mi/43 km W of FLINT; 43°00'N 84°10'W. Railroad junction; Airport to NE. Manufacturing (printing, building materials, auto parts, corrugated containers, boats, auto seats, machinery); grain, soybeans; livestock. Thomas E. Dewey born here. Incorporated 1859.

Owsa, INDIA: see AUSA.

Owsley (OUZ-lee), county (□ 198 sq mi/514.8 sq km; 2006 population 4,690), E central KENTUCKY, ⊙ BOONEVILLE; 37°24'N 83°41'W. In the Cumberland Mountains. Drained by South Fork of KENTUCKY RIVER and several creeks. Part of Daniel Boone National Forest in NW; separate section bounds county on S. Mountainous agricultural region (livestock; burley tobacco, hay); bituminous-coal mines, timber. Formed 1843.

Owston Ferry (OW-stuhn FER-ee) village (2001 population 1,128), North Lincolnshire, NE ENGLAND, on TRENT RIVER, 6 mi/9.7 km N of GAINSBOROUGH; 53°29'N 00°46'W. Farming.

Owyhee (oh-WEI-hee), county (□ 7,697 sq mi/20,012.2 sq km; 2006 population 11,104), SW IDAHO; ⊙ MURPHY; 42°34'N 116°10'W. Hilly region bordering on OREGON (W) and NEVADA (S) and bounded N by SNAKE RIVER. Irrigated areas are along Snake and in E along BRUNEAU RIVER, and in SW along forks of OWYHEE RIVER. Sheep, cattle; dairying; agriculture (hay, alfalfa; sugar beets, potatoes, beans; oats, barley, wheat); mining (gold, silver, zinc). Part of Duck Valley Indian Reservation is in S on Nevada state line. Pacific/Mountain time zone boundary follows S boundary and S part of W boundary. Bruneau Dunes State Park in N; SNAKE RIVER PLAIN in N. Formed 1863.

Owyhee (O-wee-hee), unincorporated village (□ 225 sq mi/585 sq km; 2000 population 1,017), ELKO county, N NEVADA, on EAST FORK OWYHEE RIVER, near IDAHO state line, and c.80 mi/129 km NNW of ELKO; 41°54'N 116°11'W. Elev. 5,400 ft/1,646 m. Cattle, sheep. Wild Horse Reserve and State Recreation Area to SE. Part of Humboldt National Forest to E and S.

Owyhee, river, c.200 mi/322 km long; rises in N NEVADA, S IDAHO, SE OREGON. Main stream begins in SW OWYHEE county, SW Idaho, at confluence of South and East forks, c.95 mi/153 km SSW of BOISE; flows NW into MALHEUR county, Oregon, receiving North and Middle forks at THREE FORKS locality, continues NW, then NNE through 40-mi/64-km-long Owyhee Reservoir to SNAKE RIVER 5 mi/8 km S of NYSSA, Oregon, and 2 mi/3.2 km S of mouth of BOISE RIVER. Named in 1826 for 2 Hawaiian employees of the Hudson Bay Co. who were killed by Native Americans. After gold and silver were discovered in the region in 1863, there were many mining camps along the river. The Owyhee reclamation project of the U.S. Bureau of Reclamation irrigates a large area W of the Snake River Owyhee Dam forms a reservoir 52 mi/84 km long.

Owyhee, Hawaii: see HAWAII.

Owyhee Reservoir (o-WEI-hee), MALHEUR county, SE OREGON, on Owyhee River, c.50 mi/80 km W of BOISE, IDAHO, and 30 mi/48 km SSW of ONTARIO, Oregon; c.40 mi/64 km long, average width 1 mi/1.6 km; 43°38′N 117°15′W. Dry Creek enters from W. Formed by Owyhee Dam (417 ft/127 m high), built (1928–1932) for Owyhee Power and Irrigation Project. Lake Owyhee State Park on NE shore.

Oxapampa (ok-sah-PAHM-pah), province, PASCO region, central PERU; ⊙ OXAPAMPA; 10°20′S 75°05′W. Easternmost and largest province of Pasco region. Contains PARQUE NACIONAL YANACHAGA-CHEMILLEN, RESERVA COMUNAL EL SIRA, and the BOSQUE DE PROTECCIÓN SAN MATIAS-SAN CARLOS.

Oxapampa (ok-sah-PAHM-pah), town, ⊙ OXAPAMPA province, PASCO region, central PERU, in Cordillera ORIENTAL, 60 mi/97 km NNE of TARMA; 10°34′S 75°24′W. Coffee, cacao, fruits; livestock. Center of German settlement. Near PARQUE NACIONAL YANACHAGA-CHEMILLEN.

Oxbow (ahks-BO), town (2006 population 1,139), SE SASKATCHEWAN, Canada, on SOURIS RIVER, and 40 mi/64 km ENE of ESTEVAN; 49°14′N 102°11′W. Grain elevators; livestock; dairying.

Oxbow (AWKS-bo), village, BUTHA-BUTHE district, N LESOTHO, 80 mi/129 km NE of MASERU, on Matseng River, in MALUTI MOUNTAINS; reached via Moteng and Mahlasela passes en route to Sani Pass; 28°46′S 28°39′E. SOUTH AFRICAN border 8 mi/12.9 km to NE. Only ski resort in Lesotho; tourist haven for winter sport. Horses, sheep, goats; vegetables.

Oxbow, village, OAKLAND county, SE MICHIGAN, suburb 9 mi/14.5 km W of PONTIAC on OXBOW LAKE; 42°38′N 83°28′W.

Oxbow, region, AROOSTOOK county, NE MAINE, on the AROOSTOOK, and 30 mi/48 km SW of PRESQUE ISLE; 46°25′N 68°30′W. Hunting, fishing.

Oxbow, reservoir (□ 2 sq mi/5.2 sq km), E OREGON (BAKER county)–W IDAHO (ADAMS county) border, on Snake River, 52 mi/84 km NW of WEISER; 44°58′N 116°50′W. Maximum capacity 53,200 acre-ft. Formed by Oxbow Dam (175 ft/53 m high), built (1961) for power generation. HELLS CANYON National Recreation Area to NW; Payette National Forest on E bank.

Oxbow Lake, S HAMILTON county, E central NEW YORK, in the ADIRONDACK MOUNTAINS, 3 mi/4.8 km SW of LAKE PLEASANT village; c.2 mi/3.2 km long; 43°27′N 74°28′W. Resort.

Oxchuc (OSH-chuk), town, CHIAPAS state, S MEXICO, 18 mi/29 km ENE of SAN CRISTÓBAL DE LAS CASAS; 16°51′N 92°25′W. The majority of population is Tzeltal Maya Indians. Wheat, fruits. Also known as Media Luna, and Santo Tomás Oxchuc.

Oxeia, Greece: see OXIA.

Oxelösund (OKS-el-UH-SUND), town, SÖDERMANLAND county, E SWEDEN, on BALTIC SEA, 6 mi/9.7 km SE of NYKÖPING; 58°40′N 17°06′E. Major seaport (icefree all year); manufacturing (important iron- and glassworks; coke ovens).

Oxenhope (OX-uhn-HOP), village (2001 population 2,476), WEST YORKSHIRE, N ENGLAND, 7 mi/11.3 km NNW of HALIFAX; 53°48′N 01°57′W. Previously woolen milling, textile-machinery works. Stone quarries nearby.

Oxford (AHKS-fuhrd), county (□ 787 sq mi/2,046.2 sq km; 2001 population 99,270), SW ONTARIO, E central Canada, on THAMES River; ⊙ WOODSTOCK; 43°05′N 80°50′W. Agriculture (dairying, hogs, poultry, tobacco, specialty products); agri-tourism. Includes the municipalities of Woodstock, INGERSOLL, TILLSONBURG, Blandford Blenheim, East Zorra-TAVISTOCK, NORWICH, South-West Oxford, and Zorra.

Oxford, county (□ 2,175 sq mi/5,655 sq km; 2006 population 57,118), W MAINE, bordering on NEW HAMPSHIRE and QUEBEC, CANADA; ⊙ SOUTH PARIS; 44°29′N 70°45′W. Manufacturing at RUMFORD (on the ANDROSCOGGIN RIVER), Paris (on the Little Androscoggin River), and NORWAY includes shoes, paper, wood products; lumber mills. Agriculture includes dairying. Winter sports at NEWRY and LOCKE MILLS; summer resorts in Rangeley Lakes region. Part of White Mountain National Forest at New Hampshire state line. Formed 1805.

Oxford (OKS-for-zhe), city, NE SANTA CATARINA state, BRAZIL, 38 mi/61 km NW of JOINVILLE, on IGUAÇU River (PARANÁ border); 26°14′S 49°24′W. Rice, manioc; livestock.

Oxford (AHKS-fuhd), city and district, ⊙ OXFORDSHIRE, S central ENGLAND, c.55 m/88 km NW of LONDON; 51°45′N 01°15′W. In addition to its importance as the site of Oxford University, the city has significant industries, including the manufacturing of motor vehicles and steel products. Tourism is important. A trading town and frontier fort along the THAMES, it was raided by Danes in the 10th and 11th centuries. By the 12th century, Oxford was the site of a castle, an abbey, and the university. It had foundations of several orders, including the Dominicans and the Gray Friars. During the 13th century, frequent conflicts arose between the town and the university in which the university, with the support of the church and the king, was usually victorious. During the civil wars, Oxford was the Royalist headquarters. Among its famous historic buildings (apart from the colleges) are the Radcliffe Camera (1737), the Observatory (1772), and Sheldonian Theatre (designed by Christopher Wren), the churches of St. Mary the Virgin (13th century) and St. Michael (11th century), and several old inns. The chapel (12th century) of Christ Church College is also the cathedral church of the city. The Ashmolean Museum and the Bodleian Library are notable. Besides the university, Ruskin College (1899), Oxford Brookes University, and the public Magdalen College School (c.1480) are in Oxford. Notable suburbs include Cowley, Headington, Iffley, Littlemore, North Oxford, Summertown, and Wolvercote.

Oxford, city (2000 population 14,592), Calhoun co., E Alabama, just S of Anniston. In farm area; poultry processing; manufacturing (shirts, caskets; dairy products). Settled 1855. Originally called 'Lickskillet,' a humorous name implying that the residents were so poor that they had to lick their frying pans to survive, it was renamed in the 1840's and incorporated in 1852.

Oxford, city (2000 population 11,756), ⊙ LAFAYETTE county, N central MISSISSIPPI; 34°21′N 89°31′W. Railroad terminus. Agriculture (cotton, corn, soybeans; cattle; dairying); manufacturing (contract assembly, dairy products, flower pots, apparel, building materials, machinery, wire, printing and publishing, home appliances). Tourist center. Seat of University of Mississippi ("Ole Miss"), founded 1848. Although the town was burned by Union forces in 1864, many antebellum houses remain. Rowan Oak, antebellum home of novelist William Faulkner, is here, and the city was the setting for some of his works. In 1962,

Oxford was the scene of rioting and conflict when the first African-American student, James Meredith, was enrolled in the university. The Mary Buie Museum houses one of the largest doll collections in the U.S. Holly Springs National Forest to NE. Center for Study of Southern Culture. Sardis Waterfowl Area, on SARDIS RESERVOIR, to NW. Incorporated 1837.

Oxford (AHKS-fuhrd), city (□ 4.5 sq mi/11.7 sq km; 1990 populatin 7,965; 2000 population 8,338), ⊙ GRANVILLE county, N NORTH CAROLINA, 28 mi/45 km NE of DURHAM; 36°18′N 78°35′W. Railroad junction. Service industries; manufacturing (tobacco processing; metal, concrete, and plastic products; manufactured buildings, consumer goods, textiles, cosmetics, zippers); agriculture (tobacco, soybeans, grain; livestock; dairying). Settled 1760; laid out 1811; incorporated 1816.

Oxford (AHKS-fuhrd), city (2006 population 22,394), BUTLER county, SW OHIO, near the INDIANA state line; 39°30′N 84°45′W. In farm area. Residential college town and seat of Miami University and the Western College State park and a pioneer farm (1835; now a museum) nearby. Laid out 1810; incorporated 1830.

Oxford, (OX-furd) town (2001 population 1,332), N NOVA SCOTIA, CANADA, on Philip River, and 20 mi/32 km ESE of AMHERST; 45°44′N 63°52′W. Elevation 59 ft/17 m. Known as "Wild Blueberry Capital of Canada." Formerly a manufacturing center.

Oxford, township, Waimakariri district, CANTERBURY region, SOUTH ISLAND, NEW ZEALAND, 30 mi/48 km NW of CHRISTCHURCH; 43°18′S 172°11′E. Sheep; cereals.

Oxford, town, NEW HAVEN county, SW CONNECTICUT, 14 mi/23 km NW of NEW HAVEN; 41°25′N 73°08′W. Agriculture and light industry support the economy. Part of state forest here. Settled c.1680; incorporated 1798.

Oxford, town (2000 population 1,892), NEWTON county, N central GEORGIA, 2 mi/3.2 km NNW of COVINGTON; 33°37′N 83°52′W. Manufacturing of concrete and lubricants. Oxford College of Emory University located on original site of Emory. College listed on National Register of Historic Places.

Oxford, town (2000 population 1,271), BENTON county, W INDIANA, 9 mi/14.5 km SSE of FOWLER; 40°31′N 87°15′W. In agricultural area (grain, soybeans, hybrid corn; livestock). Manufacturing (plastic bottles, vertical blinds).

Oxford, town (2000 population 705), JOHNSON county, E IOWA, 14 mi/23 km WNW of IOWA CITY; 41°43′N 91°47′W. In livestock and grain area.

Oxford, town (2000 population 1,173), SUMNER county, S KANSAS, on ARKANSAS RIVER, and 12 mi/19 km E of WELLINGTON; 37°16′N 97°10′W. In wheat area; grain milling.

Oxford, town, OXFORD county, SW MAINE, on THOMPSON LAKE, and just S of SOUTH PARIS; 44°08′N 70°28′W. Resorts; wood products. Settled 1794; set off from HEBRON and incorporated 1829.

Oxford, fishing town (2000 population 771), TALBOT county, E MARYLAND, on EASTERN SHORE, 9 mi/14.5 km WNW of CAMBRIDGE, and on TRED AVON RIVER; 38°41′N 76°10′W. Rivaling ANNAPOLIS as an official port of entry into the colony from 1683 to the Revolutionary War, the town declined, and by 1825 municipal government was suspended. Revived as an incorporated town that year, it flourished as a shipbuilding, oystering, fishpacking, and resort community. The main street is named for Robert Morris (1734–1806), who came to Oxford in 1738 as the factor of Liverpool shipping firm and became the "Financier of the American Revolution." The Tred Avon Ferry at the foot of Morris Street., started by Elizabeth Skinner in 1760 with a rowed scow, still operates, carrying cars as well as passengers to Bellevue on the boat *Southside* every 15 minutes until dusk. The remains of the Morris House (c.1774) have been incorporated into the Robert Morris Inn. The Tred Avon Yacht Club holds an

Area is shown by the symbol □, and capital city or county seat by ⊙.

annual three-day regatta here every August 1–3. Wiley's Shipyard is more than 250 years old. Byberry, a shingled 1½ story house, is said to go back to 1695.

Oxford, town, WORCESTER county, S MASSACHUSETTS; 42°08′N 71°52′W. It is chiefly residential, with some light manufacturing. Clara Barton was born here. Includes village of North Oxford. Settled 1687 by French Protestants; incorporated 1693.

Oxford, town (2000 population 3,540), OAKLAND county, SE MICHIGAN, 14 mi/23 km N of PONTIAC; 42°49′N 83°15′W. In lake and farm area (cattle; grain); manufacturing (steel castings, urethane foam systems, plastic products, dies, metal stampings); gravel pits; resort. Mount Grampian Ski Area to E. Settled 1836; incorporated 1876.

Oxford, town (2006 population 802), FURNAS and HARLAN counties, S NEBRASKA, 18 mi/29 km SW of HOLDREGE, and on REPUBLICAN RIVER; 40°15′N 99°37′W. Railroad junction. Grain; manufacturing (dairy products, wooden pallets, fertilizers, auto parts). Incorporated 1879.

Oxford, village (1990 in interior highlands 562; 2000 in interior highlands 642), IZARD county, N ARKANSAS, 22 mi/35 km N of MELBOURNE; 36°12′N 91°55′W.

Oxford, village (2000 population 53), FRANKLIN county, SE IDAHO, 17 mi/27 km NNW of PRESTON; 42°16′N 112°01′W. Caribou National Forest to W.

Oxford, village and township, WARREN county, NW NEW JERSEY, on N slope of SCOTTS MOUNTAIN, and 4 mi/6.4 km WSW of BELVIDERE; 40°48′N 75°00′W. Once an important iron-mining region.

Oxford, village (□ 1 sq mi/2.6 sq km; 2006 population 1,568), CHENANGO county, central NEW YORK, on CHENANGO River, and 28 mi/45 km NE of BINGHAMTON; 42°26′N 75°35′W. In agricultural area; dairy products. Site of New York State Women's Relief Corps Home built in 1896 and now operated as the New York State Veterans' Home. Settled 1788; incorporated 1808.

Oxford, village (2006 population 557), MARQUETTE county, central WISCONSIN, on small Neenah Creek, and 17 mi/27 km NNW of PORTAGE; 43°46′N 89°33′W. In livestock and dairy area; manufacturing (bottled artesian water).

Oxford, borough (2000 population 4,315), CHESTER county, SE PENNSYLVANIA, 17 mi/27 km SW of COATESVILLE, near MARYLAND state boundary; 39°47′N 58°75′W. Manufacturing (printing and publishing, fabricated metal products, analytical instruments, communication systems, food products, gas fuels, compost). Agricultural area (grain, soybeans, mushrooms; poultry, livestock; dairying). Lincoln University to NE; Oxford Airport to SW; Nottingham Park to SW; Octoraro Lake reservoir to W; three covered bridges to SE. Incorporated 1833.

Oxford Dam, NORTH CAROLINA: see LOOKOUT SHOALS LAKE.

Oxford Junction, town (2000 population 573), JONES county, E IOWA, near WAPSIPINICON RIVER, 20 mi/32 km ESE of ANAMOSA; 41°58′N 90°57′W. In livestock and grain area.

Oxford Lake (AHKS-fuhrd), 38 mi/61 km long, 9 mi/14 km wide (□ 155 sq mi/403 sq km), NE central MANITOBA, W central Canada, on HAYES River; 54°52′N 95°35′W.

Oxford, Mount, peak (14,153 ft/4,314 m) in ROCKY MOUNTAINS, CHAFFEE county, central COLORADO, in COLLEGIATE RANGE; also includes mounts COLUMBIA, HARVARD, and YALE. San Isabel National Forest, E of CONTINENTAL DIVIDE.

Oxfordshire (AHKS-fuhd-shir) or **Oxon,** county (□ 1,006 sq mi/2,615.6 sq km; 2001 population 605,469), S central ENGLAND, ⊙ OXFORD; 51°45′N 01°10′W. The terrain is generally flat except for an extension of the CHILTERN HILLS in the SE. The county is drained by the THAMES RIVER (or Isis, as it is sometimes locally called) and its affluents, the Windrush, the EVENLODE,

the CHERWELL, and the THAME. The chief occupation for centuries was farming (wheat, barley, and oats), with some dairying and sheep raising, but the role of agriculture as an employment source declined in the 20th century. Previously site of ironstone and limestone. Oxford is the industrial center (motor vehicles and steel products). In the Middle Ages, Oxfordshire was a part of the Anglo-Saxon kingdom of MERCIA. During the English Civil War it was a stronghold of Royalist resistance. Near WOODSTOCK, rich in historical associations, is BLENHEIM PARK. Other significant towns in the county are Abingdon, Banbury, Bicester, Burford, Chipping Norton, Henley-on-Thames, and Witney.

Oxhey, ENGLAND: see WATFORD.

Oxia (o-KSYAH), uninhabited island (□ 2 sq mi/5.2 sq km), in IONIAN Sea, in AKARNANIA prefecture, WESTERN GREECE department, at mouth of AKHELÓOS River; 3 mi/4.8 km long, 5 mi/8 km wide; 38°18′N 21°06′E. Naval battle of Lepanto (1571) took place at mouth of Gulf of PATRAS (S). Sometimes spelled Oxya or Oxeia.

Oxilithos, Greece: see OXYLITHOS.

Oxkutzcab, town, YUCATÁN state, SE MEXICO, and 10 mi/16 km SE of TICUL; 20°18′N 89°26′W. On railroad. Agricultural center (henequen, sugarcane, tobacco, corn, tropical fruits, timber). Remarkable Maya ruins are at LABNÁ, 7 mi/11.3 km SW, and interesting grottoes are nearby.

Oxley (AHKS-lee), township, NE VICTORIA, SE AUSTRALIA, 7 mi/12 km SE of WANGARATTA, on King River; 36°27′S 146°22′E.

Oxley's Peak, AUSTRALIA: see LIVERPOOL RANGE.

Ox Mountains, IRELAND: see SLIEVE GAMPH.

Oxna (AHKS-nah), uninhabited island (1 mi/1.6 km long; rises to 115 ft/35 m) of the SHETLAND ISLANDS, extreme N Scotland, off SW coast of Mainland island, at S end of the Deeps, and 4 mi/6.4 km WSW of SCALLOWAY; 60°07′N 01°22′W.

Oxnard, city (□ 35 sq mi/91 sq km; 2000 population 170,358), VENTURA county, S CALIFORNIA, 53 mi/85 km WNW of LOS ANGELES, and 7 mi/11.3 km SE of VENTURA, on the PACIFIC OCEAN; 34°12′N 119°13′W. Its economy, formerly based on agriculture and agricultural processing, mining, and nearby military bases, has broadened to include large industrial and commercial operations. Manufacturing (dairy products, sporting goods, cosmetics, metal forging, prepared foods, paper mills, asphalt, pumps, aircraft parts, building materials, uniforms, computer equipment, concrete products); dairying, avocados, citrus fruits, vegetables, flowers, nursery products. There are also oil refineries. Point Mugu Naval Air Station to SE. Oxnard is the gateway for visitors to the SANTA BARBARA ISLANDS and to Los Padres National Forest. The city has become one of the fastest growing in the U.S., and its population doubled between 1970 and 1990. Channel Island National Area c.25 mi/40 km SW; SANTA MONICA MOUNTAINS NATIONAL RECREATION AREA to E; Oxnard Air Force Base to NE; Point Mugu State Park to S; McGrath State Beach to W; Oxnard College (two-year) is here. Incorporated 1903.

Oxon, ENGLAND: see OXFORDSHIRE.

Oxon Hill, village, PRINCE GEORGES county, central MARYLAND, S suburb of WASHINGTON, D.C.; 38°48′N 76°59′W. Oxon Hill was dominated by large estates until the 1950s. The Oxon Hill House, built in the mid-18th century by Thomas Addison, the great-great grandson of College John Addison, burned in 1895. It was once owned by the family of Sumner Welles (1892–1960), Franklin Roosevelt's undersecretary of state from 1937 to 1943. The Oxon Hill Farm is run by the National Park Service as a 19th-century farm in which hand milking and horse plowing can be observed. Two 20-ton/22-metric-ton cannons, stone gun emplacements, and reinforced earthworks can still be seen in Fort Foote Park. The fort was built during the Civil War to protect the capital.

Oxted (AHK-stid), town (2001 population 10,812), E SURREY, SE ENGLAND, 9 mi/14.5 km E of REIGATE; 51°15′N 00°00′. Located at foot of NORTH DOWNS. Former agricultural market. Has 13th–15th-century church, 15th-century inn, and other old buildings.

Oxus, AFGHANISTAN, TAJIKISTAN, Turmenistand, and UZBEKISTAN: see AMU DARYA RIVER.

Oxya, Greece: see OXIA.

Oxylithos (o-KSEE-lee-thos), town, E central ÉVVIA island, ÉVVIA prefecture, CENTRAL GREECE department, port on AEGEAN SEA, S of KIMI, and 30 mi/48 km ENE of Chaleis; 38°35′N 24°07′E. Fisheries. Also spelled Oxilithos.

Oxyrhyncus (OK-si-RIN-kuhs), excavation site, MINYA province, Upper EGYPT, at village of EL BAHNASA or Al-Bahnasa (also spelled Behnesa), on the BAHR YUSUF near the NILE River, 9 mi/14.5 km WNW of BENI MAZAR. Here in 1896–1897 and 1906–1907 were made some of the largest finds of papyri, partly Ptolemaic in date, but mostly Roman and Byzantine. They tell of a Greek colony first and a large Christian monastic center later.

Oya (O-yah), town, Yabu county, HYOGO prefecture, S HONSHU, W central JAPAN, 51 mi/83 km N of KOBE; 35°19′N 134°40′E. Bookbinding, apparel; tin; akenobe ore. Skunk cabbage (SW limit). Hyonosen mountain nearby.

Oyabe (o-YAH-be), city, TOYAMA prefecture, central HONSHU, central JAPAN, 19 mi/30 km W of TOYAMA; 36°40′N 136°52′E. Traditional tile, motor vehicles, sportswear. Seat of Kurikara Fuko temple. Well-known landscape on nearby Tonami Plains.

Oyam, administrative district, NORTHERN region, N central UGANDA, on VICTORIA NILE RIVER (to SW). As of Uganda's division into eighty districts, borders GULU (N), LIRA (NE), APAC (E and S), MASINDI (SWW, formed by Victoria Nile River), and AMURU (WNW) districts. Rural and agricultural region. Formed in 2006 from NW portion of former APAC district (current Apac district formed from S and NE portions).

Oyama (O-yah-mah), city (2005 population 160,150), TOCHIGI prefecture, central HONSHU, N central JAPAN, 19 mi/30 km S of UTSUNOMIYA; 36°18′N 139°48′E. Communications equipment, engines, aluminum products; dried gourd shavings.

Oyama (O-yah-mah), town, Hita county, OITA prefecture, E KYUSHU, SW JAPAN, 37 mi/60 km W of OITA; 33°15′N 130°58′E. Food processing (jams and sweeteners). *Enoki* mushrooms.

Oyama, town, Sunto county, SHIZUOKA prefecture, central HONSHU, E central JAPAN, 43 mi/70 km N of SHIZUOKA; 35°21′N 138°59′E. Sometimes called Koyama.

Oyama, town, Kaminikawa county, TOYAMA prefecture, central HONSHU, central JAPAN, 6 mi/10 km S of TOYAMA; 36°36′N 137°18′E. Chubu-Sangaku national park (Mount Yakushidake; skiing) nearby. Includes (since early 1940s) former town of Gohyakkoku.

Oyamada (O-yah-MAH-dah), village, Ayama county, MIE prefecture, S HONSHU, central JAPAN, 16 mi/25 km W of TSU; 34°45′N 136°13′E. Greenhouse orchids. Shindaibutsu temple has noted wooden *Daibutsu* [Japanese=Buddha sculpture] (13.5 ft/4.1m high) by Kaikei.

Oyamazaki (O-yah-mah-ZAH-kee), town, Otokuni county, KYOTO prefecture, S HONSHU, W central JAPAN, 9 mi/15 km S of KYOTO; 34°53′N 135°41′E. Motor vehicles.

Oyano (O-yah-NO), town, Amakusa county, KUMAMOTO prefecture, on OYANO-SHIMA island, between AMAKUSA KAMI-SHIMA island and mainland W KYUSHU, SW JAPAN, 14 mi/22 km S of KUMAMOTO; 32°35′N 130°25′E. Prawns. Whetstones.

Oyano-shima (o-YAH-no-SHEE-mah), island (□ 12 sq mi/31.2 sq km) of AMAKUSA, in EAST CHINA SEA, KUMAMOTO prefecture, SW JAPAN, off W coast of

KYUSHU, just N of AMAKUSA KAMI-SHIMA; 5.5 mi/8.9 km N-S, 3.5 mi/5.6 km E-W. Terrain is mountainous and fertile. Coal mining, fishing; rice, wheat. Chief town, Noboritate (E).

Oyapock River (O-yah-PO-ke), c.260 mi/418 km long, NE South America; rises on and forms BRAZIL–FRENCH GUIANA border throughout its NE course and empties into the ATLANTIC OCEAN at Cape Orange. Towns on Brazilian bank are OIAPOQUE and PONTA DOS ÎNDIOS; on French Guiana bank lies SAINT-GEORGES. Called primarily Oiapoque in Brazil.

Oyash, RUSSIA: see STANTSIONNO-OYASHINSKIY.

Oyashio (o-YAH-shyo) [Japanese=parent steam], cold ocean current flowing S from BERING SEA along E KURIL ISLANDS, HOKKAIDO, and N HONSHU; meets JAPAN CURRENT in the PACIFIC OCEAN at about 38°00′N. A branch of Okhotsk Current enters Sea of JAPAN through Sea of OKHOTSK and TATAR STRAIT and flows S along E shores of Korea, meeting TSUSHIMA CURRENT (branch of JAPAN CURRENT) at about 40°00′N. Okhotsk Current chills shores of N Japan.

Oyat' River (uh-YAHT), 150 mi/241 km long, LENINGRAD oblast, NW European Russia; rises in a lake region S of Lake ONEGA; flows generally W, past ANDRONOVSKOYE and ALEKHOVSHCHINA, to the SVIR River, 5 mi/8 km E of SVIRITSA. Timber floating.

Øye, village, MØRE OG ROMSDAL county, W NORWAY, on Hjørundfjorden, at foot of the NORANGDAL, 25 mi/40 km SE of ÅLESUND. Tourist and mountaineering center.

Oyëk (uh-YOK), settlement (2005 population 3,350), S central IRKUTSK oblast, E central Siberian Russia, on road, 19 mi/31 km W of IRKUTSK; 52°35′N 104°27′E. Elevation 1,541 ft/469 m. Forestry services.

Oyem (o-YEM), city (1993 estimated population 22,669), ⊙ WOLEU-NTEM province (□ 14,700 sq mi/38,220 sq km), N GABON, near EQUATORIAL GUINEA border, 170 mi/274 km NE of Libreville; 01°38′N 11°35′E. Cocoa, coffee, and rubber plantations; rice growing. Customs station. Has Roman Catholic mission and agriculture school.

Oyen (OI-uhn), town (□ 2 sq mi/5.2 sq km; 2001 population 1,020), SE ALBERTA, W Canada, near SASKATCHEWAN border, 60 mi/97 km W of KINDERSLEY (Saskatchewan); 51°22′N 110°28′W. Grain elevators. Established as a village in 1913, and as a town in 1965.

Oyens (OY-ens), town (2000 population 132), PLYMOUTH county, NW IOWA, 6 mi/9.7 km ENE of LE MARS; 42°49′N 96°03′W. In livestock area.

Oye-Plage (oi–plahzh), town (□ 13 sq mi/33.8 sq km; 2004 population 5,707), PAS-DE-CALAIS department, NORD-PAS-DE-CALAIS region, N FRANCE; 50°59′N 02°03′E. Resort on NORTH SEA 6 mi/10 km E of CALAIS.

Øyeren (UH-ruhn), lake (□ 34 sq mi/88.4 sq km) on GLOMMA River, AKERSHUS county, SE NORWAY, 12 mi/19 km E of OSLO; 21 mi/34 km long, 1 mi/1.6 km–4 mi/6.4 km wide.

Oykel River (OI-kuhl), 25 mi/40 km long, N HIGHLAND, N Scotland; rises on Ben More Assynt (3,274 ft/998 m); flows SE, through Loch Ailsh (1 mi/1.6 km long, 1 mi/1.6 km wide) to DORNOCH FIRTH at BONAR BRIDGE. Just above its mouth it widens into Kyle of Sutherland, a lake 2 mi/3.2 km long, 0.5 mi/0.8 km wide. Receives Shin and Carron rivers.

Oy-Mittelberg (OI-MIT-tel-berg), village, SWABIA, BAVARIA, S GERMANY, on Rottach Lake, and in the Allgäu; 47°39′N 10°25′E. Metalworking; tourism.

Oymur (uh-ee-MOOR), village (2004 population 1,340), W BURYAT REPUBLIC, S Siberian Russia, on the E shore of BAYKAL LAKE, on the coastal road, 38 mi/61 km SW of ULAN-UDE, and 16 mi/26 km S of SELENGINSK; 52°19′N 106°50′E. Elevation 1,568 ft/477 m. Agricultural cooperative (grain, livestock); furniture factory.

Oymyakon (oyee-mye-KON), village, E SAKHA REPUBLIC, central RUSSIAN FAR EAST, on the INDIGIRKA River, on road junction, 430 mi/692 km ENE of YAKUTSK. Widely recognized as the "Cold Pole" of the world for a populated place (−96°F/−71°C, recorded in 1926). Local airport. Weather observation station. Formerly spelled Oymekon.

Oymyakon, RUSSIA: see OIMYAKON.

Oimyakon Plateau (oyee-myah-KON), NE SAKHA REPUBLIC, RUSSIAN FAR EAST, at the junction of VERKHOYANSK, KOLYMA, and DZHUGDZHUR ranges. Elevation 3, 000 ft/900 m–4,000 ft/1,219 m; gives rise to KOLYMA and INDIGIRKA rivers. Lowest winter temperatures in Russia.

Oyo (aw-YAW), state (□ 10,986 sq mi/28,563.6 sq km; 2006 population 5,591,589), SW NIGERIA; ⊙ IBADAN; 08°00′N 04°00′E. Bordered N and E by KWARA state, ESE by OSUN state, S by OGUN state, and W by BENIN. Evergreen forests in S; drained by OGUN River. Agriculture includes cocoa, oil palms, tobacco, maize, yams, cassava, millet, plantains, beans; livestock. Pottery and brick manufacturing; bottling, marble processing, cocoa processing; breweries. Woodcraft and leatherwork. Main centers are Ibadan, OGBOMOSHO, and Kisi. Agodi Zoological Garden, Trans Wonderland Amusement Park, and Ibadan University Zoo are here.

Oyo (aw-YAW), city, ⊙ OYO state, SW NIGERIA. It primarily is a farming settlement, producing tobacco, yams, and cassava. Traditional artisans make textiles and leather goods and carve utensils from shells of the calabash gourd. Oyo was founded c.1835 as the successor of Old Oyo (KATUNGA), capital of the Oyo Yoruba state when the Yoruba Empire was fragmented in the 17th and 18th centuries, which was destroyed in the Yoruba civil wars of the early 19th century. Oyo is about 100 mi/161 km S of OLD OYO. The city came under British rule in 1893.

Oyo (aw-YAW), one of former WESTERN PROVINCES, SW NIGERIA, on BENIN border; capital was OYO. Now within OYO state.

Oyodo (O-YO-do), town, Yoshino district, NARA prefecture, S HONSHU, W central JAPAN, on N central KII PENINSULA, 21 mi/34 km S of NARA; 34°23′N 135°47′E. Pears, tea.

Oyolo (o-YO-lo), town, PAUCAR DEL SARA SARA province, AYACUCHO region, S PERU, in CORDILLERA OCCIDENTAL, 17 mi/27 km NE of PAUSA; 15°11′S 73°11′W. Cereals, alfalfa; livestock.

Oyón (o-YON), province, LIMA region, W central PERU, ⊙ OYÓN; 10°39′S 76°47′W. NE province of Lima region. Borders on HUÁNUCO and PASCO regions.

Oyón (o-YON), town, ⊙ OYÓN province, LIMA region, W central PERU, in CORDILLERA OCCIDENTAL, 21 mi/34 km ESE of CAJATAMBO, on highway from HUACHO to Minas Raura; 10°40′S 76°46′W.

Oyón, town, ÁVILA province, N SPAIN, 4 mi/6.4 km N of LOGROÑO; 42°30′N 02°26′W. Olive-oil processing; wine, cereals, lumber; sheep.

Oyonnax (o-yo-NAHKS), town (□ 13 sq mi/33.8 sq km), AIN department, RHÔNE-ALPES region, E FRANCE, in the W JURA foothills, 21 mi/34 km ENE of BOURG-EN-BRESSE; 46°15′N 05°40′E. It is France's leading plastics manufacturing center. It also specializes in the production of frames for eyeglasses, artificial flowers, combs, auto accessories. Known as "Plastics Valley," Oyonnax and surrounding villages house hundreds of skilled workshops. Also produced here are artificial joints for surgical implants.

Oyontún, PERU: see OYOTÚN.

Oyotún (o-yo-TOON), town, CHICLAYO province, LAMBAYEQUE region, NW PERU, on W slopes of CORDILLERA OCCIDENTAL, 38 mi/61 km E of CHICLAYO; 06°51′S 79°19′W. Rice, corn, tobacco, cereals. Variant spelling: Oyontún.

Oyrot-Tura, RUSSIA: see GORNO-ALTAYSK.

Oyskhara (oyees-hah-RAH), town (2005 population 12,810), E CHECHEN REPUBLIC, in the foothills of the NE CAUCASUS Mountains, S European Russia, on highway branch, 11 mi/18 km ESE of GUDERMES; 43°16′N 46°15′E. Elevation 390 ft/118 m. Petroleum wells. Developed during World War II. Until 1944, called Oysungur; from 1944 to 1989, Novogroznenskiy.

Oyster (OIS-tuhr), unincorporated village, NORTHAMPTON county, E VIRGINIA, 5 mi/8 km E of CAPE CHARLES town, on Mockhorn Bay, arm of ATLANTIC OCEAN; 37°17′N 75°55′W. Manufacturing (oyster processing); fish, oysters, clams.

Oyster Bay, unincorporated village (□ 5 sq mi/13 sq km; 2000 population 6,826), NASSAU county, SE NEW YORK, on N LONG ISLAND, on LONG ISLAND SOUND; 40°47′N 73°30′W. It is chiefly upper-income residential with many estates. Nearby is Theodore Roosevelt's estate, SAGAMORE HILL, which was made a national shrine in 1953 and a National Historic Site in 1963. Also of interest in Oyster Bay are several 18th-century houses, the Theodore Roosevelt Memorial Bird Sanctuary, a 12-acre/5-ha wildlife sanctuary owned by the National Audubon Society, which adjoins Roosevelt's grave, and the Oyster Bay National Wildlife Refuge (□ 5 sq mi/13 sq km). Settled 1653.

Oyster Bay (OI-stuhr BAI), inlet of TASMAN SEA formed by E coast of TASMANIA (W) and FREYCINET PENINSULA and SCHOUTEN ISLAND (E); 17 mi/27 km long, 14 mi/23 km wide; opens N into lagoons; 42°40′S 148°03′E. SWANSEA on NW shore.

Oyster Bay, irregular inlet of LONG ISLAND SOUND, SE NEW YORK, indenting N shore of W LONG ISLAND, with entrance (c.2 mi/3.2 km wide) between ROCKY POINT (W) and LLOYD POINT (E); 40°54′N 73°32′W. COLD SPRING HARBOR (c.3 mi/4.8 km long) is SE arm. Oyster Bay Harbor, with irregular branches, is W arm; OYSTER BAY village is on its S shore.

Oyster Bay, AUSTRALIA: see STANSBURY.

Oyster Bay Cove, residential village (□ 4 sq mi/10.4 sq km; 2006 population 2,250), NASSAU county, SE NEW YORK, on NW LONG ISLAND, just ESE of OYSTER BAY village; 40°51′N 73°30′W. In shore recreation area. Incorporated 1931.

Oyster Creek (OI-stuhr), town (2006 population 1,231), BRAZORIA county, SE TEXAS, suburb 4 mi/6.4 km NE of FREEPORT, in Brazosport Area, on OYSTER CREEK; 29°00′N 95°19′W. Oil and natural gas. Manufacturing (pump seals). Brazoria National Wildlife Refuge to NE.

Oyster Creek Nuclear Power Plant, NEW JERSEY: see OCEAN county.

Oyster Harbors, private island, Osterville, MASSACHUSETTS, in BARNSTABLE town, CAPE COD. Summer estates and golf course.

Oyster Island, 1 mi/1.6 km long, N SLIGO county, NW IRELAND, in SLIGO BAY, 4 mi/6.4 km NW of SLIGO. Lighthouse.

Oystermouth, Wales: see SWANSEA.

Oyster Point, AUSTRALIA: see WYNNUM.

Oyster River, c.12 mi/19 km long, STRAFFORD county, SE NEW HAMPSHIRE; rises 9 mi/14.5 km W of DOVER; flows ESE, in serpentine course, past Durham, to entrance of GREAT BAY into Little Bay.

Øystese (UH-STAH-suh), village, HORDALAND county, SW NORWAY, on N shore of HARDANGERFJORDEN, 8 mi/12.9 km WSW of ALVIK.

Oysungur, RUSSIA: see OYSKHARA.

Oytal, KAZAKHSTAN: see OITAL.

Oyten (OI-ten), suburb of BREMEN, LOWER SAXONY, NW GERMANY, 8 mi/12.9 km ESE of city center; 53°04′N 09°01′E. Livestock.

Ozalj, CROATIA: see KUPA RIVER.

Ozalp, Turkish *Özlap*, village, SE TURKEY, 45 mi/72 km ENE of VAN, 15 mi/24 km W of IRAN border; c.38°39′N 43°59′E. Grain. Also called Karakalli; formerly called Kazim Pasa and Saray.

Ozama River (o-ZAH-mah), c. 65 mi/105 km long, S central and S DOMINICAN REPUBLIC; rises in the Cordillera CENTRAL SE of BONAO; flows E and S to the CARIBBEAN SEA at SANTO DOMINGO; 18°28′N 69°53′W. Navigable for c. 15 mi/24 km upstream. Sometimes called OZUMA RIVER.

Ozamis (o-SAH-mees), city (□ 56 sq mi/145.6 sq km; 2000 population 110,420), MISAMIS OCCIDENTAL province, N MINDANAO, PHILIPPINES, port on PANGUIL BAY, 35 mi/56 km NE of PAGADIAN; 08°09′N 123°50′E. Chief port for CEBU trade and the main commercial center for Misamis Occidental province. Trade center for agricultural area (corn, coconuts); ships copra. Airport at LABO, 3.7 mi/6 km S. Founded as Fort Santiago, an 18th-century religious center and fortification in Muslim territory. Was called Misamis until the late 1940s, when it was renamed in honor of Jose Ozamiz, an executed World War II resistance leader.

Ozan (o-ZAN), village (2000 population 81), HEMPSTEAD county, SW ARKANSAS, 14 mi/23 km NNW of HOPE; 33°51′N 93°43′W.

Ozarichi (o-ZUH-ri-chi), urban settlement, central GOMEL oblast, BELARUS, 28 mi/45 km N of MOZYR, 52°28′N 29°12′E. Manufacturing (furniture). Formerly also spelled Azarichi.

Ozark, county (□ 756 sq mi/1,965.6 sq km; 2006 population 9,393), S MISSOURI; ⊙ GAINESVILLE, borders ARKANSAS in S, in the OZARKS; drained by North Fork of WHITE RIVER; 36°38′N 92°26′W. Hay; cattle; timber (cedar, oak, pine); tourism, canoeing, fishing. Parts of Mark Twain National Forest in NE and NW corners. N end of NORFORK LAKE in SE (dammed in Arkansas); arms of BULL SHOALS LAKE Reservoir in SW (also dammed in Arkansas). Formed 1841.

Ozark, city (2000 population 15,119), ⊙ Dale co., SE Alabama, c.20 mi/32 km NW of Dothan. Shipping center for agr. products, such as nuts and timber; manufacturing (aircraft, railcars, farm machinery, clothing, wood products, fertilizer). The Alabama Aviation and Technical College is located in Ozark. Fort Rucker, the U.S. Army's aviation center, is nearby. Settled 1820; incorporated 1870. First called 'Merricks' in honor of John Merrick, a revolutionary war soldier who owned an early grocery store, it was also called 'Woodshop' before finally being named Ozark, after the French 'Aux Arks' which means 'in the country of the Arkansa.'

Ozark, city (2000 population 9,665), ⊙ CHRISTIAN county, SW MISSOURI, in the OZARKS, near JAMES RIVER, 14 mi/23 km S of SPRINGFIELD; 37°01′N 93°12′W. Fruits; cattle; dairying; manufacturing (electric motors, boat trailers); limestone quarry.

Ozark (O-zahrk), town (2000 population 3,525), shares ⊙ functions with CHARLESTON, FRANKLIN county, NW ARKANSAS, 33 mi/53 km ENE of FORT SMITH, and on ARKANSAS RIVER (here bridged); 35°30′N 93°50′W. Diversified agricultural area. Manufacturing (turkey processing [Cargill], concrete, shirts, air-conditioner parts). Ozark National Forest to N; OZARK JETA TAYLOR LOCK AND DAM (Arkansas River) to SE; forms Ozark Lake. Settled 1836; incorporated as city 1938.

Ozark Jeta Taylor Lock and Dam (O-zahrk), FRANKLIN county, W ARKANSAS, on the ARKANSAS RIVER, 1 mi/1.6 km SE of OZARK; 65 ft/20 m high; 35°27′N 93°48′W. Built (1969) by the Army Corps of Engineers for navigation and power generation. Forms Ozark Lake, a reservoir that is a raised navigation channel, with a maximum capacity of 148,000 acre-ft.; extends c.45 mi/72 km to JAMES W. TRIMBLE LOCK AND DAM.

Ozark Mountains, MISSOURI: see OZARKS, THE.

Ozark National Scenic Riverways, 134 mi/216 km along the CURRENT and Jack's Fork rivers, SE MISSOURI. Covers 80,791 acres/32,695 ha. Authorized 1964 as the first national riverway; established 1972. Many large springs flow into the rivers; Big Springs is the largest single-outlet spring in the U.S. Many large caves with interesting dripstone formations are found along the rivers. Forests cover about 75% of the riverways. Wildlife and fish are abundant in the area. Canoeing, floating, fishing. Park Headquarters at VAN BUREN.

Ozarks, Lake of the, MISSOURI: see LAKE OF THE OZARKS.

Ozarks, the or **Ozark Highland**, upland region, actually a dissected plateau, and sometimes called OZARK MOUNTAINS, c.50,000 sq mi/129,500 sq km, chiefly in S MISSOURI and N ARKANSAS, but partly in OKLAHOMA and KANSAS, between the ARKANSAS and MISSOURI rivers. The Ozarks, which rise from the surrounding plains, are, in a few places, locally referred to as mountains. Composed of igneous rock over 1 billion years old overlain mostly by limestone and dolomite, the ancient landform has been worn down by erosion. Summits (knobs) are found wherever there is a resistant igneous rock outcrop, as in the SAINT FRANCOIS MOUNTAINS of SE Missouri. The BOSTON MOUNTAINS are the highest and most rugged section, with several peaks more than 2,000 ft/610 m high. The Ozarks are metalliferous, especially in lead, zinc, and iron, especially in the Saint Francois Mountains and the JOPLIN areas. Cattle raising and wood products are major activities; fruit-growing areas are prevalent. Traditional household crafts have been maintained and promoted for the tourist industry. The Ozarks have many large lakes that were created by dams across numerous rivers; the dams generate electricity. The scenic Ozarks, with forests, caves, lakes, streams, and springs, are a popular tourist region, and the construction of summer homes and large retirement communities there has grown.

Ozatlán (o-saht-LAHN), municipality and town, USULUTÁN department, EL SALVADOR, WNW of USULUTÁN city; 13°23′N 88°30′W.

Ozato (O-zah-to), village, Shimajiri county, Okinawa prefecture, SW JAPAN, 4.3 mi/7 km S of NAHA; 26°10′N 127°44′E.

Ozaukee (o-ZAH-kee), county (□ 1,116 sq mi/2,901.6 sq km; 2006 population 86,321), E WISCONSIN; ⊙ PORT WASHINGTON; 43°15′N 87°30′W. Bounded E by LAKE MICHIGAN; drained by MILWAUKEE RIVER. Dairying, livestock, poultry, and vegetable farming area (barley, wheat, soybeans, peas), with dairy product processing. Manufacturing at MEQUON, CEDARBURG, and Port Washington. Fisheries. S part of county has become urbanized, extension of MILWAUKEE metropolitan area. Huntington Beach State Park in NE, on Lake Michigan. Formed 1853.

Ózd (OZD), city (2001 population 38,405), NE HUNGARY, near the SLOVAK border; 48°13′N 20°18′E. Manufacturing (apparel, steel, farinaceous foods) contracted sharply by mid-1995. Old iron and steel center dating to before World War I. Lignite mines. Enlarged after World War II. Ironworks closed in 1992. City and district severely depressed with 40% of the population unemployed. Very poor prospects for recovery.

Oździutycze, UKRAINE: see OZYUTYCHI.

Ozea, Greece: see PARNES.

Ozeblin, mountain peak (5,436 ft/1,657 m high), on Plješivica Mountain, W CROATIA, in LIKA region, on BOSNIA AND HERZEGOVINA border, 5 mi/8 km WNW of DONJI LAPAC, and SE of PLITVICE LAKES.

Özen (ooz-YEN), oil town, MANGYSTAU region, KAZAKHSTAN, 93 mi/150 km E of SHEVCHENKO; 43°30′N 53°00′E. In the MANGYSHLAK oil and gas area. Local industry supports energy ventures. Also called Uzen.

Ozernovskiy (uh-zeer-NOF-skeeye), town (2005 population 2,565), SW KAMCHATKA oblast, RUSSIAN FAR EAST, near the S tip of KAMCHATKA PENINSULA, on the Sea of OKHOTSK, 137 mi/220 km SW of PETROPAVLOVSK-KAMCHATSKIY; 51°30′N 156°31′E. Fish cannery. Incorporated a settlement of Ozërnoye (less than 2 mi/3.2 km to the N).

Ozërnoye (uh-ZYOR-nuh-ye), village, W central SARATOV oblast, SE European Russia, on the MEDVEDITSA River (tributary of the VOLGA RIVER), on railroad and terminus of local highway, 13 mi/21 km SSW of ATKARSK; 51°39′N 44°55′E. Elevation 547 ft/166 m. In agricultural area. Formerly called Durasovka.

Ozërnoye (uh-ZYOR-nuh-ye), former settlement, SW KAMCHATKA oblast, RUSSIAN FAR EAST, near the S tip of KAMCHATKA PENINSULA, on the Sea of OKHOTSK, 135 mi/217 km SW of PETROPAVLOVSK-KAMCHATSKIY. Fisheries. Now the N part of the town of OZERNOVSKIY.

Ozërnoye, RUSSIA: see BENOY-YURT.

Ozërnyy (uh-ZYOR-niyee), town (2005 population 820), central IVANOVO oblast, central European Russia, on railroad, 11 mi/18 km N of IVANOVO, and 7 mi/11 km WSW of FURMANOV; 57°10′N 40°59′E. Elevation 410 ft/124 m. In agricultural area.

Ozërnyy (uh-ZYOR-niyee), town (2006 population 6,160), N SMOLENSK oblast, W European Russia, on road and near railroad, 57 mi/92 km NNE of SMOLENSK, and 27 mi/43 km N of DUKHOVSHCHINA, to which it is administratively subordinate; 55°34′N 32°27′E. Elevation 666 ft/202 m. Hydroelectric power station. Peat works.

Ozero Bol'shoy Manych, RUSSIA: see MANYCH-GUDILO, LAKE.

Ozeros, Lake (O-ze-ros) (□ 4 sq mi/10.4 sq km), in AKARNANIA prefecture, WESTERN GREECE department, off AKHELÓOS River, 11 mi/18 km NW of MESOLONGI; 3 mi/4.8 km long, 1.5 mi/2.4 km wide; 38°39′N 21°13′E.

Ozërsk (uh-ZYORSK), city (2005 population 90,900), NW CHELYABINSK oblast, in the NE foothills of the S URALS, SW Siberian Russia, near a small lake, on road and railroad spur, 37 mi/60 km WNW of CHELYABINSK, and 6 mi/10 km E of KYSHTYM; 55°44′N 60°43′E. Elevation 748 ft/227 m. Steel production; manufacturing (electrical installation and wiring equipment); power station. Uranium enrichment and plutonium production facilities, the establishment of which gave birth to the city itself in 1948. Until 1991, a closed city with a secret designation of Chelyabinsk-65 (initially Chelyabinsk-40, changed in the 1960s, when nuclear research and development expanded into non-military areas).

Ozërsk (uh-ZYORSK), city (2005 population 5,675), S KALININGRAD oblast, W European Russia, on the Angrap River (left headstream of the PREGEL RIVER), approximately 5 mi/8 km N of the Polish border, on road, 75 mi/121 km SE of KALININGRAD, and 18 mi/29 km SE of CHERNYAKHOVSK; 54°24′N 22°00′E. Elevation 354 ft/107 m. Cheese factory. Made city in 1724. Until 1945, in East Prussia, where it was called Darkehmen and, after 1938, Angerapp.

Ozërskiy (uh-ZYOR-skeeye), town (2006 population 1,645), S SAKHALIN oblast, RUSSIAN FAR EAST, on the ANIVA GULF (SW part of the Sea of OKHOTSK), surrounded by small lakes, on coastal highway, 17 mi/27 km E of KORSAKOV; 46°36′N 143°08′E. Fisheries. Under Japanese rule (1905-1945), called Nagahama.

Ozëry (uh-ZYO-ri), city (2006 population 24,985), SE MOSCOW oblast, central European Russia, port on the OKA River, on highway and terminus of a railroad spur, 95 mi/153 km SE of MOSCOW, and 19 mi/31 km SSW of KOLOMNA; 54°51′N 38°33′E. Elevation 423 ft/128 m. Cotton milling, oil industry equipment, vegetable canning. Became city in 1925.

Ozery (o-ZE-ree), Polish *Jeziory*, town, NW GRODNO oblast, BELARUS, on small lake, 13 mi/21 km E of GRODNO; railroad- spur terminus; sawmilling center.

Ozgön, KYRGYZSTAN: see UZGEN.

Ozherel'ye (uh-zhi-RYEL-ye), city (2006 population 10,105), SE MOSCOW oblast, central European Russia, on road, 77 mi/124 km SSE of MOSCOW, and 5 mi/8 km SE of KASHIRA; 54°48′N 38°16′E. Railroad junction; railroad enterprises. Made city in 1958.

Ozinki (uh-ZEEN-kee), town (2006 population 9,710), E SARATOV oblast, SE European Russia, in the OBSHCHI SYRT hills, in the Bol'shoy IRGIZ RIVER basin, near the border with KAZAKHSTAN, on highway and railroad, 65 mi/105 km E of YERSHOV; 51°10′N 49°40′E. Elevation 377 ft/114 m. Building materials; dairy products.

Ozivs'ke, UKRAINE: see AZOVS'KE.

Ozivs'ke More, UKRAINE: see AZOV, SEA OF.

Oznachennoye, RUSSIA: see SAYANOGORSK.

Ozoir-la-Ferrière (o-ZWAHR–lah–fer-YER), residential town (□ 6 sq mi/15.6 sq km), SEINE-ET-MARNE department, ÎLE-DE-FRANCE region, N central FRANCE, an outer ESE suburb of PARIS, 16 mi/26 km from Notre Dame Cathedral; 48°46′N 02°40′E. Surrounded on three sides by wooded estates (Ferrières and Armainvilliers Forests). Has an industrial park and a golf course.

Ozolaean Locris, Greece: see LOCRIS.

Ozona (O-zon-ah), town (2000 population 3,436), ⊙ CROCKETT county, W TEXAS, c.70 mi/113 km SW of SAN ANGELO; 30°42′N 101°12′W. Trading center for oil and natural-gas and livestock region (sheep, Angora goats, cattle); manufacturing (butane, propane, and natural-gas processing).

Ozone Park, neighborhood, S central section of borough of QUEENS, NEW YORK CITY, SE NEW YORK, S of Atlantic Avenue, NW of JFK INTERNATIONAL AIRPORT, S of WOODHAVEN, a community from which it is virtually inseparable, both historically and culturally; 40°41′N 73°51′W. Both Ozone Park and Woodhaven began in June 1882 as the brainchild of Benjamin Hitchcock and Charles Denton, two real-estate developers. The idea, popular during the 1880s and 1890s, was to build nine "parks" that could be pastoral residential areas for people who wanted to escape the older, crowded districts of MANHATTAN, BROOKLYN, and Queens. The word *ozone* was drawn from the phrase "taking in the ozone" (i.e., fresh air). Ozone Park and Woodhaven flourished during the 1920s; though residential, they also attracted some light manufacturing. Today the two communities are generally quiet, residential neighborhoods of unvarying frame and brick houses. Some mafia activities are known to have taken place in and around Ozone Park. The city's only racetrack, Aqueduct ("The Big A"), is here.

Ozorkow (o-ZOR-koov), Polish *Ozorków*, Russian *Ozyurkov*, town (2002 population 21,022), ŁÓDŹ province, central POLAND, on BZURA RIVER, and 15 mi/24 km NNW of ŁÓDŹ; 51°58′N 19°17′E. Manufacturing (textiles, agricultural tools, flour). Before World War II, population was 50% Jewish.

Ozren Mountains (OZ-ren), in DINARIC ALPS, central BOSNIA, BOSNIA AND HERZEGOVINA; c. 10 mi/16 km long; highest peak (5,025 ft/1,532 m) is 5 mi/8 km N of SARAJEVO.

Ozren Mountains, in DINARIC ALPS, NE BOSNIA, BOSNIA AND HERZEGOVINA, between BOSNA and lower Spreča Rivers; highest point (3,008 ft/917 m) is 10 mi/16 km SE of DOBOJ. A major Serb stronghold during World War II and the 1992–1995 civil war.

Ozu (OZ), city, EHIME prefecture, NW SHIKOKU, W JAPAN, 28 mi/45 km S of MATSUYAMA; 33°30′N 132°32′E. Machinery. Chestnuts; pigs. Sometimes called Iyo-ozu.

Ozu (OZ), town, Kikuchi county, KUMAMOTO prefecture, W central KYUSHU, SW JAPAN, 9 mi/15 km N of KUMAMOTO; 32°52′N 130°52′E. Motor vehicles. Sometimes spelled Otu.

Ozuakoli, NIGERIA: see UZUAKOLI.

Ozuluama, MEXICO: see OZULUAMA DE MASCAREÑAS.

Ozuluama de Mascareñas, city and township, ⊙ Ozuluama municipio, VERACRUZ state, E MEXICO, in GULF lowland, 38 mi/61 km S of TAMPICO; 21°40′N 97°54′W. Cereals, coffee, sugarcane, fruits; livestock.

Ozuma River, DOMINICAN REPUBLIC: see OZAMA RIVER.

Ozumba de Alzate (o-ZOOM-bah dai ahl-SAH-te), town, ⊙ Ozumba municipio, MEXICO state, central MEXICO, at W foot of POPOCATÉPETL, and part of the ZONA METROPOLITANA DE LA CIUDAD DE MÉXICO, 35 mi/56 km SE of MEXICO CITY; 19°03′N 98°48′W. Agricultural center (cereals, fruits; livestock). Has old Franciscan church with seventeenth-century historical painting. Chimal Falls nearby.

Ozun (o-ZOON), Hungarian *Uzon*, village, COVASNA county, central ROMANIA, 5 mi/8 km SE of SFÎNTU GHEORGHE; 45°48′N 25°51′E. Manufacturing of alcohol; lumbering. In HUNGARY, 1940–1945.

Ozurgeti (o-zoor-GE-tee) or **Makharadze**, city (2002 population 18,705), SW GEORGIA, on railroad spur, and 40 mi/64 km SW of KUTAISI; 41°56′N 42°00′E. Silk winding and twisting; bentonite grinding; tire repair; cannery. Nearby are scientific research institutes of tea and subtropical crops and the tea industry.

Ozyutichi, UKRAINE: see OZYUTYCHI.

Ozyutychi (oz-YOO-ti-chee), (Russian *Ozyutichi*) (Polish *Oździutycze*), village (2004 population 3,300), S central VOLYN' oblast, UKRAINE, 17 mi/27 km E of VOLODYMYR-VOLYNS'KYY; 50°51′N 24°43′E. Elevation 679 ft/206 m. Flour milling, brick manufacturing; cattle trade. Jewish community since the 17th century, severely reduced by the killings, disease, and relocations during the first half of the 20th century; numbering 740 in 1939; eliminated by the Nazis in 1941.

Area is shown by the symbol □, and capital city or county seat by ⊙.